CHILTON®

ASIAN
MECHANICAL SERVICE
2006 EDITION
VOLUME III
Lexus
Scion
Subaru
Suzuki
Toyota

THOMSON
DELMAR LEARNING

Australia • Canada • Mexico • Singapore • Spain • United Kingdom • United States

THOMSON
™
DELMAR LEARNING

Chilton®

Asian
Mechanical Service
2006 Edition
Volume III

Lexus, Scion, Subaru, Suzuki, Toyota

Vice President,
Technology Professional Business Unit:
Gregory L. Clayton

Publisher,
Technology Prosfessional Business Unit:
David Koontz

Director of Marketing:
Beth A. Lutz

Production Director:
Patty Stephan

Editorial Assistant:
Rebecca Rokitowski

Production Manager:
Andrew Crouth

Marketing Manager:
Brian McGrath

Marketing Specialist:
Marissa Maiella

Marketing Coordinator:
Jennifer Stall

Publishing Coordinator:
Paula Baillie

Sr. Content Project Manager:
Elizabeth C. Hough

Managing Editor:
Terry Blomquist

Senior Editor:
Rich Rivele

Editors:
Nick D'Andrea
Matt Frederick
Thomas A. Mellon
Jon Wallace

Cover Design:
Melinda Possinger

NOTICE TO THE READER

Table of Contents

Sections

Lexus	1	ES 300, GS 300, GS 430, IS 300, LS 430
Toyota and Lexus	2	GX470/LX470, Land Cruiser, Sequoia
Toyota and Lexus	3	Highlander, RX300/330
Toyota and Lexus	4	Highlander Hybrid, RX 400h
Scion	5	tC, xA, xB
Subaru	6	Baja, Impreza, Impreza Outback, Impreza Outback Sport, Legacy, Legacy Outback, Legacy SUS, WRX
Subaru	7	Forester
Subaru	8	B9 Tribeca
Suzuki	9	Aerio, Verona
Suzuki	10	Forenza, Reno
Suzuki	11	Grand Vitara, Vitara, XL-7
Toyota	12	4Runner
Toyota	13	Celica, Corolla, ECHO, MR2
Toyota	14	Avalon, Camry, Camry Solara
Toyota	15	Matrix
Toyota	16	Prius
Toyota	17	RAV4
Toyota	18	Sienna
Toyota	19	Tacoma
Toyota	20	Tundra

Model Index

Model	Section No.	Model	Section No.	Model	Section No.
4Runner	12-1	GS 430	1-1	**R**	
A		GX470	2-1	RAV4	17-1
Aerio	9-1	**H**		Reno	10-1
Avalon	14-1	Highlander	3-1	RX300/330	3-1
B		Highlander Hybrid	4-1	RX 400h	4-1
B9 Tribeca	8-1	**I**		**S**	
Baja	6-1	Impreza	6-1	Sequoia	2-1
C		Impreza Outback	6-1	Sienna	18-1
Camry	14-1	Impreza Outback Sport	6-1	**T**	
Camry Solara	14-1	IS 300	1-1	Tacoma	19-1
Celica	13-1	**L**		tC	5-1
Corolla	13-1	Land Cruiser	2-1	Tundra	20-1
E		Legacy	6-1	**V**	
ECHO	13-1	Legacy Outback	6-1	Verona	9-1
ES 300	1-1	Legacy SUS	6-1	Vitara	11-1
F		LS 430	1-1	**W**	
Forenza	10-1	LX470	2-1	WRX	6-1
Forester	7-1	**M**		**X**	
G		Matrix	15-1	xA	5-1
Grand Vitara	11-1	MR2	13-1	xB	5-1
GS 300	1-1	**P**		XL-7	11-1
		Prius	16-1		

USING THIS INFORMATION

Organization

To find where a particular model section or procedure is located, look in the Table of Contents. Main topics are listed with the page number on which they may be found. Following the main topics is an alphabetical listing of all of the procedures within the section and their page numbers.

Manufacturer and Model Coverage

This product covers 2002–2006 Asian models that are produced in sufficient quantities to warrant coverage, and which have technical content available from the vehicle manufacturers before our publication date. Although this information is as complete as possible at the time of publication, some manufacturers may make changes which cannot be included here. While striving for total accuracy, the publisher cannot assume responsibility for any errors, changes, or omissions that may occur in the compilation of this data.

Part Numbers & Special Tools

Part numbers and special tools are recommended by the publisher and vehicle manufacturer to perform specific jobs. Before substituting any part or tool for the one recommended, you must be completely satisfied that neither your personal safety, nor the performance of the vehicle will be endangered.

ACKNOWLEDGEMENT

The publisher would like to express appreciation to the following vehicle manufacturers for their assistance in producing this product. No further reproduction or distribution of the material in this manual is allowed without the expressed written permission of the vehicle manufacturers and the publisher. Suzuki Motor Corporation, Fuji Heavy Industries Ltd., including Subaru Motors Ltd., Toyota Motor Sales USA, including Lexus, Scion, and Toyota Divisions.

PRECAUTIONS

Before servicing any vehicle, please be sure to read all of the following precautions, which deal with personal safety, prevention of component damage, and important points to take into consideration when servicing a motor vehicle:

- Always wear safety glasses or goggles when drilling, cutting, grinding or prying.
- Steel-toed work shoes should be worn when working with heavy parts. Pockets should not be used for carrying tools. A slip or fall can drive a screwdriver into your body.
- Work surfaces, including tools and the floor should be kept clean of grease, oil or other slippery material.
- When working around moving parts, don't wear loose clothing. Long hair should be tied back under a hat or cap, or in a hair net.
- Always use tools only for the purpose for which they were designed. Never pry with a screwdriver.
- Keep a fire extinguisher and first aid kit handy.
- Always properly support the vehicle with approved stands or lift.
- Always have adequate ventilation when working with chemicals or hazardous material.
- Carbon monoxide is colorless, odorless and dangerous. If it is necessary to operate the engine with vehicle in a closed area such as a garage, always use an exhaust collector to vent the exhaust gases outside the closed area.
- When draining coolant, keep in mind that small children and some pets are attracted by ethylene glycol antifreeze, and are quite likely to drink any left in an open container, or in puddles on the ground. This will prove fatal in sufficient quantity. Always drain the coolant into a sealable container.
- To avoid personal injury, do not remove the coolant pressure relief cap while the engine is operating or hot. The cooling system is under pressure; steam and hot liquid can come out forcefully when the cap is loosened slightly. Failure to follow these instructions may result in personal injury. The coolant must be recovered in a suitable, clean container for reuse. If the coolant is contaminated it must be recycled or disposed of correctly.
- When carrying out maintenance on the starting system be aware that heavy gauge leads are connected directly to the battery. Make sure the protective caps are in place when maintenance is completed. Failure to follow these instructions may result in personal injury.
- Do not remove any part of the engine emission control system. Operating the engine without the engine emission control system will reduce fuel economy and engine ventilation. This will weaken engine performance and shorten engine life. It is also a violation of Federal law.
- Due to environmental concerns, when the air conditioning system is drained, the refrigerant must be collected using refrigerant recovery/recycling equipment. Federal law requires that refrigerant be recovered into appropriate recovery equipment and the process be conducted by qualified technicians who have been certified by an approved organization, such as MACS, ASI, etc. Use of a recovery machine dedicated to the appropriate refrigerant is necessary to reduce the possibility of oil and refrigerant incompatibility concerns. Refer to the instructions provided by the equipment manufacturer when removing refrigerant from or charging the air conditioning system.
- Always disconnect the battery ground when working on or around the electrical system.
- Batteries contain sulfuric acid. Avoid contact with skin, eyes, or clothing. Also, shield your eyes when working near batteries to protect against possible splashing of the acid solution. In case of acid contact with skin or eyes, flush immediately with water for a minimum of 15 minutes and get prompt medical attention. If acid is swallowed, call a physician immediately. Failure to follow these instructions may result in personal injury.
- Batteries normally produce explosive gases. Therefore, do not allow flames, sparks or lighted substances to come near the battery. When charging or working near a battery, always shield your face and protect your eyes. Always provide ventilation. Failure to follow these instructions may result in personal injury.
- When lifting a battery, excessive pressure on the end walls could cause acid to spew through the vent caps, resulting in personal injury, damage to the vehicle or battery. Lift with a battery carrier or with your hands on opposite corners. Failure to follow these instructions may result in personal injury.
- Observe all applicable safety precau-

tions when working around fuel. Whenever servicing the fuel system, always work in a well-ventilated area. Do not allow fuel spray or vapors to come in contact with a spark, open flame, or excessive heat (a hot drop light, for example). Keep a dry chemical fire extinguisher near the work area. Always keep fuel in a container specifically designed for fuel storage; also, always properly seal fuel containers to avoid the possibility of fire or explosion. Do not smoke or carry lighted tobacco or open flame of any type when working on or near any fuel-related components.

• Fuel injection systems often remain pressurized, even after the engine has been turned OFF. The fuel system pressure must be relieved before disconnecting any fuel lines. Failure to do so may result in fire and/or personal injury.

• The evaporative emissions system contains fuel vapor and condensed fuel vapor. Although not present in large quantities, it still presents the danger of explosion or fire. Disconnect the battery ground cable from the battery to minimize the possibility of an electrical spark occurring, possibly causing a fire or explosion if fuel vapor or liquid fuel is present in the area. Failure to follow these instructions can result in personal injury.

• The EPA warns that prolonged contact with used engine oil may cause a number of skin disorders, including cancer! You should make every effort to minimize your exposure to used engine oil. Protective gloves should be worn when changing oil. Wash your hands and any other exposed skin areas as soon as possible after exposure to used engine oil. Soap and water, or waterless hand cleaner should be used.

• Some vehicles are equipped with an air bag system, often referred to as a Supple-mental Restraint System (SRS) or Supplemental Inflatable Restraint (SIR) system. The system must be disabled before performing service on or around system components, steering column, instrument panel components, wiring and sensors. Failure to follow safety and disabling procedures could result in accidental air bag deployment, possible personal injury and unnecessary system repairs.

• Always wear safety goggles when working with, or around, the air bag system. When carrying a non-deployed air bag, be sure the bag and trim cover are pointed away from your body. When placing a non-deployed air bag on a work surface, always face the bag and trim cover upward, away from the surface. This will reduce the motion of the module if it is accidentally deployed.

• Electronic modules are sensitive to electrical charges. The ABS module can be damaged if exposed to these charges.

• Brake pads and shoes may contain asbestos, which has been determined to be a cancer-causing agent. Never clean brake surfaces with compressed air. Avoid inhaling brake dust. Clean all brake surfaces with a commercially available brake cleaning fluid.

• When replacing brake pads, shoes, discs or drums, replace them as complete axle sets.

• When servicing drum brakes, disassemble and assemble one side at a time, leaving the remaining side intact for reference.

• Brake fluid often contains polyglycol ethers and polyglycols. Avoid contact with the eyes and wash your hands thoroughly after handling brake fluid. If you do get brake fluid in your eyes, flush your eyes with clean, running water for 15 minutes. If eye irritation persists, or if you have taken brake fluid internally, immediately seek medical assistance.

• Clean, high quality brake fluid from a sealed container is essential to the safe and proper operation of the brake system. You should always buy the correct type of brake fluid for your vehicle. If the brake fluid becomes contaminated, completely flush the system with new fluid. Never reuse any brake fluid. Any brake fluid that is removed from the system should be discarded. Also, do not allow any brake fluid to come in contact with a painted or plastic surface; it will damage the paint.

• Never operate the engine without the proper amount and type of engine oil; doing so will result in severe engine damage.

• Timing belt maintenance is extremely important! Many models utilize an interference-stype, non-freewheeling engine. If the timing belt breaks, the valves in the cylinder head may strike the pistons, causing potentially serious (also time-consuming and expensive) engine damage.

• Disconnecting the negative battery cable on some vehicles may interfere with the functions of the on-board computer system (s) and may require the computer to undergo a relearning process once the negative battery cable is reconnected.

• Steering and suspension fasteners are critical parts because they affect performance of vital components and systems and their failure can result in major service expense. They must be replaced with the same grade or part number or an equivalent part if replacement is necessary. Do not use a replacement part of lesser quality or substitute design. Torque values must be used as specified during reassembly to ensure proper retention of these parts.

1

LEXUS

ES 300 • GS 300 • GS 430 • IS 300 • LS 430

DRIVE TRAIN 1-154
ENGINE REPAIR 1-18
FRONT BRAKES 1-221
FRONT SUSPENSION 1-186
FUEL SYSTEM 1-146
REAR BRAKES 1-227
REAR SUSPENSION 1-205
SPECIFICATIONS AND
MAINTENANCE CHARTS 1-2
STEERING 1-180
Engine and Vehicle Identification 1-2
General Engine Specifications 1-3
Engine Tune-Up Specifications 1-4
Firing Order 1-5
Accessory Drive Belt Routing 1-6
Capacities 1-7
Valve Specifications 1-8
Crankshaft and Connecting
 Rod Specifications 1-9
Piston and Ring Specifications 1-10
Torque Specifications 1-11
Wheel Alignment 1-12
Tire, Wheel and Ball Joint
 Specifications 1-14
Brake Specifications 1-16
Scheduled Maintenance
 Intervals 1-17

A
Air Bag 1-180
 Arming 1-180
 Disarming 1-180
 Precautions 1-180
Alternator 1-18
 Installation 1-19
 Removal 1-18
Axle Shaft, Bearing and Seal 1-177
 Removal & Installation 1-177

B
Brake Caliper (Front) 1-221
 Removal & Installation 1-221
Brake Caliper (Rear) 1-227
 Removal & Installation 1-227

C
Camshaft and Valve Lifters 1-110
 Removal & Installation 1-110
Clutch 1-167
 Bleeding 1-167
 Removal & Installation 1-167

CV-Joints 1-175
 Overhaul 1-175
Cylinder Head 1-88
 Removal & Installation 1-88

D
Disc Brake Pads (Front) 1-226
 Removal & Installation 1-226
Disc Brake Pads (Rear) 1-231
 Removal & Installation 1-231

E
Engine Assembly 1-19
 Removal & Installation 1-19
Exhaust Manifold 1-109
 Removal & Installation 1-109

F
Fuel Filter 1-146
 Removal & Installation 1-146
Fuel Injector 1-149
 Removal & Installation 1-149
Fuel Pump 1-146
 Removal & Installation 1-146
Fuel System Pressure 1-146
 Relieving 1-146
Fuel System Service
 Precautions 1-146

H
Halfshaft 1-170
 Removal & Installation 1-170
Heater Core 1-59
 Removal & Installation 1-59

I
Ignition Timing 1-19
 Adjustment 1-19
Intake Manifold 1-101
 Removal & Installation 1-101

L
Lower Control Arm 1-193
 Control Arm Bushing
  Replacement 1-201
 Removal & Installation 1-193
Lower Control Arm 1-215
 Control Arm Bushing
  Replacement 1-215

O
Oil Pan 1-127
 Removal & Installation 1-127
Oil Pump 1-135
 Removal & Installation 1-135

P
Pinion Seal 1-179
 Removal & Installation 1-179
Piston and Ring 1-145
 Positioning 1-145
Power Rack and Pinion
 Steering Gear 1-180
 Removal & Installation 1-180

S
Starter Motor 1-126
 Removal & Installation 1-126
Strut and Coil Spring 1-186
 Removal & Installation 1-186
Strut and Coil Spring 1-205
 Removal & Installation 1-205

T
Timing Belt 1-141
 Removal & Installation 1-141
Timing Chain 1-139
 Removal & Installation 1-139
Transaxle Assembly 1-164
 Removal & Installation 1-164
Transfer Case 1-167
 Removal & Installation 1-167
Transmission Assembly 1-154
 Removal & Installation 1-154

U
Upper Ball Joint 1-192
 Removal & Installation 1-192
Upper Control Arm 1-192
 Control Arm Bushing
  Replacement 1-193
 Removal & Installation 1-192

V
Valve Lash 1-123
 Adjustment 1-123

W
Water Pump 1-51
 Removal & Installation 1-51
Wheel Bearings 1-201
 Adjustment 1-201
 Removal & Installation 1-201
Wheel Bearings 1-215
 Adjustment 1-215
 Removal & Installation 1-215

SPECIFICATIONS AND MAINTENANCE CHARTS

ENGINE AND VEHICLE IDENTIFICATION

Code ①	Liters (cc)	Cu. In.	Cyl.	Fuel Sys.	Engine Type	Eng. Mfg.
4GR-FSE	2.5 (2500)	153	V6	SFI	DOHC	Toyota
1MZ-FE	3.0 (2995)	183	V6	SFI	DOHC	Toyota
2JZ-GE	3.0 (2997)	183	I6	SFI	DOHC	Toyota
3GR-FSE	3.0 (2995)	183	V6	SFI	DOHC	Toyota
3MZ-FE	3.3 (3311)	202	V6	SFI	DOHC	Toyota
2GR-FSE	3.5 (3456)	211	V6	DI	DOHC	Toyota
3UZ-FE	4.3 (4293)	262	V8	SFI	DOHC	Toyota

Code ②	Year
2	2002
3	2003
4	2004
5	2005
6	2006

SFI: Sequential Multi-port Fuel Injection

DI: Direct Injection

DOHC: Double Overhead Camshaft

① Located on the timing belt cover

② 10th digit of the VIN

09490-LEXU-C0001

GENERAL ENGINE SPECIFICATIONS

All measurements are given in inches.

Year	Model	Engine Displacement Liters (cc)	Engine ID/VIN	Fuel System Type	Net Horsepower @ rpm	Net Torque @ rpm (ft. lbs.)	Bore x Stroke (in.)	Compression Ratio	Oil Pressure @ rpm
2002	ES 300	3.0 (2995)	1MZ-FE	SFI	210@5800	220@4400	3.44x3.27	10.5:1	43-78@3000
	GS 300	3.0 (2997)	2JZ-GE	SFI	220@5800	220@3800	3.39x3.39	10.0:1	47-84@3000
	GS 430	4.3 (4293)	3UZ-FE	SFI	300@5600	325@3400	3.58x3.25	10.4:1	43-85@3000
	IS 300	3.0 (2997)	2JZ-GE	SFI	215@5800	218@3800	3.39x3.39	10.0:1	47-84@3000
	LS 430	4.3 (4293)	3UZ-FE	SFI	290@5600	320@3400	3.58x3.25	10.4:1	43-85@3000
	SC 430	4.3 (4293)	3UZ-FE	SFI	300@5600	325@3400	3.58x3.25	10.4:1	43-85@3000
2003	ES 300	3.0 (2995)	1MZ-FE	SFI	210@5800	220@4400	3.44x3.27	10.5:1	43-78@3000
	GS 300	3.0 (2997)	2JZ-GE	SFI	220@5800	220@3800	3.39x3.39	10.0:1	47-84@3000
	GS 430	4.3 (4293)	3UZ-FE	SFI	300@5600	325@3400	3.58x3.25	10.4:1	43-85@3000
	IS 300	3.0 (2997)	2JZ-GE	SFI	215@5800	218@3800	3.39x3.39	10.0:1	47-84@3000
	LS 430	4.3 (4293)	3UZ-FE	SFI	290@5600	320@3400	3.58x3.25	10.4:1	43-85@3000
	SC 430	4.3 (4293)	3UZ-FE	SFI	300@5600	325@3400	3.58x3.25	10.4:1	43-85@3000
2004	ES 330	3.3 (3311)	3MZ-FE	SFI	225@5600	240@3600	3.62x3.27	10.8:1	36-78@3000
	GS 300	3.0 (2997)	2JZ-GE	SFI	220@5800	220@3800	3.39x3.39	10.0:1	47-84@3000
	GS 430	4.3 (4293)	3UZ-FE	SFI	300@5600	325@3400	3.58x3.25	10.4:1	43-85@3000
	IS 300	3.0 (2997)	2JZ-GE	SFI	215@5800	218@3800	3.39x3.39	10.0:1	47-84@3000
	LS 430	4.3 (4293)	3UZ-FE	SFI	290@5600	320@3400	3.58x3.25	10.4:1	43-85@3000
	SC 430	4.3 (4293)	3UZ-FE	SFI	300@5600	325@3400	3.58x3.25	10.4:1	43-85@3000
2005	ES 330	3.3 (3311)	3MZ-FE	SFI	225@5600	240@3600	3.62x3.27	10.8:1	36-78@3000
	GS 300	3.0 (2997)	2JZ-GE	SFI	220@5800	220@3800	3.39x3.39	10.0:1	47-84@3000
	GS 430	4.3 (4293)	3UZ-FE	SFI	300@5600	325@3400	3.58x3.25	10.4:1	43-85@3000
	IS 300	3.0 (2997)	2JZ-GE	SFI	215@5800	218@3800	3.39x3.39	10.0:1	47-84@3000
	LS 430	4.3 (4293)	3UZ-FE	SFI	290@5600	320@3400	3.58x3.25	10.4:1	43-85@3000
	SC 430	4.3 (4293)	3UZ-FE	SFI	300@5600	325@3400	3.58x3.25	10.4:1	43-85@3000
2006	ES 330	3.3 (3311)	3MZ-FE	SFI	218@5600	236@3600	3.62x3.27	10.8:1	36-78@3000
	GS 300	3.0 (2995)	3GR-FSE	SFI	245@6200	230@3600	3.44x3.26	11.5:1	55.5@6000
	GS 430	4.3 (4293)	3UZ-FE	SFI	300@5600	325@3400	3.58x3.25	10.5:1	43-85@3000
	IS 250	2.5 (2500)	4GR-FSE	SFI	204@6400	185@4800	3.27x3.03	12.0:1	55.5@6000
	IS 350	3.5 (3456)	2GR-FSE	DI	306@6400	277@4800	3.70x3.27	11.8:1	55.5@6000
	LS 430	4.3 (4293)	3UZ-FE	SFI	278@5600	312@3400	3.58x3.25	10.5:1	43-85@3000
	SC 430	4.3 (4293)	3UZ-FE	SFI	288@5600	317@3400	3.58x3.25	10.5:1	43-85@3000

SFI : Sequential Multi-port Fuel Injection

DI : Direct Injection

09490-LEXU-C0002

ENGINE TUNE-UP SPECIFICATIONS

Year	Engine Displacement Liters (cc)	Engine ID/VIN	Spark Plug Gap (in.)	Ignition Timing (deg.)	Fuel Pump (psi)	Idle Speed (rpm)	Valve Clearance Intake	Valve Clearance Exhaust
2002	3.0 (2995)	1MZ-FE	0.043	8-12B ①	44-50	550-650	0.006-0.010	0.010-0.014
	3.0 (2997)	2JZ-GE	0.043	8-12B ②	44-50	650-750	0.006-0.010	0.010-0.014
	4.3 (4293)	3UZ-FE	0.043	8-12B ③	44-50	700-800	0.006-0.010	0.010-0.014
2003	3.0 (2995)	1MZ-FE	0.043	8-12B ①	44-50	650-750	0.006-0.010	0.010-0.014
	3.0 (2997)	2JZ-GE	0.043	8-12B ②	44-50	650-750	0.006-0.010	0.010-0.014
	4.3 (4293)	3UZ-FE	0.043	8-12B ③	44-50	700-800	0.006-0.010	0.010-0.014
2004	3.0 (2997)	2JZ-GE	0.043	8-12B ②	44-50	650-750	0.006-0.010	0.010-0.014
	3.3 (3311)	3MZ-FE	0.043	8-12B ①	44-50	630-730	0.006-0.010	0.010-0.014
	4.3 (4293)	3UZ-FE	0.043	8-12B ④	44-50	700-800	0.006-0.010	0.010-0.014
2005	3.0 (2997)	2JZ-GE	0.043	8-12B ②	44-50	650-750	0.006-0.010	0.010-0.014
	3.3 (3311)	3MZ-FE	0.043	8-12B ①	44-50	630-730	0.006-0.010	0.010-0.014
	4.3 (4293)	3UZ-FE	0.043	8-12B ④	44-50	700-800	0.006-0.010	0.010-0.014
2006	2.5 (2500)	4GR-FSE	0.043	8-12B ①	28-85	650-750	HYD	HYD
	3.0 (2995)	3GR-FSE	0.043	8-12B ①	28-85	600-700	HYD	HYD
	3.3 (3311)	3MZ-FE	0.043	8-12B ①	44-50	630-730	0.006-0.010	0.010-0.014
	3.5 (3456)	2GR-FSE	0.043	8-12B ①	28-85	600-700	HYD	HYD
	4.3 (4293)	3UZ-FE	0.043	8-12B	44-50	700-800	0.006-0.010	0.010-0.014

NOTE: The Vehicle Emission Control Information label often reflects specification changes made during production.

The label figures must be used if they differ from those in this chart.

B: Before top dead center

HYD: Hydraulic Valve Lifters

① Terminals TC and CG of check connector must be connected

② Terminals TE and E1 of check connector must be connected

③ LS 430: Terminals TC and CG of check connector must be connected
   GS 430: Terminals TC and E1 of check connector must be connected

④ GS 430: Terminals TC and E1 of check connector must be connected

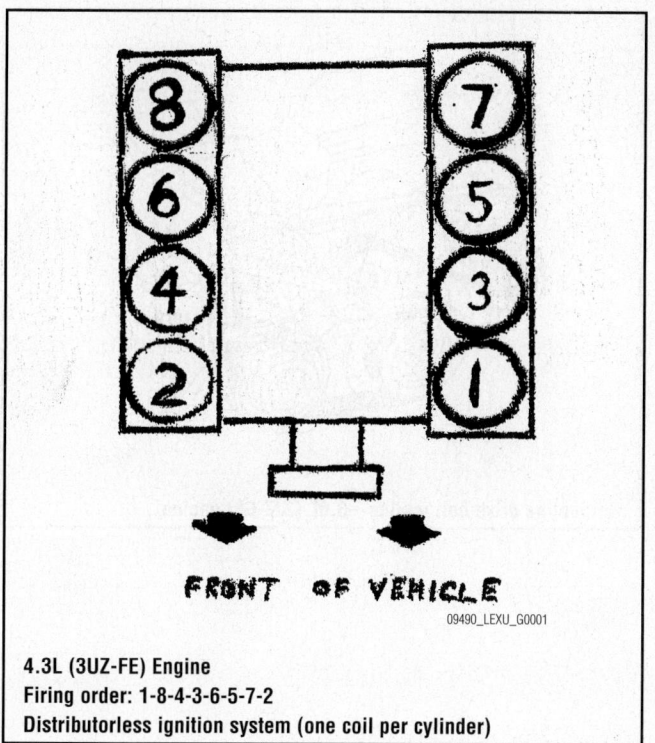

09490_LEXU_G0001

4.3L (3UZ-FE) Engine
Firing order: 1-8-4-3-6-5-7-2
Distributorless ignition system (one coil per cylinder)

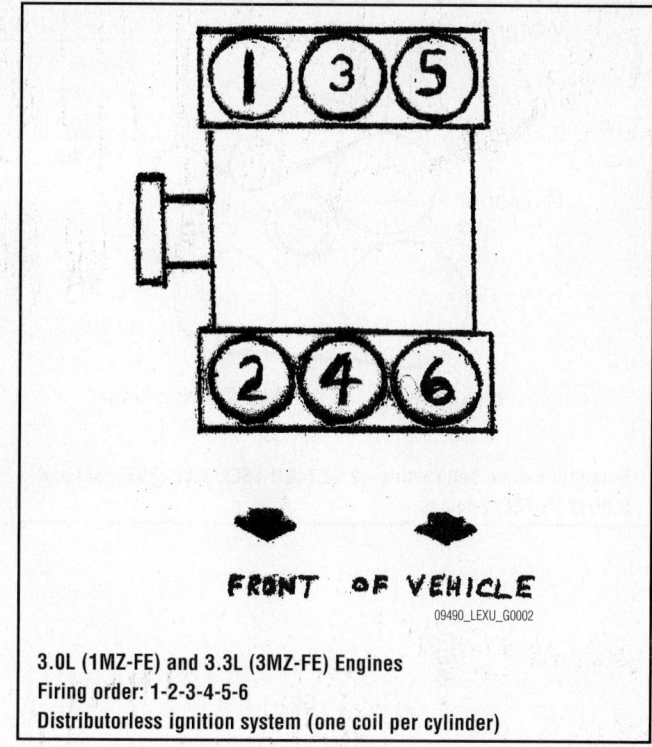

09490_LEXU_G0002

3.0L (1MZ-FE) and 3.3L (3MZ-FE) Engines
Firing order: 1-2-3-4-5-6
Distributorless ignition system (one coil per cylinder)

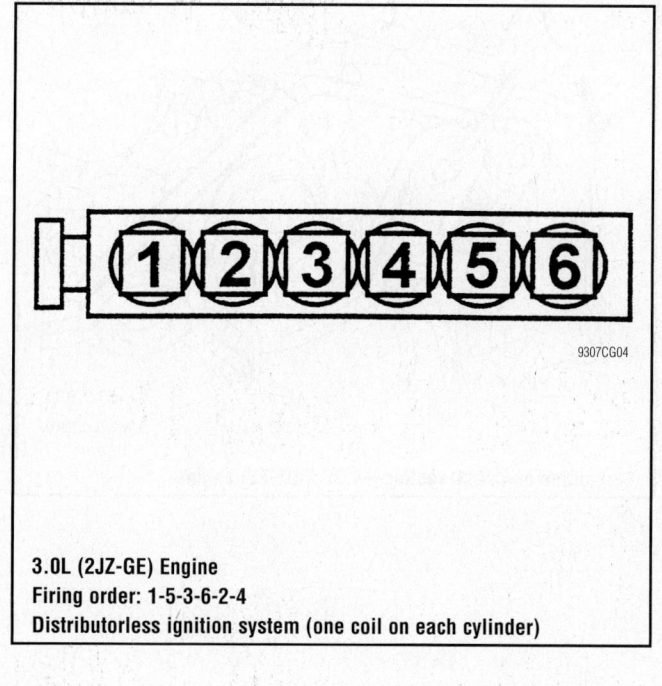

9307CG04

3.0L (2JZ-GE) Engine
Firing order: 1-5-3-6-2-4
Distributorless ignition system (one coil on each cylinder)

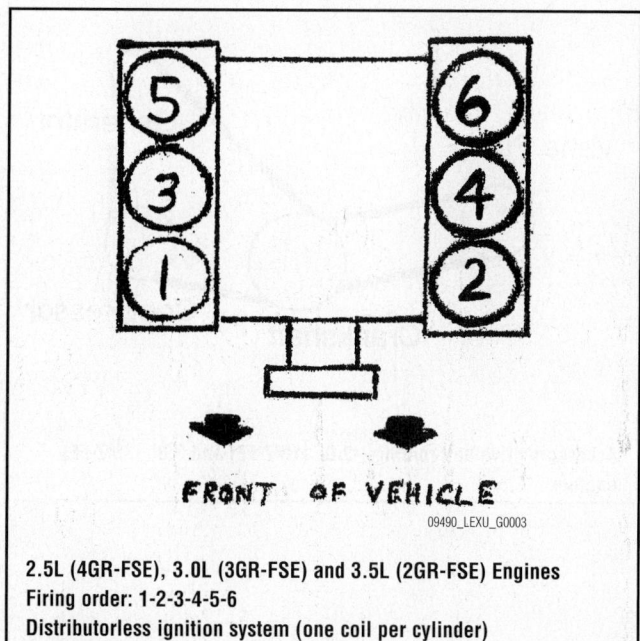

09490_LEXU_G0003

2.5L (4GR-FSE), 3.0L (3GR-FSE) and 3.5L (2GR-FSE) Engines
Firing order: 1-2-3-4-5-6
Distributorless ignition system (one coil per cylinder)

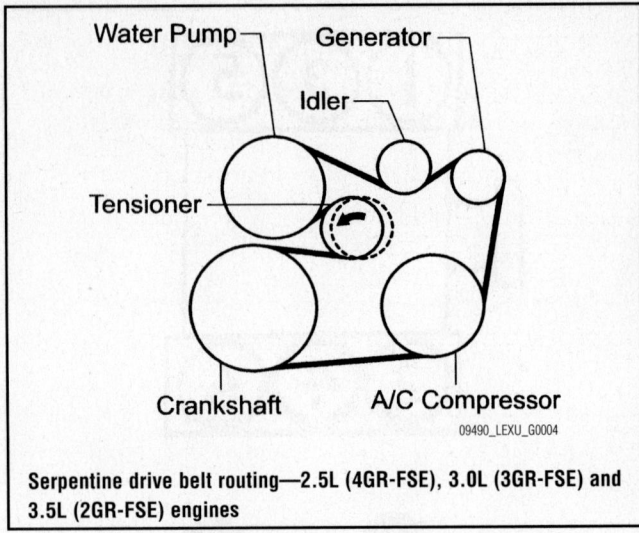

Serpentine drive belt routing—2.5L (4GR-FSE), 3.0L (3GR-FSE) and 3.5L (2GR-FSE) engines

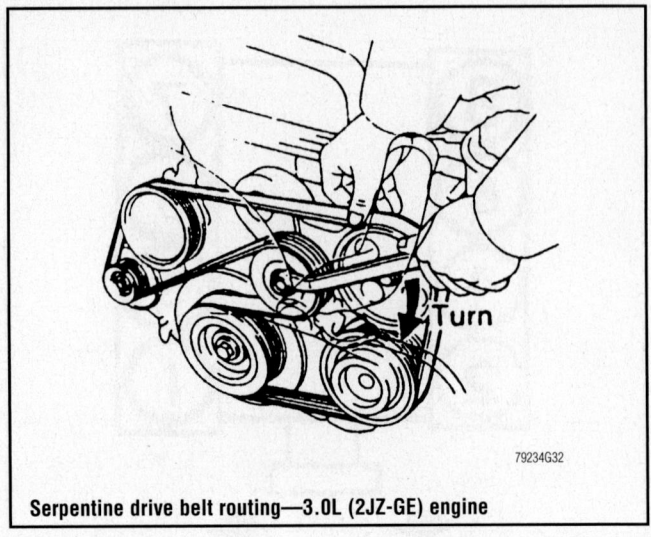

Serpentine drive belt routing—3.0L (2JZ-GE) engine

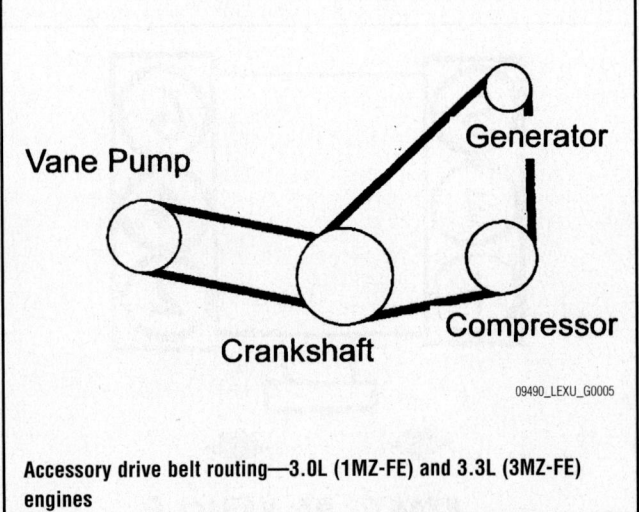

Accessory drive belt routing—3.0L (1MZ-FE) and 3.3L (3MZ-FE) engines

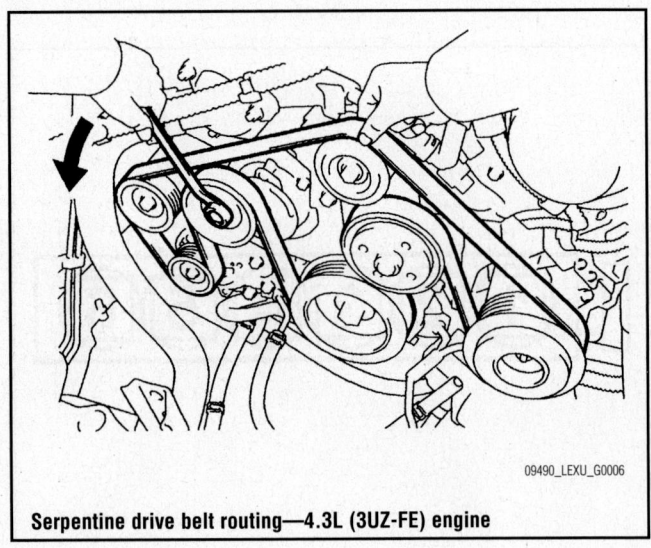

Serpentine drive belt routing—4.3L (3UZ-FE) engine

CAPACITIES

Year	Model	Engine Displacement Liters (cc)	Engine ID/VIN	Engine Oil with Filter	Transmission (pts.) ①		Drive Axle (pts.)	Fuel Tank (gal.)	Cooling System (qts.)
					Auto.	Manual			
2002	ES 300	3.0 (2995)	1MZ-FE	5.0	6.6	—	—	18.5	9.7
	GS 300	3.0 (2997)	2JZ-GE	5.7	4.2	—	2.8	19.8	8.1
	GS 430	4.3 (4293)	3UZ-FE	5.5	3.8	—	2.8	19.8	9.5
	IS 300	3.0 (2997)	2JZ-GE	5.7	4.2	5.4	2.4	17.5	7.9
	LS 430	4.3 (4293)	3UZ-FE	4.8	3.8	—	2.8	22.2	10.0
	SC 430	4.3 (4293)	3UZ-FE	5.5	3.8	—	2.8	19.8	10.5
2003	ES 300	3.0 (2995)	1MZ-FE	5.0	6.6	—	—	18.5	9.7
	GS 300	3.0 (2997)	2JZ-GE	5.7	4.2	—	2.8	19.8	8.1
	GS 430	4.3 (4293)	3UZ-FE	5.5	3.8	—	2.8	19.8	9.5
	IS 300	3.0 (2997)	2JZ-GE	5.7	4.2	5.4	2.4	17.5	7.9
	LS 430	4.3 (4293)	3UZ-FE	4.8	3.8	—	2.8	22.2	10.0
	SC 430	4.3 (4293)	3UZ-FE	5.5	3.8	—	2.8	19.8	10.5
2004	ES 330	3.3 (3311)	3MZ-FE	5.0	7.4	—	—	18.5	9.7
	GS 300	3.0 (2997)	2JZ-GE	5.7	4.2	—	2.8	19.8	8.1
	GS 430	4.3 (4293)	3UZ-FE	5.5	3.8	—	2.8	19.8	9.5
	IS 300	3.0 (2997)	2JZ-GE	5.7	4.2	5.4	2.4	17.5	7.9
	LS 430	4.3 (4293)	3UZ-FE	4.8	3.8	—	2.8	22.2	10.0
	SC 430	4.3 (4293)	3UZ-FE	5.5	3.8	—	2.8	19.8	10.5
2005	ES 330	3.3 (3311)	3MZ-FE	5.0	7.4	—	—	18.5	9.7
	GS 300	3.0 (2997)	2JZ-GE	5.7	4.2	—	2.8	19.8	8.1
	GS 430	4.3 (4293)	3UZ-FE	5.5	3.8	—	2.8	19.8	9.5
	IS 300	3.0 (2997)	2JZ-GE	5.7	4.2	5.4	2.4	17.5	7.9
	LS 430	4.3 (4293)	3UZ-FE	4.8	3.8	—	2.8	22.2	10.0
	SC 430	4.3 (4293)	3UZ-FE	5.5	3.8	—	2.8	19.8	10.5
2006	ES 330	3.3 (3311)	3MZ-FE	5.0	7.4	—	—	18.5	9.7
	GS 300	3.0 (2995)	3GR-FSE	②	③	—	④	18.7	9.6
	GS 430	4.3 (4293)	3UZ-FE	5.4	3.0	—	2.8	18.7	10.7
	IS 250	2.5 (2500)	4GR-FSE	⑤	⑥	3.8	⑦	17.2	9.6
	IS 350	3.5 (3456)	2GR-FSE	6.6	3.6	3.8	2.8	17.2	9.6
	LS 430	4.3 (4293)	3UZ-FE	4.8	3.6	—	2.8	22.2	10.0
	SC 430	4.3 (4293)	3UZ-FE	5.5	3.6	—	2.8	19.8	10.5

NOTE: All capacities are approximate. Add fluid gradually and check to be sure a proper fluid level is obtained.

① Specification is for transmission drain and refill, not overhaul.

② 2WD models: 6.6 qts.

AWD models: 6.7 qts.

③ 2WD models: 3.2 pts.

AWD models: 5.6 pts.

④ 2WD models: 2.4 pts.

AWD models front: 1.4 pts.

AWD models rear: 2.2 pts.

⑤ 2WD models: 6.2 qts.

AWD models: 6.3 qts.

⑥ 2WD models: 3.2 pts.

AWD models: 5.6 pts.

⑦ 2WD models: 2.4 pts.

AWD models: 2.2 pts. for the rear; 1.48 pts. for the front

VALVE SPECIFICATIONS

Year	Engine Displacement Liters (cc)	Engine ID/VIN	Seat Angle (deg.)	Face Angle (deg.)	Spring Test Pressure (lbs. @ in.)	Spring Free-Length (in.)	Stem-to-Guide Clearance (in.)		Stem Diameter (in.)	
							Intake	Exhaust	Intake	Exhaust
2002	3.0 (2995)	1MZ-FE	NA	44.5	41.9-46.3@ 1.331	1.791	0.0010-0.0024	0.0012-0.0026	0.2154-0.2159	0.2152-0.2157
	3.0 (2997)	2JZ-GE	NA	44.5	41.9-46.3@ 1.358	①	0.0010-0.0024	0.0012-0.0026	0.2350-0.2356	0.2348-0.2354
	4.3 (4293)	3UZ-FE	45	44.5	45.9-50.7@ 1.3795	2.130	0.0010-0.0024	0.0012-0.0026	0.2154-0.2159	0.2152-0.2157
2003	3.0 (2995)	1MZ-FE	NA	44.5	41.9-46.3@ 1.331	1.791	0.0010-0.0024	0.0012-0.0026	0.2154-0.2159	0.2152-0.2157
	3.0 (2997)	2JZ-GE	NA	44.5	41.9-46.3@ 1.358	①	0.0010-0.0024	0.0012-0.0026	0.2350-0.2356	0.2348-0.2354
	4.3 (4293)	3UZ-FE	45	44.5	45.9-50.7@ 1.3795	2.130	0.0010-0.0024	0.0012-0.0026	0.2154-0.2159	0.2152-0.2157
2004	3.0 (2997)	2JZ-GE	NA	44.5	41.9-46.3@ 1.358	①	0.0010-0.0024	0.0012-0.0026	0.2350-0.2356	0.2348-0.2354
	3.3 (3311)	3MZ-FE	45	NA	41.9-46.3@ 1.331	1.791	0.0010-0.0024	0.0012-0.0026	0.2154-0.2159	0.2152-0.2157
	4.3 (4293)	3UZ-FE	45	44.5	45.9-50.7@ 1.3795	2.130	0.0010-0.0024	0.0012-0.0026	0.2154-0.2159	0.2152-0.2157
2005	3.0 (2997)	2JZ-GE	NA	44.5	41.9-46.3@ 1.358	①	0.0010-0.0024	0.0012-0.0026	0.2350-0.2356	0.2348-0.2354
	3.3 (3311)	3MZ-FE	45	NA	41.9-46.3@ 1.331	1.791	0.0010-0.0024	0.0012-0.0026	0.2154-0.2159	0.2152-0.2157
	4.3 (4293)	3UZ-FE	45	44.5	45.9-50.7@ 1.3795	2.130	0.0010-0.0024	0.0012-0.0026	0.2154-0.2159	0.2152-0.2157
2006	2.5 (2500)	4GR-FSE	45	NA	NA	1.831	0.0010-0.0024	0.0012-0.0026	0.2154-0.2159	0.2151-0.2157
	3.3 (3311)	3MZ-FE	45	NA	41.9-46.3@ 1.331	1.791	0.0010-0.0024	0.0012-0.0026	0.2154-0.2159	0.2152-0.2157
	3.0 (2995)	3GR-FSE	45	NA	NA	1.831	0.0010-0.0024	0.0012-0.0026	0.2154-0.2159	0.2152-0.2158
	3.5 (3456)	2GR-FSE	45	NA	NA	2.035	0.0010-0.0024	0.0012-0.0026	0.2154-0.2159	0.2151-0.2157
	4.3 (4293)	3UZ-FE	45	44.5	45.9-50.7@ 1.3795	2.130	0.0010-0.0024	0.0012-0.0026	0.2154-0.2159	0.2152-0.2157

NA: Not Available
① Pink: 1.7209
   Yellow: 1.7362

CRANKSHAFT AND CONNECTING ROD SPECIFICATIONS

All measurements are given in inches.

Year	Engine Displacement Liters (cc)	Engine ID/VIN	Crankshaft				Connecting Rod		
			Main Brg. Journal Dia.	Main Brg. Oil Clearance	Shaft End-play	Thrust on No.	Journal Diameter	Oil Clearance	Side Clearance
2002	3.0 (2995)	1MZ-FE	2.4011-2.4016	①	0.0016-0.0094	2	2.0863-2.0866	0.0015-0.0026	0.0059-0.0118
	3.0 (2997)	2JZ-GE	2.4403-2.4409	0.0010-0.0016	0.0008-0.0087	4	2.0465-2.0472	0.0009-0.0016	0.0098-0.0158
	4.3 (4293)	3UZ-FE	2.6373-2.6378	②	0.0008-0.0087	3	2.0465-2.0472	0.0008-0.0019	0.0063-0.0138
2003	3.0 (2995)	1MZ-FE	2.4011-2.4016	①	0.0016-0.0095	2	2.0863-2.0866	0.0015-0.0026	0.0059-0.0118
	3.0 (2997)	2JZ-GE	2.4403-2.4409	0.0010-0.0016	0.0008-0.0087	4	2.0465-2.0472	0.0009-0.0016	0.0098-0.0158
	4.3 (4293)	3UZ-FE	2.6373-2.6378	②	0.0008-0.0087	3	2.0465-2.0472	0.0008-0.0019	0.0063-0.0138
2004	3.0 (2997)	2JZ-GE	2.4403-2.4409	0.0010-0.0016	0.0008-0.0087	4	2.0465-2.0472	0.0009-0.0016	0.0098-0.0158
	3.3 (3311)	3MZ-FE	2.4011-2.4016	③	0.0016-0.0094	4	2.0863-2.0866	0.0015-0.0026	0.0059-0.0118
	4.3 (4293)	3UZ-FE	2.6373-2.6378	②	0.0008-0.0087	3	2.0465-2.0472	0.0008-0.0019	0.0063-0.0138
2005	3.0 (2997)	2JZ-GE	2.4403-2.4409	0.0010-0.0016	0.0008-0.0087	4	2.0465-2.0472	0.0009-0.0016	0.0098-0.0158
	3.3 (3311)	3MZ-FE	2.4011-2.4016	③	0.0016-0.0094	4	2.0863-2.0866	0.0015-0.0026	0.0059-0.0118
	4.3 (4293)	3UZ-FE	2.6373-2.6378	②	0.0008-0.0087	3	2.0465-2.0472	0.0008-0.0019	0.0063-0.0138
2006	2.5 (2500)	4GR-FSE	2.4011-2.4016	0.0010-0.0019	0.0008-0.0087	2	1.8894-1.8898	0.0017-0.0027	0.0098-0.0157
	3.0 (2995)	3GR-FSE	2.4011-2.4016	0.0005-0.0014	0.0016-0.0094	NA	2.0863-2.0866	0.0018-0.0026	0.0059-0.0157
	3.3 (3311)	3MZ-FE	2.4011-2.4016	③	0.0016-0.0094	4	2.0863-2.0866	0.0015-0.0026	0.0059-0.0118
	3.5 (3456)	2GR-FSE	2.4011-2.4016	④	0.0016-0.0094	NA	2.0863-2.0866	0.0018-0.0026	0.0059-0.0157
	4.3 (4293)	3UZ-FE	2.6373-2.6378	②	0.0008-0.0087	3	2.0465-2.0472	0.0008-0.0019	0.0063-0.0138

NA: Not Available

① Journal No. 1 and 4: 0.0006 - 0.0013 inch
　Journal No. 2 and 3: 0.0010 - 0.0018 inch

② Journal No. 1 and 5: 0.0007 - 0.0013 inch
　Remaining journals: 0.0011 - 0.0018 inch

③ Journal No. 1 and 2: 0.0006 - 0.0013 inch
　Journal No. 3 and 4: 0.0010 - 0.0018 inch

④ Journal No. 1 and 4: 0.0010 - 0.0019 inch
　Journal No. 2 and 3: 0.0013 - 0.0021 inch

PISTON AND RING SPECIFICATIONS

All measurements are given in inches.

Year	Engine Displacement Liters (cc)	Engine ID/VIN	Piston Clearance	Ring Gap			Ring Side Clearance		
				Top Compression	Bottom Compression	Oil Control	Top Compression	Bottom Compression	Oil Control
2002	3.0 (2995)	1MZ-FE	0.0033-0.0042	0.0098-0.0138	0.0138-0.0177	0.0059-0.0157	0.0008-0.0028	0.0008-0.0024	0.0011-0.0043
	3.0 (2997)	2JZ-GE	0.0022-0.0031	0.0118-0.0185	0.0138-0.0205	0.0051-0.0177	0.0004-0.0028	0.0012-0.0028	SNUG
	4.3 (4293)	3UZ-FE	0.0031-0.0040	0.0118-0.0197	0.0157-0.0236	0.0059-0.0197	0.0012-0.0031	0.0008-0.0024	SNUG
2003	3.0 (2995)	1MZ-FE	0.0033-0.0042	0.0098-0.0138	0.0138-0.0177	0.0059-0.0157	0.0008-0.0028	0.0008-0.0024	0.0011-0.0043
	3.0 (2997)	2JZ-GE	0.0022-0.0031	0.0118-0.0185	0.0138-0.0205	0.0051-0.0177	0.0004-0.0028	0.0012-0.0028	SNUG
	4.3 (4293)	3UZ-FE	0.0031-0.0040	0.0118-0.0197	0.0157-0.0236	0.0059-0.0197	0.0012-0.0031	0.0008-0.0024	SNUG
2004	3.0 (2997)	2JZ-GE	0.0022-0.0031	0.0118-0.0185	0.0138-0.0205	0.0051-0.0177	0.0004-0.0028	0.0012-0.0028	SNUG
	3.3 (3311)	3MZ-FE	0.0013-0.0023	0.0118-0.0157	0.0197-0.0236	0.0059-0.0157	0.0012-0.0031	0.0008-0.0024	0.0012-0.0043
	4.3 (4293)	3UZ-FE	0.0031-0.0040	0.0118-0.0197	0.0157-0.0236	0.0059-0.0197	0.0012-0.0031	0.0008-0.0024	SNUG
2005	3.0 (2997)	2JZ-GE	0.0022-0.0031	0.0118-0.0185	0.0138-0.0205	0.0051-0.0177	0.0004-0.0028	0.0012-0.0028	SNUG
	3.3 (3311)	3MZ-FE	0.0013-0.0023	0.0118-0.0157	0.0197-0.0236	0.0059-0.0157	0.0012-0.0031	0.0008-0.0024	0.0012-0.0043
	4.3 (4293)	3UZ-FE	0.0023-0.0040	0.0118-0.0197	0.0157-0.0236	0.0059-0.0197	0.0012-0.0031	0.0008-0.0024	SNUG
2006	2.5 (2500)	4GR-FSE	0.0006-0.0014	0.0087-0.0126	0.0236-0.0276	0.0039-0.0138	0.0008-0.0028	0.0012-0.0028	0.0008-0.0026
	3.0 (2995)	3GR-FSE	0.0006-0.0014	0.0091-0.0130	0.0197-0.0236	0.0039-0.0157	0.0008-0.0028	0.0008-0.0024	0.0008-0.0026
	3.3 (3311)	3MZ-FE	0.0013-0.0023	0.0118-0.0157	0.0197-0.0236	0.0059-0.0157	0.0012-0.0031	0.0008-0.0024	0.0012-0.0043
	3.5 (3456)	2GR-FSE	0.0008-0.0020	0.0091-0.0130	0.0138-0.0177	0.0039-0.0157	0.0008-0.0028	0.0008-0.0024	0.0008-0.0028
	4.3 (4293)	3UZ-FE	0.0023-0.0040	0.0118-0.0197	0.0157-0.0236	0.0059-0.0197	0.0012-0.0031	0.0008-0.0024	SNUG

09490-LEXU-C0007

TORQUE SPECIFICATIONS
All readings in ft. lbs.

Year	Engine Displacement Liters (cc)	Engine ID/VIN	Cylinder Head Bolts	Main Bearing Bolts	Rod Bearing Bolts	Crankshaft Damper Bolts	Flywheel Bolts	Manifold Intake	Manifold Exhaust	Spark Plugs	Lug Nuts
2002	3.0 (2995)	1MZ-FE	①	②	③	159	61	11	36	13	76
	3.0 (2997)	2JZ-GE	④	⑤	⑥	243	61	21	30	13	76
	4.3 (4293)	3UZ-FE	⑦	⑧	③	181	⑨	13	32	13	76
2003	3.0 (2995)	1MZ-FE	①	②	③	159	61	11	36	13	76
	3.0 (2997)	2JZ-GE	④	⑤	⑥	243	61	21	30	13	76
	4.3 (4293)	3UZ-FE	⑦	⑧	③	181	⑨	13	32	13	76
2004	3.0 (2997)	2JZ-GE	④	⑤	⑥	243	61	21	30	13	76
	3.3 (3311)	3MZ-FE	①	②	③	162	61	11	36	18	76
	4.3 (4293)	3UZ-FE	⑦	⑧	③	181	⑨	13	32	13	76
2005	3.0 (2997)	2JZ-GE	④	⑤	⑥	243	61	21	30	13	76
	3.3 (3311)	3MZ-FE	①	②	③	162	61	11	36	18	76
	4.3 (4293)	3UZ-FE	⑦	⑧	③	181	⑨	13	32	13	76
2006	2.5 (2500)	4GR-FSE	⑩	⑪	⑫	184	⑬	15	15	18	76
	3.0 (2995)	3GR-FSE	⑩	⑪	⑫	184	30	15	15	13	76
	3.3 (3311)	3MZ-FE	①	②	③	162	61	11	36	18	76
	3.5 (3456)	2GR-FSE	⑩	⑪	⑭	184	61	15	15	18	76
	4.3 (4293)	3UZ-FE	⑦	⑧	③	181	⑨	13	32	13	76

① Head bolt:
Step 1: 40 ft. lbs.
Step 2: Plus 90 degrees
Recessed head bolt: 14 ft. lbs.

② 6-point bolts: 20 ft. lbs.
12-point bolts:
Step 1: 16 ft. lbs.
Step 2: Plus an additional 90 degrees

③ Step 1: 18 ft. lbs.
Step 2: Plus 90 degrees

④ Step 1: 26 ft. lbs.
Step 2: Tighten an additional 90 degrees
Step 3: Tighten an additional 90 degrees

⑤ Step 1: 33 ft. lbs.
Step 2: Plus 90 degrees

⑥ Step 1: 22 ft. lbs.
Step 2: Plus 90 degrees

⑦ Step 1: 44 ft. lbs.
Step 2: Plus 90 degrees

⑧ Nuts:
Step 1: 20 ft. lbs.
Step 2: Plus 90 degrees
Bolts: 36 ft. lbs.

⑨ Step 1: 36 ft. lbs.
Step 2: Plus 90 degrees

⑩ Step 1: 27 ft. lbs.
Step 2: Tighten an additional 90 degrees
Step 3: Tighten an additional 90 degrees
14mm head bolt: 22 ft. lbs.

⑪ 12mm head bolts: 19 ft. lbs.
16-point bolts:
Step 1: 45 ft. lbs.
Step 2: Plus an additional 90 degrees

⑫ Step 1: 28 ft. lbs.
Step 2: Plus 90 degrees

⑬ Automatic: 61 ft. lbs.
Manual: 54 ft. lbs.

⑭ Step 1: 30 ft. lbs.
Step 2: Plus 90 degrees

09490-LEXU-C0008

WHEEL ALIGNMENT

Year	Model		Caster Range (+/-Deg.)	Caster Preferred Setting (Deg.)	Camber Range (+/-Deg.)	Camber Preferred Setting (Deg.)	Toe-in (in.)	Steering Axis Inclination (Deg.)
2002	ES 300	F	0.75	+2.77	0.75	-0.72	0 +/- 0.08	11.45
		R	—	—	0.75	-1.38	0.16 +/- 0.08	—
	GS 300	F	0.50	+7.55	0.50	-0.27	0.06 +/- 0.08	8.87
		R	—	—	0.50	-0.78	0.06 +/- 0.08	—
	GS 430	F	0.50	+7.55	0.50	-0.27	0.06 +/- 0.08	8.87
		R	—	—	0.50	-0.78	0.06 +/- 0.08	—
	IS 300	F	0.50	+6.12	0.50	-0.50	0.04 +/- 0.08	9.42
		R	—	—	0.50	-0.92	0.08 +/- 0.08	—
	LS 430 ①	F	0.75	+6.75	0.75	-0.08	0.04 +/- 0.08	9.00
		R	—	—	0.75	-1.00	0.12 +/- 0.08	—
	LS 430 ②	F	0.75	+7.25	0.75	-0.25	0.04 +/- 0.08	9.25
		R	—	—	0.75	-1.55	0.12 +/- 0.08	—
	SC 430	F	0.75	+7.92	0.75	-0.58	0.06 +/- 0.08	9.16
		R	—	—	0.50	-1.17	0.06 +/- 0.08	—
2003	ES 300	F	0.75	+2.77	0.75	-0.72	0 +/- 0.08	11.45
		R	—	—	0.75	-1.38	0.16 +/- 0.08	—
	GS 300	F	0.50	+7.55	0.50	-0.27	0.06 +/- 0.08	8.87
		R	—	—	0.50	-0.78	0.06 +/- 0.08	—
	GS 430	F	0.50	+7.55	0.50	-0.27	0.06 +/- 0.08	8.87
		R	—	—	0.50	-0.78	0.06 +/- 0.08	—
	IS 300	F	0.50	+6.12	0.50	-0.50	0.04 +/- 0.08	9.42
		R	—	—	0.50	-0.92	0.08 +/- 0.08	—
	LS 430 ①	F	0.75	+6.75	0.75	-0.08	0.04 +/- 0.08	9.00
		R	—	—	0.75	-1.00	0.12 +/- 0.08	—
	LS 430 ②	F	0.75	+7.25	0.75	-0.25	0.04 +/- 0.08	9.25
		R	—	—	0.75	-1.55	0.12 +/- 0.08	—
	SC 430	F	0.75	+7.92	0.75	-0.58	0.06 +/- 0.08	9.16
		R	—	—	0.50	-1.17	0.06 +/- 0.08	—
2004	ES 330	F	0.75	+2.77	0.75	-0.72	0 +/- 0.08	11.45
		R	—	—	0.75	-1.38	0.16 +/- 0.08	—
	GS 300	F	0.50	+7.55	0.50	-0.27	0.06 +/- 0.08	8.87
		R	—	—	0.50	-0.78	0.06 +/- 0.08	—
	GS 430	F	0.50	+7.55	0.50	-0.27	0.06 +/- 0.08	8.87
		R	—	—	0.50	-0.78	0.06 +/- 0.08	—
	IS 300	F	0.50	+6.12	0.50	-0.50	0.04 +/- 0.08	9.42
		R	—	—	0.50	-0.92	0.08 +/- 0.08	—
	LS 430 ①	F	0.75	+6.75	0.75	-0.08	0.04 +/- 0.08	9.00
		R	—	—	0.75	-1.25	0.12 +/- 0.08	—
	LS 430 ②	F	0.75	+7.25	0.75	-0.25	0.04 +/- 0.08	9.25
		R	—	—	0.75	-1.58	0.12 +/- 0.08	—
	SC 430	F	0.75	+7.92	0.75	-0.58	0.06 +/- 0.08	9.16
		R	—	—	0.50	-1.17	0.06 +/- 0.08	—

09490-LEXU-C0009

WHEEL ALIGNMENT

Year	Model		Caster Range (+/-Deg.)	Caster Preferred Setting (Deg.)	Camber Range (+/-Deg.)	Camber Preferred Setting (Deg.)	Toe-in (in.)	Steering Axis Inclination (Deg.)
2005	ES 330	F	0.75	+2.77	0.75	-0.72	0 +/- 0.08	11.45
		R	—	—	0.75	-1.38	0.16 +/- 0.08	—
	GS 300	F	0.50	+7.55	0.50	-0.27	0.06 +/- 0.08	8.87
		R	—	—	0.50	-0.78	0.06 +/- 0.08	—
	GS 430	F	0.50	+7.55	0.50	-0.27	0.06 +/- 0.08	8.87
		R	—	—	0.50	-0.78	0.06 +/- 0.08	—
	IS 300	F	0.50	+6.12	0.50	-0.50	0.04 +/- 0.08	9.42
		R	—	—	0.50	-0.92	0.08 +/- 0.08	—
	LS 430 ①	F	0.75	+6.75	0.75	-0.08	0.04 +/- 0.08	9.00
		R	—	—	0.75	-1.25	0.12 +/- 0.08	—
	LS 430 ②	F	0.75	+7.25	0.75	-0.25	0.04 +/- 0.08	9.25
		R	—	—	0.75	-1.58	0.12 +/- 0.08	—
	SC 430	F	0.75	+7.92	0.75	-0.58	0.06 +/- 0.08	9.16
		R	—	—	0.50	-1.17	0.06 +/- 0.08	—
2006	ES 330	F	0.75	③	0.75	-0.72	0 +/- 0.08	11.45
		R	—	—	0.75	-1.38	0.16 +/- 0.08	—
	GS 300 ①	F	0.75	④	0.75	⑤	0 +/- 0.04	⑥
		R	—	—	0.75	⑦	0.12 +/- 0.08	—
	GS 300 ②	F	0.75	+7.43	0.75	-0.43	0 +/- 0.04	9.43
		R	—	—	0.75	⑦	0.12 +/- 0.08	—
	GS 430	F	0.75	+7.38	0.75	-0.38	0 +/- 0.04	9.38
		R	—	—	0.75	-1.32	0.12 +/- 0.08	—
	IS 250 2WD	F	0.75	⑧	0.75	-0.38	0.04 +/- 0.08	10.68
		R	—	—	0.75	-1.23	0.12 +/- 0.08	—
	IS 250 AWD	F	0.75	+4.63	0.75	-0.42	0.04 +/- 0.08	11.27
		R	—	—	0.75	-0.83	0.12 +/- 0.08	—
	IS 350	F	0.75	⑧	0.75	-0.38	0.04 +/- 0.08	10.68
		R	—	—	0.75	-1.23	0.12 +/- 0.08	—
	LS 430 ①	F	0.75	+6.75	0.75	-0.08	0.04 +/- 0.08	9.00
		R	—	—	0.75	-1.25	0.12 +/- 0.08	—
	LS 430 ②	F	0.75	+7.25	0.75	-0.25	0.04 +/- 0.08	9.25
		R	—	—	0.75	-1.58	0.12 +/- 0.08	—
	SC 430	F	0.75	+7.92	0.75	-0.58	0.06 +/- 0.08	9.16
		R	—	—	0.50	-1.17	0.06 +/- 0.08	—

① Except air suspension
② With air suspension
③ Tire size 215/60R16: +2.77
   Tire size 215/55R17: +2.78
④ 2WD: +7.38
   AWD: +4.88
⑤ 2WD: -0.38
   AWD: -0.42

⑥ 2WD: 9.38
   AWD: 11.18
⑦ 2WD: -1.17
   AWD: -1.05
⑧ 16 inch wheels: +8.12
   17 inch wheels: +8.18
   18 inch wheels: +8.07

TIRE, WHEEL AND BALL JOINT SPECIFICATIONS

| Year | Model | OEM Tires | | Tire Pressures (psi) | | Wheel Size | Ball Joint Inspection |
		Standard	Optional	Front	Rear		
2002	GS 300/ GS 430	225/55R16 94V	235/45ZR17	Std: 32 Opt: 33	Std: 32 Opt: 33	Std: 7.5-JJ Opt: 8-JJ	U: 9-30 in. ①
	LS 430	P225/60R16 97H	P225/55R17 95H 225/55R17 97W	Std: 30 Opt: 32 Opt: 35	Std: 30 Opt: 32 Opt: 35	Std: 7-JJ Opt: 7.5-JJ	U: 9-30 in. ①
	SC 430	245/40ZR18	None	33	33	8-JJ	U: 9-30 in. ①
	IS 300	215/45ZR17	P205/55R16 89V	33	33	Std: 7-JJ Opt: 6.5-JJ	U: 9-30 in. ①
	ES 300	P215/60R16 94V	None	29	29	6.5-JJ	L: 9-30 in. ①
2003	GS 300/ GS 430	225/55R16 94V	235/45ZR17	Std: 32 Opt: 33	Std: 32 Opt: 33	Std: 7.5-JJ Opt: 8-JJ	U: 9-30 in. ①
	LS 430	P225/55R17 95H	225/55R17 97W	Std: 32 Opt: 35	Std: 32 Opt: 35	7.5-JJ	U: 9-30 in. ①
	SC 430	245/40ZR18	245/40ZR18 93Y	33	33	8-JJ	U: 9-30 in. ①
	IS 300	215/45ZR17 {F} 225/45ZR17 {R}	215/45ZR17 P205/55R16 89V	Std: 33 Opt: 33	Std: 35 Opt: 33	Std: 7-JJ {F} Std: 7.5-JJ {R} Opt: 7-JJ Opt: 6.5-JJ	U: 9-30 in. ①
	ES 300	P215/60R16 94V	None	29	29	6.5-JJ	L: 9-30 in. ①
2004	GS 300/ GS 430	225/55R16 94V	235/45ZR17	Std: 32 Opt: 33	Std: 32 Opt: 33	Std: 7.5-JJ Opt: 8-JJ	U: 9-30 in. ①
	LS 430	P225/55R17 95H	245/45R18 96W	Std: 32 Opt: 33	Std: 32 Opt: 33	7.5-JJ	L: 4.5-31 in. ①
	SC 430	245/40ZR18	245/40ZR18 93Y	33	33	8-JJ	U: 9-30 in. ①
	IS 300	215/45ZR17 {F} 225/45ZR17 {R}	215/45ZR17 P205/55R16 89V	Std: 33 Opt: 33	Std: 35 Opt: 33	Std: 7-JJ {F} Std: 7.5-JJ {R} Opt: 7-JJ Opt: 6.5-JJ	U: 9-30 in. ①
	ES 330	P215/60R16 94V	None	29	29	6.5-JJ	L: 9-30 in. ①
2005	GS 300/ GS 430	225/55R16 94V	235/45ZR17	Std: 32 Opt: 33	Std: 32 Opt: 33	Std: 7.5-JJ Opt: 8-JJ	U: 9-30 in. ①
	LS 430	P225/55R17 95H	245/45R18 96W P245/45R18 96V	Std: 32 Opt: 33	Std: 32 Opt: 33	7.5-JJ	L: 4.5-31 in. ①
	SC 430	P245/40R18 93V	245/40ZR18 245/40ZR18 93Y	33	33	8-JJ	U: 9-30 in. ①
	IS 300	215/45ZR17 {F} 225/45ZR17 {R}	215/45ZR17 P205/55R16 89V	Std: 33 Opt: 33	Std: 35 Opt: 33	Std: 7-JJ {F} Std: 7.5-JJ {R} Opt: 7-JJ Opt: 6.5-JJ	U: 9-30 in. ①
	ES 330	P215/60R16 94V	None	29	29	6.5-JJ	L: 9-30 in. ①

TIRE, WHEEL AND BALL JOINT SPECIFICATIONS

Year	Model	OEM Tires		Tire Pressures (psi)		Wheel Size	Ball Joint Inspection
		Standard	Optional	Front	Rear		
2006	GS 300/ GS 430	245/40R18 93Y	P245/40R18 93V	33	33	8-JJ	L: 4.5 in. ①
	LS 430	P225/55R17 95H	245/45R18 96W P245/45R18 96V	Std: 32 Opt: 33	Std: 32 Opt: 33	7.5-JJ	L: 4.5-31 in. ①
	SC 430	245/40R18 93Y	P245/40R18 93V 245/40ZR18	33	33	8-JJ	U: 9-30 in. ①
	IS 250/ IS 350	205/55R16 89W	225/45R17 90W 245/45R17 95W 225/45R17 91V 225/45R17 95V 245/45R17 95V 225/40R18 88Y {F} 255/40R18 95Y {R}	35	38	Std: 7-JJ Opt: 8-JJ Opt: 8-J {F} Opt: 8.5-J {R}	L: 4.5-53 in. ①
	ES 330	P215/60R16 94V	None	29	29	6.5-JJ	L: 9-30 in. ①

OEM: Original Equipment Manufacturer

PSI: Pounds Per Square Inch

STD: Standard

OPT: Optional

L: Lower

U: Upper

① Torque required in inch lbs. to rotate ball joint when removed from the knuckle

09490-LEXU-C0012

BRAKE SPECIFICATIONS
All measurements in inches unless noted

| Year | Model | Front Brake Disc | | | Rear Brake Disc | | | Minimum Lining Thickness | Brake Caliper | |
		Original Thickness	Minimum Thickness	Maximum Run-out	Original Thickness	Minimum Thickness	Maximum Run-out		Bracket Bolts (ft. lbs.)	Mounting Bolts (ft. lbs.)
2002	ES 300	1.102	1.024	0.0020	0.472	0.413	0.0059	0.0390	①	②
	GS 300	1.260	1.181	0.0020	0.472	0.413	0.0020	0.0390	87	③
	GS 430	1.260	1.181	0.0020	0.472	0.413	0.0020	0.0390	87	③
	IS 300	1.260	1.181	0.0020	0.472	0.413	0.0020	0.0390	87	③
	LS 430	1.181	1.102	0.0020	0.630	0.571	0.0020	0.0390	—	④
	SC 430	1.260	1.181	0.0020	0.472	0.413	0.0020	0.0390	87	③
2003	ES 300	1.102	1.024	0.0020	0.472	0.413	0.0059	0.0390	①	②
	GS 300	1.260	1.181	0.0020	0.472	0.413	0.0020	0.0390	87	③
	GS 430	1.260	1.181	0.0020	0.472	0.413	0.0020	0.0390	87	③
	IS 300	1.260	1.181	0.0020	0.472	0.413	0.0020	0.0390	87	③
	LS 430	1.181	1.102	0.0020	0.630	0.571	0.0020	0.0390	—	④
	SC 430	1.260	1.181	0.0020	0.472	0.413	0.0020	0.0390	87	③
2004	ES 330	1.102	1.024	0.0020	0.472	0.413	0.0059	0.0390	79	⑤
	GS 300	1.260	1.181	0.0020	0.472	0.413	0.0020	0.0390	87	③
	GS 430	1.260	1.181	0.0020	0.472	0.413	0.0020	0.0390	87	③
	IS 300	1.260	1.181	0.0020	0.472	0.413	0.0020	0.0390	87	③
	LS 430	1.102	1.024	0.0020	0.394	0.335	0.0059	0.0390	⑥	⑦
	SC 430	1.260	1.181	0.0020	0.472	0.413	0.0020	0.0390	87	③
2005	ES 330	1.102	1.024	0.0020	0.472	0.413	0.0059	0.0390	79	⑤
	GS 300	1.260	1.181	0.0020	0.472	0.413	0.0020	0.0390	87	③
	GS 430	1.260	1.181	0.0020	0.472	0.413	0.0020	0.0390	87	③
	IS 300	1.260	1.181	0.0020	0.472	0.413	0.0020	0.0390	87	③
	LS 430	1.102	1.024	0.0020	0.394	0.335	0.0059	0.0390	⑥	⑦
	SC 430	1.260	1.181	0.0020	0.472	0.413	0.0020	0.0390	87	③
2006	ES 330	1.102	1.024	0.0020	0.472	0.413	0.0059	0.0390	79	⑤
	GS 300	1.260	1.181	0.0020	0.413	0.039	0.0020	⑧	58	⑨
	GS 430	1.181	1.063	0.0020	0.413	0.039	0.0020	⑧	58	⑨
	IS 250	1.102	0.983	0.0020	0.413	0.039	0.0020	0.0390	58	⑩
	IS 350	1.181	1.063	0.0020	0.709	0.650	0.0020	0.0390	—	⑩
	LS 430	1.102	1.024	0.0020	0.394	0.335	0.0059	0.0390	⑥	⑦
	SC 430	1.260	1.181	0.0020	0.472	0.413	0.0020	0.0390	87	③

① Front: 79 ft. lbs.
   Rear: 34 ft. lbs.

② Front: 25 ft. lbs.
   Rear: 14 ft. lbs.

③ Front: 25 ft. lbs.
   Rear: 77 ft. lbs.

④ Front: 81 ft. lbs.
   Rear: 58 ft. lbs.

⑤ Front: 25 ft. lbs.
   Rear: 46 ft. lbs.

⑥ Front: 77 ft. lbs.
   Rear: 58 ft. lbs.

⑦ Front: 25 ft. lbs.
   Rear: 32 ft. lbs.

⑧ Front: 0.039 inches
   Rear: 0.690 inches

⑨ Front: 25 ft. lbs.
   Rear: 40 ft. lbs.

⑩ Front: 58 ft. lbs.
   Rear: 40 ft. lbs.

09490-LEXU-C0013

SCHEDULED MAINTENANCE INTERVALS
Lexus—ES300, ES330, IS250, IS300, IS350, GS300, GS430, SC430 & LS430

TO BE SERVICED	TYPE OF SERVICE	VEHICLE MILEAGE INTERVAL (x1000)												
		7.5	15	22.5	30	37.5	45	52.5	60	67.5	75	82.5	90	97.5
Engine oil & filter	R	✓	✓	✓	✓	✓	✓	✓	✓	✓	✓	✓	✓	✓
A/C filter (if equipped) ①	S/I	✓	✓	✓	✓	✓	✓	✓	✓	✓	✓	✓	✓	✓
Automatic transaxle fluid & filter	S/I		✓		✓		✓		✓		✓		✓		
Ball joints & dust covers	S/I		✓		✓		✓		✓		✓		✓		
Bolts & nuts on chassis & body	S/I		✓		✓		✓		✓		✓		✓		
Brake fluid ②	S/I		✓		✓		✓		✓		✓		✓		
Brake line pipes & hoses	S/I		✓		✓		✓		✓		✓		✓		
Brake linings & drums	S/I		✓		✓		✓		✓		✓		✓		
Brake pads & discs (front & rear)	S/I		✓		✓		✓		✓		✓		✓		
Differential oil	S/I		✓		✓		✓		✓		✓		✓		
Driveshaft boots (if equipped)	S/I		✓		✓		✓		✓		✓		✓		
Steering gear housing oil	S/I		✓		✓		✓		✓		✓		✓		
Steering linkage	S/I		✓		✓		✓		✓		✓		✓		
Air filter	R				✓				✓				✓		
Exhaust pipes & mountings	S/I				✓				✓				✓		
Fuel lines & connections	S/I				✓				✓				✓		
Engine coolant	R						✓						✓		
Fuel tank cap gasket	R								✓						
Spark plugs	R								✓						
Charcoal canister	S/I								✓						
Drive belts	S/I								✓						
Valve clearance	S/I								✓						

R: Replace   S/I: Service or Inspect

① Replace at 15,000 miles.

② Replace at 30,000 miles (unless previously replaced).

FREQUENT OPERATION MAINTENANCE (SEVERE SERVICE)

If a vehicle is operated under any of the following conditions it is considered severe service

- **Extremely dusty areas.**
- **50% or more of the vehicle operation is in 32°C (90°F) or higher temperatures, or constant operation in temperatures below 0°C (32°F).**
- **Prolonged idling (vehicle operation in stop and go traffic).**
- **Frequent short running periods (engine does not warm to normal operating temperatures).**
- **Police, taxi, delivery usage or trailer towing usage.**

Oil & oil filter: change every 3750 miles.

Ball joints & dust covers: service or inspect every 7500 miles.

Bolts & nuts on chassis & body: service or inspect every 7500 miles.

Brake linings & drums: service or inspect every 7500 miles.

Brake pads & discs (front & rear): service or inspect every 7500 miles.

Driveshaft boots (if equipped): service or inspect every 7500 miles.

Brake linings & drums: service or inspect every 7500 miles.

Steering linkage: service or inspect every 7500 miles.

Air filter: service or inspect every 15,000 miles.

Automatic transmission fluid & filter: replace every 15,000 miles.

Differential oil: replace every 15,000 miles.

Exhaust pipes & mountings: service or inspect every 15,000 miles.

Drive belts: service or inspect at 60,000 miles & every 7500 miles thereafter.

Timing belts: replace every 60,000 miles.

ENGINE REPAIR

Alternator

REMOVAL

2.5L (4GR-FSE) and 3.5L (2GR-FSE) Engine

1. Before servicing the vehicle, refer to the Precautions Section.
2. Remove or disconnect the following:
   - Negative battery cable. Wait at least 90 seconds before performing any other work.
   - Cool air intake duct seal
   - Left engine room side cover
   - V-bank cover sub-assembly
   - No. 1 inlet air cleaner
   - Serpentine drive belt
   - Engine coolant
   - Radiator inlet hose
   - No. 2 engine cover
   - No. 2 idler pulley sub-assembly
   - Wiring harness
   - Terminal cap
   - Alternator wiring harnesses
   - Bolt and clamp bracket
   - Alternator connector, and 2 clamps
   - Nut and alternator bracket at the engine
   - 2 mounting bolts and the alternator
   - 2 bolts and 2 alternator mounting brackets

3.0L (3GR-FSE) Engine

1. Before servicing the vehicle, refer to the Precautions Section.
2. Remove or disconnect the following:
   - Negative battery cable. Wait at least 90 seconds before performing any other work.
   - V-bank cover sub-assembly
   - Cool air intake duct seal
   - Left engine room side cover
   - Engine under cover
   - Serpentine drive belt
3. Remove or disconnect the with pulley compressor by performing the following:
   - Bolt, nut and bracket (AWD)
   - Magnetic clutch connector
   - Nut and 3 bolts (2WD); or 2 bolts (AWD)
   - Stud bolt using an E8 "torx" socket and with pulley compressor

➡**It is not necessary to completely remove the compressor. With the hoses connected to the compressor, hang the compressor on the vehicle body with a rope.**

- No. 2 idler pulley sub-assembly
- Clamp and bolt from alternator
- Alternator connector
- Rubber cap
- Nut and battery cable.
- Nut, bolt and alternator bracket
- 2 mounting bolts and alternator

3.0L (1MZ-FE) Engine

1. Before servicing the vehicle, refer to the Precautions Section.
2. Remove or disconnect the following:
   - Negative battery cable. Wait at least 90 seconds before performing any other work.
   - Accessory drive belt
   - Alternator harness connectors
   - Alternator

3.0L (2JZ-GE) Engine

1. Before servicing the vehicle, refer to the Precautions Section.
2. Remove or disconnect the following:
   - Negative battery cable. Wait at least 90 seconds before performing any other work.
   - Engine under cover
   - Accessory drive belt
   - Alternator connector
   - Cap and nut
   - Alternator wire
   - Alternator wire clamp from the wire clip on the alternator
   - Bolt and pipe clamp
   - 2 automatic transmission oil cooler pipes from the alternator
   - Bolt, nut, pipe bracket and alternator

4.3L (3UZ-FE) Engine

1. Before servicing the vehicle, refer to the Precautions Section.
2. Remove or disconnect the following, as equipped:
   - Negative battery cable. Wait at least 90 seconds before performing any other work.
   - Air cleaner inlet
   - Accessory drive belt
   - Oil pan protector
   - Engine under cover
   - Power steering pump pulley
   - Alternator harness connectors
   - Power steering oil hose from oil pan
   - Heated Oxygen (HO_2S) sensor wiring
   - Alternator

3.3L (3MZ-FE) Engine

1. Before servicing the vehicle, refer to the Precautions Section.
2. Remove or disconnect the following:
   - Negative battery cable. Wait at least 90 seconds before performing any other work.
   - Remove two wire harness clamps.
   - Disconnect the alternator connector.
   - Open the terminal clamp.
   - Remove the nut, then disconnect the alternator wire.
   - Loosen mounting bolts to lessen tension of the alternator belt

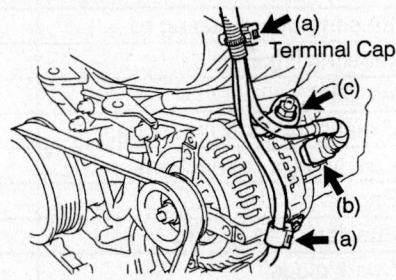

Removing wire harness clamps—3.3L (3MZ-FE)

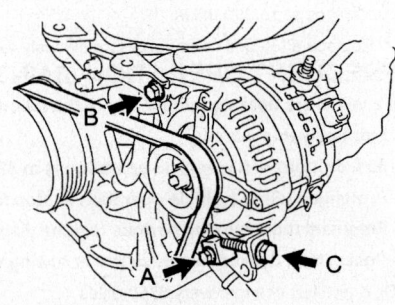

Loosening belt tension bolt—3.3L (3MZ-FE)

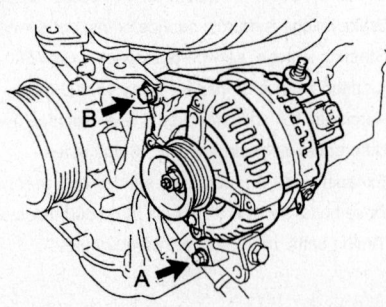

Removing alternator bolts—3.3L (3MZ-FE)

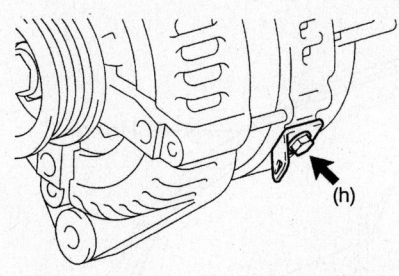

67162-LEXU-G04

Removing wiring harness clamp from alternator—3.3L (3MZ-FE)

- Remove mounting bolts
- Remove the wiring harness clamp bolt.

INSTALLATION

2.5L (4GR-FSE) and 3.5L (2GR-FSE) Engine

1. Install or connect the following:
   - 2 alternator mounting brackets and 2 bolts. Torque to 15 ft. lbs. (20 Nm).
   - Alternator and 2 mounting bolts. Torque to 32 ft. lbs. (43 Nm).
   - Alternator bracket and nut at the engine. Torque to 15 ft. lbs. (20 Nm).
   - Alternator wire to terminal. Torque nut to 87 inch lbs. (10 Nm).
   - Terminal cap
   - Clamp bracket and bolt. Torque to 7 ft. lbs. (10 Nm).
   - 2 clamps, and connect alternator connector
   - Wire harness with the 3 nuts. Torque to 7 ft. lbs. (10 Nm).
   - 2 alternator connectors.
   - No. 2 idler pulley sub-assembly
   - No. 2 engine cover
   - Radiator inlet hose
   - Serpentine drive belt
   - No. 1 inlet air cleaner
   - V-bank cover sub-assembly
   - Left engine room side cover
   - Cool air intake duct seal
   - Negative battery cable
2. Add engine coolant.

3.0L (3GR-FSE) Engine

1. Install or connect the following:
   - Alternator and 2 mounting bolts. Torque to 32 ft. lbs. (43 Nm).
   - Alternator bracket with bolt and nut at the engine. Torque to 15 ft. lbs. (20 Nm).

- Battery cable. Tighten the nut to 87 inch lbs. (10 Nm).
- Wire harness bracket and clamp to the alternator
- Alternator wire to terminal.
- Terminal cap
- No. 2 idler pulley and cover plate with the bolt. Tighten bolt to 32 ft. lbs. (43 Nm).
- With pulley compressor assembly
- Serpentine drive belt
- Engine under cover
- Left engine room side cover
- Cool air intake duct seal
- V-bank cover sub-assembly
- Negative battery cable

3.0L (1MZ-FE) Engine

1. Install or connect the following:
   - Alternator
   - Alternator harness connectors
   - Accessory drive belt. Tighten the adjusting lock bolt to 13 ft. lbs. (18 Nm) and the pivot bolt to 41 ft. lbs. (56 Nm).
   - Negative battery cable

3.0L (2JZ-GE) Engine

1. Install or connect the following:
   - Bolt, nut, pipe bracket and alternator. Tighten the fasteners to 30 ft. lbs. (40 Nm).
   - 2 automatic transmission oil cooler pipes to the alternator
   - Bolt and pipe clamp
   - Alternator wire clamp to the wire clip on the alternator
   - Alternator wire
   - Cap and nut
   - Alternator connector
   - Accessory drive belt
   - Engine under cover
   - Negative battery cable

4.3L (3UZ-FE) Engine—2002–03

1. Install or connect the following, as equipped:
   - Alternator. Tighten the fasteners to 29 ft. lbs. (39 Nm).
   - HO2S sensor wiring
   - Power steering pump oil hose to oil pan
   - Alternator harness connectors
   - Power steering pump pulley
   - Engine under cover
   - Oil pan protector
   - Accessory drive belt
   - Air cleaner inlet
   - Negative battery cable

4.3L (3UZ-FE) Engine—2004–06

- The 2 clamps from cord clips on alternator
- The nut, bolt, stay and alternator, torque to 29 ft lbs. (39Nm)
- PS oil hose to #1 oil pan bolt
- The alternator wire connector from the clamp on cord clip
- Alternator wire from the clamp on the cord clip
- The rubber cap and nut and alternator wire
- The alternator connector
- PS pulley or pump, torque to 32 ft. lbs (43Nm)
- Air cleaner inlet
- Negative battery cable
- Drive belt
- Engine undercover (if necessary)

3.3L (3MZ-FE) Engine

1. Install or connect the following:
   - Wiring harness clamp bracket with bolt
   - Temporarily install mounting bolts
   - Adjust belt with tensioner
   - Tighten mounting bolts to lower mounting bolt 13ft.lbs. (18 Nm) upper 43 ft. lbs.(58 Nm)
   - Alternator wire nut, tighten to 7 ft. lbs. (9.8 Nm)
   - Alternator wire connector
   - Wiring harness clamp
   - Battery negative cable

Ignition Timing

ADJUSTMENT

The engines covered in this section are equipped with a Distributorless Ignition System (DIS). No timing adjustments are possible.

Engine Assembly

REMOVAL & INSTALLATION

ES 300

1. Before servicing the vehicle, refer to the Precautions Section.
2. Release the fuel pressure.
3. Drain the engine coolant and engine oil.
4. Remove or disconnect the following:

   - Battery and tray
   - Hood

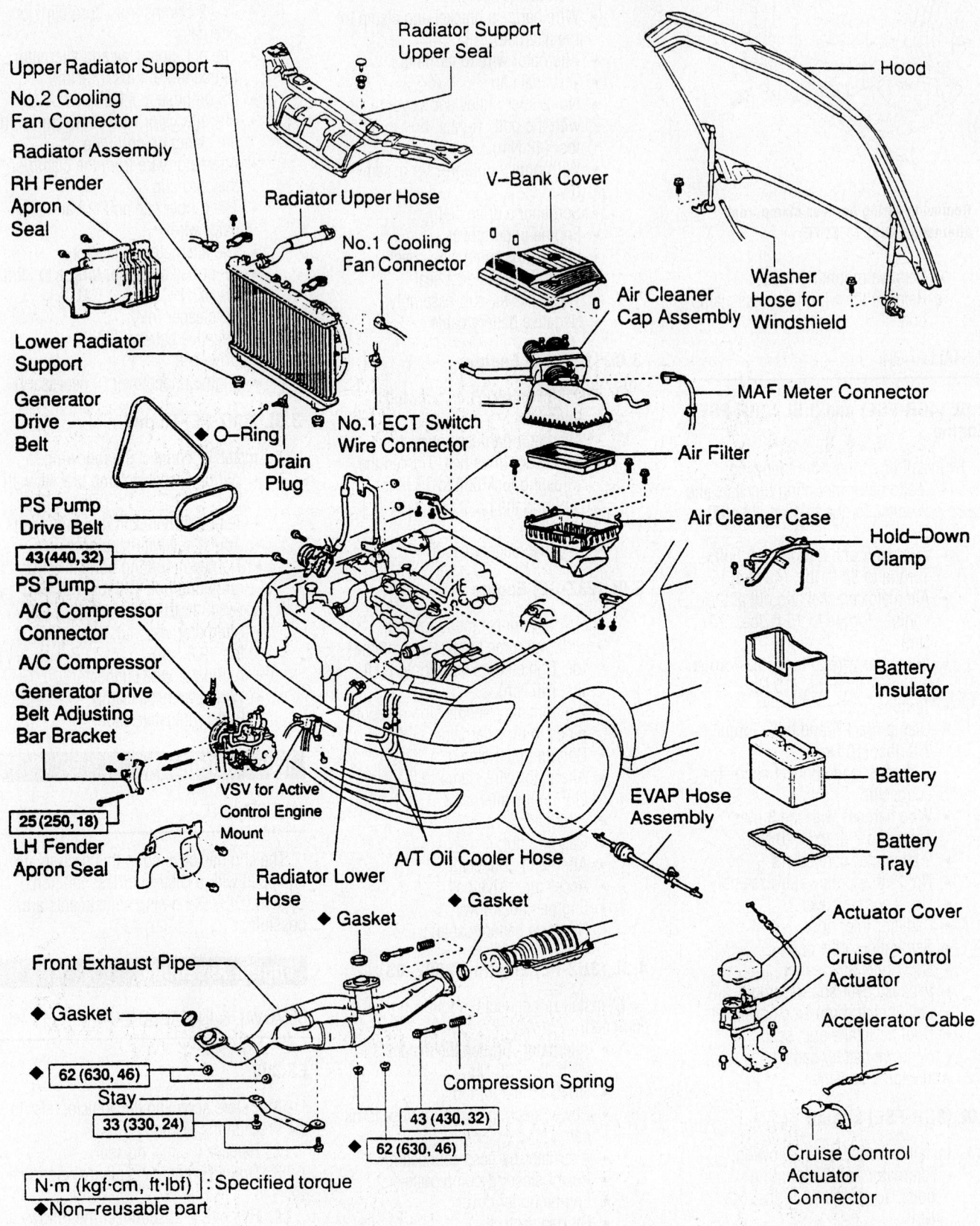

Upper Radiator Support

No.2 Cooling Fan Connector

Radiator Assembly

RH Fender Apron Seal

Radiator Support Upper Seal

Radiator Upper Hose

No.1 Cooling Fan Connector

V–Bank Cover

Hood

Air Cleaner Cap Assembly

Washer Hose for Windshield

MAF Meter Connector

Lower Radiator Support

Generator Drive Belt

◆ O–Ring

Drain Plug

No.1 ECT Switch Wire Connector

Air Filter

PS Pump Drive Belt

Air Cleaner Case

Hold–Down Clamp

43 (440, 32)

PS Pump

A/C Compressor Connector

A/C Compressor

Generator Drive Belt Adjusting Bar Bracket

Battery Insulator

Battery

25 (250, 18)

LH Fender Apron Seal

VSV for Active Control Engine Mount

A/T Oil Cooler Hose

Radiator Lower Hose

◆ Gasket

◆ Gasket

EVAP Hose Assembly

Battery Tray

Actuator Cover

Cruise Control Actuator

Accelerator Cable

Front Exhaust Pipe

◆ Gasket

62 (630, 46)

Stay

33 (330, 24)

Compression Spring

43 (430, 32)

62 (630, 46)

Cruise Control Actuator Connector

N·m (kgf·cm, ft·lbf) : Specified torque
◆ Non–reusable part

9301LG01

Exploded view of the engine removal and related components—ES 300

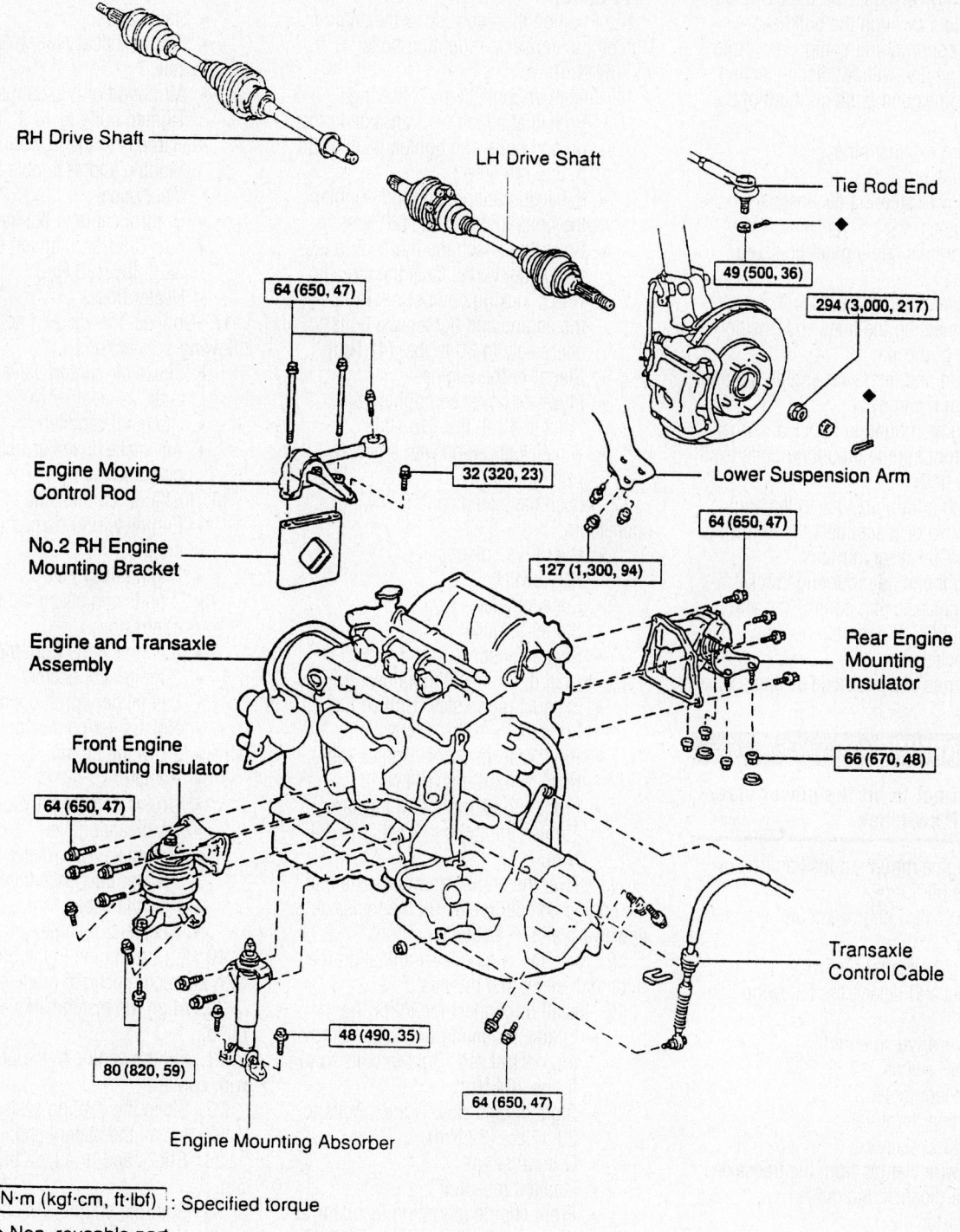

RH Drive Shaft

LH Drive Shaft

Tie Rod End

64 (650, 47)

49 (500, 36)

294 (3,000, 217)

Engine Moving
Control Rod

32 (320, 23)

Lower Suspension Arm

No.2 RH Engine
Mounting Bracket

64 (650, 47)

Engine and Transaxle
Assembly

127 (1,300, 94)

Rear Engine
Mounting
Insulator

Front Engine
Mounting Insulator

66 (670, 48)

64 (650, 47)

Transaxle
Control Cable

48 (490, 35)

80 (820, 59)

64 (650, 47)

Engine Mounting Absorber

N·m (kgf·cm, ft·lbf) : Specified torque

◆ Non–reusable part

9301LG02

Exploded view of the engine removal and related components (cont.)—ES 300

- Accelerator cable and the throttle cable
- Air cleaner cover,
- Volume air flow meter and air cleaner duct as an assembly
- Cruise control actuator, if equipped
- Radiator
- Engine relay box and 2 bolts

- 5 connections from the relay box
- 2 igniter connectors
- Left fender apron connector
- Noise filter connector
- 2 ground straps
- Engine wiring harness from the engine.
- Vacuum hoses from the following

connections: intake air control valve vacuum tank, charcoal canister, brake booster vacuum hose from the intake chamber

- 2 heater hoses from the bulk-head
- Fuel feed and return lines
- Control cable from the transaxle

- Wiring harness from the PCM and route it through the bulkhead.
- Air conditioning compressor from the engine without disconnecting the lines and position it out of the way
- Front exhaust pipe
- Halfshafts
- 2 power steering air hoses from the engine
- Hydraulic cooling fan pressure hose
- Power steering pump without disconnecting the lines and position it out of the way
- Right and left lower engine mounts from the body
- Engine mounting shock absorber
- 3 front engine mounting bolts from the body

5. Attach a lifting device to the engine.
6. Remove or disconnect the following:
   - Coolant reservoir tank.
   - Right engine mounting bracket
   - Engine moving control rod and right No. 2 engine mounting bracket
   - Engine and transaxle as an assembly

✳✳ WARNING

Be careful not to hit the power steering or PNP switches.

- Engine mounting insulator below the oil filter
- Right rear engine-mounting insulator
- Front exhaust pipe stay

7. Label and detach the following connectors:
   - Overdrive solenoid
   - PNP switch
   - Speedometer
   - Starter terminal
   - Speed sensor.
   - 2 wire clamps from the transaxle
   - Oil dipstick and guide
   - Starter
   - Flywheel housing cover

8. Turn the crankshaft pulley to gain access to the 8 torque converter bolts. Secure the crankshaft and remove them as they become accessible.
9. Install or connect the following:
   - 2 exhaust manifold stays and plate
   - 2 bolts attaching the transaxle to the oil pan
   - 6 transaxle mounting bolts
   - Transaxle

To install:

10. Position the transaxle to the engine. Tighten the transaxle mounting bolts: 47 ft. lbs. (64 Nm).
11. Install or connect the following:
    - Bolts that attach the transaxle to the oil pan bolts and tighten them to 34 ft. lbs. (46 Nm).
    - Exhaust manifold support. Tighten the bolts to 14 ft. lbs. (20 Nm).
    - Bolts that attach the flywheel to the torque converter. Coat the threads with a locking compound. Rotate the engine and tighten the bolts alternately to 30 ft. lbs. (41 Nm).
    - Starter to the engine
    - Flywheel cover and tighten the bolts to 13 ft. lbs. (18 Nm)
    - Dipstick and tube with a new O-ring

12. Attach the clamps and following connectors:
    - Overdrive solenoid
    - PNP switch
    - Speedometer
    - Starter terminal
    - Speed sensor

13. Install or connect the following:
    - Exhaust pipe's stay. Tighten the bolts to 15 ft. lbs. (21 Nm).
    - Right rear insulator. Tighten the bolts to 47 ft. lbs. (64 Nm).
    - Front engine mounting insulator. Tighten the bolts to 47 ft. lbs. (64 Nm).

14. Lower the engine and transaxle into the engine compartment. Tilt the transaxle downward and clear the left mount.
15. Keep the engine level and align the right and left engine mounts.
16. Install or connect the following:
    - Engine mounting bracket and moving control rod. Tighten bolts to 47 ft. lbs. (64 Nm).
    - Right engine stay. Tighten bolts to 23 ft. lbs. (32 Nm).
    - Ground straps
    - Coolant reservoir
    - Front engine mounting insulator to the body. Tighten bolts to 59 ft. lbs. (81 Nm).
    - Engine mounting shock absorber. Tighten bolts to 35 ft. lbs. (48 Nm).
    - Right engine mount. Tighten bolts to 48 ft. lbs. (66 Nm).
    - Left engine mount. Tighten bolts to 47 ft. lbs. (64 Nm).
    - Engine lifting device
    - Power steering pump and belt
    - Hydraulic cooling fan pressure hose. Tighten the fitting to 33 ft. lbs. (44 Nm).

- Power steering air tube and hoses
- Halfshafts
- Front exhaust pipe with new gaskets
- Air conditioning compressor. Tighten bolts to 18 ft. lbs. (25 Nm).
- Harness to the Powertrain Control Module and assemble the instrument panel
- Control cable to the transaxle
- Fuel lines and tighten the fittings to 22 ft. lbs. (30 Nm)
- Heater hoses

17. Connect the vacuum hoses to the following connections:
    - Intake air control valve vacuum tank
    - Charcoal canister
    - Air intake chamber from the brake booster

18. Install or connect the following:
    - Engine wiring harness to the engine
    - Engine relay box
    - 2 bolts and attach the following connectors:
    - 5 connections from the relay box
    - 2 igniter connectors
    - Left fender apron connector
    - Noise filter connector
    - 2 ground straps
    - Radiator
    - Cruise control actuator, if equipped
    - Air cleaner cover
    - Volume airflow meter and air cleaner duct assembly
    - Throttle cable
    - Accelerator cable

19. Fill the engine to the proper level with the recommended grade of oil.
20. Align the matchmarks and install the hood.
21. Fill the engine to the proper level with coolant.
22. Bleed the cooling system.
23. Install the battery and tray.
24. Check and/or adjust the ignition timing.
25. Start the engine and check for leaks.
26. Road test the vehicle.
27. Recheck the engine oil and coolant levels.

GS 300 (2002–05)

1. Before servicing the vehicle, refer to the Precautions Section.
2. Release the fuel pressure.
3. Drain the fuel from the tank.
4. Remove or disconnect the following:

   - Negative battery cable. Wait at least

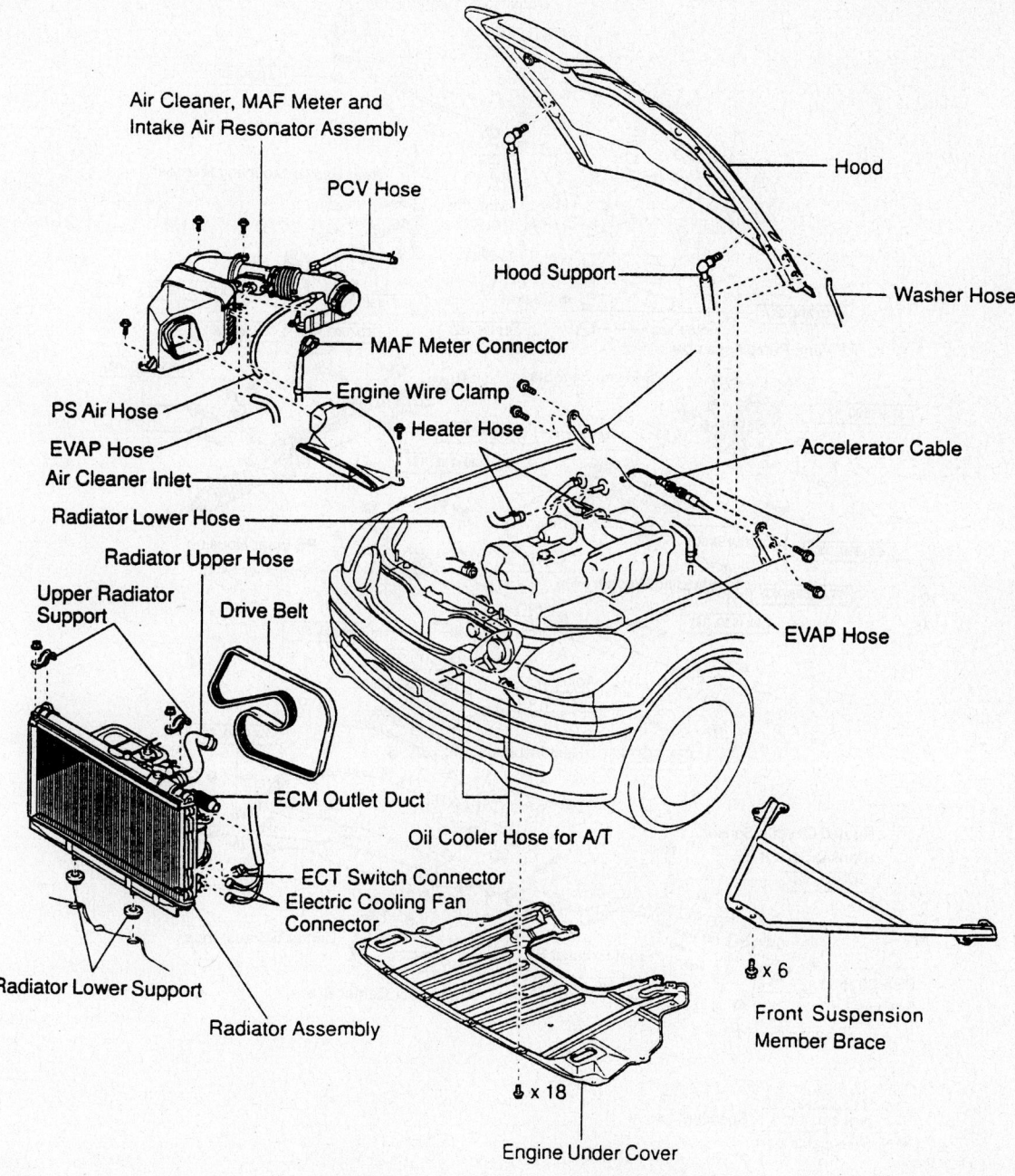

Exploded view of the related engine compartment components for 2JZ-GE engine removal—GS 300

90 seconds before performing any other work.
- Hood insulator pad and the hood
- Engine undercover, then drain the engine coolant and oil
- Front suspension member brace, if equipped
- Engine cover, if equipped
- Accelerator cable, cruise control actuator cable and the automatic transmission throttle control cable from the throttle body
- Air cleaner assembly

- Volume air flow meter (or MAF meter) and the air intake hose
- Air cleaner duct
- Accelerator cable from engine
- Drive belt
- Radiator

5. Label and detach the following wires and electrical connectors, as equipped:
- Ground strap from floor
- Fuel inlet hose
- Igniter
- Ground strap from dash panel
- Oxygen sensor connectors

- Ignition coil
- Wiring harness from the wire clamp and coolant tank
- Alternator wiring
- Ground strap from engine block and from left engine mount
- Starter
- DLC1 connector

6. Remove or disconnect the following:
- Fuel lines from the intake and return lines
- Power steering pump without dis-

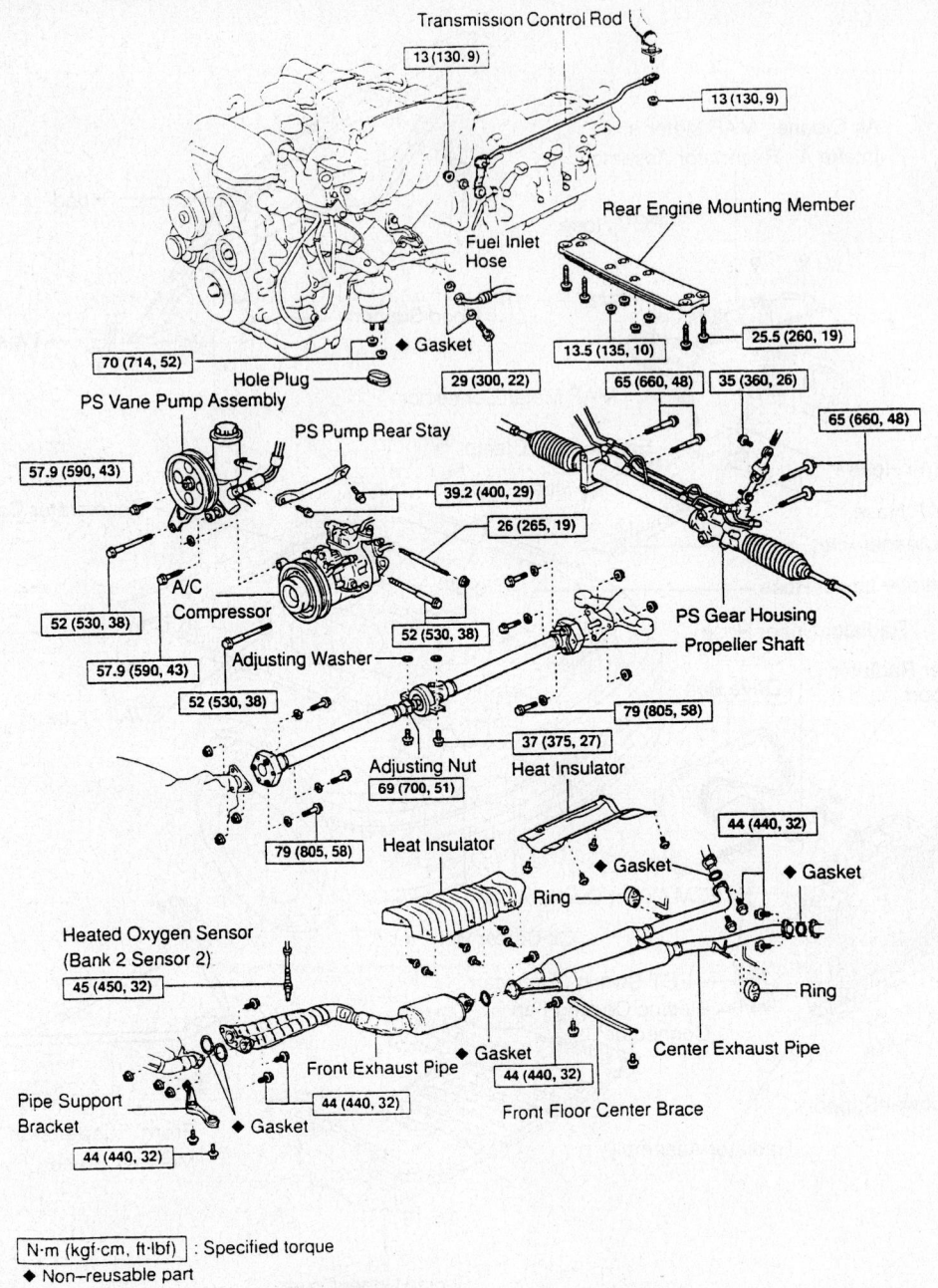

Transmission Control Rod

13 (130. 9)

13 (130, 9)

Rear Engine Mounting Member

Fuel Inlet Hose

◆ Gasket

70 (714, 52)

Hole Plug

PS Vane Pump Assembly

13.5 (135, 10)

25.5 (260, 19)

29 (300, 22)

65 (660, 48)

35 (360, 26)

65 (660, 48)

PS Pump Rear Stay

57.9 (590, 43)

39.2 (400, 29)

26 (265, 19)

A/C Compressor

52 (530, 38)

57.9 (590, 43)

Adjusting Washer

52 (530, 38)

52 (530, 38)

PS Gear Housing

Propeller Shaft

79 (805, 58)

Adjusting Nut

69 (700, 51)

37 (375, 27)

Heat Insulator

44 (440, 32)

79 (805, 58)

Heat Insulator

◆ Gasket

Ring

◆ Gasket

Heated Oxygen Sensor
(Bank 2 Sensor 2)

45 (450, 32)

Ring

Center Exhaust Pipe

◆ Gasket

Front Exhaust Pipe

44 (440, 32)

44 (440, 32)

Front Floor Center Brace

Pipe Support Bracket

◆ Gasket

44 (440, 32)

N·m (kgf·cm, ft·lbf) : Specified torque
◆ Non–reusable part

9301LG04

Exploded view of the related underbody components for 2JZ-GE engine removal (cont.)—GS 300

connecting the lines and position it aside

• Air conditioning compressor without disconnecting the air conditioning lines and position it aside
• Brake booster vacuum hose
• Evaporative Emissions (EVAP) hose
• Heater hoses from the firewall
• Heater valve and engine wire from the firewall
• Electrical harness from the PCM and route it through the firewall
• Connectors from ECM box

• Oxygen (O_2) sensor (if equipped) from the front exhaust pipe
• Front exhaust pipes and heat insulator

7. Remove the rear center floor crossmember brace.

8. Disconnect the power steering pump from it mounting and disconnect the A/C compressor from its mounting, without disconnecting hoses from either component. Position these out of the way.

9. Remove the transmission shift control rod.

10. Remove the driveshaft.

11. Disconnect the power steering gear housing from the steering column, then remove the gear housing assembly from the mounting brackets and suspend it securely.

12. Support the transmission with a jack. Use a piece of wood to prevent damage to the transmission oil pan.

13. Attach a lifting device to the engine.

14. Remove or disconnect the following:
• Rear transmission crossmember
• 2 hole plugs in the front crossmember
• 2 nuts holding the engine insulators to the front crossmember

15. Slowly and carefully remove the engine and transmission from the engine compartment as an assembly.

To install:

16. Position the engine assembly into the vehicle.

17. Install or connect the following:
- Stud bolts for the front engine mount into their bores in the front engine crossmember. Temporarily install the 2 nuts.

18. Remove the engine hoist

19. Install or connect the following:
- Temporarily, the rear engine support with the 4 nuts
- The 4 support bolts and tighten them to 19 ft. lbs. (25 Nm). Tighten the nuts to 10 ft. lbs. (13 Nm).
- Front engine crossmember to mount nuts to 54 ft. lbs. (74 Nm) and the hole plugs
- Driveshaft

20. Shift the transmission control shift rod into **N** (neutral) by shifting the lever all the way back and returning it 2 notches. Connect the shift rod to the lever and tighten it to 108 inch lbs. (13 Nm).

21. Install the power steering gear to its mountings. Torque the bolts to 48 ft. lbs. (65 Nm). Install and tighten the gear-to-steering column yoke bolt to 26 ft. lbs. (35 Nm). Connect the power steering switch connector.

22. Install or connect the following:
- Transmission control rod; torque nuts to 9 ft. lbs. (13 Nm)
- Rear center floor crossmember brace and tighten the bolts to 108 inch lbs. (13 Nm)
- Exhaust pipe heat insulator
- Front exhaust pipes
- Oxygen sensor connectors, if equipped
- Engine wiring harness to the PCM
- PCM and its cover
- The lower portion of the passenger side instrument panel, the vent, the carpet and the scuff panel
- Heater water valve and engine wire to the cowl panel
- Heater hoses
- EVAP hose
- Brake booster hose
- The air conditioning compressor and connectors; tighten the Torx® bolt to 19 ft. lbs. (26 Nm). Tighten the nut and bolts to 38 ft. lbs. (52 Nm).
- Power steering pump and connector; tighten the upper and lower P/S bracket bolts to 43 ft. lbs. (58 Nm)

and the side bracket bolt to 38 ft. lbs. (52 Nm)
- Power steering pump rear stay; torque bolts to 29 ft. lbs. (39 Nm)
- Wiring to EMC box
- Fuel lines with new gaskets and tighten the union bolts to 22 ft. lbs. (29 Nm)
- Igniter
- Ignition coil
- Wiring harness from the wire clamp and coolant tank
- Alternator and ground strap from the left engine mount
- Starter
- Radiator
- Drive belt
- Air cleaner
- Volume air flow meter and intake air connector pipe as an assembly
- Air cleaner duct
- Accelerator, cruise control and the automatic transmission throttle control cables
- Fuel
- Engine oil
- Coolant
- Negative battery cable

23. Start the engine and check for leaks.

24. Check the automatic transmission fluid level.

25. Check and/or adjust the ignition timing.

26. If equipped, install the front suspension brace and tighten the mounting bolts to 43 ft. lbs. (58 Nm).

27. Install the engine cover, hood and the hood insulator pad.

28. Refill and check all fluids.

29. Road test the vehicle.

30. Recheck all fluids.

GS 300 (2006)

1. Before servicing the vehicle, refer to the Precautions Section.

2. Release the fuel pressure.

3. Drain the engine coolant and engine oil.

4. Drain transmission fluid.

5. Drain differential fluid (AWD).

6. Remove or disconnect the following:
- Negative battery cable. Wait at least 90 seconds before performing any other work.
- Front wheels
- Hood insulator pad and the hood
- 7 clips and cool air intake duct seal
- Nut, 2 clips and right engine room side cover
- 3 clips and left engine room side cover

- 2 nuts and V-bank cover
- Engine under cover
- 4 screws, 2 grommets, 2 spacers and No. 2 engine under cover (2WD)
- Front exhaust pipe assembly
- Propeller with center bearing shaft assembly
- Front propeller shaft assembly (AWD)
- Bolt and air cleaner inlet
- Ventilation hose from the cylinder head
- Mass AirFlow (MAF) meter connector
- Clamp from the air cleaner
- EVAP VSV
- Hose clamp
- 3 bolts and air cleaner case
- Radiator inlet, outlet and reserve tank hoses
- Clamp and union-to-check valve hose
- Heater water inlet and outlet hoses
- Fuel main tube
- Engine room ECM cover
- ECM connectors and connector holder
- Positive battery cable
- Engine room No. 1 relay block, junction block cover
- Wire from engine room No. 1 relay block, junction block
- Ground cable
- Serpentine drive belt

7. Remove or disconnect the with pulley compressor by performing the following:
- Bolt, nut and bracket (AWD)
- Magnetic clutch connector
- Nut and 3 bolts (2WD); or 2 bolts (AWD)
- Stud bolt using an E8 "torx" socket and with pulley compressor

➡**It is not necessary to completely remove the compressor. With the hoses connected to the compressor, hang the compressor on the vehicle body with a rope.**

- Height control sensor link
- Front suspension member lower protector (2WD)
- Left and right front axle hub nuts (AWD)
- Left and right tie rod assemblies
- Left and right front shock absorbers
- Left and right lower ball joints (AWD)
- No. 2 steering intermediate shaft assembly
- Power steering link wire harness (2WD)

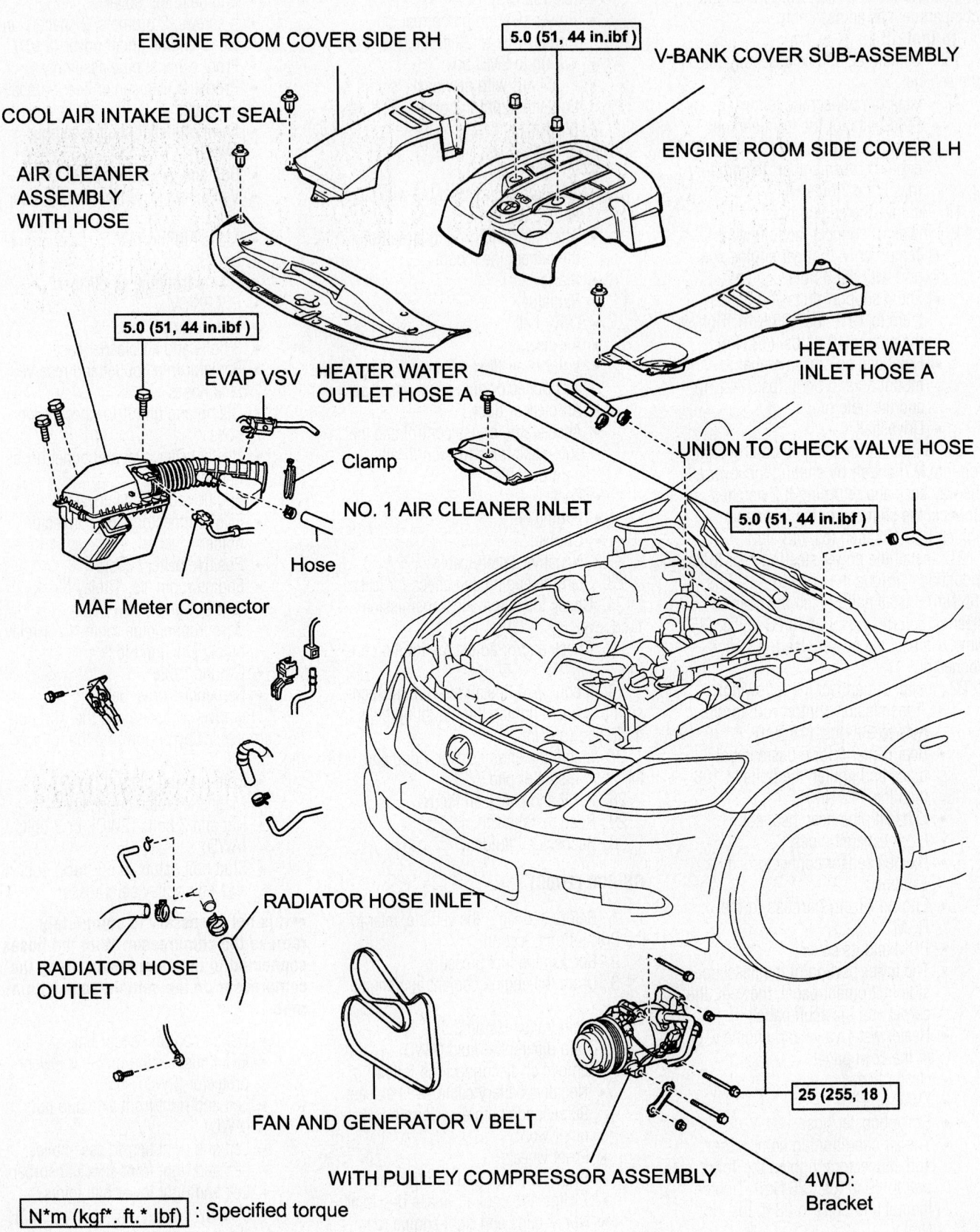

ENGINE ROOM COVER SIDE RH

5.0 (51, 44 in.ibf)

V-BANK COVER SUB-ASSEMBLY

COOL AIR INTAKE DUCT SEAL

ENGINE ROOM SIDE COVER LH

AIR CLEANER ASSEMBLY WITH HOSE

5.0 (51, 44 in.ibf)

EVAP VSV

HEATER WATER OUTLET HOSE A

HEATER WATER INLET HOSE A

Clamp

NO. 1 AIR CLEANER INLET

UNION TO CHECK VALVE HOSE

Hose

5.0 (51, 44 in.ibf)

MAF Meter Connector

RADIATOR HOSE INLET

RADIATOR HOSE OUTLET

FAN AND GENERATOR V BELT

25 (255, 18)

WITH PULLEY COMPRESSOR ASSEMBLY

4WD: Bracket

N*m (kgf*. ft.* lbf) : Specified torque

09490_LEXU_G0007

Exploded view of the related engine compartment components for 3GR-FSE engine removal—2006 GS 300

2WD:

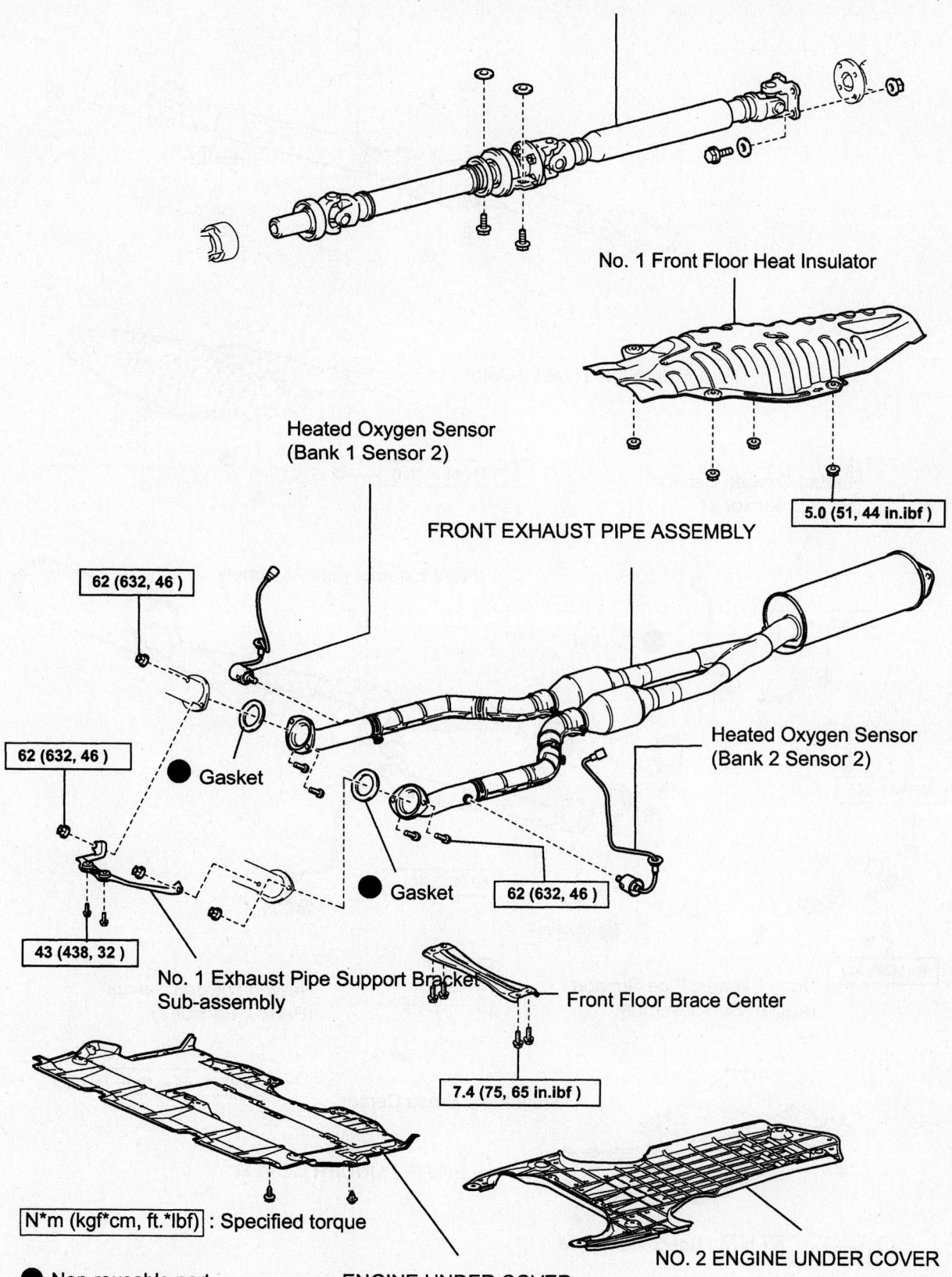

PROPELLER WITH CENTER BEARING SHAFT ASSEMBLY

No. 1 Front Floor Heat Insulator

Heated Oxygen Sensor
(Bank 1 Sensor 2)

FRONT EXHAUST PIPE ASSEMBLY

5.0 (51, 44 in.ibf)

62 (632, 46)

62 (632, 46)

● Gasket

Heated Oxygen Sensor
(Bank 2 Sensor 2)

● Gasket

62 (632, 46)

43 (438, 32)

No. 1 Exhaust Pipe Support Bracket
Sub-assembly

Front Floor Brace Center

7.4 (75, 65 in.ibf)

N*m (kgf*cm, ft.*lbf) : Specified torque

● Non-reusable part

ENGINE UNDER COVER

NO. 2 ENGINE UNDER COVER

09490_LEXU_G0008

Exploded view of the related underbody components for 3GR-FSE engine removal (cont.)—2006 GS 300 (2WD)

4WD:

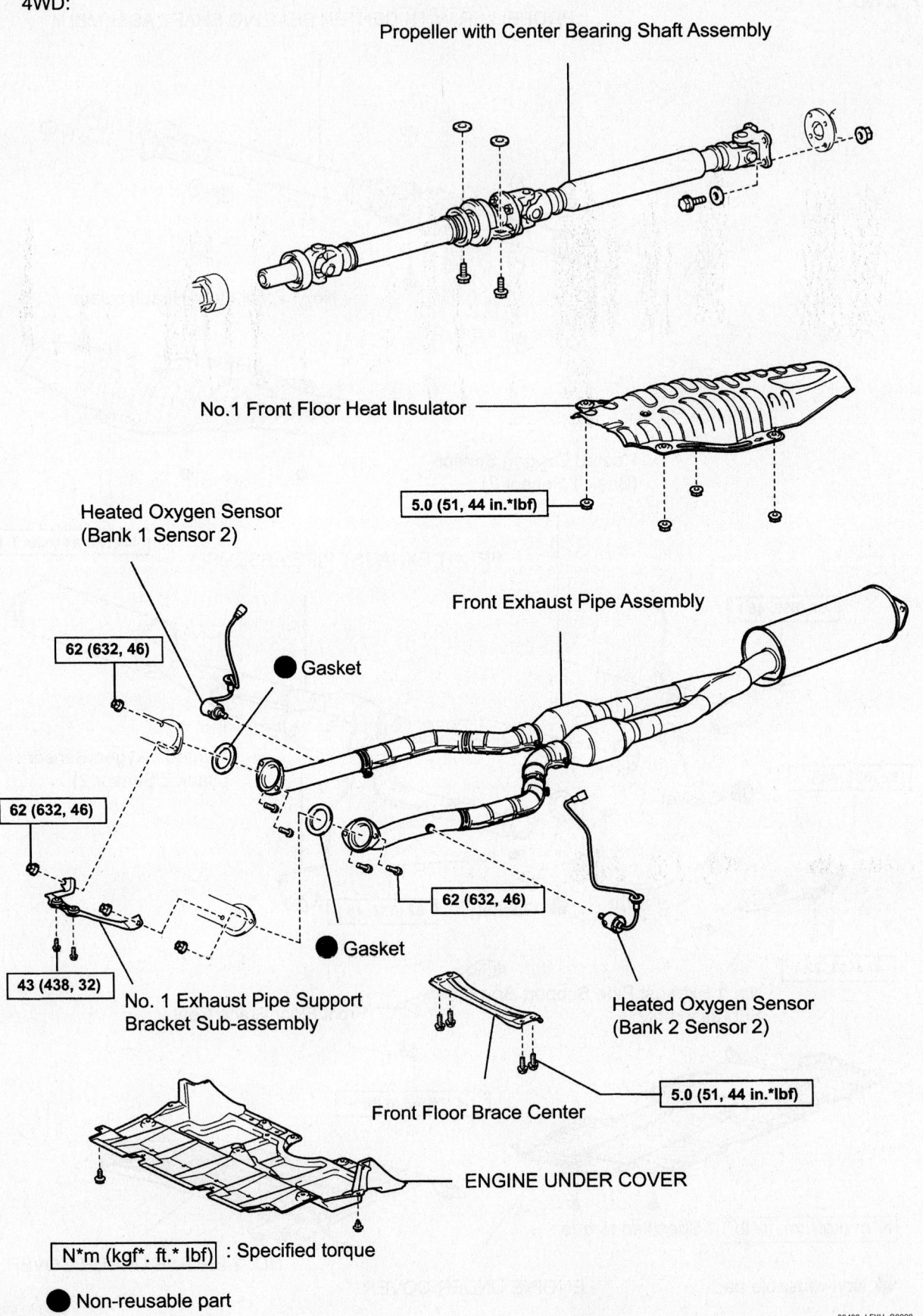

Propeller with Center Bearing Shaft Assembly

No.1 Front Floor Heat Insulator

5.0 (51, 44 in.*lbf)

Heated Oxygen Sensor
(Bank 1 Sensor 2)

62 (632, 46)

● Gasket

Front Exhaust Pipe Assembly

62 (632, 46)

62 (632, 46)

43 (438, 32)

No. 1 Exhaust Pipe Support
Bracket Sub-assembly

● Gasket

Heated Oxygen Sensor
(Bank 2 Sensor 2)

Front Floor Brace Center

5.0 (51, 44 in.*lbf)

ENGINE UNDER COVER

N*m (kgf*. ft.* lbf) : Specified torque

● Non-reusable part

09490_LEXU_G0009

Exploded view of the related underbody components for 3GR-FSE engine removal (cont.)—2006 GS 300 (AWD)

- Floor shift gear shifting rod sub-assembly

8. Remove the engine and transmission from the vehicle together as follows:

a. Set the engine lifting device.

b. Remove the 4 bolts, then separate the engine rear mounting member.

c. Remove the 12 front suspension crossmember sub-assembly bolts (2WD); or the 8 front suspension crossmember sub-assembly bolts (AWD) where they fasten to the underside of the vehicle

d. Operate the engine lifter, then slowly remove the engine and transmission with crossmember sub-assembly from the vehicle.

➡**Make sure the engine is clear of all wiring and hoses.**

e. Install engine hangers on each side of the engine using 4 bolts and tighten to 24 ft. lbs. (33 Nm).

f. Attach an engine sling device and hang the engine with a chain block.

9. Remove right and left side front drive shaft assemblies (AWD).

10. Separate the transmission assembly from the engine.

11. While holding the crankshaft, remove the 8 bolts, front spacer, drive plate and ring gear sub-assembly and rear spacer.

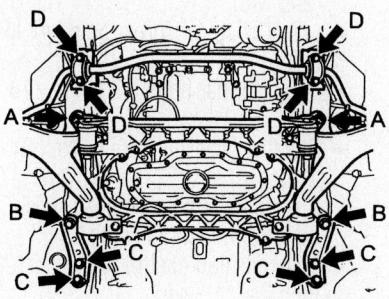

09490_LEXU_G0010

Crossmember sub-assembly mounting bolt location (2WD)—2006 GS 300

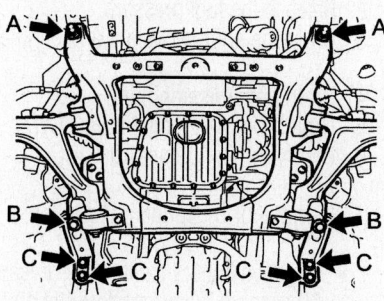

09490_LEXU_G0011

Crossmember sub-assembly mounting bolt location (AWD)—2006 GS 300

12. Remove the 2 bolts, then remove the front suspension crossmember sub-assembly from the engine.

13. Fix the engine onto engine stand with the bolts.

To install:

14. Remove the bolts and engine from the engine stand.

15. Install the front suspension crossmember sub-assembly with the 2 bolts and torque to 26 ft. lbs. (35 Nm).

16. Install the drive plate and ring gear sub-assembly and rear spacer to the crankshaft. Apply adhesive to 2 or 3 threads of the mounting bolts and install finger tight. While holding the crankshaft, tighten the 8 mounting bolts uniformly in several steps to 61 ft. lbs. (83 Nm).

17. Install the transmission assembly to the engine.

18. Install the right and left side front drive shafts (AWD).

19. Using the engine lifting device, carefully install the engine and transmission with crossmember sub-assembly into the vehicle.

20. Install the crossmember sub-assembly mounting bolts and tighten as follows:

- Bolt A: 123 ft. lbs. (167 Nm)
- Bolt B (2WD): 150 ft. lbs. (204 Nm)
- Bolt B (AWD): 121 ft. lbs. (165 Nm)
- Bolt C: 37 ft. lbs. (50 Nm)
- Bolt D (2WD): 36 ft. lbs. (49 Nm)

21. Install or connect the following:

- Engine rear mounting member with the 4 bolts. Torque the 4 bolts to 19 ft. lbs. (26 Nm).
- Floor shift gear shifting rod sub-assembly
- Power steering link wire harness (2WD)
- No. 2 steering intermediate shaft assembly
- Left and right lower ball joints (AWD)
- Left and right front shock absorbers
- Left and right tie rod assemblies
- Left and right front axle hub nuts (AWD)
- Front suspension member lower protector (2WD). Torque the 4 bolts to 71 inch lbs. (8 Nm).
- Height control sensor link and tighten the nut to 48 inch lbs. (5 Nm).

22. Install or connect to the with pulley compressor by performing the following:

- Stud bolt to the with pulley compressor using an E8 "torx" socket. Torque the stud bolt to 89 inch lbs. (10 Nm).

- With pulley compressor mounting nut and bolts. Torque to 18 ft. lbs. (25 Nm).
- Magnetic clutch connector
- Bolt, nut and bracket (AWD)
- Serpentine drive belt
- Ground cable
- Wire to engine room No. 1 relay block, junction block and cover
- Positive battery cable
- ECM connectors, connector holder and cover
- Fuel main tube
- Heater water inlet and outlet hoses
- Union-to-check valve hose and clamp
- Radiator inlet, outlet and reserve tank hoses
- Air cleaner case and hose clamp
- EVAP VSV
- MAF meter connector
- Ventilation hose to the cylinder head cover with the clamp
- Air cleaner inlet
- Front propeller shaft assembly (AWD)
- Propeller with center bearing shaft assembly
- Front exhaust pipe assembly
- No. 2 engine under cover (2WD)
- Engine under cover
- V-bank cover
- Left and right engine room side covers
- Cool air intake duct seal
- Hood
- Front wheels
- Negative battery cable

23. Add engine oil, automatic transmission fluid and coolant

24. Add front differential oil (AWD)

25. Perform system initialization (which includes power window control system, sliding roof system, clearance sonar system and variable gear ratio steering system) procedure as follows:

- Power window control system

a. Turn the ignition switch on.

b. Open power window halfway by pressing power window switch.

c. Fully pull up the switch until the power window is fully closed and continue to hold the switch for at least 1 second.

d. Check that the AUTO UP / DOWN function operates normally.

➡**If the remote UP / DOWN function does not operate after the conditions 1, 2, or 3 is satisfied, the power window regulator master switch may have a malfunction.**

- Sliding roof system
  e. Turn the ignition switch on.
  f. If the sliding roof is opened, close it fully.
  g. Push the open switch of the slide switch, or the up switch of the tilt switch on the personal light, making the sliding roof tilt up approximately 1 second, tilt down, slide open, slide close.
  h. Sliding roof stops at the fully closed position.
  i. Finish the initialization.
  j. Check that the operation works normally with AUTO operation.
- Clearance sonar system
  k. Turn the ignition switch on.
  l. Turn the clearance sonar main switch ON.
  m. Turn the steering wheel to the full left and right lock position.

➡**Make sure to completely turn the steering wheel to the left and right full lock position.**

  n. Confirm that the learning operation has been completed by checking the multi-information display.
  o. At an area with few turns and curves, and minimal traffic, drive at 20 km/h or more for 5 minutes or more.
- Variable gear ratio steering system
  p. Turn the ignition switch on, and check that the master warning light and VSC/ABS warning lights illuminate for a few seconds.

➡**If the warning lights remain on or blink, repair the applicable system.**

  q. Drive the vehicle on a straight road at 35 km/h (22 mph) or more for 5 seconds or longer.
  r. Confirm that steering angle sensor initialization is completed by doing the following:
- Drive the vehicle on a straight road at 60 km/h (37 mph) or more for 30 seconds or longer.
- Stop the vehicle (engine running).
- Slowly turn the steering wheel from lock to lock.
- If it turns approximately 2.7 turns, steering angle sensor initialization is completed. If it turns approximately 3.2 turns, steering angle sensor initialization is not completed.
26. Check shift lever position as follows:
- Remove the nut and disconnect the shifting rod.
- Turn the control shaft lever of the neutral start switch counterclockwise until it stops, and turn it

clockwise 2 notches to set it to the N position.
- Move the shift lever to the N position and tighten the nut while lightly pushing the lever toward the R position.
- After adjustment, check that the shift lever moves smoothly and the shift lever and gear operate correctly.
27. Inspect and adjust front end alignment
28. Check idle speed and ignition timing.

➡**Do not start the engine for at least 1 hour after installing.**

29. Recheck all fluid levels and add if necessary.
30. Start engine and check for fluid, fuel and exhaust leaks.
31. Road test the vehicle.

IS 300

1. Before servicing the vehicle, refer to the Precautions Section.
2. Drain the cooling system.
3. Relieve the fuel system pressure.
4. Drain the engine oil.
5. Remove or disconnect the following:
- Negative battery cable
- Engine under cover
- Air cleaner inlet
- Brake booster vacuum hose
- Radiator hoses
- Mass Air Flow (MAF) sensor connector
- Positive Crankcase Ventilation (PCV) hose
- Air intake resonator
- Accelerator cable
- Accessory drive belt
- Front subframe brace
- Floor ground strap
- Starter motor wiring harness
- Fuel inlet hose support
- Dash panel ground strap
- Heater hoses
- Evaporative Emissions (EVAP) canister hose
- Heated Oxygen (HO2S) sensor connectors
- Alternator wiring harness and clamp
- Cylinder block ground cable bracket
- Igniter connector
- Data link connector and harness clamps
- Powertrain Control Module (PCM) harness connectors

- Power steering pump
- A/C compressor
- Drive shaft
- Transmission control rod
- Exhaust front pipe
- Exhaust center pipe
- Stabilizer bar
- Front shock absorbers
- Lower ball joints from the steering knuckles
- Transmission mount crossmember. Support the powertrain from below.
- Steering intermediate shaft
- Front subframe
6. Lower the engine, transmission and subframe away from the vehicle.

To install:

7. Installation is the reverse of the removal procedure, while using the following torque values:
- Transmission flange bolts: 53 ft. lbs. (72 Nm)
- Transmission flange-to-oil pan bolts: 27 ft. lbs. (37 Nm)
- Torque converter bolts: 74 ft. lbs. (100 Nm)
- Left and right motor mount nuts: 52 ft. lbs. (70 Nm)
- Front subframe bolts: 52 ft. lbs. (70 Nm)
- Transmission mount crossmember: 19 ft. lbs. (26 Nm)
- Lower ball joint bolts: 181 ft. lbs. (245 Nm)
- Lower shock absorber bolts: 47 ft. lbs. (64 Nm)
- Stabilizer bar bolts: 13 ft. lbs. (18 Nm)
- Stabilizer bar nuts: 36 ft. lbs. (49 Nm)
- Steering intermediate shaft pinch bolt: 26 ft. lbs. (35 Nm)
- Transmission control rod nuts: 108 inch lbs. (13 Nm)

IS 250 and IS 350

1. Before servicing the vehicle, refer to the Precautions Section.
2. Release the fuel pressure.
3. Drain the engine coolant and engine oil.
4. Drain transmission fluid.
5. Drain front differential fluid (AWD).
6. Discharge refrigerant from A/C system.
7. Remove or disconnect the following:
- Negative battery cable. Wait at least 90 seconds before performing any other work.
- Hood insulator pad and the hood

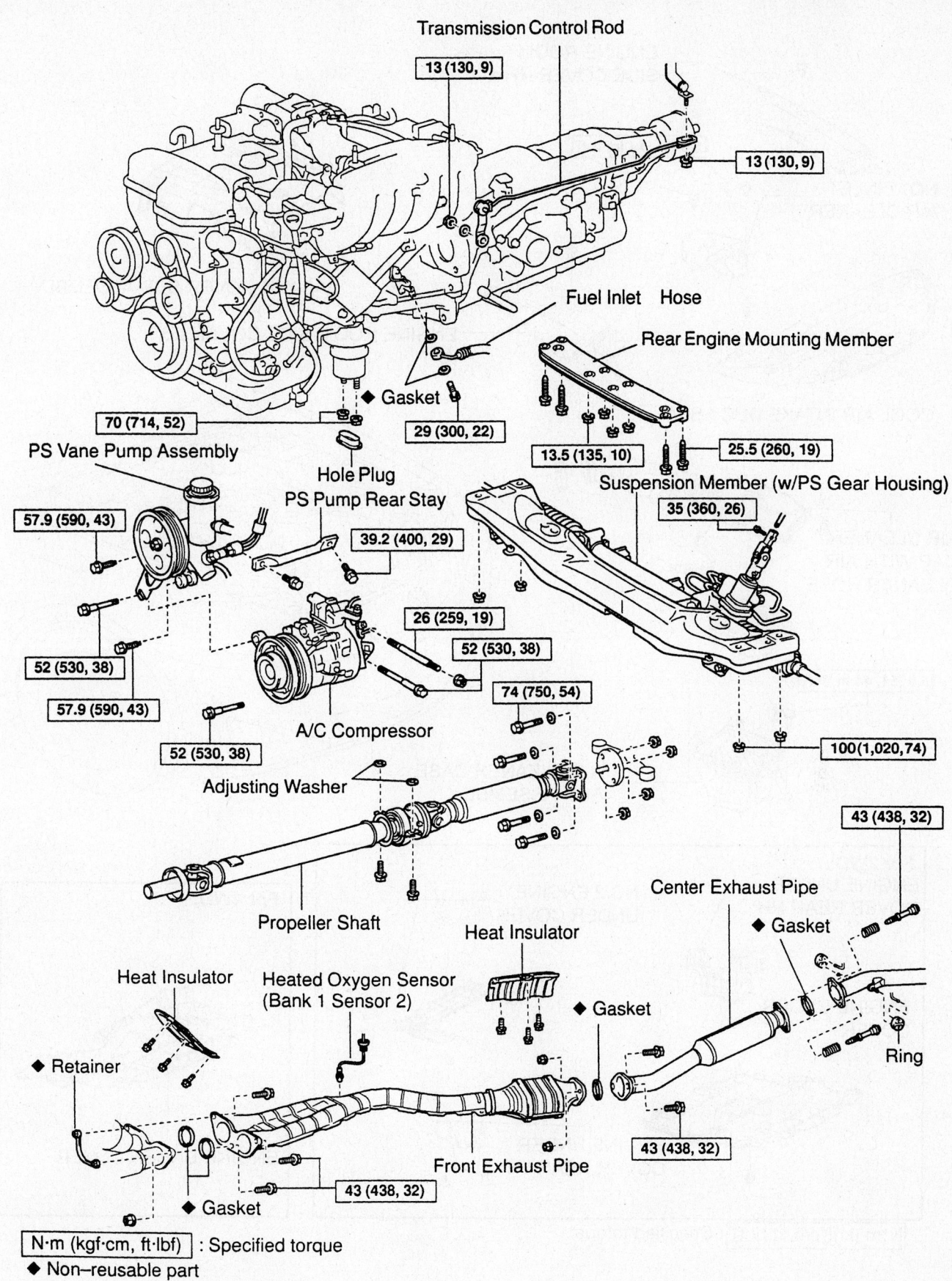

Transmission Control Rod

13 (130, 9)

13 (130, 9)

Fuel Inlet Hose

Rear Engine Mounting Member

◆ Gasket

70 (714, 52)

29 (300, 22)

PS Vane Pump Assembly

13.5 (135, 10)

25.5 (260, 19)

Hole Plug
PS Pump Rear Stay

Suspension Member (w/PS Gear Housing)

57.9 (590, 43)

35 (360, 26)

39.2 (400, 29)

26 (259, 19)

52 (530, 38)

52 (530, 38)

57.9 (590, 43)

A/C Compressor

74 (750, 54)

52 (530, 38)

100 (1,020, 74)

Adjusting Washer

43 (438, 32)

Center Exhaust Pipe

◆ Gasket

Propeller Shaft

Heat Insulator

◆ Gasket

Heat Insulator

◆ Gasket

Ring

Heated Oxygen Sensor
(Bank 1 Sensor 2)

Heat Insulator

◆ Retainer

Front Exhaust Pipe

43 (438, 32)

43 (438, 32)

◆ Gasket

N·m (kgf·cm, ft·lbf) : Specified torque

◆ Non–reusable part

9347LG03

Exploded view of the engine mounting—IS 300

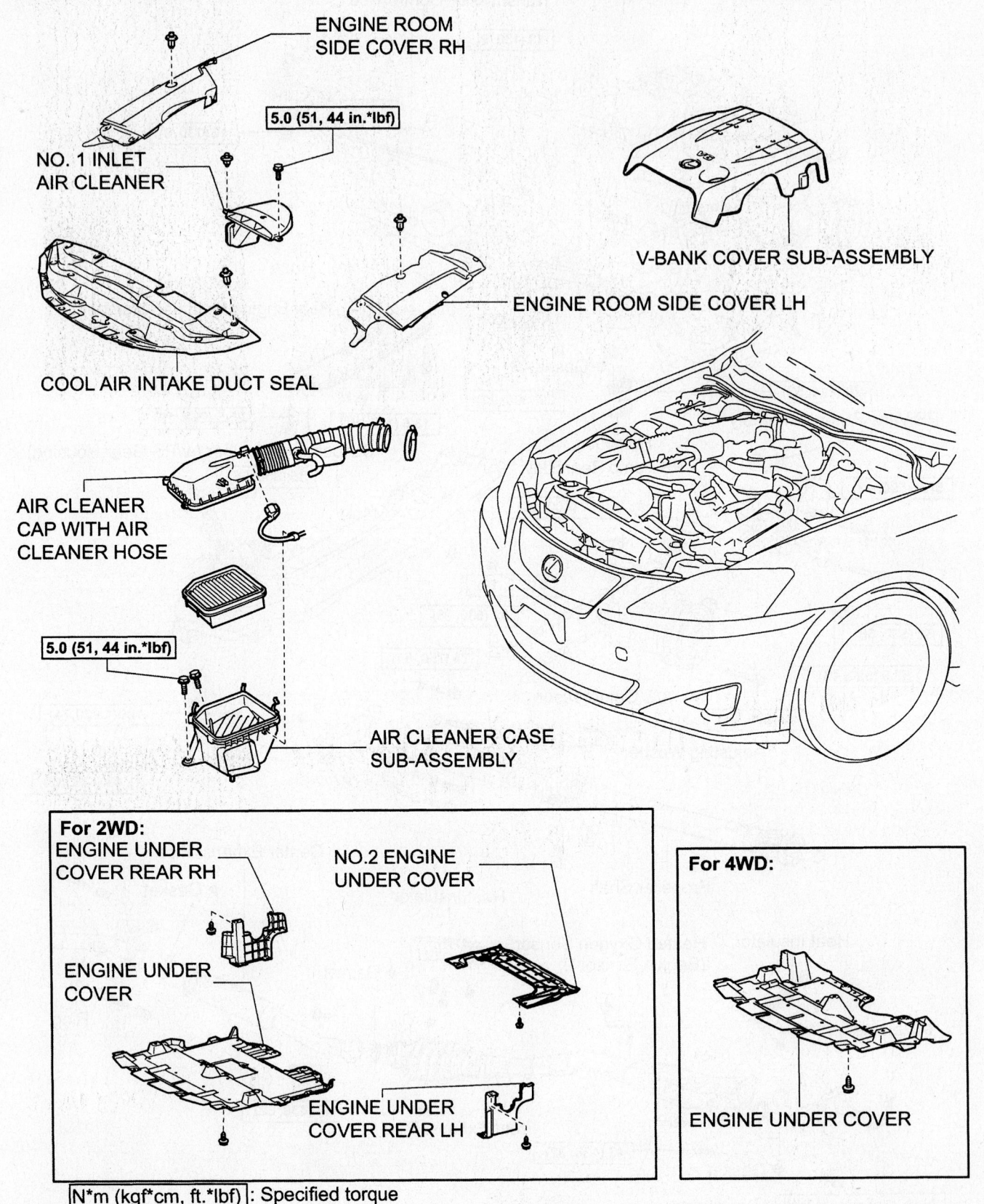

ENGINE ROOM
SIDE COVER RH

5.0 (51, 44 in.*lbf)

NO. 1 INLET
AIR CLEANER

V-BANK COVER SUB-ASSEMBLY

ENGINE ROOM SIDE COVER LH

COOL AIR INTAKE DUCT SEAL

AIR CLEANER
CAP WITH AIR
CLEANER HOSE

5.0 (51, 44 in.*lbf)

AIR CLEANER CASE
SUB-ASSEMBLY

For 2WD:
ENGINE UNDER
COVER REAR RH

NO.2 ENGINE
UNDER COVER

For 4WD:

ENGINE UNDER
COVER

ENGINE UNDER
COVER REAR LH

ENGINE UNDER COVER

N*m (kgf*cm, ft.*lbf) : Specified torque

09490_LEXU_G0012

Exploded view of the related engine compartment components for 4GR-FSE and 2GR-FSE engine removal—IS 250/IS 350

For 2WD:

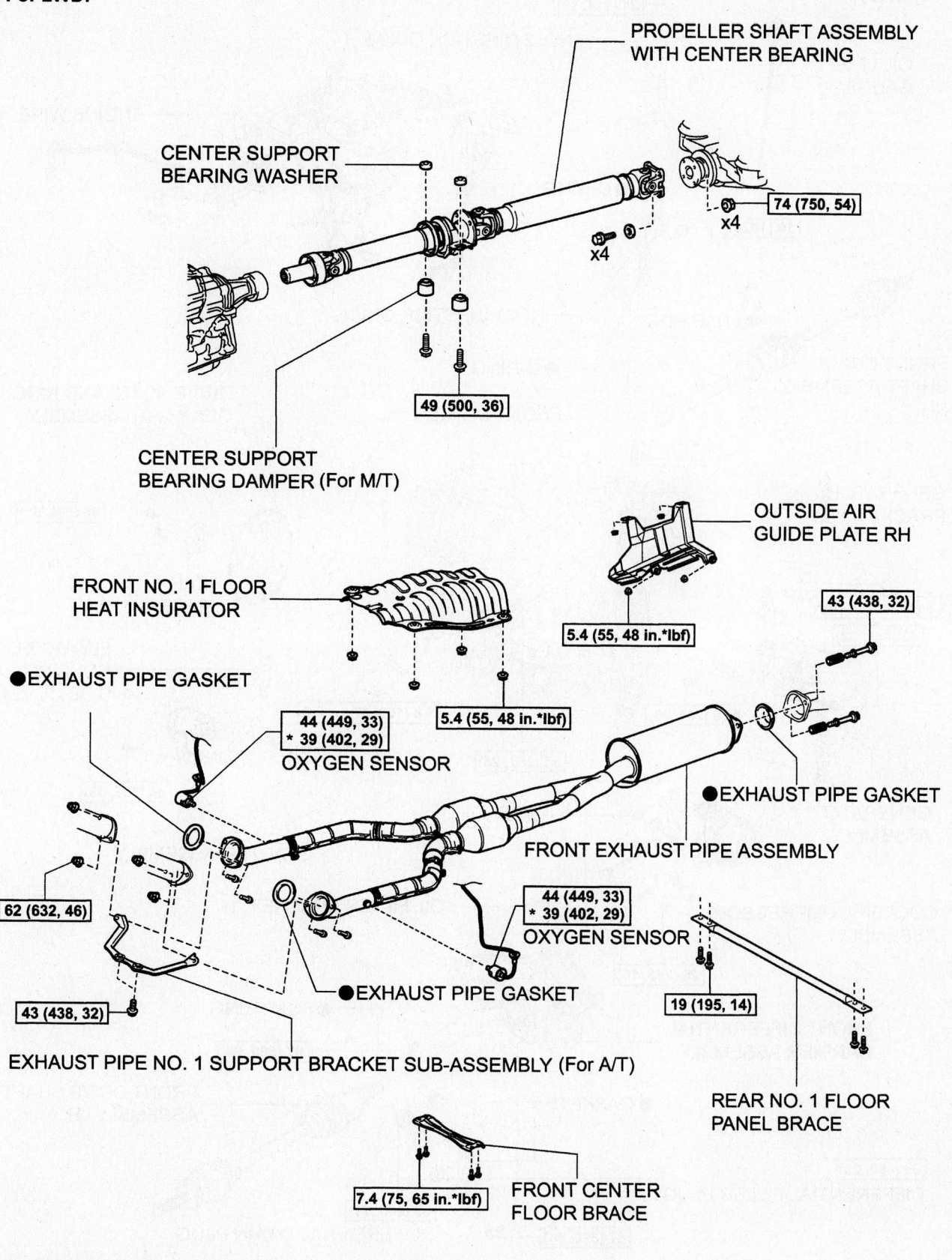

PROPELLER SHAFT ASSEMBLY
WITH CENTER BEARING

CENTER SUPPORT
BEARING WASHER

74 (750, 54)

x4

x4

49 (500, 36)

CENTER SUPPORT
BEARING DAMPER (For M/T)

OUTSIDE AIR
GUIDE PLATE RH

FRONT NO. 1 FLOOR
HEAT INSURATOR

43 (438, 32)

5.4 (55, 48 in.*lbf)

●EXHAUST PIPE GASKET

44 (449, 33)
* 39 (402, 29)
OXYGEN SENSOR

5.4 (55, 48 in.*lbf)

●EXHAUST PIPE GASKET

FRONT EXHAUST PIPE ASSEMBLY

62 (632, 46)

44 (449, 33)
* 39 (402, 29)
OXYGEN SENSOR

43 (438, 32)

●EXHAUST PIPE GASKET

19 (195, 14)

EXHAUST PIPE NO. 1 SUPPORT BRACKET SUB-ASSEMBLY (For A/T)

REAR NO. 1 FLOOR
PANEL BRACE

7.4 (75, 65 in.*lbf)

FRONT CENTER
FLOOR BRACE

N*m (kgf*cm, ft.*lbf) : Specified torque   ● Non-reusable part   * For use with SST

09490_LEXU_G0013

Exploded view of the related underbody components for 4GR-FSE and 2GR-FSE engine removal (cont.)—IS 250/IS 350 (2WD)

For 4WD:

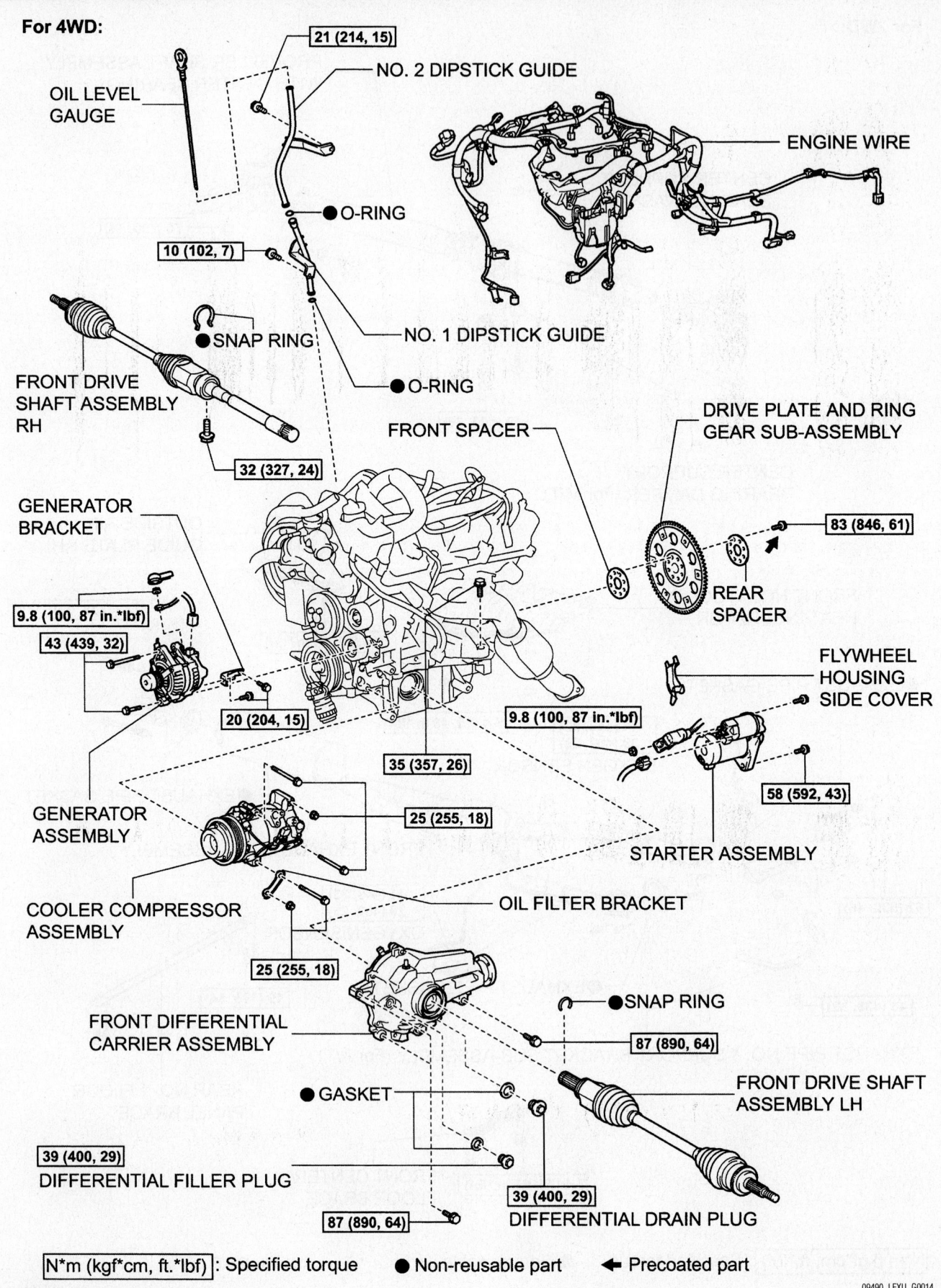

21 (214, 15)

NO. 2 DIPSTICK GUIDE

OIL LEVEL GAUGE

ENGINE WIRE

●O-RING

10 (102, 7)

NO. 1 DIPSTICK GUIDE

●SNAP RING

FRONT DRIVE SHAFT ASSEMBLY RH

●O-RING

FRONT SPACER

DRIVE PLATE AND RING GEAR SUB-ASSEMBLY

32 (327, 24)

GENERATOR BRACKET

83 (846, 61)

REAR SPACER

9.8 (100, 87 in.*lbf)

43 (439, 32)

FLYWHEEL HOUSING SIDE COVER

20 (204, 15)

9.8 (100, 87 in.*lbf)

35 (357, 26)

58 (592, 43)

GENERATOR ASSEMBLY

25 (255, 18)

STARTER ASSEMBLY

COOLER COMPRESSOR ASSEMBLY

OIL FILTER BRACKET

25 (255, 18)

●SNAP RING

FRONT DIFFERENTIAL CARRIER ASSEMBLY

87 (890, 64)

●GASKET

FRONT DRIVE SHAFT ASSEMBLY LH

39 (400, 29)

DIFFERENTIAL FILLER PLUG

39 (400, 29)

87 (890, 64)

DIFFERENTIAL DRAIN PLUG

N*m (kgf*cm, ft.*lbf): Specified torque   ● Non-reusable part   ◄ Precoated part

09490_LEXU_G0014

Exploded view of the related engine components for 4GR-FSE engine removal (cont.)—IS 250 (AWD)

- 11 clips and cool air intake duct seal
- 2 clips and right engine room side cover
- 5 clips and left engine room side cover
- 2 front retaining clips and 1 rear retaining clip and V-bank cover
- Front wheels
- Engine under cover
- No. 2 engine under cover, left rear engine under cover and right rear engine under cover (2WD)
- Front exhaust pipe assembly
- Propeller shaft with center bearing assembly
- Bolt, clip and inlet air cleaner
- Air cleaner cap with air cleaner hose
- Air cleaner case
- Clamp and union-to-check valve hose
- No. 2 fuel vapor feed hose
- Inlet and outlet radiator hoses
- Radiator reservoir tank hose
- Inlet and outlet heater water hoses
- No. 1 cooler refrigerant suction hose
- Discharge hose sub-assembly
- ECM cover
- 6 ECM connectors and connector holder
- Engine room No. 1 relay block cover
- Wire from engine room No. 1 junction block
- Positive battery cable
- Connector and 4 clamps from body
- Bolt, clamp and engine wire No. 3
- Ground cable
- Fuel main hose
- No. 3 fuel hose
- Fuel pipe clamp. Carefully pinch and pull the fuel tube's connector to disconnect it from the fuel tube sub-assembly (IS 350).
- 4 bolts and suspension member protector (2WD)
- Steering sliding with shaft yoke sub-assembly (2WD)
- No. 2 steering intermediate shaft assembly (AWD)
- Left and right front speed sensors (AWD)
- Left and right front axle hub nuts (AWD)
- Left and right tie rods
- Height sensor control link
- Left and right front shock absorbers
- Left and right front lower ball joints
- Left and right front axle assemblies (AWD)

- 2 bolts and exhaust pipe No. 1 support bracket sub-assembly (IS 250 with automatic transmission)

8. Remove or disconnect the power steering link wire harness by performing the following:
- Ground wire from the bracket
- Wire harness from the bracket
- Wire harness clamps from the front frame (AWD)
- Connectors from the power steering link assembly
- Ground wire from the power steering link assembly
- Bracket from the front frame (AWD)

9. Disconnect the floor shift gear shifting rod sub-assembly.

10. Remove the engine and transmission from the vehicle together as follows:

a. Set the engine lifting device.

b. Remove the 4 bolts, then separate the engine rear mounting member.

c. Remove the 12 front suspension crossmember sub-assembly bolts (2WD); or the 8 front suspension crossmember sub-assembly bolts (AWD) where they fasten to the underside of the vehicle

d. Operate the engine lifter, then slowly remove the engine and transmission with crossmember sub-assembly from the vehicle.

➡**Make sure the engine is clear of all wiring and hoses.**

e. Install engine hangers on each side of the engine using 4 bolts and tighten to 24 ft. lbs. (33 Nm).

f. Attach an engine sling device and hang the engine with a chain block.

11. Remove front propeller shaft assembly (AWD).

12. Remove starter motor.

13. Separate the transmission assembly from the engine.

14. Remove oil dipstick guide.

15. Remove left and right exhaust manifolds.

16. Remove left and right front drive shaft assemblies (AWD).

17. While holding the crankshaft, remove the 8 bolts, front spacer, drive plate and ring gear sub-assembly and rear spacer.

18. Remove the 2 bolts, then remove the front suspension crossmember sub-assembly from the engine.

19. Fix the engine onto engine stand with the bolts.

To install:

20. Install engine hangers on each side of the engine using 4 bolts and tighten to 24 ft. lbs. (33 Nm).

21. Install the drive plate and ring gear sub-assembly and rear spacer to the crankshaft. Apply adhesive to 2 or 3 threads of the mounting bolts and install finger tight. While holding the crankshaft, tighten the 8 mounting bolts uniformly in several steps to 61 ft. lbs. (83 Nm) for automatic transmission equipped vehicles; or 54 ft. lbs. (73 Nm) for manual transmission equipped vehicles.

22. Install the front suspension crossmember sub-assembly with the 2 bolts and torque to 26 ft. lbs. (35 Nm).

23. Install the left and right front drive shaft assemblies (AWD).

24. Install the transmission assembly to the engine.

25. Install the starter motor.

26. Install the left and right exhaust manifolds.

27. Install oil dipstick guide along with a new O-ring coated in clean engine oil.

28. Install front propeller shaft assembly (AWD).

29. Using the engine lifting device, carefully install the engine and transmission with crossmember sub-assembly into the vehicle.

30. Install the crossmember sub-assembly mounting bolts and tighten as follows:

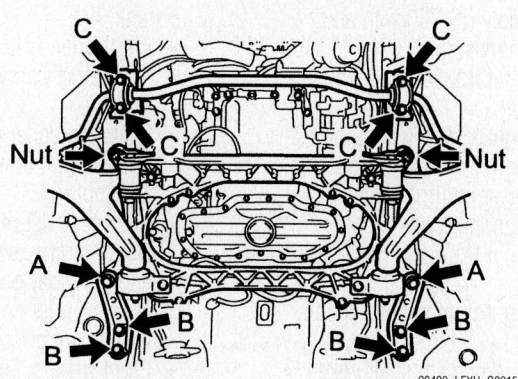

09490_LEXU_G0015
Crossmember sub-assembly mounting bolt location (2WD)—IS 250/IS 350

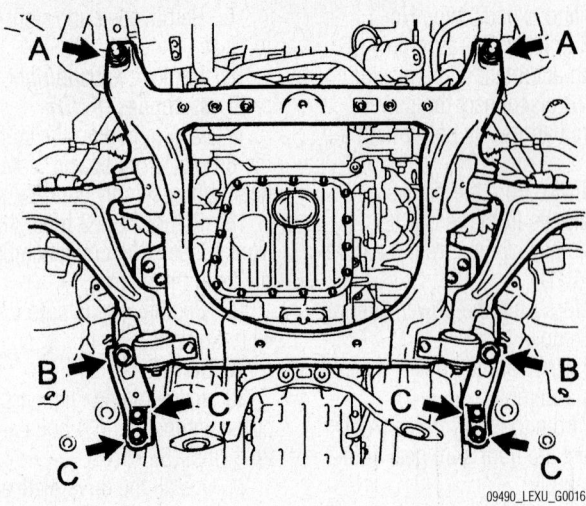

09490_LEXU_G0016

Crossmember sub-assembly mounting bolt location (AWD)—IS 250

- 2 nuts (2WD): 123 ft. lbs. (167 Nm)
- Bolt A (2WD): 150 ft. lbs. (204 Nm)
- Bolt A (AWD): 123 ft. lbs. (167 Nm)
- Bolt B (2WD): 37 ft. lbs. (50 Nm)
- Bolt B (AWD): 122 ft. lbs. (165 Nm)
- Bolt C (2WD): 36 ft. lbs. (49 Nm)
- Bolt C (AWD): 37 ft. lbs. (50 Nm)

31. Install or connect the following:
- Engine rear mounting member with the 4 bolts. Torque the 4 bolts to 19 ft. lbs. (26 Nm).

32. Connect the floor shift gear shifting rod sub-assembly.

33. Install or connect the power steering link wire harness by performing the following:
- Ground wire to the power steering link assembly
- Connectors to the power steering link assembly and secure lock connector
- Wire harness clamps to the front frame (AWD)
- Bracket to the front frame (AWD)
- Wire harness clamps to the bracket
- Ground wire to the bracket

34. Install or connect the following:
- Exhaust pipe No. 1 support bracket sub-assembly (IS 250 with automatic transmission)
- Left and right front axle assemblies (AWD)
- Left and right front lower ball joints
- Left and right front shock absorbers
- Height sensor control link
- Left and right tie rods
- Left and right front axle hub nuts (AWD)
- Left and right front speed sensors (AWD)
- Steering sliding with shaft yoke sub-assembly (2WD)

- No. 2 steering intermediate shaft assembly (AWD)
- 4 bolts and suspension member protector (2WD)
- Fuel tube sub-assembly and connector (IS 350).
- No. 3 fuel hose
- Fuel main hose
- Bolt, clamp and engine wire No. 3
- Ground cable
- Connector and 4 clamps from body
- Wire from engine room No. 1 junction block
- Engine room No. 1 relay block cover
- Positive battery cable
- 6 ECM connectors and connector holder
- ECM cover
- Discharge hose sub-assembly
- No. 1 cooler refrigerant suction hose
- Inlet and outlet heater water hoses
- Inlet and outlet radiator hoses
- Radiator reservoir tank hose
- No. 2 fuel vapor feed hose
- Union-to-check valve hose to surge tank
- Air cleaner case
- Air cleaner cap with air cleaner hose
- Inlet air cleaner with bolt and clamp
- Propeller shaft with center bearing assembly
- Front exhaust pipe assembly
- No. 2 engine under cover, left rear engine under cover and right rear engine under cover (2WD)
- Engine under cover
- Front wheels. Torque the lug nuts to 76 ft. lbs. (103 Nm)

- Engine V-bank cover
- Left and right engine room side covers
- Cool air intake duct seal and 11 clips
- Hood
- Negative battery cable

35. Add engine oil, transmission fluid and engine coolant.

36. Add front differential oil (AWD).

37. Recharge the A/C system.

➡ **Do not start the engine for at least 1 hour after installing.**

38. Start engine and check for fluid, fuel and exhaust leaks.

39. Check shift lever position as follows:
- Remove the nut and disconnect the shifting rod.
- Turn the control shaft lever of the neutral start switch counterclockwise until it stops, and turn it clockwise 2 notches to set it to the N position.
- Move the shift lever to the N position and tighten the nut while lightly pushing the lever toward the R position.
- After adjustment, check that the shift lever moves smoothly and the shift lever and gear operate correctly.

40. Inspect shift lever position as follows:
- When shifting from the P to R position with the engine switch on (IG) and the brake pedal depressed, make sure that the shift lever moves smoothly and then moves correctly into the position.
- Start the engine and make sure that the vehicle moves forward when shifting from the N to D position and moves rearward when shifting to the R position. If operation cannot be done as specified, inspect the neutral start switch assembly and check the shift lever assembly installation condition.

41. Inspect and adjust front end alignment

42. Check idle speed and ignition timing.

43. Have the CO/HC emissions inspected.

44. Recheck all fluid levels and add if necessary.

45. Road test the vehicle.

LS 430 (2002–03)

1. Before servicing the vehicle, refer to the Precautions Section.

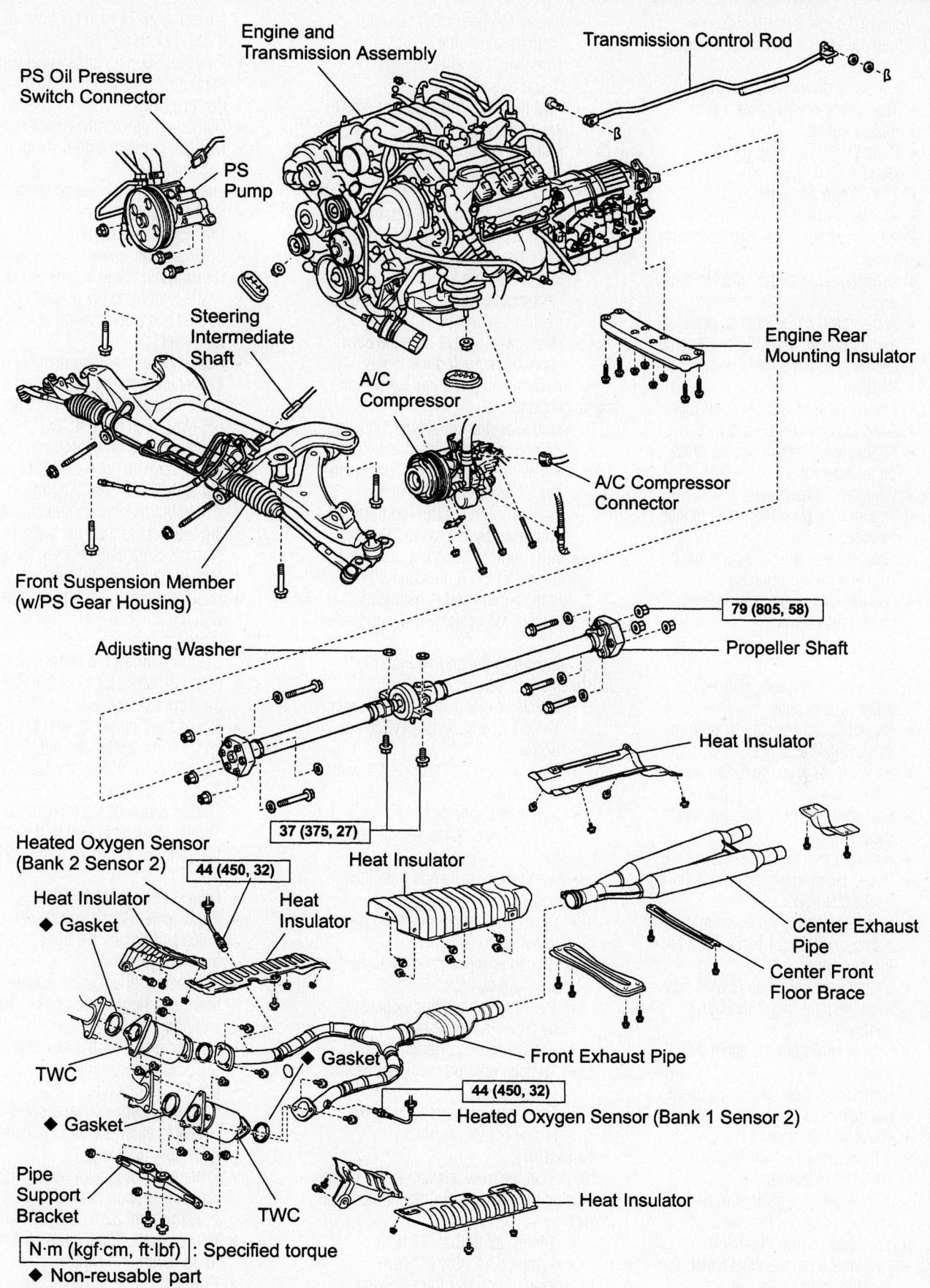

PS Oil Pressure Switch Connector

Engine and Transmission Assembly

Transmission Control Rod

PS Pump

Steering Intermediate Shaft

A/C Compressor

Engine Rear Mounting Insulator

A/C Compressor Connector

Front Suspension Member (w/PS Gear Housing)

Adjusting Washer

79 (805, 58)

Propeller Shaft

Heat Insulator

37 (375, 27)

Heated Oxygen Sensor (Bank 2 Sensor 2)

44 (450, 32)

Heat Insulator

Heat Insulator

Center Exhaust Pipe

Center Front Floor Brace

Heat Insulator

◆ Gasket

TWC

◆ Gasket

Front Exhaust Pipe

44 (450, 32)

Heated Oxygen Sensor (Bank 1 Sensor 2)

◆ Gasket

Pipe Support Bracket

TWC

Heat Insulator

N·m (kgf·cm, ft·lbf) : Specified torque

◆ Non-reusable part

09490_LEXU_G0017

Exploded view of the engine removal and related components—2002–03 LS 430

2. Relieve the fuel system pressure
3. Drain the engine coolant and engine oil
4. Remove or disconnect the following:
- The battery clamp cover
- Battery cables
- Battery
- Hood
- The oil pan protector
- Air cleaner inlet
- Air cleaner and intake air connector assembly
- Drive belt, fan clutch and fan pulley
- Accelerator, cruise control actuator and automatic transmission throttle cables from the throttle body
- Radiator
- Engine oil level sensor connector
- Alternator connector and wire
- Engine wire clamp from the bracket on the alternator
- 2 igniter connectors
- Engine wire clamp from the igniter bracket
- Ground strap from the right-hand engine mounting bracket
- Ground strap from under the left-hand fender apron
- Engine wire clamp from the cowl panel
- Radiator reservoir hose from the water bypass pipe
- Brake booster vacuum hose from the air intake chamber
- Heater hose from the heater water valve and water bypass pipe
- Fuel inlet hose from the fuel inlet pipe
- Fuel return hose to the return pipe
- Power steering air hose from the air intake chamber
- 2 power steering hoses from the clamp on the right-hand No. 3 timing belt cover
- Evaporative Emission (EVAP) hose from the pipe (from the charcoal canister).
- Engine wire from the cabin as follows:
- Undercover from under the glove compartment
- Glove compartment
- 3 Powertrain Control Module (PCM) connectors
- 2 cowl wire connectors from the connector on the bracket
- Wire clamp from the bracket
- Grommet from the cowl panel, then pull the engine wire out.
- Power steering oil cooler pipe from the oil pan

- Heated Oxygen (HO2) sensors
- Front exhaust pipe
- 2 catalytic converters
- Center exhaust pipe
- Heat insulator from the rear side of the front exhaust pipe
- Front center floor and rear center floor crossmember braces
- Driveshaft
- Air conditioning compressor without disconnecting the air conditioning lines
- Power steering pump
- Heat insulators for the front side of the front exhaust pipe
- Heater water valve from the cowl panel by removing the 2 nuts
5. Attach the engine chain hoist to the engine hangers.
6. Remove or disconnect the following:
- Engine mounting insulators from the engine suspension crossmember by removing the 2 nuts
- Transmission control rod from the shift lever by removing the nut
- Rear engine mounting member by removing the 4 nuts and 4 bolts
7. Lift the engine and transmission assembly out of the vehicle slowly and carefully.
8. Disconnect the engine from the transmission as follows:
9. Remove or disconnect the following:
- Vehicle Speed Sensor (VSS) connector
- Park/Neutral Position (PNP) switch connector
- Solenoid connector
- Direct clutch speed sensor connector
- 4 engine wire clamps from the brackets
- Oil dipstick and guide from the transmission
- Oil cooler pipes from the transmission and clamps
- The flywheel housing undercover by removing the 2 bolts
- The 6 torque converter bolts
- 10 bolts holding the transmission to the engine
- Transmission together with the torque converter clutch

To install:
10. Install the transmission to the engine and install the 10 bolts. Tighten the bolts as follows:
- 14mm: 27 ft. lbs. (37 Nm)
- 17mm: 53 ft. lbs. (72 Nm)
11. Install or connect the following:
- Torque converter clutch bolts. Apply adhesive to 2 or 3 threads of

the bolt end. Tighten the bolts to 30 ft. lbs. (41 Nm).
- Flywheel housing undercover with the 2 bolts. Tighten bolts to 14 ft. lbs. (19 Nm).
- Oil cooler pipe for the transmission
- Dipstick guide and dipstick for the transmission
- Engine wire to the transmission
- VSS connector
- PNP switch connector
- Solenoid connector
- Direct clutch speed sensor connector
- 4 wire clamps to the brackets
- Engine and transmission assembly to the vehicle
- Rear engine mounting member to the vehicle and the 4 bolts and 4 nuts; bolts tightened to 19 ft. lbs. (25 Nm), nuts to 10 ft. lbs. (14 Nm)
- The transmission control rod to the shift lever with the nut. Tighten the nut to 108 inch lbs. (13 Nm).
- 2 nuts holding the engine mounting brackets to the front suspension crossmember. Tighten the 2 nuts to 52 ft. lbs. (70 Nm).
- Heater water valve to the cowl panel with the 2 nuts
12. Remove the engine hoist.
13. Install or connect the following:
- Heat insulators for the front side of the front exhaust pipe
- Power steering pump with the nut and 3 bolts-tighten the nut to 32 ft. lbs. (43 Nm); tighten the bolts to 29 ft. lbs. (39 Nm)
- The air conditioning compressor. Tighten the bolts to 36 ft. lbs. (49 Nm) and the nut to 22 ft. lbs. (29 Nm).
- Driveshaft to the vehicle
- Front center floor crossmember brace and tighten the bolts to 108 inch lbs. (13 Nm)
- Rear center floor crossmember brace and tighten the bolts to 108 inch lbs. (13 Nm)
- Heat insulator for the rear side of the front exhaust pipe
- Center exhaust pipe
- 2 front catalytic converters with 3 new nuts each. Tighten the nuts to 46 ft. lbs. (62 Nm).
- Front exhaust pipe. 4 bolts holding the pipe support bracket to the transmission: 32 ft. lbs. (44 Nm).
- HO2 sensors. Tighten the sensors to 33 ft. lbs. (44 Nm).
- Power steering oil cooler pipe
- Engine wire harness to the passenger's compartment

- 3 PCM connectors
- 2 engine wire connectors to the connector on the bracket
- Engine wire clamp to bracket
- Glove compartment and the dash undercover
- All of the engine assembly connectors, wires, straps, clamps and hoses
- Radiator
- Accelerator and cruise control cables to the throttle body
- Throttle control cable to the throttle body, if equipped with automatic transmission
- Fan pulley
- Fan clutch and the drive belt. Tighten the 4 nuts for the fan to 16 ft. lbs. (21 Nm).
- Air cleaner and intake air connector assembly
- Air cleaner inlet
- Coolant
- Battery
- Engine oil
- Battery cables
- Battery cover
- Engine undercover
- Oil pan protector
- Hood

LS 430 (2004–06)

1. Before servicing the vehicle, refer to the Precautions Section.
2. Relieve the fuel system pressure
3. Drain the engine coolant, engine oil and automatic transmission fluid
4. Remove or disconnect the following:
   - The battery clamp cover
   - Battery cables
   - Battery
   - Hood assembly
   - V-bank covers
   - Front wheels
   - Engine undercovers
   - Radiator assembly
   - Fuel pipe No 2
   - Fan and alternator drive belts
   - Wiring from ECM box
   - Alternator wiring and wiring clamps
   - Ground strap from alternator stay
   - PS hose from oil pan bolt
   - Ground strap from the body
   - PS air hose
   - Fuel vapor feed hose
   - Heater hoses from heater core
   - PS oil reservoir
   - PS pump
   - PS air hoses
   - PS oil pressure switch connector
   - A/C compressor. Lay assembly to

the side and support. Do NOT discharge system
- Front center floor brace
- Exhaust pipe assembly
- w/Catalyst converter assembly
- Front floor heat insulator
- Parking brake cable heat insulator
- Drive shaft w/center support bearing
- Floor shift gear shifting rod
- Steering shaft with yoke
- Front disc brake calipers. Lay to the side and support. Do NOT disconnect fluid lines
- RF & LF upper suspension arms
- RF & LF shock absorbers
- Height control sensors
- Rack and Pinion assembly
- 2 hole plugs
- 4 nuts on engine mounting insulators
- 4 bolts and 4 nuts on rear engine mounting member

5. Attach engine hoist and gently remove the engine and transmission assembly
6. Disconnect transmission from the engine

To install:

7. Install the transmission to the engine and install the 10 bolts. Tighten the bolts as follows:
   - 14mm: 27 ft. lbs. (37 Nm)
   - 17mm: 53 ft. lbs. (72 Nm)
8. Install or connect the following:
   - Torque converter clutch bolts. Apply adhesive to 2 or 3 threads of the bolt end. Tighten the bolts to 30 ft. lbs. (41 Nm).
   - Flywheel housing undercover with the 2 bolts. Tighten bolts to 14 ft. lbs. (19 Nm).
   - Oil cooler pipe for the transmission
   - Dipstick guide and dipstick for the transmission
   - Engine wire to the transmission
   - VSS connector
   - PNP switch connector
   - Solenoid connector
   - Direct clutch speed sensor connector
   - 4 wire clamps to the brackets
   - Engine and transmission assembly to the vehicle
   - Rack and Pinion assembly. Tighten to 48 ft. lbs. (65 Nm)
   - Height control sensor link
   - Front shock absorbers
   - Upper suspension arms
   - Front disc brake calipers
   - Steering shaft and yoke
   - Floor gear shift rod

- Drive shaft and center bearing assembly
- Front brake cable heat insulator
- Front floor heat insulator
- Catalyst converter
- Exhaust pipe assembly
- Front floor center brace
- A/C compressor
- PS pump assembly
- PS pump reservoir
- Heater hoses to heater core
- Fuel vapor feed hose
- Air hose No 5
- Wiring to ECM, alternator, wiring clamp, ground wire on alternator stay, PS hose to oil pan bolt, body ground strap

9. reinstall remaining components in reverse order of removal.
10. Add engine oil, automatic transmission fluid and coolant
11. Add power steering fluid and bleed system
12. Start engine and check for fluid, fuel and exhaust leaks
13. Set ignition timing and engine speed
14. Inspect and adjust front end alignment
15. Check speed sensor signal

GS 430 (2002–03)

1. Before servicing the vehicle, refer to the Precautions Section.
2. Relieve the fuel pressure from the fuel lines.
3. Drain the engine coolant from the cooling system.
4. Remove or disconnect the following:
   - Battery cables and remove the battery. Wait at least 90 seconds before proceeding with any other work.
   - Hood
   - V-bank cover, if equipped
   - Engine undercover and drain the engine oil. Lower the vehicle.
   - Drive belt
   - Throttle body
   - Accelerator, transmission and cruise control cables from the throttle body.
   - Air cleaner assembly
   - Vacuum hose (from the power steering air control valve) from the air intake chamber.
   - Intake air connector
   - Coolant reservoir tank
   - Radiator
   - Igniter connectors and wire clamp
   - Engine wires located next to the relay box, which is located next to the left strut tower

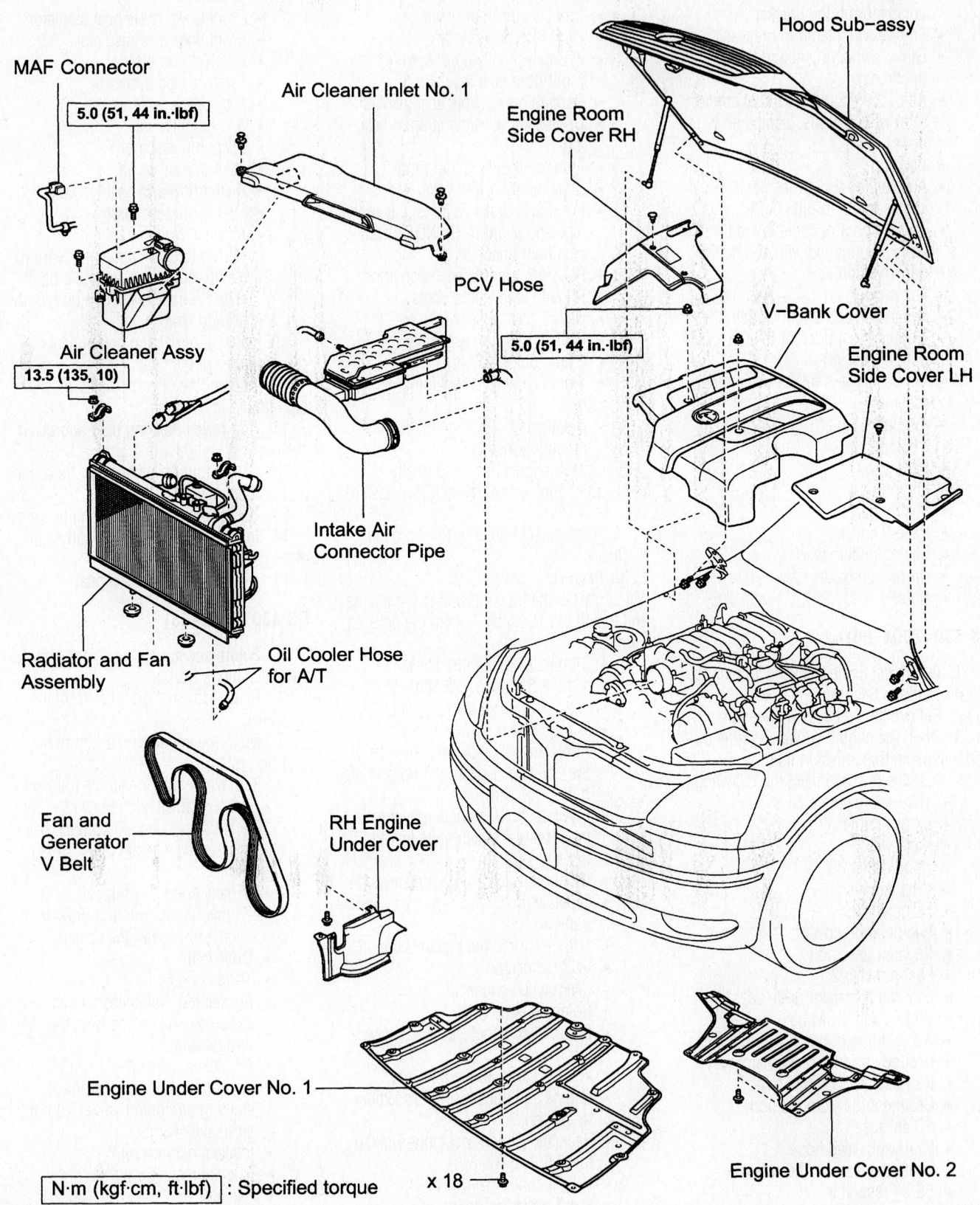

MAF Connector

5.0 (51, 44 in.·lbf)

Air Cleaner Inlet No. 1

Hood Sub-assy

Engine Room
Side Cover RH

PCV Hose

5.0 (51, 44 in.·lbf)

V-Bank Cover

Engine Room
Side Cover LH

Air Cleaner Assy

13.5 (135, 10)

Intake Air
Connector Pipe

Radiator and Fan
Assembly

Oil Cooler Hose
for A/T

Fan and
Generator
V Belt

RH Engine
Under Cover

Engine Under Cover No. 1

x 18

Engine Under Cover No. 2

N·m (kgf·cm, ft·lbf) : Specified torque

09490_LEXU_G0018

Exploded view underhood covers requiring removal—2004–06 LS 430

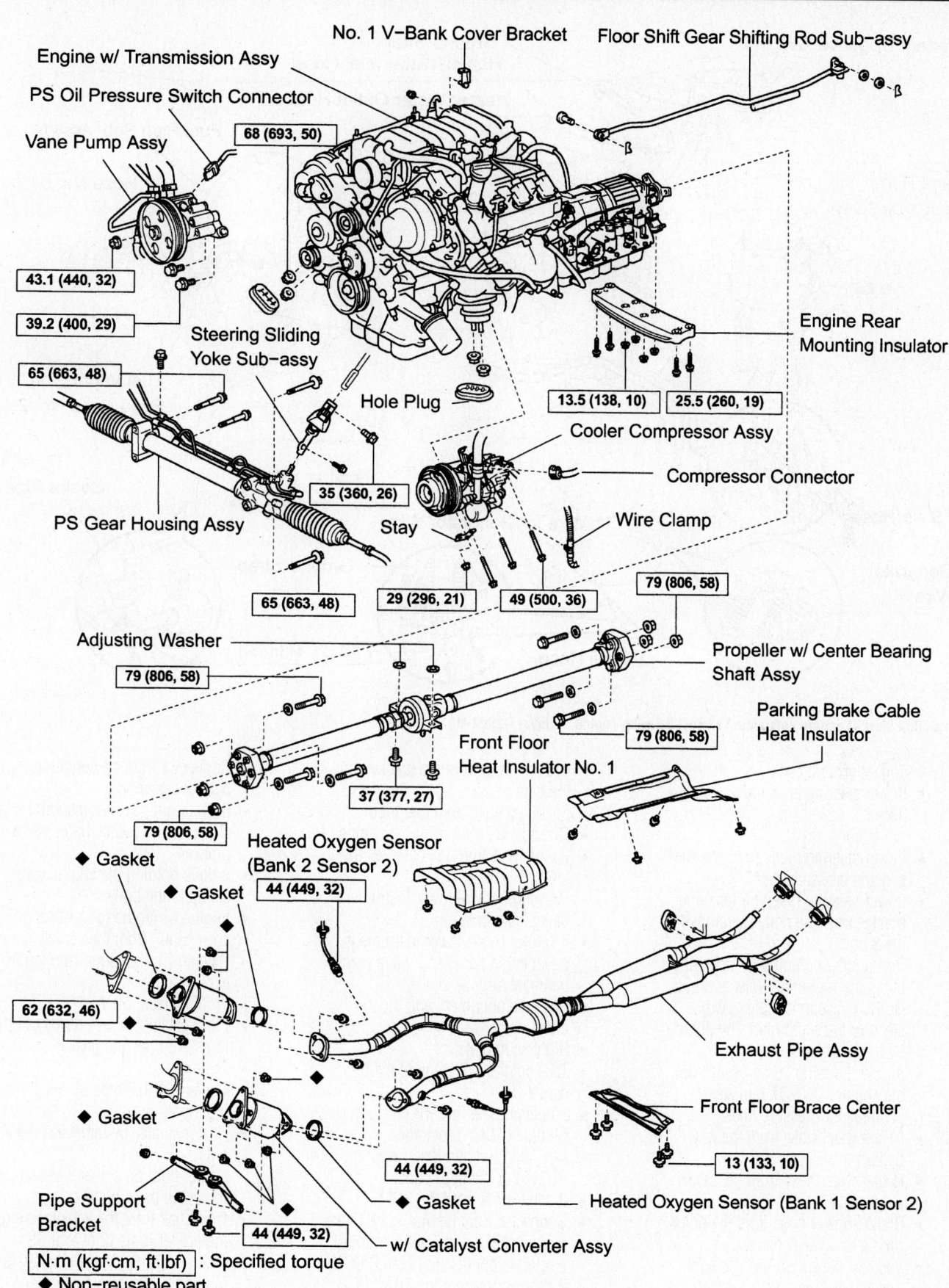

No. 1 V-Bank Cover Bracket

Floor Shift Gear Shifting Rod Sub-assy

Engine w/ Transmission Assy

PS Oil Pressure Switch Connector

Vane Pump Assy

68 (693, 50)

43.1 (440, 32)

39.2 (400, 29)

Steering Sliding Yoke Sub-assy

65 (663, 48)

Hole Plug

Engine Rear Mounting Insulator

13.5 (138, 10)   25.5 (260, 19)

Cooler Compressor Assy

Compressor Connector

35 (360, 26)

PS Gear Housing Assy

Stay

Wire Clamp

65 (663, 48)

29 (296, 21)   49 (500, 36)

79 (806, 58)

Adjusting Washer

79 (806, 58)

Propeller w/ Center Bearing Shaft Assy

Parking Brake Cable Heat Insulator

79 (806, 58)

Front Floor Heat Insulator No. 1

37 (377, 27)

79 (806, 58)

Heated Oxygen Sensor (Bank 2 Sensor 2)

◆ Gasket

◆ Gasket

44 (449, 32)

62 (632, 46)

Exhaust Pipe Assy

◆ Gasket

Front Floor Brace Center

13 (133, 10)

Pipe Support Bracket

44 (449, 32)

◆ Gasket

Heated Oxygen Sensor (Bank 1 Sensor 2)

w/ Catalyst Converter Assy

N·m (kgf·cm, ft·lbf) : Specified torque
◆ Non-reusable part

Exploded view engine attached components—2004–06 LS 430

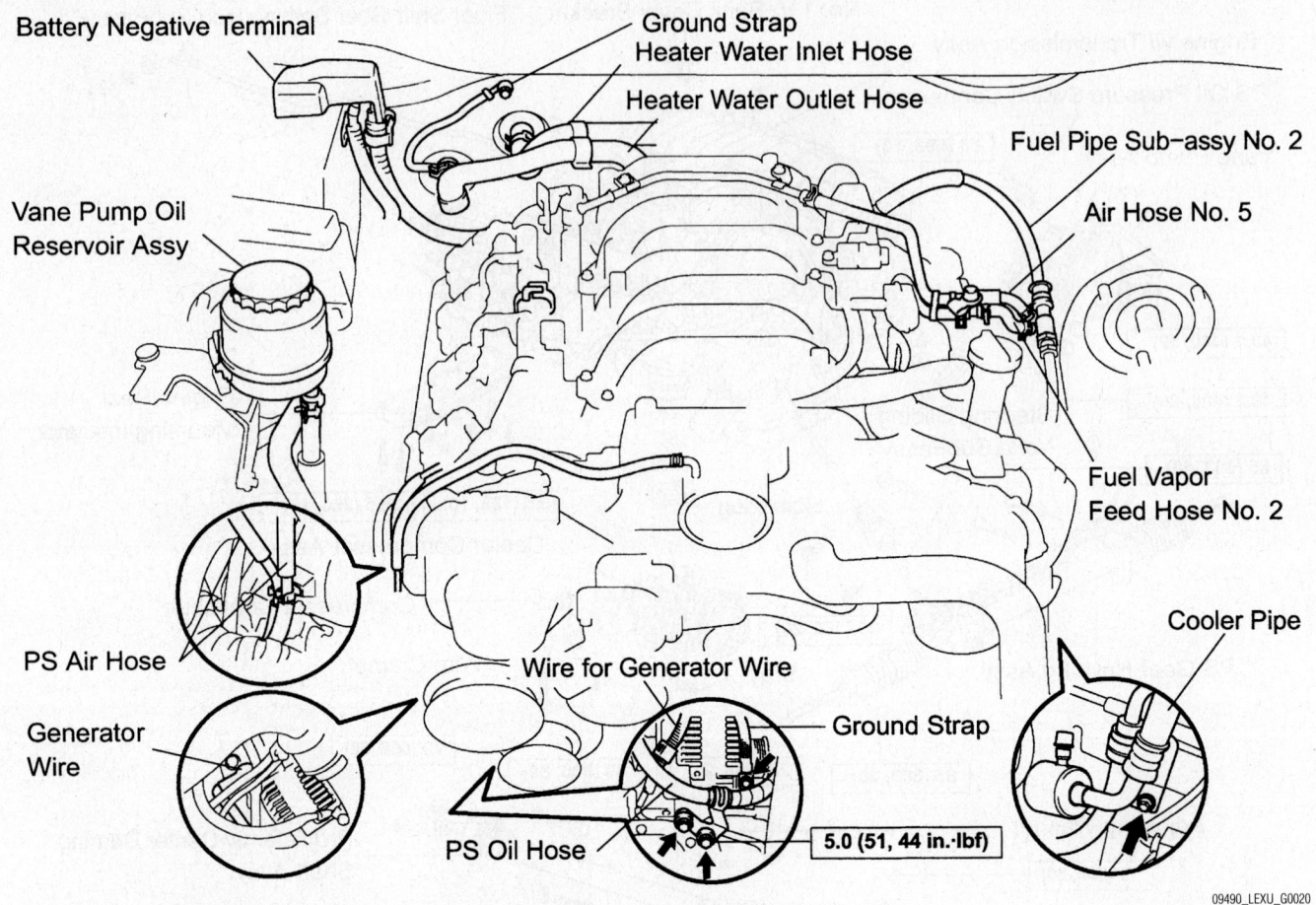

Exploded view of engine component removed during engine removal—2004–06 LS 430

- Engine ground cable
- Power steering solenoid valve connector
- Alternator
- Power steering tubes from the suspension crossmember
- Power steering reservoir tank and bracket from the body by removing the 3 bolts
- Power steering pump by removing the pump mounting bolts and nut. Do NOT disconnect the power steering lines and place the pump off to the side.
- Air conditioning compressor from the engine. Do NOT remove the compressor pressure lines.
- Heater water hose from the water bypass hose
- Heater water hose from the heater water valve
- Brake booster hose from the union on the air intake chamber
- Vacuum hose from the Vacuum Switching Valve (VSV) for the heater water valve from the air intake chamber

- Ground strap from the bracket on the body
- Fuel inlet hose from fuel tube
- Charcoal canister from the engine
- Engine wire from the cabin as follows:
- Passenger's side lower instrument panel undercover
- 4 screws to the lower instrument panel finish panel and glove compartment door assembly
- Glove compartment and finish panel.
- Right scuff plate
- Take out the front side of the floor carpet
- 2 nuts and the Powertrain Control Module (PCM) protector
- Mounting nut and disconnect the PCM from the floor panel.
- 2 connectors from the PCM
- Connector from the Anti-lock brake system (ABS) and Traction control electronic control unit (TRAC ECU)
- 2 connectors from the TRAC ECU
- 4 connectors from connector cassette

- Connector from air conditioning control assembly
- Bolt holding the engine wire clamp to the heater water valve bracket.
- 2 bolts holding the engine wire clamp to the body.
- Engine wiring harness (through the cowl panel) from the vehicle cabin
- Oxygen (O_2) sensors from the front exhaust pipe
- Front exhaust pipe
- Front catalytic converter by removing the 3 nuts and gasket
- Tailpipes
- Center exhaust pipe by disconnecting the 2 hooks
- Heat insulator by removing the 4 nuts
- Center floor crossmember brace by removing the 4 bolts
- Driveshaft from the vehicle using the proper tools (2 of tool SST 09922–10010), loosen the adjusting nut on the driveshaft. Place matchmarks on the transmission flange and the flexible coupling.

- The transmission control rod from the shift lever by removing the nut
5. Attach the engine chain hoist to the engine hangers.
6. Remove or disconnect the following:
   - 2 nuts holding the engine mounting insulators to the front suspension crossmember.
   - 4 bolts, 4 nuts and the rear engine mounting member
   - Ground strap to the rear mounting member
   - Engine out of the vehicle
   - Oil dipstick guide and dipstick for transmission
   - Oil cooler pipes for the transmission
   - All the engine wiring
   - Engine bolts holding the transmission to the engine
   - Engine from the transmission

To install:

7. Install or connect the following:
   - Transmission to the engine and install the bolts. Tighten the bolts to 42 ft. lbs. (57 Nm).
   - Engine wiring
   - Oil cooler pipe for the transmission. Tighten the unions on the pipes to 25 ft. lbs. (34 Nm).
   - Engine oil dipstick guide and the dipstick for the transmission
   - Engine and transmission to the vehicle
   - Rear engine mounting member with the 4 bolts and 4 nuts. Tighten the bolts to 19 ft. lbs. (25 Nm) and the nuts to 10 ft. lbs. (13 Nm).
   - 2 nuts holding the engine mounting brackets to the front suspension crossmember. Tighten the nuts to 43 ft. lbs. (59 Nm).
   - Engine chain hoist
   - Transmission control rod to the shift lever by installing the nut
   - Driveshaft
   - Center floor crossmember brace by installing the 4 bolts. Tighten the bolts to 108 inch lbs. (13 Nm).
   - Heat insulator for the front exhaust pipe by installing the 4 bolts
   - Center exhaust pipe by installing the 2 hooks
   - Tailpipe and tighten the 2 bolts to 14 ft. lbs. (19 Nm)
   - Front catalytic converter and tighten the nuts to 46 ft. lbs. (62 Nm)
8. Torque the front exhaust retainers as follows:
   - 4 bolts and nuts holding the catalytic converter to the front exhaust pipe to 32 ft. lbs. (43 Nm)

- 2 bolts and nuts holding the front exhaust pipe to the center exhaust pipe to 32 ft. lbs. (43 Nm)
- 4 bolts holding the pipe support bracket to the transmission to 32 ft. lbs. (43 Nm)
9. Install or reconnect the following:
   - O_2 sensors to the front exhaust and tighten the sensors to 33 ft. lbs. (44 Nm)
   - Engine wiring harness in through the cowl panel
   - Engine wire retainer with the 3 bolts
   - Reattach the connectors under the dash panel
   - PCM with the nut
   - PCM protector with the 2 nuts
   - Floor carpet
   - Scuff plate
   - Lower instrument panel finish panel and glove compartment door assembly with the 4 screws
   - Instrument panel undercover with the 2 screws
   - Charcoal canister
   - All hoses and grounds
   - Air conditioning compressor with the nut and 3 bolts. Tighten the bolts to 36 ft. lbs. (49 Nm) and the nut to 22 ft. lbs. (29 Nm).
   - Power steering pump with the nut and 3 bolts. Tighten the bolts to 29 ft. lbs. (39 Nm) and the nut to 32 ft. lbs. (43 Nm).
   - Power steering reservoir tank and bracket with the 3 bolts
   - Power steering tubes with the clamp and bolt
   - Alternator, tighten the nut and bolt to 27 ft. lbs. (37 Nm)
   - Power steering solenoid valve connector
   - Engine wire connectors
   - Theft deterrent horn connector
   - Ground cable to the body from the engine
   - Igniter connectors
   - Yellow taped connector to the igniter on the rear side
   - Radiator assembly
   - The reservoir tank and the inlet pipe to the fan shroud and tighten the 4 bolts to 43 inch lbs. (5 Nm)
   - The 2 hydraulic lines for the fan motor and tighten the bolts to 47 ft. lbs. (64 Nm)
   - Upper and lower radiator hoses to the radiator
   - 2 oil cooler hoses for the transmission to the radiator
   - Coolant tank

- Intake air connector
- Vacuum hose (from the power steering air control valve) to the air intake chamber
- Air cleaner
- Accelerator, transmission throttle control and the cruise control actuator cables to the engine
- Throttle cover and hose clamp with the cap nut and 2 bolts
- Evaporative emission control (EVAP) hose to the hose clamp
- Drive belt to the engine
- Battery to the engine compartment and attach the electrical connectors
- Engine coolant
- V-bank cover if it was removed
- Engine undercover
- Hood.
10. Fill the engine oil and check the transmission oil.
11. Start the engine, bleed the cooling system, and check for leaks.

GS 430 (2004–06)

1. Before servicing the vehicle, refer to the Precautions Section.
2. Disconnect the negative battery cable. Wait at least 90 seconds before proceeding with any other work. This provides time to disarm the airbag system.
3. Remove the hood.
4. Relieve the fuel pressure from the fuel lines.
5. Remove the engine under cover.
6. Drain the engine coolant and engine oil.
7. Remove or disconnect the following:
   - V-bank cover, if equipped
   - Air cleaner inlet, air cleaner assembly, and intake air connector pipe
   - Radiator
   - Serpentine drive belt
   - Front suspension member brace.
8. Before disconnecting any wiring, connectors, cables or hoses, be sure they are properly marked for proper reinstallation. As items are marked, remove the following:
   - Alternator wiring
   - Power steering oil hose from oil pan
   - Power steering hoses and clamp from timing belt cover
   - Starter cable from battery
   - Ground strap from body
   - Both heater hoses
   - Fuel inlet hose (rear fuel pipe) from fuel main tube
9. Disconnect all wiring connectors and grommet from ECM box

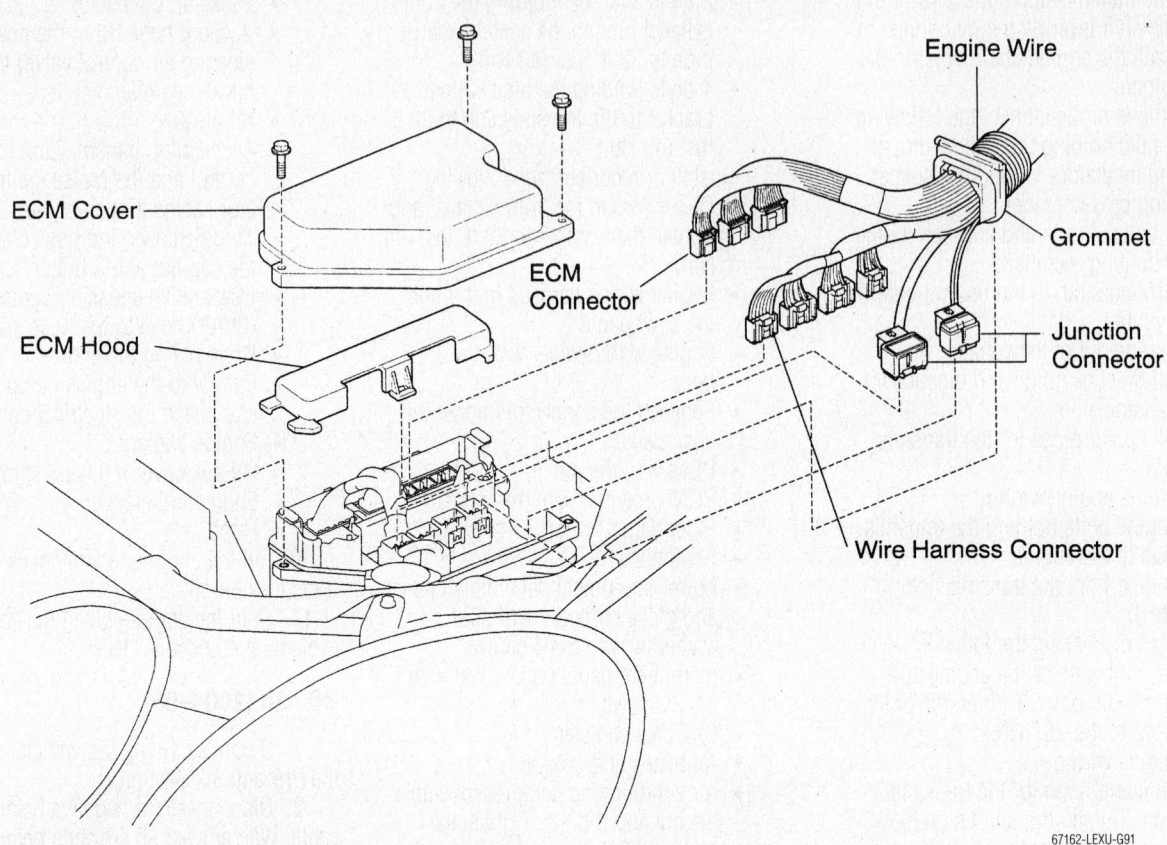

67162-LEXU-G91

Disconnecting the wiring from the ECM box—GS 430 (2004 model)

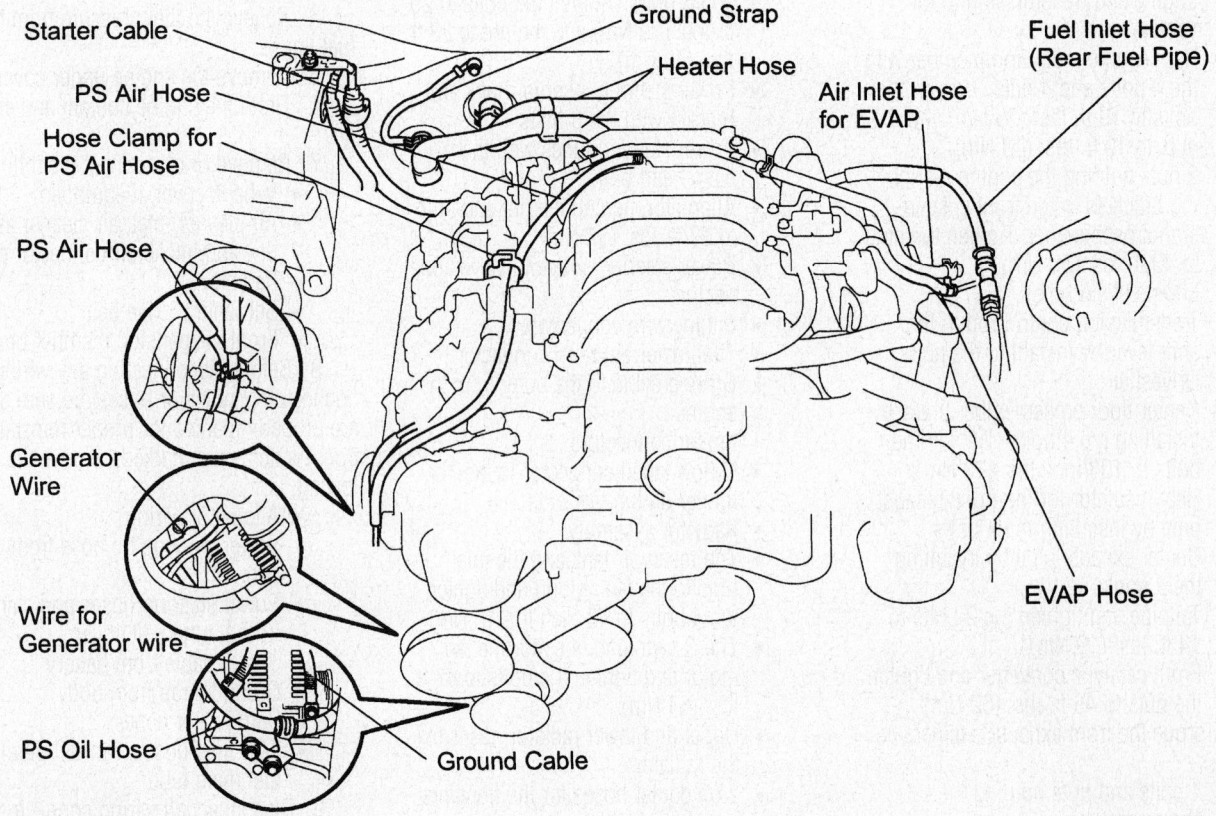

09490_LEXU_G0021

Showing the locations of wiring and other components to disconnect for engine removal—GS 430 (2004–06)

10. Disconnect the Heated Oxygen Sensors (HEGO) from the exhaust pipe, after disconnecting the wiring grommets from the floor panel.

11. Remove the front and center exhaust pipes and then remove the catalytic converters.

12. Remove or disconnect the following:
- Center front floor brace
- Heat insulators
- Driveshaft
- A/C compressor from engine mounting (Do NOT disconnect A/C hoses)
- Power steering pump oil pressure switch connector
- Power steering pump from mounting (Do NOT disconnect power steering hoses)
- Transmission shift control rod at both ends
- Power steering rack and pinion gear housing from its body mountings (Do NOT disconnect from axles or CV joints); ensure gear assembly is securely suspended

13. Attach a proper engine hangers to engine and attach a chain hoist to the hangers.

14. Remove the hole plugs to access the front engine mounts and remove the engine mount nuts.

15. Remove the rear engine mount crossmember.

16. Carefully lift the engine and transmission assembly out of the vehicle. Ensure assembly clears all wiring, hoses and components.

To install:

17. Using the chain hoist, carefully reposition the engine and transmission assembly into the vehicle.

18. Insert the stud bolts of the front engine mounting brackets into the holes of the front suspension crossmember.

➡**Ensure the engine is kept level during reattachment.**

19. Install the rear engine mount crossmember (with "V8" mark facing forward) and torque the bolts to 19 ft. lbs. (26 Nm) and the nuts to 10 ft. lbs. (14 Nm).

20. Install the 4 nuts on the front engine mount brackets and tighten to 50 ft. lbs. (68 Nm). Install the hole plugs.

21. Remove the chain hoist.

22. Reinstall the V-bank cover bracket to the engine hanger, with a nut.

23. Install or connect the following:
- Heat insulators to front side of front exhaust pipe
- Power steering gear housing to

mounting; torque nuts to 48 ft. lbs. (65 Nm)
- Power steering gear oil tube
- Steering column yoke bolt to power steering gear; torque to 26 ft. lbs. (25 Nm)
- Transmission control rod
- A/C compressor to mounting; torque bolt to 36 ft. lbs. (49 Nm) and nut to 21 ft. lbs. (29 Nm)
- A/C compressor wiring
- Power steering pump to mounting; torque bolts to 29 ft. lbs. (39 Nm) and nut to 21 ft. lbs. (29 Nm)
- Power steering oil pressure switch connector
- Driveshaft
- Front center floor brace
- Heat insulators
- Catalytic converters; torque new nuts to 46 ft. lbs. (62 Nm)
- Front and center exhaust pipes
- Oxygen sensors; tighten to 32 ft. lbs. (44 Nm)

➡**Before installing oxygen sensors, twist sensor wiring 3-1/2 turns in counterclockwise direction; hold it in this position, then install the sensors. This method prevents sensor wires from being twisted after installation.**

- Wiring connectors and grommet to ECM box
- All wiring, connectors, straps and hoses to original positions as marked
- Front suspension member brace; torque bolts to 43 ft. lbs. (58 Nm)
- Serpentine drive belt
- Radiator
- Intake air connector, air cleaner assembly and air cleaner inlet

24. Refill engine and cooling system with proper fluids and amounts.

25. Install the hood, then road test the vehicle. Recheck fluid levels after road test.

SC 430

1. Before servicing the vehicle, refer to the Precautions Section.

2. Relieve the fuel pressure from the fuel lines.

3. Drain the engine coolant from the cooling system.

4. Remove or disconnect the following:
- Battery negative cable
- Hood sub-assembly
- V-Bank cover
- Air cleaner inlet
- Air cleaner assembly and connector pipe

- Engine undercover and drain the oil
- Drain Automatic Transmission fluid
- Remove radiator
- Fuel sub-pipe assembly
- V-belts
- Engine wire from ECU
- Alternator wires
- PS hose from # 1 oil pan bolt
- Ground strap from the body
- PS Air hose
- Fuel vapor feed hose
- Heater inlet and outlet water hose
- PS pump reservoir
- RH engine under cover
- PS oil switch connector
- PS pump from the engine
- A/C Compressor electrical connector and clamp
- A/C compressor from the engine (Do NOT disconnect hoses)
- Support compressor assembly
- Front floor brace assembly
- Exhaust pipe assembly
- Front suspension member brace sub-assembly
- Catalytic converter assembly
- Front floor heat insulator
- Parking brake cable heat insulator
- Drive shaft W/Center bearing assembly
- Steering sliding yoke assembly
- The bolts and disconnect the 2 PS oil tubes from the front frame.
- The 4 bolts securing the PS gear housing from the front frame
- Suspend the PS gear housing securely
- V-bank cover bracket from number engine hanger
- Install engine chain hoist to the engine hangers and support engine
- Remove 2 hole plugs
- 4 nuts holding the engine mounting insulator to the front suspension cross member
- 4 nuts and bolts from the rear engine mounting member
- Remove engine slowly and carefully. Make sure engine is clear of all wiring, hoses and cables.

To install:

5. Lower engine and transmission assembly onto the engine compartment

6. Install or connect the following:
- front engine mounting bolts in front suspension crossmember
- Rear engine mounting member with 4 bolts and nuts. Torque to: bolts 19 ft. lbs (25.5 Nm), Nuts 10 ft. lbs (13.5 Nm)
- 4 nuts holding engine mounting

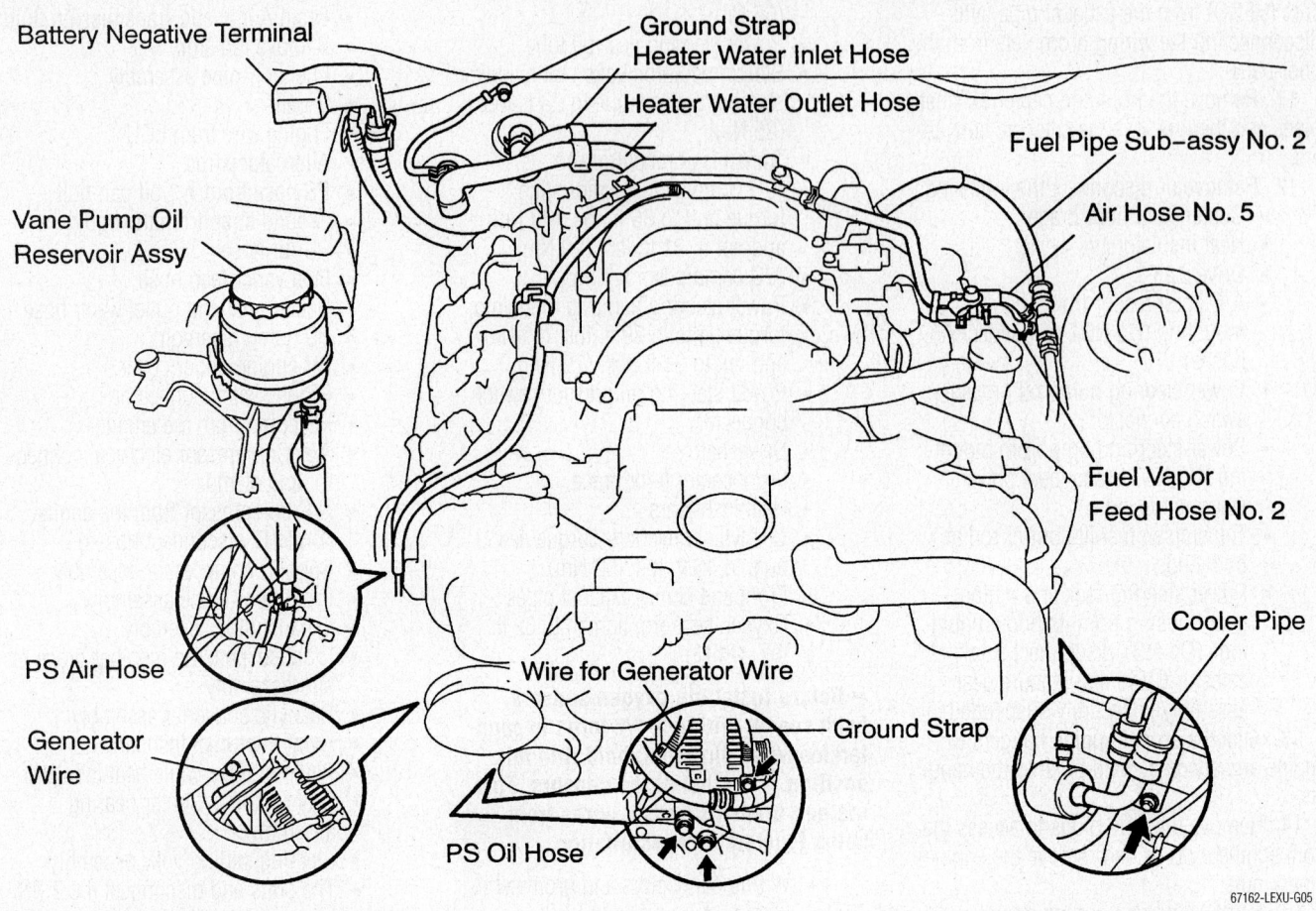

Battery Negative Terminal

Ground Strap
Heater Water Inlet Hose

Heater Water Outlet Hose

Fuel Pipe Sub–assy No. 2

Air Hose No. 5

Vane Pump Oil
Reservoir Assy

Fuel Vapor
Feed Hose No. 2

Cooler Pipe

PS Air Hose

Wire for Generator Wire

Generator
Wire

Ground Strap

PS Oil Hose

67162-LEXU-G05

Engine component locations—SC 430

brackets to front crossmember.
Torque to: 50 ft. lbs. (68 Nm)
- Front V-bank cover bracket to engine hanger with nut
- Sliding PS gear housing yoke to intermediate shaft
- PS gear housing with 4 bolts. Torque to 48 ft. lbs (65 Nm)
- PS oil tube with bolts
- Steering sliding yoke sub-assembly with 2 bolts. Torque to 26 ft. lbs (35 Nm)
- Floor gear shift rod sub-assembly
- Drive shaft W/Center bearing shaft assembly
- Parking brake cable heat insulator
- Front floor heat insulator
- Catalytic converter assembly
- Front suspension member sub-assembly. Torque to 43 ft. lbs. (58Nm)
- Exhaust pipe assembly
- Front floor brace center
- A/C compressor, stay and wire bracket with 3 bolts and nuts.

Torque bolts to 36 ft. lbs (49 Nm) and nuts to 21 ft. lbs. (29 Nm)
- A/C wiring connector and clamps
- PS pump with 2 bolts and nuts. Alternately tighten the bolts and nuts: bolts to 29 ft. lbs. (39 Nm) and nuts to 32 ft. lbs. (43 Nm)
- PS oil pressure switch connector
- PS air hoses to the PS pump
- PS pump reservoir assembly
- Heater inlet and outlet water hoses
- Vapor feed hose
- Upper PS air hose
- Engine wire from ECM box
- Alternator wires and clamp
- Ground cable to alternator
- PS hose to #1 oil pan bolt
- Ground strap to body
- V-belts
- Air cleaner assembly and connector pipes
- V-bank cover
- Hood sub-assembly
- Battery negative cable
- Automatic transmission fluid
- Engine coolant

- Engine oil
- Start engine
- Inspect for coolant, oil and transmission fluid leaks
- Adjust engine settings
- Test drive vehicle

ES 330

1. Before servicing the vehicle, refer to the Precautions Section.
2. Relieve the fuel pressure from the fuel lines.
3. Drain the engine coolant from the cooling system.
4. Remove or disconnect the following:

- Front wheels
- Engine under covers
- Front fender apron seal
- Drain engine oil
- Drain automatic transmission fluid
- V-bank cover sub-assembly
- Radiator lower air deflector
- Battery cables and remove battery
- Battery tray

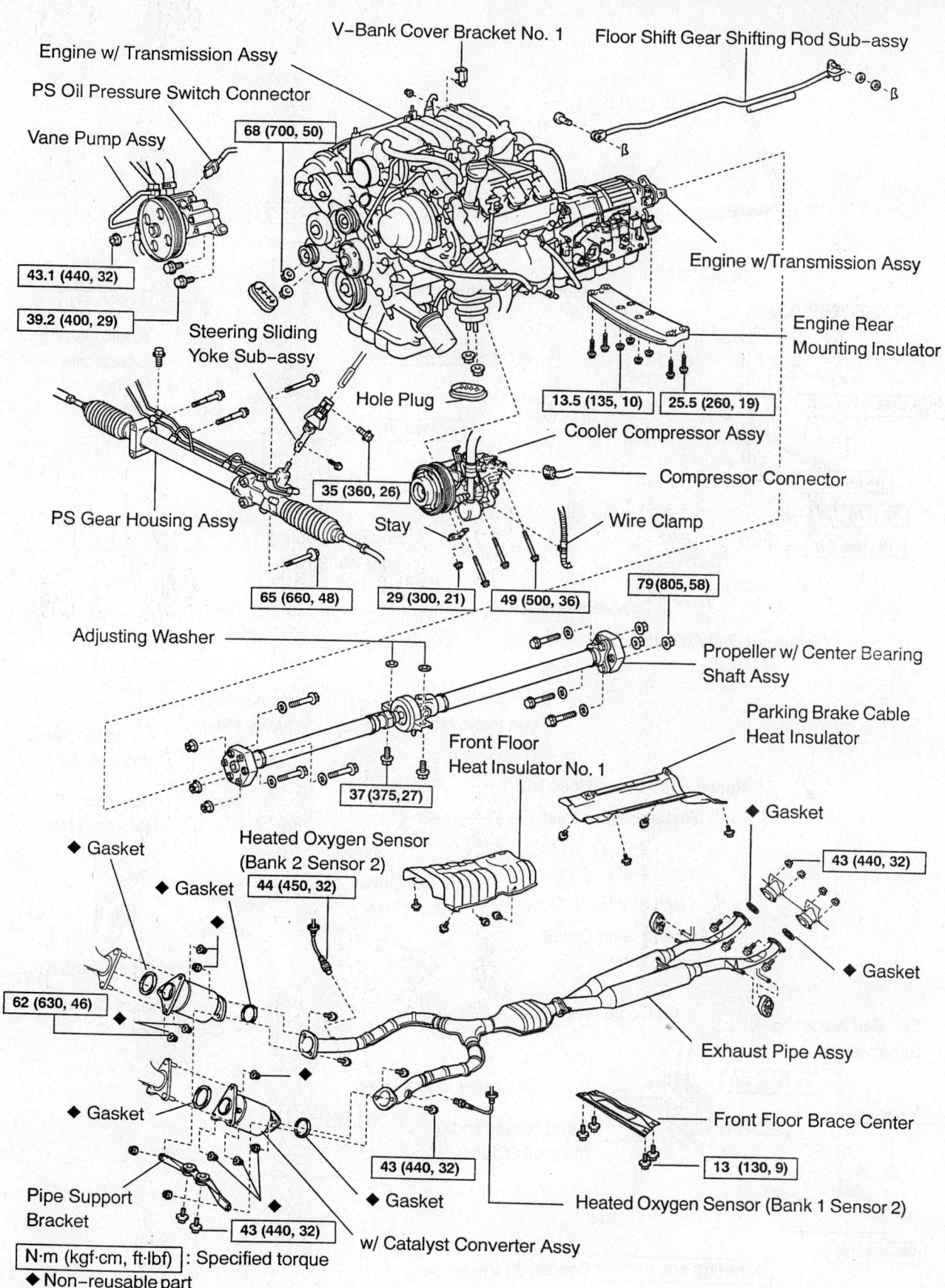

Engine w/ Transmission Assy

PS Oil Pressure Switch Connector

Vane Pump Assy

V–Bank Cover Bracket No. 1

Floor Shift Gear Shifting Rod Sub–assy

68 (700, 50)

43.1 (440, 32)

39.2 (400, 29)

Engine w/Transmission Assy

Engine Rear Mounting Insulator

Steering Sliding Yoke Sub–assy

Hole Plug

13.5 (135, 10)   25.5 (260, 19)

Cooler Compressor Assy

Compressor Connector

PS Gear Housing Assy

35 (360, 26)

Stay

Wire Clamp

65 (660, 48)   29 (300, 21)   49 (500, 36)   79 (805, 58)

Adjusting Washer

Propeller w/ Center Bearing Shaft Assy

Parking Brake Cable Heat Insulator

Front Floor Heat Insulator No. 1

37 (375, 27)

◆ Gasket

43 (440, 32)

◆ Gasket

Heated Oxygen Sensor (Bank 2 Sensor 2)

44 (450, 32)

◆ Gasket

62 (630, 46)

◆ Gasket

◆ Gasket

Exhaust Pipe Assy

Front Floor Brace Center

43 (440, 32)

Heated Oxygen Sensor (Bank 1 Sensor 2)

13 (130, 9)

Pipe Support Bracket

43 (440, 32)

◆ Gasket

w/ Catalyst Converter Assy

N·m (kgf·cm, ft·lbf) : Specified torque
◆ Non–reusable part

67162-LEXU-G06

Engine removal components, exploded view—(SC 430)

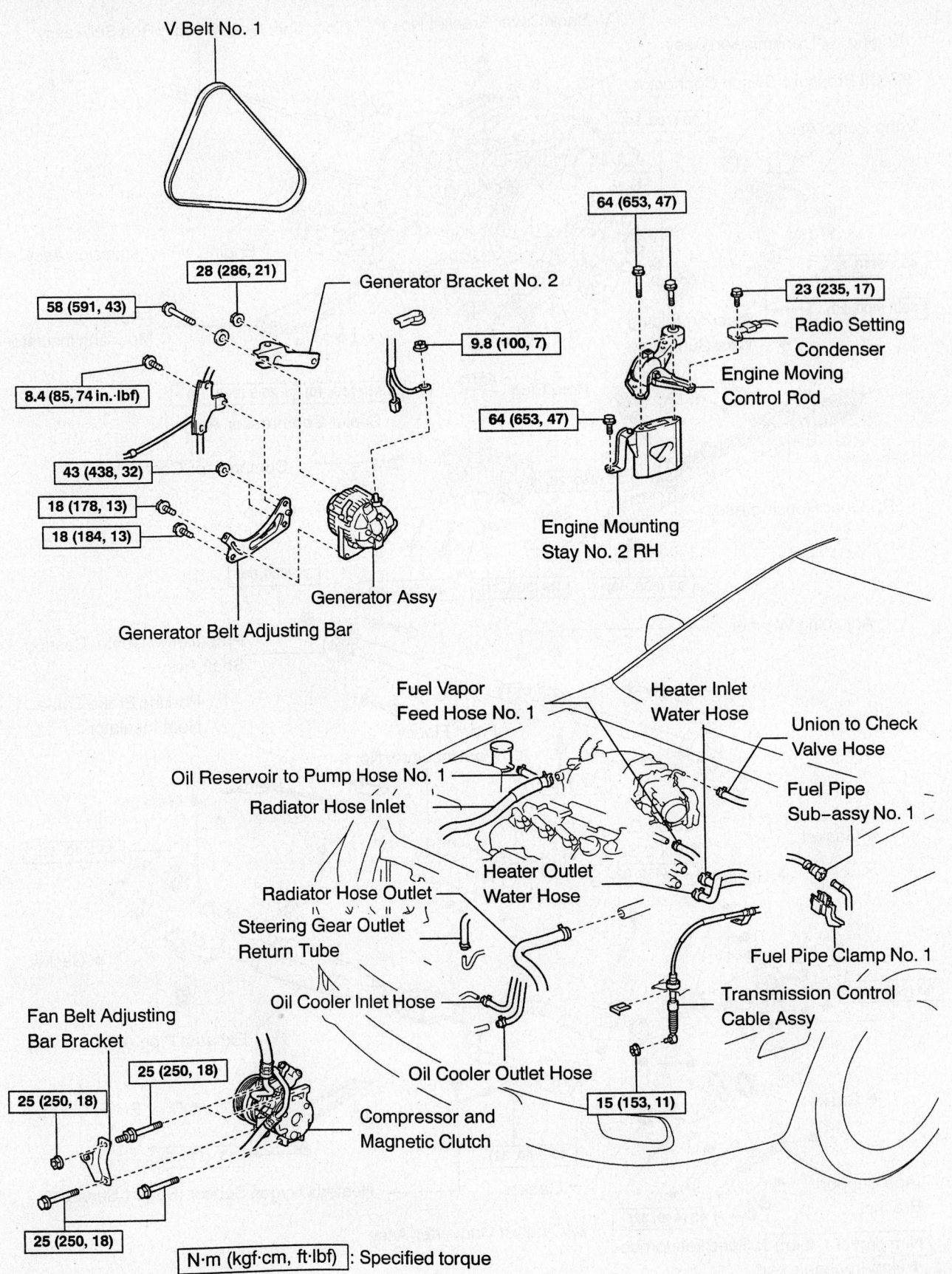

V Belt No. 1

28 (286, 21)
58 (591, 43)
Generator Bracket No. 2
8.4 (85, 74 in.·lbf)
9.8 (100, 7)
43 (438, 32)
18 (178, 13)
18 (184, 13)
Generator Assy
Generator Belt Adjusting Bar

64 (653, 47)
23 (235, 17)
Radio Setting Condenser
Engine Moving Control Rod
64 (653, 47)
Engine Mounting Stay No. 2 RH

Fuel Vapor Feed Hose No. 1
Heater Inlet Water Hose
Union to Check Valve Hose
Oil Reservoir to Pump Hose No. 1
Radiator Hose Inlet
Fuel Pipe Sub–assy No. 1
Radiator Hose Outlet
Heater Outlet Water Hose
Steering Gear Outlet Return Tube
Oil Cooler Inlet Hose
Fuel Pipe Clamp No. 1
Fan Belt Adjusting Bar Bracket
Transmission Control Cable Assy
Oil Cooler Outlet Hose
25 (250, 18)
25 (250, 18)
15 (153, 11)
Compressor and Magnetic Clutch
25 (250, 18)

N·m (kgf·cm, ft·lbf) : Specified torque

67162-LEXU-G07

Engine compartment accessory locations—ES 330

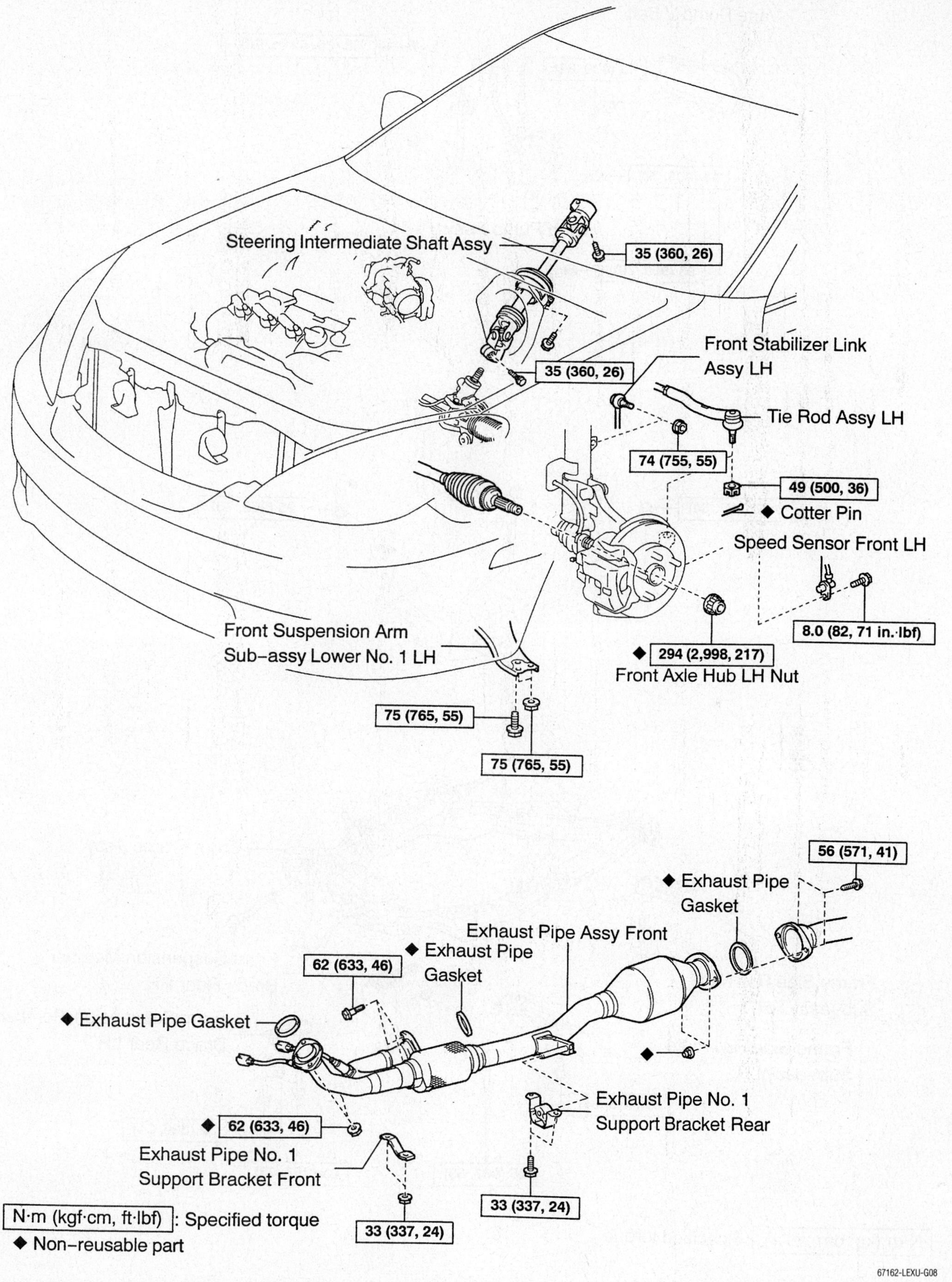

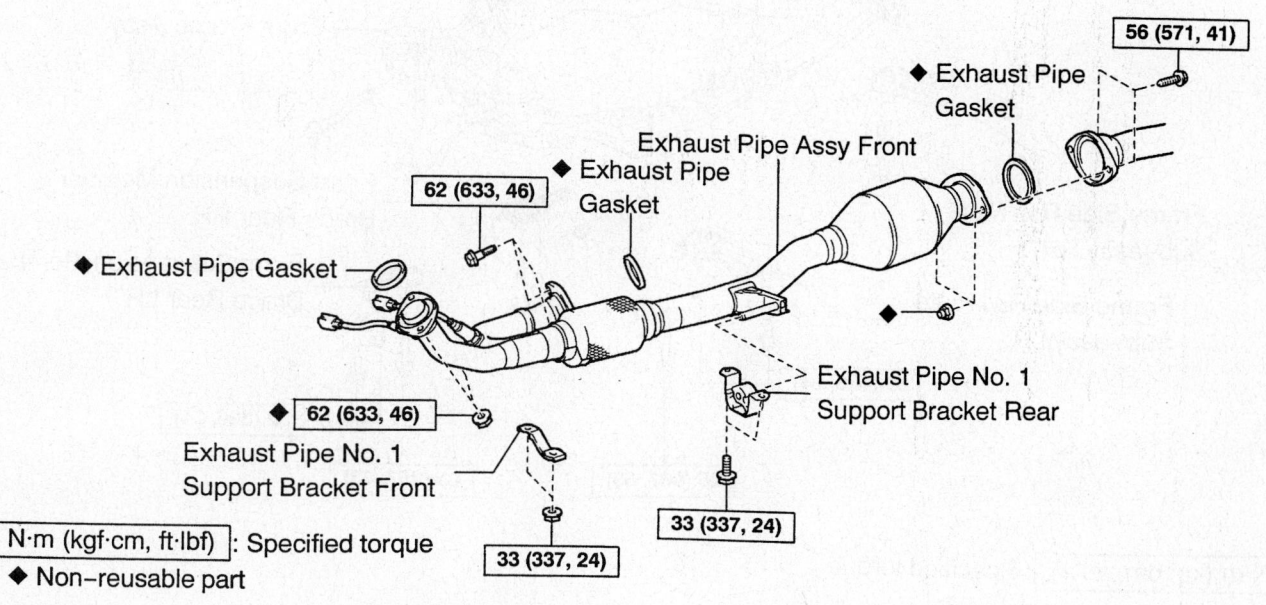

Steering Intermediate Shaft Assy

35 (360, 26)

35 (360, 26)

Front Stabilizer Link Assy LH

Tie Rod Assy LH

74 (755, 55)

49 (500, 36)

◆ Cotter Pin

Speed Sensor Front LH

8.0 (82, 71 in.·lbf)

Front Suspension Arm Sub–assy Lower No. 1 LH

◆ 294 (2,998, 217)
Front Axle Hub LH Nut

75 (765, 55)

75 (765, 55)

56 (571, 41)

◆ Exhaust Pipe Gasket

Exhaust Pipe Assy Front

62 (633, 46)

◆ Exhaust Pipe Gasket

◆ Exhaust Pipe Gasket

◆ 62 (633, 46)

Exhaust Pipe No. 1 Support Bracket Front

Exhaust Pipe No. 1 Support Bracket Rear

33 (337, 24)

N·m (kgf·cm, ft·lbf): Specified torque
◆ Non-reusable part

33 (337, 24)

67162-LEXU-G08

Steering and exhaust components removal—ES 330

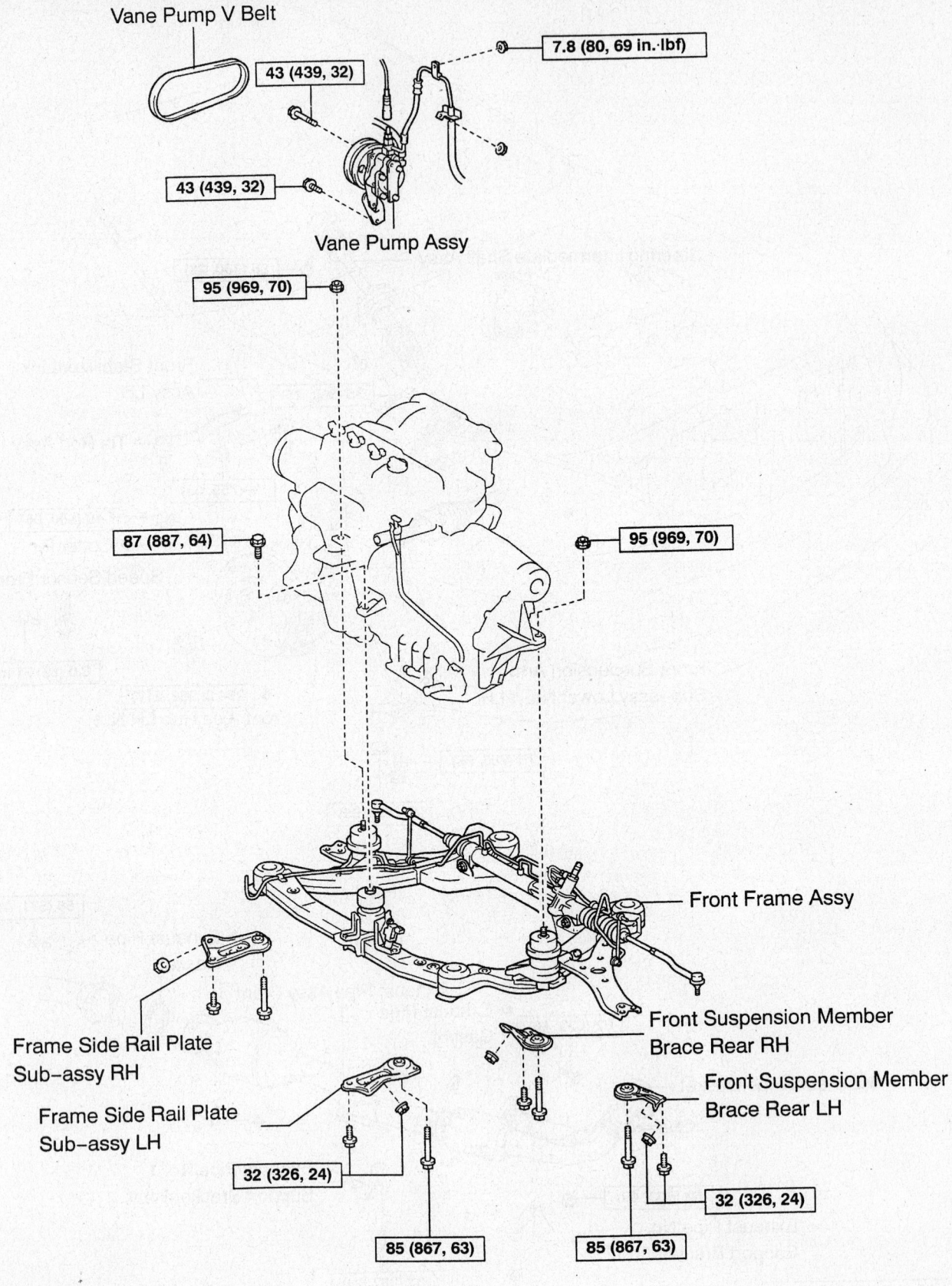

Vane Pump V Belt

43 (439, 32)

7.8 (80, 69 in.·lbf)

43 (439, 32)

Vane Pump Assy

95 (969, 70)

87 (887, 64)

95 (969, 70)

Front Frame Assy

Frame Side Rail Plate
Sub–assy RH

Front Suspension Member
Brace Rear RH

Frame Side Rail Plate
Sub–assy LH

Front Suspension Member
Brace Rear LH

32 (326, 24)

32 (326, 24)

85 (867, 63)

85 (867, 63)

N·m (kgf·cm, ft·lbf) : Specified torque

67162-LEXU-G09

Sub frame component removal—ES 330

- Air cleaner assembly, brackets and inlets
- Intake air resonator sub-assembly
- A/C compressor V-belt
- Alternator wiring and alternator
- Engine stabilizing control rod
- Engine mounting stay
- Alternator bracket
- Alternator adjusting bar
- A/C compressor (Do NOT disconnect hoses)
- Transmission control cable assembly
- Union to check valve hose
- Fuel vapor feed hose
- Fuel pipe sub-assembly
- Heater inlet and outlet hoses
- Radiator inlet and outlet hoses
- Oil cooler inlet and outlet hoses
- Steering gear outlet return tube
- Glove compartment door assembly
- Engine wire from ECU and junction box
- Engine harness from engine compartment junction block
- Front exhaust pipe support bracket
- Rear exhaust pipe support bracket
- Front exhaust pipe assembly
- Front stabilizer link assembly
- Rear stabilizer link assembly
- Both front axel hub nuts
- Both front speed sensors
- Separate both outer tie rod ends
- Separate front lower suspension arm sub assembly, both sides
- Left and Right drive axels
- Separate steering intermediate shaft assembly
- Attach engine hoist
- 4 bolts and 2 nuts from RH & LH frame side rail plate
- 4 bolts and 2 nuts from RH & LH front suspension member
- Carefully remove engine assembly

To install:

5. Lower engine and transmission assembly onto the engine compartment
6. Install or connect the following:
   - RH & LH side rail plates; torque large bolt 63 ft. lbs. (85 Nm) and small bolt and nuts 24 ft lbs (32 Nm)
   - RH & LH front suspension member brace. Torque large bolt to 63 ft. lbs (85 Nm) and small bolt and nuts 24 ft. lbs. (32 Nm)
   - Steering intermediate shaft assembly
   - LH & RH axel shaft assemblies
   - LH & RH lower suspension arm sub-assemblies
   - LH & RH tie rod assemblies

- LH & RH speed sensors
- LH & RH front axle hub nuts. Torque to 217 ft. lbs. (294 Nm)
- LH & RH stabilizer link assemblies
- Front exhaust pipe assembly
- Rear exhaust support bracket
- Front exhaust support bracket
- Fuel pipe sub-assembly
- Transmission control cable assembly
- A/C compressor assembly
- Alternator belt adjusting bar and bracket
- RH engine mounting stay
- Engine stabilizing control rod
- Alternator assembly
- A/C compressor V-belt
- Inspect drive belt deflector and tensioner
- Intake air resonator. Torque to 44 inch lbs. (5 Nm)
- Air cleaner assembly, brackets and inlets
- Verify vacuum hose connections
- Add automatic transmission fluid
- Add engine oil
- Add power steering fluid
- Inspect automatic transaxle fluid
- Check for oil, coolant, fuel and exhaust leaks
- Adjust front wheel alignment
- Adjust ignition timing and engine idle speed
- Inspect CO/HC
- Check ABS sensor signal
- System initialization

Water Pump

REMOVAL & INSTALLATION

2.5L (4GR-FSE) and 3.5L (2GR-FSE) Engines

1. Before servicing the vehicle, refer to the Precautions Section.
2. Disconnect the negative battery cable. Wait at least 90 seconds before performing any work.
3. Drain the engine coolant.
4. Drain the engine oil.
5. Discharge refrigerant from A/C system.
6. Remove or disconnect the following:
   - Cool air intake duct seal
   - V-bank engine cover
   - No. 1 inlet air cleaner
7. Loosen, but do not remove water pump pulley mounting bolts.
8. Remove or disconnect the following:

- Serpentine drive belt
- Engine under cover
- Left rear engine under cover (2WD)
- Oil filter and mounting bracket assembly (AWD)
- No. 1 cooler refrigerant suction and discharge hoses

➡ **Seal the openings of the disconnected parts using vinyl tape to prevent moisture and foreign matter from entering.**

- A/C compressor and magnetic clutch
- Serpentine belt tensioner
- Radiator inlet and outlet hoses
- No. 1 engine cover
- Bolt, 2 retaining nuts and injector driver unit
- No. 2 engine cover
- Water pump pulley
- 5 hoses from the water inlet housing
- Water inlet housing, gasket and O-ring
- No. 2 idler pulley
- 16 bolts, water pump assembly and water pump gasket

To install:

9. Install a new water pump gasket and water pump assembly with the 16 mounting bolts and tighten as follows:
   - Bolt A: 15 ft. lbs. (21 Nm)
   - Bolt B: 81 inch lbs. (9 Nm)
   - Bolt C: 81 inch lbs. (9 Nm)

➡ **Be sure to replace 2 bolts C with new ones or reuse them after applying adhesive 1344. Make sure that there is no oil on the threads of the A bolts.**

10. Install or connect the following:
   - No. 2 idler pulley, cover plate and bolt. Torque the bolt to 32 ft. lbs. (43 Nm).
   - Water inlet housing, new gasket and O-ring. Torque the 4 mounting bolts to 7 ft. lbs. (10 Nm).

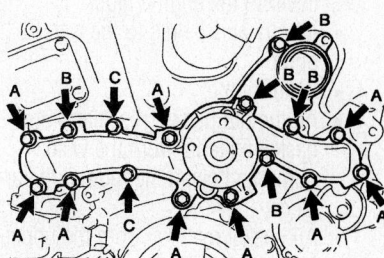

09490_LEXU_G0022

Water pump mounting bolts—2.5L (4GR-FSE), 3.0L (3GR-FSE) and 3.5L (2GR-FSE) engines

→Be careful not to allow the O-ring to get caught between parts.

- 5 hoses to the water inlet housing
- Water pump pulley with 4 bolts finger tight only
- No. 2 engine cover
- Injector driver unit
- No. 1 engine cover
- Inlet and outlet radiator hoses and new clamps
- Serpentine belt tensioner
- A/C compressor and magnetic clutch
- Oil filter and mounting bracket assembly (AWD)
- No. 1 cooler refrigerant suction and discharge hoses
- Serpentine drive belt
- Left rear engine under cover (2WD)
- Engine under cover
- No. 1 inlet air cleaner
- Negative battery cable
- V-bank engine cover
- Cool air intake duct seal

11. Torque the 4 water pump pulley bolts to 15 ft. lbs. (21 Nm)
12. Add engine oil and engine coolant.
13. Recharge the A/C system.
14. Start the engine, check for leaks and bleed the cooling system.
15. Recheck all fluid levels and add if necessary.
16. Road test the vehicle.

3.0L (1MZ-FE) Engine

1. Before servicing the vehicle, refer to the Precautions Section.
2. Remove or disconnect the following:
- Negative battery terminal
- Engine coolant
- Timing belt
- Right and left camshaft pulleys
- No. 2 idler pulley by removing the bolt
- 3 clamps and engine wire from the rear timing belt cover
- 6 bolts holding the rear timing belt cover to the engine block
- 4 bolts and 2 nuts to the water pump
- Water pump
- All the old packing (sealant) and gasket material from the water pump and clean the mounting surfaces.
- All gasket material from the upper inner timing belt cover

To install:
3. Check that the water pump turns smoothly. Also, check the air hole for coolant leakage.

Water pump mounting bolts—3.0L (1MZ-FE) engine

4. Using a new gasket, apply liquid sealer to the gasket, water pump and engine block.
5. Install or connect the following:
- Gasket and pump to the engine and install the 4 bolts and 2 nuts. Tighten the nuts and bolts to 53 inch lbs. (6 Nm).
- Rear timing belt cover and tighten the 6 bolts to 74 inch lbs. (9 Nm)
- Engine wire with the 3 clamps to the rear timing belt cover
- No. 2 idler pulley with the bolt. Tighten the bolt to 32 ft. lbs. (43 Nm). After tightening the bolt, be sure the idler pulley moves smoothly.
- Right-hand camshaft pulley, with the flange side **outward**. Be sure to align the knock pinhole on the camshaft pulley with the knock pin on the camshaft. Camshaft bolt to 65 ft. lbs. (88 Nm).
- Left-hand camshaft pulley with the flange side **inward**. Be sure to align the knock pin hole on the camshaft pulley with the knock pin on the camshaft. Camshaft bolt to 94 ft. lbs. (125 Nm).
- Timing belt
- Engine coolant
- Negative battery cable to the battery and start the engine.

3.0L (2JZ-GE) Engine

1. Before servicing the vehicle, refer to the Precautions Section.
2. Disconnect the negative battery cable. Wait at least 90 seconds before performing any work.
3. Drain the engine coolant.
4. Remove or disconnect the following:
- Radiator assembly
- Air cleaner
- Mass Air Flow (MAF) meter
- Intake air connector pipe assembly
- Serpentine drive belt
- Water pump pulley

- Timing belt
- Idler pulley
- Water bypass outlet and the No. 1 water bypass pipe
- Water inlet and the thermostat
- Alternator (position aside)
- Bolt and engine wire bracket
- Bolt and clamp bracket for the Crankshaft Sensor (CKP) connector
- Nuts and the No. 2 water bypass pipe from the water pump
- 6 bolts and the water pump and gasket
- Drain hose and the O-ring from the cylinder block

To install:
5. Install or connect the following:
- New O-ring to the cylinder block
- Drain hose
- New gasket to the water pump
- Water pump to the water bypass pipe. Do NOT install the nut yet.
- Water pump with the 2 bolts (A) and the 4 bolts (B)

→**Hand-tighten the bolts (A) first. Tighten all 6 bolts to 15 ft. lbs. (21 Nm).**

- 2 nuts holding the No. 2 water bypass pipe to the water pump. Tighten the nuts to 15 ft. lbs. (21 Nm).
- Clamp bracket for the CKP sensor connector

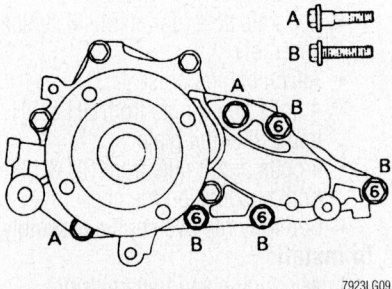

Water pump mounting bolt locations— 3.0L (2JZ-GE) engine

Be sure to use new O-rings when installing the water bypass pipe—3.0L (2JZ-GE) engine

- Engine wire bracket
- Alternator
- New O-rings to the No. 1 water bypass pipe.
- New O-ring and the water bypass outlet with the 2 bolts and tighten them to 78 inch lbs. (9 Nm)
- Thermostat and the water inlet
- Idler pulley
- Timing belt
- Serpentine drive belt
- Air cleaner, the MAF meter and the intake air connector pipe assembly
- Radiator assembly
- Negative battery cable
- Coolant. Start the engine, check for leaks and bleed the cooling system.

3.0L (3GR-FSE) Engine

1. Before servicing the vehicle, refer to the Precautions Section.

2. Disconnect the negative battery cable. Wait at least 90 seconds before performing any work.

3. Drain the engine coolant.

4. Loosen, but do not remove water pump pulley mounting bolts.

5. Remove or disconnect the following:
- Engine under cover
- Cool air intake duct seal
- No. 1 inlet air cleaner
- V-bank engine cover
- Serpentine drive belt
- Water pump pulley
- No. 2 engine cover
- No. 1 engine cover
- Mounting fasteners and injector driver unit
- Radiator inlet and outlet hoses
- 5 hoses from the water inlet housing
- Water inlet housing, gasket and O-ring
- Bolt, cover plate and No. 2 idler pulley
- Bolt, cover plate and belt tensioner pulley (marking: L)

➡**Do not turn the bolt "L" counterclockwise.**

- 16 bolts, water pump assembly and water pump gasket

To install:

6. Install a new water pump gasket and water pump assembly with the 16 mounting bolts and tighten as follows:
- Bolt A: 15 ft. lbs. (21 Nm)
- Bolt B: 81 inch lbs. (9 Nm)
- Bolt C: 81 inch lbs. (9 Nm)

➡**Be sure to replace 2 bolts C with new ones or reuse them after applying**

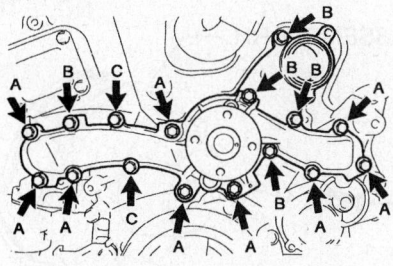

09490_LEXU_G0022

Water pump mounting bolts—2.5L (4GR-FSE), 3.0L (3GR-FSE) and 3.5L (2GR-FSE) engines

adhesive 1344. Make sure that there is no oil on the threads of the A bolts.

7. Install or connect the following:
- Belt tensioner pulley and cover plate with bolt. Torque the bolt to 32 ft. lbs. (43 Nm).

✴✴ WARNING

Be careful when tightening the bolt because it is left-hand threaded.

- No. 2 idler pulley, cover plate and bolt. Torque the bolt to 32 ft. lbs. (43 Nm).
- Water inlet housing, new gasket and O-ring. Torque the 4 mounting bolts and nut to 7 ft. lbs. (10 Nm).

➡**Be careful not to allow the O-ring to get caught between parts.**

- 5 hoses to the water inlet housing
- Inlet and outlet radiator hoses and new clamps
- Water pump pulley with 4 bolts finger tight only
- Injector driver unit
- No. 1 engine cover
- No. 2 engine cover
- Serpentine drive belt
- V-bank engine cover
- No. 1 inlet air cleaner
- Cool air intake duct seal
- Engine under cover
- Negative battery cable

8. Torque the 4 water pump pulley bolts to 15 ft. lbs. (21 Nm)

9. Add engine coolant.

10. Start the engine, check for leaks and bleed the cooling system.

11. Recheck all fluid levels and add if necessary.

12. Perform system initialization (which includes power window control system, sliding roof system, clearance sonar system and variable gear ratio steering system) procedure as follows:

- Power window control system
  a. Turn the ignition switch on.
  b. Open power window halfway by pressing power window switch.
  c. Fully pull up the switch until the power window is fully closed and continue to hold the switch for at least 1 second.
  d. Check that the AUTO UP / DOWN function operates normally.

➡**If the remote UP / DOWN function does not operate after the conditions 1), 2), or 3) is satisfied, the power window regulator master switch may have a malfunction.**

- Sliding roof system
  e. Turn the ignition switch on.
  f. If the sliding roof is opened, close it fully.
  g. Push the open switch of the slide switch, or the up switch of the tilt switch on the personal light, making the sliding roof tilt up approximately 1 second, tilt down, slide open, slide close.
  h. Sliding roof stops at the fully closed position.
  i. Finish the initialization.
  j. Check that the operation works normally with AUTO operation.
- Clearance sonar system
  k. Turn the ignition switch on.
  l. Turn the clearance sonar main switch ON.
  m. Turn the steering wheel to the full left and right lock position.

➡**Make sure to completely turn the steering wheel to the left and right full lock position.**

  n. Confirm that the learning operation has been completed by checking the multi-information display.
  o. At an area with few turns and curves, and minimal traffic, drive at 20 km/h or more for 5 minutes or more.
- Variable gear ratio steering system
  p. Turn the ignition switch on, and check that the master warning light and VSC/ABS warning lights illuminate for a few seconds.

➡**If the warning lights remain on or blink, repair the applicable system.**

  q. Drive the vehicle on a straight road at 35 km/h (22 mph) or more for 5 seconds or longer.
  r. Confirm that steering angle sensor initialization is completed by doing the following:
- Drive the vehicle on a straight road at 60 km/h (37 mph) or more for 30 seconds or longer.

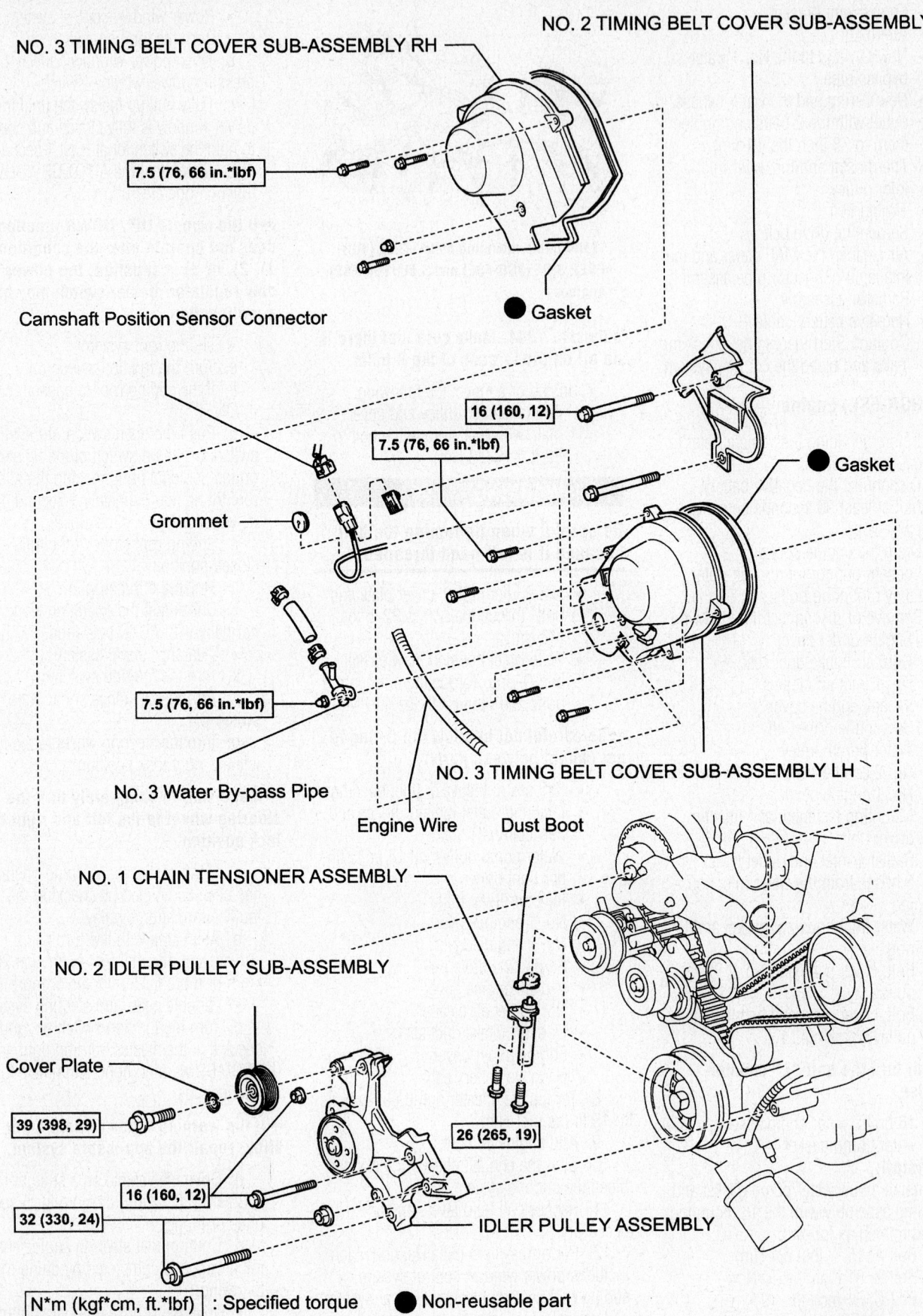

NO. 3 TIMING BELT COVER SUB-ASSEMBLY RH

NO. 2 TIMING BELT COVER SUB-ASSEMBLY

7.5 (76, 66 in.*lbf)

● Gasket

Camshaft Position Sensor Connector

16 (160, 12)

7.5 (76, 66 in.*lbf)

● Gasket

Grommet

7.5 (76, 66 in.*lbf)

No. 3 Water By-pass Pipe

NO. 3 TIMING BELT COVER SUB-ASSEMBLY LH

Engine Wire   Dust Boot

NO. 1 CHAIN TENSIONER ASSEMBLY

NO. 2 IDLER PULLEY SUB-ASSEMBLY

Cover Plate

39 (398, 29)

26 (265, 19)

16 (160, 12)

32 (330, 24)   IDLER PULLEY ASSEMBLY

N*m (kgf*cm, ft.*lbf) : Specified torque   ● Non-reusable part

09490_LEXU_G0023

Exploded view of water pump assembly—3UZ-FE Engine

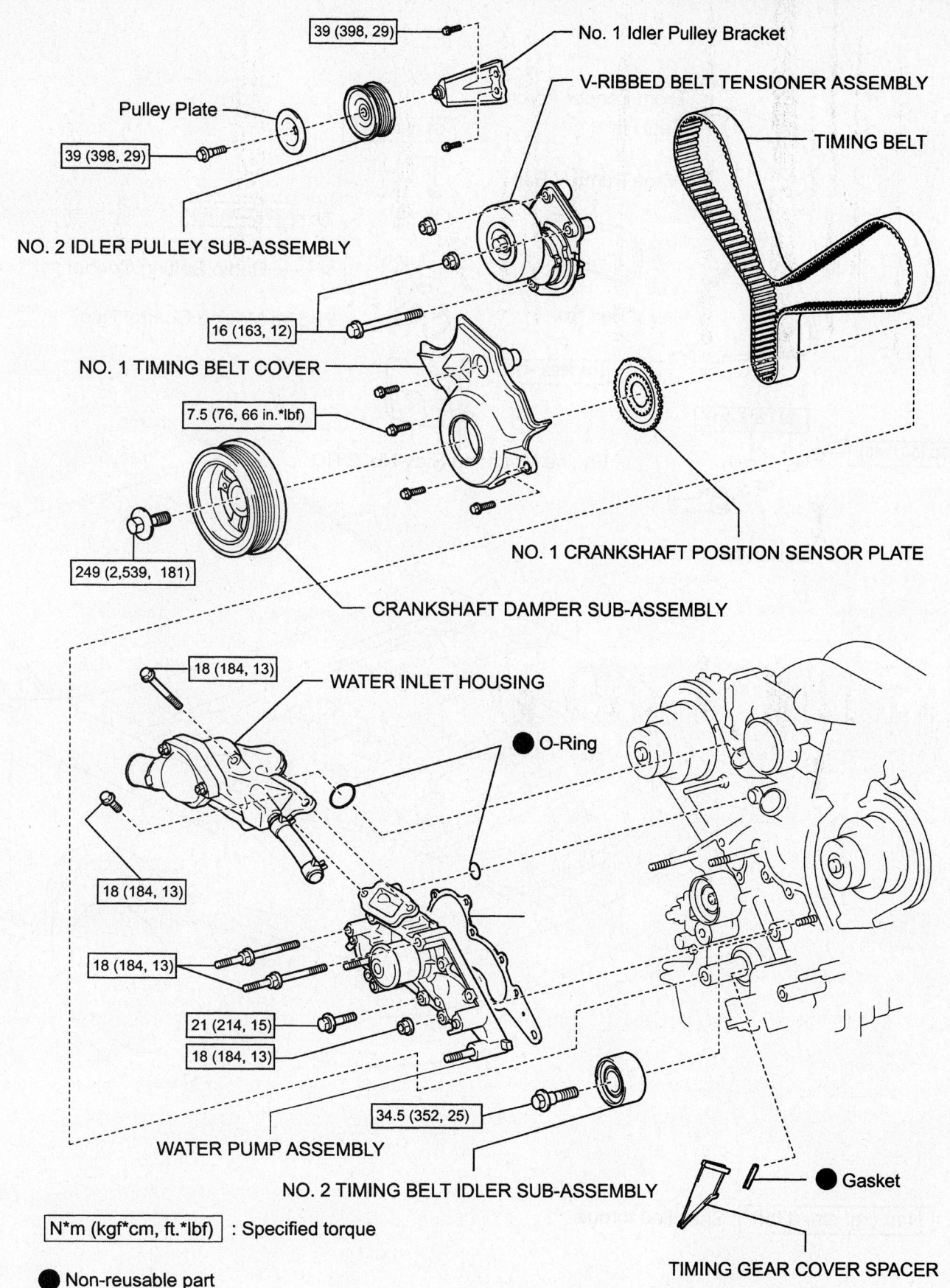

No. 1 Idler Pulley Bracket

39 (398, 29)

V-RIBBED BELT TENSIONER ASSEMBLY

Pulley Plate

TIMING BELT

39 (398, 29)

NO. 2 IDLER PULLEY SUB-ASSEMBLY

16 (163, 12)

NO. 1 TIMING BELT COVER

7.5 (76, 66 in.*lbf)

249 (2,539, 181)

NO. 1 CRANKSHAFT POSITION SENSOR PLATE

CRANKSHAFT DAMPER SUB-ASSEMBLY

18 (184, 13)

WATER INLET HOUSING

● O-Ring

18 (184, 13)

18 (184, 13)

21 (214, 15)

18 (184, 13)

34.5 (352, 25)

WATER PUMP ASSEMBLY

NO. 2 TIMING BELT IDLER SUB-ASSEMBLY

● Gasket

N*m (kgf*cm, ft.*lbf) : Specified torque

TIMING GEAR COVER SPACER

● Non-reusable part

09490_LEXU_G0024

Exploded view of water pump assembly—GS430

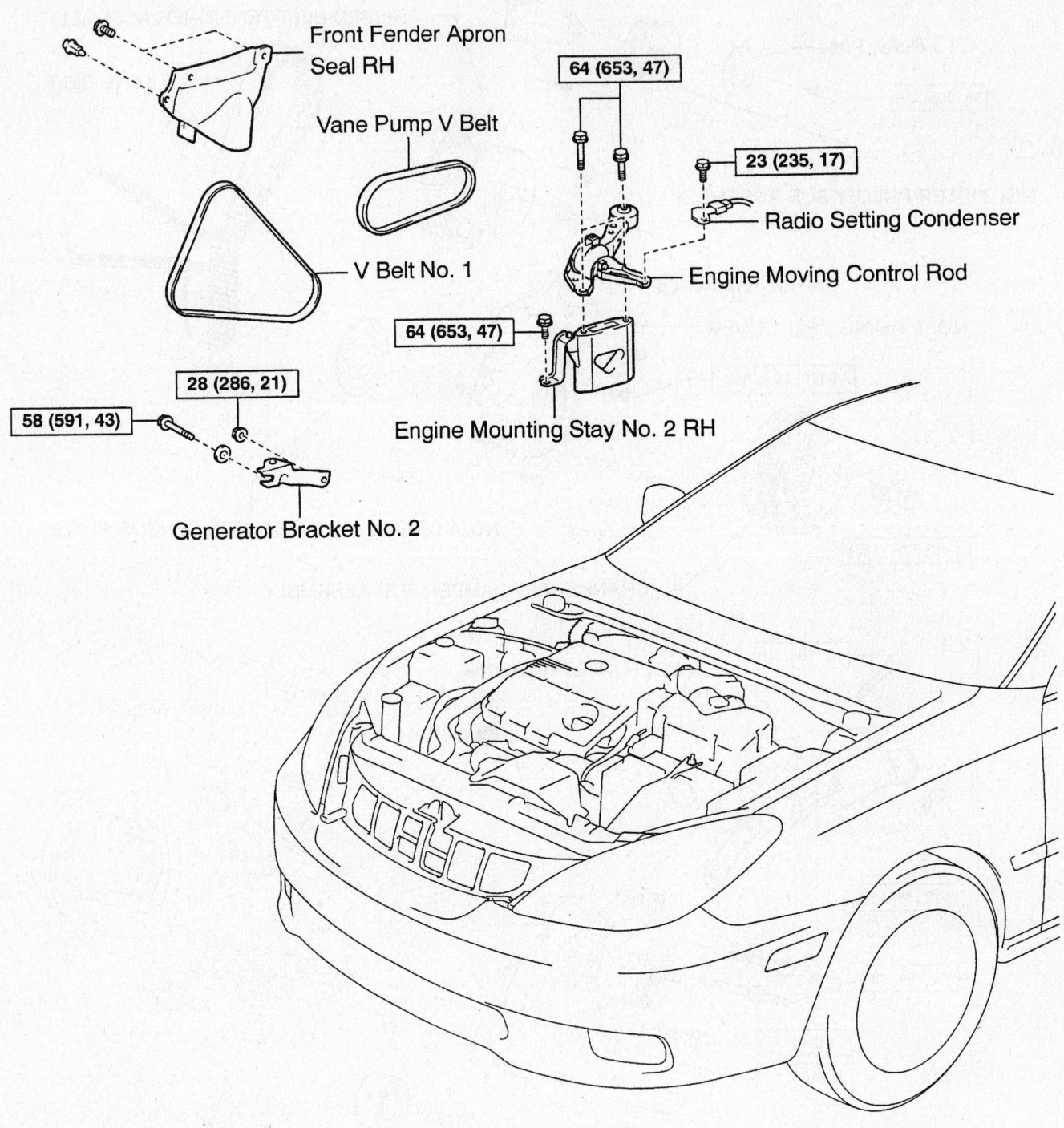

Front Fender Apron Seal RH

Vane Pump V Belt

V Belt No. 1

64 (653, 47)

23 (235, 17)

Radio Setting Condenser

Engine Moving Control Rod

28 (286, 21)

58 (591, 43)

64 (653, 47)

Engine Mounting Stay No. 2 RH

Generator Bracket No. 2

N·m (kgf·cm, ft·lbf) : Specified torque

67162-LEXU-G10

Drive belt and mount locations—3.3L (3MZ-FE)

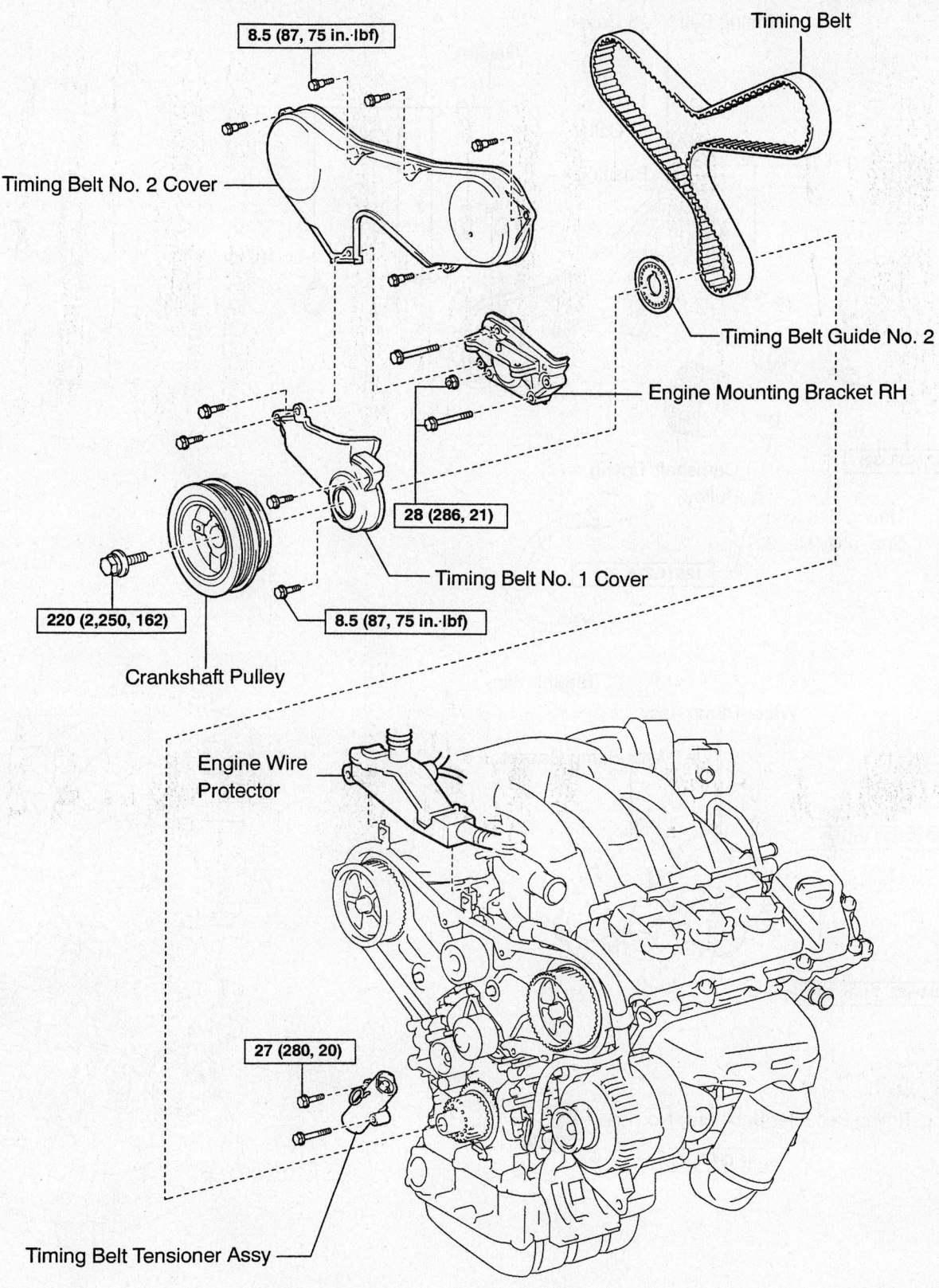

8.5 (87, 75 in.·lbf)

Timing Belt

Timing Belt No. 2 Cover

Timing Belt Guide No. 2

Engine Mounting Bracket RH

28 (286, 21)

Timing Belt No. 1 Cover

220 (2,250, 162)

8.5 (87, 75 in.·lbf)

Crankshaft Pulley

Engine Wire Protector

27 (280, 20)

Timing Belt Tensioner Assy

N·m (kgf·cm, ft·lbf) : Specified torque

Timing cover and belt components—3.3L (3MZ-FE)

67162-LEXU-G11

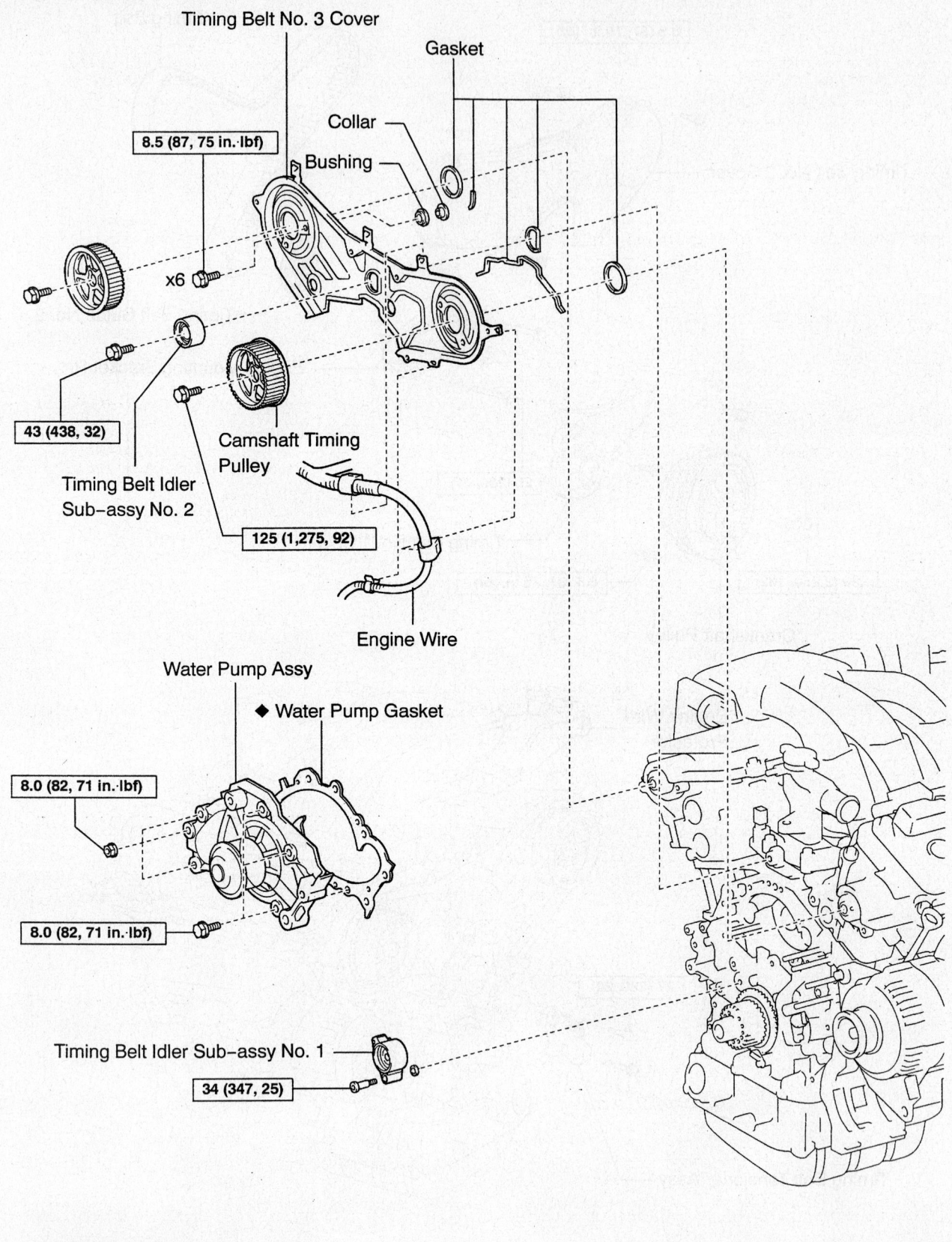

Timing Belt No. 3 Cover

Gasket

Collar

Bushing

8.5 (87, 75 in.·lbf)

x6

43 (438, 32)

Timing Belt Idler
Sub-assy No. 2

Camshaft Timing
Pulley

125 (1,275, 92)

Engine Wire

Water Pump Assy

◆ Water Pump Gasket

8.0 (82, 71 in.·lbf)

8.0 (82, 71 in.·lbf)

Timing Belt Idler Sub-assy No. 1

34 (347, 25)

N·m (kgf·cm, ft·lbf) : Specified torque

◆ Non-reusable part

67162-LEXU-G12

Camshaft pulleys, rear cover and pump components—3.3L (3MZ-FE)

- Stop the vehicle (engine running).
- Slowly turn the steering wheel from lock to lock.
- If it turns approximately 2.7 turns, steering angle sensor initialization is completed. If it turns approximately 3.2 turns, steering angle sensor initialization is not completed.

13. Road test the vehicle.

4.3L (3UZ-FE) Engine

1. Before servicing the vehicle, refer to the Precautions Section.
2. Disconnect the negative battery cable.
3. Drain the cooling system.
4. Remove or disconnect the following:
   - Engine under cover
   - Timing belt
   - Water inlet housing
   - Timing belt idler sub assembly
   - Water pump

To install:

5. Install or connect the following:
   - New gasket, water pump with 5 bolts and 2 stud bolts and nuts. Tighten bolt to 15 ft lbs (21 Nm) and stud nuts to 13 ft. lbs (18 Nm)
   - Water inlet housing with new O-ring and seal packing

➡**If O-ring contacts engine oil it must be replaced**

   - Timing belt idler sub assembly
   - Timing belt assembly
   - Radiator assembly
   - Engine coolant
   - Negative battery cable
6. Check for coolant leaks
7. Initialize the power windows and seat control systems
   - Engine under cover

3.3L (3MZ-FE)

1. Before servicing the vehicle, refer to the Precautions Section.
2. Disconnect the negative battery cable.
3. Drain the cooling system.
4. Remove or disconnect the following:
   - RH front wheel
   - RH fender apron seal
   - A/C drive belt
   - PS drive belt
   - Engine stabilizer control rod
   - RH engine stay
   - Alternator bracket #2
   - Crankshaft pulley
   - Both timing belt covers
   - RH engine mounting bracket
   - Timing belt cover 1 and 2

   - Timing belt, guide and idler pulley sub-assembly #1
   - Camshaft timing pulleys and idler pulley sub-assembly #2
   - Timing belt cover #3
   - Water pump assembly

To install:

5. Install or connect the following:
   - Water pump assembly with new gasket. Torque to 71 inch lbs. (8.0 Nm)
   - Timing belt idler #1. Torque to 25 ft. lbs (24 Nm)
   - Timing belt cover #3
   - Camshaft timing pulleys
   - Timing belt idler sub-assembly
   - Timing belt, tensioner assembly and guide
   - Engine mounting bracket
   - Upper and lower timing belts covers
   - Crankshaft pulley
   - Alternator bracket
   - Engine mounting stay
   - Engine stabilizing control rod
   - PS pump
   - A/C drive belt
   - Inspect drive belt tension
   - Right front wheel
   - Add coolant and check for leaks

Heater Core

REMOVAL & INSTALLATION

IS 250 and IS 350

1. Before servicing the vehicle, refer to the Precautions Section.
2. Discharge the A/C system.
3. Disconnect the negative battery cable. Wait 90 seconds before doing any further work while the airbag system de-energizes.
4. Drain the cooling system into a clean container for reuse.
5. Align the front wheels facing straight ahead.
6. Remove or disconnect the following:
   - Cool air intake duct seal
   - Left and right engine room side covers
7. Remove the left and right front upper fender protector as follows:
   a. Using a clip remover, separate the clip on the rubber portion of the cowl top ventilator louver subassembly from the front upper fender protector.
   b. Disengage the 3 clips and the claw to remove the front upper fender protector.

8. Remove the left and right roof drip side finish moulding as follows:
   a. Put protective tape around the roof drip side finish moulding.
   b. Using a moulding remover, disengage the 6 clips and remove the roof drip side finish moulding.

➡**Do not remove the clips. If the clips are damaged or fall off, replace them with new clips.**

9. Remove the windshield wiper motor and link assembly as follows:
   a. Using a screwdriver, remove the front wiper arm head cap.
   b. Remove the nut and the left and right front wiper arm and blade assembly.
   c. Remove the 2 clips and disengage the 11 claws, then pull out the cowl top ventilator louver sub-assembly.
   d. Disconnect the connector and the 2 bolts.
   e. Disengage the rubber pin of the windshield wiper motor from the vehicle body.
   f. Remove the windshield wiper motor and link assembly.
10. Separate suction pipe sub-assembly as follows:
   a. Remove the bolt, and slide the hook connector.
   b. Disconnect the suction pipe sub-assembly.
   c. Remove the O-ring from the suction pipe subassembly.

✳✳ WARNING

Seal the openings of the disconnected parts using vinyl tape to prevent moisture and foreign matter from entering.

11. Separate liquid tube sub-assembly as follows:
   a. Disconnect the liquid tube sub-assembly.
   b. Remove the O-ring from the liquid tube subassembly.
12. Remove or disconnect the following:
   - Inlet and outlet heater water outlet hoses
   - Shift lever knob by turning counter-clockwise
13. Remove upper No. 1 and No. 2 console panel garnishes and console panel sub-assembly (automatic transmission) as follows:
   a. Using a moulding remover, disengage the claw.
   b. Pull each upper console panel gar-

nish toward the rear of the vehicle to disengage the 2 clips and remove it.

c. Disengage the 8 clips.

d. Disconnect the connectors and remove the console panel sub-assembly.

14. Remove the front and rear console panel sub-assemblies (manual transmission) as follows:

a. Open the snap of the boot for the emergency brake lever.

b. Disengage the 7 claws and 2 clips, and then remove the rear console panel sub-assembly.

c. Open the snap of the boot for the shifter.

d. Pull the front console panel sub-assembly toward the rear of the vehicle to disengage the 6 clips and remove it.

15. Remove or disconnect the following:

- Front and rear ash receptacle assemblies
- Console box register assembly
- 6 bolts, 2 clamps, 2 claws, 2 clips, connector and console box

16. Remove the No. 3 instrument panel register assembly as follows:

a. Using a screwdriver, disengage the 4 claws. Tape the screwdriver tip before use.

b. Apply protective tape to the areas surrounding the removal.

c. Using a moulding remover, disengage the 4 claws starting from the right of the No. 3 instrument panel register assembly. Disengage the remaining 3 claws by pulling the No. 3 instrument panel register assembly by hand.

✳✳ WARNING

Do not pry the lower part of the No. 3 instrument panel register assembly. Doing so may damage the assembly.

d. Disconnect the connectors.

- Center lower instrument cluster finish panel by disengaging 4 claws

17. Remove the integration control panel with radio receiver assembly (w/o navigation system) as follows:

a. Remove the 4 bolts.

b. Pull the integration control panel w/ radio receiver assembly toward the rear of the vehicle.

c. Disconnect each connector and remove the panel.

18. Remove the multi-display with radio receiver assembly (w/ navigation system) as follows:

a. Remove the 4 bolts.

b. Pull the multi-display w/ radio receiver assembly toward the rear of the vehicle.

c. Disconnect each connector and remove the multi-display.

- The lower No. 2 and No. 3 steering wheel covers using a screwdriver. Tape up the screwdriver tip before use.

19. Remove the steering wheel airbag as follows:

a. Using a "torx" socket wrench (T30), loosen the 2 "torx" screws until the groove along the screw circumference catches on the screw case.

b. Pull out the steering airbag from the steering wheel assembly and support the airbag with one hand.

✳✳ WARNING

When removing the steering airbag, do not pull the airbag wire harness.

c. Disconnect the horn connector.

d. Disconnect the 2 connectors from the steering airbag.

✳✳ WARNING

When handling the airbag connector, take care not to damage the airbag wire harness.

e. Remove the steering wheel airbag.

20. Remove the steering wheel assembly as follows:

a. Remove the steering wheel assembly set nut.

b. Put matchmarks on the steering wheel assembly and main shaft assembly.

c. Using a steering wheel puller tool, remove the steering wheel assembly.

21. Remove the steering column cover as follows:

a. Remove the retaining screws.

b. Disengage the 2 claws to remove the lower steering column cover.

✳✳ WARNING

If equipped with the power tilt column, be careful not to damage the tilt and telescopic switch.

c. Disengage the 4 clips to separate the steering column cover upper.

d. Disengage the claw to remove the upper steering column cover.

22. Remove or disconnect the following:

- The tilt and telescopic switch (power tilt steering column)

- Turn signal switch assembly with spiral cable sub-assembly
- Left and right front door scuff plates
- Left and right front door opening trim covers
- Left and right side instrument panel
- No. 1 and 2 instrument panel under covers
- Lower instrument panel finish panel
- Driver side and passenger side knee airbag assemblies
- Instrument cluster finish panel
- Combination meter assembly
- Glove compartment door assembly
- Left and right front pillar garnishes
- No. 1 and 2 instrument panel register assemblies
- Instrument panel speaker panel
- Front stereo component speaker assembly (w/ front center speaker)
- No. 1 console box duct (automatic transmission)
- No. 2 console box duct
- Passenger airbag connector
- Instrument panel safety pad
- No. 1 air duct

23. Remove the steering sliding yoke sub-assembly (2WD) as follows:

a. Loosen upper bolt and remove lower bolt, then slide the steering sliding yoke sub-assembly.

➡ **Do not remove the upper bolt. Do not separate the steering sliding yoke sub-assembly from the power steering gear assembly.**

b. Put matchmarks on the steering sliding yoke subassembly and the power steering gear assembly.

c. Separate the steering sliding yoke sub-assembly from the power steering gear assembly.

d. Put matchmarks on the steering sliding yoke sub-assembly and the No. 2 steering intermediate shaft assembly.

e. Remove the bolt and the steering sliding yoke subassembly from the No. 2 steering intermediate shaft assembly.

24. Separate the No. 2 steering intermediate shaft assembly (AWD) as follows:

a. Remove the bolt, and then slide the No. 2 steering intermediate shaft assembly.

➡ **Do not separate the No. 2 steering intermediate shaft assembly from the power steering gear assembly.**

b. Put matchmarks on the No. 2 steering intermediate shaft assembly and the power steering gear assembly.

for Automatic Transmission:

 — NO. 3 INSTRUMENT PANEL REGISTER ASSEMBLY

with Navigation System:

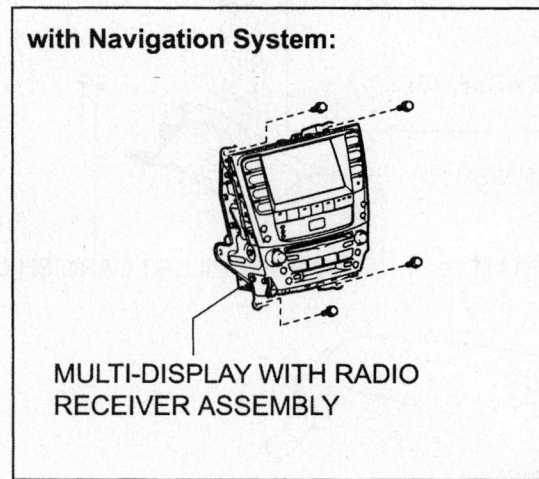

MULTI-DISPLAY WITH RADIO
RECEIVER ASSEMBLY

without Navigation System:

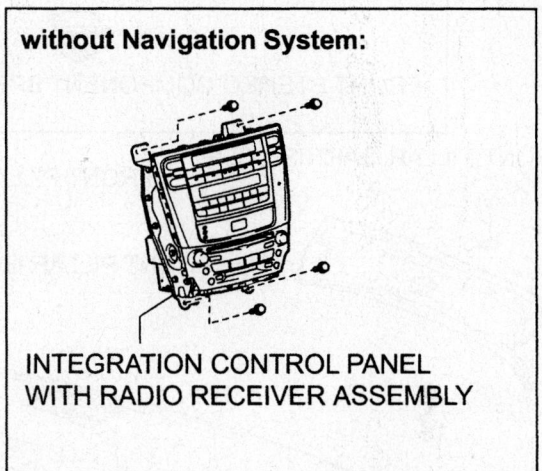

INTEGRATION CONTROL PANEL
WITH RADIO RECEIVER ASSEMBLY

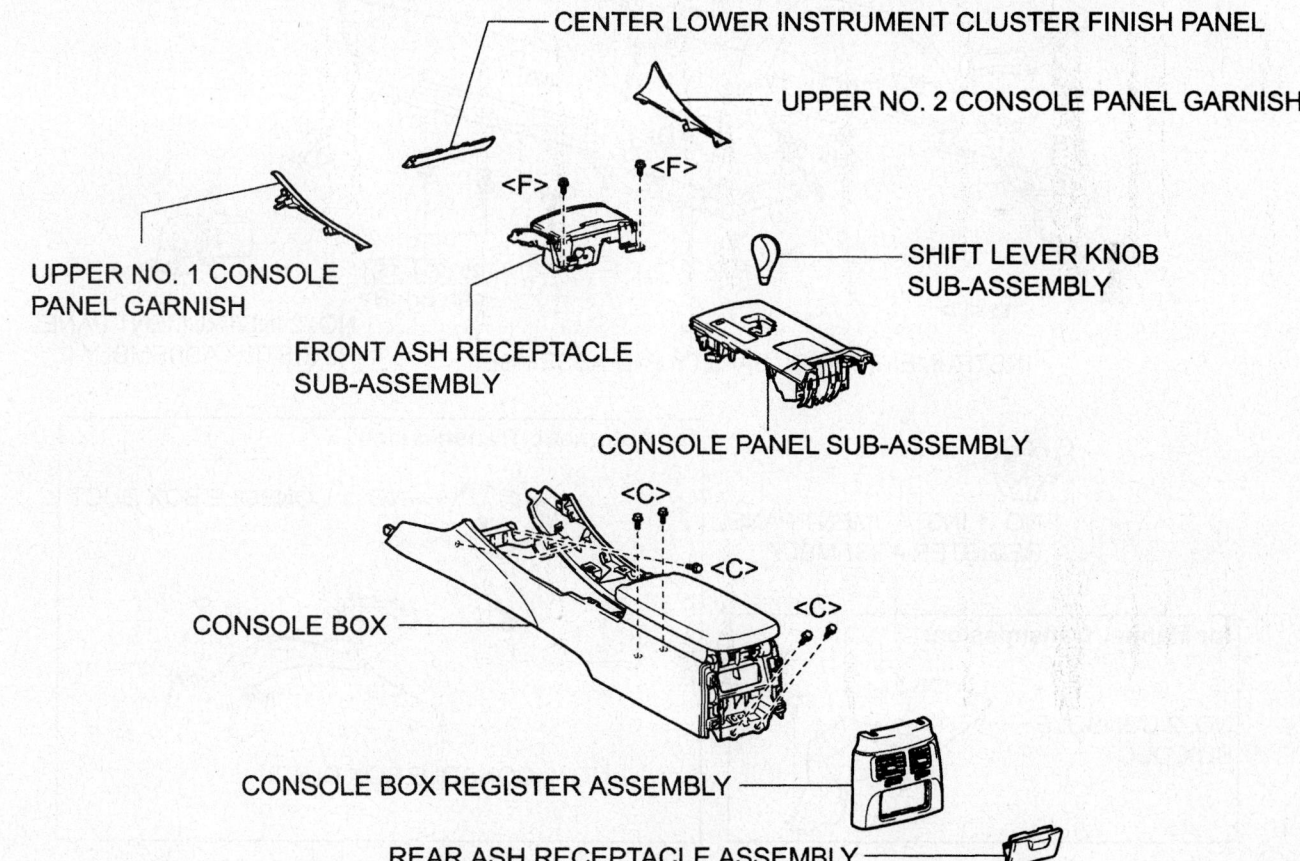

CENTER LOWER INSTRUMENT CLUSTER FINISH PANEL

UPPER NO. 2 CONSOLE PANEL GARNISH

UPPER NO. 1 CONSOLE
PANEL GARNISH

FRONT ASH RECEPTACLE
SUB-ASSEMBLY

SHIFT LEVER KNOB
SUB-ASSEMBLY

CONSOLE PANEL SUB-ASSEMBLY

CONSOLE BOX

CONSOLE BOX REGISTER ASSEMBLY

REAR ASH RECEPTACLE ASSEMBLY

09490_LEXU_G0025

Exploded view of the center console assembly and related components—IS 250 and IS 350 with automatic transmission

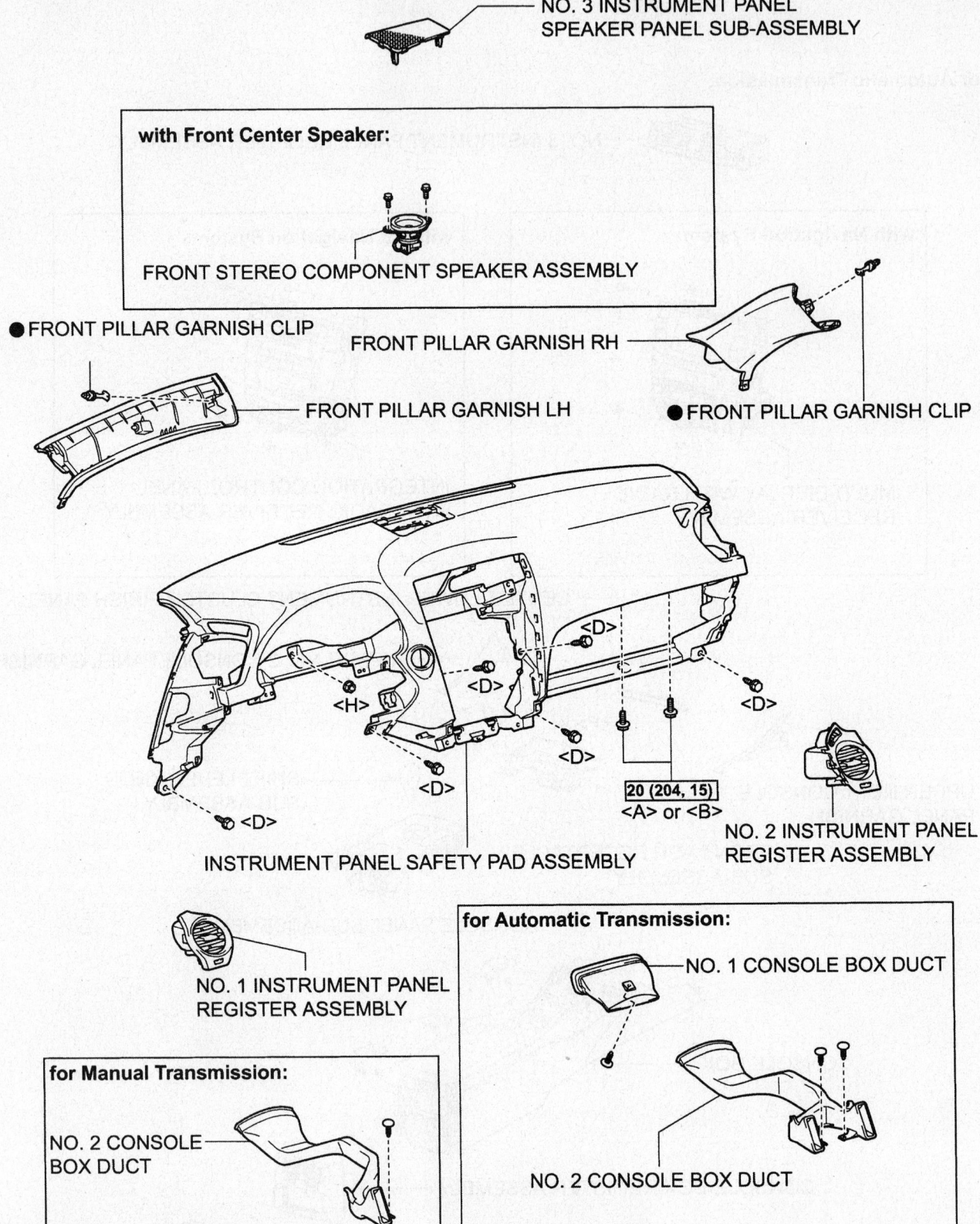

NO. 3 INSTRUMENT PANEL
SPEAKER PANEL SUB-ASSEMBLY

with Front Center Speaker:

FRONT STEREO COMPONENT SPEAKER ASSEMBLY

● FRONT PILLAR GARNISH CLIP

FRONT PILLAR GARNISH RH

FRONT PILLAR GARNISH LH

● FRONT PILLAR GARNISH CLIP

\<D\>

\<H\>

\<D\>

\<D\>

\<D\>

\<D\>

\<D\>

20 (204, 15)
\<A\> or \<B\>

NO. 2 INSTRUMENT PANEL
REGISTER ASSEMBLY

INSTRUMENT PANEL SAFETY PAD ASSEMBLY

NO. 1 INSTRUMENT PANEL
REGISTER ASSEMBLY

for Automatic Transmission:

NO. 1 CONSOLE BOX DUCT

NO. 2 CONSOLE BOX DUCT

for Manual Transmission:

NO. 2 CONSOLE
BOX DUCT

N*m (kgf*cm, ft.*lbf) : Specified torque

● Non-reusable part

09490_LEXU_G0026

Exploded view of the instrument panel, ventilation ducts and related components—IS 250 and IS 350

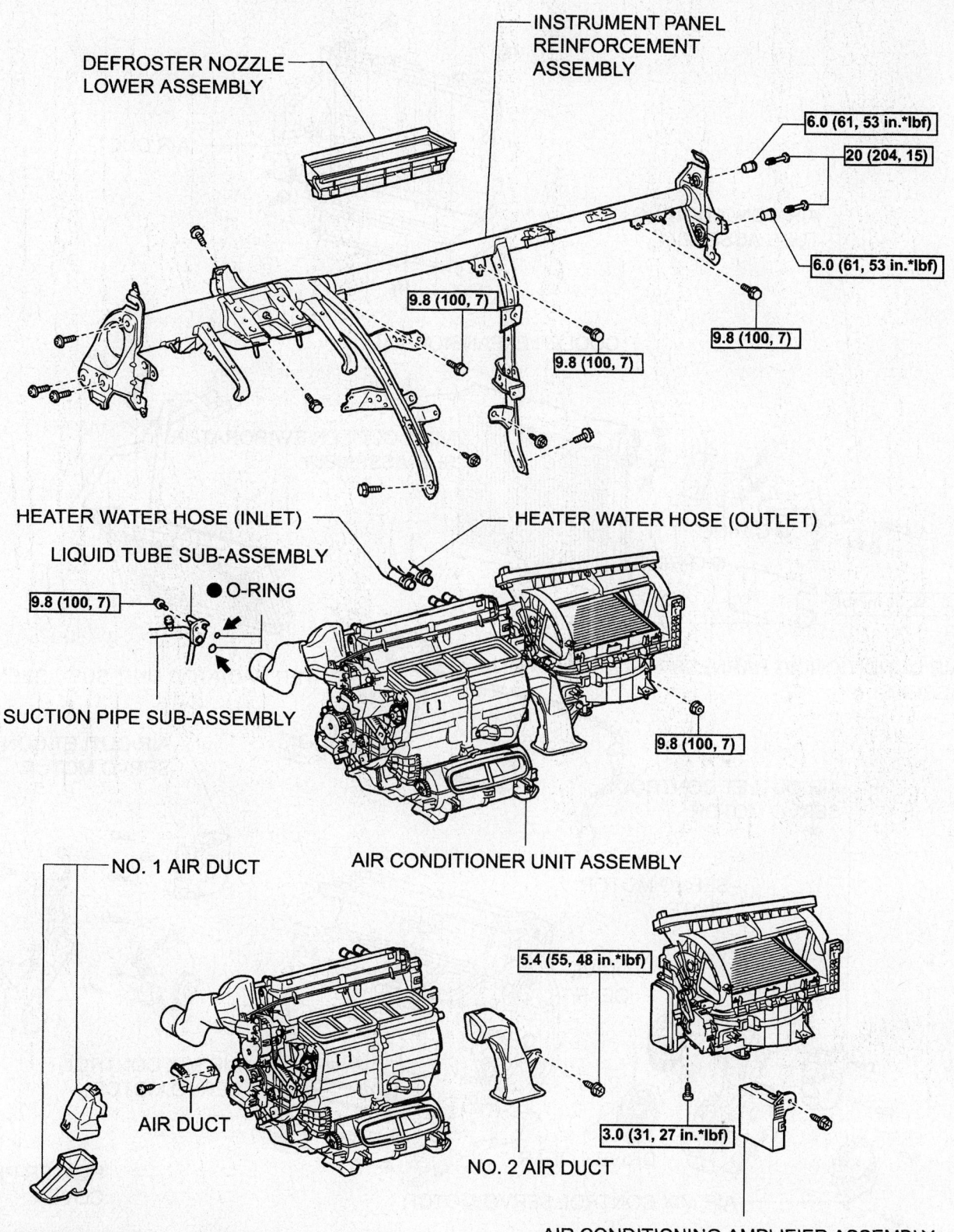

DEFROSTER NOZZLE LOWER ASSEMBLY

INSTRUMENT PANEL REINFORCEMENT ASSEMBLY

6.0 (61, 53 in.*lbf)

20 (204, 15)

6.0 (61, 53 in.*lbf)

9.8 (100, 7)

9.8 (100, 7)

9.8 (100, 7)

HEATER WATER HOSE (INLET)

HEATER WATER HOSE (OUTLET)

LIQUID TUBE SUB-ASSEMBLY

● O-RING

9.8 (100, 7)

SUCTION PIPE SUB-ASSEMBLY

9.8 (100, 7)

AIR CONDITIONER UNIT ASSEMBLY

NO. 1 AIR DUCT

5.4 (55, 48 in.*lbf)

AIR DUCT

NO. 2 AIR DUCT

3.0 (31, 27 in.*lbf)

AIR CONDITIONING AMPLIFIER ASSEMBLY

N*m (kgf*cm, ft.*lbf) : Specified torque

● Non-reusable part

◄ Compressor oil ND-OIL 8 or equivalent

09490_LEXU_G0027

Exploded view of the A/C unit assembly, instrument panel reinforcement and related components—IS 250, IS 350, GS 300 (2006) and GS 430 (2006)

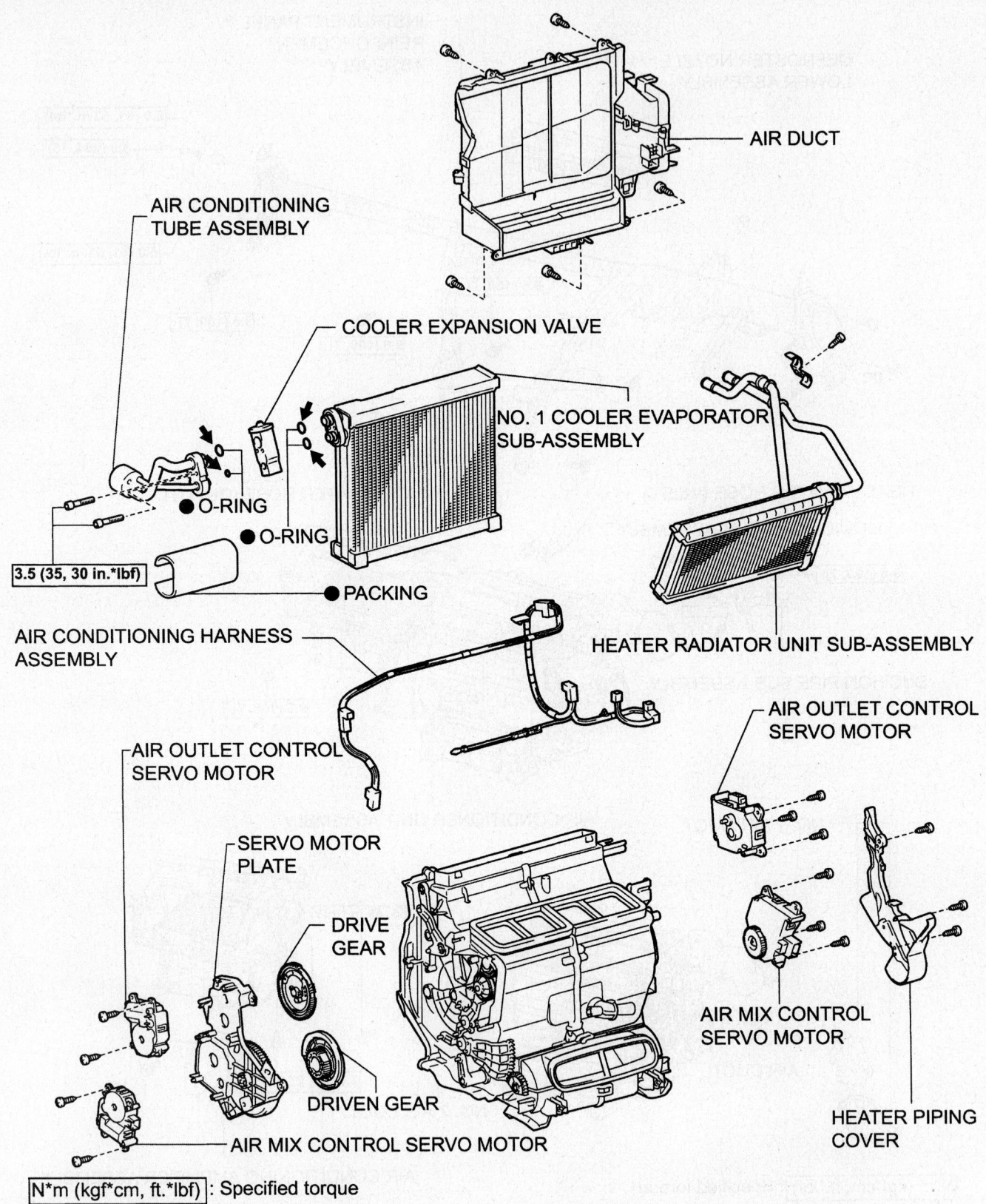

AIR DUCT

AIR CONDITIONING
TUBE ASSEMBLY

COOLER EXPANSION VALVE

NO. 1 COOLER EVAPORATOR
SUB-ASSEMBLY

● O-RING

● O-RING

3.5 (35, 30 in.*lbf)

● PACKING

HEATER RADIATOR UNIT SUB-ASSEMBLY

AIR CONDITIONING HARNESS
ASSEMBLY

AIR OUTLET CONTROL
SERVO MOTOR

AIR OUTLET CONTROL
SERVO MOTOR

SERVO MOTOR
PLATE

DRIVE
GEAR

AIR MIX CONTROL
SERVO MOTOR

DRIVEN GEAR

AIR MIX CONTROL SERVO MOTOR

HEATER PIPING
COVER

N*m (kgf*cm, ft.*lbf) : Specified torque

● Non-reusable part

◄ Compressor oil ND-OIL 8 or equivalent

09490_LEXU_G0028

Exploded view of the heater radiator unit (heater core), heater housing and related components—IS 250, IS 350, GS 300 (2006) and GS 430 (2006)

c. Separate the No. 2 intermediate shaft assembly from the power steering gear assembly.

25. Remove the steering column assembly as follows:

a. Remove the brake pedal return spring (manual tilt).

b. Remove the clamp from the steering column hole shield.

c. Disconnect the connectors and wire harness clamps from the steering column assembly.

d. Remove the 4 nuts and steering column assembly.

- 4 claws and defroster nozzle lower assembly
- Immobiliser code ECU

26. Remove the instrument panel reinforcement assembly as follows:

a. Remove the 27 clamps and 19 connectors, and then disconnect the wire harness.

b. Remove the 2 nuts, 2 bolts and 2 junction blocks.

c. Remove the cap and bolt.

d. Remove the 3 bolts and nut.

e. Using a T40 "torx" socket, remove the 5 "torx" bolts.

f. Using a 12 mm hexagon wrench, remove the 2 collars and instrument panel reinforcement.

27. Remove or disconnect the following:

- 2 screws, 3 bolts, and air conditioner unit assembly
- Mounting screw and No. 2 air duct
- Connector, mounting screw and air conditioning amplifier assembly

28. Remove the heater core from the A/C-blower assembly unit as follows:

a. Disconnect the wiring connector, remove the 3 screws and air outlet control servo motor.

b. Disengage 4 clamps, connector and wire harness to air mix control servo motor.

c. Remove the 3 screws and heater piping cover.

d. Remove the 3 screws and air mix control servo motor.

e. Remove the heater radiator unit sub-assembly (heater core)

To install:

29. Install or connect the following:

- Heater core to the A/C blower housing
- Air conditioning amplifier assembly
- No. 2 air duct
- A/C blower unit assembly. Torque the 3 bolts and 2 screws to 7 ft. lbs. (10 Nm).

- Instrument panel reinforcement assembly. Torque the 3 driver side "torx" bolts and 2 passenger side 12mm collars to 53 inch lbs. (6 Nm). Torque the 2 remaining "torx" bolts to 15 ft. lbs. (20 Nm) and the 4 remaining bolts and 1 nut to 7 ft. lbs. (10 Nm).
- Immobiliser code ECU
- Defroster nozzle lower assembly
- Steering column assembly. Torque the 4 retaining nuts to 16 ft. lbs. (21 Nm).
- No. 2 steering intermediate shaft assembly (AWD). Torque the bolt to 26 ft. lbs. (35 Nm).
- Steering sliding yoke sub-assembly (2WD). Torque the bolts to 26 ft. lbs. (35 Nm).
- No. 1 air duct
- Instrument panel safety pad
- Passenger airbag connector
- No. 2 console box duct
- No. 1 console box duct (automatic transmission)
- Front stereo component speaker assembly (w/ front center speaker)
- Instrument panel speaker panel
- No. 1 and 2 instrument panel register assemblies
- Left and right front pillar garnishes
- Glove compartment door assembly
- Combination meter assembly
- Instrument cluster finish panel
- Driver side and passenger side knee airbag assemblies
- Lower instrument panel finish panel
- No. 1 and 2 instrument panel under covers
- Left and right side instrument panel
- Left and right front door opening trim covers
- Left and right front door scuff plates
- Turn signal switch assembly with spiral cable sub-assembly

30. Adjust the spiral cable as follows:

a. Check that the engine switch is off.

b. Check that the battery negative (-) terminal is disconnected.

⁑ WARNING

After removing the terminal, wait for at least 90 seconds before starting the operation.

c. Rotate the spiral cable with steering sensor counterclockwise slowly by hand until it feels firm. Do not turn it by the wiring harness.

d. Rotate the spiral cable with steer-

ing sensor clockwise approximately 2.5 turns to align the marks.

- The tilt and telescopic switch (power tilt steering column)
- Steering column cover
- Steering wheel. Torque the steering wheel bolt to 37 ft. lbs. (50 Nm).
- Steering wheel airbag. Torque the screws to 78 inch lbs. (9 Nm).
- The lower No. 2 and No. 3 steering wheel covers
- Integration control panel with radio receiver assembly (w/o navigation system)
- Multi-display with radio receiver assembly (w/ navigation system)
- Center lower instrument cluster finish panel
- No. 3 instrument panel register assembly
- Console box
- Console box register assembly
- Front and rear ash receptacle assemblies
- Front and rear console panel sub-assemblies (manual transmission)
- Upper No. 1 and No. 2 console panel garnishes and console panel sub-assembly (automatic transmission)
- Shift lever knob
- Inlet and outlet heater water outlet hoses
- Liquid tube sub-assembly
- Suction pipe sub-assembly
- Windshield wiper motor and link assembly
- Left and right roof drip side finish mouldings
- Left and right front upper fender protectors
- Left and right engine room side covers
- Cool air intake duct seal

31. Refill the cooling system.

32. Connect the negative battery cable.

33. Evacuate, charge and leak test the air conditioning system refrigerant.

34. Operate the engine to normal operating temperatures; then, check the climate control operation and check for leaks.

IS 300 and ES 300

1. Before servicing the vehicle, refer to the Precautions Section.

2. Disconnect the negative battery cable. Wait 90 seconds before doing any further work while the airbag system de-energizes.

3. Drain the cooling system into a clean container for reuse.

4. Remove or disconnect the following:

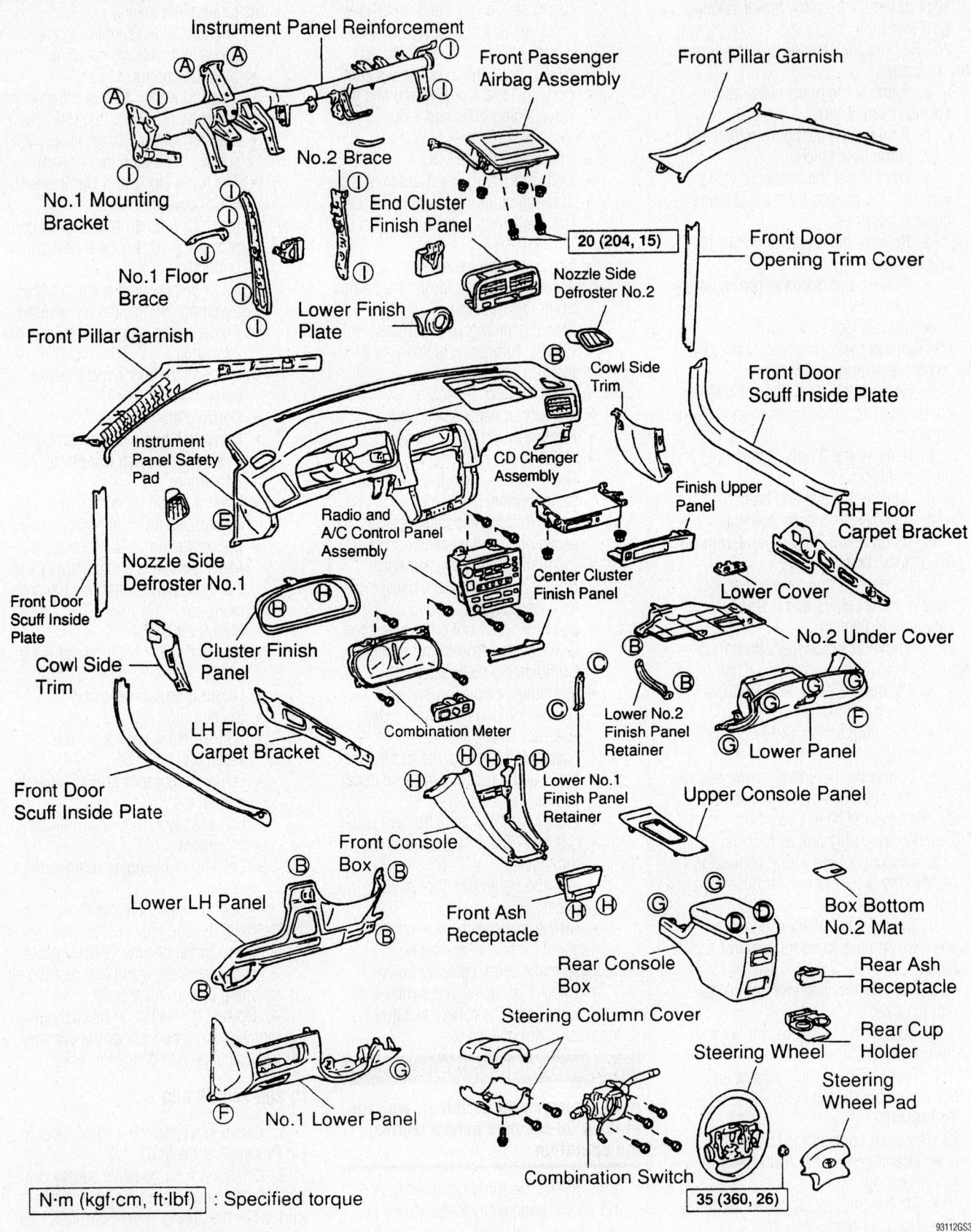

Instrument Panel Reinforcement

Front Passenger Airbag Assembly

Front Pillar Garnish

No.2 Brace

No.1 Mounting Bracket

End Cluster Finish Panel

20 (204, 15)

Nozzle Side Defroster No.2

Front Door Opening Trim Cover

No.1 Floor Brace

Lower Finish Plate

Cowl Side Trim

Front Door Scuff Inside Plate

Front Pillar Garnish

Instrument Panel Safety Pad

CD Chenger Assembly

Finish Upper Panel

Nozzle Side Defroster No.1

Radio and A/C Control Panel Assembly

Center Cluster Finish Panel

RH Floor Carpet Bracket

Front Door Scuff Inside Plate

Lower Cover

Cowl Side Trim

Cluster Finish Panel

No.2 Under Cover

Lower No.2 Finish Panel Retainer

Combination Meter

LH Floor Carpet Bracket

Lower No.1 Finish Panel Retainer

Lower Panel

Upper Console Panel

Front Door Scuff Inside Plate

Front Console Box

Lower LH Panel

Front Ash Receptacle

Box Bottom No.2 Mat

Rear Console Box

Rear Ash Receptacle

Steering Column Cover

Rear Cup Holder

Steering Wheel

No.1 Lower Panel

Combination Switch

Steering Wheel Pad

35 (360, 26)

N·m (kgf·cm, ft·lbf) : Specified torque

93112GS3

Exploded view of the instrument assembly—ES 300

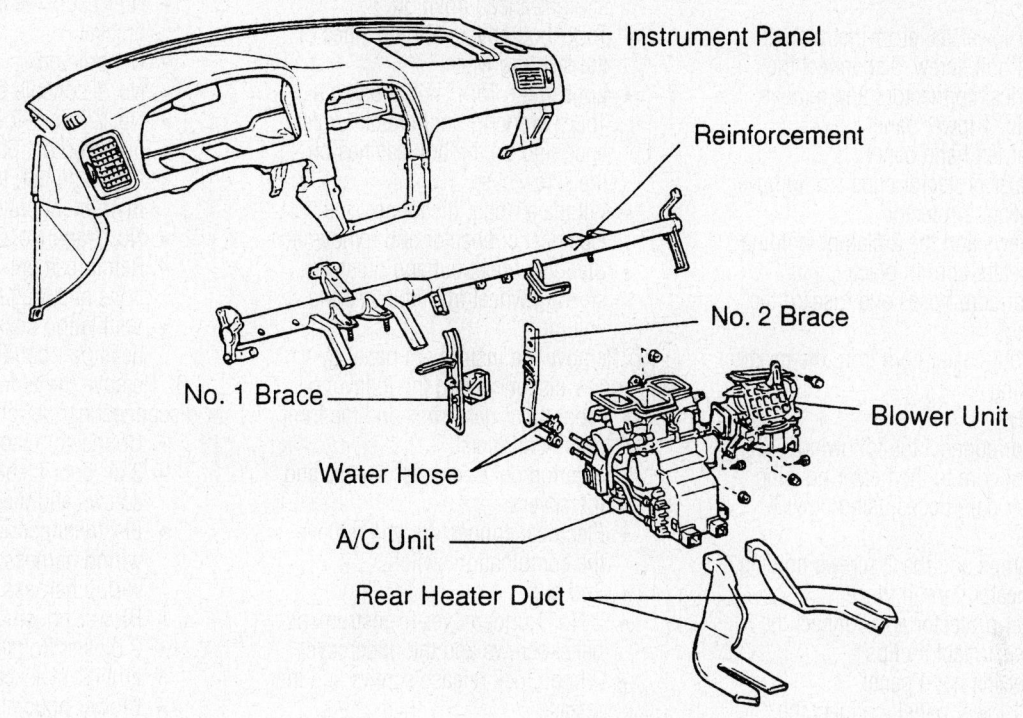

Instrument Panel

Reinforcement

No. 2 Brace

No. 1 Brace

Water Hose

A/C Unit

Rear Heater Duct

Blower Unit

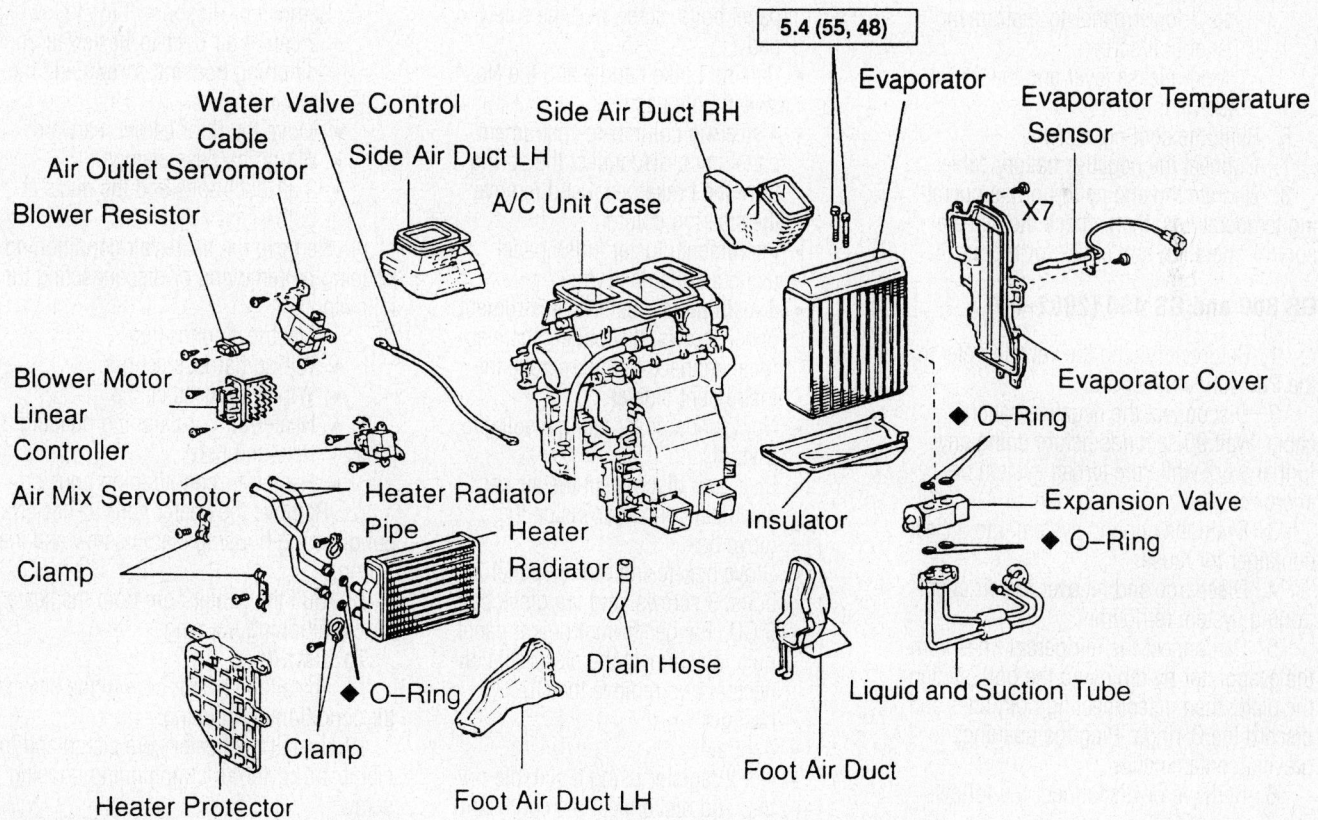

Water Valve Control Cable

Air Outlet Servomotor

Blower Resistor

Side Air Duct LH

Side Air Duct RH

A/C Unit Case

5.4 (55, 48)

Evaporator

Evaporator Temperature Sensor

X7

Blower Motor Linear Controller

Air Mix Servomotor

Clamp

Heater Radiator Pipe

Heater Radiator

Insulator

Evaporator Cover

◆ O–Ring

Expansion Valve

◆ O–Ring

◆ O–Ring

Clamp

Drain Hose

Foot Air Duct

Liquid and Suction Tube

Heater Protector

Foot Air Duct LH

N·m (kgf·cm, in.·lbf) : Specified torque

◆ Non–reusable part

93112GS4

Exploded view of the heater core, heater/air conditioning housing and related components—ES 300

- 2 hood release lever screws and the lever
- No. 1 lower panel-to-instrument panel bolt/screw, disconnect the electrical connectors and remove the No. 1 lower panel
- Lower left hand panel
- 3 heater protector clips and remove the heater protector
- 2 screws and the 2 clamps holding the heater core in place
- Heater core hoses and discard the O-rings
- Pull out heater core from the heater housing

To install:

5. Install or connect the following:
- Heater core to the heater housing
- Heater core hoses using new O-rings
- 2 clamps and the 2 screws holding the heater core in place
- Heater protector and connect the 3 heater protector clips
- Lower left hand panel
- No. 1 lower panel, connect the electrical connectors and install the No. 1 lower panel-to-instrument panel bolt/screw
- Hood release lever and the 2 lever screws

6. Refill the cooling system.
7. Connect the negative battery cable.
8. Operate the engine to normal operating temperatures; then, check the climate control operation and check for leaks.

GS 300 and GS 430 (2002–05)

1. Before servicing the vehicle, refer to the Precautions Section.
2. Disconnect the negative battery cable. Wait 90 seconds before doing any further work while the airbag system de-energizes.
3. Drain the cooling system into a clean container for reuse.
4. Discharge and recover the air conditioning system refrigerant.
5. Disconnect the refrigerant lines from the evaporator by removing the bolt, sliding the plate, then disconnecting both lines and discard the O-rings. Plug the openings to prevent contamination
6. Remove or disconnect the following:
- Heater hoses from the heater core
- No. 1 grommet, the heater pipe grommet and the drain hose grommet

7. Remove the steering wheel by removing or disconnecting the following:

- Place the front wheels in the straight-ahead position.
- Torx® bolt covers at both sides of the steering wheel
- Loosen the Torx® bolts (using a Torx® wrench), until the circumference ring on the bolt catches on the screw case
- Lift the air bag, disconnect the electrical connector and remove it
- Steering wheel nut and press the steering wheel from the steering column

8. Remove the instrument panel by removing or disconnecting the following:
- Front pillar garnishes and the front door scuff plates
- Steering column cover screws and the covers
- Electrical connectors and remove the combination switch
- End pad
- 2 No. 1 undercover-to-instrument panel screws and the undercover
- 2 hood lock release screws and the release
- 4 No. 1 safety pad-to-instrument panel bolts, screw and the safety pad
- Parking brake handle and the No. 1 switch hole base
- 4 steering column-to-instrument panel nuts, disconnect the spring from the brake pedal and remove the steering column
- Instrument cluster finish panel using a suitable prytool
- 4 instrument cluster-to-instrument panel screws, disconnect the electrical connectors and remove the instrument cluster
- No. 2 undercover using a suitable prytool
- Plate and disconnect the air bag electrical connector inside the glove box
- Glove box-to-instrument panel 2 bolts, 3 screws, and the glove box
- 3 CD changer-to-instrument panel nuts, disconnect the electrical connectors and remove the CD changer
- Ashtray
- No. 2 register using a suitable prytool and disconnect the connector
- Audio unit-to-instrument panel 2 bolts, 2 screws and the audio unit
- Cluster finish panel using a suitable prytool, disconnect the connectors and remove the panel.
- Console box carpet
- Lower rear console box

- Rear console armrest
- No. 3 console box mounting bracket
- Console box
- No. 1 console box duct
- No. 7 heater-to-register
- 5 instrument panel-to-chassis bolts, the nut, the screw and the instrument panel
- No. 1 and No. 2 brace
- Reinforcement-to-chassis 5 nuts, 4 bolts and the reinforcement
- Ventilation nozzles from the heater/air conditioning housing

9. Remove the blower unit by removing or disconnecting the following:
- Connector clamp
- 3 air duct-to-blower housing screws and the air duct
- Electrical connector bracket, the wiring harness clamps and the wiring harness
- Blower housing connectors
- 2 blower housing-to-bracket bolts and the bracket
- Blower housing-to-chassis bolt, screw, nut and the blower housing

10. Remove or disconnect the following:
- 2 center air duct-to-heater/air conditioning housing screws and the air duct
- Move the floor carpet rearward
- Wiring harness clamps
- 2 air duct bolts and the ducts at both sides

11. Remove the heater/air conditioning housing by removing or disconnecting the following:
- Electrical connector
- Wiring harness set nut
- Wiring harness clamp
- Heater/air conditioning housing 2 nuts and bolt
- Heater/air conditioning housing

12. Remove the heater core-to-heater/air conditioning housing clamp screw and the clamp.
13. Pull the heater core from the heater/air conditioning housing.

To install:

14. Install the heater core to the heater/air conditioning housing.
15. Install the heater core clamp and the clamp-to-heater/air conditioning housing screw.
16. Install the heater/air conditioning housing by installing or connecting the following:

- Heater/air conditioning housing
- Heater/air conditioning housing 2 nuts and bolt
- Wiring harness clamp

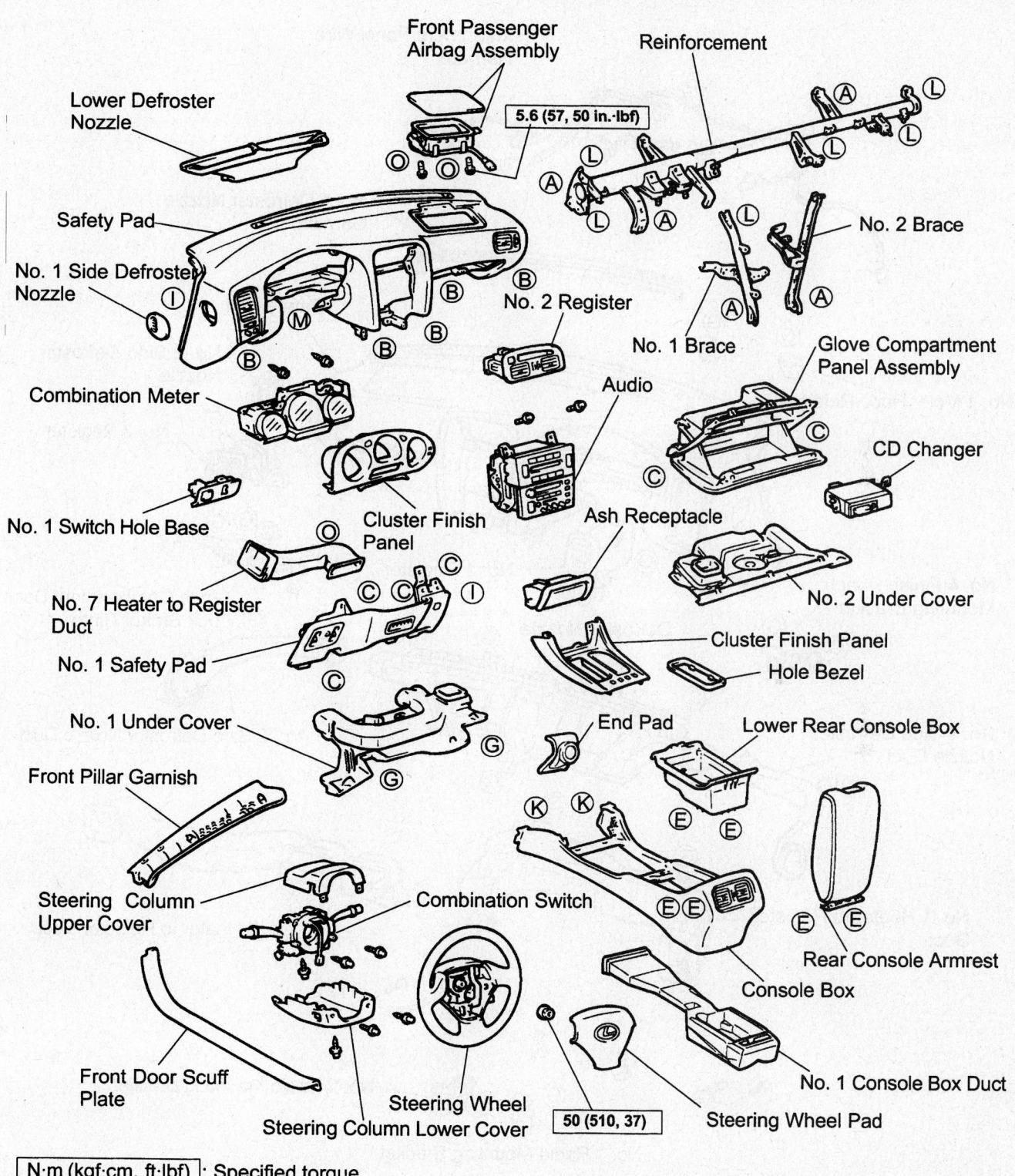

Front Passenger Airbag Assembly

Reinforcement

Lower Defroster Nozzle

5.6 (57, 50 in.·lbf)

Safety Pad

No. 2 Brace

No. 1 Side Defroster Nozzle

No. 2 Register

No. 1 Brace

Glove Compartment Panel Assembly

Audio

Combination Meter

CD Changer

No. 1 Switch Hole Base

Cluster Finish Panel

Ash Receptacle

No. 7 Heater to Register Duct

No. 2 Under Cover

Cluster Finish Panel

No. 1 Safety Pad

Hole Bezel

No. 1 Under Cover

End Pad

Lower Rear Console Box

Front Pillar Garnish

Steering Column Upper Cover

Combination Switch

Rear Console Armrest

Console Box

Front Door Scuff Plate

Steering Wheel

No. 1 Console Box Duct

Steering Column Lower Cover

50 (510, 37)

Steering Wheel Pad

N·m (kgf·cm, ft·lbf) : Specified torque

09490_LEXU_G0029

Exploded view of the instrument panel and related components—2002–05 GS 300 and GS 430

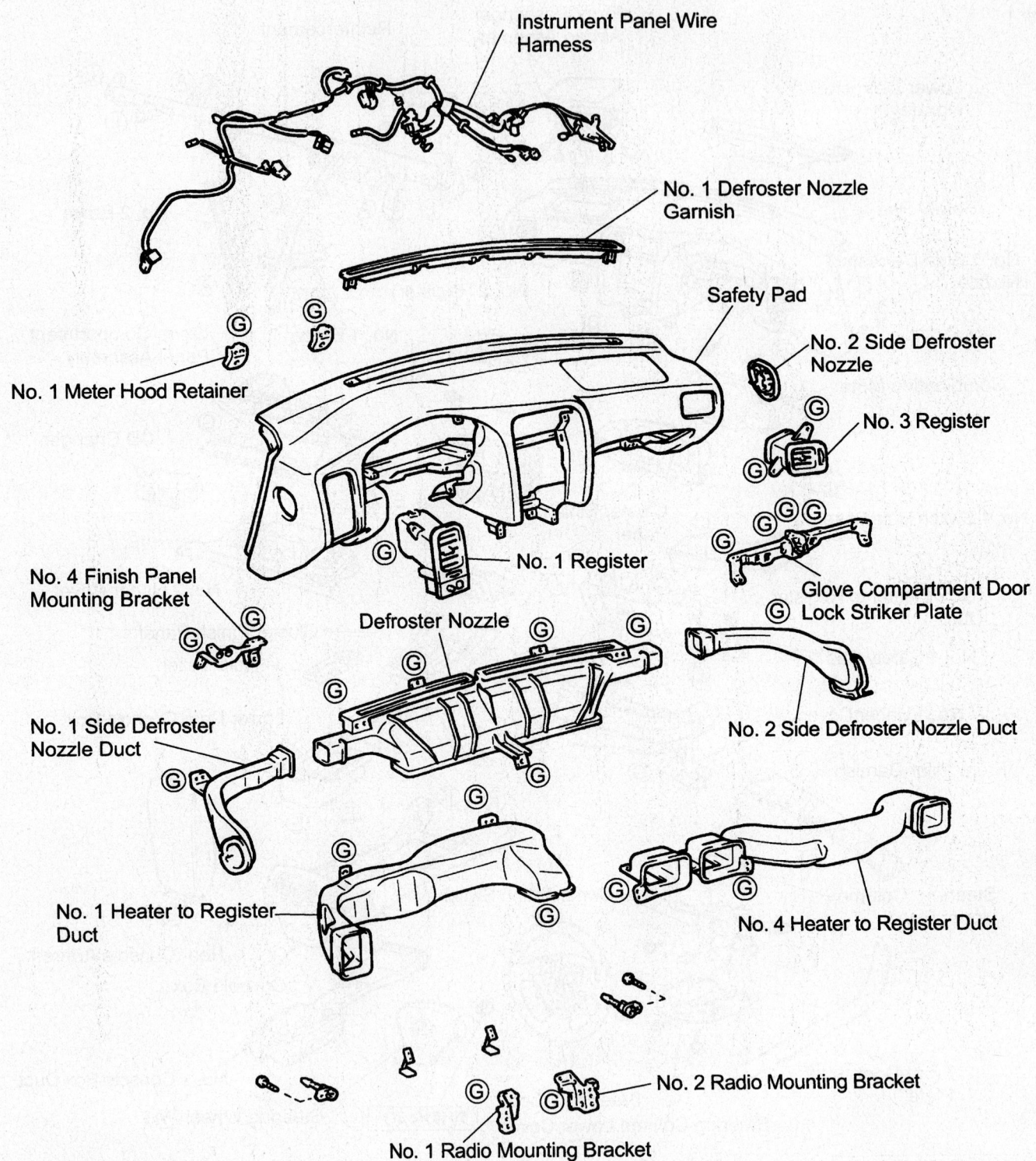

Instrument Panel Wire Harness

No. 1 Defroster Nozzle Garnish

Safety Pad

No. 2 Side Defroster Nozzle

No. 1 Meter Hood Retainer

No. 3 Register

No. 4 Finish Panel Mounting Bracket

No. 1 Register

Glove Compartment Door Lock Striker Plate

Defroster Nozzle

No. 2 Side Defroster Nozzle Duct

No. 1 Side Defroster Nozzle Duct

No. 1 Heater to Register Duct

No. 4 Heater to Register Duct

No. 2 Radio Mounting Bracket

No. 1 Radio Mounting Bracket

09490_LEXU_G0030

Exploded view of the instrument panel, ventilation ducts and related components—2002–05 GS 300 and GS 430

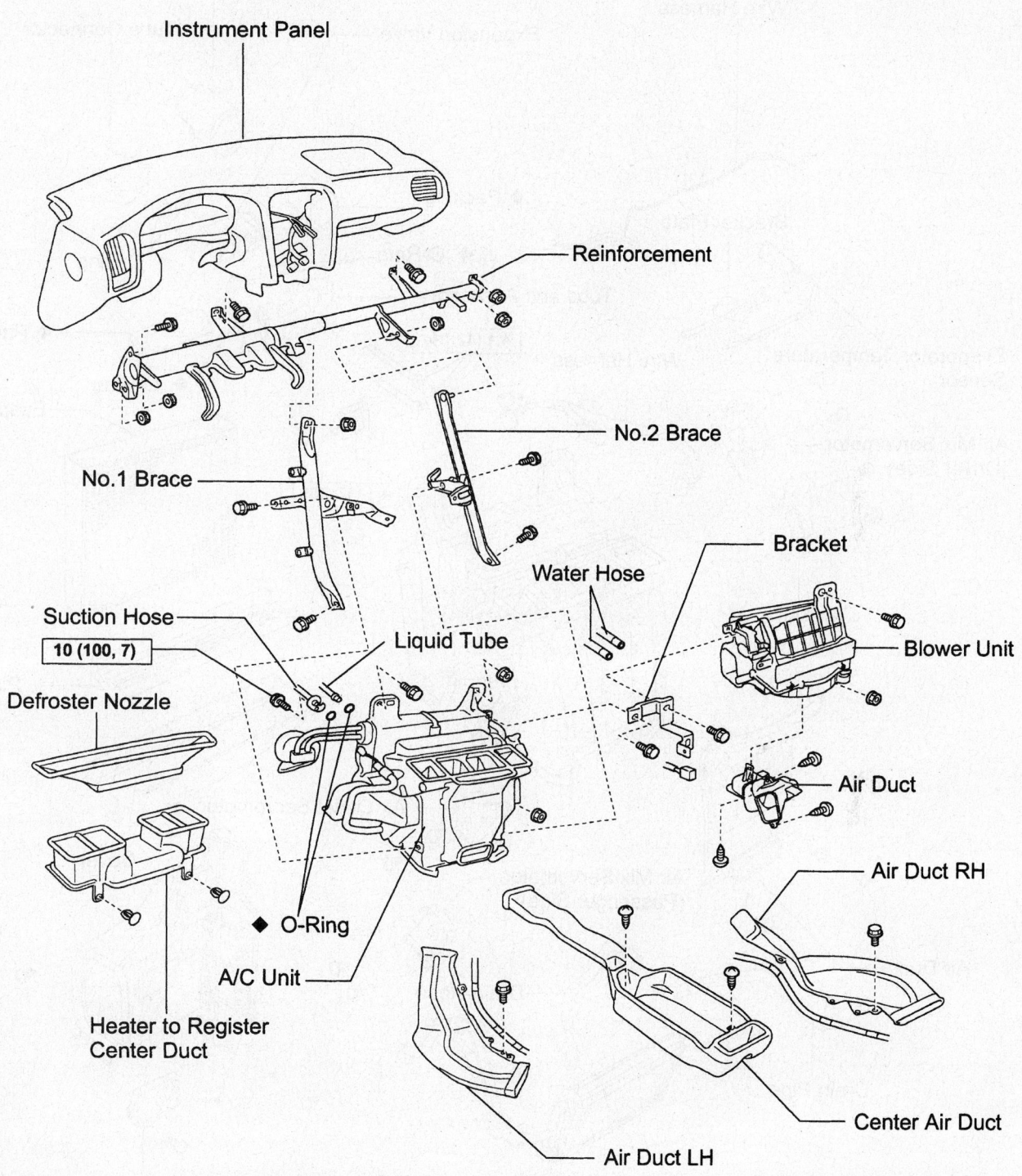

Instrument Panel

Reinforcement

No.2 Brace

No.1 Brace

Water Hose

Bracket

Suction Hose

10 (100, 7)

Liquid Tube

Blower Unit

Defroster Nozzle

Air Duct

Air Duct RH

◆ O-Ring

A/C Unit

Air Duct LH

Center Air Duct

Heater to Register
Center Duct

09490_LEXU_G0031

N·m (kgf·cm, ft·lbf) : Specified torque
◆ Non-reusable part

Exploded view of the heater/air conditioning housing and related components—2002–05 GS 300 and GS 430

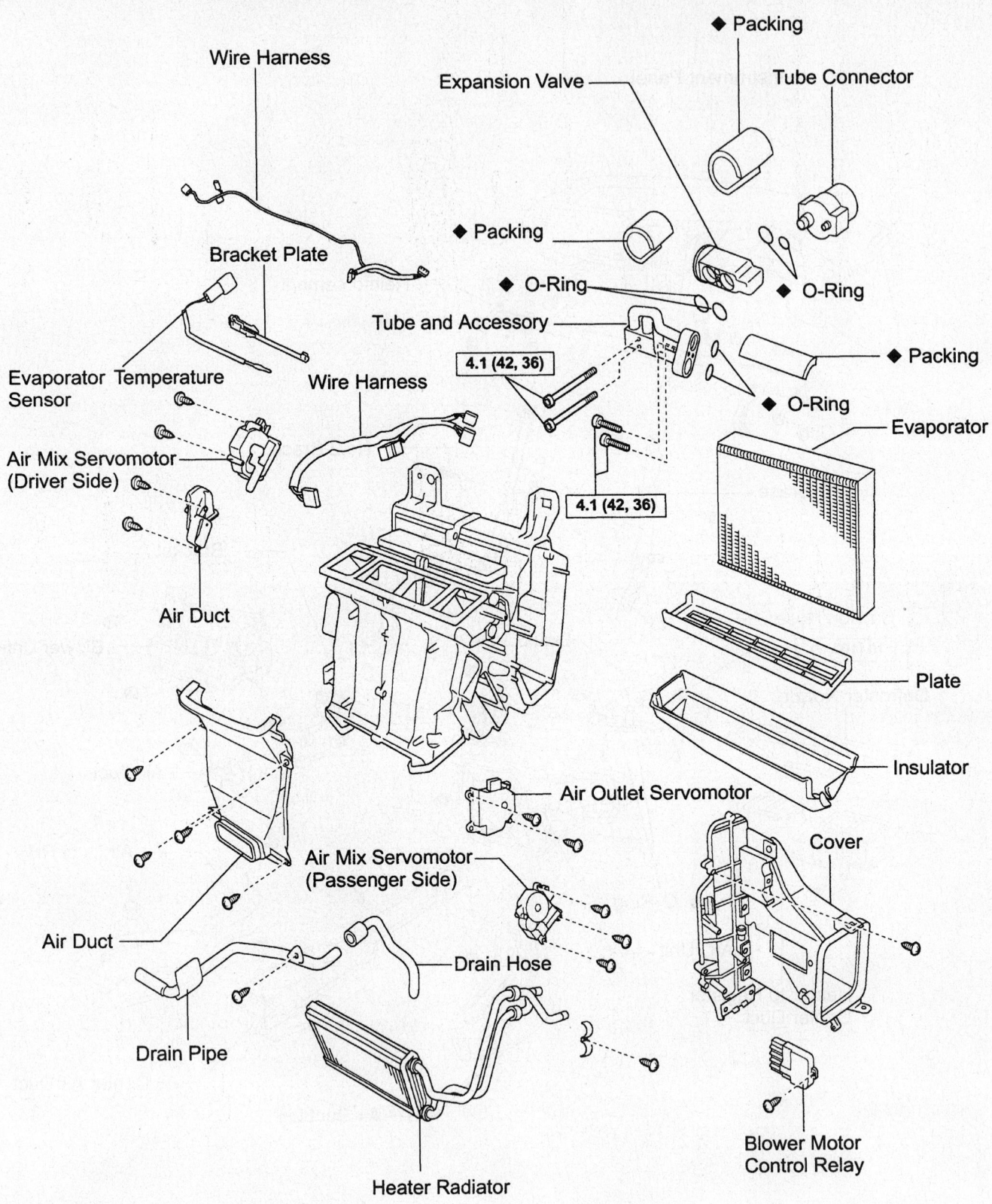

Wire Harness

◆ Packing

Expansion Valve

Tube Connector

Bracket Plate

◆ Packing

◆ O-Ring

◆ O-Ring

Evaporator Temperature Sensor

Tube and Accessory

4.1 (42, 36)

◆ Packing

Wire Harness

◆ O-Ring

Air Mix Servomotor (Driver Side)

Evaporator

4.1 (42, 36)

Air Duct

Plate

Insulator

Air Outlet Servomotor

Cover

Air Mix Servomotor (Passenger Side)

Air Duct

Drain Hose

Drain Pipe

Heater Radiator

Blower Motor Control Relay

$N \cdot m$ (kgf·cm, in.·lbf) : Specified torque

◆ Non-reusable part

09490_LEXU_G0032

Exploded view of the heater core, heater housing and related components—2002–05 GS 300 GS 430

- Wiring harness set nut
- Electrical connector

17. Install or connect the following:
- Air duct and the 2 duct bolts on both sides
- Wiring harness clamps
- Move the floor carpet forward
- Center air duct and the 2 air duct-to-heater/air conditioning housing screws

18. Install the blower unit by installing or connecting the following:
- Blower housing and the blower housing-to-chassis bolt, screw and nut
- Blower housing and the 2 bracket-to-bracket bolts
- Blower housing connectors
- Electrical connector bracket, the wiring harness clamps and the wiring harness
- Air duct and the 3 air duct-to-blower housing screws
- Connector clamp

19. Install the instrument by installing or connecting the following:
- Ventilation nozzles to the heater/air conditioning housing
- Reinforcement and the reinforce-ment-to-chassis 5 nuts and 4 bolts
- No. 1 and No. 2 brace
- Instrument panel and the 5 instru-ment panel-to-chassis bolts, the nut and the screw

20. Install or connect the following:
- No. 7 heater-to-register
- No. 1 console box duct
- Console box
- No. 3 console box mounting bracket
- Rear console armrest
- Lower rear console box
- Console box carpet
- Connectors and install the cluster finish panel
- Audio unit and the audio unit-to-instrument panel 2 bolts and 2 screws
- Connector and install the No. 2 register
- Ashtray
- CD changer, connect the electri-cal connectors and install the 3 CD changer-to-instrument panel nuts
- Glove box and the glove box-to-instrument panel 2 bolts and 3 screws
- Air bag electrical connector and install the plate inside the glove box
- No. 2 undercover

- Instrument cluster, connect the electrical connectors and install the 4 instrument cluster-to-instrument panel screws
- Instrument cluster finish panel
- Steering column, connect the spring to the brake pedal and install the 4 steering column-to-instrument panel nuts
- Parking brake handle and the No. 1 switch hole base
- No. 1 safety pad and the 4 safety pad-to-instrument panel bolts and screw
- Hood lock release and the 2 release screws
- No. 1 undercover and the 2 under-cover-to-instrument panel screws
- End pad
- Combination switch and connect the electrical connectors
- Steering column covers and the cover screws
- Front pillar garnishes and the front door scuff plates

21. Install the steering wheel by installing or connecting the following:
- Steering wheel and torque the steering wheel nut to 26 ft. lbs. (35 Nm)
- Electrical connector and install the air bag
- Tighten the Torx® bolts to 80 inch lbs. (9.0 Nm) using a Torx® wrench
- Torx® bolt covers at both sides of the steering wheel
- No. 1 grommet, the heater pipe grommet and the drain hose grom-met
- Heater hoses to the heater core
- Refrigerant lines (using new O-rings) and the refrigerant lines-to-evaporator bolt

22. Refill the cooling system.
23. Connect the negative battery cable.
24. Evacuate, charge and leak test the air conditioning system refrigerant.
25. Operate the engine to normal oper-ating temperatures; then, check the cli-mate control operation and check for leaks.

GS 300 and GS 430 (2006)

1. Before servicing the vehicle, refer to the Precautions Section.
2. Discharge the A/C system.
3. Set radio receiver assembly to ship-ment mode as follows:
   a. Be sure that all discs and tapes have been removed from the unit.

b. Be sure that the engine switch off.
c. While simultaneously pressing the "SEEK UP" and "DISC" switches, turn the engine switch on (ACC).

➡**The CD loading door indicator light blinks during mode setting and it remains lit after the setting is com-pleted.**

   d. Turn the engine switch off.
4. Disconnect the negative battery cable. Wait 90 seconds before doing any further work while the airbag system de-energizes.
5. Drain the cooling system into a clean container for reuse.
6. Align the front wheels facing straight ahead.
7. Remove or disconnect the follow-ing:
- Cool air intake duct seal
- Left and right engine room side covers
- Left and right front pillar to front side seals, using a clip remover to detach the 3 claws
- Left and right nut, windshield wiper arms and blades
- Left and right front fender to cowl side seals by moving the compo-nent toward the center of the vehicle to detach the 2 claws

8. Remove cowl top ventilator louver assembly
   a. Remove the 2 clips and detach the 5 claws.
   b. Pull the ventilator louver in the direction indicated by the arrow in the illustration to detach the 10 claws and remove the ventilator louver.
9. Remove the windshield wiper motor and link assembly as follows:
   a. Disconnect the connector. Then detach the 2 clamps and remove the wire harness from the cowl top panel.

➡**There are 6 bolts total, however, 2 bolts cannot be removed from the wiper motor and link because they are integrated into the wiper motor and link.**

   b. Remove the 4 bolts and wiper motor and link.
10. Separate suction pipe sub-assembly as follows:
   a. Remove the bolt, and slide the hook connector.
   b. Disconnect the suction pipe sub-assembly.
   c. Remove the O-ring from the suc-tion pipe sub-assembly.

> ❊❊ **WARNING**

**Seal the openings of the discon-
nected parts using vinyl tape to prevent moisture and foreign matter
from entering.**

11. Separate liquid tube sub-assembly
as follows:
   a. Disconnect the liquid tube sub-
assembly.
   b. Remove the O-ring from the liquid
tube subassembly.
- Inlet and outlet heater water outlet
hoses
- Left and right front door scuff
plates
- Left and right front door opening
trim covers
- The lower No. 2 and No. 3 steering
wheel covers using a screwdriver.
Tape up the screwdriver tip before
use.

12. Remove the steering wheel airbag
assembly as follows:
   a. Using a "torx" socket wrench
(T30), loosen the 2 "torx" screws until
the groove along the screw circumference catches on the screw case.
   b. Pull out the steering wheel airbag
from the steering wheel assembly and support the airbag carefully with one hand.

> ❊❊ **WARNING**

**When removing the steering airbag,
do not pull the airbag wire harness.**

   c. Disconnect the horn connector.
   d. Disconnect the 2 connectors from
the steering airbag.

> ❊❊ **WARNING**

**When handling the airbag connector,
take care not to damage the airbag
wire harness.**

   e. Remove the steering wheel airbag.
13. Remove the steering wheel assembly
as follows:
   a. Remove the steering wheel assembly set nut.
   b. Put matchmarks on the steering
wheel assembly and main shaft assembly.
   c. Using a steering wheel puller tool,
remove the steering wheel assembly.
14. Remove the steering column cover
as follows:
   a. Remove the 3 retaining screws.
   b. Disengage the 2 claws to remove
the lower steering column cover.

> ❊❊ **WARNING**

**Be careful not to damage the tilt and
telescopic switch.**

   c. Disengage the 4 clips to separate
the steering column cover upper.
   d. Disengage the claw to remove the
upper steering column cover.
- Spiral cable with steering sensor
- Windshield wiper switch assembly
15. Remove the headlight dimmer switch
as follows:
   a. Disconnect the connector.
   b. Detach the clamp (looks like a
hose clamp) from the headlight dimmer
switch using a set of pliers.
   c. Disengage the claw and remove the
headlight dimmer switch.
- Front console upper panel garnish
- Console upper panel
- Left and right instrument panel finish panel ends
- Shift lever knob by turning counterclockwise
- Console box plate
- Console box register assembly
- Console box
- The tilt and telescopic switch by
disengaging the claw
- Turn signal switch assembly and
spiral cable assembly
- No. 1 air duct
16. Remove the steering sliding yoke
sub-assembly (3UZ-FE) as follows:

> ❊❊ **WARNING**

**Make sure that the steering link
assembly is centered before performing this step.**

   a. Loosen upper bolt and remove
lower bolt, then slide the steering sliding
yoke sub-assembly.

➡**Do not remove the upper bolt. Do not
separate the steering sliding yoke sub-
assembly from the power steering link
assembly.**

   b. Put matchmarks on the steering
sliding yoke sub-assembly and the
power steering link assembly.
   c. Separate the steering sliding yoke
sub-assembly from the power steering
link assembly.
   d. Put matchmarks on the steering
sliding yoke sub-assembly and the steering actuator assembly.
   e. Remove the bolt and the steering
sliding yoke sub-assembly from the
steering actuator assembly.
17. Separate the No. 2 steering interme-

diate shaft assembly (3GR-FSE with 2WD)
as follows:
   a. Loosen the upper bolt and remove
the lower bolt, then slide the No. 2 steering intermediate shaft assembly.

➡**Do not remove the upper bolt. Do not
disconnect the No. 2 steering intermediate shaft assembly from the power
steering link assembly.**

   b. Put matchmarks on the steering
sliding yoke sub-assembly and the
power steering link assembly.
   c. Separate the steering sliding yoke
sub-assembly from the power steering
link assembly.
   d. Put matchmarks on the steering
sliding yoke sub-assembly and the No. 2
steering intermediate shaft assembly.
   e. Remove the bolt and the steering
sliding yoke sub-assembly from the No.
2 steering intermediate shaft assembly.
18. Separate the No. 2 steering intermediate shaft assembly (3GR-FSE with AWD)
as follows:
   a. Remove the bolt, and then slide the
No. 2 steering intermediate shaft assembly.

➡**Do not separate the No. 2 steering
intermediate shaft assembly from the
power steering link assembly.**

   b. Put matchmarks on the No. 2
steering intermediate shaft assembly and
the power steering link assembly.
   c. Separate the No. 2 intermediate
shaft assembly from the power steering
link assembly.
- Steering main shaft lower dust
cover (3UZ-FE)
19. Separate steering actuator assembly
(3UZ-FE) as follows:
   a. Disconnect the connector and separate the wire harness clamp from the
steering actuator assembly.
   b. Put matchmarks on the steering
actuator assembly and the main shaft.
   c. Remove the bolt and separate the
steering actuator assembly.
20. Remove the steering column assembly as follows:
   a. Remove the bolt and separate the
steering intermediate shaft assembly No.
2 (3GR-FSE engine).
   b. Disconnect the connectors and
wire harness clamps.
   c. Remove the 4 nuts and steering
column assembly.
- Instrument cluster finish panel
- Combination meter assembly
- Left and right side instrument panel
side panels

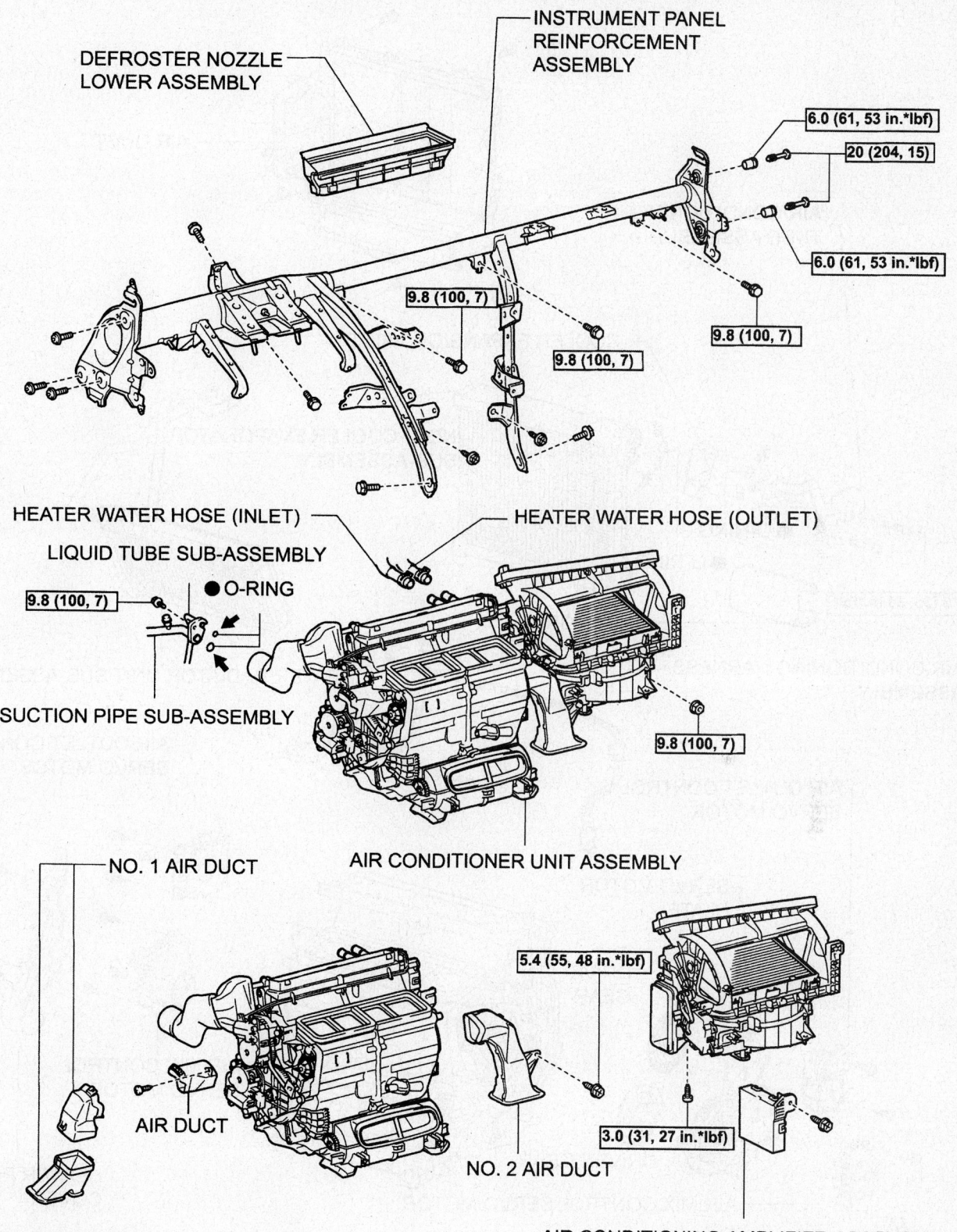

DEFROSTER NOZZLE LOWER ASSEMBLY

INSTRUMENT PANEL REINFORCEMENT ASSEMBLY

6.0 (61, 53 in.*lbf)

20 (204, 15)

6.0 (61, 53 in.*lbf)

9.8 (100, 7)

9.8 (100, 7)

9.8 (100, 7)

9.8 (100, 7)

HEATER WATER HOSE (INLET)

HEATER WATER HOSE (OUTLET)

LIQUID TUBE SUB-ASSEMBLY

9.8 (100, 7)

● O-RING

SUCTION PIPE SUB-ASSEMBLY

9.8 (100, 7)

NO. 1 AIR DUCT

AIR CONDITIONER UNIT ASSEMBLY

AIR DUCT

5.4 (55, 48 in.*lbf)

NO. 2 AIR DUCT

3.0 (31, 27 in.*lbf)

AIR CONDITIONING AMPLIFIER ASSEMBLY

N*m (kgf*cm, ft.*lbf) : Specified torque

● Non-reusable part

◄ Compressor oil ND-OIL 8 or equivalent

09490_LEXU_G0027

Exploded view of the A/C unit assembly, instrument panel reinforcement and related components—IS 250, IS 350, GS 300 (2006) and GS 430 (2006)

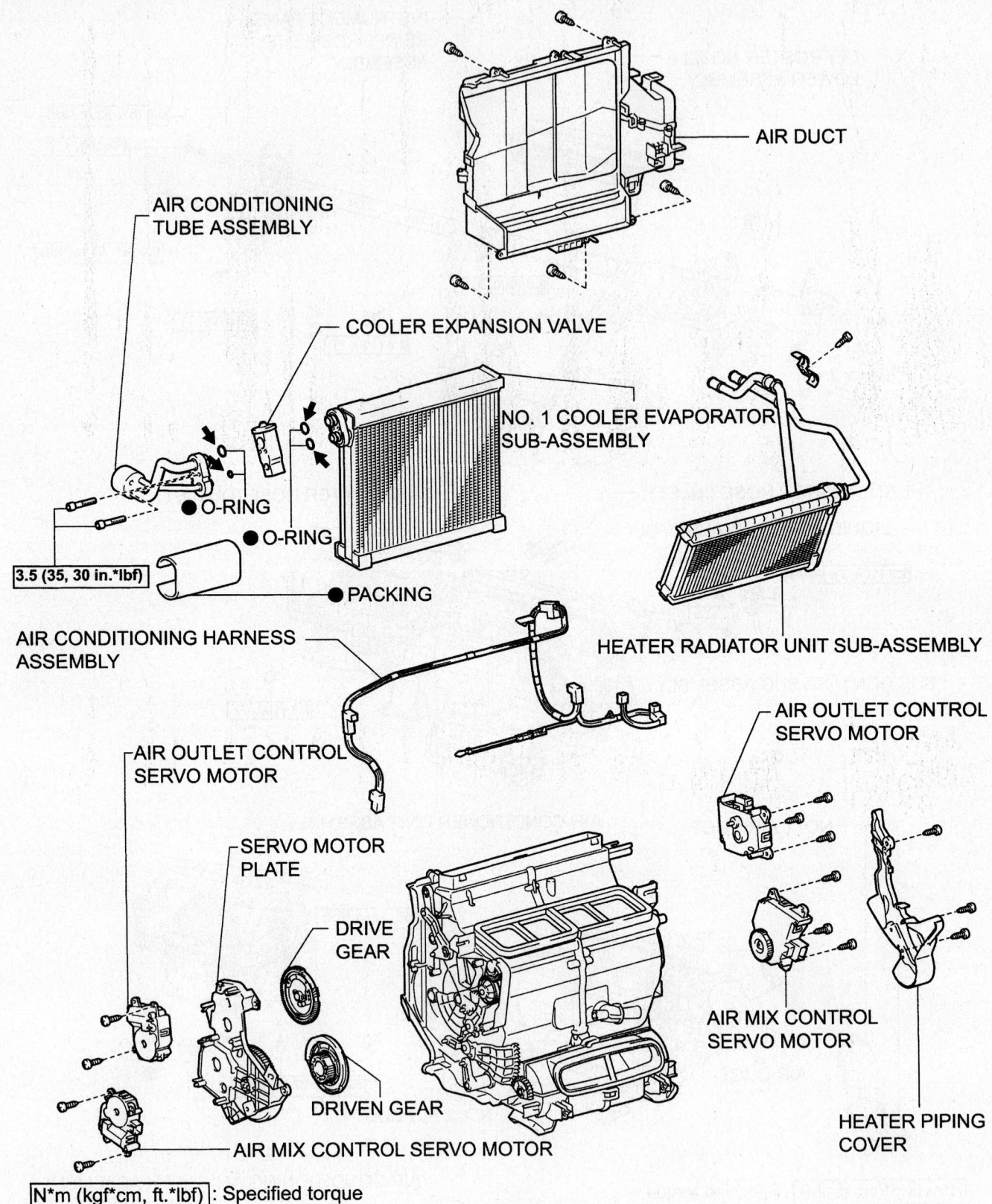

AIR DUCT

AIR CONDITIONING
TUBE ASSEMBLY

COOLER EXPANSION VALVE

NO. 1 COOLER EVAPORATOR
SUB-ASSEMBLY

● O-RING

● O-RING

3.5 (35, 30 in.*lbf)

● PACKING

HEATER RADIATOR UNIT SUB-ASSEMBLY

AIR CONDITIONING HARNESS
ASSEMBLY

AIR OUTLET CONTROL
SERVO MOTOR

AIR OUTLET CONTROL
SERVO MOTOR

SERVO MOTOR
PLATE

DRIVE
GEAR

AIR MIX CONTROL
SERVO MOTOR

DRIVEN GEAR

AIR MIX CONTROL SERVO MOTOR

HEATER PIPING
COVER

N*m (kgf*cm, ft.*lbf) : Specified torque

● Non-reusable part

◄ Compressor oil ND-OIL 8 or equivalent

09490_LEXU_G0028

Exploded view of the heater radiator unit (heater core), heater housing and related components—IS 250, IS 350, GS 300 (2006) and GS 430 (2006)

- No. 1 instrument panel under cover
- No. 1 instrument panel safety pad
- Driver side knee airbag assembly
- Integration control and panel assembly
- No. 2 instrument panel under cover
- Steering control ECU
- Passenger side knee airbag assembly
- Glove compartment door assembly
- Left and right front pillar garnishes

21. Remove the multi-display with radio receiver assembly as follows:

   a. Remove the 4 bolts.

   b. Pull the multi-display with radio receiver to detach the 2 clips on the backside of the multi-display.

   c. Disconnect each connector and remove the multi-display.

   - Instrument panel safety pad
   - No. 1 and 2 console box duct
   - 4 claws and defroster nozzle lower assembly
   - Transponder key ECU assembly

22. Remove the instrument panel reinforcement assembly as follows:

   a. Remove the 27 clamps and connectors, and then disconnect the wire harness.

   b. Remove the 3 nuts, 2 bolts and 2 junction blocks.

   c. Remove the cap and bolt.

   d. Remove the 6 bolts and 2 screws.

   e. Using a T40 "torx" socket, remove the 5 "torx" bolts.

   f. Using a 12 mm hexagon wrench, remove the 2 collars and instrument panel reinforcement.

23. Remove or disconnect the following:

   - A/C blower unit assembly
   - Mounting screw and No. 2 air duct
   - Upper and lower foot ducts
   - Connector, mounting screw and air conditioning amplifier assembly

24. Remove the heater core from the A/C-blower assembly unit as follows:

   a. Disconnect the wiring connector, remove the 3 screws and right side air outlet control servo motor.

   b. Disengage 3 clamps, connector and wire harness to right side air mix control servo motor.

   c. Remove the 2 screws and heater piping cover.

   d. Remove the 3 screws and right side air mix control servo motor.

   e. Move the A/C wiring harness out of the way of the heater radiator unit.

   f. Remove the heater radiator unit sub-assembly (heater core)

To install:

25. Install or connect the following:

   - Heater core to the A/C blower housing
   - A/C wiring harness back into position
   - Right side air mix control servo motor
   - Heater piping cover
   - Right side air outlet control servo motor
   - Air conditioning amplifier assembly
   - Upper and lower foot ducts
   - No. 2 air duct
   - A/C blower unit assembly. Torque the retaining nut to 7 ft. lbs. (10 Nm).

26. Install the instrument panel reinforcement assembly and torque the fasteners as follows:

   a. On the driver side, use a T40 "torx" socket. Install the instrument panel reinforcement with the 3 "torx" bolts and torque to 13 ft. lbs. (17 Nm).

   b. On the passenger side, use a 12 mm hexagon wrench. Install the instrument panel reinforcement with the 2 bolts and torque to 53 inch lbs. (6 Nm).

   c. On the passenger side, use a T40 "torx" socket. Install the instrument panel reinforcement with the 2 "torx" bolts and torque tp 15 ft. lbs. (20 Nm).

   d. Install the instrument panel reinforcement with the 6 bolts and 2 screws.

   e. Install the bolt and cap.

   f. Connect the 2 connectors and attach the 27 clamps.

   g. Install the 2 junction blocks with the 2 bolts and 3 nuts.

   - Transponder key ECU assembly
   - Defroster nozzle lower assembly
   - No. 1 and 2 console box duct
   - Instrument panel safety pad
   - Multi-display with radio receiver
   - Left and right front pillar garnishes
   - Glove compartment door assembly
   - Passenger side knee airbag assembly
   - Steering control ECU
   - No. 2 instrument panel under cover
   - Integration control and panel assembly
   - Driver side knee airbag assembly
   - No. 1 instrument panel safety pad
   - No. 1 instrument panel under cover
   - Left and right side instrument panel side panels
   - Combination meter assembly
   - Instrument cluster finish panel

27. Install the steering column assembly as follows:

   a. Install the No. 2 steering intermediate shaft assembly (3GR-FSE).

   b. Install the steering column assembly with the 4 nuts. Torque the 4 nuts to 19 ft. lbs. (26 Nm).

   c. Connect the connectors and wire harness clamps to the steering column assembly.

   d. Align the matchmarks on the steering sliding yoke assembly and power steering link assembly.

   e. Install the bolt and torque to 26 ft. lbs. (35 Nm).

28. Connect steering actuator assembly (3UZ-FE) as follows:

   a. Align the matchmarks on the steering actuator assembly and the main shaft.

   b. Install the bolt and torque to 26 ft. lbs. (35 Nm).

   - Steering main shaft lower dust cover (3UZ-FE)
   - No. 2 steering intermediate shaft assembly (3GR-FSE). Torque the bolt to 26 ft. lbs. (35 Nm).
   - Steering sliding yoke sub-assembly (3GR-FSE with 2WD and 3UZ-FE). Torque the bolts to 26 ft. lbs. (35 Nm).
   - No. 1 air duct
   - Turn signal switch assembly with spiral cable sub-assembly
   - The tilt and telescopic switch
   - Console box
   - Console box register assembly
   - Console box plate
   - Shift lever knob
   - Left and right instrument panel finish panel ends
   - Console upper panel
   - Front console upper panel garnish
   - Headlight dimmer switch
   - Windshield wiper switch assembly
   - Spiral cable with steering sensor

29. Adjust the spiral cable as follows:

   a. Check that the engine switch is off.

   b. Check that the battery negative (-) terminal is disconnected.

✳✳ WARNING

After removing the terminal, wait for at least 90 seconds before starting the operation.

   c. Rotate the spiral cable with steering sensor counterclockwise slowly by hand until it feels firm. Do not turn it by the wiring harness.

   d. Rotate the spiral cable with steering sensor clockwise approximately 2.5 turns to align the marks.

➡ **The spiral cable with steering sensor will rotate approximately 2.5 turns to both the left and right from the center.**

- Steering column covers
- Steering wheel. Torque the steering wheel bolt to 37 ft. lbs. (50 Nm).
- Steering wheel airbag. Torque the screws to 78 inch lbs. (9 Nm).
- The lower No. 2 and No. 3 steering wheel covers
- Left and right front door opening trim covers
- Left and right front door scuff plates
- Inlet and outlet heater water outlet hoses

30. Install the liquid tube sub-assembly as follows:

a. Remove the vinyl tape attached to the tube.

b. Sufficiently apply compressor oil to a new O-ring and the fitting surface of the liquid tube.

c. Install the O-ring on the liquid tube.

d. Install the liquid tube to the fitting hole.

31. Install the suction pipe sub-assembly as follows:

a. Remove the vinyl tape attached to the pipe.

b. Sufficiently apply compressor oil to a new O-ring and the fitting surface of the suction pipe.

c. Install the O-ring on the suction pipe.

d. Move the hook connector in a counterclockwise direction.

e. Insert the pipe joints into the fitting holes securely and tighten the bolt to 7 ft. lbs. (10 Nm).

- Wiper motor and link assembly
- Cowl top ventilator louver assembly
- Left and right front fender to cowl side seals
- Windshield wiper arms and blades
- Left and right front pillar to front side seals
- Left and right engine room side covers
- Cool air intake duct seal

32. Perform system initialization (which includes power window control system, sliding roof system, clearance sonar system and variable gear ratio steering system) procedure as follows:

- Power window control system

a. Turn the ignition switch on.

b. Open power window halfway by pressing power window switch.

c. Fully pull up the switch until the power window is fully closed and continue to hold the switch for at least 1 second.

d. Check that the AUTO UP / DOWN function operates normally.

➡ **If the remote UP / DOWN function does not operate after the conditions 1, 2, or 3 is satisfied, the power window regulator master switch may have a malfunction.**

- Sliding roof system

e. Turn the ignition switch on.

f. If the sliding roof is opened, close it fully.

g. Push the open switch of the slide switch, or the up switch of the tilt switch on the personal light, making the sliding roof tilt up approximately 1 second, tilt down, slide open, slide close.

h. Sliding roof stops at the fully closed position.

i. Finish the initialization.

j. Check that the operation works normally with AUTO operation.

- Clearance sonar system

k. Turn the ignition switch on.

l. Turn the clearance sonar main switch ON.

m. Turn the steering wheel to the full left and right lock position.

➡ **Make sure to completely turn the steering wheel to the left and right full lock position.**

n. Confirm that the learning operation has been completed by checking the multi-information display.

o. At an area with few turns and curves, and minimal traffic, drive at 20 km/h or more for 5 minutes or more.

- Variable gear ratio steering system

p. Turn the ignition switch on, and check that the master warning light and VSC/ABS warning lights illuminate for a few seconds.

➡ **If the warning lights remain on or blink, repair the applicable system.**

q. Drive the vehicle on a straight road at 35 km/h (22 mph) or more for 5 seconds or longer.

r. Confirm that steering angle sensor initialization is completed by doing the following:

- Drive the vehicle on a straight road at 60 km/h (37 mph) or more for 30 seconds or longer.
- Stop the vehicle (engine running).
- Slowly turn the steering wheel from lock to lock.
- If it turns approximately 2.7 turns,

steering angle sensor initialization is completed. If it turns approximately 3.2 turns, steering angle sensor initialization is not completed.

33. Refill the cooling system.

34. Connect the negative battery cable.

35. Evacuate, charge and leak test the air conditioning system refrigerant.

36. Operate the engine to normal operating temperatures; then, check the climate control operation and check for leaks.

LS 430

1. Before servicing the vehicle, refer to the Precautions Section.

2. Disconnect the negative battery cable. Wait 90 seconds before doing any further work while the airbag system de-energizes.

3. Disconnect the negative battery cable.

4. Drain the cooling system into a clean container for reuse.

5. Remove or disconnect the following:

- Undercover and the No. 1 safety pad-to-instrument panel screws and the panel at the driver's side
- No. 2 heater-to-register duct
- Heater core-to-heater housing screw and clamp
- Heater hoses from the heater core
- Heater core from the heater housing
- Discard the O-rings

To install:

6. Install or connect the following:

- New O-rings to the heater core
- Heater core to the heater housing
- Heater hoses to the heater core
- Heater core clamp and the heater core-to-heater housing screw
- No. 2 heater-to-register duct
- Undercover and the No. 1 safety pad-to-instrument panel and the panel screws at the driver's side

7. Refill the cooling system.

8. Connect the negative battery cable.

9. Operate the engine to normal operating temperatures; then, check the climate control operation and check for leaks.

SC 430

➡ **Removal of the heater core requires removal of the entire heater air conditioning assembly.**

1. Before servicing the vehicle, refer to the Precautions Section.

2. Drain the cooling system into a clean container for reuse.

3. Discharge and recover the air conditioning system refrigerant.

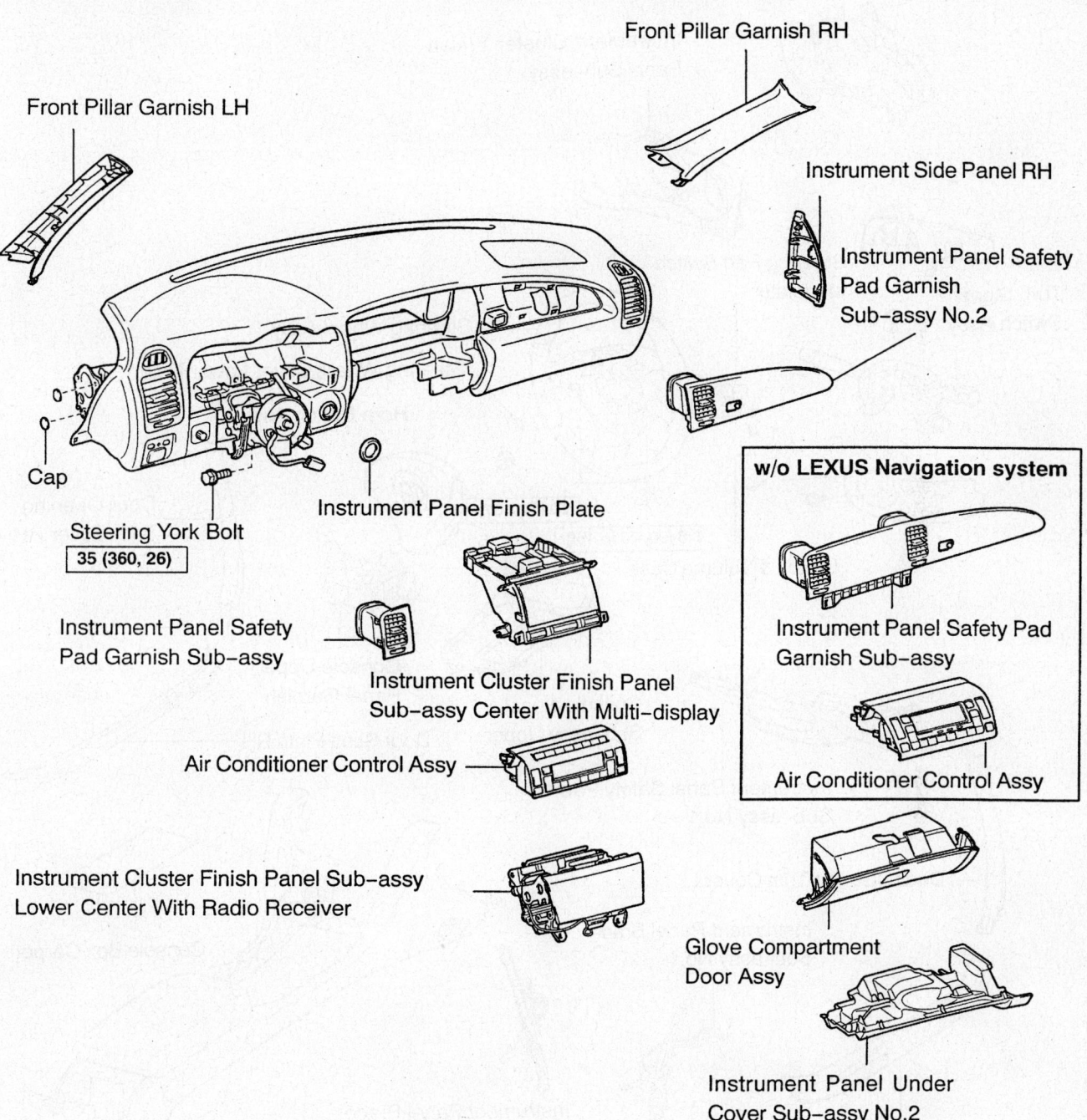

Front Pillar Garnish RH

Front Pillar Garnish LH

Instrument Side Panel RH

Instrument Panel Safety
Pad Garnish
Sub-assy No.2

Cap

Instrument Panel Finish Plate

Steering York Bolt
35 (360, 26)

w/o LEXUS Navigation system

Instrument Panel Safety Pad
Garnish Sub-assy

Air Conditioner Control Assy

Instrument Panel Safety
Pad Garnish Sub-assy

Instrument Cluster Finish Panel
Sub-assy Center With Multi-display

Air Conditioner Control Assy

Instrument Cluster Finish Panel Sub-assy
Lower Center With Radio Receiver

Glove Compartment
Door Assy

Instrument Panel Under
Cover Sub-assy No.2

N·m (kgf·cm, ft·lbf)   : Specified torque

67162-LEXU-G13

Exploded view of the dash components—SC 430

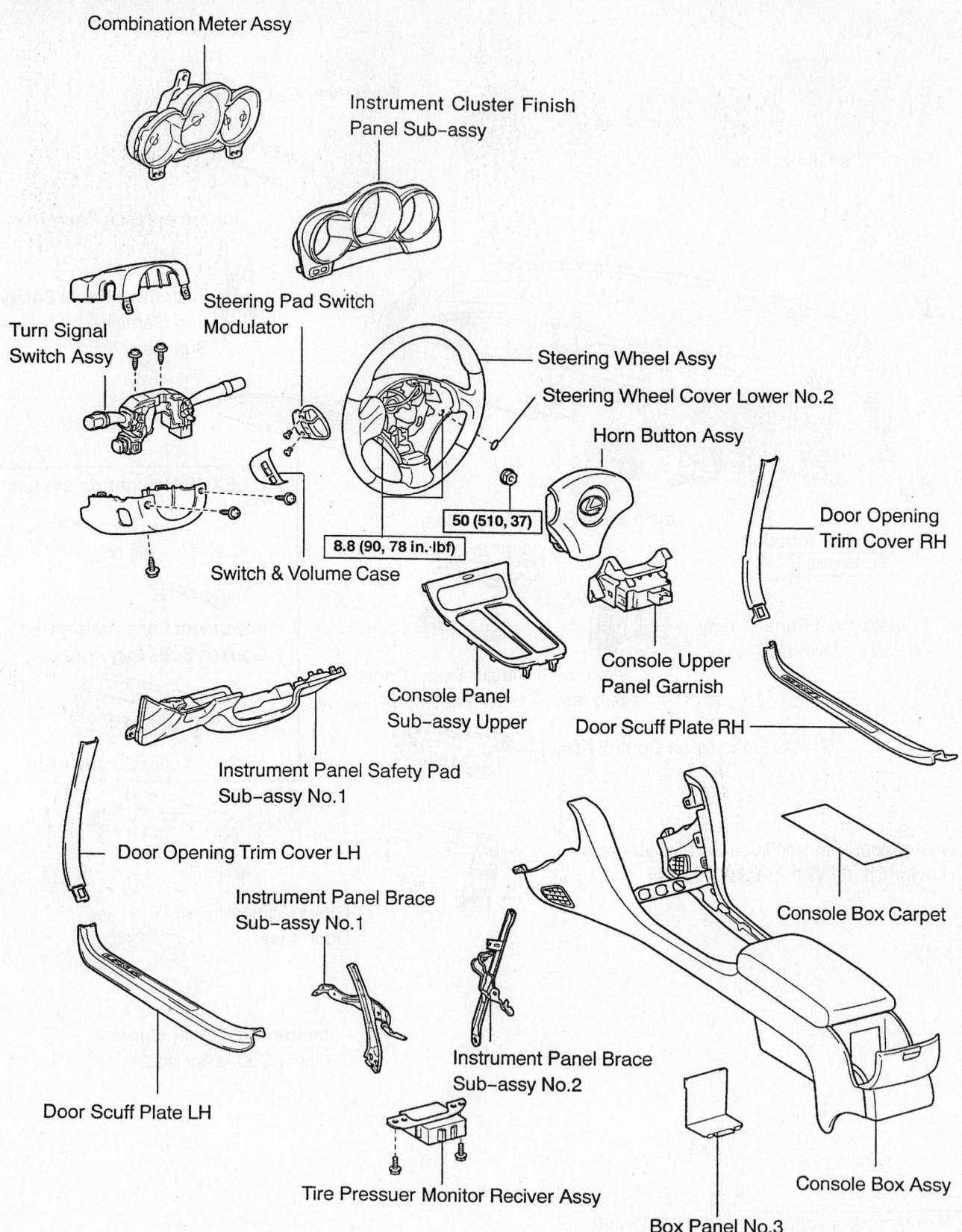

Combination Meter Assy

Instrument Cluster Finish Panel Sub-assy

Steering Pad Switch Modulator

Turn Signal Switch Assy

Steering Wheel Assy

Steering Wheel Cover Lower No.2

Horn Button Assy

Door Opening Trim Cover RH

50 (510, 37)

8.8 (90, 78 in.·lbf)

Switch & Volume Case

Console Upper Panel Garnish

Console Panel Sub-assy Upper

Door Scuff Plate RH

Instrument Panel Safety Pad Sub-assy No.1

Door Opening Trim Cover LH

Instrument Panel Brace Sub-assy No.1

Console Box Carpet

Door Scuff Plate LH

Instrument Panel Brace Sub-assy No.2

Tire Pressuer Monitor Reciver Assy

Box Panel No.3

Console Box Assy

| N·m (kgf·cm, ft·lbf) | : Specified torque |

67162-LEXU-G14

Exploded view gauge cluster, steering wheel and console—SC 430

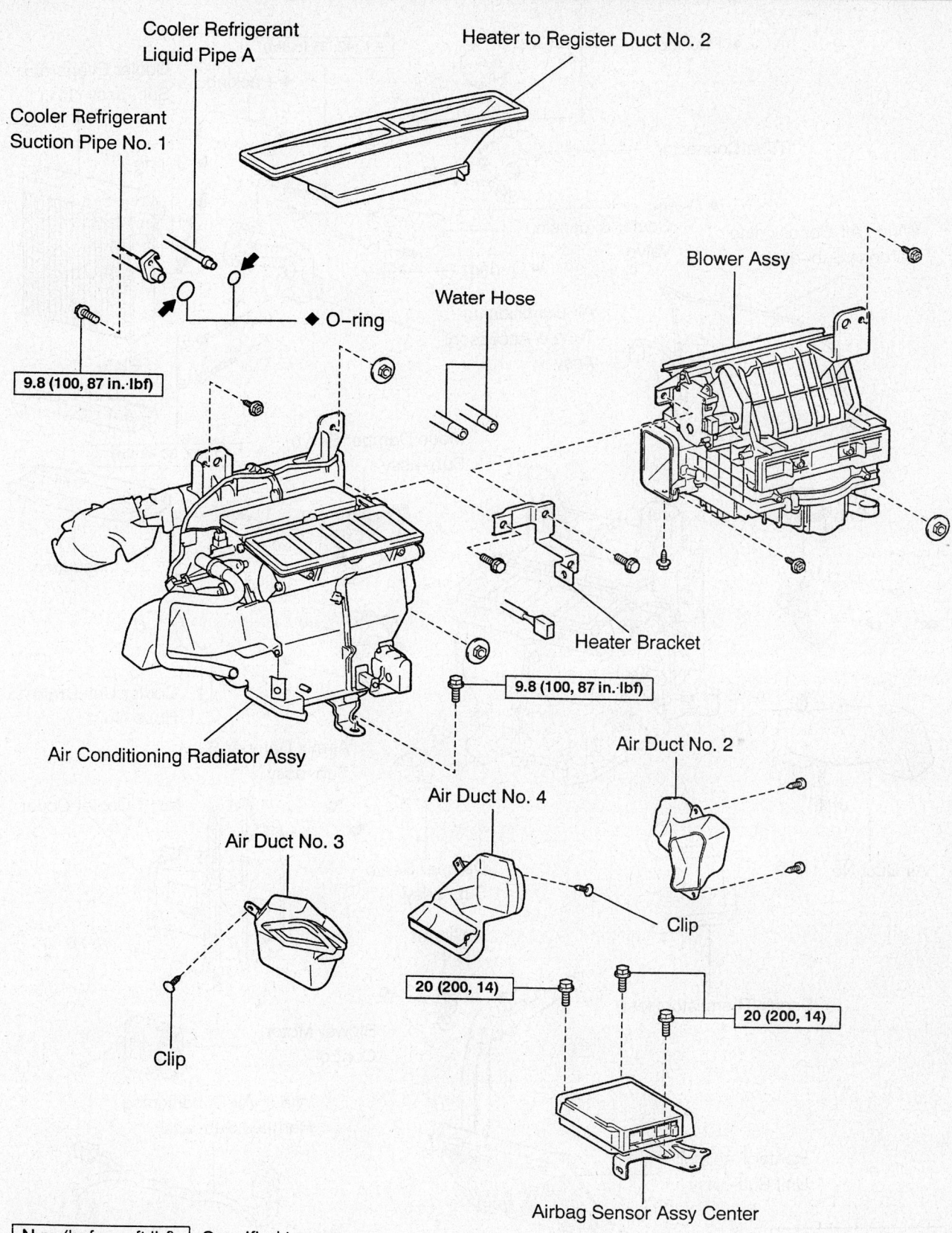

Cooler Refrigerant
Liquid Pipe A

Cooler Refrigerant
Suction Pipe No. 1

Heater to Register Duct No. 2

Water Hose

Blower Assy

◆ O-ring

9.8 (100, 87 in.·lbf)

Heater Bracket

Air Conditioning Radiator Assy

9.8 (100, 87 in.·lbf)

Air Duct No. 2

Air Duct No. 4

Air Duct No. 3

Clip

Clip

20 (200, 14)

20 (200, 14)

Clip

Airbag Sensor Assy Center

N·m (kgf·cm, ft·lbf) : Specified torque

◆ Non-resable part

◀ Compressor oil ND-OIL 8 or equivalent

View of the Heater & Air Conditioning assembly—SC 430

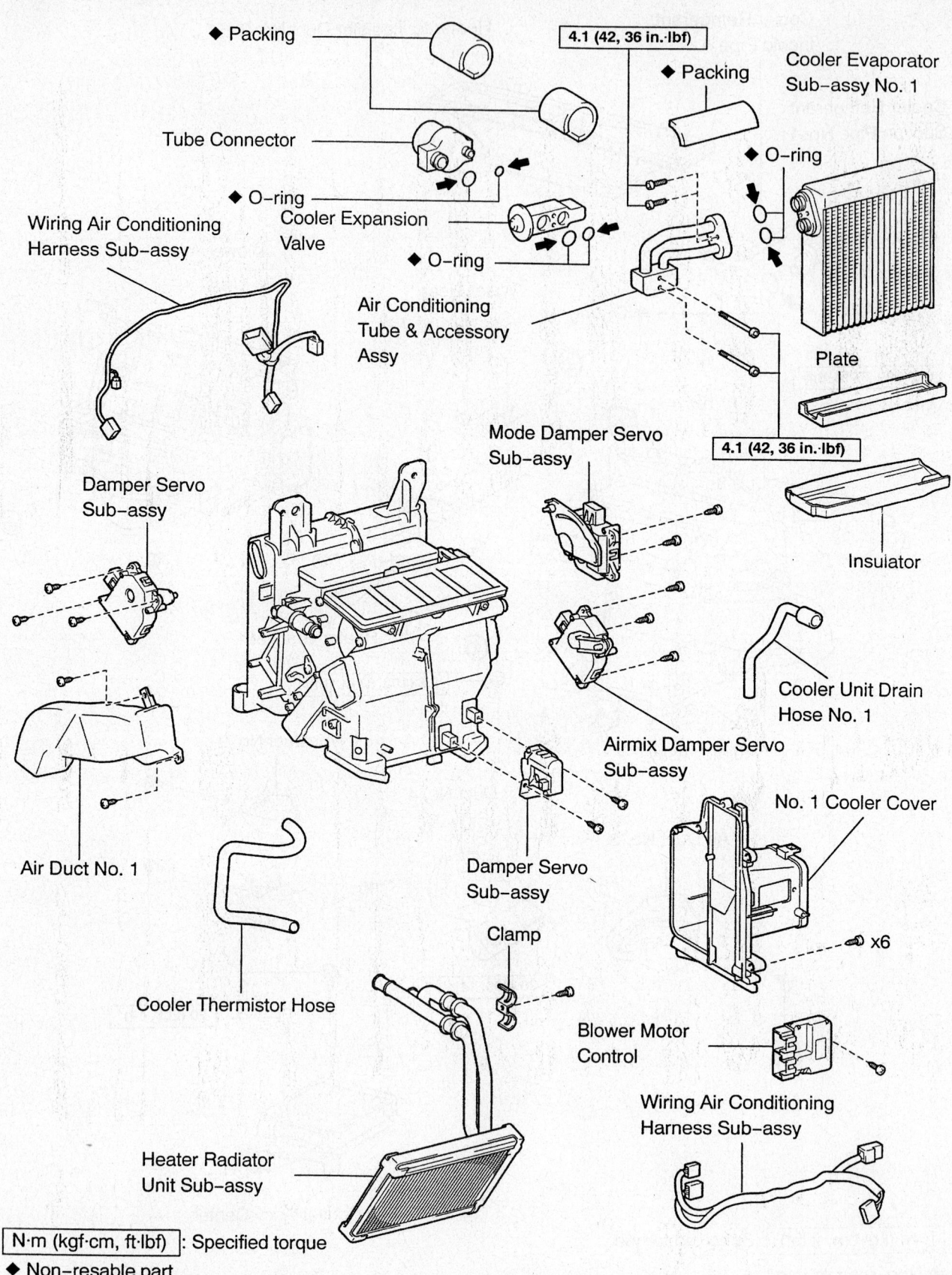

◆ Packing

4.1 (42, 36 in.·lbf)

◆ Packing

Cooler Evaporator Sub-assy No. 1

Tube Connector

◆ O-ring

Wiring Air Conditioning Harness Sub-assy

◆ O-ring

Cooler Expansion Valve

◆ O-ring

Air Conditioning Tube & Accessory Assy

Plate

4.1 (42, 36 in.·lbf)

Mode Damper Servo Sub-assy

Insulator

Damper Servo Sub-assy

Cooler Unit Drain Hose No. 1

Airmix Damper Servo Sub-assy

Air Duct No. 1

No. 1 Cooler Cover

Damper Servo Sub-assy

x6

Clamp

Cooler Thermistor Hose

Blower Motor Control

Wiring Air Conditioning Harness Sub-assy

Heater Radiator Unit Sub-assy

N·m (kgf·cm, ft·lbf) : Specified torque

◆ Non-resable part

◀ Compressor oil ND-OIL 8 or equivalent

67162-LEXU-G16

Exploded view of Heater & Air Conditioning assembly—SC 430

4. Remove or disconnect the following:
- A/C Suction and pressure hose bolt and plate at firewall

➡**Do Not use pry tools to separate**

➡**Cap the end to prevent system contamination**

- Heater hoses from the heater core

5. Set CD changer to ship mode setting using following procedure
- Remove all CDs
- Simultaneously press "Seek Up" and "Disc" while turning the ignition switch to "Acc"

➡**When mode setting is complete "Ship" appears on the display**

6. Remove or disconnect the following:
- Disconnect the negative battery cable. Wait 90 seconds before doing any further work while the airbag system de-energizes.
- Center front wheels and steering wheel
- Lower steering wheel cover
- Switch & volume case
- Steering pad switch modulator
- Horn button assembly
- Steering wheel assembly
- RH & LH door scuff plates
- RH & LH door opening trim covers
- Instrument panel safety pad. LH side under steering column
- Instrument cluster finish panel
- Upper and lower steering column covers
- Turn signal switch assembly
- Combination meter assembly
- Upper/Rear shift console garnish panel
- Shift console panel sub-assembly upper
- Air conditioning control assembly
- Instrument cluster radio finish panel assembly lower center
- Instrument panel safety pad garnish sub-assembly
- Instrument panel garnish sub-assembly No. 2
- Instrument cluster finish panel sub-assembly center
- Instrument panel under cover sub-assembly No. 2
- Console glove box carpet
- Console glove box panel No. 3
- Console glove box assembly
- Instrument panel brace sub-assembly No. 2
- Instrument panel brace sub-assembly No. 1
- RH & LH front pillar garnish

- Instrument side panel
- Instrument panel safety pad sub-assembly
- 3 bolts and separate air bag sensor assembly center
- Heater to register duct No. 2

7. Air Conditioning unit assembly
- Wire harness clamps and disconnect wire harness from air conditioning unit assembly
- 2 screws, 3 nuts, bolt and air conditioning unit

8. Remove or disconnect the following:
- Remove heater bracket

9. Air conditioning evaporator assembly

- Connector and clamp for the blower assembly
- Release claw and disconnect the connector from the bracket
- 2 screws and release the claw or the evaporator assembly
- evaporator assembly

10. Remove or disconnect the following:
- Cooler thermistor hose
- Air duct No.1, 2, 3 and 4

11. Remove heater core unit sub-assembly

- Screw and clamp
- Heater core assembly from A/C assembly

To install:

12. Install or connect the following:

13. Heater core unit sub assembly
- Heater core into A/C assembly
- Screw and clamp

14. Install or connect the following:
- Air duct No 1,2, 3 and 4
- Cooler thermistor hose

15. Install Air Conditioning Unit Assembly

- 2 screws and 3 nuts and air conditioning unit
- Bolt to air conditioning unit assembly. Torque: 87 inch lbs. (9.8 Nm)
- Clamps to the air conditioner unit assembly

16. Install or connect the following:
- Heater to register duct No. 2
- Air bag sensor assembly center
- Instrument panel safety pad sub-assembly
- Instrument side panel
- RH & LH front pillar garnish
- Instrument panel brace sub-assembly No. 1
- Instrument panel brace sub-assembly No. 2
- Console glove box assembly, panel and box carpet
- Instrument panel under cover sub-assembly No. 2

- Instrument cluster finish panel sub-assembly center
- Instrument panel garnish sub-assembly No. 2
- Instrument panel safety pad garnish sub-assembly
- Instrument cluster radio finish assembly lower center
- Air conditioning control assembly
- Shift console panel sub-assembly upper
- Upper/Rear shift console garnish panel
- Combination meter assembly
- Turn signal switch assembly
- Upper and lower steering column covers
- Instrument cluster finish panel
- Instrument panel safety pad. LH side under steering wheel
- RH & LH door opening trim covers
- RH & LH door scuff plates
- Steering wheel assembly
- Horn button assembly
- Steering pad switch modulator
- Switch and volume case
- Lower steering wheel cover
- Heater core hoses
- A/C suction and pressure hoses, attach with bolt and plate

➡**Lubricate O-rings with compressor oil**

- Negative battery cable
- Fill cooling system with coolant
- Evacuate and recharge A/C system
- Warm up engine and inspect for coolant leaks

ES 330

➡**Removal of the heater core requires removal of the entire heater air conditioning assembly.**

1. Before servicing the vehicle, refer to the Precautions Section.
2. Drain the cooling system into a clean container for reuse.
3. Set CD changer to ship mode setting using following procedure
- Remove all CDs
- Simultaneously press "Seek Up" and "Disc" while turning the ignition switch to "Acc"

➡**When mode setting is complete "Ship" appears on the display**

4. Disconnect the negative battery cable. Wait 90 seconds before doing any further work while the airbag system de-energizes.
5. Discharge and recover the air conditioning system refrigerant.

6. Disconnect A/C suction hose (No. 1) and Liquid pipe (A)

  a. Install SST to piping clamp

  b. Push down SST and release the clamp lock

✳✳ WARNING

Be careful not to deform the tube, when pushing the SST

➡ **Cap the open fittings immediately to prevent system contamination**

7. Remove or disconnect the following:

- Heater core hoses

8. Disassemble the dash components as follows

- RH & LH door scuff plates
- Instrument panel sub-assembly
- Lower LH instrument panel sub-assembly
- Lower steering column cover
- Steering column cover
- Headlamp dimmer switch assembly
- Windshield wiper switch assembly
- Instrument panel register assemblies
- Instrument cluster finish panel sub-assembly
- Combination meter assembly
- Instrument panel under cover sub-assembly
- Lower instrument panel sub-assembly
- Upper console panel sub-assembly
- Upper rear console panel sub-assembly
- Console box
- Instrument panel LH & RH end panel
- Air conditioning control assembly
- Instrument panel center cluster finish panel
- Radio receiver panel
- LH & RH front pillar garnish
- Instrument panel finish plate
- Passenger air bag connector
- Instrument panel safety pad cap
- instrument panel safety pad sub-assembly
- Instrument cluster molding
- Switch hole base
- Glove box lamp assembly
- Automatic light control sensor
- Side defroster nozzle ducts
- Defroster nozzle assembly
- Heater register ducts
- Left door instrument panel air bag assembly

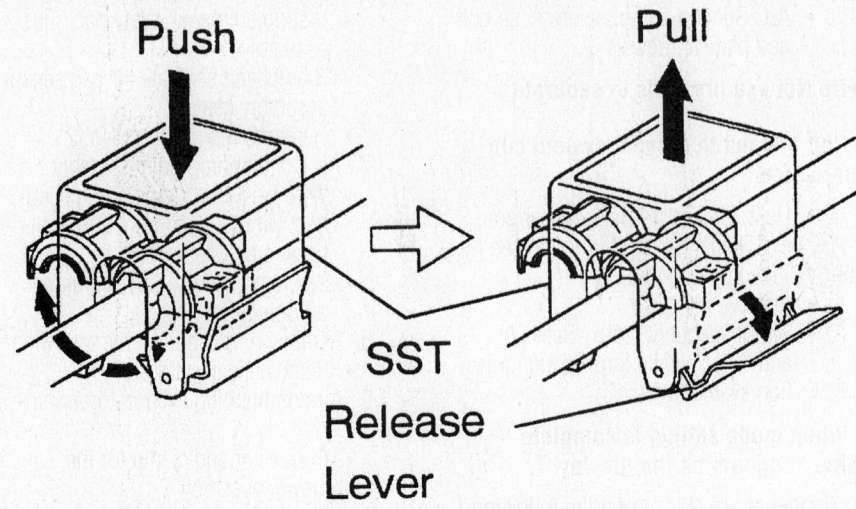

Push **Pull**

SST Release Lever

67162-LEXU-G19

Using SST tool—ES 330

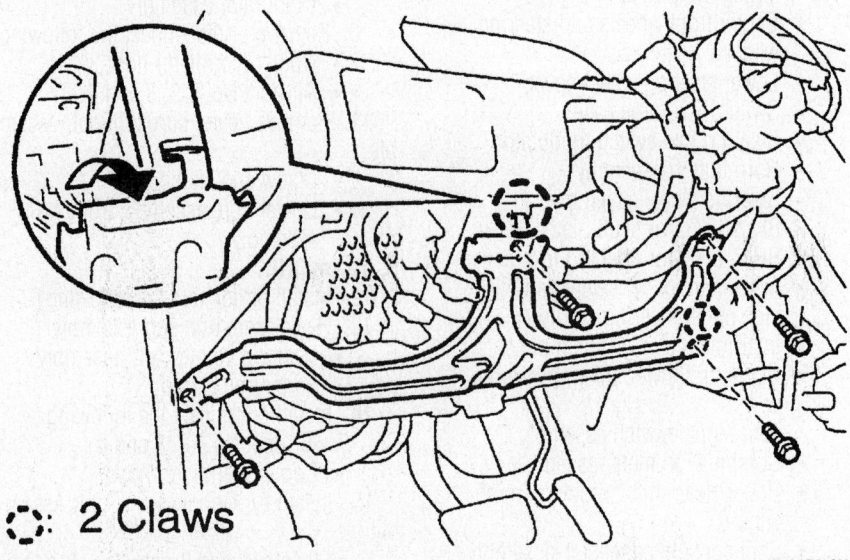

◌ **: 2 Claws**

67162-LEXU-G20

Exploded view lower instrument panel sub-assembly—ES 330

✳✳ WARNING

Follow air bag removal procedures

9. Remove or disconnect the following:

- Rear air ducts
- Console box duct
- Floor shift parking lock cable assembly
- Headlamp leveling ECU assembly
- Instrument panel brace assemblies
- Lower instrument finish panel retainer
- Heater to foot ducts
- Steering column assembly
- Instrument panel reinforcements
- Heater blower assembly

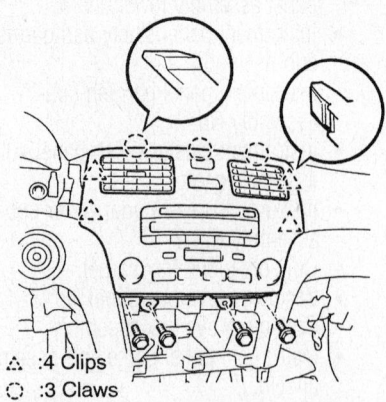

△ :4 Clips
◌ :3 Claws

67162-LEXU-G21

Exploded view center finish panel—ES 330

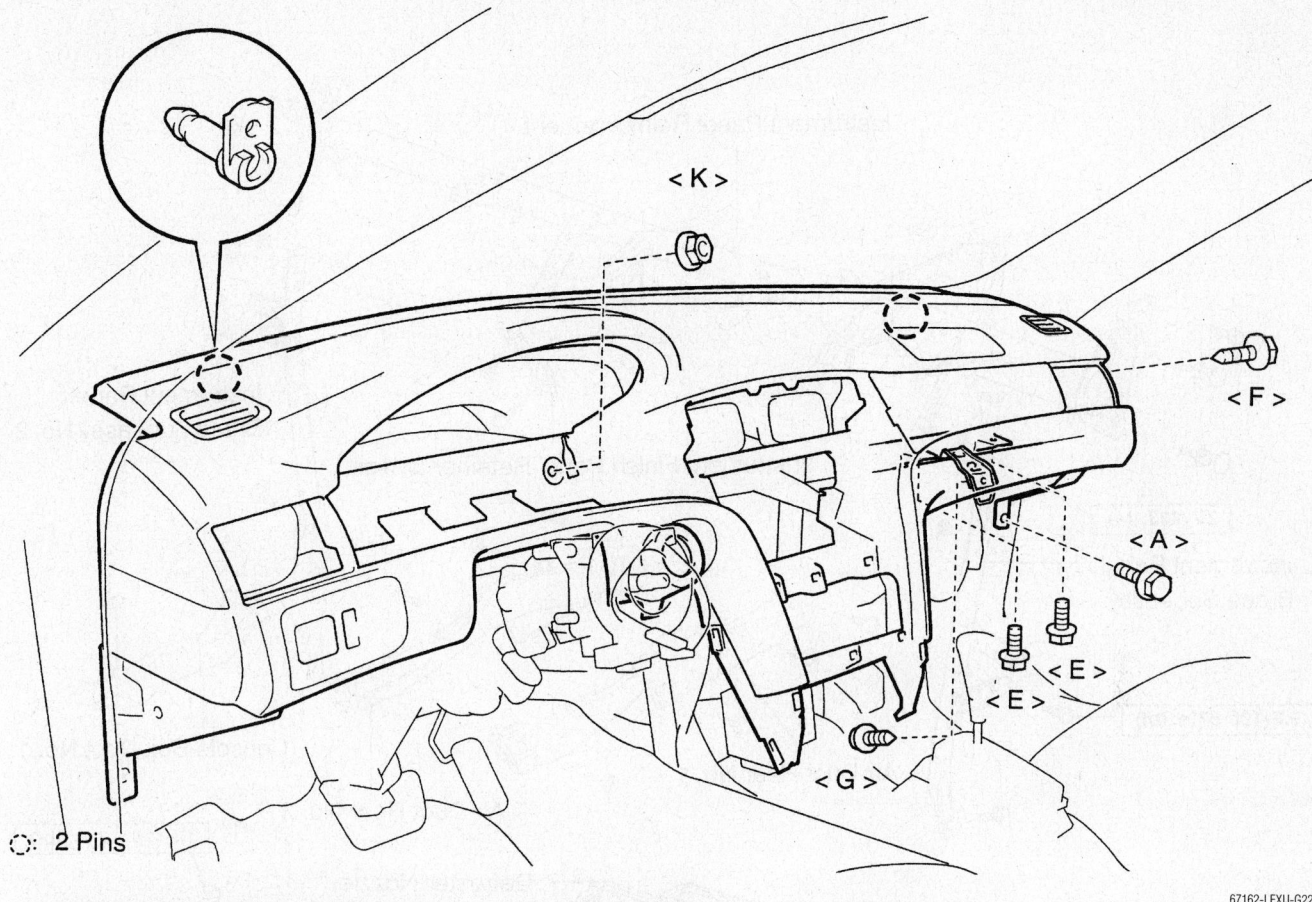

○: 2 Pins

67162-LEXU-G22

Exploded view of instrument panel safety pad sub-assembly—ES 330

- Lower defroster nozzles
- Air conditioning radiator assembly
- Mode damper servo sub-assembly
- Airmix damper servo sub-assembly
- Heater radiator (core) sub-assembly

To install:
10. Install or connect the following:
11. Heater core unit sub assembly
    - Heater core into A/C assembly
    - Screw and clamp
12. Install or connect the following:
    - Air conditioning radiator assembly
    - Lower defroster nozzles
    - Heater blower assembly
    - Instrument panel reinforcements
    - Steering column assembly
    - Heater to foot ducts
    - Lower instrument finish panel retainer
    - Instrument panel brace assemblies
    - Headlamp leveling ECU assembly
    - Floor shift parking lock cable assembly
    - Console box duct
    - Rear air ducts

13. Reassemble the dash components as follows:
    - Left door instrument panel air bag assembly
    - Heater register ducts
    - Defroster nozzle assembly
    - Side defroster nozzle ducts
    - Automatic light control sensor
    - Glove box lamp assembly
    - Instrument cluster molding
    - Instrument panel safety pad sub-assembly
    - Passenger air bag connector
    - Instrument panel finish plate
    - LH & RH front pillar garnish
    - Radio receiver panel
    - Instrument panel center cluster finish panel
    - Air conditioning control assembly
    - Instrument panel LH & RH end panel
    - Console box
    - Upper rear console panel sub-assembly
    - Upper console panel sub-assembly
    - Lower instrument panel sub-assembly

- Instrument panel under cover sub-assembly
- Combination meter assembly
- Instrument cluster finish pale sub-assembly
- Instrument panel register assemblies
- Windshield wiper switch assembly
- Headlamp dimmer switch assembly
- Steering column cover
- Lower steering column cover
- Lower LH instrument panel sub-assembly
- Instrument panel sub-assembly
- RH & LH door scuff plates
14. Install or connect the following:
    - Heater core hoses
    - A/C suction and pressure hoses, attach with bolt and plate

➡**Lubricate O-rings with compressor oil**

- Negative battery cable
- Fill cooling system with coolant
- Evacuate and recharge A/C system
- Warm up engine and inspect for coolant leaks

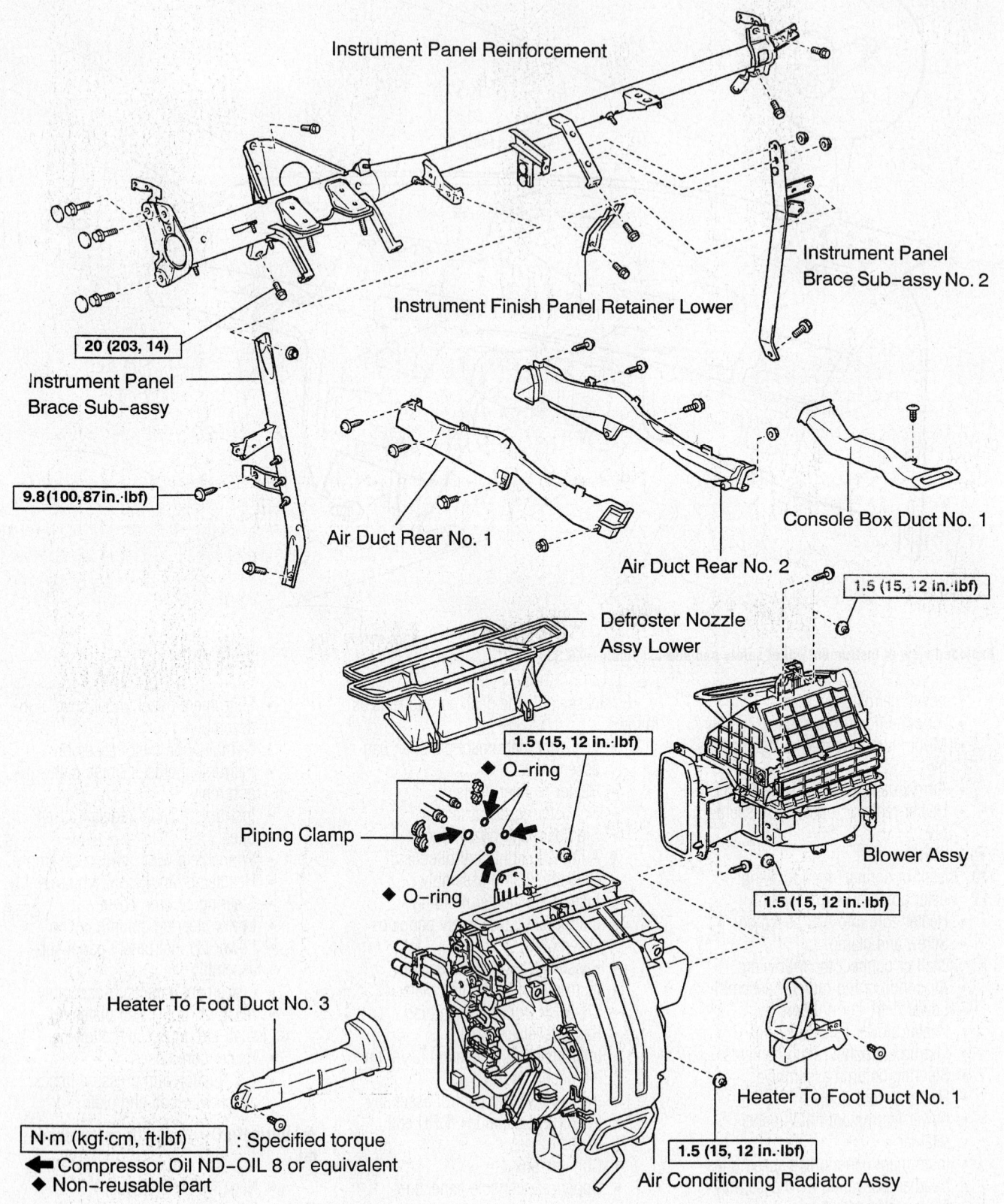

Instrument Panel Reinforcement

Instrument Panel Brace Sub–assy No. 2

Instrument Finish Panel Retainer Lower

20 (203, 14)

Instrument Panel Brace Sub–assy

9.8 (100, 87 in.·lbf)

Air Duct Rear No. 1

Air Duct Rear No. 2

Console Box Duct No. 1

Defroster Nozzle Assy Lower

1.5 (15, 12 in.·lbf)

1.5 (15, 12 in.·lbf)

◆ O–ring

Piping Clamp

◆ O–ring

Blower Assy

1.5 (15, 12 in.·lbf)

Heater To Foot Duct No. 3

Heater To Foot Duct No. 1

N·m (kgf·cm, ft·lbf) : Specified torque
◀ Compressor Oil ND–OIL 8 or equivalent
◆ Non–reusable part

1.5 (15, 12 in.·lbf)

Air Conditioning Radiator Assy

67162-LEXU-G23

Exploded view Instrument panel reinforcement and A/C assembly and sub-assembly—ES 330

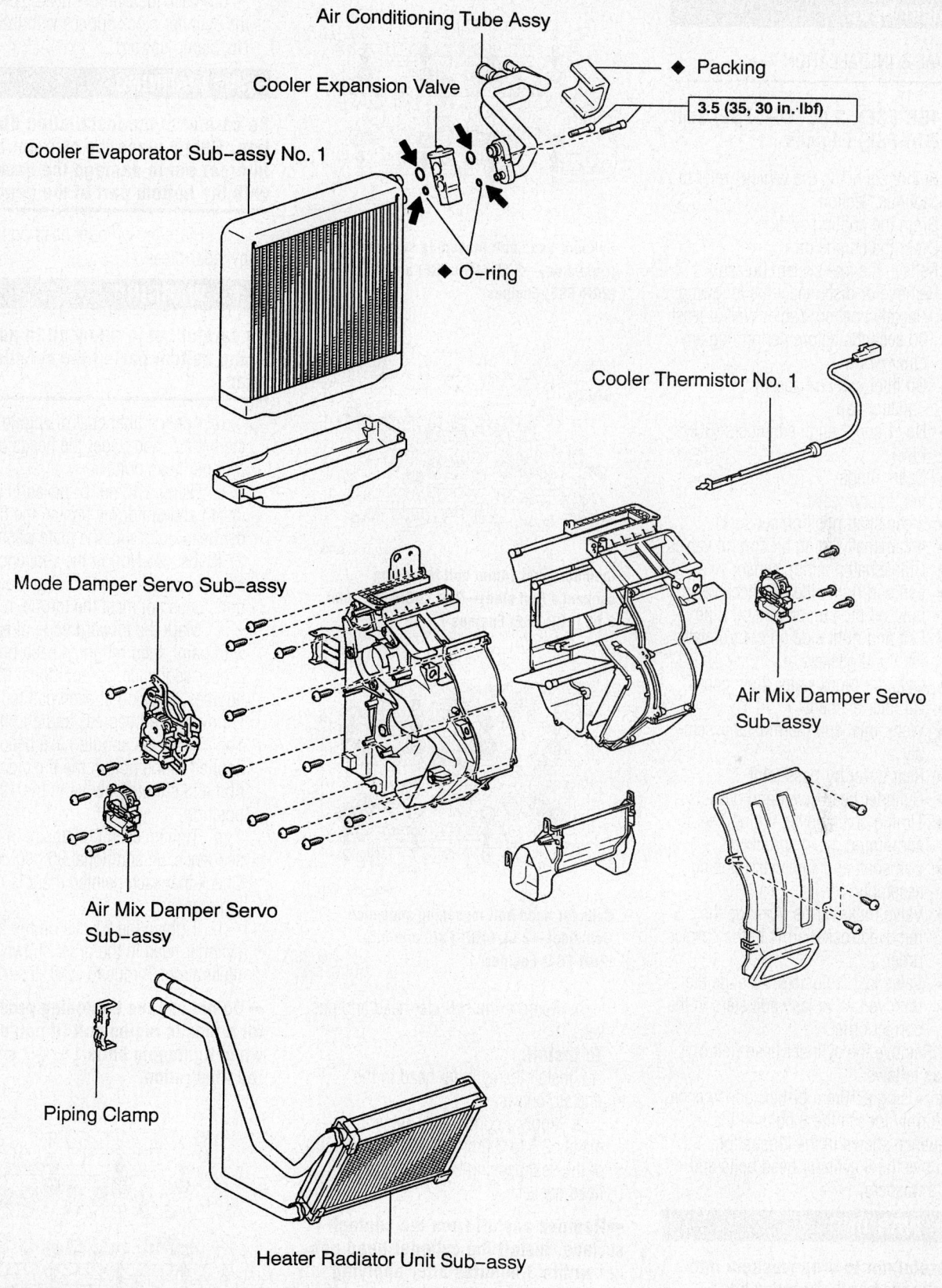

Air Conditioning Tube Assy

Cooler Expansion Valve

◆ Packing

3.5 (35, 30 in.·lbf)

Cooler Evaporator Sub–assy No. 1

◆ O–ring

Cooler Thermistor No. 1

Mode Damper Servo Sub–assy

Air Mix Damper Servo Sub–assy

Air Mix Damper Servo Sub–assy

Piping Clamp

Heater Radiator Unit Sub–assy

N·m (kgf·cm, ft·lbf) : Specified torque
◀ Compressor Oil ND–OIL 8 or equivalent
◆ Non–reusable part

67162-LEXU-G24

Exploded view of the heater case assembly—ES330

Cylinder Head

REMOVAL & INSTALLATION

2.5L (4GR-FSE), 3.0L (3GR-FSE) and 3.5L (2GR-FSE) Engines

1. Before servicing the vehicle, refer to the Precautions Section.
2. Drain the cooling system.
3. Drain the engine oil.
4. Relieve the fuel system pressure.
5. Remove or disconnect the following:
- Negative battery cable. Wait at least 90 seconds before performing any other work
- Oil filler cap and gasket
- Radiator cap
- No. 1 and 2 engine hangers (3GR-FSE)
- Spark plugs
- Ventilation valve
- 4 camshaft position sensors
- 4 camshaft timing oil control valves
- Crankshaft position sensor
- Left and right side oil check valve bolt, oil pipe union and oil pipe
- Left and right side oil control valve filter and gaskets
- Cylinder block water drain cocks
- Oil filter element
- Water inlet and thermostat assembly
- Rear water by-pass joint
- Cylinder head cover and gaskets
- Timing chain cover, timing chain and timing chain sprockets
- Camshaft and camshaft housing assembly
- Valve rocker arms. Arrange the removed rocker arms in the correct order.
- Valve lash adjusters. Arrange the removed valve lash adjusters in the correct order.
6. Remove the cylinder head (left or right) as follows:
a. Using a 10mm bi-hexagon wrench, uniformly loosen the 8 bolts in the sequence shown in the illustration. Remove the 8 cylinder head bolts and plate washers.

❈❈ WARNING

Be careful not to drop washers into the cylinder head. Cylinder head warpage or cracking could result from removing bolts in an incorrect order. Be sure to keep separate the removed parts for each installation position.

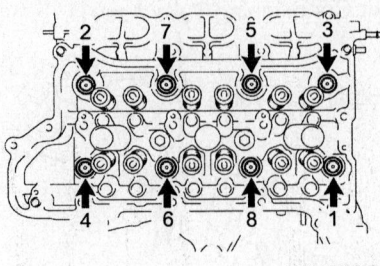

09490_LEXU_G0033

Cylinder head bolt loosening sequence (right side)—2.5L (4GR-FSE) and 3.5L (2GR-FSE) Engines

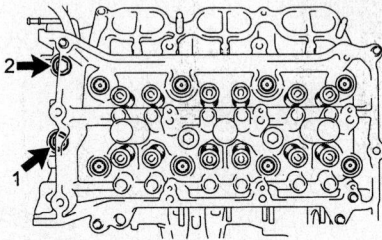

09490_LEXU_G0034

Cylinder head 14mm bolt loosening sequence (left side)—2.5L (4GR-FSE) and 3.5L (2GR-FSE) Engines

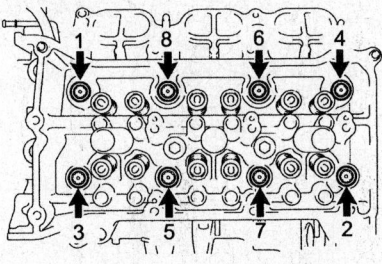

09490_LEXU_G0035

Cylinder head bolt loosening sequence (left side)—2.5L (4GR-FSE) and 3.5L (2GR-FSE) Engines

b. Remove the cylinder head and gasket.

To install:

7. Install the cylinder head to the engine as follows:
a. Apply a continuous line approximately 2.5 to 3.0mm (0.098 to 0.118 in.) of the seal packing to a new cylinder head gasket.

➡**Remove any oil from the contact surface. Install the cylinder head gasket within 3 minutes after applying the seal packing. Install the cylinder head bolt within 15 minutes after applying the seal packing. Do not apply engine oil within 2 hours of installation.**

b. Place the cylinder head gasket on the cylinder block surface with the Lot No. stamp upward.

❈❈ WARNING

Be careful of the installation direction. Gently place the cylinder head in order not to damage the gasket with the bottom part of the head.

c. Place the cylinder head on the cylinder block.

❈❈ WARNING

Be careful not to allow oil to adhere to the bottom part of the cylinder head.

d. Apply a light coat of engine oil to the threads and under the heads of the cylinder head bolts.
e. Using a 10mm bi-hexagon wrench, install and uniformly tighten the 8 cylinder head bolts with the plate washers to 27 ft. lbs. (36 Nm) in the sequence shown in the illustration. If any of the bolts does not meet the torque, replace it.
f. Mark the forward edge of each bolt with paint, then retighten each bolt, in proper sequence, an additional 90 degrees. Check that each painted mark is now at a 90 degrees angle to the front. The paint mark should have been applied to the bolt in the 9 o'clock position and should now be in the 12 o'clock position.
g. Tighten each bolt again, in proper sequence, an additional 90 degrees. Check that each painted mark is now facing rearward.
h. Tighten the 2 bolts on the left cylinder head in the order shown in the illustration. Torque to 22 ft. lbs. (30 Nm).

➡**Do not use the tightening procedure for a plastic region bolt (if equipped) when tightening bolts 1 and 2 shown in the illustration.**

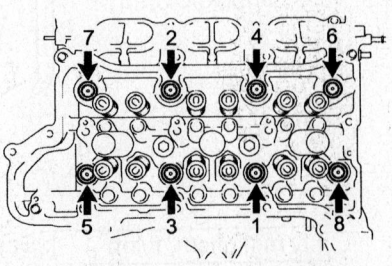

09490_LEXU_G0036

Cylinder head bolt tightening sequence (right side)—2.5L (4GR-FSE) and 3.5L (2GR-FSE) Engines

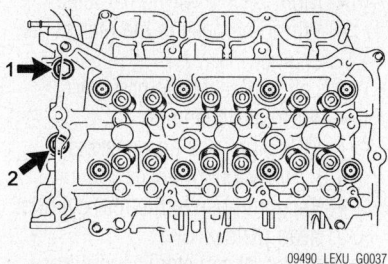

09490_LEXU_G0037

Cylinder head 14mm bolt tightening sequence (left side)—2.5L (4GR-FSE) and 3.5L (2GR-FSE) Engines

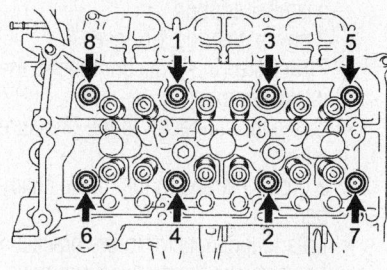

09490_LEXU_G0038

Cylinder head bolt tightening sequence (left side)—2.5L (4GR-FSE) and 3.5L (2GR-FSE) Engines

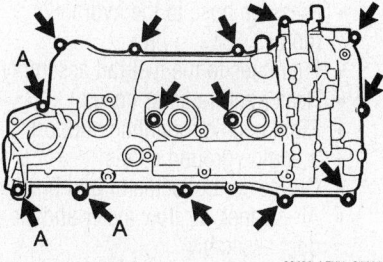

09490_LEXU_G0039

Cylinder head cover bolt tightening sequence (right side)—2.5L (4GR-FSE) and 3.5L (2GR-FSE) Engines

i. Seal packing will seep out on the engine's front side. Thoroughly wipe off seeped out seal packing.

8. Install or connect the following:
- Valve lash adjusters
- Valve rocker arms
- Camshaft and camshaft housing assembly
- Timing chain cover, timing chain and timing chain sprockets

9. Install the cylinder head cover and gaskets as follows:

a. Apply seal packing where the cylinder head meets the timing cover.

➡**Remove any oil from the contact surface. Install the head cover within 3**

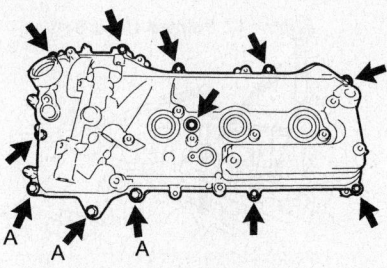

09490_LEXU_G0040

Cylinder head cover bolt tightening sequence (left side)—2.5L (4GR-FSE) and 3.5L (2GR-FSE) Engines

minutes after applying seal packing. Do not start the engine for at least 2 hours after installing.

b. Install all new gaskets.

c. Install the head cover along with the retaining bolts. Torque the A bolts to 15 ft. lbs. (21 Nm); and remaining bolts to 7 ft. lbs. (10 Nm).
- New gaskets, O-ring and rear water by-pass joint. Torque to 7 ft. lbs. (10 Nm).
- New gasket, water inlet and thermostat assembly. Torque to 7 ft. lbs. (10 Nm).
- Oil filter element
- Adhesive sealer, left and right side cylinder block water drain cocks. Torque to 22 ft. lbs. (30 Nm).
- Left and right side oil control valve filter and gaskets
- Left and right side oil check valve bolt, oil pipe union and oil pipe. Torque to 44 ft. lbs. (60 Nm).
- Crankshaft position sensor. Torque to 7 ft. lbs. (10 Nm).
- 4 camshaft timing oil control valves
- 4 camshaft position sensors
- Adhesive and ventilation valve
- Spark plugs. Torque to 18 ft. lbs. (24 Nm) for 2.5L and 3.5L engines; 13 ft. lbs. (18 Nm) for 3.0L engine.
- Engine hangers (if equipped). Torque to 24 ft. lbs. (33 Nm).
- Radiator cap
- Oil filler cap and gasket
- Negative battery cable

10. Refill the coolant and engine oil. Start the engine and check for leaks or abnormal conditions. Perform and road test. Then, recheck for leaks and recheck fluid levels.

3.0L (1MZ-FE) Engine

1. Before servicing the vehicle, refer to the Precautions Section.
2. Drain the cooling system.

3. Relieve the fuel system pressure.
4. Remove or disconnect the following:
- Negative battery cable
- Accelerator and the throttle cables
- Air cleaner cover, air flow meter and the air duct
- Cruise control actuator and bracket, if equipped
- 2 engine ground straps
- Right engine mounting support
- Radiator hoses
- 2 heater hoses

5. Plug the fuel feed and return lines from the fuel rail assembly.

6. Plug the pressure hose from the hydraulic motor

7. Remove or disconnect the following:
- V-bank cover
- Fuel pressure control Vacuum Switching Valve (VSV)
- Fuel pressure regulator
- Cylinder head rear plate
- Intake air control valve VSV
- Exhaust Gas Recirculation (EGR) vacuum modulator
- EGR valve
- Intake Air Control (IAC) valve
- Fuel pressure regulator
- EGR VSV
- 2 nuts and the emission control valve set
- Brake booster vacuum hose
- Positive Crankcase Ventilation (PCV) hose
- Intake air control valve vacuum hose
- Data link connector from the mounting bracket
- 2 ground straps from the intake chamber
- Hydraulic motor pressure hose from the intake chamber
- Right Oxygen (O2) sensor connector from the power steering pressure tube
- 2 nuts and the power steering pressure tube from the intake chamber
- 2 power steering air hoses
- Engine hanger and the intake chamber support
- EGR pipe and gaskets
- Throttle Pressure (TP) sensor connector
- Idle Air Control (IAC) valve connector
- EGR gas temperature connector
- air conditioning idle up connector
- 2 vacuum hoses from the Thermal Vacuum Valve (TVV)
- Vacuum hose from the cylinder head rear plate

- Vacuum hose from the charcoal canister
- Air assist hose and the 2 water bypass hoses
- Air intake chamber
- Left engine wiring harness and position it out of the way
- Wiring harness from the rear of the engine
- Right engine wiring harness and position it out of the way
- Ignition coils and the spark plugs
- Timing belt
- Camshaft pulleys and the timing belt rear cover
- Cylinder head rear plate
- Water inlet pipe
- Air assist hose and vacuum hose
- Intake manifold and fuel rail assembly
- Water outlet
- EGR pipe from the right exhaust manifold
- Exhaust manifolds
- Dipstick assembly and the power steering pump bracket
- Valve covers and the Camshaft Position (CMP) sensor
- Camshafts

8. Be sure the engine is at or near ambient temperature and remove the 2 (one on each head) 8mm recessed hex bolts. Loosen and remove the 8 head bolts evenly, in 3 passes, in the reverse order of the tightening sequence. Carefully lift the head from the engine; if it is necessary to pry the head loose, take great care not to damage the mating surfaces. Place the head on wood blocks in a clean work area.

❋❋ WARNING

If the cylinder head bolts are loosened out of sequence, warpage or cracking could result.

9. Remove the cylinder head gasket. With a gasket scraper, carefully remove all the old gasket material from the cylinder head and engine block surfaces.

To install:

10. Place the new cylinder head gasket onto the cylinder block. Place the cylinder head onto the gasket.

11. Coat the threads of the 8 cylinder head bolts (12-sided) with clean engine oil and install the bolts into the cylinder head. Uniformly tighten the bolts in sequence in 3 steps to an ultimate tighten of 40 ft. lbs. (54 Nm). If any of the bolts does not meet the torque, replace it.

12. Mark the forward edge of each bolt

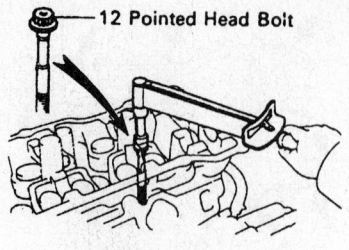

— 12 Pointed Head Bolt

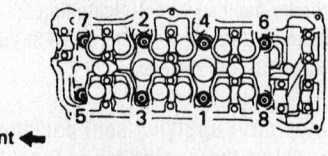

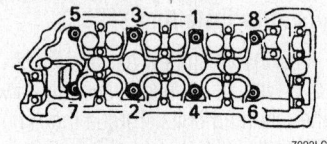

7923LG13

Cylinder head torque sequence—3.0L (1MZ-FE) engine

with paint, then retighten each bolt, in proper sequence, an additional 90 degrees. Check that each painted mark is now at a 90 degrees angle to the front. The paint mark should have been applied to the bolt in the 9 o'clock position and should now be in the 12 o'clock position.

13. Coat the threads of the 2 remaining 8mm bolts with engine oil and install them. Tighten to 14 ft. lbs. (18 Nm).

14. Install or connect the following:

- Camshafts and adjust the valves

➡ **Apply sealant to the cylinder heads where the camshaft supports meet the cylinder heads.**

- Cylinder head covers. Use new gaskets.
- Dipstick and power steering pump bracket
- Exhaust manifolds. Tighten the nuts to 36 ft. lbs. (49 Nm).
- EGR pipe to the right exhaust manifold
- Water outlet
- Intake manifold and the fuel rail assembly. Tighten the intake manifold nuts and bolts to 11 ft. lbs. (15 Nm).
- Air assist hose and the 2 water bypass hoses
- Water inlet pipe and the cylinder head rear plate
- Timing belt rear cover and the camshaft pulleys
- Timing belt
- Spark plugs and the ignition coils

- Right engine wiring harness
- Wiring harness to the rear of the engine
- Left engine wiring harness
- Air intake chamber
- EGR pipe. Use new gaskets
- 2 TVV vacuum hoses
- Vacuum hose to the rear cylinder head plate
- Charcoal canister vacuum hose
- TP sensor connector
- IAC valve connector
- EGR gas temperature connector
- Air conditioning idle up connector
- Engine hanger and the intake chamber support
- 2 power steering air hoses
- Power steering pressure tube to the intake chamber
- O_2 sensor connector to the pressure tube
- 2 ground straps to the intake chamber
- Data link connector to the bracket
- Power brake booster vacuum hose
- PCV hose
- IAC valve vacuum hose
- Emission control valve set and related vacuum hoses and connectors
- V-bank cover
- Pressure hose to the hydraulic motor
- Fuel lines to the fuel rail assembly
- Heater and radiator hoses
- Right engine mounting support
- 2 engine ground straps
- Cruise control actuator and bracket
- Air cleaner, air flow meter and air duct assembly
- Accelerator and the throttle cables
- Negative battery cable

15. Fill the cooling system to the proper level with coolant.

16. Start the engine and check for leaks. Bleed the air from the cooling system.

17. Adjust the ignition timing.

18. Road test the vehicle and check for unusual noise, shock, slippage, correct shift points and smooth operation.

19. Recheck the coolant and engine oil levels.

3.0L (2JZ-GE) Engine (2002–03)

1. Before servicing the vehicle, refer to the Precautions Section.

2. Disconnect the negative battery cable. Wait at least 90 seconds before performing any other work.

3. Relieve the fuel pressure from the fuel lines.

4. Remove or disconnect the following:
- Coolant
- Undercovers
- Accelerator, throttle control (automatic transmission only) and cruise control cables from the throttle body
- Cleaner duct
- Air cleaner, airflow meter and the intake air pipe
- Drive belt, the fan and fluid coupling and the water pump pulley
- No. 2 front exhaust pipe
- Exhaust manifold cover
- 2 Heated Oxygen (HO₂) sensor connector(s)
- Exhaust manifolds and gaskets by removing the 8 bolts
- Water bypass outlet and the No. 1 water bypass pipe
- Power steering air hose from the No. 4 timing belt cover
- Power steering hose from the air intake chamber
- 2 bolts and the vane pump from the pump bracket
- 2 bolts, the pump rear stay. Put aside the vane pump and suspend it.
- Fuel return hose from the fuel return pipe. Plug the hose end.
- Fuel return hose from the oil dipstick guide
- Bolt and bracket
- Engine wire from the intake manifold stay
- Throttle body and intake air connector assembly
- Bolt, pull out the oil dipstick guide with the dipstick and remove the O-ring from the dipstick guide
- Transmission dipstick and guide, if equipped with automatic transmission
- Connector from the No. 2 vacuum pipe
- Exhaust Gas Recirculation (EGR) gas temperature sensor wiring harness
- 2 nuts and the vacuum pipe from the air intake chamber and intake manifold
- No. 2 vacuum pipe and Vacuum Switching Valve (VSV) assembly
- Nuts and the vacuum tank from the intake manifold
- VSV connector and hoses
- Vacuum hose (from the air intake chamber) from port B of the vacuum tank
- Vacuum hose (from actuator) from the VSV

- Vacuum control valve set
- Data Link Connector (DLC1) bracket and VSV assembly
- Vacuum hose from the brake booster union and the Evaporative Emission (EVAP) hose from the No. 2 vacuum pipe
- Bolt holding the engine wire protector to the air intake chamber
- 5 bolts, nut, air intake chamber and gasket
- No. 3 (top) timing belt cover by removing the oil filler cap and the 6 bolts using a 5mm hexagon wrench.
- 4 bolts, using a 5mm hexagon wrench, and the rear cylinder head cover
- Spark plugs
- Drive belt tensioner by removing the 3 bolts

5. Set the engine to Top Dead Center (TDC)/compression for cylinder No. 1 piston

6. Remove or disconnect the following:

- Timing belt tensioner and dust boot. Remove the timing belt from the camshaft pulleys. Support the belt so that it remains in contact with the crankshaft pulley.
- Wire clamp from the bracket.
- HO₂ and the Crankshaft Position (CKP) sensors.
- 2 ground straps from the intake manifold.
- Engine Coolant Temperature (ECT) sender gauge
- Knock Sensor (KS)
- Oil pressure switch
- Oil level sensor
- air conditioning compressor
- 6 injector electrical connectors
- 3 nuts and the engine wire protector from the intake manifold
- Water bypass hose from the clamp on the oil filter bracket
- Water outlet, 2 nuts, and the bolt with the water bypass hose
- 2 bolts and the intake manifold stay
- Fuel pressure pulsation damper
- Clamp bolt from the intake manifold
- Union bolt and gaskets
- Fuel inlet pipe
- 6 bolts, 2 nuts, the intake manifold
- Delivery pipe assembly and gasket
- Cylinder head covers (valve covers)
- Camshaft timing pulleys
- Rear (No. 4) timing belt cover
- Camshafts
- Cylinder head bolts in several

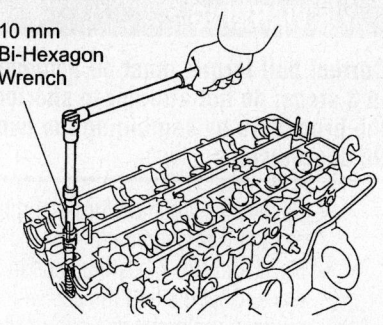

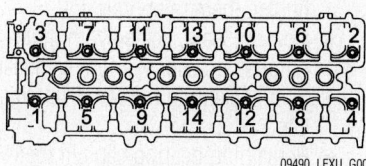

09490_LEXU_G0041

Showing the cylinder head bolt removal sequence—3.0L (2JZ-GE) engine

passes and in the reverse order of the tightening sequence
- Head from the engine

7. Clean the head and block of all gasket material.

To install:

8. Install or connect the following:
- New gasket and cylinder head on the block
- Head bolts, lightly coated with engine oil and plate washers. Uniformly tighten the head bolts in several passes, in sequence to 26 ft. lbs. (35 Nm). Following the correct order, tighten each bolt an additional 90 degrees. Again following the correct order, tighten the bolts another 90 degrees of rotation.

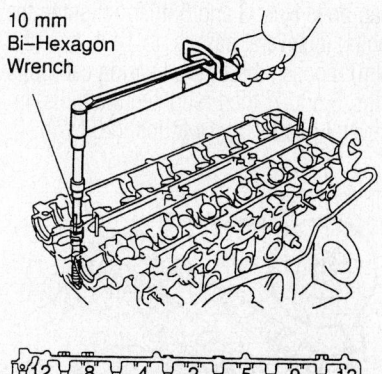

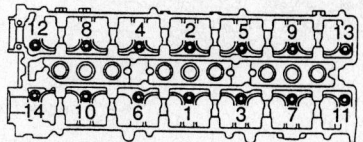

9301LG08

Cylinder head torque sequence—3.0L (2JZ-GE) engine

✳✳ WARNING

Correct bolt torque must be achieved in 3 steps; do not attempt to shorten the procedure by combining the two 90 degree steps.

- Camshafts. Coat the thrust portions of each with engine oil.
- No. 3 and No. 7 bearing caps in place. Coat the bolt threads with oil, then uniformly and alternately tighten them temporarily.
- New oil seals, coated with multi-purpose grease, over the camshafts
- Seal packing to the No. 1 bearing cap
- Remaining bearing caps in their proper locations. Coat the threads of each bolt with clean oil, then tighten them, in several passes, in the correct sequence, to 14 ft. lbs. (20 Nm).
  - The 2 oil seals in as far as it will go

9. Rotate each camshaft until the forward straight (knock) pin is straight up. Loosen the exhaust Nos. 1, 2 and 6 bearing cap bolts until they can be turned by hand; retighten the bolts, in several passes, to 14 ft. lbs. (20 Nm). Loosen the intake Nos. 1, 2 and 5 bearing cap bolts and retighten the bolts, in several passes, to 14 ft. lbs. (20 Nm).

10. Turn each camshaft ⅓ of a revolution (120 degrees). Loosen the exhaust Nos. 4 and 7 bearing cap bolts; retighten the bolts, in several passes, to 14 ft. lbs. (20 Nm). Loosen the intake Nos. 4 and 6 bearing cap bolts; retighten the bolts, in several passes, to 14 ft. lbs. (20 Nm).

11. Turn each camshaft an additional ⅓ of a revolution, loosen the exhaust bearing cap bolts Nos. 3 and 5, then retighten the bolts, in several passes, to 14 ft. lbs. (20 Nm). Loosen the intake bearing cap bolts Nos. 3 and 7, then retighten the bolts, in several passes, to 14 ft. lbs. (20 Nm).

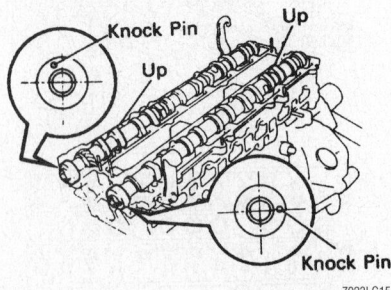

Position the knock pins as shown when installing the camshafts—3.0L (2JZ-GE) engine

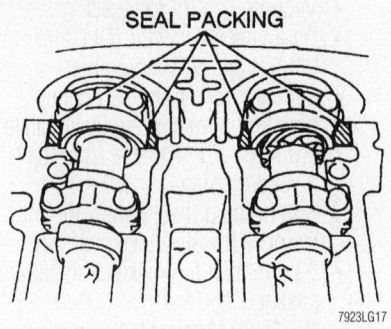

Apply sealant to the areas indicated on the cylinder head before installing the cover—3.0L (2JZ-GE) engine

12. Check and adjust the valve clearance.

13. Install or connect the following:
- Rear (No. 4) timing belt cover. Tighten the bolts to 78 inch lbs. (9 Nm).
- Camshaft timing pulleys. Align the shaft pin with the pulley groove and slide the pulley on. Install the bolt temporarily. Hold the hex portion of the camshaft with a wrench and tighten the pulley bolt to 59 ft. lbs. (79 Nm).
- Cylinder head covers
- Intake manifold and delivery pipe with a new gasket. Tighten the 6 bolts and 2 nuts to 20 ft. lbs. (27 Nm).
- Fuel inlet pipe to the fuel rail. Tighten the union bolt to 30 ft. lbs. (42 Nm).
- Clamp bolt to the intake manifold
- Fuel pressure pulsation damper

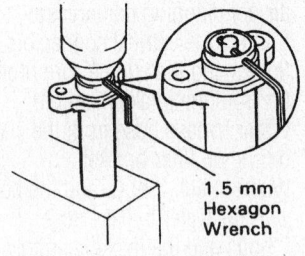

1.5 mm Hexagon Wrench

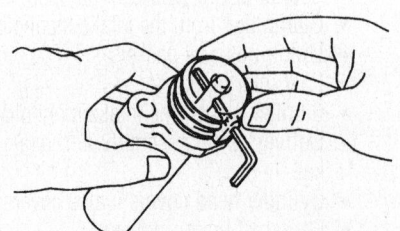

Compressing the timing belt tensioner—3.0L (2JZ-GE) engine

- Intake manifold stay and tighten the bolts to 29 ft. lbs. (39 Nm)
- Water outlet and the bypass hose. Tighten the bolts to 15 ft. lbs. (21 Nm).
- Engine wiring harness. Secure the wiring in all clamps and retainers.
- Wiring leads to the proper sender, sensor or switch
- Injector leads

14. Compress the timing belt tensioner in a vise and retain the pin with a 1.5mm hex wrench. Install the dust boot onto the tensioner.

15. Install the tensioner. Alternately tighten the bolts to 20 ft. lbs. (26 Nm). Remove the hex wrench with a pair of pliers, allowing the tensioner to be applied to the timing belt.

16. Turn the crankshaft 2 full revolutions clockwise. Check that all timing marks align as before. If the marks (cam and crankshaft) do not align, remove the timing belt and reinstall it.

17. Install the accessory drive belt tensioner. Take great care not to drop the bolts inside the lower timing cover. Tighten the bolts to 15 ft. lbs. (21 Nm).

18. Double check that the engine is still set to TDC/compression for cylinder No. 1. Check the alignment of both the crank and camshaft timing marks. Install the timing belt.

19. Install or connect the following:
- Spark plugs
- Wiring to the spark plugs
- No. 3 timing belt cover
- Cylinder head rear cover
- Air intake chamber with a new gasket. Tighten the bolts to 20 ft. lbs. (27 Nm). Install the bolt to hold the engine wire protector to the air intake chamber.
- Vacuum hose to the brake booster union and the EVAP hose to the No. 2 vacuum pipe.
- DLC connector and bracket and VSV connector
- Vacuum control set
- No. 2 vacuum pipe assembly and connect the hoses. Tighten the nuts to 20 ft. lbs. (27 Nm).
- EGR gas temperature sensor. Tighten it to 14 ft. lbs. (20 Nm).
- Vacuum hoses
- Dipstick tubes. Always use a new O-ring on each tube.
- Intake chamber supports and tighten the bolts to 13 ft. lbs. (18 Nm). The supports are marked **F** and **R** for the front and rear positions.

- Throttle body and intake air connector assembly
- Engine wire bracket
- Fuel return hose
- Vane pump to the pump bracket
- Power steering air hose to the No. 4 timing belt cover and intake chamber.
- Water bypass outlet and the bypass pipe. Always use new O-rings.
- Exhaust manifolds with new gaskets. Tighten the bolts to 29 ft. lbs. (39 Nm).
- O_2 sensor leads
- Front exhaust pipe. Tighten the bolts to 46 ft. lbs. (62 Nm).
- Manifold cover
- Water pump pulley
- Fan and coupling and the drive belt. Tighten the 4 nuts to 12 ft. lbs. (16 Nm).
- Air cleaner, airflow meter and the intake air connector pipe
- Air cleaner duct
- Control and accelerator cables to the throttle body
- Coolant
- Negative battery cable. Start the engine and check for leaks.
- Engine undercovers

3.0L (2JZ-GE) Engine (2004–05)

1. Before servicing the vehicle, refer to the Precautions Section.

2. Disconnect the negative battery cable. Wait at least 90 seconds before performing any other work.

3. Relieve the fuel pressure from the fuel lines.

4. Remove or disconnect the following:
- Undercover(s)
- Coolant
- Upper radiator hose from water outlet
- Engine cover
- Air cleaner inlet, air cleaner assembly, MAF meter, and intake air resonator assembly
- Serpentine drive belt
- Power steering pump from mounting (Do NOT disconnect hoses)
- Oxygen sensor wiring and grommet for bank 2, No. 2 sensor
- Front exhaust pipe from exhaust manifold
- Exhaust manifold
- Vacuum control valve set and No. 2 vacuum pipe from lower side of engine
- No. 3 timing belt cover

- Ignition coils and high-tension wires
- Spark plugs

5. Properly mark all engine wiring before disconnecting as needed for access to cylinder head, then disconnect the following:

- Ground strap from cylinder head
- 2 water bypass hoses from cylinder head and oil filter bracket
- Bank 2, No. 1 and bank 1, No. 1 oxygen sensor connectors and wire clamp
- Alternator wiring
- Wiring from water pump clamp
- 2 ground terminals from intake manifold
- 2 engine wire clamps from No. 1 oil pipe and clamp on intake manifold
- ECT, knock, oil pressure switch, and oil level sensor connectors
- Starter connector
- Injector connectors
- Camshaft timing oil control valve connector
- Camshaft position sensor connector
- Wiring protector from No. 2 cylinder head cover
- Engine wiring protector from intake manifold

6. Remove the fuel pressure pulsation damper.

7. Remove the intake manifold assembly, including stays.

8. Remove both cylinder head covers and gaskets.

9. While holding the timing belt so its position to timing marks does not change, remove the timing belt from the camshaft pulleys.

10. Remove both camshaft pulleys (exhaust pulley first, while holding the hex portion of the camshaft in order to loosen the pulley bolt).

➡See CAMSHAFT & VALVE LIFTERS for detailed removal procedure, if needed.

11. Remove the No. 4 timing belt cover.

12. Uniformly remove the No. 3 camshaft bearing cap bolts from both camshafts. Use a screwdriver to pry out the No. 1 and No. 3 camshaft bearing caps and oil seals. Then, evenly loosen all remaining 12 camshaft bearing cap bolts, in the order shown. Remove the No. 2 camshaft bearing caps and remove the camshafts.

13. Loosen and remove all 14 cylinder head bolts in the sequence shown, using several passes during loosening sequence.

14. Lift the cylinder head from the block

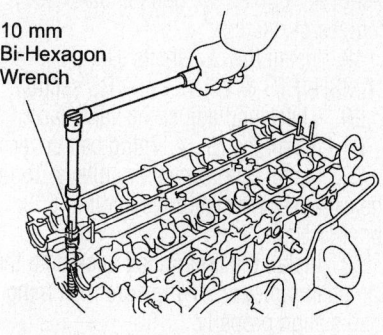

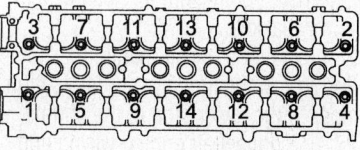

Showing the cylinder head bolt removal sequence—3.0L (2JZ-GE) engine

dowels. Disconnect the heater hose from the union. Remove the cylinder head from the vehicle and place on wooden blocks on the workbench.

To install:

15. With a new gasket, position the cylinder head onto the block.

16. Apply a light coat of engine oil to the cylinder head bolt threads and install the washers.

17. Install the cylinder head bolts, following the bolt tightening sequence shown. Torque bolts to 26 ft. lbs. (35 Nm).

18. Mark the front of each cylinder head bolt with paint, then retighten each bolt, in the same sequence, an additional 90 degrees. The paint mark should now face 90 degrees to the side. Again following the cor-

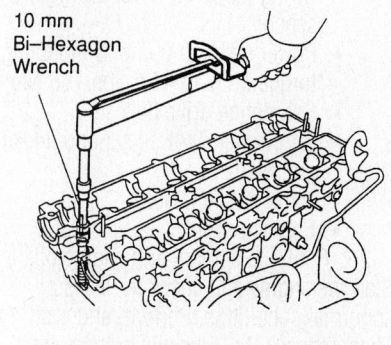

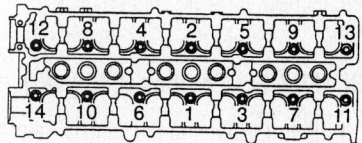

Cylinder head torque sequence—3.0L (2JZ-GE) engine

rect order, tighten the bolts another 90 degrees of rotation.

19. Install the camshafts. See CAMSHAFTS & VALVE LIFTERS section.

20. Check and adjust the valve lash.

21. Install the No. 4 timing belt cover.

22. Install both camshaft pulleys. Torque the exhaust camshaft pulley bolt to 60 ft. lbs. (81 Nm).

23. Install the timing belt back onto the camshaft pulleys, making sure the timing marks align properly.

24. Install both cylinder head covers, with new gaskets.

25. Install the intake manifold assembly.

26. Install the fuel pressure pulsation damper to the lower side of the engine.

27. Reconnect all engine wiring and hoses in reverse order of removal. Ensure components are reinstalled in original positions, as marked.

28. Install or reconnect the following:
- Spark plugs
- Ignition coils and wires
- No. 3 timing belt cover
- Vacuum control valve set and No. 2 vacuum pipe; torque nuts to 15 ft. lbs. (21 Nm)
- Air intake chamber
- Oil dipstick and guide
- Throttle body and intake air connector
- Water bypass outlet and No. 1 water bypass pipe
- Exhaust manifold, with new gaskets; torque nuts to 30 ft. lbs. (40 Nm)
- Front exhaust pipe to manifold; torque bolts and nuts to 32 ft. lbs. (44 Nm)
- Reconnect wire grommet and wiring for bank 2, No. 2 oxygen sensor
- Power steering pump to mounting; torque bolts to 43 ft. lbs. (58 Nm)
- Serpentine drive belt
- Air cleaner, MAF meter and intake air resonator
- Upper radiator hose
- Engine cover

29. Refill the coolant and engine oil. Start the engine and check for leaks or abnormal conditions. Perform and road test. Then, recheck for leaks and recheck fluid levels.

4.3L (3UZ-FE) Engines (2002–03)

1. Before servicing the vehicle, refer to the Precautions Section.

2. Relieve the fuel system pressure.

3. Remove or disconnect the following:

- Negative battery cable. Wait at least 90 seconds before performing any other work.
- Oil pan protector
- Engine undercover
- Coolant
- Battery clamp cover
- Air cleaner inlet
- V bank cover by removing the bolt and 2 cap nuts
- Air cleaner and intake air connector assembly
- Drive belt, fluid coupling and the fan pulley. The drive belt tension may be slackened by turning the tensioner counterclockwise. The pulley bolt for the drive belt tensioner has a left-handed thread.
- Radiator
- Right-hand No. 3 timing belt cover
- Left-hand No. 3 timing belt cover
- Drive belt idler pulley by removing the pulley bolt and cover plate
- Right-hand No. 2 timing belt cover
- Left-hand No. 2 timing belt cover
- No. 1 ignition coil
- Air conditioning compressor from the engine
- Fan bracket by removing the 2 bolts and 2 nuts

4. Set the engine to Top Dead Center (TDC) on cylinder No. 1.

✷✷ WARNING

Since the thrust clearance of the camshaft is small, the camshaft must be kept level while it is being removed. If the camshaft is not kept level, the portion of the cylinder head receiving the shaft thrust may crack or be damaged, causing the camshaft to seize or break.

5. Turn the crankshaft pulley approximately 50 degrees clockwise and put the timing mark of the crankshaft pulley in line with the centers of the crankshaft pulley bolt and the idler pulley bolt.

✷✷ WARNING

If the timing belt is disengaged, having the crankshaft pulley at the wrong angle can cause the piston head and valve head to come into contact with each other when you remove the camshaft timing pulley. Always set the crankshaft pulley at the correct angle before removing the timing belt.

6. If the timing belt is to be reused, turn the crank pulley slowly; check that the 3 installation marks are present on the belt. If the marks are not present, make new installation marks before removing the belt. The marks should align with the timing marks on each camshaft pulley and the crank pulley.

7. Remove the timing belt tensioner. Alternately loosen the 2 bolts; remove the bolts, the tensioner and the dust protector.

8. Loosen the tension between the left side and the right side timing pulleys by slightly turning the left side camshaft clockwise.

9. Remove or disconnect the following:
- Timing belt from the camshaft timing pulleys. Using the proper tool, remove the bolt and the camshaft timing pulleys.
- Power steering pump from the engine. Do NOT disconnect the hoses or lines from the power steering pump. Support the power steering pump with a piece of wire. Do NOT allow the pump to hang.
- Front catalytic converter
- High tension spark plug wires, wire clamps and the wire cover assembly
- No. 2 ignition coil by removing the connector and the 2 bolts
- 2 bolts and the rear timing belt plate. Remove both plates
- Intake chamber assembly
- Throttle Position Sensor (TPS) connector
- With TRAC system, sub TP sensor connector
- With TRAC system, sub TP connector
- Idle Air Control (IAC) valve connector
- Exhaust Gas Recirculation (EGR) valve connector
- Vacuum Switching Valve (VSV) connector for fuel pressure control
- VSV connector for Evaporative Emissions (EVAP) system
- EGR gas temperature sensor connector
- Brake booster vacuum hose from the union on the air intake chamber
- Positive Crankcase Ventilation (PCV) hose from the PCV valve on the left-hand cylinder head
- Water bypass hose (from the EGR valve) from the rear water bypass joint
- Water bypass hose (from the throttle body) from the rear water bypass joint

- Vacuum hose (from the VSV for fuel pressure control) from the fuel pressure regulator
- EVAP hose (from charcoal canister) from the VSV for EVAP
- Heater hose from the water bypass pipe
- Fuel inlet hose from the delivery pipe
- Fuel return hose from the fuel return pipe
- Engine wire from the delivery pipes and rear water bypass joint
- Fuel hose from the fuel pressure regulator
- 2 bolts and fuel return pipe from the intake manifold
- 8 injector connectors
- 6 bolts, 4 nuts, the intake manifold assembly and the 2 gaskets
- Water inlet and inlet housing
- Front water bypass joint
- Rear water bypass joint and No. 1 EGR pipe assembly
- Oil dipstick and guide for the automatic transmission
- Oil dipstick and guide for the engine
- Engine hangers
- Right and left cylinder head covers by removing the 8 bolts, seal washers and gaskets
- Semi-circular plug, if necessary

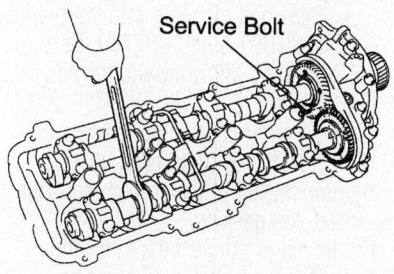

Securing the exhaust camshaft on the right cylinder head—4.3L (3UZ-FE) Engine

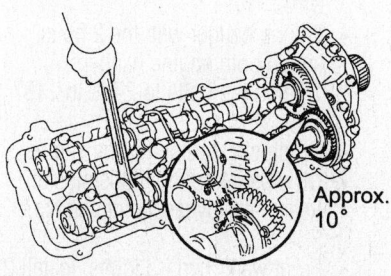

Turning the exhaust camshaft 10 degrees on the right cylinder head—4.3L (3UZ-FE) Engine

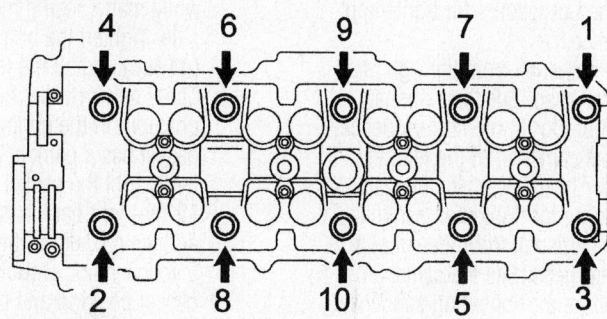

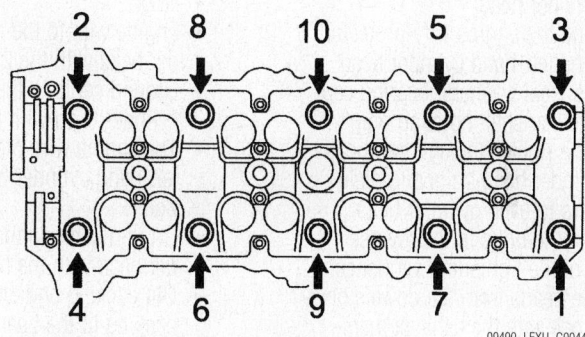

Cylinder head bolt loosening sequence—4.3L (3UZ-FE) Engine

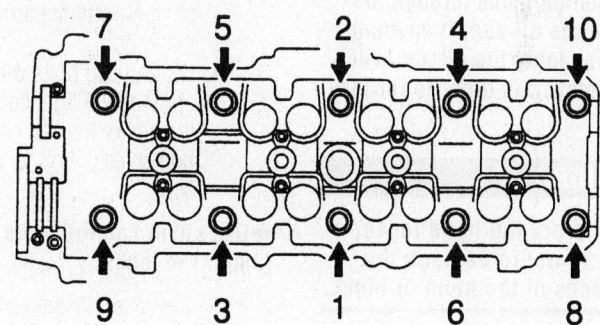

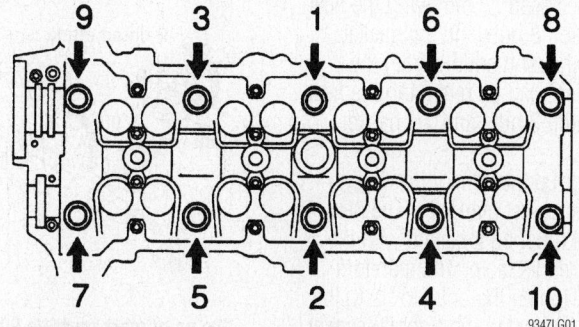

Cylinder head torque sequence—4.3L (3UZ-FE) engine

- Exhaust camshaft from the right side cylinder head. See the camshaft procedure for tightening sequence.
- Intake camshaft from the right side cylinder head. See the camshaft procedure for tightening sequence.
- Exhaust camshaft of the left side cylinder head. See the camshaft procedure for tightening sequence.
- Intake camshaft from the left side cylinder head. See the camshaft procedure for tightening sequence.
- Main Heated Oxygen (HO_2) sensor
- Bolt and HO_2 the ground cable from the right cylinder head
- Bolt and the ground strap from the left cylinder head
- Bolt and the engine wire protector from the left-hand cylinder head
- 2 bolts, seal washers, bearing cap and the camshaft housing plug from the right-hand cylinder head
- 10 cylinder head bolts and plate washers to each cylinder head. Loosen the bolts in the reverse order of the tightening sequence. Lift the heads from the dowels on the block with the exhaust manifolds attached. Place the heads on blocks of wood on the workbench.

➡ **Do NOT drop anything in the opening in the front of the right side cylinder head. The opening leads through the block and into the oil pan. If anything falls into the opening the oil pan will have to be removed in order to retrieve it.**

✳✳ WARNING

If necessary to pry the head loose, take great care not to damage the contact surfaces of the head or block.

- 2 bolts, seal washers, bearing cap and camshaft housing plug from the right-hand cylinder head
- Right exhaust manifold from the cylinder head by removing the heat insulator, 8 nuts and the gasket
- Left exhaust manifold from the cylinder head by removing the heat insulator, 8 nuts and the gasket

To install:
10. Install or connect the following:
- Right exhaust manifold. The new gasket must be installed with the white marks facing the manifold side. Tighten the bolts to 32 ft. lbs. (44 Nm). Install the right O_2 sensor.

- Left exhaust manifold. The new gasket must be installed with the white marks facing the manifold side. Tighten the bolts to 32 ft. lbs. (44 Nm). Install the left O_2 sensor.
- The 2 new cylinder head gaskets in position on the engine block. Each gasket has a painted mark denoting the rear of the gasket. The gasket for the right bank has a white mark and the gasket for the left bank has a yellow mark. Double check the gasket position and placement.

11. Install the cylinder heads and tighten the bolts in sequence as follows:
   a. Step 1: 44 ft. lbs. (59 Nm)
   b. Step 2: Plus 90 degrees
- HO_2.
- Engine wire to the right-hand cylinder head with the 2 bolt
- Ground cable to the right-hand cylinder head with the bolt
- The engine wire protector to the left-hand cylinder head with the bolt
- Ground cable to the left-hand cylinder head with the bolt
- Old packing and apply new seal packing to the bearing caps
- Bearing cap on the right side cylinder head, marked **I1**, in position with the arrow mark facing the rear. Install the bearing cap on the left side cylinder head, marked **I6**, in position with the arrow mark facing the front.
- Bearing cap bolts with new washers. Apply a light coat of oil on the threads of the cap bolts. Alternately tighten each bolt to 12 ft. lbs. (16 Nm).

➡ **Use silver colored bolts 1.50 in. (38mm) in length.**

- Camshaft housing plugs on the cylinder heads. Be sure to face the cupped side forward.

12. Turn the crankshaft pulley clockwise or counterclockwise and put the timing

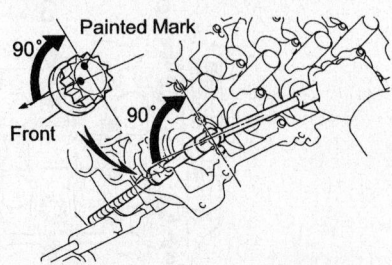

09490_LEXU_G0045

The paint mark must be 90 degrees from the starting point—4.3L (3UZ-FE) Engine

mark of the crankshaft pulley in line with the centers of the crankshaft pulley bolt and the idler pulley bolt.

✳✳ WARNING

Since the thrust clearance of the camshaft is small, the camshaft must be kept level while it is being installed. If the camshaft is not kept level, the portion of the cylinder head receiving the shaft thrust may crack or be damaged, causing the camshaft to seize or break.

13. Install or connect the following:
- Right side cylinder head intake camshaft. Tighten the bracket bolt in the reverse order of the loosening sequence.
- Right side cylinder head exhaust camshaft. Tighten the bracket bolt in the reverse order of the loosening sequence.
- Left side cylinder head intake camshaft. Tighten the bracket bolt in the reverse order of the loosening sequence.
- Left side cylinder head exhaust camshaft. Tighten the bracket bolt in the reverse order of the loosening sequence.

14. Check and adjust the valve clearance.
15. Install or connect the following:
- Camshaft oil seals with the proper tool (SST 09223-46011). Be sure to apply MP grease to the new oil seal lip.
- Semi-circular plugs, if removed

16. Clean the cylinder head covers. Apply new sealant in the correct locations and install the gaskets.

17. Install or connect the following:
- Right cylinder head cover and bolts. Tighten the bolts to 52 inch lbs. (6 Nm).
- Left cylinder head cover and bolts. Tighten the bolts to 52 inch lbs. (6 Nm).
- Engine hanger with the 2 bolts. Install both engine hangers. Tighten the bolts to 27 ft. lbs. (37 Nm).
- Oil dipstick guide for the engine
- Oil dipstick for the transmission
- Rear water bypass joint and No. 1 EGR pipe
- Front water bypass joints. Install 2 gaskets and alternately tighten the nuts to 13 ft. lbs. (18 Nm).
- Water inlet and inlet housing, alter-

nately tighten the bolts to 13 ft. lbs. (18 Nm)

- Delivery pipe and intake manifold
- Return pipe with 2 new gaskets. Tighten the union bolt to 26 ft. lbs. (35 Nm).
- Engine wire to the delivery pipes and rear water bypass joint
- Fuel return hose to the fuel return pipe
- Fuel inlet hose to the left-hand delivery pipe
- Fuel hose to the fuel pressure regulator
- Air intake chamber assembly
- Brake booster vacuum hose to the union on the air intake chamber
- PCV hose to the PCV valve on the left-hand cylinder head
- Water bypass hose (from EGR valve) to the rear water bypass joint
- Water bypass hose (from throttle body) to the rear water bypass joint
- Vacuum hose (from VSV for fuel pressure control) to the fuel pressure regulator
- EVAP hose (from charcoal canister) from the VSV for EVAP
- TPS connector
- With TRAC system, sub TPS connector
- With TRAC system, sub throttle actuator connector
- IAC valve connector
- EGR valve connector
- EGR gas temperature sensor connector
- VSV connector for fuel pressure control
- VSV connector for EVAP
- Accelerator bracket with the 2 bolts
- Accelerator, automatic transmission throttle control and the cruise control actuator cable
- Spark plug wires and clamps to the right and left cylinder head cover
- Belt rear plates by installing the bolts. Tighten the bolts to 66 inch lbs. (8 Nm).
- No. 2 ignition coil
- A new gasket to the exhaust manifold and install the catalytic converters. Tighten the 3 nuts to each converter to 46 ft. lbs. (62 Nm).
- Front exhaust pipe, tighten the bolts and nuts to 32 ft. lbs. (44 Nm). Tighten the 4 bolts holding the pipe support bracket to the transmission. Tighten the bolts to 32 ft. lbs. (44 Nm).
- Power steering pump with the nut and 3 bolts. Tighten the nut to 32

ft. lbs. (43 Nm) and the bolts to 29 ft. lbs. (39 Nm).

18. Align the knock pin on the right side camshaft with the knock pin of the timing pulley. Slide on the timing pulley with the right side mark facing forward. Tighten the bolt to 80 ft. lbs. (108 Nm).

19. Align the knock pin on the left side camshaft with the knock pin of the timing pulley. Slide on the timing pulley with the left side mark facing forward. Tighten the bolt to 80 ft. lbs. (108 Nm).

20. Install the timing belt to the left side camshaft timing pulley as follows:

a. Using the proper tool, slightly turn the left side timing pulley clockwise. Align the installation mark of the timing belt with the timing mark of the camshaft timing pulley and hang the timing belt on the left side camshaft pulley.

b. Align the timing marks of the left side camshaft pulley and the timing belt rear plate.

c. Check that the timing belt has tension between crankshaft timing pulley and the left side camshaft pulley.

21. Install the timing belt to the right side camshaft timing pulley as follows:

a. Using the proper tool, slightly turn the right side timing pulley clockwise. Align the installation mark of the timing belt with the timing mark of the camshaft timing pulley and hang the timing belt on the right side camshaft pulley.

b. Align the timing marks of the right side camshaft pulley and the timing belt rear plate.

c. Check that the timing belt has tension between crankshaft timing pulley and the right side camshaft pulley.

22. The timing belt tensioner must be set prior to installation. The tensioner can be set by:

a. Place a plate washer between the tensioner and a block. Using a press, press in the pushrod using 220–225 lbs. of pressure.

b. Align the holes of the pushrod and housing, pass a 0.05 inch (1.27mm) rod through the holes to keep the setting position of the pushrod.

c. Release the press and install the dust boot to the tensioner.

23. Loosely install the tensioner. Evenly and alternately tighten the bolts to 20 ft. lbs. (26 Nm). Remove the tool from the tensioner.

24. Turn the crankshaft pulley 2 complete revolutions from TDC to TDC. Always turn the crankshaft clockwise. Check that all belt and pulley marks align with their reference marks. If any mark is out of perfect

alignment, the timing belt must be removed and reinstalled.

25. Install or connect the following:
- Drive belt tensioner and tighten the bolt and nuts to 12 ft. lbs. (16 Nm).

26. Install the fan bracket by installing the 2 bolts and 2 nuts. Tighten as follows:
  a. 12mm: 12 ft. lbs. (16 Nm)
  b. 14mm: 24 ft. lbs. (32 Nm)

27. Remove or disconnect the following:
- Air conditioning compressor. Tighten the bolts to 36 ft. lbs. (49 Nm) and the nut to 22 ft. lbs. (29 Nm).
- No. 1 ignition coil
- Right side No. 2 timing belt cover
- Left side No. 2 timing belt cover
- Drive belt idler pulley and cover plate. Tighten the bolt to 27 ft. lbs. (37 Nm).
- Secure the ignition wires. Make certain that all clips and retainers are securely engaged and that the wires are properly routed.
- Right side No. 3 timing belt
- Left-hand No. 3 timing belt cover
- Radiator assembly
- Fan pulley, fan, fluid coupling and the drive belt
- The air cleaner and intake air connector assembly
- V bank cover
- Coolant
- Negative battery cable to the battery
- Air cleaner inlet
- Battery clamp cover
- Engine undercover
- Oil pan protector

28. Start the engine and check for leaks

29. Bleed the cooling system and recheck the engine coolant level.

30. Make all the necessary engine adjustments.

4.3L (3UZ-FE) Engine (2004–06)

1. Before servicing the vehicle, refer to the Precautions Section.

2. Relieve the fuel system pressure.

3. Remove or disconnect the following, as applicable to each engine:
- Engine under cover
- Drain engine coolant
- V-bank cover
- Air cleaner assembly
- Intake air pipe
- Radiator (if necessary)
- Throttle body
- Upper and lower intake manifold assembly
- Camshaft position sensor and LH timing belt rear plates
- RH timing belt rear plates

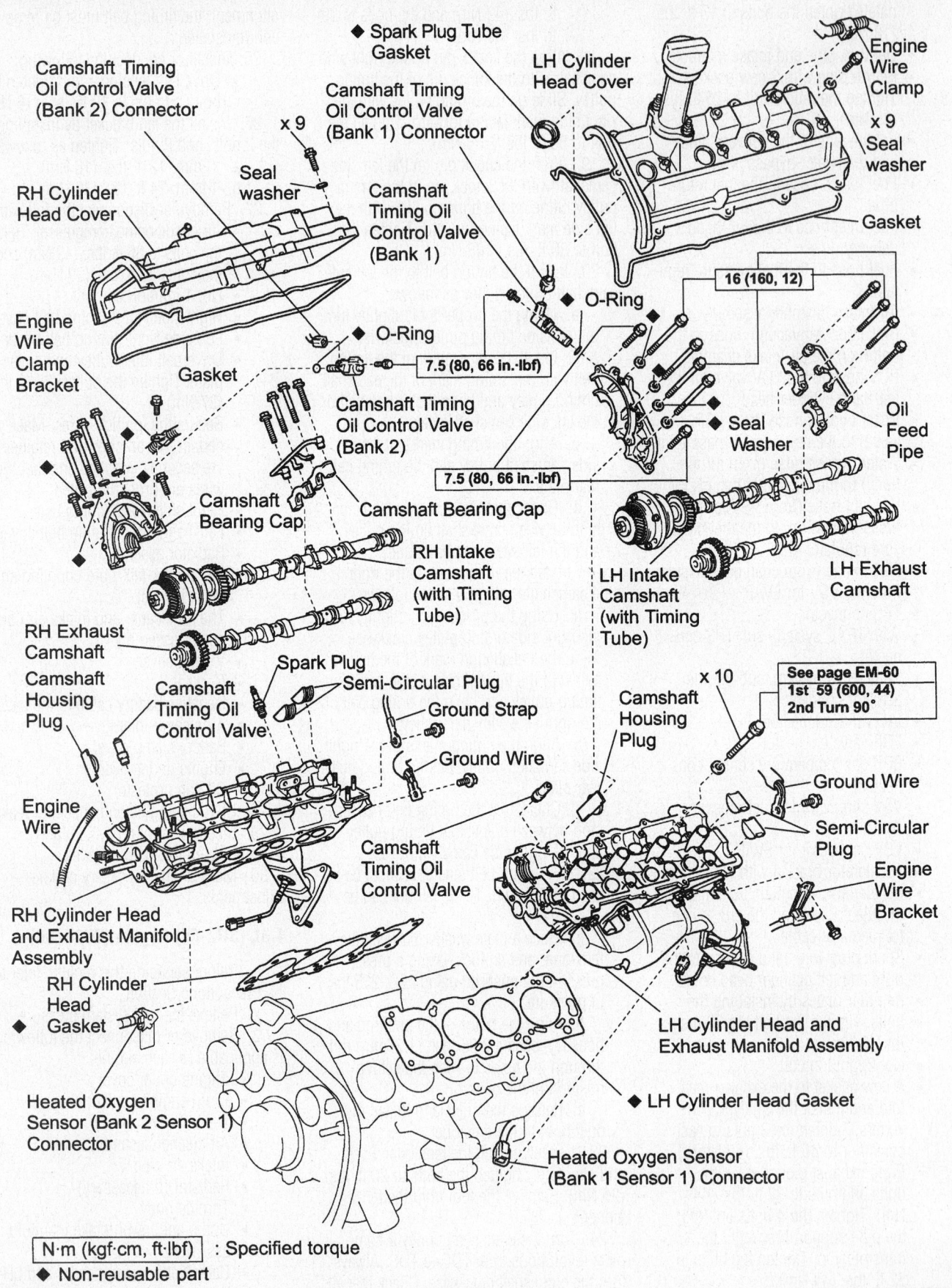

Camshaft Timing Oil Control Valve (Bank 2) Connector

Camshaft Timing Oil Control Valve (Bank 1) Connector

Spark Plug Tube Gasket

LH Cylinder Head Cover

Engine Wire Clamp

x 9

RH Cylinder Head Cover

Seal washer

x 9

Seal washer

Gasket

Camshaft Timing Oil Control Valve (Bank 1)

Engine Wire Clamp Bracket

Gasket

O-Ring

O-Ring

16 (160, 12)

7.5 (80, 66 in.·lbf)

Camshaft Timing Oil Control Valve (Bank 2)

Seal Washer

Oil Feed Pipe

7.5 (80, 66 in.·lbf)

Camshaft Bearing Cap

Camshaft Bearing Cap

RH Intake Camshaft (with Timing Tube)

LH Intake Camshaft (with Timing Tube)

LH Exhaust Camshaft

RH Exhaust Camshaft

Camshaft Housing Plug

Camshaft Timing Oil Control Valve

Spark Plug

Semi-Circular Plug

Ground Strap

Ground Wire

Camshaft Housing Plug

x 10

See page EM-60
1st 59 (600, 44)
2nd Turn 90°

Ground Wire

Semi-Circular Plug

Engine Wire

Camshaft Timing Oil Control Valve

Engine Wire Bracket

RH Cylinder Head and Exhaust Manifold Assembly

RH Cylinder Head Gasket

◆

LH Cylinder Head and Exhaust Manifold Assembly

◆ LH Cylinder Head Gasket

Heated Oxygen Sensor (Bank 2 Sensor 1) Connector

Heated Oxygen Sensor (Bank 1 Sensor 1) Connector

N·m (kgf·cm, ft·lbf) : Specified torque

◆ Non-reusable part

Exploded view of the cylinder heads and related components—4.3L (3UZ-FE) Engine

Do NOT drop anything inside timing belt cover during this procedure. Keep oil, water and dust from timing belt.

- Power steering pump from engine mount (Do NOT disconnect hoses)
- Catalytic converters
- Water inlet housing assembly
- Water bypass pipe, front bypass joint, and rear bypass joint
- Ignition coils
- Variable valve timing (VVT) sensors
- Engine hangers
- Oil dipsticks and guides for engine oil and transmission fluid
- Cylinder head covers
- Spark plugs
- Camshafts

Since the thrust clearance of the camshaft is small, the camshaft must be kept level during removal. If not, the portion of the cylinder head receiving the camshaft thrust may crack or be otherwise damaged, causing the camshaft to later seize or break. Follow the camshaft removal procedure carefully as given in this section.

- Both oxygen sensor connectors
- Ground wire from LH cylinder head
- Engine wire bracket for oxygen sensor on LH cylinder head

4. Uniformly loosen the 10 cylinder head bolts on each cylinder head, in several passes, following loosening sequence as shown.

Use care so that no bolts or washers are dropped into the recesses or enclosed portions of the cylinder head or block.

5. Carefully lift the cylinder head from the locating dowels on the engine block. Place cylinder heads on wooden blocks on the workbench.

6. If necessary, exhaust manifolds may be removed from the cylinder heads at this time.

To install:

7. If removed, install exhaust manifolds to cylinder heads, with new gaskets. Ensure the white mark on the gasket is facing the manifold side.

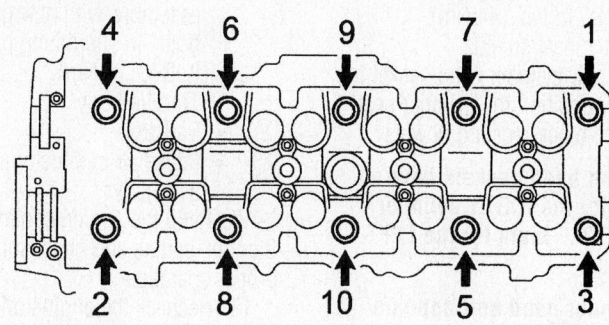

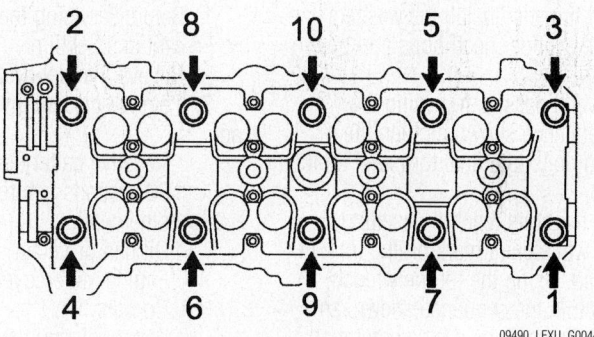

Cylinder head bolt loosening sequence—4.3L (3UZ-FE) Engine

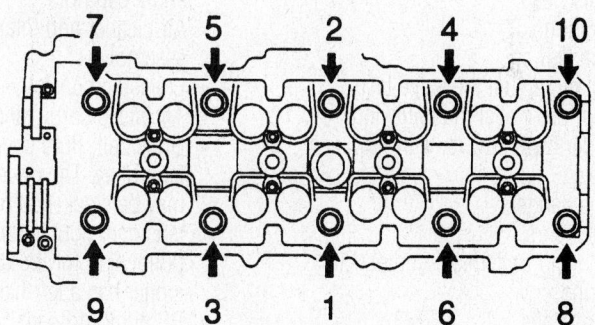

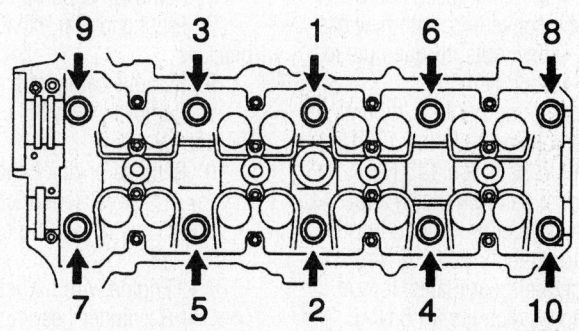

Cylinder head torque sequence—4.3L (3UZ-FE) engine

8. Install and tighten the exhaust manifold bolts, in an alternating pattern, to a final torque of 32 ft. lbs. (44 Nm).

9. Install the heat shields.

10. With a new cylinder head gasket in place, carefully position the cylinder head onto the engine block locating dowels.

➡ **The cylinder head gaskets have a "3R" marks for the RIGHT cylinder head, and a "3L" mark for the LEFT cylinder head.**

➡ **If any cylinder head bolt appears stretched or damaged, replace it. If a bolt will not reach final torque setting, replace it.**

11. Apply a light coat of oil to the cylinder head bolt threads. Install the washers and insert the cylinder head bolts into position.

12. In several passes, following the tightening sequence shown, tighten the cylinder head bolts to a final torque of 44 ft. lbs. (59 Nm).

13. Once the bolts reach this setting, then place a white paint mark on the front of each bolt head. Using the torque wrench, turn each bolt, in the sequence shown, an additional 90 degrees, using the paint mark as a reference.

14. Install or reconnect the following:
- Engine wiring and ground straps
- Oxygen sensor wire bracket on LH cylinder head
- Spark plugs
- Camshafts

15. Inspect and adjust the valve lash.

16. Install or reconnect the following:
- Cylinder head covers, with new gaskets
- Engine hangers
- VVT sensors
- Engine and transmission dipsticks and tubes
- Ignition coils
- Water bypass joints and pipe; torque nuts and bolt to 13 ft. lbs. (18 Nm)
- Water inlet housing assembly
- Catalytic converters, with new gaskets and new nuts; torque nuts to 46 ft. lbs. (62 Nm)
- Power steering pump to mounting; torque bolts to 29 ft. lbs. (39 Nm) and nut to 32 ft. lbs. (43 Nm)
- LH timing belt rear plates and camshaft position sensor; torque bolts to 66 inch lbs. (7.5 Nm)
- RH right belt rear plates; torque bolts to 66 inch lbs. (7.5 Nm)
- Camshaft pulleys

- Timing belt to camshaft pulleys
- Upper and lower intake manifold assembly, with new gaskets; torque bolts, in alternating pattern, to 13 ft. lbs. (18 Nm)
- Throttle body
- Radiator
- Intake air connector and air cleaner assembly

17. Refill the engine cooling system. Start the engine and check for leaks and proper operation.

18. Recheck the engine oil level.

19. Install the V-bank cover and the engine under cover.

3.3L (3MZ-FE) Engine

1. Before servicing the vehicle, refer to the Precautions Section.

2. Relieve the fuel system pressure.

3. Remove or disconnect the following:

- Negative battery cable. Wait at least 90 seconds before performing any other work
- Oil pan protector
- Engine undercover
- Coolant
- Battery clamp cover
- Air cleaner inlet
- Lower radiator air deflector
- RF wheel
- V bank cover by removing the bolt and 2 cap nuts
- Air cleaner and intake air connector assembly
- Emission control valve
- Air intake surge tank
- Drive belt, fluid coupling and the fan pulley. The drive belt tension may be slackened by turning the tensioner counterclockwise. The pulley bolt for the drive belt tensioner has a left-handed thread.
- PS pump drive belt
- Radiator

4. Remove intake manifold

5. Remove timing belt

6. Remove PS pump assembly

7. Front and rear exhaust pipe and brackets

8. Remove camshafts

9. Remove LH or RH cylinder head assemblies

10. Remove or disconnect the following:
- The VVT Sensor connector
- Camshaft timing oil valve connector
- Engine wire harness clamp
- 8 cylinder head bolts uniformly in the sequence

← Front

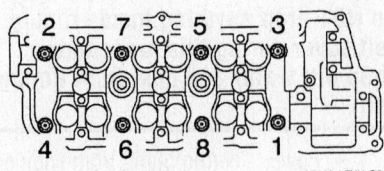

67162-LEXU-G31

Cylinder head bolt removal sequence— 3.3L (3MZ-FE) Engine

☀☀ WARNING

Head warpage or cracking could result from removing the bolts in an incorrect order.

11. Inspect the cylinder head set bolts. Ensure they match the following:
- Outside diameter is .3524 to .3563 in. (8.95 to 9.05mm)
- Minimum diameter is .3445 in. (8.75mm)

12. If diameter is lees than minimum, replace the bolts.

To install:

13. Install new head gasket with R mark upward

14. Install cylinder head assembly
- Apply light oil to the cylinder head bolts
- Install plate washers on the cylinder head bolts

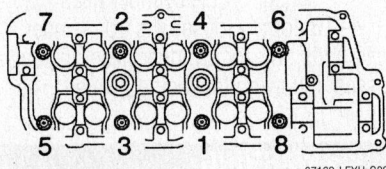
← Front

67162-LEXU-G32

Cylinder head bolt tightening sequence— 3.3L (3MZ-FE) Engine

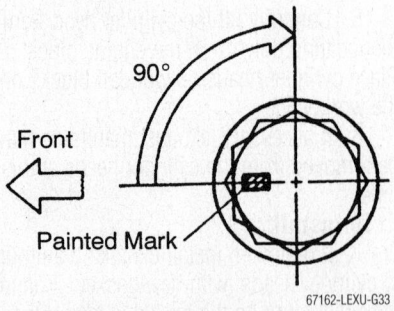

67162-LEXU-G33

Head bolt marking procedure—3.3L (3MZ-FE) Engine

➡**Cylinder head bolts are tightened in two successive steps. Install and tighten 8 cylinder head bolts in required sequence.**

    a. Tighten to 40 ft. lbs. (54 Nm)
    b. Mark the front side of each head bolt with paint.
    c. Retighten cylinder head bolts 90 degrees in the same sequence
    d. Check that each painted mark is now at a 90 degree angle to the front
15. Install or connect the following:
- Wiring harness clamp
- Camshaft timing oil valve connector
- Camshaft assemblies
- Valve cover assemblies
- Exhaust manifold assemblies and support brackets Torque to 36 ft. lbs. (49 Nm)
- Exhaust manifold heat insulators
- PS pump assembly
- Timing belt inner cover
- Camshaft timing pulleys
- Timing belt and idler assemblies
- RH engine mounting bracket
- Timing belt covers
- Alternator bracket
- Engine mounting stay No 2
- Engine stabilizer rod
- PS drive belt
- A/C compressor drive belt
- Water outlet
- Intake manifold assembly
- Intake air surge tank
- Emission control valve set
- Air cleaner assembly
- Vacuum hoses
- V-bank cover sub-assembly
- Front suspension upper brace
- RF wheel
- Install engine oil
- Installed engine coolant
16. Inspect for fuel leaks
17. Inspect for oil leaks
18. Inspect for exhaust leaks
19. Check engine timing and idle speed
20. Run system initialization

Intake Manifold

REMOVAL & INSTALLATION

2.5L (4GR-FSE) Engine

1. Before servicing the vehicle, refer to the Precautions Section.
2. Relieve fuel system pressure.
3. Remove or disconnect the following:
- Negative battery cable. Wait at least 90 seconds before performing any other work.
- Coolant
- Cool air intake duct seal
- Left and right engine room side covers
- V-bank cover
- Left and right front upper fender protectors
- Left and right roof drip side finish mouldings using a moulding removal tool
- Front wiper arm head cap
- Left and right windshield wiper arm and blade assemblies
- Cowl top ventilator louver assembly
- No. 2 ventilation hose
- Air cleaner cap with air cleaner hose
- Throttle body assembly
- Cold start injector

4. Remove intake air surge tank as follows:
    a. Disconnect the vacuum hose from the intake air surge tank.
    b. Remove the bolt and disconnect the No. 1 vacuum switching valve assembly from the intake air surge tank.
    c. Disconnect the wire harness and hose from the surge tank.
    d. Disconnect the ventilation hose from the intake air surge tank.
    e. Remove the bolt and water hose joint from the intake air surge tank.
    f. Remove the bolt and disconnect the surge tank stay.
    g. Using a 5mm hexagon socket wrench, remove the 7 bolts, 2 nuts and gasket.
5. Remove intake manifold as follows:
    a. Disconnect the connector for the SCV.
    b. Disconnect the SCV position sensor connector.
    c. Remove the 4 bolts, 4 nuts, intake manifold and gasket.

To install:
6. Place new gaskets onto the intake manifold and position the intake manifold between the cylinder heads. Tighten the nuts and bolts to 15 ft. lbs. (21 Nm).
7. Install or connect the following:
- SCV position sensor connector
- DC motor connector for the SCV
8. Install the intake air surge tank as follows:
    a. Install a new gasket to the intake air surge tank.
    b. Using a 5mm hexagon socket wrench, install the 6 bolts. Torque all bolts, except A, to 13 ft. lbs. (18 Nm).

    c. Install the bolt and 2 nuts to the intake air surge tank. Torque bolt A to 15 ft. lbs. (21 Nm) and the nuts to 12 ft. lbs. (16 Nm).
    d. Install the surge tank stay to the intake air surge tank. Torque the bolt to 15 ft. lbs. (21 Nm).
    e. Connect the water hose joint with the bolt and torque to 7 ft. lbs. (10 Nm).
    f. Connect the ventilation hose to the intake air surge tank.
    g. Connect the wire harness and hose to the intake air surge tank.
    h. Connect the No. 1 vacuum switching valve assembly to the intake air surge tank. Torque the fasteners to 13 ft. lbs. (18 Nm).
    i. Connect the vacuum hose to the intake air surge tank.
- Cold start injector. Torque the bolts to 7 ft. lbs. (10 Nm).
- New gasket and throttle body assembly. Torque the 4 bolts to 7 ft. lbs. (10 Nm)
- Air cleaner cap with air cleaner hose
- No. 2 ventilation hose
- Cowl top ventilator louver assembly
- Left and right windshield wiper arm and blade assemblies
- Front wiper arm head cap
- Left and right roof drip side finish mouldings
- Left and right front upper fender protectors
- V-bank cover
- Left and right engine room side covers
- Cool air intake duct seal
- Negative battery cable

9. Refill the cooling system. Start the engine and check for leaks and proper operation.

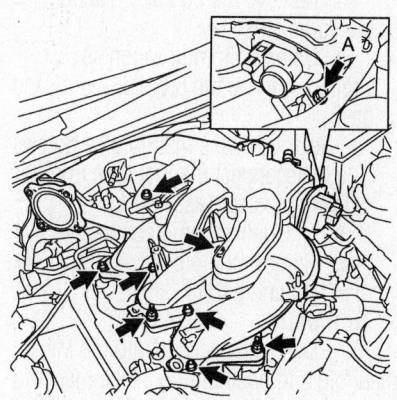

09490_LEXU_G0048

Intake air surge tank mounting bolt locations—2.5L (4GR-FSE) engine

10. Check the function of the throttle body unit

11. System initialization

3.5L (2GR-FSE) Engine

1. Before servicing the vehicle, refer to the Precautions Section.

2. Relieve fuel system pressure.

3. Remove or disconnect the following:
- Negative battery cable. Wait at least 90 seconds before performing any other work.
- Coolant
- Cool air intake duct seal
- Left and right engine room side covers
- V-bank cover
- Left and right front upper fender protectors
- Left and right roof drip side finish mouldings using a moulding removal tool
- Front wiper arm head cap
- Left and right windshield wiper arm and blade assemblies
- Cowl top ventilator louver assembly
- No. 2 ventilation hose
- Air cleaner cap with air cleaner hose
- Throttle body assembly

4. Remove intake air surge tank as follows:

a. Disconnect the vacuum hose from the intake air surge tank.

b. Remove the bolt and disconnect the No. 1 vacuum switching valve assembly from the intake air surge tank.

c. Disconnect the ventilation hose, union to check valve hose and water by-pass hose from the surge tank.

d. Disconnect the 4 wire harness clamps from the intake air surge tank.

e. Remove the bolt and water hose joint from the intake air surge tank.

f. Remove the bolt and disconnect the surge tank stay.

g. Using a 5mm hexagon socket wrench, remove the 6 bolts, 2 nuts and gasket.

5. Remove intake manifold as follows:

a. Disconnect the fuel tube from the delivery pipe sub-assembly.

b. Disconnect the 4 connectors.

c. Remove the 4 bolts, 4 nuts, intake manifold and gasket.

To install:

6. Place new gaskets onto the intake manifold and position the intake manifold between the cylinder heads. Tighten the nuts and bolts to 15 ft. lbs. (21 Nm).

7. Install or connect the following:

- 4 connectors
- Fuel tube to the delivery pipe sub-assembly

8. Install the intake air surge tank as follows:

a. Install a new gasket to the intake air surge tank.

b. Install the intake air surge tank with the 2 nuts. Torque the 2 nuts to 12 ft. lbs. (16 Nm).

c. Using a 5mm hexagon socket wrench, install the 6 bolts. Torque the bolts to 13 ft. lbs. (18 Nm).

d. Install the surge tank stay to the intake air surge tank. Torque the bolt to 15 ft. lbs. (21 Nm).

e. Connect the water hose joint with the bolt and torque to 7 ft. lbs. (10 Nm).

f. Connect the 4 wire harness clamps to the intake air surge tank.

g. Connect the ventilation hose, union to check valve hose and water by-pass hose to the intake air surge tank.

h. Connect the No. 1 vacuum switching valve assembly to the intake air surge tank. Torque the bolt to 13 ft. lbs. (18 Nm).

i. Connect the vacuum hose to the intake air surge tank.

- New gasket and throttle body assembly. Torque the 4 bolts to 7 ft. lbs. (10 Nm)
- Air cleaner cap with air cleaner hose
- No. 2 ventilation hose
- Cowl top ventilator louver assembly
- Left and right windshield wiper arm and blade assemblies
- Front wiper arm head cap
- Left and right roof drip side finish mouldings
- Left and right front upper fender protectors
- V-bank cover
- Left and right engine room side covers
- Cool air intake duct seal
- Negative battery cable

9. Refill the cooling system. Start the engine and check for leaks and proper operation.

10. Check the function of the throttle body unit

11. System initialization

3.0L (3GR-FSE) Engine

1. Before servicing the vehicle, refer to the Precautions Section.

2. Relieve fuel system pressure.

3. Remove or disconnect the following:

- Negative battery cable. Wait at least 90 seconds before performing any other work.
- Coolant
- Cool air intake duct seal
- Left and right engine room side covers
- V-bank cover
- No. 1 air cleaner inlet

4. Remove the air cleaner assembly with hose as follows:

a. Disconnect the ventilation hose from the cylinder head.

b. Disconnect the MAF meter connector.

c. Disconnect the clamp from the air cleaner.

d. Disconnect the EVAP VSV.

e. Loosen the hose clamp.

f. Remove the 3 bolts and air cleaner case.

- No. 1 water by-pass hose
- No. 2 water by-pass hose
- Rear engine cover
- No. 3 water by-pass pipe
- No. 4 water by-pass pipe

5. Remove intake air surge tank as follows:

a. Remove the 2 bolts and intake manifold stay.

b. Remove the 4 bolts and 2 surge tank stays.

c. Using a 5mm hexagon socket wrench, remove the 7 bolts, 2 nuts and gasket.

6. Remove the 4 bolts, 2 nuts, intake manifold and gasket.

❋❋ WARNING

Cover the cylinder head intake port to prevent foreign matter from entering it.

To install:

7. Place new gaskets onto the intake manifold and position the intake manifold between the cylinder heads. Tighten the nuts and bolts to 15 ft. lbs. (21 Nm).

8. Install the intake air surge tank as follows:

a. Install a new gasket to the intake air surge tank.

b. Install the intake air surge tank with the 2 nuts. Torque the 2 nuts to 13 ft. lbs. (18 Nm).

c. Using a 5mm hexagon socket wrench, install the 7 bolts. Torque the bolts to 12 ft. lbs. (16 Nm).

d. Install the 2 surge tank stay to the intake air surge tank. Torque the 4 bolt to 15 ft. lbs. (21 Nm).

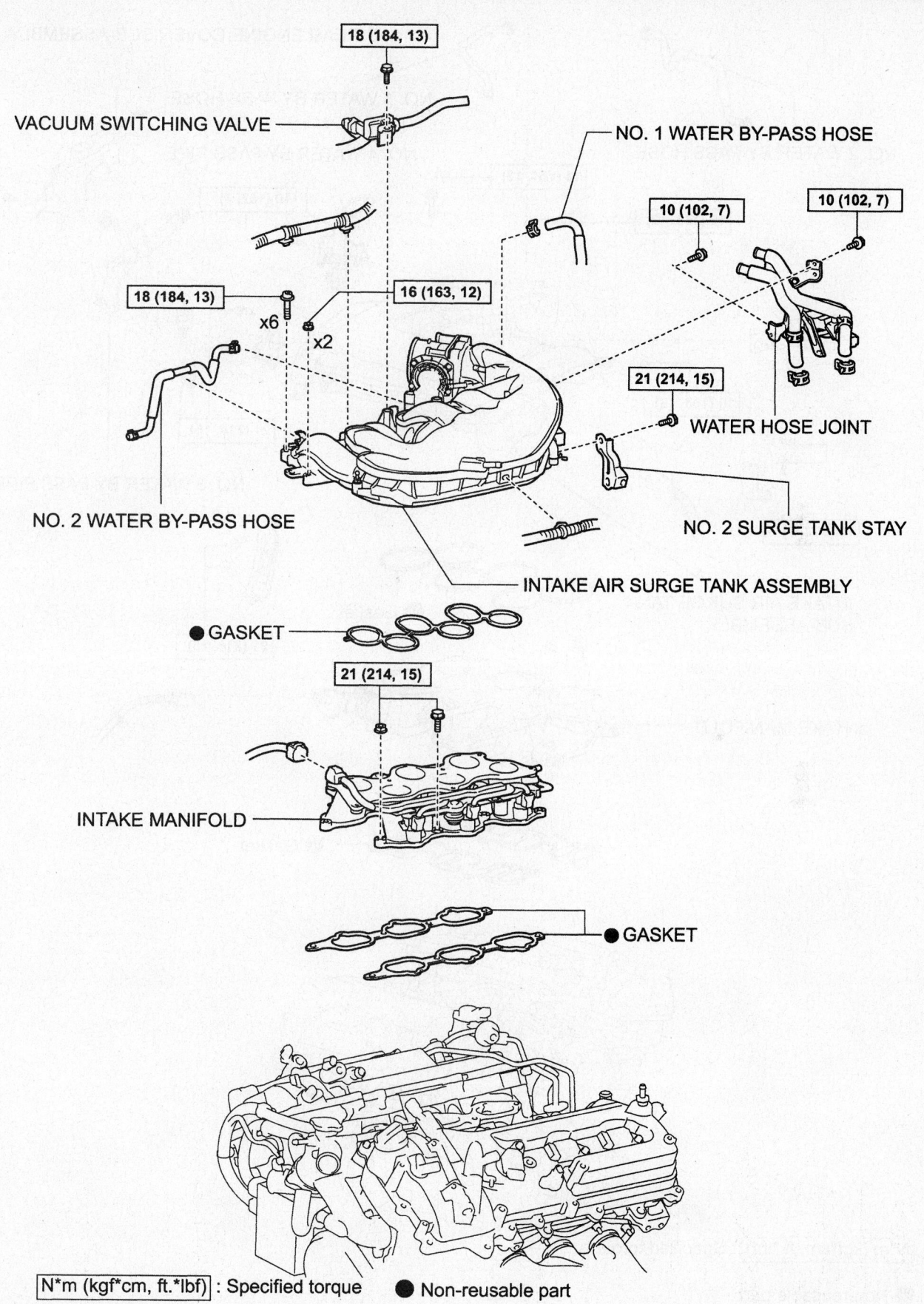

18 (184, 13)

VACUUM SWITCHING VALVE

NO. 1 WATER BY-PASS HOSE

10 (102, 7)   10 (102, 7)

18 (184, 13)   16 (163, 12)

x6   x2

21 (214, 15)

WATER HOSE JOINT

NO. 2 WATER BY-PASS HOSE

NO. 2 SURGE TANK STAY

INTAKE AIR SURGE TANK ASSEMBLY

● GASKET

21 (214, 15)

INTAKE MANIFOLD

● GASKET

N*m (kgf*cm, ft.*lbf) : Specified torque   ● Non-reusable part

Exploded view of the intake manifold mounting and related components—3.5L (2GR-FSE) engine

09490_LEXU_G0049

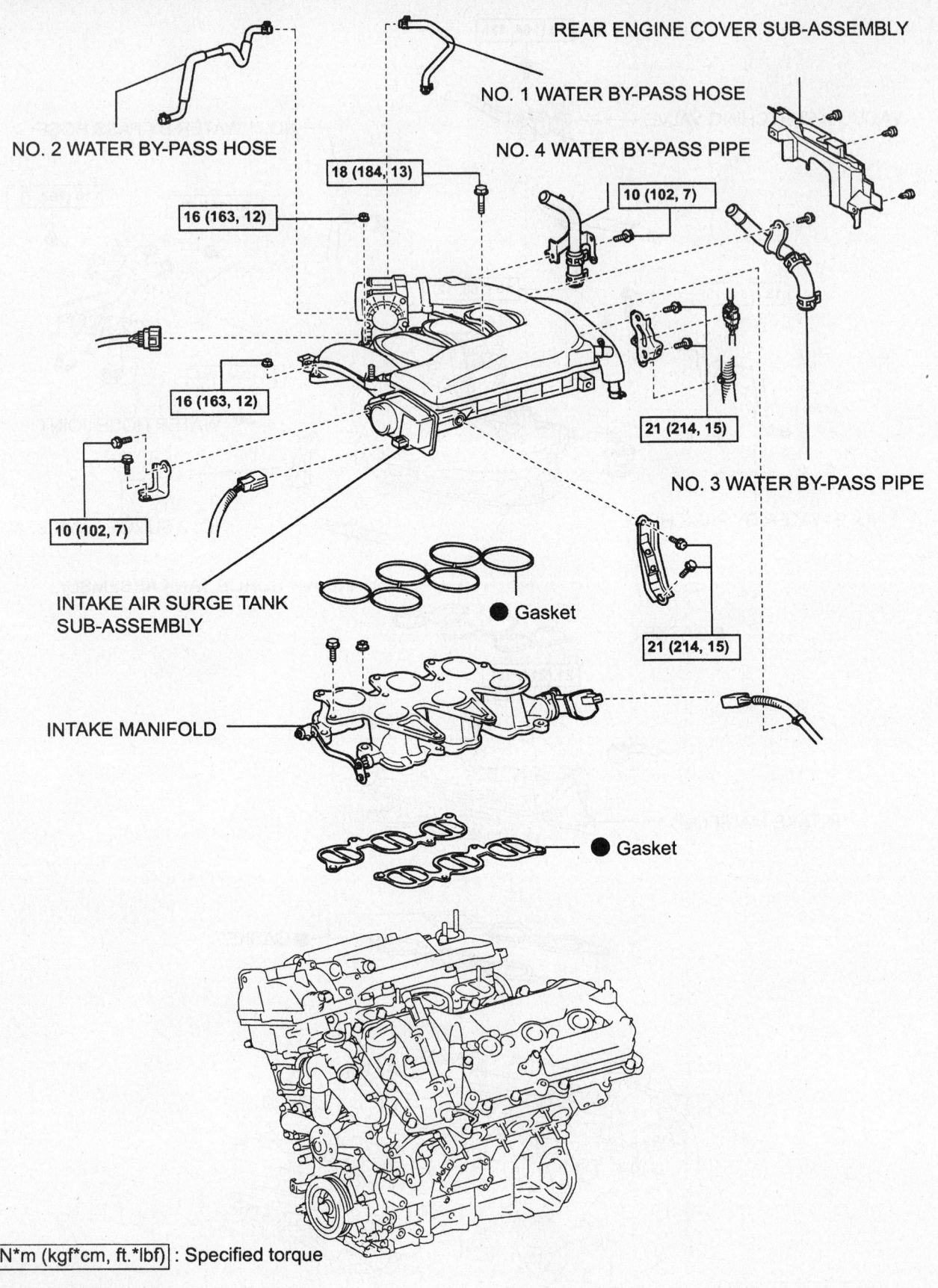

REAR ENGINE COVER SUB-ASSEMBLY

NO. 1 WATER BY-PASS HOSE

NO. 2 WATER BY-PASS HOSE

NO. 4 WATER BY-PASS PIPE

18 (184, 13)

16 (163, 12)

10 (102, 7)

16 (163, 12)

21 (214, 15)

10 (102, 7)

NO. 3 WATER BY-PASS PIPE

INTAKE AIR SURGE TANK SUB-ASSEMBLY

● Gasket

21 (214, 15)

INTAKE MANIFOLD

● Gasket

N*m (kgf*cm, ft.*lbf) : Specified torque

● Non-reusable part

09490_LEXU_G0050

Exploded view of the intake manifold mounting and related components—3.0L (3GR-FSE) engine

e. Install the intake manifold stay with 2 bolts and torque to 7 ft. lbs. (10 Nm).

9. Install or connect the following:
- No. 4 water by-pass pipe
- No. 3 water by-pass pipe
- Rear engine cover
- No. 2 water by-pass hose
- No. 1 water by-pass hose
- Air cleaner assembly
- No. 1 air cleaner inlet
- V-bank cover
- Left and right engine room side covers
- Cool air intake duct seal
- Negative battery cable

10. Refill the cooling system. Start the engine and check for leaks and proper operation.

11. Check the function of the throttle body unit

12. System initialization

3.0L (1MZ-FE) Engine

1. Before servicing the vehicle, refer to the Precautions Section.

2. Remove or disconnect the following:
- Negative battery cable
- Coolant
- Throttle/accelerator cable from the throttle body
- Air cleaner hose at the air intake chamber and remove it
- All lines and hoses. Tag them for installation.
- Idle Speed Control (ISC) valve and the throttle body
- Exhaust Gas Recirculation (EGR) valve and vacuum modulator
- Cylinder head rear plate
- Intake chamber stays, any wires, then, the air intake chamber
- Fuel injection delivery pipe and the injectors
- Water outlet and the bypass outlet
- 2 bolts and the No. 2 idler pulley bracket stay
- 8 bolts and 4 nuts, then lift out the intake manifold

To install:

3. Thoroughly clean the intake manifold and cylinder head surfaces. Using a machinist's straight edge and a feeler gauge, check the surface of the intake manifold for warpage. If the warpage is greater than 0.0039 in. (0.10mm), replace the intake manifold.

4. Place new gaskets onto the intake manifold and position the intake manifold between the cylinder heads. Tighten the nuts and bolts to 11 ft. lbs. (17 Nm). Tighten the No. 2 pulley bracket bolts to 13 ft. lbs. (18 Nm).

5. Install or connect the following:
- Water bypass outlet and tighten the bolts to 74 inch lbs. (8.3 Nm). Tighten the water outlet to 74 inch lbs. (8 Nm).
- Injectors and delivery pipe
- Air intake chamber and tighten the 2 bolts and 2 nuts to 32 ft. lbs. (43 Nm); use an 8mm hex wrench
- Chamber stays and tighten the mounting bolts to 29 ft. lbs. (39 Nm)
- Remaining components. Tighten the emission control valve set to 73 inch lbs. (8 Nm).
- All hoses
- Accelerator cable, if equipped with automatic transaxle
- Coolant
- Negative battery cable

3.0L (2JZ-GE) Engine (2002–03)

1. Before servicing the vehicle, refer to the Precautions Section.

2. Remove or disconnect the following:
- Negative battery cable
- Coolant
- Spark plug wires at the spark plugs
- Spark plugs
- Radiator
- Water pump pulley
- Timing belt
- No. 2 front exhaust pipe
- 2 Oxygen (O_2) sensor leads
- 4 nuts, and the manifold heat shield
- Exhaust manifolds
- Water bypass outlet and the No. 1 bypass pipe. Remove the 3 O-rings.
- Water outlet
- No. 1 bypass hose
- Vacuum Control Valve (VCV) set and the No. 2 vacuum pipe
- Fuel return hose from the oil dipstick guide; remove the mounting bolt and pull the guide and dipstick from the pan. Plug the hole.
- Air intake chamber
- Fuel delivery pipe, then pull out the injectors
- No. 1 and 2 fuel pipes
- Engine harness from the intake manifold
- Intake manifold stay
- Loosen the 6 bolts and 2 nuts, then lift out the intake manifold

To install:

3. Install or connect the following:
- Install the intake manifold, with a new gasket, and tighten the bolts and nuts to 21 ft. lbs. (28 Nm)
- Mounting stay and tighten the bolts to 29 ft. lbs. (39 Nm)
- Engine harness to the manifold
- 2 fuel pipes and tighten the bolts to 78 inch lbs. (9 Nm)
- Delivery pipe and injectors. Tighten the pipe bolts to 15 ft. lbs. (21 Nm).
- Air intake chamber and tighten it to 15 ft. lbs. (21 Nm)
- 2 stays and tighten them to 13 ft. lbs. (18 Nm); The No. 1 stay is marked with an **F** and the No. 2 stay is marked with an **R**.
- The oil dipstick and guide, using a new O-ring
- VCV set and the vacuum pipe. Tighten the set mounting bolts to 15 ft. lbs. (21 Nm).
- Water bypass outlet and the pipe, tighten the bolts to 78 inch lbs. (9 Nm)
- Exhaust manifolds. Tighten the bolts to 30 ft. lbs. (40 Nm).
- Heat shield and tighten it to 13 ft. lbs. (18 Nm)
- No. 2 front pipe
- Timing belt
- Radiator and water pump pulley
- Plug wires to the plugs
- Coolant
- Negative battery cable

3.0L (2JZ-GE) Engine (2004–05)

1. Before servicing the vehicle, refer to the Precautions Section.

2. Remove or disconnect the following:
- Negative battery cable
- Engine under cover
- Engine coolant
- Engine cover
- Air cleaner assembly
- Throttle body assembly
- Water bypass pipe running through intake manifold
- Engine and transmission dipsticks and tubes
- Air intake chamber
- Drive belt
- Vacuum control valve set
- Power steering pump from mounting (Do NOT disconnect hoses)
- Engine wiring, and hoses, as need for access to intake manifold

➡**Mark each wire, connector or hose, as it is removed, for proper reinstallation reference.**

- Fuel pressure pulsation damper

3. Disconnect the starter wire from the

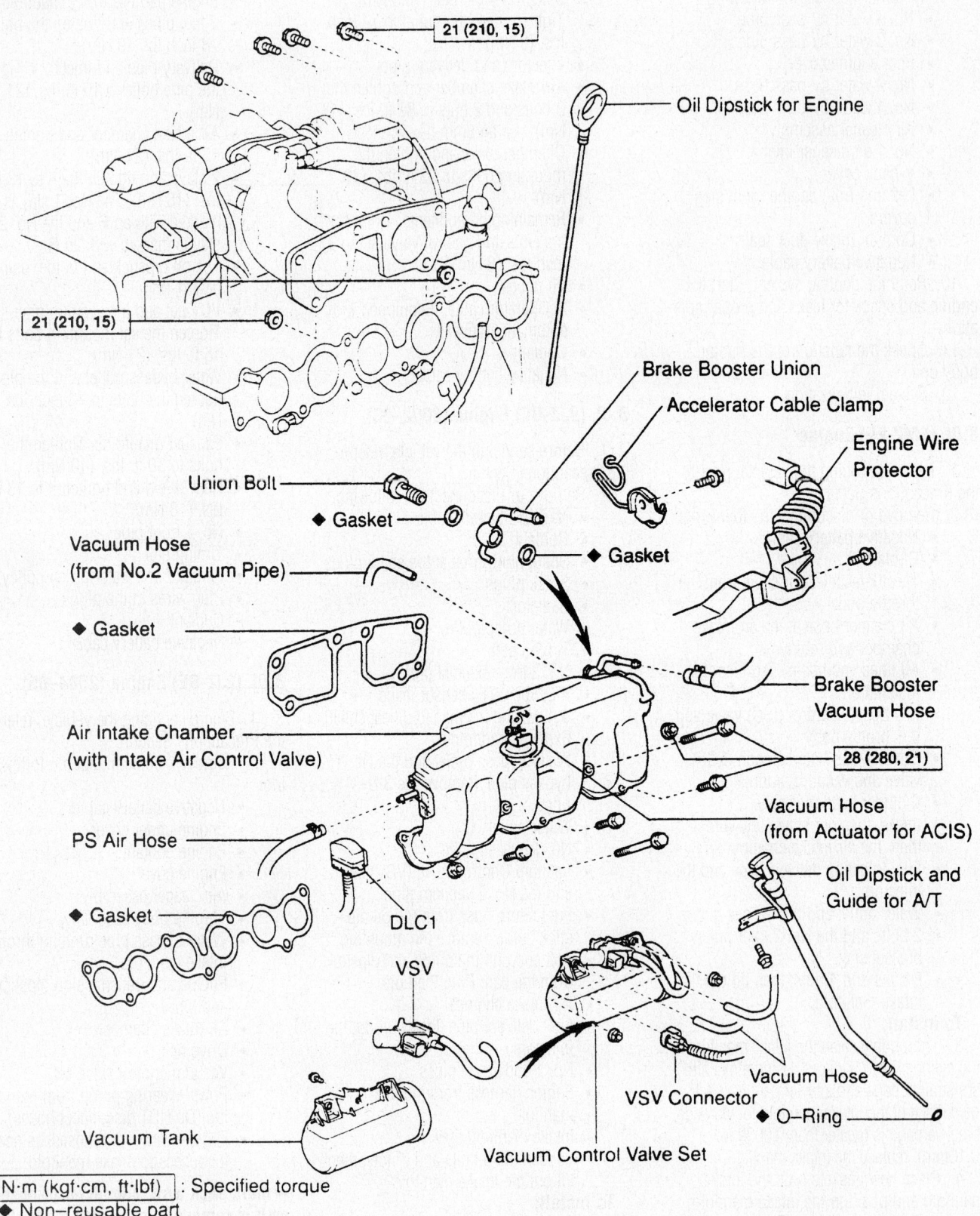

21 (210, 15)

21 (210, 15)

Oil Dipstick for Engine

Brake Booster Union

Accelerator Cable Clamp

Engine Wire Protector

Union Bolt

◆ Gasket

Vacuum Hose
(from No.2 Vacuum Pipe)

◆ Gasket

◆ Gasket

Brake Booster
Vacuum Hose

Air Intake Chamber
(with Intake Air Control Valve)

28 (280, 21)

Vacuum Hose
(from Actuator for ACIS)

PS Air Hose

◆ Gasket

DLC1

Oil Dipstick and
Guide for A/T

VSV

Vacuum Hose

VSV Connector

◆ O-Ring

Vacuum Tank

Vacuum Control Valve Set

N·m (kgf·cm, ft·lbf) : Specified torque
◆ Non-reusable part

9301LG09

Exploded view of the intake manifold mounting and related components—3.0L (2JZ-GE) engine

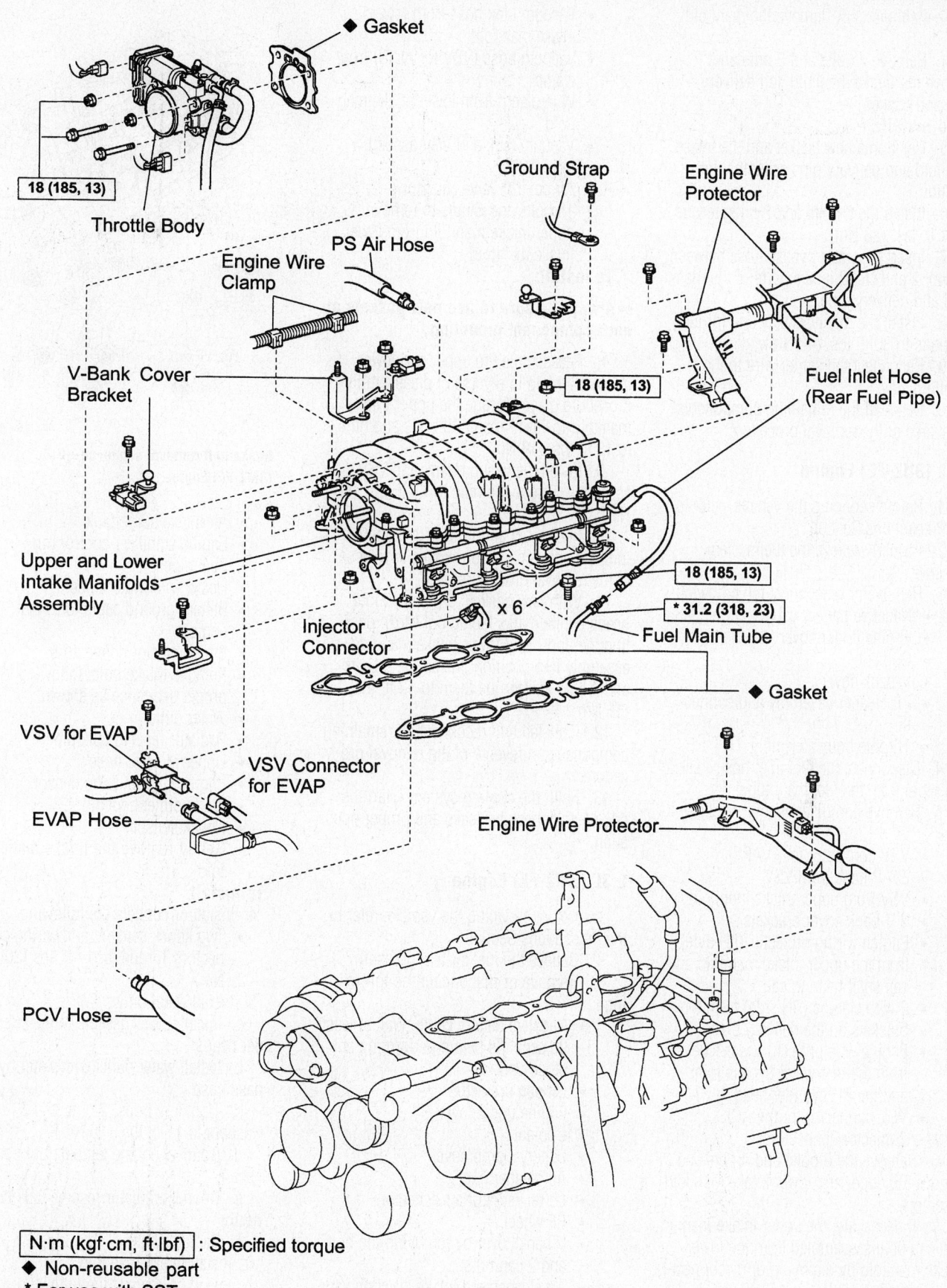

◆ Gasket

Throttle Body

18 (185, 13)

Ground Strap

Engine Wire Protector

PS Air Hose

Engine Wire Clamp

V-Bank Cover Bracket

18 (185, 13)

Fuel Inlet Hose (Rear Fuel Pipe)

Upper and Lower Intake Manifolds Assembly

Injector Connector

x 6

18 (185, 13)

* **31.2 (318, 23)**

Fuel Main Tube

◆ Gasket

VSV for EVAP

VSV Connector for EVAP

EVAP Hose

Engine Wire Protector

PCV Hose

N·m (kgf·cm, ft·lbf) : Specified torque

◆ Non-reusable part

* For use with SST

Exploded view of the intake manifold mounting and related components—4.3L (3UZ-FE) Engine

09490_LEXU_G0047

intake manifold stay. Remove the manifold stay.

4. Remove 7 bolts and 2 nuts and remove the intake manifold and delivery pipe and gasket.

To install:

5. Position a new gasket and the intake manifold and delivery pipe assembly into position.

6. Install the 7 bolts and 2 nuts; torque to 21 ft. lbs. (28 Nm).

7. Insert the water bypass hose between the No. 2 and No. 3 intake ports of the manifold and delivery pipe.

8. Install the manifold stay and torque the bolts to 30 ft. lbs. (40 Nm).

9. Reconnect the starter wire to the manifold stay.

10. Reinstall the remaining components in reverse of the removal procedure.

4.3L (3UZ-FE) Engine

1. Before servicing the vehicle, refer to the Precautions Section.

2. Properly relieve the fuel system pressure.

3. Remove or disconnect the following:
- Negative battery cable
- Engine under cover
- Coolant
- V-bank cover
- Air cleaner assembly and connectors
- Throttle body assembly

4. Disconnect the fuel inlet hose (rear fuel pipe) from the fuel main tube.

5. Remove and disconnect the following:
- VSV connector for EVAP
- EVAP hose from VSV
- VSV from upper intake manifold
- 4 V-bank cover brackets
- Engine wiring protector (LH side) from the upper intake manifold and camshaft bearing cap
- 2 wire clamps (RH side) from the brackets on the delivery pipe
- Engine wire protector (rear side) from the rear water bypass joint and the RH cylinder head
- VSV connector for the ACIS
- 8 injector connectors

6. Remove the 6 bolts and 4 nuts and remove the upper and lower intake manifold assembly.

7. If necessary, the upper intake manifold can be disassembled from the lower intake manifold by removing or disconnecting the following:
- Vacuum hose for VSV from air control valve actuator

- Vacuum tank hose from lower intake manifold
- Vacuum hose (VSV for ACIS) from clamp
- Wire clamp from lower intake manifold
- Vacuum tank and VSV assembly from ACIS
- Air control valve actuator
- 15 bolts and 5 nuts to remove upper intake manifold from lower intake manifold

To install:

➡ **Always be sure to use new gaskets at each component mounting.**

8. Reassemble the upper and lower intake manifold in reverse of disassembly procedure given. Torque the upper intake manifold-to-lower manifold bolts and nuts to 13 ft. lbs. (18 Nm).

9. Install the vacuum tank and VSV assembly. Torque the nuts to 13 ft. lbs. (18 Nm).

10. Reconnect all of the wiring connectors, clamps and vacuum hoses in reverse of the removal procedure.

11. With new gaskets on the cylinder heads (white marks facing outward), position the upper and lower intake manifold assembly into position. Install the 6 bolts and 4 nuts and torque them to 13 ft. lbs. (18 Nm).

12. Reinstall and reconnect all remaining components in reverse of the removal procedure.

13. Refill the cooling system. Start the engine and check for leaks and proper operation.

3.3L (3MZ-FE) Engine

1. Before servicing the vehicle, refer to the Precautions Section.

2. Relieve the fuel system pressure.

3. Remove or disconnect the following:
- Negative battery cable. Wait at least 90 seconds before performing any other work
- Oil pan protector
- Engine undercover
- Coolant
- Battery clamp cover
- Air cleaner inlet
- Lower radiator air deflector
- RF wheel
- V bank cover by removing the bolt and 2 cap nuts
- Air cleaner and intake air connector assembly
- Emission control valve

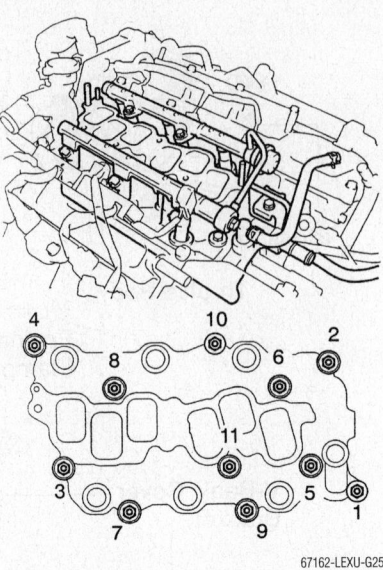

67162-LEXU-G25

Intake bolt removal sequence—3.3L (3MZ-FE) Engine

- Air intake surge tank
- Engine stabilizer control rod
- Fuel pipe
- Heater inlet water hose
- Battery ground cable form the engine
- 6 fuel injector connectors
- Remove intake bolts following the proper sequence as shown.
- Water outlet
- Radiator reserve tank pipe
- Upper radiator hose
- Engine coolant temp sensor
- Water outlet housing
- Knock sensor
- Gently remove the intake manifold

To install:

4. Install or connect the following:
- Two knock sensors and wiring connectors Torque to: 14 ft. lbs (20 Nm)

5. Install water outlet
   a. Install 2 new gaskets to the cylinder heads
   b. Install water outlet with water bypass hose
   c. Tighten 2 bolts, 2 nuts and 2 washers to 11 ft. lbs. (15 Nm)
   d. Connect engine temperature connector
   e. Connect radiator reserve tank connector
   f. Radiator hose

6. Install intake manifold

7. Install or connect the following:
- 9 bolts, 2 nuts and 2 washers; torque to 11 ft. lbs. (15 Nm)

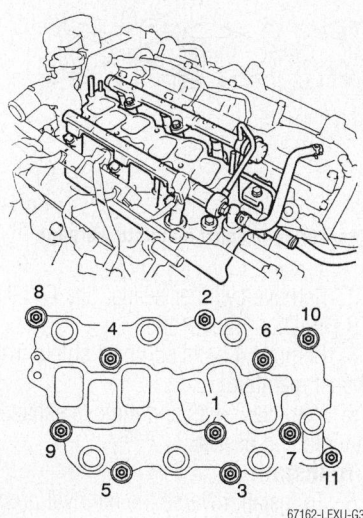

Intake manifold tightening sequence—3.3L (3MZ-FE) Engine

67162-LEXU-G34

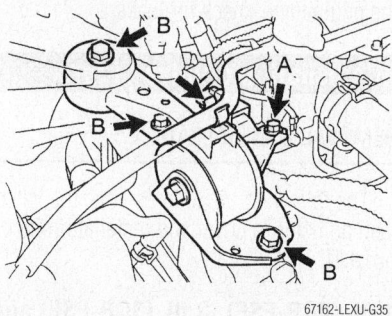

67162-LEXU-G35

Tightening procedure for Engine Stabilizing Rod —3.3L (3MZ-FE) Engine

➡**Using several steps, tighten the bolts and nuts uniformly in sequence**

- Retighten 9 bolts and 2 nuts to 11 ft. lbs. (15 Nm)
- Fuel pipe making sure the connector is seated until it clicks.

8. Install or connect the following:

- Engine stabilizing control rod; torque bolt "A" to 17 ft. lbs. (23 Nm) and bolt "B" to 47 ft. lbs. (64 Nm)
- Intake air surge tank
- Emission control valve set
- Air Cleaner assembly
- Vacuum hoses
- Battery negative wire
- Front center suspension upper brace
- Engine coolant
- V-bank cover sub-assembly

9. Run engine and check for coolant and fuel leaks

10. System initialization

Exhaust Manifold

REMOVAL & INSTALLATION

2.5L (4GR-FSE), 3.0L (3GR-FSE) and 3.5L (2GR-FSE) Engines

1. Before servicing the vehicle, refer to the Precautions Section.

2. Remove or disconnect the following:

- Negative battery cable. Wait at least 90 seconds before performing any other work
- Cool air intake duct seal
- Left and right engine room side covers
- V-bank cover
- Engine under cover
- No. 2 engine under cover (2WD)
- Left and right rear engine under cover (2WD)
- Front exhaust pipe assembly
- Front lower suspension member protector (2WD)
- Exhaust pipe No. 1 support bracket (4GR-FSE with automatic transmission)

3. Remove the oil dipstick guide assembly as follows:

a. Remove the oil level gauge.

b. Remove the bolt, then remove the No. 2 oil dipstick guide.

c. Remove the O-ring from the No. 2 oil dipstick guide.

d. Remove the bolt and clamp (2WD), then remove the No. 1 oil dipstick guide.

e. Remove the O-ring from the No. 1 oil dipstick guide.

- Oxygen sensor wiring connectors
- 6 nuts, exhaust manifold and gasket (left and right sides)

To install:

4. Install the exhaust manifold(s) as follows:

a. Install a new exhaust manifold gasket.

b. Install the exhaust manifold with 6 new nuts and torque to 15 ft. lbs. (21 Nm).

✳✳ WARNING

Do not damage the stud bolt when installing the exhaust manifold. Be sure to tighten either of nuts A first as shown in the illustration.

5. Install the oil dipstick guide assembly as follows:

a. Install a new O-ring to the oil dipstick guide.

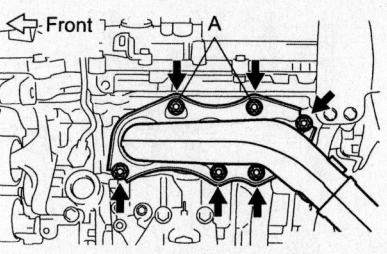

09490_LEXU_G0051

When tightening the exhaust manifold nuts, start with nuts A as shown(left side illustrated, right side the same—2.5L (4GR-FSE), 3.0L (3GR-FSE) and 3.5L (2GR-FSE) engines

b. Apply a light coat of engine oil to the O-ring.

c. Push in the oil dipstick guide end into the guide hole.

d. Install the No. 1 oil dipstick guide with the bolt and torque to 7 ft. lbs. (10 Nm).

e. Connect the clamp (2WD).

f. Install the No. 2 oil dipstick guide with the bolt and torque to 15 ft. lbs. (21 Nm).

g. Install the oil dipstick.

6. Install or connect the following:

- Oxygen sensor wiring connectors
- Front lower suspension member protector (2WD) and 4 bolts. Torque the 4 bolts to 71 inch lbs. (8 Nm).
- Front exhaust pipe assembly
- Left and right rear engine under cover (2WD)
- No. 2 engine under cover (2WD)
- Engine under cover
- V-bank cover
- Left and right engine room side covers
- Cool air intake duct seal
- Negative battery cable

7. Start the engine and check for exhaust leaks and proper operation.

8. System initialization

3.3L (3MZ-FE) Engine

1. Before servicing the vehicle, refer to the Precautions Section.

2. Remove or disconnect the following:

- Negative battery cable. Wait at least 90 seconds before performing any other work
- V-bank cover
- Air cleaner assembly (if needed for access to exhaust manifold)
- Front and rear exhaust pipe No. 1 support brackets
- Front exhaust pipe assembly

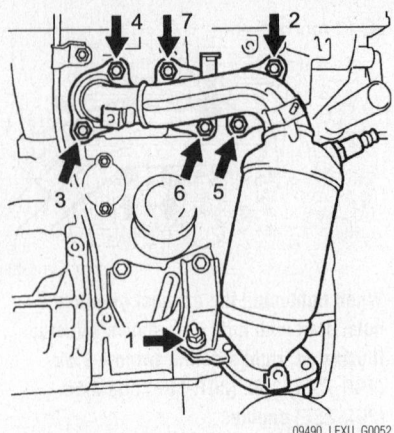

Exhaust manifold nut loosening sequence (left side shown, right side the same)—3.3L (3MZ-FE) engine

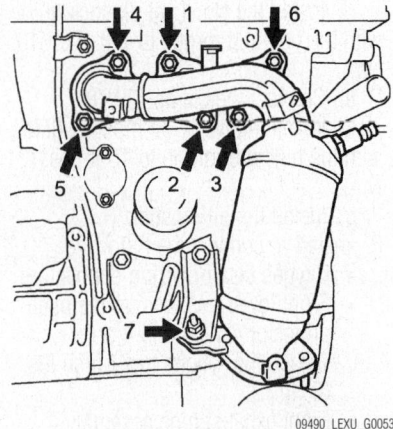

Exhaust manifold nut tightening sequence (left side shown, right side the same)—3.3L (3MZ-FE) engine

- Exhaust pipe No. 1 support bracket
- Oxygen sensor wiring connectors
- Exhaust manifold heat insulator No. 1
- Manifold converter insulator No. 3
- Exhaust manifold heat insulator No. 2
- 7 nuts in sequential order and lift off exhaust manifold and gasket

To install:

3. Position the exhaust manifold in place, with new gaskets. Torque the nuts, in sequential order, to 36 ft. lbs. (49 Nm). Retighten nuts 1 and 2 as illustrated.

4. Reinstall the remaining components in reverse of the removal procedure.

3.0L (1MZ-FE) Engine

1. Before servicing the vehicle, refer to the Precautions Section.
2. Remove or disconnect the following:
   - Negative battery cable

- Engine undercovers
- 2 front exhaust pipe stay bolts
- Front pipe from the center pipe
- 3 nuts and the front pipe
- Oxygen (O₂) sensor at the right side manifold
- 3 mounting nuts and lift off the outside heat insulator
- 6 nuts and lift off the right side manifold and gasket
- Left side heat insulator
- 6 nuts and lift off the left side manifold and gaskets

To install:

3. Install or connect the following:
   - Right manifold with a new gasket. Tighten the nuts to 36 ft. lbs. (46 Nm).
   - Outer insulator
   - Left manifold. Use a new gasket. Tighten the nuts to 36 ft. lbs. (46 Nm).
   - Outer insulator
   - Front exhaust pipe and tighten the manifold-to-pipe nuts to 46 ft. lbs. (62 Nm). Tighten the pipe-to-converter nuts to 32 ft. lbs. (43 Nm).
   - O₂ sensor
   - Undercovers
   - Battery cable

3.0L (2JZ-GE) Engine

1. Before servicing the vehicle, refer to the Precautions Section.
2. Remove or disconnect the following:
   - Negative battery cable
   - Engine cover
   - Air cleaner assembly, as needed for access
   - Oxygen sensor wiring connectors
   - Front exhaust pipe support bracket
   - Front exhaust pipe from manifold
   - Exhaust manifold (8 nuts accessible with a deep-socket wrench)

To install:

3. Position the exhaust manifold in place, with new gaskets. Torque the nuts, in an alternating pattern, to 30 ft. lbs. (40 Nm).

4. Reinstall the remaining components in reverse of the removal procedure.

4.3L (3UZ-FE) Engine

1. Before servicing the vehicle, refer to the Precautions Section.
2. Remove or disconnect the following:
   - Negative battery cable
   - Engine under cover
   - Coolant
   - V-bank cover
   - Air cleaner assembly (if needed for access to exhaust manifold)

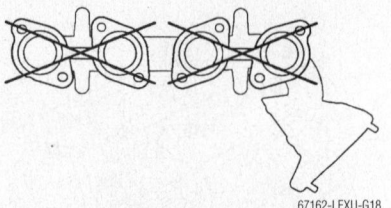

Measuring exhaust manifold warpage

3. Remove cylinder heads. See CYLINDER HEADS in this section.
4. Remove 4 bolts and heat shield from exhaust manifold.
5. Remove 8 nuts and remove exhaust manifold and gasket.

To install:

6. To install, reverse the removal procedure. Install new manifold gasket and new retaining nuts. Torque the exhaust manifold nuts to 32 ft. lbs. (44 Nm).

7. Install the manifold heat shields.

8. Refill the engine cooling system. Start the engine and check for leaks.

Camshaft and Valve Lifters

REMOVAL & INSTALLATION

The following procedures have the valve lash adjuster removal and installation incorporated.

2.5L (4GR-FSE), 3.0L (3GR-FSE) and 3.5L (2GR-FSE) Engines

1. Before servicing the vehicle, refer to the Precautions Section.
2. Drain the cooling system.
3. Drain the engine oil.
4. Relieve the fuel system pressure.
5. Remove or disconnect the following:
   - Negative battery cable. Wait at least 90 seconds before performing any other work
   - Oil filler cap and gasket
   - Radiator cap
   - No. 1 and 2 engine hangers (3GR-FSE)
   - Spark plugs
   - Ventilation valve
   - 4 camshaft position sensors
   - 4 camshaft timing oil control valves
   - Crankshaft position sensor
   - Left and right side oil check valve bolt, oil pipe union and oil pipe
   - Left and right side oil control valve filter and gaskets
   - Cylinder block water drain cocks
   - Oil filter element
   - Water inlet and thermostat assembly

- Rear water by-pass joint
- Cylinder head cover and gaskets
- Timing chain cover, timing chain and timing chain sprockets
- No. 2 chain tensioner assembly

6. Remove the camshaft timing gears and No. 2 chain (right) as follows:

a. While raising the No. 2 chain tensioner, insert a pin of 1.0mm (0.039 in.) into the hole to fix the No. 2 chain tensioner.

b. Hold the hexagonal portion of the camshaft with a wrench, and remove the 2 bolts and 2 camshaft timing gear assemblies.

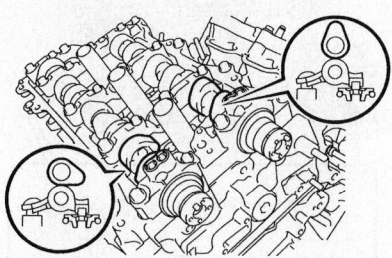

09490_LEXU_G0054

Check that the camshafts are positioned as shown (right side)—2.5L (4GR-FSE), 3.0L (3GR-FSE) and 3.5L (2GR-FSE) Engines

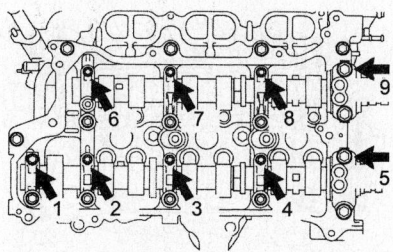

09490_LEXU_G0055

Right side camshaft bearing inner bolt loosening sequence—2.5L (4GR-FSE), 3.0L (3GR-FSE) and 3.5L (2GR-FSE) Engines

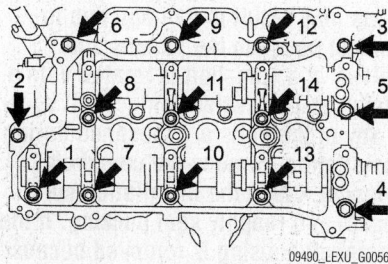

09490_LEXU_G0056

Right side camshaft bearing outer bolt loosening sequence—2.5L (4GR-FSE), 3.0L (3GR-FSE) and 3.5L (2GR-FSE) Engines

c. Remove the No. 2 chain.
- No. 2 chain tensioner

7. Remove the camshaft bearing cap (right) as follows:

a. Check that the camshafts are positioned as shown in the illustration.

b. Uniformly loosen and remove the 9 bearing cap bolts in the sequence shown in the illustration.

c. Uniformly loosen and remove the 14 bearing cap bolts in the sequence shown in the illustration.

d. Remove the 6 bearing caps.
- Camshaft
- No. 2 camshaft
- Camshaft housing sub-assembly (right) by prying between the cylinder head and camshaft housing sub-assembly (right) with a screwdriver.

❋❋ **WARNING**

Be careful not to damage the contact surfaces of the cylinder head and camshaft housing.

8. Remove the camshaft timing gears and No. 2 chain (left) as follows:

a. While pushing down the No. 3 chain tensioner, insert a pin of 1.0mm (0.039 in.) into the hole to fix the No. 3 chain tensioner.

b. Hold the hexagonal portion of the camshaft with a wrench, and remove the 2 bolts and 2 camshaft timing gear assemblies.

c. Remove the No. 2 chain.
- No. 3 chain tensioner

9. Remove the camshaft bearing cap (left) as follows:

a. Check that the camshafts are positioned as shown in the illustration.

b. Uniformly loosen and remove the 8 bearing cap bolts in the sequence shown in the illustration.

c. Uniformly loosen and remove the 13 bearing cap bolts in the sequence shown in the illustration.

d. Remove the 5 bearing caps.
- No. 3 camshaft
- No. 4 camshaft
- Camshaft housing sub-assembly (left) by prying between the cylinder head and camshaft housing sub-assembly (left) with a screwdriver.

❋❋ **WARNING**

Be careful not to damage the contact surfaces of the cylinder head and camshaft housing.

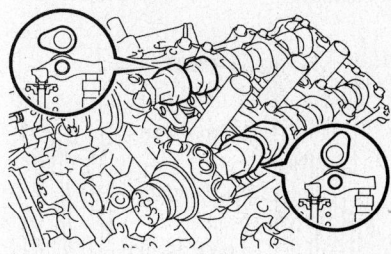

09490_LEXU_G0057

Check that the camshafts are positioned as shown (left side)—2.5L (4GR-FSE), 3.0L (3GR-FSE) and 3.5L (2GR-FSE) Engines

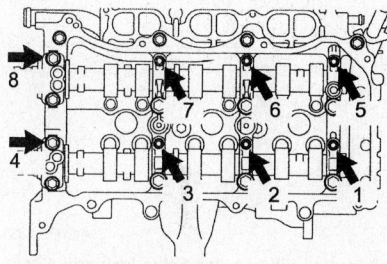

09490_LEXU_G0058

Left side camshaft bearing inner bolt loosening sequence—2.5L (4GR-FSE), 3.0L (3GR-FSE) and 3.5L (2GR-FSE) Engines

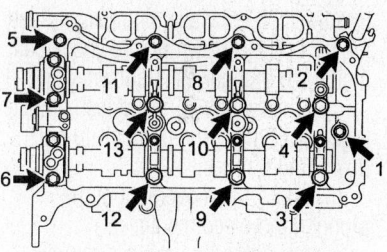

09490_LEXU_G0059

Left side camshaft bearing outer bolt loosening sequence—2.5L (4GR-FSE), 3.0L (3GR-FSE) and 3.5L (2GR-FSE) Engines

- Valve rocker arms. Arrange the removed rocker arms in the correct order.
- Valve lash adjusters. Arrange the removed valve lash adjusters in the correct order.

To install:

10. Install valve lash adjusters as follows:

❋❋ **WARNING**

Keep the lash adjuster free of dirt and foreign objects. Only use clean engine oil.

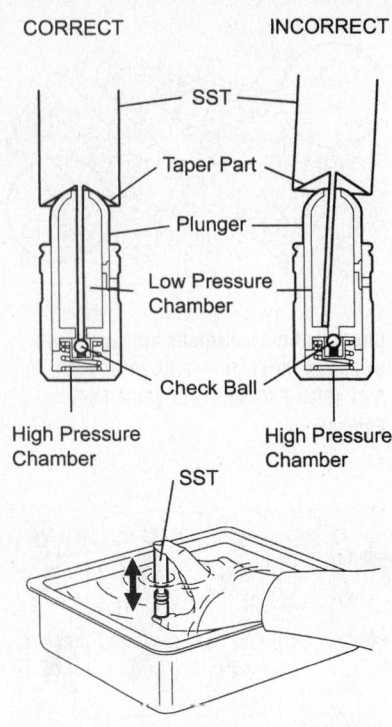

09490_LEXU_G0060

Bleeding air from the valve lash adjuster assembly—2.5L (4GR-FSE), 3.0L (3GR-FSE) and 3.5L (2GR-FSE) Engines

   a. Place the lash adjuster into a container filled with engine oil.

   b. Insert the SST's tip (09276-75010) into the lash adjuster's plunger and use the tip to press down on the check ball inside the plunger as is shown.

   c. Squeeze the SST and lash adjuster together to move the plunger up and down 5 to 6 times.

   d. Check the movement of the plunger and bleed the air.

   e. After bleeding the air, remove SST. Then, try to quickly and firmly press the plunger with a finger. If the result is not as specified, replace the lash adjuster.

   f. Install the lash adjusters.

➡ **Install the lash adjuster to the same place it was removed from.**

11. Install No. 1 valve rocker arm assembly as follows:

   a. Apply engine oil to the lash adjuster tip and valve stem cap end.

   b. Make sure that the valve rocker arms are installed as shown in the illustration.

12. Install right side camshaft bearing cap as follows:

   a. Apply engine oil to the camshaft journals, camshaft housing and bearing caps.

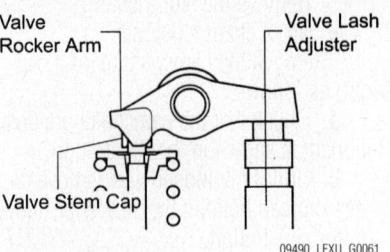

09490_LEXU_G0061

Correct installation of the valve rocker arm assembly—2.5L (4GR-FSE), 3.0L (3GR-FSE) and 3.5L (2GR-FSE) Engines

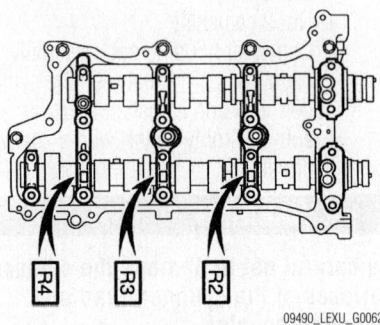

09490_LEXU_G0062

Make sure of the marks and numbers on the right camshaft bearing caps and place them in each proper position and direction.—2.5L (4GR-FSE), 3.0L (3GR-FSE) and 3.5L (2GR-FSE) Engines

   b. Install the camshaft and camshaft No. 2 to the right camshaft housing.

   c. Make sure of the marks and numbers on the camshaft bearing caps and place them in each proper position and direction.

   d. Temporarily tighten the 9 bolts to 7 ft. lbs. (10 Nm) in the order shown in the illustration.

13. Install the right side camshaft housing assembly as follows:

   a. Make sure that the valve rocker arm is installed as shown in the illustration.

   b. Apply seal packing in a continuous line approximately 3.5 to 4.0mm (0.138 to 0.158 inches) wide.

➡ **Remove any oil from the contact surface. Install the camshaft housing assembly within 3 minutes. Do not start the engine for at least 2 hours after installing.**

   c. Install the right camshaft housing and tighten the 14 bolts to 21 ft. lbs. (28 Nm) in the order shown in the illustration.

✳✳ WARNING

When installing the camshaft housing, it is necessary to correctly position the camshafts as shown in the illustration.

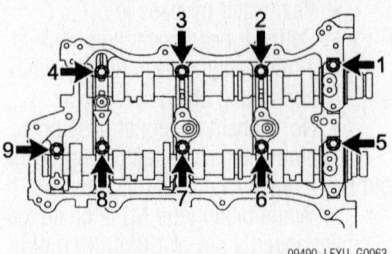

09490_LEXU_G0063

Temporarily tighten the 9 bolts in the order shown (right side).—2.5L (4GR-FSE), 3.0L (3GR-FSE) and 3.5L (2GR-FSE) Engines

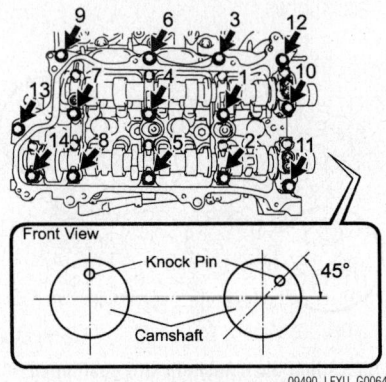

09490_LEXU_G0064

Right camshaft housing assembly bolt tightening sequence and camshaft positioning.—2.5L (4GR-FSE), 3.0L (3GR-FSE) and 3.5L (2GR-FSE) Engines

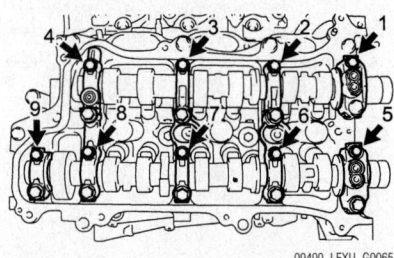

09490_LEXU_G0065

Right side camshaft bearing inner bolt tightening sequence—2.5L (4GR-FSE), 3.0L (3GR-FSE) and 3.5L (2GR-FSE) Engines

Failure to correctly position these parts may result in damage due to contact between the pistons and valves. If a camshaft is rotated with a piston at TDC, valve contact will occur. If any of the bolts are loosened during installation, remove the camshaft housing, clean the installation surfaces, and reapply seal packing. If the camshaft housing is removed because any of the bolts are loosened during installation, make sure that the previously applied seal packing does not enter any oil passages.

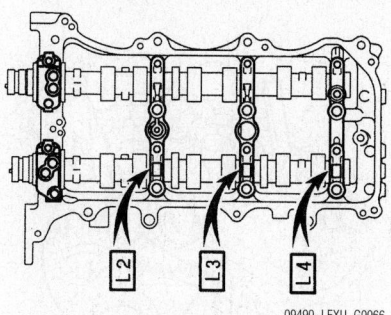

09490_LEXU_G0066

Make sure of the marks and numbers on the left camshaft bearing caps and place them in each proper position and direction.—2.5L (4GR-FSE), 3.0L (3GR-FSE) and 3.5L (2GR-FSE) Engines

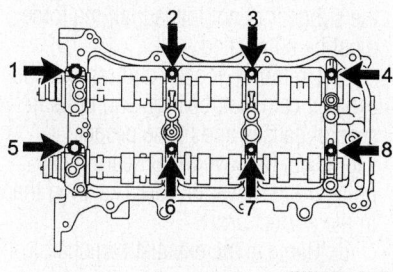

09490_LEXU_G0067

Temporarily tighten the 8 bolts in the order shown (left side).—2.5L (4GR-FSE), 3.0L (3GR-FSE) and 3.5L (2GR-FSE) Engines

d. Tighten the 9 bolts to 12 ft. lbs. (16 Nm) in the order shown in the illustration.

14. Install left side camshaft bearing cap as follows:

a. Apply engine oil to the camshaft journals, camshaft housing and bearing caps.

b. Install camshaft No. 3 and camshaft No. 4 to the left camshaft housing.

c. Make sure the marks and numbers on the camshaft bearing caps and place them in each proper position and direction.

d. Temporarily tighten the 8 bolts to 7 ft. lbs. (10 Nm) in the order shown in the illustration.

15. Install the left side camshaft housing assembly as follows:

a. Make sure that the valve rocker arm is installed as shown in the illustration.

b. Apply seal packing in a continuous line approximately 5.0 to 5.5mm (0.197 to 0.217 inches) wide.

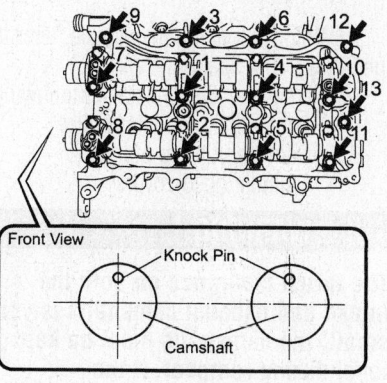

09490_LEXU_G0068

Left camshaft housing assembly bolt tightening sequence and camshaft positioning.—2.5L (4GR-FSE), 3.0L (3GR-FSE) and 3.5L (2GR-FSE) Engines

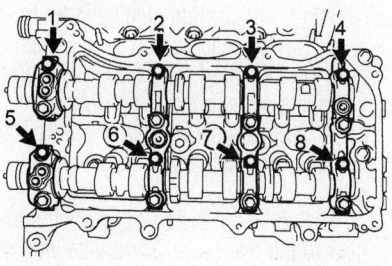

09490_LEXU_G0069

Left side camshaft bearing inner bolt tightening sequence—2.5L (4GR-FSE), 3.0L (3GR-FSE) and 3.5L (2GR-FSE) Engines

➡Remove any oil from the contact surface. Install the camshaft housing assembly within 3 minutes. Do not start the engine for at least 2 hours after installing.

c. Install the left camshaft housing and tighten the 13 bolts to 21 ft. lbs. (28 Nm) in the order shown in the illustration.

✳✳ WARNING

When installing the camshaft housing, it is necessary to correctly position the camshafts as shown in the illustration. Failure to correctly position these parts may result in damage due to contact between the pistons and valves. If a camshaft is rotated with a piston at TDC, valve contact will occur. If any of the bolts are loosened during installation, remove the camshaft housing, clean the installation surfaces, and reapply seal packing. If the camshaft housing is removed because any of the bolts

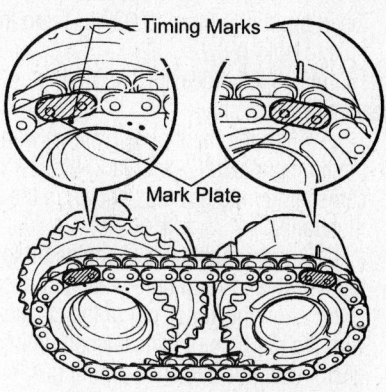

09490_LEXU_G0070

Align the mark plate with the timing marks of the camshaft timing gears (right side shown, left side similar)—2.5L (4GR-FSE), 3.0L (3GR-FSE) and 3.5L (2GR-FSE) Engines

are loosened during installation, make sure that the previously applied seal packing does not enter any oil passages.

d. Tighten the 8 bolts to 12 ft. lbs. (16 Nm) in the order shown in the illustration.

16. Install No. 2 chain tensioner assembly as follows:

a. Install the No. 2 chain tensioner with the bolt. Torque the bolt to 15 ft. lbs. (21 Nm).

b. While pushing in the tensioner, insert a pin of 1.0mm (0.039 in.) into the hole to fix it.

17. Install right side camshaft timing gears and No. 2 chain.

a. Align the mark plate (yellow) with the timing marks (1-dot mark) of the camshaft timing gears as shown in the illustration.

b. Apply a light coat of engine oil to the bolt threads and bolt-seating surface.

c. Align the knock pin of the camshaft with the pin hole of the camshaft timing gear. Install the camshaft timing gear and camshaft timing exhaust gear (right) with the No. 2 chain installed.

d. Hold the hexagonal portion of the camshaft with a wrench, and tighten the 2 bolts to 74 ft. lbs. (100 Nm)

e. Remove the pin from the chain tensioner.

18. Install No. 3 chain tensioner assembly as follows:

a. Install the chain tensioner with the bolt. Torque the bolt to 15 ft. lbs. (21 Nm).

b. While pushing in the tensioner, insert a pin of 1.0mm (0.039 in.) into the hole to hold it.

19. Install left side camshaft timing gears and No. 2 chain.

a. Align the mark plate (yellow) with the timing marks (2-dot mark) of the camshaft timing gears as shown in the illustration.

b. Apply a light coat of engine oil to the bolt threads and bolt-seating surface.

c. Align the knock pin of the camshaft with the pin hole of the camshaft timing gear. Install the camshaft timing gear and camshaft timing exhaust gear (left) with the No. 2 chain installed.

d. Hold the hexagonal portion of the camshaft with a wrench, and tighten the 2 bolts to 74 ft. lbs. (100 Nm).

e. Remove the pin from the chain tensioner.

20. Install or connect the following:

• Timing chain cover, timing chain and timing chain sprockets
• Cylinder head cover and gaskets
• New gaskets, O-ring and rear water by-pass joint. Torque to 7 ft. lbs. (10 Nm).
• New gasket, water inlet and thermostat assembly. Torque to 7 ft. lbs. (10 Nm).
• Oil filter element
• Adhesive sealer, left and right side cylinder block water drain cocks. Torque to 22 ft. lbs. (30 Nm).
• Left and right side oil control valve filter and new gaskets
• Left and right side oil check valve bolt, oil pipe union and oil pipe. Torque to 44 ft. lbs. (60 Nm).
• Crankshaft position sensor. Torque to 7 ft. lbs. (10 Nm).
• 4 camshaft timing oil control valves. Torque to 7 ft. lbs. (10 Nm).
• 4 camshaft position sensors. Torque to 7 ft. lbs. (10 Nm).
• Adhesive and ventilation valve. Torque to 20 ft. lbs. (27 Nm).
• Spark plugs. Torque to 18 ft. lbs. (24 Nm) for 2.5L and 3.5L engines; 13 ft. lbs. (18 Nm) for 3.0L engine.
• Engine hangers (if equipped). Torque to 24 ft. lbs. (33 Nm).
• Radiator cap
• Oil filler cap and gasket
• Negative battery cable

21. Refill the coolant and engine oil. Start the engine and check for leaks or abnormal conditions. Perform and road test. Then, recheck for leaks and recheck fluid levels.

3.0L (1MZ-FE) Engine

1. Before servicing the vehicle, refer to the Precautions Section.

2. Remove or disconnect the following:
• Timing belt and idler pulley
• Camshaft timing pulleys
• Cylinder head covers

❊❊ WARNING

The thrust clearance on both the intake and exhaust camshafts is very small, the camshafts must be kept level during removal. If the camshafts are removed without being kept level, the camshaft may be caught in the cylinder head causing the head to break or the camshaft to seize.

3. To remove the exhaust and intake camshafts from the right side cylinder head:

a. Turn the camshaft with a wrench until the 2 pointed marks drive and driven gears are aligned. (The right camshaft gears have 2 marks apiece; the left side camshaft gears have 1 mark each.)

b. Secure the exhaust camshaft subgear to the main gear using a service bolt. A bolt 0.63–0.79 in. (16–20mm)

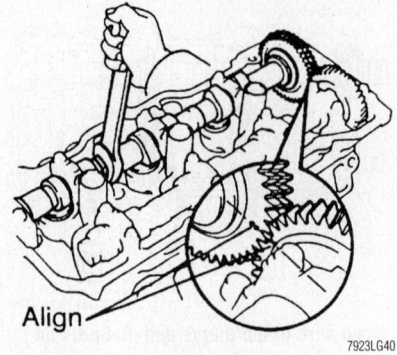

Align

Aligning the right side camshaft timing marks—3.0L (1MZ-FE) engine

long with a 6mm thread diameter and a 1mm pitch is recommended. When removing the exhaust camshaft be sure the subgear is not loaded; all the force must be eliminated.

c. Uniformly loosen and remove the exhaust camshaft bearing cap bolts in several passes and in the proper sequence. Remove the 8 bearing cap bolts and remove the caps, keeping them in the correct order.

d. Remove the exhaust camshaft from the engine.

e. Uniformly loosen and remove the

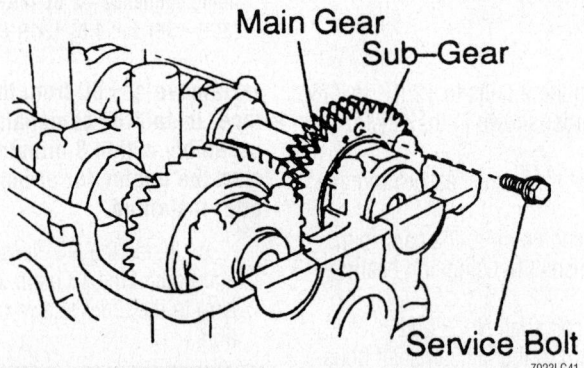

Main Gear
Sub-Gear
Service Bolt

Securing the subgear and driven gear, right side—3.0L (1MZ-FE) engine

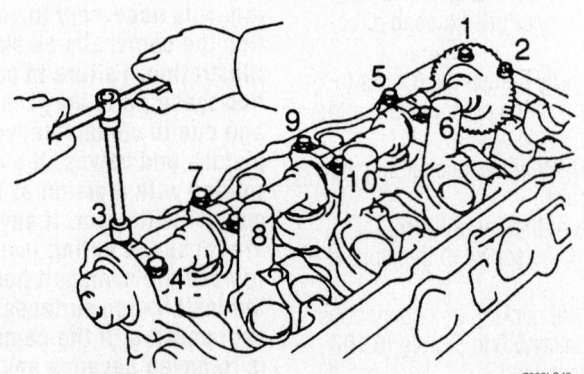

Right exhaust camshaft bearing loosening sequence—3.0L (1MZ-FE) engine

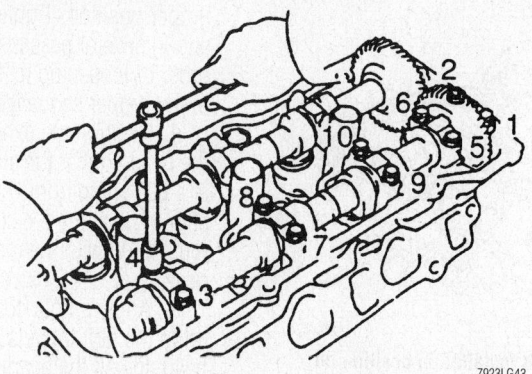

Right intake camshaft bearing loosening sequence—3.0L (1MZ-FE) engine

7923LG43

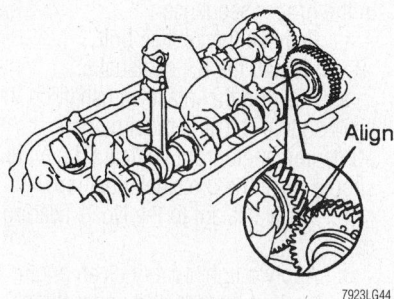

Aligning the left side camshaft timing marks—3.0L (1MZ-FE) engine

7923LG44

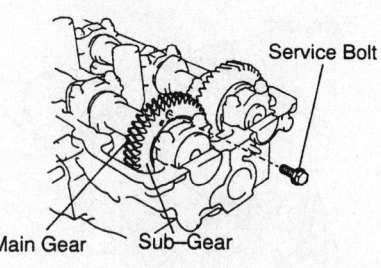

Securing the subgear and driven gear, left side—3.0L (1MZ-FE) engine

7923LG45

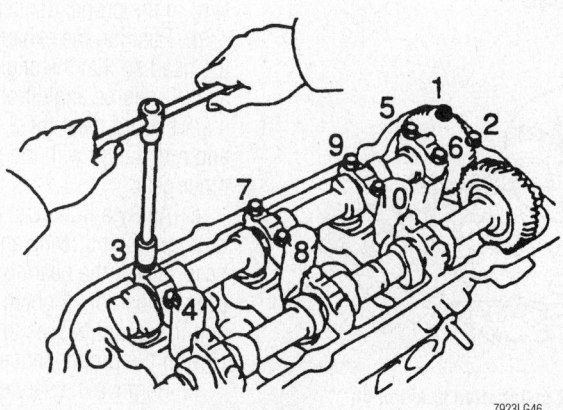

Left intake camshaft bearing cap bolt loosening sequence—3.0L (1MZ-FE) engine

7923LG46

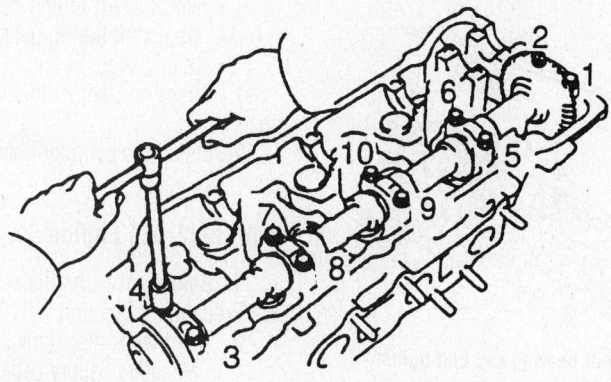

Left exhaust camshaft bearing cap bolt loosening sequence—3.0L (1MZ-FE) engine

7923LG47

10 bearing cap bolts in several passes, in the proper sequence. Remove the bearing caps, keeping them in order, remove the oil seal, then lift out the intake camshaft.

4. To remove the exhaust and intake camshafts from the left side cylinder head:

a. Turn the camshaft with a wrench until the pointed marks on the drive and driven gears are aligned. (The right camshaft gears have 2 marks apiece; the left side camshaft gears have 1 mark each.)

b. Secure the exhaust camshaft sub-gear to the main gear using a service bolt. A bolt 0.63–0.79 in. (16–20mm) long with a 6mm thread diameter and a 1mm pitch is recommended. When removing the exhaust camshaft be sure the subgear is not loaded; all the force must be eliminated.

c. Uniformly loosen and remove the exhaust camshaft bearing cap bolts in several passes and in the proper sequence. Remove the 8 bearing cap bolts and remove the caps. Keep the caps in the correct order.

d. Remove the exhaust camshaft from the engine.

e. Uniformly loosen and remove the 10 bearing cap bolts in several passes, in the proper sequence. Remove the bearing caps, keeping them in order, remove the oil seal, then lift out the intake camshaft.

5. Remove the valve lash adjuster shims and hydraulic lash adjusters. Identify each lash adjuster and shim as it is removed so it can be reinstalled in the same position. If the lash adjusters are to be reused, store them upside down in a sealed container.

To install:

6. Install or connect the following:

• Valve lash adjusters and shims. Check valve clearance and replace the shims as necessary.

➡**Before installing the camshafts in either cylinder head, apply multi-purpose grease to the thrust portions of each camshaft.**

7. To install the right camshafts:

a. Position the intake camshaft on the head so that the alignment marks are at a 90° angle from vertical. The mark should be at the 3 o'clock position.

b. Apply sealant to the No. 1 bearing cap.

c. Apply a light coat of clean engine oil to the bolt threads and under the bolt head. Install the bearing caps to their

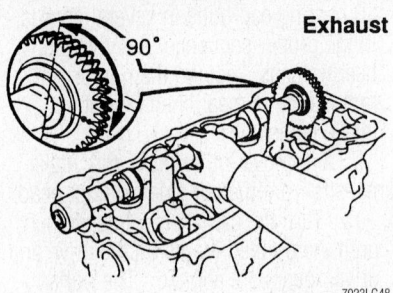

Exhaust camshaft installation position on the right cylinder head—3.0L (1MZ-FE) engine

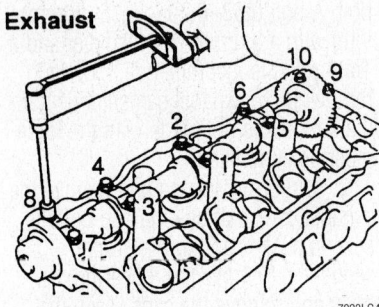

Exhaust camshaft bearing cap bolt tightening sequence on the right cylinder head—3.0L (1MZ-FE) engine

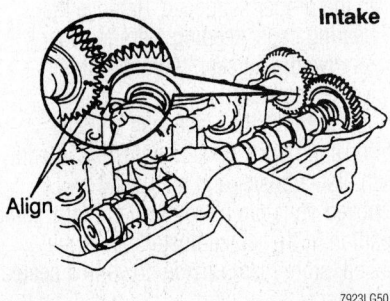

Intake camshaft installation position on the right cylinder head—3.0L (1MZ-FE) engine

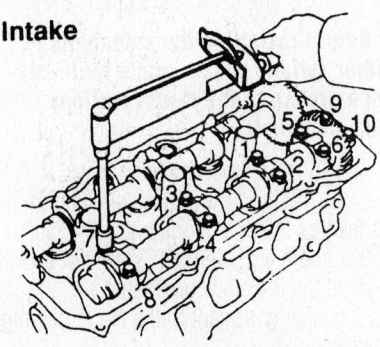

Intake camshaft bearing cap bolt tightening sequence on the right cylinder head—3.0L (1MZ-FE) engine

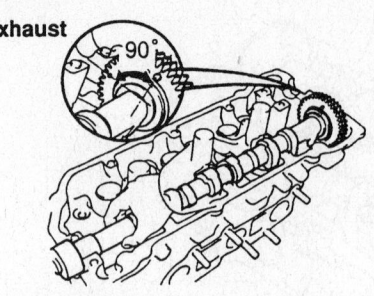

Exhaust camshaft installation position on the left cylinder head—3.0L (1MZ-FE) engine

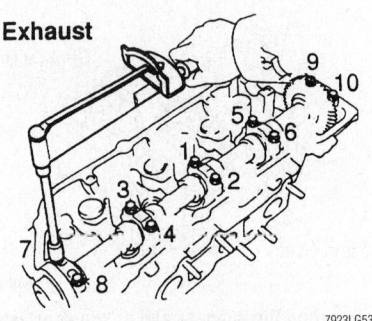

Exhaust camshaft bearing cap bolt tightening sequence on the left cylinder head—3.0L (1MZ-FE) engine

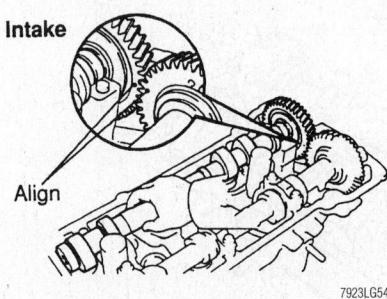

Intake camshaft installation position on the left cylinder head—3.0L (1MZ-FE) engine

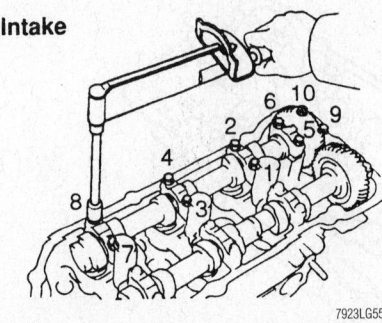

Intake camshaft bearing cap bolt tightening sequence on the left cylinder head—3.0L (1MZ-FE) engine

proper position. Tighten the bolts evenly and in several passes in the reverse order of loosening to 12 ft. lbs. (16 Nm) in the proper sequence.

d. Position the exhaust camshaft on the head so that the alignment marks are at a 90° angle from vertical. The mark should be at the 9 o'clock position and must align with the marks on the other gear.

e. Apply a light coat of clean engine oil to the bolt threads and under the bolt head. Install the bearing caps to their proper position. Tighten the bolts evenly and in several passes in the reverse order of loosening to 12 ft. lbs. (16 Nm) in the proper sequence.

f. Remove the service bolt.

8. To install the left camshafts:

a. Position the intake camshaft on the head so that the alignment mark is at a 90 degree angle from vertical. The mark should be at the 9 o'clock position.

b. Apply sealant to the No. 1 bearing cap.

c. Apply a light coat of clean engine oil to the bolt threads and under the bolt head. Install the bearing caps to their proper position. Tighten the bolts evenly and in several passes to 12 ft. lbs. (16 Nm) in the proper sequence.

d. Position the exhaust camshaft on the head so that the alignment marks are at a 90-degree angle from vertical. The mark should be at the 3 o'clock position and must align with the marks on the other gear.

e. Apply a light coat of clean engine oil to the bolt threads and under the bolt head. Install the bearing caps to their proper position. Tighten the bolts evenly and in several passes to 12 ft. lbs. (16 Nm) in the proper sequence.

f. Remove the service bolt.

9. Apply multi-purpose grease to new camshaft oil seals. Install the seals.

10. Install or connect the following:
- No. 3 (rear) timing belt cover
- Camshaft timing gears
- Idler pulley, timing belt and covers

11. Check and adjust the valve clearance.

12. Install the cylinder head (valve) covers.

3.0L (2JZ-GE) Engine

1. Before servicing the vehicle, refer to the Precautions Section.

2. Remove or disconnect the following:
- Negative battery cable from the battery

- Timing belt from the engine
- Cylinder head covers

3. Make reference marks on the timing belt to the crankshaft and idler pulley timing marks.

4. Remove the timing belt. See TIMING BELT.

✳✳ WARNING

Do NOT allow anything to drop into the lower part of the timing belt cover. Do NOT allow the timing belt to come in contact with any oil, water or dust.

5. Remove both the exhaust and intake camshaft pulleys. See procedure under TIMING BELT.

✳✳ WARNING

Since the thrust clearance of the camshaft is small, the camshaft must be kept level during removal. If not, the portion of the cylinder head receiving the camshaft thrust may crack or be otherwise damaged, causing the camshaft to later seize or break. Follow the camshaft removal procedure carefully as given in this section.

6. With a 5mm hex wrench, remove the No. 3 camshaft bearing cap bolts. Then, uniformly loosen and remove the 4 camshaft bearing cap bolts, located just rear of the camshaft pulley locations.

7. Carefully pry out the No. 1 and No. 3 camshaft bearing caps and oil seals.

8. Uniformly loosen and remove the 12 camshaft bearing cap bolts, in several passes, and following the sequence illustrated. Remove the 6 No. 2 bearing caps and the camshafts from the engine.

To install:

9. Apply engine oil to the thrust portion of the camshaft.

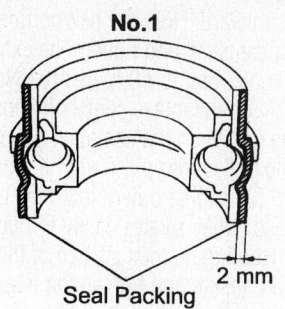

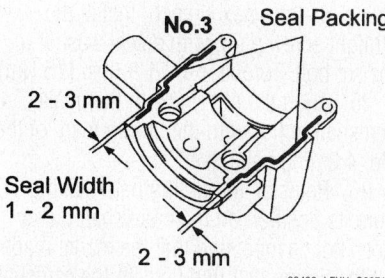

09490_LEXU_G0071

Installing packing onto camshaft bearing caps—3.0L (2JZ-GE) Engine

10. Place the camshaft into the cylinder head, with the cam lobe facing up (cam lobe mark on the exhaust camshaft will be at about 11 o'clock and the cam lobe mark on the intake camshaft will be at 3 o'clock).

11. Place the No. 3 and No. 7 camshaft journal bearing caps into their positions. Apply engine oil onto thread and under the heads of the bearing cap bolts. Temporarily tighten these bearing cap bolts, uniformly and alternately, until they are snug with the cylinder head.

7923LG60

Camshaft bearing cap bolt tightening sequence—3.0L (2JZ-GE) engine

12. Apply grease to new camshaft oil seal lip. Install both oil seals onto the camshafts.

13. Clean the installed surface of the No. 1 and No. 3 camshaft bearing caps and cylinder head. Apply seal packing along outer edge of bearing cap, outside of the bolt hole.

14. Install the other bearing caps in proper locations. Apply a light coat of engine oil to bolt threads and under the bolt heads. Install and uniformly tighten the 14 bearing cap bolts on one side, in several passes, following sequence shown. Torque these bolts to 15 ft. lbs. (20 Nm).

15. Using a 5mm hex wrench, tighten bearing cap bolts for caps No. 2 and No. 3 to 44 inch lbs. (5 Nm).

16. With a special oil seal driver, install the 2 oil seals as far as possible by pushing from front end of camshafts.

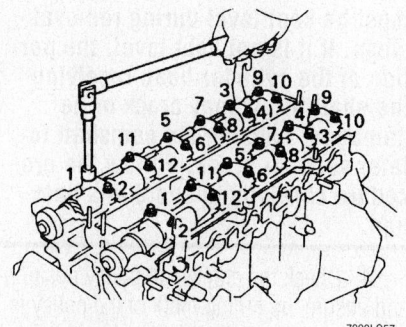

7923LG57

Camshaft bearing cap bolt removal sequence—3.0L (2JZ-GE) engine

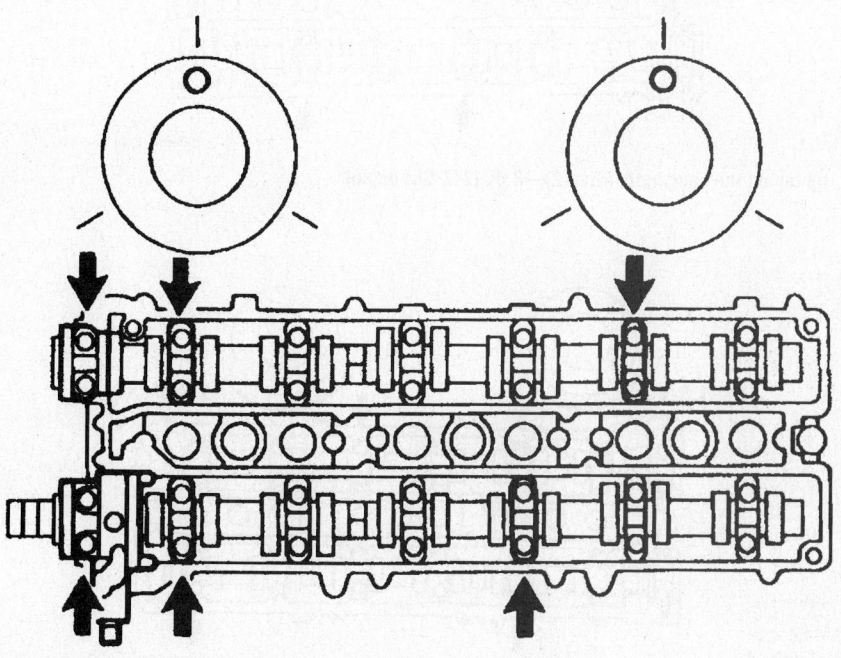

09490_LEXU_G0072

Tightening the camshafts (Step 1)—3.0L (2JZ-GE) engine

17. Rotate camshaft with a wrench on the hex section and bring the forward straight pin up to the 12 o'clock position. Loosen the 12 bearing cap bolts, highlighted in the Step 1 figure, until bolts are just finger-tight. Now, retighten these bolts, in several passes, to 15 ft. lbs. (20 Nm).

18. Turn camshaft 1/3 of a revolution clockwise, so straight pin is at about the 4 o'clock position. Now, loosen the 8 bolts, as highlighted in Step 2 illustration, until they are just finger-tight. Now, retighten these bolts, in several passes, to 15 ft. lbs. (20 Nm).

19. Turn the camshaft an additional 1/3 of a revolution, so the straight pin in now at about the 8 o'clock position. Loosen the 8 bearing cap bolts, as highlighted in Step 3 illustration, until they are just finger-tight. Now, retighten these bolts, in several passes, to 15 ft. lbs. (20 Nm).

20. Check and adjust the valve lash. See VALVE LASH

21. Install the No. 4 timing belt cover.

22. Align the exhaust camshaft knock pin with the groove in the camshaft pulley and slide on the pulley. Ensure the front mark is forward. Hold the hex portion of the exhaust camshaft and tighten the exhaust pulley center bolt to 60 ft. lbs. (81 Nm).

23. Align the intake camshaft knock pin with the groove in the camshaft pulley and slide the pulley into place until it touches bottom. Check that outer circumference of the intake pulley rotates easily through 30°.

24. Holding the hex portion of the intake camshaft, install and tighten the intake camshaft pulley bolt to 60 ft. lbs. (81 Nm). Using a 14mm hex wrench, install the straight screw plug, with a seal washer to the set bolt. Torque it to 11 ft. lbs. (15 Nm).

25. Align the dot mark on the intake camshaft pulley with the timing mark of the No. 4 timing belt cover.

26. Reinstall the timing belt, making sure its position on the crankshaft pulley does not change, and that the marks made on the timing belt line up with the camshaft pulley timing marks.

27. Clean any old packing material from outside of the 2 rearmost bearing caps and install new packing in its place.

28. Reinstall remaining components in reverse of the removal procedure.

4.3L (3UZ-FE) Engine

1. Before servicing the vehicle, refer to the Precautions Section.

2. Relieve the fuel pressure from the fuel lines.

3. Remove or disconnect the following:

- Engine under cover
- Drain engine coolant and engine oil
- V-bank cover
- Air cleaner assembly
- Intake air pipe
- Radiator
- Throttle body
- Upper and lower intake manifold assembly
- Timing belt from camshaft pulleys
- Camshaft position sensor and LH timing belt rear plates
- RH timing belt rear plates

✳✳ WARNING

Do NOT drop anything inside timing belt cover during this procedure. Keep oil, water and dust from timing belt.

- Power steering pump from engine mount (Do NOT disconnect hoses)
- Catalytic converters
- Water inlet housing assembly
- Water bypass pipe, front bypass joint, and rear bypass joint
- Ignition coils
- Variable valve timing (VVT) sensors
- Engine hangers
- Oil dipsticks and guides for engine oil and transmission fluid
- Cylinder head covers
- Spark plugs
- Semi-circular plugs and camshaft housing plugs
- Camshaft timing oil control valve

✳✳ WARNING

Since the thrust clearance of the camshaft is small, the camshaft must be kept level during removal steps. If it is not kept level, the portion of the cylinder head receiving the shaft thrust may crack or be damaged, causing the camshaft to later seize or break. Follow the procedure carefully to avoid this damage.

4. Check the crankshaft pulley position and ensure the timing mark of the pulley is aligned with the centers of the crankshaft pulley bolt and the No. 2 timing belt idler pulley bolt.

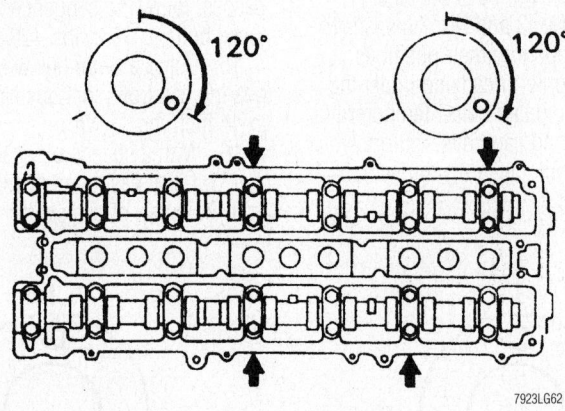

Tightening the camshafts (Step 2)—3.0L (2JZ-GE) engine

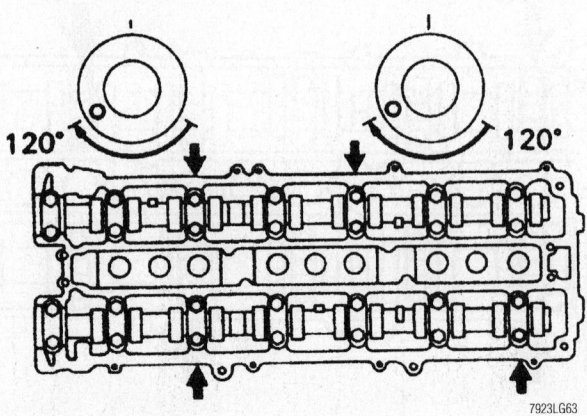

Tightening the camshafts (Step 3)—3.0L (2JZ-GE) engine

✳✳ WARNING

Having the crankshaft pulley at the wrong angle can cause the piston head and valve head to come into contact with each other during camshaft removal. Always set the crankshaft pulley at the described angle.

5. Using a special wrench, rotate the camshaft timing tube from left to right about 2-3 times, within only a 25° range of movement. Use a waste cloth to collect oil from the camshaft timing oil control valve installation hole.

6. Remove the LH camshafts first. With a hex wrench on the hex portion of the camshaft, rotate so that a 6mm service bolt can be inserted into the bolt hole in the rear face of the camshaft pulley into order to secure the camshaft in place. The bolt should be about 0.63-0.79 in. (16-20mm) long.

7. Align the timing mark (2 dots) of the camshaft drive gear by turning the camshaft with a hex wrench until the timing mark aligns.

8. Now, uniformly loosen the 22 camshaft bearing cap bolts, in several passes, following the sequence shown.

9. Remove the 22 bearing cap bolts, 4 seal washers, oil feed pipe, 9 bearing caps, the camshaft housing plug, the oil control valve filter, and both LH camshafts. Keep all parts in order for proper reinstallation.

10. Remove the RH camshafts. With a hex wrench on the hex portion of the camshaft, rotate so that a 6mm service bolt can be inserted into the bolt hole in the rear face of the camshaft pulley into order to secure the camshaft in place. The bolt should be about 0.63-0.79 in. (16-20mm) long.

11. Align the timing mark (1 dot) of the camshaft main gear about 10° angle by turning the camshaft with a hex wrench until the timing mark aligns.

12. Now, uniformly loosen the 22 camshaft bearing cap bolts, in several passes, following the sequence shown.

13. Remove the 22 bearing cap bolts, 4 seal washers, oil feed pipe, 9 bearing caps, the camshaft housing plug, the oil control valve filter, and both LH camshafts. Keep all parts in order for proper reinstallation.

To install:

✳✳ WARNING

Since the thrust clearance of the camshaft is small, the camshaft must be kept level during removal steps. If it is not kept level, the portion of the cylinder head receiving the shaft thrust may crack or be damaged, causing the camshaft to later seize or break. Follow the procedure carefully to avoid this damage.

14. Ensure the crankshaft pulley is in position so that its timing mark is in line with the centers of the pulley bolt and idler pulley bolt.

✳✳ WARNING

Having the crankshaft pulley at the wrong angle can cause the piston head and valve head to come into contact with each other during camshaft removal. Always set the crankshaft pulley at the described angle.

15. Apply grease to the thrust portion of the LH intake and exhaust camshafts. Align the timing marks (2 dots) of the camshaft drive and driven main gears. Place the camshafts into the LH cylinder head.

16. Apply new seal packing material around the opening of the camshaft housing plug. Install the camshaft housing plug and the oil control valve filter into the cylinder head, as shown.

17. Remove any old packing material, then install new packing material around the

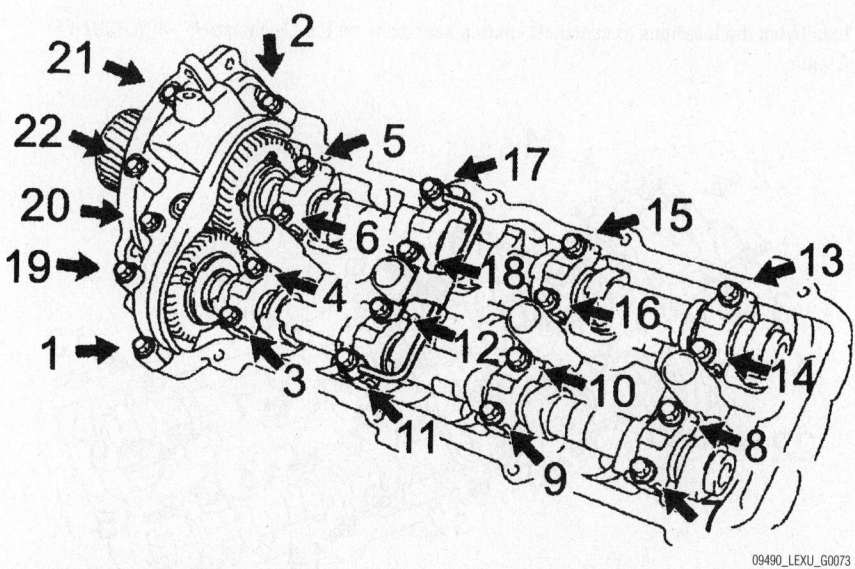

09490_LEXU_G0073

LH camshaft bearing cap bolt loosening sequence—4.3L (3UZ-FE) Engine

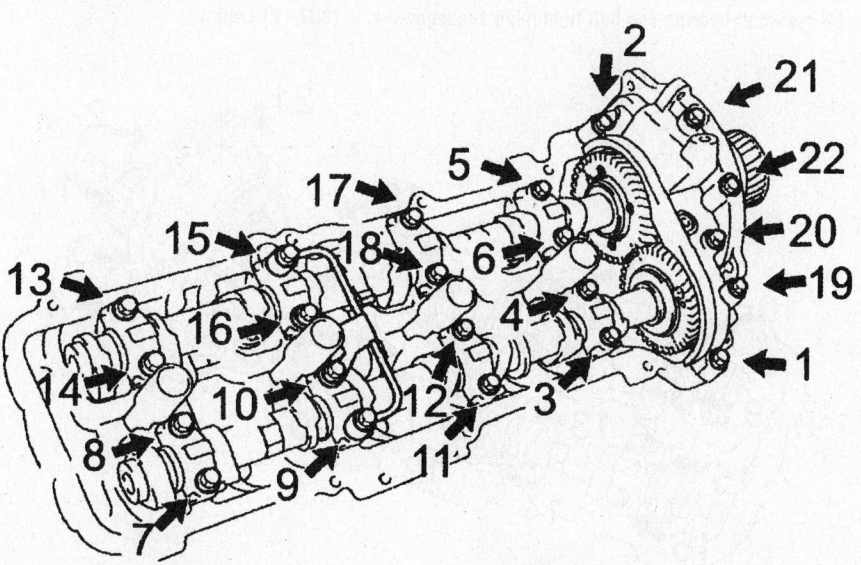

09490_LEXU_G0074

RH camshaft bearing cap bolt loosening sequence—4.3L (3UZ-FE) Engine

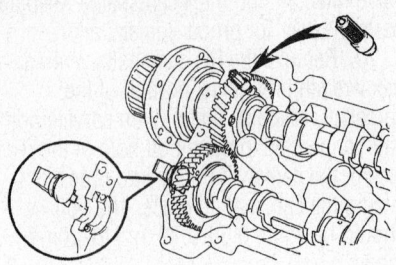

09490_LEXU_G0075

Installing the camshaft housing plug and oil control valve filter for the LH camshaft—4.3L (3UZ-FE) Engine

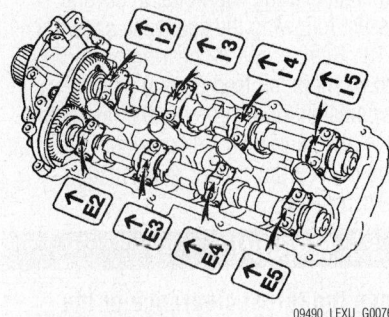

09490_LEXU_G0076

Installing the camshaft bearing caps in sequence on the camshaft (left shown)—4.3L (3UZ-FE) Engine

mounting edge (not in the grooves) of the front bearing cap.

18. Position the front bearing cap in place. This will determine the thrust portion of the camshaft.

    a. Install the other bearing caps, in sequence shown, with the arrow marks facing forward.

    b. Push the camshaft oil seal into place by pushing from the front of the engine. Install a new seal washer to the front bearing cap bolts.

    c. Apply a light coat of oil to bearing cap bolt threads and under the heads of the bearing cap bolts "D" and "E", as shown. Do NOT apply engine oil under the heads of bearing cap bolts "A", "B" and "C".

    d. Bolt lengths vary for each bearing cap. Refer to the illustration for each of the following bolts:

- Bolt "A" with seal washer is 3.70 in. (94mm)
- Bolt "B" with seal washer is 2.83 in. (72mm)
- Bolt "C" is 0.98 in. (25mm)
- Bolt "D" is 2.05 in. (52mm)
- Bolt "E" is 1.50 in. (38mm)

19. Install the oil feed pipe and the 22 bearing cap bolts in their respective loca-

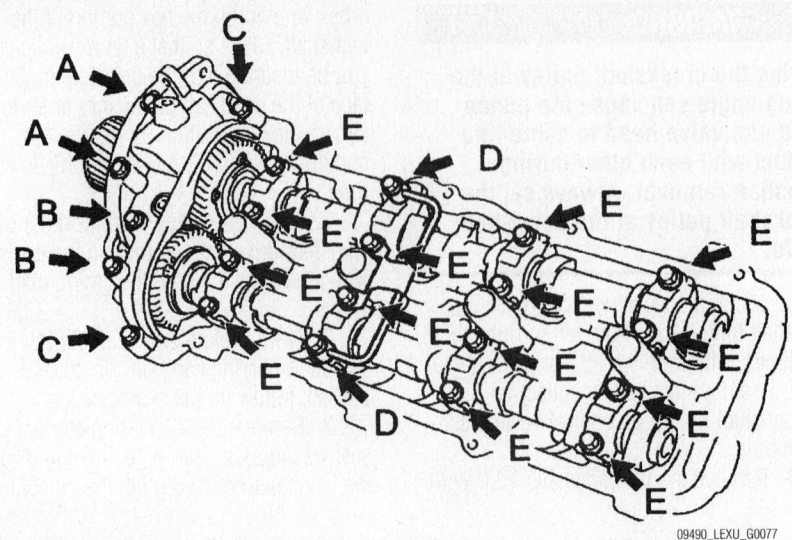

09490_LEXU_G0077

Identifying the locations of camshaft bearing caps bolts on the LH camshaft—4.3L (3UZ-FE) Engine

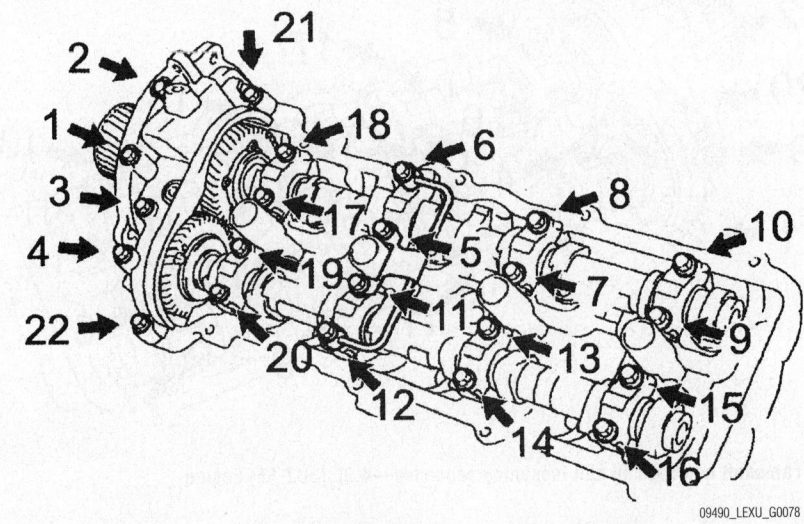

09490_LEXU_G0078

LH camshaft bearing cap bolt tightening sequence—4.3L (3UZ-FE) Engine

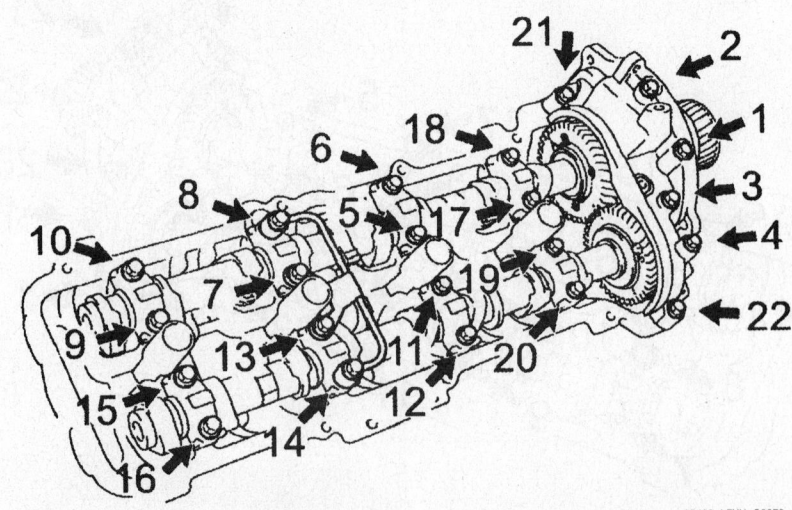

09490_LEXU_G0079

RH camshaft bearing cap bolt tightening sequence—4.3L (3UZ-FE) Engine

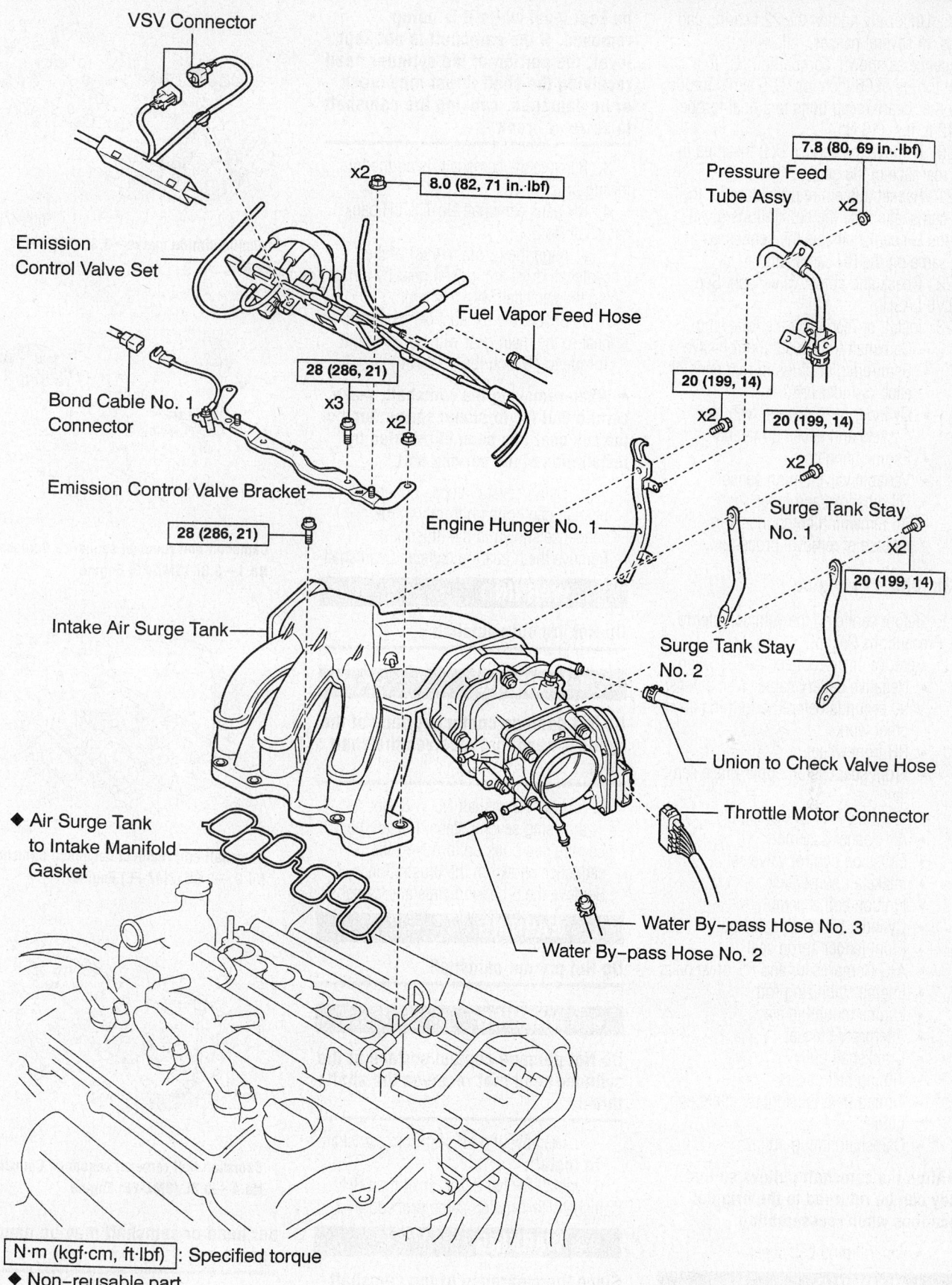

VSV Connector

7.8 (80, 69 in.·lbf)

Pressure Feed
Tube Assy

x2

8.0 (82, 71 in.·lbf)

x2

Emission
Control Valve Set

Fuel Vapor Feed Hose

28 (286, 21)

Bond Cable No. 1
Connector

x3

x2

20 (199, 14)

x2

20 (199, 14)

x2

Emission Control Valve Bracket

Engine Hunger No. 1

Surge Tank Stay
No. 1

x2

28 (286, 21)

20 (199, 14)

Intake Air Surge Tank

Surge Tank Stay
No. 2

Union to Check Valve Hose

Throttle Motor Connector

◆ Air Surge Tank
to Intake Manifold
Gasket

Water By-pass Hose No. 3

Water By-pass Hose No. 2

N·m (kgf·cm, ft·lbf) : Specified torque

◆ Non–reusable part

67162-LEXU-G36

Exploded view Intake Manifold—3.3L (3MZ-FE) Engine

tions. Uniformly tighten the 22 bearing cap bolts, in several passes, following the sequence as shown. Torque bolt "C" to a final torque of 66 inch lbs. (7.5 Nm). Torque all other bearing cap bolts to a final torque of 12 ft. lbs. (16 Nm).

20. Remove the service bolt installed in the rear face of the gear.

21. Repeat this entire procedure for the RH camshafts. Use the illustrations given for the LH camshafts, as the sequences are the same on the RH camshafts.

22. Check and adjust valve lash. See VALVE LASH.

23. Install or reconnect the following:
- Camshaft timing oil control valve.
- Semi-circular plugs in rear ends of each cylinder head
- Cylinder head covers, with new gaskets and packing material
- Engine hangers
- Variable valve timing sensors
- Oil dipsticks and tubes
- All remaining components in reverse of removal procedure.

3.3L (3MZ-FE) Engine

1. Before servicing the vehicle, refer to the Precautions Section.

2. Remove or disconnect the following:
- Negative battery cable. Wait at least 90 seconds before performing any other work
- RH front wheel
- Front suspension upper brace center
- V-bank cover sub-assembly
- Air cleaner assembly
- Emission control valve set
- Intake air surge tank
- Ignition coil assembly
- Cylinder head valve covers
- Front fender apron seal
- A/C Compressor and PS drive belts
- Engine stabilizing rod
- Engine mounting stay
- Alternator bracket
- Crankshaft pulley
- Timing belt covers
- Timing belt, tensioners, idlers and guides
- Camshaft timing pulleys

➡**Align the camshaft pulleys so that they can be returned to the original locations when reassembling.**

- Inner timing belt cover

✱✱ WARNING

Since the thrust clearance of the camshaft is small, the camshaft must

be kept level while it is being removed. If the camshaft is not kept level, the portion of the cylinder head receiving the shaft thrust may crack or be damaged, causing the camshaft to seize or break.

3. Remove the camshaft using the following procedures

4. RH bank camshaft No.1 & LH bank camshaft No.2

a. Align the (2 dot marks) of the camshaft drive and driven gear by turning the camshaft with a wrench.

b. Secure the exhaust camshaft sub gear to the main gear with service bolt. Torque to 48 inch lbs. (5.4 Nm)

➡**When removing the camshaft, make certain that the torsional spring force of the sub gear has been eliminated by installation of the service bolt.**

c. Using several steps, loosen the 10 bearing cap bolts uniformly in the sequence shown in the illustration. Remove the 5 bearing caps and camshaft

✱✱ WARNING

Do Not pry out camshaft

✱✱ WARNING

Do Not damage contact surface of the cylinder head that receives the shaft thrust.

5. LH bank camshaft No. 3 & No. 4

a. Using several steps, loosen the 10 bearing cap bolts uniformly in the sequence shown in the illustration. Remove the 5 bearing caps and camshaft

✱✱ WARNING

Do Not pry out camshaft

✱✱ WARNING

Do Not damage contact surface of the cylinder head that receives the shaft thrust.

b. Remove the oil seal from camshaft
To install:

6. Install RH No 2 Camshaft then RH Camshaft No. 1 using same procedure

✱✱ WARNING

Since the clearance of the camshaft is small, the camshaft must be kept level while bring installed. If the camshaft is not kept level, the cylin-

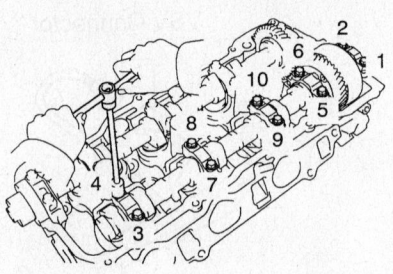

Aligning timing marks—3.3L (3MZ-FE) Engine

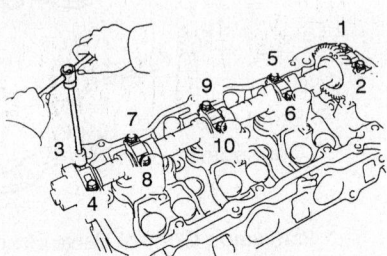

Camshaft bolt removal sequence Camshaft No.1—3.3L (3MZ-FE) Engine

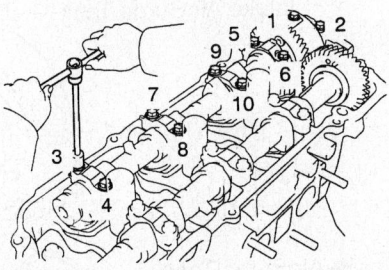

Camshaft bolt removal sequence Camshaft No.2 —3.3L (3MZ-FE) Engine

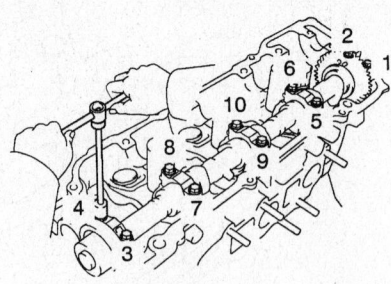

Camshaft bolt removal sequence Camshaft No.4 —3.3L (3MZ-FE) Engine

der head or camshaft may be damaged.

a. Apply engine oil to the thrust portion and journal of camshaft

b. No. 2 camshaft at 90 degree angle

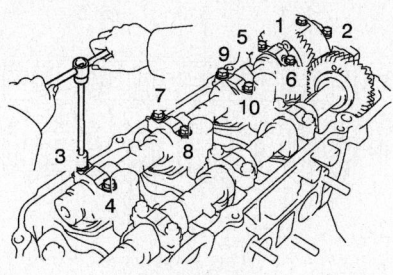

Camshaft bolt removal sequence Camshaft No. 3—3.3L (3MZ-FE) Engine

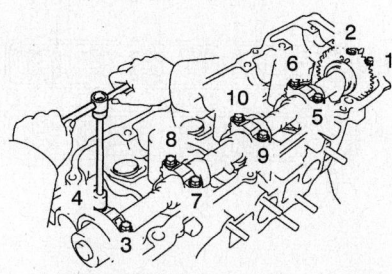

Camshaft bolt removal sequence Camshaft No. 4—3.3L (3MZ-FE) Engine

to the timing mark (2 dot marks) on the head.

   c. Multi-purpose grease to new oil seal

   d. Oil seal to camshaft

➡**Do NOT turn over the oil seal lip**

➡**Insert oil seal until it stops**

   e. Seal packing to bearing cap No. 1

➡**Install bearing cap No. 1 within 5 minutes after applying seal packing**

➡**Do NOT expose seal packing to engine oil within 2 hours after installation**

   f. 5 bearing caps in their proper locations

   g. Apply light coat of oil to the threads of the bearing caps

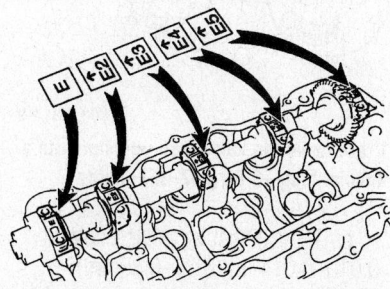

Bolt torque procedure RH camshaft No. 2—3.3L (3MZ-FE) Engine

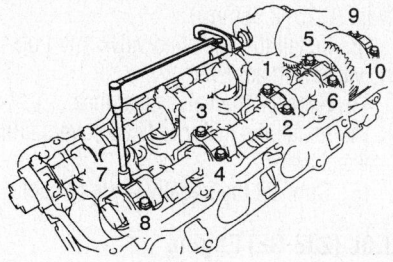

Bolt torque procedure RH camshaft No. 1—3.3L (3MZ-FE) Engine

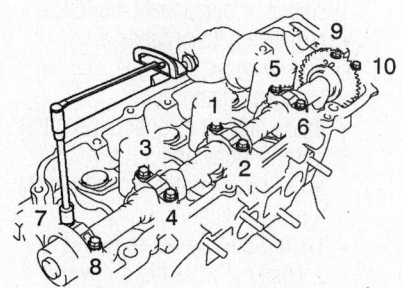

Bolt torque procedure LH camshaft No. 3—3.3L (3MZ-FE) Engine

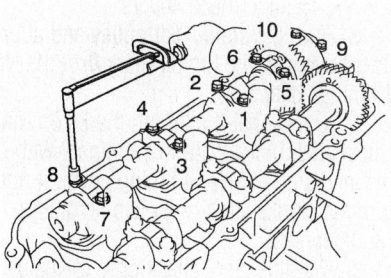

Bolt torque procedure LH camshaft No. 4—3.3L (3MZ-FE) Engine

   h. Tighten the 10 bearing cap bolts in required sequence. Torque to: 12 ft. lbs. (16 Nm)

7. Install LH camshaft No.3 and LH camshaft No. 4 using the above procedure

8. Install or connect the following:
- Inner timing belt cover
- RH then LH camshaft timing pulleys. Torque to: 92 ft. lbs. (125 Nm)
- Timing belt idlers, tensioners, guide and belt
- RH engine mounting bracket
- Timing belt covers
- Crankshaft pulley
- Alternator bracket
- RH engine mounting stay
- Engine stabilizing rod
- Inspect or adjust valve lash

- A/C compressor and PS drive belts
- Cylinder head valve covers
- Ignition coil assembly
- Intake air surge tank
- Emission control valve set
- Air cleaner assembly
- Vacuum hoses
- V-bank cover sub-assembly
- Front suspension upper center brace
- RF wheel
- Engine coolant. Check for leaks
- System initialization

Valve Lash

ADJUSTMENT

2.5L (4GR-FSE), 3.0L (3GR-FSE) and 3.5L (2GR-FSE) Engines

The 2.5L (4GR-FSE), 3.0L (3GR-FSE) and 3.5L (2GR-FSE) engines are equipped with hydraulic valves which are not adjustable.

3.0L (1MZ-FE) Engine

1. Before servicing the vehicle, refer to the Precautions Section.

➡**Adjust the valve clearance when the engine is cold.**

2. Remove or disconnect the following:
- Negative battery cable
- Accelerator/throttle cable from the throttle linkage
- Air intake chamber
- Cylinder head covers

3. Turn the crankshaft pulley and align it's groove with the timing mark **0** of the No. 1 timing cover.

4. Check that the valve lash adjusters on the No. 1 intake are loose and the exhaust are tight. If not, turn the crankshaft on complete revolution (360 degrees).

5. Measure the clearance between the

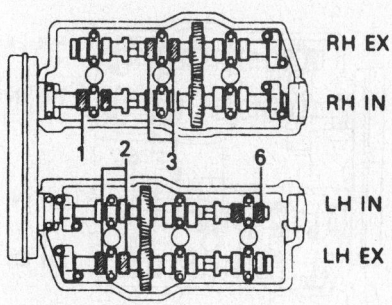

Adjust these valves FIRST—3.0L (1MZ-FE) engine

valve lash adjuster and the camshaft. Record the measurements on valves No. 1, 2, 3 and 6.

6. The intake valve clearance cold is 0.006–0.010 in. (0.15–0.25mm).

7. The exhaust valve clearance cold is 0.010–0.014 in. (0.25–0.35mm).

8. Turn the crankshaft ⅔ of a revolution (240 degrees) and check the clearance on valves No. 2, 3, 4 and 5 and record.

9. Turn the crankshaft another ⅔ of a revolution and check valves; No. 1, 4, 5 and 6 and record.

10. Remove or disconnect the following:
- Adjusting shim and turn the crankshaft to position the cam lobe of the camshaft on the adjusting valve upward. Press down the valve lash adjuster with the proper tool and place the proper tool between the camshaft and the valve lash adjuster. Remove the tool.
- Adjusting shim with the proper tool.

11. Determine the thickness of the replacement shim as follows:
a. T: Thickness of the used shim
b. A: Measured valve clearance
c. N: Thickness of new shim
d. Intake: $N = T + (A - 0.006-0.010$ in. (0.15–0.25mm))

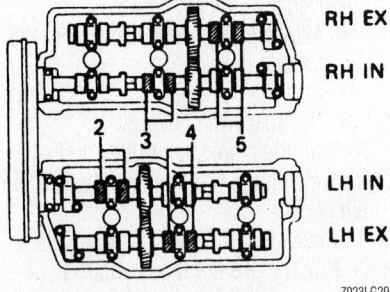

Adjust these valves SECOND—3.0L (1MZ-FE) engine

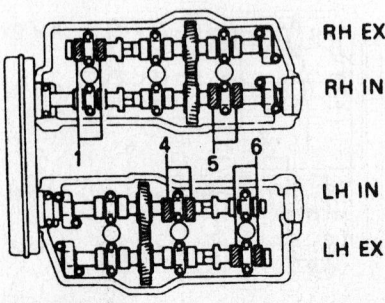

Adjust these valves THIRD—3.0L (1MZ-FE) engine

e. Exhaust: $N = T + (A - 0.010-0.014$ in. (0.25–0.35mm))

12. Install the specified valve shim on the valve lash adjuster

13. Recheck the valve clearance.

14. Install the cylinder head covers and intake chamber.

15. Connect the negative battery cable.

3.0L (2JZ-GE) Engine

➡**Adjust the valve lash when the engine is cold.**

1. Before servicing the vehicle, refer to the Precautions Section.

2. Remove or disconnect the following:
- Negative battery cable
- Engine cover
- Engine coolant
- Intake and air cleaner assemblies
- Throttle body and accelerator/throttle cable from the throttle linkage
- Wiring and hoses required to access cylinder head covers (mark each one for reinstallation)
- No. 3 timing belt cover
- Ignition coils and wires
- Spark plugs
- Cylinder head covers

3. Turn the crankshaft pulley and align its groove with the timing mark **0** of the No. 1 timing cover.

4. Check that the timing marks on the camshaft sprockets are in alignment with the marks on the No. 4 timing cover. If not, turn the crankshaft 1 complete revolution (360 degrees).

5. For this step, check only valves as indicated in appropriate illustration. Adjust these valves first.

6. Measure the clearance between the valve lash adjuster and the camshaft. Record the measurements.
a. The intake valve lash cold is 0.006–0.010 in. (0.15–0.25mm).
b. The exhaust valve lash cold is 0.010–0.014 in. (0.25–0.35mm).

7. For models except 2004, turn the crankshaft ⅔ of a revolution (240 degrees) and check the clearance on other valves as indicated.

8. Turn the crankshaft pulley one full revolution (360°) and align the groove with the "0" timing marks on the No. 1 timing belt cover.

9. Measure the clearance between the valve lash adjuster and the camshaft. Record the measurements.
a. The intake valve lash cold is 0.006–0.010 in. (0.15–0.25mm).

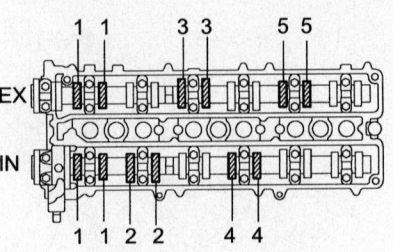

Identifying the valves to be adjusted first—3.0L (2JZ-GE) Engine

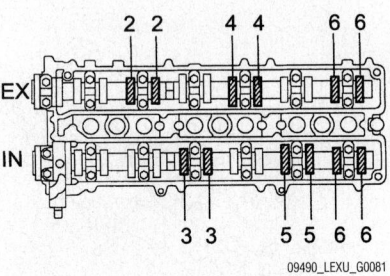

Identifying the valves to be adjusted second—3.0L (2JZ-GE) engine

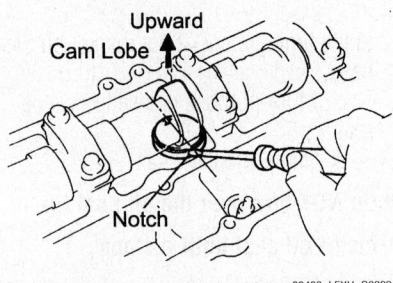

Removing the adjusting shim—3.0L (2JZ-GE) engine

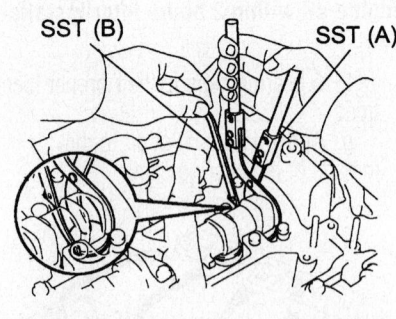

Press down the valve lash adjuster with a special tool—3.0L (2JZ-GE) engine

b. The exhaust valve lash cold is 0.010–0.014 in. (0.25–0.35mm).

10. Remove the adjusting shim and turn the crankshaft to position the cam lobe of the camshaft on the adjusting valve upward.

The notches should be perpendicular to the camshaft. Press down the valve lash adjuster with the proper tool and place the proper tool between the camshaft and the valve lash adjuster. Remove the tool.

11. Remove the adjusting shim with the proper tool (a magnetic finger).

12. Determine the thickness of the replacement shim as follows:

    a. T = Thickness of the used shim

    b. A = Measured valve lash

    c. N = Thickness of new shim

    d. Intake: N = T + (A – 0.006–0.010 in. (0.15–0.25mm))

    e. Exhaust: N = T + (A – 0.010–0.014 in. (0.25–0.35mm))

13. Install the specified valve shim on the valve lash adjuster.

14. Recheck the valve lash.

15. Reinstall all components in reverse of removal sequence.

16. Refill with engine coolant.

17. Start engine and check for leaks.

4.3L (3UZ-FE) Engine

1. Before servicing the vehicle, refer to the Precautions Section.

2. Remove or disconnect the following, as applicable:

- Negative battery cable
- V-bank cover
- Intake air connector pipe
- Ignition coils
- No. 3 timing belt covers
- Spark plug wires
- Cylinder head covers

3. Turn the crankshaft pulley and align its groove with the timing mark **0** of the No. 1 timing cover. Check that the timing marks of the camshaft timing pulleys and timing belt rear plates are aligned. If not, turn the crankshaft 1 revolution (360 degrees) and align the mark.

4. Measure the clearance between the valve lash adjuster and the camshaft on the valves, as illustrated, in the first sequence. Record the measurements.

    a. The intake valve lash cold is 0.006–0.010 in. (0.15–0.25mm).

    b. The exhaust valve lash cold is 0.010–0.014 in. (0.25–0.35mm).

5. Turn the crankshaft 1 full revolution (360 degrees) and align the mark.

6. Measure the clearance between the valve lash adjuster and the camshaft on the valves, as illustrated, in the second sequence. Record these measurements.

7. If necessary, remove the camshafts.

8. Remove the adjusting shim and turn the crankshaft to position the cam lobe of the camshaft on the adjusting valve upward.

Position the hole in the shim toward the outside of the cylinder head. Press down the valve lash adjuster with the proper tool and place the proper tool between the camshaft and the valve lash adjuster. Remove the tool.

9. Remove the adjusting shim with the proper tool.

10. Determine the thickness of the replacement shim as follows:

    a. T = Thickness of the used shim

    b. A = Measured valve lash

    c. N = Thickness of new shim

    d. Intake: N = T + (A – 0.006–0.010 in. (0.15–0.25mm))

    e. Exhaust: N = T + (A – 0.010–0.014 in. (0.25–0.35mm))

11. Recheck the valve lash. Install the cylinder head covers.

RH Bank

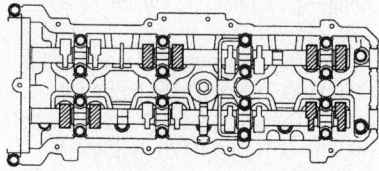

09490_LEXU_G0084

Adjust these valves FIRST—4.3L (3UZ-FE) Engine

RH Bank

LH Bank

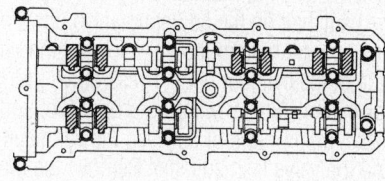

09490_LEXU_G0085

Adjust these valves SECOND—4.3L (3UZ-FE) Engine

12. Connect the spark plug wires and install the No. 3 timing belt covers.

13. Install or reconnect all other components in reverse of removal procedure.

14. Connect the negative battery cable.

3.3L (3MZ-FE) Engine

1. Before servicing the vehicle, refer to the Precautions Section.

2. Remove or disconnect the following:

- Negative battery cable. Wait at least

RH Bank:

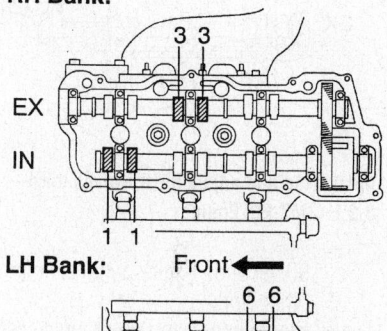

67162-LEXU-G43

Measuring and adjust these valves first—3.3 L (3MZ-FE) Engine

RH Bank:

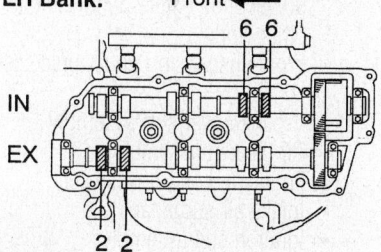

LH Bank:

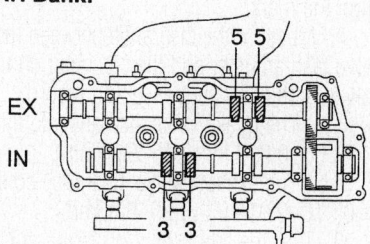

67162-LEXU-G44

Measuring and adjust these valves second—3.3 L (3MZ-FE) Engine

RH Bank:

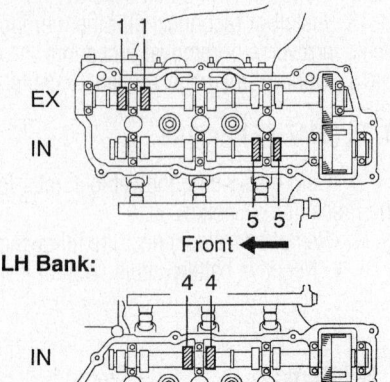

LH Bank:

67162-LEXU-G45

Measuring and adjust these valves third—3.3 L (3MZ-FE) Engine

90 seconds before performing any other work
- Front suspension upper brace center
- V-bank cover sub-assembly
- Air cleaner assembly
- Emission control valve set
- Intake air surge tank
- Ignition coil assembly
- Cylinder head valve covers

3. Turn the crankshaft pulley and align its groove with the timing mark **0** of the No. 1 timing cover. Check that the timing marks of the camshaft timing pulleys and timing belt rear plates are aligned. If not, turn the crankshaft 1 revolution (360 degrees) and align the mark.

4. Measure the clearance between the valve lash adjuster and the camshaft on the valves in the first sequence and record.
   a. The intake valve clearance cold is 0.006–0.010 in. (0.15–0.25mm).
   b. The exhaust valve clearance cold is 0.010–0.014 in. (0.25–0.35mm).

5. Turn the crankshaft 2/3 revolution and (240 degrees).

6. Measure the clearance between the valve lash adjuster and the camshaft on the valves in the second sequence and record.
   a. The intake valve clearance cold is 0.006–0.010 in. (0.15–0.25mm).
   b. The exhaust valve clearance cold is 0.010–0.014 in. (0.25–0.35mm).

7. Turn the crankshaft 2/3 revolution and (240 degrees).

8. Measure the clearance between the valve lash adjuster and the camshaft on the valves in the third sequence and record.

Front of No. 1 and No. 2 Cylinders:

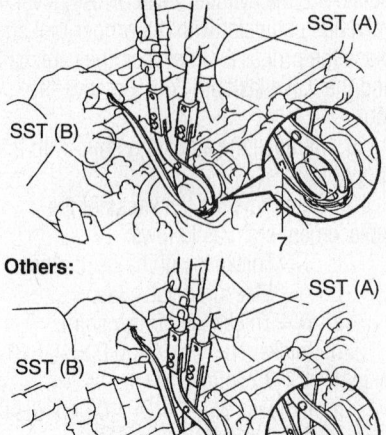

Others:

67162-LEXU-G46

Using special tool to remove the adjusting shim—3.3 L (3MZ-FE) Engine

   a. The intake valve clearance cold is 0.006–0.010 in. (0.15–0.25mm).
   b. The exhaust valve clearance cold is 0.010–0.014 in. (0.25–0.35mm).

9. Adjust the valve lash using the following procedure

10. Remove the adjusting shim and turn the crankshaft to position the cam lobe of the camshaft on the adjusting valve upward. Position the hole in the shim toward the outside of the cylinder head. Press down the valve lash adjuster with the proper tool and place the proper tool between the camshaft and the valve lash adjuster. Remove the tool.

11. Remove the adjusting shim with the proper tool.

12. Determine the thickness of the replacement shim as follows:
   a. T = Thickness of the used shim
   b. A = Measured valve clearance
   c. N = Thickness of new shim
   d. Intake: N = T + (A − 0.006–0.010 in. (0.15–0.25mm))
   e. Exhaust: N = T + (A − 0.010–0.014 in. (0.25–0.35mm))

➡**Place the adjusting shim on the valve lifter with the imprinted number facing down.**

13. Reinstall the following
- Cylinder head valve covers
- Ignition coil assembly
- Intake air surge tank
- Emission control valve set
- Air cleaner assembly
- Vacuum hoses
- Front suspension upper center brace
- Negitive battery cable

Starter Motor

REMOVAL & INSTALLATION

2.5L (4GR-FSE) and 3.5L (2GR-FSE) Engines

1. Before servicing the vehicle, refer to the Precautions Section.
2. Remove or disconnect the following:
- Negative battery cable. Wait at least 90 seconds before performing any other work.
- Cool air intake duct seal
- Left engine room side cover
- Rear No. 1 floor panel brace
- Front center floor brace
- Heated oxygen sensor
- Front exhaust pipe assembly
- Front propeller shaft assembly (AWD)
- No. 1 exhaust pipe support bracket (2.5L with automatic transmission)
- Left exhaust manifold
- Terminal 50 connector from starter
- Terminal cap
- Nut and wire harness from terminal 30
- 2 bolts and starter motor

To install:
3. Install or connect the following:
- Starter assembly with 2 bolts. Torque to 43 ft. lbs. (58 Nm).
- Starter wires
- Left exhaust manifold
- No. 1 exhaust pipe support bracket (2.5L with automatic transmission) with 2 bolts. Torque to 32 ft. lbs. (43 Nm).
- Front propeller shaft assembly (AWD)
- Front exhaust pipe assembly
- Heated oxygen sensor
- Front center floor brace. Torque to 65 inch lbs. (7 Nm).
- Rear No. 1 floor panel brace. Torque to 14 ft. lbs. (19 Nm).
- Left engine room side cover
- Cool air intake duct seal
- Negative battery cable
- System initialization

4. Start the vehicle and check for exhaust leaks.

3.0L (3GR-FSE) Engine

1. Before servicing the vehicle, refer to the Precautions Section.
2. Remove or disconnect the following:
- Negative battery cable. Wait at least 90 seconds before performing any other work.

- Front console upper panel garnish
- Front console upper panel assembly
- Left and right instrument panel finish panel ends
- Console box plate
- Console box register assembly
- Console box
- Left front seat
- Engine under cover
- Front center floor brace
- Heated oxygen sensor
- Front exhaust pipe
- Front propeller shaft (AWD)
- Engine V-bank cover
- Cool air intake duct seal
- Left engine room side cover
- Left exhaust manifold
- Starter connector
- Terminal cap
- Nut and starter cable
- 2 bolts and starter motor

To install:
3. Install or connect the following:
- Starter assembly with 2 bolts. Torque to 43 ft. lbs. (58 Nm).
- Starter wires
- Left exhaust manifold
- Front exhaust pipe assembly
- Heated oxygen sensor
- Front center floor brace. Torque to 65 inch lbs. (7 Nm).
- Negative battery cable
- Front propeller shaft assembly (AWD)
- Engine under cover
- Left engine room side cover
- Cool air intake duct seal
- Engine V-bank cover
- Left front seat
- Console box
- Console box register assembly
- Console box plate
- Left and right instrument panel finish panel ends
- Front console upper panel assembly
- Front console upper panel garnish
- System initialization
4. Start the vehicle and check for exhaust leaks.

3.0L (1MZ-FE) Engine

1. Before servicing the vehicle, refer to the Precautions Section.
2. Remove or disconnect the following:
- Negative battery cable. Wait at least 90 seconds before performing any other work
- Automatic transmission shift control cable

- Engine wire
- Starter connector
- Nut, and disconnect the starter wire
- 2 bolts, automatic transmission shift control cable clamp and starter

To install:
3. Installation is the reversal of the removal process.
4. Torque the starter bolts to 27 ft. lbs. (37 Nm).

3.0L (2JZ-GE) Engine

1. Before servicing the vehicle, refer to the Precautions Section.
2. Remove or disconnect the following:
- Negative battery cable. Wait at least 90 seconds before performing any other work
- Starter connector
- Nut, and disconnect the starter wire
- 2 bolts and starter

To install:
3. Installation is the reversal of the removal procedure.
4. Tighten the starter bolts to 27 ft. lbs. (37 Nm).

4.3L (3UZ-FE) Engine

1. Before servicing the vehicle, refer to the Precautions Section.
2. Remove or disconnect the following:
- Negative battery cable. Wait at least 90 seconds before performing any other work
- Drain engine coolant
- V-bank cover
- Accelerator cable
- Intake air connector
- Throttle Body
- Upper and lower intake manifold assembly
- Rear water bypass joint
- Water bypass pipe
- Water bypass pipe from the water pump
- Wire clamp from the bracket on the water bypass pipe
- O-ring from the water bypass pipe
- Water bypass pipe bracket from the water bypass pipe
- 2 bolts holding the starter to the cylinder block
- Starter connector
- Starter from the cylinder block
- Nut, and disconnect the starter wire
- Starter

To install:
3. Install or connect the following:
- Wire clamp to the wire bracket with the bolt. Tighten to 87 inch lbs.

- Starter wire with the nut. Tighten to 87 inch lbs.
- Starter connector
- Starter with the 2 bolts. Torque the bolts to 29 ft. lbs. (39 Nm).
- Water bypass pipe bracket to the water bypass pipe
- O-ring to the water bypass pipe
- Water bypass pipe
- Wire clamp to the bracket on the water bypass pipe
- Water bypass pipe bolts. Torque the bolts to 13 ft. lbs. (18 Nm).
- Rear water bypass joint
- Intake manifold assembly
- Throttle body
- Intake air connector
- Accelerator cable
- V-bank cover

3.3L (3MZ-FE) Engine

1. Before servicing the vehicle, refer to the Precautions Section.
2. Remove or disconnect the following:
- Negative battery cable. Wait at least 90 seconds before performing any other work
- RH radiator side deflector
- Air cleaner inlet assembly, air cleaner assembly and bracket
- Battery and battery tray
- Remove starter connector
- Starter wire
- Two bolts securing starter to engine
- Starter assembly

To install:
3. Install or connect the following:
- Starter assembly with 2 bolts; torque to 26 ft. lbs. (37 Nm)
- Starter wires
- Battery and battery tray
- Air cleaner assembly, bracket and inlet
- Radiator side deflector
4. System initialization

Oil Pan

REMOVAL & INSTALLATION

2.5L (4GR-FSE), 3.0L (3GR-FSE) and 3.5L (2GR-FSE) Engines

➡The No. 1 oil pan cannot be removed with the engine in the vehicle. The engine and transmission must be removed as a unit, then separated. See ENGINE ASSEMBLY section. It may be possible to remove the No. 2 oil pan from the vehicle while the engine is still in the vehicle.

For 2WD:

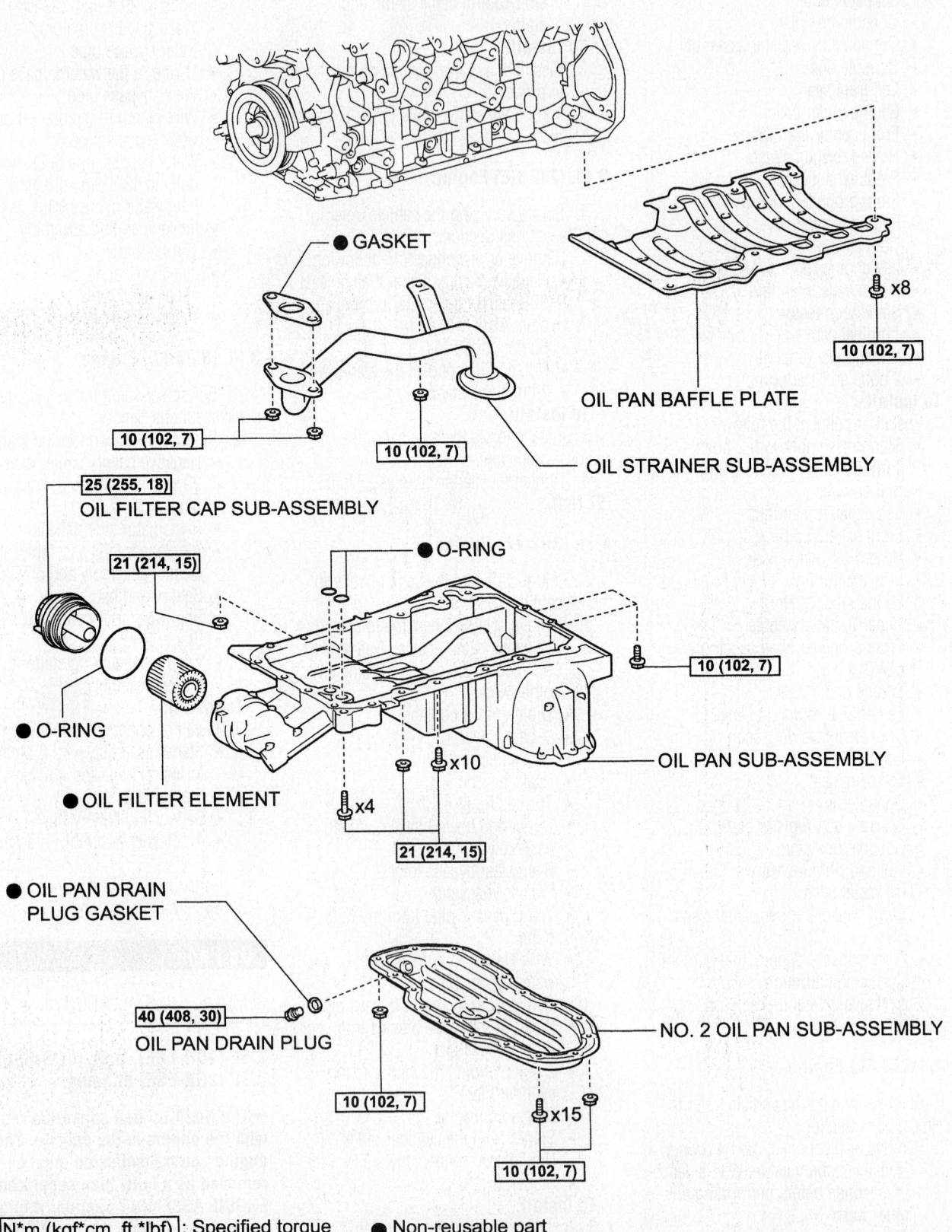

GASKET

OIL PAN BAFFLE PLATE

OIL STRAINER SUB-ASSEMBLY

10 (102, 7)

×8

10 (102, 7)

10 (102, 7)

25 (255, 18)

OIL FILTER CAP SUB-ASSEMBLY

21 (214, 15)

O-RING

10 (102, 7)

O-RING

OIL FILTER ELEMENT

OIL PAN SUB-ASSEMBLY

×10

×4

21 (214, 15)

OIL PAN DRAIN PLUG GASKET

40 (408, 30)

OIL PAN DRAIN PLUG

NO. 2 OIL PAN SUB-ASSEMBLY

10 (102, 7)

×15

10 (102, 7)

N*m (kgf*cm, ft.*lbf) : Specified torque ● Non-reusable part

09490_LEXU_G0086

Exploded view of the oil pan and related components (2WD)—2.5L (4GR-FSE), 3.0L (3GR-FSE) and 3.5L (2GR-FSE) Engines

For 4WD:

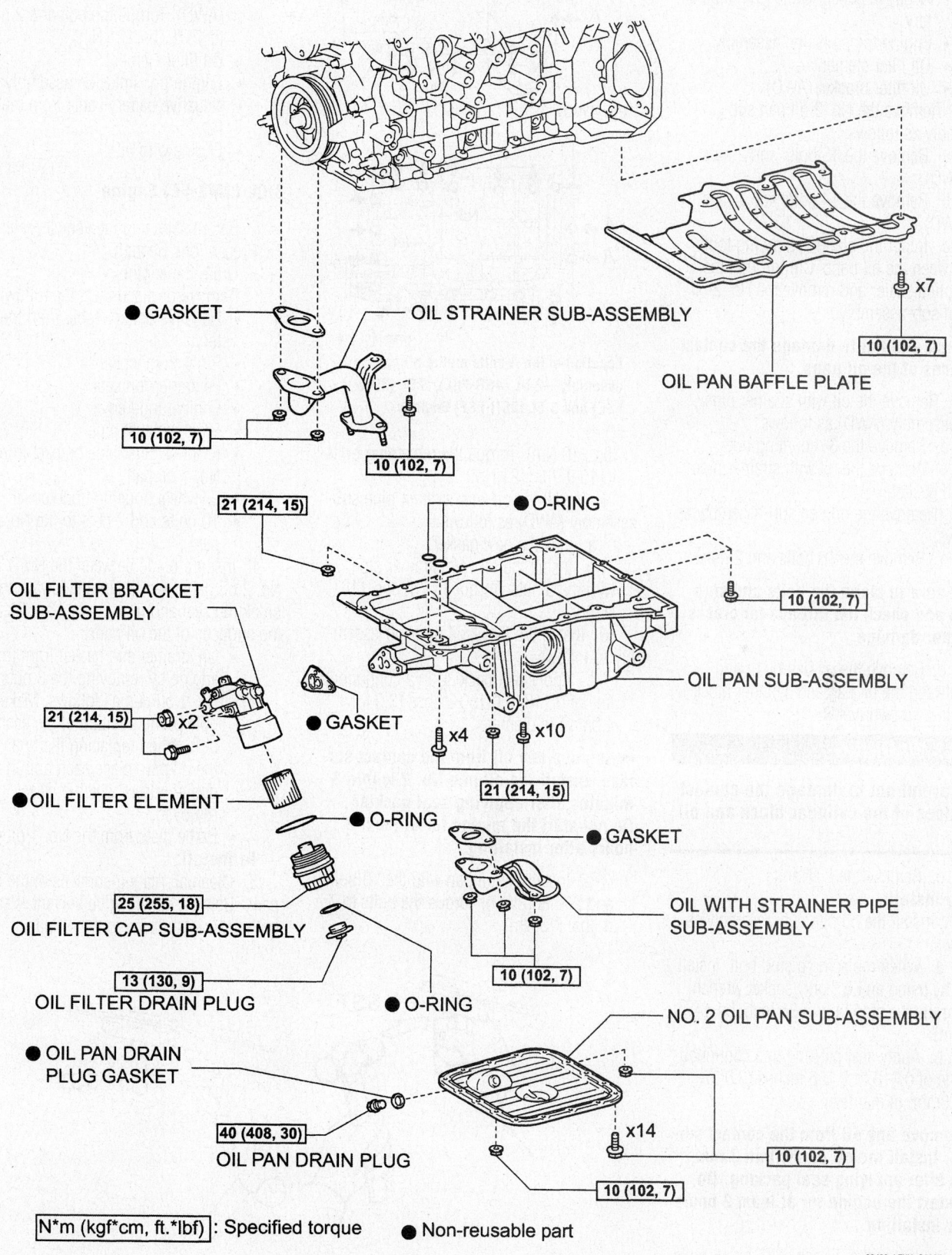

● GASKET ── OIL STRAINER SUB-ASSEMBLY

OIL PAN BAFFLE PLATE

x7

10 (102, 7)

10 (102, 7)

10 (102, 7)

21 (214, 15)

● O-RING

OIL FILTER BRACKET SUB-ASSEMBLY

10 (102, 7)

21 (214, 15) x2

OIL PAN SUB-ASSEMBLY

● GASKET

● OIL FILTER ELEMENT

x4   x10

21 (214, 15)

● O-RING

● GASKET

OIL WITH STRAINER PIPE SUB-ASSEMBLY

25 (255, 18)

OIL FILTER CAP SUB-ASSEMBLY

10 (102, 7)

13 (130, 9)

OIL FILTER DRAIN PLUG   ● O-RING

NO. 2 OIL PAN SUB-ASSEMBLY

● OIL PAN DRAIN PLUG GASKET

40 (408, 30)

OIL PAN DRAIN PLUG

x14

10 (102, 7)

10 (102, 7)

N*m (kgf*cm, ft.*lbf) : Specified torque   ● Non-reusable part

09490_LEXU_G0087

Exploded view of the oil pan and related components (AWD)—2.5L (4GR-FSE) and 3.0L (3GR-FSE) Engines

1. Before servicing the vehicle, refer to the Precautions Section.
2. Drain the engine oil.
3. Remove or disconnect the following:
   - Negative battery cable from the battery
   - Engine/transmission assembly
   - Oil filter element
   - Oil filter bracket (AWD)
4. Remove the No. 2 oil pan sub-assembly as follows:
   a. Remove the 15 bolts and 2 nuts (2WD).
   b. Remove the 14 bolts and 2 nuts (AWD).
   c. Insert the blade of a prying tool between the oil pans. Cut through the applied sealer and remove the No. 2 oil pan sub-assembly.

➡**Be careful not to damage the contact surfaces of the oil pans.**

5. Remove the oil with strainer pipe sub-assembly (AWD) as follows:
   a. Remove the 3 mounting nuts.
   b. Remove the oil with strainer pipe and gasket.
6. Remove the oil pan sub-assembly as follows:
   a. Remove the 16 bolts and 2 nuts.

➡**Be sure to clean the bolts and stud bolts and check the threads for cracks or other damage.**

   b. Remove the oil pan by prying between the oil pan and cylinder block with a screwdriver.

✳✳ WARNING

Be careful not to damage the contact surfaces of the cylinder block and oil pan.

   c. Remove the 2 O-rings.
   To install:
7. Install the oil pan sub-assembly as follows:
   a. When replacing a stud bolt, install it by using an E6 "torx" socket wrench. Torque the stud bolt to 35 inch lbs. (4 Nm).
   b. Apply seal packing in a continuous line of 0.118 to 0.156 inches (3.0 to 4.0mm) in diameter.

➡**Remove any oil from the contact surface. Install the oil pan within 3 minutes after applying seal packing. Do not start the engine for at least 2 hours after installing.**

   c. Install the oil pan with the 16 bolts and 2 nuts. Torque the A bolts to 7 ft.

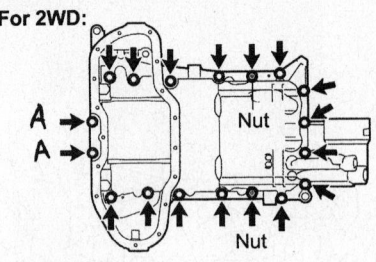

For 2WD:

For 4WD:

09490_LEXU_G0088

Location of the A bolts on the oil pan assembly—2.5L (4GR-FSE), 3.0L (3GR-FSE) and 3.5L (2GR-FSE) Engines

lbs. (10 Nm). Torque the remaining bolts to 15 ft. lbs. (21 Nm).
8. Install the oil with strainer pipe sub-assembly (AWD) as follows:
   a. Install a new gasket.
   b. Install the oil with strainer pipe with the 3 nuts. Torque to 7 ft. lbs. (10 Nm).
9. Install the No. 2 oil pan sub-assembly as follows:
   a. Apply seal packing in a continuous line of 0.118 to 0.156 inches (3.0 to 4.0mm) in diameter.

➡**Remove any oil from the contact surface. Install the oil pan No. 2 within 3 minutes after applying seal packing. Do not start the engine for at least 2 hours after installing.**

   b. Install the oil pan with the 15 bolts and 2 nuts (2WD). Torque the bolts to 7 ft. lbs. (10 Nm).

   c. Install the oil pan with the 14 bolts and 2 nuts (AWD). Torque the bolts to 7 ft. lbs. (10 Nm).
10. Install or connect the following:
   - New gasket and oil filter bracket (AWD). Torque the bolt and 2 nuts to 15 ft. lbs. (21 Nm).
   - Oil filter element
   - Engine/transmission assembly
   - Negative battery cable from the battery
   - Engine with oil

3.0L (1MZ-FE) Engine

1. Before servicing the vehicle, refer to the Precautions Section.
2. Drain the engine oil.
3. Remove or disconnect the following:
   - Negative battery cable from the battery.
   - Right front wheel
   - Fender apron seal
   - Engine undercover
   - Front exhaust pipe
   - Front exhaust pipe bracket from the No. 1 oil pan
   - Flywheel housing undercover
   - 10 bolts and 2 nuts to the No. 2 oil pan
4. Insert a blade between the No. 1 and No. 2 oil pans. Tap the tool sideways to break the seal and remove the pan. Clean the surfaces of the oil pans.
   - Oil strainer and gasket from the engine by removing the 3 nuts.
   - No. 1 oil pan as follows. Make a note of the position of the each bolt. When replacing the bolts into the oil pan, place each bolt in the position from which it was removed.
   - Baffle plate from the No. 1 oil pan
   To install:
5. Clean all mating surfaces of the oil pans. Using a non-residue solvent, clean both sealing surfaces to the oil pan.

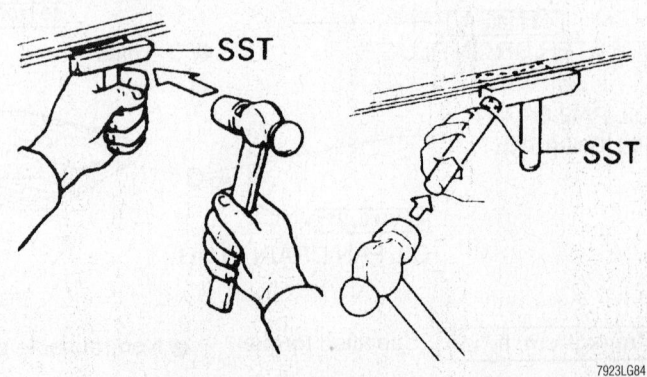

7923LG84

Use the special tool to break the seal and remove the oil pan—3.0L (1MZ-FE) engine

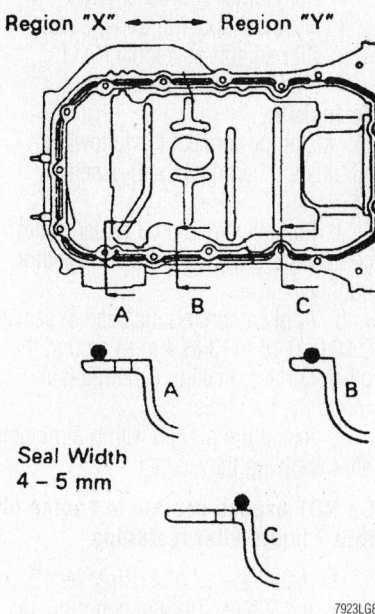

Seal Width
4 – 5 mm

7923LG85

Apply sealant as shown to the No. 1 (upper) oil pan—3.0L (1MZ-FE) engine

6. Install or connect the following:
   - Baffle plate to the No. 1 oil pan and tighten to 69 inch lbs. (8 Nm)
   - No. 1 oil pan. Apply RTV sealant to the oil pan and engine block. Uniformly tighten the bolts and nuts in several passes to: 10mm: 69 inch lbs. (8 Nm); 12mm: 14 ft. lbs. (20 Nm); 14mm: 27 ft. lbs. (37 Nm)
   - Flywheel housing undercover with the 2 bolts. Tighten the bolts to 69 inch lbs. (8 Nm).
   - Oil strainer with the 3 nuts. Tighten the nuts to 69 inch lbs. (8 Nm).
   - No. 2 oil pan. Apply RTV sealant to the oil pan and engine block. Uniformly tighten the bolts and nuts in several passes, to 69 inch lbs. (8 Nm).
   - Flywheel housing undercover
   - Front exhaust pipe bracket to the No. 1 oil pan. Tighten the bolts to 15 ft. lbs. (21 Nm).
   - Front exhaust pipe. 4 pipe-to-pipe nuts: 46 ft. lbs. (62 Nm); front exhaust pipe-to-the center exhaust pipe bolts and nuts: 41 ft. lbs. (56 Nm); bracket bolts: 14 ft. lbs. (19 Nm); support stay bolts: 22 ft. lbs. (30 Nm).
   - Engine undercover
   - Right fender apron seal
   - The right front wheel and lower the vehicle
   - Engine with oil

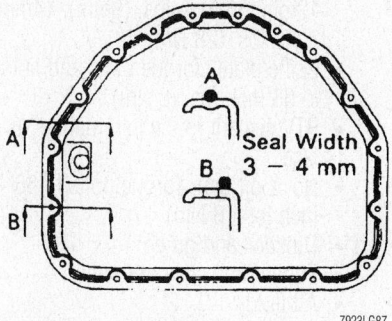

7923LG87

Lower oil pan sealant application—3.0L (2JZ-GE) engine

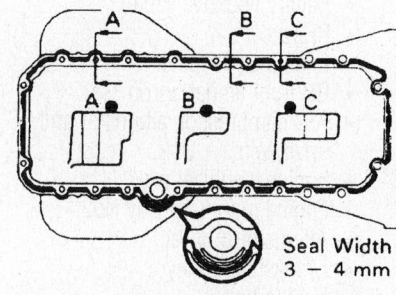

Seal Width
3 – 4 mm

7923LG86

Upper oil pan sealant application—3.0L (2JZ-GE) engine

3.0L (2JZ-GE) Engine

➡The No. 1 oil pan cannot be removed with the engine in the vehicle. The engine/transmission assembly must be removed. See ENGINE ASSEMBLY section. The manufacturer does not provide any on vehicle information for the No. 2 oil pan removal and installation. If only the No. 2 oil pan is being serviced, the engine/transmission assembly can remain in the vehicle.

1. Before servicing the vehicle, refer to the Precautions Section.
2. Remove or disconnect the following:
   - Engine/transmission assembly
   - Timing belt
   - Idler pulley
   - Crankshaft timing pulley
   - Oil dipstick and guide
   - Oil sensor lead
   - 4 attaching bolts and lift off the oil level sensor. Be careful not to drop this sensor.
   - 14 bolts (16 bolts for GS 300) and 2 nuts and pry off the lower (No. 2) oil pan. Be careful not to damage the No. 1 pan while performing this procedure.

- Bolt and 2 nuts and drop down the oil strainer and gasket
- 5 bolts and 2 nuts and drop down the baffle plate
- 22 bolts and the carefully pry off the upper (No. 1) oil pan
- O-ring from the cylinder block

To install:
3. Install or connect the following:
   - New O-ring in the block and scrape off any old sealant
   - A ⅛ inch (3–4mm) bead of RTV sealant to the pan mating surface
   - Upper pan. 12mm bolts: 15 ft. lbs. (21 Nm); 14mm bolts to 29 ft. lbs. (39 Nm)
   - Baffle plate and oil strainer. Tighten them both to 78 inch lbs. (9 Nm).
   - Lower pan in the same manner as the upper pan and tighten the bolts to 78 inch lbs. (9 Nm)
   - Oil level sensor and tighten it to 48 inch lbs. (5 Nm)
   - Oil dipstick and guide
   - Timing pulleys and belt
   - Transmission to the engine
   - Engine and transmission
   - All fluids

4.3L (3UZ-FE) Engine

LS 430

1. Before servicing the vehicle, refer to the Precautions Section.
2. Remove or disconnect the following:
   - Engine/transmission assembly
   - Remove the timing belt
   - Idler pulleys
   - Crankshaft timing pulley
   - Oil dipstick and guide
   - Oil level sensor lead
   - 4 bolts and lift off the oil level sensor. Be careful not to drop this sensor.
   - Oil filter and the bracket assembly by removing the stud bolt and 2 nuts
   - Engine Crankshaft Position (CKP) sensor connector
   - Sensor by removing the bolt
   - 12 bolts and 2 nuts to the No. 2 oil pan. Use a gasket cutting tool to separate the No. 2 (lower) oil pan. Be careful not to damage the No. 1 pan while performing this procedure.
   - 2 bolts and 3 nuts and drop down the baffle plate
   - Oil strainer by removing the bolts and nuts
   - Bolts, then carefully pry off the No. 1 oil pan. There are slots for inserting the prybar.

To install:

3. Install or connect the following:
- No. 1 pan. Apply a ⅛inch (3–4mm) bead sealant to the pan mating surface. Bolts: 10mm: 66 inch lbs. (8 Nm); 12mm: 21 ft. lbs. (28 Nm)
- Oil strainer. Bolts and nuts: 66 inch lbs. (8 Nm)
- Baffle plate. Bolts and nuts: 66 inch lbs. (8 Nm)
- No. 2 pan in the same manner as the No. 1 oil pan and tighten the bolts to 66 inch lbs. (8 Nm). Be sure the bolts are 14mm in length.
- CKP sensor. Tighten the bolt to 56 inch lbs. (6 Nm).
- New O-ring in position on the oil filter bracket
- Bracket and tighten the bolt and nuts to 13 ft. lbs. (18 Nm)
- Wiring to the pressure switch
- Oil level sensor and tighten the 4 bolts to 48 inch lbs. (5 Nm). Use a new gasket.
- Dipstick and guide
- Timing belt pulleys and the timing belt components
- Transaxle to the engine
- Engine and transaxle
- All fluids

GS 430 & SC 430

➡ **The No. 1 oil pan cannot be removed with the engine in the vehicle. The engine and transmission must be removed as a unit, then separated. It may be possible to remove the No. 2 oil pan from the vehicle while the engine is still in the vehicle.**

1. Before servicing the vehicle, refer to the Precautions Section.
2. Remove or disconnect the following:
- Engine/transmission assembly
- Oil dipstick and guide
- 12 bolts and 2 nuts. Use a gasket-cutting tool to separate the No. 2 (lower) oil pan. Be careful not to damage the No. 1 pan while performing this procedure.
- 6 bolts and 2 nuts; remove the baffle plate
- 16 bolts, then carefully pry off the No. 1 oil pan

➡ **There are slots for inserting the pry-bar.**

To install:

3. Install or connect the following:
- No. 1 pan. Apply a ⅛ inch (3–4mm) bead on RTV sealant to

the pan mating surface. Bolts: 12mm: 66 inch lbs. (8mm); 14mm: 21 ft. lbs. (28 Nm)
- Baffle plate. Torque bolts and nuts to 66 inch lbs. (8 Nm).
- RTV sealant to the pan mating surface
- No. 2 oil pan. Torque bolts to 66 inch lbs. (8 Nm)
- Dipstick and guide
- Engine/transaxle assembly
- All fluids

3.3L (3MZ-FE) Engine

1. Before servicing the vehicle, refer to the Precautions Section.
2. Remove or disconnect the following:
- Battery negative cable
- RF wheel
- Engine under covers
- RH front fender apron seal
- A/C compressor, alternator and PS drive belts
- Engine stabilizer rod
- Engine mounting stay No2.
- Alternator bracket
- Crankshaft pulley
- Timing belt
- Crankshaft timing pulley
- Exhaust pipe and support brackets
- Oil gauge guide
- Alternator belt adjusting bar
- A/C compressor/clutch assembly
- A/C compressor mounting bracket

3. Separate FR engine mounting insulator

➡ **Do NOT remove the FR engine mounting at this time**

a. Remove bolt and disconnect the power steering return hose
b. Remove the 4 nuts
c. Place a wooden block underneath the engine
d. Jack up the engine and remove the engine mounting insulator

☀☀ WARNING

Be careful not to damage the oil pan

4. Remove or disconnect the following:
- RH engine mounting bracket
- 10 bolts and 2 nuts, gently pry off the oil pan sub assembly No.2.

☀☀ WARNING

Be careful not to damage the oil pan flange area or the contact surface of the engine block.

- Oil strainer sub-assembly
- Flywheel housing under cover.
- Oil pan sub-assembly No. 1
- Engine oil level sensor

To install:

5. Install or connect the following:
6. Install the oil pan sub-assembly No. 1

a. Remove any old oil sealant from contact surface. Clean the surface thoroughly.
b. Apply a continuous bead of sealant 0.12 to 0.16 in (3 to 4 mm) around the block surface, making certain to surround the bolt holes.
c. Install the oil pan within 3 minutes after applying the sealant

➡ **Do NOT expose sealant to engine oil within 2 hours after installing**

d. Install the oil pan using the 17 bolts and 2 nuts. Tighten uniformly in several steps.
Torque to:
- 10 mm head 71 in. lbs (8.0 Nm)
- 12 mm head 14 ft. lbs.(20 Nm)
- 14 mm head 27 ft. lbs (37 Nm)
a. Engine oil level sensor
7. Oil strainer assembly
8. Oil pan sub assembly No. 2
a. Remove any old oil sealant from contact surface. Clean the surface thoroughly.
b. Apply a continuous bead of sealant 0.16 to 0.20 in. (3 to 4 mm) around the block surface, making certain to surround the bolt holes.
c. Install the oil pan within 3 minutes after applying the sealant

➡ **Do NOT expose sealant to engine oil within 2 hours after installing**

d. Install the oil pan using the 10 bolts and 2 nuts. Torque to 71 inch lbs. (8.0 Nm)
9. Install or connect the following
- RH engine mounting bracket; torque bolts "A" & "B" to 40 ft. lbs (54 Nm) and bolt "C" to 32 ft. lbs (43 Nm)
- RH engine mounting insulator; torque nut "A" to 70 ft. lbs. (95 Nm) and nut "B" to 64 ft. lbs (87 Nm)
- FR engine mounting insulator; torque bolt to 64 ft. lbs. (87 Nm) and nut to 38 ft. lbs (52 Nm)
- A/C compressor mounting bracket; torque to 18 ft. lbs (25 Nm)
- A/C compressor/clutch assembly; torque to 18 ft. lbs. (25 Nm)
- Alternator belt adjusting bar; torque

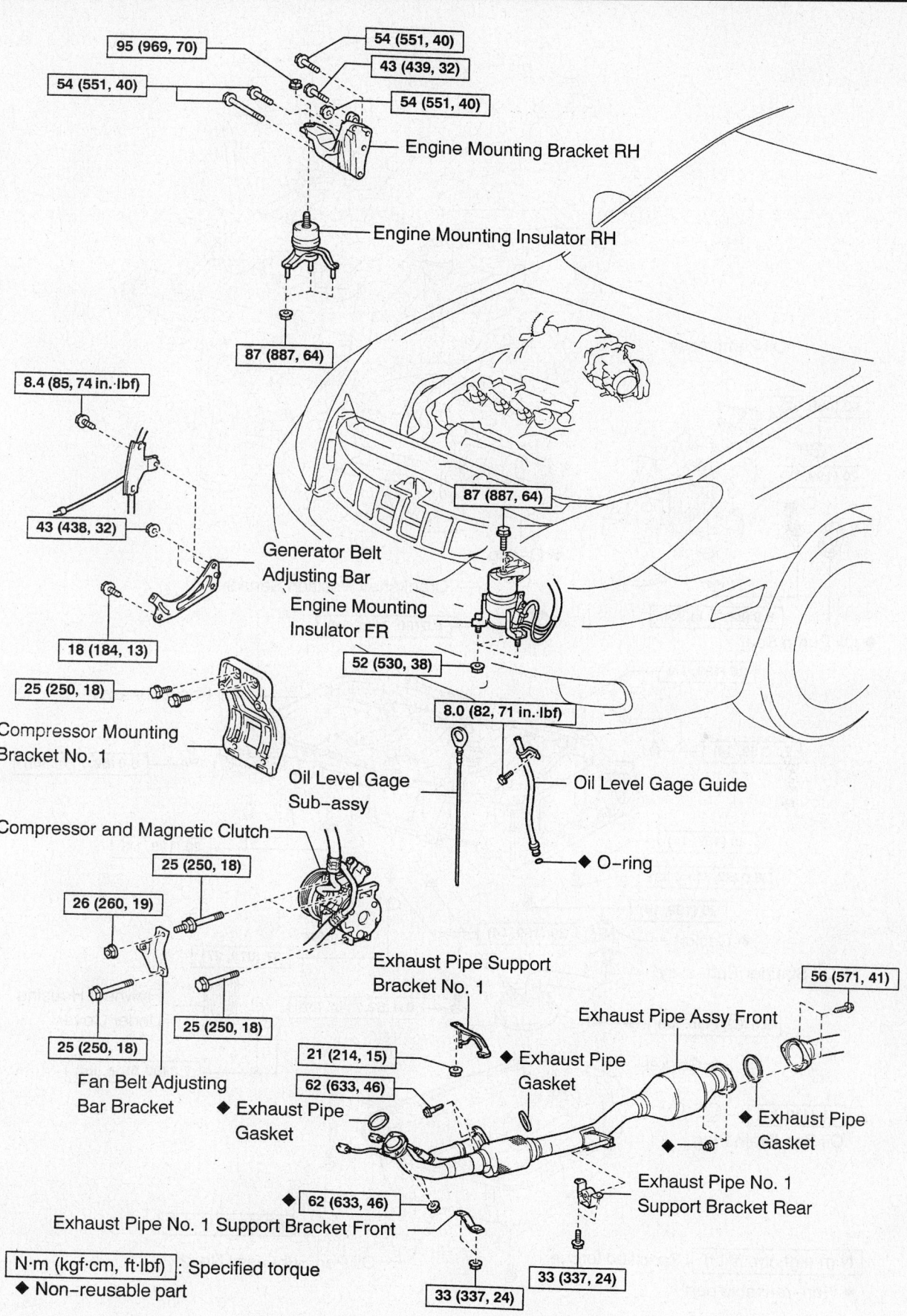

95 (969, 70)

54 (551, 40)

54 (551, 40)

43 (439, 32)

54 (551, 40)

Engine Mounting Bracket RH

Engine Mounting Insulator RH

87 (887, 64)

8.4 (85, 74 in.·lbf)

43 (438, 32)

18 (184, 13)

25 (250, 18)

Generator Belt
Adjusting Bar

Engine Mounting
Insulator FR

87 (887, 64)

52 (530, 38)

8.0 (82, 71 in.·lbf)

Compressor Mounting
Bracket No. 1

Oil Level Gage
Sub–assy

Oil Level Gage Guide

Compressor and Magnetic Clutch

◆ O–ring

25 (250, 18)

26 (260, 19)

Exhaust Pipe Support
Bracket No. 1

Exhaust Pipe Assy Front

56 (571, 41)

25 (250, 18)

25 (250, 18)

21 (214, 15)

62 (633, 46)

◆ Exhaust Pipe
Gasket

◆ Exhaust Pipe
Gasket

Fan Belt Adjusting
Bar Bracket

◆ Exhaust Pipe
Gasket

62 (633, 46)

Exhaust Pipe No. 1
Support Bracket Rear

Exhaust Pipe No. 1 Support Bracket Front

33 (337, 24)

33 (337, 24)

N·m (kgf·cm, ft·lbf) : Specified torque
◆ Non-reusable part

67162-LEXU-G47

Exploded view, component removal for oil pan—3.3L (3MZ-FE) Engine

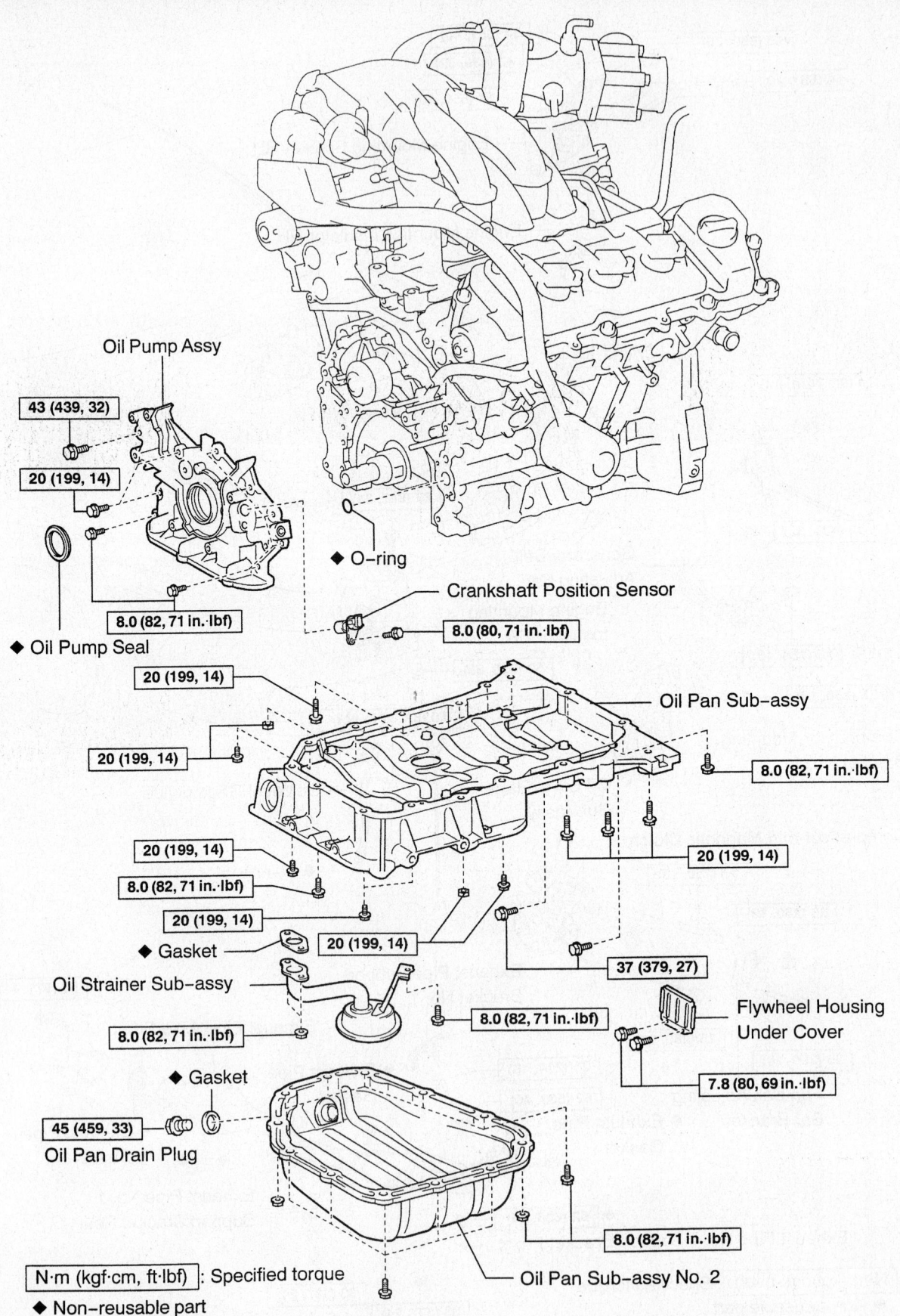

Oil Pump Assy

43 (439, 32)

20 (199, 14)

8.0 (82, 71 in.·lbf)

◆ Oil Pump Seal

◆ O-ring

Crankshaft Position Sensor

8.0 (80, 71 in.·lbf)

20 (199, 14)

20 (199, 14)

Oil Pan Sub-assy

8.0 (82, 71 in.·lbf)

20 (199, 14)

8.0 (82, 71 in.·lbf)

20 (199, 14)

20 (199, 14)

◆ Gasket

Oil Strainer Sub-assy

8.0 (82, 71 in.·lbf)

20 (199, 14)

8.0 (82, 71 in.·lbf)

37 (379, 27)

Flywheel Housing
Under Cover

7.8 (80, 69 in.·lbf)

◆ Gasket

45 (459, 33)

Oil Pan Drain Plug

8.0 (82, 71 in.·lbf)

N·m (kgf·cm, ft·lbf) : Specified torque

◆ Non-reusable part

Oil Pan Sub-assy No. 2

67162-LEXU-G48

Exploded view, removal of oil pan from engine—3.3L (3MZ-FE) Engine

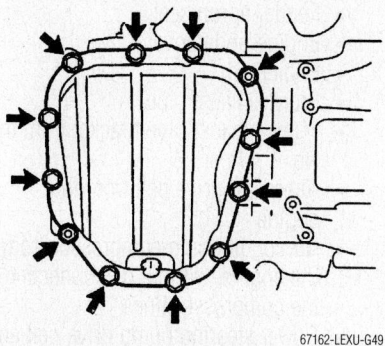

Oil pan sub-assembly No. 2—3.3L (3MZ-FE) Engine

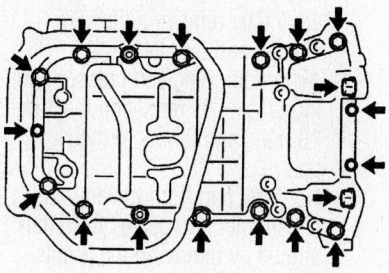

Oil pan sub-assembly No 1.—3.3L (3MZ-FE) Engine

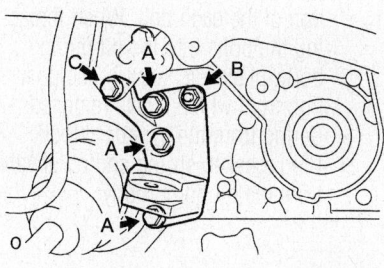

Engine mounting bracket tightening procedure—3.3L (3MZ-FE) Engine

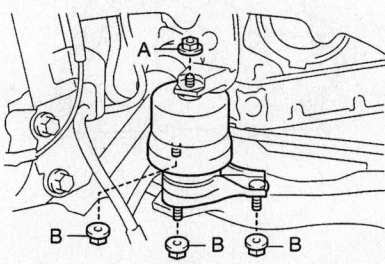

Engine mounting insulator tightening procedure—3.3L (3MZ-FE) Engine

bolt to 18 ft. lbs (25 Nm) and nut to 19 ft. lbs. (26 Nm)
- Oil level gage guide
- Exhaust pipes and support brackets
- Crankshaft timing pulley

- Timing belt assembly
- Crankshaft pulley
- Alternator bracket
- Engine mounting stay
- Engine stabilizer rod
- Alternator, PS pump and A/C drive belts
- RH fender apron seal
- RF wheel
- Engine under covers
- Negative battery cable
10. System initialization

Oil Pump

REMOVAL & INSTALLATION

2.5L (4GR-FSE), 3.0L (3GR-FSE) and 3.5L (2GR-FSE) Engines

➡The oil pump cannot be removed with the engine in the vehicle. The engine and transmission must be removed as a unit, then separated.

1. Before servicing the vehicle, refer to the Precautions Section.
2. Remove or disconnect the following:
   - Engine/transmission assembly
   - Front differential assembly (AWD)
   - Serpentine drive belt
   - No. 2 idler pulley
   - Alternator
   - A/C compressor unit, if necessary
   - Left and right engine mounting brackets
   - Serpentine belt tensioner
   - Water pump pulley
   - Fuel injector driver
   - Intake air surge tank assembly and No. 2 surge tank stay
   - Water hose joint
   - Crankshaft pulley
   - Water inlet
   - Oil pan assembly
   - Oil strainer
   - No. 1 and 2 fuel pipes
   - High pressure side fuel pump
   - Ignition coil assembly
   - No. 1 and 2 oil pipes
   - Left and right cylinder head covers
3. Remove the timing chain cover assembly as follows:
   a. Remove bolt and wiring harness clamp bracket.
   b. Remove 25 mounting bolts and 2 mounting nuts.
   c. Remove the timing chain cover by prying between the timing chain cover and cylinder head or cylinder block with a screwdriver.
   d. Remove the gasket.

➡The oil pump assembly is incorporated into the back of the timing chain cover. The oil pump assembly can be disassembled from the back of the timing chain cover for inspection purposes.

To install:
4. Install the timing chain cover assembly as follows:
   a. Apply seal packing in a continuous line of 0.197–0.217 inches (5.0–5.5mm) in diameter to the engine at the seem where the cylinder head meets the camshaft bearing cap assembly and the cylinder head meets the cylinder block

➡Be sure to clean, degrease and dry the contact surfaces before applying the seal packing. Install the component within 3 minutes after applying seal packing. Do not start the engine for at least 2 hours after installing.

   b. Apply seal packing in a continuous line of 0.138–0.158 inches (3.5–4.0mm) in diameter to the timing chain cover
   c. Install a new gasket
   d. Align the oil pump drive rotor spline and the crankshaft. Install the spline and chain cover to the crankshaft.

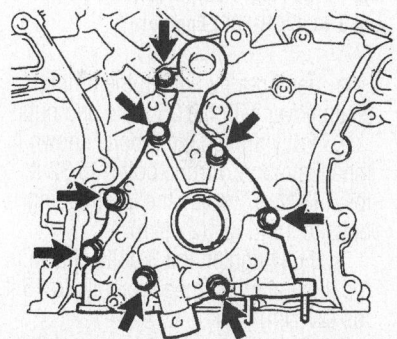

Location of mounting bolts for oil pump cover behind the timing chain cover—2.5L (4GR-FSE), 3.0L (3GR-FSE) and 3.5L (2GR-FSE) Engines

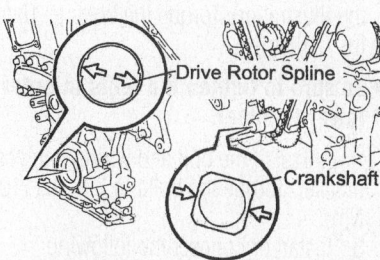

Align the oil pump drive rotor spline and the crankshaft—2.5L (4GR-FSE), 3.0L (3GR-FSE) and 3.5L (2GR-FSE) Engines

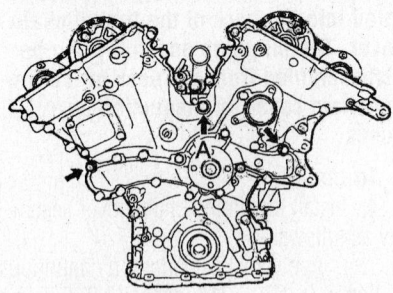

09490_LEXU_G0091

Location of 3 bolts to be tightened first (location of bolt A shown)—2.5L (4GR-FSE), 3.0L (3GR-FSE) and 3.5L (2GR-FSE) Engines

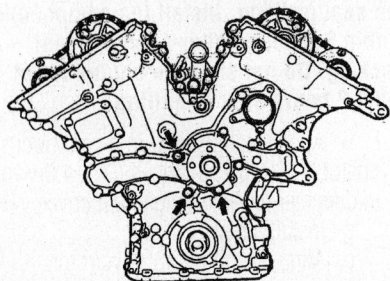

09490_LEXU_G0092

Location of 3 bolts to be tightened second—2.5L (4GR-FSE), 3.0L (3GR-FSE) and 3.5L (2GR-FSE) Engines

e. Temporarily tighten the timing chain cover with the 25 bolts and nuts.

f. Fully tighten the 3 bolts shown in the illustration. Torque bolt A to 32 ft. lbs. (43 Nm). Torque the 2 remaining bolts to 15 ft. lbs (21 Nm).

g. Fully tighten the 3 bolts shown in the illustration. Torque the bolts to 15 ft. lbs (21 Nm).

h. Fully tighten the 7 bolts and 2 nuts shown in the illustration. Torque the bolts to 15 ft. lbs (21 Nm).

➡ **Be sure to tighten the bolts and nuts in order of upper to lower.**

i. Fully tighten the 12 bolts shown in the illustration. Torque the bolts to 15 ft. lbs (21 Nm).

➡ **Be sure to tighten the bolts in order of lower to upper.**

j. Install the bolt and wiring harness bracket. Torque the bolts to 7 ft. lbs (10 Nm).

5. Install or connect the following:
- Left and right cylinder head covers
- No. 1 and 2 oil pipes
- Ignition coil assembly
- High pressure side fuel pump

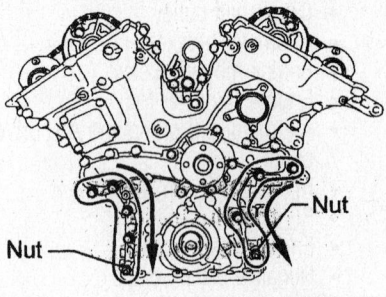

09490_LEXU_G0093

Location of 7 bolts and 2 nuts to be tightened third—2.5L (4GR-FSE), 3.0L (3GR-FSE) and 3.5L (2GR-FSE) Engines

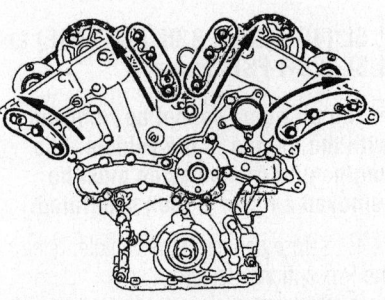

09490_LEXU_G0094

Location of 12 bolts to be tightened fourth—2.5L (4GR-FSE), 3.0L (3GR-FSE) and 3.5L (2GR-FSE) Engines

- No. 1 and 2 fuel pipes
- Oil strainer
- Oil pan assembly
- Water inlet
- Crankshaft pulley. Torque the bolt to 192 ft. lbs (260 Nm).
- Water hose joint
- Intake air surge tank assembly and No. 2 surge tank stay
- Fuel injector driver
- Water pump pulley. Torque the bolt to 15 ft. lbs (21 Nm).
- Serpentine belt tensioner
- Left and right engine mounting brackets
- A/C compressor unit, if necessary
- Alternator
- No. 2 idler pulley
- Serpentine drive belt
- Front differential assembly (AWD)
- Engine/transmission assembly

3.0L (1MZ-FE) Engine

1. Before servicing the vehicle, refer to the Precautions Section.

2. Remove or disconnect the following:
- Negative battery cable from the battery
- Right front wheel

- Fender apron seal
- Engine undercover
- Engine oil
- Front exhaust pipe
- Front exhaust pipe bracket from the No. 1 oil pan
- Alternator drive belt from the engine
- Air conditioning compressor from the engine, without disconnecting the compressor lines
- Power steering pump drive belt and adjusting strut
- Timing belt from the engine
- Timing belt pulleys
- Rear timing belt cover from the engine by removing the wire clamps and 6 bolts
- Air conditioning compressor housing bracket by removing the 3 bolts.
- 10 bolts and 2 nuts to the No. 2 oil pan
- No. 2 oil pan from the engine
- Oil strainer and gasket from the engine by removing the 3 nuts
- No. 1 oil pan
- Baffle plate from the No. 1 oil pan
- Crankshaft Position (CKP) sensor by removing the connector and bolt
- Oil pump. Make a note of the position of the each bolt. When replacing the bolts into the oil pump body, place each bolt in the position from which it was removed.
- O-ring from the cylinder block
- Plug, gasket, spring and relief valve from the oil pump body

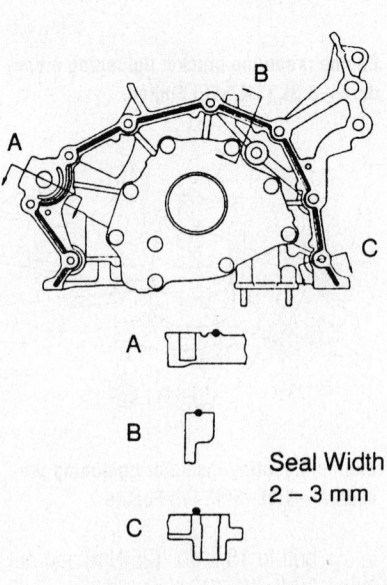

7923LG89

Apply sealant to the mounting surface of the oil pump in the areas shown—3.0L (1MZ-FE) engine

To install:

3. Install or connect the following:
- Driven rotors, drive, pump body cover, then install the 9 screws
- Relief valve, spring, gasket and the plug to the oil pump body
- New O-ring on the cylinder block
- RTV sealant to the oil pump as shown
- Pump on the engine block. Be sure to engage the spline teeth of the oil pump drive gear with the large teeth of the crankshaft.
- The 9 bolts to the oil pump and uniformly tighten the bolts in several passes. Tighten the bolts to:

10mm: 69 inch lbs. (8 Nm); 12mm: 14 ft. lbs. (20 Nm).
- CKP sensor and bolt. Tighten the bolt to 69 inch lbs. (8 Nm).
- Baffle plate to the No. 1 oil pan and tighten to 69 inch lbs. (8 Nm).
- No. 1 oil pan Uniformly tighten the bolts and nuts in several passes: 10mm–69 inch lbs. (8 Nm); 12mm–14 ft. lbs. (20 Nm); 14mm–27 ft. lbs. (37 Nm)
- Flywheel housing undercover with the 2 bolts. Tighten the bolts to 69 inch lbs. (8 Nm).
- Oil strainer with the 3 nuts. Tighten the nuts to 69 inch lbs. (8 Nm).

- No. 2 oil pan
- RTV sealant to the oil pan and engine block
- No. 2 oil pan. Uniformly tighten the bolts and nuts in several passes to 69 inch lbs. (8 Nm).
- Remaining components
- Right front wheel and lower the vehicle
- Engine with oil
- Negative battery cable to the battery

3.0L (2JZ-GE) Engine

1. Before servicing the vehicle, refer to the Precautions Section.
2. Remove or disconnect the following:

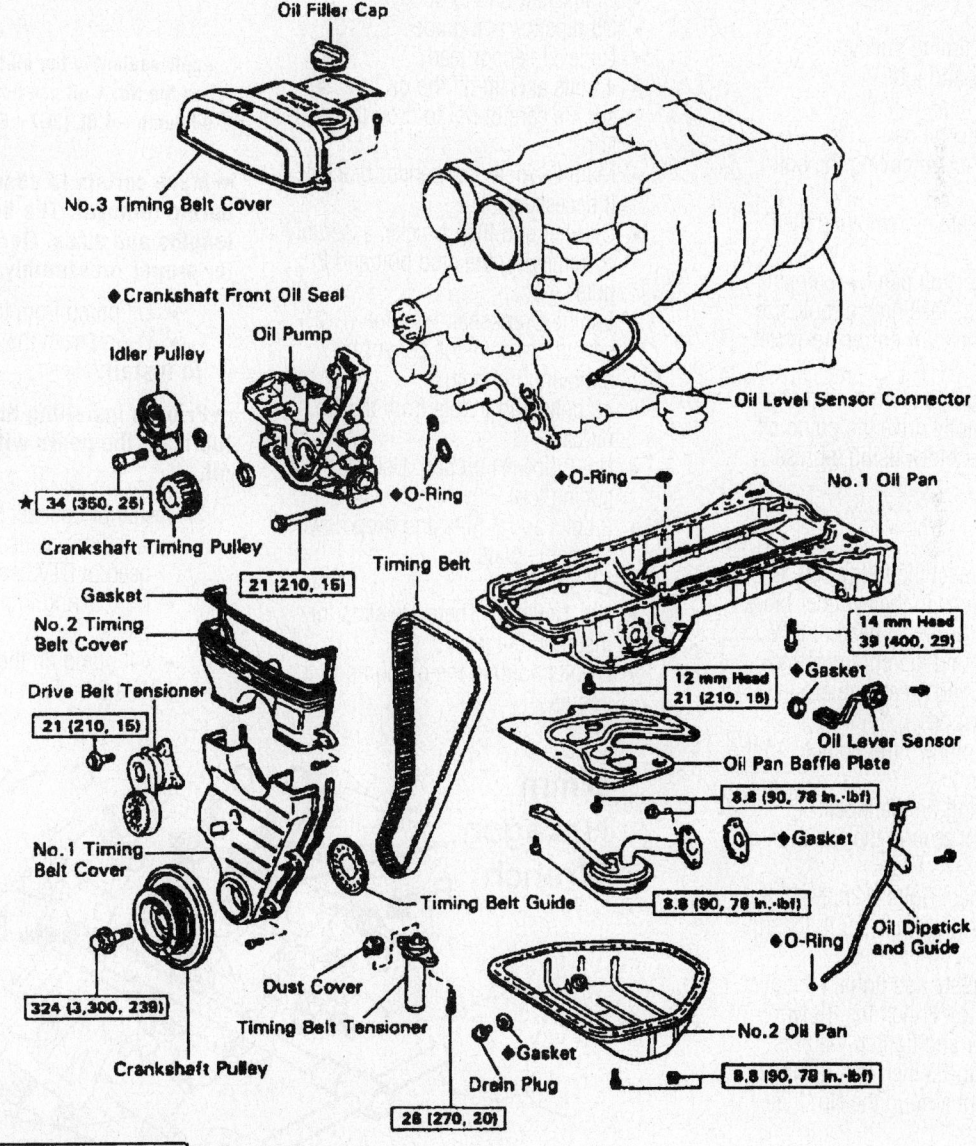

N·m (kgf·cm, ft·lbf) : Specified torque
♦ Non-reusable part
★ Precoated part

Exploded view of the oil pump and related component mountings—3.0L (2JZ-GE) engine

7923LG90

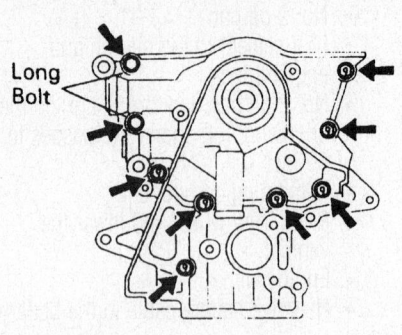

Oil pump mounting bolt installation locations—3.0L (2JZ-GE) engine

- Engine and transmission
- Timing belt
- Idler pulley
- Crankshaft timing pulley
- Oil dipstick and tube
- Oil level sensor
- No. 2 (lower) oil pan
- Oil strainer by removing the bolt and 2 nuts
- Oil baffle plate by removing the 6 bolts
- No. 1 (upper) oil pan by removing the 22 bolts. Take note of bolt size and placement for correct re-installation.
- 9 mounting bolts to the oil pump body. Carefully drive the pump off the cylinder block using a brass drift.
- 2 O-rings

To install:

3. Install or connect the following:
- 2 new O-rings in the cylinder block
- A 1/8 inch (3–4mm) bead of RTV sealant around the pump mating surface, taking great care around the oil passages
- Pump and tighten the bolts to 15 ft. lbs. (21 Nm)
- A new O-ring on the block
- RTV sealant around the No. 1 oil pan
- No. 1 oil pan. Bolts: 12mm–15 ft. lbs. (21 Nm); 14mm–29 ft. lbs. (39 Nm)
- Oil baffle plate and tighten the nuts and bolts to 78 inch lbs. (9 Nm)
- Oil strainer and tighten the nuts and bolts to 78 inch lbs. (9 Nm)
- RTV sealant around the No. 2 oil pan
- No. 2 oil pan and tighten the bolts to 78 inch lbs. (9 Nm)
- Oil lever sensor with a new gasket and tighten the bolts to 48 inch lbs. (6 Nm)

- Oil dipstick with a new O-ring
- Remaining components
- All fluids
- Negative battery cable

4.3L (3UZ-FE) Engine

➡The oil pump cannot be removed with the engine in the vehicle. The engine and transmission must be removed as a unit, then separated.

1. Before servicing the vehicle, refer to the Precautions Section.
2. Remove or disconnect the following:
- Engine/transmission assembly
- Timing belt
- Idler pulleys
- Crankshaft timing pulley
- Oil dipstick and guide
- Oil level sensor lead
- 4 bolts and lift off the oil level sensor. Be careful not to drop this sensor.
- Main Oxygen (O_2) sensor bracket, if necessary
- Oil filter and filter bracket assembly by removing the stud bolt and 2 nuts
- Engine Crankshaft Position (CKP) sensor. Remove the sensor by removing the bolt.
- 12 bolts and 2 nuts from the No. 2 oil pan
- No. 2 (lower) oil pan. Use a gasket-cutting tool
- 2 bolts and 3 nuts and drop down the baffle plate
- Oil strainer
- No. 1 oil pan. There are slots for inserting the prybar.
- 8 bolts holding the oil pump to the engine

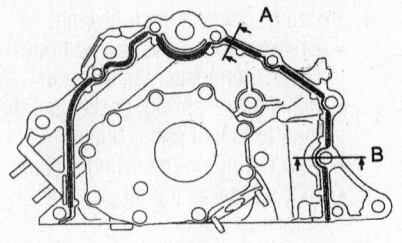

Seal Diameter: 2 to 3 mm

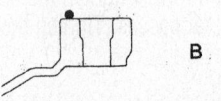

Apply sealant to the oil pump (as shown) and the No. 1 oil pan before installing the oil pump—4.3L (3UZ-FE) Engine

➡Make certain to observe bolt position during removal. The bolts are different lengths and sizes. Record their position for proper reassembly.

- Oil pump from the engine block
- O-ring from the block

To install:

➡Prior to installing the oil pump, lubricate the gears with clean engine oil.

3. Install or connect the following:
- A 2–3mm wide (0.08–0.12 in.) bead of RTV sealant to the oil pump
- New O-ring in position on the block
- Oil pump on the engine
- The 8 bolts in their correct loca-

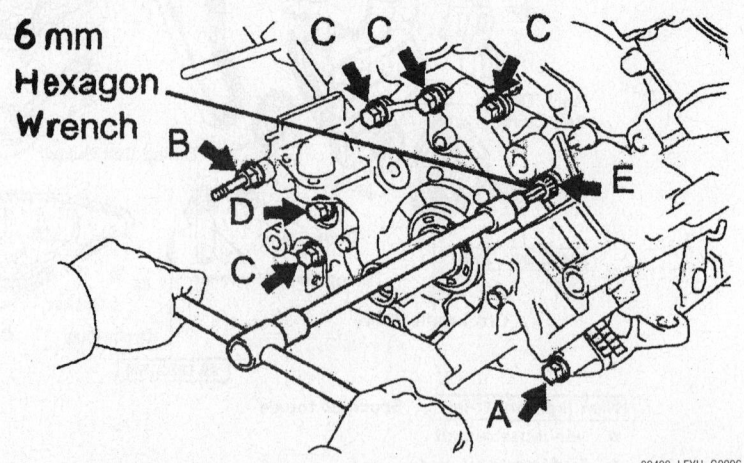

4.3L engine oil pump mounting bolt locations, according to bolt lengths—(A) 1.97 in. (50mm), (B) 4.17 in. (106mm), (C) 1.18 in. (30mm), (D) 1.73 in. (44mm) and (E) 1.10 in. (28mm)

tions. Tighten the bolts with 12mm or 6mm heads to 12 ft. lbs. (16 Nm) and the bolts with 14mm heads to 22 ft. lbs. (30 Nm).

- A ⅛ inch (3–4mm) bead of RTV sealant to the pan mating surface.
- No. 1 pan. Bolts–10mm: 66 inch lbs. (8 Nm); 12mm: 21 ft. lbs. (28 Nm)
- Oil strainer and tighten the bolts to 66 inch lbs. (8 Nm)
- Baffle plate and tighten the bolts and nuts to 66 inch lbs. (8 Nm)
- Remaining components
- Engine/transaxle

3.3L (3MZ-FE) Engine

1. Follow engine oil pan removal procedure then perform the following steps.
2. Remove or disconnect the following:
- Crankshaft position sensor
- Oil pump assembly
- 9 bolts
3. Remove oil pump by prying between the oil pump and bearing cap
4. Remove the oil ring

To install:

5. Install the oil pump seal using proper driver.
   a. Tap in the seal until it is flush with the oil pump body
   b. Apply multi-purpose grease to the seal lip.

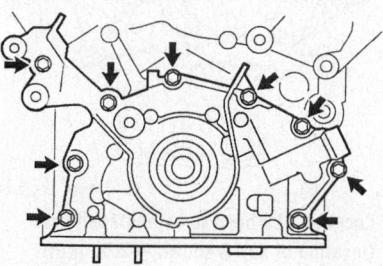

Oil pump removal bolt procedure—3.3L (3MZ-FE) Engine

67162-LEXU-G53

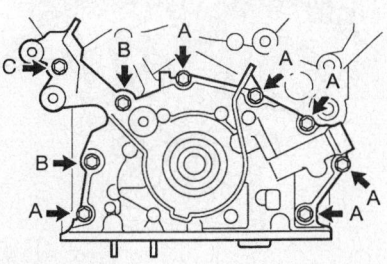

Oil pump bolt installation procedure— 3.3L (3MZ-FE) Engine

67162-LEXU-G54

6. Install oil pump assembly
   a. Remove any old sealant from the mating surfaces
   b. Apply a light coat of clean engine oil to the O-ring, then place it on the engine block.
   c. Thoroughly clean the mating surface of any oil or old sealant
   d. Apply a continuous bead of sealant on the oil pump body, making certain to surround the bolt holes.
   e. Install the oil pump within 3 minutes after applying the sealant.

➡**Do NOT expose the sealant to engine oil within 2 hours after installing**

   f. Align the key of the oil pump drive gear with the keyway located on the crankshaft, then slide the oil pump into place.
   g. Install the oil pump with the 9 bolts. Tighten the bolts uniformly in several steps. Torque to: Bolt A 71 in. lbs (8.0 Nm), Bolt B14 ft lbs. (20 Nm), Bolt C 32 ft. lbs. (43 Nm)
7. Install crankshaft position sensor
8. Install oil pans using oil pan installation procedure

Timing Chain

REMOVAL & INSTALLATION

2.5L (4GR-FSE), 3.0L (3GR-FSE) and 3.5L (2GR-FSE) Engines

➡**The timing chain cannot be removed with the engine in the vehicle. The engine and transmission must be removed as a unit, then separated.**

1. Before servicing the vehicle, refer to the Precautions Section.
2. Remove or disconnect the following:
- Engine/transmission assembly
- Front differential assembly (AWD)
- Serpentine drive belt
- No. 2 idler pulley
- Alternator
- A/C compressor unit, if necessary
- Left and right engine mounting brackets
- Serpentine belt tensioner
- Water pump pulley
- Fuel injector driver
- Intake air surge tank assembly and No. 2 surge tank stay
- Water hose joint
- Crankshaft pulley
- Water inlet
- Oil pan assembly
- Oil strainer

- No. 1 and 2 fuel pipes
- High pressure side fuel pump
- Ignition coil assembly
- No. 1 and 2 oil pipes
- Left and right cylinder head covers
3. Remove the timing chain cover assembly as follows:
   a. Remove bolt and wiring harness clamp bracket.
   b. Remove 25 mounting bolts and 2 mounting nuts.
   c. Remove the timing chain cover by

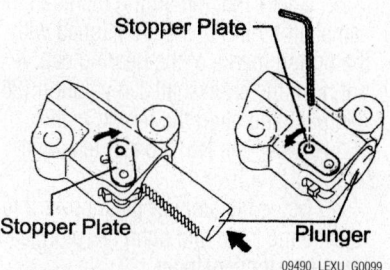

09490_LEXU_G0099

Aligning the timing marks at the block bore centerline and camshaft bearing caps—2.5L (4GR-FSE), 3.0L (3GR-FSE) and 3.5L (2GR-FSE) Engines

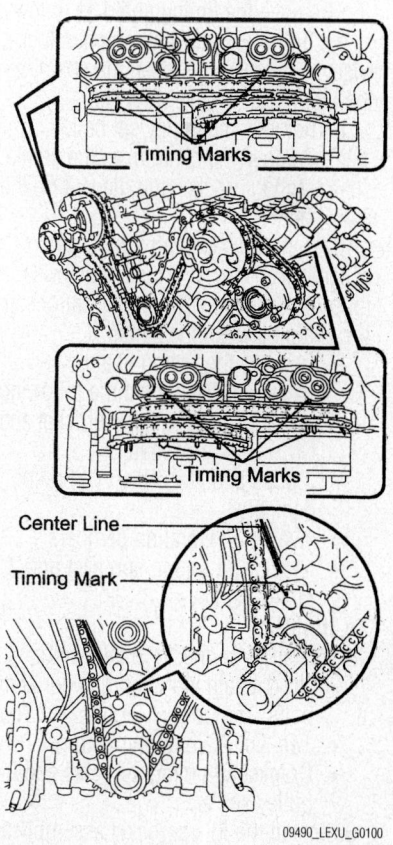

09490_LEXU_G0100

Chain tensioner component showing stopper plate and plunger—2.5L (4GR-FSE), 3.0L (3GR-FSE) and 3.5L (2GR-FSE) Engines

prying between the timing chain cover and cylinder head or cylinder block with a screwdriver.

d. Remove the gasket.

e. Remove the timing chain case oil seal.

4. Set the No. 1 cylinder to TDC/compression as follows:

a. Temporarily tighten the pulley set bolt.

b. Set the timing mark on the crank angle sensor plate to the right block bore center line (TDC/compression).

c. Check that the timing marks of the camshaft timing gears are aligned with the timing marks of the bearing cap. If not, turn the crankshaft 1 revolution (360 degrees) and align the timing marks.

5. Remove the No. 1 chain tensioner assembly as follows:

a. Move the stopper plate upward to release the lock, and push the plunger deep into the tensioner.

b. Move the stopper plate downward to set the lock, and insert a hexagon wrench into the stopper plate's hole.

c. Remove the 2 bolts and chain tensioner.

• Chain tensioner slipper

6. Remove the timing chain as follows:

a. Turn the crankshaft counterclockwise 10 degrees to loosen the chain of the crankshaft timing sprocket.

b. Remove the pulley set bolt.

c. Remove the chain from the crankshaft timing sprocket and place it on the crankshaft.

d. Turn the camshaft timing gear assembly on the right bank clockwise (approx. 60 degrees). Be sure to loosen the chain between the banks.

e. Remove the timing chain.

7. Remove or disconnect the following:

• No. 2 idle gear shaft, sprocket and No. 1 idle gear shaft

• 2 bolts and No. 1 chain vibration damper

• Two No. 2 vibration dampers

• Crankshaft timing sprocket and 2 pulley set keys

To install:

8. Install or connect the following:

• No. 1 chain vibration damper and 2 bolts. Torque to 17 ft. lbs. (23 Nm).

• Two No. 2 vibration dampers

• Crankshaft timing sprocket and 2 pulley set keys

9. Install the idle sprocket assembly as follows:

a. Apply a light coat of engine oil to the rotating surface of the No. 1 idle gear shaft.

b. Temporarily install the No. 1 idle gear shaft and idle sprocket with the No. 2 idle gear shaft while aligning the knock pin of the No. 1 idle gear with the knock pin groove of the cylinder block.

➡ **Be careful of the idle gear direction.**

c. Using a 10mm hexagon wrench, tighten the No. 2 idle gear shaft to 44 ft. lbs. (60 Nm). Check that the idle sprocket turns smoothly.

10. Install the timing chain as follows:

a. Align the mark plate and timing mark and install the chain.

b. Do not pass the chain over the crankshaft, just put it on it.

c. Turn the camshaft timing gear assembly on the right bank counterclockwise to tighten the chain between the banks.

➡ **When the idle sprocket is reused, align the chain plate with the mark where the plate had been in order to tighten the chain between the banks.**

d. Align the mark plate and timing mark and install the chain onto the crankshaft timing sprocket.

e. Temporarily tighten the pulley set bolt.

f. Turn the crankshaft clockwise to set it to the right block bore center line (TDC/compression).

• Chain tensioner slipper

11. Install the No. 1 chain tensioner assembly as follows:

a. Move the stopper plate upward to release the lock, and push the plunger deep into the tensioner.

b. Move the stopper plate downward to set the lock, and insert a hexagon wrench into the hole of the stopper plate.

c. Install the chain tensioner with the 2 bolts and torque to 7 ft. lbs. (10 Nm).

d. Remove the lock pin of the chain tensioner. Check that each timing mark is aligned with the crankshaft at the TDC/compression.

e. Remove the pulley set bolt.

12. Install the timing chain cover assembly as follows:

a. Install a new timing chain case oil seal.

b. Apply seal packing in a continuous line of 0.197–0.217 inches (5.0–5.5mm) in diameter to the engine at the seem where the cylinder head meets the camshaft bearing cap assembly and the cylinder head meets the cylinder block

➡ **Be sure to clean, degrease and dry the contact surfaces before applying the seal packing. Install the component within 3 minutes after applying seal packing. Do not start the engine for at least 2 hours after installing.**

c. Apply seal packing in a continuous line of 0.138–0.158 inches (3.5–4.0mm) in diameter to the timing chain cover

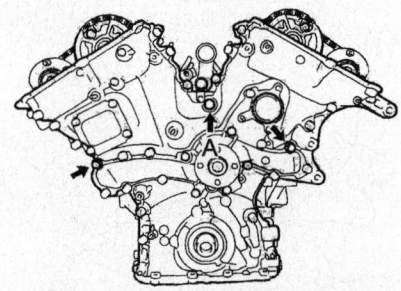

09490_LEXU_G0091

Location of 3 bolts to be tightened first (location of bolt A shown)—2.5L (4GR-FSE), 3.0L (3GR-FSE) and 3.5L (2GR-FSE) Engines

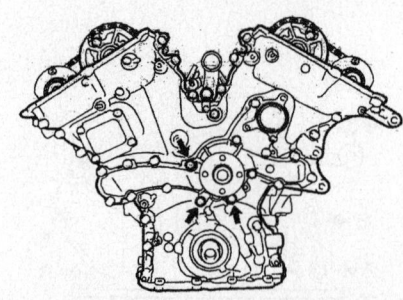

09490_LEXU_G0092

Location of 3 bolts to be tightened second—2.5L (4GR-FSE), 3.0L (3GR-FSE) and 3.5L (2GR-FSE) Engines

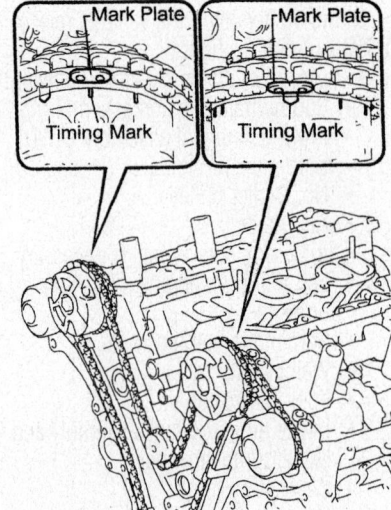

09490_LEXU_G0101

Aligning the mark plates to the timing marks—2.5L (4GR-FSE), 3.0L (3GR-FSE) and 3.5L (2GR-FSE) Engines

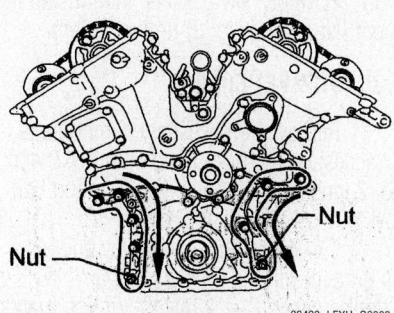

Location of 7 bolts and 2 nuts to be tightened third—2.5L (4GR-FSE), 3.0L (3GR-FSE) and 3.5L (2GR-FSE) Engines

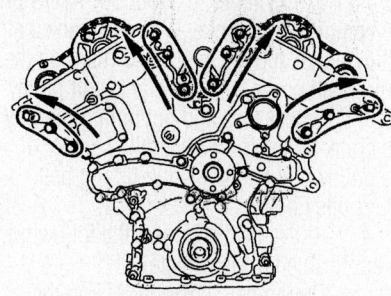

09490_LEXU_G0094

Location of 12 bolts to be tightened fourth—2.5L (4GR-FSE), 3.0L (3GR-FSE) and 3.5L (2GR-FSE) Engines

d. Install a new gasket

e. Align the oil pump drive rotor spline and the crankshaft. Install the spline and chain cover to the crankshaft.

f. Temporarily tighten the timing chain cover with the 25 bolts and nuts.

g. Fully tighten the 3 bolts shown in the illustration. Torque bolt A to 32 ft. lbs. (43 Nm). Torque the 2 remaining bolts to 15 ft. lbs (21 Nm).

h. Fully tighten the 3 bolts shown in the illustration. Torque the bolts to 15 ft. lbs (21 Nm).

i. Fully tighten the 7 bolts and 2 nuts shown in the illustration. Torque the bolts to 15 ft. lbs (21 Nm).

➡ **Be sure to tighten the bolts and nuts in order of upper to lower.**

j. Fully tighten the 12 bolts shown in the illustration. Torque the bolts to 15 ft. lbs (21 Nm).

➡ **Be sure to tighten the bolts in order of lower to upper.**

k. Install the bolt and wiring harness bracket. Torque the bolts to 7 ft. lbs (10 Nm).

13. Install or connect the following:
• Left and right cylinder head covers
• No. 1 and 2 oil pipes

• Ignition coil assembly
• High pressure side fuel pump
• No. 1 and 2 fuel pipes
• Oil strainer
• Oil pan assembly
• Water inlet
• Crankshaft pulley. Torque the bolt to 192 ft. lbs (260 Nm).
• Water hose joint
• Intake air surge tank assembly and No. 2 surge tank stay
• Fuel injector driver
• Water pump pulley. Torque the bolt to 15 ft. lbs (21 Nm).
• Serpentine belt tensioner
• Left and right engine mounting brackets
• A/C compressor unit, if necessary
• Alternator
• No. 2 idler pulley
• Serpentine drive belt
• Front differential assembly (AWD)
• Engine/transmission assembly

Timing Belt

REMOVAL & INSTALLATION

3.0L (1MZ-FE) Engine

1. Before servicing the vehicle, refer to the Precautions Section.

2. Remove all necessary components for access to the upper timing belt cover. Remove the 8 bolts and lift off the upper (No. 2) cover.

3. Paint matchmarks on the timing belt at all points where it meshes with the pulleys and the lower timing cover.

4. Set the No. 1 cylinder to Top Dead Center (TDC) of the compression stroke and check that the timing marks on the camshaft timing pulleys are aligned with those on the

No. 3 timing cover. If not, turn the engine 1 complete revolution (360 degrees) and check again.

5. Remove or disconnect the following:
• Timing belt tensioner and the dust boot

6. Turn the right camshaft pulley clockwise slightly to release tension, then remove the timing belt from the pulleys.
• Upper (No. 3) and lower (No. 1) timing belt covers
• Timing belt guide
• Timing belt from the engine

➡ **If the timing belt is to be reused, draw a directional arrow on the timing belt in the direction of engine rotation (clockwise) and place matchmarks on the timing belt and crankshaft gear to match the drilled mark on the pulley.**

To install:

➡ **If the old timing belt is being reinstalled, be sure the directional arrow is facing in the original direction and that the belt and crankshaft gear matchmarks are properly aligned.**

7. Install the lower (No. 1) timing cover and tighten the bolts.

8. Set the No. 1 cylinder to TDC again. Turn the right camshaft until the knock pin hole is aligned with the timing mark on the No. 3 belt cover. Turn the left pulley until the marks on the pulley are aligned with the mark on the No. 3 timing cover.

9. Check that the mark on the belt matches with the edge of the lower cover. If not, shift it on the crank pulley until it does. Turn the left pulley clockwise a bit and align the mark on the timing belt with the timing mark on the pulley. Slide the belt over the left pulley. Now move the pulley until the marks on it align with the one on the No. 3 cover. There should be tension on the belt

GA16DE engine

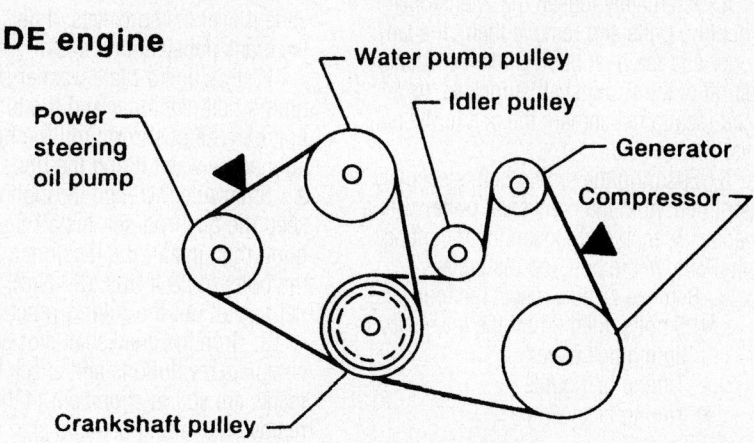

Camshaft and crankshaft pulley positioning for timing belt installation—3.0L (1MZ-FE) engine

between the crankshaft pulley and the left camshaft pulley.

10. Align the installation mark on the timing belt with the mark on the right side camshaft pulley. Hang the belt over the pulley with the flange facing inward. Align the timing marks on the right pulley with the one on the No. 3 cover and slide the pulley onto the end of the camshaft. Move the pulley until the camshaft knock pin hole is aligned with the groove in the pulley, then install the knock pin. Tighten the bolt to 55 ft. lbs. (75 Nm).

11. Position a plate washer between the timing belt tensioner and the block, then press in the pushrod until the holes are aligned between it and the housing. Slide a 0.05 in. Allen wrench through the hole to keep the pushrod set. Install the dust boot, and then install the tensioner. Tighten the bolts to 20 ft. lbs. (26 Nm). Don't forget to pull out the Allen wrench.

12. Turn the crankshaft clockwise 2 complete revolutions and check that all marks are still in alignment. If they aren't, remove the timing belt and start over again.

13. Install the remaining components.

3.0L (2JZ-GE) Engine

1. Before servicing the vehicle, refer to the Precautions Section.

2. Remove all necessary components for access to the upper timing belt covers. Using a 5mm Allen wrench, remove the 9 bolts and lift off the two upper (No. 2 and No. 3) timing belt covers.

3. Rotate the crankshaft pulley clockwise so its groove is aligned with the **0** mark in the No. 1 (lower) timing cover. Check that the timing marks on the camshaft timing sprockets are aligned with the marks on the No. 4 (inner) cover. If not, rotate the crankshaft 1 complete revolution (360 degrees).

4. Alternately loosen the 2 tensioner mounting bolts and remove them, the tensioner and the dust boot. Slide the timing belt off of the 2 camshaft sprockets. Its a good idea to matchmark the belt to the pulleys.

5. Ensuring the timing belt is securely supported, hold the crankshaft pulley with a spanner wrench and loosen the mounting bolt. Remove the bolt and the pulley.

6. Remove or disconnect the following:
- 5 bolts, then lift off the lower No. 1 timing belt cover
- Timing belt guide
- Timing belt

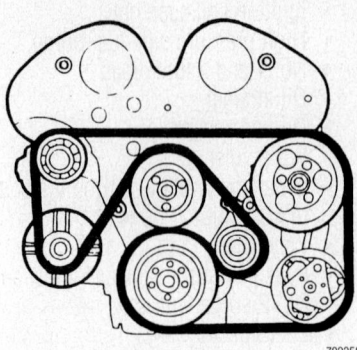

79235G46

Set the engine to TDC by aligning the marks before removing the lower timing cover—Lexus 3.0L (2JZ-GE) engine

➡️**If the timing belt is to be reused, draw a directional arrow on the timing belt in the direction of engine rotation (clockwise) and place matchmarks on the timing belt and crankshaft gear to match the drilled mark on the pulley.**

To install:

7. Install the timing belt on the crankshaft timing pulley and the idler pulleys.

➡️**If the old timing belt is being reinstalled, be sure the directional arrow is facing in the original direction and that the belt and crankshaft gear matchmarks are properly aligned.**

8. Install the timing belt guide. Install the lower (No. 1) timing cover and tighten the bolts.

9. Align the crankshaft pulley set key with the key groove on the pulley and slide the pulley on. Tighten the bolt to 239 ft. lbs. (324 Nm).

10. Set the No. 1 cylinder to TDC again. Turn the camshaft until the sprocket timing marks are aligned with the timing marks on the No. 4 belt cover.

11. Check that the marks on the belt matches with those on the sprockets, then slide it over the sprockets. If not, shift it on the crank pulley until it does.

12. Position a plate washer between the timing belt tensioner and the block, then press in the pushrod until the holes are aligned between it and the housing. Slide a 1.5mm Allen wrench through the hole to keep the pushrod set. Install the dust boot, then install the tensioner. Tighten the bolts to 20 ft. lbs. (26 Nm). Don't forget to pull out the Allen wrench.

13. Turn the crankshaft clockwise two complete revolutions and check that all marks are still in alignment. If they aren't, remove the timing belt and start over again.

14. Position new gaskets, then install the upper (No. 2 and No. 3) timing covers.

4.3L (3UZ-FE) Engine

1. Remove all necessary components for access to the right-hand side No. 3 and No. 2, and left-hand side No. 2 timing belt covers, then remove the covers.

2. Turn the crankshaft pulley and align it's groove with the timing mark **0** of the No. 1 timing cover. Check that the timing marks of the camshaft timing pulleys and timing belt rear plates are aligned. If not, turn the crankshaft 1 full revolution (360 degrees).

3. Loosen crankshaft pulley bolt, then set the No. 1 cylinder to about 50° ATDC on the compression stroke. With the crankshaft pulley notch aligned with the **0** mark on the timing belt cover, turn the crankshaft pulley about 50° clockwise and put timing mark of the crank pulley in line with the centers of the pulley bolt and the No. 2 timing belt idler pulley bolt.

4. Remove or disconnect the following, as applicable for the vehicle:
- Timing belt tensioner. Using the proper tool, loosen the tension between the left side and right side timing pulleys by slightly turning the left side camshaft clockwise.
- Timing belt from the camshaft timing pulleys
- Power steering pump pulley
- Alternator
- Drive belt tensioner
- Bolt and timing pulleys, using the proper tool
- Bolt and the crankshaft pulley with the proper tool.
- Fan bracket
- Mounting bolts and the No. 1 timing belt cover
- 2 upper and lower timing belt covers
- Timing belt guide (No. 1 crank position sensor plate)
- Timing belt
- No. 1 and No. 2 timing belt idler pulleys, if necessary.

➡️**If the timing belt is to be reused, draw a directional arrow on the timing belt in the direction of engine rotation (clockwise) and place matchmarks on the timing belt and crankshaft gear to match the drilled mark on the pulley.**

To install:

5. Align the installation mark on the timing belt with the drilled mark of the crankshaft timing pulley. Install the timing

belt on the crankshaft timing pulley, No. 1 idler pulley and the No. 2 idler pulley.

➡️**If the old timing belt is being reinstalled, be sure the directional arrow is facing in the original direction and that the belt and crankshaft gear matchmarks are properly aligned.**

6. Install the timing belt guide (No. 1 crank angle sensor plate) with the cup side facing forward. Replace the timing belt cover spacer.

7. Install the No. 1 timing belt cover and tighten the mounting bolts.

8. Align the pulley set key on the crankshaft with the key groove of the pulley.

Install the pulley, using the proper tool to tap in the pulley. Tighten the pulley bolt to 181 ft. lbs. (245 Nm).

9. Align the knock pin on the right side camshaft with the knock pin of the timing pulley. Slide on the timing pulley with the right side mark facing forward. Tighten the bolt to 80 ft. lbs. (108 Nm).

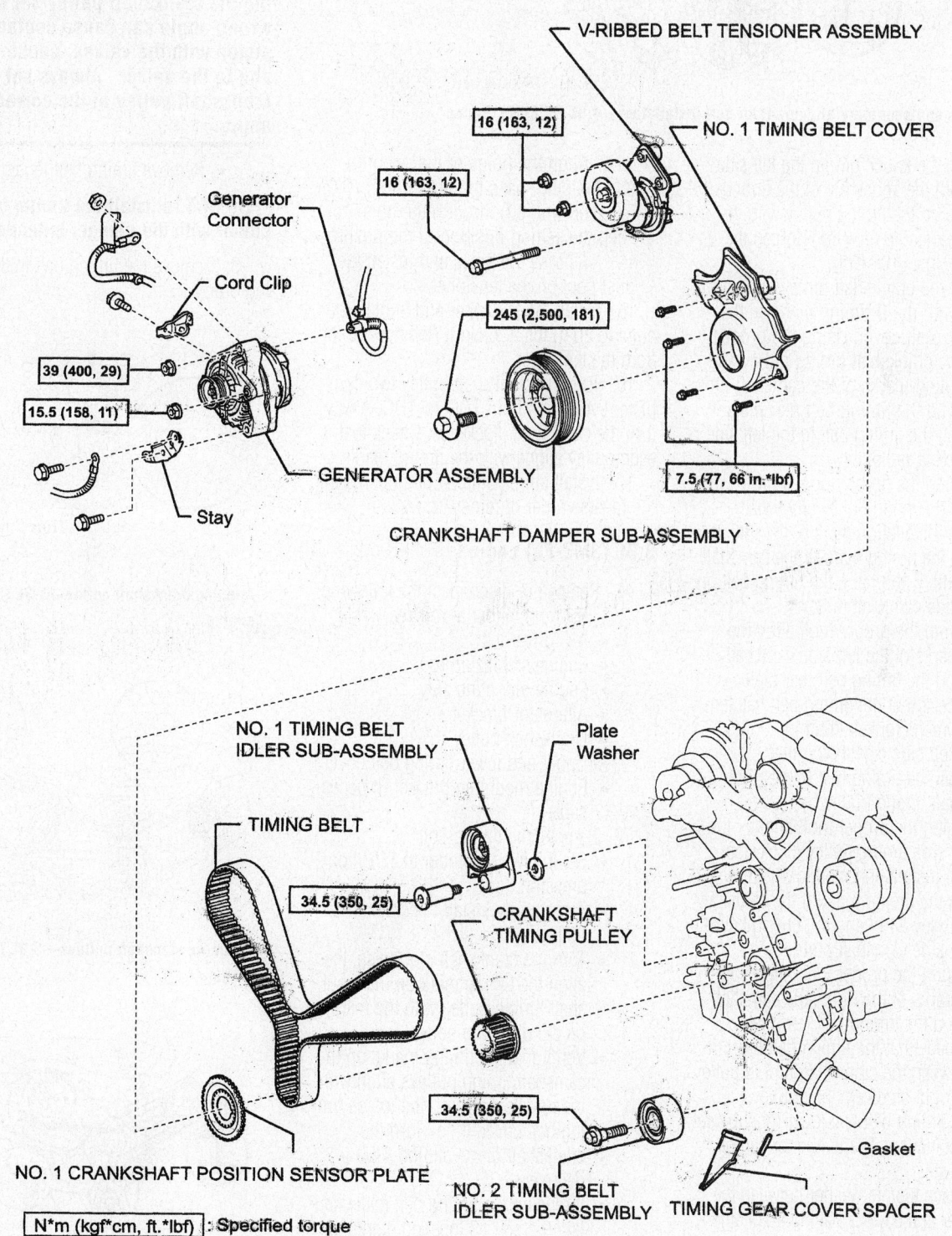

Exploded view of the timing belt and cover assembly and related components—4.3L (3UZ-FE) Engine

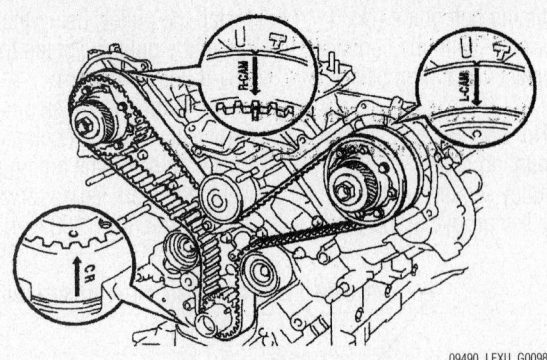

09490_LEXU_G0098

Timing belt sprocket mark alignment for belt installation—4.3L (3UZ-FE) Engine

10. Align the knock pin on the left side camshaft with the knock pin of the timing pulley. Slide on the timing pulley with the left side mark facing forward. Tighten the bolt to 80 ft. lbs. (108 Nm).

11. Turn the crankshaft pulley and align its groove with the **0** timing mark on the No. 1 timing belt cover. Using the proper tool, turn the crankshaft timing pulley and align the timing marks of the camshaft timing pulley and the timing belt rear plate.

12. Install the timing belt to the left side camshaft timing pulley by:

a. Using the proper tool, slightly turn the left side timing pulley clockwise. Align the installation mark of the timing belt with the timing mark of the camshaft timing pulley and hang the timing belt on the left side camshaft pulley.

b. Using the proper tool, align the timing marks of the left side camshaft pulley and the timing belt rear plate.

c. Check that the timing belt has tension between crankshaft timing pulley and the left side camshaft pulley.

13. Install the timing belt to the right side camshaft timing pulley by:

a. Using the proper tool, slightly turn the right side timing pulley clockwise. Align the installation mark of the timing belt with the timing mark of the camshaft timing pulley and hang the timing belt on the right side camshaft pulley.

b. Using the proper tool, align the timing marks of the right side camshaft pulley and the timing belt rear plate.

c. Check that the timing belt has tension between the crankshaft timing pulley and the right side camshaft pulley.

14. The timing belt tensioner must be set prior to installation. The tensioner can be set as follows:

a. Place a plate washer between the tensioner and a block. Using a suitable press, press in the pushrod using 220–2205 lbs. (100–1000kg) of pressure.

b. Align the holes of the pushrod and housing, pass the proper tool (0.05 in. Allen wrench) through the holes to keep the setting position of the pushrod.

c. Release the press and install the dust boot on the tensioner.

15. Install the tensioner and tighten the bolts to 20 ft. lbs. (26 Nm). Remove the tool from the tensioner.

16. Turn the crankshaft pulley two complete revolutions from TDC-to-TDC. Always turn the crankshaft clockwise. Check that each pulley aligns with the timing marks.

17. Install all remaining components in the reverse order of removal.

3.3L (3MZ-FE) Engine

1. Remove or disconnect the following:
- RH front fender apron seal
- A/C drive belt
- Engine stabilizing rod
- Engine mounting stay
- Alternator bracket
- Crankshaft pulley
- Upper and lower timing belt covers
- Engine mounting bracket (If necessary)

2. Remove the timing belt
- Set the No 1 cylinder to TDC/compression
- Temporarily install the crankshaft pulley bolt
- Turn the crankshaft clockwise, the align the timing mark on the crankshaft timing pulley with the mark on the oil pump body
- Verify that the timing marks on the camshaft timing pulleys align with the timing marks on the inside timing belt cover. If not, turn the crankshaft on revolution (360 degrees).
- If reusing the timing belt make certain there are 3 location marks on the timing belt corresponding with the marks on the timing gears.

- Set the No 1 cylinder to approximately 60 degrees BTDC/compression. Turn the crankshaft counterclockwise by approximately 60 degrees.

✲✲ WARNING

If the timing belt is disengaged, having the crankshaft pulley set at the wrong angle can cause contact of the piston with the valves, causing damage to the valves. Always set the crankshaft pulley at the correct angle.

- Remove timing belt tensioner

➡**Do NOT reinstall the timing belt tensioner with the plunger extended.**

3. Remove the timing belt in the following order.

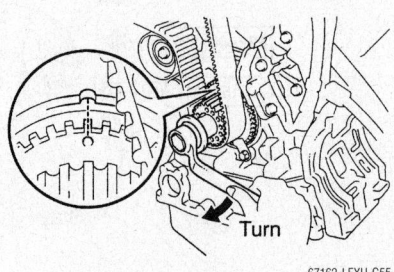

67162-LEXU-G55

Aligning crankshaft pulley—3.3L (3MZ-FE) engine

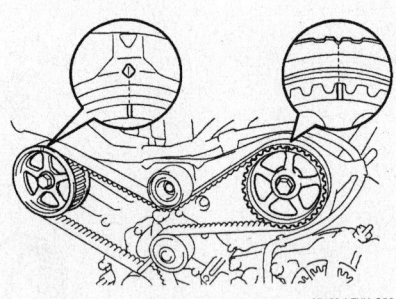

67162-LEXU-G56

Aligning camshaft pulleys—3.3L (3MZ-FE) engine

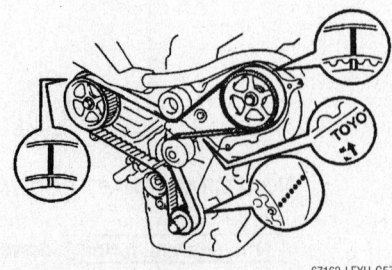

67162-LEXU-G57

Marking the timing belt for reuse—3.3L (3MZ-FE) engine

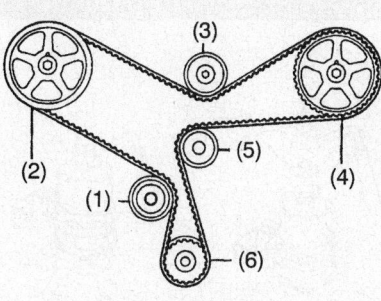

67162-LEXU-G58

Removal of timing belt from pulleys—3.3L (3MZ-FE) engine

- No 1 idler pulley
- RH camshaft timing pulley
- No. 2 idler pulley
- LH camshaft timing pulley
- Water pump pulley
- Crankshaft timing pulley
4. Inspect the timing belt

➡ **do not reuse the timing belt if there is evidence of fraying, oil contamination, cracking, tooth damage or wear, or belt distortion. If there is any doubt in the belt condition, replace the belt.**

To install:

5. After inspecting the pulleys for wear and checking for oil leaks install the timing belt using the following procedure.
- Temporarily install the crankshaft pulley bolt
- Make sure the crankshaft pulley is set at 60° BTDC
- Align the timing marks on the camshaft pulley with the respective marks on the inside timing cover.
- If reusing the old belt, align the marks previously made on the belt with the marks on the timing gears.
- Install the belt in the reverse order used when removing the belt
6. Install the timing belt tensioner using the following procedure

➡ **Keep the tensioner in an upright position**

- Slowly depress the push rod and align the hole with the hole in the housing
- Insert a 1.5mm hexagon wrench through the holes to maintain the setting position of the push rod.

- Install the tensioner with the 2 bolts and torque to 20 ft. lbs (27 Nm).
- Remove the hexagon wrench
7. Slowly turn the crankshaft 2 revolutions clockwise.
8. Check to see that all timing marks are in alignment.
9. Remove the crankshaft bolt
10. Install the timing belt guide
11. Install all remaining components in the reverse order of removal.

Piston and Ring

POSITIONING

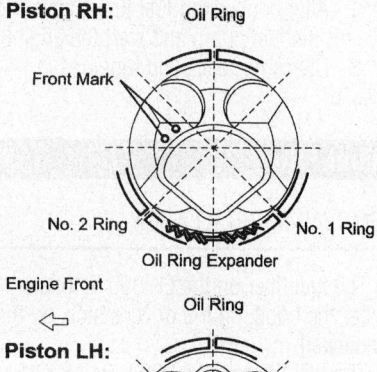

09490_LEXU_G0102

Piston ring positioning—2.5L (4GR-FSE), 3.0L (3GR-FSE) and 3.5L (2GR-FSE) Engines

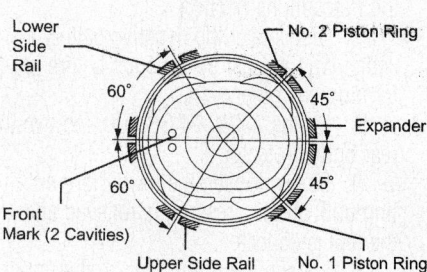

09490_LEXU_G0103

Piston ring positioning—4.3L (3UZ-FE) engine

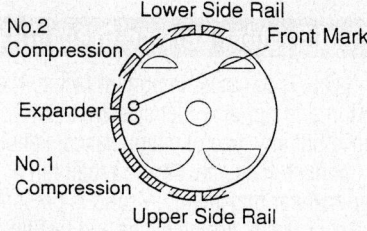
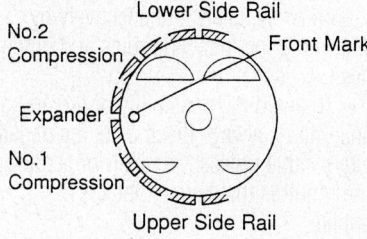

9307LG06

Piston ring positioning—3.0L (1MZE-FE) engine

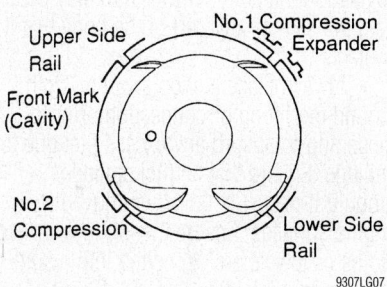

9307LG07

Piston ring positioning—3.0L (2JZ-GE) engine

FUEL SYSTEM

Fuel System Service Precautions

Safety is the most important factor when performing not only fuel system maintenance but any type of maintenance. Failure to conduct maintenance and repairs in a safe manner may result in serious personal injury or death. Maintenance and testing of the vehicle's fuel system components can be accomplished safely and effectively by adhering to the following rules and guidelines.

• To avoid the possibility of fire and personal injury, always disconnect the negative battery cable unless the repair or test procedure requires that battery voltage be applied.

• Always relieve the fuel system pressure prior to disconnecting any fuel system component (injector, fuel rail, pressure regulator, etc.), fitting or fuel line connection. Exercise extreme caution whenever relieving fuel system pressure, to avoid exposing skin, face and eyes to fuel spray. Please be advised that fuel under pressure may penetrate the skin or any part of the body that it contacts.

• Always place a shop towel or cloth around the fitting or connection prior to loosening to absorb any excess fuel due to spillage. Ensure that all fuel spillage (should it occur) is quickly removed from engine surfaces. Ensure that all fuel soaked cloths or towels are deposited into a suitable waste container.

• Always keep a dry chemical (Class B) fire extinguisher near the work area.

• Do NOT allow fuel spray or fuel vapors to come into contact with a spark or open flame.

• Always use a back-up wrench when loosening and tightening fuel line connection fittings. This will prevent unnecessary stress and torsion to fuel line piping.

• Always replace worn fuel fitting O-rings with new. Do NOT substitute fuel hose or equivalent, where a fuel pipe is installed.

Fuel System Pressure

RELIEVING

1. Before servicing the vehicle, refer to the Precautions Section.
2. Remove the fuse for the electronic fuel pump.
3. Start the engine until the engine stalls.

4. Disconnect the negative battery terminal.
5. Place a catch-pan under the joint to be disconnected. A large quantity of fuel may be released when the joint is opened.
6. Wear eye or full-face protection.
7. Place a shop towel over the area and slowly release the joint using a wrench of the correct size.
8. Allow the any fuel left in the line to bleed off slowly before fully disconnecting the joint.
9. Plug the opened lines immediately to prevent fuel spillage or the entry of dirt.
10. Dispose of the released fuel properly.
11. After connecting fuel lines, install the fuse for the fuel pump and start the engine.
12. Check for leaks and repair as needed.

Fuel Filter

REMOVAL & INSTALLATION

The fuel filter on the ES 300 is located under the hood, on the driver's side, by the fenderwell.

The fuel filter on the 2002–04 LS 430 is located under the vehicle on the left side before the rear axle.

The fuel filter on the 2002–04 GS 300/GS 430 is located under the vehicle, next to the left rear exhaust resonator.

The fuel filter for ES 330, 2005–06 GS 300/GS 430, IS 250/IS 350, IS 300, 2005–06 LS 430 and 2005–06 SC 430 models are in integral component of the in-tank fuel pump assembly. Refer to the Fuel Pump Removal procedure later in this section.

1. Before servicing the vehicle, refer to the Precautions Section.
2. Disconnect the negative battery cable. Wait at least 90 seconds before performing any other work.
3. On the 2002–04 GS 300, remove the rear body protector.
4. Slowly loosen the lower flare nut fitting until all the pressure is relieved and all the fuel is collected.
5. Loosen the union bolt on the upper portion of the filter and remove the banjo fitting and 2 metal gaskets. Discard the gaskets.
6. Loosen the fuel filter bracket bolt, remove the fuel line with the flared nut from the filter and pull the filter from the mounting bracket.

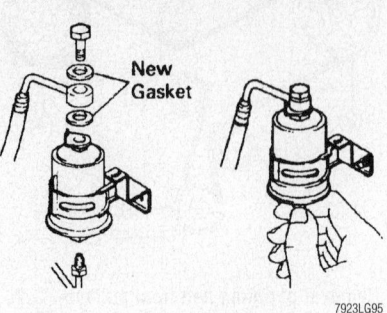

Exploded view of a typical fuel line connection at the filter

To install:

7. Install or connect the following:
• A new fuel filter to the vehicle and tighten the bracket bolt
• Banjo fitting with a new metal gasket on each side
• Union bolt. Tighten the union bolt to 22 ft. lbs. (30 Nm).
• Flare nut to the lower connection. Tighten the flare nut to 22 ft. lbs. (30 Nm).
8. On the 2002–04 GS 300, install the body protector.
9. Lower the vehicle if raised.
10. Remove the drain pan and/or rags and connect the negative battery cable.
11. Start the engine and visually inspect the upper and lower connections for leaks.

Fuel Pump

REMOVAL & INSTALLATION

IS 250 and IS 350

1. Before servicing the vehicle, refer to the Precautions Section.
2. Relieve the fuel system pressure.
3. Disconnect the negative battery cable. Wait at least 90 seconds before performing any other work.
4. Remove or disconnect the following:
• Rear seat cushion
• Floor service hole cover
• Fuel pump electrical connector
• Retaining clip and fuel tank return vent tube
• Retaining clip and fuel tank main tube
• 8 screws and lift out the pump/bracket assembly with gasket

➡**Do NOT lift the fuel pump assembly up using the wiring harness.**

- Fuel tube from fuel pump

5. Separate the fuel pump assembly from the fuel pump/sending unit

6. Separate the fuel filter from the fuel pump/sending unit

To install:

7. Install or connect the following:
- Filter and rubber cushion on the pump/sending unit
- New pump on the pump/sending unit
- Fuel tube to the fuel pump
- Pump/sending unit, using a new gasket and tighten the 8 screws to 53 inch lbs. (6 Nm)
- Fuel tank main tube and retaining clip
- Fuel tank return vent tube and retaining clip
- Fuel pump electrical connector
- Floor service hole cover
- Rear seat cushion
- Negative battery cable

8. Start the engine and check for leaks.

ES 300 and IS 300

1. Before servicing the vehicle, refer to the Precautions Section.

2. Relieve the fuel system pressure.

3. With the ignition switch in the **LOCK** position, disconnect the negative battery terminal.

4. Remove or disconnect the following:
- Rear seat cushion
- Fuel pump connector
- Floor service hole cover

➡**Do NOT lift the fuel pump assembly up using the wiring harness.**

- Fuel filler cap
- Fuel outlet pipe and the return hose from the pump bracket
- 8 screws and lift out the pump/bracket assembly with gasket
- Fuel pump lead wire
- Lower end of the pump off the bracket
- Fuel hose from the pump and remove the pump
- Rubber cushion from the pump

To install:

5. Install or connect the following:
- Filter and rubber cushion on the new pump
- Pump on the bracket
- Fuel hose and the wire connector on the pump
- Pump, using a new gasket and tighten the 8 screws to 35 inch lbs. (4 Nm)
- Fuel pipe and return hose to the

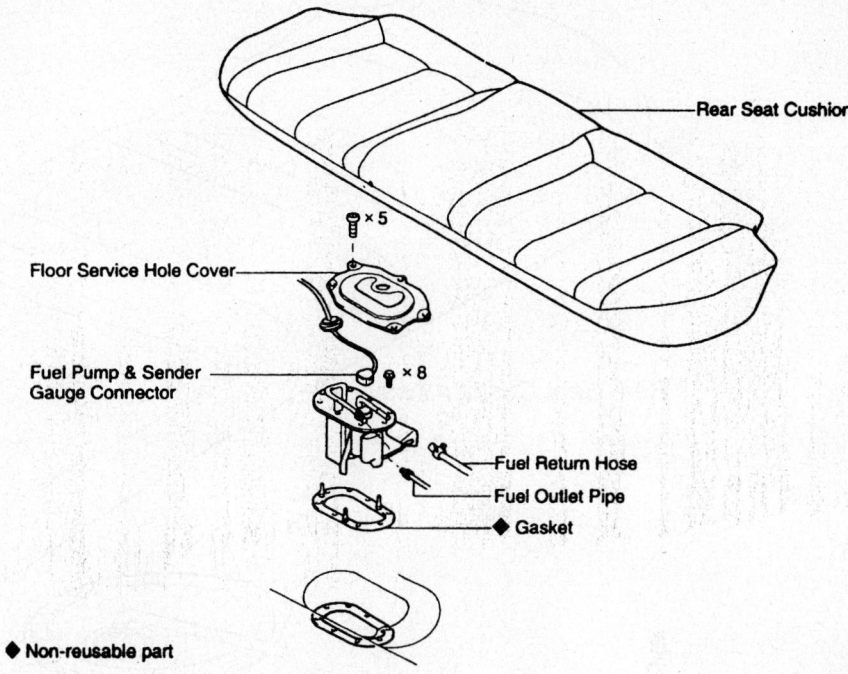

Rear Seat Cushion

Floor Service Hole Cover

× 5

Fuel Pump & Sender Gauge Connector

× 8

Fuel Return Hose
Fuel Outlet Pipe
◆ Gasket

◆ **Non-reusable part**

7923LG96

Exploded view of the fuel pump assembly—ES 300

pump and tighten the bolts to 22 ft. lbs. (29 Nm)
- Wire
- Service cover, and replace the rear seat
- Negative battery cable

6. Start the engine and check for leaks.

GS 300 and GS 430

1. Before servicing the vehicle, refer to the Precautions Section.

2. Relieve the fuel system pressure.

3. Remove or disconnect the following:

- Negative battery cable. Wait at least 90 seconds before performing any other work.
- Rear seat bottom
- Partition cover
- Floor service hole cover
- Fuel pump electrical connector
- Fuel main tube and fuel pump tube from the top of the fuel pump
- Mounting bolts, or retaining ring using SST 09808-14020
- Pump, bracket and set plate as an assembly

To install:

4. Install or connect the following:
- A new gasket on the set plate
- Fuel pump and bracket assembly. Torque the mounting bolts to 31 inch lbs. (3.5 Nm). If equipped with a retaining ring, use SST 09808-

14020 to tighten the retainer 2 full turns so that the mark on the ring lines up within the 2 marks indicated next to the ring on the fuel tank.
- Fuel main tube and fuel pump tube to the top of the fuel pump
- Fuel pump electrical connector
- Floor service hole cover
- Partition cover
- Rear seat bottom
- Negative battery cable

5. Start the engine; check the fuel system for leaks

SC 430 and LS 430

1. Before servicing the vehicle, refer to the Precautions Section.

2. Remove or disconnect the following:
- Negative battery cable. Wait at least 90 seconds before performing any other work.
- Rear seat bottom and seat back
- Fuel pump tube
- 8 bolts and fuel tank vent tube set plate
- Fuel pump and sensor gauge assembly
- Fuel suction hose and support
- Fuel pump cushion rubber
- Fuel pressure w/jet pump regulator assembly
- Fuel suction plate w/sender gauge
- Fuel pump and filter

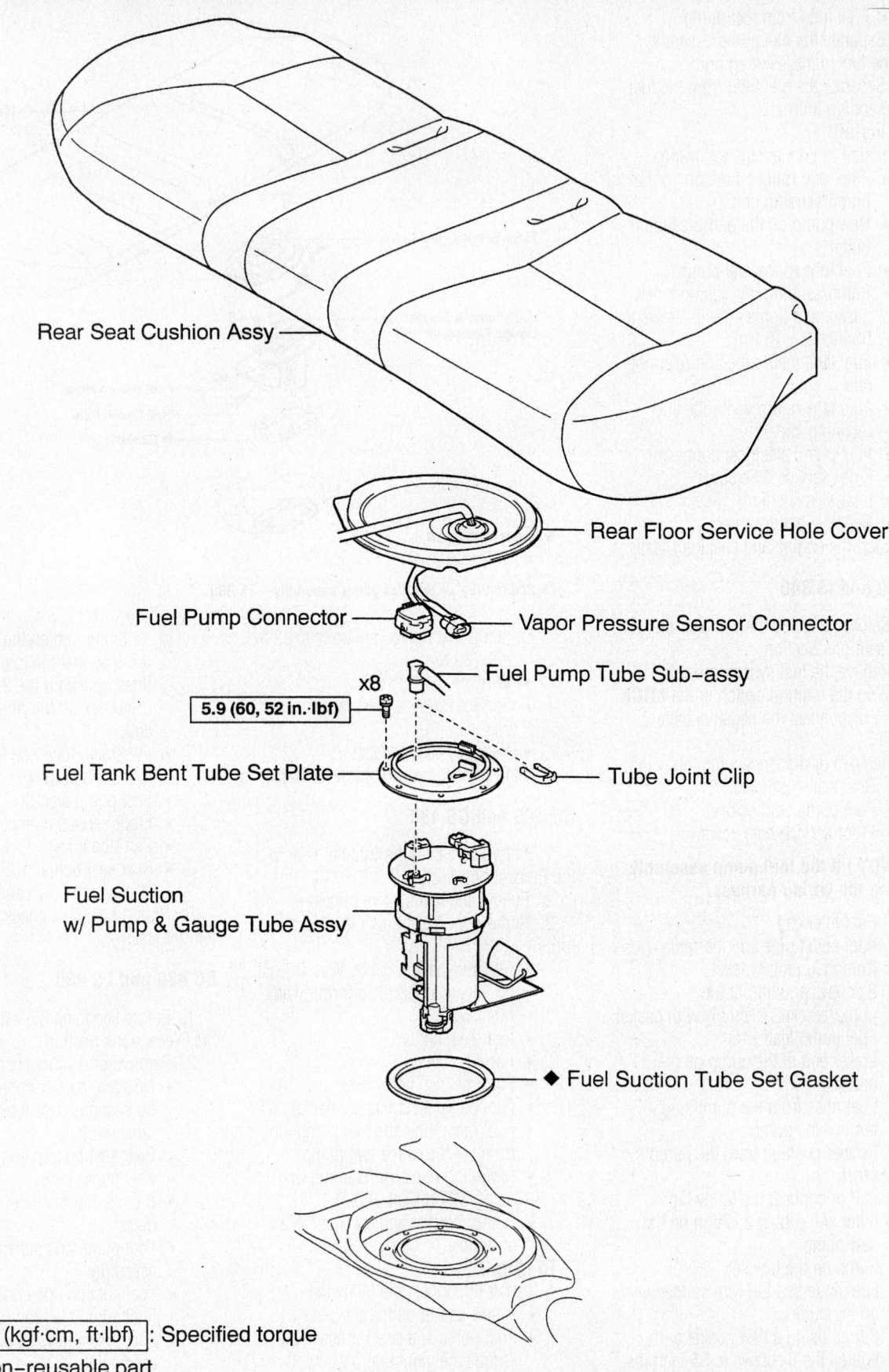

Rear Seat Cushion Assy

Rear Floor Service Hole Cover

Fuel Pump Connector

Vapor Pressure Sensor Connector

Fuel Pump Tube Sub-assy

x8

5.9 (60, 52 in.·lbf)

Fuel Tank Bent Tube Set Plate

Tube Joint Clip

Fuel Suction
w/ Pump & Gauge Tube Assy

◆ Fuel Suction Tube Set Gasket

N·m (kgf·cm, ft·lbf) : Specified torque

◆ Non-reusable part

67162-LEXU-G60

Exploded view for fuel pump removal—ES 330

To install:

3. To install, reverse the removal procedure. Install a new O-ring on the fuel jet pump regulator assembly and a new gasket on the fuel pump/sender gauge assembly. Torque the 8 bolts securing the fuel tank set plate to 52 in. lbs (6.0 Nm).

4. Inspect fuel pump operation and check for fuel leaks.

ES 330

1. Before servicing the vehicle, refer to the Precautions Section.

2. Remove or disconnect the following:
- Negative battery cable. Wait at least 90 seconds before performing any other work.
- Rear seat bottom and seat back
- Rear floor service hole cover
- Fuel suction w/pump and gauge tube assembly
- 8 bolts and fuel tank vent tube set plate
- Vapor pressure sensor assembly
- Fuel suction hose and support
- Fuel pump cushion rubber
- Fuel sender gauge assembly
- Fuel suction plate w/sender gauge
- Fuel pump harness
- Fuel pump
- Fuel pump filter. Pry out clip and remove from pump
- Fuel pressure regulator assembly

To install:

3. To install, reverse the removal procedure. Install a new O-ring on the fuel suction w/pump and gauge assembly. Torque the 8 bolts securing the fuel tank set plate to 52 inch lbs. (6.0 Nm).

4. Inspect fuel pump operation and check for fuel leaks.

▌ Fuel Injector

REMOVAL & INSTALLATION

2.5L (4GR-FSE) and 3.5L (2GR-FSE) Engines

1. Before servicing the vehicle, refer to the Precautions Section.

2. Relieve fuel system pressure.

3. Remove or disconnect the following:
- Negative battery cable. Wait at least 90 seconds before performing any other work.
- Coolant
- Cool air intake duct seal
- Left and right engine room side covers
- V-bank cover

- Left and right front upper fender protectors
- Left and right roof drip side finish mouldings using a moulding removal tool
- Front wiper arm head cap
- Left and right windshield wiper arm and blade assemblies
- Cowl top ventilator louver assembly
- No. 2 ventilation hose
- Air cleaner cap with air cleaner hose
- Throttle body assembly
- Cold start injector
- Intake air surge tank
- Intake manifold
- Water hose joint
- Fuel pressure pulsation damper
- No. 3 fuel hose from the No. 1 fuel pipe and remove fuel pipe
- No. 2 and 3 fuel pipes
- No. 1 and 2 fuel delivery pipes
- Fuel injectors, O-rings and seals

To install:

4. Install or connect the following:
- 2 new seals to each injector

5. Install the fuel injectors as follows:

a. Install a new O-ring, new backup rings (No. 1, No. 2, No. 3) and new E-ring to the fuel injector.

❋❋ **WARNING**

Check that there is no foreign matter or damaged areas in the injector's O-ring groove. Check that the installation direction of the No. 1 and No. 2 backup ring are correct. Make sure the backup rings and O-ring are installed in the correct order. Check that the alignment openings of the backup rings are not overlapped or stretched. After installing the O-ring, check that it is not contaminated with foreign matter and is not damaged.

b. Install the injector nozzle holder clamp.

c. Apply gasoline to the O-ring. Install the nozzle holder clamp by aligning the protruding part of the clamp to the notch of the delivery pipe.

❋❋ **WARNING**

Make sure there is no gap between the delivery pipe and clamp. Check that there is no foreign matter or damage in the injector insertion hole of the delivery pipe. Insert the injector straight into the delivery pipe without tilting it.

- No. 1 and 2 fuel delivery pipes. Torque the fasteners to 15 ft. lbs. (21 Nm).
- No. 2 and 3 fuel pipes and new O-rings. Torque No. 3 fuel pipe fastener to 7 ft. lbs. (10 Nm).
- No. 1 fuel pipe. Torque fasteners to 7 ft. lbs. (10 Nm).
- Fuel pressure pulsation damper and new gasket. Torque to 28 ft. lbs. (40 Nm).
- Water hose joint. Torque fasteners to 7 ft. lbs. (10 Nm).
- Intake manifold
- Intake air surge tank
- Cold start injector. Torque the bolts to 7 ft. lbs. (10 Nm).
- New gasket and throttle body assembly. Torque the 4 bolts to 7 ft. lbs. (10 Nm)
- Air cleaner cap with air cleaner hose
- No. 2 ventilation hose
- Cowl top ventilator louver assembly
- Left and right windshield wiper arm and blade assemblies
- Front wiper arm head cap
- Left and right roof drip side finish mouldings
- Left and right front upper fender protectors
- V-bank cover
- Left and right engine room side covers
- Cool air intake duct seal
- Negative battery cable

6. Refill the cooling system. Start the engine and check for coolant and fuel leaks and proper operation.

7. Check the function of the throttle body unit

8. System initialization

3.0L (3GR-FSE) Engine

1. Before servicing the vehicle, refer to the Precautions Section.

2. Relieve fuel system pressure.

3. Remove or disconnect the following:
- Negative battery cable. Wait at least 90 seconds before performing any other work.
- Coolant
- Cool air intake duct seal
- Engine under cover
- Right engine room side cover
- V-bank cover
- No. 2 ventilation hose
- Air cleaner cap with air cleaner hose
- Cold start injector
- Water by-pass hose

- Inlet and outlet heater water hoses
- Engine rear cover
- Union to check valve hose
- Ventilation hose
- Fuel main tube
- Intake air surge tank
- Intake manifold
- Fuel pressure pulsation damper
- No. 1 and 2 fuel pipes
- High pressure side fuel pump
- No. 2 and 3 fuel pipes
- No. 1 and 2 fuel delivery pipes
- Fuel injectors, O-rings and seals

To install:

4. Install or connect the following:
   - 2 new seals to each injector
5. Install the fuel injectors as follows:
   a. Install a new O-ring, new backup rings (No. 1, No. 2, No. 3) and new E-ring to the fuel injector.

✳ WARNING

Check that there is no foreign matter or damaged areas in the injector's O-ring groove. Check that the installation direction of the No. 1 and No. 2 backup ring are correct. Make sure the backup rings and O-ring are installed in the correct order. Check that the alignment openings of the backup rings are not overlapped or stretched. After installing the O-ring, check that it is not contaminated with foreign matter and is not damaged.

   b. Install the injector nozzle holder clamp.
   c. Apply gasoline to the O-ring. Install the nozzle holder clamp by aligning the protruding part of the clamp to the notch of the delivery pipe.

✳ WARNING

Make sure there is no gap between the delivery pipe and clamp. Check that there is no foreign matter or damage in the injector insertion hole of the delivery pipe. Insert the injector straight into the delivery pipe without tilting it.

6. Install the No. 1 and 2 fuel delivery pipes as follows:
   a. Install a new injector vibration insulator to the cylinder head.
   b. Apply lubricant to the installation hole of the injector.
   c. Insert the stud bolt into the fuel delivery pipe until the screw threads protrude enough so that a nut can be attached.

➡**If an injector is dropped, replace it with a new one. Check that there is no foreign matter or damage in the injector insertion hole of the delivery pipe. Be extremely careful not to touch or strike the tips of the injectors. When inserting the fuel delivery pipe, push it in evenly without tilting it.**

   d. Install the fuel delivery pipe by uniformly tightening the 2 bolts and 2 nuts in several passes to 15 ft. lbs. (21 Nm).
   e. Connect the 3 connectors and 2 clamps.
   - No. 2 and 3 fuel pipes and all new rings and seals. Torque No. 3 fuel pipe fastener to 7 ft. lbs. (10 Nm).
   - High pressure side fuel pump
   - No. 1 and 2 fuel pipes. Torque fasteners to 7 ft. lbs. (10 Nm).
   - Fuel pressure pulsation damper and new gasket. Torque to 28 ft. lbs. (40 Nm).
   - Intake manifold
   - Intake air surge tank
   - Fuel main tube
   - Ventilation hose
   - Union to check valve hose
   - Engine rear cover
   - Inlet and outlet heater water hoses
   - Water by-pass hose
   - Cold start injector. Torque the bolts to 7 ft. lbs. (10 Nm).
   - Air cleaner cap with air cleaner hose
   - No. 2 ventilation hose
   - V-bank cover
   - Right engine room side cover
   - Cool air intake duct seal
   - Engine under cover
   - Negative battery cable

7. Refill the cooling system. Start the engine and check for coolant and fuel leaks and proper operation.
8. Check the function of the throttle body unit
9. System initialization

3.0L (1MZ-FE) Engine

1. Before servicing the vehicle, refer to the Precautions Section.
2. Remove or disconnect the following:
   - 3 cap nuts, using a 5mm hexagon wrench. Loosen the V-bank cover fastener counterclockwise.
   - V-Bank cover
   - Air cleaner hose with resonator
   - air intake chamber assembly
   - Injector connectors
   - Air assist hoses and pipe
   - No. 1 fuel pipe and remove the fuel hose clamp

   - No. 1 fuel pipe (fuel tube connector) from the fuel filter outlet
   - 5 bolts and delivery pipes together with the 6 injectors and No. 1 fuel pipe.
   - 4 spacers from the intake manifold
   - 6 injectors from the delivery pipes
   - 2 O-rings and 2 grommets from each injector

To install:

3. Install or connect the following:
   - New insulator and grommet to each injector
   - New O-rings, coated with gasoline, to each injector
   - A light coat of gasoline on the place where a delivery pipe touches an O-ring of the injector
   - Injector, while turning it clockwise, into the delivery pipe

➡**Position the injector connector outward.**

   - The 4 spacers in position on the intake manifold
   - A light coat of gasoline on the place where the intake manifold touches an O-ring
   - The delivery pipe and fuel pipe together with the 6 injectors in position on the intake manifold. Position the injector connector outward.
   - 4 bolts holding the delivery pipes to the intake manifold, temporarily
   - Bolt holding the No. 1 fuel pipe to the intake manifold, temporarily

➡**Check that the injectors rotate smoothly. If the injectors do not rotate smoothly, the probable cause is incorrect installation of the O-rings. Replace the O-rings.**

   - 4 bolts holding the delivery pipes to the intake manifold. Torque to 14 ft. lbs. (19.5 Nm)
   - No. 1 fuel pipe (fuel tube connector) to the fuel filter
   - Fuel hose clamp to the fuel filter with a "click" sound. After installing the clamp, check that the clamp is fixed by pulling up the clamp.
   - Air assist hoses
   - Injector connectors
   - Air intake chamber assembly
   - Air cleaner hose with resonator
   - V-bank cover, using a 5mm hexagon wrench with the 3 cap nuts
   - Press down the V-bank cover.

3.0L (2JZ-GE) Engine

1. Before servicing the vehicle, refer to the Precautions Section.
2. Remove or disconnect the following:
   - Air intake chamber
   - Fuel pressure pulsation damper
   - Engine wire from intake manifold
   - Bolt holding the engine wire protector to the body
   - 6 injector connectors
   - Camshaft Position (CMP) sensor connector
   - Throttle Position (TP) sensor connector
   - Vacuum Switching Valve (VSV) connector for Evaporative Emission (EVAP) control
   - VSV connector for Acoustic Control Induction System (ACIS)
   - 3 nuts holding the engine wire protector to the intake manifold

➡**Be careful not to drop the injectors when removing the delivery pipe.**

   - 3 bolts and delivery pipe together with the 6 injectors
   - The injectors from the delivery pipe
   - O-rings, insulator and grommet from each injector
   - 3 spacers from the intake manifold

To install:

3. Install or connect the following:
   - A new insulator and grommet to each injector
   - A light coat of gasoline on the place where a delivery pipe touches an O-ring of the injector
   - Injector, while turning clockwise and counterclockwise, into the delivery pipe

➡**Position the injector connector outward.**

   - The 3 spacers in position on the intake manifold
   - A light coat of gasoline on the place where an intake manifold touches an O-ring
   - Injectors together with the delivery pipe and 3 bolts in position on the intake manifold. Check that the injectors rotate smoothly. Position the injector connector upward.

➡**If the injectors do not rotate smoothly, the probable cause is incorrect installation of the O-rings. Replace the O-rings.**

4. Tighten the 3 bolts holding the delivery pipe to the intake manifold. Tighten the bolts to 15 ft. lbs. (21 Nm).

5. Install or connect the following:
   - Engine wire protector with the 3 nuts
   - 6 injector connectors

➡**The No. 1, No. 3 and No. 5 injector connectors are dark gray, and the No. 2, No. 4, and the No. 6 injectors connectors are brown.**

6. Install or connect the following:
   - Camshaft position sensor connector
   - Throttle position sensor connector
   - VSV connector for EVAP
   - Bolt holding the engine wire protector to the body

4.3L (3UZ-FE) Engine

1. Before servicing the vehicle, refer to the Precautions Section.
2. Remove or disconnect the following:
   - V-bank cover
   - Intake air connector
   - Accelerator cable
   - Fuel pressure pulsation dampers.
   - VVT sensor connectors
   - Vacuum Switching Valve (VSV) for Evaporative Emissions (EVAP)
   - 2 nuts and accelerator cable bracket
   - 2 nuts and accelerator cable bracket
   - 3 V-bank cover brackets
   - VSV connector for Acoustic Control Induction System (ACIS) from the No. 1 V-bank cover bracket
   - 4 bolts and 3 V-bank cover brackets
   - Engine wire from the delivery pipe
   - 2 wire clamps from the wire clamp bracket on the right-hand deliver pipe
   - 8 injector connectors
   - 4 nuts holding the delivery pipe to the intake manifold
   - 2 delivery pipes and 8 injectors assembly and 4 spacers
   - 2 O-rings, grommet and insulator from each injector

To install:

3. Install or connect the following:
   - A new insulator and grommet to each injector
   - A light coat of gasoline to new O-rings and install them to each injector
   - A light coat of gasoline on the place where a delivery pipe touches an O-ring of the injector
   - Injector, while turning the clockwise and counterclockwise, into the delivery pipe

➡**Position the injector connector outward.**

   - The 4 spacers in position on the intake manifold
   - A light coat of gasoline on the place where an intake manifold touches an O-ring
   - The delivery pipes in position on the intake manifold
   - Temporarily, the 3 bolts holding the delivery pipe to the intake manifold

➡**Check that the injectors rotate smoothly. If the injectors do not rotate smoothly, the probable cause is incorrect installation of the O-rings. Replace the O-rings.**

4. Tighten the 3 bolts holding the delivery pipe to the intake manifold. Tighten the bolts to 15 ft. lbs. (21 Nm).
5. Install or connect the following:
   - Engine wire protector with the 3 nuts
   - Injector connectors
   - Remaining components

3.3L (3MZ-FE) Engine

1. Before servicing the vehicle, refer to the Precautions Section.
2. Remove or disconnect the following:
   - Negative battery cable
   - Suspension upper center brace
   - V-bank cover sub assembly
   - Air cleaner assembly
   - Emission control valve set
3. Remove intake air surge tank
4. Remove or disconnect the following:
   - Throttle motor connector
   - Water by-pass hoses
   - Union check valve hose
   - Ventilation hose
   - Pressure feed hose
   - Engine hangers
   - Surge tank stays
   - Bond cable connector
   - Emission control valve bracket
   - Intake air surge tank
   - Gasket from intake air surge tank
5. Remove fuel pipe assembly
6. Remove or disconnect the following:
   - Fuel pulsation damper and gasket
   - Fuel pipe union bolt and 2 gaskets
   - Bolt and separate the fuel pipe
7. Remove the fuel injector assembly
8. Remove or disconnect the following:
   - 6 fuel injector connectors
   - 4 bolts, then remove the fuel injector delivery pipes

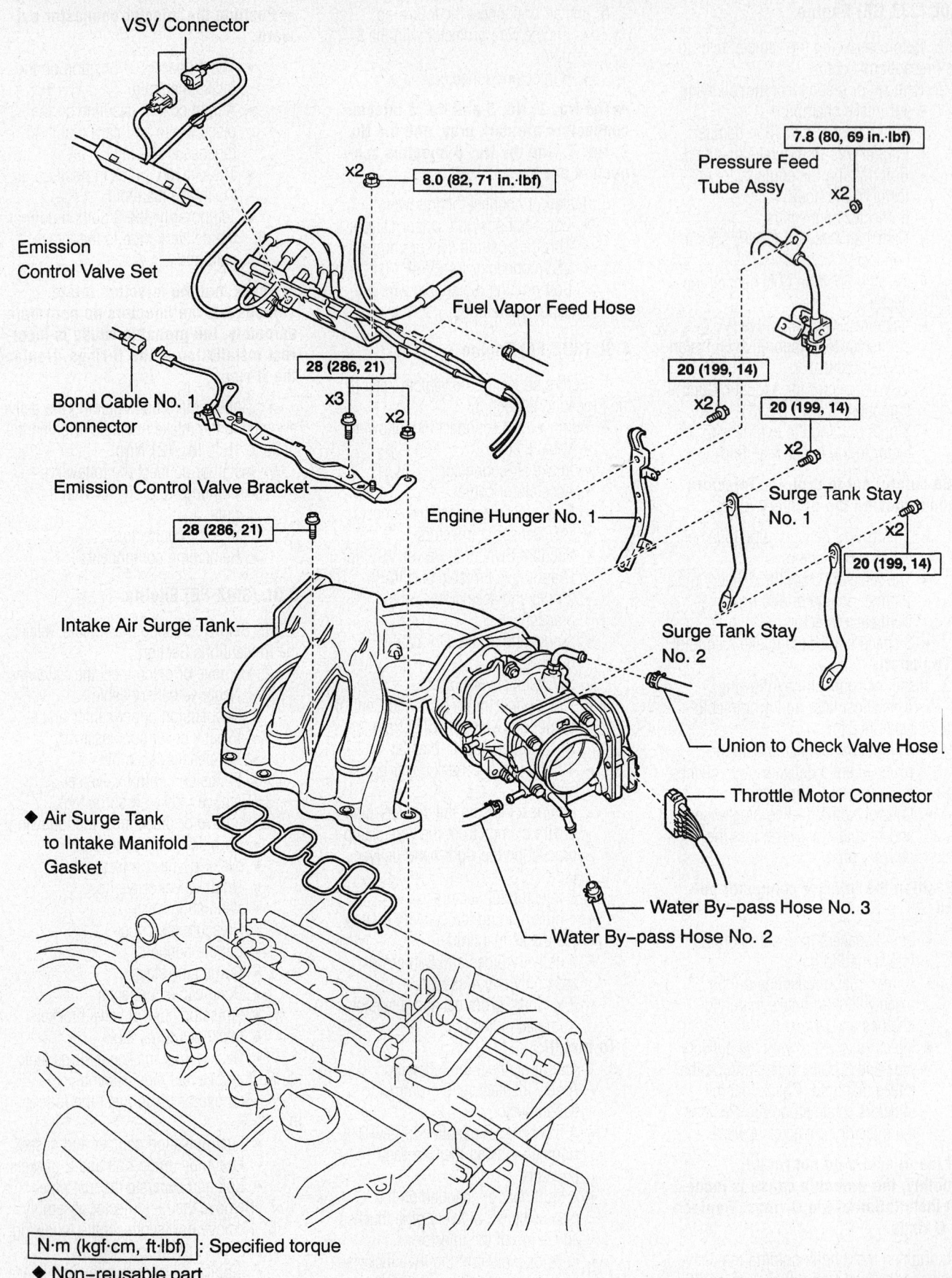

VSV Connector

Emission Control Valve Set

Bond Cable No. 1 Connector

Emission Control Valve Bracket

x2 8.0 (82, 71 in.·lbf)

28 (286, 21)
x3
x2

Fuel Vapor Feed Hose

Pressure Feed Tube Assy

7.8 (80, 69 in.·lbf)
x2

20 (199, 14)
x2

20 (199, 14)
x2

20 (199, 14)
x2

Engine Hunger No. 1

Surge Tank Stay No. 1

Surge Tank Stay No. 2

28 (286, 21)

Intake Air Surge Tank

◆ Air Surge Tank to Intake Manifold Gasket

Union to Check Valve Hose

Throttle Motor Connector

Water By-pass Hose No. 3

Water By-pass Hose No. 2

N·m (kgf·cm, ft·lbf) : Specified torque
◆ Non-reusable part

67162-LEXU-G61

Exploded view of Intake Air Surge tank—3.3L (3MZ-FE) Engine

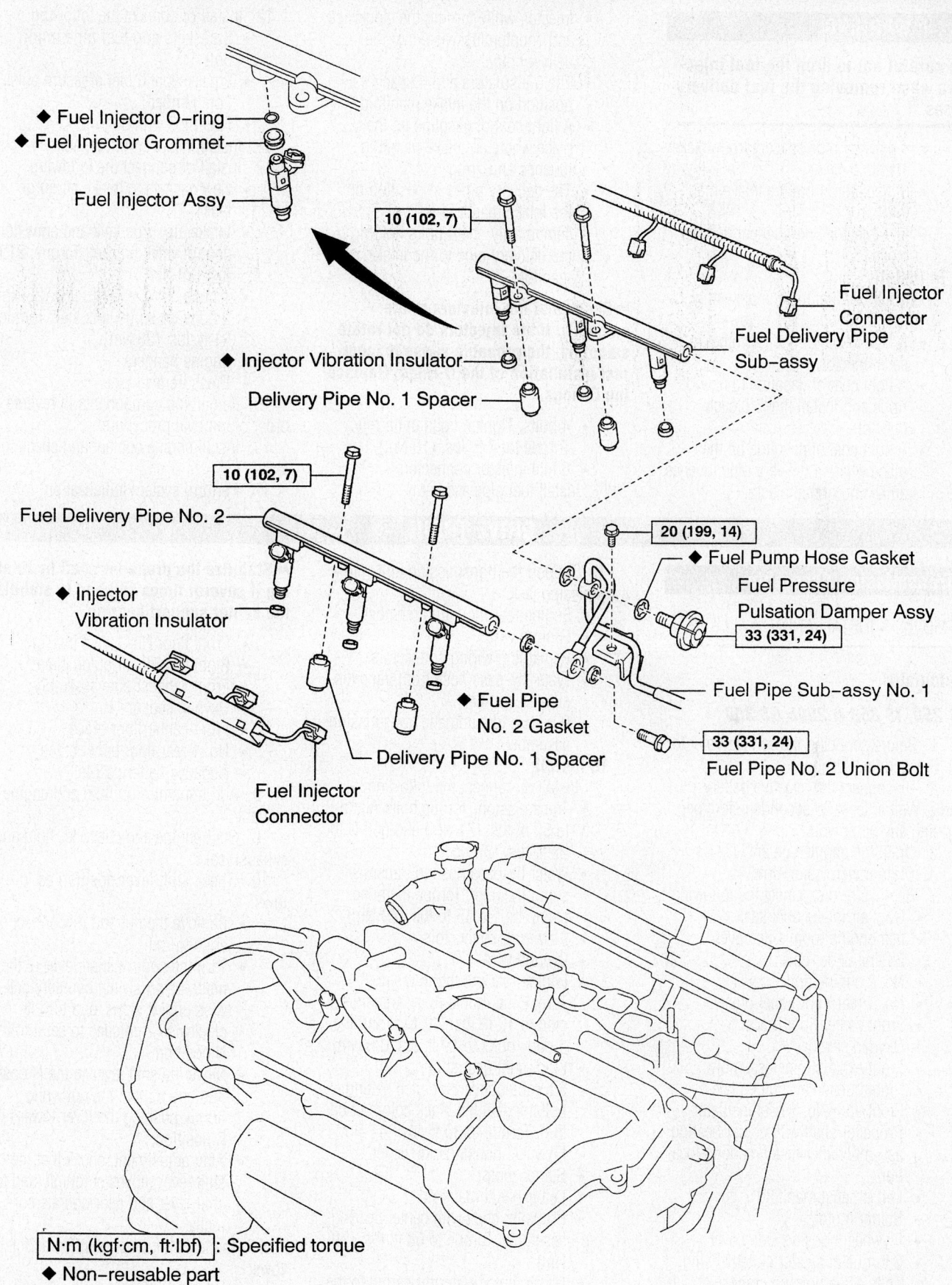

◆ Fuel Injector O-ring
◆ Fuel Injector Grommet

Fuel Injector Assy

10 (102, 7)

Fuel Injector Connector

Fuel Delivery Pipe Sub-assy

◆ Injector Vibration Insulator

Delivery Pipe No. 1 Spacer

10 (102, 7)

Fuel Delivery Pipe No. 2

◆ Injector Vibration Insulator

20 (199, 14)

◆ Fuel Pump Hose Gasket

Fuel Pressure Pulsation Damper Assy

33 (331, 24)

Fuel Pipe Sub-assy No. 1

◆ Fuel Pipe No. 2 Gasket

Delivery Pipe No. 1 Spacer

Fuel Injector Connector

33 (331, 24)

Fuel Pipe No. 2 Union Bolt

N·m (kgf·cm, ft·lbf) : Specified torque

◆ Non-reusable part

67162-LEXU-G62

Exploded view of fuel injector delivery pipe and injectors—3.3L (3MZ-FE) Engine

✳✳ WARNING

Be careful not to drop the fuel injectors when removing the fuel delivery pipes.

- 4 delivery pipe spacers from intake manifold
- 6 insulators from the intake manifold
- fuel injector from the fuel delivery pipes.

To install:

9. Install new fuel injector assembly
10. Install or connect the following
    - A new insulator and grommet to each injector
    - A light coat of gasoline to new O-rings and install them to each injector
    - A light coat of gasoline on the place where a delivery pipe touches an O-ring of the injector
- Injector, while turning the clockwise and counterclockwise, into the delivery pipe
- The 6 insulators and 4 spacers in position on the intake manifold
- A light coat of gasoline on the place where an intake manifold touches an O-ring
- The delivery pipes in position on the intake manifold
- Temporarily, the 4 bolts holding the delivery pipe to the intake manifold

➡ **Check that the injectors rotate smoothly. If the injectors do not rotate smoothly, the probable cause is incorrect installation of the O-rings. Replace the O-rings.**

- 4 bolts. Tighten bolts uniformly. Torque to: 7 ft. lbs. (10 Nm)
- 6 fuel injector connectors
11. Install fuel pipe assembly

12. Install or connect the following
    - 2 gaskets and fuel pipe union bolt
    - 2 gaskets and fuel pressure pulsation damper
    - Fuel pipe with bolt
13. Install intake air surge tank
14. Install or connect the following
    - New gaskets to intake air surge tank
    - Intake air surge tank and emission control valve bracket. Torque: 21 ft. lbs. (28 Nm)
    - 4 bolts. Torque: 21 ft. lbs. (28 Nm)
    - 2 bolts on surge tank stay. Torque: 21 ft. lbs. (28 Nm)
    - Engine hangers
    - Pressure feed tube
15. Remaining components in reverse order of removal procedure
16. Install engine coolant and check for leaks
17. Perform system initialization

DRIVE TRAIN

Transmission Assembly

REMOVAL & INSTALLATION

Automatic

IS 250, IS 350 & 2006 GS 300

1. Before servicing the vehicle, refer to the Precautions Section.
2. Disconnect the negative battery cable. Wait at least 90 seconds before performing any other work.
3. Drain the engine coolant.
4. Drain transmission fluid.
5. Remove or disconnect the following:
   - Cool air intake duct seal
   - Left engine room side cover
   - Engine under cover
   - No. 2 engine under cover
   - No. 1 rear floor panel brace
   - Front center floor brace
   - Oxygen sensor
   - Front exhaust pipe assembly
   - Right outside air guide plate
   - Front floor No. 1 heat insulator
   - Propeller shaft with center bearing assembly after matchmarking location
   - Left exhaust manifold
   - Starter motor
   - Flywheel housing side cover
   - 6 torque converter clutch setting bolts while holding crankshaft pulley nut with wrench
   - Floor shift gear shifting rod

6. Support the transmission unit with a transmission jack.
   - Engine rear mounting member
   - Ground wire
   - Remaining wiring connectors
   - Water by-pass hose from transmission oil cooler
   - 9 bolts and automatic transmission assembly

To install:

7. Install or connect the following:
   - Transmission. Torque bolts A and B to 52 ft. lbs. (71 Nm) and bolt C to 27 ft. lbs. (37 Nm)
   - Water by-pass hose to transmission oil cooler. Torque the hose clamp bolt to 15 ft. lbs. (20 Nm)
   - All wiring connectors
   - Ground wire
   - Engine rear mounting member. Torque the 4 nuts to the transmission to 10 ft. lbs. (13 Nm) and the 4 body bolts to 19 ft. lbs. (26 Nm)
   - Floor shift gear shifting rod
   - 6 torque converter clutch setting bolts. Install the black colored bolt first. Torque to 30 ft. lbs. (41 Nm)
   - Flywheel housing side cover
   - Starter motor
   - Left exhaust manifold
   - Propeller shaft with center bearing assembly. Torque to 58 ft. lbs. (79 Nm)
   - No. 1 center support bearing to the matchmark. Torque to 51 ft. lbs. (69 Nm)

➡ **Stabilize the propeller shaft by rotating it several times by hand to stabilize the center support bearing.**

   - Front floor No. 1 heat insulator
   - Right outside air guide plate
   - Front exhaust pipe assembly
   - Oxygen sensor
   - Front center floor brace
   - No. 1 rear floor panel brace
   - Negative battery cable
8. Add transmission fluid and engine coolant.
9. Start engine and check for fluid and exhaust leaks.
10. Check shift lever position as follows:
    - Remove the nut and disconnect the shifting rod.
    - Turn the control shaft lever of the neutral start switch counterclockwise until it stops, and turn it clockwise 2 notches to set it to the N position.
    - Move the shift lever to the N position and tighten the nut while lightly pushing the lever toward the R position.
    - After adjustment, check that the shift lever moves smoothly and the shift lever and gear operate correctly.
11. Inspect shift lever position as follows:
    - When shifting from the P to R position with the engine switch on

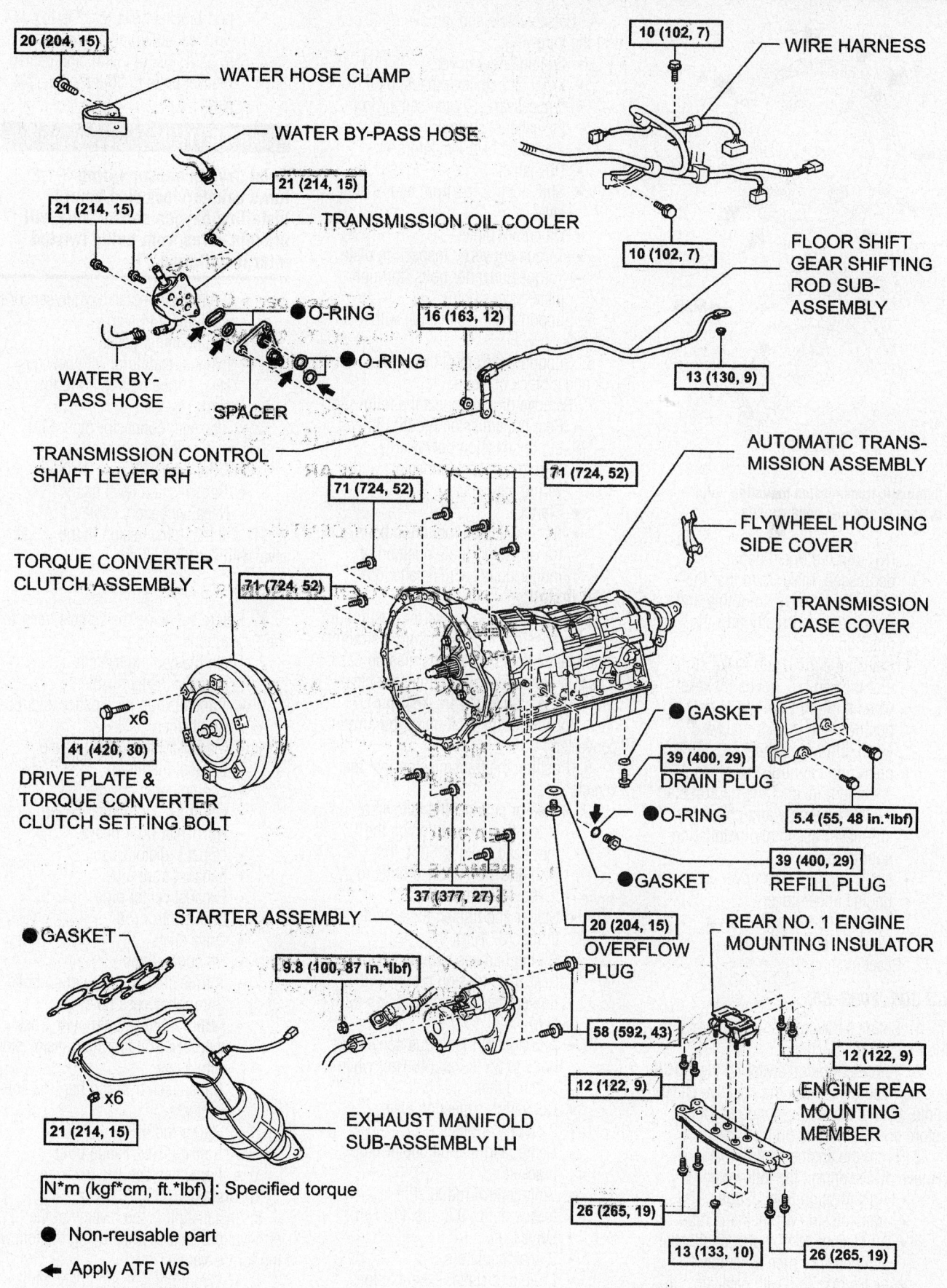

20 (204, 15) — WATER HOSE CLAMP

— WATER BY-PASS HOSE

10 (102, 7) — WIRE HARNESS

21 (214, 15)

21 (214, 15) — TRANSMISSION OIL COOLER

10 (102, 7)

FLOOR SHIFT GEAR SHIFTING ROD SUB-ASSEMBLY

● O-RING

16 (163, 12)

● O-RING

13 (130, 9)

WATER BY-PASS HOSE

SPACER

TRANSMISSION CONTROL SHAFT LEVER RH

AUTOMATIC TRANS-MISSION ASSEMBLY

71 (724, 52)

71 (724, 52)

FLYWHEEL HOUSING SIDE COVER

71 (724, 52)

TORQUE CONVERTER CLUTCH ASSEMBLY

71 (724, 52)

TRANSMISSION CASE COVER

x6

● GASKET

41 (420, 30)

39 (400, 29) DRAIN PLUG

DRIVE PLATE & TORQUE CONVERTER CLUTCH SETTING BOLT

● O-RING

5.4 (55, 48 in.*lbf)

● GASKET

39 (400, 29) REFILL PLUG

37 (377, 27)

● GASKET

STARTER ASSEMBLY

20 (204, 15) OVERFLOW PLUG

REAR NO. 1 ENGINE MOUNTING INSULATOR

9.8 (100, 87 in.*lbf)

58 (592, 43)

12 (122, 9)

12 (122, 9)

ENGINE REAR MOUNTING MEMBER

x6

21 (214, 15)

EXHAUST MANIFOLD SUB-ASSEMBLY LH

26 (265, 19)

N*m (kgf*cm, ft.*lbf) : Specified torque

● Non-reusable part

◄ Apply ATF WS

13 (133, 10)

26 (265, 19)

Exploded view of automatic transmission components—IS 250, IS 350 and 2006 GS 300

09490_LEXU_G0104

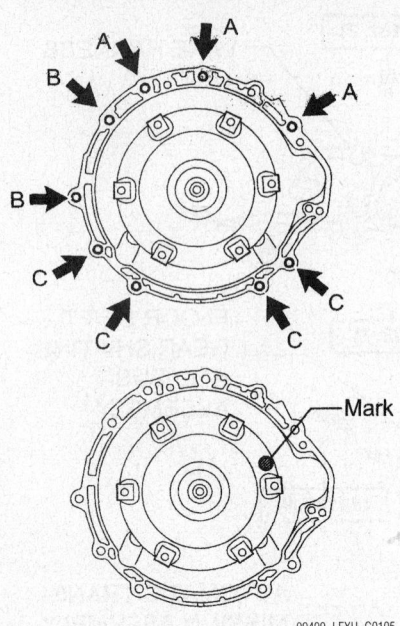

09490_LEXU_G0105

Automatic transmission mounting bolts— IS 250, IS 350 and 2006 GS 300

(IG) and the brake pedal depressed, make sure that the shift lever moves smoothly and then moves correctly into the position.

- Start the engine and make sure that the vehicle moves forward when shifting from the N to D position and moves rearward when shifting to the R position. If operation cannot be done as specified, inspect the neutral start switch assembly and check the shift lever assembly installation condition.
- No. 2 engine under cover
- Engine under cover
- Left engine room side cover
- Cool air intake duct seal

12. Reset memory.

GS 300 (2002–05)

1. Before servicing the vehicle, refer to the Precautions Section.

2. Turn the ignition switch to the **LOCK** position and disconnect the negative battery cable. Wait at least 90 seconds or longer before doing any work on the vehicle.

3. From the engine compartment, remove or disconnect the following:

- Transmission level gauge
- Transmission dipstick and tube
- Air cleaner, MAF meter, and intake air connector pipe assembly
- Intake manifold with catalytic converters

4. Raise vehicle and remove or disconnect the following:

- Engine under cover
- Front and center exhaust pipe
- If necessary, oxygen sensor from the exhaust system.
- 2 exhaust heat insulators
- Driveshaft
- Shift control rod from end of shift lever
- Oil cooler pipes
- Torque converter inspection plate
- Torque converter bolts (through plate access hole)

5. Support the transmission with a suitable jack.

6. Support the front of the engine with a jack and a block of wood.

7. Remove or disconnect the following:

- Rear transmission mount (tilt end of transmission downward)
- Wiring harness connectors and wiring harness
- Starter
- 9 transmission mounting bolts and transmission (note position of longer bolts for reinstallation)

To install:

8. Inspect torque converter position in the transmission housing, using a straight-edge placed across the transmission case. If torque converter-to-transmission clearance is less than 0.004 in. (0.1 mm), check for proper installation of the torque converter.

9. Position the transmission into the vehicle.

10. Install or connect the following:

- Transmission and tighten the 17 mm head bolts to 53 ft. lbs. (72 Nm) and 14 mm head bolts to 27 ft. lbs. (37 Nm)
- Starter and tighten the bolts to 27 ft. lbs. (37 Nm)
- Wiring harness and connectors
- Rear transmission mount and tighten the bolts to 20 ft. lbs. (26 Nm)
- And tighten the torque converter bolts to 35 ft. lbs. (48 Nm) while rotating the crankshaft
- Converter inspection plate
- 2 oil cooler pipes
- Center and rear oil cooler pipe brackets
- Shift control rod to shift lever; torque nut to 9 ft. lbs. (13 Nm)
- Driveshaft
- 2 heat insulators
- Front and center exhaust pipes, with new gaskets; torque pipe sup-

port bracket bolts to 33 ft. lbs. (44 Nm), exhaust pipe connector nuts to 33 ft. lbs. (44 Nm), and center bracket bolts to 65 inch lbs. (7.4 Nm).

✷✷ CAUTION

Twist oxygen sensor wiring 3-1/2 turns counterclockwise before installing oxygen sensor; this will prevent wires from being twisted after installation.

- If removed, install oxygen sensor to 33 ft. lbs. (44 Nm)
- Engine under cover
- Exhaust manifold, with new gaskets; torque nuts to 29 ft. lbs. (39 Nm)
- Intake air connector pipe, MAF meter, and air cleaner
- Filler pipe with new O-ring
- Transmission level gauge
- Negative battery cable

11. Fill the transmission to the proper level with Type T-IV fluid.

IS 300

1. Before servicing the vehicle, refer to the Precautions Section.

2. Drain the cooling system.

- Negative battery cable
- Transmission oil dipstick and tube
- Air cleaner
- Mass Air Flow (MAF) sensor
- Exhaust manifold
- Engine under covers
- Left front floor center cover
- No. 1 rear floor board
- Upper radiator hose
- Exhaust front pipe
- Exhaust center pipe
- Shift control rod
- Drive shaft
- Oil cooler lines
- Torque converter mounting bolts (through access hole)
- Rear transmission mount crossmember. Support the transmission with a jack.
- Transmission wiring harness connectors
- Starter motor
- Transmission flange bolts
- Transmission

To install:

3. Installation is the reverse of the removal procedure, while using the following torque values:

- 17mm transmission flange bolts: 53 ft. lbs. (72 Nm)

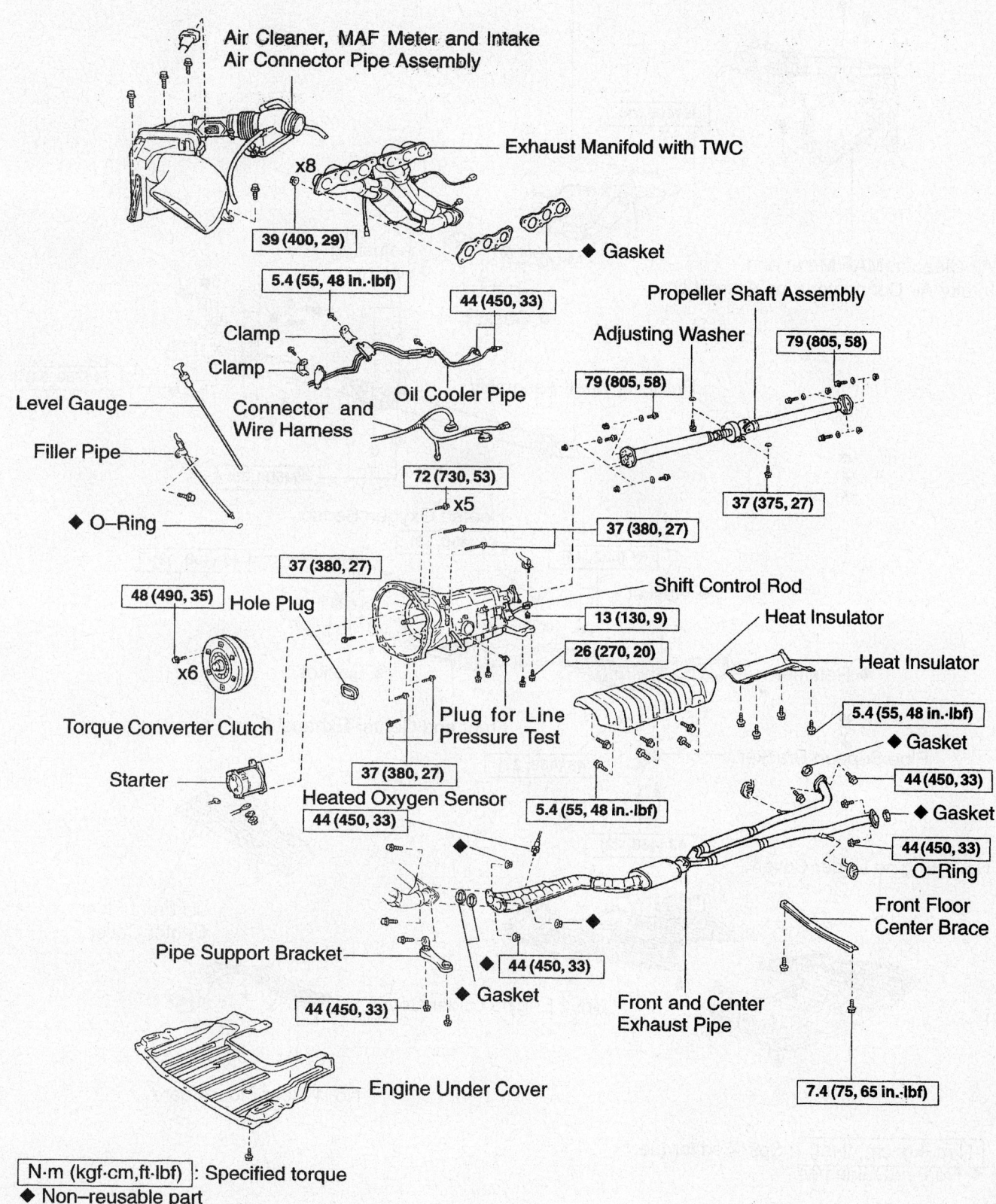

Air Cleaner, MAF Meter and Intake Air Connector Pipe Assembly

Exhaust Manifold with TWC

x8

39 (400, 29)

◆ Gasket

5.4 (55, 48 in.·lbf)

44 (450, 33)

Clamp

Clamp

Oil Cooler Pipe

Propeller Shaft Assembly

Adjusting Washer

79 (805, 58)

79 (805, 58)

Level Gauge

Connector and Wire Harness

Filler Pipe

72 (730, 53)

x5

37 (380, 27)

37 (375, 27)

◆ O–Ring

37 (380, 27)

48 (490, 35)

Hole Plug

Shift Control Rod

Heat Insulator

13 (130, 9)

Heat Insulator

x6

26 (270, 20)

5.4 (55, 48 in.·lbf)

Torque Converter Clutch

Plug for Line Pressure Test

◆ Gasket

44 (450, 33)

Starter

5.4 (55, 48 in.·lbf)

◆ Gasket

Heated Oxygen Sensor

44 (450, 33)

44 (450, 33)

O–Ring

Front Floor Center Brace

Pipe Support Bracket

44 (450, 33)

◆ Gasket

Front and Center Exhaust Pipe

44 (450, 33)

◆ Gasket

Engine Under Cover

7.4 (75, 65 in.·lbf)

N·m (kgf·cm,ft·lbf) : Specified torque
◆ Non–reusable part

Exploded view of automatic transmission components—2002–05 GS 300

09490_LEXU_G0106

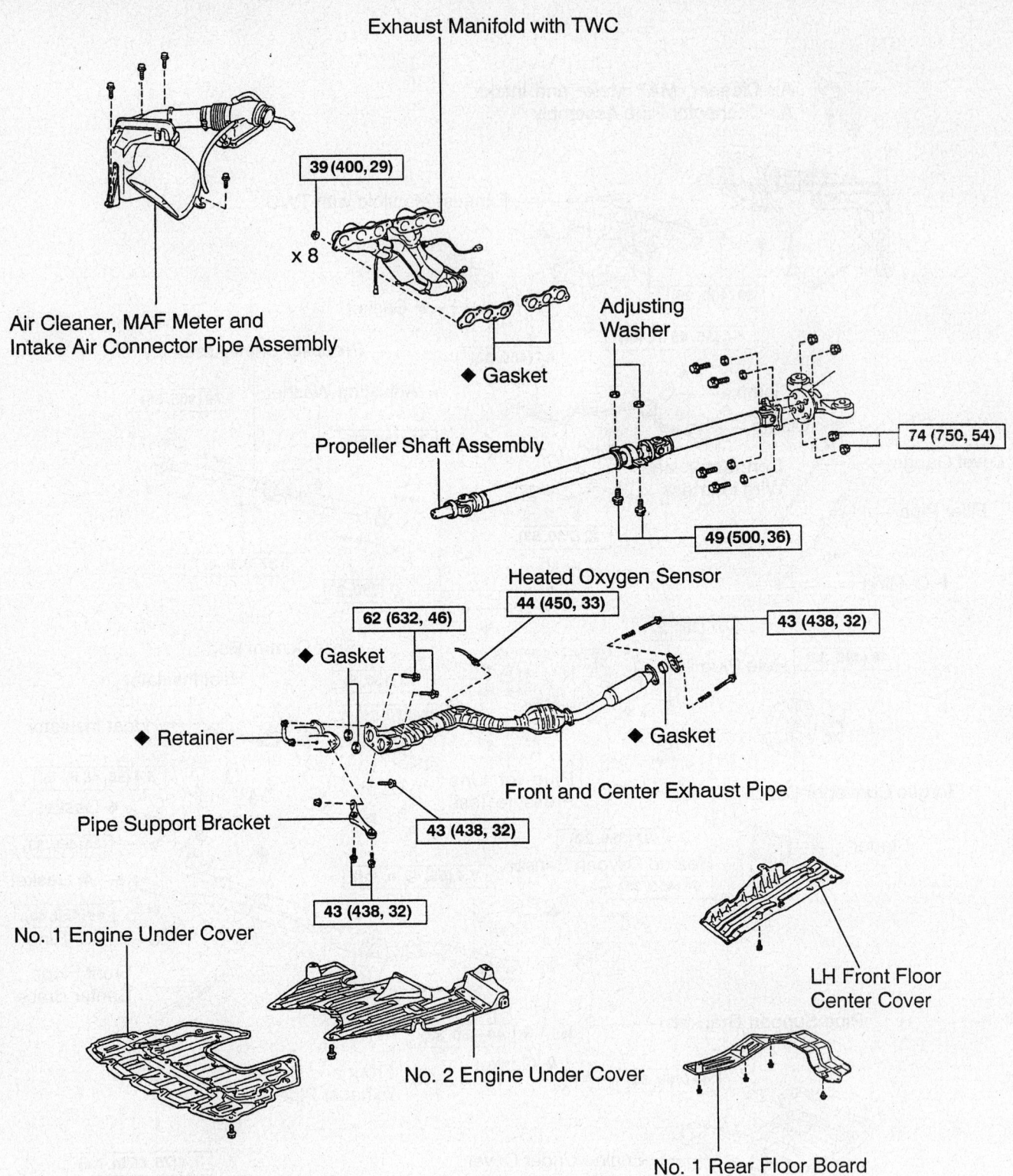

Exhaust Manifold with TWC

39 (400, 29)

x 8

Air Cleaner, MAF Meter and
Intake Air Connector Pipe Assembly

◆ Gasket

Adjusting
Washer

Propeller Shaft Assembly

74 (750, 54)

49 (500, 36)

Heated Oxygen Sensor
44 (450, 33)

62 (632, 46)

43 (438, 32)

◆ Gasket

◆ Retainer

◆ Gasket

Front and Center Exhaust Pipe

Pipe Support Bracket

43 (438, 32)

43 (438, 32)

No. 1 Engine Under Cover

LH Front Floor
Center Cover

No. 2 Engine Under Cover

No. 1 Rear Floor Board

N·m (kgf·cm, ft·lbf) : Specified torque
◆ Non-reusable part

67162-LEXU-G64

Exploded view of automatic transmission components for removal—IS300 (1 of 2)

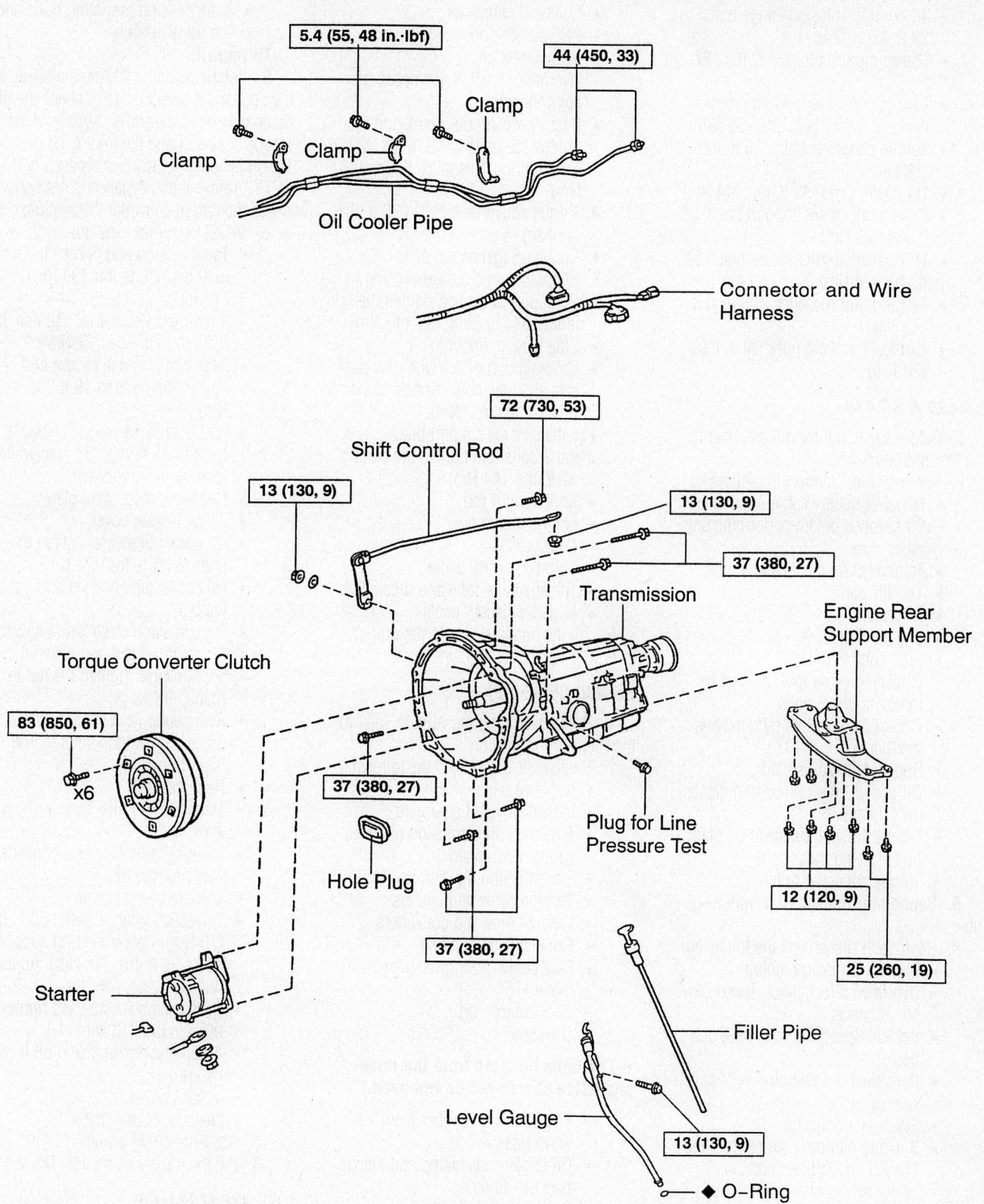

5.4 (55, 48 in.·lbf)

44 (450, 33)

Clamp

Clamp

Clamp

Clamp

Oil Cooler Pipe

Connector and Wire Harness

72 (730, 53)

Shift Control Rod

13 (130, 9)

13 (130, 9)

37 (380, 27)

Transmission

Engine Rear Support Member

Torque Converter Clutch

83 (850, 61)

x6

37 (380, 27)

Hole Plug

Plug for Line Pressure Test

12 (120, 9)

25 (260, 19)

Starter

37 (380, 27)

Filler Pipe

Level Gauge

13 (130, 9)

◆ O-Ring

N·m (kgf·cm, ft·lbf) : Specified torque
◆ Non-reusable part

67162-LEXU-G65

Exploded view of automatic transmission components for removal—IS300 (2 of 2)

- 14mm transmission flange bolts: 27 ft. lbs. (37 Nm)
- Starter motor bolts: 27 ft. lbs. (37 Nm)
- Rear transmission mount cross-member bolts: 19 ft. lbs. (25 Nm)
- Torque converter bolts: 35 ft. lbs. (48 Nm)
- Oil cooler lines: 33 ft. lbs. (44 Nm)
- Drive shaft center support bolts: 36 ft. lbs. (49 Nm)
- Drive shaft U-joint flange bolts: 54 ft. lbs. (74 Nm)
- Shift control rod nuts: 108 inch lbs. (13 Nm)
- Exhaust manifold nuts: 29 ft. lbs. (39 Nm)

LS 430 & SC 430

1. Before servicing the vehicle, refer to the Precautions Section.
2. Remove or disconnect the following:
   - Negative battery cable. Wait at least 90 seconds before performing any other work.
   - Transmission dipstick and tube
   - Throttle cable
   - Driveshaft
   - Engine undercover
   - Shift control rod
   - Exhaust pipe support bracket by removing the 2 bolts
   - Catalytic converters by removing the 6 nuts
   - Both side heat insulators
   - Oil cooler tube clamps and disconnect the tubes
   - Torque converter inspection plate by removing the 2 bolts
   - Torque converter bolts
3. Support the transmission with a suitable jack.
4. Remove or disconnect the following:
   - Rear transmission mount
   - Overdrive direct clutch speed sensor connector
   - Vehicle Speed Sensor (VSS) connector
   - Park/Neutral Position (PNP) switch connector
   - Solenoid connector
   - 3 wiring harness clamps from the bracket on the transmission.
   - 10 transmission mounting bolts and the transmission.

To install:

5. Install or connect the following:
   - Transmission and tighten the bolts as follows: 14mm–27 ft. lbs. (37 Nm); 17mm–53 ft. lbs. (72 Nm)
   - 3 wiring harness clamp to the bracket on the transmission

- Solenoid connector
- PNP switch connector
- VSS connector
- Overdrive direct clutch speed sensor connector
- Rear transmission mount and tighten the bolts to 19 ft. lbs. (20 Nm) and the nuts to 10 ft. lbs. (13 Nm)
- Torque converter bolts to 30 ft. lbs. (41 Nm)
- Converter inspection plate
- Support from the transmission
- Oil cooler pipes and tighten the union nuts to 32 ft. lbs. (44 Nm)
- Side heat insulators
- Catalytic converters with new gaskets and new nuts. Tighten the nuts to 46 ft. lbs. (62 Nm).
- Exhaust pipe support bracket with the 2 bolts and tighten the bolts to 32 ft. lbs. (44 Nm)
- Shift control rod
- Engine undercover
- Driveshaft
- Throttle control cable
- Transmission tube and dipstick
- Negative battery cable

6. Fill the transmission to the proper level with recommended type fluid.

GS 430 (2002–03)

1. Before servicing the vehicle, refer to the Precautions Section.
2. Remove or disconnect the following:
   - Negative battery cable.
   - V-bank cover, if equipped
   - Automatic transmission oil level gauge if equipped
   - Transmission dipstick and tube
   - Throttle cable and clamps
   - Exhaust pipe and converters
   - Exhaust heat insulator
   - Rear center floor crossmember brace
   - Shift control rod
   - Driveshaft

➡ **The bolts inserted from the drive-shaft side should not be removed.**

   - The electrical harness from the transmission.
   - Oil cooler tube clamp and disconnect the tubes
   - Lower engine cover
   - Torque converter inspection plate
   - Torque converter bolts
3. Support the transmission with a suitable jack.
4. Remove or disconnect the following:
   - Starter, if necessary
   - Rear transmission mount

- Transmission mounting bolts and the transmission.

To install:

5. Before installing the transmission, use calipers and a straightedge to check the distance between the installed surface of the torque converter and the front edge of the transmission case. Correct distance is 0.673 in. (17.1mm). If this distance is not correct, check the torque converter installation.
6. Install or connect the following:
   - Transmission and tighten the bolts to 14mm: 29 ft. lbs. (39 Nm); 17mm: 42 ft. lbs. (57 Nm)
   - If removed, the starter. Tighten the bolts to 27 ft. lbs. (37 Nm).
   - Rear transmission mount and tighten the bolts to 19 ft. lbs. (20 Nm)
   - And tighten the torque converter bolts to 25 ft. lbs. (33 Nm) while rotating the crankshaft
   - Converter inspection plate
   - Lower engine cover
   - Oil cooler lines and tighten the lines to 25 ft. lbs. (34 Nm)
   - Oil cooler pipe bracket and tighten the bolt
   - Transmission electrical connectors
   - Shift control rod and adjust the shift linkage. Tighten the nut to 12 ft. lbs. (16 Nm)
   - Rear center floor crossmember brace. Tighten the bolts to 108 inch lbs. (13 Nm).
   - Heat insulator
   - Transmission filler tube and dipstick
   - Front exhaust pipe and converters with new gaskets
   - Throttle control cable
   - Driveshaft. Flange bolts: 58 ft. lbs. (79 Nm). Center bearing support bolts: 36 ft. lbs. (49 Nm). Adjusting nut: 35 ft. lbs. (48 Nm).
   - Crossmember brace and tighten to 96 inch lbs. (13 Nm)
   - Automatic transmission oil level gauge
   - V-bank cover
   - Negative battery cable
7. Adjust the PNP switch
8. Fill the transmission with Dexron®II.

GS 430 (2004–06)

1. Before servicing the vehicle, refer to the Precautions Section.
2. Remove or disconnect the following:
   - Negative battery cable.
   - Transmission oil level dipstick
   - Front and center exhaust pipes (disconnect oxygen sensors)

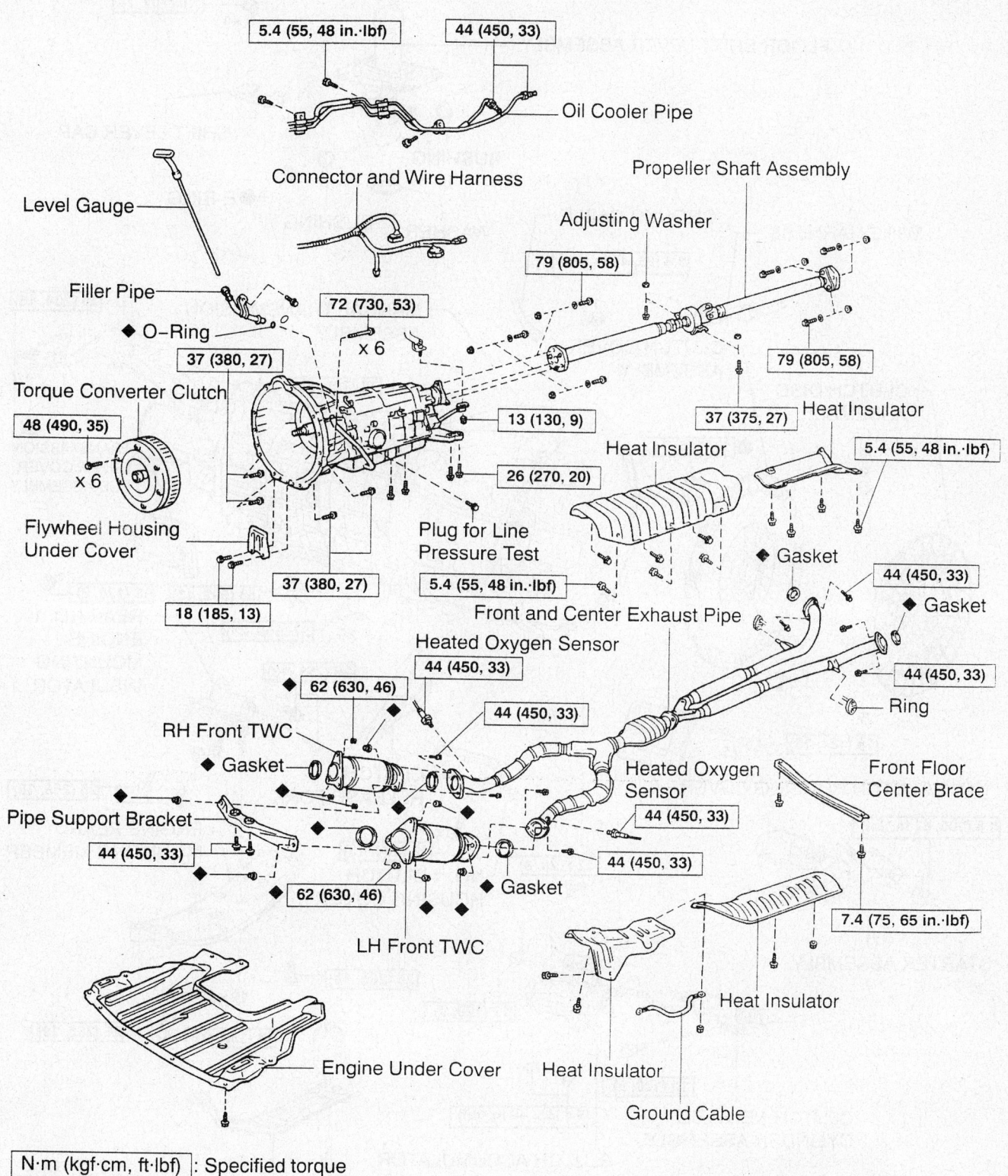

5.4 (55, 48 in.·lbf)

44 (450, 33)

Oil Cooler Pipe

Connector and Wire Harness

Propeller Shaft Assembly

Adjusting Washer

Level Gauge

79 (805, 58)

Filler Pipe

◆ O-Ring

72 (730, 53)

×6

79 (805, 58)

37 (380, 27)

37 (375, 27)

Heat Insulator

Torque Converter Clutch

48 (490, 35)

13 (130, 9)

Heat Insulator

5.4 (55, 48 in.·lbf)

×6

26 (270, 20)

Flywheel Housing
Under Cover

Plug for Line
Pressure Test

◆ Gasket

44 (450, 33)

37 (380, 27)

5.4 (55, 48 in.·lbf)

◆ Gasket

18 (185, 13)

Front and Center Exhaust Pipe

44 (450, 33)

Heated Oxygen Sensor

44 (450, 33)

62 (630, 46)

Ring

RH Front TWC

44 (450, 33)

Heated Oxygen
Sensor

Front Floor
Center Brace

◆ Gasket

44 (450, 33)

Pipe Support Bracket

44 (450, 33)

44 (450, 33)

62 (630, 46)

◆ Gasket

7.4 (75, 65 in.·lbf)

LH Front TWC

Heat Insulator

Engine Under Cover

Heat Insulator

Ground Cable

N·m (kgf·cm, ft·lbf) : Specified torque

◆ Non–reusable part

67162-LEXU-G66

Exploded view of the transmission components for removal—GS 430 (2004 application shown)

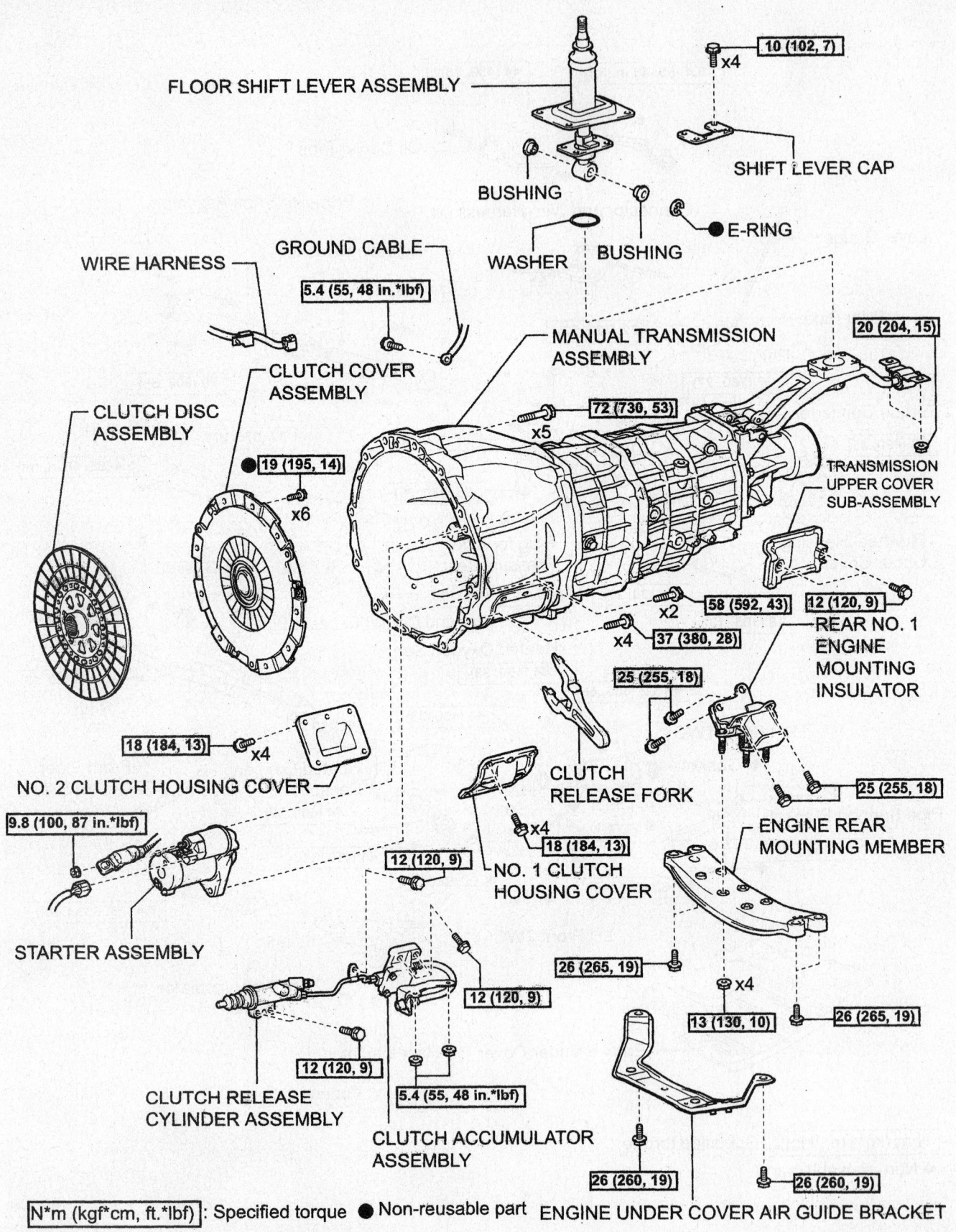

FLOOR SHIFT LEVER ASSEMBLY

10 (102, 7) x4

SHIFT LEVER CAP

BUSHING

E-RING

WASHER   BUSHING

WIRE HARNESS   GROUND CABLE

5.4 (55, 48 in.*lbf)

MANUAL TRANSMISSION ASSEMBLY

20 (204, 15)

CLUTCH COVER ASSEMBLY

72 (730, 53) x5

CLUTCH DISC ASSEMBLY

● 19 (195, 14) x6

TRANSMISSION UPPER COVER SUB-ASSEMBLY

58 (592, 43) x2

12 (120, 9)

37 (380, 28) x4

REAR NO. 1 ENGINE MOUNTING INSULATOR

25 (255, 18)

25 (255, 18)

18 (184, 13) x4

NO. 2 CLUTCH HOUSING COVER

CLUTCH RELEASE FORK

ENGINE REAR MOUNTING MEMBER

9.8 (100, 87 in.*lbf)

18 (184, 13) x4

NO. 1 CLUTCH HOUSING COVER

STARTER ASSEMBLY

12 (120, 9)

26 (265, 19)

x4

26 (265, 19)

13 (130, 10)

12 (120, 9)

12 (120, 9)

5.4 (55, 48 in.*lbf)

CLUTCH RELEASE CYLINDER ASSEMBLY

CLUTCH ACCUMULATOR ASSEMBLY

26 (260, 19)

26 (260, 19)

ENGINE UNDER COVER AIR GUIDE BRACKET

N*m (kgf*cm, ft.*lbf) : Specified torque ● Non-reusable part

Exploded view of manual transmission components—IS 250

09490_LEXU_G0107

- Both catalytic converters
- Exhaust heat insulator
- Transmission fluid filler pipe
- Driveshaft
- Shift control rod from end of shift lever
- Oil cooler pipes
- Torque converter inspection plate
- Torque converter bolts

3. Support the transmission with a suitable jack.

4. Remove or disconnect the following:
- Engine rear mount (4 bolts)
- Wiring and connectors (tilt transmission rear downward)
- Transmission mounting bolts and the transmission (place jack and wood block under front end of engine)

To install:

5. Before installing the transmission, use calipers and a straightedge to check the distance between the installed surface of the torque converter and the front edge of the transmission case. Correct distance is 0.673 in. (17.1mm). If this distance is not correct, check the torque converter installation.

6. Position transmission to the vehicle and install or connect the following:
- Transmission and tighten the bolts to 14mm: 27 ft. lbs. (37 Nm); 17mm: 53 ft. lbs. (72 Nm)
- Wiring and connectors
- Rear engine mount and tighten the bolts to 20 ft. lbs. (26 Nm)
- Torque converter bolts to 35 ft. lbs. (48 Nm) while rotating the crankshaft
- Converter inspection plate; torque bolts to 13 ft. lbs. (18 Nm)
- Oil cooler lines and tighten the clamp bolts to 48 inch lbs. (5.4 Nm)
- Oil cooler pipe union nuts to 33 ft. lbs. (44 Nm)
- Shift control rod; torque nut to 9 ft. lbs. (13 Nm).
- Driveshaft
- Transmission fluid filler pipe, with new O-ring
- Heat insulator
- Both catalytic converters, with new gaskets; torque nuts to 46 ft. lbs. (62 Nm)
- Tail pipe to center exhaust pipe, with new gaskets; torque bolts to 33 ft. lbs. (44 Nm)
- Exhaust pipe support bracket bolts to 33 ft. lbs. (44 Nm)
- Exhaust pipe-to-catalytic converters, with new gaskets, to 33 ft. lbs. (44 Nm)

- Front floor center brace bolts to 65 inch lbs. (7.4 Nm)
- If removed, oxygen sensors to 33 ft. lbs. (44 Nm)

➡**Before installation of the oxygen sensor, twist wiring 3½ turns in counterclockwise position; this will prevent wires being twisted after sensors are installed.**

- Engine under cover
- Transmission dipstick
- Negative battery cable

7. Fill the transmission with recommended fluid type.

Manual

IS 250

1. Before servicing the vehicle, refer to the Precautions Section.

2. Disconnect the negative battery cable. Wait at least 90 seconds before performing any other work.

3. Drain manual transmission fluid.

4. Remove or disconnect the following:

- Shift lever knob
- Front and rear console panel sub-assemblies
- Front ashtray
- Console box register assembly
- Console box
- No. 2 console box duct
- Shift lever boot
- 4 bolts and shift lever cap
- Engine under cover
- No. 2 engine under cover
- No. 1 rear floor panel brace
- Front center floor brace
- Oxygen sensor
- Front exhaust pipe assembly
- Front floor No. 1 heat insulator
- Right outside air guide plate
- Propeller shaft with center bearing assembly after matchmarking location
- Engine under cover air guide bracket
- Starter motor
- 2 bolts, 2 nuts and clutch accumulator assembly
- 2 bolts and clutch release cylinder
- Ground wire
- Remaining wiring connectors
- No. 1 and 2 clutch housing covers
- 6 bolts and clutch cover assembly. Matchmark clutch cover to the flywheel.

➡**Loosen each bolt one turn at a time until spring tension is released.**

- Floor shift control shift lever retainer
- E-ring and floor shift lever assembly

5. Support the transmission unit with a transmission jack.
- Engine rear mounting member
- 9 bolts and manual transmission assembly

To install:

6. Install or connect the following:
- Manual transmission. Torque bolt A to 53 ft. lbs. (72 Nm) and bolt B to 28 ft. lbs. (37 Nm)

➡**Temporarily tighten bolt *1, fully tighten bolt *2, lift the rear end of the transmission, and fully tighten bolt *1 and the other bolts.**

- Engine rear mounting member. Torque the 4 nuts to the transmission to 10 ft. lbs. (13 Nm) and the 4 body bolts to 19 ft. lbs. (26 Nm)
- Floor shift lever assembly with a new E-ring
- Shift lever boot
- Floor shift control shift lever retainer. Torque the 2 nuts to 15 ft. lbs. (20 Nm).
- Align matchmarks of the clutch cover to the flywheel. Install 6 new bolts and torque them to 14 ft. lbs. (19 Nm) a little at a time in a diagonal pattern.
- No. 1 and 2 clutch housing covers. Torque to 13 ft. lbs. (18 Nm)
- Wiring connectors
- Ground wire
- Clutch release cylinder. Torque to 9 ft. lbs. (12 Nm)
- Clutch accumulator assembly. Torque the bolts to 9 ft. lbs. (12 Nm) and the nuts to 48 inch lbs. (5 Nm).
- Starter motor
- Engine under cover air guide

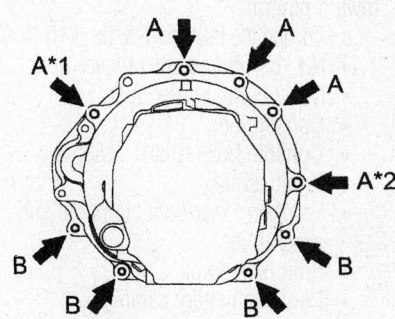

09490_LEXU_G0108

Automatic transmission mounting bolts—IS 250

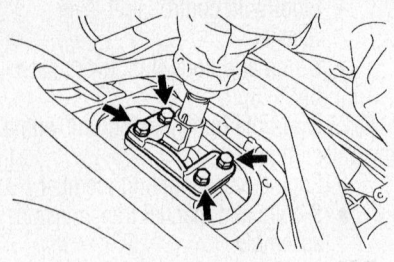

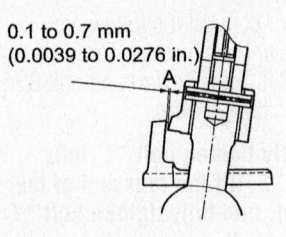

0.1 to 0.7 mm
(0.0039 to 0.0276 in.)

09490_LEXU_G0109

Adjusting the shift lever guide plate—IS 250

bracket. Torque to 19 ft. lbs. (26 Nm).
- Propeller shaft with center bearing assembly. Torque to 58 ft. lbs. (79 Nm)
- No. 1 center support bearing to the matchmark. Torque to 51 ft. lbs. (69 Nm)

➡ **Stabilize the propeller shaft by rotating it several times by hand to stabilize the center support bearing.**

- Front floor No. 1 heat insulator
- Right outside air guide plate
- Front exhaust pipe assembly
- Oxygen sensor
- Front center floor brace
- No. 1 rear floor panel brace
- No. 2 engine under cover
- Engine under cover

7. Install the shift lever cap as follows:
a. Move the shift lever to the second gear position.
b. Adjust the guide plate so that dimension A is as shown in the illustration with the shift lever cap pushed toward neutral.
c. Torque the bolts to 7 ft. lbs. (10 Nm).
- No. 1 shift and select lever boot
- No. 2 console box duct
- Console box
- Console box register assembly
- Front ashtray
- Front and rear console panel sub-assemblies
- Shift lever knob
- Negative battery cable

8. Add manual transmission fluid.
9. Start engine and check for fluid and exhaust leaks.

Transaxle Assembly

REMOVAL & INSTALLATION

ES 300

1. Before servicing the vehicle, refer to the Precautions Section.
2. Turn the ignition switch to the **LOCK** position and disconnect the negative battery cable. Wait at least 90 seconds or longer before doing any work on the vehicle.
3. Remove or disconnect the following:
- Battery
- Air cleaner assembly
- Throttle cable from the throttle body
- Cruise control actuator cover and detach the connector
- Ground wire
- Starter
- Vehicle Speed Sensor (VSS) connectors
- Direct clutch speed sensor
- Park/Neutral Position (PNP) switch connector on the transaxle
- Solenoid connector on the transaxle
- Shift control cable
- Oil cooler hoses
- 2 front side transaxle mounting bolts
- 2 front engine mounting bolts
- Oil cooler line mounting bolts from the front frame
- 3 upper transaxle to engine mounting bolts

4. Install an engine support fixture. Tie steering gear housing to engine support fixture.
5. Raise and safely support the vehicle.
6. Drain the transaxle/differential fluid.
7. Remove or disconnect the following:
- Front wheels
- Front exhaust pipe
- Engine side covers and undercovers
- Both halfshafts
- Front side engine mounting nut
- Rear side engine mounting bolts (remove hole plugs)
- 4 left side transaxle mounting bolts
- Steering gear housing
- Front frame assembly

8. Properly support the transaxle assembly.
9. Remove or disconnect the following:
- Rear end plate mounting bolts
- Torque converter cover
- Torque converter retaining bolts
- Remaining transaxle mounting bolts

10. Carefully remove the transaxle assembly from the vehicle.

To install:

11. Position the transaxle, aligning the 2 dowel pins on the block with the converter housing. Tighten the bolts as follows: 10mm–34 ft. lbs. (46 Nm); 12mm–47 ft. lbs. (64 Nm).
12. Install or connect the following:
- Torque converter bolts. Coat the threads with sealer. Install the bolts starting with the green bolt followed by the rest and tighten the bolts evenly to 20 ft. lbs. (27 Nm).
- End plate and tighten the bolts to 27 ft. lbs. (37 Nm)
- Front frame assembly and tighten the fasteners as follows: 12mm–24 ft. lbs. (32 Nm); 19mm–134 ft. lbs. (181 Nm); nut–27 ft. lbs. (36 Nm).
- 2 fender liner set screws
- Steering gear to the frame and tighten the bolts and nuts to 134 ft. lbs. (181 Nm)
- Sway bar brackets and toque the bolts to 14 ft. lbs. (19 Nm)
- Left transaxle mounting bolts and tighten them to 38 ft. lbs. (52 Nm)
- Rear side mounting bolts and nuts and tighten them to 48 ft. lbs. (66 Nm). Install the plugs.
- Front engine mounting nut and tighten it to 59 ft. lbs. (80 Nm)
- Halfshafts
- Right and left engine side covers
- Lower engine cover
- Exhaust pipe to the engine with new gaskets and tighten the nuts to 46 ft. lbs. (62 Nm). Connect the exhaust pipe to the converter with a new gasket and tighten the nuts and bolts to 32 ft. lbs. (43 Nm).
- Wheel
- Engine support
- 4 upper transaxle mounting bolts and tighten them to 47 ft. lbs. (64 Nm)
- Oil cooler clamping bolts to the front frame
- 2 front side engine mounting bolts and tighten them to 59 ft. lbs. (80 Nm)
- 2 front side transaxle mounting bolts and tighten them to 59 ft. lbs. (80 Nm)
- Remaining components
- Battery and connect the battery cables

13. Fill the transaxle/differential to the proper level with Dexron® II, or equivalent.

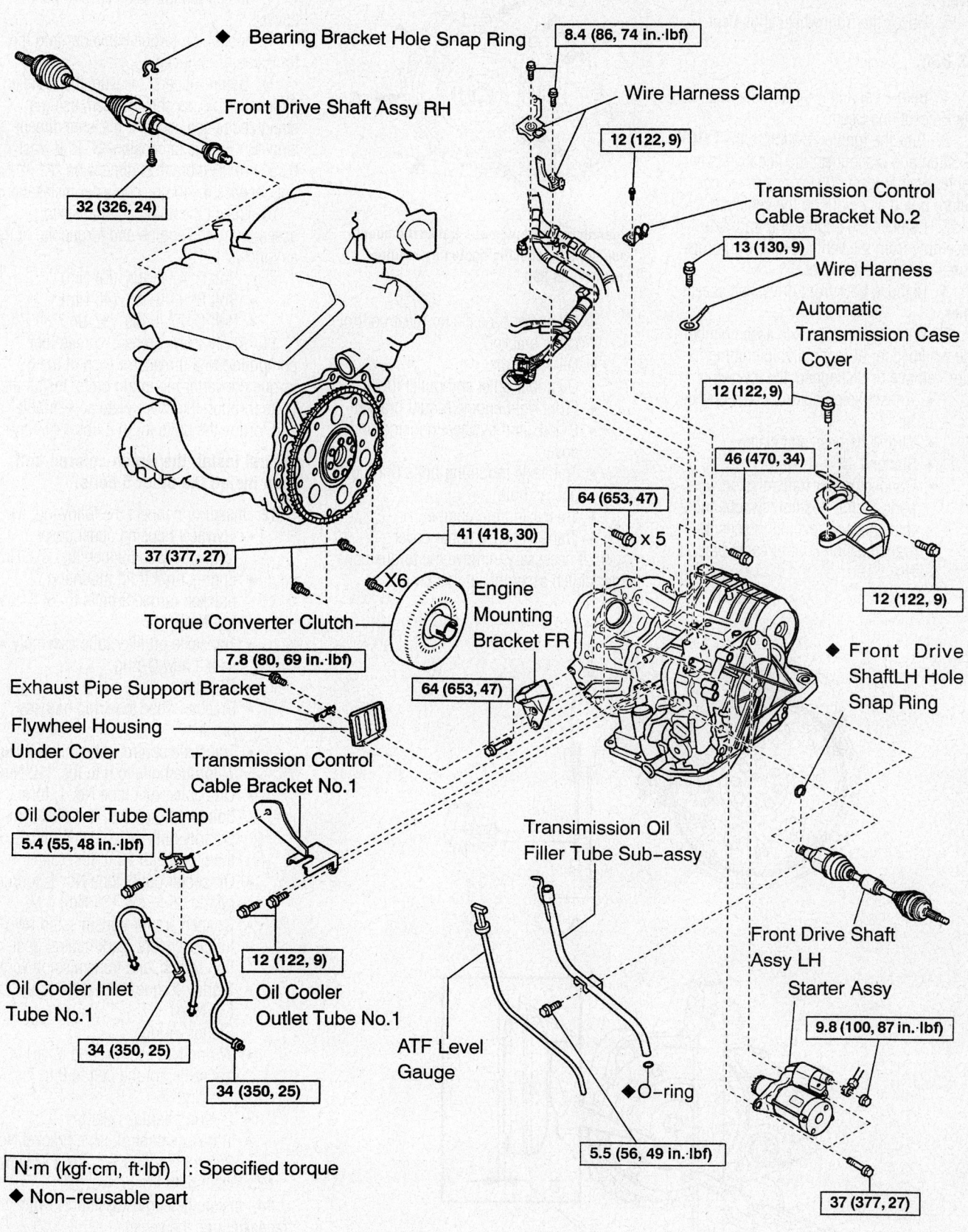

◆ Bearing Bracket Hole Snap Ring

Front Drive Shaft Assy RH

8.4 (86, 74 in.·lbf)

Wire Harness Clamp

12 (122, 9)

32 (326, 24)

Transmission Control Cable Bracket No.2

13 (130, 9)

Wire Harness

Automatic Transmission Case Cover

12 (122, 9)

46 (470, 34)

37 (377, 27)

41 (418, 30)

64 (653, 47) × 5

64 (653, 47)

12 (122, 9)

◆ Front Drive ShaftLH Hole Snap Ring

X6

Torque Converter Clutch

Engine Mounting Bracket FR

7.8 (80, 69 in.·lbf)

Exhaust Pipe Support Bracket

Flywheel Housing Under Cover

Transmission Control Cable Bracket No.1

Oil Cooler Tube Clamp

5.4 (55, 48 in.·lbf)

Transimission Oil Filler Tube Sub-assy

Front Drive Shaft Assy LH

12 (122, 9)

Starter Assy

9.8 (100, 87 in.·lbf)

Oil Cooler Inlet Tube No.1

Oil Cooler Outlet Tube No.1

ATF Level Gauge

◆O-ring

34 (350, 25)

34 (350, 25)

5.5 (56, 49 in.·lbf)

37 (377, 27)

N·m (kgf·cm, ft·lbf) : Specified torque

◆ Non-reusable part

67162-LEXU-G67

Exploded view of the transaxle assembly components for removal—ES 330

14. Check the transaxle/differential fluid level.

15. Check the front wheel alignment.

ES 330

1. Before servicing the vehicle, refer to the Precautions Section.

2. Turn the ignition switch to the **LOCK** position and disconnect the negative battery cable. Wait at least 90 seconds or longer before doing any work on the vehicle.

3. Remove the engine and transaxle assembly from the vehicle, following procedures for engine removal.

4. Remove both front driveshaft assemblies.

5. With engine/transaxle assembly on the workbench, or in a suitable holding fixture, remove or disconnect the following:

- Transmission control cable bracket No. 2
- Wiring harness and clamp
- Starter
- Connectors for transmission wire, park/neutral position switch, and 2 transmission revolution sensors
- Transmission control cable bracket No. 1

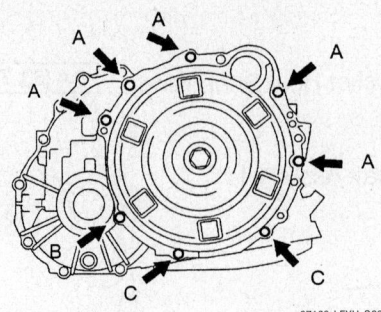

67162-LEXU-G69

Identifying the automatic transaxle mounting bolts for specific tightening requirements—ES 330

- Oil cooler tube clamp (near control cable bracket)
- Oil filler tube
- Oil cooler inlet and outlet tubes
- Front right engine mount bracket
- 2 bolts and flywheel housing under cover
- Transaxle mounting bolts (through access hole)
- Transaxle from engine
- Transaxle case upper cover

6. If necessary, remove the torque converter clutch assembly at this time.

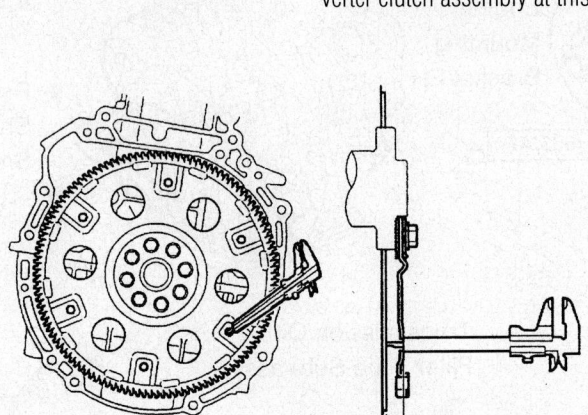

A

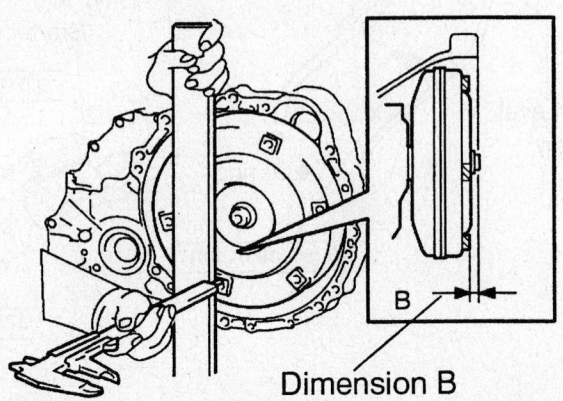

B

Dimension B

67162-LEXU-G68

Measuring the torque converter installed position—ES 330

To install:

7. Install the transaxle case upper cover.

8. Install the torque converter into the transaxle, if removed.

9. Using calipers, measure the installed dimension "A", as shown. Then, using a straightedge and calipers, measure dimension "B". Ensure dimension "B" is at least 0.25" (1mm) more than dimension "A". If not, recheck the torque converter installation.

10. Install the automatic transaxle assembly to the engine and torque the bolts as follows:

- Bolt A: 47 ft. lbs. (64 Nm)
- Bolt B: 34 ft. lbs. (46 Nm)
- Bolt C: 27 ft. lbs. (37 Nm)

11. Apply a few drops of thread lock compound to 2 threads of each of the 6 torque converter mounting bolts. Install the bolts through the cover plate access hole and torque the bolts to 30 ft. lbs. (41 Nm).

➡️**First install the Green-colored bolt, then install the other 5 bolts.**

12. Install or connect the following:

- Flywheel housing under cover; torque bolts to 69 inch lbs. (7.8 Nm)
- Engine right front mounting bracket; torque 3 bolts to 47 ft. lbs. (64 Nm)
- Transaxle oil filler tube assembly, with a new O-ring
- Dipstick
- Breather hose to wiring harness bracket
- Transaxle control cable bracket No. 1; torque bolts to 9 ft. lbs. (12 Nm)
- Oil cooler inlet tube No. 1; torque bolt to 48 inch lbs. (5.4 Nm) temporarily until assembly is installed, then torque to 25 ft. lbs. (34 Nm)
- Oil cooler outlet tube No. 1; torque bolt to 25 ft. lbs. (34 Nm)
- Connectors for transmission revolution sensors, park/neutral position switch, and transmission wire.
- Starter; torque bolts to 27 ft. lbs. (37 Nm)
- Starter wiring
- Wiring harness (ground strap) to transaxle; torque bolt to 9 ft. lbs. (12 Nm)
- 2 wiring harness clamps
- Transaxle control cable bracket No. 2; torque bolt to 9 ft. lbs. (12 Nm)

13. Install both front driveshafts.

14. Install the engine assembly, with transaxle, into the vehicle.

15. Refill the transaxle with recommended fluid type.

16. Check the front wheel alignment.

Clutch

REMOVAL & INSTALLATION

IS 250

1. Before servicing the vehicle, refer to the Precautions Section.
2. Disconnect the negative battery cable. Wait at least 90 seconds before performing any other work.
3. Drain manual transmission fluid.
4. Remove or disconnect the following:

- Manual transmission assembly
- Clutch release fork
- Clutch disc assembly
- Clutch cover assembly
- Clutch release bearing shaft snapring, using a snapring expander
- Clutch release bearing hub, wave washer and plate washer
- Clutch release hub snapring, using a snapring expander
- Clutch release hub ball bearing, thrust cone spring plate washer and thrust cone spring.
- Release fork support and release fork support spring

To install:

5. Install or connect the following:

- Release fork support and release fork support spring. Torque the 2 bolts to 19 ft. lbs. (26 Nm).
- Thrust cone spring, thrust cone spring plate washer and clutch release hub ball bearing
- Clutch release hub snapring, using a snapring expander

6. Install the clutch release bearing hub as follows:

   a. Fill the groove inside the release bearing hub with grease.

   b. Apply grease to the contact surface between the release hub bearing and clutch cover. Be sure to apply the grease in an even, thin coat.

   c. Using SST 09950-60020, secure the clutch release hub ball bearing

   d. Clutch cover assembly, release bearing plate washer and release bearing wave washer.

   e. Install snapring, using a snapring expander.

- Clutch cover assembly. Apply clutch spline grease to the input shaft spline.
- Clutch disc assembly
- Clutch release fork with a new E-ring. Apply grease to the pivot

points and contact surfaces.
- Manual transmission assembly
- Manual transmission fluid.
- Negative battery cable

BLEEDING

IS 250

1. Remove bleeder plug cap.
2. Connect a vinyl tube to the bleeder plug.
3. Depress clutch pedal several times, and then loosen the bleeder plug with the pedal depressed.
4. When fluid no longer comes out, tighten the bleeder plug, and then release the clutch pedal.
5. Repeat the previous 2 steps until all the air in the fluid is completely bled.
6. Tighten the bleeder plug to 8 ft. lbs. (11 Nm).
7. Install the bleeder plug cap.
8. Check that all of the air has been bled from the clutch line.
9. Check the fluid level and add as necessary.

Transfer Case

REMOVAL & INSTALLATION

IS 250 and 2006 GS 300 with AWD

1. Before servicing the vehicle, refer to the Precautions Section.
2. Disconnect the negative battery cable. Wait at least 90 seconds before performing any other work.
3. Remove or disconnect the following:

- Automatic transmission assembly
- Transfer case oil pan

4. Remove the transfer valve body assembly as follows:

   a. Remove the 4 bolts and the transfer valve body assembly

   b. Disconnect the transfer wire connector from the transfer control solenoid.

   c. Remove the bolt and the transfer wire.

- Transfer extension housing using SST 09950-40011
- 8 bolts (in order), clamp and transfer case. Use a plastic hammer to tap the transfer case to remove it.
- Transfer rear output shaft assembly

5. Remove the flange yoke as follows:

   a. Using a hammer and SST 09930-00010, release the staked part of the nut.

   b. Using SST 09330-00021 and 09213-54015, hold the yoke.

   c. Remove the nut and the flange yoke.

- 10 bolts and rear transfer chain case. Use a plastic hammer to tap the rear transfer chain case to remove it.
- Transfer front drive chain, transfer drive sprocket and transfer front driven clutch sleeve
- 6 bolts (in order) and front transfer chain case. Use a plastic hammer to tap the front transfer chain case to remove it.

To install:

6. Install front transfer chain case as follows:

   a. Clean the front mating surface of any residual sealant.

   b. Apply new sealant in a width of 0.04–0.06 inches (1.0–1.5mm) all around to the front transfer chain case front mating surface.

   c. Temporarily tighten the 6 bolts in several steps, and then tighten them to 25 ft. lbs. (34 Nm).

7. Install or connect the following:

- Transfer front drive chain, transfer drive sprocket and transfer front driven clutch sleeve

8. Install rear transfer chain case as follows:

   a. Clean the front mating surface of any residual sealant.

   b. Apply new sealant in a width of 0.04–0.06 inches (1.0–1.5mm) all around to the rear transfer chain case front mating surface.

   c. Temporarily tighten the 10 bolts in several steps, and then tighten them to 25 ft. lbs. (34 Nm). Tighten the bolt in the middle of the case cover to 42 ft. lbs. (57 Nm).

9. Install the flange yoke as follows:

   a. Install the flange yoke.

   b. Using SST 09330-00021 and 09213-54015, hold the yoke.

   c. Torque the new nut to 91 ft. lbs. (123 Nm).

- Transfer rear output shaft assembly
- Transfer extension housing using SST 09950-40011

10. Install transfer case as follows:

   a. Clean the front mating surface of any residual sealant.

   b. Apply new sealant in a width of 0.04–0.06 inches (1.0–1.5mm) all around to the transfer case front mating surface.

   c. Temporarily tighten the 8 bolts and clamp in several steps, and then tighten them to 25 ft. lbs. (34 Nm)

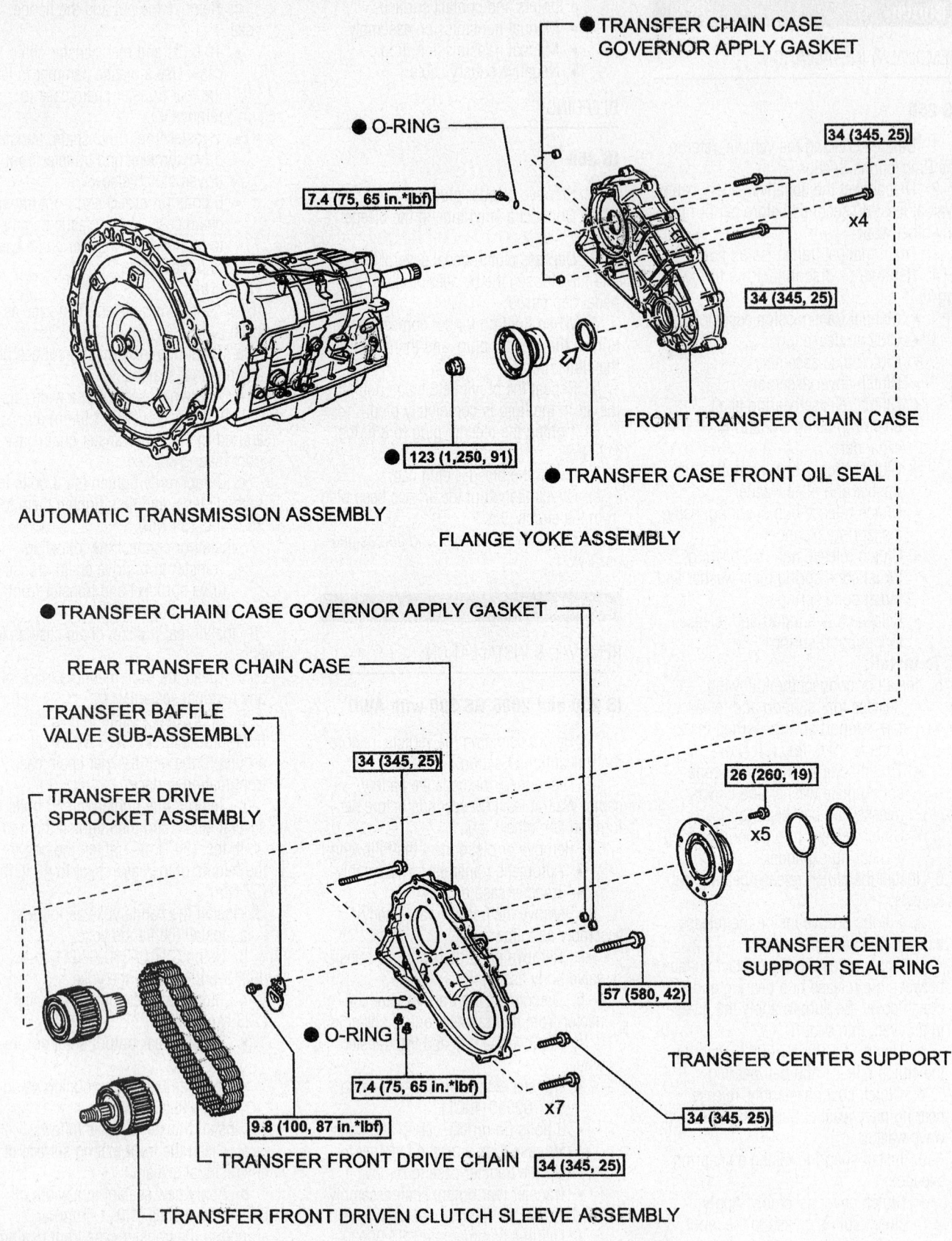

● TRANSFER CHAIN CASE
GOVERNOR APPLY GASKET

● O-RING

34 (345, 25)

7.4 (75, 65 in.*lbf)

x4

34 (345, 25)

FRONT TRANSFER CHAIN CASE

123 (1,250, 91)

● TRANSFER CASE FRONT OIL SEAL

AUTOMATIC TRANSMISSION ASSEMBLY

FLANGE YOKE ASSEMBLY

● TRANSFER CHAIN CASE GOVERNOR APPLY GASKET

REAR TRANSFER CHAIN CASE

TRANSFER BAFFLE
VALVE SUB-ASSEMBLY

34 (345, 25)

26 (260, 19)

TRANSFER DRIVE
SPROCKET ASSEMBLY

x5

TRANSFER CENTER
SUPPORT SEAL RING

57 (580, 42)

● O-RING

TRANSFER CENTER SUPPORT

7.4 (75, 65 in.*lbf)

9.8 (100, 87 in.*lbf)

x7

34 (345, 25)

TRANSFER FRONT DRIVE CHAIN

34 (345, 25)

TRANSFER FRONT DRIVEN CLUTCH SLEEVE ASSEMBLY

N*m (kgf*cm, ft.*lbf) : Specified torque ● Non-reusable part ⇐ Apply MP grease

09490_LEXU_G0110

Exploded view of the transfer case assembly—IS 250 and 2006 GS 300 with AWD

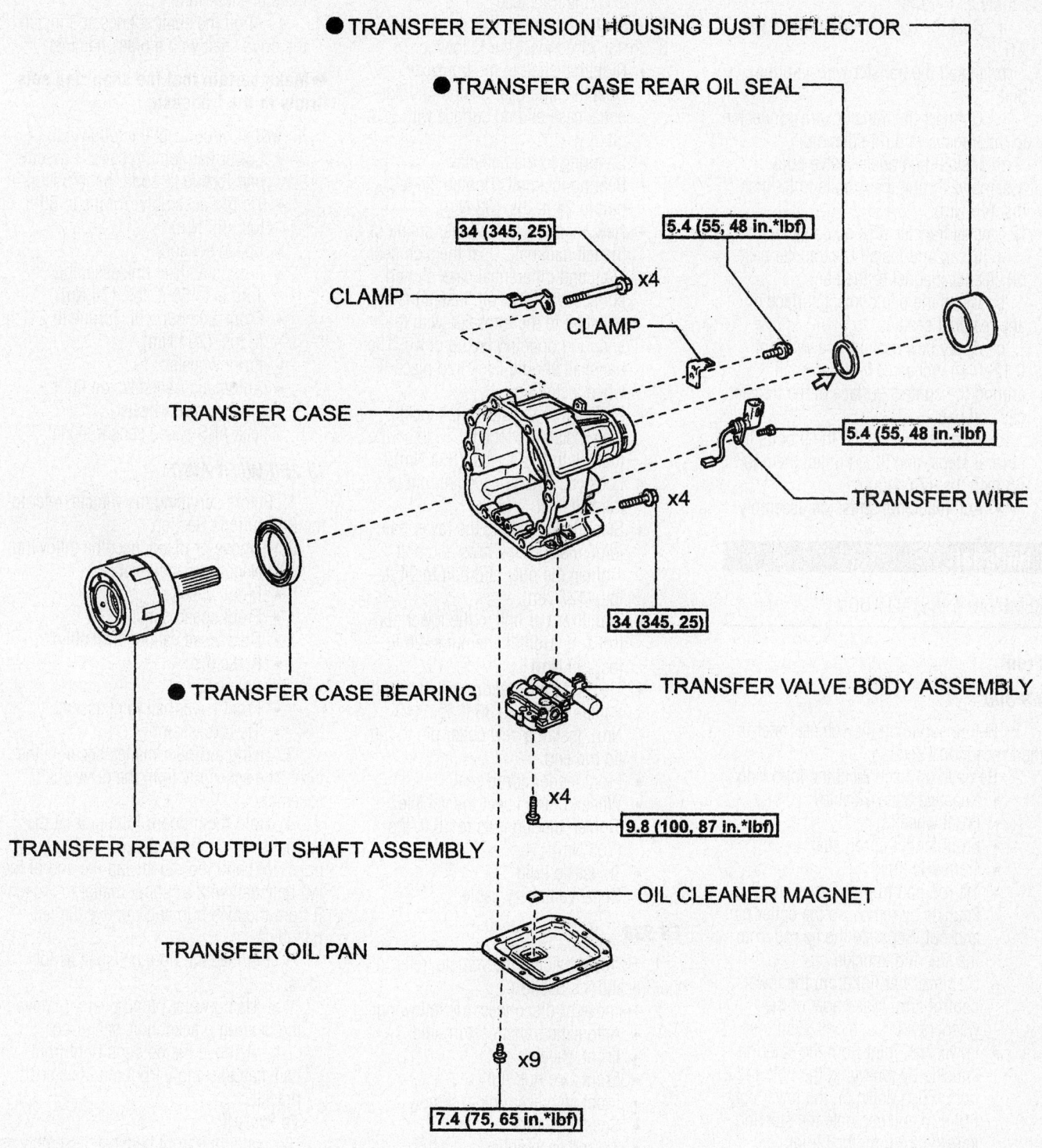

● TRANSFER EXTENSION HOUSING DUST DEFLECTOR

● TRANSFER CASE REAR OIL SEAL

34 (345, 25)

5.4 (55, 48 in.*lbf)

CLAMP

CLAMP

TRANSFER CASE

x4

5.4 (55, 48 in.*lbf)

TRANSFER WIRE

x4

34 (345, 25)

● TRANSFER CASE BEARING

TRANSFER VALVE BODY ASSEMBLY

x4

9.8 (100, 87 in.*lbf)

TRANSFER REAR OUTPUT SHAFT ASSEMBLY

OIL CLEANER MAGNET

TRANSFER OIL PAN

x9

7.4 (75, 65 in.*lbf)

N*m (kgf*cm, ft.*lbf) : Specified torque   ● Non-reusable part   ⬅ Apply MP grease

Exploded view of the transfer case assembly (cont.)—IS 250 and 2006 GS 300 with AWD

11. Install the transfer valve body assembly as follows:

a. Coat the transfer wire O-ring with ATF.

b. Install the transfer wire with the bolt.

c. Connect the transfer wire connector to the transfer control solenoid.

d. Install the transfer valve body assembly. Torque the 4 bolts to 87 inch lbs. (10 Nm).

12. Install transfer case oil pan as follows:

a. Clean and install the transfer case oil cleaner magnet to the pan.

b. Clean the pan contact surface of any residual sealant.

c. Apply new sealant in a width of 0.12–0.14 inches (3.0–3.5mm) all around the contact surface of the transfer case oil pan.

d. Temporarily tighten the 9 bolts in several steps, and then tighten them to 65 inch lbs. (7.5 Nm).

• Automatic transmission assembly

Halfshaft

REMOVAL & INSTALLATION

Front

ES 300

1. Before servicing the vehicle, refer to the Precautions Section.

2. Remove or disconnect the following:
   • Negative battery cable
   • Front wheel(s)
   • Front fender apron seal
   • Transaxle fluid
   • Tie rod end from the steering knuckle by removing the cotter pin and nut. Separate the tie rod from the steering knuckle.
   • Stabilizer bar link from the lower control arm. Make note of the washers and cushions positions.
   • Lower ball joint from the steering knuckle by removing the bolt and 2 nuts. Push down on the lower control arm and separate the steering knuckle from the ball joint.
   • Cotter pin, lock cap and locknut holding the halfshaft to the steering knuckle
   • Left halfshaft from the steering knuckle
   • Halfshaft from the transaxle
   • Snapring from the halfshaft
   • Right halfshaft bearing lockbolt. The lockbolt is located in the center of the halfshaft, near the dampener.

• Snapring and pull the halfshaft from the transaxle

To install:

3. Install or connect the following:
   • Right halfshaft to the transaxle. Coat the side gear shaft and differential case sliding surface with gear oil.
   • Snapring to the halfshaft
   • Bearing lockbolt. Tighten the lockbolt to 24 ft. lbs. (32 Nm).
   • New snapring to the inner spline of the left halfshaft. Coat the side gear shaft and differential case sliding surface with gear oil. Install the halfshaft to the transaxle with the snapring opening facing down. The halfshaft should click into place when installing.
   • Halfshaft to the steering knuckle, then install the locknut. Tighten the locknut to 217 ft. lbs. (294 Nm).
   • Lock cap and a new cotter pin to the halfshaft
   • Steering knuckle to the lower ball joint. Install the 2 nuts and bolt. Tighten the nuts and bolt to 94 ft. lbs. (127 Nm).
   • Stabilizer bar link to the lower control arm. Tighten the nut to 29 ft. lbs. (39 Nm).
   • Tie rod to the steering knuckle and tighten the nut to 36 ft. lbs. (49 Nm). Install a new cotter pin to the tie rod end.
   • Front fender apron seal
   • Wheel(s) and lower the vehicle. Tighten the lug nuts to 76 ft. lbs. (103 Nm).
   • Transaxle fluid
   • Negative battery cable

ES 330

1. Before servicing the vehicle, refer to the Precautions Section.

2. Remove or disconnect the following:
   • Automatic transmission fluid
   • Front wheels
   • Front axle hub nut
   • Front stabilizer link assembly
   • Speed sensors
   • Tie rod assemblies
   • Lower ball joint
   • Drive shaft from axle hub
   • Drive shaft from transaxle

➡**Be careful not to damage the drive shaft transaxle case oil seal.**

To install:

3. Install front drive shaft assembly

a. New snap ring with opening side facing down

b. Coat inboard spline with clean transmission fluid

c. Align the shaft splines and install the drive shaft with a brass hammer

➡**Make certain that the snap ring sets firmly in the transaxle**

4. Install or connect the following:
   • Lower ball joint to spindle assembly. Torque to 55 ft. lbs. (75 Nm)
   • Tie rod assembly. Torque to 36 ft. lbs. (49 Nm)
   • Speed sensors
   • Front stabilizer link assembly Torque to 55 ft. lbs. (74 Nm)
   • Front axle hub nut. Torque to 217 ft. lbs. (294 Nm)
   • Front wheels
   • Automatic transmission fluid

5. Align the front wheels

6. Check ABS speed sensor signal

IS 250 WITH AWD

1. Before servicing the vehicle, refer to the Precautions Section.

2. Remove or disconnect the following:
   • Negative battery cable
   • Front wheel(s)
   • Front speed sensor
   • Disc brake caliper assembly
   • Brake disc
   • Axle hub nut
   • Front lower ball joint assembly
   • Tie rod assembly

3. Using a rubber mallet, separate the front axle assembly from the drive shaft assembly.

4. Hold the inboard joint side of the left halfshaft so the outboard joint side does not bend too much. Tap the end of the left halfshaft with a rubber mallet to loosen it from the axle hub and remove the left halfshaft.

5. Remove the right halfshaft as follows:

a. Using water pump pliers, remove the bearing bracket hole snap ring.

b. Remove the bolt and right front halfshaft assembly from the drive shaft bracket.

To install:

6. Install left front halfshaft assembly as follows:

a. Coat inboard spline with clean gear oil.

➡**Make certain that the snap ring sets firmly in the transaxle**

b. New snap ring with opening side facing down.

c. Align the shaft splines and install the drive shaft with a brass hammer.

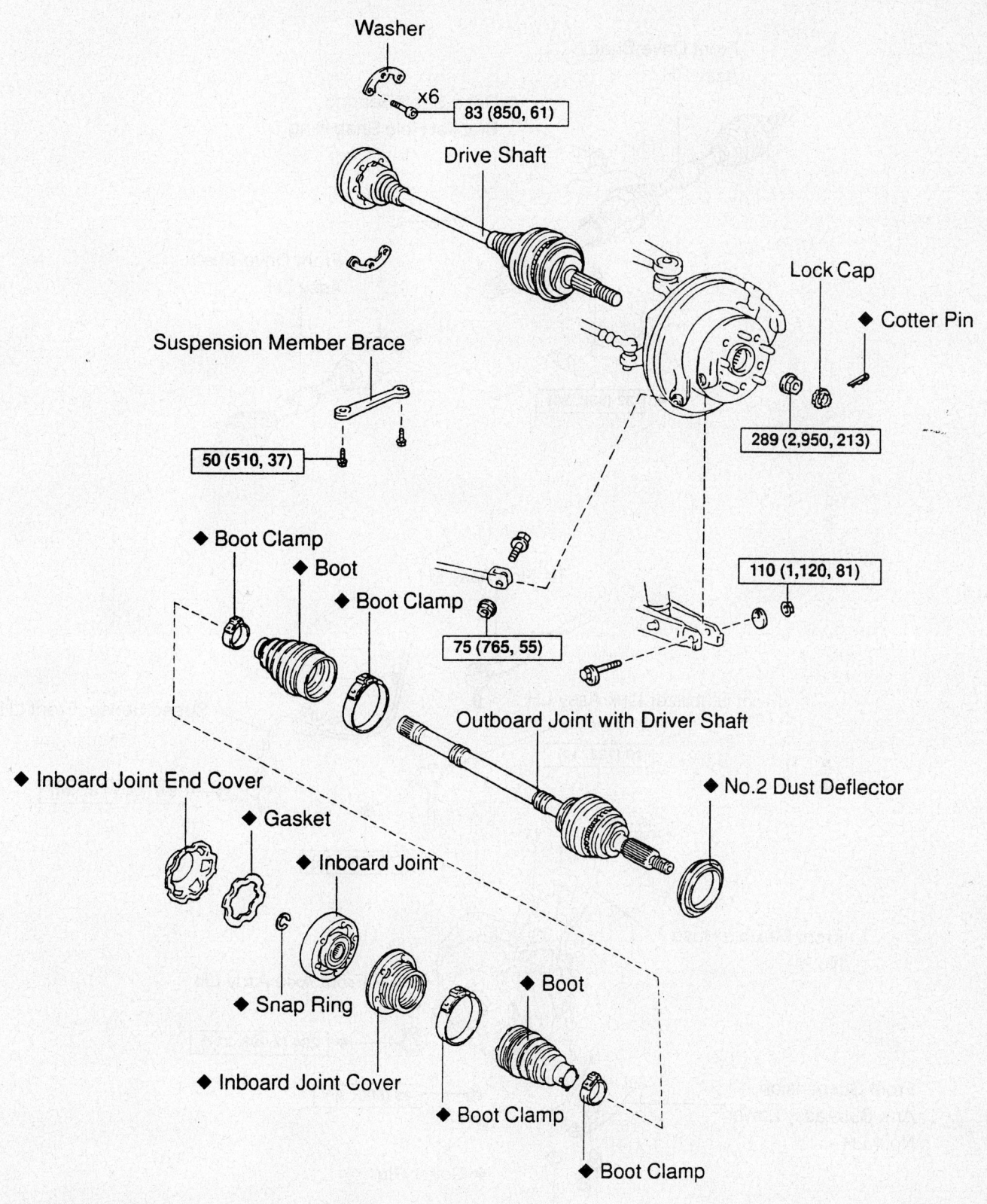

Washer

x6 | 83 (850, 61) |

Drive Shaft

Lock Cap

◆ Cotter Pin

Suspension Member Brace

| 50 (510, 37) |

| 289 (2,950, 213) |

◆ Boot Clamp

◆ Boot

◆ Boot Clamp

| 75 (765, 55) |

| 110 (1,120, 81) |

Outboard Joint with Driver Shaft

◆ Inboard Joint End Cover

◆ Gasket

◆ No.2 Dust Deflector

◆ Inboard Joint

◆ Snap Ring

◆ Boot

◆ Inboard Joint Cover

◆ Boot Clamp

◆ Boot Clamp

| N·m (kgf·cm, ft·lbf) | : Specified torque
◆ Non-reusable part

Exploded view of the rear halfshaft and related components—except ES 300

9301LG12

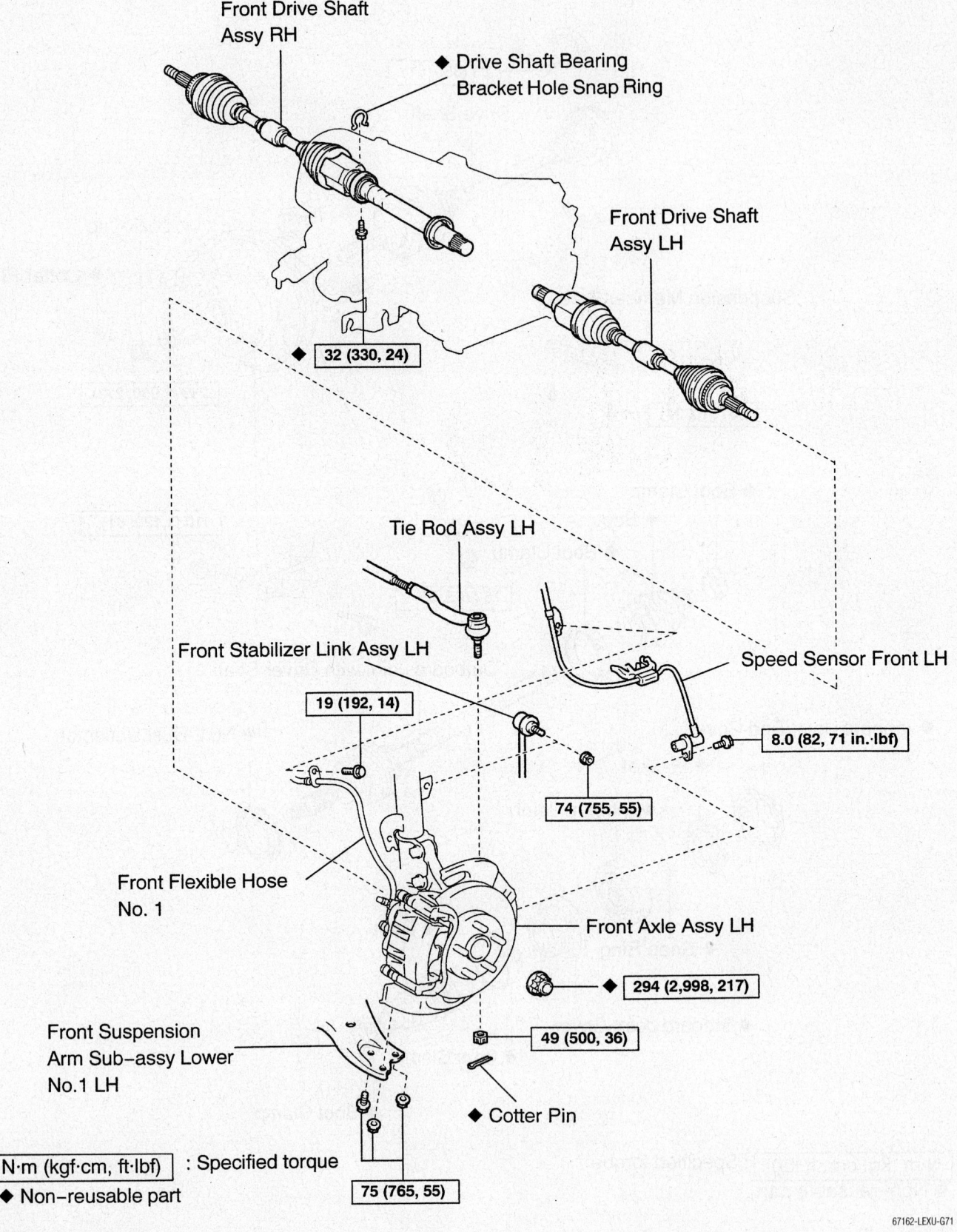

Front Drive Shaft
Assy RH

◆ Drive Shaft Bearing
Bracket Hole Snap Ring

Front Drive Shaft
Assy LH

◆ 32 (330, 24)

Tie Rod Assy LH

Front Stabilizer Link Assy LH

Speed Sensor Front LH

19 (192, 14)

8.0 (82, 71 in.·lbf)

74 (755, 55)

Front Flexible Hose
No. 1

Front Axle Assy LH

◆ 294 (2,998, 217)

49 (500, 36)

Front Suspension
Arm Sub–assy Lower
No.1 LH

◆ Cotter Pin

N·m (kgf·cm, ft·lbf) : Specified torque

◆ Non–reusable part

75 (765, 55)

67162-LEXU-G71

Exploded view drive shaft assembly—ES 330

7. Install right front halfshaft assembly as follows:

a. Coat inboard spline with clean gear oil.

b. Install the halfshaft assembly.

c. Using water pump pliers, install a new bearing bracket hole snap ring.

d. Install a new bolt and torque it to 24 ft. lbs. (32 Nm).

8. Install the front halfshaft assembly to the front axle assembly.

- Tie rod assembly. Torque to 48 ft. lbs. (65 Nm) and install a new cotter pin.
- Front lower ball joint assembly. Torque to 89 ft. lbs. (120 Nm).
- Brake disc
- Disc brake caliper assembly
- New axle hub nut. Torque the nut to 217 ft. lbs. (294 Nm). Stake the nut using a hammer and chisel.
- Front speed sensor. Torque the mounting bolt to 75 inch lbs. (8.5 Nm). Torque the wiring bracket bolt to 10 ft. lbs. (14 Nm).
- Front wheel(s)
- Negative battery cable

9. Inspect and adjust front wheel alignment.

10. Check for ABS speed sensor signal.

2006 GS 300 WITH AWD

1. Before servicing the vehicle, refer to the Precautions Section.

2. Remove or disconnect the following:
- Negative battery cable
- Front wheel(s)
- Front speed sensor
- Axle hub nut
- Front lower ball joint assembly
- Tie rod assembly

3. Using a rubber mallet, separate the front axle assembly from the drive shaft assembly.

4. Hold the inboard joint side of the left halfshaft so the outboard joint side does not bend too much. Tap the end of the left halfshaft with a rubber mallet to loosen it from the axle hub and remove the left halfshaft.

5. Remove the right halfshaft as follows:

a. Using water pump pliers, remove the bearing bracket hole snap ring.

b. Remove the bolt and right front halfshaft assembly from the drive shaft bearing bracket.

To install:

6. Install left front halfshaft assembly as follows:

a. Coat inboard spline with clean gear oil.

➡**Make certain that the snap ring sets firmly in the transaxle**

b. New snap ring with opening side facing down.

c. Align the shaft splines and install the drive shaft with a brass hammer.

7. Install right front halfshaft assembly as follows:

a. Coat inboard spline with clean gear oil.

b. Install the halfshaft assembly.

c. Using water pump pliers, install a new bearing bracket hole snap ring.

d. Install a new bolt and torque it to 24 ft. lbs. (32 Nm).

8. Install the front halfshaft assembly to the front axle assembly.

- Tie rod assembly. Torque to 48 ft. lbs. (65 Nm) and install a new cotter pin.
- Front lower ball joint assembly. Torque to 89 ft. lbs. (120 Nm).
- New axle hub nut. Torque the nut to 217 ft. lbs. (294 Nm). Stake the nut using a hammer and chisel.
- Front speed sensor. Torque the mounting bolt to 75 inch lbs. (8.5 Nm). Torque the wiring bracket bolt to 53 inch lbs. (6 Nm).
- Front wheel(s)
- Negative battery cable

9. Inspect and adjust front wheel alignment.

10. Check for ABS speed sensor signal.

Rear

IS 250 & IS 350

1. Before servicing the vehicle, refer to the Precautions Section.

2. Remove or disconnect the following:
- Negative battery cable
- Rear tire and wheel assembly
- Stabilizer link
- No.2 differential support protector
- No. 3 parking brake cable assembly
- Axle shaft nut
- Rear speed sensor
- Disc brake caliper assembly
- Brake disc
- No. 2 rear upper control arm assembly
- No. 1 rear upper control arm assembly
- No. 1 and 2 rear suspension arm assembly

3. Hold the inboard joint side of the halfshaft so the outboard joint side does not bend too much. Tap the end of the halfshaft with a rubber mallet to loosen it from the axle hub and remove the halfshaft.

To install:

4. Install rear halfshaft assembly as follows:

a. Coat inboard spline with clean gear oil.

➡**Make certain that the snap ring sets firmly in the transmission**

b. New snap ring with opening side facing down.

c. Align the shaft splines and install the halfshaft with a brass hammer.

d. Install the rear halfshaft assembly to the rear axle carrier.

5. Install or connect the following:
- No. 2 rear upper control arm assembly. Do not fully tighten.
- No. 1 rear upper control arm assembly. Do not fully tighten.
- No. 1 and 2 rear suspension arm assembly. Do not fully tighten.
- Brake disc
- Disc brake caliper assembly
- Rear speed sensor
- Axle shaft nut. Torque the nut to 214 ft. lbs. (290 Nm). Stake the nut using a hammer and chisel.
- No. 3 parking brake cable assembly. Torque the 2 bolts to 14 ft. lbs. (19 Nm).
- Stabilizer link assembly. Torque to 20 ft. lbs. (27 Nm).

6. Stabilize the rear suspension.

7. Fully tighten the No. 1 rear upper control arm assembly. Torque to 119 ft. lbs. (161 Nm).

8. Fully tighten the No. 1 rear suspension arm assembly. Torque to 70 ft. lbs. (95 Nm).

9. Fully tighten the No. 2 rear suspension arm assembly. Torque to 119 ft. lbs. (161 Nm).

10. Install the No.2 differential support protector.

11. Install the rear tire and wheel assembly.

12. Connect the negative battery cable.

13. Inspect and adjust rear wheel alignment.

14. Check for ABS speed sensor signal.

2006 GS 300 & GS 430

1. Before servicing the vehicle, refer to the Precautions Section.

2. Remove or disconnect the following:
- Negative battery cable
- Rear tire and wheel assembly
- Load sensing valve sensor bracket
- No.2 differential support protector
- Rear suspension member brace
- No. 3 parking brake cable assembly
- Axle shaft nut

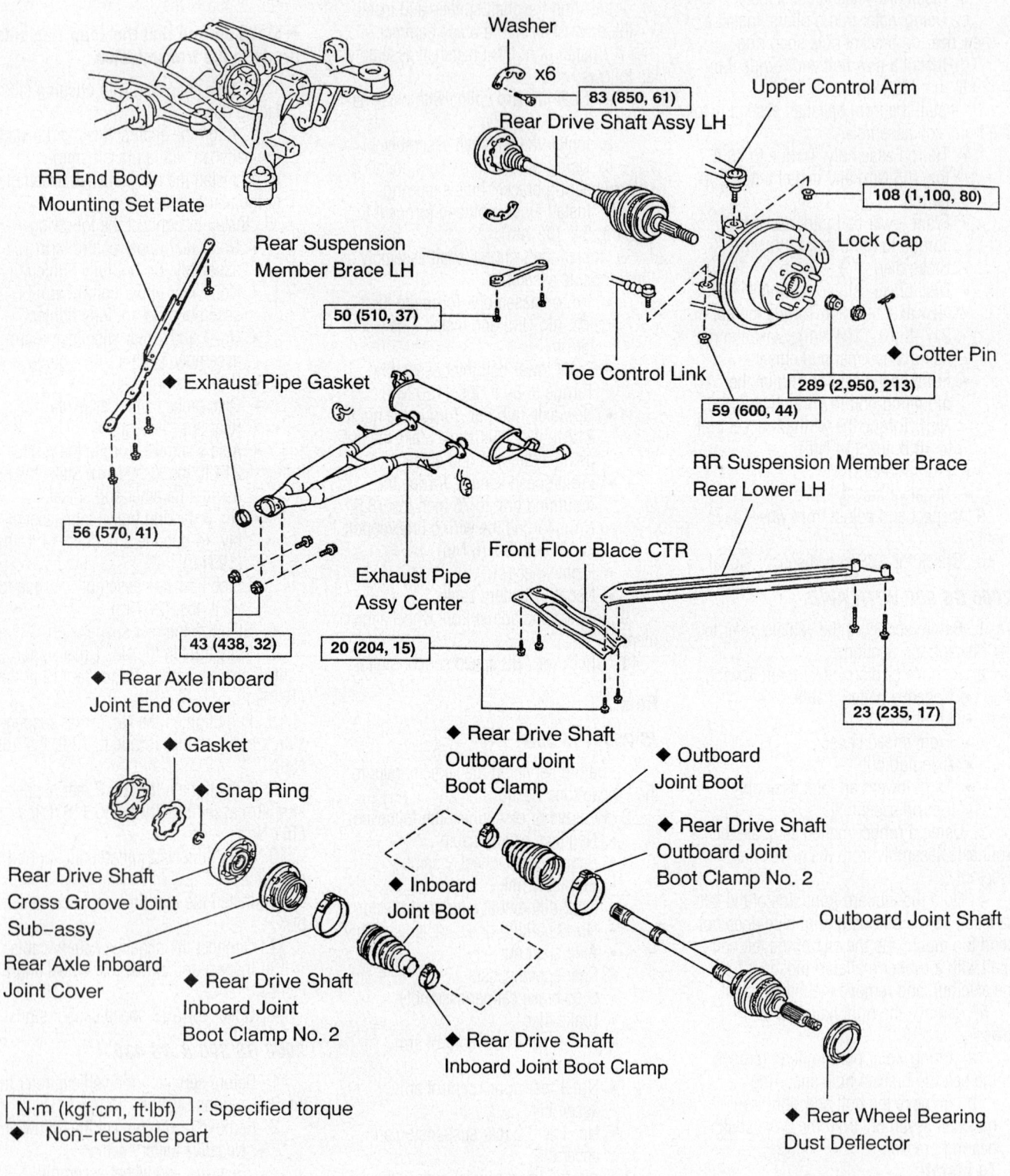

Washer
x6
83 (850, 61)
Rear Drive Shaft Assy LH

Upper Control Arm

108 (1,100, 80)

RR End Body
Mounting Set Plate

Lock Cap

Rear Suspension
Member Brace LH

Cotter Pin

50 (510, 37)

289 (2,950, 213)

Toe Control Link

◆ Exhaust Pipe Gasket

59 (600, 44)

RR Suspension Member Brace
Rear Lower LH

56 (570, 41)

Exhaust Pipe
Assy Center

Front Floor Blace CTR

43 (438, 32)

20 (204, 15)

23 (235, 17)

◆ Rear Axle Inboard
Joint End Cover

◆ Rear Drive Shaft
Outboard Joint
Boot Clamp

◆ Outboard
Joint Boot

◆ Gasket

◆ Snap Ring

◆ Rear Drive Shaft
Outboard Joint
Boot Clamp No. 2

Rear Drive Shaft
Cross Groove Joint
Sub-assy

◆ Inboard
Joint Boot

Outboard Joint Shaft

Rear Axle Inboard
Joint Cover

◆ Rear Drive Shaft
Inboard Joint
Boot Clamp No. 2

◆ Rear Drive Shaft
Inboard Joint Boot Clamp

N·m (kgf·cm, ft·lbf) : Specified torque
◆ Non-reusable part

◆ Rear Wheel Bearing
Dust Deflector

67162-LEXU-G70

Exploded view rear drive shaft assembly—SC 430

- Rear speed sensor
- Disc brake caliper assembly
- Brake disc
- No. 2 rear upper control arm assembly
- No. 1 rear upper control arm assembly
- No. 1 and 2 rear suspension arm assembly

3. Hold the inboard joint side of the halfshaft so the outboard joint side does not bend too much. Tap the end of the halfshaft with a rubber mallet to loosen it from the axle hub and remove the halfshaft.

To install:

4. Install rear halfshaft assembly as follows:

a. Coat inboard spline with clean gear oil.

➡**Make certain that the snap ring sets firmly in the transmission**

b. New snap ring with opening side facing down.

c. Align the shaft splines and install the halfshaft with a brass hammer.

d. Install the rear halfshaft assembly to the rear axle carrier.

5. Install or connect the following:

- No. 2 rear upper control arm assembly. Do not fully tighten.
- No. 1 rear upper control arm assembly. Do not fully tighten.
- No. 1 and 2 rear suspension arm assemblies. Do not fully tighten.
- Load sensing valve sensor bracket. Torque to 20 ft. lbs. (27 Nm).
- Brake disc
- Disc brake caliper assembly
- Rear speed sensor
- Axle shaft nut. Torque the nut to 214 ft. lbs. (290 Nm). Stake the nut using a hammer and chisel.
- No. 3 parking brake cable assembly. Torque the 2 bolts to 14 ft. lbs. (19 Nm).

6. Stabilize the rear suspension.

7. Fully tighten the No. 1 rear upper control arm assembly. Torque to 119 ft. lbs. (161 Nm).

8. Fully tighten the No. 1 rear suspension arm assembly. Torque to 70 ft. lbs. (95 Nm).

9. Fully tighten the No. 2 rear suspension arm assembly. Torque to 119 ft. lbs. (161 Nm).

10. Install the rear suspension member brace. Torque the 2 mounting bolts to 37 ft. lbs. (50 Nm).

11. Install the No.2 differential support protector.

12. Install the rear tire and wheel assembly.

13. Connect the negative battery cable.

14. Inspect and adjust rear wheel alignment.

15. Check for ABS speed sensor signal.

SC 430

1. Before servicing the vehicle, refer to the Precautions Section.

2. Remove or disconnect the following:

- Negative battery cable
- Rear tire and wheel assembly
- Cotter pin, locknut cap, and locknut
- Front center floor brace
- 2 exhaust pipe support brackets, if necessary
- Rear lower suspension member brace
- End body mounting set plate
- Rear suspension member brace
- Rear wheel speed sensor
- Toe control link assembly
- Upper control arm assembly
- Rear drive shaft assembly

To install:

3. Insert the outboard joint side of the halfshaft through the axle hub. Align the matchmarks on the side gear shaft and the halfshaft.

4. Coat the threads with clean oil and install the hex bolts.

5. Install the remaining components reversing the removal procedure.

6. Perform speed sensor signal test.

CV-Joints

OVERHAUL

IS 250 and IS 350

1. Before servicing the vehicle, refer to the Precautions Section.

2. Remove or disconnect the following:

- Inboard joint boot clamps using a screwdriver
- Inboard joint boot

3. Remove the inboard joint assembly as follows:

a. Put matchmarks on the inboard joint assembly and the outboard joint shaft. Do NOT use a punch to make the marks.

b. Remove the inboard joint assembly from the outboard joint shaft.

c. Put matchmarks on the outboard joint shaft and the tripod joint. Do NOT use a punch to make the marks.

d. Remove the snapring using a snapring expander.

e. Using a brass bar and a hammer, remove the tripod joint from the outboard joint shaft. Do not tap the roller.

- Outboard joint boot clamps using pliers
- Outboard joint boot
- Left drive shaft hole snapring, using a screwdriver
- Dust cover from the left drive shaft, using a press
- Right drive shaft bearing snapring, using a snapring expander
- Right drive shaft bearing, using the press
- No. 1 wheel bearing dust deflector, using a hammer and suitable chisel

To install:

4. Install or connect the following:

- No. 1 wheel bearing dust deflector, using a press
- New right drive shaft bearing, using the press and steel plate
- New right drive shaft bearing snapring, using a snapring expander
- New dust cover to the left drive shaft, using a press and steel plate
- New left drive shaft hole snapring
- Outboard joint boot and new boot clamps as a temporary measure. Before installing the boot, place 2 new clamps to the small boot end and large end (wheel side) and install it to the driveshaft. Coat the outboard joint and boot with grease.

➡**Before installing the boots, wrap the spline of the shaft with vinyl tape to prevent the boots from being damaged.**

5. Using SST 09521-24010, tighten the new boot clamps.

6. Install the inboard joint assembly as follows:

a. Temporarily install a new inboard joint boot with 2 new clamps to the drive shaft.

b. Align the matchmarks and place the beveled side of the tripod joint axial spline toward the outboard joint shaft.

c. Using a brass bar and hammer, tap the tripod joint into the outboard joint shaft. Do not tap the roller.

d. Install a new snapring using a snapring expander.

e. Coat the outboard joint and boot with grease.

f. Align the matchmarks and install

the inboard joint assembly to the outboard joint shaft assembly.
- Inboard joint boot
- New inboard joint boot clamps using a screwdriver.

ES 300

1. Before servicing the vehicle, refer to the Precautions Section.

2. Once the driveshaft is removed from the vehicle, place matchmarks on the outboard and inboard joints and the shaft. Do NOT use a punch to make the marks.

3. Remove or disconnect the following:
- Boot clamps. Use a side cutter or pliers
- Outboard joint shaft expanding the snapring
- 2 boots
- Dust cover on the left-hand driveshaft, using a hammer and suitable chisel
- Dust cover from the inboard joint shaft, using a press
- Dust cover
- Snapring. Use a snapring expander.
- Bearing, using the press
- Snapring
- No. 2 dust deflector, using a hammer and suitable chisel

➡ **Be careful not to damage the Antilock Brake System (ABS) speed sensor rotor.**

To install:

4. Install or connect the following:
- No. 2 dust deflector
- New snapring to the inboard joint shaft
- New bearing
- New dust cover
- Dust cover on the left-hand driveshaft
- A new dust cover, using a press
- Outboard and inboard joint boots and new boot clamps as a temporary measure. Before installing the boot, place 3 new clamps to the small boot ends and large end (wheel side) and install it to the driveshaft.
- Inboard joint shaft to outboard joint shaft
- Using a snapring expander, put in the inboard joint shaft expanding the snapring
- Boot to outboard joint, before assembling the boot, pack the outboard joint and boot with grease in the boot kit
- Boot to inboard joint shaft. Pack the inboard joint, and boot with

grease in the boot kit, and install the boot to the inboard joint shaft.
- Boot clamps to both boots, make sure that the 2 boots are on the shaft groove. Hold the clamp near the clamp's free end over the closing hooks.

5. Secure the clamp by drawing the closing hooks together. Secure the clamp onto the boot.

GS 300, GS 430 and IS 300

1. Before servicing the vehicle, refer to the Precautions Section.

2. Remove or disconnect the following:
- End cover. Use nuts and bolts to keep the inboard joint together. Hand tighten only.
- 4 Boot clamps, using a side cutter or pliers
- Inboard joint, place matchmarks on the inboard joint and driveshaft; do not use punch marks.
- Snapring, using a snapring expander
- Using a press, the inboard joint from the driveshaft
- Inboard and outboard boot
- Inboard joint cover from the inboard joint

To install:

3. Install or connect the following:
- Inboard and outboard joint boots
- New No. 2 Dust deflector, using a press

➡ **Be careful not to damage the Anti-lock Brake System (ABS) speed sensor rotor.**

- Inboard joint
- Inner race to the cage so that the indented beveled part of the inner race is on the opposite side to the beveled top of the cage
- Outer race so that the indented side of the outer race is facing the same side as the beveled surface of the cage

4. Match the narrow projections of the inner race with the wide projections of the outer race.

5. Tilt the cage and inner race to the side and insert the balls one by one.

6. Install or connect the following:
- New boots and new boot clamps, temporarily
- 4 new boot clamps to the boots
- 2 boots to the driveshaft
- Inboard joint cover and apply RTV to the inboard joint cover

7. Remove grease from the surface of the inboard joint facing the cover.

8. Align the bolt holes of the cover with those of the inboard joint, then insert the hexagon bolts.

9. Use a plastic hammer to tap the rim of the inboard joint cover into place

10. To install the inboard joint, align the matchmarks placed before removal.

11. Using a brass bar and hammer, tap the inboard joint onto the driveshaft.

12. Install or connect the following:
- New snapring
- Boots to joints, pack with the proper grease. 3.5–3.7 oz. (100–105g)
- New boot clamps to both boots
- 6 hexagon bolts and washers from the end cover side, install the 6 nuts to the boot side.

LS 430 and SC 430

1. Before servicing the vehicle, refer to the Precautions Section.

2. Using a suitable prytool, remove the end cover.

3. Use nuts and bolts to keep the inboard joint together. Hand tighten only.

4. Remove or disconnect the following:
- 4 Boot clamps, using a side cutter or pliers
- Inboard joint, place matchmarks on the inboard joint and driveshaft; do not use punch marks.
- Snapring, using a snapring expander
- Using a press, the inboard joint from the driveshaft
- Inboard joint cover from the inboard joint
- Inboard and outboard boot
- No. 2 dust deflector, using a suitable chisel and hammer
- New No. 3 dust deflector, using a press

To install:

5. If the joint has come apart, reassemble it in the following order.
   a. Align the matchmarks placed before removal
   b. Inner race to the cage so that the indented beveled part of the inner race is on the opposite side to the beveled top of the cage.
   c. Outer race so that the indented side of the outer race is facing the same side as the beveled surface of the cage.

6. Match the narrow projections of the inner race with the wide projections of the outer race.

7. Tilt the cage and inner race to the side and insert the balls one by one.

8. Install or connect the following:

- New boots and new boot clamps, temporarily
- 4 new boot clamps to the boots
- 2 boots to the driveshaft
- Inboard joint cover and apply Formed In Place Gasket (FIPG) to the inboard joint cover. Avoid applying an excessive amount to the surface

9. Remove grease from the surface of the inboard joint facing the cover

10. Align the bolt holes of the cover with those of the inboard joint, then insert the hexagon bolts.

11. Use a plastic hammer to tap the rim of the inboard joint cover into place

12. To install the inboard joint, align the matchmarks placed before removal.

13. Using a brass bar and hammer, tap the inboard joint onto the driveshaft.

14. Install or connect the following:

- New snapring
- Boots to joints, pack with 3.5–3.7 oz. (100–105g) of grease
- A new gasket, with the side with adhesive on it facing toward the outer race side of the inboard joint
- 6 hexagon bolts an washer from the end cover side
- 6 nuts to the boot side

15. Check that the claw of the end cover touches the inboard joint

ES 330

1. Remove drive shaft
2. Remove inboard drive axle clamps and slide boot off the inboard joint.
   - a. Mark the joint and housing for proper alignment during installation.
   - b. Remove the snap clip with snap ring expander
   - c. Using and brass bar and hammer remove the tripod joint from the shaft
   - d. Remove the boot from the shaft
3. Remove outboard drive axle clamps and slide boot off the inboard joint.
   - a. Mark the joint and housing for proper alignment during installation.
   - b. Remove the snap clip with snap ring expander
   - c. Using and brass bar and hammer remove the tripod joint from the shaft
   - d. Remove the boot from the shaft

➡ **Clean all the old grease from the joints.**

To install:

4. Install new outboard joint
   - a. Slip new boot with clamps over the shaft

b. Pack the boot with grease
c. Install the new joint over the shaft
d. Install new snap rings on the shaft

➡ **Do NOT reuse the old snap rings**

e. Install joint housing over the joint making certain to align the marks
f. Slide the boot over the joint and position on joint housing
g. Crimp the clamps using special tool

5. Install new inboard joint
   a. Slip new boot with clamps over the shaft
   b. Pack the boot with grease
   c. Install the new joint over the shaft
   d. Install new snap rings on the shaft

➡ **Do NOT reuse the old snap rings**

e. Install joint housing over the joint making certain to align the marks
f. Slide the boot over the joint and position on joint housing
g. Crimp the clamps using special tool

6. Install new snap ring on inboard shat before installing in transaxle
7. Install drive shaft assembly

Axle Shaft, Bearing and Seal

REMOVAL & INSTALLATION

IS 300, 2002–05 GS 300 and 2002–05 GS 430

1. Drain the gear oil.
2. Remove or disconnect the following:
   - Rear driveshaft
   - Side gear shaft
   - Snapring from the side gear, using a suitable tool
   - Side gear shaft oil seal

To install:
   - Side gear shaft oil seal
   - New oil seal

3. Check installation of side gear shaft. There should be 0.08–0.12 inch (2–3mm) of play in the axial direction. Check that the side gear shaft will not come out by pulling on it.

LS 430

1. Remove or disconnect the following:
   - Gear oil
   - Rear driveshaft
   - Side gear shaft
   - Snapring from the side gear
   - Side gear shaft oil seal

To install:
2. Install or connect the following:
   - Side gear shaft oil seal

- New oil seal
- Multi-Purpose (MP) grease to the oil seal lip
- Side gear shaft
- Snapring from the side gear, using a suitable tool
- Side gear shaft oil seal
- Gear oil

SC 430

1. Remove or disconnect the following:
   - Rear wheel
   - Differential oil
   - Rear suspension member brace
   - Rear suspension arm
   - Speed sensor
   - Matchmark and remove drive shaft
   - Snap ring
   - Differential side gear shaft
   - Side gear seal using proper puller

To install:
2. Install or connect the following:
   - New snap ring on shaft
   - New seal using proper seal driver
   - Rear differential side gear shaft

➡ **Take care not to damage the seal**

- Rear drive shaft assembly making certain to align the matchmarks
- Speed sensor
- Temporarily tighten rear suspension arm No. 2

➡ **Do NOT torque nut**

- Temporarily tighten rear suspension arm No. 1

➡ **Do NOT torque nut**

- Rear wheels
- Differential oil. Synthetic gear oil GL-5 75W-90 or equivalent

➡ **Stabilize the rear suspension**

- Cross member brace

3. Fully tighten suspension arm No. 2. Torque: 81 ft. lbs. (110 Nm)
4. Fully tighten suspension arm No. 1. Torque: 55 ft. lbs (75 Nm)
5. Inspect and adjust rear wheel alignment
6. Speed sensor check.

ES 330

1. Remove or disconnect the following:
   - Front wheels
   - Engine under cover
   - Transaxle fluid
   - Drive shaft assembly
   - Transaxle housing oil seal using proper puller
   - Differential side bearing retainer oil seal using proper puller

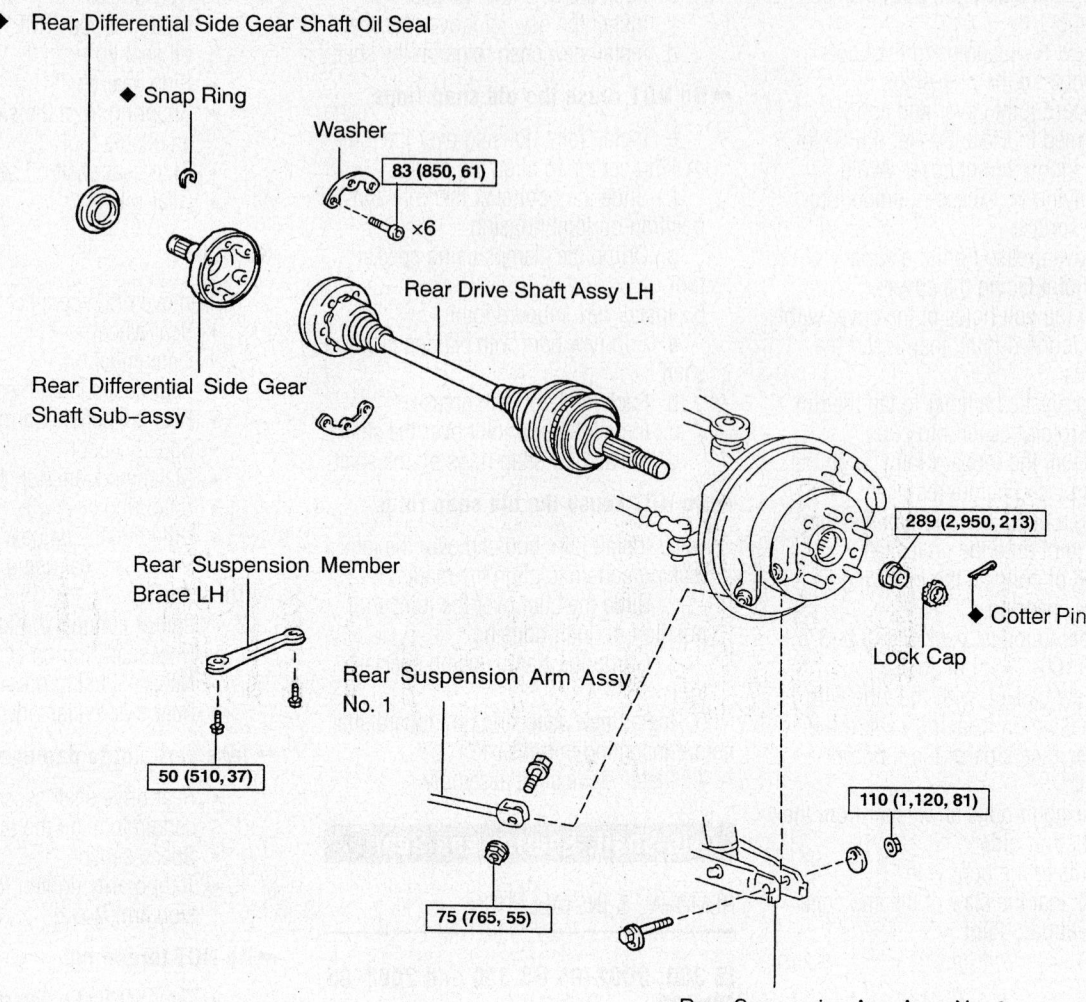

◆ Rear Differential Side Gear Shaft Oil Seal

◆ Snap Ring

Washer

83 (850, 61)

×6

Rear Drive Shaft Assy LH

Rear Differential Side Gear Shaft Sub-assy

289 (2,950, 213)

◆ Cotter Pin

Lock Cap

Rear Suspension Member Brace LH

Rear Suspension Arm Assy No. 1

50 (510, 37)

110 (1,120, 81)

75 (765, 55)

Rear Suspension Arm Assy No. 2

67162-LEXU-G72

N·m (kgf·cm, ft·lbf) : Specified torque

◆ Non-reusable part

Expanded view, side gear shaft oil seal—SC 430

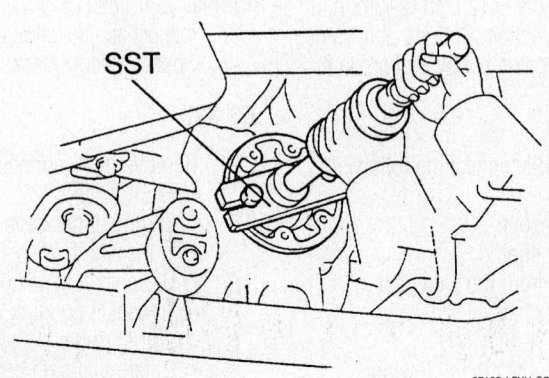

SST

67162-LEXU-G73

Removing differential side shaft—SC 430

To install:
2. Install or connect the following:
- Differential side bearing oil seal using proper driver

➡**Coat seal lip with MP grease**

- Transaxle housing oil seal using proper driver

➡**Coat seal lip with MP grease**

- Install drive shaft assembly
- Front wheels
- Automatic Transaxle fluid
3. Check ABS speed sensor signal

Pinion Seal

REMOVAL & INSTALLATION

IS 250 and IS 350

1. Drain the gear oil.
2. Remove the propeller shaft.
3. Remove the companion flange as follows:

   a. Using a hammer and SST 09930-00010, release the staked part of the drive pinion nut.

   b. Using SST 09330-00021 and 09950-30012, hold the flange and remove the drive pinion nut.

   c. Using SST 09950-30012, remove the companion flange.
4. Remove or disconnect:
   • Differential dust deflector (only if damaged) using a press
   • Carrier oil seal
5. Check oil slinger.

To install:
6. Install or connect:
   • Oil slinger
   • New oil seal

➡**Front carrier oil seal drive-in depth: 0.028–0.051 inches (0.7–1.3mm). Rear carrier oil seal drive-in depth: -0.020–0.020 inches (-0.5–0.5mm).**

   • Multi-Purpose (MP) grease to the oil seal lip
   • Companion flange on the drive pinion
   • Gear oil on the threads of a new nut. Torque to 80–173 ft. lbs. (108–235 Nm) for the front differential. Torque to 361 ft. lbs. (490 Nm) for the rear differential.
7. Adjust the drive pinion preload as necessary.
8. The pinion bearing preload specifications for the front differential is as follows:

   a. Used bearings: 4.3–6.9 inch lbs. (0.49–0.78 Nm).

   b. New bearings: 8.7–13.9 inch lbs. (0.98–1.57 Nm).
9. The pinion bearing preload specifications for the rear differential is as follows:

   a. Used bearings: 4.4–7.1 inch lbs. (0.50–0.80 Nm).

   b. New bearings: 10.7–16.0 inch lbs. (1.22–1.80 Nm).
10. Stake the drive pinion nut using a chisel and hammer, and install the propeller shaft.
11. Fill the differential with hypoid gear oil

GS 300 and GS 430

1. Drain the gear oil.
2. Remove the driveshaft and the companion flange
3. Remove the oil seal
4. Check oil slinger
To install:
5. Installation is the reversal of the removal procedure.

LS 430, and SC 430

1. Remove or disconnect:
   • Gear oil
   • Driveshaft
   • Companion flange
   • Oil seal and slinger
To install:
2. Install or connect:
   • Oil slinger
   • New oil seal

➡**Oil seal drive-in depth: 0.079 inch (2.0mm).**

   • Multi-Purpose (MP) grease to the oil seal lip
   • Companion flange on the shaft
   • Gear oil on the threads of a new nut. Torque to 80 ft. lbs. (108 Nm).
3. Adjust the drive pinion preload as necessary, stake drive the pinion nut, and install the driveshaft.

4. Fill the differential with hypoid gear oil

IS 300

1. Before servicing the vehicle, refer to the Precautions Section.
2. Remove or disconnect the following:
   • Driveshaft
   • Rear wheels
   • Rear brake calipers
   • Pinion flange
   • Pinion seal

➡**The rear brake calipers must be removed so that there is no additional drag when measuring pinion bearing preload.**

To install:
3. Install or connect the following:
   • Pinion seal and flange
   • New pinion flange nut
4. Rotate the pinion flange occasionally while tightening the flange nut to make sure the pinion bearings seat correctly. Do NOT exceed 249 ft. lbs. (338 Nm).
5. Take frequent bearing preload torque readings.
6. The pinion bearing preload specifications are as follows:

   a. Used bearings: 4.3–6.9 inch lbs. (0.49–0.78 Nm).

   b. New bearings: 8.7–13.9 inch lbs. (0.98–1.57 Nm).

✳✳ CAUTION

Never loosen the pinion nut to reduce bearing preload. If it is necessary to reduce bearing preload, install a new collapsible spacer and pinion nut.

7. Install or connect the following:
   • Driveshaft
   • Brake calipers
   • Rear wheels
8. Fill the differential with gear lubricant and check for leaks.

STEERING

Air Bag

✳✳ CAUTION

These vehicles are equipped with an air bag system. The system must be disabled before performing service on or around system components, steering column, instrument panel components, wiring and sensors. Failure to follow safety and disabling procedures could result in accidental air bag deployment, possible personal injury and unnecessary system repairs.

PRECAUTIONS

Several precautions must be observed when handling the inflator module to avoid accidental deployment and possible personal injury.

• Never carry the inflator module by the wires or connector on the underside of the module.

• When carrying a live inflator module, hold securely with both hands and ensure that the bag and trim cover are pointed away.

• Place the inflator module on a bench or other surface with the bag and trim cover facing up.

• With the inflator module on the bench, never place anything on or close to the module which may be thrown in the event of an accidental deployment.

DISARMING

To avoid personal injury when working on vehicles equipped with an air bag, the negative battery cable must be disconnected and at least 90 seconds must elapse before working on the system. Failure to do so may result in deployment of the air bag.

ARMING

To rearm the air bag system, simply reconnect the battery cable(s).

Power Rack and Pinion Steering Gear

REMOVAL & INSTALLATION

IS 250 and IS 350

1. Before servicing the vehicle, refer to the Precautions Section.

2. Remove or disconnect the following:
• Negative battery cable and wait at least 90 seconds before working on the vehicle to disarm the air bag.
• Front wheels
• Engine under cover
• No. 2 engine under cover
• Front lower suspension member protector

3. Separate the steering sliding yoke sub-assembly as follows:

a. To prevent steering wheel rotation and possible damage to the spiral cable, fix the steering wheel with the seat belt.

b. Loosen the bolts and slide, but do not separate the steering sliding yoke sub-assembly.

c. Matchmark the power steering gear assembly and the steering sliding yoke sub-assembly.

d. Separate the steering sliding yoke sub-assembly from the power steering gear assembly.
• Left and right tie rods from the steering knuckles

4. Remove the bolts and remove the power steering gear assembly.

To install:

5. Install the steering gear assembly to its mounting position. Install the 2 bolts, washers and 2 nuts and torque to 87 ft. lbs. (118 Nm).

6. Reconnect the ground wire and other wiring connectors.

7. Install or connect the following:
• Tie rods to the steering knuckles and the nuts. Tighten the nut to 50 ft. lbs. (65 Nm) and install a new cotter pin. The prongs of the cotter pin should be firmly wrapped around the flats of the nut.
• Steering sliding yoke sub-assembly to the power steering gear assembly by aligning the matchmarks. Torque the 2 bolts to 26 ft. lbs. (35 Nm).
• Front lower suspension member protector
• No. 2 engine under cover
• Engine under cover
• Front wheels
• Negative battery cable

8. Inspect and adjust front wheel alignment.

ES 300

1. Before servicing the vehicle, refer to the Precautions Section.

2. Remove or disconnect the following:

• Negative battery cable and wait at least 90 seconds before working on the vehicle to disarm the air bag.
• Front wheels
• Left and right front fender apron seals
• Cotter pin and nut holding the steering knuckle to the tie rod end. Using a tie rod puller, disconnect the tie rod end from the steering knuckle.

3. Place matchmarks on the intermediate shaft and the control valve shaft.

4. Loosen the upper bolt and remove the lower bolt holding the control valve shaft to the intermediate shaft. Disconnect the intermediate shaft from steering rack housing.

5. Remove or disconnect the following:
• Tube clamp
• Return line and the pressure line from the control valve housing
• 4 stabilizer bar bolts and 2 nuts. Position the stabilizer bar out of the way. Do NOT remove the sway bar from the vehicle.
• Heated Oxygen (HO2) sensor (bank 1 sensor 1)
• 2 steering gear mounting bolts and nuts. Remove the steering gear through the left side of the vehicle.

To install:

6. Install or connect the following:
• Steering gear on the vehicle and install the 2 mounting bolts and nuts. Tighten the nuts and bolts to 134 ft. lbs. (181 Nm).
• HO2 sensor. Tighten the sensor to 33 ft. lbs. (44 Nm).
• Stabilizer bar bolts and nuts and tighten as follows: Bolts: 14 ft. lbs. (19 Nm); Nuts: 29 ft. lbs. (39 Nm).
• Pressure and return lines and tighten the connectors to 18 ft. lbs. (25 Nm)
• Tube clamp and tighten the nut to 84 inch lbs. (10 Nm)
• Intermediate shaft to the steering rack and tighten the retaining bolts to 26 ft. lbs. (35 Nm)
• Tie rods to the steering knuckles with the castellated nuts. Tighten the nut to 36 ft. lbs. and install a new cotter pin. The prongs of the cotter pin should be firmly wrapped around the flats of the nut.
• Front fender apron seals by installing the 2 bolts
• Front wheels and lower the vehicle
• Power steering fluid
• Negative battery cable

7. Release the steering wheel
8. Bleed the system
9. Check for leaks, adjust the toe-in and check the steering wheel center point

GS 300 and GS 430

1. Before servicing the vehicle, refer to the Precautions Section.

2. Set front wheels in straight-ahead position, then remove the front wheels.

3. Disconnect the negative battery cable and wait for at least 90 seconds before proceeding.

4. Remove the steering wheel pad:

   a. Remove the 2 lower steering wheel covers (3.0L) or the No. 2 cover and the pad switch modulator cover (4.3L).

   b. Loose the Torx® head bolts on either side of the steering wheel until the grooves along the bolt circumference catches on the bolt case.

   c. Pull the wheel pad out and disconnect the airbag connector. Store the airbag face up and in a safe place.

5. Remove the steering wheel:

   a. Disconnect the wiring connector.

   b. Remove the steering wheel set nut.

   c. Make matchmarks across steering wheel and main shaft assembly.

6. Disconnect the left and right tie rod ends, placing matchmarks on screw portion for proper reinstallation.

7. Disconnect the intermediate shaft assembly after making matchmarks as shown.

8. Remove the front suspension member brace.

9. Disconnect the pressure feed and return power steering pipes.

10. Disconnect the tube clamp(s).

11. Remove the bolts and remove the power steering gear assembly.

To install:

12. Center the rack and pinion gear assembly as shown.

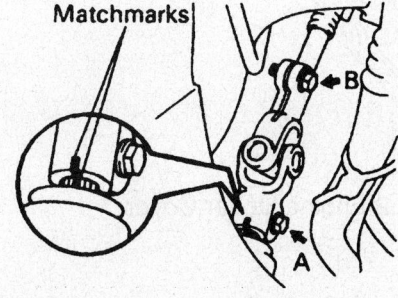

Matchmarking the intermediate shaft to the control valve shaft—GS 300 and GS 430

7923LGA3

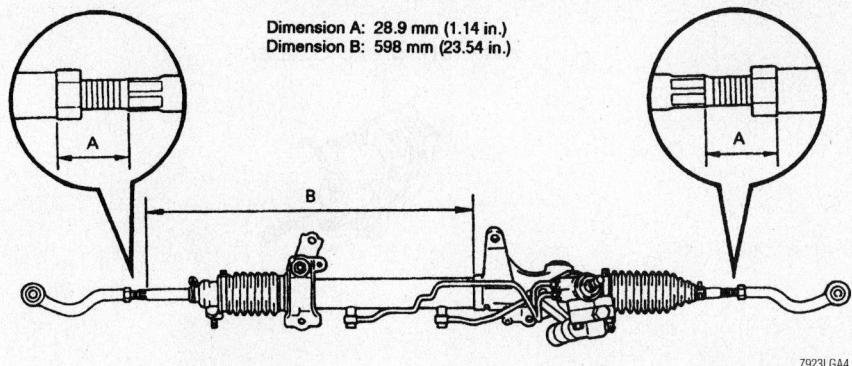

Dimension A: 28.9 mm (1.14 in.)
Dimension B: 598 mm (23.54 in.)

7923LGA4

Centering the rack and pinion—GS 300 and GS 430

13. Install the steering gear assembly to its mounting position. Install and torque the bolts as follows:
   • 2002–03: 72 ft. lbs. (98 Nm)
   • 2004–06: 48 ft. lbs. (65 Nm)

14. Reconnect the tube clamp(s); torque tube clamps and tighten the bolt to 12 ft. lbs. (17 Nm).

15. Connect the steering gear to the intermediate shaft assembly and torque the bolts above and below the U-joint to 26 ft. lbs. (35 Nm).

16. Reconnect the pressure feed tube and the return tube, with a new gasket on the union bolts. Torque the bolts to 36 ft. lbs. (49 Nm).

17. Reconnect the tie rod ends, installing them to their original positions as marked.

18. Install the front wheels and position them in a straight-ahead position.

19. Install and center the spiral cable:

   a. Turn the spiral cable counterclockwise by hand until it becomes hard to turn.

   b. Then, rotate the cable clockwise about 3 turns to align the marks at the bottom of the assembly.

20. Install the steering wheel, aligning the matchmarks made during removal. Temporarily install and tighten the wheel set nut. Connect the connector.

21. Bleed the power steering system.

22. Check the steering wheel center point; if okay, torque the wheel set nut to 37 ft. lbs. (50 Nm).

23. Install the steering wheel pad:

   a. Connect the airbag connector.

   b. Install the pad, after confirming that the circumference groove of the Torx bolts is caught on the screw case. Then, torque the Torx bolts to 78 inch lbs. (8.8 Nm).

   c. Install the lower steering wheel covers.

24. Reconnect the negative battery cable.

25. Check the front wheel alignment.

IS 300

1. Before servicing the vehicle, refer to the Precautions Section.

2. Remove or disconnect the following:
   • Negative battery cable
   • Steering wheel
   • Front wheels
   • Brake calipers
   • Outer tie rod ends
   • Engine under cover
   • Intermediate shaft
   • Front subframe brace
   • Pressure and return lines
   • Steering gear

To install:

3. Installation is the reverse of the removal procedure, while using the following torque values:
   • Steering gear mounting bracket bolts: 54 ft. lbs. (74 Nm)
   • Fluid return line: 30 ft. lbs. (40 Nm)
   • Fluid pressure line: 31 ft. lbs. (42 Nm)
   • Front subframe brace large bolts: 88 ft. lbs. (119 Nm)
   • Front subframe brace small bolts: 43 ft. lbs. (58 Nm)
   • Tie rod end nuts: 40 ft. lbs. (54 Nm)
   • Steering wheel nut: 26 ft. lbs. (35 Nm)

2002–03 LS 430

1. Before servicing the vehicle, refer to the Precautions Section.

2. Remove or disconnect the following:
   • Wheel(s)
   • Engine undercover by removing the 8 bolts and 5 screws
   • Cotter pin and nut holding each tie rod to the steering knuckle
   • Tie rod end from the steering knuckle with a tie rod end puller

3. Place matchmarks on the sliding yoke and control valve shaft.

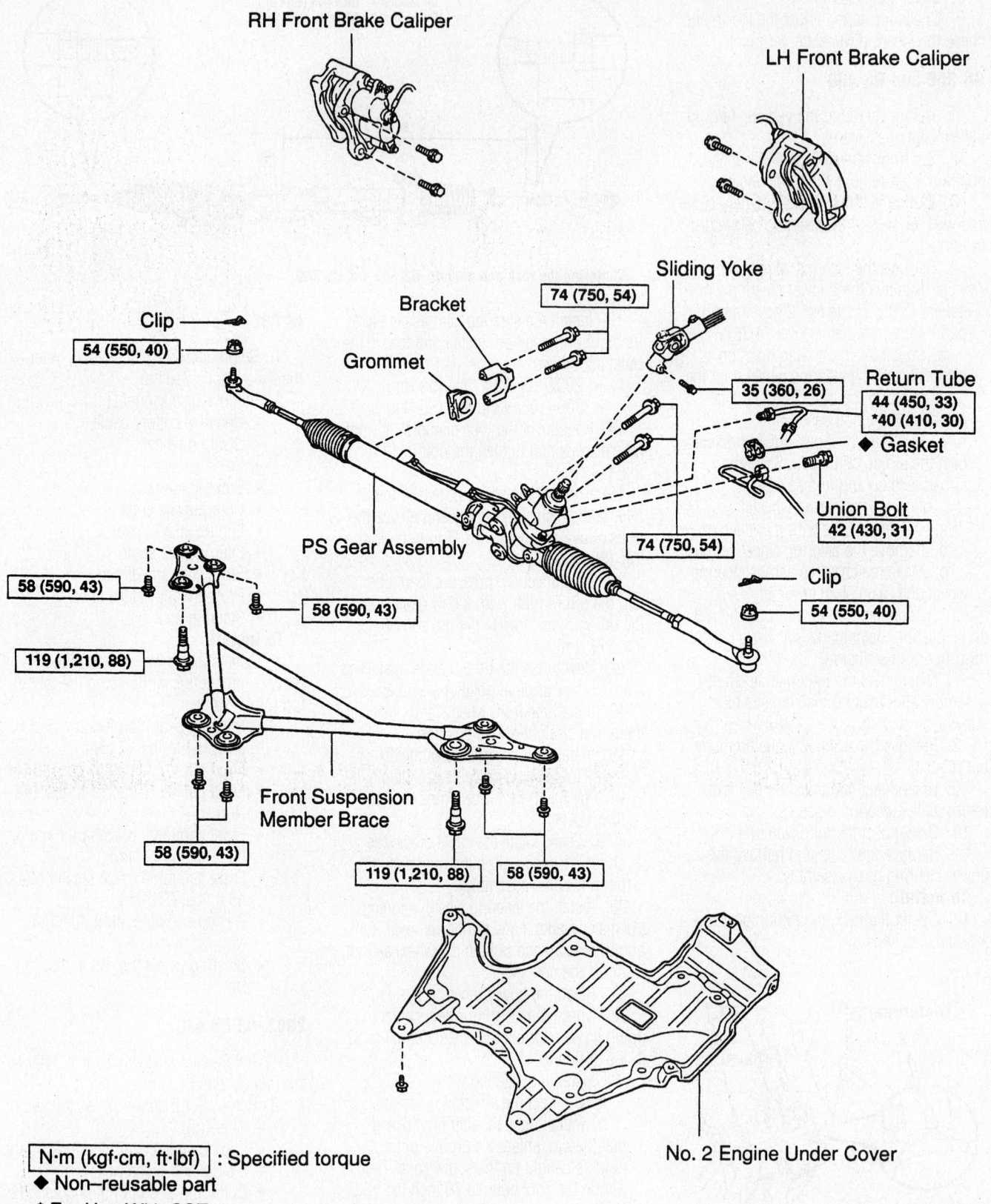

RH Front Brake Caliper

LH Front Brake Caliper

Sliding Yoke

74 (750, 54)

Clip

54 (550, 40)

Bracket

Grommet

35 (360, 26)

Return Tube
44 (450, 33)
*40 (410, 30)

◆ Gasket

Union Bolt
42 (430, 31)

74 (750, 54)

PS Gear Assembly

58 (590, 43)

58 (590, 43)

119 (1,210, 88)

Clip

54 (550, 40)

Front Suspension
Member Brace

58 (590, 43)

119 (1,210, 88)   58 (590, 43)

No. 2 Engine Under Cover

| N·m (kgf·cm, ft·lbf) | : Specified torque
◆ Non–reusable part
* For Use With SST

9347LG02

Exploded view of the steering gear mounting—IS 300

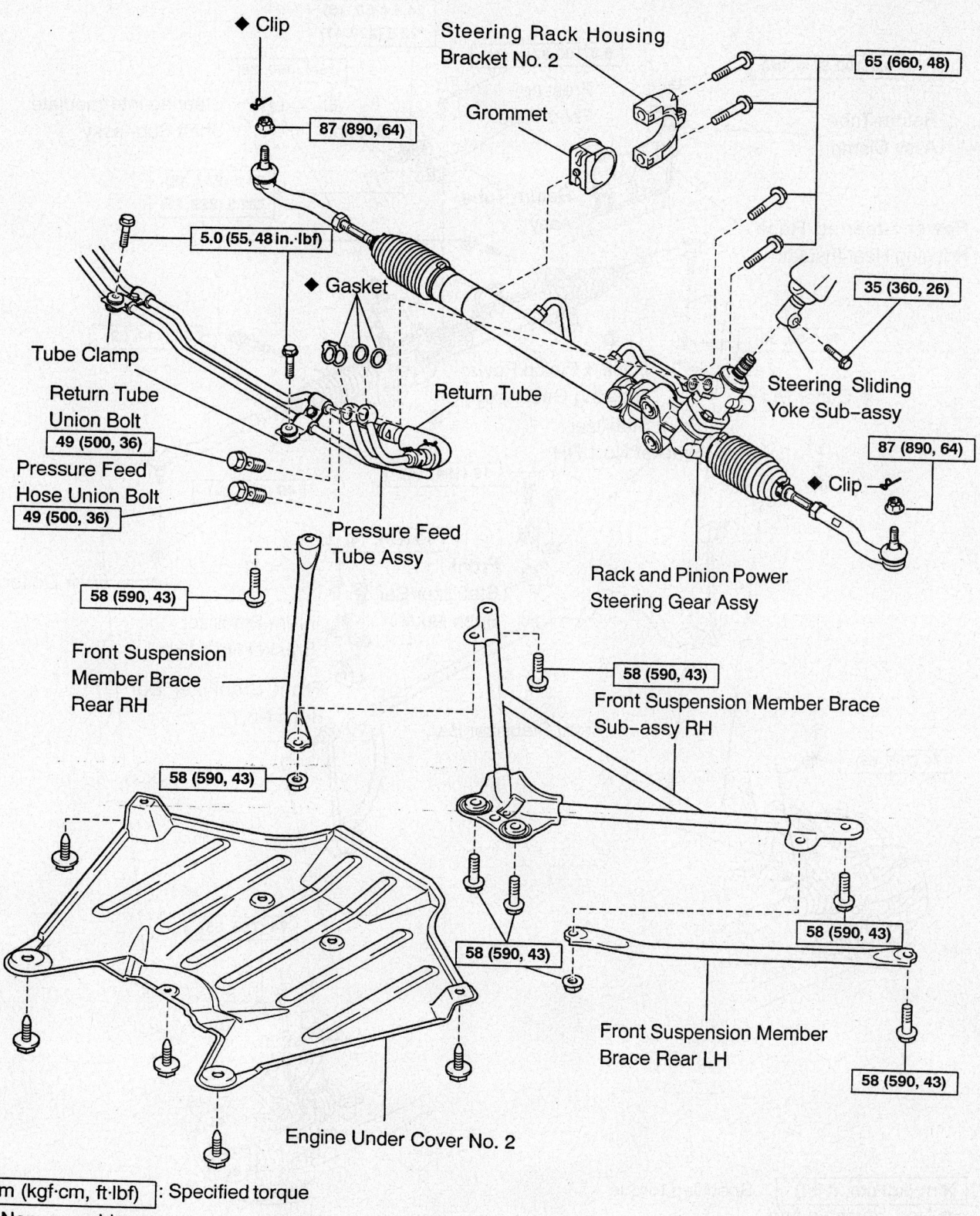

◆ Clip

Steering Rack Housing
Bracket No. 2

Grommet

65 (660, 48)

87 (890, 64)

5.0 (55, 48 in.·lbf)

◆ Gasket

Return Tube

35 (360, 26)

Tube Clamp

Return Tube
Union Bolt
49 (500, 36)

Pressure Feed
Hose Union Bolt
49 (500, 36)

Steering Sliding
Yoke Sub-assy

87 (890, 64)

◆ Clip

Pressure Feed
Tube Assy

Rack and Pinion Power
Steering Gear Assy

58 (590, 43)

Front Suspension
Member Brace
Rear RH

58 (590, 43)

Front Suspension Member Brace
Sub-assy RH

58 (590, 43)

58 (590, 43)

58 (590, 43)

58 (590, 43)

Front Suspension Member
Brace Rear LH

58 (590, 43)

Engine Under Cover No. 2

N·m (kgf·cm, ft·lbf) : Specified torque

◆ Non-reusable part

67162-LEXU-G74

Exploded view power steering gear assembly—SC 430

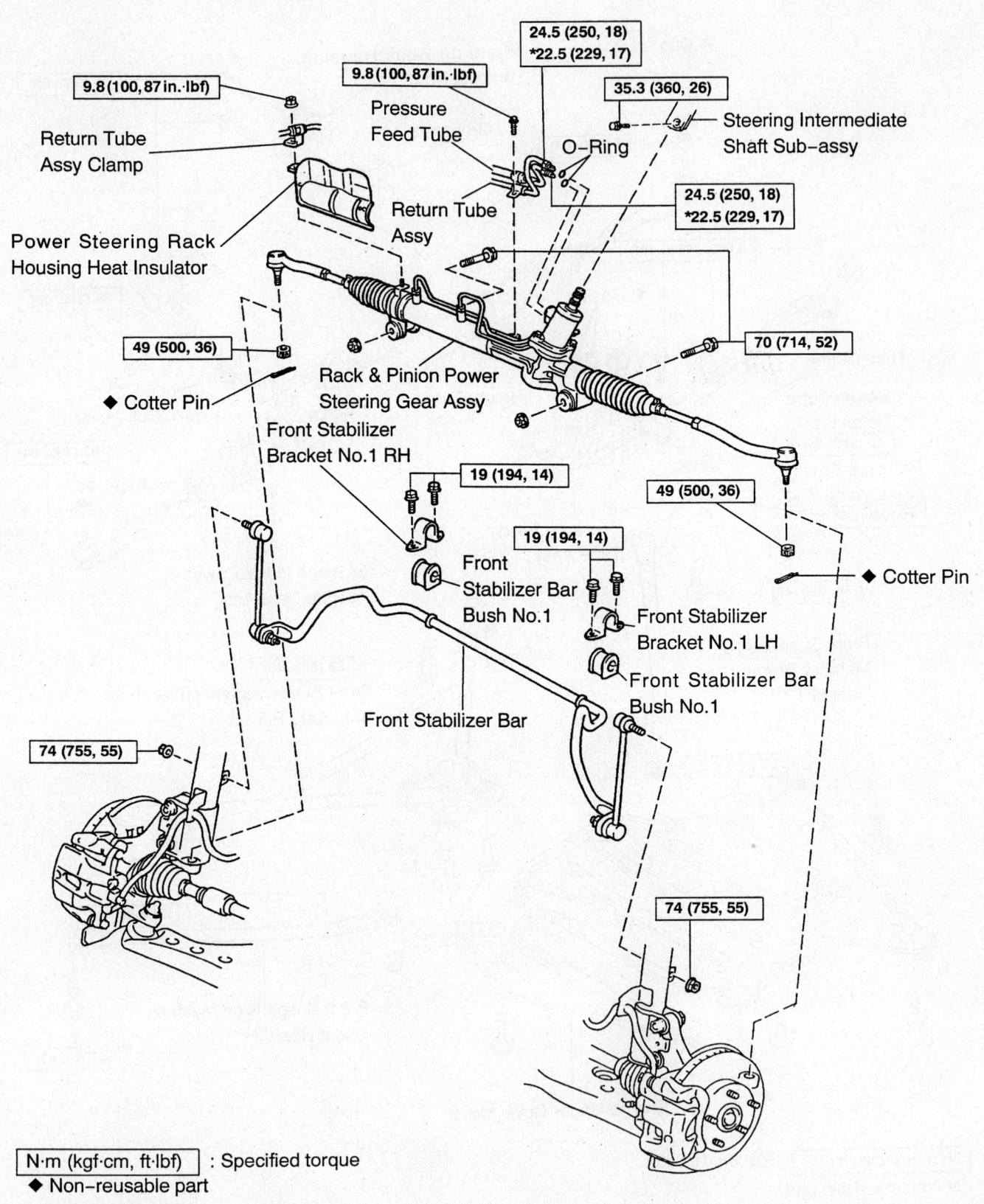

9.8 (100, 87 in. lbf)

Return Tube
Assy Clamp

Power Steering Rack
Housing Heat Insulator

49 (500, 36)

◆ Cotter Pin

9.8 (100, 87 in. lbf)

Pressure
Feed Tube

Return Tube
Assy

24.5 (250, 18)
*22.5 (229, 17)

35.3 (360, 26)

O-Ring

Steering Intermediate
Shaft Sub-assy

24.5 (250, 18)
*22.5 (229, 17)

70 (714, 52)

Rack & Pinion Power
Steering Gear Assy

Front Stabilizer
Bracket No.1 RH

19 (194, 14)

49 (500, 36)

Front
Stabilizer Bar
Bush No.1

19 (194, 14)

◆ Cotter Pin

Front Stabilizer
Bracket No.1 LH

Front Stabilizer Bar
Bush No.1

74 (755, 55)

Front Stabilizer Bar

74 (755, 55)

N·m (kgf·cm, ft·lbf) : Specified torque

◆ Non-reusable part

* For use with SST

67162-LEXU-G75

Exploded view power steering gear assembly—ES 330

4. Loosen the top bolt holding the sliding yoke to the intermediate shaft. Remove the bottom bolt holding the sliding yoke to the steering rack.

5. Remove or disconnect the following:
- Pressure feed and return lines to the rack and pinion
- Power steering connector
- 4 mount bolts and nuts to the power steering rack
- 2 brackets and grommets
- Power steering rack from the vehicle

To install:

6. Install or connect the following:
- Power steering rack to the vehicle.
- 2 brackets and grommets to the power steering rack.
- 4 bolts and tighten the bolts to 56 ft. lbs. (76 Nm).
- Power steering solenoid connector
- Pressure feed and return tubes. Tighten the union bolt to 36 ft. lbs. (49 Nm).

7. Align the matchmarks on the sliding yoke and control valve shaft.

8. Tighten the bolt holding the sliding yoke to the steering rack to 26 ft. lbs. (35 Nm).

9. Tighten the bolt holding the sliding yoke to the intermediate shaft to 26 ft. lbs. (35 Nm).

10. Install or connect the following:
- Tie rod end to the steering knuckle. Tighten the nut to 48 ft. lbs. (65 Nm). Install a new cotter pin.
- Engine undercover
- Wheel(s)

11. Bleed the power steering system and check the front end alignment.

SC 430 AND 2004–06 LS 430

1. Before servicing the vehicle, refer to the Precautions Section.

2. Place the front wheels facing straight ahead.

3. Remove or disconnect the following:
- Negative battery cable. Wait at least 90 seconds before performing any work.
- Front wheels
- Steering wheel switch & volume case
- Steering wheel pad
- Steering wheel column lower cover
- Horn button

- Steering wheel
- Brake caliper
- Tie rods from lower ball joint
- Engine under cover
- Front suspension member braces
- Steering slide yoke. Match-mark with control valve shaft before removal
- Oil feed tubes

4. Remove Rack & Pinion Assembly

5. Remove or disconnect the following:
- Tube clamps on rack assembly
- Connector assembly
- 4 gear assembly set bolts
- Steering gear
- Steering rack housing bracket

6. Match-mark and remove tie rod assemblies

To install:

7. Install or connect the following:
- Tie rod assemblies. Align match marks
- Oil feed tubes
- Rack & Pinion gear assembly with 4 set bolts; torque to 48 ft. lbs (65 Nm)
- Steering sliding yoke assembly
- Pressure and return tubes
- Tie rod assemblies to lower ball joint; torque to 64 ft. lbs. (87 Nm)
- Brake calipers
- Front suspension member braces
- Engine under cover
- Center spiral cable
- Temporarily tighten steering wheel assembly
- Center steering wheel and fully tighten set nut
- Horn button
- Steering wheel covers
- Steering pad modulator switch
- Negative battery cable
- Bleed power steering system
- Front wheels

8. Inspect toe in and adjust as necessary

9. Test drive

ES 330

1. Before servicing the vehicle, refer to the Precautions Section.

2. Place the front wheels facing straight ahead.

3. Remove or disconnect the following:
- Negative battery cable. Wait at least 90 seconds before performing any work.

- Front wheels
- Steering wheel switch & volume case
- Steering wheel pad
- Steering wheel column lower cover
- Horn button
- Steering wheel
- Tie rods from steering knuckle
- Steering intermediate shaft Match-mark with control valve shaft before removal
- Front stabilizer link assembly
- Front stabilizer brackets
- Oil feed tubes

4. Remove Rack & Pinion Assembly

5. Remove or disconnect the following:
- Tube clamps on rack assembly
- Connector assembly
- 2 bolts and 2 nuts
- Steering gear

6. Power steering rack housing heat insulator

7. Right & Left turn pressure tubes from the rack assembly

8. Match-mark and remove tie rod assemblies

To install:

9. Install or connect the following:
- Tie rod assemblies. Align match marks
- Oil feed tubes
- Steering rack heat insulator
- Rack and pinion gear assembly with 2 bolts and 2 nuts; torque to 52 ft. lbs (70 Nm)
- Pressure and return tubes
- Front stabilizer brackets
- Front stabilizer link assembly
- Steering intermediate shaft
- Tie rod assemblies to 36 ft. lbs. (49 Nm)
- Center spiral cable
- Temporarily tighten steering wheel assembly
- Center steering wheel and fully tighten set nut
- Horn button
- Steering wheel covers
- Steering pad modulator switch
- Negative battery cable
- Bleed power steering system
- Front wheels

10. Inspect toe in and adjust as necessary

11. Test drive

FRONT SUSPENSION

Strut and Coil Spring

REMOVAL & INSTALLATION

IS 250 and IS 350

1. Before servicing the vehicle, refer to the Precautions Section.
2. Remove or disconnect the following:
   - Negative battery cable
   - Tire and wheel assembly
   - Front speed sensor
   - Stabilizer link assembly
   - Front upper control arm at the steering knuckle
3. Loosen, but do not remove the bolt securing the strut to the lower suspension arm. Support lower arm with a floor jack and block of wood.
   - Engine room side cover
   - Loosen the top strut lock nut
   - 3 upper mounting nuts from the strut tower
   - Spring support reinforcement
4. Slowly lower the jack.
   - Lower strut bolt
   - Strut assembly

✳✳ CAUTION

Do NOT remove the center nut to the strut at this time. The spring on the strut is under high pressure and can cause serious injury.

5. Secure the strut in a vise.
6. Compress the coil spring
7. Remove or disconnect the following:
   - Upper strut retaining nut

- Suspension support
- Upper insulator
- Bumper
- Coil spring
- Insulator

To install:

8. Install or connect the following:
   - Lower insulator
   - Coil spring end into the step of the lower seat
   - Bumper to the piston rod
   - Upper insulator
   - Upper support to the piston rod, aligning it with the groove in the strut rod
9. Adjust the front suspension support assembly so that the bolts come to the positions shown in the illustration.
   - Front spring support reinforcement. Tighten the 3 new upper strut retaining nuts to 49 ft. lbs. (67 Nm).
   - Lower strut to lower suspension arm. Temporarily tighten.
   - Front upper control arm at the steering knuckle. Torque to 64 ft. lbs. (87 Nm) and install a new clip.
   - Front speed sensor. Torque to 4 ft. lbs. (6 Nm).
   - Stabilizer link assembly. Torque to 48 inch lbs. (5 Nm).
10. Stabilize the suspension as follows:
    a. Install the front wheel(s).
    b. Lower the vehicle and bounce it up and down several times to stabilize the front suspension.
    c. Raise the vehicle.
    d. Remove the front wheel.
    e. Jack up the front lower suspension

armplacing a wood block in between. Apply a load to the front suspension so that the front lower suspension arm is plkaced in a horizontal position.
11. Fully tighten the lower strut to lower suspension arm nut and bolt to 116 ft. lbs. (157 Nm).
    - Engine room side cover
    - Tire and wheel assembly
    - Negative battery cable
12. Check the front alignment.

ES 300

1. Before servicing the vehicle, refer to the Precautions Section.
2. Remove or disconnect the following:
   - Negative battery cable
   - Tire and wheel assembly
   - If equipped with an Anti-lock Brake System (ABS), the ABS speed sensor connector
   - Brake line from the strut housing
   - Strut assembly from the steering knuckle
   - 3 upper mounting nuts from the strut tower
   - Strut assembly

✳✳ CAUTION

Do NOT remove the center nut to the strut at this time. The spring on the strut is under high pressure and can cause serious injury.

3. Temporarily install the bolt and nuts to the lower bracket of the strut to support it and secure the strut in a vise.
4. Compress the coil spring
5. Remove or disconnect the following:
   - Spring seat
   - Upper strut retaining nut
   - Suspension support
   - Upper insulator

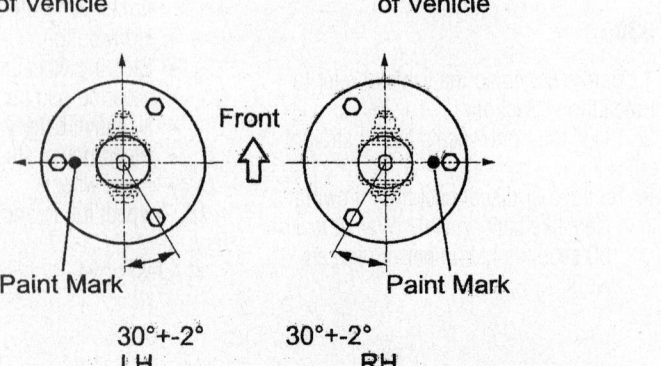

Front Suspension Upper Side View

09490_LEXU_G0112

Align the out mark of the upper spring seat with the mark on the upper insulator—IS 250 and IS 350

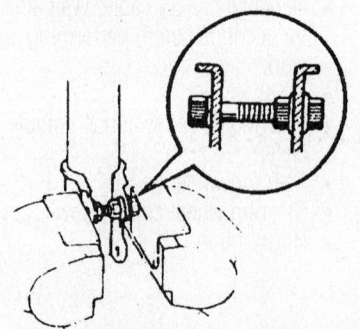

7923LGA5

Temporarily install the support nuts and bolt to the strut—ES 300

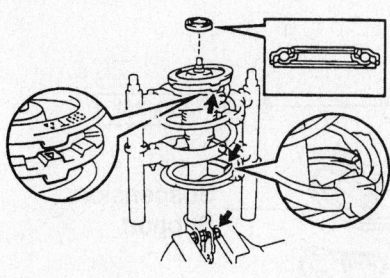

7923LGA6

Align the out mark of the upper spring seat with the mark on the upper insulator—ES 300

- Spring
- Bumper
- Insulator

To install:

6. Install or connect the following:
   - Lower insulator
   - Bumper to the piston rod
   - Coil spring end into the gap of the lower seat
   - Upper insulator
   - Upper support to the piston rod, aligning it with the groove in the strut rod
7. Install or connect the following:
   - Spring seat. Tighten the new upper strut retaining nut to 36 ft. lbs. (49 Nm).
8. Remove the strut from the vise and disassemble the securing nuts and bolt.
9. Rotate the upper support so the lowest bolt on the support aligns with the projection part of the lower spring.
10. Install or connect the following:
    - Strut and tighten the strut to body bolts to 59 ft. lbs. (80 Nm)
    - Strut to the steering knuckle and tighten the bolts to 156 ft. lbs. (211 Nm)
11. Run the brake hose through the brake hose bracket and install the clip.
12. Install or connect the following:
    - ABS speed sensor and tighten the mounting bolt to 48 inch lbs. (5 Nm)
    - Brake line to the strut housing and tighten the bolt to 22 ft. lbs. (29 Nm)
    - Wheel
    - Negative battery cable
13. Check the front alignment.

GS 300 and GS 430

1. Before servicing the vehicle, refer to the Precautions Section.
2. Remove or disconnect the following:
   - Negative battery cable.

- Front wheel
- Brake caliper, leaving the line attached
3. Loosen the 3 upper strut mounting nuts.
4. Loosen, but do not remove, the upper strut rod nut.

✳✳ CAUTION

Do NOT remove the upper strut nut at this time.

5. Remove or disconnect the following:
   - Anti-lock Brake System (ABS) speed sensor and harness
   - Upper suspension arm from the steering knuckle
   - Stabilizer bar from the link and remove the bracket
   - Strut from the lower suspension arm.
   - 3 upper strut mounting nuts and remove the strut
6. Compress the coil spring.
7. Remove or disconnect the following:
   - Piston rod locknut
   - Suspension support, coil spring and bumper
8. If disposing the strut, perform the following procedure:
   a. Fully extend the strut rod.
   b. Drill a hole near the bottom of the shock to remove the gas inside.

✳✳ CAUTION

The gas is harmless, but be careful of chips that may fly up when the gas is released.

To install:
9. Install or connect the following:
   - Spring bumper
   - Coil spring
   - Suspension support to the rod and temporarily install a new nut
10. Turn the suspension support so one of the bolts on the support faces the same direction as shown in the illustration.

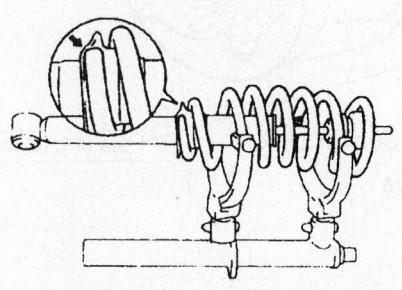

7923LGA7

Matching the spring to the seat

➡**Align the bolt so a line drawn between the rod and bolt would be at 90° to the direction of the lower bushing.**

11. Install or connect the following:
    - Spring compressor
    - Strut and tighten the upper retaining nuts to 41 ft. lbs. (56 Nm)
    - New upper strut rod nut to 20 ft. lbs. (27 Nm)
    - Strut to the lower arm and temporarily tighten the nut and bolt
    - Stabilizer bar bracket and tighten the bolts to 21 ft. lbs. (28 Nm)
    - The stabilizer bar to the link and tighten the bolts to 29 ft. lbs. (39 Nm)
    - Upper suspension arm to the steering knuckle. Tighten the nut to 64 ft. lbs. (87 Nm) and install a new cotter pin.
    - ABS speed sensor and tighten the bolt to 69 inch lbs. (8 Nm)
    - Caliper
    - Wheel
12. Bounce the vehicle several times to stabilize the suspension.
13. Tighten the lower strut bolt and nut to 116 ft. lbs. (157 Nm).
14. Check the front wheel alignment.

IS 300

1. Before servicing the vehicle, refer to the Precautions Section.
2. Remove or disconnect the following:
   - Front wheel
   - Wheel speed sensor and harness clamp
   - Upper ball joint
   - Level control sensor link
   - Stabilizer bar link
   - Lower strut bolt
   - Upper strut mount cap
   - Upper strut mount nuts
   - Strut assembly
3. Install a suitable spring compressor and remove the center nut.
4. Remove the upper strut mount and the coil spring.

To install:
5. Installation is the reverse of the removal procedure, while using the following torque values:
   - Upper strut mount center nut: 25 ft. lbs. (34 Nm)
   - Upper strut mounting nuts: 26 ft. lbs. (35 Nm)
   - Lower strut mount bolt: 47 ft. lbs. (64 Nm)
   - Upper ball joint nut: 50 ft. lbs. (65 Nm)

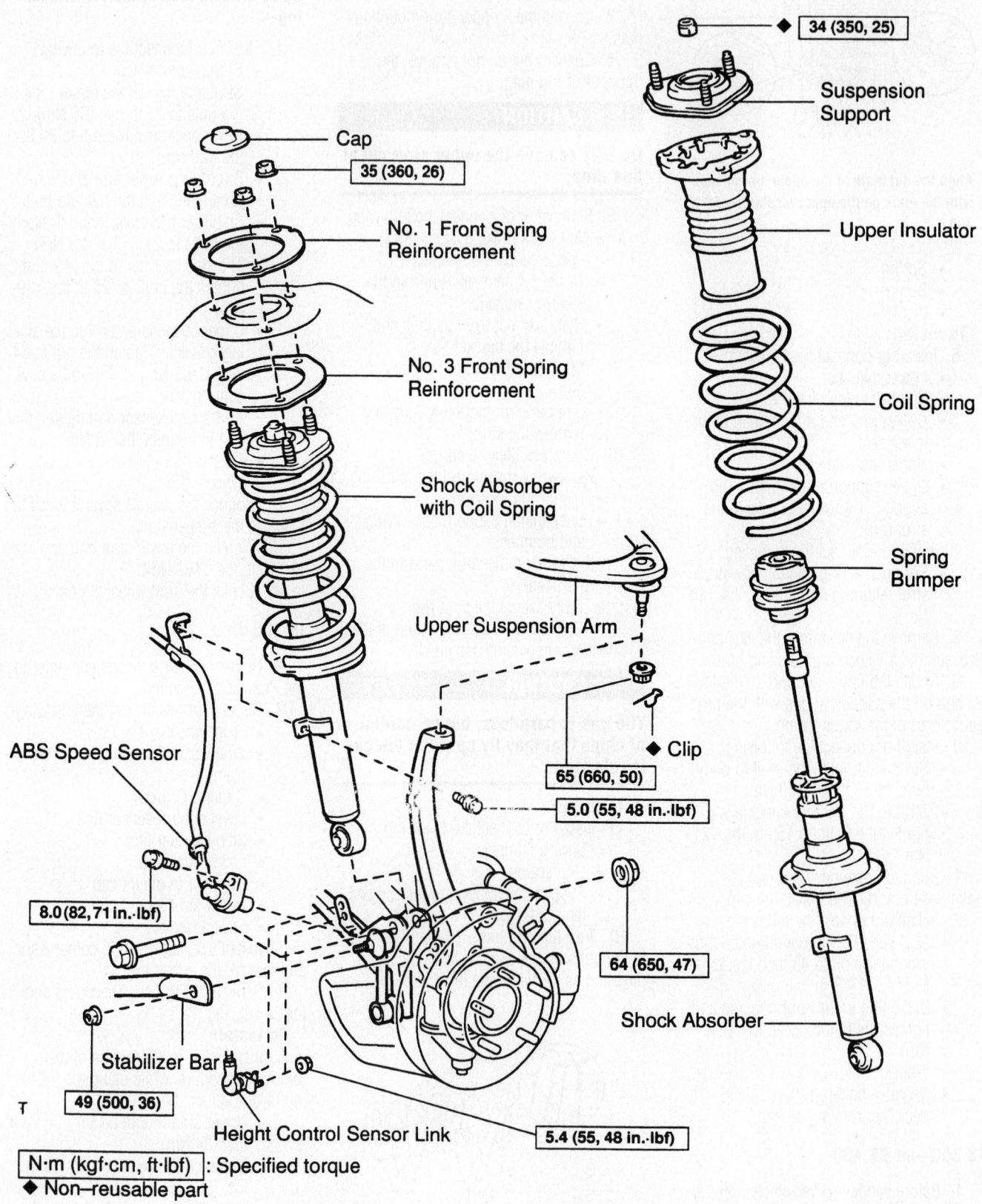

◆ 34 (350, 25)

Suspension Support

Cap

35 (360, 26)

No. 1 Front Spring Reinforcement

Upper Insulator

No. 3 Front Spring Reinforcement

Coil Spring

Shock Absorber with Coil Spring

Upper Suspension Arm

Spring Bumper

ABS Speed Sensor

◆ Clip

65 (660, 50)

5.0 (55, 48 in.·lbf)

8.0 (82, 71 in.·lbf)

64 (650, 47)

Shock Absorber

Stabilizer Bar

49 (500, 36)

Height Control Sensor Link

5.4 (55, 48 in.·lbf)

N·m (kgf·cm, ft·lbf) : Specified torque
◆ Non–reusable part

9347LG05

Exploded view of the front strut assembly mounting—IS 300

- Stabilizer bar link nut: 36 ft. lbs. (49 Nm)

LS 430—WITHOUT AIR SUSPENSION

1. Before servicing the vehicle, refer to the Precautions Section.
2. Remove or disconnect the following:
   - Tire and wheel assembly
   - Steering knuckle from the upper ball joint
   - Strut assembly from the lower strut bracket
   - Strut cover from the upper strut mount
   - 3 mounting nuts and remove the strut assembly with the coil spring from the vehicle.

✳✳ CAUTION

Do NOT remove the center nut to the strut at this time.

3. Compress the coil spring.
4. Remove or disconnect the following:
   - Piston rod locknut
   - Suspension support, coil spring and the bumper
5. If disposing the strut, perform the following procedure:
   a. Fully extend the strut rod.
   b. Drill a hole within the shaded area shown in the illustration to remove the gas inside.

✳✳ CAUTION

The gas is harmless, but be careful of chips that may fly up when drilling.

   c. Properly dispose of the strut assembly.
To install:
6. Install or connect the following:
   - Spring bumper
   - Coil spring. Match the end of the coil into the recess of the strut spring seat.
   - Suspension support to the rod and temporarily install a new nut
7. Turn the suspension support so one of the bolts on the support faces the same direction as shown in the illustration.

➡**Align the bolt so a line drawn between the rod and bolt would be at 90˚ to the direction of the lower bushing.**

8. Tighten the strut rod nut to 20 ft. lbs. (27 Nm) and install the cap.
9. Remove the spring compressor
10. Install or connect the following:

- Strut and tighten the upper retaining nuts to 43 ft. lbs. (58 Nm)
- Strut to the lower bracket and temporarily install the nut and bolt
- Upper control arm to the steering knuckle. Tighten the nut to 48 ft. lbs. (65 Nm) and install a new cotter pin.
- Wheel
11. Lower the vehicle.
12. Bounce the vehicle several times to stabilize the suspension.
13. Tighten the lower strut bolt and nut to 116 ft. lbs. (157 Nm).
14. Check the front wheel alignment.

LS 430—WITH AIR SUSPENSION

1. Before servicing the vehicle, refer to the Precautions Section.
2. Move the height control switch to **OFF**.
3. Bleed the air from the suspension.
4. Remove or disconnect the following:
   - Wheel
   - Height control sensor link from the lower strut bracket
   - Cotter pin and nut holding the upper control arm to the steering knuckle
   - Upper ball joint from the steering knuckle
   - Pneumatic cylinder from the lower bracket by removing the through-bolt
   - Air tube from the strut
   - The actuator cover

✳✳ CAUTION

Do NOT remove the center nut from the pneumatic cylinder.

   - Actuator electrical connector
   - 2 bolts to the suspension control actuator and position the actuator aside
   - 3 upper mounting nuts and the strut from the vehicle
5. If disposing the strut perform the following procedure:
   a. Using a screwdriver, remove the air from inside the cylinder.
   b. Fully extend the cylinder.
   c. Drill a hole in the cylinder at a point above 1.57 in. (40mm) from the bottom of the strut assembly. This will release the gas charge in the strut. Do NOT puncture the pneumatic cylinder.

✳✳ CAUTION

The gas coming out is harmless, but be careful of chips that may fly up while drilling.

To install:
6. Install or connect the following:
   - Strut and tighten the upper mounting nuts to 43 ft. lbs. (58 Nm)
   - Suspension control actuator and tighten the bolts. Tighten the 2 nuts to 13 ft. lbs. (17 Nm).
   - Suspension control actuator cover and tighten the nuts to 43 ft. lbs. (58 Nm)
   - 2 new O-rings to the air tube. Install the tube and tighten it to 13 ft. lbs. (17 Nm). Install the grommet.
   - The strut to the lower strut bracket and temporarily install the nut and bolt
   - Steering knuckle to the upper ball joint. Tighten the nut to 48 ft. lbs. (65 Nm) and install a new cotter pin.
   - Height control sensor link and tighten a new nut to 48 inch lbs. (5 Nm)
   - Wheel
7. Turn the height control switch **ON**.
8. Start the engine to fill the strut with air.
9. Bounce the vehicle several times to normalize the suspension.
10. Support the lower control arm with a jack.
11. Install or connect the following:
   - Front wheel
   - Lower strut mounting nut and bolt to 76 ft. lbs. (106 Nm)
   - Wheel
12. Check the front end alignment.

SC 430

1. Before servicing the vehicle, refer to the Precautions Section.
2. Remove or disconnect the following:
   - Tire and wheel assembly
   - Speed sensor assembly
   - Upper suspension arm from the steering knuckle
   - Shock absorber from the mounting bracket

✳✳ CAUTION

Loosen the piston rod lock nut. Do Not remove at this time

   - 3 nuts front spring support reinforcement and strut assembly
3. Remove the shock absorber from the spring assembly
4. Remove or disconnect the following:
   a. Compress the coil spring with proper spring compressor

SHOCK ABSORBER ASSY FRONT:

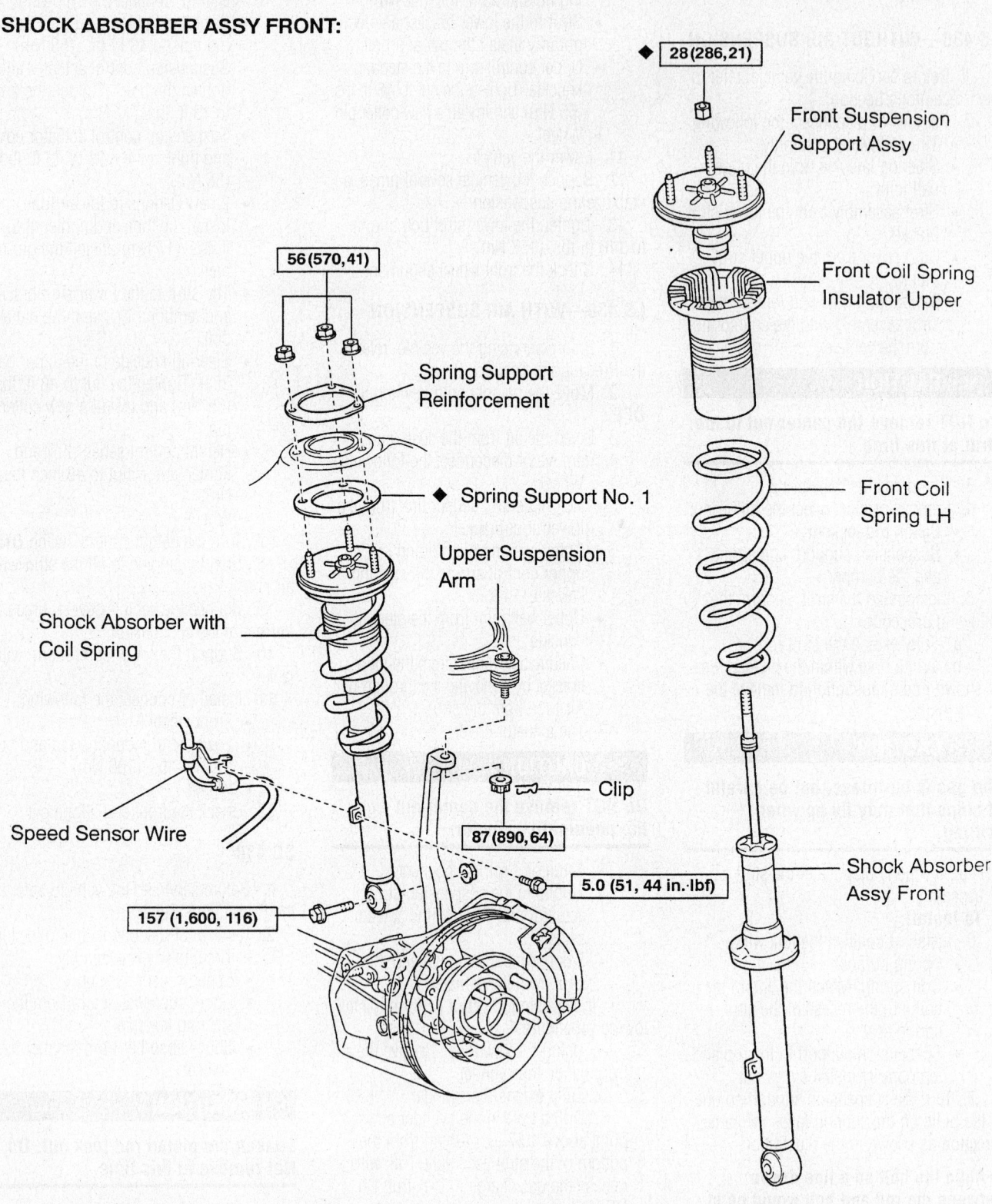

28 (286, 21)

Front Suspension Support Assy

Front Coil Spring Insulator Upper

Front Coil Spring LH

Shock Absorber Assy Front

56 (570, 41)

Spring Support Reinforcement

◆ Spring Support No. 1

Upper Suspension Arm

Shock Absorber with Coil Spring

Clip

87 (890, 64)

Speed Sensor Wire

5.0 (51, 44 in.·lbf)

157 (1,600, 116)

N·m (kgf·cm, ft·lbf) : Specified torque
◆ Non−reusable part

67162-LEXU-G76

Exploded view of front strut assembly—SC 430

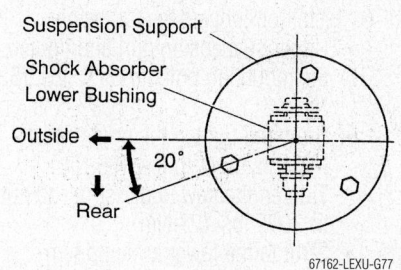

67162-LEXU-G77

Strut to suspension support alignment—SC 430

   b. Remove the piston lock nut
   c. Remove the front suspension support assembly, front coil spring insulator and coil spring.

To install:
   5. Install the shock into the spring assembly
      a. Install the spring insulator
      b. Compress the coil spring using the proper spring compressor
      c. Install spring making sure that the spring seats properly
      d. Temporarily install a new lock nut
      e. Align the suspension support with the shock absorber lower bolt
   6. Install or connect the following:
- 3 bolts attaching the strut assembly to the support assembly to 41 ft. lbs (56 Nm)
- Fully tighten piston lock nut to 21 ft. lbs. (28 Nm)

- Strut assembly to mounting bracket to 116 ft lbs. (157 Nm)
- Upper suspension arm to steering knuckle to 64 ft. lbs. (87 Nm)
- Speed sensor
- Tire & wheel

ES 330

   1. Before servicing the vehicle, refer to the Precautions Section.
   2. Remove or disconnect the following:
- Tire and wheel assembly
- Stabilizer link assembly

   3. Equipped with H-Tems® suspension
      a. Disconnect the wiring connector
      b. Disconnect the 3 nuts, harness clamp and shock absorber cap

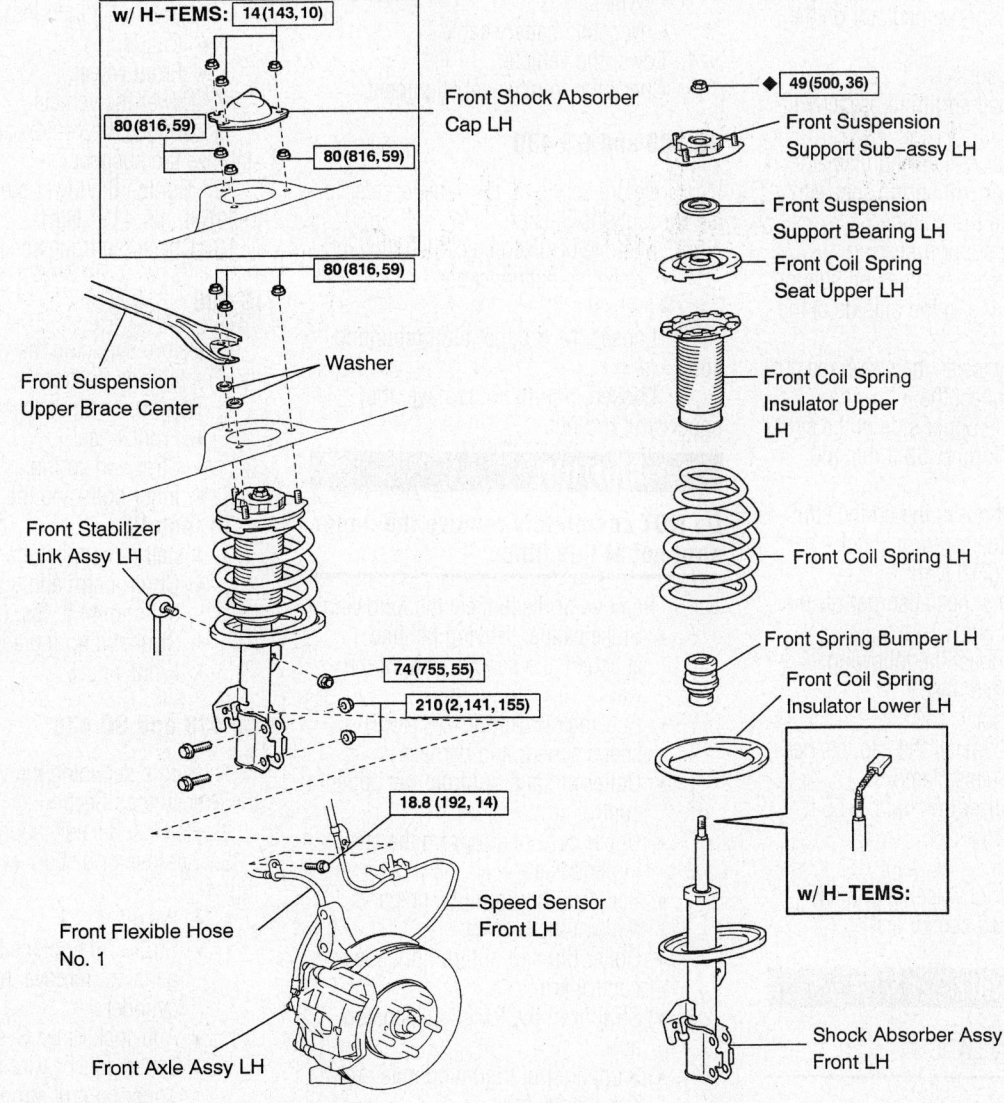

N·m (kgf·cm, ft·lbf) : Specified torque
◆ Non-reusable part

67162-LEXU-G78

Exploded view front suspension—ES 330

c. With special tool loosen the piston rod nut. Do NOT remove nut at this time.

4. W/out H-Tems®

5. Remove or disconnect the following:
- Loosen lock nut on piston rod. Do NOT remove at this time
- Brake hose from strut
- Speed sensor
- 2 bolts and nuts on strut mount on steering knuckle
- Strut assembly

6. Secure strut in a vise and compress the spring with a spring compressor

7. Remove the piston rod nut and remove the support assembly

8. Remove the shock absorber from the spring assembly

To install:

9. Install shock absorber into the spring coil assembly.

a. Install coil spring insulator on the piston rod

b. Install coil spring

c. Align the coil spring to properly fit in lower seat

d. Install upper coil spring insulator

e. Install upper coil spring seat with the mark facing to the outside

f. Install new support bearing

g. Install the front suspension support with the mark facing to the outside of the vehicle.

h. Temporarily install the piston rod nut

10. Install or connect the following:
- 3 nuts to the upper side of the strut assembly. Torque: 59 ft lbs. (80 Nm)
- 2 bolts and nut to the on the strut mount on the steering knuckle to 155 ft. lbs (210 Nm)

11. Fully tighten shock absorber piston nut to 36 ft. lbs (49 Nm)

12. Install or connect the following:
- Flexible brake hose
- Speed sensor
- W/H-Tems, Install the cap, connector and harness clamp
- Front stabilizer link nuts to 55 ft. lbs. (74 Nm)
- Front wheel

13. Adjust the front wheel alignment

14. Perform speed sensor test

Upper Ball Joint

REMOVAL & INSTALLATION

The upper ball joint is an integral part of the upper arm and is not replaced separately. The upper ball joint replacement is accomplished by replacing the upper arm.

Upper Control Arm

REMOVAL & INSTALLATION

IS 250 and IS 350

1. Before servicing the vehicle, refer to the Precautions Section.

2. Remove or disconnect the following:
- Negative battery cable
- Wheel
- Front strut assembly
- Front upper control arm

To install:

3. Install or connect the following:
- Upper suspension arm and tighten the mounting bolts to 36 ft. lbs. (49 Nm)
- Front strut assembly
- Wheel
- Negative battery cable

4. Lower the vehicle.

5. Check the front wheel alignment.

GS 300 and GS 430

1. Before servicing the vehicle, refer to the Precautions Section.

2. Remove or disconnect the following:
- Negative battery cable
- Wheel

3. Loosen the 3 upper strut mounting nuts.

4. Loosen, but do not remove, the upper strut rod nut.

✳✳ CAUTION

Do NOT completely remove the upper strut nut at this time.

5. Remove or disconnect the following:
- Brake caliper, leaving the line attached and secure it out of the way
- Anti-lock Brake System (ABS) speed sensor and harness
- Cotter pin and nut from the upper control arm
- Upper control arm from the steering knuckle
- Stabilizer bar from the link and remove the bracket
- Cotter pin and nut from the lower control arm
- Strut from the lower suspension arm
- 3 upper strut mounting nuts and remove the strut
- Mounting bolts holding the upper control arm to the frame
- Upper control arm from the vehicle

To install:

6. Install or connect the following:
- Upper suspension arm and tighten the mounting bolts to 39 ft. lbs. (53 Nm)
- Strut and tighten the upper retaining nuts to 41 ft. lbs. (56 Nm). Tighten the new upper strut rod nut to 20 ft. lbs. (27 Nm).
- Strut to the lower arm and temporarily tighten the nut and bolt
- Stabilizer bar bracket and tighten the bolts to 21 ft. lbs. (28 Nm)
- Stabilizer bar to the link and tighten the bolts to 29 ft. lbs. (39 Nm)
- Upper suspension arm to the steering knuckle. Tighten the nut to 64 ft. lbs. (87 Nm) and install a new cotter pin.
- ABS speed sensor and tighten the bolt to 69 inch lbs. (8 Nm)
- Caliper
- Front wheel

7. Lower the vehicle.

8. Bounce the vehicle several times to stabilize the suspension.

9. Tighten the lower strut bolt and nut to 116 ft. lbs. (157 Nm).

10. Check the front wheel alignment.

IS 300

1. Before servicing the vehicle, refer to the Precautions Section.

2. Remove or disconnect the following:
- Front wheel
- Strut and spring assembly
- Inner bolts and the control arm

To install:

3. Install or connect the following:
- Control arm and tighten the inner bolts to 44 ft. lbs. (59 Nm)
- Strut and spring assembly
- Front wheel

LS 430 and SC 430

1. Before servicing the vehicle, refer to the Precautions Section.

2. Raise and safely support the vehicle.

3. Remove or disconnect the following:
- Wheel
- Strut or if equipped with air suspension, remove the pneumatic cylinder
- Anti-lock Brake System (ABS) speed sensor wire harness from the upper control arm by removing the bolt.
- Mounting bolts holding the upper control arm to the vehicle
- Upper control arm

To install:

4. Install or connect the following:
- Upper control arm and tighten the 2 mounting bolts to (except SC 430) 83 ft. lbs. (113 Nm), or (SC 430) 39 ft. lbs. (53 Nm)
- ABS speed sensor wire harness to the upper control arm with the attaching bolt
- Strut, or if equipped with air suspension, install the pneumatic cylinder
- Wheel

5. Lower the vehicle.

6. Check and adjust the wheel alignment as necessary.

CONTROL ARM BUSHING REPLACEMENT

The control arm bushings are serviced with the control arm as an assembly.

Lower Control Arm

REMOVAL & INSTALLATION

IS 250 and IS 350

1. Before servicing the vehicle, refer to the Precautions Section.

2. Remove or disconnect the following:
- Negative battery cable
- Front wheel(s)
- Front speed sensor
- Tie rod assembly
- Height control sensor link assembly
- Lower part of strut from lower control arm
- Stabilizer link assembly
- Engine under cover

3. Remove the lower control arm as follows:

a. Support front suspension crossmember with a transmission jack and a block of wood

b. Remove the 2 bolts from the front lower ball joint.

c. Loosen, but do not remove the nut of the lower No. 2 arm bracket assembly.

d. Remove the bolt, washer and nut on the front of the front lower control arm.

e. Remove the 4 bolts, side rail plate and front lower suspension arm with the lower No. 2 arm bracket assembly.

To install:

4. Install and temporarily tighten the front lower control arm as follows:

a. Torque bolt 1 to 150 ft. lbs. (204 Nm); bolt 2 to 63 ft. lbs. (86 Nm); bolt 3 to 37 ft. lbs. (50 Nm).

b. Install bolt from the front and temporarily tighten the bolt, washer and nut.

c. Install the front lower ball joint, with 2 bolts and tighten to 89 ft. lbs. (120 Nm).

5. Install or connect the following:
- Lower part of strut to lower control arm and temporarily tighten the bolt
- Stabilizer link assembly. Torque to 48 inch lbs. (5 Nm).
- Tie rods to the steering knuckles and the nuts. Tighten the nut to 50 ft. lbs. (65 Nm) and install a new cotter pin. The prongs of the cotter pin should be firmly wrapped around the flats of the nut.
- Front speed sensor. Torque to 4 ft. lbs. (6 Nm).
- Height control sensor link assembly. Torque to 48 inch lbs. (5 Nm).

6. Stabilize the suspension as follows :

a. Install the front wheel(s).

b. Lower the vehicle and bounce it up and down several times to stabilize the front suspension.

c. Raise the vehicle.

d. Remove the front wheel.

e. Jack up the front lower suspension armplacing a wood block in between. Apply a load to the front suspension so that the front lower suspension arm is plkaced in a horizontal position.

7. Fully tighten the lower strut to lower suspension arm nut and bolt to 116 ft. lbs. (157 Nm).

8. Fully tighten the bolt on the front of the front lower control arm to 100 ft. lbs. (135 Nm).

9. Fully tighten the nut on the lower No. 2 arm bracket assembly to 83 ft. lbs. (113 Nm).

- Engine under cover
- Front wheel

10. Inspect and adjust front wheel alignment.

ES 300

1. Before servicing the vehicle, refer to the Precautions Section.

2. Remove or disconnect the following:
- Negative battery cable
- Front wheel(s)
- Side fender apron seal
- Steering knuckle with the axle hub, from the vehicle
- Dust deflector from the knuckle
- Cotter pin and the nut from the ball joint stud

3. Remove the lower ball joint from the steering knuckle.

To install:

4. Install the lower ball joint onto the steering knuckle and tighten nut to 90 ft. lbs. (123 Nm). Install new cotter pin.

5. Align the hole in the dust deflector with the ABS speed sensor. Using the appropriate driver, install a new dust deflector.

6. Install or connect the following:
- Steering knuckle and hub onto the vehicle
- Fender apron seal
- Front wheel(s)
- Negative battery cable

GS 300 and GS 430 (Except 2004)

1. Before servicing the vehicle, refer to the Precautions Section.

2. Remove or disconnect the following:
- Negative battery cable
- Wheel(s)
- Caliper, leaving the brake line connected and suspend it out of the way

✺✺ WARNING

Never allow the brake caliper to hang freely from the brake hose.

- Rotor
- Anti-lock Brake System (ABS) speed sensor and harness

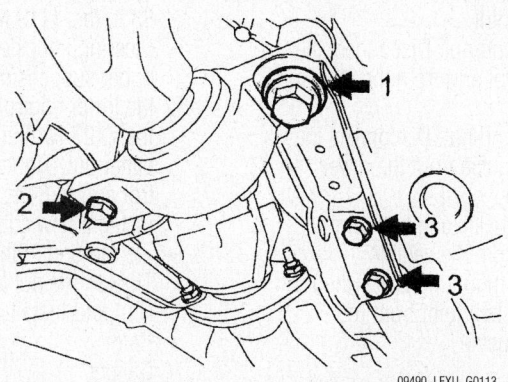

09490_LEXU_G0113

Lower control arm and side rail plate bolt locations—IS 250 and IS 350

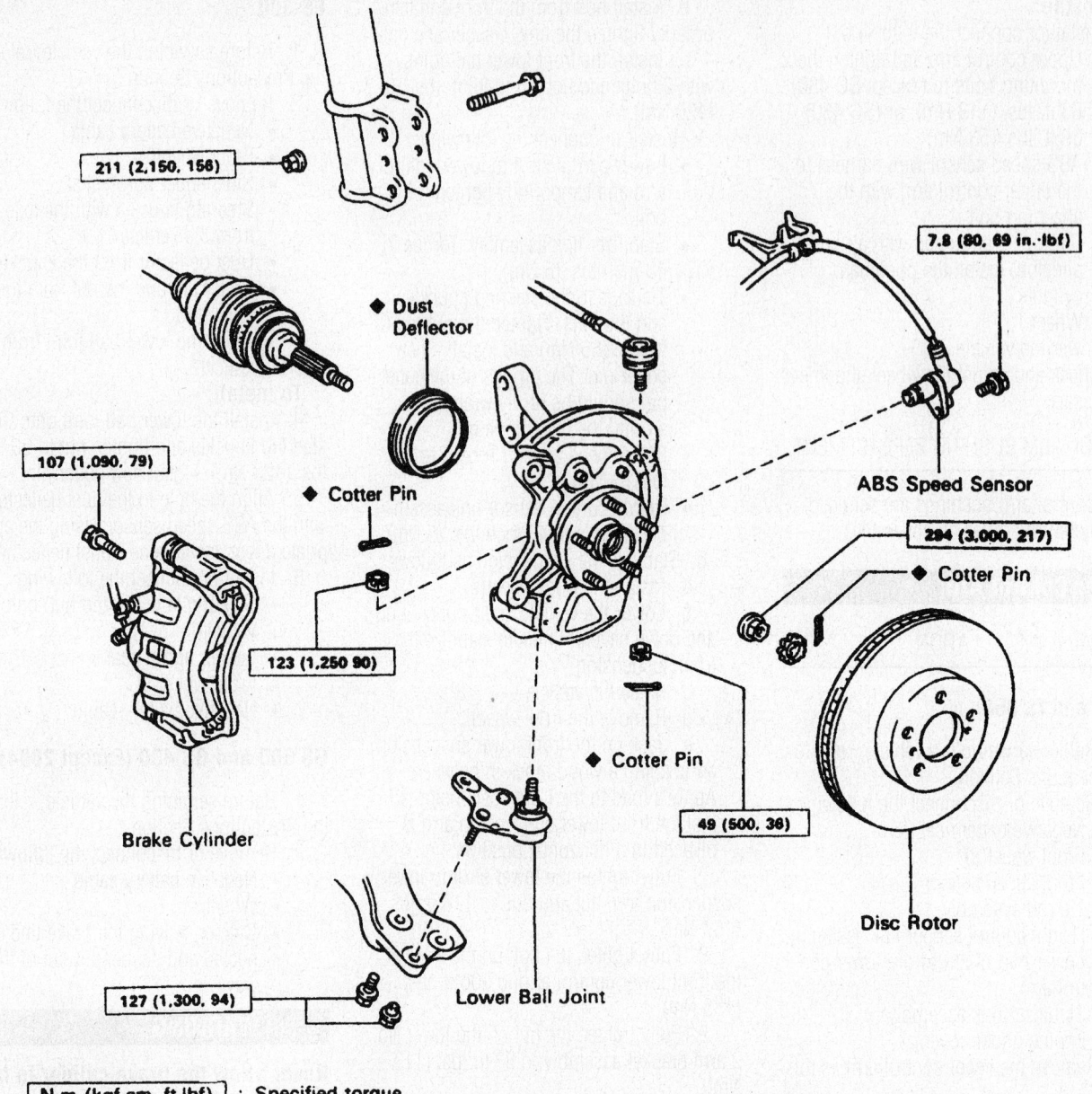

211 (2,150, 156)

107 (1,090, 79)

7.8 (80, 69 in.·lbf)

◆ Dust Deflector

◆ Cotter Pin

ABS Speed Sensor

294 (3,000, 217)

◆ Cotter Pin

123 (1,250 90)

Brake Cylinder

◆ Cotter Pin

49 (500, 36)

Disc Rotor

127 (1,300, 94)

Lower Ball Joint

N·m (kgf·cm, ft·lbf) : Specified torque
◆ Non-reusable part

7923LGB3

Exploded view of the lower suspension—ES 300

- Tie rod end from the arm on the lower ball joint
- Cotter pin and nut. Disconnect the upper control arm from the steering knuckle.
- Cotter pin and nut. Disconnect the steering knuckle from the lower control arm.
- Steering knuckle and ball joint assembly from the vehicle
- 2 ball joint mounting bolts, then remove the ball joint from the steering knuckle

To install:

3. Install or connect the following:

- Ball joint and tighten the bolts to 83 ft. lbs. (113 Nm)
- Steering knuckle to the lower and upper suspension arms. Tighten the lower control arm nut to 95 ft. lbs. (127 Nm) and install a new cotter pin. Tighten the upper control arm to 64 ft. lbs. (87 Nm) and install a new cotter pin.
- Tie rod end to the ball joint arm. Tighten the nut to 64 ft. lbs. (87 Nm) and install a new cotter pin.
- Rotor
- Caliper
- ABS speed sensor and harness.

Tighten the sensor retaining bolt to 69 inch lbs. (8 Nm).
- Wheel(s)
- Negative battery cable

4. Check the front wheel alignment.

GS 300 and GS 430 (2004)

1. Before servicing the vehicle, refer to the Precautions Section.

2. Raise the vehicle on a hoist, so that front suspension components are hanging and accessible.

3. Remove or disconnect the following:
- Front wheels.

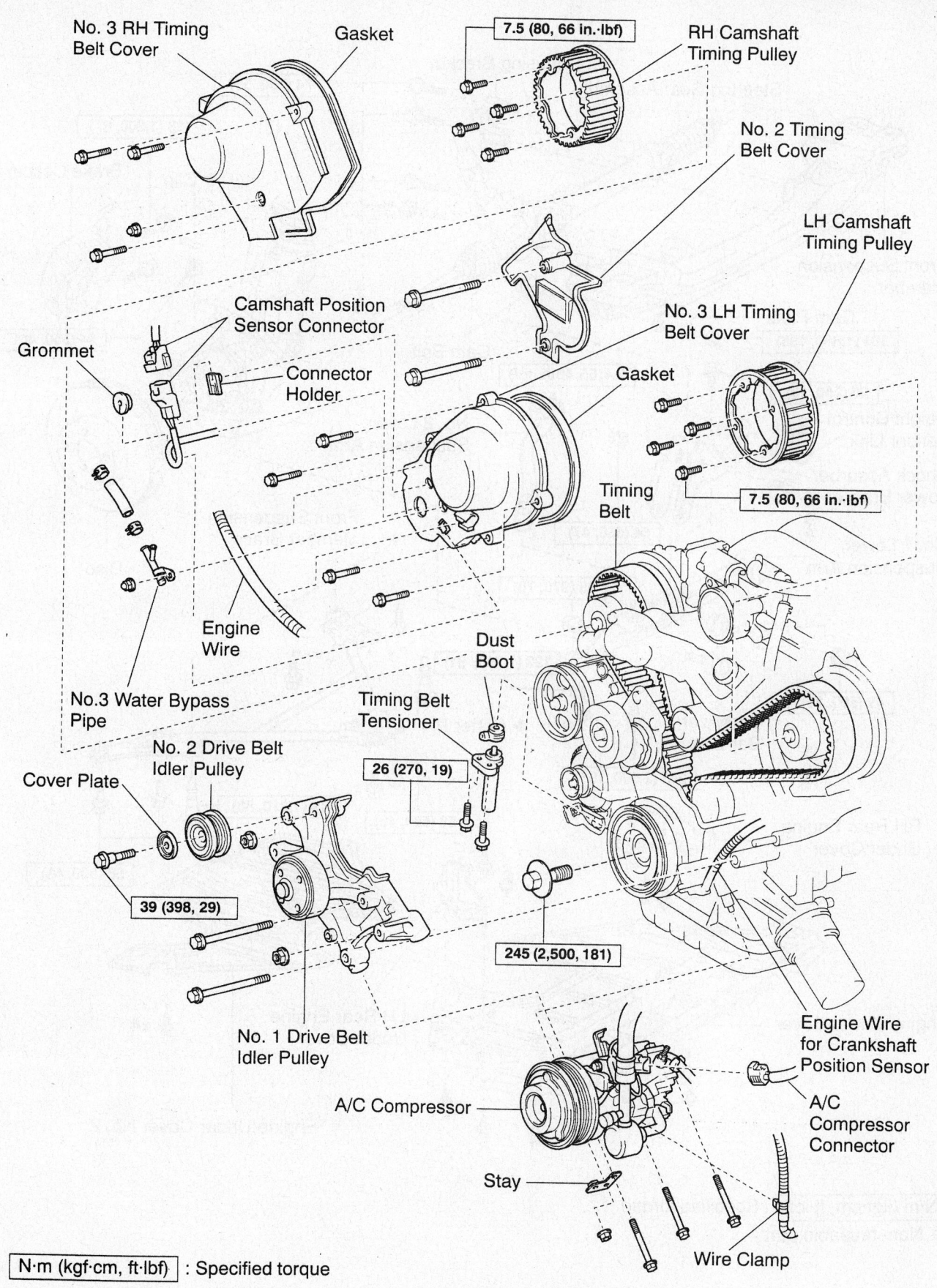

No. 3 RH Timing Belt Cover

Gasket

7.5 (80, 66 in.·lbf)

RH Camshaft Timing Pulley

No. 2 Timing Belt Cover

LH Camshaft Timing Pulley

Camshaft Position Sensor Connector

Connector Holder

No. 3 LH Timing Belt Cover

Gasket

Grommet

Timing Belt

7.5 (80, 66 in.·lbf)

Engine Wire

No.3 Water Bypass Pipe

Dust Boot

Timing Belt Tensioner

26 (270, 19)

Cover Plate

No. 2 Drive Belt Idler Pulley

39 (398, 29)

245 (2,500, 181)

No. 1 Drive Belt Idler Pulley

A/C Compressor

Engine Wire for Crankshaft Position Sensor

A/C Compressor Connector

Stay

Wire Clamp

N·m (kgf·cm, ft·lbf) : Specified torque

Exploded view of the front lower control arm and related components—GS 300 and GS 430 (2004 models)

67162-LEXU-G116

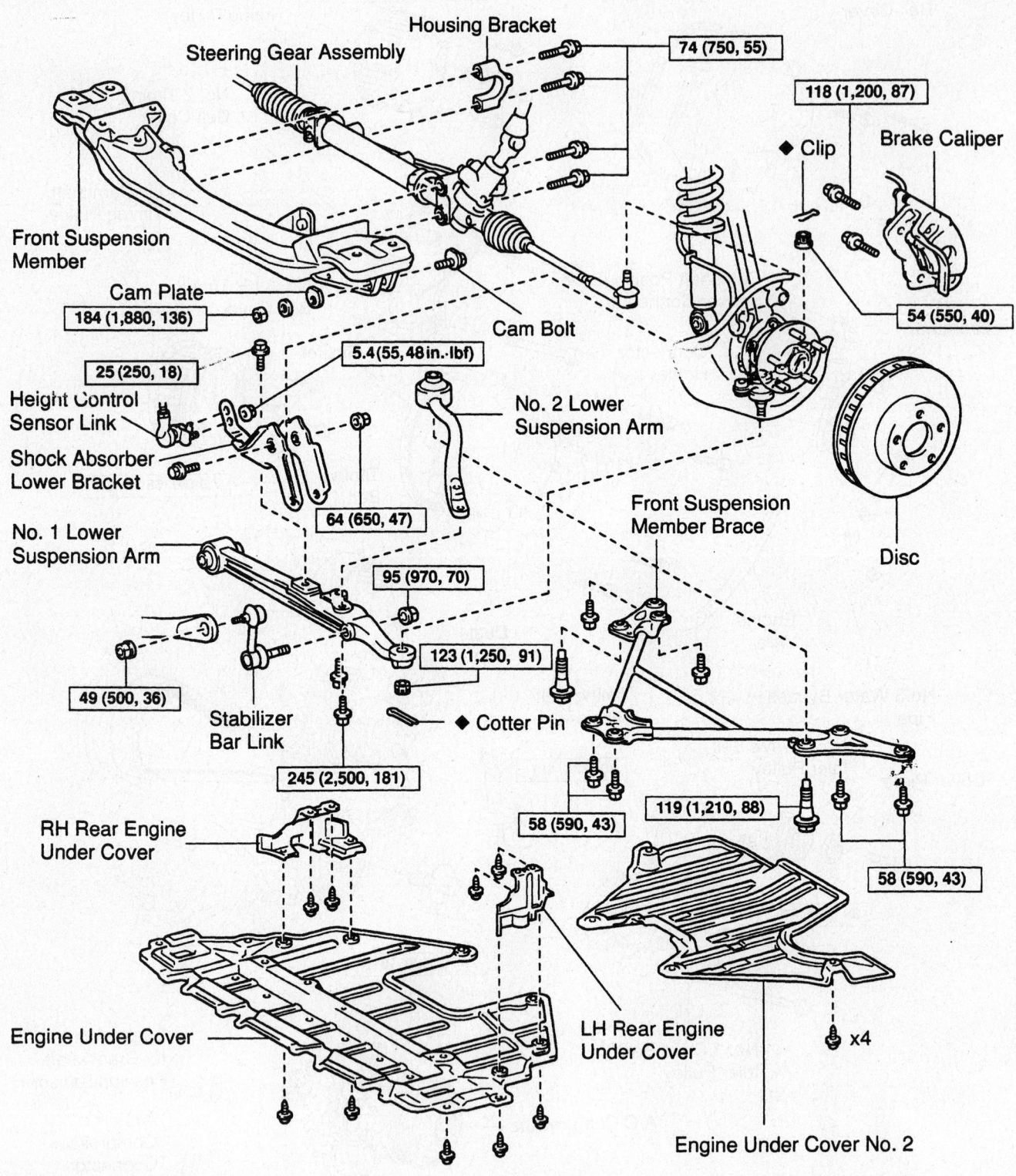

Steering Gear Assembly

Housing Bracket

74 (750, 55)

118 (1,200, 87)

◆ Clip

Brake Caliper

Front Suspension Member

Cam Plate

184 (1,880, 136)

Cam Bolt

54 (550, 40)

25 (250, 18)

5.4 (55, 48 in.·lbf)

No. 2 Lower Suspension Arm

Height Control Sensor Link

Shock Absorber Lower Bracket

64 (650, 47)

Front Suspension Member Brace

Disc

No. 1 Lower Suspension Arm

95 (970, 70)

49 (500, 36)

123 (1,250, 91)

Stabilizer Bar Link

◆ Cotter Pin

119 (1,210, 88)

58 (590, 43)

245 (2,500, 181)

RH Rear Engine Under Cover

58 (590, 43)

LH Rear Engine Under Cover

× 4

Engine Under Cover

Engine Under Cover No. 2

N·m (kgf·cm, ft·lbf) : Specified torque

◆ Non–reusable part

9347LG06

Exploded view of the front suspension—IS 300

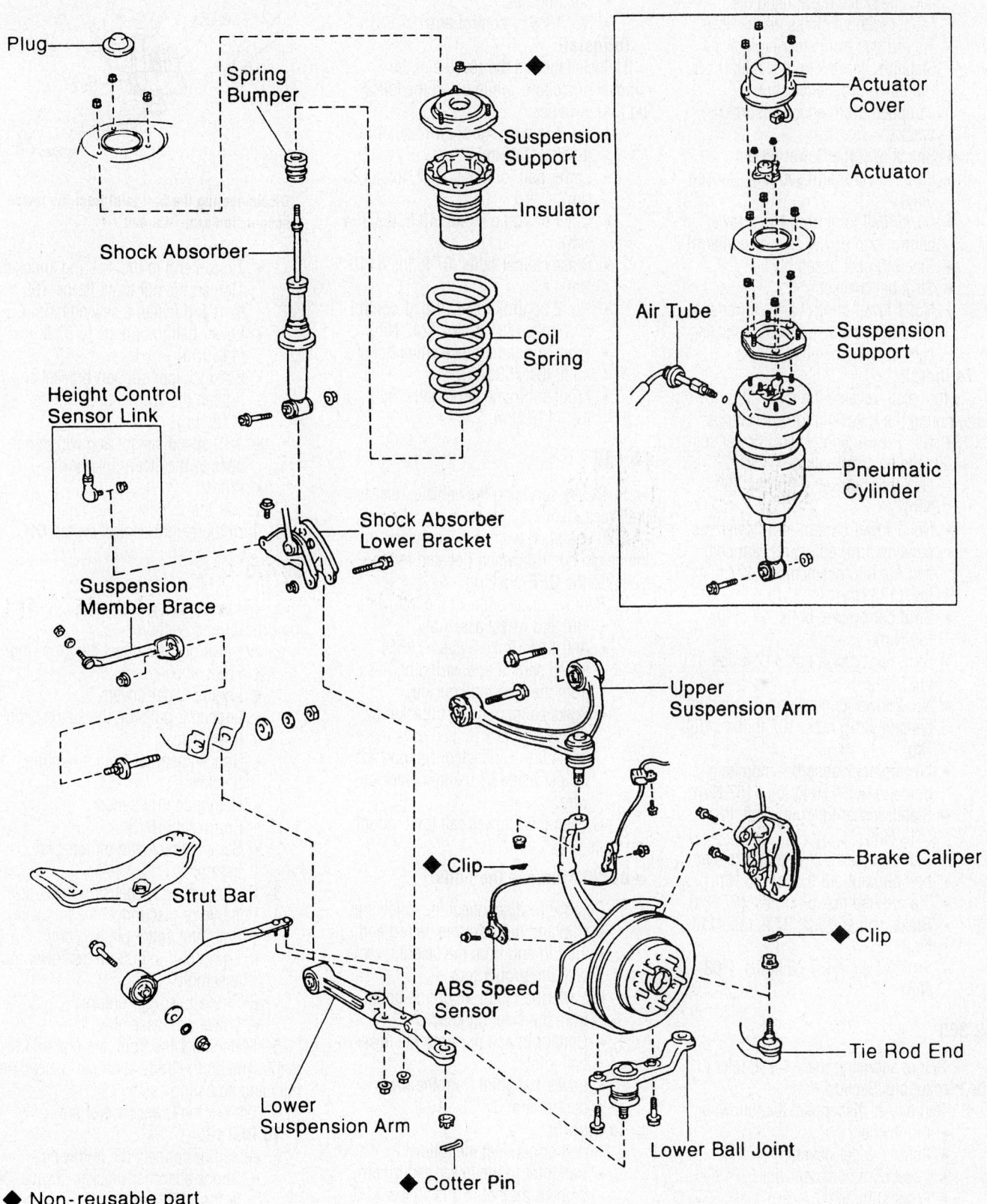

Plug

Spring Bumper

Shock Absorber

Suspension Support

Insulator

Coil Spring

Actuator Cover

Actuator

Air Tube

Suspension Support

Pneumatic Cylinder

Height Control Sensor Link

Shock Absorber Lower Bracket

Suspension Member Brace

Upper Suspension Arm

◆ Clip

Brake Caliper

◆ Clip

Strut Bar

ABS Speed Sensor

Tie Rod End

Lower Suspension Arm

Lower Ball Joint

◆ Cotter Pin

◆ Non-reusable part

7923LGB4

Exploded view of the lower ball joint mounting—LS 430

- Engine under covers.
- Brake caliper(s); Do NOT disconnect the brake hose; hang the caliper without stress on the hose
- Tie rod end from steering knuckle
- Stabilizer bar link from stabilizer bar
- Height control sensor link, if equipped, from shock absorber bracket
- Shock absorber lower mount
- Lower control arm set bolts (loosen only)
- Lower ball joint from No. 2 lower control arm (lower suspension arm)
- Steering gear assembly
- Strut bar bracket
- No. 1 lower control arm (lower suspension arm); matchmark adjusting cam to crossmember

To install:

4. To install, reverse the removal procedure, noting the following torque settings:
- No. 1 lower control arm (lower suspension arm) bolt to shock absorber bracket: 44 ft. lbs. (59 Nm)
- No. 1 lower control arm (lower suspension arm) adjusting cam bolt and nut to crossmember: 127 ft. lbs. (172 Nm)
- Strut bar bracket bolts: 43 ft. lbs. (58 Nm)
- Strut bar bracket nut: 112 ft. lbs. (152 Nm)
- No. 2 lower control arm (lower suspension arm) nuts: 122 ft. lbs. (164 Nm)
- Lower shock absorber mounting bolt and nut: 116 ft. lbs. (157 Nm)
- Stabilizer bar link nut: 83 ft. lbs. (113 Nm)
- Stabilizer bar link-to-stabilizer bar bolt and nut: 43 ft. lbs. (55 Nm)
- Tie rod end nut: 64 ft. lbs. (87 Nm)
- Brake caliper bolts: 87 ft. lbs. (118 Nm)
- Front wheel nuts: 76 ft. lbs. (103 Nm)

IS 300

1. Before servicing the vehicle, refer to the Precautions Section.
2. Remove or disconnect the following:
- Front wheel
- Engine under covers
- Level control sensor link
- Front subframe brace
- No. 2 lower control arm
- Brake caliper and rotor
- Outer tie rod end
- Stabilizer bar link

- Lower strut bolt
- Lower ball joint
- Steering gear
- No. 1 lower control arm

To install:

3. Installation is the reverse of the removal procedure, while using the following torque values:
- No. 1 lower control arm bolt: 136 ft. lbs. (184 Nm)
- Lower ball joint nut: 91 ft. lbs. (123 Nm)
- Outer tie rod end nut: 40 ft. lbs. (54 Nm)
- Brake caliper bolts: 87 ft. lbs. (118 Nm)
- No. 2 control arm-to-No. 1 control arm bolts: 181 ft. lbs. (245 Nm)
- Front subframe brace small bolts: 43 ft. lbs. (58 Nm)
- Front subframe large bolts: 88 ft. lbs. (119 Nm)

LS 430

1. Before servicing the vehicle, refer to the Precautions Section.
2. If equipped with air suspension, move the height control switch (located in the trunk) to the **OFF** position.
3. Remove or disconnect the following:
- Tire and wheel assembly
- Anti-lock Brake System (ABS) speed sensor and wiring harness from the steering knuckle
- Brake caliper support bracket by removing the 2 bolts. Leave the brake line connected. Support the caliper aside by using a piece of wire.
4. Loosen the 2 lower ball joint mounting bolts.

➡**Do NOT remove the bolts.**

5. Remove or disconnect the following:
- Clip and nut from the tie rod end
- Tie rod end from the steering arm with the proper tool
- Lower ball joint mounting bolts from the steering knuckle
- Cotter pin and nut from the lower ball joint
- Lower ball joint from the lower control arm

To install:

6. Install or connect the following:
- Ball joint to the lower control arm. Tighten the nut to 112 ft. lbs. (152 Nm) and install a new cotter pin.
- Mounting bolts, temporarily, holding the ball joint to the steering knuckle

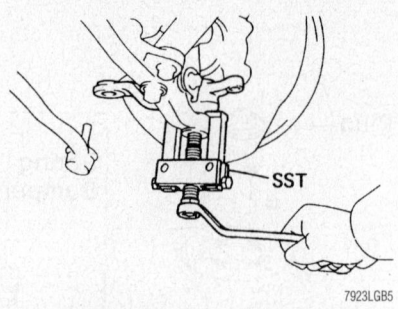

7923LGB5

Disconnecting the ball joint from the lower suspension arm—LS 430

- Tie rod end to the steering knuckle. Tighten the nut to 48 ft. lbs. (65 Nm) and install a new cotter pin.
- Lower ball joint bolts to 83 ft. lbs. (113 Nm)
- Brake caliper support bracket and tighten the 2 bolts to 87 ft. lbs. (118 Nm)
- ABS speed sensor and wiring harness to the steering knuckle
- Wheel

7. Lower the vehicle.
8. Turn the height control switch **ON**.

SC 430

1. Before servicing the vehicle, refer to the Precautions Section.
2. Remove or disconnect the following:
- Front wheel
- Engine under covers
- Loosen 2 bolts on the suspension lower arm
- Shock absorber from mounting bracket
- Height control sensor
- Front stabilizer link
- Separate rack and pinion gear assembly

3. Remove front suspension lower arm
4. Remove or disconnect the following:
- Ball joint cotter pin and bolt
- Lower ball joint from the lower arm assembly
- Shock absorber bracket
- Lower arm assembly

5. Matchmark the front and rear adjustment cams to the body and then remove the nuts and adjusting cams.
6. Lift out the lower control arm.

To install:

7. Install or connect the following:
- Shock absorber bracket. Torque: 44 ft. lbs. (59 Nm)
- Lower control arm to the body and temporarily install the adjusting cams and nuts. Do NOT tighten the nuts at this time.

- Lower control arm to the knuckle and tighten the ball joint nut to 119 ft. lbs. (162 Nm). Install a new cotter pin.
- 2 bolts on lower suspension arm to 121 ft. lbs. (164 Nm)
- Rack and pinion steering gear assembly nuts to 48 ft. lbs. (65 Nm)
- Front stabilizer link bolts and nuts

(to stabilizer bar) to 38 ft. lbs. (51 Nm) and (to stabilizer link) to 116 ft. lbs. (157 Nm)
- Shock absorber assembly bolt to 116 ft. lbs. (157 Nm)
- Height control sensor
- Front wheel
8. Inspect and adjust front wheel alignment
9. Adjust height control sensor

ES 330

➡**Removal of the lower control arm requires the removal of the engine and transaxle.**

1. Before servicing the vehicle, refer to the Precautions Section.
2. Remove or disconnect the following:
- Transverse engine mounting insulator

FRONT SUSPENSION LOWER ARM ASSY:

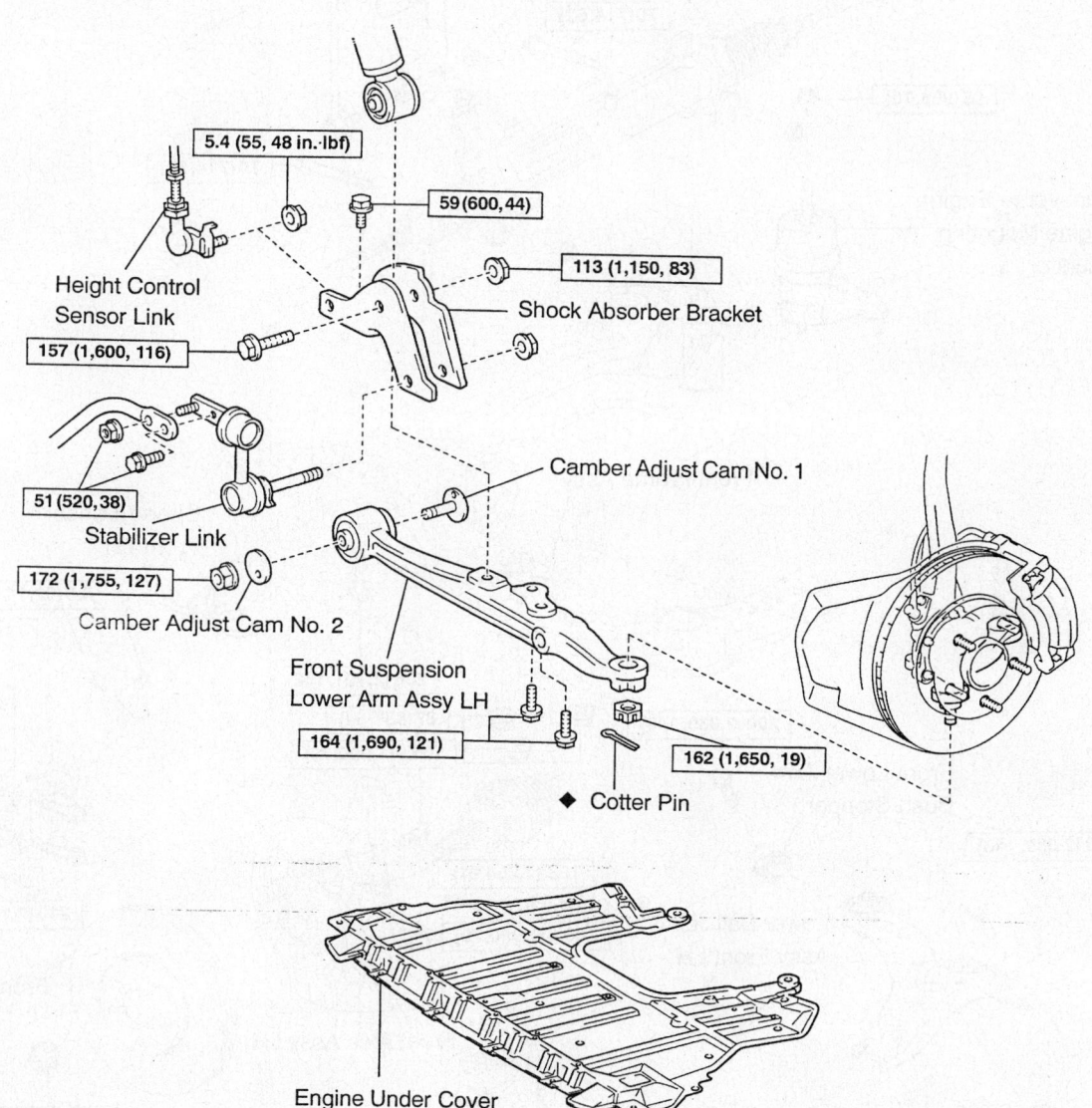

Expanded view, lower front suspension arm—SC 430

N·m (kgf·cm, ft·lbf) : Specified torque
◆ Non-reusable part

67162-LEXU-G85

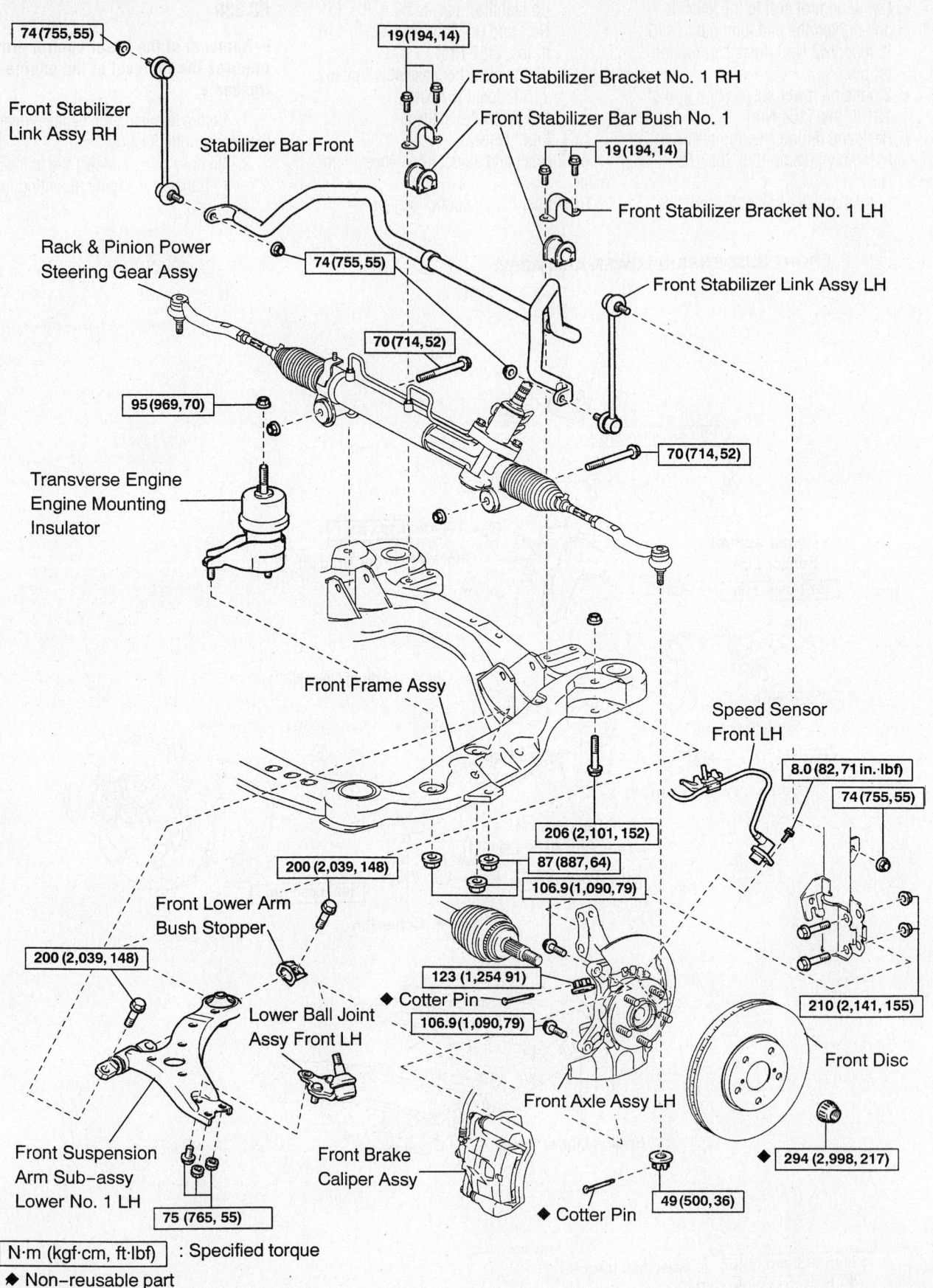

74 (755, 55)

Front Stabilizer
Link Assy RH

19 (194, 14)

Front Stabilizer Bracket No. 1 RH

Stabilizer Bar Front

Front Stabilizer Bar Bush No. 1

19 (194, 14)

Front Stabilizer Bracket No. 1 LH

Rack & Pinion Power
Steering Gear Assy

Front Stabilizer Link Assy LH

70 (714, 52)

95 (969, 70)

70 (714, 52)

Transverse Engine
Engine Mounting
Insulator

Front Frame Assy

Speed Sensor
Front LH

8.0 (82, 71 in.·lbf)

74 (755, 55)

206 (2,101, 152)

200 (2,039, 148)

87 (887, 64)

Front Lower Arm
Bush Stopper

106.9 (1,090, 79)

200 (2,039, 148)

123 (1,254 91)

◆ Cotter Pin

Lower Ball Joint
Assy Front LH

106.9 (1,090, 79)

210 (2,141, 155)

Front Disc

Front Axle Assy LH

Front Suspension
Arm Sub–assy
Lower No. 1 LH

Front Brake
Caliper Assy

75 (765, 55)

294 (2,998, 217)

49 (500, 36)

◆ Cotter Pin

N·m (kgf·cm, ft·lbf) : Specified torque

◆ Non–reusable part

67162-LEXU-G86

Expanded view, front suspension—ES 330

- Engine and transaxle assembly
- 2 bolts on front side of suspension arm
- Bolt and nut on rear side of suspension arm and lower arm
- Lower bush stopper

To install:

3. Install or connect the following:
- Lower bush arm stopper
- 2 bolt on the front side to 148 ft. lbs. (200 Nm)
- Rear side bolt and nut to 152 ft. lbs. (206 Nm)
- Transverse engine mounting insulator to 64 ft. lbs. (87 Nm)
- Engine and transaxle assembly

CONTROL ARM BUSHING REPLACEMENT

The control arm bushings are serviced with the control arm as an assembly.

Wheel Bearings

ADJUSTMENT

Check the backlash in bearing shaft direction and the axle hub deviation. Maximum for backlash should be 0.0020 in. (0.05mm) and for axle hub deviation 0.020 in. (0.05mm).

➡**The front wheel bearings are non-adjustable. If the wheel bearing is out of specifications, replace the wheel bearing.**

REMOVAL & INSTALLATION

IS 250 and IS 350

1. Before servicing the vehicle, refer to the Precautions Section.
2. Remove or disconnect the following:
- Negative battery cable
- Front wheels
- Speed sensor

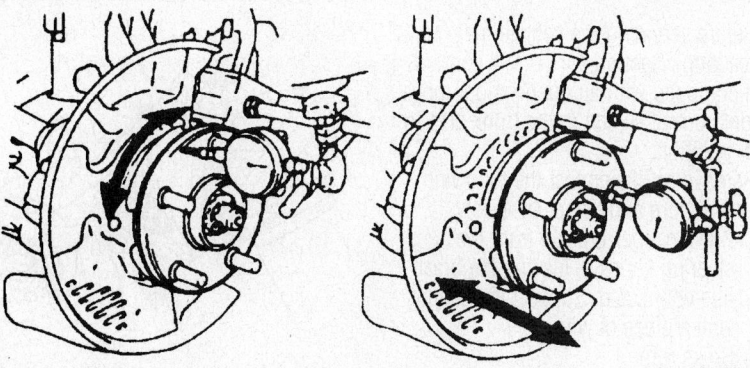

Checking wheel bearings for excessive play

7923LGB6

- Disc brake caliper
- Brake disc
- Front axle hub nut (AWD)
- 4 bolts and front axle hub

➡**On AWD models, use a plastic hammer to tap the hub unit away from the drive shaft.**

To install:

3. Install or connect the following:
- Front axle hub. Torque the mounting bolts to 51 ft. lbs. (69 Nm).
- New front axle hub nut (AWD). Torque to 217 ft. lbs. (294 Nm).
- Brake disc
- Disc brake caliper
4. Stake the front axle hub nut using a hammer and chisel.
- Speed sensor
- Front wheels
- Negative battery cable

ES 300

1. Before servicing the vehicle, refer to the Precautions Section.
2. Remove or disconnect the following:
- Negative battery cable
- Front wheels
- Fender apron seal
- Cotter pin and lock cap from the end of the halfshaft
- Halfshaft locknut.
- Brake caliper and use a wire to support it out of the way

❊❊ WARNING

Never allow the caliper to hang freely from the brake hose.

- Rotor
- Anti-lock Brake System (ABS) speed sensor from the steering knuckle
- Nuts on the lower end of the strut
- Tie rod end from the steering knuckle

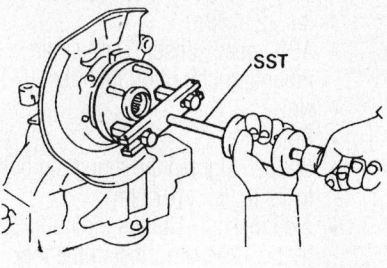

Removing the axle hub from the steering knuckle—ES 300

7923LGB7

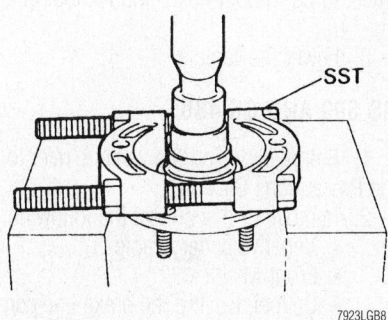

Remove the inner race from the hub—ES 300

7923LGB8

- Lower control arm from the ball joint
- Driveshaft from the axle hub
- 2 nuts on the lower end of the strut
- Steering knuckle
3. Clamp the steering knuckle in a vise with soft jaws to protect the knuckle.
- Dust deflector from the hub
- Ball joint from the steering knuckle
- Hub from the knuckle
- Inner race from the hub
- Dust cover
- Snapring
- Bearing from the steering knuckle

To install:

4. Install or connect the following:
- Bearing into the knuckle
- Snapring
- Dust cover. Tighten the 4 bolts to 74 inch lbs. (8.3 Nm).
- Hub into the steering knuckle
- Lower ball joint to the steering knuckle. Tighten the nut to 90 ft. lbs. (123 Nm) and install a new cotter pin.
- Dust deflector
- Knuckle on the lower strut
- Lower ball joint to the lower arm. Tighten the bolts to 94 ft. lbs. (127 Nm).
- Tie rod end to the steering knuckle. Tighten the nut to 36 ft. lbs. (49 Nm).

- Nuts on the lower strut to 156 ft. lbs. (211 Nm)
- ABS speed sensor. Tighten the mounting bolt to 69 inch lbs. (8 Nm).
- Rotor
- Caliper. Tighten the mounting bolts to 79 ft. lbs. (107 Nm).
- Axle locknut. Tighten the nut to 217 ft. lbs. (294 Nm). Install the lock cap and a new cotter pin.
- Front fender apron seal
- Wheel

5. Turn the wheel by hand, verify that the wheel turns without noise and without binding.
6. Lower the vehicle.

GS 300 AND GS 430

1. Before servicing the vehicle, refer to the Precautions Section.
2. Remove or disconnect the following:
- Negative battery cable
- Front wheel
- Caliper, leaving the brake line connected and suspend it out of the way

✴✴ WARNING

Never allow the brake caliper to hang freely from the brake hose.

- Rotor
- Anti-lock Brake System (ABS) speed sensor and harness
- Tie rod from the arm on the lower ball joint
- Upper suspension arm from the steering knuckle
- Steering knuckle from the lower control arm
- Ball joint from the steering knuckle
- Front hub grease cap

3. Clamp the hub in a soft jaw vise.
4. Using a hammer and chisel, loosen the staked part of the locknut.
5. Remove or disconnect the following:
- Locknut
- ABS speed sensor rotor

➡**Do NOT scratch the serrations of the sensor rotor.**

- Brake dust cover bolts and shift the cover toward the outside.
- Hub from the steering knuckle
- Inner bearing race from the hub shaft
- Oil seal from the knuckle
- Bearing snapring from the steering knuckle
- Bearing from the steering knuckle

To install:
6. Install or connect the following:
- New bearing into the steering knuckle

➡**If the inner race and balls come loose from the bearing outer race, be sure to install them on the same side as before.**

- Snapring
- New outside inner race and tap in the new seal. Tap the seal until it is flush with the end surface of the steering knuckle.
- Brake dust cover to the knuckle and tighten the bolts to 74 inch lbs. (8 Nm)
- Hub into the steering knuckle
- ABS speed sensor rotor
- Axle hub locknut. Tighten the nut to 147 ft. lbs. (199 Nm) and stake it.
- Grease cap to the steering knuckle by tapping lightly around the circumference of the cap with a hammer
- Ball joint to the steering knuckle. Tighten the 2 bolts to 83 ft. lbs. (113 Nm).
- Steering knuckle to the upper and lower suspension arms. Tighten the upper nut to 64 ft. lbs. (87 Nm) and the lower nut to 95 ft. lbs. (127 Nm). Install a new cotter pin on the lower nut. Install the clip on the upper suspension arm nut.
- Tie rod end to the steering knuckle. Tighten the nut to 64 ft. lbs. (87 Nm) and install a new cotter pin.
- Rotor, disc brake pads and the brake caliper
- ABS speed sensor and harness. Tighten the sensor retaining bolt to 69 inch lbs. (8 Nm).
- Wheel

7. Lower the vehicle and connect the negative battery cable.
8. Check the front wheel alignment.

SC 430 and 2002–03 LS 430

1. Before servicing the vehicle, refer to the Precautions Section.
2. If equipped with air suspension, move the height control switch in the trunk area to the **OFF** position.
3. Remove or disconnect the following:
- Front tire and wheel assembly
- Brake caliper bracket from the steering knuckle, leaving the brake line connected. Support the caliper with a piece of wire.
- Brake rotor
- Anti-lock Brake System (ABS) speed sensor from the steering knuckle
- Steering knuckle from the lower ball joint by removing the 2 bolts
- Steering knuckle from the upper ball joint
- Steering knuckle with the axle hub from the vehicle
- Grease cap from the hub
- Nut and the speed sensor rotor
- 4 bolts and shift the brake dust cover towards the hub side
- Axle hub from the steering knuckle
- Outside inner race from the axle
- Oil seal from the steering knuckle
- Snapring and bearing from the steering knuckle

To install:
4. Install or connect the following:
- Bearing in the steering knuckle
- Snapring
- Inner race (outside)
- New oil seal until it is flush with the end surface of the steering knuckle
- Brake dust cover to the steering knuckle and tighten the bolts to 74 inch lbs. (8.4 Nm)
- Axle hub to the steering knuckle
- ABS speed sensor
- New nut on the axle shaft. Tighten the nut to 147 ft. lbs. (199 Nm). Stake the nut and install the grease cap.
- Steering knuckle to the lower ball joint and tighten the bolts to 83 ft. lbs. (113 Nm)
- Steering knuckle to the upper ball joint and tighten the nut to 48 ft. lbs. (65 Nm)
- Brake rotor
- Brake caliper and tighten the 2 bolts to 87 ft. lbs. (118 Nm)
- Speed sensor to the steering knuckle
- Front tire and wheel assembly

5. If equipped with air suspension, turn the height control switch to the **ON** position.

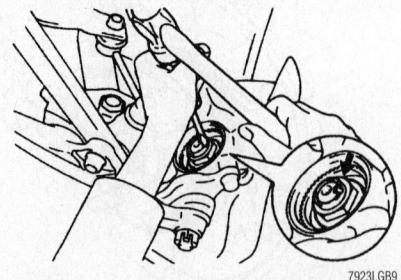

Axle hub nut is located on the inboard side of the knuckle assembly—LS 430

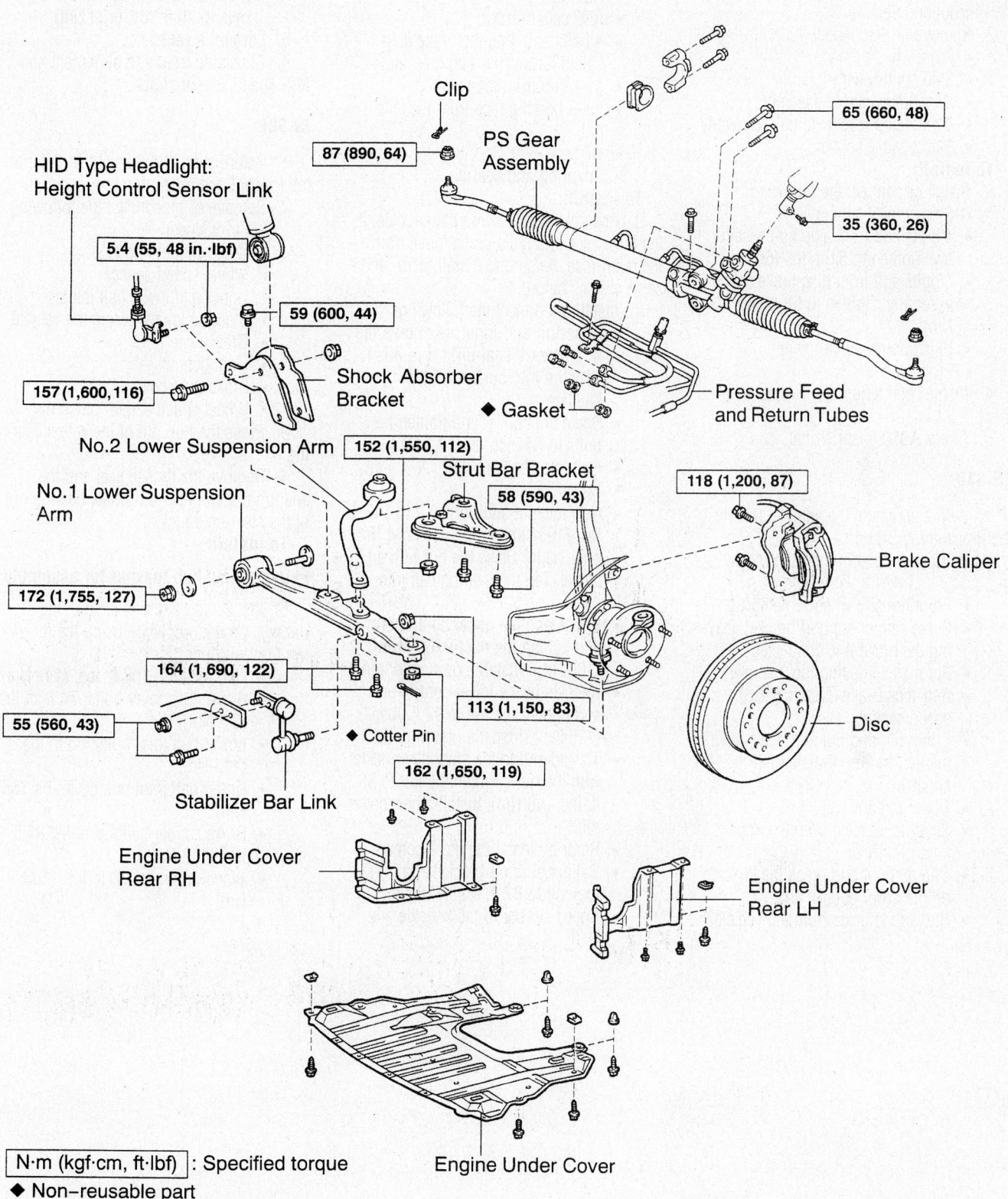

Clip

PS Gear Assembly

65 (660, 48)

87 (890, 64)

HID Type Headlight: Height Control Sensor Link

5.4 (55, 48 in.·lbf)

35 (360, 26)

59 (600, 44)

157 (1,600, 116)

Shock Absorber Bracket

Pressure Feed and Return Tubes

◆ Gasket

No.2 Lower Suspension Arm   152 (1,550, 112)

Strut Bar Bracket

No.1 Lower Suspension Arm

58 (590, 43)

118 (1,200, 87)

Brake Caliper

172 (1,755, 127)

164 (1,690, 122)

113 (1,150, 83)

55 (560, 43)

Disc

◆ Cotter Pin

162 (1,650, 119)

Stabilizer Bar Link

Engine Under Cover Rear RH

Engine Under Cover Rear LH

N·m (kgf·cm, ft·lbf) : Specified torque
◆ Non-reusable part

Engine Under Cover

67162-LEXU-G117

Exploded view front hub and bearing assembly—LS 430 (2004)

2004–06 LS 430

1. Before servicing the vehicle, refer to the Precautions Section.
2. Remove or disconnect the following:
   - Front wheels
   - Skid control wire
   - Disc brake caliper assembly
   - 4 bolts and front axle hub assembly
   - Skid control sensor

To install:

3. Install or connect the following:
   - Skid control sensor
   - 4 bolts and front axle hub assembly. Tighten to 51 ft. lbs (69 Nm)
   - 2 bolts and front disc brake caliper assembly. Tighten to 58 ft. lbs. (78 Nm)
   - Skid control sensor wire
   - Front wheel
4. Inspect and adjust front wheel alignment
5. Check ABS sensor signal

ES 330

1. Before servicing the vehicle, refer to the Precautions Section.
2. Remove or disconnect the following:
   - Front tire and wheel assembly
   - Brake caliper support bracket, leaving the brake line connected
   - Rotor by removing the 2 screws
   - Anti-lock Brake System (ABS) speed sensor
   - Cotter pin and nut and disconnect the tie rod from the steering knuckle
   - Cotter pin and nut
   - Steering knuckle from the upper control arm
   - Clip and nut and press the knuckle off the lower control arm
   - Steering knuckle from the vehicle

- Hub bearing cap from the steering knuckle
- Hub nut
- ABS sensor rotor
- 4 bolts and shift the brake dust shield toward the hub (outside)
- Axle hub from the knuckle
- Inner bearing race from the axle hub
- Oil seal
- Snapring and bearing

To install:

3. Press the bearing into the knuckle. If the inner race and balls come loose from the outer race, be sure to install them on the same side as before.
4. Install or connect the following:
   - Snapring and inner race, then tap in a new oil seal until it is flush with the end surface of the knuckle
   - Brake dust cover and tighten the bolts to 74 inch lbs. (8.3 Nm)
   - Hub into the knuckle
   - Speed sensor
   - New locknut and tighten it to 147 ft. lbs. (199 Nm), 217 ft lbs (294 Nm) on ES 330. Stake the nut with a chisel. Tap the bearing cap into place.
   - Knuckle to the upper control arm and tighten the nut to 76 ft. lbs. (103 Nm). Install a new cotter pin
   - Knuckle to the lower control arm and tighten the nut to 92 ft. lbs. (125 Nm). Install a new clip.
   - Tie rod end to the steering knuckle with the nut. Tighten the nut to 36 ft. lbs. (49 Nm). Install a new cotter pin.
   - Rotor by installing the 2 screws
   - Caliper support bracket and tighten the bolt to 87 ft. lbs. (118 Nm)
   - Speed sensor to the knuckle and

tighten the bolt to 69 inch lbs. (8 Nm)
   - Front wheel and tighten the lug nuts to 76 ft. lbs. (103 Nm)
5. Lower the vehicle.
6. Check the front end alignment and ABS speed sensor signal.

IS 300

1. Before servicing the vehicle, refer to the Precautions Section.
2. Remove or disconnect the following:
   - Front wheel
   - Brake caliper and rotor
   - Wheel speed sensor
   - Upper and lower ball joints
   - Steering knuckle from the vehicle
   - Grease cap
   - Hub locknut
   - Brake dust cover
   - Wheel speed sensor pulse ring
3. Press the hub out of the wheel bearing.
4. Remove the grease seal and the snapring, then press the wheel bearing out of the steering knuckle.

To install:

➡**Use a new hub locknut for assembly.**

5. Installation is the reverse of the removal procedure, while using the following torque values:
   - Hub locknut: 108 ft. lbs. (147 Nm)
   - Brake dust cover bolts: 74 inch lbs. (8.3 Nm)
   - Lower ball joint bolts: 83 ft. lbs. (113 Nm)
   - Upper ball joint nut: 50 ft. lbs. (65 Nm)
   - Brake caliper support bolts: 87 ft. lbs. (118 Nm)
   - Wheel lug nuts: 76 ft. lbs. (103 Nm)

REAR SUSPENSION

Strut and Coil Spring

REMOVAL & INSTALLATION

IS 250 and IS 350

1. Before servicing the vehicle, refer to the Precautions Section.
2. Remove or disconnect the following:
   - No. 2 luggage compartment trim cover
   - Left and right deck side trim box
   - Rear luggage compartment trim cover
   - Front luggage compartment trim cover
   - Left and right side luggage compartment trim covers
   - Rear wheel
   - Rear wheel fender liner
   - No. 2 differential support protector
   - Rear suspension member brace
   - Rear No. 2 lower control arm
   - Mounting bolts and strut assembly

✳✳ CAUTION

Do NOT remove the center nut to the strut at this time. The spring on the strut is under high pressure and can cause serious injury.

3. Secure the strut in a vise.
4. Compress the coil spring
5. Remove or disconnect the following:
   - Upper strut retaining nut
   - Suspension support
   - Upper insulator
   - Bumper
   - Coil spring
   - Insulator

To install:
6. Install or connect the following:
   - Lower insulator
   - Coil spring end into the step of the lower seat
   - Bumper to the piston rod
   - Upper insulator
   - Upper support to the piston rod, aligning it with the groove in the strut rod
   - Strut assembly into vehicle. Tighten the 3 new upper strut retaining nuts to 55 ft. lbs. (74 Nm). Tighten the 2 body bolts to 15 ft. lbs. (21 Nm).
   - Rear No. 2 lower control arm
7. Stabilize the suspension as follows :
   a. Install the wheel(s).
   b. Lower the vehicle and bounce it up and down several times to stabilize the rear suspension.
   c. Raise the vehicle.
   d. Remove the wheel.
   e. Jack up the rear lower suspension armplacing a wood block in between. Apply a load to the suspension so that the lower suspension arm is plkaced in a horizontal position.
8. Fully tighten the No. 2 rear lower control arm as follows:
   a. Torque bolt A to 118 ft. lbs. (161 Nm); bolt B to 80 ft. lbs. (110 Nm); nut C to 20 ft. lbs. (27 Nm).
   b. Using a 19mm ball joint lock nut wrench, tighten to 75 ft. lbs. (102 Nm).
   - Rear suspension member brace. Torque to 37 ft. Lbs. (50 Nm).
   - No. 2 differential support protector
   - Rear wheel fender liner
   - Rear wheel
   - Left and right side luggage compartment trim covers
   - Front luggage compartment trim cover
   - Rear luggage compartment trim cover
   - Left and right deck side trim box
   - No. 2 luggage compartment trim cover
9. Check the front alignment.

ES 300

1. Before servicing the vehicle, refer to the Precautions Section.
2. Remove or disconnect the following:
   - Tire and wheel assembly
   - Load sensing proportioning valve spring assembly from the lower arm
   - Anti-lock Brake System (ABS) speed sensor harness and brake line from the strut assembly
   - Stabilizer bar link from the strut
3. Loosen the 2 nuts attaching the strut to the axle carrier.
4. Support the axle carrier.
5. Remove or disconnect the following:
   - Rear seat back and package tray trim
   - Upper mounting nuts
   - 2 lower mounting bolts and remove the strut assembly
6. Compress the coil spring.
7. Temporarily install a bolt and 2 nuts on the bracket at the lower end of the strut and secure it in a vise.
8. Secure the upper support and remove the strut rod retaining nut.
9. Remove or disconnect the following:
   - Upper suspension support
   - Upper insulator
   - Coil spring
   - Spring bumper
   - Lower insulator
10. If discarding the strut, perform the following:
    a. Fully extend the strut rod.
    b. Drill a hole in the side of the strut to release the gas.

✳✳ WARNING

The gas coming out is harmless, but be careful of chips which may fly up while drilling.

To install:
11. Install or connect the following:
    - Lower insulator to the strut
    - Spring bumper to the strut piston rod
    - Compressed coil spring
    - Coil spring with the end butted against the gap in the lower seat
    - Upper insulator and support matching the bolt of the support with the cut-off part of the insulator

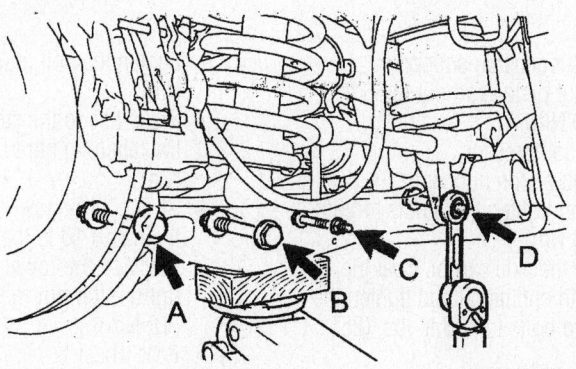

09490_LEXU_G0114

Rear No. 2 lower control arm nut and bolt locations—IS 250 and IS 350

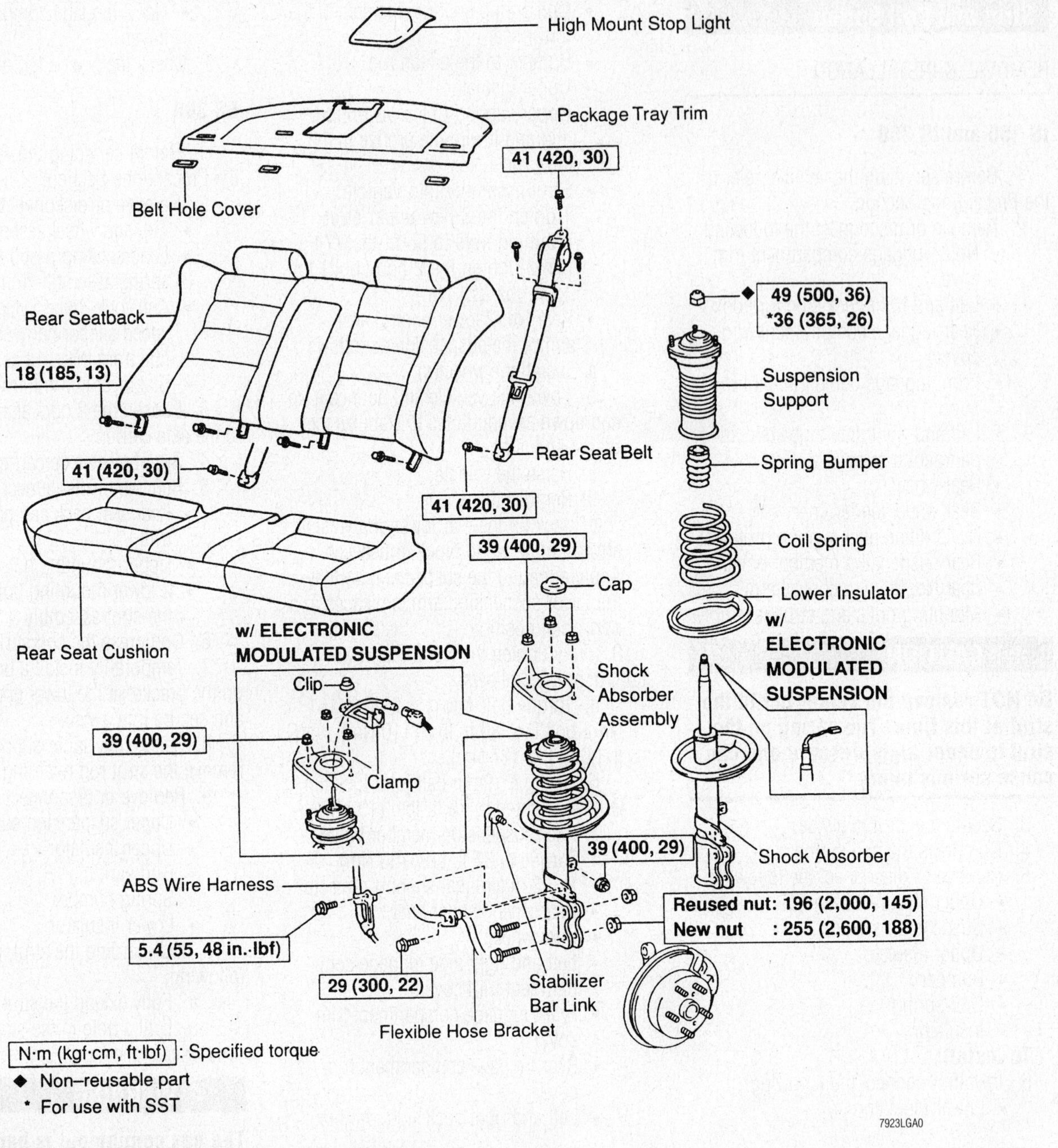

High Mount Stop Light

Package Tray Trim

Belt Hole Cover

41 (420, 30)

Rear Seatback

18 (185, 13)

49 (500, 36)
*36 (365, 26)

Suspension Support

Spring Bumper

Rear Seat Belt

41 (420, 30)

41 (420, 30)

Coil Spring

Lower Insulator

39 (400, 29)

Cap

w/ ELECTRONIC MODULATED SUSPENSION

w/ ELECTRONIC MODULATED SUSPENSION

Rear Seat Cushion

Clip

39 (400, 29)

Clamp

Shock Absorber Assembly

Shock Absorber

ABS Wire Harness

39 (400, 29)

5.4 (55, 48 in.·lbf)

Reused nut: 196 (2,000, 145)
New nut   : 255 (2,600, 188)

29 (300, 22)

Stabilizer Bar Link

Flexible Hose Bracket

N·m (kgf·cm, ft·lbf) : Specified torque

◆ Non–reusable part

* For use with SST

7923LGA0

Exploded view of the rear strut and coil spring mounting—ES 300

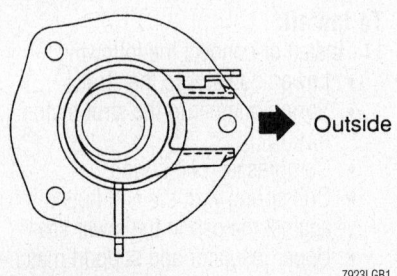

7923LGB1

Position the upper suspension support as shown when assembling the strut—ES 300

Outside

- Upper suspension support
- New strut piston rod nut to 36 ft. lbs. (49 Nm)
- Spring compressor
- Strut rod piston nut cap
- Strut and tighten the 3 nuts to 29 ft. lbs. (39 Nm)
- Strut to the axle carrier. Coat the nuts with engine oil and tighten the nuts and bolts to 188 ft. lbs. (255 Nm)
- ABS harness to the strut and

- tighten the bolt to 48 inch lbs. (6 Nm)
- Brake line to the strut and tighten the retaining nut to 22 ft. lbs. (29 Nm)
- Spring to the lower arm and tighten the nut to 10 ft. lbs. (13 Nm)
- LSPV to the lower arm and tighten the nut to 108 inch lbs. (12 Nm)
- Rear wheel
- Rear seat and package tray

GS 300 and GS 430

1. Before servicing the vehicle, refer to the Precautions Section.

2. Remove the corresponding rear wheel(s).

3. Remove the luggage compartment trim front cover.

4. Remove the rear fender apron seal.

5. Remove the rear lower suspension arm:

    a. Remove 2 bolts, nuts and the No. 1 lower suspension arm (trailing arm).

    b. Remove the bolt and nuts and disconnect the stabilizer bar link (and height control link, if equipped) from the No. 2 lower suspension arm.

    c. Remove the bolt and nut and disconnect the strut from the No. 2 lower suspension arm.

    d. Place matchmarks on the adjusting cam and on the No. 2 lower suspension arm.

    e. Remove the nut and adjusting cams.

    f. Remove the bolt, nut and washer and the No. 2 lower suspension arm.

6. Remove the 3 upper retaining nuts. Loosen, but Do NOT remove the center stud nut.

7. Remove the lower strut mounting bolt.

8. Remove the strut assembly, with the coil spring.

9. Compress the coil spring in a suitable spring compressor.

➡**Do NOT remove these items with an impact wrench; it will damage the spring compressor.**

10. Remove or disconnect the following:

- Suspension support nut
- Washer
- 2 cushions
- Collar
- Suspension support
- Upper insulator
- Lower cup
- Spring bumper

11. Carefully release the spring compressor and remove the coil spring.

12. If the shock absorber is being replaced, fully extend the shock absorber rod and drill a hole to discharge the gas from the cylinder. Drill this hole about 5-8 in. (130-185mm) from lower mounting bolt hole.

To install:

13. With a new shock absorber in place, install the suspension support and compress the coil spring.

14. With a non-impact wrench, install the coil spring to the shock absorber, fitting the lower end of the spring into the recent of the spring seat on the shock absorber.

15. Install the spring bumper, lower cup, cushion, collar, upper insulator, suspension support, cushion and washer onto the shock absorber. Temporarily tighten a new nut.

16. Rotate the suspension support so the rod and one of the bolts on the suspension support are aligned with the lower shock mounting hole such that while the lower mounting bolt hole is in proper position, the 3 studs on the top of the strut assembly will align with the holes in the body.

17. Carefully remove the spring compressor.

18. Install the strut assembly into the vehicle. Torque the 2 bolts on top of the coil spring to 13 ft. lbs. (18 Nm), the 3 nuts on the top mounting studs to 47 ft. lbs. (64 Nm), and the nut in the center of the upper strut mounting to 20 ft. lbs. (27 Nm).

19. Install the bolt, nut and washer and the No. 2 lower suspension arm. Torque bolt to 81 ft. lbs. (110 Nm).

20. Position both adjusting cams, referencing the matchmarks made during removal. Install and torque the retaining nuts to 81 ft. lbs. (110 Nm).

21. Install the strut assembly lower mounting bolt to the No. 2 lower suspension arm. Torque the nut to 81 ft. lbs. (110 Nm).

22. Install the stabilizer bar link (and height control link, if equipped) to the No. 2 lower suspension arm. Torque the bolt and nut to 22 ft. lbs. (30 Nm).

23. Install the No. 1 lower suspension arm (trailing arm) and torque the bolts and nuts to 55 ft. lbs. (75 Nm).

24. Install the rear fender apron seal.

25. Install the luggage compartment trim front cover.

26. Install the rear wheel(s). Torque the wheel nuts to 76 ft. lbs. (103 Nm).

LS 430—Without Air Suspension

1. Before servicing the vehicle, refer to the Precautions Section.

2. Remove or disconnect the following:

- Rear seat cushion and seat back.
- Tray trim
- Tire and wheel assembly
- Rear halfshaft
- Stabilizer bar link from the stabilizer bar
- Anti-lock Brake System (ABS) speed sensor and wiring harness
- Brake caliper bracket from the axle carrier, leaving the brake line connected. Suspend the brake caliper aside with a piece of wire.
- Nut on the lower side of the strut. Do NOT remove the bolt.
- Rear axle assembly with a lifting device
- Strut cap by removing the 3 nuts
- 3 mounting nuts holding the strut assembly to the strut tower. Do NOT remove the center bolt.

✳✳ CAUTION

Do NOT remove the center nut to the strut at this time.

- Bolt on the lower side of the strut assembly
- Strut assembly with the coil spring

3. Compress the coil spring.

4. Secure the strut housing in a vise.

5. Remove or disconnect the following:

- Strut rod retaining nut
- Upper suspension support
- Upper insulator
- Coil spring
- Spring bumper
- Lower insulator

6. If discarding the strut, perform the following:

    a. Fully extend the strut rod.

    b. Drill a hole in the strut (about 1 in. above the strut lower mount) and drain the gas inside

✳✳ CAUTION

The gas coming out is harmless, but be careful of chips which may fly up while drilling.

To install:

7. Install or connect the following:

- Lower insulator to the strut.
- Spring bumper to the strut piston rod
- Coil spring
- Upper insulator and support
- Upper suspension support

8. Temporarily install the upper strut rod retaining nut.

9. Rotate the suspension support so that the rod and one of the bolts on the suspension support are aligned with the lower bushing.

10. Remove the spring compressor.

11. Install or connect the following:

- Strut assembly to the vehicle and tighten the 3 nuts to 47 ft. lbs. (64 Nm). Tighten the strut rod retaining nut to 20 ft. lbs. (27 Nm).
- Strut assembly cap and install the 3 nuts

- Strut to the rear axle carrier. Install the bolt from the rear of the vehicle and temporarily tighten the nut.
- Brake caliper and tighten the mounting bolts to 77 ft. lbs. (104 Nm)
- ABS speed sensor and wiring harness
- Stabilizer link to the stabilizer bar and tighten the nut to 48 ft. lbs. (65 Nm)
- Rear halfshaft
- Tire and wheel assembly

12. Bounce the vehicle up and down to stabilize the suspension.

13. Support the rear axle assembly with a lifting device. Tighten the lower strut bolt to 101 ft. lbs. (137 Nm).

14. Install or connect the following:
- Rear seat cushion and rear seat back.
- Package tray trim

15. Check the wheel alignment.

2002–03 LS 430—With Air Suspension

1. Before servicing the vehicle, refer to the Precautions Section.

2. Bleed the air system from the suspension.

3. Remove or disconnect the following:
- Rear seat cushion and seat back
- Package tray trim
- Trunk trim panel. Move the height control switch, located in the trunk area, to the **OFF** position.
- Tire and wheel assembly
- Rear halfshaft
- Stabilizer links from the stabilizer bar
- Anti-lock Brake System (ABS) speed sensor and wiring harness
- Brake caliper bracket from the rear axle carrier. Do NOT disconnect the brake line.

4. Place matchmarks on the height control sensor link and bracket. Disconnect the height control sensor link from the No. 1 lower control arm.

5. Support the rear axle assembly with a lifting device.

6. Remove or disconnect the following:
- Nut on the lower side of the shock absorber. Do NOT remove the bolt.
- Grommet and disconnect the air tube from the shock absorber
- Actuator cover from the strut tower by removing the 3 nuts
- Actuator electrical connector from the top of the strut

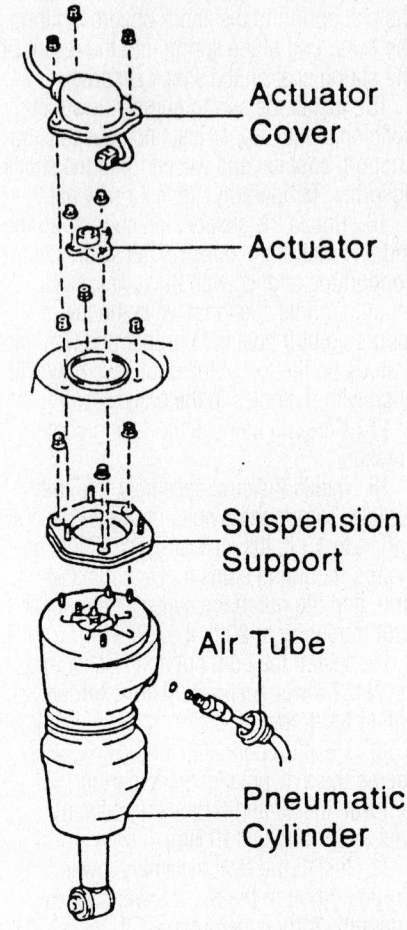

Pneumatic cylinder (strut) component overview (air suspension)

- Actuator by removing the 2 nuts
- 3 upper mounting nuts holding the strut to the strut tower

7. Lower the rear axle assembly

8. Remove or disconnect the following:
- Bolt on the lower side of the shock absorber
- Pneumatic cylinder strut assembly from the vehicle
- Suspension support from the strut assembly by removing the 3 nuts

9. If discarding the pneumatic cylinder, perform the following:
   a. Using a screwdriver, depressurize the air from inside the cylinder.
   b. Drill a hole in the shaded area shown in the illustration and remove the gas inside.

✳✳ CAUTION

The gas coming out is harmless, but be careful of chips which may fly up when drilling.

To install:

10. Install or connect the following:
- Suspension support to the pneumatic cylinder (strut) and tighten the nuts to 27 ft. lbs. (36 Nm)
- Strut assembly to the vehicle and tighten the upper mounting nuts to 47 ft. lbs. (64 Nm)

11. Match the holes in the pneumatic cylinder with the holes in the suspension control actuator.

12. Install or connect the following:
- Actuator and tighten the mounting nuts to 69 inch lbs. (8 Nm)
- Actuator cover and tighten the 3 nuts to 18 ft. lbs. (25 Nm)
- New O-rings and connect the air line to the shock absorber. Tighten the fitting to 13 ft. lbs. (18 Nm).
- Strut to the rear axle carrier. Insert the bolt from the vehicle's rear and temporarily tighten the nut.
- Height control sensor link to the No. 1 lower control arm. Mounting nut: 48 inch lbs. (5 Nm).
- Rear brake caliper to the rear axle carrier and tighten the mounting bolts to 77 ft. lbs. (104 Nm)
- ABS speed sensor and wiring harness
- Stabilizer bar link and tighten the nut to 48 ft. lbs. (65 Nm)
- Halfshaft
- Actuator electrical connector to the top of the strut
- Tire

13. Move the height control switch to the **ON** position. Start the engine and fill the pneumatic cylinder with air.

14. Bounce the vehicle up and down several times to stabilize the suspension.

15. Turn the suspension height control to the **OFF** position.

16. Remove the tire and wheel assembly.

17. Support the rear axle carrier with a lifting device. Tighten the lower strut bolt to 101 ft. lbs. (137 Nm).

18. Install or connect the following:
- Package tray trim
- Rear seat cushion and seat back

19. Turn the suspension control switch to the **ON** position.

20. Check the wheel alignment.

2004–06 LS 430—With Air Suspension

1. Before servicing the vehicle, refer to the Precautions Section.

2. Bleed the air system from the suspension.

3. Remove or disconnect the following:

Coil Spring:

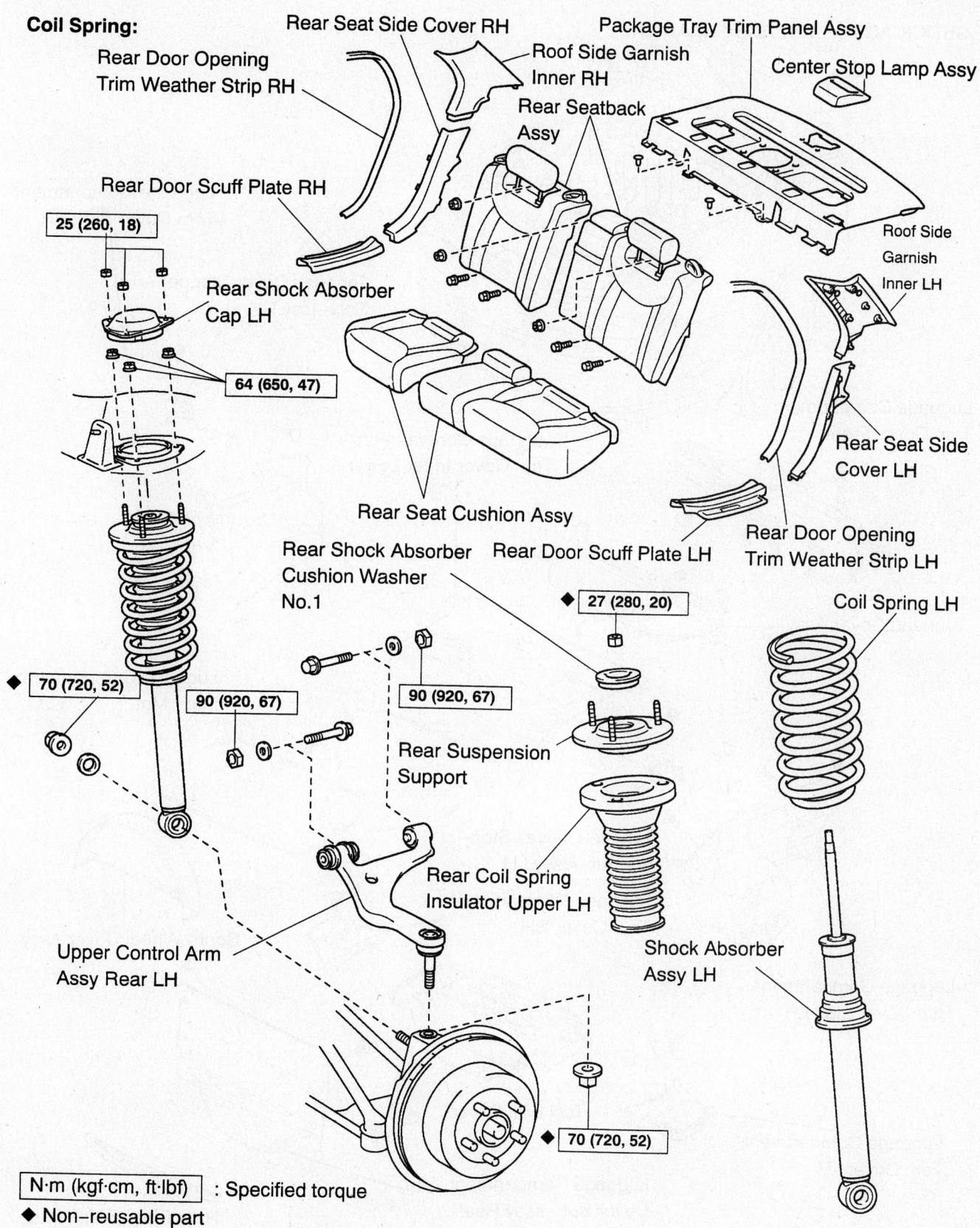

Rear Seat Side Cover RH

Package Tray Trim Panel Assy

Rear Door Opening Trim Weather Strip RH

Roof Side Garnish Inner RH

Center Stop Lamp Assy

Rear Seatback Assy

Rear Door Scuff Plate RH

25 (260, 18)

Rear Shock Absorber Cap LH

64 (650, 47)

Roof Side Garnish Inner LH

Rear Seat Side Cover LH

Rear Seat Cushion Assy

Rear Door Opening Trim Weather Strip LH

Rear Shock Absorber Cushion Washer No.1

Rear Door Scuff Plate LH

◆ 27 (280, 20)

Coil Spring LH

◆ 70 (720, 52)

90 (920, 67)

90 (920, 67)

Rear Suspension Support

Rear Coil Spring Insulator Upper LH

Upper Control Arm Assy Rear LH

Shock Absorber Assy LH

N·m (kgf·cm, ft·lbf) : Specified torque

◆ Non-reusable part

◆ 70 (720, 52)

67172-LEXU-G90

Expanded view rear suspension—LS 430

SHOCK ABSORBER ASSY REAR LH:

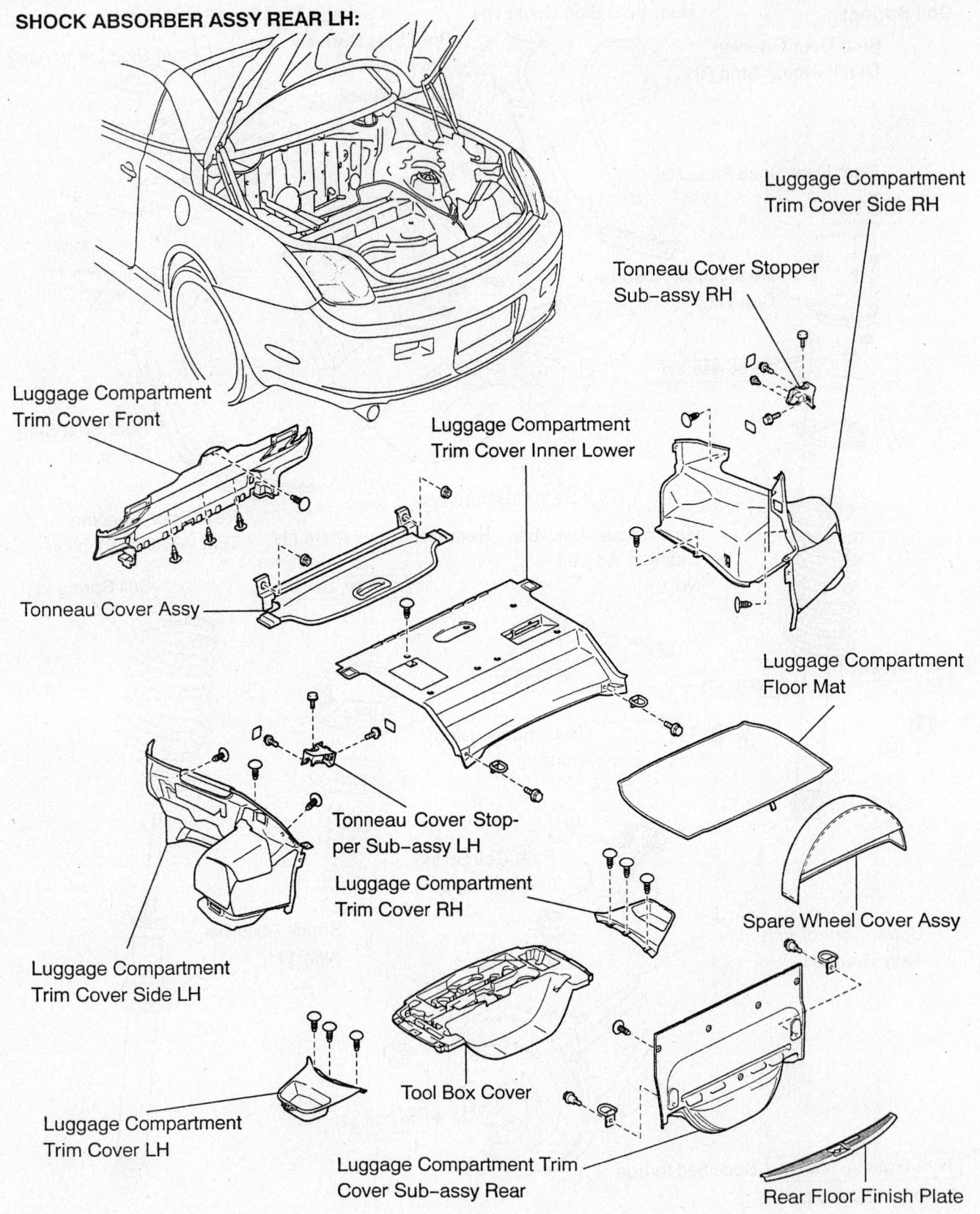

Luggage Compartment Trim Cover Side RH

Tonneau Cover Stopper Sub–assy RH

Luggage Compartment Trim Cover Front

Luggage Compartment Trim Cover Inner Lower

Tonneau Cover Assy

Luggage Compartment Floor Mat

Luggage Compartment Trim Cover Side LH

Tonneau Cover Stopper Sub–assy LH

Luggage Compartment Trim Cover RH

Spare Wheel Cover Assy

Luggage Compartment Trim Cover LH

Tool Box Cover

Luggage Compartment Trim Cover Sub–assy Rear

Rear Floor Finish Plate

67162-LEXU-G79

Exploded view of trunk access for rear struts—SC 430

SHOCK ABSORBER ASSY REAR LH:

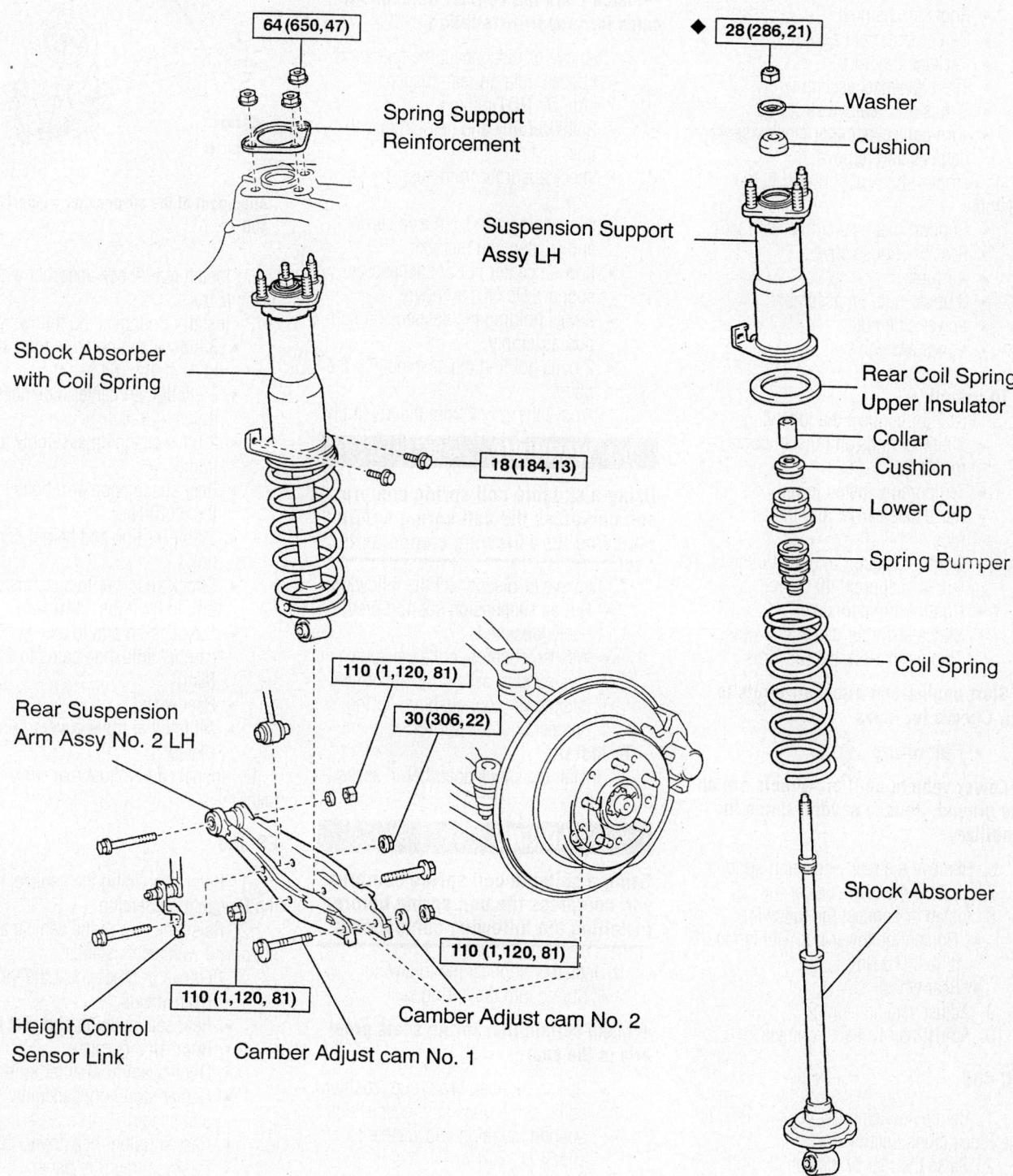

64 (650, 47)

28 (286, 21)

Spring Support
Reinforcement

Washer

Cushion

Suspension Support
Assy LH

Shock Absorber
with Coil Spring

18 (184, 13)

Rear Coil Spring
Upper Insulator

Collar

Cushion

Lower Cup

Spring Bumper

Coil Spring

110 (1,120, 81)

Rear Suspension
Arm Assy No. 2 LH

30 (306, 22)

Shock Absorber

110 (1,120, 81)

110 (1,120, 81)

Height Control
Sensor Link

Camber Adjust cam No. 2

Camber Adjust cam No. 1

N·m (kgf·cm, ft·lbf) : Specified torque
◆ Non-reusable part

Exploded view of rear strut assembly—SC 430

67162-LEXU-G80

- Rear seat cushion and seat back
- Scuff plates
- Door opening weather strip
- Rear seat side covers
- Roof side garnish
- Center stop light assembly
- Package tray trim
- Tire and wheel assembly
- 3 nuts and remove strut cap
- Turn control actuator clockwise 40 degrees and remove

4. Remove shock absorber cylinder assembly

- Support rear axle carrier with jack
- Rear tire house cover
- Air tube
- 3 upper nuts on assembly
- Lower side nut
- Lower assembly
- O-rings

To install:

5. Install or connect the following:

- 2 new O-rings and install connector No 2
- Temporarily install lower side nut
- Raise axle carrier and install air tube
- Control actuator until response is felt, turn approx. 40 degrees
- Control arm connector
- Shock absorber cap with 3 nuts. Tighten to 18 ft. lbs (25 Nm)

➡**Start engine and allow air struts to fill. Checks for leaks**

- Rear wheels

➡**Lower vehicle until all wheels are on the ground. Jounce several times to stabilize.**

6. Remove the rear wheel and support
7. Jack up the rear axle carrier
8. Install or connect the following:

- Tighten the lower nut. Tighten to 52 ft. lbs. (70 Nm)
- Rear wheel

9. Adjust vehicle height
10. Adjust rear wheel alignment.

SC 430

1. Before servicing the vehicle, refer to the Precautions Section.
2. Raise the rear of the vehicle and support it with safety stands.
3. Remove or disconnect the following:

- Tonneau stoppers
- Rear floor finish plate
- Luggage compartment floor mat
- Luggage compartment trim covers
- Spare wheel cover
- Tool box cover

- Rear wheels

4. Remove rear suspension arm assembly

➡**Match mark the camber adjustment cams for proper installation**

5. Remove or disconnect the following:

- Loosen nuts on rear suspension arm. Do NOT remove
- Stabilizer link and height control link
- shock absorber from suspension arm
- suspension arm from axle carrier and suspension member
- Loosen center nut of suspension support. Do NOT remove.
- 3 nuts holding the suspension support assembly
- 2 bolts holding the assembly to the frame

6. Remove the shock from the assembly

✳✳ WARNING

Using a suitable coil spring compressor, compress the coil spring before removing the following components

7. Remove or disconnect the following:

- Nut on suspension support previously loosened
- Washer, cushion, collar, suspension support assembly, upper insulator, lower cup, cushion, spring bumper and coil spring.

To install:

8. Install the shock absorber in assembly.

✳✳ WARNING

Using a suitable coil spring compressor, compress the coil spring before installing the following components

9. Install or connect the following:

- Spring into the spring seat

➡**Make certain that spring seats properly in the seat**

- Spring bumper, lower cup, cushion, collar, upper insulator, suspension support, cushion and washer to shock
- Temporarily install center nut

➡**Rotate the suspension support so that the rod and 1 of the bolts on the suspension support are aligned with the lower shock absorber.**

10. Release and remove the coil spring compressor

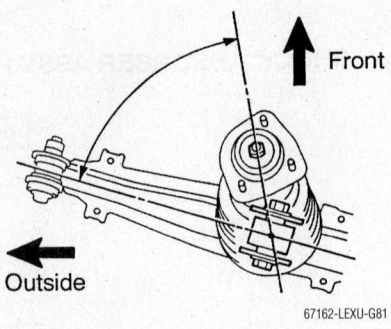

Alignment of the suspension support—SC 430

11. Install rear shock absorber with spring (Strut)
12. Install or connect the following:

- 3 nuts attaching the spring support to 47 ft. lbs (64 Nm)
- Fully tighten center lock nut to 13 ft. lbs. (18 Nm)
- 2 bolts attaching assembly to the frame
- Rear suspension arm bolts to 81 ft. lbs. (110 Nm)
- Stabilizer link and height control link
- Shock absorber to suspension arm bolt to 81 ft lbs. (110 Nm)
- Suspension arm to axle carrier with camber adjusting cams to 81 ft lbs. (Nm)
- Rear wheel
- All internal trunk covers, trims and plates

13. Inspect and adjust rear wheel alignment

ES 330

1. Before servicing the vehicle, refer to the Precautions Section.
2. Raise the rear of the vehicle and support it with safety stands.
3. Remove or disconnect the following:

- Rear wheels
- Rear seat cushion and seat back
- Rood side garnish
- Door opening weather strip trim
- Center stop light assembly (w/o sun shade)
- Rear shoulder belt cover
- Package tray trim panel
- Rear seat 3 point seat belt outer assembly
- Rear stabilizer bar link
- Rear suspension support No. 1 cover
- 2 bolts for brake flex hose and ABS wire harness
- Loosen 2 bolts on lower shock absorber

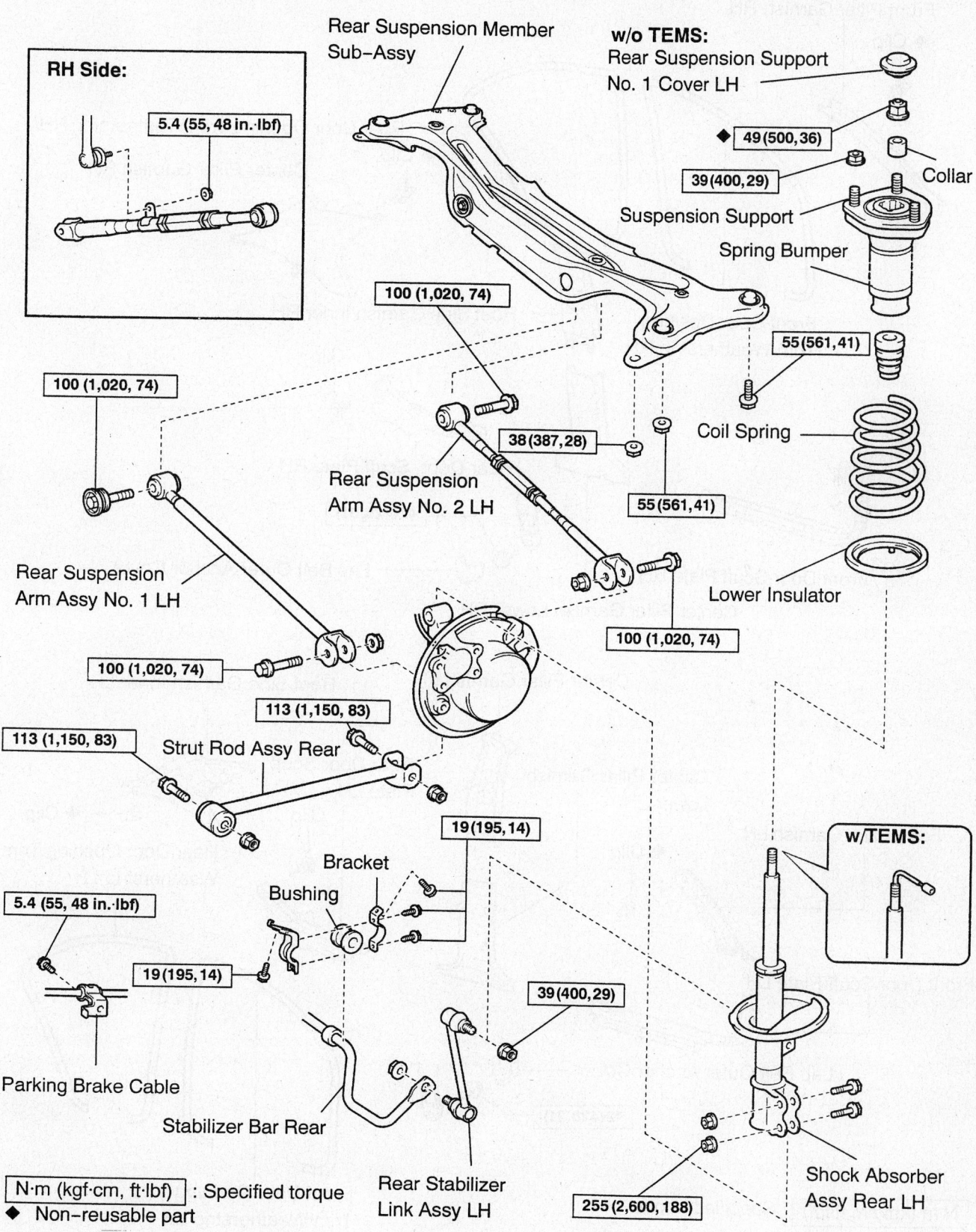

RH Side:

5.4 (55, 48 in.·lbf)

Rear Suspension Member Sub–Assy

w/o TEMS:
Rear Suspension Support No. 1 Cover LH

◆ 49 (500, 36)

39 (400, 29)

Collar

Suspension Support

Spring Bumper

55 (561, 41)

100 (1,020, 74)

38 (387, 28)

Rear Suspension Arm Assy No. 2 LH

Coil Spring

100 (1,020, 74)

Rear Suspension Arm Assy No. 1 LH

55 (561, 41)

Lower Insulator

100 (1,020, 74)

100 (1,020, 74)

113 (1,150, 83)

113 (1,150, 83)

Strut Rod Assy Rear

19 (195, 14)

w/TEMS:

5.4 (55, 48 in.·lbf)

Bracket

Bushing

19 (195, 14)

39 (400, 29)

Parking Brake Cable

Stabilizer Bar Rear

Rear Stabilizer Link Assy LH

Shock Absorber Assy Rear LH

N·m (kgf·cm, ft·lbf) : Specified torque
◆ Non–reusable part

255 (2,600, 188)

67162-LEXU-G82

Exploded view, rear suspension—ES 330

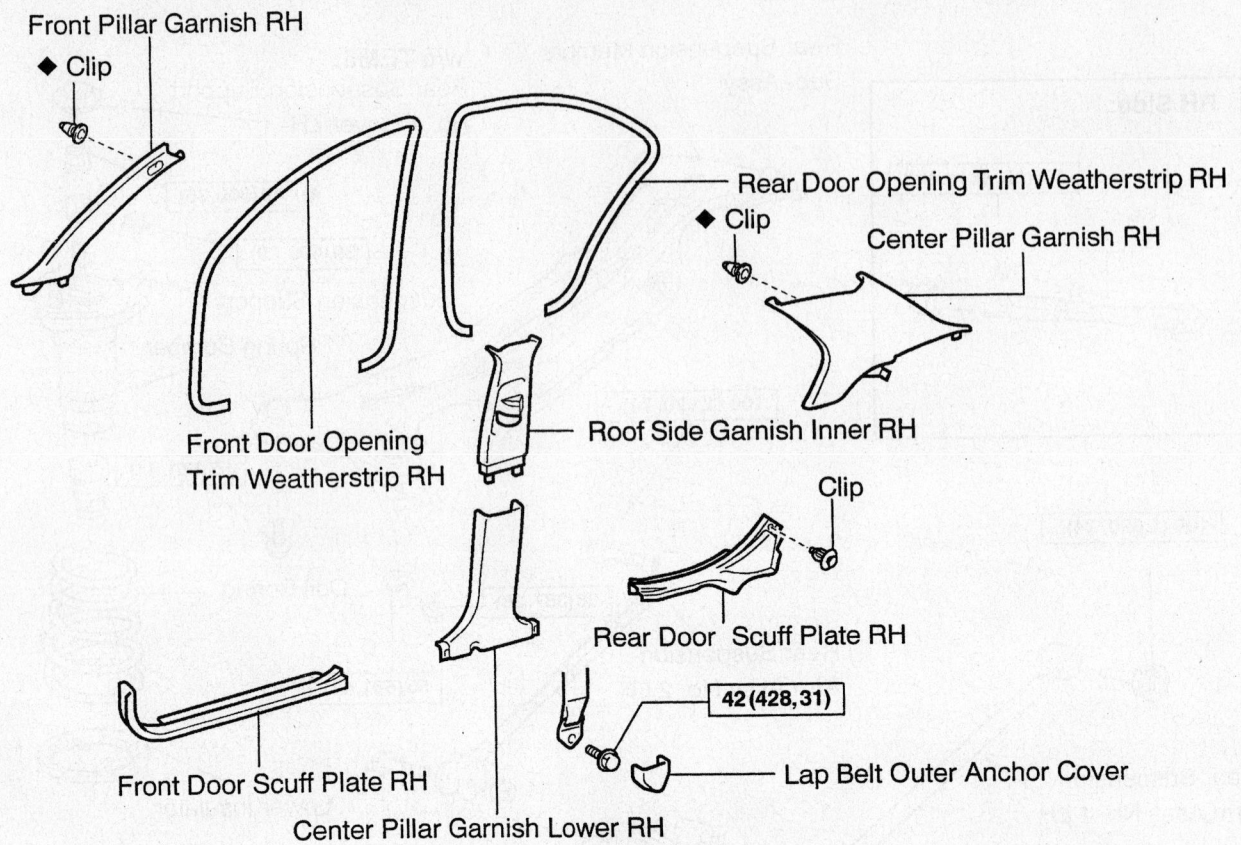

Front Pillar Garnish RH

◆ Clip

Rear Door Opening Trim Weatherstrip RH

◆ Clip

Center Pillar Garnish RH

Front Door Opening Trim Weatherstrip RH

Roof Side Garnish Inner RH

Clip

Rear Door Scuff Plate RH

42 (428, 31)

Front Door Scuff Plate RH

Center Pillar Garnish Lower RH

Lap Belt Outer Anchor Cover

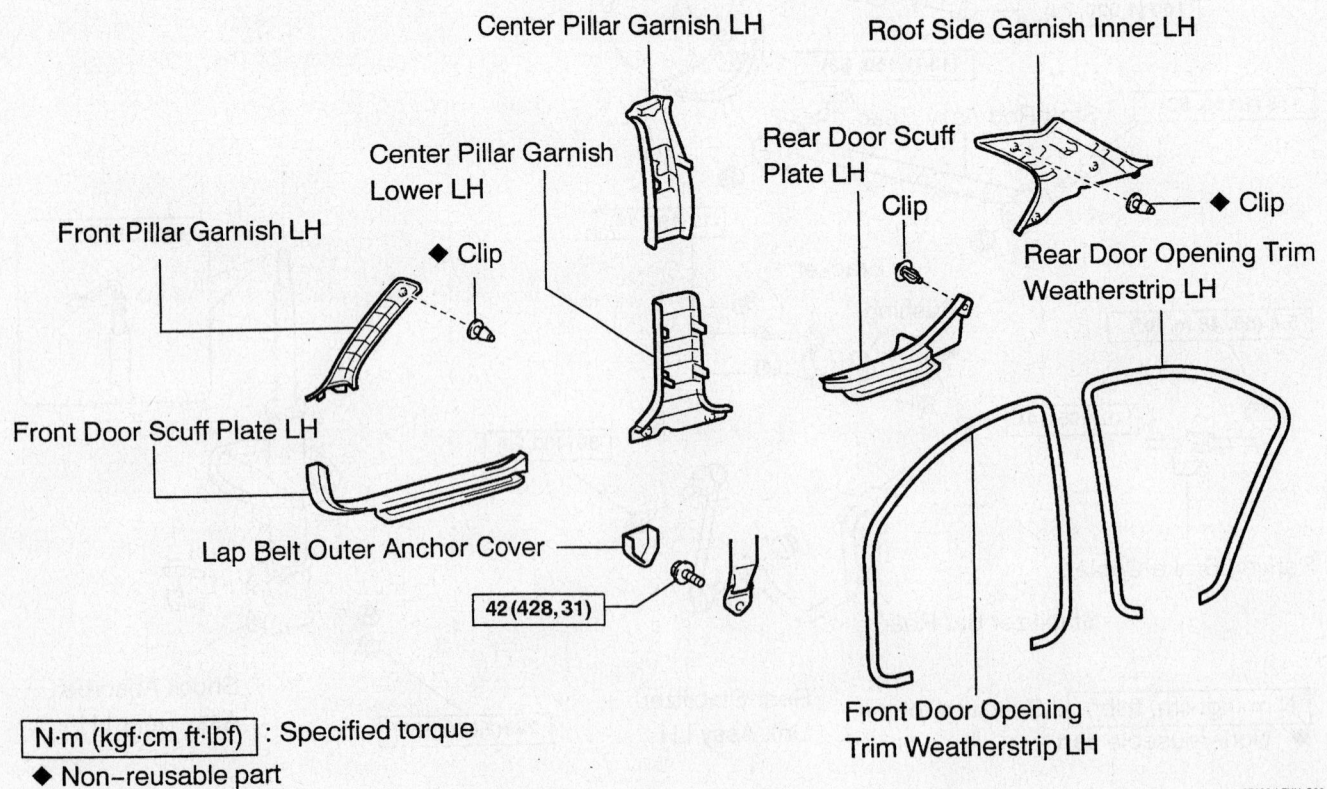

Center Pillar Garnish LH

Roof Side Garnish Inner LH

Center Pillar Garnish Lower LH

Rear Door Scuff Plate LH

Clip

◆ Clip

Front Pillar Garnish LH

◆ Clip

Rear Door Opening Trim Weatherstrip LH

Front Door Scuff Plate LH

Lap Belt Outer Anchor Cover

42 (428, 31)

Front Door Opening Trim Weatherstrip LH

N·m (kgf·cm ft·lbf) : Specified torque

◆ Non-reusable part

Exploded view, Roof headlining assembly—ES 330

67162-LEXU-G83

- Loosen center support nut. Do NOT remove
- 3 nuts from center suspension support
- Lower rear axle carrier and remove the 2 bolts loosened earlier
- Rear shock absorber and coil assembly

4. Remove the shock from the assembly

✳✳ WARNING

Using a suitable coil spring compressor, compress the coil spring before removing the following components

5. Remove or disconnect the following:
   - Nut on suspension support previously loosened
   - Collar, suspension support assembly, spring bumper and coil spring.
   - Lower insulator

To install:

6. Install the shock absorber in assembly

✳✳ WARNING

Using a suitable coil spring compressor, compress the coil spring before installing the following components

7. Install or connect the following:
   - Spring into the spring seat

➡**Make certain that spring seats properly in the seat**

   - Lower insulator, coil spring, suspension support assembly, collar
   - Temporarily install center nut

➡**Align the suspension support with the shock absorber lower bracket**

8. Release and remove the coil spring compressor

9. Install rear shock absorber with spring (Strut)

10. Install or connect the following:
    - 3 nuts attaching the spring support to 29 ft. lbs (39 Nm)

- 2 bolts attaching sock to the rear axle carrier assembly to 188 ft. lbs. (255 Nm)
- Flex hose and ABS speed sensor wire harness
- Center nut on suspension support to 36 ft. lbs. (49 Nm)
- Rear suspension support cover
- Rear stabilizer link to 29 ft. lbs. (39 Nm)
- Rear wheels
- Reinstall interior components in reverse order of removal

11. Inspect and align the rear wheels

Lower Control Arm

CONTROL ARM BUSHING REPLACEMENT

The control arm bushings are serviced with the control arm as an assembly.

Wheel Bearings

ADJUSTMENT

Check the backlash in bearing shaft direction and the axle hub deviation. Maximum for backlash should be 0.0020 in. (0.05mm) and for axle hub deviation 0.020 in. (0.05mm).

➡**The rear wheel bearings are non-adjustable. If the wheel bearing is out of specifications, replace the wheel bearing.**

REMOVAL & INSTALLATION

IS 250 and IS 350

1. Before servicing the vehicle, refer to the Precautions Section.
2. Raise and safely support the vehicle.
3. Remove or disconnect the following:
   - Rear tire and wheel assembly

- Rear stabilizer link
- Rear axle shaft nut
- Rear disc brake caliper and disc
- Speed sensor
- Parking brake assembly at the wheel hub
- No. 1 and 2 upper control arm assembly
- No. 1 and 2 rear suspension arm assembly
- Toe control link
- Rear axle from hub assembly using a plastic hammer
- No. 2 rear wheel bearing dust deflector using a screwdriver
- 4 bolts and rear axle hub and bearing assembly from the axle carrier assembly

To install:

4. Install or connect the following:
   - Hub on the carrier and tighten the bolts to 52 ft. lbs. (70 Nm)
   - No. 2 rear wheel bearing dust deflector using a screwdriver
   - Rear axle into hub assembly
   - No. 1 and 2 upper control arm assembly. Torque the new nut to the No. 2 rear upper control arm to 52 ft. lbs. (70 Nm). Temporarily tighten the No. 1 arm bolt.
   - No. 1 and 2 rear suspension arm assembly. Temporarily tighten.
   - Toe control link. Torque new nut to 52 ft. lbs. (70 Nm).
   - Rear stabilizer link. Torque to 20 ft. lbs. (27 Nm).
   - Parking brake assembly at the wheel hub
   - Rear speed sensor
   - Rear axle shaft nut. Torque to 214 ft. Lbs. (290 Nm) and stake the nut with a chisel and hammer.
   - Rear brake disc and disc brake caliper

5. Adjust the parking brake.
6. Stabilize the suspension as follows:

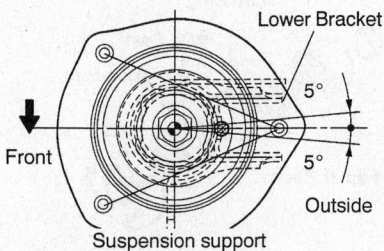

Aligning support bracket—ES 330

67162-LEXU-G84

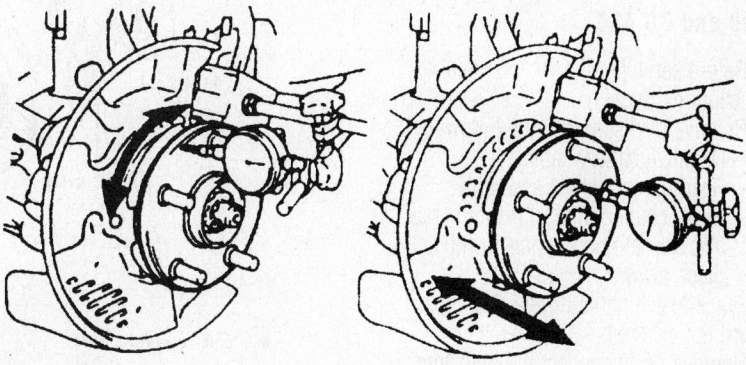

Checking wheel bearings for excessive play

7923LGB6

a. Install the wheel(s).

b. Lower the vehicle and bounce it up and down several times to stabilize the rear suspension.

c. Raise the vehicle.

d. Remove the wheel.

e. Jack up the rear lower suspension armplacing a wood block in between. Apply a load to the suspension so that the lower suspension arm is plkaced in a horizontal position.

7. Fully tighten the No. 1 rear upper control arm assembly to 119 ft. lbs. (161 Nm).

8. Fully tighten the No. 1 rear suspension arm assembly to 70 ft. lbs. (95 Nm).

9. Fully tighten the No. 2 rear suspension arm assembly to 119 ft. lbs. (161 Nm).

- Rear tire and wheel assembly

ES 300

1. Before servicing the vehicle, refer to the Precautions Section.

2. Raise and safely support the vehicle.

3. Remove or disconnect the following:
- Rear tire and wheel assembly
- If equipped with rear disc brakes, the caliper mounting bolts. Leave the brake line connected and suspend the assembly out of the way.
- Brake rotor or drum
- 4 bolts and pull off the rear axle hub
- O-ring

➡**If it is necessary to replace the hub or bearing, replace the components as an assembly.**

To install:

4. Install or connect the following:
- Hub on the carrier and tighten the bolts to 59 ft. lbs. (80 Nm)
- Rotor or drum
- Caliper, if equipped with rear disc brakes and tighten the bolts to 34 ft. lbs. (64 Nm)
- Wheel

GS 300 and GS 430

1. Before servicing the vehicle, refer to the Precautions Section.

2. Remove or disconnect the following:
- Negative battery cable.
- Rear tire and wheel assembly
- Brake caliper support from the rear axle carrier and support it with a piece of wire

3. Place matchmarks on the disc brake rotor and the axle hub.

4. Remove or disconnect the following:
- Brake rotor

- Speed sensor
- Rear halfshaft
- Parking brake shoes
- Parking brake cable
- Strut rod

5. Place matchmarks on the adjusting cam and rear control crossmember.
- Nut, adjusting cam and the washer to the No. 1 control arm
- No. 1 lower control arm from the crossmember
- Loosen the nut holding the lower control arm to the axle carrier
- No. 2 lower control arm from the axle carrier
- Nut, then remove the No. 2 lower control arm from the axle carrier
- Nut holding the upper control arm to the axle carrier
- Axle carrier
- Nut holding the No. 1 control arm to the axle carrier
- No. 1 lower control arm from the axle carrier
- Dust deflector
- Axle hub from the carrier
- Backing plate
- Inner race (outside)
- Oil seal
- Snapring
- Bearing

To install:

6. Install or connect the following:
- Bearing to the axle carrier

➡**If the inner races come loose from the bearing outer race, be sure to install them on the same side as before.**

- Snapring. Install the inner race (outside) and a new oil seal.

- Backing plate. Install the inner race (inside) and press in the axle hub with the proper tools.
- Inner oil seal. Align the holes for the speed sensor in the dust deflector and axle carrier. Install the dust deflector.
- No. 1 lower arm to the axle carrier and install a new nut. Tighten the nut to 43 ft. lbs. (59 Nm).
- Upper control arm to the axle carrier. Tighten the new nut and bolt to 80 ft. lbs. (109 Nm).
- No. 2 lower control arm to the axle carrier and tighten a new nut to 110 ft. lbs. (150 Nm)
- No. 1 lower control arm to the rear crossmember. Tighten the nut to 136 ft. lbs. (184 Nm).
- Strut rod to the axle carrier. Tighten the nuts and bolts to 134 ft. lbs. (184 Nm).
- Parking brake cable and slide the backing plate to the inside. Install the hex bolt and tighten it to 132 ft. lbs. (180 Nm).
- Shoe guide plate set bolt. Tighten the bolt to 13 ft. lbs. (18 Nm).
- 4 hub bolts and tighten them to 19 ft. lbs. (26 Nm)
- Bolts at the speed sensor and tighten them to 69 inch lbs. (8 Nm)
- Parking brake shoes
- Halfshafts. Apply the brakes and tighten the locknut to 213 ft. lbs. (289 Nm).
- Brake rotor
- Brake caliper support to the rear axle carrier. Tighten the bolts to 77 ft. lbs. (104 Nm).
- Rear tire and wheel assembly

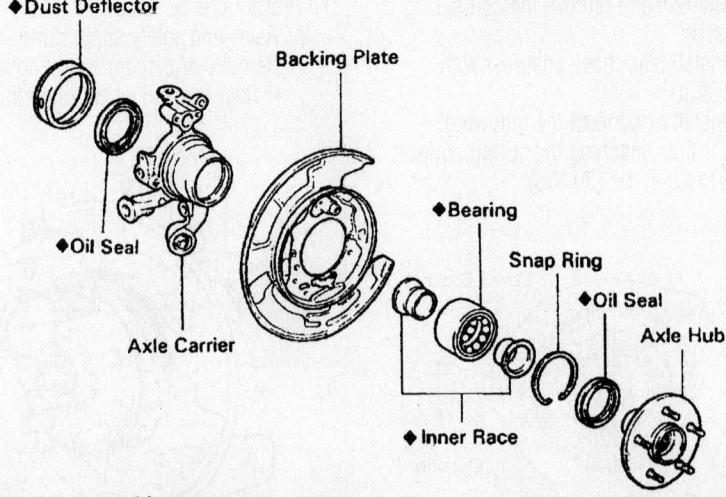

◆Dust Deflector

Backing Plate

◆Oil Seal

◆Bearing

Snap Ring

◆Oil Seal

Axle Hub

Axle Carrier

◆Inner Race

◆ Non-reusable part

7923LGC1

Exploded view of the axle carrier—GS 300

- Negative battery cable

7. Lower the vehicle and bounce it a few times to stabilize the suspension.

2002–03 LS 430

1. Before servicing the vehicle, refer to the Precautions Section.

2. If equipped with air suspension, move the height control switch in the trunk area to the **OFF** position.

3. Remove or disconnect the following:
- Negative battery cable
- Rear wheel(s)
- Height control sensor link from the lower control arm
- Anti-lock Brake System (ABS) speed sensor and wiring harness
- Brake caliper bracket from the rear axle carrier by removing the 2 bolts. Support the caliper with a piece of wire.
- Brake rotor

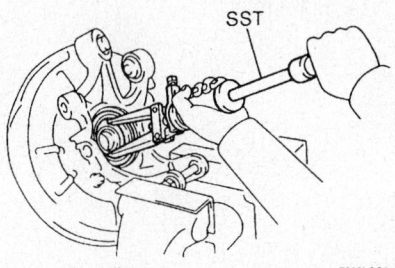

Removing the oil seal (inner)—LS 430

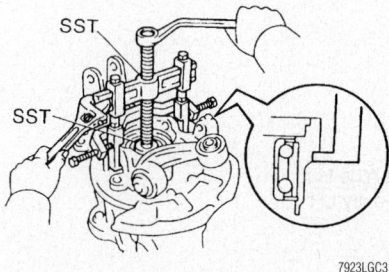

Removing the axle hub from the axle carrier—LS 430

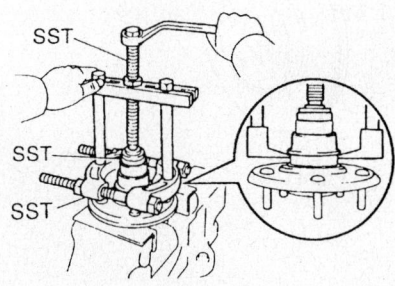

Removing the inner race (outside) from the axle hub—LS 430

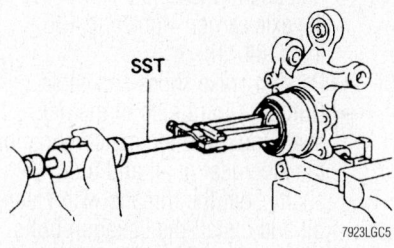

Removing the oil seal (outer)—LS 430

- Parking brake shoes and cable
- Cotter pin, lock cap and the nut holding the halfshaft to the rear axle
- Suspension member brace by removing the 2 bolts
- Halfshaft bolts and washers
- Halfshaft from the vehicle
- Strut rod

4. Place matchmarks on the adjusting cam and body for the No. 1 control arm.

5. Remove or disconnect the following:
- Nut and adjusting cam
- Nut on the axle carrier side of the No. 1 lower control arm
- Separate the control arm from the axle carrier
- No. 1 lower control arm
- Stabilizer bar link from the No. 2 lower control arm.

6. Place matchmarks on the adjusting cam and body.

7. Remove or disconnect the following:
- Nut and adjusting cam from the No. 2 lower control arm

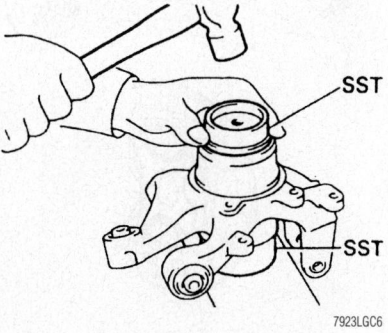

Installing the oil seal (outer)—LS 430

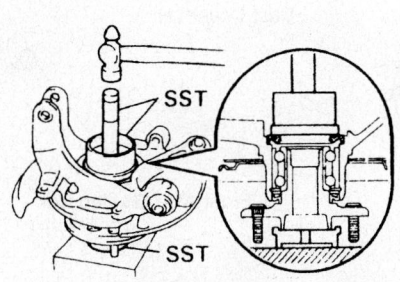

Installing the oil seal (inner)—LS 430

- Nut and bolt holding the No. 2 lower control arm to the axle carrier
- No. 2 control arm from the vehicle
- Nut and bolt on the lower side of the strut assembly
- 2 upper control arm set nuts and bolts
- Axle carrier with the upper control arm

8. Secure the axle carrier in a vise.

9. Remove or disconnect the following:
- Nut holding the upper control arm to the axle carrier and remove the control arm
- Dust deflector. Use a suitable pry-tool.
- Oil seal

10. Remove the 2 bolts and nuts and shift the backing the plate towards the hub side (outside).

11. Remove or disconnect the following:
- Axle hub
- Backing plate.
- Inner race (outside) from the axle hub
- Oil seal (outer) from the axle
- Snapring from inside the axle housing
- Bearing from the axle housing

To install:

12. Install or connect the following:
- New bearing to the axle housing
- Snapring to the axle carrier, using snapring pliers
- New outer oil seal. Coat the oil seal lip with multipurpose grease.
- Backing plate to the axle housing. Do NOT install the bolts or nuts at this time.
- Inner race (inside) to the axle housing
- Axle hub to the axle housing
- Backing plate in position. Tighten the bolts and nuts to 43 ft. lbs. (59 Nm).
- New oil seal (inner) to the axle housing. Coat the oil seal lip with multipurpose grease.
- New dust deflector. Be sure to align the hose for the ABS speed sensor in the dust deflector and axle carrier.
- Upper control arm to the axle carrier by installing the nut. Tighten the nut to 80 ft. lbs. (108 Nm).
- Axle carrier and upper control arm to the vehicle as an assembly
- 2 upper control arm set bolts and tighten the bolts to 121 ft. lbs. (164 Nm)
- Bolt and nut holding the strut to the axle carrier. Tighten to 101 ft. lbs. (137 Nm).

- Bolt and nut connecting the No. 2 lower control arm to the axle carrier. Tighten the bolt to 60 ft. lbs. (81 Nm).
- Nut and adjusting cam to hold the No. 2 lower control arm to the body. Align the adjusting cam marks and tighten the nut to 57 ft. lbs. (78 Nm).
- Stabilizer bar link to the No. 2 lower control arm and tighten the nut to 48 ft. lbs. (65 Nm)
- No. 1 lower control arm to the axle carrier and body. Install the nut to hold the No. 1 lower control arm to the axle carrier. Tighten the nut to 43 ft. lbs. (59 Nm).
- Nut and adjusting cam to hold the No. 1 lower control arm to the body. Align the matchmarks and tighten the nut to 57 ft. lbs. (78 Nm).
- Strut rod to the axle carrier and body. Install the bolt and nut to hold the strut rod to the body. Tighten to 57 ft. lbs. (78 Nm).

- Bolt and nut to hold the strut rod to the axle carrier. Tighten to 136 ft. lbs. (184 Nm).
- Parking brake shoes and cable
- Outboard joint side of the half-shaft and align the matchmarks on the side gear shaft and the half-shaft. Coat the threads with clean oil and install the hexagon bolts. Tighten bolts to 61 ft. lbs. (83 Nm).
- Suspension member brace with the 2 bolts. Tighten the 2 bolts to 37 ft. lbs. (50 Nm).
- Nut to hold the halfshaft to the rear axle. Tighten the nut to 213 ft. lbs. (289 Nm).
- Lock cap and cotter pin
- Brake disc to the axle hub with the matchmarks aligned. Install the 2 screws and tighten the screws to 48 inch lbs. (5 Nm).
- Brake caliper to the vehicle and install the 2 bolts. Tighten the bolts to 77 ft. lbs. (104 Nm).

- ABS speed sensor and wiring harness
- Height control sensor link with the matchmarks aligned. Tighten the nut to 48 inch lbs. (5 Nm).
- Rear wheel(s)
- Negative battery cable

13. Lower the vehicle and turn **ON** the air suspension switch.

2004–06 LS 430

1. Before servicing the vehicle, refer to the Precautions Section.
2. Remove or disconnect the following:
   - Rear wheel(s)
   - Height control sensor link from the lower control arm
   - Anti-lock Brake System (ABS) speed sensor and wiring harness
   - Brake caliper bracket from the rear axle carrier by removing the 2 bolts. Support the caliper with a piece of wire.
   - Brake rotor

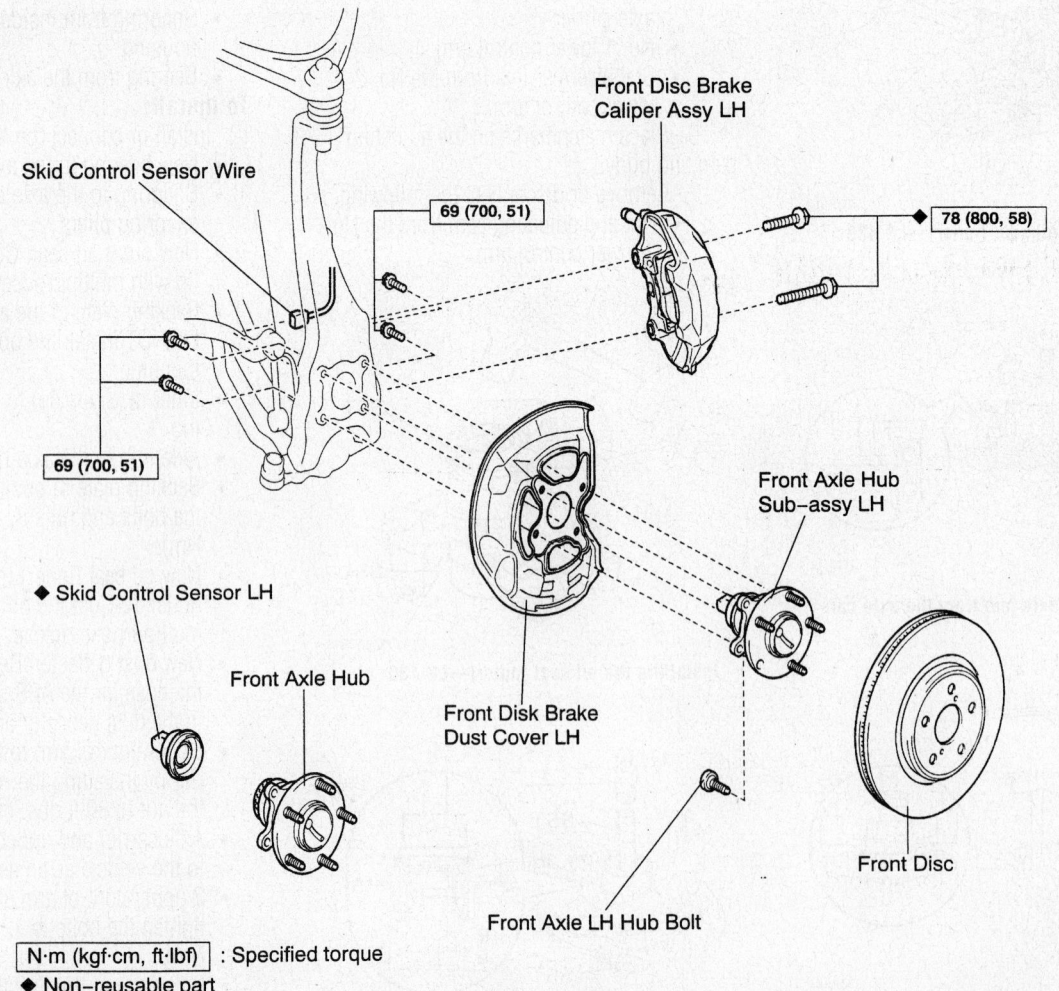

Skid Control Sensor Wire

Front Disc Brake Caliper Assy LH

69 (700, 51)

78 (800, 58)

69 (700, 51)

◆ Skid Control Sensor LH

Front Axle Hub

Front Disk Brake Dust Cover LH

Front Axle Hub Sub–assy LH

Front Axle LH Hub Bolt

Front Disc

N·m (kgf·cm, ft·lbf) : Specified torque
◆ Non–reusable part

67162-LEXU-G118

Exploded view rear carrier assembly, hub and bearing—2004-06 LS 430

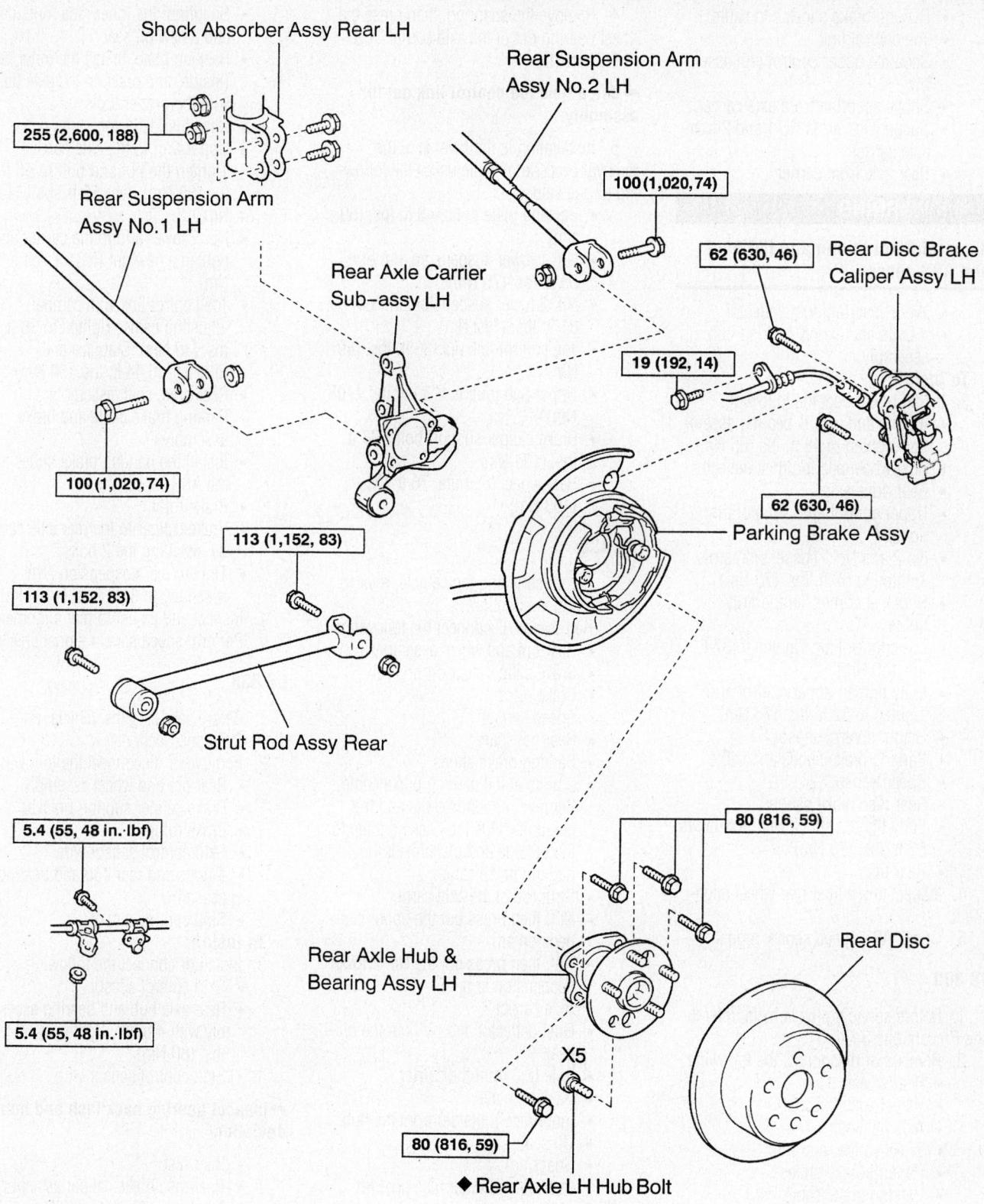

Shock Absorber Assy Rear LH

255 (2,600, 188)

Rear Suspension Arm
Assy No.2 LH

100 (1,020, 74)

Rear Suspension Arm
Assy No.1 LH

62 (630, 46)

Rear Disc Brake
Caliper Assy LH

Rear Axle Carrier
Sub–assy LH

19 (192, 14)

100 (1,020, 74)

113 (1,152, 83)

62 (630, 46)

Parking Brake Assy

113 (1,152, 83)

Strut Rod Assy Rear

5.4 (55, 48 in.·lbf)

80 (816, 59)

Rear Disc

Rear Axle Hub &
Bearing Assy LH

5.4 (55, 48 in.·lbf)

X5

80 (816, 59)

◆ Rear Axle LH Hub Bolt

N·m (kgf·cm, ft·lbf) : Specified torque

◆ Non–reusable part

Exploded view rear Axle carrier and hub assembly—ES 330

67162-LEXU-G87

- Parking brake shoes and cable
- Toe control link
- Separate upper control arm assembly
- Shock absorber from axle carrier
- Suspension arms No 1 and 2 from axle carrier
- Rear axle from carrier

WARNING

Be careful not to damage the boot and ABS sensor

- Wheel bearing dust deflector
- 4 bolts and axle & bearing hub assembly

To install:

3. Install or connect the following:
   - 4 bolts and axle & bearing assembly. Tighten to 48 ft. lbs (65 Nm)
   - Rear wheel bearing dust deflector
   - Rear drive axle
   - Upper control arm. Temporarily tighten
   - No 2 and No 1 suspension arms. Tighten to 52 ft. lbs. (70 Nm)
   - Shock absorber. Temporarily tighten
   - Toe control link. Tighten to 37 ft. lbs. (50 Nm)
   - Fully tighten upper control arm. Tighten to 52 ft. lbs. (70 Nm)
   - Height control sensor
   - Parking brake shoes and cable
   - Speed sensor
   - Rear disc brake caliper
   - Fully tighten shock with new nut to 52 ft. lbs. (70 Nm)
   - Rear tire
4. Inspect and adjust rear wheel alignment
5. Check ABS speed sensor signal

IS 300

1. Before servicing the vehicle, refer to the Precautions Section.
2. Remove or disconnect the following:
   - Rear wheel
   - Wheel speed sensor
   - Axle halfshaft
   - Brake caliper and rotor
   - Parking brake shoes
   - Parking brake cable
   - No. 1 lower suspension arm bolt
   - No. 2 lower suspension arm bolt
   - Toe control link
   - Upper ball joint
   - Axle carrier from the vehicle
3. Press the hub out of the wheel bearing, then remove the backing plate.
4. Remove the snapring, then press the wheel bearing out of the axle carrier.

To install:

➡Use a new toe control link nut for assembly.

5. Installation is the reverse of the removal procedure, while using the following torque values:
   - Backing plate bolts: 43 ft. lbs. (59 Nm)
   - No. 1 lower suspension arm bolt: 55 ft. lbs. (75 Nm)
   - No. 2 lower suspension arm bolt: 81 ft. lbs. (110 Nm)
   - Toe control link nut: 36 ft. lbs. (49 Nm)
   - Upper ball joint nut: 80 ft. lbs. (108 Nm)
   - Brake caliper support bolts: 77 ft. lbs. (104 Nm)
   - Rear wheel lug nuts: 76 ft. lbs. (103 Nm)

SC 430

1. Before servicing the vehicle, refer to the Precautions Section.
2. Remove or disconnect the following:
   - Rear tire and wheel assembly
   - Brake caliper support bracket
   - Brake rotor
   - Speed sensor
   - Rear halfshaft
   - Parking brake shoes
   - 2 bolts at the parking brake cable. Remove the 2 hub bolts and the hex bolt. Slide the backing plate to the outside and disconnect the parking brake cable.
   - Strut rod at the axle carrier
   - Nut, then press out the upper suspension arm.
   - Nut, then press out the No. 2 lower suspension arm
   - Axle carrier
   - Dust deflector and pull out the oil seal
   - Axle hub from the carrier
   - Backing plate
   - Inner race (outside) from the hub
   - Oil seal
   - Snapring
   - Bearing and inner race (inside)

To install:

3. Install or connect the following:
   - Bearing to the axle carrier

➡If the inner races come loose from the bearing outer race, be sure to install them on the same side as before.

   - Snapring, the inner race (outside) and a new oil seal
   - Backing plate. Install the inner race (inside) and press in the axle hub with the proper tools.
   - New dust deflector
   - Upper arm to the axle carrier. Tighten the nut and bolt to 65 ft. lbs (88 Nm), rear 55 ft. lbs (74 Nm)
   - No. 2 lower arm to the carrier and tighten a new nut to 81 ft. lbs. (110 Nm).
   - Toe control link with camber adjusting cams. Tighten to 36 ft. lbs. (49 Nm). Stabilize and retighten to 44 ft. lbs. (59 Nm).
   - Rear drive shaft assembly
   - Parking brake cable and brake assembly
   - Install the parking brake shoes and the ABS sensor.
   - Brake rotor
   - Brake caliper to the rear axle carrier by installing the 2 bolts.
   - Tighten rear suspension arm assembly to 81 ft. lbs. (110 Nm)
4. Inspect and adjust to rear alignment
5. Perform speed sensor signal check.

ES 330

1. Before servicing the vehicle, refer to the Precautions Section.
2. Remove or disconnect the following:
   - Rear tire and wheel assembly
   - Brake caliper support bracket
   - Brake rotor
   - Skid control sensor wire
   - 4 bolts and rear hub and bearing assembly
   - Skid control sensor

To install:

3. Install or connect the following:
   - Skid control sensor
   - Rear axle hub and bearing assembly with 4 bolts. Tighten to 59 ft. lbs. (80 Nm)
   - Skid control sensor wire

➡Inspect bearing back lash and hub deviation

   - Rear disc
   - Rear disc brake caliper assembly. Tighten to 46 ft. lbs (62 Nm)
   - Flexible brake hose
   - Rear tire and wheel
4. inspect and adjust rear wheel alignment
5. Check ABS speed sensor signal

FRONT BRAKES

Brake Caliper

REMOVAL & INSTALLATION

ES 300

1. Before servicing the vehicle, refer to the Precautions Section.
2. Remove or disconnect the following:
   - Wheels
   - Brake hose from the caliper
   - Bolts that attach the caliper to the torque plate
   - Caliper assembly by lifting the bottom

To install:

3. Grease the caliper slides and bolts with lithium grease.
4. Install or connect the following:
   - Caliper. Torque the bolts to 25 ft. lbs. (34 Nm).
   - Brake hose to the caliper using 2 new washers. Torque the union bolt to 20 ft. lbs. (29 Nm).
5. Fill the master cylinder to the proper level and bleed the brake system.

IS 300

1. Before servicing the vehicle, refer to the Precautions Section.
2. Remove or disconnect the following:
   - Wheels
   - Brake hose from the caliper
   - Bolts that attach the caliper to the torque plate
   - Caliper assembly by lifting the bottom

To install:

3. Grease the caliper slides and bolts with lithium grease.
4. Install or connect the following:
   - Caliper. Torque the bolts to 25 ft. lbs. (34 Nm).
   - Brake hose to the caliper using 2 new washers. Torque the union bolt to 22 ft. lbs. (30 Nm).
5. Fill the master cylinder to the proper level and bleed the brake system.

IS 250

1. Before servicing the vehicle, refer to the Precautions Section.
2. Remove or disconnect the following:
   - Wheels
   - Brake hose from the caliper
   - Bolts that attach the caliper to the mounting bracket

- Caliper assembly by lifting the bottom

To install:

3. Grease the caliper slides and bolts with lithium grease.
4. Install or connect the following:
   - Caliper. Hold the sliding pin and tighten the mounting bolts to 25 ft. lbs. (34 Nm).
   - Brake hose to the caliper using 2 new washers. Torque the union bolt to 22 ft. lbs. (30 Nm).
   - Wheels
5. Fill the master cylinder to the proper level and bleed the brake system.

IS 350

1. Before servicing the vehicle, refer to the Precautions Section.
2. Remove or disconnect the following:
   - Wheels
   - Hole pin hold clip
   - Front disc brake anti-rattle with hole pin while pushing on the anti-rattle spring
   - Anti-rattle spring
   - 2 pads and the 2 anti-squeal shims from each pad
   - Brake hose from the caliper
   - Bolts that attach the caliper to the steering knuckle
   - Caliper assembly by lifting the bottom

To install:

3. Install or connect the following:
   - Caliper. Torque the bolts to 58 ft. lbs. (78 Nm).
   - Brake hose to the caliper using 2 new washers. Torque the union bolt to 29 ft. lbs. (39 Nm).
   - 2 pads and the 2 anti-squeal shims to each pad
   - Anti-rattle spring
   - Hole pin while pushing on the front disc brake anti-rattle spring
   - Hole pin hold clip
   - Wheels
4. Fill the master cylinder to the proper level and bleed the brake system.

GS 300 and GS 430

1. Before servicing the vehicle, refer to the Precautions Section.
2. Remove or disconnect the following:
   - Wheels
   - Brake line at the caliper
   - Anti-squeal springs

- Mounting bolts, while holding the sliding pin with a wrench
- Caliper assembly

To install:

3. Install or connect the following:
   - Caliper. Hold the sliding pin and tighten the mounting bolts to 25 ft. lbs. (34 Nm).
   - Anti-squeal springs
   - Connect the brake line with 2 new gaskets and tighten the union bolt to 22 ft. lbs. (30 Nm).
4. Bleed the brake system.
   - Wheels
5. Check and if necessary fill the master cylinder reservoir.

LS 430 and SC 430

1. Before servicing the vehicle, refer to the Precautions Section.
2. Remove or disconnect the following:
   - Front wheels
   - Brake line at the caliper
   - 2 bolts to the holding the caliper to the steering knuckle
   - Caliper assembly

To install:

3. Install or connect the following:
   - Caliper. Tighten the 2 bolts to 87 ft. lbs. (118 Nm) for SC 430; or 77 ft. lbs. (104 Nm) for LS 430
   - Brake line with 2 new gaskets and tighten the union to 29 ft. lbs. (39 Nm)
4. Refill the reservoir as necessary and bleed the brake system.

ES 330

1. Before servicing the vehicle, refer to the Precautions Section.
2. Install or connect the following:
   - Wheels
   - Brake line at the caliper by removing the union bolt and 2 gaskets
   - Mounting bolts, while holding the sliding pin with a wrench
   - Caliper from the caliper support

To install:

3. Install or connect the following:
   - Caliper to the caliper support
   - Caliper bolts. Hold the sliding pin and tighten the mounting bolts to 25 ft. lbs. (34 Nm).
   - Brake line with (2) new gaskets and tighten the union to 22 ft. lbs. (30 Nm)
4. Bleed the brake system.
   - Wheels

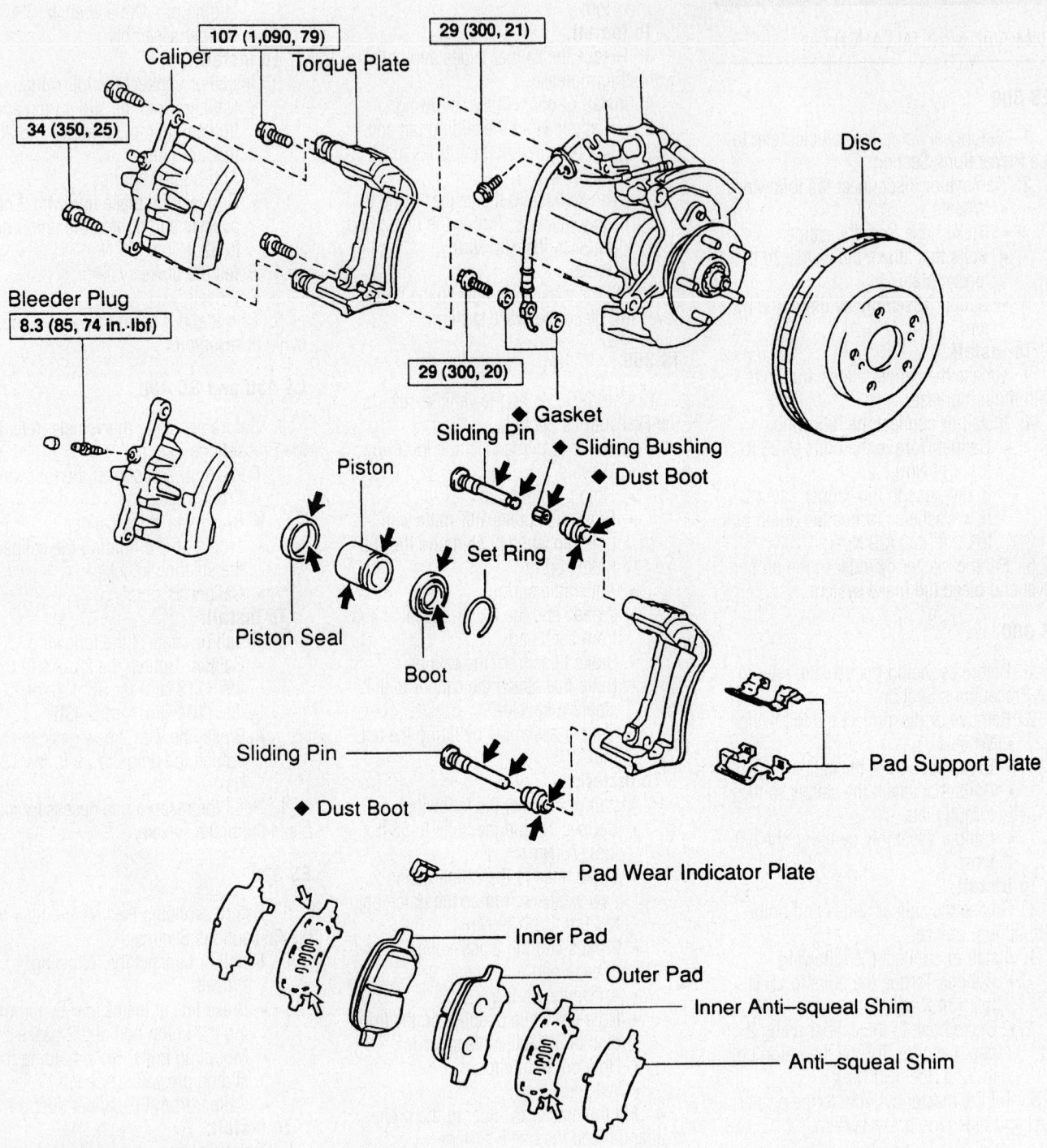

Caliper

107 (1,090, 79)   Torque Plate   29 (300, 21)

34 (350, 25)

Disc

Bleeder Plug
8.3 (85, 74 in.·lbf)

29 (300, 20)

◆ Gasket

Piston   Sliding Pin   ◆ Sliding Bushing

◆ Dust Boot

Set Ring

Piston Seal

Boot

Sliding Pin

Pad Support Plate

◆ Dust Boot

Pad Wear Indicator Plate

Inner Pad

Outer Pad

Inner Anti–squeal Shim

Anti–squeal Shim

N·m (kgf·cm, ft·lbf) : Specified torque

◆ Non–reusable part

◀ Lithium soap base glycol grease

◁ Disc brake grease

93016G18

Front disc brakes—ES 300

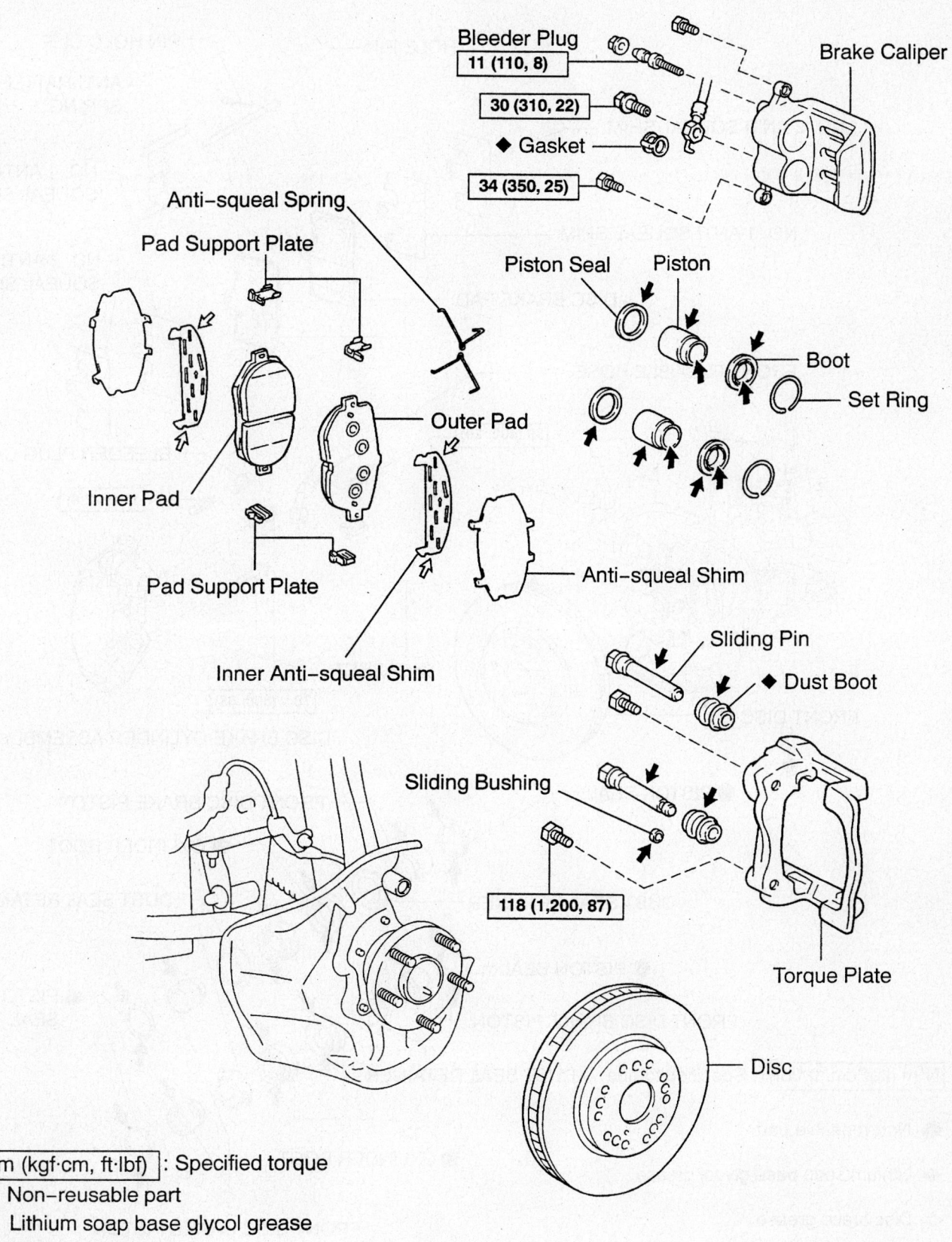

Bleeder Plug
11 (110, 8)

30 (310, 22)

◆ Gasket

34 (350, 25)

Brake Caliper

Anti-squeal Spring

Pad Support Plate

Piston Seal   Piston

Boot

Set Ring

Outer Pad

Inner Pad

Pad Support Plate

Anti-squeal Shim

Inner Anti-squeal Shim

Sliding Pin

◆ Dust Boot

Sliding Bushing

118 (1,200, 87)

Torque Plate

Disc

N·m (kgf·cm, ft·lbf) : Specified torque
◆ Non-reusable part
► Lithium soap base glycol grease
⇨ Disc brake grease

67162-LEXU-G88

Front disc brakes—IS 300

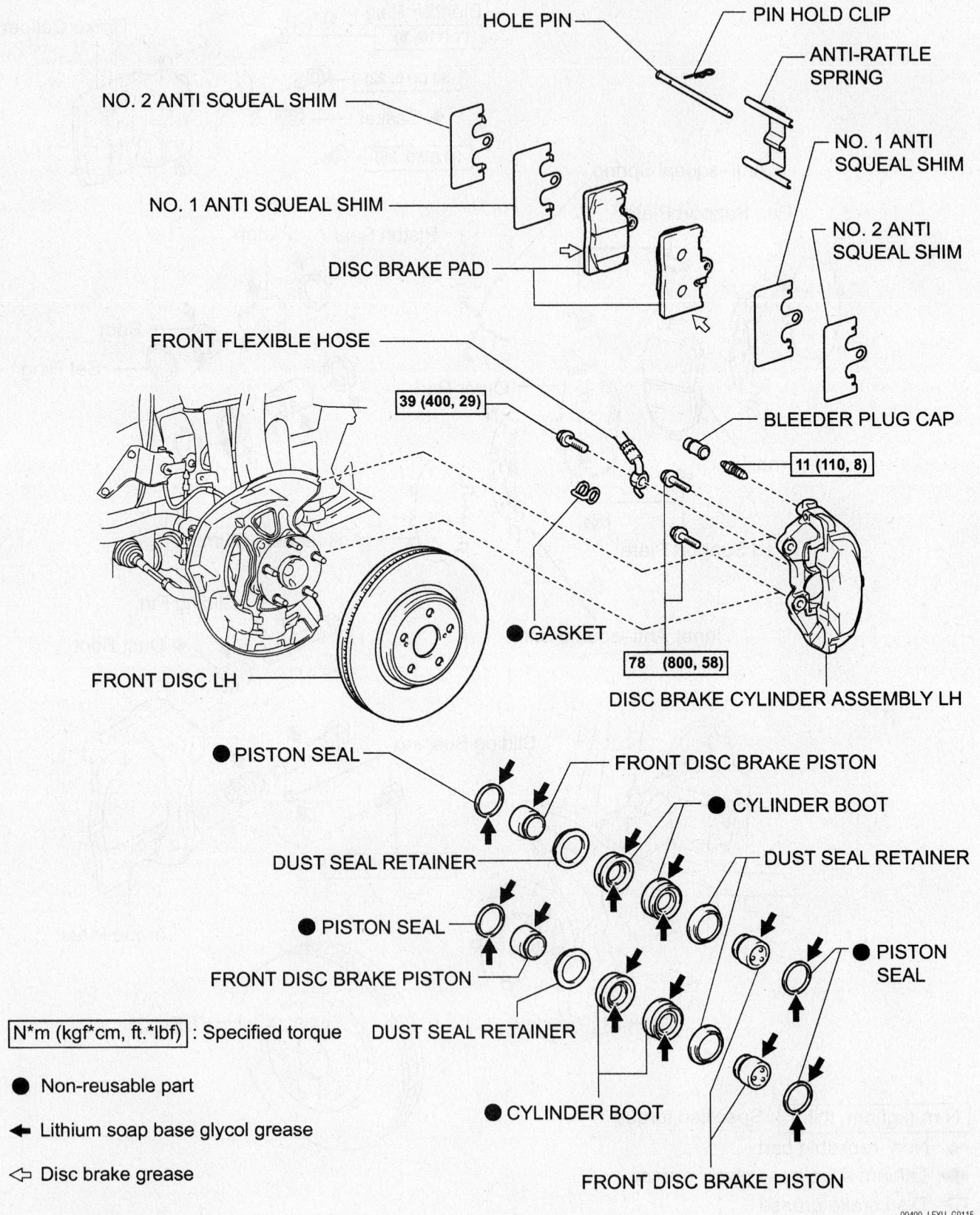

HOLE PIN

PIN HOLD CLIP

ANTI-RATTLE SPRING

NO. 2 ANTI SQUEAL SHIM

NO. 1 ANTI SQUEAL SHIM

NO. 1 ANTI SQUEAL SHIM

DISC BRAKE PAD

NO. 2 ANTI SQUEAL SHIM

FRONT FLEXIBLE HOSE

39 (400, 29)

BLEEDER PLUG CAP

11 (110, 8)

● GASKET

78   (800, 58)

FRONT DISC LH

DISC BRAKE CYLINDER ASSEMBLY LH

● PISTON SEAL

FRONT DISC BRAKE PISTON

● CYLINDER BOOT

DUST SEAL RETAINER

DUST SEAL RETAINER

● PISTON SEAL

FRONT DISC BRAKE PISTON

● PISTON SEAL

DUST SEAL RETAINER

N*m (kgf*cm, ft.*lbf) : Specified torque

● Non-reusable part

◄ Lithium soap base glycol grease

◁ Disc brake grease

● CYLINDER BOOT

FRONT DISC BRAKE PISTON

09490_LEXU_G0115

Front disc brakes—IS 350

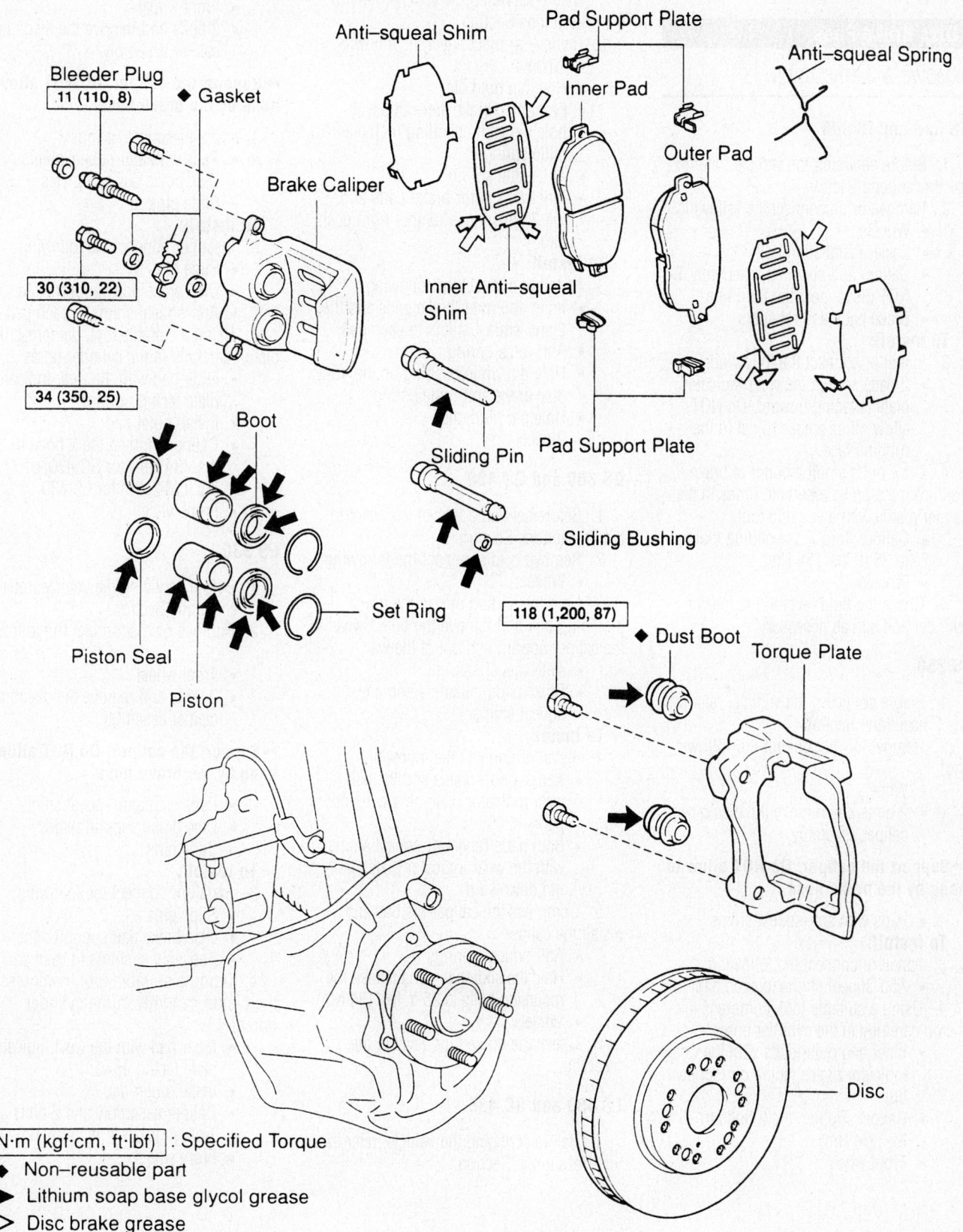

Bleeder Plug
.11 (110, 8)

◆ Gasket

Anti–squeal Shim

Pad Support Plate

Anti–squeal Spring

Inner Pad

Outer Pad

Brake Caliper

Inner Anti–squeal Shim

30 (310, 22)

34 (350, 25)

Boot

Sliding Pin

Pad Support Plate

Piston Seal

Piston

Set Ring

Sliding Bushing

118 (1,200, 87)

◆ Dust Boot

Torque Plate

Disc

N·m (kgf·cm, ft·lbf) : Specified Torque

◆ Non–reusable part

➡ Lithium soap base glycol grease

⇨ Disc brake grease

93016G20

Front disc brakes—GS 300 and GS 430

5. Check and fill the master cylinder reservoir, if needed.

Disc Brake Pads

REMOVAL & INSTALLATION

ES 300 and IS 300

1. Before servicing the vehicle, refer to the Precautions Section.
2. Remove or disconnect the following:
   - Wheels
   - Lower installation bolt
   - Caliper and suspend it securely. Do NOT disconnect the fluid line.
   - Brake pads and retainers

To install:

3. Install or connect the following:
   - 2 pads so that the wear indicator plate is facing upward. Do NOT allow oil or grease to get in the rubbing face.
4. Draw out a small amount of brake fluid from the brake reservoir. Press in the caliper piston with a suitable tool.
   - Caliper. Torque the sliding main pin to 25 ft. lbs. (34 Nm).
   - Wheels
5. Check the fluid level in the master cylinder and add as necessary.

IS 250

1. Before servicing the vehicle, refer to the Precautions Section.
2. Remove or disconnect the following:
   - Wheels
   - 2 bolts and remove the disc brake caliper assembly

➡ **Support the caliper. Do NOT allow to hang by the brake hose**

   - Pads with anti-squeal shims

To install:

3. Install or connect the following:
   - Anti-squeal shims to each pad
4. Using a suitable tool, compress the piston carefully in the cylinder bores
   - Inner and outer pads with the wear indicator plates facing correct position
   - Caliper. Tighten the 2 bolts to 25 ft. lbs. (34 Nm)
   - Front wheel

IS 350

1. Before servicing the vehicle, refer to the Precautions Section.
2. Remove or disconnect the following:
   - Wheels
   - Hole pin hold clip
   - Front disc brake anti-rattle with hole pin while pushing on the anti-rattle spring
   - Anti-rattle spring
   - Inner and outer brake pads and the 2 anti-squeal shims from each pad

To install:

3. Install or connect the following:
   - Inner and outer brake pads and the 2 anti-squeal shims to each pad
   - Anti-rattle spring
   - Hole pin while pushing on the front disc brake anti-rattle spring
   - Hole pin hold clip
   - Wheels

GS 300 and GS 430

1. Before servicing the vehicle, refer to the Precautions Section.
2. Remove or disconnect the following:
   - Wheels
3. Hold the sliding pin on the lower mounting bolt and remove the bolt. Swivel the caliper upward and out of the way.
   - Anti-squeal springs
   - Brake pads, retainers and anti-squeal shims

To install:

4. Install or connect the following:
   - Pad support plates and the pad wear indicator plate on the inside pad
   - Both pads (and anti-squeal shims) with the wear indicator plates facing downward
5. Compress the caliper pistons and install the caliper.
   - Anti-squeal springs
   - Hold the sliding pin and tighten the mounting bolts to 25 ft. lbs. (34 Nm)
   - Wheels
6. Check the brake fluid level in the reservoir.

LS 430 and SC 430

1. Before servicing the vehicle, refer to the Precautions Section.

2. Remove or disconnect the following:
   - Front wheel
   - 2 bolts and remove the disc brake caliper assembly

➡ **Support the caliper. Do NOT allow to hang by the brake hose**

   - 2 anti-squeal springs
   - Pads with anti-squeal shims
   - Disc brake support plates
   - Slide pins

To install:

3. Install or connect the following:
   - Slide pins
   - Disc brake pad support plate
   - Anti-squeal shims to each pad
4. Using a suitable tool, compress the piston carefully in the cylinder bores
   - Inner pad with the wear indicator plate facing upward
   - Install outer pad
   - Caliper. Tighten the 2 bolts to 25 ft. lbs. (34 Nm) for SC 430; or 77 ft. lbs. (104 Nm) for LS 430
   - Front wheel

ES 330

1. Before servicing the vehicle, refer to the Precautions Section.
2. Remove or disconnect the following:
   - Front wheel
   - 2 bolts and remove the disc brake caliper assembly

➡ **Support the caliper. Do NOT allow to hang by the brake hose**

   - Pads with anti-squeal shims
   - Disc brake support plates
   - Slide pins

To install:

3. Install or connect the following:
   - Slide pins
   - Disc brake pad support plate
   - Anti-squeal shims to each pad
4. Using a suitable tool, compress the piston carefully in the cylinder bores
   - Inner pad with the wear indicator plate facing upward
   - Install outer pad
   - Caliper assembly and 2 bolts. Tighten to 25 ft. lbs (34 Nm)
   - Front wheel

REAR BRAKES

Brake Caliper

REMOVAL & INSTALLATION

ES 300

1. Before servicing the vehicle, refer to the Precautions Section.
2. Remove or disconnect the following:
   • Wheels
   • Brake hose from the caliper
   • Bolts that attach the caliper to the torque plate
   • Caliper assembly by lifting the bottom

To install:

3. Grease the caliper slides and bolts with lithium grease.
4. Install or connect the following:
   • Caliper. Torque the bolts to 25 ft. lbs. (34 Nm).
   • Brake hose to the caliper using 2 new washers. Torque the union bolt to 21 ft. lbs. (29 Nm).
5. Fill the master cylinder to the proper level and bleed the brake system.

IS 300

1. Before servicing the vehicle, refer to the Precautions Section.
2. Remove or disconnect the following:

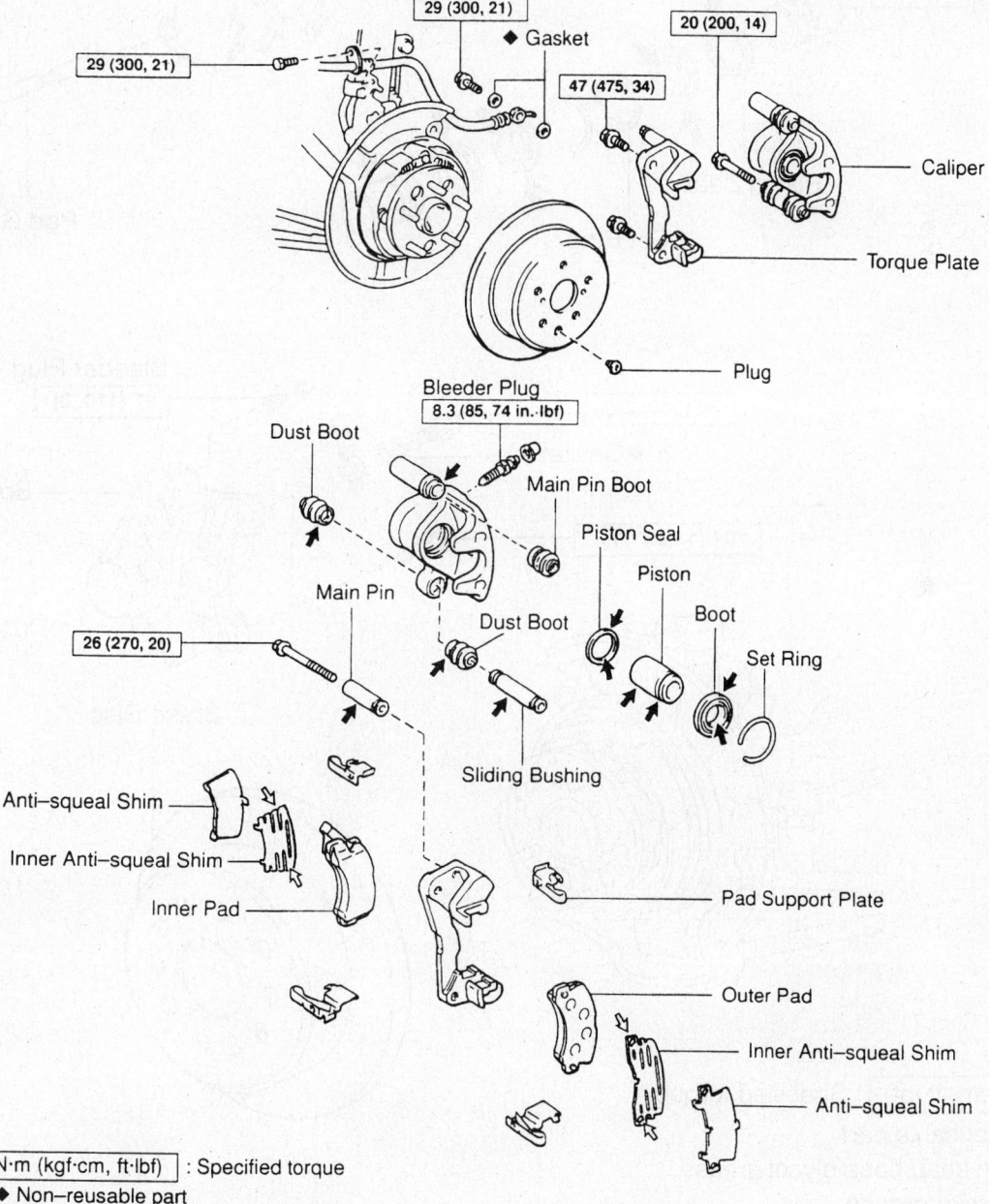

29 (300, 21)
29 (300, 21)
20 (200, 14)
◆ Gasket
47 (475, 34)
Caliper
Torque Plate
Plug
Bleeder Plug
8.3 (85, 74 in.·lbf)
Dust Boot
Main Pin Boot
Piston Seal
Piston
Boot
Main Pin
Dust Boot
Set Ring
26 (270, 20)
Sliding Bushing
Anti–squeal Shim
Inner Anti–squeal Shim
Inner Pad
Pad Support Plate
Outer Pad
Inner Anti–squeal Shim
Anti–squeal Shim

N·m (kgf·cm, ft·lbf) : Specified torque
◆ Non–reusable part
◀ Lithium soap base glycol grease
◁ Disc brake grease

93016G19

Rear disc brakes—ES 300

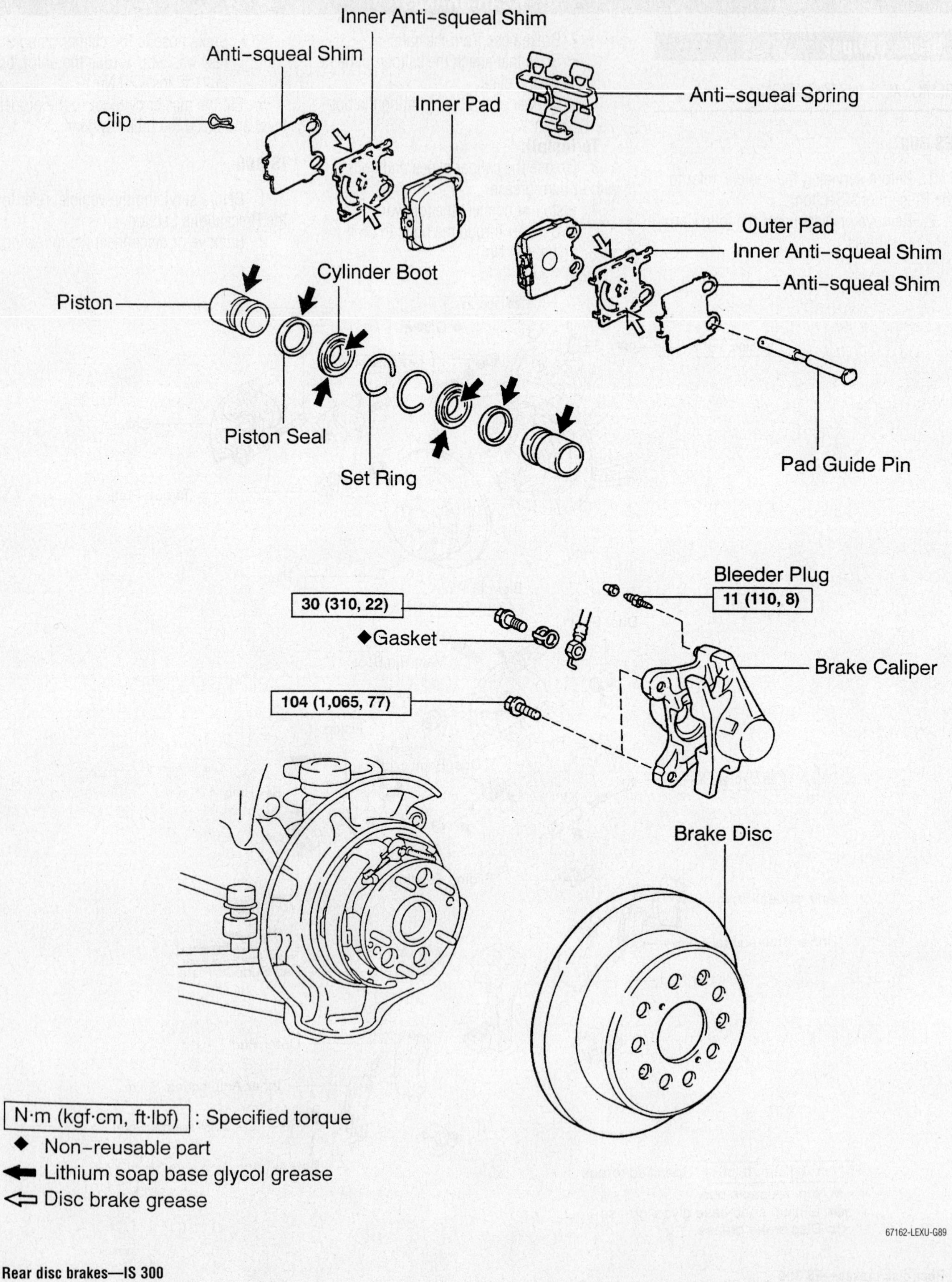

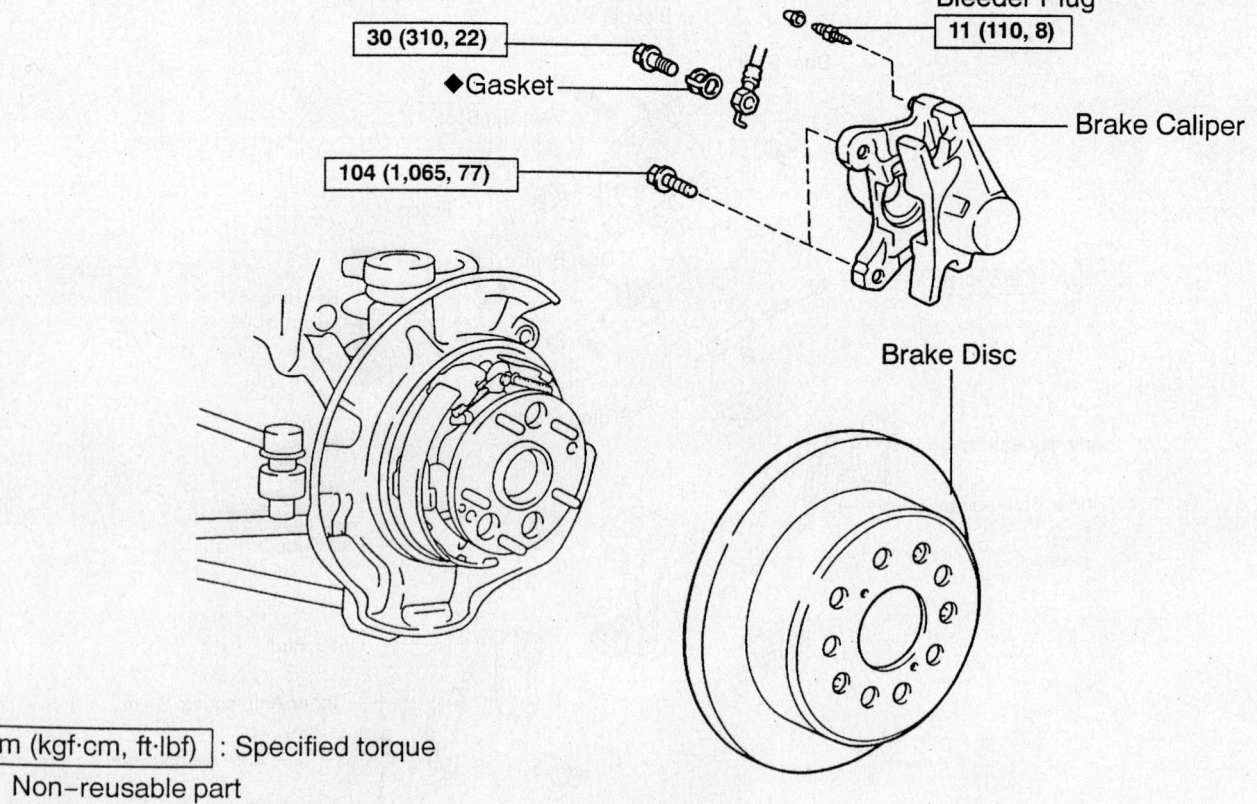

Clip

Anti-squeal Shim

Inner Anti-squeal Shim

Inner Pad

Anti-squeal Spring

Piston

Cylinder Boot

Outer Pad

Inner Anti-squeal Shim

Anti-squeal Shim

Piston Seal

Set Ring

Pad Guide Pin

Bleeder Plug

30 (310, 22)

11 (110, 8)

◆Gasket

Brake Caliper

104 (1,065, 77)

Brake Disc

N·m (kgf·cm, ft·lbf) : Specified torque

◆ Non-reusable part

◀ Lithium soap base glycol grease

◁ Disc brake grease

67162-LEXU-G89

Rear disc brakes—IS 300

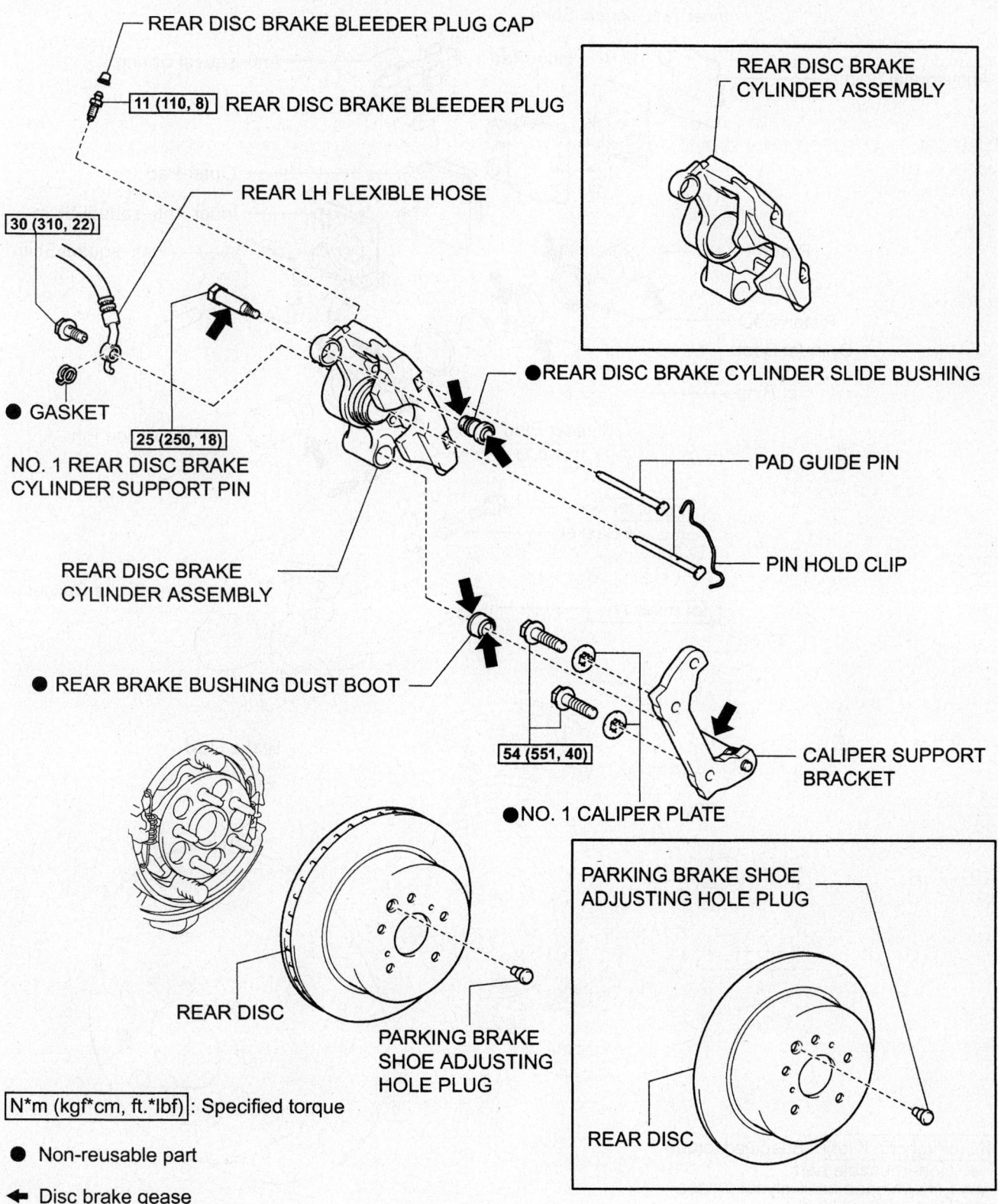

REAR DISC BRAKE BLEEDER PLUG CAP

11 (110, 8) REAR DISC BRAKE BLEEDER PLUG

REAR DISC BRAKE
CYLINDER ASSEMBLY

REAR LH FLEXIBLE HOSE

30 (310, 22)

● GASKET

25 (250, 18)

NO. 1 REAR DISC BRAKE
CYLINDER SUPPORT PIN

REAR DISC BRAKE
CYLINDER ASSEMBLY

●REAR DISC BRAKE CYLINDER SLIDE BUSHING

PAD GUIDE PIN

PIN HOLD CLIP

● REAR BRAKE BUSHING DUST BOOT

54 (551, 40)

●NO. 1 CALIPER PLATE

CALIPER SUPPORT
BRACKET

REAR DISC

PARKING BRAKE
SHOE ADJUSTING
HOLE PLUG

PARKING BRAKE SHOE
ADJUSTING HOLE PLUG

REAR DISC

N*m (kgf*cm, ft.*lbf): Specified torque

● Non-reusable part

◄ Disc brake gease

09490_LEXU_G0116

Rear disc brakes—IS 250 and IS 350

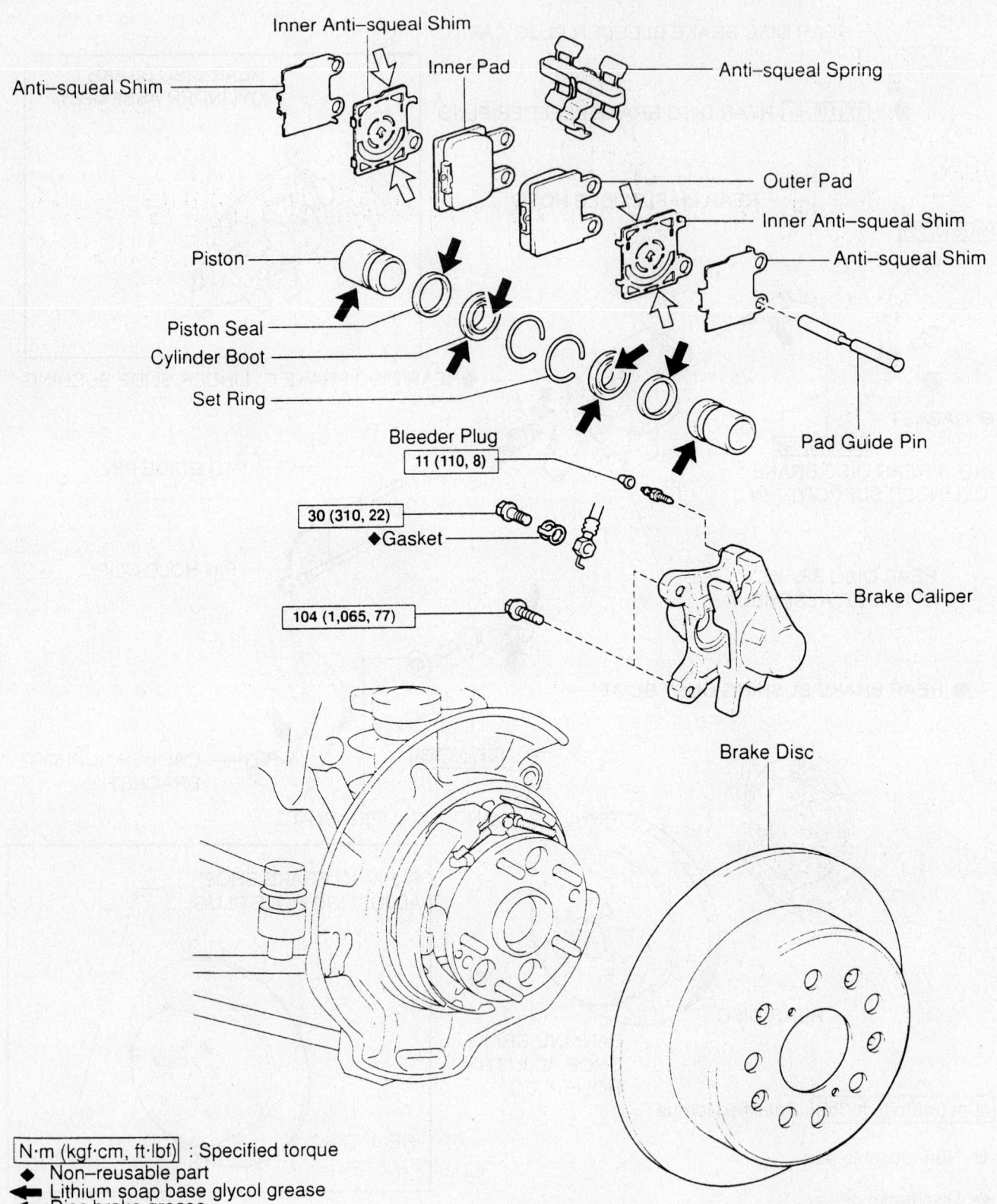

Anti-squeal Shim

Inner Anti-squeal Shim

Inner Pad

Anti-squeal Spring

Outer Pad

Inner Anti-squeal Shim

Anti-squeal Shim

Piston

Piston Seal

Cylinder Boot

Set Ring

Pad Guide Pin

Bleeder Plug
11 (110, 8)

30 (310, 22)

◆Gasket

104 (1,065, 77)

Brake Caliper

Brake Disc

N·m (kgf·cm, ft·lbf) : Specified torque
◆ Non-reusable part
◀ Lithium soap base glycol grease
◁ Disc brake grease

93016G21

Rear disc brakes—GS 300 and GS 430

- Wheels
- Brake hose from the caliper
- Bolts that attach the caliper to the torque plate
- Caliper assembly by lifting the bottom

To install:

3. Grease the caliper slides and bolts with lithium grease.

4. Install or connect the following:
  - Torque the caliper bolts to 77 ft. lbs. (34 Nm)
  - Brake hose to the caliper using 2 new washers. Torque the union bolt to 22 ft. lbs. (30 Nm)

5. Fill the master cylinder to the proper level and bleed the brake system.

IS 250 and IS 350

1. Before servicing the vehicle, refer to the Precautions Section.

2. Remove or disconnect the following:
  - Wheels
  - Pin hold clip
  - 2 pad guide pins and anti-squeal spring
  - Both brake pads with anti-squeal shims (remove the 4 anti-squeal shims from each pad)
  - Brake hose from the caliper
  - Caliper support pins
  - Caliper assembly by lifting the bottom

To install:

3. Install or connect the following:
  - Caliper assembly to the caliper support bracket
  - Caliper support pins and torque to 18 ft. lbs. (25 Nm)
  - Brake hose to the caliper using a new washer. Torque the union bolt to 22 ft. lbs. (30 Nm)
  - Both brake pads with anti-squeal shims (apply disc brake grease to both sides of the 2 antisqueal shims)
  - 2 pad guide pins and anti-squeal springs (apply disc brake grease to both sides of the 2 antisqueal springs)
  - Pin hold clip
  - Wheels

4. Fill the master cylinder to the proper level and bleed the brake system.

GS 300 and GS 430

1. Before servicing the vehicle, refer to the Precautions Section.

2. Remove or disconnect the following:
  - Wheels
  - Brake line at the caliper

- Anti-squeal springs
- Mounting bolts, while holding the sliding pin with a wrench
- Caliper assembly

To install:

3. Install or connect the following:
  - Caliper. Hold the sliding pin and tighten the mounting bolts to 25 ft. lbs. (34 Nm).
  - Anti-squeal springs
  - Connect the brake line with 2 new gaskets and tighten the union bolt to 22 ft. lbs. (30 Nm)

4. Bleed the brake system.
  - Wheels

5. Check and if necessary fill the master cylinder reservoir.

LS 430 and SC 430

1. Before servicing the vehicle, refer to the Precautions Section.

2. Remove or disconnect the following:
  - Rear wheels
  - Brake line at the caliper, then plug it
  - Mounting bolts and the caliper assembly

To install:

3. Temporarily install the caliper on the torque plate with the 2 installation bolts.

4. Hold the sliding pin and tighten the mounting bolts to (except SC 430) 25 ft. lbs. (34 Nm), SC 430 tighten to 77 ft. lbs. (104 Nm).

5. Connect the brake line with 2 new gaskets and tighten the union to 29 ft. lbs. (39 Nm).

6. Refill the reservoir as necessary and bleed the brake system.

ES 330

1. Before servicing the vehicle, refer to the Precautions Section.

2. Install or connect the following:
  - Wheels
  - Brake line at the caliper by removing the union bolt and 2 gaskets
  - Mounting bolts, while holding the sliding pin with a wrench
  - Caliper from the caliper support

To install:

3. Install or connect the following:
  - Caliper to the caliper support
  - Caliper bolts. Hold the sliding pin and tighten the mounting bolts to 25 ft. lbs. (34 Nm).
  - Brake line with (2) new gaskets and tighten the union to 22 ft. lbs. (30 Nm)

4. Bleed the brake system.
  - Wheels

5. Check and fill the master cylinder reservoir, if needed.

Disc Brake Pads

REMOVAL & INSTALLATION

ES 300 and IS 300

1. Before servicing the vehicle, refer to the Precautions Section.

2. Remove or disconnect the following:
  - Wheels
  - Lower installation bolt
  - Caliper and suspend it securely. Do NOT disconnect the fluid line.
  - Brake pads and retainers

To install:

3. Install or connect the following:
  - 2 pads so that the wear indicator plate is facing upward. Do NOT allow oil or grease to get in the rubbing face.

4. Draw out a small amount of brake fluid from the brake reservoir. Press in the caliper piston with a suitable tool.
  - Caliper. Torque the sliding main pin to 25 ft. lbs. (34 Nm).
  - Wheels

5. Check the fluid level in the master cylinder and add as necessary.

IS 250 and IS 350

1. Before servicing the vehicle, refer to the Precautions Section.

2. Remove or disconnect the following:
  - Wheels
  - Pin hold clip
  - 2 pad guide pins and anti-squeal spring
  - Both brake pads with anti-squeal shims (remove the 4 anti-squeal shims from each pad)

To install:

3. Install or connect the following:
  - Compress the caliper pistons
  - Both brake pads with anti-squeal shims (apply disc brake grease to both sides of the 2 anti-squeal shims)
  - 2 pad guide pins and anti-squeal springs (apply disc brake grease to both sides of the 2 anti-squeal springs)
  - Pin hold clip
  - Wheels

GS 300 and GS 430

1. Before servicing the vehicle, refer to the Precautions Section.

2. Remove or disconnect the following:
- Wheels

3. Hold the sliding pin on the lower mounting bolt and remove the bolt. Swivel the caliper upward and out of the way.
- Anti-squeal springs
- Brake pads, retainers and anti-squeal shims

To install:

4. Install or connect the following:
- Pad support plates and the pad wear indicator plate on the inside pad
- Both pads (and anti-squeal shims) with the wear indicator plates facing downward

5. Compress the caliper pistons and install the caliper.
- Anti-squeal springs
- Hold the sliding pin and tighten the mounting bolts to 25 ft. lbs. (34 Nm)
- Wheels

6. Check the brake fluid level in the reservoir.

LS 430 and SC 430

1. Before servicing the vehicle, refer to the Precautions Section.

2. Remove or disconnect the following:
- Rear wheel
- Anti-squeal springs
- Clip and guide pin
- Disc pads
- 4 anti-squeal shims from each pad

To install:

3. Install or connect the following:
- Apply disc brake grease to both sides of the inner anti-squeal shims
- Install 2 shims on each pad

➡**Make sure that the arrows on the anti-squeal shims face the direction of wheel rotation.**

4. Using a suitable tool, compress the piston carefully in the cylinder bores
- Install pads
- Anti-squeal springs
- Rear wheel

ES 330

1. Before servicing the vehicle, refer to the Precautions Section.

2. Remove or disconnect the following:
- Rear wheel
- Anti-squeal springs
- Clip and guide pin
- Disc pads
- 4 anti-squeal shims from each pad

To install:

3. Install or connect the following:
- Apply disc brake grease to both sides of the inner anti-squeal shims
- Install 2 shims on each pad

➡**Make sure that the arrows on the anti-squeal shims face the direction of wheel rotation.**

4. Using a suitable tool, compress the piston carefully in the cylinder bores
- Install pads
- Anti-squeal springs
- Rear wheel

TOYOTA AND LEXUS

BRAKES 2-123
DRIVE TRAIN 2-74
ENGINE REPAIR 2-11
FUEL SYSTEM 2-68
**SPECIFICATIONS AND
  MAINTENANCE CHARTS** 2-3
Engine and Vehicle
  Identification 2-3
General Engine Specifications 2-3
Engine Tune-Up Specifications 2-4
Firing Order 2-4
Accessory Drive Belt Routing 2-4
Capacities 2-5
Valve Specifications 2-5
Crankshaft and Connecting Rod
  Specifications 2-6
Piston and Ring Specifications 2-6
Torque Specifications 2-7
Wheel Alignment 2-7
Tire, Wheel and Ball Joint
  Specifications 2-8
Brake Specifications 2-9
Scheduled Maintenance
  Intervals...........................2-10
**STEERING AND
  SUSPENSION** 2-90

A

Air Bag..............................2-90
  Disarming 2-90
  Precautions 2-90
Alternator 2-11
  Removal 2-11
Automatic Transmission
  Assembly 2-74
  Removal & Installation.............2-74
Axle Shaft, Bearing and Seal.........2-85
  Removal & Installation.............2-85

C

Camshaft and Valve Lifters 2-48
  Removal & Installation.............2-48
CV-Joints...........................2-82
  Overhaul 2-82
Cylinder Head.......................2-35
  Removal & Installation.............2-35

D

Distributor...........................2-11

E

Engine Assembly 2-13
  Removal & Installation.............2-13
Exhaust Manifold 2-47
  Removal & Installation.............2-47

F

Front Brake Caliper..................2-123
  Removal & Installation.............2-123
Front Coil Spring 2-104
  Removal & Installation.............2-104
Front Crankshaft Seal 2-47
  Removal & Installation.............2-47
Front Disc Brake Pads 2-130
  Removal & Installation.............2-130
Front Shock Absorber 2-97
  Removal & Installation.............2-97
Front Wheel Bearing 2-117
  Adjustment.......................2-117
  Removal & Installation.............2-118
Fuel Filter 2-68
  Removal & Installation.............2-68
Fuel Injector........................2-71
  Removal & Installation.............2-71
Fuel Pump 2-69
  Removal & Installation.............2-69
Fuel System Pressure 2-68
  Relieving 2-68
Fuel System Service
  Precautions.......................2-68

H

Halfshaft............................2-79
  Removal & Installation.............2-79
Heater Core........................2-15
  Removal & Installation.............2-15

I

Ignition Timing 2-13
  Adjustment........................2-13
Intake Manifold 2-42
  Removal & Installation.............2-42

L

Lower Ball Joint.....................2-108
  Removal & Installation.............2-108

Lower Control Arm 2-112
  Control Arm Bushing
    Replacement.....................2-116
  Removal & Installation.............2-112

M

MacPherson Struts 2-96
  Removal & Installation.............2-96

O

Oil Pan.............................2-63
  Removal & Installation.............2-63
Oil Pump 2-66
  Removal & Installation.............2-66

P

Pinion Seal 2-88
  Removal & Installation.............2-88
Piston and Ring 2-67
  Positioning 2-67
Power Rack And Pinion Steering
  Gear..............................2-90
  Removal & Installation.............2-90

R

Rear Air Spring.....................2-105
  Removal & Installation.............2-105
Rear Brake Caliper 2-126
  Removal & Installation.............2-126
Rear Coil Spring 2-105
  Removal & Installation.............2-105
Rear Disc Brake Pads 2-130
  Removal & Installation.............2-130
Rear Main Seal 2-67
  Removal & Installation.............2-67
Rear Shock Absorber 2-97
  Removal & Installation.............2-97

S

Sequoia Front Strut.................2-103
  Removal & Installation.............2-103
Spindle Bearings 2-83
  Removal, Packing &
    Installation......................2-83
Starter Motor 2-57
  Removal & Installation.............2-57

T

Timing Belt 2-57
  Removal & Installation.............2-57

Torsion Bars 2-105
   Removal & Installation............2-105
Transfer Case Assembly.................2-77
   Removal & Installation..............2-77

U

Upper Ball Joint..........................2-107
   Removal & Installation............2-107

Upper Control Arm 2-108
  Control Arm Bushing
    Replacement.........................2-110
  Removal & Installation............2-108

V

Valve Lash 2-54
  Adjustment................................2-54

W

Water Pump 2-34
  Removal & Installation..............2-34

SPECIFICATIONS AND MAINTENANCE CHARTS

ENGINE AND VEHICLE IDENTIFICATION

Code ①	Liters (cc)	Cu. In.	Cyl.	Fuel Sys.	Engine Type	Eng. Mfg.	Code ②	Year
				Engine			**Model Year**	
2UZ-FE	4.7 (4664)	285	8	SFI	DOHC	Toyota	2	2002

SFI: Sequential Fuel Injection

DOHC: Double Overhead Camshaft

① Stamped on the left side of the engine block

② 10th digit of the Vehicle Identification Number (VIN)

Code ②	Year
2	2002
3	2003
4	2004
5	2005
6	2006

09490_LCRUIS_C0001

GENERAL ENGINE SPECIFICATIONS

Year	Model	Engine Displacement Liters	Engine Series ID	Net Horsepower @ rpm	Net Torque @ rpm (ft. lbs.)	Bore x Stroke (in.)	Compression Ratio	Oil Pressure @ rpm
2002	LX470	4.7	2UZ-FE	245@4800	315@3400	3.70x3.31	9.6:1	45-65@3000
	Land Cruiser	4.7	2UZ-FE	245@4800	315@3400	3.70x3.30	9.6:1	45-65@3000
	Sequoia	4.7	2UZ-FE	245@4800	315@3400	3.70x3.30	9.6:1	45-65@3000
2003	LX470	4.7	2UZ-FE	245@4800	315@3400	3.70x3.31	9.6:1	45-65@3000
	GX470	4.7	2UZ-FE	245@4800	320@3400	3.70x3.31	9.6:1	45-65@3000
	Land Cruiser	4.7	2UZ-FE	245@4800	315@3400	3.70x3.30	9.6:1	45-65@3000
	Sequoia	4.7	2UZ-FE	245@4800	315@3400	3.70x3.30	9.6:1	45-65@3000
2004	LX470	4.7	2UZ-FE	245@4800	315@3400	3.70x3.31	9.6:1	45-65@3000
	GX470	4.7	2UZ-FE	245@4800	315@3400	3.70x3.31	9.6:1	45-65@3000
	Land Cruiser	4.7	2UZ-FE	245@4800	315@3400	3.70x3.31	9.6:1	45-65@3000
	Sequoia	4.7	2UZ-FE	245@4800	315@3400	3.70x3.31	9.6:1	45-65@3000
2005-06	LX470	4.7	2UZ-FE	245@4800	315@3400	3.70x3.31	9.6:1	45-65@3000
	GX470	4.7	2UZ-FE	245@4800	315@3400	3.70x3.31	9.6:1	45-65@3000
	Land Cruiser	4.7	2UZ-FE	245@4800	315@3400	3.70x3.31	9.6:1	45-65@3000
	Sequoia	4.7	2UZ-FE	245@4800	315@3400	3.70x3.31	9.6:1	45-65@3000

09490_LCRUIS_C0002

ENGINE TUNE-UP SPECIFICATIONS

Year	Engine Displacement Liters	Engine ID	Spark Plug Gap (in.)	Ignition Timing (deg.)*	Fuel Pump (psi)	Idle Speed (rpm) MT	AT	Valve Clearance Intake	Exhaust
2002	4.7	2UZ-FE	0.031	5-15B	38-44	—	650-750	0.006-0.010	0.010-0.014
2003	4.7	2UZ-FE	0.031	5-15B	38-44	—	650-750	0.006-0.010	0.010-0.014
2004	4.7	2UZ-FE	0.043	5-15B	38-44	—	650-750	0.006-0.010	0.010-0.014
2005-06	4.7	2UZ-FE	0.043	5-15B	38-44	—	650-750	0.006-0.010	0.010-0.014

NOTE: The Vehicle Emission Control Information label often reflects specification changes made during production.

The label figures must be used if they differ from those in this chart.

B: Before top dead center

* With terminals TC and E1 connected to DLC1

09490_LCRUIS_C0003

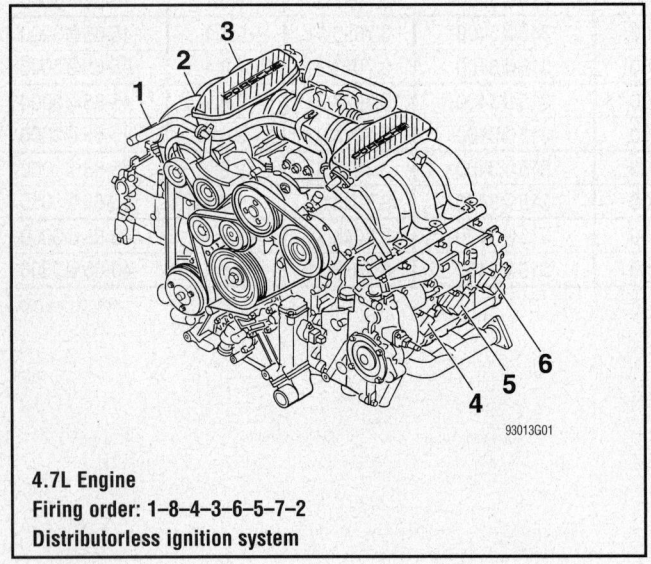

4.7L Engine
Firing order: 1–8–4–3–6–5–7–2
Distributorless ignition system

93013G01

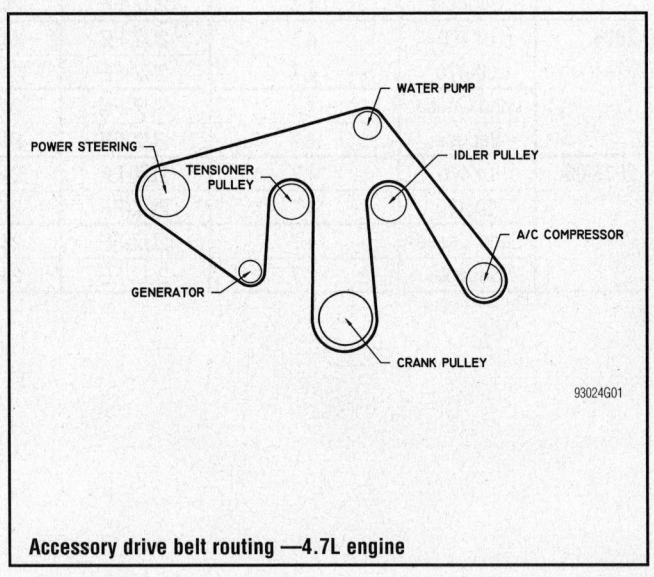

Accessory drive belt routing —4.7L engine

93024G01

CAPACITIES

Year	Model	Engine Displacement Liters	Engine ID	Engine Oil with Filter (qts.)	Transmission (pts.) 5-Spd	Transmission (pts.) Auto.*	Transfer Case (pts.)	Drive Axle Front (pts.)	Drive Axle Rear (pts.)	Fuel Tank (gal.)	Cooling System (qts.)
2002	LX470	4.7	2UZ-FE	6.6	—	4.2	2.6	2.4	7.7	25.4	12.3
	Land Cruiser	4.7	2UZ-FE	7.2	—	7.4	2.8	3.6	①	25.4	②
	Sequoia	4.7	2UZ-FE	6.6	—	4.2	2.6	2.4	7.7	26.1	12.3
2003	LX470	4.7	2UZ-FE	6.6	—	4.2	2.6	2.4	7.7	25.4	12.3
	GX470	4.7	2UZ-FE	6.5	—	6.4	3.0	3.9	6.4	23.0	13.6
	Land Cruiser	4.7	2UZ-FE	7.2	—	6.4	2.8	3.6	①	25.4	②
	Sequoia	4.7	2UZ-FE	6.6	—	4.2	2.6	2.4	7.7	26.1	12.3
2004	LX470	4.7	2UZ-FE	7.2	—	6.4	2.8	3.4	7.0	25.4	②
	GX470	4.7	2UZ-FE	6.5	—	6.4	3.0	3.9	6.4	23.0	13.6
	Land Cruiser	4.7	2UZ-FE	7.2	—	6.4	2.8	3.6	①	25.4	②
	Sequoia	4.7	2UZ-FE	6.6	—	4.2	2.6	2.4	7.7	26.1	12.3
2005-06	LX470	4.7	2UZ-FE	7.2	—	6.4	2.8	3.4	7.0	25.4	②
	GX470	4.7	2UZ-FE	6.5	—	6.4	3.0	3.9	6.4	23.0	13.6
	Land Cruiser	4.7	2UZ-FE	7.2	—	6.4	2.8	3.6	①	25.4	②
	Sequoia	4.7	2UZ-FE	6.6	—	4.2	2.6	2.4	7.7	26.1	12.3

* After draining, add the following amounts, then fill to the cold full line

① With rear heater: 9.5
   Without rear heater: 8.5

② Without rear heater: 15.6
   With rear heater: 16.2

09490_LCRUIS_C0004

VALVE SPECIFICATIONS

Year	Engine Displacement Liters	Engine ID	Seat Angle (deg.)	Face Angle (deg.)	Spring Test Pressure (lbs. @ in.)	Spring Installed Height (in.)	Stem-to-Guide Clearance (in.) Intake	Stem-to-Guide Clearance (in.) Exhaust	Stem Diameter (in.) Intake	Stem Diameter (in.) Exhaust
2002	4.7	2UZ-FE	45	44.5	45.9-50.7@ 1.378	1.380	0.0010-0.0024	0.0012-0.0026	0.2154-0.2159	0.2152-0.2157
2003	4.7	2UZ-FE	45	44.5	45.9-50.7@ 1.378	1.380	0.0010-0.0024	0.0012-0.0026	0.2154-0.2159	0.2152-0.2157
2004	4.7	2UZ-FE	45	44.5	45.9-50.7@ 1.378	1.380	0.0010-0.0024	0.0012-0.0026	0.2154-0.2159	0.2152-0.2157
2005-06	4.7	2UZ-FE	45	44.5	45.9-50.7@ 1.378	1.380	0.0010-0.0024	0.0012-0.0026	0.2154-0.2159	0.2152-0.2157

09490_LCRUIS_C0005

CRANKSHAFT AND CONNECTING ROD SPECIFICATIONS
All measurements are given in inches.

Year	Engine Displacement Liters	Engine ID	Crankshaft				Connecting Rod		
			Main Brg. Journal Dia.	Main Brg. Clearance	Shaft End-play	Thrust on No.	Journal Diameter	Oil Clearance	Side Clearance
2002	4.7	2UZ-FE	2.6373-2.6378	①	0.0008-0.0087	3	2.0465-2.0472	0.0011-0.0021	0.0063-0.0138
2003	4.7	2UZ-FE	2.6373-2.6378	①	0.0008-0.0087	3	2.0465-2.0472	0.0011-0.0021	0.0063-0.0138
2004	4.7	2UZ-FE	2.6373-2.6378	①	0.0008-0.0087	3	2.0465-2.0472	0.0011-0.0021	0.0063-0.0138
2005-06	4.7	2UZ-FE	2.6373-2.6378	①	0.0008-0.0087	3	2.0465-2.0472	0.0011-0.0021	0.0063-0.0138

① Nos. 1 and 2: 0.0011-0.0018

All others: 0.0016-0.0023

09490_LCRUIS_C0006

PISTON AND RING SPECIFICATIONS
All measurements are given in inches.

Year	Engine Displacement Liters	Engine ID	Piston Clearance	Ring Gap			Ring Side Clearance		
				Top Comp.	Bottom Comp.	Oil Control	Top Comp.	Bottom Comp.	Oil Control
2002	4.7	2UZ-FE	0.0035-0.0044	0.0118-0.0157	0.0157-0.0217	0.0051-0.0150	0.0012-0.0031	0.0012-0.0028	SNUG
2003	4.7	2UZ-FE	0.0035-0.0044	0.0118-0.0157	0.0157-0.0217	0.0051-0.0150	0.0012-0.0031	0.0012-0.0028	SNUG
2004	4.7	2UZ-FE	0.0035-0.0044	0.0118-0.0157	0.0157-0.0217	0.0051-0.0150	0.0012-0.0031	0.0012-0.0028	SNUG
2005-06	4.7	2UZ-FE	0.0035-0.0044	0.0118-0.0157	0.0157-0.0217	0.0051-0.0150	0.0012-0.0031	0.0012-0.0028	SNUG

09490_LCRUIS_C0007

TORQUE SPECIFICATIONS
All readings in ft. lbs.

Year	Engine Displacement Liters	Engine ID	Cylinder Head Bolts	Main Bearing Bolts	Rod Bearing Bolts	Crankshaft Damper Bolts	Flywheel Bolts	Manifold		Spark Plugs	Oil Pan Drain Plug
								Intake	Exhaust		
2002	4.7	2UZ-FE	①	②	③	181	④	13	33	13	29
2003	4.7	2UZ-FE	①	②	③	181	④	13	33	13	29
2004	4.7	2UZ-FE	①	②	③	181	④	13	33	13	29
2005	4.7	2UZ-FE	①	②	③	181	④	13	33	13	29
2006	4.7	2UZ-FE	⑤	②	③	181	④	13	33	13	29

① Step 1: 24
  Step 2: Plus 90 degrees
  Step 3: Plus 90 degrees

② Step 1: 20 ft. lbs.
  Step 2: Plus 90 degrees

③ Step 1: 18 ft. lbs.
  Step 2: Plus 90 degrees

④ Step 1: 35 ft. lbs.
  Step 2: Plus 90 degrees

⑤ Step 1: 30
  Step 2: Plus 90 degrees
  Step 3: Plus 90 degrees

09490_LCRUIS_C0008

WHEEL ALIGNMENT

Year	Model	Caster Range (+/-Deg.)	Caster Preferred Setting (Deg.)	Camber Range (+/-Deg.)	Camber Preferred Setting (Deg.)	Toe-in (in.)	Steering Axis Inclination (Deg.)
2002	LX470	0.75	①	0.75	+0.13	0.05+/-0.08	10.635+/-0.75
	Land Cruiser	0.75	+2.50	0.75	+0.08	0.04+/-0.08	12.17+/-0.75
	Sequoia	0.75	①	0.75	+0.13	0.05+/-0.08	10.635+/-0.75
2003	LX470	0.75	①	0.75	+0.13	0.05+/-0.08	10.635+/-0.75
	GX470	0.75	+3.28	0.75	-0.02	0.08+/-0.16	12.48+/-0.75
	Land Cruiser	0.75	+2.50	0.75	+0.08	0.04+/-0.08	12.17+/-0.75
	Sequoia	0.75	①	0.75	+0.13	0.05+/-0.08	10.635+/-0.75
2004	LX470	0.75	+3.08	0.75	0	0+/-0.08	12.25+/-0.75
	GX470	0.75	+3.28	0.75	-0.02	0.08+/-0.16	12.48+/-0.75
	Land Cruiser	0.75	+2.50	0.75	+0.08	0.04+/-0.08	12.17+/-0.75
	Sequoia	0.75	①	0.75	+0.13	0.05+/-0.08	10.635+/-0.75
2005-06	LX470	0.75	+3.08	0.75	0	0+/-0.08	12.25+/-0.75
	GX470	0.75	+3.28	0.75	-0.02	0.08+/-0.16	12.48+/-0.75
	Land Cruiser	0.75	+2.50	0.75	+0.08	0.04+/-0.08	12.17+/-0.75
	Sequoia	0.75	①	0.75	+0.13	0.05+/-0.08	10.635+/-0.75

Note: All alignment specifications are based on nominal ride height and standard tires

① P245/70R16: +2.95
  P265/70R16: +3.00

09490_LCRUIS_C0009

TIRE, WHEEL AND BALL JOINT SPECIFICATIONS

| Year | Model | OEM Tires | | Tire Pressures (psi) | | Wheel Size | Ball Joint Inspection | Lugnut Torque (ft. lbs.) |
		Standard	Optional	Front	Rear			
2002	LX470	P275/70HR16	None	32	32	8-JJ	②	97
	Land Cruiser	P275/70R16	None	29①	32①	8-JJ	③	97
	Sequoia	P245/70R16	P265/70R16	32	Std: 35; Opt: 32	7-JJ	④	83
2003	LX470	P275/60HR18	None	⑥	⑥	8-JJ	③	97
	GX470	P265/65SR17	None	⑥	⑥	7.5	④	NA
	Land Cruiser	P275/65R17	P275/60R18	⑥	⑥	8-JJ	③	97
	Sequoia	P265/70R16	P265/6R17	32	Std: 35; Opt: 32	7-JJ	④	83
2004	LX470	P275/60HR18	None	NA	NA	8-JJ	⑤	97
	GX470	P265/65SR17	None	⑥	⑥	7.5	④	NA
	Land Cruiser	P275/65R17	P275/60R18	⑥	⑥	8-JJ	②	97
	Sequoia	P265/70R16	P265/65R17	⑥	⑥	std: 7; opt: 7.5	③	83
2005-06	LX470	P275/60HR18	None	⑥	⑥	8-JJ	⑤	97
	GX470	P265/65SR17	None	⑥	⑥	7.5	④	NA
	Land Cruiser	P275/65R17	P275/60R18	⑥	⑥	8-JJ	②	97
	Sequoia	P265/70R16	P265/65R17	⑥	⑥	std: 7; opt: 7.5	③	83

OEM: Original Equipment Manufacturer

PSI: Pounds Per Square Inch

STD: Standard

OPT: Optional

NA: Not Available

① Trailer towing: front 32; rear 35

② Replace if any measurable movement is found.

③ Upper ball joint turning torque: 6-39 inch lbs.

   Lower ball joint turning torque: 1-22 inch lbs.

   Lower ball joint excessive play: 0.020 in.

④ Upper arm ball joint turning torque: 40 inch lbs. or less

   Lower arm ball joint turning torque: 27 inch lbs or less

⑤ Upper arm ball joint turning torque: 9-39 inch lbs.

   Lower arm ball joint turning torque: 2.6-26 inch lbs.

⑥ See placard on vehicle

09490_LCRUIS_C0010

BRAKE SPECIFICATIONS

All measurements in inches unless noted

| Year | Model | | Brake Disc | | | Minimum Lining Thickness | Brake Caliper | |
			Original Thickness	Minimum Thickness	Maximum Runout		Bracket Bolts (ft. lbs.)	Mounting Bolts (ft. lbs.)
2002	LX470	F	1.260	1.181	0.0028	0.039	—	90
		R	0.709	0.611	0.0040	0.039	—	20
	Land Cruiser	F	1.260	1.181	0.0028	0.039	—	90
		R	0.709	0.611	—	0.039	76	20
	Sequoia	F	1.102	1.024	0.0028	0.039	—	90
		R	0.709	0.611	0.0039	0.039	77	65
2003	LX470	F	1.260	1.181	0.0028	0.039	—	90
		R	0.709	0.611	0.0040	0.039	—	20
	GX 470	F	1.102	1.024	0.0020	0.039	—	91
		R	0.709	0.630	0.0079	0.039	77	65
	Land Cruiser	F	1.260	1.181	0.0028	0.039	—	90
		R	0.709	0.611	—	0.039	76	20
	Sequoia	F	1.102	1.024	0.0028	0.039	—	90
		R	0.709	0.611	0.0039	0.039	77	65
2004	LX470	F	1.260	1.181	0.0028	0.039	—	90
		R	0.709	0.611	0.0040	0.039	—	20
	GX 470	F	1.102	1.024	0.0020	0.039	—	91
		R	0.709	0.630	0.0079	0.039	77	65
	Land Cruiser	F	1.260	1.181	0.0028	0.039	—	90
		R	0.709	0.611	—	0.039	76	20
	Sequoia	F	1.102	1.024	0.0028	0.039	—	90
		R	0.709	0.611	0.0039	0.039	77	65
2005-06	LX470	F	1.260	1.181	0.0028	0.039	—	90
		R	0.709	0.611	0.0040	0.039	—	20
	GX 470	F	1.102	1.024	0.0020	0.039	—	91
		R	0.709	0.630	0.0079	0.039	77	65
	Land Cruiser	F	1.260	1.181	0.0028	0.039	—	90
		R	0.709	0.611	—	0.039	76	20
	Sequoia	F	1.102	1.024	0.0028	0.039	—	90
		R	0.709	0.611	0.0039	0.039	77	65

F: Front

R: Rear

09490_LCRUIS_C0011

SCHEDULED MAINTENANCE INTERVALS
Toyota Land Cruiser/Sequoia, Lexus GX470/LX470

TO BE SERVICED	TYPE OF SERVICE	VEHICLE MILEAGE INTERVAL (x1000)												
		7.5	15	22.5	30	37.5	45	52.5	60	67.5	75	82.5	90	97.5
Engine oil & filter	R	✓	✓	✓	✓	✓	✓	✓	✓	✓	✓	✓	✓	✓
Automatic transmission fluid & filter	S/I		✓		✓		✓		✓		✓		✓	
Ball joints & dust covers	S/I		✓		✓		✓		✓		✓		✓	
Bolts & nuts on chassis & body	S/I		✓		✓		✓		✓		✓		✓	
Brake line pipes & hoses	S/I		✓		✓		✓		✓		✓		✓	
Brake pads & discs	S/I		✓		✓		✓		✓		✓		✓	
Propeller shaft grease	S/I		✓		✓		✓		✓		✓		✓	
Steering knuckle & chassis grease	S/I		✓		✓		✓		✓		✓		✓	
Steering linkage	S/I		✓		✓		✓		✓		✓		✓	
Transfer and differential oil	S/I		✓		✓		✓		✓		✓		✓	
Air cleaner filter	R				✓				✓				✓	
Spark plugs ①	R				✓				✓				✓	
Drive belts	S/I				✓				✓				✓	
Exhaust pipes & mountings	S/I					✓			✓				✓	
Fuel lines & connections	S/I					✓			✓				✓	
Engine coolant	R						✓				✓			
Charcoal canister	R								✓					
Fuel tank cap gasket	R								✓					
Heated oxygen sensors (exc. Cal.) ②	R													

R: Replace   S/I: Service or Inspect

① Platinum plugs, replace every 100,000 miles

② Heated oxygen sensors (except Calif.): replace every 80,000 miles.

FREQUENT OPERATION MAINTENANCE (SEVERE SERVICE)

If a vehicle is operated under any of the following conditions it is considered severe service:

- Extremely dusty areas.

- 50% or more of the vehicle operation is in 32°C (90°F) or higher temperatures, or constant operation in temperatures below 0°C (32°F).

- Prolonged idling (vehicle operation in stop and go traffic).

- Frequent short running periods (engine does not warm to normal operating temperatures).

- Police, taxi, delivery usage or trailer towing usage.

Air cleaner filter: service or inspect every 3750 miles.

Engine oil & filter: replace every 3750 miles.

Ball joints & dust covers: service or inspect every 7500 miles.

Bolts & nuts on chassis & body: service or inspect every 7500 miles.

Brake pads & discs (front & rear): service or inspect every 7500 miles.

Steering knuckle & chassis grease: service or inspect every 7500 miles.

Steering linkage: service or inspect every 7500 miles.

Propeller shaft grease: service or inspect every 7500 miles.

Exhaust pipes & mountings: service or inspect every 15,000 miles.

ENGINE REPAIR

➡**Disconnecting the negative battery cable on some vehicles may interfere with the functions of the on board computer system. The computer may undergo a relearning process once the negative battery cable is reconnected.**

Distributor

All models are equipped with a distibutorless ignition system.

Alternator

REMOVAL

Land Cruiser

2002–05

1. Before servicing the vehicle, refer to the precautions section.

2. Drain the cooling system.
3. Remove or disconnect the following:

- Negative battery cable
- Accessory drive belt
- Engine under cover
- Radiator
- Power steering pump pulley
- Alternator harness connectors
- Alternator

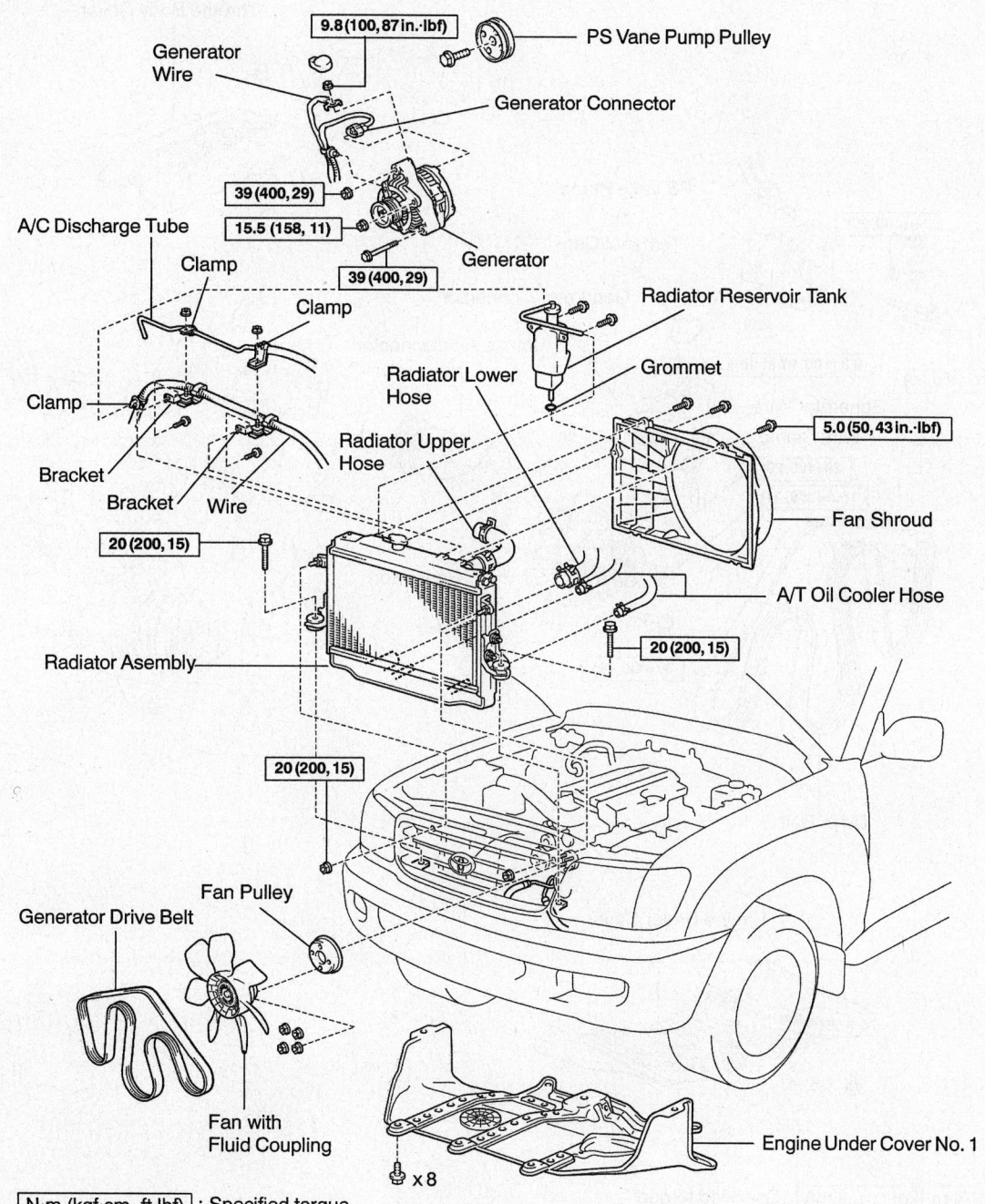

N·m (kgf·cm, ft·lbf) : Specified torque

Alternator and related parts—Land Cruiser

67170-LCSQ-G01

To install:

4. Install or connect the following:
- Alternator. Tighten the fasteners to 29 ft. lbs. (39 Nm).
- Alternator harness connectors
- Power steering pump pulley
- Radiator
- Engine under cover
- Accessory drive belt
- Negative battery cable

5. Fill the cooling system.
6. Start the engine and check for leaks.

2006

1. Before servicing the vehicle, refer to the precautions section.
2. Disconnect the negative battery cable.
3. Remove the drive belt.
4. Disconnect the 2 oil cooler lines from the fan shroud, remove the fan shroud.
5. Remove the 4 nuts and remove the fan with the fluid coupling.
6. Disconnect the vane pump assembly.
7. Disconnect the alternator wiring.

8. Remove the nuts and bolts and remove the alternator.

To install:

9. Install the alternator. Tighten the bolt to 29 ft. lbs. (39 Nm), the upper nut to 29 ft. lbs. (39 Nm) and the side nut to 12 ft. lbs. (16 Nm).
10. Attach the alternator wiring.
11. Install the vane pump assembly.
12. Install the shroud and fluid coupling together and tighten shroud bolts. Tighten the fan coupling nuts to 21 ft. lbs. (29 Nm).

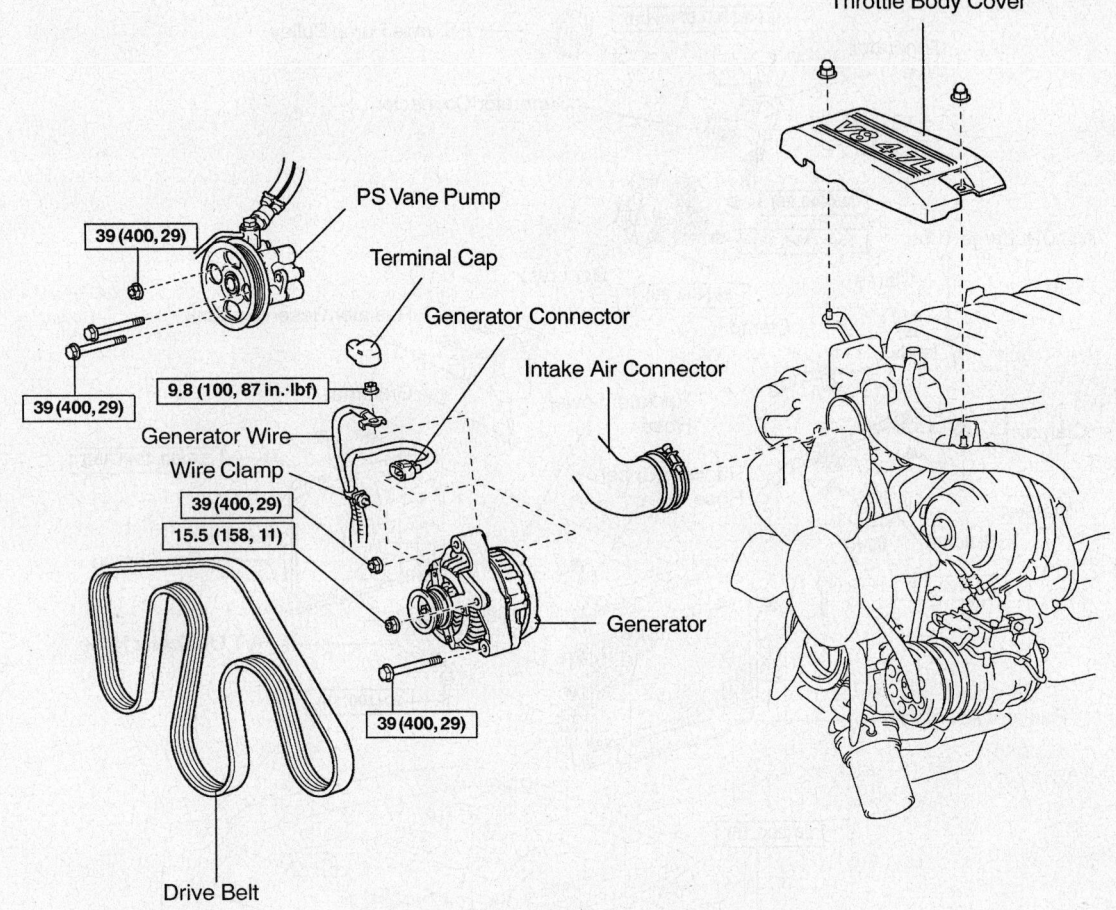

Throttle Body Cover

PS Vane Pump

Terminal Cap

Generator Connector

Intake Air Connector

39 (400, 29)

39 (400, 29)

9.8 (100, 87 in.·lbf)

Generator Wire

Wire Clamp

39 (400, 29)

15.5 (158, 11)

Generator

39 (400, 29)

Drive Belt

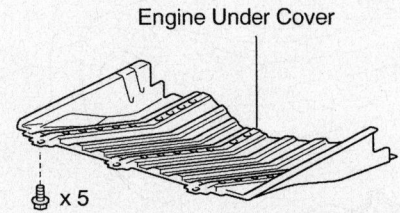

Engine Under Cover

x 5

N·m (kgf·cm, ft·lbf) : Specified torque

67170-LCSQ-G02

Alternator and related parts—Sequoia

Sequoia

1. Before servicing the vehicle, refer to the precautions section.
2. Drain the cooling system.
3. Remove or disconnect the following:
   - Negative battery cable
   - Engine under cover
   - Throttle body cover
   - Intake air connector from the throttle body
   - Accessory drive belt
   - Power steering pump
   - Alternator harness connectors
   - Alternator

To install:

4. Install or connect the following:
   - Alternator. Tighten the fasteners to 29 ft. lbs. (39 Nm).
   - Alternator harness connectors
   - Power steering pump
   - Engine under cover
   - Intake air connector to the throttle body
   - Throttle body cover
   - Accessory drive belt
   - Negative battery cable
5. Fill the cooling system.
6. Start the engine and check for leaks.

GX470 and LX470

2002–05

1. Before servicing the vehicle, refer to the precautions section.
2. Drain the cooling system.
3. Remove or disconnect the following:
   - Negative battery cable
   - Accessory drive belt
   - Engine under cover
   - Radiator
   - Power steering pump pulley
   - Alternator harness connectors
   - Alternator

To install:

4. Install or connect the following:
   - Alternator. Tighten the fasteners to 29 ft. lbs. (39 Nm).
   - Alternator harness connectors
   - Power steering pump pulley
   - Radiator
   - Engine under cover
   - Accessory drive belt
   - Negative battery cable
5. Fill the cooling system.
6. Start the engine and check for leaks.

2006

1. Before servicing the vehicle, refer to the precautions section.
2. Disconnect the negative battery cable.

3. Remove the drive belt.
4. Disconnect the 2 oil cooler lines from the fan shroud, remove the fan shroud.
5. Remove the 4 nuts and remove the fan with the fluid coupling.
6. Disconnect the vane pump assembly.
7. Disconnect the alternator wiring.
8. Remove the nuts and bolts and remove the alternator.

To install:

9. Install the alternator. Tighten the bolt to 29 ft. lbs. (39 Nm), the upper nut to 29 ft. lbs. (39 Nm) and the side nut to 12 ft. lbs. (16 Nm).
10. Attach the alternator wiring.
11. Install the vane pump assembly.
12. Install the shroud and fluid coupling together and tighten shroud bolts. Tighten the fan coupling nuts to 21 ft. lbs. (29 Nm).

Ignition Timing

ADJUSTMENT

All engines are equipped with a Distributorless Ignition System (DIS). No timing adjustment is possible.

Engine Assembly

REMOVAL & INSTALLATION

Sequoia

1. Before servicing the vehicle, refer to the precautions section.
2. Relieve the fuel system pressure.
3. Drain the cooling system.
4. Drain the engine oil.
5. Remove or disconnect the following:
   - Battery and tray
   - Hood
   - Engine appearance cover
   - Air intake pipe
   - Engine under covers
   - Coolant recovery tank
   - Radiator hoses
   - Radiator and fan shroud
   - Accessory drive belt
   - Cooling fan and pulley
   - Powertrain Control Module (PCM) harness connectors and pass the wiring harness through the firewall
   - Accelerator cable
   - Power steering vacuum hoses
   - Alternator harness connectors
   - Heater hoses
   - Engine control wiring harness and grommet at the firewall
   - Ground cable connector
   - Fuel lines

   - Evaporative Emissions (EVAP) canister hoses
   - Wire clamp at right inner fender
   - Negative battery cable at the relay box and right inner fender
   - Positive battery cable
   - Center console
   - Transmission shift lever assembly
   - Transfer case shift lever and rod
   - Exhaust front pipes
   - Stabilizer bar
   - Front and rear driveshafts
   - A/C compressor
   - Power steering pump
6. Attach a hoist to the engine lifting eyes.
7. Remove or disconnect the following:
   - Transfer case skid plate
   - Left and right motor mounts
   - Transmission mount crossmember
8. Attach a hoist to the engine lifting eyes and raise the powertrain out of the vehicle.

To install:

9. Lower the powertrain into the vehicle.
10. Install or connect the following:
    - Rear mount bracket on 2wd models. Tighten to 48 ft. lbs. (65 Nm).
    - Transmission mount crossmember. Tighten the nuts to 13 ft. lbs. (18 Nm) and the nuts to 53 ft. lbs. (72 Nm).
    - Transfer case skid plate
    - Left and right motor mounts. Tighten the fasteners to 28 ft. lbs. (38 Nm).
    - Power steering pump. Tighten the bolts to 13 ft. lbs. (17 Nm).
    - A/C compressor. Tighten the bolts to 36 ft. lbs. (49 Nm).
    - Front driveshaft. Tighten the fasteners to 59 ft. lbs. (80 Nm).
    - Rear driveshaft. Tighten the fasteners to 78 ft. lbs. (106 Nm).
    - Stabilizer bar. Tighten the bracket bolts to 13 ft. lbs. (18 Nm) and the link nuts to 18 ft. lbs. (25 Nm).
    - Exhaust front pipes
    - Transfer case shift lever and rod
    - Transmission shift lever assembly
    - Center console
    - Positive battery cable
    - Negative battery cable at the relay box and right inner fender
    - Wire clamp at right inner fender
    - EVAP canister hoses
    - Fuel lines
    - Ground cable connector
    - Engine control wiring harness and grommet at the firewall
    - Heater hoses

- Alternator harness connectors
- Power steering vacuum hoses
- Accelerator cable
- PCM harness connectors
- Cooling fan and pulley
- Accessory drive belt
- Radiator and fan shroud
- Radiator hoses
- Coolant recovery tank
- Engine under covers
- Air intake pipe
- Engine appearance cover
- Hood
- Battery and tray

11. Fill the crankcase to the correct level.
12. Fill the cooling system.
13. Start the engine and check for leaks.

Land Cruiser

1. Before servicing the vehicle, refer to the precautions section.
2. Relieve the fuel system pressure.
3. Drain the cooling system.
4. Drain the engine oil.
5. Remove or disconnect the following:
   - Battery and tray
   - Hood
   - Engine appearance cover
   - Air intake pipe
   - Engine under covers
   - Coolant recovery tank
   - Radiator hoses
   - Radiator and fan shroud
   - Accessory drive belt
   - Cooling fan and pulley
   - Powertrain Control Module (PCM) harness connectors and pass the wiring harness through the firewall
   - Accelerator cable
   - Power steering vacuum hoses
   - Alternator harness connectors
   - Heater hoses
   - Engine control wiring harness and grommet at the firewall
   - Ground cable connector
   - Fuel lines
   - Evaporative Emissions (EVAP) canister hoses
   - Wire clamp at right inner fender
   - Negative battery cable at the relay box and right inner fender
   - Positive battery cable
   - Center console
   - Transmission shift lever assembly
   - Transfer case shift lever and rod
   - Exhaust front pipes
   - Stabilizer bar
   - Front and rear driveshafts
   - A/C compressor
   - Power steering pump

6. Attach a hoist to the engine lifting eyes.
7. Remove or disconnect the following:
   - Transfer case skid plate
   - Left and right motor mounts
   - Transmission mount crossmember
8. Attach a hoist to the engine lifting eyes and raise the powertrain out of the vehicle.

To install:

9. Lower the powertrain into the vehicle.
10. Install or connect the following:
    - Transmission mount crossmember. Tighten the bolts to 37 ft. lbs. (50 Nm) and the nuts to 55 ft. lbs. (74 Nm).
    - Transfer case skid plate
    - Left and right motor mounts. Tighten the fasteners to 22 ft. lbs. (30 Nm).
    - Power steering pump. Tighten the bolts to 13 ft. lbs. (17 Nm).
    - A/C compressor. Tighten the bolts to 36 ft. lbs. (49 Nm).
    - Front driveshaft. Tighten the fasteners to 59 ft. lbs. (80 Nm).
    - Rear driveshaft. Tighten the fasteners to 78 ft. lbs. (106 Nm).
    - Stabilizer bar. Tighten the bracket bolts to 13 ft. lbs. (18 Nm) and the link nuts to 18 ft. lbs. (25 Nm).
    - Exhaust front pipes
    - Transfer case shift lever and rod
    - Transmission shift lever assembly
    - Center console
    - Positive battery cable
    - Negative battery cable at the relay box and right inner fender
    - Wire clamp at right inner fender
    - EVAP canister hoses
    - Fuel lines
    - Ground cable connector
    - Engine control wiring harness and grommet at the firewall
    - Heater hoses
    - Alternator harness connectors
    - Power steering vacuum hoses
    - Accelerator cable
    - PCM harness connectors
    - Cooling fan and pulley
    - Accessory drive belt
    - Radiator and fan shroud
    - Radiator hoses
    - Coolant recovery tank
    - Engine under covers
    - Air intake pipe
    - Engine appearance cover
    - Hood
    - Battery and tray

11. Fill the crankcase to the correct level.
12. Fill the cooling system.

13. Start the engine and check for leaks.

GX470

1. Before servicing the vehicle, refer to the precautions section.
2. Remove the transmission.
3. Remove the hood.
4. Remove the V-bank cover.
5. Remove the air cleaner assembly.
6. Remove the under-covers.
7. Remove the radiator.
8. Remove the fan shroud.
9. Tag and disconnect all hoses, pipes and wires necessary for engine removal.
10. Remove the fan.
11. Remove the power steering pump and secure it out of the way.
12. Remove the alternator and secure it out of the way.
13. Remove the compressor and secure it out of the way.
14. Remove the transmission filler tube.
15. Remove the oil level sending unit.
16. Remove the exhaust manifolds.
17. Attach a crane and equalizer to the engine.
18. Support the weight of the engine with the crane and remove the mount bolts.
19. Remove the engine.
20. Installation is the reverse of removal. Observe the following torques:
    - Engine mount bolts: 28 ft. lbs. (38 Nm)
    - Exhaust manifold nuts: 33 ft. lbs. (44 Nm)
    - Oil level sending unit: 11 ft. lbs. 15 Nm)
    - Fan bolts: 21 ft. lbs. (29 Nm)
    - Compressor: bolt, 34 ft. lbs. (47 Nm); nut, 18 ft. lbs. (25 Nm)
    - Power steering pump: 32 ft. lbs. (43 Nm)
    - Hood: 10 ft. lbs. (13 Nm)

LX470

2002–05

1. Before servicing the vehicle, refer to the precautions section.
2. Relieve the fuel system pressure.
3. Drain the cooling system.
4. Drain the engine oil.
5. Remove or disconnect the following:
   - Battery and tray
   - Hood
   - Engine appearance cover
   - Air intake pipe
   - Engine under covers
   - Coolant recovery tank
   - Radiator hoses

- Radiator and fan shroud
- Accessory drive belt
- Cooling fan and pulley
- Powertrain Control Module (PCM) harness connectors and pass the wiring harness through the firewall
- Accelerator cable
- Power steering vacuum hoses
- Alternator harness connectors
- Heater hoses
- Engine control wiring harness and grommet at the firewall
- Ground cable connector
- Fuel lines
- Evaporative Emissions (EVAP) canister hoses
- Wire clamp at right inner fender
- Negative battery cable at the relay box and right inner fender
- Positive battery cable
- Center console
- Transmission shift lever assembly
- Transfer case shift lever and rod
- Exhaust front pipes
- Stabilizer bar
- Front and rear driveshafts
- A/C compressor
- Power steering pump

6. Attach a hoist to the engine lifting eyes.

7. Remove or disconnect the following:
- Transfer case skid plate
- Left and right motor mounts
- Transmission mount crossmember

8. Attach a hoist to the engine lifting eyes and raise the powertrain out of the vehicle.

To install:

9. Lower the powertrain into the vehicle.

10. Install or connect the following:
- Transmission mount crossmember. Tighten the bolts to 37 ft. lbs. (50 Nm) and the nuts to 55 ft. lbs. (74 Nm).
- Transfer case skid plate
- Left and right motor mounts. Tighten the fasteners to 22 ft. lbs. (30 Nm).
- Power steering pump. Tighten the bolts to 13 ft. lbs. (17 Nm).
- A/C compressor. Tighten the bolts to 36 ft. lbs. (49 Nm).
- Front driveshaft. Tighten the fasteners to 59 ft. lbs. (80 Nm).
- Rear driveshaft. Tighten the fasteners to 78 ft. lbs. (106 Nm).
- Stabilizer bar. Tighten the bracket bolts to 13 ft. lbs. (18 Nm) and the link nuts to 18 ft. lbs. (25 Nm).
- Exhaust front pipes
- Transfer case shift lever and rod

- Transmission shift lever assembly
- Center console
- Positive battery cable
- Negative battery cable at the relay box and right inner fender
- Wire clamp at right inner fender
- EVAP canister hoses
- Fuel lines
- Ground cable connector
- Engine control wiring harness and grommet at the firewall
- Heater hoses
- Alternator harness connectors
- Power steering vacuum hoses
- Accelerator cable
- PCM harness connectors
- Cooling fan and pulley
- Accessory drive belt
- Radiator and fan shroud
- Radiator hoses
- Coolant recovery tank
- Engine under covers
- Air intake pipe
- Engine appearance cover
- Hood
- Battery and tray

11. Fill the crankcase to the correct level.
12. Fill the cooling system.
13. Start the engine and check for leaks.

2006

1. Before servicing the vehicle, refer to the precautions section.
2. Properly relieve the fuel system pressure.
3. Disconnect the negative battery cable.
4. Remove the hood.
5. Remove the under-covers.
6. Drain the engine oil.
7. Drain and recycle the engine coolant.
8. Drain the transmission fluid.
9. Remove the V-bank cover.
10. Remove the air cleaner assembly.
11. Remove the drive belt.
12. Disconnect the radiator hoses and the cooler tubes.
13. Remove the fan shroud.
14. Remove the radiator.
15. Tag and disconnect all hoses, pipes and wires necessary for engine removal.
16. Remove the front exhaust pipes.
17. Remove the compressor and secure it out of the way.
18. Disconnect the vane pump.
19. Remove the rear and front drive shafts.
20. Attach a crane and equalizer to the engine.
21. Support the weight of the engine with the crane and remove the mount bolts.

22. Remove the engine and transmission assembly.

23. Installation is the reverse of removal. Observe the following torques:
- Frame cross member bolts to 37 ft. lbs. (50 Nm) and the nuts to 55 ft. lbs. (74 Nm)
- Engine mount bolts: 22 ft. lbs. (30 Nm)
- Transmission bolts: 14mm 27 ft. lbs. (37 Nm) and 17mm (53 ft. lbs. (72 Nm)
- Exhaust pipes: 46 ft. lbs. (62 Nm)
- Rear drive shaft: 78 ft. lbs. (106 Nm)
- Front drive shaft: 59 ft. lbs. (80 Nm)
- Compressor: bolt, 36 ft. lbs. (49 Nm); and stay , 45 ft. lbs. (61 Nm)

Heater Core

REMOVAL & INSTALLATION

Land Cruiser

FRONT HEATER

1. Disconnect the negative battery cable.
2. Drain the cooling system into a clean container for reuse.
3. Disconnect the heater hoses from the heater core.
4. Remove the steering wheel by performing the following procedure:
   a. Position the front wheels facing straight-ahead.
   b. Remove the steering wheel side covers.
   c. Using a Torx® wrench, loosen the 2 screws located at each side of the steering wheel until the screw's circumference groove catches on the screw case.
   d. Pull the air bag module from the steering wheel and disconnect the electrical connector.

✸✸ CAUTION

Place the air bag module in a safe place with the front side facing upward.

   e. Remove the steering wheel nut.
   f. Place alignment marks on the steering wheel and the main shaft.
   g. Using a steering wheel puller, press the steering wheel from the steering column.

5. Remove the instrument panel and reinforcement by performing the following procedure:

a. Remove the front door scuff plates, the cowl side trim and the front door opening trim.

b. At the driver's side, remove the 2 assist grip plugs, the 2 screws and assist grip and the front pillar garnish.

c. At the passenger's side, remove the 4 assist grip plugs, the 4 screws, the 2 assist grips and the front pillar garnish.

d. Remove the instrument cluster finish panel.

e. Remove the 2 screws and the hood lock control cable.

f. Remove the 2 screws and the fuel lid control cable lever.

g. Remove the lower No. 1 panel screw and the panel.

h. Remove the lower left side panel.

i. Remove the 3 steering column cover screws and the covers.

j. At the steering column, disconnect the electrical connectors; then, remove the clamp, the 3 screws and the combination switch.

k. Remove the No. 2 heater-to-register duct screw and the duct.

l. Remove the steering column-to-instrument panel bolts and the steering column.

m. At the combination meter, disconnect the electrical connectors; then, remove the 4 screws and the combination meter.

n. Remove the glove compartment door stoppers, the 2 screws and the glove box door.

o. At the passenger's side air bag module, remove the No. 1 undercover, pull the air bag connector up from the undercover and disconnect it; then, remove the air bag.

✳✳ CAUTION

Place the air bag module in a safe place with the front side facing upward.

p. Remove the 3 lower No. 2 panel screws and the panel.

q. Remove the center cluster; then, pry the center cluster from the dash by prying the 8 clips in the following order:
- Left side
- Right side
- Top left side
- Top right side

r. Remove the 4 radio screws, pull the radio outward, disconnect the electrical connectors and remove the radio.

s. At the rear console panel, remove the transfer shift lever knob. Pry the panel upward disengaging the 4 clips (2 on each side) and remove the panel.

t. At the rear of the console, remove the 2 rear end panel-to-console screws; then, pry the end panel rearward disengaging the 2 clips and remove the panel.

u. If not equipped with a rear air conditioning system, disconnect the connector and control cable; then, remove the 3 rear heater control panel screws and the panel.

v. Remove the 4 rear console box-to-chassis screws/bolts and the console box.

w. Remove the center lower cluster finish panel by prying panel rearward disengaging the 5 clips; then, disconnect the electrical connector.

x. Remove the 2 front console-to-chassis bolts/screws, disengage the 2 clips and remove the console.

y. At the instrument panel, disconnect the junction connectors (the connectors can be disconnected by loosening the bolts), the instrument panel-to-chassis 8 bolts and 2 nuts. Using an assistant, remove the instrument panel.

z. Disconnect the electrical connector and remove the ECM.

aa. Remove the No. 3 and No. 4 heater-to-register ducts.

bb. Remove the floor brace, the No. 1 brace and the reinforcement.

6. Remove the evaporator housing by performing the following procedure:

a. Discharge and recover the air conditioning system refrigerant.

b. Remove the air conditioning liquid line clamp.

c. Remove the air conditioning suction line clamp.

d. Disconnect both air conditioning lines and plug the openings to prevent contamination. Discard the 4 O-rings.

e. Remove the antenna relay electrical connector, the 2 screws and the relay.

f. Remove the evaporator housing-to-chassis 4 screws/2 nuts and the housing.

7. Remove the heater housing by performing the following procedure:

a. Remove the defroster nozzle.

b. Disconnect the electrical connector.

c. Remove the 4 nuts and the heater housing.

8. Remove the heater core-to-heater housing packing, the screw, the bracket, the clamp and the heater core.

To install:

9. Install the heater core, the clamp, the bracket, the screw and the heater core-to-heater housing packing.

10. Install the heater housing by performing the following procedure:

a. Install the heater housing and the 4 nuts.

b. Connect the electrical connector.

c. Install the defroster nozzle.

11. Install the evaporator housing by performing the following procedure:

a. Install the evaporator housing and the housing-to-chassis 4 screws and 2 nuts.

b. Install the antenna relay, the 2 screws and the electrical connector.

c. Using new O-rings, connect both air conditioning lines.

d. Install the air conditioning liquid line and suction line clamp.

12. Install the instrument panel and reinforcement by performing the following procedure:

a. Install the reinforcement, the No. 1 brace and the floor brace.

b. Install the No. 3 and No. 4 heater-to-register ducts.

c. Install the ECM and connect the electrical connector.

d. Using an assistant, install the instrument panel, connect the junction connectors, the instrument panel-to-chassis 8 bolts and 2 nuts.

e. Install the front the console,

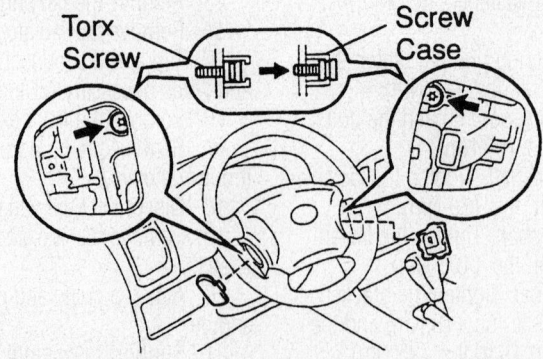

View the steering wheel's Torx® bolts—Toyota Land Cruiser

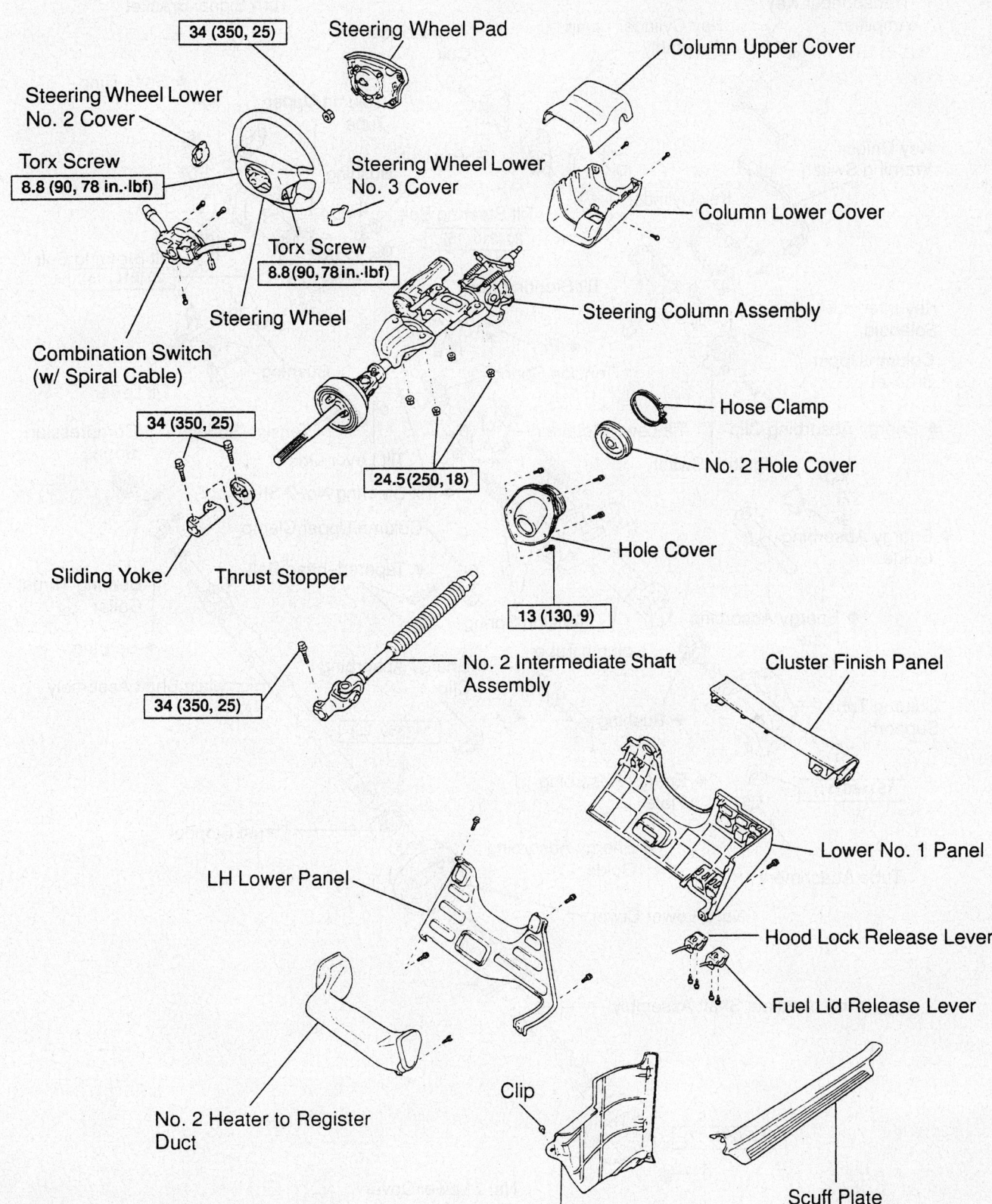

34 (350, 25)
Steering Wheel Pad
Column Upper Cover
Steering Wheel Lower No. 2 Cover
Torx Screw
8.8 (90, 78 in.·lbf)
Column Lower Cover
Steering Wheel Lower No. 3 Cover
Torx Screw
8.8 (90, 78 in.·lbf)
Steering Column Assembly
Steering Wheel
Combination Switch (w/ Spiral Cable)
Hose Clamp
No. 2 Hole Cover
34 (350, 25)
24.5 (250, 18)
Hole Cover
Sliding Yoke
Thrust Stopper
13 (130, 9)
34 (350, 25)
No. 2 Intermediate Shaft Assembly
Cluster Finish Panel
Lower No. 1 Panel
LH Lower Panel
Hood Lock Release Lever
Fuel Lid Release Lever
No. 2 Heater to Register Duct
Clip
Scuff Plate
Cowl Trim

N·m (kgf·cm, ft·lbf) : Specified torque

93113GG5

Exploded view the steering column—Toyota Land Cruiser (Part 1 of 2)

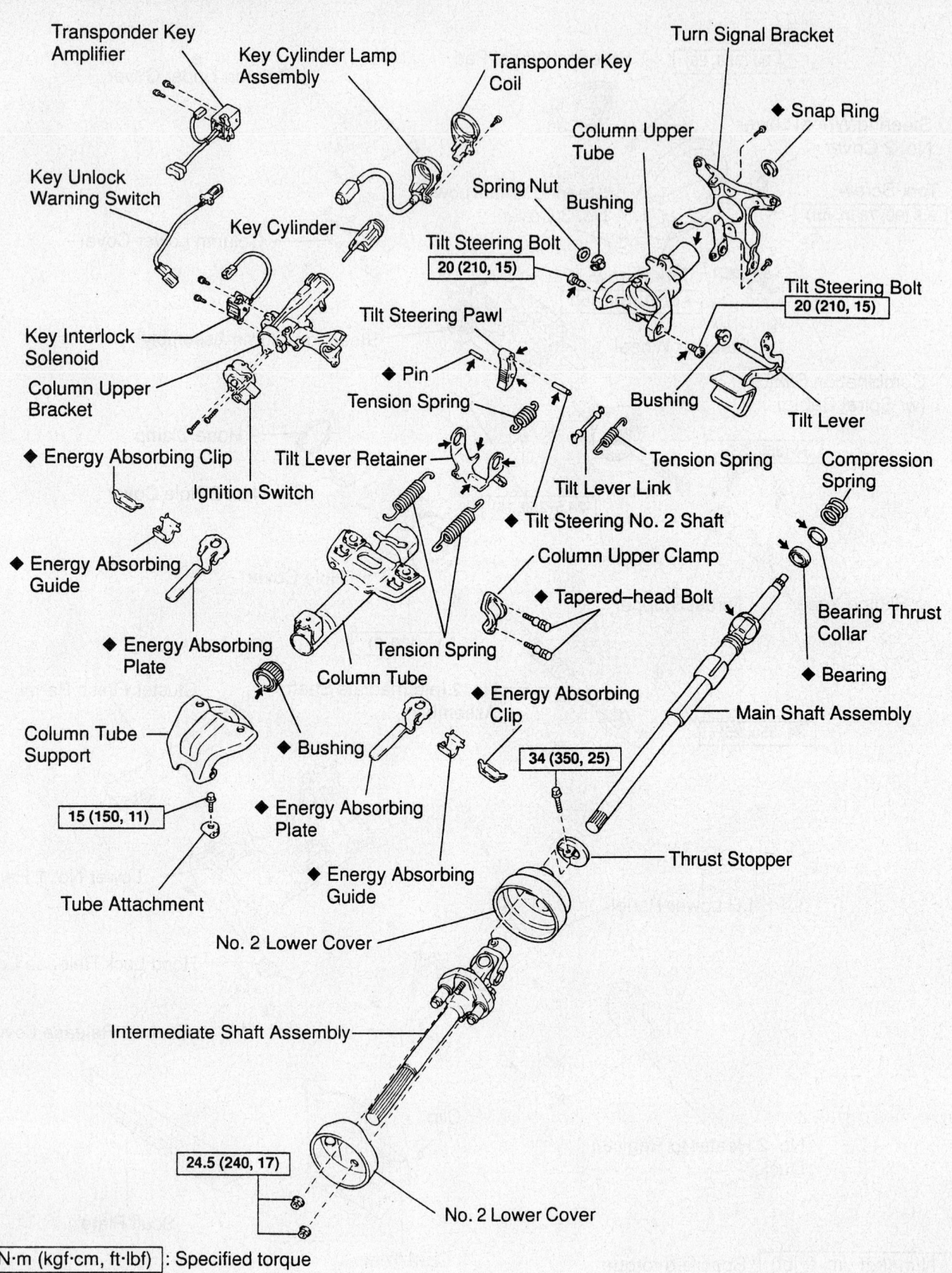

Transponder Key Amplifier
Key Cylinder Lamp Assembly
Transponder Key Coil
Turn Signal Bracket
Column Upper Tube
◆ Snap Ring
Key Unlock Warning Switch
Spring Nut
Bushing
Tilt Steering Bolt
20 (210, 15)
Key Cylinder
Tilt Steering Bolt
20 (210, 15)
Key Interlock Solenoid
Tilt Steering Pawl
Bushing
Tilt Lever
Column Upper Bracket
◆ Pin
Tension Spring
Tension Spring
Compression Spring
◆ Energy Absorbing Clip
Tilt Lever Retainer
Ignition Switch
Tilt Lever Link
◆ Tilt Steering No. 2 Shaft
◆ Energy Absorbing Guide
Column Upper Clamp
Bearing Thrust Collar
◆ Energy Absorbing Plate
◆ Tapered–head Bolt
◆ Bearing
◆ Energy Absorbing Clip
Column Tube Support
Tension Spring
Column Tube
34 (350, 25)
Main Shaft Assembly
15 (150, 11)
◆ Bushing
◆ Energy Absorbing Plate
Thrust Stopper
Tube Attachment
◆ Energy Absorbing Guide
No. 2 Lower Cover
Intermediate Shaft Assembly
24.5 (240, 17)
No. 2 Lower Cover

N·m (kgf·cm, ft·lbf) : Specified torque
◆ Non–reusable part
◀ Molybdenum disulfide lithium base grease

93113GG6

Exploded view the steering column—Toyota Land Cruiser (Part 2 of 2)

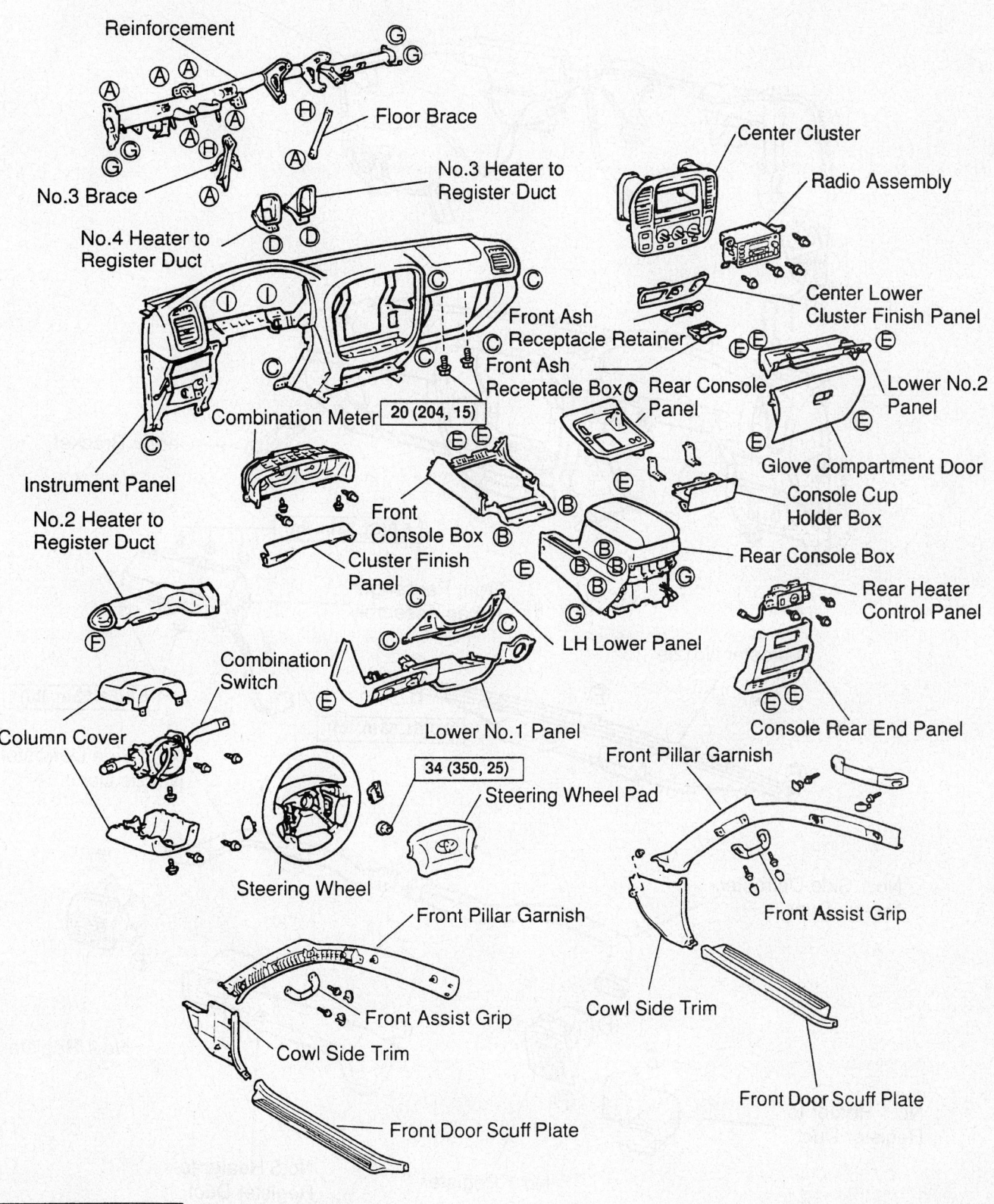

Reinforcement

Floor Brace

No.3 Brace

No.3 Heater to Register Duct

No.4 Heater to Register Duct

Center Cluster

Radio Assembly

Center Lower Cluster Finish Panel

Front Ash Receptacle Retainer

Front Ash Receptacle Box

Rear Console Panel

Lower No.2 Panel

Combination Meter 20 (204, 15)

Glove Compartment Door

Instrument Panel

Console Cup Holder Box

No.2 Heater to Register Duct

Front Console Box

Cluster Finish Panel

Rear Console Box

Rear Heater Control Panel

LH Lower Panel

Column Cover

Combination Switch

Lower No.1 Panel

Console Rear End Panel

Front Pillar Garnish

34 (350, 25)

Steering Wheel Pad

Front Assist Grip

Steering Wheel

Front Pillar Garnish

Cowl Side Trim

Front Door Scuff Plate

Front Assist Grip

Cowl Side Trim

Front Door Scuff Plate

N·m (kgf·cm, ft·lbf) : Specified torque

93113GG7

Exploded view the instrument panel and related components—Toyota Land Cruiser

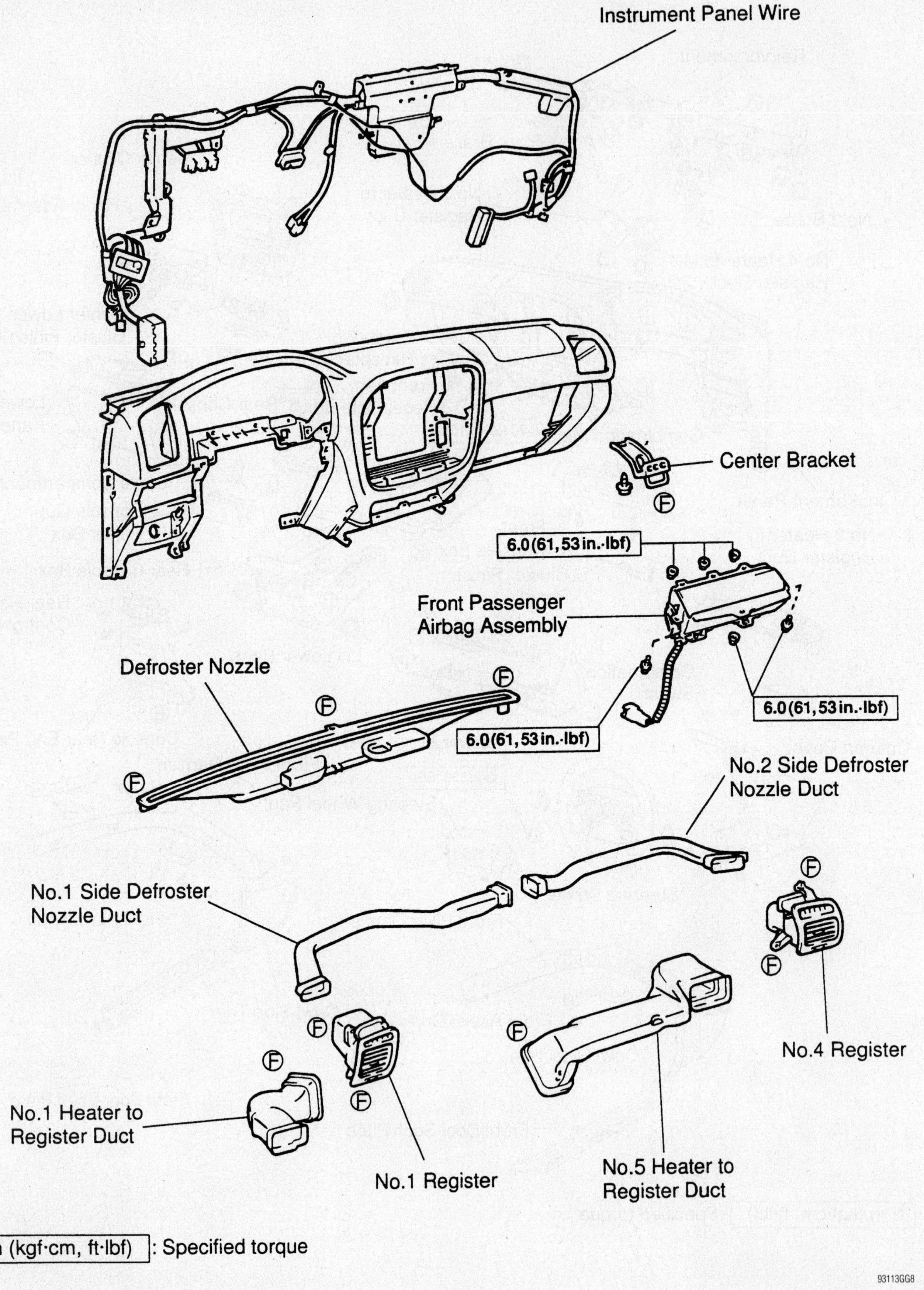

Instrument Panel Wire

Center Bracket

6.0 (61, 53 in.·lbf)

Front Passenger
Airbag Assembly

6.0 (61, 53 in.·lbf)

Defroster Nozzle

6.0 (61, 53 in.·lbf)

No.2 Side Defroster
Nozzle Duct

No.1 Side Defroster
Nozzle Duct

No.4 Register

No.1 Heater to
Register Duct

No.1 Register

No.5 Heater to
Register Duct

N·m (kgf·cm, ft·lbf) : Specified torque

Exploded view the front ventilation ducts and related components—Toyota Land Cruiser

93113GG8

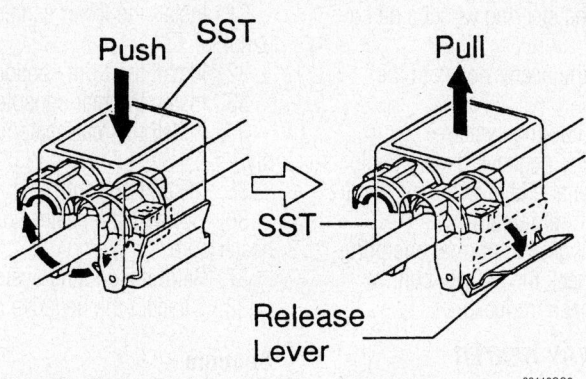

Push SST **Pull**

SST

Release Lever

93113GG9

View the air conditioning line clamp removal tool—Toyota Land Cruiser

engage the 2 clips and install the 2 console-to-chassis bolts/screws.

f. Connect the electrical connector; then, install the center lower cluster finish panel by engaging the 5 clips.

g. Install the console box and the 4 rear console box-to-chassis screws/bolts.

h. If not equipped with a rear air conditioning system, install rear heater control panel, the 3 panel screws; then, connect the connector and control cable.

i. Install the rear of the console and

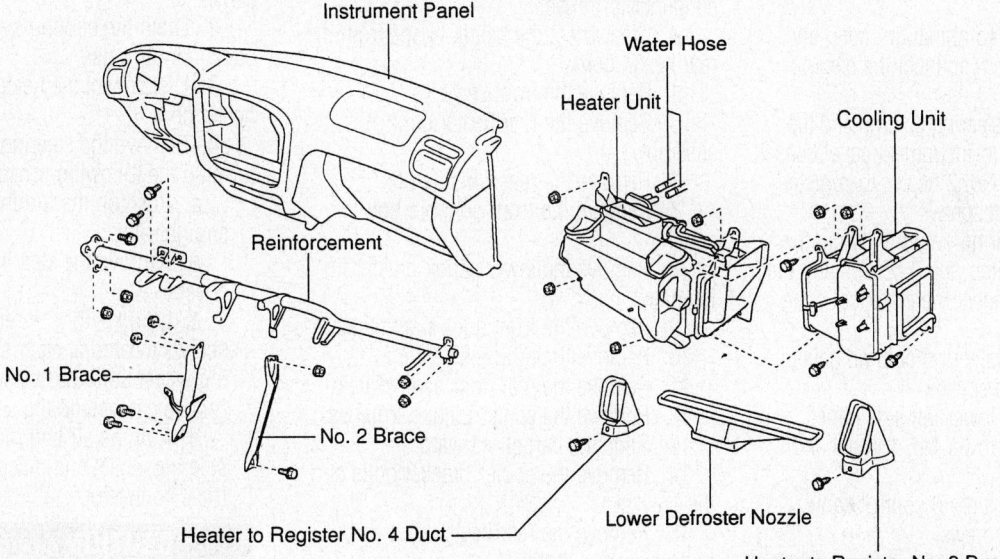

Instrument Panel

Water Hose

Heater Unit

Cooling Unit

Reinforcement

No. 1 Brace

No. 2 Brace

Heater to Register No. 4 Duct

Lower Defroster Nozzle

Heater to Register No. 3 Duct

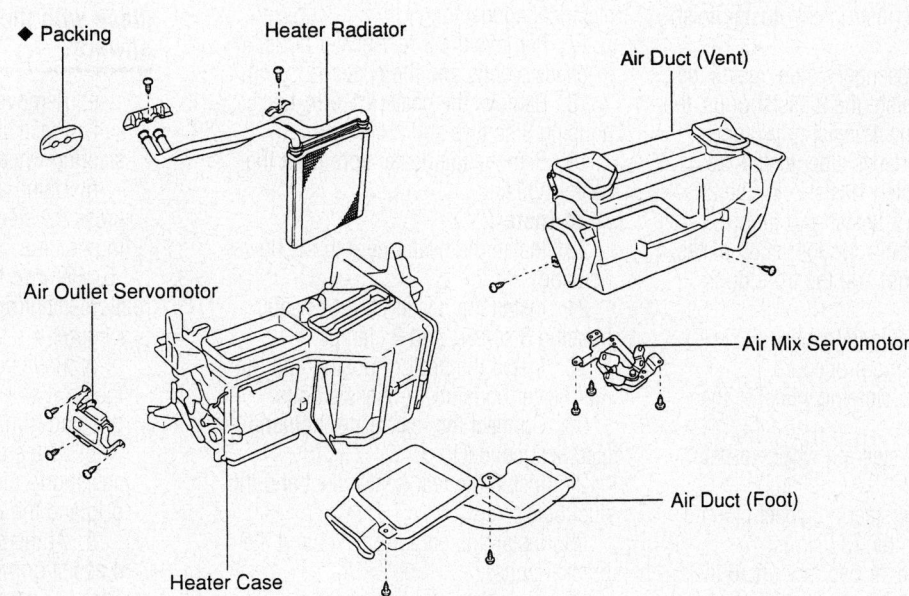

◆ Packing

Heater Radiator

Air Duct (Vent)

Air Outlet Servomotor

Air Mix Servomotor

Air Duct (Foot)

Heater Case

◆ Non–reusable part

93113GG0

Exploded view of the front heater core, heater housing, evaporator housing and related components—Toyota Land Cruiser

engage the 2 clips; then, install the 2 rear end panel-to-console screws.

j. Install the rear console panel and engage the 4 clips (2 on each side); then, install the transfer shift lever knob.

k. Install the radio, connect the electrical connectors and the 4 radio screws.

l. Install the center cluster and engage the 8 center cluster clips.

m. Install the lower No. 2 panel and the 3 panel screws.

n. Install the passenger's side air bag module, connect it and install the No. 1 undercover.

o. Install the glove box door, the 2 screws and the glove compartment door stoppers.

p. Install the combination meter and the 4 screws; then, connect the electrical connectors.

q. Install the steering column and the steering column-to-instrument panel bolts.

r. Install the No. 2 heater-to-register duct and the duct screw.

s. At the steering column, install the combination switch, the 3 screws and the clamp; then, connect the electrical connectors.

t. Install the steering column covers and the 3 covers screws.

u. Install the lower left side panel.

v. Install the lower No. 1 panel and the panel screw.

w. Install the fuel lid control cable lever and the 2 screws.

x. Install the hood lock control cable and the 2 screws.

y. Install the instrument cluster finish panel.

z. At the passenger's side, install the front pillar garnish, the 2 assist grips, the 4 screws and the 4 assist grip plugs.

aa. At the driver's side, install the front pillar garnish, assist grip, the 2 screws and the 2 assist grip plugs.

bb. Install the front door scuff plates, the cowl side trim and the front door opening trim.

13. Install the steering wheel by performing the following procedure:

a. Install the steering wheel to the steering column.

b. Align the steering wheel-to-main shaft marks.

c. Install the steering wheel nut and torque to 25 ft. lbs. (34 Nm).

d. Install the air bag module to the steering wheel and connect the electrical connector.

e. Using a Torx® wrench, tighten the 2 screws located at each side of the steering wheel to 78 inch lbs. (8.8 Nm).

f. Install the steering wheel side covers.

14. Connect the heater hoses to the heater core.

15. Refill the cooling system.

16. Connect the negative battery cable.

a. Evacuate and charge the air conditioning system refrigerant.

17. Run the engine to normal operating temperatures. Check the climate control operation and check for leaks.

REAR AUXILIARY HEATER

1. Disconnect the negative battery cable.

2. Drain the cooling system into a clean container for reuse.

3. Disconnect the heater hoses from the rear heater core.

4. Remove the front seats.

5. Remove the rear heater control assembly.

6. Remove the rear console box.

7. Remove the front console box cover.

8. Remove the lower center cluster finish panel.

9. Remove the front door scuff plates.

10. Remove the cowl side trim.

11. Remove the rear door scuff plates.

12. Remove the center pillar garnishes.

13. Slide the carpet rearward.

14. Remove the cooler bracket bolts and the bracket.

15. Remove the rear heater duct bolt/screw and the duct.

16. Disconnect the rear heater housing electrical connector.

17. Remove the 3 rear heater housing-to-chassis bolts and the heater housing.

18. Remove the heater core-to-heater housing 3 screws and 2 clamps.

19. Remove the heater core from the heater housing.

To install:

20. Install the heater core to the heater housing.

21. Install the heater core-to-heater housing 3 screws and 2 clamps.

22. Install the heater housing and the 3 rear heater housing-to-chassis bolts.

23. Connect the rear heater housing electrical connector.

24. Install the rear heater duct and the duct bolt/screw.

25. Install the cooler bracket and the bracket bolts.

26. Slide the carpet rearward.

27. Install the center pillar garnishes.

28. Install the rear door scuff plates.

29. Install the cowl side trim.

30. Install the front door scuff plates.

31. Install the lower center cluster finish panel.

32. Install the front console box cover.

33. Install the rear console box.

34. Install the rear heater control assembly.

35. Install the front seats.

36. Connect the heater hoses to the rear heater core.

37. Refill the cooling system.

38. Connect the negative battery cable.

Sequoia

FRONT HEATER

1. Disconnect the negative battery cable.

2. Drain the cooling system into a clean container for reuse.

3. Disconnect the heater hoses from the heater core.

4. Remove the steering wheel by performing the following procedure:

a. Position the front wheels facing straight-ahead.

b. Remove the steering wheel side covers.

c. Using a Torx® wrench, loosen the 2 screws located at each side of the steering wheel until the screw's circumference groove catches on the screw case.

d. Pull the air bag module from the steering wheel and disconnect the electrical connector.

✳✳ CAUTION

Place the air bag module in a safe place with the front side facing upward.

e. Remove the steering wheel nut.

f. Place alignment marks on the steering wheel and the main shaft.

g. Using a steering wheel puller, press the steering wheel from the steering column.

5. Remove the instrument panel and reinforcement by performing the following procedure:

a. Remove the front door scuff plates, the cowl side trim and the front door opening trim.

b. At the driver's side, remove the 2 assist grip plugs, the 2 screws and assist grip and the front pillar garnish.

c. At the passenger's side, remove the 4 assist grip plugs, the 4 screws, the 2 assist grips and the front pillar garnish.

d. Remove the instrument cluster finish panel.

e. Remove the 2 screws and the hood lock control cable.

f. Remove the 2 screws and the fuel lid control cable lever.

g. Remove the lower No. 1 panel screw and the panel.

h. Remove the lower left side panel.

i. Remove the 3 steering column cover screws and the covers.

j. At the steering column, disconnect the electrical connectors; then, remove the clamp, the 3 screws and the combination switch.

k. Remove the No. 2 heater-to-register duct screw and the duct.

l. Remove the steering column-to-instrument panel bolts and the steering column.

m. At the combination meter, disconnect the electrical connectors; then, remove the 4 screws and the combination meter.

n. Remove the glove compartment door stoppers, the 2 screws and the glove box door.

o. At the passenger's side air bag module, remove the No. 1 undercover, pull the air bag connector up from the undercover and disconnect it; then, remove the air bag.

✴✴ CAUTION

Place the air bag module in a safe place with the front side facing upward.

p. Remove the 3 lower No. 2 panel screws and the panel.

q. Remove the center cluster; then, pry the center cluster from the dash by prying the 8 clips in the following order:
- Left side
- Right side
- Top left side

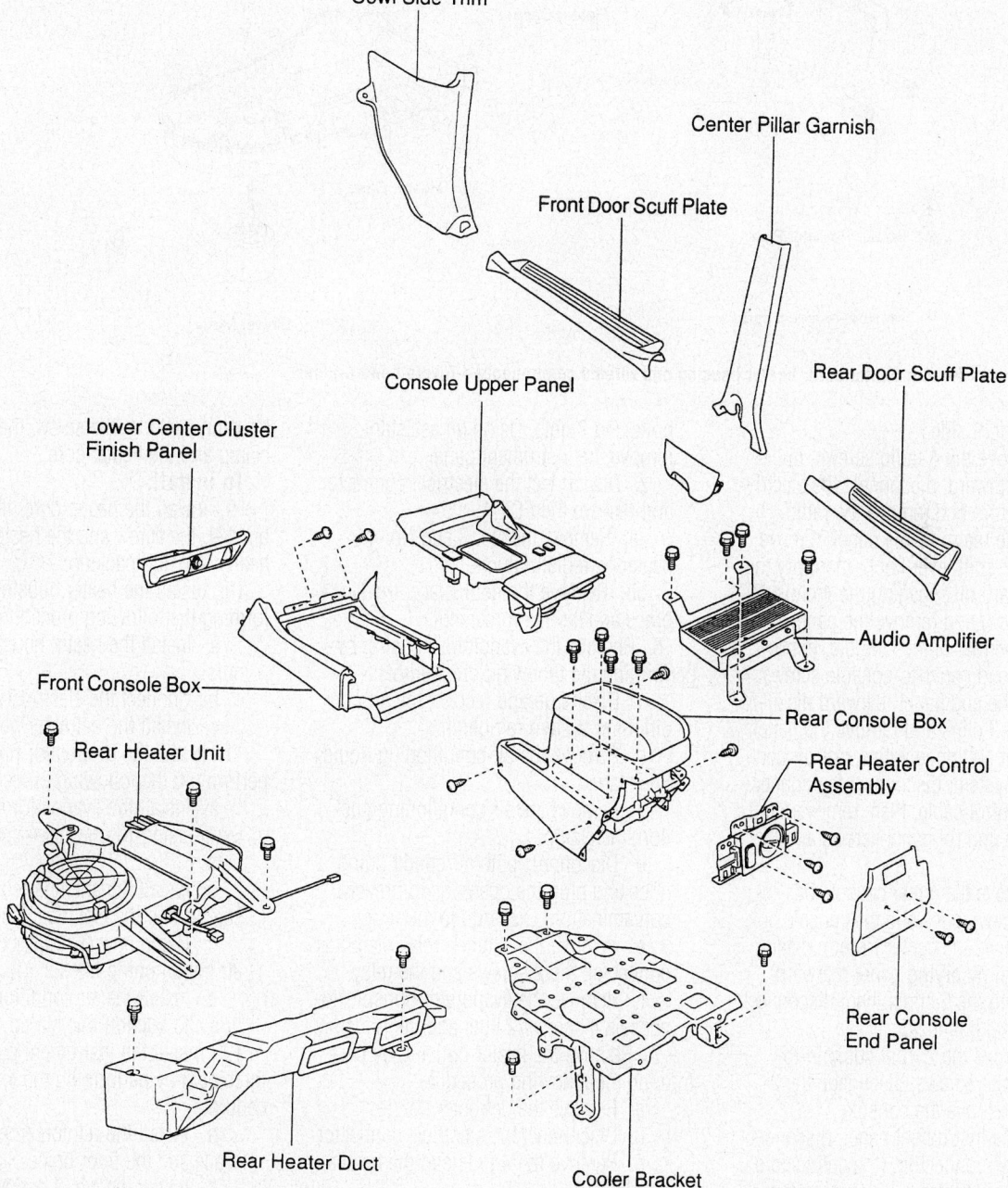

Cowl Side Trim

Center Pillar Garnish

Front Door Scuff Plate

Console Upper Panel

Rear Door Scuff Plate

Lower Center Cluster Finish Panel

Audio Amplifier

Front Console Box

Rear Console Box

Rear Heater Unit

Rear Heater Control Assembly

Rear Console End Panel

Rear Heater Duct

Cooler Bracket

93113GH1

Exploded view of the rear heater housing and related components—Toyota Land Cruiser

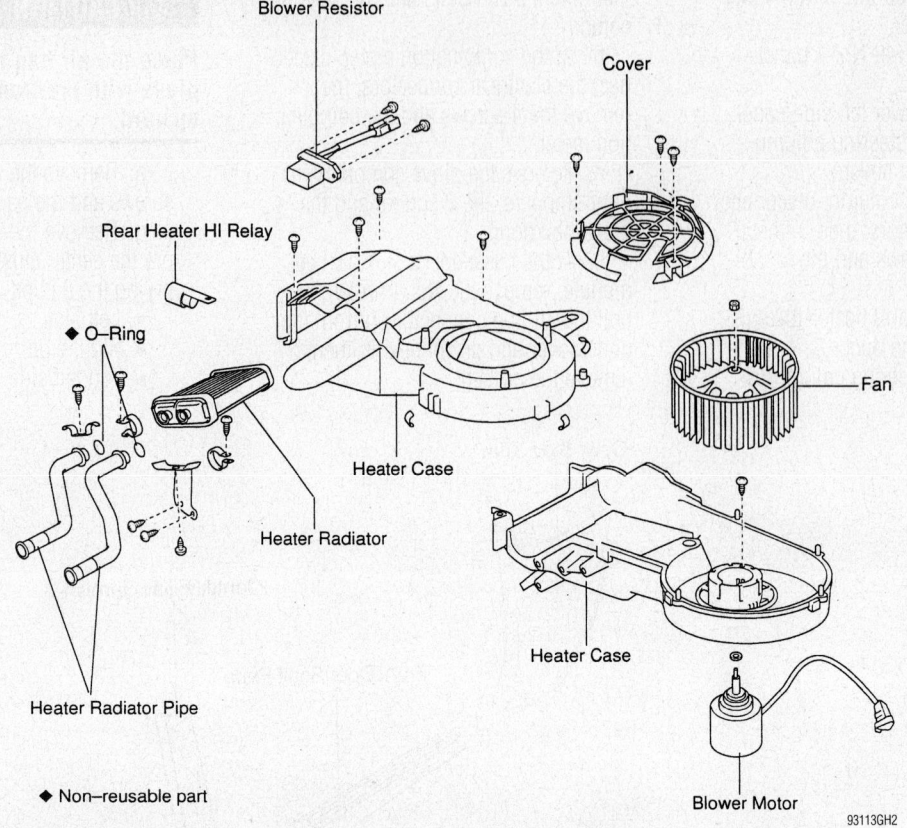

Blower Resistor

Cover

Rear Heater HI Relay

◆ O–Ring

Heater Case

Fan

Heater Radiator

Heater Case

Heater Radiator Pipe

Blower Motor

◆ Non–reusable part

93113GH2

Exploded view of the rear heater core, heater housing and related components—Toyota Land Cruiser

- Top right side

r. Remove the 4 radio screws, pull the radio outward, disconnect the electrical connectors and remove the radio.

s. At the rear console panel, remove the transfer shift lever knob; then, pry the panel upward disengaging the 4 clips (2 on each side) and remove the panel.

t. At the rear of the console, remove the 2 rear end panel-to-console screws; then, pry the end panel rearward disengaging the 2 clips and remove the panel.

u. If not equipped with a rear air conditioning system, disconnect the connector and control cable; then, remove the 3 rear heater control panel screws and the panel.

v. Remove the 4 rear console box-to-chassis screws/bolts and the console box.

w. Remove the center lower cluster finish panel by prying panel rearward disengaging the 5 clips; then, disconnect the electrical connector.

x. Remove the 2 front console-to-chassis bolts/screws, disengage the 2 clips and remove the console.

y. At the instrument panel, disconnect the junction connectors (the connectors can be disconnected by loosening the bolts), the instrument panel-to-chassis 8

bolts and 2 nuts. Using an assistant, remove the instrument panel.

z. Disconnect the electrical connector and remove the ECM.

aa. Remove the No. 3 and No. 4 heater-to-register ducts.

bb. Remove the floor brace, the No. 1 brace and the reinforcement.

6. Remove the evaporator housing by performing the following procedure:

a. Discharge and recover the air conditioning system refrigerant.

b. Remove the air conditioning liquid line clamp.

c. Remove the air conditioning suction line clamp.

d. Disconnect both air conditioning lines and plug the openings to prevent contamination. Discard the 4 O-rings.

e. Remove the antenna relay electrical connector, the 2 screws and the relay.

f. Remove the evaporator housing-to-chassis 4 screws/2 nuts and the housing.

7. Remove the heater housing by performing the following procedure:

a. Remove the defroster nozzle.

b. Disconnect the electrical connector.

c. Remove the 4 nuts and the heater housing.

8. Remove the heater core-to-heater

housing packing, the screw, the bracket, the clamp and the heater core.

To install:

9. Install the heater core, the clamp, the bracket, the screw and the heater core-to-heater housing packing.

10. Install the heater housing by performing the following procedure:

a. Install the heater housing and the 4 nuts.

b. Connect the electrical connector.

c. Install the defroster nozzle.

11. Install the evaporator housing by performing the following procedure:

a. Install the evaporator housing and the housing-to-chassis 4 screws and 2 nuts.

b. Install the antenna relay, the 2 screws and the electrical connector.

c. Using new O-rings, connect both air conditioning lines.

d. Install the air conditioning liquid line and suction line clamp.

12. Install the instrument panel and reinforcement by performing the following procedure:

a. Install the reinforcement, the No. 1 brace and the floor brace.

b. Install the No. 3 and No. 4 heater-to-register ducts.

c. Install the ECM and connect the electrical connector.

d. Using an assistant, install the instrument panel, connect the junction connectors, the instrument panel-to-chassis 8 bolts and 2 nuts.

e. Install the front the console, engage the 2 clips and install the 2 console-to-chassis bolts/screws.

f. Connect the electrical connector; then, install the center lower cluster finish panel by engaging the 5 clips.

g. Install the console box and the 4 rear console box-to-chassis screws/bolts.

h. If not equipped with a rear air conditioning system, install rear heater control panel, the 3 panel screws; then, connect the connector and control cable.

i. Install the rear of the console and engage the 2 clips; then, install the 2 rear end panel-to-console screws.

j. Install the rear console panel and engage the 4 clips (2 on each side); then, install the transfer shift lever knob.

k. Install the radio, connect the electrical connectors and the 4 radio screws.

l. Install the center cluster and engage the 8 center cluster clips.

m. Install the lower No. 2 panel and the 3 panel screws.

n. Install the passenger's side air bag module, connect it and install the No. 1 undercover.

o. Install the glove box door, the 2 screws and the glove compartment door stoppers.

p. Install the combination meter and the 4 screws; then, connect the electrical connectors.

q. Install the steering column and the steering column-to-instrument panel bolts.

r. Install the No. 2 heater-to-register duct and the duct screw.

s. At the steering column, install the combination switch, the 3 screws and the clamp; then, connect the electrical connectors.

t. Install the steering column covers and the 3 covers screws.

u. Install the lower left side panel.

v. Install the lower No. 1 panel and the panel screw.

w. Install the fuel lid control cable lever and the 2 screws.

x. Install the hood lock control cable and the 2 screws.

y. Install the instrument cluster finish panel.

z. At the passenger's side, install the front pillar garnish, the 2 assist grips, the 4 screws and the 4 assist grip plugs.

aa. At the driver's side, install the front pillar garnish, assist grip, the 2 screws and the 2 assist grip plugs.

bb. Install the front door scuff plates, the cowl side trim and the front door opening trim.

13. Install the steering wheel by performing the following procedure:

a. Install the steering wheel to the steering column.

b. Align the steering wheel-to-main shaft marks.

c. Install the steering wheel nut and torque to 25 ft. lbs. (34 Nm).

d. Install the air bag module to the steering wheel and connect the electrical connector.

e. Using a Torx® wrench, tighten the 2 screws located at each side of the steering wheel to 78 inch lbs. (8.8 Nm).

f. Install the steering wheel side covers.

14. Connect the heater hoses to the heater core.

15. Refill the cooling system.

16. Connect the negative battery cable.

a. Evacuate and charge the air conditioning system refrigerant.

17. Run the engine to normal operating temperatures; then, check the climate control operation and check for leaks.

REAR AUXILIARY HEATER

1. Disconnect the negative battery cable.
2. Drain the cooling system into a clean container for reuse.
3. Disconnect the heater hoses from the rear heater core.
4. Remove the front seats.
5. Remove the rear heater control assembly.
6. Remove the rear console box.
7. Remove the front console box cover.
8. Remove the lower center cluster finish panel.
9. Remove the front door scuff plates.
10. Remove the cowl side trim.
11. Remove the rear door scuff plates.
12. Remove the center pillar garnishes.
13. Slide the carpet rearward.
14. Remove the cooler bracket bolts and the bracket.
15. Remove the rear heater duct bolt/screw and the duct.
16. Disconnect the rear heater housing electrical connector.
17. Remove the 3 rear heater housing-to-chassis bolts and the heater housing.
18. Remove the heater core-to-heater housing 3 screws and 2 clamps.
19. Remove the heater core from the heater housing.

To install:

20. Install the heater core to the heater housing.
21. Install the heater core-to-heater housing 3 screws and 2 clamps.
22. Install the heater housing and the 3 rear heater housing-to-chassis bolts.
23. Connect the rear heater housing electrical connector.
24. Install the rear heater duct and the duct bolt/screw.
25. Install the cooler bracket and the bracket bolts.
26. Slide the carpet rearward.
27. Install the center pillar garnishes.
28. Install the rear door scuff plates.
29. Install the cowl side trim.
30. Install the front door scuff plates.
31. Install the lower center cluster finish panel.
32. Install the front console box cover.
33. Install the rear console box.
34. Install the rear heater control assembly.
35. Install the front seats.
36. Connect the heater hoses to the rear heater core.
37. Refill the cooling system.
38. Connect the negative battery cable.

LX470

FRONT HEATER

1. Disconnect the negative battery cable.
2. Drain the cooling system into a clean container for reuse.
3. Disconnect the heater hoses from the heater core.
4. Remove the steering wheel by performing the following procedure:

a. Position the front wheels facing straight-ahead.

b. Remove the steering wheel side covers.

c. Using a Torx® wrench, loosen the 2 screws located at each side of the steering wheel until the screw's circumference groove catches on the screw case.

d. Pull the air bag module from the steering wheel and disconnect the electrical connector.

⁂ CAUTION

Place the air bag module in a safe place with the front side facing upward.

e. Remove the steering wheel nut.

f. Place alignment marks on the steering wheel and the main shaft.

g. Using a steering wheel puller,

press the steering wheel from the steering column.

5. Remove the instrument panel and reinforcement by performing the following procedure:

a. Remove the front door scuff plates, the cowl side trim and the front door opening trim.

b. At the driver's side, remove the 2 assist grip plugs, the 2 screws and assist grip and the front pillar garnish.

c. At the passenger's side, remove the 4 assist grip plugs, the 4 screws, the 2 assist grips and the front pillar garnish.

d. Remove the instrument cluster finish panel.

e. Remove the 2 screws and the hood lock control cable.

f. Remove the 2 screws and the fuel lid control cable lever.

g. Remove the lower No. 1 panel screw and the panel.

h. Remove the lower left side panel.

i. Remove the 3 steering column cover screws and the covers.

j. At the steering column, disconnect the electrical connectors; then, remove the clamp, the 3 screws and the combination switch.

k. Remove the No. 2 heater-to-register duct screw and the duct.

l. Remove the steering column-to-instrument panel bolts and the steering column.

m. At the combination meter, disconnect the electrical connectors; then, remove the 4 screws and the combination meter.

n. Remove the glove compartment door stoppers, the 2 screws and the glove box door.

o. At the passenger's side air bag module, remove the No. 1 undercover, pull the air bag connector up from the undercover and disconnect it; then, remove the air bag.

✳✳ CAUTION

Place the air bag module in a safe place with the front side facing upward.

p. Remove the 3 lower No. 2 panel screws and the panel.

q. Remove the center cluster; then, pry the center cluster from the dash by prying the 8 clips in the following order:
• Left side
• Right side
• Top left side
• Top right side

r. Remove the 4 radio screws, pull the radio outward, disconnect the electrical connectors and remove the radio.

s. At the rear console panel, remove the transfer shift lever knob; then, pry the panel upward disengaging the 4 clips (2 on each side) and remove the panel.

t. At the rear of the console, remove the 2 rear end panel-to-console screws; then, pry the end panel rearward disengaging the 2 clips and remove the panel.

u. If not equipped with a rear air conditioning system, disconnect the connector and control cable; then, remove the 3 rear heater control panel screws and the panel.

v. Remove the 4 rear console box-to-chassis screws/bolts and the console box.

w. Remove the center lower cluster finish panel by prying panel rearward disengaging the 5 clips; then, disconnect the electrical connector.

x. Remove the 2 front console-to-chassis bolts/screws, disengage the 2 clips and remove the console.

y. At the instrument panel, disconnect the junction connectors (the connectors can be disconnected by loosening the bolts), the instrument panel-to-chassis 8 bolts and 2 nuts. Using an assistant, remove the instrument panel.

z. Disconnect the electrical connector and remove the ECM.

aa. Remove the No. 3 and No. 4 heater-to-register ducts.

bb. Remove the floor brace, the No. 1 brace and the reinforcement.

6. Remove the evaporator housing by performing the following procedure:

a. Discharge and recover the air conditioning system refrigerant.

b. Remove the air conditioning liquid line clamp.

c. Remove the air conditioning suction line clamp.

d. Disconnect both air conditioning lines and plug the openings to prevent contamination. Discard the 4 O-rings.

e. Remove the antenna relay electrical connector, the 2 screws and the relay.

f. Remove the evaporator housing-to-chassis 4 screws/2 nuts and the housing.

7. Remove the heater housing by performing the following procedure:

a. Remove the defroster nozzle.

b. Disconnect the electrical connector.

c. Remove the 4 nuts and the heater housing.

8. Remove the heater core-to-heater housing packing, the screw, the bracket, the clamp and the heater core.

To install:

9. Install the heater core, the clamp, the bracket, the screw and the heater core-to-heater housing packing.

10. Install the heater housing by performing the following procedure:

a. Install the heater housing and the 4 nuts.

b. Connect the electrical connector.

c. Install the defroster nozzle.

11. Install the evaporator housing by performing the following procedure:

a. Install the evaporator housing and the housing-to-chassis 4 screws and 2 nuts.

b. Install the antenna relay, the 2 screws and the electrical connector.

c. Using new O-rings, connect both air conditioning lines.

d. Install the air conditioning liquid line and suction line clamp.

12. Install the instrument panel and reinforcement by performing the following procedure:

a. Install the reinforcement, the No. 1 brace and the floor brace.

b. Install the No. 3 and No. 4 heater-to-register ducts.

c. Install the ECM and connect the electrical connector.

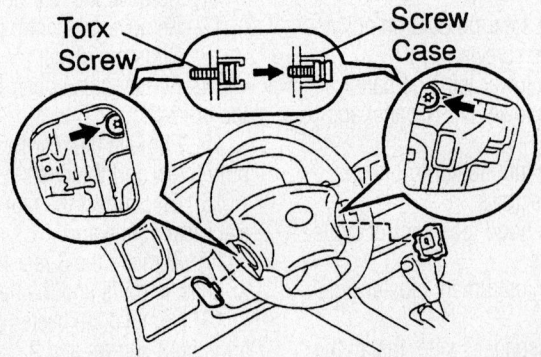

View the steering wheel's Torx® bolts—Lexus LX470

93113GG4

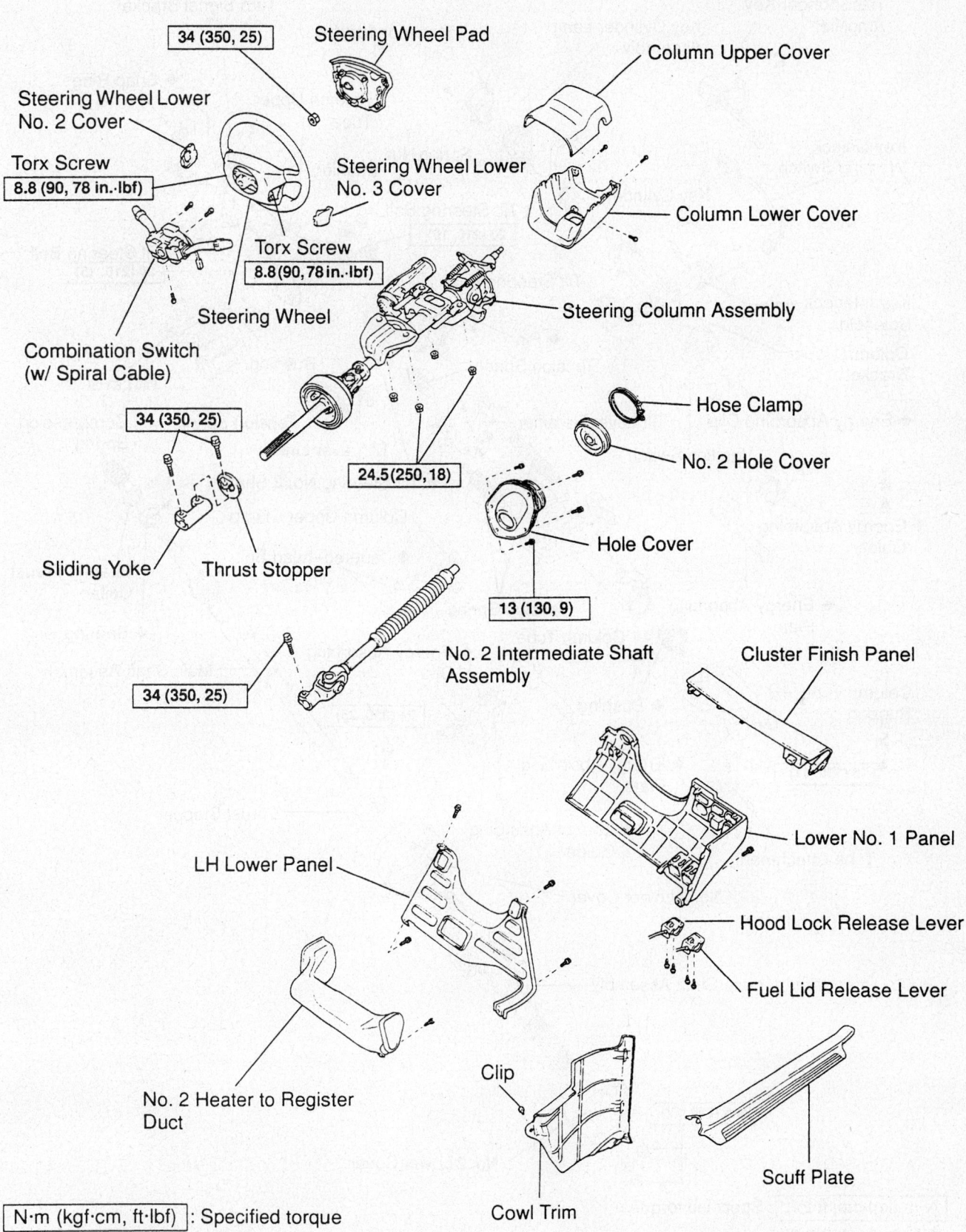

34 (350, 25)

Steering Wheel Pad

Column Upper Cover

Steering Wheel Lower
No. 2 Cover

Torx Screw
8.8 (90, 78 in.·lbf)

Steering Wheel Lower
No. 3 Cover

Column Lower Cover

Torx Screw
8.8 (90, 78 in.·lbf)

Steering Wheel

Steering Column Assembly

Combination Switch
(w/ Spiral Cable)

34 (350, 25)

Hose Clamp

No. 2 Hole Cover

24.5 (250, 18)

Hole Cover

Sliding Yoke   Thrust Stopper

13 (130, 9)

34 (350, 25)

No. 2 Intermediate Shaft
Assembly

Cluster Finish Panel

Lower No. 1 Panel

LH Lower Panel

Hood Lock Release Lever

Fuel Lid Release Lever

No. 2 Heater to Register
Duct

Clip

Scuff Plate

Cowl Trim

N·m (kgf·cm, ft·lbf) : Specified torque

93113GG5

Exploded view the steering column—Lexus LX470 (Part 1 of 2)

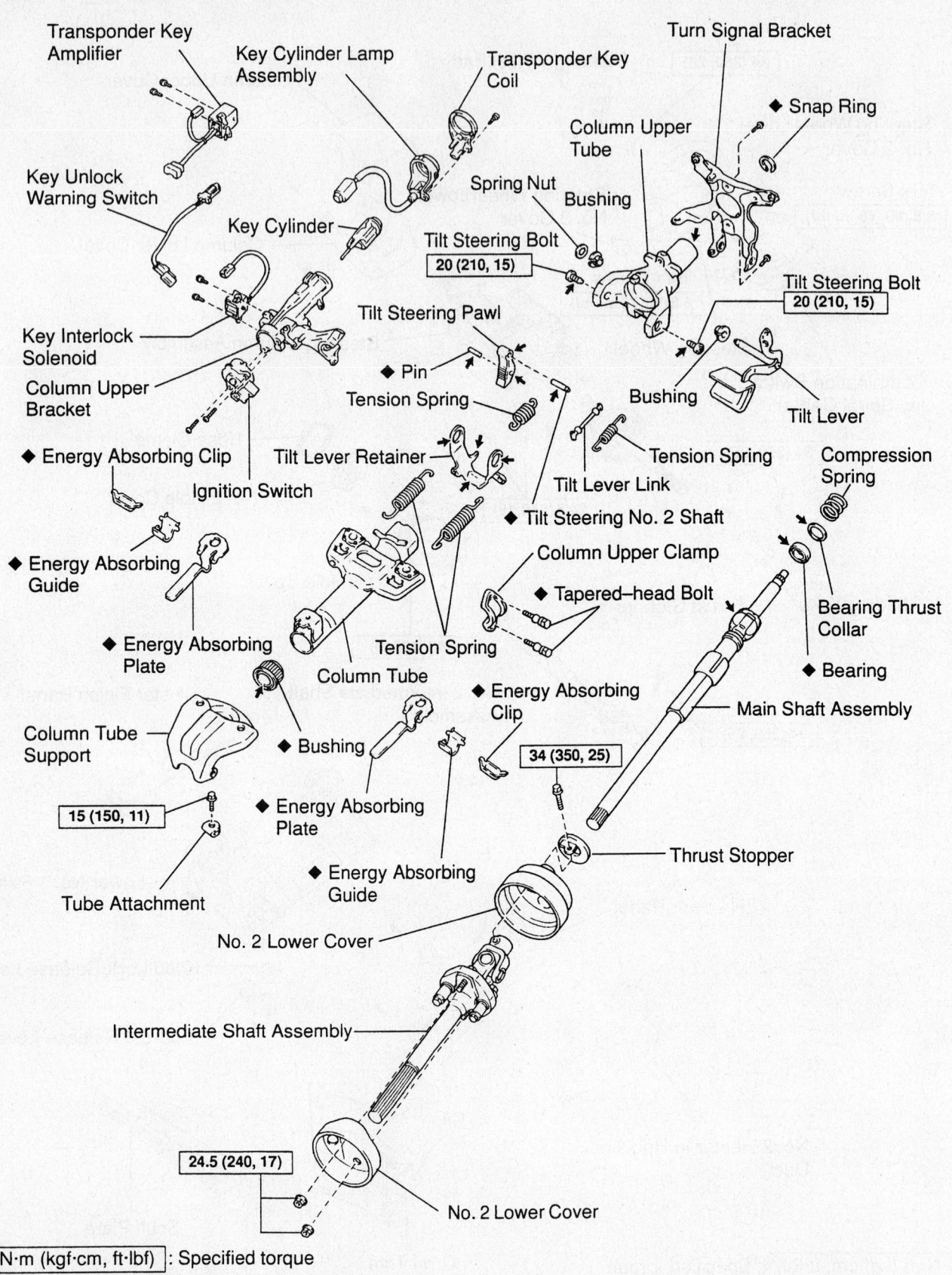

Transponder Key Amplifier

Key Cylinder Lamp Assembly

Transponder Key Coil

Turn Signal Bracket

◆ Snap Ring

Column Upper Tube

Key Unlock Warning Switch

Spring Nut

Bushing

Tilt Steering Bolt
20 (210, 15)

Key Cylinder

Tilt Steering Bolt
20 (210, 15)

Key Interlock Solenoid

Tilt Steering Pawl

◆ Pin

Bushing

Tilt Lever

Column Upper Bracket

Tension Spring

Tension Spring

◆ Energy Absorbing Clip

Tilt Lever Retainer

Tilt Lever Link

Compression Spring

Ignition Switch

◆ Tilt Steering No. 2 Shaft

◆ Energy Absorbing Guide

Column Upper Clamp

Bearing Thrust Collar

◆ Tapered-head Bolt

◆ Energy Absorbing Plate

◆ Bearing

◆ Energy Absorbing Clip

Main Shaft Assembly

Column Tube Support

◆ Bushing

34 (350, 25)

15 (150, 11)

◆ Energy Absorbing Plate

Column Tube

Tension Spring

Thrust Stopper

Tube Attachment

◆ Energy Absorbing Guide

No. 2 Lower Cover

Intermediate Shaft Assembly

24.5 (240, 17)

No. 2 Lower Cover

N·m (kgf·cm, ft·lbf) : Specified torque
◆ Non-reusable part
◀ Molybdenum disulfide lithium base grease

93113GG6

Exploded view the steering column—Lexus LX470 (Part 2 of 2)

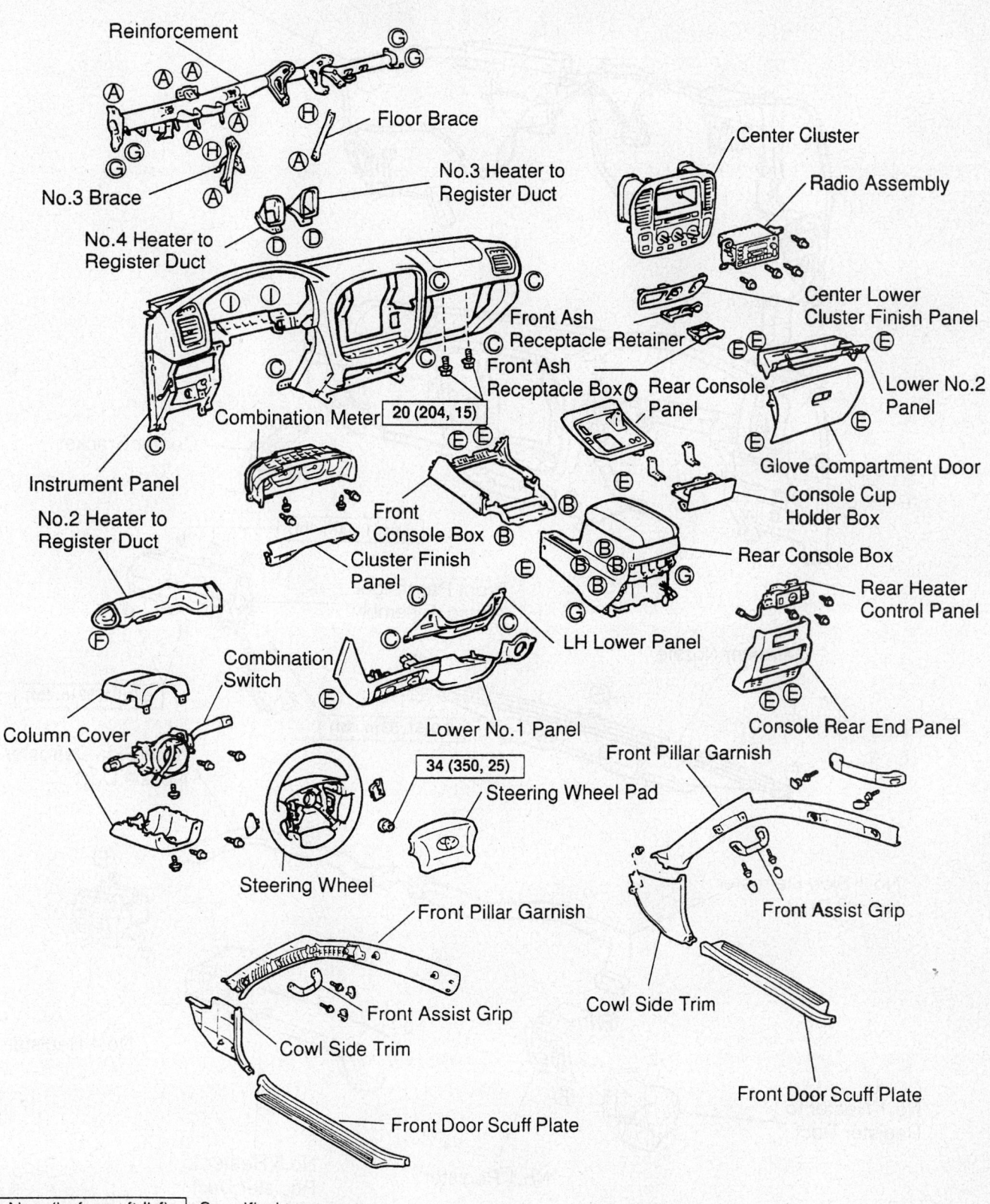

Reinforcement

Floor Brace

No.3 Brace

No.3 Heater to Register Duct

No.4 Heater to Register Duct

Center Cluster

Radio Assembly

Center Lower Cluster Finish Panel

Front Ash Receptacle Retainer

Front Ash Receptacle Box

Rear Console Panel

Lower No.2 Panel

Glove Compartment Door

Console Cup Holder Box

Rear Console Box

Combination Meter | 20 (204, 15) |

Instrument Panel

No.2 Heater to Register Duct

Front Console Box

Cluster Finish Panel

LH Lower Panel

Rear Heater Control Panel

Console Rear End Panel

Combination Switch

Column Cover

Lower No.1 Panel

| 34 (350, 25) |

Steering Wheel Pad

Front Pillar Garnish

Front Assist Grip

Steering Wheel

Front Pillar Garnish

Front Assist Grip

Cowl Side Trim

Cowl Side Trim

Front Door Scuff Plate

Front Door Scuff Plate

| N·m (kgf·cm, ft·lbf) | : Specified torque

93113GG7

Exploded view the instrument panel and related components—LX470

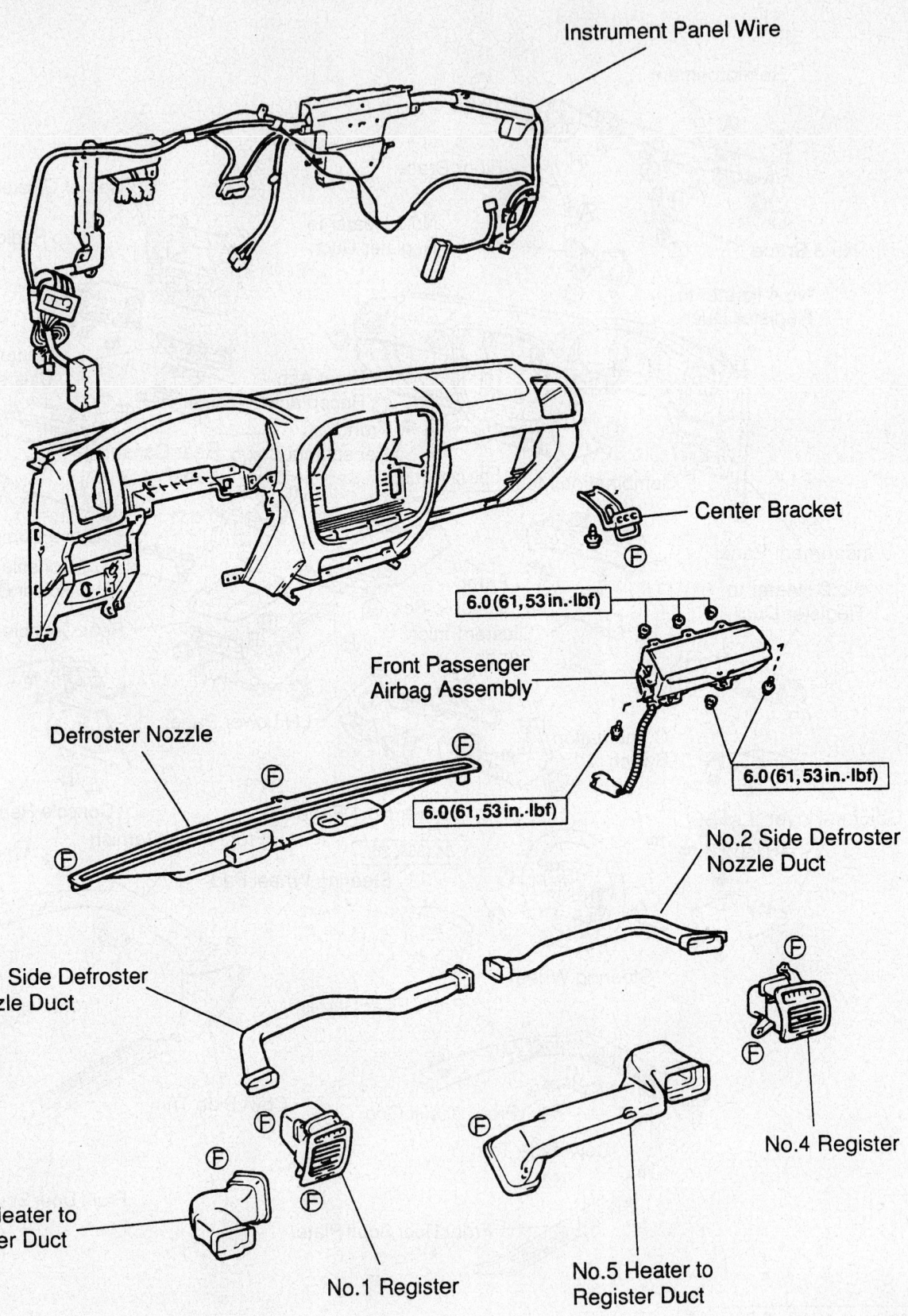

Instrument Panel Wire

Center Bracket

6.0 (61, 53 in.·lbf)

Front Passenger
Airbag Assembly

6.0 (61, 53 in.·lbf)

Defroster Nozzle

6.0 (61, 53 in.·lbf)

No.2 Side Defroster
Nozzle Duct

No.1 Side Defroster
Nozzle Duct

No.4 Register

No.1 Heater to
Register Duct

No.1 Register

No.5 Heater to
Register Duct

N·m (kgf·cm, ft·lbf) : Specified torque

93113GG8

Exploded view the front ventilation ducts and related components—Lexus LX470

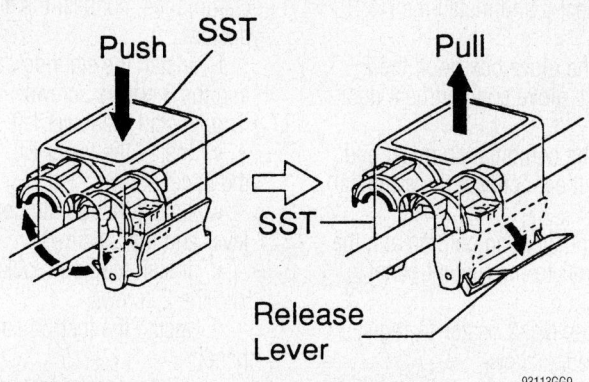

Push SST   Pull

SST

Release
Lever

93113GG9

View the air conditioning line clamp removal tool—Lexus LX470

d. Using an assistant, install the instrument panel, connect the junction connectors, the instrument panel-to-chassis 8 bolts and 2 nuts.

e. Install the front the console, engage the 2 clips and install the 2 console-to-chassis bolts/screws.

f. Connect the electrical connector; then, install the center lower cluster finish panel by engaging the 5 clips.

g. Install the console box and the 4 rear console box-to-chassis screws/bolts.

h. If not equipped with a rear air conditioning system, install rear heater control panel, the 3 panel screws; then,

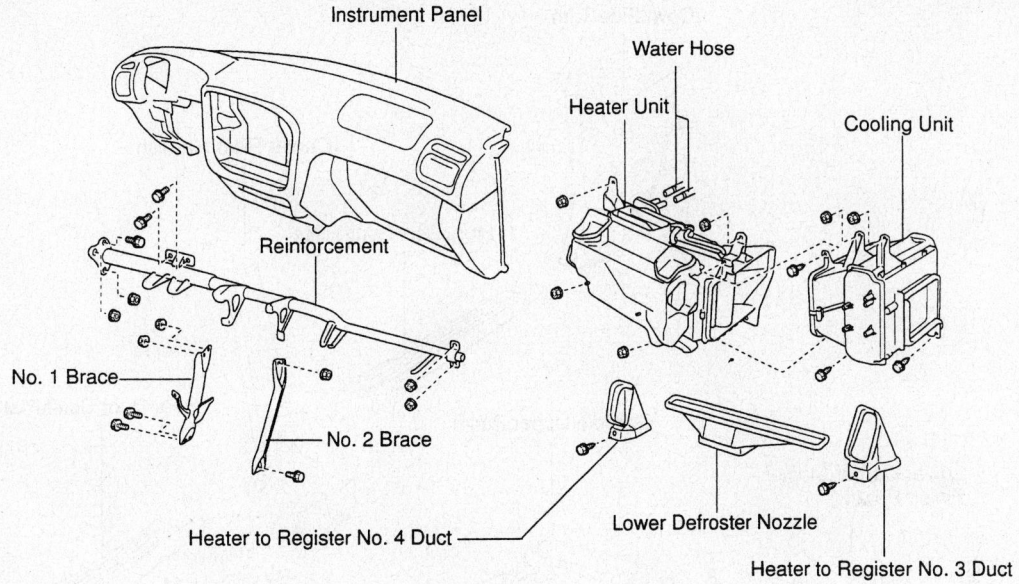

Instrument Panel

Water Hose

Heater Unit

Cooling Unit

Reinforcement

No. 1 Brace

No. 2 Brace

Heater to Register No. 4 Duct

Lower Defroster Nozzle

Heater to Register No. 3 Duct

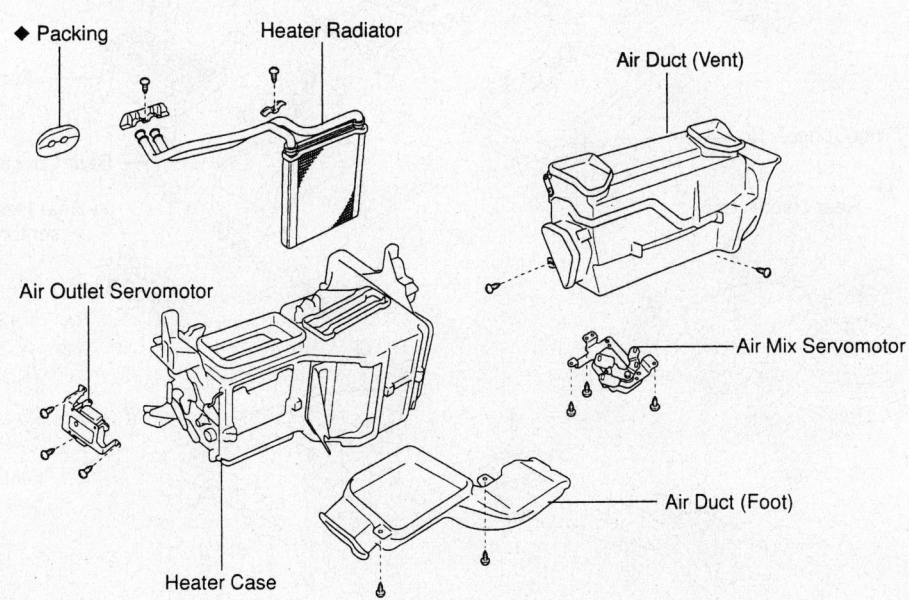

◆ Packing   Heater Radiator

Air Duct (Vent)

Air Outlet Servomotor

Air Mix Servomotor

Air Duct (Foot)

Heater Case

◆ Non–reusable part

93113GG0

Exploded view the front heater core, heater housing, evaporator housing and related components—Lexus LX470

connect the connector and control cable.

i. Install the rear of the console and engage the 2 clips; then, install the 2 rear end panel-to-console screws.

j. Install the rear console panel and engage the 4 clips (2 on each side); then, install the transfer shift lever knob.

k. Install the radio, connect the electrical connectors and the 4 radio screws.

l. Install the center cluster and engage the 8 center cluster clips.

m. Install the lower No. 2 panel and the 3 panel screws.

n. Install the passenger's side air bag

module, connect it and install the No. 1 undercover.

o. Install the glove box door, the 2 screws and the glove compartment door stoppers.

p. Install the combination meter and the 4 screws; then, connect the electrical connectors.

q. Install the steering column and the steering column-to-instrument panel bolts.

r. Install the No. 2 heater-to-register duct and the duct screw.

s. At the steering column, install the combination switch, the 3 screws and the

clamp; then, connect the electrical connectors.

t. Install the steering column covers and the 3 covers screws.

u. Install the lower left side panel.

v. Install the lower No. 1 panel and the panel screw.

w. Install the fuel lid control cable lever and the 2 screws.

x. Install the hood lock control cable and the 2 screws.

y. Install the instrument cluster finish panel.

z. At the passenger's side, install the front pillar garnish, the 2 assist grips,

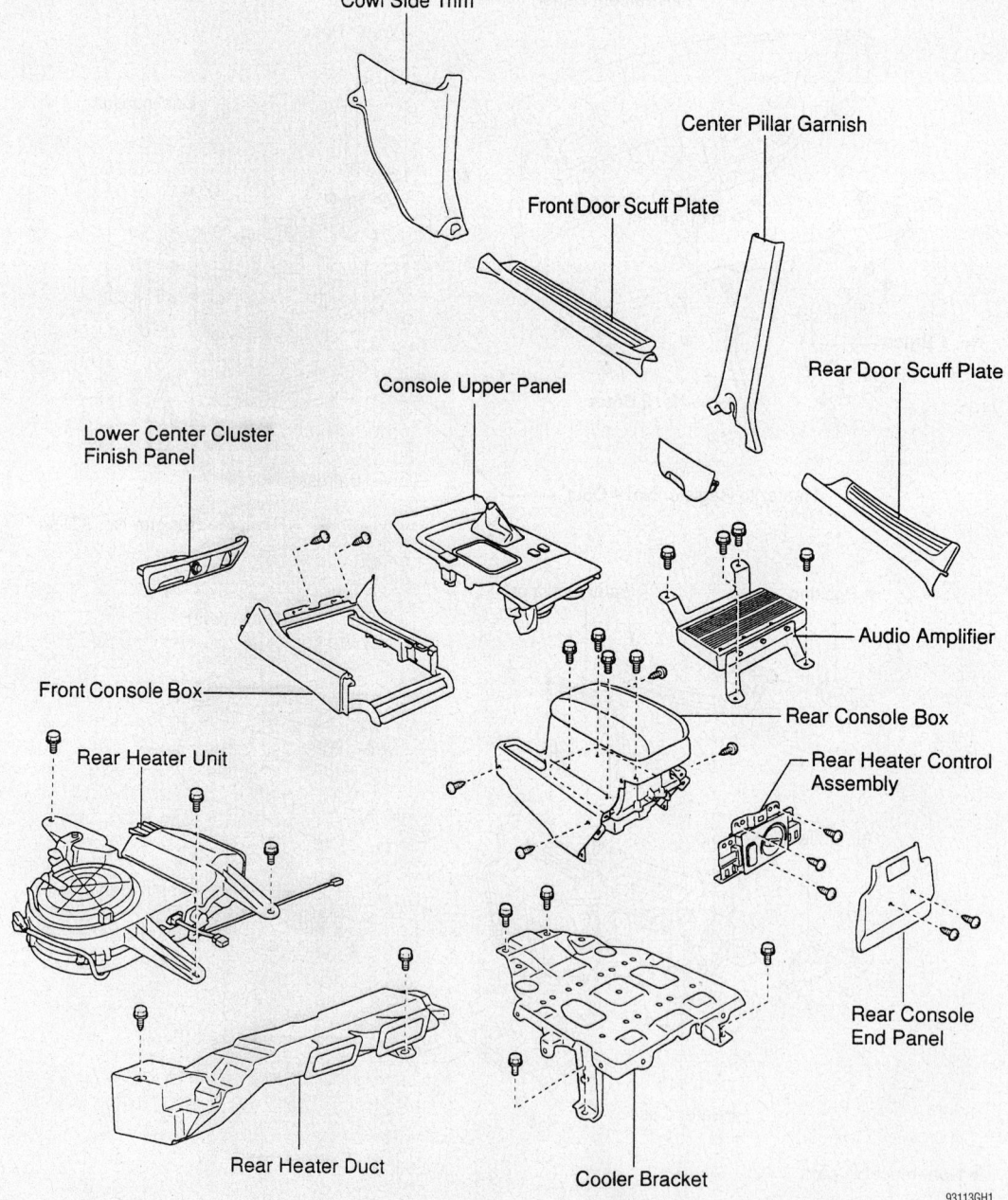

Cowl Side Trim

Center Pillar Garnish

Front Door Scuff Plate

Rear Door Scuff Plate

Console Upper Panel

Lower Center Cluster Finish Panel

Audio Amplifier

Front Console Box

Rear Console Box

Rear Heater Unit

Rear Heater Control Assembly

Rear Console End Panel

Rear Heater Duct

Cooler Bracket

93113GH1

Exploded view of the rear heater housing and related components—Lexus LX470

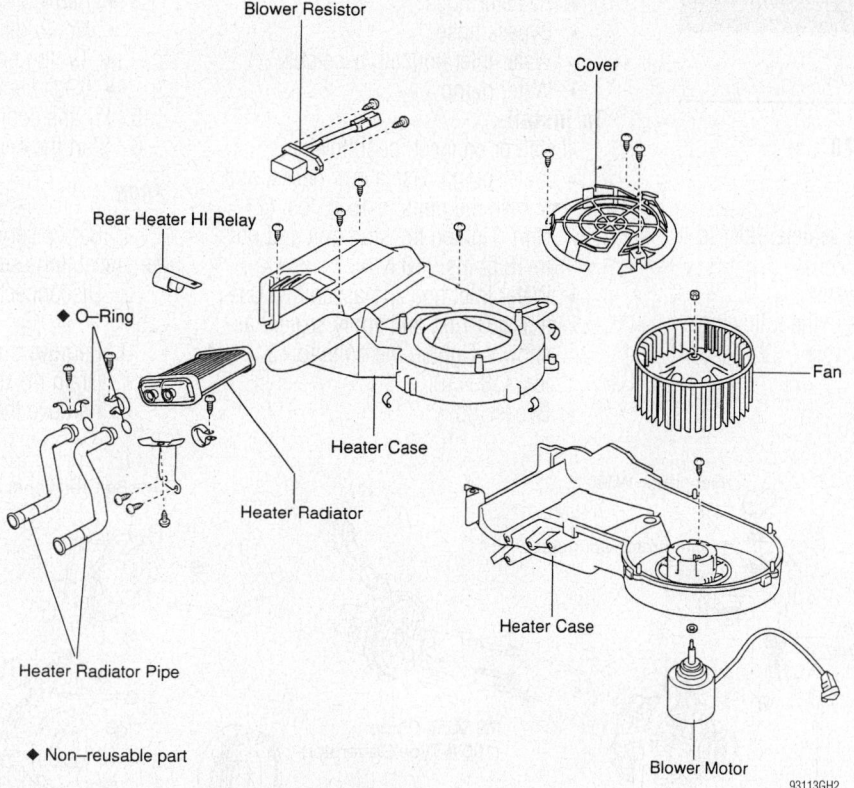

Blower Resistor

Cover

Rear Heater HI Relay

◆ O–Ring

Fan

Heater Case

Heater Radiator

Heater Case

Heater Radiator Pipe

Blower Motor

◆ Non–reusable part

93113GH2

Exploded view of the rear heater core, heater housing and related components—Lexus LX470

the 4 screws and the 4 assist grip plugs.

aa. At the driver's side, install the front pillar garnish, assist grip, the 2 screws and the 2 assist grip plugs.

bb. Install the front door scuff plates, the cowl side trim and the front door opening trim.

13. Install the steering wheel by performing the following procedure:

a. Install the steering wheel to the steering column.

b. Align the steering wheel-to-main shaft marks.

c. Install the steering wheel nut and torque to 25 ft. lbs. (34 Nm).

d. Install the air bag module to the steering wheel and connect the electrical connector.

e. Using a Torx® wrench, tighten the 2 screws located at each side of the steering wheel to 78 inch lbs. (8.8 Nm).

f. Install the steering wheel side covers.

14. Connect the heater hoses to the heater core.

15. Refill the cooling system.

16. Connect the negative battery cable.

a. Evacuate and charge the air conditioning system refrigerant.

17. Run the engine to normal operating temperatures; then, check the climate control operation and check for leaks.

REAR AUXILIARY HEATER

1. Disconnect the negative battery cable.

2. Drain the cooling system into a clean container for reuse.

3. Disconnect the heater hoses from the rear heater core.

4. Remove the front seats.

5. Remove the rear heater control assembly.

6. Remove the rear console box.

7. Remove the front console box cover.

8. Remove the lower center cluster finish panel.

9. Remove the front door scuff plates.

10. Remove the cowl side trim.

11. Remove the rear door scuff plates.

12. Remove the center pillar garnishes.

13. Slide the carpet rearward.

14. Remove the cooler bracket bolts and the bracket.

15. Remove the rear heater duct bolt/screw and the duct.

16. Disconnect the rear heater housing electrical connector.

17. Remove the 3 rear heater housing-to-chassis bolts and the heater housing.

18. Remove the heater core-to-heater housing 3 screws and 2 clamps.

19. Remove the heater core from the heater housing.

To install:

20. Install the heater core to the heater housing.

21. Install the heater core-to-heater housing 3 screws and 2 clamps.

22. Install the heater housing and the 3 rear heater housing-to-chassis bolts.

23. Connect the rear heater housing electrical connector.

24. Install the rear heater duct and the duct bolt/screw.

25. Install the cooler bracket and the bracket bolts.

26. Slide the carpet rearward.

27. Install the center pillar garnishes.

28. Install the rear door scuff plates.

29. Install the cowl side trim.

30. Install the front door scuff plates.

31. Install the lower center cluster finish panel.

32. Install the front console box cover.

33. Install the rear console box.

34. Install the rear heater control assembly.

35. Install the front seats.

36. Connect the heater hoses to the rear heater core.

37. Refill the cooling system.

38. Connect the negative battery cable.

Water Pump

REMOVAL & INSTALLATION

Land Cruiser and LX470

2002–05

1. Before servicing the vehicle, refer to the precautions section.
2. Drain the cooling system.
3. Remove or disconnect the following:
   - Negative battery cable
   - Timing belt.
   - No. 2 idler pulley
   - Radiator hose
   - Bypass hose
   - Water inlet housing assembly
   - Water pump

To install:

4. Install or connect the following:
   - Water pump. Use a new gasket and tighten the bolts to 15 ft. lbs. (21 Nm). Tighten the stud bolt and nut to 13 ft. lbs. (18 Nm).
   - Water inlet housing assembly. Use a new O-ring and apply sealant as shown. Tighten the bolts to 13 ft. lbs. (18 Nm).
   - Bypass hose
   - Radiator hose
   - No. 2 idler pulley
   - Timing belt
   - Negative battery cable
5. Fill the cooling system.
6. Start the engine and check for leaks.

2006

1. Before servicing the vehicle, refer to the precautions section.
2. Disconnect the negative battery cable.
3. Remove the engine undercover.
4. Drain the cooling system.
5. Remove the v-bank cover.

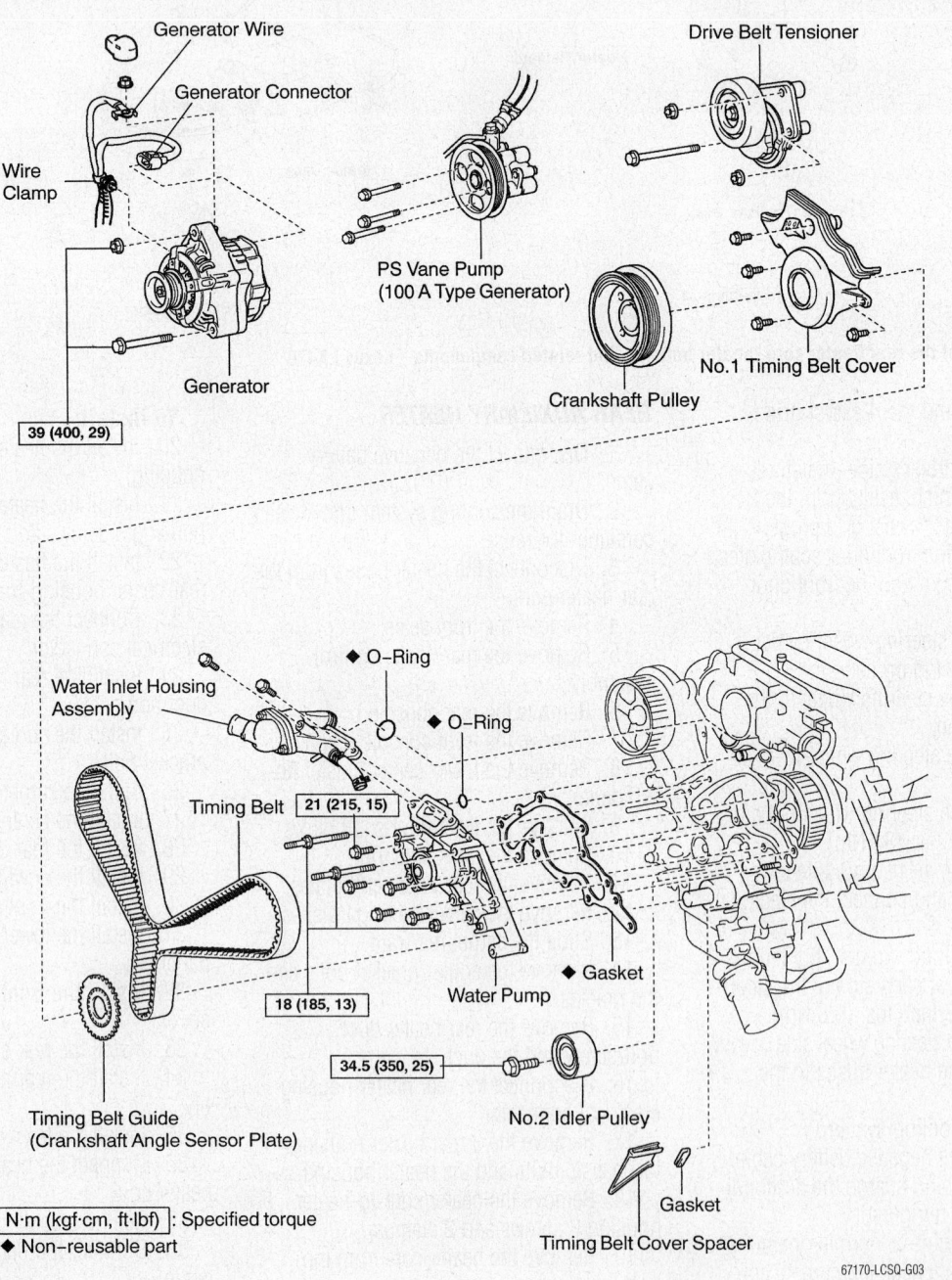

N·m (kgf·cm, ft·lbf) : Specified torque
◆ Non-reusable part

Water pump and related parts—2002–05 models

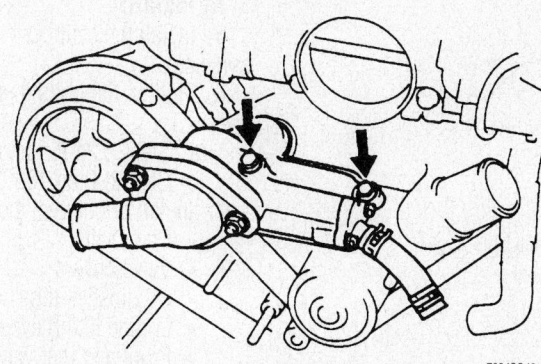

Water inlet housing attaching bolts

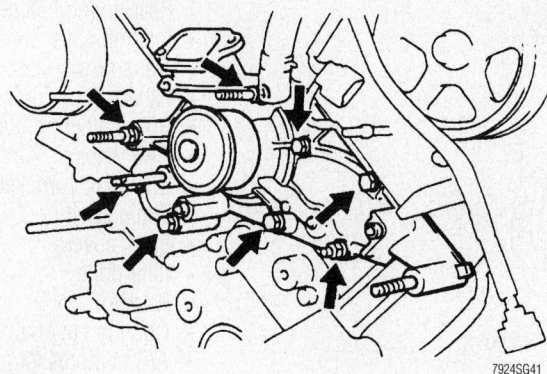

Water pump mounting bolts, stud bolts and nut locations

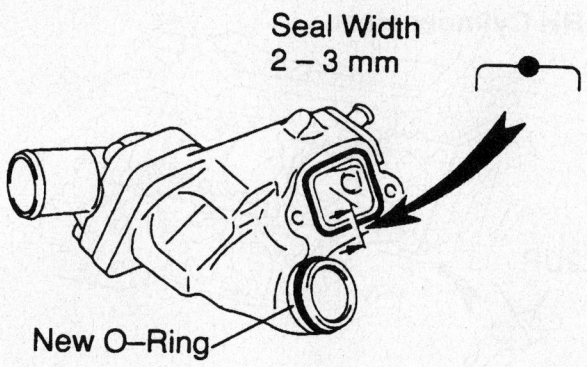

Water inlet housing sealant application

6. Remove the intake air pipe.

7. Remove the A/C compressor and set aside with the lines attached.

8. Remove the drive belt.

9. Remove the radiator.

10. Remove the oil cooler pipe.

11. Remove the timing belt.

12. Remove the number 2 timing belt idler pulley sub assembly.

13. Disconnect the water bypass hose from the water inlet housing.

14. Remove the bolts, nuts and stud attaching the inlet housing to the water pump.

15. Disconnect the hose from the inlet housing front joint and remove the inlet housing.

16. Remove the water pump bolts, gasket and the O-ring from the by-pass pipe.

To install:

17. Coat a new O-ring with soapy water, install the O-ring on the pipe end and connect the water pump to the by-pass pipe.

18. Using a new gasket, install the water pump. Tighten the bolts to 15 ft. lbs. (21 Nm), the nuts and stud to 24 ft. lbs. (32 Nm).

19. Clean all packing material from the

water inlet housing gasket surfaces and sealing groove.

20. Apply a 0.078–0.118 inch (2–3 mm) bead of sealant to the inlet housing.

21. Install a new O-ring to the inlet housing and lube the O-ring with soapy water.

22. Attach the inlet housing end to the water by-pass tube.

23. Install the assembly and tighten to 13 ft. lbs. (18 Nm).

24. Install the idler pulley sub-assembly and tighten to 25 ft. lbs. (34 Nm).

25. Install the timing belt.

26. Install the remaining components in the reverse order of removal.

27. Fill the cooling system.

28. Start the engine and check for leaks.

Sequoia and GX470

1. Before servicing the vehicle, refer to the precautions section.

2. Drain the cooling system.

3. Remove or disconnect the following:
- Negative battery cable
- Timing belt.
- No. 2 idler pulley
- Radiator hose
- Bypass hose
- Water inlet housing assembly
- Water pump

To install:

4. Install or connect the following:
- Water pump. Use a new gasket and tighten the bolts to 15 ft. lbs. (21 Nm). Tighten the stud bolt and nut to 13 ft. lbs. (18 Nm).
- Water inlet housing assembly. Use a new O-ring and apply sealant as shown. Tighten the bolts to 13 ft. lbs. (18 Nm).
- Bypass hose
- Radiator hose
- No. 2 idler pulley
- Timing belt
- Negative battery cable

5. Fill the cooling system.

6. Start the engine and check for leaks.

Cylinder Head

REMOVAL & INSTALLATION

2002 Models

LAND CRUISER AND SEQUOIA

1. Before servicing the vehicle, refer to the precautions section.

2. Drain the cooling system.

3. Relieve the fuel system pressure.

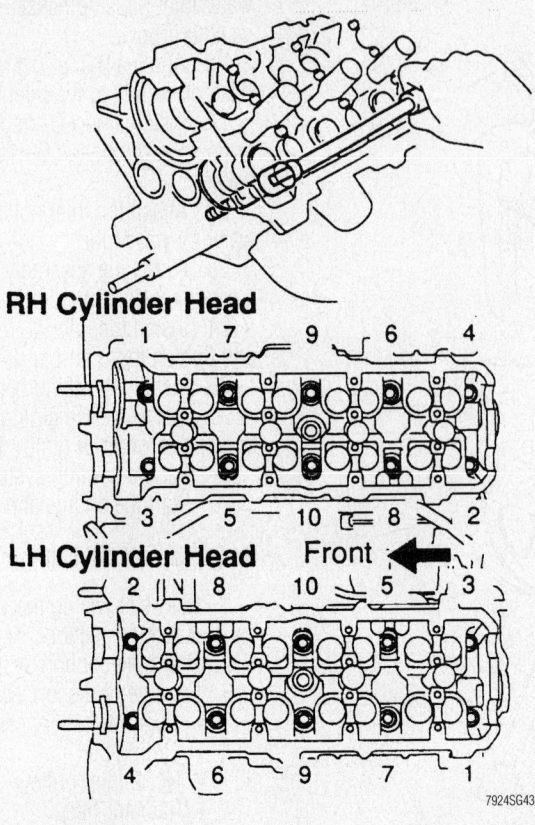

RH Cylinder Head

LH Cylinder Head Front

Cylinder head loosening sequence—2002–05 4.7L 2UZ-FE engine

To install:

5. Install the cylinder heads with new gaskets.

6. Tighten the bolts in sequence as follows:
   a. Step 1: 24 ft. lbs. (32 Nm).
   b. Step 2: Plus 180 degrees.

7. Install or connect the following:
   - Camshafts
   - Valve covers
   - Oil dipstick tube
   - Engine lifting eyes
   - Front and rear water bypass joints
   - Water inlet housing assembly
   - Intake manifold
   - Fuel lines
   - Rear timing belt covers
   - Ignition coils
   - Transmission dipstick tube
   - Exhaust front pipes
   - Power steering pump
   - CMP sensor
   - Camshaft sprockets
   - Timing belt
   - Front covers
   - Idler pulley
   - Radiator
   - Cooling fan and bracket
   - A/C compressor and bracket
   - Accessory drive belt

4. Remove or disconnect the following:
   - Battery and tray
   - Engine appearance cover
   - Engine under covers
   - Air intake assembly
   - Accessory drive belt
   - A/C compressor and bracket
   - Cooling fan and bracket
   - Radiator
   - Idler pulley
   - Front covers
   - Timing belt.
   - Camshaft sprockets
   - Camshaft Position (CMP) sensor
   - Power steering pump
   - Exhaust front pipes
   - Transmission dipstick tube
   - Ignition coils
   - Rear timing belt covers
   - Fuel lines
   - Intake manifold
   - Water inlet housing assembly
   - Front and rear water bypass joints
   - Engine lifting eyes
   - Oil dipstick tube
   - Valve covers
   - Camshafts
   - Cylinder heads with the exhaust manifolds attached. Loosen the bolts in the sequence shown.

RH Cylinder Head

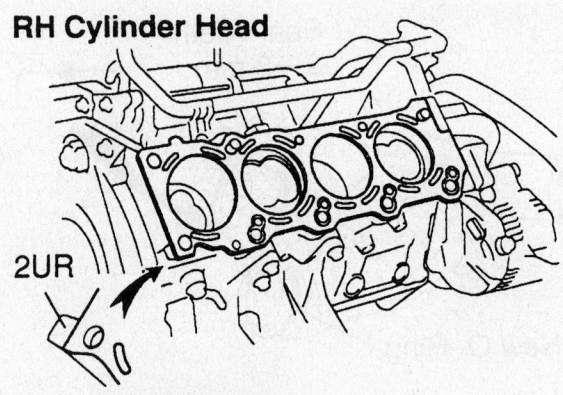

2UR

LH Cylinder Head

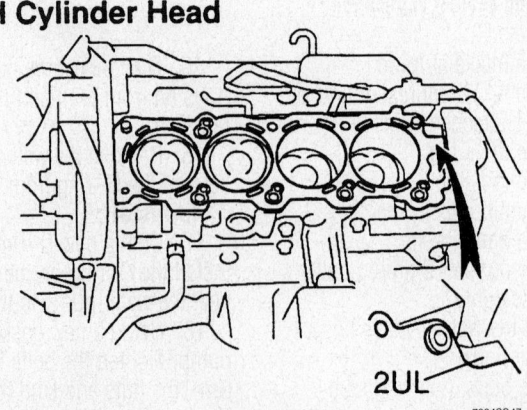

2UL

Cylinder head gasket identification—4.7L 2UZ-FE engine

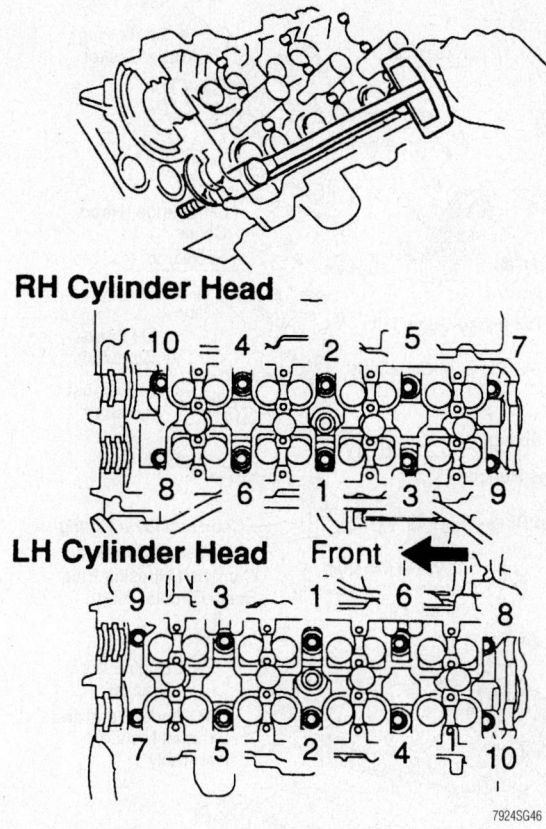

RH Cylinder Head

LH Cylinder Head   Front

Cylinder head torque sequence—2002–05 4.7L 2UZ-FE engine

RH Bank

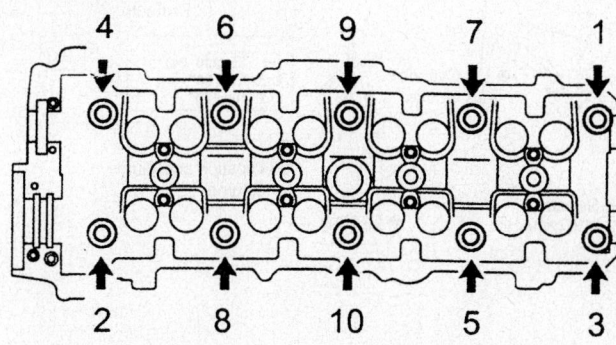

LH Bank

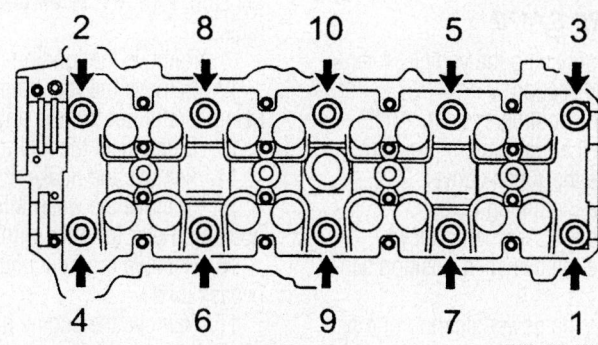

Cylinder head torque sequence—2006 4.7L 2UZ-FE engine

- Air intake assembly
- Engine under covers
- Engine appearance cover
- Battery and tray

8. Fill the cooling system.

9. Start the engine and check for leaks.

GX470 AND LX470

1. Before servicing the vehicle, refer to the precautions section.

2. Drain the cooling system.

3. Relieve the fuel system pressure.

4. Remove or disconnect the following:
- Battery and tray
- Engine appearance cover
- Engine under covers
- Air intake assembly
- Accessory drive belt
- A/C compressor and bracket
- Cooling fan and bracket
- Radiator
- Idler pulley
- Front covers
- Timing belt
- Camshaft sprockets
- Camshaft Position (CMP) sensor
- Power steering pump
- Exhaust front pipes
- Transmission dipstick tube
- Ignition coils
- Rear timing belt covers
- Fuel lines
- Intake manifold
- Water inlet housing assembly
- Front and rear water bypass joints
- Engine lifting eyes
- Oil dipstick tube
- Valve covers
- Camshafts
- Cylinder heads with the exhaust manifolds attached. Loosen the bolts in the sequence shown.

To install:

5. Install the cylinder heads with new gaskets. Tighten the bolts in sequence as follows:

   a. Step 1: 24 ft. lbs. (32 Nm).

   b. Step 2: Plus 180 degrees.

6. Install or connect the following:
- Camshafts
- Valve covers
- Oil dipstick tube
- Engine lifting eyes
- Front and rear water bypass joints
- Water inlet housing assembly
- Intake manifold
- Fuel lines
- Rear timing belt covers
- Ignition coils
- Transmission dipstick tube
- Exhaust front pipes
- Power steering pump

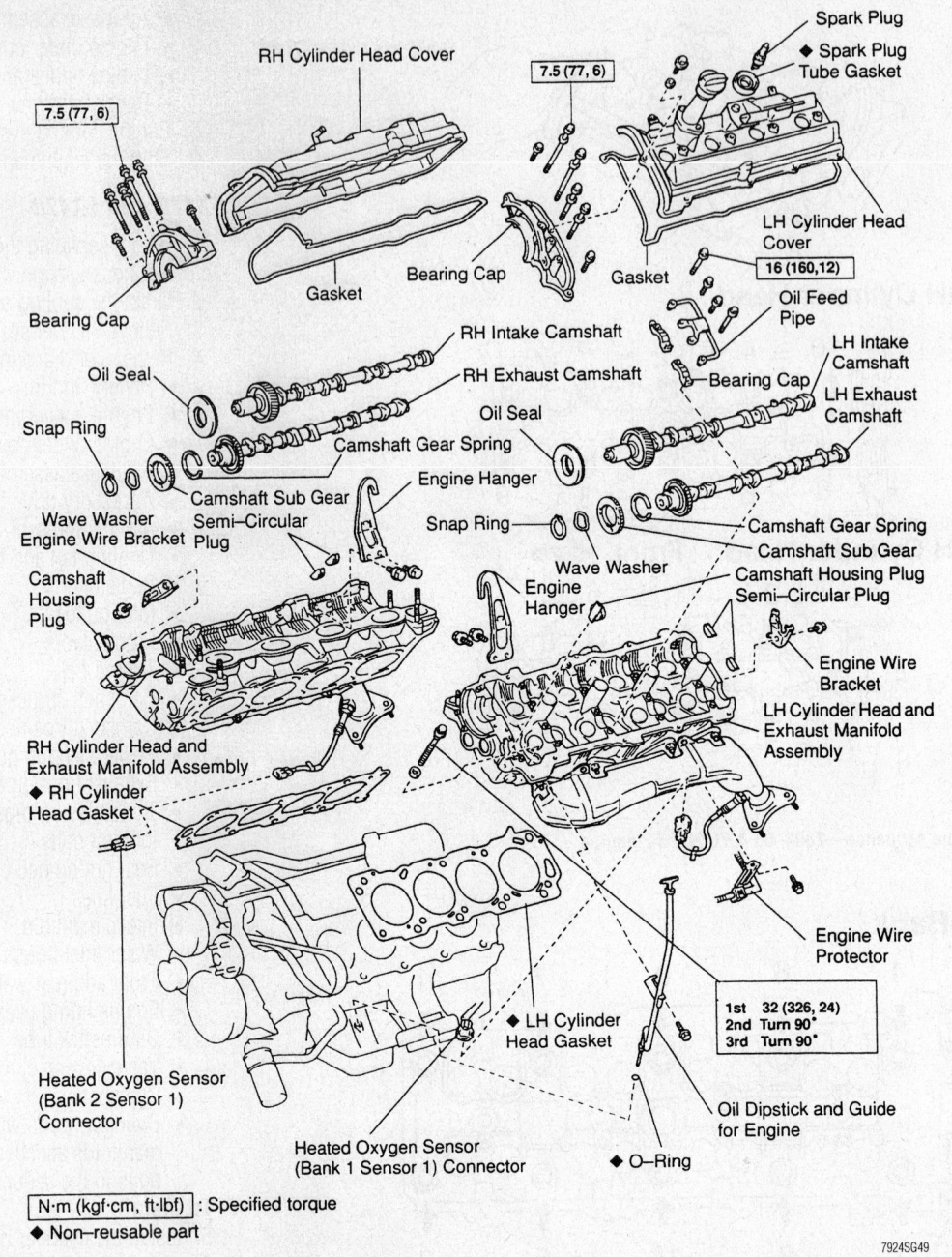

Exploded view of the cylinder head mounting—4.7L 2UZ-FE engine

- CMP sensor
- Camshaft sprockets
- Timing belt
- Front covers
- Idler pulley
- Radiator
- Cooling fan and bracket
- A/C compressor and bracket
- Accessory drive belt
- Air intake assembly
- Engine under covers
- Engine appearance cover
- Battery and tray
7. Fill the cooling system.
8. Start the engine and check for leaks.

2003–06 Models

SEQUOIA AND GX470

1. Before servicing the vehicle, refer to the precautions section.
2. Drain the cooling system.
3. Relieve the fuel system pressure.
4. Remove the V bank cover.
5. Remove the timing belt.
6. Remove the camshaft pulleys.
7. Remove the camshaft position sensor.
8. Remove the power steering pump and set it aside with the lines still attached.
9. Remove the front exhaust pipe.

10. On models with an automatic transmission, remove the oil dipstick and tube.
11. Remove the ignition coils.
12. Remove the rear timing belt plates being careful not to drop anything.
13. Disconnect the fuel inlet hose.
14. Remove the intake manifold.
15. Remove the water inlet and inlet housing. Refer to water pump removal.
16. Remove the front and rear water bypass joint.
17. Remove the engine hangers and if needed the oil dipstick and tube.
18. Remove the valve covers.

➡Since the thrust level of the camshaft is small, the camshaft must be kept level during removal. If not kept level serious damage could occur.

19. Check the timing mark of the crankshaft pulley is aligned with the center(s) of the crankshaft pulley bolt and idler pulley bolt.

➡If the crankshaft pulley is wrongly positioned, this can cause the piston to contact the head causing severe damage. Make sure the crankshaft pulley is properly positioned.

20. Release the oil from the front bearing caps using the tool illustrated. Rotate the camshaft timing tube from left to right 2 to 3 times within its VVT-I range of 25 degrees and collect the oil from the timing oil control valve installation hole using a rag.

21. Remove the left hand camshafts as follows:

a. Bring the service bolt of the sub gear up by turning the left exhaust camshaft using a wrench on the hexagon head portion of the shaft.

b. Secure the sub gear to the main gear using a 16 to 20 mm bolt with a diameter of 6mm and a thread pitch of 1mm.

c. Make sure the torsional force of the sub gear is retained by the bolt.

d. Align the 2 dot timing mark of the left side camshaft by turning the left exhaust camshaft using a wrench on the hexagon head portion of the shaft.

➡Mark the position of the caps so they can be reinstalled in their original positions.

e. Loosen the 22 bearing cap bolts in the sequence illustrated using several passes.

f. Remove the bolts, washers, oil feed pipe, bearing caps, camshaft housing plug, oil control valve filter and the camshafts.

22. Remove the right hand camshafts as follows:

a. Bring the service bolt of the sub gear up by turning the right exhaust camshaft using a wrench on the hexagon head portion of the shaft.

b. Secure the sub gear to the main gear using a 16 to 20 mm bolt with a diameter of 6mm and a thread pitch of 1mm.

c. Make sure the torsional force of the sub gear is retained by the bolt.

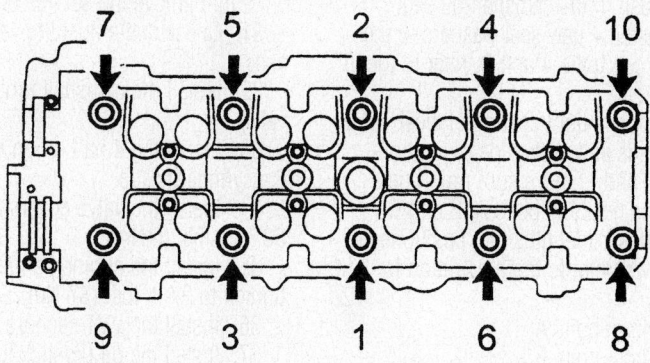

LH Bank

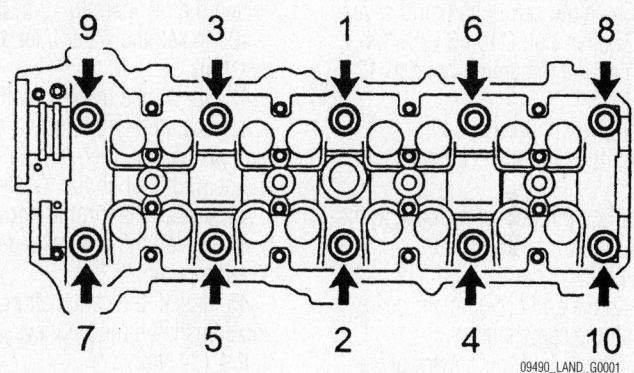

Cylinder head loosening sequence—2006 4.7L 2UZ-FE engine

d. Align the 1 dot timing mark of the camshaft main gear (about 10 degrees) angle by turning the right exhaust camshaft using a wrench on the hexagon head portion of the shaft.

➡Mark the position of the caps so they can be reinstalled in their original positions.

e. Loosen the 22 bearing cap bolts in the sequence illustrated using several passes.

f. Remove the bolts, washers, oil feed pipe, bearing caps, camshaft housing plug, oil control valve filter and the camshafts.

23. Loosen the cylinder head bolts in the sequence shown, using several passes.

24. Remove the cylinder heads and exhaust manifolds together as an assembly.

To install:

25. Install new gaskets and the cylinder heads

26. Tighten the bolts in sequence on 2003–05 models as follows:

a. Step 1: 24 ft. lbs. (32 Nm).

b. Step 2: Plus 180 degrees.

27. Tighten the bolts in sequence on 2006 models as follows:

a. Step 1: 30 ft. lbs. (40 Nm).

b. Step 2: Plus 90 degrees.

c. Step 3: Plus 90 degrees.

28. Check the timing mark of the crankshaft pulley is aligned with the center(s) of the crankshaft pulley bolt and idler pulley bolt.

➡If the crankshaft pulley is wrongly positioned, this can cause the piston to contact the head causing severe damage. Make sure the crankshaft pulley is properly positioned.

29. Install the left side camshafts as follows:

a. Apply multi purpose grease to the thrust portion of the camshafts.

b. Align the 2 dot timing mark of the camshaft drive and driven main gears and install the camshafts.

c. Apply seal packing to the camshaft housing plug.

d. Install the camshaft housing plug on the cylinder head as illustrated. Install the strainer on the head being careful it is properly positioned.

e. Apply seal packing to the front bearing cap.

f. Install the front bearing cap and then the other caps in the sequence illustrated.

g. Push in the camshaft oil seal.

h. Install 4 new seal washers to the bearing cap bolts A and B, refer to the illustration.

i. Apply a light coating of clean oil to the threads and underside of the bolt heads D and E. make sure no oil gets under the heads of bolts A, B and C.

j. The bolt lengths and positions are as follows. refer to the illustration for bolt location:

- 94mm bolts A
- 72mm bolts B
- 25mm bolts C
- 52mm bolts D
- 38mm bolts E

k. Tighten the cap bolts using several passes. Tighten bolt C to 66 inch lbs. (7.5 Nm) an the remaining bolts to 12 ft. lbs. (16 Nm).

l. Remove the service bolt.

30. Install the right side camshafts as follows:

a. Apply multi purpose grease to the thrust portion of the camshafts.

b. Align the 1 dot timing mark of the camshaft drive and driven main gears and install the camshafts.

c. Set the 1 dot timing mark of the camshaft drive and driven gears at a 10 degree angle.

d. Apply seal packing to the camshaft housing plug.

e. Install the camshaft housing plug on the cylinder head as illustrated. Install the strainer on the head being careful it is properly positioned.

f. Apply seal packing to the front bearing cap.

g. Install the front bearing cap and then the other caps in the sequence illustrated.

h. Push in the camshaft oil seal.

i. Install 4 new seal washers to the bearing cap bolts A and B, refer to the illustration.

j. Apply a light coating of clean oil to the threads and underside of the bolt heads D and E. make sure no oil gets under the heads of bolts A, B and C.

k. The bolt lengths and positions are as follows. refer to the illustration for bolt location:

- 94mm bolts A
- 72mm bolts B
- 25mm bolts C
- 52mm bolts D
- 38mm bolts E

l. Tighten the cap bolts using several passes. Tighten bolt C to 66 inch lbs. (7.5 Nm) an the remaining bolts to 12 ft. lbs. (16 Nm).

m. Remove the service bolt.

31. Check and adjust the valve clearance.

32. Install the camshaft timing control valve.

33. Install the 4 half moon plugs onto the cylinder heads.

34. Install the valve covers and tighten to 53 inch lbs. (6 Nm).

35. Install the engine hangers and tighten to 27 ft. lbs. (37 Nm).

36. Install the VVT sensors.

37. Install the oil dipstick tube and dipstick.

38. Install the ignition coils.

39. Install the water bypass joint and tighten the retainers to 13 ft. lbs. (18 Nm).

40. Install the water inlet and housing assembly.

41. Install the intake manifold.

42. Install the timing belt rear plates, right plates first, then left plates. Tighten the retainers to 66 inch lbs. (7 Nm).

43. Install the throttle body cover.

44. Install the front exhaust pipe, power steering pump.

45. Install the camshaft position sensor and camshaft timing pulleys, tighten to 25 ft. lbs. (34 Nm).

46. Install the timing belt.

47. Fill the cooling system and perform an oil change.

48. Start the vehicle and check for leaks.

LAND CRUISER AND LX470

1. Before servicing the vehicle, refer to the precautions section.

2. Relieve the fuel system pressure.

3. Drain the cooling system.

4. Drain the oil.

5. Remove the timing belt.

6. Disconnect the fuel lines.

7. Remove the front exhaust pipes from both sides.

8. Remove the intake manifold.

9. Remove the air switching valve.

10. Remove the air pump.

11. Remove the camshaft position sensor.

12. Remove the VVT sensor on banks 1 and 2.

13. Remove the knock sensor.

14. Remove the camshaft timing oil control valve.

15. Remove the starter.

16. Remove the water inlet housing and front bypass joint. Refer to water pump removal.

17. Disconnect the heater hose and remove the water bypass pipe sub assembly.

18. Remove the rear water bypass joint.

19. Remove the oil dipstick and tube.

20. Remove the ignition coils.

21. Remove the spark plugs.

22. Remove the valve covers.

23. Turn the crankshaft pulley counter-clockwise by 5 degrees until the marks are aligned as illustrated.

➡**Make sure to match the cut part by turning counterclockwise.**

24. Remove the camshaft pulleys.

25. Remove the rear timing belt plates.

➡**Since the thrust level of the camshaft is small, the camshaft must be kept level during removal. If not kept level serious damage could occur.**

26. Check the timing mark of the crankshaft pulley is aligned with the center(s) of the crankshaft pulley bolt and idler pulley bolt.

➡**If the crankshaft pulley is wrongly positioned, this can cause the piston to contact the head causing severe damage. Make sure the crankshaft pulley is properly positioned.**

27. Release the oil from the front bearing caps using the tool illustrated. Rotate the camshaft timing tube from left to right 2 to 3 times within its VVT-I range of 25 degrees and collect the oil from the timing oil control valve installation hole using a rag.

28. Remove the left hand camshafts as follows:

a. Bring the service bolt of the sub gear up by turning the left exhaust camshaft using a wrench on the hexagon head portion of the shaft.

b. Secure the sub gear to the main gear using a 16 to 20 mm bolt with a diameter of 6mm and a thread pitch of 1mm.

c. Make sure the torsional force of the sub gear is retained by the bolt.

d. Align the 2 dot timing mark of the left side camshaft by turning the left exhaust camshaft using a wrench on the hexagon head portion of the shaft.

➡**Mark the position of the caps so they can be reinstalled in their original positions.**

e. Loosen the 22 bearing cap bolts in the sequence illustrated using several passes.

f. Remove the bolts, washers, oil feed pipe, bearing caps, camshaft housing plug, oil control valve filter and the camshafts.

29. Remove the right hand camshafts as follows:

a. Bring the service bolt of the sub gear up by turning the right exhaust camshaft using a wrench on the hexagon head portion of the shaft.

b. Secure the sub gear to the main gear using a 16 to 20 mm bolt with a diameter of 6mm and a thread pitch of 1mm.

c. Make sure the torsional force of the sub gear is retained by the bolt.

d. Align the 1 dot timing mark of the camshaft main gear (about 10 degrees) angle by turning the right exhaust camshaft using a wrench on the hexagon head portion of the shaft.

➡**Mark the position of the caps so they can be reinstalled in their original positions.**

e. Loosen the 22 bearing cap bolts in the sequence illustrated using several passes.

f. Remove the bolts, washers, oil feed pipe, bearing caps, camshaft housing plug, oil control valve filter and the camshafts.

30. Remove the engine hangers.

31. Loosen the cylinder head bolts in the sequence shown, using several passes.

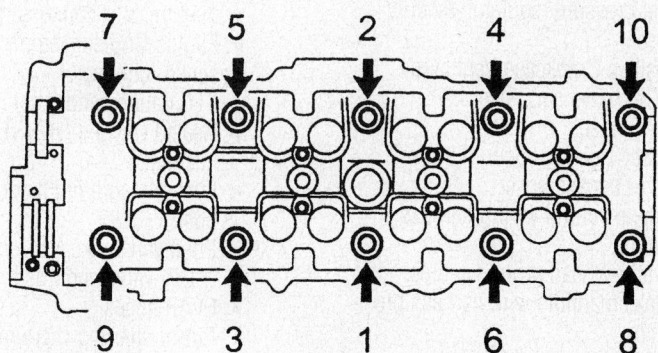

LH Bank

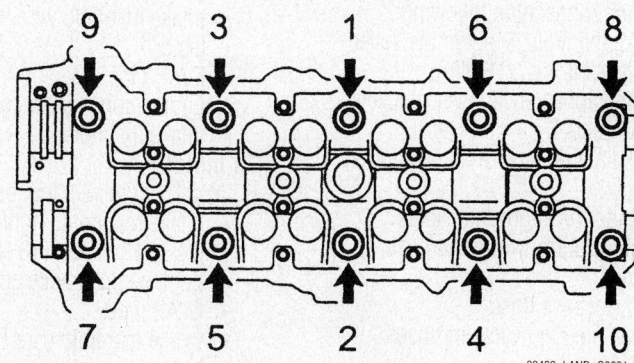

Cylinder head loosening sequence—2006 4.7L 2UZ-FE engine

32. Remove the cylinder heads and exhaust manifolds together as an assembly.

To install:

33. Install new gaskets and the cylinder heads

34. Tighten the bolts in sequence on 2003–05 models as follows:

a. Step 1: 24 ft. lbs. (32 Nm).

b. Step 2: Plus 180 degrees.

35. Tighten the bolts in sequence on 2006 models as follows:

a. Step 1: 30 ft. lbs. (40 Nm).

b. Step 2: Plus 90 degrees.

c. Step 3: Plus 90 degrees.

36. Check the timing mark of the crankshaft pulley is aligned with the center(s) of the crankshaft pulley bolt and idler pulley bolt.

➡**If the crankshaft pulley is wrongly positioned, this can cause the piston to contact the head causing severe damage. Make sure the crankshaft pulley is properly positioned.**

37. Install the left side camshafts as follows:

a. Apply multi purpose grease to the thrust portion of the camshafts.

b. Align the 2 dot timing mark of the camshaft drive and driven main gears and install the camshafts.

c. Apply seal packing to the camshaft housing plug.

d. Install the camshaft housing plug on the cylinder head as illustrated. Install the strainer on the head being careful it is properly positioned.

e. Apply seal packing to the front bearing cap.

f. Install the front bearing cap and then the other caps in the sequence illustrated.

g. Push in the camshaft oil seal.

h. Install 4 new seal washers to the bearing cap bolts A and B, refer to the illustration.

i. Apply a light coating of clean oil to the threads and underside of the bolt heads D and E. make sure no oil gets under the heads of bolts A, B and C.

j. The bolt lengths and positions are as follows. refer to the illustration for bolt location:

- 94mm bolts A
- 72mm bolts B
- 25mm bolts C
- 52mm bolts D
- 38mm bolts E

k. Tighten the cap bolts using several passes. Tighten bolt C to 66 inch lbs. (7.5 Nm) an the remaining bolts to 12 ft. lbs. (16 Nm).

l. Remove the service bolt.

38. Install the right side camshafts as follows:

a. Apply multi purpose grease to the thrust portion of the camshafts.

b. Align the 1 dot timing mark of the camshaft drive and driven main gears and install the camshafts.

c. Set the 1 dot timing mark of the camshaft drive and driven gears at a 10 degree angle.

d. Apply seal packing to the camshaft housing plug.

e. Install the camshaft housing plug on the cylinder head as illustrated. Install the strainer on the head being careful it is properly positioned.

f. Apply seal packing to the front bearing cap.

g. Install the front bearing cap and then the other caps in the sequence illustrated.

h. Push in the camshaft oil seal.

i. Install 4 new seal washers to the bearing cap bolts A and B, refer to the illustration.

j. Apply a light coating of clean oil to the threads and underside of the bolt heads D and E. make sure no oil gets under the heads of bolts A, B and C.

k. The bolt lengths and positions are

as follows. refer to the illustration for bolt location:

- 94mm bolts A
- 72mm bolts B
- 25mm bolts C
- 52mm bolts D
- 38mm bolts E

  l. Tighten the cap bolts using several passes. Tighten bolt C to 66 inch lbs. (7.5 Nm) an the remaining bolts to 12 ft. lbs. (16 Nm).

  m. Remove the service bolt.

39. Check and adjust the valve clearance.

40. Install the camshaft timing control valve.

41. Install the 4 half moon plugs onto the cylinder heads.

42. Install the valve covers and tighten to 53 inch lbs. (6 Nm).

43. Install the spark plugs.

44. Install the ignition coils.

45. Install the engine hangers and tighten to 27 ft. lbs. (37 Nm).

46. Install the VVT sensors.

47. Install the oil dipstick tube and dipstick.

48. Install the water bypass joint and tighten the retainers to 13 ft. lbs. (18 Nm).

49. Install the water inlet and housing assembly.

50. Install new intake manifold gaskets and the manifold. Tighten the bolts to 13 ft. lbs. (18 Nm) in several passes.

51. Install the throttle cover bracket, wire bracket and wire to the engine hanger bracket.

52. Install the wire to the timing belt rear plate.

53. Attach the wire protector to the intake manifold.

54. Attach the 2 ground cables the cylinder heads.

55. Connect the water bypass hoses to the throttle body.

56. Connect the wire clamps to the bracket on the right delivery pipe.

57. Attach the hoses to the intake manifold.

58. Attach the electrical connectors to the intake manifold.

59. Connect the fuel hose.

60. Install the timing belt rear plates, right plates first, then left plates. Tighten the retainers to 66 inch lbs. (7 Nm).

61. Install the throttle body cover.

62. Install the front exhaust pipe, power steering pump.

63. Install the camshaft position sensor and camshaft timing pulleys, tighten to 25 ft. lbs. (34 Nm).

64. Install the timing belt.

65. Fill the cooling system and perform an oil change.

66. Start the vehicle and check for leaks.

Intake Manifold

REMOVAL & INSTALLATION

2002

LAND CRUISER AND SEQUOIA

1. Before servicing the vehicle, refer to the precautions section.
2. Drain the cooling system.
3. Relieve the fuel system pressure.
4. Remove or disconnect the following:
   - Negative battery cable
   - Engine appearance cover
   - Accelerator cable
   - Throttle Position (TP) sensor connector
   - Accelerator pedal position sensor
   - Throttle motor connector
   - Evaporative Emissions (EVAP) vacuum switching valve connector
   - Fuel injector connectors
   - Engine Coolant Temperature (ECT) sensor connector
   - ETC gauge sender connector
   - Heated Oxygen (HO2S) sensor connectors
   - Fuel pressure regulator vacuum hose
   - Positive Crankcase Ventilation (PCV) valve and hose
   - EVAP hoses
   - Power steering vacuum hoses
   - Water bypass hose
   - Engine control wiring harness clamps
   - Cylinder head ground cables
   - Intake manifold wire harness protector
   - EVAP pipe
   - Engine appearance cover brackets
   - Intake manifold

To install:

5. Install or connect the following:
   - Intake manifold. Tighten the fasteners to 13 ft. lbs. (18 Nm).
   - Engine appearance cover brackets
   - EVAP pipe
   - Intake manifold wire harness protector
   - Cylinder head ground cables
   - Engine control wiring harness clamps
   - Water bypass hose
   - Power steering vacuum hoses
   - EVAP hoses
   - PCV valve and hose

   - Fuel pressure regulator vacuum hose
   - HO2S sensor connectors
   - ETC gauge sender connector
   - ECT sensor connector
   - Fuel injector connectors
   - EVAP vacuum switching valve connector
   - Throttle motor connector
   - Accelerator pedal position sensor
   - TP sensor connector
   - Accelerator cable
   - Engine appearance cover
   - Negative battery cable

6. Fill the cooling system.
7. Start the engine and check for leaks.

GX470 AND LX470

1. Before servicing the vehicle, refer to the precautions section.
2. Drain the cooling system.
3. Relieve the fuel system pressure.
4. Remove or disconnect the following:
   - Negative battery cable
   - Engine appearance cover
   - Accelerator cable
   - Throttle Position (TP) sensor connector
   - Accelerator pedal position sensor
   - Throttle motor connector
   - Evaporative Emissions (EVAP) vacuum switching valve connector
   - Fuel injector connectors
   - Engine Coolant Temperature (ECT) sensor connector
   - ETC gauge sender connector
   - Heated Oxygen (HO2S) sensor connectors
   - Fuel pressure regulator vacuum hose
   - Positive Crankcase Ventilation (PCV) valve and hose
   - EVAP hoses
   - Power steering vacuum hoses
   - Water bypass hose
   - Engine control wiring harness clamps
   - Cylinder head ground cables
   - Intake manifold wire harness protector
   - EVAP pipe
   - Engine appearance cover brackets
   - Intake manifold

To install:

5. Install or connect the following:
   - Intake manifold. Tighten the fasteners to 13 ft. lbs. (18 Nm).
   - Engine appearance cover brackets
   - EVAP pipe
   - Intake manifold wire harness protector
   - Cylinder head ground cables

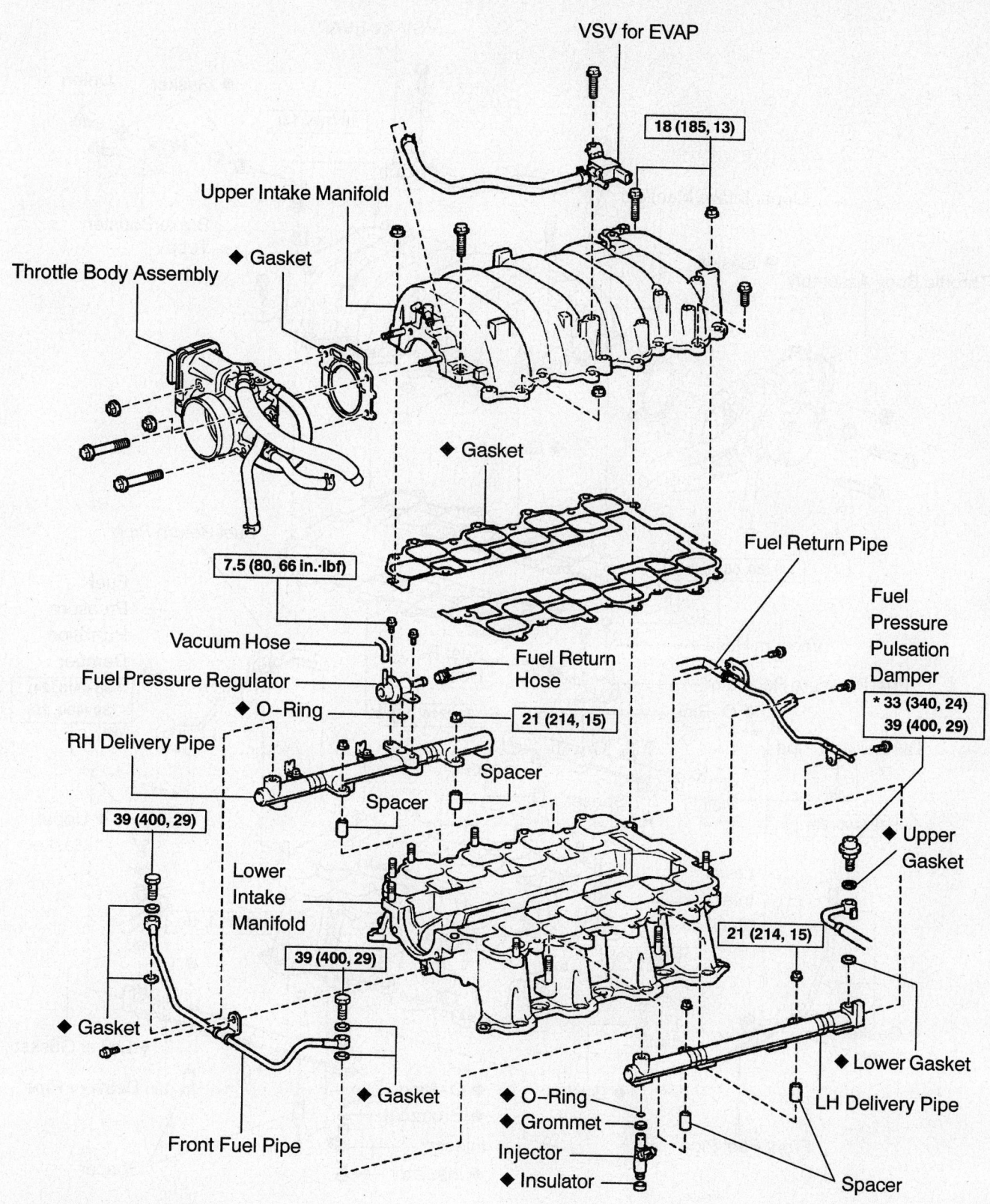

VSV for EVAP

18 (185, 13)

Upper Intake Manifold

◆ Gasket

Throttle Body Assembly

◆ Gasket

7.5 (80, 66 in.·lbf)

Vacuum Hose

Fuel Pressure Regulator

◆ O-Ring

RH Delivery Pipe

39 (400, 29)

Lower Intake Manifold

39 (400, 29)

◆ Gasket

◆ Gasket

Front Fuel Pipe

Fuel Return Hose

21 (214, 15)

Spacer

Spacer

Fuel Return Pipe

Fuel Pressure Pulsation Damper

* 33 (340, 24)
39 (400, 29)

◆ Upper Gasket

21 (214, 15)

◆ Lower Gasket

LH Delivery Pipe

Spacer

◆ O-Ring

◆ Grommet

Injector

◆ Insulator

N·m (kgf·cm, ft·lbf) : Specified torque

◆ Non-reusable part

* For use with SST

Intake manifold and related parts—early model Land Cruiser

67170-LCSQ-G04

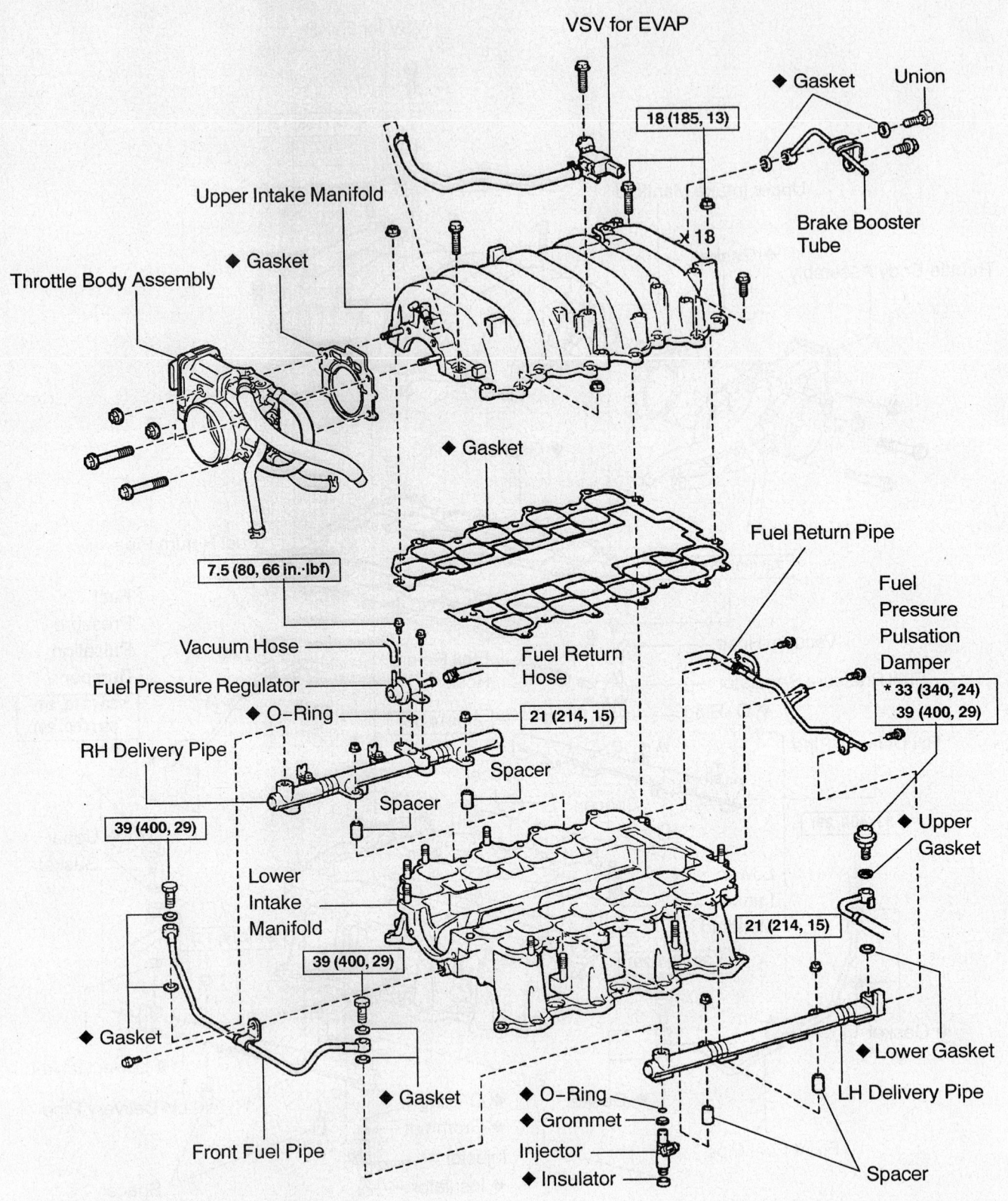

VSV for EVAP

Gasket

Union

18 (185, 13)

Upper Intake Manifold

Brake Booster Tube

× 18

◆ Gasket

Throttle Body Assembly

◆ Gasket

◆ Gasket

Fuel Return Pipe

Fuel Pressure Pulsation Damper

7.5 (80, 66 in.·lbf)

* 33 (340, 24)
39 (400, 29)

Vacuum Hose

Fuel Return Hose

Fuel Pressure Regulator

◆ O–Ring

21 (214, 15)

RH Delivery Pipe

Spacer

◆ Upper Gasket

39 (400, 29)

Spacer

Lower Intake Manifold

39 (400, 29)

21 (214, 15)

◆ Gasket

◆ Gasket

◆ O–Ring

◆ Lower Gasket

LH Delivery Pipe

Front Fuel Pipe

◆ Grommet

Injector

◆ Insulator

Spacer

N·m (kgf·cm, ft·lbf) : Specified torque

◆ Non-reusable part

* For use with SST

67170-LCSQ-G05

Intake manifold and related parts—early model Sequoia

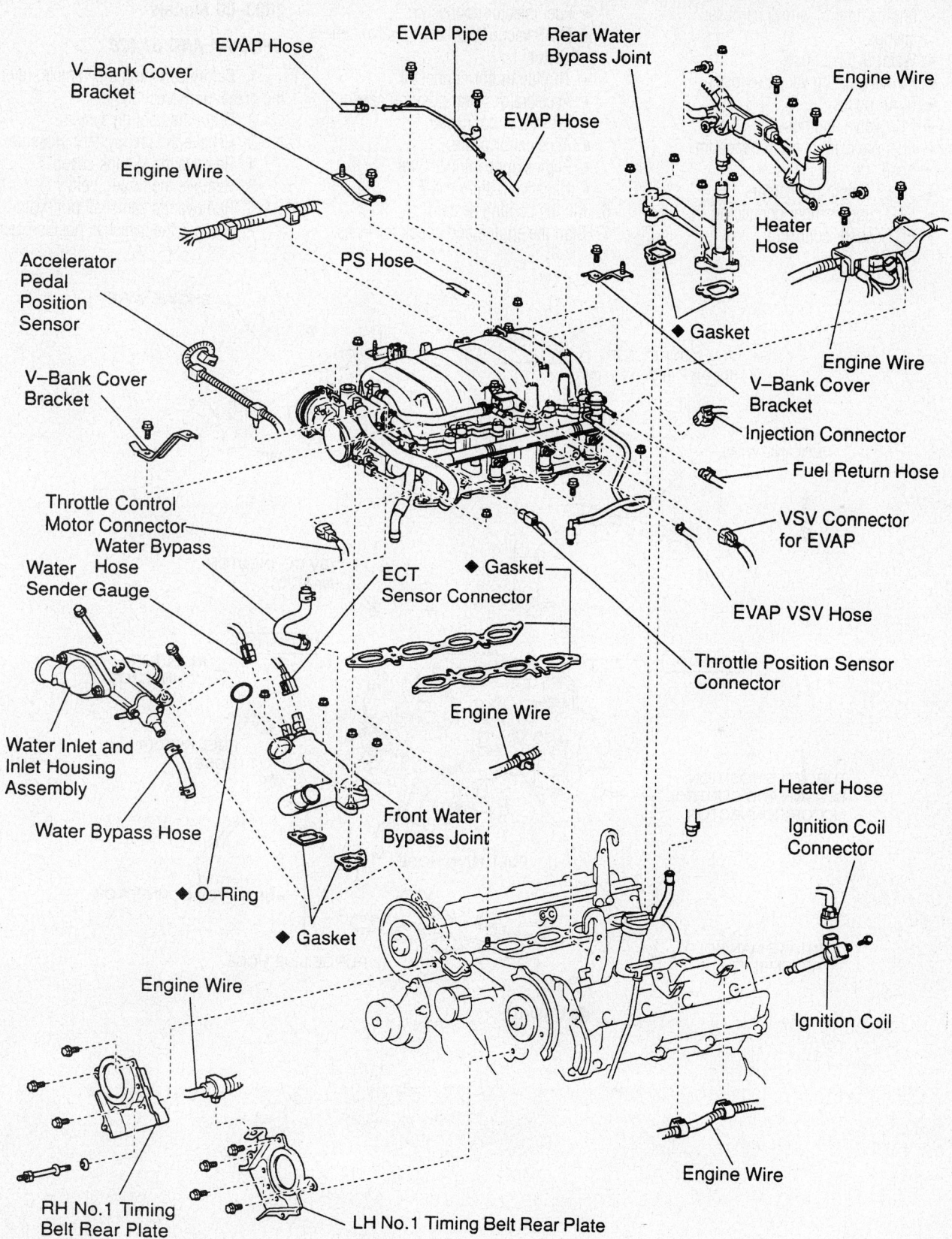

EVAP Hose

EVAP Pipe

Rear Water Bypass Joint

V-Bank Cover Bracket

Engine Wire

EVAP Hose

Engine Wire

Accelerator Pedal Position Sensor

PS Hose

Heater Hose

◆ Gasket

Engine Wire

V-Bank Cover Bracket

V-Bank Cover Bracket

Injection Connector

Fuel Return Hose

Throttle Control Motor Connector

Water Bypass Hose

VSV Connector for EVAP

Water Sender Gauge

ECT Sensor Connector

◆ Gasket

EVAP VSV Hose

Throttle Position Sensor Connector

Water Inlet and Inlet Housing Assembly

Engine Wire

Water Bypass Hose

Heater Hose

Ignition Coil Connector

◆ O-Ring

Front Water Bypass Joint

◆ Gasket

Engine Wire

Ignition Coil

Engine Wire

RH No.1 Timing Belt Rear Plate

LH No.1 Timing Belt Rear Plate

◆ Non-reusable part

Exploded of the intake manifold mounting—2002 GX470 and LX470

7924SG50

- Engine control wiring harness clamps
- Water bypass hose
- Power steering vacuum hoses
- EVAP hoses
- PCV valve and hose
- Fuel pressure regulator vacuum hose
- HO$_2$S sensor connectors
- ETC gauge sender connector
- ECT sensor connector

- Fuel injector connectors
- EVAP vacuum switching valve connector
- Throttle motor connector
- Accelerator pedal position sensor
- TP sensor connector
- Accelerator cable
- Engine appearance cover
- Negative battery cable
6. Fill the cooling system.
7. Start the engine and check for leaks.

2003–06 Models

SEQUOIA AND GX470

1. Before servicing the vehicle, refer to the precautions section.
2. Drain the cooling system.
3. Relieve the fuel system pressure.
4. Remove the V bank cover.
5. Remove the timing belt.
6. Remove the camshaft pulleys.
7. Remove the camshaft position sensor.

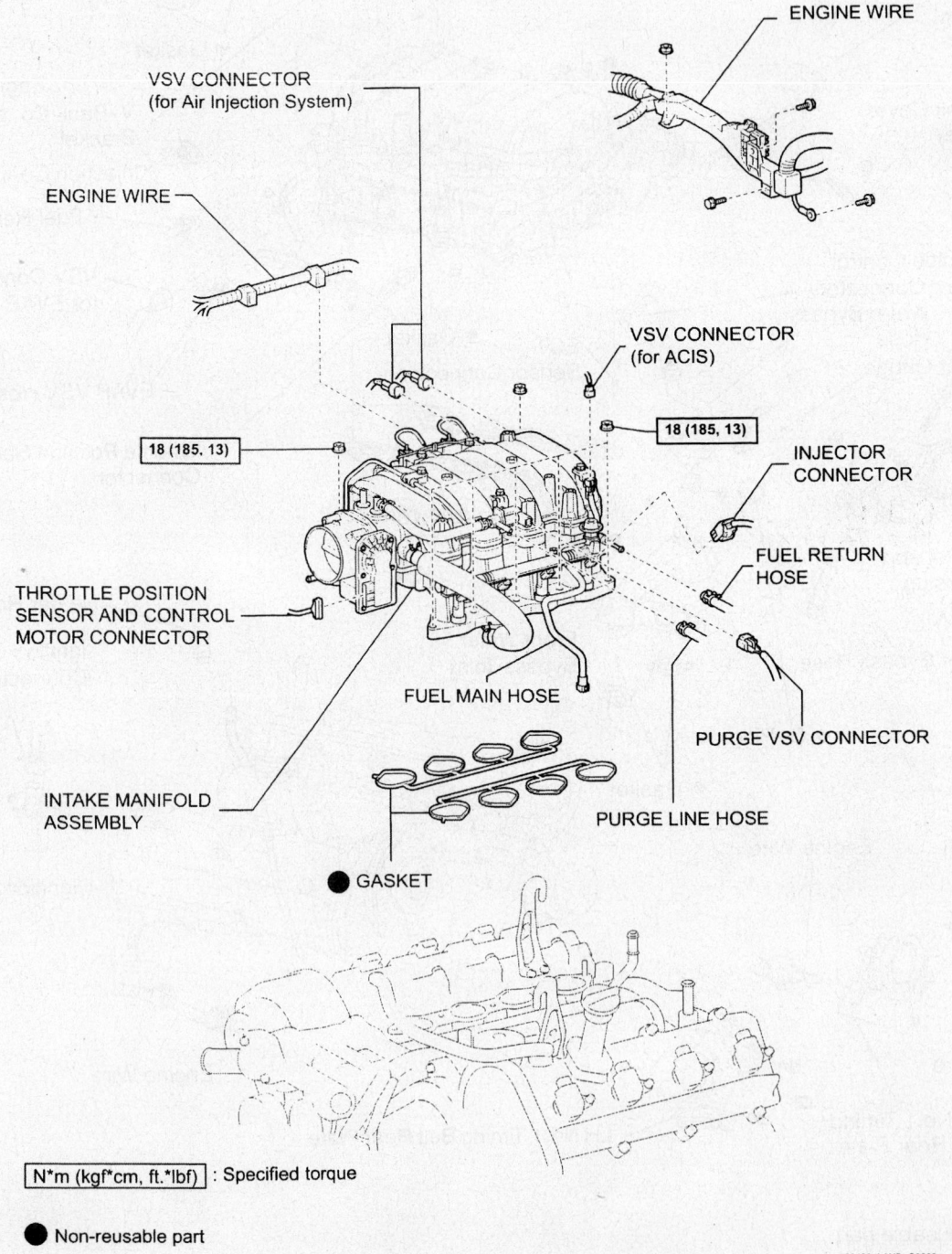

ENGINE WIRE

VSV CONNECTOR
(for Air Injection System)

ENGINE WIRE

VSV CONNECTOR
(for ACIS)

18 (185, 13)

18 (185, 13)

INJECTOR
CONNECTOR

THROTTLE POSITION
SENSOR AND CONTROL
MOTOR CONNECTOR

FUEL RETURN
HOSE

FUEL MAIN HOSE

PURGE VSV CONNECTOR

INTAKE MANIFOLD
ASSEMBLY

PURGE LINE HOSE

● GASKET

N*m (kgf*cm, ft.*lbf) : Specified torque

● Non-reusable part

09490_LAND_G0003

Intake manifold and related parts—2003–06 Land Cruiser, Sequoia, GX470 and LX470

8. Remove the power steering pump and set it aside with the lines still attached.

9. Remove the front exhaust pipe.

10. On models with an automatic transmission, remove the oil dipstick and tube.

11. Remove the ignition coils.

12. Remove the rear timing belt plates being careful not to drop anything.

13. Disconnect the fuel inlet hose.

14. Remove the intake manifold as follows:

    a. Disconnect the all electrical connectors from the manifold.

    b. Disconnect all hoses from the intake manifold.

    c. Disconnect the wire clamp bracket on the right hand delivery pipe.

    d. Remove the engine wire harness protector from the water bypass joint and right hand cylinder head.

    e. Remove the guide from the A/T bracket on the left side.

    f. Remove the two ground cable from the left and right head.

    g. Remove the bolts and disconnect the wire harness protector from the intake manifold.

    h. Remove the engine wire from the engine hanger and bracket.

    i. Remove the left and right side front V cover brackets.

    j. Remove the nuts, bolts and manifold.

    k. Remove the throttle body.

    l. Remove the bolts and nuts that attach the upper manifold to the lower manifold.

    m. Disconnect the EVAP hose from the upper manifold.

    n. Remove the accelerator cable clamp and VSV for the EVAP.

    o. Remove the bolt, union, gaskets and brake booster from the upper manifold.

    p. Remove the EVAP pipe from the manifold.

    q. Disconnect the fuel return hose from the regulator.

    r. Remove the bolts attaching the fuel return hose to the lower manifold.

    s. Remove the fuel pressure regulator, pulsatation damper gaskets.

    t. Remove the retainer and rear fuel pipe.

    u. Remove the fuel rail and injectors.

To install:

15. Install new intake manifold gaskets and the manifold. Tighten the bolts to 13 ft. lbs. (18 Nm) in several passes.

16. Install the throttle cover bracket, wire bracket and wire to the engine hanger bracket.

17. Install the wire to the timing belt rear plate.

18. Attach the wire protector to the intake manifold.

19. Attach the 2 ground cables the cylinder heads.

20. Connect the water bypass hoses to the throttle body.

21. Connect the wire clamps to the bracket on the right delivery pipe.

22. Attach the hoses to the intake manifold.

23. Attach the electrical connectors to the intake manifold.

24. Connect the fuel hose.

LAND CRUISER AND LX470

1. Before servicing the vehicle, refer to the precautions section.

2. Relieve the fuel system pressure.

3. Drain the cooling system.

4. Drain the oil.

5. Remove the timing belt.

6. Disconnect the fuel lines.

7. Remove the front exhaust pipes from both sides.

8. Remove the intake manifold as follows:

    a. Remove the V bank cover.

    b. Disconnect the all electrical connectors from the manifold.

    c. Disconnect all hoses from the intake manifold.

    d. Disconnect the water bypass hoses from the throttle body.

    e. Disconnect the wire clamp bracket on the right hand delivery pipe.

    f. Remove the engine wire harness protector from the water bypass joint and right hand cylinder head.

    g. Remove the two ground cables from the left and right head.

    h. Remove the throttle body cover.

    i. Remove the engine wire from the engine hanger and bracket.

    j. Remove the bolts and nuts that attach the upper manifold to the lower manifold.

To install:

9. Install new intake manifold gaskets and the manifold. Tighten the bolts to 13 ft. lbs. (18 Nm) in several passes.

10. Install the throttle cover bracket, wire bracket and wire to the engine hanger bracket.

11. Install the wire to the timing belt rear plate.

12. Attach the wire protector to the intake manifold.

13. Attach the 2 ground cables the cylinder heads.

14. Connect the water bypass hoses to the throttle body.

15. Connect the wire clamps to the bracket on the right delivery pipe.

16. Attach the hoses to the intake manifold.

17. Attach the electrical connectors to the intake manifold.

18. Connect the fuel hose.

19. Install the throttle body cover.

20. Install the front exhaust pipe, power steering pump.

21. Install the camshaft position sensor and camshaft timing pulleys, tighten to 25 ft. lbs. (34 Nm).

22. Install the timing belt.

23. Fill the cooling system and perform an oil change.

24. Start the vehicle and check for leaks.

Exhaust Manifold

REMOVAL & INSTALLATION

1. Before servicing the vehicle, refer to the precautions section.

2. Attach a hoist to the engine lifting eyes.

3. Remove or disconnect the following:

- Negative battery cable
- Heated Oxygen (HO_2S) sensor connectors
- Exhaust manifold heat shield
- Exhaust front pipe
- Motor mount
- Motor mount bracket
- Exhaust manifold

To install:

➡**Use new exhaust manifold nuts for assembly.**

4. Install or connect the following:

- Exhaust manifold. Tighten the nuts to 32 ft. lbs. (44 Nm).
- Motor mount bracket. Tighten the bolts to 27 ft. lbs. (36 Nm).
- Motor mount. Tighten the fasteners to 22 ft. lbs. (30 Nm).
- Exhaust front pipe. Tighten the nuts to 46 ft. lbs. (62 Nm).
- Exhaust manifold heat shield
- HO_2S sensor connectors
- Negative battery cable

5. Start the engine and check for leaks.

Front Crankshaft Seal

REMOVAL & INSTALLATION

1. Before servicing the vehicle, refer to the precautions section.

2. Drain the cooling system.

3. Remove or disconnect the following:

- Negative battery cable
- Engine under cover
- Engine appearance cover
- Air intake assembly
- Accessory drive belt
- Cooling fan and pulley
- Radiator
- Drive belt idler pulley
- Camshaft Position (CMP) sensor connector
- Upper timing covers
- Oil cooler pipe
- Center timing cover
- A/C compressor
- Cooling fan bracket
- Crankshaft pulley
- Lower timing cover
- Timing belt
- Crankshaft timing sprocket
- Front crankshaft seal

To install:

4. Install the oil seal so that it is flush with the oil pump housing.

5. Install or connect the following:
- Crankshaft timing sprocket
- Timing belt
- Lower timing cover
- Crankshaft pulley. Tighten the bolt to 181 ft. lbs. (245 Nm).
- Cooling fan bracket. Tighten the 12mm bolts to 12 ft. lbs. (16 Nm) and the 14mm bolts to 24 ft. lbs. (32 Nm).
- A/C compressor
- Center timing cover
- Oil cooler pipe
- Upper timing covers
- CMP sensor connector
- Drive belt idler pulley. Tighten the bolt to 27 ft. lbs. (37 Nm).
- Radiator
- Cooling fan and pulley. Tighten the nuts to 16 ft. lbs. (21 Nm).
- Accessory drive belt
- Air intake assembly
- Engine appearance cover
- Engine under cover
- Negative battery cable

6. Fill the cooling system.
7. Start the engine and check for leaks.

Camshaft and Valve Lifters

REMOVAL & INSTALLATION

2002 Models

1. Before servicing the vehicle, refer to the precautions section.
2. Drain the cooling system.
3. Relieve the fuel system pressure.

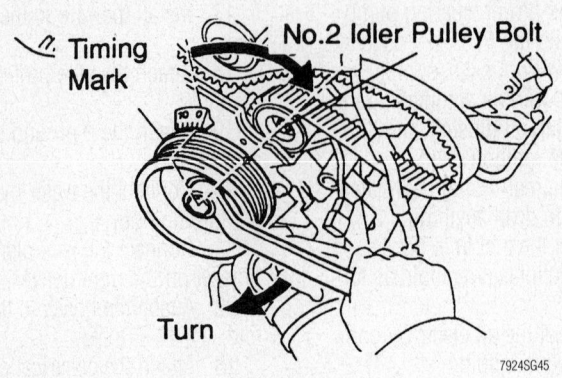

Setting the crankshaft to 50 degrees ATDC—4.7L 2UZ-FE engine

4. Remove or disconnect the following:
- Negative battery cable
- Engine under covers
- Engine appearance cover
- Air intake hose
- Accessory drive belt
- Cooling fan
- Radiator
- Idler pulley
- Upper and middle timing belt covers
- A/C compressor
- Cooling fan bracket
- Alternator
- Accessory drive belt tensioner

5. Set the engine to Top Dead Center (TDC) with the camshaft sprocket timing marks aligned with the rear cover timing marks.

6. Rotate the crankshaft to 50 degrees After TDC as shown. The crankshaft pulley timing mark should align with the center of the No. 2 idler pulley bolt.

7. Remove or disconnect the following:
- Crankshaft pulley
- Lower timing cover
- Timing belt
- Camshaft timing sprockets
- Camshaft Position (CMP) sensor
- Ignition coils

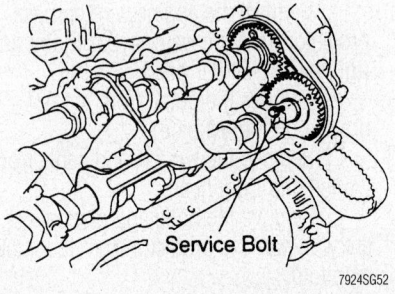

Camshaft service bolt installation—4.7L 2UZ-FE engine

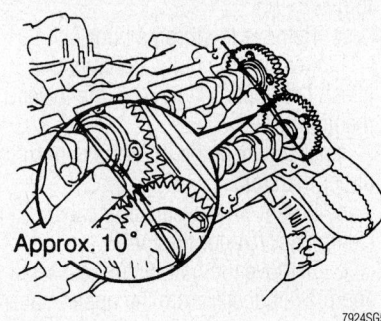

Right bank camshaft timing mark (1 dot marks) alignment —4.7L 2UZ-FE engine

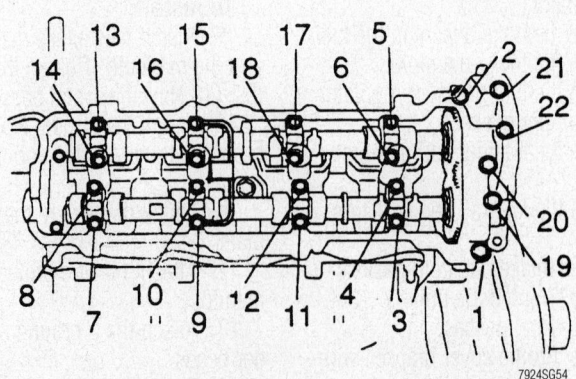

Right bank camshaft bearing cap loosening sequence—4.7L 2UZ-FE engine

- Valve cover
- Timing belt rear covers

8. Rotate the right bank camshafts as necessary to access the exhaust camshaft sub-gear service bolt hole and install a 6mm x 1.0mm bolt.

➡**Keep all valvetrain components in order for assembly.**

9. Align the right bank camshaft 1 dot timing marks to a **10** degree angle as shown.

10. Loosen the bearing cap bolts in sequence and in several passes.

11. Remove the right bank camshafts.

12. Rotate the left bank camshafts as necessary to access the exhaust camshaft sub-gear service bolt hole and install a 6mm x 1.0mm bolt.

13. Align the left bank camshaft 2 dot timing marks as shown.

14. Loosen the bearing cap bolts in sequence and in several passes.

15. Remove the left bank camshafts.

16. Remove the valve lifters and shims.

To install:

17. Ensure that the crankshaft is at 50 degrees After TDC.

18. Install or connect the following:

- Valve lifters and shims in their original positions
- Right bank camshafts with the 1 dot timing marks at 10 degrees
- Left bank camshafts with the 2 dot timing marks aligned
- Left and right bank camshaft bearing caps in their original positions. Apply sealant to the front bearing caps as shown.
- Camshaft oil seals

19. The bearing cap bolts vary in length and are identified as follows:

- A: 3.70 inches (94mm)
- B: 2.83 inches (72mm)
- C: 0.98 inches (25mm)
- D: 2.05 inches (52mm)

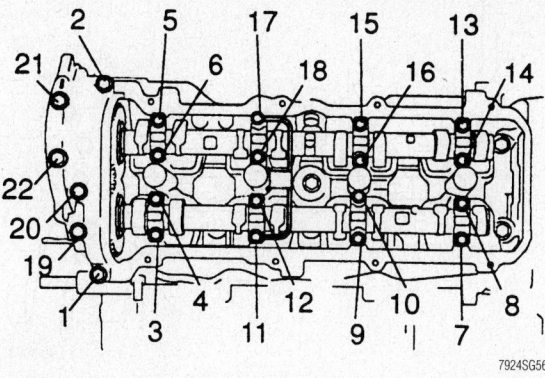

Left bank camshaft bearing cap loosening sequence—4.7L 2UZ-FE engine

- E: 1.50 inches (38mm)

20. Bolts in positions **A**, **B** and **C** are installed dry.

21. Lubricate the threads and under the contact flange for bolts in positions **D** and **E**.

22. Install oil feed pipes and the bearing cap bolts according to position in the illustrations.

23. Tighten the camshaft bearing bolts in sequence and in several passes to the following specifications:

- Bolt C: 66 inch lbs. (7.5 Nm)
- All others: 12 ft. lbs. (16 Nm)

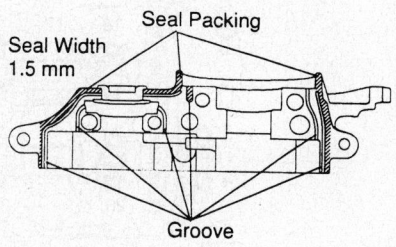

Apply a 1.5mm bead of sealant to the front bearing caps—4.7L 2UZ-FE engine

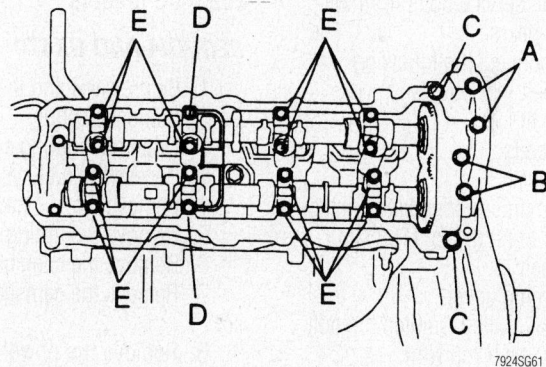

Right bank bearing cap bolt location—4.7L 2UZ-FE engine

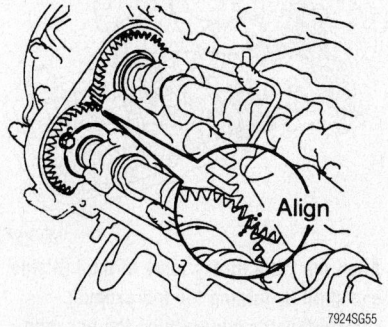

Left bank camshaft timing mark (2 dot marks) alignment—4.7L 2UZ-FE engine

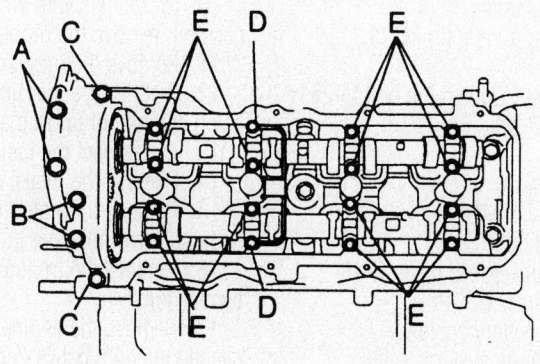

Left camshaft bearing cap bolt locations—4.7L 2UZ-FE engine

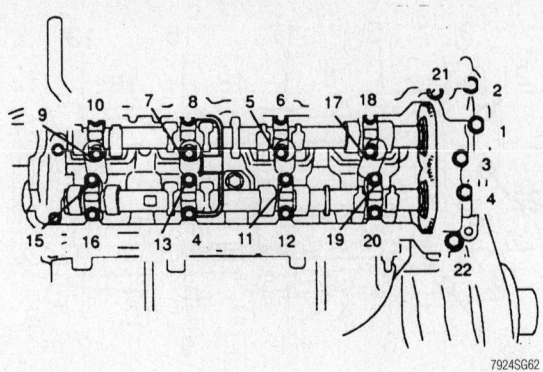

Right bank camshaft bearing cap bolt torque sequence—4.7L 2UZ-FE engine

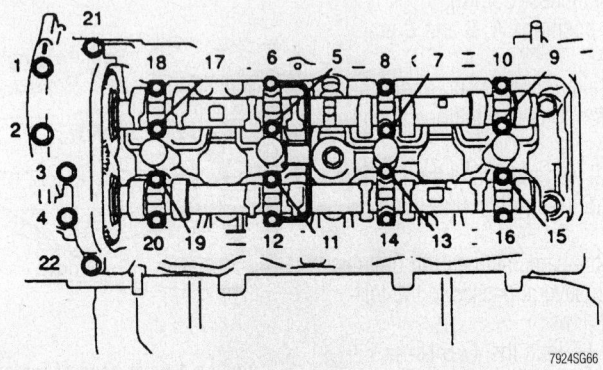

Left bank camshaft bearing cap bolt torque sequence—4.7L 2UZ-FE engine

24. Remove the service bolts from the exhaust camshaft gears.
25. Install or connect the following:
- Timing belt rear covers
- Valve cover
- Ignition coils
- CMP sensor
- Camshaft timing sprockets. Tighten the bolts to 80 ft. lbs. (108 Nm).
- Timing belt
- Lower timing cover
- Crankshaft pulley. Tighten the bolt to 181 ft. lbs. (245 Nm).
- Accessory drive belt tensioner
- Alternator
- Cooling fan bracket
- A/C compressor
- Upper and middle timing belt covers
- Idler pulley. Tighten the bolt to 27 ft. lbs. (37 Nm).
- Radiator
- Cooling fan
- Accessory drive belt
- Air intake hose
- Engine appearance cover
- Engine under covers
- Negative battery cable
26. Fill the cooling system.
27. Start the engine and check for leaks.

2003–06 Models

SEQUOIA AND GX470

1. Before servicing the vehicle, refer to the precautions section.
2. Drain the cooling system.
3. Relieve the fuel system pressure.
4. Remove the V bank cover.
5. Remove the timing belt.
6. Remove the camshaft pulleys.
7. Remove the camshaft position sensor.
8. Remove the power steering pump and set it aside with the lines still attached.
9. Remove the front exhaust pipe.
10. On models with an automatic transmission, remove the oil dipstick and tube.
11. Remove the ignition coils.
12. Remove the rear timing belt plates being careful not to drop anything.
13. Disconnect the fuel inlet hose.
14. Remove the intake manifold.
15. Remove the water inlet and inlet housing. Refer to water pump removal.
16. Remove the front and rear water bypass joint.
17. Remove the engine hangers and if needed the oil dipstick and tube.
18. Remove the valve covers.

➡Since the thrust level of the camshaft is small, the camshaft must be kept level during removal. If not kept level serious damage could occur.

19. Check the timing mark of the crankshaft pulley is aligned with the center(s) of the crankshaft pulley bolt and idler pulley bolt.

➡If the crankshaft pulley is wrongly positioned, this can cause the piston to contact the head causing severe damage. Make sure the crankshaft pulley is properly positioned.

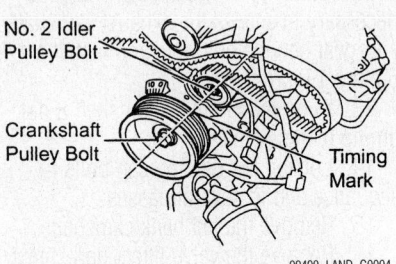

Check the timing mark of the crankshaft pulley is aligned with the center(s) of the crankshaft pulley bolt and idler pulley bolt—2003–06 Land Cruiser

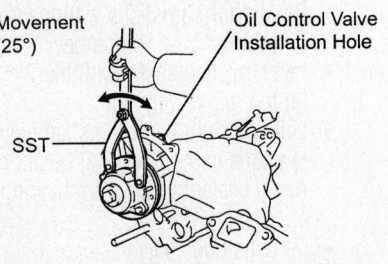

Release the oil from the front bearing caps using the tool illustrated—2003–06 Land Cruiser and Sequoia

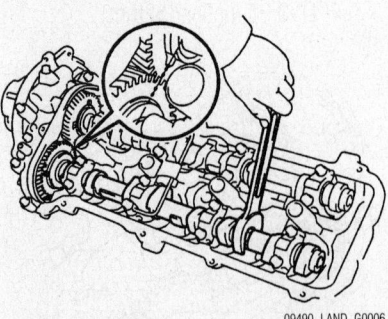

Align the 2 dot timing mark of the left side camshaft by turning the left exhaust camshaft using a wrench on the hexagon head portion of the shaft—2003–06 Land Cruiser and Sequoia

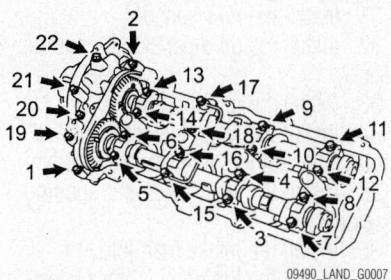

Loosen the left side 22 bearing cap bolts in the sequence illustrated using several passes—2003–06 Land Cruiser and Sequoia

20. Release the oil from the front bearing caps using the tool illustrated. Rotate the camshaft timing tube from left to right 2 to 3 times within its VVT-I range of 25 degrees and collect the oil from the timing oil control valve installation hole using a rag.

21. Remove the left hand camshafts as follows:

   a. Bring the service bolt of the sub gear up by turning the left exhaust camshaft using a wrench on the hexagon head portion of the shaft.

   b. Secure the sub gear to the main gear using a 16 to 20 mm bolt with a diameter of 6mm and a thread pitch of 1mm.

   c. Make sure the torsional force of the sub gear is retained by the bolt.

   d. Align the 2 dot timing mark of the left side camshaft by turning the left exhaust camshaft using a wrench on the hexagon head portion of the shaft.

➡**Mark the position of the caps so they can be reinstalled in their original positions.**

   e. Loosen the 22 bearing cap bolts in the sequence illustrated using several passes.

   f. Remove the bolts, washers, oil feed pipe, bearing caps, camshaft housing plug, oil control valve filter and the camshafts.

22. Remove the right hand camshafts as follows:

   a. Bring the service bolt of the sub gear up by turning the right exhaust camshaft using a wrench on the hexagon head portion of the shaft.

   b. Secure the sub gear to the main gear using a 16 to 20 mm bolt with a diameter of 6mm and a thread pitch of 1mm.

   c. Make sure the torsional force of the sub gear is retained by the bolt.

   d. Align the 1 dot timing mark of the camshaft main gear (about 10 degrees)

angle by turning the right exhaust camshaft using a wrench on the hexagon head portion of the shaft.

➡**Mark the position of the caps so they can be reinstalled in their original positions.**

   e. Loosen the 22 bearing cap bolts in the sequence illustrated using several passes.

   f. Remove the bolts, washers, oil feed pipe, bearing caps, camshaft housing plug, oil control valve filter and the camshafts.

To install:

23. Check the timing mark of the crankshaft pulley is aligned with the center(s) of the crankshaft pulley bolt and idler pulley bolt.

➡**If the crankshaft pulley is wrongly positioned, this can cause the piston to contact the head causing severe damage. Make sure the crankshaft pulley is properly positioned.**

24. Install the left side camshafts as follows:

   a. Apply multi purpose grease to the thrust portion of the camshafts.

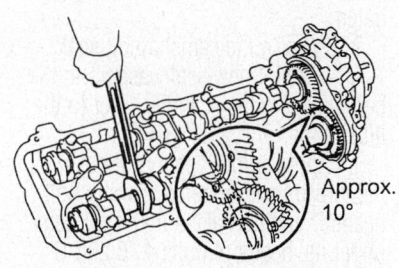

Align the 1 dot timing mark of the camshaft main gear (about 10 degrees) angle by turning the right exhaust camshaft using a wrench on the hexagon head portion of the shaft—2003–06 Land Cruiser and Sequoia

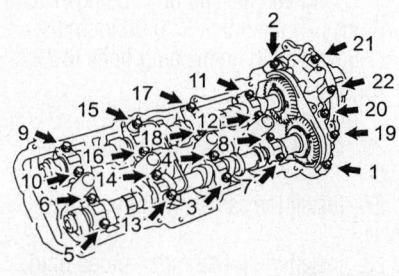

Loosen the right side 22 bearing cap bolts in the sequence illustrated using several passes—2003–06 Land Cruiser and Sequoia

   b. Align the 2 dot timing mark of the camshaft drive and driven main gears and install the camshafts.

   c. Apply seal packing to the camshaft housing plug.

   d. Install the camshaft housing plug on the cylinder head as illustrated. Install the strainer on the head being careful it is properly positioned.

   e. Apply seal packing to the front bearing cap.

   f. Install the front bearing cap and

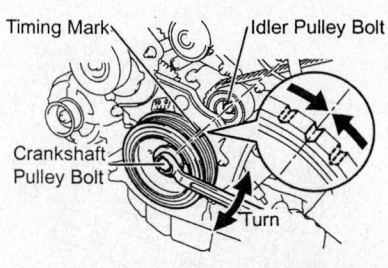

Check the timing mark of the crankshaft pulley is aligned with the center(s) of the crankshaft pulley bolt and idler pulley bolt—2003–06 Land Cruiser and Sequoia

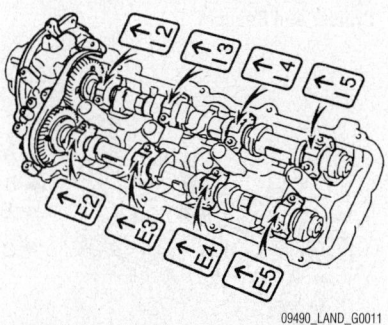

Install the front bearing cap and then the other caps in the sequence illustrated on the left side camshafts—2003–06 Land Cruiser and Sequoia

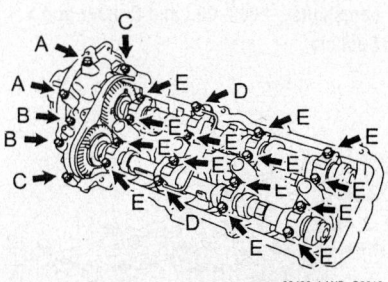

Apply a light coating of clean oil to the threads and underside of the bolt heads D and E. make sure no oil gets under the heads of bolts A, B and C on the left side camshafts—2003–06 Land Cruiser and Sequoia

then the other caps in the sequence illustrated.

g. Push in the camshaft oil seal.

h. Install 4 new seal washers to the bearing cap bolts A and B, refer to the illustration.

i. Apply a light coating of clean oil to the threads and underside of the bolt heads D and E. make sure no oil gets under the heads of bolts A, B and C.

j. The bolt lengths and positions are as follows. refer to the illustration for bolt location:

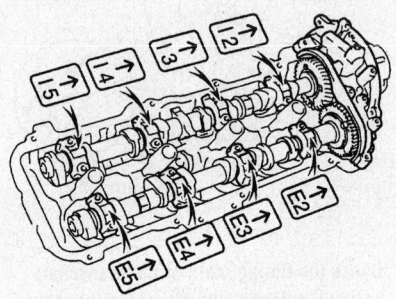

09490_LAND_G0014

Install the front bearing cap and then the other caps in the sequence illustrated on the left side camshafts—2003–06 Land Cruiser and Sequoia

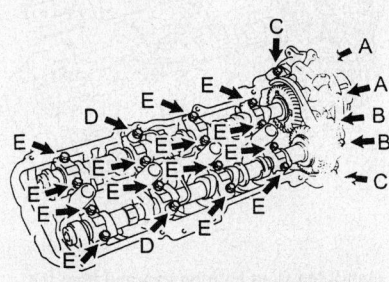

09490_LAND_G0015

Apply a light coating of clean oil to the threads and underside of the bolt heads D and E. make sure no oil gets under the heads of bolts A, B and C on the right side camshafts—2003–06 Land Cruiser and Sequoia

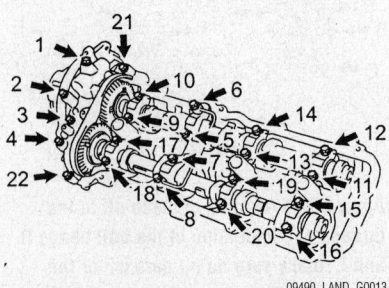

09490_LAND_G0013

Left side camshaft bolt torque sequence— 2003–06 Land Cruiser and Sequoia

• 94mm bolts A
• 72mm bolts B
• 25mm bolts C
• 52mm bolts D
• 38mm bolts E

k. Tighten the cap bolts using several passes. Tighten bolt C to 66 inch lbs. (7.5 Nm) an the remaining bolts to 12 ft. lbs. (16 Nm).

l. Remove the service bolt.

25. Install the right side camshafts as follows:

a. Apply multi purpose grease to the thrust portion of the camshafts.

b. Align the 1 dot timing mark of the camshaft drive and driven main gears and install the camshafts.

c. Set the 1 dot timing mark of the camshaft drive and driven gears at a 10 degree angle.

d. Apply seal packing to the camshaft housing plug.

e. Install the camshaft housing plug on the cylinder head as illustrated. Install the strainer on the head being careful it is properly positioned.

f. Apply seal packing to the front bearing cap.

g. Install the front bearing cap and then the other caps in the sequence illustrated.

h. Push in the camshaft oil seal.

i. Install 4 new seal washers to the bearing cap bolts A and B, refer to the illustration.

j. Apply a light coating of clean oil to the threads and underside of the bolt heads D and E. make sure no oil gets under the heads of bolts A, B and C.

k. The bolt lengths and positions are as follows. refer to the illustration for bolt location:

• 94mm bolts A
• 72mm bolts B
• 25mm bolts C
• 52mm bolts D
• 38mm bolts E

l. Tighten the cap bolts using several passes. Tighten bolt C to 66 inch lbs. (7.5 Nm) an the remaining bolts to 12 ft. lbs. (16 Nm).

m. Remove the service bolt.

26. Check and adjust the valve clearance.

27. Install the camshaft timing control valve.

28. Install the 4 half moon plugs onto the cylinder heads.

29. Install the valve covers and tighten to 53 inch lbs. (6 Nm).

30. Install the engine hangers and tighten to 27 ft. lbs. (37 Nm).

31. Install the VVT sensors.

32. Install the oil dipstick tube and dipstick.

33. Install the ignition coils.

34. Install the water bypass joint and tighten the retainers to 13 ft. lbs. (18 Nm).

35. Install the water inlet and housing assembly.

36. Install the intake manifold.

37. Install the timing belt rear plates, right plates first, then left plates. Tighten the retainers to 66 inch lbs. (7 Nm).

38. Install the throttle body cover.

39. Install the front exhaust pipe, power steering pump.

40. Install the camshaft position sensor and camshaft timing pulleys, tighten to 25 ft. lbs. (34 Nm).

41. Install the timing belt.

42. Fill the cooling system and perform an oil change.

43. Start the vehicle and check for leaks.

LAND CRUISER AND LX470

1. Before servicing the vehicle, refer to the precautions section.

2. Relieve the fuel system pressure.

3. Drain the cooling system.

4. Drain the oil.

5. Remove the timing belt.

6. Disconnect the fuel lines.

7. Remove the front exhaust pipes from both sides.

8. Remove the intake manifold.

9. Remove the air switching valve.

10. Remove the air pump.

11. Remove the camshaft position sensor.

12. Remove the VVT sensor on banks 1 and 2.

13. Remove the knock sensor.

14. Remove the camshaft timing oil control valve.

15. Remove the starter.

16. Remove the water inlet housing and front bypass joint. Refer to water pump removal.

17. Disconnect the heater hose and remove the water bypass pipe sub assembly.

18. Remove the rear water bypass joint.

19. Remove the oil dipstick and tube.

20. Remove the ignition coils.

21. Remove the spark plugs.

22. Remove the valve covers.

23. Turn the crankshaft pulley counterclockwise by 5 degrees until the marks are aligned as illustrated.

➡**Make sure to match the cut part by turning counterclockwise.**

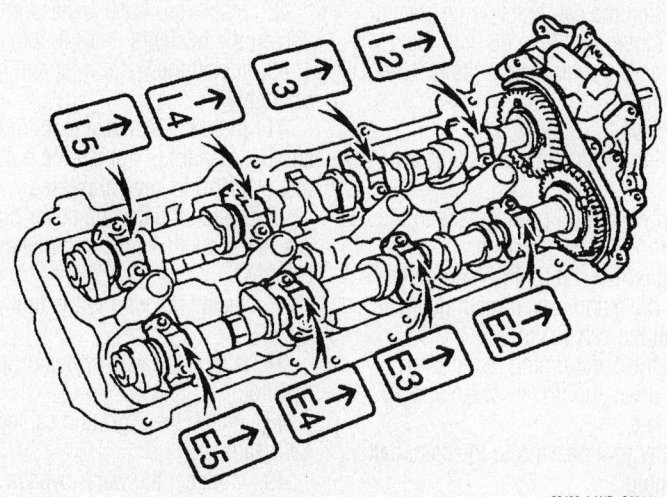

09490_LAND_G0014

Install the front bearing cap and then the other caps in the sequence illustrated on the right side camshafts—2003–06 Land Cruiser and Sequoia

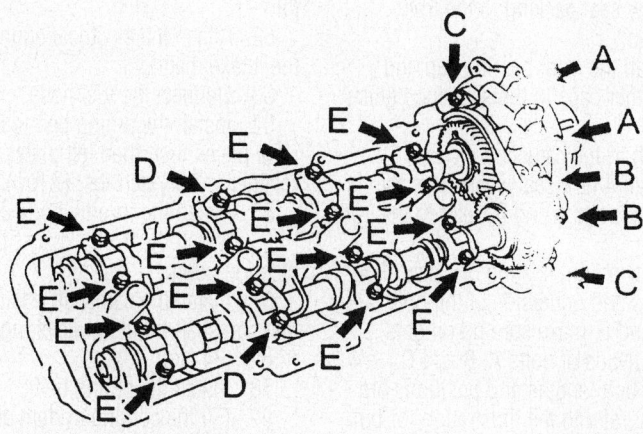

09490_LAND_G0015

Apply a light coating of clean oil to the threads and underside of the bolt heads D and E. make sure no oil gets under the heads of bolts A, B and C on the right side camshafts—2003–06 Land Cruiser and Sequoia

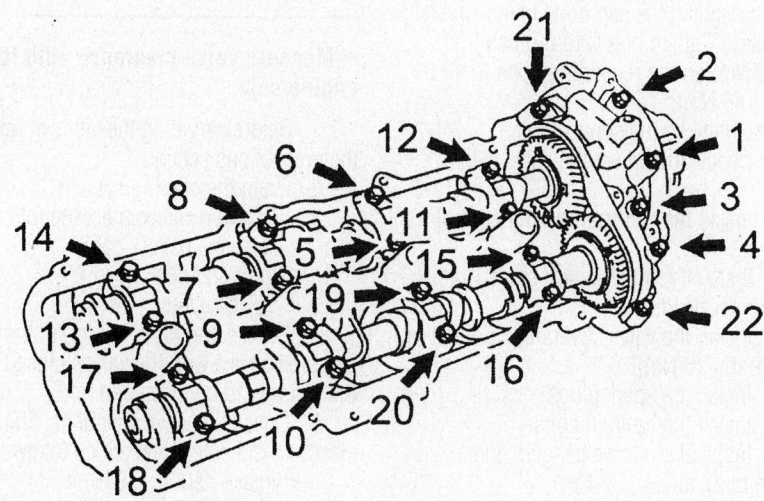

09490_LAND_G0016

Right side camshaft bolt torque sequence—2003–06 Land Cruiser and Sequoia

24. Remove the camshaft pulleys.

25. Remove the rear timing belt plates.

➡**Since the thrust level of the camshaft is small, the camshaft must be kept level during removal. If not kept level serious damage could occur.**

26. Check the timing mark of the crankshaft pulley is aligned with the center(s) of the crankshaft pulley bolt and idler pulley bolt.

➡**If the crankshaft pulley is wrongly positioned, this can cause the piston to contact the head causing severe damage. Make sure the crankshaft pulley is properly positioned.**

27. Release the oil from the front bearing caps using the tool illustrated. Rotate the camshaft timing tube from left to right 2 to 3 times within its VVT-I range of 25 degrees and collect the oil from the timing oil control valve installation hole using a rag.

28. Remove the left hand camshafts as follows:

a. Bring the service bolt of the sub gear up by turning the left exhaust camshaft using a wrench on the hexagon head portion of the shaft.

b. Secure the sub gear to the main gear using a 16 to 20 mm bolt with a diameter of 6mm and a thread pitch of 1mm.

c. Make sure the torsional force of the sub gear is retained by the bolt.

d. Align the 2 dot timing mark of the left side camshaft by turning the left exhaust camshaft using a wrench on the hexagon head portion of the shaft.

➡**Mark the position of the caps so they can be reinstalled in their original positions.**

e. Loosen the 22 bearing cap bolts in the sequence illustrated using several passes.

f. Remove the bolts, washers, oil feed pipe, bearing caps, camshaft housing plug, oil control valve filter and the camshafts.

29. Remove the right hand camshafts as follows:

a. Bring the service bolt of the sub gear up by turning the right exhaust camshaft using a wrench on the hexagon head portion of the shaft.

b. Secure the sub gear to the main gear using a 16 to 20 mm bolt with a diameter of 6mm and a thread pitch of 1mm.

c. Make sure the torsional force of the sub gear is retained by the bolt.

d. Align the 1 dot timing mark of the camshaft main gear (about 10 degrees) angle by turning the right exhaust camshaft using a wrench on the hexagon head portion of the shaft.

➡ **Mark the position of the caps so they can be reinstalled in their original positions.**

e. Loosen the 22 bearing cap bolts in the sequence illustrated using several passes.

f. Remove the bolts, washers, oil feed pipe, bearing caps, camshaft housing plug, oil control valve filter and the camshafts.

To install:

30. Check the timing mark of the crankshaft pulley is aligned with the center(s) of the crankshaft pulley bolt and idler pulley bolt.

➡ **If the crankshaft pulley is wrongly positioned, this can cause the piston to contact the head causing severe damage. Make sure the crankshaft pulley is properly positioned.**

31. Install the left side camshafts as follows:

a. Apply multi purpose grease to the thrust portion of the camshafts.

b. Align the 2 dot timing mark of the camshaft drive and driven main gears and install the camshafts.

c. Apply seal packing to the camshaft housing plug.

d. Install the camshaft housing plug on the cylinder head as illustrated. Install the strainer on the head being careful it is properly positioned.

e. Apply seal packing to the front bearing cap.

f. Install the front bearing cap and then the other caps in the sequence illustrated.

g. Push in the camshaft oil seal.

h. Install 4 new seal washers to the bearing cap bolts A and B, refer to the illustration.

i. Apply a light coating of clean oil to the threads and underside of the bolt heads D and E. make sure no oil gets under the heads of bolts A, B and C.

j. The bolt lengths and positions are as follows. refer to the illustration for bolt location:

- 94mm bolts A
- 72mm bolts B
- 25mm bolts C
- 52mm bolts D
- 38mm bolts E

k. Tighten the cap bolts using several passes. Tighten bolt C to 66 inch lbs. (7.5 Nm) an the remaining bolts to 12 ft. lbs. (16 Nm).

l. Remove the service bolt.

32. Install the right side camshafts as follows:

a. Apply multi purpose grease to the thrust portion of the camshafts.

b. Align the 1 dot timing mark of the camshaft drive and driven main gears and install the camshafts.

c. Set the 1 dot timing mark of the camshaft drive and driven gears at a 10 degree angle.

d. Apply seal packing to the camshaft housing plug.

e. Install the camshaft housing plug on the cylinder head as illustrated. Install the strainer on the head being careful it is properly positioned.

f. Apply seal packing to the front bearing cap.

g. Install the front bearing cap and then the other caps in the sequence illustrated.

h. Push in the camshaft oil seal.

i. Install 4 new seal washers to the bearing cap bolts A and B, refer to the illustration.

j. Apply a light coating of clean oil to the threads and underside of the bolt heads D and E. make sure no oil gets under the heads of bolts A, B and C.

k. The bolt lengths and positions are as follows. refer to the illustration for bolt location:

- 94mm bolts A
- 72mm bolts B
- 25mm bolts C
- 52mm bolts D
- 38mm bolts E

l. Tighten the cap bolts using several passes. Tighten bolt C to 66 inch lbs. (7.5 Nm) an the remaining bolts to 12 ft. lbs. (16 Nm).

m. Remove the service bolt.

33. Check and adjust the valve clearance.

34. Install the camshaft timing control valve.

35. Install the 4 half moon plugs onto the cylinder heads.

36. Install the valve covers and tighten to 53 inch lbs. (6 Nm).

37. Install the spark plugs.

38. Install the ignition coils.

39. Install the engine hangers and tighten to 27 ft. lbs. (37 Nm).

40. Install the VVT sensors.

41. Install the oil dipstick tube and dipstick.

42. Install the water bypass joint and tighten the retainers to 13 ft. lbs. (18 Nm).

43. Install the water inlet and housing assembly.

44. Install new intake manifold gaskets and the manifold. Tighten the bolts to 13 ft. lbs. (18 Nm) in several passes.

45. Install the throttle cover bracket, wire bracket and wire to the engine hanger bracket.

46. Install the wire to the timing belt rear plate.

47. Attach the wire protector to the intake manifold.

48. Attach the 2 ground cables the cylinder heads.

49. Connect the water bypass hoses to the throttle body.

50. Connect the wire clamps to the bracket on the right delivery pipe.

51. Attach the hoses to the intake manifold.

52. Attach the electrical connectors to the intake manifold.

53. Connect the fuel hose.

54. Install the timing belt rear plates, right plates first, then left plates. Tighten the retainers to 66 inch lbs. (7 Nm).

55. Install the throttle body cover.

56. Install the front exhaust pipe, power steering pump.

57. Install the camshaft position sensor and camshaft timing pulleys, tighten to 25 ft. lbs. (34 Nm).

58. Install the timing belt.

59. Fill the cooling system and perform an oil change.

60. Start the vehicle and check for leaks.

Valve Lash

ADJUSTMENT

➡ **Measure valve clearance with the engine cold.**

1. Before servicing the vehicle, refer to the precautions section.

2. Drain the cooling system.

3. Remove or disconnect the following:
- Negative battery cable
- Ignition coils
- Valve covers

4. Set the engine to the top of the compression stroke with the valves closed for the cylinder to be measured.

5. Check the valve clearance. The valve clearance specifications are as follows:
- Intake: 0.006–0.010 in. (0.15–0.25mm)
- Exhaust: 0.010–0.014 in. (0.25–0.35mm)

New shim thickness mm (in.)

Shim No.	Thickness	Shim No.	Thickness	Shim No.	Thickness
00	2.000 (0.0787)	28	2.280 (0.0898)	56	2.560 (0.1008)
02	2.020 (0.0795)	30	2.300 (0.0906)	58	2.580 (0.1016)
04	2.040 (0.0803)	32	2.320 (0.0913)	60	2.600 (0.1024)
06	2.060 (0.0811)	34	2.340 (0.0921)	62	2.620 (0.1031)
08	2.080 (0.0819)	36	2.360 (0.0929)	64	2.640 (0.1039)
10	2.100 (0.0827)	38	2.380 (0.0937)	66	2.660 (0.1047)
12	2.120 (0.0835)	40	2.400 (0.0945)	68	2.680 (0.1055)
14	2.140 (0.0843)	42	2.420 (0.0953)	70	2.700 (0.1063)
16	2.160 (0.0850)	44	2.440 (0.0961)	72	2.720 (0.1071)
18	2.180 (0.0858)	46	2.460 (0.0969)	74	2.740 (0.1079)
20	2.200 (0.0866)	48	2.480 (0.0976)	76	2.760 (0.1087)
22	2.220 (0.0874)	50	2.500 (0.0984)	78	2.780 (0.1094)
24	2.240 (0.0882)	52	2.520 (0.0992)	80	2.800 (0.1102)
26	2.260 (0.0890)	54	2.540 (0.1000)		

Intake valve clearance (Cold):
0.15 – 0.25 mm (0.006 – 0.010 in.)

EXAMPLE:
The 2.300 mm (0.0906 in.) shim is installed, and the measured clearance is 0.440 mm (0.0173 in.). Replace the 2.300 mm (0.0906 in.) shim with a No. 54 shim.

Intake valve clearance shim selection chart—4.7L 2UZ-FE engine

7924SG71

New shim thickness

mm (in.)

Shim No.	Thickness	Shim No.	Thickness	Shim No.	Thickness
00	2.000 (0.0787)	28	2.280 (0.0898)	56	2.560 (0.1008)
02	2.020 (0.0795)	30	2.300 (0.0906)	58	2.580 (0.1016)
04	2.040 (0.0803)	32	2.320 (0.0913)	60	2.600 (0.1024)
06	2.060 (0.0811)	34	2.340 (0.0921)	62	2.620 (0.1031)
08	2.080 (0.0819)	36	2.360 (0.0929)	64	2.640 (0.1039)
10	2.100 (0.0827)	38	2.380 (0.0937)	66	2.660 (0.1047)
12	2.120 (0.0835)	40	2.400 (0.0945)	68	2.680 (0.1055)
14	2.140 (0.0843)	42	2.420 (0.0953)	70	2.700 (0.1063)
16	2.160 (0.0850)	44	2.440 (0.0961)	72	2.720 (0.1071)
18	2.180 (0.0858)	46	2.460 (0.0969)	74	2.740 (0.1079)
20	2.200 (0.0866)	48	2.480 (0.0976)	76	2.760 (0.1087)
22	2.220 (0.0874)	50	2.500 (0.0984)	78	2.780 (0.1094)
24	2.240 (0.0882)	52	2.520 (0.0992)	80	2.800 (0.1102)
26	2.260 (0.0890)	54	2.540 (0.1000)		

Exhaust valve clearance (Cold):
0.25 – 0.35 mm (0.010 – 0.014 in.)

EXAMPLE:
The 2.300 mm (0.0906 in.) shim is installed, and the measured clearance is 0.440 mm (0.0173 in.). Replace the 2.300 mm (0.0906 in.) shim with a No. 44 shim.

Exhaust valve clearance shim selection chart—4.7L 2UZ-FE engine

7924SG72

6. Record the measurements for each valve.

7. When all valve clearances have been measured, remove the camshafts.

8. Remove the valve shims and measure them. Note this measurement along with the clearance measurement recorded earlier.

9. Using the valve clearance and shim thickness measurements, find replacement shims in the Adjusting Shim Selection charts.

10. Install or connect the following:
- Replacement valve shims
- Camshafts
- Valve covers
- Ignition coils
- Negative battery cable

11. Fill the cooling system.

12. Start the engine and check for leaks.

Starter Motor

REMOVAL & INSTALLATION

1. Before servicing the vehicle, refer to the precautions section.

2. Drain the cooling system.

3. Relieve the fuel system pressure.

4. Remove or disconnect the following:
- Negative battery cable
- Engine appearance cover
- Air intake tube
- Intake manifold
- Starter motor mounting bolts
- Starter wiring connectors
- Starter motor

To install:

5. Install or connect the following:
- Starter motor
- Starter wiring connectors. Tighten the cable nut to 86 inch lbs. (10 Nm).
- Starter motor mounting bolts. Tighten the bolts to 29 ft. lbs. (39 Nm).
- Intake manifold
- Air intake tube
- Engine appearance cover
- Negative battery cable

6. Fill the cooling system.

7. Start the engine and check for leaks.

Timing Belt

REMOVAL & INSTALLATION

1. Disconnect the negative battery cable.

2. Raise and safely support the vehicle.

3. Remove the oil pan protector and the engine under cover.

4. Drain the cooling system and store the coolant for refilling purposes.

5. Lower the vehicle and remove the battery clamp cover.

6. From the top of the engine, remove the fuel return hose, the engine cover nuts/bolts and the cover.

7. Remove the air cleaner and the intake air connector assembly.

8. Remove the cooling fan pulley by performing the following procedures:

   a. Loosen the 4 fan clutch-to-fan pulley nuts.

   b. Using a box-end wrench on the serpentine drive belt tensioner bolt, rotate the tensioner counterclockwise and remove the drive belt.

➡**The serpentine drive belt tensioner bolt is a left-hand thread.**

   c. Remove the fan clutch-to-fan pulley nuts, the fan, the clutch assembly and the fan pulley.

9. Remove the radiator by performing the following procedures:

   a. Disconnect the upper, lower and reservoir hoses from the radiator.

   b. Disconnect and plug the automatic transmission oil cooler at the radiator. Disconnect the automatic transmission oil cooler hoses from the fan shroud clamp.

   c. Remove the radiator reservoir tank.

   d. Remove the fan shroud-to-radiator bolts and the shroud.

   e. Remove the 2 upper radiator-to-chassis nuts.

   f. Remove the middle radiator-to-chassis nut/bolts and brackets.

   g. Carefully, lift the radiator from the vehicle.

10. Remove the serpentine drive belt idler pulley bolt, cover plate and pulley.

11. Remove the right side (No. 3) timing belt cover.

12. Remove the left side (No. 3) timing belt cover by performing the following procedures:

   a. Disconnect the engine wire from both wire clamps.

   b. Disconnect the camshaft position sensor wire from the wire clamp on the left-side (No.3) timing belt cover.

   c. Disconnect the sensor connector from the connector bracket.

   d. Disconnect the sensor connector.

   e. Remove the wire grommet from the left-side (No. 3) timing belt cover.

   f. Remove the oil cooler tube bolts and tube.

13. Remove the middle (No. 2) timing belt cover bolts and cover.

14. Remove the cooling fan bracket nuts/bolts and bracket.

➡**If reusing the timing belt, make sure that there are 3 installation marks on the belt; if there are none, install them.**

15. Using the Crankshaft Pulley Holding tool 09213-70010, Bolt tool 90105-08076 and Companion Flange Holding tool 09330-00021, or equivalent, loosen the crankshaft pulley bolt.

16. Position the No. 1 cylinder to approximately 50 degrees After Top Dead Center (ATDC) of the compression stroke by performing the following procedures:

   a. Rotate the crankshaft pulley (CLOCKWISE) to align its groove with the timing mark "0" on the lower (No. 1) timing belt cover.

   b. Check that the camshaft sprocket timing marks are aligned with the rear timing belt plate marks; if not, rotate the crankshaft 1 revolution (360 degrees).

   c. Rotate the crankshaft pulley approximately 50 degrees (CLOCKWISE) and align the crankshaft pulley timing mark between the centers of the crankshaft pulley bolt and the idler pulley bolt.

✳✳ WARNING

If the timing belt is disengaged, having the crankshaft pulley in the wrong angle can cause the valve to come into contact with the piston when removing the camshaft pulley.

17. Remove the crankshaft pulley bolt.

➡**If reusing the timing belt and the installation marks have disappeared, place new installation marks on the timing belt to match the camshaft timing sprocket marks.**

➡**To avoid meshing the timing sprocket and the timing belt, secure one with a string; then, place matchmarks on the timing belt and the right-side camshaft timing sprocket.**

18. Remove the timing belt tensioner bolts and the tensioner.

19. Using the Camshaft Holding tool 09960-10010, or equivalent, slightly turn the left-side camshaft sprocket clockwise to loosen the tension spring. Then, disconnect the timing belt from the camshaft sprockets.

20. Remove the alternator by performing the following procedures:

   a. Disconnect the electrical connector from the alternator.

b. Remove the rubber cap/nut and disconnect the battery wire from the alternator.

c. Disconnect the wire clamp from the alternator cord clip.

d. Remove the alternator-to-engine nuts/bolts and the alternator.

21. Remove the serpentine drive belt tensioner nuts/bolts and the tensioner.

22. Using the Crankshaft Puller Assembly tool 09950-50012, or equivalent, press the crankshaft pulley from the crankshaft.

⁂ WARNING

DO NOT rotate the crankshaft pulley.

23. Remove the lower (No. 1) timing belt cover bolts and the cover.

24. Remove the timing belt guide, spacer and the timing belt.

To install:

➡ **With the timing belt removed, this is a perfect opportunity to inspect and/or replace the water pump.**

25. Inspect the timing belt tensioner by performing the following procedures:

a. Inspect the seal for leakage; if leakage is suspected, replace the tensioner.

b. Using both hands to hold the ten-

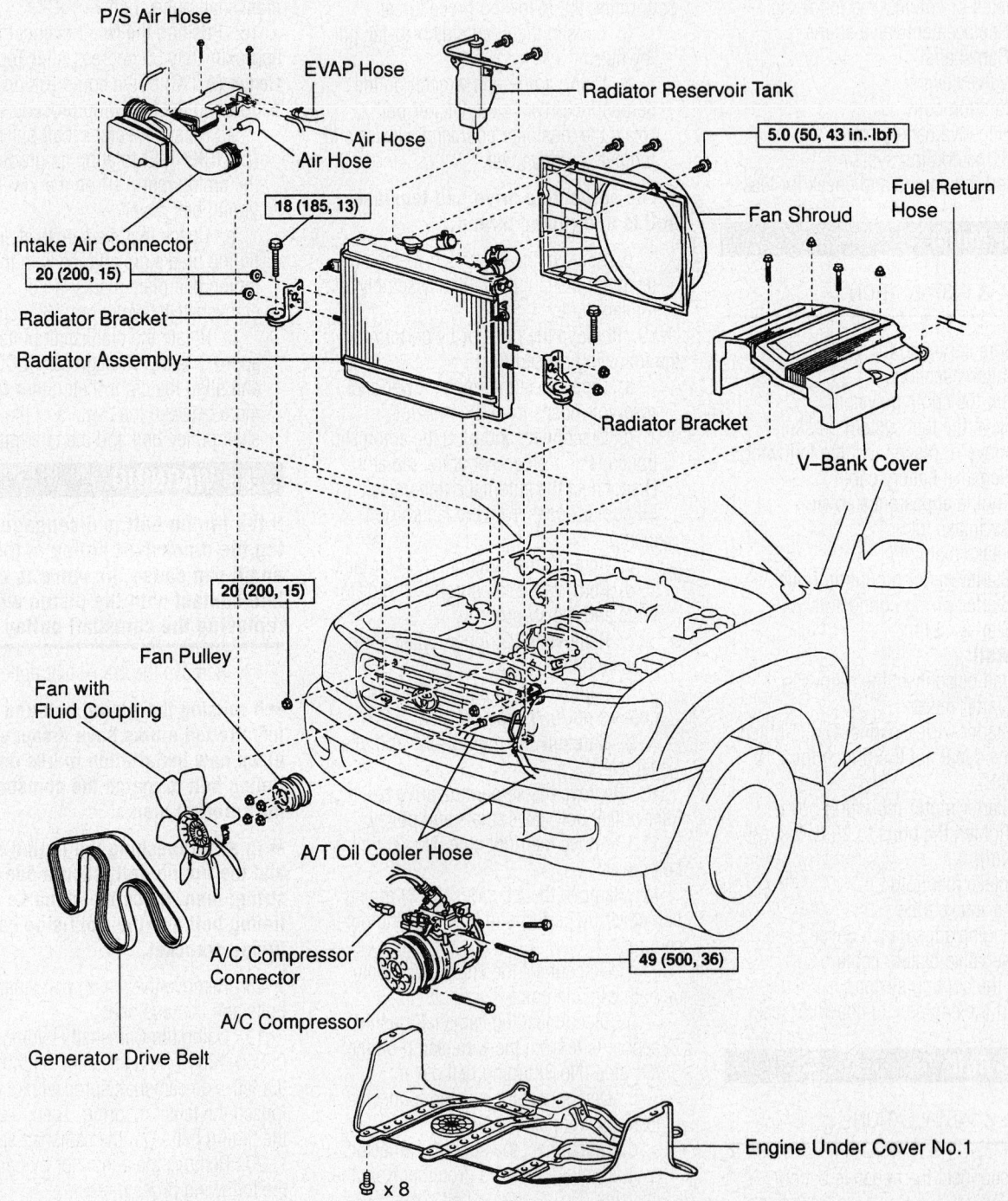

Exploded view of vehicle components for timing belt replacement—Land Cruiser

93025G24

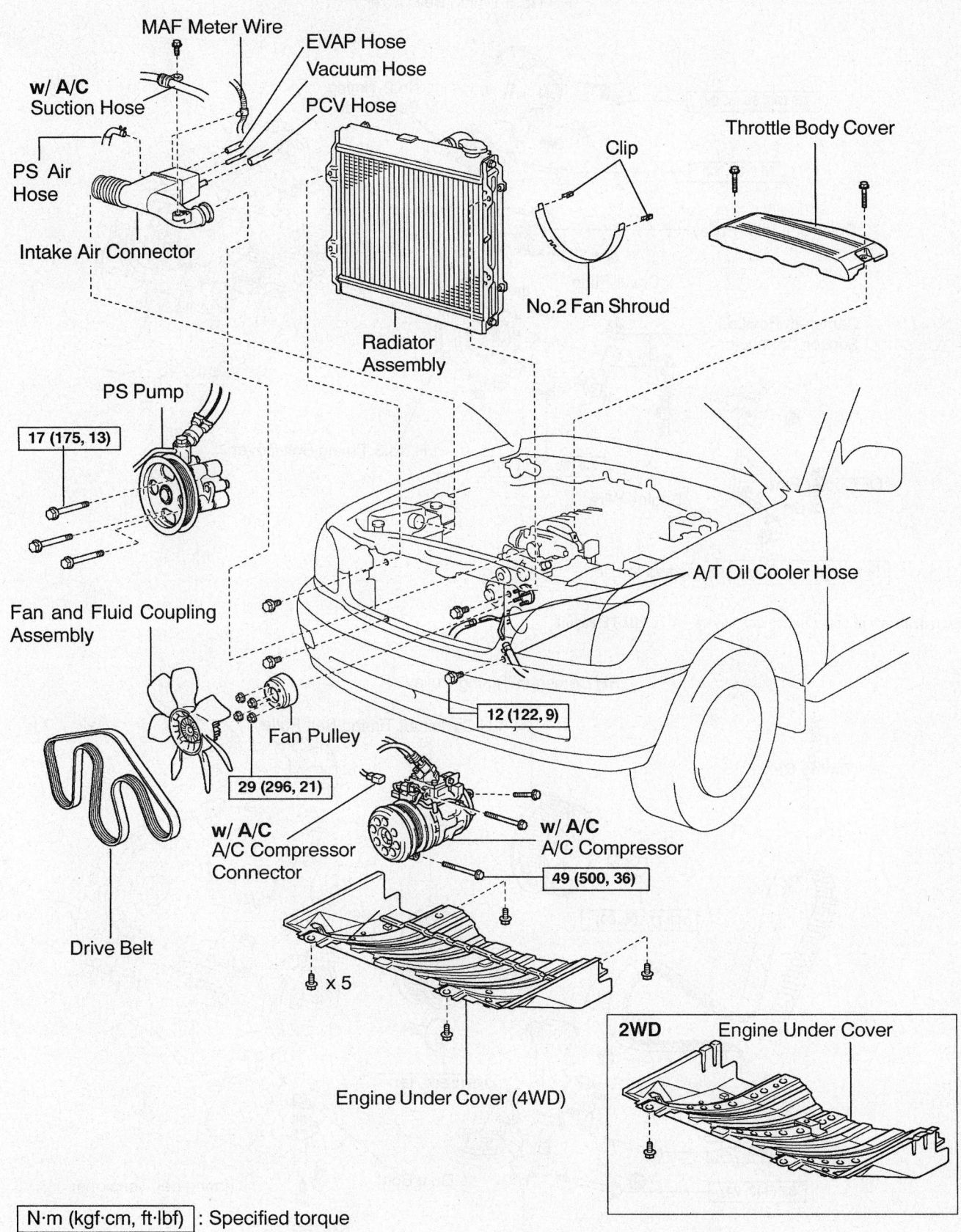

MAF Meter Wire

EVAP Hose
Vacuum Hose

w/ A/C
Suction Hose

PCV Hose

Clip

Throttle Body Cover

PS Air
Hose

Intake Air Connector

No.2 Fan Shroud

Radiator
Assembly

PS Pump

17 (175, 13)

A/T Oil Cooler Hose

Fan and Fluid Coupling
Assembly

12 (122, 9)

Fan Pulley

29 (296, 21)

w/ A/C
A/C Compressor
Connector

w/ A/C
A/C Compressor

Drive Belt

49 (500, 36)

x 5

2WD   Engine Under Cover

Engine Under Cover (4WD)

N·m (kgf·cm, ft·lbf) : Specified torque

Exploded view of vehicle components for timing belt replacement—Sequoia

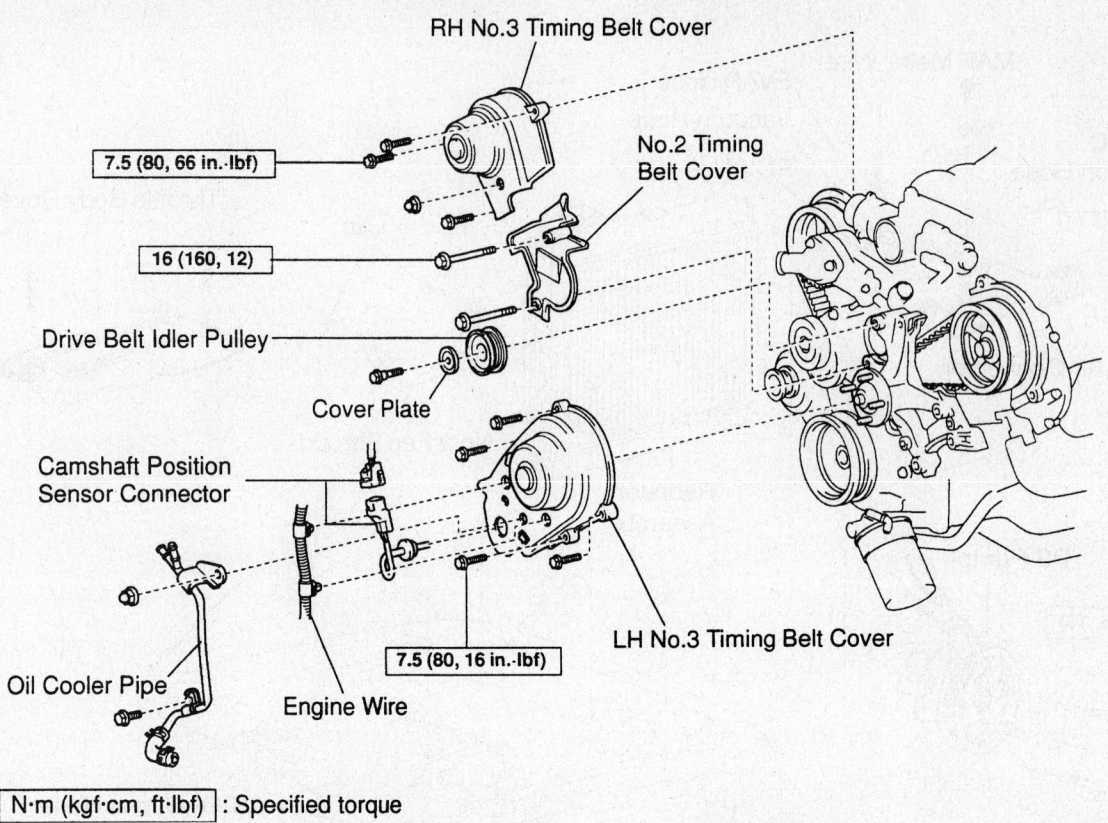

RH No.3 Timing Belt Cover

No.2 Timing Belt Cover

7.5 (80, 66 in.-lbf)

16 (160, 12)

Drive Belt Idler Pulley

Cover Plate

Camshaft Position Sensor Connector

LH No.3 Timing Belt Cover

7.5 (80, 16 in.-lbf)

Oil Cooler Pipe

Engine Wire

N·m (kgf·cm, ft·lbf) : Specified torque

93025G25

Exploded view of upper timing belt covers—4.7L 2UZ-FE engine

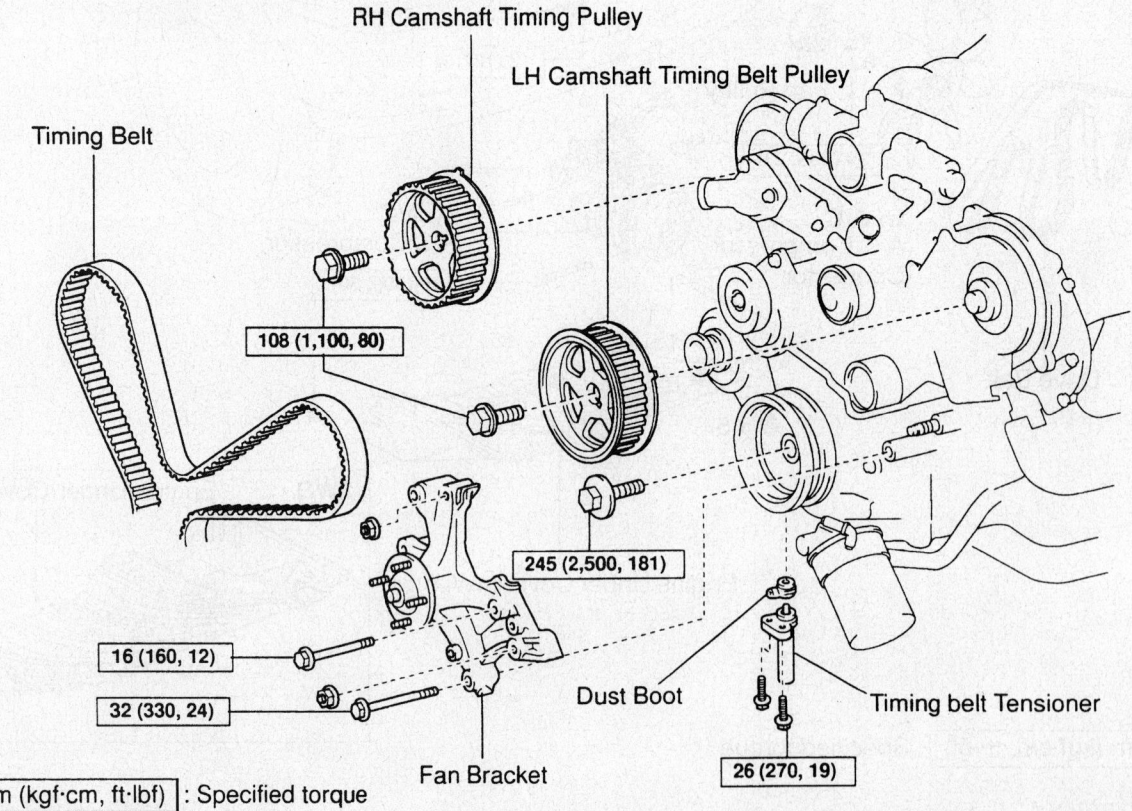

RH Camshaft Timing Pulley

LH Camshaft Timing Belt Pulley

Timing Belt

108 (1,100, 80)

245 (2,500, 181)

16 (160, 12)

32 (330, 24)

Dust Boot

Timing belt Tensioner

Fan Bracket

26 (270, 19)

N·m (kgf·cm, ft·lbf) : Specified torque

93025G26

Exploded view of upper timing sprockets and components—4.7L 2UZ-FE engine

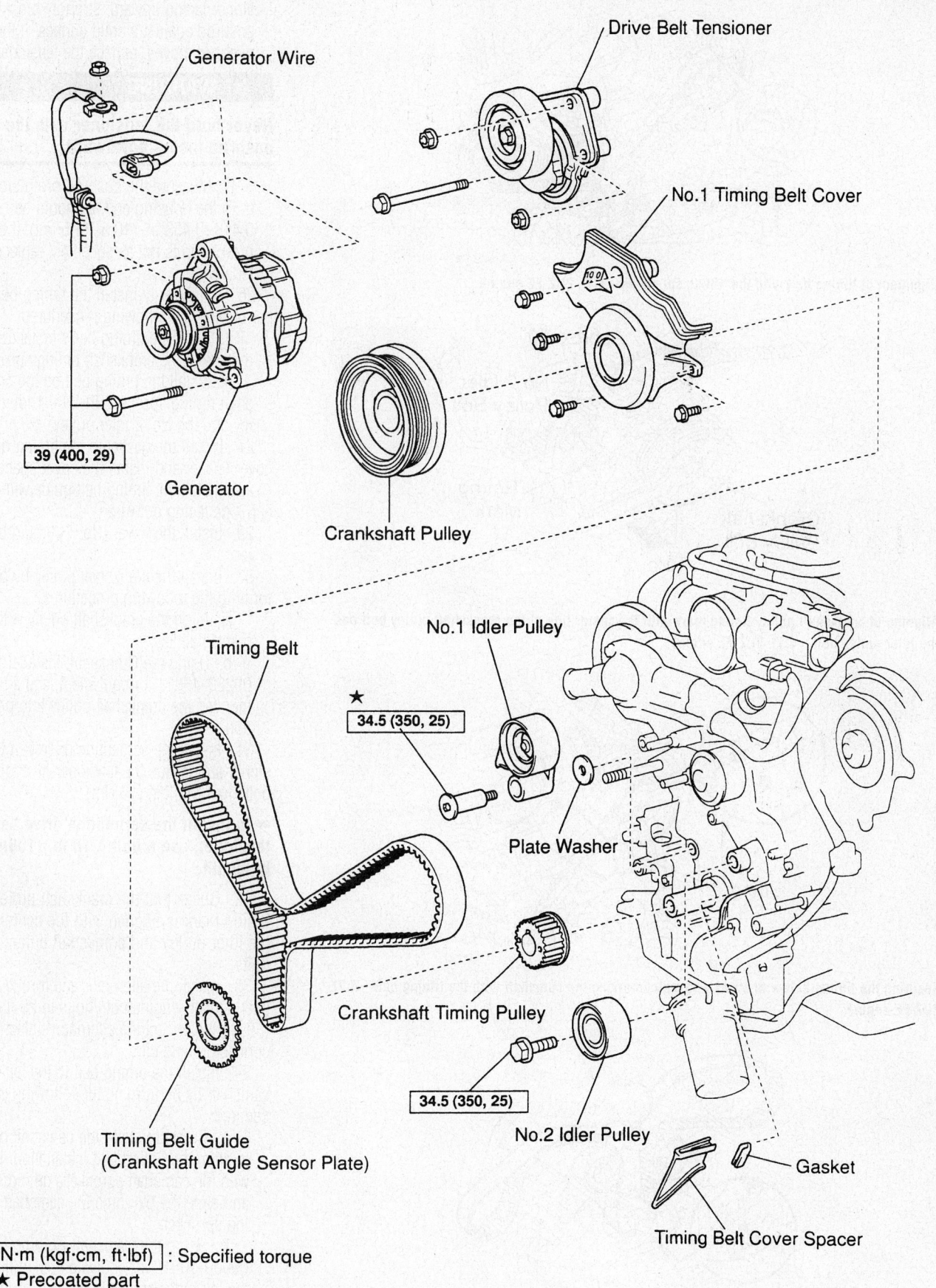

Generator Wire

Drive Belt Tensioner

No.1 Timing Belt Cover

39 (400, 29)

Generator

Crankshaft Pulley

Timing Belt

No.1 Idler Pulley

★

34.5 (350, 25)

Plate Washer

Crankshaft Timing Pulley

Timing Belt Guide
(Crankshaft Angle Sensor Plate)

34.5 (350, 25)

No.2 Idler Pulley

Gasket

Timing Belt Cover Spacer

N·m (kgf·cm, ft·lbf) : Specified torque
★ Precoated part

93025G27

Exploded view of lower timing belt cover, sprockets and components—4.7L 2UZ-FE engine

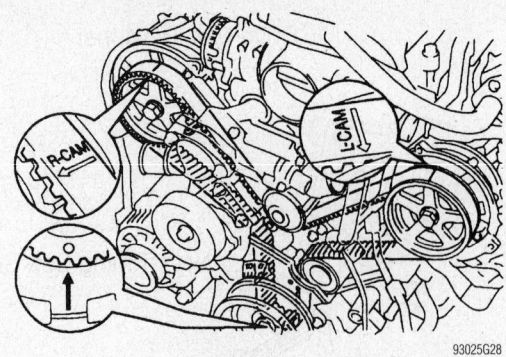

Alignment of timing belt with the timing sprockets—4.7L 2UZ-FE engine

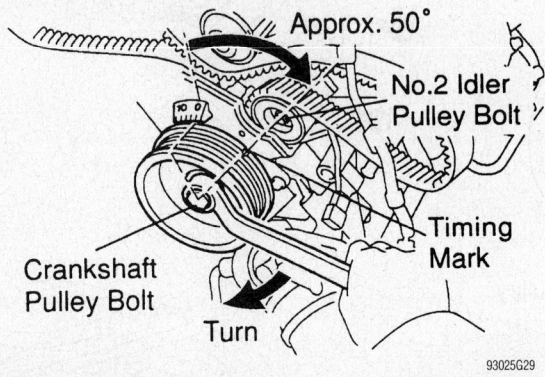

Aligning of crankshaft pulley timing mark with the center line of the crankshaft pulley bolt and the idler pulley bolt—4.7L 2UZ-FE engine

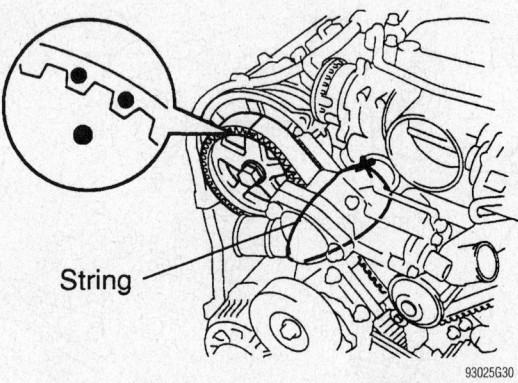

Securing the timing belt with string and matchmarking the camshaft with the timing belt—4.7L 2UZ-FE engine

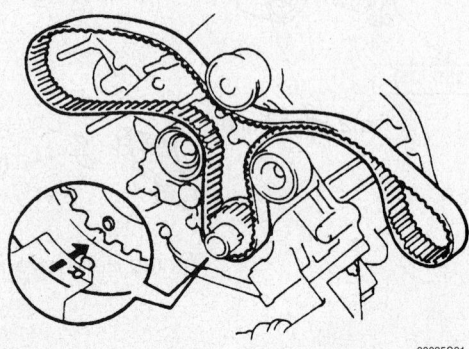

Installing the timing belt on the crankshaft sprocket—4.7L 2UZ-FE engine

sioner facing upward, strongly press the pushrod against a solid surface. If the pushrod moves, replace the tensioner.

✻✻ WARNING

Never hold the tensioner with the pushrod facing downward.

    c. Measure the pushrod protrusion from the housing end, it should be 0.413–0.453 in. (10.5–11.5mm). If the protrusion is not as specified, replace the tensioner.

26. Temporarily install the timing belt by performing the following procedures:

    a. Align the timing belt's installation mark with the crankshaft timing sprocket.

    b. Install the timing belt on the crankshaft timing sprocket, the No. 1 idler pulley and the No. 2 idler pulley.

27. Install the gasket to the timing belt cover spacer and install the cover spacer.

28. Install the timing belt guide with the cup side facing outward.

29. Install the lower (No. 1) timing belt cover.

30. Install the crankshaft pulley by performing the following procedures:

    a. Align the crankshaft pulley with the crankshaft key.

    b. Using the Crankshaft Installer tool 09223-46011, or equivalent, and a hammer, tap the crankshaft pulley into position.

31. Install the serpentine drive belt tensioner and torque the tensioner-to-engine bolts to 12 ft. lbs. (16 Nm).

➡ **To install the serpentine drive belt tensioner, use a bolt 4.18 in. (106mm) in length.**

32. Check that the crankshaft pulley's timing mark is aligned with the centers of the idler pulley and crankshaft pulley bolts.

33. Install the alternator and torque the alternator-to-engine nuts/bolts to 29 ft. lbs. (39 Nm). Connect the alternator's electrical connectors and clip.

34. Install the timing belt to the left-side camshaft by performing the following procedures:

    a. Rotate the left-side camshaft pulley to align the timing belt installation mark with the camshaft sprocket's timing mark and slide the belt onto the camshaft timing sprocket.

    b. Using the Camshaft Holding tool 09960-10010, or equivalent, slightly turn the left-side camshaft sprocket counterclockwise to place tension on the timing

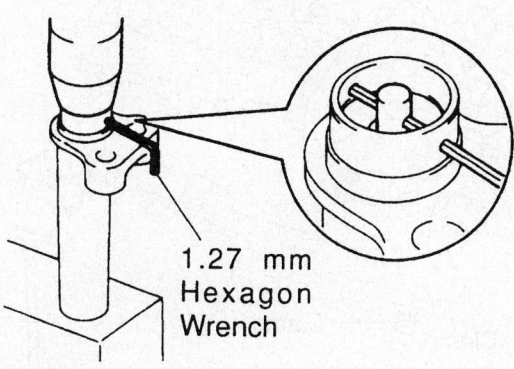

Securing the timing belt tensioner pushrod—4.7L 2UZ-FE engine

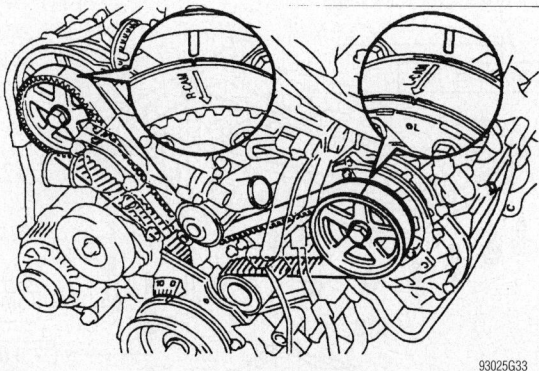

Checking the TDC alignment marks after rotating the crankshaft 2 revolutions—4.7L 2UZ-FE engine

belt between the crankshaft sprocket and the camshaft sprocket.

35. Rotate the right-side camshaft pulley to align the timing belt installation mark with the camshaft sprocket's timing mark and slide the belt onto the camshaft timing sprocket.

36. Using a vertical press, slowly press the pushrod into the housing using 200–2205 lbs. (981–9807 N) until the holes align, then, install a 1.27mm Allen® wrench to secure the pushrod and release the press. Install the dust boot on the tensioner housing.

37. Install the timing belt tensioner and torque the bolts to 19 ft. lbs. (26 Nm).

38. Using a pair of pliers, remove the Allen® wrench from the tensioner housing.

39. Check the valve timing by performing the following procedure:

a. Temporarily install the crankshaft pulley bolt.

b. Slowly, rotate the crankshaft pulley 2 revolutions (CLOCKWISE) and realign the TDC marks.

➡**If the pulley/sprocket timing marks do not realign, remove the timing belt and reinstall it.**

40. Using the Crankshaft Pulley Holding tool 09213-70010, Bolt tool 90105-08076 and Companion Flange Holding tool

09330-00021, or equivalent, torque the crankshaft pulley bolt to 181 ft. lbs. (245 Nm).

41. Install the cooling fan bracket and torque the 12mm (head size) bolt to 12 ft. lbs. (16 Nm) and the 14mm (head size) bolt to 24 ft. lbs. (32 Nm).

42. Install the air conditioning compressor.

43. Install the middle (No. 2) timing belt cover and torque the bolts to 12 ft. lbs. (16 Nm).

44. Install the upper right-side (No. 3) timing belt cover and torque the bolts to 66 inch lbs. (7.5 Nm).

45. Install the upper left-side (No. 3) timing belt cover by performing the following procedures:

a. Install the oil cooler tube and bolt.

b. Feed the Camshaft Position Sensor (CPS) through the left-side (No. 3) timing belt cover hole.

c. Install the left-side (No. 3) timing belt cover and torque the bolts to 66 inch lbs. (7.5 Nm).

d. Install the wire grommet to the left-side (No. 3) timing belt cover.

e. Install the sensor connector to the connector bracket and connect the sensor connector.

f. Install the sensor wire and the engine wire to the clamps on the left-side (No. 3) timing belt cover.

46. Install the drive belt idler pulley and cover plate; then, torque the pulley bolt to 27 ft. lbs. (37 Nm).

47. To complete the installation, reverse the removal procedures.

48. Refill the cooling system and connect the negative battery cable.

Oil Pan

REMOVAL & INSTALLATION

1. Before servicing the vehicle, refer to the precautions section.

2. Remove the engine from the vehicle and mount it on a stand.

3. Remove or disconnect the following:
- Oil dipstick tube
- Lower oil pan
- Oil pan baffle
- Upper oil pan

To install:

4. The upper oil pan bolts are different lengths and are identified as follows:
- A: 0.79 inch (20mm) w/10mm head
- B: 0.98 inch (25mm) w/12mm head
- C: 2.36 inch (60mm) w/12mm head

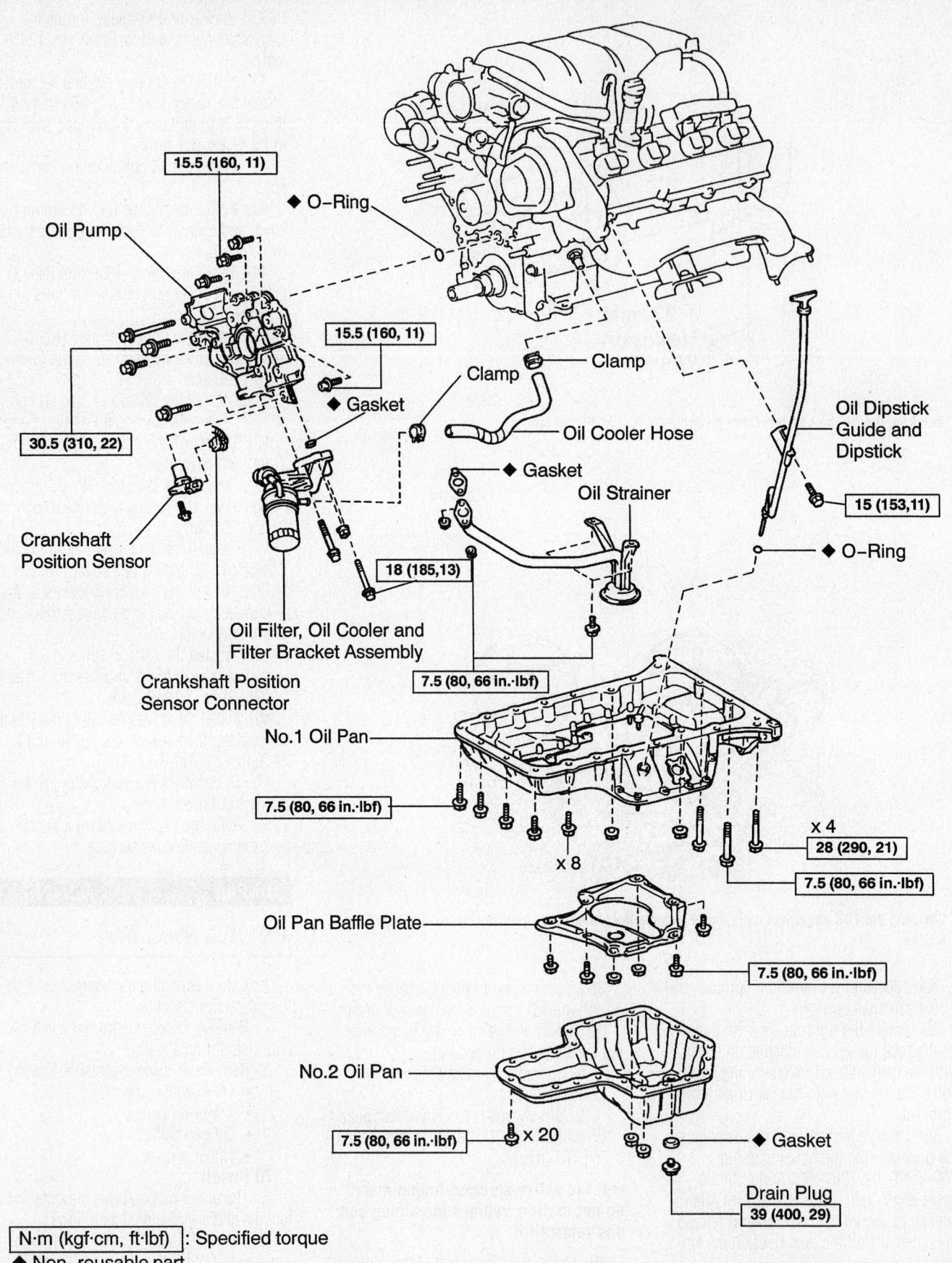

15.5 (160, 11)

◆ O–Ring

Oil Pump

15.5 (160, 11)

◆ Gasket

Clamp

Clamp

Oil Cooler Hose

Oil Dipstick
Guide and
Dipstick

30.5 (310, 22)

◆ Gasket

Oil Strainer

15 (153,11)

Crankshaft
Position Sensor

◆ O–Ring

18 (185,13)

Oil Filter, Oil Cooler and
Filter Bracket Assembly

Crankshaft Position
Sensor Connector

7.5 (80, 66 in.·lbf)

No.1 Oil Pan

7.5 (80, 66 in.·lbf)

x 4

28 (290, 21)

x 8

7.5 (80, 66 in.·lbf)

Oil Pan Baffle Plate

7.5 (80, 66 in.·lbf)

No.2 Oil Pan

7.5 (80, 66 in.·lbf)   x 20

◆ Gasket

Drain Plug

39 (400, 29)

N·m (kgf·cm, ft·lbf) : Specified torque
◆ Non–reusable part

Oil pan and pump—Land Cruiser

67170-LCSQ-G09

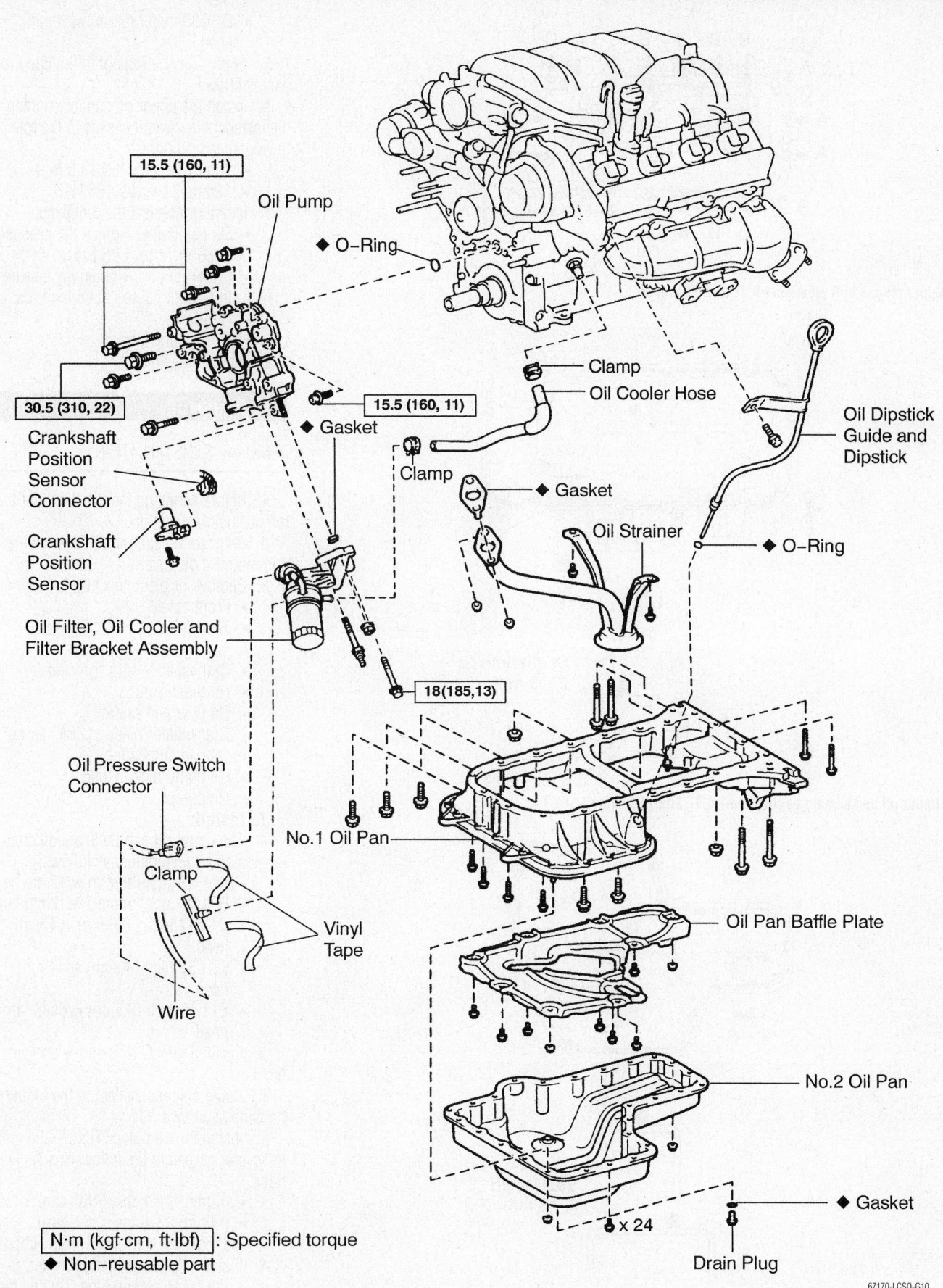

15.5 (160, 11)

Oil Pump

◆ O–Ring

30.5 (310, 22)

Crankshaft Position Sensor Connector

Crankshaft Position Sensor

Oil Filter, Oil Cooler and Filter Bracket Assembly

15.5 (160, 11)

◆ Gasket

Clamp

Clamp
Oil Cooler Hose

◆ Gasket

Oil Strainer

Oil Dipstick Guide and Dipstick

◆ O–Ring

Oil Pressure Switch Connector

Clamp

Vinyl Tape

Wire

18(185,13)

No.1 Oil Pan

Oil Pan Baffle Plate

No.2 Oil Pan

◆ Gasket

x 24

Drain Plug

N·m (kgf·cm, ft·lbf) : Specified torque
◆ Non-reusable part

Oil pan and pump—Sequoia

67170-LCSQ-G10

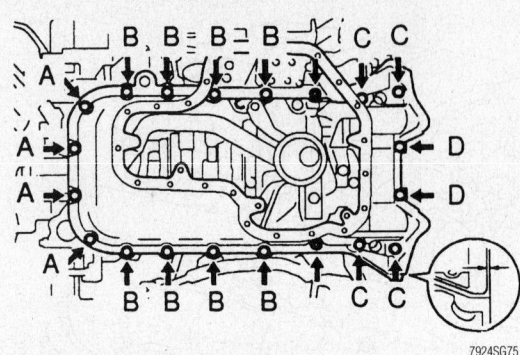

Upper oil pan bolt location—4.7L 2UZ-FE engine

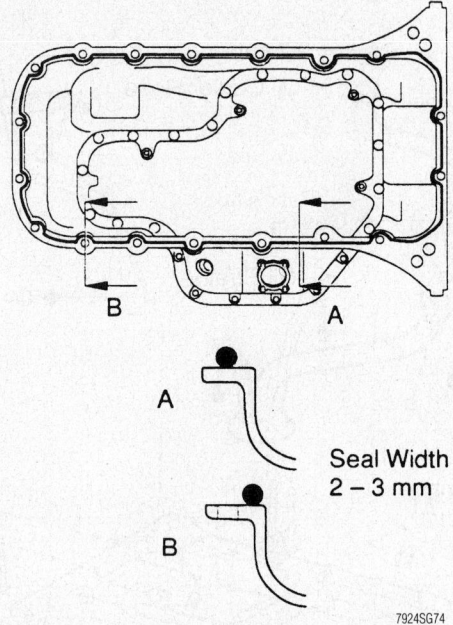

Seal Width
2 – 3 mm

Upper oil pan sealant application—4.7L 2UZ-FE engine

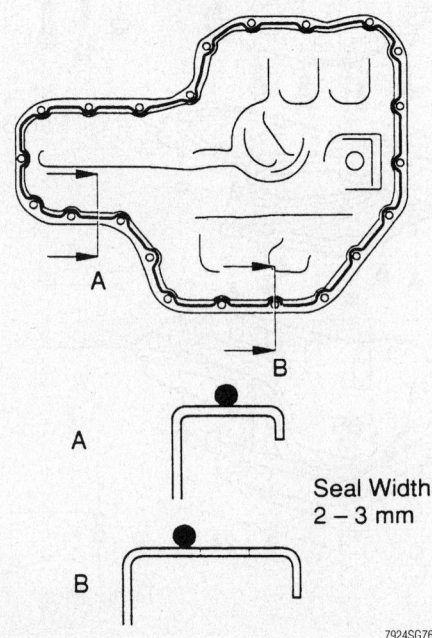

Seal Width
2 – 3 mm

Lower oil pan sealant application—4.7L 2UZ-FE engine

- D: 1.38 inch (35mm) w/10mm head

5. Apply silicone sealant to the upper oil pan as shown.

6. Install the upper oil pan and tighten the fasteners in several passes to the following specifications:
- 10mm: 66 inch lbs. (7.5 Nm)
- 12mm: 21 ft. lbs. (28 Nm)

7. Install or connect the following:
- Oil pan baffle. Tighten the fasteners to 66 inch lbs. (7.5 Nm).
- Lower oil pan. Tighten the fasteners in several passes to 66 inch lbs. (7.5 Nm).
- Oil dipstick tube

8. Install the engine.

Oil Pump

REMOVAL & INSTALLATION

1. Before servicing the vehicle, refer to the precautions section.

2. Remove the engine from the vehicle and mount it on a stand.

3. Remove or disconnect the following:
- Front cover
- Timing belt.
- Timing belt idler pulleys
- Crankshaft timing sprocket
- Oil dipstick tube
- Oil filter and bracket
- Crankshaft Position (CKP) sensor
- Oil pan and baffle
- Oil pump pickup tube
- Oil pump

To install:

4. The upper oil pan bolts are different lengths and are identified as follows:
- A: 1.38 inch (35mm) w/12mm head
- B: 1.97 inch (50mm) w/12mm head
- C: 4.17 inch (106mm) w/12mm head
- D: 1.57 inch (40mm) w/14mm head
- E: 1.18 inch (30mm) w/6mm hex head

5. Install a new O-ring on the engine block.

6. Apply silicone sealant to the oil pump housing as shown.

7. Install the oil pump. Tighten the bolts in several passes to the following specifications:
- 12mm: 11 ft. lbs. (15.5 Nm)
- 14mm: 22 ft. lbs. (30.5 Nm)
- 6mm Hex: 11 ft. lbs. (15.5 Nm)

8. Install or connect the following:
- Oil pump pickup tube. Tighten the bolts to 66 inch lbs. (7.5 Nm).

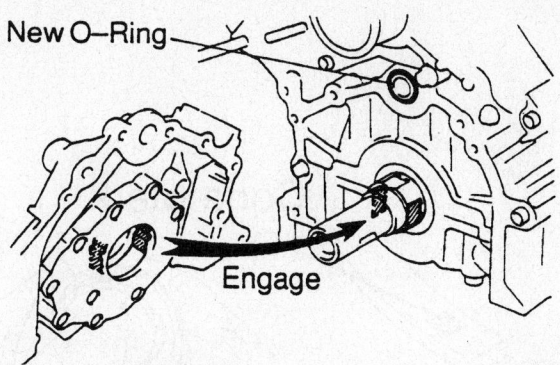

Location of the O-ring seal—4.7L 2UZ-FE engine

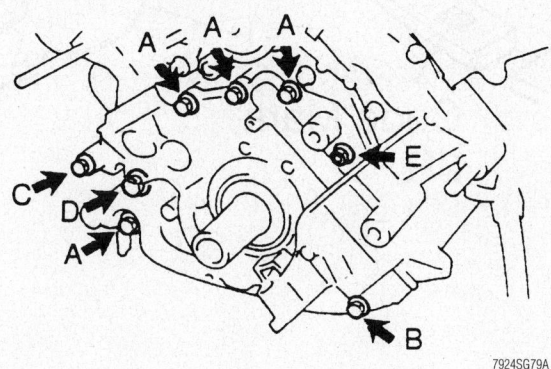

Oil pump bolt location—4.7L 2UZ-FE engine

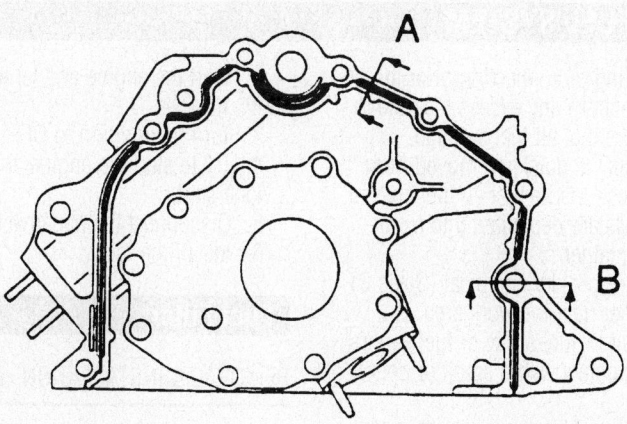

Seal Width
2 – 3 mm

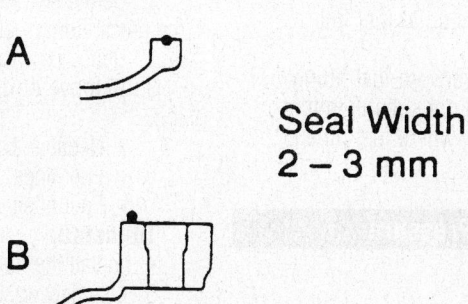

Oil pump housing sealant application—4.7L 2UZ-FE engine

- Oil pan and baffle
- CKP sensor
- Oil filter and bracket. Tighten the bolts to 13 ft. lbs. (18 Nm).
- Oil dipstick tube
- Crankshaft timing sprocket
- Timing belt idler pulleys
- Timing belt
- Front cover

9. Install the engine.

Rear Main Seal

REMOVAL & INSTALLATION

1. Before servicing the vehicle, refer to the precautions section.
2. Remove the transmission and flywheel from the vehicle.
3. Cut off the rubber lip portion of the seal with a sharp knife.
4. Pry out the oil seal.

To install:

5. Install the rear main seal so that it is flush with the seal retainer housing.
6. Install or connect the following:

- Flywheel/driveplate. Tighten the bolts to 35 ft. lbs. (48 Nm) plus a 90 degree turn.
- Transmission

Piston and Ring

POSITIONING

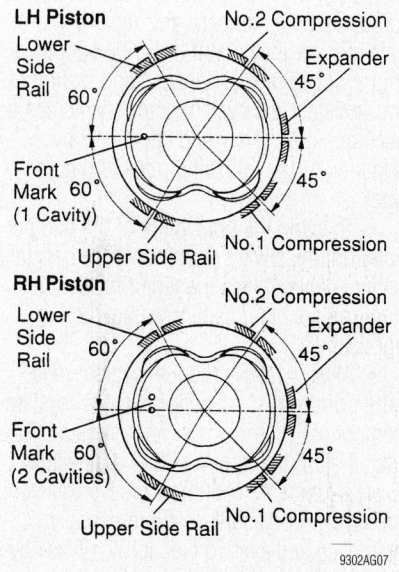

Piston ring positioning—4.7L 2UZ-FE engine

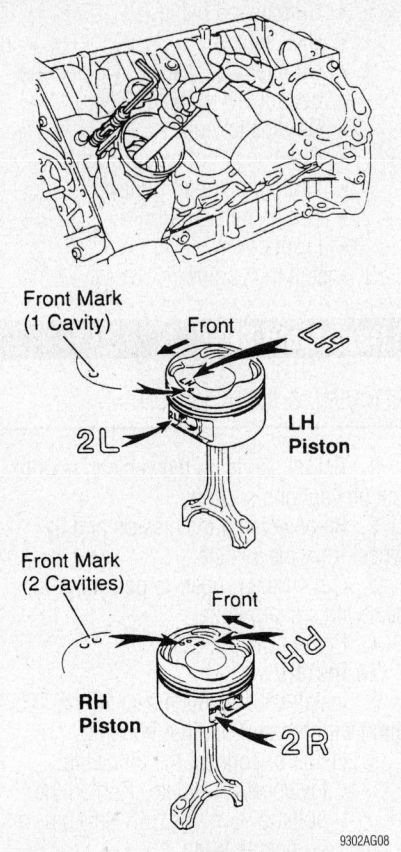

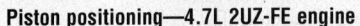

Piston positioning—4.7L 2UZ-FE engine

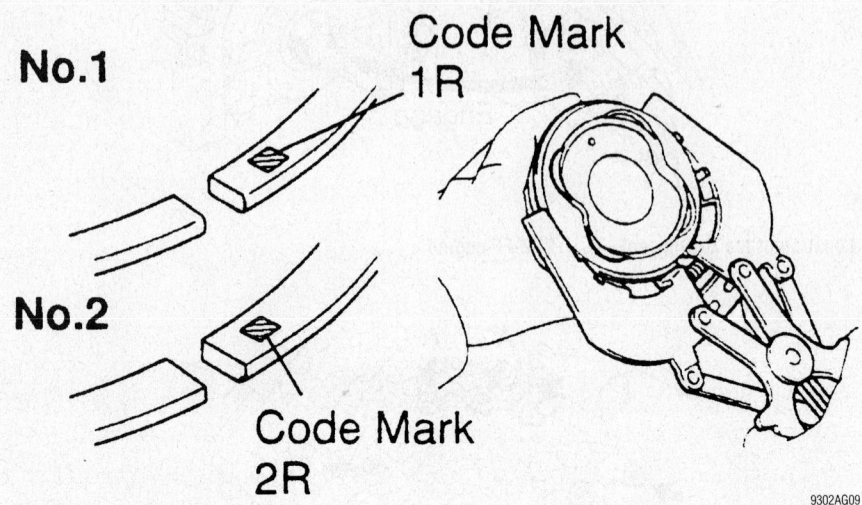

Piston ring identification—4.7L 2UZ-FE engine

FUEL SYSTEM

Fuel System Service Precautions

Safety is the most important factor when performing not only fuel system maintenance but any type of maintenance. Failure to conduct maintenance and repairs in a safe manner may result in serious personal injury or death. Maintenance and testing of the vehicle's fuel system components can be accomplished safely and effectively by adhering to the following rules and guidelines.

• To avoid the possibility of fire and personal injury, always disconnect the negative battery cable unless the repair or test procedure requires that battery voltage be applied.

• Always relieve the fuel system pressure prior to disconnecting any fuel system component (injector, fuel rail, pressure regulator, etc.), fitting or fuel line connection. Exercise extreme caution whenever relieving fuel system pressure, to avoid exposing skin, face and eyes to fuel spray. Please be advised that fuel under pressure may penetrate the skin or any part of the body that it contacts.

• Always place a shop towel or cloth around the fitting or connection prior to loosening to absorb any excess fuel due to spillage. Ensure that all fuel spillage (should it occur) is quickly removed from engine surfaces. Ensure that all fuel soaked cloths or towels are deposited into a suitable waste container.

• Always keep a dry chemical (Class B) fire extinguisher near the work area.

• Do not allow fuel spray or fuel vapors to come into contact with a spark or open flame.

• Always use a back-up wrench when loosening and tightening fuel line connection fittings. This will prevent unnecessary stress and torsion to fuel line piping.

• Always replace worn fuel fitting O-rings with new. Do not substitute fuel hose or equivalent, where fuel pipe is installed.

Fuel System Pressure

RELIEVING

1. Remove the fuel pump relay from the engine compartment relay block.

2. Start the engine and let it run until it shuts off.
3. Turn the ignition to OFF.
4. Try to start the engine and make sure it won't start.
5. Disconnect the negative battery cable.
6. Install the relay.

Fuel Filter

REMOVAL & INSTALLATION

Except GX470

1. Before servicing the vehicle, refer to the precautions section.
2. Relieve the fuel system pressure.
3. Remove or disconnect the following:

• Negative battery cable
• Fuel lines
• Fuel filter

To install:
4. Install the fuel filter.
5. Use new washers and tighten the fuel line bolts to the following specifications:

• Banjo bolt fittings: 21 ft. lbs. (29 Nm)

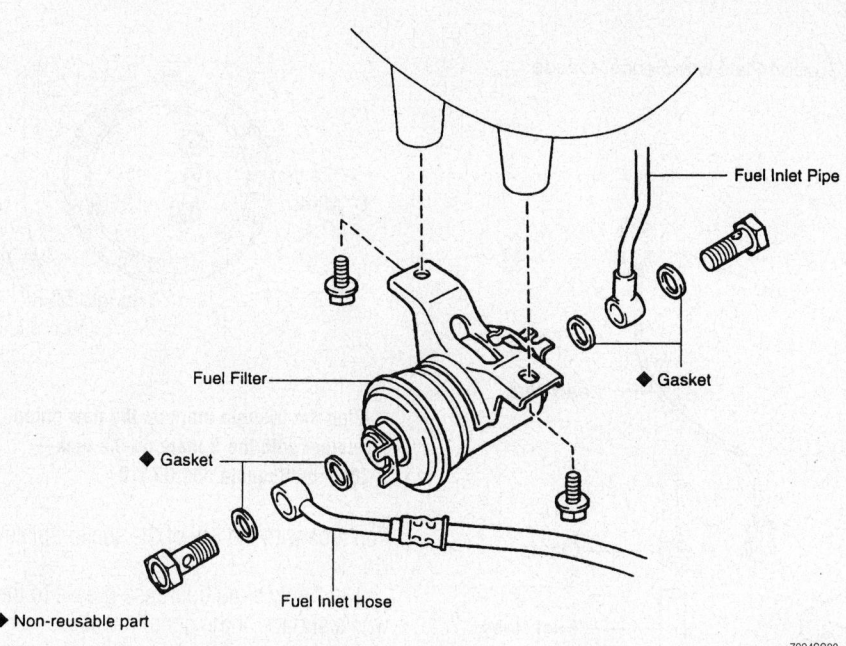

Always use new gaskets when replacing the fuel filter

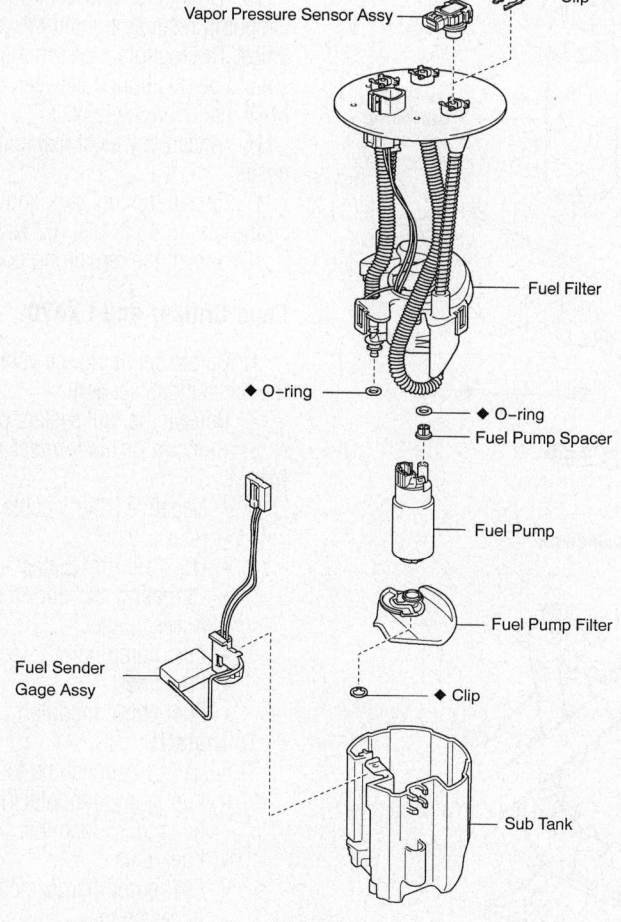

Fuel pump components—GX470

- Flare nut fitting: 28 ft. lbs. (38 Nm)
6. Connect the negative battery cable.
7. Start the engine and check for leaks.

GX470

The fuel filter is part of the fuel pump module unit and is not a normally replaced item.

Fuel Pump

REMOVAL & INSTALLATION

2002–04 Sequoia and GX470

1. Before servicing the vehicle, refer to the precautions section.
2. Relieve the fuel system pressure.
3. Remove or disconnect the following:

- Negative battery cable
- Fuel tank
- Fuel pump harness connector
- Fuel lines
- Fuel pump module

To install:
4. Install or connect the following:
- Fuel pump module. Tighten the bolts to 35 inch lbs. (4 Nm).
- Fuel lines
- Fuel pump harness connector
- Fuel tank
- Negative battery cable
5. Start the engine and check for leaks.

2005–06 Sequoia and GX470

1. Before servicing the vehicle, refer to the precautions section.
2. Relieve the fuel system pressure.
3. Remove the spare tire.
4. Disconnect the fuel pump connector and remove the fuel tank protector.
5. Disconnect the main and fuel return tubes.
6. Disconnect the fuel tank vent hose.
7. Disconnect the inlet and breather hoses.
8. Support the fuel tank with a jack, loosen the tank strap bolts remove the straps and lower the tank.
9. Disconnect any necessary hoses and wiring from the pump.
10. Using the tool illustrated, loosen the pump retainer.
11. Remove the pump and gasket.
To install:
12. Install a new gasket and the pump. Make sure to align the keyway of the suc-

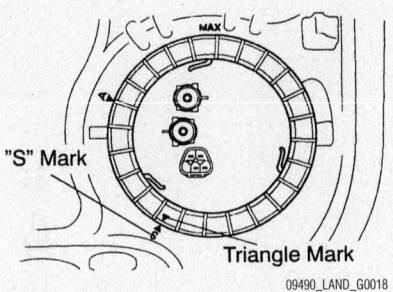

"S" Mark

Triangle Mark

09490_LAND_G0018

Align the triangle mark on the new pump retainer with the S mark on the tank—2005–06 Sequoia and GX470

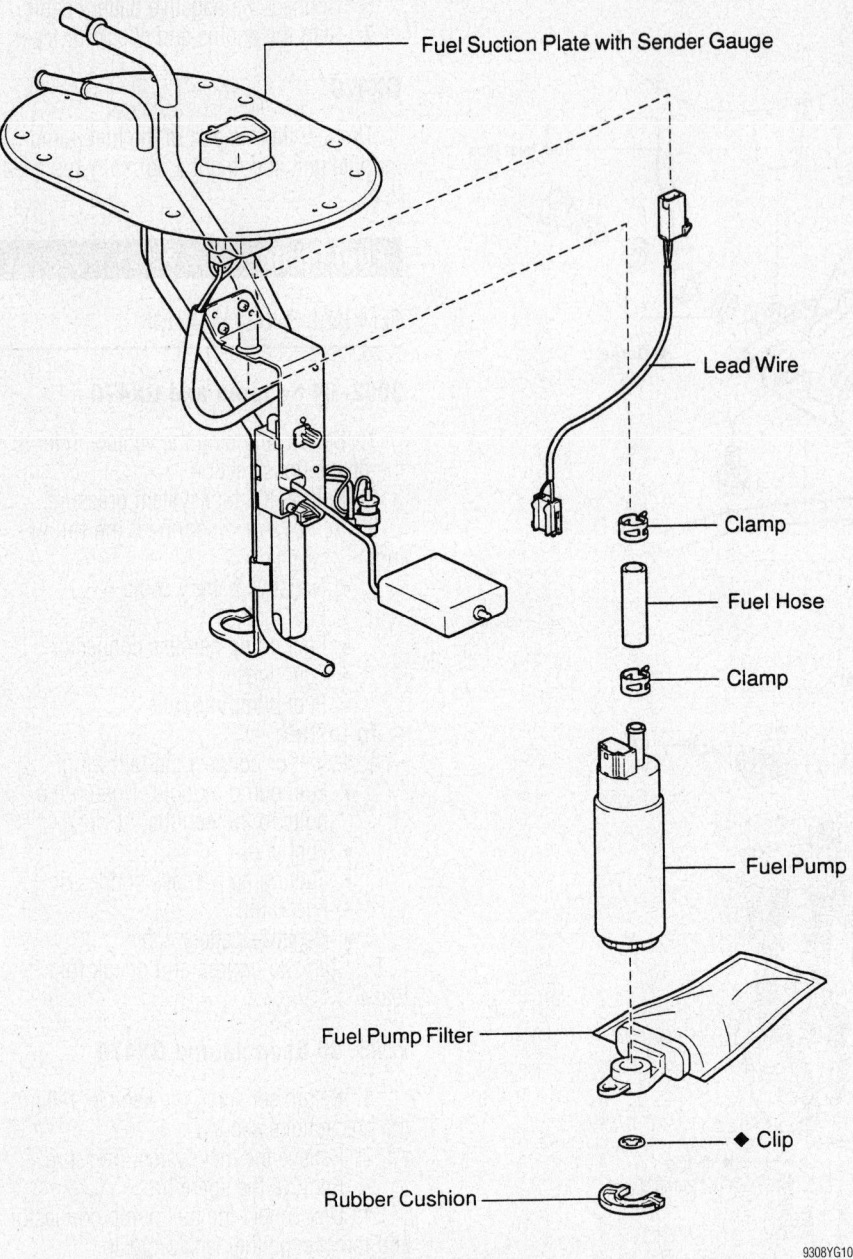

Fuel Suction Plate with Sender Gauge

Lead Wire

Clamp

Fuel Hose

Clamp

Fuel Pump

Fuel Pump Filter

◆ Clip

Rubber Cushion

9308YG10

Exploded view of the fuel pump and related components—2002–04 Sequoia

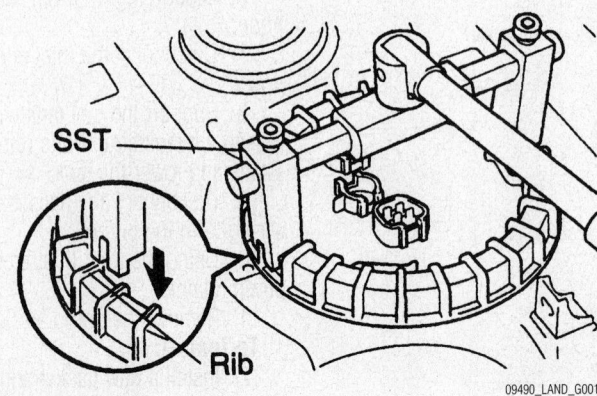

SST

Rib

09490_LAND_G0017

Use the tool illustrated to remove the fuel pump retainer—2005–06 Sequoia and GX470

tion tube with the key of the suction plate No. 1.

13. Apply a multipurpose grease to the whole surface of the pump retainer.

14. Align the triangle mark on the new pump retainer with the S mark on the tank while pushing the suction tube down and attach the gauge retainer.

15. Using the same tool used to remove the pump retainer, tighten the retainer 1 ½ times. The triangle mark on the pump should be positioned between the A and MAX marks on the tank.

16. Attach any electrical connections and hoses.

17. Install the fuel tank and tighten the strap bolts to 45 ft. lbs. (62 Nm).

18. Install the remaining components.

Land Cruiser and LX470

1. Before servicing the vehicle, refer to the precautions section.

2. Relieve the fuel system pressure.

3. Remove or disconnect the following:

- Negative battery cable
- Rear seats
- Door sill trim plates
- Carpeting and floor mats
- Access panel
- Fuel pump harness connector
- Fuel lines
- Fuel pump module

To install:

4. Install or connect the following:

- Fuel pump module. Tighten the bolts to 35 inch lbs. (4 Nm).
- Fuel lines
- Fuel pump harness connector
- Access panel
- Carpeting and floor mats
- Door sill trim plates
- Rear seats

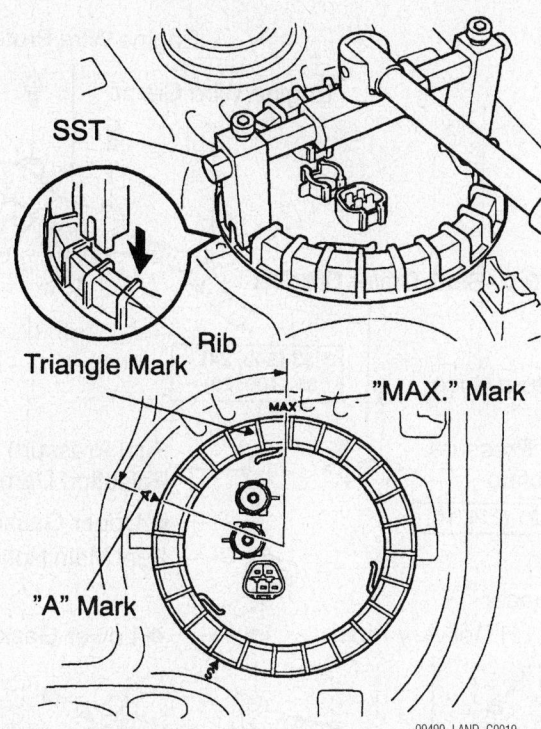

09490_LAND_G0019

The triangle mark on the pump should be positioned between the A and MAX marks on the tank when properly tightened—2005–06 Sequoia

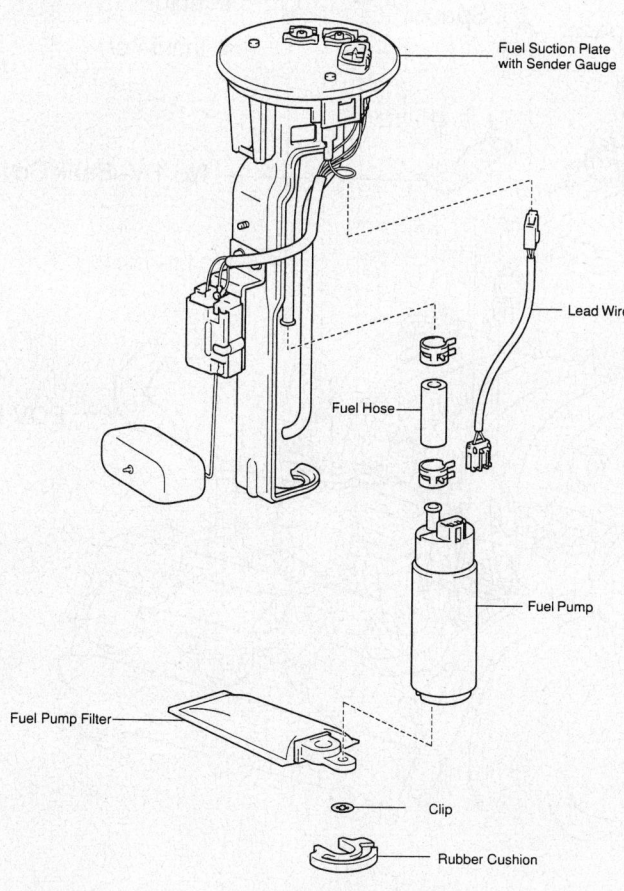

Fuel Suction Plate with Sender Gauge

Lead Wire

Fuel Hose

Fuel Pump

Fuel Pump Filter

Clip

Rubber Cushion

◆ Non–reusable part

7924SG81

Exploded view of the fuel pump and related components— Land Cruiser and LX470

• Negative battery cable
5. Start the engine and check for leaks.

Fuel Injector

REMOVAL & INSTALLATION

1. Before servicing the vehicle, refer to the precautions section.
2. Relieve the fuel system pressure.
3. Remove or disconnect the following:

- Negative battery cable
- Engine appearance cover
- Air intake tube
- Fuel lines
- Fuel pulsation damper
- Fuel pressure regulator vacuum line
- Accelerator cable and bracket
- Positive Crankcase Ventilation (PCV) valve and hose
- Evaporative Emissions (EVAP) vacuum switching valve
- Engine appearance cover brackets
- Fuel injector harness connectors
- Engine harness protector
- Fuel supply manifold crossover pipe
- Fuel supply manifolds with injectors attached
- Fuel injectors

To install:

4. Install the fuel injectors to the supply manifold with new O-ring seals and new grommets.
5. Install new injector insulators to the intake manifold.
6. Install or connect the following:

- Fuel supply manifolds with injectors attached. Tighten the bolts to 66 inch lbs. (7.5 Nm).
- Fuel supply manifold crossover pipe. Tighten the bolts to 29 ft. lbs. (39 Nm).
- Engine harness protector
- Fuel injector harness connectors
- Engine appearance cover brackets
- EVAP vacuum switching valve
- PCV valve and hose
- Accelerator cable and bracket
- Fuel pressure regulator vacuum line
- Fuel pulsation damper
- Fuel lines
- Air intake tube
- Engine appearance cover
- Negative battery cable

7. Start the engine and check for leaks.

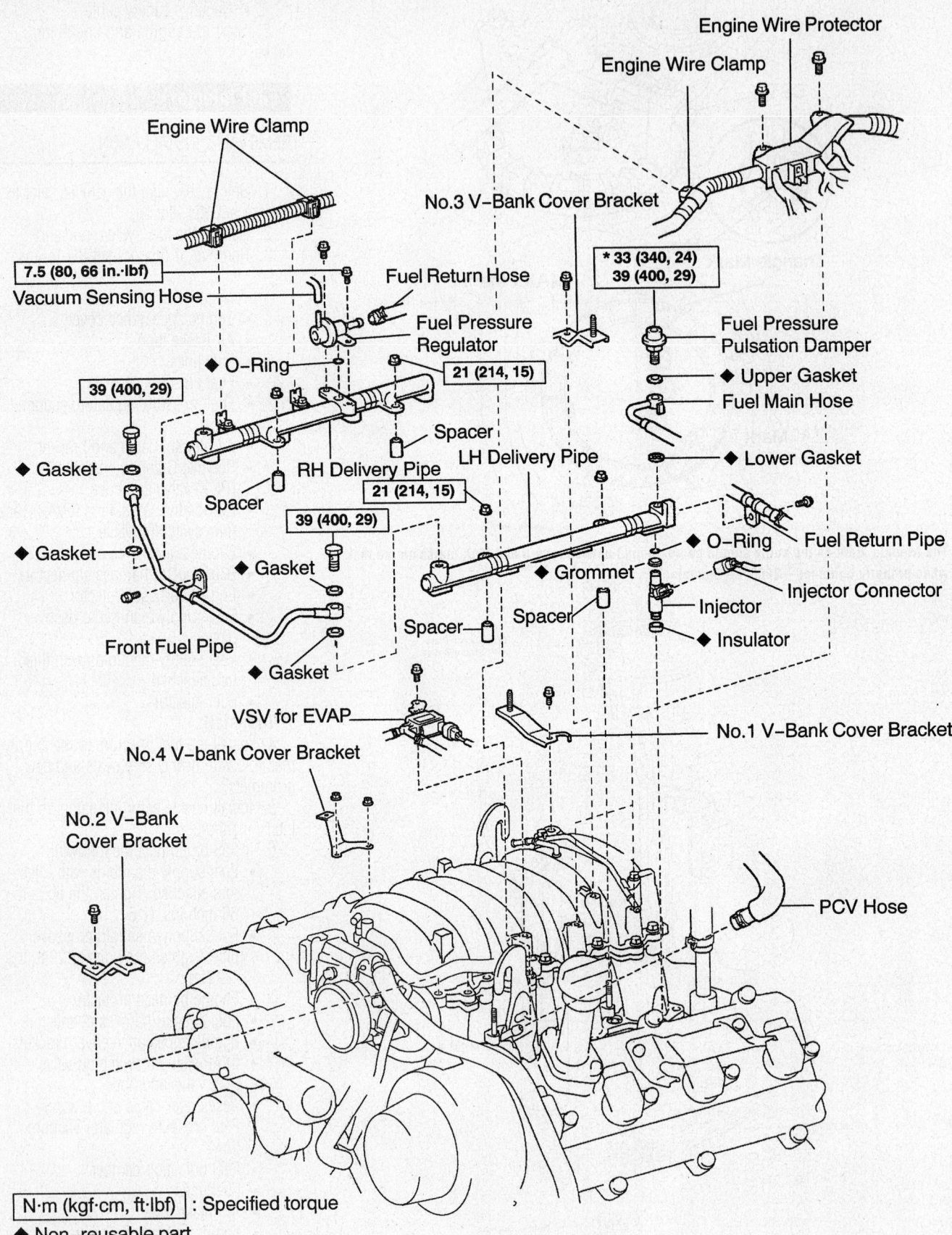

Engine Wire Protector

Engine Wire Clamp

Engine Wire Clamp

No.3 V–Bank Cover Bracket

Engine Wire Clamp

7.5 (80, 66 in.·lbf)

Vacuum Sensing Hose

Fuel Return Hose

Fuel Pressure Regulator

* 33 (340, 24)
39 (400, 29)

Fuel Pressure Pulsation Damper

◆ O–Ring

◆ Upper Gasket

Fuel Main Hose

39 (400, 29)

21 (214, 15)

◆ Lower Gasket

Spacer

◆ Gasket

RH Delivery Pipe

LH Delivery Pipe

◆ Gasket

Spacer

21 (214, 15)

Fuel Return Pipe

39 (400, 29)

◆ O–Ring

◆ Gasket

◆ Gasket

◆ Grommet

Injector Connector

Spacer

Spacer

Injector

Front Fuel Pipe

◆ Gasket

◆ Insulator

VSV for EVAP

No.1 V–Bank Cover Bracket

No.4 V–bank Cover Bracket

No.2 V–Bank Cover Bracket

PCV Hose

N·m (kgf·cm, ft·lbf) : Specified torque

◆ Non–reusable part

* For use with SST

Fuel injectors and related parts—Land Cruiser

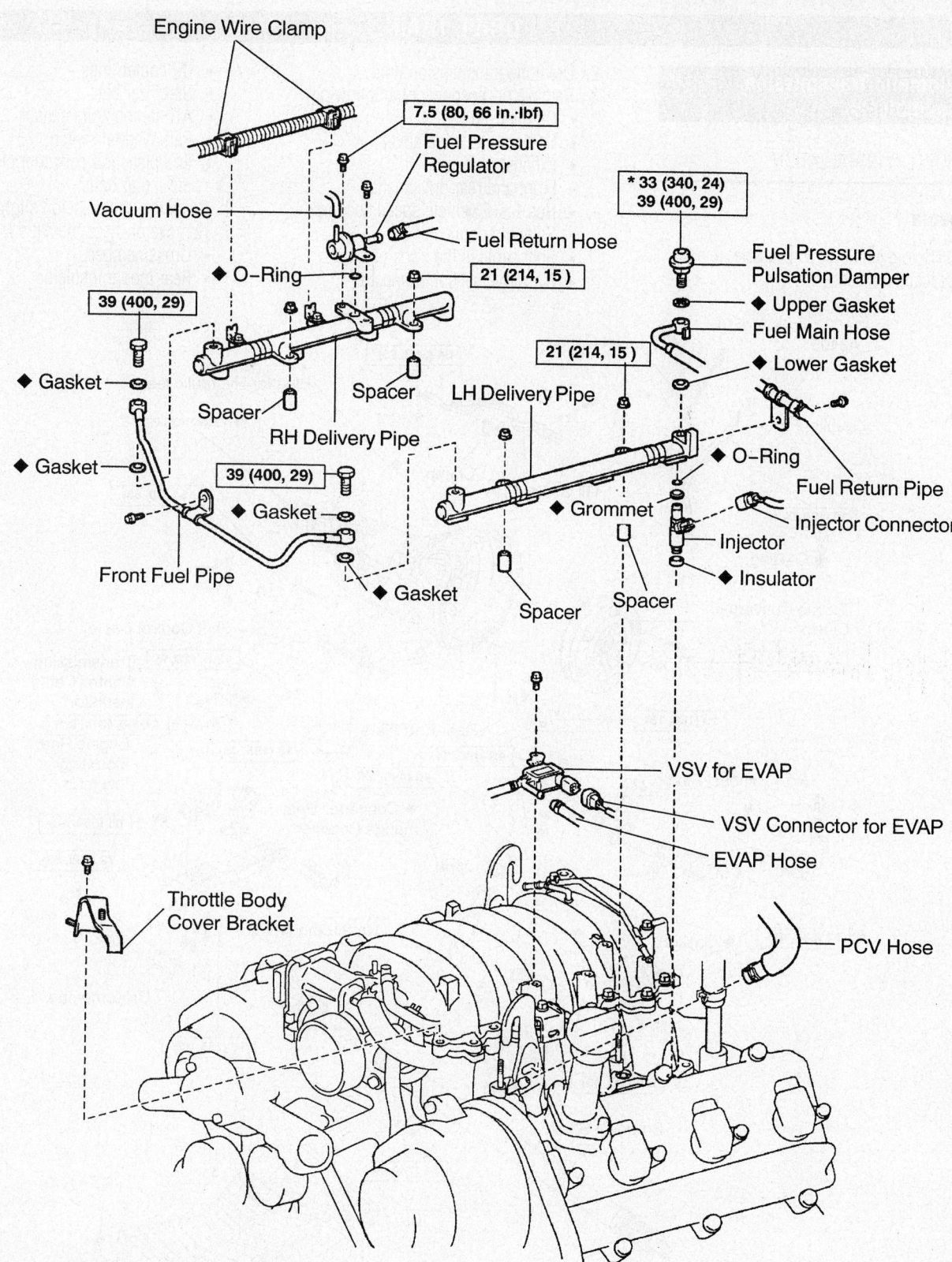

Engine Wire Clamp

7.5 (80, 66 in.·lbf)

Fuel Pressure Regulator

Vacuum Hose

Fuel Return Hose

* 33 (340, 24)
39 (400, 29)

Fuel Pressure Pulsation Damper

◆ O–Ring

◆ Upper Gasket

Fuel Main Hose

21 (214, 15)

39 (400, 29)

◆ Lower Gasket

◆ Gasket

Spacer

Spacer

LH Delivery Pipe

21 (214, 15)

◆ O–Ring

◆ Gasket

RH Delivery Pipe

Fuel Return Pipe

39 (400, 29)

◆ Grommet

Injector Connector

◆ Gasket

Injector

Front Fuel Pipe

◆ Gasket

Spacer

Spacer

◆ Insulator

VSV for EVAP

VSV Connector for EVAP

EVAP Hose

Throttle Body Cover Bracket

PCV Hose

N·m (kgf·cm, ft·lbf) : Specified torque

◆ Non–reusable part

* For use with SST

Fuel injectors and related parts—Sequoia

67170-LCSQ-G12

DRIVE TRAIN

Automatic Transmission Assembly

REMOVAL & INSTALLATION

Sequoia

1. Before servicing the vehicle, refer to the precautions section.

2. Drain the transmission fluid.
3. Remove or disconnect the following:
   - Oil filler pipe
   - No.1 engine undercover
   - Exhaust pipes
   - Front and rear driveshafts
   - Nos.1 & 2 vehicle speed sensors
   - Solenoid connector
   - Shift cable at the transmission
   - Overdrive sensor connector
   - Oil cooler lines
   - Stabilizer bar
   - ATF temperature sensor
   - Park/Neutral switch
   - End plate and converter clutch mounting bolts
4. Raise the transmission slightly.
5. Remove or disconnect the following:
   - Crossmember
   - Rear mount insulator

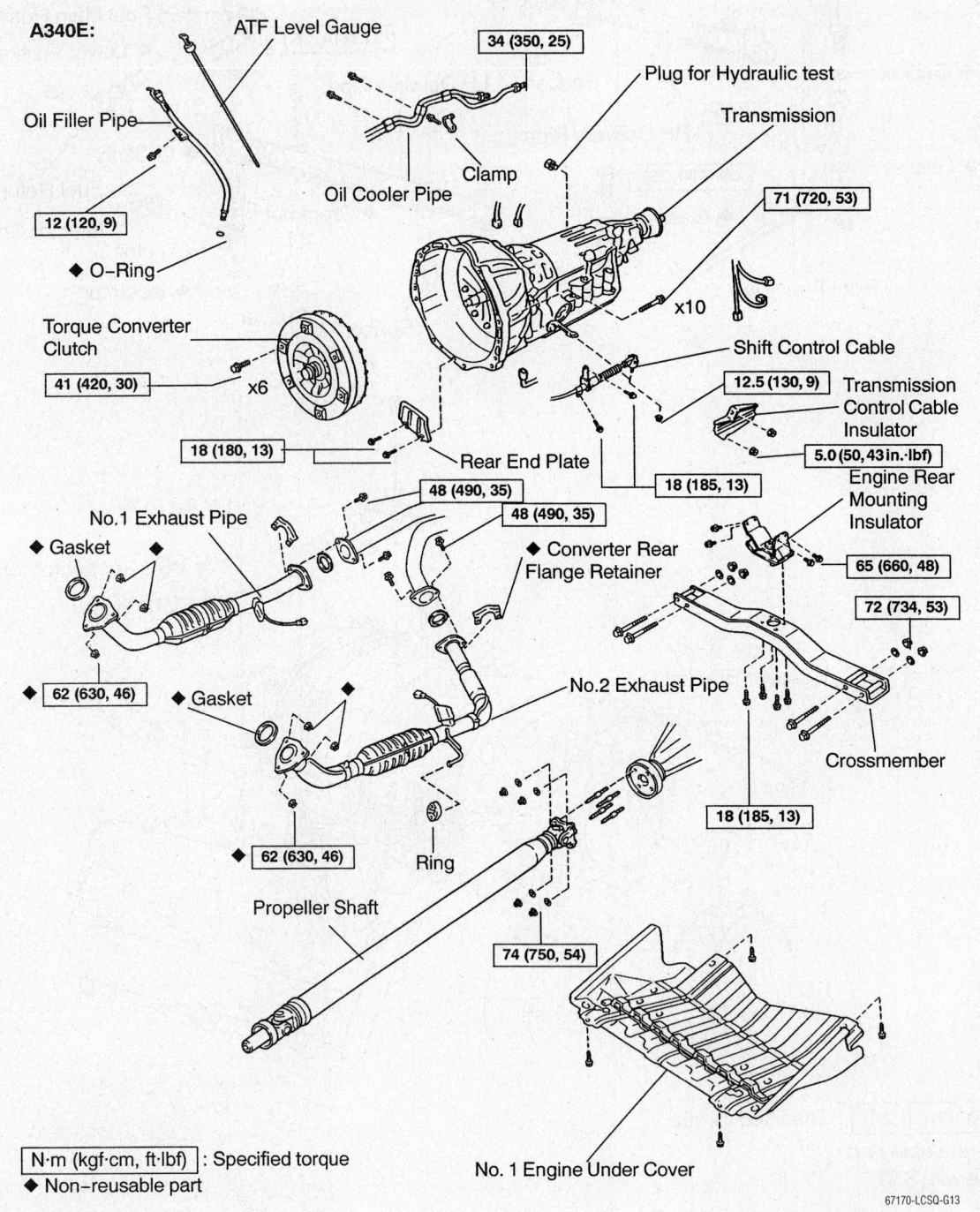

A340E:

ATF Level Gauge

Oil Filler Pipe

34 (350, 25)

Plug for Hydraulic test

Transmission

Oil Cooler Pipe

Clamp

71 (720, 53)

12 (120, 9)

◆ O–Ring

x10

Torque Converter Clutch

Shift Control Cable

41 (420, 30)

x6

12.5 (130, 9) Transmission Control Cable Insulator

18 (180, 13)

Rear End Plate

18 (185, 13)

5.0 (50, 43 in.·lbf)

Engine Rear Mounting Insulator

48 (490, 35)

No.1 Exhaust Pipe

48 (490, 35)

◆ Converter Rear Flange Retainer

65 (660, 48)

◆ Gasket   ◆

72 (734, 53)

62 (630, 46)

◆ Gasket

No.2 Exhaust Pipe

Crossmember

18 (185, 13)

62 (630, 46)

Ring

Propeller Shaft

74 (750, 54)

N·m (kgf·cm, ft·lbf) : Specified torque
◆ Non–reusable part

No. 1 Engine Under Cover

67170-LCSQ-G13

- Transmission and transfer case as a unit
6. Installation is the reverse of removal. Note the following torques:
  - Transmission-to-transfer case bolts: 17mm to 53 ft. lbs. (71Nm) and 14mm to 27 ft. lbs. (37Nm)
  - Rear mount insulator-to-transmission: 48 ft. lbs. (65Nm)
  - Rear mount insulator-to-cross-member: 13 ft. lbs. (18Nm)

- Cross member-to-frame: 53 ft. lbs. (71Nm)
- Torque converter clutch: 30 ft. lbs. (41Nm)

➡**Install the green bolt first.**

- Rear end plate: 13 ft. lbs. (18Nm)
- Stabilizer bar: 27 ft. lbs. (37Nm)
- Oil cooler lines: 25 ft. lbs. (34Nm)
- Shift control bracket: 13 ft. lbs. (18Nm)

Land Cruiser

1. Before servicing the vehicle, refer to the precautions section.
2. Drain the transmission fluid.
3. Remove or disconnect the following:
   - Battery and tray
   - Air intake assembly
   - Cooling fan and shroud
   - Coolant recovery reservoir
   - Transmission dipstick tube

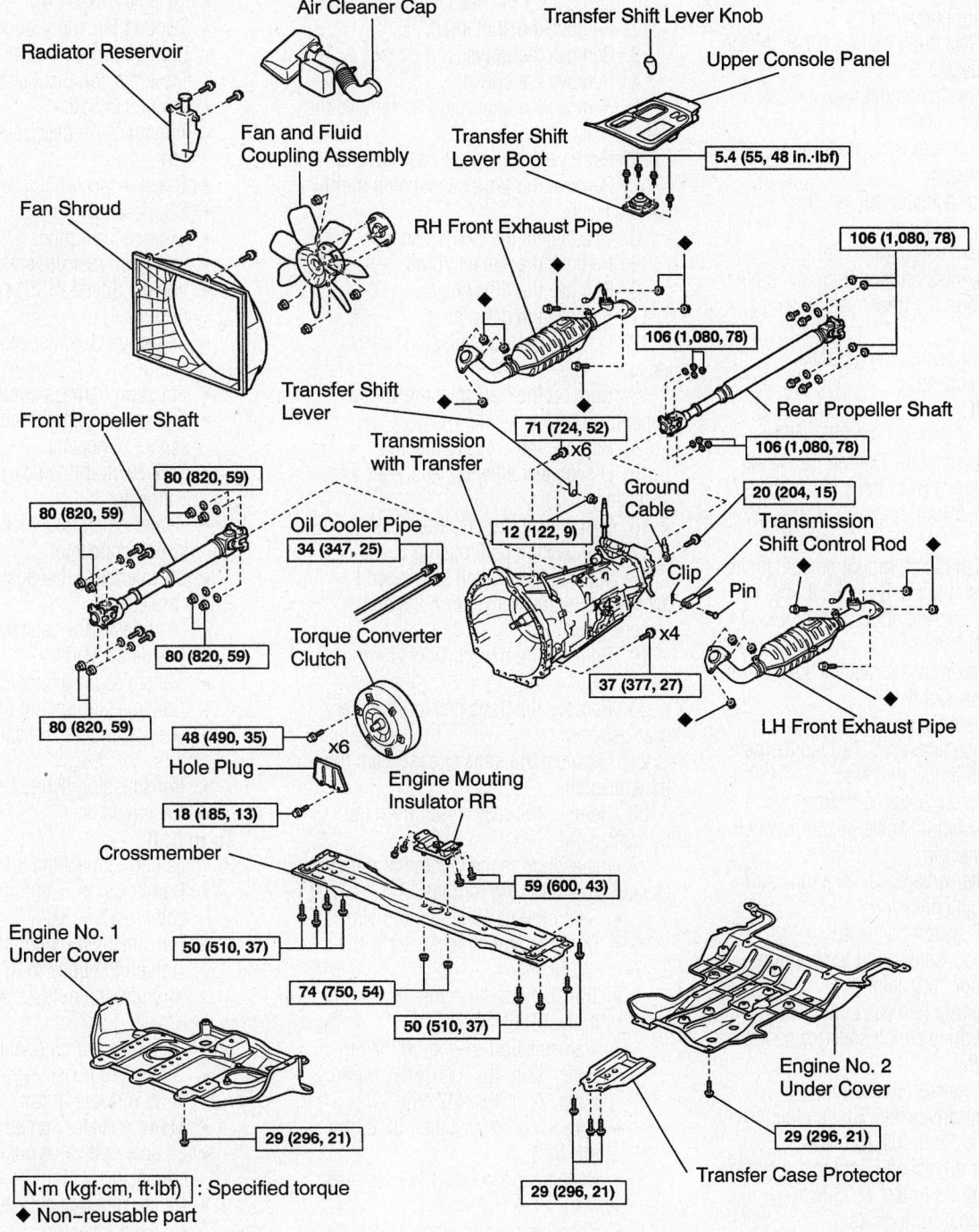

N·m (kgf·cm, ft·lbf) : Specified torque
◆ Non–reusable part

67170-LCSQ-G14

Transmission and related parts—Land Cruiser

- Center console
- Transmission gear select lever and rod
- Transfer case shift lever and rod
- Engine under covers
- Exhaust front pipes
- Front and rear driveshafts
- Vehicle Speed (VSS) sensor connectors
- Overdrive clutch speed sensor connector
- Solenoid harness connector
- Transmission fluid temperature sensor connector
- Park/Neutral Position (PNP) switch connector
- Center differential lock indicator switch connector
- L4 solenoid valve position switch connector
- Motor actuator connector
- Torque converter
- Transmission oil cooler lines
- Transmission mount crossmember. Support the transmission with a jack.
- Transmission flange bolts
- Transmission

To install:

4. Install or connect the following:
- Transmission. Tighten the flange bolts to 17mm bolts, 53 ft. lbs. (71 Nm); 14mm bolts, 27 ft. lbs. (37 Nm)
- Transmission mount crossmember. Tighten the bolts to 37 ft. lbs. (50 Nm) and the nuts to 54 ft. lbs. (74 Nm).
- Transfer case protector bolts to 21 ft. lbs. (29 Nm).
- Transmission oil cooler lines
- Torque converter. Tighten the bolts to 35 ft. lbs. (48 Nm).
- Motor actuator connector
- L4 solenoid valve position switch connector
- Center differential lock indicator switch connector
- PNP switch connector
- Transmission fluid temperature sensor connector
- Solenoid harness connector
- Overdrive clutch speed sensor connector
- VSS sensor connectors
- Front driveshaft. Tighten the fasteners to 59 ft. lbs. (80 Nm).
- Rear driveshaft. Tighten the fasteners to 78 ft. lbs. (106 Nm).
- Exhaust front pipes
- Engine under covers
- Transfer case shift lever and rod

- Transmission gear select lever and rod
- Center console
- Transmission dipstick tube
- Coolant recovery reservoir
- Cooling fan and shroud
- Air intake assembly
- Battery and tray

5. Check the transmission and transfer case fluid levels and adjust as necessary.

GX470

1. Disconnect the negative battery cable.
2. Remove the shift knob.
3. Remove the upper trim panels.
4. Remove the console.
5. Remove the snapring and remove the transfer case lever.
6. Remove the engine under-covers.
7. Remove the front suspension member brackets.
8. Disconnect the oxygen sensor.
9. Remove the exhaust pipe.
10. Remove the driveshafts.
11. Remove the drain plug.
12. Remove the transmission control cable.
13. Support the transmission with a transmission jack.
14. Remove the crossmember.
15. Disconnect all wires and lines as necessary.
16. Disconnect the breather hose.
17. Remove the bellhousing cover.
18. Turn the crankshaft as needed to access the torque converter bolts and remove them.
19. Remove the transmission-to-engine bolts.
20. Remove the transmission/transfer case assembly.
21. Separate the transfer case from the transmission.
22. Remove the rear mount from the transmission.
23. Installation is the reverse of removal. Observe the following torques:
- Rear mount: 48 ft. lbs. (65 Nm)
- Control cable bracket: 19 ft. lbs. (25 Nm)
- Transfer case-to-transmission: 17 ft. lbs. (24 Nm)
- Transmission-to-engine: 17mm bolts, 53 ft. lbs. (71 Nm); 14mm bolts, 27 ft. lbs. (37 Nm)
- Torque converter bolts: 35 ft. lbs. (48 Nm)
- Bellhousing cover: 13 ft. lbs. (18 Nm)
- Crossmember-to-frame: 53 ft. lbs. (72 Nm)

- Transmission-to-crossmember: 13 ft. lbs. (18 Nm)
- Front and rear driveshaft flanges: 65 ft. lbs. (88 Nm)
- Suspension member brackets: 24 ft. lbs. (33 Nm)

LX470

1. Before servicing the vehicle, refer to the precautions section.
2. Remove or disconnect the following:
- Battery and tray
- Air intake assembly
- Cooling fan and shroud
- Coolant recovery reservoir
- Transmission dipstick tube
- Center console
- Transmission gear select lever and rod
- Transfer case shift lever and rod
- Engine under covers
- Exhaust front pipes
- Front and rear driveshafts
- Vehicle Speed (VSS) sensor connectors
- Overdrive clutch speed sensor connector
- Solenoid harness connector
- Transmission fluid temperature sensor connector
- Park/Neutral Position (PNP) switch connector
- Center differential lock indicator switch connector
- L4 solenoid valve position switch connector
- Motor actuator connector
- Torque converter
- Transmission oil cooler lines
- Transmission mount crossmember. Support the transmission with a jack.
- Transmission flange bolts
- Transmission

To install:

3. Install or connect the following:
- Transmission. Tighten the flange bolts to 53 ft. lbs. (72 Nm).
- Transmission mount crossmember. Tighten the bolts to 37 ft. lbs. (50 Nm) and the nuts to 54 ft. lbs. (74 Nm).
- Transmission oil cooler lines
- Torque converter. Tighten the bolts to 35 ft. lbs. (48 Nm).
- Motor actuator connector
- L4 solenoid valve position switch connector
- Center differential lock indicator switch connector
- PNP switch connector

- Transmission fluid temperature sensor connector
- Solenoid harness connector
- Overdrive clutch speed sensor connector
- VSS sensor connectors
- Front driveshaft. Tighten the fasteners to 59 ft. lbs. (80 Nm).
- Rear driveshaft. Tighten the fasteners to 78 ft. lbs. (106 Nm).
- Exhaust front pipes
- Engine under covers

- Transfer case shift lever and rod
- Transmission gear select lever and rod
- Center console
- Transmission dipstick tube
- Coolant recovery reservoir
- Cooling fan and shroud
- Air intake assembly
- Battery and tray

4. Check the transmission and transfer case fluid levels and adjust as necessary.

Transfer Case Assembly

REMOVAL & INSTALLATION

Sequoia

1. Before servicing the vehicle, refer to the precautions section.
2. Drain the transfer case oil.
3. Place the shift lever in the **H** position and the one-touch 2-4 switch **OFF**.

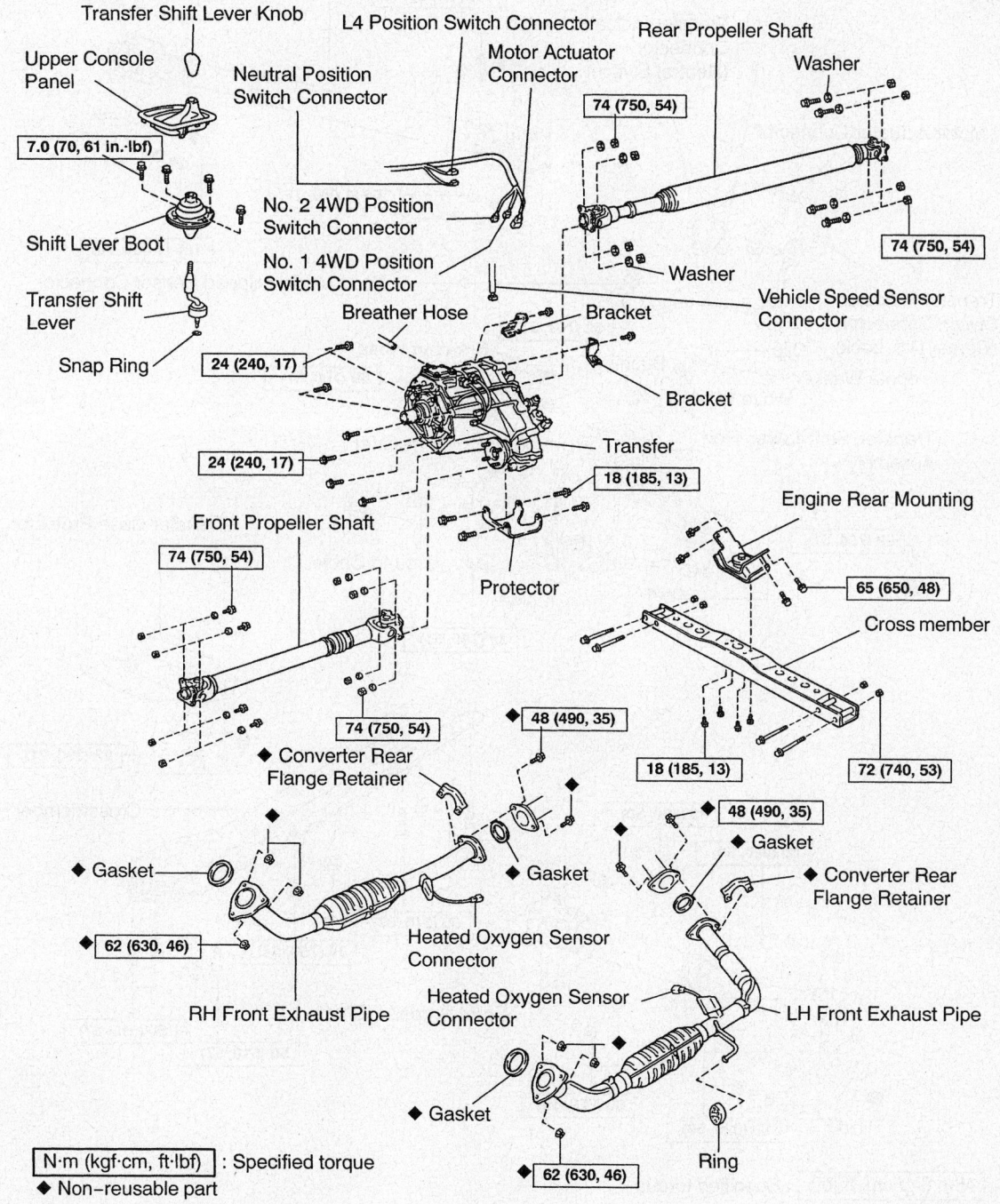

Transfer case and related parts—Sequoia

4. Remove or disconnect the following:
- Shift lever
- Skid plate
- Oil from the transfer case
- Exhaust pipes
- Front and rear driveshafts
- Crossmember
- Rear engine mount
- All wiring connectors

5. Support the transfer case, remove the case-to-adapter bolts and lower the transfer case.

6. Installation is the reverse of removal. Note the following torques:
- Transfer case-to-adapter bolts: 17 ft. lbs. (24Nm)
- Engine rear mount-to-adapter: 48 ft. lbs. (65Nm)

- Crossmember: 53 ft. lbs. (72Nm)
- Engine rear mount set bolts: 13 ft. lbs. (18Nm)
- Skid plate: 13 ft. lbs. (18Nm)

Land Cruiser

1. Before servicing the vehicle, refer to the precautions section.

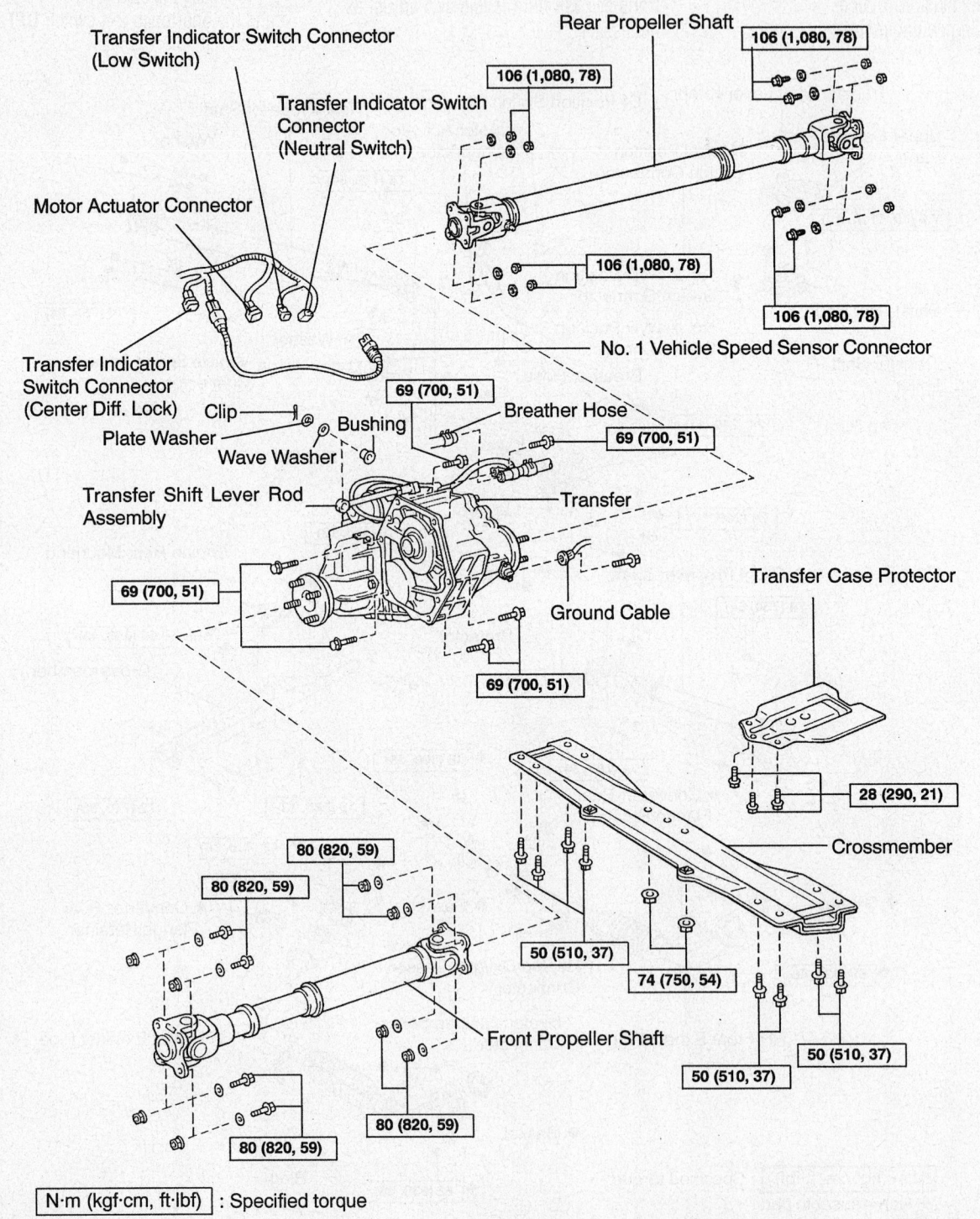

N·m (kgf·cm, ft·lbf) : Specified torque

Transfer case and related parts—Land Cruiser

67170-LCSQ-G16

2. Drain the transfer case oil.
3. Remove or disconnect the following:
- Transfer case protector
- Front and rear driveshafts
- Transfer case shift lever rod
- Ground cable
- Transmission mount crossmember. Support the transmission with a jack.
- Transfer case vent hose
- Vehicle Speed (VSS) sensor connector
- Center differential lock indicator switch connector
- Motor actuator connectors
- Transfer case adapter bolts
- Transfer case

To install:
4. Install or connect the following:
- Transfer case. Tighten the adapter bolts to 51 ft. lbs. (69 Nm).
- Motor actuator connectors
- Center differential lock indicator switch connector
- VSS sensor connector
- Transfer case vent hose
- Transmission mount crossmember. Tighten the bolts to 37 ft. lbs. (50 Nm) and the nuts to 54 ft. lbs. (74 Nm).
- Ground cable
- Transfer case shift lever rod
- Front driveshaft. Tighten the fasteners to 59 ft. lbs. (80 Nm).
- Rear driveshaft. Tighten the fasteners to 78 ft. lbs. (106 Nm).
- Transfer case protector
5. Fill the transfer case to the correct level.

GX470

1. Drain the fluid.
2. Remove the skid plate.
3. Remove the transmission.
4. Separate the transfer case from the transmission.
5. Installation is the reverse of removal. Torque the bolts to 17 ft. lbs. (24 Nm).

LX470

1. Before servicing the vehicle, refer to the precautions section.
2. Drain the transfer case oil.
3. Remove or disconnect the following:
- Transfer case protector
- Front and rear driveshafts
- Transfer case shift lever rod
- Ground cable
- Transmission mount crossmember. Support the transmission with a jack.
- Transfer case vent hose
- Vehicle Speed (VSS) sensor connector

- Center differential lock indicator switch connector
- Motor actuator connectors
- Transfer case adapter bolts
- Transfer case

To install:
4. Install or connect the following:
- Transfer case. Tighten the adapter bolts to 51 ft. lbs. (69 Nm).
- Motor actuator connectors
- Center differential lock indicator switch connector
- VSS sensor connector
- Transfer case vent hose
- Transmission mount crossmember. Tighten the bolts to 37 ft. lbs. (50 Nm) and the nuts to 54 ft. lbs. (74 Nm).
- Ground cable

- Transfer case shift lever rod
- Front driveshaft. Tighten the fasteners to 59 ft. lbs. (80 Nm).
- Rear driveshaft. Tighten the fasteners to 78 ft. lbs. (106 Nm).
- Transfer case protector
5. Fill the transfer case to the correct level.

Halfshaft

REMOVAL & INSTALLATION

Sequoia

1. Before servicing the vehicle, refer to the precautions section.
2. Remove or disconnect the following:

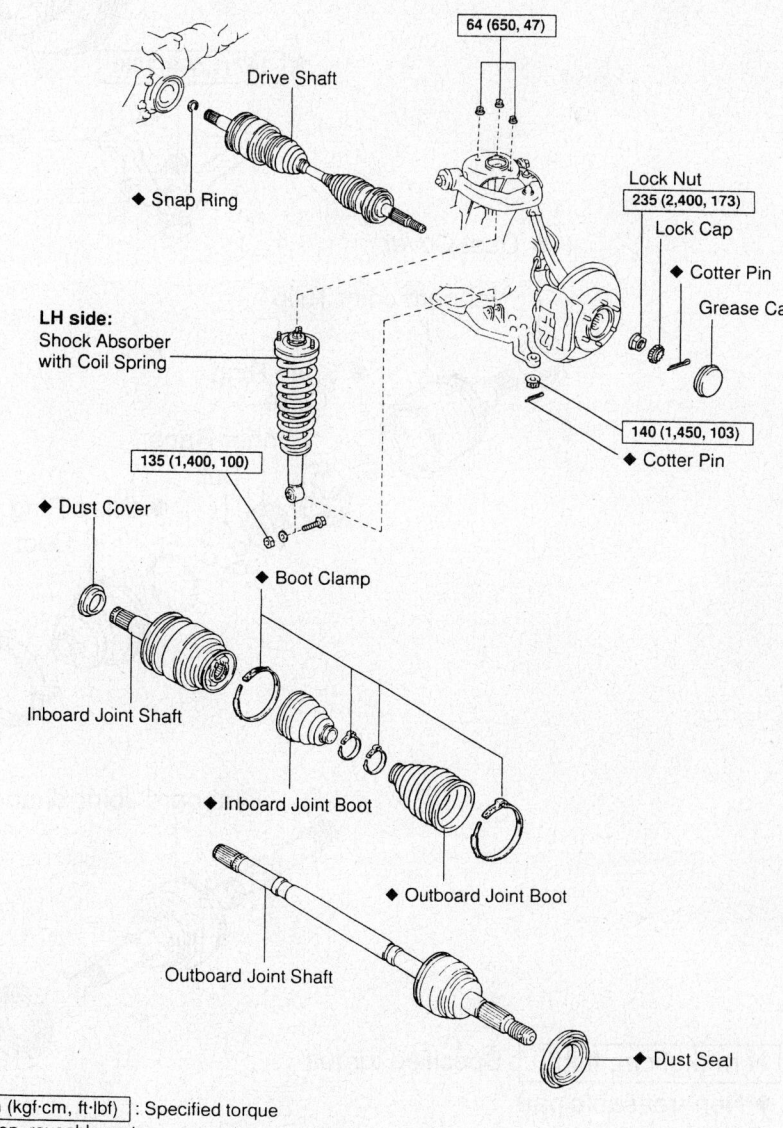

64 (650, 47)

Drive Shaft

◆ Snap Ring

Lock Nut
235 (2,400, 173)

Lock Cap

◆ Cotter Pin

Grease Cap

LH side:
Shock Absorber with Coil Spring

135 (1,400, 100)

◆ Dust Cover

140 (1,450, 103)

◆ Cotter Pin

◆ Boot Clamp

Inboard Joint Shaft

◆ Inboard Joint Boot

◆ Outboard Joint Boot

Outboard Joint Shaft

◆ Dust Seal

N·m (kgf·cm, ft·lbf) : Specified torque
◆ Non−reusable part

View of the halfshaft and related components—Sequoia

9308YG12

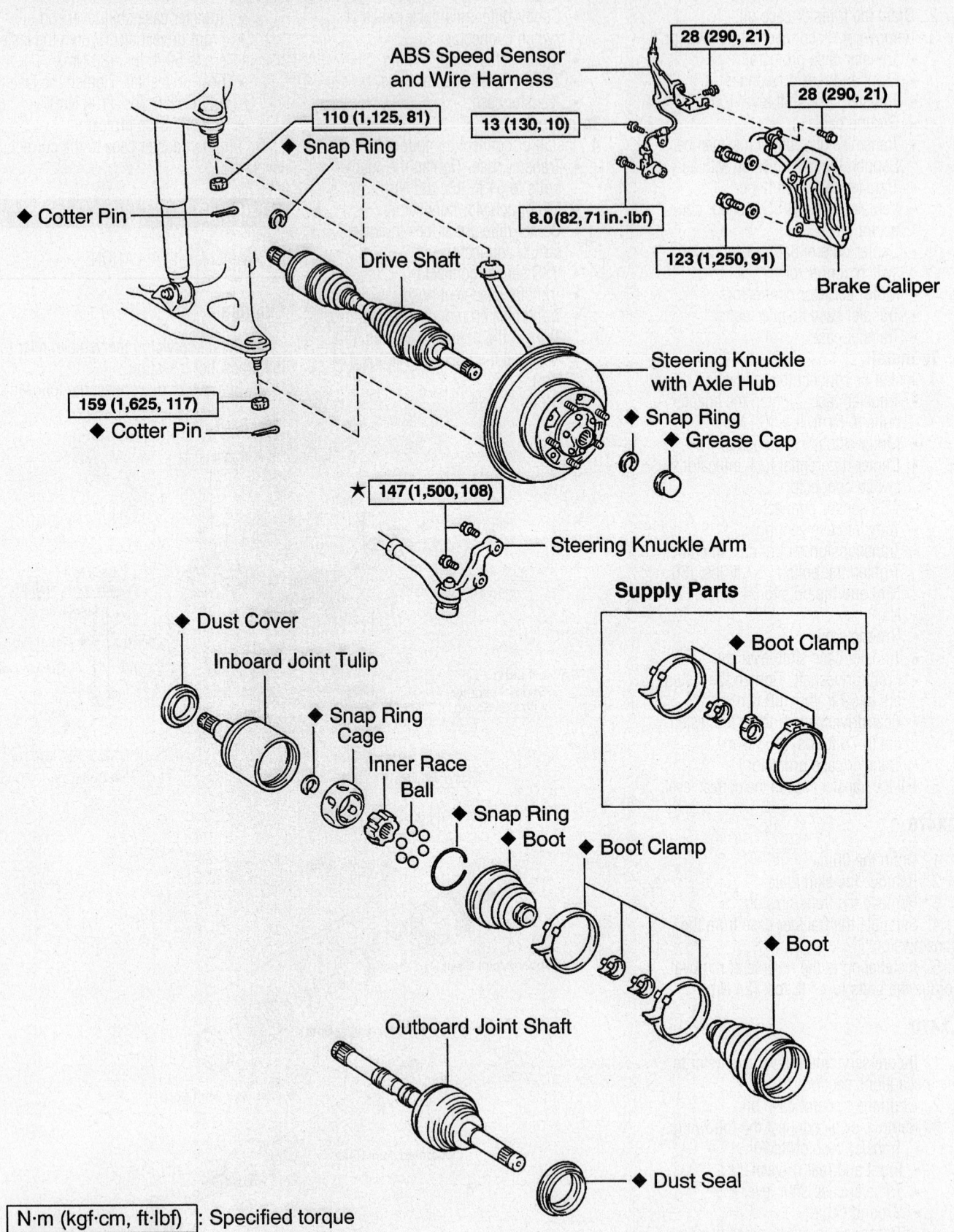

ABS Speed Sensor and Wire Harness

28 (290, 21)

28 (290, 21)

110 (1,125, 81)

◆ Snap Ring

13 (130, 10)

8.0 (82, 71 in.·lbf)

123 (1,250, 91)

Brake Caliper

◆ Cotter Pin

Drive Shaft

Steering Knuckle with Axle Hub

◆ Snap Ring

◆ Grease Cap

159 (1,625, 117)

◆ Cotter Pin

★ 147 (1,500, 108)

Steering Knuckle Arm

Supply Parts

◆ Boot Clamp

◆ Dust Cover

Inboard Joint Tulip

◆ Snap Ring

Cage

Inner Race

Ball

◆ Snap Ring

◆ Boot

◆ Boot Clamp

◆ Boot

Outboard Joint Shaft

◆ Dust Seal

N·m (kgf·cm, ft·lbf) : Specified torque
◆ Non-reusable part
★ Precoated part

Halfshaft and related components—Land Cruiser

67170-LCSQ-G17

- Front wheel
- Under cover
3. Drain the differential oil.
4. Remove or disconnect the following:
- Grease cap
- Cotter pin and lock cap
- Halfshaft locknut by applying the brakes
- Lower control arm from the lower ball joint
- Halfshaft from the steering knuckle, using a plastic hammer
- Left strut, for the left halfshaft
- Right halfshaft, using a brass bar and a hammer
- Left halfshaft, using tools 09520-01010 and 09520-24010
- Snapring from the inboard joint shaft

To install:
5. Install or connect the following:
- New snapring, onto the inboard joint shaft with the opening facing downward
- Halfshafts to the differential using a brass bar and a hammer
- Halfshafts to the steering knuckles

✳✳ WARNING

Be careful not to damage the oil seal, boot or dust seal.

- Lower control arm to the lower ball joint using a new cotter pin. Torque the ball joint nut to 103 ft. lbs. (140 Nm).
- Halfshaft locknut by applying the

LH side:

RH side:

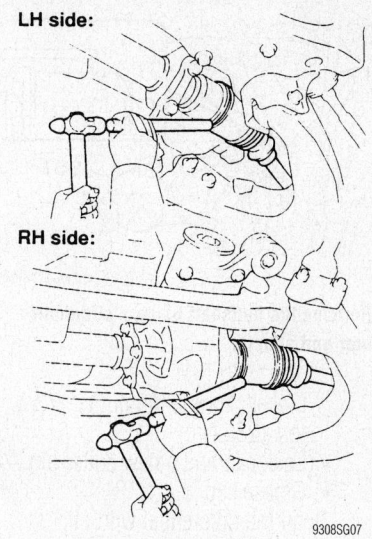

9308SG07

Axle halfshaft removal—Land Cruiser

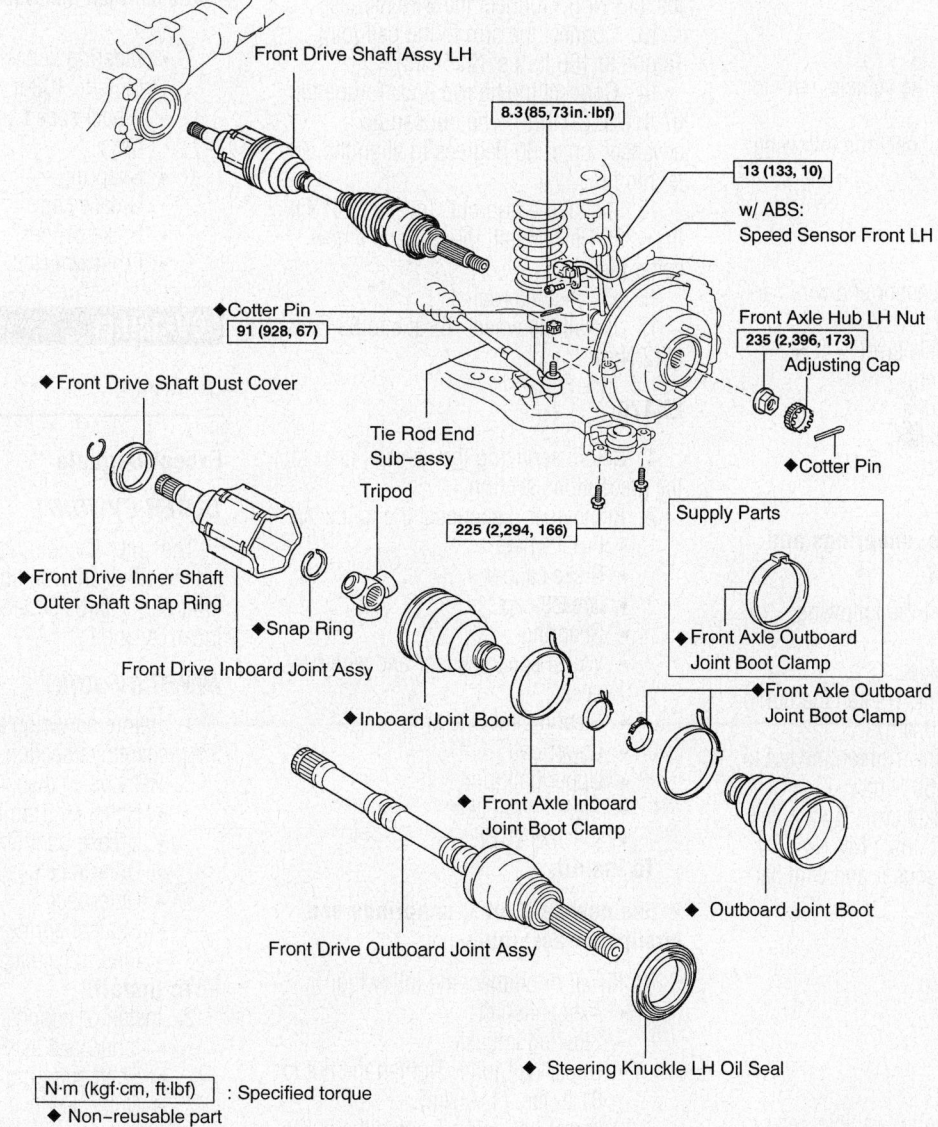

Front Drive Shaft Assy LH

8.3 (85, 73 in.·lbf)

13 (133, 10)

w/ ABS:
Speed Sensor Front LH

◆Cotter Pin
91 (928, 67)

Front Axle Hub LH Nut
235 (2,396, 173)
Adjusting Cap

◆Front Drive Shaft Dust Cover

Tie Rod End
Sub-assy

Tripod

◆Cotter Pin

225 (2,294, 166)

Supply Parts

◆Front Drive Inner Shaft
Outer Shaft Snap Ring

◆Snap Ring

Front Drive Inboard Joint Assy

◆Inboard Joint Boot

◆Front Axle Outboard
Joint Boot Clamp

◆Front Axle Outboard
Joint Boot Clamp

◆ Front Axle Inboard
Joint Boot Clamp

◆ Outboard Joint Boot

Front Drive Outboard Joint Assy

◆ Steering Knuckle LH Oil Seal

N·m (kgf·cm, ft·lbf) : Specified torque
◆ Non-reusable part

67162-GX470-G08

Front halfshaft, left side shown—GX470

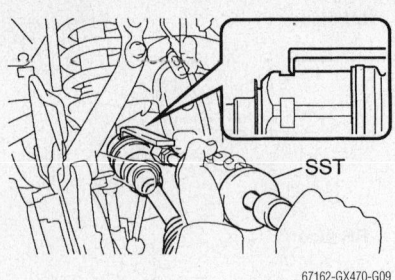

67162-GX470-G09

Remove the halfshaft using a slidehammer and adapter

brakes. Torque the nut to 173 ft. lbs. (235 Nm).
- Lock cap and a new cotter pin
- Grease cap
6. Refill the differential with oil.
7. Install or connect the following:
- Under cover
- Front wheel

Land Cruiser

1. Before servicing the vehicle, refer to the precautions section.
2. Remove or disconnect the following:
- Front wheel
- Brake caliper
- Grease cap
- Snapring
- Wheel speed sensor and wire harness
- Steering knuckle arm
- Lower ball joint
- Upper ball joint
- Steering knuckle
- Axle halfshaft

To install:

➡**Use new split pins, snaprings and circlips for assembly.**

3. Install or connect the following:
- Axle halfshaft
- Steering knuckle
- Upper ball joint. Tighten the nut to 81 ft. lbs. (110 Nm).
- Lower ball joint. Tighten the nut to 117 ft. lbs. (159 Nm).
- Steering knuckle arm. Tighten the bolts to 108 ft. lbs. (147 Nm).
- Wheel speed sensor and wire harness
- Snapring
- Grease cap
- Brake caliper
- Front wheel

GX470

1. Before servicing the vehicle, refer to the precautions section.

2. Remove the wheel.
3. Drain the differential oil.
4. Remove the cotter pin and cap, then remove the hub nut.
5. Remove the speed sensor wiring harness. Remove the sensor.
6. Remove the tie rod end from the knuckle.
7. Remove the 2 bolts and separate the lower arm from the ball joint.
8. Remove the halfshaft using a slidehammer and adapter. Keep the halfshaft level when carrying it.

To install:

9. Coat the inboard end splines of the halfshaft with clean ATF.
10. Align the splines and drive the halfshaft into place with a brass drift.
11. Install a new snapring with the opening facing down.
12. Install the sensor. Torque to 10 ft. lbs. (13 Nm). Connect the wire harness.
13. Connect the arm to the ball joint. Torque to 166 ft. lbs. (225 Nm).
14. Connect the tie rod end. Torque to 67 ft. lbs. (91 Nm). The nut can be advanced up to 60 degrees to align the cotter pin hole.
15. Install the hub nut. Torque to 173 ft. lbs. (235 Nm). Install the cap and a new cotter pin.
16. Fill the differential.
17. Install the wheel. Torque to 83 ft. lbs. (112 Nm).

LX470

1. Before servicing the vehicle, refer to the precautions section.
2. Remove or disconnect the following:
- Front wheel
- Brake caliper
- Grease cap
- Snapring
- Wheel speed sensor and wire harness
- Steering knuckle arm
- Lower ball joint
- Upper ball joint
- Steering knuckle
- Axle halfshaft

To install:

➡**Use new split pins, snaprings and circlips for assembly.**

3. Install or connect the following:
- Axle halfshaft
- Steering knuckle
- Upper ball joint. Tighten the nut to 81 ft. lbs. (110 Nm).
- Lower ball joint. Tighten the nut to 117 ft. lbs. (159 Nm).

LH side:

RH side:

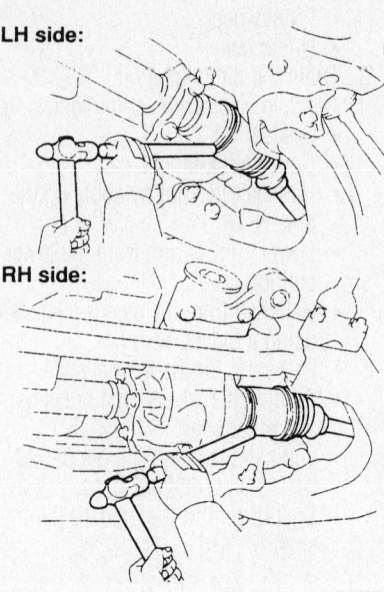

9308SG07

Axle halfshaft removal—LX470

- Steering knuckle arm. Tighten the bolts to 108 ft. lbs. (147 Nm).
- Wheel speed sensor and wire harness
- Snapring
- Grease cap
- Brake caliper
- Front wheel

CV-Joints

OVERHAUL

Except Sequoia

OUTER CV-JOINT

The outer CV-joint is serviced with the axle shaft as an assembly. The outer CV-joint boot can be serviced by removing the inner CV-joint.

INNER CV-JOINT

1. Before servicing the vehicle, refer to the precautions section.
2. Remove or disconnect the following:
- Halfshaft from the vehicle
- Grease boot clamps
- Outer race snapring
- Outer race
- Shaft snapring
- Inner race, cage and balls

To install:

3. Install or connect the following:
- Inner race, cage and balls
- Shaft snapring
- Outer race
- Outer race snapring
4. Fill the outer race and the grease boot

with CV-joint grease and tighten the boot clamps.

5. Install the axle halfshaft.

Sequoia

OUTER CV-JOINT

The outer CV-joint is serviced with the axle shaft as an assembly. The outer CV-joint boot can be serviced by removing the inner CV-joint.

INNER CV-JOINT

1. Before servicing the vehicle, refer to the precautions section.
2. Remove or disconnect the following:
   - Halfshaft from the vehicle
   - Large boot clamps
   - Small boot clamps
3. Matchmark inboard CV-joint to the shaft
4. Remove or disconnect the following:
   - Inboard CV-joint from the shaft by expanding the snapring
   - Both CV-joint boots
   - Outer dust seal, using a shop press and tool 09950-00020
   - Outer dust cover, using a shop press and tool 09950-00020

To install:

5. Install or connect the following:
   - Outer dust cover, using a screwdriver and a hammer
   - Outer dust seal, using a screwdriver and a hammer
6. Wrap the shaft splines with tape to protect the boot from damage.
7. Install or connect the following:
   - Both CV-joint boots with clamps, temporarily
   - Inboard CV-joint to the shaft by aligning the matchmarks and expanding the snapring
8. Lubricate the outboard joint with 7.23–7.94 oz. (205–225g) grease, provided in the boot kit.
9. Lubricate the inboard joint with 6.70–7.41 oz. (190–210g) grease, provided in the boot kit.
10. Install or connect the following:
   - Both joint boots making sure the boots are in the shaft groove
   - Standard halfshaft length is 20.531–20.689 in. (521.5–525.5mm) when the shaft is not expanded or contracted
   - Large inboard boot clamp
   - All other boot clamps using tool 09521-24010. Tighten the crimping tool until the clamp clearance is 0.039–0.059 in. (1.0–1.5mm)
   - Halfshaft

Spindle Bearings

REMOVAL, PACKING & INSTALLATION

Sequoia

1. Before servicing the vehicle, refer to the precautions section.
2. Remove or disconnect the following:
   - Front wheel
   - Grease cap, for 2WD
   - Cotter pin and lock cap, for 4WD
   - Locknut, by applying the brakes
   - Anti-lock Brake System (ABS) speed sensor and wiring harness clamp from the steering knuckle
   - Brake line clamp from the steering knuckle

➡**Be careful not to damage the brake tube.**

   - Brake caliper and disc
3. Support the steering knuckle
4. Remove or disconnect the following:
   - 4 lower ball joint-to-steering knuckle bolts
   - Upper ball joint cotter pin and loosen the nut
   - Upper ball joint from the steering knuckle using tool 09950-40011
   - Steering knuckle by placing it in a soft-jawed vise
   - Inside oil seal
   - 4 bolts and shift the dust cover towards the hub side (outside)
   - Axle hub from the steering knuckle using tools 09710-30021 and 09950-40011
   - Dust cover from the steering knuckle
   - Bearing spacer and ABS speed sensor (with ABS) or spacer (without ABS)

✳✳ WARNING

Be careful not to scratch the speed sensor rotor serrations

   - Outside oil seal from the steering knuckle
   - Bearing snapring from the steering knuckle
   - Bearing from the steering knuckle, using a shop press and tools 09950-60020 and 09950-70010

To install:

5. Install or connect the following:
   - Bearing to the steering knuckle, using a shop press and tools 09527-17011 and 09950-60020

   - Bearing snapring to the steering knuckle
   - New outside oil seal to the steering knuckle, using tools 09223-15030 and 09527-17011
   - Dust cover to the steering knuckle. Torque the 4 bolts to 13 ft. lbs. (18 Nm).
   - Axle hub to the steering knuckle using a shop press and tool 09649-17010
   - ABS speed sensor (with ABS) or spacer (without ABS)

✳✳ WARNING

Be careful not to scratch the speed sensor rotor serrations

   - Bearing spacer, using a shop press and tools 09950-60010 and 09950-70010
   - New inside oil seal, using tool 09527-17011 and a plastic hammer
6. For 4WD, install or connect the following:
   a. Halfshaft into the axle hub and temporarily tighten the nut

✳✳ WARNING

Be careful not to damage the oil seal or boot.

   b. Steering knuckle to the upper control arm. Tighten the nut to 77 ft. lbs. (105 Nm).
   c. New cotter pin.
7. Install or connect the following:
   - Lower ball joint to the steering knuckle. Torque the 4 bolts to 59 ft. lbs. (80 Nm).
   - Strut
   - Brake disc and caliper. Torque both caliper bolts to 90 ft. lbs. (123 Nm).
   - Brake line clamp to the steering knuckle. Torque the bolt to 21 ft. lbs. (28 Nm).
   - ABS speed sensor. Torque both bolts to 7.1 ft. lbs. (8.2 Nm).
   - Halfshaft locknut. Torque the nut to 173 ft. lbs. (235 Nm), by applying the brakes.
   - Lock cap and new cotter pin
   - Grease cap, for 2WD
   - Front wheel
8. Depress the brake pedal several times.
9. Check and/or adjust the front wheel alignment.
10. Check the ABS speed sensor signal.

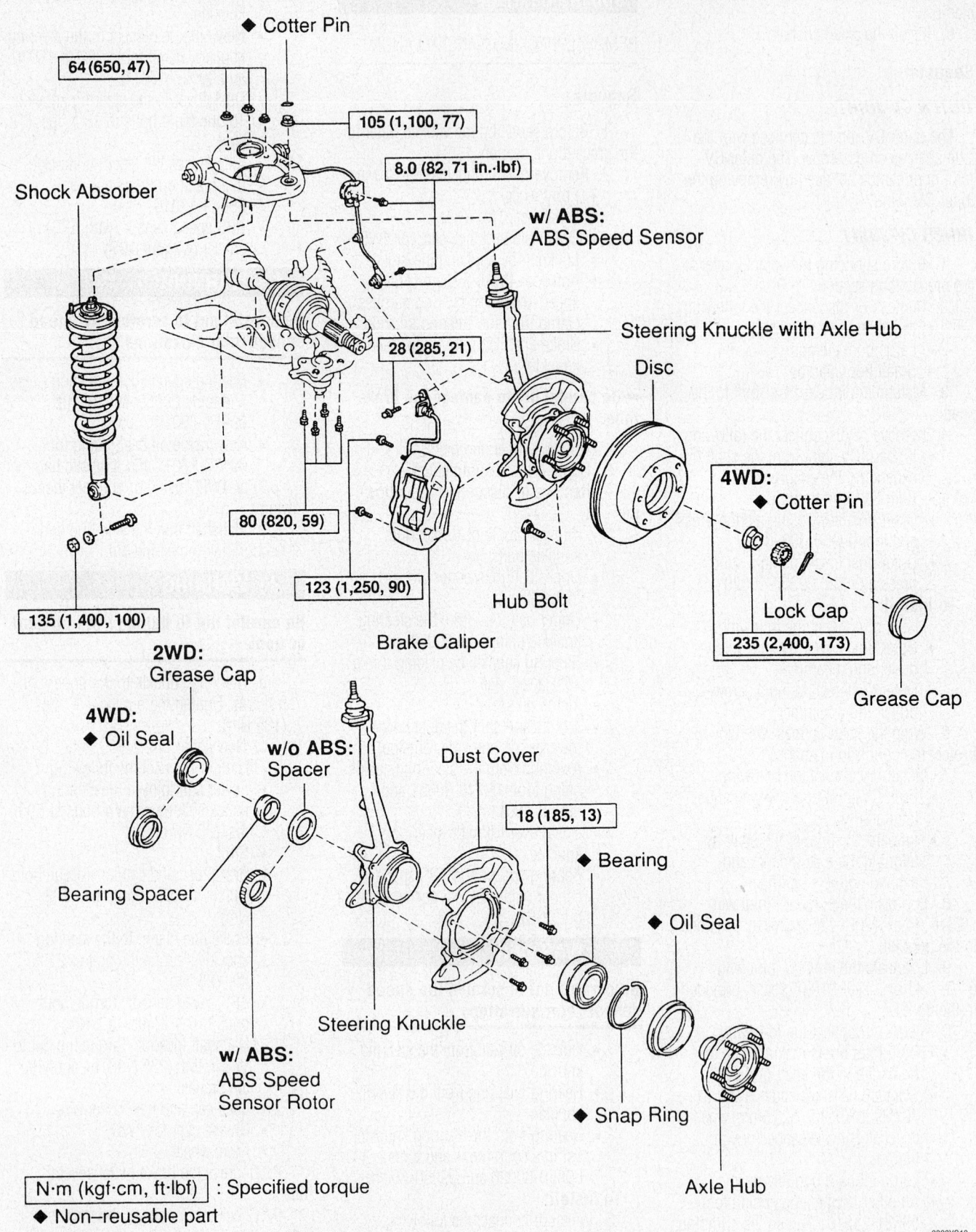

◆ Cotter Pin

64 (650, 47)

105 (1,100, 77)

8.0 (82, 71 in.·lbf)

Shock Absorber

w/ ABS:
ABS Speed Sensor

Steering Knuckle with Axle Hub

Disc

28 (285, 21)

4WD:
◆ Cotter Pin

80 (820, 59)

Lock Cap

235 (2,400, 173)

135 (1,400, 100)

123 (1,250, 90)

Hub Bolt

Brake Caliper

Grease Cap

2WD:
Grease Cap

4WD:
◆ Oil Seal

w/o ABS:
Spacer

Dust Cover

18 (185, 13)

◆ Bearing

◆ Oil Seal

Bearing Spacer

Steering Knuckle

w/ ABS:
ABS Speed
Sensor Rotor

◆ Snap Ring

Axle Hub

N·m (kgf·cm, ft·lbf) : Specified torque

◆ Non–reusable part

9308YG13

Exploded view of the front axle hub and related components—Sequoia

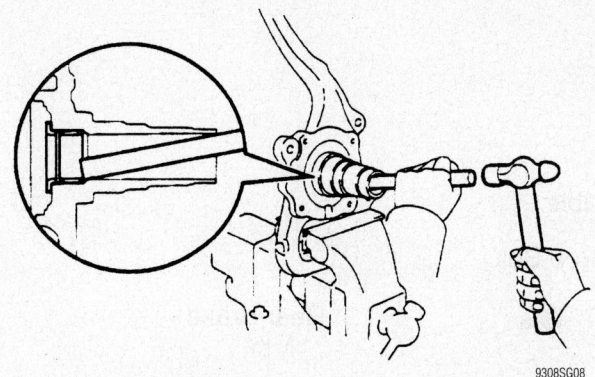

Removing the oil seal, bushing and spindle bearing—Land Cruiser and LX470

Land Cruiser and LX470

1. Before servicing the vehicle, refer to the precautions section.
2. Remove or disconnect the following:
   • Front wheel
   • Brake caliper
   • Grease cap
   • Snapring
   • Hub drive flange
   • Locknut
   • Lockwasher
   • Adjusting nut
   • Outer bearing
   • Wheel hub
   • Disc brake dust shield
   • Wheel speed sensor and harness
   • Outer tie rod end
   • Upper ball joint
   • Lower ball joint
   • Steering knuckle
   • Oil seal, bushing and spindle bearing

To install:

3. Coat the spindle bearing and bushing with lithium grease.
4. Fill the spindle cavity with lithium grease.
5. Press the spindle bearing and bushing into the spindle.
6. Install or connect the following:
   • Oil seal
   • Steering knuckle
   • Upper ball joint. Tighten the nut to 81 ft. lbs. (110 Nm).
   • Lower ball joint. Tighten the nut to 117 ft. lbs. (159 Nm).
   • Outer tie rod end. Tighten the nut to 91 ft. lbs. (122 Nm).
   • Wheel speed sensor and harness
   • Disc brake dust shield. Tighten the bolts to 13 ft. lbs. (18 Nm).
   • Wheel hub
   • Outer bearing
   • Adjusting nut. Adjust the wheel bearings.

   • Lockwasher
   • Locknut. Tighten the nut to 47 ft. lbs. (64 Nm).
   • Hub drive flange. Tighten the nuts to 26 ft. lbs. (35 Nm).
   • Snapring
   • Grease cap
   • Brake caliper
   • Front wheel

Axle Shaft, Bearing and Seal

REMOVAL & INSTALLATION

Rear

SEQUOIA

1. Before servicing the vehicle, refer to the precautions section.
2. Remove or disconnect the following:
   • Rear wheel
   • Brake drum and gasket
3. Using a dial indicator, check the bearing backlash and the axle shaft deviation. If the bearing backlash exceeds a maximum or 0.028 in. (0.7mm), replace it. If the axle shaft deviation exceeds the maximum of 0.004 in. (0.1mm), replace it.
4. Remove or disconnect the following:
   • Anti-lock Brake System (ABS) speed sensor from the rear axle housing, if equipped
   • Brake line from the wheel cylinder, using tool 09023-00100
   • Parking brake cable
   • 4 backing plate nuts
   • Axle shaft assembly, by pulling it from the axle housing

✳✳ WARNING

Be careful not to damage the oil seal.

   • O-ring from the rear axle housing
   • Inner side oil seal using tool 09308-00010

5. If equipped with ABS, perform the following:
   a. Remove and discard the 4 serration bolt nuts; then, using a hammer, drive the bolts from the backing plate.
   b. Using a grinder, grind the retainer and sensor rotor surfaces; then, chisel them out.
6. Remove the snapring from the axle shaft.
7. Remove the axle shaft from the backing plate, as follows:
   a. Position tool 09521-25011 onto the backing plate with the 4 nuts.
   b. Using a shop press, remove the axle shaft and bearing retainer from the backing plate.
8. Using tool 09308-00010, pull the oil seal from the backing plate.
9. Using a shop press and tools 09223-56010 and 09950-60010, press the bearing from the backing plate.

To install:

10. Install or connect the following:
    • Bearing into the backing plate, using a shop press and tools 09223-56010 and 09950-60010
    • New O-ring to the rear axle housing
    • New oil seal into the backing plate, using a hammer and tools 09950-70010 and 09950-60010
11. Install the axle shaft to the backing plate, as follows:
    • New outer side seal, lubricate the oil seal lip with multi-purpose grease
    • Backing plate and bearing retainer onto the rear axle shaft
    • Axle shaft onto the backing plate, by pressing it using a shop press and tool 09316-60011
    • New snapring

✳✳ WARNING

Be careful not to damage the oil seal.

12. Install or connect the following:
    • New sensor rotor and new bearing retainer onto the axle shaft, using a shop press and tool 09316-60011 to a standard length of 4.77–4.85 in. (121.2–123.2mm), if equipped with ABS
    • New inner side oil seal, using a hammer and tools 09950-60020 and 09950-70010
    • Axle shaft assembly. Torque the bolts to 51 ft. lbs. (69 Nm).

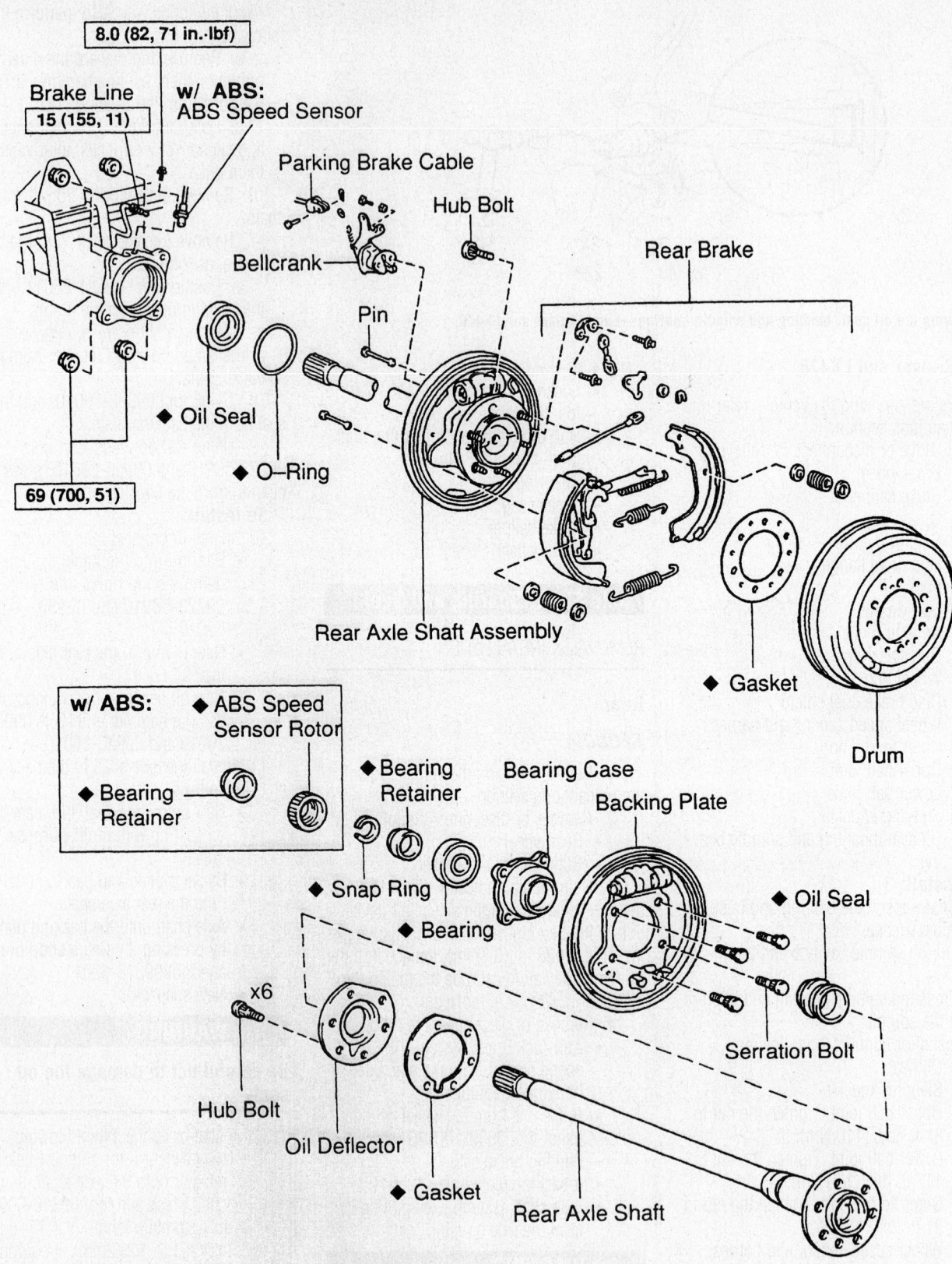

8.0 (82, 71 in.·lbf)

Brake Line
15 (155, 11)

w/ ABS:
ABS Speed Sensor

Parking Brake Cable

Hub Bolt

Rear Brake

Bellcrank

Pin

◆ Oil Seal

◆ O–Ring

69 (700, 51)

Rear Axle Shaft Assembly

◆ Gasket

Drum

w/ ABS: ◆ ABS Speed Sensor Rotor

◆ Bearing Retainer

◆ Bearing Retainer

Bearing Case

Backing Plate

◆ Snap Ring

◆ Bearing

◆ Oil Seal

Serration Bolt

x6

Hub Bolt

Oil Deflector

◆ Gasket

Rear Axle Shaft

N·m (kgf·cm, ft·lbf) : Specified torque

◆ Non–reusable part

9308YG14

Exploded view of the rear axle—Sequoia

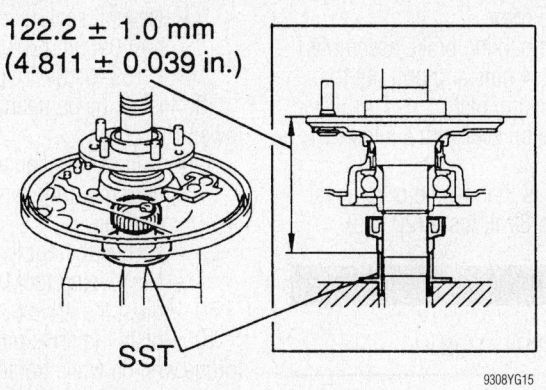

122.2 ± 1.0 mm
(4.811 ± 0.039 in.)

SST

9308YG15

Standard length of rear axle ABS speed sensor rotor and bearing retainer—Sequoia

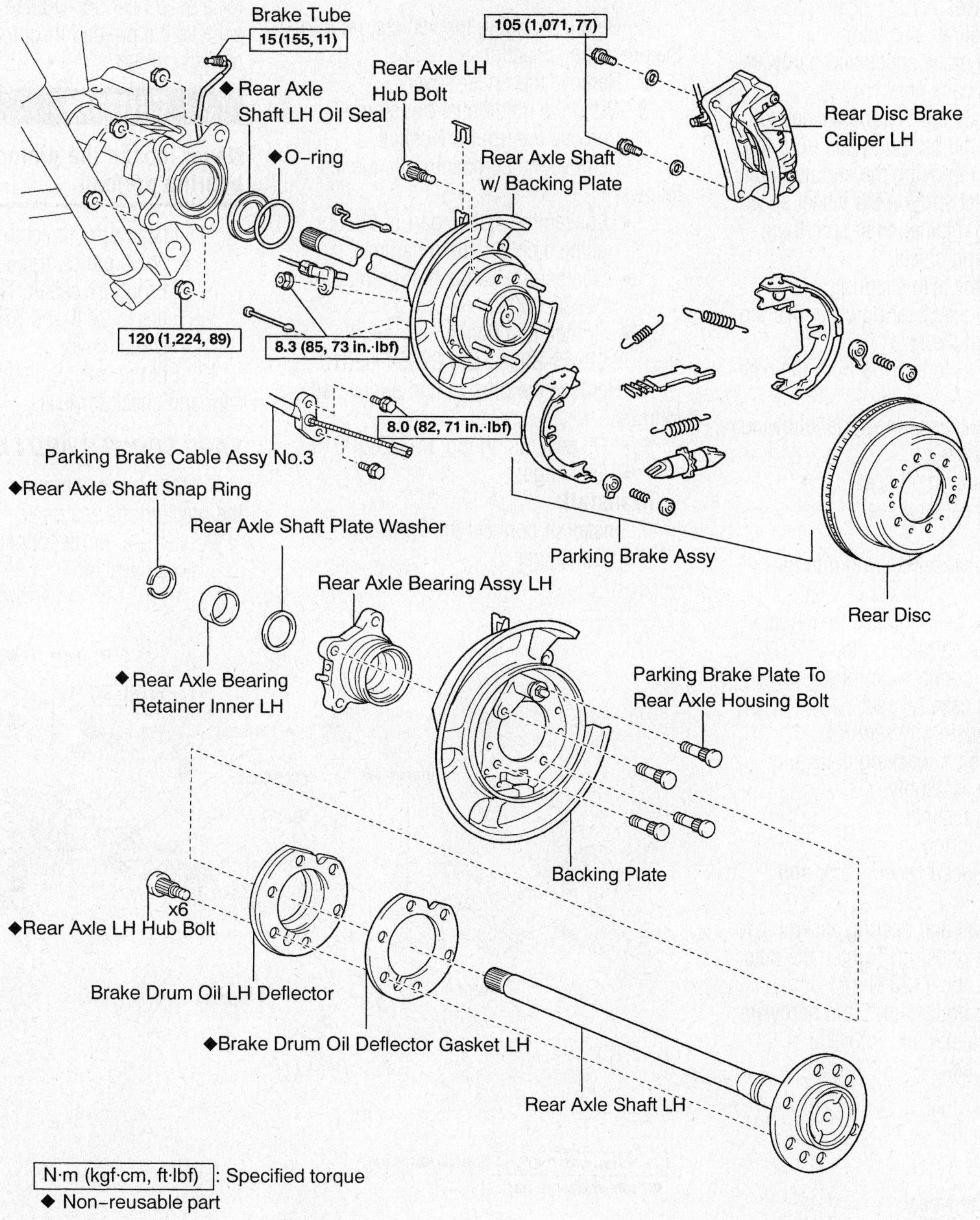

⁂ WARNING

Be careful not to damage the oil seal.

- Parking brake cable
- Brake line to the wheel cylinder, using tool 09023-00100. Torque the brake line to 11 ft. lbs. (15 Nm).
- Rear brake assembly
- ABS speed sensor to the rear axle housing. Torque it to 7.1 ft. lbs. (8.0 Nm).

13. Using a dial indicator, check the

Brake Tube
15 (155, 11)
◆ Rear Axle
Shaft LH Oil Seal
◆ O-ring
Rear Axle LH Hub Bolt
105 (1,071, 77)
Rear Axle Shaft w/ Backing Plate
Rear Disc Brake Caliper LH
120 (1,224, 89)
8.3 (85, 73 in.·lbf)
8.0 (82, 71 in.·lbf)
Parking Brake Cable Assy No.3
◆ Rear Axle Shaft Snap Ring
Rear Axle Shaft Plate Washer
Parking Brake Assy
Rear Axle Bearing Assy LH
Rear Disc
◆ Rear Axle Bearing Retainer Inner LH
Parking Brake Plate To Rear Axle Housing Bolt
Backing Plate
x6
◆ Rear Axle LH Hub Bolt
Brake Drum Oil LH Deflector
◆ Brake Drum Oil Deflector Gasket LH
Rear Axle Shaft LH

N·m (kgf·cm, ft·lbf): Specified torque
◆ Non-reusable part

67162-GX470-G11

Rear axle shaft and related parts—GX470

bearing backlash and the axle shaft deviation. If the bearing backlash exceeds a maximum or 0.028 in. (0.7mm), replace it. If the axle shaft deviation exceeds the maximum of 0.004 in. (0.1mm), replace it.

14. Install or connect the following:
  • New gasket and brake drum
  • Rear wheel. Torque the lug nuts to 81 ft. lbs. (110 Nm).
15. Bleed the brake system.
16. Check the ABS speed sensor signal.

LAND CRUISER AND LX470

1. Before servicing the vehicle, refer to the precautions section.
2. Remove or disconnect the following:
  • Rear wheel
  • Brake caliper and rotor
  • Parking brake shoes and hardware
  • Bearing case nuts
  • Axle shaft and bearing assembly
3. Separate the backing plate from the bearing case by removing the serrated bolts.
4. Grind a flat spot on the wheel speed sensor rotor and retainer, then split them with a hammer and chisel.
5. Remove the axle snapring.
6. Press the axle bearing case, bearing and retainer off of the axle.
7. Press the axle bearing from the bearing case.
8. Remove or disconnect the following:
  • Backing plate
  • Axle housing oil seal
  • Bearing case oil seal

To install:
9. Press the wheel bearing into the bearing case.
10. Install the bearing case to the backing plate with the serrated bolts.
11. Install or connect the following:
  • Bearing case oil seal
  • Axle housing oil seal
  • Axle shaft to backing plate and bearing assembly
  • Bearing retainer
  • Axle snapring
  • Wheel speed sensor rotor and retainer
  • Axle shaft and bearing assembly to the axle housing. Tighten the nuts to 91 ft. lbs. (123 Nm).
  • Parking brake shoes and hardware
  • Brake caliper and rotor
  • Rear wheel

Rear

GX470

1. Remove the wheel.
2. Remove the speed sensor.
3. Remove the caliper.

4. Remove the rotor.
5. Remove the parking brake assembly.
6. Remove the 4 nuts and pull out the axle shaft with backing plate.
7. Remove the oil seal with a slidehammer.
8. Installation is the reverse of removal. Torque the nuts to 89 ft. lbs. (120 Nm).

Pinion Seal

REMOVAL & INSTALLATION

Front

SEQUOIA

1. Before servicing the vehicle, refer to the precautions section.
2. Remove the under cover.
3. Drain the differential housing oil.
4. Remove the front driveshaft.
5. Remove the companion flange, as follows:
  • Loosen the staked part of the nut, using a chisel and a hammer
  • Companion flange nut, using tool 09330-00021
  • Companion flange, using tools 09950-30011 and 09954-03010
6. Remove the oil seal and slinger, as follows:
  • Oil seal, using tool 09308-10010
  • Oil slinger

To install:
7. Install or connect the following:
  • Oil slinger

• New oil seal, using a hammer and tool 09554-22010 to a depth of 0.153–0.189 in. (4.2–4.8mm).
8. Install the companion flange, as follows:
  • Companion flange
  • New nut, lubricated with hypoid gear oil
  • Torque the nut to 80 ft. lbs. (108 Nm), using tool 09330-00021
9. Adjust the drive pinion preload
10. Rotate the drive pinion, using a torque wrench while tightening the flange nut to make sure the bearing preload is 10.4–16.5 inch lbs. (1.2–1.9 Nm) for a new bearing or 5.2–8.7 inch lbs. (0.6–1.0 Nm) for a used bearing. Tighten the flange nut to achieve the preload torque readings originally recorded.

✳✳ CAUTION

Never loosen the pinion nut to reduce bearing preload.

11. Install or connect the following:
  • Drive pinion nut, stake it
  • Front driveshaft. Tighten the fasteners to 54 ft. lbs. (74 Nm).
  • Under cover
12. Fill the differential with gear lubricant and check for leaks.

LAND CRUISER AND LX470

1. Before servicing the vehicle, refer to the precautions section.
2. Remove or disconnect the following:
  • Driveshaft

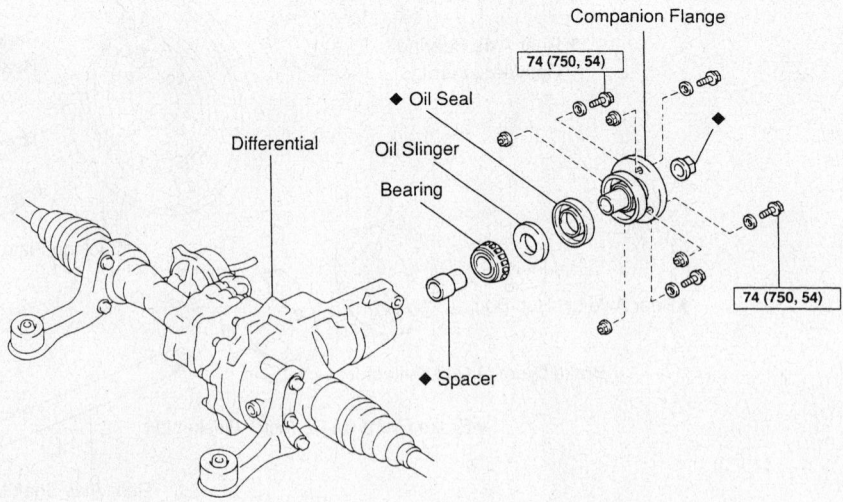

N·m (kgf·cm, ft·lbf) : Specified torque
◆ Non-reusable part

9308YG16

Exploded view of the Sequoia front differential assembly—Rear differential assembly is similar

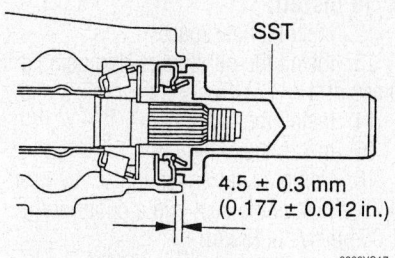

4.5 ± 0.3 mm
(0.177 ± 0.012 in.)

9308YG17

Positioning the Sequoia front pinion seal in the differential housing—Rear differential assembly is similar

- Front wheels
- Front brake calipers

➡**The front brake calipers must be removed so that there is no additional drag when measuring pinion bearing preload.**

3. Use an inch lb. torque wrench and measure the amount of torque required to maintain pinion rotation through several revolutions.
4. Remove or disconnect the following:

- Pinion flange
- Oil seal
- Oil slinger
- Pinion bearing and race
- Oil storage ring
- Collapsible spacer

To install:

➡**Use a new collapsible spacer and flange nut for assembly.**

5. Install or connect the following:

- Collapsible spacer
- Oil storage ring
- Pinion bearing and race
- Pinion seal
- Pinion flange. Tighten the nut to 80 ft. lbs. (108 Nm).

6. Rotate the pinion flange occasionally while tightening the flange nut to make sure the pinion bearings seat correctly.
7. Take frequent bearing preload torque readings. Tighten the flange nut to achieve the preload torque readings originally recorded. Do not exceed 249 ft. lbs. (338 Nm) torque when tightening the pinion flange nut.

❊❊ CAUTION

Never loosen the pinion nut to reduce bearing preload. If it is necessary to reduce bearing preload, install a new collapsible spacer and pinion nut.

8. Install or connect the following:

- Front brake calipers

- Front wheels
- Driveshaft. Tighten the fasteners to 59 ft. lbs. (80 Nm).

9. Fill the differential with gear lubricant and check for leaks.

GX470

1. Before servicing the vehicle, refer to the precautions section.
2. Remove the wheels.
3. Remove the engine under-covers.
4. Remove the front driveshaft.
5. Remove the pinion nut.
6. Remove the companion flange with a puller.
7. Remove the oil seal with a seal puller.
8. Remove the oil slinger.
9. Remove the bearing with a puller.
10. Remove the oil storage ring.
11. Remove the spacer and discard it.

To install:

12. Install a new spacer.
13. Install the oil storage ring using a brass drift.
14. Install the bearing.
15. Install the slinger.
16. Using a seal driver, install the new oil seal. Drive the seal into a depth of 4.35mm +/- 0.45mm.
17. Install the companion flange. Coat the threads of a new flange nut with gear oil. Hold the flange and torque the nut to 273 ft. lbs. (370 Nm).
18. Using an inch-pound torque wrench, check the preload. Preload for a new bearing should be 9-14 inch lbs.; for a used bearing, 4.3-7 inch lbs. If not, a new spacer must be installed.
19. When preload is correct, stake the nut.
20. Install the driveshaft. Torque the bolts to 65 ft. lbs. (88 Nm).
21. Fill the differential.
22. Install the under-covers.

Rear

SEQUOIA

1. Before servicing the vehicle, refer to the precautions section.
2. Drain the differential housing oil.
3. Remove the rear driveshaft.
4. Remove the companion flange, as follows:

- Loosen the staked part of the nut, using a chisel and a hammer
- Companion flange nut, using tool 09330-00021
- Companion flange, using tools 09950-30011 and 09954-03010
- Oil seal, using tool 09308-10010

To install:

5. Install the new oil seal until it is flush with the housing, using a plastic hammer and tools 09316-12010 and 09649-17010

➡**Use vinyl tape to connect both oil seal installation tools.**

6. Install the companion flange, as follows:

- Companion flange
- New nut, lubricated with hypoid gear oil
- Torque the nut to 109 ft. lbs. (147 Nm), using tool 09330-00021.

7. Adjust the drive pinion preload
8. Rotate the drive pinion, using a torque wrench while tightening the flange nut to make sure the bearing preload is 11.4–16.7 inch lbs. (1.3–1.9 Nm) for a new bearing or 4.3–6.9 inch lbs. (0.5–0.8 Nm) for a used bearing. Tighten the flange nut to achieve the preload torque readings originally recorded.

❊❊ CAUTION

Never loosen the pinion nut to reduce bearing preload.

9. Install or connect the following:

- Drive pinion nut, stake it
- Rear driveshaft. Tighten the fasteners to 54 ft. lbs. (74 Nm).

10. Refill the differential with gear lubricant and check for leaks; 3.33 qts. for 2WD or 3.12 qts. for 4WD.

LAND CRUISER AND LX470

1. Before servicing the vehicle, refer to the precautions section.
2. Remove or disconnect the following:

- Driveshaft
- Rear wheels
- Rear brake calipers

➡**The rear brake calipers must be removed so that there is no additional drag when measuring pinion bearing preload.**

3. Use an inch lb. torque wrench and measure the amount of torque required to maintain pinion rotation through several revolutions.
4. Remove or disconnect the following:

- Pinion flange
- Oil seal
- Oil slinger
- Pinion bearing and race
- Collapsible spacer

To install:

➡**Use a new collapsible spacer and flange nut for assembly.**

5. Install or connect the following:
- Collapsible spacer
- Pinion bearing and race
- Pinion seal
- Pinion flange. Tighten the nut to 181 ft. lbs. (245 Nm).

6. Rotate the pinion flange occasionally while tightening the flange nut to make sure the pinion bearings seat correctly.

7. Take frequent bearing preload torque readings. Tighten the flange nut to achieve the preload torque readings originally recorded. Do not exceed 326 ft. lbs. (441 Nm) torque when tightening the pinion flange nut.

❋❋ CAUTION

Never loosen the pinion nut to reduce bearing preload. If it is necessary to reduce bearing preload, install a new collapsible spacer and pinion nut.

8. Install or connect the following:
- Rear brake calipers
- Rear wheels
- Driveshaft. Tighten the fasteners to 78 ft. lbs. (106 Nm).

9. Fill the differential with gear lubricant and check for leaks.

GX470

1. Before servicing the vehicle, refer to the precautions section.
2. Remove the wheels.
3. Remove the engine under-covers.
4. Remove the front driveshaft.
5. Remove the pinion nut.
6. Remove the companion flange with a puller.
7. Remove the oil seal with a seal puller.
8. Remove the oil slinger.
9. Remove the bearing with a puller.
10. Remove the oil storage ring.
11. Remove the spacer and discard it.

To install:
12. Install a new spacer.
13. Install the oil storage ring using a brass drift.
14. Install the bearing.
15. Install the slinger.
16. Using a seal driver, install the new oil seal. Drive the seal into a depth of 1.00mm +/- 0.45mm.
17. Install the companion flange. Coat the threads of a new flange nut with gear oil. Hold the flange and torque the nut to 273 ft. lbs. (370 Nm).
18. Using an inch-pound torque wrench, check the preload. Preload for a new bearing should be 9-15 inch lbs.; for a used bearing, 5-7.5 inch lbs. If not, a new spacer must be installed.
19. When preload is correct, stake the nut.
20. Install the driveshaft. Torque the bolts to 65 ft. lbs. (88 Nm).
21. Fill the differential.
22. Install the under-covers.

STEERING AND SUSPENSION

Air Bag

❋❋ CAUTION

Some vehicles are equipped with an air bag system. The system must be disarmed before performing service on, or around, system components, the steering column, instrument panel components, wiring and sensors. Failure to follow the safety precautions and the disarming procedure could result in accidental air bag deployment, possible injury and unnecessary system repairs.

PRECAUTIONS

Several precautions must be observed when handling the inflator module to avoid accidental deployment and possible personal injury.

- Never carry the inflator module by the wires or connector on the underside of the module.
- When carrying a live inflator module, hold securely with both hands and ensure that the bag and trim cover are pointed away.
- Place the inflator module on a bench or other surface with the bag and trim cover facing up.
- With the inflator module on the bench, never place anything on or close to the module which may be thrown in the event of an accidental deployment.

DISARMING

To avoid personal injury when working on vehicles equipped with an air bag, the negative battery cable must be disconnected and at least 90 seconds must elapse before working on the system. Failure to do so may result in deployment of the air bag.

Power Rack And Pinion Steering Gear

REMOVAL & INSTALLATION

Sequoia

1. Before servicing the vehicle, refer to the precautions section.
2. Position the front wheels in the straight-ahead position.
3. Remove or disconnect the following:
- Engine under cover
- Steering wheel pad
- Steering wheel
- Left and right outer tie-rod ends from the steering knuckles

4. Matchmark the No. 2 intermediate shaft to the steering gear input shaft.
5. Remove or disconnect the following:
- Clamp plate
- Pressure feed and return tubes from the power steering gear, using tool 09631-22020
- Power steering gear assembly

To install:
6. Install or connect the following:
- Power steering gear assembly. Torque the set bolt to 123 ft. lbs. (165 Nm) and the set nut/bolt to 96 ft. lbs. (91 Nm).
- Pressure feed and return tubes to the power steering gear. Torque them to 27 ft. lbs. (32 Nm), using tool 09631-22020.
- Clamp plate. Torque the bolt to 21 ft. lbs. (29 Nm).
- No. 2 intermediate shaft to the steering gear input shaft
- Left and right outer tie-rod ends to the steering knuckles. Torque the nuts to 67 ft. lbs. (91 Nm).
- Steering wheel. Torque the nut to 26 ft. lbs. (35 Nm).
- Steering wheel pad
- Engine under cover

7. Fill and bleed the power steering system.
8. Check and/or adjust the wheel alignment, as necessary.

Land Cruiser

2002–03

1. Before servicing the vehicle, refer to the precautions section.
2. Matchmark the intermediate shaft to the steering gear input shaft.
3. Remove or disconnect the following:
- Negative battery cable

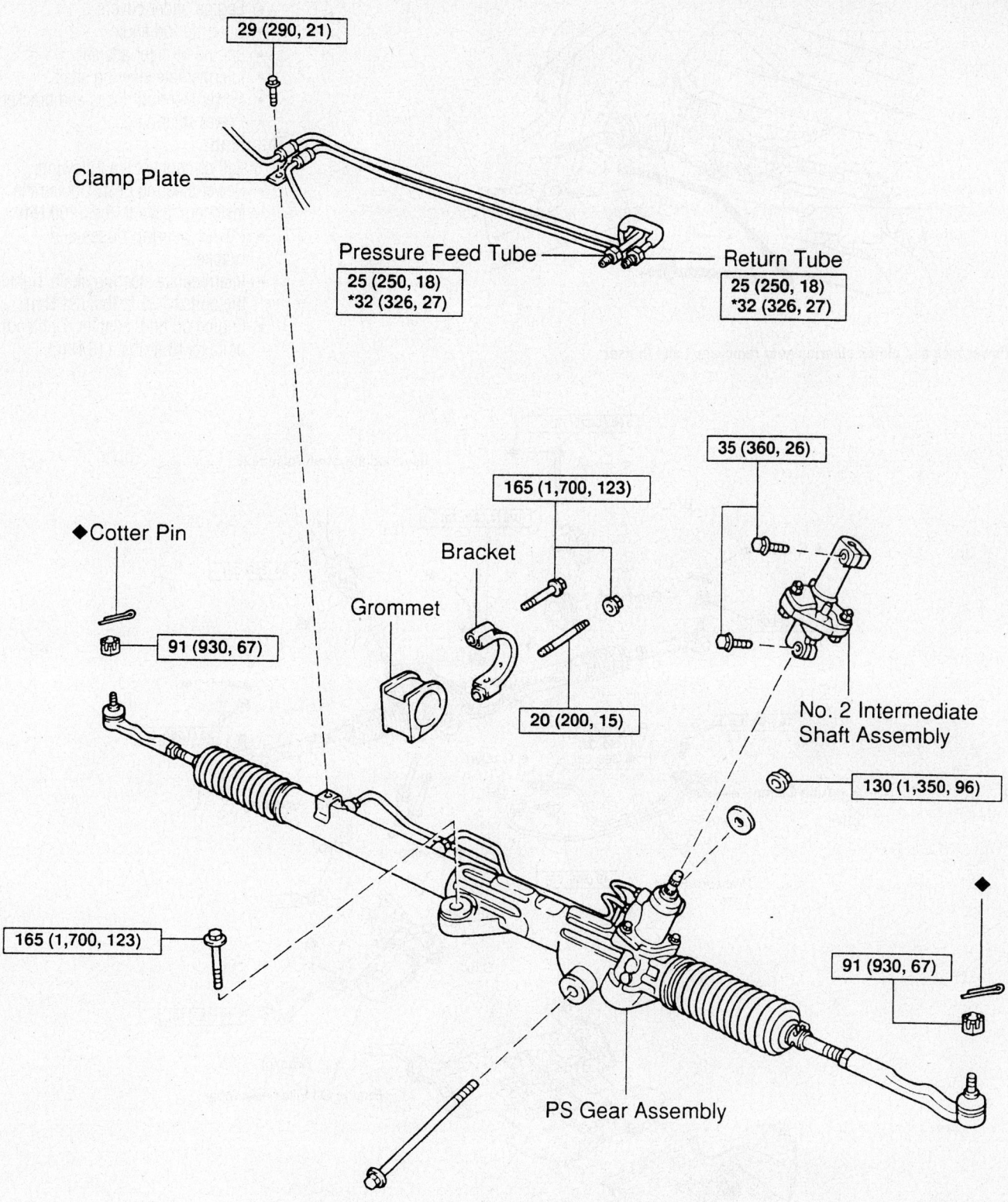

29 (290, 21)

Clamp Plate

Pressure Feed Tube — — Return Tube

25 (250, 18)
*32 (326, 27)

25 (250, 18)
*32 (326, 27)

35 (360, 26)

165 (1,700, 123)

Bracket

◆Cotter Pin

Grommet

91 (930, 67)

20 (200, 15)

No. 2 Intermediate
Shaft Assembly

130 (1,350, 96)

165 (1,700, 123)

◆

91 (930, 67)

PS Gear Assembly

N·m (kgf·cm, ft·lbf) : Specified torque
◆Non−reusable part
* For use with SST

9308YG18

Exploded view of the power rack and pinion steering gear mounting—Sequoia

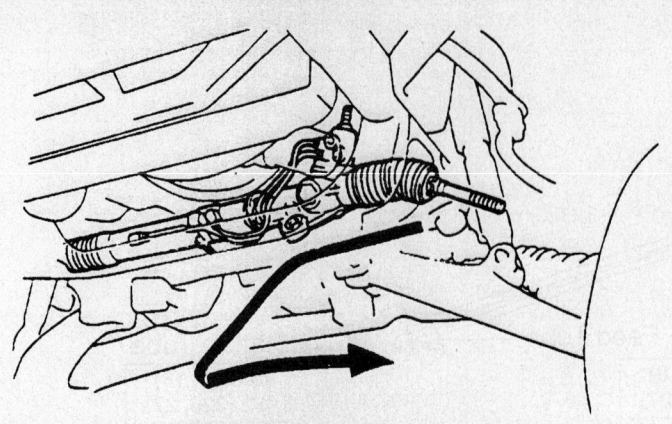

- Engine under covers
- Outer tie rod ends
- Engine oil filter adapter
- Intermediate steering shaft
- Power steering hoses and bracket
- Power steering gear

To install:
4. Install or connect the following:
- Power steering gear. Tighten the fasteners to 74 ft. lbs. (100 Nm).
- Power steering hoses and bracket
- Intermediate steering shaft. Tighten the bolts to 25 ft. lbs. (34 Nm).
- Engine oil filter adapter. Tighten the bolts to 13 ft. lbs. (18 Nm).

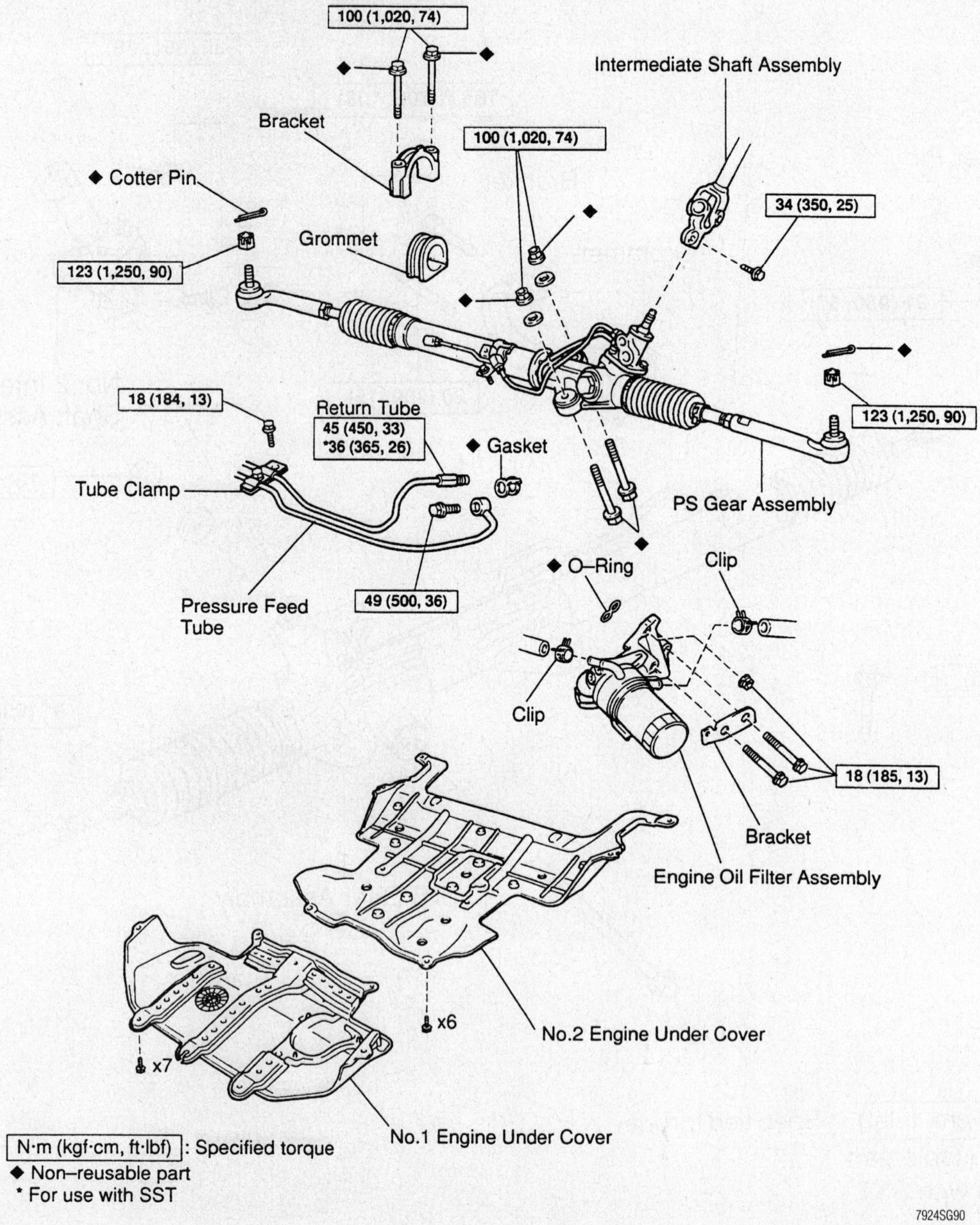

7924SG89

Power rack and pinion steering gear removal—Land Cruiser

100 (1,020, 74)

Intermediate Shaft Assembly

Bracket

100 (1,020, 74)

◆ Cotter Pin

34 (350, 25)

123 (1,250, 90)

Grommet

18 (184, 13)

Return Tube
45 (450, 33)
*36 (365, 26)

◆ Gasket

123 (1,250, 90)

Tube Clamp

PS Gear Assembly

Pressure Feed Tube

49 (500, 36)

◆ O—Ring

Clip

Clip

18 (185, 13)

Bracket

Engine Oil Filter Assembly

x6

No.2 Engine Under Cover

x7

No.1 Engine Under Cover

N·m (kgf·cm, ft·lbf) : Specified torque
◆ Non–reusable part
* For use with SST

7924SG90

Exploded view of the rack and pinion steering gear mounting—2002–03 Land Cruiser

- Outer tie rod ends. Tighten the nuts to 90 ft. lbs. (122 Nm).
- Engine under covers
- Negative battery cable

5. Fill the power steering fluid reservoir.

6. Check the wheel alignment and adjust as necessary.

2004–06

1. Place the front wheels in the straight-ahead position.

2. Remove the steering wheel.

3. Remove the engine under-covers.

4. Disconnect the left and right tie rod ends.

5. Remove the oil filter.

6. Remove the oil filter adapter.

7. Turn the steering fully to the right, matchmark and disconnect the intermediate shaft from the gear.

8. Disconnect the pressure and return lines.

9. Remove the 2 mounting bolts and remove the gear, sliding it to the right and pulling it from the left side.

10. Installation is the reverse of removal. Observe the following torques:
- Steering gear mounting bolts: 89 ft. lbs. (120 Nm)
- Return line: 37 ft. lbs. (50 Nm). Use a torque wrench with a fulcrum length of 300mm.
- Pressure line: 31 ft. lbs. (42 Nm)

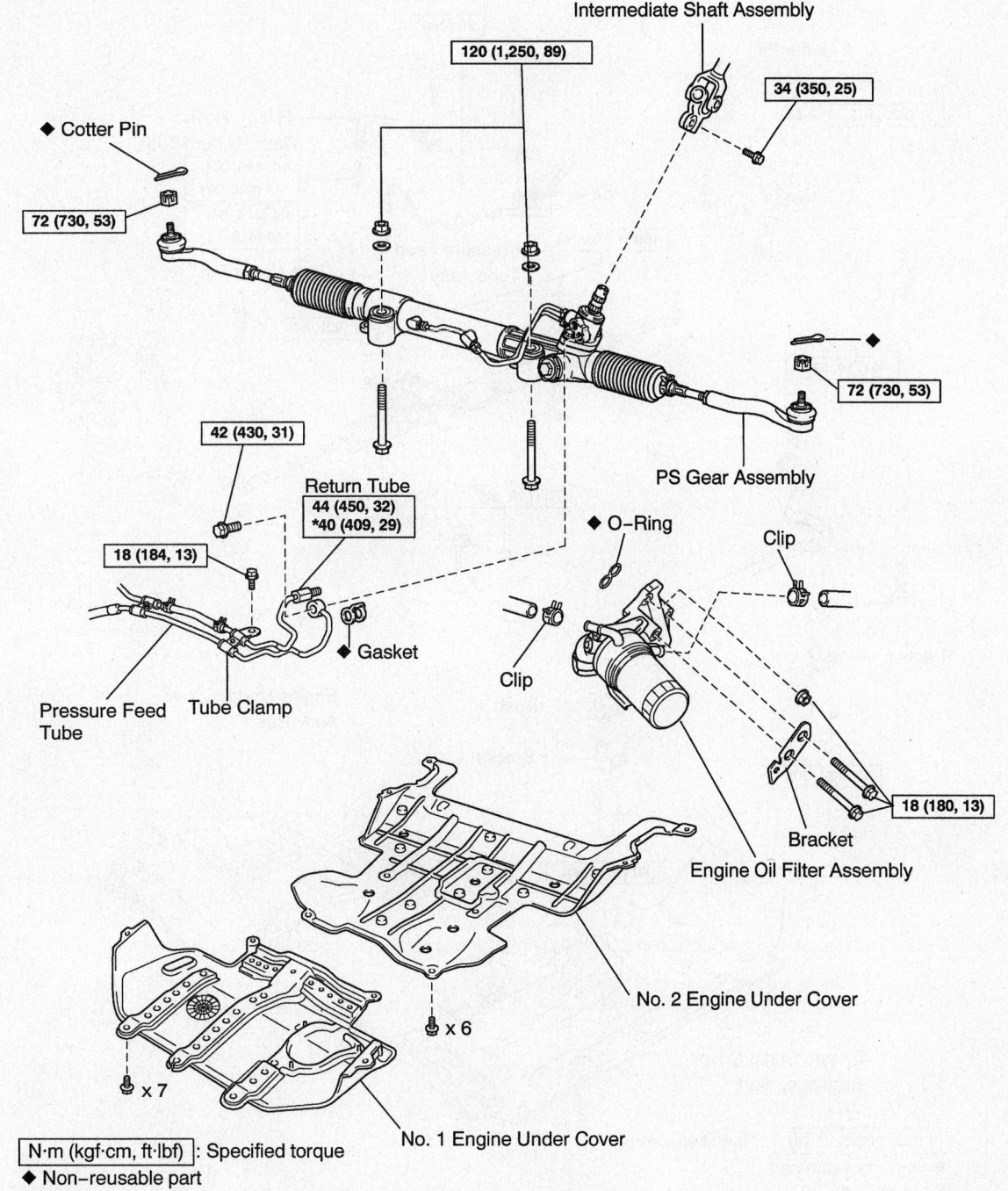

Steering gear and related components—2004–06 Land Cruiser

67170-LCSQ-G18

- Oil filter adapter: 13 ft. lbs. (18 Nm)
- Tie rod ends: 41 ft. lbs. (55 Nm)

GX470

1. Before servicing the vehicle, refer to the precautions section.
2. Disconnect the battery ground cable.
3. Place the front wheels in the straight ahead position.
4. Remove the horn pad.
5. Remove the steering wheel.
6. Remove the lower steering column cover.
7. Remove the turn signal switch.
8. Remove the spiral cable assembly.
9. Remove the front wheels.
10. Remove the engine under-covers.
11. Remove the stabilizer bar.
12. Remove the tie rod ends from the knuckle.
13. Remove the steering intermediate shaft.
14. Disconnect the pressure and return lines.
15. Remove the 2 bolts and remove the steering gear assembly.

To install:

16. Position the gear and install the 2 bolts. Torque to 74 ft. lbs. (100 Nm).

➡**The nuts have detents. Never turn the nuts, just the bolts.**

17. Install the stabilizer bar. Torque the end links to 52 ft. lbs. (70 Nm); the clamp bolts to 30 ft. lbs. (40 Nm).
18. Connect the return line. Use a torque

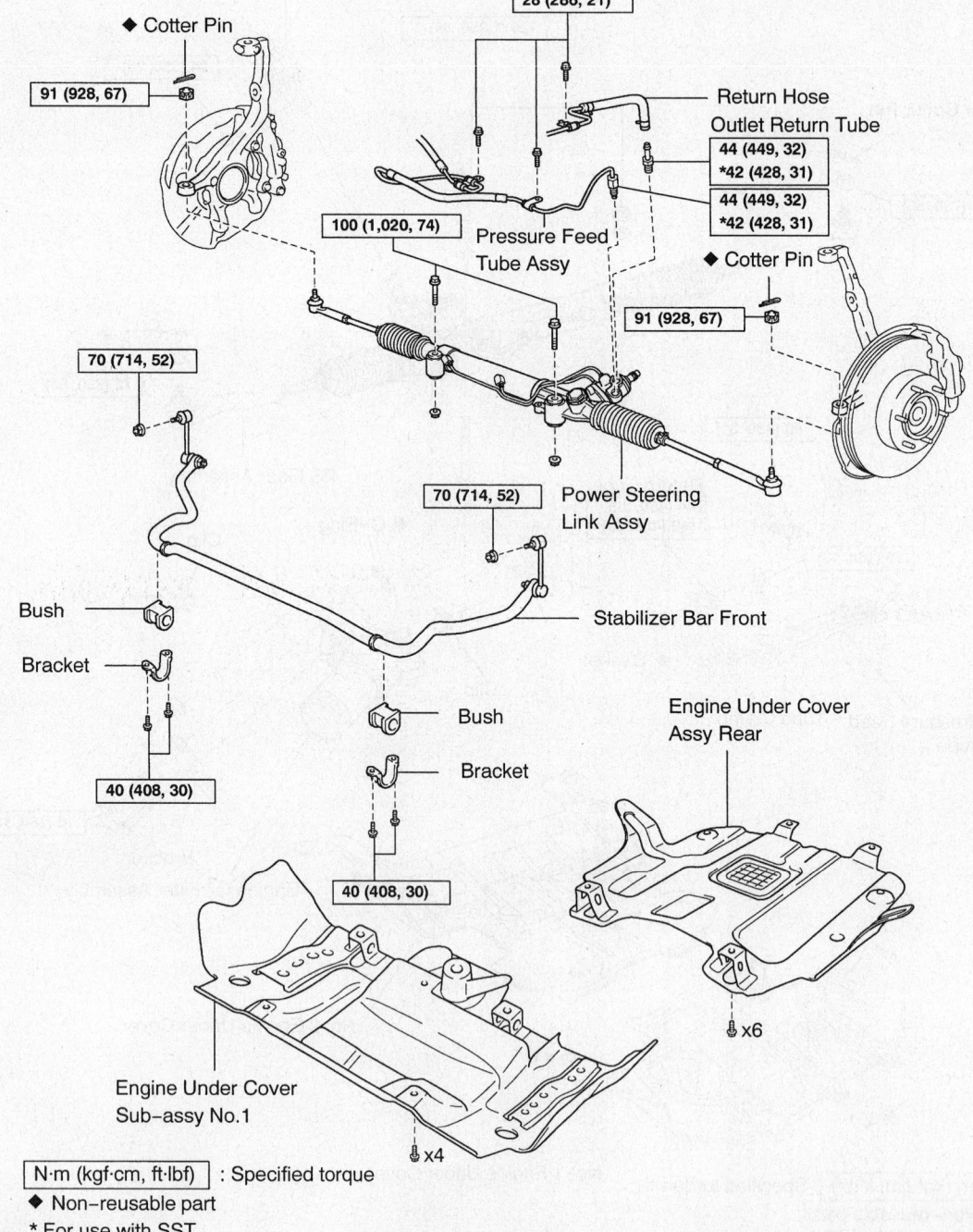

◆ Cotter Pin
91 (928, 67)
28 (286, 21)
Return Hose
Outlet Return Tube
44 (449, 32)
*42 (428, 31)
44 (449, 32)
*42 (428, 31)
100 (1,020, 74)
Pressure Feed Tube Assy
◆ Cotter Pin
91 (928, 67)
70 (714, 52)
70 (714, 52)
Power Steering Link Assy
Bush
Bracket
Stabilizer Bar Front
Bush
Bracket
Engine Under Cover Assy Rear
40 (408, 30)
40 (408, 30)
☐×6
Engine Under Cover Sub-assy No.1
☐×4

N·m (kgf·cm, ft·lbf) : Specified torque
◆ Non-reusable part
* For use with SST

67162-GX470-G14

Steering gear and related parts—GX470

wrench with SST 09023-12700, or equivalent. The torque wrench should have a fulcrum length of 300mm. Torque to 31 ft. lbs. (42 Nm).

19. Connect the pressure line at the subframe. Torque to 21 ft. lbs. (28 Nm).

20. Connect the pressure line to the gear. Use a torque wrench with SST 09023-12700, or equivalent. The torque wrench should have a fulcrum length of 300mm. Torque to 31 ft. lbs. (42 Nm).

21. Connect the intermediate shaft. Torque to 26 ft. lbs. (36 Nm).

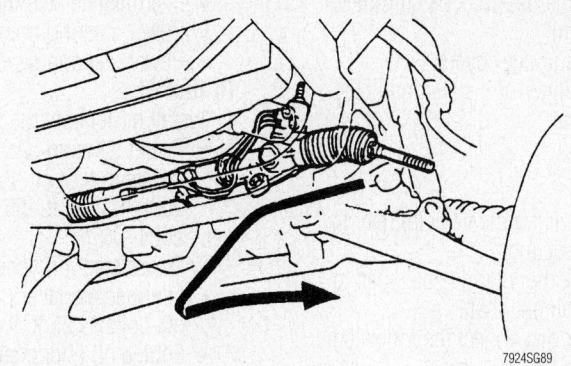

Power rack and pinion steering gear removal—LX470

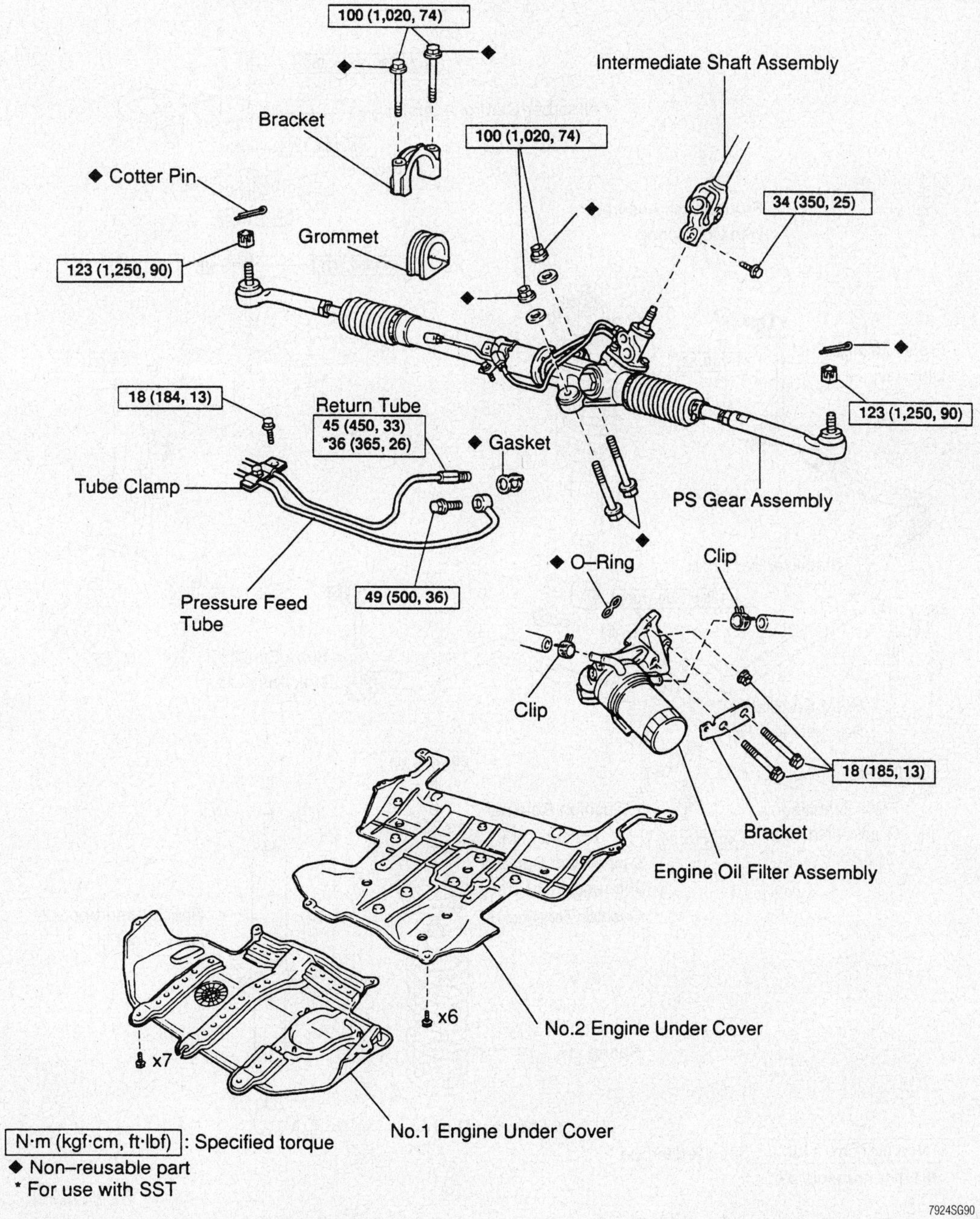

100 (1,020, 74)

Intermediate Shaft Assembly

Bracket

100 (1,020, 74)

34 (350, 25)

Cotter Pin

Grommet

123 (1,250, 90)

123 (1,250, 90)

18 (184, 13)

Return Tube
45 (450, 33)
*36 (365, 26)

Gasket

Tube Clamp

PS Gear Assembly

Pressure Feed Tube

49 (500, 36)

O-Ring

Clip

Clip

18 (185, 13)

Bracket

Engine Oil Filter Assembly

x6

No.2 Engine Under Cover

x7

No.1 Engine Under Cover

N·m (kgf·cm, ft·lbf) : Specified torque
◆ Non-reusable part
* For use with SST

Exploded view of the rack and pinion steering gear mounting—LX470

22. Connect the tie rod ends. Torque to 67 ft. lbs. (91 Nm).

23. Install the under-covers.

24. The remainder of installation is the reverse of removal.

LX470

1. Before servicing the vehicle, refer to the precautions section.

2. Matchmark the intermediate shaft to the steering gear input shaft.

3. Remove or disconnect the following:
- Negative battery cable
- Engine under covers
- Outer tie rod ends
- Engine oil filter adapter

- Intermediate steering shaft
- Power steering hoses and bracket
- Power steering gear

To install:

4. Install or connect the following:
- Power steering gear. Tighten the fasteners to 74 ft. lbs. (100 Nm) for 2002–03; 89 ft. lbs. (120 Nm) for 2004–06.
- Power steering hoses and bracket
- Intermediate steering shaft. Tighten the bolts to 25 ft. lbs. (34 Nm).
- Engine oil filter adapter. Tighten the bolts to 13 ft. lbs. (18 Nm).
- Outer tie rod ends. Tighten the nuts to 90 ft. lbs. (122 Nm) for

2002–03; 53 ft. lbs. (72 Nm) for 2004–06.
- Engine under covers
- Negative battery cable

5. Fill the power steering fluid reservoir.

6. Check the wheel alignment and adjust as necessary.

MacPherson Struts

REMOVAL & INSTALLATION

GX470

1. Before servicing the vehicle, refer to the precautions section.

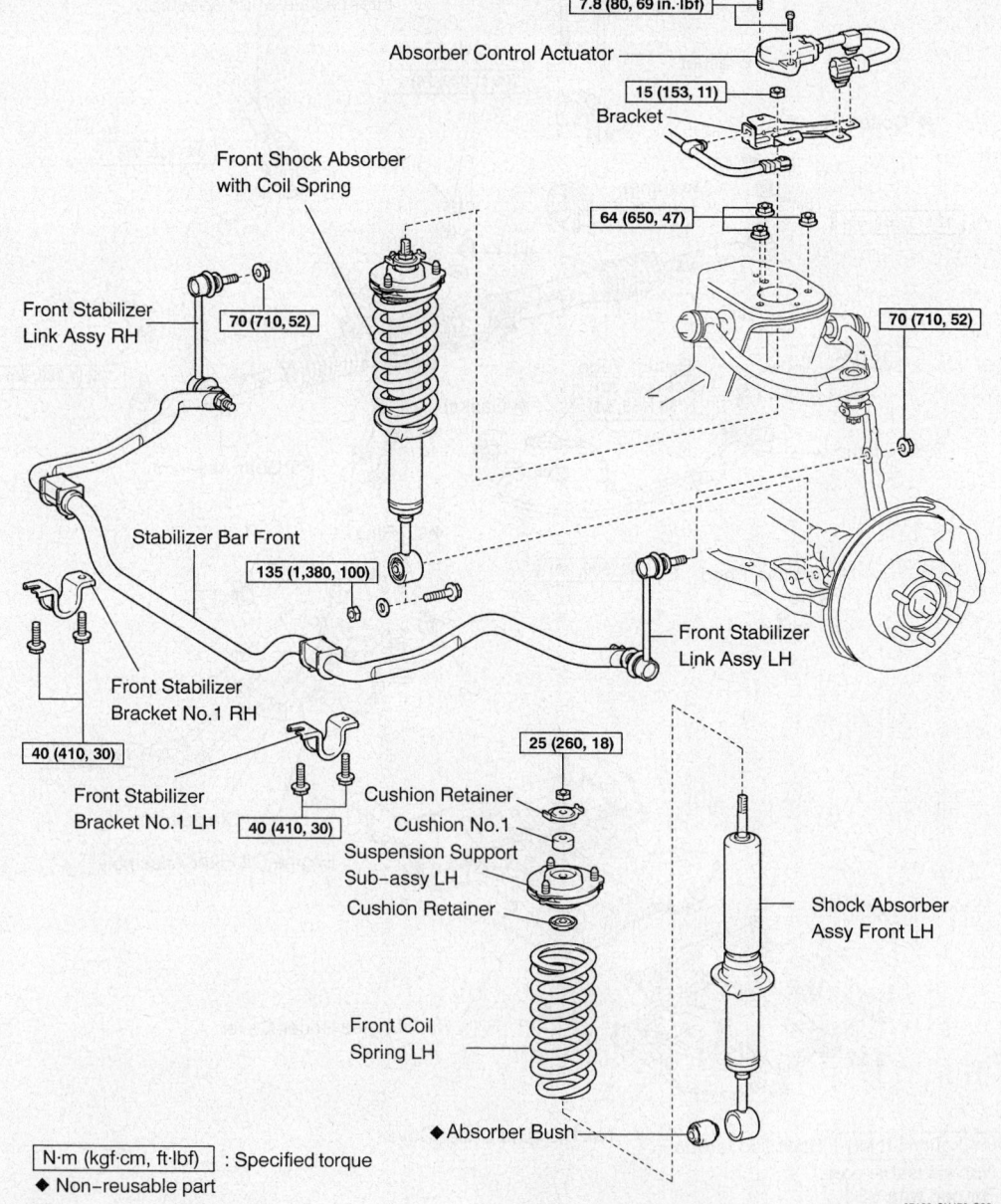

7.8 (80, 69 in.·lbf)
Absorber Control Actuator
15 (153, 11)
Bracket
64 (650, 47)
Front Shock Absorber with Coil Spring
70 (710, 52)
Front Stabilizer Link Assy RH
70 (710, 52)
Stabilizer Bar Front
135 (1,380, 100)
Front Stabilizer Link Assy LH
Front Stabilizer Bracket No.1 RH
40 (410, 30)
25 (260, 18)
Front Stabilizer Bracket No.1 LH
40 (410, 30)
Cushion Retainer
Cushion No.1
Suspension Support Sub-assy LH
Cushion Retainer
Shock Absorber Assy Front LH
Front Coil Spring LH
◆Absorber Bush

N·m (kgf·cm, ft·lbf) : Specified torque
◆ Non-reusable part

Front strut and related components—GX470

2. Remove the wheel.
3. Remove the stabilizer bar.
4. Remove the clamps and connector.
5. Remove the wire bracket.
6. Remove the lower strut bolt.
7. Remove the 3 upper strut nuts.
8. Remove the strut.
9. Installation is the reverse of removal. Do not fully tighten the lower strut bolt until the vehicle is resting on the ground and the suspension has been jounced a few times. Observe the following torques:
- Upper nuts: 47 ft. lbs. (64 Nm)
- Bracket nut: 11 ft. lbs. (15 Nm)
- Stabilizer bar links: 52 ft. lbs. (70 Nm)

- Wheel: 83 ft. lbs. (112 Nm)
- Lower strut bolt: 100 ft. lbs. (135 Nm)

Front Shock Absorber

REMOVAL & INSTALLATION

Land Cruiser

1. Before servicing the vehicle, refer to the precautions section.
2. Support the axle with a jackstand.
3. Remove or disconnect the following:
- Front wheel
- Shock absorber

To install:
4. Install or connect the following:
- Shock absorber. Tighten the nut to 50–57 ft. lbs. (68–78 Nm) and the bolt to 100 ft. lbs. (135 Nm).
- Front wheel

Rear Shock Absorber

REMOVAL & INSTALLATION

Land Cruiser

1. Before servicing the vehicle, refer to the precautions section.
2. Support the axle with a jackstand.

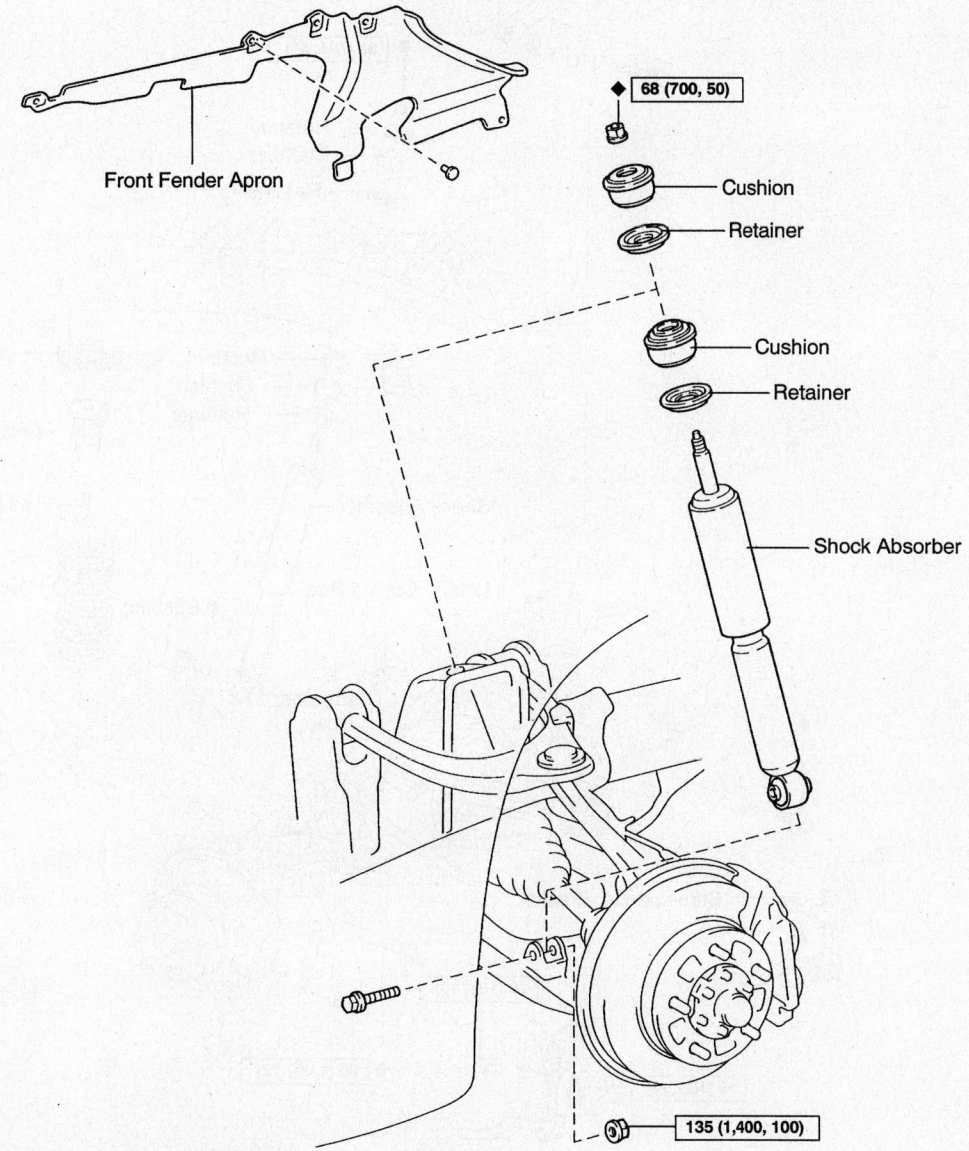

Front Fender Apron

68 (700, 50)

Cushion

Retainer

Cushion

Retainer

Shock Absorber

135 (1,400, 100)

N·m (kgf·cm, ft·lbf) : Specified torque
◆ Non–reusable part

67170-LCSQ-G19

Front shock absorber—Land Cruiser

3. Remove or disconnect the following:
- Rear wheel
- Shock absorber

To install:

4. Install or connect the following:
- Shock absorber. Tighten the nut to 51 ft. lbs. (69 Nm) and the bolt to 72 ft. lbs. (98 Nm).
- Rear wheel

Sequoia

2002–03

1. Before servicing the vehicle, refer to the precautions section.

2. Remove or disconnect the following:
- Rear wheel
- Shock absorber

To install:

3. Install or connect the following:
- Shock absorber. Tighten the upper nut to 15 ft. lbs. (20 Nm) and the lower nut/bolt to 64 ft. lbs. (87 Nm).
- Rear wheel.

2004–06

1. Remove the rear wheels.
2. Support the axle with a jack.
3. Without Auto Leveler:

  a. Disconnect the lower end of the shock from the axle.

  b. Remove the upper nut, retainer and bushings.

4. With Auto Leveler:

✳✳ CAUTION

Perform this procedure with the shock absorber stretched completely. The primary reaction force of the shock absorber is approximately 1,000 N.

  a. Remove the bolt and disconnect the shock absorber from the axle.

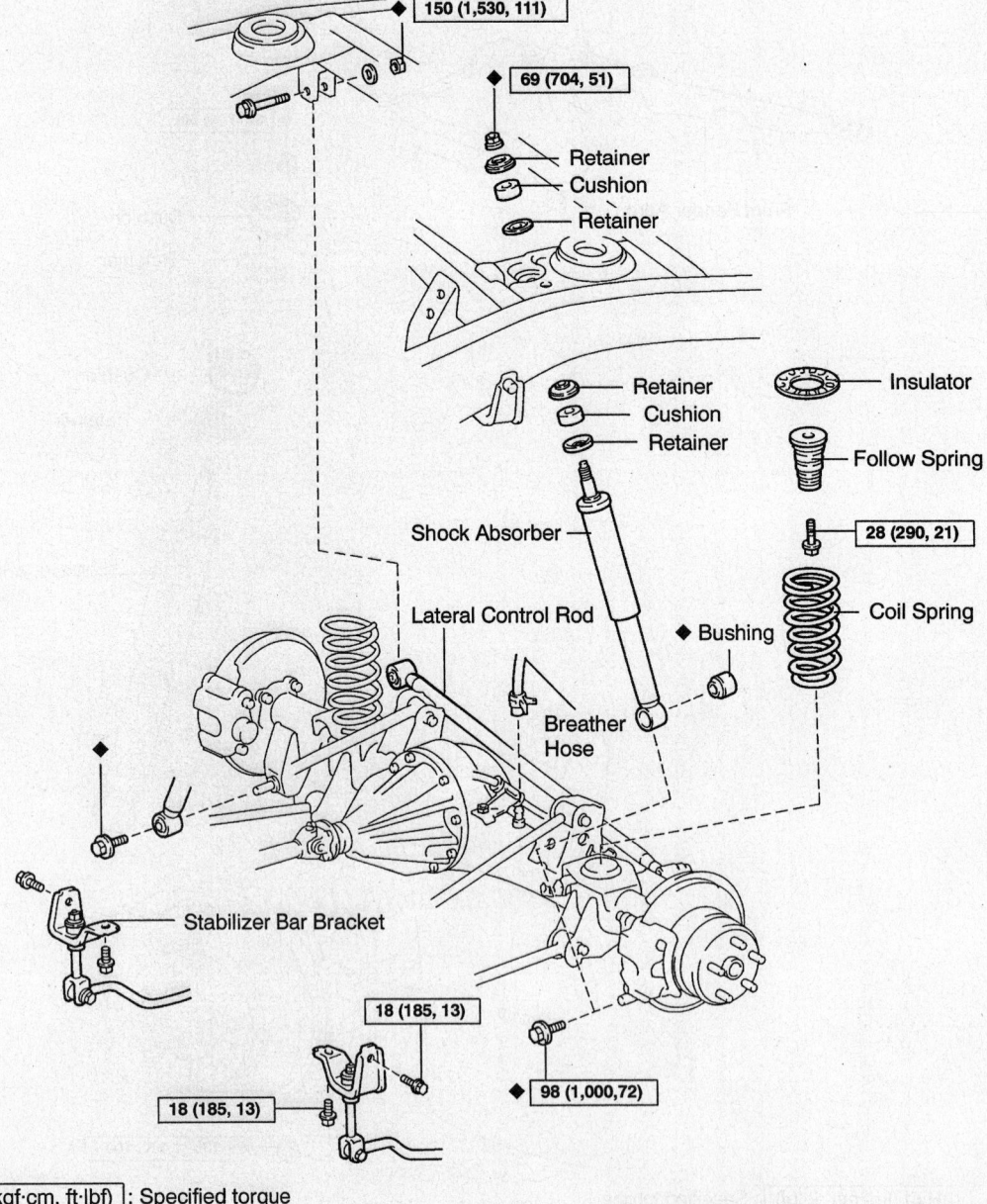

150 (1,530, 111)

69 (704, 51)

— Retainer
— Cushion
— Retainer

— Retainer — Insulator
— Cushion
— Retainer — Follow Spring

Shock Absorber

28 (290, 21)

Lateral Control Rod

◆ Bushing — Coil Spring

Breather Hose

Stabilizer Bar Bracket

18 (185, 13)

18 (185, 13)

◆ 98 (1,000, 72)

N·m (kgf·cm, ft·lbf) : Specified torque
◆ Non-reusable part

67170-LCSQ-G20

Rear suspension—Land Cruiser

b. Remove the upper nut, retainer and bushings.

5. Installation is the reverse of removal. Torque the lower end bolt to 64 ft. lbs. (87 Nm); the upper end nut to 43 ft. lbs. (58 Nm).

GX470

REAR

1. Before servicing the vehicle, refer to the precautions section.

2. Support the axle with a jackstand.

3. Disconnect the actuator at the shock absorber.

➡**Don't over-extend the pneumatic shock.**

4. Remove the lower shock bolt.

5. Remove the upper nut and remove the shock.

6. Installation is the reverse of removal. Don't fully tighten the lower bolt until the vehicle is on the ground and the suspension jounced a few times. Torque the upper nut to 18 ft. lbs. (25 Nm); the lower bolt to 72 ft. lbs. (98 Nm).

LX470 Without Active Height Control

FRONT

1. Before servicing the vehicle, refer to the precautions section.

2. Support the axle with a jackstand.

3. Remove or disconnect the following:
   • Front wheel
   • Shock absorber

To install:

4. Install or connect the following:
   • Shock absorber. Tighten the nut to 51 ft. lbs. (69 Nm) for 2002–03; 57

Normal Type:

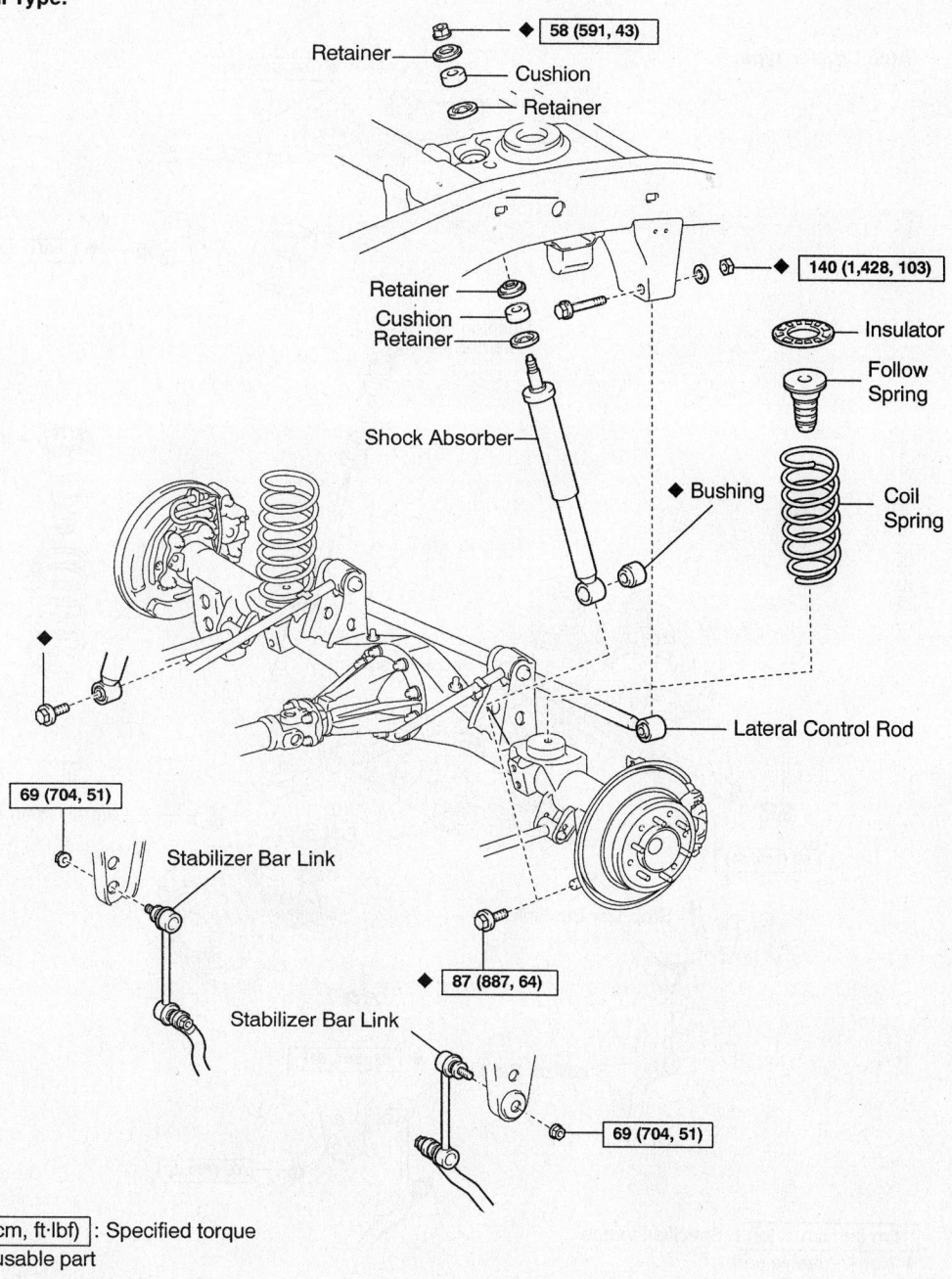

69 (704, 51)

Stabilizer Bar Link

58 (591, 43)
Retainer
Cushion
Retainer

140 (1,428, 103)
Retainer
Cushion
Retainer
Insulator
Follow Spring
Shock Absorber
Bushing
Coil Spring
Lateral Control Rod

87 (887, 64)

Stabilizer Bar Link

69 (704, 51)

N·m (kgf·cm, ft·lbf) : Specified torque
◆ Non–reusable part

2004–06 Sequoia rear suspension, without Auto Leveler

67170-LCSQ-G21

ft. lbs. (78 Nm) for 2004–06; and the bolt to 100 ft. lbs. (135 Nm).

- Front wheel

REAR

1. Before servicing the vehicle, refer to the precautions section.
2. Support the axle with a jackstand.
3. Remove or disconnect the following:

- Rear wheel
- Shock absorber

To install:

4. Install or connect the following:

- Shock absorber. Tighten the nut to

51 ft. lbs. (69 Nm) and the bolt to 72 ft. lbs. (98 Nm).

- Rear wheel

LX470 With Active Height Control

FRONT

> **⁑ CAUTION**
>
> **The vehicle ride height may change suddenly when relieving system pressure.**

1. Before servicing the vehicle, refer to the precautions section.

2. Relieve the Active Height Control (AHC) hydraulic pressure as follows:

   a. Connect a hose to the control actuator bleed screw and place the other end in a container.

   b. Open the bleed screw.

   c. When the fluid pressure has dropped and oil stops flowing, close the bleed screw.

3. Remove or disconnect the following:

- Front wheel
- Inner fender liner
- Lower shock absorber mounting bolt
- AHC pressure hose

Auto Leveler Type:

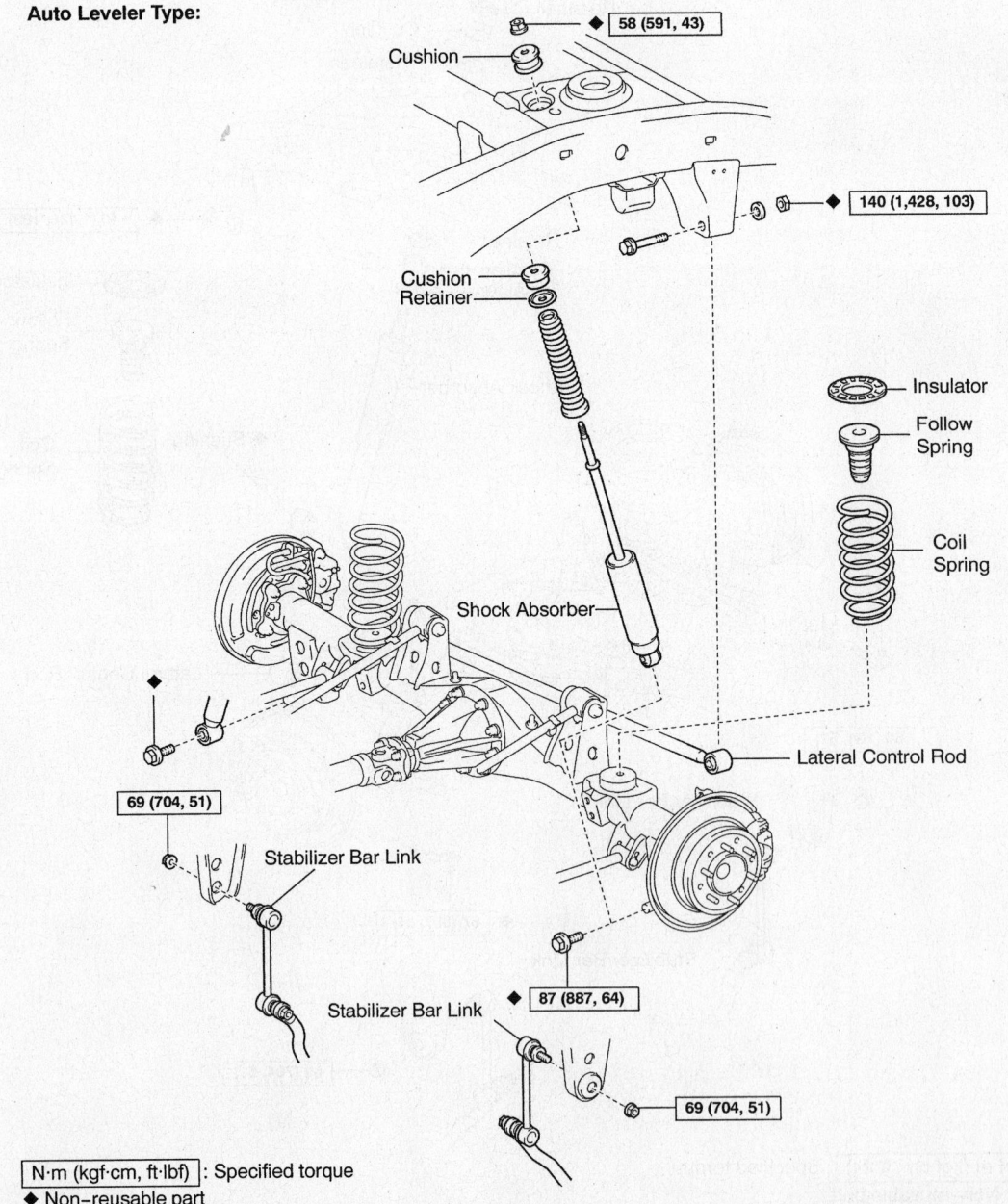

N·m (kgf·cm, ft·lbf) : Specified torque
◆ Non-reusable part

2004–06 Sequoia rear suspension, with Auto Leveler

67170-LCSQ-G22

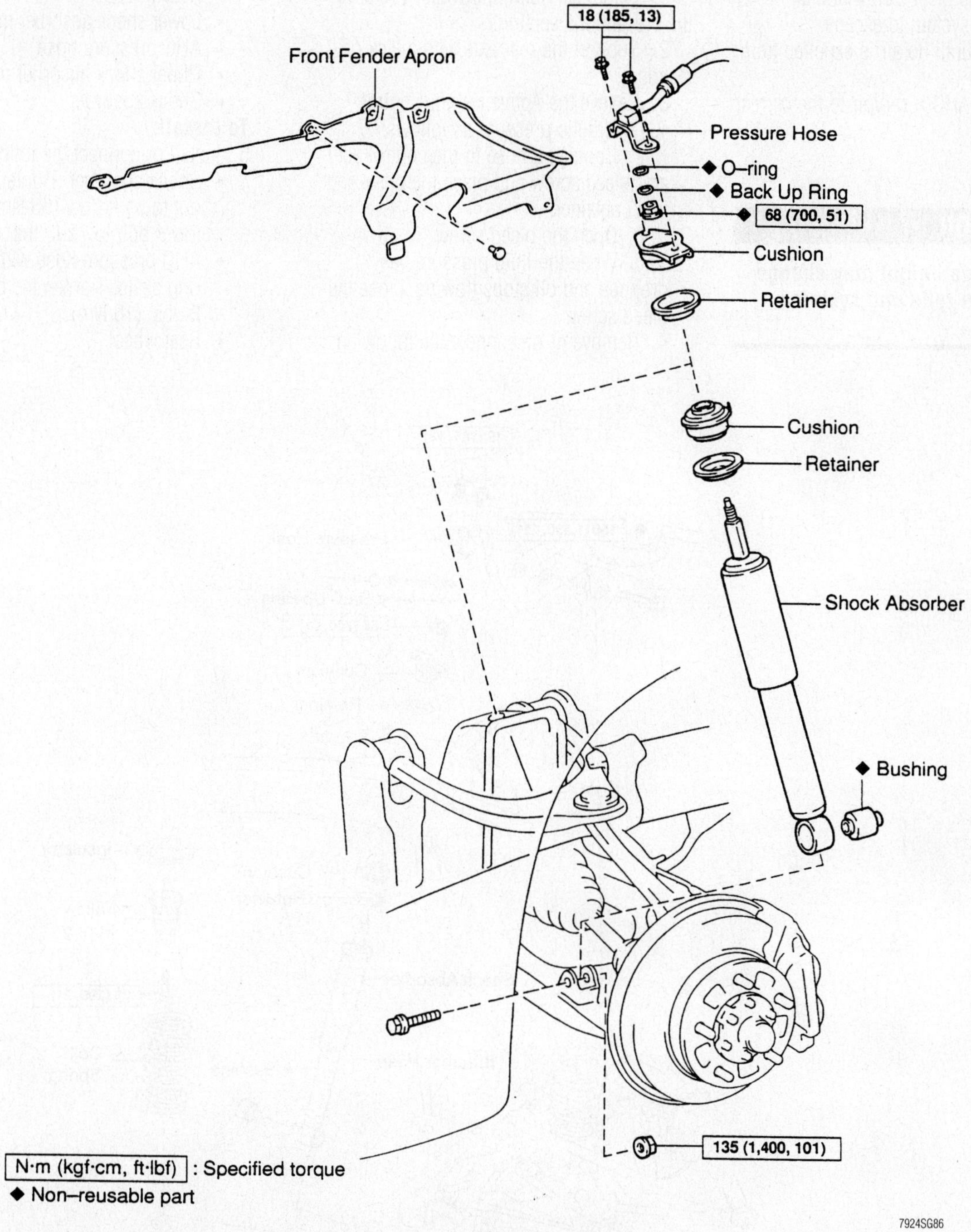

18 (185, 13)

Front Fender Apron

Pressure Hose

◆ O–ring
◆ Back Up Ring
◆ 68 (700, 51)
Cushion
Retainer

Cushion
Retainer

Shock Absorber

◆ Bushing

N·m (kgf·cm, ft·lbf) : Specified torque
◆ Non–reusable part

135 (1,400, 101)

7924SG86

Exploded view of the front shock absorber mounting—LX470 models with Active Height Control (AHC)

- Upper shock absorber mounting nut
- Shock absorber

To install:

4. Install or connect the following:
- Shock absorber. Tighten the upper nut to 51 ft. lbs. (68 Nm) for 2002–03; 57 ft. lbs. (78 Nm) for 2004–06; and the lower bolt to 101 ft. lbs. (135 Nm).
- AHC pressure hose with new O-ring seals. Tighten the bolts to 13 ft. lbs. (18 Nm).

- Inner fender liner
- Front wheel

➡**Do not let the AHC reservoir run empty during this procedure.**

5. Bleed the AHC system as follows:
a. Fill the AHC system reservoir with AHC fluid 08886-01805.
b. Start the engine and push **N** on the vehicle height select switch.
c. When the AHC pump stops, turn the engine **OFF**.

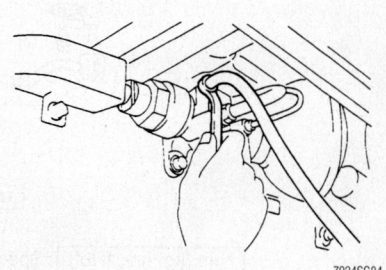

7924SG84

Relieving system pressure—LX470 with Active Height Control (AHC)

d. Open the bleed screw and allow any air in the system to escape.

e. Repeat until no air is expelled from the bleed screw.

f. Fill the AHC reservoir to the correct level.

REAR

> **⁂ CAUTION**
>
> **The vehicle ride height may change suddenly when relieving system pressure.**

1. Before servicing the vehicle, refer to the precautions section.

2. Support the rear axle with a jack or stands.

3. Relieve the Active Height Control (AHC) hydraulic pressure as follows:

a. Connect a hose to the control actuator bleed screw and place the other end in a container.

b. Open the bleed screw.

c. When the fluid pressure has dropped and oil stops flowing, close the bleed screw.

4. Remove or disconnect the following:

- Rear wheel
- Lower shock absorber mounting bolt
- AHC pressure hose
- Upper shock absorber mounting nut
- Shock absorber

To install:

5. Install or connect the following:

- Shock absorber. Tighten the upper nut to 51 ft. lbs. (68 Nm) and the lower bolt to 72 ft. lbs. (98 Nm).
- AHC pressure hose with new O-ring seals. Tighten the bolts to 13 ft. lbs. (18 Nm).
- Rear wheel

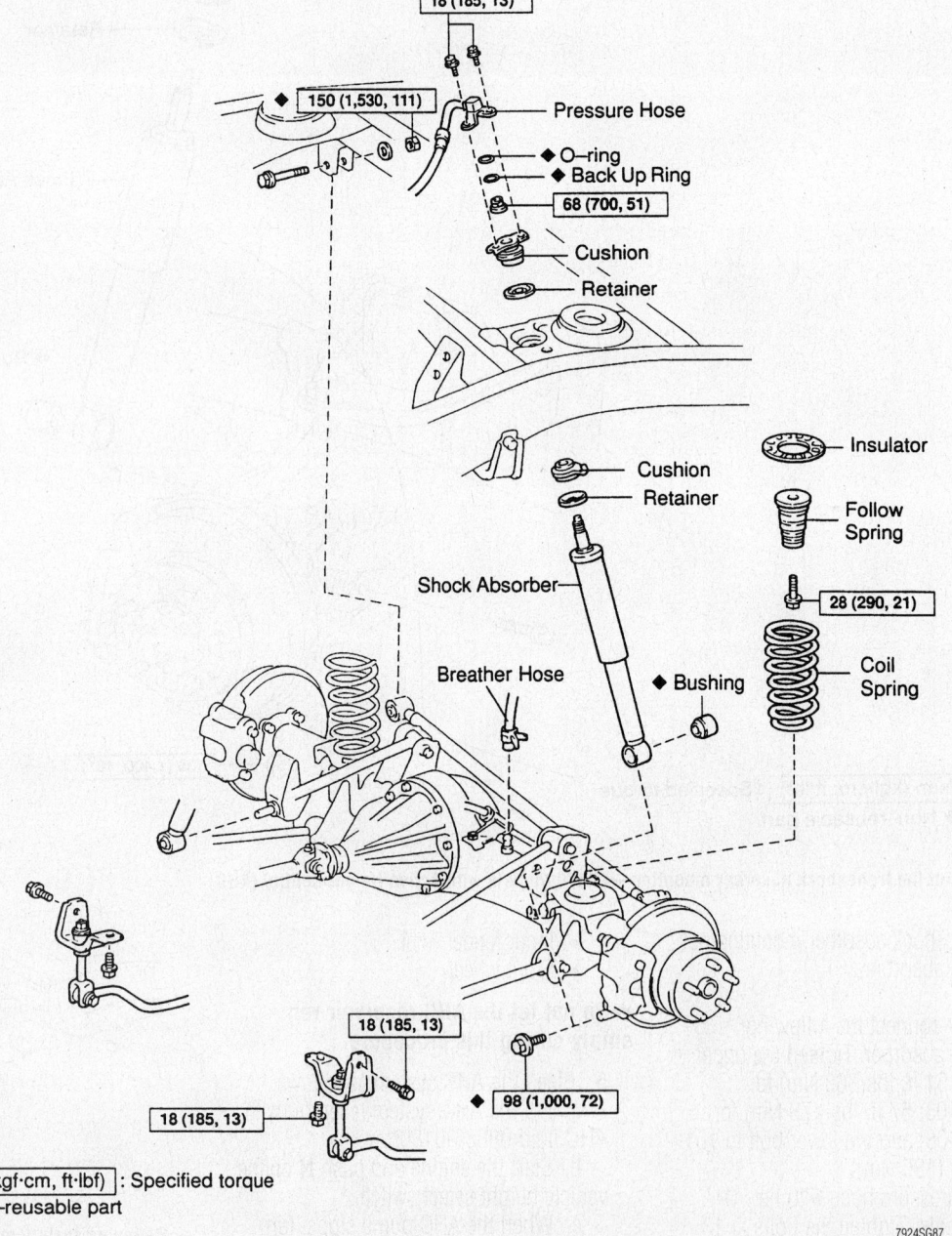

18 (185, 13)

150 (1,530, 111)

Pressure Hose

◆ O-ring

◆ Back Up Ring

68 (700, 51)

Cushion

Retainer

Insulator

Cushion

Retainer

Follow Spring

Shock Absorber

28 (290, 21)

Breather Hose

◆ Bushing

Coil Spring

18 (185, 13)

18 (185, 13)

◆ 98 (1,000, 72)

N·m (kgf·cm, ft·lbf) : Specified torque
◆ Non-reusable part

7924SG87

Exploded view of the rear shock absorber mounting—LX470 with Active Height Control (AHC)

➡**Do not let the AHC reservoir run empty during this procedure.**

6. Bleed the AHC system as follows:

a. Fill the AHC system reservoir with AHC fluid 08886-01805.

b. Start the engine and push **N** on the vehicle height select switch.

c. When the AHC pump stops, turn the engine **OFF**.

d. Open the bleed screw and allow any air in the system to escape.

e. Repeat until no air is expelled from the bleed screw.

f. Fill the AHC reservoir to the correct level.

Sequoia Front Strut

REMOVAL & INSTALLATION

1. Before servicing the vehicle, refer to the precautions section.

2. Remove or disconnect the following:

- Front wheel
- Strut-to-lower control arm nut/bolt and the strut
- Strut-to-chassis 3 nuts/bolts and the strut

To install:

3. Install or connect the following:

- Strut to the chassis. Torque the 3 nuts/bolts to 47 ft. lbs. (64 Nm).

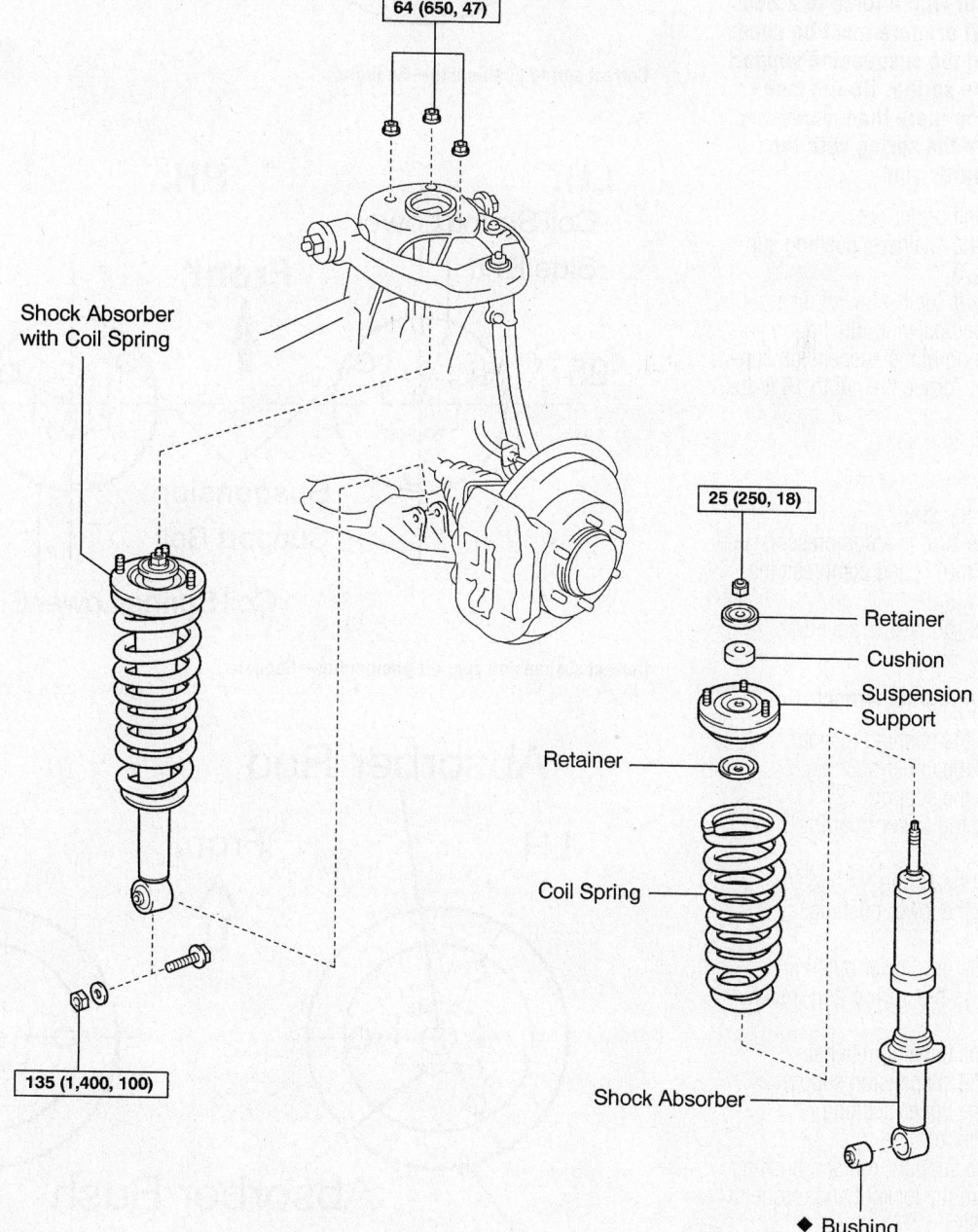

64 (650, 47)

Shock Absorber with Coil Spring

135 (1,400, 100)

25 (250, 18)

Retainer

Cushion

Suspension Support

Retainer

Coil Spring

Shock Absorber

◆ Bushing

N·m (kgf·cm, ft·lbf) : Specified torque

67170-LCSQ-G23

Front strut—Sequoia

- Strut to the lower control arm. Torque the nut/bolt to 100 ft. lbs. (135 Nm).
- Front wheel

Front Coil Spring

REMOVAL & INSTALLATION

1. Remove the strut.
2. Using a compressor, compress the coil spring.

➡A compressor with a force of 2,860 lbs. (12,740 N) or more must be used. Make sure that the suspension support is free from the spring. Do not compress the spring more than necessary. Do not position the spring with the upper end towards you.

3. Remove the center nut.
4. Remove the retainers, bushing support and spring.
5. Assembly is the reverse of disassembly. See the accompanying illustration for correct positioning of the suspension support and spring. Torque the nut to 18 ft. lbs. (25 Nm).

GX470

1. Remove the strut.
2. Place the strut in a compressor, such as SST 09727-30021, and compress the spring.
3. Hold the rod and remove the nut.

➡Don't use an impact wrench.

4. Remove the bushing retainer
5. Remove the upper bushing.
6. Remove the support.
7. Remove the lower bushing retainer.
8. Remove the spring.
9. Remove the lower bushing.
To install:
10. Install the new lower bushing.
11. Compress the spring and install it.
12. Install the bushing retainer.
13. Install the suspension support.
14. Install the upper bushing.
15. Install the retainer.
16. Align the support, rod and bushing as shown. Install the locknut and torque to 18 ft. lbs. (25 Nm).
17. Release the spring from the compressor and check the alignment of the parts.
18. Install the strut.

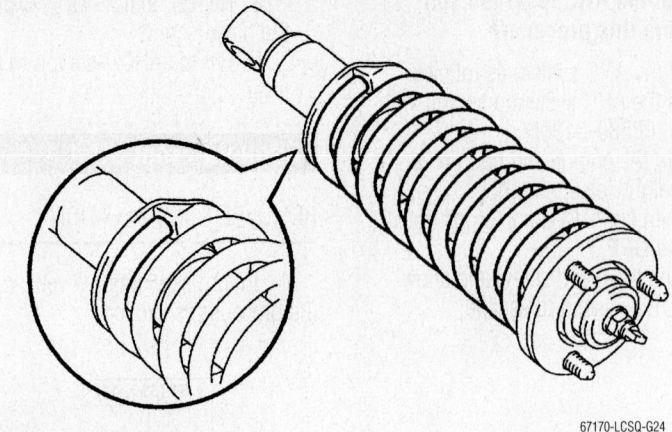

Correct spring positioning—Sequoia

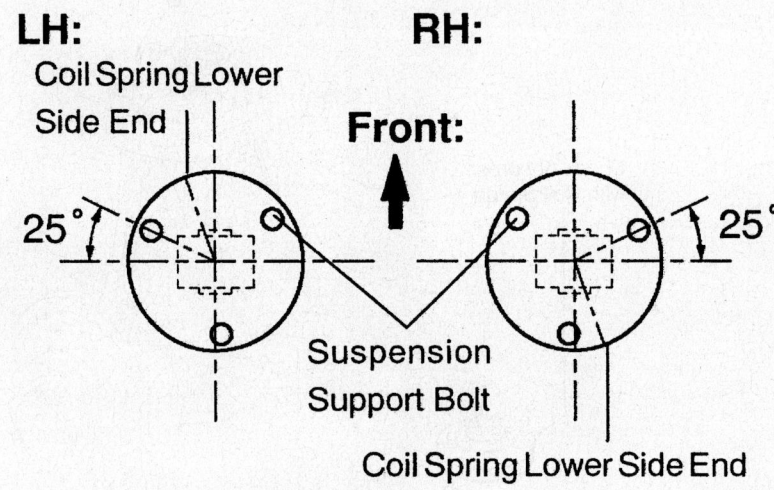

Correct suspension support positioning—Sequoia

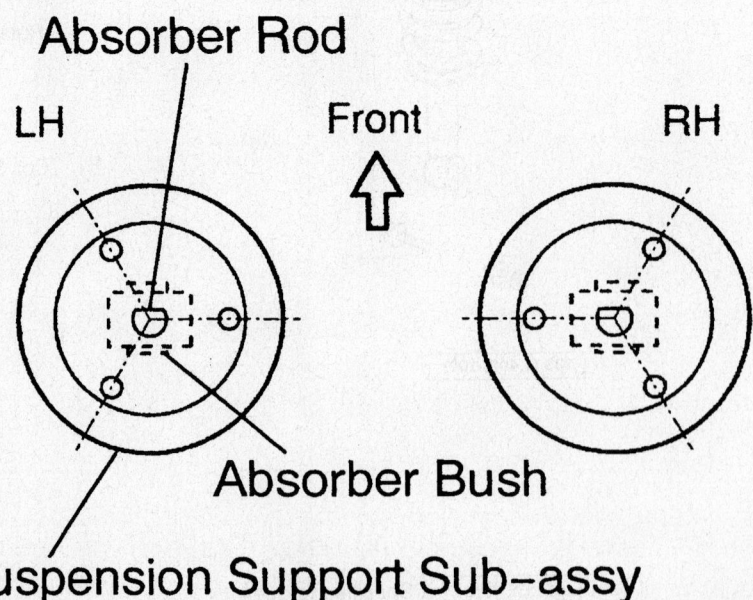

Align the support, rod and bushing as shown

Rear Coil Spring

REMOVAL & INSTALLATION

➡️**The front coil springs on the Sequoia are part of the front strut. The Land Cruiser employs front torsion bars.**

Land Cruiser

1. Before servicing the vehicle, refer to the precautions section.
2. Support the vehicle at the frame.
3. Support the axle with a floor jack.
4. Remove or disconnect the following:

- Rear wheel
- Shock absorber
- Stabilizer bar brackets
- Lateral control rod
- Breather hose
- Coil spring

To install:

5. Install or connect the following:

- Coil spring
- Lateral control rod. Tighten to 111 ft. lbs. (150 Nm).
- Stabilizer bar brackets. Tighten the bolts to 13 ft. lbs. (18 Nm)
- Shock absorber
- Rear wheel

Sequoia

1. Remove the shock absorber.
2. Disconnect the left and right stabilizer bar links.
3. Disconnect the lateral rod.
4. Lower the axle slowly and remove the spring.
5. Installation is the reverse of removal. Make sure that the lower end of the coil spring is correctly installed, against the stop. Torque the later rod to 103 ft. lbs. (140 Nm); the stabilizer bar links to 51 ft. lbs. (69 Nm).

LX470

1. Before servicing the vehicle, refer to the precautions section.
2. Support the vehicle at the frame.
3. Support the axle with a floor jack.
4. Remove or disconnect the following:

- Rear wheel
- Shock absorber
- Stabilizer bar brackets
- Lateral control rod
- Coil spring

To install:

5. Install or connect the following:

- Coil spring
- Lateral control rod. Tighten the axle housing bolt to 181 ft. lbs. (245 Nm) for 2002–03; 110 ft. lbs. (149 Nm) for 2004–06.
- Stabilizer bar brackets. Tighten the bolts to 13 ft. lbs. (18 Nm)
- Shock absorber
- Rear wheel

Rear Air Spring

REMOVAL & INSTALLATION

GX470

1. Before servicing the vehicle, refer to the precautions section.
2. Remove the wheel.
3. Support the frame with jackstands and allow the axle to hang.
4. Disconnect the height control tube.
5. Disconnect the clip on the underside of the air spring. If the clip is difficult to remove, thread a wire through the hole and pull it. Discharge the air from the air spring to retract it.
6. Turn the unit 90 degrees and remove it from the axle.

➡️**Don't manually extend the unit.**

7. Installation is the reverse of removal. Use new O-rings on the height control tube.

Torsion Bars

REMOVAL & INSTALLATION

Land Cruiser and LX470

1. Before servicing the vehicle, refer to the precautions section.
2. Remove or disconnect the following:

- Front wheel
- Engine under cover

3. Measure dimension **A** as shown between the adjustment bolt head and the frame.
4. Loosen the adjusting bolt until all spring tension is relieved.
5. Measure dimension **B** as shown between the adjustment bolt head and the frame.
6. Remove or disconnect the following:

- Adjustment bolt, swivel and seat
- Torsion bar and anchor arm. Separate the anchor arm from the torsion bar.
- Torque arm

To install:

7. Install or connect the following:

- Torque arm. Tighten the fasteners to 166 ft. lbs. (225 Nm).
- Torsion bar and anchor arm. Align the matchmarks.

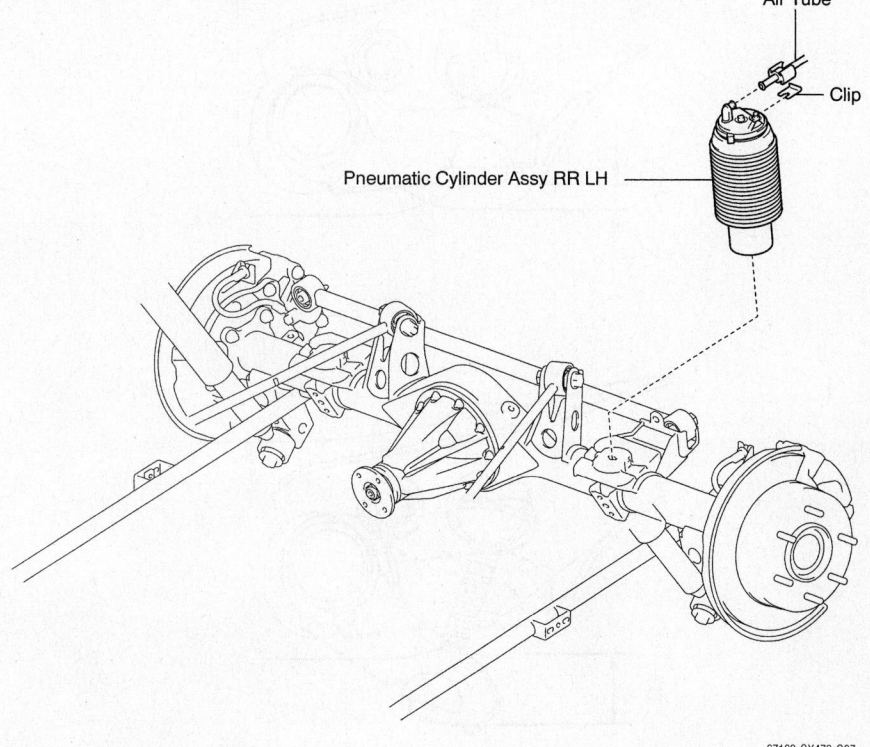

Air Tube

Clip

Pneumatic Cylinder Assy RR LH

67162-GX470-G07

Rear air spring—GX470

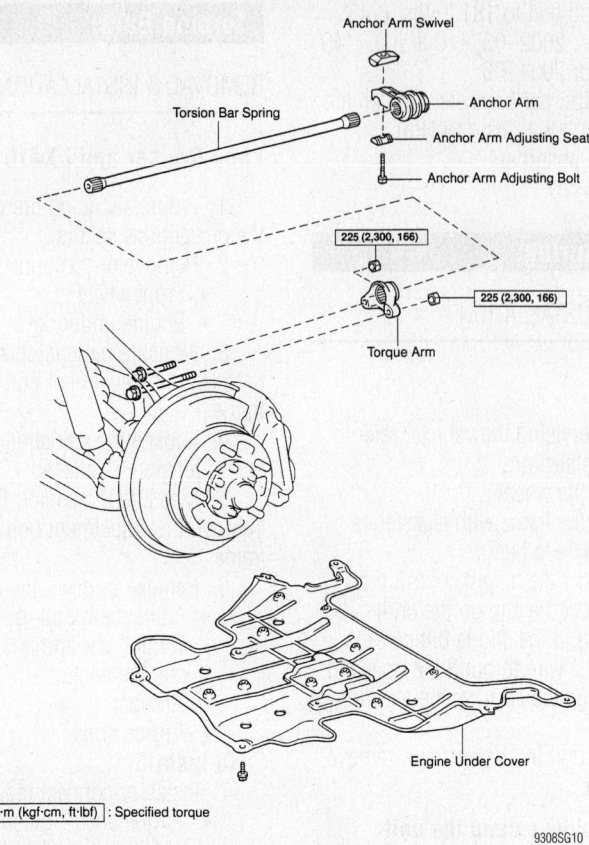

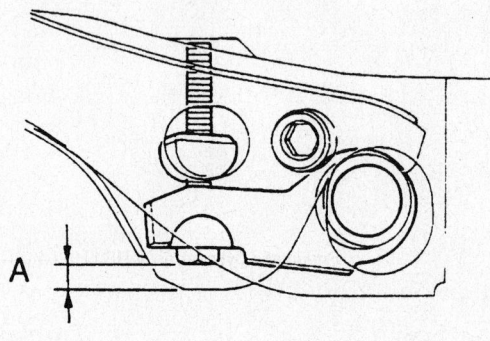

Torsion bar mounting exploded view—Land Cruiser and LX470

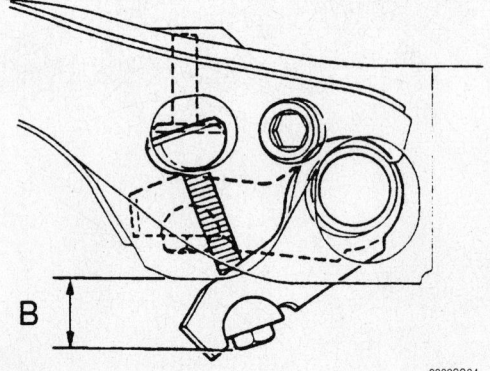

Reference measurements A and B—Land Cruiser and LX470

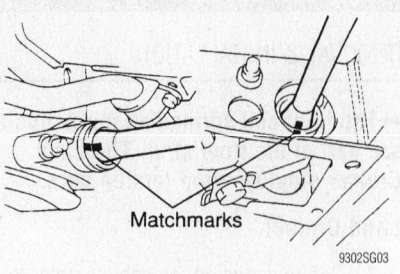

Matchmark the torsion bar to the anchor arm and torque arm—Land Cruiser and LX470

- Adjustment bolt, swivel and seat

8. Check that dimension **B** is close to the measurement made at disassembly.

9. If installing a new torsion bar, tighten the adjustment bolt until dimension **A** is as follows:

- Left torsion bar: 0.315–0.984 inches (8–25mm)
- Right torsion bar: 0.079–0.709 inches (2–18mm)

10. If installing the original torsion bar, tighten the adjustment bolt until dimension **A** is close to the measurement made at disassembly.

11. Install or connect the following:

- Engine under cover
- Front wheel

12. Place the vehicle on a flat, level surface and check the vehicle curb height as follows:

a. Step 1: Measure dimension **A** between the spindle center and the ground.

b. Step 2: Measure dimension **B** between the lower control arm front bolt center and the ground.

c. Step 3: Turn the adjusting bolt so that **A** minus **B** is equal to 2.795 inches (71mm).

Front:

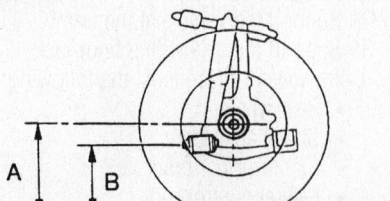

Ride height measurements A and B—Land Cruiser and LX470

Upper Ball Joint

REMOVAL & INSTALLATION

Except Sequoia

The upper ball joint is serviced with the upper control arm as an assembly.

Sequoia

1. Before servicing the vehicle, refer to the precautions section.
2. Remove or disconnect the following:

- Front wheel
- Steering knuckle with the axle hub
- Wire and boot
- Snapring
- Upper ball joint from the steering knuckle, using a deep socket wrench and tool 09050-40011, or equivalent press.

To install:
3. Install or connect the following:
- New upper ball joint to the steering knuckle, using a deep socket and tool 09309-37010, or equivalent press.
- New snapring

4. Using a torque wrench, inspect the upper ball joint rotation, as follows:

a. Flip the ball joint back-and-forth 5 times.

b. Using a torque wrench, continuously turn the nut 1 turn in 2–4 seconds.

c. Take the reading on the 5th turn; it should be 6–39 inch lbs. (0.7–4.4 Nm). If not, replace the upper ball joint.
5. Install or connect the following:
- New boot secured with a wire
- Front wheel.

6. Check and/or adjust the front wheel alignment.

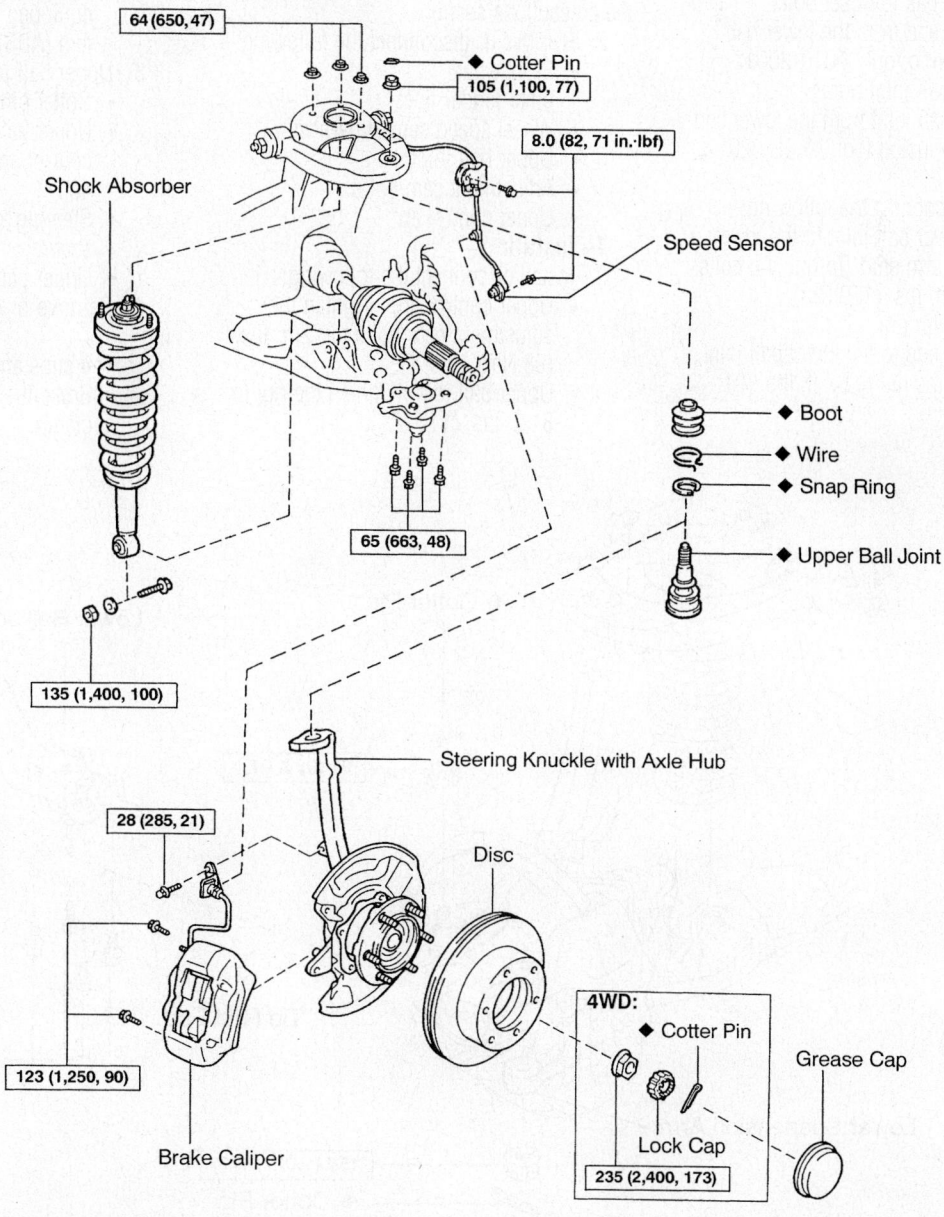

N·m (kgf·cm, ft·lbf) : Specified torque

◆ Non–reusable part

67170-LCSQ-G26

Upper ball joint and related parts—Sequoia

Lower Ball Joint

REMOVAL & INSTALLATION

Except Sequoia

The lower ball joint is serviced with the lower control arm as an assembly.

Sequoia

1. Before servicing the vehicle, refer to the precautions section.
2. Remove or disconnect the following:
   - Front wheel
   - 4 lower ball joint set bolts
   - Tie-rod end from the lower ball joint, using tool 09610-20012
   - Lower ball joint nut.
   - Lower ball joint from the lower control arm, using tool 09628-62011

To install:

3. Install or connect the following:
   - New lower ball joint to the lower control arm stud. Torque the bolts to 117 ft. lbs. (159 Nm).
   - New cotter pin
   - Tie-rod end to the lower ball joint. Torque the nut to 67 ft. lbs. (91 Nm).

- Lower ball joint set bolts. Torque the 4 bolts to 59 ft. lbs. (80 Nm) for 2002–03; 48 ft. lbs. (65 Nm) for 2004–06.
- Front wheel.

4. Check and/or adjust the front wheel alignment.

Upper Control Arm

REMOVAL & INSTALLATION

Land Cruiser

1. Before servicing the vehicle, refer to the precautions section.
2. Remove or disconnect the following:
   - Front wheel
   - Inner fender liner
   - Wheel speed sensor harness
   - Upper ball joint
   - Adjustment cam bolts
   - Upper control arm

To install:

3. Install or connect the following:
   - Upper control arm. Tighten the adjustment cam bolts to 72 ft. lbs. (98 Nm).
   - Upper ball joint. Tighten the nut to 81 ft. lbs. (110 Nm).

- Wheel speed sensor harness. Tighten the bolts to 10 ft. lbs. (13 Nm).
- Inner fender liner
- Front wheel

4. Check the wheel alignment and adjust as necessary.

Sequoia

1. Before servicing the vehicle, refer to the precautions section.
2. Remove or disconnect the following:
   - Front wheel
   - Strut
   - Wheel speed sensor harness, if equipped with Anti-lock Brake System (ABS)
3. Upper ball joint, as follows:
   - Cotter pin and loosen the nut
   - Upper ball joint from the upper control arm, using tool 09950-40011
   - Steering knuckle, support it securely
   - Upper ball joint nut
4. Remove or disconnect the following:

   - 4 clips and the fender apron seal
   - Brake/fuel line clamp nut and clamp

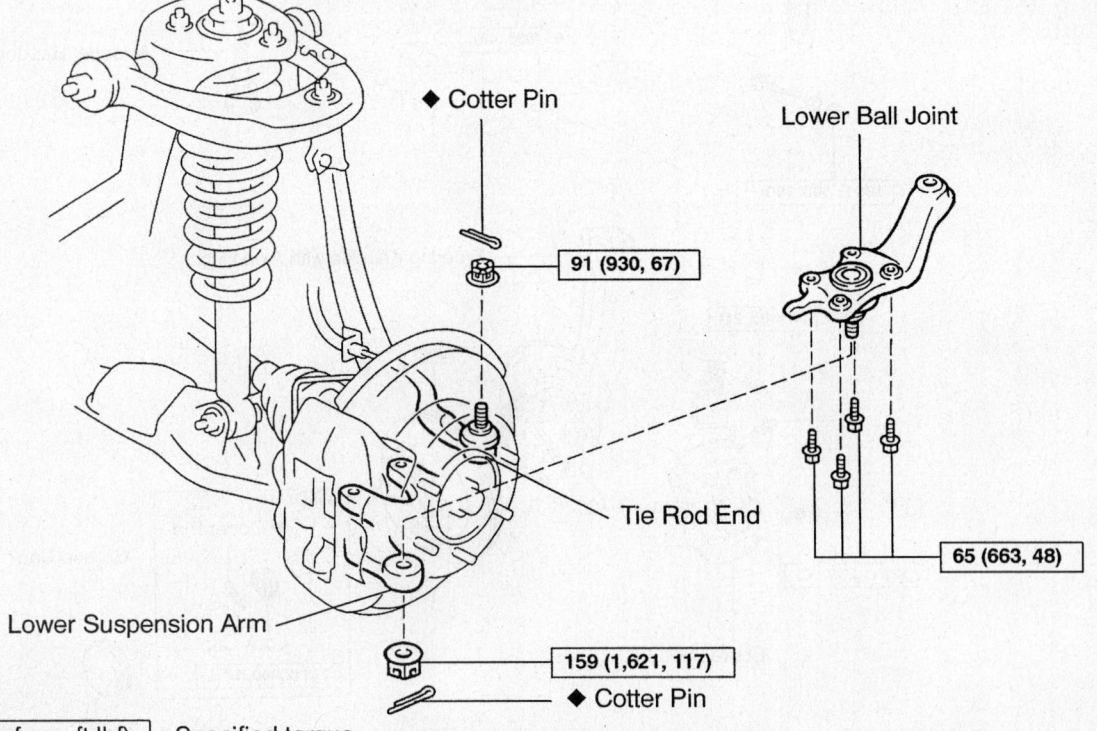

◆ Cotter Pin

91 (930, 67)

Lower Ball Joint

Tie Rod End

65 (663, 48)

Lower Suspension Arm

159 (1,621, 117)

◆ Cotter Pin

| N·m (kgf·cm, ft·lbf) | : Specified torque

◆ Non–reusable part

Lower ball joint installation—2004–06 Sequoia

67170-LCSQ-G34

- Both upper control arm-to-chassis nuts/bolts
- Upper control arm

To install:

5. Install or connect the following:
- Upper control arm. Torque both upper control arm-to-chassis nuts/bolts to 72 ft. lbs. (98 Nm).
- Brake/fuel line clamp nut and clamp. Torque the clamp nut to 49 inch lbs. (5.5 Nm).
- Fender apron seal
- Upper ball joint. Torque the nut to 77 ft. lbs. (105 Nm).
- New cotter pin

- Steering knuckle
- Wheel speed sensor harness, if equipped with Anti-lock Brake System (ABS). Torque it to 71 inch lbs. (8.0 Nm).
- Strut
- Front wheel

6. Check and/or adjust the wheel alignment.

GX470

1. Before servicing the vehicle, refer to the precautions section.
2. Remove the wheel.

3. Disconnect the skid control wire.
4. Support the lower arm with a jack.
5. Remove the cable bracket.
6. Disconnect the ball joint from the knuckle.
7. Remove the through-bolt, washers and nut.
8. Remove the arm.
9. Installation is the reverse of removal. Don't fully tighten the through-bolt until the vehicle is on the ground and the suspension is jounced a few times.
- Ball joint nut: 81 ft. lbs. (110 Nm)
- Through-bolt: 85 ft. lbs. (115 Nm)

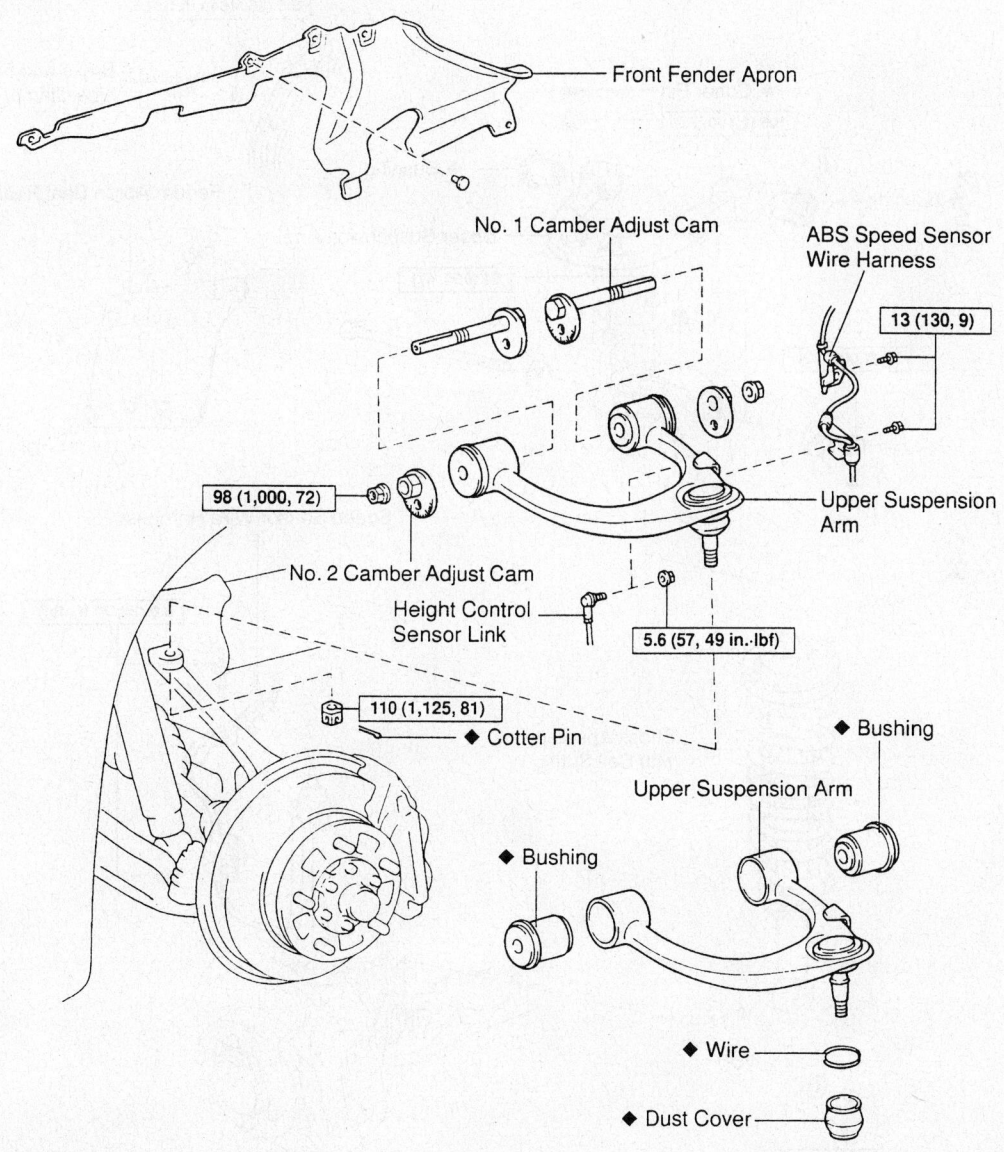

Front Fender Apron

No. 1 Camber Adjust Cam

ABS Speed Sensor Wire Harness

13 (130, 9)

Upper Suspension Arm

98 (1,000, 72)

No. 2 Camber Adjust Cam

Height Control Sensor Link

5.6 (57, 49 in.·lbf)

110 (1,125, 81)

◆ Cotter Pin

◆ Bushing

Upper Suspension Arm

◆ Bushing

◆ Wire

◆ Dust Cover

N·m (kgf·cm, ft·lbf) : Specified torque
◆ Non−reusable part

9302SG01

Exploded view of the upper control arm and related components—Land Cruiser

LX470

1. Before servicing the vehicle, refer to the precautions section.

2. Remove or disconnect the following:
- Front wheel
- Inner fender liner
- Wheel speed sensor harness
- Upper ball joint
- Adjustment cam bolts
- Upper control arm

To install:

3. Install or connect the following:
- Upper control arm. Tighten the adjustment cam bolts to 72 ft. lbs. (98 Nm).
- Upper ball joint. Tighten the nut to 81 ft. lbs. (110 Nm).
- Wheel speed sensor harness. Tighten the bolts to 10 ft. lbs. (13 Nm).
- Inner fender liner
- Front wheel

4. Check the wheel alignment and adjust as necessary.

CONTROL ARM BUSHING REPLACEMENT

Except Sequoia

1. Before servicing the vehicle, refer to the precautions section.

2. Remove the control arm from the vehicle.

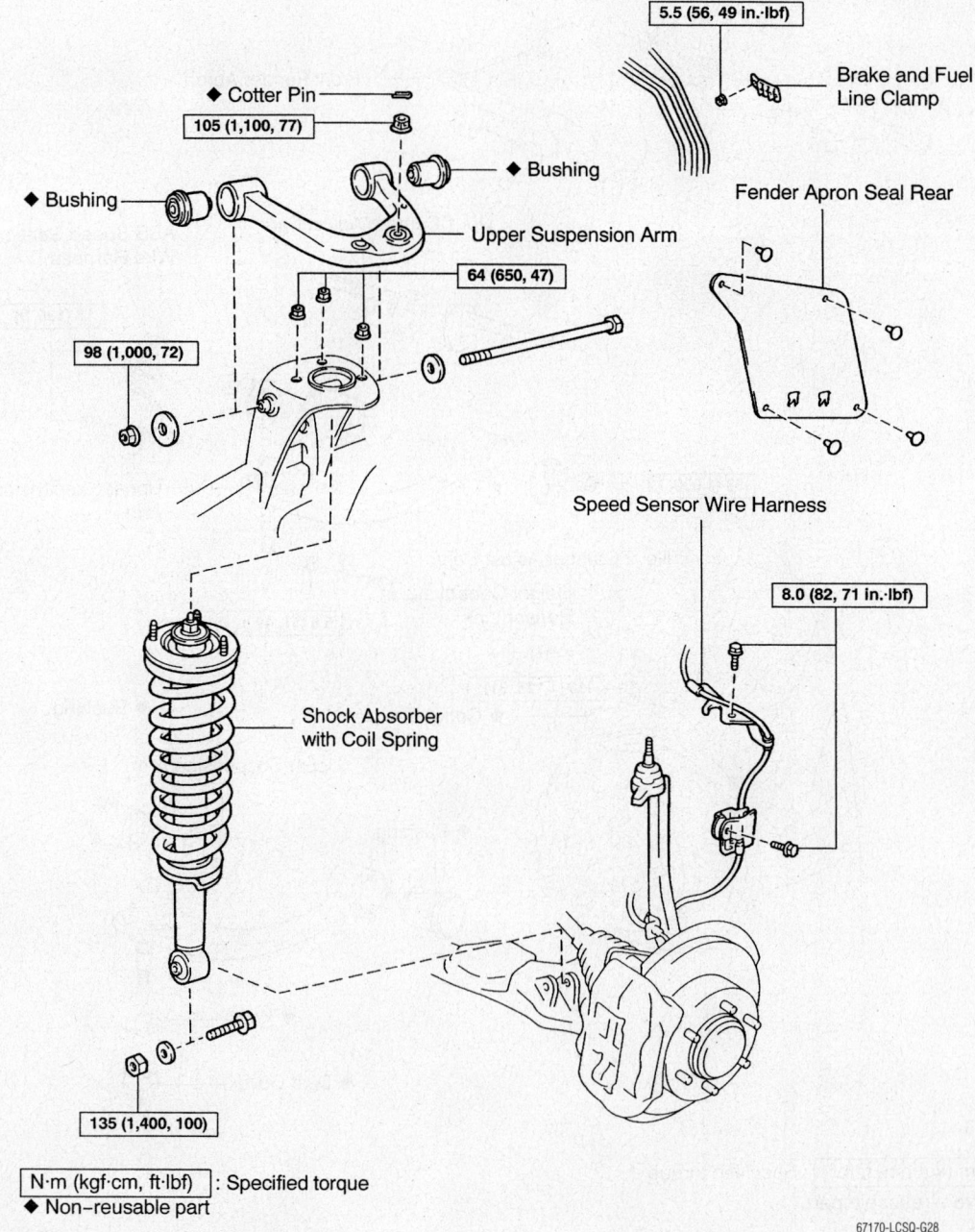

| 5.5 (56, 49 in.·lbf) |
| 105 (1,100, 77) |
| 64 (650, 47) |
| 98 (1,000, 72) |
| 8.0 (82, 71 in.·lbf) |
| 135 (1,400, 100) |

◆ Cotter Pin

◆ Bushing

◆ Bushing

Upper Suspension Arm

Brake and Fuel Line Clamp

Fender Apron Seal Rear

Speed Sensor Wire Harness

Shock Absorber with Coil Spring

| N·m (kgf·cm, ft·lbf) | : Specified torque

◆ Non-reusable part

67170-LCSQ-G28

Upper control arm and related parts—Sequoia

3. Remove the control arm bushings with a hydraulic press.

To install:

4. Lubricate the control arm bushings with liquid soap.

5. Press the bushings into the control arm until the bushing flange contacts the housing edge of the control arm.

6. Install the control arm to the vehicle.

7. Check the wheel alignment and adjust as necessary.

Sequoia

1. Before servicing the vehicle, refer to the precautions section.

2. Remove the upper control arm from the vehicle.

3. Remove the control arm bushings, as follows:

- Pry up the bushing flange, using a chisel and a hammer
- Press the bushing(s) from the

upper control arm, using a shop press and tools 09613-26010, 09631-20060 and 09950-00020

To install:

4. Lubricate the new control arm bushings with liquid soap.

5. Press the bushings into the control arm until the bushing flange contacts the housing edge of the control arm, using a shop press, a steel plate and tools 09631-12090 and 09710-30021

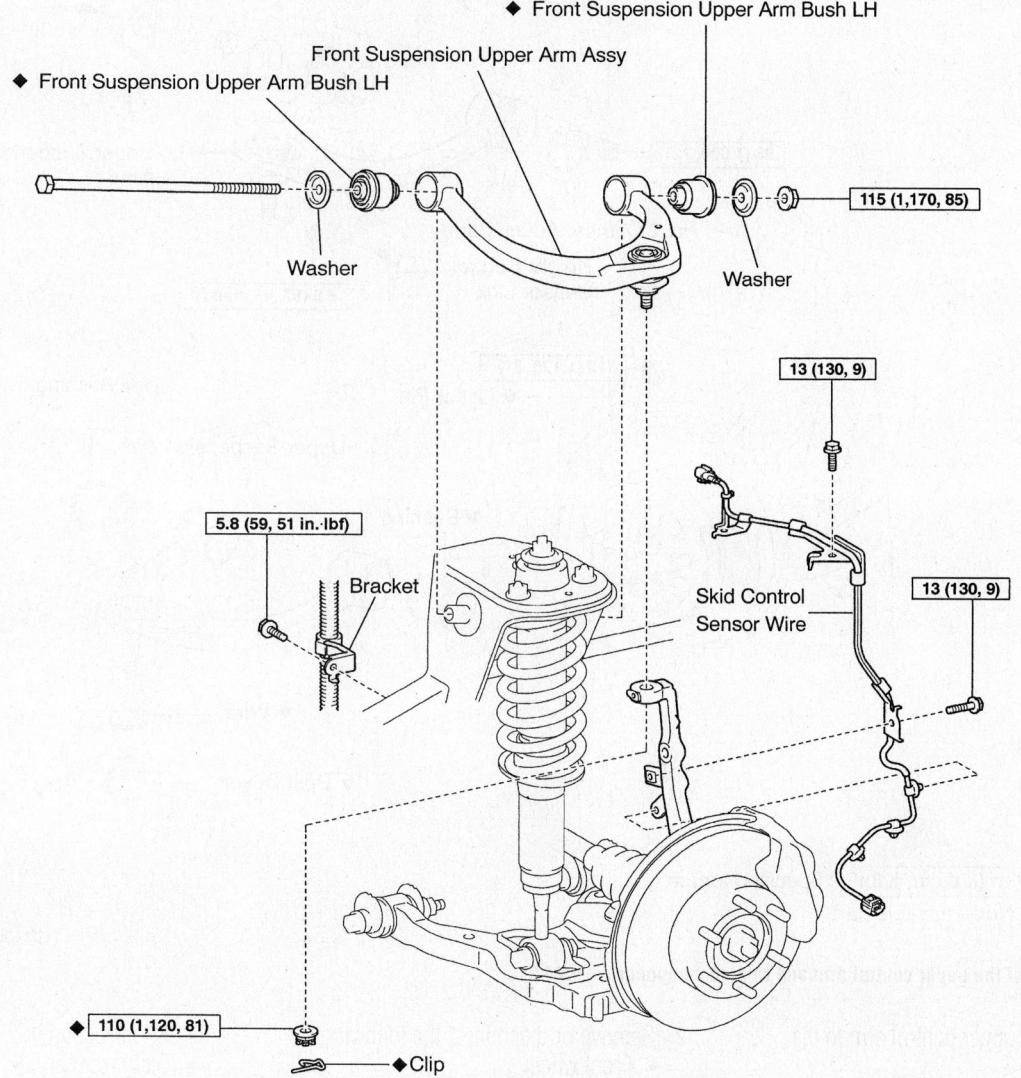

◆ Front Suspension Upper Arm Bush LH

Front Suspension Upper Arm Assy

◆ Front Suspension Upper Arm Bush LH

Washer

115 (1,170, 85)

Washer

13 (130, 9)

5.8 (59, 51 in.·lbf)

Bracket

Skid Control Sensor Wire

13 (130, 9)

110 (1,120, 81)

◆ Clip

N·m (kgf·cm, ft·lbf) : Specified torque
◆ Non-reusable part

67162-GX470-G04

Upper control arm and related parts—GX470

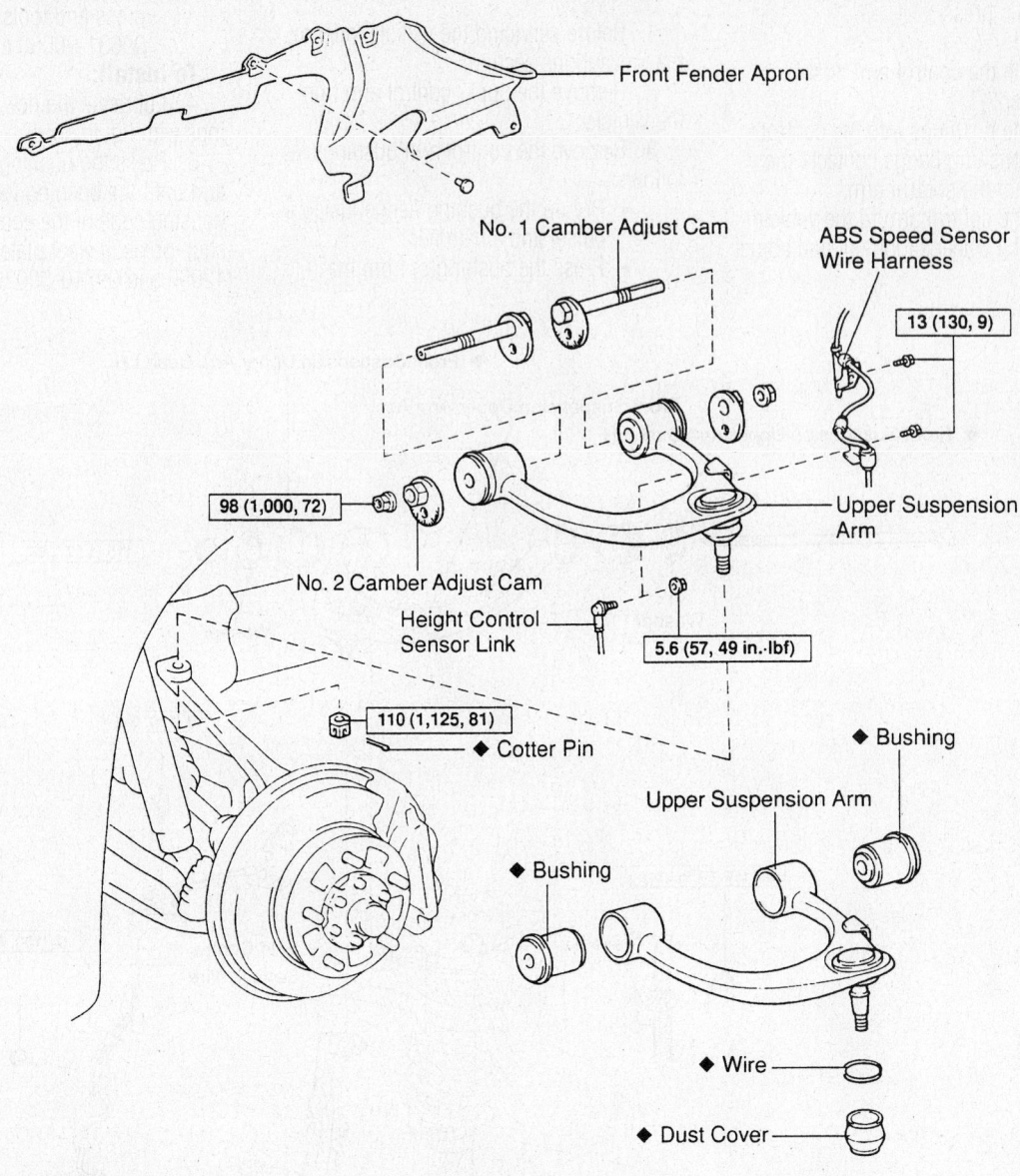

Front Fender Apron

No. 1 Camber Adjust Cam

ABS Speed Sensor
Wire Harness

13 (130, 9)

98 (1,000, 72)

Upper Suspension
Arm

No. 2 Camber Adjust Cam

Height Control
Sensor Link

5.6 (57, 49 in.-lbf)

110 (1,125, 81)

◆ Cotter Pin

Upper Suspension Arm

◆ Bushing

◆ Bushing

◆ Wire

◆ Dust Cover

N·m (kgf·cm, ft·lbf) : Specified torque
◆ Non–reusable part

9302SG01

Exploded view of the upper control arm and related components—LX470

6. Install the upper control arm to the vehicle.

7. Check and/or adjust the wheel alignment.

Lower Control Arm

REMOVAL & INSTALLATION

Land Cruiser

1. Before servicing the vehicle, refer to the precautions section.

2. Remove or disconnect the following:
- Front wheel
- Engine under cover
- Torsion bar
- Stabilizer bar link
- Shock absorber
- Lower ball joint
- Lower control arm

To install:

3. Install or connect the following:
- Lower control arm. Tighten the bolts to 170 ft. lbs. (230 Nm).
- Lower ball joint. Tighten the nut to 117 ft. lbs. (159 Nm).

- Shock absorber
- Stabilizer bar link. Tighten the bolt to 38 ft. lbs. (52 Nm).
- Torsion bar
- Engine under cover
- Front wheel

4. Check the wheel alignment and adjust as necessary.

Sequoia

1. Before servicing the vehicle, refer to the precautions section.

2. Remove front wheel.

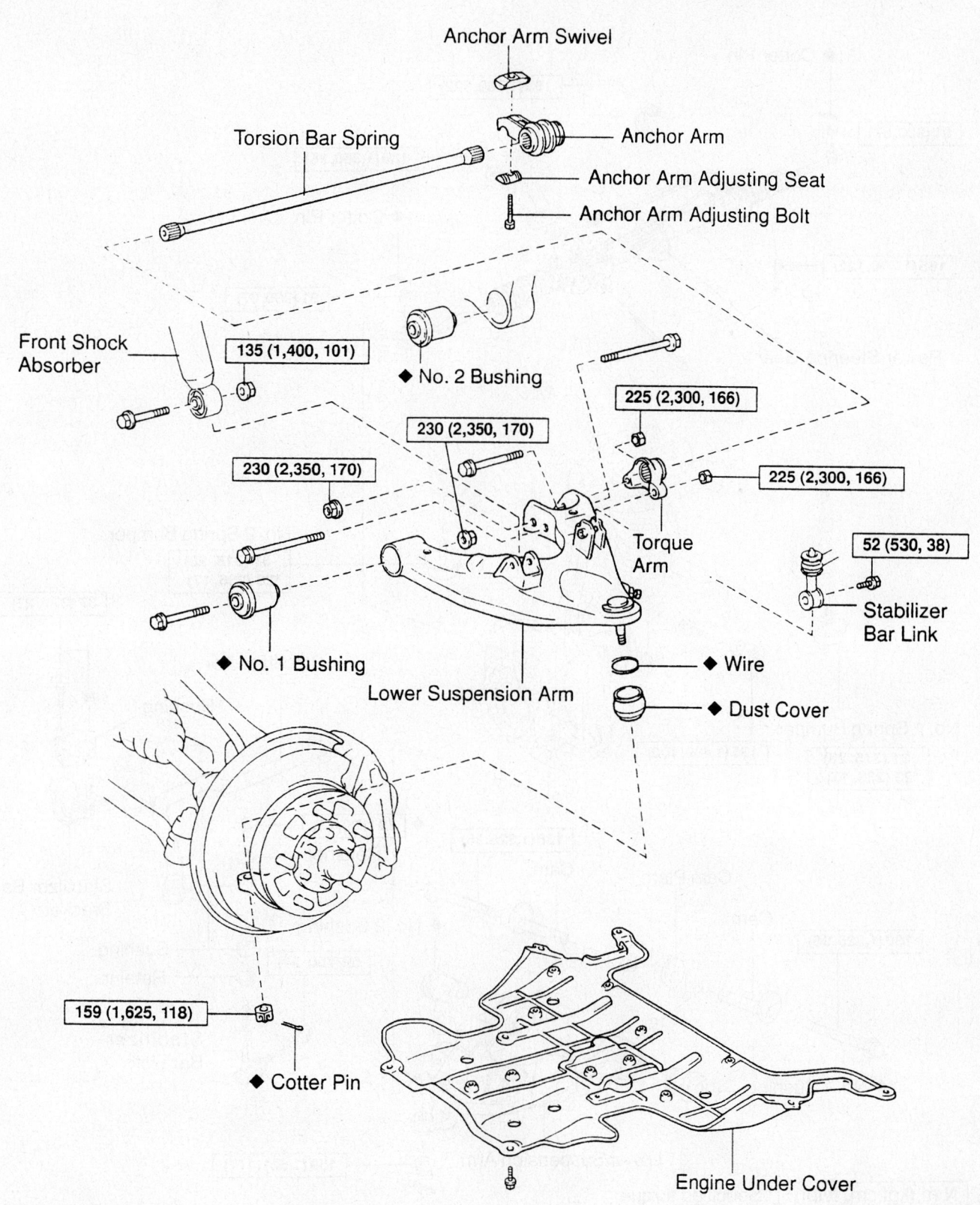

Anchor Arm Swivel

Torsion Bar Spring

Anchor Arm

Anchor Arm Adjusting Seat

Anchor Arm Adjusting Bolt

Front Shock Absorber

135 (1,400, 101)

◆ No. 2 Bushing

225 (2,300, 166)

230 (2,350, 170)

230 (2,350, 170)

225 (2,300, 166)

Torque Arm

52 (530, 38)

Stabilizer Bar Link

◆ No. 1 Bushing

◆ Wire

Lower Suspension Arm

◆ Dust Cover

159 (1,625, 118)

◆ Cotter Pin

Engine Under Cover

N·m (kgf·cm, ft·lbf) : Specified torque

◆ Non–reusable part

9302SG02

Exploded view of the lower control arm and related components—Land Cruiser

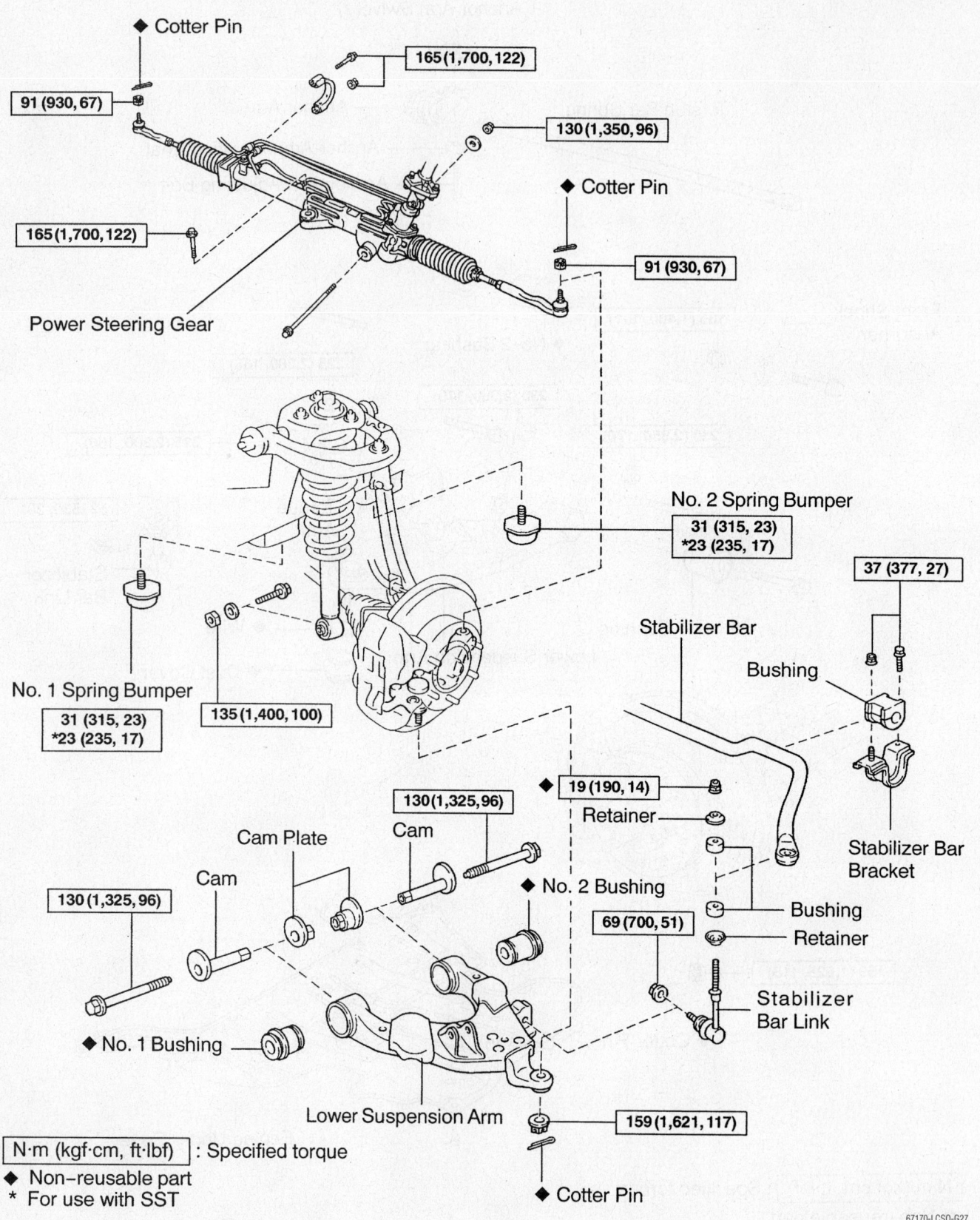

◆ Cotter Pin

165(1,700,122)

91 (930,67)

130(1,350,96)

◆ Cotter Pin

165(1,700,122)

91 (930,67)

Power Steering Gear

No. 2 Spring Bumper

31 (315, 23)
*23 (235, 17)

37 (377, 27)

Stabilizer Bar

Bushing

No. 1 Spring Bumper

31 (315, 23)
*23 (235, 17)

135 (1,400,100)

Stabilizer Bar Bracket

◆ 19 (190, 14)

Retainer

Cam Plate

130(1,325,96)

Cam

Bushing

No. 2 Bushing

69 (700, 51)

Retainer

Cam

130 (1,325,96)

Cam

Stabilizer Bar Link

◆ No. 1 Bushing

Lower Suspension Arm

159(1,621,117)

N·m (kgf·cm, ft·lbf) : Specified torque
◆ Non–reusable part
* For use with SST

◆ Cotter Pin

67170-LCSQ-G27

Lower control arm and related parts—Sequoia

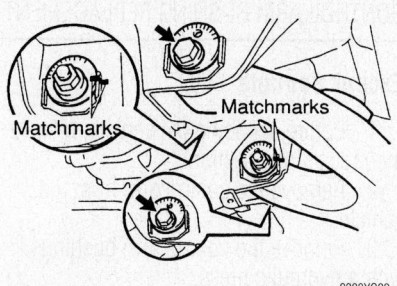

View of the lower control arm's cam plate alignment—Sequoia

9308YG22

Matchmarks

Matchmarks

3. Disconnect the tie-rod end, as follows:
- Cotter pin and nut
- Tie-rod end from the lower ball joint, using tool 09610-20012

4. Remove or disconnect the following:
- Power steering gear set bolts and nuts
- Stabilizer bar link from the lower control arm
- Strut from the lower control arm

5. Disconnect the lower ball joint, as follows:
- Cotter pin and nut
- Lower ball joint from the lower control arm

6. Matchmark both front and rear cam plates and chassis frame.

7. Remove the lower control arm while slightly shifting the power steering gear rearward.

To install:

8. Install or connect the following:
- Lower control arm while slightly shifting the power steering gear rearward

Front Shock Absorber with Coil Spring

135 (1,380, 100)

Camber Adjust Cam Assy

Toe Adjust Plate No.2

Camber Adjust Cam No.2

Toe Adjust Cam Sub-assy

135 (1,380, 100)

135 (1,380, 100)

◆ Front Lower Arm Bush No.2 LH

◆ Front Lower Arm Bush No.1 LH

Front Suspension Arm Sub-assy Lower No.1 LH

Front Lower Ball Joint Attachment LH

◆ Cotter Pin

◆ 140 (1,430, 103)

225 (2,290, 166)

N·m (kgf·cm, ft·lbf) : Specified torque

◆ Non-reusable part

67162-GX470-G05

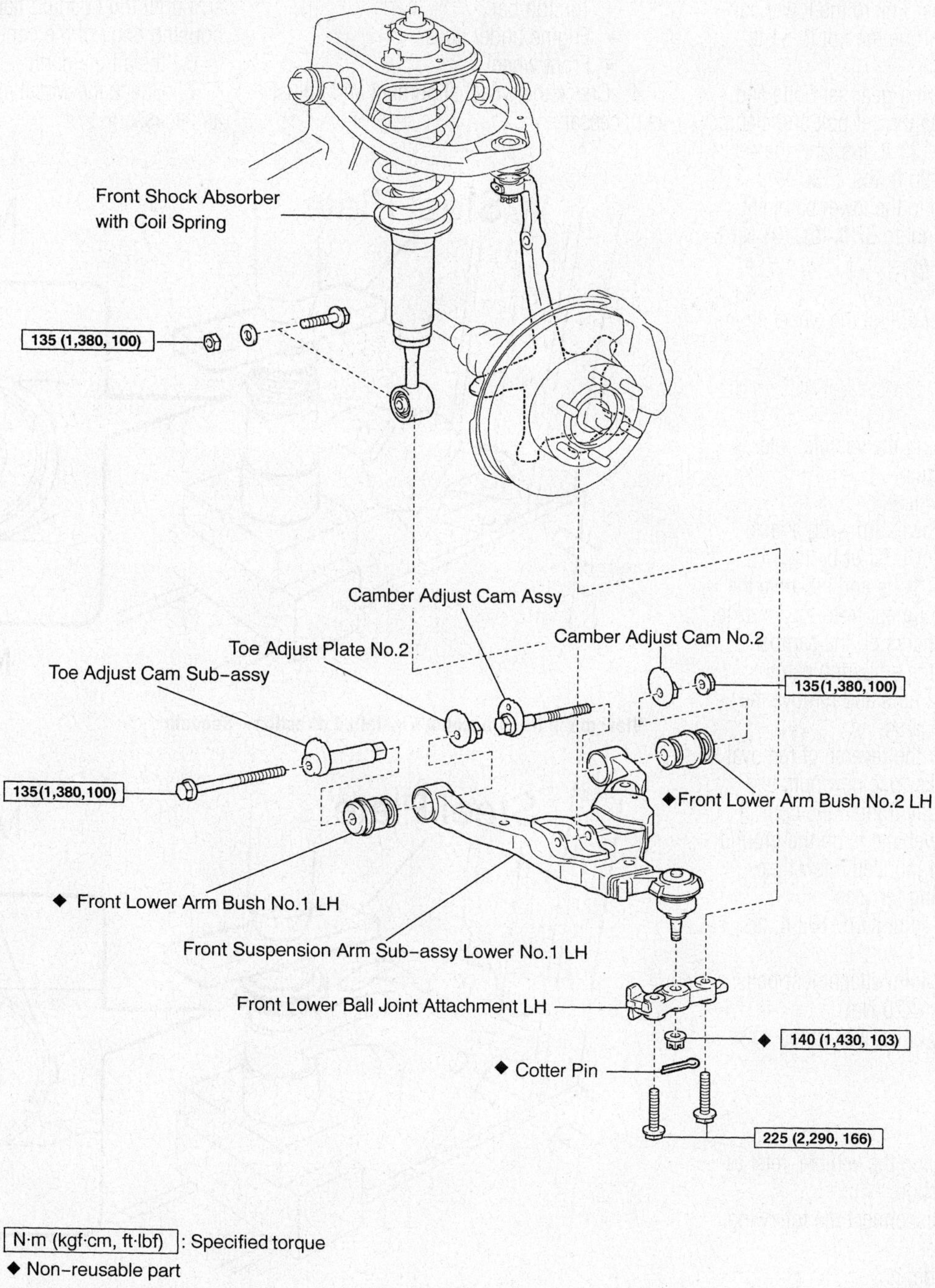

Lower control arm and related parts—GX470

- Align both front and rear cam plates and chassis frame matchmarks. Torque both bolts to 96 ft. lbs. (130 Nm).
9. Connect the lower ball joint, as follows:
   - Lower ball joint to the lower control arm. Torque the nut to 103 ft. lbs. (140 Nm).
   - New cotter pin
10. Install or connect the following:
   - Strut to the lower control arm. Torque the nut/bolt to 100 ft. lbs. (135 Nm).
   - Stabilizer bar link to the lower control arm. Torque the nut to 51 ft. lbs. (69 Nm).
   - Power steering gear set bolts and nuts. Torque the set bolt and clamp nut/bolt to 122 ft. lbs. and the set nut/bolt to 96 ft. lbs. (130 Nm)
   - Tie-rod end to the lower ball joint. Torque the nut to 67 ft. lbs. (91 Nm).
   - New cotter pin
   - Front wheel.
11. Check and/or adjust the wheel alignment.

GX470

1. Before servicing the vehicle, refer to the precautions section.
2. Remove the wheel.
3. Support the lower arm with a jack.
4. Remove the lower strut bolt.
5. Remove the 2 bolts and separate the lower ball joint attachment from the knuckle.
6. Place matchmarks on the camber adjusting cam and toe adjusting cam.
7. Remove the 2 nuts and remove the arm along with the cams.
8. Installation is the reverse of removal. Align all matchmarks. Use new nuts and cotter pins. Don't fully tighten the control arm bolts until the vehicle is on the ground and the suspension jounced a few times. Observe the following torques:
   - Lower ball joint stud: 103 ft. lbs. (140 Nm)
   - Lower ball joint attachment bolts: 166 ft. lbs. (225 Nm)
   - Lower arm bolts: 100 ft. lbs. (135 Nm)

LX470

1. Before servicing the vehicle, refer to the precautions section.
2. Remove or disconnect the following:
   - Front wheel
   - Engine under cover
   - Torsion bar
   - Stabilizer bar link

- Shock absorber
- Lower ball joint
- Lower control arm

To install:
3. Install or connect the following:
   - Lower control arm. Tighten the bolts to 170 ft. lbs. (230 Nm) for 2002–03; 123 ft. lbs. (167 Nm) for 2004–06.
   - Lower ball joint. Tighten the nut to 117 ft. lbs. (159 Nm).
   - Shock absorber
   - Stabilizer bar link. Tighten the bolt to 38 ft. lbs. (52 Nm).
   - Torsion bar
   - Engine under cover
   - Front wheel
4. Check the wheel alignment and adjust as necessary.

CONTROL ARM BUSHING REPLACEMENT

Except Sequoia

1. Before servicing the vehicle, refer to the precautions section.
2. Remove the control arm from the vehicle.
3. Remove the control arm bushings with a hydraulic press.
   To install:
4. Lubricate the control arm bushings with liquid soap.
5. Press the bushings into the control arm until the bushing flange contacts the housing edge of the control arm.
6. Install the control arm to the vehicle.
7. Check the wheel alignment and adjust as necessary.

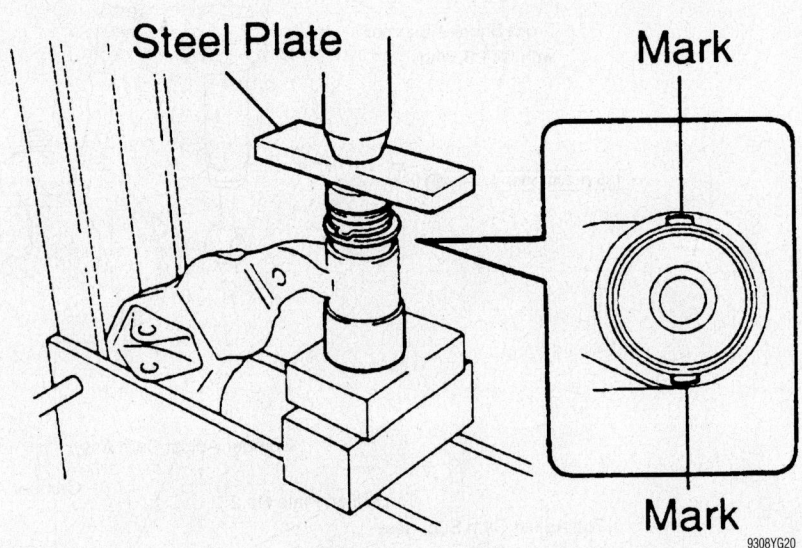

View of the No. 1 bushing's installed direction—Sequoia

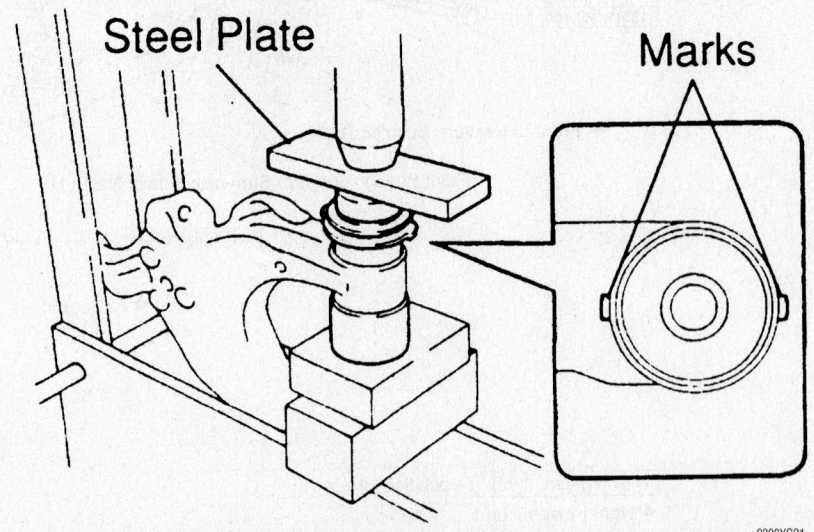

View of the No. 2 bushing's installed direction—Sequoia

Sequoia

1. Before servicing the vehicle, refer to the precautions section.
2. Remove the lower control arm from the vehicle.
3. Remove the control arm bushings, as follows:

- Pry up the bushing flange, using a chisel and a hammer
- Press the bushing(s) from the upper control arm, using a shop press and tools 09613-26010, 09632-36010 and 09950-00020

To install:

4. Lubricate the new control arm bushings with liquid soap.
5. Press the No. 1 bushing into the control arm until the bushing flange contacts the housing edge of the control arm, using a shop press, a steel plate and tools 09631-12090 and 09502-12010, facing the correct direction.
6. Press the No. 2 bushing into the control arm until the bushing flange contacts the housing edge of the control arm, using a shop press, a steel plate and tools 09631-12090 and 09950-60020, facing the correct direction.

7. Install the lower control arm to the vehicle.
8. Check and/or adjust the wheel alignment.

Front Wheel Bearing

ADJUSTMENT

Land Cruiser

1. Before servicing the vehicle, refer to the precautions section.

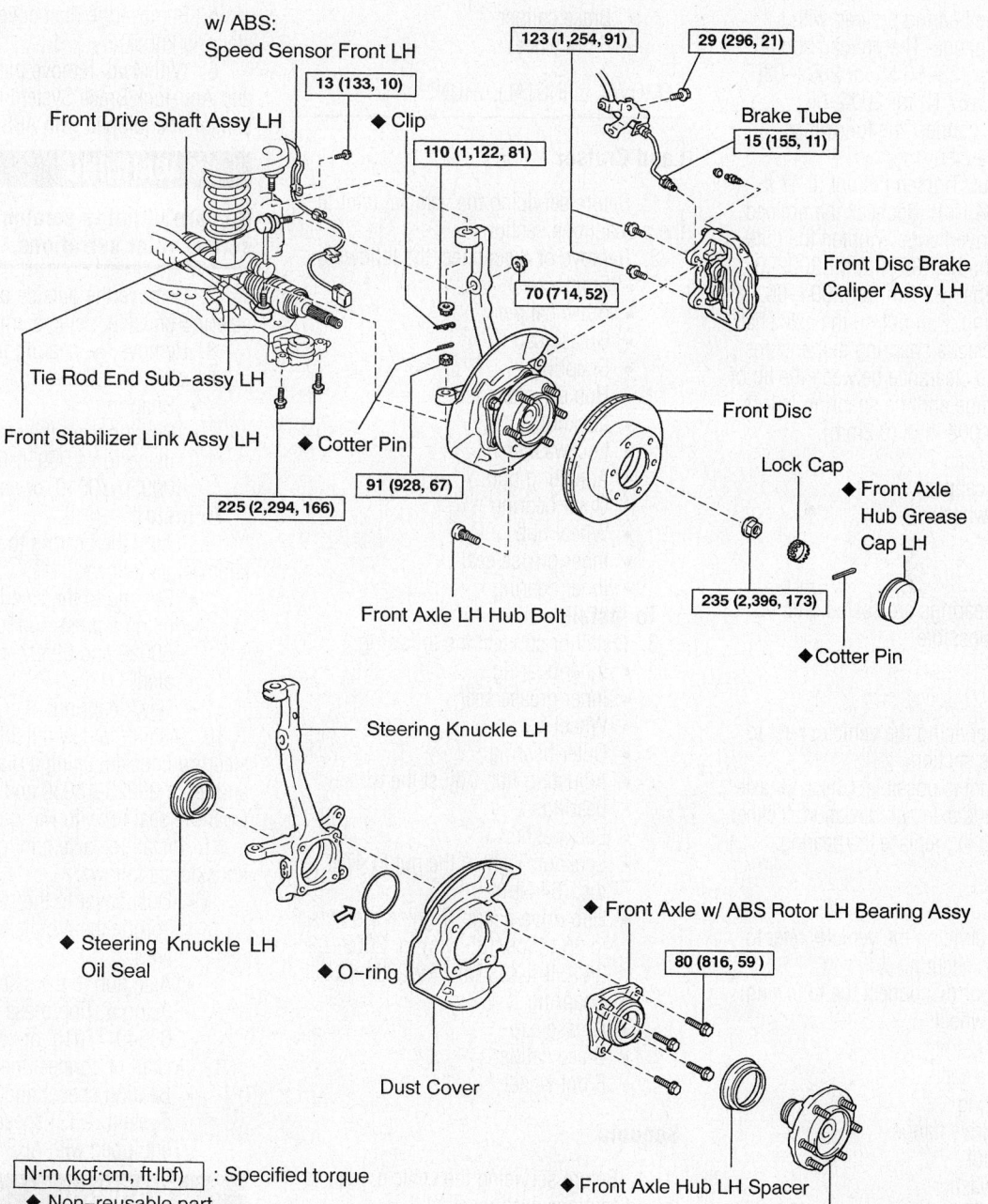

w/ ABS:
Speed Sensor Front LH
13 (133, 10)
Front Drive Shaft Assy LH
◆ Clip
123 (1,254, 91)
29 (296, 21)
Brake Tube
15 (155, 11)
110 (1,122, 81)
Front Disc Brake Caliper Assy LH
70 (714, 52)
Tie Rod End Sub-assy LH
Front Disc
Lock Cap
◆ Front Axle Hub Grease Cap LH
Front Stabilizer Link Assy LH
◆ Cotter Pin
91 (928, 67)
235 (2,396, 173)
◆ Cotter Pin
225 (2,294, 166)
Front Axle LH Hub Bolt

Steering Knuckle LH
◆ Steering Knuckle LH Oil Seal
◆ O-ring
◆ Front Axle w/ ABS Rotor LH Bearing Assy
80 (816, 59)
Dust Cover
◆ Front Axle Hub LH Spacer
Front Axle Hub Sub-assy LH

N·m (kgf·cm, ft·lbf) : Specified torque
◆ Non-reusable part
⇐ Mp Grease

67162-GX470-G10

Front hub and related parts—GX470

2. Remove or disconnect the following:
- Front wheel
- Brake caliper
- Grease cap
- Snapring
- Hub drive flange
- Locknut
- Lockwasher

3. Tighten the adjusting nut to 43 ft. lbs. (59 Nm) while rotating the hub to seat the bearings.

4. Loosen the adjusting nut.

5. Tighten the adjusting nut to 48 inch lbs. (5.4 Nm) and check that the bearing has no play.

6. Check the bearing preload with a spring tension gauge. The preload should be 6.4–12.6 lbs. (28–56 N) for 2002–03; 9.5–15 lbs. (42–67 N) for 2004–06.

7. Install or connect the following:
- Lockwasher
- Locknut. Tighten the nut to 47 ft. lbs. (64 Nm). Recheck the preload.
- Hub drive flange. Tighten the nuts to 26 ft. lbs. (35 Nm) for 2002–03; 24 ft. lbs. (33 Nm) for 2004–06.
- Snapring. Pull out on the axle shaft and select a snapring that ensures that the clearance between the tip of the flange and the snapring is less than 0.008 inch (0.2mm).
- Grease cap
- Brake caliper
- Front wheel

Sequoia

The wheel bearings are sealed unit; no adjustment is possible.

GX470

1. Before servicing the vehicle, refer to the precautions section.

No adjustment is possible. Check for axle hub backlash and axle hub deviation. If either exceeds 0.0020 in., replace the bearing.

LX470

1. Before servicing the vehicle, refer to the precautions section.

2. Remove or disconnect the following:
- Front wheel
- Brake caliper
- Grease cap
- Snapring
- Hub drive flange
- Locknut
- Lockwasher

3. Tighten the adjusting nut to 43 ft. lbs. (59 Nm) while rotating the hub to seat the bearings.

4. Loosen the adjusting nut.

5. Tighten the adjusting nut to 48 inch lbs. (5.4 Nm) and check that the bearing has no play.

6. Check the bearing preload with a spring tension gauge. The preload should be 6.4–12.6 lbs. (28–56 N).

7. Install or connect the following:
- Lockwasher
- Locknut. Tighten the nut to 47 ft. lbs. (64 Nm).
- Hub drive flange. Tighten the nuts to 26 ft. lbs. (35 Nm).
- Snapring
- Grease cap
- Brake caliper
- Front wheel

REMOVAL & INSTALLATION

Land Cruiser

1. Before servicing the vehicle, refer to the precautions section.

2. Remove or disconnect the following:
- Front wheel
- Brake caliper
- Grease cap
- Snapring
- Hub drive flange
- Locknut
- Lockwasher
- Adjusting nut
- Outer bearing
- Wheel hub
- Inner grease seal
- Inner bearing

To install:

3. Install or connect the following:
- Inner bearing
- Inner grease seal
- Wheel hub
- Outer bearing
- Adjusting nut. Adjust the wheel bearings.
- Lockwasher
- Locknut. Tighten the nut to 47 ft. lbs. (64 Nm).
- Hub drive flange. Tighten the nuts to 26 ft. lbs. (35 Nm) for 2002–03; 24 ft. lbs. (33 Nm) for 2004–06.
- Snapring
- Grease cap
- Brake caliper
- Front wheel

Sequoia

1. Before servicing the vehicle, refer to the precautions section.

2. Remove or disconnect the following:
- Front wheel

- Grease cap
- With 4wd, cotter pin, lock cap and halfshaft nut
- Speed sensor wire from the knuckle
- Caliper and rotor. Support the caliper out of the way.
- Lower ball joint
- Axle hub/steering knuckle assembly and place it in a vise
- With 2wd, the grease cap
- Inner grease seal, for 4WD

3. Remove the 4 bolts and shift the dust cover towards the hub side.

4. Using a suitable puller, remove the hub from the knuckle.

5. Remove the dust cover from the steering knuckle.

6. With 4wd, remove the bearing spacer and Anti-lock Brake System (ABS) speed sensor, if equipped with ABS

✳✳ WARNING

Be careful not to scratch the speed sensor rotor serrations.

7. Remove the outside oil seal from steering knuckle, using a small prybar

8. Remove the bearing from the steering knuckle, as follows:
- Snapring
- Bearing from the steering knuckle, using tools 09950-60020 and 09950-70010, or equivalent.

To install:

9. Install the bearing to the steering knuckle, as follows:
- Bearing to the steering knuckle, using a press and tools 09950-60020 and 09527-17011, or equivalent
- New snapring

10. Install the new outside oil seal to steering knuckle, using a plastic hammer and tools 09223-15030 and 09527-17011. Coat the seal lip with MP grease.

11. Install the axle hub to the steering knuckle, as follows:
- Dust cover to the steering knuckle. Torque the 4 bolts to 13 ft. lbs. (18 Nm).
- Axle hub to the steering knuckle, using a shop press and tool 09649-17010, or equivalent.

12. Install or connect the following:
- Bearing spacer and Anti-lock Brake System (ABS) speed sensor, if equipped with ABS

✳✳ WARNING

Be careful not to scratch the speed sensor rotor serrations.

- Bearing spacer, if not equipped with ABS, using a shop press and tools 09950-60010 and 09950-70010
- With 2wd, install a new locknut. Torque to 203 ft. lbs. (274 Nm). Stake the nut.
- With 4wd, the bearing spacer using a press
- Grease cap, for 2WD
- Inner grease seal, for 4WD, using a plastic hammer and tool 09527-17011

13. Install the hub/steering knuckle assembly. With 4wd, install the halfshaft and install the nut loosely.

14. Connect the knuckle to the upper arm. Torque the nut to 77 ft. lbs. (105 Nm). The nut can be tightened up to an additional 60 degrees to align the hole. Install a new cotter pin.

15. Connect the lower ball joint. Torque the 4 bolts to 48 ft. lbs. (65 Nm).

16. Install the shock absorber.

17. Install the rotor and caliper.

18. Attach the brake line to the knuckle.

19. Connect the speed sensor.

20. With 4wd, apply the brakes and tighten the shaft nut to 173 ft. lbs. (235 Nm). The nut can be tightened up to an additional 60 degrees to align the hole. Install a new cotter pin.

21. Install the grease cap.
22. Install the wheel.
23. Check the alignment.

GX470

1. Remove the wheel.
2. Remove the caliper.
3. Remove the hub grease cap.
4. Remove the cotter pin.
5. Remove the hub nut.
6. Remove the speed sensor.
7. Remove the stabilizer links from the knuckles.
8. Remove the tie rod end from the knuckle.

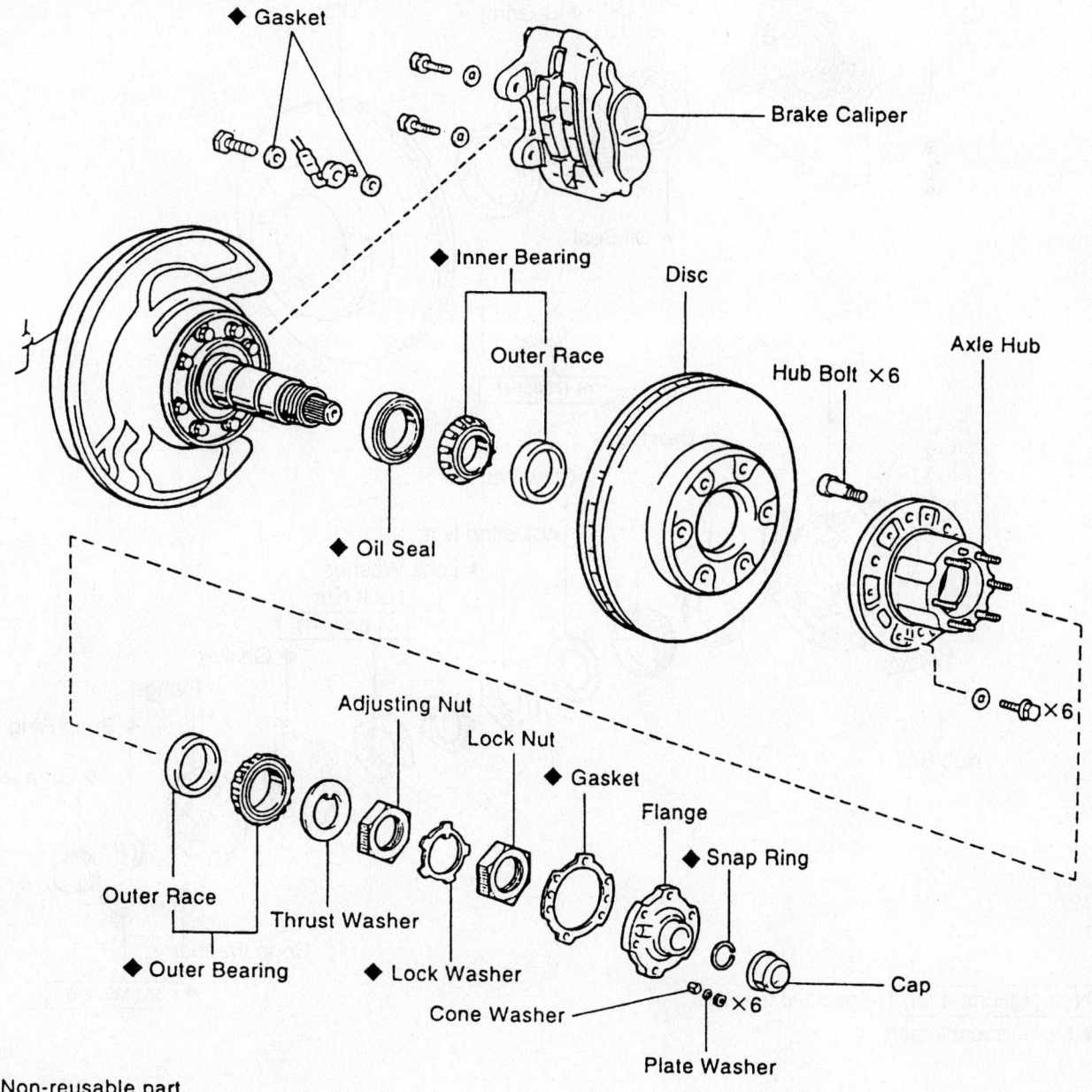

◆ Non-reusable part

Exploded view of the front hub and related components—2002 Land Cruiser

7924SG31

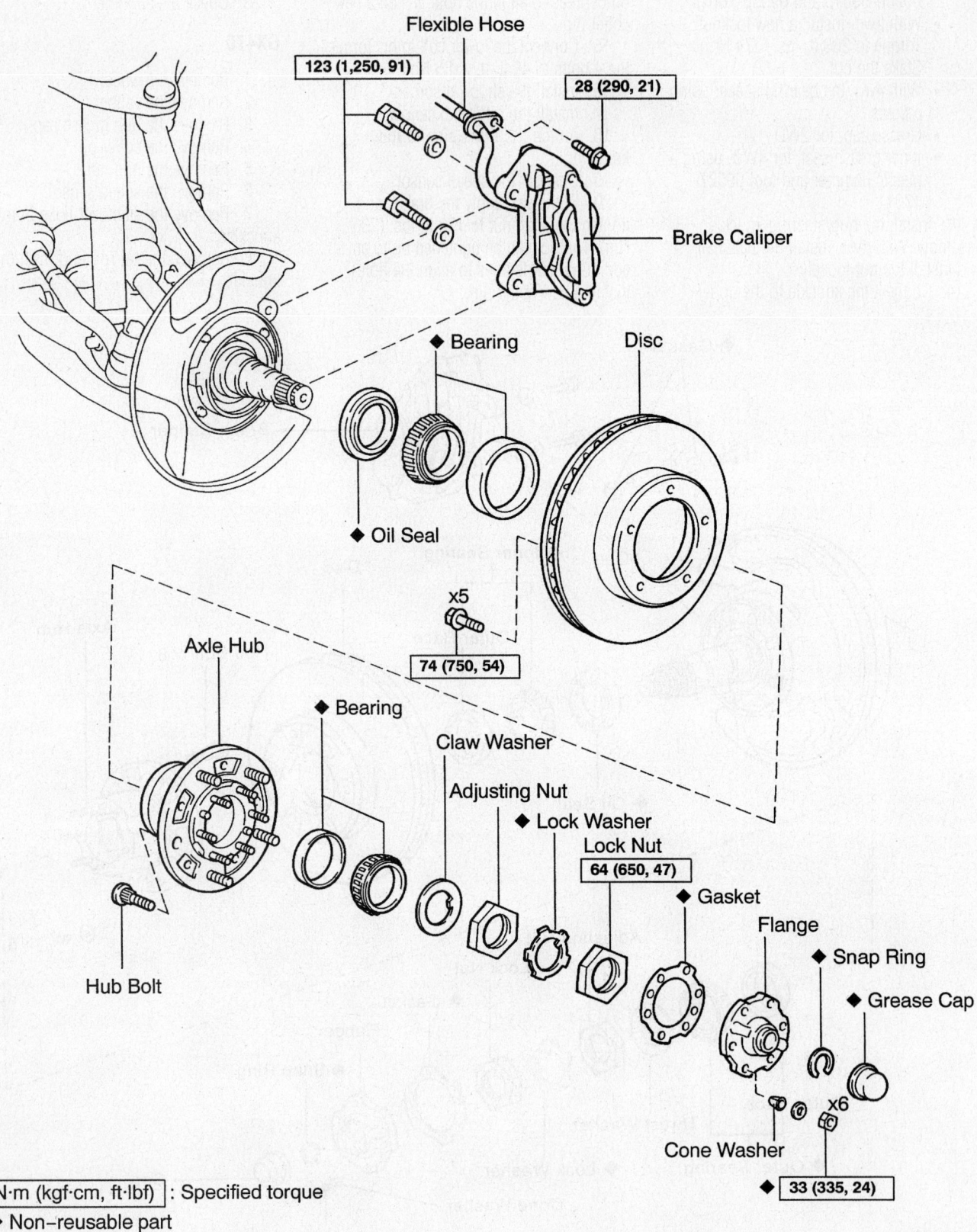

Flexible Hose

123 (1,250, 91)

28 (290, 21)

Brake Caliper

◆ Bearing

Disc

◆ Oil Seal

x5

74 (750, 54)

Axle Hub

◆ Bearing

Claw Washer

Adjusting Nut

◆ Lock Washer

Lock Nut

64 (650, 47)

◆ Gasket

Flange

◆ Snap Ring

◆ Grease Cap

Hub Bolt

Cone Washer

x6

33 (335, 24)

N·m (kgf·cm, ft·lbf) : Specified torque

◆ Non-reusable part

67170-LCSQ-G35

Exploded view of the front hub and related components—2004–06 Land Cruiser

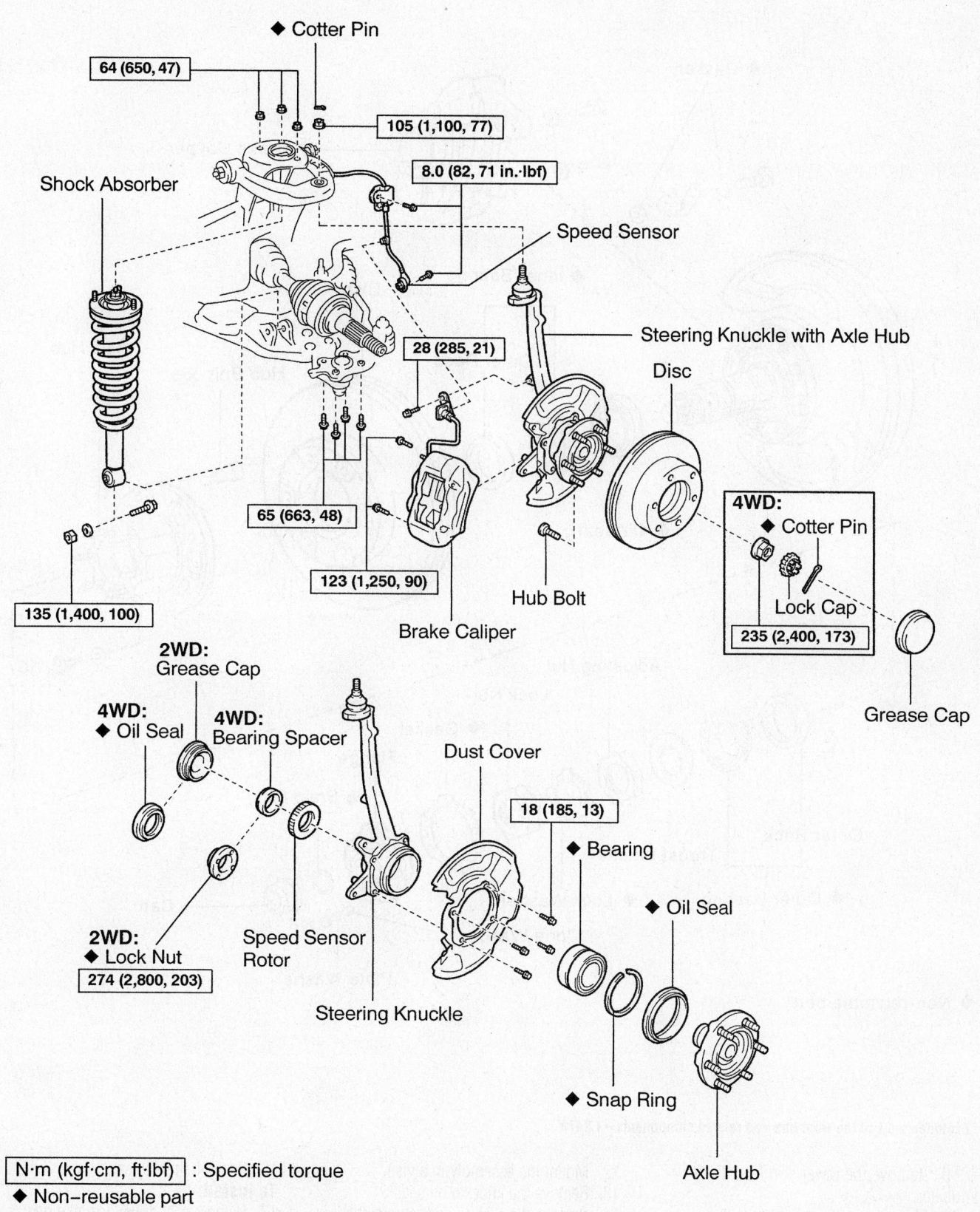

◆ Cotter Pin

64 (650, 47)

105 (1,100, 77)

8.0 (82, 71 in.·lbf)

Shock Absorber

Speed Sensor

Steering Knuckle with Axle Hub

Disc

28 (285, 21)

4WD:
◆ Cotter Pin

Lock Cap

65 (663, 48)

235 (2,400, 173)

Hub Bolt

123 (1,250, 90)

135 (1,400, 100)

Brake Caliper

Grease Cap

2WD:
Grease Cap

4WD:
◆ Oil Seal

4WD:
Bearing Spacer

Dust Cover

18 (185, 13)

◆ Bearing

◆ Oil Seal

2WD:
◆ Lock Nut

274 (2,800, 203)

Speed Sensor
Rotor

Steering Knuckle

◆ Snap Ring

Axle Hub

N·m (kgf·cm, ft·lbf) : Specified torque
◆ Non–reusable part

67170-LCSQ-G29

Front hub and related parts—Sequoia

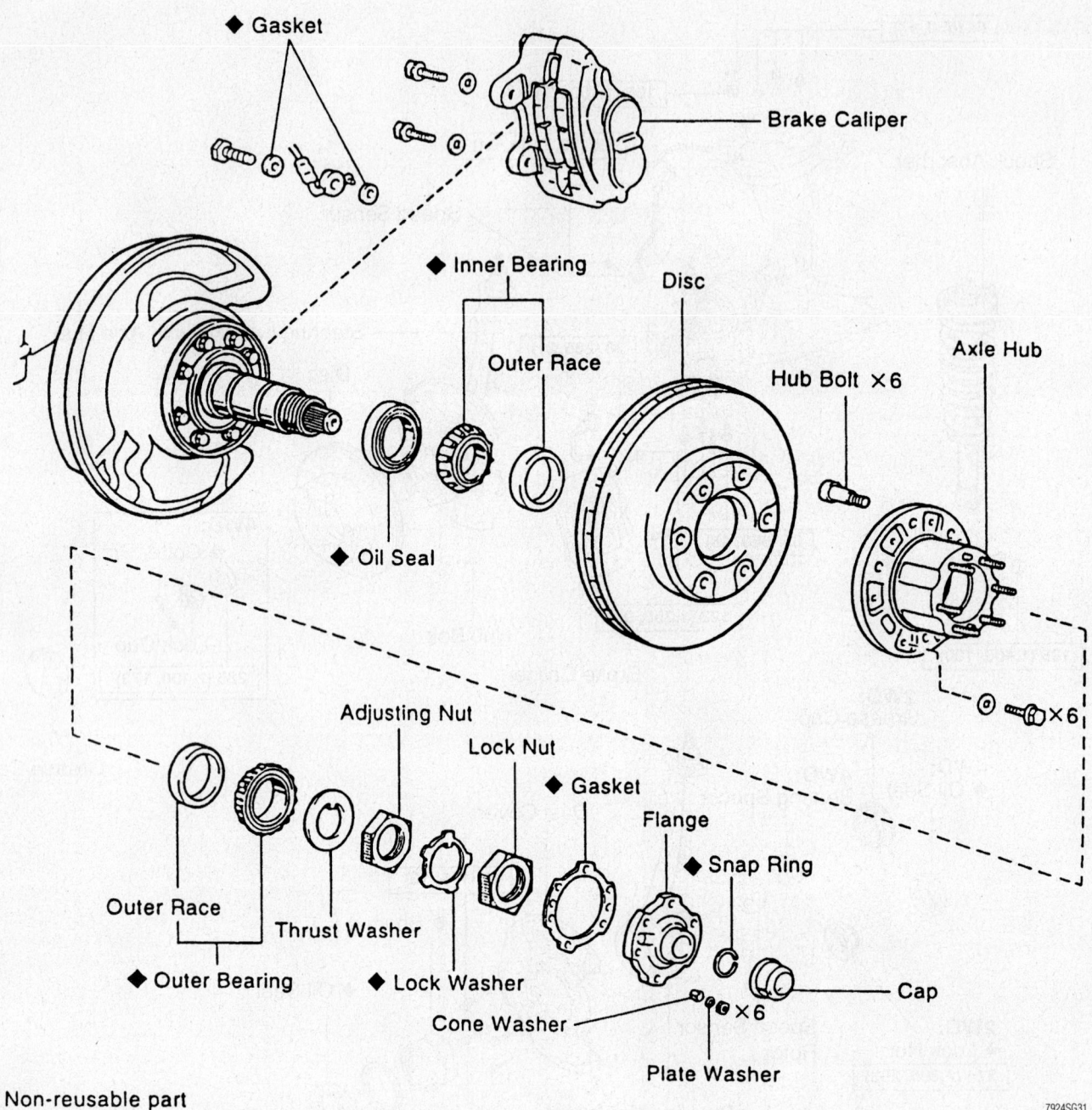

Gasket

Brake Caliper

Inner Bearing

Disc

Outer Race

Hub Bolt ×6

Axle Hub

Oil Seal

Adjusting Nut

Lock Nut

◆ Gasket

Flange

◆ Snap Ring

Outer Race

Thrust Washer

◆ Outer Bearing

◆ Lock Washer

Cone Washer

Cap

Plate Washer

◆ Non-reusable part

7924SG31

Exploded view of the front hub and related components—LX470

9. Remove the lower arm from the knuckle.

10. Remove the upper arm from the knuckle.

11. Remove the hub/knuckle assembly from the shaft.

12. Mount the assembly in a vise.

13. Remove the knuckle oil seal.

14. Remove the 4 bolts and remove the hub assembly from the knuckle.

15. Using SST 09710-30021 and its components, remove the bearing from the hub.

16. Remove the oil seal.

To install:

17. Using a seal driver, install a new seal.

➡**Take care to avoid damage to the spacer.**

18. Press a new bearing into the hub.

19. Coat a new O-ring with MP grease and install it in the hub.

20. Attach the hub to the knuckle. Torque to 59 ft. lbs. (80 Nm).

21. Install a new knuckle oil seal.

22. The remainder of installation is the reverse of removal. Observe the following torques:

- Upper arm ball stud nut: 81 ft. lbs. (110 Nm)
- Lower arm ball joint attachment bolts: 166 ft. lbs. (225 Nm)
- Tie rod end ball stud nut: 67 ft. lbs. (91 Nm)
- Stabilizer end links: 52 ft. lbs. (70 Nm)
- Hub nut: 173 ft. lbs. (235 Nm)

LX470

1. Before servicing the vehicle, refer to the precautions section.

2. Remove or disconnect the following:

- Front wheel
- Brake caliper
- Grease cap
- Snapring
- Hub drive flange
- Locknut
- Lockwasher
- Adjusting nut
- Outer bearing
- Wheel hub
- Inner grease seal
- Inner bearing

To install:

3. Install or connect the following:

- Inner bearing
- Inner grease seal
- Wheel hub
- Outer bearing
- Adjusting nut. Adjust the wheel bearings.
- Lockwasher
- Locknut. Tighten the nut to 47 ft. lbs. (64 Nm).
- Hub drive flange. Tighten the nuts to 26 ft. lbs. (35 Nm).
- Snapring
- Grease cap
- Brake caliper
- Front wheel

BRAKES

Front Brake Caliper

REMOVAL & INSTALLATION

Land Cruiser

1. Disconnect the negative battery cable from the battery.

2. Raise and support the vehicle safely.

3. Remove the wheels.

4. Disconnect the brake hose from the caliper by removing the union bolt and 2 gaskets. Plug the end of the hose to prevent loss of fluid.

5. Remove the 2 caliper mounting bolts.

6. Lift the bottom of the caliper up and remove the caliper assembly.

To install:

7. Grease the caliper slides and bolts with lithium grease or equivalent. Install the caliper and secure with the bolts. Torque the bolts to 90 ft. lbs. (123 Nm).

8. Connect the brake hose to the

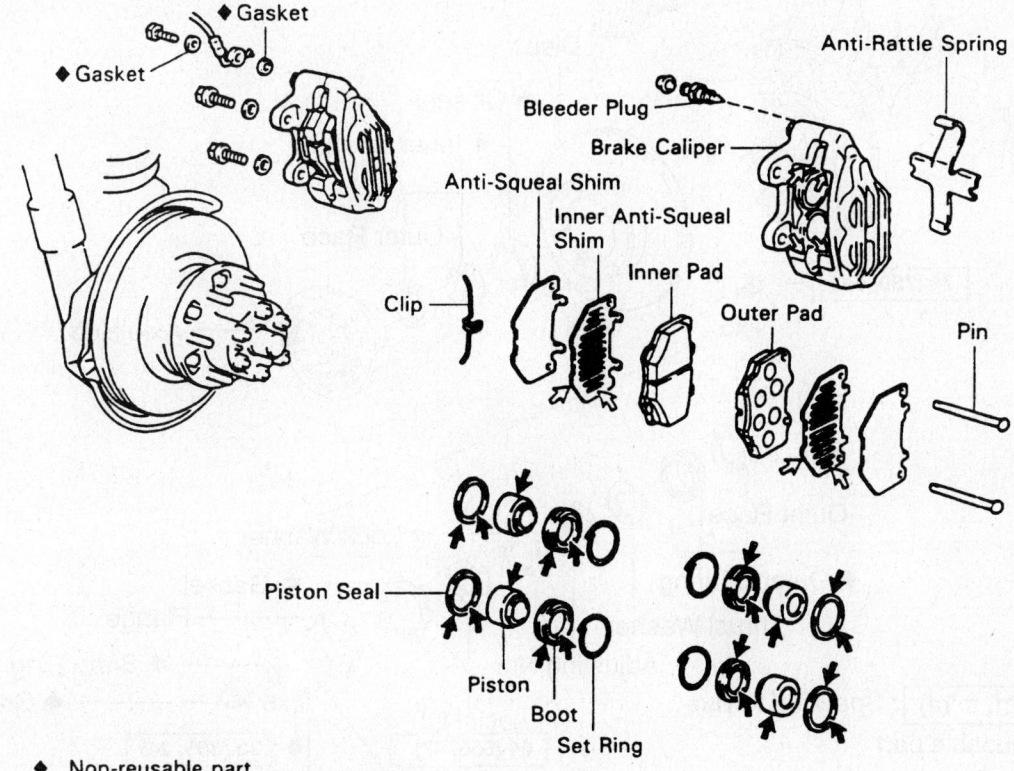

Gasket

Gasket

Anti-Rattle Spring

Bleeder Plug

Brake Caliper

Anti-Squeal Shim

Inner Anti-Squeal Shim

Clip

Inner Pad

Outer Pad

Pin

Piston Seal

Piston

Boot

Set Ring

◆ Non-reusable part

◀ Lithium soap base glycol grease

◁ Disc Brake Grease

93026G75

Exploded view of the front disc brake components—2002 Land Cruiser

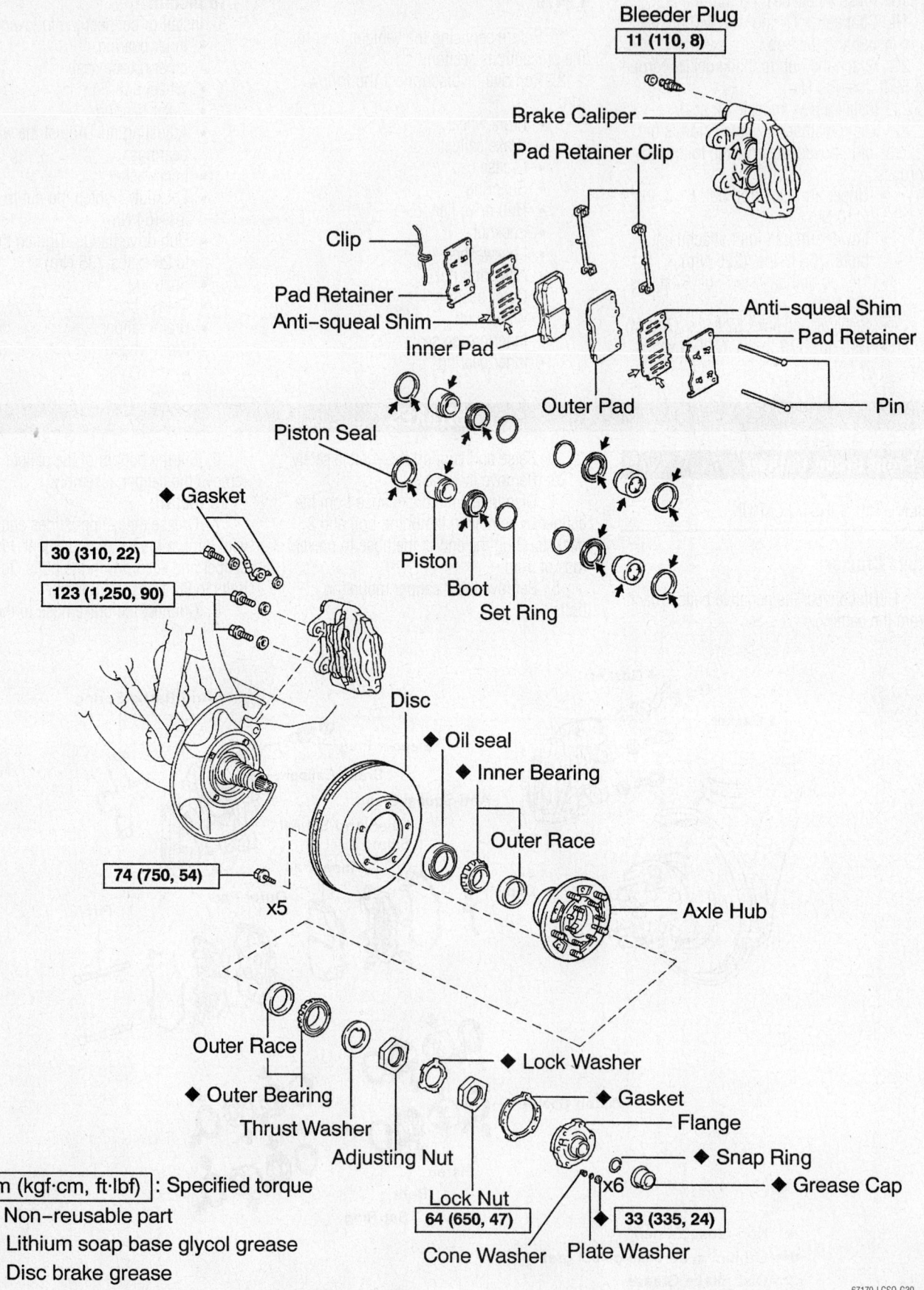

Bleeder Plug
11 (110, 8)

Brake Caliper

Pad Retainer Clip

Clip

Pad Retainer

Anti-squeal Shim

Inner Pad

Anti-squeal Shim

Pad Retainer

Outer Pad

Pin

Piston Seal

◆ Gasket

30 (310, 22)

123 (1,250, 90)

Piston

Boot

Set Ring

Disc

◆ Oil seal

◆ Inner Bearing

Outer Race

74 (750, 54)

x5

Axle Hub

Outer Race

◆ Outer Bearing

Thrust Washer

Adjusting Nut

◆ Lock Washer

◆ Gasket

Flange

◆ Snap Ring

◆ Grease Cap

Lock Nut
64 (650, 47)

Cone Washer

x6

Plate Washer

33 (335, 24)

N·m (kgf·cm, ft·lbf) : Specified torque

◆ Non-reusable part

► Lithium soap base glycol grease

⇨ Disc brake grease

67170-LCSQ-G30

Exploded view of the front disc brake components—2004–06 Land Cruiser

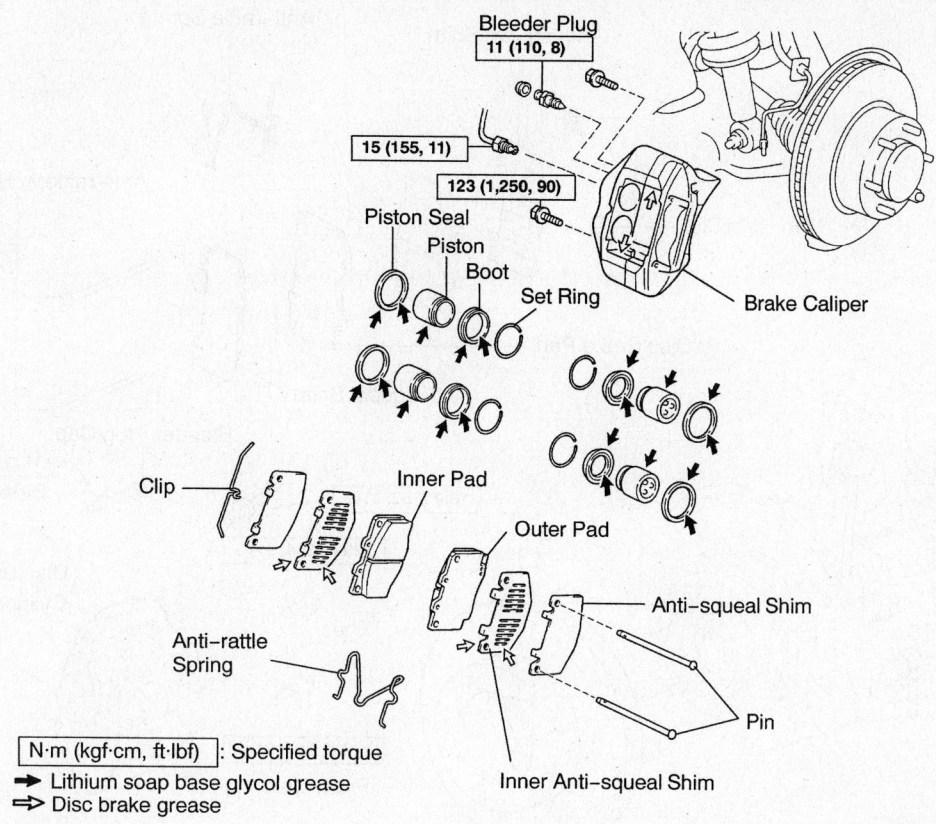

Bleeder Plug
11 (110, 8)

15 (155, 11)

123 (1,250, 90)

Piston Seal

Piston

Boot

Set Ring

Brake Caliper

Clip

Inner Pad

Outer Pad

Anti-squeal Shim

Anti-rattle
Spring

Pin

N·m (kgf·cm, ft·lbf) : Specified torque
➡ Lithium soap base glycol grease
⇨ Disc brake grease

Inner Anti-squeal Shim

67170-LCSQ-G32

Front caliper and related parts—Sequoia

caliper, using 2 new washers. Make sure the flexible hose lock is securely in the lock hole of the caliper. Torque the union bolt to 22 ft. lbs. (30 Nm).

9. Fill the brake system to the proper level and bleed the brake system.

10. Install the tire and wheel assembly.

11. Top off the brake fluid level in the master cylinder. Check for leaks and proper brake operation.

12. Connect the negative battery cable to the battery.

Sequoia

1. Disconnect the negative battery cable from the battery.

2. Raise and support the vehicle safely.

3. Remove the wheels.

4. Disconnect the brake hose from the caliper. Plug the end of the hose to prevent loss of fluid.

5. Remove the bolts that attach the caliper to the torque plate.

6. Lift the bottom of the caliper up and remove the caliper assembly.

To install:

7. Grease the caliper slides and bolts with lithium grease or equivalent. Install the caliper and secure with the bolts. Torque the bolts to 90 ft. lbs. (123 Nm).

8. Connect the brake hose to the caliper. Torque 11 ft. lbs. (15 Nm).

9. Fill the brake system to the proper level and bleed the brake system.

10. Install the tire and wheel assembly.

11. Top off the brake fluid level in the master cylinder. Check for leaks and proper brake operation.

12. Connect the negative battery cable to the battery.

GX470

1. Remove the wheel.

2. Remove the anti-rattle spring from the caliper.

3. Remove the clips and anti-rattle pins.

4. Lift out the pads and shims.

5. If the caliper is being replaced, disconnect the brake line. Plug the line to prevent fluid loss.

6. Remove the caliper mounting bolts. Lift off the caliper.

7. Installation is the reverse of removal. Bleed the brakes. Observe the following torques:

- Caliper mounting bolts: 91 ft. lbs. (123 Nm)
- Brake line-to-caliper: 11 ft. lbs. (15 Nm)

LX470

1. Disconnect the negative battery cable from the battery.

2. Raise and support the vehicle safely.

3. Remove the wheels.

4. Disconnect the brake hose from the caliper by removing the union bolt and 2 gaskets. Plug the end of the hose to prevent loss of fluid.

5. Remove the bolts that attach the caliper to the torque plate.

6. Lift the bottom of the caliper up and remove the caliper assembly.

To install:

7. Grease the caliper slides and bolts with lithium grease or equivalent. Install the caliper and secure with the bolts. Torque the bolts to 90 ft. lbs. (123 Nm).

8. Connect the brake hose to the caliper, using 2 new washers. Make sure the flexible hose lock is securely in the lock hole of the caliper. Torque the union bolt to 22 ft. lbs. (30 Nm).

9. Fill the brake system to the proper level and bleed the brake system.

10. Install the tire and wheel assembly.

11. Top off the brake fluid level in the master cylinder. Check for leaks and proper brake operation.

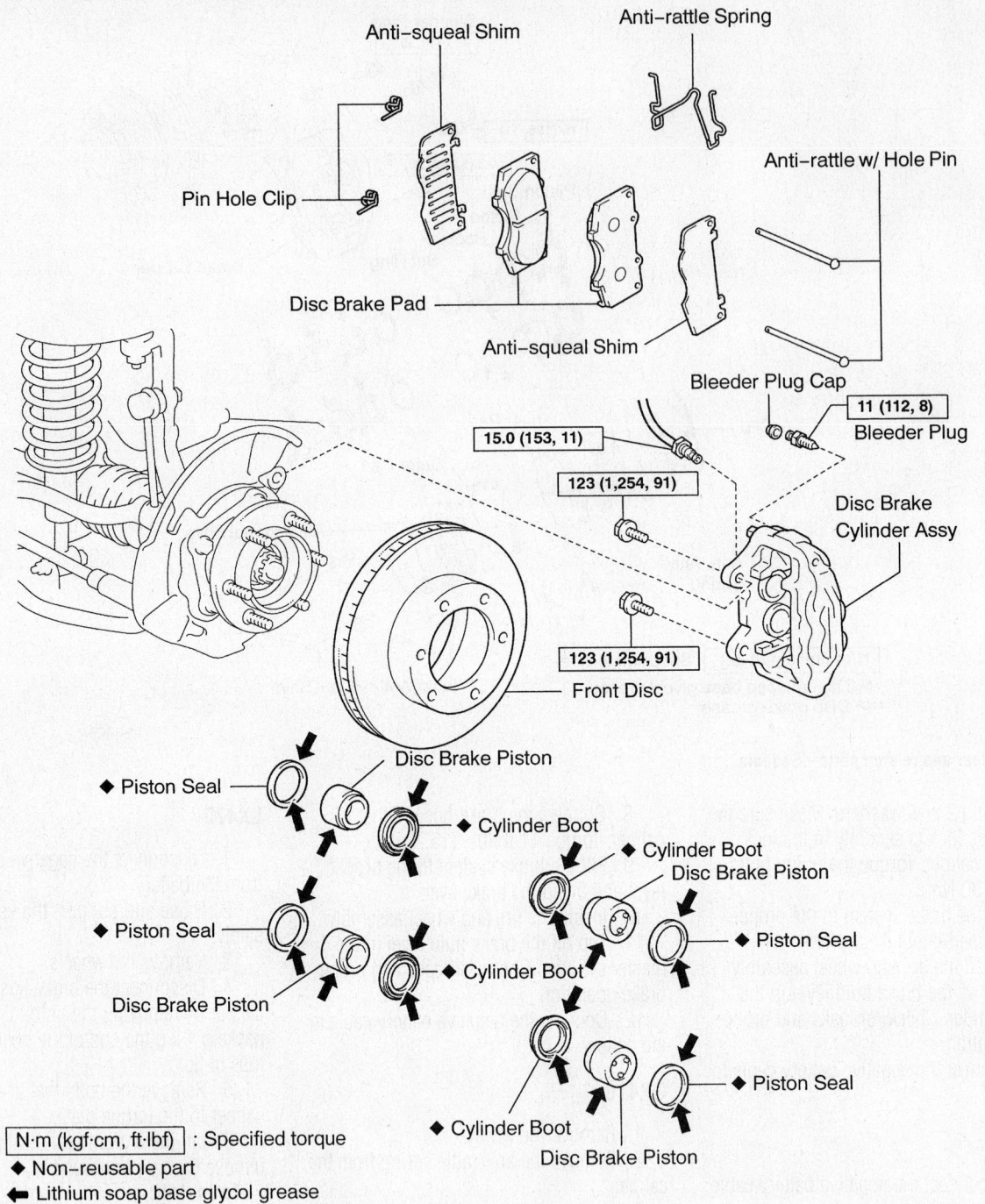

Front brake components—GX470

67162-GX470-G12

12. Connect the negative battery cable to the battery.

Rear Brake Caliper

REMOVAL & INSTALLATION

Land Cruiser

1. Remove the brake line from the caliper.

2. Hold the siding pin and remove the 2 bolts.

3. Remove the caliper from the torque plate.

4. Remove the pads and shims.

5. Remove the pad support plates.

6. Installation is the reverse of removal. Torque the caliper bolts to 20 ft. lbs. (26 Nm). Torque the brake line union bolt to 22 ft. lbs. (30 Nm).

Sequoia

1. Disconnect the negative battery cable from the battery.

2. Raise and support the vehicle safely.

3. Remove the wheels.

4. Disconnect the brake hose from the caliper by removing the union bolt and 2 gaskets. Plug the end of the hose to prevent loss of fluid.

5. Remove the 2 sliding pins.

6. Lift the bottom of the caliper up and remove the caliper assembly.

To install:

7. Grease the caliper slides and pins with silicone grease or equivalent. Install the caliper and secure with the bolts. Torque the pins to 65 ft. lbs. (883 Nm).

8. Connect the brake hose to the caliper, using 2 new washers. Torque the union bolt to 22 ft. lbs. (30 Nm).

9. Fill the brake system to the proper level and bleed the brake system.

10. Install the tire and wheel assembly.

11. Top off the brake fluid level in the master cylinder. Check for leaks and proper brake operation.

12. Connect the negative battery cable to the battery.

GX470

1. Remove the wheel.

2. Remove the anti-rattle spring from the caliper.

3. Remove the clips and anti-rattle pins.

4. Lift out the pads and shims.

5. If the caliper is being replaced, disconnect the brake line. Plug the line to prevent fluid loss.

6. Remove the caliper mounting bolts. Lift off the caliper.

7. Installation is the reverse of removal. Bleed the brakes. Observe the following torques:

- Caliper mounting bolts: 91 ft. lbs. (123 Nm)
- Brake line-to-caliper: 11 ft. lbs. (15 Nm)

LX470

1. Disconnect the negative battery cable from the battery.

2. Raise and support the vehicle safely.

3. Remove the wheels.

4. Disconnect the brake hose from the caliper by removing the union bolt and 2

gaskets. Plug the end of the hose to prevent loss of fluid.

5. Remove the bolts that attach the caliper to the torque plate.

6. Lift the bottom of the caliper up and remove the caliper assembly.

To install:

7. Grease the caliper slides and bolts with lithium grease or equivalent. Install the caliper and secure with the bolts. Torque the bolts to 20 ft. lbs. (27 Nm).

8. Connect the brake hose to the caliper, using 2 new washers. Make sure the flexible hose lock is securely in the lock hole of the caliper. Torque the union bolt to 22 ft. lbs. (30 Nm).

9. Fill the brake system to the proper level and bleed the brake system.

10. Install the tire and wheel assembly.

11. Top off the brake fluid level in the master cylinder. Check for leaks and proper brake operation.

12. Connect the negative battery cable to the battery.

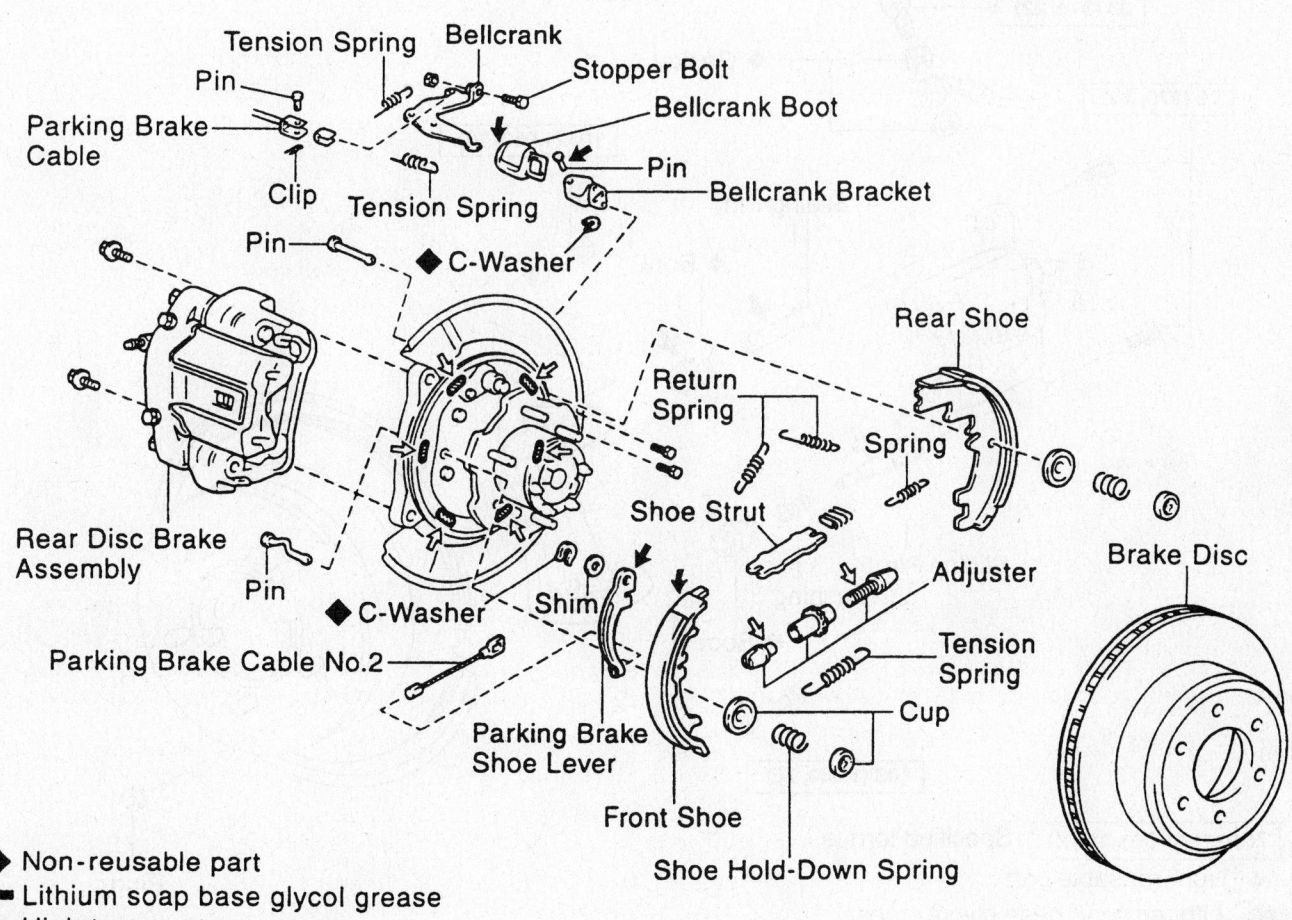

Rear disc brake components—2002 Land Cruiser

◆ Non-reusable part
◀ Lithium soap base glycol grease
◁ High temperature grease

93026G76

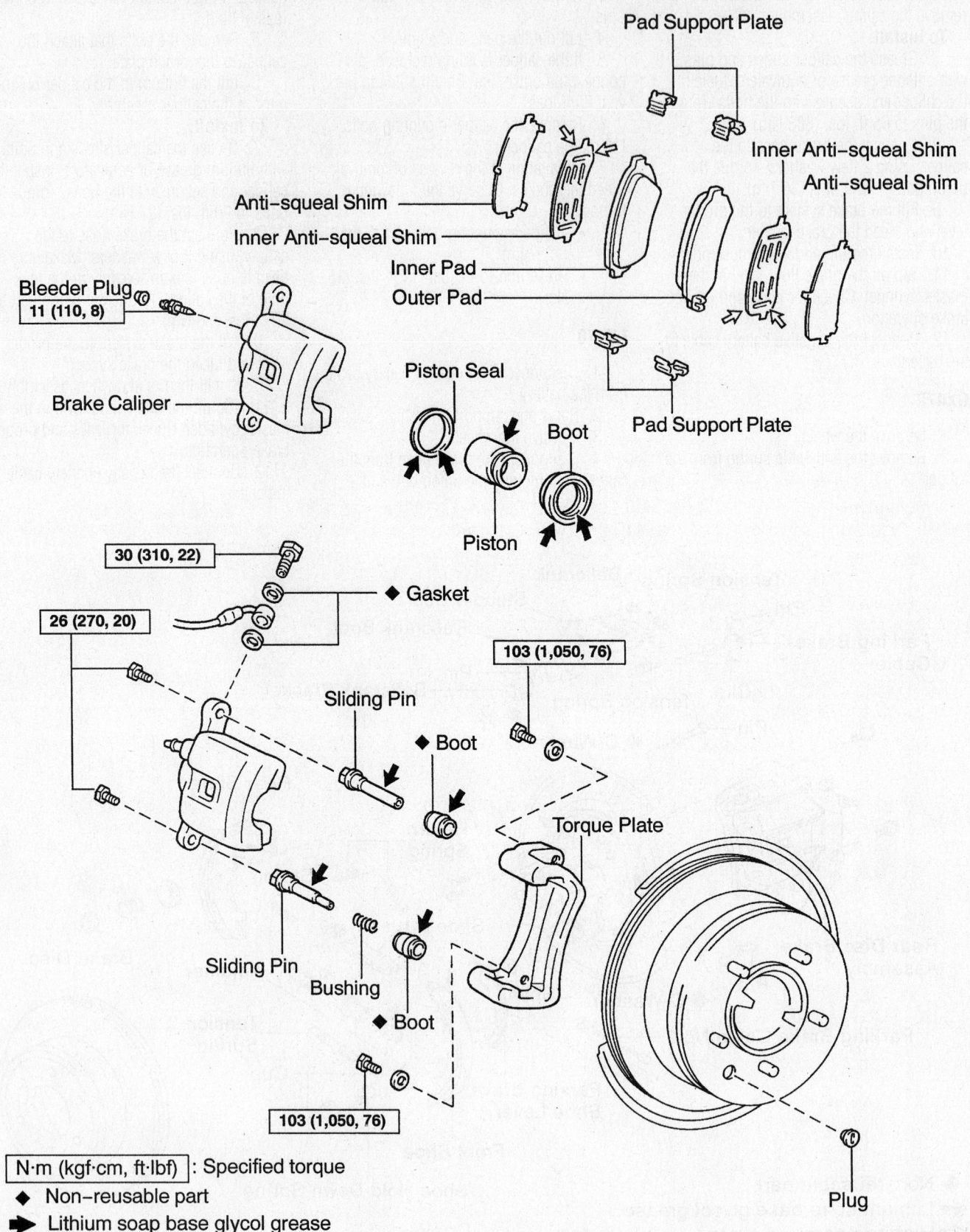

Pad Support Plate

Inner Anti-squeal Shim

Anti-squeal Shim

Anti-squeal Shim

Inner Anti-squeal Shim

Inner Pad

Outer Pad

Pad Support Plate

Bleeder Plug

11 (110, 8)

Brake Caliper

Piston Seal

Boot

Piston

30 (310, 22)

◆ Gasket

26 (270, 20)

103 (1,050, 76)

Sliding Pin

◆ Boot

Torque Plate

Sliding Pin

Bushing

◆ Boot

103 (1,050, 76)

Plug

N·m (kgf·cm, ft·lbf) : Specified torque

◆ Non-reusable part

➡ Lithium soap base glycol grease

⇨ Disc brake grease

67170-LCSQ-G31

Rear brake caliper and related parts—2004–06 Land Cruiser

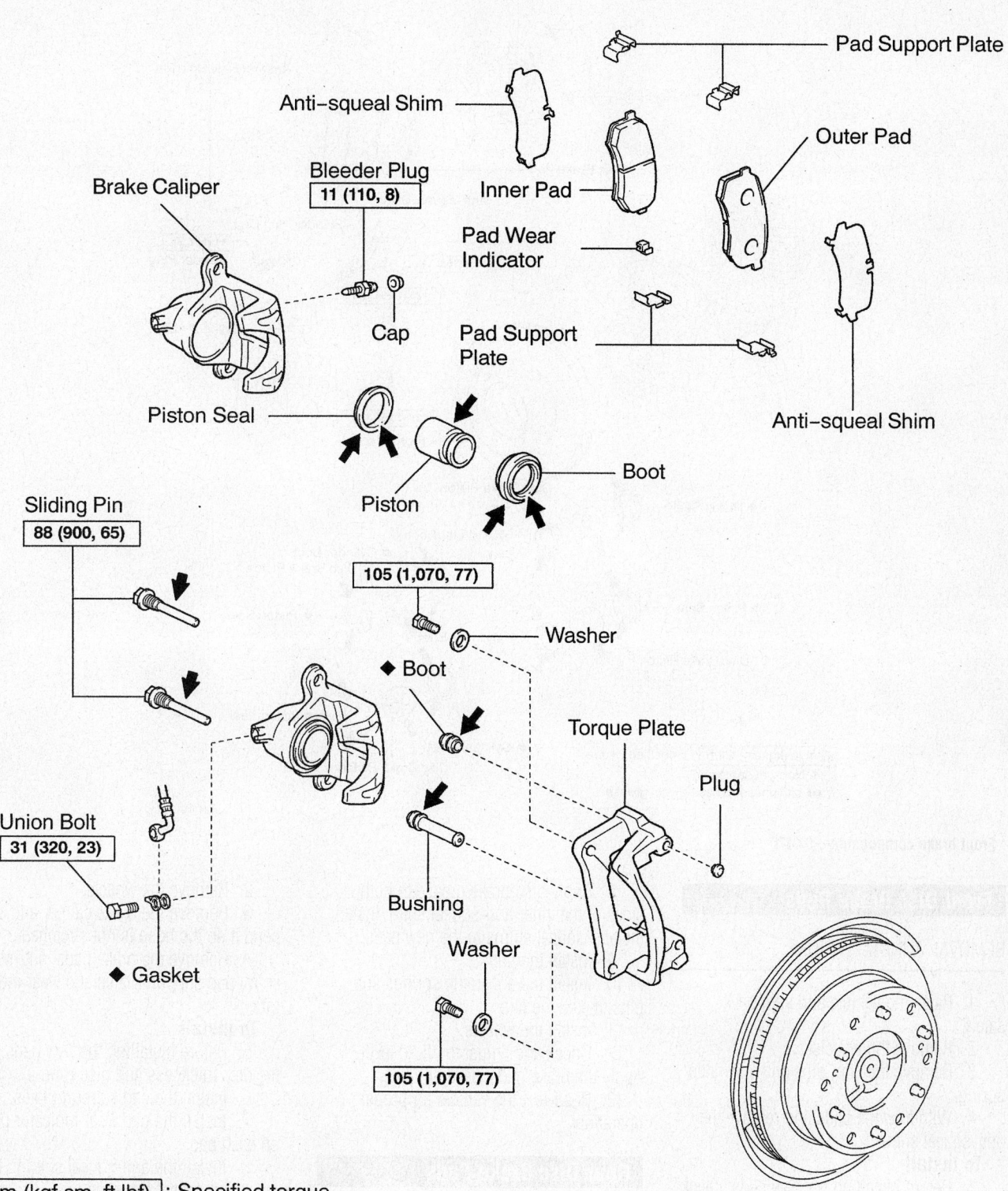

Pad Support Plate

Anti-squeal Shim

Outer Pad

Brake Caliper

Bleeder Plug
11 (110, 8)

Inner Pad

Pad Wear Indicator

Cap

Pad Support Plate

Anti-squeal Shim

Piston Seal

Boot

Piston

Sliding Pin
88 (900, 65)

105 (1,070, 77)

Washer

◆ Boot

Torque Plate

Plug

Union Bolt
31 (320, 23)

Bushing

◆ Gasket

Washer

105 (1,070, 77)

N·m (kgf·cm, ft·lbf) : Specified torque

◆ Non-reusable part

➤ Lithium soap base glycol grease

67170-LCSQ-G33

Rear brake caliper—Sequoia

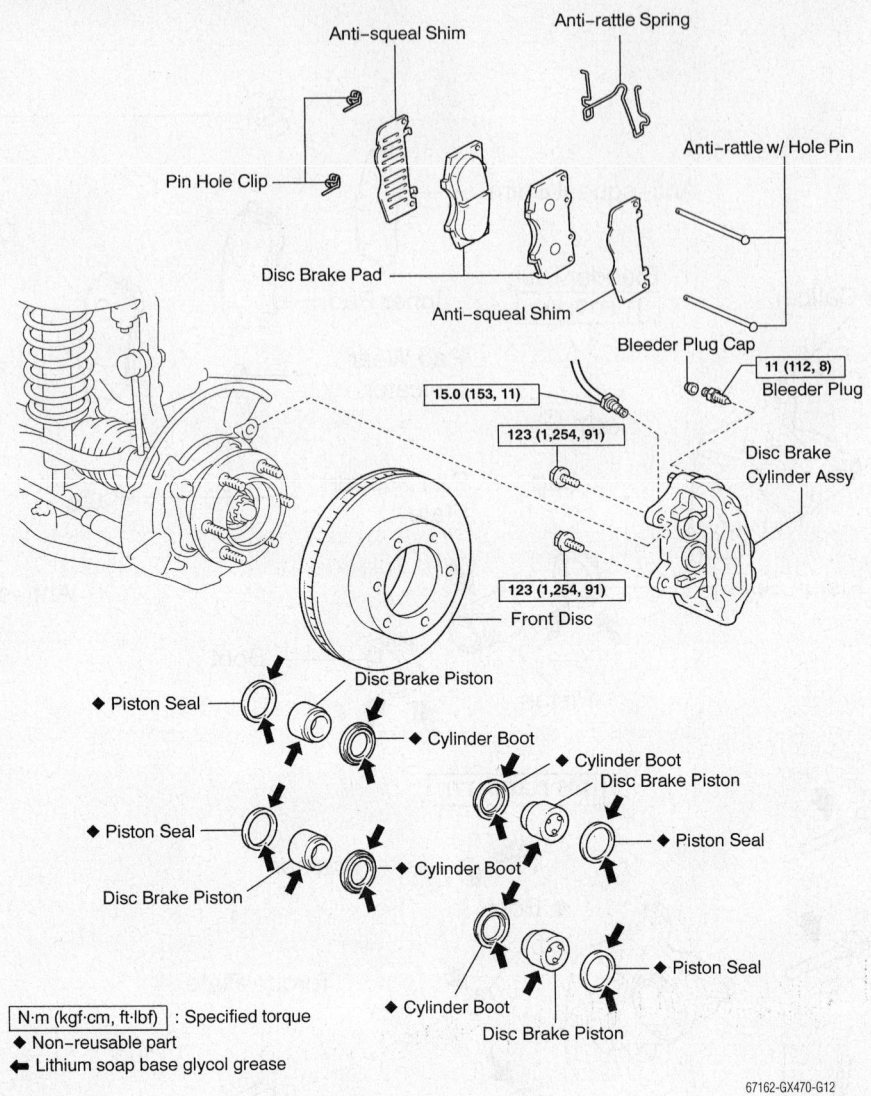

N·m (kgf·cm, ft·lbf) : Specified torque
◆ Non-reusable part
← Lithium soap base glycol grease

67162-GX470-G12

Front brake components—GX470

Front Disc Brake Pads

REMOVAL & INSTALLATION

1. Raise the vehicle and support it safely.
2. Remove the wheels.
3. Remove the clip, pins and anti-rattle spring.
4. Withdraw the pads and remove the anti-squeal shims.

To install:
5. Before installing the new pads, check the disc thickness and disc runout.
6. Siphon out a small amount of brake fluid from the reservoir.
7. Press in the pistons with a hammer handle or equivalent.

8. Apply disc brake grease to both sides of the inner anti-squeal shim. Install the anti-squeal shims to the new pads.
9. Install the pads.
10. Install the anti-rattle springs and pins. Install the clip.
11. Install the wheels.
12. Check and adjust the fluid level. Apply the brake pedal several times.
13. Road-test the vehicle for proper operation.

Rear Disc Brake Pads

REMOVAL & INSTALLATION

1. Raise the vehicle and support it safely.

2. Remove the wheels.
3. Remove the brake caliper and suspend it so the hose is not stretched.
4. Remove the brake pads, anti-squeal shim, pad support plates and wear indicators.

To install:
5. Before installing the new pads, check the disc thickness and disc runout.
6. Install the pad support plates.
7. Install the pad wear indicator plates on each pad.
8. Install the anti-squeal shim to the outer pad. Install the pads.
9. Install the brake caliper.
10. Install the wheels.
11. Apply the brake pedal several times.
12. Road-test the vehicle for proper operation.

BRAKES 3-128
DRIVE TRAIN 3-101
ENGINE REPAIR 3-10
FUEL SYSTEM 3-96
**SPECIFICATIONS AND
MAINTENANCE CHARTS** 3-2
Engine and Vehicle Identification
  Chart........................3-2
General Engine Specifications 3-2
Engine Tune-Up Specifications 3-3
Firing Order 3-3
Accessory Drive Belt Routing 3-4
Capacities 3-4
Valve Specifications 3-5
Crankshaft and Connecting Rod
  Specifications 3-5
Piston and Ring Specifications 3-6
Torque Specifications 3-6
Wheel Alignment 3-7
Tire, Wheel and Ball Joint
  Specifications 3-8
Brake Specifications 3-8
Scheduled Maintenance
  Intervals......................3-9
**STEERING AND
SUSPENSION** 3-114
A
Air Bag..........................3-114
  Disarming 3-114
  Precautions 3-114
Alternator 3-10
  Removal 3-10
Automatic Transaxle Assembly 3-101
  Removal & Installation...........3-101
C
Camshaft and Valve Lifters 3-65
  Removal & Installation...........3-65
Coil Spring 3-124
  Removal & Installation...........3-124
CV-Joints........................3-113
  Overhaul 3-113
Cylinder Head 3-49
  Removal & Installation...........3-49
E
Engine Assembly 3-10
  Removal & Installation...........3-10

Exhaust Manifold 3-62
  Removal & Installation............3-62
F
Front Brake Caliper 3-128
  Removal & Installation............3-128
Front Crankshaft Seal 3-64
  Removal & Installation............3-64
Front Disc Brake Pads 3-130
  Removal & Installation............3-130
Front Wheel Bearing 3-125
  Disassembly And Assembly....3-125
  Installation 3-125
  Removal 3-125
Fuel Filter 3-96
  Removal & Installation............3-96
Fuel Injector 3-97
  Removal & Installation............3-97
Fuel Pump 3-96
  Removal & Installation............3-96
Fuel System Pressure 3-96
  Relieving 3-96
Fuel System Service
  Precautions...................3-96
H
Halfshaft.........................3-107
  Removal & Installation............3-107
Heater Core......................3-31
  Removal & Installation............3-31
I
Ignition Timing 3-10
  Adjustment...................3-10
Intake Manifold 3-57
  Removal & Installation............3-57
L
Lower Ball Joint..................3-124
  Removal & Installation............3-124
Lower Control Arm 3-125
  Control Arm Bushing
    Replacement.................3-125
  Removal & Installation............3-125
O
Oil Pan..........................3-84
  Removal & Installation............3-84
Oil Pump 3-88
  Removal & Installation............3-88

P
Pinion Seal 3-114
  Removal & Installation............3-114
Piston and Ring 3-95
  Positioning 3-95
Pneumatic Front Strut 3-123
  Removal & Installation............3-123
Pneumatic Rear Strut 3-123
  Removal & Installation............3-123
Power Rack And Pinion Steering
  Gear.........................3-114
  Removal & Installation............3-114
R
Rear Brake Caliper 3-129
  Removal & Installation............3-129
Rear Disc Brake Pads 3-130
  Removal & Installation............3-130
Rear Main Seal 3-91
  Removal & Installation............3-91
Rear Wheel Bearings.............3-127
  Removal & Installation............3-127
S
Starter Motor 3-84
  Removal & Installation............3-84
Strut.............................3-117
  Removal & Installation............3-117
  Strut Overhaul................3-121
T
Timing Belt 3-91
  Removal & Installation............3-91
Timing Chain, Sprockets, Front
  Cover and Seal 3-92
  Removal & Installation............3-92
Transfer Case Assembly............3-107
  Removal & Installation............3-107
V
Valve Lash 3-75
  Adjustment...................3-75
W
Water Pump 3-44
  Removal & Installation............3-44

SPECIFICATIONS AND MAINTENANCE CHARTS

ENGINE AND VEHICLE IDENTIFICATION

		Engine							Model Year	
Code ①	Liters (cc)	Cu. In.	Cyl.	Fuel Sys.	Engine Type	Eng. Mfg.		Code ②	Year
1MZ-FE	3.0 (2995)	183	6	SFI	DOHC	Toyota		2	2002
2AZ-FE	2.4 (2362)	144	4	SFI	DOHC	Toyota		3	2003
3MZ-FE	3.3 (NA)	NA	6	SFI	DOHC	Toyota		4	2004
								5	2005
								6	2006

SFI: Sequential Fuel Injection

DOHC: Double Overhead Camshaft

NA: Information not available

① Stamped on the left side of the engine block

② 10th digit of the Vehicle Identification Number (VIN)

09490_HIGH_C0001

GENERAL ENGINE SPECIFICATIONS

Year	Model	Engine Displacement Liters	Engine Series ID	Net Horsepower @ rpm	Net Torque @ rpm (ft. lbs.)	Bore x Stroke (in.)	Com-pression Ratio	Oil Pressure @ rpm
2002	Highlander	2.4	2AZ-FE	155@5600	163@4000	3.48x3.78	NA	36@3000
		3.0	1MZ-FE	220@5800	222@4400	3.44x3.27	10.5:1	43-78@3000
	RX300	3.0	1MZ-FE	220@5800	222@4400	3.44x3.27	10.5:1	43-78@3000
2003	Highlander	2.4	2AZ-FE	155@5600	163@4000	3.48x3.78	NA	36@3000
		3.0	1MZ-FE	220@5800	222@4400	3.44x3.27	10.5:1	43-78@3000
	RX300	3.0	1MZ-FE	220@5800	222@4400	3.44x3.27	10.5:1	43-78@3000
2004	Highlander	2.4	2AZ-FE	155@5600	163@4000	3.48x3.78	NA	36@3000
		3.3	3MZ-FE	230@5600	242@3600	NA	NA	36-78@3000
	RX330	3.3	3MZ-FE	230@5600	242@3600	NA	NA	36-78@3000
2005-06	Highlander	2.4	2AZ-FE	155@5700	163@4000	3.48x3.78	NA	36@3000
		3.3	3MZ-FE	215@5600	222@3600	NA	NA	36-78@3000
	RX330	3.3	3MZ-FE	215@5600	222@3600	NA	NA	36-78@3000

NA: Information not available

09490_HIGH_C0002

ENGINE TUNE-UP SPECIFICATIONS

Year	Engine Displacement Liters	Engine ID	Spark Plug Gap (in.)	Ignition Timing (deg.)*	Fuel Pump (psi)	Idle Speed (rpm)	Intake	Exhaust
2002	2.4	2AZ-FE	0.041	8-12B	44-50	600-700	0.007-0.011	0.012-0.016
	3.0	1MZ-FE	0.039-0.043	8-12B	44-50	650-750	0.006-0.010	0.010-0.014
2003	2.4	2AZ-FE	0.041	8-12B	44-50	600-700	0.007-0.011	0.012-0.016
	3.0	1MZ-FE	0.039-0.043	8-12B	44-50	650-750	0.006-0.010	0.010-0.014
2004	2.4	2AZ-FE	0.041	8-12B	44-50	600-700	0.007-0.011	0.012-0.016
	3.3	3MZ-FE	0.039-0.043	8-12B	44-50	650-750	0.006-0.010	0.010-0.014
2005-06	2.4	2AZ-FE	0.041	8-12B	44-50	600-700	0.007-0.011	0.012-0.016
	3.3	3MZ-FE	0.039-0.043	8-12B	44-50	650-750	0.006-0.010	0.010-0.014

NOTE: The Vehicle Emission Control Information label often reflects specification changes made during production.

The label figures must be used if they differ from those in this chart.

B: Before top dead center

* With terminals TC and CG connected to DLC3

09490_HIGH_C0003

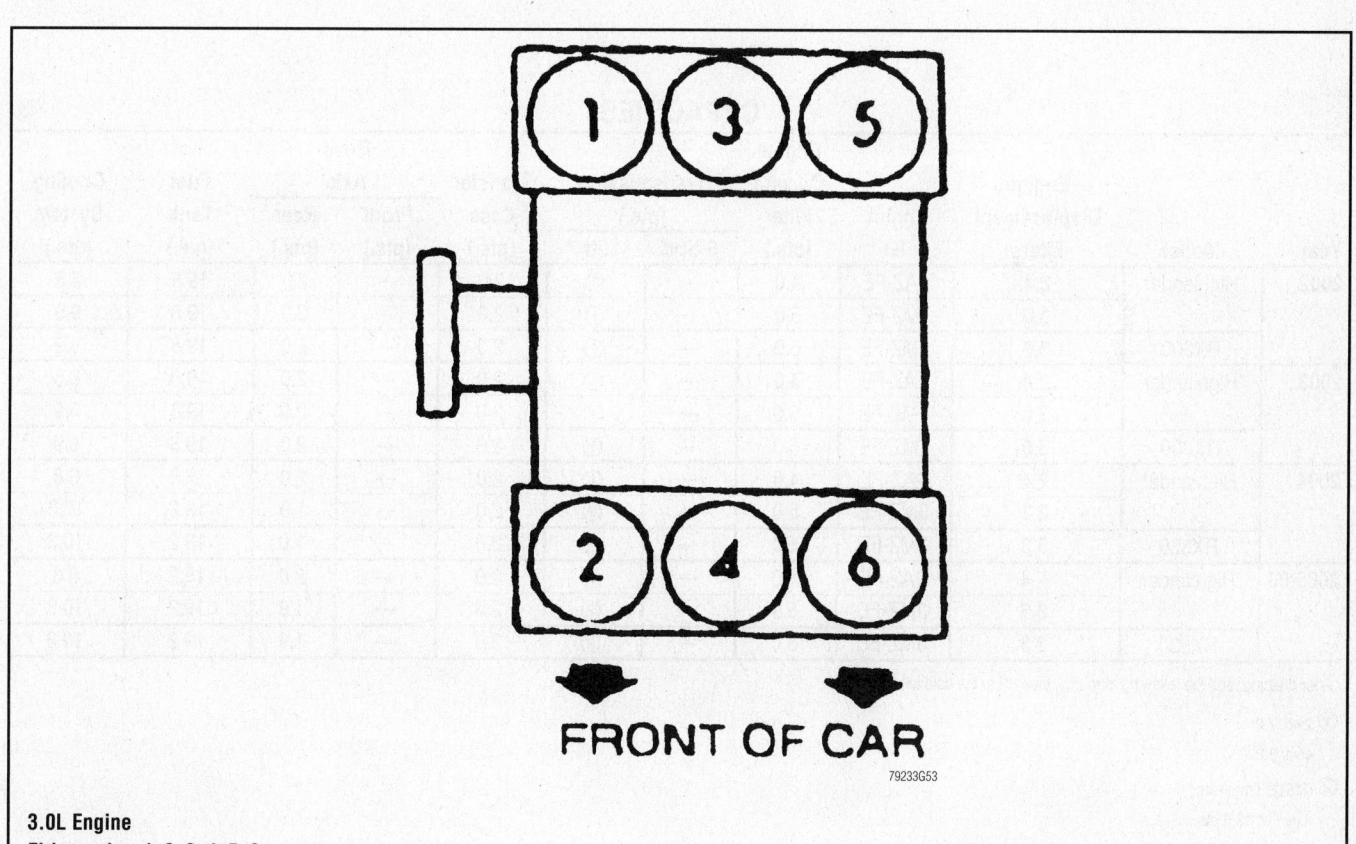

79233G53

3.0L Engine
Firing order: 1–2–3–4–5–6
Distributorless ignition system

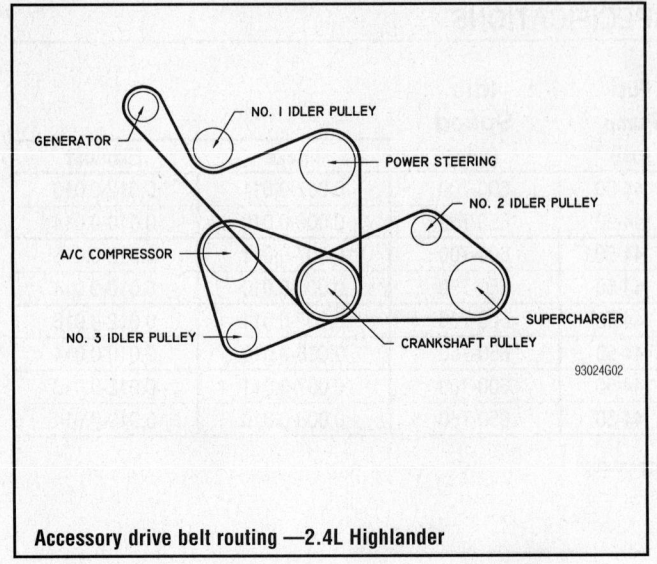

Accessory drive belt routing —2.4L Highlander

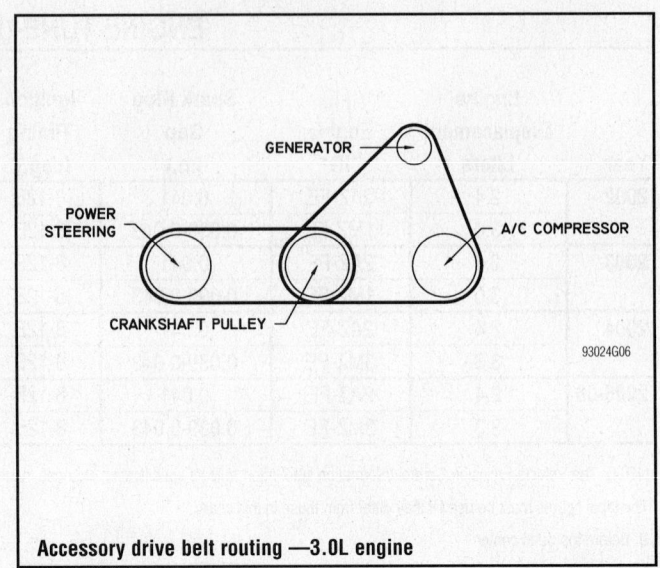

Accessory drive belt routing —3.0L engine

CAPACITIES

Year	Model	Engine Displacement Liters	Engine ID	Engine Oil with Filter (qts.)	Transmission (pts.) 5-Spd	Auto.*	Transfer Case (pts.)	Drive Axle Front (pts.)	Rear (pts.)	Fuel Tank (gal.)	Cooling System (qts.)
2002	Highlander	2.4	2AZ-FE	4.0	—	①	2.0	—	2.0	19.8	6.8
		3.0	1MZ-FE	5.0	—	①	2.0	—	2.0	19.8	9.9
	RX300	3.0	1MZ-FE	5.0	—	①	2.0	—	2.0	19.8	9.9
2003	Highlander	2.4	2AZ-FE	4.0	—	①	2.0	—	2.0	19.8	6.8
		3.0	1MZ-FE	5.0	—	①	2.0	—	2.0	19.8	9.9
	RX300	3.0	1MZ-FE	5.0	—	①	2.0	—	2.0	19.8	9.9
2004	Highlander	2.4	2AZ-FE	4.0	—	①	2.0	—	2.0	19.2	6.8
		3.3	3MZ-FE	5.0	—	②	2.0	—	1.9	19.2	10.3
	RX330	3.3	3MZ-FE	5.0	—	②	2.0	—	1.9	19.2	10.3
2005-06	Highlander	2.4	2AZ-FE	4.0	—	①	2.0	—	2.0	19.2	6.8
		3.3	3MZ-FE	5.0	—	②	2.0	—	1.9	19.2	10.3
	RX330	3.3	3MZ-FE	5.0	—	②	2.0	—	1.9	19.2	10.3

*After draining, add the following amounts, then, fill to the cold full line.

① 2wd: 7.0
  4wd: 8.2

② U151E Transaxle
  Dry Fill: 18.6 pts.
  Drain and Refill: 7.4 pts.
  U151F Transaxle
  Dry Fill: 19.0 pts.
  Drain and Refill: 7.6 pts.

VALVE SPECIFICATIONS

Year	Engine Displacement Liters	Engine ID	Seat Angle (deg.)	Face Angle (deg.)	Spring Test Pressure (lbs. @ in.)	Spring Installed Height (in.)	Stem-to-Guide Clearance (in.)		Stem Diameter (in.)	
							Intake	Exhaust	Intake	Exhaust
2002	2.4	2AZ-FE	45	44.5	NA	NA	0.0010-0.0024	0.0012-0.0026	0.2154-0.2159	0.2152-0.2157
	3.0	1MZ-FE	45	40.5	41.9-46.3@ 1.437	1.331	0.0010-0.0024	0.0012-0.0026	0.2154-0.2159	0.2152-0.2156
2003	2.4	2AZ-FE	45	44.5	NA	NA	0.0010-0.0024	0.0012-0.0026	0.2154-0.2159	0.2152-0.2157
	3.0	1MZ-FE	45	40.5	41.9-46.3@ 1.437	1.331	0.0010-0.0024	0.0012-0.0026	0.2154-0.2159	0.2152-0.2156
2004	2.4	2AZ-FE	45	44.5	NA	NA	0.0010-0.0024	0.0012-0.0026	0.2154-0.2159	0.2152-0.2157
	3.3	3MZ-FE	45	40.5	41.9-46.3@ 1.437	1.331	0.0010-0.0024	0.0012-0.0026	0.2154-0.2159	0.2152-0.2156
2005-06	2.4	2AZ-FE	45	44.5	NA	NA	0.0010-0.0024	0.0012-0.0026	0.2154-0.2159	0.2152-0.2157
	3.3	3MZ-FE	45	40.5	41.9-46.3@ 1.437	1.331	0.0010-0.0024	0.0012-0.0026	0.2154-0.2159	0.2152-0.2156

NA: Information not available

09490_HIGH_C0005

CRANKSHAFT AND CONNECTING ROD SPECIFICATIONS

All measurements are given in inches.

Year	Engine Displacement Liters	Engine ID	Crankshaft				Connecting Rod		
			Main Brg. Journal Dia.	Main Brg. Oil Clearance	Shaft End-play	Thrust on No.	Journal Diameter	Oil Clearance	Side Clearance
2002	2.4	2AZ-FE	2.0654-2.1648	0.0009-0.0019	0.0016-0.0094	2	1.8894-1.8898	0.0009-0.0019	0.0063-0.0143
	3.0	1MZ-FE	2.4011-2.4016	①	0.0016-0.0095	2	2.0863-2.0866	0.0015-0.0025	0.0059-0.0188
2003	2.4	2AZ-FE	2.0654-2.1648	0.0009-0.0019	0.0016-0.0094	2	1.8894-1.8898	0.0009-0.0019	0.0063-0.0143
	3.0	1MZ-FE	2.4011-2.4016	①	0.0016-0.0095	2	2.0863-2.0866	0.0015-0.0025	0.0059-0.0188
2004	2.4	2AZ-FE	2.0654-2.1648	0.0009-0.0019	0.0016-0.0094	2	1.8894-1.8898	0.0009-0.0019	0.0063-0.0143
	3.3	3MZ-FE	2.4011-2.4016	①	0.0016-0.0095	2	2.0863-2.0866	0.0015-0.0026	0.0059-0.0118
2005-06	2.4	2AZ-FE	2.0654-2.1648	0.0009-0.0019	0.0016-0.0094	2	1.8894-1.8898	0.0009-0.0019	0.0063-0.0143
	3.3	3MZ-FE	2.4011-2.4016	①	0.0016-0.0095	2	2.0863-2.0866	0.0015-0.0026	0.0059-0.0118

① Journals 1 and 4: 0.0006 - 0.0013 in.
   Journals 2 and 3: 0.0010 - 0.0018 in.

09490_HIGH_C0006

PISTON AND RING SPECIFICATIONS
All measurements are given in inches.

Year	Engine Displ. Liters	Engine ID	Piston Clearance	Ring Gap			Ring Side Clearance		
				Top Comp.	Bottom Comp.	Oil Control	Top Comp.	Bottom Comp.	Oil Control
2002	2.4	2AZ-FE	0.0020-0.0029	0.0087-0.0126	0.0197-0.0236	0.0039-0.0138	0.0012-0.0028	0.0012-0.0028	SNUG
	3.0	1MZ-FE	0.0033-0.0042	0.0098-0.0138	0.0138-0.0177	0.0059-0.0157	0.0008-0.0028	0.0008-0.0024	SNUG
2003	2.4	2AZ-FE	0.0020-0.0029	0.0087-0.0126	0.0197-0.0236	0.0039-0.0138	0.0012-0.0028	0.0012-0.0028	SNUG
	3.0	1MZ-FE	0.0033-0.0042	0.0098-0.0138	0.0138-0.0177	0.0059-0.0157	0.0008-0.0028	0.0008-0.0024	SNUG
2004	2.4	2AZ-FE	0.0020-0.0029	0.0087-0.0126	0.0197-0.0236	0.0039-0.0138	0.0012-0.0028	0.0012-0.0028	SNUG
	3.3	3MZ-FE	0.0013-0.0023	0.0118-0.0138	0.0197-0.0236	0.0059-0.0157	0.0012-0.0031	0.0008-0.0024	0.0012-0.0043
2005-06	2.4	2AZ-FE	0.0020-0.0029	0.0087-0.0126	0.0197-0.0236	0.0039-0.0138	0.0012-0.0028	0.0012-0.0028	SNUG
	3.3	3MZ-FE	0.0013-0.0023	0.0118-0.0138	0.0197-0.0236	0.0059-0.0157	0.0012-0.0031	0.0008-0.0024	0.0012-0.0043

09490_HIGH_C0007

TORQUE SPECIFICATIONS
All readings in ft. lbs.

Year	Engine Displacement Liters	Engine ID	Cylinder Head Bolts	Main Bearing Bolts	Rod Bearing Bolts	Crankshaft Damper Bolts	Flywheel Bolts	Manifold Intake	Manifold Exhaust	Spark Plugs	Oil Pan Drain Plug
2002	2.4	2AZ-FE	①	29	②	125	72	22	27	14	18
	3.0	1MZ-FE	④	④	②	159	61	32	36	13	33
2003	2.4	2AZ-FE	①	29	②	125	72	22	27	14	18
	3.0	1MZ-FE	③	④	②	159	61	32	36	13	33
2004	2.4	2AZ-FE	①	29	②	125	72	22	27	14	18
	3.3	3MZ-FE	⑤	④	②	162	61	11	36	18	33
2005-06	2.4	2AZ-FE	①	29	②	125	72	22	27	14	18
	3.3	3MZ-FE	⑤	④	②	162	61	11	36	18	33

① Step 1: 58
   Step 2: plus 90 degrees
② Step 1: 18 ft. lbs.
   Step 2: Plus 90 degrees
③ Step 1: 12 point bolts to 40 ft. lbs.
   Step 2: 12 point bolts plus 90 degrees
   Step 3: Hex head recessed bolt to 13 ft. lbs.
④ Step 1: 12 point cap bolts to 16 ft. lbs.
   Step 2: 12 point cap bolts plus 90 degrees
   Step 3: Hex head side bolts to 20 ft. lbs.
⑤ Step 1: 12 point bolts to 40 ft. lbs.
   Step 2: 12 point bolts plus 90 degrees
   Step 3: Hex head recessed bolt to 13 ft. lbs.

09490_HIGH_C0008

WHEEL ALIGNMENT

Year	Model		Caster Range (+/-Deg.)	Caster Preferred Setting (Deg.)	Camber Range (+/-Deg.)	Camber Preferred Setting (Deg.)	Toe-in (in.)	Steering Axis Inclination (Deg.)
2002	Highlander	2WD F	0.75	+2.75	0.75	-0.67	0+/-0.08	10.75+/-0.75
		4WD F	0.75	+2.75	0.75	-0.58	0+/-0.08	10.58+/-0.75
		2WD R	—	—	0.75	-1.33	0.12+/-0.08	—
		4WD R	—	—	0.75	-0.75	0.12+/-0.08	—
	RX300	2WD F	0.75	+2.08	0.75	-0.33	0.04+/-0.08	10.58+/-0.75
		2WD R	—	—	0.75	-0.33	0.08+/-0.08	—
		4WD R	—	—	0.75	-0.35	0.12+/-0.08	—
2003	Highlander	2WD F	0.75	+2.75	0.75	-0.67	0+/-0.08	10.75+/-0.75
		4WD F	0.75	+2.75	0.75	-0.58	0+/-0.08	10.58+/-0.75
		2WD R	—	—	0.75	-1.33	0.12+/-0.08	—
		4WD R	—	—	0.75	-0.75	0.12+/-0.08	—
	RX300	2WD F	0.75	+2.08	0.75	-0.33	0.04+/-0.08	10.58+/-0.75
		2WD R	—	—	0.75	-0.33	0.08+/-0.08	—
		4WD R	—	—	0.75	-0.35	0.12+/-0.08	—
2004	Highlander	2WD F	0.75	+2.75	0.75	-0.67	0+/-0.08	10.75+/-0.75
		4WD F	0.75	+2.75	0.75	-0.58	0+/-0.08	10.58+/-0.75
		2WD R	—	—	0.75	-1.33	0.12+/-0.08	—
		4WD R	—	—	0.75	-0.75	0.12+/-0.08	—
	RX330 without air suspension	2WD F	0.75	+2.85	0.75	-0.67	0+/-0.08	10.75+/-0.75
		4WD F	0.75	+2.83	0.75	-0.58	0+/-0.08	10.58+/-0.75
		2WD R	—	—	0.75	-1.33	0.12+/-0.08	—
		4WD R	—	—	0.75	-0.83	0.12+/-0.08	—
	with air suspension	2WD F	0.75	+2.85	0.75	-0.67	0+/-0.08	10.75+/-0.75
		4WD F	0.75	+2.67	0.75	-0.62	0+/-0.08	10.58+/-0.75
		2WD R	—	—	0.75	-1.35	0.12+/-0.08	—
		4WD R	—	—	0.75	-0.92	0.12+/-0.08	—
2005-06	Highlander	2WD F	0.75	+2.75	0.75	-0.67	0+/-0.08	10.75+/-0.75
		4WD F	0.75	+2.75	0.75	-0.58	0+/-0.08	10.58+/-0.75
		2WD R	—	—	0.75	-1.33	0.12+/-0.08	—
		4WD R	—	—	0.75	-0.75	0.12+/-0.08	—
	RX330 without air suspension	2WD F	0.75	+2.85	0.75	-0.67	0+/-0.08	10.75+/-0.75
		4WD F	0.75	+2.83	0.75	-0.58	0+/-0.08	10.58+/-0.75
		2WD R	—	—	0.75	-1.33	0.12+/-0.08	—
		4WD R	—	—	0.75	-0.83	0.12+/-0.08	—
	with air suspension	2WD F	0.75	+2.85	0.75	-0.67	0+/-0.08	10.75+/-0.75
		4WD F	0.75	+2.67	0.75	-0.62	0+/-0.08	10.58+/-0.75
		2WD R	—	—	0.75	-1.35	0.12+/-0.08	—
		4WD R	—	—	0.75	-0.92	0.12+/-0.08	—

09490_HIGH_C0009

TIRE, WHEEL AND BALL JOINT SPECIFICATIONS

Year	Model	OEM Tires Standard	OEM Tires Optional	Tire Pressures (psi) Front	Tire Pressures (psi) Rear	Wheel Size	Ball Joint Inspection	Lugnut Torque (ft. lbs.)
2002	Highlander	P225/70R16	None	30	30	6.5-JJ	①	70
	RX300	P255/70HR16	None	30	30	6.5-JJ	①	70
2003	Highlander	P225/70R16	None	30	30	6.5-JJ	①	70
	RX300	P255/70HR16	None	30	30	6.5-JJ	①	70
2004	Highlander	P225/70R16	P225/65R17	30	30	6.5-JJ	①	76
	RX330	P255/65R17	P235/55R18	30	30	6.5-JJ	①	76
2005-06	Highlander	P225/70R16	P225/65R17	30	30	6.5-JJ	①	76
	RX330	P255/65R17	P235/55R18	30	30	6.5-JJ	①	76

OEM: Original Equipment Manufacturer

PSI: Pounds Per Square Inch

STD: Standard

OPT: Optional

① Replace if any measurable movement is found.

09490_HIGH_C0010

BRAKE SPECIFICATIONS
All measurements in inches unless noted

Year	Model		Brake Disc Original Thickness	Brake Disc Minimum Thickness	Brake Disc Maximum Runout	Minimum Lining Thickness	Brake Caliper Bracket Bolts (ft. lbs.)	Brake Caliper Mounting Bolts (ft. lbs.)
2002	Highlander	F	1.102	1.024	0.0020	0.039	79	25
		R	0.394	0.354	0.0059	0.039	43	25
	RX300	F	1.020	1.024	0.0020	0.039	79	25
		R	0.394	0.354	0.0059	0.039	34	14
2003	Highlander	F	1.102	1.024	0.0020	0.039	79	25
		R	0.394	0.354	0.0059	0.039	43	25
	RX300	F	1.020	1.024	0.0020	0.039	79	25
		R	0.394	0.354	0.0059	0.039	34	14
2004	Highlander	F	1.102	1.024	0.0020	0.039	77	25
		R	0.394	0.335	0.0059	0.039	58	32
	RX330	F	1.020	1.024	0.0020	0.039	77	25
		R	0.394	0.335	0.0059	0.039	58	32
2005-06	Highlander	F	1.102	1.024	0.0020	0.039	77	25
		R	0.394	0.335	0.0059	0.039	58	32
	RX330	F	1.020	1.024	0.0020	0.039	77	25
		R	0.394	0.335	0.0059	0.039	58	32

F: Front

R: Rear

09490_HIGH_C0011

SCHEDULED MAINTENANCE INTERVALS
TOYOTA—HIGHLANDER, LEXUS—RX300 AND RX330

TO BE SERVICED	TYPE OF SERVICE	VEHICLE MILEAGE INTERVAL (x1000)												
		7.5	15	22.5	30	37.5	45	52.5	60	67.5	75	82.5	90	97.5
Engine oil & filter	R	✓	✓	✓	✓	✓	✓	✓	✓	✓	✓	✓	✓	✓
Automatic transmission fluid	S/I		✓		✓		✓		✓		✓		✓	
Ball joints & dust covers	S/I		✓		✓		✓		✓		✓		✓	
Bolts & nuts on chassis & body	S/I		✓		✓		✓		✓		✓		✓	
Brake linings & drums	S/I		✓		✓		✓		✓		✓		✓	
Brake line pipes & hoses	S/I		✓		✓		✓		✓		✓		✓	
Brake pads & discs (front & rear)	S/I		✓		✓		✓		✓		✓		✓	
Propeller shaft grease	S/I		✓		✓		✓		✓		✓		✓	
Steering knuckle & chassis grease	S/I		✓		✓		✓		✓		✓		✓	
Steering linkage	S/I		✓		✓		✓		✓		✓		✓	
Air cleaner filter	R				✓				✓				✓	
Spark plugs ①	R				✓				✓				✓	
Drive belts	S/I				✓				✓				✓	
Exhaust pipes & mountings	S/I				✓				✓				✓	
Fuel lines & connections	S/I				✓				✓				✓	
Engine coolant	R						✓				✓			
Charcoal canister	R								✓					
Fuel tank cap gasket	R								✓					
Heated oxygen sensors (except Calif.) ②	R													

R: Replace   S/I: Service or Inspect

① Platinum plugs are replaced at 100,000 mile intervals

② Heated oxygen sensors (except Calif.): replace every 80,000 miles.

FREQUENT OPERATION MAINTENANCE (SEVERE SERVICE)

If a vehicle is operated under any of the following conditions it is considered severe service:

- Extremely dusty areas.

- 50% or more of the vehicle operation is in 32°C (90°F) or higher temperatures, or constant temperatures below 0°C (32°F).

- Prolonged idling (vehicle operation in stop and go traffic).

- Frequent short running periods (engine does not warm to normal operating temperatures).

- Police, taxi, delivery usage or trailer towing usage.

Air cleaner filter: service or inspect every 3750 miles

Engine oil & filter: replace every 3750 miles.

Ball joints & dust covers: service or inspect every 7500 miles.

Bolts & nuts on chassis & body: service or inspect every 7500 miles.

Brake pads & discs (front & rear): service or inspect every 7500 miles.

Steering knuckle & chassis grease: service or inspect every 7500 miles.

Steering linkage: service or inspect every 7500 miles.

Exhaust pipes & mountings: service or inspect every 15,000 miles.

09490_HIGH_C0012

ENGINE REPAIR

→Disconnecting the negative battery cable on some vehicles may interfere with the functions of the on board computer system. The computer may undergo a relearning process once the negative battery cable is reconnected.

Alternator

REMOVAL

Highlander

2.4L ENGINE

1. Before servicing the vehicle, refer to the precautions section.
2. Remove or disconnect the following:
   • Electrical wiring from the alternator
   • Drive belt
   • 1 adjusting and 2 mounting bolts
   • Alternator
3. Installation is the reverse of removal. Observe the following torques:
   • M8 bolts: 15 ft. lbs. (21Nm)
   • M10 bolts: 38 ft. lbs. (52Nm)

3.0L ENGINE

1. Before servicing the vehicle, refer to the precautions section.
2. Remove or disconnect the following:
   • Alternator electrical connectors
   • Wiring harness from the clip
   • Pivot bolt
   • Adjuster lockbolt
   • Drive belt
   • Alternator

To install:

3. Install or connect the following:
   • Alternator
   • Drive belt
   • Adjusting lockbolt. Tighten the bolt to 13 ft. lbs. (18 Nm).
   • Pivot bolt. Tighten the bolt to 43 ft. lbs. (58 Nm).
   • Wiring harness from the clip
   • Alternator electrical connectors

3.3L ENGINE

1. Before servicing the vehicle, refer to the precautions section.
2. Remove or disconnect the following:
   • Alternator electrical connectors
   • Wiring harness from the clip
   • Pivot bolt
   • Plate washer
   • Adjusting lockbolt
   • Drive belt
   • Alternator

To install:

3. Install or connect the following:
   • Alternator
   • Drive belt. Tension the belt to 170–180 lbs. for a new belt or 95–135 lbs. for a used belt.
   • Adjusting lockbolt. Tighten the bolt to 13 ft. lbs. (18 Nm).
   • Plate washer
   • Pivot bolt. Tighten the bolt to 41 ft. lbs. (56 Nm) for the 3.0L and 43 ft. lbs. (58 Nm) for the 3.3L.
   • Wiring harness from the clip
   • Alternator electrical connectors

RX300 and RX330

On some models, the alternator is mounted very low on the engine. On these models, it may be necessary to remove the gravel shield and work from beneath the vehicle in order to gain access to the alternator. Replacing the alternator while the engine is cold is recommended. A hot engine can result in personal injury.

1. Before servicing the vehicle, refer to the precautions section.
2. Remove or disconnect the following:
   • Alternator electrical connectors
   • Wiring harness from the clip
   • Pivot bolt
   • Plate washer
   • Adjusting lockbolt
   • Drive belt
   • Alternator

To install:

3. Install or connect the following:
   • Alternator
   • Drive belt. Tension the belt to 170–180 lbs. for a new belt or 95–135 lbs. for a used belt.
   • Adjusting lockbolt. Tighten the bolt to 13 ft. lbs. (18 Nm).
   • Plate washer
   • Pivot bolt. Tighten the bolt to 41 ft. lbs. (56 Nm) for the 3.0L and 43 ft. lbs. (58 Nm) for the 3.3L.
   • Wiring harness from the clip
   • Alternator electrical connectors

Ignition Timing

ADJUSTMENT

All engines are equipped with a Distributorless Ignition System (DIS). No timing adjustment is possible.

Engine Assembly

REMOVAL & INSTALLATION

Highlander

2.4L ENGINE

1. Before servicing the vehicle, refer to the precautions section.
2. Matchmark the hood position.
3. Remove or disconnect the following:
   • Front wheels
   • No.1 engine undercover
   • Right and left fender splash shields
   • Right fender apron seal
   • Engine oil
   • Coolant
   • Transaxle fluid
   • Transfer case oil
   • Battery
   • Air cleaner
   • Radiator hoses
   • Oil cooler hoses
   • Upper engine stay
   • Upper engine mount bracket
   • Accessory drive belts
   • Steering pump reservoir
   • Steering pump hoses
   • All cables and wires connected to the engine
   • Exhaust pipe
   • Front drive shaft
   • Stabilizer links
   • Left and right axle hub nuts
   • Left and right speed sensors

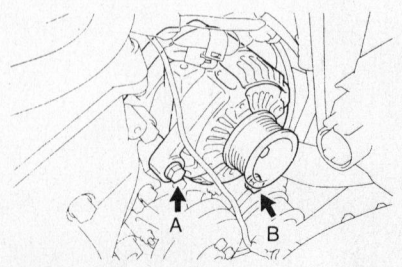

67170-HIGH-G01
Alternator bolt locations—2.4L Highlander

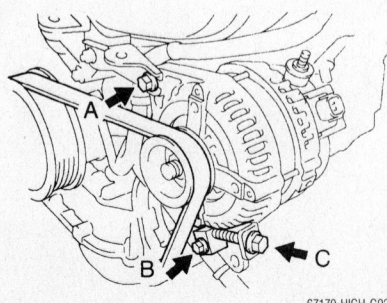

67170-HIGH-G02
Alternator bolt locations—3.3L Highlander

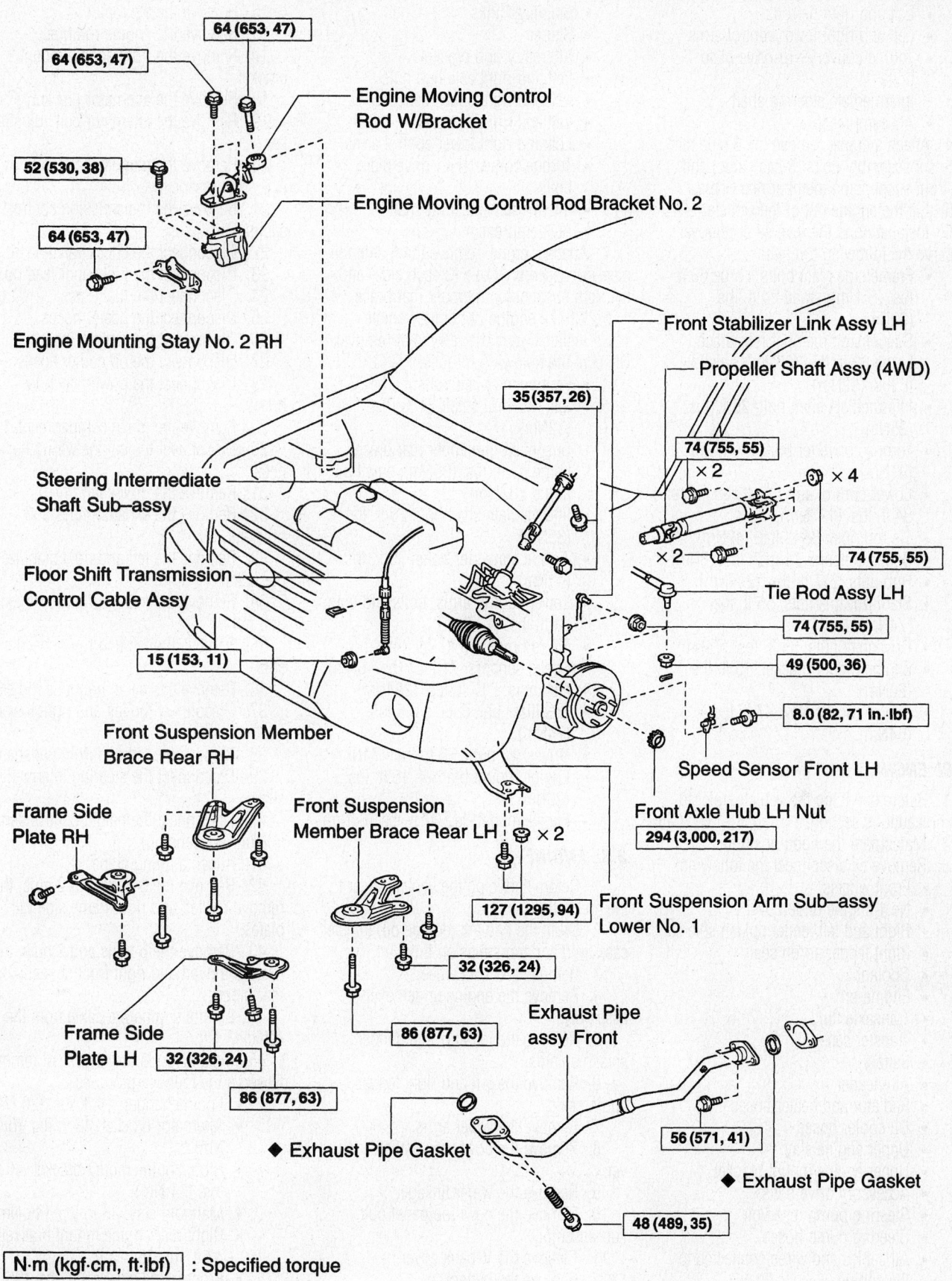

64 (653, 47)

64 (653, 47)

Engine Moving Control
Rod W/Bracket

52 (530, 38)

Engine Moving Control Rod Bracket No. 2

64 (653, 47)

Engine Mounting Stay No. 2 RH

Front Stabilizer Link Assy LH

Propeller Shaft Assy (4WD)

35 (357, 26)

74 (755, 55)
× 2   × 4

Steering Intermediate
Shaft Sub–assy

× 2

74 (755, 55)

Floor Shift Transmission
Control Cable Assy

Tie Rod Assy LH

74 (755, 55)

15 (153, 11)

49 (500, 36)

8.0 (82, 71 in.·lbf)

Front Suspension Member
Brace Rear RH

Speed Sensor Front LH

Frame Side
Plate RH

Front Suspension
Member Brace Rear LH   × 2

Front Axle Hub LH Nut
294 (3,000, 217)

127 (1295, 94)

Front Suspension Arm Sub–assy
Lower No. 1

32 (326, 24)

Frame Side
Plate LH   32 (326, 24)

86 (877, 63)

Exhaust Pipe
assy Front

86 (877, 63)

◆ Exhaust Pipe Gasket

56 (571, 41)

◆ Exhaust Pipe Gasket

48 (489, 35)

N·m (kgf·cm, ft·lbf) : Specified torque

◆ Non–reusable part

Engine mounts and related parts—2.4L Highlander

67170-HIGH-G03

- Left and right tie rods
- Left and right lower control arms
- Torque converter-to-drive plate bolts
- Intermediate steering shaft
- AC compressor

4. Attach a crane, remove the 6 side rail plate subassembly bolts (3 each side) and the front suspension member rear brace.

5. Lift the engine out of the vehicle.

6. Installation is the reverse of removal. Observe the following torques:

- Frame side plate bolts: Large 63 ft. lbs. (85Nm); small 24 ft. lbs. (32Nm)
- Suspension member rear brace: Large 63 ft. lbs. (85Nm); small 24 ft. lbs. (32Nm)
- Intermediate shaft bolt: 26 ft. lbs. (35Nm)
- Torque converter bolts: 30 ft. lbs. (41Nm)
- Lower control arms, bolts and nuts: 94 ft. lbs. (127Nm)
- Tie rod nuts: 36 ft. lbs. (49Nm)
- Speed sensors: 71 inch lbs. (8Nm)
- Hub nuts: 217 ft. lbs. (294Nm)
- Stabilizer link nuts: 55 ft. lbs. (74Nm)
- Driveshaft nuts: 55 ft. lbs. (74Nm)
- Engine mount bracket: 15 ft. lbs. (20Nm)
- Engine mount stay: 47 ft. lbs. (64Nm)

3.0L ENGINE

1. Before servicing the vehicle, refer to the precautions section.

2. Matchmark the hood position.

3. Remove or disconnect the following:

- Front wheels
- No.1 engine undercover
- Right and left fender splash shields
- Right fender apron seal
- Coolant
- Engine oil
- Transaxle fluid
- Transfer case oil
- Battery
- Air cleaner
- Radiator and heater hoses
- Oil cooler hoses
- Upper engine stay
- Upper engine mount bracket
- Accessory drive belts
- Steering pump reservoir
- Steering pump hoses
- All cables and wires connected to the engine
- Exhaust pipes
- Front drive shaft with center bearing

- Stabilizer links
- Starter
- Alternator and brackets
- Left and right axle hub nuts
- Left and right speed sensors
- Left and right tie rods
- Left and right lower control arms
- Torque converter-to-drive plate bolts
- Intermediate steering shaft
- AC compressor

4. Attach a crane, remove the 6 side rail plate subassembly bolts (3 each side) and the front suspension member rear brace.

5. Lift the engine out of the vehicle.

6. Installation is the reverse of removal. Observe the following torques:

- Frame side plate bolts: Large 63 ft. lbs. (85Nm); small 24 ft. lbs. (32Nm)
- Suspension member rear brace: Large 63 ft. lbs. (85Nm); small 24 ft. lbs. (32Nm)
- Intermediate shaft bolt: 26 ft. lbs. (35Nm)
- Torque converter bolts: 30 ft. lbs. (41Nm)
- Lower control arms, bolts and nuts: 94 ft. lbs. (127Nm)
- Tie rod nuts: 36 ft. lbs. (49Nm)
- Speed sensors: 71 inch lbs. (8Nm)
- Hub nuts: 217 ft. lbs. (294Nm)
- Stabilizer link nuts: 55 ft. lbs. (74Nm)
- Driveshaft nuts: 55 ft. lbs. (74Nm)
- Engine mount bracket: 15 ft. lbs. (20Nm)
- Engine mount stay: 47 ft. lbs. (64Nm)

3.3L ENGINE

1. Before servicing the vehicle, refer to the precautions section.

2. Drain the coolant, engine oil, transfer case fluid and transmission fluid.

3. Remove the front wheels.

4. Remove the engine undercover assembly.

5. Remove the left and right fender splash shields.

6. Remove the left and right fender apron seals.

7. Remove the wiper arms.

8. Remove the cowl top ventilator louver.

9. Remove the wiper linkage.

10. Remove the cowl top panel outer sub-assembly.

11. Remove the V-bank cover.

12. Remove the battery.

13. Remove the air cleaner assembly.

14. Remove the A/C compressor drive belt.

15. Remove the alternator.

16. Remove the engine roll brace.

17. Remove the front engine mount bracket.

18. Remove the alternator bracket.

19. Remove the alternator belt adjusting bar.

20. Remove the magnetic clutch from the A/C compressor.

21. Remove the transmission control cable.

22. Disconnect the check valve hose.

23. Disconnect the fuel vapor feed hose.

24. Disconnect the fuel pipes.

25. Disconnect the heater hoses.

26. Disconnect the radiator hoses.

27. Disconnect the oil cooler hoses.

28. Disconnect the power steering hoses.

29. Remove the glove compartment door.

30. Disconnect the engine wiring harness.

31. Remove the driveshaft (4wd).

32. Remove the exhaust pipes and brackets.

33. Remove the left and right stabilizer bar links.

34. Remove the left and right front axle hub nuts.

35. Remove the left and right speed sensors.

36. Remove the left and right tie rod ends.

37. Disconnect the left and right lower control arms.

38. Remove the left and right halfshafts.

39. Disconnect the steering intermediate shaft.

40. Disconnect the height control sensor link (air suspension).

41. Attach a lifting crane.

42. Remove the 6 bolts and 2 nuts, then, remove the left and right frame side rail plates.

43. Remove the 6 bolts and 2 nuts, then, remove the left and right front suspension rear braces.

44. Lift the engine/transaxle from the vehicle.

45. Installation is the reverse of removal. Observe the following torques:

- Engine hanger: 14 ft. lbs. (20 Nm)
- Alternator bracket: 43 ft. lbs. (58 Nm)
- Right engine mount bracket: 40 ft. lbs. (54 Nm)
- Manifold stay: 36 ft. lbs. (49 Nm)
- Right rear engine mount bracket: 47 ft. lbs. (64 Nm)
- Front frame nuts: 70 ft. lbs. (95 Nm)
- Front right engine mount insulator nut: 64 ft. lbs. (87 Nm)

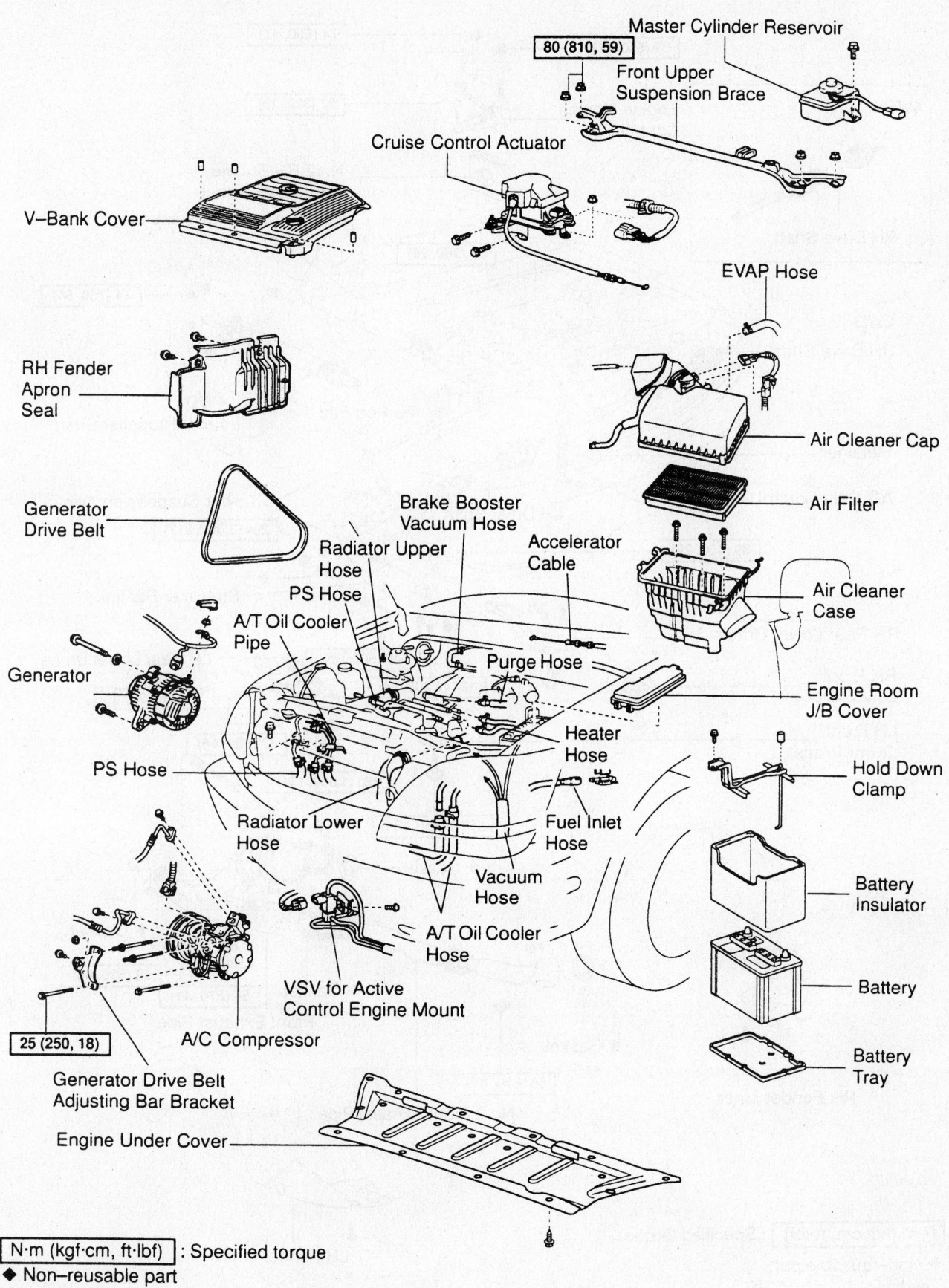

Master Cylinder Reservoir

80 (810, 59)

Front Upper
Suspension Brace

Cruise Control Actuator

V–Bank Cover

EVAP Hose

RH Fender
Apron
Seal

Air Cleaner Cap

Air Filter

Generator
Drive Belt

Brake Booster
Vacuum Hose

Radiator Upper
Hose

Accelerator
Cable

Air Cleaner
Case

PS Hose

A/T Oil Cooler
Pipe

Purge Hose

Generator

Heater
Hose

Engine Room
J/B Cover

PS Hose

Hold Down
Clamp

Radiator Lower
Hose

Fuel Inlet
Hose

Battery
Insulator

Vacuum
Hose

A/T Oil Cooler
Hose

Battery

25 (250, 18)

VSV for Active
Control Engine Mount

A/C Compressor

Battery
Tray

Generator Drive Belt
Adjusting Bar Bracket

Engine Under Cover

N·m (kgf·cm, ft·lbf) : Specified torque

◆ Non–reusable part

7924ZG84

Exploded view of engine pre-removal components—3.0L Highlander

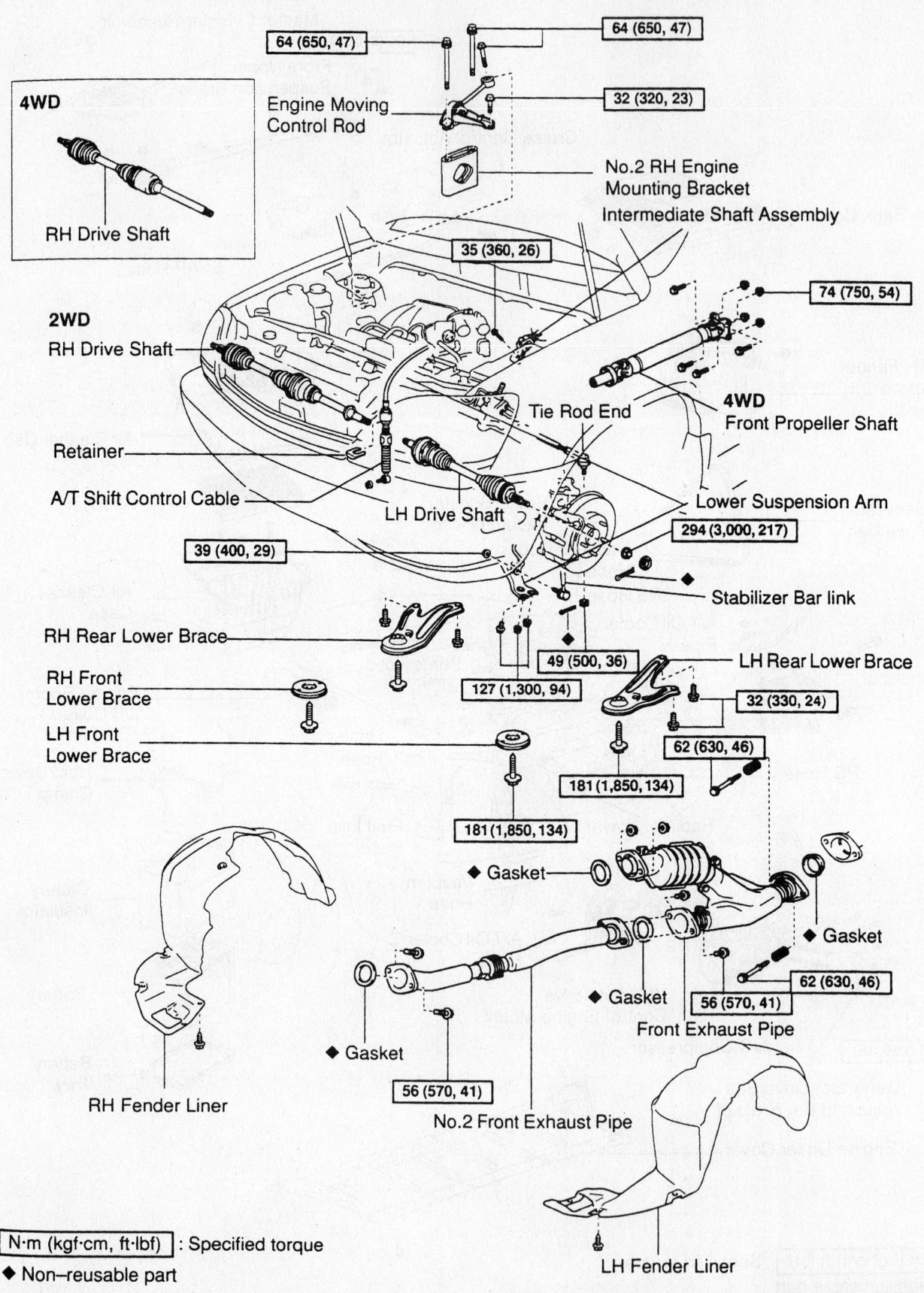

4WD

RH Drive Shaft

64 (650, 47)

64 (650, 47)

32 (320, 23)

Engine Moving
Control Rod

No.2 RH Engine
Mounting Bracket

Intermediate Shaft Assembly

35 (360, 26)

74 (750, 54)

2WD

RH Drive Shaft

4WD
Front Propeller Shaft

Tie Rod End

Retainer

A/T Shift Control Cable

LH Drive Shaft

Lower Suspension Arm

294 (3,000, 217)

39 (400, 29)

Stabilizer Bar link

RH Rear Lower Brace

49 (500, 36)

LH Rear Lower Brace

RH Front
Lower Brace

127 (1,300, 94)

32 (330, 24)

LH Front
Lower Brace

62 (630, 46)

181 (1,850, 134)

181 (1,850, 134)

Gasket

Gasket

Gasket

62 (630, 46)

Gasket

56 (570, 41)

Front Exhaust Pipe

RH Fender Liner

Gasket

56 (570, 41)

No.2 Front Exhaust Pipe

LH Fender Liner

N·m (kgf·cm, ft·lbf) : Specified torque

◆ Non–reusable part

7924ZG85

Exploded view of engine removal and installation tightening specifications of the related components—3.0L Highlander

2WD

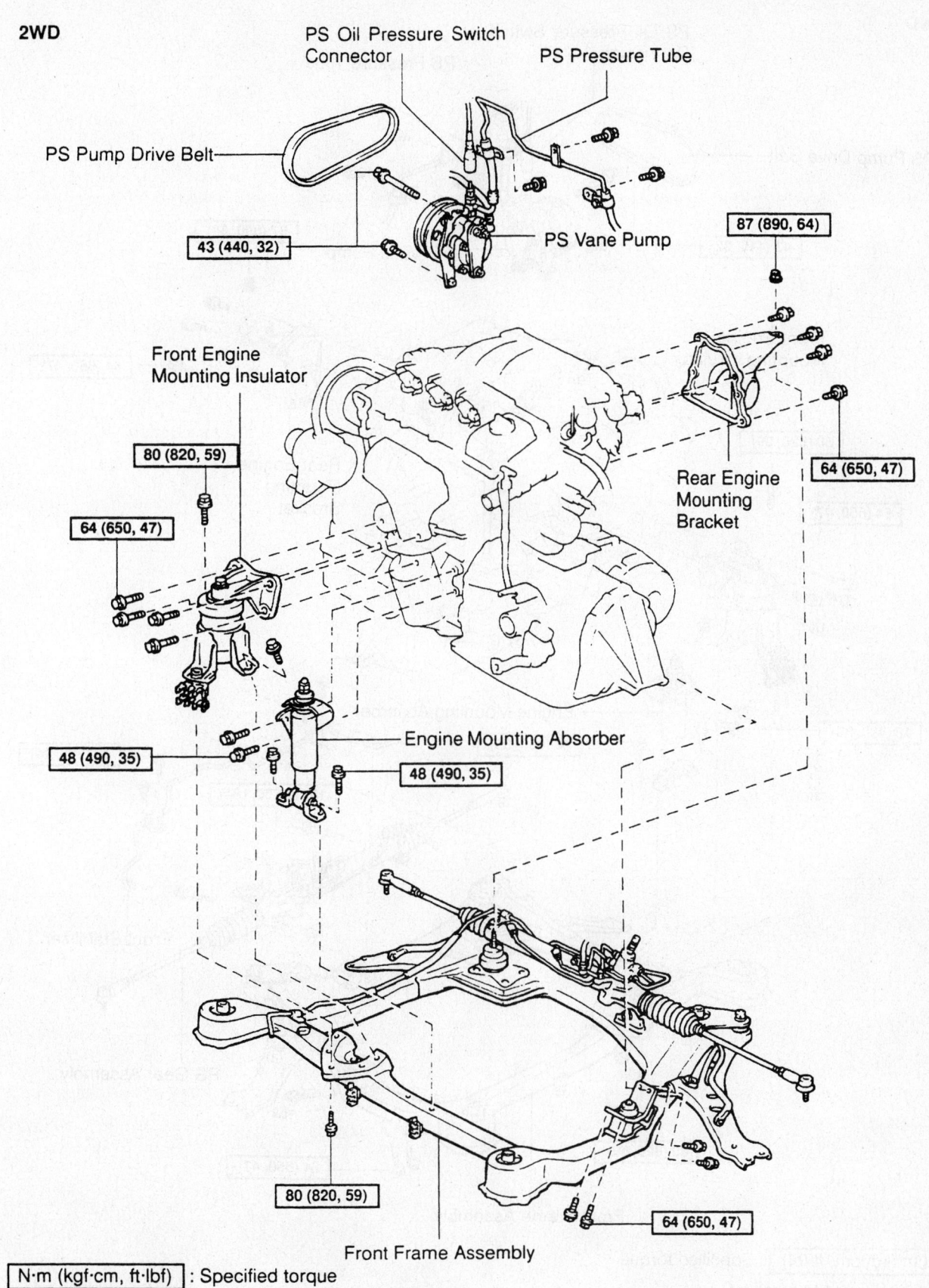

PS Oil Pressure Switch
Connector

PS Pressure Tube

PS Pump Drive Belt

PS Vane Pump

43 (440, 32)

87 (890, 64)

64 (650, 47)

Front Engine
Mounting Insulator

80 (820, 59)

64 (650, 47)

Rear Engine
Mounting
Bracket

Engine Mounting Absorber

48 (490, 35)

48 (490, 35)

80 (820, 59)

64 (650, 47)

Front Frame Assembly

N·m (kgf·cm, ft·lbf) : Specified torque

◆ Non–reusable part

7924ZG86

Exploded view of the suspension component removal and installation for engine removal—2WD Highlander

4WD

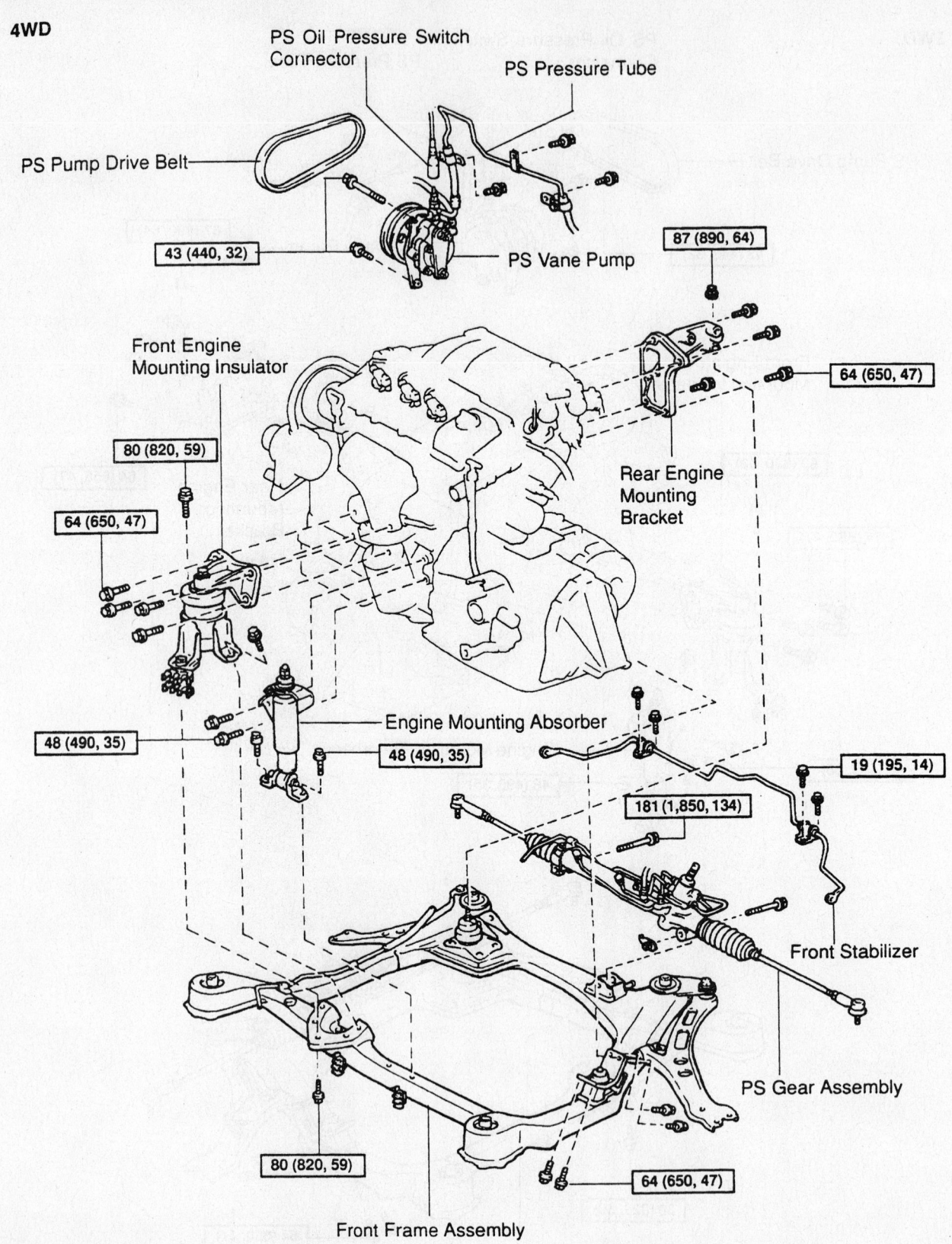

PS Oil Pressure Switch Connector

PS Pressure Tube

PS Pump Drive Belt

43 (440, 32)

87 (890, 64)

PS Vane Pump

Front Engine Mounting Insulator

80 (820, 59)

64 (650, 47)

64 (650, 47)

Rear Engine Mounting Bracket

Engine Mounting Absorber

48 (490, 35)

48 (490, 35)

19 (195, 14)

181 (1,850, 134)

Front Stabilizer

PS Gear Assembly

80 (820, 59)

64 (650, 47)

Front Frame Assembly

N·m (kgf·cm, ft·lbf) : Specified torque

◆ Non–reusable part

7924ZG87

Exploded view of the suspension component removal and installation for engine removal—4WD Highlander

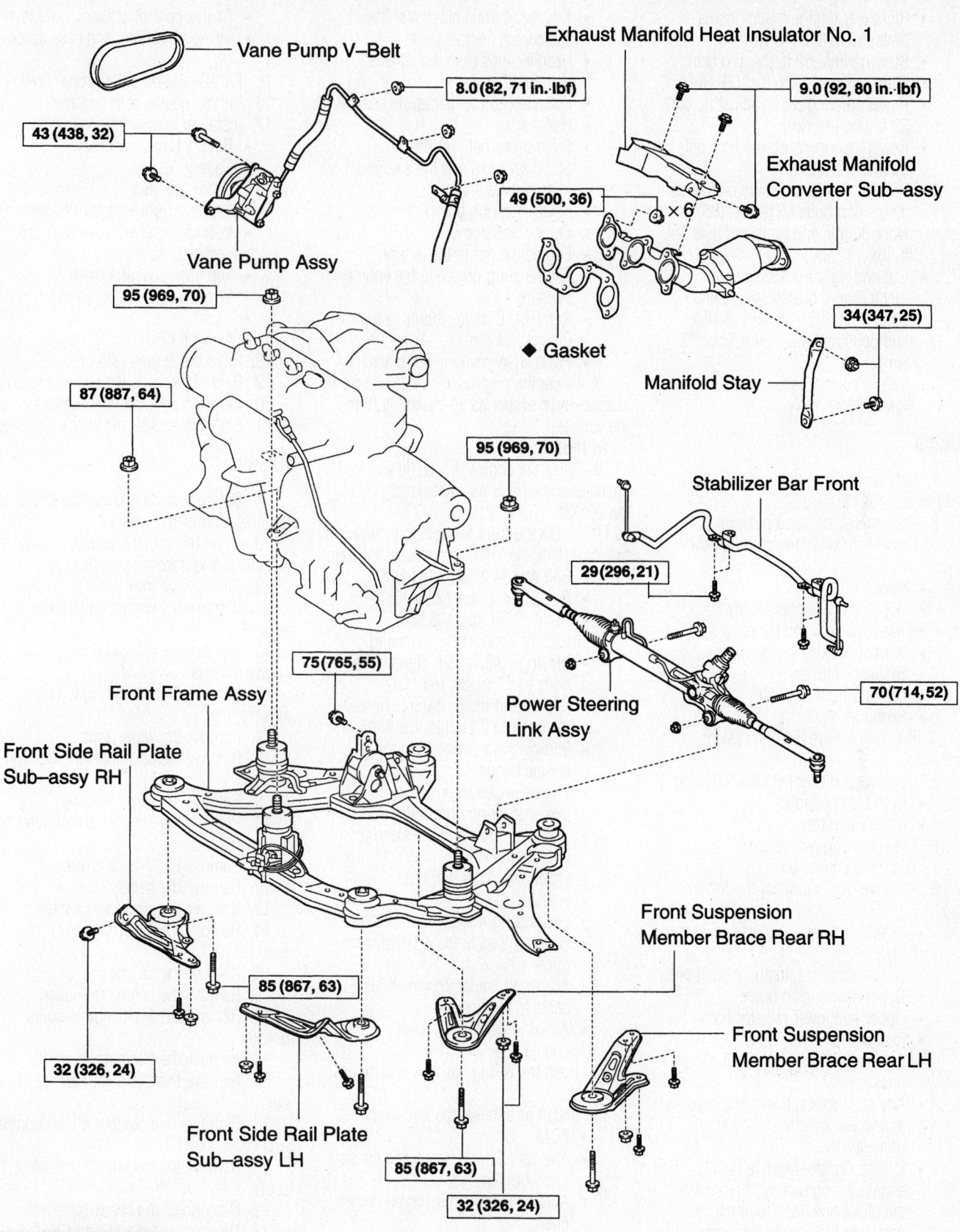

Vane Pump V–Belt

Exhaust Manifold Heat Insulator No. 1

8.0 (82, 71 in.·lbf)

9.0 (92, 80 in.·lbf)

43 (438, 32)

Exhaust Manifold
Converter Sub–assy

49 (500, 36) × 6

Vane Pump Assy

34 (347, 25)

95 (969, 70)

◆ Gasket

Manifold Stay

87 (887, 64)

95 (969, 70)

Stabilizer Bar Front

29 (296, 21)

Front Frame Assy

75 (765, 55)

Power Steering
Link Assy

70 (714, 52)

Front Side Rail Plate
Sub–assy RH

Front Suspension
Member Brace Rear RH

85 (867, 63)

32 (326, 24)

Front Suspension
Member Brace Rear LH

Front Side Rail Plate
Sub–assy LH

85 (867, 63)

32 (326, 24)

N·m (kgf·cm, ft·lbf) : Specified torque

◆ Non–reusable part

Engine mounting points and related parts—3.3L Highlander

67170-HIGH-G04

- Right rear engine mount insulator bolts: 55 ft. lbs. (75 Nm)
- Steering link: 52 ft. lbs. (70 Nm)
- Stabilizer bar: 21 ft. lbs. (29 Nm)
- Power steering pump adjusting bar: 32 ft. lbs. (43 Nm)
- Power steering pressure tube nuts: 69 inch lbs. (8 Nm)
- Left and right frame side rail plates: single end bolts 63 ft. lbs. (85 Nm); double end bolts and nuts: 24 ft. lbs. (32 Nm)
- Left and right front suspension member rear braces: single end bolts 63 ft. lbs. (85 Nm); double end bolts and nuts: 24 ft. lbs. (32 Nm)
- Height control sensor link: 48 inch lbs. (5 Nm)

RX300

1. Before servicing the vehicle, refer to the precautions section.
2. Matchmark the hood position.
3. Remove or disconnect the following:
- Hood
- Wiper and blade assembly
- Top cowl seal and panel
- Window washer hoses from the ventilator louvers
- Left and right ventilator louvers
- Heater air duct
4. Properly relieve the fuel system pressure.
5. Remove or disconnect the following:
- Both battery cables
- Battery and tray
6. Drain the engine coolant.
7. Drain the engine oil.
8. Remove or disconnect the following:
- Intake air cleaner and case assembly
- Cruise control actuator, if equipped
- Upper suspension brace
- Upper and lower radiator hoses
- Radiator
- Automatic transmission oil cooler lines
- Any connectors, hoses and sensors that would interfere with engine removal
- Engine Control Module (ECM) engine wiring harness from inside the glove box; then, pull the harness into the engine compartment
- Compressor

➡**It may be necessary to remove the air conditioning compressor lines in order to remove the engine.**

- Automatic transmission shifter cable from the transaxle
- Header pipes from the exhaust manifolds
- Left and right fender apron seals
- Halfshafts
- Front driveshaft, for 4WD
- Stabilizer links and the steering intermediate shaft
- Power steering pump
- Engine undercover
- Engine hanger to the engine
- Engine sling device to the engine hangers
- Right-hand motor mount and moving control rod
- Front suspension lower braces
9. Lower the engine, transaxle and front suspension member as an assembly from the vehicle.

To install:
10. Raise the engine, transaxle and front suspension member as an assembly into the vehicle.
11. Install the front suspension lower braces, and tighten the fasteners, as follows:
- Bolt A: 134 ft. lbs. (181 Nm)
- Bolt B: 24 ft. lbs. (32 Nm)
- Nut C: 27 ft. lbs. (36 Nm)
12. Install or connect the following:
- Moving control rod. Tighten the bolts to 47 ft. lbs. (64 Nm).
- Right-hand motor mount. Tighten the bolts to 23 ft. lbs. (32 Nm).
- Engine sling device from the engine hangers
- Engine undercover
- Power steering pump hoses
- Stabilizer links and the steering intermediate shaft
- Front driveshaft, for 4WD
- Halfshafts
- Left and right fender apron seals
- Header pipes to the exhaust manifolds
- Automatic transmission shifter cable to the transaxle
- Air conditioning compressor to the engine
13. Push the wiring harness into the glove box.
14. Install or connect the following:
- ECM
- Any connectors, hoses and sensors that were removed
- Automatic transmission oil cooler lines
- Upper and lower radiator hoses and fit the radiator
- Front upper suspension brace. Tighten the nuts to 59 ft. lbs. (80 Nm)

- Cruise control actuator, if removed
- Intake air cleaner and case assembly
15. Fill the engine oil to proper level.
16. Fill the engine with coolant.
17. Install or connect the following:
- Battery tray and battery
- Battery cables
- Heater air duct
- Left and right ventilator louvers
- Window washer hoses from the ventilator louvers
- Top cowl seal and panel
- Wiper and blade assembly
- Hood
- New oil filter
18. Refill the engine with oil.
19. Refill the engine with engine coolant.
20. Install the engine undercovers.
21. Start the engine and check for leaks.

RX330

1. Before servicing the vehicle, refer to the precautions section.
2. Drain the coolant, engine oil, transfer case fluid and transmission fluid.
3. Remove the front wheels.
4. Remove the engine undercover assembly.
5. Remove the left and right fender splash shields.
6. Remove the left and right fender apron seals.
7. Remove the wiper arms.
8. Remove the cowl top ventilator louver.
9. Remove the wiper linkage.
10. Remove the cowl top panel outer sub-assembly.
11. Remove the V-bank cover.
12. Remove the battery.
13. Remove the air cleaner assembly.
14. Remove the A/C compressor drive belt.
15. Remove the alternator.
16. Remove the engine roll brace.
17. Remove the front engine mount bracket.
18. Remove the alternator bracket.
19. Remove the alternator belt adjusting bar.
20. Remove the magnetic clutch from the A/C compressor.
21. Remove the transmission control cable.
22. Disconnect the check valve hose.
23. Disconnect the fuel vapor feed hose.
24. Disconnect the fuel pipes.
25. Disconnect the heater hoses.
26. Disconnect the radiator hoses.
27. Disconnect the oil cooler hoses.

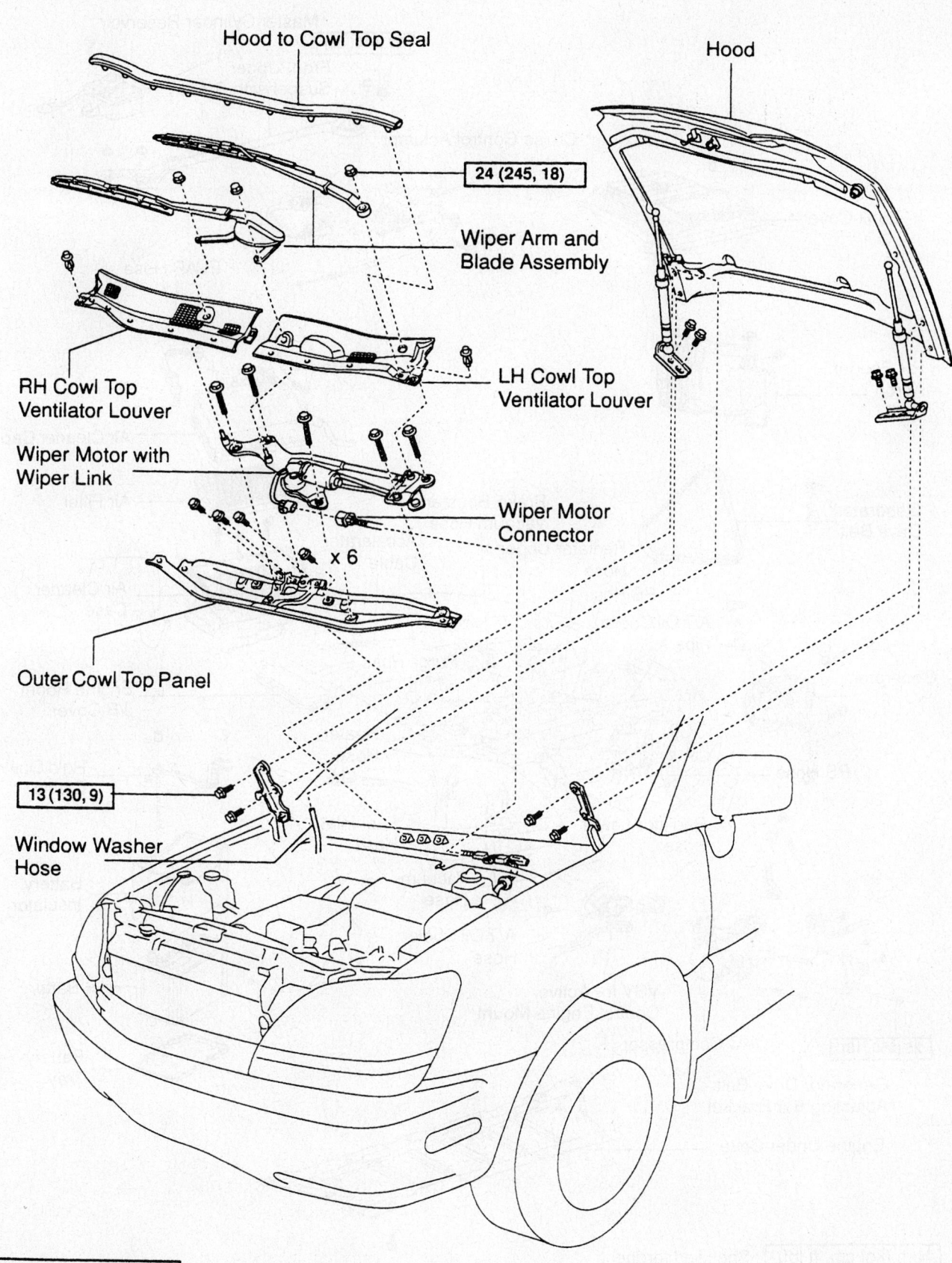

Hood to Cowl Top Seal

Hood

24 (245, 18)

Wiper Arm and
Blade Assembly

RH Cowl Top
Ventilator Louver

LH Cowl Top
Ventilator Louver

Wiper Motor with
Wiper Link

Wiper Motor
Connector

x 6

Outer Cowl Top Panel

13 (130, 9)

Window Washer
Hose

N·m (kgf·cm, ft·lbf) : Specified torque

7924ZG83

Exploded view of the top cowl and related components—RX300

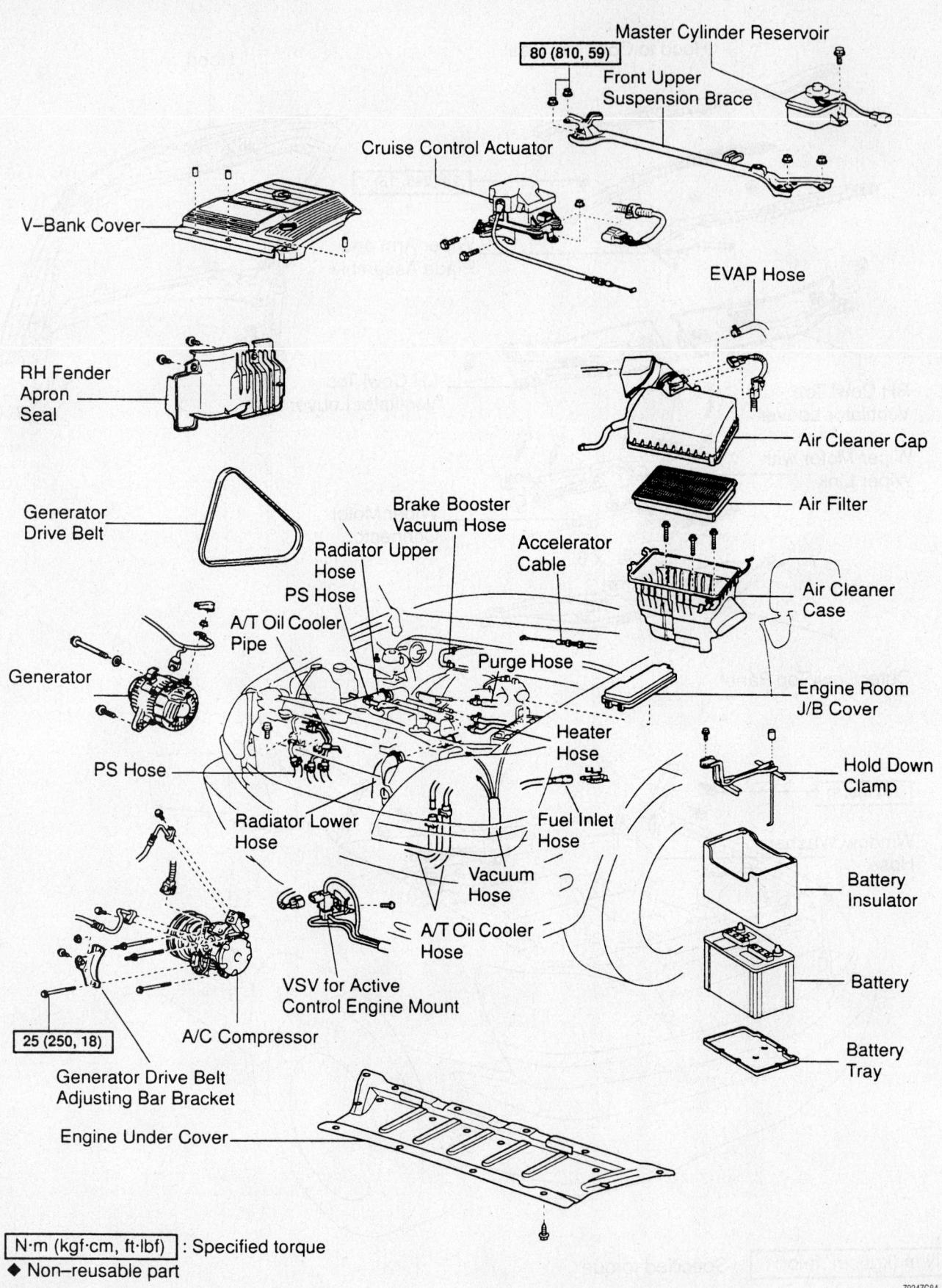

Master Cylinder Reservoir

80 (810, 59)

Front Upper
Suspension Brace

Cruise Control Actuator

V−Bank Cover

EVAP Hose

RH Fender
Apron
Seal

Air Cleaner Cap

Air Filter

Generator
Drive Belt

Brake Booster
Vacuum Hose

Radiator Upper
Hose

Accelerator
Cable

PS Hose

A/T Oil Cooler
Pipe

Purge Hose

Air Cleaner
Case

Generator

Heater
Hose

Engine Room
J/B Cover

PS Hose

Hold Down
Clamp

Radiator Lower
Hose

Fuel Inlet
Hose

Vacuum
Hose

Battery
Insulator

A/T Oil Cooler
Hose

VSV for Active
Control Engine Mount

Battery

25 (250, 18)

A/C Compressor

Battery
Tray

Generator Drive Belt
Adjusting Bar Bracket

Engine Under Cover

N·m (kgf·cm, ft·lbf) : Specified torque

◆ Non−reusable part

7924ZG84

Exploded view of engine pre-removal components—3.0L

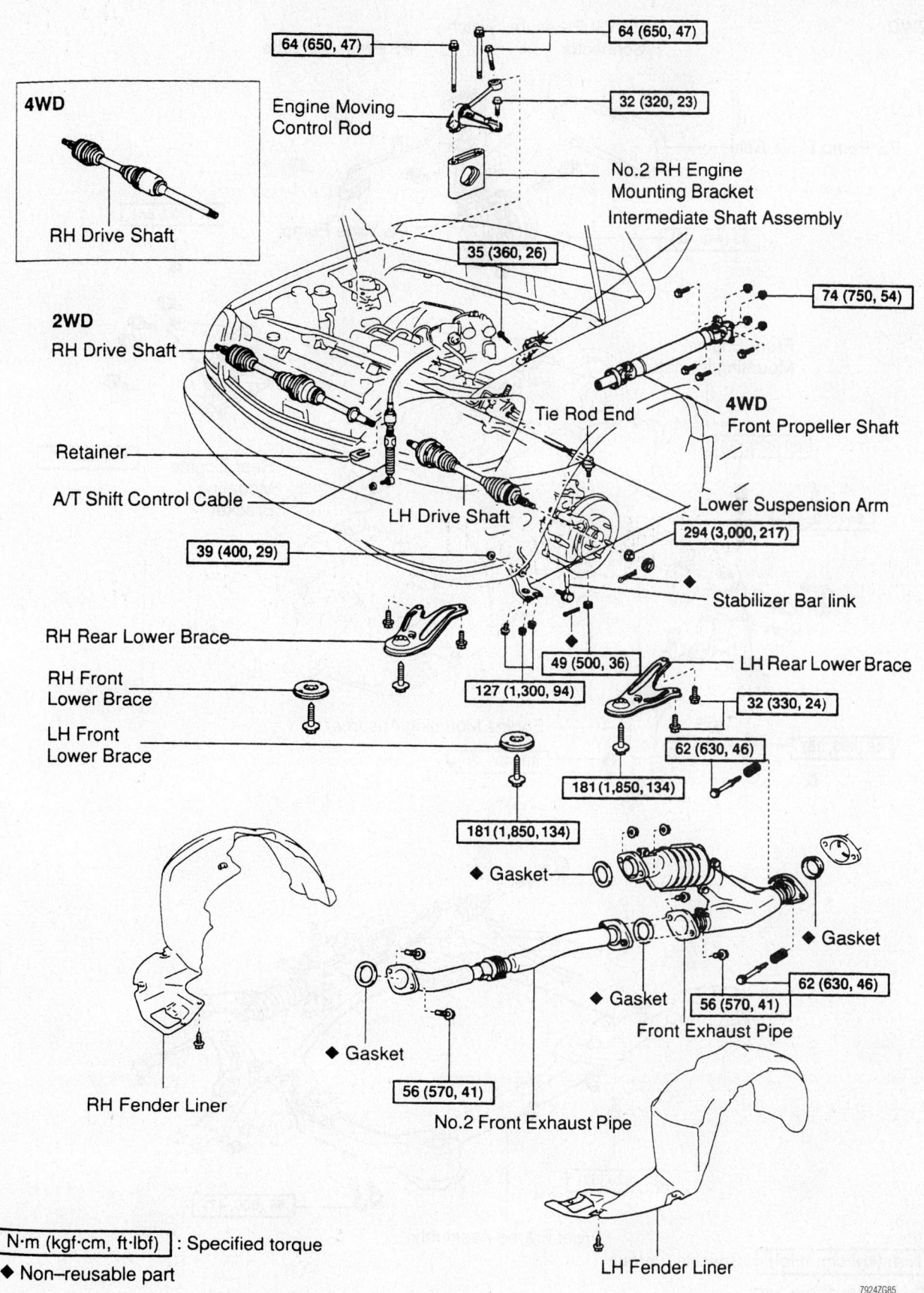

4WD

RH Drive Shaft

64 (650, 47)

64 (650, 47)

Engine Moving
Control Rod

32 (320, 23)

No.2 RH Engine
Mounting Bracket

Intermediate Shaft Assembly

35 (360, 26)

74 (750, 54)

2WD
RH Drive Shaft

4WD
Front Propeller Shaft

Tie Rod End

Retainer

A/T Shift Control Cable

LH Drive Shaft

Lower Suspension Arm

294 (3,000, 217)

39 (400, 29)

Stabilizer Bar link

RH Rear Lower Brace

49 (500, 36)

LH Rear Lower Brace

RH Front
Lower Brace

127 (1,300, 94)

32 (330, 24)

LH Front
Lower Brace

62 (630, 46)

181 (1,850, 134)

181 (1,850, 134)

◆ Gasket

62 (630, 46)

◆ Gasket

◆ Gasket

56 (570, 41)

Front Exhaust Pipe

RH Fender Liner

◆ Gasket

56 (570, 41)

No.2 Front Exhaust Pipe

LH Fender Liner

N·m (kgf·cm, ft·lbf) : Specified torque

◆ Non-reusable part

7924ZG85

Exploded view of engine removal and installation tightening specifications of the related components—3.0L

2WD

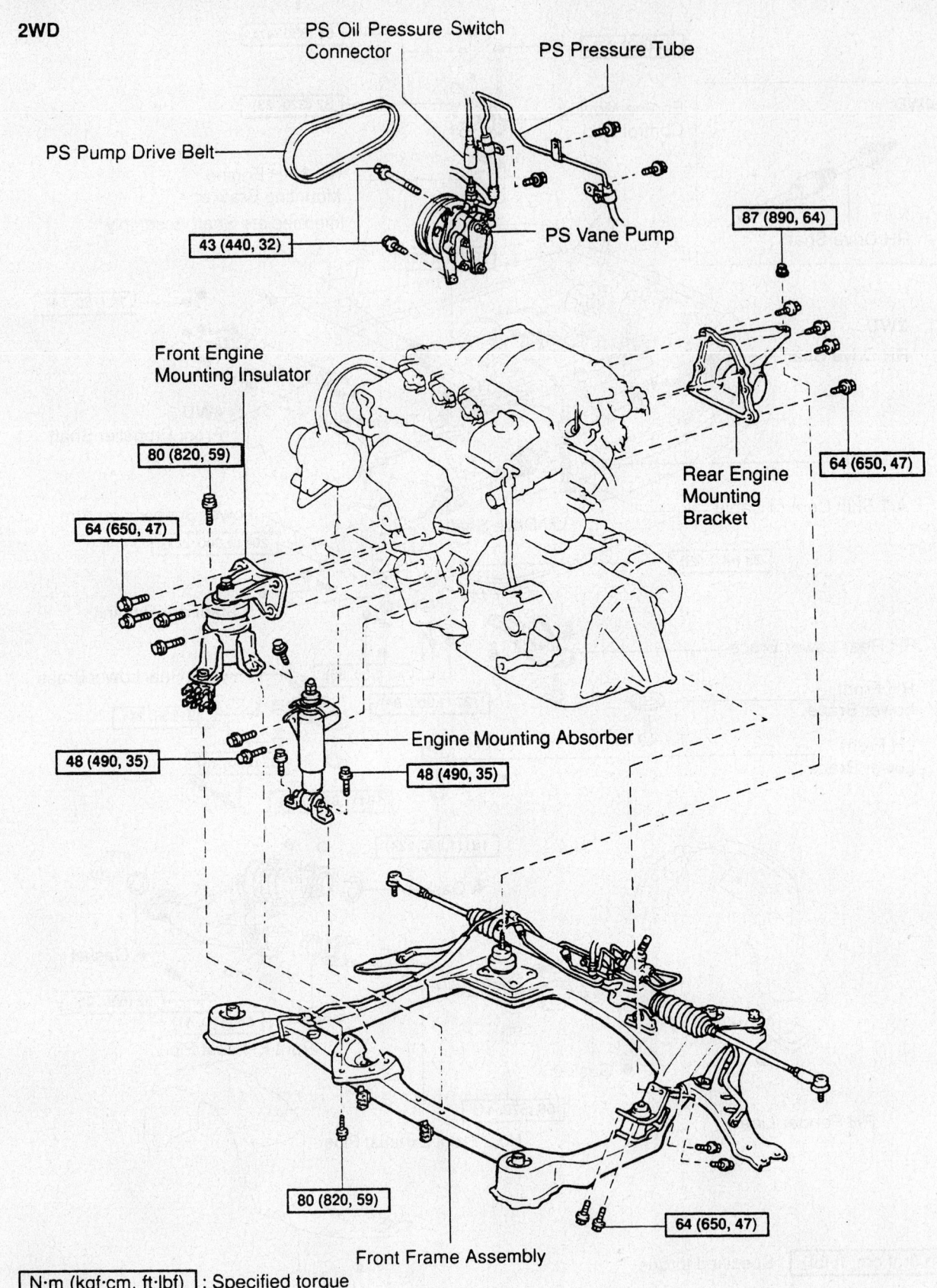

PS Oil Pressure Switch Connector

PS Pressure Tube

PS Pump Drive Belt

43 (440, 32)

PS Vane Pump

87 (890, 64)

Front Engine Mounting Insulator

80 (820, 59)

64 (650, 47)

Rear Engine Mounting Bracket

64 (650, 47)

48 (490, 35)

Engine Mounting Absorber

48 (490, 35)

80 (820, 59)

64 (650, 47)

Front Frame Assembly

N·m (kgf·cm, ft·lbf) : Specified torque

◆ Non−reusable part

7924ZG86

Exploded view of the suspension component removal and installation for engine removal—2WD RX300

4WD

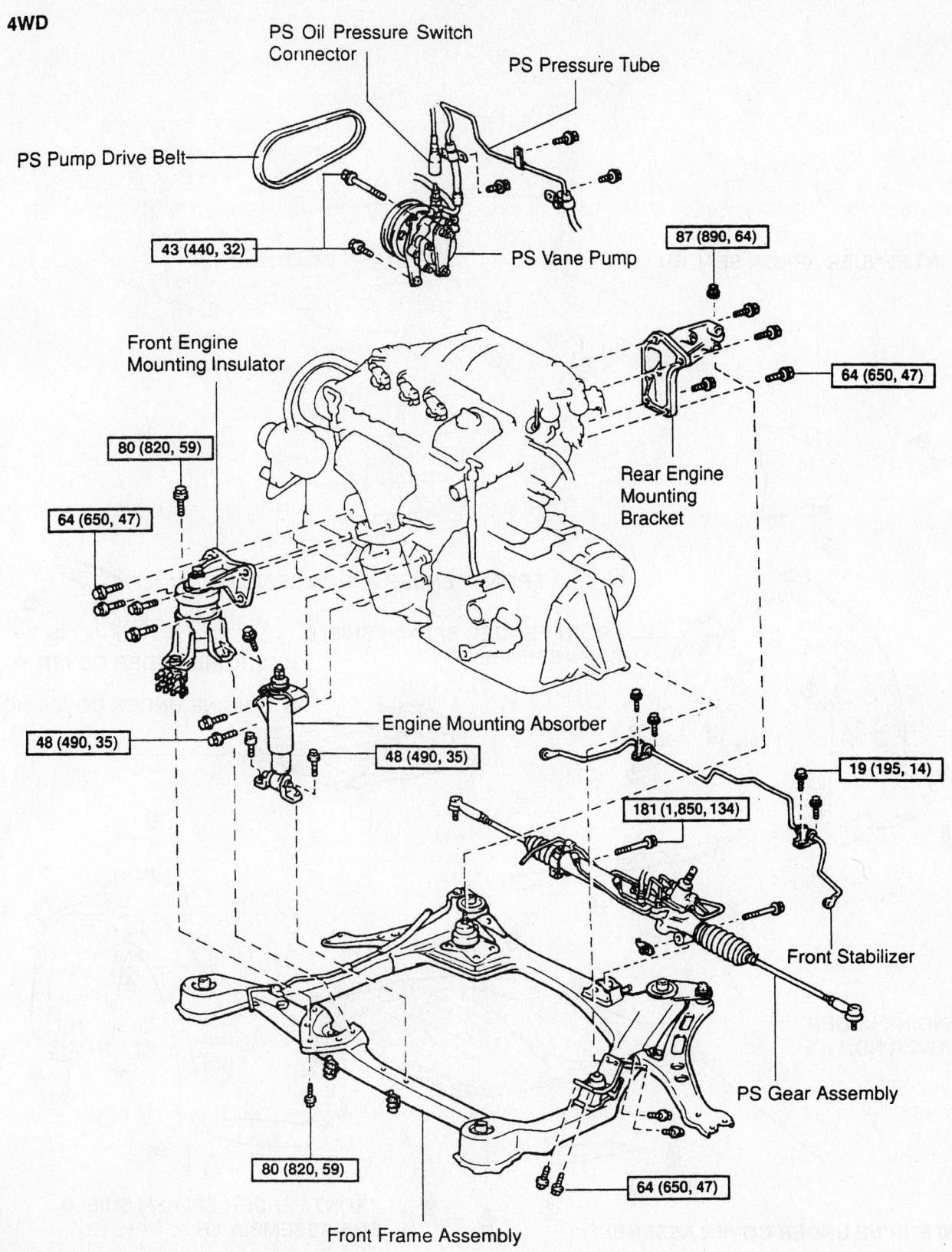

PS Oil Pressure Switch Connector

PS Pressure Tube

PS Pump Drive Belt

43 (440, 32)

PS Vane Pump

87 (890, 64)

64 (650, 47)

Front Engine Mounting Insulator

Rear Engine Mounting Bracket

80 (820, 59)

64 (650, 47)

48 (490, 35)

Engine Mounting Absorber

48 (490, 35)

19 (195, 14)

181 (1,850, 134)

Front Stabilizer

PS Gear Assembly

80 (820, 59)

64 (650, 47)

Front Frame Assembly

N·m (kgf·cm, ft·lbf) : Specified torque

◆ Non–reusable part

7924ZG87

Exploded view of the suspension component removal and installation for engine removal—4WD RX300

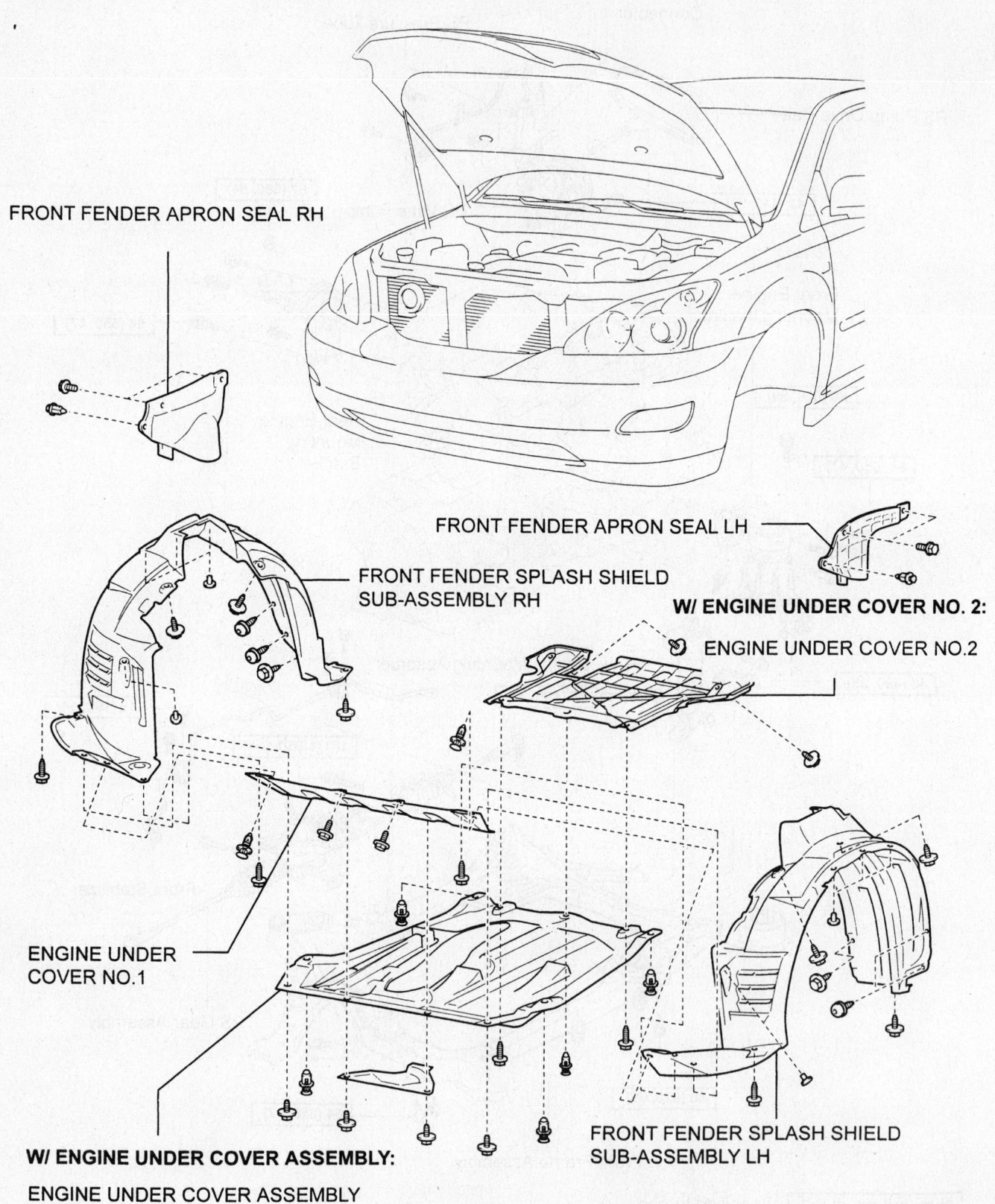

FRONT FENDER APRON SEAL RH

FRONT FENDER APRON SEAL LH

FRONT FENDER SPLASH SHIELD
SUB-ASSEMBLY RH

W/ ENGINE UNDER COVER NO. 2:

ENGINE UNDER COVER NO.2

ENGINE UNDER
COVER NO.1

FRONT FENDER SPLASH SHIELD
SUB-ASSEMBLY LH

W/ ENGINE UNDER COVER ASSEMBLY:

ENGINE UNDER COVER ASSEMBLY

09490_HIGH_G0001

Exploded view of the undercover and apron component removal and installation for engine removal—RX330

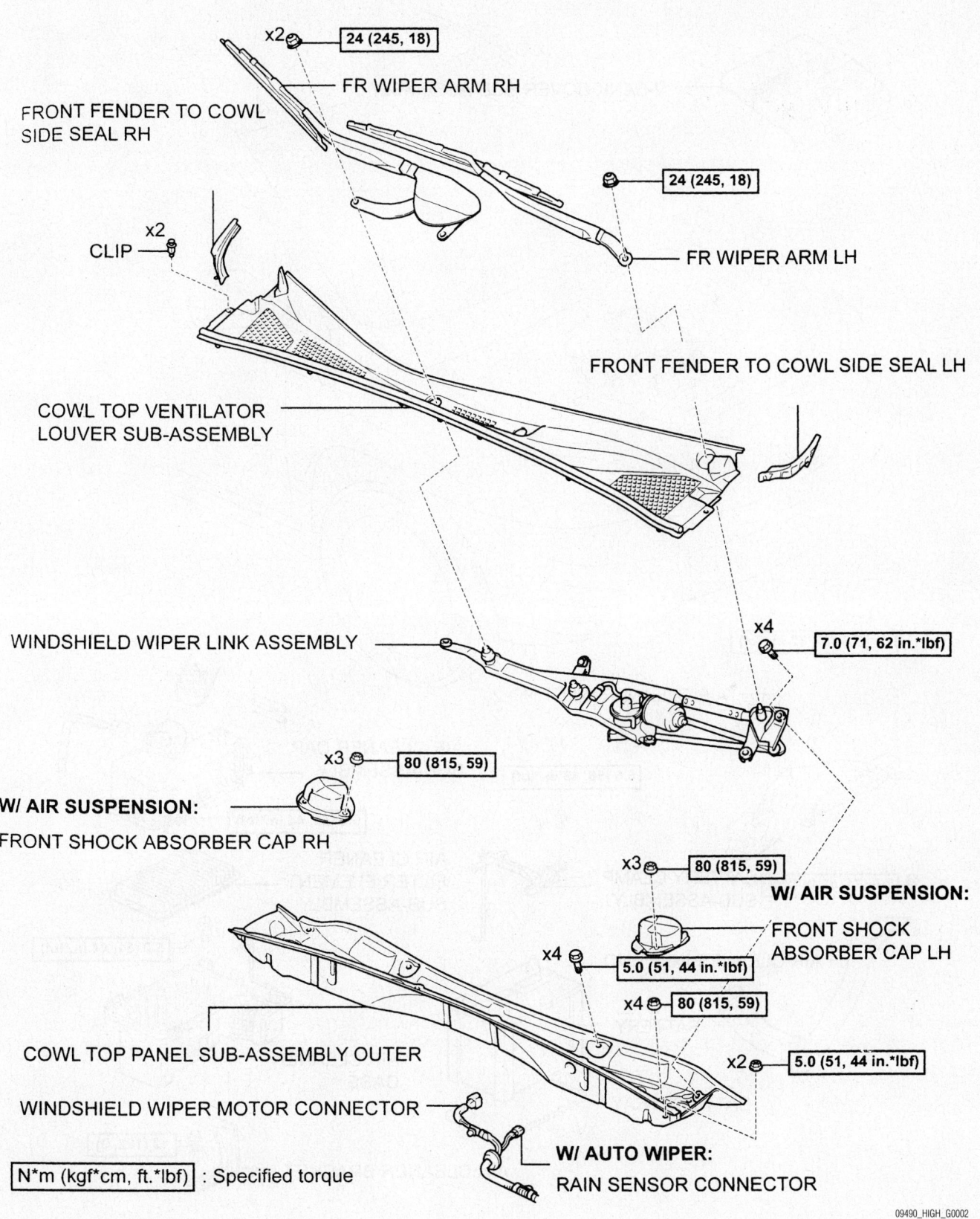

x2 — 24 (245, 18)

FR WIPER ARM RH

FRONT FENDER TO COWL
SIDE SEAL RH

24 (245, 18)

CLIP — x2

FR WIPER ARM LH

FRONT FENDER TO COWL SIDE SEAL LH

COWL TOP VENTILATOR
LOUVER SUB-ASSEMBLY

WINDSHIELD WIPER LINK ASSEMBLY

x4 — 7.0 (71, 62 in.*lbf)

x3 — 80 (815, 59)

W/ AIR SUSPENSION:
FRONT SHOCK ABSORBER CAP RH

x3 — 80 (815, 59)

W/ AIR SUSPENSION:
FRONT SHOCK
ABSORBER CAP LH

x4 — 5.0 (51, 44 in.*lbf)

x4 — 80 (815, 59)

COWL TOP PANEL SUB-ASSEMBLY OUTER

x2 — 5.0 (51, 44 in.*lbf)

WINDSHIELD WIPER MOTOR CONNECTOR

W/ AUTO WIPER:
RAIN SENSOR CONNECTOR

N*m (kgf*cm, ft.*lbf) : Specified torque

09490_HIGH_G0002

Exploded view of the cowl and wiper component removal and installation for engine removal—RX330

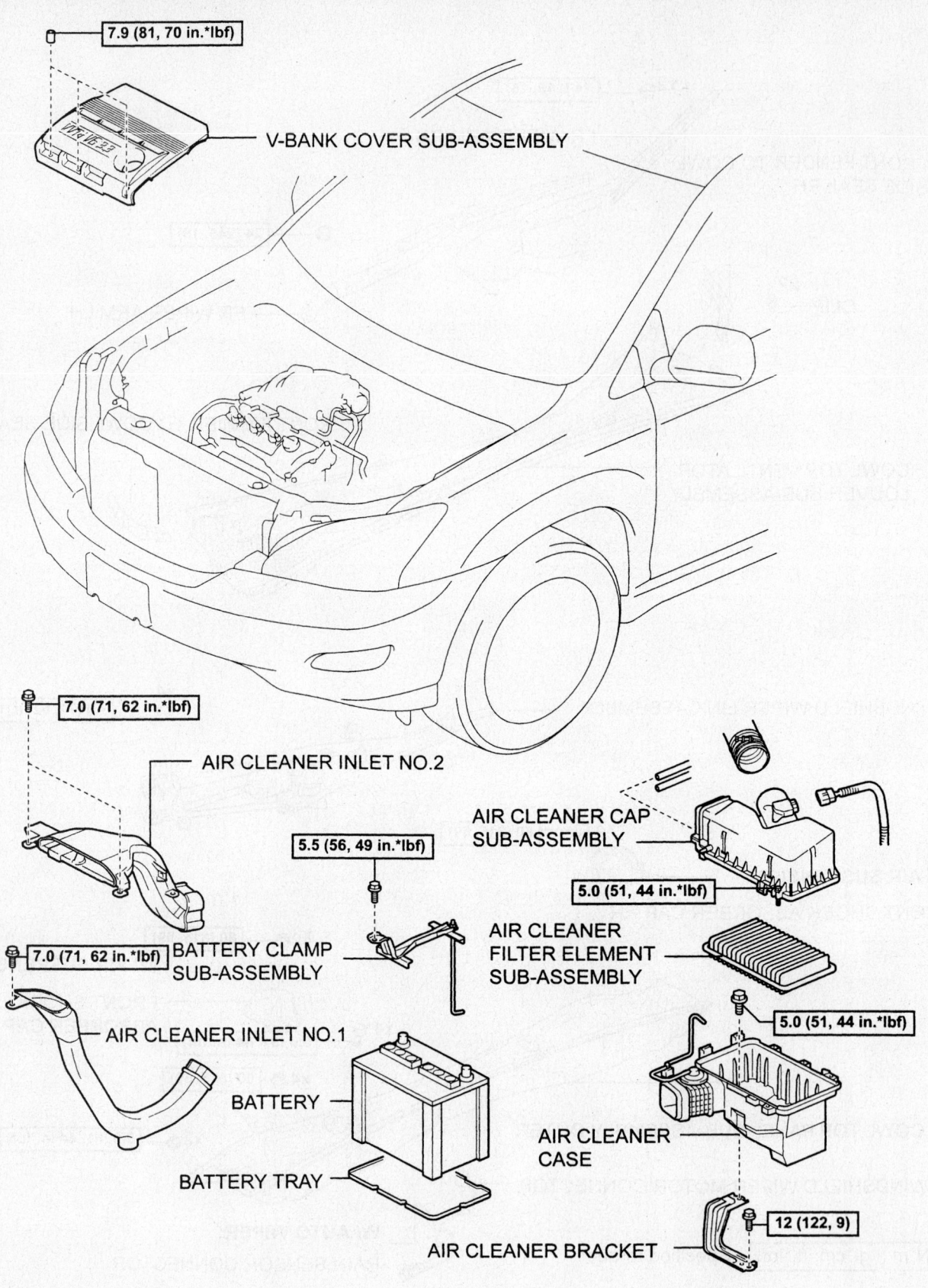

7.9 (81, 70 in.*lbf)

V-BANK COVER SUB-ASSEMBLY

7.0 (71, 62 in.*lbf)

AIR CLEANER INLET NO.2

AIR CLEANER CAP
SUB-ASSEMBLY

5.5 (56, 49 in.*lbf)

5.0 (51, 44 in.*lbf)

AIR CLEANER
FILTER ELEMENT
SUB-ASSEMBLY

7.0 (71, 62 in.*lbf)

BATTERY CLAMP
SUB-ASSEMBLY

AIR CLEANER INLET NO.1

5.0 (51, 44 in.*lbf)

BATTERY

AIR CLEANER
CASE

BATTERY TRAY

12 (122, 9)

AIR CLEANER BRACKET

N*m (kgf*cm, ft.*lbf) : Specified torque

Exploded view of the battery and air cleaner component removal and installation for engine removal—RX330

09490_HIGH_G0003

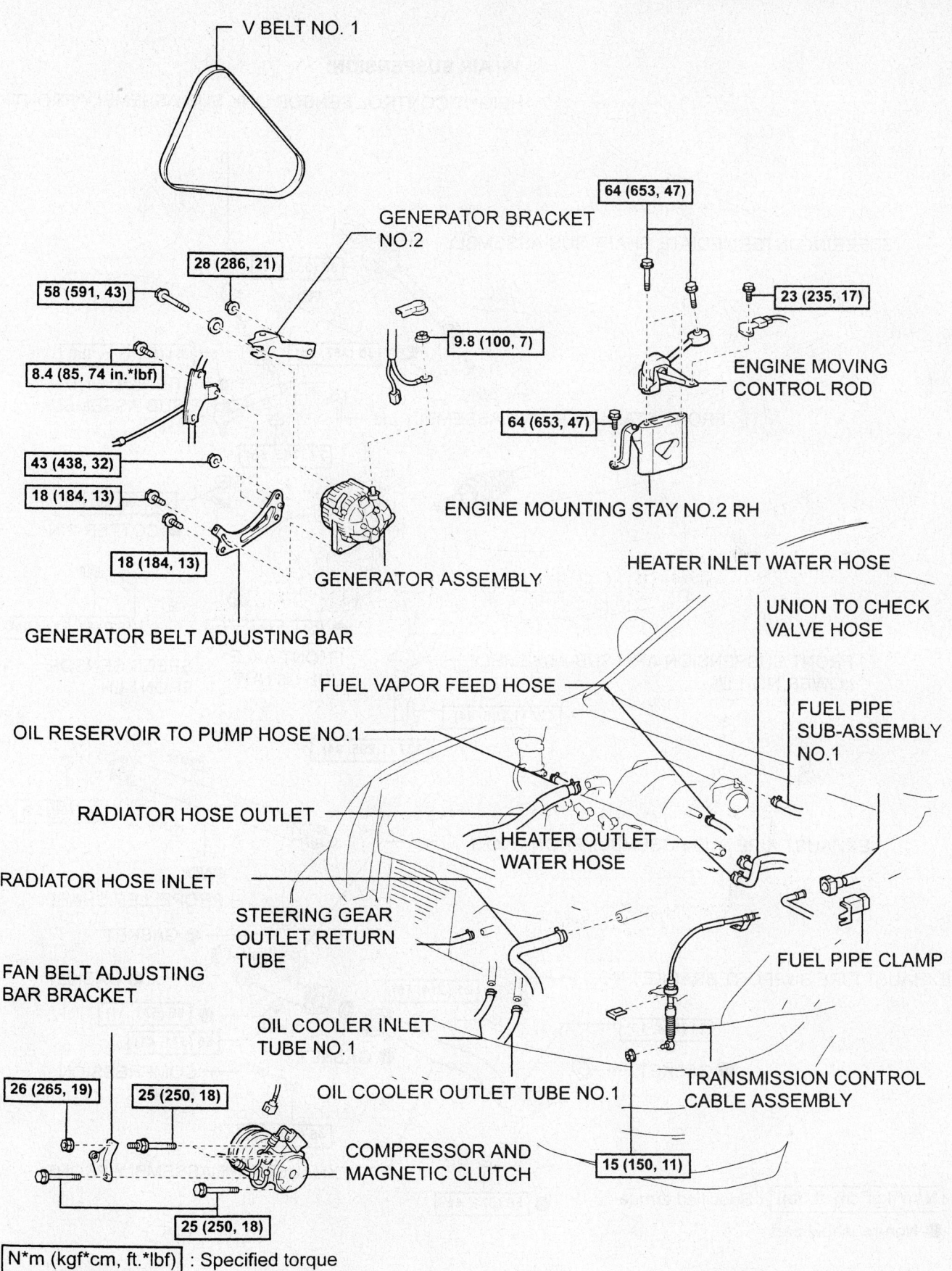

V BELT NO. 1

GENERATOR BRACKET NO.2

28 (286, 21)

58 (591, 43)

64 (653, 47)

23 (235, 17)

9.8 (100, 7)

8.4 (85, 74 in.*lbf)

ENGINE MOVING CONTROL ROD

64 (653, 47)

43 (438, 32)

18 (184, 13)

18 (184, 13)

ENGINE MOUNTING STAY NO.2 RH

GENERATOR ASSEMBLY

GENERATOR BELT ADJUSTING BAR

HEATER INLET WATER HOSE

UNION TO CHECK VALVE HOSE

FUEL VAPOR FEED HOSE

OIL RESERVOIR TO PUMP HOSE NO.1

FUEL PIPE SUB-ASSEMBLY NO.1

RADIATOR HOSE OUTLET

HEATER OUTLET WATER HOSE

RADIATOR HOSE INLET

STEERING GEAR OUTLET RETURN TUBE

FAN BELT ADJUSTING BAR BRACKET

FUEL PIPE CLAMP

OIL COOLER INLET TUBE NO.1

26 (265, 19)

25 (250, 18)

OIL COOLER OUTLET TUBE NO.1

TRANSMISSION CONTROL CABLE ASSEMBLY

COMPRESSOR AND MAGNETIC CLUTCH

15 (150, 11)

25 (250, 18)

N*m (kgf*cm, ft.*lbf) : Specified torque

09490_HIGH_G0004

Exploded view of some of the various hoses, belts and brackets that must be disconnected or removed to facilitate engine removal—RX330

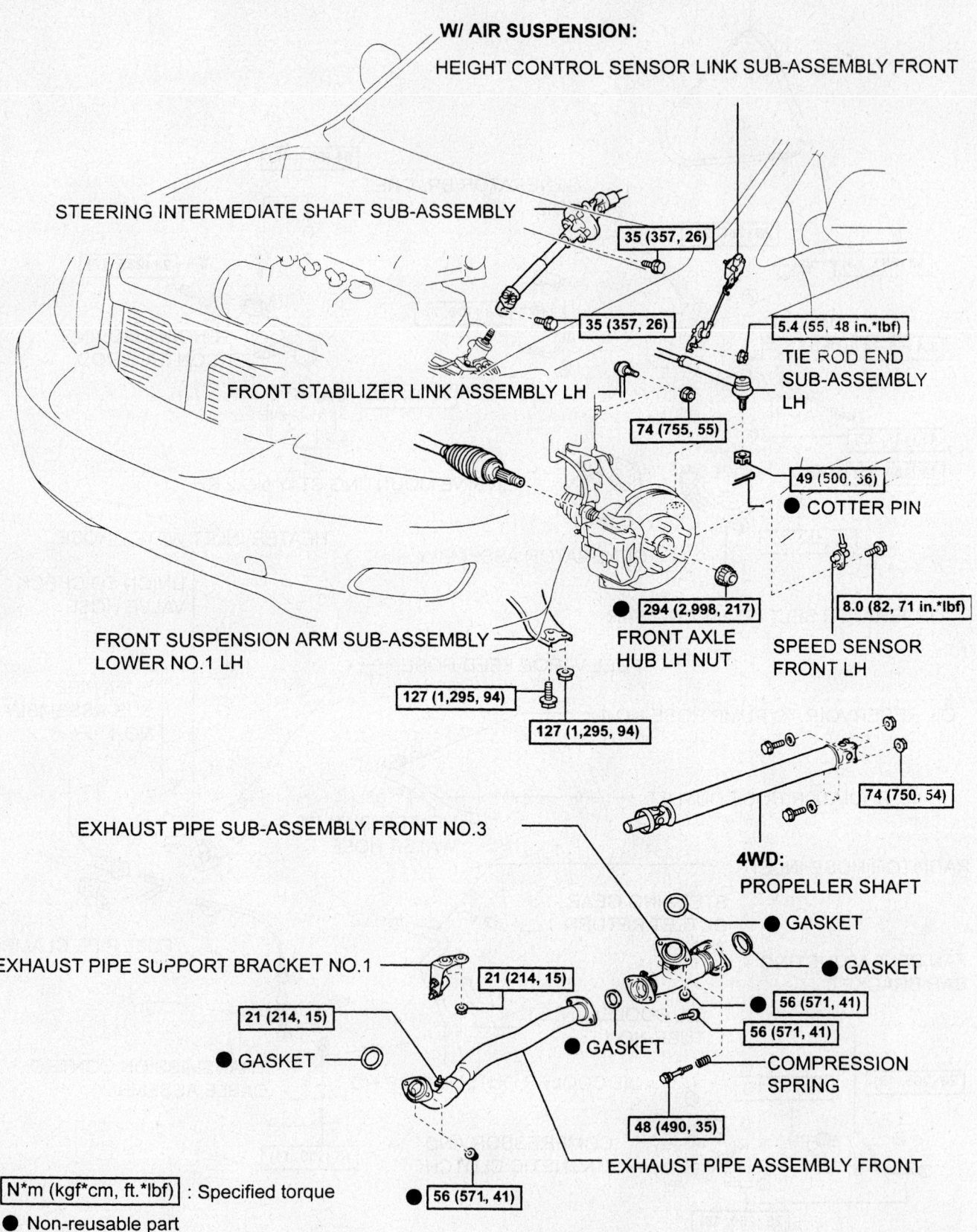

W/ AIR SUSPENSION:

HEIGHT CONTROL SENSOR LINK SUB-ASSEMBLY FRONT

STEERING INTERMEDIATE SHAFT SUB-ASSEMBLY

35 (357, 26)

35 (357, 26)

5.4 (55, 48 in.*lbf)

TIE ROD END
SUB-ASSEMBLY
LH

FRONT STABILIZER LINK ASSEMBLY LH

74 (755, 55)

49 (500, 36)

● COTTER PIN

8.0 (82, 71 in.*lbf)

● 294 (2,998, 217)

FRONT AXLE
HUB LH NUT

SPEED SENSOR
FRONT LH

FRONT SUSPENSION ARM SUB-ASSEMBLY
LOWER NO.1 LH

127 (1,295, 94)

127 (1,295, 94)

74 (750, 54)

EXHAUST PIPE SUB-ASSEMBLY FRONT NO.3

4WD:
PROPELLER SHAFT

● GASKET

● GASKET

EXHAUST PIPE SUPPORT BRACKET NO.1

21 (214, 15)

21 (214, 15)

● 56 (571, 41)

56 (571, 41)

● GASKET

● GASKET

COMPRESSION
SPRING

48 (490, 35)

EXHAUST PIPE ASSEMBLY FRONT

N*m (kgf*cm, ft.*lbf) : Specified torque
● Non-reusable part

● 56 (571, 41)

09490_HIGH_G0005

Exploded view of some of the steering, driveshaft and exhaust component removal and installation for engine removal—RX330

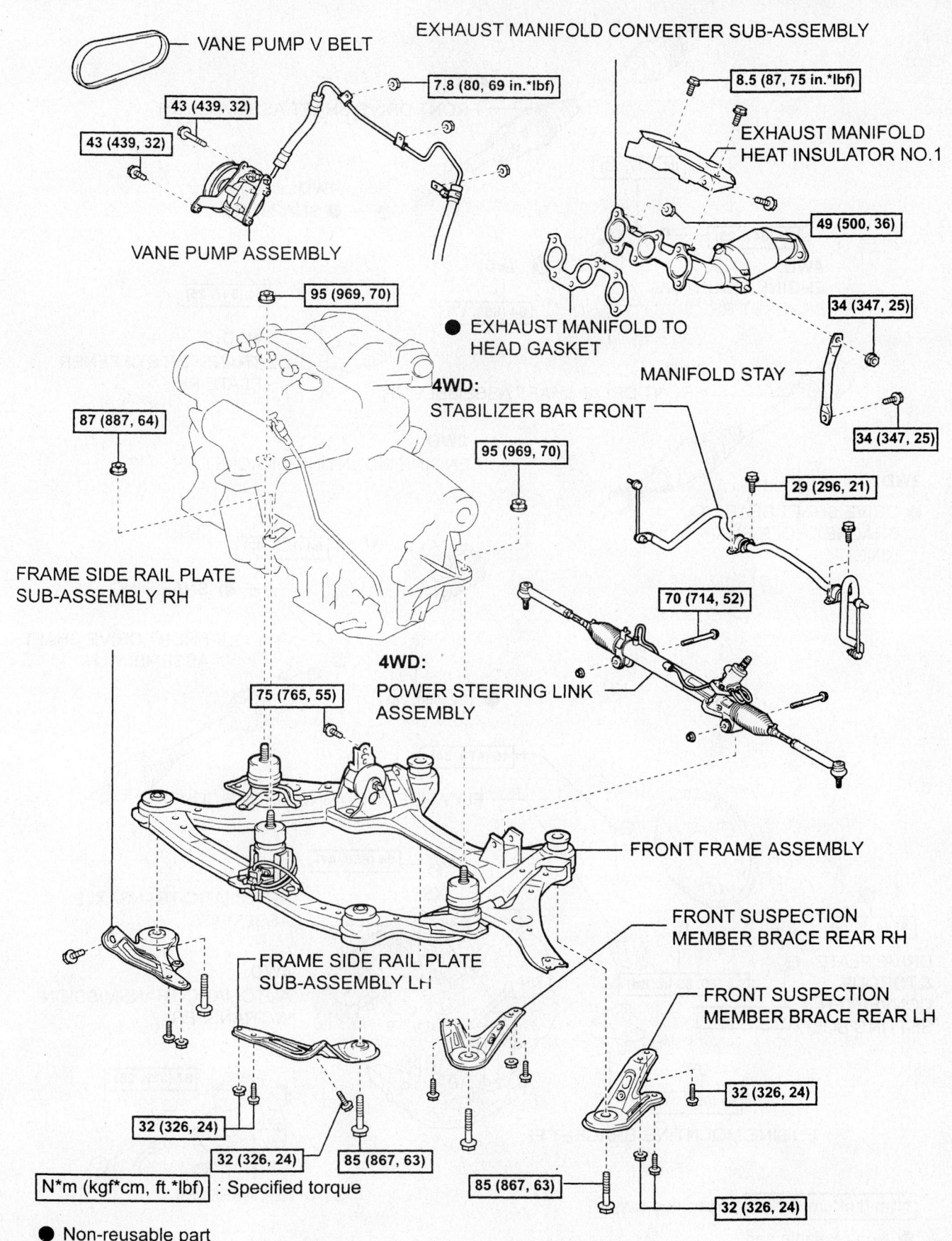

VANE PUMP V BELT

EXHAUST MANIFOLD CONVERTER SUB-ASSEMBLY

7.8 (80, 69 in.*lbf)

8.5 (87, 75 in.*lbf)

43 (439, 32)

43 (439, 32)

EXHAUST MANIFOLD HEAT INSULATOR NO.1

49 (500, 36)

VANE PUMP ASSEMBLY

34 (347, 25)

● EXHAUST MANIFOLD TO HEAD GASKET

MANIFOLD STAY

95 (969, 70)

4WD:
STABILIZER BAR FRONT

34 (347, 25)

87 (887, 64)

95 (969, 70)

29 (296, 21)

FRAME SIDE RAIL PLATE SUB-ASSEMBLY RH

70 (714, 52)

4WD:
POWER STEERING LINK ASSEMBLY

75 (765, 55)

FRONT FRAME ASSEMBLY

FRONT SUSPENSION MEMBER BRACE REAR RH

FRAME SIDE RAIL PLATE SUB-ASSEMBLY LH

FRONT SUSPENSION MEMBER BRACE REAR LH

32 (326, 24)

32 (326, 24)

32 (326, 24)

32 (326, 24)

85 (867, 63)

N*m (kgf*cm, ft.*lbf) : Specified torque

85 (867, 63)

32 (326, 24)

● Non-reusable part

Exploded view of some of the suspension component removal and installation for engine removal—RX330

09490_HIGH_G0006

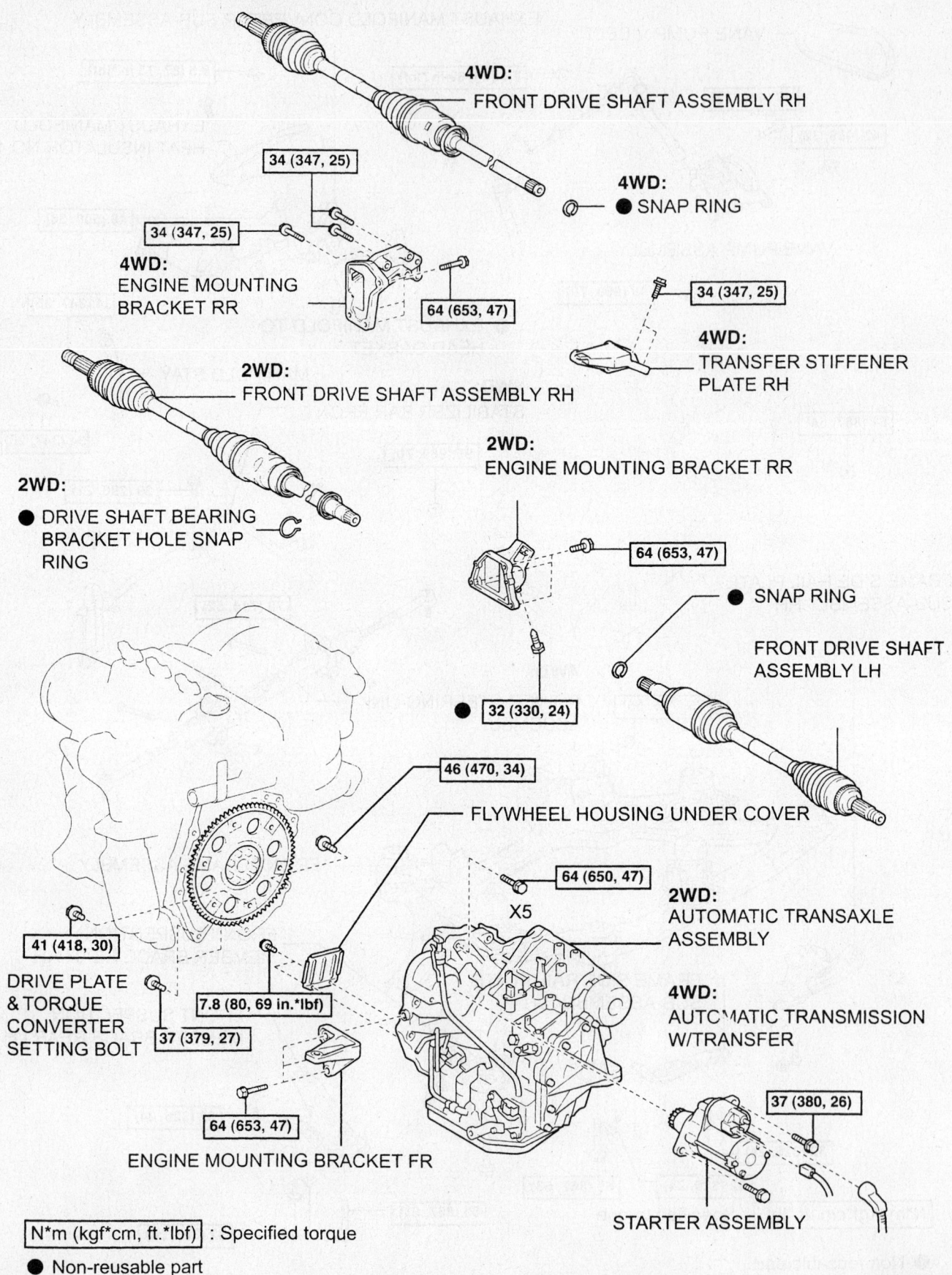

4WD:
FRONT DRIVE SHAFT ASSEMBLY RH

34 (347, 25)

34 (347, 25)

4WD:
ENGINE MOUNTING
BRACKET RR

64 (653, 47)

4WD:
● SNAP RING

34 (347, 25)

4WD:
TRANSFER STIFFENER
PLATE RH

2WD:
FRONT DRIVE SHAFT ASSEMBLY RH

2WD:
● DRIVE SHAFT BEARING
BRACKET HOLE SNAP
RING

2WD:
ENGINE MOUNTING BRACKET RR

64 (653, 47)

● SNAP RING

FRONT DRIVE SHAFT
ASSEMBLY LH

● 32 (330, 24)

46 (470, 34)

FLYWHEEL HOUSING UNDER COVER

64 (650, 47)

X5

2WD:
AUTOMATIC TRANSAXLE
ASSEMBLY

41 (418, 30)

DRIVE PLATE
& TORQUE
CONVERTER
SETTING BOLT

7.8 (80, 69 in.*lbf)

37 (379, 27)

4WD:
AUTOMATIC TRANSMISSION
W/TRANSFER

64 (653, 47)

ENGINE MOUNTING BRACKET FR

37 (380, 26)

STARTER ASSEMBLY

N*m (kgf*cm, ft.*lbf) : Specified torque

● Non-reusable part

09490_HIGH_G0007

Exploded view of some of the suspension and engine-to-transmission components—RX330

28. Disconnect the power steering hoses.
29. Remove the glove compartment door.
30. Disconnect the engine wiring harness.
31. Remove the driveshaft (4wd).
32. Remove the exhaust pipes and brackets.
33. Remove the left and right stabilizer bar links.
34. Remove the left and right front axle hub nuts.
35. Remove the left and right speed sensors.
36. Remove the left and right tie rod ends.
37. Disconnect the left and right lower control arms.
38. Remove the left and right halfshafts.
39. Disconnect the steering intermediate shaft.
40. Disconnect the height control sensor link (air suspension).
41. Attach a lifting crane.
42. Remove the 6 bolts and 2 nuts, then, remove the left and right frame side rail plates.
43. Remove the 6 bolts and 2 nuts, then, remove the left and right front suspension rear braces.
44. Lift the engine/transaxle from the vehicle.
45. Installation is the reverse of removal. Observe the following torques:
- Engine hanger: 14 ft. lbs. (20 Nm)
- Alternator bracket: 43 ft. lbs. (58 Nm)
- Right engine mount bracket: 40 ft. lbs. (54 Nm)
- Manifold stay: 36 ft. lbs. (49 Nm)
- Right rear engine mount bracket: 47 ft. lbs. (64 Nm)
- Front frame nuts: 70 ft. lbs. (95 Nm)
- Front right engine mount insulator nut: 64 ft. lbs. (87 Nm)
- Right rear engine mount insulator bolts: 55 ft. lbs. (75 Nm)
- Steering link: 52 ft. lbs. (70 Nm)
- Stabilizer bar: 21 ft. lbs. (29 Nm)
- Power steering pump adjusting bar: 32 ft. lbs. (43 Nm)
- Power steering pressure tube nuts: 69 inch lbs. (8 Nm)
- Left and right frame side rail plates: single end bolts 63 ft. lbs. (85 Nm); double end bolts and nuts: 24 ft. lbs. (32 Nm)
- Left and right front suspension member rear braces: single end bolts 63 ft. lbs. (85 Nm); double end bolts and nuts: 24 ft. lbs. (32 Nm)

- Height control sensor link: 48 inch lbs. (5 Nm)

Heater Core

REMOVAL & INSTALLATION

Highlander

FRONT HEATER

1. Before servicing the vehicle, refer to the precautions section.
2. Disconnect the negative battery cable.
3. Drain the cooling system into a clean container for reuse.
4. Disconnect the heater hoses from the heater core.
5. Remove the steering wheel by performing the following procedure:
    a. Position the front wheels facing straight-ahead.
    b. Remove the steering wheel side covers.
    c. Using a Torx® wrench, loosen the 2 screws located at each side of the steering wheel until the screw's circumference groove catches on the screw case.
    d. Pull the air bag module from the steering wheel and disconnect the electrical connector.

✳✳ CAUTION

Place the air bag module in a safe place with the front side facing upward.

    e. Remove the steering wheel nut.
    f. Place alignment marks on the steering wheel and the main shaft.
    g. Using a steering wheel puller, press the steering wheel from the steering column.
6. Remove the instrument panel and reinforcement by performing the following procedure:
    a. Remove the front door scuff plates.
    b. Remove the cowl side boards.
    c. Remove the front door trim covers.
    d. Remove the front pillar garnish by disengaging the 5 clips. If equipped with a tweeter speaker, disconnect the electrical connector.
    e. Remove the steering column covers-to-steering column screws and the covers.
    f. Remove the combination switch-to-steering column screws, disconnect the electrical connector(s) and remove the combination switch.

    g. Remove the 2 hood open lever screws and the hood open lever.
    h. Remove the 2 lower finish panel bolts and disengage the panel from the 3 clips.
    i. Remove the 2 No. 1 safety pad insert bolts and the insert.
    j. Remove the 2 No. 2 finish panel bolts and disengage the panel from the 4 clips.
    k. In the left side of the glove compartment, pry out the glove box door finish plate and disconnect the air bag module connector.
    l. Remove the glove box 3 nuts and 2 screws and the glove box.
    m. Remove the center cluster finish panel by disengaging the claw (bottom center) and 4 clips (1 at each corner).
    n. Remove the ashtray, the 2 ashtray receptacle box screws.
    o. Remove the 4 lower center cluster finish panel screws and disconnect the connector.
    p. Remove the clock, the No. 1 and No. 2 registers from the panel.
    q. Remove the 3 cluster finish panel screws, disengage the 8 clips and remove the panel.
    r. Remove the combination meter.
    s. Remove the radio assembly.
    t. Remove the heater control assembly.
    u. Remove 2 passenger's side air bag module bolts; then, disconnect and remove the air bag module.

✳✳ CAUTION

Place the air bag module in a safe place with the front side facing upward.

    v. Remove the instrument panel-to-chassis 5 bolts and nut.
    w. Remove the audio amplifier.
    x. Remove the No. 1 and No. 2 braces.
    y. Remove the No. 2 cowl brace.
    z. Remove the instrument panel reinforcement.
7. Remove the evaporator housing by performing the following procedure:
    a. Discharge and recover the air conditioning system refrigerant.
    b. In the engine compartment, remove the refrigerant lines-to-cowl connector bolts; then, disconnect the lines and discard the O-rings.
    c. Disconnect the electrical connector at the evaporator housing.
    d. Disconnect the wiring harness clamp.

e. Remove the evaporator housing-to-chassis 2 rivets, 3 bolts and nut.

f. Remove the evaporator housing.

8. Remove the 4 defroster nozzle nuts and the nozzle.

9. Disconnect and remove the theft deterrent and the wireless door lock ECUs.

10. Release the 2 air duct claws and the air duct.

11. Remove the 2 heater housing-to-chassis rivets and the heater housing.

➡**When installing the heater housing, use new screws in place of the rivets.**

12. Remove the heater core-to-heater housing cover.

13. Remove both heater core screws and clamps; then, remove the heater core.

To install:

14. Install the heater core and both heater core screws and clamps.

15. Install the heater core-to-heater housing cover.

➡**When installing the heater housing, use new screws in place of the rivets.**

16. Install the heater housing-to-chassis and the 2 heater housing screws.

17. Release the air duct and the air duct claws.

18. Connect and install the theft deterrent and the wireless door lock ECUs.

19. Install the defroster nozzle and the 4 nozzle nuts.

20. Install the evaporator housing by performing the following procedure:

a. Install the evaporator housing.

b. Install the evaporator housing-to-chassis 2 rivets, 3 bolts and nut.

c. Connect the wiring harness clamp.

d. Connect the electrical connector at the evaporator housing.

e. In the engine compartment, use new O-rings and install the refrigerant lines-to-cowl connector and install the bolts.

21. Install the instrument panel and reinforcement by performing the following procedure:

a. Install the instrument panel reinforcement.

b. Install the No. 2 cowl brace.

c. Install the No. 1 and No. 2 braces.

d. Install the audio amplifier.

e. Install the instrument panel-to-chassis 5 bolts and nut.

f. Connect and install the air bag module and the 2 passenger's side air bag module bolts.

g. Install the heater control assembly.

h. Install the radio assembly.

i. Install the combination meter.

j. Install the cluster finish panel, engage the 8 clips and install the panel screws.

k. Install the No. 1 and No. 2 registers and the clock to the panel.

l. Connect the lower center cluster finish panel connector and install the 4 lower center cluster finish panel screws.

m. Install the 2 ashtray receptacle box screws and the ashtray.

n. Install the center cluster finish panel by engaging the 4 clips (1 at each corner) and the claw (bottom center).

o. Install the glove box and the glove box 3 nuts and 2 screws.

p. In the left side of the glove compartment, connect the air bag module connector and install the glove box door finish plate.

q. Install the No. 2 finish panel, engage the 4 panel clips and install the 3 panel bolts.

r. Install the No. 1 safety pad insert and the 2 insert bolts.

s. Install the finish panel, engage the 3 finish panel clips and install 2 lower finish panel bolts.

t. Install the hood open lever and the 2 hood open lever screws.

u. Install the combination switch, connect the electrical connector(s) and install the combination switch-to-steering column screws.

v. Install the steering column covers and the covers-to-steering column screws.

w. Install the front pillar garnish by engaging the 5 clips. If equipped with a tweeter speaker, connect the electrical connector.

x. Install the front door trim covers.

y. Install the cowl side boards.

z. Install the front door scuff plates.

22. Install the steering wheel by performing the following procedure:

a. Install the steering wheel to the steering column.

b. Align the steering wheel-to-main shaft marks.

c. Install the steering wheel nut and torque the nut to 25 ft. lbs. (34 Nm).

d. Install the air bag module to the steering wheel and connect the electrical connector.

e. Using a Torx® wrench, tighten the steering wheel screws to 78 inch lbs. (8.8 Nm).

f. Install the steering wheel side covers.

23. Connect the heater hoses to the heater core.

24. Refill the cooling system.

25. Connect the negative battery cable.

26. Evacuate and charge the air conditioning system.

27. Run the engine to normal operating temperatures; then, check the climate control operation and check for leaks.

REAR AUXILIARY HEATER

1. Before servicing the vehicle, refer to the precautions section.

2. Disconnect the negative battery cable.

3. Drain the cooling system into a clean container for reuse.

4. Disconnect the heater hoses from the rear heater core.

5. Remove the front seats.

6. Remove the front door scuff plates.

7. Remove the cowl side trim.

8. Remove the rear door scuff plates.

9. Remove the lower door scuff plates.

10. Remove the rear console box.

11. Remove the left side air outlet grille.

12. Pull the carpet rearward.

13. Remove the 3 clips and the air outlet grille.

14. Remove the rear air duct 2 bolts, 2 clips and the duct.

15. Disconnect the electrical connectors.

16. Remove the 3 rear heater housing bolts and the housing.

17. Remove both heater core-to-heater housing screws and clamps.

18. Remove the heater core-to-heater housing screw and plate.

19. Remove the heater core.

To install:

20. Install the heater core.

21. Install the heater core-to-heater housing screw and plate.

22. Install both heater core-to-heater housing screws and clamps.

23. Install the rear heater housing and the 3 housing bolts.

24. Connect the electrical connectors.

25. Install the rear air duct and the duct 2 bolts and 2 clips.

26. Install the 3 clips and the air outlet grille.

27. Move the carpet forward.

28. Install the left side air outlet grille.

29. Install the rear console box.

30. Install the lower door scuff plates.

31. Install the rear door scuff plates.

32. Install the cowl side trim.

33. Install the front door scuff plates.

34. Install the front seats.

35. Connect the heater hoses to the rear heater core.

36. Refill the cooling system.

37. Connect the negative battery cable.

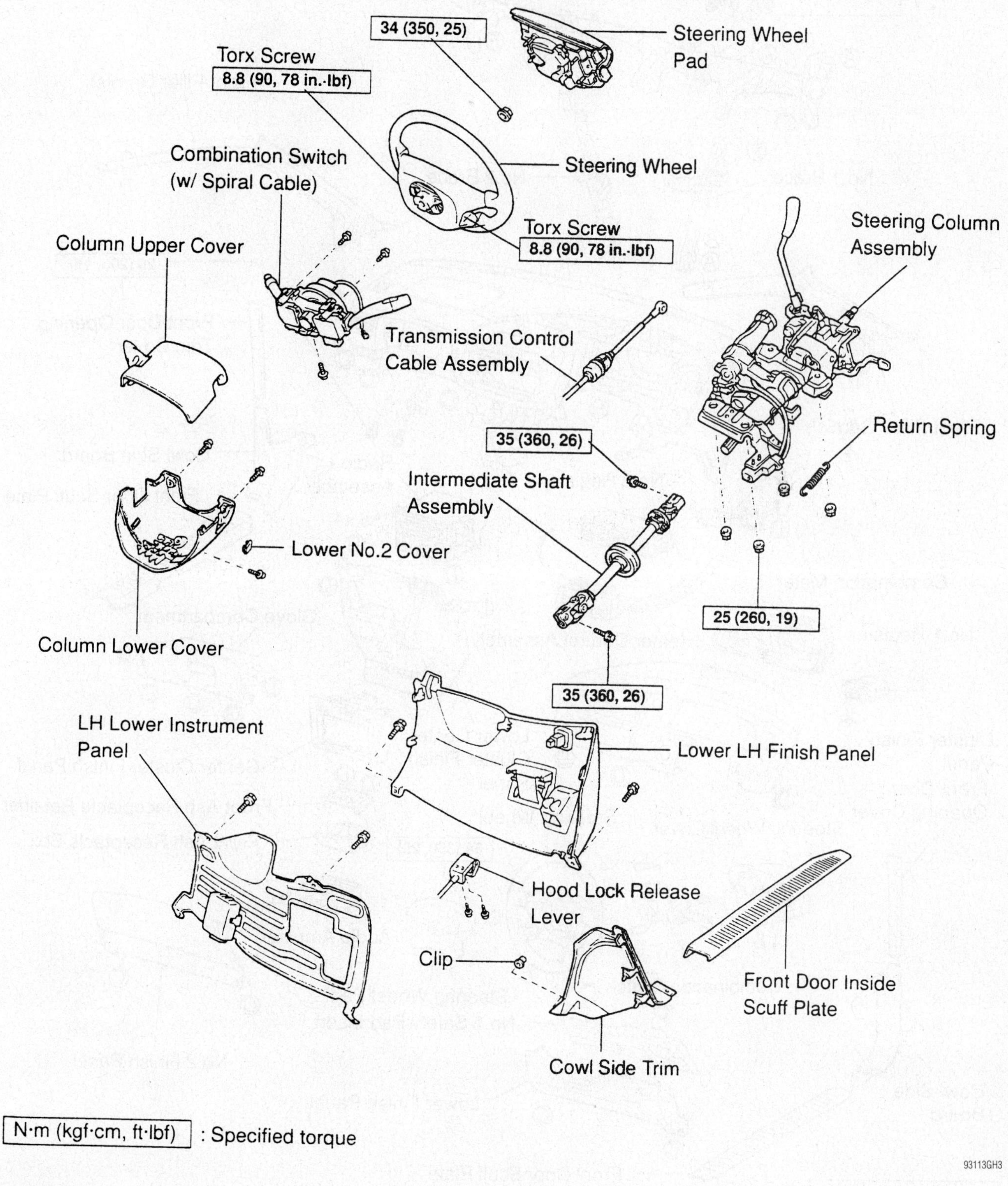

34 (350, 25)

Steering Wheel Pad

Torx Screw
8.8 (90, 78 in.·lbf)

Combination Switch
(w/ Spiral Cable)

Column Upper Cover

Steering Wheel

Torx Screw
8.8 (90, 78 in.·lbf)

Steering Column Assembly

Transmission Control Cable Assembly

Return Spring

35 (360, 26)

Intermediate Shaft Assembly

Lower No.2 Cover

25 (260, 19)

Column Lower Cover

35 (360, 26)

LH Lower Instrument Panel

Lower LH Finish Panel

Hood Lock Release Lever

Clip

Cowl Side Trim

Front Door Inside Scuff Plate

N·m (kgf·cm, ft·lbf) : Specified torque

93113GH3

Exploded view of the steering wheel, steering column and related components—Highlander

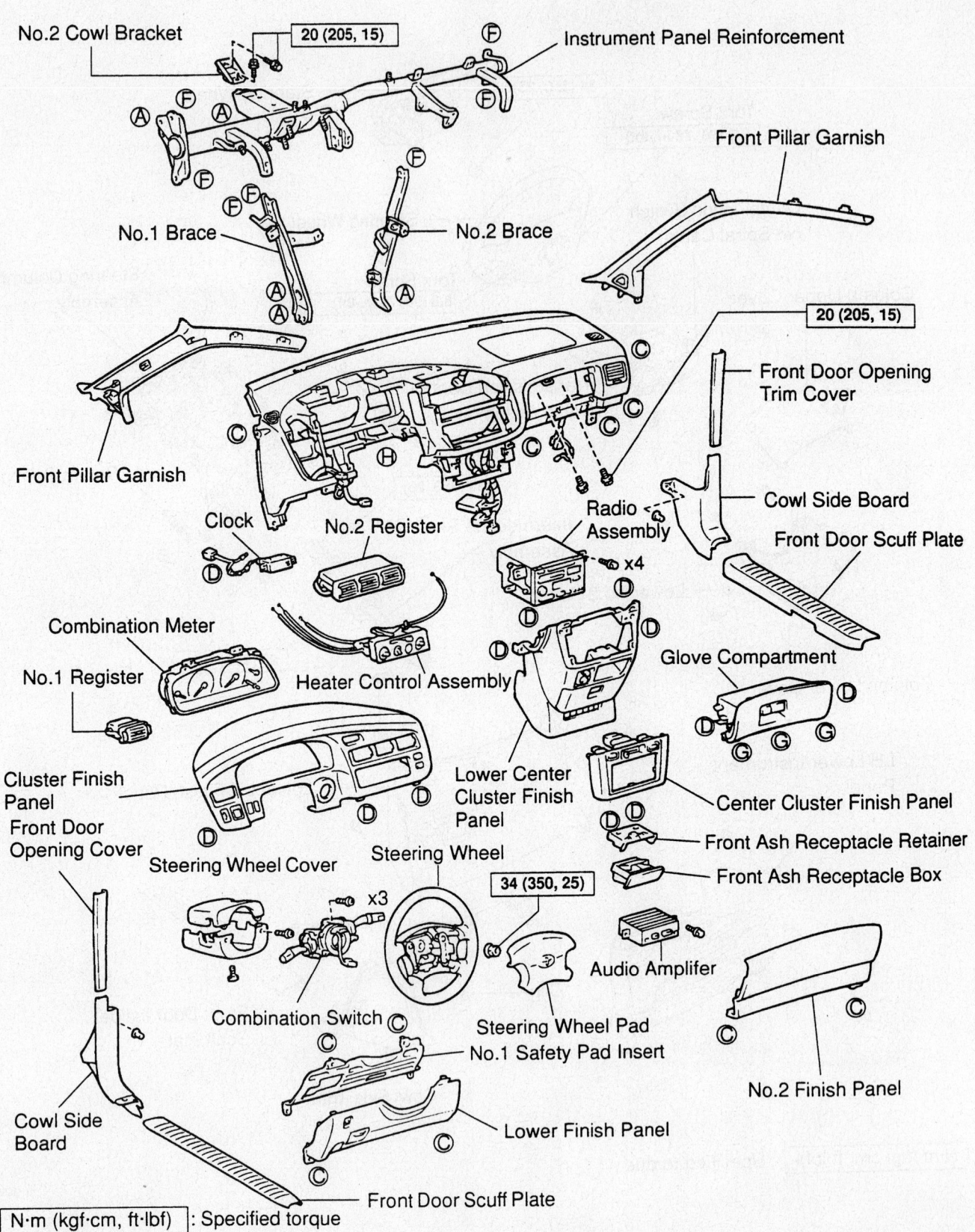

No.2 Cowl Bracket

20 (205, 15)

Instrument Panel Reinforcement

Front Pillar Garnish

No.1 Brace

No.2 Brace

20 (205, 15)

Front Door Opening Trim Cover

Front Pillar Garnish

Cowl Side Board

Front Door Scuff Plate

Clock

No.2 Register

Radio Assembly

x4

Combination Meter

Heater Control Assembly

Glove Compartment

No.1 Register

Cluster Finish Panel

Front Door Opening Cover

Steering Wheel Cover

Steering Wheel

Lower Center Cluster Finish Panel

Center Cluster Finish Panel

Front Ash Receptacle Retainer

Front Ash Receptacle Box

34 (350, 25)

x3

Audio Amplifer

Combination Switch

Steering Wheel Pad

No.1 Safety Pad Insert

No.2 Finish Panel

Cowl Side Board

Lower Finish Panel

Front Door Scuff Plate

N·m (kgf·cm, ft·lbf) : Specified torque

93113GH4

Exploded view of the instrument panel and related components–Highlander

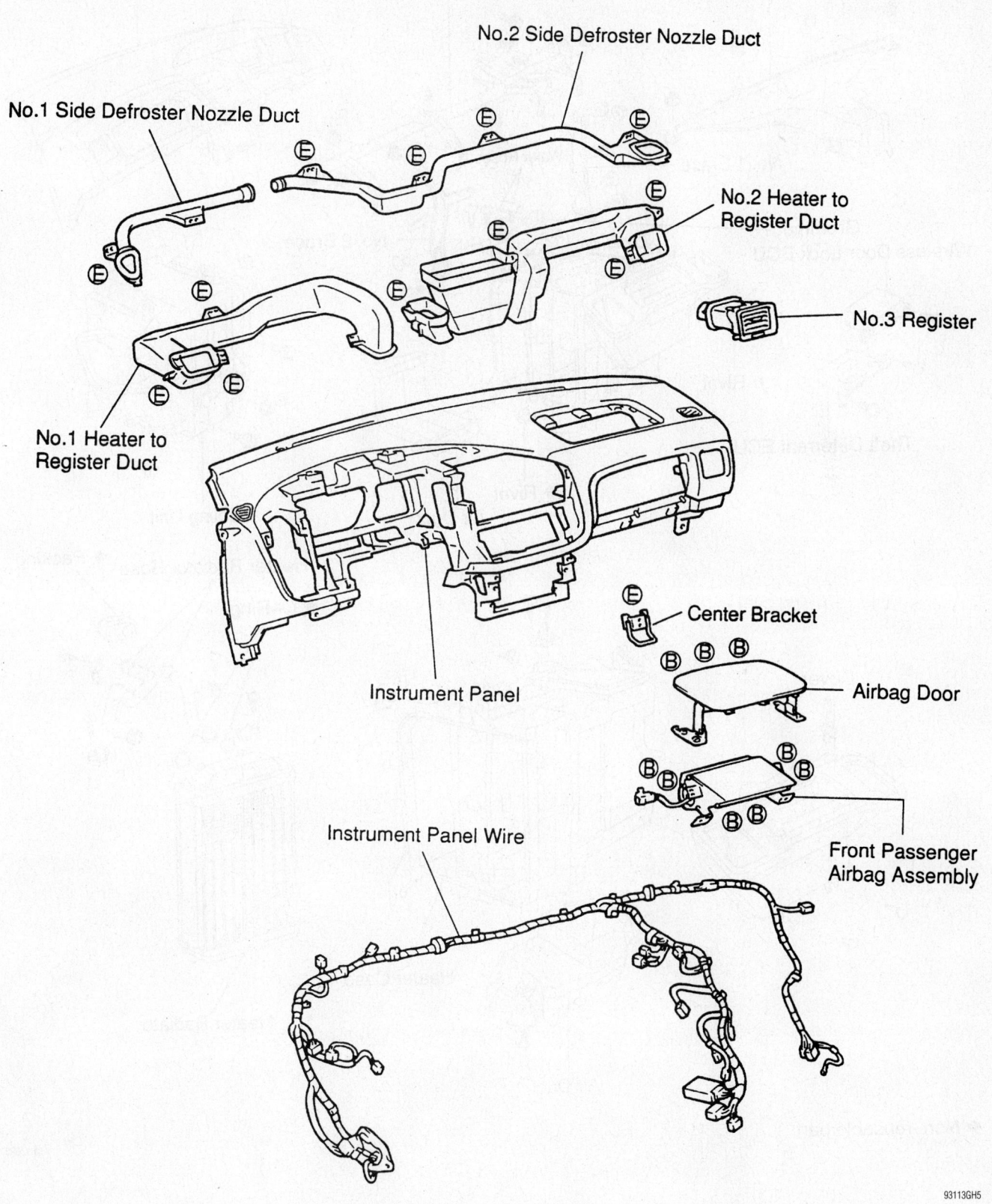

No.2 Side Defroster Nozzle Duct

No.1 Side Defroster Nozzle Duct

No.2 Heater to Register Duct

No.3 Register

No.1 Heater to Register Duct

Instrument Panel

Center Bracket

Airbag Door

Front Passenger Airbag Assembly

Instrument Panel Wire

93113GH5

Exploded view of the ventilation system and related components–Highlander

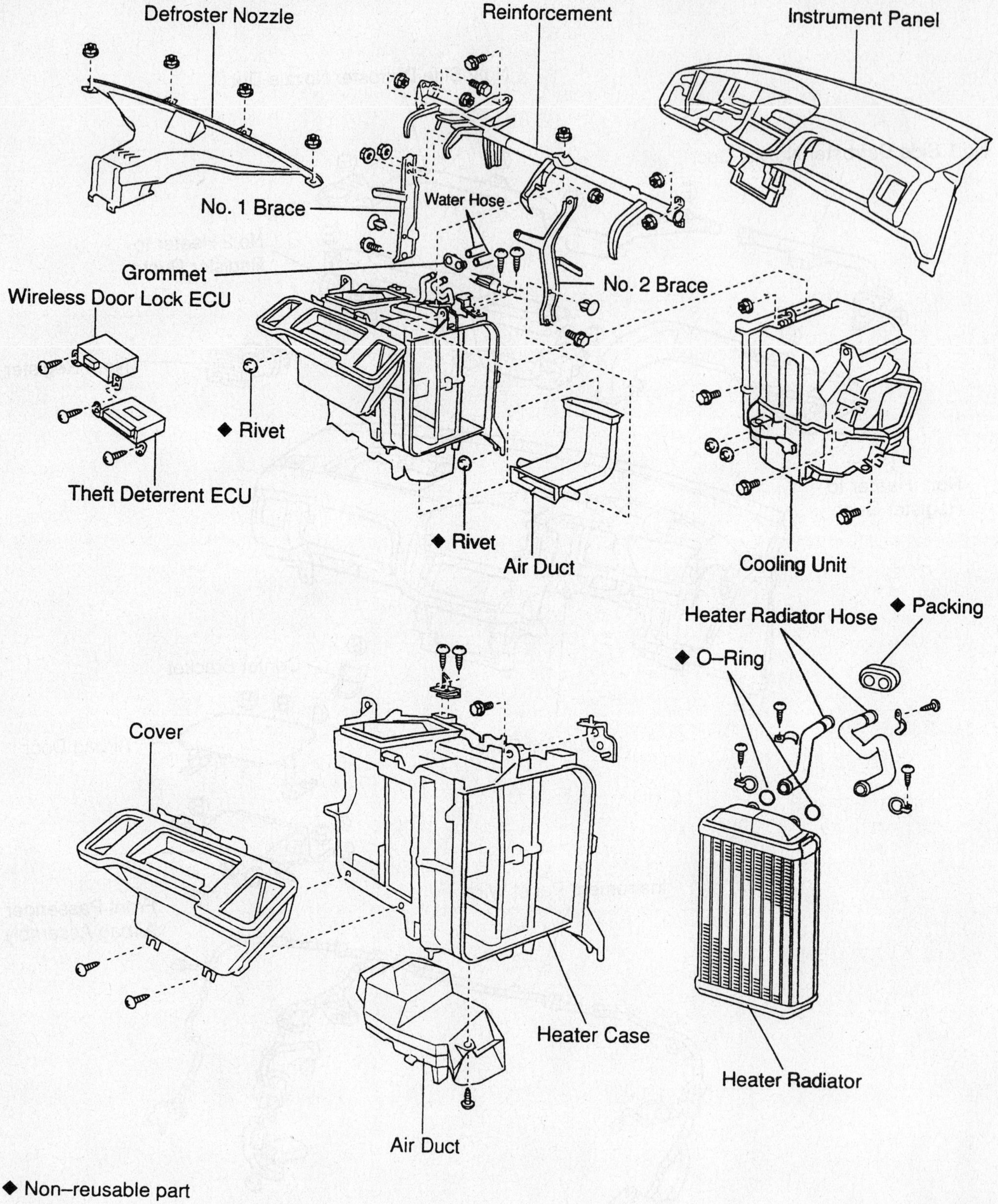

Defroster Nozzle

Reinforcement

Instrument Panel

No. 1 Brace

Water Hose

No. 2 Brace

Grommet

Wireless Door Lock ECU

◆ Rivet

Theft Deterrent ECU

◆ Rivet

Air Duct

Cooling Unit

Heater Radiator Hose   ◆ Packing

◆ O–Ring

Cover

Heater Case

Heater Radiator

Air Duct

◆ Non–reusable part

93113GH6

Exploded view of the heater core, heater housing, evaporator housing and related components–Highlander

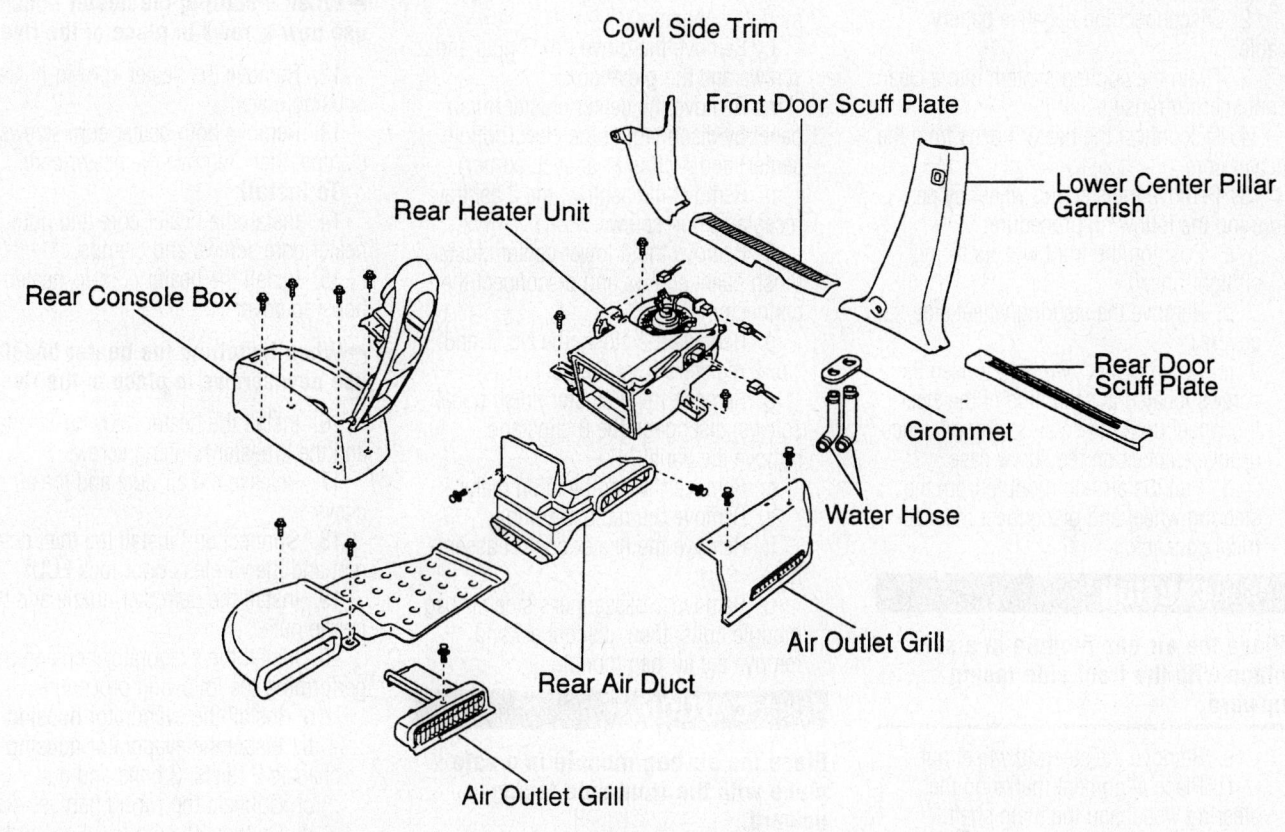

Cowl Side Trim

Front Door Scuff Plate

Lower Center Pillar Garnish

Rear Heater Unit

Rear Console Box

Rear Door Scuff Plate

Grommet

Water Hose

Air Outlet Grill

Rear Air Duct

Air Outlet Grill

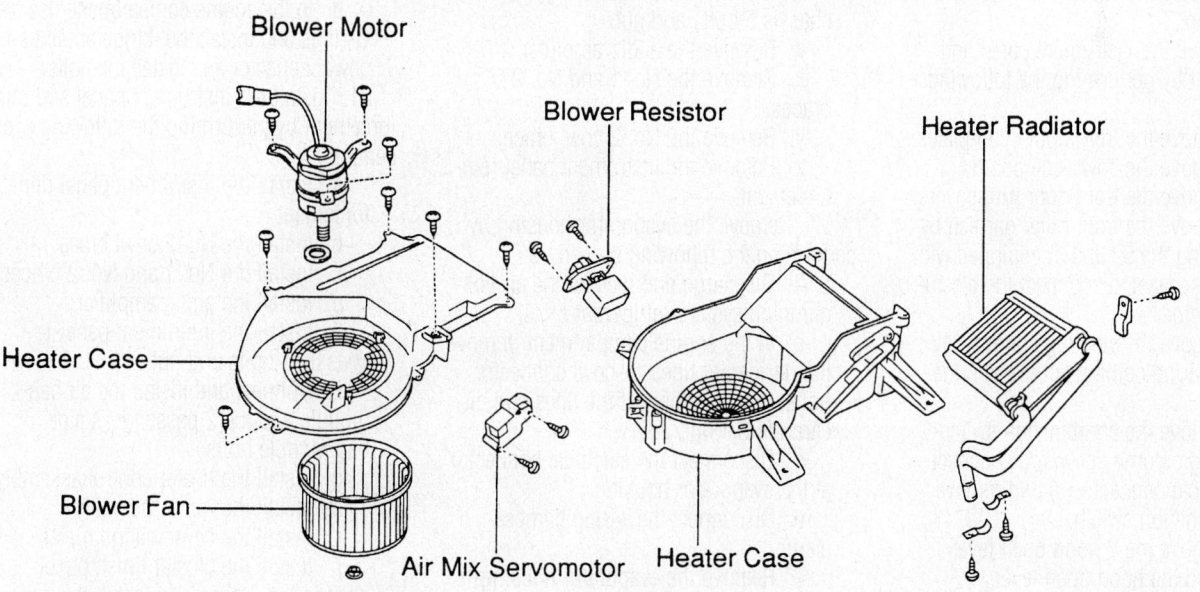

Blower Motor

Blower Resistor

Heater Radiator

Heater Case

Heater Case

Air Mix Servomotor

Blower Fan

93113GH7

Exploded view of the rear heater core, the rear heater housing and related components–Highlander

RX300

FRONT HEATER

1. Before servicing the vehicle, refer to the precautions section.

2. Disconnect the negative battery cable.

3. Drain the cooling system into a clean container for reuse.

4. Disconnect the heater hoses from the heater core.

5. Remove the steering wheel by performing the following procedure:

    a. Position the front wheels facing straight-ahead.

    b. Remove the steering wheel side covers.

    c. Using a Torx® wrench, loosen the 2 screws located at each side of the steering wheel until the screw's circumference groove catches on the screw case.

    d. Pull the air bag module from the steering wheel and disconnect the electrical connector.

✷✷ CAUTION

Place the air bag module in a safe place with the front side facing upward.

    e. Remove the steering wheel nut.

    f. Place alignment marks on the steering wheel and the main shaft.

    g. Using a steering wheel puller, press the steering wheel from the steering column.

6. Remove the instrument panel and reinforcement by performing the following procedure:

    a. Remove the front door scuff plates.

    b. Remove the cowl side boards.

    c. Remove the front door trim covers.

    d. Remove the front pillar garnish by disengaging the 5 clips. If equipped with a tweeter speaker, disconnect the electrical connector.

    e. Remove the steering column covers-to-steering column screws and the covers.

    f. Remove the combination switch-to-steering column screws, disconnect the electrical connector(s) and remove the combination switch.

    g. Remove the 2 hood open lever screws and the hood open lever.

    h. Remove the 2 lower finish panel bolts and disengage the panel from the 3 clips.

    i. Remove the 2 No. 1 safety pad insert bolts and the insert.

    j. Remove the 2 No. 2 finish panel bolts and disengage the panel from the 4 clips.

    k. In the left side of the glove compartment, pry out the glove box door finish plate and disconnect the air bag module connector.

    l. Remove the glove box 3 nuts and 2 screws and the glove box.

    m. Remove the center cluster finish panel by disengaging the claw (bottom center) and 4 clips (1 at each corner).

    n. Remove the ashtray, the 2 ashtray receptacle box screws.

    o. Remove the 4 lower center cluster finish panel screws and disconnect the connector.

    p. Remove the clock, the No. 1 and No. 2 registers from the panel.

    q. Remove the 3 cluster finish panel screws, disengage the 8 clips and remove the panel.

    r. Remove the combination meter.

    s. Remove the radio assembly.

    t. Remove the heater control assembly.

    u. Remove 2 passenger's side air bag module bolts; then, disconnect and remove the air bag module.

✷✷ CAUTION

Place the air bag module in a safe place with the front side facing upward.

    v. Remove the instrument panel-to-chassis 5 bolts and nut.

    w. Remove the audio amplifier.

    x. Remove the No. 1 and No. 2 braces.

    y. Remove the No. 2 cowl brace.

    z. Remove the instrument panel reinforcement.

7. Remove the evaporator housing by performing the following procedure:

    a. Discharge and recover the air conditioning system refrigerant.

    b. In the engine compartment, remove the refrigerant lines-to-cowl connector bolts; then, disconnect the lines and discard the O-rings.

    c. Disconnect the electrical connector at the evaporator housing.

    d. Disconnect the wiring harness clamp.

    e. Remove the evaporator housing-to-chassis 2 rivets, 3 bolts and nut.

    f. Remove the evaporator housing.

8. Remove the 4 defroster nozzle nuts and the nozzle.

9. Disconnect and remove the theft deterrent and the wireless door lock ECUs.

10. Release the 2 air duct claws and the air duct.

11. Remove the 2 heater housing-to-chassis rivets and the heater housing.

➡ **When installing the heater housing, use new screws in place of the rivets.**

12. Remove the heater core-to-heater housing cover.

13. Remove both heater core screws and clamps; then, remove the heater core.

To install:

14. Install the heater core and both heater core screws and clamps.

15. Install the heater core-to-heater housing cover.

➡ **When installing the heater housing, use new screws in place of the rivets.**

16. Install the heater housing-to-chassis and the 2 heater housing screws.

17. Release the air duct and the air duct claws.

18. Connect and install the theft deterrent and the wireless door lock ECUs.

19. Install the defroster nozzle and the 4 nozzle nuts.

20. Install the evaporator housing by performing the following procedure:

    a. Install the evaporator housing.

    b. Install the evaporator housing-to-chassis 2 rivets, 3 bolts and nut.

    c. Connect the wiring harness clamp.

    d. Connect the electrical connector at the evaporator housing.

    e. In the engine compartment, use new O-rings and install the refrigerant lines-to-cowl connector and install the bolts.

21. Install the instrument panel and reinforcement by performing the following procedure:

    a. Install the instrument panel reinforcement.

    b. Install the No. 2 cowl brace.

    c. Install the No. 1 and No. 2 braces.

    d. Install the audio amplifier.

    e. Install the instrument panel-to-chassis 5 bolts and nut.

    f. Connect and install the air bag module and the 2 passenger's side air bag module bolts.

    g. Install the heater control assembly.

    h. Install the radio assembly.

    i. Install the combination meter.

    j. Install the cluster finish panel, engage the 8 clips and install the panel screws.

    k. Install the No. 1 and No. 2 registers and the clock to the panel.

    l. Connect the lower center cluster finish panel connector and install the 4 lower center cluster finish panel screws.

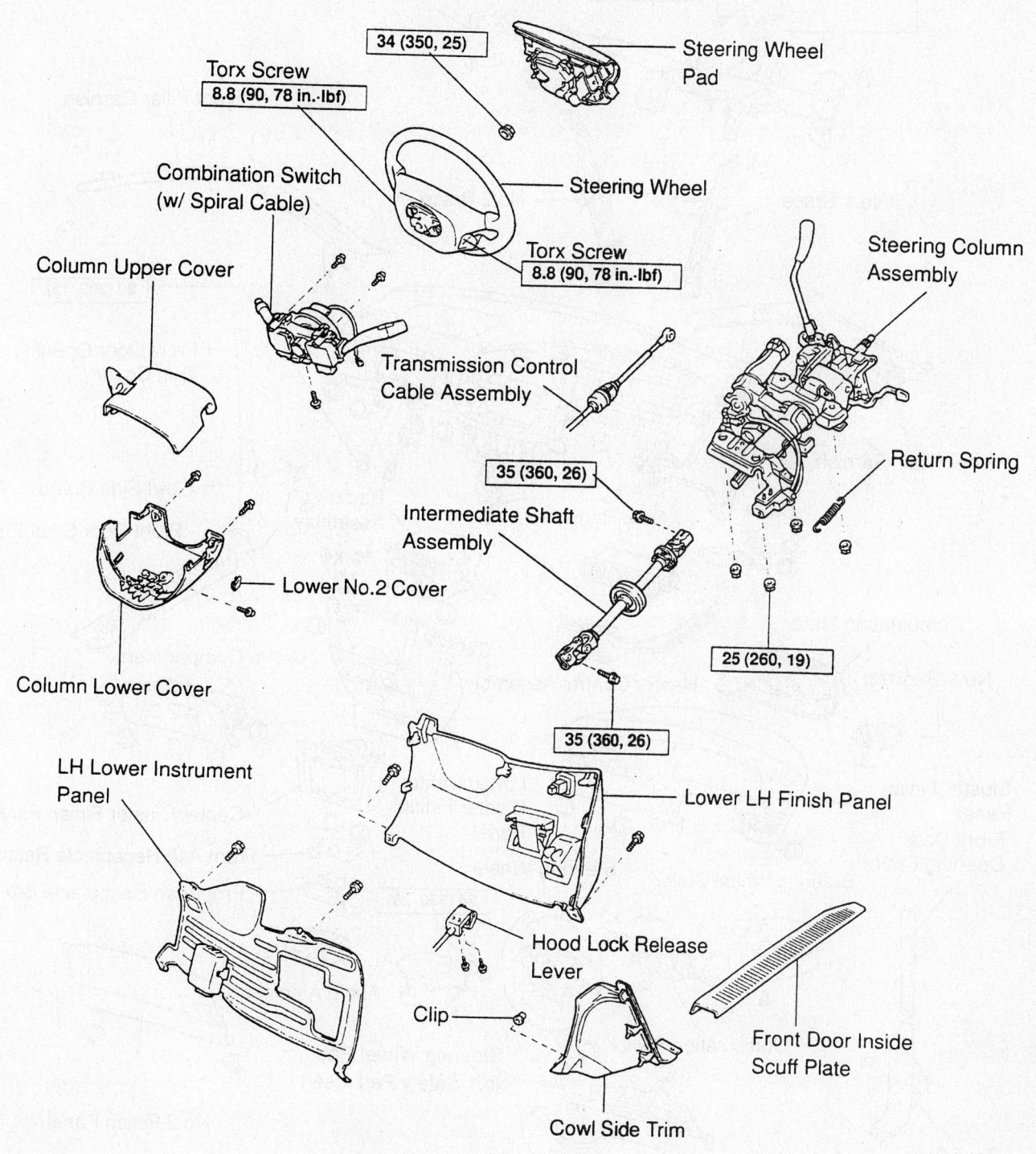

Torx Screw
8.8 (90, 78 in.·lbf)

34 (350, 25)

Steering Wheel
Pad

Combination Switch
(w/ Spiral Cable)

Steering Wheel

Steering Column
Assembly

Column Upper Cover

Torx Screw
8.8 (90, 78 in.·lbf)

Transmission Control
Cable Assembly

Return Spring

35 (360, 26)

Intermediate Shaft
Assembly

Lower No.2 Cover

25 (260, 19)

Column Lower Cover

35 (360, 26)

LH Lower Instrument
Panel

Lower LH Finish Panel

Hood Lock Release
Lever

Clip

Front Door Inside
Scuff Plate

Cowl Side Trim

N·m (kgf·cm, ft·lbf) : Specified torque

Exploded view of the steering wheel, steering column and related components—Lexus RX300

93113GH3

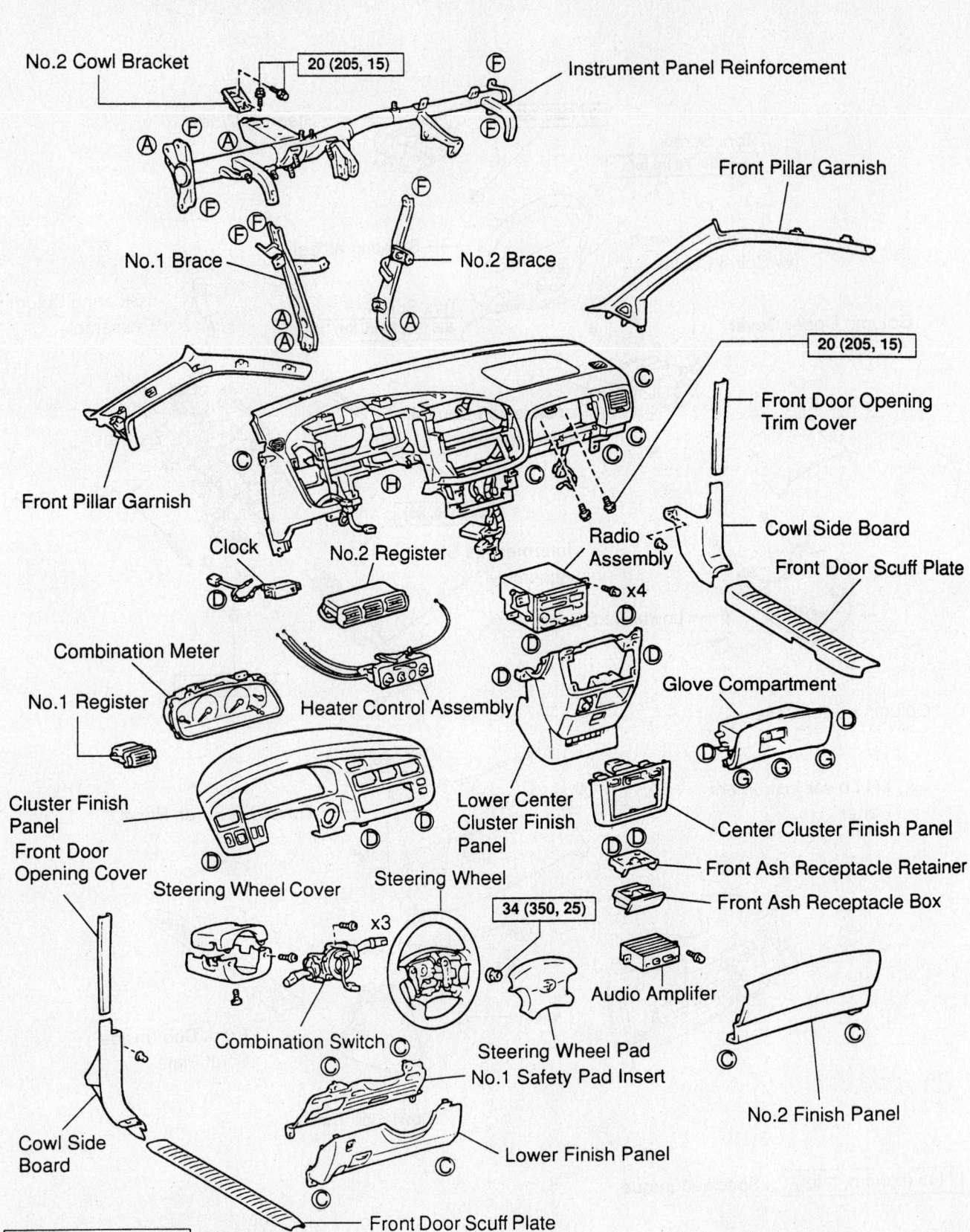

No.2 Cowl Bracket

20 (205, 15)

Instrument Panel Reinforcement

Front Pillar Garnish

No.1 Brace

No.2 Brace

20 (205, 15)

Front Door Opening Trim Cover

Front Pillar Garnish

Cowl Side Board

Front Door Scuff Plate

Clock

No.2 Register

Radio Assembly

x4

Combination Meter

Heater Control Assembly

Glove Compartment

No.1 Register

Lower Center Cluster Finish Panel

Center Cluster Finish Panel

Cluster Finish Panel

Front Door Opening Cover

Steering Wheel Cover

Steering Wheel

Front Ash Receptacle Retainer

Front Ash Receptacle Box

34 (350, 25)

x3

Audio Amplifer

Combination Switch

Steering Wheel Pad

No.1 Safety Pad Insert

No.2 Finish Panel

Cowl Side Board

Lower Finish Panel

Front Door Scuff Plate

N·m (kgf·cm, ft·lbf) : Specified torque

93113GH4

Exploded view of the instrument panel and related components—Lexus RX300

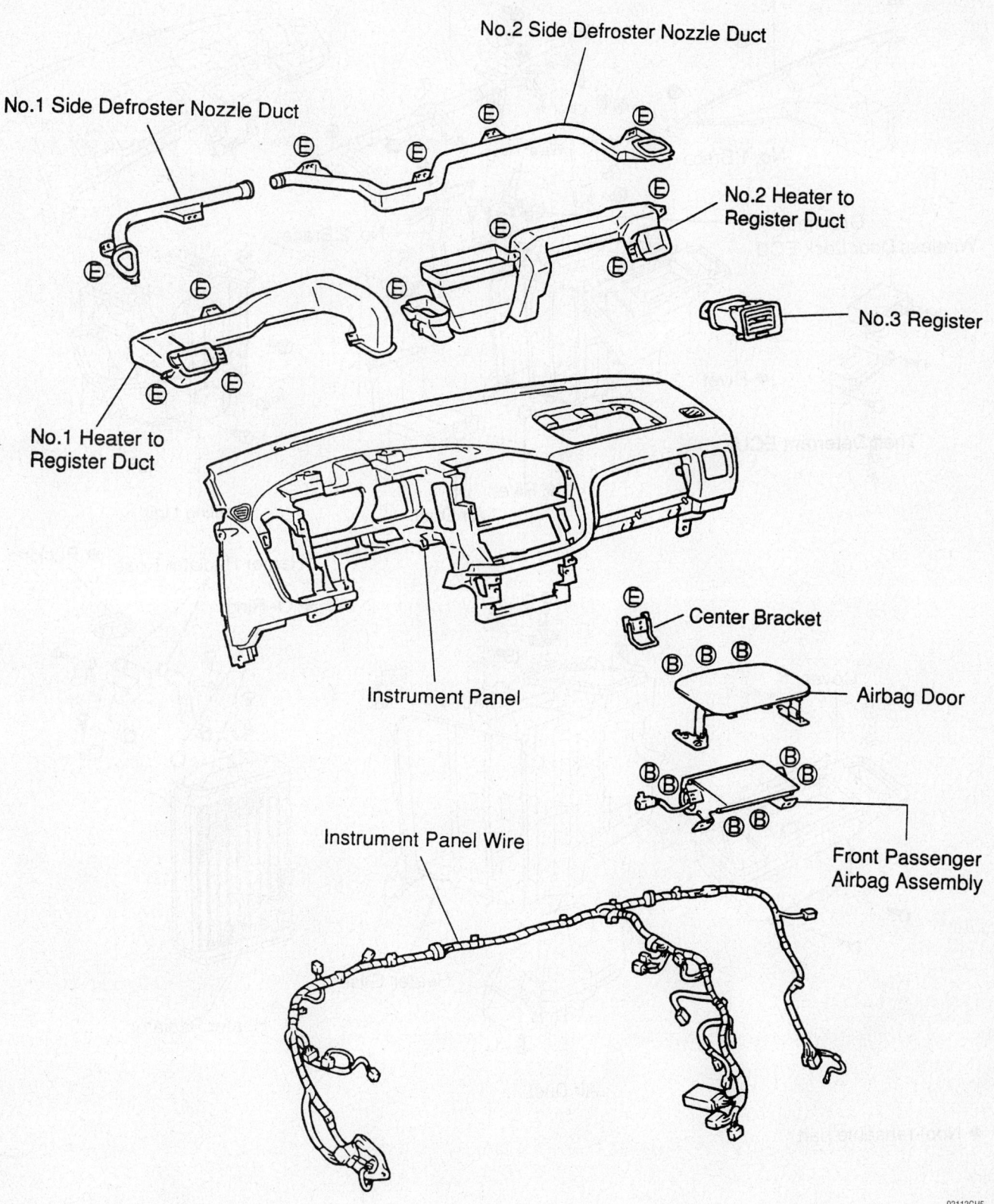

No.2 Side Defroster Nozzle Duct

No.1 Side Defroster Nozzle Duct

No.2 Heater to Register Duct

No.3 Register

No.1 Heater to Register Duct

Instrument Panel

Center Bracket

Airbag Door

Instrument Panel Wire

Front Passenger Airbag Assembly

93113GH5

Exploded view of the ventilation system and related components—Lexus RX300

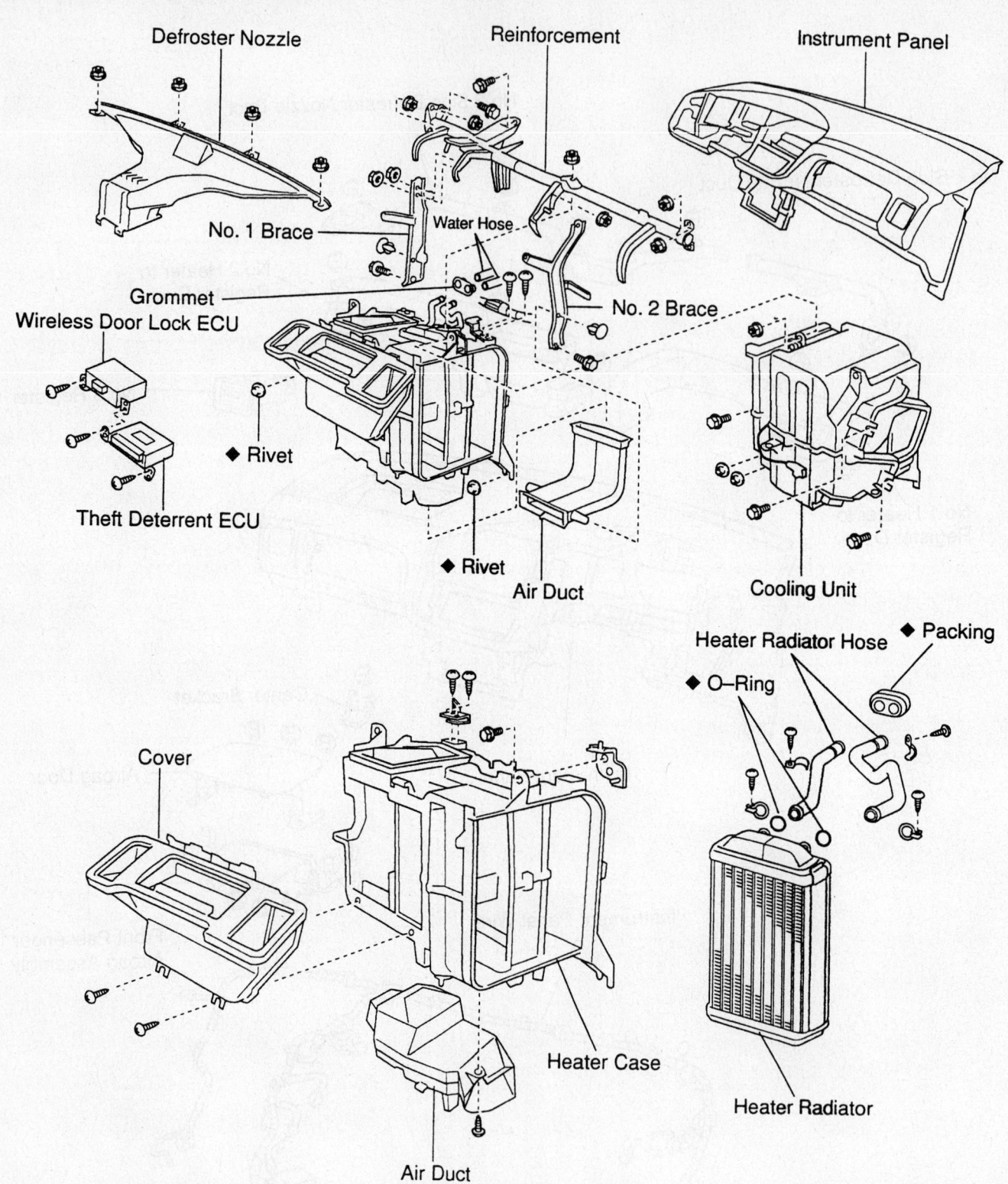

Defroster Nozzle

Reinforcement

Instrument Panel

No. 1 Brace

Water Hose

No. 2 Brace

Grommet

Wireless Door Lock ECU

◆ Rivet

Theft Deterrent ECU

◆ Rivet

Air Duct

Cooling Unit

Heater Radiator Hose

◆ Packing

◆ O–Ring

Cover

Heater Case

Heater Radiator

Air Duct

◆ Non–reusable part

93113GH6

Exploded view of the heater core, heater housing, evaporator housing and related components—Lexus RX300

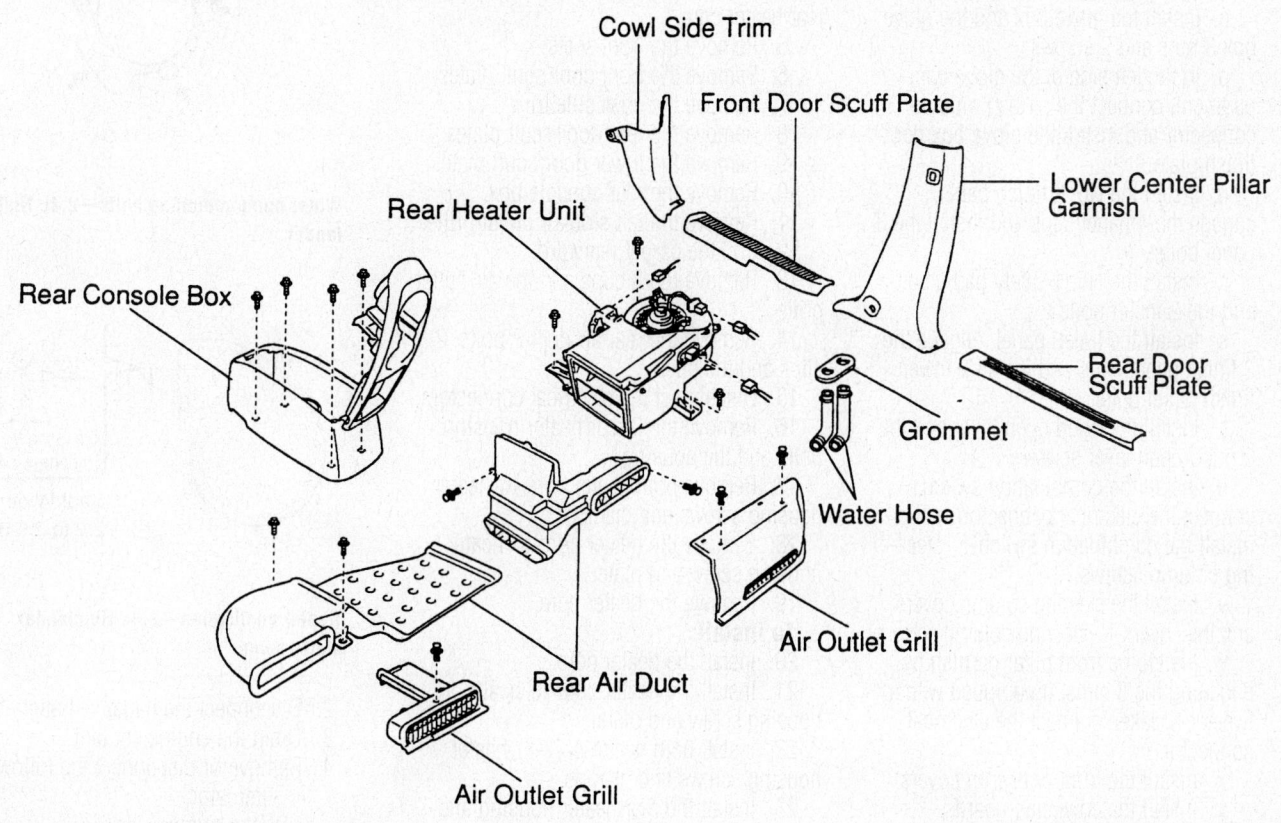

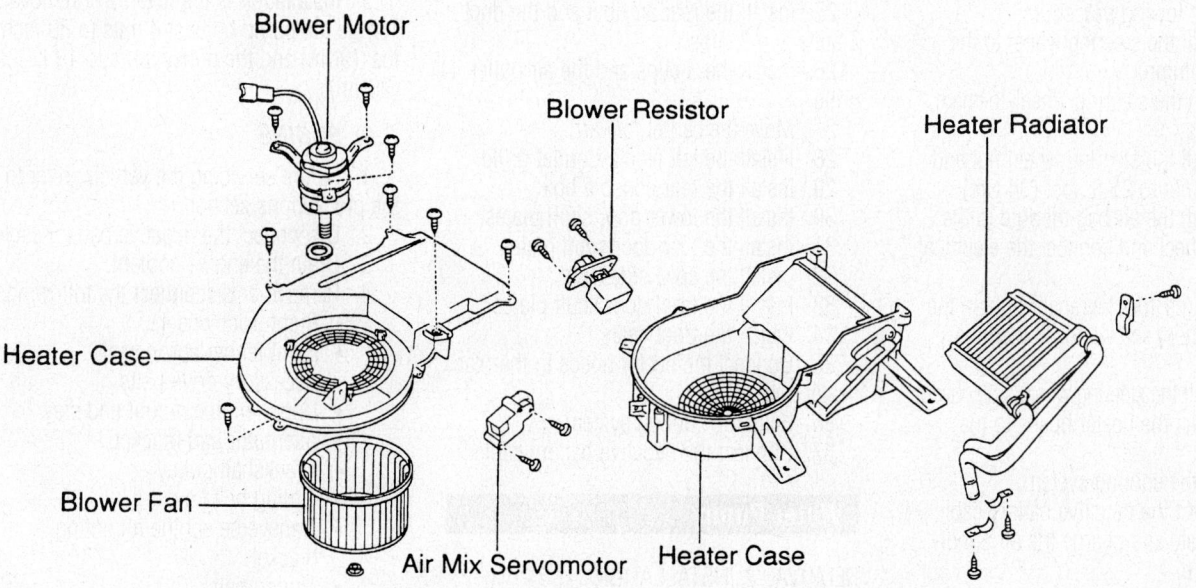

Exploded view of the rear heater core, the rear heater housing and related components—Lexus RX300

m. Install the 2 ashtray receptacle box screws and the ashtray.

n. Install the center cluster finish panel by engaging the 4 clips (1 at each corner) and the claw (bottom center).

o. Install the glove box and the glove box 3 nuts and 2 screws.

p. In the left side of the glove compartment, connect the air bag module connector and install the glove box door finish plate.

q. Install the No. 2 finish panel, engage the 4 panel clips and install the 3 panel bolts.

r. Install the No. 1 safety pad insert and the 2 insert bolts.

s. Install the finish panel, engage the 3 finish panel clips and install 2 lower finish panel bolts.

t. Install the hood open lever and the 2 hood open lever screws.

u. Install the combination switch, connect the electrical connector(s) and install the combination switch-to-steering column screws.

v. Install the steering column covers and the covers-to-steering column screws.

w. Install the front pillar garnish by engaging the 5 clips. If equipped with a tweeter speaker, connect the electrical connector.

x. Install the front door trim covers.

y. Install the cowl side boards.

z. Install the front door scuff plates.

22. Install the steering wheel by performing the following procedure:

a. Install the steering wheel to the steering column.

b. Align the steering wheel-to-main shaft marks.

c. Install the steering wheel nut and torque the nut to 25 ft. lbs. (34 Nm).

d. Install the air bag module to the steering wheel and connect the electrical connector.

e. Using a Torx® wrench, tighten the steering wheel screws to 78 inch lbs. (8.8 Nm).

f. Install the steering wheel side covers.

23. Connect the heater hoses to the heater core.

24. Refill the cooling system.

25. Connect the negative battery cable.

26. Evacuate and charge the air conditioning system.

27. Run the engine to normal operating temperatures; then, check the climate control operation and check for leaks.

REAR AUXILIARY HEATER

1. Before servicing the vehicle, refer to the precautions section.

2. Disconnect the negative battery cable.

3. Drain the cooling system into a clean container for reuse.

4. Disconnect the heater hoses from the rear heater core.

5. Remove the front seats.

6. Remove the front door scuff plates.

7. Remove the cowl side trim.

8. Remove the rear door scuff plates.

9. Remove the lower door scuff plates.

10. Remove the rear console box.

11. Remove the left side air outlet grille.

12. Pull the carpet rearward.

13. Remove the 3 clips and the air outlet grille.

14. Remove the rear air duct 2 bolts, 2 clips and the duct.

15. Disconnect the electrical connectors.

16. Remove the 3 rear heater housing bolts and the housing.

17. Remove both heater core-to-heater housing screws and clamps.

18. Remove the heater core-to-heater housing screw and plate.

19. Remove the heater core.

To install:

20. Install the heater core.

21. Install the heater core-to-heater housing screw and plate.

22. Install both heater core-to-heater housing screws and clamps.

23. Install the rear heater housing and the 3 housing bolts.

24. Connect the electrical connectors.

25. Install the rear air duct and the duct 2 bolts and 2 clips.

26. Install the 3 clips and the air outlet grille.

27. Move the carpet forward.

28. Install the left side air outlet grille.

29. Install the rear console box.

30. Install the lower door scuff plates.

31. Install the rear door scuff plates.

32. Install the cowl side trim.

33. Install the front door scuff plates.

34. Install the front seats.

35. Connect the heater hoses to the rear heater core.

36. Refill the cooling system.

37. Connect the negative battery cable.

Water Pump

REMOVAL & INSTALLATION

Highlander

2.4L ENGINE

1. Before servicing the vehicle, refer to the precautions section.

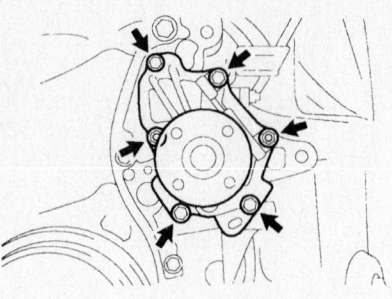

Water pump mounting bolts—2.4L Highlander

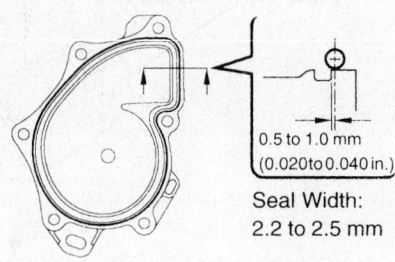

Sealer application—2.4L Highlander water pump

2. Disconnect the negative battery cable.

3. Drain the engine coolant.

4. Remove or disconnect the following:
   - Alternator
   - Water pump pulley
   - Water pump

5. Installation is the reverse of removal. Torque the pump bolts and nuts to 80 inch lbs. (9Nm) and the pulley bolts to 19 ft. lbs. (26Nm).

3.0L ENGINE

1. Before servicing the vehicle, refer to the precautions section.

2. Disconnect the negative battery cable.

3. Drain the engine coolant.

4. Remove or disconnect the following:
   - Right front wheel
   - Right fender apron seal
   - Accessory drive belts
   - Upper engine mount and stay
   - Alternator and bracket
   - Crankshaft pulley
   - Timing belt covers
   - Transverse engine mounting bracket
   - Timing belt
   - Timing belt idler
   - Camshaft pulley
   - Water pump

5. Installation is the reverse of removal. Torque the water pump bolts and nuts to 71 inch lbs. (8Nm).

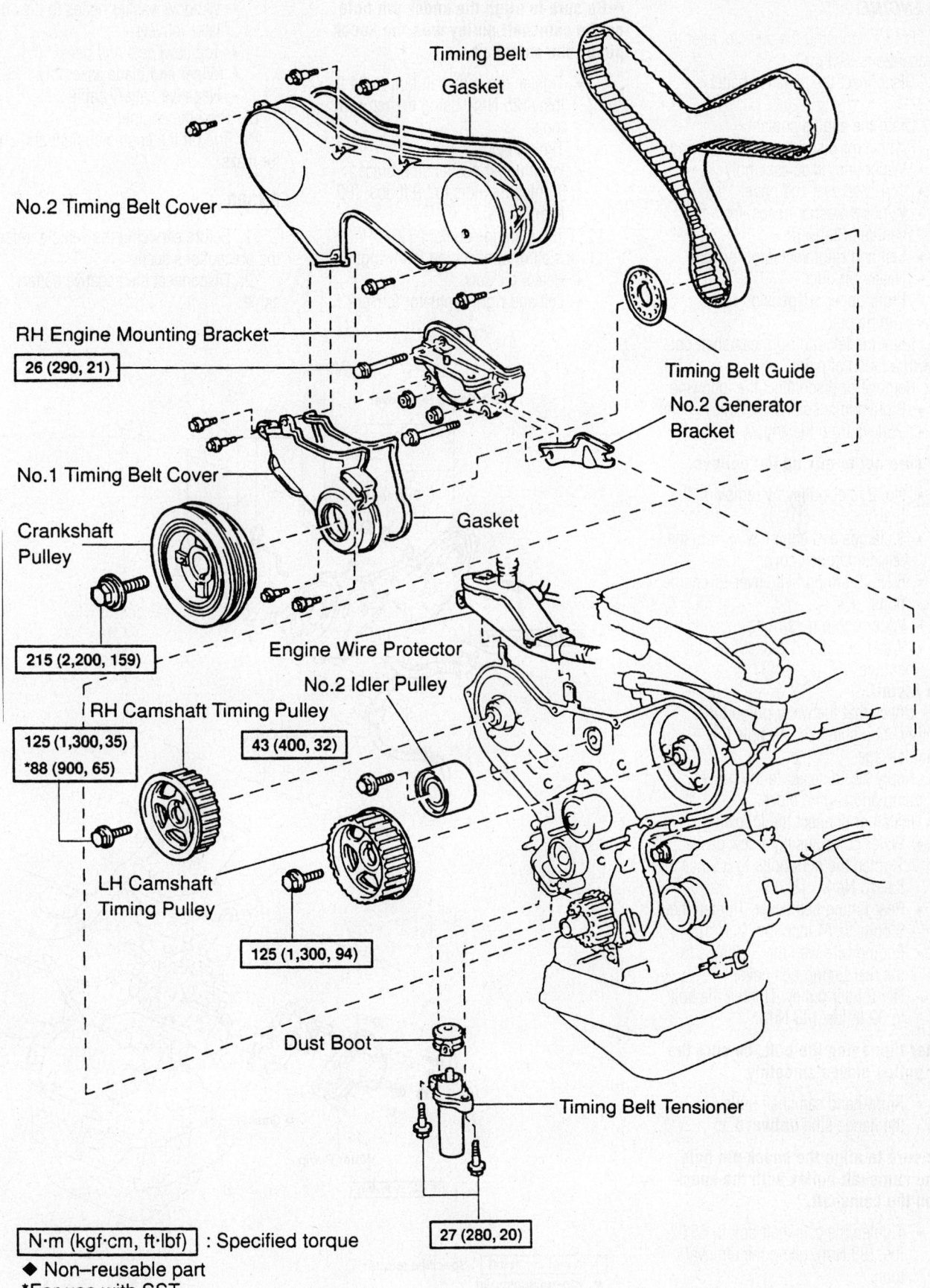

Timing Belt

Gasket

No.2 Timing Belt Cover

RH Engine Mounting Bracket

26 (290, 21)

No.1 Timing Belt Cover

Crankshaft Pulley

Gasket

Timing Belt Guide

No.2 Generator Bracket

215 (2,200, 159)

Engine Wire Protector

No.2 Idler Pulley

RH Camshaft Timing Pulley

43 (400, 32)

125 (1,300, 35)
*88 (900, 65)

LH Camshaft Timing Pulley

125 (1,300, 94)

Dust Boot

Timing Belt Tensioner

27 (280, 20)

N·m (kgf·cm, ft·lbf) : Specified torque

◆ Non–reusable part
*For use with SST

7924ZG15

Exploded view of the components to gain access to the water pump—3.3L Highlander shown; RX330 similar

3.3L ENGINE

1. Before servicing the vehicle, refer to the precautions section.
2. Disconnect the negative battery cable.
3. Drain the engine coolant.
4. Remove or disconnect the following:
   - Wiper and blade assembly
   - Top cowl seal and panel
   - Window washer hoses, from the ventilator louvers
   - Left and right ventilator louvers
   - Heater air duct
   - Front upper suspension brace
   - Timing belt
5. Mark the left and right camshaft pulleys with a touch of paint.
6. Remove or disconnect the following:
   - Right and left camshaft pulleys bolts
   - Pulleys from the engine

➡ **Be sure not to mix up the pulleys.**

   - No. 2 idler pulley by removing the bolt
   - 3 clamps and engine wire from the rear timing belt cover
   - 6 No. 3 timing belt cover-to-engine bolts
   - Water pump nuts/bolts
   - Water pump and gasket from the engine

To install:

7. Check that the water pump turns smoothly. Also check the air hole for coolant leakage.
8. Apply liquid sealer to the gasket, water pump and engine block.
9. Install or connect the following:
   - Water pump, using a new gasket. Tighten the nuts/bolts to 53 inch lbs. (6 Nm).
   - Rear timing belt cover. Tighten the 6 bolts to 74 inch lbs. (9 Nm).
   - Engine wire with the 3 clamps to the rear timing belt cover
   - No. 2 idler pulley. Tighten the bolt to 32 ft. lbs. (43 Nm).

➡ **After tightening the bolt, be sure the idler pulley moves smoothly.**

   - Right-hand camshaft pulley, with the flange side **outward**.

➡ **Be sure to align the knock pin hole on the camshaft pulley with the knock pin on the camshaft.**

   - Tighten the camshaft bolt to 65 ft. lbs. (88 Nm), using the removal tools
   - Left-hand camshaft pulley, with the flange side **inward**.

➡ **Be sure to align the knock pin hole on the camshaft pulley with the knock pin on the camshaft.**

   - Tighten the camshaft bolt to 94 ft. lbs. (125 Nm), using the removal tools
   - Timing belt
   - Front upper suspension brace. Tighten the nuts to 59 ft. lbs. (80 Nm).
10. Fill the engine coolant.
11. Install or connect the following:
   - Heater air duct
   - Left and right ventilator louvers
   - Window washer hoses to the ventilator louvers
   - Top cowl seal and panel
   - Wiper and blade assembly
   - Negative battery cable
12. Start the engine.
13. Top off the engine coolant and check for leaks.

RX300

1. Before servicing the vehicle, refer to the precautions section.
2. Disconnect the negative battery cable.

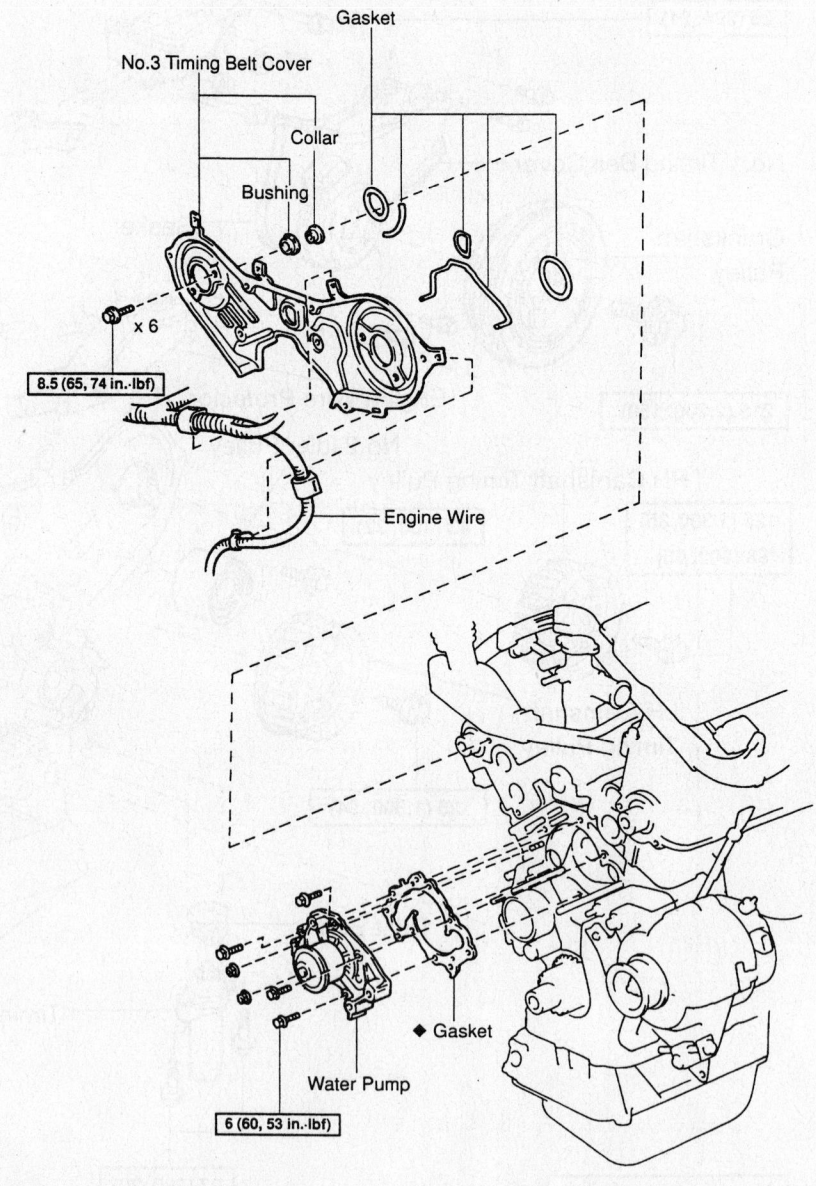

N·m (kgf·cm, ft·lbf) : Specified torque
◆ Non—reusable part

Exploded view of the water pump and related components—3.3L Highlander shown; RX330 similar

7924ZG16

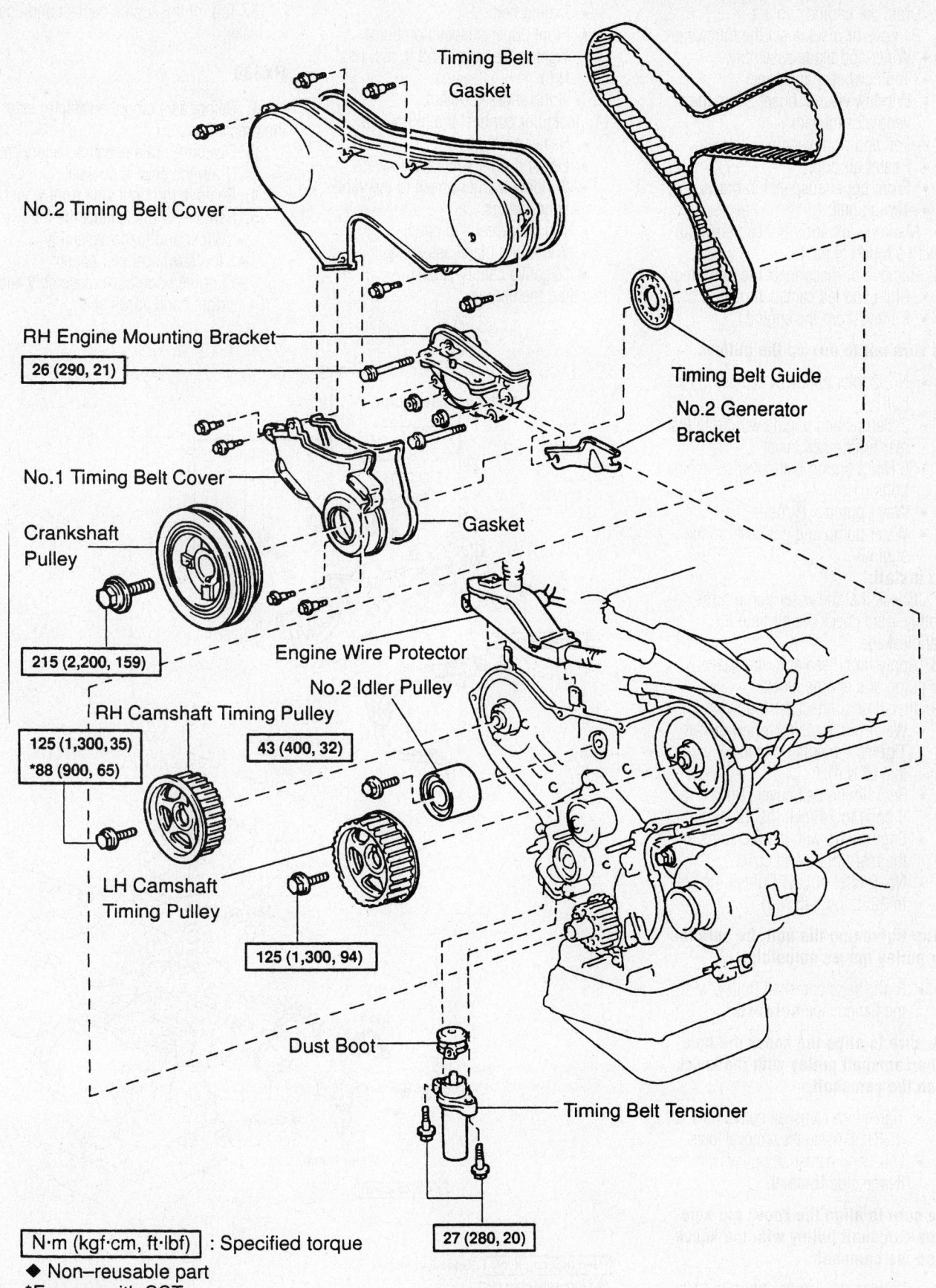

Timing Belt

Gasket

No.2 Timing Belt Cover

RH Engine Mounting Bracket

26 (290, 21)

Timing Belt Guide

No.2 Generator Bracket

No.1 Timing Belt Cover

Gasket

Crankshaft Pulley

215 (2,200, 159)

Engine Wire Protector

No.2 Idler Pulley

125 (1,300, 35)
*88 (900, 65)

43 (400, 32)

RH Camshaft Timing Pulley

LH Camshaft Timing Pulley

125 (1,300, 94)

Dust Boot

Timing Belt Tensioner

27 (280, 20)

N·m (kgf·cm, ft·lbf) : Specified torque

◆ Non–reusable part
*For use with SST

7924ZG15

Exploded view of the components to gain access to the water pump—3.0L RX300

3. Drain the engine coolant.
4. Remove or disconnect the following:
   • Wiper and blade assembly
   • Top cowl seal and panel
   • Window washer hoses, from the ventilator louvers
   • Left and right ventilator louvers
   • Heater air duct
   • Front upper suspension brace
   • Timing belt
5. Mark the left and right camshaft pulleys with a touch of paint.
6. Remove or disconnect the following:
   • Right and left camshaft pulleys bolts
   • Pulleys from the engine

➡ **Be sure not to mix up the pulleys.**

   • No. 2 idler pulley by removing the bolt
   • 3 clamps and engine wire from the rear timing belt cover
   • 6 No. 3 timing belt cover-to-engine bolts
   • Water pump nuts/bolts
   • Water pump and gasket from the engine

To install:

7. Check that the water pump turns smoothly. Also check the air hole for coolant leakage.
8. Apply liquid sealer to the gasket, water pump and engine block.
9. Install or connect the following:
   • Water pump, using a new gasket. Tighten the nuts/bolts to 53 inch lbs. (6 Nm).
   • Rear timing belt cover. Tighten the 6 bolts to 74 inch lbs. (9 Nm).
   • Engine wire with the 3 clamps to the rear timing belt cover
   • No. 2 idler pulley. Tighten the bolt to 32 ft. lbs. (43 Nm).

➡ **After tightening the bolt, be sure the idler pulley moves smoothly.**

   • Right-hand camshaft pulley, with the flange side **outward**.

➡ **Be sure to align the knock pin hole on the camshaft pulley with the knock pin on the camshaft.**

   • Tighten the camshaft bolt to 65 ft. lbs. (88 Nm), using the removal tools
   • Left-hand camshaft pulley, with the flange side **inward**.

➡ **Be sure to align the knock pin hole on the camshaft pulley with the knock pin on the camshaft.**

   • Tighten the camshaft bolt to 94 ft. lbs. (125 Nm), using the removal tools

   • Timing belt
   • Front upper suspension brace. Tighten the nuts to 59 ft. lbs. (80 Nm).
10. Fill the engine coolant.
11. Install or connect the following:
   • Heater air duct
   • Left and right ventilator louvers
   • Window washer hoses to the ventilator louvers
   • Top cowl seal and panel
   • Wiper and blade assembly
   • Negative battery cable
12. Start the engine.
13. Top off the engine coolant and check for leaks.

RX330

1. Before servicing the vehicle, refer to the precautions section.
2. Disconnect the negative battery cable.
3. Drain the engine coolant.
4. Remove the right side front wheel.
5. Remove or disconnect the following:
   • Wiper and blade assembly
   • Top cowl seal and panel
   • Engine undercover assembly and right hand apron seal

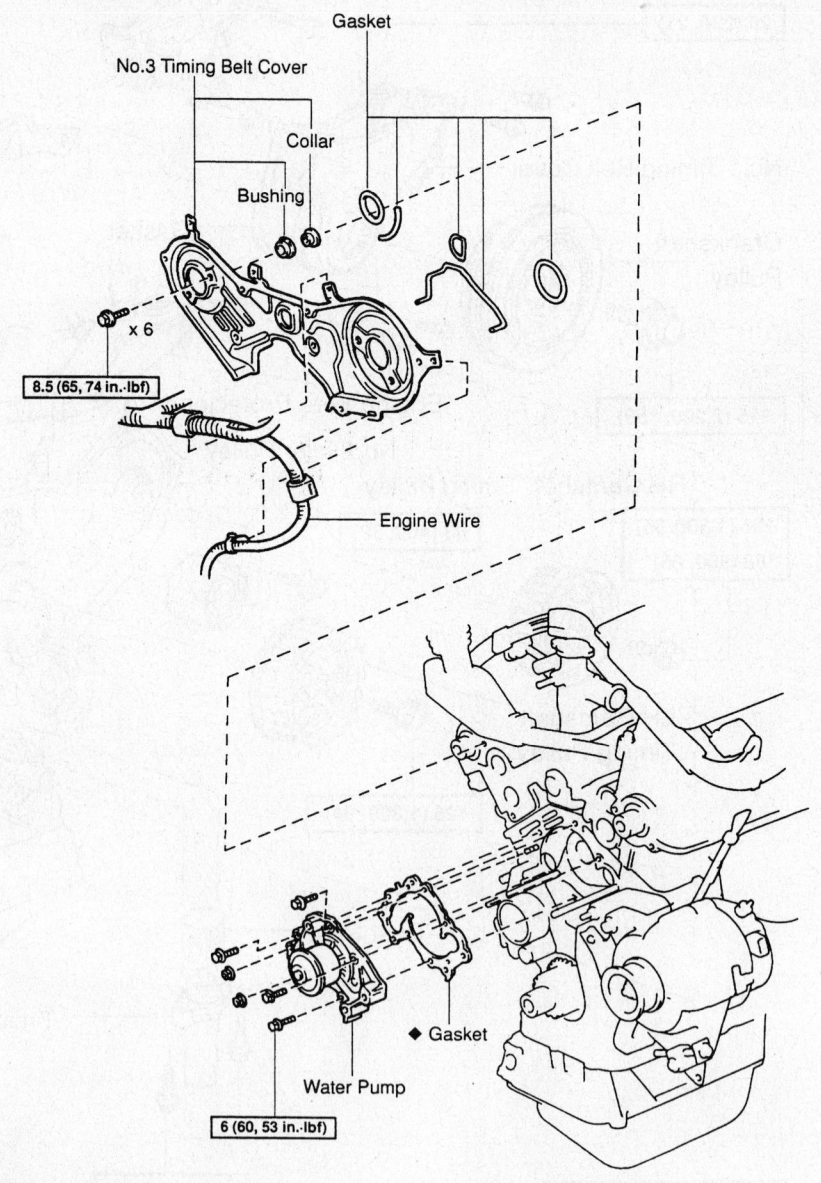

Gasket

No.3 Timing Belt Cover

Collar

Bushing

8.5 (65, 74 in.·lbf)

x 6

Engine Wire

◆ Gasket

Water Pump

6 (60, 53 in.·lbf)

N·m (kgf·cm, ft·lbf) : Specified torque
◆ Non–reusable part

7924ZG16

Exploded view of the water pump and related components—3.0L RX300

- Drive belts
- Engine moving control rod
- Right hand engine mount stay No: 2 and alternator bracket No: 2
- Crankshaft pulley
- Timing belt

6. Mark the left and right camshaft pulleys with a touch of paint.

7. Remove or disconnect the following:
- Right and left camshaft pulleys bolts
- Pulleys from the engine

➡ **Be sure not to mix up the pulleys.**

- No. 3 timing belt cover
- No. 1 idler subassembly
- Water pump nuts/bolts
- Water pump and gasket from the engine

To install:

8. Check that the water pump turns smoothly. Also check the air hole for coolant leakage.

9. Apply liquid sealer to the gasket, water pump and engine block.

10. Install or connect the following:
- Water pump, using a new gasket. Tighten the nuts/bolts to 71 inch lbs. (8 Nm).
- Rear timing belt cover. Tighten the 6 bolts to 74 inch lbs. (9 Nm).

11. Install the remaining components in the reverse order of removal, but note the following:

a. Tighten the camshaft bolt to 92 ft. lbs. (125 Nm), using the removal tools

b. Tighten the crankshaft pulley to 162 ft. lbs. (220 Nm)

12. Fill the engine coolant.

13. Start the engine.

14. Top off the engine coolant and check for leaks.

Cylinder Head

REMOVAL & INSTALLATION

Highlander

2.4L ENGINE

1. Before servicing the vehicle, refer to the precautions section.

2. Remove or disconnect the following:
- Front center suspension brace
- Timing chain
- Coolant
- Transfer case oil
- Radiator hoses
- Power steering hoses
- Heater hoses
- Fuel rail lines
- Camshaft timing oil control valve
- Front driveshaft
- Rear engine mount insulator (4wd)
- Transverse engine mount bracket (4wd)
- Intake manifold
- All wires and cables connected to the head
- Exhaust manifold
- Camshafts

3. Loosen the 10 head bolts evenly, a little at a time in several passes and lift off the head. Check the head bolt length. Any bolt longer than 6.465 in. (164.2mm) should be replaced.

4. Installation is the reverse of removal.

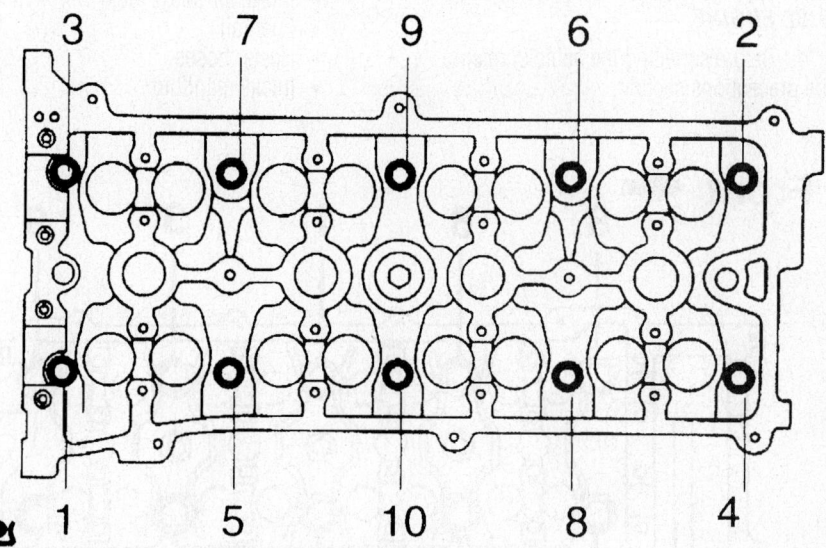

Cylinder head bolt loosening sequence—2.4L Highlander

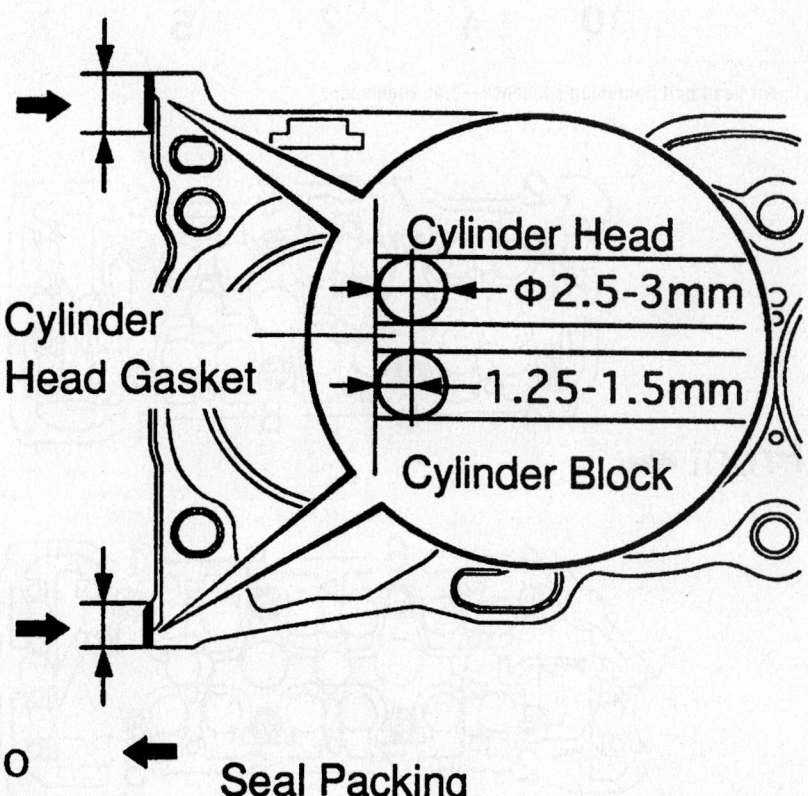

Apply a bead of RTV sealant as shown— Highlander 2.4L engine cylinder head

Install the head gasket with the lot number stamp upward. Apply a bead of RTV sealer as shown. The head must be installed within 3 minutes of applying the sealer, and the head bolts must be tightened with 15 minutes. The head bolts must be tightened in sequence, in several passes, to 58 ft. lbs. (79Nm), then, an additional 90 degree turn each.

3.0L ENGINE

1. Before servicing the vehicle, refer to the precautions section.

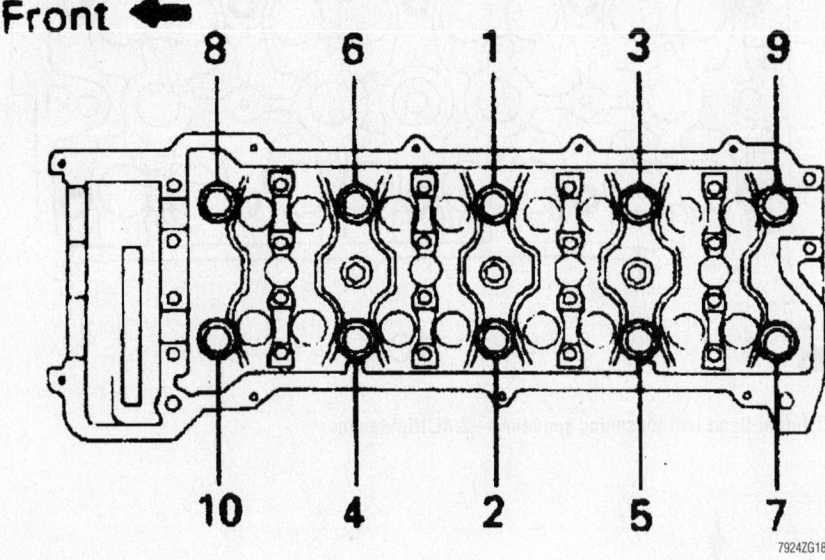

Cylinder head bolt tightening sequence—2.4L Highlander

7924ZG18

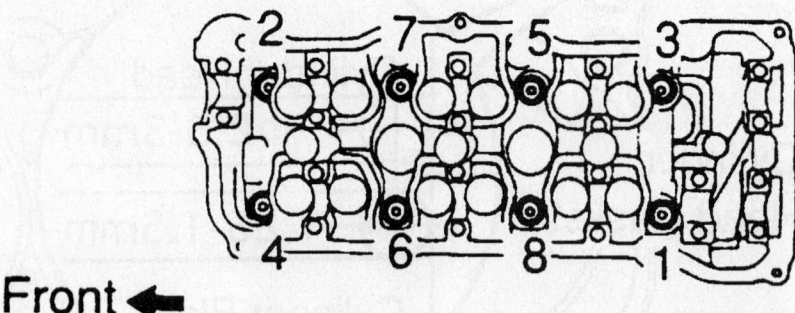

Front ←

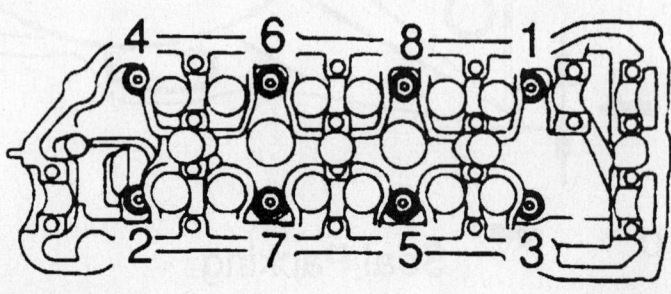

Cylinder head bolt loosening sequence—3.0L Highlander

7924ZG19

2. Remove or disconnect the following:
  - Coolant
  - Engine oil
  - Exhaust pipes
  - Exhaust manifold
  - Camshaft cover
  - Upper center front suspension brace
  - Air cleaner
  - Intake air surge tank
  - Fuel rail
  - Heater hoses
  - Intake manifold

  - Radiator hose
  - Water outlet
  - Right front wheel
  - Right fender apron seal
  - Accessory drive belts
  - Engine roll stopper rod
  - Right engine mount
  - Alternator and brackets
  - Crankshaft pulley
  - Timing belt covers
  - Transverse engine mounting bracket
  - Timing belt
  - Timing belt tensioner
  - Power steering pump
  - Ignition coil pack
  - Camshafts
  - The hexagonal bolt, using an 8mm hex wrench, then the 8 head bolts
  - Cylinder head

3. Installation is the reverse of removal. Measure the head bolts. The minimum diameter of the stretch portion of each bolt should be at least 8.75mm (0.3775 in.). Replace any bolt that does not measure up. The head gasket is installed with the **R** mark upwards. Install the 8 head bolts first, tightened in sequence, in 2 equal steps, to 40 ft. lbs. (54Nm). Then tighten each an additional 90 degrees. Finally, install the hex bolt, torqued to 14 ft. lbs. (19 Nm).

3.3L ENGINES

1. Before servicing the vehicle, refer to the precautions section.
2. Remove or disconnect the following:
  - Wiper and blade assembly
  - Top cowl seal and panel
  - Window washer hoses from the ventilator louvers
  - Left and right ventilator louvers
  - Heater air duct
3. Relieve the fuel pressure.
4. Remove or disconnect the following:
  - Turn the ignition key to the **OFF** position
  - Negative battery cable

➡ **Wait at least 90 seconds from the time the negative battery was disconnected to start work.**

5. Drain the cooling system.
6. Remove or disconnect the following:
  - Accelerator and throttle cables, if equipped with an automatic transaxle
  - Air cleaner cover, air flow meter and the air duct
  - Front upper suspension brace
  - Cruise control actuator and bracket, if equipped
  - 2 engine ground straps

- Right engine mounting support
- Radiator hoses
- 2 heater hoses
- Fuel feed and return lines from the fuel rail assembly
- Pressure hose from the hydraulic motor
- V-bank cover

7. Disconnect the following vacuum hoses:

- Fuel pressure control Vacuum Switching Valve (VSV)
- Fuel pressure regulator
- Cylinder head rear plate

- Intake air control valve VSV
- Exhaust Gas Recirculation (EGR) vacuum modulator
- EGR valve

8. Disconnect the following wiring and hoses:

- Intake air control valve
- Fuel pressure regulator
- EGR VSV

9. Remove the 2 nuts and the emission control valve set.

10. Disconnect the following hoses;

- Brake booster vacuum hose
- PCV hose

- Intake air control valve vacuum hose

11. Remove or disconnect the following:

- Data Link Connector (DLC) from the mounting bracket
- 2 ground straps from the intake chamber
- Hydraulic motor pressure hose from the intake chamber
- Right Oxygen (O_2) sensor connector from the power steering pressure tube
- 2 nuts and the power steering pressure tube from the intake chamber
- Both power steering air hoses
- Engine hanger and the intake chamber support
- EGR pipe and gaskets

12. Disconnect the following wiring:

- Throttle Position (TP) sensor connector
- Idle Air Control (IAC) valve connector
- EGR gas temperature connector
- Air conditioning idle up connector

13. Disconnect the following vacuum hoses:

- 2 vacuum hoses from the Thermal Vacuum Valve (TVV)
- Vacuum hose from the cylinder head rear plate
- Vacuum hose from the charcoal canister

14. Remove or disconnect the following:

- Air assist hose and the 2 water bypass hoses
- Air intake chamber
- Left engine wiring harness and move it aside
- Wiring harness from the rear of the engine
- Right engine wiring harness and move it aside
- Ignition coils and move them aside
- Timing belt
- Camshaft pulleys and the timing belt rear cover
- Cylinder head rear plate
- Water inlet pipe
- Air assist hose and vacuum hose
- Intake manifold and fuel rail assembly
- Water outlet
- EGR pipe from the right exhaust manifold
- Front exhaust pipe and exhaust manifolds
- Dipstick assembly and the power steering pump bracket
- Valve covers and the Camshaft Position (CMP) sensor
- Camshafts

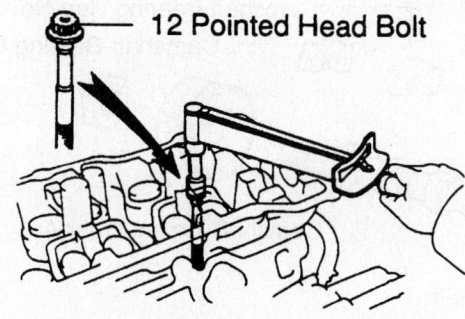

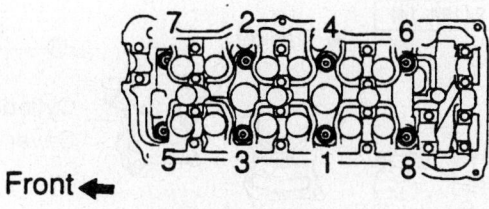

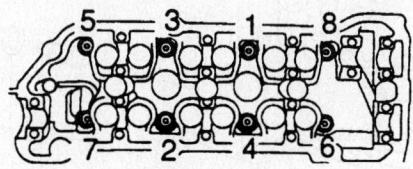

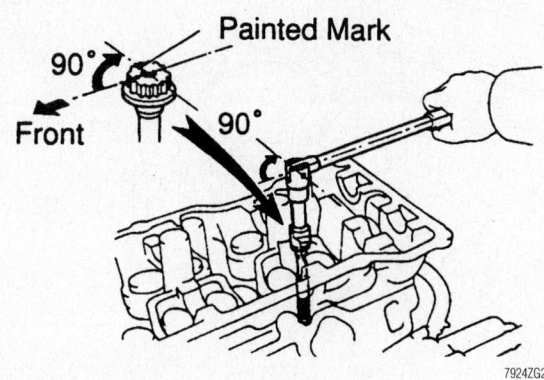

7924ZG20

Cylinder head bolt tightening sequence—3.0L Highlander

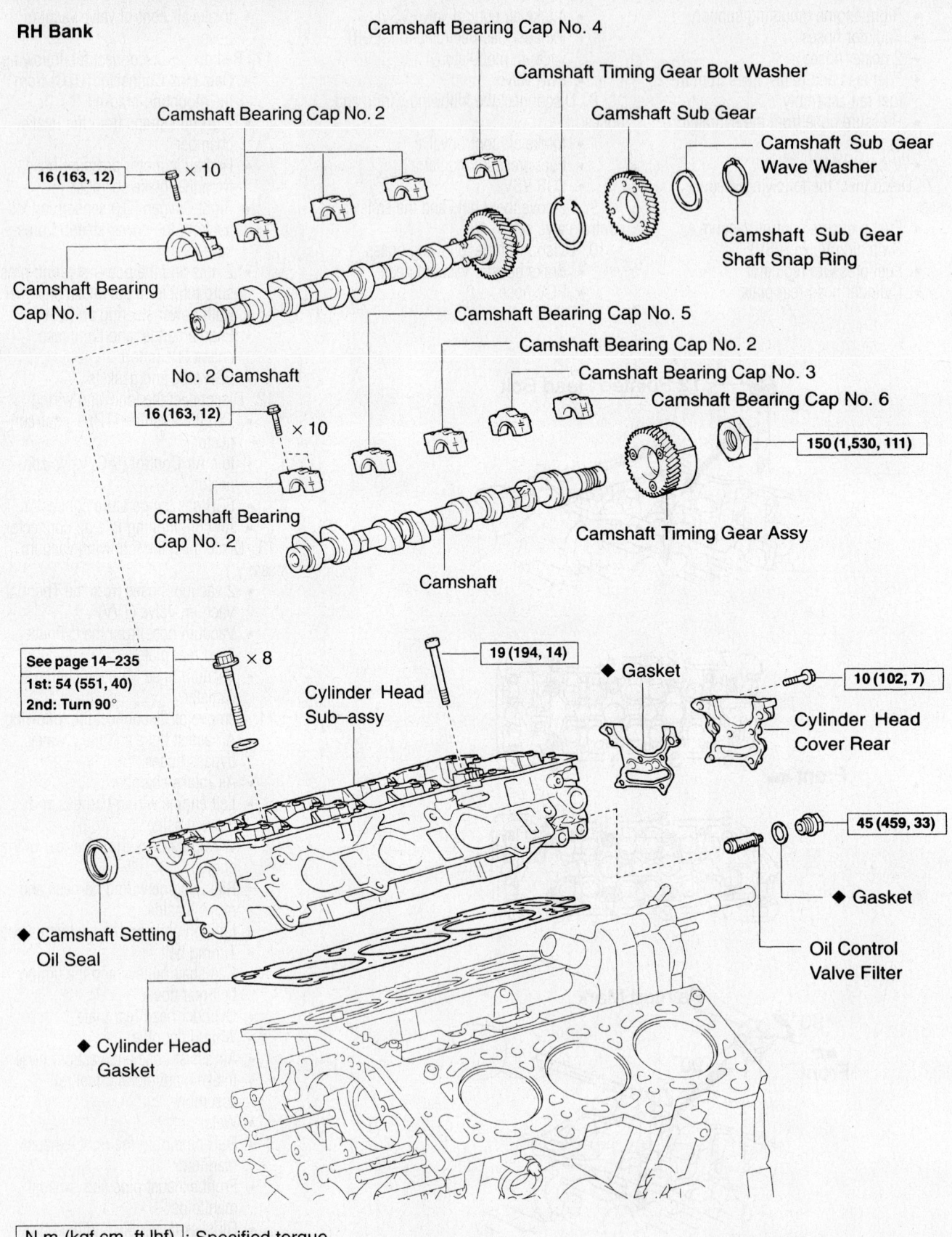

RH Bank

Camshaft Bearing Cap No. 4

Camshaft Timing Gear Bolt Washer

Camshaft Sub Gear

Camshaft Bearing Cap No. 2

Camshaft Sub Gear
Wave Washer

16 (163, 12) × 10

Camshaft Sub Gear
Shaft Snap Ring

Camshaft Bearing
Cap No. 1

No. 2 Camshaft

Camshaft Bearing Cap No. 5

Camshaft Bearing Cap No. 2

16 (163, 12) × 10

Camshaft Bearing Cap No. 3

Camshaft Bearing Cap No. 6

Camshaft Bearing
Cap No. 2

150 (1,530, 111)

Camshaft

Camshaft Timing Gear Assy

See page 14–235
1st: 54 (551, 40)
2nd: Turn 90°

× 8

19 (194, 14)

◆ Gasket

10 (102, 7)

Cylinder Head
Sub–assy

Cylinder Head
Cover Rear

45 (459, 33)

◆ Gasket

◆ Camshaft Setting
Oil Seal

Oil Control
Valve Filter

◆ Cylinder Head
Gasket

N·m (kgf·cm, ft·lbf) : Specified torque

◆ Non–reusable part

67170-HIGH-G08

Right cylinder head and related parts—3.3L Highlander

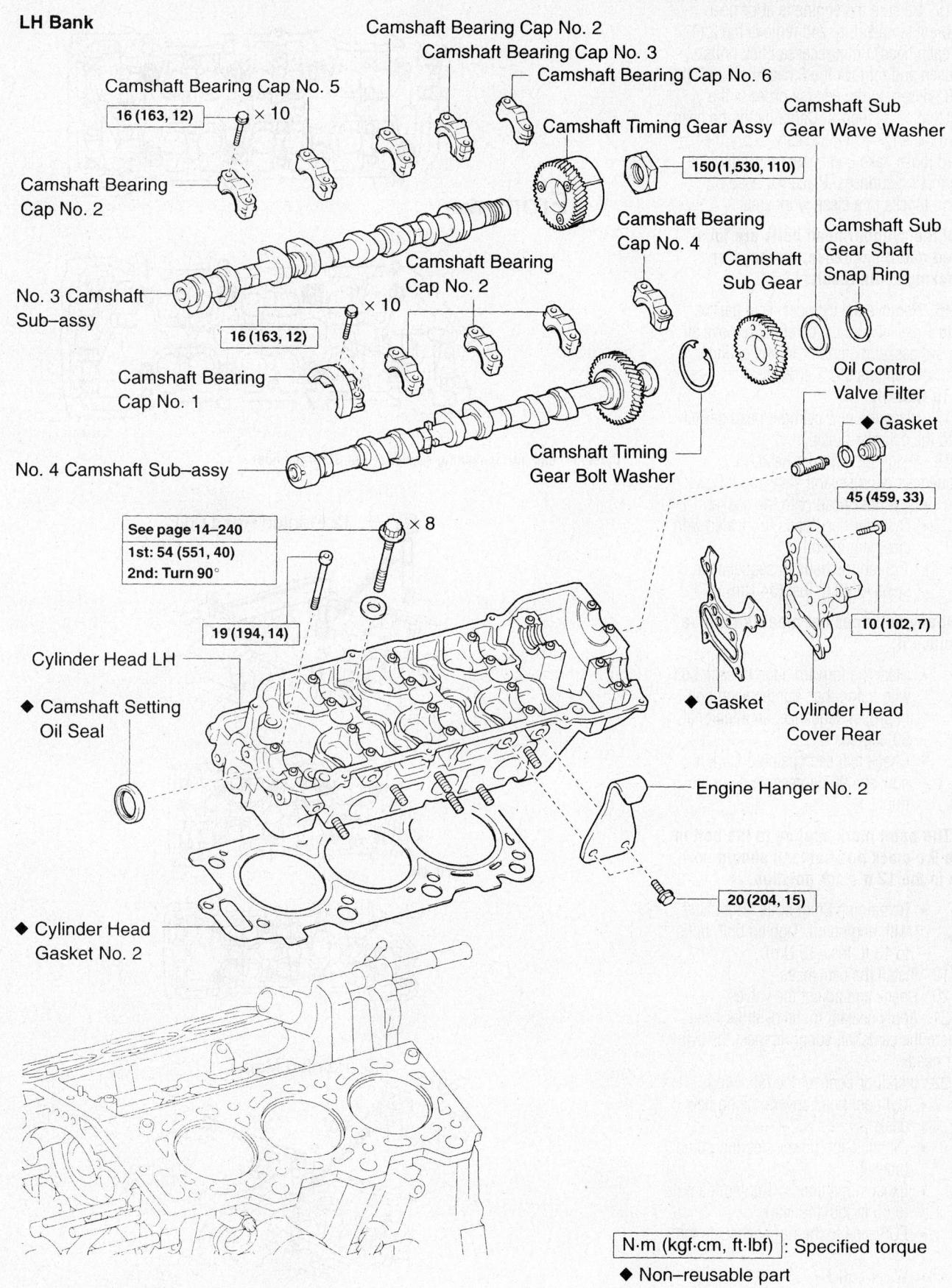

LH Bank

Camshaft Bearing Cap No. 2

Camshaft Bearing Cap No. 3

Camshaft Bearing Cap No. 6

Camshaft Bearing Cap No. 5

16 (163, 12)   ×10

Camshaft Timing Gear Assy

Camshaft Sub Gear Wave Washer

150 (1,530, 110)

Camshaft Bearing Cap No. 2

Camshaft Bearing Cap No. 4

Camshaft Sub Gear Shaft Snap Ring

Camshaft Bearing Cap No. 2

Camshaft Sub Gear

No. 3 Camshaft Sub–assy

16 (163, 12)   ×10

Camshaft Bearing Cap No. 1

Oil Control Valve Filter

◆ Gasket

No. 4 Camshaft Sub–assy

Camshaft Timing Gear Bolt Washer

45 (459, 33)

See page 14–240
1st: 54 (551, 40)
2nd: Turn 90°   ×8

19 (194, 14)

10 (102, 7)

Cylinder Head LH

◆ Gasket   Cylinder Head Cover Rear

◆ Camshaft Setting Oil Seal

Engine Hanger No. 2

20 (204, 15)

◆ Cylinder Head Gasket No. 2

N·m (kgf·cm, ft·lbf) : Specified torque

◆ Non–reusable part

67170-HIGH-G09

Left cylinder head and related parts—3.3L Highlander

15. Be sure the engine is at/or near ambient temperature and remove the 2 (1 on each head) 8mm recessed hex bolts. Loosen and remove the 8 head bolts evenly, in 3 passes, in the reverse order of the installation sequence. Carefully lift the head from the engine; if necessary to pry the head loose, take great care not to damage the mating surfaces. Place the head on wood blocks in a clean work area.

➡**If the cylinder head bolts are loosened out of sequence, warpage or cracking could result.**

16. Remove the cylinder head gasket. With a gasket scraper, carefully remove all the old gasket material from the cylinder head and engine block surfaces.

To install:

17. Place the new cylinder head gasket onto the cylinder block.

18. Install the cylinder head, in sequence, using several steps, as follows:
- Cylinder head onto the gasket
- Cylinder head bolts lubricated with clean engine oil
- Tighten the bolts in sequence in 3 steps to 40 ft. lbs. (54 Nm).

➡**If any bolt does not meet the torque, replace it.**

- Mark the forward edge of each bolt with paint, then tighten each bolt, in proper sequence, an additional 90 degrees.
- Check that each painted mark is now at a 90 degrees angle to the front

➡**The paint mark applied to the bolt in the 9 o'clock position and should now be in the 12 o'clock position.**

- Remaining 8mm bolts, lubricated with engine oil. Tighten both bolts to 13 ft. lbs. (18 Nm).

19. Install the camshafts.

20. Check and adjust the valves.

21. Apply sealant to the cylinder heads where the camshaft supports meet the cylinder heads.

22. Install or connect the following:
- Cylinder head covers, using new gaskets
- Dipstick and power steering pump bracket
- Exhaust manifolds. Tighten the nuts to 36 ft. lbs. (49 Nm).
- EGR pipe to the right exhaust manifold
- Water outlet
- Intake manifold and the fuel rail assembly. Tighten the intake mani-

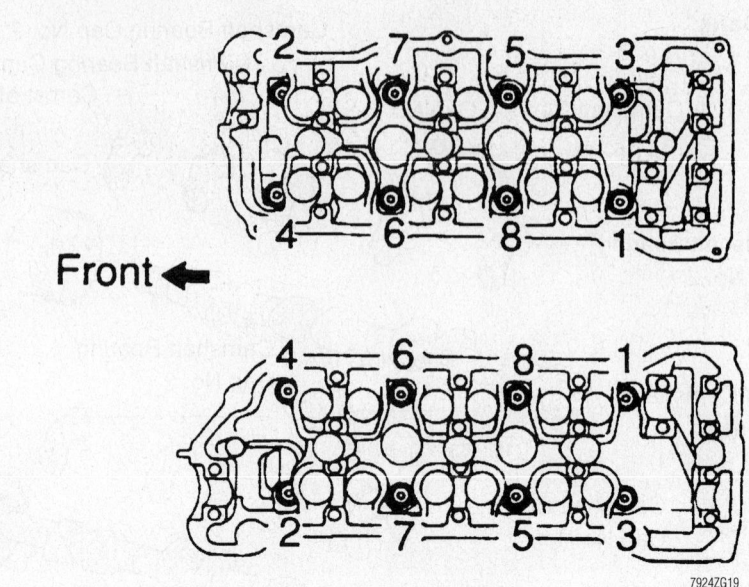

Front ←

7924ZG19

Cylinder head bolt loosening sequence—3.3L Highlander

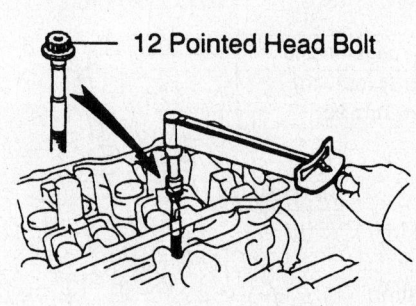

12 Pointed Head Bolt

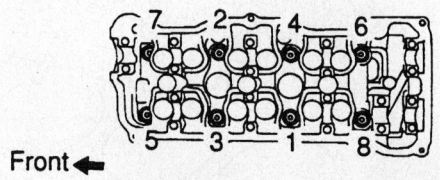

Front ←

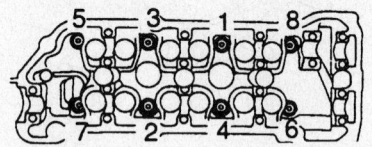

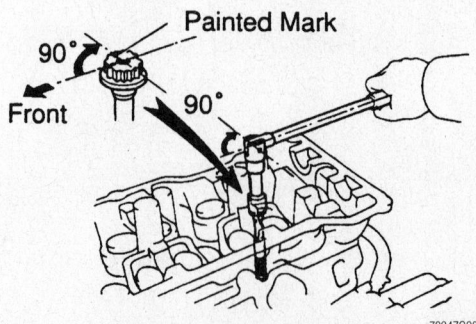

Painted Mark

90° **Front** 90°

7924ZG20

Cylinder head bolt tightening sequence—3.3L Highlander

fold nuts/bolts to 11 ft. lbs. (15 Nm).
- Air assist hose and the 2 water bypass hoses
- Water inlet pipe and cylinder head rear plate
- Timing belt rear cover and camshaft pulleys
- Timing belt
- Spark plugs and ignition coils
- Right engine wiring harness
- Wiring harness to the rear of the engine
- Left engine wiring harness
- Air intake chamber
- EGR pipe, using new gaskets

23. Connect the following vacuum hoses:
   - The 2 TVV vacuum hoses
   - The vacuum hose to the rear cylinder head plate
   - Charcoal canister vacuum hose

24. Connect the following electrical wiring:
   - TP sensor connector
   - IAC valve connector
   - EGR gas temperature connector
   - Air conditioning idle up connector

25. Install or connect the following:
   - Engine hanger and the intake chamber support
   - Both power steering air hoses
   - Power steering pressure tube to the intake chamber
   - O₂ sensor connector to the pressure tube.
   - Both ground straps, to the intake chamber
   - DLC to the bracket

26. Connect the following hoses:
   - Power brake booster vacuum hose
   - PCV hose
   - IAC valve vacuum hose

27. Install or connect the following:
   - Emission control valve set and related vacuum hoses and connectors
   - V-bank cover
   - Pressure hose to the hydraulic motor
   - Fuel lines to the fuel rail assembly
   - Heater and radiator hoses
   - Right engine mounting support
   - Both engine ground straps
   - Upper front suspension brace, if removed. Tighten the nuts to 59 ft. lbs. (80 Nm).
   - Cruise control actuator and bracket
   - Air cleaner, air flow meter and air duct assembly
   - Accelerator and throttle cables

28. Fill the cooling system.

29. Install or connect the following:
   - Negative battery cable
   - Heater air duct
   - Left and right ventilator louvers
   - Window washer hoses from the ventilator louvers
   - Top cowl seal and panel
   - Wiper and blade assembly

30. Start the engine and check for leaks.
31. Bleed the air from the cooling system.
32. Road test the vehicle and check for unusual noise, shock, slippage, correct shift points and smooth operation.
33. Recheck the coolant and engine oil levels.

RX300 and RX330

1. Before servicing the vehicle, refer to the precautions section.
2. Remove or disconnect the following:
   - Wiper and blade assembly
   - Top cowl seal and panel
   - Window washer hoses from the ventilator louvers
   - Left and right ventilator louvers
   - Heater air duct
3. Relieve the fuel pressure.
4. Remove or disconnect the following:
   - Turn the ignition key to the **OFF** position
   - Negative battery cable

➡ **Wait at least 90 seconds from the time the negative battery was disconnected to start work.**

5. Drain the cooling system.
6. Remove or disconnect the following:
   - Accelerator and throttle cables, if equipped with an automatic transaxle
   - Air cleaner cover, air flow meter and the air duct
   - Front upper suspension brace
   - Cruise control actuator and bracket, if equipped
   - 2 engine ground straps
   - Right engine mounting support
   - Radiator hoses
   - 2 heater hoses
   - Fuel feed and return lines from the fuel rail assembly
   - Pressure hose from the hydraulic motor
   - V-bank cover
7. Disconnect the following vacuum hoses:
   - Fuel pressure control Vacuum Switching Valve (VSV)
   - Fuel pressure regulator
   - Cylinder head rear plate

- Intake air control valve VSV
- Exhaust Gas Recirculation (EGR) vacuum modulator
- EGR valve
8. Disconnect the following wiring and hoses:
   - Intake air control valve
   - Fuel pressure regulator
   - EGR VSV
9. Remove the 2 nuts and the emission control valve set.
10. Disconnect the following hoses;
   - Brake booster vacuum hose
   - PCV hose
   - Intake air control valve vacuum hose
11. Remove or disconnect the following:
   - Data Link Connector (DLC) from the mounting bracket
   - 2 ground straps from the intake chamber
   - Hydraulic motor pressure hose from the intake chamber
   - Right Oxygen (O₂) sensor connector from the power steering pressure tube
   - 2 nuts and the power steering pressure tube from the intake chamber
   - Both power steering air hoses
   - Engine hanger and the intake chamber support
   - EGR pipe and gaskets
12. Disconnect the following wiring:
   - Throttle Position (TP) sensor connector
   - Idle Air Control (IAC) valve connector
   - EGR gas temperature connector
   - Air conditioning idle up connector
13. Disconnect the following vacuum hoses:
   - 2 vacuum hoses from the Thermal Vacuum Valve (TVV)
   - Vacuum hose from the cylinder head rear plate
   - Vacuum hose from the charcoal canister
14. Remove or disconnect the following:
   - Air assist hose and the 2 water bypass hoses
   - Air intake chamber
   - Left engine wiring harness and move it aside
   - Wiring harness from the rear of the engine
   - Right engine wiring harness and move it aside
   - Ignition coils and move them aside
   - Timing belt
   - Camshaft pulleys and the timing belt rear cover
   - Cylinder head rear plate

- Water inlet pipe
- Air assist hose and vacuum hose
- Intake manifold and fuel rail assembly
- Water outlet
- EGR pipe from the right exhaust manifold
- Front exhaust pipe and exhaust manifolds
- Dipstick assembly and the power steering pump bracket
- Valve covers and the Camshaft Position (CMP) sensor
- Camshafts

15. Be sure the engine is at/or near ambient temperature and remove the 2 (1 on each head) 8mm recessed hex bolts. Loosen and remove the 8 head bolts evenly, in 3 passes, in the reverse order of the installation sequence. Carefully lift the head from the engine; if necessary to pry the head loose, take great care not to damage the mating surfaces. Place the head on wood blocks in a clean work area.

➡**If the cylinder head bolts are loosened out of sequence, warpage or cracking could result.**

16. Remove the cylinder head gasket. With a gasket scraper, carefully remove all the old gasket material from the cylinder head and engine block surfaces.

To install:

17. Place the new cylinder head gasket onto the cylinder block.

18. Install the cylinder head, in sequence, using several steps, as follows:
- Cylinder head onto the gasket
- Cylinder head bolts lubricated with clean engine oil
- Tighten the bolts in sequence in 3 steps to 40 ft. lbs. (54 Nm).

➡**If any bolt does not meet the torque, replace it.**

- Mark the forward edge of each bolt with paint, then tighten each bolt, in proper sequence, an additional 90 degrees.
- Check that each painted mark is now at a 90 degrees angle to the front

➡**The paint mark applied to the bolt in the 9 o'clock position and should now be in the 12 o'clock position.**

- Remaining 8mm bolts, lubricated with engine oil. Tighten both bolts to 13 ft. lbs. (18 Nm).

19. Install the camshafts.
20. Check and adjust the valves.
21. Apply sealant to the cylinder heads

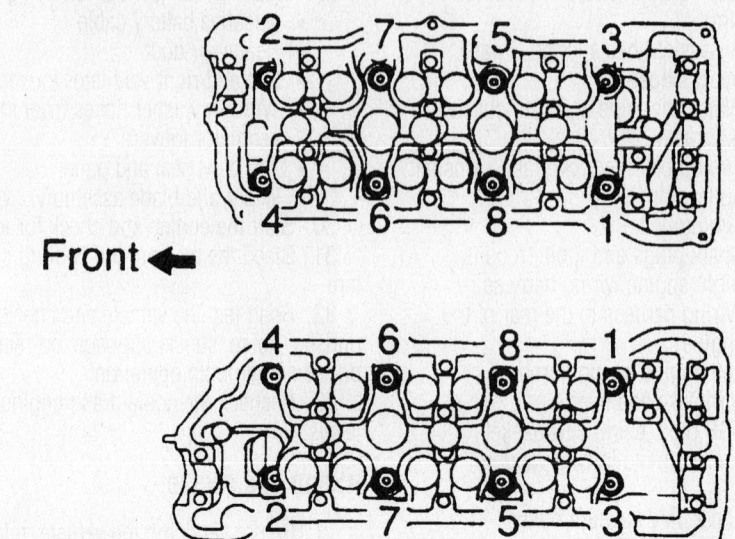

Cylinder head bolt loosening sequence—3.0L and 3.3L engines

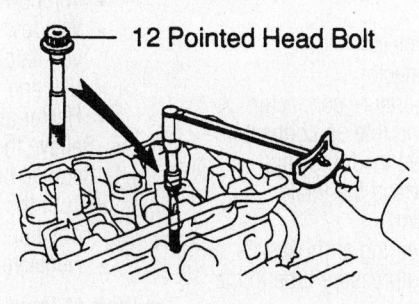

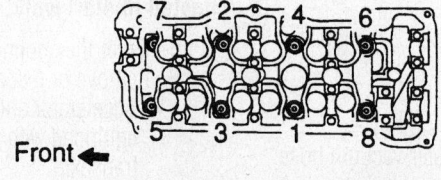

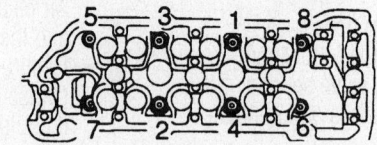

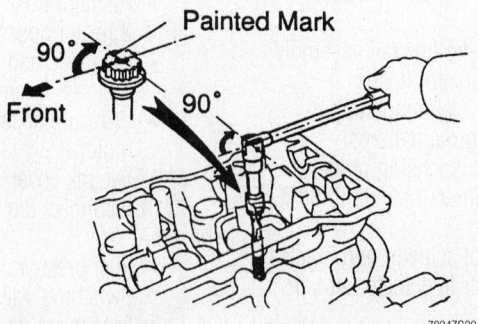

Cylinder head bolt tightening sequence—3.0L and 3.3L engines engine

where the camshaft supports meet the cylinder heads.

22. Install or connect the following:
    - Cylinder head covers, using new gaskets
    - Dipstick and power steering pump bracket
    - Exhaust manifolds. Tighten the nuts to 36 ft. lbs. (49 Nm).
    - EGR pipe to the right exhaust manifold
    - Water outlet
    - Intake manifold and the fuel rail assembly. Tighten the intake manifold nuts/bolts to 11 ft. lbs. (15 Nm).
    - Air assist hose and the 2 water bypass hoses
    - Water inlet pipe and cylinder head rear plate
    - Timing belt rear cover and camshaft pulleys
    - Timing belt
    - Spark plugs and ignition coils
    - Right engine wiring harness
    - Wiring harness to the rear of the engine
    - Left engine wiring harness
    - Air intake chamber
    - EGR pipe, using new gaskets
23. Connect the following vacuum hoses:
    - The 2 TVV vacuum hoses
    - The vacuum hose to the rear cylinder head plate
    - Charcoal canister vacuum hose
24. Connect the following electrical wiring:
    - TP sensor connector
    - IAC valve connector
    - EGR gas temperature connector
    - Air conditioning idle up connector
25. Install or connect the following:
    - Engine hanger and the intake chamber support
    - Both power steering air hoses
    - Power steering pressure tube to the intake chamber
    - O_2 sensor connector to the pressure tube.
    - Both ground straps, to the intake chamber
    - DLC to the bracket
26. Connect the following hoses:
    - Power brake booster vacuum hose
    - PCV hose
    - IAC valve vacuum hose
27. Install or connect the following:
    - Emission control valve set and related vacuum hoses and connectors
    - V-bank cover

- Pressure hose to the hydraulic motor
- Fuel lines to the fuel rail assembly
- Heater and radiator hoses
- Right engine mounting support
- both engine ground straps
- Upper front suspension brace, if removed. Tighten the nuts to 59 ft. lbs. (80 Nm).
- Cruise control actuator and bracket
- Air cleaner, air flow meter and air duct assembly
- Accelerator and throttle cables
28. Fill the cooling system.
29. Install or connect the following:
    - Negative battery cable
    - Heater air duct
    - Left and right ventilator louvers
    - Window washer hoses from the ventilator louvers
    - Top cowl seal and panel
    - Wiper and blade assembly
30. Start the engine and check for leaks.
31. Bleed the air from the cooling system.
32. Road test the vehicle and check for unusual noise, shock, slippage, correct shift points and smooth operation.
33. Recheck the coolant and engine oil levels.

Intake Manifold

REMOVAL & INSTALLATION

Highlander

2.4L ENGINE

1. Before servicing the vehicle, refer to the precautions section.
2. Disconnect the negative battery cable.
3. Release the fuel system pressure.
4. Drain the engine coolant.
5. Remove or disconnect the following:
    - Air cleaner cap
    - Mass Air Flow (MAF) meter and the resonator
    - Accelerator cable from the throttle body, if equipped with a manual transmission
    - Accelerator and throttle cables from the throttle body, if equipped with an automatic transmission
    - Cruise control cable from the actuator, if equipped with cruise control
    - Intake air connector
    - Air hose for Idle Air Control (IAC)
    - Vacuum sensing hose
    - Wire clamp for the engine wiring harness

- Positive Crankcase Ventilation (PCV) hoses.
- Engine wiring harness
- Air conditioning compressor connector, if equipped with air conditioning
- Oil pressure sensor connector
- Engine Coolant Temperature (ECT) sensor connector
- ECT sender gauge connector
- Exhaust Gas Recirculation (EGR) gas temperature sensor connector
- Vacuum Switching Valve (VSV) connector, for the EGR
- 2 vacuum hoses, from the VSV for the EGR
- Ground strap, from the cowl top panel
- Engine wiring harness, from the air intake chamber
- Throttle Position (TP) sensor connector
- IAC valve connector
- Crankshaft Position (CKP) sensor connector
- Knock (KS) sensor connector
- Data Link Connector 1 (DLC1), from the bracket
- Engine wiring harness clamp
- EGR pipe
- Intake chamber stay
- Air intake chamber assembly
6. Disconnect the following hoses:
    - Evaporative Emission (EVAP) hose, from the throttle body
    - Brake booster vacuum hose, from the union
    - Water bypass hose, from the water bypass pipe
    - Water bypass hose, from the cylinder head rear cover
    - Injector connectors
    - Fuel inlet pipe
    - Hoses and the fuel return pipe.
7. Remove the delivery pipe and injectors, as follows:
    - Delivery pipe, together with the 4 injectors
    - 4 insulators from the 4 spacers
    - 4 injectors, from the delivery pipe
    - O-ring and grommets, from each injector
    - 4 spacers, by carefully prying them out
8. Remove the intake manifold.
To install:
9. Install or connect the following:
    - Intake manifold. Tighten the bolts to 22 ft. lbs. (29 Nm).
    - Injectors and the delivery pipe
    - Fuel return pipe
    - Fuel inlet pipe, with a new gasket.

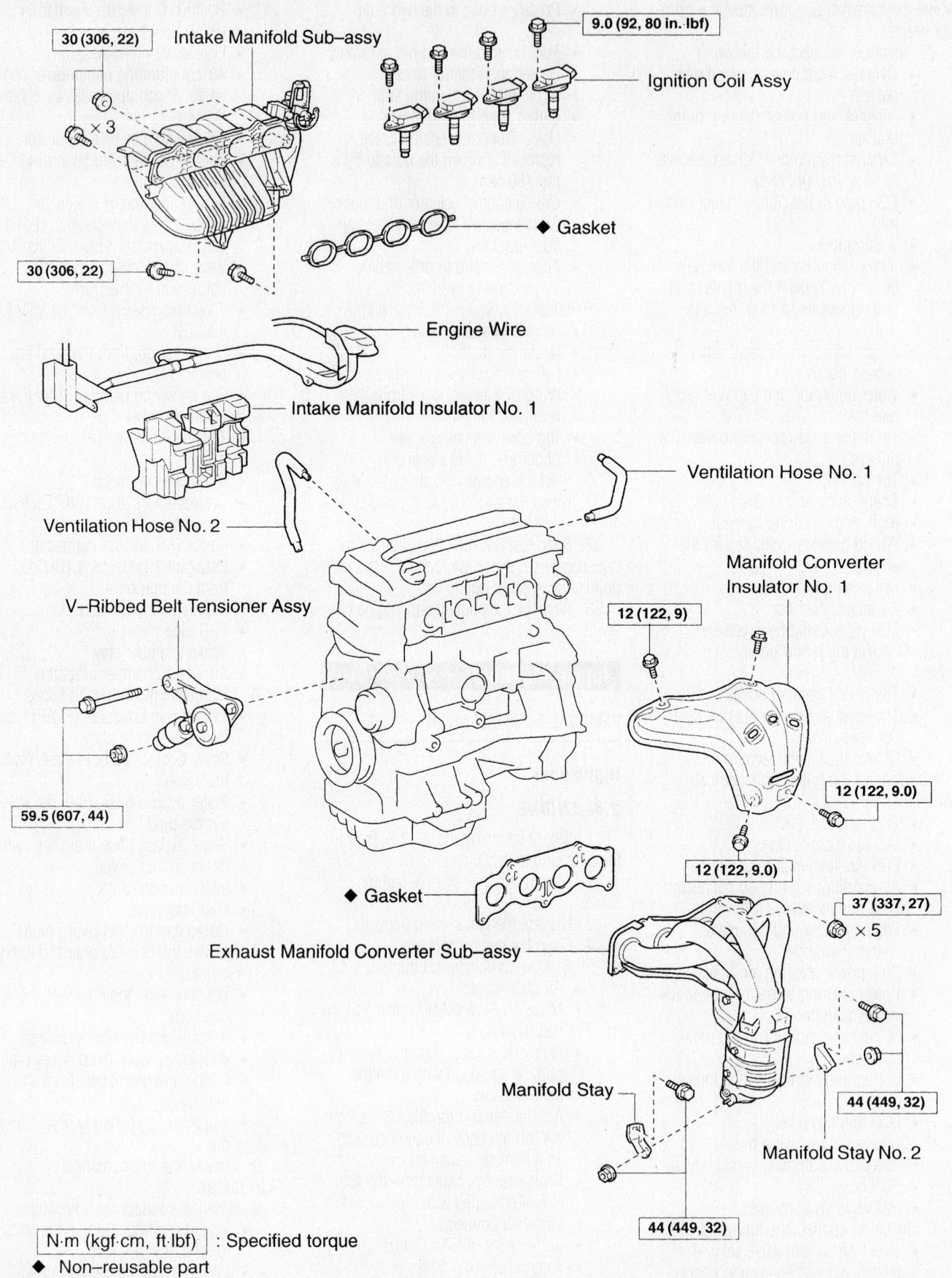

30 (306, 22)

Intake Manifold Sub–assy

9.0 (92, 80 in. lbf)

Ignition Coil Assy

◆ Gasket

30 (306, 22)

Engine Wire

Intake Manifold Insulator No. 1

Ventilation Hose No. 1

Ventilation Hose No. 2

Manifold Converter Insulator No. 1

V–Ribbed Belt Tensioner Assy

12 (122, 9)

12 (122, 9.0)

59.5 (607, 44)

12 (122, 9.0)

37 (337, 27) × 5

◆ Gasket

Exhaust Manifold Converter Sub–assy

Manifold Stay

Manifold Stay No. 2

44 (449, 32)

44 (449, 32)

N·m (kgf·cm, ft·lbf) : Specified torque

◆ Non–reusable part

67170-HIGH-G11

Intake manifold and related components—2004 2.4L Highlander

Tighten the bolts to 22 ft. lbs. (29 Nm).
- Injector connectors
- Air intake chamber assembly. Tighten the bolts to 15 ft. lbs. (21 Nm).
10. Connect the following hoses:
- Evaporative Emissions (EVAP) hose, to the throttle body
- Brake booster vacuum hose, to the union
- Water bypass hose, to water bypass pipe
- Water bypass hose, to cylinder head rear cover
11. Install or connect the following:
- Air intake chamber stay. Tighten the bolts to 15 ft. lbs. (20 Nm).
- EGR pipe. Tighten bolts to 13 ft. lbs. (18 Nm), nut "A" to 14 ft. lbs. (19 Nm) and nut B to 15 ft. lbs. (20 Nm).
- Air conditioning compressor connector
- Oil pressure sensor connector
- ECT sensor connector
- ECT sender gauge connector
- EGR gas temperature sensor connector
- VSV connector for the EGR
- 2 vacuum hose to the VSV for the EGR
- Ground strap to the cowl top panel
- Engine wiring harness to the air intake chamber
- TP sensor connector
- IAC valve connector
- CKP sensor connector
- KS sensor connector
- DLC1 to the bracket
- Engine wiring harness clamp
- PCV hoses
- Intake air connector. Tighten the bolts to 13 ft. lbs. (18 Nm).
- Cruise control cable to the actuator, if equipped with cruise control
- Accelerator cable to the throttle body, if equipped with a manual transmission
- Accelerator and throttle cables to the throttle body, if equipped with an automatic transmission
12. Fill the engine and radiator with engine coolant.
13. Install or connect the following:
- Air cleaner cap, MAF meter and resonator assembly
- Negative battery cable
14. Start the engine and check for leaks.
15. Road test the vehicle for proper operation.
16. Recheck all fluid levels.

3.0L ENGINE

1. Before servicing the vehicle, refer to the precautions section.
2. Remove or disconnect the following:
3. Properly relieve the fuel system pressure.
4. Drain and recycle the engine coolant.
5. Remove or disconnect the following:
- Accelerator cable
- Throttle cable
- Air cleaner
- Any wiring or hoses interfering with removal
- Right side engine mount stay
- Radiator and heater hoses in the way of the intake manifold removal
- V-bank cover
- All the vacuum hose and wiring for the emission control valve set
- Air intake chamber and discard the gasket
- Exhaust Gas Recirculation (EGR) pipe and discard the gaskets
- Hydraulic motor pressure hose from the air intake chamber
- Engine wiring harnesses from the left side, right side, rear and No. 3 timing belt cover
- Front exhaust pipe, if necessary
- Timing belt, camshaft timing pulleys, No. 2 idler pulley and No. 3 timing belt cover
- Cylinder head rear plate
- 2 bolts, nuts and plate washers with the intake manifold

➡**The delivery pipes with injectors will be attached to the manifold.**

- Other fuel related components such as the No. 2 fuel pipe and pulsation damper, if needed
- Delivery pipes from the intake manifold
6. Clean and inspect the intake manifold mating surfaces. Scrape all old gasket material off.
   To install:
7. Install or connect the following:
- Delivery pipes with injectors to the intake manifold.

➡**Be sure to place 4 spacers in position on the manifold. Temporarily install 4 bolts to retain the delivery pipes to the manifold. Inspect the injectors for smooth rotation.**

- Tighten the delivery pipes bolts to 84 inch lbs. (10 Nm), once the injectors are properly seated
- No. 2 fuel pipe with union bolts

and gaskets. Tighten the bolts to 24 ft. lbs. (32 Nm).
- No. 1 fuel pipe with pulsation damper, using 4 new gaskets. Tighten the damper to 35 ft. lbs. (32 Nm) and the bolt to 11 ft. lbs. (15 Nm).
- Fuel pressure regulator, if removed
- Intake manifold. Tighten the 9 bolts and 2 nuts in a crisscross pattern to 11 ft. lbs. (15 Nm).

➡**Be sure the gasket is in place properly prior to tightening.**

8. Retighten the water outlet mounting nuts/bolts to 11 ft. lbs. (15 Nm), if loosened.
9. Install or connect the following:
- Air assist hose and water inlet pipe, using a new O-ring, by applying a small amount of soapy water. Tighten the fastener(s) to 14 ft. lbs. (20 Nm).
- Ground strap
- Vacuum hoses removed to the air intake chamber and vacuum tank
- Any remaining components, using new gaskets. Tighten the air intake chamber nuts/bolts to 32 ft. lbs. (43 Nm), the EGR pipe nuts to 108 inch lbs. (12 Nm) and the emission control valve set to 69 inch lbs. (8 Nm).
- Air cleaner assembly
- Heater hoses
- Throttle cable with bracket onto the throttle body
- Accelerator cable, by adjusting it, if equipped with an automatic transaxle
10. Refill the cooling system
11. Install or connect the following:
- Negative battery cable
- Heater air duct
12. Start the engine and inspect for leaks.

3.3L ENGINES

1. Before servicing the vehicle, refer to the precautions section.
2. Remove or disconnect the following:
- Wiper and blade assembly
- Top cowl seal and panel
- Window washer hoses from the ventilator louvers
- Left and right ventilator louvers
- Heater air duct
- Front upper suspension brace
3. Properly relieve the fuel system pressure.
4. Remove the battery and battery tray.
5. Drain and recycle the engine coolant.

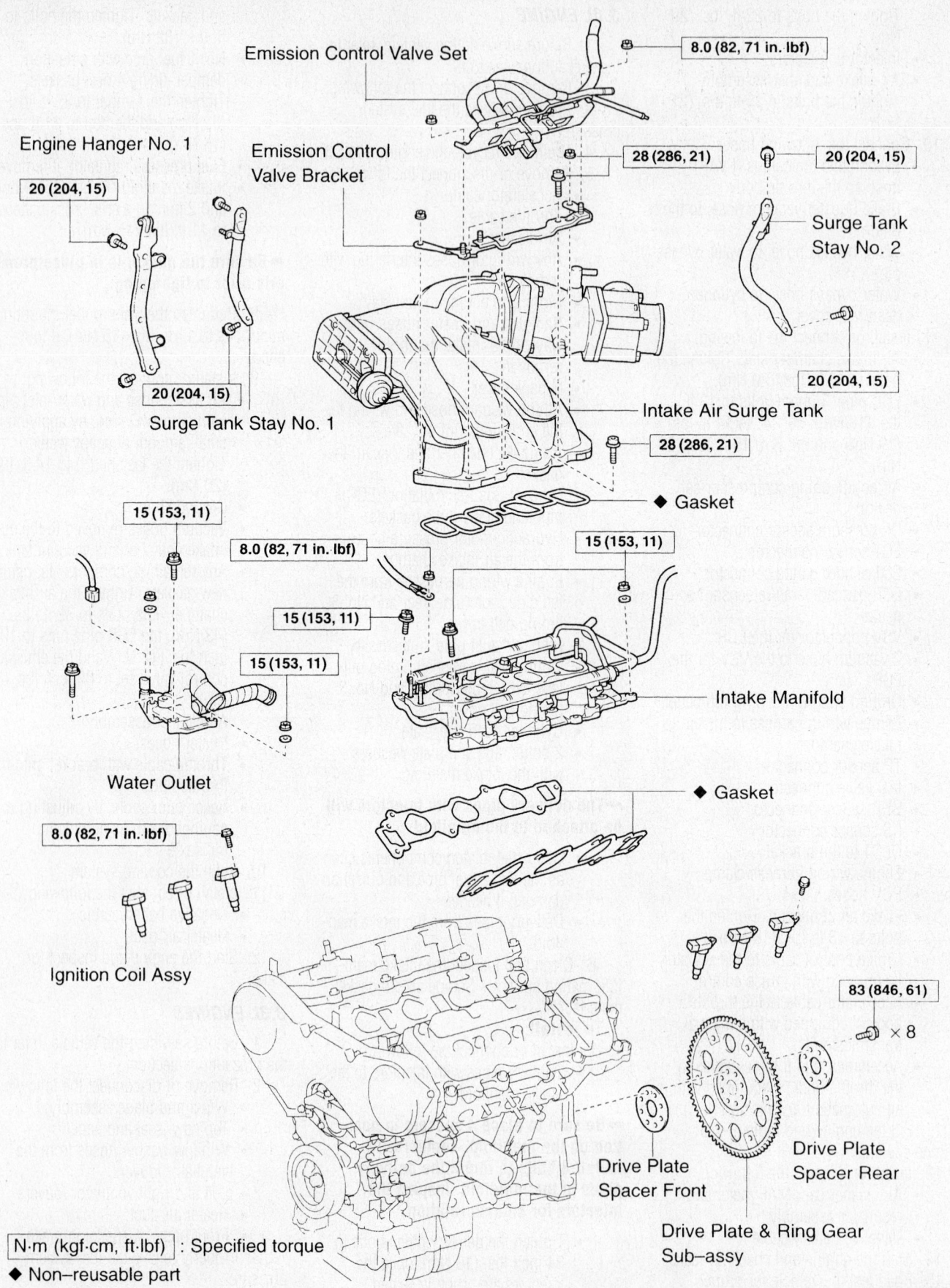

Emission Control Valve Set

8.0 (82, 71 in. lbf)

Engine Hanger No. 1

Emission Control Valve Bracket

20 (204, 15)

28 (286, 21)

20 (204, 15)

Surge Tank Stay No. 2

20 (204, 15)

Surge Tank Stay No. 1

20 (204, 15)

Intake Air Surge Tank

28 (286, 21)

◆ Gasket

15 (153, 11)

8.0 (82, 71 in. lbf)

15 (153, 11)

15 (153, 11)

15 (153, 11)

Intake Manifold

Water Outlet

◆ Gasket

8.0 (82, 71 in. lbf)

Ignition Coil Assy

83 (846, 61)

× 8

Drive Plate Spacer Front

Drive Plate Spacer Rear

Drive Plate & Ring Gear Sub–assy

N·m (kgf·cm, ft·lbf) : Specified torque

◆ Non–reusable part

67170-HIGH-G10

Intake manifold and related parts—3.3L Highlander

6. Remove or disconnect the following:
- Accelerator cable
- Throttle cable
- Air cleaner cap assembly
- Any wiring or hoses interfering with removal
- Right side engine mount stay
- Radiator and heater hoses in the way of the intake manifold removal
- V-bank cover
- All the vacuum hose and wiring for the emission control valve set
- Air intake chamber and discard the gasket
- Exhaust Gas Recirculation (EGR) pipe and discard the gaskets
- Hydraulic motor pressure hose from the air intake chamber
- Engine wiring harnesses from the left side, right side, rear and No. 3 timing belt cover
- Front exhaust pipe, if necessary
- Timing belt, camshaft timing pulleys, No. 2 idler pulley and No. 3 timing belt cover
- Cylinder head rear plate
- 2 bolts, nuts and plate washers with the intake manifold.

➡**The delivery pipes with injectors will be attached to the manifold.**

- Other fuel related components such as the No. 2 fuel pipe and pulsation damper, if needed
- Delivery pipes from the intake manifold

7. Clean and inspect the intake manifold mating surfaces. Scrape all old gasket material off.

To install:

8. Install or connect the following:
- Delivery pipes with injectors to the intake manifold.

➡**Be sure to place 4 spacers in position on the manifold. Temporarily install 4 bolts to retain the delivery pipes to the manifold. Inspect the injectors for smooth rotation.**

- Tighten the delivery pipes bolts to 84 inch lbs. (10 Nm), once the injectors are properly seated
- No. 2 fuel pipe with union bolts and gaskets. Tighten the bolts to 24 ft. lbs. (32 Nm).
- No. 1 fuel pipe with pulsation damper, using 4 new gaskets. Tighten the damper to 35 ft. lbs. (32 Nm) and the bolt to 11 ft. lbs. (15 Nm).
- Fuel pressure regulator, if removed
- Intake manifold. Tighten the 9 bolts

and 2 nuts in a crisscross pattern to 11 ft. lbs. (15 Nm).

➡**Be sure the gasket is in place properly prior to tightening.**

9. Retighten the water outlet mounting nuts/bolts to 11 ft. lbs. (15 Nm), if loosened.

10. Install or connect the following:
- Air assist hose and water inlet pipe, using a new O-ring, by applying a small amount of soapy water. Tighten the fastener(s) to 14 ft. lbs. (20 Nm).
- Ground strap
- Vacuum hoses removed to the air intake chamber and vacuum tank
- Any remaining components, using new gaskets. Tighten the air intake chamber nuts/bolts to 32 ft. lbs. (43 Nm), the EGR pipe nuts to 108 inch lbs. (12 Nm) and the emission control valve set to 69 inch lbs. (8 Nm).
- Air cleaner assembly
- Heater hoses
- Battery and tray
- Throttle cable with bracket onto the throttle body
- Accelerator cable, by adjusting it, if equipped with an automatic transaxle
- Front upper suspension brace. Tighten the nuts to 59 ft. lbs. (80 Nm).

11. Refill the cooling system
12. Install or connect the following:
- Negative battery cable
- Heater air duct
- Left and right ventilator louvers
- Window washer hoses from the ventilator louvers
- Top cowl seal and panel
- Wiper and blade assembly

13. Start the engine and inspect for leaks.

RX300 and RX330

1. Before servicing the vehicle, refer to the precautions section.
2. Remove or disconnect the following:
- Wiper and blade assembly
- Top cowl seal and panel
- Window washer hoses from the ventilator louvers
- Left and right ventilator louvers
- Heater air duct
- Front upper suspension brace

3. Properly relieve the fuel system pressure.
4. Remove the battery and battery tray.

5. Drain and recycle the engine coolant.
6. Remove or disconnect the following:
- Accelerator cable
- Throttle cable
- Air cleaner cap assembly
- Any wiring or hoses interfering with removal
- Right side engine mount stay
- Radiator and heater hoses in the way of the intake manifold removal
- V-bank cover
- All the vacuum hose and wiring for the emission control valve set
- Air intake chamber and discard the gasket
- Exhaust Gas Recirculation (EGR) pipe and discard the gaskets
- Hydraulic motor pressure hose from the air intake chamber
- Engine wiring harnesses from the left side, right side, rear and No. 3 timing belt cover
- Front exhaust pipe, if necessary
- Timing belt, camshaft timing pulleys, No. 2 idler pulley and No. 3 timing belt cover
- Cylinder head rear plate
- 2 bolts, nuts and plate washers with the intake manifold.

➡**The delivery pipes with injectors will be attached to the manifold.**

- Other fuel related components such as the No. 2 fuel pipe and pulsation damper, if needed
- Delivery pipes from the intake manifold

7. Clean and inspect the intake manifold mating surfaces. Scrape all old gasket material off.

To install:

8. Install or connect the following:
- Delivery pipes with injectors to the intake manifold.

➡**Be sure to place 4 spacers in position on the manifold. Temporarily install 4 bolts to retain the delivery pipes to the manifold. Inspect the injectors for smooth rotation.**

- Tighten the delivery pipes bolts to 84 inch lbs. (10 Nm), once the injectors are properly seated
- No. 2 fuel pipe with union bolts and gaskets. Tighten the bolts to 24 ft. lbs. (32 Nm).
- No. 1 fuel pipe with pulsation damper, using 4 new gaskets. Tighten the damper to 35 ft. lbs. (32 Nm) and the bolt to 11 ft. lbs. (15 Nm).
- Fuel pressure regulator, if removed

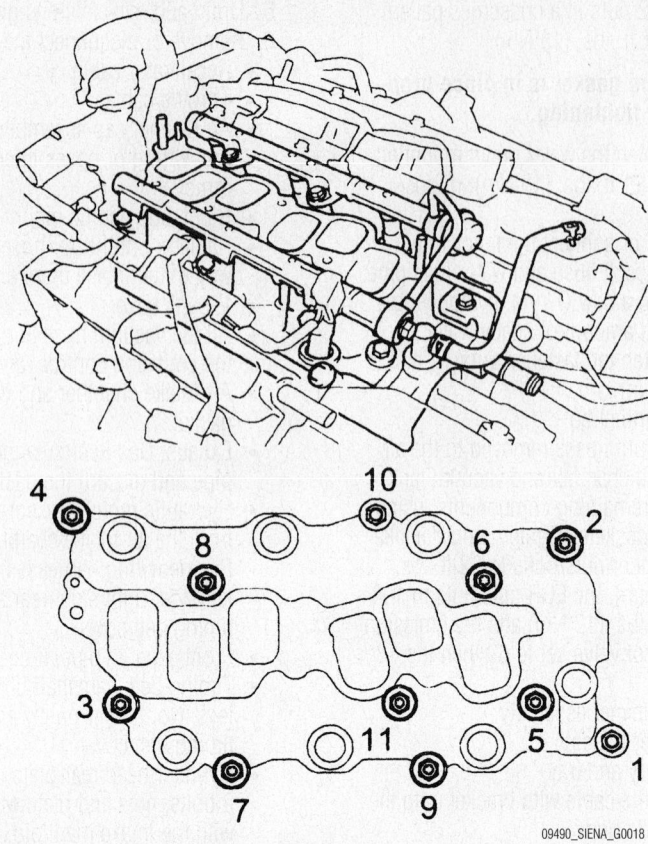

Remove the intake manifold nuts and bolts in the sequence shown—3.3L engine

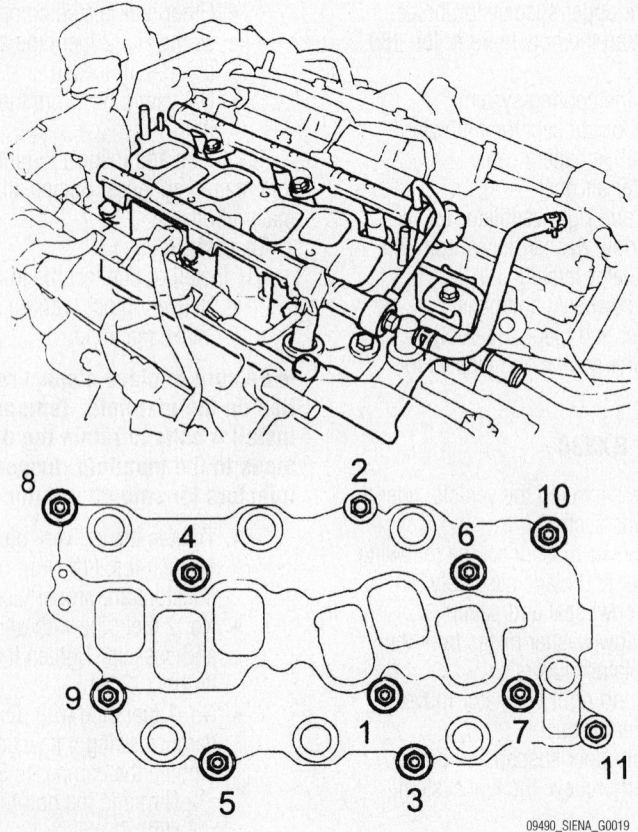

Tighten the intake manifold nuts and bolts in the sequence shown—3.3L engine

- Intake manifold. Tighten the 9 bolts and 2 nuts in a crisscross pattern to 11 ft. lbs. (15 Nm).

➡**Be sure the gasket is in place properly prior to tightening.**

9. Retighten the water outlet mounting nuts/bolts to 11 ft. lbs. (15 Nm), if loosened.
10. Install or connect the following:
   - Air assist hose and water inlet pipe, using a new O-ring, by applying a small amount of soapy water. Tighten the fastener(s) to 14 ft. lbs. (20 Nm).
   - Ground strap
   - Vacuum hoses removed to the air intake chamber and vacuum tank
   - Any remaining components, using new gaskets. Tighten the air intake chamber nuts/bolts to 32 ft. lbs. (43 Nm), the EGR pipe nuts to 108 inch lbs. (12 Nm) and the emission control valve set to 69 inch lbs. (8 Nm).
   - Air cleaner assembly
   - Heater hoses
   - Battery and tray
   - Throttle cable with bracket onto the throttle body
   - Accelerator cable, by adjusting it, if equipped with an automatic transaxle
   - Front upper suspension brace. Tighten the nuts to 59 ft. lbs. (80 Nm).
11. Refill the cooling system
12. Install or connect the following:
   - Negative battery cable
   - Heater air duct
   - Left and right ventilator louvers
   - Window washer hoses from the ventilator louvers
   - Top cowl seal and panel
   - Wiper and blade assembly
13. Start the engine and inspect for leaks.

Exhaust Manifold

REMOVAL & INSTALLATION

Highlander

2.4L ENGINE

1. Before servicing the vehicle, refer to the precautions section.
2. Remove or disconnect the following:
   - Clamp from the support bracket
   - Support bracket
   - Front exhaust pipe and gaskets from the exhaust manifold

- Heat insulator
- Exhaust manifold and gasket

To install:

3. Install or connect the following:
   - Exhaust manifold and gasket. Tighten the nuts to 36 ft. lbs. (49 Nm).
   - Heat insulator. Tighten the bolts and nuts to 48 inch lbs. (5.5 Nm).
   - Front exhaust pipe assembly to the exhaust manifold. Tighten the nuts to 46 ft. lbs. (62 Nm).
   - Support bracket. Tighten the bolts to 29 ft. lbs. (39 Nm).
   - Clamp. Tighten the bolt to 14 ft. lbs. (19 Nm).
4. Start the engine.
5. Check for exhaust leaks.

3.0L ENGINE–FRONT MANIFOLD

➡**Removing the oil filter helps gain access to a lower bolt in the front exhaust manifold.**

1. Before servicing the vehicle, refer to the precautions section.
2. Remove or disconnect the following:
   - Negative battery cable
   - Engine undercovers
   - Front exhaust pipe from the exhaust manifolds, by removing the nuts

➡**Check for access to some of the manifold lower bolts, if so remove any possible.**

   - Heated Oxygen (HO2) sensor
   - Exhaust manifold stay, by removing the bolt and nut
   - Remaining exhaust manifold nuts; then, separate the exhaust manifold from the engine

To install:

3. Install or connect the following:
   - Exhaust manifold, using a new gasket. Uniformly, tighten the bolts to 36 ft. lbs. (49 Nm).
   - Exhaust manifold stay. Tighten the nut/bolt to 15 ft. lbs. (20 Nm).
   - Heated Oxygen (HO2) sensor to the exhaust manifold
   - Front exhaust pipe to the exhaust manifold, using a new gasket. Tighten both nuts to 46 ft. lbs. (62 Nm).
   - Engine undercovers
   - Negative battery cable

3.0L ENGINE–REAR MANIFOLD

1. Before servicing the vehicle, refer to the precautions section.
2. Remove or disconnect the following:

- Negative battery cable
- Engine undercovers
- Front exhaust pipe from both exhaust manifolds, from below the engine
- Exhaust Gas Recirculation (EGR) pipe from the rear exhaust manifold, by removing the 4 nuts
- Heated Oxygen (HO2) sensor wiring, from the right exhaust manifold
- Exhaust manifold stay
- 6 exhaust manifold nuts and the exhaust manifold

To install:

3. Install or connect the following:
   - Exhaust manifold to the engine, using a new gasket. Tighten the 6 nuts to 36 ft. lbs. (49 Nm).
   - Exhaust manifold stay. Tighten the nut/bolt to 15 ft. lbs. (20 Nm).
   - HO2 sensor wiring to the exhaust manifold
   - EGR pipe to the exhaust manifold and the engine, using new gaskets. Tighten the 4 nuts to 108 inch lbs. (12 Nm).
   - Front exhaust pipe to the exhaust manifold, use a new gasket. Tighten both nuts to 46 ft. lbs. (62 Nm).
   - Engine undercovers
   - Negative battery cable

3.3L ENGINE–FRONT MANIFOLD

➡**Removing the oil filter helps gain access to a lower bolt in the front exhaust manifold.**

1. Before servicing the vehicle, refer to the precautions section.
2. Remove or disconnect the following:
   - Negative battery cable
   - Engine undercovers
   - Front exhaust pipe from the exhaust manifolds, by removing the nuts

➡**Check for access to some of the manifold lower bolts, if so remove any possible.**

   - Heated Oxygen (HO2) sensor
   - Exhaust manifold stay, by removing the bolt and nut
   - Remaining exhaust manifold nuts; then, separate the exhaust manifold from the engine

To install:

3. Install or connect the following:
   - Exhaust manifold, using a new gasket. Uniformly, tighten the bolts to 36 ft. lbs. (49 Nm).
   - Exhaust manifold stay. Tighten the nut/bolt to 15 ft. lbs. (20 Nm).

- Heated Oxygen (HO2) sensor to the exhaust manifold
- Front exhaust pipe to the exhaust manifold, using a new gasket. Tighten both nuts to 46 ft. lbs. (62 Nm).
- Engine undercovers
- Negative battery cable

3.3L ENGINE–REAR MANIFOLD

1. Before servicing the vehicle, refer to the precautions section.
2. Remove or disconnect the following:

   - Negative battery cable
   - Engine undercovers
   - Front exhaust pipe from both exhaust manifolds, from below the engine
   - Exhaust Gas Recirculation (EGR) pipe from the rear exhaust manifold, by removing the 4 nuts
   - Heated Oxygen (HO2) sensor wiring, from the right exhaust manifold
   - Exhaust manifold stay
   - 6 exhaust manifold nuts and the exhaust manifold

To install:

3. Install or connect the following:
   - Exhaust manifold to the engine, using a new gasket. Tighten the 6 nuts to 36 ft. lbs. (49 Nm).
   - Exhaust manifold stay. Tighten the nut/bolt to 15 ft. lbs. (20 Nm).
   - HO2 sensor wiring to the exhaust manifold
   - EGR pipe to the exhaust manifold and the engine, using new gaskets. Tighten the 4 nuts to 108 inch lbs. (12 Nm).
   - Front exhaust pipe to the exhaust manifold, use a new gasket. Tighten both nuts to 46 ft. lbs. (62 Nm).
   - Engine undercovers
   - Negative battery cable

RX300 and RX330

FRONT MANIFOLD

➡**Removing the oil filter helps gain access to a lower bolt in the front exhaust manifold.**

1. Before servicing the vehicle, refer to the precautions section.
2. Remove or disconnect the following:
   - Negative battery cable
   - Engine undercovers
   - Front exhaust pipe from the exhaust manifolds, by removing the nuts

➡**Check for access to some of the manifold lower bolts, if so remove any possible.**

- Heated Oxygen (HO2) sensor
- Exhaust manifold stay, by removing the bolt and nut
- Remaining exhaust manifold nuts; then, separate the exhaust manifold from the engine

To install:

3. Install or connect the following:
- Exhaust manifold, using a new gasket. Uniformly, tighten the bolts to 36 ft. lbs. (49 Nm).
- Exhaust manifold stay. Tighten the nut/bolt to 15 ft. lbs. (20 Nm).
- Heated Oxygen (HO2) sensor to the exhaust manifold
- Front exhaust pipe to the exhaust manifold, using a new gasket. Tighten both nuts to 46 ft. lbs. (62 Nm).
- Engine undercovers
- Negative battery cable

REAR MANIFOLD

1. Before servicing the vehicle, refer to the precautions section.
2. Remove or disconnect the following:
- Negative battery cable
- Engine undercovers
- Front exhaust pipe from both exhaust manifolds, from below the engine
- Exhaust Gas Recirculation (EGR) pipe from the rear exhaust manifold, by removing the 4 nuts
- Heated Oxygen (HO2) sensor wiring, from the right exhaust manifold
- Exhaust manifold stay
- 6 exhaust manifold nuts and the exhaust manifold

To install:

3. Install or connect the following:
- Exhaust manifold to the engine, using a new gasket. Tighten the 6 nuts to 36 ft. lbs. (49 Nm).
- Exhaust manifold stay. Tighten the nut/bolt to 15 ft. lbs. (20 Nm).
- HO2 sensor wiring to the exhaust manifold
- EGR pipe to the exhaust manifold and the engine, using new gaskets. Tighten the 4 nuts to 108 inch lbs. (12 Nm).
- Front exhaust pipe to the exhaust manifold, use a new gasket. Tighten both nuts to 46 ft. lbs. (62 Nm).
- Engine undercovers
- Negative battery cable

Front Crankshaft Seal

REMOVAL & INSTALLATION

Highlander

2.4L ENGINES

For 2.4L engines, see the Timing Chain procedure later in this chapter.

3.0L ENGINE

1. Before servicing the vehicle, refer to the precautions section.
2. Remove or disconnect the following:

- Engine coolant reservoir tank and the alternator belt
- Right front wheel and the splash shield
- Power steering pump drive belt, by loosening both bolts
- Both ground wire connectors
- Right engine mounting stay
- Engine moving control rod and the No. 2 right engine mount bracket

➡**To extract the engine bracket and control rod, raise the engine slightly.**

- No. 2 alternator bracket
- Crankshaft pulley bolt, using a pry-bar and wrench or Crankshaft Pulley Holding tool 09213-54015 and Flange Holding tool 09330-00021
- Crankshaft pulley, using a puller
- No. 1 timing belt cover

3. Remove the No. 2 timing belt cover, as follows:
- Engine wire protector from the No. 3 (rear) timing belt cover
- Engine wire protector clamp from the No. 3 timing belt cover
- 5 bolts from the No. 2 timing belt cover
- No. 2 cover

To install:

4. Install or connect the following:
- No. 2 timing belt cover, using a new gasket

➡**Install it evenly to the part of the belt cover shaded black. After installation, press down on it so that the adhesive sticks to the belt cover firmly.**

- No. 2 timing belt cover. Tighten the 5 bolts to 74 inch lbs. (8 Nm).
- Engine wire protector clamp to the No. 3 timing belt cover
- Engine wire protector to the No. 3 timing belt cover with the bolt
- No. 3 timing belt cover, using a new gasket

- Tighten the 4 No. 1 timing belt cover bolts to 74 inch lbs. (8 Nm).
- Crankshaft pulley. Tighten the bolt to 159 ft. lbs. (215 Nm).
- No. 2 alternator bracket. Tighten the nut to 21 ft. lbs. (28 Nm). Do not tighten the pivot bolt at this time.
- No. 2 right engine mounting bracket and the moving control rod
- Right engine mount stay
- Both ground wire connectors
- Drive belts by adjusting them
- Coolant reservoir
- Right front splash shield and wheel
- Negative battery cable

5. Start the vehicle and check for any leaks.
6. Recheck the ignition timing.

3.3L ENGINES

1. Before servicing the vehicle, refer to the precautions section.
2. Remove or disconnect the following:

- Engine coolant reservoir tank and the alternator belt
- Right front wheel and the splash shield
- Power steering pump drive belt, by loosening both bolts
- Both ground wire connectors
- Right engine mounting stay
- Engine moving control rod and the No. 2 right engine mount bracket

➡**To extract the engine bracket and control rod, raise the engine slightly.**

- No. 2 alternator bracket
- Crankshaft pulley bolt, using a pry-bar and wrench or Crankshaft Pulley Holding tool 09213-54015 and Flange Holding tool 09330-00021
- Crankshaft pulley, using a puller
- No. 1 timing belt cover

3. Remove the No. 2 timing belt cover, as follows:
- Engine wire protector from the No. 3 (rear) timing belt cover
- Engine wire protector clamp from the No. 3 timing belt cover
- 5 bolts from the No. 2 timing belt cover
- No. 2 cover

To install:

4. Install or connect the following:
- No. 2 timing belt cover, using a new gasket

➡**Install it evenly to the part of the belt cover shaded black. After installation, press down on it so that the adhesive sticks to the belt cover firmly.**

- No. 2 timing belt cover. Tighten the 5 bolts to 74 inch lbs. (8 Nm).
- Engine wire protector clamp to the No. 3 timing belt cover
- Engine wire protector to the No. 3 timing belt cover with the bolt
- No. 3 timing belt cover, using a new gasket
- Tighten the 4 No. 1 timing belt cover bolts to 74 inch lbs. (8 Nm).
- Crankshaft pulley. Tighten the bolt to 159 ft. lbs. (215 Nm).
- No. 2 alternator bracket. Tighten the nut to 21 ft. lbs. (28 Nm). Do not tighten the pivot bolt at this time.
- No. 2 right engine mounting bracket and the moving control rod
- Right engine mount stay
- Both ground wire connectors
- Drive belts by adjusting them
- Coolant reservoir
- Right front splash shield and wheel
- Negative battery cable

5. Start the vehicle and check for any leaks.

6. Recheck the ignition timing.

RX300 and RX330

1. Before servicing the vehicle, refer to the precautions section.

2. Remove or disconnect the following:
- Engine coolant reservoir tank and the alternator belt
- Right front wheel and the splash shield
- Power steering pump drive belt, by loosening both bolts
- Both ground wire connectors
- Right engine mounting stay
- Engine moving control rod and the No. 2 right engine mount bracket

➡ **To extract the engine bracket and control rod, raise the engine slightly.**

- No. 2 alternator bracket
- Crankshaft pulley bolt, using a prybar and wrench or Crankshaft Pulley Holding tool 09213-54015 and Flange Holding tool 09330-00021
- Crankshaft pulley, using a puller
- No. 1 timing belt cover

3. Remove the No. 2 timing belt cover, as follows:
- Engine wire protector from the No. 3 (rear) timing belt cover
- Engine wire protector clamp from the No. 3 timing belt cover
- 5 bolts from the No. 2 timing belt cover
- No. 2 cover

To install:

4. Install or connect the following:
- No. 2 timing belt cover, using a new gasket

➡ **Install it evenly to the part of the belt cover shaded black. After installation, press down on it so that the adhesive sticks to the belt cover firmly.**

- No. 2 timing belt cover. Tighten the 5 bolts to 74 inch lbs. (8 Nm).
- Engine wire protector clamp to the No. 3 timing belt cover
- Engine wire protector to the No. 3 timing belt cover with the bolt
- No. 3 timing belt cover, using a new gasket
- Tighten the 4 No. 1 timing belt cover bolts to 74 inch lbs. (8 Nm).
- Crankshaft pulley. Tighten the bolt to 159 ft. lbs. (215 Nm).
- No. 2 alternator bracket. Tighten the nut to 21 ft. lbs. (28 Nm). Do not tighten the pivot bolt at this time.
- No. 2 right engine mounting bracket and the moving control rod
- Right engine mount stay
- Both ground wire connectors
- Drive belts by adjusting them
- Coolant reservoir
- Right front splash shield and wheel
- Negative battery cable

5. Start the vehicle and check for any leaks.

6. Recheck the ignition timing.

Camshaft and Valve Lifters

REMOVAL & INSTALLATION

Highlander

2.4L ENGINE

1. Before servicing the vehicle, refer to the precautions section.

2. Disconnect the negative battery cable.

3. Drain the engine coolant.

4. Remove or disconnect the following:
- Right front wheel
- Right fender splash shield
- Right fender apron seal
- No.1 engine undercover
- Coil pack
- Cylinder head cover

5. Set the No.1 piston at TDC compression.

6. Remove or disconnect the following:
- Timing chain tensioner No.1

7. Loosen the camshaft timing gear set bolt.

8. Raise the camshaft and remove the set bolt.

9. Remove or disconnect the following:
- Timing gear and chain
- Exhaust camshaft

10. Loosen the intake camshaft cap bolts in several passes, in the sequence shown. Remove the caps. Remove the camshaft.

11. Installation is the reverse of removal.

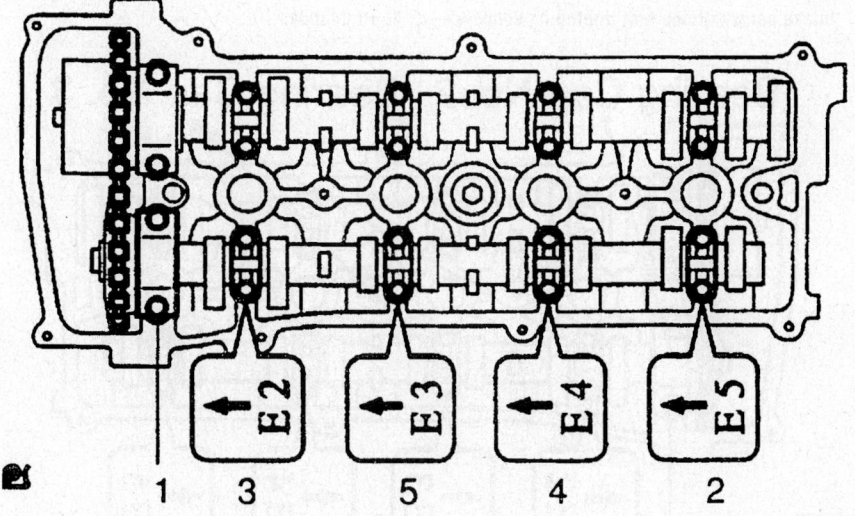

Exhaust camshaft cap bolt loosening sequence—2.4L Highlander

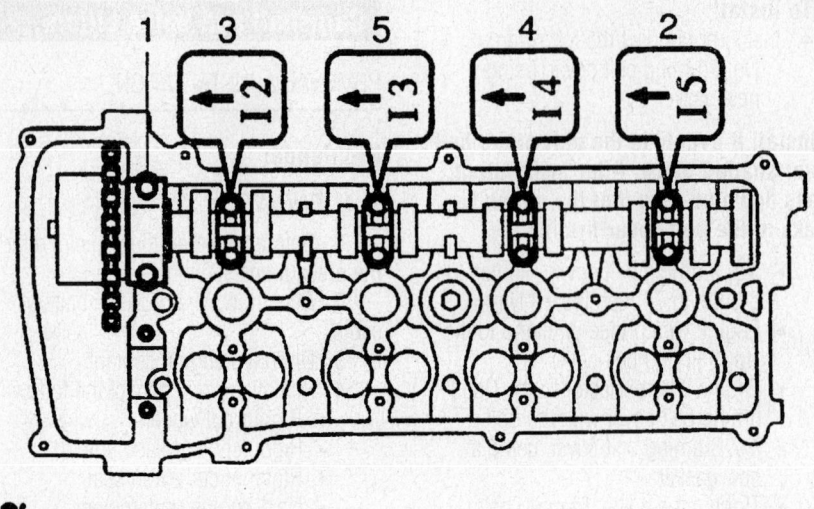

9355YG04

Intake camshaft cap bolt loosening sequence—2.4L Highlander

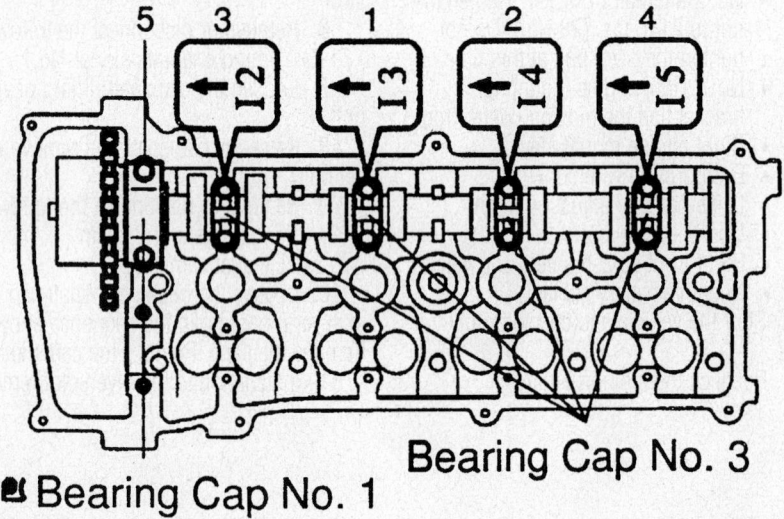

Bearing Cap No. 3

Bearing Cap No. 1

9355YG05

Intake camshaft cap bolt tightening sequence—2.4L Highlander

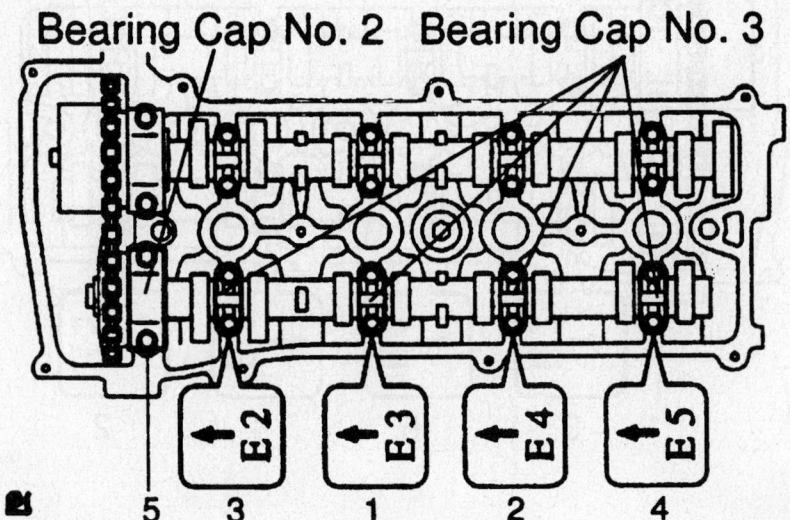

Bearing Cap No. 2 **Bearing Cap No. 3**

9355YG06

Exhaust camshaft cap bolt tightening sequence—2.4L Highlander

Tighten the cap bolt, in several passes, in the sequences shown, to:
- Front caps: 22 ft. lbs. (30Nm)
- All other caps: 80 inch lbs. (9Nm)

12. See the Timing Chain Removal and Installation procedure.

3.0L ENGINE

1. Before servicing the vehicle, refer to the precautions section.

2. Remove or disconnect the following:
- Timing belt and idler pulley
- Camshaft timing pulleys
- Cylinder head covers

➡ **The thrust clearance on both the intake and exhaust camshafts is very small; the camshafts must be kept level during removal. If the camshafts are removed without being kept level, the camshaft may be caught in the cylinder head, causing the head to break or the camshaft to seize.**

3. Remove the exhaust and intake camshafts from the right side cylinder head, as follows:

a. Turn the camshaft with a wrench until the 2 pointed marks drive and driven gears are aligned. (The right camshaft gears have 2 marks apiece; the left side camshaft gears have 1 mark each.)

b. Secure the exhaust camshaft sub-gear to the main gear using a service bolt. A bolt 0.63–0.79 in. (16–20mm) long with a 6mm thread diameter and a 1mm pitch is recommended. When removing the exhaust camshaft be sure the sub-gear is not loaded; all the force must be eliminated.

c. Uniformly loosen and remove the exhaust camshaft bearing cap bolts in several passes and in the proper sequence. Remove the 8 bearing cap bolts and remove the caps, keeping them in the correct order.

d. Remove the exhaust camshaft from the engine.

e. Uniformly loosen and remove the 10 bearing cap bolts in several passes, in the proper sequence. Remove the bearing caps, keeping them in order, remove the oil seal, then lift out the intake camshaft.

4. Remove the exhaust and intake camshafts from the left side cylinder head, as follows:

a. Turn the camshaft with a wrench until the pointed marks on the drive and driven gears are aligned. (The right camshaft gears have 2 marks apiece; the left side camshaft gears have 1 mark each.)

Intake

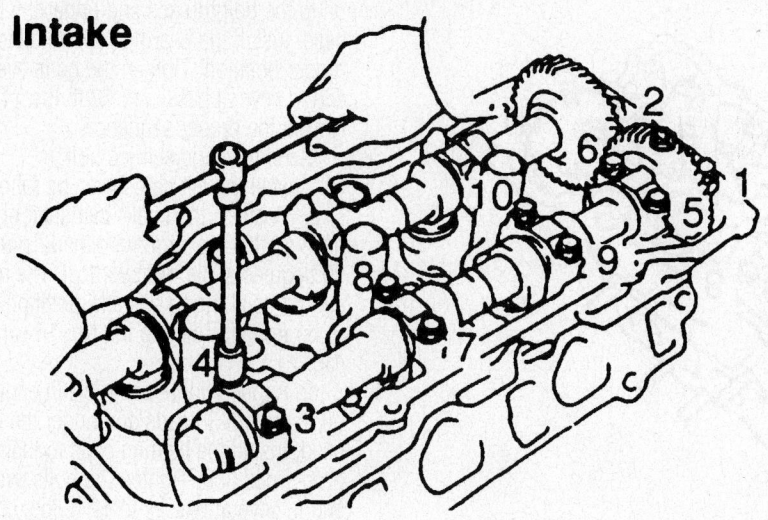

Right intake camshaft bearing cap bolt loosening sequence—3.0L Highlander

Exhaust

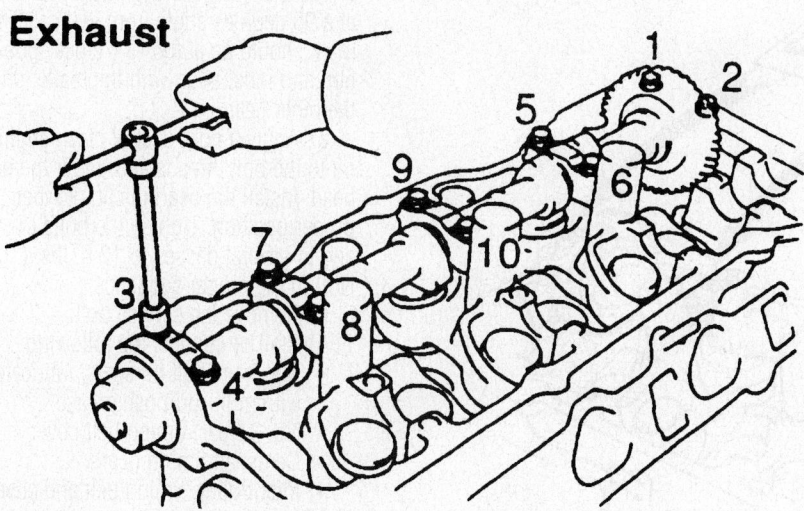

Right side exhaust camshaft bearing cap bolt loosening sequence—3.0L Highlander

b. Secure the exhaust camshaft sub-gear to the main gear using a service bolt. A bolt 16–20mm long with a 6mm thread diameter and a 1mm pitch is recommended. When removing the exhaust camshaft be sure the sub-gear is not loaded; all the force must be eliminated.

c. Uniformly loosen and remove the exhaust camshaft bearing cap bolts in several passes and in the proper sequence. Remove the 8 bearing cap bolts and remove the caps. Keep the caps in the correct order.

d. Remove the exhaust camshaft from the engine.

e. Uniformly loosen and remove the 10 bearing cap bolts in several passes, in the reverse order of the installation sequence. Remove the bearing caps,

keeping them in order, remove the oil seal, then lift out the intake camshaft.

5. Remove the valve lifter shims and hydraulic lifters. Identify each lifter and shim as it is removed so it can be reinstalled in the same position. If the lifters are to be reused, store them upside down in a sealed container.

To install:

6. Install the valve lifters into their original positions and install the shims. Check valve clearance and replace the shims as necessary.

7. When reinstalling, remember that the camshafts must be handled carefully and kept straight and level to avoid damage.

8. Before installing the camshafts in either cylinder head, apply multi-purpose grease to each camshaft.

9. Install the right camshafts, as follows:

a. Position the intake camshaft on the head so that the alignment marks are at a 90 degrees angle from vertical. The mark should be at the "3 o'clock" position.

b. Apply sealant to the No. 1 bearing cap.

c. Apply a light coat of clean engine oil to the bolt threads and under the bolt head. Install the bearing caps to their proper position. Tighten the bolts evenly and in several passes to 12 ft. lbs. (16 Nm) in the proper sequence.

d. Position the exhaust camshaft on the head so that the alignment marks are at a 90 degrees angle from vertical. The mark should be at the "9 o'clock" position and must align with the marks on the other gear.

e. Apply a light coat of clean engine

Intake

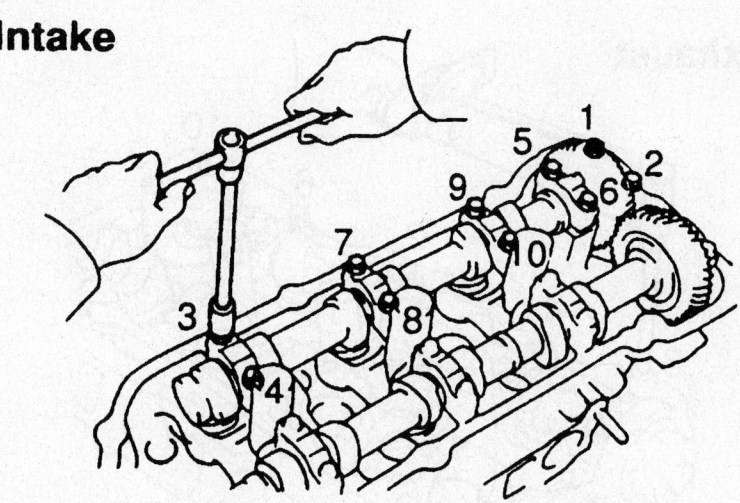

Left intake camshaft bearing cap bolt loosening sequence—3.0L Highlander

Exhaust

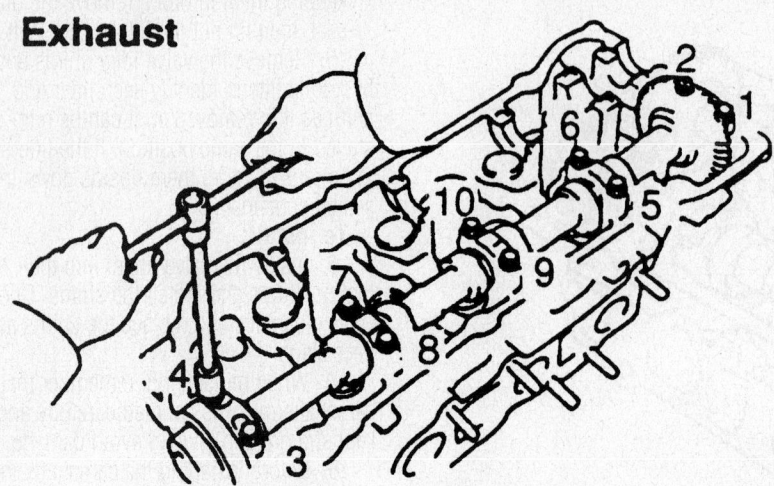

7924ZG47

Left side exhaust camshaft bearing cap bolt loosening sequence—3.0L Highlander

Exhaust

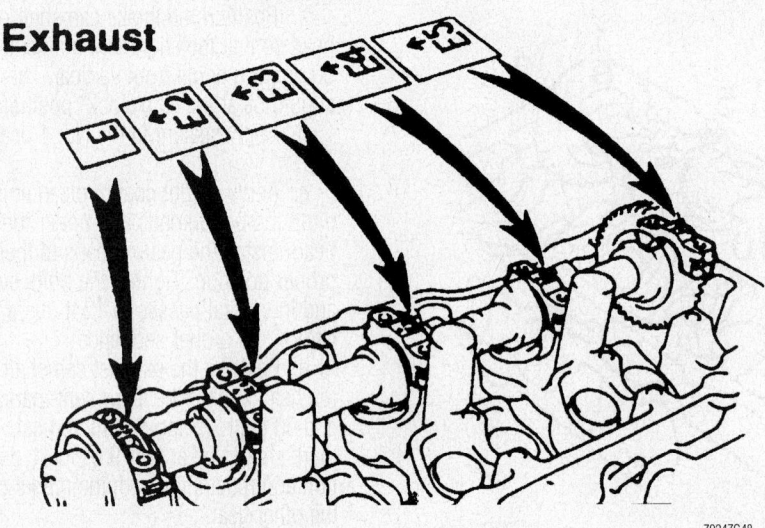

7924ZG48

Right exhaust bearing caps must be placed in their proper locations—3.0L Highlander

Exhaust

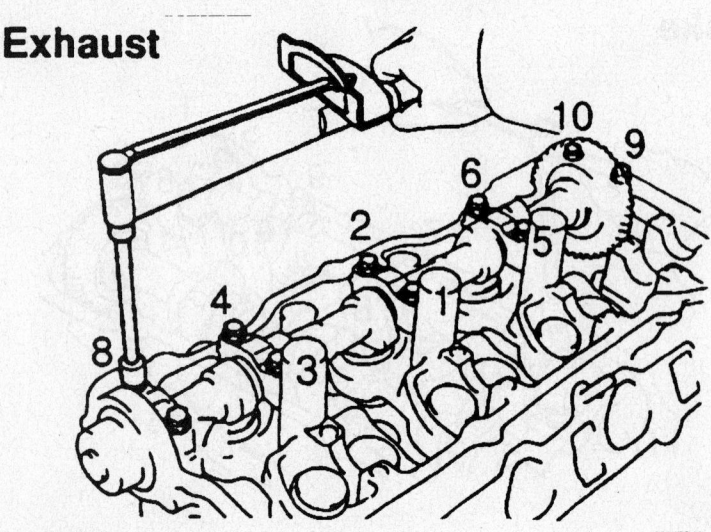

7924ZG49

Right exhaust camshaft bearing cap bolt tightening sequence—3.0L Highlander

oil to the bolt threads and under the bolt head. Install the bearing caps to their proper position. Tighten the bolts evenly and in several passes to 12 ft. lbs. (16 Nm) in the proper sequence.

f. Remove the service bolt.

10. Install the left camshafts, as follows:

a. Position the intake camshaft on the head so that the alignment mark is at a 90 degrees angle from vertical. The mark should be at the "9 o'clock" position.

b. Apply sealant to the No. 1 bearing cap.

c. Apply a light coat of clean engine oil to the bolt threads and under the bolt head. Install the bearing caps to their proper position. Tighten the bolts evenly and in several passes to 12 ft. lbs. (16 Nm) in the proper sequence.

d. Position the exhaust camshaft on the head so that the alignment marks are at a 90 degrees angle from vertical. The mark should be at the "3 o'clock" position and must align with the marks on the other gear.

e. Apply a light coat of clean engine oil to the bolt threads and under the bolt head. Install the bearing caps to their proper position. Tighten the bolts evenly and in several passes to 12 ft. lbs. (16 Nm) in the proper sequence.

f. Remove the service bolt.

11. Install or connect the following:

• New camshaft oil seals, lubricated with multi-purpose grease
• No. 3 (rear) timing belt cover
• Camshaft timing gears
• Idler pulley, timing belt and covers

12. Check and adjust the valve clearance.

13. Install the cylinder head (valve) covers.

14. Start the engine. Check the ignition timing.

15. Test drive the vehicle.

16. Check all fluid levels.

3.3L ENGINES

1. Before servicing the vehicle, refer to the precautions section.

2. Remove or disconnect the following:

• Timing belt and idler pulley
• Camshaft timing pulleys
• Cylinder head covers

➡**The thrust clearance on both the intake and exhaust camshafts is very small; the camshafts must be kept level during removal. If the camshafts are removed without being kept level, the camshaft may be caught in the cylinder head, causing the head to break or the camshaft to seize.**

Intake

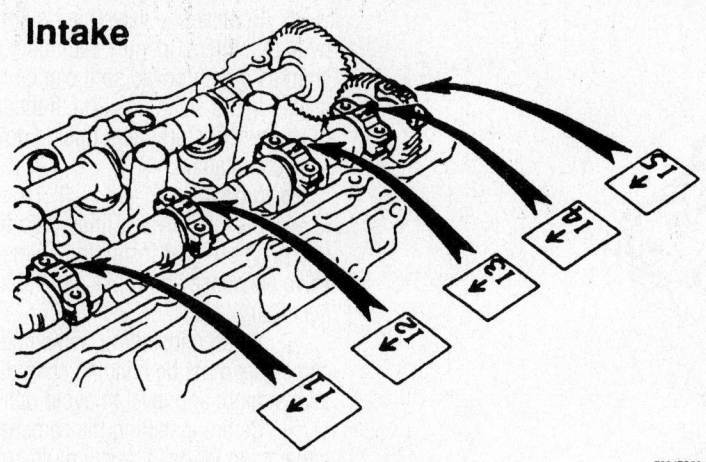

7924ZG50

Right intake bearing caps must be placed in their proper locations—3.0L Highlander

Intake

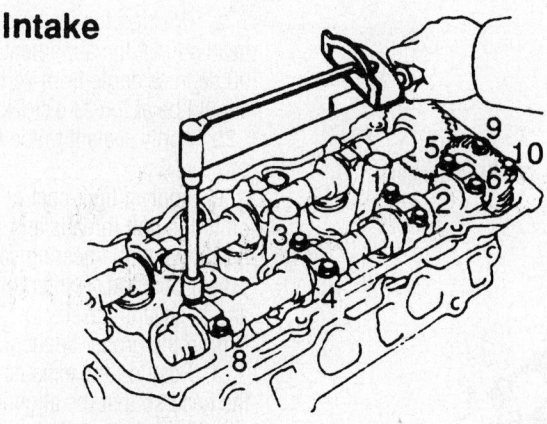

7924ZG51

Right intake camshaft bearing cap bolt tightening sequence—3.0L Highlander

3. Remove the exhaust and intake camshafts from the right side cylinder head, as follows:

a. Turn the camshaft with a wrench until the 2 pointed marks drive and driven gears are aligned. (The right camshaft gears have 2 marks apiece; the left side camshaft gears have 1 mark each.)

b. Secure the exhaust camshaft sub-gear to the main gear using a service bolt. A bolt 0.63–0.79 in. (16–20mm) long with a 6mm thread diameter and a 1mm pitch is recommended. When removing the exhaust camshaft be sure the sub-gear is not loaded; all the force must be eliminated.

c. Uniformly loosen and remove the exhaust camshaft bearing cap bolts in several passes and in the proper sequence. Remove the 8 bearing cap bolts and remove the caps, keeping them in the correct order.

d. Remove the exhaust camshaft from the engine.

e. Uniformly loosen and remove the 10 bearing cap bolts in several passes, in the proper sequence. Remove the bearing caps, keeping them in order, remove the oil seal, then lift out the intake camshaft.

4. Remove the exhaust and intake camshafts from the left side cylinder head, as follows:

a. Turn the camshaft with a wrench until the pointed marks on the drive and driven gears are aligned. (The right camshaft gears have 2 marks apiece; the left side camshaft gears have 1 mark each.)

b. Secure the exhaust camshaft sub-gear to the main gear using a service bolt. A bolt 16–20mm long with a 6mm thread diameter and a 1mm pitch is recommended. When removing the exhaust camshaft be sure the sub-gear is not loaded; all the force must be eliminated.

c. Uniformly loosen and remove the exhaust camshaft bearing cap bolts in several passes and in the proper sequence. Remove the 8 bearing cap bolts and remove the caps. Keep the caps in the correct order.

d. Remove the exhaust camshaft from the engine.

e. Uniformly loosen and remove the 10 bearing cap bolts in several passes, in the reverse order of the installation sequence. Remove the bearing caps, keeping them in order, remove the oil seal, then lift out the intake camshaft.

Exhaust

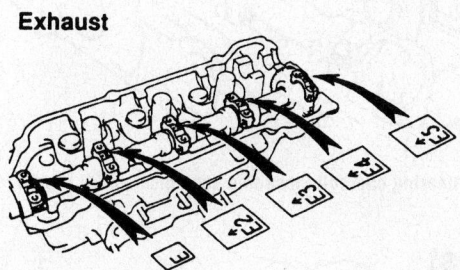

Exhaust

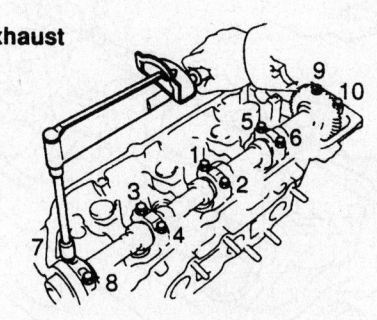

7924ZG52

Left exhaust bearing caps locations and bolt tightening sequence—3.0L Highlander

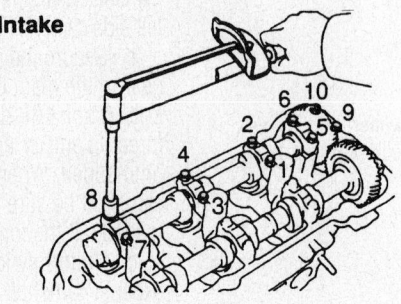

Left intake camshaft bearing cap locations and bolt tightening sequence—3.0L Highlander

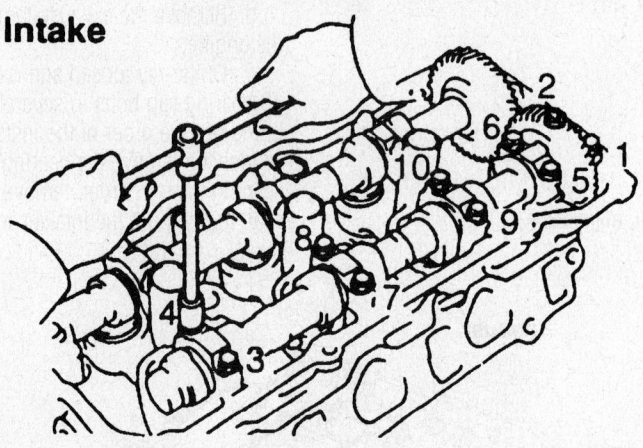

Right intake camshaft bearing cap bolt loosening sequence—3.3L Highlander

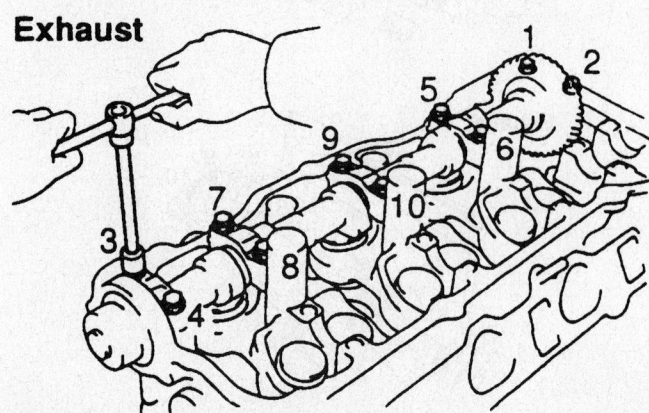

Right side exhaust camshaft bearing cap bolt loosening sequence—3.3L Highlander

5. Remove the valve lifter shims and hydraulic lifters. Identify each lifter and shim as it is removed so it can be reinstalled in the same position. If the lifters are to be reused, store them upside down in a sealed container.

To install:

6. Install the valve lifters into their original positions and install the shims. Check valve clearance and replace the shims as necessary.

7. When reinstalling, remember that the camshafts must be handled carefully and kept straight and level to avoid damage.

8. Before installing the camshafts in either cylinder head, apply multi-purpose grease to each camshaft.

9. Install the right camshafts, as follows:

a. Position the intake camshaft on the head so that the alignment marks are at a 90 degrees angle from vertical. The mark should be at the "3 o'clock" position.

b. Apply sealant to the No. 1 bearing cap.

c. Apply a light coat of clean engine oil to the bolt threads and under the bolt head. Install the bearing caps to their proper position. Tighten the bolts evenly and in several passes to 12 ft. lbs. (16 Nm) in the proper sequence.

d. Position the exhaust camshaft on the head so that the alignment marks are at a 90 degrees angle from vertical. The mark should be at the "9 o'clock" position and must align with the marks on the other gear.

e. Apply a light coat of clean engine oil to the bolt threads and under the bolt head. Install the bearing caps to their proper position. Tighten the bolts evenly and in several passes to 12 ft. lbs. (16 Nm) in the proper sequence.

f. Remove the service bolt.

10. Install the left camshafts, as follows:

a. Position the intake camshaft on the head so that the alignment mark is at a 90 degrees angle from vertical. The mark should be at the "9 o'clock" position.

b. Apply sealant to the No. 1 bearing cap.

c. Apply a light coat of clean engine oil to the bolt threads and under the bolt head. Install the bearing caps to their proper position. Tighten the bolts evenly and in several passes to 12 ft. lbs. (16 Nm) in the proper sequence.

d. Position the exhaust camshaft on the head so that the alignment marks are at a 90 degrees angle from vertical. The mark should be at the "3 o'clock" posi-

Intake

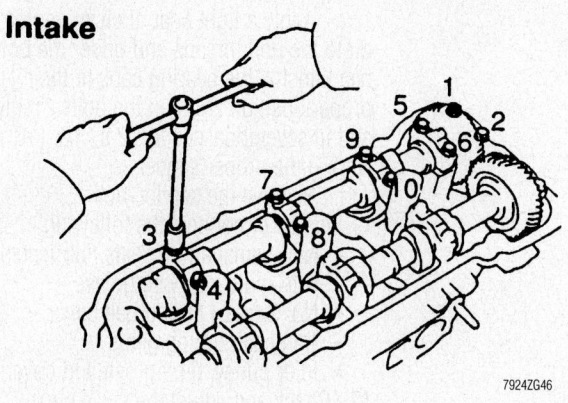

7924ZG46

Left intake camshaft bearing cap bolt loosening sequence—3.3L Highlander

Exhaust

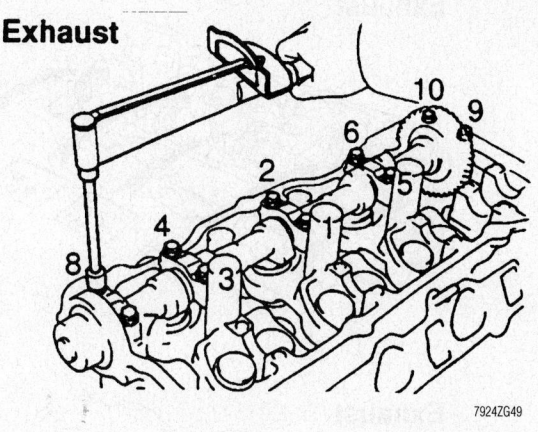

7924ZG49

Right exhaust camshaft bearing cap bolt tightening sequence—3.3L engines

Exhaust

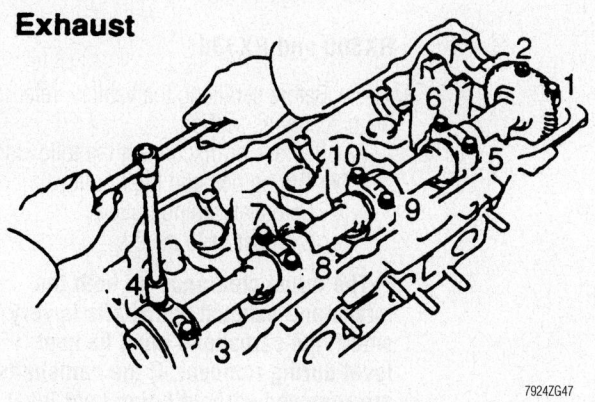

7924ZG47

Left side exhaust camshaft bearing cap bolt loosening sequence—3.3L Highlander

Intake

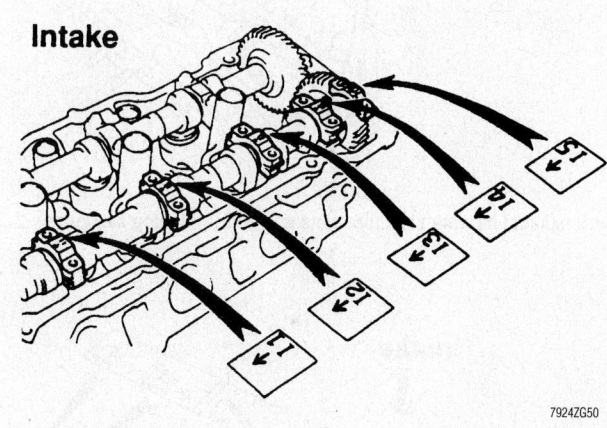

7924ZG50

Right intake bearing caps must be placed in their proper locations—3.3L Highlander

Exhaust

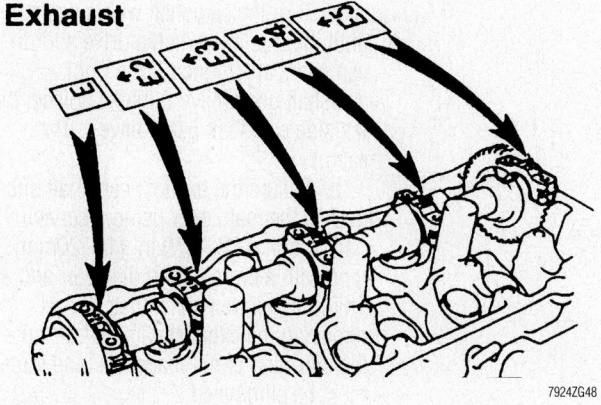

7924ZG48

Right exhaust bearing caps must be placed in their proper locations—3.3L Highlander

Intake

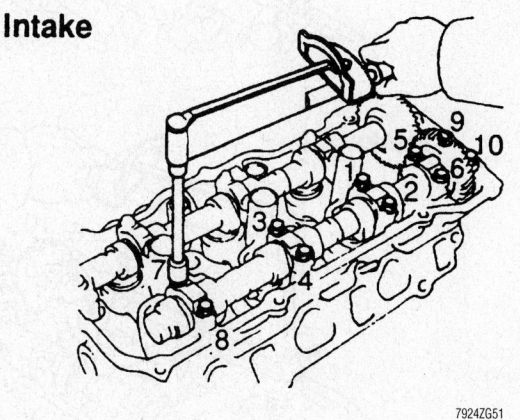

7924ZG51

Right intake camshaft bearing cap bolt tightening sequence—3.3L Highlander

Exhaust

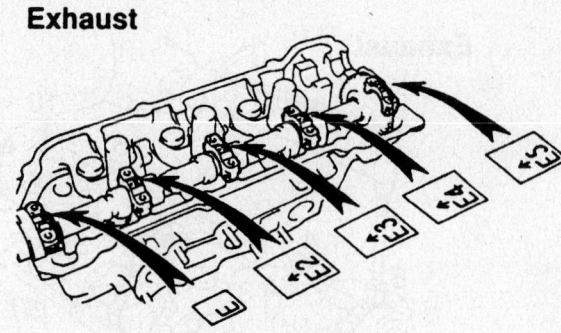

Exhaust

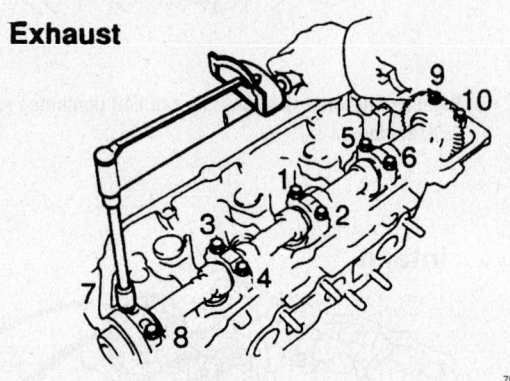

Left exhaust bearing caps locations and bolt tightening sequence—3.3L Highlander

Intake

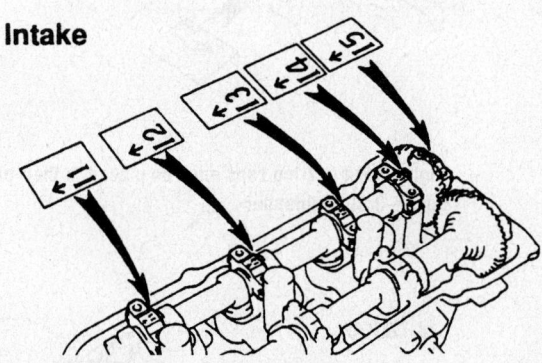

Intake

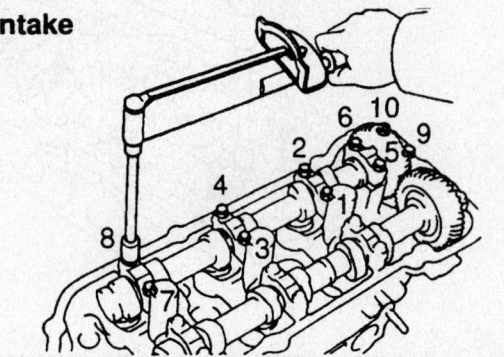

Left intake camshaft bearing cap locations and bolt tightening sequence—3.3L Highlander

tion and must align with the marks on the other gear.

 e. Apply a light coat of clean engine oil to the bolt threads and under the bolt head. Install the bearing caps to their proper position. Tighten the bolts evenly and in several passes to 12 ft. lbs. (16 Nm) in the proper sequence.

 f. Remove the service bolt.

11. Install or connect the following:
- New camshaft oil seals, lubricated with multi-purpose grease
- No. 3 (rear) timing belt cover
- Camshaft timing gears
- Idler pulley, timing belt and covers

12. Check and adjust the valve clearance.

13. Install the cylinder head (valve) covers.

14. Start the engine. Check the ignition timing.

15. Test drive the vehicle.

16. Check all fluid levels.

RX300 and RX330

1. Before servicing the vehicle, refer to the precautions section.

2. Remove or disconnect the following:
- Timing belt and idler pulley
- Camshaft timing pulleys
- Cylinder head covers

➡**The thrust clearance on both the intake and exhaust camshafts is very small; the camshafts must be kept level during removal. If the camshafts are removed without being kept level, the camshaft may be caught in the cylinder head, causing the head to break or the camshaft to seize.**

3. Remove the exhaust and intake camshafts from the right side cylinder head, as follows:

 a. Turn the camshaft with a wrench until the 2 pointed marks drive and driven gears are aligned. (The right camshaft gears have 2 marks apiece; the left side camshaft gears have 1 mark each.)

 b. Secure the exhaust camshaft sub-gear to the main gear using a service bolt. A bolt 0.63–0.79 in. (16–20mm) long with a 6mm thread diameter and a 1mm pitch is recommended. When removing the exhaust camshaft be sure the sub-gear is not loaded; all the force must be eliminated.

 c. Uniformly loosen and remove the exhaust camshaft bearing cap bolts in several passes and in the proper sequence. Remove the 8 bearing cap

Intake

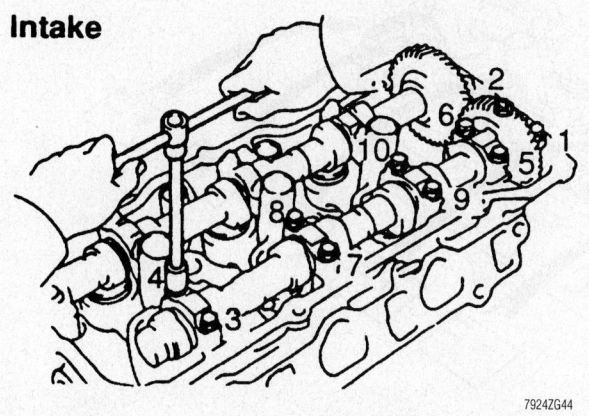

7924ZG44

Right intake camshaft bearing cap bolt loosening sequence

Exhaust

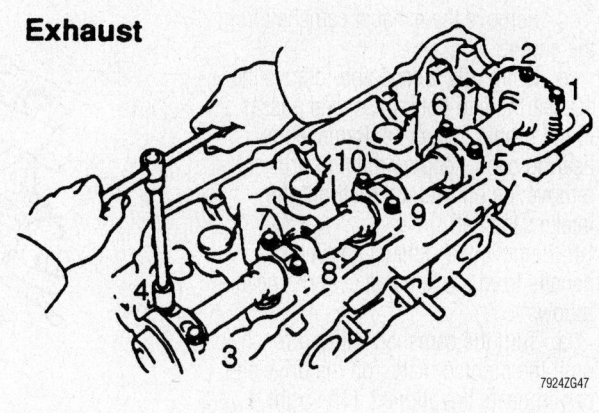

7924ZG47

Left side exhaust camshaft bearing cap bolt loosening sequence

Exhaust

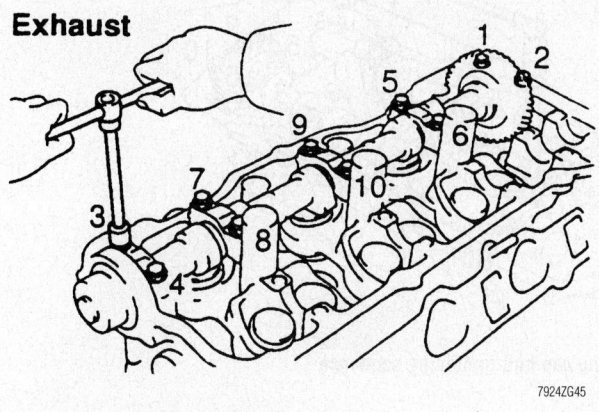

7924ZG45

Right side exhaust camshaft bearing cap bolt loosening sequence

Exhaust

7924ZG48

Right exhaust bearing caps must be placed in their proper locations

Intake

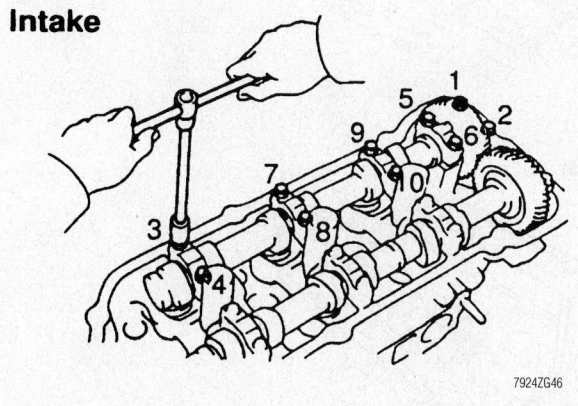

7924ZG46

Left intake camshaft bearing cap bolt loosening sequence

Exhaust

7924ZG49

Right exhaust camshaft bearing cap bolt tightening sequence

bolts and remove the caps, keeping them in the correct order.

d. Remove the exhaust camshaft from the engine.

e. Uniformly loosen and remove the 10 bearing cap bolts in several passes, in the proper sequence. Remove the bearing caps, keeping them in order, remove the oil seal, then lift out the intake camshaft.

4. Remove the exhaust and intake camshafts from the left side cylinder head, as follows:

a. Turn the camshaft with a wrench until the pointed marks on the drive and driven gears are aligned. (The right camshaft gears have 2 marks apiece; the left side camshaft gears have 1 mark each.)

b. Secure the exhaust camshaft sub-gear to the main gear using a service bolt. A bolt 16–20mm long with a 6mm thread diameter and a 1mm pitch is recommended. When removing the exhaust camshaft be sure the sub-gear is not loaded; all the force must be eliminated.

c. Uniformly loosen and remove the exhaust camshaft bearing cap bolts in several passes and in the proper sequence. Remove the 8 bearing cap bolts and remove the caps. Keep the caps in the correct order.

d. Remove the exhaust camshaft from the engine.

e. Uniformly loosen and remove the 10 bearing cap bolts in several passes, in the reverse order of the installation sequence. Remove the bearing caps, keeping them in order, remove the oil seal, then lift out the intake camshaft.

5. Remove the valve lifter shims and hydraulic lifters. Identify each lifter and shim as it is removed so it can be reinstalled in the same position. If the lifters are to be reused, store them upside down in a sealed container.

To install:

6. Install the valve lifters into their original positions and install the shims. Check valve clearance and replace the shims as necessary.

7. When reinstalling, remember that the camshafts must be handled carefully and kept straight and level to avoid damage.

8. Before installing the camshafts in either cylinder head, apply multi-purpose grease to each camshaft.

9. Install the right camshafts, as follows:

a. Position the intake camshaft on the head so that the alignment marks are at a 90 degrees angle from vertical. The mark should be at the "3 o'clock" position.

Intake

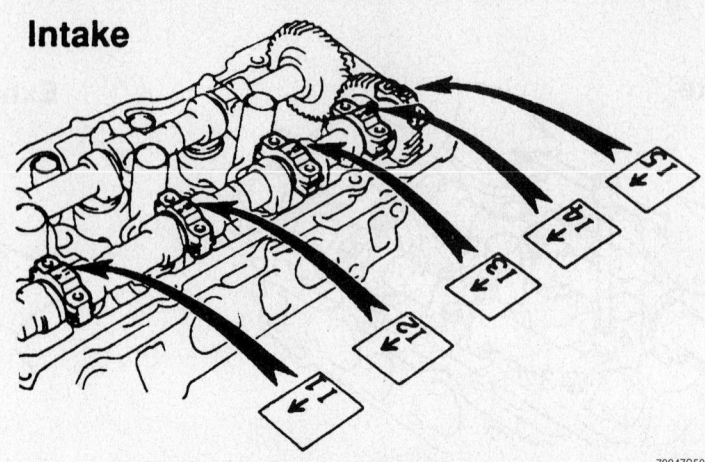

Right intake bearing caps must be placed in their proper locations

Intake

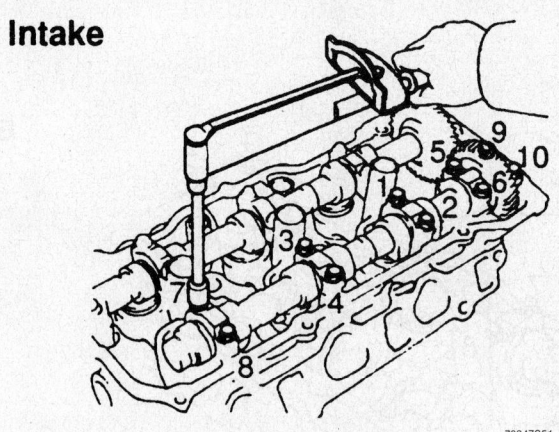

Right intake camshaft bearing cap bolt tightening sequence

Exhaust

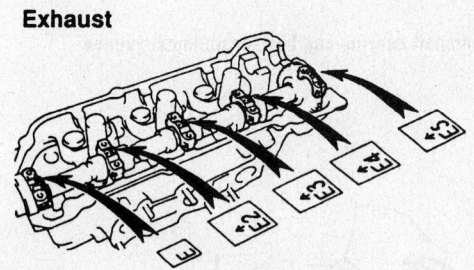

Exhaust

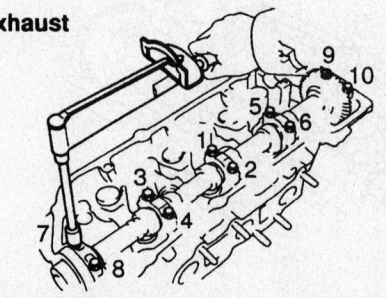

Left exhaust bearing caps locations and bolt tightening sequence

Intake

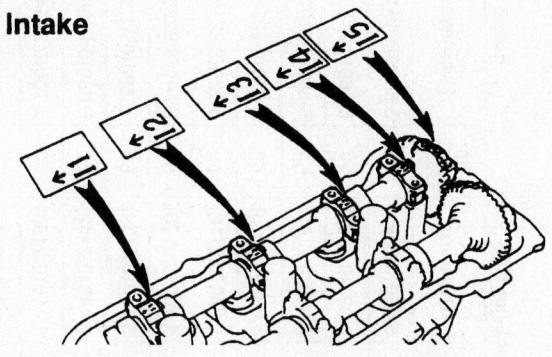

Intake

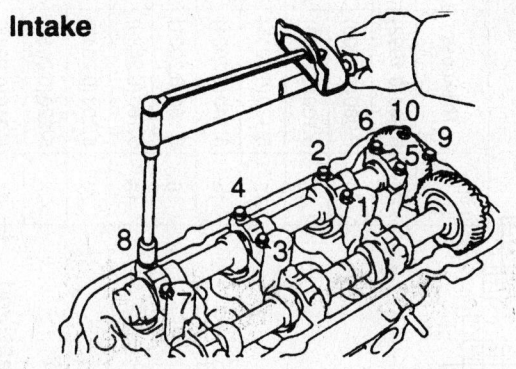

Left intake camshaft bearing cap locations and bolt tightening sequence

b. Apply sealant to the No. 1 bearing cap.

c. Apply a light coat of clean engine oil to the bolt threads and under the bolt head. Install the bearing caps to their proper position. Tighten the bolts evenly and in several passes to 12 ft. lbs. (16 Nm) in the proper sequence.

d. Position the exhaust camshaft on the head so that the alignment marks are at a 90 degrees angle from vertical. The mark should be at the "9 o'clock" position and must align with the marks on the other gear.

e. Apply a light coat of clean engine oil to the bolt threads and under the bolt head. Install the bearing caps to their proper position. Tighten the bolts evenly and in several passes to 12 ft. lbs. (16 Nm) in the proper sequence.

f. Remove the service bolt.

10. Install the left camshafts, as follows:

a. Position the intake camshaft on the head so that the alignment mark is at a 90 degrees angle from vertical. The mark should be at the "9 o'clock" position.

b. Apply sealant to the No. 1 bearing cap.

c. Apply a light coat of clean engine oil to the bolt threads and under the bolt

head. Install the bearing caps to their proper position. Tighten the bolts evenly and in several passes to 12 ft. lbs. (16 Nm) in the proper sequence.

d. Position the exhaust camshaft on the head so that the alignment marks are at a 90 degrees angle from vertical. The mark should be at the "3 o'clock" position and must align with the marks on the other gear.

e. Apply a light coat of clean engine oil to the bolt threads and under the bolt head. Install the bearing caps to their proper position. Tighten the bolts evenly and in several passes to 12 ft. lbs. (16 Nm) in the proper sequence.

f. Remove the service bolt.

11. Install or connect the following:
- New camshaft oil seals, lubricated with multi-purpose grease
- No. 3 (rear) timing belt cover
- Camshaft timing gears
- Idler pulley, timing belt and covers

12. Check and adjust the valve clearance.

13. Install the cylinder head (valve) covers.

14. Start the engine. Check the ignition timing.

15. Test drive the vehicle.

16. Check all fluid levels.

ADJUSTMENT

Highlander

2.4L ENGINE

➡**Adjust the valve clearance when the engine is cold.**

1. Before servicing the vehicle, refer to the precautions section.

2. Remove or disconnect the following:
- Negative battery cable. If equipped with an air bag, wait at least 90 seconds before proceeding.
- Right front wheel, splash shield and apron seal
- engine undercover

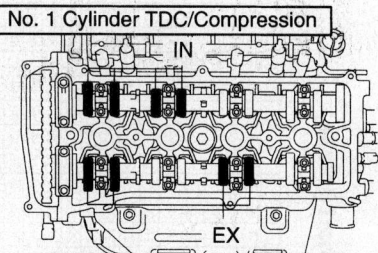

Check the clearance on these valves with the engine at No.1 TDC compression—2.4L Highlander

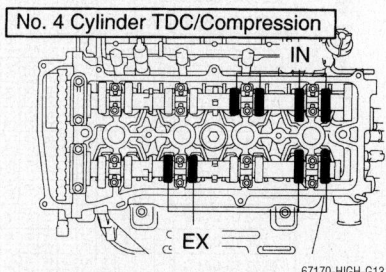

Check the clearance on these valves with the engine at No.4 TDC compression—2.4L Highlander

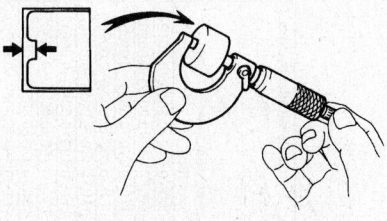

Measure the lifters at the point shown—2.4L Highlander

Valve Lifter Selection Chart (Intake)

New Lifter Thickness — mm (in.)

Lifter No.	Thickness	Lifter No.	Thickness	Lifter No.	Thickness
06	5.060 (0.1992)	30	5.300 (0.2087)	54	5.540 (0.2181)
08	5.080 (0.2000)	32	5.320 (0.2094)	56	5.560 (0.2189)
10	5.100 (0.2008)	34	5.340 (0.2102)	58	5.580 (0.2197)
12	5.120 (0.2016)	36	5.360 (0.2110)	60	5.600 (0.2205)
14	5.140 (0.2024)	38	5.380 (0.2118)	62	5.620 (0.2213)
16	5.160 (0.2031)	40	5.400 (0.2126)	64	5.640 (0.2220)
18	5.180 (0.2039)	42	5.420 (0.2134)	66	5.660 (0.2228)
20	5.200 (0.2047)	44	5.440 (0.2142)	68	5.680 (0.2236)
22	5.220 (0.2055)	46	5.460 (0.2150)	70	5.700 (0.2244)
24	5.240 (0.2063)	48	5.480 (0.2157)	72	5.720 (0.2252)
26	5.260 (0.2071)	50	5.500 (0.2165)	74	5.740 (0.2260)
28	5.280 (0.2079)	52	5.520 (0.2173)		

Intake valve clearance (Cold):
0.19 to 0.29 mm (0.008 to 0.011 in.)

EXAMPLE:

The 5.250 mm (0.2067 in.) lifter is installed, and the measured clearance is 0.400 mm (0.0157 in.). Replace the 5.250 mm (0.2067 in.) lifter with a new No. 42 lifter.

The valve lifter selection chart cross-references installed lifter thickness (mm/in., across the top, from 5.060 (0.1992) to 5.740 (0.2260)) against measured clearance (mm/in., down the left side), yielding the new lifter number to install:

Measure clearance mm (in.)
0.000–0.030 (0.0000–0.0012)
0.031–0.050 (0.0012–0.0020)
0.051–0.070 (0.0020–0.0028)
0.071–0.090 (0.0028–0.0035)
0.091–0.110 (0.0036–0.0043)
0.111–0.130 (0.0044–0.0051)
0.131–0.150 (0.0052–0.0059)
0.171–0.170 (0.0059–0.0067)
0.171–0.189 (0.0067–0.0074)
0.190–0.290 (0.0075–0.0114)
0.291–0.310 (0.0115–0.0122)
0.311–0.330 (0.0122–0.0130)
0.331–0.350 (0.0130–0.0138)
0.351–0.370 (0.0138–0.0146)
0.371–0.390 (0.0146–0.0154)
0.391–0.410 (0.0154–0.0161)
0.411–0.430 (0.0162–0.0169)
0.431–0.450 (0.0170–0.0177)
0.451–0.470 (0.0178–0.0185)
0.471–0.490 (0.0185–0.0193)
0.491–0.510 (0.0193–0.0201)
0.511–0.530 (0.0201–0.0209)
0.531–0.550 (0.0209–0.0217)
0.551–0.570 (0.0217–0.0224)
0.571–0.590 (0.0225–0.0232)
0.591–0.610 (0.0233–0.0240)
0.611–0.630 (0.0241–0.0248)
0.631–0.650 (0.0248–0.0256)
0.651–0.670 (0.0256–0.0264)
0.671–0.690 (0.0264–0.0272)
0.691–0.710 (0.0272–0.0280)
0.711–0.730 (0.0280–0.0287)
0.731–0.750 (0.0288–0.0295)
0.751–0.770 (0.0296–0.0303)
0.771–0.790 (0.0304–0.0311)
0.791–0.810 (0.0311–0.0319)
0.811–0.830 (0.0319–0.0327)
0.831–0.850 (0.0327–0.0335)
0.851–0.870 (0.0335–0.0343)
0.871–0.890 (0.0343–0.0350)
0.891–0.910 (0.0351–0.0358)
0.911–0.930 (0.0359–0.0366)

Intake valve lifter selection chart—2.4L Highlander

67170-HIGH-G15

Valve Lifter Selection Chart (Exhaust)

Measure clearance mm (in.) \ Installed lifter thickness mm (in.)	5.060 (0.1992)	5.080 (0.2000)	5.100 (0.2008)	5.120 (0.2016)	5.140 (0.2024)	5.160 (0.2031)	5.180 (0.2039)	5.200 (0.2047)	5.210 (0.2051)	5.220 (0.2055)	5.230 (0.2059)	5.240 (0.2063)	5.250 (0.2067)	5.260 (0.2071)	5.270 (0.2075)	5.280 (0.2079)	5.290 (0.2083)	5.300 (0.2087)	5.310 (0.2091)	5.320 (0.2094)	5.330 (0.2098)	5.340 (0.2102)	5.350 (0.2106)	5.360 (0.2110)	5.370 (0.2114)	5.380 (0.2118)	5.390 (0.2122)
0.000–0.030 (0.0000–0.0012)																							06	06	08	08	10
0.031–0.050 (0.0012–0.0020)																					06	06	08	08	10	10	12
0.051–0.070 (0.0020–0.0028)																			06	06	08	08	10	10	12	12	14
0.071–0.090 (0.0028–0.0035)																	06	06	08	08	10	10	12	12	14	14	16
0.091–0.110 (0.0036–0.0043)															06	06	08	08	10	10	12	12	14	14	16	16	18
0.111–0.130 (0.0044–0.0051)													06	06	08	08	10	10	12	12	14	14	16	16	18	18	20
0.131–0.150 (0.0052–0.0059)											06	06	08	08	10	10	12	12	14	14	16	16	18	18	20	20	22
0.151–0.170 (0.0059–0.0067)									06	06	08	08	10	10	12	12	14	14	16	16	18	18	20	20	22	22	24
0.171–0.190 (0.0067–0.0075)								06	08	08	10	10	12	12	14	14	16	16	18	18	20	20	22	22	24	24	26
0.191–0.210 (0.0075–0.0083)							06	08	10	10	12	12	14	14	16	16	18	18	20	20	22	22	24	24	26	26	28
0.211–0.230 (0.0083–0.0091)						06	08	10	12	12	14	14	16	16	18	18	20	20	22	22	24	24	26	26	28	28	30
0.231–0.250 (0.0091–0.0098)					06	08	10	12	14	14	16	16	18	18	20	20	22	22	24	24	26	26	28	28	30	30	32
0.251–0.270 (0.0099–0.0106)				06	08	10	12	14	16	16	18	18	20	20	22	22	24	24	26	26	28	28	30	30	32	32	34
0.271–0.290 (0.0107–0.0114)			06	08	10	12	14	16	18	18	20	20	22	22	24	24	26	26	28	28	30	30	32	32	34	34	36
0.291–0.299 (0.0115–0.0118)		06	08	10	12	14	16	18	20	20	22	22	24	24	26	26	28	28	30	30	32	32	34	34	36	36	38
0.300–0.400 (0.0118–0.0157)																											
0.401–0.420 (0.0158–0.0165)	12	14	16	18	20	22	24	26	26	28	28	30	30	32	32	34	34	36	36	38	38	40	40	42	42	44	44
0.421–0.440 (0.0166–0.0173)	14	16	18	20	22	24	26	28	28	30	30	32	32	34	34	36	36	38	38	40	40	42	42	44	44	46	46
0.441–0.460 (0.0174–0.0181)	16	18	20	22	24	26	28	30	30	32	32	34	34	36	36	38	38	40	40	42	42	44	44	46	46	48	48
0.461–0.480 (0.0181–0.0189)	18	20	22	24	26	28	30	32	32	34	34	36	36	38	38	40	40	42	42	44	44	46	46	48	48	50	50
0.481–0.500 (0.0189–0.0197)	20	22	24	26	28	30	32	34	34	36	36	38	38	40	40	42	42	44	44	46	46	48	48	50	50	52	52
0.501–0.520 (0.0197–0.0205)	22	24	26	28	30	32	34	36	36	38	38	40	40	42	42	44	44	46	46	48	48	50	50	52	52	54	54
0.521–0.540 (0.0205–0.0213)	24	26	28	30	32	34	36	38	38	40	40	42	42	44	44	46	46	48	48	50	50	52	52	54	54	56	56
0.541–0.560 (0.0213–0.0220)	26	28	30	32	34	36	38	40	40	42	42	44	44	46	46	48	48	50	50	52	52	54	54	56	56	58	58
0.561–0.580 (0.0221–0.0228)	28	30	32	34	36	38	40	42	42	44	44	46	46	48	48	50	50	52	52	54	54	56	56	58	58	60	60
0.581–0.600 (0.0229–0.0236)	30	32	34	36	38	40	42	44	44	46	46	48	48	50	50	52	52	54	54	56	56	58	58	60	60	62	62
0.601–0.620 (0.0237–0.0244)	32	34	36	38	40	42	44	46	46	48	48	50	50	52	52	54	54	56	56	58	58	60	60	62	62	64	64
0.621–0.640 (0.0244–0.0252)	34	36	38	40	42	44	46	48	48	50	50	52	52	54	54	56	56	58	58	60	60	62	62	64	64	66	66
0.641–0.660 (0.0252–0.0260)	36	38	40	42	44	46	48	50	50	52	52	54	54	56	56	58	58	60	60	62	62	64	64	66	66	68	68
0.661–0.680 (0.0260–0.0268)	38	40	42	44	46	48	50	52	52	54	54	56	56	58	58	60	60	62	62	64	64	66	66	68	68	70	70
0.681–0.700 (0.0268–0.0276)	40	42	44	46	48	50	52	54	54	56	56	58	58	60	60	62	62	64	64	66	66	68	68	70	70	72	72
0.701–0.720 (0.0276–0.0283)	42	44	46	48	50	52	54	56	56	58	58	60	60	62	62	64	64	66	66	68	68	70	70	72	72	74	74
0.721–0.740 (0.0284–0.0291)	44	46	48	50	52	54	56	58	58	60	60	62	62	64	64	66	66	68	68	70	70	72	72	74	74	74	74
0.741–0.760 (0.0292–0.0299)	46	48	50	52	54	56	58	60	60	62	62	64	64	66	66	68	68	70	70	72	72	74	74	74	74	74	74
0.761–0.780 (0.0300–0.0307)	48	50	52	54	56	58	60	62	62	64	64	66	66	68	68	70	70	72	72	74	74	74	74	74	74		
0.781–0.800 (0.0307–0.0315)	50	52	54	56	58	60	62	64	64	66	66	68	68	70	70	72	72	74	74	74	74	74	74				
0.801–0.820 (0.0315–0.0323)	52	54	56	58	60	62	64	66	66	68	68	70	70	72	72	74	74	74	74	74	74						
0.821–0.840 (0.0323–0.0331)	54	56	58	60	62	64	66	68	68	70	70	72	72	74	74	74	74	74	74								
0.841–0.860 (0.0331–0.0339)	56	58	60	62	64	66	68	70	70	72	72	74	74	74	74	74	74										
0.861–0.880 (0.0339–0.0346)	58	60	62	64	66	68	70	72	72	74	74	74	74	74	74												
0.881–0.900 (0.0347–0.0354)	60	62	64	66	68	70	72	74	74	74	74	74	74														
0.901–0.920 (0.0355–0.0362)	62	64	66	68	70	72	74	74	74	74	74																
0.921–0.940 (0.0363–0.0370)	64	66	68	70	72	74	74	74	74																		
0.941–0.960 (0.0370–0.0378)	66	68	70	72	74	74	74																				
0.961–0.980 (0.0378–0.0386)	68	70	72	74	74	74																					
0.981–1.000 (0.0386–0.0394)	70	72	74	74	74																						
1.001–1.020 (0.0394–0.0402)	72	74	74	74																							
1.021–1.040 (0.0402–0.0409)	74	74	74																								
1.041–1.060 (0.0410–0.0417)	74	74																									
1.061–1.080 (0.0418–0.0425)	74																										

Measure clearance mm (in.) \ Installed lifter thickness mm (in.)	5.400 (0.2126)	5.410 (0.2130)	5.420 (0.2134)	5.430 (0.2138)	5.440 (0.2142)	5.450 (0.2146)	5.460 (0.2150)	5.470 (0.2154)	5.480 (0.2157)	5.490 (0.2161)	5.500 (0.2165)	5.510 (0.2169)	5.520 (0.2173)	5.530 (0.2177)	5.540 (0.2181)	5.550 (0.2185)	5.560 (0.2189)	5.570 (0.2193)	5.580 (0.2197)	5.590 (0.2201)	5.600 (0.2205)	5.620 (0.2213)	5.640 (0.2220)	5.660 (0.2228)	5.680 (0.2236)	5.700 (0.2244)	5.720 (0.2252)	5.740 (0.2260)
0.000–0.030 (0.0000–0.0012)	10	12	12	14	14	16	16	18	18	20	20	22	22	24	24	26	26	28	28	30	30	32	34	36	38	40	42	44
0.031–0.050 (0.0012–0.0020)	12	14	14	16	16	18	18	20	20	22	22	24	24	26	26	28	28	30	30	32	32	34	36	38	40	42	44	46
0.051–0.070 (0.0020–0.0028)	14	16	16	18	18	20	20	22	22	24	24	26	26	28	28	30	30	32	32	34	34	36	38	40	42	44	46	48
0.071–0.090 (0.0028–0.0035)	16	18	18	20	20	22	22	24	24	26	26	28	28	30	30	32	32	34	34	36	36	38	40	42	44	46	48	50
0.091–0.110 (0.0036–0.0043)	18	20	20	22	22	24	24	26	26	28	28	30	30	32	32	34	34	36	36	38	38	40	42	44	46	48	50	52
0.111–0.130 (0.0044–0.0051)	20	22	22	24	24	26	26	28	28	30	30	32	32	34	34	36	36	38	38	40	40	42	44	46	48	50	52	54
0.131–0.150 (0.0052–0.0059)	22	24	24	26	26	28	28	30	30	32	32	34	34	36	36	38	38	40	40	42	42	44	46	48	50	52	54	56
0.151–0.170 (0.0059–0.0067)	24	26	26	28	28	30	30	32	32	34	34	36	36	38	38	40	40	42	42	44	44	46	48	50	52	54	56	58
0.171–0.190 (0.0067–0.0075)	26	28	28	30	30	32	32	34	34	36	36	38	38	40	40	42	42	44	44	46	46	48	50	52	54	56	58	60
0.191–0.210 (0.0075–0.0083)	28	30	30	32	32	34	34	36	36	38	38	40	40	42	42	44	44	46	46	48	48	50	52	54	56	58	60	62
0.211–0.230 (0.0083–0.0091)	30	32	32	34	34	36	36	38	38	40	40	42	42	44	44	46	46	48	48	50	50	52	54	56	58	60	62	64
0.231–0.250 (0.0091–0.0098)	32	34	34	36	36	38	38	40	40	42	42	44	44	46	46	48	48	50	50	52	52	54	56	58	60	62	64	66
0.251–0.270 (0.0099–0.0106)	34	36	36	38	38	40	40	42	42	44	44	46	46	48	48	50	50	52	52	54	54	56	58	60	62	64	66	68
0.271–0.290 (0.0107–0.0114)	36	38	38	40	40	42	42	44	44	46	46	48	48	50	50	52	52	54	54	56	56	58	60	62	64	66	68	70
0.291–0.299 (0.0115–0.0118)	38	40	40	42	42	44	44	46	46	48	48	50	50	52	52	54	54	56	56	58	58	60	62	64	66	68	70	72
0.300–0.400 (0.0118–0.0157)																												
0.401–0.420 (0.0158–0.0165)	46	46	48	48	50	50	52	52	54	54	56	56	58	58	60	60	62	62	64	64	66	68	70	72	74	74	74	
0.421–0.440 (0.0166–0.0173)	48	48	50	50	52	52	54	54	56	56	58	58	60	60	62	62	64	64	66	66	68	70	72	74	74	74		
0.441–0.460 (0.0174–0.0181)	50	50	52	52	54	54	56	56	58	58	60	60	62	62	64	64	66	66	68	68	70	72	74	74	74			
0.461–0.480 (0.0181–0.0189)	52	52	54	54	56	56	58	58	60	60	62	62	64	64	66	66	68	68	70	70	72	74	74	74				
0.481–0.500 (0.0189–0.0197)	54	54	56	56	58	58	60	60	62	62	64	64	66	66	68	68	70	70	72	72	74	74	74					
0.501–0.520 (0.0197–0.0205)	56	56	58	58	60	60	62	62	64	64	66	66	68	68	70	70	72	72	74	74	74	74						
0.521–0.540 (0.0205–0.0213)	58	58	60	60	62	62	64	64	66	66	68	68	70	70	72	72	74	74	76	76	74							
0.541–0.560 (0.0213–0.0220)	60	60	62	62	64	64	66	66	68	68	70	70	72	72	74	74	74	74	74	74								
0.561–0.580 (0.0221–0.0228)	62	62	64	64	66	66	68	68	70	70	72	72	74	74	74	74	74	74										
0.581–0.600 (0.0229–0.0236)	64	64	66	66	68	68	70	70	72	72	74	74	74	74	74	74												
0.601–0.620 (0.0237–0.0244)	66	66	68	68	70	70	72	72	74	74	74	74	74	74														
0.621–0.640 (0.0244–0.0252)	68	68	70	70	72	72	74	74	74	74	74	74																
0.641–0.660 (0.0252–0.0260)	70	70	72	72	74	74	74	74	74	74																		
0.661–0.680 (0.0260–0.0268)	72	72	74	74	74	74	74	74																				
0.681–0.700 (0.0268–0.0276)	74	74	74	74	74	74																						
0.701–0.720 (0.0276–0.0283)	74	74	74	74																								
0.721–0.740 (0.0284–0.0291)	74	74																										

New lifter thickness mm (in.)

Lifter No.	Thickness	Lifter No.	Thickness	Lifter No.	Thickness
06	5.060 (0.1992)	30	5.300 (0.2087)	54	5.540 (0.2181)
08	5.080 (0.2000)	32	5.320 (0.2094)	56	5.560 (0.2189)
10	5.100 (0.2008)	34	5.340 (0.2102)	58	5.580 (0.2197)
12	5.120 (0.2016)	36	5.360 (0.2110)	60	5.600 (0.2205)
14	5.140 (0.2024)	38	5.380 (0.2118)	62	5.620 (0.2213)
16	5.160 (0.2031)	40	5.400 (0.2126)	64	5.640 (0.2220)
18	5.180 (0.2039)	42	5.420 (0.2134)	66	5.660 (0.2228)
20	5.200 (0.2047)	44	5.440 (0.2142)	68	5.680 (0.2236)
22	5.220 (0.2055)	46	5.460 (0.2150)	70	5.700 (0.2244)
24	5.240 (0.2063)	48	5.480 (0.2157)	72	5.720 (0.2252)
26	5.260 (0.2071)	50	5.500 (0.2165)	74	5.740 (0.2260)
28	5.280 (0.2079)	52	5.520 (0.2173)		

Exhaust valve clearance (Cold):
0.30 to 0.40 mm (0.012 to 0.016 in.)
EXAMPLE:
The 5.340 mm (0.2102 in.) lifter is installed, and the measured clearance is 0.440 mm (0.0173 in.). Replace the 5.340 mm (0.2102 in.) lifter with a new No. 44 lifter.

67170-HIGH-G16

Exhaust valve lifter selection chart—2.4L Highlander

- Coil pack
- Air intake hoses
- Cylinder head cover

3. Place the No.1 piston at TDC compression. Check only those valves shown. Record the clearance. If out of clearance, the measurement will be used to calculate the adjusting shims.

4. Place the No.4 piston at TDC compression. Check only those valves shown. Record the clearance. If out of clearance, the measurement will be used to calculate the adjusting shims.

Clearance range is:
- Intake: 0.19–0.29mm (0.008–0.011 in.)
- Exhaust: 0.30–0.40mm (0.012–0.016 in.)

To adjust the valves:

5. Turn the crankshaft 1 complete revolution (360 degrees) clockwise and set the No.1 piston at TDC compression. Place matchmarks on the chain and camshaft sprocket.

6. Remove the tensioner.

7. Loosen the camshaft sprocket bolt.

8. Remove the exhaust camshaft bearing caps, raise the camshaft and remove the sprocket. Tie the chain out of the way.

9. Remove the camshaft.

10. Remove the lifters and keep them in order.

11. Measure the thickness of any lifter on which the clearance was out of range. Calculate the thickness of the necessary replacement lifter. Lifters are available in 0.020mm increments from 5.060mm to 5.740mm.

12. For Camshaft and Timing Chain installation, see the respective procedures in this section.

3.0L ENGINE

➡Adjust the valve clearance when the engine is cold.

1. Before servicing the vehicle, refer to the precautions section.

2. Remove or disconnect the following:
- Negative battery cable. If equipped with an air bag, wait at least 90 seconds before proceeding.
- Accelerator/throttle cable from the throttle linkage
- Air cleaner cover, air flow meter and air duct assembly
- V-bank cover
- Emission control valve set
- Air intake chamber
- Engine harness from the injectors and the ignition coils
- Ignition coils and keep them in order for reassembly
- Spark plugs

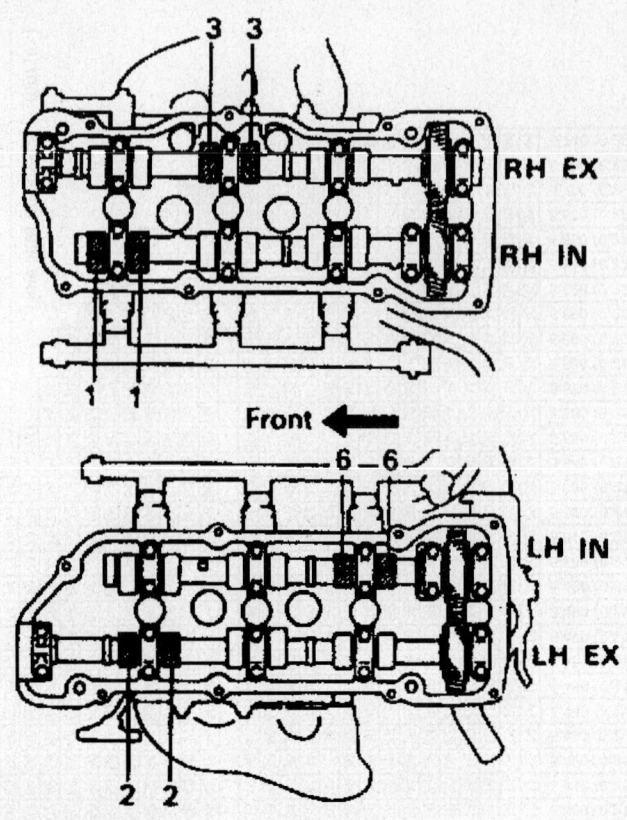

Adjust these valves during the 1st step—3.0L Highlander

7923VG65

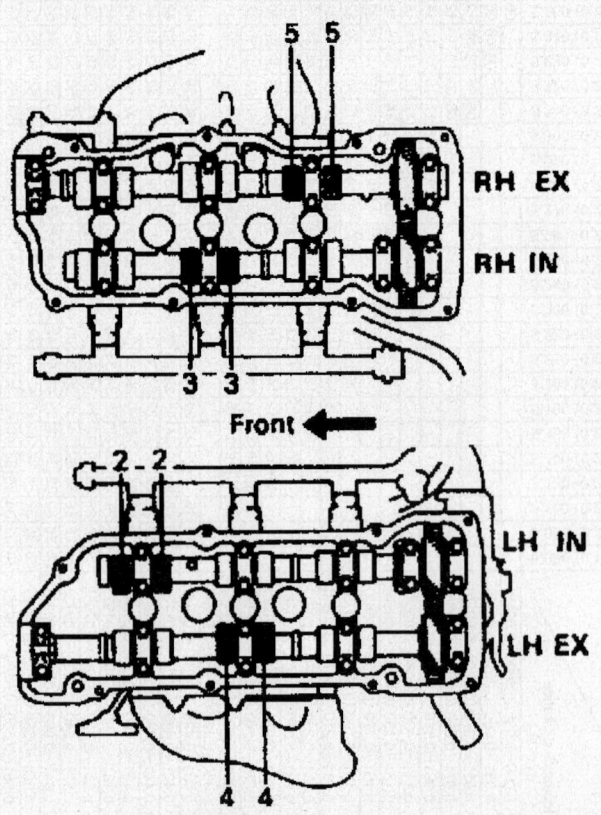

Adjust these valves during the 2nd step—3.0L Highlander

7923VG66

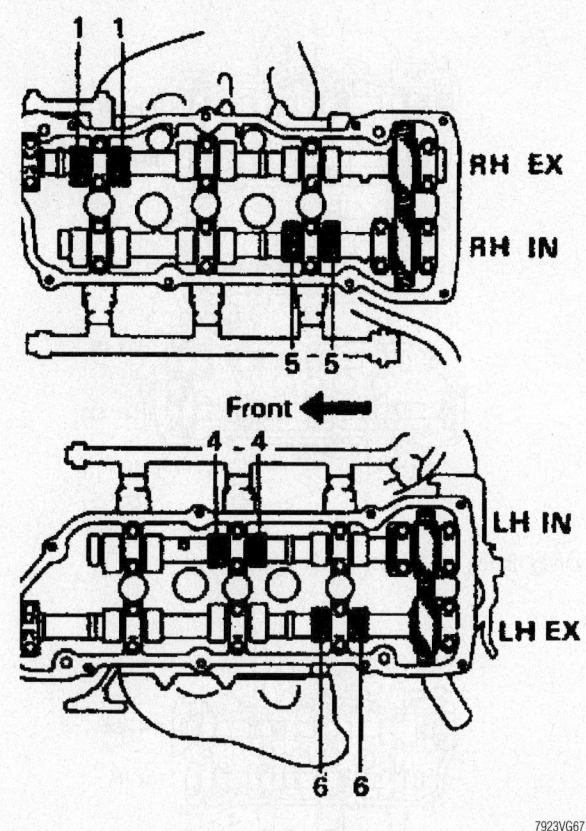

Adjust these valves during the 3rd step—3.0L Highlander

- Cylinder head covers

3. Turn the crankshaft pulley and align its groove with the timing mark **0** of the No. 1 timing cover.

4. Check that the valve lifters on the No. 1 intake are loose and the No. 1 exhaust are tight. If not, turn the crankshaft 1 complete revolution (360 degrees).

➡**All measurements should be written down. These recorded measurements will need to be used in conjunction with a mathematical formula to determine the thickness of the replacement shims.**

5. Measure the clearance between the valve lifters and the camshaft. Record the measurements on valves No. 1 and 6 intake; No. 2 and 3 exhaust.

  a. The intake valve clearance cold is 0.006–0.010 in. (0.15–0.25mm).

  b. The exhaust valve clearance cold is 0.010–0.014 in. (0.25–0.35mm).

6. Turn the crankshaft ⅔ of a revolution (240 degrees). Record the measurements on valves No. 2 and 3 intake; No. 4 and 5 exhaust.

7. Turn the crankshaft another ⅔ of a revolution. Record the measurements on valves No. 4 and 5 intake; No. 1 and 6 exhaust.

8. Remove the adjusting shim by turning the crankshaft to position the cam lobe of the camshaft in the up position on the valve to be adjusted. Using a small thin flat bladed tool, turn the valve lifter so that the notches are perpendicular to the camshaft. Press down the valve lifter with tool 09248-55010 part A. Place too 09248-55010 part B between the camshaft and the valve lifter; remove part A.

9. Remove the adjusting shim with a magnet and a small screwdriver.

10. Determine the replacement adjusting shim size by either using the charts or the following formulas:

- Intake: $N = T + (A - 0.008$ in./0.020mm$)$
- Exhaust: $N = T + (A - 0.012$ in./0.30mm$)$
- T = Thickness of removed shim
- A = Measured valve clearance
- N = Thickness of new shim

11. Select a new shim with a thickness as close as possible to the calculated value. Install the new replacement shim.

➡**Shims are available in 17 sizes in increments of 0.0020 in. (0.050mm), from 0.0984 in. (2.500mm) to 0.1299 in. (3.300mm).**

12. Recheck the valve clearance.

13. Install or connect the following:

- Cylinder head covers

- Spark plugs and the ignition coils
- Engine wiring harness to the injectors and the coils
- Intake chamber
- Emission control valve set
- V-bank cover
- Air flow meter, air duct and air cleaner cover
- Negative battery cable

3.3L ENGINE

➡**Adjust the valve clearance when the engine is cold.**

1. Before servicing the vehicle, refer to the precautions section.

2. Remove or disconnect the following:

- Negative battery cable. If equipped with an air bag, wait at least 90 seconds before proceeding.
- Accelerator/throttle cable from the throttle linkage
- Air cleaner cover, air flow meter and air duct assembly
- V-bank cover
- Emission control valve set
- Air intake chamber
- Engine harness from the injectors and the ignition coils
- Ignition coils and keep them in order for reassembly
- Spark plugs
- Cylinder head covers

3. Turn the crankshaft pulley and align its groove with the timing mark **0** of the No. 1 timing cover.

4. Check that the valve lifters on the No. 1 intake are loose and the No. 1 exhaust are tight. If not, turn the crankshaft 1 complete revolution (360 degrees).

➡**All measurements should be written down. These recorded measurements will need to be used in conjunction with a mathematical formula to determine the thickness of the replacement shims.**

5. Measure the clearance between the valve lifters and the camshaft. Record the

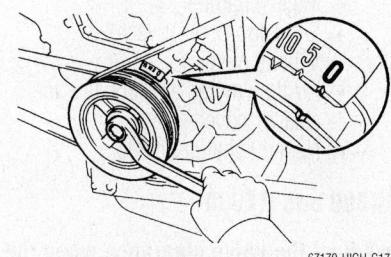

Turn the crankshaft pulley and align its groove with the timing mark 0 of the No. 1 timing cover—3.3L Highlander

measurements on valves No. 1 and 6 intake; No. 2 and 3 exhaust.

  a. The intake valve clearance cold is 0.006–0.010 in. (0.15–0.25mm).

  b. The exhaust valve clearance cold is 0.010–0.014 in. (0.25–0.35mm).

6. Turn the crankshaft ⅔ of a revolution (240 degrees). Record the measurements on valves No. 2 and 3 intake; No. 4 and 5 exhaust.

7. Turn the crankshaft another ⅔ of a revolution. Record the measurements on valves No. 4 and 5 intake; No. 1 and 6 exhaust.

8. Remove the adjusting shim by turning the crankshaft to position the cam lobe of the camshaft in the up position on the valve to be adjusted. Using a small thin flat bladed tool, turn the valve lifter so that the notches are perpendicular to the camshaft. Press down the valve lifter with tool 09248-55010 part A. Place too 09248-55010 part B between the camshaft and the valve lifter; remove part A.

9. Remove the adjusting shim with a magnet and a small screwdriver.

10. Determine the replacement adjusting shim size by either using the charts or the following formulas:

- Intake: N = T + (A—0.008 in./0.020mm)
- Exhaust: N = T + (A—0.012 in./0.30mm)
- T = Thickness of removed shim
- A = Measured valve clearance
- N = Thickness of new shim

11. Select a new shim with a thickness as close as possible to the calculated value. Install the new replacement shim.

➡Shims are available in 17 sizes in increments of 0.0020 in. (0.050mm), from 0.0984 in. (2.500mm) to 0.1299 in. (3.300mm).

12. Recheck the valve clearance.
13. Install or connect the following:
- Cylinder head covers
- Spark plugs and the ignition coils
- Engine wiring harness to the injectors and the coils
- Intake chamber
- Emission control valve set
- V-bank cover
- Air flow meter, air duct and air cleaner cover
- Negative battery cable

RX300 and RX330

➡Adjust the valve clearance when the engine is cold.

1. Before servicing the vehicle, refer to the precautions section.

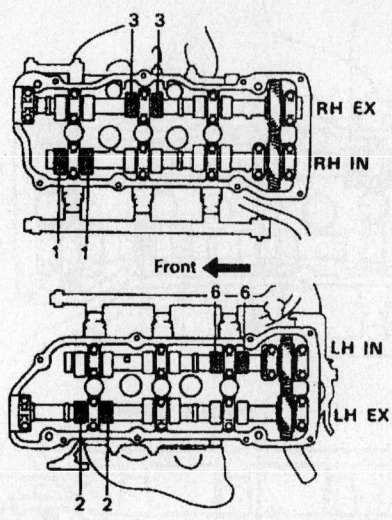

Adjust these valves during the 1st step—3.3L Highlander

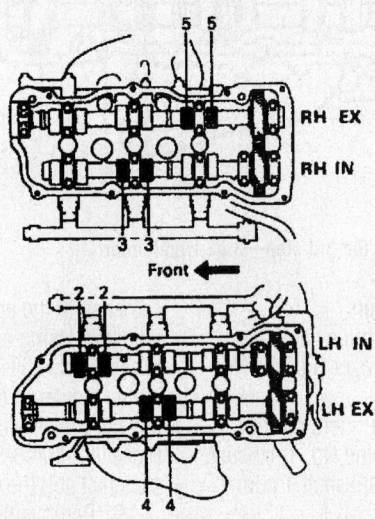

Adjust these valves during the 2nd step—3.3L Highlander

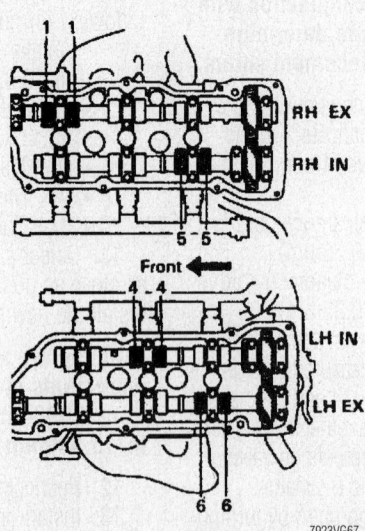

Adjust these valves during the 3rd step—3.3L Highlander

Adjusting Shim Selection Chart (Intake)

New shim thickness mm (in.)

Shim No.	Thickness	Shim No.	Thickness
1	2.500 (0.0984)	10	2.950 (0.1161)
2	2.550 (0.1004)	11	3.000 (0.1181)
3	2.600 (0.1024)	12	3.050 (0.1201)
4	2.650 (0.1043)	13	3.100 (0.1220)
5	2.700 (0.1063)	14	3.150 (0.1240)
6	2.750 (0.1083)	15	3.200 (0.1260)
7	2.800 (0.1102)	16	3.250 (0.1280)
8	2.850 (0.1122)	17	3.300 (0.1299)
9	2.900 (0.1142)		

HINT:
A shim's thickness is written on its face in millimeters.

Intake valve clearance (Cold):
0.15 to 0.25 mm (0.006 to 0.010 in.)

EXAMPLE:
The 2.800 mm (0.1102 in.) shim is installed, and the measured clearance is 0.450 mm (0.0177 in.). Replace the 2.800 mm (0.1102 in.) shim with a new No. 12 shim.

Intake valve shim selection chart—3.3L Highlander

67170-HIGH-G18

Adjusting Shim Selection Chart (Exhaust)

Exhaust valve shim selection chart—3.3L Highlander

New shim thickness

Shim No.	Thickness mm (in.)	Shim No.	Thickness mm (in.)
1	2.500 (0.0984)	10	2.950 (0.1161)
2	2.550 (0.1004)	11	3.000 (0.1181)
3	2.600 (0.1024)	12	3.050 (0.1201)
4	2.650 (0.1043)	13	3.100 (0.1220)
5	2.700 (0.1063)	14	3.150 (0.1240)
6	2.750 (0.1083)	15	3.200 (0.1260)
7	2.800 (0.1102)	16	3.250 (0.1280)
8	2.850 (0.1122)	17	3.300 (0.1299)
9	2.900 (0.1142)		

HINT:
A shim's thickness is written on its face in millimeters.

Exhaust valve clearance (Cold):
0.25 to 0.35 mm (0.010 to 0.014 in.)

EXAMPLE:
The 2.800 mm (0.1102 in.) shim is installed, and the measured clearance is 0.450 mm (0.0177 in.). Replace the 2.800 mm (0.1102 in.) shim with a new No. 10 shim.

Measured clearance mm (in.) — row ranges (left axis):

0.000 – 0.020 (0.0000 – 0.0008); 0.021 – 0.040 (0.0008 – 0.0016); 0.041 – 0.060 (0.0016 – 0.0024); 0.061 – 0.080 (0.0024 – 0.0031); 0.081 – 0.100 (0.0032 – 0.0039); 0.101 – 0.120 (0.0040 – 0.0047); 0.121 – 0.140 (0.0048 – 0.0055); 0.141 – 0.160 (0.0056 – 0.0063); 0.161 – 0.180 (0.0063 – 0.0071); 0.181 – 0.200 (0.0071 – 0.0079); 0.201 – 0.220 (0.0079 – 0.0087); 0.221 – 0.240 (0.0087 – 0.0094); 0.241 – 0.249 (0.0095 – 0.0098); 0.250 – 0.350 (0.0098 – 0.0138); 0.351 – 0.360 (0.0138 – 0.0142); 0.361 – 0.380 (0.0142 – 0.0150); 0.381 – 0.400 (0.0150 – 0.0157); 0.401 – 0.420 (0.0158 – 0.0165); 0.421 – 0.440 (0.0166 – 0.0173); 0.441 – 0.460 (0.0174 – 0.0181); 0.461 – 0.480 (0.0181 – 0.0189); 0.481 – 0.500 (0.0189 – 0.0197); 0.501 – 0.520 (0.0197 – 0.0205); 0.521 – 0.540 (0.0205 – 0.0213); 0.541 – 0.560 (0.0213 – 0.0220); 0.561 – 0.580 (0.0221 – 0.0228); 0.581 – 0.600 (0.0229 – 0.0236); 0.601 – 0.620 (0.0237 – 0.0244); 0.621 – 0.640 (0.0244 – 0.0252); 0.641 – 0.660 (0.0252 – 0.0260); 0.661 – 0.680 (0.0260 – 0.0268); 0.681 – 0.700 (0.0268 – 0.0276); 0.701 – 0.720 (0.0276 – 0.0283); 0.721 – 0.740 (0.0284 – 0.0291); 0.741 – 0.760 (0.0292 – 0.0299); 0.761 – 0.780 (0.0300 – 0.0307); 0.781 – 0.800 (0.0307 – 0.0315); 0.801 – 0.820 (0.0315 – 0.0323); 0.821 – 0.840 (0.0323 – 0.0331); 0.841 – 0.860 (0.0331 – 0.0339); 0.861 – 0.880 (0.0339 – 0.0346); 0.881 – 0.900 (0.0347 – 0.0354); 0.901 – 0.920 (0.0354 – 0.0362); 0.921 – 0.940 (0.0362 – 0.0370); 0.941 – 0.960 (0.0370 – 0.0378); 0.961 – 0.980 (0.0378 – 0.0386); 0.981 – 1.000 (0.0386 – 0.0394); 1.001 – 1.020 (0.0394 – 0.0402); 1.021 – 1.040 (0.0402 – 0.0409); 1.041 – 1.060 (0.0410 – 0.0417); 1.061 – 1.080 (0.0418 – 0.0425); 1.081 – 1.100 (0.0426 – 0.0433); 1.101 – 1.120 (0.0433 – 0.0441); 1.121 – 1.140 (0.0441 – 0.0449); 1.141 – 1.150 (0.0449 – 0.0453)

Installed shim thickness mm (in.) — column headings (top axis), left to right:

2.500 (0.0984); 2.520 (0.0992); 2.540 (0.1000); 2.560 (0.1008); 2.580 (0.1016); 2.600 (0.1024); 2.620 (0.1031); 2.640 (0.1039); 2.650 (0.1043); 2.660 (0.1047); 2.670 (0.1051); 2.680 (0.1055); 2.690 (0.1059); 2.700 (0.1063); 2.710 (0.1067); 2.720 (0.1071); 2.730 (0.1075); 2.740 (0.1079); 2.750 (0.1083); 2.760 (0.1087); 2.770 (0.1091); 2.780 (0.1094); 2.790 (0.1098); 2.800 (0.1102); 2.810 (0.1106); 2.820 (0.1110); 2.830 (0.1114); 2.840 (0.1118); 2.850 (0.1122); 2.860 (0.1126); 2.870 (0.1130); 2.880 (0.1134); 2.890 (0.1138); 2.900 (0.1142); 2.910 (0.1146); 2.920 (0.1150); 2.930 (0.1154); 2.940 (0.1157); 2.950 (0.1161); 2.960 (0.1165); 2.970 (0.1169); 2.980 (0.1173); 2.990 (0.1177); 3.000 (0.1181); 3.010 (0.1185); 3.020 (0.1189); 3.030 (0.1193); 3.040 (0.1197); 3.050 (0.1201); 3.060 (0.1205); 3.080 (0.1213); 3.100 (0.1220); 3.120 (0.1228); 3.140 (0.1236); 3.150 (0.1240); 3.160 (0.1244); 3.180 (0.1252); 3.200 (0.1260); 3.220 (0.1268); 3.240 (0.1276); 3.250 (0.1280); 3.260 (0.1283); 3.280 (0.1291); 3.300 (0.1299)

2. Remove or disconnect the following:
- Negative battery cable. If equipped with an air bag, wait at least 90 seconds before proceeding.
- Accelerator/throttle cable from the throttle linkage
- Air cleaner cover, air flow meter and air duct assembly
- V-bank cover
- Emission control valve set
- Air intake chamber
- Engine harness from the injectors and the ignition coils
- Ignition coils and keep them in order for reassembly
- Spark plugs
- Cylinder head covers

3. Turn the crankshaft pulley and align its groove with the timing mark **0** of the No. 1 timing cover.

4. Check that the valve lifters on the No. 1 intake are loose and the No. 1 exhaust are tight. If not, turn the crankshaft 1 complete revolution (360 degrees).

➡**All measurements should be written down. These recorded measurements will need to be used in conjunction with a mathematical formula to determine the thickness of the replacement shims.**

5. Measure the clearance between the valve lifters and the camshaft. Record the measurements on valves No. 1 and 6 intake; No. 2 and 3 exhaust.
   a. The intake valve clearance cold is 0.006–0.010 in. (0.15–0.25mm).
   b. The exhaust valve clearance cold is 0.010–0.014 in. (0.25–0.35mm).

6. Turn the crankshaft ⅔ of a revolution (240 degrees). Record the measurements on valves No. 2 and 3 intake; No. 4 and 5 exhaust.

7. Turn the crankshaft another ⅔ of a revolution. Record the measurements on valves No. 4 and 5 intake; No. 1 and 6 exhaust.

8. Remove the adjusting shim by turning the crankshaft to position the cam lobe of the camshaft in the up position on the valve to be adjusted. Using a small thin flat bladed tool, turn the valve lifter so that the notches are perpendicular to the camshaft. Press down the valve lifter with tool 09248-55010 part A. Place too 09248-55010 part B between the camshaft and the valve lifter; remove part A.

9. Remove the adjusting shim with a magnet and a small screwdriver.

10. Determine the replacement adjusting shim size by either using the charts or the following formulas:

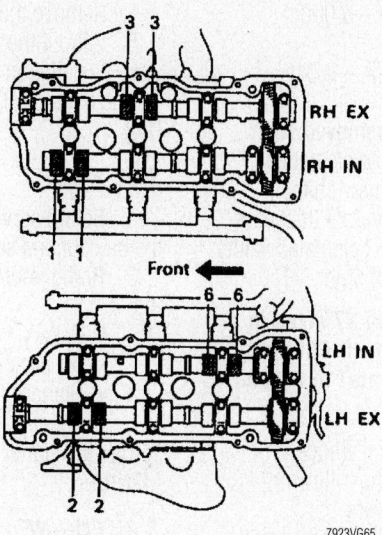

Adjust these valves during the 1st step—RX300 and RX330

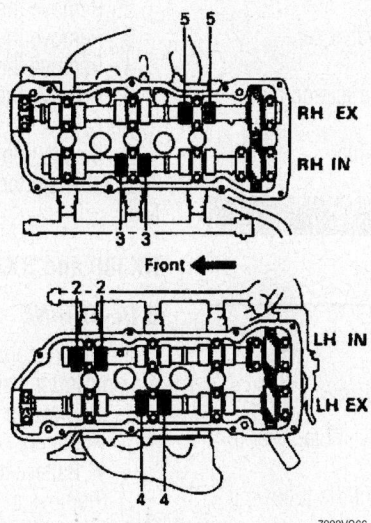

Adjust these valves during the 2nd step—RX300 and RX330

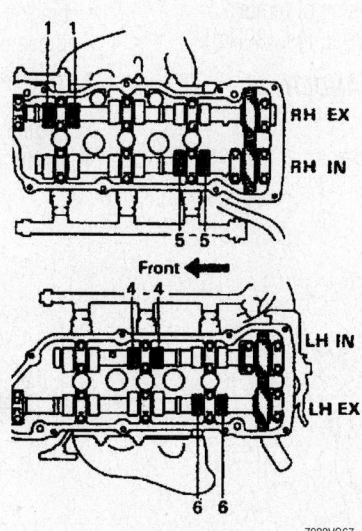

Adjust these valves during the 3rd step—RX300 and RX330

- Intake: N = T + (A—0.008 in./0.020mm)
- Exhaust: N = T + (A—0.012 in./0.30mm)
- T = Thickness of removed shim
- A = Measured valve clearance
- N = Thickness of new shim

11. Select a new shim with a thickness as close as possible to the calculated value. Install the new replacement shim.

➡ Shims are available in 17 sizes in increments of 0.0020 in. (0.050mm), from 0.0984 in. (2.500mm) to 0.1299 in. (3.300mm).

12. Recheck the valve clearance.
13. Install or connect the following:
- Cylinder head covers
- Spark plugs and the ignition coils
- Engine wiring harness to the injectors and the coils
- Intake chamber
- Emission control valve set
- V-bank cover
- Air flow meter, air duct and air cleaner cover
- Negative battery cable

Starter Motor

REMOVAL & INSTALLATION

Highlander

2002—2.4L HIGHLANDER

1. Before servicing the vehicle, refer to the precautions section.
2. Remove or disconnect the following:
- Air cleaner
- Starter wiring
- Starter

3. Installation is the reverse of removal. Torque the starter bolts to 29 ft. lbs. (39Nm).

2003–06—2.4L HIGHLANDER

1. Remove the battery.
2. Remove the battery tray.
3. Disconnect the starter wiring.

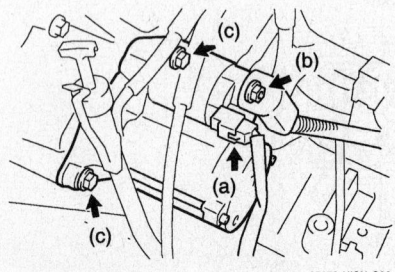

67170-HIGH-G20

Starter mounting—2.4L Highlander

4. Remove the mounting bolts.
5. Lift out the starter.
6. Installation is the reverse of removal. Torque the mounting bolts to 27 ft. lbs. (37 Nm).

3.0L ENGINE

1. Before servicing the vehicle, refer to the precautions section.
2. Remove or disconnect the following:
- Battery
- Air cleaner
- Starter

3. Installation is the reverse of removal. Torque the starter bolts to 31 ft. lbs. (42Nm).

3.3L ENGINE

1. Before servicing the vehicle, refer to the precautions section.
2. Remove the battery.
3. Remove the battery tray.
4. Remove the wiring from the starter.
5. Remove the 2 bolts and lower the starter.
6. Installation is the reverse of removal. Torque the starter bolts to 27 ft. lbs. (37 Nm).

RX300 and RX330

3.0L ENGINE

1. Before servicing the vehicle, refer to the precautions section.
2. Remove or disconnect the following:
- Battery
- Battery tray

3. Remove or disconnect the cruise control actuator, if equipped, as follows:
- Actuator connector and clamp
- 3 bolts and the actuator with the bracket

4. Remove or disconnect the following:
- Automatic transaxle shift control cable
- Engine wiring
- Starter electrical connectors

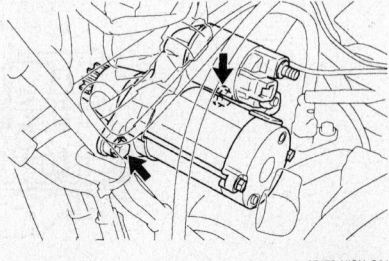

67170-HIGH-G21

Starter mounting bolt locations—3.3L Highlander

- Both bolts, shift control cable clamp and the starter

To install:

5. Install or connect the following:
- Starter and the shift control cable clamp. Tighten the bolts to 27 ft. lbs. (37 Nm).
- Starter electrical connectors
- Engine wiring
- Automatic transaxle shift control cable

6. Install or connect the following, if equipped with cruise control:
- 3 bolts and the actuator with the bracket
- Actuator connector and clamp

7. Install or connect the following:
- Battery tray
- Battery

3.3L ENGINE

1. Before servicing the vehicle, refer to the precautions section.
2. Remove the air cleaner assembly and inlet tubes.
3. Remove the air cleaner bracket.
4. Remove the wiring from the starter.
5. Remove the 2 bolts and lower the starter.
6. Installation is the reverse of removal. Torque the starter bolts to 26 ft. lbs. (37 Nm).

Oil Pan

REMOVAL & INSTALLATION

Highlander

2.4L ENGINE

1. Before servicing the vehicle, refer to the precautions section.
2. Remove or disconnect the following:
- Engine undercover
- Engine oil
- Oil pan bolts, nuts and pan

3. Installation is the reverse of removal. Torque the 12 bolts and 2 nuts to 80 inch lbs. (9Nm).

3.0L ENGINE

1. Before servicing the vehicle, refer to the precautions section.
2. Remove or disconnect the following:
- Right front wheel
- Fender apron seal
- Engine undercover

3. Drain the engine oil from the engine.
4. Remove or disconnect the following:
- Front exhaust pipe

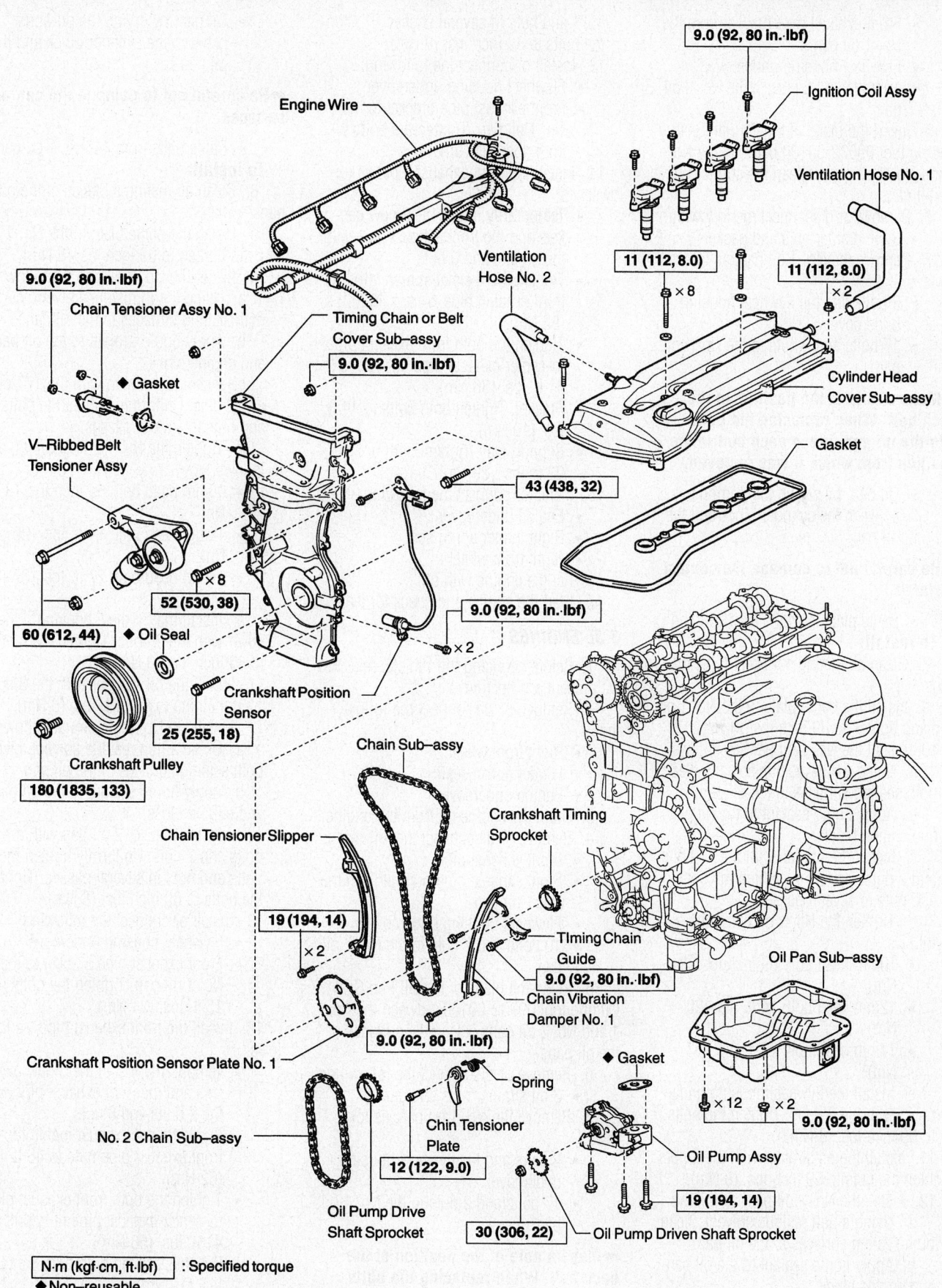

9.0 (92, 80 in.·lbf)
Ignition Coil Assy
Engine Wire
Ventilation Hose No. 1
Ventilation Hose No. 2
11 (112, 8.0) ×8
11 (112, 8.0) ×2
9.0 (92, 80 in.·lbf)
Chain Tensioner Assy No. 1
Timing Chain or Belt Cover Sub–assy
Cylinder Head Cover Sub–assy
◆ Gasket
9.0 (92, 80 in.·lbf)
Gasket
V–Ribbed Belt Tensioner Assy
43 (438, 32)
52 (530, 38) ×8
9.0 (92, 80 in.·lbf) ×2
60 (612, 44)   ◆ Oil Seal
Crankshaft Position Sensor
25 (255, 18)
Crankshaft Pulley
180 (1835, 133)
Chain Sub–assy
Crankshaft Timing Sprocket
Chain Tensioner Slipper
19 (194, 14) ×2
Timing Chain Guide
9.0 (92, 80 in.·lbf)
Chain Vibration Damper No. 1
Oil Pan Sub–assy
×12 ×2
9.0 (92, 80 in.·lbf)
9.0 (92, 80 in.·lbf)
Crankshaft Position Sensor Plate No. 1
Spring
◆ Gasket
No. 2 Chain Sub–assy
Chin Tensioner Plate
12 (122, 9.0)
Oil Pump Assy
Oil Pump Drive Shaft Sprocket
30 (306, 22)   Oil Pump Driven Shaft Sprocket
19 (194, 14)

N·m (kgf·cm, ft·lbf) : Specified torque
◆ Non–reusable

Oil pan, pump and related parts—2.4L Highlander

67170-HIGH-G22

- Front exhaust pipe bracket from the No. 1 oil pan
- Flywheel housing undercover
- 10 bolts and 2 nuts to the No. 2 oil pan

5. Insert the blade of the Oil Pan Seal Cutting tool 09032-00100 between the No. 1 and No. 2 oil pans. Clean the surfaces of the oil pans.

6. Remove or disconnect the following:
- 3 oil strainer nuts and gasket

7. Remove the No. 1 oil pan, as follows:

- 2 bolts and the flywheel housing undercover
- 17 bolts and 2 nuts to the No. 1 oil pan

➡**Make a note of the position of the each bolt. When replacing the bolts into the oil pan, place each bolt in the position from which it was removed.**

- Oil pan, by prying the portions between the cylinder block and the oil pan

➡**Be careful not to damage the contact surfaces.**

- Baffle plate from the No. 1 oil pan

To install:

8. Clean all mating surfaces of the oil pans.

9. Install the baffle plate to the No. 1 oil pan and tighten to 69 inch lbs. (8 Nm).

10. Install the No. 1 oil pan, as follows:

a. Using a non residue solvent, clean both sealing surfaces to the oil pan.

b. Apply liquid sealant to the oil pan and engine block.

c. Install the oil pan with the 17 bolts and 2 nuts. Uniformly tighten the bolts and nuts in several passes.

d. Tighten the No. 1 oil pan bolts, as follows:

- 10mm head bolt: 69 inch lbs. (8 Nm)
- 12mm head bolt: 14 ft. lbs. (20 Nm)
- 14mm head bolt: 27 ft. lbs. (37 Nm)

e. Install the flywheel housing undercover with the 2 bolts. Tighten the bolts to 69 inch lbs. (8 Nm).

11. Install the oil strainer with the 3 nuts. Tighten the nuts to 69 inch lbs. (8 Nm).

12. Install the No. 2 oil pan, as follows:

a. Using a non residue solvent, clean both sealing surfaces to the oil pan.

b. Apply liquid sealant to the oil pan and engine block.

c. Install the No. 2 oil pan with the 10 bolts and 2 nuts. Uniformly tighten the

bolts and nuts in several passes. Tighten the bolts to 69 inch lbs. (8 Nm).

13. Install or connect the following:
- Flywheel housing undercover
- Front exhaust pipe bracket to the No. 1 oil pan. Tighten the bolts to 15 ft. lbs. (21 Nm).

14. Install the front exhaust pipe, as follows:

- Temporarily install the 3 new gaskets and the front exhaust pipe with the 2 bolts and 6 nuts
- Tighten the 4 exhaust manifolds-to-front exhaust pipe nuts to 46 ft. lbs. (62 Nm).
- Tighten the both front exhaust pipe-to-center exhaust pipe nuts/bolts to 41 ft. lbs. (56 Nm).
- Bracket. Tighten both bolts to 14 ft. lbs. (19 Nm).
- Support stay. Tighten both bolts to 22 ft. lbs. (29 Nm).

15. Install or connect the following:
- Engine undercover
- Right fender apron seal
- Right front wheel

16. Fill the engine with oil.

17. Start the engine and check for leaks.

3.3L ENGINES

1. Before servicing the vehicle, refer to the precautions section.

2. Remove or disconnect the following:

- Right front wheel
- Fender apron seal
- Engine undercover

3. Drain the engine oil from the engine.

4. Remove or disconnect the following:
- Front exhaust pipe
- Front exhaust pipe bracket from the No. 1 oil pan
- Flywheel housing undercover
- 10 bolts and 2 nuts to the No. 2 oil pan

5. Insert the blade of the Oil Pan Seal Cutting tool 09032-00100 between the No. 1 and No. 2 oil pans. Clean the surfaces of the oil pans.

6. Remove or disconnect the following:
- 3 oil strainer nuts and gasket

7. Remove the No. 1 oil pan, as follows:

- 2 bolts and the flywheel housing undercover
- 17 bolts and 2 nuts to the No. 1 oil pan

➡**Make a note of the position of the each bolt. When replacing the bolts into the oil pan, place each bolt in the position from which it was removed.**

- Oil pan, by prying the portions between the cylinder block and the oil pan

➡**Be careful not to damage the contact surfaces.**

- Baffle plate from the No. 1 oil pan

To install:

8. Clean all mating surfaces of the oil pans.

9. Install the baffle plate to the No. 1 oil pan and tighten to 69 inch lbs. (8 Nm).

10. Install the No. 1 oil pan, as follows:

a. Using a non residue solvent, clean both sealing surfaces to the oil pan.

b. Apply liquid sealant to the oil pan and engine block.

c. Install the oil pan with the 17 bolts and 2 nuts. Uniformly tighten the bolts and nuts in several passes.

d. Tighten the No. 1 oil pan bolts, as follows:

- 10mm head bolt: 69 inch lbs. (8 Nm)
- 12mm head bolt: 14 ft. lbs. (20 Nm)
- 14mm head bolt: 27 ft. lbs. (37 Nm)

e. Install the flywheel housing undercover with the 2 bolts. Tighten the bolts to 69 inch lbs. (8 Nm).

11. Install the oil strainer with the 3 nuts. Tighten the nuts to 69 inch lbs. (8 Nm).

12. Install the No. 2 oil pan, as follows:

a. Using a non residue solvent, clean both sealing surfaces to the oil pan.

b. Apply liquid sealant to the oil pan and engine block.

c. Install the No. 2 oil pan with the 10 bolts and 2 nuts. Uniformly tighten the bolts and nuts in several passes. Tighten the bolts to 69 inch lbs. (8 Nm).

13. Install or connect the following:
- Flywheel housing undercover
- Front exhaust pipe bracket to the No. 1 oil pan. Tighten the bolts to 15 ft. lbs. (21 Nm).

14. Install the front exhaust pipe, as follows:

- Temporarily install the 3 new gaskets and the front exhaust pipe with the 2 bolts and 6 nuts
- Tighten the 4 exhaust manifolds-to-front exhaust pipe nuts to 46 ft. lbs. (62 Nm).
- Tighten the both front exhaust pipe-to-center exhaust pipe nuts/bolts to 41 ft. lbs. (56 Nm).
- Bracket. Tighten both bolts to 14 ft. lbs. (19 Nm).
- Support stay. Tighten both bolts to 22 ft. lbs. (29 Nm).

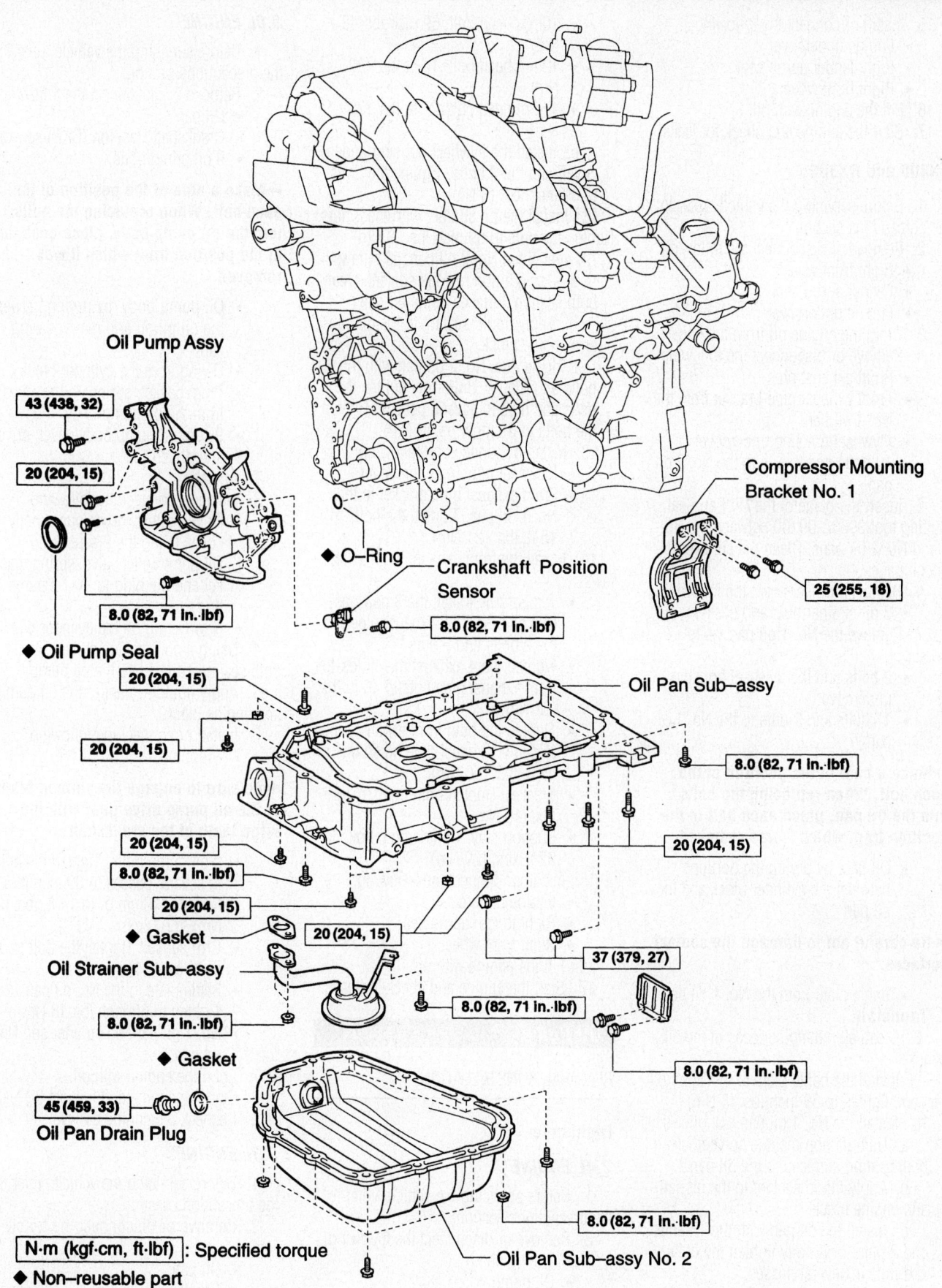

Oil Pump Assy

43 (438, 32)

20 (204, 15)

8.0 (82, 71 in. lbf)

◆ **Oil Pump Seal**

◆ **O–Ring**

Crankshaft Position Sensor

8.0 (82, 71 in. lbf)

Compressor Mounting Bracket No. 1

25 (255, 18)

20 (204, 15)

20 (204, 15)

Oil Pan Sub–assy

8.0 (82, 71 in. lbf)

20 (204, 15)

8.0 (82, 71 in. lbf)

20 (204, 15)

20 (204, 15)

20 (204, 15)

◆ **Gasket**

37 (379, 27)

Oil Strainer Sub–assy

8.0 (82, 71 in. lbf)

8.0 (82, 71 in. lbf)

8.0 (82, 71 in. lbf)

◆ **Gasket**

45 (459, 33)

Oil Pan Drain Plug

8.0 (82, 71 in. lbf)

N·m (kgf·cm, ft·lbf) : Specified torque

◆ Non–reusable part

Oil Pan Sub–assy No. 2

67170-HIGH-G23

Oil pan, pump and related parts—3.3L Highlander shown; RX330 similar

15. Install or connect the following:
- Engine undercover
- Right fender apron seal
- Right front wheel

16. Fill the engine with oil.

17. Start the engine and check for leaks.

RX300 and RX330

1. Before servicing the vehicle, refer to the precautions section.

2. Remove or disconnect the following:
- Right front wheel
- Fender apron seal
- Engine undercover

3. Drain the engine oil from the engine.

4. Remove or disconnect the following:
- Front exhaust pipe
- Front exhaust pipe bracket from the No. 1 oil pan
- Flywheel housing undercover
- 10 bolts and 2 nuts to the No. 2 oil pan

5. Insert the blade of the Oil Pan Seal Cutting tool 09032-00100 between the No. 1 and No. 2 oil pans. Clean the surfaces of the oil pans.

6. Remove or disconnect the following:
- 3 oil strainer nuts and gasket

7. Remove the No. 1 oil pan, as follows:
- 2 bolts and the flywheel housing undercover
- 17 bolts and 2 nuts to the No. 1 oil pan

➡**Make a note of the position of the each bolt. When replacing the bolts into the oil pan, place each bolt in the position from which it was removed.**

- Oil pan, by prying the portions between the cylinder block and the oil pan

➡**Be careful not to damage the contact surfaces.**

- Baffle plate from the No. 1 oil pan

To install:

8. Clean all mating surfaces of the oil pans.

9. Install the baffle plate to the No. 1 oil pan and tighten to 69 inch lbs. (8 Nm).

10. Install the No. 1 oil pan, as follows:

   a. Using a non residue solvent, clean both sealing surfaces to the oil pan.

   b. Apply liquid sealant to the oil pan and engine block.

   c. Install the oil pan with the 17 bolts and 2 nuts. Uniformly tighten the bolts and nuts in several passes.

   d. Tighten the No. 1 oil pan bolts, as follows:

- 10mm head bolt: 69 inch lbs. (8 Nm)
- 12mm head bolt: 14 ft. lbs. (20 Nm)
- 14mm head bolt: 27 ft. lbs. (37 Nm)

   e. Install the flywheel housing undercover with the 2 bolts. Tighten the bolts to 69 inch lbs. (8 Nm).

11. Install the oil strainer with the 3 nuts. Tighten the nuts to 69 inch lbs. (8 Nm).

12. Install the No. 2 oil pan, as follows:

   a. Using a non residue solvent, clean both sealing surfaces to the oil pan.

   b. Apply liquid sealant to the oil pan and engine block.

   c. Install the No. 2 oil pan with the 10 bolts and 2 nuts. Uniformly tighten the bolts and nuts in several passes. Tighten the bolts to 69 inch lbs. (8 Nm).

13. Install or connect the following:
- Flywheel housing undercover
- Front exhaust pipe bracket to the No. 1 oil pan. Tighten the bolts to 15 ft. lbs. (21 Nm).

14. Install the front exhaust pipe, as follows:

- Temporarily install the 3 new gaskets and the front exhaust pipe with the 2 bolts and 6 nuts
- Tighten the 4 exhaust manifolds-to-front exhaust pipe nuts to 46 ft. lbs. (62 Nm).
- Tighten the both front exhaust pipe-to-center exhaust pipe nuts/bolts to 41 ft. lbs. (56 Nm).
- Bracket. Tighten both bolts to 14 ft. lbs. (19 Nm).
- Support stay. Tighten both bolts to 22 ft. lbs. (29 Nm).

15. Install or connect the following:
- Engine undercover
- Right fender apron seal
- Right front wheel

16. Fill the engine with oil.

17. Start the engine and check for leaks.

Oil Pump

REMOVAL & INSTALLATION

Highlander

2.4L ENGINE

1. Before servicing the vehicle, refer to the precautions section.

2. Remove or disconnect the following:
- Timing chain
- Oil pump

3. Installation is the reverse of removal. Torque the pump bolts to 14 ft. lbs. (19Nm).

3.0L ENGINE

1. Before servicing the vehicle, refer to the precautions section.

2. Remove or disconnect the following:
- Oil pan
- Crankshaft Position (CKP) sensor
- 9 oil pump bolts

➡**Make a note of the position of the each bolt. When replacing the bolts into the oil pump body, place each bolt in the position from which it was removed.**

- Oil pump body, by prying between the oil pump and main bearing cap
- O-ring from the cylinder block
- Plug, gasket, spring and relief valve from the oil pump body
- 9 screws, pump body cover, drive and driven rotors

To install:

3. Install or connect the following:
- Driven rotors, drive, pump body cover, using the 9 screws
- Oil pump relief valve, spring, gasket and the plug to the oil pump body
- New O-ring on the cylinder block

4. Using a non residue solvent, clean both sealing surfaces to the oil pump.

5. Apply liquid sealant to the oil pump and engine block.

6. Install or connect the following:
- Oil pump

➡**Be sure to engage the splined teeth of the oil pump drive gear with the large teeth of the crankshaft.**

- 9 oil pump bolts. Tighten the bolts in several passes to 69 inch lbs. (8 Nm), for 10mm or to 14 ft. lbs. (20 Nm), for 12mm.
- CKP sensor. Tighten the bolt to 69 inch lbs. (8 Nm).
- Baffle plate to the No. oil pan. Tighten to 69 inch lbs. (8 Nm).
- No. 1 oil pan, oil strainer and No. 2 oil pan

7. Refill the engine with oil.

8. Start the engine and inspect for leaks.

9. Recheck the engine oil level.

3.3L ENGINES

1. Before servicing the vehicle, refer to the precautions section.

2. Remove or disconnect the following:
- Oil pan
- Crankshaft Position (CKP) sensor
- 9 oil pump bolts

➡Make a note of the position of the each bolt. When replacing the bolts into the oil pump body, place each bolt in the position from which it was removed.

- Oil pump body, by prying between the oil pump and main bearing cap
- O-ring from the cylinder block
- Plug, gasket, spring and relief valve from the oil pump body
- 9 screws, pump body cover, drive and driven rotors

To install:

3. Install or connect the following:
   - Driven rotors, drive, pump body cover, using the 9 screws
   - Oil pump relief valve, spring, gasket and the plug to the oil pump body
   - New O-ring on the cylinder block
4. Using a non residue solvent, clean both sealing surfaces to the oil pump.

5. Apply liquid sealant to the oil pump and engine block.
6. Install or connect the following:
   - Oil pump

➡Be sure to engage the splined teeth of the oil pump drive gear with the large teeth of the crankshaft.

- 9 oil pump bolts. Tighten the bolts in several passes to 69 inch lbs. (8 Nm), for 10mm; 14 ft. lbs. (20 Nm), for 12mm; 32 ft. lbs. (43 Nm) for 14mm
- CKP sensor. Tighten the bolt to 69 inch lbs. (8 Nm).
- Baffle plate to the No. oil pan. Tighten to 69 inch lbs. (8 Nm).
- No. 1 oil pan, oil strainer and No. 2 oil pan
7. Refill the engine with oil.
8. Start the engine and inspect for leaks.
9. Recheck the engine oil level.

RX300 and RX330

1. Before servicing the vehicle, refer to the precautions section.
2. Remove or disconnect the following:
   - Oil pan
   - Crankshaft Position (CKP) sensor
   - 9 oil pump bolts

➡Make a note of the position of the each bolt. When replacing the bolts into the oil pump body, place each bolt in the position from which it was removed.

- Oil pump body, by prying between the oil pump and main bearing cap
- O-ring from the cylinder block
- Plug, gasket, spring and relief valve from the oil pump body
- 9 screws, pump body cover, drive and driven rotors

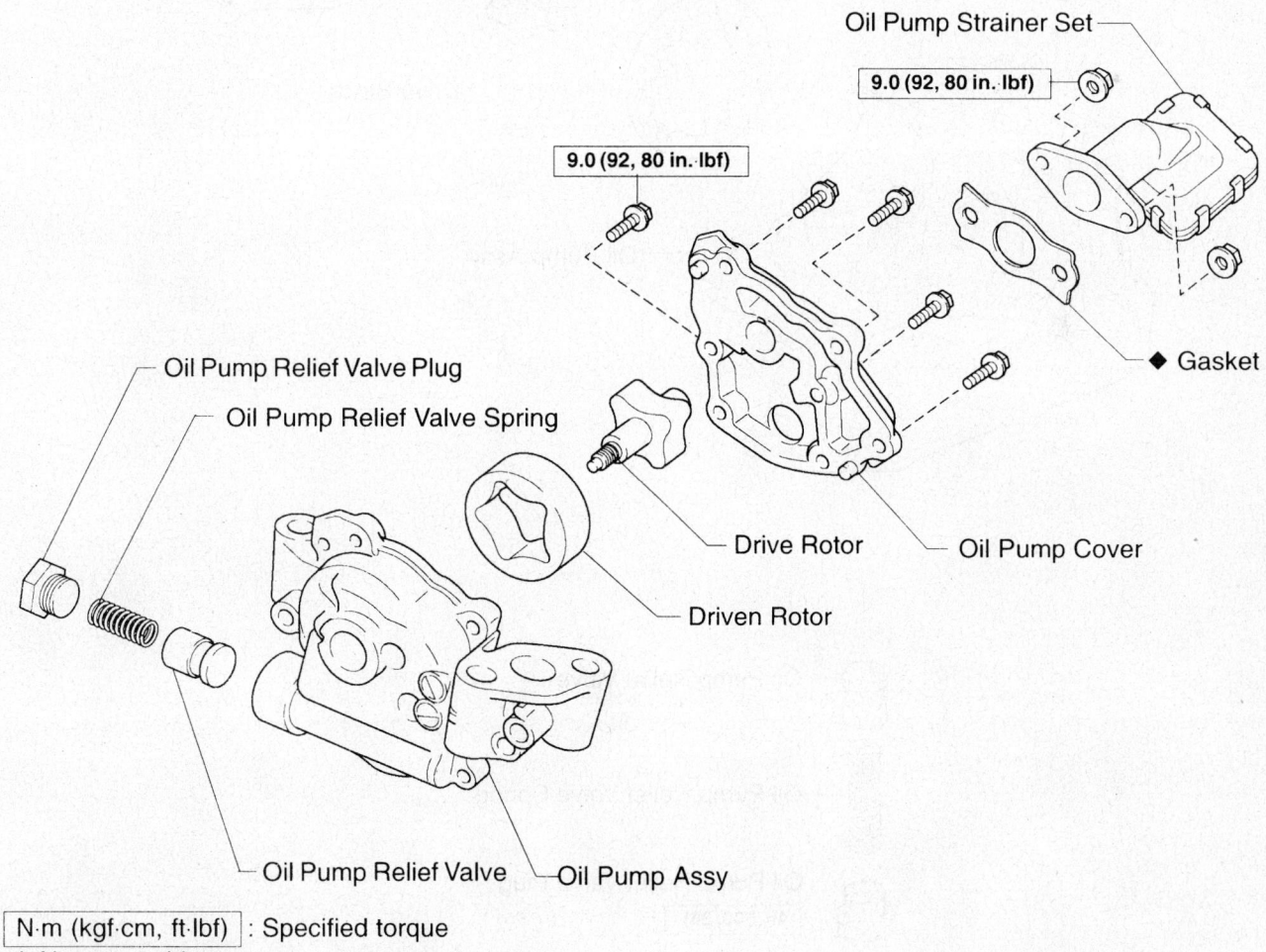

Oil Pump Strainer Set

9.0 (92, 80 in. lbf)

9.0 (92, 80 in. lbf)

◆ Gasket

Oil Pump Relief Valve Plug

Oil Pump Relief Valve Spring

Drive Rotor

Oil Pump Cover

Driven Rotor

Oil Pump Relief Valve

Oil Pump Assy

N·m (kgf·cm, ft·lbf) : Specified torque
◆ Non–reusable part

67170-HIGH-G24

Oil pump exploded view—2.4L Highlander

To install:

3. Install or connect the following:
   - Driven rotors, drive, pump body cover, using the 9 screws
   - Oil pump relief valve, spring, gasket and the plug to the oil pump body
   - New O-ring on the cylinder block

4. Using a non residue solvent, clean both sealing surfaces to the oil pump.

5. Apply liquid sealant to the oil pump and engine block.

6. Install or connect the following:
   - Oil pump

➡️**Be sure to engage the splined teeth of the oil pump drive gear with the large teeth of the crankshaft.**

   - 9 oil pump bolts. Tighten the bolts in several passes to 69 inch lbs. (8 Nm), for 10mm; 14 ft. lbs. (20 Nm), for 12mm; 32 ft. lbs. (43 Nm) for 14mm
   - CKP sensor. Tighten the bolt to 69 inch lbs. (8 Nm).
   - Baffle plate to the No. oil pan. Tighten to 69 inch lbs. (8 Nm).
   - No. 1 oil pan, oil strainer and No. 2 oil pan

7. Refill the engine with oil.

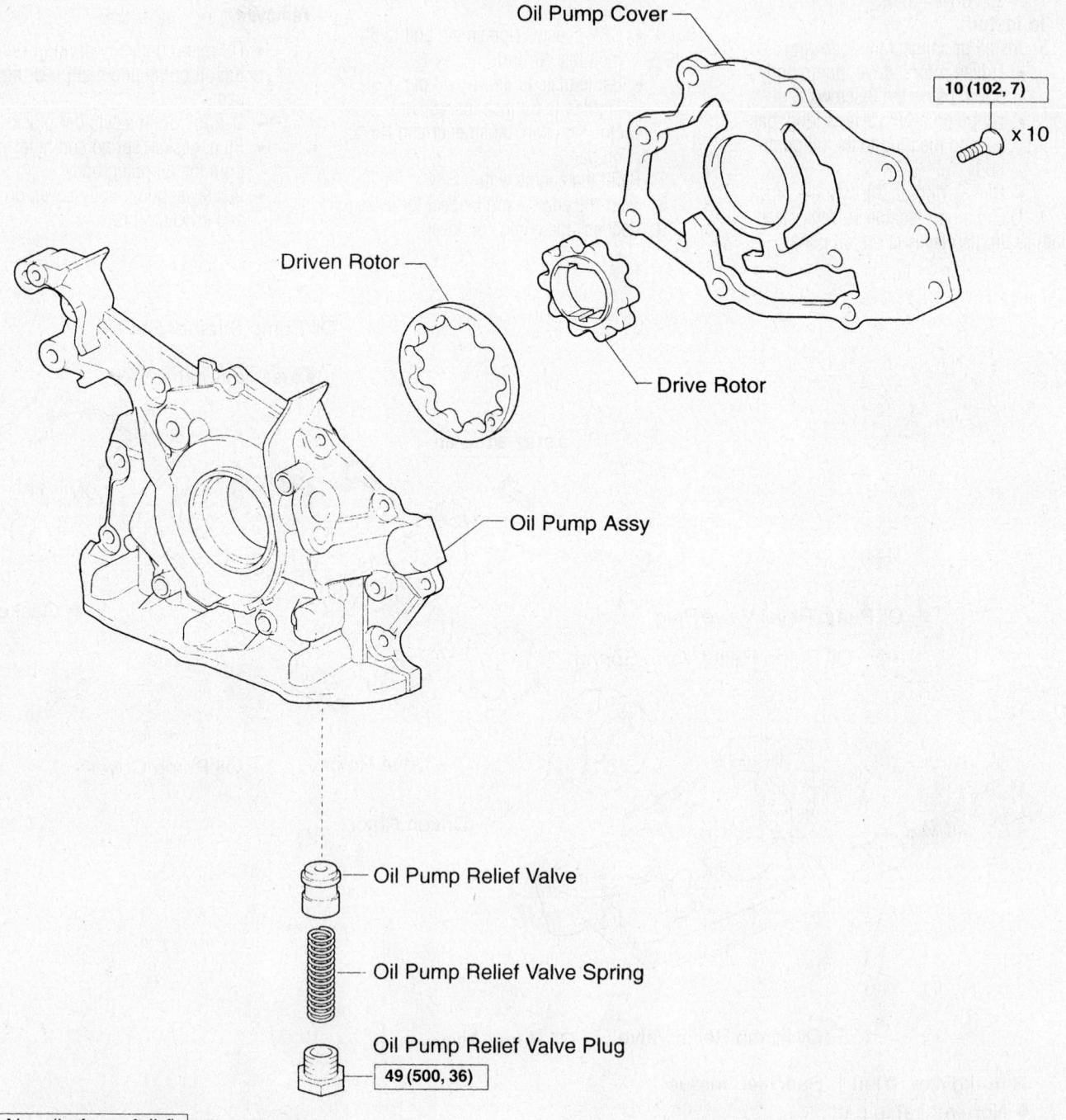

Oil Pump Cover

10 (102, 7) × 10

Driven Rotor

Drive Rotor

Oil Pump Assy

Oil Pump Relief Valve

Oil Pump Relief Valve Spring

Oil Pump Relief Valve Plug

49 (500, 36)

N·m (kgf·cm, ft·lbf) : Specified torque

Oil pump exploded view—3.3L Highlander shown, RX330 similar

8. Start the engine and inspect for leaks.
9. Recheck the engine oil level.

Rear Main Seal

REMOVAL & INSTALLATION

All Engines

If the rear oil seal retainer is not installed to the block, use a tapered ended screwdriver and hammer to remove the oil seal. Apply multi-purpose grease to the new oil seal lip. Using a seal driver, tap the seal into place. Be careful not to install it slantwise.

1. Before servicing the vehicle, refer to the precautions section.

If the rear oil seal retainer is installed on

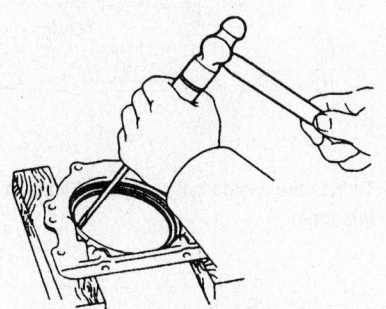

Carefully tap the old seal from the retainer

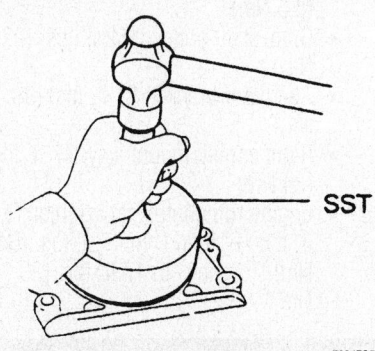

Use the proper sized driver to seat the seal

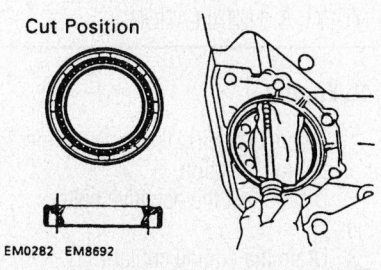

Cut off the oil seal lip, then pry the seal out of the retaining plate

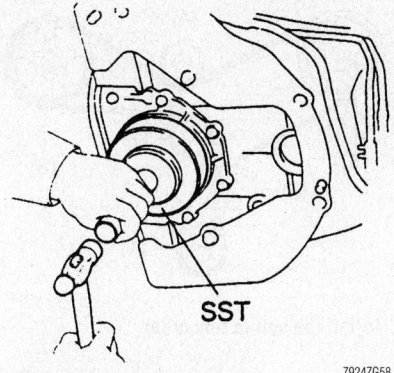

Tap a new seal into place

the cylinder block, using a knife, cut off the lip of the seal. Using a taped ended prytool, pry the old seal out of the retainer. Inspect the oil seal lip contacting surface of the crankshaft for cracks or damage. Apply multipurpose grease to the new oil seal, then tap the seal in place with a seal installer. Be careful not to install the seal slantwise.

Timing Belt

REMOVAL & INSTALLATION

1. Before servicing the vehicle, refer to the precautions section.
2. Remove the right front wheel.
3. Remove the wiper arms.
4. Remove the wiper linkage.
5. Remove the top cowl panel.
6. Remove the engine undercovers.
7. Remove the right front fender apron seal.
8. Remove the A/C compressor drive belt.
9. Remove the power steering pump belt.
10. Remove the engine roll control rod.
11. Remove the right side engine mount stay.
12. Remove the alternator bracket.

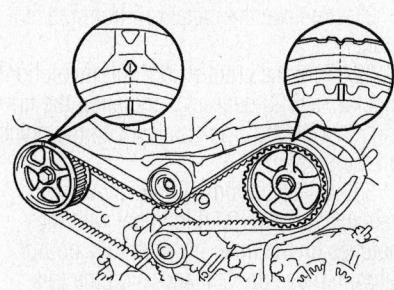

Check that the timing marks on the camshaft pulleys are aligned with the notches on the inner belt cover

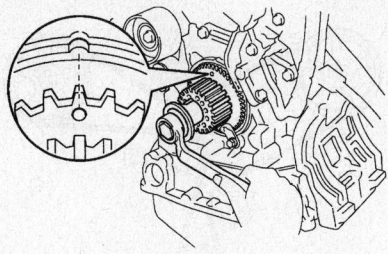

Turn the crankshaft clockwise to align the timing mark on the crankshaft timing pulley with the notch in the oil pump body

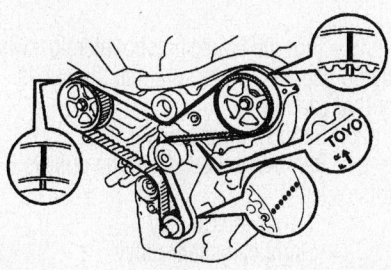

If the timing belt is re-used, check that the 3 original installation marks are visible on the belt as shown

13. Remove the crankshaft pulley.
14. Remove the upper belt cover.
15. Remove the right engine mount bracket.
16. Remove the no. 2 timing belt guide.
17. Set the no.1 cylinder to TDC compression.
18. Temporarily install the crank pulley bolt. Turn the crankshaft clockwise to align the timing mark on the crankshaft timing pulley with the notch in the oil pump body. Check that the timing marks on the camshaft pulleys are aligned with the notches on the inner belt cover. If not, rotate the crankshaft 360 degrees clockwise.

➡ If the timing belt is re-used, check that the 3 original installation marks are visible on the belt as shown. If not, paint three new marks on the belt.

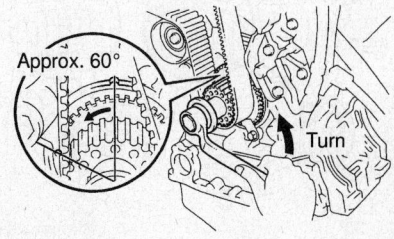

Turn the crankshaft counterclockwise by 60 degrees

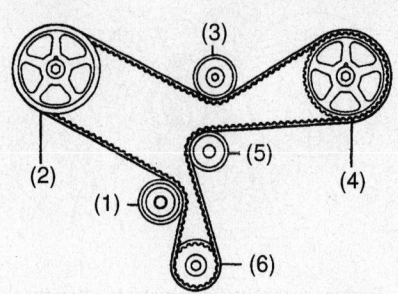

67162-X300-G04

Remove the belt from the pulleys in this order

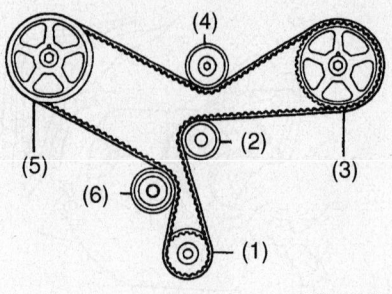

67162-X300-G06

Install the belt in this order

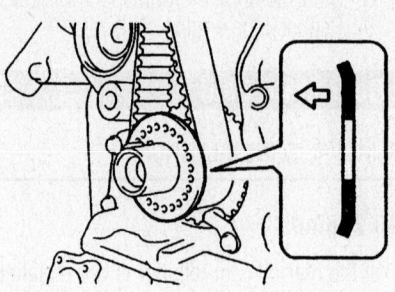

67162-X300-G08

Install the timing belt guide with the cupped side facing front

19. Turn the crankshaft counterclockwise by 60 degrees. Make sure that the belt is still engaged.

20. Remove the tensioner.

21. Remove the belt from the pulleys in this order:

- Lower idler pulley
- Right camshaft pulley
- Upper idler pulley
- Left camshaft pulley
- Water pump pulley
- Crankshaft timing pulley

22. If the belt is being re-used, check it for wear or damage; don't twist it or turn it inside-out. If there is any doubt as to its condition, replace it.

To install:

23. Clean all the pulleys.

24. Turn the crankshaft another 60 degrees counterclockwise.

25. Turn the camshaft pulleys back into alignment so the marks align with the notches on the inner cover.

26. Turn the crankshaft back so that the timing mark aligns with the notch on the oil pump.

27. Align the installation marks on the belt with the timing marks on the pulleys.

28. Install the belt in this order:

- Crankshaft
- Water pump
- Left camshaft

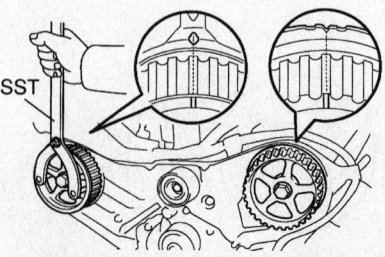

67162-X300-G05

Turn the camshaft pulleys back into alignment so the marks align with the notches on the inner cover

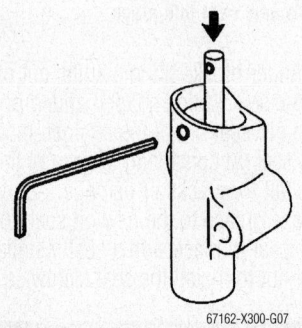

67162-X300-G07

Set the tensioner in a press and collapse the plunger. Do not apply more that 2,205 lbs (9.8 kN) of force. Insert a suitable metal rod through the holes to hold the plunger in position

- Upper idler
- Right camshaft
- Lower idler

29. Set the tensioner in a press and collapse the plunger. Do not apply more that 2,205 lbs (9.8 kN) of force. Insert a suitable metal rod through the holes to hold the plunger in position.

30. Install the tensioner and torque the 2 bolts alternately to 20 ft. lbs. (27 Nm).

☀☀ WARNING

Be sure to tighten to bolts alternately and evenly so the tensioner seats flat.

31. Remove the metal rod from the tensioner.

32. Turn the crankshaft 2 full revolutions clockwise (720 degrees), and align the timing mark on the crank pulley with the notch on the oil pump.

33. Check the timing marks on the camshaft pulleys for alignment with the notches on the inner cover. If they do not align, remove the belt and align the mismatched mark(s).

34. The remainder of installation is the reverse of removal. Observe the following torques:

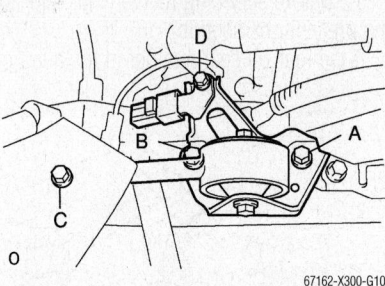

67162-X300-G10

Tighten the engine roll control rod bolts in this order

- Right engine mount bracket: 21 ft. lbs. (28 Nm)
- Right engine mount insulator: 70 ft. lbs. (95 Nm)
- Timing belt covers: 75 inch lbs. (8.5 Nm)
- Crankshaft pulley: 162 ft. lbs. (220 Nm)
- Alternator bracket: 21 ft. lbs. (28 Nm)
- Right engine mount stay: 47 ft. lbs. (64 Nm)
- Engine roll control rod: tighten first A, then B, then C to 47 ft. lbs. (64 Nm). Torque D to 17 ft. lbs. (23 Nm)

Timing Chain, Sprockets, Front Cover and Seal

REMOVAL & INSTALLATION

2.4L Engine

1. Before servicing the vehicle, refer to the precautions section.

2. Disconnect the negative battery cable.

3. Drain the engine coolant.

4. Remove or disconnect the following:

- Hood
- Engine oil

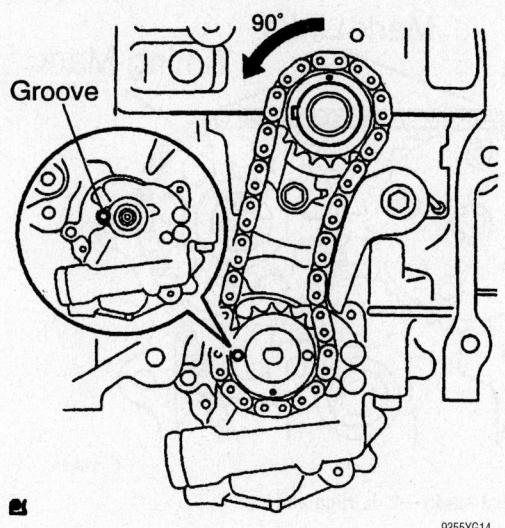

Aligning the adjusting hole and groove—2.4L Highlander

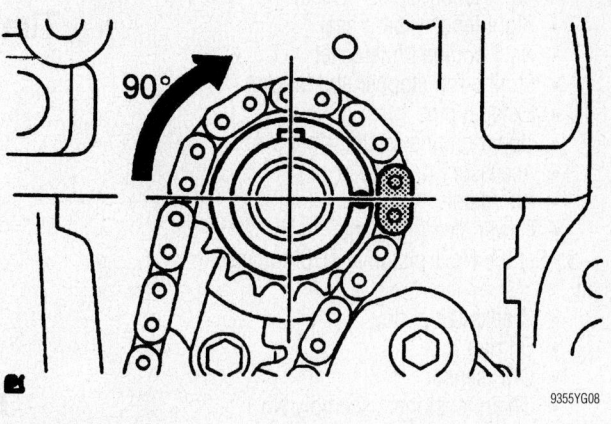

Aligning the crankshaft with the key in the 12 o'clock position—2.4L Highlander

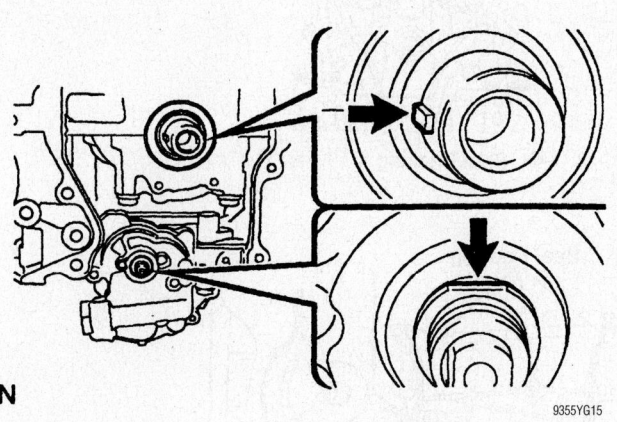

Aligning the crankshaft with the key in the left horizontal position—2.4L Highlander

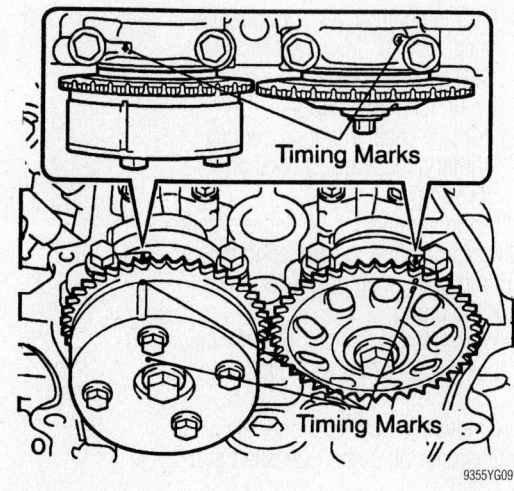

Aligning the timing marks with the No.1 piston at TDC compression—2.4L Highlander

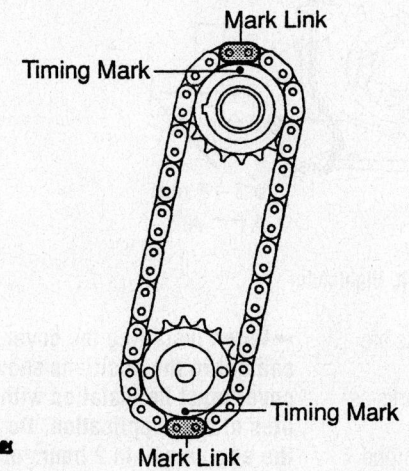

Install the secondary chain and gears with the timing marks aligned as shown—2.4L Highlander

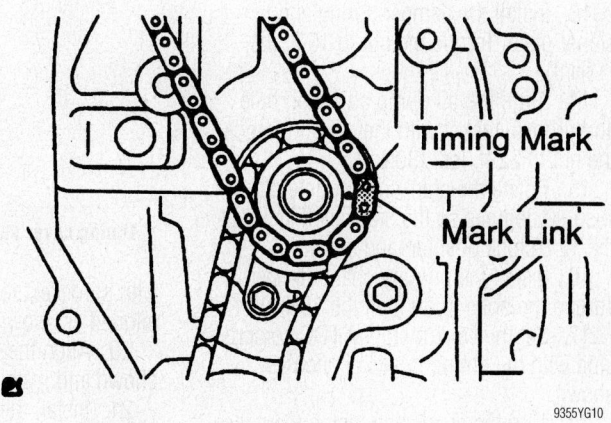

Aligning the timing chain bottom end marks—2.4L Highlander

- Right front wheel
- Right fender splash shield
- Right fender apron seal
- No.1 engine undercover
- Engine roll stopper and bracket
- Exhaust pipe
- Upper engine mount, right side
- Accessory drive belts
- Alternator
- Power steering pump

5. Set the No.1 piston at TDC compression.

- Crankshaft pulley
- Oil pan
- CKP sensor
- Chain tensioner assembly No.1
- V-belt tensioner

6. Take up the weight of the engine with a crane.

7. Remove or disconnect the following:
- Transverse engine mount insulator
- Transverse engine mount bracket
- Timing chain cover (14 bolts and 2 nuts
- CKP sensor plate
- Chain tensioner slipper
- Primary chain vibration damper
- Primary chain and crankshaft sprocket

8. Turn the crankshaft 90 degrees counterclockwise and align the adjusting hole of the oil pump drive shaft gear with the groove in the pump.

9. Insert a 4mm diameter bar into the hole to lock the gear in position and remove the nut.

10. Remove the bolt, tensioner plate, spring, tensioner oil pump driveshaft gear and the chain.

To install:

11. Turn the crankshaft so that the key is in the left horizontal position.

12. Install the secondary chain and gears with the timing marks aligned as shown.

13. Install the damper spring and tensioner plate. Torque the nut to 10 ft. lbs. (13Nm).

14. Align the oil pump adjusting hole and groove, lock it with the bar and torque the nut to 22 ft. lbs. (30Nm).

15. Rotate the crankshaft counterclockwise 90 degrees so the crankshaft key is at the 12 o'clock position and shown.

16. Install the primary chain damper. Torque the bolts to 80 inch lbs. (9Nm).

17. Set the No.1 piston at TDC compression with the timing marks aligned as shown.

18. Turn the crankshaft, using the pulley bolt, until the key is at the 12 o'clock position.

19. Install the bottom end of the chain,

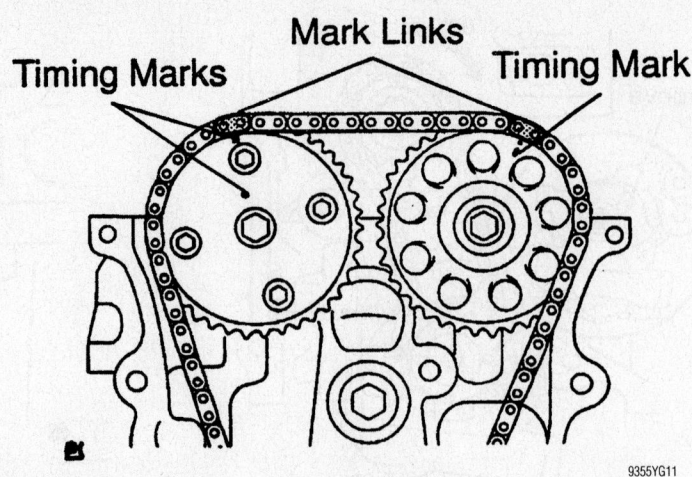

Aligning the timing chain upper end marks—2.4L Highlander

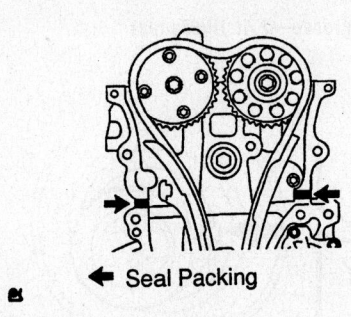

← Seal Packing

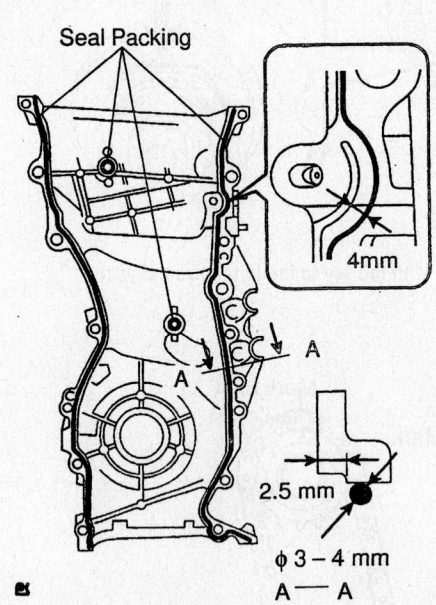

Timing cover sealant application—2.4L Highlander

with sprocket, so that the colored links are aligned as shown.

20. Align the upper end timing marks as shown and install the chain.

21. Install the tensioner slipper. Torque the bolt to 14 ft. lbs. (19Nm).

22. Install the CKP sensor plate with the **F** mark outwards.

➡**When installing the cover, use RTV sealant in the positions shown. The cover must be installed within 3 minutes of seal application. Do not start the engine within 2 hours of seal application.**

23. Apply the sealant and install the cover. Torque the cover bolts as follows:

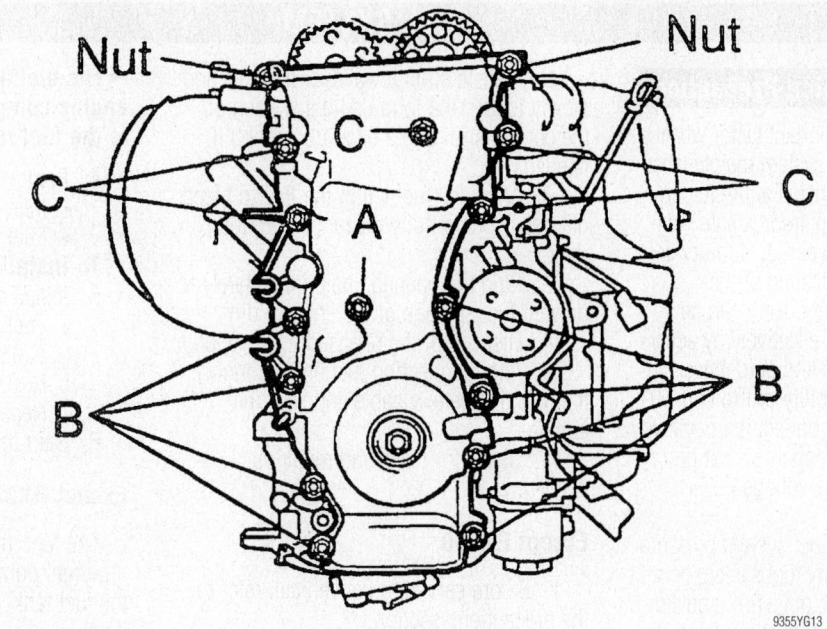

Timing cover bolt positions—2.4L Highlander

- Bolt A: 80 inch lbs. (9Nm)
- Bolts B: 15 ft. lbs. (21 Nm)
- Bolts C: 32 ft. lbs. (43Nm)
- Nuts: 80 inch lbs. (9Nm)

24. The remainder of installation is the reverse of removal.

Piston and Ring

POSITIONING

RH Piston

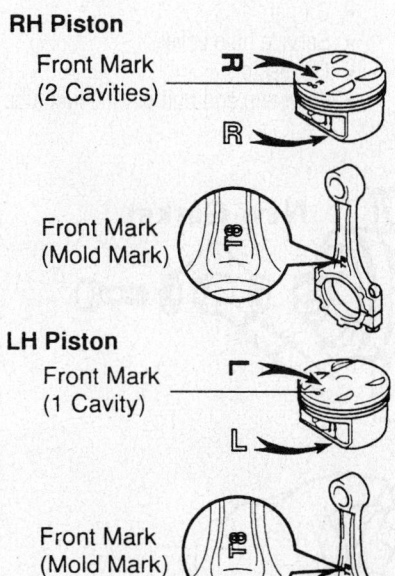

Front Mark (2 Cavities)

Front Mark (Mold Mark)

LH Piston

Front Mark (1 Cavity)

Front Mark (Mold Mark)

Piston/connecting rod-to-engine positioning–3.0L and 3.3L engines

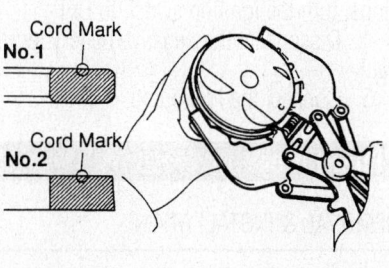

Cord Mark No.1

Cord Mark No.2

Piston ring positioning–3.0L and 3.3L engines

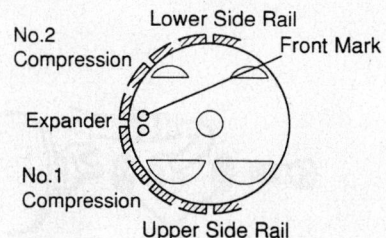

RH Piston

No.2 Compression
Lower Side Rail
Front Mark
Expander
No.1 Compression
Upper Side Rail

LH Piston

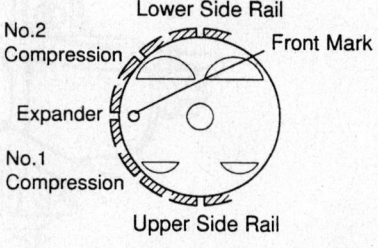

No.2 Compression
Lower Side Rail
Front Mark
Expander
No.1 Compression
Upper Side Rail

Piston ring identification–3.0L and 3.3L engines

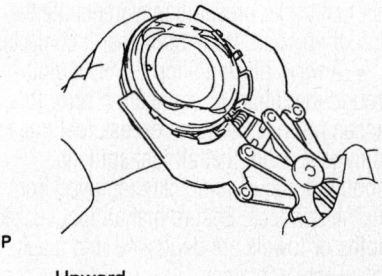

Upward

No. 1

No. 2

Painted Mark

Code Mark (2N)

Piston ring identification—2.4L engine

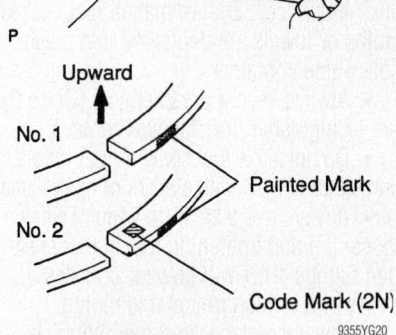

No. 1 and Expander
Side Rail Lower
Front
Side Rail Upper
No. 2 Compression

Piston ring installation—2.4L engine

FUEL SYSTEM

Fuel System Service Precautions

Safety is the most important factor when performing not only fuel system maintenance but any type of maintenance. Failure to conduct maintenance and repairs in a safe manner may result in serious personal injury or death. Maintenance and testing of the vehicle's fuel system components can be accomplished safely and effectively by adhering to the following rules and guidelines.

• To avoid the possibility of fire and personal injury, always disconnect the negative battery cable unless the repair or test procedure requires that battery voltage be applied.

• Always relieve the fuel system pressure prior to disconnecting any fuel system component (injector, fuel rail, pressure regulator, etc.), fitting or fuel line connection. Exercise extreme caution whenever relieving fuel system pressure, to avoid exposing skin, face and eyes to fuel spray. Please be advised that fuel under pressure may penetrate the skin or any part of the body that it contacts.

• Always place a shop towel or cloth around the fitting or connection prior to loosening to absorb any excess fuel due to spillage. Ensure that all fuel spillage (should it occur) is quickly removed from engine surfaces. Ensure that all fuel soaked cloths or towels are deposited into a suitable waste container.

• Always keep a dry chemical (Class B) fire extinguisher near the work area.

• Do not allow fuel spray or fuel vapors to come into contact with a spark or open flame.

• Always use a back-up wrench when loosening and tightening fuel line connection fittings. This will prevent unnecessary stress and torsion to fuel line piping.

• Always replace worn fuel fitting O-rings with new. Do not substitute fuel hose or equivalent, where fuel pipe is installed.

Fuel System Pressure

RELIEVING

RX300

1. Before servicing the vehicle, refer to the precautions section.
2. Disconnect the negative battery terminal.
3. Place a catch-pan under the joint to be disconnected. A large quantity of fuel may be released when the joint is opened.
4. Wear eye or full face protection.

5. Place a shop towel over the area and slowly loosen the joint using a wrench of the correct size. Use a back-up wrench if needed.
6. Allow the fuel left in the line to bleed off slowly before fully disconnecting the joint.
7. Plug the opened lines immediately to prevent fuel spillage or the entry of dirt.
8. Dispose of the released fuel properly.
9. After connecting fuel lines, connect the negative battery cable and start the engine.
10. Check for leaks and repair as needed.

Except RX300

1. Before servicing the vehicle, refer to the precautions section.
2. Disconnect the fuel pump wire at the pump.
3. Start the engine. After the engine stops, turn the ignition switch to OFF.
4. Disconnect the negative battery terminal.
5. Connect the fuel pump.

Fuel Filter

REMOVAL & INSTALLATION

RX300

1. Before servicing the vehicle, refer to the precautions section.
2. Disconnect the negative battery cable.
3. Relieve the fuel system pressure.

➡ The fuel filter is located in the engine compartment, at the inlet line to the fuel rail.

4. Remove or disconnect the following:
   • Inlet and outlet lines from the filter
   • Fuel filter

To install:
5. Install or connect the following:
   • Fuel filter, using new O-rings. Tighten the lines to 22 ft. lbs. (29 Nm).
   • Negative battery cable
6. Start the engine and check for leaks.

Except RX330

The fuel filter is part of the fuel suction tube/fuel pump assembly and is located in the fuel tank. It is not a normally serviced item.

Fuel Pump

REMOVAL & INSTALLATION

Highlander

2002-03

1. Before servicing the vehicle, refer to the precautions section.
2. Relieve the fuel pressure.
3. Remove or disconnect the following:
   • Both rear seats
   • Carpet
   • Service hole cover
   • Connector
   • Joint clip and pull out the fuel tube

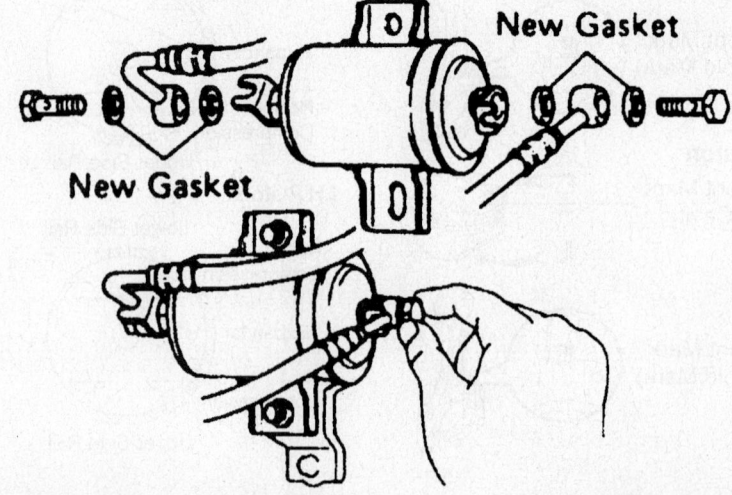

Exploded view of the fuel filter—RX300

7924ZG59

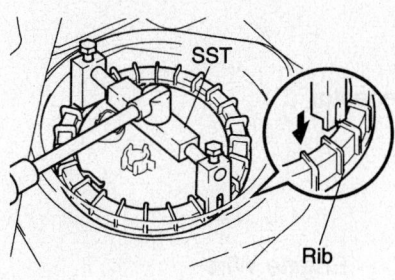

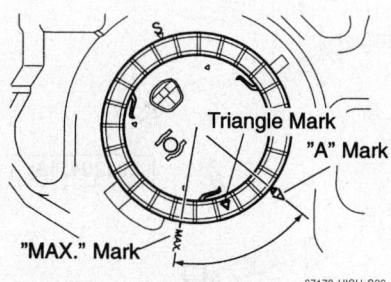

67170-HIGH-G26

Lock ring positioning during installation—2004–05 Highlander

- Vent tube set plate
- Pump and gauge assembly

4. Installation is the reverse of removal.

2004–05

1. Before servicing the vehicle, refer to the precautions section.
2. Relieve the fuel system pressure.
3. Remove the deck board assembly.
4. Remove the rear seats.
5. Remove the left side door scuff plate.
6. Remove the left side trim cover.
7. Remove the left rear seat side cover.
8. Remove the carpet.
9. Remove the rear floor service cover.
10. Disconnect the fuel pump wiring.
11. Disconnect the fuel hose.
12. Remove the fuel pump locking ring. Special tool 09808-14020 is available for this job.
13. Remove the pump assembly from the tank.
14. Installation is the reverse of removal. When installing the locking ring, tighten it first by hand, then, using the special tool turn the ring 1½ turn. The triangle mark on the lockring must be positioned between the **A** and **MAX** marks on the fuel tank.

> ✳✳ **WARNING**
>
> **No other type of tool should be used for this operation.**

2006

1. Before servicing the vehicle, refer to the precautions section.
2. Relieve the fuel system pressure.

3. Disconnect the negative battery cable.
4. Remove the center exhaust pipe.
5. Remove the front floor heat insulator No: 3.
6. Remove the fuel tank protector assembly.
7. Separate the parking brake assembly.
8. Disconnect the fuel tank wire.
9. Remove the charcoal canister protector and hose.
10. Disconnect the fuel tank main tube sub assembly.
11. Disconnect the breather lower tube and remove the fuel tank.
12. Disconnect all wiring and hoses from the pump.
13. Remove the pump cover bolts and remove the cover and pump.

To install:

14. Install the pump assembly, cover and bolts. Tighten the bolts to 53 inch lbs. (6 Nm).
15. Install the fuel tank and tighten the strap bolts to 29 ft. lbs. (39 Nm).
16. Install the remaining components in the reverse order of removal.

RX300

1. Before servicing the vehicle, refer to the precautions section.
2. Relieve the fuel system pressure.
3. Remove or disconnect the following:
   - Negative battery cable
   - Left-hand rear seat assembly
   - Floor service hole by pulling back the carpet; then, remove the 4 screws
   - Fuel pump and sender gauge connector

➡ **Loosen the fuel cap to relieve any fuel pressure within the tank.**

   - Fuel pipe union bolt and both gaskets
   - Fuel pump outlet pipe
   - Return vent hose from the fuel pump
   - 8 fuel pump bolts and the pump assembly from the tank

To install:

4. Install or connect the following:
   - Fuel pump to the fuel tank. Tighten the 8 bolts to 31 inch lbs. (3.5 Nm).
   - Return vent hose to the fuel pump
   - Outlet pipe to the fuel pump, using new gaskets. Tighten the union bolts to 22 ft. lbs. (29 Nm).
   - Fuel pump and sender gauge connector

   - Floor hole cover with the 4 screws
   - Carpet
   - Left rear seat assembly
   - Negative battery cable
   - Fuel cap

5. Start the vehicle and check for leaks.

RX330

1. Before servicing the vehicle, refer to the precautions section.
2. Relieve the fuel system pressure.
3. Remove the deck board assembly.
4. Remove the rear seats.
5. Remove the left side door scuff plate.
6. Remove the left side trim cover.
7. Remove the left rear seat side cover.
8. Remove the carpet.
9. Remove the rear floor service cover.
10. Disconnect the fuel pump wiring.
11. Disconnect the fuel hose.
12. Remove the fuel pump locking ring. Special tool 09808-14020 is available for this job.
13. Remove the pump assembly from the tank.
14. Installation is the reverse of removal. When installing the locking ring, tighten it first by hand, then, using the special tool turn the ring 1½ turn. The triangle mark on the lockring must be positioned between the **A** and **MAX** marks on the fuel tank.

> ✳✳ **WARNING**
>
> **No other type of tool should be used for this operation.**

Fuel Injector

REMOVAL & INSTALLATION

Highlander

2.4L ENGINE

1. Before servicing the vehicle, refer to the precautions section.
2. Relieve the fuel system pressure.
3. Remove or disconnect the following:

   - Air cleaner and hoses
   - Fuel line from the rail
   - Injector connectors
   - Injector wiring harness
   - Fuel rail with injectors
   - Injector spacers from the head

4. Installation is the reverse of removal. Coat the new o-rings with clean fuel. Before tightening the fuel rail bolts, make sure that each injector rotates smoothly. Tighten the bolts to 15 ft. lbs. (20 Nm).

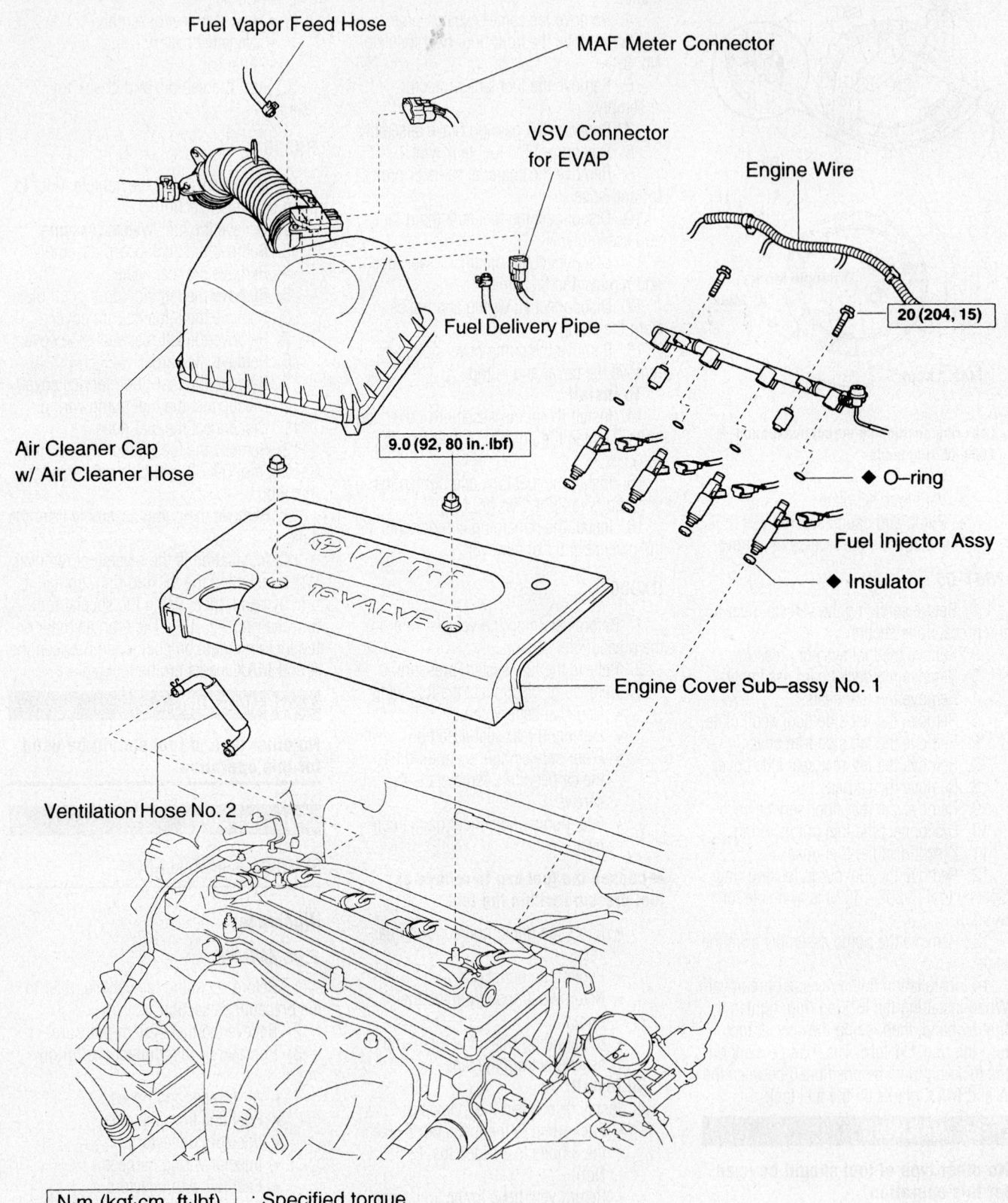

Fuel Vapor Feed Hose

MAF Meter Connector

VSV Connector
for EVAP

Engine Wire

20 (204, 15)

Fuel Delivery Pipe

9.0 (92, 80 in.·lbf)

◆ O–ring

Air Cleaner Cap
w/ Air Cleaner Hose

Fuel Injector Assy

◆ Insulator

Engine Cover Sub–assy No. 1

Ventilation Hose No. 2

| **N·m (kgf·cm, ft·lbf)** | : Specified torque |

◆ Non–reusable part

67170-HIGH-G27

Fuel injectors and related parts—2004 2.4L Highlander model shown

3.0L ENGINE

1. Before servicing the vehicle, refer to the precautions section.
2. Relieve the fuel system pressure.
3. Remove or disconnect the following:
   - Coolant
   - V-bank cover
   - Battery
   - Air cleaner assembly
   - Upper front suspension brace
   - Intake air surge tank
   - Fuel supply line
   - Injector connectors
   - Fuel rail with injectors
4. Installation is the reverse of removal. Coat the new o-rings with clean fuel. Before tightening the fuel rail bolts, make sure that each injector rotates smoothly. Tighten the bolts to 84 inch lbs. (10Nm).

3.3L ENGINE

1. Before servicing the vehicle, refer to the precautions section.
2. Relieve the fuel system pressure.
3. Drain the coolant.
4. Remove the wiper arms.
5. Remove the wiper linkage.
6. Remove the fender-to-cowl side seals.
7. Remove the rain sensor.
8. Remove the front shock absorber caps (air suspension).
9. Remove the 4 set nuts from the strut (w/o air suspension).
10. Remove the cowl top outer panel.
11. Remove the 6 set nuts from the shock absorber.
12. Remove the V-bank cover.
13. Remove the air cleaner assembly and inlet tubes.
14. Remove the emission control valve set.
15. Remove the upper intake manifold (intake air surge tank). Discard the gasket.
16. Remove the fuel pipe sub-assembly.
17. Disconnect the wiring at the injectors.
18. Remove the 4 bolts and 2 delivery pipe along with the injectors.
19. Remove the delivery pipe spacers and insulators from the manifold.
20. Pull each injector from the pipe.

To install:

21. Install new O-rings on each injector. Apply a light coating of gasoline to the O-rings and mating points on the pipes.
22. Using a twisting motion, install the injectors on the pipes.

➡**Be careful to avoid twisting the O-rings. After installation, check that the injectors turn smoothly. If not, use new O-rings.**

23. Install the pipes and injectors.
24. Loosely install the bolts and make sure that the injectors still turn freely. If not, replace the O-rings.
25. Torque the bolts to 84 inch lbs. (10 Nm).
26. The remainder of installation is the reverse of removal. Observe the following torques:
   - Fuel line union bolt: 24 ft. lbs. (33 Nm)
   - Pulsation damper: 24 ft. lbs. (33 Nm)
   - Fuel feed pipe: 14 ft. lbs. (20 Nm)
   - Upper intake manifold (air surge tank): 21 ft. lbs. (28 Nm)
   - Upper intake manifold stays: 14 ft. lbs. (20 Nm)

RX300

1. Before servicing the vehicle, refer to the precautions section.
2. Remove or disconnect the following:
   - Outer front cowl top panel assembly
   - Air cleaner cap with hose
   - Negative battery cable. Work must be started approximately 90 seconds or longer after the negative battery cable has been disconnected, if equipped with an air bag.
   - Coolant
   - Accelerator and throttle cables
   - V-bank cover
   - Emission valve control set
   - No. 2 EGR pipe
   - Hydraulic motor pressure pipe from the water inlet and air inlet chamber
   - Air intake chamber assembly
   - Injector wiring
   - Air assist pipe from the bracket on the No. 1 fuel pipe
   - Air assist hoses from the intake manifold
   - Fuel return hose from the No. 1 fuel pipe
   - Fuel inlet hose for the fuel filter
   - 2 union bolts holding the No. 2 fuel pipe to the delivery pipes
   - Fuel return hose from the fuel pressure regulator
   - Union bolt for the right hand delivery pipe, 2 gaskets, 2 bolts, left hand delivery pipe together with the 3 injectors and the No. 2 fuel pipe
   - Union bolt for the delivery pipe and 2 gaskets from the No. 2 fuel pipe
   - The 3 bolts, right hand delivery pipe together with the 3 injectors and the No. 1 fuel pipe
   - The 4 spacers from the intake manifold

   - The 6 injectors from the delivery pipes
   - The two O-rings and two grommets from each injector

To install:

3. Install or connect the following:
   - 2 new grommets to each injector
   - New O-rings, with a light coat of fuel, to each injector
   - Injectors
   - The 4 spacers on the intake manifold
   - Right hand delivery pipe and the No. 1 fuel pipe together with the 3 injectors in position on the intake manifold
   - Bolt holding the right side delivery pipe, temporarily, to the intake manifold
   - Left hand delivery pipe and the No. 2 fuel pipe together with the 3 injectors in position on the intake manifold
   - Fuel return hose to the fuel pressure regulator
4. Temporarily install the 2 bolts holding the left hand delivery pipe to the intake manifold.
5. Temporarily install the No. 2 fuel pipe to the left side delivery pipe with the union bolt and 2 new gaskets.
6. Check that the injectors rotate smoothly. If they do not, replace the O-rings.
7. Position the injector connector outward. Tighten the 4 bolts holding the delivery pipes to the intake manifold and tighten to 7 ft. lbs. (10 Nm). Tighten the bolt holding the No. 1 fuel pipe to the intake manifold to 14 ft. lbs. (20 Nm). Tighten the 2 union bolts holding the no. 2 fuel pipe to the delivery pipes to 24 ft. lbs. (32 Nm).
8. Install or connect the following:
   - Fuel inlet and return hoses. Union bolt: 22 ft. lbs. (30 Nm)
   - Fuel return hose to the No. 1 fuel pipe. Pass the fuel return hose under the heater hoses.
   - Air assist hoses to the intake manifold
   - Air assist pipe to the bracket on the No. 1 fuel pipe
   - Fuel injector wiring connectors
   - Air intake chamber assembly
   - Hydraulic motor pressure pipe to the intake chamber. Bolts: 69 inch lbs. (8 Nm)
   - No. 2 EGR pipe with new gaskets, tighten to 9 ft. lbs. (12 Nm)
   - Emission control valve set
   - V-bank cover

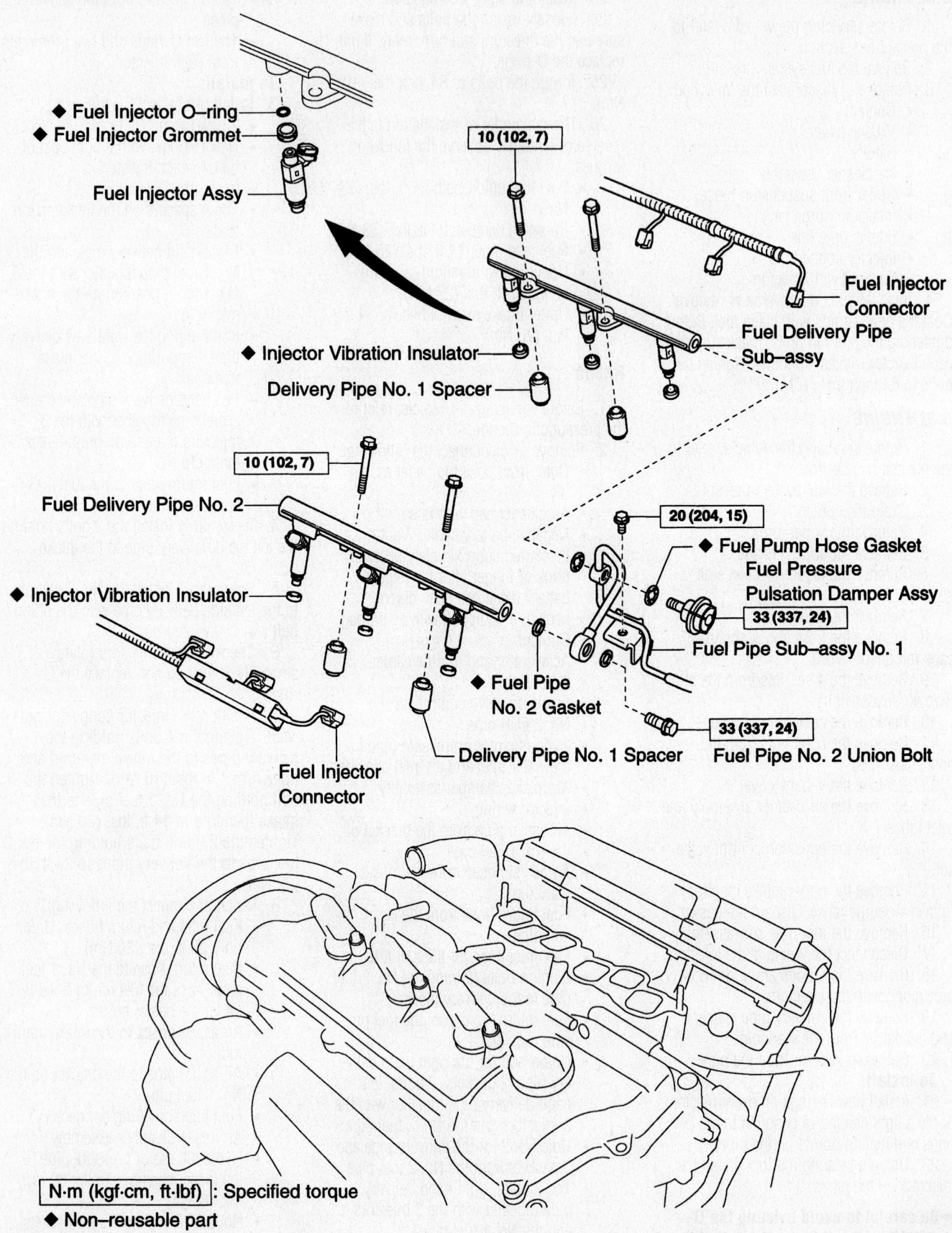

◆ Fuel Injector O–ring

◆ Fuel Injector Grommet

Fuel Injector Assy

10 (102, 7)

Fuel Injector Connector

Fuel Delivery Pipe Sub–assy

◆ Injector Vibration Insulator

Delivery Pipe No. 1 Spacer

10 (102, 7)

Fuel Delivery Pipe No. 2

◆ Injector Vibration Insulator

20 (204, 15)

◆ Fuel Pump Hose Gasket
Fuel Pressure
Pulsation Damper Assy

33 (337, 24)

Fuel Pipe Sub–assy No. 1

◆ Fuel Pipe No. 2 Gasket

Delivery Pipe No. 1 Spacer

33 (337, 24) Fuel Pipe No. 2 Union Bolt

Fuel Injector Connector

N·m (kgf·cm, ft·lbf) : Specified torque

◆ Non–reusable part

67170-HIGH-G28

Fuel injectors and related parts—3.3L Highlander

- Air cleaner hose
- Throttle and accelerator cables
- Coolant
- Air cleaner cap with hose
- Outer front cowl top panel assembly
- Negative battery cable

RX330

1. Before servicing the vehicle, refer to the precautions section.
2. Relieve the fuel system pressure.
3. Drain the coolant.
4. Remove the wiper arms.
5. Remove the wiper linkage.
6. Remove the fender-to-cowl side seals.
7. Remove the rain sensor.
8. Remove the front shock absorber caps (air suspension).
9. Remove the 4 set nuts from the strut (w/o air suspension).
10. Remove the cowl top outer panel.

11. Remove the 6 set nuts from the shock absorber.
12. Remove the V-bank cover.
13. Remove the air cleaner assembly and inlet tubes.
14. Remove the emission control valve set.
15. Remove the upper intake manifold (intake air surge tank). Discard the gasket.
16. Remove the fuel pipe sub-assembly.
17. Disconnect the wiring at the injectors.
18. Remove the 4 bolts and 2 delivery pipe along with the injectors.
19. Remove the delivery pipe spacers and insulators from the manifold.
20. Pull each injector from the pipe.

To install:

21. Install new O-rings on each injector. Apply a light coating of gasoline to the O-rings and mating points on the pipes.
22. Using a twisting motion, install the injectors on the pipes.

➡**Be careful to avoid twisting the O-rings. After installation, check that the injectors turn smoothly. If not, use new O-rings.**

23. Install the pipes and injectors.
24. Loosely install the bolts and make sure that the injectors still turn freely. If not, replace the O-rings.
25. Torque the bolts to 84 inch lbs. (10 Nm).
26. The remainder of installation is the reverse of removal. Observe the following torques:

- Fuel line union bolt: 24 ft. lbs. (33 Nm)
- Pulsation damper: 24 ft. lbs. (33 Nm)
- Fuel feed pipe: 14 ft. lbs. (20 Nm)
- Upper intake manifold (air surge tank): 21 ft. lbs. (28 Nm)
- Upper intake manifold stays: 14 ft. lbs. (20 Nm)

DRIVE TRAIN

Automatic Transaxle Assembly

REMOVAL & INSTALLATION

Highlander

2.4L ENGINE

1. Remove the engine/transaxle assembly.
2. Remove the halfshafts.
3. Remove the engine mounting bracket (4wd).
4. Remove the transfer case (4wd).
5. Disconnect the wiring.
6. Remove the starter.
7. Remove the cables and hoses.
8. Remove the filler tube.
9. Remove the front engine mount bracket.
10. Remove the flywheel housing undercover.
11. Turn the crankshaft to gain access to the torque converter bolts. There is one green bolt.
12. Remove the 9 engine-to-transaxle bolts. Separate the transaxle from the engine.
13. Installation is the reverse of removal. Observe the following torques:

- Transaxle-to-engine: bolts A 47 ft. lbs (64 Nm); bolts B 34 ft. lbs. (46 Nm); bolts C 27 ft. lbs. (37 Nm)
- Torque converter bolts (use a thread locking compound such as

Three Bond 1324): 30 ft. lbs. (41 Nm)

➡**Install the green bolt first.**

- Undercover: 69 inch lbs. (8 Nm)
- Engine mount bracket: 47 ft. lbs. (64 Nm)
- Transfer case-to-transaxle: 51 ft. lbs. (69 Nm)

3.0L ENGINE

1. Before servicing the vehicle, refer to the precautions section.
2. Remove or disconnect the following:

- Hood
- Wiper and blade assembly
- Top cowl seal and panel
- Window washer hoses, from the ventilator louvers
- Left and right ventilator louvers
- Heater air duct
- Battery and tray
- Throttle cable
- Front upper suspension brace
- Cruise control actuator with its bracket, if equipped
- Starter
- Shift control cable
- Driveshaft, for 4WD
- Body-to-engine ground strap
- Park/Neutral Position (PNP) switch, solenoid and ATF temperature connectors
- 5 upper transaxle-to-engine mounting bolts

- Front wheel
- Engine undercover
- Halfshafts
- Front exhaust pipe
- Stabilizer bar
- Both steering gear mounting bolts and support it in the vehicle
- Shift control cable from its bracket
- Power steering pipe and the oil cooler clamps from the frame

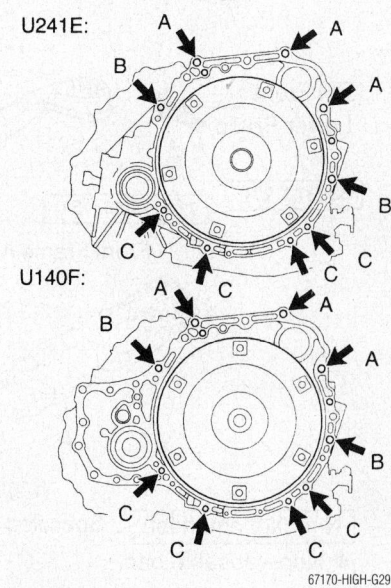

U241E:

U140F:

67170-HIGH-G29

Engine-to-transaxle bolt identification— 2.4L Highlander

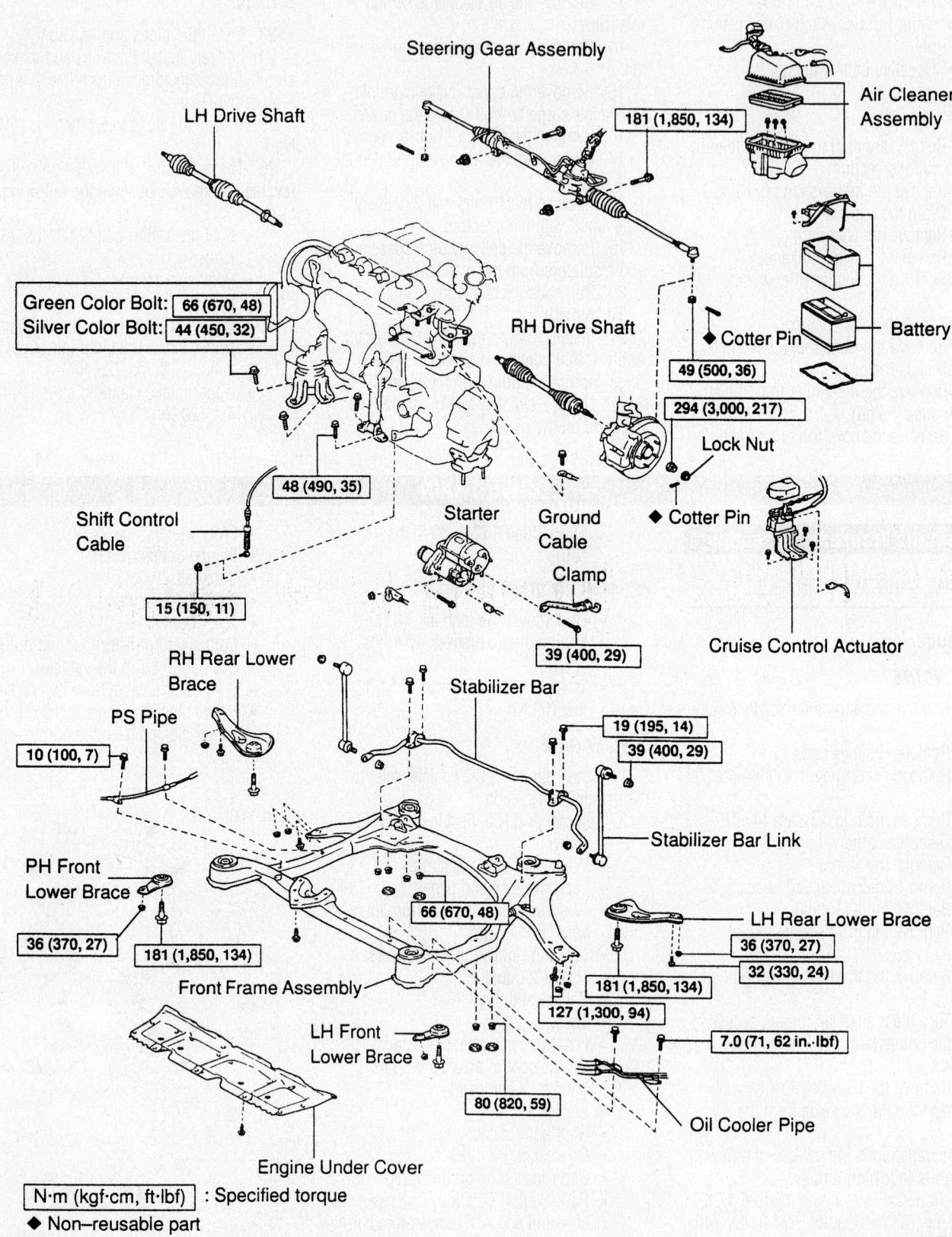

Steering Gear Assembly

LH Drive Shaft

181 (1,850, 134)

Air Cleaner Assembly

Green Color Bolt: 66 (670, 48)
Silver Color Bolt: 44 (450, 32)

RH Drive Shaft

◆ **Cotter Pin**

Battery

49 (500, 36)

294 (3,000, 217)

Lock Nut

Shift Control Cable

48 (490, 35)

Starter

Ground Cable

◆ **Cotter Pin**

Clamp

15 (150, 11)

39 (400, 29)

Cruise Control Actuator

RH Rear Lower Brace

Stabilizer Bar

19 (195, 14)

39 (400, 29)

PS Pipe

10 (100, 7)

Stabilizer Bar Link

PH Front Lower Brace

66 (670, 48)

LH Rear Lower Brace

36 (370, 27)

36 (370, 27)

181 (1,850, 134)

32 (330, 24)

181 (1,850, 134)

Front Frame Assembly

127 (1,300, 94)

LH Front Lower Brace

7.0 (71, 62 in.·lbf)

80 (820, 59)

Oil Cooler Pipe

Engine Under Cover

N·m (kgf·cm, ft·lbf) : Specified torque

◆ Non-reusable part

7924ZG65

Exploded view of the transaxle removal and installation components—3.0L Highlander

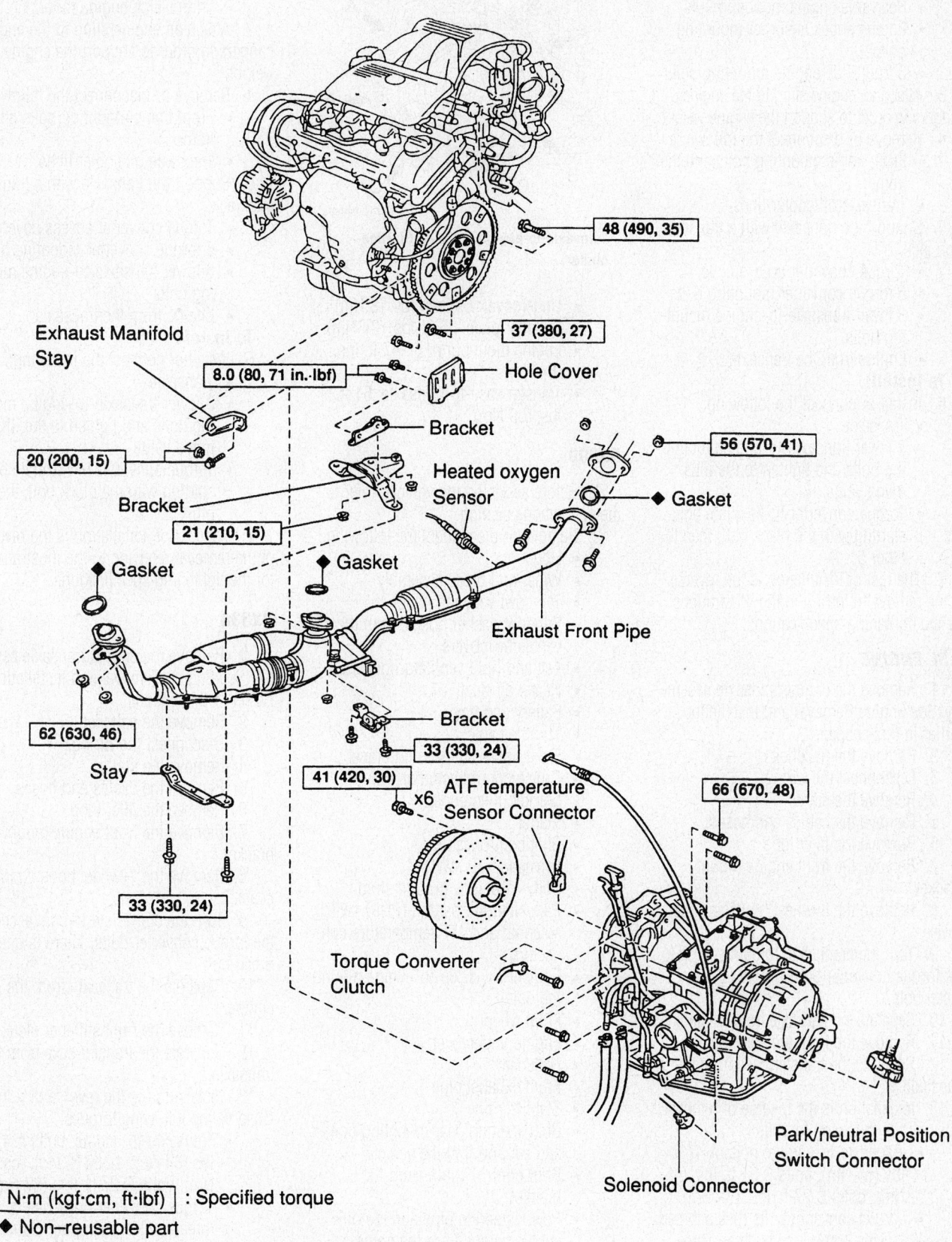

Exhaust Manifold Stay

8.0 (80, 71 in.·lbf)

Hole Cover

48 (490, 35)

37 (380, 27)

Bracket

56 (570, 41)

20 (200, 15)

Heated oxygen Sensor

Bracket

◆ Gasket

21 (210, 15)

◆ Gasket

◆ Gasket

Exhaust Front Pipe

62 (630, 46)

Bracket

Stay

33 (330, 24)

41 (420, 30)

x6

ATF temperature Sensor Connector

66 (670, 48)

33 (330, 24)

Torque Converter Clutch

Solenoid Connector

Park/neutral Position Switch Connector

N·m (kgf·cm, ft·lbf) : Specified torque

◆ Non–reusable part

7924ZG66

Exploded view of the transaxle removal and installation components—3.0L Highlander, Cont.

- Both left-side transaxle mounting nuts
- Rear-side engine mounting nuts
- Engine shock absorber mounting bolts
- 3 front-side engine mounting bolts

3. Attach an engine sling to the engine hangers in order to support the engine weight.

4. Remove or disconnect the following:
- Front frame mounting bolts and the frame
- Transaxle oil cooler lines

5. Support the transaxle with a transmission jack.
- Torque converter access cover
- 6 torque converter mounting bolts
- 3 lower transaxle-to-engine mounting bolts
- Engine from the transaxle

To install:

6. Install or connect the following:
- Transaxle
- 3 lower transaxle-to-engine mounting bolts and tighten to the illustrated value.
- Torque converter-to-flexplate bolts, starting with the black bolt, then the other 5.

7. The rest of installation is the reverse of the removal referring to the illustrations for the tightening specifications.

3.3L ENGINE

1. Remove the engine/transaxle assembly. See Engine Removal and Installation, earlier in this chapter.

2. Remove the halfshafts.

3. Disconnect the wiring.

4. Remove the starter.

5. Remove the cables and hoses.

6. Remove the filler tube.

7. Remove the front engine mount bracket.

8. Remove the flywheel housing undercover.

9. Turn the crankshaft to gain access to the torque converter bolts. There is one green bolt.

10. Separate the transaxle from the engine.

11. Remove the right stiffener plate.

12. Separate the transfer case from the transaxle.

13. Installation is the reverse of removal. Observe the following torques:
- Transaxle-to-engine: bolts A 47 ft. lbs (64 Nm); bolts B 34 ft. lbs. (46 Nm); bolts C 27 ft. lbs. (37 Nm);
- Torque converter bolts (use a thread locking compound such as Three Bond 1324): 30 ft. lbs. (41 Nm)

➡**Install the green bolt first.**

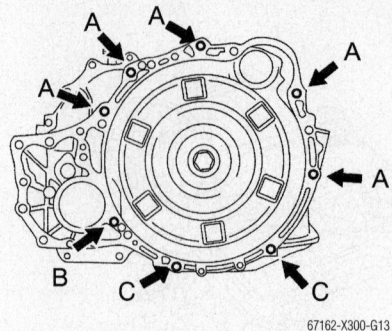

67162-X300-G13

Transaxle-to-engine bolts—3.3L Highlander

- Undercover: 69 inch lbs. (8 Nm)
- Stiffener plate: 25 ft. lbs. (34 Nm)
- Engine mount bracket: 47 ft. lbs. (64 Nm)
- Transfer case-to-transaxle: 51 ft. lbs. (69 Nm)

RX300

1. Before servicing the vehicle, refer to the precautions section.

2. Remove or disconnect the following:
- Hood
- Wiper and blade assembly
- Top cowl seal and panel
- Window washer hoses, from the ventilator louvers
- Left and right ventilator louvers
- Heater air duct
- Battery and tray
- Throttle cable
- Front upper suspension brace
- Cruise control actuator with its bracket, if equipped
- Starter
- Shift control cable
- Driveshaft, for 4WD
- Body-to-engine ground strap
- Park/Neutral Position (PNP) switch, solenoid and ATF temperature connectors
- 5 upper transaxle-to-engine mounting bolts
- Front wheel
- Engine undercover
- Halfshafts
- Front exhaust pipe
- Stabilizer bar
- Both steering gear mounting bolts and support it in the vehicle
- Shift control cable from its bracket
- Power steering pipe and the oil cooler clamps from the frame
- Both left-side transaxle mounting nuts
- Rear-side engine mounting nuts

- Engine shock absorber mounting bolts
- 3 front-side engine mounting bolts

3. Attach an engine sling to the engine hangers in order to support the engine weight.

4. Remove or disconnect the following:
- Front frame mounting bolts and the frame
- Transaxle oil cooler lines

5. Support the transaxle with a transmission jack.
- Torque converter access cover
- 6 torque converter mounting bolts
- 3 lower transaxle-to-engine mounting bolts
- Engine from the transaxle

To install:

6. Install or connect the following:
- Transaxle
- 3 lower transaxle-to-engine mounting bolts and tighten to the illustrated value.
- Torque converter-to-flexplate bolts, starting with the black bolt, then the other 5.

7. The rest of installation is the reverse of the removal referring to the illustrations for the tightening specifications.

RX330

1. Remove the engine/transaxle assembly. See Engine Removal and Installation, earlier in this chapter.

2. Remove the halfshafts.

3. Disconnect the wiring.

4. Remove the starter.

5. Remove the cables and hoses.

6. Remove the filler tube.

7. Remove the front engine mount bracket.

8. Remove the flywheel housing undercover.

9. Turn the crankshaft to gain access to the torque converter bolts. There is one green bolt.

10. Separate the transaxle from the engine.

11. Remove the right stiffener plate.

12. Separate the transfer case from the transaxle.

13. Installation is the reverse of removal. Observe the following torques:
- Transaxle-to-engine: bolts A 47 ft. lbs (64 Nm); bolts B 34 ft. lbs. (46 Nm); bolts C 27 ft. lbs. (37 Nm)
- Torque converter bolts (use a thread locking compound such as Three Bond 1324): 30 ft. lbs. (41 Nm)

➡**Install the green bolt first.**

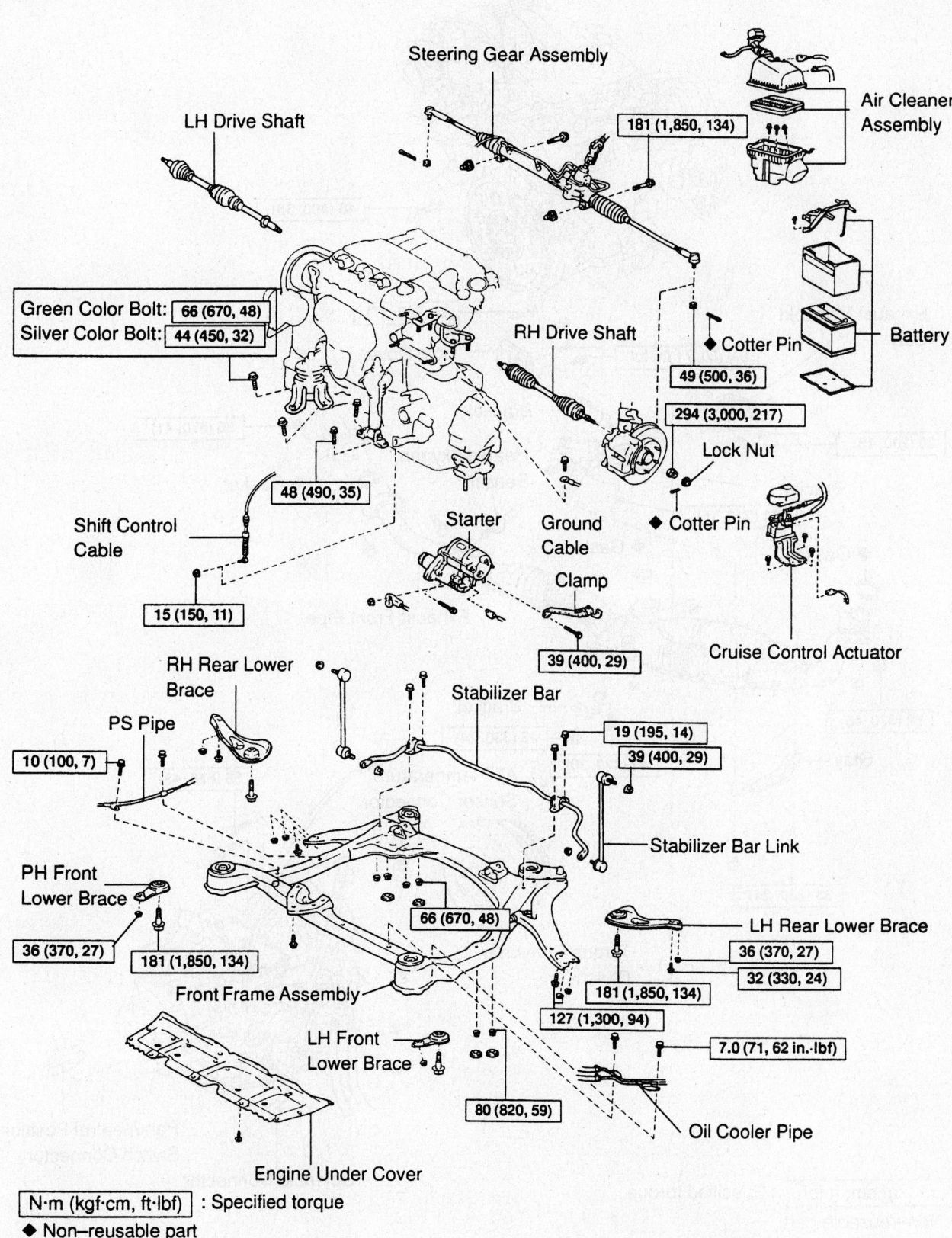

Steering Gear Assembly

LH Drive Shaft

181 (1,850, 134)

Air Cleaner Assembly

Green Color Bolt: 66 (670, 48)
Silver Color Bolt: 44 (450, 32)

RH Drive Shaft

◆ Cotter Pin

Battery

49 (500, 36)

294 (3,000, 217)

Lock Nut

48 (490, 35)

Shift Control Cable

Starter

Ground Cable

◆ Cotter Pin

Clamp

Cruise Control Actuator

15 (150, 11)

39 (400, 29)

RH Rear Lower Brace

Stabilizer Bar

PS Pipe

19 (195, 14)

39 (400, 29)

10 (100, 7)

Stabilizer Bar Link

PH Front Lower Brace

LH Rear Lower Brace

66 (670, 48)

36 (370, 27)

36 (370, 27)

181 (1,850, 134)

32 (330, 24)

Front Frame Assembly

181 (1,850, 134)

127 (1,300, 94)

LH Front Lower Brace

7.0 (71, 62 in.·lbf)

80 (820, 59)

Oil Cooler Pipe

Engine Under Cover

N·m (kgf·cm, ft·lbf) : Specified torque
◆ Non–reusable part

7924ZG65

Exploded view of the transaxle removal and installation components—RX300 models

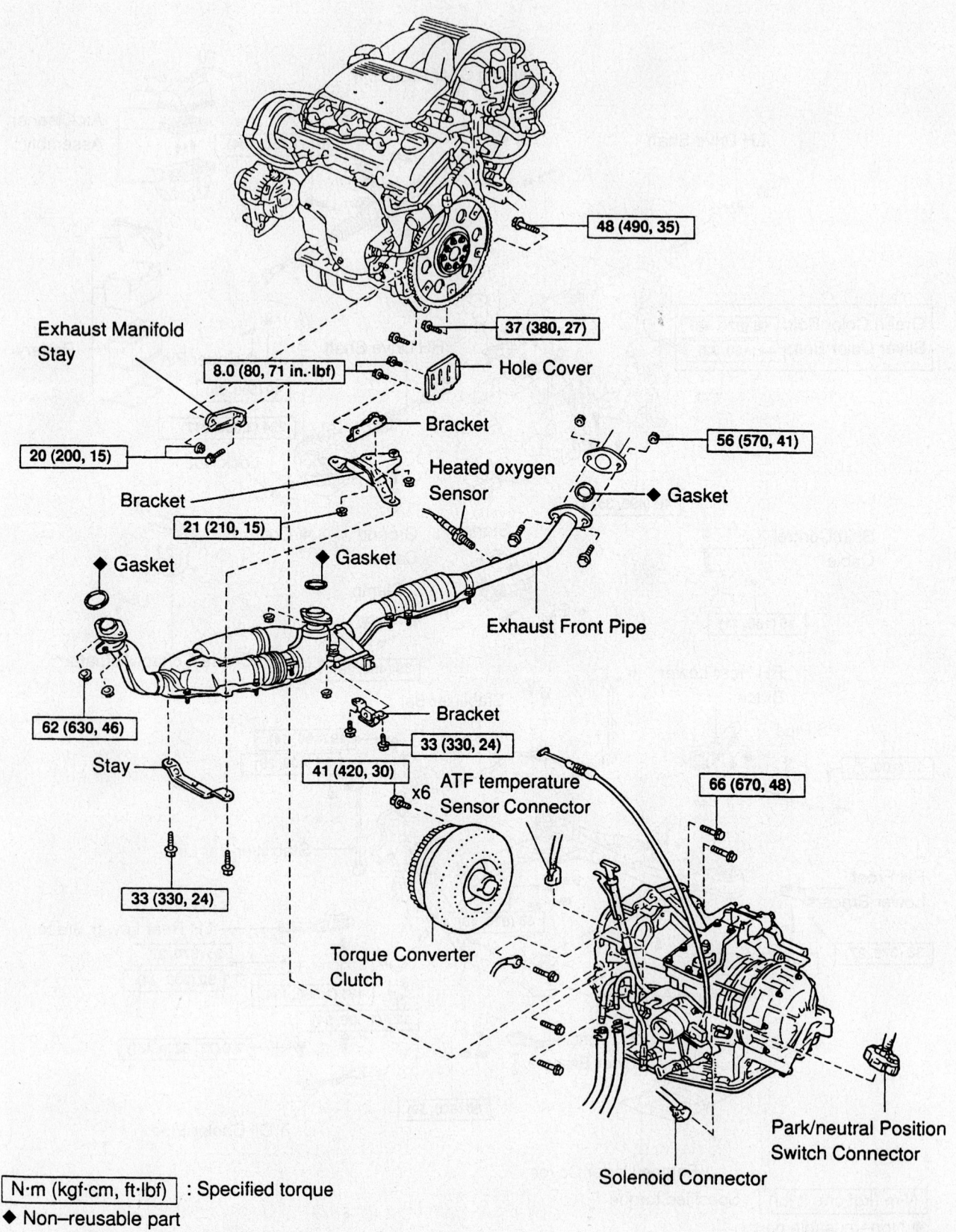

Exhaust Manifold Stay

48 (490, 35)

37 (380, 27)

8.0 (80, 71 in..lbf)

Hole Cover

20 (200, 15)

Bracket

Bracket

56 (570, 41)

Heated oxygen Sensor

21 (210, 15)

◆ Gasket

◆ Gasket

◆ Gasket

Exhaust Front Pipe

62 (630, 46)

Stay

Bracket

33 (330, 24)

41 (420, 30)

x6

ATF temperature Sensor Connector

66 (670, 48)

33 (330, 24)

Torque Converter Clutch

Park/neutral Position Switch Connector

Solenoid Connector

N·m (kgf·cm, ft·lbf) : Specified torque

◆ Non–reusable part

7924ZG66

Exploded view of the transaxle removal and installation components—RX300 models, Cont.

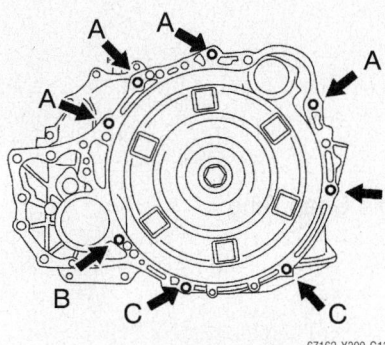

Transaxle-to-engine bolts—RX330

- Undercover: 69 inch lbs. (8 Nm)
- Stiffener plate: 25 ft. lbs. (34 Nm)
- Engine mount bracket: 47 ft. lbs. (64 Nm)
- Transfer case-to-transaxle: 51 ft. lbs. (69 Nm)

Transfer Case Assembly

REMOVAL & INSTALLATION

Highlander

1. Remove the engine/transaxle assembly.
2. Drain the transaxle.
3. With the 2.4L engine, remove the stiffener plate (5 bolts).
4. With the 2.4L engine, remove the right rear engine mount bracket (3 bolts).
5. Separate the engine and transaxle.

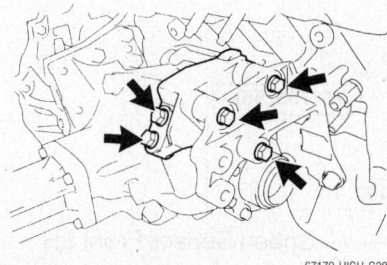

Stiffener plate—Highlander

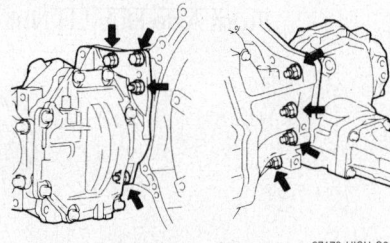

Transfer case fastener locations—Highlander

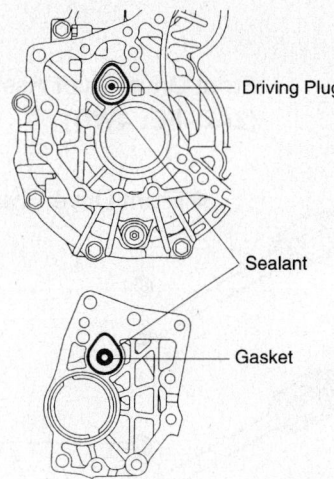

Gasket material application—Highlander

6. Remove the 2 bolts and 6 nuts and separate the transfer case from the transaxle. It will be necessary to break it loose with a plastic mallet.

➡**Keep the transfer case level during removal. Don't grasp the oil seals.**

To install:

7. Clean all grease from the mating surfaces.
8. Apply a continuous 1.2mm diameter bead of silicone gasket material to the transaxle and transfer case as shown.
9. Join the transfer case to the transaxle within 10 minutes of gasket material application. If not, remove the material and start again.
10. Torque the nuts and bolts to 51 ft. lbs. (69 Nm).
11. The remainder of installation is the reverse of removal. Observe the following torques:
    - Engine mount bracket: 47 ft. lbs. (64 Nm)
    - Stiffener plate: 25 ft. lbs. (34 Nm)
    - Drain plug: 36 ft. lbs. (49 Nm)

RX300 and RX330

The transfer case is part of the transmission/transaxle assembly and is serviced with those units.

Halfshaft

REMOVAL & INSTALLATION

Highlander

FRONT

1. Before servicing the vehicle, refer to the precautions section.

2. Remove or disconnect the following:
   - Front wheels
   - Fender apron seal
   - Transaxle fluid
   - Transfer case oil (4wd)
   - Hub nut
   - Stabilizer bar link
   - Speed sensor
   - Tie rod end
   - Lower arm from the ball joint
3. Slide the halfshaft from the hub, then, carefully, pry the shaft from the transaxle.
4. Installation is the reverse of removal. Torque the hub nut to 217 ft. lbs. (294Nm).

REAR—2002–03

1. Before servicing the vehicle, refer to the precautions section.
2. Remove or disconnect the following:
   - Negative battery cable
   - Rear wheels
   - Anti-lock Brake System (ABS) speed sensor from the axle assembly by removing the bolt, if equipped
   - Cotter pin, lock cap and the nut holding the halfshaft to the axle carrier
3. Place matchmarks on the halfshaft and differential side gear shaft.
4. Remove or disconnect the following:
   - 4 nuts, washers and the halfshaft from the differential
   - Halfshaft from the axle carrier

To install:

5. Install or connect the following:
   - Halfshaft into the axle carrier. Tighten the 4 nuts to 51 ft. lbs. (69 Nm).
   - Halfshaft. Tighten the locknut to 159 ft. lbs. (216 Nm).
   - ABS sensor
   - Rear wheels
   - Negative battery cable

REAR—2004–06

1. Before servicing the vehicle, refer to the precautions section.
2. Remove the rear wheel.
3. Disconnect and remove the speed sensor.
4. Remove the axle shaft nut.
5. Disconnect the height control sensor.
6. Disconnect the rear control arms.
7. Disconnect the strut rod.
8. Push the rear axle carrier sub-assembly towards the outside and separate the shaft from the carrier.
9. Remove the shaft, keeping it level.

To install

10. Align the shaft splines and install the shaft with a brass bar and hammer.

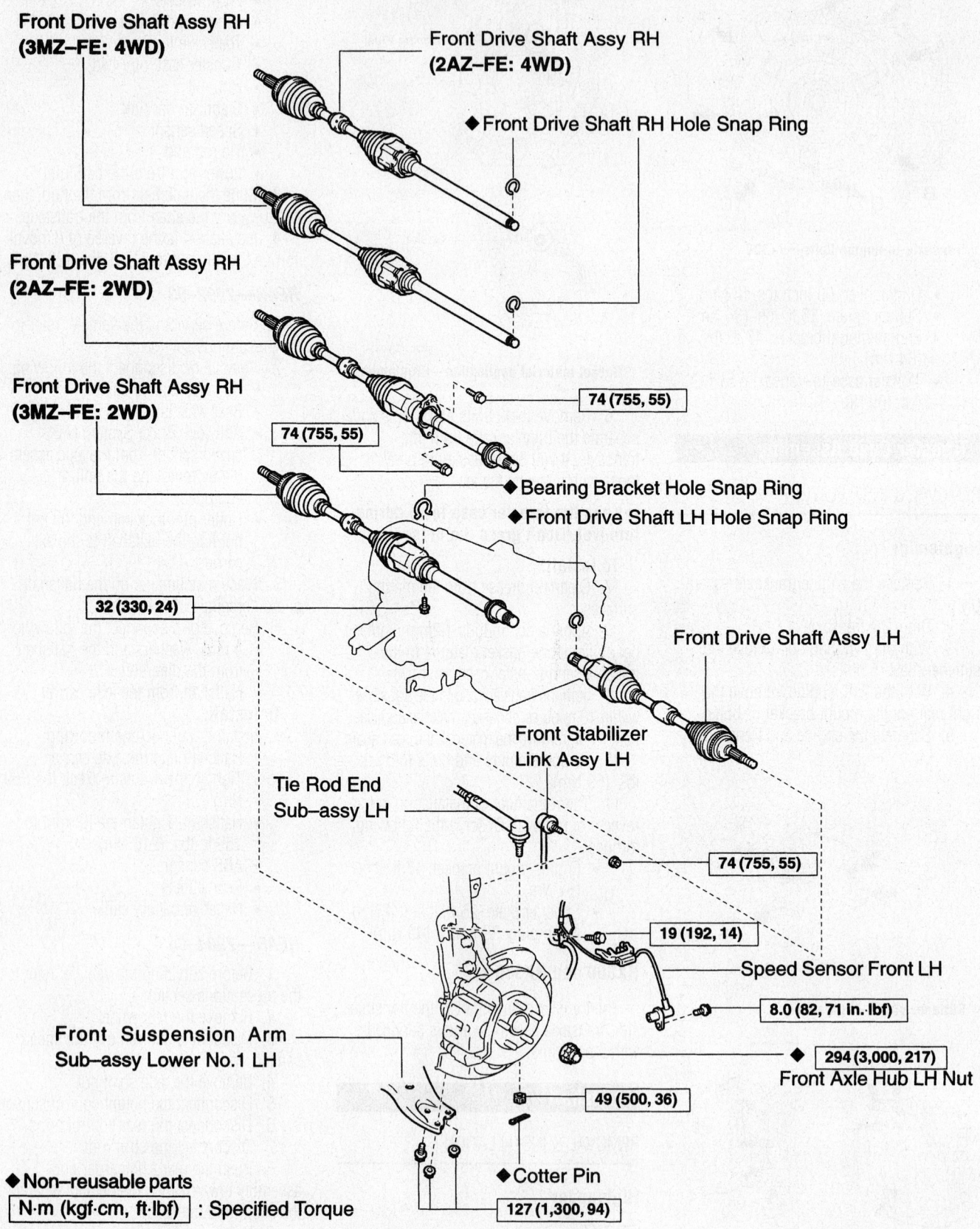

Front Drive Shaft Assy RH
(3MZ–FE: 4WD)

Front Drive Shaft Assy RH
(2AZ–FE: 4WD)

◆ Front Drive Shaft RH Hole Snap Ring

Front Drive Shaft Assy RH
(2AZ–FE: 2WD)

Front Drive Shaft Assy RH
(3MZ–FE: 2WD)

74 (755, 55)

74 (755, 55)

◆ Bearing Bracket Hole Snap Ring

◆ Front Drive Shaft LH Hole Snap Ring

32 (330, 24)

Front Drive Shaft Assy LH

Front Stabilizer
Link Assy LH

Tie Rod End
Sub–assy LH

74 (755, 55)

19 (192, 14)

Speed Sensor Front LH

8.0 (82, 71 in.·lbf)

Front Suspension Arm
Sub–assy Lower No.1 LH

◆ 294 (3,000, 217)
Front Axle Hub LH Nut

49 (500, 36)

◆ Non–reusable parts

N·m (kgf·cm, ft·lbf) : Specified Torque

◆ Cotter Pin

127 (1,300, 94)

Front halfshaft and related parts—Highlander

67170-HIGH-G33

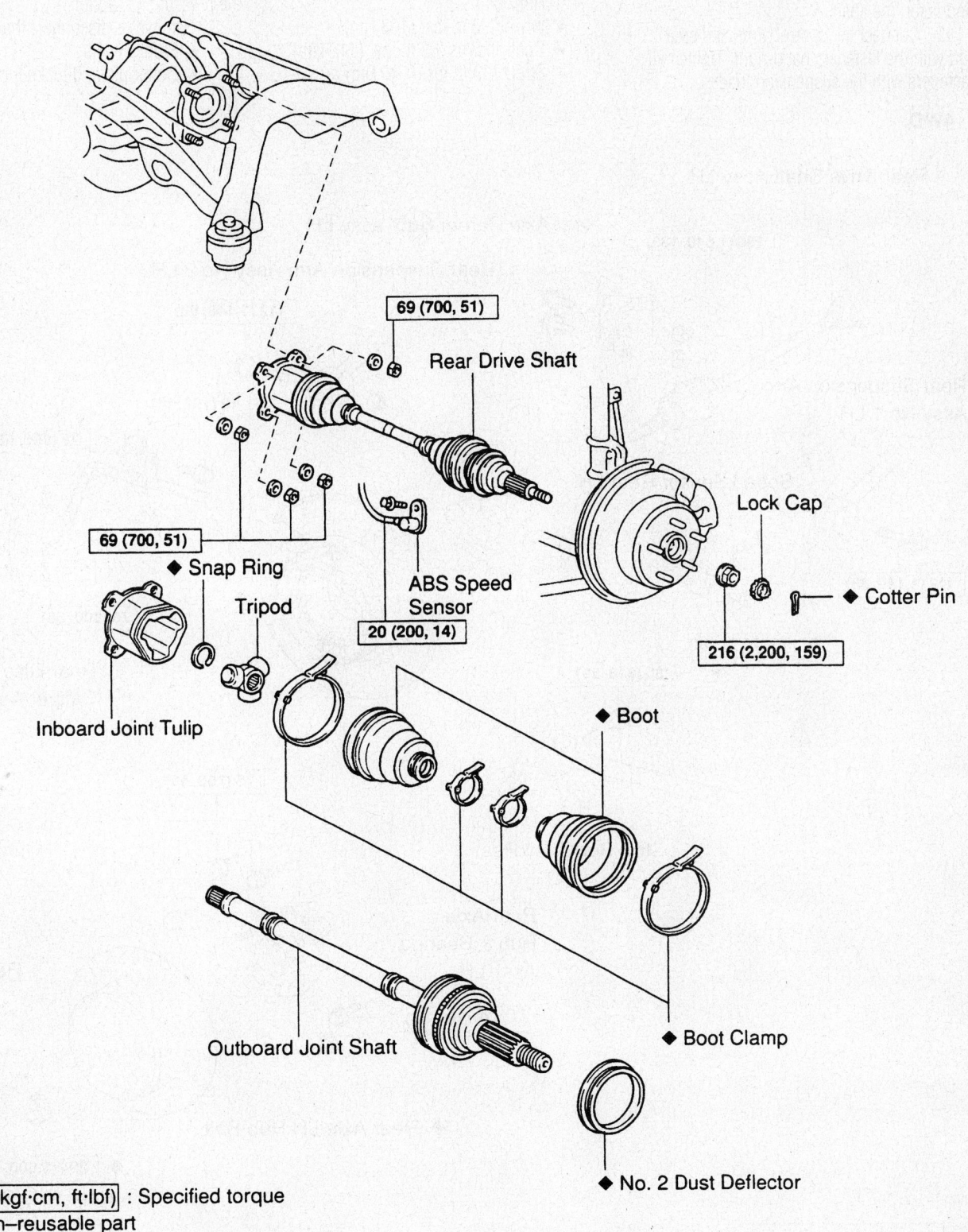

69 (700, 51)

Rear Drive Shaft

69 (700, 51)

◆ Snap Ring

Tripod

Inboard Joint Tulip

ABS Speed
Sensor

20 (200, 14)

Lock Cap

◆ Cotter Pin

216 (2,200, 159)

◆ Boot

◆ Boot Clamp

Outboard Joint Shaft

◆ No. 2 Dust Deflector

N·m (kgf·cm, ft·lbf) : Specified torque
◆ Non–reusable part

7924ZG88

Exploded view of the rear halfshaft—2002–03 Highlander

➡**Set the snapring with the opening side facing downward. Keep the shaft level.**

11. Push the carrier towards the inside and insert the shaft.

12. Connect the control arms and strut rod with the fasteners hand-tight. Tighten all fasteners with the suspension loaded.

13. The remainder of installation is the reverse of removal. Observe the following torques:
  - Axle shaft nut: 217 ft. lbs. (294 Nm)
  - Wheel: 76 ft. lbs. (103 Nm)
  - Control arms: 83 ft. lbs. (112 Nm)
  - Strut: 133 ft. lbs. (180 Nm)

RX300 and RX330

FRONT

1. Before servicing the vehicle, refer to the precautions section.
2. Remove or disconnect the following:
  - Front wheels
  - Cotter pin and locknut cap

4WD:

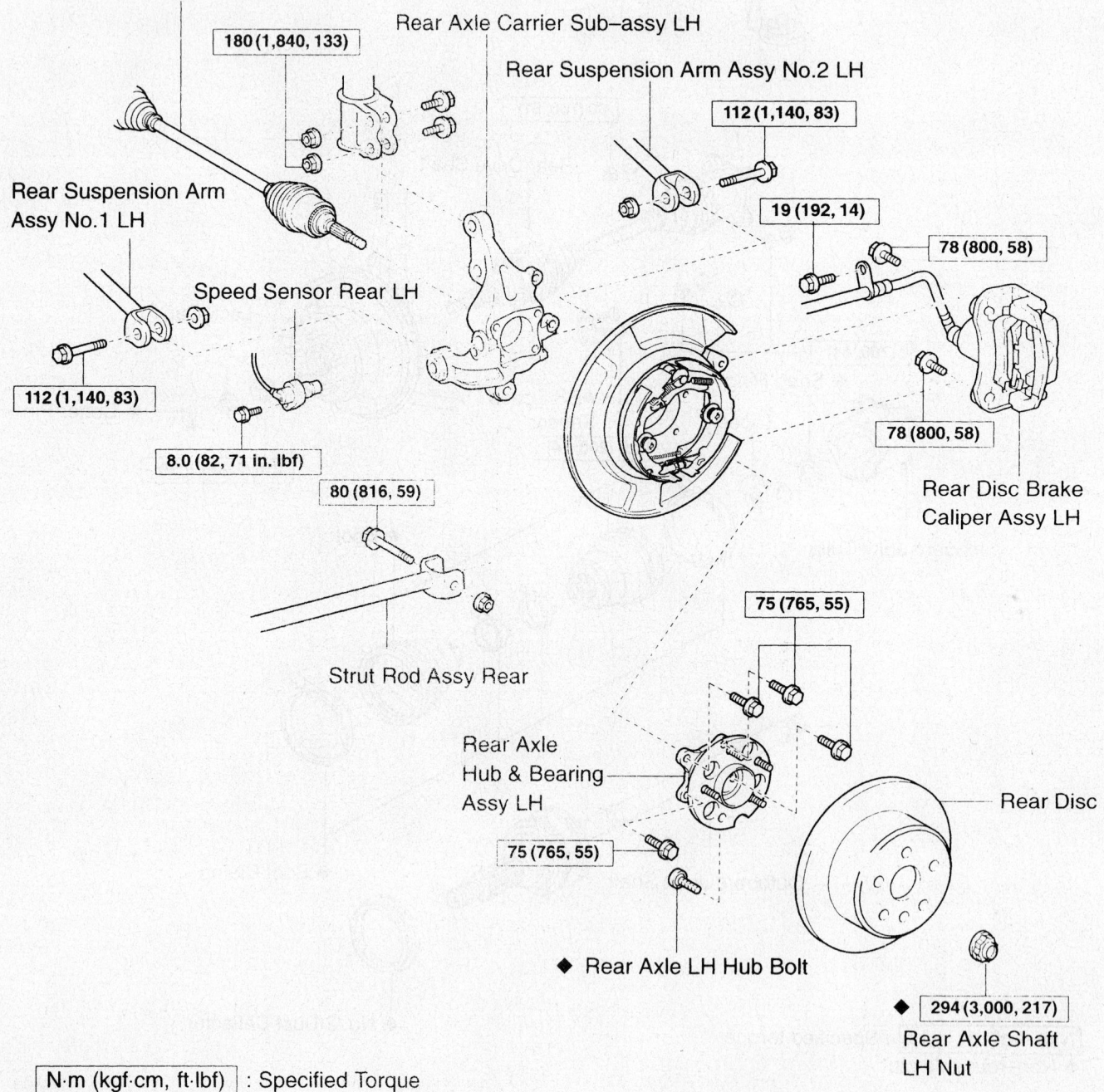

Rear Drive Shaft Assy LH

180 (1,840, 133)

Rear Axle Carrier Sub–assy LH

Rear Suspension Arm Assy No.2 LH

112 (1,140, 83)

Rear Suspension Arm Assy No.1 LH

Speed Sensor Rear LH

19 (192, 14)

78 (800, 58)

112 (1,140, 83)

8.0 (82, 71 in. lbf)

80 (816, 59)

78 (800, 58)

Rear Disc Brake Caliper Assy LH

Strut Rod Assy Rear

75 (765, 55)

Rear Axle Hub & Bearing Assy LH

Rear Disc

75 (765, 55)

◆ Rear Axle LH Hub Bolt

◆ 294 (3,000, 217)

Rear Axle Shaft LH Nut

N·m (kgf·cm, ft·lbf) : Specified Torque
◆ Non–reusable parts

67170-HIGH-G34

Exploded view of the rear halfshaft—2004–06 Highlander

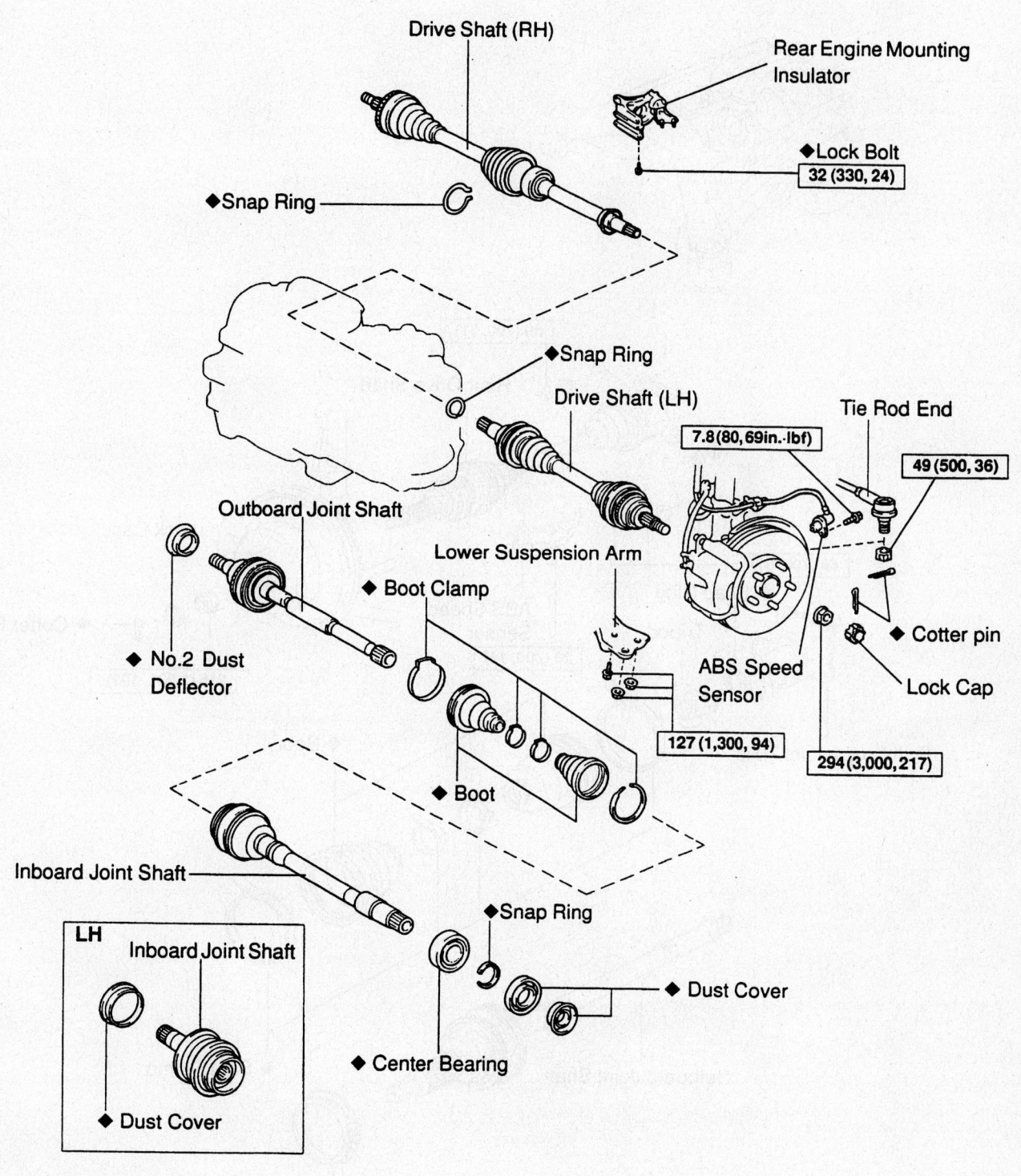

Drive Shaft (RH)

Rear Engine Mounting Insulator

◆Snap Ring

◆Lock Bolt
32 (330, 24)

◆Snap Ring

Drive Shaft (LH)

Tie Rod End

7.8 (80, 69in.·lbf)

49 (500, 36)

Outboard Joint Shaft

Lower Suspension Arm

◆ Boot Clamp

◆ No.2 Dust Deflector

◆ Cotter pin

ABS Speed Sensor

Lock Cap

◆ Boot

127 (1,300, 94)

294 (3,000, 217)

Inboard Joint Shaft

◆Snap Ring

LH

Inboard Joint Shaft

◆ Dust Cover

◆ Center Bearing

◆ Dust Cover

N·m (kgf·cm, ft·lbf) : Specified torque

◆ Non–reusable part

7924ZG73

Exploded view of front halfshaft—RX300 and RX330

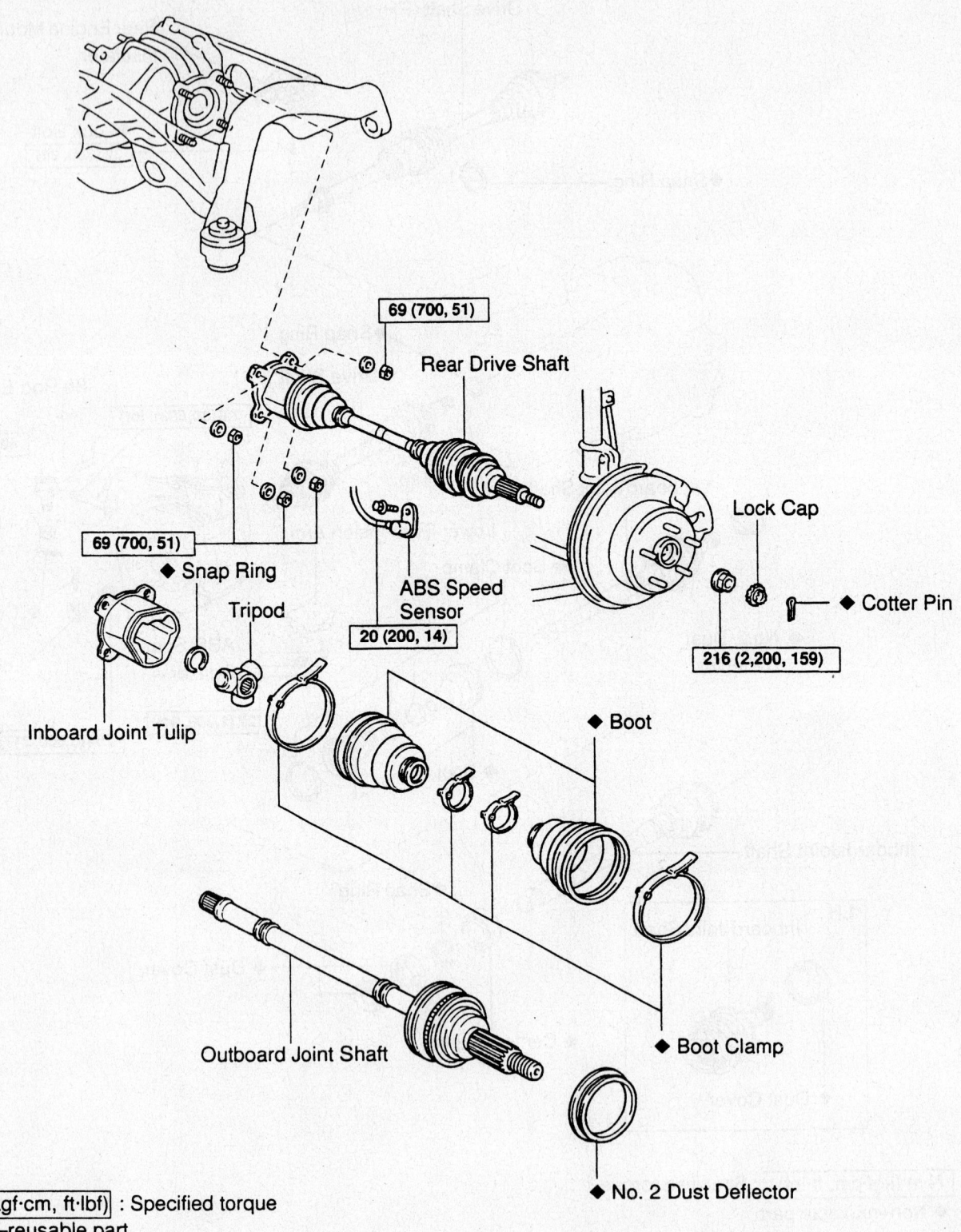

69 (700, 51)

Rear Drive Shaft

69 (700, 51)

Lock Cap

◆ Snap Ring

Tripod

◆ Cotter Pin

ABS Speed Sensor

20 (200, 14)

Inboard Joint Tulip

216 (2,200, 159)

◆ Boot

Outboard Joint Shaft

◆ Boot Clamp

◆ No. 2 Dust Deflector

N·m (kgf·cm, ft·lbf) : Specified torque
◆ Non–reusable part

7924ZG88

Exploded view of the rear halfshaft—RX300 model with 4WD

→**Have an assistant depress the brake pedal and loosen the bearing locknut.**

- Engine undercover
- Fender apron seal
- Tie rod end, from the steering knuckle
- Steering knuckle, from the lower control arm
- Halfshaft from the axle hub, using a plastic hammer
- Cover the outer boot with a rag
- Halfshaft from the transaxle, using the proper tools

To install:

3. Reverse the removal procedures to complete installation, tightening fasteners to specifications.

4. Fill the transaxle with gear oil, install the fender apron, check front end alignment and test drive.

→**If the cotter pin holes do not align, always correct by tightening the nut until the next hole aligns.**

5. Install a new cotter pin.

REAR—RX300

1. Before servicing the vehicle, refer to the precautions section.
2. Remove or disconnect the following:
   - Negative battery cable
   - Rear wheels
   - Anti-lock Brake System (ABS) speed sensor from the axle assembly by removing the bolt, if equipped
   - Cotter pin, lock cap and the nut holding the halfshaft to the axle carrier
3. Place matchmarks on the halfshaft and differential side gear shaft.
4. Remove or disconnect the following:
   - 4 nuts, washers and the halfshaft from the differential
   - Halfshaft from the axle carrier

To install:

5. Install or connect the following:
   - Halfshaft into the axle carrier. Tighten the 4 nuts to 51 ft. lbs. (69 Nm).
   - Halfshaft. Tighten the locknut to 159 ft. lbs. (216 Nm).
   - ABS sensor
   - Rear wheels
   - Negative battery cable

REAR—RX330

1. Before servicing the vehicle, refer to the precautions section.
2. Remove the rear wheel.
3. Disconnect and remove the speed sensor.

4. Remove the axle shaft nut.
5. Disconnect the height control sensor.
6. Disconnect the rear control arms.
7. Disconnect the strut rod.
8. Push the rear axle carrier sub-assembly towards the outside and separate the shaft from the carrier.
9. Remove the shaft, keeping it level.

To install

10. Align the shaft splines and install the shaft with a brass bar and hammer.

→**Set the snapring with the opening side facing downward. Keep the shaft level.**

11. Push the carrier towards the inside and insert the shaft.
12. Connect the control arms and strut rod with the fasteners hand-tight. Tighten all fasteners with the suspension loaded.
13. The remainder of installation is the reverse of removal. Observe the following torques:
   - Axle shaft nut: 217 ft. lbs. (294 Nm)
   - Wheel: 76 ft. lbs. (103 Nm)
   - Control arms: 83 ft. lbs. (112 Nm)
   - Strut: 133 ft. lbs. (180 Nm)

CV-Joints

OVERHAUL

1. Before servicing the vehicle, refer to the precautions section.
2. Remove the inboard and outboard joint boot clamps.
3. Disassemble the inboard joint tulip, as follows:
   - Matchmark the tri-pot, inboard joint tulip or center driveshaft to the driveshaft

❊❊ WARNING

Do not use punch marks.

   - Inboard joint tulip from the driveshaft
4. Remove the inboard and outboard joint clamps.
5. Remove the tri-pot joint, as follows:
   - Snapring
   - Matchmark the tri-pot joint to the driveshaft
   - Tri-pot joint, using a brass bar and hammer

❊❊ WARNING

Do not tap the roller.

6. Remove or disconnect the following:
   - Inboard and outboard joint boots

→**Do not disassemble the outboard joint.**

   - Dust cover from the center drive-shaft, using a press, for 2WD on the right side
   - Dust cover from the inboard joint tulip, using tool 09950-00020 and a press, for 2WD on the left side and 4WD
7. Disassemble the center driveshaft, as follows:
   - Snapring
   - Bearing case, using a press
   - Straight pin from the bearing case, using a pin punch and hammer
   - Dust cover, using tool 09950-00020 and a press
   - Snapring
   - Bearing, using a press
8. Remove the No. 2 dust deflector, using a screwdriver and hammer.

To assemble:

9. Install a new No. 2 dust deflector, using a press.
10. Assemble the center driveshaft, as follows:
   - Straight pin into the bearing case, using a pin punch and hammer
   - New bearing, using tools 09959-60010, 09950-70010 and a press
   - New snapring
   - Bearing with the bearing case assembly to the center driveshaft, using tool 09710-30021 and a press
   - New snapring
   - New dust cover, until the clearance between the dust cover and the bearing is 0.039 in. (1.0mm)
11. Install or connect the following:
   - Right dust cover (2WD), until the distance from the tip of the center drive is 3.39–3.34 in. (86–87mm) to the inner edge of the dust cover
   - Left side dust cover (2WD and 4WD), using a press
12. Temporarily install new outboard/inboard joint boots using new clamps, as follows:
   a. Warp tape around the driveshaft splines.
   b. Install the new outboard joint boot onto the driveshaft.
   c. Install the new inboard joint boot onto the driveshaft.
13. Install the tri-pot joint, as follows:
   - Tri-pot joint, face the beveled side toward the outboard joint and align the matchmarks
   - Tri-pot joint onto the driveshaft, using a press

☀☀ WARNING

Be careful not to tap the roller.

- New snapring

14. Install the outboard joint boot packed with grease from the boot kit.

15. Install the inboard joint tulip, as follows:

- Pack the inboard joint boot with grease from the boot kit
- Inboard joint tulip, by aligning the matchmarks
- Temporarily, install the inboard joint boot packed with grease from the kit

16. Install the boot clamps to both boots, as follows:

- Both boots to the shaft grooves
- Halfshaft length should be 33.055–33.449 in. (839.6–849.6mm) for the right side on 2WD with A/T, 21.397–21.791 in.

(543.5–553.5mm) for the left side on 2WD with A/T, 19.929–20.323 in. (506.2–516.2mm) for the right side on 4WD or 19.803–20.197 in. (503–511mm) for the left side on 4WD

- Both new boot clamps boot
- Bend the band and lock it using a screwdriver

Pinion Seal

REMOVAL & INSTALLATION

Rear

1. Before servicing the vehicle, refer to the precautions section.
2. Drain the differential oil.
3. Remove or disconnect the following:

- Exhaust pipe
- Driveshaft by matchmarking it

- Companion flange nut, by loosen the staked portion
- Companion flange, using a screw-type extractor
- Oil seal, using an extractor
- Slinger
- Front bearing
- Spacer

To install:

4. Install or connect the following

- New spacer
- Bearing
- Slinger
- New seal

➡**Seal installation depth: 2.0mm +/- 0.3mm**

- Companion flange
- New nut. Coat the threads with clean differential oil. Torque the nut to 80 ft. lbs. (108Nm).

5. The remainder of installation is the reverse of removal.

STEERING AND SUSPENSION

Air Bag

☀☀ CAUTION

Some vehicles are equipped with an air bag system. The system must be disarmed before performing service on, or around, system components, the steering column, instrument panel components, wiring and sensors. Failure to follow the safety precautions and the disarming procedure could result in accidental air bag deployment, possible injury and unnecessary system repairs.

PRECAUTIONS

Several precautions must be observed when handling the inflator module to avoid accidental deployment and possible personal injury.

- Never carry the inflator module by the wires or connector on the underside of the module.
- When carrying a live inflator module, hold securely with both hands and ensure that the bag and trim cover are pointed away.
- Place the inflator module on a bench or other surface with the bag and trim cover facing up.
- With the inflator module on the bench, never place anything on or close to the

module which may be thrown in the event of an accidental deployment.

DISARMING

To avoid personal injury when working on vehicles equipped with an air bag, the negative battery cable must be disconnected and at least 90 seconds must elapse before working on the system. Failure to do so may result in deployment of the air bag.

Power Rack And Pinion Steering Gear

REMOVAL & INSTALLATION

Highlander

1. Before servicing the vehicle, refer to the precautions section.
2. Remove or disconnect the following:

- Negative battery cable

➡**Wait at least 90 seconds before working on the vehicle to allow the Supplemental Restraint System (SRS) system to disarm.**

- Steering wheel
- Front wheels
- Tie rod ends
- Intermediate shaft

➡**Matchmark the shaft and gear.**

- Stabilizer bar end links

- Pressure and return lines
- Steering gear
- Installation is the reverse of removal. Observe the following torques:
- Rack mounting bolts: 52 ft. lbs. (70Nm)
- Stabilizer bar end links: 55 ft. lbs. (74Nm)
- Intermediate shaft bolt: 26 ft. lbs. (35Nm)
- Tie rod end nuts: 36 ft. lbs. (49Nm)

RX300

1. Before servicing the vehicle, refer to the precautions section.
2. Remove or disconnect the following:

- Negative battery cable

➡**Wait at least 90 seconds before working on the vehicle to allow the Supplemental Restraint System (SRS) system to disarm.**

- Right and left side fender apron seals
- Right and left tie rod ends

3. Place matchmarks on the intermediate shaft.
4. Remove or disconnect the following:

- Pinch bolt and the intermediate shaft out from under the vehicle
- Power steering line clamp
- Pressure and feed lines
- Stabilizer bar, unbolt it but do not remove it

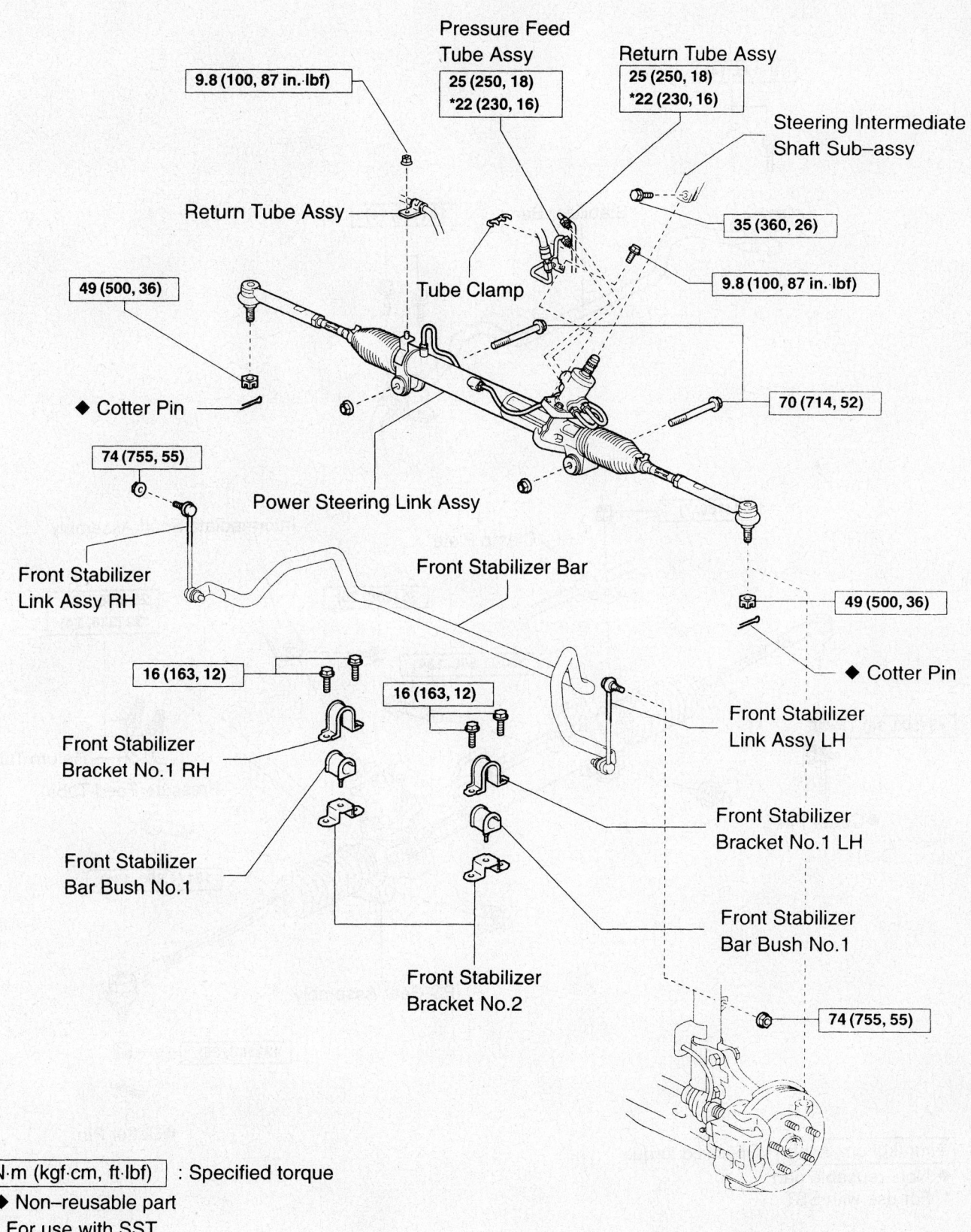

9.8 (100, 87 in. lbf)

Pressure Feed
Tube Assy
25 (250, 18)
***22 (230, 16)**

Return Tube Assy
25 (250, 18)
***22 (230, 16)**

Steering Intermediate
Shaft Sub–assy

Return Tube Assy

Tube Clamp

35 (360, 26)

9.8 (100, 87 in. lbf)

49 (500, 36)

◆ Cotter Pin

70 (714, 52)

74 (755, 55)

Power Steering Link Assy

Front Stabilizer
Link Assy RH

Front Stabilizer Bar

49 (500, 36)

◆ Cotter Pin

Front Stabilizer
Link Assy LH

16 (163, 12)

16 (163, 12)

Front Stabilizer
Bracket No.1 RH

Front Stabilizer
Bracket No.1 LH

Front Stabilizer
Bar Bush No.1

Front Stabilizer
Bar Bush No.1

Front Stabilizer
Bracket No.2

74 (755, 55)

N·m (kgf·cm, ft·lbf) : Specified torque
◆ Non–reusable part
* For use with SST

67170-HIGH-G35

Steering rack and related parts—Highlander

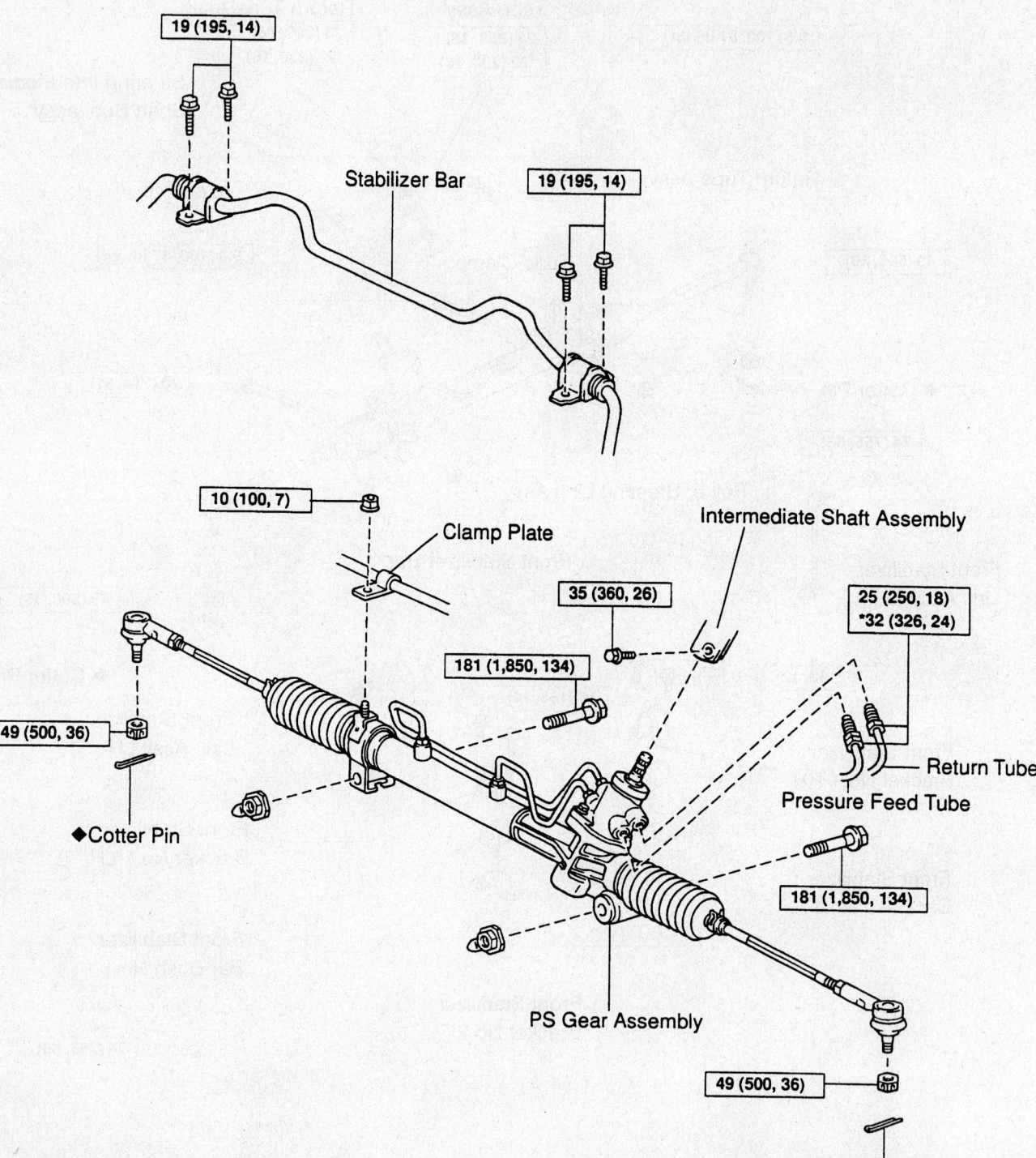

19 (195, 14)

Stabilizer Bar

19 (195, 14)

10 (100, 7)

Clamp Plate

Intermediate Shaft Assembly

35 (360, 26)

25 (250, 18)
*32 (326, 24)

181 (1,850, 134)

Return Tube

Pressure Feed Tube

49 (500, 36)

◆Cotter Pin

181 (1,850, 134)

PS Gear Assembly

49 (500, 36)

◆Cotter Pin

N·m (kgf·cm, ft·lbf) : Specified torque
◆ Non–reusable part
* For use with SST

7924ZG76

Exploded view of the power steering gear and related components—RX300 models

- Heated Oxygen (HO₂) sensor
- Both gear assembly set bolts and nuts, by lifting the stabilizer bar
- Gear assembly from the left side of the vehicle

To install:

5. Install or connect the following:
- Gear assembly to the left side of the vehicle

✳✳ WARNING

Be careful not to damage the power steering lines.

- Tighten the gear assembly set bolts and nuts to 134 ft. lbs. (181 Nm), by lifting the stabilizer bar
- HO₂ sensor
- Stabilizer bar. Tighten the bolt to 14 ft. lbs. (19 Nm) and the nut to 29 ft. lbs. (39 Nm).
- Pressure and feed return lines. Tighten them to 18 ft. lbs. (25 Nm).
- Line clamps. Tighten the nut to 84 inch lbs. (10 Nm).
- Intermediate shaft, by aligning the joint and main shaft matchmarks. Tighten to 26 ft. lbs. (35 Nm).
- Tie rod ends
- Fender apron seals. Securely tighten the bolts.

6. Remove or disconnect the following:
- Steering wheel pad
- Steering wheel

7. Position the front wheels facing straight-ahead. Do this with the front of the vehicle on jackstands.

8. Center the spiral cable.

9. Install the steering wheel at the straight-ahead position. Temporarily tighten the wheel set nut. Attach the wiring.

10. Bleed the power steering system.

11. Check the steering wheel center point. Tighten the steering nut to 26 ft. lbs. (35 Nm).

12. Check and/or adjust the front wheel alignment.

RX330

1. Before servicing the vehicle, refer to the precautions section.

2. Center the steering wheel.

3. Matchmark and disconnect the intermediate shaft.

4. Remove the wheels.

5. Separate the tie rods.

6. Disconnect the stabilizer bar end links.

7. Remove the front exhaust pipe.

8. Remove the stabilizer bar brackets.

9. Remove the height control sensor.

10. Disconnect the pressure and return lines.

11. Remove the bolts and nuts and remove the steering rack assembly.

To install:

12. Position the assembly and install the bolts and nuts. Torque the nuts to 52 ft. lbs. (70 Nm).

13. Connect the pressure and lines. Torque to 16 ft. lbs. (22 Nm). Torque the clamp bolt to 87 inch lbs. (10 Nm).

14. The remainder of installation is the reverse of removal. Observe the following torques:
- Tie rod end nuts: 36 ft. lbs. (49 Nm)
- Stabilizer bar end links: 55 ft. lbs. (74 Nm)
- Stabilizer bar bracket bolts: 12 ft. lbs. (16 Nm)
- Intermediate shaft bolt: 26 ft. lbs. (35 Nm)

Strut

REMOVAL & INSTALLATION

Highlander

FRONT—2002–03

1. Before servicing the vehicle, refer to the precautions section.

➡**Do not support the weight of the vehicle on the suspension arm; the arm will deform under its weight.**

2. Remove or disconnect the following:
- Wheel
- Stabilizer bar link
- Brake hose and the Anti-lock Brake System (ABS) speed sensor wire from the strut
- Strut lower end from the steering knuckle lower arm
- 3 upper strut mounting plate-to-upper wheel arch nuts
- Strut

To install:

3. Install or connect the following:
- Tighten the 3 suspension support-to-wheel arch nuts to 59 ft. lbs. (80 Nm).
- Tighten the strut-to-steering knuckle arm bolts to 155 ft. lbs. (210 Nm).
- Sway bar link to the strut. Tighten the nut to 55 ft. lbs. (74 Nm).
- ABS speed sensor and the brake hose to the strut, if equipped
- Wheel

4. Check and/or adjust the front wheel alignment.

FRONT—2004–06

1. Remove the wheel.

2. Disconnect the stabilizer bar link.

3. Loosen, don't remove, the strut locknut.

4. Disconnect the brake hose from the strut.

5. Remove the lower mounting bolts.

6. Remove the upper retaining nuts.

To install:

7. Position the strut and install the upper nuts. Torque to 59 ft. lbs. (80 Nm).

8. Install the lower bolts and torque to 170 ft. lbs. (230 Nm).

9. Connect the brake line.

10. Tighten the strut locknut to 36 ft. lbs. (49 Nm).

11. Connect the stabilizer links and torque to 55 ft. lbs. (74 Nm).

12. Install the wheel. Torque to 76 ft. lbs. (103 Nm).

REAR—2002–03

1. Before servicing the vehicle, refer to the precautions in the beginning of this section.

➡**Do not support the weight of the vehicle on the suspension arm; the arm will deform under its weight.**

2. Remove or disconnect the following:
- Wheel
- Brake hose and the Anti-lock Brake System (ABS) speed sensor wire from the strut
- Sway bar link from the strut

3. Loosen, but do not remove the 2 lower bolts.

4. Support the axle carrier with a jack and remove cap.

5. If the strut is being disassembled, loosen the center nut.

6. Remove the 3 mounting nuts.

7. Lower the carrier and remove the 2 lower nuts and bolts.

8. Installation is the reverse of removal. Observe the following torques:
- 3 mounting nuts: 29 ft. lbs. (39Nm)
- 2 lower nuts: 188 ft. lbs. (255Nm)
- Center nut: 36 ft. lbs. (49Nm)
- Stabilizer link: 29 ft. lbs. (39Nm)

FRONT—2004–06

1. Remove the tonneau cover.

2. Remove the left side deck trim cover.

3. Remove the rear wheels.

4. Remove the stabilizer link from the strut.

5. On 2-wheel drive models, disconnect the skid control sensor wire and brake hose from the strut and carrier.

6. On 4-wheel drive models, disconnect

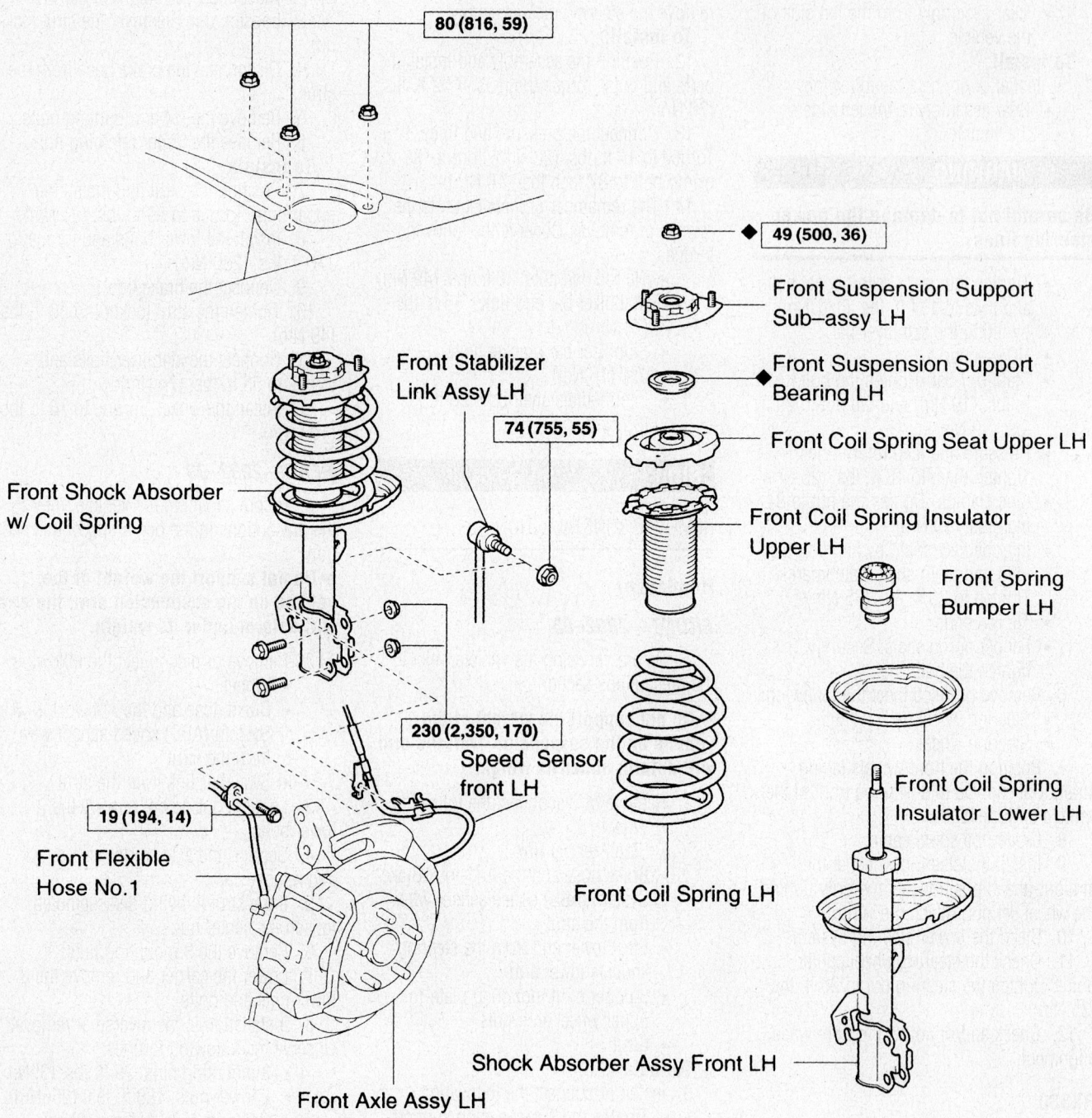

80 (816, 59)

49 (500, 36)

Front Suspension Support Sub–assy LH

Front Suspension Support Bearing LH

Front Coil Spring Seat Upper LH

74 (755, 55)

Front Stabilizer Link Assy LH

Front Shock Absorber w/ Coil Spring

Front Coil Spring Insulator Upper LH

Front Spring Bumper LH

230 (2,350, 170)

Speed Sensor front LH

Front Coil Spring Insulator Lower LH

19 (194, 14)

Front Flexible Hose No.1

Front Coil Spring LH

Shock Absorber Assy Front LH

Front Axle Assy LH

N·m (kgf·cm, ft·lbf) : Specified torque

◆ Non–reusable part

67170-HIGH-G36

Front strut and related parts—2004–06 Highlander models

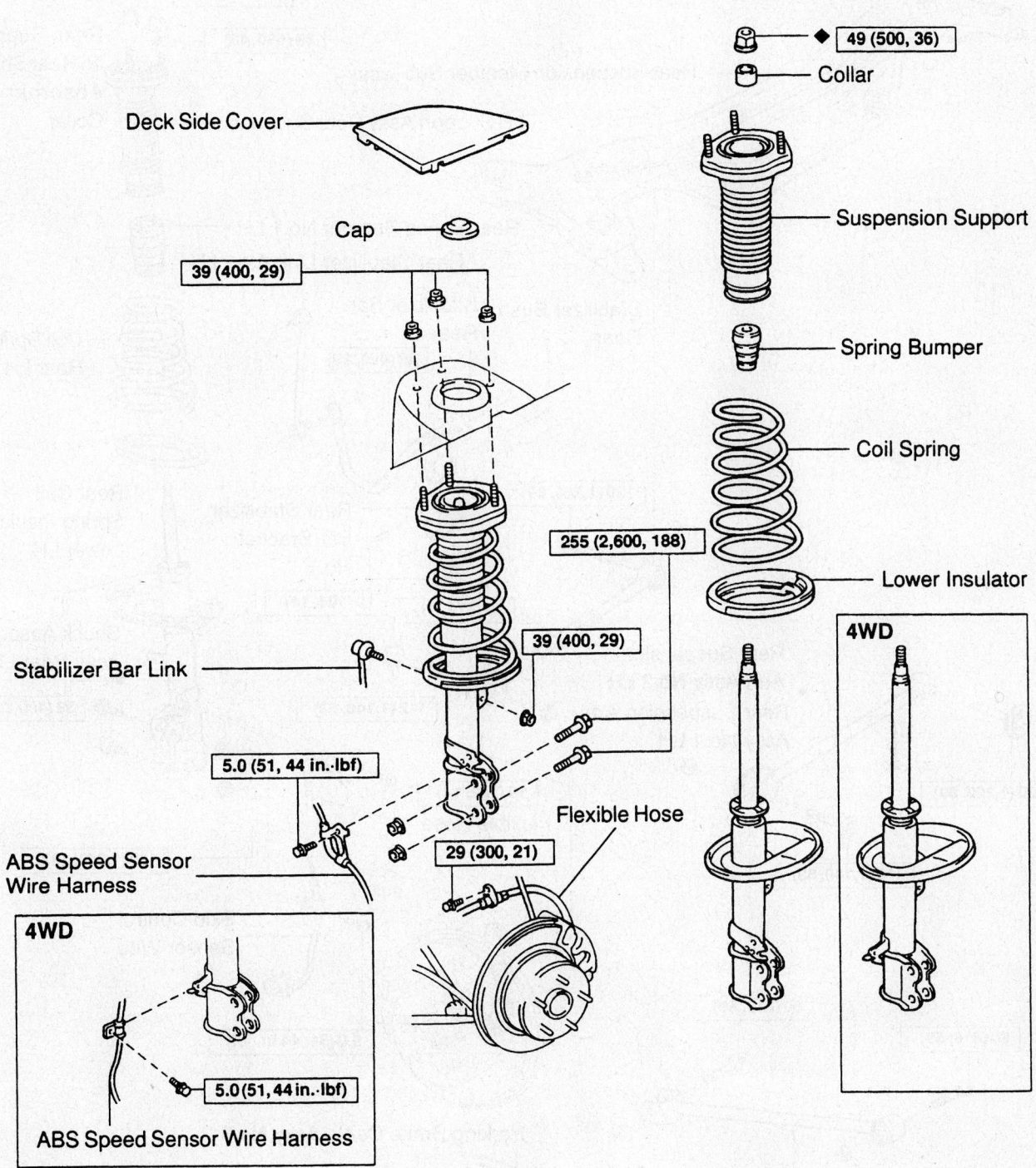

Deck Side Cover

Cap

39 (400, 29)

49 (500, 36)

Collar

Suspension Support

Spring Bumper

Coil Spring

255 (2,600, 188)

Lower Insulator

39 (400, 29)

4WD

Stabilizer Bar Link

5.0 (51, 44 in.·lbf)

ABS Speed Sensor
Wire Harness

Flexible Hose

29 (300, 21)

4WD

5.0 (51, 44 in.·lbf)

ABS Speed Sensor Wire Harness

N·m (kgf·cm, ft·lbf) : Specified torque

◆ Non-reusable part

7924ZG89

Rear strut and related parts—2002–03 Highlander models

FF:

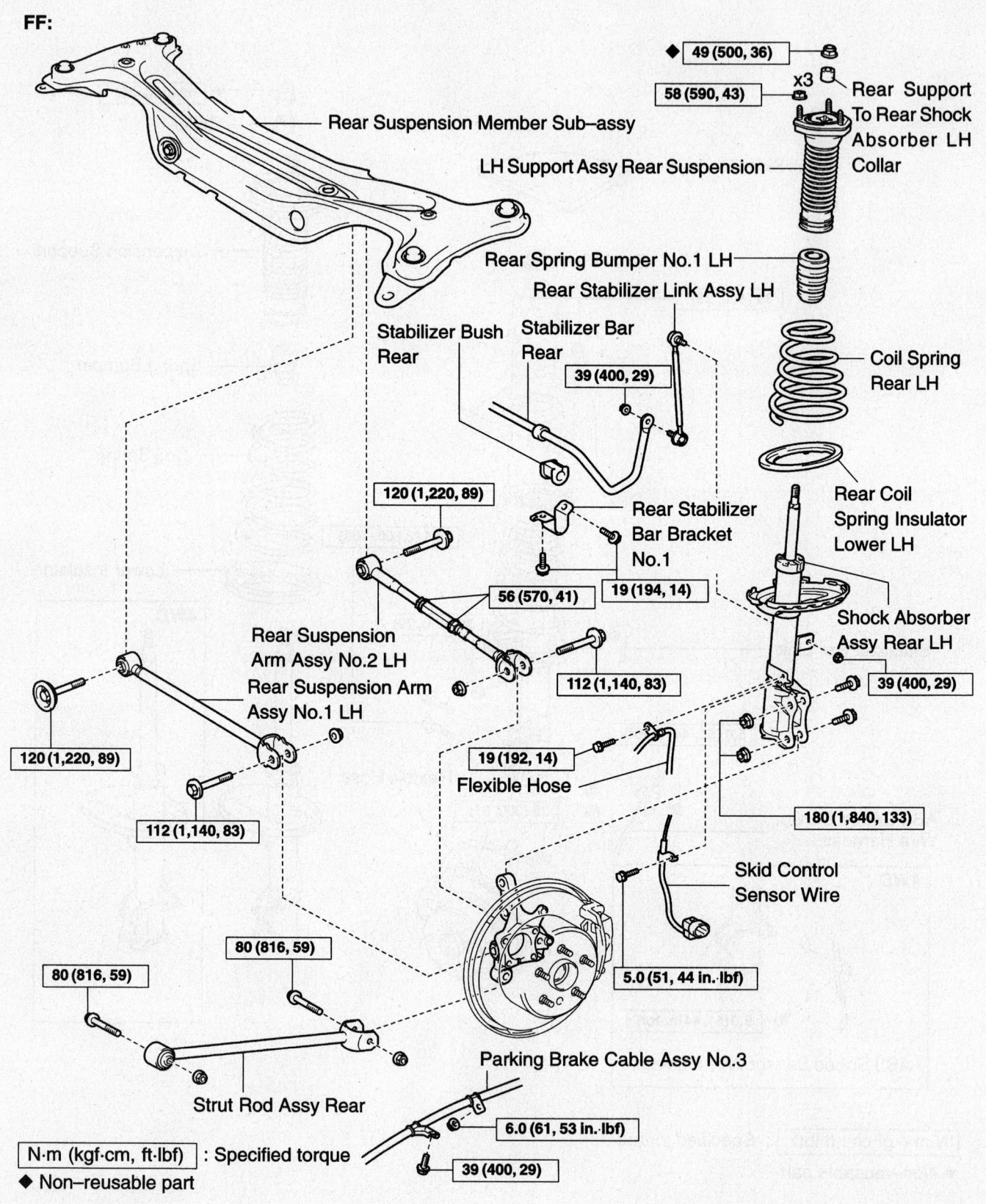

Rear Suspension Member Sub–assy

LH Support Assy Rear Suspension

◆ 49 (500, 36)

58 (590, 43)   x3

Rear Support To Rear Shock Absorber LH Collar

Rear Spring Bumper No.1 LH

Rear Stabilizer Link Assy LH

Stabilizer Bush Rear

Stabilizer Bar Rear

39 (400, 29)

Coil Spring Rear LH

120 (1,220, 89)

Rear Stabilizer Bar Bracket No.1

Rear Coil Spring Insulator Lower LH

56 (570, 41)

19 (194, 14)

Rear Suspension Arm Assy No.2 LH

Rear Suspension Arm Assy No.1 LH

112 (1,140, 83)

Shock Absorber Assy Rear LH

39 (400, 29)

120 (1,220, 89)

19 (192, 14)

Flexible Hose

180 (1,840, 133)

112 (1,140, 83)

Skid Control Sensor Wire

80 (816, 59)

80 (816, 59)

5.0 (51, 44 in.·lbf)

Parking Brake Cable Assy No.3

Strut Rod Assy Rear

N·m (kgf·cm, ft·lbf) : Specified torque

◆ Non–reusable part

6.0 (61, 53 in.·lbf)

39 (400, 29)

67170-HIGH-G37

Rear strut and related parts—2004–06 Highlander models

the brake hose and speed sensor from the strut and carrier.

7. Loosen the 2 nuts at the lower end of the strut, but don't remove them.

8. Support the rear axle carrier with a jack.

9. Remove the 3 upper strut nuts and lower the axle.

10. Remove the lower strut bolts and nuts. Lift out the strut.

To install:

11. Position the strut and install the 3 upper nuts. Torque to 43 ft. lbs. (58 Nm).

12. Lift the axle and install the 2 lower bolts and nuts. Torque to 133 ft. lbs. (180 Nm).

13. The remainder of installation is the reverse of removal. Observe the following torques:

- Brake hose clamp: 14 ft. lbs. (19 Nm)
- Wire-to-strut clamp: 44 inch lbs. (5 Nm)
- Sensor clamp: 71 inch lbs. (8 Nm)
- Stabilizer link: 29 ft. lbs. (39 Nm)

RX300

FRONT

1. Before servicing the vehicle, refer to the precautions section.

➡**Do not support the weight of the vehicle on the suspension arm; the arm will deform under its weight.**

2. Remove or disconnect the following:
- Wheel
- Brake hose and the Anti-lock Brake System (ABS) speed sensor wire from the strut
- Sway bar link from the strut

3. Matchmark on the strut lower bracket and camber adjust cam, if equipped.

4. Remove or disconnect the following:
- Strut lower end from the steering knuckle lower arm
- 3 upper strut mounting plate-to-upper wheel arch nuts
- Strut

To install:

5. Align the upper suspension support hole with the strut piston or end, so they fit properly.

6. Install or connect the following:
- Strut piston rod end to the upper suspension support. Tighten the new nut to 29–40 ft. lbs. (39–54 Nm).

➡**Do not use an impact wrench to tighten the nut.**

- Lubricate the suspension support bearing with multi-purpose grease.
- Pack the upper support space with

multi-purpose grease, also, after installation.
- Tighten the 3 suspension support-to-wheel arch nuts to 47 ft. lbs. (64 Nm).
- Tighten the strut-to-steering knuckle arm bolts to 156 ft. lbs. (211 Nm).
- Sway bar link to the strut. Tighten the nut to 29 ft. lbs. (39 Nm).
- ABS speed sensor and the brake hose to the strut, if equipped
- Wheel

7. Check and/or adjust the front wheel alignment.

REAR

1. Before servicing the vehicle, refer to the precautions section.

2. Remove or disconnect the following:
- Negative battery cable
- Deck side cover
- Rear wheels
- Anti-lock Brake System (ABS) sensor from the strut bracket
- Flexible brake hose from the strut
- Sway bar link from the strut
- Loosen the 2 lower strut mounting bolts

3. Support the rear axle carrier with a jack.

4. Remove or disconnect the following:
- 3 upper strut mounting nuts
- Strut, by lower the rear axle

To install:

5. Install or connect the following:
- Strut
- Both lower strut mounting bolts, but do not tighten
- Axle carrier by aligning the 3 upper mounting studs. Tighten the nuts to 29 ft. lbs. (39 Nm).

6. Lower the axle carrier.

7. Install or connect the following:
- Tighten both lower mounting bolts to 188 ft. lbs. (255 Nm).
- Sway bar link. Tighten the nut to 29 ft. lbs. (39 Nm).
- Flexible brake hose and the ABS sensor to the strut
- Rear wheels and the deck side cover
- Negative battery cable

RX330

FRONT

1. Remove the wheel.
2. Disconnect the stabilizer bar link.
3. Loosen, don't remove, the strut locknut.
4. Disconnect the brake hose from the strut.
5. Remove the lower mounting bolts.
6. Remove the upper retaining nuts.

To install:

7. Position the strut and install the upper nuts. Torque to 59 ft. lbs. (80 Nm).

8. Install the lower bolts and torque to 170 ft. lbs. (230 Nm).

9. Connect the brake line.

10. Tighten the strut locknut to 36 ft. lbs. (49 Nm).

11. Connect the stabilizer links and torque to 55 ft. lbs. (74 Nm).

12. Install the wheel. Torque to 76 ft. lbs. (103 Nm).

REAR

1. Before servicing the vehicle, refer to the precautions section.

2. Remove the tonneau cover.

3. Remove the left side deck trim cover.

4. Remove the rear wheels.

5. Remove the stabilizer link from the strut.

6. On 2-wheel drive models, disconnect the skid control sensor wire and brake hose from the strut and carrier.

7. On 4-wheel drive models, disconnect the brake hose and speed sensor from the strut and carrier.

8. Loosen the 2 nuts at the lower end of the strut, but don't remove them.

9. Support the rear axle carrier with a jack.

10. Remove the 3 upper strut nuts and lower the axle.

11. Remove the lower strut bolts and nuts. Lift out the strut.

To install:

12. Position the strut and install the 3 upper nuts. Torque to 43 ft. lbs. (58 Nm).

13. Lift the axle and install the 2 lower bolts and nuts. Torque to 133 ft. lbs. (180 Nm).

14. The remainder of installation is the reverse of removal. Observe the following torques:

- Brake hose clamp: 14 ft. lbs. (19 Nm)
- Wire-to-strut clamp: 44 inch lbs. (5 Nm)
- Sensor clamp: 71 inch lbs. (8 Nm)
- Stabilizer link: 29 ft. lbs. (39 Nm)

STRUT OVERHAUL

Highlander

1. Before servicing the vehicle, refer to the precautions section.

2. Remove or disconnect the following:
- Wheel

➡**If equipped, be careful not to damage the oil seal, driveshaft boot and/or speed sensor rotor when removing the steering knuckle.**

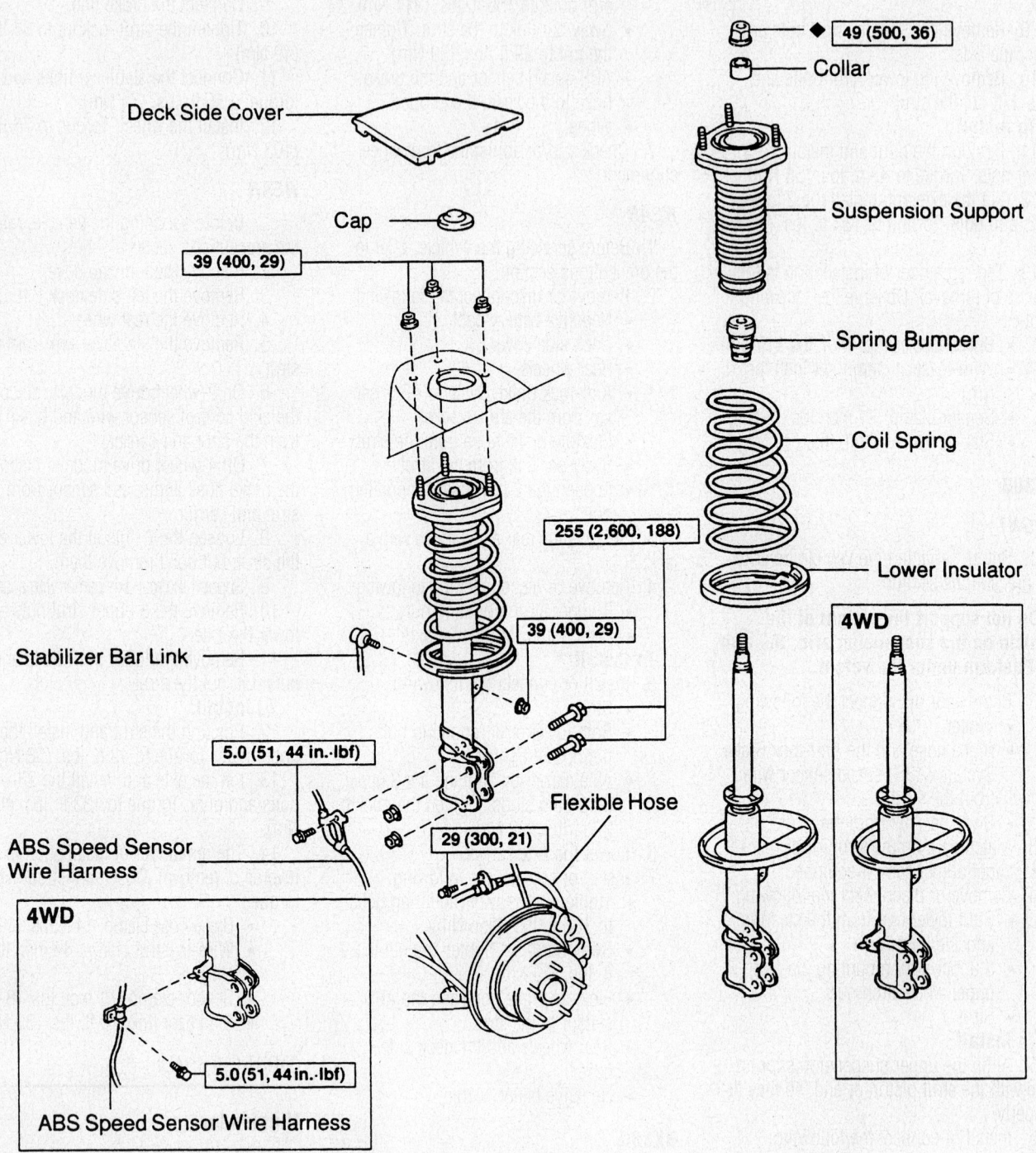

49 (500, 36) ◆

Collar

Suspension Support

Spring Bumper

Coil Spring

Lower Insulator

Deck Side Cover

Cap

39 (400, 29)

255 (2,600, 188)

39 (400, 29)

Stabilizer Bar Link

5.0 (51, 44 in.·lbf)

ABS Speed Sensor Wire Harness

Flexible Hose

29 (300, 21)

4WD

4WD

5.0 (51, 44 in.·lbf)

ABS Speed Sensor Wire Harness

N·m (kgf·cm, ft·lbf) : Specified torque

◆ Non–reusable part

7924ZG89

Exploded view of the rear strut assembly—RX300

- Shock absorber (strut assembly)

3. Install a nut/bolt to the bracket at the lower portion of the strut assembly and secure it in a vise.

4. Compress the coil spring with a spring compressor.

✳✳ CAUTION

The proper tools must be used for this procedure. The spring on the strut is under high pressure and can cause serious injury if not properly removed and installed.

5. Remove or disconnect the following:
- Center retaining nut, by holding the spring seat
- Support, dust seal, spring seat, insulator and spring from the strut assembly

To install:

6. Install the spring bumper and lower insulator to the strut assembly.

7. Compress the coil spring and fit the lower end of the spring into the spring seat gap.

8. Install or connect the following:
- Upper insulator, spring seat, dust seal, support and spring seat. Tighten the new retaining nut to 36 ft. lbs. (49Nm).
- Strut
- Wheel

9. If required, bleed the brake system and check for leaks.

10. Check and/or adjust the front wheel alignment.

RX300 and RX330

1. Before servicing the vehicle, refer to the precautions section.

2. Remove or disconnect the following:
- Wheel

➡**If equipped, be careful not to damage the oil seal, driveshaft boot and/or speed sensor rotor when removing the steering knuckle.**

- Shock absorber (strut assembly)

3. Install a nut/bolt to the bracket at the lower portion of the strut assembly and secure it in a vise.

4. Compress the coil spring with a spring compressor.

✳✳ CAUTION

The proper tools must be used for this procedure. The spring on the strut is under high pressure and can cause serious injury if not properly removed and installed.

5. Remove or disconnect the following:
- Center retaining nut, by holding the spring seat
- Support, dust seal, spring seat, insulator and spring from the strut assembly

To install:

6. Install the spring bumper and lower insulator to the strut assembly.

7. Compress the coil spring and fit the lower end of the spring into the spring seat gap.

8. Install or connect the following:
- Upper insulator, spring seat, dust seal, support and spring seat. Tighten the new retaining nut to 34 ft. lbs. (47 Nm) for RX300; 36 ft. lbs. (49 Nm) for RX330.
- Strut
- Wheel

9. If required, bleed the brake system and check for leaks.

10. Check and/or adjust the front wheel alignment.

Pneumatic Front Strut

REMOVAL & INSTALLATION

RX300 and RX330

1. Remove the wheels.

➡**Before disconnecting the air tube, press the height control OFF SW to disable the system.**

2. Remove the cowl top silencer pad.

3. Remove the strut cap.

4. Remove the height control tube clamp and disconnect the tube by loosening the nut.

➡**Keep the chamber of the strut from moving.**

5. Support the lower arm with a jack.

6. Remove the stabilizer link from the strut.

7. Disconnect the height control sensor sub-assembly from the lower arm.

8. Remove the brake hose and speed sensor wire from the strut.

9. Remove the nuts from the 2 lower strut bolts, but leave the bolts in place.

10. Remove the 3 upper strut nuts.

11. Lower the jack slowly until the strut is free, then remove the bolts and lift out the strut.

To install:

12. Position the strut and install the bolts from the front side. Torque the nuts to 170 ft. lbs. (230 Nm).

13. Raise the arm and position the upper end. Install the nuts and torque to 59 ft. lbs. (80 Nm).

14. Install the brake hose and sensor wire and torque the bolt to 14 ft. lbs. (19 Nm).

15. Connect the stabilizer link. Torque the nut to 55 ft. lbs. (74 Nm).

16. The remainder of installation is the reverse of removal.

Pneumatic Rear Strut

REMOVAL & INSTALLATION

RX300 and RX330

1. Press the height control switch to disable the system.

2. Support the axle carrier with a jack.

3. Remove the deck side trim cover.

4. Remove the wheel.

5. Disconnect the stabilizer link from the strut.

6. Disconnect the height control sensor at the strut.

7. Disconnect the height control tube at the strut.

8. On 2-wheel drive models, disconnect the skid control sensor wire and brake hose from the strut and carrier.

9. On 4-wheel drive models, disconnect the brake hose and speed sensor from the strut and carrier.

10. Loosen the 2 nuts at the lower end of the strut, but don't remove them.

11. Remove the 3 upper strut nuts and lower the axle.

12. Remove the lower strut bolts and nuts. Lift out the strut.

To install:

13. Coat new O-rings and plate with multi-purpose grease and install them on the tube.

14. Connect the height control tube. It helps to push the tube into place with a piece of rolled up cardboard. Push the connector into place until a click is heard. Turn the connection 90 degrees and lightly pull on the tube to make sure it's secure.

15. Position the strut and install the 3 upper nuts. Torque to 43 ft. lbs. (58 Nm).

16. Raise the axle and install the lower bolts and nuts. Torque to 133 ft. lbs. (180 Nm).

17. The remainder of installation is the reverse of removal. Observe the following torques:
- Brake hose clamp: 14 ft. lbs. (19 Nm)

- Wire-to-strut clamp: 44 inch lbs. (5 Nm)
- Sensor clamp: 71 inch lbs. (8 Nm)
- Stabilizer link: 29 ft. lbs. (39 Nm)

Coil Spring

REMOVAL & INSTALLATION

➡**See Strut Overhaul in this section.**

Lower Ball Joint

REMOVAL & INSTALLATION

Highlander

1. Before servicing the vehicle, refer to the precautions section.
2. Remove or disconnect the following:
   - Wheel
   - Hub nut

- Caliper and rotor
- Lower control arm from the ball joint
- Tie rod end
- Halfshaft
- Ball joint from the knuckle
3. Installation is the reverse of removal. Observe the following torques:
   - Ball stud nut: 90 ft. lbs. (123Nm)
   - Lower arm-to-ball joint: 94 ft. lbs. (127Nm)
   - Tie rod end: 36 ft. lbs. (49Nm)

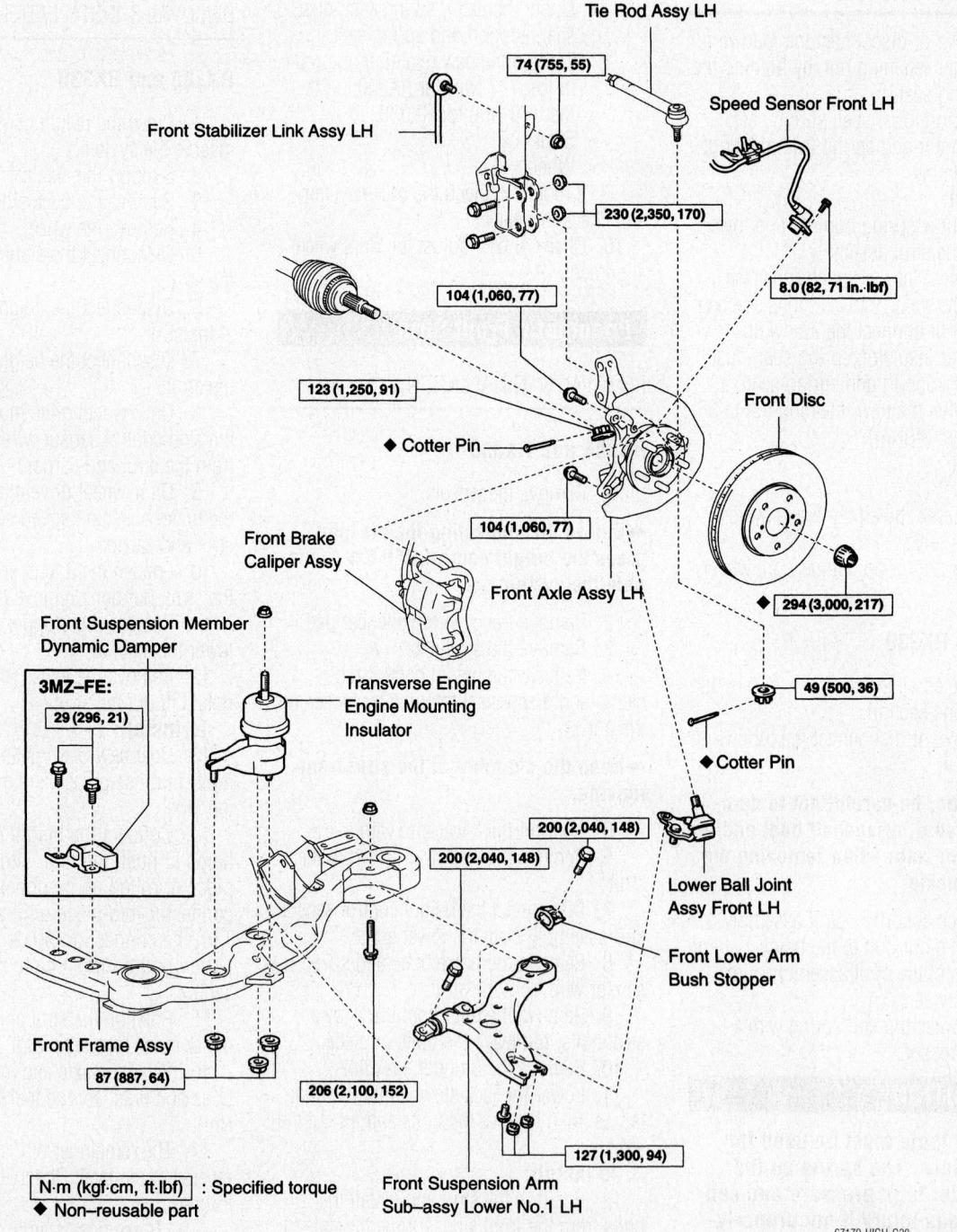

N·m (kgf·cm, ft·lbf) : Specified torque
◆ Non-reusable part

Lower control arm and related parts—Highlander

67170-HIGH-G38

RX300

1. Before servicing the vehicle, refer to the precautions section.
2. Remove or disconnect the following:
   - Wheel
   - Steering knuckle with the axle hub
   - Dust deflector, by prying it from the knuckle
   - Cotter pin and nut from the ball joint
   - Ball joint from the steering knuckle, by removing the 2 bolts
   - Lower ball joint, using a Ball Joint Separator tool 09628-62011

To install:

3. Install or connect the following:
   - Lower ball joint. Tighten the nut to 76 ft. lbs. (103 Nm) and both bolts to 94 ft. lbs. (127 Nm).
   - New cotter pin
   - Wheel

RX330

1. Remove the wheel.
2. Remove the axle hub nut.
3. Disconnect the speed sensor.
4. Remove the caliper and suspend it out of the way.
5. Remove the rotor.
6. Disconnect the tie rod end.
7. Remove the lower arm.
8. Pull the knuckle from the halfshaft.
9. Remove the bolts securing the ball joint to the arm.
10. Installation is the reverse of removal. Observe the following torques:
    - Ball joint-to-arm: 94 ft. lbs. (127 Nm)
    - Ball joint-to-knuckle: 91 ft. lbs. (123 Nm)
    - Arm-to-frame: 148 ft. lbs. (200 Nm)
    - Stabilizer bar link: 55 ft. lbs. (74 Nm)
    - Caliper support bolts: 77 ft. lbs. (104 Nm)
    - Caliper pins: 25 ft. lbs. (34 Nm)
    - Hub nut: 217 ft. lbs. (294 Nm)

Lower Control Arm

REMOVAL & INSTALLATION

Highlander

2002–03

1. Before servicing the vehicle, refer to the precautions section.
2. Remove or disconnect the following:
   - Engine/transaxle assembly

   - Transverse engine mount insulator
   - 2 front and 1 rear lower arm mount bolts
   - Lower arm
3. Installation is the reverse of removal. Observe the following torques:
   - Front side bolts: 148 ft. lbs. (200Nm)
   - Rear arm bolt/nut: 152 ft. (206Nm)
   - Insulator: 64 ft. lbs. (87Nm)

2004–06

1. Remove the engine/transaxle assembly.
2. Remove the transverse engine mounting insulator.
3. Remove the 3 bolts securing the arm to the engine support member.
4. Remove the front lower arm bush stopper.
5. Remove the ball joint-to-arm bolts.
6. Installation is the reverse of removal. Observe the following torques:
   - 2 short arm-to-support bolts: 148 ft. lbs. (200 Nm)
   - 1 long arm-to-support bolt: 152 ft. lbs. (206 Nm)
   - Ball joint-to-arm: 94 ft. lbs. (127 Nm)
   - Transverse engine mounting insulator: 64 ft. lbs. (87 Nm)

RX300

1. Before servicing the vehicle, refer to the precautions section.
2. Remove or disconnect the following:
   - Engine/transaxle assembly
   - Transverse engine mount insulator
   - 2 front and 1 rear lower arm mount bolts
   - Lower arm
3. Installation is the reverse of removal. Observe the following torques:
   - Front side bolts: 148 ft. lbs. (200Nm)
   - Rear arm bolt/nut: 152 ft. (206Nm)
   - Insulator: 64 ft. lbs. (87Nm)

RX330

1. Remove the engine/transaxle assembly.
2. Remove the transverse engine mounting insulator.
3. Remove the 3 bolts securing the arm to the engine support member.
4. Remove the front lower arm bush stopper.
5. Remove the ball joint-to-arm bolts.
6. Installation is the reverse of removal. Observe the following torques:
   - 2 short arm-to-support bolts: 148 ft. lbs. (200 Nm)

   - 1 long arm-to-support bolt: 152 ft. lbs. (206 Nm)
   - Ball joint-to-arm: 94 ft. lbs. (127 Nm)
   - Transverse engine mounting insulator: 64 ft. lbs. (87 Nm)

CONTROL ARM BUSHING REPLACEMENT

➡**These vehicles do not have replaceable bushings.**

Front Wheel Bearing

REMOVAL

Highlander

1. Remove the wheel.
2. Remove the hub nut.
3. Remove the caliper and suspend it out of the way.
4. Remove the brake rotor.
5. Disconnect the tie rod end from the knuckle.
6. Disconnect the control arm from the ball joint.
7. Remove the halfshaft from the knuckle.
8. Remove the lower strut-to-knuckle bolts and remove the hub/knuckle assembly.
9. Remove the ball joint.

DISASSEMBLY AND ASSEMBLY

Highlander

1. Remove the inner seal from the hub.
2. Remove the snap ring from the hub.
3. Mount the knuckle in a vise and, using a slidehammer, remove the hub from the knuckle.
4. Remove the outer seal.
5. Using a press, the bearings and races can now be replaced.
6. Replace the snap ring and seals.
7. Press the knuckle onto the hub assembly.

INSTALLATION

Highlander

1. Install the ball joint. Torque to 91 ft. lbs. (123 Nm). Advance the nut as much as 60 degrees to align the cotter pin hole. Use a new cotter pin.
2. Attach the knuckle assembly to the strut. Torque to 170 ft. lbs. (230 Nm).
3. The remainder of installation is the

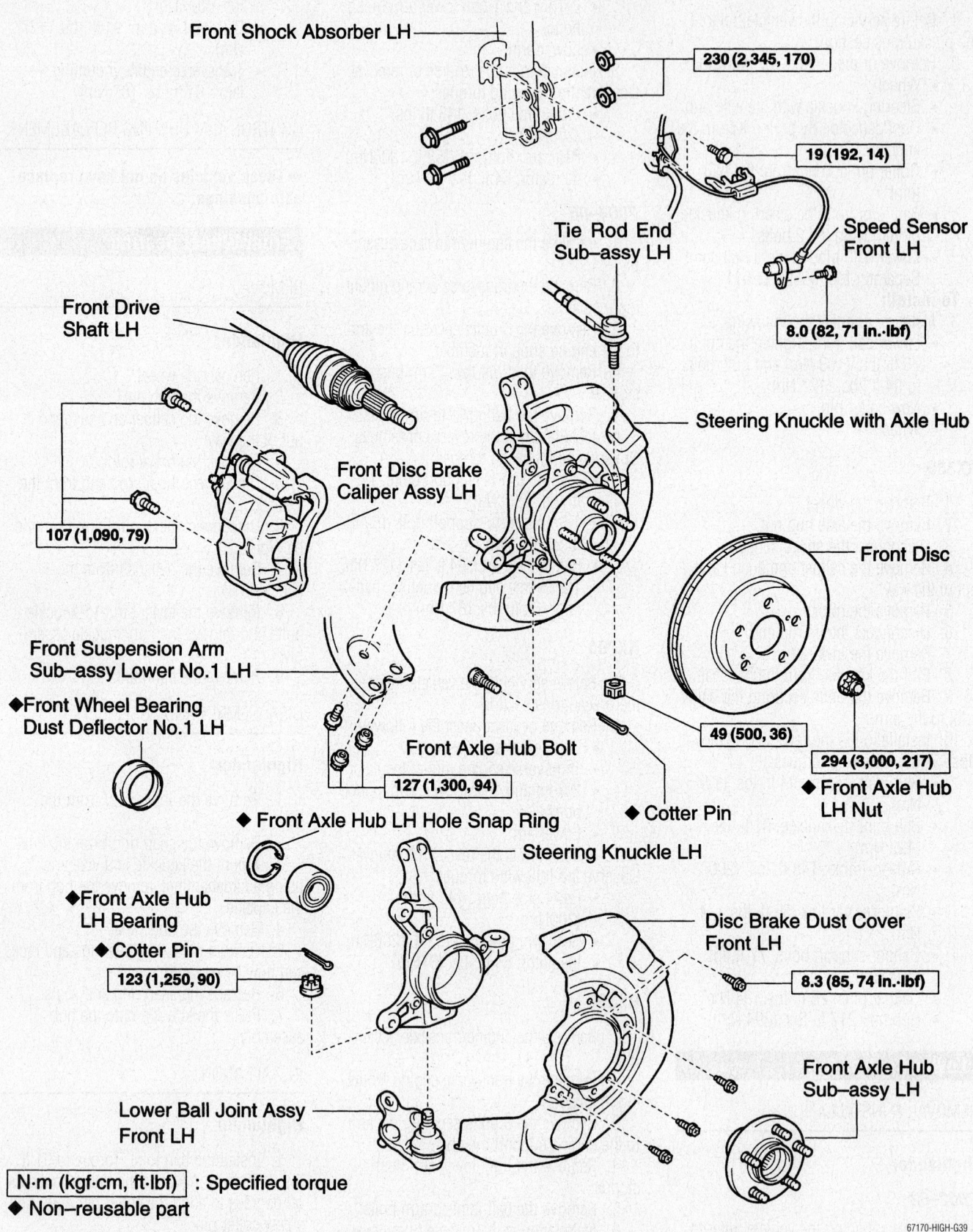

Front Shock Absorber LH

230 (2,345, 170)

19 (192, 14)

Tie Rod End Sub–assy LH

Speed Sensor Front LH

8.0 (82, 71 in.·lbf)

Front Drive Shaft LH

Steering Knuckle with Axle Hub

Front Disc Brake Caliper Assy LH

Front Disc

107 (1,090, 79)

Front Suspension Arm Sub–assy Lower No.1 LH

◆Front Wheel Bearing Dust Deflector No.1 LH

Front Axle Hub Bolt

49 (500, 36)

294 (3,000, 217)

127 (1,300, 94)

◆ Front Axle Hub LH Nut

◆ Front Axle Hub LH Hole Snap Ring

Steering Knuckle LH

◆ Cotter Pin

◆Front Axle Hub LH Bearing

Disc Brake Dust Cover Front LH

◆ Cotter Pin

8.3 (85, 74 in.·lbf)

123 (1,250, 90)

Front Axle Hub Sub–assy LH

Lower Ball Joint Assy Front LH

N·m (kgf·cm, ft·lbf) : Specified torque

◆ Non–reusable part

67170-HIGH-G39

Front hub and related parts—Highlander

reverse of removal. Observe the following torques:

- Caliper support: 77 ft. lbs. (104 Nm)
- Hub nut: 217 ft. lbs. (294 Nm)
- Caliper pins: 25 ft. lbs. (34 Nm)

RX300

The front wheel bearings and 4wd rear wheel bearings are serviced as a unit with the hub and are not replaceable. See the Halfshaft Removal and Installation procedure.

RX330

REMOVAL

1. Remove the wheel.
2. Remove the hub nut.
3. Remove the caliper and suspend it out of the way.
4. Remove the brake rotor.
5. Disconnect the tie rod end from the knuckle.
6. Disconnect the control arm from the ball joint.
7. Remove the halfshaft from the knuckle.
8. Remove the lower strut-to-knuckle bolts and remove the hub/knuckle assembly.
9. Remove the ball joint.

DISASSEMBLY

1. Remove the inner seal from the hub.
2. Remove the snap ring from the hub.
3. Mount the knuckle in a vise and, using a slidehammer, remove the hub from the knuckle.
4. Remove the outer seal.
5. Using a press, the bearings and races can now be replaced.
6. Replace the snap ring and seals.
7. Press the knuckle onto the hub assembly.

INSTALLATION

1. Install the ball joint. Torque to 91 ft. lbs. (123 Nm). Advance the nut as much as 60 degrees to align the cotter pin hole. Use a new cotter pin.
2. Attach the knuckle assembly to the strut. Torque to 170 ft. lbs. (230 Nm).
3. The remainder of installation is the reverse of removal. Observe the following torques:

- Caliper support: 77 ft. lbs. (104 Nm)
- Hub nut: 217 ft. lbs. (294 Nm)
- Caliper pins: 25 ft. lbs. (34 Nm)

Rear Wheel Bearings

REMOVAL & INSTALLATION

Highlander

The 4wd rear wheel bearings are serviced as a unit with the hub and are not replaceable. See the Halfshaft Removal and Installation procedure.

2002–03 WITH 2WD

1. Before servicing the vehicle, refer to the precautions section.
2. Remove or disconnect the following:
- Rear wheel
- Flexible brake hose from the rear strut assembly
- Brake caliper and support it on using wire
- Brake rotor
3. Remove or disconnect the following:
- Anti-lock Brake System (ABS) speed sensor connector
- 4 rear axle hub assembly nuts
- Hub assembly

To install:
4. Install or connect the following:
- New hub assembly. Tighten the nuts to 59 ft. lbs. (80 Nm).
- ABS speed sensor
- Brake rotor
- Brake caliper. Tighten the mounting bolts to 34 ft. lbs. (47 Nm).
- Flexible brake hose to the rear strut assembly. Tighten the mounting bolt to 21 ft. lbs. (29 Nm).
- Wheels
5. Test drive the vehicle.

2004–06 WITH 2WD

1. Remove the wheel.
2. Remove the speed sensor from the axle carrier.
3. Remove the axle shaft nut.
4. Remove the caliper and support assembly and suspend it out of the way.
5. Check the bearing backlash. It should not exceed 0.0020 in. (0.05mm). If it does, it must be replaced.
6. Remove the 4 bolts and remove the hub/bearing assembly.
7. Installation is the reverse of removal. Observe the following torques:

- Hub/bearing assembly bolts: 55 ft. lbs. (75 Nm)
- Caliper support: 58 ft. lbs. (78 Nm)
- Brake hose clamp: 14 ft. lbs. (19 Nm)
- Hub nut: 217 ft. lbs. (294 Nm)

RX300

The 4wd rear wheel bearings are serviced as a unit with the hub and are not replaceable. See the Halfshaft Removal and Installation procedure.

WITH 2WD

1. Before servicing the vehicle, refer to the precautions section.
2. Remove or disconnect the following:

- Rear wheel
- Flexible brake hose from the rear strut assembly
- Brake caliper and support it on using wire
- Brake rotor
3. Remove or disconnect the following:

- Anti-lock Brake System (ABS) speed sensor connector
- 4 rear axle hub assembly nuts
- Hub assembly

To install:
4. Install or connect the following:
- New hub assembly. Tighten the nuts to 59 ft. lbs. (80 Nm).
- ABS speed sensor
- Brake rotor
- Brake caliper. Tighten the mounting bolts to 34 ft. lbs. (47 Nm).
- Flexible brake hose to the rear strut assembly. Tighten the mounting bolt to 21 ft. lbs. (29 Nm).
- Wheels
5. Test drive the vehicle.

RX330

1. Remove the wheel.
2. Remove the speed sensor from the axle carrier.
3. Remove the axle shaft nut.
4. Remove the caliper and support assembly and suspend it out of the way.
5. Check the bearing backlash. It should not exceed 0.0020 in. (0.05mm). If it does, it must be replaced.
6. Remove the 4 bolts and remove the hub/bearing assembly.
7. Installation is the reverse of removal. Observe the following torques:

- Hub/bearing assembly bolts: 55 ft. lbs. (75 Nm)
- Caliper support: 58 ft. lbs. (78 Nm)
- Brake hose clamp: 14 ft. lbs. (19 Nm)
- Hub nut: 217 ft. lbs. (294 Nm)

BRAKES

Front Brake Caliper

REMOVAL & INSTALLATION

2002–03

1. Before servicing the vehicle, refer to the precautions section.

2. Remove or disconnect the following:
   • Front wheel
   • 2 mounting bolts and caliper

3. If the caliper is being replaced, disconnect the brake line and plug both openings.

➡**Depending on the brake type, there may be either 1 or 2 sealing washers.**

4. Installation is the reverse of removal. Torque the mounting bolts to 25 ft. lbs. (34Nm). If the brake hose was removed, torque the union bolt to 21 ft. lbs. (29Nm).

2004–06

1. Disconnect the brake line from the caliper and plug it.

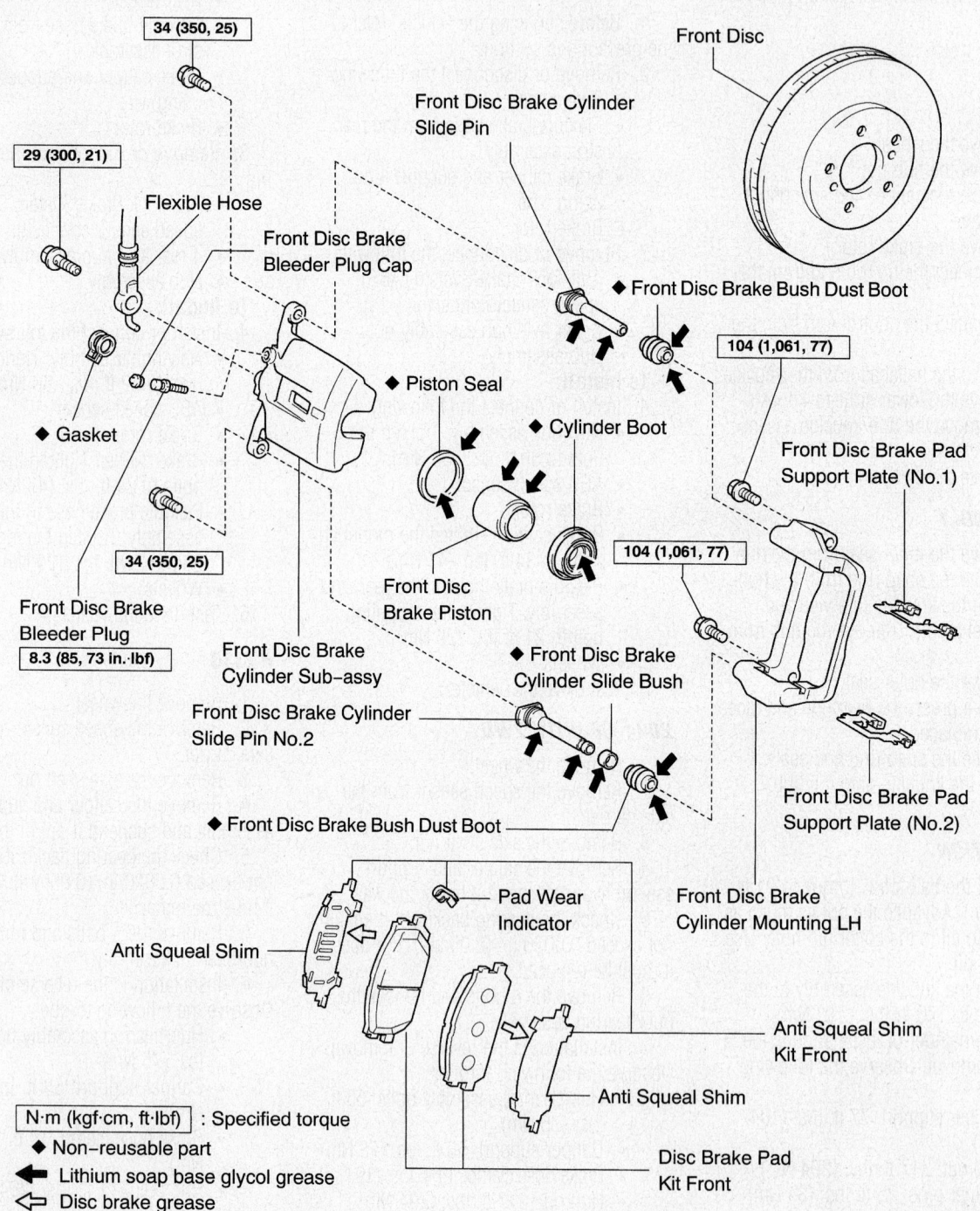

34 (350, 25)

29 (300, 21)

Flexible Hose

Front Disc Brake Cylinder Slide Pin

Front Disc Brake Bleeder Plug Cap

Front Disc Brake Bush Dust Boot

104 (1,061, 77)

◆ Gasket

◆ Piston Seal

◆ Cylinder Boot

Front Disc Brake Pad Support Plate (No.1)

Front Disc Brake Piston

104 (1,061, 77)

34 (350, 25)

Front Disc Brake Bleeder Plug

8.3 (85, 73 in.·lbf)

Front Disc Brake Cylinder Sub-assy

◆ Front Disc Brake Cylinder Slide Bush

Front Disc Brake Cylinder Slide Pin No.2

◆ Front Disc Brake Bush Dust Boot

Front Disc Brake Cylinder Mounting LH

Front Disc Brake Pad Support Plate (No.2)

Front Disc

Pad Wear Indicator

Anti Squeal Shim

Anti Squeal Shim Kit Front

Anti Squeal Shim

Disc Brake Pad Kit Front

N·m (kgf·cm, ft·lbf) : Specified torque

◆ Non-reusable part

◀ Lithium soap base glycol grease

◁ Disc brake grease

67162-X300-G11

Front disc brake components—2004–06

2. Hold the caliper slide pins and remove the mounting bolts.

3. Lift off the caliper.

4. Remove the pads and anti-squeal shims.

5. Remove the wear indicator from the inner pad.

6. Installation is the reverse of removal. Grease the caliper slides and bolts with lithium grease or equivalent. Apply disc brake grease to the anti-squeal shims. Torque the caliper bolts to 25 ft. lbs. (34

Nm); the brake line union bolt to 21 ft. lbs. (29 Nm).

Rear Brake Caliper

REMOVAL & INSTALLATION

2002–03

1. Before servicing the vehicle, refer to the precautions section.

2. Remove or disconnect the following:

• Rear wheel
• Slide pin and caliper

3. If the caliper is being replaced, disconnect the brake line and plug both openings.

➡**Depending on the brake type, there may be either 1 or 2 sealing washers.**

4. Installation is the reverse of removal. Torque the slide pin to 25 ft. lbs. (34Nm). If

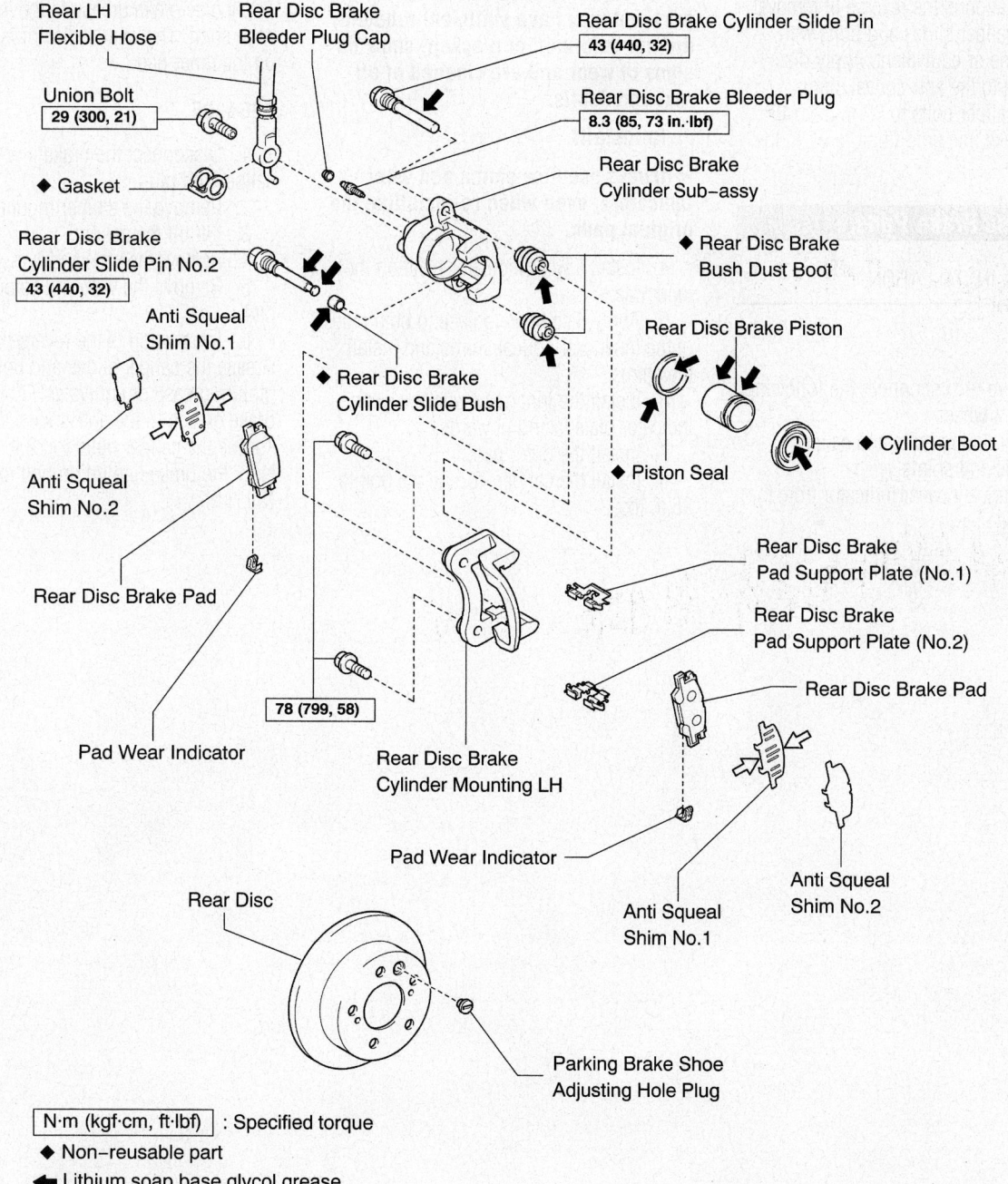

Rear LH Flexible Hose

Rear Disc Brake Bleeder Plug Cap

Rear Disc Brake Cylinder Slide Pin
43 (440, 32)

Union Bolt
29 (300, 21)

Rear Disc Brake Bleeder Plug
8.3 (85, 73 in.·lbf)

◆ Gasket

Rear Disc Brake Cylinder Sub–assy

Rear Disc Brake Cylinder Slide Pin No.2
43 (440, 32)

◆ Rear Disc Brake Bush Dust Boot

Anti Squeal Shim No.1

◆ Rear Disc Brake Cylinder Slide Bush

Rear Disc Brake Piston

Anti Squeal Shim No.2

◆ Cylinder Boot

◆ Piston Seal

Rear Disc Brake Pad

Rear Disc Brake Pad Support Plate (No.1)

Rear Disc Brake Pad Support Plate (No.2)

Rear Disc Brake Pad

78 (799, 58)

Pad Wear Indicator

Rear Disc Brake Cylinder Mounting LH

Pad Wear Indicator

Anti Squeal Shim No.2

Rear Disc

Anti Squeal Shim No.1

Parking Brake Shoe Adjusting Hole Plug

N·m (kgf·cm, ft·lbf) : Specified torque
◆ Non–reusable part
◀ Lithium soap base glycol grease
⇦ Disc brake grease

Rear disc brake components—2004–06

67162-X300-G12

the brake hose was removed, torque the union bolt to 21 ft. lbs. (29Nm).

2004–06

1. Disconnect the brake line from the caliper and plug it.
2. Remove the caliper mounting bolts.
3. Lift off the caliper.
4. Remove the pads and anti-squeal shims.
5. Remove the wear indicators from each pad.
6. Installation is the reverse of removal. Grease the caliper slides and bolts with lithium grease or equivalent. Apply disc brake grease to the anti-squeal shims. Torque the caliper bolts to 32 ft. lbs. (43 Nm); the brake line union bolt to 21 ft. lbs. (29 Nm).

Front Disc Brake Pads

REMOVAL & INSTALLATION

2002–03

1. Remove or disconnect the following:
   - Front wheel
   - 2 mounting bolts and caliper
   - Pads and shims
   - Shims and wear indicator from the pads

2. Installation is the reverse of removal. Apply disc brake grease to the inside of each shim. The wear indicator is installed on the inner pad.

2004–06

1. Hold the sliding pin and remove the lower bolt.
2. Lift the caliper up and secure it.
3. Remove the pads, 4 shims and wear indicator plate. Remove the 2 pad support plates.

➡The support plates can be reused, provided they have sufficient rebound, are not deformed or cracked, show no signs of wear and are cleaned of all rust and debris.

To install:

➡Always use new shims and wear indicators, even when re-installing the original pads.

4. Install a wear indicator plate on the inner pad.
5. Apply disc brake grease to both sides of the inner anti-squeal shims and install the shims.
6. Install the inner pad with the wear indicator plate facing upwards.
7. Install the outer pad.
8. Install the caliper. Torque the bolt to 25 ft. lbs.

Rear Disc Brake Pads

REMOVAL & INSTALLATION

2002–03

1. Remove or disconnect the following:
   - Rear wheel
   - Caliper
   - Pads and shims
   - Shims and wear indicator from the pads
2. Installation is the reverse of removal. Apply disc brake grease to the inside of each shim. The wear indicator is installed on the inner pad.

2004–06

1. Disconnect the brake line from the caliper and plug it.
2. Remove the caliper mounting bolts.
3. Lift off the caliper.
4. Remove the pads and anti-squeal shims.
5. Remove the wear indicators from each pad.
6. Installation is the reverse of removal. Grease the caliper slides and bolts with lithium grease or equivalent. Apply disc brake grease to the anti-squeal shims. Torque the caliper bolts to 32 ft. lbs. (43 Nm); the brake line union bolt to 21 ft. lbs. (29 Nm).

TOYOTA AND LEXUS

Highlander Hybrid • RX 400h

4

BRAKES **4-30**
DRIVE TRAIN **4-22**
ENGINE REPAIR................**4-7**
FUEL SYSTEM **4-21**
SPECIFICATIONS AND
MAINTENANCE CHARTS **4-2**
Engine and Vehicle Identification
  Chart.....................................4-2
General Engine Specifications 4-2
Engine Tune-Up Specifications........4-2
Capacities 4-3
Valve Specifications.....................4-3
Crankshaft and Connecting
  Rod Specifications.....................4-3
Camshaft and Bearing
  Specifications...........................4-4
Piston and Ring Specifications........4-4
Torque Specifications 4-4
Wheel Alignment 4-5
Tire, Wheel and Ball Joint
  Specifications...........................4-5
Brake Specifications 4-5
Scheduled Maintenance
  Intervals..................................4-6
STEERING AND
SUSPENSION **4-28**
A
Air Bag.....................................4-28
  Disarming 4-28
  Precautions 4-28
Alternator 4-7
B
Brake Caliper 4-30
  Removal And Installation 4-30
C
Camshaft and Valve Lifters 4-14
  Inspection 4-16
  Removal & Installation 4-14
CV-Joints...................................4-24
  Overhaul 4-24

Cylinder Head.............................4-10
  Removal & Installation..............4-10
D
Disc Brake Pads..........................4-32
  Removal And Installation 4-32
E
Engine Assembly 4-7
  Removal & Installation..............4-7
Exhaust Manifold 4-12
  Removal & Installation..............4-12
F
Front Crankshaft Seal 4-14
  Removal & Installation..............4-14
Front Halfshaft 4-23
  Removal & Installation..............4-23
Front Strut 4-28
  Removal & Installation..............4-28
Front Wheel Bearing 4-29
Fuel Filter 4-21
Fuel Injector 4-22
  Removal & Installation..............4-22
Fuel Pump 4-21
  Removal & Installation..............4-21
Fuel System Pressure 4-21
  Relieving.................................4-21
H
Heater Core...............................4-7
  Removal & Installation..............4-7
Hybrid Transaxle Assembly...........4-22
  Removal & Installation..............4-22
I
Ignition Timing 4-7
  Adjustment..............................4-7
Intake Manifold 4-11
  Removal & Installation..............4-11
L
Lower Ball Joint...........................4-29
  Removal & Installation..............4-29
Lower Control Arm 4-29
  Removal & Installation..............4-29

O
Oil Pan.....................................4-18
  Removal & Installation..............4-18
Oil Pump 4-18
  Removal & Installation..............4-18
P
Pinion Seal 4-27
  Removal & Installation..............4-27
Piston and Ring 4-21
  Positioning 4-21
Power Rack And Pinion
  Steering Gear...........................4-28
  Removal & Installation..............4-28
R
Rear Halfshaft 4-24
  Removal And Installation 4-24
Rear Main Seal 4-19
  Removal & Installation..............4-19
Rear Strut..................................4-28
  Conventional Strut Overhaul.....4-28
  Removal And Installation 4-28
Rear Traction Motor 4-22
  Removal & Installation..............4-22
Rear Wheel Bearing 4-29
  Removal And Installation 4-29
T
Timing Belt 4-19
  Removal & Installation..............4-19
V
Valve Lash 4-17
  Adjustment..............................4-17
W
Water Pump 4-10
  Removal & Installation..............4-10

SPECIFICATIONS AND MAINTENANCE CHARTS

ENGINE AND VEHICLE IDENTIFICATION

Engine							Model Year	
Code ①	Liters (cc)	Cu. In.	Cyl.	Fuel Sys.	Engine Type	Eng. Mfg.	Code ②	Year
3MZ-FE	3.3 (3311)	202.1	6	SFI	DOHC	Toyota	6	2006

SFI: Sequential Fuel Injection

DOHC: Double Overhead Camshaft

① Stamped on the left side of the engine block

② 10th digit of the Vehicle Identification Number (VIN)

09490_RX400H_C0001

GENERAL ENGINE SPECIFICATIONS

Year	Model	Engine Displacement Liters	Engine Series ID	Net Horsepower @ rpm	Net Torque @ rpm (ft. lbs.)	Bore x Stroke (in.)	Com- pression Ratio	Oil Pressure @ rpm
2006	RX 400h	3.3	3MZ-FE	268@5600	212@4400	3.62x3.27	10.8:1	36-78@3000
	Highlander Hybrid	3.3	3MZ-FE	268@5600	212@4400	3.62x3.27	10.8:1	36-78@3000

09490_RX400H_C0002

ENGINE TUNE-UP SPECIFICATIONS

Year	Engine Displacement Liters	Engine ID	Spark Plug Gap (in.)	Ignition Timing (deg.)	Fuel Pump (psi)	Idle Speed (rpm)	Valve Clearance	
							Intake	Exhaust
2006	3.3	3MZ-FE	0.039-0.043	8-12B①	44-50	850-950	0.006-0.010	0.010-0.014

NOTE: The Vehicle Emission Control Information label often reflects specification changes made during production.

The label figures must be used if they differ from those in this chart.

B: Before top dead center

① With terminals TC and CG of DLC3 connected

09490_RX400H_C0003

CAPACITIES

Year	Model	Engine Displacement Liters	Engine ID	Engine Oil with Filter (qts.)	Transaxle (pts)	Rear Transaxle (pts.)	Rear Drive Axle (pts.)	Fuel Tank (gal.)	Cooling System (qts.)
2006	RX 400h	3.3	3MZ-FE	5.0	①	4.2	N/A	17.2	10.9
	Highlander Hybrid	3.3	3MZ-FE	5.0	①	4.2	N/A	17.2	②

③ With towing package: 8.8 pts.

Without towing package: 8.2 pts.

② With Rear Heater: 12.3 qts.

Without Rear Heater: 10.9 qts.

09490_RX400H_C0004

VALVE SPECIFICATIONS

Year	Engine Displacement Liters	Engine ID	Seat Angle (deg.)	Face Angle (deg.)	Spring Test Pressure (lbs. @ in.)	Spring Installed Height (in.)	Stem-to-Guide Clearance (in.) Intake	Stem-to-Guide Clearance (in.) Exhaust	Stem Diameter (in.) Intake	Stem Diameter (in.) Exhaust
2006	3.3	3MZ-FE	45	40.5	41.9-46.3@ 1.331	1.331	0.0010- 0.0024	0.0012- 0.0026	0.2154- 0.2159	0.2152 0.2157

09490_RX400H_C0005

CRANKSHAFT AND CONNECTING ROD SPECIFICATIONS

All measurements are given in inches.

Year	Engine Displacement Liters	Engine ID	Main Brg. Journal Dia.	Main Brg. Oil Clearance	Shaft End-play	Thrust on No.	Journal Diameter	Oil Clearance	Side Clearance
2006	3.3	3MZ-FE	2.4011- 2.4016	①	0.0016- 0.0094	2	2.0863- 2.0866	0.0015- 0.0026	0.0059- 0.0118

① Journals 1 and 4: 0.0006 - 0.0013 in.

Journals 2 and 3: 0.0010 - 0.0018 in.

09490_RX400H_C0006

CAMSHAFT AND BEARING SPECIFICATIONS CHART

All measurements are given in inches.

Year	Engine Displ. Liters	Engine ID/VIN	Journal Dia.	Brg. Oil Clearance	Shaft End-play	Runout	Journal Bore	Lobe Height Intake	Lobe Height Exhaust
2006	3.3	3MZ-FE	1.0614-1.0620	①	0.0016-0.0035	0.0024	NA	1.6981-1.7020	1.6933-1.6972

① Intake Journals 4 and 5: 0.0010 - 0.0022 in.　　　　NA: Not Available
　All Others: 0.0010 - 0.0024 in.

09490_RX400H_C0007

PISTON AND RING SPECIFICATIONS

All measurements are given in inches.

Year	Engine Displ. Liters	Engine ID	Piston Clearance	Ring Gap Top Comp.	Ring Gap Bottom Comp.	Ring Gap Oil Control	Ring Side Clearance Top Comp.	Ring Side Clearance Bottom Comp.	Ring Side Clearance Oil Control
2006	3.3	3MZ-FE	0.0013-0.0023	0.0118-0.0157	0.0197-0.0236	0.0059-0.0157	0.0012-0.0031	0.0008-0.0024	0.0012-0.0043

09490_RX400H_C0008

TORQUE SPECIFICATIONS

All readings in ft. lbs.

Year	Engine Displacement Liters	Engine ID	Cylinder Head Bolts	Main Bearing Bolts	Rod Bearing Bolts	Crankshaft Damper Bolts	Flywheel Bolts	Manifold Intake	Manifold Exhaust	Spark Plugs	Oil Pan Drain Plug
2006	3.3	3MZ-FE	①	②	③	162	61	11	36	18	33

① Step 1: 12 point bolts to 40 ft. lbs.　　　　　　　　　③ Step 1: 18 ft. lbs.
　Step 2: 12 point bolts plus 90 degrees　　　　　　　　　Step 2: Plus 90 degrees
　Step 3: Hex head recessed bolt to 14 ft. lbs.

② Step 1: 12 point cap bolts to 16 ft. lbs.
　Step 2: 12 point cap bolts plus 90 degrees
　Step 3: Hex head side bolts to 20 ft. lbs.

09490_RX400H_C0009

WHEEL ALIGNMENT

Year	Model		Caster Range (+/-Deg.)	Caster Preferred Setting (Deg.)	Camber Range (+/-Deg.)	Camber Preferred Setting (Deg.)	Toe-in (in.)
2006	RX 400h	2WD F	0.75	+2.75	0.75	-0.58	0+/-0.08
		4WD F	0.75	+2.50	0.75	-0.58	0+/-0.08
		2WD R	—	—	0.75	-1.17	0.12+/-0.08
		4WD R	—	—	0.75	-0.67	0.12+/-0.08
	Highlander Hybrid	2WD F	0.75	+2.75	0.75	-0.58	0+/-0.08
		4WD F	0.75	+2.50	0.75	-0.58	0+/-0.08
		2WD R	—	—	0.75	-1.17	0.12+/-0.08
		4WD R	—	—	0.75	-0.67	0.12+/-0.08

F: Front

R: Rear

09490_RX400H_C0010

TIRE, WHEEL AND BALL JOINT SPECIFICATIONS

Year	Model	OEM Tires Standard	OEM Tires Optional	Tire Pressures (psi) Front	Tire Pressures (psi) Rear	Wheel Size	Ball Joint Inspection	Lugnut Torque (ft. lbs.)
2006	RX 400h	P225/65R17	P235/55VR18	32	32	6.5-JJ	①	76
	Highlander Hybrid	P225/65R17	N/A	32	32	6.5-J	①	76

OEM: Original Equipment Manufacturer

PSI: Pounds Per Square Inch

STD: Standard

OPT: Optional

① Replace if any measurable movement is found.

09490_RX400H_C0011

BRAKE SPECIFICATIONS

All measurements in inches unless noted

Year	Model		Brake Disc Original Thickness	Brake Disc Minimum Thickness	Brake Disc Maximum Runout	Minimum Lining Thickness	Brake Caliper Bracket Bolts (ft. lbs.)	Brake Caliper Mounting Bolts (ft. lbs.)
2006	RX 400h	F	1.102	1.024	0.0020	0.039	78	25
		R	0.394	0.335	0.0059	0.039	56	32
	Highlander Hybrid	F	1.102	1.024	0.0020	0.039	78	25
		R	0.394	0.335	0.0059	0.039	56	32

F: Front

R: Rear

09490_RX400H_C0012

SCHEDULED MAINTENANCE INTERVALS

LEXUS—RX 400h, TOYOTA Highlander Hybrid

TO BE SERVICED	TYPE OF SERVICE	VEHICLE MILEAGE INTERVAL (x1000)												
		5	10	15	20	25	30	35	40	45	50	55	60	65
Engine oil & filter	R	✓	✓	✓	✓	✓	✓	✓	✓	✓	✓	✓	✓	✓
Automatic transmission fluid	R												✓	
Ball joints & dust covers	S/I	✓	✓	✓	✓	✓	✓	✓	✓	✓	✓	✓	✓	✓
Bolts & nuts on chassis & body	S/I	✓	✓		✓	✓		✓	✓		✓		✓	✓
Brake line pipes & hoses	S/I			✓			✓			✓			✓	
Brake fluid	R						✓						✓	
Brake pads & discs (front & rear)	S/I	✓	✓	✓	✓	✓	✓	✓	✓	✓	✓	✓	✓	✓
Propeller shaft grease	S/I	✓	✓	✓	✓	✓	✓	✓	✓	✓	✓	✓	✓	✓
Steering knuckle & chassis grease	S/I	✓	✓	✓	✓	✓	✓	✓	✓	✓	✓	✓	✓	✓
Steering linkage	S/I	✓	✓	✓	✓	✓	✓	✓	✓	✓	✓	✓	✓	✓
Air cleaner filter	R						✓						✓	
Air conditioner filter	R						✓							
Spark plugs	R	Replace at 120,000 miles												
Exhaust pipes & mountings	S/I			✓			✓			✓			✓	
Fuel lines & connections	S/I												✓	
Engine coolant	R	Replace at 120,000 miles												
Timing belt	R	Replace at 90,000 miles												
Rear differential fluid	R												✓	
Rotate Tires	S/I	✓	✓	✓	✓	✓	✓	✓	✓	✓	✓	✓	✓	✓

R: Replace   S/I: Service or Inspect

FREQUENT OPERATION MAINTENANCE (SEVERE SERVICE)

 If a vehicle is operated under any of the following conditions it is considered severe service:

- Extremely dusty areas.

- 50% or more of the constant operation is in 32°C (90°F) or higher temperatures, or in temperatures below 0°C (32°F).

- Prolonged idling (vehicle operation in stop and go traffic).

- Frequent short running periods (engine does not warm to normal operating temperatures).

- Police, taxi, delivery usage or trailer towing usage.

Air cleaner filter: service or inspect every 3750 miles

Engine oil & filter: replace every 3750 miles.

Ball joints & dust covers: service or inspect every 7500 miles.

Bolts & nuts on chassis & body: service or inspect every 7500 miles.

Brake pads & discs (front & rear): service or inspect every 7500 miles.

Steering knuckle & chassis grease: service or inspect every 7500 miles.

Steering linkage: service or inspect every 7500 miles.

Exhaust pipes & mountings: service or inspect every 15,000 miles.

09490_RX400H_C00013

ENGINE REPAIR

→**Disconnecting the negative battery cable on some vehicles may interfere with the functions of the on board computer system. The computer may undergo a relearning process once the negative battery cable is reconnected.**

Alternator

The 3.3L engine has a DC electric converter and therefore does not use a standard alternator.

Ignition Timing

ADJUSTMENT

All engines are equipped with a Distributorless Ignition System (DIS). No timing adjustment is possible.

Engine Assembly

REMOVAL & INSTALLATION

1. Before servicing the vehicle, refer to the Precautions Section.
2. Drain the coolant from the engine and hybrid assemblies.
3. Drain the engine oil.
4. Drain the transfer case fluid, if equipped.
5. Drain the hybrid transaxle fluid.
6. Discharge the fuel system pressure.
7. Remove the service plug grip, found underneath the Battery Service cover on the rear seat. Wait 5 minutes to discharge the high voltage capacitor.
8. Remove or disconnect the following:
   - Engine room side covers
   - Front wheels
   - Engine under cover assembly
   - Left and right fender splash shields
   - Left and right fender apron seals
   - Wiper arms
   - Cowl top ventilator louver
   - Wiper linkage
   - Cowl top panel outer sub-assembly
   - Battery and battery tray
   - Air intake assembly
   - Inverter support bracket No. 5
   - Power steering ECU assembly
9. Remove the inverter with converter assembly as follows:
   a. Disconnect circuit breaker sensor No. 1.
   b. Disconnect engine room wire No. 2.

c. Remove the inverter reserve tank sub-assembly.
d. Disconnect the coolant hose.
e. Disconnect the power steering ECU bracket.
f. Remove the inverter cover.
g. Verify the voltage of w/ converter inverter assembly is 0 volts.
h. Disconnect engine wire No. 4.
i. Disconnect the high voltage cable of the front motor.
j. Disconnect the Motor Generator ECU connector.
k. Disconnect No. 3 wire frame.
l. Install the inverter cover.
m. Separate the engine room relay block assembly.
n. Remove the inverter bracket No. 4.
o. Remove the w/ converter inverter assembly.
10. Remove or disconnect the following:
   - Brake master cylinder reservoir sub-assembly
   - Master cylinder reservoir bracket
   - Air intake assembly bracket
   - Engine moving control rod
   - Right-hand engine mount
   - Compressor with the motor assembly
   - Inverter bracket No. 1
   - Transmission control cable assembly
   - Fuel vapor feed hose
   - Fuel pipe sub-assembly No. 1
   - Heater hoses
   - Radiator hoses
   - Oil cooler hoses
   - Water pump assembly
   - Glove compartment door assembly
   - Engine wire harnesses from the ECU
   - Front exhaust pipe assembly
   - Front stabilizer link assembly
   - Front axle hub nut, both sides
   - Wheel speed sensors
   - Tie rod end sub-assemblies
   - Lower control arms
11. Separate the halfshafts from the axle hub.
12. Separate the steering intermediate shaft sub-assembly.
13. Attach a lifting crane.
14. Remove the left and right hand frame side rail plates.
15. Remove the front suspension member rear braces.
16. Lift the engine/transaxle from the vehicle.

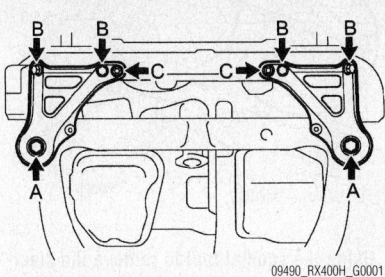
Frame side rail plates

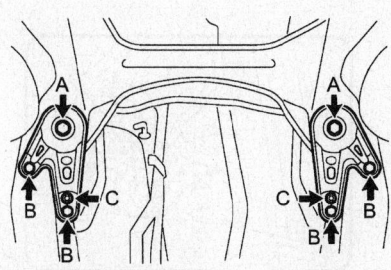
Front suspension member braces

17. Installation is the reverse of removal. Observe the following torques:
   - Frame side rail plates—Bolt A: 63 ft. lbs. (85 Nm). Bolt B: 24 ft. lbs. (32 Nm). Nut C: 24 ft. lbs. (32 Nm).
   - Front suspension member brace— Bolt A: 63 ft. lbs. (85 Nm). Bolt B: 24 ft. lbs. (32 Nm). Nut C: 24 ft. lbs. (32 Nm).
   - Front exhaust pipe assembly: 41 ft. lbs. (56 Nm).
   - Front stabilizer link assembly: 55 ft. lbs. (74 Nm).
   - Steering link: 26 ft. lbs. (35 Nm).

Heater Core

REMOVAL & INSTALLATION

Front Heater Assembly

1. Before servicing the vehicle, refer to the Precautions Section.
2. Discharge and recover the refrigerant from the A/C system.
3. Disconnect the negative battery cable.

✳✳ CAUTION

Wait 90 seconds after disconnect the cable to allow the airbag to discharge.

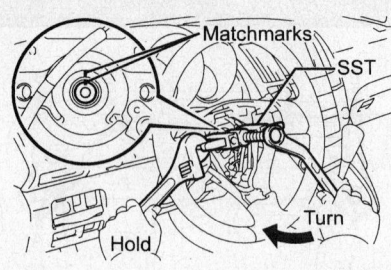

Matchmarks

SST

Turn

Hold

09490_RX400H_G0003

Using the special tool to remove the steering wheel assembly—RX 400h, Highlander Hybrid similar

4. Drain the cooling system.
5. Remove or disconnect the following:
   - Wiper arms
   - Cowl top ventilator louver assembly
   - Wiper link assembly
   - Air conditioning tube and accessory assembly
   - Cooler refrigerant suction hose No.1
   - Heater hoses from the heater core.
6. Remove the instrument panel assembly as follows:
   a. Remove the shift knob and console panel.

b. Remove the lower instrument panel.
c. Remove the radio.
d. Remove the multi-display unit.
e. Remove the steering wheel lower covers.
f. Using a Torx® wrench, loosen the 2 screws located at each side of the steering wheel until the screw's circumference groove catches on the screw case
g. Pull the air bag module from the steering wheel and disconnect the electrical connector.

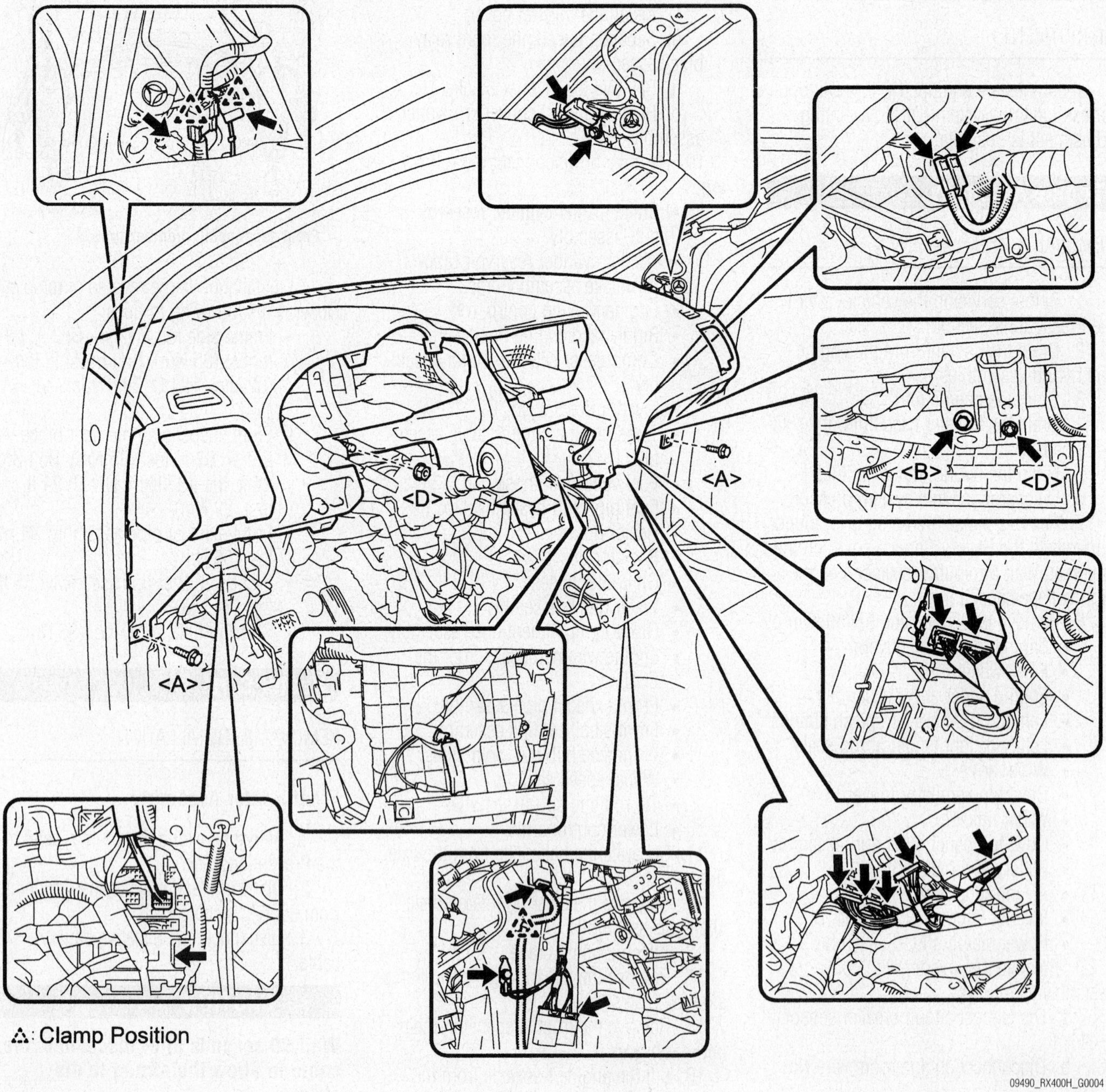

<A>

<D>

<A>

<B>   <D>

⚠: Clamp Position

09490_RX400H_G0004

Disconnect each of the clamps and connectors to remove the instrument panel assembly.

※※ **CAUTION**

Place the air bag module in a safe place with the front side facing upward.

h. Disconnect the horn connector.

i. Remove the steering pad.

j. Matchmark the steering wheel and main shaft assembly.

k. Using Special Tool 00950-50013, remove the steering wheel assembly.

l. Remove the steering column cover.

m. Remove the tilt and telescopic switch.

n. Remove the turn signal switch assembly with spiral cable assembly.

o. Remove the left-hand cowl side trim sub-assembly.

p. Remove the lower instrument panel.

q. Remove the driver side knee airbag assembly.

r. Remove the gauge cluster trim.

s. Remove the instrument gauge assembly.

t. Remove the right-hand front door scuff plate.

u. Remove the right-hand cowl side trim assembly.

v. Remove the instrument panel No. 2 under cover.

w. Remove the glove compartment door assembly.

x. Remove the center console carpet.

y. Remove the lower center instrument panel.

z. Remove the A-pillar trim.

aa. Remove the instrument panel finish plate.

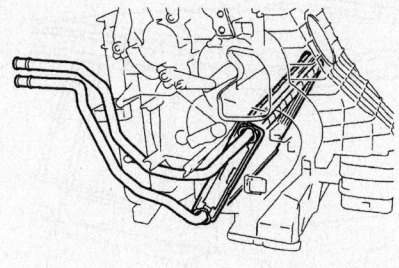

09490_RX400H_G0006

Remove the heater core from the air conditioner unit.

bb. Remove the No.1 register panel.

cc. Remove the instrument panel assembly.

7. Remove the hybrid vehicle control ECU.

8. Remove the air conditioner amplifier assembly.

9. Remove the rear air ducts.

10. Remove air duct No.1.

11. Remove the No.4 instrument panel bracket.

12. Remove the No.1 instrument panel brace.

13. Remove the No.2 instrument panel brace.

14. Matchmark and separate the steering intermediate shaft sub-assembly.

15. Remove the steering column assembly.

16. Remove the No.1 air duct sub-assembly.

17. Remove the instrument panel reinforcement assembly.

18. Remove the air conditioner unit.

19. Remove the No.2 air duct.

20. Remove the blower assembly.

21. Remove the heater to register center duct.

22. Remove the air outlet control servo motor.

23. Remove the air mix control servo motor.

24. Remove the evaporator temperature sensor.

25. Remove the heater core.

26. Installation is the reverse order of removal.

Rear Heater Assembly

1. Before servicing the vehicle, refer to the Precautions Section.

2. Remove or disconnect the following:

- Negative battery cable
- Rear No. 2 deck board
- Rear floor finishing plate
- Backdoor weatherstrip
- Right-hand deck side trim box
- Rear door scuff plate
- Rear door right-hand weatherstrip
- Right-hand deck trim side panel

3. Using pliers, disconnect the rear heater hoses.

➡**Prepare a drain pan to catch any coolant overflow.**

4. Remove or disconnect the following:

- Rear foot air duct
- Rear air duct No. 5
- Rear heater assembly
- Heater core from the rear heater assembly

5. Installation is the reverse order of removal. Tighten the rear heater assembly mounting bolts to 87 inch lbs. (10 Nm).

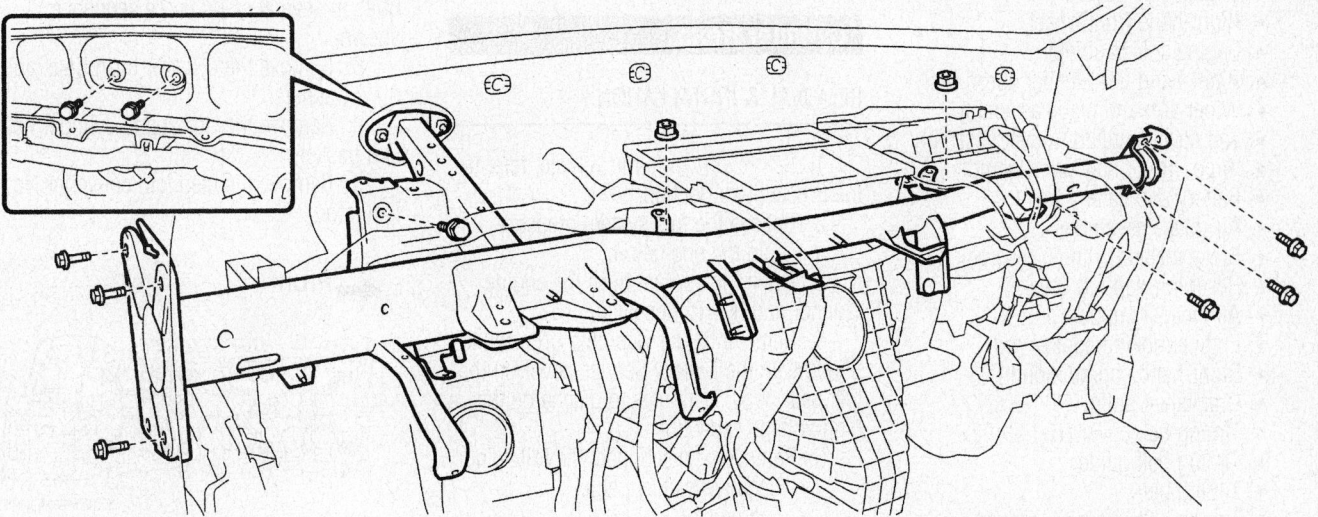

09490_RX400H_G0005

Remove the 9 bolts and 2 nuts to remove the instrument panel reinforcement assembly.

Heater Water Inlet Hose

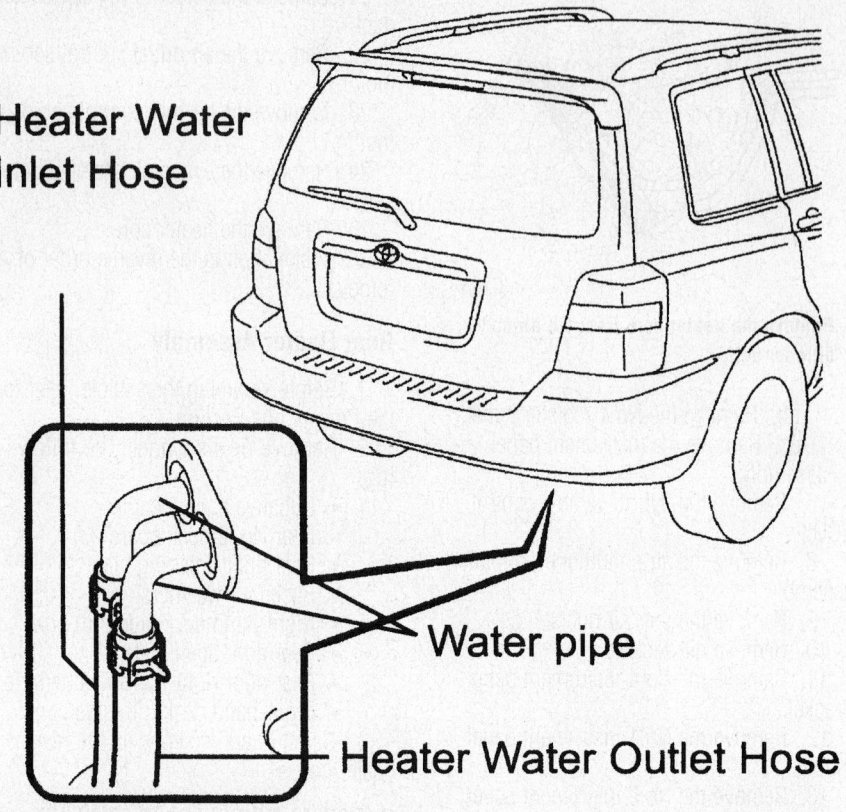

Water pipe

Heater Water Outlet Hose

09490_RX400H_G0038

Disconnect the heater hoses from the rear heater assembly.

Water Pump

REMOVAL & INSTALLATION

1. Before servicing the vehicle, refer to the Precautions Section.
2. Drain the engine coolant.
3. Remove or disconnect the following:
   - Negative battery cable
   - Engine side covers
   - Right-hand front wheel
   - Engine splash shield
   - Right-hand front fender apron seal
   - Wiper arm and blade assembly
   - Top cowl ventilator louver assembly
   - Wiper motor and link assembly
   - Battery and battery tray
   - Air intake assembly
   - Brake master cylinder reservoir
   - Reservoir support bracket
   - Air cleaner support bracket
   - Engine moving control rod
   - Right-hand engine mount
   - Crankshaft pulley
   - Timing belt cover No.1 and 2
   - Timing belt guide
   - Timing belt
   - Timing belt idler sub-assembly No.1

   - Camshaft timing pulley
   - Timing belt cover No.3
   - Timing belt idler sub-assembly No.2
   - Water pump

4. Installation is the reverse order of removal. Tighten the water pump with a new gasket to 71 inch lbs. (8 Nm).
5. Refill the coolant to the correct level.
6. Start the engine and check for leaks.

Cylinder Head

REMOVAL & INSTALLATION

1. Before servicing the vehicle, refer to the Precautions Section.
2. Relieve the fuel system pressure.
3. Drain the engine oil.
4. Drain the coolant from the engine radiator and hybrid transaxle.
5. Remove the service plug grip, found underneath the Battery Service cover on the rear seat. Wait 5 minutes to discharge the high voltage capacitor.
6. Remove or disconnect the following:
   - Negative battery cable
   - Engine cover
   - Right-hand front wheel

   - Engine splash shields
   - Right-hand front fender apron seal
   - Wiper and blade assembly
   - Top cowl ventilator louver assembly
   - Wiper motor and link assembly
   - Battery and battery tray
   - Air intake assembly
   - Converter with Inverter assembly
   - Emission control valve set
   - Intake air surge tank
   - Fuel supply hose
   - Heater inlet hose
   - Intake manifold
   - Radiator hoses
   - Water outlet from the cylinder heads
   - Brake master cylinder reservoir
   - Air cleaner bracket
   - Engine moving control rod
   - Right-hand No.2 engine mounting stay
   - Crankshaft pulley
   - Timing belt No.1 and No.2 covers
   - Right-hand engine mounting bracket
   - Timing belt guide No.2
   - Timing belt
   - Timing belt No.2 idler
   - Camshaft timing pulley
   - Timing belt No.3 cover
   - Front exhaust pipe assembly
   - Exhaust manifold heat insulator
   - Exhaust manifold stay
   - Right-hand exhaust manifold and gasket
   - Ignition coil
   - Right-hand cylinder head cover
   - Camshaft
   - VVT sensor connector
   - Camshaft timing oil control valve connecter

7. Loosen the right-hand cylinder head bolts in several steps in the sequence shown.
8. Remove the cylinder head bolts and plate washers.
9. Remove the right-hand cylinder head and gasket.
10. Remove the manifold converter No.3 insulator.

 Front

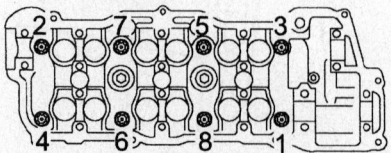

09490_RX400H_G0007

Right-hand cylinder head loosening sequence

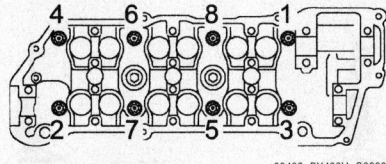

Left-hand cylinder head loosening sequence

11. Remove the exhaust manifold No.2 heat insulator.

12. Separate the cooling fan ECU and hang securely with mechanic's wire.

13. Remove or disconnect the following:
- Left-hand exhaust manifold
- Oil level gauge guide
- Water inlet pipe
- Left-hand cylinder head cover
- Camshaft

14. Loosen the right-hand cylinder head bolts in several steps in the sequence shown.

15. Remove the cylinder head bolts and plate washers.

16. Remove the left-hand cylinder head and gasket.

To install:

17. Install the left-hand cylinder head with a new gasket. Tighten the cylinder head bolts as follows:

a. Step 1: Tighten the 8 cylinder head bolts to 40 ft. lbs. (54 Nm)

b. Step 2: Tighten each bolt 90°

c. Step 3: Tighten each bolt an additional 90°

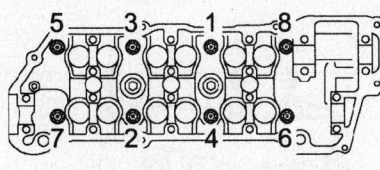

Left-hand cylinder head tightening sequence

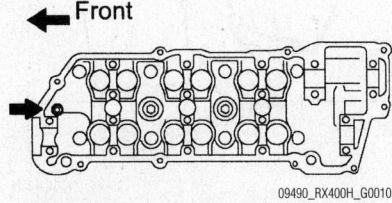

8mm hexagon bolt on the cylinder head

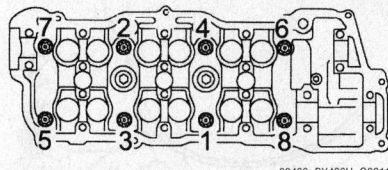

Right-hand cylinder head tightening sequence

d. Step 4: Tighten the single 8mm hexagon bolt to 14 ft. lbs. (19 Nm)

18. Install or connect the following:
- Camshaft
- Left-hand cylinder head cover. Tighten to 71 inch lbs. (8 Nm).
- Water inlet pipe
- Oil level gauge guide. Tighten to 71 inch lbs. (8 Nm).
- Exhaust manifold No. 2 converter
- Cooling fan ECU
- Exhaust manifold No.2 heat insulator
- Manifold converter No.2 insulator

19. Install the right cylinder head with a new gasket. Tighten the cylinder head bolts as follows:

a. Step 1: Tighten the 8 cylinder head bolts to 40 ft. lbs. (54 Nm)

b. Step 2: Tighten each bolt 90°

c. Step 3: Tighten each bolt an additional 90°

d. Step 4: Tighten the single 8mm hexagon bolt to 14 ft. lbs. (19 Nm)

20. The remainder of installation is the reverse order of removal.

21. Refill the engine oil to the correct level.

22. Refill the coolant to the engine radiator and hybrid transaxle to the correct level.

23. Replace the service plug grip.

24. Start the engine and check for leaks.

Intake Manifold

REMOVAL & INSTALLATION

1. Before servicing the vehicle, refer to the Precautions Section.

2. Relieve the fuel system pressure.

3. Drain the engine oil.

4. Drain the coolant from the engine radiator and hybrid transaxle.

5. Remove the service plug grip, found underneath the Battery Service cover on the rear seat. Wait 5 minutes to discharge the high voltage capacitor.

6. Remove or disconnect the following:

- Negative battery cable
- Engine cover
- Right-hand front wheel
- Engine splash shields
- Right-hand front fender apron seal
- Wiper and blade assembly
- Top cowl ventilator louver assembly
- Wiper motor and link assembly
- Battery and battery tray
- Air intake assembly
- Converter with Inverter assembly
- Emission control valve set
- Intake air surge tank
- Fuel supply hose
- Heater inlet hose
- Intake manifold ground cable
- Fuel injector connectors

7. Loosen the intake manifold mounting

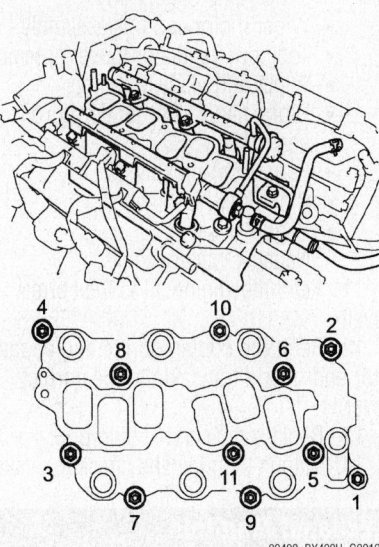

Intake manifold removal sequence

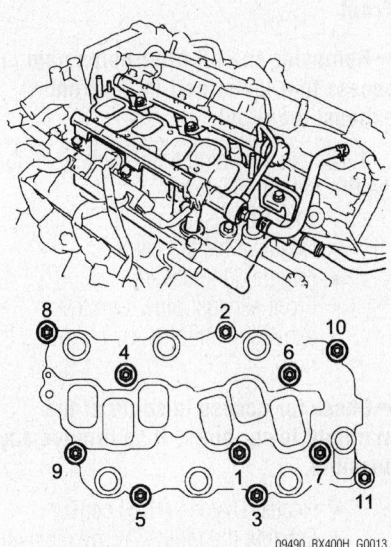

Intake manifold installation sequence

bolts in several steps, in sequence as shown.

8. Remove the intake manifold and gaskets.

To install:

9. Install the intake manifold and gaskets. Tighten the bolts in sequence to 11 ft. lbs. (15 Nm).

10. Install or connect the following:
- Fuel injector connectors
- Intake manifold ground cable. Tighten to 11 ft. lbs. (15 Nm).
- Heater inlet hose. Tighten to 74 inch lbs. (8.4 Nm).
- Fuel supply hose
- Intake air surge tank
- Emission control valve set
- Converter with Inverter assembly
- Air intake assembly
- Battery and battery tray
- Wiper motor and link assembly
- Top cowl ventilator louver assembly
- Wiper and blade assembly
- Right-hand front fender apron seal
- Engine splash shields
- Right-hand front wheel
- Engine cover
- Negative battery cable

11. Refill the engine oil to the correct level.

12. Refill the coolant to the engine radiator and hybrid transaxle to the correct level.

13. Replace the service plug grip.

14. Start the engine and check for leaks.

Exhaust Manifold

REMOVAL & INSTALLATION

Front

➡**Removing the oil filter helps gain access to a lower bolt in the front exhaust manifold.**

1. Before servicing the vehicle, refer to the Precautions Section.

2. Remove or disconnect the following:
- Negative battery cable
- Engine undercovers
- Front exhaust pipe from the exhaust manifolds, by removing the nuts

➡**Check for access to some of the manifold lower bolts, if so remove any possible.**

- Heated Oxygen (HO₂) sensor
- Exhaust manifold stay, by removing the bolt and nut

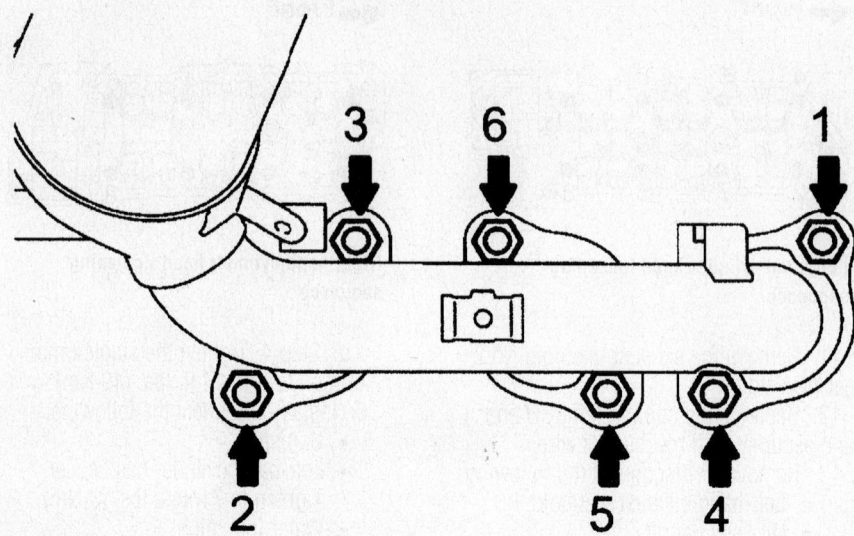

Front manifold nut locations

09490_RX400H_G0014

- Remaining exhaust manifold nuts; then, separate the exhaust manifold from the engine

To install:

3. Install or connect the following:
- Exhaust manifold, using a new gasket. Uniformly, tighten the bolts to 36 ft. lbs. (49 Nm).
- Exhaust manifold stay. Tighten the nut/bolt to 15 ft. lbs. (20 Nm).
- Heated Oxygen (HO₂) sensor to the exhaust manifold
- Front exhaust pipe to the exhaust manifold, using a new gasket. Tighten both nuts to 41 ft. lbs. (56 Nm).
- Engine undercovers
- Negative battery cable

Rear

1. Before servicing the vehicle, refer to the Precautions Section.

2. Remove or disconnect the following:

- Negative battery cable
- Engine undercovers
- Front exhaust pipe from both exhaust manifolds, from below the engine
- Exhaust Gas Recirculation (EGR) pipe from the rear exhaust manifold, by removing the 4 nuts
- Heated Oxygen (HO₂) sensor wiring, from the right exhaust manifold
- Exhaust manifold stay
- 6 exhaust manifold nuts and the exhaust manifold

To install:

3. Install or connect the following:
- Exhaust manifold to the engine, using a new gasket. Tighten the 6 nuts to 36 ft. lbs. (49 Nm).
- Exhaust manifold stay. Tighten the nut/bolt to 25 ft. lbs. (34 Nm).
- HO₂ sensor wiring to the exhaust manifold
- EGR pipe to the exhaust manifold and the engine, using new gaskets. Tighten the 4 nuts to 108 inch lbs. (12 Nm).
- Front exhaust pipe to the exhaust manifold, use a new gasket. Tighten both nuts to 41 ft. lbs. (56 Nm).
- Engine undercovers
- Negative battery cable

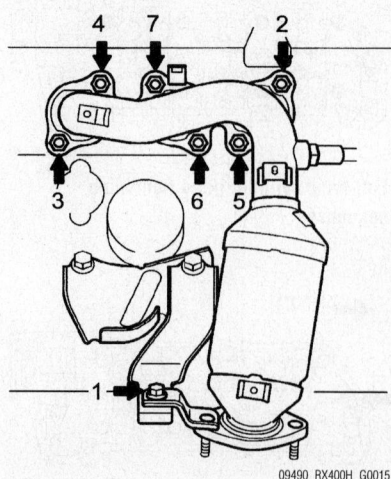

09490_RX400H_G0015

Rear manifold nut locations

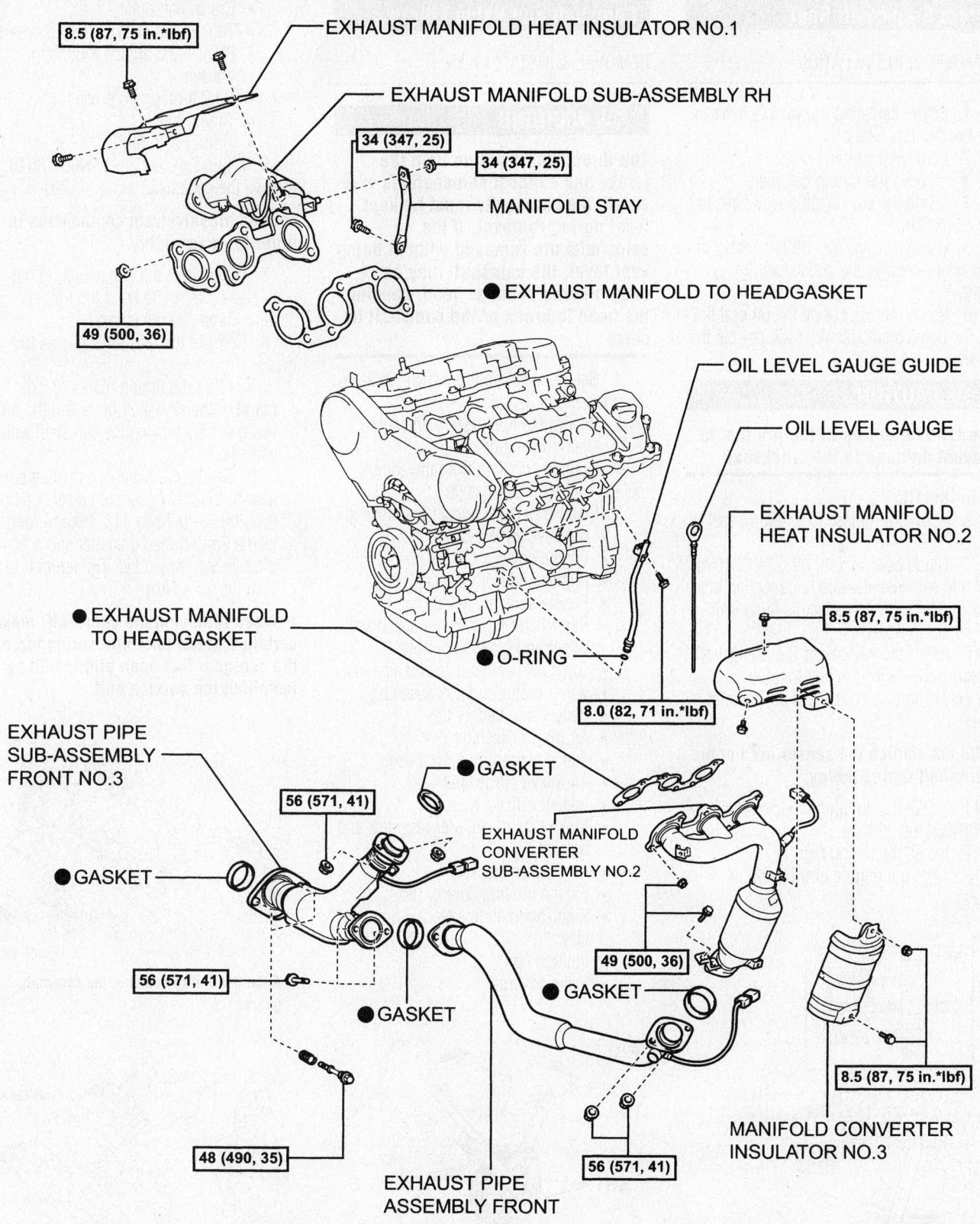

8.5 (87, 75 in.*lbf) — EXHAUST MANIFOLD HEAT INSULATOR NO.1

EXHAUST MANIFOLD SUB-ASSEMBLY RH

34 (347, 25)

34 (347, 25)

MANIFOLD STAY

● EXHAUST MANIFOLD TO HEADGASKET

49 (500, 36)

OIL LEVEL GAUGE GUIDE

OIL LEVEL GAUGE

EXHAUST MANIFOLD HEAT INSULATOR NO.2

● EXHAUST MANIFOLD TO HEADGASKET

● O-RING

8.5 (87, 75 in.*lbf)

8.0 (82, 71 in.*lbf)

EXHAUST PIPE SUB-ASSEMBLY FRONT NO.3

● GASKET

56 (571, 41)

EXHAUST MANIFOLD CONVERTER SUB-ASSEMBLY NO.2

● GASKET

56 (571, 41)

● GASKET

● GASKET

49 (500, 36)

8.5 (87, 75 in.*lbf)

48 (490, 35)

56 (571, 41)

EXHAUST PIPE ASSEMBLY FRONT

MANIFOLD CONVERTER INSULATOR NO.3

N*m (kgf*cm, ft.*lbf) : Specified torque   ● Non-reusable part

09490_RX400H_G0016

Exploded view of the exhaust system—RX 400h, Highlander Hybrid similar

Front Crankshaft Seal

REMOVAL & INSTALLATION

1. Before servicing the vehicle, refer to the Precautions Section.
2. Remove the timing belt.
3. Remove the timing belt plate.
4. Install the crankshaft pulley bolts to the crankshaft.
5. Using Special Tool 09950-50013 or equivalent, remove the crankshaft timing pulley.
6. Using a knife, cut off the oil seal lip.
7. Using a suitable pry tool, pry out the oil seal.

✷ CAUTION

Use tape on the tip of the pry tool to prevent damage to the crankshaft.

To install:

8. Apply MP grease to a new oil seal lip.
9. Using Special Tool 09223-00010 or suitable seal installer and hammer, tap in a new oil seal until its surface is flush with the oil pump edge.
10. Align the keyway of the crankshaft timing pulley with the key located on the crankshaft and slide the pulley into place.

➡**Do not scratch the sensor area of the crankshaft timing pulley.**

11. Install the timing belt plate. Tighten to 71 inch lbs. (8 Nm).
12. Install the timing belt.
13. Start the engine and check for leaks.

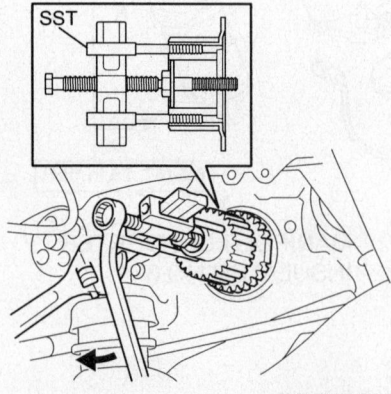

Using the Special Tool or equivalent puller, remove the crankshaft timing pulley.

Camshaft and Valve Lifters

REMOVAL & INSTALLATION

✷✷ WARNING

The thrust clearance on both the intake and exhaust camshafts is very small; the camshafts must be kept level during removal. If the camshafts are removed without being kept level, the camshaft may be caught in the cylinder head, causing the head to break or the camshaft to seize.

1. Before servicing the vehicle, refer to the Precautions Section.
2. Relieve the fuel system pressure.
3. Drain the engine oil.
4. Drain the coolant from the engine radiator and hybrid transaxle.
5. Remove or disconnect the following:
   - Negative battery cable
   - Engine cover
   - Right-hand front wheel
   - Engine splash shields
   - Right-hand front fender apron seal
   - Wiper and blade assembly
   - Top cowl ventilator louver assembly
   - Wiper motor and link assembly
   - Battery and battery tray
   - Air intake assembly
   - Emission control valve hoses
   - Air intake surge tank
   - Radiator intake hose
   - Brake master cylinder reservoir and bracket
   - Air cleaner bracket
   - Engine moving control rod
   - Right-hand engine No.2 mounting stay
   - Ignition coil
   - Valve covers

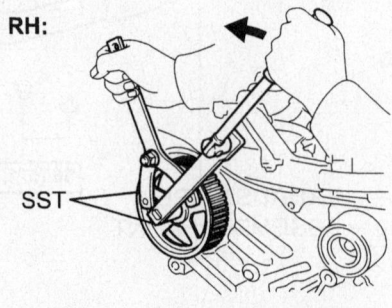

RH:

Removing the right-hand camshaft timing pulley, left-hand similar.

   - Crankshaft pulley
   - Timing belt No.1 and No.2 covers
   - Right-hand engine mounting bracket
   - No.2 timing belt guide
   - Timing belt
   - Timing belt idler

6. Using Special Tool 09960-10010, remove the camshaft timing pulleys.

➡**Keep all valvetrain components in order for reassembly.**

7. Disconnect the engine wiring harness clamps from the No.3 timing belt cover and remove the cover.
8. Remove the left camshafts as follows:
   a. Align the timing marks (2-dot mark) of the camshaft drive and the driven gears by turning the camshaft with a wrench.
   b. Secure the exhaust camshaft sub-gear to the main gear with a service bolt. A bolt 0.63–0.79 in. (16–20mm) long with a 6mm thread diameter and a 1mm pitch is recommended. Tighten bolt to 48 inch lbs. (5.4 Nm).

➡**When removing the camshaft, make certain that the torsional spring force of the sub-gear has been eliminated by installing the service bolt.**

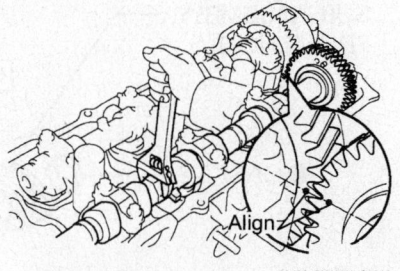

Align the timing marks of the camshaft gears.

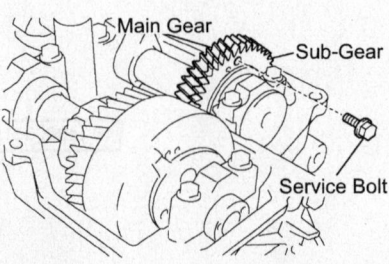

Install a service bolt to secure the camshaft gears.

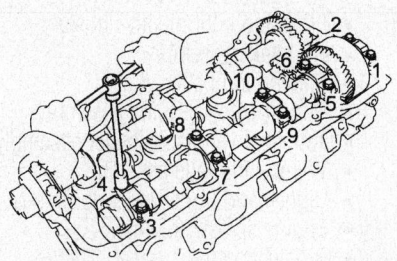

09490_RX400H_G0021

Intake camshaft bearing cap loosening sequence

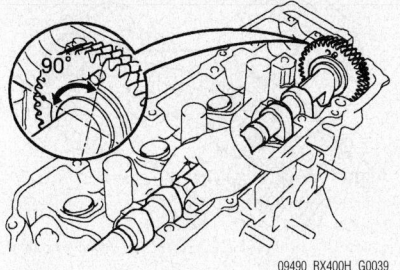

09490_RX400H_G0039

Install the right exhaust camshaft with the alignment marks in the correct position.

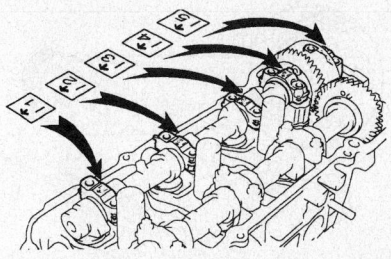

09490_RX400H_G0042

Right intake camshaft bearing caps must be placed in their proper locations

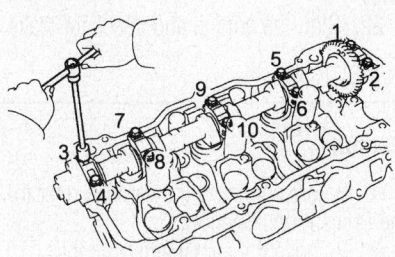

09490_RX400H_G0028

Exhaust camshaft bearing cap loosening sequence

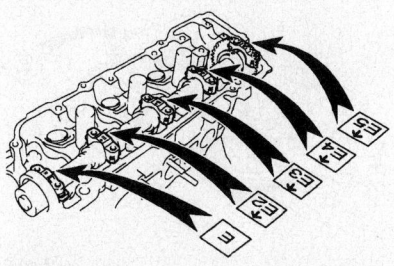

09490_RX400H_G0040

Right exhaust camshaft bearing caps must be placed in their proper locations.

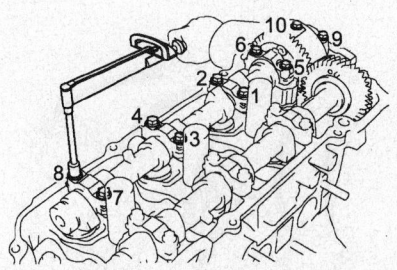

09490_RX400H_G0043

Right intake camshaft bearing cap bolt tightening sequence

c. Using several steps, loosen and remove the 10 bearing cap bolts uniformly in the sequence shown.

d. Remove the 5 bearing caps and the exhaust camshaft.

e. Using several steps, loosen and remove the 10 bearing cap bolts uniformly in the sequence shown.

f. Remove the 5 bearing caps and the intake camshaft.

g. Remove the oil seal from the intake camshaft.

9. Repeat the same process to remove the right-side camshafts, beginning with the intake camshaft.

10. Remove the valve lifter shims and hydraulic lifters. Identify each lifter and shim as it is removed so it can be reinstalled in the same position. If the lifters are to be reused, store them upside down in a sealed container.

To install:

11. Install the valve lifters into their original positions and install the shims. Check valve clearance and replace the shims as necessary.

12. When reinstalling, remember that the camshafts must be handled carefully and kept straight and level to avoid damage.

13. Install the right camshafts, as follows:

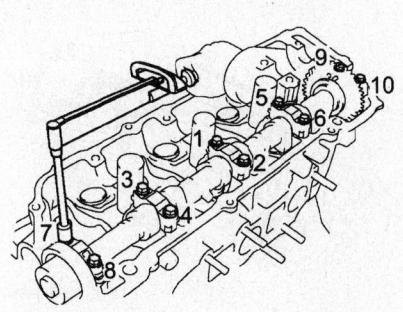

09490_RX400H_G0041

Right exhaust camshaft bearing cap torque sequence

a. Apply new engine oil to the thrust portion and journal of the camshaft.

b. Position the exhaust camshaft on the head so that the alignment marks are at a 90 degrees angle from vertical.

c. Apply multi-purpose grease to the lip of a new oil seal.

d. Install the oil seal to the camshaft.

e. Apply sealant to the No. 1 bearing cap.

f. Apply a light coat of clean engine oil to the bolt threads and under the bolt head. Install the bearing caps to their proper position. Tighten the bolts evenly and in several passes to 12 ft. lbs. (16 Nm) in the proper sequence.

g. Position the intake camshaft on the

head so that the alignment marks are at a 90 degrees angle from vertical. The mark should be at the "9 o'clock" position and must align with the marks on the other gear.

h. Apply a light coat of clean engine oil to the bolt threads and under the bolt head. Install the bearing caps to their proper position. Tighten the bolts evenly and in several passes to 12 ft. lbs. (16 Nm) in the proper sequence.

i. Remove the service bolt.

14. Install the left camshafts, as follows:

a. Apply new engine oil to the thrust portion and journal of the camshaft

b. Position the exhaust camshaft on the head so that the alignment mark is at a 90 degrees angle from vertical. The mark should be at the "9 o'clock" position.

c. Apply multi-purpose grease to the oil seal lip and install the new oil seal to the camshaft.

d. Apply sealant to the No. 1 bearing cap.

e. Apply a light coat of clean engine oil to the bolt threads and under the bolt head. Install the bearing caps to their proper position. Tighten the bolts evenly and in several passes to 12 ft. lbs. (16 Nm) in the proper sequence.

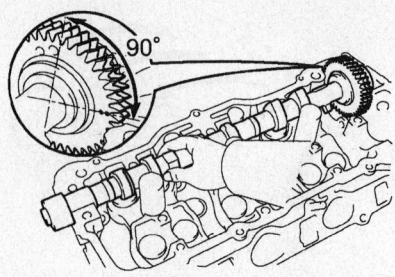

09490_RX400H_G0044

Install the left exhaust camshaft with the alignment mark in the correct position.

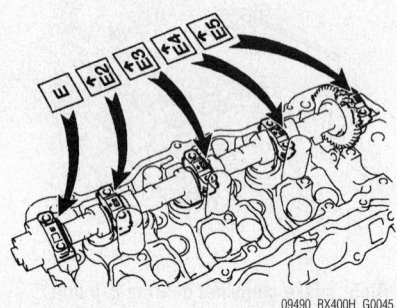

09490_RX400H_G0045

Left exhaust camshaft bearing caps must be placed in their proper locations

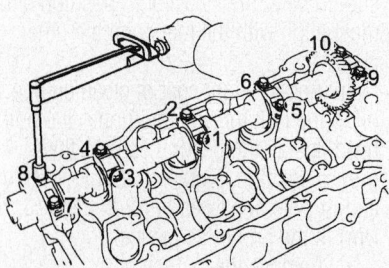

09490_RX400H_G0046

Left exhaust camshaft bearing cap torque sequence

 f. Position the intake camshaft on the head so that the alignment marks are at a 90 degrees angle from vertical. The mark should be at the "3 o'clock" position and must align with the marks on the exhaust camshaft gear.

 g. Apply a light coat of clean engine oil to the bolt threads and under the bolt head. Install the bearing caps to their proper position. Tighten the bolts evenly and in several passes to 12 ft. lbs. (16 Nm) in the proper sequence.

 h. Remove the service bolt.

15. Install or connect the following:

16. Install the timing belt cover No. 3. Tighten to 76 inch lbs. (8.5 Nm)

17. Using Special Tool 09960-10010,

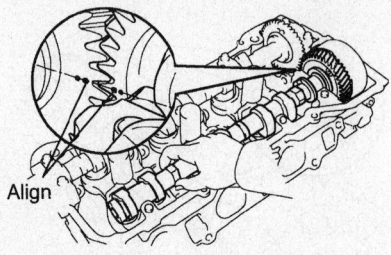

09490_RX400H_G0047

Install the left intake camshaft with the alignment mark in the correct position.

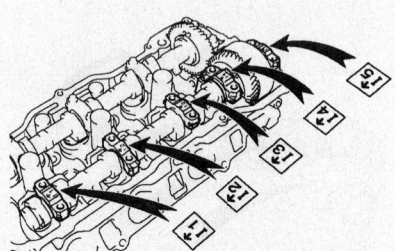

09490_RX400H_G0048

Left exhaust camshaft bearing caps must be placed in their proper locations.

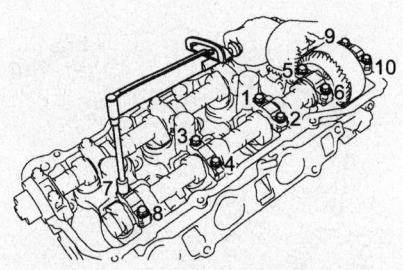

09490_RX400H_G0049

Left exhaust bearing cap torque sequence

install the camshaft timing pulleys. Tighten to 92 ft. lbs. (125 Nm).

18. Install the timing belt idler and tighten to 32 ft. lbs. (43 Nm).

19. Install or connect the following:
- Timing belt
- No. 2 Timing belt guide
- Right-hand engine mounting bracket
- Timing belt covers Nos. 1 and 2
- Crankshaft pulley. Tighten to 162 ft. lbs. (220 Nm).
- Right-hand engine mounting stay No. 2
- Engine moving control rod
- Air cleaner bracket
- Brake master cylinder reservoir and bracket
- Valve covers
- Ignition coil
- Radiator intake hose

- Air intake surge tank
- Emission control valve hoses
- Air intake assembly
- Battery and battery tray
- Wiper motor and link assembly
- Top cowl ventilator louver assembly
- Wiper and blade assembly
- Right-hand front fender apron seal
- Engine splash shields
- Right-hand front wheel
- Engine covers
- Negative battery cable

20. Refill the cooling system to the correct level.

21. Refill the engine oil to the correct level.

22. Start the engine and check for leaks.

INSPECTION

Runout

1. Before servicing the vehicle, refer to the Precautions Section.

2. Remove the camshafts from the vehicle.

3. Place the camshaft on V-blocks

4. Using a dial indicator, measure the amount of runout at the center journal.

5. If the runout is greater than 0.0024 in. (0.06 mm), replace the camshaft.

Camshaft Lobe Height

1. Before servicing the vehicle, refer to the Precautions Section.

2. Remove the camshafts from the vehicle.

3. Using a micrometer, measure the cam lobe height.

4. If the lobe height on the intake camshaft is less than 1.6921 in. (42.98 mm), the camshaft must be replaced. If the

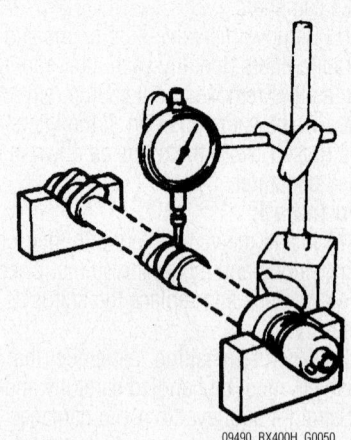

09490_RX400H_G0050

Use a dial indicator to inspect the camshaft runout.

lobe height on the exhaust camshaft is less than 1.6874 in. (42.86 mm), the camshaft must be replaced.

Camshaft Journal Diameter

1. Before servicing the vehicle, refer to the Precautions Section.
2. Remove the camshafts from the vehicle.
3. Using a micrometer, measure the journal diameter.
4. If the journal diameter is less than 1.0614 in. (26.959 mm), replace the camshaft.

Valve Lash

ADJUSTMENT

➡**Adjust the valve clearance when the engine is cold.**

1. Before servicing the vehicle, refer to the Precautions Section.
2. Relieve the fuel system pressure.
3. Drain the engine oil.
4. Drain the coolant from the engine radiator and hybrid transaxle.
5. Remove or disconnect the following:
   • Negative battery cable
   • Engine cover
   • Right-hand front wheel
   • Engine splash shields
   • Right-hand front fender apron seal
   • Wiper and blade assembly
   • Top cowl ventilator louver assembly
   • Wiper motor and link assembly
   • Battery and battery tray
   • Air intake assembly
   • Emission control valve hoses
   • Air intake surge tank
   • Radiator intake hose
   • Brake master cylinder reservoir and bracket
   • Air cleaner bracket
   • Engine moving control rod
   • Right-hand engine No.2 mounting stay
   • Ignition coil
   • Valve covers
6. Turn the crankshaft pulley and align its groove with the timing mark **0** of the No. 1 timing cover.
7. Check that the valve lifters on the No. 1 cylinder (intake and exhaust) are loose. If not, turn the crankshaft 1 complete revolution (360 degrees).

➡**All measurements should be written down. These recorded measurements will need to be used in conjunction with a mathematical formula to deter-**

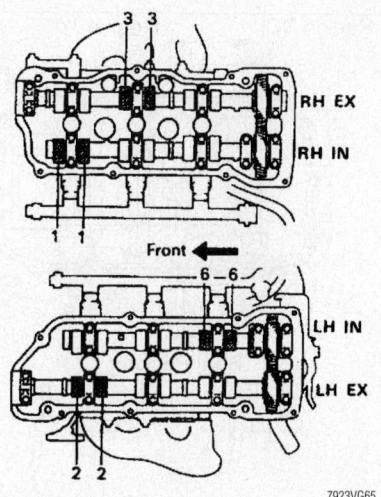

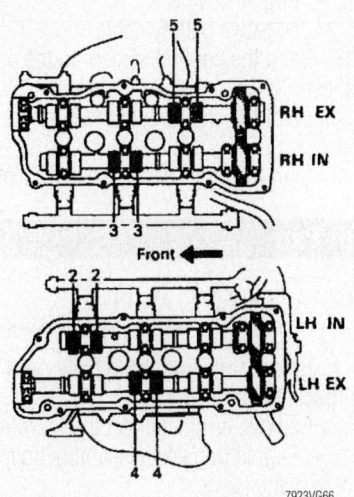

7923VG65

7923VG66

Adjust these valves during the 1st step

Adjust these valves during the 2nd step

mine the thickness of the replacement shims.

8. Measure the clearance between the valve lifters and the camshaft. Record the measurements on valves No. 1 and 6 intake; No. 2 and 3 exhaust.
   a. The intake valve clearance cold is 0.006–0.010 in. (0.15–0.25mm).
   b. The exhaust valve clearance cold is 0.010–0.014 in. (0.25–0.35mm).
9. Turn the crankshaft ⅔ of a revolution (240 degrees). Record the measurements on valves No. 2 and 3 intake; No. 4 and 5 exhaust.
10. Turn the crankshaft another ⅔ of a revolution (240 degrees). Record the measurements on valves No. 4 and 5 intake; No. 1 and 6 exhaust.
11. Remove the adjusting shim by turning the crankshaft to position the cam lobe of the camshaft in the up position on the

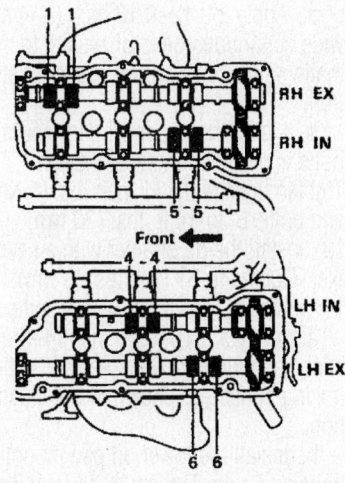

7923VG67

Adjust these valves during the 3rd step

valve to be adjusted. Using a small thin flat bladed tool, turn the valve lifter so that the notches are perpendicular to the camshaft. Press down the valve lifter with tool 09248-55010 part A. Place too 09248-55010 part B between the camshaft and the valve lifter; remove part A.
12. Remove the adjusting shim with a magnet and a small screwdriver.
13. Determine the replacement adjusting shim size by either using the charts or the following formulas:
   • Intake: N = T + (A−0.008 in./0.020mm)
   • Exhaust: N = T + (A−0.012 in./0.30mm)
   • T = Thickness of removed shim
   • A = Measured valve clearance
   • N = Thickness of new shim
14. Select a new shim with a thickness as close as possible to the calculated value. Install the new replacement shim.

➡**Shims are available in 17 sizes in increments of 0.0020 in. (0.050mm), from 0.0984 in. (2.500mm) to 0.1299 in. (3.300mm).**

15. Recheck the valve clearance.
16. Install or connect the following:
   • Valve covers
   • Ignition coil
   • Radiator intake hose
   • Air intake surge tank
   • Emission control valve hoses
   • Air intake assembly
   • Battery and battery tray
   • Wiper motor and link assembly
   • Top cowl ventilator louver assembly
   • Wiper and blade assembly
   • Right-hand front fender apron seal
   • Engine splash shields
   • Right-hand front wheel

- Engine covers
- Negative battery cable

17. Refill the cooling system to the correct level.

18. Refill the engine oil to the correct level.

19. Start the engine and check for leaks.

Oil Pan

REMOVAL & INSTALLATION

1. Before servicing the vehicle, refer to the Precautions Section.

2. Remove or disconnect the following:
- Engine/transaxle assembly from the vehicle
- Right-hand exhaust manifold
- Transaxle mass damper
- Front frame assembly
- Halfshafts
- Flywheel housing undercover
- Front engine mounting bracket
- Transaxle assembly from the engine
- Transmission input damper assembly
- Flywheel

3. Install the engine to a suitable engine stand.

4. Remove or disconnect the following:
- Remaining exhaust manifold heat shields
- Right-hand engine mounting bracket
- Compressor mounting bracket
- Crankshaft pulley
- Timing belt
- Timing belt idler
- Crankshaft timing pulley
- Oil level gauge assembly

5. Remove the lower oil pan as follows:

a. Remove the mounting bolts and nuts

b. Using Special Tool 09032-00100 or suitable seal cutter, cut the sealant between the upper and lower oil pans.

c. Remove the lower oil pan.

6. Remove the oil strainer and gasket.

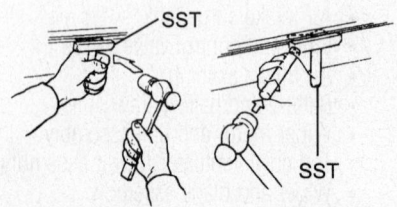

09490_RX400H_G0022

Use a suitable tool to cut the sealant between the oil pans.

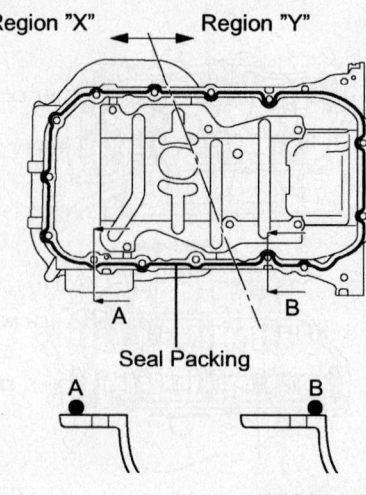

09490_RX400H_G0023

Apply the sealant to the upper oil pan as shown.

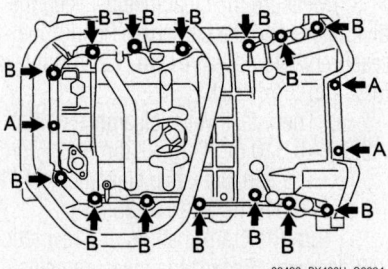

09490_RX400H_G0024

Upper oil pan bolt locations

7. Remove the upper oil pan as follows:

a. Uniformly loosen and remove the mounting bolts.

b. Using a suitable pry tool, pry the upper oil pan from the cylinder block.

To install:

8. Remove any old sealant from the mating surface of the oil pans.

9. Install the upper oil pan as follows:

a. Apply a 0.12–0.16 inch (3–4 mm) wide continuous bead of sealant to the mating surface as shown in the illustration.

b. Install the upper oil pan mounting bolts and tighten in several steps. Tighten bolts 'A' to 71 inch lbs. (8 Nm) and bolts 'B' to 14 ft. lbs. (20 Nm).

10. Install the oil strainer with a new gasket. Tighten to 71 inch lbs. (8 Nm).

11. Install the lower oil pan as follows:

a. Apply a 0.16–0.20 inch (4–5 mm) wide continuous bead of sealant to the mating surface as shown in the illustration.

b. Install the lower oil pan mounting bolts and nuts. Tighten to 71 inch lbs. (8 Nm).

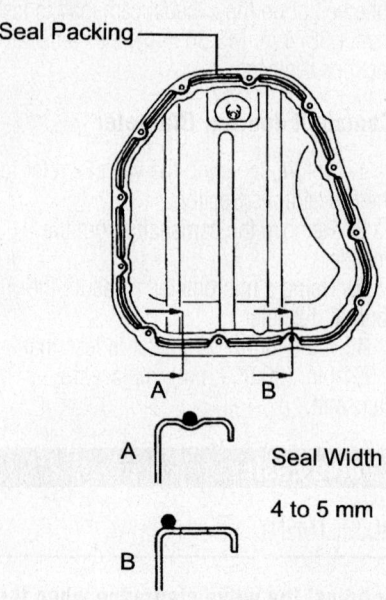

09490_RX400H_G0025

Apply sealant to the lower oil pan as shown.

12. The remainder of the installation is the reverse order of removal.

13. Refill the engine with oil to the correct level.

14. Start the engine and check for leaks.

Oil Pump

REMOVAL & INSTALLATION

1. Before servicing the vehicle, refer to the Precautions Section.

2. Remove or disconnect the following:
- Upper and lower oil pans
- Crankshaft Position (CKP) sensor
- 9 oil pump bolts

➡**Make a note of the position of the each bolt. When replacing the bolts into the oil pump body, place each bolt in the position from which it was removed.**

- Oil pump body, by prying between the oil pump and main bearing cap
- O-ring from the cylinder block
- Plug, gasket, spring and relief valve from the oil pump body
- 9 screws, pump body cover, drive and driven rotors

To install:

3. Install or connect the following:
- Driven rotors, drive, pump body cover, using the 9 screws
- Oil pump relief valve, spring, gas-

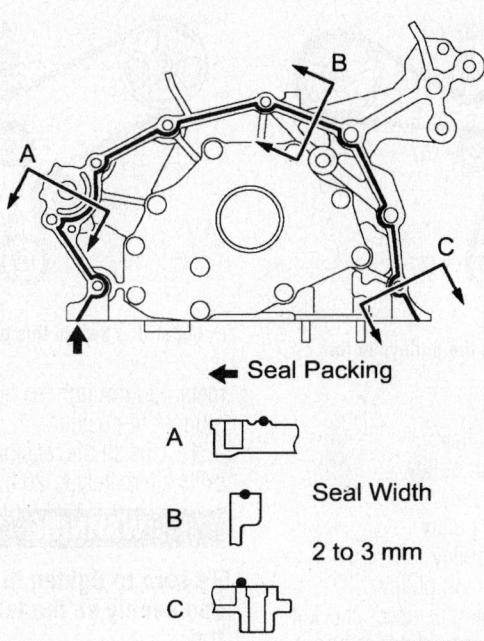

← Seal Packing

A | Seal Width

B | 2 to 3 mm

C

09490_RX400H_G0026

Apply the sealant to the oil pump as shown

ket and the plug to the oil pump body
- New O-ring on the cylinder block

4. Using a non residue solvent, clean both sealing surfaces to the oil pump.

5. Apply liquid sealant to the oil pump and engine block.

6. Install or connect the following:
- Oil pump

➡ **Be sure to engage the splined teeth of the oil pump drive gear with the large teeth of the crankshaft.**

- 9 oil pump bolts. Tighten the bolts in several passes to 71 inch lbs. (8 Nm) for bolt 'A'; 14 ft. lbs. (20 Nm), for bolts 'B'; 32 ft. lbs. (43 Nm) for bolt 'C'
- CKP sensor. Tighten the bolt to 71 inch lbs. (8 Nm).
- Upper and lower oil pans

7. Refill the engine with oil to the correct level.

8. Start the engine and inspect for leaks.

9. Recheck the engine oil level.

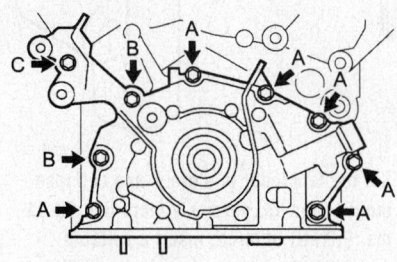

09490_RX400H_G0027

Oil pump bolt locations

Rear Main Seal

REMOVAL & INSTALLATION

1. Before servicing the vehicle, refer to the Precautions Section.

2. Remove or disconnect the following:

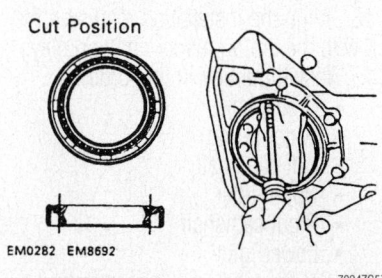

Cut Position

EM0282 EM8692

7924ZG57

Cut off the oil seal lip, then pry the seal out of the retaining plate

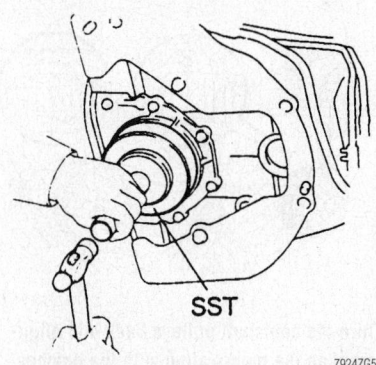

SST

7924ZG58

Tap a new seal into place

- Transaxle assembly
- Transmission input damper
- Flywheel
- Rear main seal

To install:

3. Using Special Tool 09223-15030 or equivalent, tap the new seal into place until the surface is flush with the retainer edge.

4. Install or connect the following:
- Flywheel
- Transmission input damper
- Transaxle assembly

Timing Belt

REMOVAL & INSTALLATION

1. Before servicing the vehicle, refer to the Precautions Section.

2. Remove or disconnect the following:
- Negative battery cable
- Engine covers
- Right front wheel
- Fender splash shields
- Wiper arms
- Top cowl ventilator louver
- Wiper motor and linkage assembly
- Battery and battery tray
- Air intake assembly
- Brake master cylinder reservoir and bracket
- Air cleaner bracket
- Engine moving control rod
- Right-hand engine mounting stay No. 2

3. Use Special Tool 09213-54015 to hold the crankshaft pulley in order to loosen the pulley bolt.

4. Use Special Tool 09950-50013 to remove the crankshaft pulley.

5. Remove or disconnect the following:
- Timing belt cover No. 1
- Timing belt cover No. 2
- Right-hand engine mounting bracket

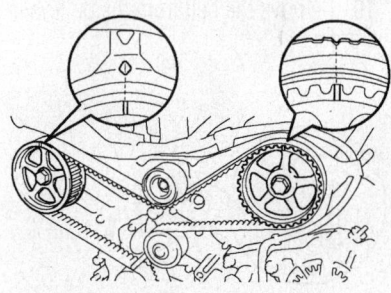

67162-RX300-G01

Check that the timing marks on the camshaft pulleys are aligned with the notches on the inner belt cover

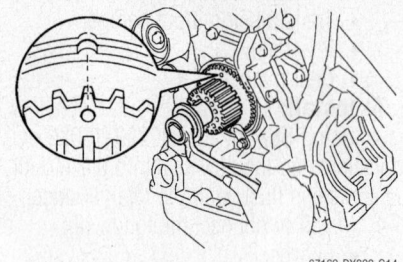

67162-RX300-G14

Turn the crankshaft clockwise to align the timing mark on the crankshaft timing pulley with the notch in the oil pump body

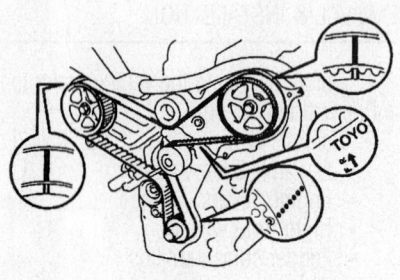

67162-RX300-G02

If the timing belt is re-used, check that the 3 original installation marks are visible on the belt as shown

• Timing belt guide No. 2

6. Temporarily install the crank pulley bolt. Turn the crankshaft clockwise to align the timing mark on the crankshaft timing pulley with the notch in the oil pump body.

7. Check that the timing marks on the camshaft pulleys are aligned with the notches on the inner belt cover. If not, rotate the crankshaft 360 degrees clockwise.

➡ **If the timing belt is re-used, check that the 3 original installation marks are visible on the belt as shown. If not, paint three new marks on the belt.**

8. Turn the crankshaft counterclockwise by 60 degrees. Make sure that the belt is still engaged.

9. Remove the timing belt tensioner.

10. Remove the belt from the pulleys in this order:

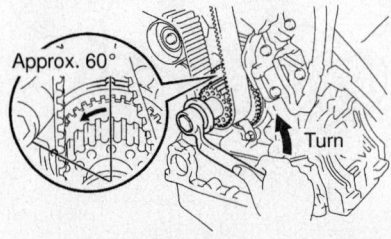

Approx. 60°

Turn

67162-RX300-G03

Turn the crankshaft counterclockwise by 60 degrees

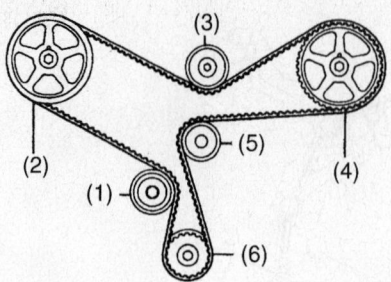

67162-RX300-G04

Remove the belt from the pulleys in this order

• Lower idler pulley
• Right camshaft pulley
• Upper idler pulley
• Left camshaft pulley
• Water pump pulley
• Crankshaft timing pulley

11. If the belt is being re-used, check it for wear or damage; don't twist it or turn it inside-out. If there is any doubt as to it's condition, replace it.

To install:

12. Clean all the pulleys.

13. Turn the crankshaft another 60 degrees counterclockwise.

14. Turn the camshaft pulleys back into alignment so the marks align with the notches on the inner cover.

15. Turn the crankshaft back so that the timing mark aligns with the notch on the oil pump.

16. Align the installation marks on the belt with the timing marks on the pulleys.

17. Install the belt in this order:

• Crankshaft
• Water pump
• Left camshaft
• Upper idler
• Right camshaft
• Lower idler

18. Set the tensioner in a press and collapse the plunger. Do not apply more that 2,205 lbs (9.8 kN) of force. Insert a suitable

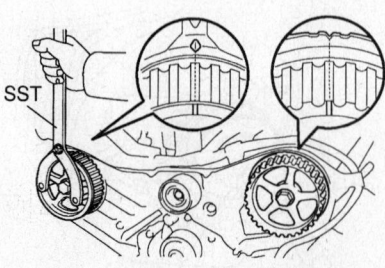

SST

67162-RX300-G05

Turn the camshaft pulleys back into alignment so the marks align with the notches on the inner cover

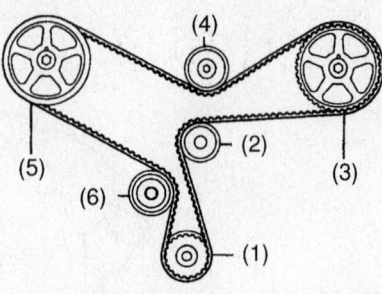

67162-RX300-G06

Install the belt in this order

metal rod through the holes to hold the plunger in position.

19. Install the tensioner and torque the 2 bolts alternately to 20 ft. lbs. (27 Nm).

✳✳ WARNING

Be sure to tighten to bolts alternately and evenly so the tensioner seats flat.

20. Remove the metal rod from the tensioner.

21. Turn the crankshaft 2 full revolutions clockwise (720 degrees), and align the timing mark on the crank pulley with the notch on the oil pump.

22. Check the timing marks on the camshaft pulleys for alignment with the notches on the inner cover. If they do not align, remove the belt and align the mismatched mark(s).

23. The remainder of installation is the reverse of removal. Observe the following torques:

• Right engine mount bracket: 21 ft. lbs. (28 Nm)

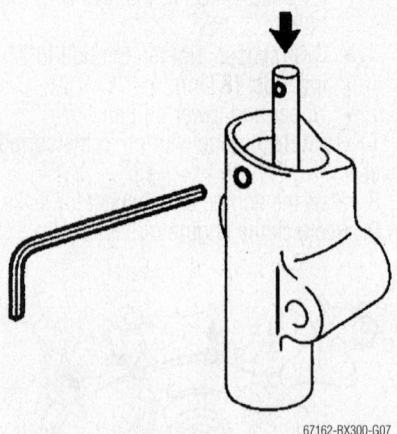

67162-RX300-G07

Set the tensioner in a press and collapse the plunger. Do not apply more that 2,205 lbs (9.8 kN) of force. Insert a suitable metal rod through the holes to hold the plunger in position

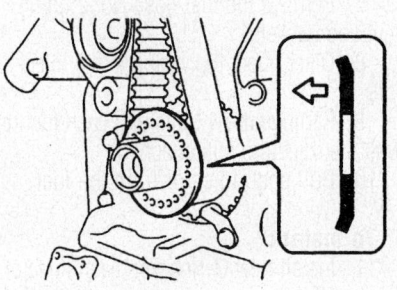

Install the timing belt guide with the cupped side facing front

67162-RX300-G08

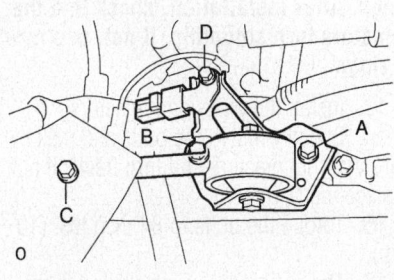

Tighten the engine roll control rod bolts in this order

67162-RX300-G10

- Right engine mount insulator: 70 ft. lbs. (95 Nm)
- Timing belt covers: 75 inch lbs. (8.5 Nm)
- Crankshaft pulley: 162 ft. lbs. (220 Nm)
- Alternator bracket: 21 ft. lbs. (28 Nm)

- Right engine mount stay: 47 ft. lbs. (64 Nm)
- Engine roll control rod: tighten first A, then B, and then C to 47 ft. lbs. (64 Nm). Torque D to 17 ft. lbs. (23 Nm)

Piston and Ring

POSITIONING

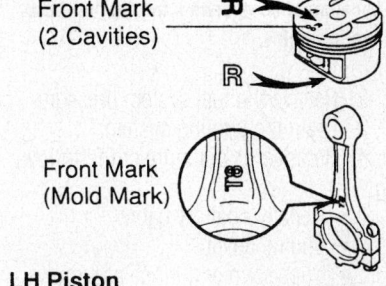

RH Piston
Front Mark (2 Cavities)

Front Mark (Mold Mark)

LH Piston
Front Mark (1 Cavity)

Front Mark (Mold Mark)

9302AG10

Piston/connecting rod-to-engine positioning

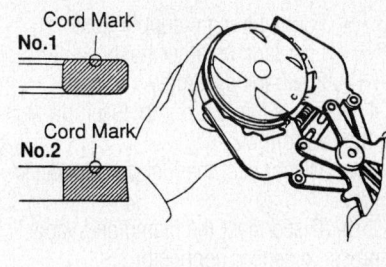

Cord Mark No.1

Cord Mark No.2

9302AG11

Piston ring positioning

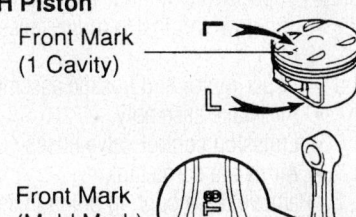

RH Piston
- No.2 Compression
- Lower Side Rail
- Front Mark
- Expander
- No.1 Compression
- Upper Side Rail

LH Piston
- No.2 Compression
- Lower Side Rail
- Front Mark
- Expander
- No.1 Compression
- Upper Side Rail

9302AG12

Piston ring identification

FUEL SYSTEM

Fuel System Pressure

RELIEVING

1. Before servicing the vehicle, refer to the Precautions Section.
2. Disconnect the No. 3 relay block.
3. Remove the No. 2 junction block cover.
4. Remove the C/OPN RLY.
5. Put the vehicle in Inspection Mode and start the engine.
6. Turn the ignition switch to **OFF** immediately after the engine comes to 'rough idle state'.

➡**The hybrid system has a complicated process from an 'out of gas' to 'engine stall' condition. Therefore, 'rough idle' is regarded as 'stop'.**

7. Disconnect the negative battery cable.
8. Reinstall the C/OPN RLY.

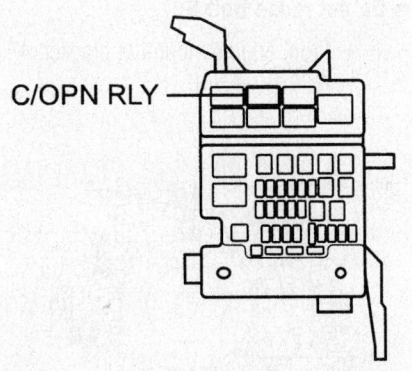

C/OPN RLY

09490_RX400H_G0029

Location of the C/OPN relay in the junction box.

Fuel Filter

The fuel filter is part of the fuel suction tube/fuel pump assembly and is located in the fuel tank.

Fuel Pump

REMOVAL & INSTALLATION

1. Before servicing the vehicle, refer to the Precautions Section.
2. Relieve the fuel system pressure.
3. Remove or disconnect the following:
- Negative battery cable
- Center exhaust pipe assembly
- Front floor heat insulation
- Fuel tank skid plate
- Parking brake cable support
- Fuel tank electrical connection

- Charcoal canister protector and fuel hose
- Fuel tank main supply hose
- Fuel tank breather hose
- Fuel tank assembly

4. Remove the fuel pump assembly from the tank as follows:

a. Disconnect the fuel pump connector.

b. Disconnect the clamp and vapor pressure sensor connector.

c. Remove the hose joint clip and clamp. Pull out the fuel hose.

d. Remove the mounting bolts and pull the fuel pump assembly from the tank.

To install:

5. Install a the fuel pump assembly into the tank with a new gasket.

➡**Align the fuel tank vent tube set plate with the cutout on the fuel pump assembly.**

6. Install the fuel connector plate and tighten the bolts to 53 inch lbs. (6 Nm).

7. The remainder of installation is the reverse order of removal.

Fuel Injector

REMOVAL & INSTALLATION

1. Before servicing the vehicle, refer to the Precautions Section.

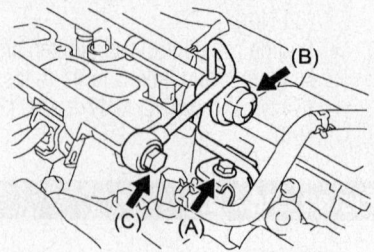

A) FUEL HOSE NO. 1
B) FUEL PRESSURE PULSATION DAMPER
C) FUEL HOSE NO. 2 UNION BOLT

09490_RX400H_G0030

Location of components for the fuel hose sub-assembly

2. Relieve the fuel system pressure.
3. Drain the cooling system.
4. Remove or disconnect the following:

- Negative battery cable
- Engine cover
- Wiper arm and blade assembly
- Top cowl ventilator louver assembly
- Wiper motor and linkage assembly
- Air intake assembly
- Emission control valve hoses
- Air intake surge tank

5. Remove the mounting bolt and separate fuel hose No. 1

6. Remove the fuel pressure pulsation damper and gaskets.

7. Remove the fuel hose No. 2 union bolt and gaskets.

8. Disconnect the wiring at the injectors.

9. Remove the 4 bolts and each fuel rail with the injectors still attached.

10. Pull each injector from the fuel rail..

To install:

11. Install new O-rings on each injector. Apply a light coating of gasoline to the O-rings and mating points on the pipes.

12. Using a twisting motion, install the injectors on the pipes.

➡**Be careful to avoid twisting the O-rings. After installation, check that the injectors turn smoothly. If not, use new O-rings.**

13. Install the pipes and injectors.

14. Loosely install the bolts and make sure that the injectors still turn freely. If not, replace the O-rings.

15. Torque the bolts to 84 inch lbs. (10 Nm).

16. The remainder of installation is the reverse of removal. Observe the following torques:

- Fuel hose No. 2 union bolt: 24 ft. lbs. (33 Nm)
- Pulsation damper: 24 ft. lbs. (33 Nm)
- Fuel hose No.1: 14 ft. lbs. (20 Nm)

DRIVE TRAIN

Hybrid Transaxle Assembly

REMOVAL & INSTALLATION

1. Before servicing the vehicle, refer to the Precautions Section.

2. Remove or disconnect the following:

- Engine/transaxle assembly. See Engine Removal and Installation, earlier in this chapter.
- Manifold stay
- Transaxle damper
- Front frame assembly
- Halfshafts
- Flywheel housing undercover
- Engine wiring harnesses
- Transaxle case cover
- Coolant hose
- Front engine mounting bracket
- Transaxle oil cooler assembly
- Transmission control cable bracket

3. Remove the 8 mounting bolts and separate the transaxle assembly from the vehicle.

4. Installation is the reverse of removal. Observe the following torques:

- Transaxle-to-engine: Bolts A to 47 ft. lbs (64 Nm); Bolt B to 34 ft. lbs. (46 Nm); Bolts C to 47 ft. lbs. (64 Nm); Bolts D to 27 ft. lbs. (37 Nm)

➡**Do not reuse Bolt B.**

- Front engine mounting bracket: 47 ft. lbs. (64 Nm)

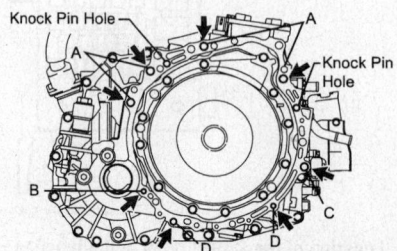

09490_RX400H_G0031

Transaxle-to-engine bolts

- Transaxle case cover: 74 inch lbs. (8.4 Nm)
- Undercover: 69 inch lbs. (8 Nm)

Rear Traction Motor

REMOVAL & INSTALLATION

1. Before servicing the vehicle, refer to the Precautions Section.

2. Remove the service plug grip, found underneath the Battery Service cover on the rear seat. Wait 5 minutes to discharge the high voltage capacitor.

3. Remove or disconnect the following:

- Negative battery cable
- Left-hand engine side cover
- No. 5 inverter bracket
- Power steering ECU
- Inverter reserve tank
- Power steering ECU bracket
- Inverter cover

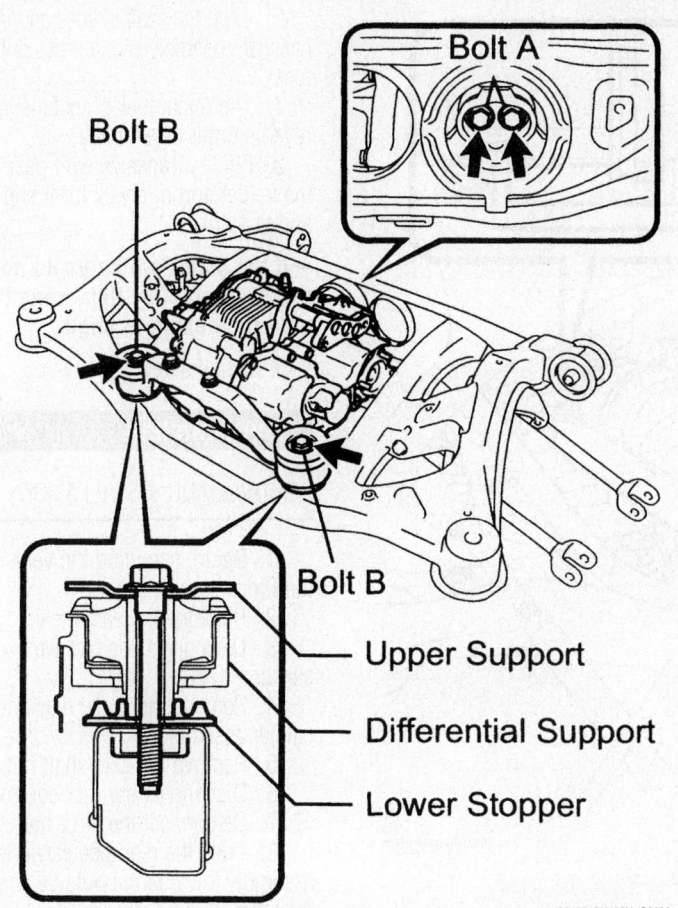

Rear traction motor mounting bolts

.09490_RX400H_G0032

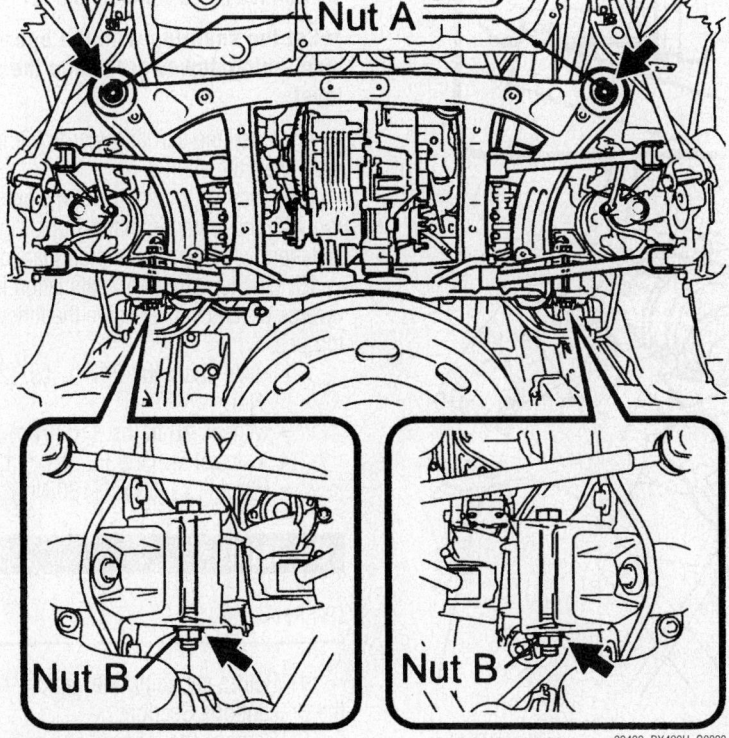

Rear suspension member mounting nuts

09490_RX400H_G0033

4. Verify that the voltage of the W/ Converter Inverter assembly is 0 volts.

5. Install the inverter cover.

6. Drain the hybrid transaxle fluid.

7. Remove or disconnect the following:
- Rear wheels
- Exhaust pipe assembly
- Hub nut
- Height control sensor, RX 400h only

8. Support the rear axle carrier with a suitable jack.

9. Separate the rear strut rod assembly from the knuckle

10. Separate the No. 1 and No. 2 rear suspension arm assemblies.

11. Remove or disconnect the following:
- Rear halfshafts
- Parking brake cable
- No. 3 wire frame
- Rear suspension member
- Rear traction motor assembly

To install:

12. Install the rear traction motor assembly. If motor assembly was removed from the rear suspension member, tighten the mounting bolts as follows:

    a. Bolt A to 70 ft. lbs. (95 Nm).

    b. Bolt B to 76 ft. lbs. (103 Nm).

13. Slowly raise the rear suspension member into place. Tighten Nuts A to 85 ft. lbs. (115 Nm) and Nuts B to 134 ft. lbs. (181 Nm).

14. The remainder of installation is the reverse order of removal.

Front Halfshaft

REMOVAL & INSTALLATION

1. Before servicing the vehicle, refer to the Precautions Section.

2. Remove or disconnect the following:
- Engine splash shields
- Front wheels
- Cotter pin and hub nut
- Front speed sensor
- Brake caliper
- Brake disc
- Tie rod end, from the steering knuckle
- Steering knuckle, from the lower control arm
- Halfshaft from the axle hub, using a plastic hammer
- Stabilizer link
- Front strut

3. Using Special Tool 095020-01010, remove the halfshaft from the transaxle.

To install:

4. Install a new halfshaft hole snapring.

5. Coat the splines of the inboard joint shaft assembly with ATF.

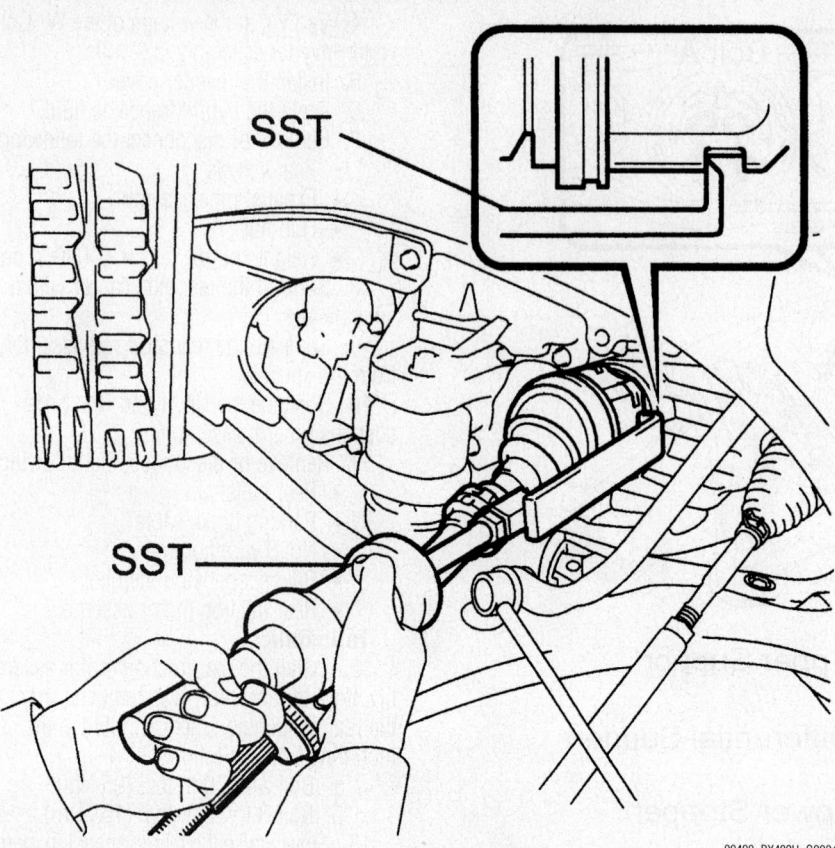

Use the Special Tool to remove the halfshaft from the transaxle.

09490_RX400H_G0034

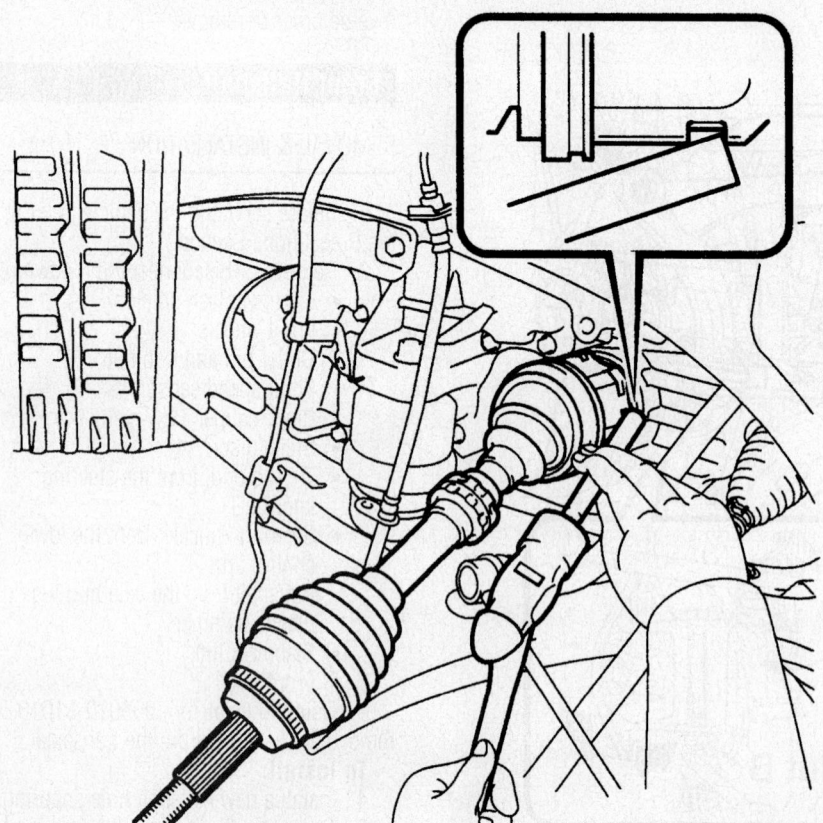

Insert a brass drift into the groove to install the halfshaft.

09490_RX400H_G0035

6. Align the shaft splines and install the halfshaft assembly with a brass drift and hammer.

7. The remainder of installation is the reverse order of removal.

8. Fill the transaxle with gear oil, install the fender apron, check front end alignment and test drive.

➡**If the cotter pin holes do not align, always correct by tightening the nut until the next hole aligns.**

9. Install a new cotter pin.

Rear Halfshaft

REMOVAL AND INSTALLATION

1. Before servicing the vehicle, refer to the Precautions Section.

2. Remove the rear wheel.

3. Disconnect and remove the speed sensor.

4. Remove the brake disc and brake caliper assembly.

5. Remove the axle shaft nut.

6. Disconnect the rear control arms.

7. Disconnect the strut rod.

8. Push the rear axle carrier sub-assembly towards the outside and separate the shaft from the carrier.

9. Remove the shaft, keeping it level.

To install

10. Align the shaft splines and install the shaft with a brass bar and hammer.

➡**Set the snapring with the opening side facing downward. Keep the shaft level.**

11. Push the carrier towards the inside and insert the shaft.

12. Connect the control arms and strut rod with the fasteners hand-tight. Tighten all fasteners with the suspension loaded.

13. The remainder of installation is the reverse of removal. Observe the following torques:

- Axle shaft nut: 217 ft. lbs. (294 Nm)
- Wheel: 76 ft. lbs. (103 Nm)
- Control arms: 83 ft. lbs. (112 Nm)
- Strut: 133 ft. lbs. (180 Nm)

CV-Joints

OVERHAUL

1. Before servicing the vehicle, refer to the Precautions Section.

2. Remove the inboard and outboard joint boot clamps.

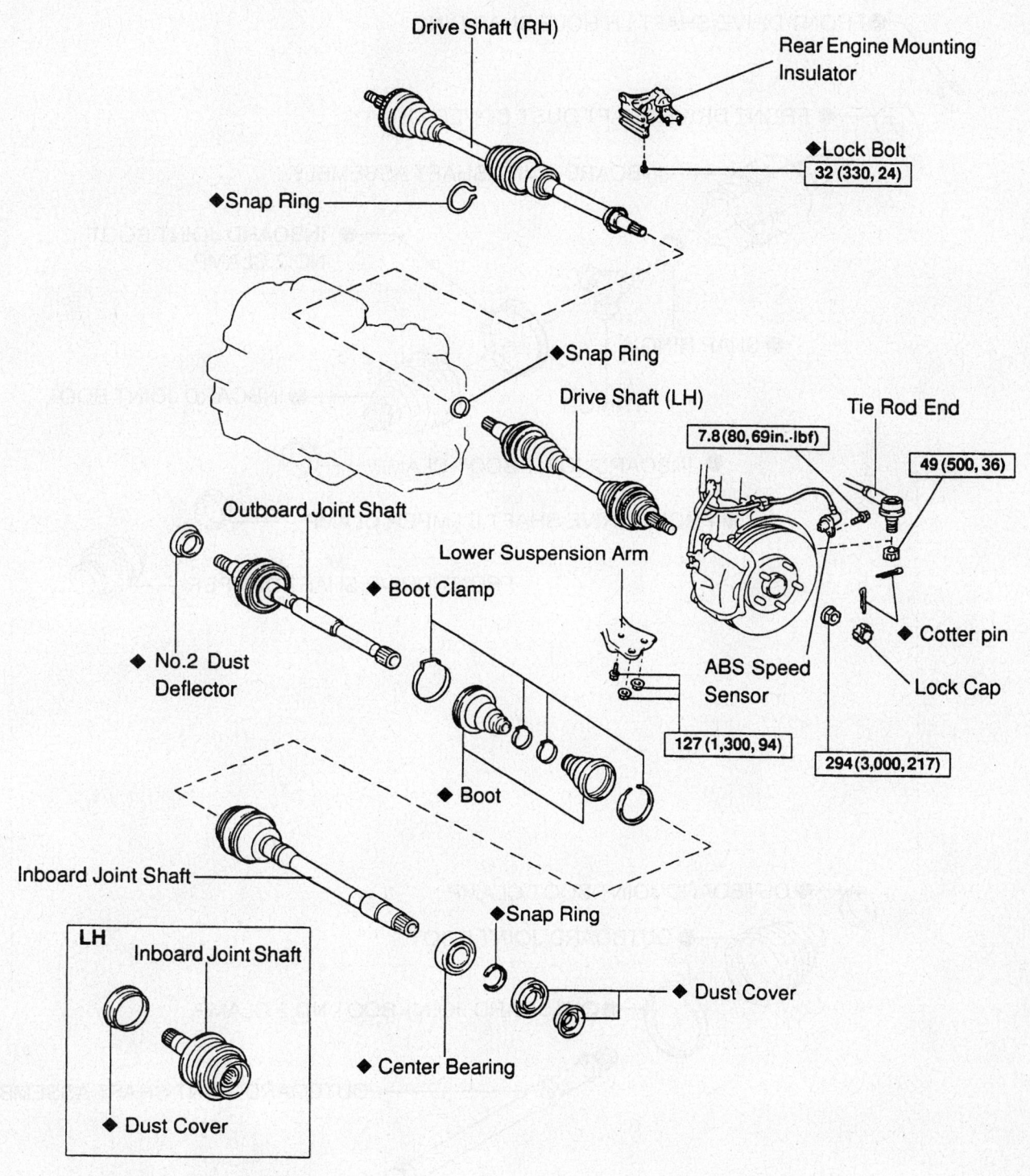

Drive Shaft (RH)

Rear Engine Mounting Insulator

◆Lock Bolt
32 (330, 24)

◆Snap Ring

◆Snap Ring

Drive Shaft (LH)

Tie Rod End

7.8 (80, 69 in.·lbf)

49 (500, 36)

Outboard Joint Shaft

Lower Suspension Arm

◆ Boot Clamp

◆ No.2 Dust Deflector

◆ Boot

ABS Speed Sensor

◆ Cotter pin

Lock Cap

127 (1,300, 94)

294 (3,000, 217)

Inboard Joint Shaft

◆Snap Ring

◆ Dust Cover

LH Inboard Joint Shaft

◆ Center Bearing

◆ Dust Cover

N·m (kgf·cm, ft·lbf) : Specified torque
◆ Non–reusable part

7924ZG73

Exploded view of front right halfshaft

FRONT DRIVE SHAFT ASSEMBLY LH :

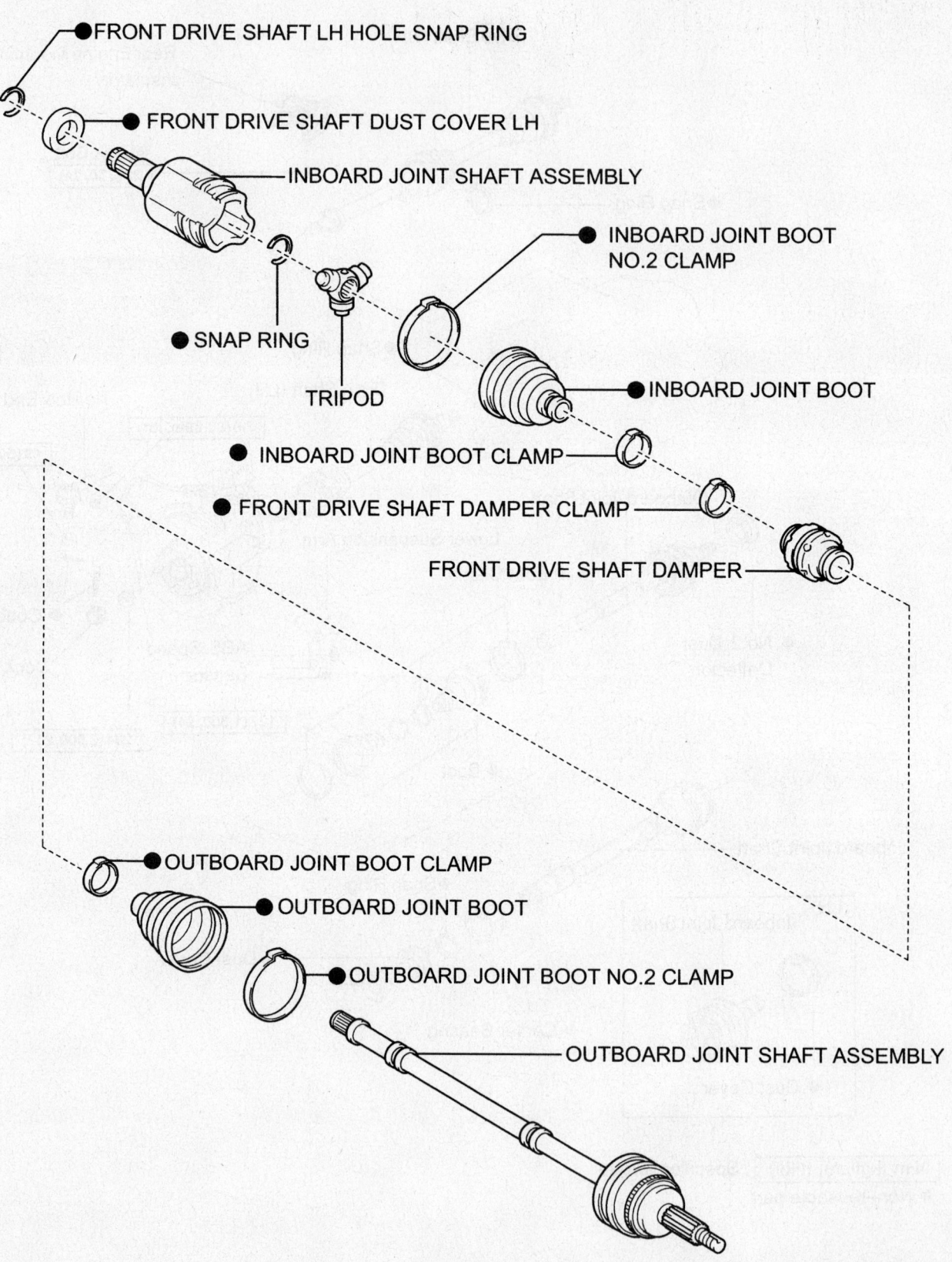

●FRONT DRIVE SHAFT LH HOLE SNAP RING

●FRONT DRIVE SHAFT DUST COVER LH

INBOARD JOINT SHAFT ASSEMBLY

●INBOARD JOINT BOOT NO.2 CLAMP

●SNAP RING

TRIPOD

●INBOARD JOINT BOOT

●INBOARD JOINT BOOT CLAMP

●FRONT DRIVE SHAFT DAMPER CLAMP

FRONT DRIVE SHAFT DAMPER

●OUTBOARD JOINT BOOT CLAMP

●OUTBOARD JOINT BOOT

●OUTBOARD JOINT BOOT NO.2 CLAMP

OUTBOARD JOINT SHAFT ASSEMBLY

09490_RX400H_G0036

Exploded view of front left halfshaft

3. Disassemble the inboard joint tulip, as follows:
- Matchmark the tri-pot, inboard joint tulip or center driveshaft to the driveshaft

✱✱ WARNING

Do not use a punch to make the marks.

- Inboard joint tulip from the drive-shaft

4. Remove the inboard and outboard joint clamps.

5. Remove the tri-pot joint, as follows:
- Snapring
- Matchmark the tri-pot joint to the driveshaft
- Tri-pot joint, using a brass bar and hammer

✱✱ WARNING

Do not tap the roller.

6. Remove or disconnect the following:
- Inboard and outboard joint boots

➡ **Do not disassemble the outboard joint.**

- Dust cover from the center drive-shaft, using a press, for 2WD on the right side
- Dust cover from the inboard joint tulip, using tool 09950-00020 and a press, for 2WD on the left side and 4WD

7. Disassemble the center driveshaft, as follows:
- Snapring
- Bearing case, using a press
- Straight pin from the bearing case, using a pin punch and hammer
- Dust cover, using tool 09950-00020 and a press
- Snapring
- Bearing, using a press

8. Remove the No. 2 dust deflector, using a screwdriver and hammer.

To assemble:

9. Install a new No. 2 dust deflector, using a press.

10. Assemble the center driveshaft, as follows:
- Straight pin into the bearing case, using a pin punch and hammer
- New bearing, using tools 09959-60010, 09950-70010 and a press
- New snapring
- Bearing with the bearing case assembly to the center driveshaft, using tool 09726-40010 and a press
- New snapring
- New dust cover, until the clearance between the dust cover and the bearing is 0.039 in. (1.0mm)

11. Install or connect the following:
- Right dust cover (2WD), until the distance from the tip of the center drive 4.311 in. (109.5 mm) to the inner edge of the dust cover
- Left side dust cover (2WD and 4WD), using a press

12. Temporarily install new outboard/inboard joint boots using new clamps, as follows:
   a. Warp tape around the driveshaft splines.
   b. Install the new outboard joint boot onto the driveshaft.
   c. Install the new inboard joint boot onto the driveshaft.

13. Install the tri-pot joint, as follows:
- Tri-pot joint, face the beveled side toward the outboard joint and align the matchmarks
- Tri-pot joint onto the driveshaft, using a press

✱✱ WARNING

Be careful not to tap the roller.

- New snapring

14. Install the outboard joint boot packed with grease from the boot kit.

15. Install the inboard joint tulip, as follows:
- Pack the inboard joint boot with grease from the boot kit
- Inboard joint tulip, by aligning the matchmarks
- Temporarily, install the inboard joint boot packed with grease from the kit

16. Install the boot clamps to both boots, as follows:
- Both boots to the shaft grooves
- Halfshaft length should be 33.055–33.449 in. (839.6–849.6mm) for the right side on 2WD with A/T, 21.397–21.791 in. (543.5–553.5mm) for the left side on 2WD with A/T, 19.929–20.323 in. (506.2–516.2mm) for the right side on 4WD or 19.803–20.197 in. (503–511mm) for the left side on 4WD
- Both new boot clamps boot
- Bend the band and lock it using a screwdriver

Pinion Seal

REMOVAL & INSTALLATION

1. Before servicing the vehicle, refer to the Precautions Section.

2. Drain the differential oil.

3. Remove or disconnect the following:
- Exhaust pipe
- Driveshaft by matchmarking it
- Companion flange nut, by loosen the staked portion
- Companion flange, using a screw-type extractor
- Oil seal, using an extractor
- Slinger
- Front bearing
- Spacer

To install:

4. Install or connect the following
- New spacer
- Bearing
- Slinger
- New seal

➡ **Seal installation depth: 2.0mm +/- 0.3mm**

- Companion flange
- New nut. Coat the threads with clean differential oil. Torque the nut to 80 ft. lbs. (108Nm).

5. The remainder of installation is the reverse of removal.

STEERING AND SUSPENSION

Air Bag

✳✳ CAUTION

Some vehicles are equipped with an air bag system. The system must be disarmed before performing service on, or around, system components, the steering column, instrument panel components, wiring and sensors. Failure to follow the safety precautions and the disarming procedure could result in accidental air bag deployment, possible injury and unnecessary system repairs.

PRECAUTIONS

Several precautions must be observed when handling the inflator module to avoid accidental deployment and possible personal injury.

• Never carry the inflator module by the wires or connector on the underside of the module.

• When carrying a live inflator module, hold securely with both hands and ensure that the bag and trim cover are pointed away.

• Place the inflator module on a bench or other surface with the bag and trim cover facing up.

• With the inflator module on the bench, never place anything on or close to the module which may be thrown in the event of an accidental deployment.

DISARMING

To avoid personal injury when working on vehicles equipped with an air bag, the negative battery cable must be disconnected and at least 90 seconds must elapse before working on the system. Failure to do so may result in deployment of the air bag.

Power Rack And Pinion Steering Gear

REMOVAL & INSTALLATION

1. Before servicing the vehicle, refer to the Precautions Section.
2. Center the steering wheel to place the front wheels facing straight ahead.
3. Disconnect the negative battery cable.
4. Matchmark and disconnect the intermediate shaft.

5. Remove the wheels.
6. Separate the tie rods.
7. Remove the engine/transaxle assembly.
8. Remove the power steering gear from the front frame assembly.
To install:
9. Install the power steering gear to the front frame assembly. Tighten the mounting bolts to 52 ft. lbs. (70 Nm).
10. Install the engine/transaxle assembly.
11. The remainder of installation is the reverse of removal. Observe the following torques:

• Tie rod end nuts: 36 ft. lbs. (49 Nm)
• Intermediate shaft bolt: 26 ft. lbs. (35 Nm)

Front Strut

REMOVAL & INSTALLATION

1. Before servicing the vehicle, refer to the Precautions Section.
2. Remove the wheel.
3. Disconnect the stabilizer bar link.
4. Loosen, don't remove, the strut locknut.
5. Disconnect the brake hose from the strut.
6. Remove the lower mounting bolts.
7. Remove the upper retaining nuts.
To install:
8. Position the strut and install the upper nuts. Torque to 59 ft. lbs. (80 Nm).
9. Install the lower bolts and torque to 214 ft. lbs. (290 Nm).
10. Connect the brake line.
11. Tighten the strut locknut to 36 ft. lbs. (49 Nm).
12. Connect the stabilizer links and torque to 55 ft. lbs. (74 Nm).
13. Install the wheel. Torque to 76 ft. lbs. (103 Nm).

Rear Strut

REMOVAL AND INSTALLATION

1. Before servicing the vehicle, refer to the Precautions Section.
2. Remove the rear interior trim for RX 400h as follows:
   a. Remove the front and rear deck floor box.
   b. Remove the No. 2 and No. 3 deck board assemblies.

   c. Remove rear floor finishing plate.
   d. Remove jack assembly.
   e. Remove left and right deck side trim box.
   f. Remove rear door scuff plate.
   g. Remove deck side trim cover.
   h. Remove rear floor finishing side plate.
   i. Remove deck trim side panel.
3. Remove the rear interior trim for Highlander Hybrid as follows:
   a. Remove the left side deck trim cover.
4. Remove the rear wheels.
5. Remove the stabilizer link from the strut.
6. On 2-wheel drive models, disconnect the skid control sensor wire and brake hose from the strut and carrier.
7. On 4-wheel drive models, disconnect the brake hose and speed sensor from the strut and carrier.
8. Loosen the 2 nuts at the lower end of the strut, but don't remove them.
9. Support the rear axle carrier with a jack.
10. Remove the 3 upper strut nuts and lower the axle.
11. Remove the lower strut bolts and nuts. Lift out the strut.
To install:
12. Position the strut and install the 3 upper nuts. Torque to 43 ft. lbs. (58 Nm).
13. Lift the axle and install the 2 lower bolts and nuts. Torque to 133 ft. lbs. (180 Nm).
14. Tighten the strut locknut to 36 ft. lbs. (49 Nm).
15. The remainder of installation is the reverse of removal. Observe the following torques:

• Brake hose clamp: 14 ft. lbs. (19 Nm)
• Wire-to-strut clamp: 44 inch lbs. (5 Nm)
• Sensor clamp: 71 inch lbs. (8 Nm)
• Stabilizer link: 29 ft. lbs. (39 Nm)

CONVENTIONAL STRUT OVERHAUL

1. Before servicing the vehicle, refer to the Precautions Section.
2. Remove or disconnect the following:
   • Wheel

➡**If equipped, be careful not to damage the oil seal, driveshaft boot and/or speed sensor rotor when removing the steering knuckle.**

- Shock absorber (strut assembly)

3. Install a nut/bolt to the bracket at the lower portion of the strut assembly and secure it in a vise.

4. Compress the coil spring with a spring compressor.

✳✳ CAUTION

The proper tools must be used for this procedure. The spring on the strut is under high pressure and can cause serious injury if not properly removed and installed.

5. Remove or disconnect the following:
- Center retaining nut, by holding the spring seat
- Support, dust seal, spring seat, insulator and spring from the strut assembly

To install:

6. Install the spring bumper and lower insulator to the strut assembly.

7. Compress the coil spring and fit the lower end of the spring into the spring seat gap.

8. Install or connect the following:
- Upper insulator, spring seat, dust seal, support and spring seat. Tighten the new retaining nut to 36 ft. lbs. (49 Nm).
- Strut
- Wheel

9. If required, bleed the brake system and check for leaks.

10. Check and/or adjust the front wheel alignment.

Lower Ball Joint

REMOVAL & INSTALLATION

1. Remove the wheel.
2. Remove the axle hub nut.

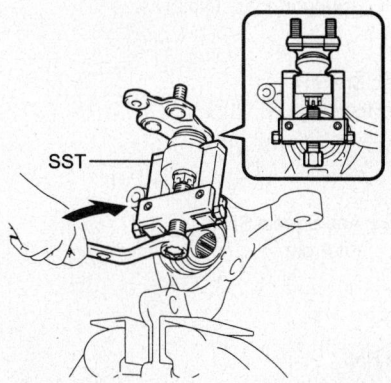

09490_RX400H_G0037

Remove the lower ball joint from the knuckle with a suitable puller.

3. Disconnect the speed sensor.
4. Remove the caliper and suspend it out of the way.
5. Remove the rotor.
6. Disconnect the tie rod end.
7. Remove the lower arm.
8. Remove the disc brake dust cover.
9. Pull the knuckle from the half-shaft.
10. Secure the knuckle in a vise.
11. Remove the cotter pin and nut.
12. Using Special Tool 09628-62011 or suitable ball joint puller, remove the lower ball joint.
13. Installation is the reverse of removal. Observe the following torques:
- Ball joint-to-arm: 94 ft. lbs. (127 Nm)
- Ball joint-to-knuckle: 91 ft. lbs. (123 Nm)
- Hub nut: 217 ft. lbs. (294 Nm)

Lower Control Arm

REMOVAL & INSTALLATION

1. Remove the engine/transaxle assembly.
2. Remove the transverse engine mounting insulator.
3. Remove the 3 bolts securing the arm to the engine support member.
4. Remove the front lower arm bush stopper.
5. Remove the ball joint-to-arm bolts.
6. Installation is the reverse of removal. Observe the following torques:
- 2 short arm-to-support bolts: 148 ft. lbs. (200 Nm)
- 1 long arm-to-support bolt: 152 ft. lbs. (206 Nm)
- Ball joint-to-arm: 94 ft. lbs. (127 Nm)
- Transverse engine mounting insulator: 64 ft. lbs. (87 Nm)

Front Wheel Bearing

Removal

1. Remove the wheel.
2. Remove the hub nut.
3. Remove the caliper and suspend it out of the way.
4. Remove the brake rotor.
5. Disconnect the tie rod end from the knuckle.
6. Disconnect the control arm from the ball joint.
7. Remove the halfshaft from the knuckle.

8. Remove the lower strut-to-knuckle bolts and remove the hub/knuckle assembly.
9. Remove the ball joint.

Disassembly

1. Remove the inner seal from the hub.
2. Remove the snap ring from the hub.
3. Mount the knuckle in a vise and, using a slidehammer, remove the hub from the knuckle.
4. Remove the outer seal.
5. Using a press, the bearings and races can now be replaced.
6. Replace the snap ring and seals.
7. Press the knuckle onto the hub assembly.

Installation

1. Install the ball joint. Torque to 91 ft. lbs. (123 Nm). Advance the nut as much as 60 degrees to align the cotter pin hole. Use a new cotter pin.
2. Attach the knuckle assembly to the strut. Torque to 170 ft. lbs. (230 Nm).
3. The remainder of installation is the reverse of removal. Observe the following torques:
- Caliper support: 77 ft. lbs. (104 Nm)
- Hub nut: 217 ft. lbs. (294 Nm)
- Caliper pins: 25 ft. lbs. (34 Nm)

Rear Wheel Bearing

REMOVAL AND INSTALLATION

1. Remove the wheel.
2. Remove the speed sensor from the axle carrier.
3. Remove the axle shaft nut.
4. Remove the caliper and support assembly and suspend it out of the way.
5. Check the bearing backlash. It should not exceed 0.0020 in. (0.05mm). If it does, it must be replaced.
6. Remove the 4 bolts and remove the hub/bearing assembly.
7. Installation is the reverse of removal. Observe the following torques:
- Hub/bearing assembly bolts: 55 ft. lbs. (75 Nm)
- Caliper support: 58 ft. lbs. (78 Nm)
- Brake hose clamp: 14 ft. lbs. (19 Nm)
- Hub nut: 217 ft. lbs. (294 Nm)

BRAKES

Brake Caliper

REMOVAL AND INSTALLATION

FRONT

1. Disconnect the brake line from the caliper and plug it.
2. Hold the caliper slide pins and remove the mounting bolts.
3. Lift off the caliper.

4. Remove the pads and anti-squeal shims.
5. Remove the wear indicator from the inner pad.
6. Installation is the reverse of removal. Grease the caliper slides and bolts with lithium grease or equivalent. Apply disc brake grease to the anti-squeal shims. Torque the caliper bolts to 25 ft. lbs. (34 Nm); the brake line union bolt to 21 ft. lbs. (29 Nm).

REAR

1. Disconnect the brake line from the caliper and plug it.
2. Remove the caliper mounting bolts.
3. Lift off the caliper.
4. Remove the pads and anti-squeal shims.
5. Remove the wear indicators from each pad.
6. Installation is the reverse of removal.

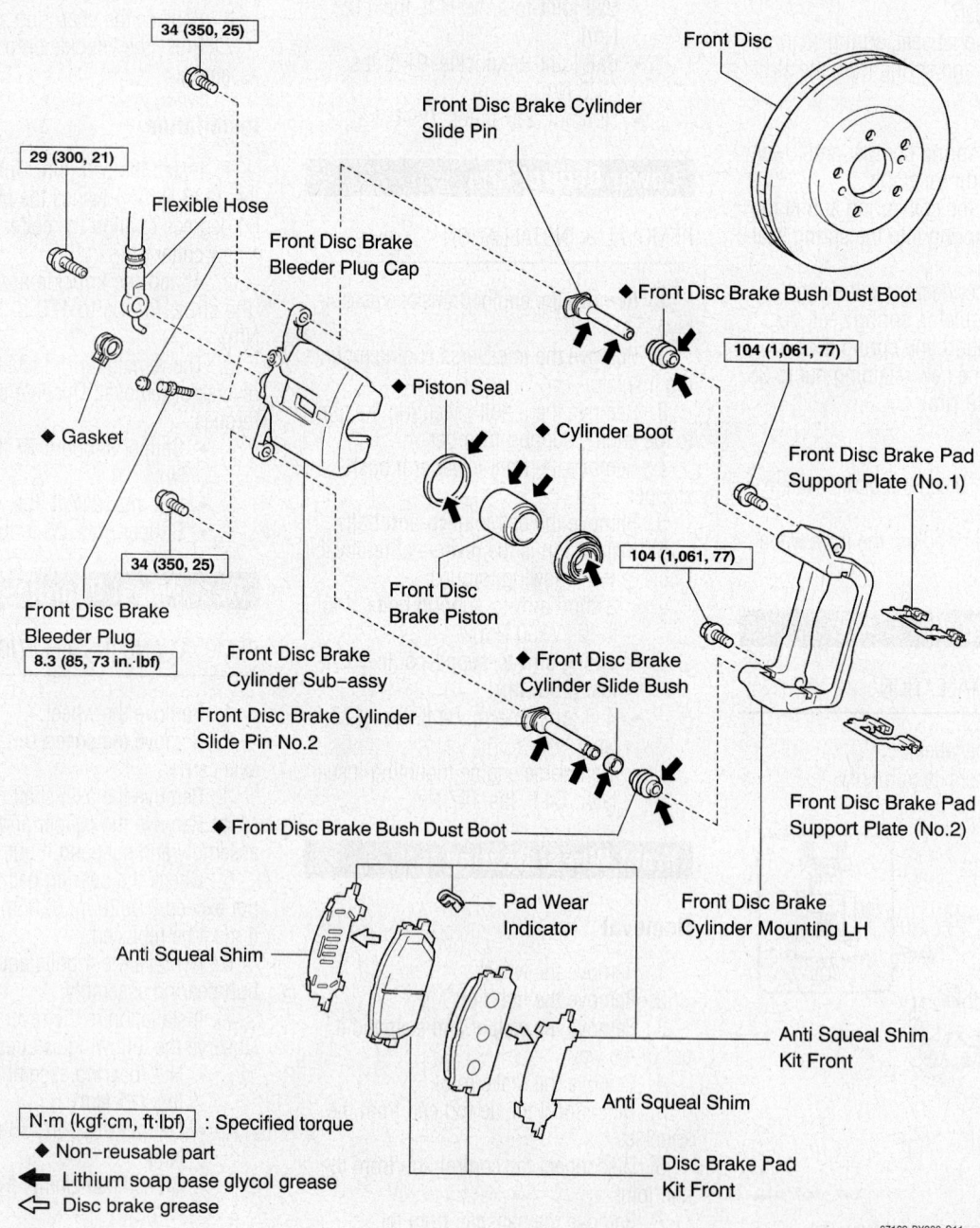

34 (350, 25)

Front Disc Brake Cylinder Slide Pin

Front Disc

29 (300, 21)

Flexible Hose

Front Disc Brake Bleeder Plug Cap

◆ Front Disc Brake Bush Dust Boot

104 (1,061, 77)

◆ Piston Seal

◆ Gasket

Cylinder Boot

Front Disc Brake Pad Support Plate (No.1)

Front Disc Brake Piston

104 (1,061, 77)

34 (350, 25)

Front Disc Brake Bleeder Plug
8.3 (85, 73 in.·lbf)

Front Disc Brake Cylinder Sub-assy

◆ Front Disc Brake Cylinder Slide Bush

Front Disc Brake Cylinder Slide Pin No.2

◆ Front Disc Brake Bush Dust Boot

Pad Wear Indicator

Front Disc Brake Pad Support Plate (No.2)

Front Disc Brake Cylinder Mounting LH

Anti Squeal Shim

Anti Squeal Shim Kit Front

Anti Squeal Shim

N·m (kgf·cm, ft·lbf) : Specified torque
◆ Non-reusable part
← Lithium soap base glycol grease
⇐ Disc brake grease

Disc Brake Pad Kit Front

67162-RX300-G11

Front disc brake components

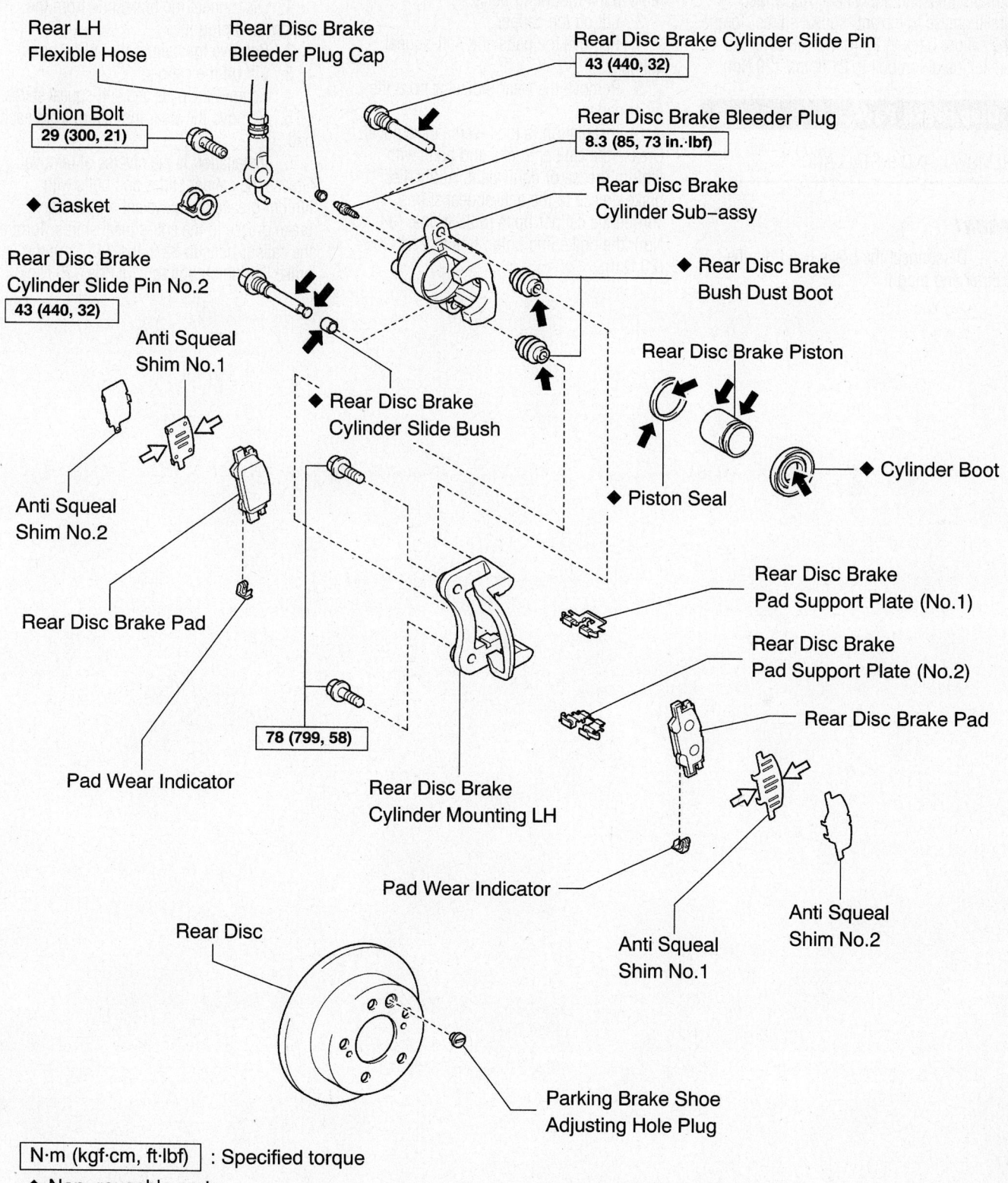

Rear LH Flexible Hose

Rear Disc Brake Bleeder Plug Cap

Rear Disc Brake Cylinder Slide Pin
43 (440, 32)

Union Bolt
29 (300, 21)

Rear Disc Brake Bleeder Plug
8.3 (85, 73 in.·lbf)

◆ Gasket

Rear Disc Brake Cylinder Sub-assy

Rear Disc Brake Cylinder Slide Pin No.2
43 (440, 32)

◆ Rear Disc Brake Bush Dust Boot

Anti Squeal Shim No.1

Rear Disc Brake Piston

◆ Rear Disc Brake Cylinder Slide Bush

Anti Squeal Shim No.2

◆ Cylinder Boot

◆ Piston Seal

Rear Disc Brake Pad

Rear Disc Brake Pad Support Plate (No.1)

Rear Disc Brake Pad Support Plate (No.2)

Pad Wear Indicator

Rear Disc Brake Pad

78 (799, 58)

Rear Disc Brake Cylinder Mounting LH

Pad Wear Indicator

Anti Squeal Shim No.1

Anti Squeal Shim No.2

Rear Disc

Parking Brake Shoe Adjusting Hole Plug

N·m (kgf·cm, ft·lbf) : Specified torque
◆ Non-reusable part
← Lithium soap base glycol grease
⇐ Disc brake grease

67162-RX300-G12

Rear disc brake components

Grease the caliper slides and bolts with lithium grease or equivalent. Apply disc brake grease to the anti-squeal shims. Torque the caliper bolts to 32 ft. lbs. (43 Nm); the brake line union bolt to 21 ft. lbs. (29 Nm).

Disc Brake Pads

REMOVAL AND INSTALLATION

FRONT

1. Disconnect the brake line from the caliper and plug it.

2. Hold the caliper slide pins and remove the mounting bolts.
3. Lift off the caliper.
4. Remove the pads and anti-squeal shims.
5. Remove the wear indicator from the inner pad.
6. Installation is the reverse of removal. Grease the caliper slides and bolts with lithium grease or equivalent. Apply disc brake grease to the anti-squeal shims. Torque the caliper bolts to 25 ft. lbs. (34 Nm); the brake line union bolt to 21 ft. lbs. (29 Nm).

REAR

1. Disconnect the brake line from the caliper and plug it.
2. Remove the caliper mounting bolts.
3. Lift off the caliper.
4. Remove the pads and anti-squeal shims.
5. Remove the wear indicators from each pad.
6. Installation is the reverse of removal. Grease the caliper slides and bolts with lithium grease or equivalent. Apply disc brake grease to the anti-squeal shims. Torque the caliper bolts to 32 ft. lbs. (43 Nm); the brake line union bolt to 21 ft. lbs. (29 Nm).

SCION

tC • xA • xB

BRAKES 5-46
DRIVE TRAIN 5-30
ENGINE REPAIR................5-9
FUEL SYSTEM 5-29
**SPECIFICATIONS AND
  MAINTENANCE CHARTS** 5-2
Engine and Vehicle Identification
  Chart..............................5-2
General Engine Specifications 5-2
Engine Tune-Up Specifications.......5-2
Firing Order 5-3
Accessory Drive Belt Routing 5-3
Capacities 5-3
Valve Specifications.....................5-4
Crankshaft and Connecting Rod
  Specifications 5-4
Camshaft Specifications 5-5
Piston and Ring Specifications 5-5
Torque Specifications 5-6
Wheel Alignment 5-6
Tire, Wheel and Ball Joint
  Specifications 5-7
Brake Specifications 5-7
Scheduled Maintenance Intervals 5-8
**STEERING AND
  SUSPENSION** 5-37

A

Air Bag.................................5-37
  Disarming 5-37
Alternator..............................5-9
  Removal & Installation................5-9

B

Brake Caliper 5-46
  Removal & Installation...............5-46
Brake Drums 5-48
  Removal & Installation...............5-48
Brake Shoes 5-49
  Removal & Installation...............5-49

C

Camshaft and Valve Lifters 5-18
  Inspection 5-20
  Removal & Installation.............5-18
Clutch 5-32
  Adjustments 5-32
  Removal & Installation.............5-33
Coil Spring 5-43
  Removal & Installation.............5-43
CV-Joints...............................5-34
  Overhaul 5-34
Cylinder Head 5-15
  Removal & Installation.............5-15

D

Disc Brake Pads........................5-47
  Removal & Installation.............5-47
Distributor...............................5-9

E

Engine Assembly 5-9
  Removal & Installation.............5-9
Exhaust Manifold.......................5-17
  Removal & Installation.............5-17

F

Fuel Filter 5-29
Fuel Injector 5-30
  Removal & Installation..............5-30
Fuel Pump 5-29
  Removal & Installation..............5-29
Fuel System Pressure 5-29
  Relieving 5-29

H

Halfshaft..............................5-34
  Removal & Installation.............5-34
Heater Core...........................5-11
  Removal And Installation 5-11

I

Ignition Timing 5-9
  Adjustment.........................5-9

Intake Manifold........................5-16
  Removal & Installation..............5-16

L

Lower Ball Joint.........................5-43
  Removal & Installation..............5-43
Lower Control Arm 5-44
  Removal & Installation..............5-44

O

Oil Pan.................................5-22
  Removal & Installation.............5-22
Oil Pump 5-23
  Removal & Installation.............5-23

P

Piston and Ring 5-28
  Positioning 5-28

R

Rack and Pinion Steering Gear 5-37
  Removal & Installation.............5-37
Rear Main Seal 5-24
  Removal & Installation.............5-24

S

Starter Motor 5-22
  Removal & Installation.............5-22
Struts.................................5-40
  Removal & Installation.............5-40

T

Timing Chain..........................5-24
  Removal & Installation.............5-24
Transaxle Assembly 5-30
  Removal & Installation.............5-30

V

Valve Lash 5-20
  Adjustment.........................5-20

W

Water Pump...........................5-10
  Removal & Installation.............5-10
Wheel Bearings........................5-44
  Removal & Installation.............5-44

SPECIFICATIONS AND MAINTENANCE CHARTS

ENGINE AND VEHICLE IDENTIFICATION

	Engine							Model Year	
Code ①	Liters (cc)	Cu. In.	Cyl.	Fuel Sys.	Engine Type	Eng. Mfg.		Code ②	Year
1NZ-FE	1.5 (1496)	91.3	4	EFI	DOHC	Toyota		4	2004
2AZ-FE	2.4 (2398)	144.1	4	EFI	DOHC	Toyota		5	2005
								6	2006

SFI: Sequential Fuel Injection

DOHC: Double Overhead Camshaft

① Stamped on the left side of the engine block

② 10th digit of the Vehicle Identification Number (VIN)

09490_SION_C0001

GENERAL ENGINE SPECIFICATIONS

Year	Model	Engine Displacement Liters	Engine Series ID	Net Horsepower @ rpm	Net Torque @ rpm (ft. lbs.)	Bore x Stroke (in.)	Com-pression Ratio	Oil Pressure @ rpm
2004	xA	1.5	1NZ-FE	108@6000	105@4200	2.95x3.33	10.5:1	22-80@3000
	xB	1.5	1NZ-FE	108@6000	105@4200	2.95x3.33	10.5:1	22-80@3000
2005	xA	1.5	1NZ-FE	108@6000	105@4200	2.95x3.33	10.5:1	22-80@3000
	xB	1.5	1NZ-FE	108@6000	105@4200	2.95x3.33	10.5:1	22-80@3000
	tC	2.4	2AZ-FE	160@5700	163@4000	3.48x3.78	9.6:1	36-78@3000
2006	xA	1.5	1NZ-FE	103@6000	101@4200	2.95x3.33	10.5:1	22-80@3000
	xB	1.5	1NZ-FE	103@6000	101@4200	2.95x3.33	10.5:1	22-80@3000
	tC	2.4	2AZ-FE	160@5700	163@4000	3.48x3.78	9.6:1	36-78@3000

09490_SION_C0002

ENGINE TUNE-UP SPECIFICATIONS

Year	Engine Displacement Liters	Engine ID	Spark Plug Gap (in.)	Ignition Timing (deg.)	Fuel Pump (psi)	Idle Speed (rpm)	Valve Clearance Intake	Valve Clearance Exhaust
2004	1.5	1NZ-FE	0.031	8-12B	44-50	①	0.0060-0.0100	0.0100-0.0140
2005	1.5	1NZ-FE	0.031	8-12B	44-50	①	0.0060-0.0100	0.0100-0.0140
	2.4	2AZ-FE	0.043	5-15B	44-50	②	0.0070-0.0120	0.0120-0.0160
2006	1.5	1NZ-FE	0.031	8-12B	44-50	①	0.0060-0.0100	0.0100-0.0140
	2.4	2AZ-FE	0.043	5-15B	44-50	②	0.0070-0.0120	0.0120-0.0160

NOTE: The Vehicle Emission Control Information label often reflects specification changes made during production.

The label figures must be used if they differ from those in this chart.

B: Before top dead center

① M/T: 600-700
A/T: 650-750

② A/T: 610-710
M/T: 650-750

09490_SION_C0003

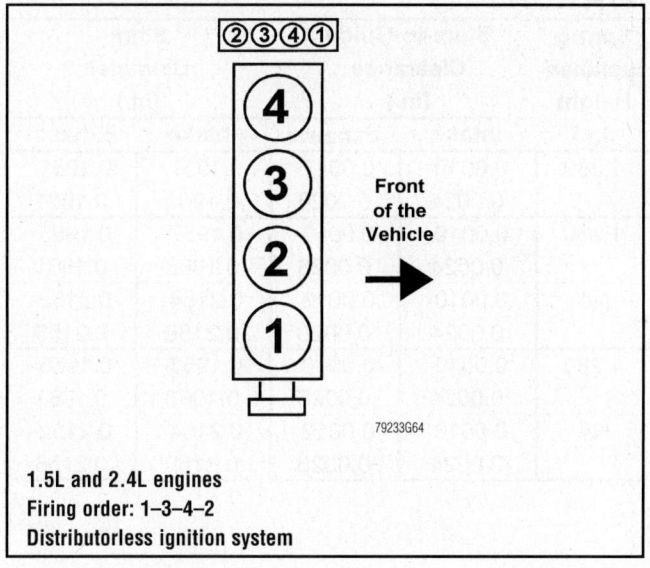

②③④①

4
3
2
1

Front
of the
Vehicle →

79233G64

1.5L and 2.4L engines
Firing order: 1–3–4–2
Distributorless ignition system

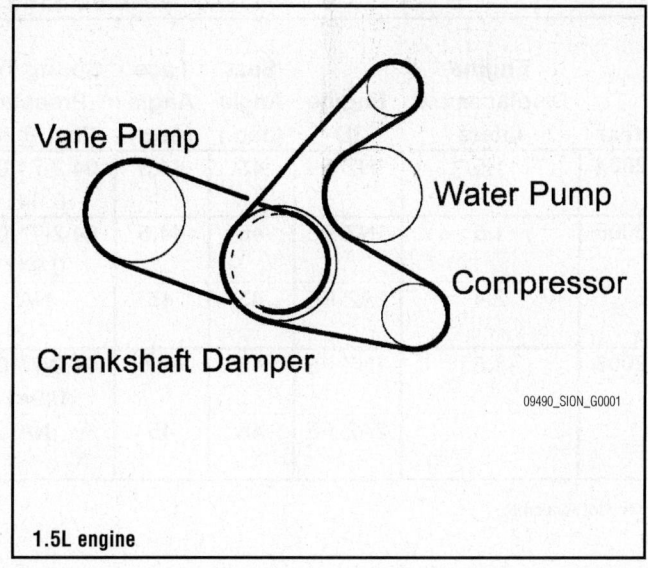

Vane Pump

Water Pump

Compressor

Crankshaft Damper

09490_SION_G0001

1.5L engine

CAPACITIES

Year	Model	Engine Displacement Liters	Engine ID	Engine Oil with Filter (qts.)	Transmission (pts.)		Front Drive Axle (pts.)	Fuel Tank (gal.)	Cooling System (qts.)
					Manual	Auto			
2004	xA	1.5	1NZ-FE	3.9	4.0	6.2	③	11.9	①
	xB	1.5	1NZ-FE	3.9	4.0	6.2	③	11.9	①
2005	xA	1.5	1NZ-FE	3.9	4.0	6.2	③	11.9	①
	xB	1.5	1NZ-FE	3.9	4.0	6.2	③	11.9	①
	tC	2.4	2AZ-FE	4.0	5.2	7.4	③	14.5	②
2006	xA	1.5	1NZ-FE	3.9	4.0	6.2	③	11.9	①
	xB	1.5	1NZ-FE	3.9	4.0	6.2	③	11.9	①
	tC	2.4	2AZ-FE	4.0	5.2	7.4	③	14.5	②

① With Manual Transmission: 4.7 qts.
  With Automatic Transmission: 4.5 qts.

② With Manual Transmission: 7.2 qts.
  With Automatic Transmission: 7.1 qts.

③ Included in transmission capacity

09490_SION_C0004

VALVE SPECIFICATIONS

Year	Engine Displacement Liters	Engine ID	Seat Angle (deg.)	Face Angle (deg.)	Spring Test Pressure (lbs. @ in.)	Spring Installed Height (in.)	Stem-to-Guide Clearance (in.)		Stem Diameter (in.)	
							Intake	Exhaust	Intake	Exhaust
2004	1.5	1NZ-FE	45	44.5	64.2-71.0@ 0.941	1.280	0.0010- 0.0024	0.0012- 0.0026	0.1957- 0.1963	0.1955- 0.1961
2005	1.5	1NZ-FE	45	44.5	64.2-71.0@ 0.941	1.280	0.0010- 0.0024	0.0012- 0.0026	0.1957- 0.1963	0.1955- 0.1961
	2.4	2AZ-FE	45	45	NA	NA	0.0010- 0.0024	0.0012- 0.0026	0.2154- 0.2159	0.2152- 0.2158
2006	1.5	1NZ-FE	45	44.5	64.2-71.0@ 0.941	1.280	0.0010- 0.0024	0.0012- 0.0026	0.1957- 0.1963	0.1955- 0.1961
	2.4	2AZ-FE	45	45	NA	NA	0.0010- 0.0024	0.0012- 0.0026	0.2154- 0.2159	0.2152- 0.2158

NA: Not Available

09490_SION_C0005

CRANKSHAFT AND CONNECTING ROD SPECIFICATIONS

All measurements are given in inches.

Year	Engine Displacement Liters	Engine ID	Crankshaft				Connecting Rod		
			Main Brg. Journal Dia.	Main Brg. Oil Clearance	Shaft End-play	Thrust on No.	Journal Diameter	Oil Clearance	Side Clearance
2004	1.5	1NZ-FE	1.8106- 1.8110	0.0004- 0.0009	0.0035- 0.0075	3	1.5745- 1.5748	0.0006- 0.0016	0.0063- 0.0142
2005	1.5	1NZ-FE	1.8106- 1.8110	0.0004- 0.0009	0.0035- 0.0075	3	1.5745- 1.5748	0.0006- 0.0016	0.0063- 0.0142
	2.4	2AZ-FE	2.1649- 2.1654	0.0007- 0.0016	0.0016- 0.0095	3	1.8894- 1.8898	0.0009- 0.0019	0.0063- 0.0143
2006	1.5	1NZ-FE	1.8106- 1.8110	0.0004- 0.0009	0.0035- 0.0075	3	1.5745- 1.5748	0.0006- 0.0016	0.0063- 0.0142
	2.4	2AZ-FE	2.1649- 2.1654	0.0007- 0.0016	0.0016- 0.0095	3	1.8894- 1.8898	0.0009- 0.0019	0.0063- 0.0143

09490_SION_C0006

CAMSHAFT AND BEARING SPECIFICATIONS CHART

All measurements are given in inches.

Year	Engine Displ. Liters	Engine ID/VIN	Journal Dia.	Brg. Oil Clearance	Shaft End-play	Runout	Lobe Height Intake	Lobe Height Exhaust
2004	1.5	1NZ-FE	①	0.0016-0.0037	0.0016-0.0037	0.0012	1.7566-1.7605	1.7585-1.7624
2005	1.5	1NZ-FE	①	0.0016-0.0037	0.0016-0.0037	0.0012	1.7566-1.7605	1.7585-1.7624
	2.4	2AZ-FE	②	③	④	0.0012	1.8305-1.8345	1.8104-1.8143
2006	1.5	1NZ-FE	①	0.0016-0.0037	0.0016-0.0037	0.0012	1.7566-1.7605	1.7585-1.7624
	2.4	2AZ-FE	②	③	④	0.0012	1.8305-1.8345	1.8104-1.8143

① No. 1 Journal: 1.3563-1.3569 in.
All Others: 0.9035-0.9041 in.

② No. 1 Journal: 1.4162-1.4167 in.
All Others: 0.9039-0.9045 in.

NA: Not Available

③ Intake: 0.0016-0.0037 in.
Exhaust: 0.0032-0.0053 in.

④ No. 1 Intake: 0.0016-0.0037 in.
No. 1 Exhaust: 0.0016-0.0031 in.
All Others: 0.0010-0.0024 in.

09490_SION_C0007

PISTON AND RING SPECIFICATIONS

All measurements are given in inches.

Year	Engine Displ. Liters	Engine ID	Piston Clearance	Ring Gap Top Comp.	Ring Gap Bottom Comp.	Ring Gap Oil Control	Ring Side Clearance Top Comp.	Ring Side Clearance Bottom Comp.	Ring Side Clearance Oil Control
2004	1.5	1NZ-FE	0.0018-0.0027	0.0098-0.0138	0.0138-0.0197	0.0039-0.0138	0.0012-0.0028	0.0008-0.0024	SNUG
2005	1.5	1NZ-FE	0.0018-0.0027	0.0098-0.0138	0.0138-0.0197	0.0039-0.0138	0.0012-0.0028	0.0008-0.0024	SNUG
	2.4	2AZ-FE	0.0020-0.0029	0.0087-0.0126	0.0197-0.0236	0.0039-0.0138	0.0012-0.0028	0.0012-0.0028	0.0012-0.0043
2006	1.5	1NZ-FE	0.0018-0.0027	0.0098-0.0138	0.0138-0.0197	0.0039-0.0138	0.0012-0.0028	0.0008-0.0024	SNUG
	2.4	2AZ-FE	0.0020-0.0029	0.0087-0.0126	0.0197-0.0236	0.0039-0.0138	0.0012-0.0028	0.0012-0.0028	0.0012-0.0043

09490_SION_C0008

TORQUE SPECIFICATIONS

All readings in ft. lbs.

Year	Engine Displacement Liters	Engine ID	Cylinder Head Bolts	Main Bearing Bolts	Rod Bearing Bolts	Crankshaft Damper Bolts	Flywheel Bolts	Manifold		Spark Plugs	Oil Pan Drain Plug
								Intake	Exhaust		
2004	1.5	1NZ-FE	①	②	③	95	④	22	20	13	28
2005	1.5	1NZ-FE	①	②	③	95	④	22	20	13	28
	2.4	2AZ-FE	⑤	⑤	⑦	133	96	22	27	18	30
2006	1.5	1NZ-FE	①	②	③	95	④	22	20	13	28
	2.4	2AZ-FE	⑤	⑤	⑦	133	96	22	27	18	30

① Step 1: 22 ft. lbs.
    Step 2: Plus 90 degrees
    Step 3: Plus 90 degrees again

② Step 1: 16 ft. lbs.
    Step 2: Plus 90 degrees

③ Step 1: 11 ft. lbs.
    Step 2: Plus 90 degrees

④ Step 1: 36 ft. lbs.
    Step 2: Plus 90 degrees

⑤ Step 1: 58 ft. lbs.
    Step 2: Plus 90 degrees

⑥ Step 1: 30 ft. lbs.
    Step 2: Plus 90 degrees

⑦ Step 1: 18 ft. lbs.
    Step 2: Plus 90 degrees

09490_SION_C0009

WHEEL ALIGNMENT

Year	Model		Caster		Camber		Toe-in (in.)
			Range (+/-Deg.)	Preferred Setting (Deg.)	Range (+/-Deg.)	Preferred Setting (Deg.)	
2004	xA	F	0.75	+1.78	0.75	-0.65	0+/-0.08
		R	—	—	0.75	-0.95	0.13+/-0.12
	xB	F	0.75	+1.75	0.75	-0.57	0+/-0.08
		R	—	—	0.42	-0.95	0.13+/-0.12
2005	xA	F	0.75	+1.78	0.75	-0.65	0+/-0.08
		R	—	—	0.75	-0.95	0.13+/-0.12
	xB	F	0.75	+1.75	0.75	-0.57	0+/-0.08
		R	—	—	0.42	-0.95	0.13+/-0.12
	tC	F	0.75	+3.03	0.75	-0.52	0+/-0.08
		R	—	—	0.75	-0.90	0.12+/-0.08
2006	xA	F	0.75	+1.78	0.75	-0.65	0+/-0.08
		R	—	—	0.75	-0.95	0.13+/-0.12
	xB	F	0.75	+1.75	0.75	-0.57	0+/-0.08
		R	—	—	0.42	-0.95	0.13+/-0.12
	tC	F	0.75	+3.03	0.75	-0.52	0+/-0.08
		R	—	—	0.75	-0.90	0.12+/-0.08

F: Front

R: Rear

09490_SION_C0010

TIRE, WHEEL AND BALL JOINT SPECIFICATIONS

Year	Model	OEM Tires Standard	OEM Tires Optional	Tire Pressures (psi) Front	Tire Pressures (psi) Rear	Wheel Size	Ball Joint Inspection	Lugnut Torque (ft. lbs.)
2004	xA	P185/60R15	N/A	29	29	5.5-JJ	44 in. ①	76
	xB	P185/60R15	N/A	29	29	6.0-JJ	44 in. ①	76
2005	xA	P185/60R15	N/A	29	29	5.5-JJ	44 in. ①	76
	xB	P185/60R15	N/A	29	29	6.0-JJ	44 in. ①	76
	tC	P215/45ZR17	N/A	32	29	7.0-JJ	44 in. ①	76
2006	xA	P185/60R15	N/A	29	29	5.5-JJ	44 in. ①	76
	xB	P185/60R15	N/A	29	29	6.0-JJ	44 in. ①	76
	tC	P215/45ZR17	N/A	32	29	7.0-JJ	44 in. ①	76

OEM: Original Equipment Manufacturer

PSI: Pounds Per Square Inch

STD: Standard

OPT: Optional

① Torque required (in inch lbs.) to rotate ball joint when removed from the knuckle

09490_SION_C0011

BRAKE SPECIFICATIONS

All measurements in inches unless noted

Year	Model		Brake Disc Original Thickness	Brake Disc Minimum Thickness	Brake Disc Maximum Run-out	Brake Drum Original Inside Diameter	Brake Drum Max. Wear Limit	Brake Drum Maximum Machine Diameter	Minimum Lining Thickness	Brake Caliper Bracket Bolts (ft. lbs.)	Brake Caliper Mounting Bolts (ft. lbs.)
2004	xA	F	0.866	0.787	0.0020	-	-	-	0.039	65	25
		R	-	-	-	7.874	7.913	NA	0.039	-	-
	xB	F	1.100	1.039	0.0010	-	-	-	0.039	130	26
		R	-	-	-	11.50	7.913	NA	0.039	-	-
2005	xA	F	0.866	0.787	0.0020	-	-	-	0.039	65	25
		R	-	-	-	7.874	7.913	NA	0.039	-	-
	xB	F	1.100	1.039	0.0010	-	-	-	0.039	130	26
		R	-	-	-	11.50	7.913	NA	0.039	-	-
	tC	F	0.984	0.906	0.0020	-	-	-	0.039	79	25
		R	0.354	0.295	0.0059	-	-	-	0.039	34	29
2006	xA	F	0.866	0.787	0.0020	-	-	-	0.039	65	25
		R	-	-	-	7.874	7.913	NA	0.039	-	-
	xB	F	1.100	1.039	0.0010	-	-	-	0.039	130	26
		R	-	-	-	11.50	A	NA	0.039	-	-
	tC	F	0.984	0.906	0.0020	-	-	-	0.039	79	25
		R	0.354	0.295	0.0059	-	-	-	0.039	34	29

F: Front

R: Rear

09490_SION_C0012

SCHEDULED MAINTENANCE INTERVALS

SCION - xA, xB and tC

TO BE SERVICED	TYPE OF SERVICE	VEHICLE MILEAGE INTERVAL (x1000)												
		5	10	15	20	25	30	35	40	45	50	55	60	65
Air cleaner filter	R						✓						✓	
Transmission fluid	S/I						✓						✓	
Ball joints & dust covers	S/I			✓			✓			✓			✓	
Bolts & nuts on chassis & body	S/I													
Brake line pipes & hoses	S/I			✓			✓			✓			✓	
Brake pads & discs/linings & drums (front & rear)	S/I	✓	✓	✓	✓	✓	✓	✓	✓	✓	✓	✓	✓	✓
Drive belts	S/I												✓	
Driveshaft boots	S/I			✓			✓			✓			✓	
Engine coolant	S/I			✓			✓			✓			✓	
Engine coolant	R	Replace at 100,000 miles												
Engine oil & filter	R	✓	✓	✓	✓	✓	✓	✓	✓	✓	✓	✓	✓	✓
Exhaust pipes & mountings	S/I			✓			✓			✓			✓	
Fuel lines & connections	S/I						✓						✓	
Propeller shaft bolt	S/I			✓			✓			✓			✓	
Radiator core & condenser	S/I			✓			✓			✓			✓	
Front differential fluid	S/I						✓						✓	
Rotate Tires	S/I	✓	✓	✓	✓	✓	✓	✓	✓	✓	✓	✓	✓	✓
Spark plugs (tC)	R	Replace at 120,000 miles												
Spark plugs (xA & xB)	R						✓						✓	
Steering linkage & gear box	S/I			✓			✓			✓			✓	

R: Replace   S/I: Service or Inspect

Drivebelts: After initial inspection at 60,000 miles, inspect every 15,000 miles thereafter.

FREQUENT OPERATION MAINTENANCE (SEVERE SERVICE)

If a vehicle is operated under any of the following conditions it is considered severe service:

- Desert/Extremely dusty areas.

- Trailer towing usage.

Air cleaner filter: service or inspect every 5000 miles.

Ball joints & dust covers: service or inspect every 5000 miles.

Bolts & nuts on chassis & body: service or inspect every 5000 miles.

Driveshaft boots: service or inspect every 5000 miles.

Steering linkage: service or inspect every 5000 miles.

Transmission and Front differential fluid: replace every 30,000 miles.

09490_SION_C0013

ENGINE REPAIR

➡Disconnecting the negative battery cable on some vehicles may interfere with the functions of the on board computer system. The computer may undergo a relearning process once the negative battery cable is reconnected. On tC models, the power window and sliding room systems will have to be initialized.

Distributor

All engines are equipped with a Distributorless Ignition System (DIS).

Alternator

REMOVAL & INSTALLATION

1.5L Engine

1. Before servicing the vehicle, refer to the Precautions Section.
2. Disconnect the negative battery cable.
3. Disconnect the alternator wiring connections.
4. Loosen, but do not remove the alternator mounting bolts.
5. Release the alternator drive belt tension and remove the belt.
6. Remove the top alternator mounting bolt and fan belt adjusting bar.
7. Remove the bottom alternator mounting bolt and alternator.
To install:
8. Temporarily install the alternator with the bottom mounting bolt.
9. Install the fan belt adjusting bar with the top alternator mounting bolt. Tighten fan belt adjusting bar nuts to 8 ft. lbs. (11 Nm).
10. Install the alternator drive belt and push the adjusting bar towards the vehicle front to adjust the tension.

✲✲ WARNING

Do not insert the adjusting bar between the oil control valve and alternator as it could damage the oil control valve.

11. Tighten the top mounting bolt to 14 ft. lbs. (19 Nm). Tighten the bottom mounting bolt to 40 ft. lbs. (54 Nm).
12. Connect the alternator wiring connections.
13. Connect the negative battery cable.

2.4L Engine

1. Before servicing the vehicle, refer to the Precautions Section.
2. Remove or disconnect the following:
   • Negative battery cable
   • Right-hand front fender apron seal
   • Alternator drive belt
   • Alternator wiring harnesses
   • Alternator mounting bolts and alternator
   To install:

➡Confirm the crankshaft position sensor wiring harnesses is secured in the clamp bracket on the timing chain cover.

3. Install or connect the following:
   • Alternator. Tighten upper bolt to 16 ft. lbs. (21 Nm) and lower bolt to 38 ft. lbs. (52 Nm).
   • Alternator wiring harnesses
   • Alternator drive belt
   • Right-hand front fender apron seal
   • Negative battery cable

Ignition Timing

ADJUSTMENT

These engines are equipped with a Distributorless Ignition System (DIS). No adjustment is necessary.

Engine Assembly

REMOVAL & INSTALLATION

1.5L Engine

1. Before servicing the vehicle, refer to the Precautions Section.
2. Relieve the fuel system pressure.
3. Drain the engine cooling system.
4. Drain the engine oil.
5. Remove or disconnect the following:
   • Negative battery cable
   • Engine undercover
   • Front wheels
   • Engine appearance cover
   • Battery
   • Air intake assembly
   • Fuel hose
   • Accelerator cable
   • Radiator hoses
   • Transmission oil cooler hoses, if equipped
   • Grille assembly
   • Hood lock assembly

• Radiator supports
• Radiator assembly
• Transmission control cables
• Clutch release cylinder assembly, if equipped
• Brake booster check valve hose
• Heater hoses
• Engine wire from engine room junction block
• Engine body ground wire
• Alternator/fan drive belt
• A/C compressor
• Front floor panel brace
• Front exhaust pipe assembly
6. Separate the intermediate steering shaft assembly.

➡Matchmark the main shaft and intermediate shaft for reassembly.

7. Using Special Tool 09930-00010, remove the left-hand front axle hub nut.
8. Separate the following left-hand front suspension components:
   • Wheel speed sensor
   • Tie rod end
   • Lower control arm
   • Halfshaft
9. Support the engine with a suitable lifting device.
10. Remove the engine mounting bolts from the left side, right side and crossmember.
11. Lift the engine/transaxle assembly from the vehicle.
12. Separate the transaxle from the engine assembly.
13. Installation is the reverse order of removal. Note the following torques:
   a. Left-hand engine mounting bolts: 36 ft. lbs. (49 Nm).
   b. Right-hand engine mounting bolts: Bolts A to 33 ft. lbs. (45 Nm). Bolts B to 38 ft. lbs. (52 Nm). Bolt C to 7 ft. lbs. (10 Nm).
   c. Crossmember mounting bolts:

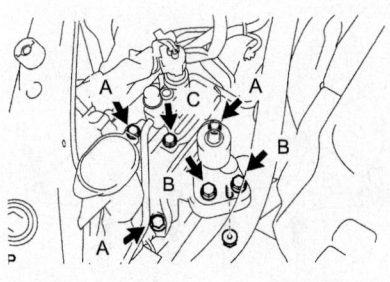

09490_SION_G0002

Right-hand engine mount bolt identification—1.5L engine

Front bolts to 52 ft. lbs. (70 Nm). Rear bolts to 86 ft. lbs. (116 Nm).

   d. Crossmember support bars: 35 ft. lbs. (47 Nm).

14. Refill the engine cooling system to the correct level.

15. Refill the engine with oil to the correct level.

16. Start the engine and check for leaks.

2.4L Engine

1. Before servicing the vehicle, refer to the Precautions Section.

2. Relieve the fuel system pressure.

3. Drain the engine cooling system.

4. Drain the engine oil.

5. Drain the transmission fluid.

6. Remove or disconnect the following:

- Negative battery cable
- Front wheels
- Engine undercover
- Right-hand front fender apron
- Engine appearance cover
- Radiator hoses
- Transmission oil cooler hoses, if equipped
- Radiator
- Air intake assembly
- Battery and battery tray
- Battery carrier
- Brake booster hose
- Heater hoses
- Transmission control cables
- Clutch release cylinder assembly, if equipped
- Accessory drive belt
- Alternator
- A/C compressor
- Glove compartment door
- Engine wire from the ECM and junction block
- Engine ground wire
- Front exhaust pipe assembly

7. Separate the intermediate steering shaft assembly.

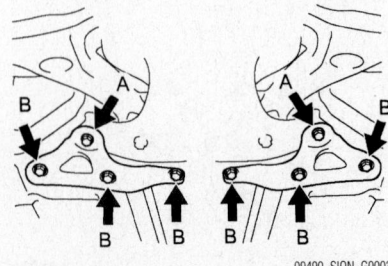

Bolt identification for front suspension braces.

09490_SION_G0003

➥**Matchmark the main shaft and intermediate shaft for reassembly.**

8. Remove or disconnect the the following from both sides of the front suspension:

   a. Hub nut
   b. Stabilizer link
   c. Wheel speed sensor
   d. Tie rod assembly
   e. Lower control arm

9. Separate the axle assembly from the halfshaft.

➥**Matchmark the halfshaft and axle hub for reassembly.**

10. Disconnect the oil pump reservoir and return hose.

11. Secure the engine with a suitable engine lifting device.

12. Remove the mounting bolts from the left and right engine mounts.

13. Remove the mounting bolts from the left and right front suspension braces.

14. Remove the engine/transaxle assembly from the vehicle.

15. Installation is the reverse order of removal. Note the following torques:

   a. Left-hand engine mount bolts: 64 ft. lbs. (87 Nm).

   b. Right-hand engine mount bolts: 38 ft. lbs. (52 Nm).

   c. Front suspension braces: Bolts A to 98 ft. lbs. (133 Nm). Bolts B to 59 ft. lbs. (80 Nm).

16. Refill the transmission with fluid to the correct level.

17. Refill the engine with oil.

18. Refill the cooling system to the correct level.

19. Start the engine and check for leaks.

Water Pump

REMOVAL & INSTALLATION

1.5L Engine

1. Before servicing the vehicle, refer to the Precautions Section.

2. Drain the cooling system.

3. Disconnect the negative battery cable.

4. Remove the accessory drive belt.

5. Support the engine with a suitable jack.

6. Remove the right-hand engine mount.

7. Using Special Tool 09960-10010, remove the water pump pulley.

8. Remove the water pump and gasket.

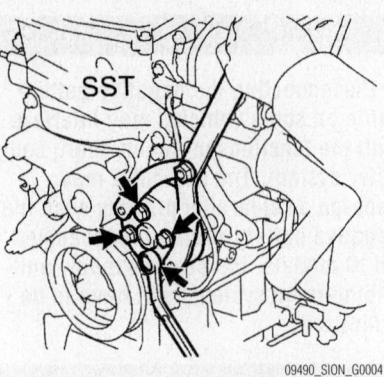

09490_SION_G0004

Use Special Tool 09960-10010 to hold the water pump pulley while removing the mounting bolts.

To install:

9. Install the water pump with a new gasket. Tighten the bolts and nuts to 96 inch lbs. (11 Nm).

10. Using Special Tool 09960-10010, install the water pump pulley and tighten to 11 ft. lbs. (15 Nm).

11. Install or connect the following:

- Right-hand engine mount
- Accessory drive belt
- Negative battery cable

12. Refill the cooling system to the correct level.

13. Start the engine and check for leaks.

2.4L Engine

1. Before servicing the vehicle, refer to the Precautions Section.

2. Drain the cooling system.

3. Remove or disconnect the following:

- Negative battery cable
- Right-hand front fender apron
- Engine undercover
- Accessory drive belt
- Alternator

4. Using Special Tool 09960-10010, remove the water pump pulley.

5. Remove the water pump assembly.

To install:

6. Install the water pump with a new gasket. Tighten to 80 inch lbs. (9 Nm).

7. Using Special Tool 09960-10010, install the water pump pulley and tighten to 19 ft. lbs. (26 Nm).

8. Install or connect the following:

- Alternator
- Accessory drive belt
- Engine undercover
- Right-hand front fender apron
- Negative battery cable

9. Refill the engine cooling system to the correct level.

10. Start the engine and check for leaks.

Heater Core

REMOVAL AND INSTALLATION

xA

1. Before servicing the vehicle, refer to the Precautions Section.
2. Discharge and recover the air conditioning system refrigerant.
3. Drain the cooling system.
4. Place front wheel facing straight ahead.
5. Remove or disconnect the following:
   • Negative battery cable

✳✳ CAUTION

Wait 90 seconds after disconnecting the cable to allow the airbag to discharge.

   • Heater hoses from the heater unit
   • A-pillar trim

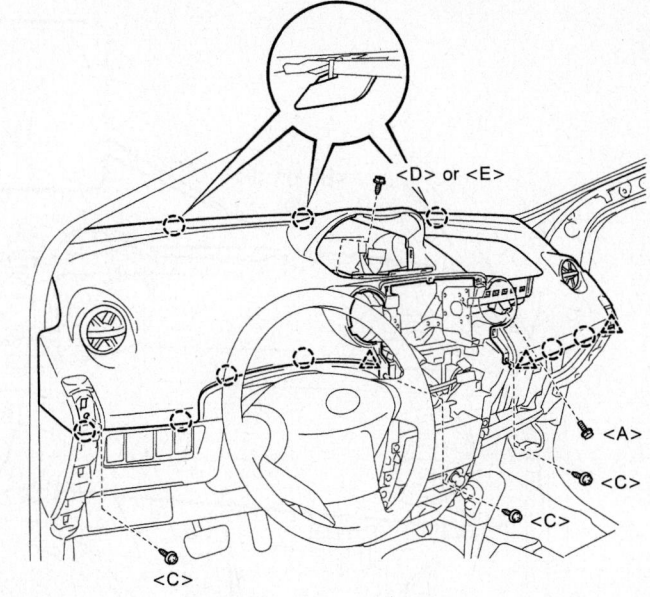

○: Claw
△: Clip

09490_SION_G0005

Remove the instrument panel assembly—xA models

<H>

<H>

<F>

<C> <C>

<F>

09490_SION_G0006

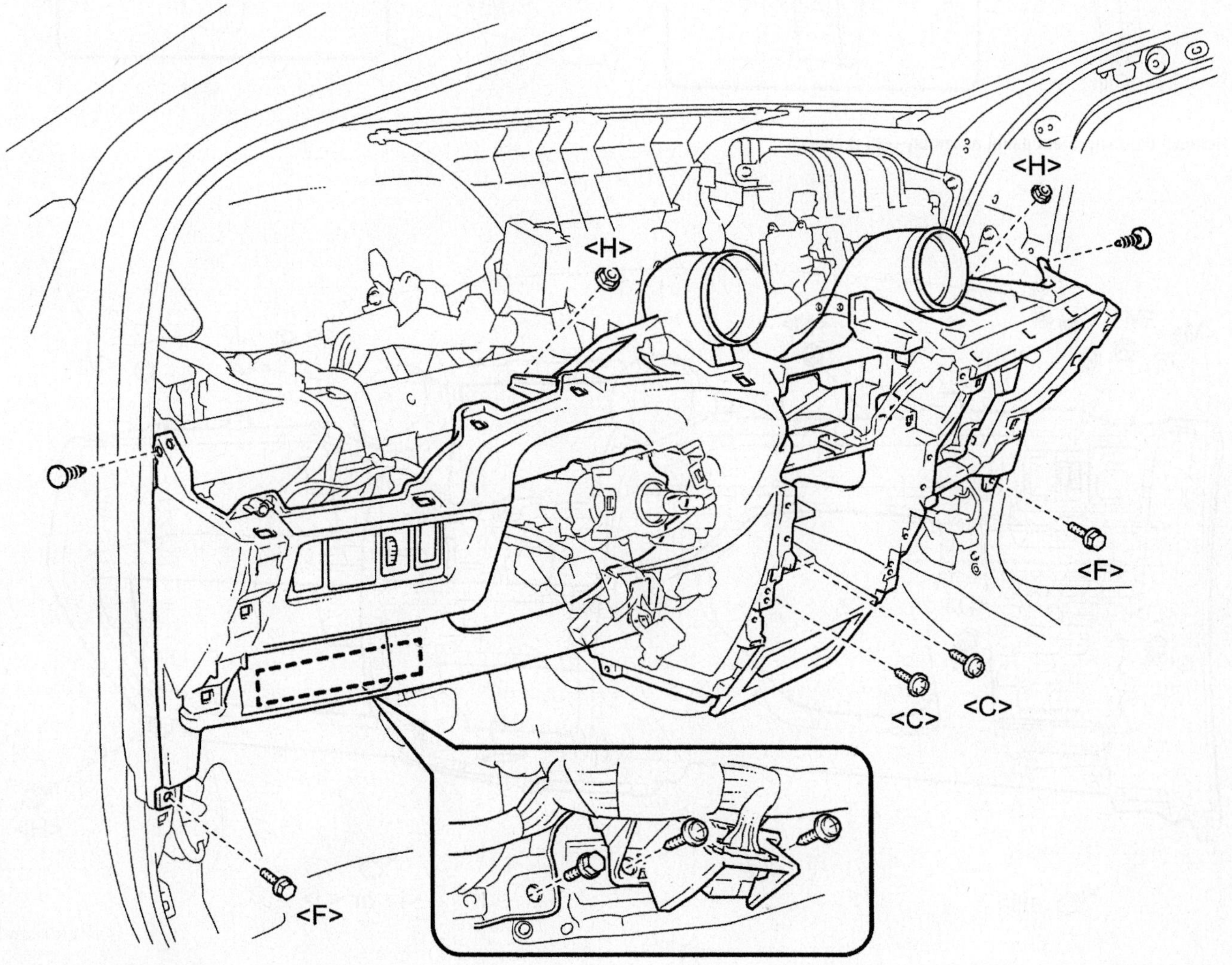

Removing the lower instrument panel assembly—xA models

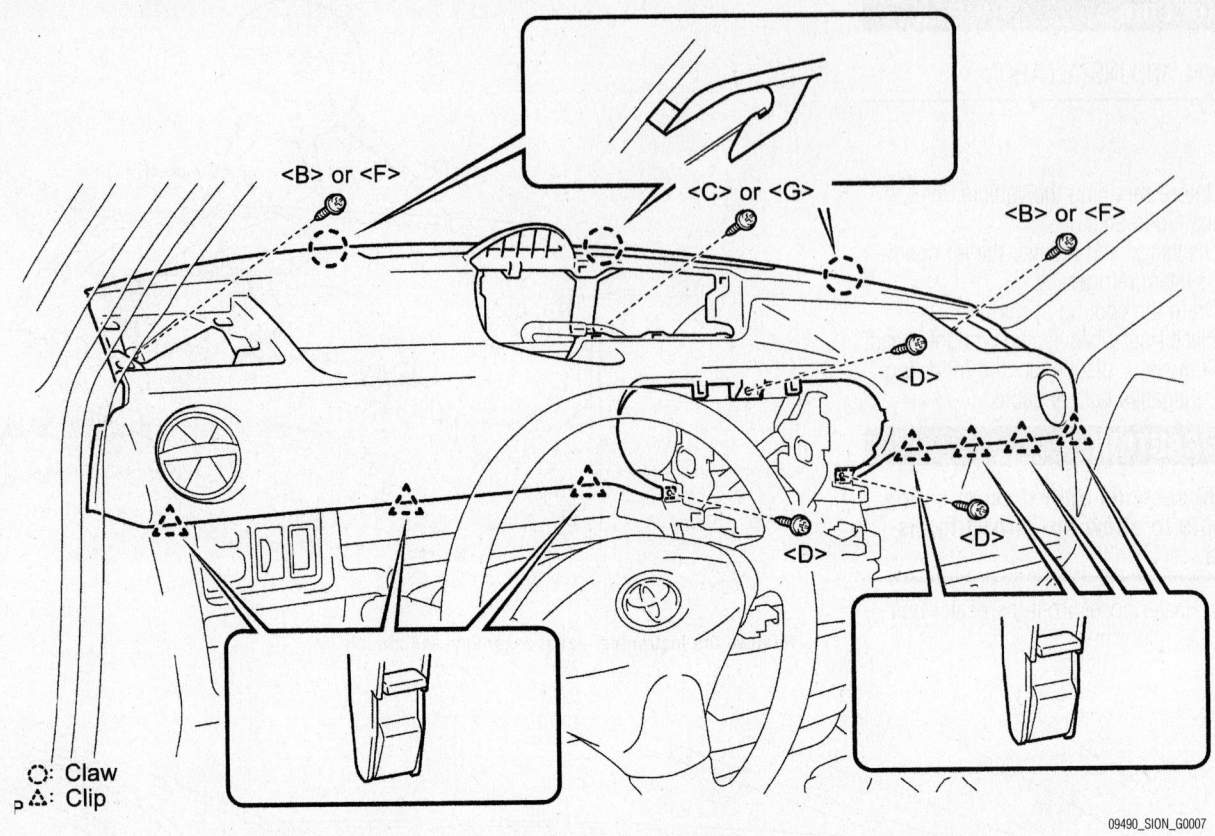

Remove the instrument panel assembly—xB models

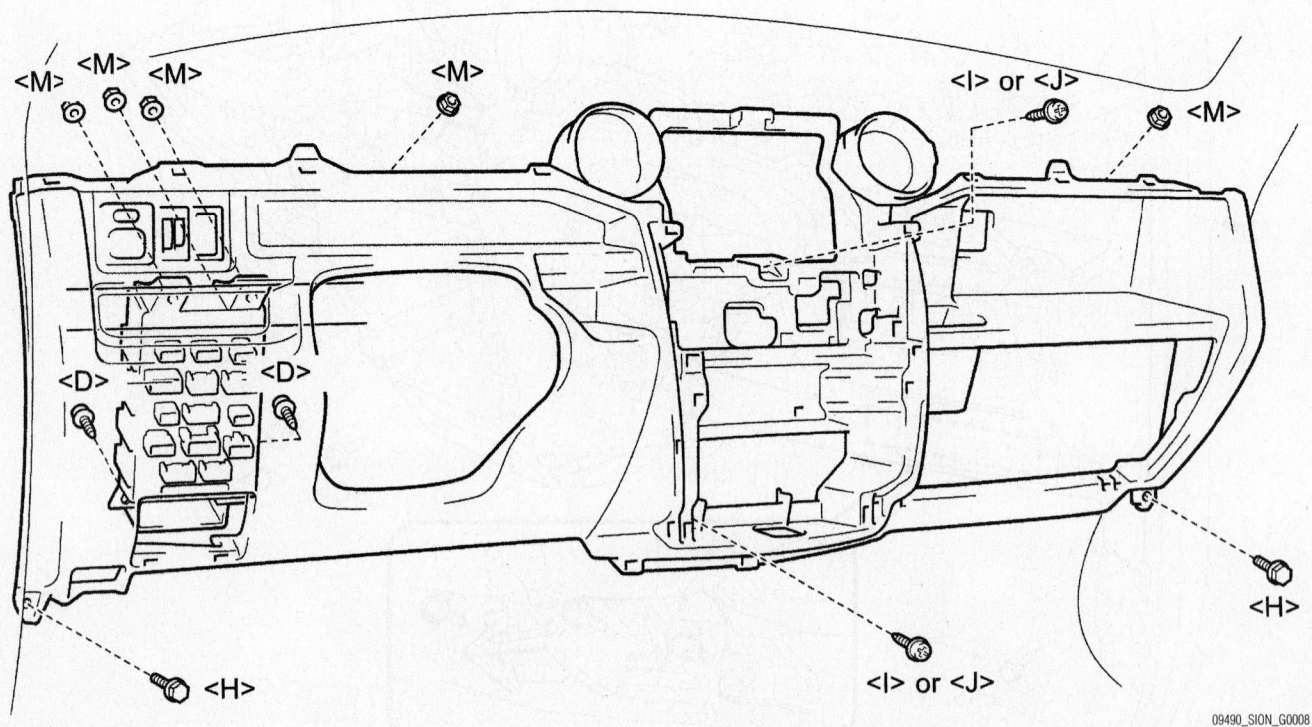

Removing the lower instrument panel assembly—xB models (71-25)

- Lower instrument panel finishing trim
- Gauge cluster cover
- Gauge cluster assembly
- Instrument cluster finishing trim
- Heater control knobs
- Center instrument cluster trim panel
- Stereo opening cover and bracket
- Radio assembly
- Glove compartment door
- Passenger airbag connector
- Instrument panel assembly with passenger airbag assembly attached
- Horn button assembly
- Steering wheel
- Steering wheel column cover
- Headlight dimmer switch
- Windshield wiper switch
- Front door scuff plates
- Cowl side trim boards
- Heater control assembly
- Center instrument cluster trim panel
- Front ash receptacle assembly
- Center lower instrument cover

- Center lower instrument panel trim panel
- Shift lever knob, if equipped with manual transmission
- Floor console
- Fuse box opening cover
- Lower instrument panel assembly
- ECM
- Air conditioner blower assembly
- Air conditioner amplifier assembly
- Defroster nozzle
- Instrument panel brace
- Defroster damper control cable
- Air-mix damper control cable
- Rear air ducts from heater unit
- Heater unit assembly
- Air duct assembly from the heater unit
- Heater core cover
- Thermistor assembly
- Heater core

6. Installation is the reverse order of removal.

7. Refill the cooling system to the correct level.

8. Start the engine and check for leaks.

xB

1. Before servicing the vehicle, refer to the Precautions Section.

2. Discharge and recover the air conditioning system refrigerant.

3. Drain the cooling system.

4. Place front wheel facing straight ahead.

5. Remove or disconnect the following:
- Negative battery cable

✳✳ CAUTION

Wait 90 seconds after disconnecting the cable to allow the airbag to discharge.

- Heater hoses from the heater unit
- Heater control knobs
- Instrument cluster cover panel
- Instrument cluster
- Instrument panel speaker panel
- Front speaker
- Center instrument cluster trim panel
- Radio assembly

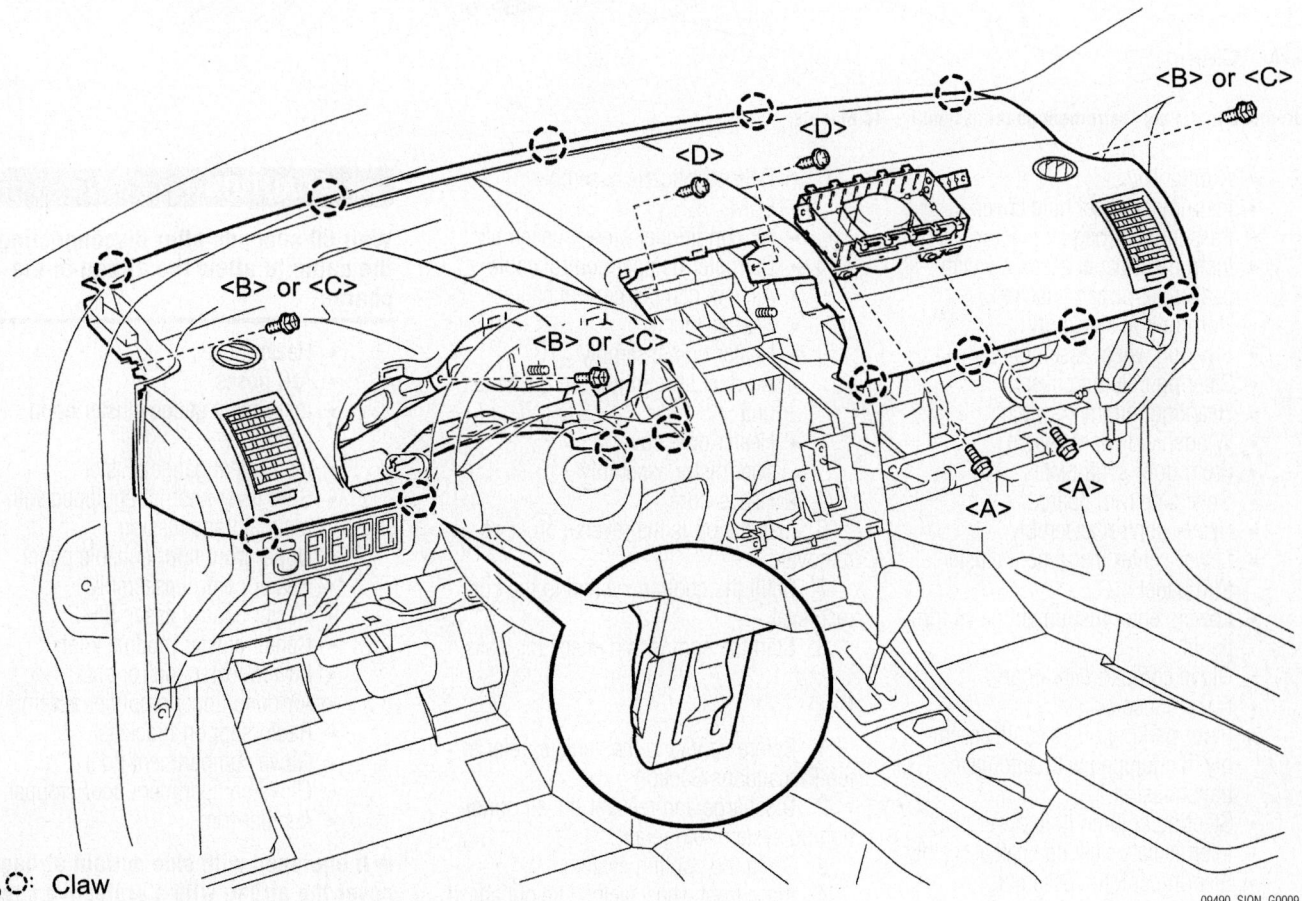

⟜⃝ Claw

09490_SION_G0009

Remove the instrument panel assembly—tC Models

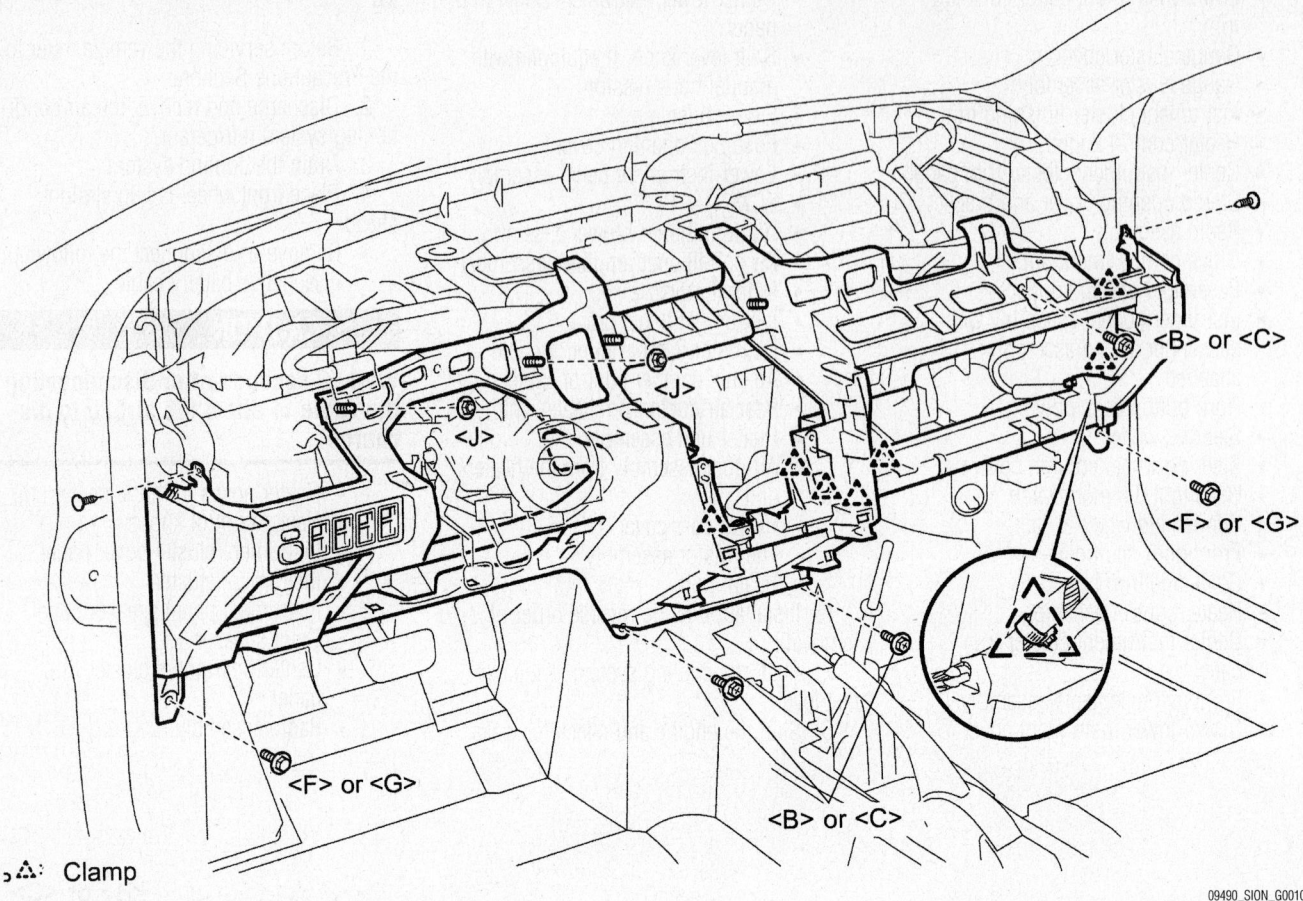

<B> or <C>

<F> or <G>

<F> or <G>

<B> or <C>

⟳△ Clamp

09490_SION_G0010

Remove the lower instrument panel assembly—tC Models

- A-pillar trim
- Instrument panel hole cover
- Passenger airbag connector
- Instrument panel assembly with passenger airbag attached
- Horn button assembly
- Steering wheel assembly
- Steering column cover
- Headlight dimmer switch
- Windshield wiper switch
- Front door scuff plates
- Cowl side trim boards
- Heater control assembly
- Lower center instrument cluster trim panel
- Lower center instrument panel trim panel
- Glove compartment door
- Floor console
- Floor parking brake cable assembly, if equipped with automatic transmission
- Steering column hole cover plate
- Intermediate steering shaft assembly
- Steering column assembly
- Antenna cord
- Lower instrument panel assembly with panel reinforcement attached

- Defroster nozzle assembly
- ECM
- Air conditioner blower assembly
- Defroster damper control cable
- Air-mix damper control cable
- Rear air ducts
- Heater unit assembly
- Air duct assembly from the heater unit
- Heater core cover
- Thermistor assembly
- Heater core

6. Installation is the reverse order of removal.

7. Refill the cooling system to the correct level.

8. Start the engine and check for leaks.

tC

1. Before servicing the vehicle, refer to the Precautions Section.

2. Discharge and recover the air conditioning system refrigerant.

3. Drain the cooling system.

4. Place front wheel facing straight ahead.

5. Remove or disconnect the following:
- Negative battery cable

✳✳ CAUTION

Wait 90 seconds after disconnecting the cable to allow the airbag to discharge.

- Heater hoses
- A/C hoses
- Instrument gauge cluster hood panel
- Instrument gauge cluster
- Shift lever knob, if equipped with manual transmission
- Upper front floor console panel
- Heater control assembly
- Heater control assembly
- Center cluster module knob
- Lower heater control base
- Air conditioner amplifier assembly
- Radio support brackets
- Glove compartment door
- Glove compartment door stopper
- A-pillar trim

➡**If equipped with side curtain airbags, cover the airbag with a protective cover as soon as the trim pieces are removed.**

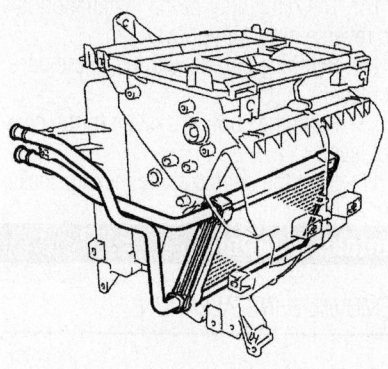

Remove the heater core from the air condition unit—tC models

- Passenger airbag connector
- Instrument panel assembly with passenger airbag attached.
- Steering wheel assembly
- Steering column cover
- Headlight dimmer switch
- Windshield wiper switch assembly
- Floor console top panel
- Floor console
- Front door scuff plates
- Cowl side trim panels
- Center heater to register duct
- Lower instrument panel assembly
- ECM
- Center heater to register duct
- Lower defroster nozzle assembly
- Instrument panel-to-cowl brace assembly
- Blower assembly
- Air conditioning unit assembly

6. Remove the following from the air conditioning unit:
- Drain hose
- Air duct
- Air outlet control motor
- Air-mix control motor
- Heater piping cover
- Heater core

7. Installation is the reverse order of removal.

8. Refill the cooling system to the correct level.

9. Start the engine and check for leaks.

Cylinder Head

REMOVAL & INSTALLATION

1.5L Engines

1. Before servicing the vehicle, refer to the Precautions Section.
2. Drain the cooling system.
3. Drain the engine oil.

4. Relieve the fuel system pressure.
5. Remove or disconnect the following:
- Negative battery cable
- Air intake assembly
- Engine appearance cover
- Right-hand front wheel
- Engine undercover
- Accelerator control cable
- Fuel tube from the fuel supply hose
- Brake booster check valve hose
- Radiator hose
- Heater hoses
- Radiator filler hose
- Front exhaust pipe
- Exhaust manifold
- Intake manifold
- Oil dipstick guide
- Ignition coil
- Accessory drive belts
- Alternator
- Camshaft timing oil control valve
- Ventilation hoses from the cylinder head cover
- Cylinder head cover
- Water pump
- Front cover
- Crankshaft pulley
- Crankshaft position sensor

6. Support the engine with a suitable jack.

7. Remove or disconnect the following:
- Right-hand engine mount
- Timing chain
- Fuel supply hose
- Fuel rail and injectors
- Camshaft and sprockets

8. Loosen the cylinder head mounting bolts, in several steps, in the sequence shown.

9. Remove the cylinder head and gasket.

To install:

10. Install the cylinder head with a new gasket. The Lot Number on the gasket should face upward.

11. Apply a light coat of new engine oil to the threads of the cylinder head bolts.

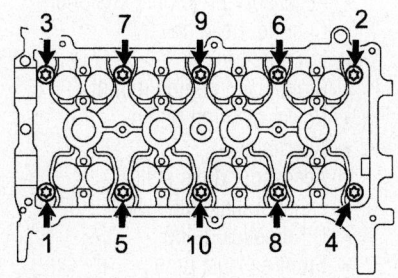

Cylinder head bolt removal sequence—1.5L engine

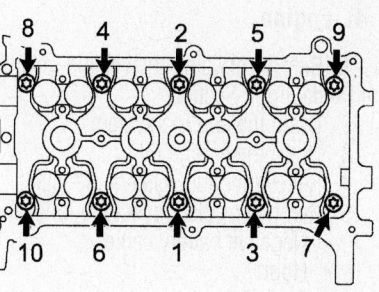

Cylinder head torque sequence—1.5L engine

Tighten the bolts in the sequence shown as follows:
a. Step 1: Tighten to 22 ft. lbs. (29 Nm).
b. Step 2: Plus 90 degrees.
c. Step 3: Plus an additional 90 degrees.

12. Install or connect the following:
- Camshaft and sprockets
- Fuel rail and injectors
- Fuel supply hose
- Timing chain
- Right-hand engine mount
- Camshaft timing oil control valve
- Crankshaft position sensor
- Front cover
- Water pump
- Cylinder head cover
- Alternator
- Accessory drive belts
- Ventilation hoses to the cylinder head cover
- Ignition coil. Tighten bolts to 80 inch lbs. (9 Nm).
- Oil dipstick guide
- Intake manifold
- Exhaust manifold
- Front exhaust pipe. Tighten to 32 ft. lbs. (43 Nm).
- Radiator filler hose. Tighten support bracket to 66 inch lbs. (7.5 Nm).
- Heater hose
- Radiator hose
- Brake booster check valve hose
- Fuel tube to the fuel supply hose
- Accelerator control cable
- Engine undercover
- Front wheel
- Engine appearance cover
- Air intake assembly
- Negative battery cable

13. Refill the engine with oil to the correct level.

14. Refill the cooling system to the correct level.

15. Start the engine and check for leaks.

2.4L Engine

1. Before servicing the vehicle, refer to the Precautions Section.
2. Drain the cooling system.
3. Drain the engine oil.
4. Relieve the fuel system pressure.
5. Remove or disconnect the following:
   • Negative battery cable
   • Hood
   • Right-hand front wheel
   • Engine undercover
   • Right-hand front fender apron
   • Engine appearance cover
   • Wiper arm assembly
   • Top cowl seal
   • Left-hand top cowl ventilator louver
   • Wiper linkage assembly
   • Top outer cowl panel
   • Left-hand cowl body mounting bracket
   • Air intake assembly
   • Throttle body
   • Fuel rail
   • Intake manifold
   • Front exhaust pipe
   • Oil dipstick guide
   • Exhaust manifold
   • Accessory drive belts
   • Alternator
   • Power steering pump
6. Support the engine with a suitable jack.
7. Remove or disconnect the following:
   • Right-hand engine mount
   • Ignition coil
   • Cylinder head cover
   • Accessory drive belt tensioner
   • Crankshaft position sensor
   • Oil pan
8. Turn the crankshaft pulley until its groove and the timing mark on the front cover are aligned to set the No. 1 cylinder to TDC.
9. Remove or disconnect the following:
   • Camshafts
   • Crankshaft pulley
   • Front cover

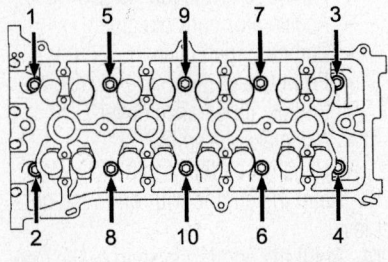

Cylinder head bolt removal sequence— 2.4L Engine

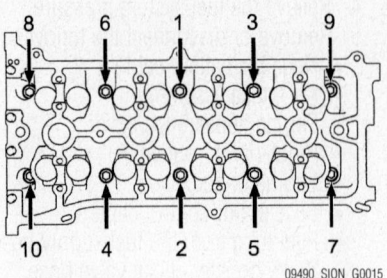

09490_SION_G0015

Cylinder head bolt torque sequence—2.4L Engine

   • Timing chain
   • Timing chain vibration damper
   • Camshaft timing oil control valve
   • Radiator inlet hose
   • All remaining sensors connectors
   • Ground wire
10. Loosen the cylinder head bolts in the sequence shown and remove the cylinder head and gasket.

To install:

11. Install the cylinder head with a new gasket. The Lot Number on the gasket should face upward.
12. Apply a light coat of new engine oil to the threads of the cylinder head bolts. Tighten the bolts in the sequence shown as follows:
   a. Step 1: Tighten to 58 ft. lbs. (79 Nm).
   b. Step 2: Plus 90 degrees.
13. Install or connect the following:
   • Ground wire
   • Sensor connectors to the cylinder head
   • Radiator inlet hose
   • Camshafts
   • Camshaft timing oil control valve
   • Exhaust manifold
   • Oil dipstick guide
   • Timing chain vibration damper
   • Timing chain
   • Front cover
   • Crankshaft pulley. Tighten to 133 ft. lbs. (180 Nm).
   • Oil pan
   • Crankshaft position sensor
   • Accessory drive belt tensioner
   • Cylinder head cover
   • Ignition coil
   • Right-hand engine mount
   • Power steering pump
   • Alternator
   • Accessory drive belts
   • Exhaust manifold
   • Oil dipstick guide
   • Front exhaust pipe
   • Intake manifold
   • Fuel rail
   • Throttle body

14. The remainder of the installation is the reverse order of removal.
15. Refill the engine with oil to the correct level.
16. Refill the cooling system to the correct level.
17. Start the engine and check for leaks.

Intake Manifold

REMOVAL & INSTALLATION

1.5L Engine

1. Before servicing the vehicle, refer to the Precautions Section.
2. Drain the cooling system.
3. Relieve the fuel system pressure.
4. Remove or disconnect the following:
   • Negative battery cable
   • Air intake assembly
   • Engine appearance cover
   • Right-hand front wheel
   • Engine undercover
   • Accelerator control cable
   • Fuel tube from the fuel supply hose
   • Brake booster check valve hose
   • Radiator hose
   • Heater hose
   • Radiator filler hose
   • Front exhaust pipe
   • Exhaust manifold
5. Loosen the intake manifold bolts and nuts, using several passes, in the sequence shown.
6. Remove the intake manifold and gasket.

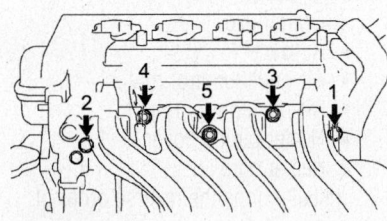

09490_SION_G0016

Intake manifold removal sequence—1.5L Engine

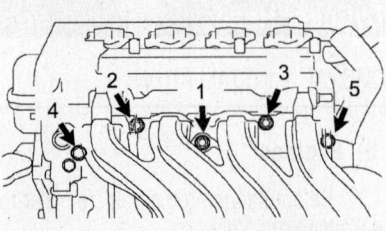

09490_SION_G0017

Intake manifold torque sequence—1.5L Engine

To install:

7. Install the intake manifold with a new gasket.

8. In several steps, tighten the mounting bolts and nuts in the sequence shown to 22 ft. lbs. (30 Nm).

9. Install or connect the following:
- Exhaust manifold
- Front exhaust pipe. Tighten to 32 ft. lbs. (43 Nm).
- Radiator filler hose. Tighten support bracket to 66 inch lbs. (7.5 Nm).
- Heater hose
- Radiator hose
- Brake booster check valve hose
- Fuel tube to the fuel supply hose
- Accelerator control cable
- Engine undercover
- Front wheel
- Engine appearance cover
- Air intake assembly
- Negative battery cable

10. Refill the cooling system to the correct level.

11. Start the engine and check for leaks.

2.4L Engine

1. Before servicing the vehicle, refer to the Precautions Section.

2. Drain the cooling system.

3. Relieve the fuel system pressure.

4. Remove or disconnect the following:
- Negative battery cable
- Hood
- Engine appearance cover
- Wiper arm assembly
- Top cowl seal
- Left-hand top cowl ventilator louver
- Wiper linkage assembly
- Top outer cowl panel
- Left-hand cowl body mounting bracket
- Air intake assembly
- Throttle body
- Fuel rail
- Water by-pass hoses from the throttle body
- Intake manifold and gasket

To install:

5. Install a new gasket into the intake manifold.

6. Install the intake manifold and tighten fasteners in sequence to 22 ft. lbs. (30 Nm).

7. Install or connect the following:
- Water bypass hoses from the throttle body
- Fuel rail

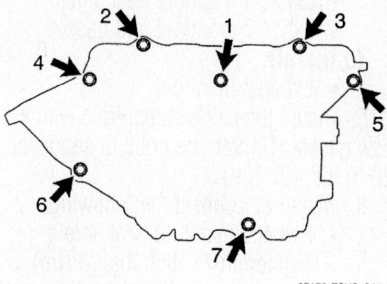

Intake manifold fastener torque sequence—2.4L Engine

- Throttle body. Tighten bolts to 22 ft lbs. (30 Nm).
- Air intake assembly

8. The remainder of the installation is the reverse order of removal.

9. Refill the cooling system to the correct level.

10. Start the engine and check for leaks.

Exhaust Manifold

REMOVAL & INSTALLATION

1.5L Engine

1. Before servicing the vehicle, refer to the Precautions Section.

2. Drain the cooling system.

3. Relieve the fuel system pressure.

4. Remove or disconnect the following:
- Negative battery cable
- Air intake assembly
- Engine appearance cover
- Right-hand front wheel
- Engine undercover
- Accelerator control cable
- Fuel tube from the fuel supply hose
- Brake booster check valve hose
- Radiator hose
- Heater hose
- Radiator filler hose
- Front exhaust pipe
- Exhaust manifold support bracket

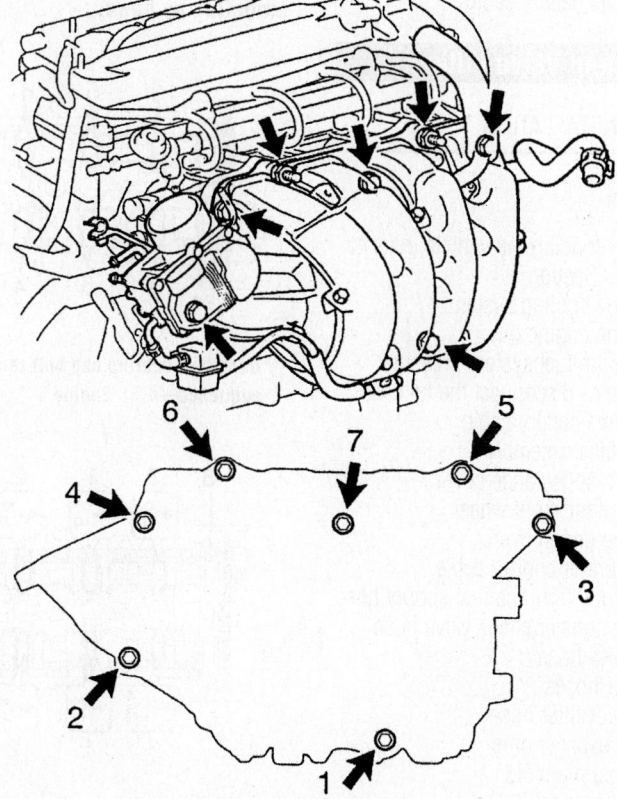

Intake manifold fastener location and loosening sequence—2.4L Engine

67170-TOYO-G24

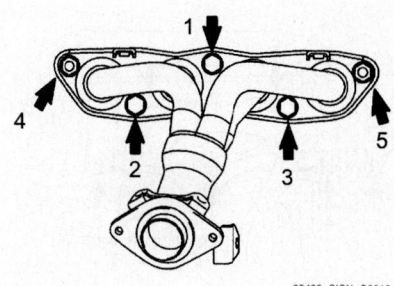

Exhaust manifold torque sequence—1.5L Engine

09490_SION_G0018

- Exhaust manifold heat shield
- Exhaust manifold and gasket

To install:

- Exhaust manifold

5. Install the exhaust manifold with a new gasket. Tighten the bolts in sequence to 20 ft. lbs. (27 Nm).

6. Install or connect the following:
- Exhaust manifold heat shield. Tighten to 71 inch lbs. (8 Nm).
- Exhaust manifold support bracket. Tighten to 32 ft. lbs. (44 Nm).
- Front exhaust pipe. Tighten to 32 ft. lbs. (43 Nm).
- Radiator filler hose. Tighten support bracket to 66 inch lbs. (7.5 Nm).
- Heater hose
- Radiator hose
- Brake booster check valve hose
- Fuel tube to the fuel supply hose
- Accelerator control cable
- Engine undercover
- Front wheel
- Engine appearance cover
- Air intake assembly
- Negative battery cable

7. Refill the engine with oil to the correct level.

8. Refill the cooling system to the correct level.

9. Start the engine and check for leaks.

2.4L Engine

1. Before servicing the vehicle, refer to the Precautions Section.

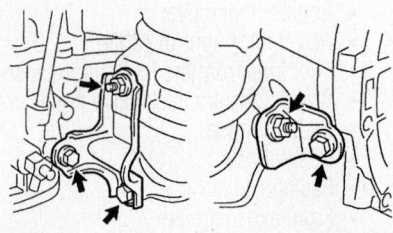

Exhaust manifold stays—2.4L Engine

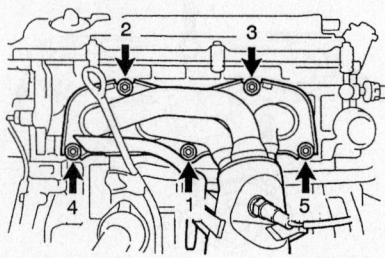

Exhaust manifold torque sequence—2.4L Engine

2. Drain the cooling system.
3. Relieve the fuel system pressure.
4. Remove or disconnect the following:
- Negative battery cable
- Engine appearance cover
- Air intake assembly
- Throttle body
- Fuel rail
- Intake manifold
- Front exhaust pipe
- Oil dipstick guide
- Exhaust manifold stays
- Exhaust manifold heat shield
- Exhaust manifold and gasket

To install:

5. Install the exhaust manifold with a new gasket. Tighten the bolts in sequence to 27 ft. lbs. (37 Nm).

6. Install or connect the following:
- Exhaust manifold heat shield. Tighten to 9 ft. lbs. (12 Nm).
- Exhaust manifold stays. Tighten to 32 ft. lbs. (44 Nm).
- Oil dipstick guide. Tighten to 80 inch lbs. (9 Nm).
- Front exhaust pipe. Tighten to 32 ft. lbs. (44 Nm).
- Intake manifold
- Fuel rail
- Throttle body
- Air intake assembly
- Engine appearance cover
- Negative battery cable

Camshaft and Valve Lifters

REMOVAL & INSTALLATION

1.5L Engines

1. Before servicing the vehicle, refer to the Precautions Section.
2. Drain the cooling system.
3. Drain the engine oil.
4. Relieve the fuel system pressure.
5. Remove or disconnect the following:
- Negative battery cable
- Air intake assembly
- Engine appearance cover
- Right-hand front wheel
- Engine undercover
- Accelerator control cable
- Fuel tube from the fuel supply hose
- Brake booster check valve hose
- Radiator hose
- Heater hoses
- Radiator filler hose
- Front exhaust pipe
- Exhaust manifold
- Intake manifold
- Oil dipstick guide

- Ignition coil
- Accessory drive belts
- Alternator
- Camshaft timing oil control valve
- Ventilation hoses from the cylinder head cover
- Cylinder head cover
- Water pump
- Front cover
- Crankshaft pulley
- Crankshaft position sensor

6. Support the engine with a suitable jack.

7. Remove or disconnect the following:
- Right-hand engine mount
- Timing chain
- Fuel supply hose
- Fuel rail and injectors

8. In several passes, uniformly loosen and remove the camshaft bearing cap bolts in the sequence shown.

9. Remove the bearing caps and camshafts.

10. Remove the valve lifters.

➡**Keep all valvetrain components in order for reassembly.**

To install:

11. Apply a light coat of clean engine oil to each valve lifter.

12. Install the valve lifters.

➡**Ensure that each lifter rotates smoothly by hand.**

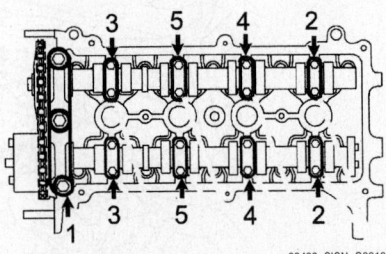

Camshaft bearing cap bolt removal sequence—1.5L Engine

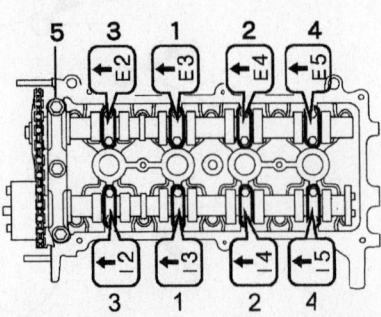

Camshaft bearing cap bolt torque sequence—1.5L Engine

13. Apply a light coat of clean engine oil to the camshaft journals.

14. Place the camshaft on the cylinder head with the timing mark on the camshaft gear facing upward.

15. Tighten the camshaft intake and exhaust bearing cap bolts in sequence to 9 ft. lbs. (13 Nm).

16. Install the No. 1 bearing cap and tighten to 17 ft. lbs. (23 Nm).

17. The remainder of the installation is the reverse order of removal.

18. Refill the engine with oil to the correct level.

19. Refill the cooling system to the correct level.

20. Start the engine and check for leaks.

2.4L Engine

1. Before servicing the vehicle, refer to the Precautions Section.
2. Drain the cooling system.
3. Drain the engine oil.
4. Relieve the fuel system pressure.
5. Remove or disconnect the following:
- Negative battery cable
- Hood
- Right-hand front wheel
- Engine undercover
- Right-hand front fender apron
- Engine appearance cover
- Wiper arm assembly
- Top cowl seal
- Left-hand top cowl ventilator louver
- Wiper linkage assembly
- Top outer cowl panel
- Left-hand cowl body mounting bracket
- Air intake assembly
- Throttle body
- Fuel rail
- Intake manifold
- Front exhaust pipe
- Oil dipstick guide
- Exhaust manifold
- Accessory drive belts
- Alternator
- Power steering pump

6. Support the engine with a suitable jack.
7. Remove or disconnect the following:
- Right-hand engine mount
- Ignition coil
- Cylinder head cover
- Accessory drive belt tensioner
- Crankshaft position sensor
- Oil pan

8. Turn the crankshaft pulley until its groove and the timing mark on the front cover are aligned to set the No. 1 cylinder to TDC.

9. Matchmark the timing chain and camshaft sprockets.

10. Holding the exhaust camshaft with a wrench, loosen the camshaft timing set bolt.

11. Using several steps, loosen the bearing cap bolts in the sequence shown. Remove the bearing caps.

12. Remove the camshaft timing set bolt while holding the exhaust camshaft in place.

13. Remove the exhaust camshaft, leaving the camshaft sprocket wrapped in the timing chain.

14. Remove the camshaft sprocket.

15. Holding the intake camshaft with a wrench, loosen the camshaft timing set bolt.

16. Using several steps, loosen the intake camshaft bearing cap bolts in the sequence shown. Remove the bearing caps.

17. Remove the intake camshaft, with

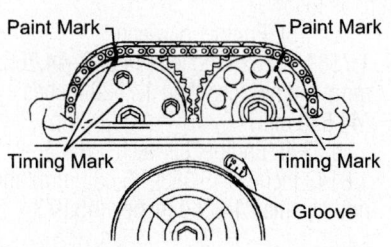

Match the groove on the crankshaft pulley to the front cover timing mark. Place matchmarks on the camshaft sprocket and timing chain—2.4L Engine

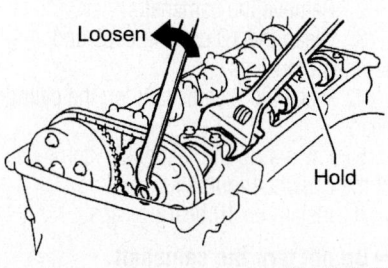

Secure the camshaft with a wrench when removing the set bolt—2.4L Engine

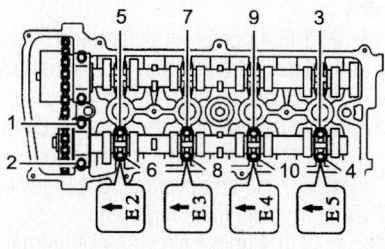

Exhaust camshaft bearing cap bolt removal sequence—2.4L Engine

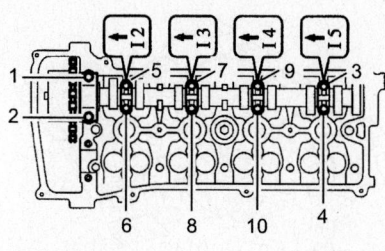

Intake camshaft bearing cap bolt removal sequence—2.4L Engine

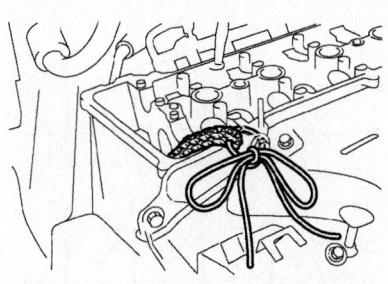

Secure the timing chain with string after camshafts are removed—2.4L Engine

sprocket attached, while holding the timing chain by hand.

18. Secure the timing chain with string to prevent it from falling down into the front cover.

19. Remove the valve lifters.

➡ **Keep all valvetrain components in order for reassembly.**

To install:

20. Apply a light coat of clean engine oil to each valve lifter.

21. Install the valve lifters in their original places.

22. Apply a light coat of clean engine oil to the journals of the camshafts.

23. Install the camshafts on the cylinder head with the No. 1 cam lobes facing the directions shown.

24. Examine the camshaft markings to

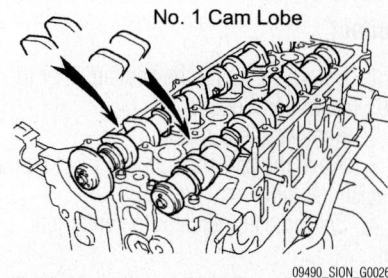

Ensure the No. 1 cam lobes are facing the correct direction during installation—2.4L Engine

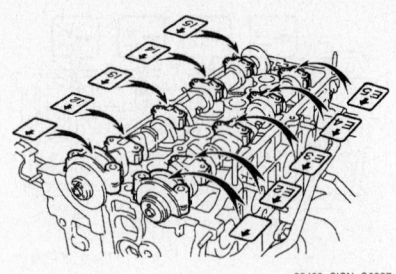

Compare the markings on the camshafts to the illustration for correct installation orientation—2.4L Engine

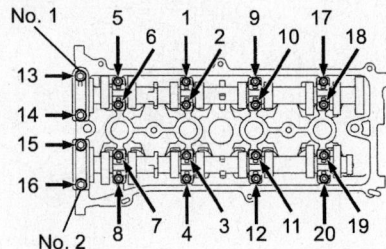

Camshaft bearing cap bolt torque sequence—2.4L Engine

ensure correct orientation of the camshafts for installation.

25. Apply a light coat of clean engine oil to the threads and under the heads of the bearing cap bolts.

26. Tighten the bearing cap bolts in sequence using the following torque values:

a. No. 1 and No. 2 bearing cap to 22 ft. lbs. (30 Nm).

b. Remaining bolts to 80 inch lbs. (9 Nm).

27. The remainder of the installation is the reverse order of removal.

28. Refill the engine with oil to the correct level.

29. Refill the cooling system to the correct level.

30. Start the engine and check for leaks.

INSPECTION

Runout

1. Before servicing the vehicle, refer to the Precautions Section.

2. Remove the camshafts.

3. Place the camshaft on a V-block, on a precise flat table.

4. Set the dial indicator to center journal.

5. Turn the camshaft to one direction by hand and measure the camshaft runout.

6. Runout should measure less than 0.0012 inches (0.03 mm).

7. Camshaft should be replaced if it exceeds the limit.

Cam Height

1. Before servicing the vehicle, refer to the Precautions Section.

2. Remove the camshafts.

3. Measure the cam height with a micrometer.

4. The intake camshaft should measure as follows:

a. 1.5L Engine: between 1.7566–1.7605 inches (44.617–44.717 mm) and not less than 1.7508 inches (44.470 mm).

b. 2.4L Engine: between 1.8305–1.8345 inches (46.495–46.595 mm) and not less than 1.8262 inches (46.385 mm).

5. The exhaust camshaft should measure as follows:

a. 1.5L Engine: between 1.7585–1.7624 inches (44.666–44.766 mm) and not less than 1.7528 inches (44.520 mm).

b. 2.4L Engine: between 1.8104–1.8143 inches (45.983–46.083 mm) and not less than 1.8060 inches (45.873 mm).

6. Camshaft should be replaced if it exceeds the limit.

Journal Oil Clearance

1. Before servicing the vehicle, refer to the Precautions Section.

2. Remove the camshafts.

3. Clean the 10 bearing caps and camshaft journals.

4. Place the 2 camshafts on the cylinder head.

5. Lay a strip of Plastigage® across each of the camshaft journals.

6. Install the 10 bearing caps.

➡ Do not turn the camshaft.

7. Remove the 10 bearing caps.

8. Measure the Plastigage® at its widest point.

9. Oil clearance should measure as follows:

a. 1.5L Engine: All other journals should measure between 0.0016–0.0037 inches (0.040–0.095 mm). If any clearance measurement is more than 0.0045 inches (0.115 mm), the camshaft or cylinder head and bearing caps together (or both) need to be replaced.

b. 2.4L Engine: No. 1 intake journal should measure between 0.0003–0.0015 inches (0.007–0.038 mm). No. 1 exhaust journal should measure between

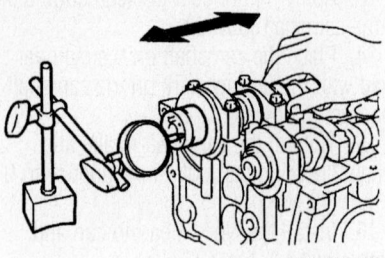

Measuring for camshaft endplay

0.0016–0.0031 inches (0.040–0.079 mm). All other journals should be between 0.0010–0.0024 inches (0.025–0.062 mm). If any clearance measurement is more than 0.0028 inches (0.07 mm) for the No. 1 intake journal, or more than 0.0039 inches (0.10 mm) for all other journals, the camshaft or cylinder head sub-assembly (or both) needs to be replaced. If the No. 1 journal clearance is greater than the maximum, replace the bearing cap.

End Play

1. Before servicing the vehicle, refer to the Precautions Section.

2. Install a dial indicator in the thrust direction on the front end of the camshaft. Measure the end play of the dial indicator when the camshaft is moved back and forth. The dial indicator should measure as follows:

a. 1.5L Engine: between 0.0016–0.0037 inches (0.040–0.095 mm) and not exceed 0.0043 inches (0.11 mm).

b. 2.4L Engine: Intake camshaft between 0.0016–0.0037 inches (0.040–0.095 mm) and not exceed 0.0043 inches (0.11 mm). Exhaust camshaft between 0.0032–0.0053 inches (0.080–0.135 mm) and not exceed 0.0059 inches (0.15 mm).

3. Replace the cylinder head assembly if the measurement is exceeded. Replace the camshaft if damage is found on the thrust surfaces.

Valve Lash

ADJUSTMENT

1.5L Engine

1. Before servicing the vehicle, refer to the Precautions Section.

2. Remove or disconnect the following:

- Negative battery cable
- Engine appearance cover

- Ignition coil
- Ventilation hoses from the cylinder head cover
- Cylinder head cover

→**Inspect the valve clearance when the engine is cold.**

3. Turn the crankshaft pulley to set the No. 1 cylinder at TDC.

4. Using a feeler gauge, measure the valve clearance shown in the illustration. Record the measurements.

5. Turn the crankshaft 1 complete revolution until the timing mark 0 of the chain cover are aligned to set the No. 4 cylinder at TDC.

6. Using a feeler gauge, measure the valve clearance shown in the illustration. Record the measurements.

   a. Intake valve clearance is 0.006–0.010 inches (0.15–0.25 mm).

   b. Exhaust valve clearance is 0.010–0.014 inches (0.25–0.35 mm).

7. To adjust the valve clearance:

   a. Set the No. 1 cylinder at TDC compression. Matchmark the timing chain and cam shaft sprockets.

   b. Remove the screw plug and insert a screwdriver into the service hole of the chain tensioner to hold the stopper plate

No. 1 Cylinder TDC/Compression

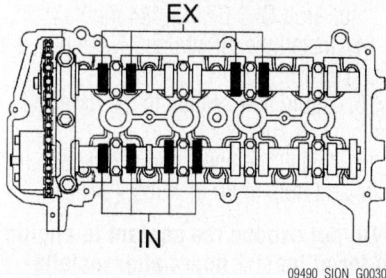

09490_SION_G0030

Measure indicated valve lifters with No. 1 cylinder at TDC—1.5L Engine

No. 4 Cylinder TDC/Compression

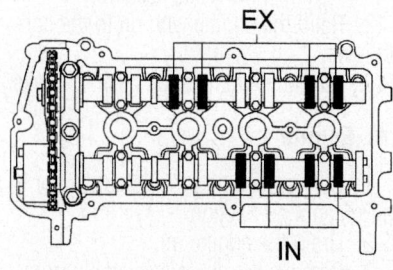

09490_SION_G0031

Measure indicated valve lifters with No. 4 cylinder at TDC—1.5L Engine

of the chain tensioner at an upward position.

   c. Using a wrench, rotate the exhaust camshaft clockwise to push in the plunger of the chain tensioner.

   d. Remove the screwdriver from the service hole, and then align the hole of the stopper plate with the service hole and insert a 0.08–0.12 in. (2 –3 mm) diameter bar into the holes to hold the stopper plate.

   e. Using Special Tool 09023-38400, hold the camshaft with a wrench on the hexagonal lobe, and remove the flange bolt.

   f. Remove the exhaust camshaft bearing caps.

→**Loosen each bolt uniformly, keeping the camshaft level.**

   g. Remove the flange bolt with the exhaust camshaft lifted up. Then detach the exhaust camshaft and the camshaft timing sprocket.

   h. Remove the intake camshaft bearing caps.

→**Loosen each bolt uniformly, keeping the camshaft level.**

   i. Hold the timing chain by hand, and remove the camshaft and the camshaft timing gear assembly.

   j. Remove the valve lifters.

   k. Using a micrometer, measure the thickness of the removed lifter.

8. Calculate the thickness of a new lifter so that the valve clearance comes within the specified value.

- Intake: New lifter thickness = Used lifter thickness + Recorded valve clearance measurement–0.008 in. (0.20 mm)
- Exhaust: New lifter thickness = Used lifter thickness + Recorded valve clearance measurement– 0.012 in. (0.30 mm)

9. Select a new lifter with the thickness as close to the calculated value as possible.

10. Reinstall the camshafts and check the valve clearance.

11. Install the cylinder head cover and remaining components.

12. Start the engine and check for leaks.

2.4L Engine

1. Support the engine with a suitable jack.

2. Remove or disconnect the following:
- Negative battery cable
- Right-hand front wheel
- Engine undercover
- Engine appearance cover

- Accessory drive belt
- Power steering pump
- Right-hand engine mount
- Ignition coil
- Cylinder head cover

3. Turn the crankshaft to set the No. 1 cylinder at TDC compression.

→**Inspect the valve clearance when the engine is cold.**

4. Turn the crankshaft pulley to set the No. 1 cylinder at TDC.

5. Using a feeler gauge, measure the valve clearance shown in the illustration. Record the measurements.

6. Turn the crankshaft 1 complete revolution and set the No. 4 cylinder to TDC.

7. Using a feeler gauge, measure the valve clearance shown in the illustration. Record the measurements.

   a. Intake valve clearance is 0.0075–0.0114 inches (0.19–0.29 mm).

   b. Exhaust valve clearance is 0.0118–0.0158 inches (0.30–0.40 mm).

8. To adjust the valve clearance:

9. Remove the camshafts.

10. Remove the valve lifters.

   a. Using a micrometer, measure the thickness of the removed lifter.

11. Calculate the thickness of a new lifter so that the valve clearance comes within the specified value.

- Intake: New lifter thickness = Used lifter thickness + Recorded valve

No. 1 Cylinder TDC/Compression

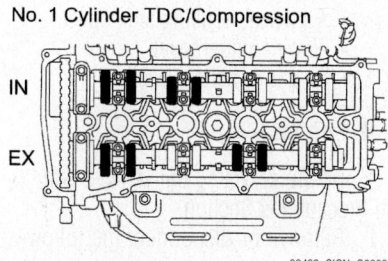

09490_SION_G0032

Measure indicated valve lifters with No. 1 cylinder at TDC—2.4L Engine

N. 4 Cylinder TDC/Compression:

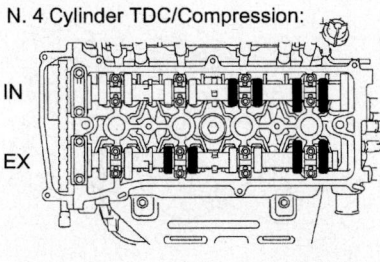

09490_SION_G0033

Measure indicated valve lifters with No. 4 cylinder at TDC—2.4L Engine

clearance measurement—0.0095 in. (0.24 mm)
- Exhaust: New lifter thickness = Used lifter thickness + Recorded valve clearance measurement—0.0138 in. (0.35 mm)

12. Select a new lifter with the thickness as close to the calculated value as possible.

13. Reinstall the camshafts and check the valve clearance.

14. Install the cylinder head cover and remaining components.

15. Start the engine and check for leaks.

Starter Motor

REMOVAL & INSTALLATION

1.5L Engine

1. Before servicing the vehicle, refer to the Precautions Section.
2. Remove or disconnect the following:
   - Negative battery cable
   - Engine undercover
   - Flywheel housing side cover
   - Starter electrical connections
   - Starter mounting bolts
   - Starter

To install:

3. Install or connect the following:
   - Starter. Tighten the mounting bolts to 27 ft. lbs. (37 Nm).
   - Starter electrical connections. Tighten the starter wire nut to 7 ft. lbs. (10 Nm).
   - Flywheel housing side cover
   - Engine undercover
   - Negative battery cable

2.4L Engine

1. Before servicing the vehicle, refer to the Precautions Section.
2. Remove or disconnect the following:
   - Negative battery cable
   - Starter electrical connections

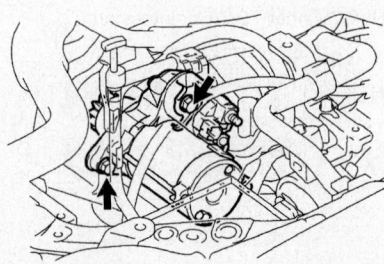

Location of the starter mounting bolts— 2.4L Engine

- Starter mounting bolts
- Starter

To install:

3. Install or connect the following:
   - Starter. Tighten the mounting bolts to 27 ft. lbs. (37 Nm).
   - Starter electrical connections. Tighten the starter wire nut to 7 ft. lbs. (10 Nm).
   - Negative battery cable

Oil Pan

REMOVAL & INSTALLATION

1.5L Engine

1. Before servicing the vehicle, refer to the Precautions Section.
2. Drain the engine oil.
3. Remove the engine/transaxle assembly.
4. Separate the transaxle from the engine assembly.
5. Remove or disconnect from the engine unit:
   - Front cover
   - Timing chain
   - Oil filter
6. Using Special Tool 09032-00100, cut the sealer between the upper and lower oil pans.
7. Remove the lower oil pan.
8. Remove the oil strainer and gasket.
9. Remove the mounting mount for the upper oil pan.
10. Using a suitable prying tool, pry the upper oil pan from the cylinder block.

❋❋ WARNING

Be careful not to damage the contact surfaces of the oil and cylinder block.

To install:

11. Clean the contact surfaces of all old sealant.
12. Apply a continuous line of sealant to the upper oil pan mating surfaces shown.
13. Install the upper oil pan and tighten

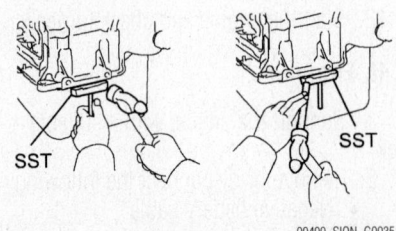

Cut off the applied sealer and remove the lower oil pan—1.5L Engine

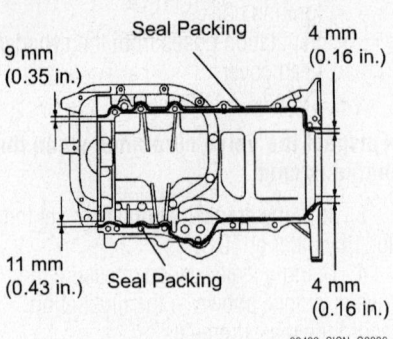

Apply sealant to the upper oil pan—1.5L Engine

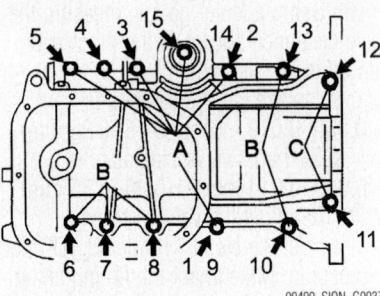

Upper oil pan torque sequence—1.5L Engine

the mounting bolts in sequence to 18 ft. lbs. (24 Nm). Each bolt length is as follows:
   a. Bolt A: 1.93 in. (49 mm)
   b. Bolt B: 3.47 in. (88 mm)
   c. Bolt C: 5.67 in. (144 mm)

14. Install the oil strainer. Tighten to 8 ft. lbs. (11 Nm).

15. Apply new sealant to the lower oil pan contact surfaces.

16. Install the lower pan and tighten the bolts and nuts to 80 inch lbs. (9 Nm).

➡ **Do not expose the sealant to engine oil for at least 2 hours after installation.**

17. Install or connect the following:
   - Oil filter
   - Timing chain
   - Front cover
   - Engine/Transaxle assembly

18. Refill the engine with oil to the correct level.

19. Start the engine and check for leaks.

2.4L Engine

1. Before servicing the vehicle, refer to the Precautions Section.
2. Drain the engine oil.
3. Remove the engine assembly from the vehicle and secure to a suitable chain block and sling device.

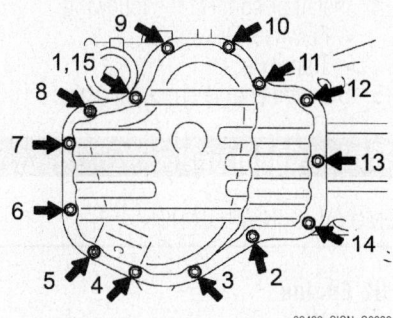

Oil pan torque sequence—2.4L Engine

4. Remove the oil pan mounting bolts.
5. Using Special Tool 09032-00100, cut off the sealant between the front cover, cylinder block and oil pan.
6. Remove the oil pan.
To install:
7. Remove any old sealant from the oil pan.
8. Apply a continuous bead of sealant to the contact surfaces of the oil pan.

➡**Do not expose the sealant to engine oil for at least 2 hours after installation.**

9. Install the oil pan. Tighten the bolts in sequence to 80 inch lbs. (9 Nm).
10. Install the engine assembly.
11. Fill the engine with oil to the correct level.
12. Start the engine and check for leaks.

Oil Pump

REMOVAL & INSTALLATION

1.5L Engine

1. Before servicing the vehicle, refer to the Precautions Section.
2. Drain the engine oil.
3. Drain the cooling system.
4. Remove or disconnect the following:
   • Negative battery cable
   • Front cover
   • Oil pump seal
   • Oil pump relief valve
   • 3 bolts and 2 screws and the oil pump cover from the front cover.
   • Oil pump
To install:
5. Coat the oil pump rotors with clean engine oil and install into the pump body with the marks facing the pump body cover side. Tighten the mounting bolts to 78 inch lbs. (9 Nm) and 95 inch lbs. (10 Nm) for the screws.
6. Install or connect the following:
   • Oil pump relief valve. Tighten to 18 ft. lbs. (25 Nm).

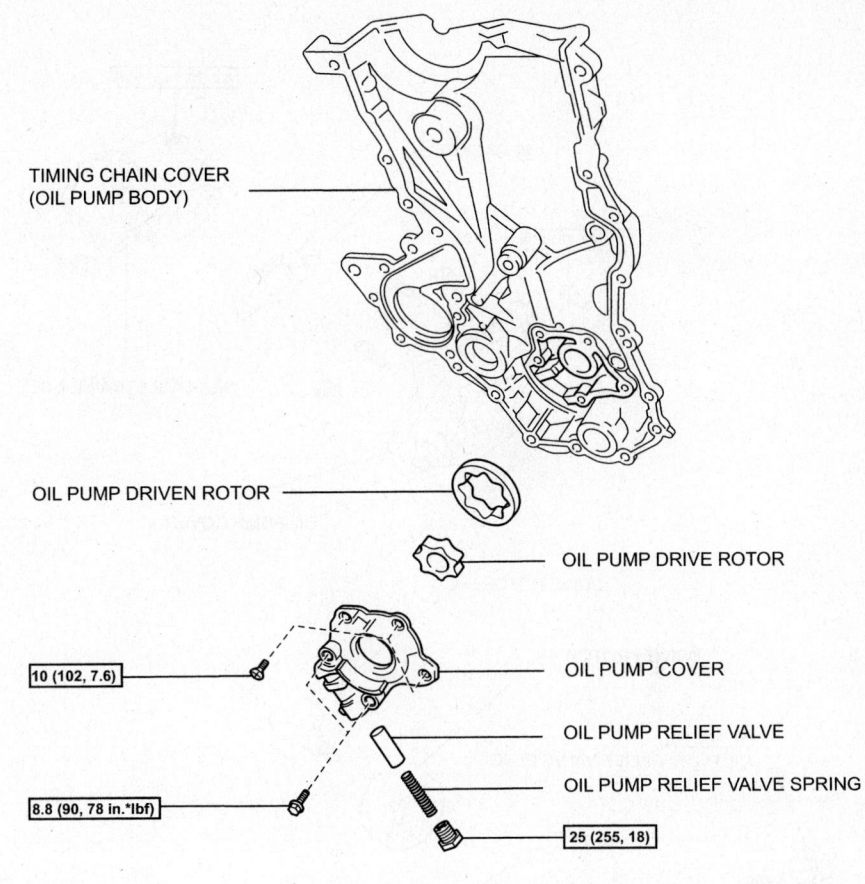

TIMING CHAIN COVER (OIL PUMP BODY)

OIL PUMP DRIVEN ROTOR

OIL PUMP DRIVE ROTOR

10 (102, 7.6)

OIL PUMP COVER

OIL PUMP RELIEF VALVE

8.8 (90, 78 in.*lbf)

OIL PUMP RELIEF VALVE SPRING

25 (255, 18)

N*m (kgf*cm, ft.*lbf) : Specified torque

09490_SION_G0042

Exploded view of the oil pump components—1.5L Engine

   • Oil pump seal using Special Tool 09950-60010.
   • Front cover
   • Negative battery cable
7. Refill the engine with oil to the correct level.
8. Start the engine and check for leaks.

2.4L Engine

1. Before servicing the vehicle, refer to the Precautions Section.
2. Drain the engine oil.
3. Remove or disconnect the following:
   • Negative battery cable
   • Front cover
   • Timing chain
   • Oil pump assembly
To install:
4. Install the oil pump assembly with a new gasket. Tighten the mounting bolts to 14 ft. lbs. (19 Nm).

09490_SION_G0039

Install the oil pump rotor set with the marks facing the pump body cover side— 1.5L Engine

5. Install or connect the following:
   • Timing chain
   • Front cover
   • Negative battery cable
6. Refill the engine with oil to the correct level.
7. Start the engine and check for leaks.

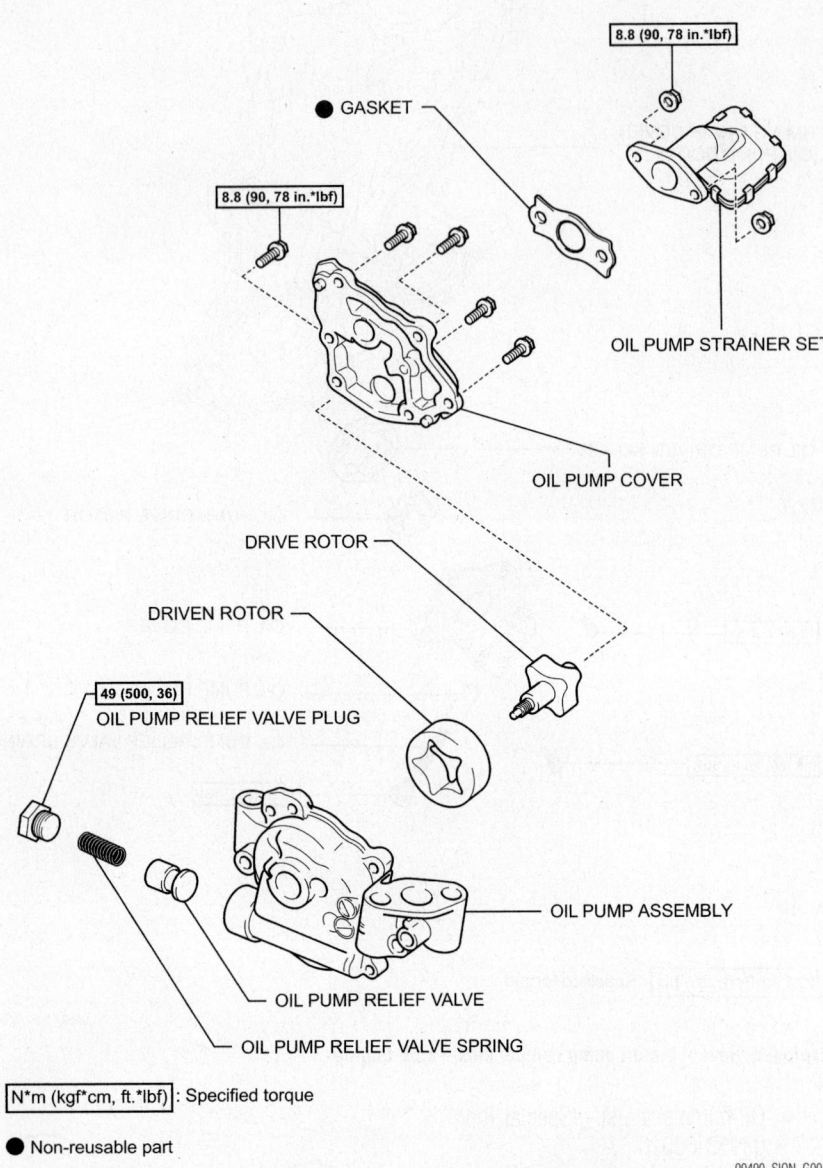

8.8 (90, 78 in.*lbf)

● GASKET

8.8 (90, 78 in.*lbf)

OIL PUMP STRAINER SET

OIL PUMP COVER

DRIVE ROTOR

DRIVEN ROTOR

49 (500, 36)
OIL PUMP RELIEF VALVE PLUG

OIL PUMP ASSEMBLY

OIL PUMP RELIEF VALVE

OIL PUMP RELIEF VALVE SPRING

N*m (kgf*cm, ft.*lbf) : Specified torque

● Non-reusable part

09490_SION_G0040

Exploded view of the oil pump components—2.4L Engine

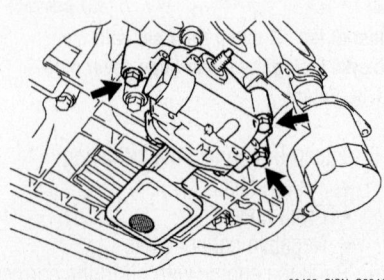

09490_SION_G0041

Remove the three mounting bolts to remove the oil pump assembly—2.4L Engine

Rear Main Seal

REMOVAL & INSTALLATION

1. Before servicing the vehicle, refer to the Precautions Section.
2. Remove or disconnect the following:
   • Transaxle
   • Flywheel/Driveplate
   • Oil seal

To install:

3. Using Special Tool 09223-56010 or suitable seal installer, tap in the oil until its surface is flush with the seal retainer edge.

4. Install or connect the following:
   • Flywheel/Driveplate
   • Transaxle
5. Start the engine check for leaks.

Timing Chain

REMOVAL & INSTALLATION

1.5L Engine

1. Before servicing the vehicle, refer to the Precautions Section.
2. Drain the engine oil.
3. Drain the cooling system.
4. Support the engine with a suitable jack.

5. Remove or disconnect the following:
   • Negative battery cable
   • Right-hand front wheel
   • Ignition coil
   • Ventilation hoses from the cylinder head cover
   • Cylinder head cover
   • Accessory drive belts
   • Alternator
   • Right-hand engine undercover
   • Water pump
6. Set the No. 1 cylinder to TDC.
7. Using Special Tool 09213-58012, remove the crankshaft pulley bolt.
8. Remove the crankshaft pulley. Use Special Tool 09950-50013 and the pulley bolt if necessary.
9. Remove or disconnect the following:
   • Camshaft timing oil control valve assembly
   • Crankshaft position sensor
   • Right-hand engine mount
   • Front cover

➡**Be careful not to damage the contact surfaces of the cover, cylinder head and engine block when prying the cover.**

   • Oil pump seal
   • Timing chain tensioner

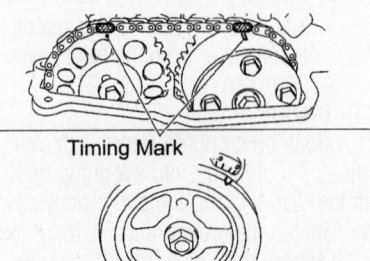

Timing Mark

09490_SION_G0043

Align the timing marks on the camshaft timing gears and timing chain to ensure the No. 1 cylinder is at TDC compression.

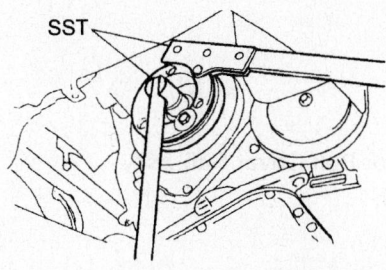

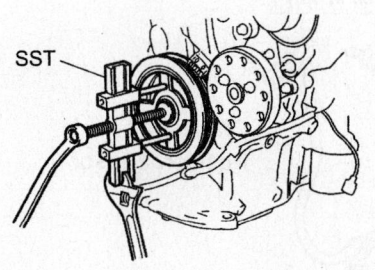

09490_SION_G0044

Remove crankshaft pulley with Special Tool 09213-58012 and 09950-50013.

- Timing chain tensioner slipper
- Timing chain vibration damper
- Timing chain

To install:

10. Set the crankshaft at 40–140° ATDC. Set the cams of intake and exhaust timing sprockets at 20° ATDC. Reset the crankshaft to 20° ATDC.

11. Install the chain vibration damper with the 2 bolts. Tighten to 80 inch lbs. (9 Nm).

12. Align the matchmarks of the camshaft timing sprocket, the camshaft timing gear and the crankshaft timing sprocket with each mark plate (colored in yellow) of the timing chain.

➡**To prevent the exhaust camshaft from spring back, turn it using a wrench and set it at the mark on the chain.**

13. Install the chain tensioner slipper with the bolt. Tighten to 80 inch lbs. (9 Nm).

14. Install the chain tensioner with the 2 bolts. Tighten to 80 inch lbs. (9 Nm).

15. Remove the bar from the chain tensioner.

16. Install or connect the following:
- Oil pump seal
- Oil pump
- Water pump
- Right-hand engine mount
- Camshaft timing oil control valve. Tighten to 66 inch lbs. (8 Nm).
- Crankshaft position sensor
- Water pump pulley
- Crankshaft. Tighten to 95 ft. lbs. (128 Nm).
- Cylinder head cover. Tighten to 7 ft. lbs. (10 Nm).
- Ignition coil
- Engine appearance cover
- Alternator
- Accessory drive belts
- Negative battery cable

17. Refill the engine with oil to the correct level.

18. Refill the cooling system to the correct level.

19. Start the engine and check for leaks.

2.4L Engine

1. Before servicing the vehicle, refer to the Precautions Section.

2. Drain the engine oil.

3. Drain the cooling system.

4. Support the engine with a suitable jack.

5. Remove or disconnect the following:
- Negative battery cable
- Hood
- Right-hand front wheel
- Engine undercover
- Front fender apron
- Engine appearance cover
- Front exhaust pipe
- Accessory drive belt
- Alternator
- Power steering pump and reservoir
- Right-hand engine mount
- Ignition coil
- Cylinder head cover
- Accessory drive belt tensioner

➡**Lift the engine with a suitable jack to gain access to the tensioner mounting bolts.**

- Crankshaft position sensor
- Oil pan

6. Set the No. 1 cylinder to TDC.

7. Using Special Tool 09213-54015, remove the crankshaft pulley bolt.

8. Remove the crankshaft pulley. Use

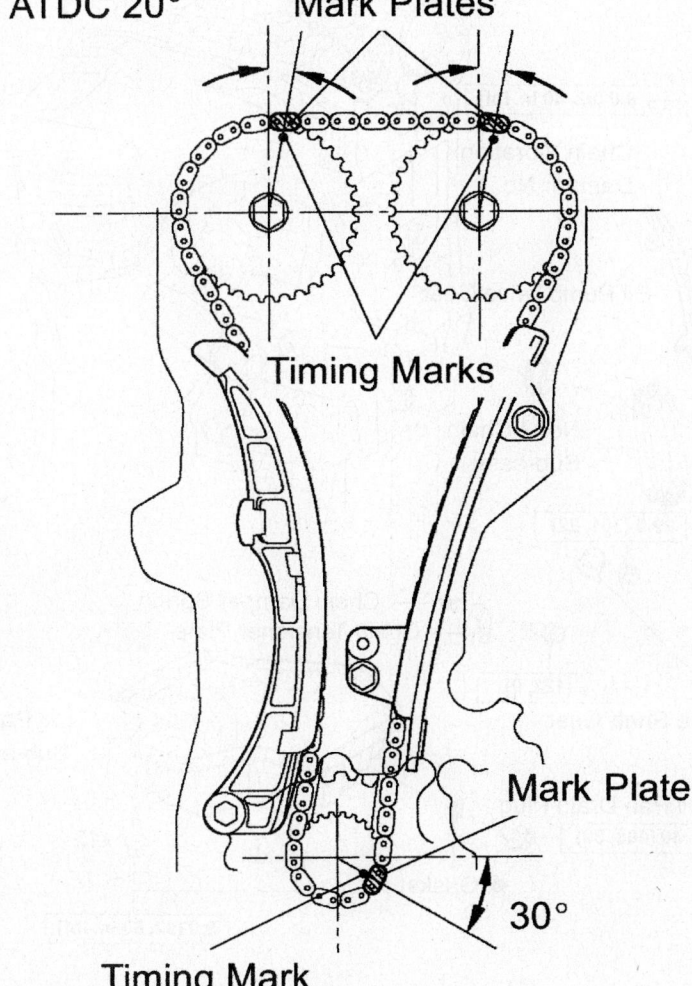

ATDC 20°   **Mark Plates**

Timing Marks

Mark Plate

30°

Timing Mark

09490_SION_G0045

Setting the crankshaft to 20° ATDC—1.5L Engine

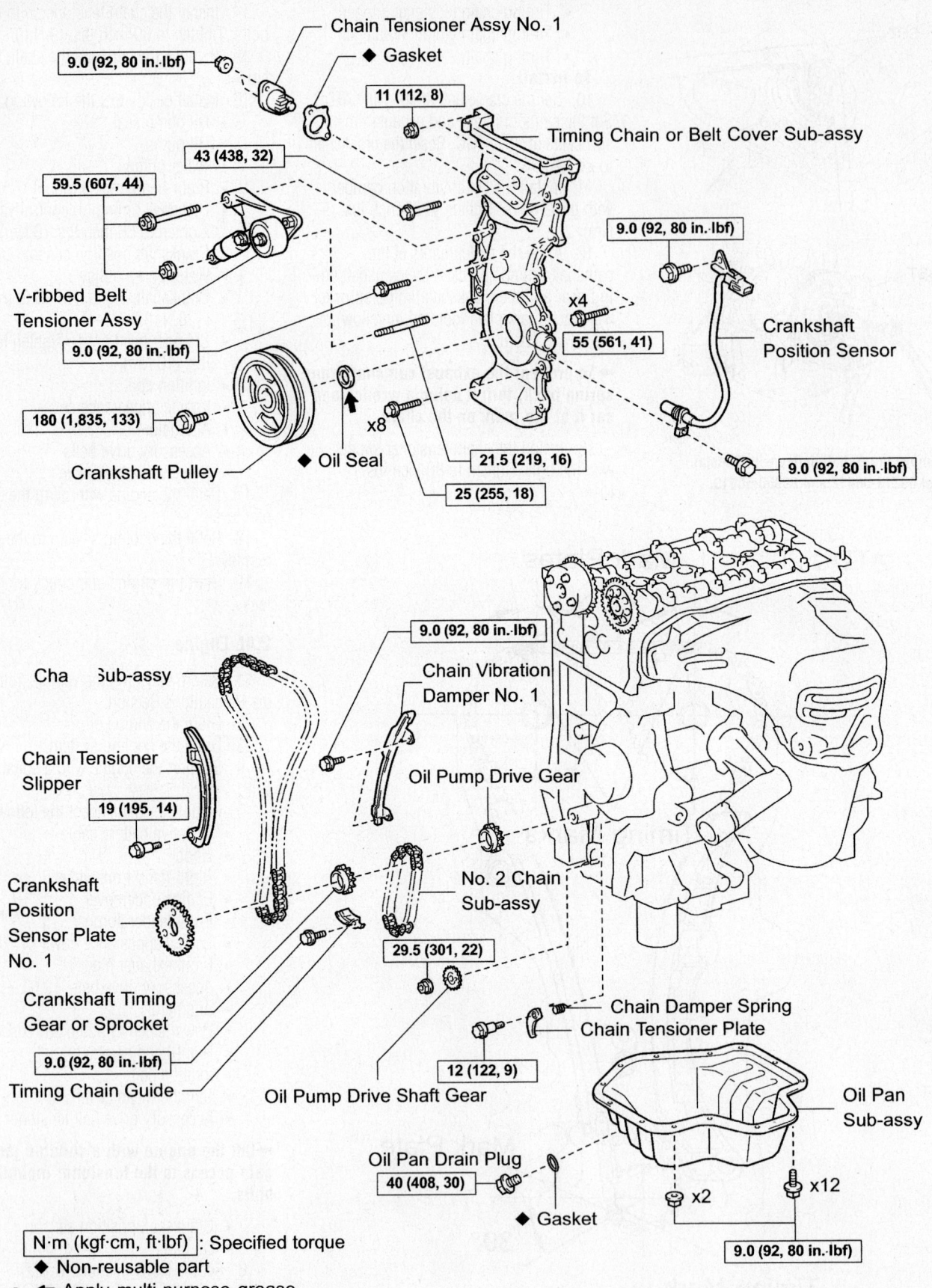

9.0 (92, 80 in.·lbf)

Chain Tensioner Assy No. 1

◆ Gasket

11 (112, 8)

Timing Chain or Belt Cover Sub-assy

43 (438, 32)

59.5 (607, 44)

9.0 (92, 80 in.·lbf)

V-ribbed Belt Tensioner Assy

x4

55 (561, 41)

Crankshaft Position Sensor

9.0 (92, 80 in.·lbf)

180 (1,835, 133)

x8

◆ Oil Seal

Crankshaft Pulley

21.5 (219, 16)

25 (255, 18)

9.0 (92, 80 in.·lbf)

9.0 (92, 80 in.·lbf)

Cha   Sub-assy

Chain Vibration Damper No. 1

Oil Pump Drive Gear

Chain Tensioner Slipper

19 (195, 14)

No. 2 Chain Sub-assy

Crankshaft Position Sensor Plate No. 1

29.5 (301, 22)

Crankshaft Timing Gear or Sprocket

9.0 (92, 80 in.·lbf)

Timing Chain Guide

Oil Pump Drive Shaft Gear

Chain Damper Spring
Chain Tensioner Plate

12 (122, 9)

Oil Pan Sub-assy

Oil Pan Drain Plug

40 (408, 30)

x2

x12

◆ Gasket

9.0 (92, 80 in.·lbf)

N·m (kgf·cm, ft·lbf) : Specified torque

◆ Non-reusable part

o ◄ Apply multi-purpose grease

09490_SION_G0046

Exploded view of the timing chain components—2.4L Engine

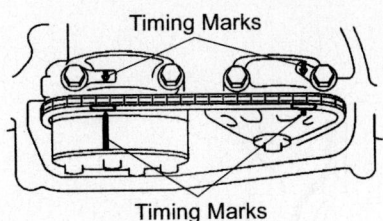

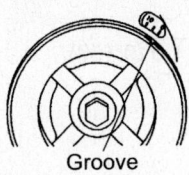

09490_SION_G0047

Align the timing marks on the camshaft timing gears and Nos. 1 and 2 camshaft bearing caps to ensure the No. 1 cylinder is at TDC compression.

Special Tool 09950-50013 and the pulley bolt if necessary.

9. Remove or disconnect the following:
- Timing chain tensioner
- Front cover
- Crankshaft position sensor plate
- Timing chain guide

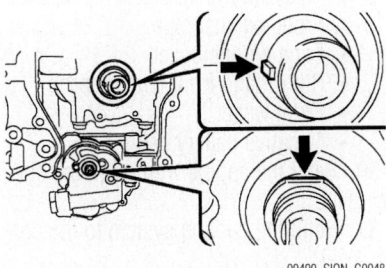

09490_SION_G0048

Set the crankshaft key and drive shaft cutout as shown—2.4L Engine

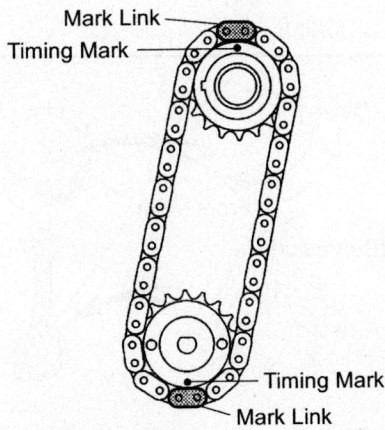

09490_SION_G0049

Align the marked links with the timing gear marks—2.4L Engine

- Timing chain tensioner slipper
- Upper timing chain
- Timing chain vibration damper
- Crankshaft sprocket

10. Remove the lower timing chain as follows:

a. Turn the crankshaft by 90° counter-clockwise to align the adjusting hole of the oil pump drive shaft gear with the groove of the oil pump.

b. Insert a 4 mm diameter bar into the adjusting hole of the oil pump drive shaft gear to lock in position, then remove the nut.

c. Remove the bolt, then remove the chain tensioner plate and spring.

d. Remove the oil pump drive gear, oil pump drive shaft gear and lower timing chain.

To install:

11. Set the crankshaft key in the left horizontal position. Turn the cutout of the drive shaft to the top.

12. Align the yellow mark links with the timing marks of the each gear as shown in the illustration.

13. Install the gears onto the crankshaft and oil pump shaft with the lower chain wrapped.

14. Temporarily tighten the oil pump drive shaft gear with the nut.

15. Insert the damper spring into the adjusting hole, then install the chain tensioner plate with the bolt. Tighten to 9 ft. lbs. (12 Nm).

16. Align the adjusting hole of the oil pump drive shaft gear with the groove of the oil pump.

17. Insert a 4 mm diameter bar into the adjusting hole of the oil pump drive shaft gear to lock in position, then tighten the nut to 22 ft. lbs. (30 Nm).

18. Turn the crankshaft clockwise by 90° to position the crankshaft key upward.

19. Install the crankshaft sprocket.

20. Turn the camshafts with a wrench on the hexagonal lobe to align the timing

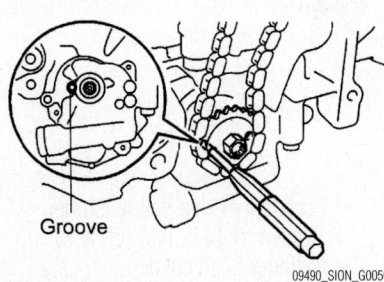

09490_SION_G0050

Insert a bar into the adjusting hole to lock the drive shaft gear in position—2.4L Engine

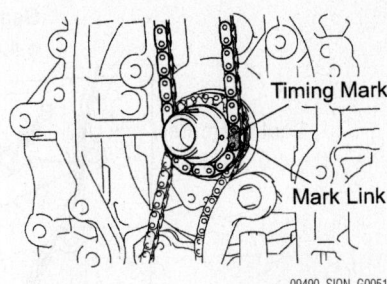

09490_SION_G0051

Align the timing marks installing the upper timing chain on the crankshaft sprocket—2.4L Engine

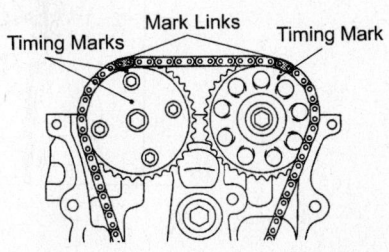

09490_SION_G0052

Align the timing marks with the camshaft timing gears when installing the upper timing chain—2.4L Engine

marks of the camshaft timing gear with each timing mark located on the No. 1 and No. 2 bearing caps as shown in the illustration.

21. Using the crankshaft pulley bolt, turn the crankshaft to position the key on the crankshaft upward.

22. Install the upper timing chain onto the crankshaft timing gear with the gold or orange mark link aligned with the timing mark on the crankshaft sprocket.

23. Using Special Tool 09309-37010, tap in the crankshaft timing gear.

24. Align the gold or yellow mark links with each timing mark located on the camshaft timing gears, then install the upper timing chain.

25. Install or connect the following:
- Timing chain tensioner slipper. Tighten to 14 ft. lbs. (19 Nm).
- Timing chain guide. Tighten to 80 inch lbs. (9 Nm).
- Crankshaft position sensor plate

26. Remove any old sealant from the front cover.

27. Apply a continuous bead of sealant to the front cover contact surfaces as shown.

28. Install the front cover. Tighten the bolts as follows:

a. Bolt A to 80 inch lbs. (9 Nm).

b. Bolts B to 18 ft. lbs. (25 Nm).

c. Bolts C to 41 ft. lbs. (55 Nm).

d. Bolt D to 32 ft. lbs. (43 Nm).

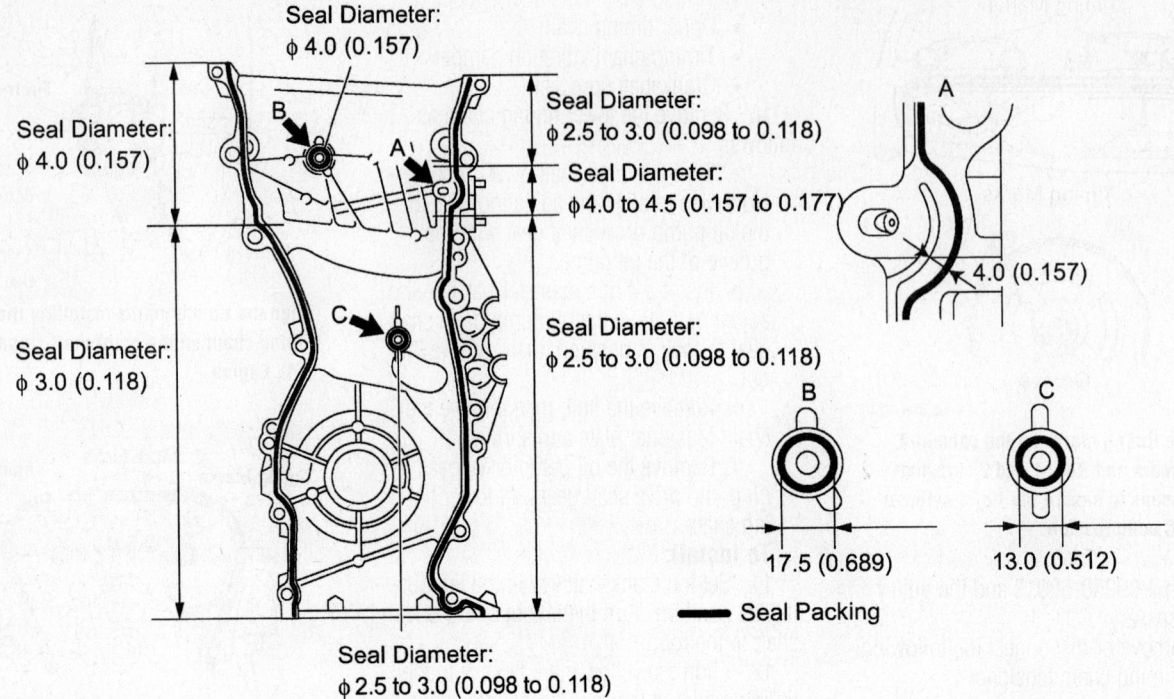

Seal Diameter:
φ 4.0 (0.157)

Seal Diameter:
φ 4.0 (0.157)

Seal Diameter:
φ 2.5 to 3.0 (0.098 to 0.118)

Seal Diameter:
φ 4.0 to 4.5 (0.157 to 0.177)

Seal Diameter:
φ 3.0 (0.118)

Seal Diameter:
φ 2.5 to 3.0 (0.098 to 0.118)

Seal Diameter:
φ 2.5 to 3.0 (0.098 to 0.118)

4.0 (0.157)

17.5 (0.689)   13.0 (0.512)

━ Seal Packing

09490_SION_G0053

Apply sealant to the front cover contact surfaces shown—2.4L Engine

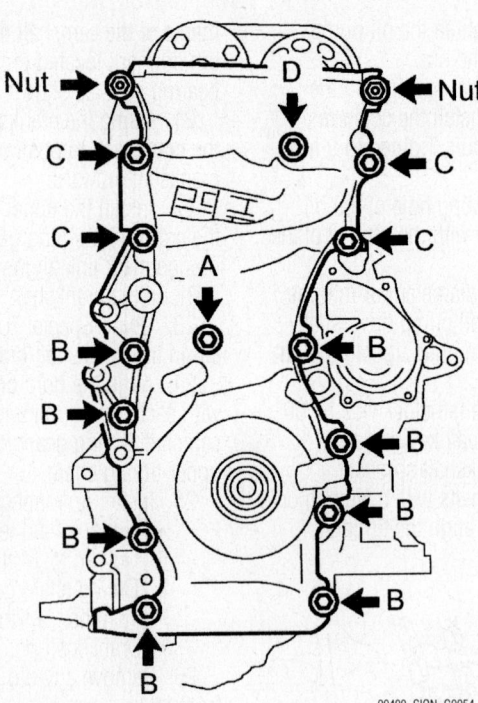

Nut   D   Nut

09490_SION_G0054

Front cover bolt identification—2.4L Engine

e. Nuts to 8 ft. lbs. (11 Nm).

29. Install or connect the following:
- Timing chain tensioner. Tighten to 80 inch lbs. (9 Nm).
- Crankshaft pulley. Tighten to 133 ft. lbs. (180 Nm).
- Accessory drive belt tensioner. Tighten to 44 ft. lbs. (60 Nm).
- Cylinder head cover
- Ignition coil
- Right-hand engine mount
- Power steering pump and reservoir
- Alternator
- Accessory drive belts
- Front exhaust pipe
- Engine undercover
- Right-hand front fender apron
- Front wheel
- Negative battery cable

30. Refill the engine with oil to the correct level.

31. Refill the cooling system to the correct level.

32. Start the engine and check for leaks.

Piston and Ring

POSITIONING

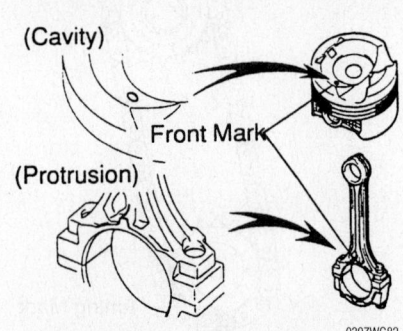

(Cavity)

Front Mark

(Protrusion)

9307WG82

Align the cavity of the piston with protruding portion of the connecting rod—1.5L and 2.4L Engines

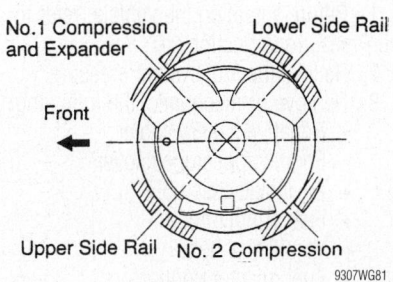

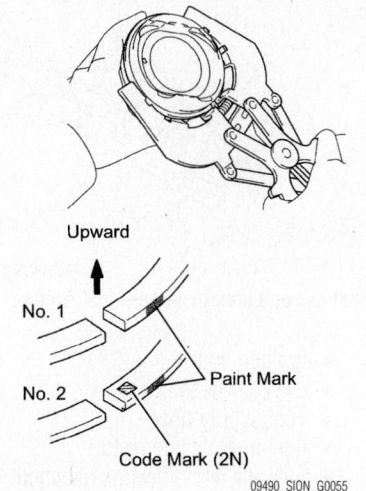

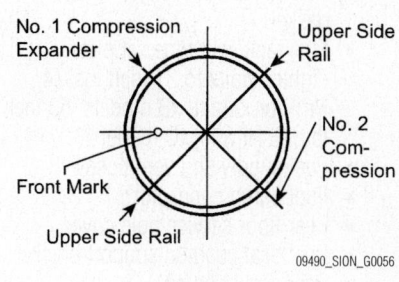

Piston ring end-gap spacing—1.5L Engine

Install the two compression rings with the paint mark as shown—2.4L Engine

Piston ring end-gap spacing—2.4L Engine

FUEL SYSTEM

Fuel System Pressure

RELIEVING

1. Before servicing the vehicle, refer to the Precautions Section.
2. Remove or disconnect the following:
   • Rear seat cushion
   • Access panel
   • Fuel pump module connector
3. Start the engine and allow it to run until it stalls.
4. Turn the ignition switch to the **OFF**-position.
5. Disconnect the negative battery cable.
6. Attach the fuel pump connector.

Fuel Filter

The fuel filter is in the tank as part of the fuel pump assembly.

Fuel Pump

REMOVAL & INSTALLATION

1. Before servicing the vehicle, refer to the Precautions Section.
2. Relieve the fuel system pressure.
3. Remove or disconnect the following:
   • Negative battery cable
   • Rear seat cushion assembly
   • Rear seat cushion support bracket, xA and xB only
   • Rear floor service hole cover
   • Fuel pump connector
   • Fuel supply and vent hoses

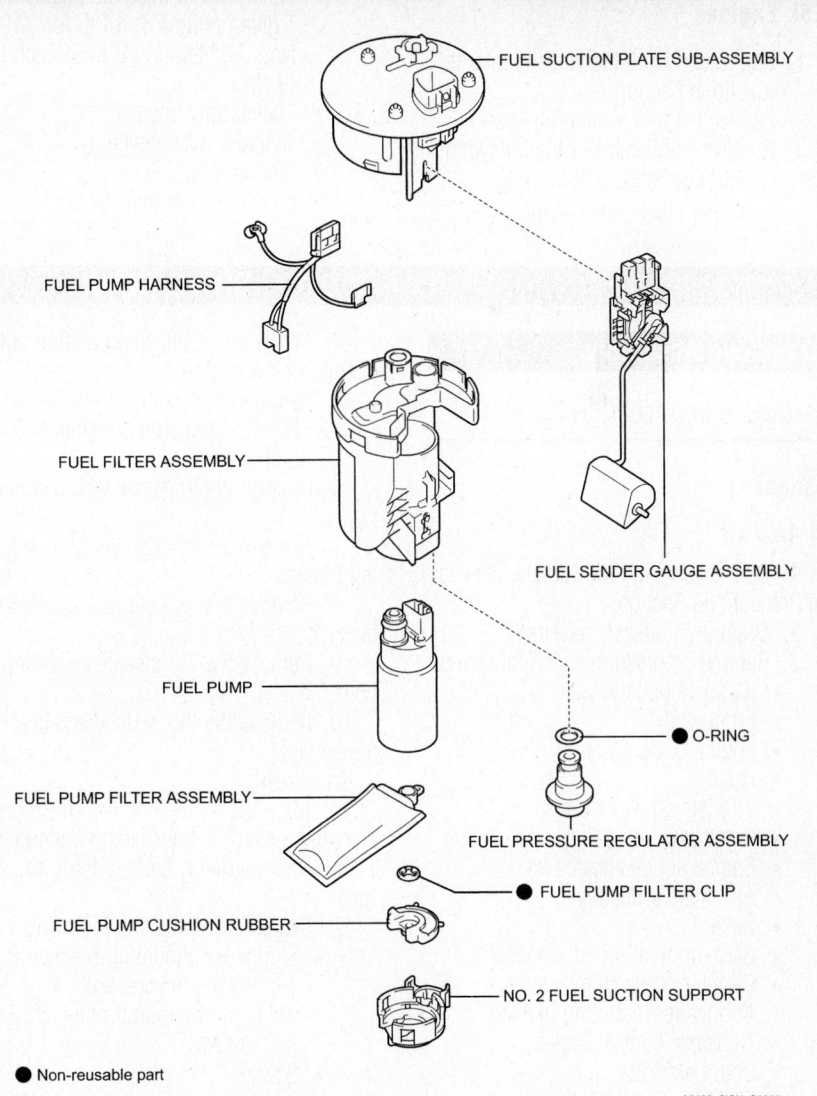

● Non-reusable part

Exploded view of the fuel pump assembly

- Fuel tank vent tube set plate
- Fuel pump assembly

To install:

4. Install or connect the following:
- Fuel pump assembly with a new gasket.
- Fuel tank vent tube set plate. Tighten bolts to 31 inch lbs. (4 Nm) for xA and xB models. 53 inch lbs. (6 Nm) for tC models.
- Fuel supply and vent hoses
- Fuel pump connector
- Rear floor service hole cover
- Rear seat cushion support bracket
- Rear seat cushion
- Negative battery cable

5. Start the engine and check for leaks.

Fuel Injector

REMOVAL & INSTALLATION

1.5L Engine

1. Before servicing the vehicle, refer to the Precautions Section.
2. Relieve the fuel system pressure.
3. Remove or disconnect the following:
- Negative battery cable
- Engine appearance cover

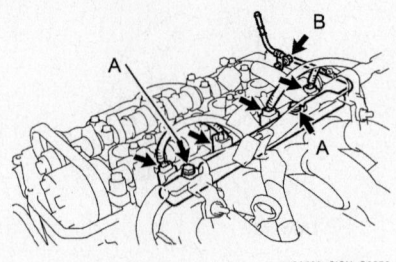

09490_SION_G0058

Fuel rail bolt identification—1.5L Engine

- Ignition coil
- Cylinder head cover
- Fuel supply hose
- Fuel injector connectors
- Fuel rail with injectors attached
- Fuel injector from the fuel rail

To install:

4. Install or connect the following:
- Injectors to the fuel rail using new O-rings
- Fuel rail with injectors attached. Tighten bolts A to 14 ft. lbs. (19 Nm). Tighten bolt B to 80 inch lbs. (9 Nm).
- Fuel supply hose
- Cylinder head cover
- Ignition coil
- Engine appearance cover

- Negative battery cable

5. Start the engine and check for leaks.

2.4L Engine

1. Before servicing the vehicle, refer to the Precautions Section.
2. Relieve the fuel system pressure.
3. Remove or disconnect the following:
- Negative battery cable
- Engine appearance cover
- Air intake assembly
- Fuel supply hose
- Ventilation hose
- Fuel injector connectors
- Fuel rail with the injectors attached
- Fuel injector from the fuel rail

To install:

4. Install or connect the following:
- Injectors to the fuel rail using new O-rings
- Fuel rail with injectors attached and torque the bolts to 15 ft. lbs. (20 Nm).
- Fuel injector connectors
- Ventilation hose
- Fuel supply hose
- Air intake assembly
- Engine appearance cover
- Negative battery cable

5. Start the engine and check for leaks.

DRIVE TRAIN

Transaxle Assembly

REMOVAL & INSTALLATION

Manual

xA AND xB

1. Before servicing the vehicle, refer to the Precautions Section.
2. Drain the transaxle assembly.
3. Remove or disconnect the following:
- Negative battery cable
- Front wheel
- Engine undercover
- Hood
- Windshield wiper arms
- Outer cowl top panel
- Engine appearance cover
- Air intake assembly
- Battery
- Back-up light switch connector
- Clutch release cylinder hose
- Air cleaner mounting bracket
- Transaxle control cables
- Both halfshafts
- Front floor panel brace
- Front exhaust pipe

4. At this point, attach an engine hoist to support the engine.
5. Remove or disconnect the following:
- Front suspension crossmember
- Starter
6. Support the transaxle with a suitable jack.
7. Remove the left-hand engine mounting bracket.
8. Remove the rear engine mounting bracket.
9. Remove the 7 transaxle mounting bolts.
10. Remove the transaxle assembly from the engine.

To install:

11. Align the input shaft with the clutch disc and install the transaxle to the engine. Tighten the mounting bolts to 24 ft. lbs. (33 Nm).
12. Install or connect the following:
- Engine left mounting bracket. Tighten the bracket bolts 47 ft. lbs. (64 Nm); the mount bolts to 36 ft. lbs. (49 Nm).
- Starter
- Front suspension crossmember
- Halfshafts

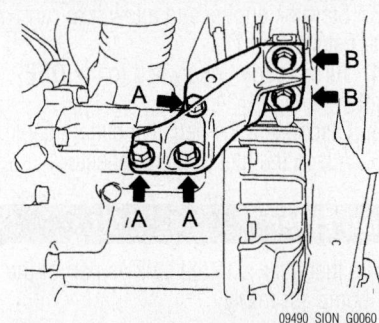

09490_SION_G0060

Rear engine mount—Bracket-to-transaxle bolts (A) and engine mounting bolts (B)—xA and xB

- Front exhaust pipe
- Front floor panel brace
- Transaxle control cables
- Air cleaner mounting bracket. Tighten to 14 ft. lbs. (19 Nm).
- Clutch release cylinder hose
- Back-up light switch connector
- Battery
- Air intake assembly
- Engine appearance cover
- Outer cowl top panel
- Windshield wiper arms

- Hood
- Engine undercover
- Front wheel
- Negative battery cable

13. Refill the transaxle with fluid to the correct level.

14. Start the engine and check for leaks.

tC

1. Before servicing the vehicle, refer to the Precautions Section.

2. Drain the transaxle assembly.

3. Place the front wheels facing straight ahead.

4. Remove or disconnect the following:
- Negative battery cable
- Steering intermediate shaft
- Front wheel
- Engine undercover
- Center exhaust pipe
- Hood
- Engine appearance cover
- Air intake assembly
- Battery and battery tray
- Starter
- Back-up light switch
- Clutch release cylinder assembly
- Transaxle control cables
- Both halfshafts

5. At this point, attach an engine hoist to support the engine.

6. Remove or disconnect the following:
- Front suspension crossmember
- Left engine mount and bracket
- Transaxle

To install:

7. Install the transaxle to the engine. Tighten the bolts as follows:
   a. Bolts A to 47 ft. lbs. (64 Nm).
   b. Bolts B to 34 ft. lbs. (46 Nm).
   c. Bolts C to 32 ft. lbs. (44 Nm).

8. Install or connect the following:
- Left engine mount and bracket. Tighten the bracket bolts to 38 ft. lbs. (52 Nm). Tighten the mount bolt to 64 ft. lbs. (87 Nm).
- Front suspension crossmember

![Manual transaxle mounting bolt identification—tC]

09490_SION_G0062

Manual transaxle mounting bolt identification—tC

- Halfshafts
- Transaxle control cables
- Clutch release cylinder assembly
- Back-up light switch
- Starter
- Battery tray and battery
- Air intake assembly
- Engine appearance cover
- Hood
- Center exhaust pipe
- Front wheel
- Negative battery cable

9. Refill the transaxle with fluid to correct level.

10. Start the engine and check for leaks.

Automatic

xA AND xB

1. Before servicing the vehicle, refer to the Precautions Section.

2. Drain the transaxle assembly.

3. Remove or disconnect the following:
- Negative battery cable
- Hood
- Windshield wiper arms
- Outer cowl top panel
- Engine appearance cover
- Air intake assembly
- Transmission control cable and bracket
- Transmission wiring connectors
- Front wheel
- Engine undercover
- Front exhaust pipe assembly
- Both halfshafts
- Transaxle oil cooler hose

4. At this point, attach an engine crane to support the engine.

5. Remove the halfshaft heat insulator.

6. Remove the starter.

7. Support the transaxle assembly with a suitable jack.

8. Remove or disconnect the following:
- Left-hand engine mounting bracket
- Rear engine mount and bracket

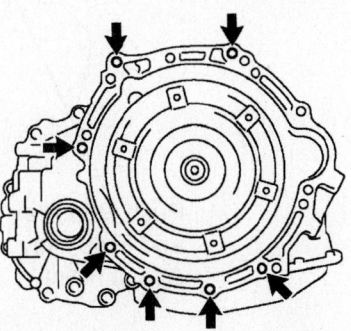

09490_SION_G0059

Automatic transaxle mounting bolt location—xA and xB

- Front suspension crossmember
- Flywheel housing cover

9. Turn the crankshaft to gain access and remove the 6 torque converter mounting bolts.

10. Remove the 7 transaxle mounting bolts.

11. Remove the automatic transaxle from the engine.

To install:

12. Install the transaxle assembly to the engine. Tighten the 7 mounting bolts to 22 ft. lbs. (30 Nm).

13. Install the 6 torque converter mounting bolts. Tighten to 20 ft. lbs. (27 Nm).

14. Install or connect the following:
- Flywheel housing cover
- Rear engine mounting bracket. Tighten to 36 ft. lbs. (49 Nm).
- Front suspension crossmember
- Rear engine mount. Tighten to 47 ft. lbs. (64 Nm).
- Left-hand engine mounting bracket. Tighten to 36 ft. lbs. (49 Nm).
- Starter
- Halfshaft heat insulator. Tighten to 13 ft. lbs. (18 Nm).
- Halfshafts
- Front exhaust pipe assembly
- Transaxle oil cooler hose
- Engine undercover
- Front wheel
- Transaxle wiring connectors
- Transaxle control cable
- Air intake assembly
- Engine appearance cover
- Outer cowl top panel
- Wiper arms
- Hood
- Negative battery

15. Refill the transaxle with fluid to correct level.

16. Start the engine and check for leaks.

tC

1. Before servicing the vehicle, refer to the Precautions Section.

2. Drain the transaxle assembly.

3. Remove or disconnect the following:
- Negative battery cable
- Hood
- Engine appearance cover
- Battery and battery carrier
- Air intake assembly
- Transaxle control cables
- Transaxle wiring connectors
- Breather plug hose
- Transaxle control cable brackets
- Transaxle oil filler tube
- Front wheel
- Engine undercover
- Transaxle oil cooler hoses

4. At this point, attach an engine crane to support the engine.

5. Remove or disconnect the following:
- Both halfshafts
- Starter

6. Support the transaxle with a suitable jack.

7. Remove or disconnect the following:
- Front engine mount
- Rear engine mount
- Center engine crossmember
- Left engine mount and bracket
- Front engine mounting bracket from the transaxle
- Rear engine mounting bracket from the transaxle
- Flywheel housing cover

8. Turn the crankshaft to remove the 6 torque converter mounting bolts.

9. Remove the 7 transaxle mounting bolts.

10. Remove the transaxle assembly from the engine.

To install:

11. Install the transaxle. Tighten the bolts as follows:
   a. Bolt A: 47 ft. lbs. (64 Nm).
   b. Bolt B: 34 ft. lbs. (46 Nm).
   c. Bolts C: 32 ft. lbs. (44 Nm).

12. Install or connect the following:
- Torque converter mounting bolts. Tighten to 30 ft. lbs. (41 Nm).
- Flywheel housing cover
- Rear engine mounting bracket. Tighten to 47 ft. lbs. (64 Nm).
- Front engine mounting bracket. Tighten to 47 ft. lbs. (64 Nm).
- Left engine mounting bracket. Tighten to 47 ft. lbs. (64 Nm).
- Left engine mount
- Center engine crossmember
- Rear engine mount
- Front engine mount
- Starter
- Halfshafts
- Front exhaust pipe

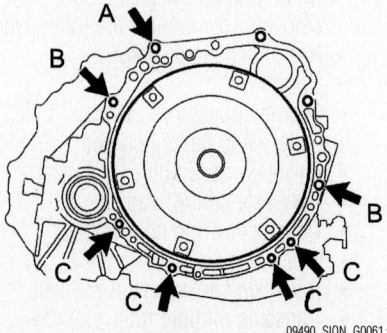

Automatic transaxle mounting bolt identification—tC

- Engine undercover
- Front wheel
- Transaxle oil cooler hoses
- Transaxle control cable support brackets
- Transaxle oil filler tube
- Breather plug hose
- Transaxle wiring connectors
- Transaxle control cable
- Air intake assembly
- Battery carrier and battery
- Engine appearance cover
- Hood
- Negative battery cable

13. Refill the transaxle with fluid to the correct level.

14. Start the engine and check for leaks.

Clutch

ADJUSTMENTS

Pedal Height

1. Loosen the lock nut and turn the stopper bolt until the pedal height is 5.287–5.681 in. (134.3–144.3 mm).

2. Tighten the lock nut to 13 ft. lbs. (18 Nm).

Pedal Free Play

1. Depress the pedal until clutch resistance begins to be felt. Standard pedal free play is 0.197–0.591 in. (5–15 mm).

2. Gently depress the pedal until the resistance begins to increase a little. Standard push rod play at pedal top is 0.039–0.197 in. (1–5 mm).

3. Loosen the lock nut and turn the push rod until correct free play and push rod play are obtained. Tighten the lock nut to 9 ft. lbs. (12 Nm).

➡After adjusting the pedal free play, check the pedal height.

Clutch Release Point

1. Pull the parking brake lever and install wheel chocks.

2. Start the engine and run it at idle.

3. Without depressing the clutch pedal, slowly move the shift lever into reverse until the gears contact.

4. Gradually depress the clutch pedal and measure the stroke distance from the point that the gear noise stops (release

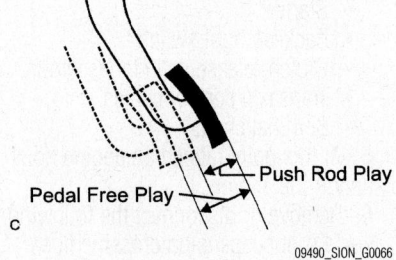

Measuring pedal and push rod free play.

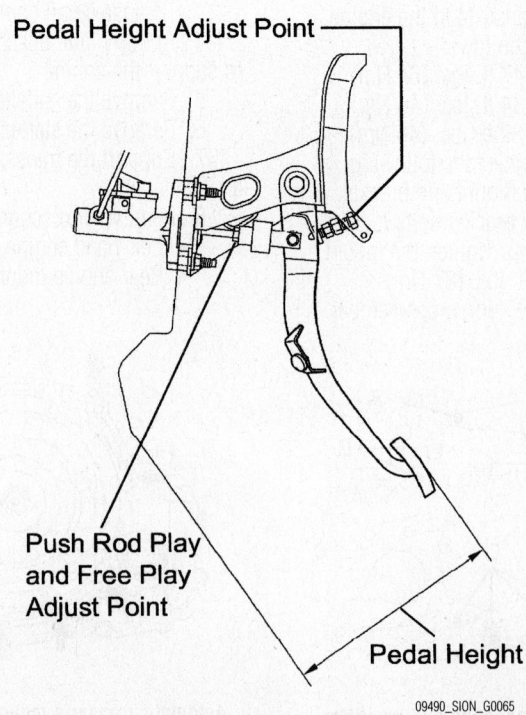

Adjusting the pedal height.

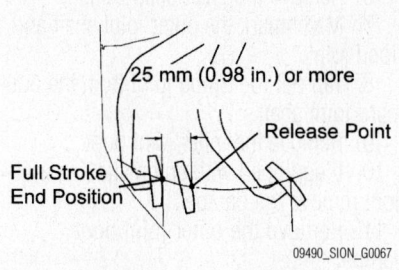

25 mm (0.98 in.) or more

Release Point

Full Stroke
End Position

09490_SION_G0067

Measuring the clutch release point.

point) up to the full stroke end position. Standard distance is 0.98 in. (25 mm) or more.

REMOVAL & INSTALLATION

1. Before servicing the vehicle, refer to the Precautions Section.
2. Remove or disconnect the following:
   • Transaxle
   • Release fork and boot
   • Release bearing assembly
   • Release fork support
3. Matchmark the clutch cover and flywheel. Loosen the clutch cover bolts one turn at a time until the spring tension is released.
4. Remove the clutch cover.
5. Remove the clutch disc.
To install:
6. Install the clutch disc on the flywheel.
7. Align the matchmarks on the clutch cover and flywheel.

130 (1,325,96)

CLUTCH DISC ASSEMBLY

CLUTCH COVER ASSEMBLY

19 (195,14)

FLYWHEEL SUB-ASSEMBLY

CLUTCH RELEASE BEARING ASSEMBLY

CLUTCH RELEASE FORK SUB-ASSEMBLY

RELEASE BEARING HUB CLIP

47 (480,35)

RELEASE FORK SUPPORT

N*m (kgf*cm, ft.*lbf) : Specified torque

⇦ Clutch spline grease

◄ Release hub grease

CLUTCH RELEASE FORK BOOT

09490_SION_G0064

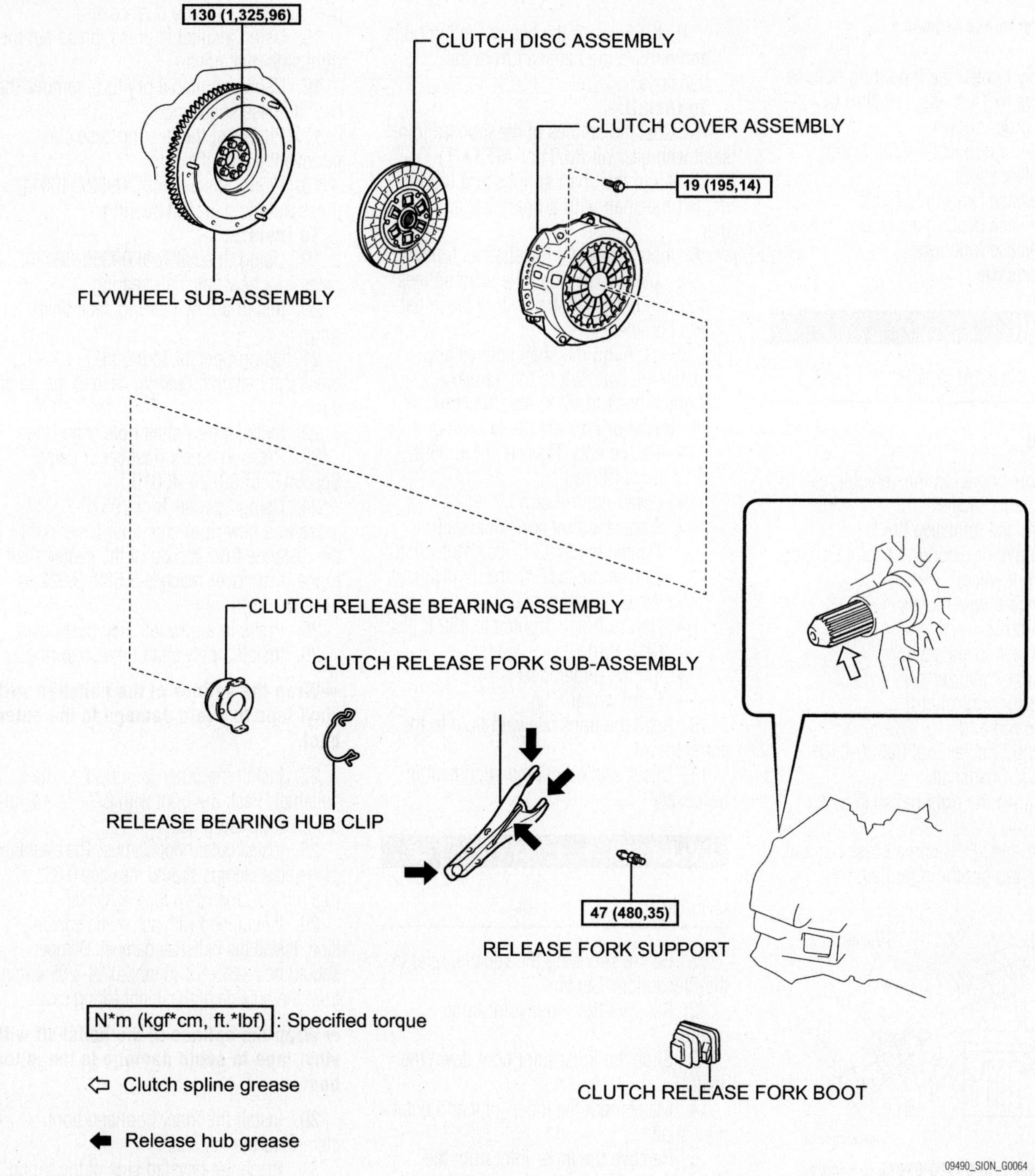

Exploded view of the clutch components.

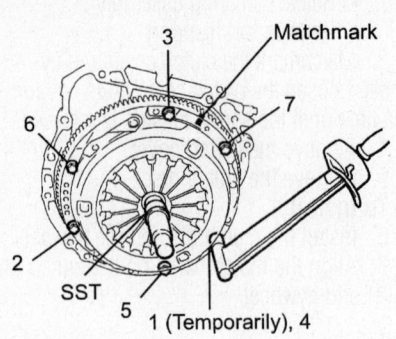

Clutch cover torque sequence

09490_SION_G0063

8. Evenly tighten the mounting bolts in several steps to 14 ft. lbs. (19 Nm) by following the order shown.

9. Install or connect the following:
- Release fork
- Release bearing hub clip
- Release bearing assembly
- Release fork boot
- Transaxle

Halfshaft

REMOVAL & INSTALLATION

xA and xB

1. Before servicing the vehicle, refer to the Precautions Section.

2. Drain the transaxle fluid.

3. Remove or disconnect the following:
- Front wheel
- Engine undercover
- Hub nut
- Wheel speed sensor
- Front stabilizer bar
- Lower control arm
- Tie rod end

4. Using Special Tool 09520-01010, tap out the left halfshaft.

5. Remove the right halfshaft as follows:

a. xA and xB: Using a brass bar and hammer, tap out the right halfshaft.

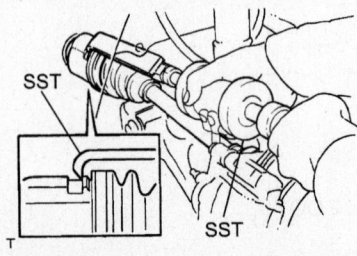

Use Special Tool 09520-01010 to remove the left halfshaft—xA and xB

09490_SION_G0068

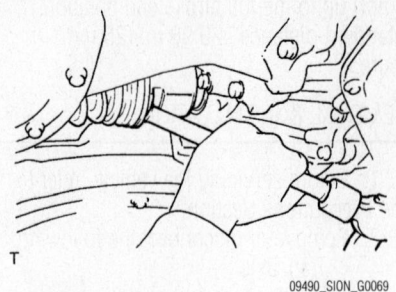

Tap out the right halfshaft with a brass bar and hammer—xA and xB

09490_SION_G0069

b. tC: Remove the two mounting bolts and remove the halfshaft from the transaxle.

To install:

6. Coat the splines of the inboard joint shaft with gear oil (M/T) or ATF (A/T).

7. Align the shaft splines and tap in the left halfshaft with a brass bar and hammer.

8. Install the right halfshaft as follows:

a. xA and xB: Align the shaft splines and tap in the halfshaft with a brass bar and hammer.

b. tC: Align the shaft splines and install the halfshaft to the transaxle. Tighten bolts to 47 ft. lbs. (64 Nm).

9. Install or connect the following:
- Tie rod end. Tighten nut to 36 ft. lbs. (49 Nm).
- Lower control arm
- Front stabilizer arm. xA and xB: Tighten nut to 13 ft. lbs. (18 Nm). tC: Tighten nut to 55 ft. lbs. (74 Nm).
- Wheel speed sensor
- New hub nut. Tighten to 159 ft. lbs. (216 Nm).
- Engine undercover
- Front wheel

10. Refill the transaxle with fluid to the correct level.

11. Check and adjust the alignment if necessary.

CV-Joints

OVERHAUL

1. Before servicing the vehicle, refer to the Precautions Section.

2. Remove the inner joint boot clamps.

3. Slide the inner joint boot down the shaft.

4. Matchmark the inner joint and outer joint shaft.

5. Remove the inner joint from the outer joint shaft.

6. Remove the shaft snap ring.

7. Matchmark the outer joint shaft and tripod joint.

8. Tap out the tripod joint from the outboard joint shaft.

9. Remove the inner joint boot.

10. If equipped with automatic transmission, remove the damper.

11. Remove the outer joint boot clamps.

12. Remove the outer joint boot.

13. Remove the shaft hole snap ring.

14. Using Special Tool 09950-00020, press out the left side dust cover.

15. Using a suitable press, press out the right side dust cover.

16. Using a suitable pry tool, remove the bearing case snap ring.

17. Press out the bearing case and remove the straight pin.

18. Using Special Tool, 09527-10011, press out the halfshaft bearing.

To install:

19. Using Special Tool 09950-60020, press in a new halfshaft bearing.

20. Install a new bearing case snap ring.

21. Using Special Tool 09527-10011, press the halfshaft bearing case to the inner joint.

22. Install a new shaft hole snap ring.

23. Press in a new dust cover using Special Tool 09726-40010.

24. Using Special Tool 09527-10011, press in a new right side dust cover until the distance from the tip of the center shaft to the dust cover reaches 3.582–3.622 in. (91–92 mm).

25. Press in a new left side dust cover.

26. Install a new shaft hole snap ring.

➡ **Wrap the splines of the halfshaft with vinyl tape to avoid damage to the outer boot.**

27. Install the outer joint boot to the halfshaft. Pack the boot with 6.7–7.1 oz. of the kit supplied grease.

28. Install outer boot clamps. The clearance of the boot clamps should measure 0.031 in. (0.8 mm) or less when fully tightened.

29. If equipped with automatic transmission, install the halfshaft damper. Damper should be 7.969–8.126 in. (202.4–206.4 mm) from the outside of the joint bearing case.

➡ **Wrap the splines of the halfshaft with vinyl tape to avoid damage to the inner boot.**

30. Install the inner boot and boot clamps to the halfshaft.

31. Place the beveled side of the tripod axial spline toward the outer joint.

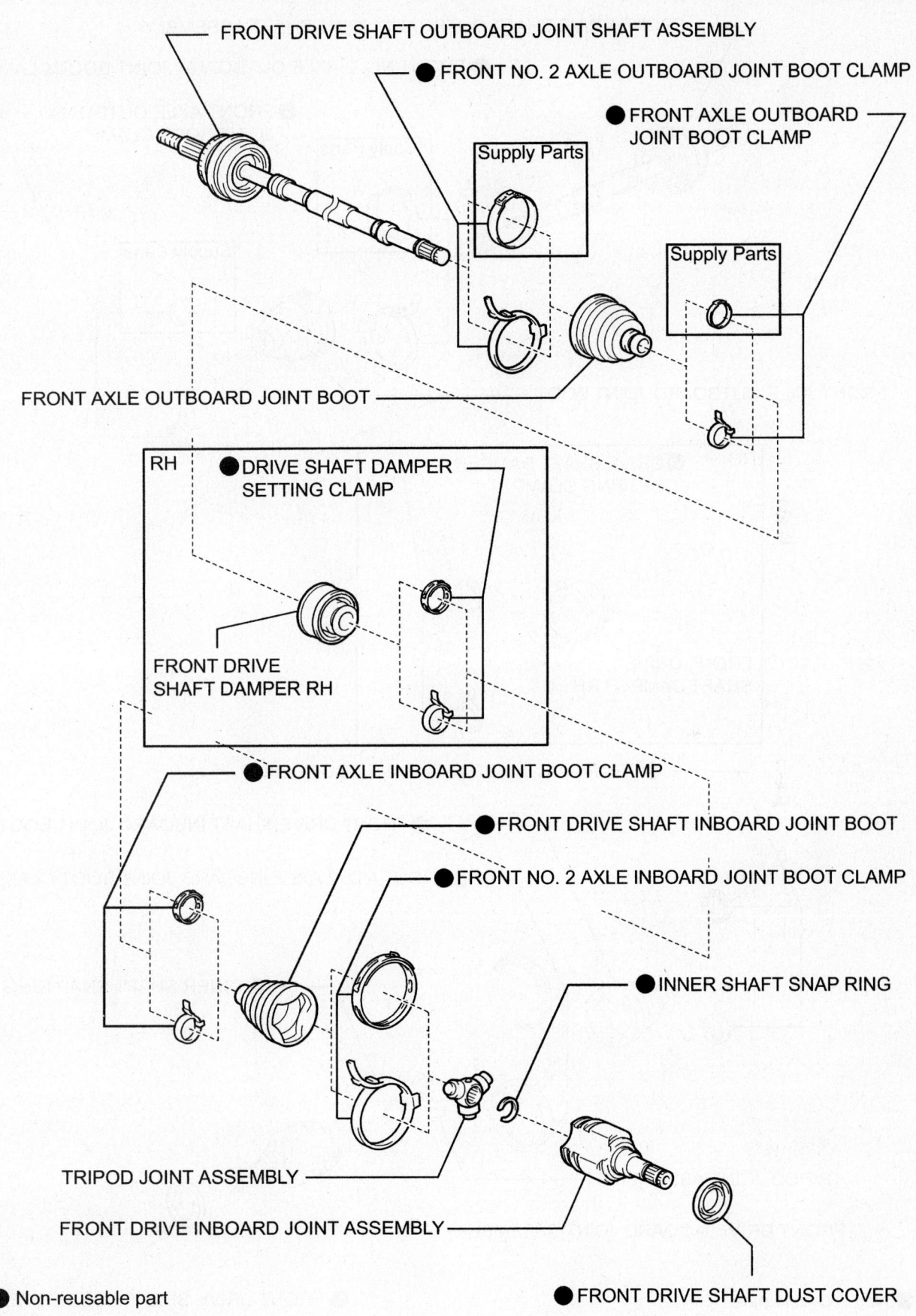

FRONT DRIVE SHAFT OUTBOARD JOINT SHAFT ASSEMBLY

●FRONT NO. 2 AXLE OUTBOARD JOINT BOOT CLAMP

●FRONT AXLE OUTBOARD JOINT BOOT CLAMP

Supply Parts

Supply Parts

FRONT AXLE OUTBOARD JOINT BOOT

RH

●DRIVE SHAFT DAMPER SETTING CLAMP

FRONT DRIVE SHAFT DAMPER RH

●FRONT AXLE INBOARD JOINT BOOT CLAMP

●FRONT DRIVE SHAFT INBOARD JOINT BOOT

●FRONT NO. 2 AXLE INBOARD JOINT BOOT CLAMP

●INNER SHAFT SNAP RING

TRIPOD JOINT ASSEMBLY

FRONT DRIVE INBOARD JOINT ASSEMBLY

● Non-reusable part

●FRONT DRIVE SHAFT DUST COVER

09490_SION_G0070

Exploded view of the halfshaft—tC

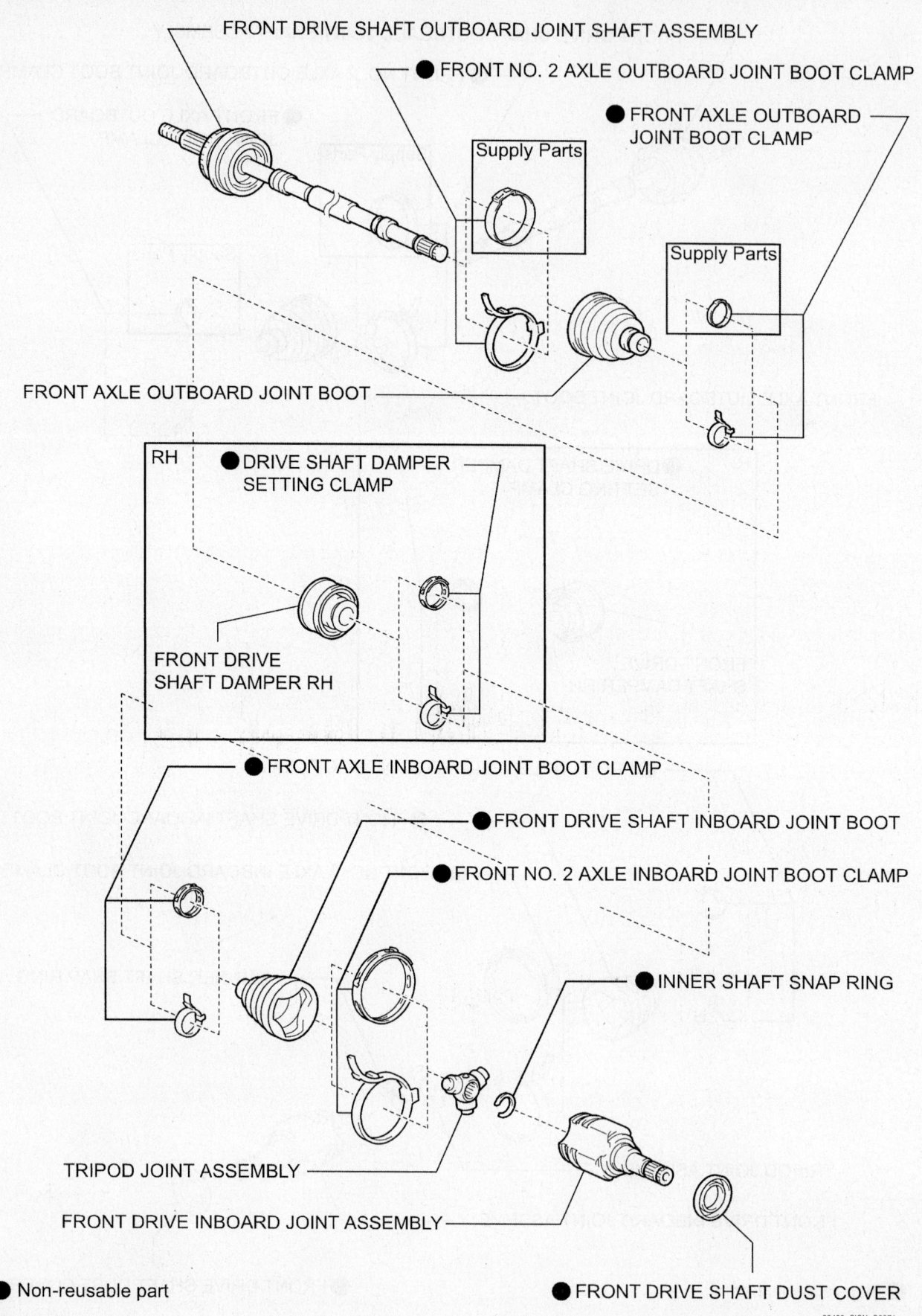

FRONT DRIVE SHAFT OUTBOARD JOINT SHAFT ASSEMBLY

● FRONT NO. 2 AXLE OUTBOARD JOINT BOOT CLAMP

● FRONT AXLE OUTBOARD JOINT BOOT CLAMP

Supply Parts

Supply Parts

FRONT AXLE OUTBOARD JOINT BOOT

RH ● DRIVE SHAFT DAMPER SETTING CLAMP

FRONT DRIVE SHAFT DAMPER RH

● FRONT AXLE INBOARD JOINT BOOT CLAMP

● FRONT DRIVE SHAFT INBOARD JOINT BOOT

● FRONT NO. 2 AXLE INBOARD JOINT BOOT CLAMP

● INNER SHAFT SNAP RING

TRIPOD JOINT ASSEMBLY

FRONT DRIVE INBOARD JOINT ASSEMBLY

● Non-reusable part

● FRONT DRIVE SHAFT DUST COVER

09490_SION_G0071

Exploded view of the halfshaft components—xA and xB

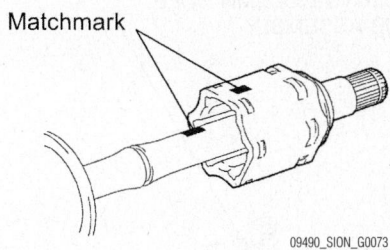

Matchmark the inner joint and joint shaft.

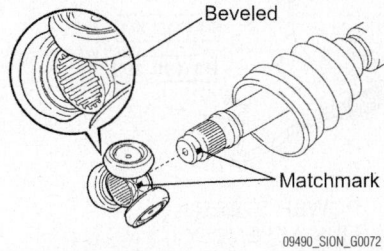

Matchmark the inner tripod joint to the shaft.

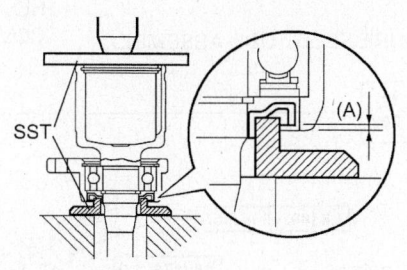

When installing the new dust cover, distance (A) should be less than 0.039 in. (1 mm).

32. Align the matchmarks made during removal.

33. Using a brass bar and hammer, tap the tripod joint onto the halfshaft.

34. Pack the inner boot with 6.2–6.5 oz. of the kit supplied grease.

35. Install a new shaft snap ring.

36. Align the matchmarks of the inner joint case to the shalfshaft.

37. Install the inner joint boot clamps.

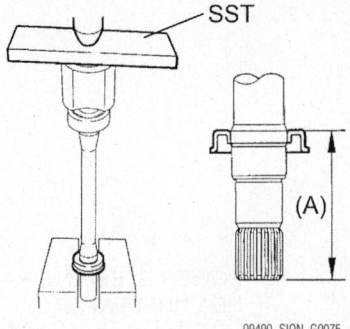

Ensure of the correct installation distance of the right side dust cover.

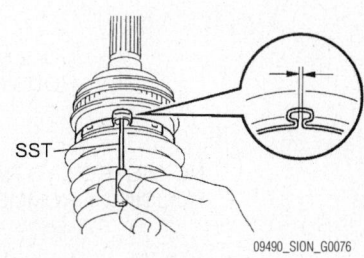

Boot clamp clearance should be less than 0.031 in. when installed.

STEERING AND SUSPENSION

Air Bag

✳ CAUTION

Some vehicles are equipped with an air bag system. The system must be disarmed before performing service on, or around, system components, the steering column, instrument panel components, wiring and sensors. Failure to follow the safety precautions and the disarming procedure could result in accidental air bag deployment, possible injury and unnecessary system repairs.

DISARMING

Disconnect and isolate the negative battery cable. Wait 90 seconds for the system capacitor to discharge before performing any service.

Rack and Pinion Steering Gear

REMOVAL & INSTALLATION

xA and xB

1. Before servicing the vehicle, refer to the Precautions Section.

2. Face the front wheels straight ahead.

3. Drain the power steering fluid.

4. Remove or disconnect the following:
   - Wiper arms
   - Top cowl ventilator louvers
   - Wiper motor and linkage
   - Steering column hole cover plate

5. Matchmark the steering sliding yoke and intermediate shaft.

6. Disconnect the sliding yoke from the intermediate shaft.

7. Remove or disconnect the following:
   - Hood
   - Front wheels
   - Tie rod ends
   - Engine undercover
   - Power steering hoses
   - Lower control arms

8. At this point, attach an engine crane to support the engine.

9. Remove or disconnect the following:
   - Front suspension crossmember
   - Power steering rack heat insulator
   - Intermediate shaft
   - Steering gear assembly attached to the front crossmember.

10. Remove the steering rack clamps and remove the steering gear from the crossmember.

To install:

11. Install the steering gear to the crossmember. Tighten to the nuts 54 ft. lbs. (74 Nm).

12. Install or connect the following:
   - Intermediate shaft
   - Power steering rack heat insulator. Tighten to 26 ft. lbs. (35 Nm).
   - Front suspension crossmember. Tighten the front outside bolts to 52 ft. lbs. (70 Nm). Tighten the rear outside bolts to 86 ft. lbs. (116 Nm). Tighten to inside nuts to 53 ft. lbs. (72 Nm).
   - Lower control arms
   - Power steering hoses
   - Tie rod ends
   - Front wheels
   - Sliding yoke to the intermediate shaft
   - Hood
   - Engine undercover
   - Steering column hole cover plate
   - Wiper motor and linkage
   - Top cowl ventilator louvers
   - Wiper arms

13. Refill the power steering reservoir to the correct level.

14. Bleed the air from the power steering system.

15. Start the engine and check for leaks.

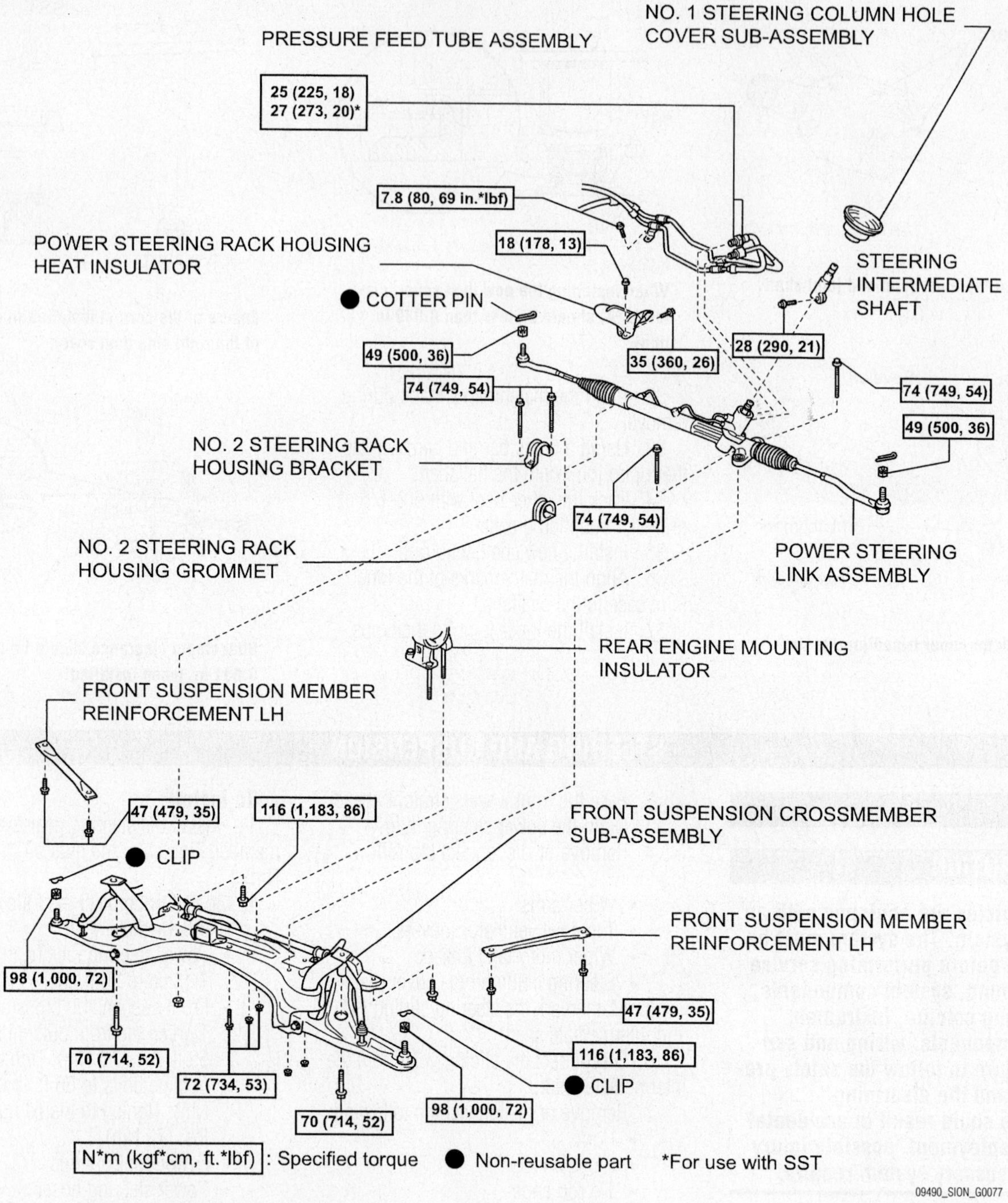

PRESSURE FEED TUBE ASSEMBLY

NO. 1 STEERING COLUMN HOLE
COVER SUB-ASSEMBLY

25 (225, 18)
27 (273, 20)*

7.8 (80, 69 in.*lbf)

POWER STEERING RACK HOUSING
HEAT INSULATOR

18 (178, 13)

STEERING
INTERMEDIATE
SHAFT

● COTTER PIN

49 (500, 36)

35 (360, 26)

28 (290, 21)

74 (749, 54)

74 (749, 54)

49 (500, 36)

NO. 2 STEERING RACK
HOUSING BRACKET

74 (749, 54)

NO. 2 STEERING RACK
HOUSING GROMMET

POWER STEERING
LINK ASSEMBLY

REAR ENGINE MOUNTING
INSULATOR

FRONT SUSPENSION MEMBER
REINFORCEMENT LH

47 (479, 35)

116 (1,183, 86)

FRONT SUSPENSION CROSSMEMBER
SUB-ASSEMBLY

● CLIP

FRONT SUSPENSION MEMBER
REINFORCEMENT LH

98 (1,000, 72)

47 (479, 35)

70 (714, 52)

72 (734, 53)

116 (1,183, 86)

70 (714, 52)

98 (1,000, 72)

● CLIP

N*m (kgf*cm, ft.*lbf) : Specified torque   ● Non-reusable part   *For use with SST

09490_SION_G0077

Exploded view of the steering linkage components—xA

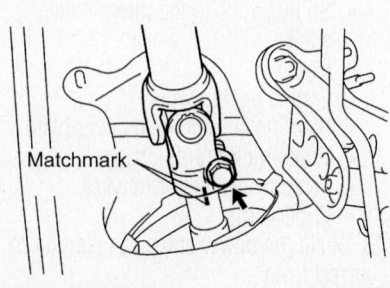

Matchmark

09490_SION_G0078

Place matchmarks on the yoke and inter-mediate shaft.

16. Check and adjust the alignment if necessary.

tC

1. Before servicing the vehicle, refer to the Precautions Section.
2. Face the front wheels straight ahead.
3. Drain the power steering fluid.
4. Remove or disconnect the following:
   • Steering column hole cover plate
   • Intermediate steering shaft
   • Front wheel

• Engine undercover
• Tie rod ends
• Power steering hoses
• Front stabilizer links
• Lower control arms
• Front floor panel brace
• Center exhaust pipe
• Hood
• Engine appearance cover

5. Attach an engine crane to support the engine.
6. Remove or disconnect the following:

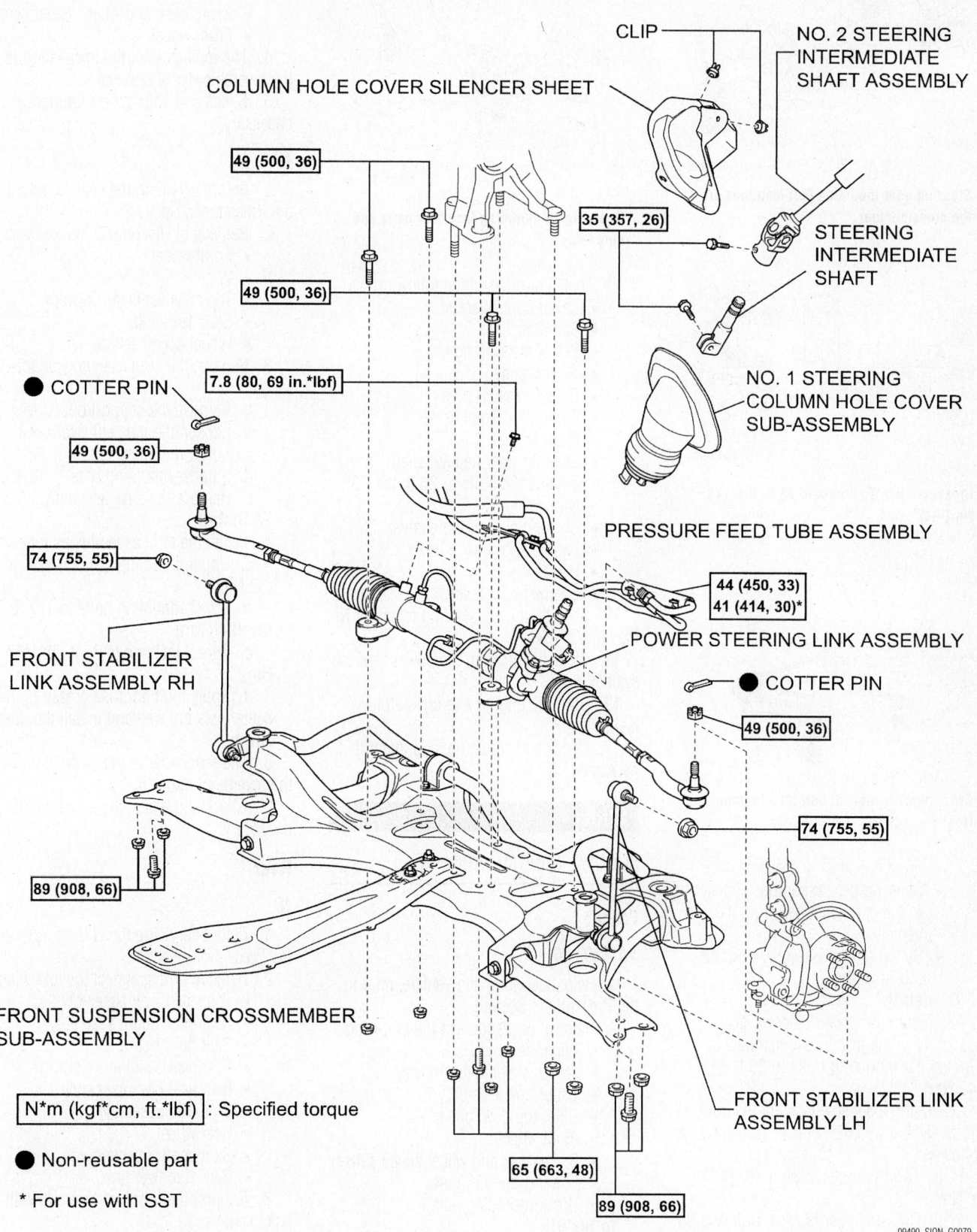

COLUMN HOLE COVER SILENCER SHEET

CLIP

NO. 2 STEERING INTERMEDIATE SHAFT ASSEMBLY

49 (500, 36)

35 (357, 26)

STEERING INTERMEDIATE SHAFT

49 (500, 36)

7.8 (80, 69 in.*lbf)

● COTTER PIN

NO. 1 STEERING COLUMN HOLE COVER SUB-ASSEMBLY

49 (500, 36)

74 (755, 55)

PRESSURE FEED TUBE ASSEMBLY

44 (450, 33)
41 (414, 30)*

FRONT STABILIZER LINK ASSEMBLY RH

POWER STEERING LINK ASSEMBLY

● COTTER PIN

49 (500, 36)

74 (755, 55)

89 (908, 66)

FRONT SUSPENSION CROSSMEMBER SUB-ASSEMBLY

N*m (kgf*cm, ft.*lbf) : Specified torque

● Non-reusable part

* For use with SST

FRONT STABILIZER LINK ASSEMBLY LH

65 (663, 48)

89 (908, 66)

09490_SION_G0079

Exploded view of the power steering assembly—tC

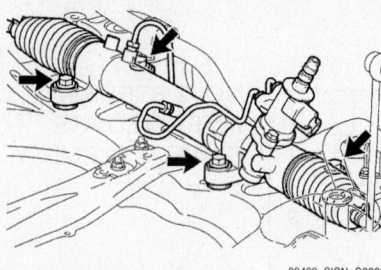

Steering gear mounting bolt locations on the crossmember

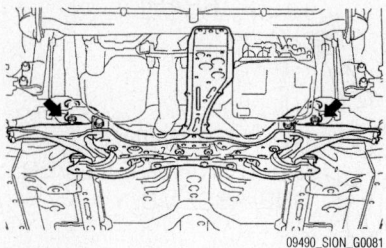

Crossmember front nuts to 98 ft. lbs. (133 Nm)—tC

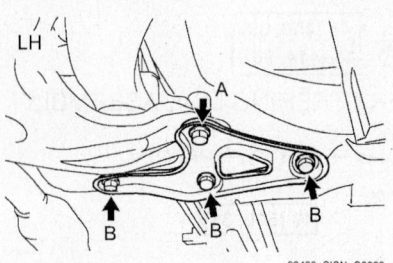

Crossmember bracket bolt identification, right-hand similar—tC

- Front suspension crossmember
- Steering gear assembly, attached to the crossmember, from the vehicle
- Steering gear assembly from the crossmember

To install:

7. Install the power steering gear assembly to the front crossmember. Tighten the mounting bolts to 36 ft. lbs. (49 Nm).

8. Using a suitable jack, lift the front crossmember into the vehicle. Tighten as follows:

a. Two front nuts to 98 ft. lbs (133 Nm)

b. Crossmember bracket: Bolt A to 98 ft. lbs. (133 Nm). Bolts B to 59 ft. lbs. (80 Nm)

c. Rear engine mounting insulator bolt and nuts to 48 ft. lbs. (65 Nm).

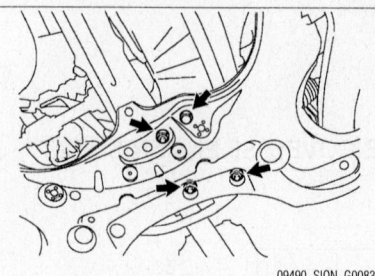

Rear engine mounting insulator bolts and nuts—tC

9. Install or connect the following:
- Lower control arms
- Stabilizer links
- Power steering hoses
- Tie rod ends
- Center exhaust pipe
- Front floor panel brace
- Front wheels
- Steering intermediate shaft. Tighten the bolts to 26 ft. lbs. (35 Nm).
- Steering column hole cover
- Engine appearance cover
- Hood
- Engine undercover

10. Refill the power steering reservoir to the correct level.

11. Bleed the air from the power steering system.

12. Start the engine and check for leaks.

13. Check and adjust the alignment if necessary.

Struts

REMOVAL & INSTALLATION

Front

xA AND xB

1. Before servicing the vehicle, refer to the Precautions Section.

2. Remove or disconnect the following:
- Wiper arms
- Top cowl ventilator louvers
- Wiper link assembly
- Outer top cowl panel
- Front wheel
- Brake hose and wheel speed sensor
- Lower mounting bolts
- Upper mounting nuts

To install:

3. Install or connect the following:
- Strut assembly. Tighten the upper mounting nuts to 29 ft. lbs. (39 Nm). Lower mounting bolts to 97 ft. lbs. (132 Nm).
- Brake hose and wheel speed sensor
- Front wheel

4. The remainder of the installation is the reverse order of removal.

5. Check and adjust the alignment if necessary.

tC

1. Before servicing the vehicle, refer to the Precautions Section.

2. Remove or disconnect the following:
- Front wheel
- Wiper arms
- Top cowl ventilator louvers
- Stabilizer links
- Wheel speed sensor

3. Remove the strut assembly as follows:
a. Remove the support dust cover
b. Loosen the top center lock nut
c. Lower mounting nuts
d. 3 upper mounting nuts
e. Remove the strut assembly

To install:

4. Install the strut assembly as follows:
a. Upper mounting nuts to 38 ft. lbs. (52 Nm)
b. Lower mounting bolts to 177 ft. lbs. (240 Nm)
c. Center lock nut to 35 ft. lbs. (47 Nm)
d. Apply multipurpose grease to the center lock nut well and install the dust cover.

5. The remainder of the installation is the reverse of removal.

6. Check and adjust the alignment if necessary.

Rear

tC

1. Before servicing the vehicle, refer to the Precautions Section.

2. Remove or disconnect the following:
- Tonneau cover assembly
- Rear deck board assembly
- Rear floor board
- Rear seat cushion assembly
- Rear seat back assembly
- Side trim assembly
- Rear wheel
- Skid control sensor wire
- Rear stabilizer link

3. Support the lower control arm with a suitable jack.

4. Remove or disconnect the following:
- Strut lower mounting bolt
- Upper mounting nuts
- Lower suspension brace

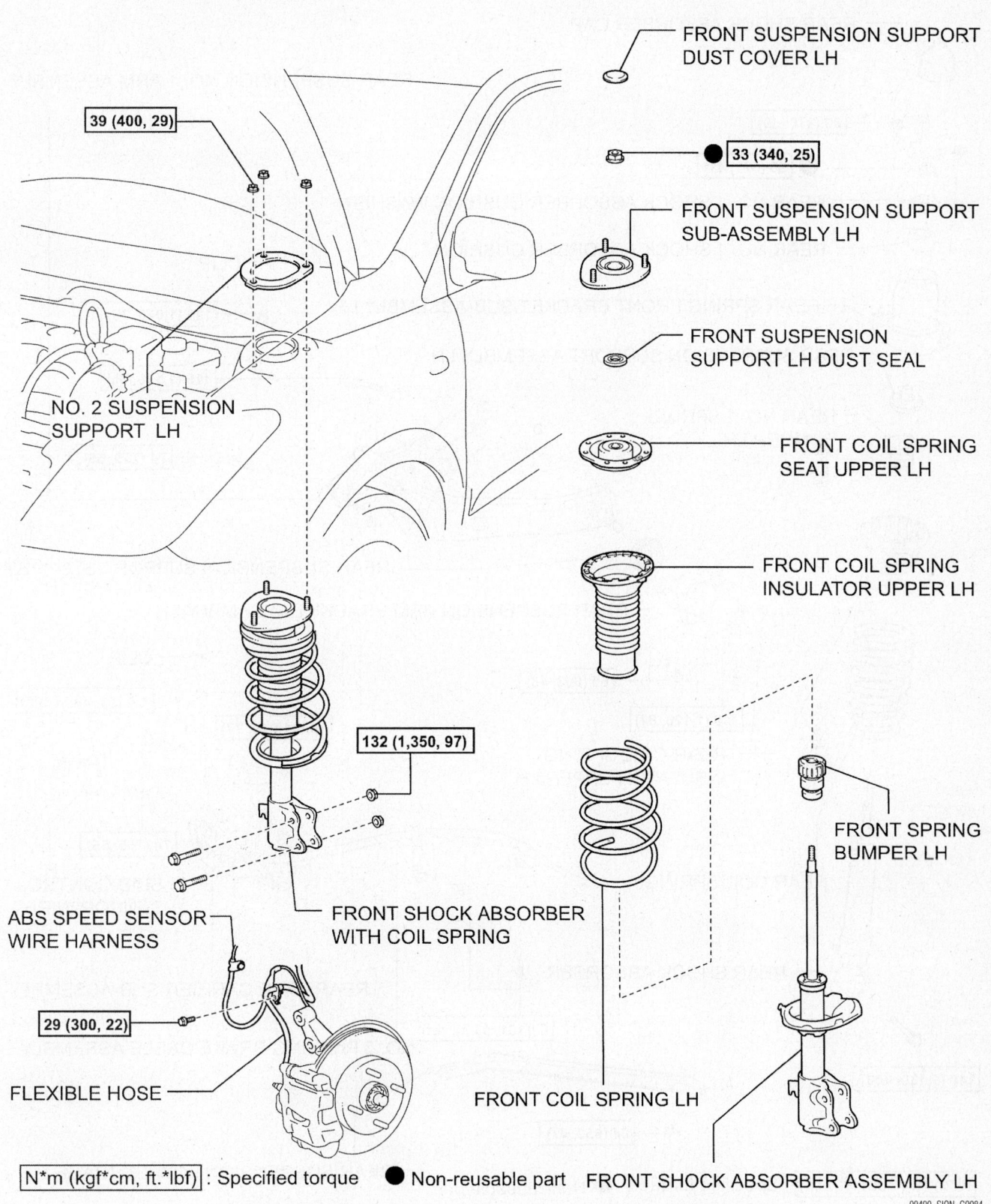

FRONT SUSPENSION SUPPORT
DUST COVER LH

39 (400, 29)

● 33 (340, 25)

FRONT SUSPENSION SUPPORT
SUB-ASSEMBLY LH

FRONT SUSPENSION
SUPPORT LH DUST SEAL

NO. 2 SUSPENSION
SUPPORT LH

FRONT COIL SPRING
SEAT UPPER LH

FRONT COIL SPRING
INSULATOR UPPER LH

132 (1,350, 97)

FRONT SPRING
BUMPER LH

ABS SPEED SENSOR
WIRE HARNESS

FRONT SHOCK ABSORBER
WITH COIL SPRING

29 (300, 22)

FLEXIBLE HOSE

FRONT COIL SPRING LH

N*m (kgf*cm, ft.*lbf) : Specified torque   ● Non-reusable part

FRONT SHOCK ABSORBER ASSEMBLY LH

09490_SION_G0084

Exploded view of the front strut components—xA and xB

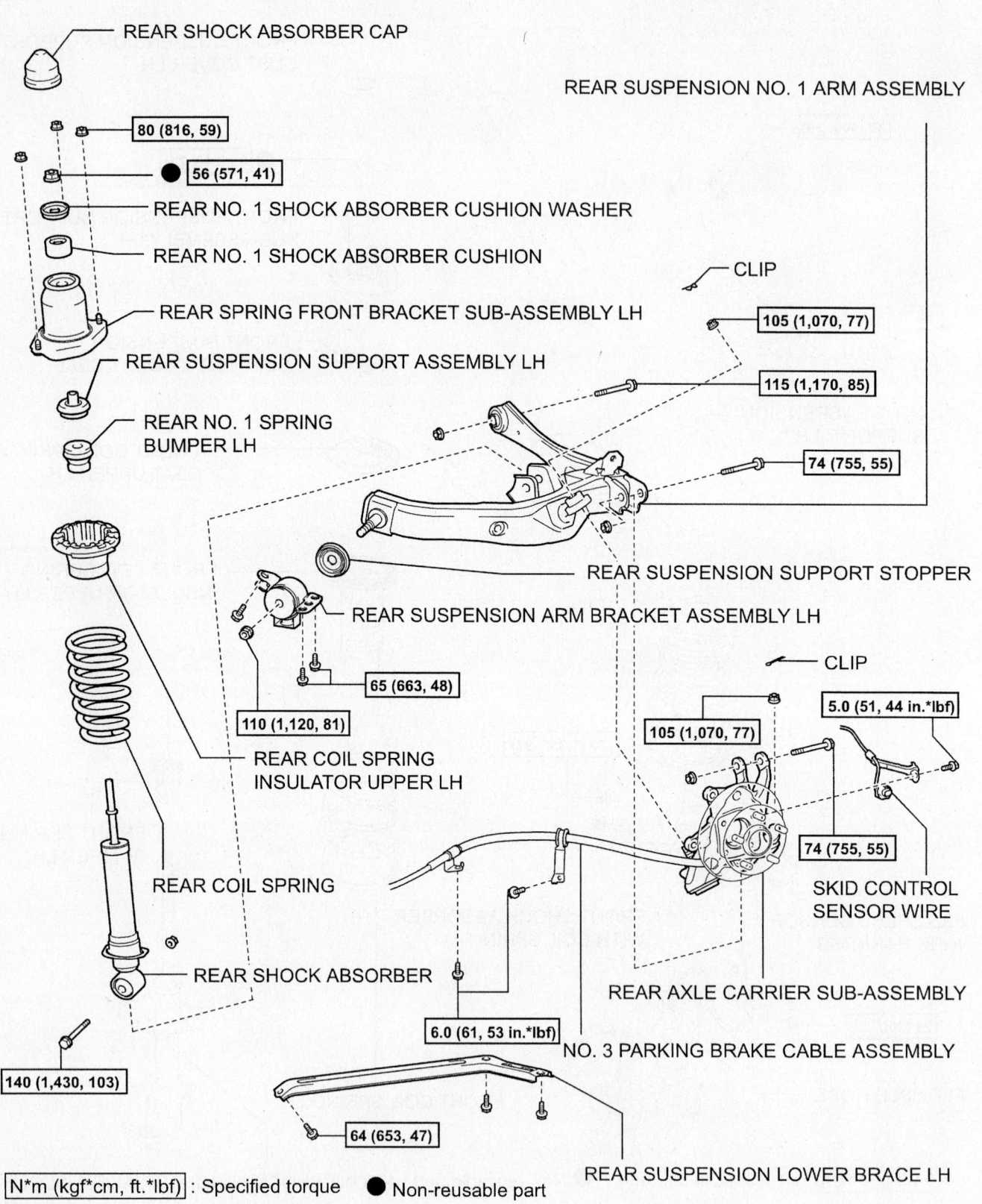

REAR SHOCK ABSORBER CAP

REAR SUSPENSION NO. 1 ARM ASSEMBLY

80 (816, 59)

● 56 (571, 41)

REAR NO. 1 SHOCK ABSORBER CUSHION WASHER

REAR NO. 1 SHOCK ABSORBER CUSHION

CLIP

REAR SPRING FRONT BRACKET SUB-ASSEMBLY LH

105 (1,070, 77)

REAR SUSPENSION SUPPORT ASSEMBLY LH

115 (1,170, 85)

REAR NO. 1 SPRING
BUMPER LH

74 (755, 55)

REAR SUSPENSION SUPPORT STOPPER

REAR SUSPENSION ARM BRACKET ASSEMBLY LH

CLIP

65 (663, 48)

5.0 (51, 44 in.*lbf)

110 (1,120, 81)

105 (1,070, 77)

REAR COIL SPRING
INSULATOR UPPER LH

74 (755, 55)

REAR COIL SPRING

SKID CONTROL
SENSOR WIRE

REAR SHOCK ABSORBER

REAR AXLE CARRIER SUB-ASSEMBLY

6.0 (61, 53 in.*lbf)

NO. 3 PARKING BRAKE CABLE ASSEMBLY

140 (1,430, 103)

64 (653, 47)

REAR SUSPENSION LOWER BRACE LH

N*m (kgf*cm, ft.*lbf) : Specified torque ● Non-reusable part

09490_SION_G0085

Exploded view of the rear suspension components—tC

- Parking brake cable

5. Lower the jack and remove the strut assembly.

To install:

6. Install the strut assembly. Tighten the upper mounting nuts to 59 ft. lbs. (80 Nm). Temporarily tighten the lower mounting bolt.

7. Install or connect the following:
- Parking brake cable
- Lower suspension brace. Tighten to 47 ft. lbs. (64 Nm).
- Rear stabilizer link
- Skid control sensor wire
- Rear wheel
- Lower strut mounting bolt. Fully tighten to 103 ft. lbs. (140 Nm).

8. The remainder of the installation is the reverse order of removal.

9. Check and adjust the rear wheel alignment if necessary.

Coil Spring

REMOVAL & INSTALLATION

Front

xA AND xB

1. Before servicing the vehicle, refer to the Precautions Section.

2. Remove the strut from the vehicle and install a spring compressor.

3. Compress the coil spring so that the end of the spring comes away from the spring seat.

4. Remove or disconnect the following:
- Top center dust cover
- Upper strut nut
- Upper spring seat
- Dust seal
- Coil spring insulator
- Compressed spring from the strut
- Spring from the spring compressor

To install:

5. Compress the spring and install it on the strut.

6. Install or connect the following:
- Coil spring insulator
- Dust seal
- Upper spring seat and the upper strut mount. Torque the nut to 25 ft. lbs. (33 Nm)
- Strut to the vehicle

7. Check and/or adjust the wheel alignment.

tC

1. Before servicing the vehicle, refer to the Precautions Section.

2. Remove the strut from the vehicle and install a Spring Compressor Tool.

3. Compress the coil spring so that the end of the spring comes away from the spring seat.

4. Remove or disconnect the following:
- Lock nut
- Suspension support
- Dust seal
- Upper seat
- Coil spring insulator
- Coil spring

To install:

5. Install the lower coil spring insulator to the shock absorber so both recessed parts are aligned.

6. Compress the coil spring with Spring Compress tool.

7. Install or connect the following:
- Coil spring to the strut. Fit the lower end of the spring into the recessed part of the lower spring seat.
- Upper seat, dust seal and lock nut.

8. Install the strut assembly.

Rear

xA AND xB

1. Before servicing the vehicle, refer to the Precautions Section.

2. Remove or disconnect the following:
- Rear wheel
- Skid control sensor wire
- Brake hose
- Parking brake cable

3. Support the rear axle with a suitable jack stand.

4. Loosen the rear axle beam assembly.

5. Remove the lower shock absorber mounting nut.

6. Remove the coil spring.

To install:

7. Install the rear coil spring.

8. Temporarily install the lower shock absorber mounting nut.

9. Install or connect the following:
- Parking brake cable
- Brake hose
- Skid control sensor wire

10. Lower the rear suspension.

11. Tighten the rear axle beam assembly to 60 ft. lbs. (82 Nm).

12. Fully tighten the lower shock absorber nut to 36 ft. lbs (49 Nm).

13. Install the wheel.

14. Check and adjust the rear wheel alignment if necessary.

tC

1. Before servicing the vehicle, refer to the Precautions Section.

2. Remove the strut from the vehicle and install a Spring Compressor Tool.

3. Compress the coil spring so that the end of the spring comes away from the spring seat.

4. Remove or disconnect the following:
- Upper mounting nut
- Shock cushion washer
- Suspension support
- Front spring bracket
- Spring bumper
- Upper spring insulator
- Coil spring

To install:

5. Compress the coil spring with Spring Compress tool.

6. Install the coil spring on the strut assembly.

7. Install or connect the following:
- Spring bumper
- Suspension support
- Upper spring insulator
- Shock cushion washer
- Upper mounting nut. Tighten to 41 ft. lbs. (56 Nm)

Lower Ball Joint

REMOVAL & INSTALLATION

tC

1. Before servicing the vehicle, refer to the Precautions Section.

2. Remove or disconnect the following:
- Front wheel
- Hub nut
- Wheel speed sensor
- Tie rod end
- Lower control arm mounting bolts

3. Disconnect the halfshaft from the axle hub.

4. Remove the lower shock absorber mounting bolts.

5. Using Special Tool 09628-62011 or suitable puller, remove the lower ball joint from the steering knuckle.

To install:

➡**Use a new split pin for assembly.**

6. Install the ball joint and torque the mounting nut to 76 ft. lbs. (103 Nm)

7. Install the halfshaft to the front axle. Tighten the steering knuckle mounting bolts to 76 ft. Lbs (103 Nm).

8. Install or connect the following :
- Lower control arm mounting bolts. Tighten to 66 ft. lbs. (89 Nm).

- Tie rod ends. Tighetn to 36 ft. lbs. (49 Nm).
- Hub nut. Tighten to 159 ft. lbs. (216 Nm).
- Front wheel

9. Check and/or adjust the wheel alignment.

Lower Control Arm

REMOVAL & INSTALLATION

xA and xB

1. Before servicing the vehicle, refer to the Precautions Section.
2. Remove or disconnect the following:
   - Front wheel
   - Hood
3. Attach an engine crane to support the engine.
4. Disconnect the lower control arm from the steering knuckle.
5. Disconnect the front stabilizer bar from the lower control arm.
6. Disconnect the power steering gear assembly.
7. Disconnect the power steering hoses from the control arm.
8. Support the front crossmember with a suitable jack.
9. Disconnect the front suspension crossmember.
10. Remove the lower control arm from the front suspension crossmember.

To install:

11. Install the lower control arm to the crossmember and temporarily tighten the bolts.
12. Using a suitable jack, lift the front crossmember into the vehicle. Tighten as follows:
   a. Two front nuts to 98 ft. lbs (133 Nm)
   b. Crossmember bracket: Bolt A to 98 ft. lbs. (133 Nm). Bolts B to 59 ft. lbs. (80 Nm)
   c. Rear engine mounting insulator bolt and nuts to 48 ft. lbs. (65 Nm).
13. Install or connect the following:
   - Power steering hoses
   - Steering gear assembly
   - Front stabilizer bar. Tighten to 13 ft. lbs. (18 Nm).
   - Lower control arm to the steering knuckle. Tighten to 72 ft. lbs. (98 Nm).
   - Front wheel
14. Fully tighten the lower control arm mounting on the crossmember to 97 ft. lbs. (132 Nm).

tC

1. Before servicing the vehicle, refer to the Precautions Section.
2. Face the front wheels straight ahead.
3. Drain the power steering fluid.
4. Remove or disconnect the following:
   - Steering column hole cover plate
   - Intermediate steering shaft
   - Front wheel
   - Engine undercover
   - Tie rod ends
   - Power steering hoses
   - Front stabilizer links
   - Lower control arms
   - Front floor panel brace
   - Center exhaust pipe
   - Hood
   - Engine appearance cover
5. Attach an engine crane to support the engine.
6. Remove or disconnect the following:
   - Front suspension crossmember
   - Lower control arm from the crossmember

To install:

7. Install the lower control arm to the crossmember, and temporarily tighten the bolts.
8. Using a suitable jack, lift the front crossmember into the vehicle. Tighten as follows:
   a. Two front nuts to 98 ft. lbs (133 Nm)
   b. Crossmember bracket: Bolt A to 98 ft. lbs. (133 Nm). Bolts B to 59 ft. lbs. (80 Nm)
   c. Rear engine mounting insulator bolt and nuts to 48 ft. lbs. (65 Nm).
9. Install or connect the following:
   - Lower control arms to the steering knuckle. Tighten to 66 ft. lbs. (89 Nm).
   - Stabilizer links. Tighten to 55 ft. lbs. (74 Nm).
   - Power steering hoses
   - Tie rod ends
   - Center exhaust pipe
   - Front floor panel brace
   - Front wheels
   - Steering intermediate shaft. Tighten the bolts to 26 ft. lbs. (35 Nm).
   - Steering column hole cover
   - Engine appearance cover
   - Hood
   - Engine undercover
10. Start the engine and check for leaks.
11. Check and adjust the alignment if necessary.

Wheel Bearings

REMOVAL & INSTALLATION

1. Before servicing the vehicle, refer to the Precautions Section.
2. Remove or disconnect the following:
   - Front wheel
   - Hub nut
   - Wheel speed sensor
   - Brake caliper
   - Front disc
   - Tie rod ends
   - Lower control arm
   - Axle from the hub assembly
   - Steering knuckle from strut
   - Lower ball joint
   - Axle hub snap ring
3. Mount the steering knuckle assembly in a vise.
4. Using Special Tool 09520-00031 to

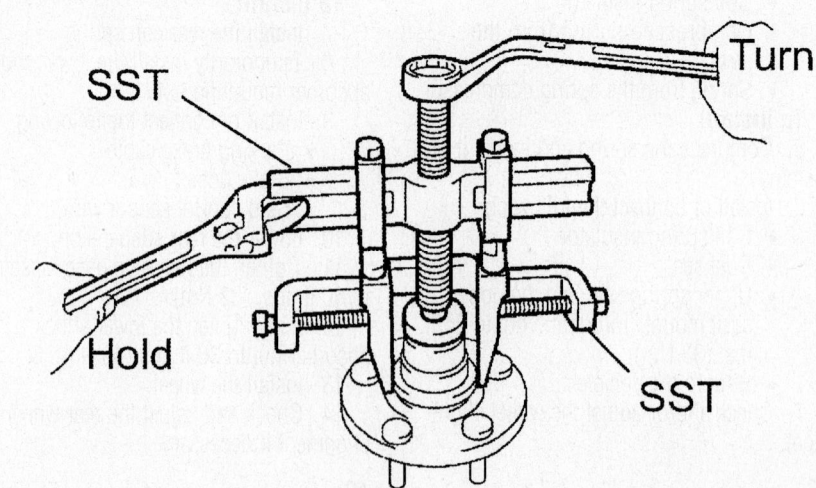

Use Special Tool 09950-40011 to remove the wheel bearing from the hub—tC

09490_SION_G0088

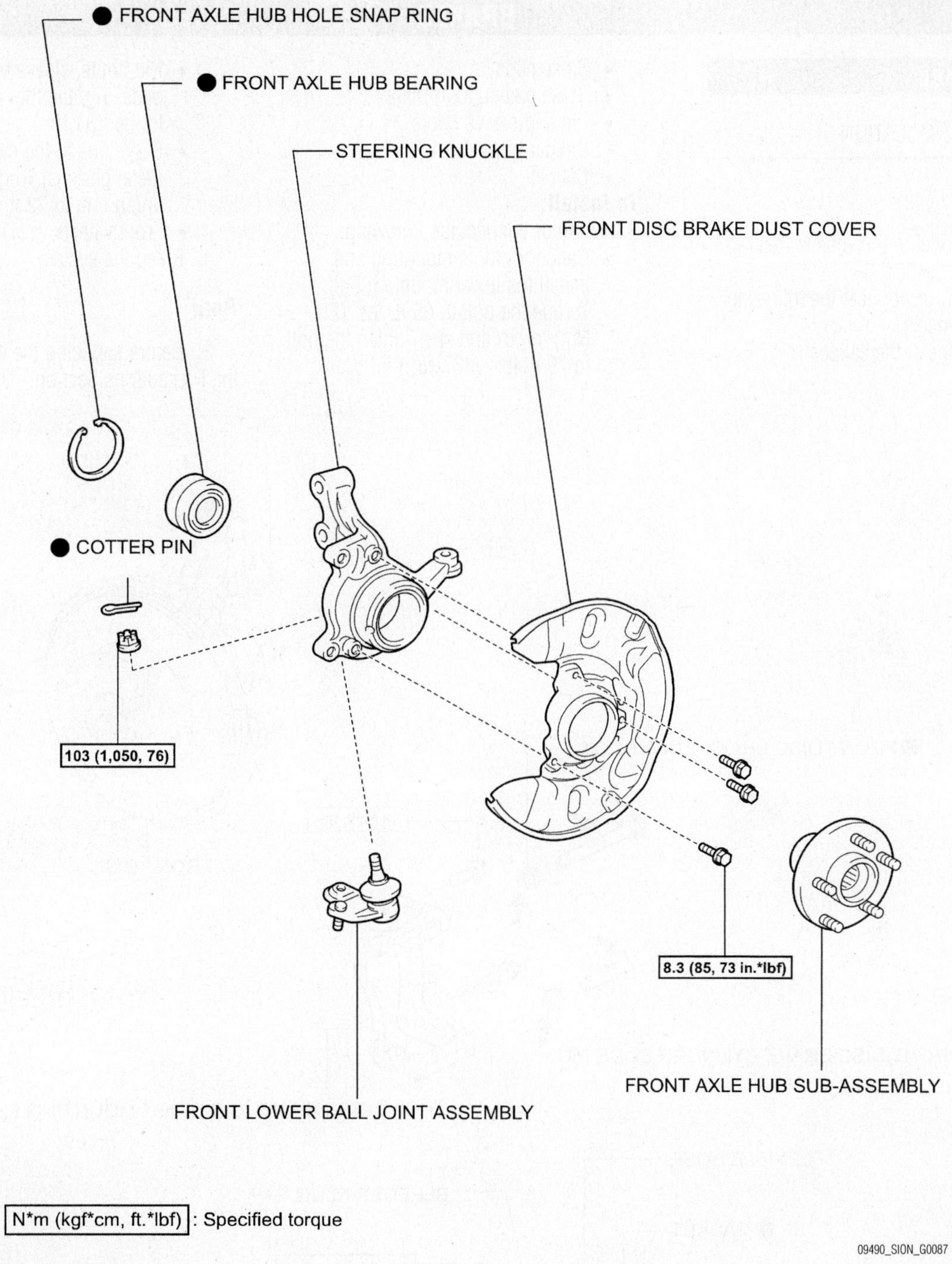

● FRONT AXLE HUB HOLE SNAP RING

● FRONT AXLE HUB BEARING

STEERING KNUCKLE

FRONT DISC BRAKE DUST COVER

● COTTER PIN

103 (1,050, 76)

8.3 (85, 73 in.*lbf)

FRONT LOWER BALL JOINT ASSEMBLY

FRONT AXLE HUB SUB-ASSEMBLY

N*m (kgf*cm, ft.*lbf) : Specified torque

09490_SION_G0087

Exploded view of the hub components—tC

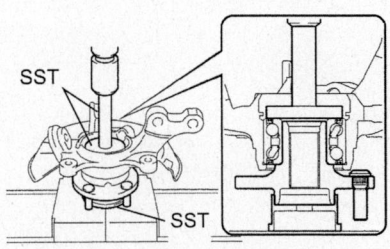

SST

SST

09490_SION_G0089

Press the axle hub assembly into the steering knuckle.

remove the axle hub from the steering knuckle.

5. Using Special Tool 09950-40011, remove the wheel bearing from the axle hub.

To install:

6. Use Special Tool 09950-60020 to press a new wheel bearing into the axle hub.

7. Using Special Tool 09608-32010, press the axle hub assembly into the steering knuckle.

8. Install or connect the following:
   • Axle hub snap ring
   • Lower ball joint
   • Steering knuckle mounting bolts
   • Axle into the hub assembly
   • Lower control arm
   • Tie rod ends
   • Front disc
   • Brake caliper
   • Wheel speed sensor
   • Hub nut
   • Front wheel

BRAKES

Brake Caliper

REMOVAL & INSTALLATION

Front

1. Before servicing the vehicle, refer to the Precautions Section.
2. Remove or disconnect the following:
   - Front wheels
   - Brake line at the caliper
   - Brake pads
   - Brake pad support plate
   - Pin and sleeve boots
   - Caliper bolts
   - Caliper

To install:

3. Install or connect the following:
   - Caliper onto its mounting and install the lower mounting bolt. Torque the bolt to 65 ft. lbs. (88 Nm) for xA and xB. Tighten the bolt to 79 ft. lbs. (107 Nm) for tC.
   - Pin boots, sleeve boots and brake pads. Tighten the slide pins to 25 ft. lbs. (34 Nm).
   - Brake line to the caliper with 2 new metal gaskets. Torque the brake line union bolt to 22 ft. lbs. (30 Nm).
   - Front wheels
4. Bleed the system.

Rear

1. Before servicing the vehicle, refer to the Precautions Section.

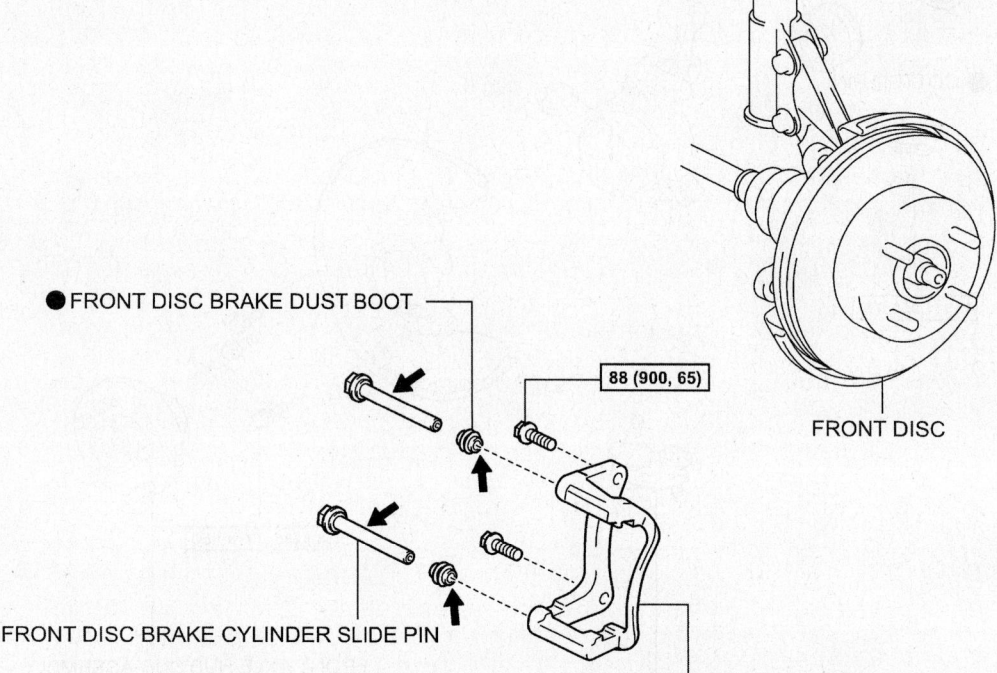

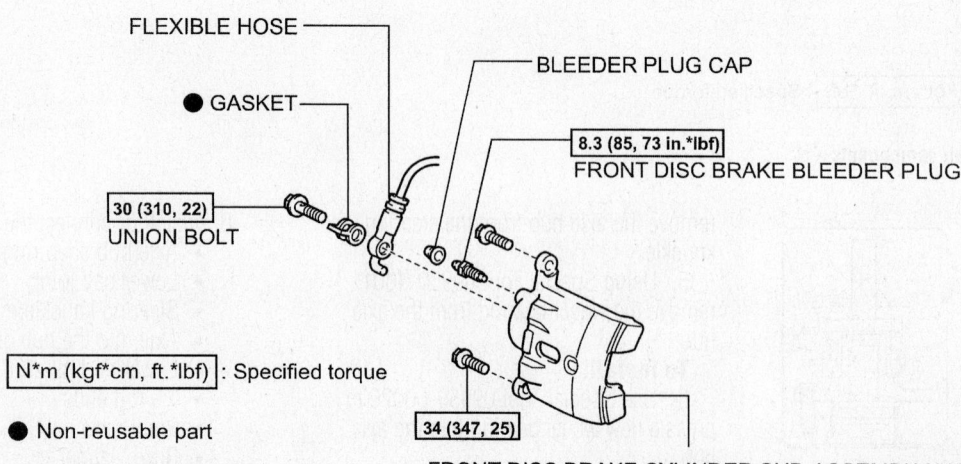

- FRONT DISC BRAKE DUST BOOT

88 (900, 65)

FRONT DISC

FRONT DISC BRAKE CYLINDER SLIDE PIN

FRONT DISC BRAKE CYLINDER MOUNTING LH

FLEXIBLE HOSE

BLEEDER PLUG CAP

- GASKET

8.3 (85, 73 in.*lbf)
FRONT DISC BRAKE BLEEDER PLUG

30 (310, 22)
UNION BOLT

N*m (kgf*cm, ft.*lbf) : Specified torque

- Non-reusable part

← Lithium soap base glycol grease

34 (347, 25)
FRONT DISC BRAKE CYLINDER SUB-ASSEMBLY LH

09490_SION_G0090

Exploded view of the brake components—Front

2. Remove or disconnect the following:
- Wheel
- Brake hose
- Caliper assembly mounting bolts
- Caliper

To install:

3. Install or connect the following:
- Caliper. Tighten the mounting bolts to 29 ft. lbs. (39 Nm).
- Brake hose
- Wheel

Disc Brake Pads

REMOVAL & INSTALLATION

1. Before servicing the vehicle, refer to the Precautions Section.

2. Remove or disconnect the following:
- Front wheels
- Slide pins and boots

- Brake pads from the caliper mounting

To install:

3. Install or connect the following:
- Brake pads to the caliper mounting
- Pin boots, sleeve boots and brake pads. Tighten the slide pins to 25 ft. lbs. (34 Nm) for Front. Tighten to 29 ft. lbs. (39 Nm) for Rear brakes.
- Wheels

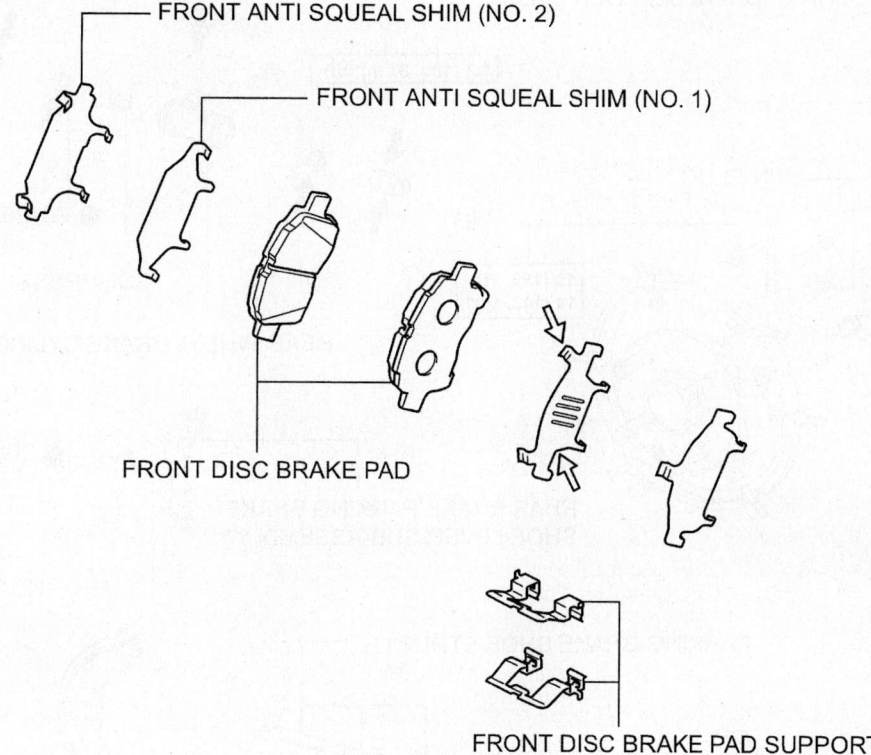

FRONT ANTI SQUEAL SHIM (NO. 2)

FRONT ANTI SQUEAL SHIM (NO. 1)

FRONT DISC BRAKE PAD

FRONT DISC BRAKE PAD SUPPORT PLATE

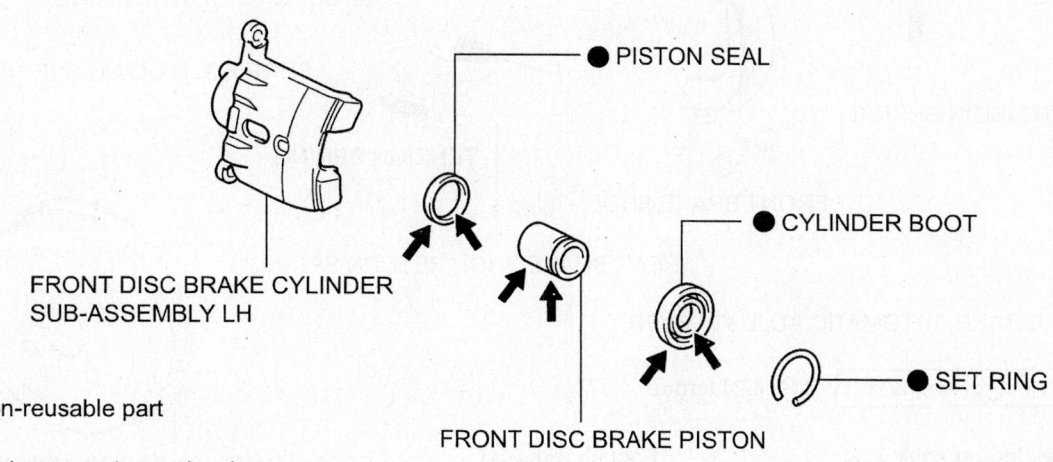

● PISTON SEAL

● CYLINDER BOOT

FRONT DISC BRAKE CYLINDER SUB-ASSEMBLY LH

● SET RING

FRONT DISC BRAKE PISTON

● Non-reusable part

◄ Lithium soap base glycol grease

◁ Disc brake grease

09490_SION_G0091

Exploded view of the brake pad components—Front

Brake Drums

REMOVAL & INSTALLATION

1. Before servicing the vehicle, refer to the Precautions Section.

2. Remove or disconnect the following:
- Wheels
- Brake drum

➡ If drum cannot be removed easily, remove the pin plug, insert a screwdriver into the hole on the backing plate, and hold the automatic adjust lever away from the adjuster. Using another screwdriver, compress the brake shoe adjuster by turning the adjusting wheel.

To install:

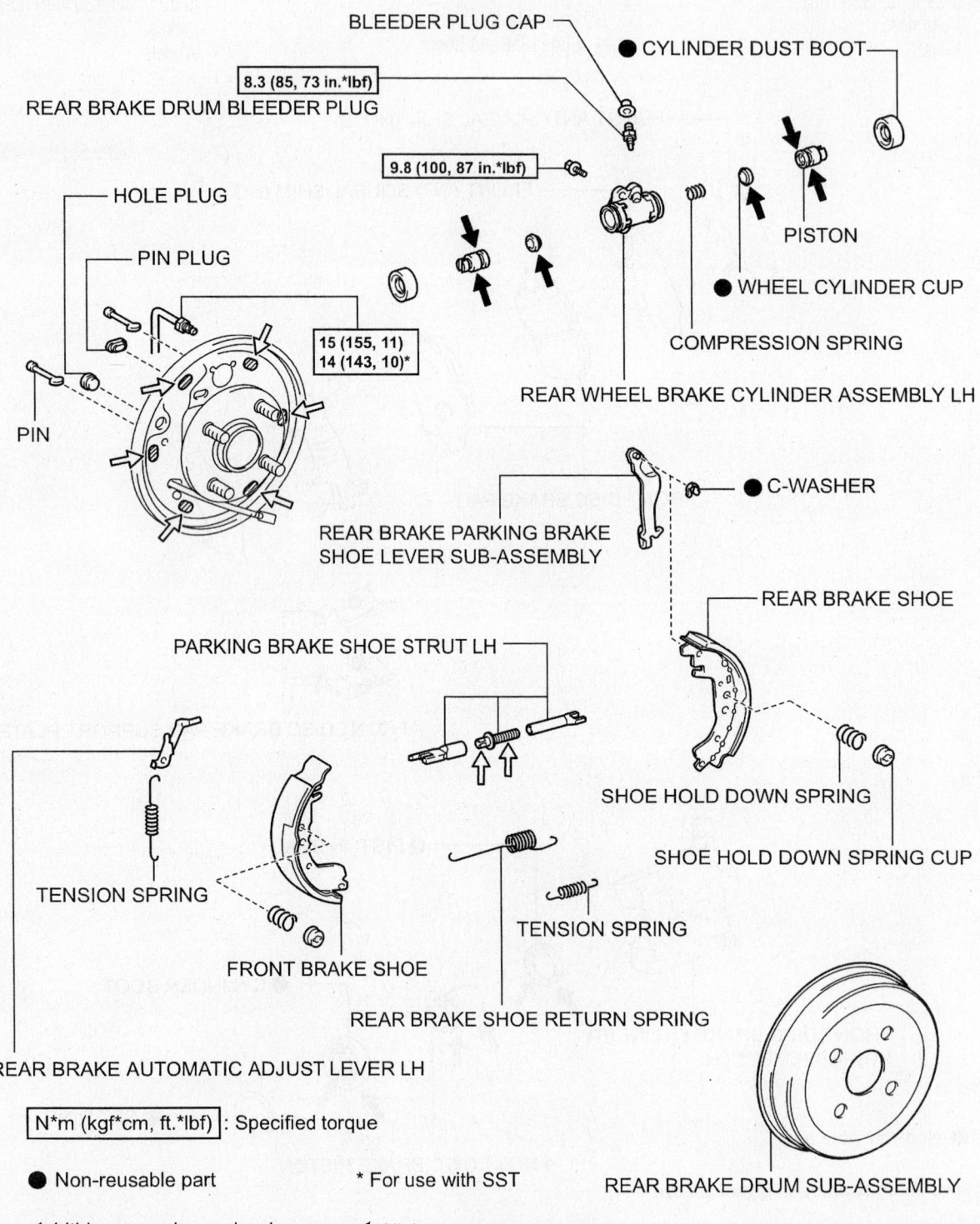

BLEEDER PLUG CAP

8.3 (85, 73 in.*lbf)

REAR BRAKE DRUM BLEEDER PLUG

9.8 (100, 87 in.*lbf)

● CYLINDER DUST BOOT

PISTON

● WHEEL CYLINDER CUP

COMPRESSION SPRING

REAR WHEEL BRAKE CYLINDER ASSEMBLY LH

HOLE PLUG

PIN PLUG

15 (155, 11)
14 (143, 10)*

PIN

● C-WASHER

REAR BRAKE PARKING BRAKE SHOE LEVER SUB-ASSEMBLY

REAR BRAKE SHOE

PARKING BRAKE SHOE STRUT LH

SHOE HOLD DOWN SPRING

SHOE HOLD DOWN SPRING CUP

TENSION SPRING

TENSION SPRING

FRONT BRAKE SHOE

REAR BRAKE SHOE RETURN SPRING

REAR BRAKE AUTOMATIC ADJUST LEVER LH

N*m (kgf*cm, ft.*lbf) : Specified torque

● Non-reusable part   * For use with SST

REAR BRAKE DRUM SUB-ASSEMBLY

◄ Lithium soap base glycol grease   ◁ High temperature grease

09490_SION_G0092

Exploded view of the rear drum brake system

3. Install or connect the following:
- Brake drum
- Wheel

Brake Shoes

REMOVAL & INSTALLATION

1. Before servicing the vehicle, refer to the Precautions Section.
2. Remove or disconnect the following:

- Wheels
- Brake drum
- Shoe return spring from the front and rear brake shoe.
- Front shoe hold down spring cup, shoe hold down spring and pin
- Front brake shoe and tension spring
- Parking brake shoe strut from the rear brake shoe
- Adjust lever tension spring and automatic adjust lever

- Rear shoe hold down spring cup, shoe hold down spring and pin
- Parking brake cable from parking brake shoe lever
- C-washer and rear brake shoe.

To install:

3. Apply high temperature grease to the contact surfaces of the backing plate.
4. Install or connect the following:
- Parking brake lever with a new C-washer

LH   RH

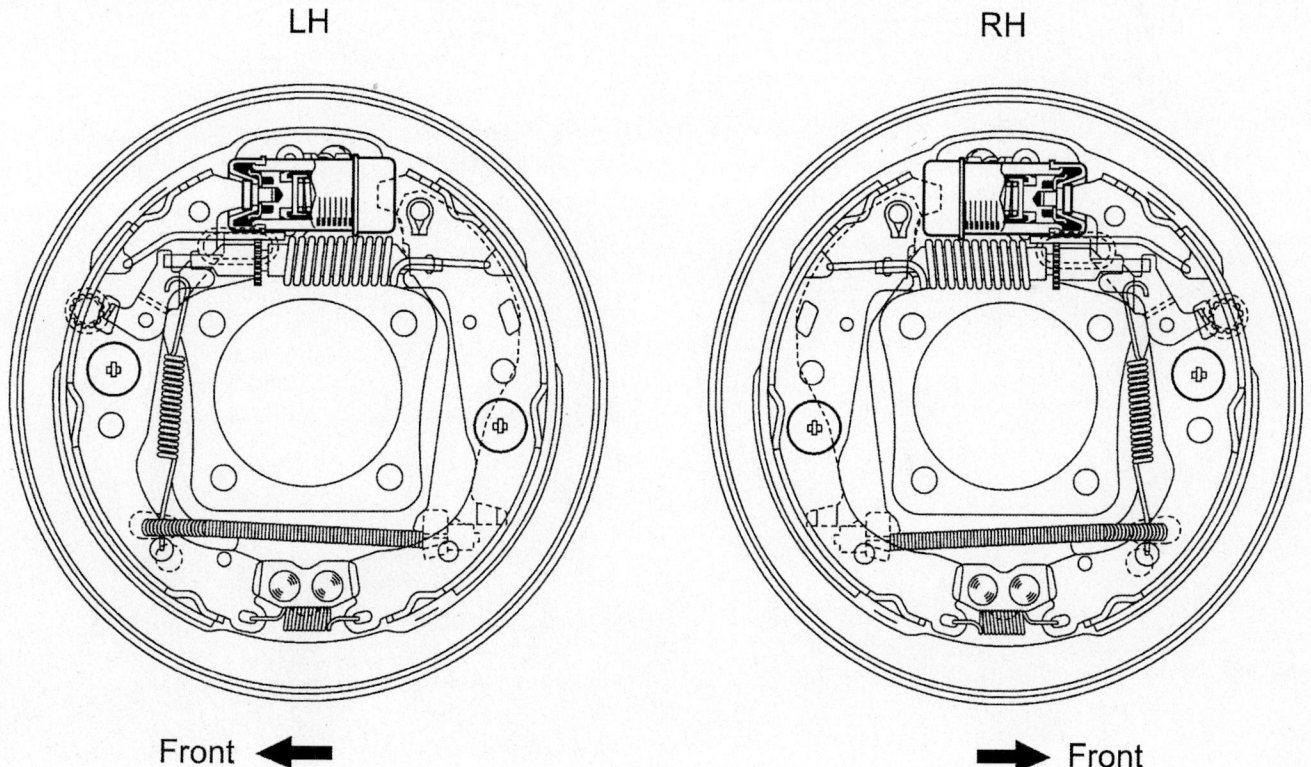

Front ⬅   ➡ Front

09490_SION_G0093

Proper installation of Brake Drum components

- Rear brake shoe, pin, shoe hold down spring and shoe hold down spring cup
- Automatic adjust lever and automatic lever tension spring to front brake shoe

- Shoe return spring to the shoe strut
- Tension spring to the front and rear brake shoe
- Front brake shoe, pin, shoe hold down spring and shoe hold down spring cup

- Tension spring to the front and rear brake shoe
- Brake drum

5. Adjust the brake shoes
- Install the wheels.

SUBARU

Baja • Impreza • Impreza Outback • Impreza Outback Sport • Legacy •
Legacy Outback • Legacy SUS • WRX

BRAKES **6-150**
DRIVE TRAIN **6-103**
ENGINE REPAIR **6-28**
FRONT SUSPENSION **6-128**
FUEL SYSTEM **6-99**
REAR SUSPENSION **6-139**
SPECIFICATION AND
MAINTENANCE CHARTS **6-3**
Engine and Vehicle Identification
  Chart.................................6-3
General Engine Specifications 6-4
Engine Tune-Up Specifications........6-5
Firing Order 6-7
Accessory Drive Belt Routing 6-7
Capacities 6-8
Valve Specifications..................6-9
Camshaft Specifications Chart.......6-10
Crankshaft and Connecting
  Rod Specifications.................6-11
Piston and Ring
  Specifications 6-12
Torque Specifications 6-13
Wheel Alignment 6-22
Tire, Wheel and Ball Joint
  Specifications 6-24
Brake Specifications 6-25
Scheduled Maintenance
  Intervals............................6-27
STEERING **6-124**
A
Air Bag..............................6-124
  Disarming 6-124
  Precautions 6-124
Alternator 6-28
  Removal & Installation.............6-28
Automatic Transmission 6-106
  Removal & Installation...........6-106
B
Brake Caliper 6-150
  Removal & Installation.............6-150
Brake Drums 6-160
  Removal & Installation.............6-160

Brake Pads...........................6-159
  Removal & Installation.............6-159
Brake Shoes..........................6-161
  Removal & Installation.............6-161
C
Camshaft and Valve Lifters 6-59
  Inspection 6-68
  Removal & Installation.............6-59
Clutch 6-110
  Adjustment........................6-110
  Removal & Installation.............6-111
CV-Joints.............................6-122
  Overhaul 6-122
Cylinder Head 6-42
  Removal & Installation.............6-42
E
Engine Assembly 6-28
  Removal & Installation.............6-28
Exhaust Manifold 6-58
  Removal & Installation.............6-58
F
Front Crankshaft Seal 6-59
  Removal & Installation.............6-59
Front Lower Control Arm............6-136
  Control Arm Bushing
    Replacement.....................6-138
  Removal & Installation.............6-136
Fuel Filter 6-99
  Removal & Installation.............6-99
Fuel Injector.........................6-101
  Removal & Installation...........6-101
Fuel Pump 6-99
  Removal & Installation.............6-99
Fuel System Pressure 6-99
  Relieving 6-99
Fuel System Service
  Precautions........................6-99
H
Halfshaft.............................6-115
  Removal & Installation.............6-115
Heater Core.........................6-34
  Removal & Installation.............6-34

Hydraulic Clutch System 6-114
  Bleeding............................6-114
I
Ignition Timing 6-28
  Adjustment........................6-28
Intake Manifold 6-50
  Removal & Installation.............6-50
L
Lower Ball Joint.....................6-135
  Removal & Installation...........6-135
M
Manual Transmission 6-103
  Removal & Installation...........6-103
O
Oil Pan..............................6-77
  Removal & Installation.............6-77
Oil Pump 6-80
  Removal & Installation.............6-80
P
Piston and Ring 6-97
  Positioning 6-97
Power Steering Gear 6-124
  Removal & Installation...........6-124
R
Rear Lower Control Arm 6-145
  Control Arm Bushing
    Replacement.....................6-147
  Removal & Installation...........6-145
Rear Main Seal 6-82
  Removal & Installation.............6-82
Rocker Arms/Shafts 6-49
  Removal & Installation.............6-49
S
Shock Absorber 6-144
  Removal & Installation...........6-144
Stabilizer Bar (Front)................6-145
  Removal & Installation...........6-145
Stabilizer Bar (Rear)................6-135
  Removal & Installation...........6-135
Starter Motor 6-77
  Removal & Installation.............6-77

Strut (Front Suspension) 6-128
    Disassembly & Assembly 6-134
    Removal & Installation...........6-128
Strut (Rear Suspension)...............6-139
    Disassembly & Assembly 6-139
    Removal & Installation...........6-139

T

Timing Belt, Cover and Crankshaft
seal...6-82
    Removal & Installation.............6-82

Timing Chain, Sprockets, Front
Cover and Seal 6-93
    Removal & Installation..............6-93
Transfer Case Assembly..............6-115
    Removal & Installation...........6-115
Turbocharger.................................6-59
    Removal & Installation..............6-59

V

Valve Lash 6-68
    Adjustment.................................6-68

W

Water Pump...................................6-33
    Removal & Installation..............6-33
Wheel Bearings
(Front Suspension).....................6-138
    Adjustment...............................6-138
    Removal & Installation...........6-138
Wheel Bearings
(Rear Suspension).....................6-148
    Adjustment...............................6-148
    Removal & Installation...........6-148

SPECIFICATION AND MAINTENANCE CHARTS

ENGINE AND VEHICLE IDENTIFICATION CHART

Engine Code							Model Year	
Code ①	Liters (cc)	Cu. In.	Cyl.	Fuel Sys.	Type	Eng. Mfg.	Code ②	Year
2	2.0 (1994)	121	4	MFI	DOHC	Subaru	2	2002
6	2.5 (2457)	150	4	MFI	DOHC	Subaru	3	2003
6	2.5 (2457)	150	4	MFI	SOHC	Subaru	4	2004
7	2.5 (2457)	150	4	MFI	DOHC	Subaru	5	2005
8	3.0 (3000)	183	6	MFI	DOHC	Subaru	6	2006

MFI: Multiport Fuel Injection

SOHC: Single Overhead Camshaft

DOHC: Double Overhead Camshaft

① 6th digit of the VIN

② 10th digit of the VIN

09490_SBCR_C0001

GENERAL ENGINE SPECIFICATIONS

Year	Model		Engine Displacement Liters (VIN)	Net Horsepower @ rpm	Net Torque @ rpm (ft. lbs.)	Bore x Stroke (in.)	Compression Ratio	Oil Pressure @ rpm
2002	Impreza		2.5 (6)	165@5600	166@4000	3.92x3.11	10.0:1	14@600
	WRX		2.0 (2)	227@6000	217@4000	3.62x2.95	8.0:1	14@800
	Outback		2.5 (6)	165@5600	166@4000	3.92x3.11	10.0:1	14@600
	Outback		3.0 (8)	212@6000	210@4400	3.51x3.15	10.7:1	14@600
	Legacy		2.5 (6)	165@5600	166@4000	3.92x3.11	10.0:1	14@600
2003	Impreza		2.5 (6)	165@5600	166@4000	3.92x3.11	10.0:1	14@600
	WRX		2.0 (2)	227@6000	217@4000	3.62x2.95	8.0:1	14@800
	Outback		2.5 (6)	165@5600	166@4000	3.92x3.11	10.0:1	14@600
	Outback		3.0 (8)	212@6000	210@4400	3.51x3.15	10.7:1	14@600
	Baja		2.5 (6)	165@5600	166@4000	3.92x3.11	10.0:1	14@600
	Legacy		2.5 (6)	165@5600	166@4000	3.92x3.11	10.0:1	14@600
2004	Impreza		2.5 (6)	165@5600	166@4000	3.92x3.11	10.0:1	14@600
	Impreza	③	2.5 (7)	230@5600	235@3600	3.92x3.11	8.4:1	14@600
	WRX		2.0 (2)	227@6000	217@4000	3.62x2.95	8.0:1	14@800
	Outback		2.5 (6)	165@5600	166@4000	3.92x3.11	10.0:1	14@600
	Outback		3.0 (8)	212@6000	210@4400	3.51x3.15	10.7:1	14@600
	Baja	①	2.5 (6)	165@5600	166@4000	3.92x3.11	10.0:1	14@600
	Baja	②	2.5 (6)	210@5600	235@3600	3.92x3.11	8.2:1	14@600
	Legacy		2.5 (6)	165@5600	166@4000	3.92x3.11	10.0:1	14@600
2005	Impreza		2.5 (6)	173@6000	166@4400	3.92x3.11	10.0:1	14@600
	Impreza	③	2.5 (7)	230@5600	235@3600	3.92x3.11	8.4:1	14@600
	WRX	④	2.5 (7)	300@6000	300@4000	3.92x3.11	8.2:1	14@600
	Outback	①	2.5 (6)	175@6000	169@4400	3.92x3.11	10.0:1	14@600
	Outback	②	2.5 (6)	250@6000	250@3600	3.92x3.11	8.2:1	14@600
	Outback		3.0 (8)	250@6600	219@4400	3.51x3.15	10.7:1	20@600
	Baja	①	2.5 (6)	165@5600	166@4000	3.92x3.11	10.0:1	14@600
	Baja	②	2.5 (6)	210@5600	235@3600	3.92x3.11	8.2:1	14@600
	Legacy	①	2.5 (6)	175@6000	169@4400	3.92x3.11	10.0:1	14@600
	Legacy	②	2.5 (6)	250@6000	250@3600	3.92x3.11	8.2:1	14@600
2006	Impreza		2.5 (6)	173@6000	166@4400	3.92x3.11	10.0:1	14@600
	Impreza	③	2.5 (7)	230@5600	235@3600	3.92x3.11	8.4:1	14@600
	WRX	④	2.5 (7)	300@6000	300@4000	3.92x3.11	8.2:1	14@600
	Outback	①	2.5 (6)	175@6000	169@4400	3.92x3.11	10.0:1	14@600
	Outback	②	2.5 (6)	250@6000	250@3600	3.92x3.11	8.2:1	14@600
	Outback		3.0 (8)	250@6600	219@4400	3.51x3.15	10.7:1	20@600
	Baja	①	2.5 (6)	165@5600	166@4000	3.92x3.11	10.0:1	14@600
	Baja	②	2.5 (6)	210@5600	235@3600	3.92x3.11	8.2:1	14@600
	Legacy	①	2.5 (6)	175@6000	169@4400	3.92x3.11	10.0:1	14@600
	Legacy	②	2.5 (6)	250@6000	250@3600	3.92x3.11	8.2:1	14@600

① SOHC

② DOHC/Turbo

③ Impreza WRX sedan and sport wagon

④ STI

ENGINE TUNE-UP SPECIFICATIONS

Year	Liters (VIN)	Engine Displacement Liters (VIN)	Spark Plug Gap (in.)	Ignition Timing (deg.) ①		Fuel Pump (psi)	Idle Speed (rpm)		Valve Clearance ②	
				MT	AT		MT	AT	In.	Ex.
2002	Impreza	2.5 (6)	0.039-0.043	7-23	7-23	34-38	600-800	600-800	0.0071-0.0087	0.0090-0.0106
	WRX	2.0 (2)	0.028-0.031	2-22	2-22	33-38	650-850	650-850	0.0071-0.0087	0.0090-0.0106
	Outback/SUS ③	2.5 (6)	0.039-0.043	7-23	7-23	34-38	600-800	600-800	0.0071-0.0087	0.0090-0.0106
	Outback/SUS ③	3.0 (8)	0.039-0.043	2-18 ④	2-18 ④	34-38	600	600	0.0063-0.0095	0.0078-0.0118
	Legacy	2.5 (6)	0.039-0.043	7-23	7-23	34-38	600-800	600-800	0.0071-0.0087	0.0090-0.0106
2003	Impreza	2.5 (6)	0.039-0.043	7-23	7-23	34-38	600-800	600-800	0.0071-0.0087	0.0090-0.0106
	WRX	2.0 (2)	0.028-0.031	2-22	2-22	33-38	650-850	650-850	0.0071-0.0087	0.0090-0.0106
	Outback/SUS ③	2.5 (6)	0.039-0.043	7-23	7-23	34-38	600-800	600-800	0.0071-0.0087	0.0090-0.0106
	Outback/SUS ③	3.0 (8)	0.039-0.043	7-23	7-23	34-38	600-800	600-800	0.0063-0.0095	0.0078-0.0118
	Baja	2.5 (6)	0.039-0.043	7-23	7-23	34-38	600-800	600-800	0.0071-0.0087	0.0090-0.0106
	Legacy	2.5 (6)	0.039-0.043	7-23	7-23	34-38	600-800	600-800	0.0071-0.0087	0.0090-0.0106
2004	Impreza	2.5 (6)	0.039-0.043	7-23	7-23	34-38	600-800	600-800	0.0071-0.0087	0.0090-0.0106
	Impreza ④	2.5 (7)	0.039-0.043	7-27	7-27	⑫	600-800	600-800	0.0071-0.0087	0.0128-0.0144
	WRX	2.0 (2)	0.028-0.031	2-22	2-22	33-38	650-850	650-850	0.0071-0.0087	0.0090-0.0106
	Outback/SUS ③	2.5 (6)	0.039-0.043	7-23	7-23	34-38	600-800	600-800	0.0071-0.0087	0.0090-0.0106
	Outback/SUS ③	3.0 (8)	0.039-0.043	7-23	7-23	34-38	600-800	600-800	0.0063-0.0095	0.0078-0.0118
	Baja ⑥	2.5 (6)	0.039-0.043	⑧	⑨	34-38	600-800	600-800	0.0071-0.0087	0.0090-0.0106
	Baja ⑦	2.5 (6)	0.028-0.031	⑩	⑪	⑫	600-800	600-800	0.0071-0.0087	0.0130-0.0146
	Legacy	2.5 (6)	0.039-0.043	7-23	7-23	34-38	600-800	600-800	0.0071-0.0087	0.0090-0.0106
2005	Impreza	2.5 (6)	0.039-0.043	⑧	⑨	⑬	550-750	600-800	0.0063-0.0095	0.0082-0.0114
	Impreza ④	2.5 (7)	0.028-0.031	⑪	⑪	⑫	600-800	600-800	0.0071-0.0087	0.0128-0.0144
	WRX ⑤	2.5 (7)	0.028-0.031	⑪	⑪	⑫	600-800	600-800	0.0071-0.0087	0.0130-0.0146
	Outback ⑥	2.5 (6)	0.039-0.043	13	13	⑬	650	700	0.0063-0.0095	0.0082-0.0114
	Outback ⑦	2.5 (6)	0.028-0.031	12	17	⑫	750	750	0.0055-0.0095	0.0118-0.0158

09490_SBCR_C0003A

ENGINE TUNE-UP SPECIFICATIONS

Year	Liters (VIN)		Engine Displacement Liters (VIN)	Spark Plug Gap (in.)	Ignition Timing (deg.) ①		Fuel Pump (psi)	Idle Speed (rpm)		Valve Clearance ②	
					MT	AT		MT	AT	In.	Ex.
2005 cont.	Outback		3.0 (8)	0.028-0.031	15	15	⑭	650	650	0.0055-0.0095	0.0118-0.0158
	Baja	⑥	2.5 (6)	0.039-0.043	⑧	⑨	⑬	650	700	0.0063-0.0095	0.0082-0.0114
	Baja	⑦	2.5 (6)	0.028-0.031	⑩	⑪	⑫	600-800	600-800	0.0071-0.0087	0.0130-0.0146
	Legacy	⑥	2.5 (6)	0.039-0.043	13	13	⑫	650	700	0.0063-0.0095	0.0082-0.0114
	Legacy	⑦	2.5 (6)	0.028-0.031	12	17	⑭	750	750	0.0055-0.0095	0.0118-0.0158
2006	Impreza		2.5 (6)	0.039-0.043	⑧	⑨	49-52	550-750	600-800	0.0063-0.0095	0.0082-0.0114
	Impreza	④	2.5 (7)	0.028-0.031	⑪	⑪	⑫	600-800	600-800	0.0071-0.0087	0.0128-0.0144
	WRX	⑤	2.5 (7)	0.028-0.031	⑪	⑪	⑫	600-800	600-800	0.0071-0.0087	0.0130-0.0146
	Outback	⑥	2.5 (6)	0.039-0.043	10	15	49-52	650	700	0.0063-0.0095	0.0082-0.0114
	Outback	⑦	2.5 (6)	0.028-0.031	12	17	⑫	750	750	0.0071-0.0087	0.0130-0.0146
	Outback		3.0 (8)	0.028-0.031	15	15	⑭	650-800	650-800	0.0055-0.0095	0.0118-0.0158
	Baja	⑥	2.5 (6)	0.039-0.043	⑧	⑨	⑬	650	700	0.0063-0.0095	0.0082-0.0114
	Baja	⑦	2.5 (6)	0.028-0.031	⑩	⑪	⑫	600-800	600-800	0.0071-0.0087	0.0130-0.0146
	Legacy	⑥	2.5 (6)	0.039-0.043	13	13	49-52	650	700	0.0063-0.0095	0.0082-0.0114
	Legacy	⑦	2.5 (6)	0.028-0.031	12	17	⑭	750	750	0.0071-0.0087	0.0130-0.0146

Note: The Vehicle Emission Control Information label often reflects specification changes made during production.

The lable figures must be used if they differ from those in this chart.

① Before Top Dead Center
② Measured with engine cold
③ Sport Utility Sedan
④ Impreza WRX sedan and sport wagon
⑤ STI
⑥ SOHC
⑦ DOHC/Turbo
⑧ 10 degrees +/- 8 degrees at 650 rpm
⑨ 15 degrees +/- 8 degrees at 700 rpm
⑩ 13 degrees +/- 10 degrees at 700 rpm
⑪ 17 degrees +/- 10 degrees at 700 rpm
⑫ 41-46 while disconnecting pressure regulator vacuum hose from intake manifold
33-38 after connecting pressure regulator vacuum hose
⑬ 41-46 while disconnecting pressure regulator vacuum hose from intake manifold
30-34 after connecting pressure regulator vacuum hose
⑭ 48-53 while disconnecting pressure regulator vacuum hose from intake manifold
40-45 after connecting pressure regulator vacuum hose

79233G37

2.5L Engine
Firing order: 1–3–2–4
Distributorless ignition system

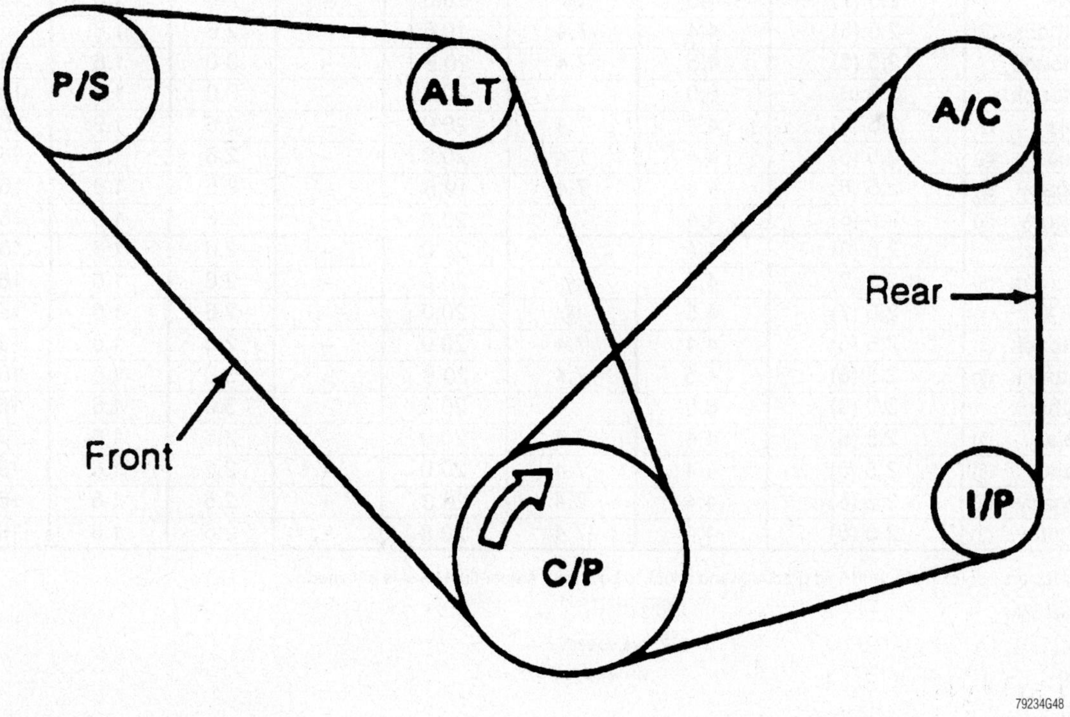

Accessory drive belt routing—2.5L engine

79234G48

CAPACITIES

Year	Liters (VIN)		Engine Displacement Liters (VIN)	Engine Oil with Filter (qts.)	Transmission (pts.) 5-Spd	Transmission (pts.) Auto.	Transfer Case (pts.)	Drive Axle Front ① (pts.)	Drive Axle Rear (pts.)	Fuel Tank (gal.)	Cooling System (qts.)
2002	Impreza		2.5 (6)	4.2	7.4	20.0	–	2.6	1.6	13.2	6.6
	WRX		2.0 (2)	4.8	7.4	19.6	–	2.6	1.6	15.9	8.0
	Outback		2.5 (6)	4.4	7.4	20.0	–	2.6	1.6	15.9	6.9
	Outback		3.0 (8)	5.9	7.4	20.2	–	2.6	1.6	16.9	8.4
	Legacy		2.5 (6)	4.2	7.4	20.0	–	2.6	1.6	16.9	7.1
2003	Impreza		2.5 (6)	4.2	7.4	20.0	–	2.6	1.6	13.2	6.6
	WRX		2.0 (2)	4.8	7.4	19.6	–	2.6	1.6	15.9	8.0
	Outback		2.5 (6)	4.4	7.4	20.0	–	2.6	1.6	15.9	6.9
	Outback		3.0 (8)	5.9	7.4	20.2	–	2.6	1.6	16.9	8.4
	Baja		2.5 (6)	4.2	7.4	20.0	–	2.6	1.6	16.9	7.1
	Legacy		2.5 (6)	4.2	7.4	20.0	–	2.6	1.6	16.9	7.1
2004	Impreza		2.5 (6)	4.2	7.4	20.0	–	2.6	1.6	15.9	6.6
	Impreza	③	2.5 (7)	4.5	⑤	20.0	–	2.6	1.6	15.9	8.1
	WRX		2.0 (2)	4.8	7.4	19.6	–	2.6	1.6	15.9	8.0
	Outback		2.5 (6)	4.4	7.4	19.6	–	2.6	1.6	16.9	6.9
	Outback		3.0 (8)	5.9	7.4	20.2	–	2.6	1.6	16.9	8.4
	Baja	②	2.5 (6)	4.2	7.4	20.0	–	2.6	1.6	16.9	7.1
	Baja	③	2.5 (6)	4.2	7.4	20.0	–	2.6	1.6	16.9	7.1
	Legacy		2.5 (6)	4.2	7.4	20.0	–	2.6	1.6	16.9	7.1
2005	Impreza		2.5 (6)	4.4	⑤	20.0	–	2.6	1.6	15.9	7.3
	Impreza	③	2.5 (7)	4.5	⑤	20.0	–	2.6	1.6	15.9	8.1
	WRX	④	2.5 (7)	4.5	⑤	20.0	–	2.6	1.6	15.9	8.1
	Outback	②	2.5 (6)	4.4	7.4	19.6	–	2.6	1.6	16.9	6.9
	Outback	③	2.5 (6)	4.5	7.4	20.8	–	3.0	1.6	16.9	7.7
	Outback		3.0 (8)	6.0	–	20.8	–	3.0	1.6	16.9	7.8
	Baja	②	2.5 (6)	4.4	7.4	20.0	–	2.6	1.6	16.9	7.2
	Baja	③	2.5 (6)	4.4	7.4	20.0	–	2.6	1.6	16.9	8.1
	Legacy	②	2.5 (6)	4.4	7.4	19.6	–	2.6	1.6	16.9	6.9
	Legacy	③	2.5 (6)	4.4	7.4	20.8	–	2.6	1.6	16.9	7.7
2006	Impreza		2.5 (6)	4.4	⑤	20.0	–	2.6	1.6	15.9	7.3
	Impreza	③	2.5 (7)	4.5	⑤	20.0	–	2.6	1.6	15.9	8.1
	WRX	④	2.5 (7)	4.5	⑤	20.0	–	2.6	1.6	15.9	8.1
	Outback	②	2.5 (6)	4.4	7.4	20.0	–	2.6	1.6	16.9	6.9
	Outback	③	2.5 (6)	4.5	7.4	20.8	–	3.0	1.6	16.9	7.7
	Outback		3.0 (8)	6.0	–	20.8	–	3.0	1.6	16.9	7.8
	Baja	②	2.5 (6)	4.4	7.4	20.0	–	2.6	1.6	16.9	7.2
	Baja	③	2.5 (6)	4.4	7.4	20.0	–	2.6	1.6	16.9	8.1
	Legacy	②	2.5 (6)	4.4	7.4	20.0	–	2.6	1.6	16.9	6.9
	Legacy	③	2.5 (6)	4.4	7.4	20.8	–	2.6	1.6	16.9	7.7

Note: All capacities are approximate. Add fluid gradually and check to be sure a proper fluid level is obtained.

① A/T differential only
② SOHC
③ DOHC/Turbo
④ STI
⑤ five speed: 7.4
   six speed: 8.6

09490_SBCR_C0004

VALVE SPECIFICATIONS

Year	Engine Displacement Liters (VIN)	Seat Angle (deg.)	Face Angle (deg.)	Spring Test Pressure (lbs. @ in.)	Spring Installed Height (in.)	Stem-to-Guide Clearance (in.)		Stem Diameter (in.)	
						Intake	Exhaust	Intake	Exhaust
2002	2.0 (2)	90	NA	91 - 103@ 1.110	①	0.0014-0.0059	0.0016-0.0059	0.2343-0.2348	0.2341-0.2346
	2.5 (6)	90	NA	102 - 118@ 1.315	②	0.0014-0.0024	0.0016-0.0026	0.2343-0.2348	0.2341-0.2346
	3.0 (8)	90	NA	102 - 118@ 1.315	③	0.0012-0.0022	0.0016-0.0026	002148-0.2154	0.2148-0.2150
2003	2.0 (2)	90	NA	91 - 103@ 1.110	①	0.0014-0.0059	0.0016-0.0059	0.2343-0.2348	0.2341-0.2346
	2.5 (6)	90	NA	102 - 118@ 1.315	②	0.0014-0.0024	0.0016-0.0026	0.2343-0.2348	0.2341-0.2346
	3.0 (8)	90	NA	102 - 118@ 1.315	③	0.0012-0.0022	0.0016-0.0026	0.2148-0.2154	0.2148-0.2150
2004	2.0 (2)	90	NA	91 - 103@ 1.110	①	0.0014-0.0059	0.0016-0.0059	0.2343-0.2348	0.2341-0.2346
	2.5 (6)	90	NA	102 - 118@ 1.315	②	0.0014-0.0024	0.0016-0.0026	0.2343-0.2348	0.2341-0.2346
	2.5 (7)	90	NA	④	⑤	0.0012-0.0022	0.0016-0.0026	0.2344-0.2350	0.2341-0.2346
	3.0 (8)	90	NA	102 - 118@ 1.315	③	0.0012-0.0022	0.0016-0.0026	002148-0.2154	0.2148-0.2150
2005	2.5 (6)	90	NA	⑥	⑦	0.0014-0.0024	0.0016-0.0026	0.2343-0.2348	0.2341-0.2346
	2.5 (7)	90	NA	④	⑤	0.0012-0.0022	0.0016-0.0026	0.2344-0.2350	0.2341-0.2346
	3.0 (8)	90	NA	NA	⑧	0.0012-0.0022	0.0016-0.0026	002148-0.2154	0.2148-0.2150
2006	2.5 (6)	90	NA	⑨	⑩	0.0014-0.0024	0.0016-0.0026	0.2343-0.2348	0.2341-0.2346
	2.5 (7)	90	NA	④	⑤	0.0012-0.0022	0.0016-0.0026	0.2344-0.2350	0.2341-0.2346
	3.0 (8)	90	NA	NA	⑧	0.0012-0.0022	0.0016-0.0026	002148-0.2154	0.2148-0.2150

NA: Not Available

① Free length: 1.7587 in.

② Free length: 1.8913 in.

③ Free length: 1.8421 in.

④ Set: 46.1-52.8@1.417
   Lift: 95.8-110.0@1.043

⑤ Free length: 1.863 in.

⑥ Set: 48.0-55.0@1.772
   Lift: 119.0-130.0@1.366

⑦ Free length: 2.1378 in.

⑧ Free length: Intake inner 1.557 in., outer 1.621 in. exhaust 1.824 in.

⑨ Set: 52.9-60.8@1.772
   Lift: 130.3-143.9@1.366

⑩ Free length: 2.173 in.

CAMSHAFT SPECIFICATIONS
All measurements in inches unless noted

Year	Engine Displacement Liters (VIN)	Journal Dia.	Brg. Oil Clearance	Shaft End-play ①	Runout	Lobe Height Intake	Lobe Height Exhaust
2002	2.0 (2)	②	0.0015-0.0028	0.0006-0.0028	NA	1.8210-1.8250	1.8170-1.8210
	2.5 (6)	1.2570-1.2577	0.0022-0.0035	0.0012-0.0035	NA	1.5545-1.5585	1.5455-1.5495
	3.0 (8)	③	0.0015-0.0028	④	NA	1.8130-1.8169	1.7933-1.7972
2003	2.0 (2)	②	0.0015-0.0028	0.0006-0.0028	NA	1.8210-1.8250	1.8170-1.8210
	2.5 (6)	1.2570-1.2577	0.0022-0.0035	0.0012-0.0035	NA	1.5545-1.5585	1.5455-1.5495
	3.0 (8)	③	0.0015-0.0028	④	NA	1.8130-1.8169	1.7815-1.7854
2004	2.0 (2)	②	0.0015-0.0028	0.0016-0.0031	NA	1.8210-1.8250	1.8170-1.8210
	2.5 (6)	1.2570-1.2577	0.0022-0.0035	0.0012-0.0035	NA	1.5545-1.5585	⑤
	2.5 (7)	②	0.0015-0.0028	0.0027-0.0047	NA	1.8330-1.8370	1.8410-1.8440
	3.0 (8)	③	0.0015-0.0028	④	NA	1.8012-1.8051	1.7815-1.7854
2005	2.5 (6)	1.2570-1.2577	0.0022-0.0035	0.0012-0.0035	NA	1.5545-1.5585	⑥
	2.5 (7)	②	0.0015-0.0028	0.0027-0.0047	NA	1.8330-1.8370	1.8410-1.8440
	3.0 (8)	⑦	0.0015-0.0028	⑧	NA	⑨	1.6398-1.6437
2006	2.5 (6)	1.2570-1.2577	0.0022-0.0035	0.0012-0.0035	NA	⑩	1.5783-1.5822
	2.5 (7)	②	0.0015-0.0028	0.0027-0.0047	NA	1.8330-1.8370	1.8410-1.8440
	3.0 (8)	⑦	0.0015-0.0028	⑧	NA	⑪	1.6398-1.6437

NA: Not Available

① Side clearance

② Front: 1.4939-1.4946
Except front: 1.1790-1.1796

③ Front: 1.4939-1.4946
Except front: 1.1002-1.1009

④ Intake: 0.0030-0.0053
Exhaust: 0.0019-0.0043

⑤ Exhaust #1: 1.5455-1.5495
Exhaust #1: 1.5686-1.5726

⑥ California: 1.5686-1.5726
Except California: 1.5638-1.5677

⑦ Front: 1.4939-1.4946
Except front: 1.0215-1.0222

⑧ Intake: 0.0030-0.0053
Exhaust: 0.0012-0.0035

⑨ High: 1.2598
Low 1: 1.2535
Low 2: 1.2598

⑩ Constant: 1.5778-1.5817
Low speed: 1.3851-1.3891
High speed: 1.5872-1.5911

⑪ High: 1.6571-1.6610
Low 1: 1.5016-1.5055
Low 2: 1.3756-1.3795

09490_SBCR_C0006

CRANKSHAFT AND CONNECTING ROD SPECIFICATIONS

All measurements are given in inches.

Year	Engine Displacement Liters (VIN)	Crankshaft				Connecting Rod		
		Main Brg. Journal Dia.	Main Brg. Oil Clearance	Shaft End-play	Thrust on No.	Journal Diameter	Oil Clearance	Side Clearance
2002	2.0 (2)	2.3619-2.3625	0.0004-0.0012	0.0012-0.0048	3	1.8891-1.8898	0.0008-0.0018	0.0028-0.0130
	2.5 (6)	2.5194-2.5200	①	0.0012-0.0098	3	1.8891-1.8898	0.0005-0.0015	0.0028-0.0130
	3.0 (8)	2.3619-2.3625	0.0006-0.0012	0.0012-0.0098	3	1.8891-1.8898	0.0009-0.0020	0.0028-0.0130
2003	2.0 (2)	2.3619-2.3625	0.0004-0.0012	0.0012-0.0048	3	1.8891-1.8898	0.0008-0.0018	0.0028-0.0130
	2.5 (6)	2.3619-2.3625	①	0.0012-0.0098	3	1.8891-1.8898	0.0005-0.0015	0.0028-0.0130
	3.0 (8)	2.3619-2.3625	0.0006-0.0012	0.0012-0.0098	3	1.8891-1.8898	0.0009-0.0020	0.0028-0.0130
2004	2.0 (2)	2.3619-2.3625	0.0004-0.0012	0.0012-0.0048	3	1.8891-1.8898	0.0008-0.0018	0.0028-0.0130
	2.5 (6)	2.3619-2.3625	①	0.0012-0.0098	3	1.8891-1.8898	0.0005-0.0015	0.0028-0.0130
	2.5 (7)	2.3619-2.3625	0.0004-0.0012	0.0012-0.0045	NA	NA	0.0007-0.0018	0.0028-0.0130
	3.0 (8)	2.3619-2.3625	0.0006-0.0012	0.0012-0.0098	3	1.8891-1.8898	0.0009-0.0020	0.0028-0.0130
2005	2.5 (6)	2.3619-2.3625	0.0004-0.0012	0.0012-0.0045	NA	NA	0.0006-0.0017	0.0028-0.0130
	2.5 (7)	2.3619-2.3625	0.0004-0.0012	0.0012-0.0045	NA	NA	0.0007-0.0018	0.0028-0.0130
	3.0 (8)	2.5194-2.5200	0.0004-0.0012	0.0012-0.0045	NA	NA	0.0006-0.0017	0.0028-0.0130
2006	2.5 (6)	2.3619-2.3625	0.0004-0.0012	0.0012-0.0045	NA	NA	0.0006-0.0017	0.0028-0.0130
	2.5 (7)	2.3619-2.3625	0.0004-0.0012	0.0012-0.0045	NA	NA	0.0007-0.0018	0.0028-0.0130
	3.0 (8)	2.5194-2.5200	0.0004-0.0012	0.0012-0.0045	NA	NA	0.0006-0.0017	0.0028-0.0130

NA: Not Available

① Journals 1 and 5: 0.0001 - 0.0016 in.
  Journals 2 and 4: 0.0004 - 0.0018 in.
  Journal 3: 0.0004 - 0.0016 in.

PISTON AND RING SPECIFICATIONS

All measurements are given in inches.

Year	Engine Displacement Liters (VIN)	Piston Clearance	Ring Gap			Ring Side Clearance		
			Top Compression	Bottom Compression	Oil Control	Top Compression	Bottom Compression	Oil Control
2002	2.0 (2)	0.0004-0.0012	0.0079-0.0138	0.0138-0.0197	0.0079-0.0276	0.0016-0.0031	0.0012-0.0028	NA
	2.5 (6)	0.0004-0.0012	0.0079-0.0138	0.0146-0.0250	0.0079-0.0197	0.0016-0.0031	0.0012-0.0028	NA
	3.0 (8)	0.0004-0.0012	0.0079-0.0138	0.0138-0.0197	0.0079-0.0236	0.0016-0.0031	0.0012-0.0028	NA
2003	2.0 (2)	0.0004-0.0012	0.0079-0.0138	0.0138-0.0197	0.0079-0.0276	0.0016-0.0031	0.0012-0.0028	NA
	2.5 (6)	0.0004-0.0012	0.0079-0.0138	0.0146-0.0250	0.0079-0.0197	0.0016-0.0031	0.0012-0.0028	NA
	3.0 (8)	0.0004-0.0012	0.0079-0.0138	0.0138-0.0197	0.0079-0.0236	0.0016-0.0031	0.0012-0.0028	NA
2004	2.0 (2)	0.0004-0.0012	0.0079-0.0138	0.0138-0.0197	0.0079-0.0276	0.0016-0.0031	0.0012-0.0028	NA
	2.5 (6)	0.0004-0.0012	0.0079-0.0138	0.0146-0.0250	0.0079-0.0197	0.0016-0.0031	0.0012-0.0028	NA
	2.5 (7)	NA	0.0079-0.0098	0.0150-0.0203	0.0079-0.0197	0.0016-0.0031	0.0012-0.0028	NA
	3.0 (8)	0.0004-0.0012	0.0079-0.0138	0.0138-0.0197	0.0079-0.0236	0.0016-0.0031	0.0012-0.0028	NA
2005	2.5 (6)	NA	0.0079-0.0138	0.0144-0.0203	0.0079-0.0197	0.0016-0.0031	0.0012-0.0028	NA
	2.5 (7)	NA	0.0079-0.0098	0.0150-0.0203	0.0079-0.0197	0.0016-0.0031	0.0012-0.0028	NA
	3.0 (8)	NA	0.0079-0.0138	0.0138-0.0197	0.0079-0.0236	0.0016-0.0031	0.0012-0.0028	0.0018-0.0049
2006	2.5 (6)	NA	0.0079-0.0138	0.0144-0.0203	0.0079-0.0197	0.0016-0.0031	0.0012-0.0028	NA
	2.5 (7)	NA	0.0079-0.0098	0.0150-0.0203	0.0079-0.0197	0.0016-0.0031	0.0012-0.0028	NA
	3.0 (8)	NA	0.0079-0.0138	0.0138-0.0197	0.0079-0.0236	0.0016-0.0031	0.0012-0.0028	0.0018-0.0049

NA: Not Available

09490_SBCR_C0008

TORQUE SPECIFICATIONS

All readings in ft. lbs.

Year	Engine Displacement Liters (VIN)	Cylinder Head Bolts	Main ① Bearing Bolts	Rod Bearing Bolts	Crankshaft Damper Bolts	Flywheel Bolts	Manifold		Spark Plugs	Oil Pan Drain Plug
							Intake	Exhaust		
2002	2.0 (2)	②	③	31 - 44	94	51 - 55	18	26	13 - 17	33
	2.5 (6)	②	③	31 - 44	123 - 137	51 - 55	17 - 20	19 - 26	13 - 17	33
	3.0 (8)	④	14-18	39	131	60	18	22	15	33
2003	2.0 (2)	②	③	31 - 44	94	51 - 55	18	26	13 - 17	33
	2.5 (6)	②	③	31 - 44	123 - 137	51 - 55	17 - 20	19 - 26	13 - 17	33
	3.0 (8)	④	14-18	39	131	60	18	22	15	33
2004	2.0 (2)	②	③	31 - 44	94	51 - 55	18	26	13 - 17	33
	2.5 (6)	②	③	31 - 44	123 - 137	51 - 55	17 - 20	19 - 26	13 - 17	33
	2.5 (7)	④	⑤	38.4	132.7	⑥	18	⑦	13-17	33
	3.0 (8)	⑧	14-18	39	131	60	18	22	15	33
2005	2.5 (6)	⑨	⑤	33.3	132.7	52.8	18	19-26	13-17	33
	2.5 (7)	④	⑤	38.4	132.7	⑥	18	⑦	13-17	33
	3.0 (8)	⑩	⑪	39	131	60	18	22	15	33
2006	2.5 (6)	⑨	⑤	33.3	132.7	52.8	18	19-26	13-17	33
	2.5 (7)	④	⑤	38.4	132.7	⑥	18	⑦	13-17	33
	3.0 (8)	⑩	⑪	39	131	60	18	22	15	33

① Engine block connecting bolts

② Step 1: Tighten all bolts to 22 ft. lbs.
Step 2: Tighten all bolts to 51 ft. lbs.
Step 3: Loosen all bolts 180 degrees.
Step 4: Repeat Step 3.
Step 5: Tighten bolts 1 and 2 to 25 ft. lbs.
Step 6: Tighten bolts 3, 4, 5 and 6 to 11 ft. lbs.
Step 7: Tighten all bolts 80 to 90 degrees.
Step 8: Repeat Step 7. Do not exceed 180 degrees total tightening.

③ Split engine case connecting bolts:
Short bolts: 17-20 ft. lbs.
Long bolts: 33-37 ft. lbs.
Smaller short bolts (if used) 5 ft. lbs.

④ Step 1: Tighten all bolts to 21.4 ft. lbs.
Step 2: Tighten all bolts to 51 ft. lbs.
Step 3: Loosen all bolts 180 degrees, and again 180 degrees
Step 4: Tighten all bolts to 36 ft. lbs.
Step 5: + 80-90 degrees
Step 6: + 40-45 degrees

⑤ Step 1: Left side 10mm bolts (A-D) 7.2 ft. lbs.
Step 2: Right side 10mm bolts (E-J) 7.2 ft. lbs.
Step 3: Left side 10mm bolts (A-D) 13 ft. lbs.
Step 4: Right side 10mm bolts (E-J) 13 ft. lbs.
Step 5: Left side bolts (A and C) +90 degrees, bolts (B and D) 29.5 ft. lbs.
Step 6: Right side 10mm bolts (E-J) + 90 degrees
Step 7: Left side cylinder block connecting bolts (A-G) 18.1 ft. lbs., bolt (H) 4.7 ft. lbs.

⑥ Automatic: 53.1 ft. lbs.
Manual: 38.4 ft. lbs.

⑦ Exhaust pipe to manifold 26 ft. lbs.
Right upper cover 5.5 ft. lbs.
Front exhaust pipe 13.7 ft. lbs.
Right manifold to turbocharger joint pipe 26-28 ft. lbs.
Lower covers 13.7 ft. lbs.

⑧ Step 1: Tighten all bolts to 14 ft. lbs.
Step 2: Tighten all bolts to 37 ft. lbs.
Step 3: Loosen all bolts 180 degrees the an additional 180 degrees in two steps in the reverse order of tightening sequence.
Step 4: Tighten all bolts to 18 ft. lbs.
Step 5: Tighten all bolts to 18 ft. lbs.
Step 6: Tighten bolts 90 degrees
Step 7: Tighten bolts 1, 2, 3 and 4 90 degrees.
Step 8: Tighten bolts 5, 6, 7 and 8 45 degrees.

⑨ Step 1: Tighten all bolts to 22 ft. lbs.
Step 2: Tighten all bolts to 51 ft. lbs.
Step 3: Loosen all bolts 180 degrees, and again 180 degrees
Step 4: Tighten all bolts to 36 ft. lbs.
Step 5: + 80-90 degrees
Step 6: + 40-45 degrees

⑩ Step 1: Tighten all bolts to 14 ft. lbs.
Step 2: Tighten all bolts to 37 ft. lbs.
Step 3: Loosen all bolts 180 degrees, and again 180 degrees
Step 4: Tighten all bolts to 14 ft. lbs.
Step 5: Bolts (1-4) 35.4 ft. lbs.
Step 6: Bolts (5-8) 33 ft. lbs.
Step 7: + 90 degrees
Step 8: Bolts (1-4) 45 degrees

⑪ Step 1: Bolts (1-11 and 13) 18 ft. lbs.
Bolts (12 and 14) 14. ft. lbs.
Step 2: Retighten bolts (1-11 and 13) 18. ft. lbs.
Retighten bolts (12 and 14) 14. ft. lbs.
Step 3: + 90 degrees
Step 4: Upper bolt to cylinder block 18 ft. lbs.
Step 7: + 90 degrees
Step 8: Bolts (1-4) 45 degrees

CAUTION:
Remove oil in the mating surface of bearing and cylinder block before installation. Also apply a coat of engine oil to crankshaft pins.

1) Position the crankshaft on the #2 and #4 cylinder block.

2) Apply fluid packing to the mating surface of #1 and #3 cylinder block, and position it on #2 and #4 cylinder block.

CAUTION:
Do not allow fluid packing to jut into O-ring grooves, oil passages, bearing grooves, etc.

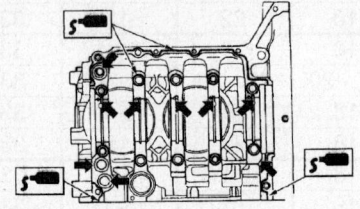

3) Temporarily tighten the 10 mm cylinder block connecting bolts in alphabetical sequence shown in the figure.

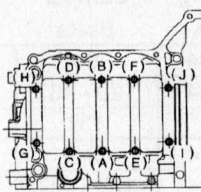

4) Tighten the 10 mm cylinder block connecting bolts in alphabetical sequence.

Tightening torque:
47 N·m (4.8 kgf-m, 34.7 ft-lb)

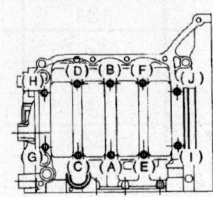

5) Tighten the 8 mm and 6 mm cylinder block connecting bolts in alphabetical sequence shown in the figure.

Tightening torque:
(A) — (G): 25 N·m (2.5 kgf-m, 18.1 ft-lb)
(H): 6.4 N·m (0.65 kgf-m, 4.7 ft-lb)

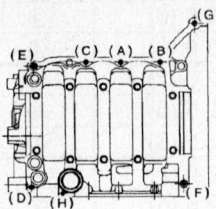

09490_SBCR_G0001

Main bearing torque sequence—2002–2004 2.0L engine

CAUTION:
Remove oil in the mating surface of bearing and cylinder block before installation. Also apply a coat of engine oil to crankshaft pins.

1) Position the crankshaft on the #2 and #4 cylinder block.

2) Apply fluid packing to the mating surface of #1 and #3 cylinder block, and position it on #2 and #4 cylinder block.

CAUTION:
Do not allow fluid packing to jut into O-ring grooves, oil passages, bearing grooves, etc.

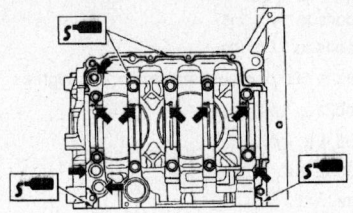

3) Temporarily tighten the 10 mm cylinder block connecting bolts in alphabetical sequence shown in the figure.

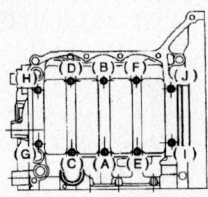

4) Tighten the 10 mm cylinder block connecting bolts in alphabetical sequence.

Tightening torque:
47±3 N·m (4.8±0.3 kgf-m, 34.7±2.2 ft-lb)

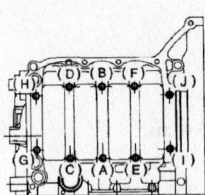

5) Tighten the 8 mm and 6 mm cylinder block connecting bolts in alphabetical sequence shown in the figure.

Tightening torque:
(A) — (G): 25 N·m (2.5 kgf-m, 18.1 ft-lb)
(H): 6.4 N·m (0.65 kgf-m, 4.7 ft-lb)

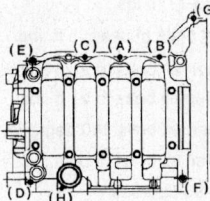

09490_SBCR_G0002

Main bearing torque sequence—2002 2.5L SOHC engine

CAUTION:
Remove oil in the mating surface of bearing and cylinder block before installation. Also apply a coat of engine oil to crankshaft pins.

1) Position the crankshaft on the #2 and #4 cylinder block.

2) Apply fluid packing to the mating surface of #1 and #3 cylinder block, and position it on #2 and #4 cylinder block.

NOTE:
Do not allow fluid packing to jut into O-ring grooves, oil passages, bearing grooves, etc.

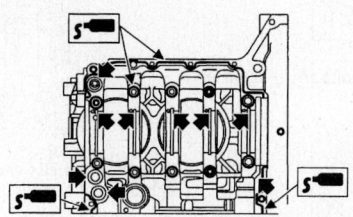

3) Tighten 10 mm cylinder block connecting bolts in alphabetical sequence shown in figure. (LH side)

Tightening torque:
15 N·m (1.5 kgf-m, 11 ft-lb)

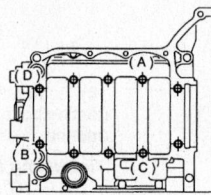

4) Tighten 10 mm cylinder block connecting bolts in alphabetical sequence shown in figure. (RH side)

Tightening torque:
15 N·m (1.5 kgf-m, 11 ft-lb)

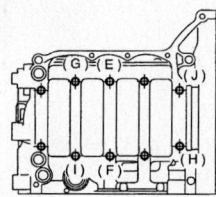

5) Tighten bolts (A to D) on left side of cylinder block more 90° in alphabetical sequence.

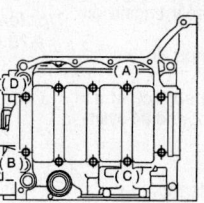

6) Tighten bolts (E to J) on right side of cylinder block more 90° in alphabetical sequence.

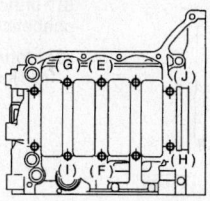

7) Tighten 8 mm and 6 mm cylinder block connecting bolts in alphabetical sequence shown in figure.

Tightening torque:
(A) — (G): 25 N·m (2.5 kgf-m, 18.1 ft-lb)
(H): 6.4 N·m (0.65 kgf-m, 4.7 ft-lb)

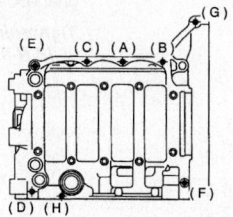

09490_SBCR_G0003

Main bearing torque sequence—2003–2004 2.5L SOHC engine

NOTE:
Remove oil on the mating surface of bearing and cylinder block before installation. Apply engine oil to crankshaft pins.

1) Position the crankshaft and O-rings on the #1 and #3 cylinder block.

2) Apply liquid gasket to the mating surface of #1 and #3 cylinder block, and position #2 and #4 cylinder block.

Liquid gasket:
Three bond 1215 (Part No. 004403007) or equivalent

NOTE:
Do not allow liquid gasket to jut into O-ring grooves, oil passages, bearing grooves, etc.

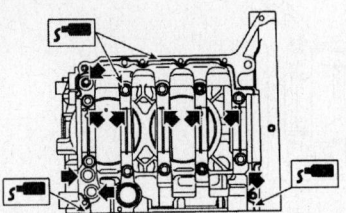

3) Apply a coat of engine oil to the washer and bolt thread.

4) Tighten the 10 mm cylinder block connecting bolts on LH side (A — D) in alphabetical sequence.

Tightening torque:
9.75 N·m (1.0 kgf-m, 7.2 ft-lb)

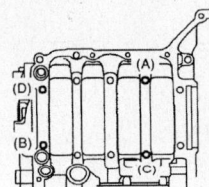

5) Tighten the 10 mm cylinder block connecting bolts on RH side (E — J) in alphabetical sequence.

Tightening torque:
9.75 N·m (1.0 kgf-m, 7.2 ft-lb)

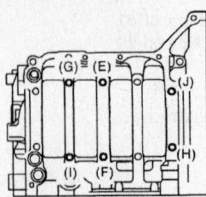

6) Further tighten the LH side bolts (A — D) in alphabetical sequence.

Tightening torque:
18 N·m (1.8 kgf-m, 13 ft-lb)

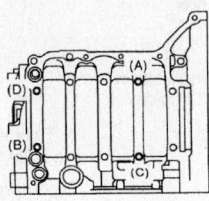

7) Further tighten the RH side bolts (E — J) in alphabetical sequence.

Tightening torque:
18 N·m (1.8 kgf-m, 13 ft-lb)

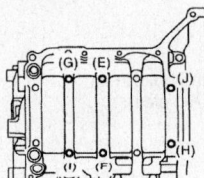

8) Further tighten the LH side bolts (A — D) in alphabetical sequence.
• (A), (C): Angle tightening

Tightening angle:
90°
• (B), (D): Torque tightening

Tightening torque:
40 N·m (4.1 kgf-m, 29.6 ft-lb)

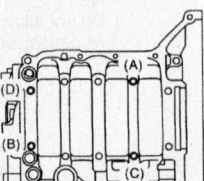

9) Tighten the RH side bolts (E — J) 90° further in alphabetical sequence.

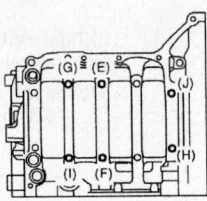

10) Tighten the 8 mm and 6 mm cylinder block connecting bolts on LH side (A — H) in alphabetical sequence.

Tightening torque:
(A) — (G): 25 N·m (2.5 kgf-m, 18.1 ft-lb)
(H): 6.4 N·m (0.65 kgf-m, 4.7 ft-lb)

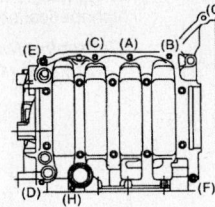

09490_SBCR_G0004

Main bearing torque sequence—2005–2006 2.5L SOHC engine

CAUTION:
Remove oil in the mating surface of bearing and cylinder block before installation. Also apply a coat of engine oil to crankshaft pins.

1) Position the crankshaft on the #2 and #4 cylinder block.

2) Apply fluid packing to the mating surface of #1 and #3 cylinder block, and position it on #2 and #4 cylinder block.

CAUTION:
Do not allow fluid packing to jut into O-ring grooves, oil passages, bearing grooves, etc.

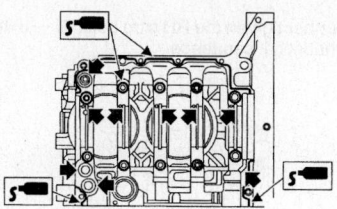

3) Tighten the 10 mm cylinder block connecting bolts in alphabetical sequence shown in the figure (LH side).

Tightening torque:
 15 N·m (1.5 kgf-m, 10.8 ft-lb)

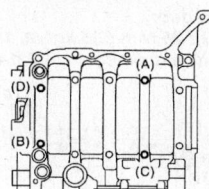

4) Tighten the 10 mm cylinder block connecting bolts in alphabetical sequence shown in the figure (RH side).

Tightening torque:
 15 N·m (1.5 kgf-m, 10.8 ft-lb)

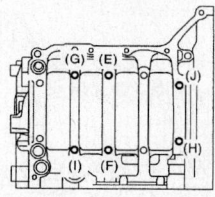

5) Tighten LH side bolts (A) — (D) for a further 90° in alphabetical order.

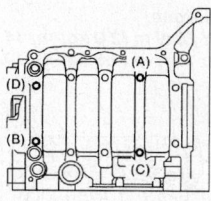

6) Tighten RH side bolts (E) — (J) for a further 90° in alphabetical order.

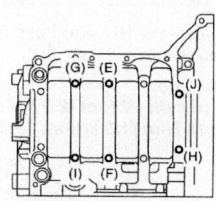

7) Tighten the 8 mm and 6 mm cylinder block connecting bolts in alphabetical sequence shown in the figure.

Tightening torque:
 (A) — (G): 25 N·m (2.5 kgf-m, 18.1 ft-lb)
 (H): 6.4 N·m (0.65 kgf-m, 4.7 ft-lb)

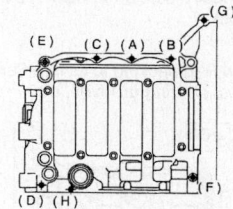

09490_SBCR_G0005

Main bearing torque sequence—2003 2.5L DOHC engine

1) Remove oil in the mating surface of bearing and cylinder block before installation. Also apply a coat of engine oil to crankshaft pins.
2) Position the crankshaft on #2 and #4 cylinder block.
3) Apply fluid packing to the mating surface of #1 and #3 cylinder block, and position it on #2 and #4 cylinder block.

NOTE:
Do not allow fluid packing to jut into O-ring grooves, oil passages, bearing grooves, etc.

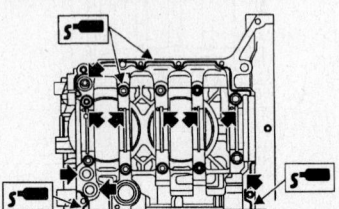

4) Apply engine oil to washers and thread of bolts.
5) Tighten the 10 mm cylinder block connecting bolts in alphabetical sequence shown in the figure. (LH side)

Tightening torque:
10 N·m (1.0 kgf-m, 7.4 ft-lb)

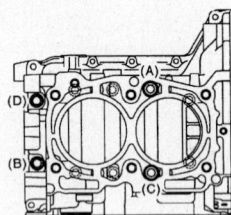

6) Tighten the 10 mm cylinder block connecting bolts in alphabetical sequence shown in the figure. (RH side)

Tightening torque:
10 N·m (1.0 kgf-m, 7.4 ft-lb)

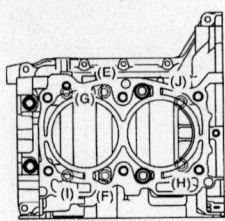

7) Further tighten the LH side bolts (A — D) in alphabetical sequence.

Tightening torque:
(A), (C): 20 N·m (2.0 kgf-m, 14.8 ft-lb)
(B), (D): 15 N·m (1.5 kgf-m, 10.8 ft-lb)

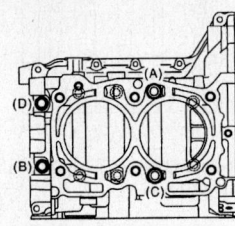

8) Further tighten the RH side bolts (E — J) in alphabetical sequence.

Tightening torque:
(E), (F), (G), (I): 20 N·m (2.0 kgf-m, 14.8 ft-lb)
(H), (J): 18 N·m (1.8 kgf-m, 13.3 ft-lb)

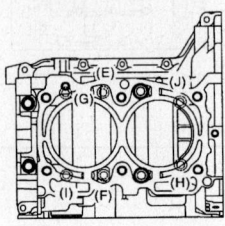

9) Further tighten the LH side bolts (A — D) by 90° in alphabetical sequence.

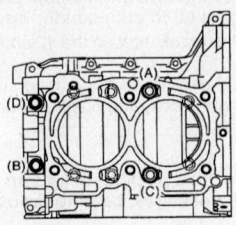

10) Further tighten the RH side bolts (E — J) by 90° in alphabetical sequence.

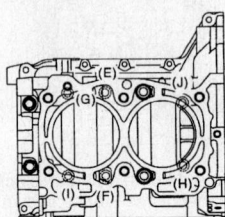

11) Tighten the 8 mm and 6 mm cylinder block connecting bolts in alphabetical sequence shown in the figure.

Tightening torque:
(A) — (G): 25 N·m (2.5 kgf-m, 18.1 ft-lb)
(H): 6.4 N·m (0.65 kgf-m, 4.7 ft-lb)

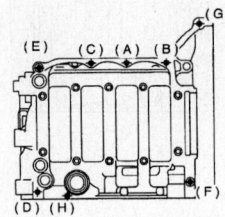

09490_SBCR_G0006

Main bearing torque sequence—2004 2.5L DOHC engine

1) Remove oil on the mating surface of bearing and cylinder block before installation. Apply engine oil to crankshaft pins.

2) Position the crankshaft and O-rings on the #1 and #3 cylinder block.

3) Apply liquid gasket to the mating surface of #1 and #3 cylinder blocks, and position cylinder block #2 and #4.

NOTE:
Do not allow liquid gasket to run over to O-ring grooves, oil passages, bearing grooves, etc.

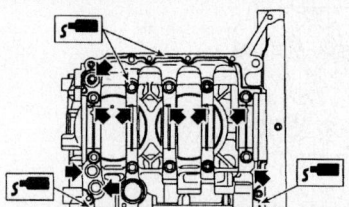

4) Apply a coat of engine oil to the washer and bolt thread.

5) Tighten the 10 mm cylinder block connecting bolts on LH side (A — D) in alphabetical sequence.

Tightening torque:
 10 N·m (1.0 kgf-m, 7.2 ft-lb)

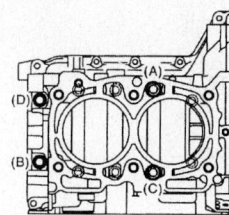

6) Tighten the 10 mm cylinder block connecting bolts on RH side (E — J) in alphabetical sequence.

Tightening torque:
 10 N·m (1.0 kgf-m, 7.2 ft-lb)

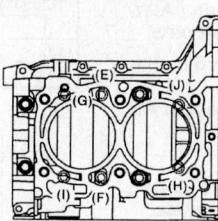

7) Further tighten the LH side bolts (A — D) in alphabetical sequence.

Tightening torque:
 18 N·m (1.8 kgf-m, 13.0 ft-lb)

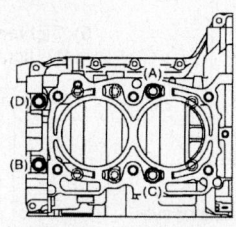

8) Further tighten the RH side bolts (E — J) in alphabetical sequence.

Tightening torque:
 18 N·m (1.8 kgf-m, 13.0 ft-lb)

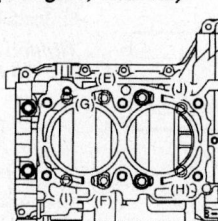

9) Further tighten the LH side bolts (A — D) in alphabetical sequence.

 (A), (C): 90°
 (B), (D): 40 N·m (4.1 kgf-m, 29.5 ft-lb)

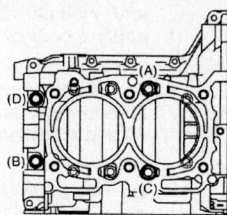

10) Tighten the RH side bolts (E — J) 90° further in alphabetical sequence.

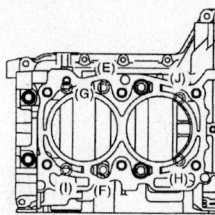

11) Tighten the 8 mm and 6 mm cylinder block connecting bolts on LH side (A — H) in alphabetical sequence.

Tightening torque:
 (A) — (G): 25 N·m (2.5 kgf-m, 18.1 ft-lb)
 (H): 6.4 N·m (0.65 kgf-m, 4.7 ft-lb)

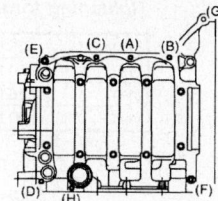

09490_SBCR_G0007

Main bearing torque sequence—2005–2006 2.5L DOHC engine

1) Install ST to cylinder block, then install crankshaft bearing.

CAUTION:
Remove oil in the mating surface of bearing and cylinder block before installation. Also apply a coat of engine oil to crankshaft pins.

2) Position crankshaft and connecting rod on the #2, #4 and #6 cylinder.

3) Apply fluid packing to the mating surface of #1, #3 and #5 cylinder block.

CAUTION:
Do not allow fluid packing to jut into O-ring grooves, oil passages, bearing grooves, etc.

Fluid packing application diameter:
1.0±0.2 mm (0.039±0.008 in)

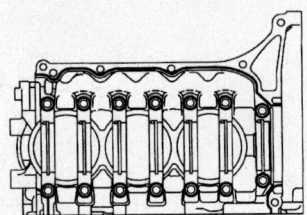

4) Apply engine oil to washers and threads of cylinder block connecting bolts. Tighten the bolts following the steps below.
   (1) Tighten all the bolts in the numerical order shown in the figure.

Tightening torque:

(1) to (11)	25 N·m (2.5 kgf-m, 18 ft-lb)
(12)	20 N·m (2.0 kgf-m, 14 ft-lb)
(13)	25 N·m (2.5 kgf-m, 18 ft-lb)
(14)	20 N·m (2.0 kgf-m, 14 ft-lb)

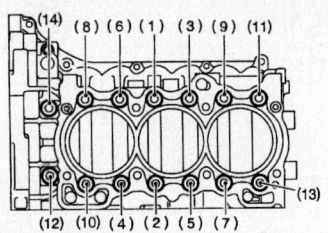

(2) Tighten all the bolts again in the order shown in the figure.

Tightening torque:

(1) to (11)	25 N·m (2.5 kgf-m, 18 ft-lb)
(12)	20 N·m (2.0 kgf-m, 14 ft-lb)
(13)	25 N·m (2.5 kgf-m, 18 ft-lb)
(14)	20 N·m (2.0 kgf-m, 14 ft-lb)

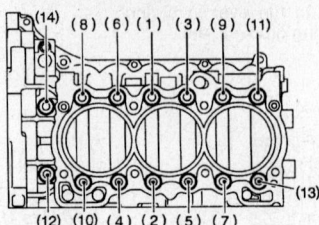

5) Tighten all the bolts by 90° in the order shown in the figure.

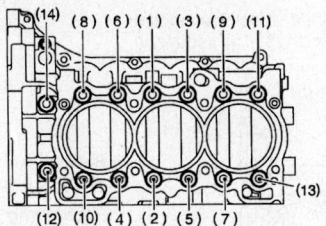

6) Install upper bolts on cylinder block.

Tightening torque:
25 N·m (2.5 kgf-m, 18 ft-lb)

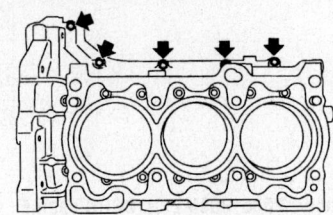

09490_SBCR_G0008

Main bearing torque sequence—2002–2004 3.0L engine

1) After setting the cylinder block to ST, install the crankshaft bearing.

NOTE:
Remove oil on the mating surface of bearing and cylinder block before installation. Apply a coat of engine oil to crankshaft pins.

2) Position the crankshaft and connecting rod on #2, #4 and #6 cylinder block.

3) Apply liquid gasket to the mating surface of the #1, #3 and #5 cylinder blocks, and position it on #2, #4 and #6 cylinder blocks.

NOTE:
Do not allow liquid gasket to run over to oil passages, bearing grooves, etc.

Applying liquid gasket diameter:
  1.0±0.2 mm (0.039±0.008 in)

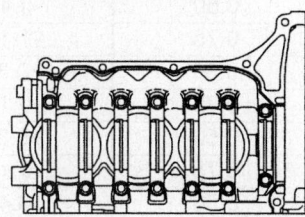

4) Apply a coat of engine oil to the washer and bolt thread.

5) Tighten all bolts in the numerical order as shown in the figure.

Tightening torque:
  (1) — (11), (13): 25 N·m (2.5 kgf-m, 18 ft-lb)
  (12), (14): 20 N·m (2.0 kgf-m, 14 ft-lb)

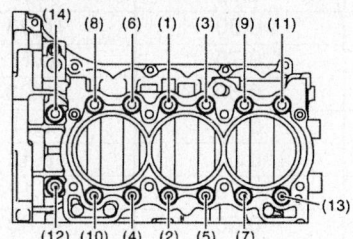

6) Retighten all bolts in the numerical order as shown in the figure.

Tightening torque:
  (1) — (11), (13): 25 N·m (2.5 kgf-m, 18.4 ft-lb)
  (12), (14): 20 N·m (2.0 kgf-m, 14 ft-lb)

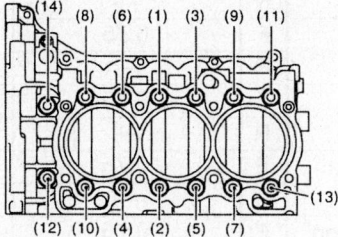

7) Tighten all bolts 90° — 110° in the numerical order as shown in the figure.

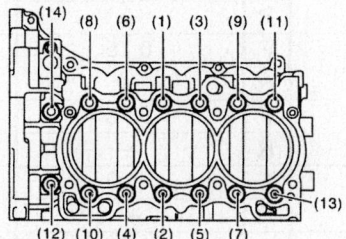

8) Install the upper bolt to cylinder block.

Tightening torque:
  25 N·m (2.5 kgf-m, 18 ft-lb)

NOTE:
Remove the liquid gasket which is running over to sealing surface between cylinder block and rear chain cover, cylinder block and oil pan upper, after tightening the bolts which combine the cylinder block.

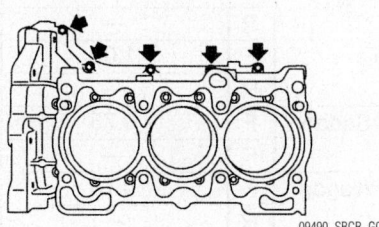

09490_SBCR_G0009

Main bearing torque sequence—2005–2006 3.0L engine

WHEEL ALIGNMENT

Year	Model		Caster		Camber		Toe-in
			Range (+/-Deg.)	Preferred Setting (Deg.)	Range (+/-Deg.)	Preferred Setting (Deg.)	(in.)
2002	Impreza	F	1.00	+3.05	0.50	-0.42	0+/-0.12
		R	—	—	0.75	-1.17	0+/-0.12
	WRX	F	0.75	+2.40	0.50	+0.33	0+/-0.12
		R	—		0.75	-0.17	0+/-0.12
	Outback	F	0.75	+2.40	0.50	+0.33	0+/-0.12
		R	—		0.75	-0.17	0+/-0.12
	Legacy Sedan	F	0.75	+3.05	0.50	-0.05	0+/-0.12
		R	—		0.75	-0.50	0+/-0.12
	Legacy Wagon	F	0.75	+2.05	0.50	-0.05	0+/-0.12
		R	—		0.75	-0.20	0+/-0.12
2003	Impreza	F	1.00	+3.05	0.50	-0.42	0+/-0.12
		R	—	—	0.75	-1.17	0+/-0.12
	WRX	F	0.75	+2.40	0.50	+0.33	0+/-0.12
		R	—		0.75	-0.17	0+/-0.12
	Outback	F	0.75	+2.40	0.50	+0.33	0+/-0.12
		R	—		0.75	-0.17	0+/-0.12
	Baja	F	1.00	+3.05	0.50	-0.42	0+/-0.12
		R	—	—	0.75	-1.17	0+/-0.12
	Legacy Sedan	F	0.75	+3.05	0.50	-0.05	0+/-0.12
		R	—		0.75	-0.50	0+/-0.12
	Legacy Wagon	F	0.75	+2.05	0.50	-0.05	0+/-0.12
		R	—		0.75	-0.20	0+/-0.12
2004	Impreza	F	1.00	+3.05	0.50	-0.42	0+/-0.12
		R	—	—	0.75	-1.17	0+/-0.12
	WRX	F	0.75	+2.40	0.50	+0.33	0+/-0.12
		R	—		0.75	-0.17	0+/-0.12
	Outback	F	0.75	+2.40	0.50	+0.33	0+/-0.12
		R	—		0.75	-0.17	0+/-0.12
	Baja	F	1.00	+3.05	0.50	-0.42	0+/-0.12
		R	—	—	0.75	-1.17	0+/-0.12
	Legacy Sedan	F	0.75	+3.05	0.50	-0.05	0+/-0.12
		R	—		0.75	-0.50	0+/-0.12
	Legacy Wagon	F	0.75	+2.05	0.50	-0.05	0+/-0.12
		R	—		0.75	-0.20	0+/-0.12
2005	Impreza	F	1.00	+3.05	0.50	-0.42	0+/-0.12
		R	—	—	0.75	-1.17	0+/-0.12
	WRX	F	0.75	+2.40	0.50	+0.33	0+/-0.12
		R	—		0.75	-0.17	0+/-0.12
	Outback	F	0.75	+2.40	0.50	+0.33	0+/-0.12
		R	—		0.75	-0.17	0+/-0.12
	Baja	F	1.00	+3.05	0.50	-0.42	0+/-0.12
		R	—	—	0.75	-1.17	0+/-0.12
	Legacy Sedan	F	0.75	+3.05	0.50	-0.05	0+/-0.12
		R	—		0.75	-0.50	0+/-0.12
	Legacy Wagon	F	0.75	+2.05	0.50	-0.05	0+/-0.12
		R	—		0.75	-0.20	0+/-0.12

09490_SBCR_C0010A

WHEEL ALIGNMENT

Year	Model		Caster Range (+/-Deg.)	Caster Preferred Setting (Deg.)	Camber Range (+/-Deg.)	Camber Preferred Setting (Deg.)	Toe-in (in.)
2006	Impreza	F	1.00	+3.05	0.50	-0.42	0+/-0.12
		R	—	—	0.75	-1.17	0+/-0.12
	WRX	F	0.75	+2.40	0.50	+0.33	0+/-0.12
		R	—		0.75	-0.17	0+/-0.12
	Outback	F	0.75	+2.40	0.50	+0.33	0+/-0.12
		R	—		0.75	-0.17	0+/-0.12
	Baja	F	1.00	+3.05	0.50	-0.42	0+/-0.12
		R	—	—	0.75	-1.17	0+/-0.12
	Legacy Sedan	F	0.75	+3.05	0.50	-0.05	0+/-0.12
		R	—		0.75	-0.50	0+/-0.12
	Legacy Wagon	F	0.75	+2.05	0.50	-0.05	0+/-0.12
		R	—		0.75	-0.20	0+/-0.12

09490_SBCR_C0010B

TIRE, WHEEL AND BALL JOINT SPECIFICATIONS

Year	Model	OEM Tires Standard	OEM Tires Optional	Tire Pressures (psi) Front	Tire Pressures (psi) Rear	Wheel Size	Ball Joint Inspection	Lug Nut
2002	Impreza	P195/60R15	None	③	③	6-JJ	0.012 in.	58 - 72
	Impreza RS	P205/55R16	None	③	③	7-JJ	0.012 in.	58 - 72
	Impreza Outback	P225/60R16	T145/80R16	③	③	16x16 1/2JJ	0.012 in.	65
	Impreza WRX	P195/60R15	P205/55R16 P215/45R17	③	③	7-JJ	0.012 in.	58 - 72
	Legacy	P185/70R14	P195/60R15 P205/55R16 P205/70R15	③	③	7-JJ	0.012 in.	58 - 72
2003	Impreza	P195/60R15	None	③	③	6-JJ	0.012 in.	58 - 72
	Impreza RS	P205/55R16	None	③	③	7-JJ	0.012 in.	58 - 72
	Impreza Outback	P225/60R16	T145/80R16	③	③	16x16 1/2JJ	0.012 in.	65
	Impreza WRX	P195/60R15	P205/55R16 P215/45R17	③	③	6-JJ 7-JJ	0.012 in.	58 - 72
	Baja	P185/70R14	P195/60R15 P205/55R16 P205/70R15	③	③	6.5-JJ	0.012 in.	58 - 72
	Legacy	P185/70R14	P195/60R15 P205/55R16 P205/70R15	③	③	6.5-JJ	0.012 in.	58 - 72
2004	Impreza	P195/60R15	None	③	③	6-JJ	0.012 in.	58 - 72
	Impreza RS	P205/55R16	None	③	③	7-JJ	0.012 in.	58 - 72
	Impreza Outback	P225/60R16	T145/80R16	③	③	16x16 1/2JJ	0.012 in.	65
	Impreza WRX	P195/60R15	P205/55R16 P215/45R17	③	③	6-JJ 7-JJ	0.012 in.	58 - 72
	Baja	P185/70R14	P195/60R15 P205/55R16 P205/70R15	③	③	6.5-JJ	0.012 in.	58 - 72
	Legacy	P185/70R14	P195/60R15 P205/55R16 P205/70R15	③	③	6.5-JJ	0.012 in.	58 - 72
2005	Impreza ①	②	None	③	③	④	0.012 in.	66
	Outback	⑤	None	③	③	⑥	0.012 in.	81
	Baja	P215/60R16	P225/60R16	③	③	NA	0.012 in.	81
	Legacy	⑦	None	③	③	⑧	0.012 in.	81
2006	Impreza ①	⑨	None	③	③	⑩	0.012 in.	66
	Outback	P205/55R17	None	③	③	7JJ	0.012 in.	81
	Baja	P225/60R16	None	③	③	NA	0.012 in.	81
	Legacy	⑪	None	③	③	7JJ	0.012 in.	81

OEM: Original Equipment Manufacturer

PSI: Pounds Per Square Inch

NA: Not Available

① Includes WRX

② Except STI: P205/55R16
STI: P225/45R17

③ See vehicle placard on driver's door

④ Except STI: 6.5JJ
STI: 8JJ

⑤ 2.5L SOHC: P225/60R16
2.5L DOHC and 3.0L P225/55R17

⑥ 2.5L SOHC: 6.5JJ
2.5L DOHC and 3.0L: 7JJ

⑦ 2.5L SOHC: 205/55R16
2.5L DOHC: P215/45R17

⑧ 2.5L SOHC: 6.5JJ
2.5L DOHC: 7JJ

⑨ 2.5i, Outback: P205/55R16
WRX: P215/45R17
STI: P225/45R17

⑩ 2.5i, Outback: 6.5JJ
WRX: 7JJ
STI: 8JJ

⑪ 2.5L SOHC: P205/55R17
2.5L DOHC: P215/ZR17

BRAKE SPECIFICATIONS

All measurements in inches unless noted

Year	Model		Brake Disc Original Thickness	Brake Disc Minimum Thickness	Brake Disc Maximum Runout	Brake Drum Diameter Original Inside Diameter	Brake Drum Diameter Max. Wear Limit	Brake Drum Diameter Maximum Machine Diameter	Minimum Lining Thickness Front	Minimum Lining Thickness Rear	Brake Caliper Bracket Bolts (ft. lbs.)	Brake Caliper Mounting Bolts (ft. lbs.)
2002	Impreza	F	0.940	0.870	0.0030	—	—	—	0.059	—	59	23-30
		R	0.390	0.335	0.0040	9.000 ①	9.079 ②	NA	—	0.059	—	23-30
	WRX	F	0.940	0.870	0.0030	—	—	—	0.295	—	59	19.5
		R	0.390	0.335	0.0040	9.000 ①	9.080 ②	9.000 ①	—	0.059	59	27.5
	Outback	F	0.945	0.866	0.0030	—	—	—	0.295	—	59	29
		R	0.390	0.335	0.0030	9.000 ①	9.079 ②	NA	—	0.256	59	29
	Legacy	F	0.940	0.870	0.0030	—	—	—	0.059	—	59	23-30
		R	0.390	0.335	0.0040	9.000 ①	9.079 ②	NA	—	0.059	—	23-30
2003	Impreza	F	0.940	0.870	0.0030	—	—	—	0.059	—	59	23-30
		R	0.390	0.335	0.0040	9.000 ①	9.079 ②	NA	—	0.059	—	23-30
	WRX	F	0.940	0.870	0.0030	—	—	—	0.295	—	59	19.5
		R	0.390	0.335	0.0040	9.000 ①	9.080 ②	9.000 ①	—	0.059	59	27.5
	Outback	F	0.945	0.866	0.0030	—	—	—	0.295	—	59	29
		R	0.390	0.335	0.0030	9.000 ①	9.079 ②	NA	—	0.256	59	29
	Baja	F	0.940	0.870	0.0030	—	—	—	0.059	—	59	23-30
		R	0.390	0.335	0.0040	9.000 ①	9.079 ②	NA	—	0.059	—	23-30
	Legacy	F	0.940	0.870	0.0030	—	—	—	0.059	—	59	23-30
		R	0.390	0.335	0.0040	9.000 ①	9.079 ②	NA	—	0.059	—	23-30
2004	Impreza	F	0.940	0.870	0.0030	—	—	—	0.059	—	59	23-30
		R	0.390	0.335	0.0040	9.000 ①	9.079 ②	NA	—	0.059	—	23-30
	WRX	F	0.940	0.870	0.0030	—	—	—	0.295	—	59	19.5
		R	0.390	0.335	0.0040	9.000 ①	9.080 ②	9.000 ①	—	0.059	59	27.5
	Outback	F	0.945	0.866	0.0030	—	—	—	0.295	—	59	29
		R	0.390	0.335	0.0030	9.000 ①	9.079 ②	NA	—	0.256	59	29
	Baja	F	0.940	0.870	0.0030	—	—	—	0.059	—	59	23-30
		R	0.390	0.335	0.0040	9.000 ①	9.079 ②	NA	—	0.059	—	23-30
	Legacy	F	0.940	0.870	0.0030	—	—	—	0.059	—	59	23-30
		R	0.390	0.335	0.0040	9.000 ①	9.079 ②	NA	—	0.059	—	23-30
2005	Impreza	F	③	④	0.0030	—	—	—	⑤	—	59	19.5
		R	⑥	⑦	0.0028	—	—	—	—	⑧	—	⑨
	WRX	F	③	④	0.0030	—	—	—	⑤	—	59	19.5
		R	⑥	⑦	0.0028	—	—	—	—	⑧	—	⑨
	Outback	F	③	④	0.0020	—	—	—	0.590	—	59	19.9
		R	⑩	⑪	0.0020	—	—	—	—	0.059	—	⑫
	Baja	F	0.940	0.870	0.0030	—	—	—	0.295	—	59	27.5
		R	0.390	0.335	0.0028	—	—	—	—	0.256	—	28.9
	Legacy	F	③	④	0.0020	—	—	—	0.590	—	59	19.9
		R	⑩	⑪	0.0020	—	—	—	—	0.059	—	⑫
2006	Impreza	F	③	④	0.0030	—	—	—	⑤	—	⑬	⑭
		R	⑮	⑯	0.0028	—	—	—	—	⑧	—	⑰
	WRX	F	③	④	0.0030	—	—	—	⑤	—	⑬	⑭
		R	⑮	⑯	0.0028	—	—	—	—	⑧	—	⑰
	Outback	F	③	④	0.0020	—	—	—	0.590	—	59	19.9
		R	⑩	⑪	0.0020	—	—	—	—	0.059	—	⑫
	Baja	F	0.940	0.870	0.0030	—	—	—	0.295	—	59	27.5
		R	0.390	0.335	0.0028	—	—	—	—	0.256	—	28.9
	Legacy	F	③	④	0.0020	—	—	—	0.590	—	59	19.9
		R	⑩	⑪	0.0020	—	—	—	—	0.059	—	⑫

09490_SBCR_C0012A

BRAKE SPECIFICATIONS
All measurements in inches unless noted

Year	Model	Brake Disc			Brake Drum Diameter			Minimum Lining Thickness		Brake Caliper	
		Original Thickness	Minimum Thickness	Maximum Runout	Original Inside Diameter	Max. Wear Limit	Maximum Machine Diameter	Front	Rear	Bracket Bolts (ft. lbs.)	Mounting Bolts (ft. lbs.)

NA: Not Available

① Parking brake drum on vehicles with rear disc brake: 6.69 inches

② Specification for parking brake drum

③ 15 and 16 inch wheels: 0.940
   17 inch wheel: 1.180

④ 15 and 16 inch wheels: 0.870
   17 inch wheel: 1.100

⑤ 15 and 16 inch wheels: 0.059
   17 inch wheel: 0.047

⑥ 15 and 16 inch wheels: 0.039
   17 inch wheel: 0.790

⑦ 14 inch wheel: 0.335
   17 inch wheel: 0.790

⑧ 14 inch wheel: 0.059
   17 inch wheel: 0.047

⑨ 14 inch wheel: 27.5
   17 inch wheel: 47.9

⑩ 15 and 16 inch wheels: 0.390
   17 inch wheel: 0.710

⑪ Solid disc: 0.335
   Ventilated disc: 27.2

⑫ Solid disc: 19.9
   Ventilated disc: 27.2

⑬ 15 and 16 inch wheels: 59
   17 inch wheel: 114.3

⑭ 15 inch wheel: 19.5

⑮ 15 inch wheel: 0.390
   16 inch wheel: 0.710
   17 inch wheel: 10.790

⑯ 14 inch wheel: 0.335
   15 inch wheel: 0.630
   17 inch wheel: 0.710

⑰ 14 inch wheel: 27.5
   15 inch wheel: 39.1
   17 inch wheel: 47.9

09490_SBCR_C0012B

SCHEDULED MAINTENANCE INTERVALS

SUBARU—IMPREZA, OUTBACK, WRX, LEGACY & BAJA

TO BE SERVICED	TYPE OF SERVICE	VEHICLE MILEAGE INTERVAL (x1000)												
		7.5	15	22.5	30	37.5	45	52.5	60	67.5	75	82.5	90	97.5
Engine oil & filter	R	✓	✓	✓	✓	✓	✓	✓	✓	✓	✓	✓	✓	✓
Brake lines	S/I		✓		✓		✓		✓		✓		✓	
Clutch & hill holder system	S/I		✓		✓		✓		✓		✓		✓	
Disc brake pads & discs, front & rear axle boots & axle shaft joint portions	S/I		✓		✓		✓		✓		✓		✓	
Parking brake	S/I		✓		✓		✓		✓		✓		✓	
Steering & suspension	S/I		✓		✓		✓		✓		✓		✓	
Air filter element	R				✓				✓				✓	
Engine coolant	R				✓				✓				✓	
Fuel filter	R				✓				✓				✓	
Spark plugs (2002-2004)	R								✓					
Spark plugs: non turbo (2005-2006)	R				✓				✓				✓	
Spark plugs: turbo (2005-2006)	R								✓					
Automatic transmission fluid & filter	S/I				✓				✓				✓	
Brake fluid	R				✓				✓				✓	
Brake linings & drums	S/I				✓				✓				✓	
Camshaft drive belt (2002-2004) ①	S/I				✓				✓				✓	
Camshaft drive belt (2005-2006)	S/I				✓				✓					
Coolant level, hoses & clamps	S/I				✓				✓				✓	
Drive belts	S/I				✓				✓				✓	
Fuel system, hoses & connections	S/I				✓				✓				✓	
Transmission and/or differential gear fluid	S/I				✓								✓	
Tires (rotate)	S/I	✓	✓	✓	✓	✓	✓	✓	✓	✓	✓	✓	✓	✓
Front & rear wheel bearing repack	S/I								✓					

R: Replace   S/I: Service or Inspect

① Non-California vehicles: replace every 60,000 miles.

FREQUENT OPERATION MAINTENANCE (SEVERE SERVICE)

If a vehicle is operated under any of the following conditions it is considered severe service:

- Extremely dusty areas.

- 50% or more of the vehicle operation is in 32°C (90°F) or higher temperatures, or constant operation in temperatures below 0°C (32°F).

- Prolonged idling (vehicle operation in stop and go traffic).

- Frequent short running periods (engine does not warm to normal operating temperatures).

- Police, taxi, delivery usage or trailer towing usage.

Oil & oil filter change: change every 3750 miles.

Clutch & hill holder system: service or inspect every 7500 miles.

Air filter element: service or inspect every 15,000 miles.

Automatic transmission fluid: service or inspect every 15,000 miles.

Brake linings & drums: service or inspect every 15,000 miles.

Coolant level, hoses & clamps: service or inspect every 15,000 miles.

Drive belts: service or inspect every 15,000 miles.

Transmission/differential gear oil (except SVX): service or inspect every 15,000 miles.

Front & rear wheel bearing repack: service or inspect every 30,000 miles.

ENGINE REPAIR

➡ **Disconnecting the negative battery cable on some vehicles may interfere with the functions of the on board computer systems and may require the computer to undergo a relearning process, once the negative battery cable is reconnected.**

Alternator

REMOVAL & INSTALLATION

1. Before servicing the vehicle, refer to the Precautions Section.
2. Remove or disconnect the following:
   - Negative battery cable
   - Connector and terminal from the alternator
   - V-belt cover, if equipped
   - Front side V-belt
   - Alternator to bracket bolts
   - Alternator from the vehicle

To install:

3. Install or connect the following:
   - Alternator into the vehicle
   - Alternator to bracket bolts
   - Front side V-belt
   - V-belt cover, if equipped
   - Connector and terminal to the alternator
   - Negative battery cable
4. Check and adjust the belt tension.

Ignition Timing

ADJUSTMENT

All Subaru's are equipped with Distributorless Ignition System (DIS). The ignition timing is controlled by the Powertrain Control Module (PCM) and is not adjustable.

Engine Assembly

❋❋ CAUTION

Whenever working near any of the Supplemental Restraint System (SRS) components, such as the impact sensors, the air bag module, steering column and instrument panel, properly disable the SRS.

REMOVAL & INSTALLATION

2.0L Engine

1. Before servicing the vehicle, refer to the Precautions Section.

2. Properly relieve the fuel system pressure.

3. Disconnect the fuel pump relay connector, start the engine and let it stall. Once the engine stalls, crank it for a further 5 seconds to ensure the fuel system is properly relieved.

4. Disconnect the negative battery cable.

5. Drain the engine oil and coolant into suitable containers.

6. Raise the rear seat and turn the floor mat up.

7. Remove the fuel filler cap.

8. Remove or disconnect the following:
   - Air cleaner cover and element
   - Radiator
   - Coolant filler tank

9. If equipped with air conditioning, discharge the system using an approved recovery/recycling machine. Disconnect and cap the lines from the compressor.
   - Intercooler

10. Disconnect the following electrical connections:
    - Engine harness connector
    - Engine ground terminal
    - Alternator connector, terminal and A/C compressor connections

11. Remove or disconnect the following:
    - Accelerator cable
    - Clutch release spring
    - Brake booster hose
    - Heater inlet and outlet hoses

12. Remove the power steering pump from the bracket by performing the following steps:
    a. Loosen the lock and slider bolts.
    b. Remove the V-belt.
    c. Disconnect the power steering switch connection.
    d. Remove the pipe with bracket from the intake manifold.
    e. Remove the power steering pump from the engine.
    f. Remove the power steering tank from the bracket by pulling it upwards.

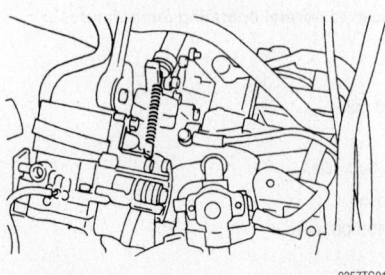

Clutch release spring location—2.0L engine

g. Place the power steering pump on the wheel apron on the right.

13. Remove or disconnect the following:
    - Center exhaust pipe
    - Nuts that attach the lower side of the engine to the transmission
    - Nuts that attach the front cushion rubber onto the crossmember

14. Disconnect the clutch release fork from the release bearing as follows:
    a. Remove the clutch cylinder from the transmission.
    b. Using a 10mm wrench, remove the plug.
    c. Screw a 6mm diameter bolt into the release fork and remove it.
    d. Raise the release fork and unfasten the release tabs to free the release fork.

15. Disconnect the torque converter

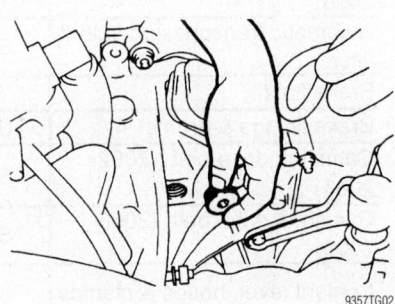

Using a 10mm wrench, remove the plug—2.0L engine

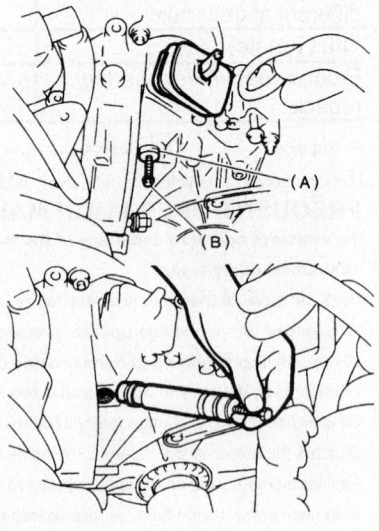

(A) Shaft
(B) Bolt

Screw a 6mm diameter bolt into the release fork and remove it—2.0L engine

clutch from the drive plate if equipped with automatic transmission as follows:

    a. Remove the service hole plug.

    b. Remove the torque converter clutch-to-drive plate bolts.

    c. Remove the remaining bolts while rotating the engine using a crankshaft pulley wrench.

16. Remove or disconnect the following:

- Pitching stopper
- Fuel delivery, return and evaporation hoses
- Fuel filter and bracket

17. Attach a lifting device to the engine.

18. Using a floor jack, support the transmission.

19. Remove the starter.

20. Separate the engine from the transmission.

21. Remove the upper right transmission-to-engine bolts.

22. Remove the engine as follows:

    a. Raise the engine slightly.

    b. Using the floor jack, raise the transmission.

    c. Move the engine horizontally until the mainshaft is withdrawn from the clutch cover.

    d. Remove the engine.

To install:

23. Installation is the reverse of removal, please note the following torques:

- Clutch release fork plug: 32 ft. lbs. (44 Nm)
- Front cushion rubbers: 25 ft. lbs. (34 Nm)
- Bolts attaching the upper right side of the transmission to the engine: 37 ft. lbs. (50 Nm)
- Pitching stopper-to-fender bolt: 37 ft. lbs. (50 Nm)
- Pitching stopper-to-engine bolt: 43 ft. lbs. (58 Nm)
- Torque converter clutch-to-drive plate bolts, while rotating the engine: 18 ft. lbs. (25 Nm)
- Power steering pump bolts: 15 ft. lbs. (20 Nm)
- Bolts attaching the lower side of the transmission to the engine: 37 ft. lbs. (50 Nm)
- Front cushion rubber-to-crossmember bolts: 61 ft. lbs. (83 Nm)

24. Fill the engine with the recommended oil.

25. Fill and bleed the cooling system.

26. Charge the air conditioning system using an approved recovery/recycling machine.

27. Adjust the clutch cable.

28. If equipped, check the automatic transmission fluid level and add Dexron®II if necessary.

29. Start the engine and allow it to reach normal operating temperature. Check for leaks.

2.5L Engine

2002–2004

1. Before servicing the vehicle, refer to the Precautions Section.

2. Relieve the fuel system pressure.

3. Drain the engine oil and coolant into suitable containers.

4. Remove or disconnect the following:

- Battery cables
- Battery from the vehicle
- Radiator hoses
- Fan motor harness
- Radiator

5. If equipped with air conditioning, discharge the system using an approved recovery/recycling machine. Disconnect and cap the lines from the compressor.

- Air intake duct
- Air cleaner element and upper cover
- Evaporator canister and bracket
- Oxygen Sensor (O$_2$S) connector
- Engine ground terminal
- Crankshaft Position (CKP) sensor connector
- Camshaft Position (CMP) sensor connector
- Knock Sensor (KS) connector
- Alternator connector and terminal
- Air conditioning compressor connectors, if equipped
- Accelerator cable
- Cruise control cable, if equipped
- Clutch release spring, clutch cable and hill holder cable, if equipped with a manual transmission
- Brake booster hose(s)
- Heater inlet and outlet hoses
- Alternator drive belt
- Spark plug wires from left side of engine
- Power steering pump line bracket
- Power steering pump, leave the lines connected and position aside
- Exhaust Y-pipe
- Lower starter nuts
- Lower engine-to-transmission nuts
- Front engine mount-to-crossmember nuts
- Starter

6. If equipped with an automatic transmission, perform the following:

    a. Remove the torque converter service hole plug.

    b. Rotate the engine. Remove the torque converter-to-drive plate bolts as they become accessible.

7. Remove or disconnect the following:

- Pitching stopper
- Fuel delivery, return and evaporation hoses

8. Support the engine with a suitable lifting device attached to the engine lifting eyes.

9. Slightly raise the engine.

10. Raise the transmission with a floor jack.

11. Slowly remove the engine from the vehicle.

To install:

12. Apply a small amount of grease to the splines of the mainshaft.

13. Position the engine in the engine compartment and align it with the transmission.

14. Install or connect the following:

- Engine upper bolts and tighten to 34–40 ft. lbs. (44–54 Nm)

15. Remove the lifting device and floor jack.

16. Install the pitching stopper and tighten the bolts to the following specifications:

    a. Body side: 49 ft. lbs. (67 Nm).

    b. Bracket side: 40 ft. lbs. (54 Nm).

17. If equipped with an automatic transmission, perform the following:

    a. Install the torque converter-to-drive plate bolts while rotating the engine, and tighten to 20 ft. lbs. (26 Nm).

    b. Install the service hole cover.

18. Install or connect the following:

- Evaporator canister and bracket
- Power steering pump. Torque retainers to 22–36 ft. lbs. (29–47 Nm).
- Drive belt, adjust tension
- Starter. Tighten bolts to 34–40 ft. lbs. (44–52 Nm).
- Lower engine-to-transmission nuts. Tighten to 34–40 ft. lbs. (44–52 Nm).
- Lower engine mounting nuts. Tighten to 61 ft. lbs. (83 Nm) in the inner most elliptical hole in the front crossmember so the clearance is 0.16–0.24 in. (4–6mm).
- Exhaust Y-pipe with new gaskets and nuts
- Brake booster hose
- Heater inlet and outlet hoses
- Accelerator and the cruise control cables, if equipped

19. If equipped with a manual transmission, install the following:

- Clutch release spring
- Clutch cable

- Hill holder cable
20. Install or connect the following:
  - Engine harness connectors
  - O₂ sensor connector
  - Engine ground terminal
  - CKP sensor connector
  - CMP sensor connector
  - Knock sensor connector
  - Alternator connector and terminal
  - Air conditioning compressor connectors, if equipped
  - Air cleaner element and cover
  - Air conditioning lines, if equipped, with new O-rings. Tighten the bolts to 23 ft. lbs. (31 Nm).
  - Radiator
  - Engine cover
  - Battery
21. Fill the engine with the recommended oil.
22. Fill and bleed the cooling system.
23. Charge the air conditioning system using an approved recovery/recycling machine.
24. Adjust the clutch cable.
25. If equipped, check the automatic transmission fluid level and add Dexron®II if necessary.
26. Start the engine and allow it to reach normal operating temperature. Check for leaks.

2005–2006 IMPREZA AND WRX

1. Before servicing the vehicle, refer to the Precautions Section.
2. Open the hood to the full open position.
3. Properly relieve the fuel system pressure. Remove the fuel cap.
4. Properly disarm the SRS system.
5. Discharge the air conditioning system.
6. Disconnect the negative battery cable.
7. Drain the engine oil. Drain the cooling system.
8. Remove the radiator. Remove the coolant filler tank.
9. On STI, remove the secondary air pump.
10. Disconnect the air conditioning pressure hoses from the compressor.
11. Disconnect the engine harness connector, ground cable, alternator connector and terminal and air conditioning compressor electrical connector.
12. If equipped, disconnect the clutch release spring.
13. Disconnect the brake booster hose and the heater hoses.
14. Remove the front side V-belt. Remove the power steering pipe and

bracket. Remove the reservoir tank. Remove the power steering pump and position it aside.
15. Raise and support the vehicle safely.
16. Remove the automatic transmission fluid line from the frame, if equipped.
17. Remove the center exhaust pipe.
18. Remove the lower transmission to engine retaining nuts.
19. Remove the front cushion rubber onto front crossmember retaining nuts.
20. Separate the clutch release fork from the release bearing, if equipped.
21. Lower the vehicle.
22. If equipped with automatic transmission, remove the service plug hole cap. Remove the torque converter to drive plate retaining bolts.
23. Remove the pitching stopper. Disconnect the fuel hoses from the fuel pipes.
24. Position the engine lifting fixture in place. Support the transmission using a suitable jack.

➡**Before separating the engine from the transmission, check to be sure nothing has been overlooked that will stop the engine from being removed. This is very important in order to facilitate reinstallation and because the transmission lowers under its own weight.**

25. Separate the engine from the transmission.
26. Remove the starter.
27. If equipped with automatic transmission, install tool ST498277200, with will hold the torque converter in place.
28. Remove the upper right side bolts that retain the transmission to the engine.
29. Slightly raise the engine. Raise the transmission, using the suitable jack. Move the engine horizontally until the mainshaft is withdrawn from the clutch cover. Carefully remove the engine from the vehicle.

To install:
30. Installation is the reverse of the removal procedure.
31. Torque the upper transmission to engine retaining bolts to 37 ft. lbs.
32. Torque the lower transmission to engine retaining bolts to 37 ft. lbs.
33. If equipped, tighten the torque converter to drive plate bolts to 18.4 ft. lbs.
34. Fill the engine with the proper grade and type engine oil.
35. Fill and bleed the cooling system.
36. Charge the air conditioning system.
37. If equipped, adjust the clutch cable as required.
38. If equipped, check the automatic transmission fluid level, correct as required

using the proper grade and type transmission fluid.
39. Start the engine and allow it to reach normal operating temperature. Check for leaks and correct as required.

2005–2006 OUTBACK AND LEGACY

1. Before servicing the vehicle, refer to the Precautions Section.
2. Open the hood to the full open position.
3. Properly relieve the fuel system pressure. Remove the fuel cap.
4. Properly disarm the SRS system.
5. Discharge the air conditioning system.
6. Disconnect the negative battery cable.
7. Drain the engine oil. Drain the cooling system.
8. Remove the collector cover.
9. Remove the radiator. Remove the coolant filler tank.
10. Disconnect the air conditioning pressure hoses from the compressor.
11. If equipped, remove the turbocharger.
12. Disconnect the engine harness connector, ground cable, alternator connector and terminal and air conditioning compressor electrical connector.
13. Disconnect the brake booster hose and the heater hoses.
14. Remove the hose between the intake manifold and the pressure regulator valve.
15. Loosen the lock bolt and slider bolt and remove the front side V-belt. Disconnect the power steering pressure switch connector. Remove the power steering pump and position it aside.
16. Raise and support the vehicle safely.
17. Remove the center exhaust pipe.
18. Remove the lower transmission to engine retaining nuts.
19. Remove the front cushion rubber onto front crossmember retaining nuts.
20. Separate the clutch release fork from the release bearing, if equipped.
21. Lower the vehicle.
22. If equipped with automatic transmission, remove the service plug hole cap. Remove the torque converter to drive plate retaining bolts.
23. Remove the pitching stopper. Disconnect the fuel hoses from the fuel pipes.
24. Position the engine lifting fixture in place. Support the transmission using a suitable jack.

➡**Before separating the engine from the transmission, check to be sure nothing has been overlooked that will** stop the engine from being removed. This is very important in order to facilitate reinstallation and because the transmission lowers under its own weight.

25. Separate the engine from the transmission.
26. Remove the starter.
27. If equipped with automatic transmission, install tool ST498277200, with will hold the torque converter in place.
28. Remove the upper right side bolts that retain the transmission to the engine.
29. Slightly raise the engine. Raise the transmission, using the suitable jack. Move the engine horizontally until the mainshaft is withdrawn from the clutch cover. Carefully remove the engine from the vehicle.

To install:
30. Installation is the reverse of the removal procedure.
31. Torque the upper transmission to engine retaining bolts to 36.9 ft. lbs.
32. Torque the lower transmission to engine retaining bolts to 36.9 ft. lbs.
33. If equipped, tighten the torque converter to drive plate bolts to 18.1 ft. lbs.
34. Fill the engine with the proper grade and type engine oil.
35. Fill and bleed the cooling system.
36. Charge the air conditioning system.
37. If equipped. Adjust the clutch cable as required.
38. If equipped, check the automatic transmission fluid level, correct as required using the proper grade and type transmission fluid.
39. Start the engine and allow it to reach normal operating temperature. Check for leaks and correct as required.

2005–2006 BAJA

1. Before servicing the vehicle, refer to the Precautions Section.
2. Open the hood to the full open position.
3. Properly relieve the fuel system pressure. Remove the fuel cap.
4. Properly disarm the SRS system.
5. Discharge the air conditioning system.
6. Disconnect the negative battery cable.
7. Drain the engine oil. Drain the cooling system.
8. Remove the radiator. Remove the coolant filler tank.
9. Disconnect the air conditioning pressure hoses from the compressor.
10. If equipped, remove the turbocharger.

11. Disconnect the engine harness connector, left and right side ground cables, alternator connector and terminal and air conditioning compressor electrical connector.
12. Disconnect the brake booster hose and the heater hoses.
13. If equipped, disconnect the clutch release spring.
14. Remove the hose between the intake manifold and the pressure regulator valve.
15. Loosen the lock bolt and slider bolt and remove the front side V-belt. Disconnect the power steering pressure switch connector. Remove the pipe with the bracket from the intake manifold. Remove the power steering pump from the engine. Remove the reservoir tank from the bracket, by pulling it upward. Position the power steering pump on the right side of the wheel apron.
16. Raise and support the vehicle safely.
17. Remove the automatic transmission fluid line from the frame, if equipped.
18. Remove the center exhaust pipe.
19. Remove the lower transmission to engine retaining nuts.
20. Remove the front cushion rubber onto front crossmember retaining nuts.
21. If equipped with manual transmission, remove the clutch slave cylinder from the transmission. Remove the plug and screw a 6mm bolt into the release fork shaft and remove it. Raise the release fork and then unfasten the release bearing tabs to free the fork.

➡This is being doe to prevent interference with the engine when removing the engine from the transmission.

22. Lower the vehicle.
23. If equipped with automatic transmission, remove the service plug hole cap. Remove the torque converter to drive plate retaining bolts.
24. Remove the pitching stopper. Disconnect the fuel hoses from the fuel pipes. Remove the fuel filter and bracket.
25. Position the engine lifting fixture in place. Support the transmission using a suitable jack.

➡Before separating the engine from the transmission, check to be sure nothing has been overlooked that will stop the engine from being removed. This is very important in order to facilitate reinstallation and because the transmission lowers under its own weight.

26. Separate the engine from the transmission.
27. Remove the starter.

28. If equipped with automatic transmission, install tool ST498277200, with will hold the torque converter in place.
29. Remove the upper right side bolts that retain the transmission to the engine.
30. Slightly raise the engine. Raise the transmission, using the suitable jack. Move the engine horizontally until the mainshaft is withdrawn from the clutch cover. Carefully remove the engine from the vehicle.

To install:
31. Installation is the reverse of the removal procedure.
32. Torque the upper transmission to engine retaining bolts to 36.9 ft. lbs.
33. Torque the lower transmission to engine retaining bolts to 36.9 ft. lbs.
34. If equipped, tighten the torque converter to drive plate bolts to 18.1 ft. lbs.
35. Fill the engine with the proper grade and type engine oil.
36. Fill and bleed the cooling system.
37. Charge the air conditioning system.
38. If equipped. Adjust the clutch cable as required.
39. If equipped, check the automatic transmission fluid level, correct as required using the proper grade and type transmission fluid.
40. Start the engine and allow it to reach normal operating temperature. Check for leaks and correct as required.

3.0L Engine

2002–2004

1. Before servicing the vehicle, refer to the Precautions Section.
2. Relieve the fuel system pressure.
3. Drain the engine oil and coolant into suitable containers.
4. Remove or disconnect the following:
   - Battery cables
   - Battery
   - Air intake duct
   - Engine undercover
   - Radiator
   - Drive belt
5. If equipped with air conditioning, discharge the system using an approved recovery/recycling machine. Disconnect and cap the lines from the compressor.
   - Engine ground terminal
   - Engine harness connectors
   - Alternator connector and terminal
   - Air conditioning compressor connectors, if equipped
   - Accelerator and the cruise control cables, if equipped
   - Brake booster hose
   - Heater inlet and outlet hoses
   - Power steering pump line bracket

- Power steering pump, leave the lines connected and position aside
- Exhaust Y-pipe
- Lower engine-to-transmission nuts
- Front engine mount-to-crossmember nuts

6. If equipped with an automatic transmission, perform the following:

a. Remove the torque converter service hole plug.

b. Rotate the engine. Remove the torque converter-to-drive plate bolts as they become accessible.

7. Remove or disconnect the following:

- Pitching stopper
- Fuel delivery, return and evaporation hoses

8. Support the engine with a suitable lifting device attached to the engine lifting eyes.

9. Support the transmission with a floor jack.

10. Remove or disconnect the following:
- Starter
- Upper engine-to-transmission bolts

11. Slightly raise the engine.

12. Raise the transmission with a floor jack.

13. Slowly remove the engine from the vehicle.

To install:

14. Apply a small amount of grease to the splines of the mainshaft.

15. Position the engine in the engine compartment and align it with the transmission.

16. Install or connect the following:
- Engine upper bolts and tighten to 36 ft. lbs. (50 Nm)

17. Remove the lifting device and floor jack.

18. Install the pitching stopper and tighten the bolts to the following specifications:

a. Body side: 42 ft. lbs. (57 Nm).

b. Bracket side: 42 ft. lbs. (49 Nm).

19. Install the starter. Tighten bolts to 37 ft. lbs. (50 Nm).

20. If equipped with an automatic transmission, perform the following:

a. Install the torque converter-to-drive plate bolts while rotating the engine, and tighten to 18 ft. lbs. (25 Nm).

b. Install the service hole cover.

21. Install or connect the following:
- Power steering pump. Torque retainers to 14 ft. lbs. (20 Nm).
- Lower engine-to-transmission nuts. Tighten to 36 ft. lbs. (50 Nm).
- Lower engine mounting nuts. Tighten to 54 ft. lbs. (74 Nm).

- Exhaust Y-pipe with new gaskets and nuts
- Fuel delivery, return and evaporation hoses
- Heater inlet and outlet hoses
- Brake booster hose
- Engine ground terminal
- Engine harness connectors
- Alternator connector and terminal
- Air conditioning compressor connectors, if equipped
- Accelerator and the cruise control cables, if equipped
- Air conditioning lines, if equipped, with new O-rings. Tighten the bolts to 10 ft. lbs. (15 Nm).
- Radiator
- Drive belt, adjust tension
- Air cleaner element and cover
- Engine under cover
- Battery
- Battery cables

22. Fill the engine with the recommended oil.

23. Fill and bleed the cooling system.

24. Charge the air conditioning system using an approved recovery/recycling machine.

25. Check the automatic transmission fluid level and add Dexron®II if necessary.

26. Start the engine and allow it to reach normal operating temperature. Check for leaks.

2005–2006

1. Before servicing the vehicle, refer to the Precautions Section.

2. Open the hood to the full open position.

3. Properly relieve the fuel system pressure. Remove the fuel cap.

4. Properly disarm the SRS system.

5. Discharge the air conditioning system.

6. Disconnect the negative battery cable.

7. Drain the engine oil. Drain the cooling system.

8. Disconnect the air conditioning pressure hoses from the compressor.

9. If equipped, remove the turbocharger.

10. Remove the radiator. Remove the coolant filler tank.

11. Disconnect the engine harness connector, ground cable, power steering switch connector, alternator connector and terminal and air conditioning compressor electrical connector.

12. Disconnect the brake booster hose, pressure regulator vacuum hose and the heater hoses.

13. Remove the power steering pump from the bracket.

➡️**Do not disconnect the hoses from the pump body. Position the pump assembly aside.**

14. Raise and support the vehicle safely.

15. Remove the under cover.

16. Remove the center exhaust pipe.

➡️**Do not let the front exhaust pipe interfere with the coolant pipes on the engine side. Remove the ground cable.**

17. Remove the lower transmission to engine retaining nuts.

18. Remove the front cushion rubber onto front crossmember retaining nuts.

19. Lower the vehicle.

20. Remove the service plug hole cap. Remove the torque converter to drive plate retaining bolts.

21. Remove the pitching stopper. Disconnect the fuel hoses from the fuel pipes.

22. Position the engine lifting fixture in place. Support the transmission using a suitable jack.

➡️**Before separating the engine from the transmission, check to be sure nothing has been overlooked that will stop the engine from being removed. This is very important in order to facilitate reinstallation and because the transmission lowers under its own weight.**

23. Separate the engine from the transmission.

24. Remove the starter.

25. If equipped with automatic transmission, install tool ST498277200, with will hold the torque converter in place.

26. Remove the upper right side bolts that retain the transmission to the engine.

27. Slightly raise the engine. Raise the transmission, using the suitable jack. Move the engine horizontally until the mainshaft is withdrawn from the clutch cover. Carefully remove the engine from the vehicle.

➡️**Be careful not to damage adjacent body parts or panels with the crank pulley, oil level gauge etc.**

28. Remove the front cushion rubber mounts.

To install:

29. Installation is the reverse of the removal procedure.

30. Torque the upper transmission to engine retaining bolts to 37 ft. lbs.

31. Torque the lower transmission to engine retaining bolts to 37 ft. lbs.

32. If equipped, tighten the torque converter to drive plate bolts to 18 ft. lbs.

33. Fill the engine with the proper grade and type engine oil.

34. Fill and bleed the cooling system.

35. Charge the air conditioning system.

36. If equipped. Adjust the clutch cable as required.

37. If equipped, check the automatic transmission fluid level, correct as required using the proper grade and type transmission fluid.

38. Start the engine and allow it to reach normal operating temperature. Check for leaks and correct as required.

Water Pump

REMOVAL & INSTALLATION

Except 3.0L Engine

1. Before servicing the vehicle, refer to the Precautions Section.
2. Remove or disconnect the following:
   • Negative battery cable
   • Engine undercover, if equipped
3. Drain the coolant into a suitable container.
4. Remove or disconnect the following:
   • Radiator fan connector(s)
   • Radiator outlet and heater hoses
   • Heater bypass hose or overflow hose, if equipped

• Reservoir tank, on Legacy
• Radiator fan motor assembly(ies)
• Accessory drive belts
• Timing belt
• Belt tension adjuster
• Belt idler No. 2
• Camshaft Position (CMP) sensor
• Left side camshaft pulley(s)
• Left side rear timing belt cover
• Tensioner bracket
• Radiator and heater hoses from water pump
• Water pump retainer bolts
• Water pump

5. Inspect the radiator hoses for deterioration and replace as necessary.

To install:

6. Clean the gasket mating surfaces thoroughly. Always use new gaskets during installation.

7. Install or connect the following:
   • On 2002–2004 engines, tighten the pump bolts in sequence to 10 ft. lbs. (13 Nm). After tightening the bolts once, retighten to the same specification again.
   • On 2005–2006 engines, tighten the pump bolts in sequence to 8.9 ft. lbs. (13 Nm). After tightening the bolts once, retighten to the same specification again.
   • Radiator heater hoses to water pump
   • Tensioner bracket and tighten to 18 ft. lbs. (25 Nm)

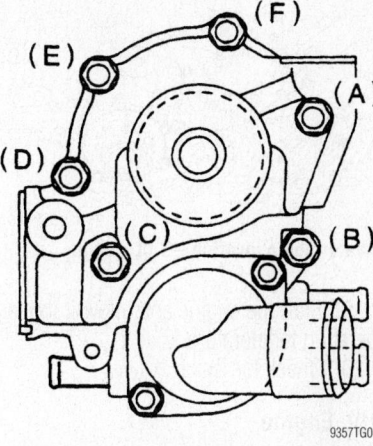

Tighten the water pump bolts in two steps using the following sequence—2.0L engine

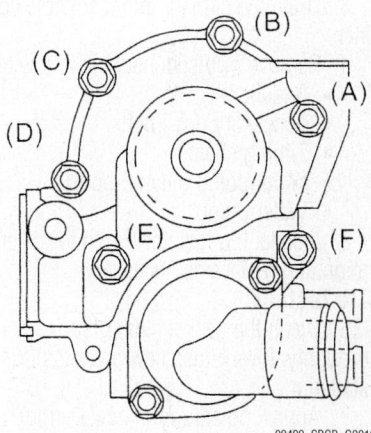

Tighten the water pump bolts in two steps using the following sequence—2005–2006 2.5L engine

• Left side rear timing belt cover
• Left side camshaft pulley(s). Tighten to 58 ft. lbs. (78 Nm) on non turbocharged engine and 72 ft. lbs. (98 Nm) on turbocharged engine.
• CMP sensor
• Belt idler No. 2 and tighten to 29 ft. lbs. (39 Nm)
• Belt tension adjuster
• Timing belt
• Accessory drive belts
• Radiator fan assembly(ies)
• Reservoir tank, if removed
• Heater bypass hose or overflow hose, if equipped
• Air intake duct
• Radiator outlet and heater hoses
• Radiator fan connector(s)
• Engine undercover, if removed

8. Fill the system with coolant and connect the negative battery cable.

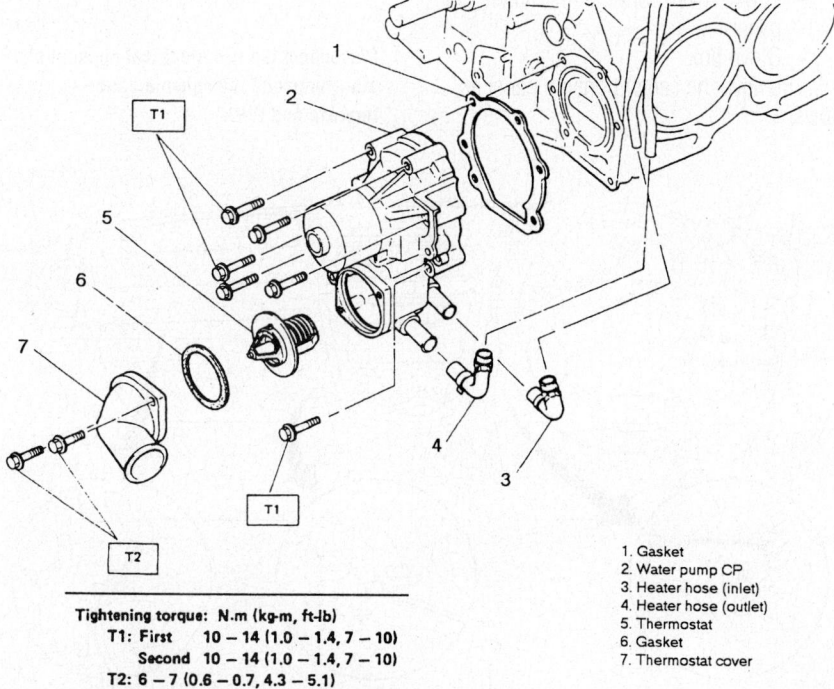

Tightening torque: N.m (kg-m, ft-lb)
T1: First   10 – 14 (1.0 – 1.4, 7 – 10)
    Second 10 – 14 (1.0 – 1.4, 7 – 10)
T2: 6 – 7 (0.6 – 0.7, 4.3 – 5.1)

1. Gasket
2. Water pump CP
3. Heater hose (inlet)
4. Heater hose (outlet)
5. Thermostat
6. Gasket
7. Thermostat cover

Water pump and related components—except 3.0L engine

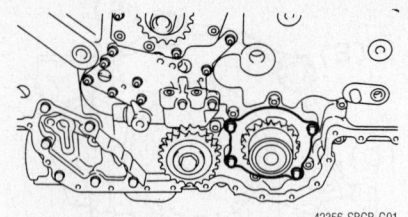

Water pump location—3.0L engine

9. Start the engine and allow it to reach operating temperature.

10. Check for leaks.

3.0L Engine

1. Before servicing the vehicle, refer to the Precautions Section.

2. Remove or disconnect the following:
- Negative battery cable
- Engine undercover, if equipped

3. Drain the coolant into a suitable container.

4. Remove or disconnect the following:
- Radiator
- Accessory drive belts
- Timing chain
- Water pump retainer bolts
- Water pump

5. Inspect the radiator hoses for deterioration and replace as necessary.

To install:

6. Clean the gasket mating surfaces thoroughly. Always use new gaskets during installation.

7. Apply coolant to the new O-ring before installation

8. Install or connect the following:
- Water pump with a new O-ring, tighten the bolts to 5 ft. lbs. (7 Nm).
- Timing chain
- Front chain cover
- Accessory drive belts
- Radiator
- Engine undercover, if removed

9. Fill the system with coolant and connect the negative battery cable.

10. Start the engine and allow it to reach operating temperature.

11. Check for leaks.

Heater Core

REMOVAL & INSTALLATION

2002–2004

WRX

1. Before servicing the vehicle, refer to the Precautions Section.

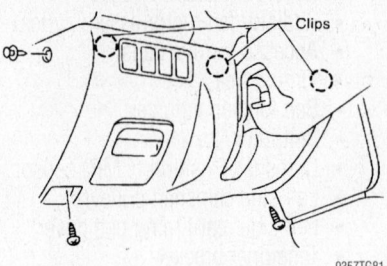

Location of the lower cover retainers— Impreza and WRX

2. Disconnect the negative battery cable.

✳✳ CAUTION

If equipped with an air bag system, wait 10 minutes after disconnecting the negative battery cable before performing any further work while the system fully de-energizes in order to avoid accidental deployment. All air bag system wiring is yellow. Do not use electrically powered test equipment on these circuits.

3. Drain the cooling system into a clean container for reuse.

4. Remove or disconnect the following:
- Bolts retaining the expansion valve and pipe
- Heater hoses from the heater core. Plug the heater core and heater hoses.
- Lower cover from the instrument panel
- Glove box

5. Remove the center console panel as follows:

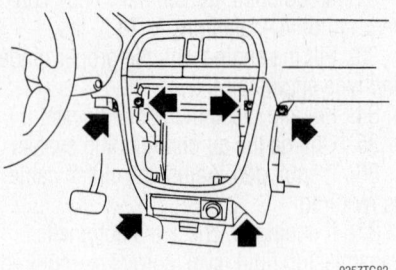

Location of the lower console panel retainers—Impreza and WRX

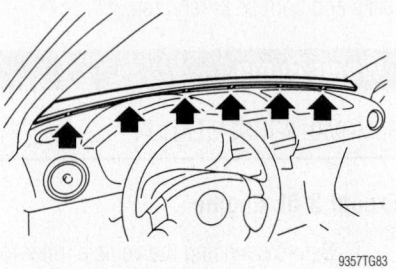

Loosen the hooks that retain the defroster panel—Impreza and WRX

Disconnect the two electrical connections after removing the console panel— Impreza and WRX

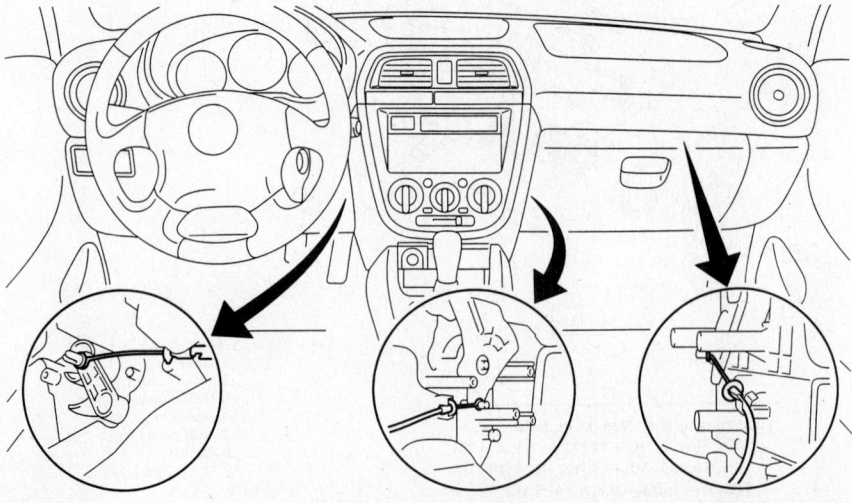

Location of the control unit control wires—Impreza and WRX

a. Remove lower panel and control wires.

b. Pull the control panel out and disconnect the connections.

6. Remove the passenger side air bag module as follows:

a. Disconnect the airbag connector from the support beam bracket and then unplug the connector.

b. Unfasten the 3 airbag module bolts and remove the module.

7. Remove or disconnect the following:
- 4 screws and two nuts then remove the lower console panel

- Hooks that retain the defroster panel
- Nuts and remove the electrical connections as shown in the illustration
- Instrument panel bolts and remove the panel
- Support beam
- Blower motor
- Heater core retainers
- Heater core

To install:

8. Install or connect the following:
- Heater core

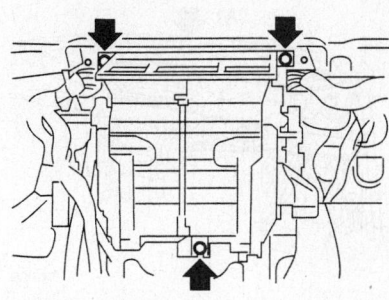

9357TG80

Location of the heater core retainers—Impreza and WRX

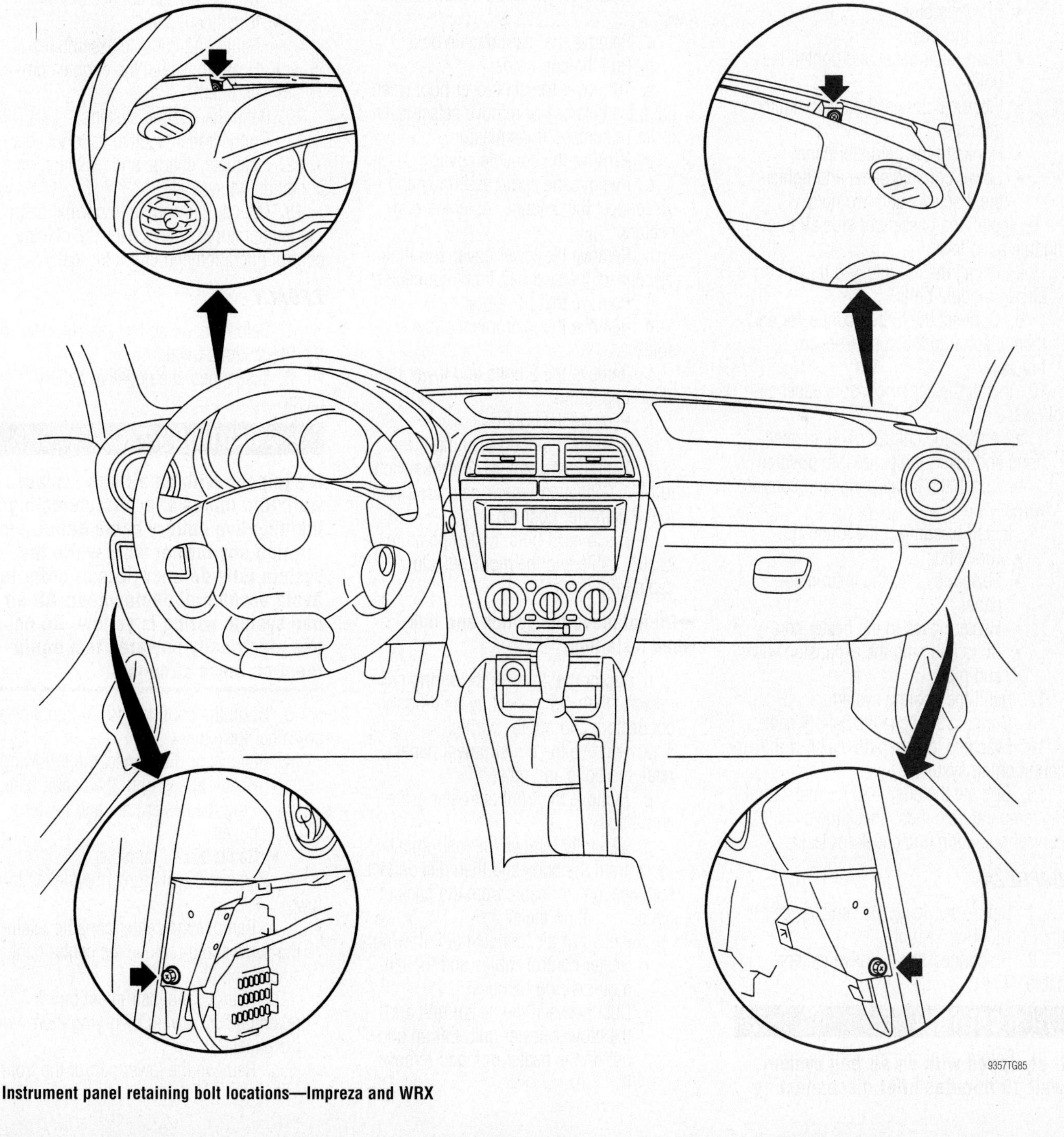

Instrument panel retaining bolt locations—Impreza and WRX

9357TG85

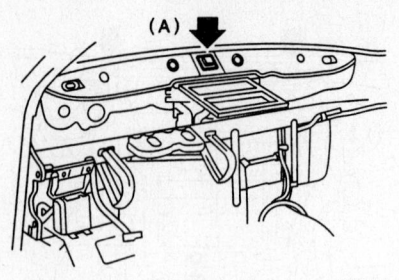

9357TG86

When installing the instrument panel, push the hook into grommet (A) on the body panel—Impreza and WRX

- Heater core retainers
- Blower motor
- Support beam
- Instrument panel and tighten the bolts
- Electrical connections and tighten the nuts
- Hooks to the defroster panel
- Lower console panel and tighten the 4 screws and two nuts

9. Install the passenger side air bag module as follows:

a. Install the module and tighten the 3 airbag module bolts.

b. Connect the airbag connector and then attach it to the support beam bracket.

10. Install the center console panel as follows:

a. Attach the control panel connections and insert the panel into position.

b. Install the lower panel and control wires.

11. Install or connect the following:

- Glove box
- Lower cover to the instrument panel
- Heater hoses to the heater core
- Bolts retaining the expansion valve and pipe

12. Refill the cooling system.

13. Connect the negative battery cable.

14. Evacuate, charge and leak test the air conditioning system.

15. Operate the engine to normal operating temperatures; then, check the climate control operation and check for leaks.

IMPREZA

1. Before servicing the vehicle, refer to the Precautions Section.

2. Disconnect the negative battery cable.

☀✳ CAUTION

If equipped with an air bag system, wait 10 minutes after disconnecting

the negative battery cable before performing any further work while the system fully de-energizes in order to avoid accidental deployment. All air bag system wiring is yellow. Do not use electrically powered test equipment on these circuits.

3. Drain the cooling system into a clean container for reuse.

4. Remove or disconnect the following:

- Heater hoses from the heater core. Plug the heater core and heater hoses.
- Radio box or console

5. Remove the instrument panel as follows:

a. Remove the rear console box.

b. Pull the cup holder.

c. Turn over the shift lever boot (manual transmission) or remove select lever cover (automatic transmission).

d. Remove the console cover.

e. Remove the audio assembly and disconnect the antenna cable and connectors.

f. Remove the lower cover and then disconnect the seat belt timer connector.

g. Remove the glove box.

h. Remove the instrument panel console.

i. Remove the 2 bolts and lower the steering column.

j. Remove the column cover.

k. Remove the hood opening lever.

l. Set the temperature control switch to MAX COLD, and mode selector switch to the defroster position.

m. Disconnect both the temperature control cable and the mode selector cable from the link.

➡ **Do not move the switch and link when installing.**

n. Tag or match mark the wiring connectors, then disconnect by holding the connectors and not the wiring.

o. Remove the 6 instrument panel retaining bolts and nuts.

p. Remove the front defroster grille and 2 bolts.

q. Carefully remove the instrument panel from the body and then disconnect the speedometer cable from the back of the combination meter.

6. Remove or disconnect the following:

- Heater control cables and the fan motor wiring harness
- Duct between the heater unit and the blower heater unit. Lift up and out on the heater unit and remove it.

7. With the heater assembly out of the vehicle, remove the heater core tube retaining clamps and lift the core from the heater case.

8. Remove the heater core.

To install:

9. Install or connect the following:

- Heater core into the heater case. Secure it in place with the retaining clamps and screws.
- Heater assembly to its mounting position under the dash. Torque the mounting bolts to 48–84 inch lbs. (5–9 Nm).
- Heater control cables and fan motor wiring harness connectors
- Instrument panel
- Radio and console assemblies
- Heater hoses in the engine compartment

10. Refill the cooling system.

11. Connect the negative battery cable.

12. Evacuate, charge and leak test the air conditioning system.

13. Operate the engine to normal operating temperatures; then, check the climate control operation and check for leaks.

LEGACY

1. Before servicing the vehicle, refer to the Precautions Section.

2. Disconnect the negative battery cable.

☀✳ CAUTION

If equipped with an air bag system, wait 10 minutes after disconnecting the negative battery cable before performing any further work while the system fully de-energizes in order to avoid accidental deployment. All air bag system wiring is yellow. Do not use electrically powered test equipment on these circuits.

3. Drain the cooling system into a clean container for reuse.

4. Remove or disconnect the following:

- Heater hoses from the heater core. Plug the heater core and heater hoses.
- Radio box or console

5. Remove the instrument panel as follows:

a. Remove the center console retaining screws and remove the center console assembly.

b. Remove the instrument panel retaining bolt covers by prying them from the panel.

c. Remove the lower part of the front

A pillar trim. Remove the instrument panel under covers from the driver's and passenger's sides.

d. Remove the hood release cable from the hood release lever.

e. Disconnect the wiring harness connectors under the instrument panel.

f. Remove the instrument cluster assembly. Remove the glove box assembly

g. Disconnect the ventilation control cables and electrical connectors at the heater unit. Disconnect the vacuum line at the blower housing.

h. Disconnect the radio antenna feeder wire. Disconnect the main harness connector at the fuse box.

i. Remove the lower steering column

covers. Remove the steering column retaining bolts and allow the column to hang down.

j. Remove the instrument panel retaining bolts.

➡**When removing the instrument panel, check that all wiring and cables are disconnected before pulling it completely away from the firewall.**

k. With the help of an assistant, lift and remove the instrument panel from the vehicle.

6. Remove or disconnect the following:
• Heater control cables and the fan motor wiring harness
• Duct between the heater unit and the blower heater unit. Lift up and

out on the heater unit and remove it.

7. Remove the evaporator case assembly by performing the following procedure:

a. Discharge and recover the air conditioning system refrigerant.

b. Disconnect the low and high pressure line from the evaporator outlet and cap the fittings.

c. Remove the inlet and outlet pipe grommets.

d. Remove the glove box and support bracket.

e. Disconnect the air conditioning wiring harness from the evaporator. Disconnect the drain hose from the evaporator.

f. Remove the evaporator mounting nut and bolt.

1. Heater case	8. Mix door
2. Heater core	9. Sub mix door
3. Vent duct	10. Heat door
4. Heat duct	11. Defroster lever
5. Defroster door	12. Vent lever 1
6. Vent door 1	13. Vent lever 2
7. Vent door 2	14. Mix lever

15. Heat lever
16. Main link
17. Screw
18. Spring
19. Motor actuator
20. Motor actuator bracket
21. Rod motor actuator
22. Mix rod 1
23. Mix rod 2
24. Rod hold
25. Clip
26. Clamp
27. Clamp
28. Bracket
29. Mix rod 3
30. Mix link 1
31. Mix link 2

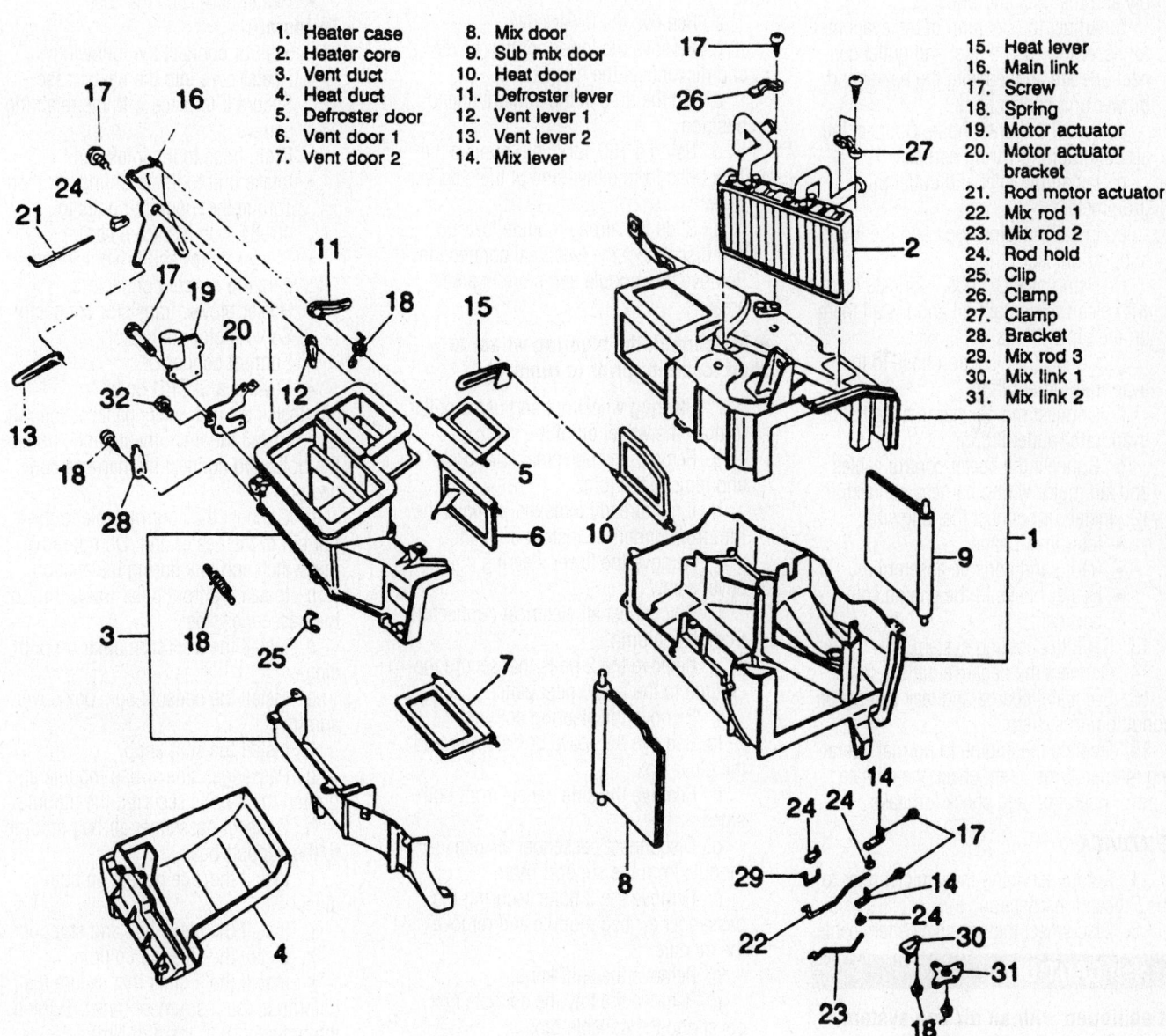

93112G93

Heater assembly and related components—2002–2004 Legacy

g. Remove the evaporator case assembly from the vehicle.

8. With the heater assembly out of the vehicle, remove the heater core tube retaining clamps and lift the core from the heater case.

9. Remove the heater core.

To install:

10. Install or connect the following:

- Heater core into the heater case. Secure it in place with the retaining clamps and screws.
- Heater assembly to its mounting position under the dash Torque the mounting bolts to 48–84 inch lbs. (5–9 Nm).

11. Install the evaporator by performing the following procedure:

a. Install the evaporator case assembly and the nuts and bolts.

b. Adjust the position of the evaporator assembly so the inlet and outlet connections are aligned with the heater and blower unit connections.

c. Install the drain hose. Connect the air conditioning wiring harness.

d. Install the inlet and outlet pipe grommets.

e. Install the glove box and the lower support bracket.

f. Using new O-rings, lubricate them with clean refrigerant oil and install them on the pipe fittings.

g. Connect the suction hose to the evaporator inlet fitting.

h. Connect the discharge hose to the evaporator outlet fitting.

i. Connect the heater control cables and fan motor wiring harness connectors.

12. Install or connect the following:

- Instrument panel
- Radio and console assemblies
- Heater hoses in the engine compartment

13. Refill the cooling system.

14. Connect the negative battery cable.

15. Evacuate, charge and leak test the air conditioning system.

16. Operate the engine to normal operating temperatures; then, check the climate control operation and check for leaks.

OUTBACK

1. Before servicing the vehicle, refer to the Precautions Section.

2. Disconnect the negative battery cable.

✳✳ CAUTION

If equipped with an air bag system, wait 10 minutes after disconnecting the negative battery cable before performing any further work while the system fully de-energizes in order to avoid accidental deployment. All air bag system wiring is yellow. Do not use electrically powered test equipment on these circuits.

3. Drain the cooling system into a clean container for reuse.

4. Remove or disconnect the following:

- Negative battery cable
- LLC
- Heater hoses from the heater core. Plug the heater core and heater hoses.
- Bolts attaching the expansion valve and pipe in the engine compartment

5. Remove the instrument panel as follows:

a. Remove the lower cover.

b. Remove the lower column cover and disconnect the harness.

c. Set the tires in the straight ahead position.

d. Using a T30 Torx® bit, remove the Torx bolts from either side of the steering wheel.

e. Slide the airbag module forward and disconnect the electrical connection. Remove the module and store in a safe area.

➡**Matchmark the steering wheel to shaft location, prior to removal.**

f. Steering wheel nut and use a puller to draw the wheel off of the shaft.

g. Remove the universal joint bolts and remove the joint.

h. If not already removed, remove the trim from under the instrument panel.

i. Remove the lower steering column cover screw.

j. Disconnect all electrical connectors from the column.

k. Remove the 2 bolts that secure the column to the instrument panel.

l. Remove the steering column.

m. Remove the glove box stopper and the glove box.

n. Remove the side panels from both sides.

o. Disconnect passenger air bag module from the support beam.

p. Remove the 3 bolts securing the passenger air bag module and remove the module.

q. Remove the shift knob.

r. Remove the tray the console box cover and the console box.

s. Remove the front trim pillar from both sides.

t. Remove the front pillar lower trim from the passenger side.

u. Set the temperature control switch to **FULL HOT** and disconnect the control cable from the bottom of the heater unit. Do not move the switch and link.

v. Remove the instrument panel bolts, harness connectors and the panel.

6. Remove or disconnect the following:

- Keyless and CRU units
- Sunroof connector
- Servo motor connector
- Heater blower transistor connector
- Blower motor and in-vehicle temperature sensor connectors
- Intake unit bolts and nuts
- Drain hose from the intake unit
- Intake unit
- Heater unit case screws
- Heater core from the case

To install:

7. Install or connect the following:

- Heater core into the heater case. Secure it in place with the retaining screws.
- Drain hose to the intake unit
- Intake unit to its mounting position. Torque the mounting bolts to 48–84 inch lbs. (5–9 Nm).
- Blower motor and in-vehicle temperature sensor connectors
- Heater blower transistor connector
- Servo motor connector
- Sunroof connector
- Keyless and CRU units

8. Install the instrument panel as follows:

a. Install the instrument panel, tighten the bolts and connect the harness connectors

b. Connect the control cable to the bottom of the heater unit. Do not move the switch and link during installation.

c. Install the front pillar lower trim to the passenger side.

d. Install the front trim pillar on both sides.

e. Install the console box, box cover and tray.

f. Install the shift knob.

g. Passenger side airbag module and tighten the 3 bolts securing the module.

h. Connect passenger air bag module to the support beam.

i. Install the side panels on both sides.

j. Install the glove box and stopper.

k. Install the steering column.

l. Install the 2 bolts that secure the column to the instrument panel. Tighten the bolts to 18 ft. lbs. (25 Nm).

m. Connect all electrical connectors to the column.

n. Install the lower steering column cover with the tilt lever held in the lowered position.

o. Install the universal joint. Align the bolt hole on the long yoke side of the joint with the cutout at the serrated section of the shaft end and insert the joint. Align the bolt hole on the short yoke side of the joint at the serrated section of the gearbox assembly and lower the joint completely. Temporarily tighten the bolt on the short yoke side and raise the joint to ensure the bolt is passing properly through the cutout. Tighten the bolt on the long yoke side, then the short yoke side to 17 ft. lbs. (24 Nm).

❋❋ CAUTION

Make sure the joint bolt is tightened through the notch in the shaft serration. Do not over tighten the joint bolts as this can lead to heavy steering wheel operation. Make sure the clearance between the gearbox is over 0.59 inch (15mm).

p. Making sure the wheels are still in the straight ahead position, turn the roll connector (A) clockwise until it stops. Then turn the roll connector pin (A) counterclockwise approximately 2.65 turns until the marks are aligned. Refer to the accompanying illustration for more detail.

q. Steering wheel, align the matchmarks made during removal and install the nut. Tighten the nut to 32 ft. lbs. (45 Nm). make sure the column cover-to-steering wheel clearance is 0.08–0.16 inch (2–4mm).

r. Drivers side airbag, engage the connector and tighten the fasteners.

❋❋ CAUTION

Insert the roll connector guide pin into the guide hole on the lower end of the surface of the steering wheel to prevent any damage. Draw out the airbag, horn and cruise control connectors from the guide hole of the steering wheel lower end.

s. Connect any remaining connections and install the lower column cover.

t. Install the lower cover.

9. Install or connect the following:
• Bolts attaching the expansion valve and pipe in the engine compartment
• Heater hoses to the heater core after removing the plugs.

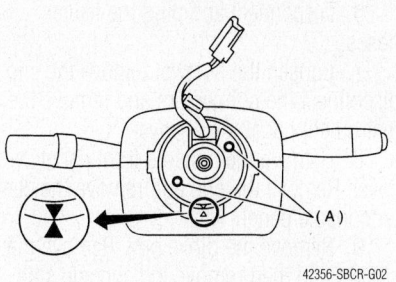

Make sure the roll pin is properly aligned—2002–2004

42356-SBCR-G02

• LLC
• Negative battery cable
10. Refill the cooling system.
11. Connect the negative battery cable.
12. Evacuate, charge and leak test the air conditioning system.
13. Operate the engine to normal operating temperatures; then, check the climate control operation and check for leaks.

2002–2003

BAJA

1. Before servicing the vehicle, refer to the Precautions Section.
2. Disconnect the negative battery cable.

❋❋ CAUTION

If equipped with an air bag system, wait 10 minutes after disconnecting the negative battery cable before performing any further work while the system fully de-energizes in order to avoid accidental deployment. All air bag system wiring is yellow. Do not use electrically powered test equipment on these circuits.

3. Drain the cooling system into a clean container for reuse.
4. Remove or disconnect the following:
• Heater hoses from the heater core. Plug the heater core and heater hoses.
• Radio box or console
5. Remove the instrument panel as follows:
a. Remove the center console retaining screws and remove the center console assembly.
b. Remove the instrument panel retaining bolt covers by prying them from the panel.
c. Remove the lower part of the front A pillar trim. Remove the instrument panel under covers from the driver's and passenger's sides.

d. Remove the hood release cable from the hood release lever.
e. Disconnect the wiring harness connectors under the instrument panel.
f. Remove the instrument cluster assembly. Remove the glove box assembly
g. Disconnect the ventilation control cables and electrical connectors at the heater unit. Disconnect the vacuum line at the blower housing.
h. Disconnect the radio antenna feeder wire. Disconnect the main harness connector at the fuse box.
i. Remove the lower steering column covers. Remove the steering column retaining bolts and allow the column to hang down.
j. Remove the instrument panel retaining bolts.

➡**When removing the instrument panel, check that all wiring and cables are disconnected before pulling it completely away from the firewall.**

k. With the help of an assistant, lift and remove the instrument panel from the vehicle.
6. Remove or disconnect the following:
• Heater control cables and the fan motor wiring harness
• Duct between the heater unit and the blower heater unit. Lift up and out on the heater unit and remove it.
7. Remove the evaporator case assembly by performing the following procedure:
a. Discharge and recover the air conditioning system refrigerant.
b. Disconnect the low and high pressure line from the evaporator outlet and cap the fittings.
c. Remove the inlet and outlet pipe grommets.
d. Remove the glove box and support bracket.
e. Disconnect the air conditioning wiring harness from the evaporator. Disconnect the drain hose from the evaporator.
f. Remove the evaporator mounting nut and bolt.
g. Remove the evaporator case assembly from the vehicle.
8. With the heater assembly out of the vehicle, remove the heater core tube retaining clamps and lift the core from the heater case.
9. Remove the heater core.

To install:
10. Install or connect the following:
• Heater core into the heater case.

Secure it in place with the retaining clamps and screws.

- Heater assembly to its mounting position under the dash
- Torque the mounting bolts to 48–84 inch lbs. (5–9 Nm).

11. Install the evaporator by performing the following procedure:

a. Install the evaporator case assembly and the nuts and bolts.

b. Adjust the position of the evaporator assembly so the inlet and outlet connections are aligned with the heater and blower unit connections.

c. Install the drain hose. Connect the air conditioning wiring harness.

d. Install the inlet and outlet pipe grommets.

e. Install the glove box and the lower support bracket.

f. Using new O-rings, lubricate them with clean refrigerant oil and install them on the pipe fittings.

g. Connect the suction hose to the evaporator inlet fitting.

h. Connect the discharge hose to the evaporator outlet fitting.

i. Connect the heater control cables and fan motor wiring harness connectors.

12. Install or connect the following:

- Instrument panel
- Radio and console assemblies
- Heater hoses in the engine compartment

13. Refill the cooling system.

14. Connect the negative battery cable.

15. Evacuate, charge and leak test the air conditioning system.

16. Operate the engine to normal operating temperatures; then, check the climate control operation and check for leaks.

2005–2006

IMPREZA AND WRX

1. Before servicing the vehicle, refer to the Precautions Section.

2. Disarm the SRS system. Wait at least 20 seconds before starting any repair work.

➡**Air bag connectors are colored yellow. Do not use electrical equipment on these circuits. Be careful not to damage the airbag system harness when servicing.**

3. Disconnect the negative battery cable.

4. Drain the engine coolant. Discharge the air conditioning system.

5. Remove the bolts securing the expansion valve and pipe.

6. Disconnect and plug the heater hoses.

7. Loosen the screws, remove the clip, disconnect the connectors and remove the instrument panel lower cover.

8. Remove the console front panel cover. Remove the clip and remove the center console panel.

9. Remove the glove box. Remove the screws and then remove the console side panel.

10. Detach the air bag connector from the support beam bracket. Disconnect the airbag connector.

11. Remove the three bolts and carefully remove the passenger's side air bag module.

12. Loosen the four screws and two nuts and then remove the lower console panel.

13. Remove the hook and remove the defroster panel.

14. Remove the front pillar upper trim. Disconnect the two connectors and then loosen the nuts.

15. Remove the instrument panel retaining bolts.

16. Be sure that all the electrical harnesses are disconnected, and then remove the instrument panel from the vehicle.

➡**Put alignment marks as necessary, in order to facilitate reassembly.**

17. Remove the support beam.

18. Remove the blower unit motor assembly from the vehicle.

19. Loosen the bolt and nuts and remove the heater/cooling unit from the vehicle.

20. Loosen the screws and remove the heater core cover.

21. Remove the heater core from its mounting.

To install:

22. Installation is the reverse of the removal procedure.

23. Refill the cooling system with the proper grade and type coolant.

24. Evacuate, charge and leak test the air conditioning system.

25. Start the engine and check for leaks, correct as required.

LEGACY AND OUTBACK

1. Before servicing the vehicle, refer to the Precautions Section.

2. Disarm the SRS system. Wait at least 20 seconds before starting any repair work.

3. Be sure the tires are pointed in the straight ahead position.

➡**Air bag connectors are colored yellow. Do not use electrical equipment on these circuits. Be careful not to damage the airbag system harness when servicing.**

4. Disconnect the negative battery cable.

5. Drain the engine coolant. Discharge the air conditioning system.

6. Remove the bolts securing the expansion valve and pipe.

7. Disconnect and plug the heater hoses.

8. Remove the front pillar upper trim. Remove the console box. Remove the center console.

9. Remove the instrument panel lower cover. Remove the glove box.

10. Remove the combination meter assembly. Remove the screws. Remove the driver's side instrument panel side cover.

11. Remove the screws at the side of the center console. Remove the center air vent grille of the instrument panel.

12. Remove the screws at the side of the passenger's side instrument panel.

13. Remove the bolts securing the passenger's side airbag module to the steering support beam.

14. Be sure that all the electrical harnesses are disconnected, and then remove the instrument panel from the vehicle.

➡**Put alignment marks as necessary, in order to facilitate reassembly.**

15. Remove the bolts on the under side of the steering wheel pad. Disconnect the horn harness.

16. Disconnect the air bag connector on the back of the airbag module. Remove the airbag module from the steering wheel.

17. Put alignment marks on the steering wheel and steering shaft. Remove the steering wheel nut. Using a suitable puller, remove the steering wheel.

18. Matchmark the universal joint. Remove the universal joint bolt. Remove the universal joint.

19. Remove all of the connectors on the steering column. Remove the steering column retaining bolts. Remove the steering column.

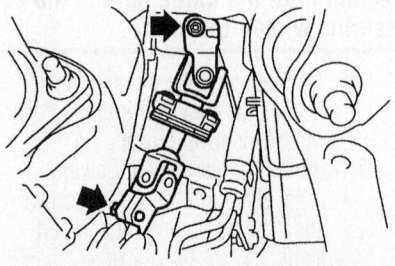

09490_SBCR_G0012

Universal joint and connection points (arrows)—2005–2006 Legacy and Outback

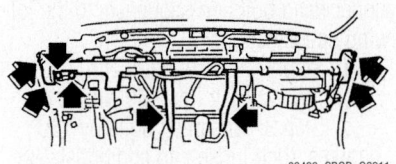

Support beam bolt locations (arrows)—
2005–2006 Legacy and Outback

➡**Be sure to remove the universal joint before removing the steering column assembly. Do not loosen the tilt lever when the steering column is not secured to the vehicle.**

20. Remove the harness clips and remove the harness from the steering support beam. Make matchmarks, as required for reassembly.

21. Remove the retaining bolts and remove the steering support beam from the vehicle.

22. Remove the glove box. Disconnect the connectors from the air conditioning control module, intake door actuator, blower motor, power transistor and blower resistor.

23. Loosen the bolt and nut and remove the blower motor assembly.

24. Loosen the bolt and nuts and remove the heater/cooling unit from the vehicle.

25. Loosen the screws and remove the heater core cover.

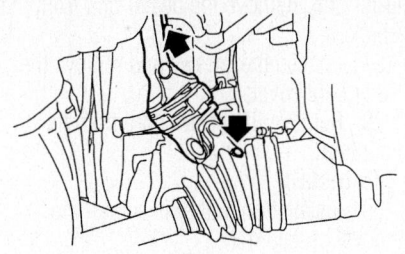

Universal joint and connection points (arrows)—2005–2006 Baja

26. Remove the heater core from its mounting.

To install:

27. Installation is the reverse of the removal procedure.

28. When installing the steering column be sure to align the cutout portion at the serrated section on the column shaft and yoke, and then install the universal joint into the column shaft. Torque the bolt to 17.4 ft. lbs.

29. Refill the cooling system with the proper grade and type coolant.

30. Evacuate, charge and leak test the air conditioning system.

31. Start the engine and check for leaks, correct as required.

BAJA

1. Before servicing the vehicle, refer to the Precautions Section.

2. Disarm the SRS system. Wait at least 20 seconds before starting any repair work.

3. Be sure the tires are pointed in the straight ahead position.

➡**Air bag connectors are colored yellow. Do not use electrical equipment on these circuits. Be careful not to damage the airbag system harness when servicing.**

4. Disconnect the negative battery cable.

5. Drain the engine coolant. Discharge the air conditioning system.

6. Pull out the LLC. Remove the air cleaner case. Disconnect and plug the heater hoses.

7. Remove the bolts securing the expansion valve and pipe.

8. Remove the lower instrument panel cover. Remove the lower steering column cover. Disconnect the harness connectors to the steering column.

9. Remove the bolts on the under side of the steering wheel pad. Disconnect the horn harness.

10. Disconnect the air bag connector on

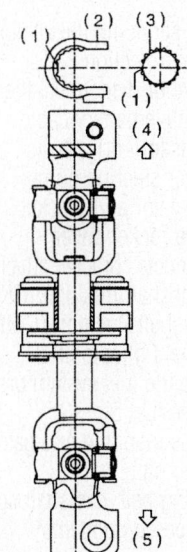

(1) Cutout portion
(2) Yoke
(3) Column shaft
(4) Column shaft side
(5) Gearbox side

Steering column joint alignment—
2005–2006 Legacy and Outback

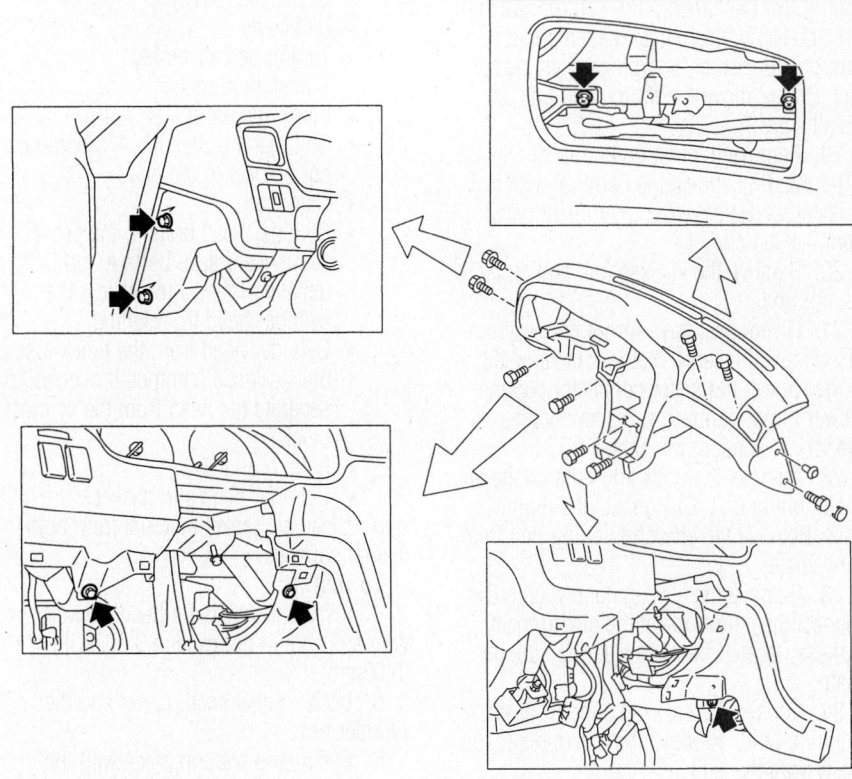

Instrument panel retaining bolt locations—2005–2006 Baja

the back of the airbag module. Remove the airbag module from the steering wheel.

11. Put alignment marks on the steering wheel and steering shaft. Remove the steering wheel nut. Using a suitable puller, remove the steering wheel.

12. Matchmark the universal joint. Remove the universal joint bolt. Remove the universal joint.

13. Remove all of the connectors on the steering column. Remove the steering column retaining bolts. Remove the steering column.

➡ **Be sure to remove the universal joint before removing the steering column assembly. Do not loosen the tilt lever when the steering column is not secured to the vehicle.**

14. Remove the glovebox. Remove the side panel on both sides of the console.

15. Detach the air bag connector from the support beam bracket. Disconnect the airbag connector.

16. Remove the three bolts and carefully remove the passenger's side air bag module.

17. Remove the center console box. Remove the front pillar upper trim from both sides. Remove the front pillar lower trim from the passenger's side of the vehicle.

18. If equipped with manual air conditioning, set the temperature control switch to FULL HOT. Disconnect the temperature control cable from the bottom of the heater unit. Do not move the switch and link on installation.

19. Remove the instrument panel retaining bolts. Disconnect the harness electrical connectors. Remove the instrument panel from the vehicle.

20. Remove the keyless unit and cruise control unit.

21. Disconnect the sunroof connector. Disconnect the servo motor connector, the blower power transistor connector and the blower motor connector. Disconnect the evaporator sensor connector.

22. Remove the bolts and nuts on the air conditioning unit. Disconnect the drain hose. Remove the air conditioning unit from the vehicle.

23. Remove the air bag control unit. Disconnect the connector of the airbag main harness, located near the steering support beam.

24. Loosen the steering support beam retaining bolts. Remove the steering support beam from the vehicle.

25. Disconnect the servo connector.

26. Loosen the bolts and nuts of the heater unit. Remove the heater unit from the vehicle.

27. Loosen the screws and remove the heater core cover.

28. Remove the heater core from its mounting.

To install:

29. Installation is the reverse of the removal procedure.

30. Refill the cooling system with the proper grade and type coolant.

31. Evacuate, charge and leak test the air conditioning system.

32. Start the engine and check for leaks, correct as required.

Cylinder Head

➡ **On some vehicles, engine compartment room is limited, so it may be necessary to remove the engine to service the cylinder heads.**

REMOVAL & INSTALLATION

2.0L Engine

1. Before servicing the vehicle, refer to the Precautions Section.

2. Remove or disconnect the following:

- Negative battery cable
- Drive belt
- Crankshaft pulley
- Belt cover
- Timing belt assembly
- Camshaft sprocket
- Intake manifold
- Bolt that attaches the A/C compressor bracket to the head
- Camshaft
- Cylinder head bolts in the proper sequence. Leave bolts A and D installed loosely to prevent the cylinder head from falling.
- Cylinder head from the block, use a plastic-faced hammer, if needed, to separate the head from the cylinder block
- Bolts A and D
- Cylinder head and gasket

3. Clean all gasket material from both mating surfaces.

To install:

4. Inspect the cylinder head for warpage. Warpage should not exceed 0.0020 in. (0.05mm).

5. Install a new head gasket and the cylinder head.

6. Secure the head in place with the mounting bolts. Coat each bolts with clean engine oil, and hand-tighten. Tighten the cylinder head bolts, in sequence, to the following specifications:

   a. Step 1: 22 ft. lbs. (29 Nm).

   b. Step 2: 51 ft. lbs. (69 Nm).

   c. Step 3: loosen all bolts by 180 degrees, then loosen an additional 180 degrees.

   d. Step 4: bolts 1 and 2 to 25 ft. lbs. (34 Nm).

   e. Step 5: bolts 3, 4, 5 and 6 to 11 ft. lbs. (15 Nm).

   f. Step 6: all bolts plus 80–90 degrees.

   g. Step 7: all bolts plus 80–90 degrees.

➡ **Do not exceed 180 degrees total tightening.**

7. Install or connect the following:

- Camshaft
- Bolt that attaches the A/C compressor bracket to the head
- Camshaft sprocket
- Intake manifold
- Timing belt assembly
- Belt cover
- Crankshaft pulley
- Drive belt
- Negative battery cable

8. Start the engine and allow it to reach operating temperature. Check for leaks.

2.5L Engine

2002–2004

1. Before servicing the vehicle, refer to the Precautions Section.

2. Remove or disconnect the following:

- Negative battery cable
- Accessory drive belts
- Power steering pump
- Alternator and bracket
- Valve rocker cover
- Connector bracket attaching bolt
- Crankshaft Position (CKP) and Camshaft Position (CMP) sensors
- Coolant filler tank

3. Relieve the fuel system pressure and disconnect the fuel pipes.

- Intake manifold and gasket
- Water pipe
- Timing belt, camshaft sprocket and related components
- Oil level gauge guide attaching bolt on the left cylinder head
- Valve covers
- Camshafts, refer to the camshaft procedure in this section
- Cylinder head bolts, in the proper sequence. On the DOHC engine leave bolts A and D installed loosely to prevent the cylinder head

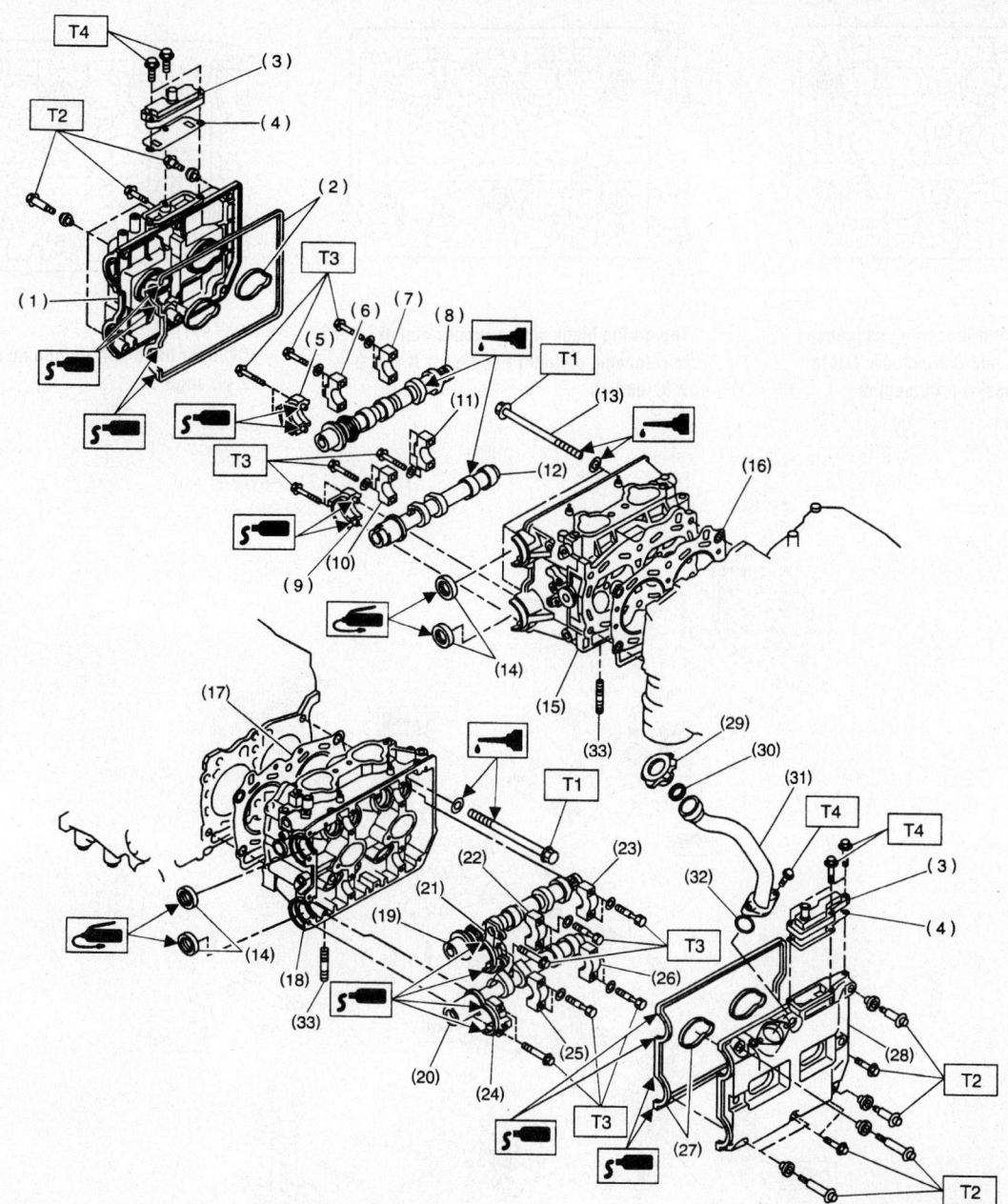

(1)	Rocker cover (RH)	(15)	Cylinder head (RH)	(29)	Oil filler cap
(2)	Rocker cover gasket (RH)	(16)	Cylinder head gasket (RH)	(30)	Gasket
(3)	Oil separator cover	(17)	Cylinder head gasket (LH)	(31)	Oil filler duct
(4)	Gasket	(18)	Cylinder head (LH)	(32)	O-ring
(5)	Intake camshaft cap (Front RH)	(19)	Intake camshaft (LH)	(33)	Stud bolt
(6)	Intake camshaft cap (Center RH)	(20)	Exhaust camshaft (LH)		
(7)	Intake camshaft cap (Rear RH)	(21)	Intake camshaft cap (Front LH)		
(8)	Intake camshaft (RH)	(22)	Intake camshaft cap (Center LH)		
(9)	Exhaust camshaft cap (Front RH)	(23)	Intake camshaft cap (Rear LH)		
(10)	Exhaust camshaft cap (Center RH)	(24)	Exhaust camshaft (Front LH)		
(11)	Exhaust camshaft cap (Rear RH)	(25)	Exhaust camshaft cap (Center LH)		
(12)	Exhaust camshaft (RH)	(26)	Exhaust camshaft cap (Rear LH)		
(13)	Cylinder head bolt	(27)	Rocker cover gasket (LH)		
(14)	Oil seal	(28)	Rocker cover (LH)		

Tightening torque: N·m (kgf-m, ft-lb)

T1: *<Ref. to ME(DOHC TURBO)-64, INSTALLATION, Cylinder Head Assembly.>*

T2: *5 (0.5, 3.6)*

T3: *10 (1.0, 7)*

T4: *6.4 (0.65, 4.7)*

9357TG05

Cylinder head and related components—2.0L engine

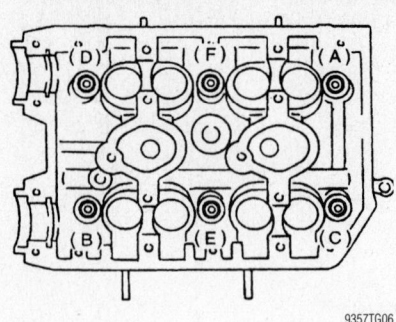

Cylinder head bolt loosening sequence (except bolts A and D which are left in place at this time)—2.0L engine

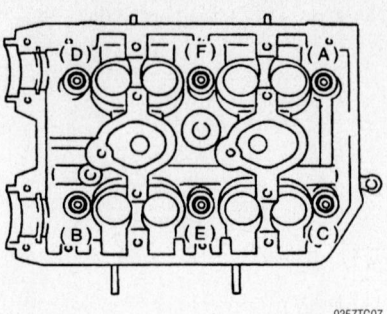

Tap on the block with a rubber mallet prior to removing cylinder head bolts A and D—2.0L engine

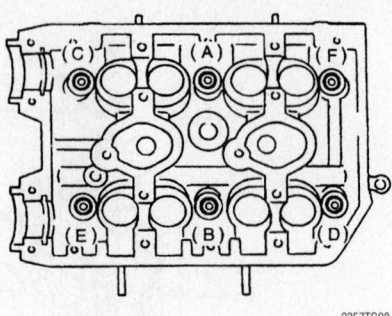

Cylinder head bolt tightening sequence—2.0L engine

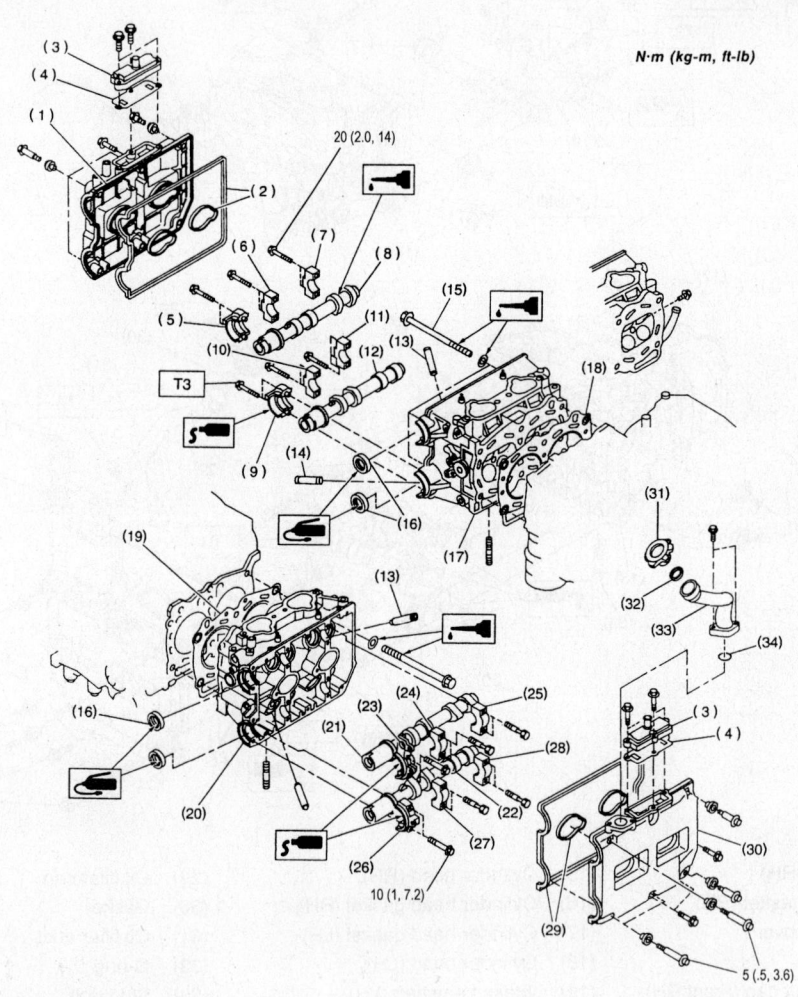

N·m (kg-m, ft-lb)

(1) Rocker cover (RH)
(2) Rocker cover gasket (RH)
(3) Oil separator cover
(4) Gasket
(5) Intake camshaft cap (Front RH)
(6) Intake camshaft cap (Center RH)
(7) Intake camshaft cap (Rear RH)
(8) Intake camshaft (RH)
(9) Exhaust camshaft cap (Front RH)
(10) Exhaust camshaft cap (Center RH)
(11) Exhaust camshaft cap (Rear RH)
(12) Exhaust camshaft (RH)
(13) Intake valve guide
(14) Exhaust valve guide

(15) Cylinder head bolt
(16) Oil seal
(17) Cylinder head (RH)
(18) Cylinder head gasket (RH)
(19) Cylinder head gasket (LH)
(20) Cylinder head (LH)
(21) Intake camshaft (LH)
(22) Exhaust camshaft (LH)
(23) Intake camshaft cap (Front LH)
(24) Intake camshaft cap (Center LH)
(25) Intake camshaft cap (Rear LH)
(26) Exhaust camshaft cap (Front LH)
(27) Exhaust camshaft cap (Center LH)
(28) Exhaust camshaft cap (Rear LH)

(29) Rocker cover gasket (LH)
(30) Rocker cover (LH)
(31) Oil filler cap
(32) Gasket
(33) Oil filler duct
(34) O-ring

Cylinder head and related components—2.5L DOHC engine

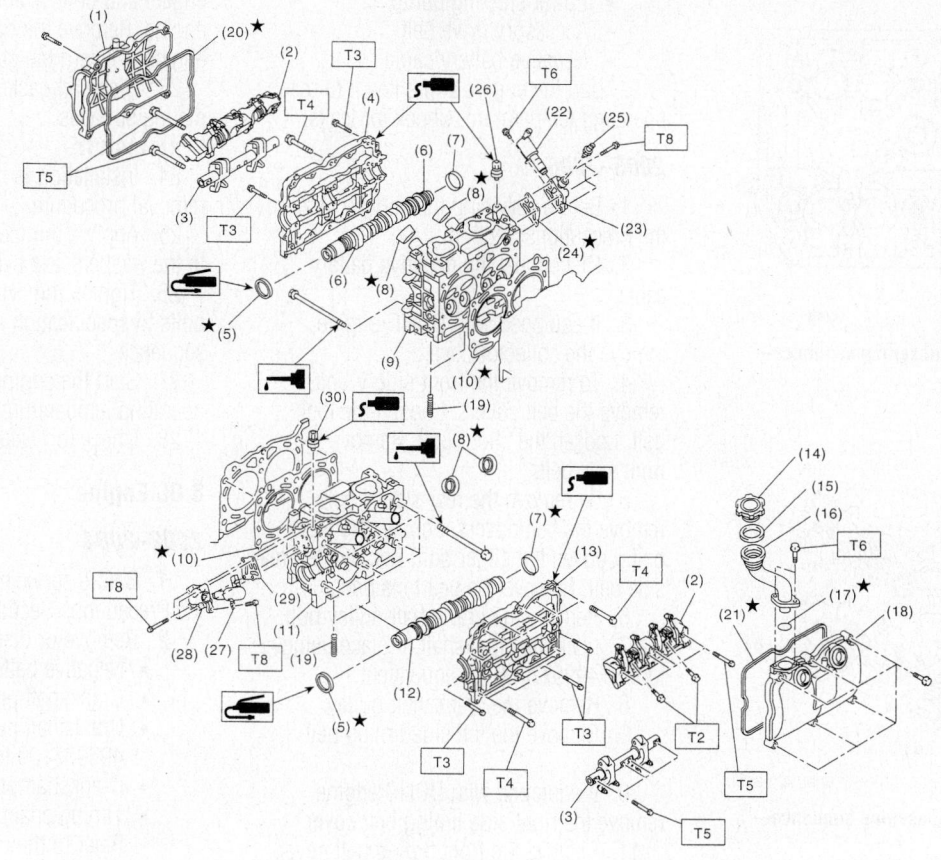

(1)	Rocker cover (RH)	(17)	O-ring	(30)	Variable valve lift diagnosis oil
(2)	Intake valve rocker assembly	(18)	Rocker cover (LH)		pressure switch (LH)
(3)	Exhaust valve rocker assembly	(19)	Stud bolt		
(4)	Camshaft cap (RH)	(20)	Rocker cover gasket (RH)		
(5)	Oil seal	(21)	Rocker cover gasket (LH)		
(6)	Camshaft (RH)	(22)	Oil switching solenoid valve (RH)		
(7)	Plug	(23)	Oil switching solenoid valve holder (RH)		
(8)	Spark plug pipe gasket				
(9)	Cylinder head (RH)	(24)	Gasket		
(10)	Cylinder head gasket	(25)	Oil temperature sensor		
(11)	Cylinder head (LH)	(26)	Variable valve lift diagnosis oil pressure switch (RH)		
(12)	Camshaft (LH)				
(13)	Camshaft cap (LH)	(27)	Oil switching solenoid valve (LH)		
(14)	Oil filler cap	(28)	Oil switching solenoid valve holder (LH)		
(15)	Gasket				
(16)	Oil filler duct	(29)	Gasket		

Tightening torque: N·m (kgf-m, ft-lb)

T3:	9.75 (1.0, 7.2)
T4:	18 (1.8, 13.0)
T5:	25 (2.5, 18.1)
T6:	6.4 (0.65, 4.7)
T7:	8 (0.8, 5.9)
T8:	10 (1.0, 7.4)

09490_SBCR_G0016

Cylinder head and related components—2.5L SOHC engine

from falling. On the SOHC engine leave bolts A and C installed loosely to prevent the cylinder head from falling.
• Cylinder head from the block using a plastic-faced hammer, if needed
• Remove bolts A and D on the DOHC engine and bolts A and C on the SOHC engine.
• Cylinder head and gasket
4. Clean all gasket material from both mating surfaces.

To install:
5. Inspect the cylinder head for warpage.

Warpage should not exceed 0.0020 in. (0.05mm).
6. Install a new head gasket and the cylinder head.
7. Secure the head in place with the mounting bolts. Coat each bolts with clean engine oil, and hand-tighten. Tighten the cylinder head bolts to specification.
8. Install or connect the following:
• Camshafts, refer to the procedure in this section
• Valve covers
• Oil level gauge guide attaching bolt on the left cylinder head

• Timing belt, camshaft sprockets and related components
• Water pipe
• Intake manifold and tighten the bolts to specification
• Fuel delivery pipes
• Blow-by hose
• Knock sensor
• CKP and CMP sensors
• Connector bracket attaching bolt
• Spark plug wires
• Valve rocker cover and tighten the bolts to 48 inch lbs. (9 Nm)
• Alternator

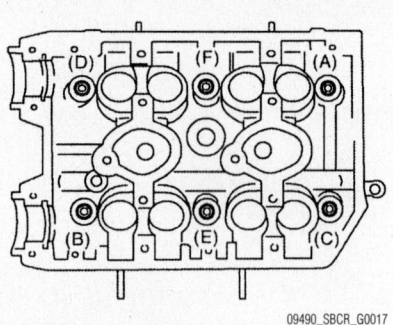

Cylinder head bolt loosening sequence—2.5L DOHC engine

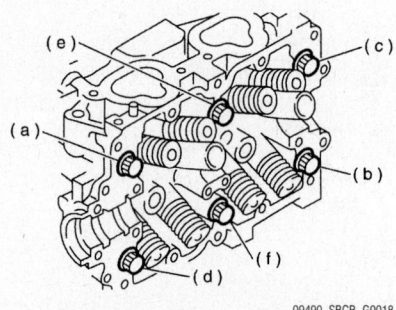

Cylinder head bolt loosening sequence—2.5L SOHC engine

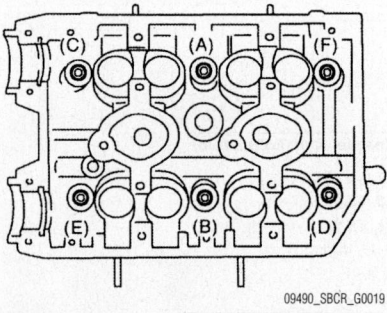

Cylinder head bolt tightening sequence—2.5L DOHC engine

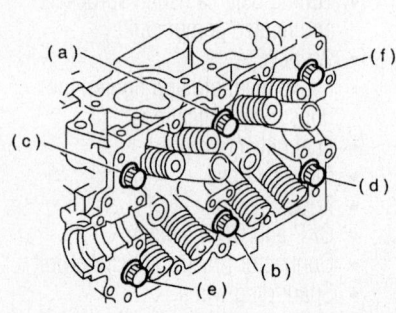

Cylinder head bolt tightening sequence—2.5L SOHC engine

- Power steering pump
- Accessory drive belt
- Negative battery cable

9. Start the engine and allow it to reach operating temperature. Check for leaks.

2005–2006

1. Before servicing the vehicle, refer to the Precautions Section.

2. Disconnect the negative battery cable.

3. If equipped with DOHC engine, remove the collector cover.

4. To remove the front side V belt, remove the belt covers. Loosen the lock bolt. Loosen the slider bolt. Remove the front side belt.

5. To remove the rear side V belt, remove the belt covers. Loosen the lock bolt. Loosen the slider bolt. Remove the rear side belt. Remove the belt tensioner.

6. Remove the crankshaft pulley bolt.

7. Lock the crankshaft in place using tool ST499977100, or equivalent.

8. Remove the crankshaft pulley.

9. Remove the left side timing belt cover.

10. If equipped with DOHC engine, remove the right side timing belt cover.

11. Remove the front timing belt cover.

12. Remove the timing belt.

13. Remove the camshaft position sensor.

14. Remove the camshaft sprockets.

➡**Be sure to lock the camshaft in place using tool ST18231AA010 (SOHC engine) and ST499207400 (DOHC engine), or equivalent.**

15. Remove the intake manifold.

16. If equipped with SOHC engine, remove the bolt that retains the air conditioning compressor bracket to the cylinder head.

17. Remove the rocker cover retaining bolts. Remove the rocker cover.

18. If equipped with SOHC engine, remove the rocker arm assembly.

19. If equipped with SOHC engine remove the camshaft. If equipped with DOHC engine, remove the camshafts.

20. Remove the cylinder head, in the proper sequence. On the DOHC engine leave bolts A and D installed loosely to prevent the cylinder head from falling. On the SOHC engine leave bolts A and C installed loosely to prevent the cylinder head from falling.

21. Loosen the cylinder head from the block using a plastic-faced hammer, if needed.

22. Remove bolts A and D on the DOHC

engine and bolts A and C on the SOHC engine. Remove the cylinder head from the engine. Discard the gasket

23. Clean all gasket material from both mating surfaces.

To install:

24. Installation is the reverse of the removal procedure.

25. Apply a thin coat of clean engine oil to the washers and cylinder head bolts.

26. Tighten the cylinder head retaining bolts to specification and in the proper sequence.

27. Start the engine and allow it to reach operating temperature.

28. Check for leaks, correct as required.

3.0L Engine

2002–2004

1. Before servicing the vehicle, refer to the Precautions Section.

2. Remove or disconnect the following:

- Negative battery cable
- Crankshaft pulley cover
- Crankshaft pulley bolt using tool 499977100 to hold the crankshaft
- Crankshaft pulley
- Timing chain cover and the chain. Refer to the procedure in this section.
- Camshaft and crankshaft sprockets.

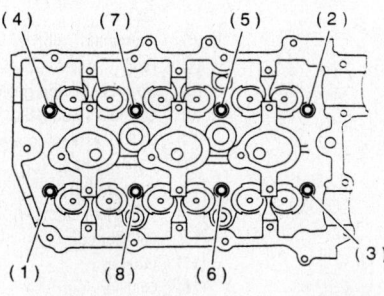

Cylinder head bolt loosening sequence—2002–2004 3.0L engine

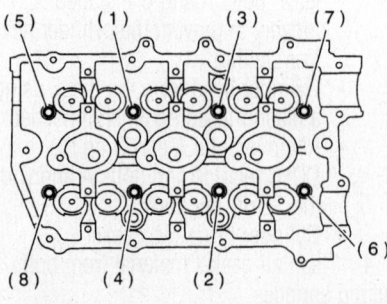

Cylinder head bolt tightening sequence—2002–2004 3.0L engine

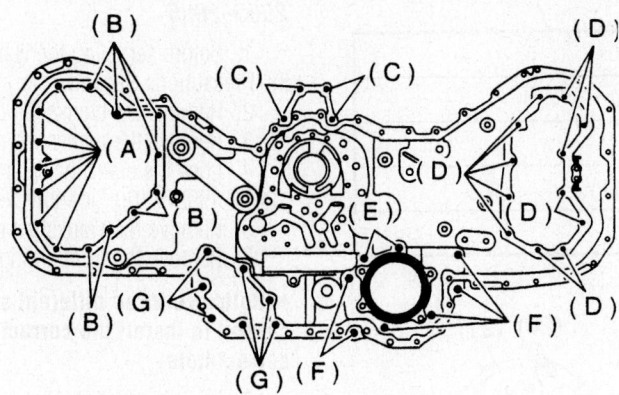

(A) 6 × 14
(B) 6 × 18 (Silver)
(C) 6 × 30
(D) 6 × 18
(E) 6 × 40
(F) 6 × 30
(G) 6 × 22

42356-SBCR-G05

Rear timing chain bolt sizes and locations—2002–2004 3.0L engine

Fluid gasket application diameter:
(A) 1.0±0.5 mm (0.039±0.020 in)
(B) 3.0±1.0 mm (0.118±0.039 in)

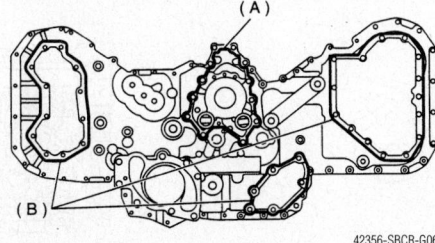

42356-SBCR-G06

Apply liquid gasket maker to the mating surfaces of the cover as shown—3.0L engine

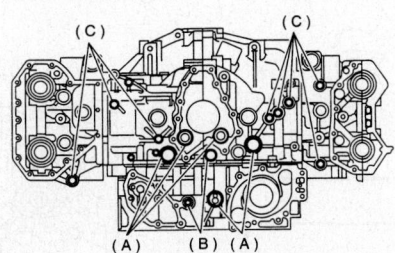

(A) O-ring (Large)
(B) O-ring (Medium)
(C) O-ring (Small)

42356-SBCR-G07

Rear cover O-ring sizes and locations—2002–2004 3.0L engine

Refer to the procedure in this section.
• Oil pump. Refer to the procedure in this section.
• Oil pump relief valve
• Water pump. Refer to the procedure in this section.
• Rear chain cover
• Camshafts. Refer to the procedure in this section.
• Cylinder head bolts in the sequence illustrated. Leave bolts 2 and 4 connected by a few threads to prevent the head from falling. Tap the head with a plastic mallet to separate it from the block.
• Bolts 2 and 4, the cylinder head and gasket

3. Clean all gasket material from both mating surfaces.

To install:
4. Inspect the cylinder head for warpage. Warpage should not exceed 0.0020 in. (0.05mm).
5. Make sure not to scratch or damage the mating surfaces of the cylinder head, block and oil pump.
6. Install a new head gasket and the cylinder head.
7. Secure the head in place with the mounting bolts. Coat each bolts with clean engine oil, and hand-tighten. Tighten the cylinder head bolts to specification.
8. Install or connect the following:
• Camshafts. Refer to the procedure in this section.

9. Install the rear chain cover as follows:

➡**There are several size bolts used, refer to the illustration for size and locations.**

a. Rear chain cover gasket and clean the mating surfaces.
b. Apply liquid gasket maker to the mating surfaces of the cover. Refer to the illustration for gasket maker application and diameter.
c. Install new O-rings. Refer to the illustration for O-ring location and size
d. Install the rear chain cover and temporarily tighten the bolts, refer to the illustration for size and locations.
e. On 2002–2004 engine, replace mounting bolts **G** with new bolts, refer to the illustration for location.
f. Tighten the cover bolts in the sequence illustrated to the specifications shown in the illustration.

10. Install or connect the following:
• Water pump
• Oil pump relief valve and tighten

(1) to (11)	9 N·m (0.9 kgf-m, 6.5 ft-lb)
(12) to (19)	20 N·m (2.0 kgf-m, 14 ft-lb)
(20) to (31)	9 N·m (0.9 kgf-m, 6.5 ft-lb)
(32) to (39)	12 N·m (1.2 kgf-m, 8.7 ft-lb)
(40) to (46)	9 N·m (0.9 kgf-m, 6.5 ft-lb)

Rear cover bolt tightening sequence and torque specifications—2002–2004 3.0L engine

the bolts to 4.7 ft. lbs. (6.4 Nm) in the sequence illustrated.
- Oil pump
- Crankshaft sprocket. Refer to the procedure in this section.
- Camshaft sprocket. Refer to the procedure in this section.
- Timing chain. Refer to the procedure in this section.
- Crankshaft pulley, apply clean oil to the bolt threads and tighten the bolt to 131 ft. lbs. (178 Nm)
- Camshaft pulley cover and tighten the bolts to 5 ft. lbs. (7 Nm)
- Negative battery cable

11. Start the engine and allow it to reach operating temperature. Check for leaks.

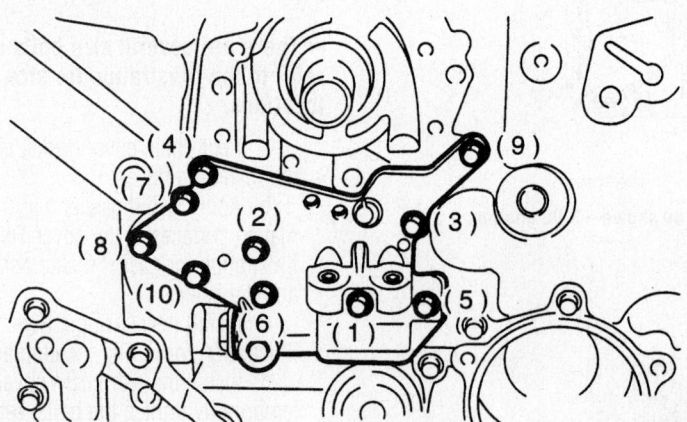

Bolt installation position	Bolt dimension
(1) and (5)	6 x 26
(2), (3), (4) and (9)	6 x 35
(6), (7), (8) and (10)	6 x 16

Oil pump relief valve bolt tightening sequence and bolt specifications—2002–2004 3.0L engine

2005–2006

1. Before servicing the vehicle, refer to the Precautions Section.
2. Remove the crankshaft pulley cover.
3. Remove the crankshaft pulley bolt.
4. Lock the crankshaft in place using tool ST499977100, or equivalent.
5. Remove the crankshaft pulley.
6. Remove the front timing chain cover.

➡**Bolts are three different sizes. Be careful to install the correct bolt in the correct hole.**

7. Remove the timing chain.
8. Remove the camshaft sprocket.

➡**Be sure to lock the camshaft in place using tool ST499977500, or equivalent.**

9. Remove the crankshaft sprocket.
10. Remove the oil pump.
11. Remove the water pump.
12. Remove the rear timing chain cover retaining bolts. Remove the rear timing chain cover.

➡**There are seven different size bolts. Be sure not to confuse them on installation.**

13. Remove the camshafts.
14. Remove the cylinder head bolts in the proper sequence. Leave bolts 2 and 4 connected by a few threads to prevent the head from falling. Tap the head with a plastic mallet to separate it from the block.

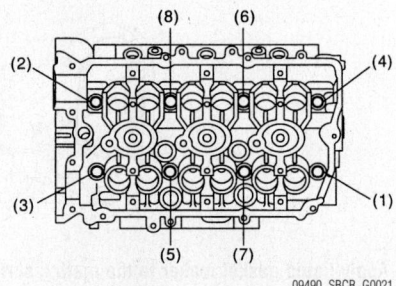

Cylinder head bolt loosening sequence—2005–2006 3.0L engine

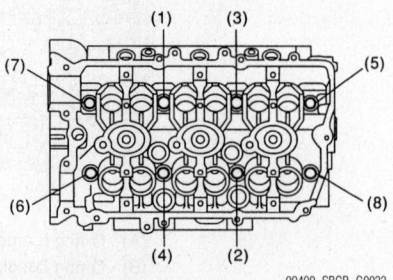

Cylinder head bolt tightening sequence—2005–2006 3.0L engine

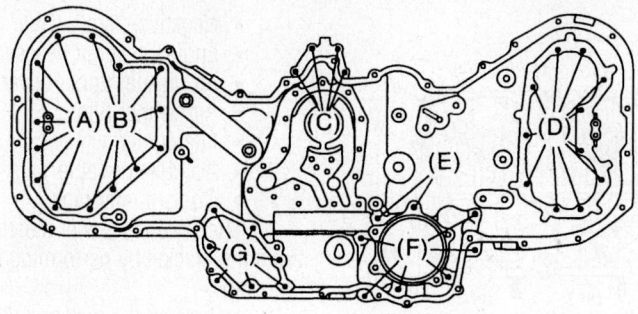

(A)	M6 × 14	(E)	M8 × 40
(B)	M6 × 18 (Silver)	(F)	M8 × 30
(C)	M6 × 30	(G)	M6 × 22
(D)	M6 × 18		

09490_SBCR_G0023

Rear timing chain bolt sizes and locations—2005–2006 3.0L engine

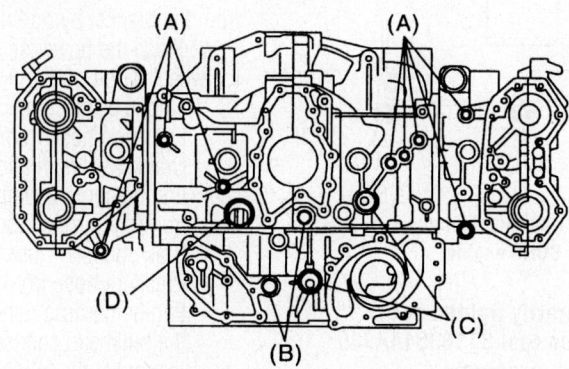

(A)	14.2 × 1.9	(C)	25 × 2
(B)	19.2 × 2.4	(D)	31.2 × 1.9

09490_SBCR_G0024

Rear cover O-ring sizes and locations—2005–2006 3.0L engine

(1) — (11)	9 N·m (0.9 kgf-m, 6.5 ft-lb)
(12) — (19)	20 N·m (2.0 kgf-m, 14 ft-lb)
(20) — (30)	9 N·m (0.9 kgf-m, 6.5 ft-lb)
(31) — (38)	12 N·m (1.2 kgf-m, 8.7 ft-lb)
(39) — (45)	9 N·m (0.9 kgf-m, 6.5 ft-lb)

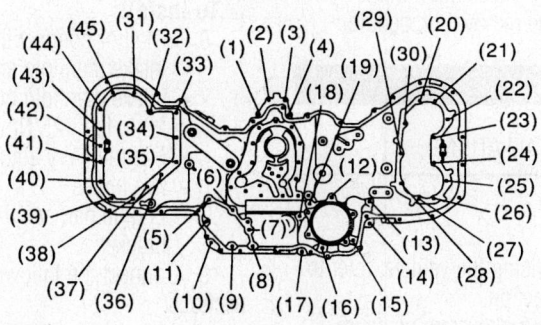

09490_SBCR_G0025

Rear cover bolt tightening sequence and torque specifications—2005–2006 3.0L engine

15. Remove bolts 2 and 4 from the cylinder head. Remove the cylinder head from the engine. Discard the gasket.

16. Clean all gasket material from both mating surfaces.

To install:

17. Installation is the reverse of the removal procedure.

18. Apply a thin coat of clean engine oil to the washers and cylinder head bolts.

19. Tighten the cylinder head retaining bolts to specification and in the proper sequence.

20. Install the rear chain cover as follows:

➡ **There are several size bolts used, refer to the illustration for size and locations.**

  a. Rear chain cover gasket and clean the mating surfaces.

  b. Apply liquid gasket maker to the mating surfaces of the cover. Refer to the illustration for gasket maker application and diameter.

  c. Install new O-rings. Refer to the illustration for O-ring location and size.

  d. Install the rear chain cover and temporarily tighten the bolts, refer to the illustration for size and locations.

  e. Tighten the cover bolts in the sequence illustrated to the specifications shown in the illustration.

21. Start the engine and allow it to reach operating temperature.

22. Check for leaks, correct as required.

Rocker Arms/Shafts

REMOVAL & INSTALLATION

2.5L SOHC Engine

1. Before servicing the vehicle, refer to the Precautions Section.

2. Disconnect the negative battery cable.

3. To remove the front side V belt, remove the belt covers. Loosen the lock bolt. Loosen the slider bolt. Remove the front side belt.

4. To remove the rear side V belt, remove the belt covers. Loosen the lock bolt. Loosen the slider bolt. Remove the rear side belt. Remove the belt tensioner.

5. Remove the crankshaft pulley bolt.

6. Lock the crankshaft in place using tool ST499977100, or equivalent.

7. Remove the crankshaft pulley.

8. Remove the left side timing belt cover.

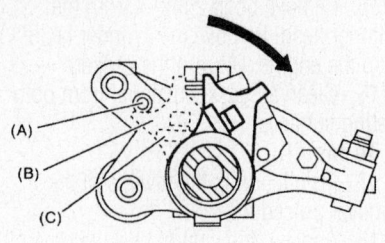

(A) Adjuster pin
(B) Spring stopper
(C) Spring

09490_SBCR_G0030

Rocker arm spring stopper removal—2.5L SOHC engine

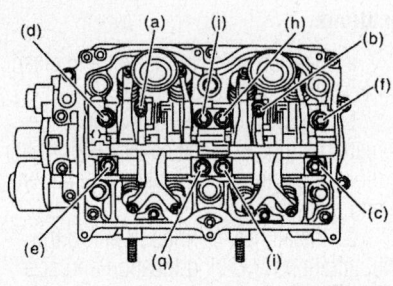

09490_SBCR_G0031

Rocker arm assembly bolt loosening sequence—2.5L SOHC engine

9. Remove the front timing belt cover.
10. Remove the timing belt.
11. Remove the camshaft position sensor.
12. Remove the camshaft sprockets, No.1 and No.2.

➡**Be sure to lock the camshaft in place using tool ST18231AA010, or equivalent.**

13. Disconnect the PCV valve hose. Remove the rocker cover retaining bolts. Remove the rocker cover from the engine.
14. Use tool St18258AA000, or equivalent, to rotate the spring stopper in direction of the arrow, see illustration, and remove the spring stopper from the adjuster spring.
15. Remove bolts "A" through "J" in sequence. Remove the assembly from the engine.

➡**Leave two or three threads on bolts "I" and "J" engaged in order to retain the rocker arm assembly.**

16. Position tool ST18354AA000 on the rocker arm assembly. Remove the assembly from the engine.
To install:
17. Install the rocker arm assembly on its mounting.
18. Temporarily tighten the bolts equally in the proper sequence.

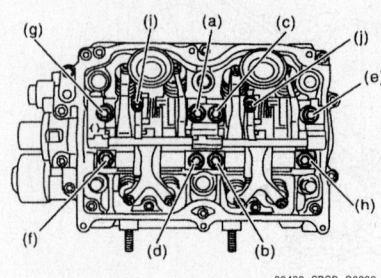

09490_SBCR_G0032

Rocker arm assembly bolt tightening sequence—2.5L SOHC engine

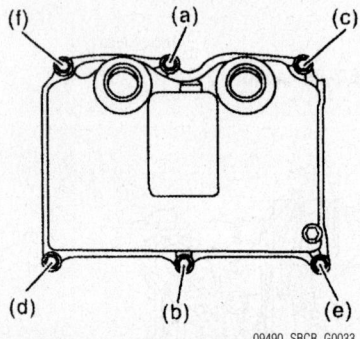

09490_SBCR_G0033

Rocker arm cover bolt tightening sequence—2.5L SOHC engine

➡**Do not temporarily tighten bolts "I" and "J". Position tool ST18354AA000 to the rocker arm assembly.**

19. Tighten bolts "A" through "H" to specification (18.1 ft. lbs.).
20. Tighten bolts "I" through "J" to specification (5.9 ft. lbs.).
21. Use tool ST18258AA000 to rotate the spring stopper in the opposite direction of the arrow, see illustration, to fasten the adjuster pin.
22. Adjust the valve clearance.
23. Use a new gasket and install the rocker arm cover. Torque the retaining bolts to 4.7 ft. lbs and in the proper sequence. Recheck and retorque to 4.7 ft. lbs in the proper sequence.
24. Continue the installation in the reverse order of the removal procedure.

Intake Manifold

REMOVAL & INSTALLATION

2.0L Engine

1. Before servicing the vehicle, refer to the Precautions Section.
2. Release the fuel system pressure.
3. Remove or disconnect the following:

- Negative battery cable
- Engine cover, if necessary
- Air cleaner upper cover and boot
- Air cleaner filter
- Intercooler
- Accelerator cable
- Coolant filler tank

4. Remove the power steering pump from the bracket by performing the following steps:
   a. Loosen the lock and slider bolts.
   b. Remove the V-belt.
   c. Disconnect the power steering switch connection.
   d. Remove the bolts that attach the power steering pump pipe brackets to the intake manifold. Do not disconnect the hose.
   e. Bolts that attach the power steering pump bracket.
   f. Remove the power steering tank from the bracket by pulling it upwards.
   g. Place the power steering pump on the wheel apron on the right.

5. Remove or disconnect the following:
- Emission hose from the Positive Crankcase Ventilation (PCV) valve
- Engine coolant temperature hoses from the throttle body
- Brake booster hose
- Pressure hose from the intake duct
- Engine harness connectors from the bulkhead connections

6. Disconnect the following electrical connections:
- Engine Coolant Temperature (ECT) sensor
- Oil pressure switch
- Crankshaft Position (CKP) sensor
- Knock Sensor (KS)
- Camshaft Position (CMP) sensor
- Ignition coil

7. Remove or disconnect the following:
- Engine harness fixed clip from the bracket
- Fuel delivery, return and evaporative hoses
- Intake manifold bolts
- Intake manifold and gasket

To install:
8. Install or connect the following:
- Intake manifold and gasket
- Intake manifold bolts and tighten to 18 ft. lbs. (25 Nm)
- Fuel delivery, return and evaporative hoses
- Engine harness fixed clip from the bracket

9. Connect the following electrical connections:
- Oil pressure switch
- CKP sensor

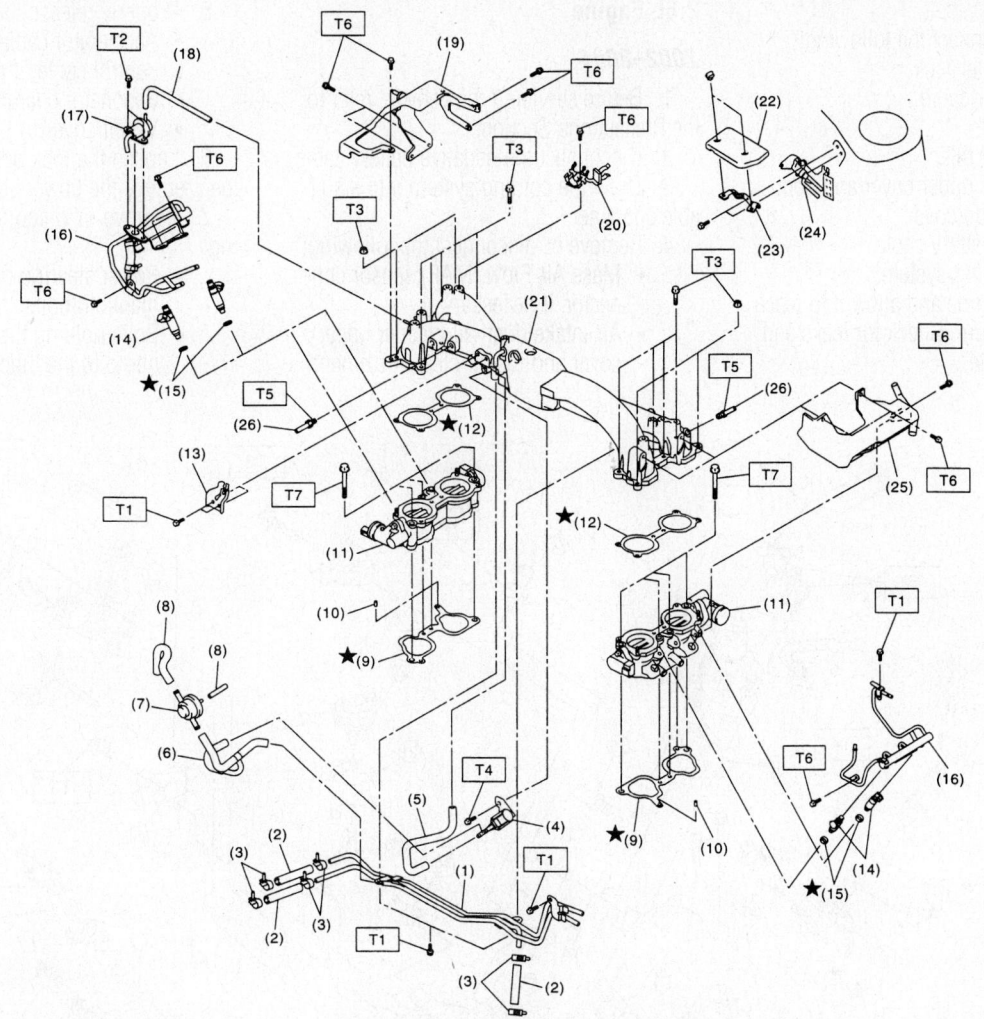

(1)	Fuel pipe ASSY	(13)	Accelerator cable bracket	(25)	Fuel pipe protector LH
(2)	Fuel hose	(14)	Fuel injector	(26)	Nipple
(3)	Clip	(15)	Insulator		
(4)	Purge control solenoid valve	(16)	Fuel injector pipe		
(5)	Vacuum hose	(17)	Pressure regulator		
(6)	Vacuum control hose	(18)	Pressure regulator hose		
(7)	Purge valve	(19)	Fuel pipe protector RH		
(8)	Purge hose	(20)	Blow-by hose stay		
(9)	Intake manifold gasket	(21)	Intake manifold		
(10)	Guide pin	(22)	Solenoid valve cover		
(11)	Tumble generator valve ASSY	(23)	Solenoid valve cover stay		
(12)	Tumble generator valve gasket	(24)	Wastegate control solenoid valve ASSY		

Tightening torque: N·m (kgf-m, ft-lb)

T1:	**4.9 (0.5, 3.6)**
T2:	**6.4 (0.65, 4.7)**
T3:	**8.25 (0.84, 6.1)**
T4:	**16 (1.6, 11.8)**
T5:	**17 (1.73, 12.5)**
T6:	**19 (1.94, 13.7)**
T7:	**25 (2.5, 18.1)**

9357TG09

Intake manifold and related components—2.0L engine

- ECT sensor
- KS sensor
- CMP sensor
- Ignition coil
- Engine harness connectors to the bulkhead connections

10. Install or connect the following:
- Brake booster hose
- Engine coolant temperature hoses to the throttle body
- Emission hose to the PCV valve
- Pressure hose to the intake duct

11. Install the power steering pump on the bracket by performing the following steps:

 a. Install the power steering tank on the bracket.

 b. Attach the power steering switch connection.

 c. Install the bolts that attach the power steering pump bracket and tighten to 16 ft. lbs. (22 Nm).

 d. Attach the power steering pump pipe brackets to the intake manifold.

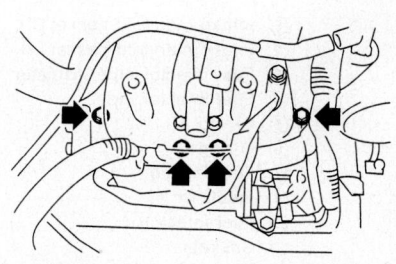

9357TG10

Intake manifold bolt location—2.0L engine

e. Install the V-belt.
12. Install or connect the following:
- Coolant filler tank
- Accelerator cable
- Intercooler
- Air cleaner filter
- Air cleaner upper cover and boot
- Engine undercover
- Negative battery cable
13. Fill the cooling system.
14. Start the engine and allow it to reach operating temperature. Check for leaks and test drive the vehicle.

2.5L Engine

2002–2004

1. Before servicing the vehicle, refer to the Precautions Section.
2. Disconnect the negative battery cable.
3. Drain the cooling system into a suitable container.
4. Remove or disconnect the following:
- Mass Air Flow (MAF) sensor connector, if necessary
- Air intake duct, air cleaner upper cover and the air cleaner element

5. Properly release the fuel pressure.
- Accelerator cable and the cruise control cable, if equipped
- Resonator chamber, if equipped
- V-belt cover(s)
6. Loosen the lock bolt and slider bolt, then remove the power steering belt
7. Remove or disconnect the following:
- Power steering pipe bracket-to-manifold bolts
- Bolts holding the power steering pump to the bracket

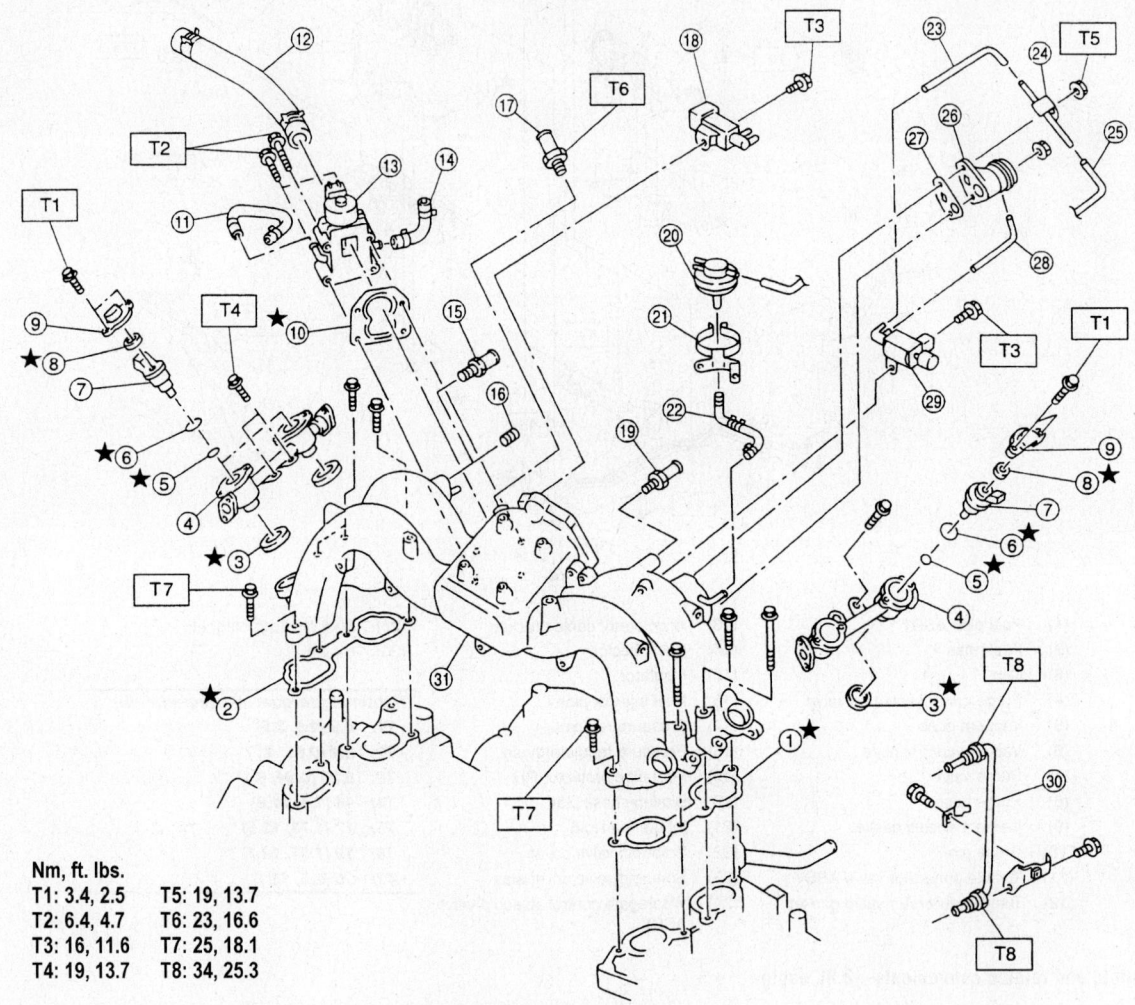

Nm, ft. lbs.

T1: 3.4, 2.5	T5: 19, 13.7
T2: 6.4, 4.7	T6: 23, 16.6
T3: 16, 11.6	T7: 25, 18.1
T4: 19, 13.7	T8: 34, 25.3

① Intake manifold gasket RH
② Intake manifold gasket LH
③ Fuel injector pipe insulator
④ Fuel injector pipe
⑤ O-ring A
⑥ O-ring B
⑦ Fuel injector
⑧ Insulator
⑨ Fuel injector cap
⑩ Gasket
⑪ Engine coolant hose B

⑫ Air by-pass hose
⑬ Idle air control solenoid valve
⑭ Engine coolant hose A
⑮ Nipple (AT model)
⑯ Plug
⑰ PCV valve
⑱ Purge control solenoid valve
⑲ Nipple
⑳ BPT
㉑ BPT holder bracket

㉒ Back pressure hose
㉓ EGR vacuum hose A
㉔ EGR vacuum pipe
㉕ EGR vacuum hose C
㉖ EGR valve
㉗ Gasket
㉘ EGR vacuum hose B
㉙ EGR solenoid valve
㉚ EGR pipe
㉛ Intake manifold

Intake manifold and related components—2002–2004 2.5L SOHC engine

7923TG17

- Connector from the power steering pump switch, if equipped
- Power steering pump, and place on the right side wheel apron. Do NOT disconnect the fluid lines.
- Spark plug wires from the spark plugs
- Positive Crankcase Ventilation (PCV) hose and vacuum hose from the intake manifold

- Engine coolant hoses from the throttle body
- Brake booster hose
- Air cleaner case stay (right side) and engine harness bracket, if necessary
- Engine harness connectors form the bulkhead harness connectors
- Engine Coolant Temperature (ECT) sensor connector

- Knock sensor, Camshaft Position (CMP) sensor and Crankshaft Position (CKP) sensor electrical connectors
- Oil pressure switch
- Fuel hoses from the fuel pipes
- Intake manifold mounting bolts
- Intake manifold and discard the gaskets

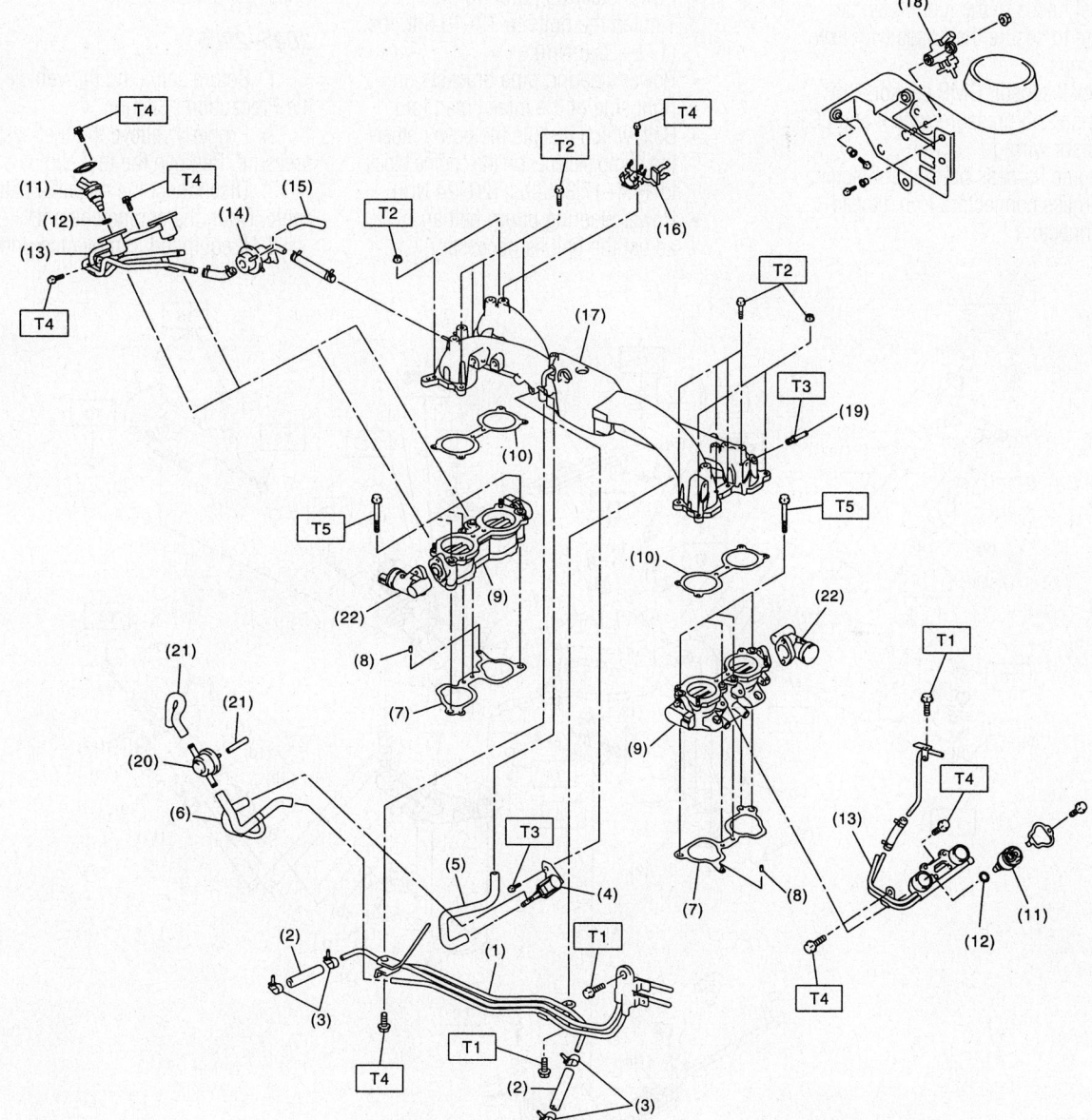

(1) Fuel pipe ASSY	(9) Tumble generator valve ASSY	(17) Intake manifold
(2) Fuel hose	(10) Tumble generator valve gasket	(18) Wastegate control solenoid valve ASSY
(3) Clip	(11) Fuel injector	
(4) Purge control solenoid valve	(12) O-ring	(19) Nipple
(5) Vacuum hose	(13) Fuel injector pipe	(20) Purge valve
(6) Vacuum control hose	(14) Pressure regulator	(21) Purge hose
(7) Intake manifold gasket	(15) Pressure regulator hose	(22) Tumble generator valve actuator
(8) Guide pin	(16) Blow-by hose stay	

09490_SBCR_G0026

Intake manifold and related components—2002–2004 2.5L DOHC engine

➡The intake manifold sits on pins that protrude from the cylinder heads. Be sure the pins remain in the cylinder heads.

To install:

8. Install or connect the following:
- New gaskets
- Intake manifold to the engine. Tighten the mounting bolts to 16.7–19.5 ft. lbs. (23–27 Nm).
- Fuel hoses to the fuel pipes, be sure to secure the hoses with new clamps
- Knock sensor, CMP sensor, CKP sensor, oil pressure switch and ECT sensor wiring
- Engine harness bracket and engine harness connectors to bulkhead connectors

- EGR pipe, if removed
- Brake booster hose
- Engine coolant hose and air bypass hose to idle air control solenoid valve, if removed
- Engine coolant hoses to the throttle body
- PCV hoses to intake manifold
- Spark plug wires to ignition coil or plugs, as applicable
- Power steering pump to bracket. Tighten the bolts to 13–16.6 ft. lbs. (17.6–22.6 Nm).
- Power steering pipe brackets on right side of the intake manifold
- Bolt, which installs the power steering pump stiffener on the engine block, to 14.4–17.3 ft. lbs. (20–24 Nm).
- Power steering pump belt and adjust the belt as necessary

- V-belt cover(s)
- Resonator chamber, if equipped
- Accelerator cable and the cruise control cable, if equipped
- Air cleaner assembly
- MAF sensor connector, if disconnected

9. Connect the negative battery cable and refill the cooling system. Start the engine, and bleed the cooling system. Check for leaks.

2005–2006

1. Before servicing the vehicle, refer to the Precautions Section.
2. Properly relieve the fuel system pressure. Remove the fuel cap.
3. Disconnect the negative battery cable. Drain the engine coolant.
4. If equipped, remove the under cover.

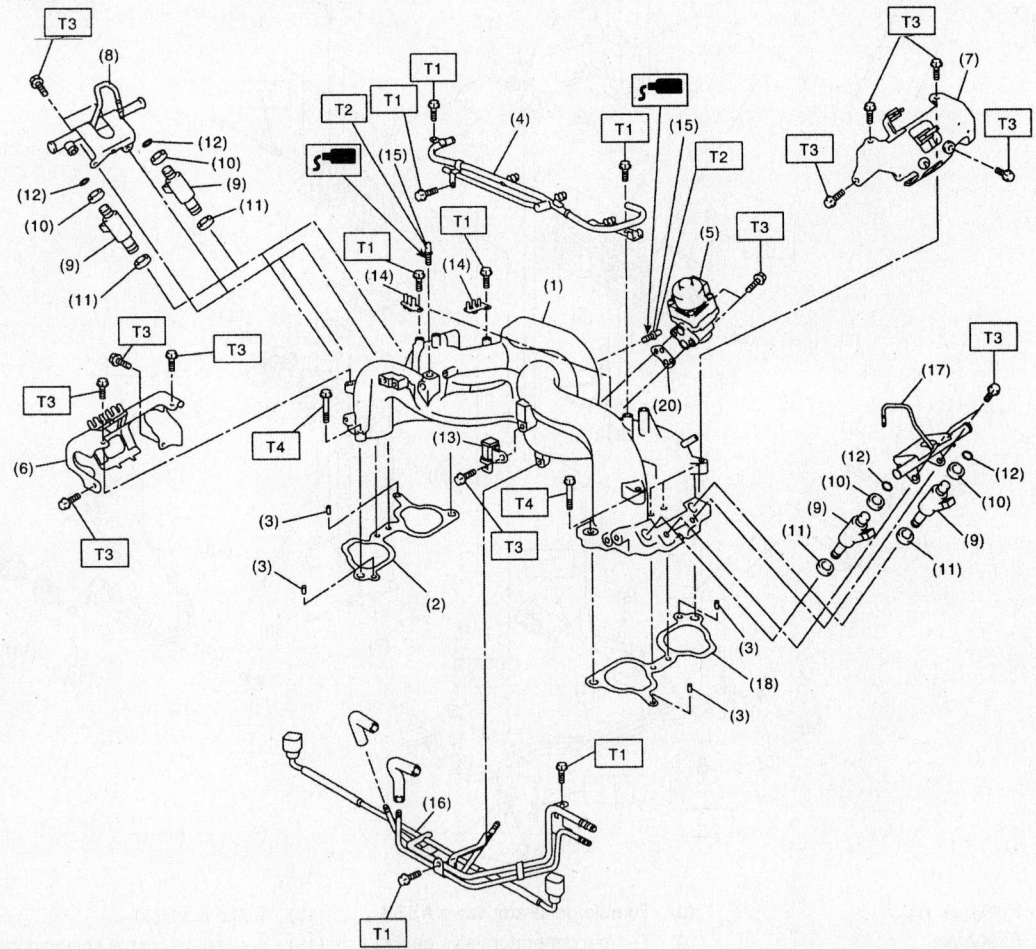

(1)	Intake manifold	(7)	Fuel pipe protector LH	(13)	Purge control solenoid valve
(2)	Gasket (RH)	(8)	Fuel injector pipe RH	(14)	Plug cord holder
(3)	Guide pin	(9)	Fuel injector	(15)	Nipple
(4)	PCV pipe	(10)	O-ring	(16)	Fuel pipe
(5)	EGR valve	(11)	O-ring	(17)	Fuel injector pipe LH
(6)	Fuel pipe protector RH	(12)	O-ring	(18)	Gasket (LH)

09490_SBCR_G0027

Intake manifold and related components—2005–2006 2.5L SOHC engine

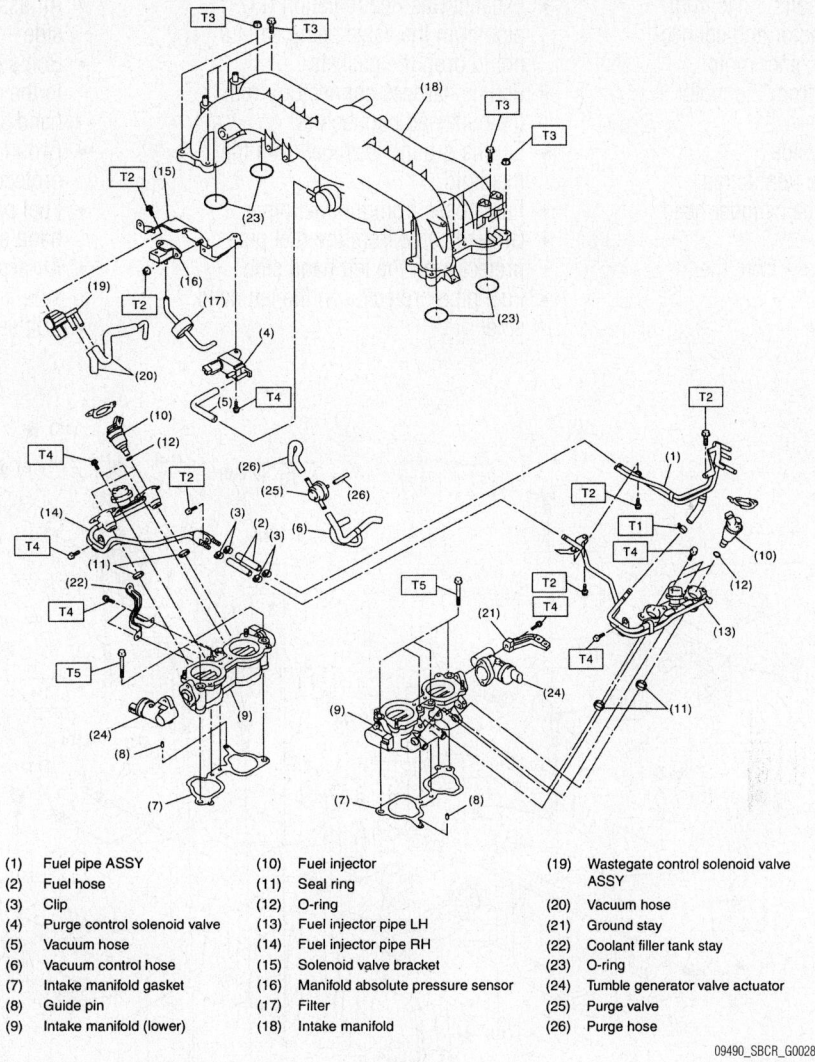

(1) Fuel pipe ASSY
(2) Fuel hose
(3) Clip
(4) Purge control solenoid valve
(5) Vacuum hose
(6) Vacuum control hose
(7) Intake manifold gasket
(8) Guide pin
(9) Intake manifold (lower)
(10) Fuel injector
(11) Seal ring
(12) O-ring
(13) Fuel injector pipe LH
(14) Fuel injector pipe RH
(15) Solenoid valve bracket
(16) Manifold absolute pressure sensor
(17) Filter
(18) Intake manifold
(19) Wastegate control solenoid valve ASSY
(20) Vacuum hose
(21) Ground stay
(22) Coolant filler tank stay
(23) O-ring
(24) Tumble generator valve actuator
(25) Purge valve
(26) Purge hose

09490_SBCR_G0028

Intake manifold and related components—2005–2006 2.5L DOHC engine

5. Remove the air intake duct, air cleaner case and air intake chamber.

6. If equipped, remove the intercooler.

7. Remove the alternator.

8. If equipped with DOHC engine, remove the coolant filler tank.

9. If equipped with SOHC engine, remove the spark plug wires.

10. Disconnect the engine coolant hoses from the throttle body. Disconnect the brake booster hose.

11. Disconnect the PCV hose from the intake manifold. Disconnect the engine harness electrical connectors from the bulkhead harness connectors.

12. Disconnect the engine coolant temperature sensor electrical connector, knock sensor electrical connector and crankshaft position sensor connector.

13. Disconnect the power steering pump switch electrical connector, oil pressure switch connector and camshaft sensor connector.

14. If equipped with DOHC engine, disconnect the oil flow solenoid valve electrical connector and the ignition coil connector.

15. If equipped with SOHC engine, remove the EGR pipe from the intake manifold and disconnect the fuel lines from the fuel pipe.

16. If equipped with DOHC engine, disconnect the fuel delivery hose, return hose and evaporation hose.

17. Remove the intake manifold retaining bolts. Remove the intake manifold from the engine.

To install:

18. Installation is the reverse of the removal procedure.

19. Be sure to use new intake manifold gaskets. Torque the manifold retaining bolts to specification and in alternating sequence.

20. Be sure to fill the cooling system with the proper grade and type engine coolant.

21. Start the engine and check for leaks, correct as required.

3.0L Engine

2002–2004

1. Before servicing the vehicle, refer to the Precautions Section.

2. Disconnect the negative battery cable.

3. Properly release the fuel pressure.

4. Remove or disconnect the following:

- Air intake duct, air cleaner upper cover and the air cleaner element
- Resonator chamber
- Accelerator cable and the cruise control cable, if equipped
- Power steering pipe bracket-to-manifold bolts
- V-belt
- Bolts holding the power steering pump to the bracket
- Connector from the power steering pump switch
- Power steering pump, and place on the right side wheel apron. Do NOT disconnect the fluid lines.

- Washer reservoir bolts, hose from the front washer motor and connector from the rear washer motor.
- Rear washer hose from the motor and plug the hose
- Washer reservoir aside
- Positive Crankcase Ventilation (PCV) hose from the cylinder head cover
- Engine coolant hoses from the throttle body
- Brake booster hose

- Exhaust Gas Recirculation (EGR) pipe from the valve being careful not to drop the gaskets
- Engine harness connectors from the bulkhead connectors
- Engine ground terminal from the manifold
- Fuel hoses from the fuel pipes
- Ground cable from the fuel pipe protector on the left hand side
- Fuel pipe protector on the left hand side

- Air assist hose on the left hand side
- Bolt securing the fuel injector pipe to the cylinder head on the left hand side
- Ground cable from the fuel pipe protector on the right hand side
- Fuel pipe protector on the right hand side
- Air assist hose on the right hand side
- Bolt securing the fuel injector pipe

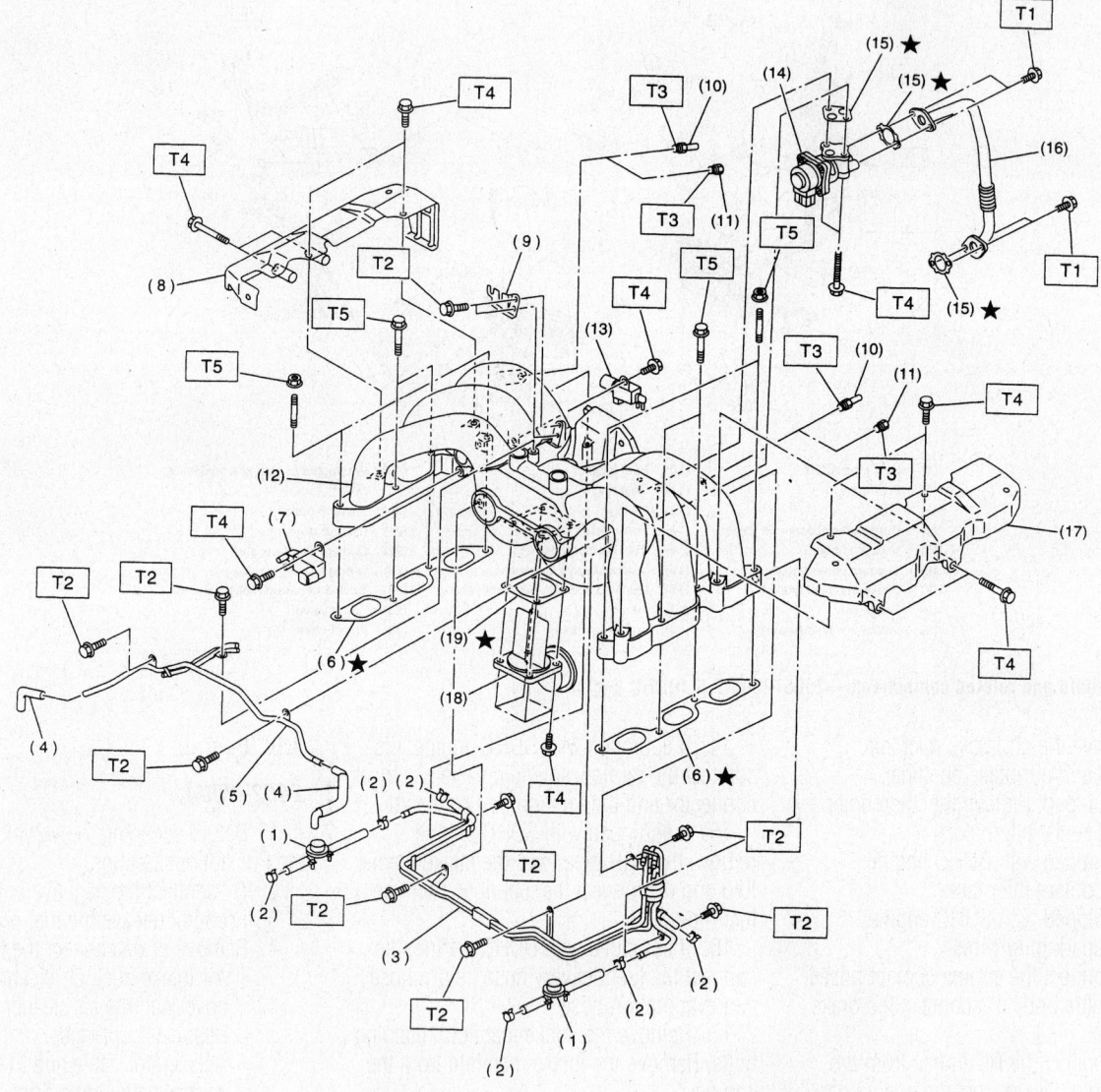

(1)	Fuel damper valve	(10)	Nipple	(19)	Gasket
(2)	Clamp	(11)	Plug		
(3)	Fuel pipe ASSY	(12)	Intake manifold		
(4)	Air assist hose	(13)	Induction valve control solenoid		
(5)	Air assist and purge pipe ASSY	(14)	EGR valve		
(6)	Gasket	(15)	Gasket		
(7)	Purge control solenoid valve	(16)	EGR pipe		
(8)	Fuel pipe protector RH	(17)	Fuel pipe protector LH		
(9)	Accelerator cable bracket	(18)	Induction valve		

Tightening torque: N·m (kgf-m, ft-lb)

T1:	6.4 (0.65, 4.7)
T2:	5.0 (0.51, 3.7)
T3:	17 (1.7, 12)
T4:	19 (1.9, 14)
T5:	25 (2.5, 18)

42356-SBCR-G10

Intake manifold and related components—2002-2004 3.0L engine

to the cylinder head on the right hand side
- Intake manifold mounting bolts
- Intake manifold and discard the gaskets

To install:

5. Install or connect the following:
- New gaskets
- Intake manifold to the engine. Tighten the mounting bolts to 18 ft. lbs. (25 Nm).
- Bolt securing the fuel injector pipe to the cylinder head on the right

hand side. Tighten to 14 ft. lbs. (19 Nm).
- Air assist hose on the right hand side
- Fuel pipe protector on the right hand side. Tighten to 14 ft. lbs. (19 Nm).
- Ground cable from the fuel pipe protector on the right hand side
- Bolt securing the fuel injector pipe to the cylinder head on the left hand side. Tighten to 14 ft. lbs. (19 Nm).

- Air assist hose on the left hand side
- Fuel pipe protector on the left hand side. Tighten to 14 ft. lbs. (19 Nm).
- Ground cable to the fuel pipe protector on the left hand side
- EGR pipe to the valve using new gaskets and tighten to 5 ft. lbs. (7 Nm)
- Fuel hoses to the fuel pipes
- Engine ground terminal to the manifold

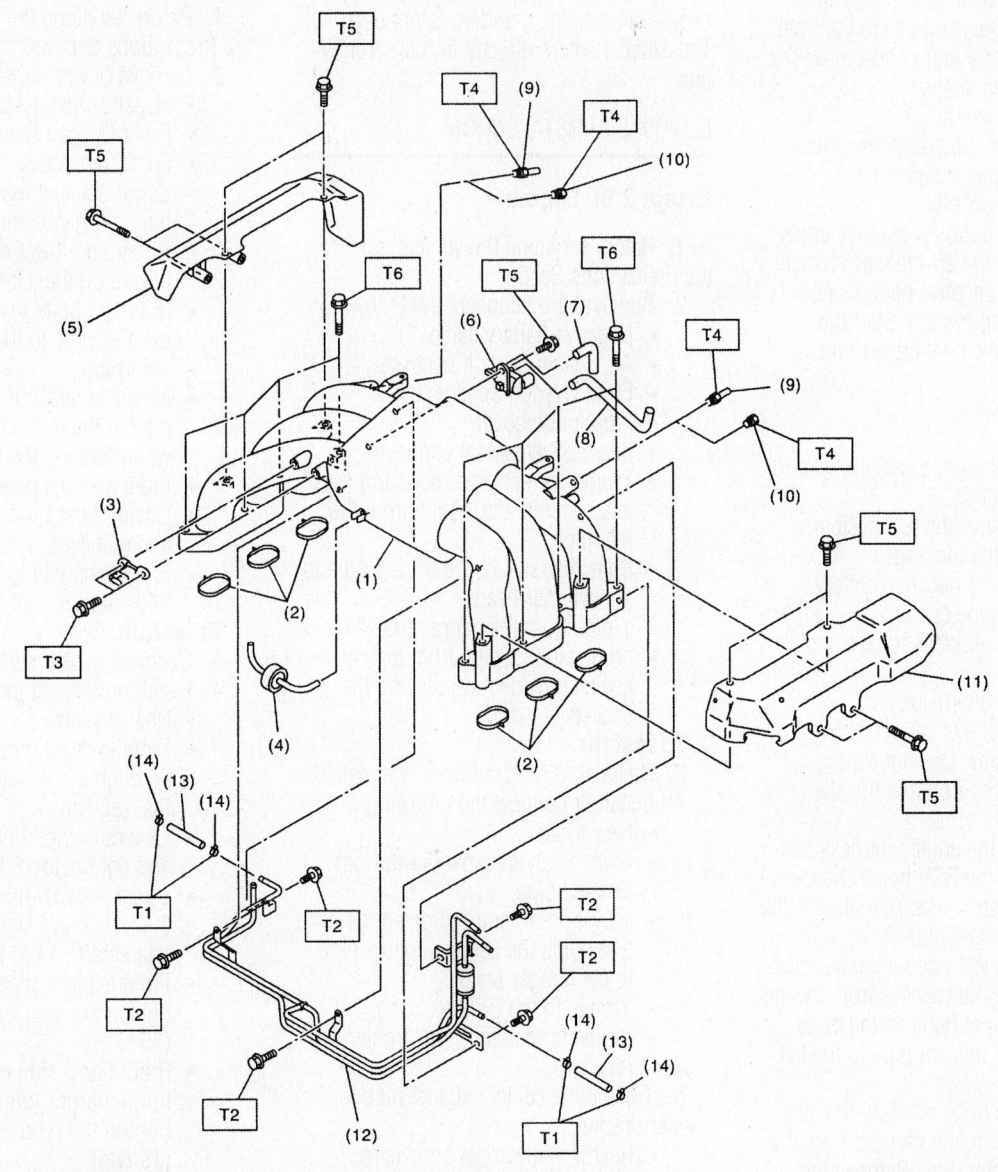

(1)	Intake manifold	(6)	Purge control solenoid valve	(11)	Fuel pipe protector LH
(2)	O-ring	(7)	Hose	(12)	Fuel pipe ASSY
(3)	Manifold absolute pressure sensor	(8)	Hose	(13)	Hose
(4)	Filter	(9)	Nipple	(14)	Clamp
(5)	Fuel pipe protector RH	(10)	Plug		

09490_SBCR_G0029

Intake manifold and related components—2005–2006 3.0L engine

- Engine harness connectors to the bulkhead connectors
- Brake booster hose
- Engine coolant hoses to the throttle body
- PCV hose to the cylinder head cover
- Power steering pump
- Connector to the power steering pump switch
- Bolts holding the power steering pump to the bracket and tighten to 15 ft. lbs. (20 Nm)
- V-belt
- Rear washer hose to the motor
- Washer reservoir hose to the front washer motor and connector to the rear washer motor.
- Washer reservoir
- Accelerator cable and the cruise control cable, if equipped
- Resonator chamber
- Air intake duct, air cleaner upper cover and the air cleaner element

6. Connect the negative battery cable and refill the cooling system. Start the engine, and bleed the cooling system. Check for leaks.

2005–2006

1. Before servicing the vehicle, refer to the Precautions Section.

2. Properly relieve the fuel system pressure. Remove the fuel cap.

3. Disconnect the negative battery cable. Drain the engine coolant.

4. Remove the air cleaner case and air intake chamber.

5. Remove the alternator.

6. Disconnect the electrical connector from the throttle body. Disconnect the engine coolant hoses from the throttle body.

7. Disconnect the engine harness connector. Disconnect the PCV hose. Disconnect the brake booster hose. Disconnect the fuel hoses from the fuel pipe.

8. Remove the left side fuel line protector. Remove the engine harness from the left side fuel injector pipe. Remove the bolts which hold the fuel injector pipe to the left cylinder head.

9. Remove the right side fuel line protector. Remove the engine harness from the right side fuel injector pipe. Remove the bolts which hold the fuel injector pipe to the right cylinder head.

10. Remove the right and left intake manifold retaining bolts. Remove the intake manifold from the engine.

To install:

11. Installation is the reverse of the removal procedure.

12. Be sure to use new intake manifold O-rings. Torque the manifold retaining bolts to specification and in alternating sequence.

13. Be sure to fill the cooling system with the proper grade and type engine coolant.

14. Start the engine and check for leaks, correct as required.

Exhaust Manifold

Due to the unique design of the Subaru engine, an exhaust manifold is not used. The exhaust enters directly into the front Y-pipe.

REMOVAL & INSTALLATION

Except 2.0L Engine

1. Before servicing the vehicle, refer to the Precautions Section.

2. Remove or disconnect the following:
- Negative battery cable
- Air cleaner case, if necessary
- Front Oxygen Sensor (O$_2$S)
- Front undercover
- Rear O$_2$S electrical connector
- Y-pipe-to-rear pipe mounting nuts and separate the Y-pipe from the rear pipe
- Bolts that secure the front Y-pipe to the cylinder head
- Y-pipe from the hanger bracket
- Front exhaust pipe from the catalytic converter and discard the gaskets

To install:

3. Clean all gasket surfaces completely.

4. Install or connect the following:
- New gaskets
- Front catalytic converter to front exhaust pipe.
- Y-pipe. Temporarily tighten the bolt that holds the center exhaust pipe to the hanger bracket.
- Y-pipe, to the cylinder head.
- Y-pipe to the rear pipe. Tighten the retainers.

5. Tighten the center exhaust pipe to hanger bracket
- Rear O$_2$S electrical connector
- Front O$_2$S electrical connector
- Front undercover, if equipped
- Air cleaner case, if removed
- Negative battery cable

6. Start the engine and check for exhaust leaks.

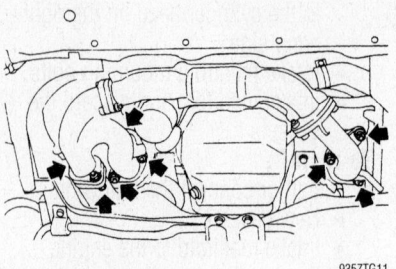

Location of the front exhaust pipe retainers—2.0L engine

2.0L Engine

1. Before servicing the vehicle, refer to the Precautions Section.

2. Remove or disconnect the following:
- Negative battery cable
- Front Oxygen Sensor (O$_2$S)
- Front undercover, if equipped
- Lower exhaust manifold cover on the right hand side
- Upper and lower exhaust manifold covers on the left hand side
- Nuts and bolts that attach the front exhaust pipe to the turbocharger joint pipe
- Nuts that attach the front exhaust pipe to the to the cylinder head while holding the front pipe
- Front exhaust pipe assembly
- Covers from the front exhaust pipe and manifold
- Front exhaust pipe from the manifolds and discard the gaskets

To install:

3. Clean all gasket surfaces completely.

4. Install or connect the following:
- New gaskets
- Front exhaust pipe to the manifolds and tighten the retainers to 26 ft. lbs. (35 Nm)
- Covers to the front exhaust pipe and tighten to 18 ft. lbs. (25 Nm)
- Upper exhaust manifold cover on the right hand side and tighten the retainers to 13 ft. lbs. (19 Nm)
- Front exhaust pipe assembly and tighten the retainers to 26 ft. lbs. (35 Nm)
- Right hand side manifold to the turbocharger joint pipe and tighten the retainers to 13 ft. lbs. (19 Nm)
- Upper and lower manifold covers on the left hand side to 13 ft. lbs. (19 Nm)
- Front O$_2$S
- Front undercover, if equipped
- Negative battery cable

Turbocharger

REMOVAL & INSTALLATION

2.0L Engine

1. Before servicing the vehicle, refer to the Precautions Section.
2. Remove or disconnect the following:
   - Negative battery cable
   - Center exhaust pipe
   - Turbocharger joint pipe from the turbocharger
   - Engine coolant hose from the filler tank
   - Clamp that attaches the turbocharger to the inlet duct
   - Bolt that attaches the bracket of the oil pipe to the turbocharger
   - Oil pipe from the turbocharger
   - Turbocharger bracket
   - Oil outlet hose from the pipe
   - Turbocharger.

To install:

3. Installation is the reverse of removal.
4. Tighten the front pipe to the turbocharger retainers to 22 ft. lbs. (30 Nm).

2.5L Engine

1. Before servicing the vehicle, refer to the Precautions Section.
2. Disconnect the negative battery cable.
3. Raise and support the vehicle safely.
4. Remove the collector cover, if equipped.
5. On 2005–2006 engines remove the intercooler and intercooler bracket.
6. Remove the center exhaust pipe.
7. Lower the vehicle.
8. Separate the turbocharger joint pipe from the turbocharger.
9. Disconnect the engine coolant hose from the coolant filler tank.
10. Loosen the clamp that attaches the turbocharger to the intake duct.
11. Remove the oil pipe–to–turbocharger bolt and remove the oil pipe from the turbocharger.
12. Remove the turbocharger bracket and disconnect the oil outlet hose from the pipe.
13. Remove the turbocharger.

To install:

14. Installation is the reverse of removal, please note the following torque specs:
    a. Turbocharger–to–intake duct: 2 ft. lbs. (3 Nm).
    b. Oil pipe–to–turbocharger: 11 ft. lbs. (16 Nm).
    c. Joint pipe–to–turbocharger using a new gasket: 25 ft. lbs. (35 Nm).
    d. Turbocharger bracket: 24 ft. lbs. (33 Nm).

Front Crankshaft Seal

REMOVAL & INSTALLATION

2.0L Engine

1. Before servicing the vehicle, refer to the Precautions Section.
2. Remove or disconnect the following:
   - Negative battery cable
   - Drive belt
3. Secure the crankshaft pulley with tool No. 499977300.
4. Remove or disconnect the following:
   - Crankshaft pulley bolt and pulley
   - Left, right and center timing belt cover(s) mounting bolts
   - Belt covers
   - Timing belt
   - Crankshaft seal from the oil pump housing

To install:

5. Using a suitable seal driver, install a new crankshaft seal.
6. Install or connect the following:
   - Timing belt crankshaft sprocket and timing belt
   - Belt covers and tighten the bolts to 36–48 inch lbs. (4–5 Nm)
   - Crankshaft pulley and tighten the bolt to 33 ft. lbs. (44 Nm)
   - Drive belt
   - Negative battery cable

2.5L Engine

2002–2004

1. Before servicing the vehicle, refer to the Precautions Section.
2. Remove or disconnect the following:
   - Negative battery cable
   - Radiator electric fan motor wiring connectors
   - Coolant reservoir tank
   - 4 bolts that secure the radiator shroud, and then remove the fan assembly
3. Position the No. 1 piston to Top Dead Center (TDC) of its compression stroke.
   - Accessory drive belt cover
   - Air conditioning compressor drive belt and tensioner
4. Secure the crankshaft pulley with tool No. ST499977000.
   - Crankshaft pulley bolt and pulley
   - Left timing belt cover mounting bolts and the left cover
   - Right timing belt cover mounting bolts and the right cover
   - Center timing belt cover mounting bolts and the center cover

   - Timing belt
   - Timing belt crankshaft sprocket
   - Crankshaft seal from the oil pump housing

To install:

5. Install or connect the following:
   - New crankshaft seal, using a suitable seal driver
   - Timing belt crankshaft sprocket and the timing belt
   - Center, right, then the left timing belt covers. Tighten the bolts to 44 inch lbs. (5 Nm).
   - Crankshaft pulley and tighten the bolt to 94 ft. lbs. (127 Nm)
   - Air conditioning compressor drive belt tensioner and the drive belts
   - Fan shroud and fan motor assembly
   - Accessory drive belt cover
   - Negative battery cable

3.0L Engine

1. Before servicing the vehicle, refer to the Precautions Section.
2. Remove or disconnect the following:
   - Negative battery cable
   - Drive belt
3. Secure the crankshaft pulley with tool No. 499977100.
4. Remove or disconnect the following:
   - Crankshaft pulley bolt and pulley
   - Front chain cover. Refer to the timing chain procedure in this section.
   - Timing chain
   - Crankshaft seal

To install:

5. Using a suitable seal driver, install a new crankshaft seal.
6. Install or connect the following:
   - Timing chain
   - Front chain cover. Refer to the timing chain procedure in this section.
   - Crankshaft pulley and tighten the bolt to 131 ft. lbs. (178 Nm)
   - Drive belt
   - Negative battery cable

Camshaft and Valve Lifters

On some vehicles, it may be necessary to remove the engine from the vehicle to perform this service.

REMOVAL & INSTALLATION

2.0L Engine

1. Before servicing the vehicle, refer to the Precautions Section.
2. Remove or disconnect the following:
   - Negative battery cable

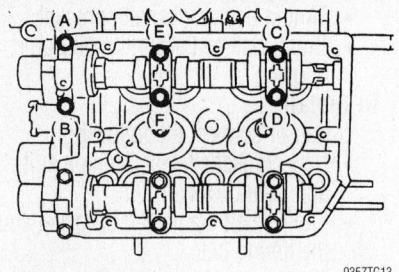

9357TG13

Intake camshaft cap bolt loosening sequence—2.0L engine and 2005–2006 2.5L DOHC engine

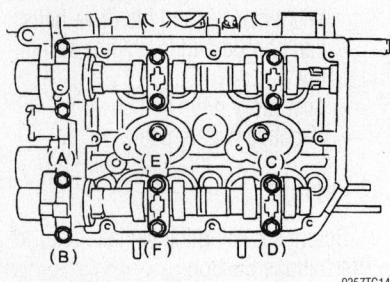

9357TG14

Exhaust camshaft cap bolt loosening sequence—2.0L engine and 2005–2006 2.5L DOHC engine

- Drive belt

3. Secure the crankshaft pulley with tool No. 499977300.

4. Remove or disconnect the following:
- Crankshaft pulley bolt and pulley
- Belt covers
- Timing belt
- Camshaft Position (CMP) sensor
- Camshaft sprockets using locking tool 499207400 to lock them in place while loosening the bolt
- Crankshaft sprocket
- Right hand belt cover No. 2
- Dipstick tube
- Spark plug cord
- Rocker cover and gasket
- Intake camshaft bolts as illustrated a little at a time
- Camshaft caps and the camshaft
- Exhaust camshaft bolts as illustrated a little at a time
- Camshaft caps and the camshaft
- Camshafts

To install:

5. Apply clean engine oil to the bearings on the head.

6. Install the camshaft so that the valves are closed or in contact with the "base circle" of the cam lobe.

7. If the camshafts are positioned as shown in the accompanying illustration, the camshafts need to be rotated at a mini-

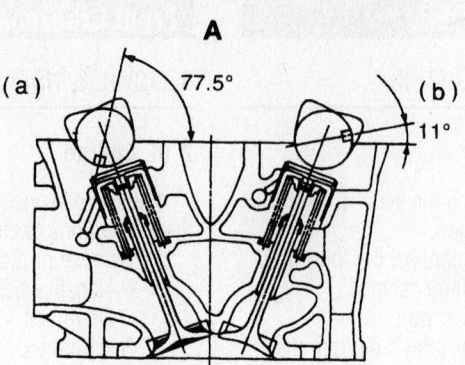

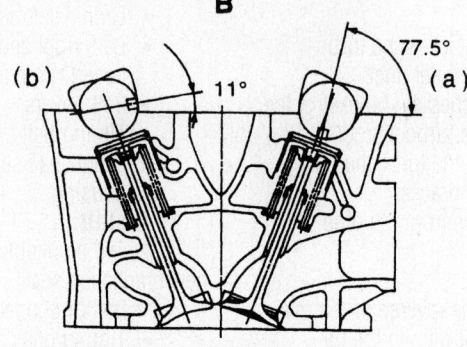

A   Left side cylinder head
B   Right side cylinder head
(a)   Intake camshaft
(b)   Exhaust camshaft

9357TG12

The right hand camshaft need not be rotated when set at the position as shown in the illustration, the left hand intake camshaft should be rotated 80 degrees clockwise. The left hand exhaust camshaft should be rotated 45 degrees counterclockwise—2.0L engine and 2005–2006 2.5L DOHC engine

mum to align the timing belt during installation.

8. The right hand camshaft need not be rotated when set at the position illustrated, the left hand camshaft should be rotated 80 degrees clockwise. The left hand camshaft should be rotated 45 degrees counterclockwise.

9. Apply fluid packing (three bond 1215) sparingly to the cap mating surface.

✳✳ WARNING

Do not apply fluid packing excessively. Failure to do so may cause excess packing to come out and flow towards oil seal, resulting in leaks.

10. Apply engine oil to the cap on the camshaft as shown by mark "A" in the accompanying illustration.

11. Tighten the cap in two stages to 14.5 ft. lbs. (20 Nm) in the sequence illustrated.

12. Apply oil to the lip of the new

camshaft seal and using guide tool 499597200 and installer 499587600, install the new seal.

13. Install or connect the following:
- Rocker cover, making sure to apply fluid packing (three bond 1215) to the four front open edges of the gasket
- Spark plug cord

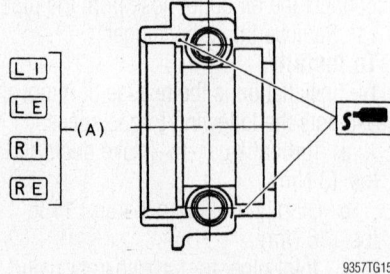

9357TG15

Apply engine oil to the cap on the camshaft as shown by mark "A"—2.0L engine

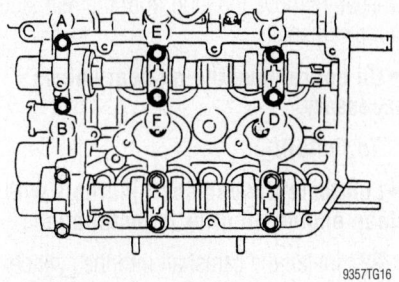

Camshaft cap bolt tightening sequence—
2.0L engine

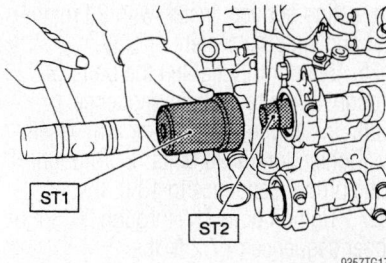

Using guide tool 499597200 and installer
499587600, install the new camshaft
seal—2.0L engine and 2005–2006 2.5L
DOHC engine

- Right hand belt cover No. 2 and
  tighten to 3 ft. lbs. (4 Nm)
- Tensioner bracket and tighten to 18
  ft. lbs. (25 Nm)
- Left hand belt cover No. 2 and
  tighten to 3 ft. lbs. (4 Nm)
- Crankshaft sprocket
- Camshaft sprockets using locking
  tool 499207400 to lock them in
  place while tightening the bolt to
  72 ft. lbs. (98 Nm)
- Timing belt
- Belt cover(s)
- Crankshaft pulley and tighten the
  bolt to 33 ft. lbs. (44 Nm)
- Drive belt
- Negative battery cable

2.5L Engine 2002–2004

1. Before servicing the vehicle, refer to
the Precautions Section.
2. Remove or disconnect the following:
- Negative battery cable
- Timing belt covers, timing belt and
  camshaft sprockets
- Valve rocker covers
- Rocker arm assemblies. Refer to
  the rocker arms/shafts procedure in
  this section.
- Camshaft cap bolts in the proper
  sequence
- Camshaft cap

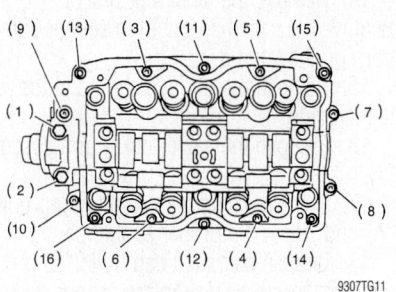

Camshaft cap bolt loosening sequence—
2002–2004 2.5L engine

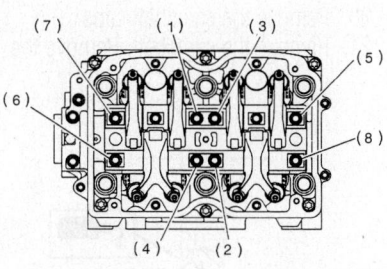

Camshaft cap bolt locations 1–8—
2002–2004 2.5L engine

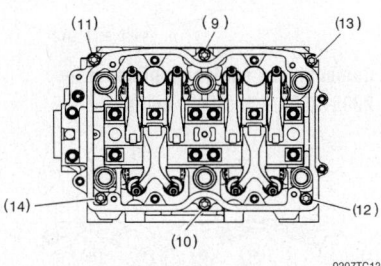

Camshaft cap bolt locations 9–14—
2002–2004 2.5L engine

- Camshaft and rear seal
- Oil seal from the rear side of the
  camshaft

To install:

➡Lubricate the camshaft journals with
clean engine oil prior to installation.

3. Install the camshaft into the cylinder
head
4. Apply a bead of sealant in the
camshaft cap. Position the camshaft cap,
then tighten the bolts 7–10, in sequence,
temporarily
5. Install the rocker arm assemblies.
Refer to the rocker arms/shafts procedure in
this section.
6. Tighten the remaining camshaft cap
bolts, in sequence, as follows:
  a. Step 1: bolts 1–8 to 17–20 ft. lbs.
(23–27 Nm).

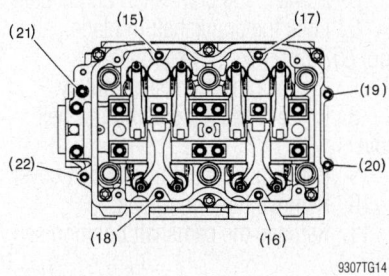

Camshaft cap bolt locations 15–22—
2002–2004 2.5L engine

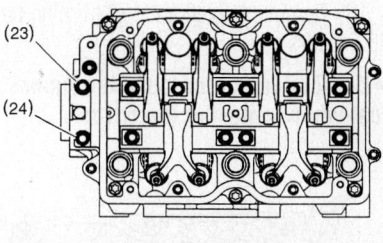

Camshaft cap bolt locations 23–24—
2002–2004 2.5L engine

  b. Step 2: bolts 9–14 to 12–15 ft. lbs.
(16–20 Nm).
  c. Step 3: bolts 15–22 to 6–9 ft. lbs.
(8–12 Nm), using SST 499497000.
  d. Step 4: bolts 23–24 to 6–9 ft. lbs.
(8–12 Nm).
7. Install or connect the following:

➡Lubricate the seals lips with clean
engine oil prior to installation

- Oil seal and plug with suitable
  tools
- Camshaft sprockets, timing belt,
  timing belt covers, and related
  components
- Negative battery cable

8. Check the fluid levels and start the
engine.
9. Allow the engine to reach normal
operating temperature and check for leaks.

2.5L Engine 2005–2006

SOHC ENGINE

1. Before servicing the vehicle, refer to
the Precautions Section.
2. Disconnect the negative battery cable.
3. To remove the front side V belt,
remove the belt covers. Loosen the lock
bolt. Loosen the slider bolt. Remove the
front side belt.
4. To remove the rear side V belt,
remove the belt covers. Loosen the lock
bolt. Loosen the slider bolt. Remove the rear
side belt. Remove the belt tensioner.

5. Remove the crankshaft pulley bolt.

6. Lock the crankshaft in place using tool ST499977100, or equivalent.

7. Remove the crankshaft pulley.

8. Remove the left side timing belt cover.

9. Remove the front timing belt cover.

10. Remove the timing belt.

11. Remove the camshaft position sensor.

12. Remove the camshaft sprockets.

➡**Be sure to lock the camshaft in place using tool ST18231AA010, or equivalent.**

13. Remove the left and right timing belt N02 covers.

➡**Do not damage or loose the rubber seal when removing the covers.**

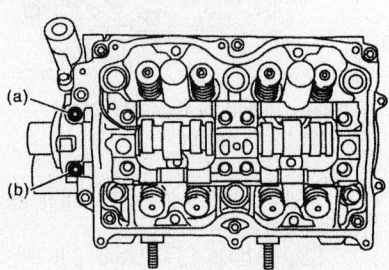

Camshaft bolt loosening sequence—2005–2006 2.5L SOHC engine

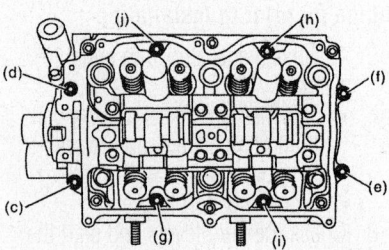

Camshaft bolt loosening sequence—2005–2006 2.5L SOHC engine

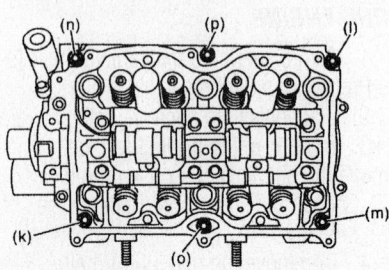

Camshaft bolt loosening sequence—2005–2006 2.5L SOHC engine

14. Remove the tensioner bracket. Remove the camshaft position sensor support, on the left side.

15. Remove the oil level gauge guide, on the left side.

16. Remove the valve rocker arm assembly.

17. Remove camshaft cap retaining bolts "A" and "B" in the proper sequence.

18. Loosen camshaft cap bolts "C" through "J" all the way in the proper sequence.

19. Remove camshaft cap bolts "K" through "P" in the proper sequence using a torx head bit.

20. Remove the camshaft caps.

21. Remove the camshaft. Remove the

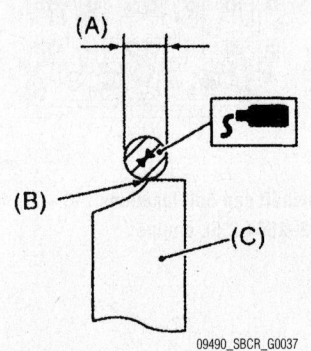

Camshaft cap sealant application—2005–2006 2.5L SOHC engine

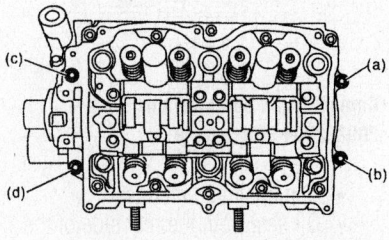

Camshaft bolt tightening sequence—2005–2006 2.5L SOHC engine

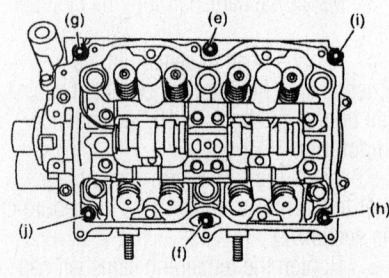

Camshaft bolt tightening sequence—2005–2006 2.5L SOHC engine

oil seal. Remove the plug from the rear side of the camshaft.

➡**Do not remove the oil seal unless necessary.**

To install:

➡**Lubricate the camshaft journals with clean engine oil prior to installation.**

22. Install the camshaft into the cylinder head.

23. Apply liquid gasket to the mating surfaces of the camshaft cap.

24. Apply a bead of sealant (0.12 inch in diameter) along the edge of the camshaft cap mating surface. Install with 20 minutes after applying the sealant.

25. Temporarily tighten the bolts "A" through "D" in the proper sequence.

26. Install the valve rocker arm assembly. Tighten torx head bolts "E" through "J" in the proper sequence to 13 ft. lbs.

27. Tighten bolts "K" through "R" in the proper sequence to 7.2 ft. lbs.

28. Tighten bolts "S" through "T" in the proper sequence to 7.2 ft. lbs.

➡**Be sure to use a new seal washer.**

29. Using tools ST499597000 and ST499587500, install a new seal on the camshaft. Be sure to coat the seal with clean engine oil before installation. Use tool ST499587700 and install the plug.

30. Adjust the valve clearance.

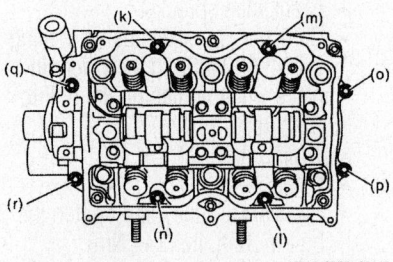

Camshaft bolt tightening sequence—2005–2006 2.5L SOHC engine

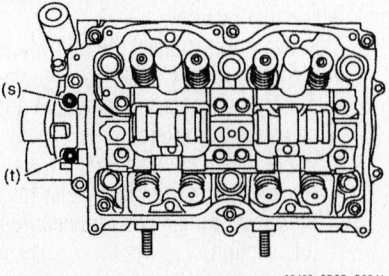

Camshaft bolt tightening sequence—2005–2006 2.5L SOHC engine

31. Continue the installation in the reverse order of the removal procedure.

DOHC ENGINE

1. Before servicing the vehicle, refer to the Precautions Section.

2. Disconnect the negative battery cable.

3. Remove the collector cover.

4. To remove the front side V belt, remove the belt covers. Loosen the lock bolt. Loosen the slider bolt. Remove the front side belt.

5. To remove the rear side V belt, remove the belt covers. Loosen the lock bolt. Loosen the slider bolt. Remove the rear side belt. Remove the belt tensioner.

6. Remove the crankshaft pulley bolt.

7. Lock the crankshaft in place using tool ST499977100, or equivalent.

8. Remove the crankshaft pulley.

9. Remove the left side timing belt cover.

10. Remove the right side timing belt cover.

11. Remove the front timing belt cover.

12. Remove the timing belt.

13. Remove the camshaft position sensor.

14. Remove the camshaft sprockets.

➡**Be sure to lock the camshaft in place using tool ST499207400, or equivalent.**

15. Lock the crankshaft in place using tool ST499977100, or equivalent.

16. Remove the crankshaft pulley.

17. Remove the tensioner bracket. Remove the right and left timing belt No.2 covers.

18. Remove the spark plug wires. Remove the oil level gauge, on the left side.

19. Remove the rocker cover and gasket. Remove the oil pipe.

20. Loosen the oil flow control solenoid valve assembly and the intake camshaft cap bolts equally and in the proper sequence.

21. Loosen the exhaust camshaft cap bolts equally and in the proper sequence.

22. Remove the oil flow control solenoid valve assembly, intake camshaft cap and camshaft.

23. Remove the exhaust camshaft caps and camshaft.

➡**Arrange the camshafts caps so that they can be installed in their original positions.**

To install:

➡**Lubricate the camshaft journals with clean engine oil prior to installation.**

24. Install the camshaft so that the

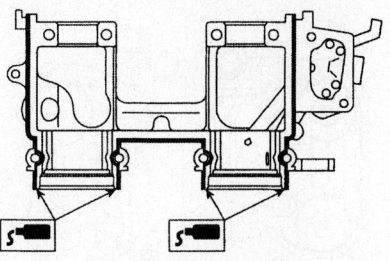

Camshaft cap liquid gasket application points—2005–2006 2.5L DOHC engine

valves are closed or in contact with the "base circle" of the cam lobe.

25. If the camshafts are positioned as shown in the, the camshafts need to be rotated at a minimum to align the timing belt during installation.

26. The right hand camshaft need not be rotated when set at the position illustrated, the left hand intake camshaft should be rotated 80 degrees clockwise. The left hand exhaust camshaft should be rotated 45 degrees counterclockwise.

27. To install the camshaft cap and oil flow control solenoid valve, apply a small amount of liquid gasket to the mating surface of the cap.

➡**Do not apply an excessive amount of sealant, as it will squish out and flow toward the seal resulting in an oil leak.**

28. Apply a thin coat of engine oil to the cap bearing surface and install the cap according to the cap identification mark. Gradually tighten the cap bolts in two stages, first to 7.2 ft. lbs. and than to 14.5 ft. lbs. in the proper sequence.

➡**After tightening the camshaft cap, ensure that the camshaft rotates slightly while holding it at base circle.**

29. Using tools ST49587600 and ST499597200, install a new seal on the camshaft. Be sure to coat the seal with clean engine oil before installation. Use tool ST499587700 and install the plug.

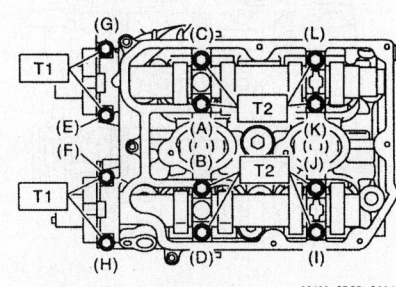

Camshaft bolt tightening sequence— 2005–2006 2.5L DOHC engine

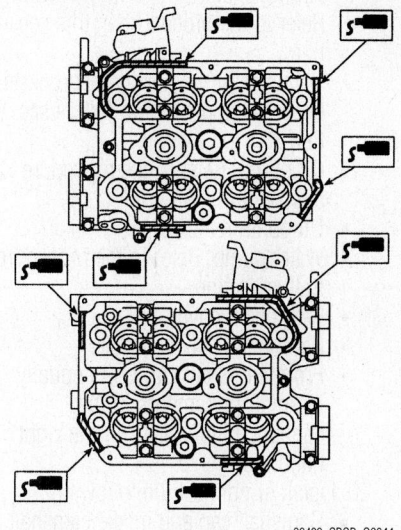

Rocker cover liquid gasket application points—2005–2006 2.5L DOHC engine

30. Install a new gasket on the rocker cover. Apply liquid gasket to the cylinder head (see illustration).

➡**Apply an extra amount of liquid gasket around the semicircular plugs, 5mm or more.**

31. Temporarily tighten the rocker cover retaining bolts, in the proper sequence, and then tighten to 4.7 ft. lbs. Install the oil pipe.

32. Continue the installation in the reverse order of the removal procedure.

3.0L Engine

2002–2004

1. Before servicing the vehicle, refer to the Precautions Section.

2. Remove or disconnect the following:
- Negative battery cable
- Crankshaft pulley cover
- Crankshaft pulley bolt using tool 499977100 to hold the crankshaft
- Crankshaft pulley

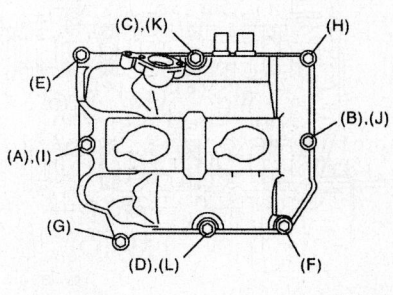

Rocker cover tightening sequence— 2005–2006 2.5L DOHC engine

- Timing chain cover and the chain. Refer to the procedure in this section.
- Camshaft and crankshaft sprockets. Refer to the procedure in this section.
- Oil pump. Refer to the procedure in this section.
- Oil pump relief valve
- Water pump. Refer to the procedure in this section.
- Rear chain cover
- Right hand valve cover
- Front camshaft cap bolts equally in small increments in the sequence illustrated on the right hand side

3. Install or connect the following:
- Camshaft cap and intake camshaft on the right hand side
- Camshaft cap bolts equally in small increments in the sequence illustrated
- Camshaft cap and exhaust camshaft of the right hand side

➡ **Mark the camshaft caps so that they can be reinstalled in their original positions.**

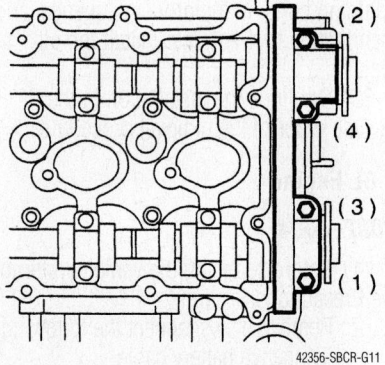

Remove the front camshaft cap bolts equally in small increments in the sequence shown—2002–2004 3.0L engine

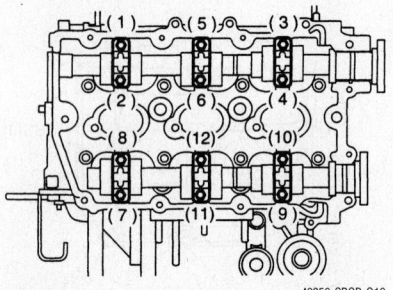

Remove the camshaft cap bolts equally in small increments in the sequence shown—2002–2004 3.0L engine

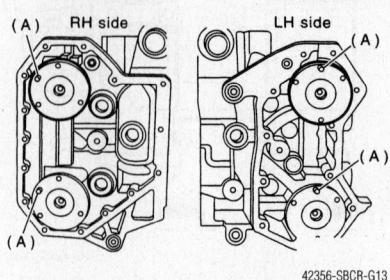

Install the camshafts so the flange knock pin (A) is positioned as shown—2002–2004 3.0L engine

- Plug from the left hand side
- Left hand camshaft components in the same manner as the right hand components

To install:

4. Apply a coat of engine oil to the journals on the camshafts and place the camshafts into position.

5. When installing the camshaft, adjust the camshaft front flange knock pin (A) at the 12 O'clock position on the left hand side and the 10 O'clock position on the right hand side. Refer to the illustration for more detail.

6. Apply 0.059–0.099 inch (1.5–

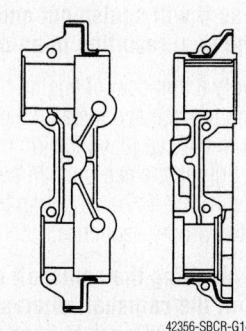

Apply 0.059–0.099 inch (1.5–2.5mm) of fluid gasket sparingly to the front and back of the camshaft cap where indicated—2002–2004 3.0L engine

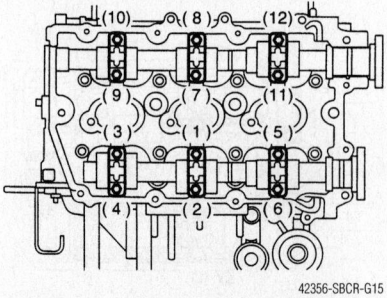

Tighten the camshaft cap bolts in the sequence shown to the proper specification—2002–2004 3.0L engine

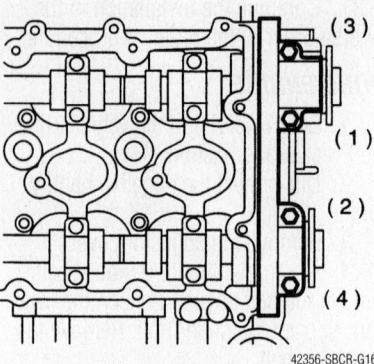

Tighten the front camshaft cap bolts in the sequence shown—2002–2004 3.0L engine

2.5mm) of fluid packing sparingly to the front and back of the camshaft cap as illustrated. Subaru recommends Three Bond 1280B for this purpose.

> ✳✳ **CAUTION**

Do not apply too much gasket maker. This may cause excess fluid gasket maker to come out and flow towards the camshaft journal resulting in engine damage.

7. Apply oil to the cap bearing surface.
8. Install or connect the following:
- Camshaft cap in the their original location
- Camshaft cap bolts and tighten to 11.6 ft. lbs. (16 Nm) in the sequence illustrated
- Front camshaft cap bolts and tighten to 7 ft. lbs. (10 Nm) in the sequence illustrated

9. Apply fluid gasket maker of the of the cylinder heads and valve covers as illustrated. Subaru recommends Three Bond 1280B for this purpose.

> ✳✳ **CAUTION**

Do not apply too much gasket maker. This may cause excess fluid gasket maker to come out and flow towards the camshaft journal resulting in engine damage.

10. Install or connect the following:
- Valve cover bolts and tighten to 5 ft. lbs. (7 Nm) in the sequence illustrated.

11. Install the rear chain cover as follows:

➡ **There are several size bolts used, refer to the illustration for size and locations.**

a. Rear chain cover gasket and clean the mating surfaces.

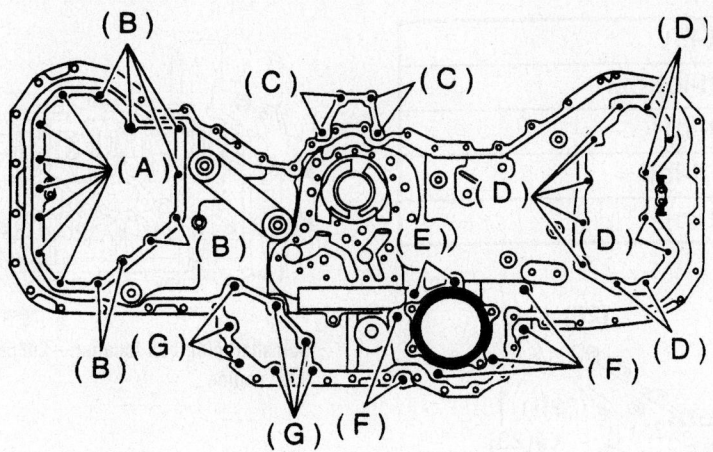

(A) 6 × 14
(B) 6 × 18 (Silver)
(C) 6 × 30
(D) 6 × 18
(E) 6 × 40
(F) 6 × 30
(G) 6 × 22

42356-SBCR-G05

Rear timing chain bolt sizes and locations—2002–2004 3.0L engine

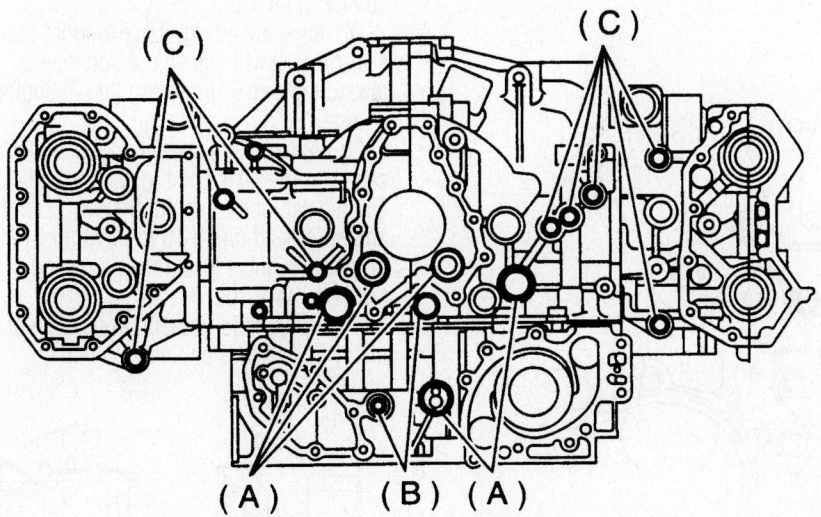

(A)   O-ring (Large)
(B)   O-ring (Medium)
(C)   O-ring (Small)

42356-SBCR-G07

Rear cover O-ring sizes and locations—2002–2004 3.0L engine

b. Apply liquid gasket maker to the mating surfaces of the cover. Refer to the illustration for gasket maker application and diameter.

c. Install new O-rings. Refer to the illustration for O-ring location and size

d. Install the rear chain cover and temporarily tighten the bolts, refer to the illustration for size and locations.

e. Replace mounting bolts **G** with new bolts, refer to the illustration for location.

f. Tighten the cover bolts in the sequence illustrated to the specifications shown in the illustration.

12. Install or connect the following:
- Water pump
- Oil pump relief valve and tighten the bolts to 4.7 ft. lbs. (6.4 Nm) in the sequence illustrated.
- Oil pump
- Crankshaft sprocket. Refer to the procedure in this section.
- Camshaft sprocket. Refer to the procedure in this section.
- Timing chain. Refer to the procedure in this section.
- Crankshaft pulley, apply clean oil to the bolt threads and tighten the bolt to 131 ft. lbs. (178 Nm)
- Camshaft pulley cover and tighten the bolts to 5 ft. lbs. (7 Nm)
- Negative battery cable

13. Start the engine and allow it to reach operating temperature. Check for leaks.

2005–2006

1. Before servicing the vehicle, refer to the Precautions Section.

2. Disconnect the negative battery cable.

3. Remove the crankshaft pulley cover.

4. Remove the crankshaft pulley bolt.

5. Lock the crankshaft in place using tool ST499977100, or equivalent.

6. Remove the crankshaft pulley.

7. Remove the front timing chain cover.

➡**Bolts are three different sizes. Be careful to install the correct bolt in the correct hole.**

8. Remove the timing chain.

9. Remove the camshaft sprocket.

➡**Be sure to lock the camshaft in place using tool ST499977500, or equivalent.**

10. Remove the crankshaft sprocket.

11. Remove the oil pump.

12. Remove the water pump.

13. Remove the rear timing chain cover retaining bolts. Remove the rear timing chain cover.

(1) to (11)	9 N·m (0.9 kgf-m, 6.5 ft-lb)
(12) to (19)	20 N·m (2.0 kgf-m, 14 ft-lb)
(20) to (31)	9 N·m (0.9 kgf-m, 6.5 ft-lb)
(32) to (39)	12 N·m (1.2 kgf-m, 8.7 ft-lb)
(40) to (46)	9 N·m (0.9 kgf-m, 6.5 ft-lb)

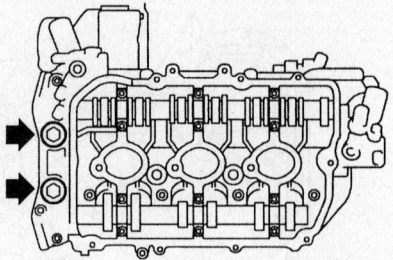

Camshaft plug bolt location—2005–2006 3.0L engine

Rear cover bolt tightening sequence and torque specifications—2002–2004 3.0L engine

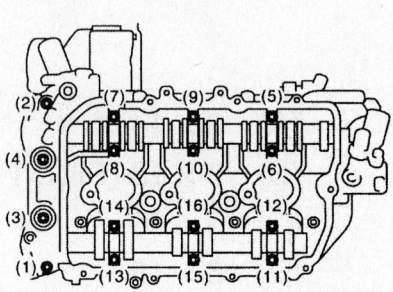

Camshaft bolt loosening sequence—2005–2006 3.0L engine

➡ **There are seven different size bolts. Be sure not to confuse them on installation.**

14. Disconnect the oil pipe.

15. Remove the rocker cover retaining bolts. Remove the rocker cover. Discard the gasket.

16. Remove the plugs, see illustration for location.

17. Loosen the camshaft cap bolts in the proper sequence. Remove the camshaft caps and remove the camshaft.

To install:

18. Apply a coat of engine oil to the

journals on the camshafts and place the camshafts into position.

19. To install the camshaft cap, apply a small amount of liquid gasket to the mating surface of the cap.

Do not apply an excessive amount of sealant, as it will squish out and flow toward the cam journal resulting in engine seizure.

20. Apply a thin coat of engine oil to the cap bearing surface and install the cap. Tighten cap bolts 1 through 12 to 12 ft. lbs. and bolts 13 through 16 to 7.2 ft. lbs. in the proper sequence. Install the plugs and torque to 44 ft. lbs.

21. Apply fluid gasket maker of the of the cylinder heads and valve covers.

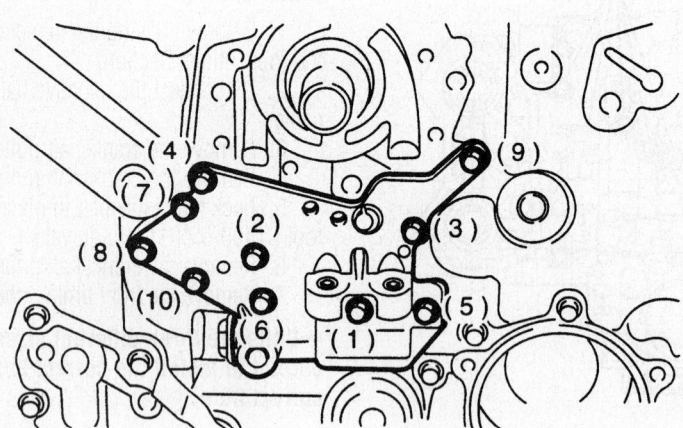

Bolt installation position	Bolt dimension
(1) and (5)	6 x 26
(2), (3), (4) and (9)	6 x 35
(6), (7), (8) and (10)	6 x 16

Oil pump relief valve bolt tightening sequence and bolt specifications—2002–2004 3.0L engine

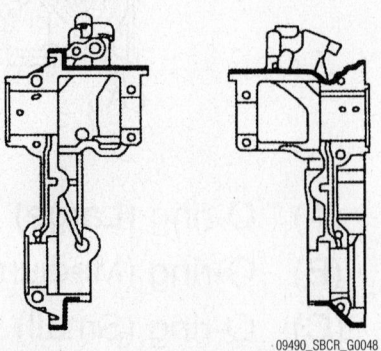

Camshaft cap liquid gasket application points—2005–2006 3.0L engine

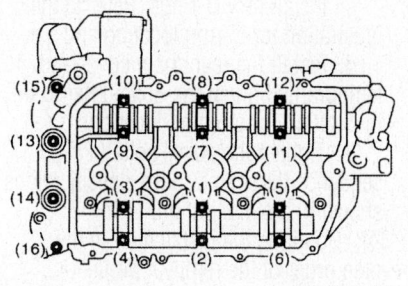

Camshaft bolt tightening sequence—
2005–2006 3.0L engine

❊❊ CAUTION

Do not apply too much gasket maker. This may cause excess fluid gasket maker to come out and flow towards the camshaft journal resulting in engine damage.

22. Tighten the valve cover bolts to 4.7 ft. lbs. and in the proper sequence.

23. Install the rear chain cover as follows:

➡There are several size bolts used, refer to the illustration for size and locations.

    a. Rear chain cover gasket and clean the mating surfaces.

    b. Apply liquid gasket maker to the mating surfaces of the cover. Refer to the illustration for gasket maker application and diameter.

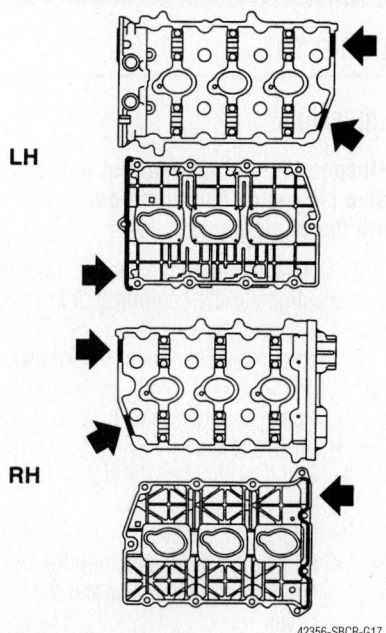

Apply fluid gasket maker of the of the cylinder heads and valve covers as shown—3.0L engine

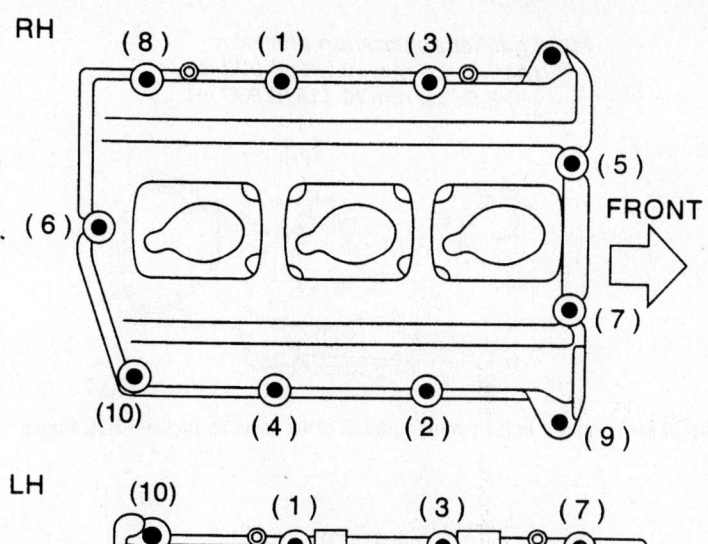

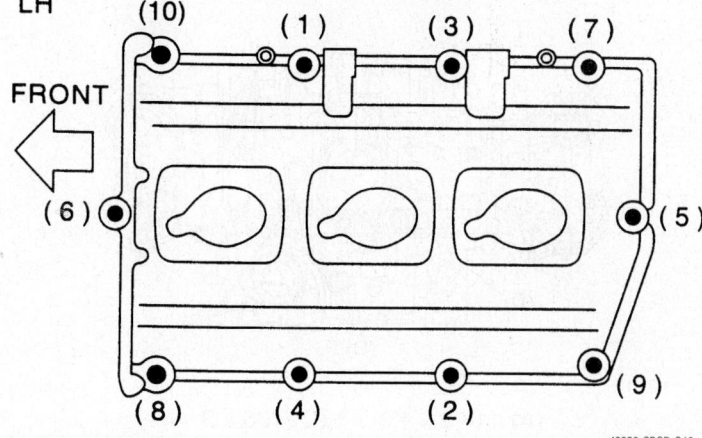

Tighten the valve cover bolts in this sequence—3.0L engine

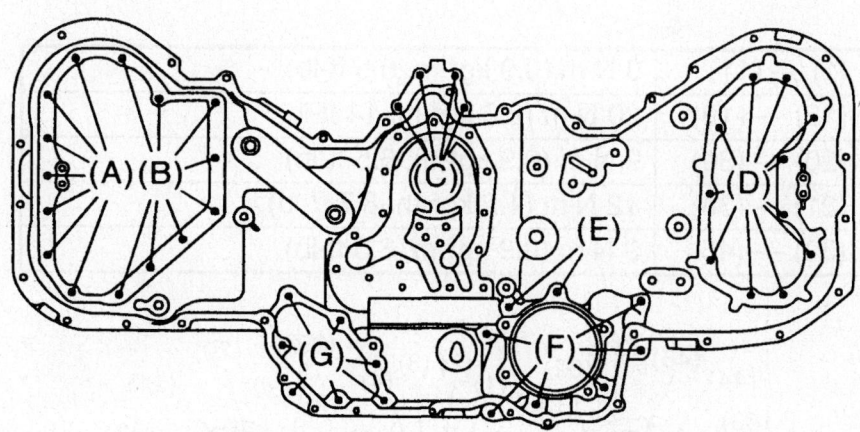

(A)	M6 × 14	(E) M8 × 40
(B)	M6 × 18 (Silver)	(F) M8 × 30
(C)	M6 × 30	(G) M6 × 22
(D)	M6 × 18	

Rear timing chain bolt sizes and locations—2005–2006 3.0L engine

Fluid gasket application diameter:
(A) 1.0±0.5 mm (0.039±0.020 in)
(B) 3.0±1.0 mm (0.118±0.039 in)

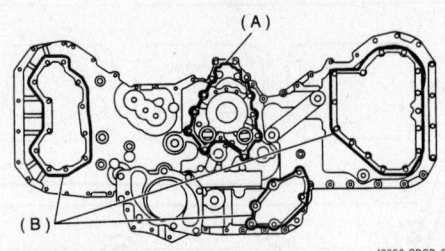

42356-SBCR-G06

Apply liquid gasket maker to the mating surfaces of the cover as shown—3.0L engine

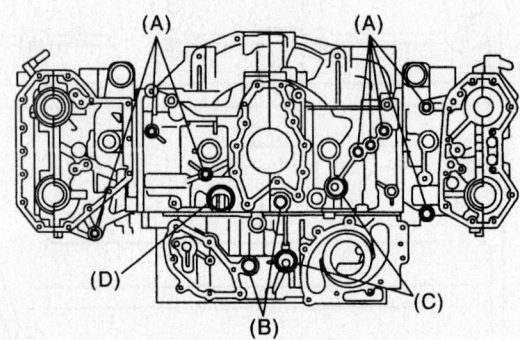

(A)	14.2 × 1.9	(C)	25 × 2
(B)	19.2 × 2.4	(D)	31.2 × 1.9

09490_SBCR_G0024

Rear cover O-ring sizes and locations—2005–2006 3.0L engine

(1) — (11)	9 N·m (0.9 kgf-m, 6.5 ft-lb)
(12) — (19)	20 N·m (2.0 kgf-m, 14 ft-lb)
(20) — (30)	9 N·m (0.9 kgf-m, 6.5 ft-lb)
(31) — (38)	12 N·m (1.2 kgf-m, 8.7 ft-lb)
(39) — (45)	9 N·m (0.9 kgf-m, 6.5 ft-lb)

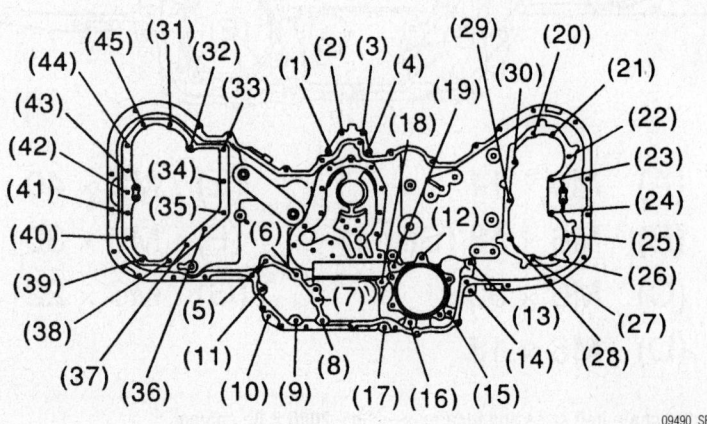

09490_SBCR_G0025

Rear cover bolt tightening sequence and torque specifications—2005–2006 3.0L engine

c. Install new O-rings. Refer to the illustration for O-ring location and size

d. Install the rear chain cover and temporarily tighten the bolts, refer to the illustration for size and locations.

e. Tighten the cover bolts in the sequence illustrated to the specifications shown in the illustration.

24. Continue the installation in the reverse order of the removal procedure.

INSPECTION

1. Before servicing the vehicle, refer to the Precautions Section.

2. Remove the camshaft from the engine.

3. Check the camshaft bearing journals for damage and binding.

4. If the journals are binding, check the cylinder head for damage.

5. Check the cylinder head for clogged oil holes.

6. Check the camshaft surface for abnormal wear and damage. Replace the camshaft, as required.

7. Measure the camshaft lobe surface and replace the camshaft if not within specification.

8. Measure the camshaft journal diameter and replace the camshaft if not within specification.

9. Measure the camshaft run out and replace the camshaft if not within specification.

Valve Lash

ADJUSTMENT

2.0L Engine

→**Inspection and adjustment of the valve clearance should be performed with the engine cold.**

1. Before servicing the vehicle, refer to the precautions in the beginning of this section.

2. Remove or disconnect the following:

- Negative battery cable
- Air intake duct
- Bolt that attaches the right hand timing cover
- Engine undercover
- Remaining bolts attaching the right hand timing belt cover and the cover

3. When inspecting the # 1 and # 3 cylinders:

a. Pull out the engine harness con-

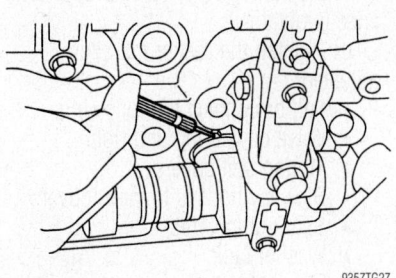

9357TG27

Remove the shim from the lifter—2.0L engine

nector with the bracket from the air cleaner upper cover.

b. Remove the air cleaner case.

c. Disconnect the spark plug wires from the # 1 and # 3 cylinders.

d. Disconnect the Positive Crankcase Ventilation (PCV) hose from the right hand rocker cover.

e. Remove the right hand rocker cover.

4. When inspecting the # 2 and # 4 cylinders:

a. Remove the battery and tray.

b. Remove the bolt that attaches the engine harness onto the body.

c. Disconnect the washer motor connectors.

d. Remove the washer tank bolts and lift the tank upwards.

e. Disconnect the spark plug wires from the # 2 and # 4 cylinders.

f. Disconnect the PCV hose from the left hand rocker cover.

g. Remove the left hand rocker cover.

h. Turn the crankshaft pulley clockwise until the arrow mark on the camshaft is positioned as shown in the illustration to measure the # 1 intake and # 3 exhaust valves.

5. Using a suitable feeler gauge, measure the # 1 and # 3 cylinder exhaust valve clearance. Insert the gauge in as horizontal a direction with respect to the shim. Make sure to measure the exhaust valve clearances while lifting up the vehicle.

6. The intake valve clearance should be 0.0071–0.0087 inch (0.18–0.22mm). The exhaust valve clearance should be 0.0090–0.0106 inch (0.23–0.27mm).

	Unit: mm
Intake valve:S =(V + T) - 0.20	
Exhaust valve:S =(V + T) - 0.25	
S: Shim thickness to be used	
V: Measured valve clearance	
T: Shim thickness required	

9357TG28

Use this table to help you select a suitable shim—2.0L engine

Part No.	Thickness mm (in)
13218 AK010	2.00 (0.0787)
13218 AK020	2.02 (0.0795)
13218 AK030	2.04 (0.0803)
13218 AK040	2.06 (0.0811)
13218 AK050	2.08 (0.0819)
13218 AK060	2.10 (0.0827)
13218 AK070	2.12 (0.0835)
13218 AK080	2.14 (0.0843)
13218 AK090	2.16 (0.0850)
13218 AK100	2.18 (0.0858)
13218 AK110	2.20 (0.0866)
13218 AE710	2.22 (0.0874)
13218 AE730	2.24 (0.0882)
13218 AE750	2.26 (0.0890)
13218 AE770	2.28 (0.0898)
13218 AE790	2.30 (0.0906)
13218 AE810	2.32 (0.0913)
13218 AE830	2.34 (0.0921)
13218 AE850	2.36 (0.0929)
13218 AE870	2.38 (0.0937)
13218 AE890	2.40 (0.0945)
13218 AE910	2.42 (0.0953)
13218 AE920	2.43 (0.0957)
13218 AE930	2.44 (0.0961)
13218 AE940	2.45 (0.0965)
13218 AE950	2.46 (0.0969)
13218 AE960	2.47 (0.0972)
13218 AE970	2.48 (0.0976)
13218 AE980	2.49 (0.0980)
13218 AE990	2.50 (0.0984)
13218 AF000	2.51 (0.0988)
13218 AF010	2.52 (0.0992)
13218 AF020	2.53 (0.0996)
13218 AF030	2.54 (0.1000)
13218 AF040	2.55 (0.1004)
13218 AF050	2.56 (0.1008)
13218 AF060	2.57 (0.1012)
13218 AF070	2.58 (0.1016)
13218 AF090	2.60 (0.1024)
13218 AF110	2.62 (0.1031)
13218 AF130	2.64 (0.1039)
13218 AF150	2.66 (0.1047)
13218 AF170	2.68 (0.1055)
13218 AF190	2.70 (0.1063)

9357TG30

Valve adjusting shim chart—2.0L engine

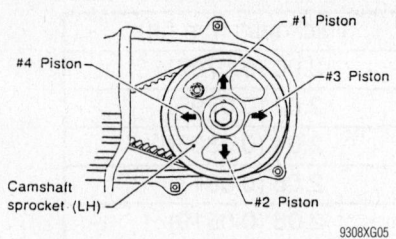

Position the camshaft for adjustment to valves—2.0L engine

7. If not within specification, adjust the valve as outlined in the adjustment steps.

8. Turn the crankshaft pulley clockwise to measure the valve clearance for the # 2 exhaust and # 3 intake valves as shown in the illustration.

9. Turn the crankshaft pulley clockwise to measure the valve clearance for the # 2 intake and # 4 exhaust valves as shown in the illustration.

10. Turn the crankshaft pulley clockwise to measure the valve clearance for the # 1 exhaust and # 4 intake valves as shown in the illustration.

11. Adjust the valve clearance as follows:

a. Measure and record all valve clearances using the procedures outlined in the inspection steps in this section.

b. Prepare shim replacer tool 498187200.

c. Rotate the notch of the valve lifter outwards 45 degrees.

d. Adjust the shim replacer tool notch to the lifter and set it.

➡**Make sure when setting the tool that the edge does not touch the shim.**

e. Tighten bolt "A" and attach it to the cylinder head. Refer to the illustration for bolt locations.

f. Tighten bolt "B" and insert the lifter. Refer to the illustration for bolt locations.

g. Use tweezers and remove the shim from the lifter. A magnet can also be used to remove the shim.

h. Measure the shim thickness using a micrometer.

i. Using the table supplied, select a suitable shim using measured valve clearance and shim thickness.

j. Install the replacement shim to the lifter.

k. After all shims have been adjusted, inspect the valve clearances again.

l. After completion, install all removed components.

2.5L Engine 2002–2004

➡**The valve adjustment should be performed while the engine is cold. A Shim Replace Kit 498187100 will be needed to perform the valve adjustment.**

1. Before servicing the vehicle, refer to the precautions in the beginning of this section.

2. Adjustment should be performed when engine is cold.

3. Remove or disconnect the following:
- Negative battery cable
- Engine coolant reservoir tank
- Timing belt cover on the left hand side

4. When inspecting the No. 1 and 3 cylinders remove the following:
- Air intake duct as a unit
- Resonator chamber
- Spark plug wires from the No. 1 and 3 cylinders
- Blow-by house from valve cover
- Engine undercover
- Timing belt cover on the right hand side
- Valve cover on the right hand side

5. When inspecting the No. 2 and 4 cylinders remove the following:
- Battery and battery tray
- Window washer motor connectors front and rear
- Rear gate glass washer hose from the washer motor
- Washer tank mounting bolts and secure out of the way
- Spark plug wires from the No. 2 and 4 cylinders
- Blow by house from valve cover
- Timing belt cover on the right hand side
- Valve cover on the left hand side

6. Set No. 1 cylinder to Top Dead Center (TDC).

➡**When arrow mark on the left hand side comes exactly to the top, No. 1 cylinder piston is brought to TDC of the compression stroke.**

7. Check the valve clearance:
- Intake valve: 0.0071–0.0087 in. (0.18–0.22mm)
- Exhaust valve: 0.0090–0.0106 in. (0.23–0.27mm)

8. If any valve needs adjustment, perform the following:

a. Loosen the valve rocker nut and screw.

b. Place a thickness gage in at as horizontal a direction as possible with respect to the valve stem and face.

c. Adjust the screw until proper clearance is obtained.

d. Tighten the rocker nut after adjusted.

9. Install or connect the following:
- Valve covers left and right
- Timing belt covers
- Blow-by houses to valve covers
- Spark plug wires
- Washer tank
- Rear gate glass washer hose to the washer motor
- Washer motor connectors
- Battery and battery tray
- Engine undercover
- Resonator chamber
- Air intake duct unit
- Engine coolant reservoir tank

2.5L Engine 2005–2006

SOHC ENGINE

➡**The valve adjustment should be performed while the engine is cold.**

1. Before servicing the vehicle, refer to the precautions in the beginning of this section.

2. Raise and support the vehicle safely.

3. Remove the under cover.

4. Lower the vehicle.

5. Disconnect the negative battery cable.

6. To remove the front side V belt, remove the belt covers. Loosen the lock bolt. Loosen the slider bolt. Remove the front side belt.

7. To remove the rear side V belt, remove the belt covers. Loosen the lock bolt. Loosen the slider bolt. Remove the rear side belt. Remove the belt tensioner.

8. Remove the crankshaft pulley bolt.

9. Lock the crankshaft in place using tool ST499977100, or equivalent.

10. Remove the crankshaft pulley.

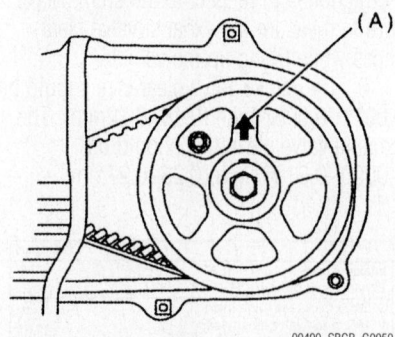

TDC alignment—2005–2006 2.5L SOHC engine

11. Remove the left side timing belt cover.

12. Remove the fuel injector.

13. Remove the rocker cover.

14. Position the number one piston at TDC of the compression stroke.

➡️**When the arrow (see illustration) on the camshaft sprocket (left side) comes exactly to the top, number one cylinder piston is at TDC of the compression stroke.**

15. Measure the valve clearance, using a feeler gauge.

16. If adjustment is needed, loosen the valve rocker nut and screw. Position the feeler gauge.

➡️**Insert the feeler gauge in a horizontally as possible with respect to the valve stem end face. Adjust the exhaust valve clearance while lifting up the vehicle.**

17. While noting the valve clearance, tighten the rocker adjusting screw.

18. When the proper valve clearance is obtained, tighten the valve rocker nut to 7.2 ft. lbs.

19. Adjust the valve clearance on the remaining cylinders, following the above procedure.

➡️**Be sure to position the pistons to their respective TDC positions on the compression stroke, before checking and adjusting the valves. By rotating the crankshaft pulley clockwise every 180 degrees from the state that number one piston is on TDC of the compression stroke, the remaining pistons come to TDC of the compression stroke in the following order, #3, #2, and #4.**

20. After adjustment, replace any removed components.

21. Be sure to use new gaskets and seals, as required.

DOHC ENGINE

➡️**The valve adjustment should be performed while the engine is cold.**

1. Before servicing the vehicle, refer to the precautions in the beginning of this section.

2. Raise and support the vehicle safely.

3. Remove the under cover.

4. Lower the vehicle.

5. Remove the collector cover.

6. Disconnect the negative battery cable.

7. Remove the air intake duct.

8. Remove the bolt that retains the right

side timing belt cover. Remove the remaining bolts and remove the right side timing belt cover.

9. Disconnect the ignition coil electrical connector. Remove the ignition coil.

10. Position a suitable container under the vehicle.

11. Disconnect the PCV hose from the rocker cover. Remove the rocker cover retaining bolts. Remove the rocker cover from the vehicle.

12. Position the number one piston at TDC of the compression stroke.

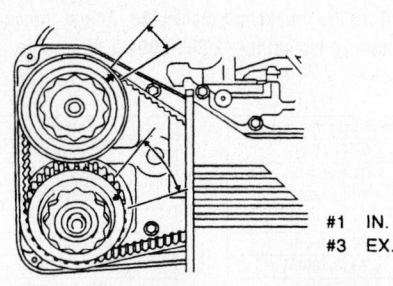

Turn the crankshaft pulley clockwise until the arrow mark on the camshaft is positioned as shown to measure the # 1 intake and # 3 exhaust valves—2.0L engine and 2005–2006 2.5L DOHC engine

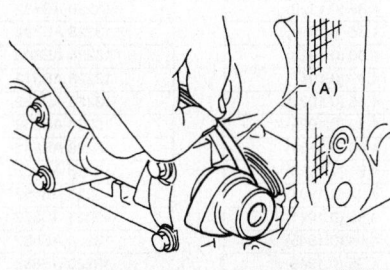

Use a feeler gauge to inspect the valve clearance—2.0L engine and 2005–2006 2.5L DOHC engine

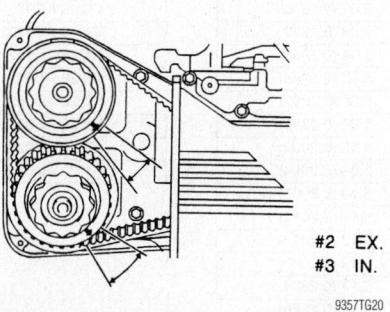

Turn the crankshaft pulley clockwise until the arrow mark on the camshaft is positioned as shown to measure the # 2 exhaust and # 3 intake valves—2.0L engine and 2005–2006 2.5L DOHC engine

13. Using a feeler gauge, measure and record the clearance of the number one cylinder intake and the number three cylinder exhaust valves.

➡️**Insert the feeler gauge in a horizontally as possible with respect to the valve lifter. Measure and record the exhaust valve clearance while lifting up the vehicle.**

14. Rotate the crankshaft pulley clockwise until the arrow mark on the camshaft is positioned as shown to measure and record the clearance on the number two exhaust and number three intake valves.

15. Rotate the crankshaft pulley clockwise until the arrow mark on the camshaft is positioned as shown to measure and record the number two intake and number four exhaust valves.

16. Rotate the crankshaft pulley clockwise until the arrow mark on the camshaft is positioned as shown to measure and record the number one exhaust and number four intake valves.

17. If adjustment is required, remove the camshafts.

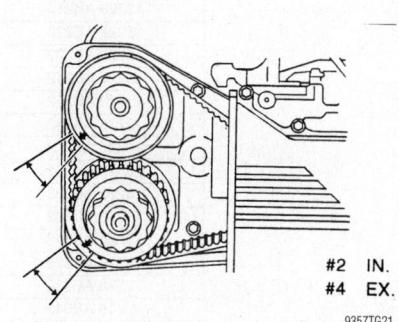

Turn the crankshaft pulley clockwise until the arrow mark on the camshaft is positioned as shown to measure the # 2 intake and # 4 exhaust valves—2.0L engine and 2005–2006 2.5L DOHC engine

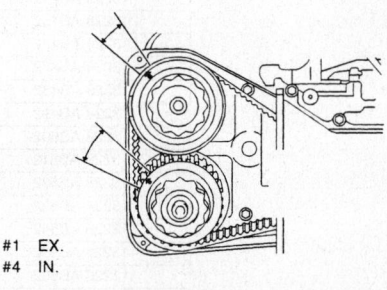

Turn the crankshaft pulley clockwise until the arrow mark on the camshaft is positioned as shown to measure the # 1 exhaust and # 4 intake valves—2.0L engine and 2005–2006 2.5L DOHC engine

	Unit: (mm)
Intake valve: S = (V + T) − 0.20	
Exhaust valve: S = (V + T) − 0.35	
S: Valve lifter thickness required	
V: Measured valve clearance	
T: Valve lifter thickness to be used	

09490_SBCR_G0051

Use this table to help you select a suitable shim—2005–2006 2.5L DOHC engine

18. Remove and measure the thickness of the valve lifter. Select a suitable shim, using the shim selection chart.

19. Install the replacement shim to the lifter.

20. After all shims have been adjusted, inspect the valve clearances again.

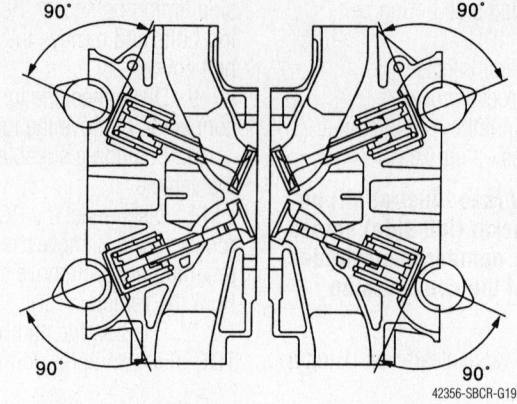

42356-SBCR-G19

Turn the crankshaft clockwise. Adjust the camshaft position so the camshaft lobe is perpendicular to the shim —2002–2004 3.0L engine

Part No.	Thickness mm (in)	Part No.	Thickness mm (in)
13228 AB102	4.68 (0.1843)	13228 AB622	5.20 (0.2047)
13228 AB112	4.69 (0.1846)	13228 AB632	5.21 (0.2051)
13228 AB122	4.70 (0.1850)	13228 AB642	5.22 (0.2055)
13228 AB132	4.71 (0.1854)	13228 AB652	5.23 (0.2059)
13228 AB142	4.72 (0.1858)	13228 AB662	5.24 (0.2063)
13228 AB152	4.73 (0.1862)	13228 AB672	5.25 (0.2067)
13228 AB162	4.74 (0.1866)	13228 AB682	5.26 (0.2071)
13228 AB172	4.75 (0.1870)	13228 AB692	5.27 (0.2075)
13228 AB182	4.76 (0.1874)	13228 AB702	4.38 (0.1724)
13228 AB192	4.77 (0.1878)	13228 AB712	4.40 (0.1732)
13228 AB202	4.78 (0.1882)	13228 AB722	4.42 (0.1740)
13228 AB212	4.79 (0.1886)	13228 AB732	4.44 (0.1748)
13228 AB222	4.80 (0.1890)	13228 AB742	4.46 (0.1756)
13228 AB232	4.81 (0.1894)	13228 AB752	4.48 (0.1764)
13228 AB242	4.82 (0.1898)	13228 AB762	4.50 (0.1771)
13228 AB252	4.83 (0.1902)	13228 AB772	4.52 (0.1780)
13228 AB262	4.84 (0.1906)	13228 AB782	4.54 (0.1787)
13228 AB272	4.85 (0.1909)	13228 AB792	4.56 (0.1795)
13228 AB282	4.86 (0.1913)	13228 AB802	4.58 (0.1803)
13228 AB292	4.87 (0.1917)	13228 AB812	4.60 (0.1811)
13228 AB302	4.88 (0.1921)	13228 AB822	4.62 (0.1819)
13228 AB312	4.89 (0.1925)	13228 AB832	4.64 (0.1827)
13228 AB322	4.90 (0.1929)	13228 AB842	4.66 (0.1835)
13228 AB332	4.91 (0.1933)	13228 AB852	5.29 (0.2083)
13228 AB342	4.92 (0.1937)	13228 AB862	5.31 (0.2091)
13228 AB352	4.93 (0.1941)	13228 AB872	5.33 (0.2098)
13228 AB362	4.94 (0.1945)	13228 AB882	5.35 (0.2106)
13228 AB372	4.95 (0.1949)	13228 AB892	5.37 (0.2114)
13228 AB382	4.96 (0.1953)	13228 AB902	5.39 (0.2122)
13228 AB392	4.97 (0.1957)	13228 AB912	5.41 (0.2123)
13228 AB402	4.98 (0.1961)	13228 AB922	5.43 (0.2138)
13228 AB412	4.99 (0.1965)	13228 AB932	5.45 (0.2146)
13228 AB422	5.00 (0.1969)	13228 AB942	5.47 (0.2154)
13228 AB432	5.01 (0.1972)	13228 AB952	5.49 (0.2161)
13228 AB442	5.02 (0.1976)	13228 AB962	5.51 (0.2169)
13228 AB452	5.03 (0.1980)	13228 AB972	5.53 (0.2177)
13228 AB462	5.04 (0.1984)	13228 AB982	5.55 (0.2185)
13228 AB472	5.05 (0.1988)	13228 AB992	5.57 (0.2193)
13228 AB482	5.06 (0.1992)	13228 AC002	5.59 (0.2201)
13228 AB492	5.07 (0.1996)	13228 AC012	5.61 (0.2209)
13228 AB502	5.08 (0.2000)	13228 AC022	5.63 (0.2217)
13228 AB512	5.09 (0.2004)	13228 AC032	5.65 (0.2224)
13228 AB522	5.10 (0.2008)		
13228 AB532	5.11 (0.2012)		
13228 AB542	5.12 (0.2016)		
13228 AB552	5.13 (0.2020)		
13228 AB562	5.14 (0.2024)		
13228 AB572	5.15 (0.2028)		
13228 AB582	5.16 (0.2031)		
13228 AB592	5.17 (0.2035)		
13228 AB602	5.18 (0.2039)		
13228 AB612	5.19 (0.2043)		

Valve adjusting shim chart—2005–2006 2.5L DOHC engine

09490_SBCR_G0052

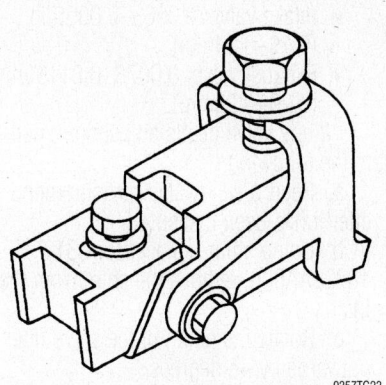

Shim replacer tool is required to adjust the valves—2.0L engine and 2002–2004 3.0L engine

9357TG23

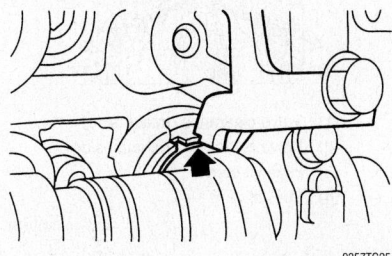

Rotate the notch of the valve lifter outwards 45 degrees—2.0L engine and 2002–2004 3.0L engine

9357TG24

Adjust the shim replacer tool notch to the lifter and set it—2.0L engine and 2002–2004 3.0L engine

9357TG25

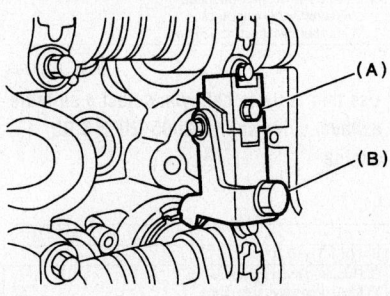

Location of bolts "A" and "B" on the shim replacer tool—2.0L engine and 2002–2004 3.0L engine

9357TG26

	Unit: mm
Intake valve: S = (V + T) - 0.20	
Exhaust valve: S = (V + T) - 0.25	
S: Shim thickness to be used	
V: Measured valve clearance	
T: Shim thickness required	

9357TG28

Use this table to help you select a suitable shim—2002–2004 3.0L engine

21. After completion, install all removed components.

3.0L Engine

2002–2004

➡**The valve adjustment should be performed while the engine is cold.**

1. Before servicing the vehicle, refer to the precautions in the beginning of this section.

2. Remove or disconnect the following:
- Negative battery cable
- Engine undercover

3. To inspect the right hand side perform remove the following:
- Drive belt
- Power steering hose from the bracket
- Power steering pump bracket bolts and place the pump assembly on the right side wheel apron with the lines still attached
- Fuel pipe protector from the right hand side

Part No.	Thickness mm (in)
13218 AK010	2.00 (0.0787)
13218 AK020	2.02 (0.0795)
13218 AK030	2.04 (0.0803)
13218 AK040	2.06 (0.0811)
13218 AK050	2.08 (0.0819)
13218 AK060	2.10 (0.0827)
13218 AK070	2.12 (0.0835)
13218 AK080	2.14 (0.0843)
13218 AK090	2.16 (0.0850)
13218 AK100	2.18 (0.0858)
13218 AK110	2.20 (0.0866)
13218 AE710	2.22 (0.0874)
13218 AE720	2.23 (0.0878)
13218 AE730	2.24 (0.0882)
13218 AE740	2.25 (0.0886)
13218 AE750	2.26 (0.0890)
13218 AE760	2.27 (0.0894)
13218 AE770	2.28 (0.0898)
13218 AE780	2.29 (0.0902)
13218 AE790	2.30 (0.0906)
13218 AE800	2.31 (0.0909)
13218 AE810	2.32 (0.0913)
13218 AE820	2.33 (0.0917)
13218 AE830	2.34 (0.0921)
13218 AE840	2.35 (0.0925)
13218 AE850	2.36 (0.0929)
13218 AE860	2.37 (0.0933)
13218 AE870	2.38 (0.0937)
13218 AE880	2.39 (0.0941)
13218 AE890	2.40 (0.0945)
13218 AE900	2.41 (0.0949)
13218 AE910	2.42 (0.0953)

42356-SBCR-G24

Valve adjusting shim chart (1 of 2)—2002–2004 3.0L engine

- Fuel injector connections
- Front Oxygen Sensor (O_2S) connector
- Oil pressure switch connector
- Ignition coils
- Valve cover on the right hand side

4. To inspect the left hand side perform remove the following:

- Battery
- Window washer motor connectors front and rear
- Rear gate glass washer hose from the washer motor
- Washer tank mounting bolts and secure out of the way

- Positive Crankcase Ventilation (PCV) and blow–by hose from the left hand rocker cover
- Fuel pipe protector from the left hand side
- Fuel injector connections
- Front O_2S connector
- Ignition coils
- Valve cover on the left hand side

5. Using crankshaft socket tool ST 18252AA000, turn the crankshaft clockwise. Adjust the camshaft position so the camshaft lobe is perpendicular to the shim as illustrated.

6. Check the valve clearance:

- Intake valve: 0.0063–0.0095 in. (0.16–0.24mm).
- Exhaust valve: 0.0078–0.0118 in. (0.20–0.25mm).

7. If any valve needs adjustment, perform the following:

a. Record each valve measurement after it has been measured.

b. Using shim replacer tool ST 18329AA000, remove the shim from the lifter.

c. Rotate the notch of the valve lifter outwards by 45 degrees.

d. Adjust the shim replacer notch to the valve lifter and set it as illustrated. Make sure when setting the replacer, the edge does not touch the shim.

e. Tighten bolt **A** on the tool and attach it to the cylinder head, then tighten bolt **B** and insert the valve lifter. Refer to the accompanying illustration.

f. Use tweezers to remove the shim. A magnet may be used as well to remove the shim without dropping it.

g. Use a micrometer to measure the shim.

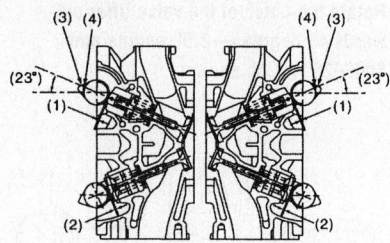

(1) Valve clearance (Intake side)
(2) Valve clearance (Exhaust side)
(3) High lift cam
(4) Low lift cam

09490_SBCR_G0053

Valve adjustment crankshaft positioning— 2005–2006 3.0L engine

	Unit: (mm)
S = (V + T) – 0.35	
S: Valve lifter thickness required	
V: Measured valve clearance	
T: Valve lifter thickness to be used	

09490_SBCR_G0054

Use this table to help you select a suitable exhaust valve shim—2005–2006 3.0L engine

	Unit: (mm)
S = (V + T) – 0.20	
S: Required shim thickness	
V: Measured valve clearance	
T: Shim thickness to be used	

09490_SBCR_G0055

Use this table to help you select a suitable intake valve shim—2005–2006 3.0L

Part No.	Thickness mm (in)
13218 AE920	2.43 (0.0957)
13218 AE930	2.44 (0.0961)
13218 AE940	2.45 (0.0965)
13218 AE950	2.46 (0.0969)
13218 AE960	2.47 (0.0972)
13218 AE970	2.48 (0.0976)
13218 AE980	2.49 (0.0980)
13218 AE990	2.50 (0.0984)
13218 AF000	2.51 (0.0988)
13218 AF010	2.52 (0.0992)
13218 AF020	2.53 (0.0996)
13218 AF030	2.54 (0.1000)
13218 AF040	2.55 (0.1004)
13218 AF050	2.56 (0.1008)
13218 AF060	2.57 (0.1012)
13218 AF070	2.58 (0.1016)
13218 AF090	2.60 (0.1024)
13218 AF100	2.61 (0.1028)
13218 AF110	2.62 (0.1031)
13218 AF120	2.63 (0.1035)
13218 AF130	2.64 (0.1039)
13218 AF140	2.65 (0.1043)
13218 AF150	2.66 (0.1047)
13218 AF160	2.67 (0.1051)
13218 AF170	2.68 (0.1055)
13218 AF180	2.69 (0.1059)
13218 AF190	2.70 (0.1063)
13218 AF200	2.71 (0.1067)
13218 AF210	2.72 (0.1071)
13218 AF220	2.73 (0.1075)
13218 AF230	2.74 (0.1079)
13218 AF240	2.75 (0.1083)
13218 AF250	2.76 (0.1087)
13218 AF260	2.77 (0.1091)
13218 AF270	2.78 (0.1094)
13218 AF280	2.79 (0.1098)
13218 AF290	2.80 (0.1102)
13218 AF300	2.81 (0.1106)

42356-SBCR-G25

Valve adjusting shim chart (2 of 2)—2002–2004 3.0L engine

Part No.	Thickness mm (in)	Part No.	Thickness mm (in)
13228AD180	4.32 (0.1701)	13228AC860	4.90 (0.1929)
13228AD190	4.34 (0.1709)	13228AC870	4.91 (0.1933)
13228AD200	4.36 (0.1717)	13228AC880	4.92 (0.1937)
13228AD210	4.38 (0.1724)	13228AC890	4.93 (0.1941)
13228AD220	4.40 (0.1732)	13228AC900	4.94 (0.1945)
13228AD230	4.42 (0.1740)	13228AC910	4.95 (0.1949)
13228AD240	4.44 (0.1748)	13228AC920	4.96 (0.1953)
13228AD250	4.46 (0.1756)	13228AC930	4.97 (0.1957)
13228AD260	4.48 (0.1764)	13228AC940	4.98 (0.1961)
13228AD270	4.50 (0.1772)	13228AC950	4.99 (0.1965)
13228AD280	4.52 (0.1780)	13228AC960	5.00 (0.1969)
13228AD290	4.54 (0.1787)	13228AC970	5.01 (0.1972)
13228AD300	4.56 (0.1795)	13228AC980	5.02 (0.1976)
13228AD310	4.58 (0.1803)	13228AC990	5.03 (0.1980)
13228AD320	4.60 (0.1811)	13228AD000	5.04 (0.1984)
13228AC580	4.62 (0.1819)	13228AD010	5.05 (0.1988)
13228AC590	4.63 (0.1823)	13228AD020	5.06 (0.1992)
13228AC600	4.64 (0.1827)	13228AD030	5.07 (0.1996)
13228AC610	4.65 (0.1831)	13228AD040	5.08 (0.2000)
13228AC620	4.66 (0.1835)	13228AD050	5.09 (0.2004)
13228AC630	4.67 (0.1839)	13228AD060	5.10 (0.2008)
13228AC640	4.68 (0.1843)	13228AD070	5.11 (0.2012)
13228AC650	4.69 (0.1846)	13228AD080	5.12 (0.2016)
13228AC660	4.70 (0.1850)	13228AD090	5.13 (0.2020)
13228AC670	4.71 (0.1854)	13228AD100	5.14 (0.2024)
13228AC680	4.72 (0.1858)	13228AD110	5.15 (0.2028)
13228AC690	4.73 (0.1862)	13228AD120	5.16 (0.2032)
13228AC700	4.74 (0.1866)	13228AD130	5.17 (0.2035)
13228AC710	4.75 (0.1870)	13228AD140	5.18 (0.2039)
13228AC720	4.76 (0.1874)	13228AD150	5.19 (0.2043)
13228AC730	4.77 (0.1878)	13228AD160	5.20 (0.2047)
13228AC740	4.78 (0.1882)	13228AD170	5.21 (0.2051)
13228AC750	4.79 (0.1886)	13228AD330	5.23 (0.2059)
13228AC760	4.80 (0.1890)	13228AD340	5.25 (0.2067)
13228AC770	4.81 (0.1894)	13228AD350	5.27 (0.2075)
13228AC780	4.82 (0.1898)	13228AD360	5.29 (0.2083)
13228AC790	4.83 (0.1902)	13228AD370	5.31 (0.2091)
13228AC800	4.84 (0.1906)	13228AD380	5.33 (0.2098)
13228AC810	4.85 (0.1909)	13228AD390	5.35 (0.2106)
13228AC820	4.86 (0.1913)	13228AD400	5.37 (0.2114)
13228AC830	4.87 (0.1917)	13228AD410	5.39 (0.2122)
13228AC840	4.88 (0.1921)	13228AD420	5.41 (0.2130)
13228AC850	4.89 (0.1925)	13228AD430	5.43 (0.2138)
		13228AD440	5.45 (0.2146)
		13228AD450	5.47 (0.2154)
		13228AD460	5.49 (0.2161)
		13228AD470	5.51 (0.2169)
		13228AD480	5.53 (0.2177)
		13228AD490	5.55 (0.2185)
		13228AD500	5.57 (0.2193)
		13228AD510	5.59 (0.2201)

09490_SBCR_G0056

Exhaust valve adjusting shim chart—2005–2006 3.0L engine

8. Measure the shim thickness using a micrometer.

9. Using the table supplied, select a suitable shim using measured valve clearance and shim thickness.

10. Install the replacement shim to the lifter.

11. After all shims have been adjusted, inspect the valve clearances again.

12. After completion, install all removed components.

2005–2006

➡The valve adjustment should be performed while the engine is cold.

1. Before servicing the vehicle, refer to the precautions in the beginning of this section.

2. Raise and support the vehicle safely.

3. Remove the under cover.

4. Lower the vehicle.

5. Remove the collector cover.

6. Disconnect the negative battery cable.

7. On the right side, remove the air intake duct and air cleaner case. Remove the fuel tank protector. Disconnect the oil pressure switch electrical connector. Remove the ignition coil.

8. On the left side, remove the battery and battery carrier. Disconnect the PCV hose from the rocker cover. Remove the ignition coil.

9. Remove the rocker cover retaining bolts. Remove the rocker covers from the engine.

10. Rotate the crankshaft clockwise until the cam is set in position, see illustration.

11. Using a feeler gauge measure and record the clearance of the intake and exhaust valve.

➡Measure it within the range of +/- 30 degrees from the specified position,

Part No.	Thickness mm (in)
13218AK890	1.92 (0.0756)
13218AK900	1.94 (0.0764)
13218AK910	1.96 (0.0772)
13218AK920	1.98 (0.0780)
13218AK930	2.00 (0.0787)
13218AK940	2.02 (0.0795)
13218AK950	2.04 (0.0803)
13218AK960	2.06 (0.0811)
13218AK970	2.07 (0.0815)
13218AK980	2.08 (0.0819)
13218AK990	2.09 (0.0823)
13218AL000	2.10 (0.0827)
13218AL010	2.11 (0.0831)
13218AL020	2.12 (0.0835)
13218AL030	2.13 (0.0839)
13218AL040	2.14 (0.0843)
13218AL050	2.15 (0.0846)
13218AL060	2.16 (0.0850)
13218AL070	2.17 (0.0854)
13218AL080	2.18 (0.0858)
13218AL090	2.19 (0.0862)
13218AL100	2.20 (0.0866)
13218AL110	2.21 (0.0870)
13218AL120	2.22 (0.0874)
13218AL130	2.23 (0.0878)
13218AL140	2.24 (0.0882)
13218AL150	2.25 (0.0886)
13218AL160	2.26 (0.0890)
13218AL170	2.27 (0.0894)
13218AL180	2.28 (0.0898)
13218AL190	2.29 (0.0902)
13218AL200	2.30 (0.0906)
13218AL210	2.31 (0.0909)
13218AL220	2.32 (0.0913)
13218AL230	2.33 (0.0917)
13218AL240	2.34 (0.0921)
13218AL250	2.35 (0.0925)
13218AL260	2.36 (0.0929)
13218AL270	2.37 (0.0933)
13218AL280	2.38 (0.0937)
13218AL290	2.39 (0.0941)
13218AL300	2.40 (0.0945)
13218AL310	2.41 (0.0949)

Part No.	Thickness mm (in)
13218AL320	2.42 (0.0953)
13218AL330	2.43 (0.0957)
13218AL340	2.44 (0.0961)
13218AL350	2.45 (0.0965)
13218AL360	2.46 (0.0969)
13218AL370	2.47 (0.0972)
13218AL380	2.48 (0.0976)
13218AL390	2.49 (0.0980)
13218AL400	2.50 (0.0984)
13218AL410	2.51 (0.0988)
13218AL420	2.52 (0.0992)
13218AL430	2.53 (0.0996)
13218AL440	2.54 (0.1000)
13218AL450	2.55 (0.1004)
13218AL460	2.56 (0.1008)
13218AL470	2.57 (0.1012)
13218AL480	2.58 (0.1016)
13218AL490	2.59 (0.1020)
13218AL500	2.60 (0.1024)
13218AL510	2.61 (0.1028)
13218AL520	2.62 (0.1032)
13218AL530	2.64 (0.1039)
13218AL540	2.66 (0.1047)
13218AL550	2.68 (0.1055)
13218AL560	2.70 (0.1063)
13218AL570	2.72 (0.1071)
13218AL580	2.74 (0.1079)
13218AL590	2.76 (0.1087)

Intake valve adjusting shim chart—2005–2006 3.0L engine

shown in the illustration. Measure it in the low cam for the intake side. Insert the feeler gauge in a horizontally as possible with respect to the valve lifter.

12. Further turn the crankshaft pulley clockwise and then measure and record the valve clearance again.

13. If adjustment is required, remove the camshafts.

14. Remove and measure the thickness of the valve lifter. Select a suitable shim, using the shim selection chart.

15. Install the replacement shim to the lifter.

16. After all shims have been adjusted, inspect the valve clearances again.

17. After completion, install all removed components.

Starter Motor

REMOVAL & INSTALLATION

2002–2004

1. Before servicing the vehicle, refer to the Precautions Section.
2. Remove or disconnect the following:
   - Negative battery cable
   - Intake Air Temperature (IAT) connector, on Legacy equipped with a manual transmission
   - Air cleaner case and duct
   - Air cleaner case stay, on Legacy
   - Connector and terminal from starter
   - Retaining bolts and/or nuts
   - Starter from transmission

To install:

3. Install or connect the following:
   - Starter to the transmission
   - Starter retaining bolts and/or nuts and tighten to 34–40 ft. lbs. (46–54 Nm)
   - Connector and terminal to starter
   - Air cleaner case stay, on Legacy
   - Air cleaner case and duct
   - IAT connector, on Legacy equipped with a manual transmission
   - Negative battery cable

2005–2006

1. Before servicing the vehicle, refer to the Precautions Section.
2. Disconnect the negative battery cable.
3. Remove the air intake chamber, on non turbocharged engine.
4. Remove the intercooler, on turbocharged engine.
5. Remove the air intake chamber stay, on non turbocharged engine.

6. Disconnect the electrical connectors from the starter.

7. Remove the starter retaining bolts. Remove the starter from the vehicle.

To install:

8. Installation is the reverse of the removal procedure.

9. Torque the starter retaining bolts to 37 ft. lbs.

Oil Pan

REMOVAL & INSTALLATION

2002–2004

1. Before servicing the vehicle, refer to the Precautions Section.
2. Remove or disconnect the following:
   - Negative battery cable
   - Air intake duct
   - Mass Air Flow (MAF) sensor on turbocharged engine
   - Air intake boot and air cleaner upper cover on turbocharged engine
   - Intercooler on turbocharged engine
   - Front Oxygen Sensor (O2S) electrical connector
   - Pitching stopper
   - Upper radiator brackets
3. Support the engine with a suitable lifting device.
   - Front wheel and tire assemblies
4. Lift up the engine slightly.
   - Engine undercover
5. Drain the oil from the engine into a suitable container.
6. Install the drain plug with a new gasket and tighten it to 33–36 ft. lbs. (43–47 Nm).
7. Remove or disconnect the following:
   - Rear O2S electrical connector
   - Exhaust Y-pipe
   - Nuts that secure the front engine mounts to the front crossmember
   - Oil pan mounting bolts
8. Insert an oil pan cutter blade between the upper and lower pans, on 3.0L engine
9. While supporting the oil pan, use a rubber mallet and tap the oil pan to free it from the engine.
10. Clean all gasket material from both mating surfaces.

To install:

11. Apply a continuous bead of sealer to a new oil pan gasket.
12. Install the oil pan assembly. Tighten the bolts to 36–48 inch lbs. (4–5 Nm) on all except 3.0L engine. On 3.0L

engine tighten the bolts to 5 ft. lbs. (7 Nm).

13. Lower the engine onto the front crossmember.

14. Install or connect the following:
   - Front engine mount nuts and tighten to 61 ft. lbs. (83 Nm)
   - Y-pipe with new gaskets. Tighten the pipe-to-engine nuts to 23 ft. lbs. (30 Nm)
   - Rear O2S electrical connector
   - Engine undercover
   - Front wheel and tire assemblies
15. Remove the engine lifting device.
   - Front O2S sensor electrical connector
   - Pitching stopper. Tighten the front bolt to 40 ft. lbs. (54 Nm) and the rear bolt to 49 ft. lbs. (67 Nm).
   - Upper radiator brackets
   - MAF sensor on turbocharged engine
   - Air intake boot and air cleaner upper cover on turbocharged engine
   - Intercooler on turbocharged engine
   - Air intake duct
   - Negative battery cable
16. Fill the engine to the proper level with the recommended oil and run the engine. Check for leaks.

2005–2006

2.5L SOHC ENGINE

1. Before servicing the vehicle, refer to the Precautions Section.
2. Disconnect the negative battery cable.
3. Raise and support the vehicle safely.
4. Remove the front tires and wheels.
5. Lower the vehicle.
6. Remove the air intake duct and the air cleaner case. Remove the air intake chamber.
7. Remove the pitching stopper.
8. Remove the hood stay holder and the radiator upper brackets.
9. Properly support the engine with a lifting device and wire ropes.
10. Lift the vehicle and support it safely.

➡ **When lifting the vehicle, raise the wire ropes at the same time.**

11. Remove the under cover.
12. Drain the engine oil.
13. Remove the front and center exhaust pipes.
14. Remove the nuts which retain the front cushion rubber onto the front crossmember.
15. Remove the bolts that retain the oil

pan to the cylinder block, with the engine in the raised position.

16. Insert an oil pan gasket cutter tool into the gap between the cylinder block and the oil pan. Remove the oil pan from the engine.

➡**Do not use a screwdriver or similar tool in place of the cutter tool.**

17. Remove the oil strainer, if required. Remove the baffle plate, if required.

To install:

18. Be sure to clean the old gasketing material from the mating surfaces.

19. Apply a continuous bead of sealer to a new oil pan gasket.

20. Make sure that the seals (A) are installed securely on the baffle plate and in the direction shown in the illustration. Install the baffle plate; tighten the retaining bolts to 4.7 ft. lbs.

21. Replace the O-ring and install the oil strainer. Tighten the bolt to 7.2 ft. lbs.

22. Apply liquid gasket, part number 004403012 or equivalent, to the oil pan mating surface. Install the oil pan. Torque the retaining bolts to specification.

23. Continue the installation in the reverse order of the removal procedure.

24. Torque the front cushion mounting bolts to 63 ft. lbs.

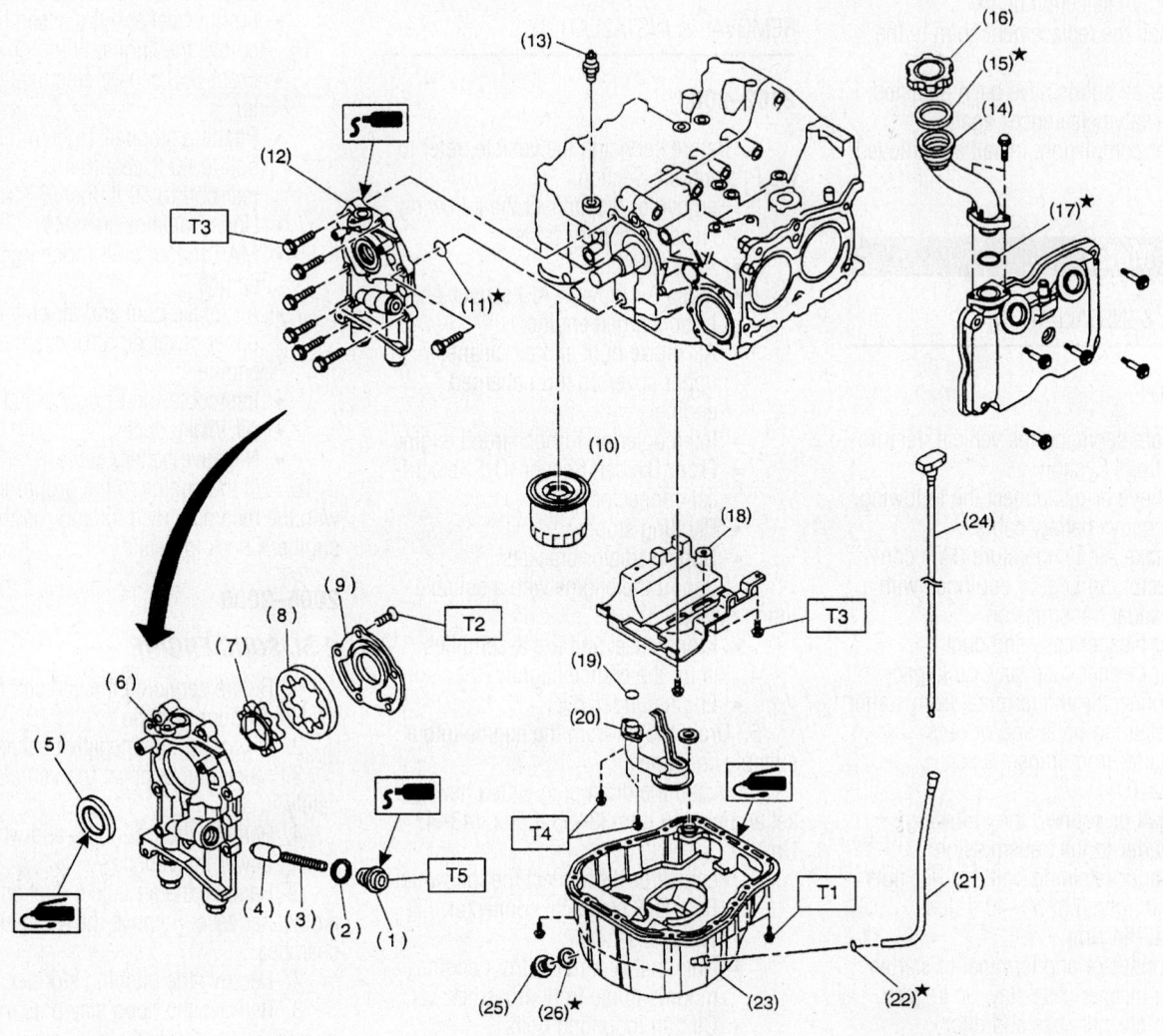

1) Plug	13) Oil pressure switch	25) Drain plug
2) Washer	14) Oil filler duct	26) Metal gasket
3) Relief valve spring	15) O-ring	
4) Relief valve	16) Oil filler cap	
5) Oil seal	17) O-ring	
6) Oil pump case	18) Baffle plate	
7) Inner rotor	19) O-ring	
8) Outer rotor	20) Oil strainer	
9) Oil pump cover	21) Oil level gauge guide	
10) Oil filter	22) O-ring	
11) O-ring	23) Oil pan	
12) Oil pump ASSY	24) Oil level gauge	

Tightening torque: N·m (kg-m, ft-lb)

T1: 5 (0.5, 3.6)

T2: $5^{+1}/_{-0}$ ($0.5^{+0.1}/_{-0}$, $3.6^{+0.7}/_{-0}$)

T3: 6.4 (0.65, 4.7)

T4: 10 (1.0, 7.2)

T5: 44.1±3.4 (4.5±0.35, 32.5±2.5)

Oil pan and related components—2.5L engine

9307TG03

25. Be sure to fill the engine with the correct grade and type engine oil.

26. Start the engine and check for leaks. Correct as required.

2.5L DOHC ENGINE

1. Before servicing the vehicle, refer to the Precautions Section.

2. Disconnect the negative battery cable.

3. Raise and support the vehicle safely.

4. Remove the front tires and wheels.

5. Remove the collector cover.

6. Lower the vehicle.

7. Disconnect the connector from the MAF sensor.

8. Remove the air intake boot and air cleaner upper cover.

9. Remove the intercooler.

10. Remove the pitching stopper.

11. Remove the radiator upper brackets.

12. Properly support the engine with a lifting device and wire ropes.

13. Lift the vehicle and support it safely.

➤**When lifting the vehicle, raise the wire ropes at the same time.**

14. Remove the under cover.

15. Drain the engine oil.

16. Remove the front exhaust pipe.

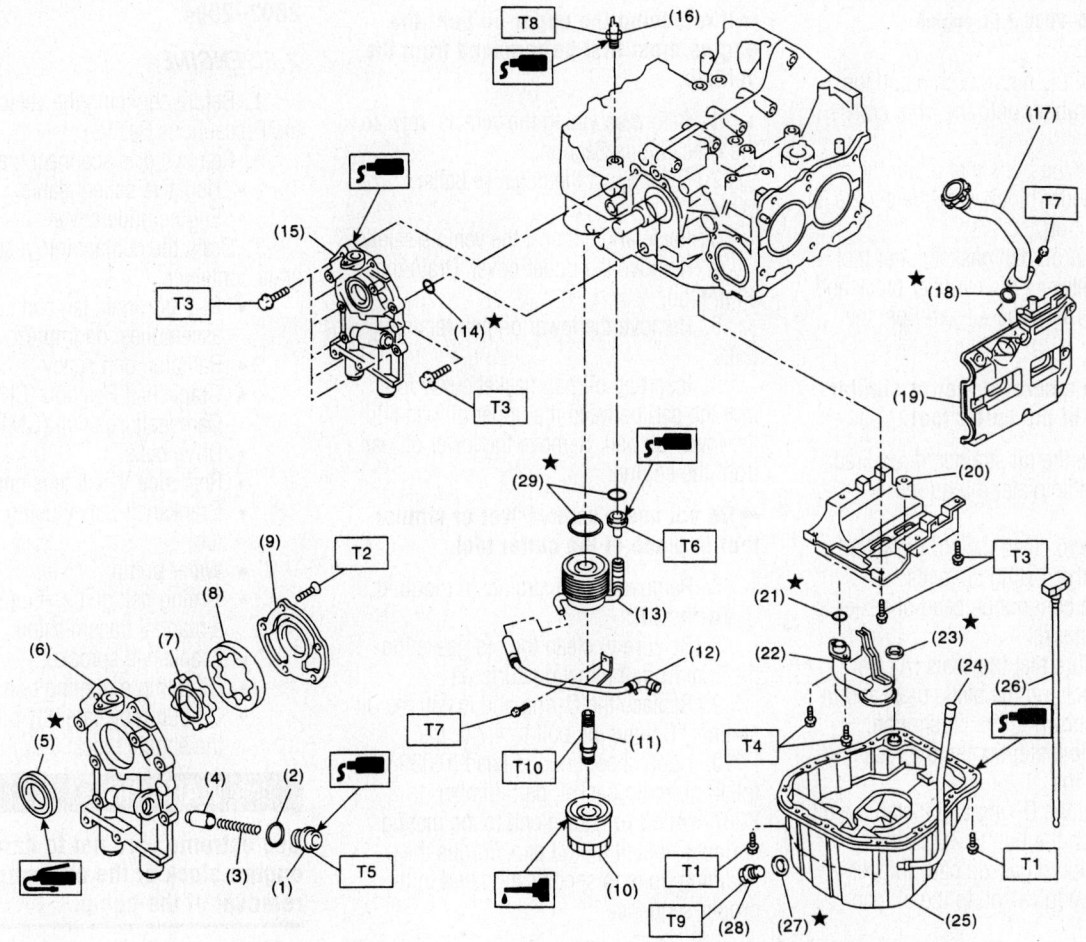

(1)	Plug	(15)	Oil pump ASSY	(29)	O-ring
(2)	Washer	(16)	Oil pressure switch		
(3)	Relief valve spring	(17)	Oil filler duct		
(4)	Relief valve	(18)	O-ring		
(5)	Oil seal	(19)	Cylinder head cover		
(6)	Oil pump case	(20)	Baffle plate		
(7)	Inner rotor	(21)	O-ring		
(8)	Outer rotor	(22)	Oil strainer		
(9)	Oil pump cover	(23)	Gasket		
(10)	Oil filter	(24)	Oil level gauge guide		
(11)	Connector	(25)	Oil pan		
(12)	Water by-pass pipe	(26)	Oil level gauge		
(13)	Oil cooler	(27)	Metal gasket		
(14)	O-ring	(28)	Drain plug		

Tightening torque: N·m (kgf-m, ft-lb)

T1:	**5 (0.5, 3.6)**
T2:	**5 (0.5, 3.6)**
T3:	**6.4 (0.65, 4.7)**
T4:	**10 (1.0, 7.0)**
T5:	**44.1 (4.5, 32.5)**
T6:	**69 (7.0, 4.7)**
T7:	**6.4 (0.65, 50.6)**
T8:	**25 (2.5, 18.1)**
T9:	**44 (4.5, 33)**
T10:	**54 (5.5, 40)**

Oil pan and related components—2.0L engine

9357TG31

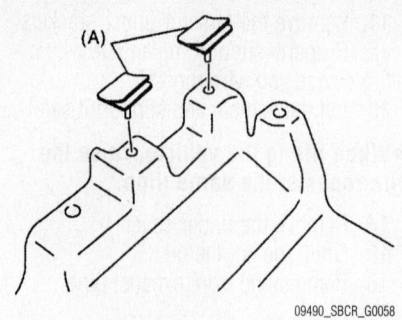

09490_SBCR_G0058

Oil pan baffle plate seal location and positioning—2005–2006 2.5L engine

17. Remove the nuts which retain the front cushion rubber onto the front crossmember.

18. Remove the bolts that retain the oil pan to the cylinder block, with the engine in the raised position.

19. Insert an oil pan gasket cutter tool into the gap between the cylinder block and the oil pan. Remove the oil pan from the engine.

➡**Do not use a screwdriver or similar tool in place of the cutter tool.**

20. Remove the oil strainer, if required. Remove the baffle plate, if required.

To install:

21. Be sure to clean the old gasketing material from the mating surfaces.

22. Apply a continuous bead of sealer to a new oil pan gasket.

23. Make sure that the seals (A) are installed securely on the baffle plate and in the direction shown in the illustration. Install the baffle plate; tighten the retaining bolts to 4.7 ft. lbs.

24. Replace the O-ring and install the oil strainer. Tighten the bolt to 7.2 ft. lbs.

25. Apply liquid gasket, part number 004403012 or equivalent, to the oil pan

mating surface. Install the oil pan. Torque the retaining bolts to specification.

26. Continue the installation in the reverse order of the removal procedure.

27. Torque the front cushion mounting bolts to 61 ft. lbs.

28. Be sure to fill the engine with the correct grade and type engine oil.

29. Start the engine and check for leaks. Correct as required.

3.0L ENGINE

➡**If removing the upper oil pan, the engine must first be removed from the vehicle.**

1. Before servicing the vehicle, refer to the Precautions Section.

2. Disconnect the negative battery cable.

3. Raise and support the vehicle safely.

4. Remove the under cover. Drain the engine oil.

5. Remove the lower oil pan retaining bolts.

6. Insert an oil pan gasket cutter tool into the gap between the upper oil pan and the lower oil pan. Remove the lower oil pan from the engine.

➡**Do not use a screwdriver or similar tool in place of the cutter tool.**

7. Remove the oil strainer, if required.

To install:

8. Be sure to clean the old gasketing material from the mating surfaces.

9. Replace the O-ring and install the oil strainer. Tighten the bolt to 4.7 ft. lbs.

10. Apply a continuous bead (0.039 inch thick) of liquid gasket, part number K0877YA018 or equivalent, to the mating surfaces. Install the oil pan. Torque the retaining bolts to specification and in the proper sequence.

11. Continue the installation in the reverse order of the removal procedure.

12. Be sure to fill the engine with the correct grade and type engine oil.

13. Start the engine and check for leaks. Correct as required.

Oil Pump

REMOVAL & INSTALLATION

2002–2004

2.5L ENGINE

1. Before servicing the vehicle, refer to the Precautions Section.

2. Remove or disconnect the following:
   • Negative battery cable
   • Engine undercover

3. Drain the coolant into a suitable separate container.
   • Radiator main fan and sub fan assemblies, on Impreza
   • Radiator, on Legacy
   • Crankshaft Position (CKP) and Camshaft Position (CMP) sensors
   • Drive belts
   • Rear side V-belt tensioner
   • Crankshaft pulley using a suitable tool
   • Water pump
   • Timing belt guide, if equipped with a manual transmission
   • Crankshaft sprocket
   • Oil pump mounting bolts
   • Oil pump by carefully prying it from the engine block

✳ WARNING

Use extreme care not to damage the engine block or the oil pump during removal of the pump.

To install:

4. Measure the tip clearance of the rotors. If clearance is greater than 0.0071 in. (0.18mm), replace the rotors.

5. Measure the clearance between the outer rotor and the cylinder block rotor housing. If clearance exceeds 0.0079 in. (0.20mm), replace the rotor.

6. Measure the side clearance between the oil pump inner rotor and the pump cover. If clearance exceeds 0.0059 in. (0.15mm), replace the rotor or pump body.

7. Assemble the oil pump.

8. Apply sealant and a new O-ring to the oil pump.

9. Install or connect the following:
   • Oil pump and tighten the bolts to 60 inch lbs. (7 Nm)

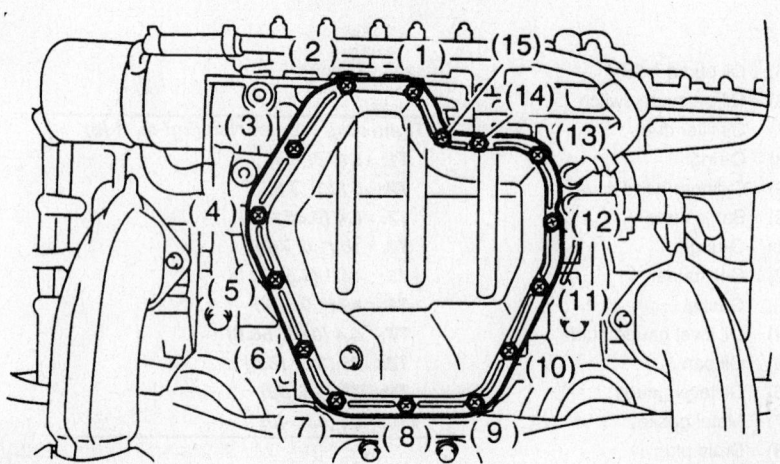

42356-SBCR-G26

Oil pan bolt torque sequence—3.0L engine

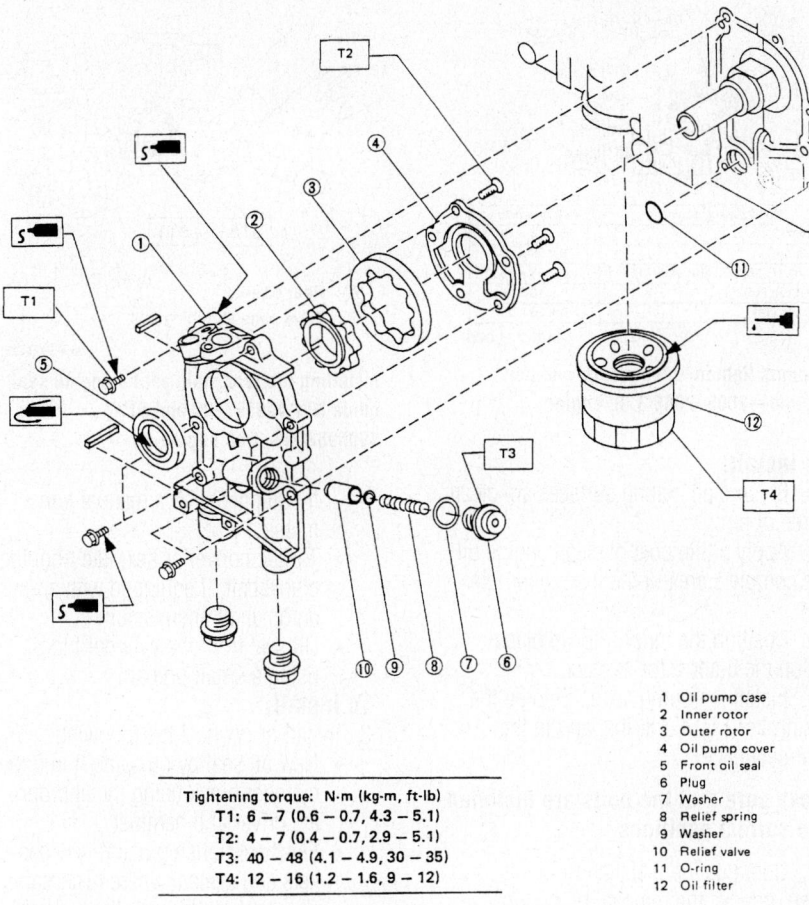

Tightening torque: N·m (kg-m, ft-lb)
T1: 6 − 7 (0.6 − 0.7, 4.3 − 5.1)
T2: 4 − 7 (0.4 − 0.7, 2.9 − 5.1)
T3: 40 − 48 (4.1 − 4.9, 30 − 35)
T4: 12 − 16 (1.2 − 1.6, 9 − 12)

1	Oil pump case
2	Inner rotor
3	Outer rotor
4	Oil pump cover
5	Front oil seal
6	Plug
7	Washer
8	Relief spring
9	Washer
10	Relief valve
11	O-ring
12	Oil filter

7923TG37

Oil pump and related components—2.5L engine

• Crankshaft sprocket
• Timing belt guide, if equipped with a manual transmission
• Water pump
• Crankshaft pulley using a suitable tool
• Rear side V-belt tensioner
• Drive belts
• CKP and CMP sensors
• Radiator, on Legacy
• Radiator main fan and sub fan assemblies, on Impreza
• Engine undercover
• Negative battery cable
10. Fill and bleed the cooling system.
11. Start the engine and check for leaks.

3.0L ENGINE

1. Before servicing the vehicle, refer to the Precautions Section.
2. Remove or disconnect the following:
• Negative battery cable
• Engine under cover
3. Drain the coolant.
• Radiator
• Drive belt
• Front timing chain cover and chain.

Refer to the timing chain removal procedure in this section.
• Oil pump cover and crankshaft sprocket
• Inner and outer rotor

To install:

4. Apply engine oil to the entire surface area of the inner and outer rotor
5. Install or connect the following:
• Inner rotor by fitting it into the groove on the crankshaft and then assemble the outer rotor

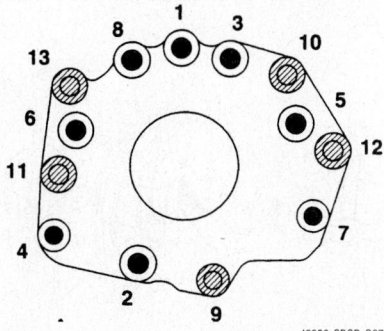

Oil pump cover torque sequence—2002–2004 3.0L engine

42356-SBCR-G27

• Oil pump cover and tighten the bolts in the sequence illustrated to 5 ft. lbs. (7 Nm)
• Crankshaft sprocket
• Timing chain and cover. Refer to the timing chain removal procedure in this section.
• Drive belt
• Radiator
• Engine undercover
• Negative battery cable
6. Fill and bleed the cooling system.
7. Start the engine and check for leaks.

2005–2006

2.5L ENGINE

1. Before servicing the vehicle, refer to the Precautions Section.
2. Disconnect the negative battery cable. Drain the cooling system.
3. On the DOHC engine, remove the collector cover.
4. Raise and support the vehicle safely.
5. Remove the under cover.
6. On the DOHC engine, remove the bolts which retain the water pipe of the oil cooler to the oil pump. Remove the water pipe and hoses between the oil cooler and the water pump.
7. Lower the vehicle. Remove the radiator.
8. To remove the front side V belt, remove the belt covers. Loosen the lock bolt. Loosen the slider bolt. Remove the front side belt.
9. To remove the rear side V belt, remove the belt covers. Loosen the lock bolt. Loosen the slider bolt. Remove the rear side belt.
10. On the SOHC engine remove the belt tensioner.
11. On the DOHC engine remove the rear side V belt tensioner
12. Remove the crankshaft position sensor.
13. Remove the crankshaft pulley bolt.
14. Lock the crankshaft in place using tool ST499977100 for the SOHC engine and tool ST499207400 for the DOHC engine, or equivalent.
15. Remove the crankshaft pulley.
16. Remove the water pump.
17. If equipped, remove the timing belt guide. Remove the crankshaft sprocket.
18. Remove the oil pump retaining bolts.

➡**When disassembling and checking the oil pump, loosen the relief valve plug before removing the oil pump from its mounting.**

19. Using a flat tip tool remove the oil pump from the engine.

To install:

20. Be sure all mating surfaces are clean and free of dirt.

21. Apply liquid gasket part number 004403007, or equivalent to the mating surfaces of the oil pump.

22. Be sure to replace the O-ring with a new one.

23. Apply a thin coat of clean engine oil to the inside of the oil seal.

24. Position the oil pump to its mounting, aligning the notched area with the crankshaft and push the pump straight.

➡**Be sure that the oil seal lip is not folded.**

25. Install the oil pump. Apply liquid gasket part number 004403042, or equivalent to the three retaining bolt threads. Install the bolts and tighten to 4.7 ft. lbs.

26. Continue the installation in the reverse order of the removal procedure.

27. Be sure to fill the cooling system with the proper grade and type coolant.

28. Start the engine and check for leaks. Correct, as required.

3.0L ENGINE

1. Before servicing the vehicle, refer to the Precautions Section.

2. Disconnect the negative battery cable. Drain the cooling system.

3. Remove the collector cover.

4. Raise and support the vehicle safely.

5. Remove the under cover.

6. Lower the vehicle. Remove the radiator.

7. To remove the front side V belt, remove the belt covers. Loosen the lock bolt. Loosen the slider bolt. Remove the front side belt.

8. To remove the rear side V belt, remove the belt covers. Loosen the lock bolt. Loosen the slider bolt. Remove the rear side belt.

9. Remove the crankshaft pulley cover.

10. Remove the crankshaft pulley bolt.

11. Lock the crankshaft in place using tool ST499977100, or equivalent.

12. Remove the crankshaft pulley.

13. Remove the front timing chain cover.

➡**Bolts are three different sizes. Be careful to install the correct bolt in the correct hole.**

14. Remove the timing chain.

15. Remove the crankshaft sprocket.

16. Remove the oil pump cover retaining bolts. Remove the oil pump cover.

17. Remove the inner and outer rotors.

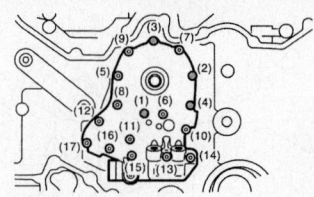

Bolt installing position	Bolt dimension
(1) and (3)	6 × 14 × 14
(2) and (4)	6 × 35 × 18
(5), (6), (7), (8), (9), (10) and (11)	6 × 35 × 15
(12), (15), (16) and (17)	6 × 16 × 16
(13) and (14)	6 × 26 × 15

09490_SBCR_G0059

Oil pump tightening sequence and bolt location—2005–2006 3.0L engine

To install:

18. Be sure all mating surfaces are clean and free of dirt.

19. Apply a thin coat of clean engine oil to the complete area of the inner and outer rotors.

20. Position the inner rotor in place. Position the outer rotor in place.

21. Install the pump cover. Tighten the retaining bolts to 4.7 ft. lbs. and in the proper sequence.

➡**Make sure that the bolts are installed in the correct positions.**

22. Continue the installation in the reverse order of the removal procedure.

23. Be sure to fill the cooling system with the proper grade and type coolant.

24. Start the engine and check for leaks. Correct, as required.

Rear Main Seal

REMOVAL & INSTALLATION

1. Before servicing the vehicle, refer to the Precautions Section.

2. Remove or disconnect the following:

- Engine from the vehicle
- Clutch assembly/flywheel using the Clutch Disc Guide tool 499747000,

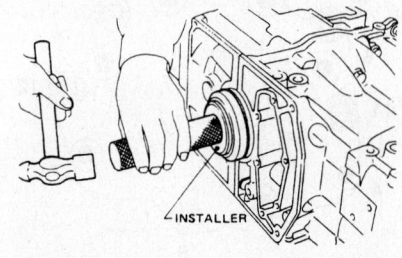

Installing the rear main seal—except 3.0L engine

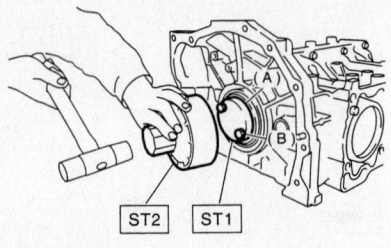

ST2	ST1

(A) Rear oil seal
(B) Drive plate attaching bolt

42356-SBCR-G28

Installing the rear main seal using oil seal guide ST1 499597100 and ST2 499598200—3.0L engine

if equipped with a manual transmission

- Torque converter flexplate from the crankshaft, if equipped with an automatic transmission
- Oil seal from the cylinder block using a small prybar

To install:

3. Install or connect the following:

- New oil seal by pressing it into the cylinder block using the appropriate driver and hammer
- Flywheel housing using new gaskets and sealant where necessary.
- Flywheel and tighten the bolts to specification.
- Engine

Timing Belt, Cover and Crankshaft seal

REMOVAL & INSTALLATION

2.0L Engine

1. Before servicing the vehicle, refer to the Precautions Section.

2. Disconnect the negative battery cable.

3. Remove the V-belt.

4. Remove the crankshaft pulley.

5. Remove the belt cover as follows:
   a. Crankshaft pulley.
   b. Left hand belt cover.
   c. Right hand belt cover.
   d. Front hand belt cover.

6. Remove the timing belt guides on vehicles equipped with a manual transmission.

7. If the alignment marks that indicate rotation are faded, put new marks on the belt before removal as follows:
   a. Turn the crankshaft using crankshaft sprocket tool 499987500 and a

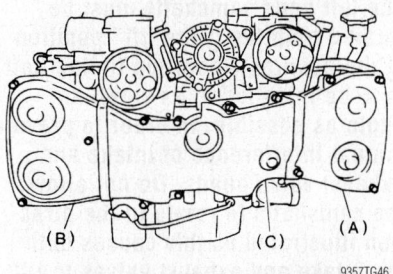

Remove the left hand (A), right hand (B) and front (C) belt covers–2.0L engine

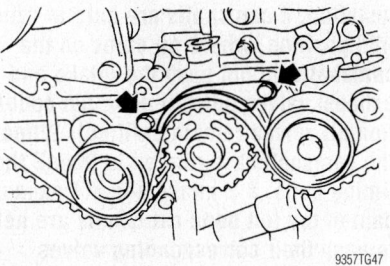

Location of the upper belt guide–2.0L engine

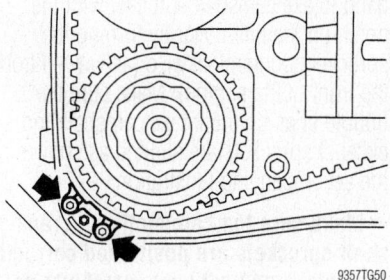

Location of the lower left belt guide–2.0L engine

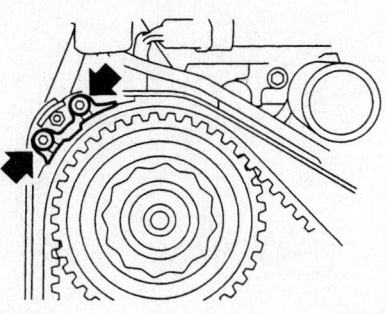

Location of the upper left belt guide–2.0L engine

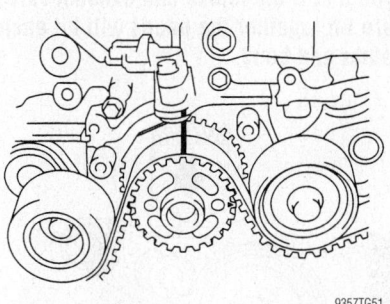

Mark the upper belt-to-sprocket alignment–2.0L engine

breaker bar to align the crankshaft sprocket, left hand intake camshaft sprocket, left hand exhaust camshaft, right hand intake camshaft sprocket and right hand exhaust camshaft sprocket with the on the cover and cylinder block.

b. Using white paint such as white out, place alignment marks on the belts in relation to the sprockets.

8. Remove the belt idler (A), illustrated in the accompanying illustration.

9. Remove the timing belt.

10. If necessary, remove belt idlers (B) and (C).

11. Remove the belt idler 2.

12. Remove the automatic belt tension adjuster assembly.

To install:

13. To prepare the automatic belt tensioner for assembly, perform the following steps:

a. Always use a vertical type pressing tool to move the adjuster rod down.

b. Do not use a lateral type vise.

c. Always push the adjuster rod vertically.

d. Make sure to slowly move the adjuster rod down applying a pressure of 66 lbs. (294 N).

e. Press in the push adjuster rod gradually taking more than 3 minutes.

f. Never allow the press pressure to exceed 2,205 lbs. (9,807 N).

g. Press the adjuster rod as far as the end surface of the cylinder. Do not press the rod into the cylinder as doing so may damage the cylinder.

h. Never release the press pressure until the stopper pin has been fully inserted.

14. Attach the automatic belt tension adjuster assembly to the vertical pressing tool.

15. Move the adjuster rod down slowly using a pressure of 66 lbs. (294 N) until the rod is aligned with the stopper pin hole in the cylinder.

16. Insert a 0.08 inch (2mm) stopper pin

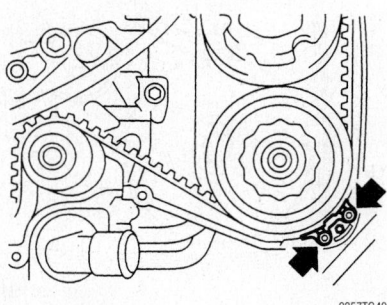

Location of the lower right belt guide–2.0L engine

or diameter Allen wrench into the stopper pin hole in the cylinder to retain the rod.

17. Install the adjuster assembly and tighten the retainers to 29 ft. lbs. (39 Nm).

18. Install belt idle 2 and tighten the retainers to 29 ft. lbs. (39 Nm).

19. Install the belt idler and tighten to 29 ft. lbs. (39 Nm).

20. Align the mark on the crankshaft sprocket with the mark on the oil pump.

21. Align the single line mark on the right hand exhaust camshaft sprocket with the notch on the belt cover.

22. Align the single line mark on the right hand intake camshaft with the notch on the belt cover. Make sure the double lines on the intake camshaft and exhaust sprock-

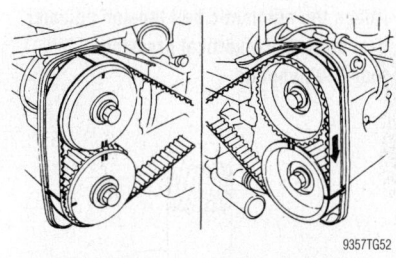

Mark the lower belt-to-sprocket alignment–2.0L engine

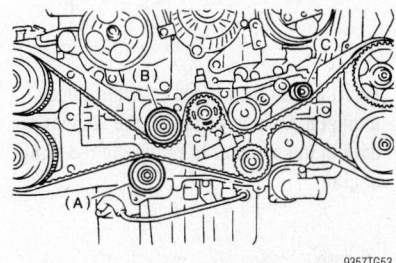

Remove the belt idler (A)–2.0L engine

ets are aligned as shown in the accompanying illustration.

23. Align the single line mark on the left hand exhaust camshaft sprocket with the notch on the belt cover by turning the sprocket counterclockwise (as viewed from the front of the engine).

24. Align the single line mark on the left hand intake camshaft sprocket with the notch on the belt cover by turning the sprocket counterclockwise (as viewed from the front of the engine). Make sure the double lines on the intake camshaft and exhaust sprockets are aligned as shown in the accompanying illustration.

➡ Make sure the camshaft and crankshaft sprockets are positioned correctly. The intake and exhaust camshafts on this engine can be rotated independently with the timing belt removed. By looking at the illustration it will show you that if the intake and exhaust valve are lift together the heads will hit each other and bend.

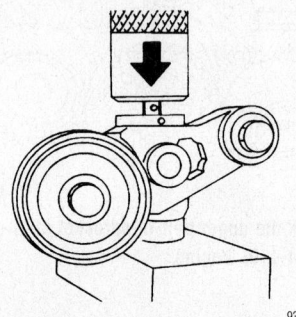

Attach the automatic belt tension adjuster assembly to the vertical pressing tool–2.0L engine

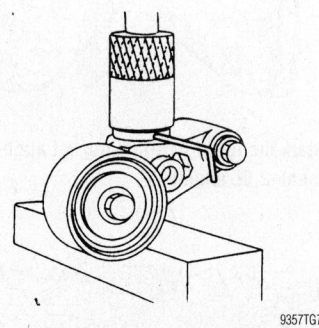

Insert a 0.08 inch (2mm) stopper pin or diameter Allen wrench into the stopper pin hole in the cylinder to retain the rod–2.0L engine

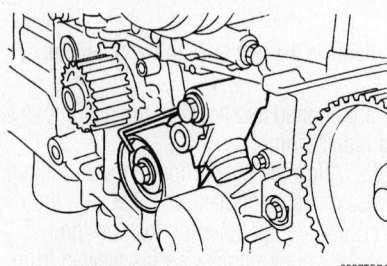

Install the adjuster assembly–2.0L engine

➡ When the timing belts are not installed, 4 camshafts are held at "zero lift" position, where all cams on the camshafts do not push the intake and exhaust valves down (under this condition all valves remain unlifted). When the camshafts are rotated to install the timing belts, # 2 intake and # 4 exhaust cam of the left hand camshafts are held to push their corresponding valves down. Under this condition these valves are held lifted. The right side camshafts are held in so that their cams do not push the valves down.

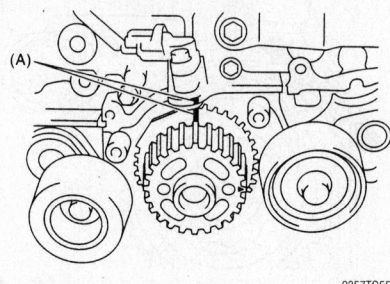

Align the mark on the crankshaft sprocket with the mark on the oil pump–2.0L engine

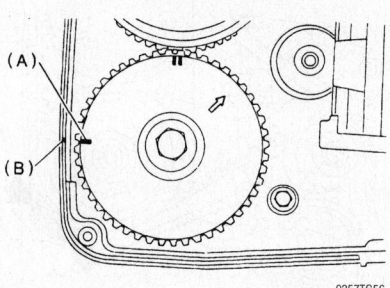

Align the single line mark on the right hand exhaust camshaft sprocket with the notch on the belt cover–2.0L engine

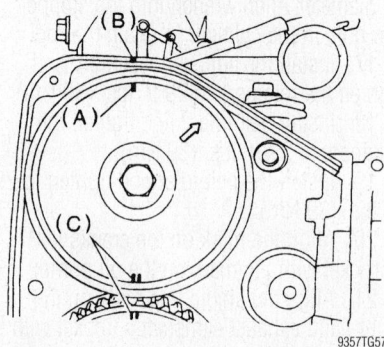

Align the single line mark on the right hand intake camshaft with the notch on the belt cover. Make sure the double lines on the intake camshaft and exhaust sprockets are aligned–2.0L engine

The left hand camshafts must be rotated from the "zero lift" position to the position where the timing belt is to be installed at as small an angle as possible, in order to prevent mutual interference of intake and exhaust valve heads. Do not allow the camshafts to rotate in the direction illustrated as this causes both the intake and exhaust valves to lift off at the same time with will cause valve damage.

25. When installing the belt, make sure to align the marks made during removal or if using a new belt, align the in alphabetical order as shown in the illustration.

❋❋ WARNING

Disengagement of more than 3 timing belt teeth may result in contact between the valve and piston. Always make sure the belts rotation is correct.

26. Install the belt idlers and tighten to 29 ft. lbs. (39 Nm).

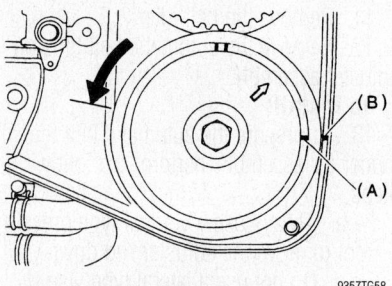

Align the single line mark on the left hand exhaust camshaft sprocket with the notch on the belt cover by turning the sprocket counterclockwise (as viewed from the front of the engine)–2.0L engine

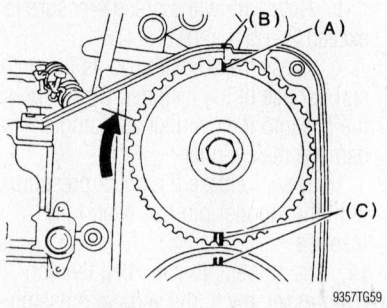

Align the single line mark on the left hand intake camshaft sprocket with the notch on the belt cover by turning the sprocket counterclockwise (as viewed from the front of the engine). Make sure the double lines on the intake camshaft and exhaust sprockets are aligned–2.0L engine

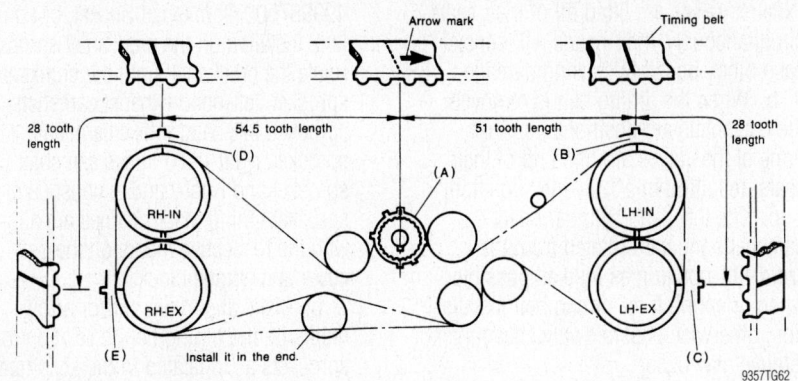

Align the marks in alphabetical order as shown if using a new belt–2.0L engine

Make sure the marks on the belt and sprockets are properly aligned.

27. Once the marks on the belt and sprockets are aligned, remove the stopper pin from the tensioner adjuster.

28. Install the timing belt guide on vehicles with manual transmission. Measure the clearance between the belt and guide. the clearance should be 0.019–0.059 inch (0.5–1.5mm) and tighten the retainers to 7 ft. lbs. (10 Nm).

Nm, ft. lbs.

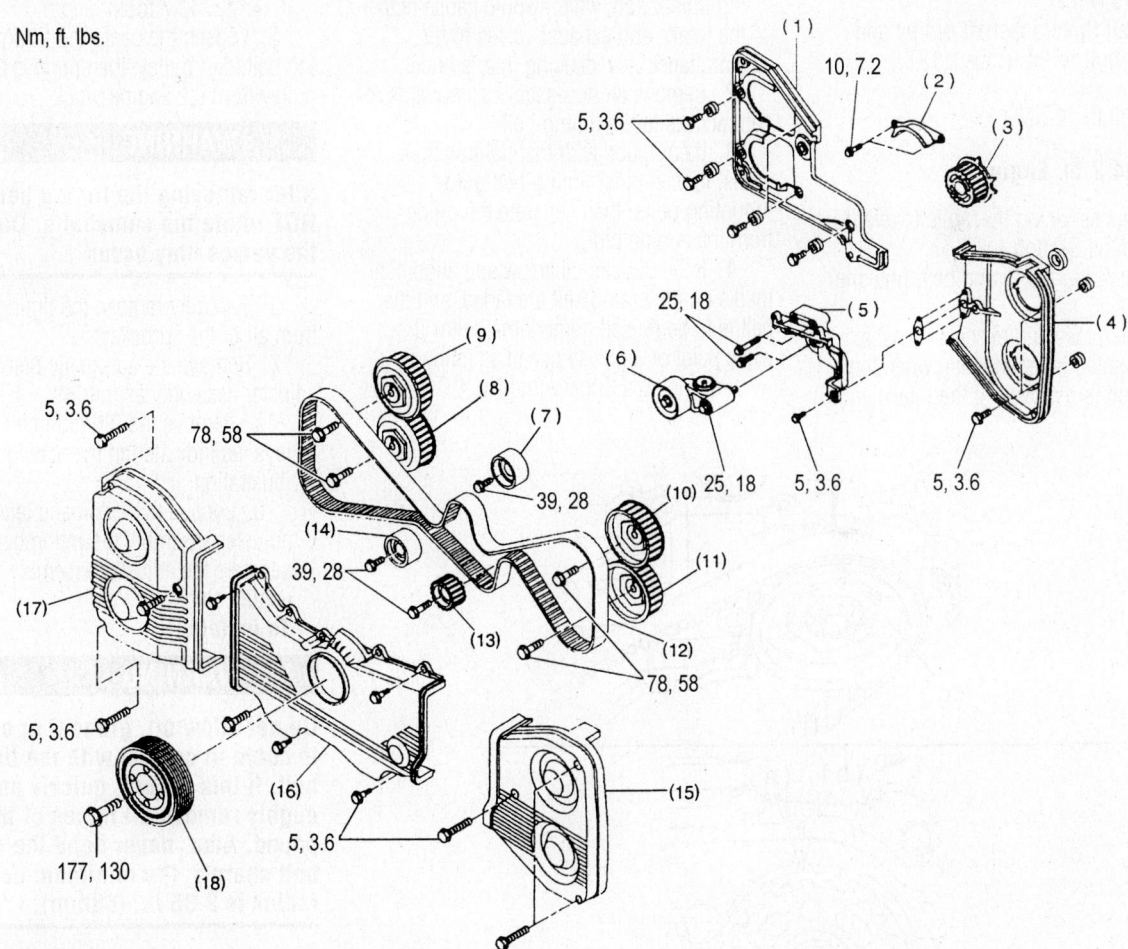

(1) Right-hand belt cover No. 2	(9) Right-hand intake camshaft sprocket	(17) Right-hand belt cover
(2) Timing belt guide (MT vehicles only)	(10) Left-hand intake camshaft sprocket	(18) Crankshaft pulley
(3) Crankshaft sprocket	(11) Left-hand exhaust camshaft sprocket	
(4) Left-hand belt cover No. 2	(12) Timing belt	
(5) Tensioner bracket	(13) Belt idler No. 2	
(6) Automatic belt tension adjuster ASSY	(14) Belt idler	
(7) Belt idler	(15) Left-hand belt cover	
(8) Right-hand exhaust camshaft sprocket	(16) Front belt cover	

Exploded view of the timing belt covers and components—2002–2004 2.5L engine

42356-SBFR-G50

Remove the timing belt guide on vehicles equipped with manual transmission— 2002–2004 2.5L engine

29. Install the belt covers and tighten to 3.5 ft. lbs. (5 Nm).

30. Install the crankshaft pulley and tighten the bolt to 94 ft. lbs. (127 Nm).

31. Install the V-belt.

2002–2004 2.5L Engine

1. Before servicing the vehicle, refer to the Precautions Section.

When servicing the timing belt, note the following:

a. The intake and exhaust camshafts can be rotated independently when the timing belt is removed. If the intake and exhaust valves are lifted off of their seats simultaneously, their heads will contact each other, possibly causing damage.

b. When the timing belt is removed, the camshafts are positioned so that none of the valves are lifted off of their seats, resulting in a "zero-lift" position.

c. The left-hand cylinder head camshafts must be rotated from the "zero-lift" position as little as possible when orienting it for timing belt installation, otherwise possible valve head interference may occur.

d. Never allow the camshafts to rotate in the direction shown in the accompanying illustration, which would cause both the intake and exhaust valves to lift simultaneously, causing interference.

2. Remove all necessary components to gain access to the timing belt.

3. If equipped with manual transmissions, loosen the 2 timing belt guide mounting bolts, then separate the guide from the engine block.

4. If the directional arrow and alignment marks on the timing belt are faded, and the belt is to be reused, remark the belt with white paint or a grease pencil as follows:

a. Using a Subaru tool No. ST-499987500 Crankshaft Socket, or equivalent, installed on the crankshaft sprocket, rotate the crankshaft until the crankshaft sprocket, left-hand exhaust camshaft sprocket, left-hand intake camshaft sprocket, right-hand intake camshaft sprocket and right-hand exhaust camshaft sprocket timing mark notches are aligned with the respective marks on the belt cover and engine block.

b. Make alignment and/or arrow marks on the timing belt in relation to the sprockets as indicated in the accompanying illustration.

- Z1: 46.8 tooth length
- Z2: 43.7 tooth length

5. Loosen the center bolt from the timing belt idler pulley, then remove the idler pulley from the engine block.

❋❋ WARNING

After removing the timing belt, DO NOT rotate the camshafts. Damage to the valves may occur.

6. Carefully remove the timing belt from all of the sprockets.

7. Remove the automatic belt tension adjuster assembly as follows:

a. Remove the 2 timing belt idler pulleys, as indicated in the accompanying illustration.

b. Loosen the automatic tension adjuster assembly mounting bolts, then separate the adjuster assembly from the engine block.

To install:

❋❋ WARNING

Do not allow oil, grease, or coolant to come in contact with the timing belt. If this occurs, quickly and thoroughly remove all traces of the compound. Also, never bend the timing belt sharply; the minimum bending radius is 2.36 in. (60mm).

8. Inspect the camshaft and crankshaft sprocket teeth for abnormal or excessive wear or scratches. Ensure there is no free-play between the sprocket and the key. Inspect the crankshaft sprocket sensor notch for damage or contamination with debris or dirt.

➡**When preparing the automatic tension adjuster assembly for installation, adhere to the following points:**

- Always use a vertical press, rather than a horizontal press or vise, to depress the adjuster assembly rod

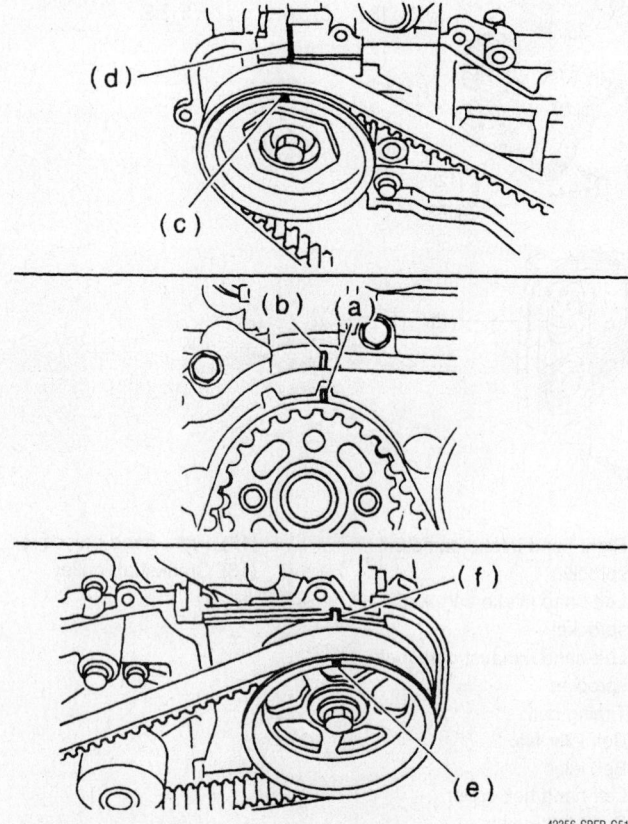

42356-SBFR-G51

Before removing the timing belt, turn the crankshaft sprocket until all of the alignment marks are aligned as indicated—2002–2004 2.5L engine

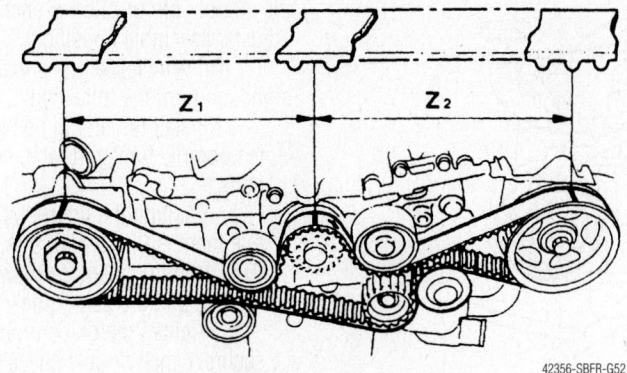

42356-SBFR-G52

If the original marks on the timing belt are worn or faded, make new alignment marks in the positions indicated—2002–2004 2.5L engine

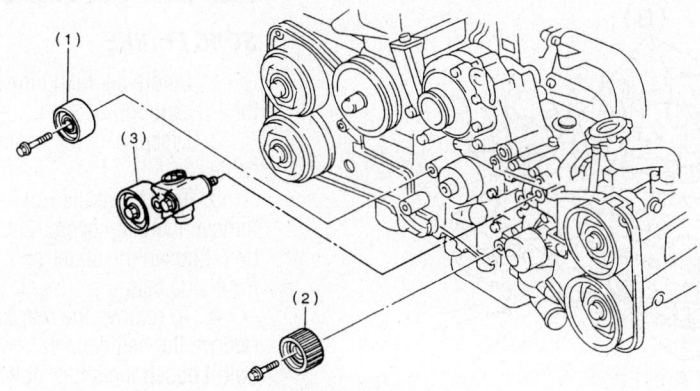

(1)	Belt idler	(3)	Automatic belt tension adjuster
(2)	Belt idler No. 2		ASSY

79245G52

It is necessary to remove the automatic adjuster assembly and reset the pushrod for timing belt installation—2002–2004 2.5L engine

- Depress the adjuster rod in a vertical position ONLY
- Depress the adjuster rod slowly (taking more than 3 minutes) with a force of 66 lbs. (30 kg)
- Do not allow the press force to exceed 2205 lbs. (1000 kg)
- Press the adjuster rod in as far as the end surface of the cylinder — do not press the rod into the cylinder, which may cause damage to the assembly

- Do not release the press force from the rod until the stopper pin is completely inserted in the cylinder

9. Prepare the automatic timing belt tension adjuster assembly for installation as follows:

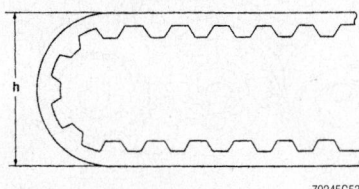

79245G53

Never bend the timing belt into a radius tighter than 2.36 in./60mm (h), otherwise it will be damaged beyond—2002–2004 2.5L engine

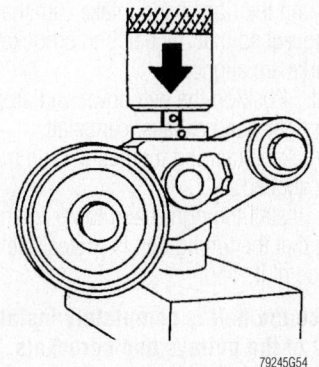

79245G54

Use a vertical press to push the adjuster rod into its housing until it is flush with the assembly's outer surface—2002–2004 2.5L engine

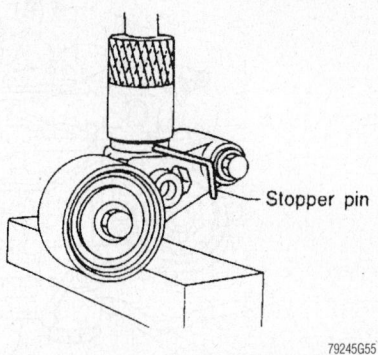

— Stopper pin

79245G55

Insert a 0.08 in. (2mm) diameter pin or Allen wrench into the housing and rod holes to hold it in position—2002–2004 2.5L engine

a. Position the adjuster assembly in a vertical press.

b. Slowly depress the adjuster rod with a force of 66 lbs. (30 kg) until the hole in the rod is aligned with the hole in the adjuster cylinder housing.

c. Insert a 0.08 in. (2mm) diameter stopper pin or Allen wrench through the hole in the cylinder housing and rod, then slowly release the press force from the adjuster rod.

10. Install the adjuster assembly onto the engine block.

11. Install timing belt idler pulley No. 2 on the engine block. Tighten the bolts to 28 ft. lbs. (39 Nm).

12. Install the timing belt idler pulley No. 1 on the engine block. Tighten the bolts to 28 ft. lbs. (39 Nm).

13. If the camshaft and crankshaft timing marks are no longer aligned, perform the following:

a. Position the crankshaft sprocket so that its mark is aligned with the mark on the oil pump cover on the engine block.

b. Align the single line mark on the right-hand exhaust camshaft sprocket with the notch on the belt cover.

c. Rotate the right-hand intake camshaft so that the single line mark is aligned with the notch on the belt cover.

➡ **At this point, the double line marks on both right-hand camshaft sprockets should be aligned.**

d. Turn the left-hand exhaust (lower) camshaft counterclockwise (as viewed from the front of the engine) until the single line mark is aligned with the notch on the belt cover.

e. Position the single line mark on the left-hand intake camshaft sprocket so that it is aligned with the notch on the belt cover. When rotating the camshaft,

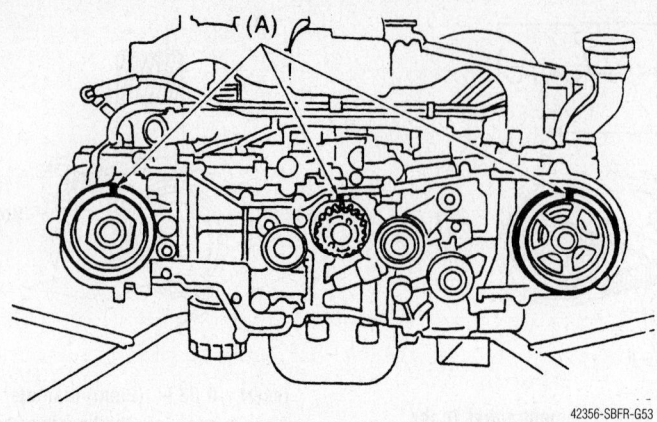

Make sure the sprockets are aligned as shown—2002–2004 2.5L engine

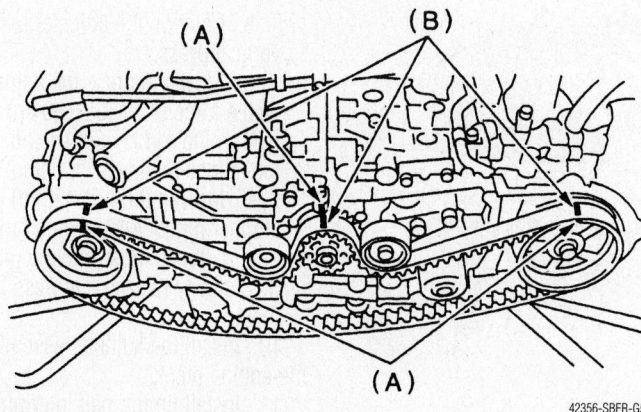

After installing the timing belt, the alignment marks should be positioned as shown. If not, remove the belt and align the sprockets and reinstall the belt—2002–2004 2.5L engine

do so only in a clockwise direction (as viewed from the front of the engine).

➡**At this point, the double line marks on both left-hand camshaft sprockets should be aligned.**

  f. Ensure the timing marks are aligned as shown in the accompanying illustration. If they are not, repeat Sub steps 12a through 12e until they are properly aligned.

14. Install the timing belt around the camshaft, crankshaft and idler pulleys so that the positioning marks on the timing belt are aligned with the marks on the sprockets as follows:

  a. Position the timing belt on the crankshaft sprocket so that the marks are aligned.

  b. Route the belt down and under the left-hand, upper idler pulley, then up and around the left-hand intake camshaft sprocket, ensuring the camshaft sprocket mark is aligned with the mark on the belt.

  c. Route the belt down and around the left-hand exhaust camshaft sprocket, making sure the marks are properly

aligned, then up and over the first lower idler pulley and down and around the second lower idler pulley.

  d. While holding the timing belt on the inner, left-hand, lower idler pulley, route the other side of the timing belt (from the crankshaft sprocket) down and under the right-hand upper idler pulley.

  e. Route the timing belt up and around the right-hand intake camshaft sprocket so that the belt and sprocket marks are aligned.

  f. Position the belt down and around the right-hand exhaust camshaft sprocket, ensuring the positioning marks are aligned.

15. Install the right-hand lower idler pulley so that the timing belt is routed over the top side of it.

➡**Once the belt is completely installed on all of the pulleys and sprockets, ensure that the positioning marks are still all aligned.**

16. After ensuring all of the marks are still aligned, use a pair of pliers to withdraw

the stopper pin or Allen wrench from the adjuster assembly housing.

17. On vehicles with manual transmissions, perform the following:

  a. Install the timing belt guide by temporarily tightening the mounting bolts.

  b. Position the timing belt guide so that there is 0.019–0.059 in. (0.5–1.5mm) clearance between the timing belt and the belt guide.

  c. Tighten the guide mounting bolts securely, then double check the guide clearance.

18. Install the timing belt covers and all remaining engine components.

2005–2006 2.5L Engine

SOHC ENGINE

1. Before servicing the vehicle, refer to the Precautions Section.

2. Disconnect the negative battery cable.

3. To remove the front side V belt, remove the belt covers. Loosen the lock bolt. Loosen the slider bolt. Remove the front side belt.

4. To remove the rear side V belt, remove the belt covers. Loosen the lock bolt. Loosen the slider bolt. Remove the rear side belt. Remove the belt tensioner.

5. Remove the crankshaft pulley bolt.

6. Lock the crankshaft in place using tool ST499977100, or equivalent.

7. Remove the crankshaft pulley.

8. Remove the left side timing belt cover.

9. Remove the front timing belt cover.

10. If equipped with manual transmission, remove the timing belt guide.

➡**If the belt is going to be reused and the alignment mark on the belt is not readable, put a new mark on the belt to indicate the direction of rotation. Using tool ST499987500, turn the crankshaft to align the mark of the sprocket "A" to**

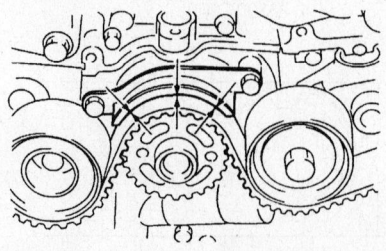

On manual transmission, ensure the timing belt-to-guide clearance (arrows) is correct before tightening the mounting bolts—2002–2004 2.5L engine

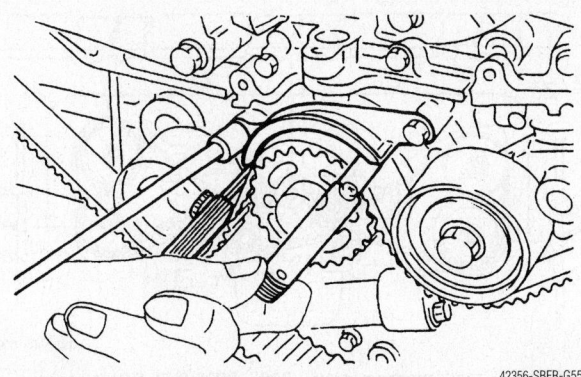

On manual transmission, use a feeler gauges to adjust the clearance between the timing belt and the belt guide—2002–2004 2.5L engine

the cylinder mark notch "B". Ensure that the right side cam sprocket mark "C", cam cap and cylinder head matching surface "D" or left side cam sprocket mark "E", timing belt cover notch "F" are properly aligned. Paint an alignment mark on the belt in relation to the crankshaft sprocket and

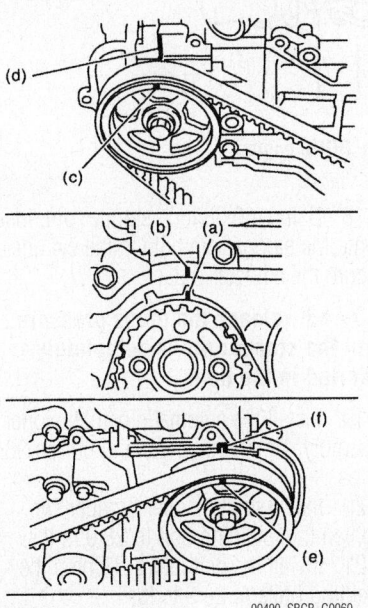

Timing belt alignment—2005–2006 2.5L SOHC engine

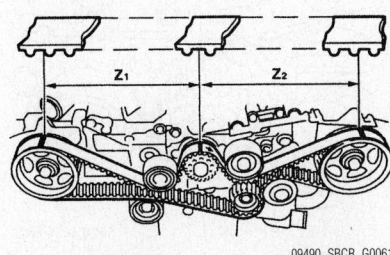

Timing belt Z1 and Z2 teeth measurement—2005–2006 2.5L SOHC engine

camshaft sprockets. Z1 measurement is 46.8 teeth. Z2 measurement is 43.7 teeth.

11. Remove both the number two belt idlers. Remove the timing belt from the engine.

12. Remove the number one belt idler. Remove the automatic belt tension adjuster assembly.

To install:

13. Attach the automatic belt tension adjuster assembly to a vertical pressing tool.

➡**Always use a vertical type pressing tool to move the adjuster rod downward. Do not use a lateral type vise. Push the adjuster rod vertically. Press in the push adjuster rod gradually, which should take three minutes or more. Do not allow pressure to exceed 2,205 lb. force.**

14. Slowly move the adjuster rod down until the adjuster rod is aligned with the stopper pin hole in the cylinder.

➡**Press the adjuster rod as far as the end surface of the cylinder. Do not press the adjuster rod into the cylinder. Doing so may damage the cylinder.**

15. Using a 0.08 inch stopper pin, insert it into the stopper pin hole in the cylinder. Secure the adjuster rod.

➡**Do not release the press pressure until the stopper pin is completely inserted in the hole.**

16. Install the automatic belt tensioner assembly. Tighten the retaining bolt to 28.9 ft. lbs.

17. Install the belt idler number one. Tighten the retaining bolt to 28.9 ft. lbs.

18. Turn the number one and number two camshaft sprockets, using tool ST499207100 or tool ST18231AA010 and

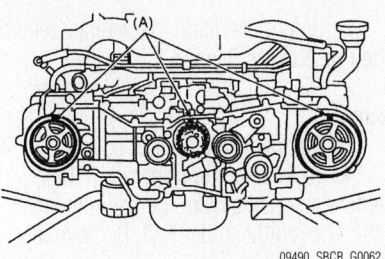

Timing mark alignment position A—2005–2006 2.5L SOHC engine

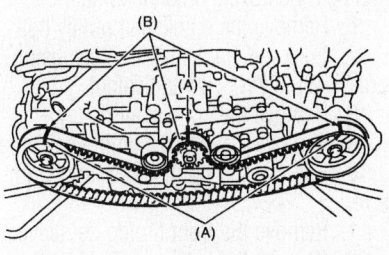

Timing mark alignment position B—2005–2006 2.5L SOHC engine

position the alignment marks "A" on each at the highest position.

19. While aligning the alignment mark "B" on the timing belt with mark "A" on the sprockets, position the timing belt properly.

20. Install both belt idler number two's. Tighten the retaining bolt to 28.9 ft. lbs.

21. After checking to be sure the marks on the timing belt and the camshaft sprockets are aligned remove the stopper pin from the belt tension adjuster.

22. Install the timing belt guide, if equipped with manual transmission. Temporarily tighten the bolts. Check and adjust the clearance between the belt and the guide. It should be 0.039 +/- 0.020 inch. Tighten the bolts to 7.2 ft. lbs.

23. Continue the installation in the reverse order of the removal procedure.

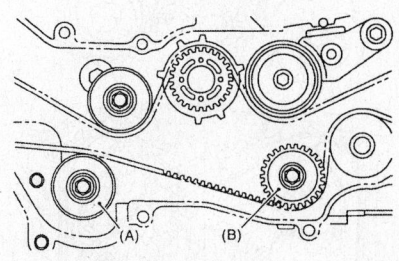

(A) Belt idler (No. 2)
(B) Belt idler No. 2

Belt idler number two locations—2005–2006 2.5L SOHC engine

DOHC ENGINE

1. Before servicing the vehicle, refer to the Precautions Section.

2. Disconnect the negative battery cable.

3. Remove the collector cover.

4. To remove the front side V belt, remove the belt covers. Loosen the lock bolt. Loosen the slider bolt. Remove the front side belt.

5. To remove the rear side V belt, remove the belt covers. Loosen the lock bolt. Loosen the slider bolt. Remove the rear side belt. Remove the belt tensioner.

6. Remove the crankshaft pulley bolt.

7. Lock the crankshaft in place using tool ST499977100, or equivalent.

8. Remove the crankshaft pulley.

9. Remove the left side timing belt cover.

10. Remove the right side timing belt cover.

11. Remove the front timing belt cover.

12. Remove the timing belt guide, if equipped.

➡If the belt is going to be reused and the alignment mark on the belt is not readable, put a new mark on the belt to indicate the direction of rotation. Using tool ST499987500, turn the crankshaft to align the mark on the crankshaft sprocket, intake camshaft sprocket (left), exhaust camshaft sprocket (left), intake camshaft sprocket (right), exhaust camshaft sprocket (right) with the notches of the timing belt cover and cylinder block. Paint an alignment mark on the belts in relation to the camshaft sprockets. Z1 measurement is 54.4 teeth. Z2 measurement is 51.0 teeth and Z3 measurement is 28.0 teeth.

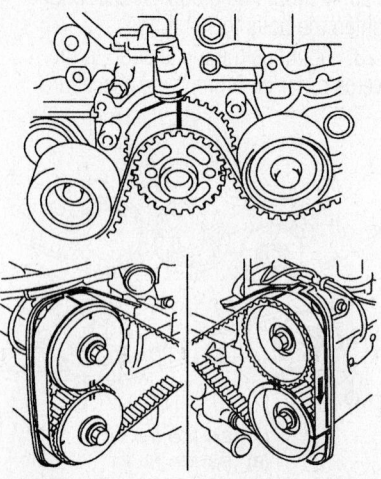

Timing belt alignment—2005–2006 2.5L DOHC engine

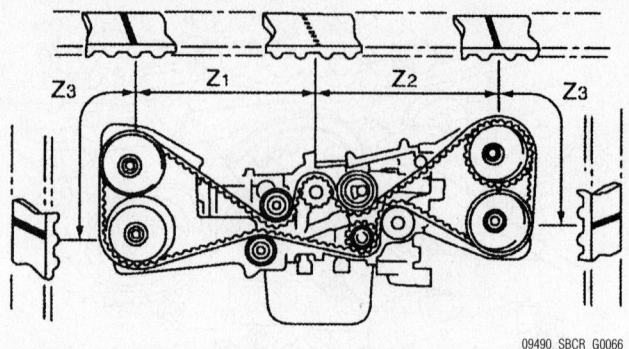

Timing belt Z1, Z2 and Z3 teeth measurement—2005–2006 2.5L DOHC engine

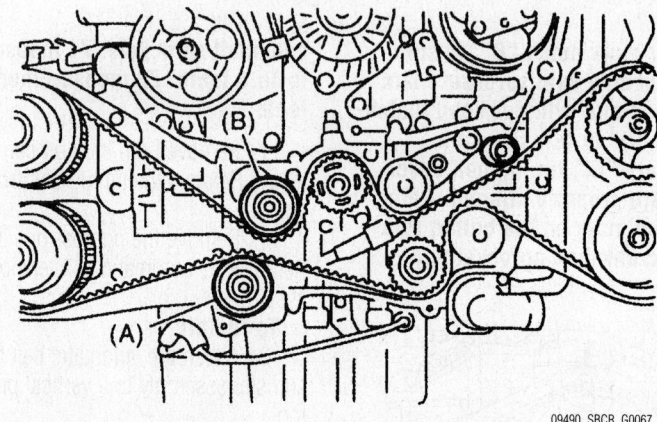

Belt idler identification and location—2005–2006 2.5L DOHC engine

13. Remove the belt idler belt "A". Remove the timing belt.

14. Remove the belt idlers belt "B" and "C".

15. Remove the belt idler number two. Remove the automatic belt tension adjuster assembly.

To install:

16. Attach the automatic belt tension adjuster assembly to a vertical pressing tool.

➡Always use a vertical type pressing tool to move the adjuster rod downward. Do not use a lateral type vise. Push the adjuster rod vertically. Press in the push adjuster rod gradually, which should take three minutes or more. Do not allow pressure to exceed 2,205 lb. force.

17. Slowly move the adjuster rod down until the adjuster rod is aligned with the stopper pin hole in the cylinder.

➡Press the adjuster rod as far as the end surface of the cylinder. Do not press the adjuster rod into the cylinder. Doing so may damage the cylinder.

18. Using a 0.08 inch stopper pin, insert it into the stopper pin hole in the cylinder. Secure the adjuster rod.

➡Do not release the press pressure until the stopper pin is completely inserted in the hole.

19. Install the automatic belt tensioner assembly. Tighten the retaining bolt to 28.9 ft. lbs.

20. Install the belt idler number two. Tighten the retaining bolt to 28.9 ft. lbs.

21. Install the belt idlers. Tighten the retaining bolts to 28.9 ft. lbs.

22. Align the mark "A" on the crankshaft

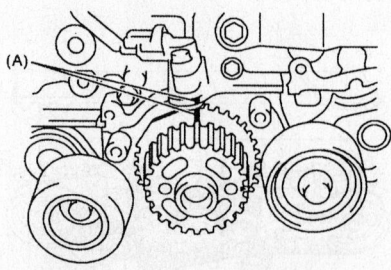

Crankshaft sprocket mark A to oil pump cover alignment—2005–2006 2.5L DOHC engine

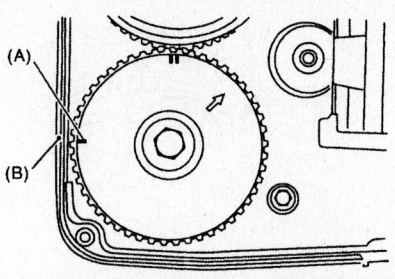

Right exhaust camshaft sprocket alignment mark A with timing belt cover B alignment mark—2005–2006 2.5L DOHC engine

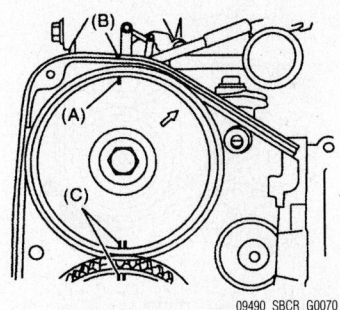

Right intake camshaft sprocket alignment mark A with timing belt cover B alignment mark an double line C alignment mark—2005–2006 2.5L DOHC engine

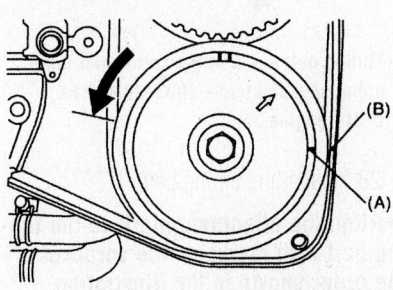

Left exhaust camshaft sprocket alignment mark A with timing belt cover B alignment mark—2005–2006 2.5L DOHC engine

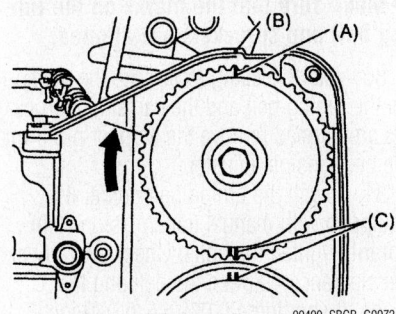

Left intake camshaft sprocket alignment mark A with timing belt cover B alignment mark an double line C alignment mark—2005–2006 2.5L DOHC engine

sprocket with the mark on the oil pump at the cylinder block.

Align the single line mark "A" on the right exhaust camshaft sprocket with the notch "B" on the timing belt cover.

23. Align single line mark "A" on the right intake camshaft sprocket with the notch "B" on the timing cover. Ensure that the double lines "C" on the intake and exhaust camshaft sprockets are aligned.

24. Align single line mark "A" on the left exhaust camshaft sprocket with the notch "B" on the timing cover by turning the sprocket counterclockwise as viewed from the front of the engine.

25. Align single line mark "A" on the left intake camshaft sprocket with the notch "B" on the timing cover, by turning the sprocket clockwise as viewed from the front of the engine. Ensure that the double lines "C" on the intake and exhaust camshaft sprockets are aligned.

26. Make sure that the camshaft and crankshaft sprockets are positioned properly.

→The intake and exhaust camshafts on this engine can be rotated independently with the timing belt removed. By looking at the illustration it will show

you that if the intake and exhaust valve are lift together the heads will hit each other and bend.

→When the timing belts are not installed, 4 camshafts are held at "zero lift" position, where all cams on the camshafts do not push the intake and exhaust valves down (under this condition all valves remain unlifted). When the camshafts are rotated to install the timing belts, # 2 intake and # 4 exhaust cam of the left hand camshafts are held to push their corresponding valves down. Under this condition these valves are held lifted. The right side camshafts are held in so that their cams do not push the valves down. The left hand camshafts must be rotated from the "zero lift" position to the position where the timing belt is to be installed at as small an angle as possible, in order to prevent mutual interference of intake and exhaust valve heads. Do not allow the camshafts to rotate in the direction illustrated as this causes both the intake and exhaust valves to lift off at the same time with will cause valve damage.

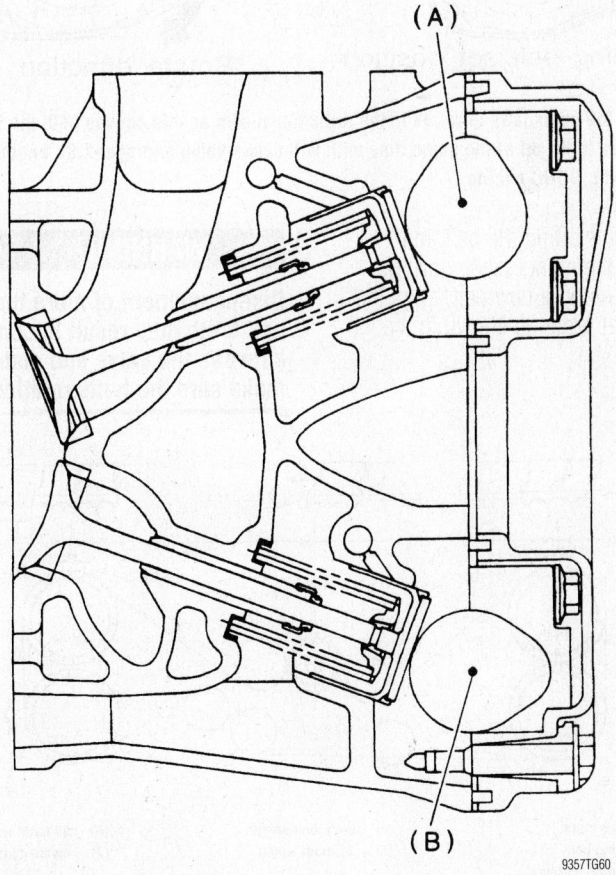

If the intake and exhaust valve are lift together the heads will hit each other and bend –2.0L engine and 2005–2006 2.5L DOHC engine

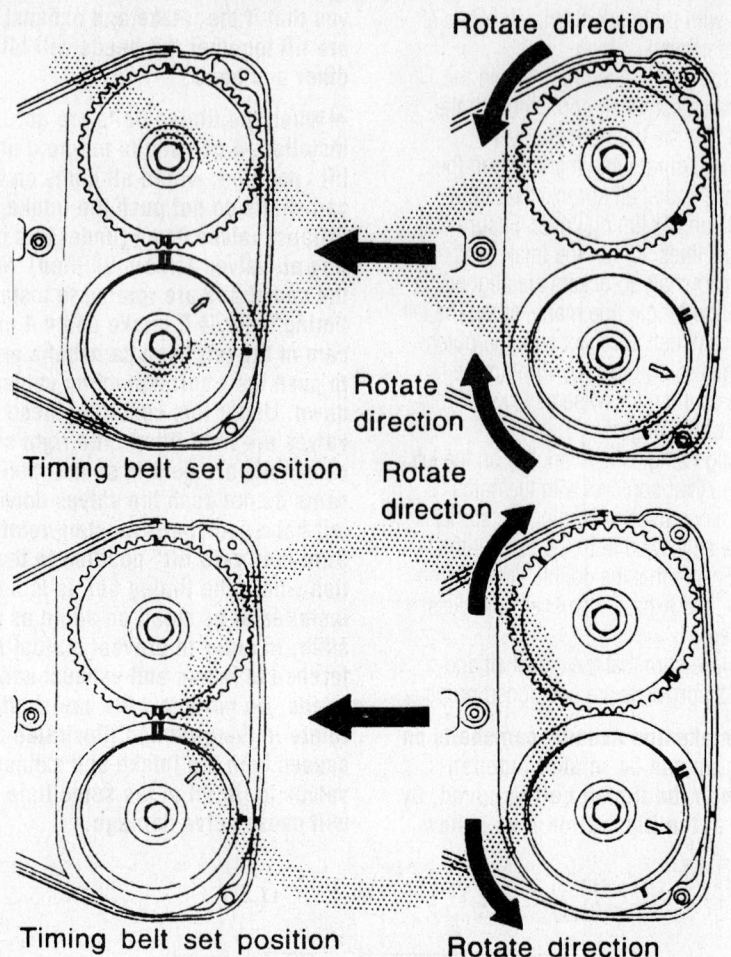

Timing belt set position

Timing belt set position

Do not allow the camshafts to rotate in the direction shown as this causes both the intake and exhaust valves to lift off at the same time with will cause valve damage–2.0L engine and 2005–2006 2.5L DOHC engine

27. When installing the belt, make sure to align the marks made during removal or if using a new belt, align the in alphabetical order as shown in the illustration.

※※ WARNING

Disengagement of more than 3 timing belt teeth may result in contact between the valve and piston. Always make sure the belts rotation is correct.

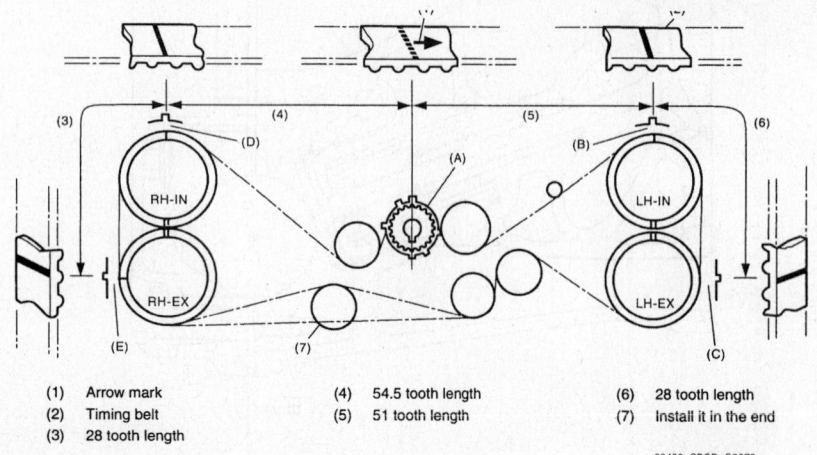

(1)	Arrow mark	(4)	54.5 tooth length	(6)	28 tooth length
(2)	Timing belt	(5)	51 tooth length	(7)	Install it in the end
(3)	28 tooth length				

09490_SBCR_G0073

Timing belt alignment and installation sequence—2005–2006 2.5L DOHC engine

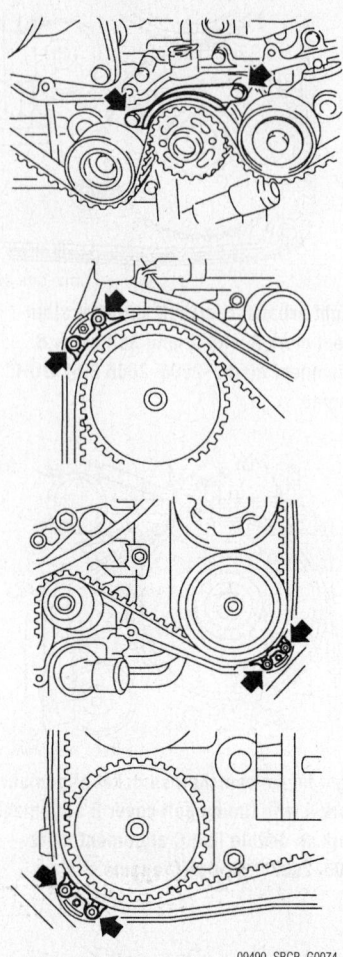

09490_SBCR_G0074

Timing belt guide bolt location and tightening specification—2005–2006 2.5L DOHC engine

28. Install the timing belt.

➡Align the alignment mark on the timing belt with marks on the sprocket in the order shown in the illustration. While aligning the timing marks, position the timing belt properly.

29. Install the belt idlers. Tighten the retaining bolts to 28.9 ft. lbs.

➡Make sure that the marks on the timing belt and sprockets are aligned.

30. After checking to be sure the marks on the timing belt and the camshaft sprockets are aligned remove the stopper pin from the belt tension adjuster.

31. Install the timing belt guide, if equipped with manual transmission. Temporarily tighten the bolts. Check and adjust the clearance between the belt and the guide. It should be 0.039 +/- 0.020 inch. Tighten the guide bolts to specification, see illustration.

32. Continue the installation in the reverse order of the removal procedure.

Timing Chain, Sprockets, Front Cover and Seal

REMOVAL & INSTALLATION

3.0L Engine

2002–2004

1. Before servicing the vehicle, refer to the Precautions Section.
2. Remove or disconnect the following:
   • Negative battery cable
   • Crankshaft pulley cover
   • Crankshaft pulley bolt using tool 499977100 to hold the crankshaft
   • Crankshaft pulley

➡**There are 4 different types of front cover bolts. Note their sizes and keep them separate to avoid a problem during installation.**

   • Front timing cover bolts. Note the location and the size of the bolts as you remove them, this will help during installation
   • Chain tensioner on the right hand side. The plunger **A** does not come out. Refer to illustration for plunger location.
   • Chain guide on the right hand side located between the cams
   • Chain guide from the right hand side
   • Chain tensioner lever from the right hand side
   • Timing chain from the right hand side
   • Chain tensioner on the left hand side. The plunger **A** does not come out. Refer to illustration for plunger location.
   • Chain tensioner lever from the left hand side
   • Chain guide on the left hand side located between the cams
   • Chain guide from the left hand side
   • Center chain guide
   • Upper idler sprocket
   • Timing chain from the left hand side
   • Lower idler sprocket

To install:

3. Make sure all components are clean. Apply oil to the chain guide, tensioner lever and idler sprockets.
4. Place the screw, spring, pin and tension rod into the tensioner body.
5. While pressing the tensioner onto a rubber mat, twist it to the left and right to shorten the rod. Place a thin pin into the holes between the rod and body to hold it in place. Always perform this task on a rubber mat.
6. Using the crankshaft socket tool, align the **TOP MARK** on the crankshaft sprocket to the 9 O'clock position as shown in the illustration.
7. Using camshaft sprocket wrench ST 18231AA000, align the four key grooves on the camshaft sprockets to the 12 O'clock position as illustrated.
8. Rotate the crankshaft sprocket clockwise to align the **TOP MARK** to the 12 O'clock position as shown in the illustration. Piston 1 is now at Top Dead Center (TDC).

❊❊ CAUTION

Do not rotate the camshaft or crankshaft sprockets until the chain is completely routed or damage will occur.

9. Install the lower idler sprocket and tighten the bolt to 50 ft. lbs. (69 Nm).

Front timing cover bolt sizes and locations—2002–2004 3.0L engine

42356-SBCR-G29

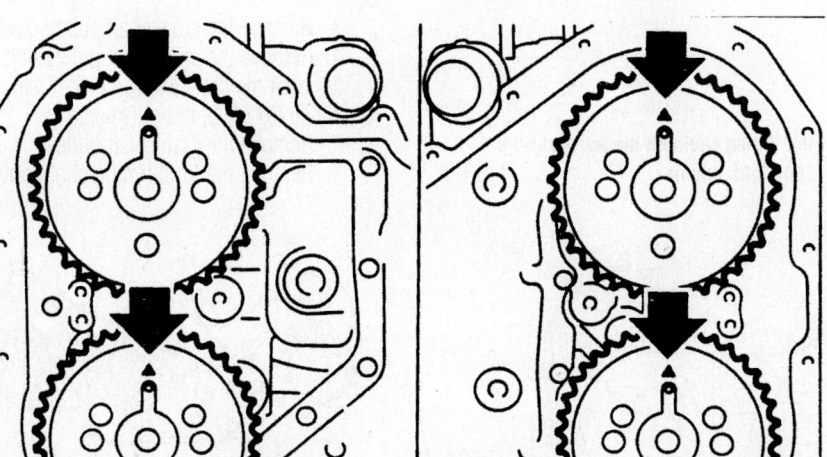

Align the four key grooves on the camshaft sprockets to the 12 O'clock position—2002–2004 3.0L engine

42356-SBCR-G37

(A) Gold
(B) Mark

42356-SBCR-G38

Align the mark B on the crankshaft sprocket with the matching mark A on the left hand timing chain—2002–2004 3.0L engine

10. Install the left hand timing chain, align the mark **B** on the crankshaft sprocket with the matching mark **A** on the timing chain. Refer to the illustration for more detail

11. Route the left hand timing chain onto the lower idler sprocket, water pump, exhaust cam sprocket and the intake cam sprocket in that order.

12. Make sure the mark **A** on the chain and the mark **B** camshaft sprocket are aligned the same way as the one on the crankshaft sprocket or damage will occur.

13. Install or connect the following:
- Upper chain idler and tighten the bolt to 50 ft. lbs. (69 Nm)
- Chain guide on the left hand side between the cams and tighten the bolt to 4.6 ft. lbs. (6 Nm) using a NEW bolt
- Chain guide on the left hand side and tighten the bolts to 11 ft. lbs. (16 Nm)
- Tensioner lever on the left hand

side and tighten the bolt to 11 ft. lbs. (16 Nm)
- Chain tensioner on the left hand side and tighten the bolts to 11 ft. lbs. (16 Nm)
- Right hand timing chain. On the lower idler sprocket align the match marks on the timing chain on the left and right hand sides as illustrated. Route the chain onto the intake cam sprocket and the exhaust cam sprocket.

14. Make sure the mark **A** on the chain and the mark **B** camshaft sprocket are aligned the same way as the one on the crankshaft sprocket or damage will occur.

15. Install or connect the following:
- Right hand chain guide
- Right hand chain tensioner lever and tighten the bolts to 11 ft. lbs. (16 Nm)
- Right hand chain guide and tighten the NEW bolt to 4.6 ft. lbs. (6 Nm)
- Right hand chain tensioner and

tighten the bolts to 11 ft. lbs. (16 Nm)

16. Adjust the clearance between the chain guide on the right hand side and the center chain guide so that there is range between 0.331–0.339 inch (8.4–8.6mm).
- Center chain guide and tighten the NEW bolt to 6 ft. lbs. (8 Nm)

17. Check the match marks on each sprocket and corresponding timing chain are correct, remove the stopper from the tensioner.

18. Clean the mating surfaces on the front timing cover. Apply a 0.078–0.126 inch (2–5mm) bead of gasket maker to the mating surface of the front cover. Subaru recommends Three Bond 1280B for this procedure.

19. Install the front chain cover and temporarily tighten the bolts. Use the illustration showing the bolt sizes and location for proper installation.

20. Tighten the front timing cover bolts in the proper sequence to 5 ft. lbs. (7 Nm).

21. Install or connect the following:
- Crankshaft pulley, apply clean oil to the bolt threads and tighten the bolt to 131 ft. lbs. (178 Nm)
- Camshaft pulley cover and tighten the bolts to 5 ft. lbs. (7 Nm)
- Negative battery cable

22. Start the engine and allow it to reach operating temperature. Check for leaks.

2005–2006

1. Before servicing the vehicle, refer to the Precautions Section.

2. Disconnect the negative battery cable.

3. Remove the crankshaft pulley cover.

4. Remove the crankshaft pulley bolt.

5. Lock the crankshaft in place using tool ST499977100, or equivalent.

6. Remove the crankshaft pulley.

7. Remove the front timing chain cover.

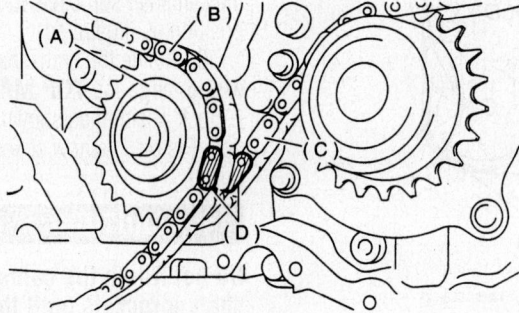

(A) Lower idler sprocket
(B) Timing chain RH
(C) Timing chain LH
(D) Dark gray

42356-SBCR-G40

On the lower idler sprocket align the match marks on the timing chain on the left and right hand sides–right hand side chain installation—2002–2004 3.0L engine

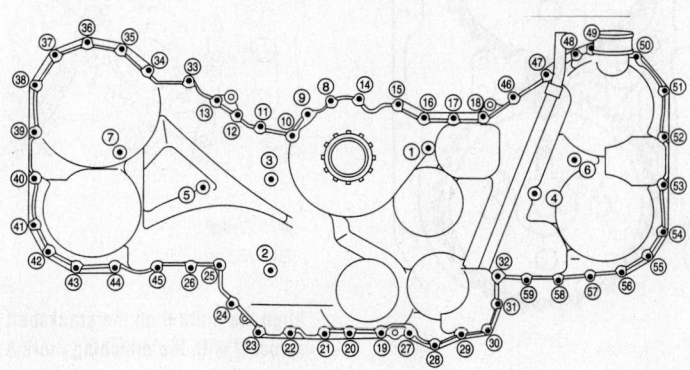

42356-SBCR-G42

Front timing cover bolt tightening sequence—2002–2004 3.0L engine

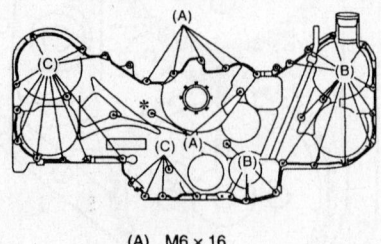

(A) M6 × 16
(B) M6 × 30
(C) M6 × 45
*: Sealing washer

09490_SBCR_G0075

Front cover bolt sizes and locations—2005–2006 3.0L engine

30. Install the right side chain guide. Install the right side chain tensioner lever and tighten the bolts to 12 ft. lbs. (16 Nm). Install the right side chain guide and tighten the NEW bolt to 4.7 ft. lbs. (6 Nm). Install the right side chain tensioner and tighten the bolts to 12 ft. lbs. (16 Nm).

31. Adjust the clearance between the chain guide on the right side and the center chain guide so that there is range between 0.331–0.339 inch (8.4–8.6mm).

32. Install the center chain guide and tighten the NEW bolt to 5.8 ft. lbs. (8 Nm).

33. Check the match marks on each sprocket and corresponding timing chain are correct, remove the stopper from the tensioner.

34. Clean the mating surfaces on the front timing cover. Apply a bead of liquid gasket 0.020 inch in diameter to the mating surface of the front timing chain cover. Install the timing chain cover. Torque the retaining bolts to 4.8 ft. lbs. and in the proper sequence.

35. Continue the installation in the reverse order of the removal procedure.

Piston and Ring

POSITIONING

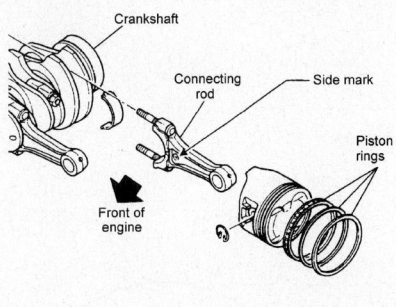

7923AG83

Piston and connecting rod assembly positioning

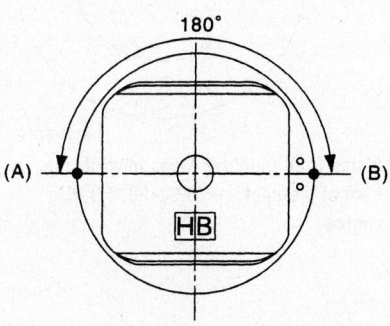

9357TG32

Top ring end-gap spacing—2.0L engine

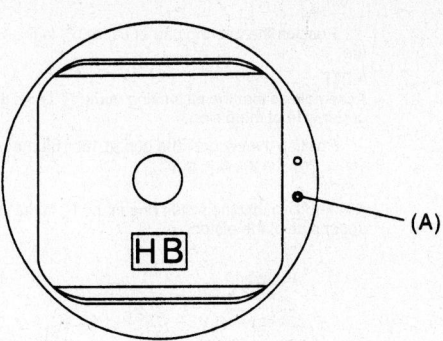

9357TG34

(A) Front mark

Piston front mark faces towards the front of the engine—2.0L engine

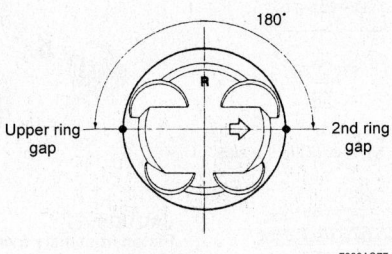

7923AG77

Compression ring end-gap spacing—2002–2004 2.5L engine

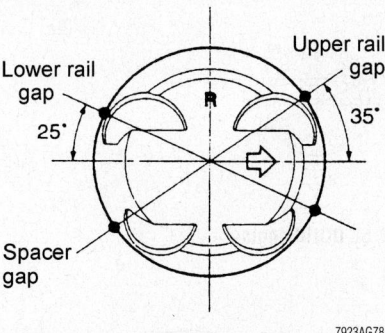

7923AG78

Upper, spacer and lower oil ring end-gap spacing—2002–2004 2.5L engine

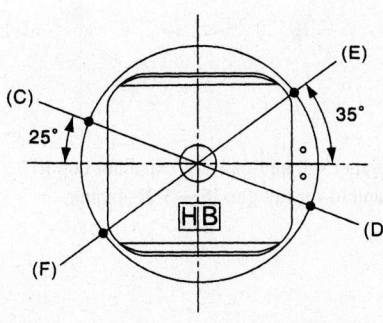

9357TG33

Upper rail end-gap spacing—2.0L engine

Position the top ring gap at (A) or (B) in the figure.

Position the second ring gap at 180° on the reverse side the top ring gap.

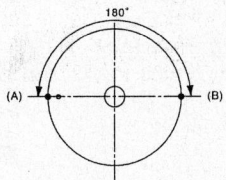

Position the upper rail gap at (C) in the figure.

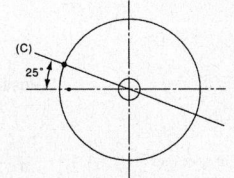

Align the upper rail spin stopper (D) to the side hole (E) on the piston.

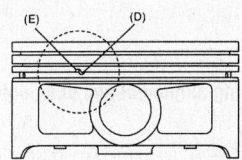

Position the expander gap at (F) in the figure on the 180° opposite direction of (C).

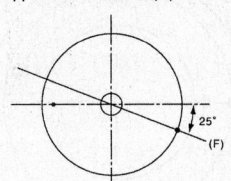

Position the lower rail gap at (G) in the figure.

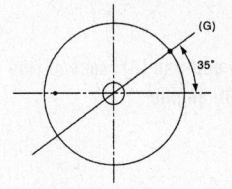

09490_SBCR_G0079

Piston ring alignment and positioning—2005–2006 2.5L SOHC engine

Position the top ring gap at (A) or (B) in the figure.

NOTE:
Assemble so that the piston ring mark "R" faces the upper side of the piston.

Position the second ring gap at 180° on the reverse side the top ring gap.

NOTE:
Assemble so that the piston ring mark "R" faces the upper side of the piston.

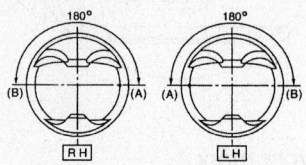

Position the upper rail gap at (C) in the figure.

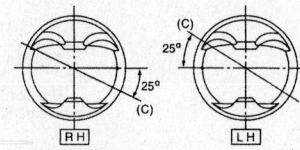

Align the upper rail spin stopper (E) to the side hole (D) on the piston.

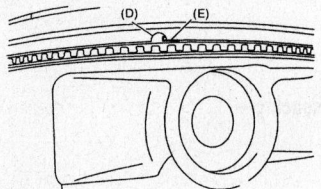

Position the expander gap at (F) in the figure.

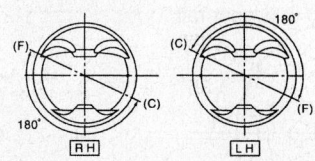

Position the lower rail gap at (G) in the figure.

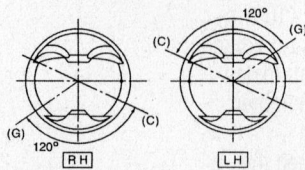

CAUTION:
• **Make sure ring gaps do not face the same direction.**
• **Make sure ring gaps are not within the piston skirt area.**
Install the snap ring.
Install the snap rings in the piston holes located opposite to the service holes in cylinder block when positioning all pistons in corresponding cylinders.

NOTE:
Use new snap rings.

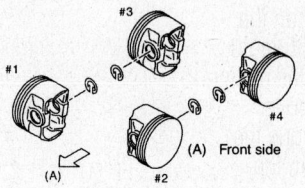

CAUTION:
Piston front mark faces towards the front of engine.

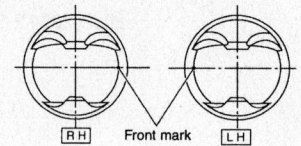

09490_SBCR_G0080

Piston ring alignment and positioning—2005–2006 2.5L DOHC engine

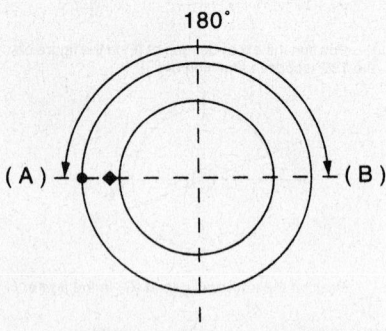

42356-SBCR-G43

Top ring end-gap (A), second ring gap (B)—3.0L engine

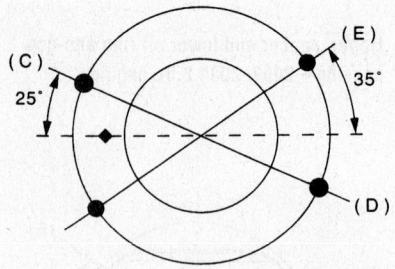

42356-SBCR-G44

Upper rail end-gap (C), expander gap (D) and lower rail gap (E)—3.0L engine

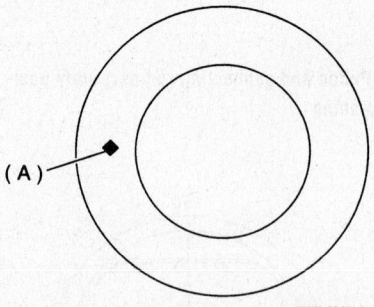

42356-SBCR-G45

Piston front mark (A) faces towards the front of the engine—2002–2004 3.0L engine

FUEL SYSTEM

Fuel System Service Precautions

Safety is the most important factor when performing not only fuel system maintenance, but any type of maintenance. Failure to conduct maintenance and repairs in a safe manner may result in serious personal injury or death. Maintenance and testing of the vehicle's fuel system components can be accomplished safely and effectively by adhering to the following rules and guidelines.

• To avoid the possibility of fire and personal injury, always disconnect the negative battery cable unless the repair or test procedure requires that battery voltage be applied.

• Always relieve the fuel system pressure prior to disconnecting any fuel system component (injector, fuel rail, pressure regulator, etc.), fitting or fuel line connection. Exercise extreme caution whenever relieving fuel system pressure, to avoid exposing skin, face and eyes to fuel spray. Please be advised that fuel under pressure may penetrate the skin or any part of the body that it contacts.

• Always place a shop towel or rag around the fitting or connection prior to loosening to absorb any excess fuel due to spillage. Ensure that all fuel spillage (should it occur) is quickly removed from engine surfaces. Ensure that all fuel soaked cloths or towels are deposited into a suitable waste container.

• Always keep a dry chemical (Class B) fire extinguisher near the work area.

• Do not allow fuel spray or fuel vapors to come into contact with a spark or open flame.

• Always use a back-up wrench when loosening and tightening fuel line connection fittings. This will prevent unnecessary stress and torsion to fuel line piping.

• Always replace worn fuel fitting O-rings with new. Do not substitute fuel hose or equivalent, where fuel pipe is installed.

Fuel System Pressure

RELIEVING

2002–2004

➡**This procedure must be performed prior to servicing any component of the fuel injection system.**

1. Before servicing the vehicle, refer to the Precautions Section.
2. Disconnect the fuel pump connector from the fuel pump relay.
3. Start the engine and let it stall.
4. Crank the engine for 5 seconds or more to ensure the fuel pressure is properly relieved. If the engine starts during this time, allow it to run until it stalls.
5. After performing the required service, connect the fuel pump harness.

2005–2006

➡**This procedure must be performed prior to servicing any component of the fuel injection system.**

1. Before servicing the vehicle, refer to the Precautions Section.
2. On Legacy, Outback and Baja Remove the fuel pump fuse from the main fuse box.
3. On Impreza and WRX disconnect the fuel pump connector from the fuel pump relay.
4. Start the engine and run until it stalls.
5. Crank the engine for 5 seconds or more to ensure the fuel pressure is properly relieved. If the engine starts during this time, allow it to run until it stalls.
6. Turn the ignition switch to the OFF position. Remove the key.
7. Disconnect the negative battery cable.
8. Remove the fuel cap.

Fuel Filter

REMOVAL & INSTALLATION

2002–2004

1. Before servicing the vehicle, refer to the Precautions Section.
2. Locate the fuel filter in the engine compartment.
3. Properly relieve the fuel system pressure.
4. Remove or disconnect the following:

• Negative battery cable
• Hose clamp screws and slide the hoses off the filter
• Filter from the bracket

To install:
5. Inspect the hoses for wear or cracks, and replace if needed.
6. Install or connect the following:

• New filter into the bracket and tighten the hose clamp screws
• Negative battery cable
7. Start the engine and check for leaks.

2005–2006

➡**On Legacy and Outback the fuel filter is an integral part of the fuel pump, no filter replacement procedures are given by the manufacturer.**

1. Before servicing the vehicle, refer to the Precautions Section.
2. Properly relieve the fuel system pressure. Disconnect the negative battery cable.
3. Remove the fuel pump assembly.
4. On Impreza and WRX, separate the fuel filter from the fuel pump. Turn the filter holder around and remove the fuel filter.
5. On Baja, Disconnect the ground cable from the fuel filter holder. Remove the fuel filter holder by turning it to the left from the body pawls. Remove the fuel filter.

To install:
6. Installation is the reverse of the removal procedure.

Fuel Pump

REMOVAL & INSTALLATION

2002–2004

1. Before servicing the vehicle, refer to the Precautions Section.
2. Properly relieve the fuel system pressure.
3. Disconnect the negative battery cable.
4. On the 2002–2003 Legacy, Outback and turbocharged vehicles, perform the following:

a. Raise and safely support the vehicle.

b. Remove the front side fuel tank cover.

c. Drain the fuel tank into a suitable container.

d. Tighten the drain plug to 14–24 ft. lbs. (19–33 Nm).

e. Install the front side fuel tank cover Tighten the retainers to 9.4–16.6 ft. lbs. (13–23 Nm).

5. Remove or disconnect the following:

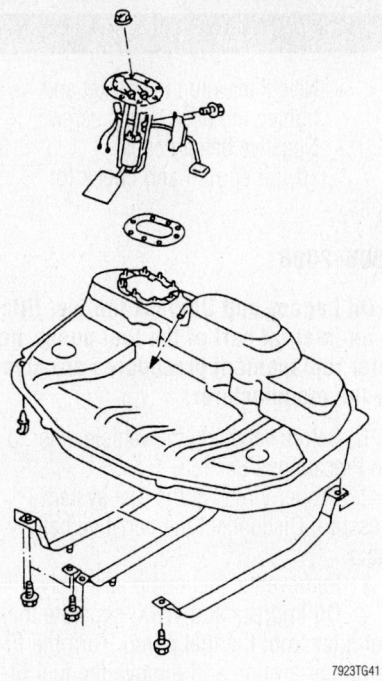

Fuel pump and related components— 2002–2004 vehicles

- Rear seat bottom, to reach the fuel pump access cover, if not already done
6. On Legacy, fold the seat back, then roll the floor mat back.
- Fuel pump cover mounting bolts and the cover
- Electrical harness from the pump assembly

➡**Label the fuel lines before disconnecting them from the pump.**

- Fuel lines from the fuel pump
- Fuel pump mounting nuts
- Fuel pump assembly from the tank

To install:

7. Install or connect the following:
- New gasket
- Fuel pump assembly into the fuel tank and secure with the mounting nuts. Tighten the nuts to 24–48 inch lbs. (3–6 Nm) on non turbocharged engine and 3 ft. lbs. (4 Nm) on turbocharged engine.
- Electrical harness to the fuel pump assembly
- Fuel lines to the pump assembly, then tighten the clamps and fittings
- Fuel pump service cover and cover mounting bolts
- Rear seat bottom
- Negative battery cable
8. Start the engine and check for leaks.

2005–2006

IMPREZA AND WRX

1. Before servicing the vehicle, refer to the Precautions Section.
2. Properly relieve the fuel system pressure.
3. Disconnect the negative battery cable.
4. Remove the fuel filler cap.
5. Raise and support the vehicle safely.
6. Drain the fuel from the fuel tank into a suitable container.
7. Remove the luggage floor mat.
8. Remove the fuel pump access cover retaining screws. Remove the access cover.
9. Disconnect the electrical connector from the fuel pump.
10. Disconnect and plug the fuel line hoses at the fuel pump. Remove the clip and disconnect the jet pump hose.
11. Remove the fuel pump assembly retaining nuts. Remove the fuel pump from the vehicle.

To install:

12. Installation is the reverse of the removal procedure.
13. Be sure to use a new gasket and retainer.
14. Tighten the fuel pump retaining nuts to 3.3 ft. lbs. in an alternating sequence pattern.
15. Continue the installation in the reverse order of the removal procedure.
16. Start the engine and check for leaks, correct as required.

LEGACY AND OUTBACK

1. Before servicing the vehicle, refer to the Precautions Section.
2. Properly relieve the fuel system pressure.
3. Disconnect the negative battery cable.
4. Remove the fuel filler cap.
5. Raise and support the vehicle safely.
6. Drain the fuel from the fuel tank into a suitable container.
7. Lower the vehicle.
8. Remove the rear seat.
9. Remove the fuel pump access cover retaining screws. Remove the access cover.
10. Disconnect the electrical connector from the fuel pump.
11. Disconnect and plug the fuel line hoses at the fuel pump. Remove the clip and disconnect the jet pump hose.
12. Remove the fuel pump assembly retaining nuts. Remove the fuel pump from the vehicle.

To install:

13. Installation is the reverse of the removal procedure.
14. Be sure to use a new gasket and retainer.

➡**When installing, point the protrusion of the gasket toward the front of the vehicle. Insert the protrusion of the gasket into the upper plate (three locations). Align the protrusion of the fuel pump assembly to the cutout in the upper plate.**

15. Tighten the fuel pump retaining nuts to 3.2 ft. lbs. in an alternating sequence pattern.
16. Continue the installation in the reverse order of the removal procedure.
17. Start the engine and check for leaks, correct as required.

BAJA

1. Before servicing the vehicle, refer to the Precautions Section.
2. Properly relieve the fuel system pressure.
3. Disconnect the negative battery cable.
4. Remove the fuel filler cap.
5. Raise and support the vehicle safely.
6. On the DOHC engine, remove the front side fuel tank cover.
7. Drain the fuel from the fuel tank into a suitable container.

➡**Vehicles equipped with the DOHC engine are equipped with a fuel tank drain plug. After draining the fuel tank, use a new gasket and replace the plug. Tighten the plug bolt to 19.2 ft. lbs.**

8. Lower the vehicle.
9. Remove the rear seat. Turn the floor mat up.
10. Remove the fuel pump access cover retaining screws. Remove the access cover.
11. Disconnect the electrical connector from the fuel pump.
12. Disconnect and plug the fuel line hoses at the fuel pump.
13. Remove the fuel pump assembly retaining nuts. Remove the fuel pump from the vehicle.

To install:

14. Installation is the reverse of the removal procedure.
15. Be sure to use a new gasket and retainer.

16. Tighten the fuel pump retaining nuts to 4.3 ft. lbs. in an alternating sequence pattern.

17. Continue the installation in the reverse order of the removal procedure.

18. Start the engine and check for leaks, correct as required.

Fuel Injector

REMOVAL & INSTALLATION

2002–2004

2.0L ENGINE

1. Before servicing the vehicle, refer to the Precautions Section.
2. Properly relieve the fuel system pressure.
3. Disconnect the negative battery cable.
4. Remove or disconnect the following:

- Intake manifold
- Fuel pipe protector
- Electrical connector from the fuel injector
- Bolts that attach the injector pipe to the intake manifold
- Fuel injector while lifting up the fuel injector pipe

To install:
- Install or connect the following:
- New injector O-rings
- Remaining components in the reverse order of removal. Tighten the injector pipe bolts to 13 ft. lbs. (19 Nm).

EXCEPT 2.0L ENGINE

1. Before servicing the vehicle, refer to the Precautions Section.
2. Properly relieve the fuel system pressure.
3. Disconnect the negative battery cable.
4. To remove the right side injectors, remove or disconnect the following:

- Air cleaner ducts and resonator chamber, on California vehicles
- Mass Air Flow (MAF) sensor connector and air intake duct and air cleaner upper cover as a unit, on non-California vehicles
- Air cleaner element
- Spark plug wires from the right side spark plugs
- V-belt covers and power steering pump belt
- Power steering pump brackets-to-intake manifold bolts
- Power steering pump-to-bracket

bolts, then position the pump on the right side wheel apron
5. To remove the injectors on the left side, remove or disconnect the following:

- Windshield washer motor electrical connector
- Electrical connector from the rear window washer, on station wagon only
- Rear window washer hose from the washer motor and plug or cap the line
- Two bolts that secure the washer tank to the body
- Washer tank and secure it out of the way
- Spark plug wires from the left side spark plugs
- Fuel pipe protector

6. Remove or disconnect the following (for either side):

- Band that secures the engine harness to the fuel injector pipe, if equipped
- Intake manifold protector, if equipped
- Fuel injector electrical connector(s)
- Bolts that hold the fuel injector pipe (fuel rail) to the intake manifold, if applicable

➡**Automatic transmission equipped Legacy's may have a retaining clip that must be removed before the injector can be removed.**

7. Pull up on the injector pipe (fuel rail), then remove the fuel injector(s) from the intake manifold. Remove and discard the injector O-rings.

To install:
- Install or connect the following:
- New injector O-rings
- Fuel injector(s) into the intake manifold
- Retaining clips, if applicable
- Injector pipe (fuel rail) and secure with the retaining bolts. Tighten the bolts to 14 ft. lbs. (19 Nm).
- Fuel injector electrical connector(s)
- Intake manifold protector, if equipped
- Band that secures the engine harness to the fuel injector pipe, if equipped

8. To install the injectors on the left side, install or connect the following:

- Fuel pipe protector
- Spark plug wires to the left side spark plugs
- Washer tank and secure with the two mounting bolts

- Rear window washer hose to the washer motor
- Electrical connector to the rear window washer, on station wagon only
- Windshield washer motor electrical connector

9. To install the right side injectors, install or connect the following:

- Power steering pump into position
- Pump-to-bracket bolts
- Power steering pump brackets-to-intake manifold bolts
- Power steering pump belt and V-belt covers
- Spark plug wires to the right side spark plugs
- Air cleaner element
- Air intake duct and upper cover and the MAF sensor connector
- Air cleaner ducts and resonator chamber, on California vehicles

2005–2006

IMPREZA AND WRX (2.5L SOHC)

1. Before servicing the vehicle, refer to the Precautions Section.
2. Properly relieve the fuel system pressure. Remove the fuel cap.
3. Disconnect the negative battery cable.
4. If removing the right side injectors, remove the air intake chamber and air cleaner case. To remove the front side V belt, remove the belt covers. Loosen the lock bolt. Loosen the slider bolt. Remove the front side belt.
5. If removing the right side injectors, remove the bolts that retain the power steering hoses to the intake manifold protector. Do not disconnect the power steering hoses. Remove the bolts which retain the power steering pump to the bracket. Disconnect the power steering pump switch connector. Remove the reservoir tank from the bracket, by pulling upward. Place the power steering pump and tank assembly to the side.
6. Remove the spark plug wires.
7. Remove the fuel line protector cover(s).
8. Disconnect the electrical connector from the fuel injector.
9. Remove the harness band that holds the engine harness to the fuel injector line.
10. Remove the bolts that hold the fuel injector line to the intake manifold.
11. Remove the fuel injector while lifting up the fuel injector line.

To install:
12. Installation is the reverse of the removal procedure.

13. Be sure to use new O-rings.

14. Start the engine and check for leaks, correct as required.

IMPREZA AND WRX (2.5L DOHC)

1. Before servicing the vehicle, refer to the Precautions Section.

2. Properly relieve the fuel system pressure. Remove the fuel cap.

3. Disconnect the negative battery cable. Drain the radiator, as required.

4. If removing the right side injectors, remove the air cleaner upper cover and air intake boot. Remove the air cleaner element. Remove the coolant filler tank.

5. If removing the right side injectors, remove the front side V belt, remove the belt covers. Loosen the lock bolt. Loosen the slider bolt. Remove the front side belt.

6. If removing the right side injectors, disconnect the power steering switch connector. Remove the bolts that retain the power steering hoses to the intake manifold protector. Do not disconnect the power steering hoses. Remove the bolts which retain the power steering pump to the bracket. Remove the reservoir tank from the bracket, by pulling upward. Place the power steering pump and tank assembly to the side.

7. If removing the left side injectors, remove the intake manifold.

8. Remove the fuel line protector cover(s).

9. Disconnect the electrical connector from the fuel injector.

10. If removing the left side injectors, disconnect the connector from the purge control solenoid valve and remove the valve.

11. Remove the harness band that holds the engine harness to the fuel injector line.

12. Remove the bolts that hold the fuel injector line to the intake manifold.

13. Remove the fuel injector while lifting up the fuel injector line.

To install:

14. Installation is the reverse of the removal procedure.

15. Be sure to use new O-rings and insulators.

16. Start the engine and check for leaks, correct as required.

LEGACY AND OUTBACK (2.5L SOHC)

1. Before servicing the vehicle, refer to the Precautions Section.

2. Properly relieve the fuel system pressure. Remove the fuel cap.

3. Disconnect the negative battery cable.

4. If removing the right side injectors, remove the air intake duct and air cleaner case.

5. If removing the left side injectors, remove the battery.

6. Remove the spark plug wires.

7. Remove the fuel line protector cover(s).

8. Disconnect the electrical connector from the fuel injector.

9. Remove the harness band that holds the engine harness to the fuel injector line.

10. Remove the bolts that hold the fuel injector line to the intake manifold.

11. Remove the fuel injector while lifting up the fuel injector line.

To install:

12. Installation is the reverse of the removal procedure.

13. Be sure to use new O-rings.

14. Start the engine and check for leaks, correct as required.

LEGACY AND OUTBACK (2.5L DOHC)

1. Before servicing the vehicle, refer to the Precautions Section.

2. Properly relieve the fuel system pressure. Remove the fuel cap.

3. Disconnect the negative battery cable.

4. Remove the collector cover.

5. If removing the right side injectors, drain the radiator, as required and the coolant filler tank.

6. Disconnect the electrical connector from the fuel injector.

7. Remove the screw and then remove the fuel injector.

To install:

8. Installation is the reverse of the removal procedure.

9. Be sure to use new O-rings and insulators.

10. Start the engine and check for leaks, correct as required.

LEGACY AND OUTBACK (3.L DOHC)

1. Before servicing the vehicle, refer to the Precautions Section.

2. Properly relieve the fuel system pressure. Remove the fuel cap.

3. Disconnect the negative battery cable.

4. Remove the collector cover.

5. If removing the right side injectors, remove the air cleaner case.

6. If removing the left side injectors, remove the battery. Remove the alternator

harness from the left side fuel line protector cover.

7. Remove the fuel line protector cover(s).

8. Disconnect the electrical connector from the fuel injector.

9. Remove the engine harness from the fuel injector line.

10. Remove the bolt that hold the fuel injector line to the intake manifold.

11. Remove the fuel injector while lifting up the fuel injector line.

To install:

12. Installation is the reverse of the removal procedure.

13. Be sure to use new O-rings and insulators.

14. Start the engine and check for leaks, correct as required.

BAJA (2.5L SOHC)

1. Before servicing the vehicle, refer to the Precautions Section.

2. Properly relieve the fuel system pressure. Remove the fuel cap.

3. Disconnect the negative battery cable.

4. If removing the right side injectors, remove the resonator chamber. Remove the number one and three spark plug wires.

5. If removing the right side injectors, remove the front side V belt, remove the belt covers. Loosen the lock bolt. Loosen the slider bolt. Remove the front side belt. Remove the bolts that retain the power steering hoses to the intake manifold protector. Do not disconnect the power steering hoses. Remove the bolts which retain the power steering pump to the bracket. Disconnect the power steering pump switch connector. Remove the reservoir tank from the bracket, by pulling upward. Place the power steering pump and tank assembly to the side.

6. If removing the left side injectors, remove the two bolts that retain the washer fluid tank to its mounting. Disconnect the electrical connector from the front window washer motor. Disconnect the electrical connector and the washer hose from the rear gate glass washer motor. Position the tank to the side. Remove the number two and four spark plug wires.

7. Remove the fuel line protector cover(s).

8. Disconnect the electrical connector from the fuel injector.

9. Remove the bolts that hold the fuel injector line to the intake manifold.

10. Remove the fuel injector retaining clip. Remove the fuel injector while lifting up the fuel injector line.

To install:

11. Installation is the reverse of the removal procedure.

12. Be sure to use new O-rings.

13. Start the engine and check for leaks, correct as required.

BAJA (2.5L DOHC)

1. Before servicing the vehicle, refer to the Precautions Section.

2. Properly relieve the fuel system pressure. Remove the fuel cap.

3. Disconnect the negative battery cable.

4. If removing the left side injectors, remove the intake manifold.

5. Disconnect the electrical connector from the fuel injector.

6. Remove the screw and then remove the fuel injector.

To install:

7. Installation is the reverse of the removal procedure.

8. Be sure to use new O-rings.

9. Start the engine and check for leaks, correct as required.

DRIVE TRAIN

Manual Transmission

REMOVAL & INSTALLATION

2002–2004

1. Before servicing the vehicle, refer to the Precautions Section.

2. Remove or disconnect the following:

- Negative battery cable
- Air intake duct and cleaner case
- Air cleaner stay, if equipped
- Intercooler on turbocharged engine
- Front Oxygen Sensor (O$_2$S) connector
- Neutral position switch connector
- Back-up light switch connector
- Vehicle Speed Sensor (VSS) connector, if equipped
- Transmission ground cable, if necessary
- Clutch cable, if equipped
- Clutch release spring, if equipped
- Starter
- Operating cylinder from the transmission
- Pitching stopper

3. On turbocharged engine, perform the

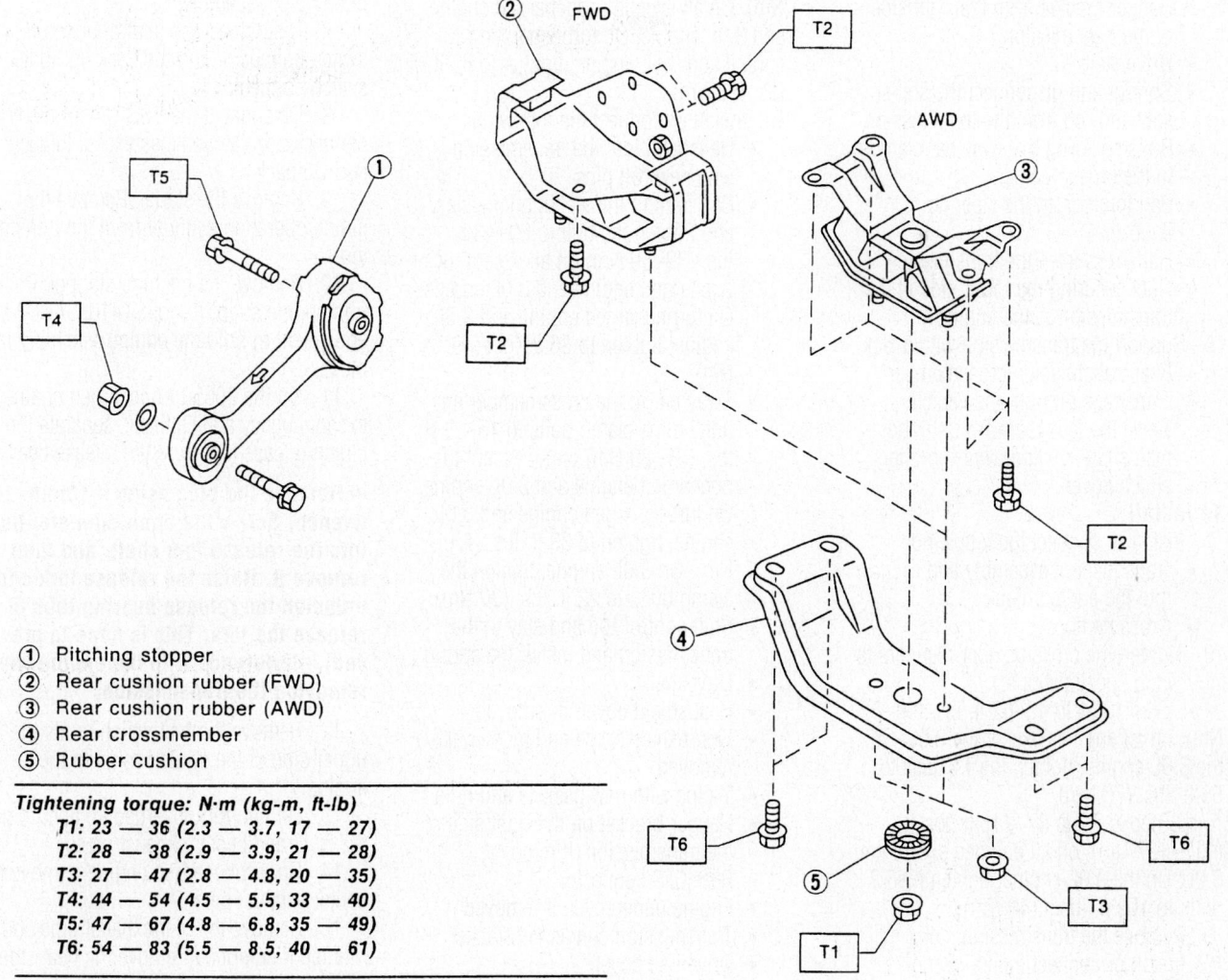

① Pitching stopper
② Rear cushion rubber (FWD)
③ Rear cushion rubber (AWD)
④ Rear crossmember
⑤ Rubber cushion

Tightening torque: N·m (kg-m, ft-lb)
T1: 23 — 36 (2.3 — 3.7, 17 — 27)
T2: 28 — 38 (2.9 — 3.9, 21 — 28)
T3: 27 — 47 (2.8 — 4.8, 20 — 35)
T4: 44 — 54 (4.5 — 5.5, 33 — 40)
T5: 47 — 67 (4.8 — 6.8, 35 — 49)
T6: 54 — 83 (5.5 — 8.5, 40 — 61)

Manual transmission crossmember and related components—2002–2004 Impreza

7923TG44

following to disconnect the clutch release fork from the release bearing:

    a. Remove the clutch cylinder from the transmission.

    b. Using a 10mm wrench, remove the plug.

    c. Screw a 6mm diameter bolt into the release fork and remove it.

    d. Raise the release fork and unfasten the release tabs to free the release fork.

4. Remove or disconnect the following:
- Drive belt cover, if necessary
- Slave (operating) cylinder, on 2.5L engine

5. Install engine support assembly 927670000, on 3.0L engine, install support assembly 41099AA00.
- Bolt securing the right upper side of the transmission to the engine
- Engine undercover, if equipped
- Rear O2S connector
- Front Y-pipe
- Rear exhaust pipe and muffler, on Legacy
- Heat shield cover
- Hanger bracket from the right side of the transmission
- Driveshaft
- Spring, and disconnect the shifter stay and rod from the transmission
- Bolts securing the sway bar clamps to the crossmember
- Ball joints from the steering knuckle
- Halfshafts from the transmission
- Nuts securing the lower side of the transmission to the engine

6. Support the transmission with a jack.
- Rear transmission crossmember
- Transmission from the vehicle. Move the jack rearward until the mainshaft is withdrawn from the clutch cover.

To install:

7. Install or connect the following:
- Transmission assembly and secure it to the engine block
- Crossmember

8. Tighten the crossmember retainers to the following specifications:

    a. Step 1: T1 to 40–62 ft. lbs. (54–84 Nm), on all engines except the 3.0L. On the 3.0L engine, tighten the T1 bolts to 55 ft. lbs. (75 Nm)

    b. Step 2: T2 to 87–115 ft. lbs. (117–157 Nm), on all engines except the 3.0L. On the 3.0L engine, tighten the T2 bolts to 103 ft. lbs. (140 Nm)

9. Remove the transmission jack.

10. Install or connect the following:
- Nuts securing the lower portion of the engine to the transmission and

tighten to 40 ft. lbs. (54 Nm) on all except turbocharged engine and 3.0L engine. On turbocharged engine and 3.0L engine, tighten to 37 ft. lbs. (50 Nm).
- Bolt securing the right upper side of the transmission to the engine and tighten it to 40 ft. lbs. (54 Nm) on all except turbocharged engine and 3.0L engine. On turbocharged engine and 3.0L engine, tighten to 37 ft. lbs. (50 Nm).

11. Remove the engine support.
- Drive belt cover
- Slave (operating) cylinder, on 2.5L engine

12. Install the pitching stopper and tighten the bolts to the following specifications:

    a. Step 1: T1 to 33–40 ft. lbs. (44–54 Nm) on all except turbocharged engine and 3.0L engine. On turbocharged engine and 3.0L engine, tighten to 37 ft. lbs. (50 Nm).

    b. Step 2: T2 to 35–49 ft. lbs. (47–67 Nm). On all except turbocharged engine and 3.0L engine. On turbocharged engine and 3.0L engine, tighten to 43 ft. lbs. (58 Nm).

13. Install or connect the following:
- Halfshafts into the transmission with new roll pins
- Ball joint to the steering knuckle and tighten the bolt to 29–43 ft. lbs. (39–59 Nm) on all except turbocharged engine and 3.0L engine. On turbocharged engine and 3.0L engine, tighten to 36 ft. lbs. (49 Nm).
- Sway bar to the crossmember and tighten the clamp bolts to 15–21 ft. lbs. (21–29 Nm) on all except turbocharged engine and 3.0L engine. On turbocharged engine and 3.0L engine, tighten to 33 ft. lbs. (45 Nm). On 3.0L engine, tighten the clamp bolts to 22 ft. lbs. (30 Nm).
- Shift control rod and stay to the transmission and install the spring
- Driveshaft
- Heat shield cover, if removed
- Rear exhaust pipe and muffler, if removed
- Y-pipe with new gaskets and nuts
- Hanger bracket on the right side of the transmission, if removed
- Rear O2S connector
- Engine undercover, if removed
- Transmission connector bracket
- Drive belt cover
- Pitching stopper
- Starter

- Front O2S connector
- VSS connector, if equipped
- Neutral position switch connector
- Back-up light switch connector
- Clutch cable (if equipped)
- Clutch release spring
- Air cleaner case stay and case
- Air intake duct and attach the air-flow sensor connector
- Negative battery cable

2005–2006

FIVE SPEED

1. Before servicing the vehicle, refer to the Precautions Section.

2. Open the hood to the full open position.

3. Disconnect the negative battery cable.

4. Drain the transmission fluid.

5. On non turbocharged engine, remove the air intake duct and cleaner case. Remove the air cleaner case stay.

6. On turbocharged engine, remove the intercooler assembly.

7. Disconnect the neutral position switch connector and the back up light switch connector.

8. Disconnect the VSS sensor electrical connector, on all vehicles except Legacy and Outback.

9. Remove the starter. Remove the clutch operating cylinder from the transmission.

10. Remove the pitching stopper. Position engine support tool ST41099AC000, or equivalent to hold the engine assembly in place.

11. On the 2005 turbocharged engine, except Legacy and Outback, separate the clutch release fork from the release bearing.

➡ **Remove the plug using a 10mm wrench. Screw the 6mm diameter bolt into the release fork shaft, and then remove it. Raise the release fork and unfasten the release bearing tabs to release the fork. This is done to prevent interference with the engine when removing the transmission.**

12. Remove the bolt which holds the upper side of the transmission to the engine.

13. On non turbocharged engine, remove the front and center exhaust pipes.

14. On turbocharged engine, remove the center exhaust pipe.

15. Remove the rear exhaust pipe and muffler. If equipped, remove the heat shield cover.

16. Remove the hanger bracket from the

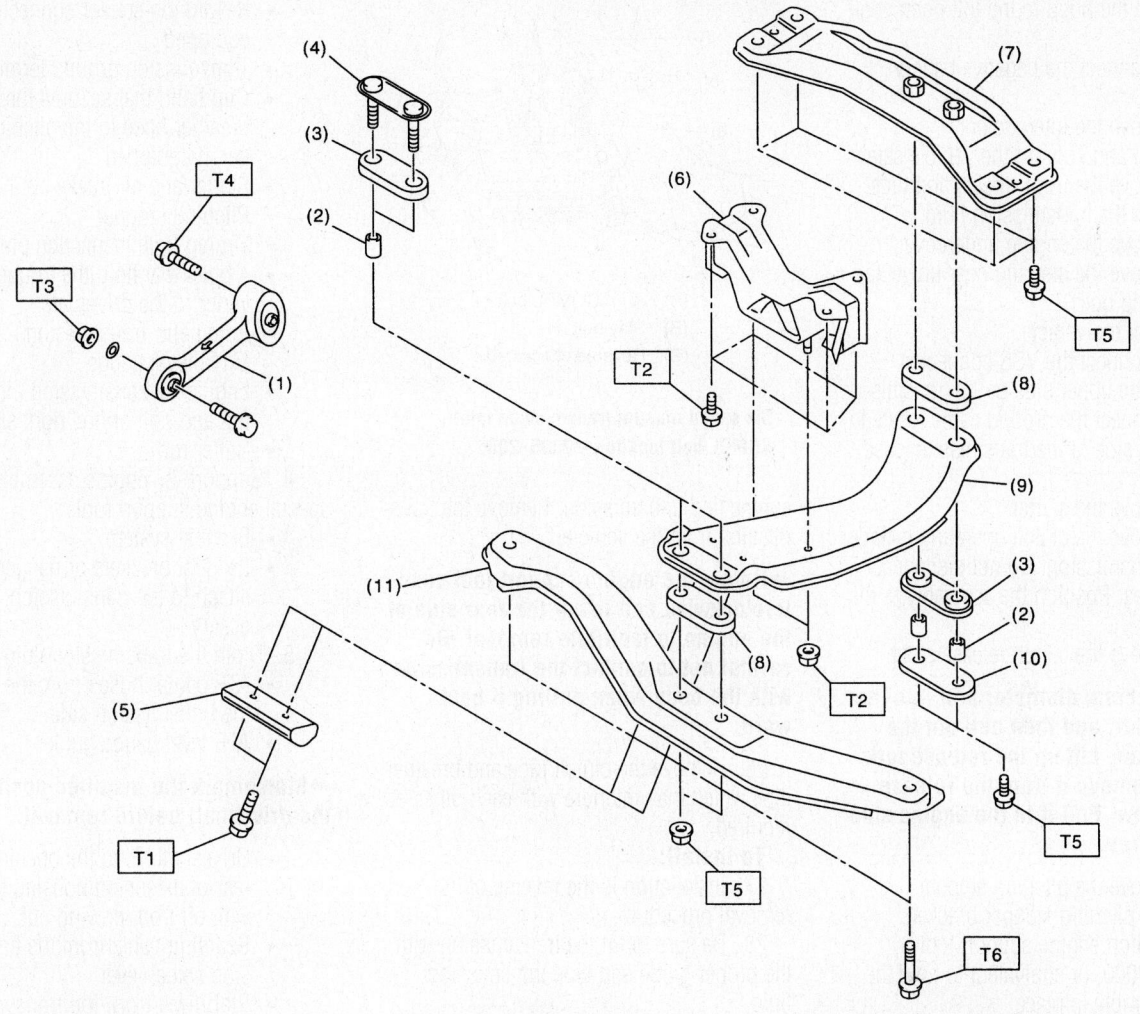

(1)	Pitching stopper	(8)	Cushion D		
(2)	Spacer	(9)	Center crossmember		
(3)	Cushion C	(10)	Rear plate		
(4)	Front plate	(11)	Front crossmember		
(5)	Dynamic damper				
(6)	Rear cushion rubber				
(7)	Rear crossmember				

Tightening torque: N·m (kgf-m, ft-lb)
T1: 7.5 (0.76, 5.5)
T2: 35 (3.6, 26)
T3: 50 (5.1, 37)
T4: 58 (5.9, 43)
T5: 70 (7.1, 51)
T6: 140 (14.3, 103)

09490_SBCR_G0081

Manual transmission crossmember and related components—2005–2006

right side of the transmission, except on Legacy and Outback.

17. Remove the drive shaft.

18. Remove the gear shift rod and the stay from the transmission.

19. On all except Legacy and Outback, disconnect the stabilizer link from the transverse link. Remove the bolt securing the ball joint of the transverse link to housing.

20. On Legacy and Outback, disconnect the stabilizer link from the front arm. Remove the bolt securing the ball joint of the front arm to housing.

21. Remove the halfshafts from the transmission.

22. Remove the nuts that hold the lower side of the transmission to the engine.

23. Position a transmission jack under the transmission assembly.

24. Remove the transmission rear crossmember retaining bolts. Remove the rear crossmember from the vehicle.

25. Carefully remove the transmission from the vehicle.

➡**Move the transmission jack rearward until the main shaft is withdrawn from the clutch cover.**

26. Separate the transmission assembly and rear cushion rubber.

27. On turbocharged engine, remove the release bearing from the engine side.

To install:

28. Installation is the reverse of the removal procedure.

29. Be sure to fill the transmission with the proper grade and type transmission fluid.

30. Start the engine and check for leaks, correct as required.

31. Roadtest the vehicle.

SIX SPEED

1. Before servicing the vehicle, refer to the Precautions Section.

2. Open the hood to the full open position.

3. Disconnect the negative battery cable.

4. Remove the intercooler.

5. Raise and support the vehicle safely.

6. Remove the front wheels and tires.

7. Drain the transmission fluid.

8. Remove the engine undercover.

9. Remove the steering gear universal joint retaining bolt.

10. Lower the vehicle.

11. Disconnect the VSS connector located on the upper side of the transmission. Disconnect the ground cable located at the upper side of the transmission case and body.

12. Remove the starter.

13. Remove the clutch operating cylinder from the transmission. Do not disconnect the fluid lines. Position the assembly to the side.

14. Remove the clutch release shaft.

➡**Screw a 6mm diameter bolt into the release shaft, and then pull out the release shaft. Lift up the release fork and then remove it from the release bearing claw. Pull it to the engine side and set it free.**

15. Remove the pitching stopper. Remove the pitching stopper bracket.

16. Position engine support tool ST41099AC000, or equivalent to hold the engine assembly in place.

17. Remove the center and rear exhaust pipes.

18. Remove the drive shaft.

19. Remove the front stabilizer bolt. Disconnect the ball joint of the transverse link from the housing.

20. On Legacy and Outback, disconnect the stabilizer link from the front arm. Remove the bolt securing the ball joint of the front arm to housing.

21. Using a pry bar, separate halfshafts from the transmission. Remove the front half shaft, then the front halfshaft.

22. Position a transmission jack under the transmission assembly.

23. Remove the transmission front and rear crossmember retaining bolts. Remove the front and rear crossmembers from the vehicle.

24. Move the transmission to the right side. Remove the joint COMPL bolt, stay bolt and reverse check cable.

➡**If the transmission is not moved, the joint COMPL and stay bolt will contact the body and damage may occur.**

25. Remove the retaining bolts from the

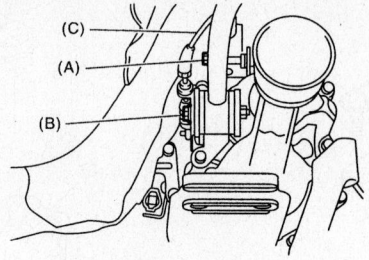

(A) Joint COMPL bolt
(B) Stay bolt
(C) Reverse check cable

09490_SBCR_G0082

Six speed manual transmission joint COMPL bolt location—2005–2006

engine and transmission. Remove the transmission from the vehicle.

➡**Rotate the engine support tool counterclockwise and lower the rear side of the engine to facilitate removal. Be careful not to contact the transmission with the body when pulling it backward.**

26. Remove the clutch pipe and breather pipe which may interfere with each other, as required.

To install:

27. Installation is the reverse of the removal procedure.

28. Be sure to fill the transmission with the proper grade and type transmission fluid.

29. Start the engine and check for leaks, correct as required.

30. Roadtest the vehicle.

Automatic Transmission

REMOVAL & INSTALLATION

2002–2004

EXCEPT WRX

1. Before servicing the vehicle, refer to the Precautions Section.

2. Drain the transmission fluid.

3. Remove or disconnect the following:
- Negative battery cable
- Air intake duct with air cleaner case
- Air cleaner case stay
- Front Oxygen Sensor (O2S) connector
- Speedometer cable or electronic wiring connector from the speed sensor
- Transmission harness connector
- Inhibitor switch connector, if equipped

- Revolution sensor connector, if equipped
- Transmission ground terminal
- Clip band that secures the air breather hose to the pitching stopper, if equipped
- Starter and air intake boot
- Pitching stopper
- Timing hole inspection plug
- 4 bolts that hold the torque converter to the driveplate
- Automatic Transmission Fluid (ATF) level gauge
- Engine-to-transmission mounting nut and bolt on the right side
- Buffer rod

4. Support the engine assembly with special engine support tool.
- Exhaust system
- Exhaust brackets or hangers that attach to the transmission, as necessary

5. Drain the transmission fluid.
- ATF cooler hoses from the pipes of the transmission side
- ATF level gauge guide

➡**Matchmark the installed position of the driveshaft before removal.**

- Driveshaft. Plug the opening at the rear of extension housing to prevent oil from flowing out.
- Gearshift cable from the transmission select lever
- Stabilizer from the transverse link
- Parking brake cable bracket from the transverse link
- Transverse link bolts and lower the link
- Spring pins
- Halfshafts from the transmission

➡**Discard the old spring pin and always install a new pin.**

- Oil cooler hoses

6. Place a transmission jack under the transmission.
- Engine to transmission mounting nuts

➡**Do not place the jack under the oil pan otherwise the oil pan may be damaged.**

- Rear cushion rubber mounting nuts and the rear crossmember

7. Move the torque converter and transmission as a unit away from the engine and lower it from the vehicle.

To install:

8. Install or connect the following:
- Transmission to the engine and

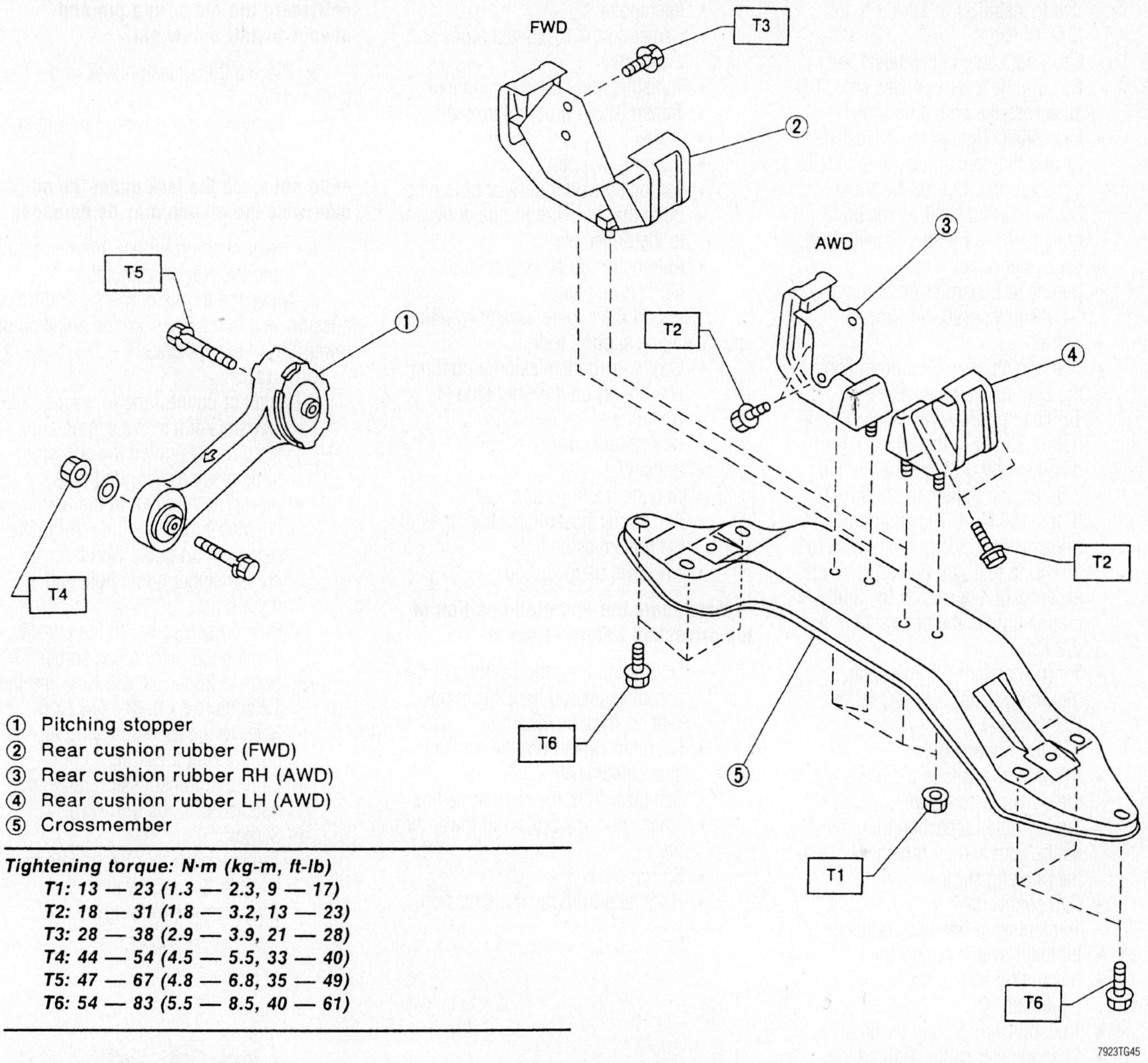

FWD

T3

②

AWD

③

T5

T2

①

④

T2

T4

T6

⑤

T1

T6

① Pitching stopper
② Rear cushion rubber (FWD)
③ Rear cushion rubber RH (AWD)
④ Rear cushion rubber LH (AWD)
⑤ Crossmember

Tightening torque: N·m (kg-m, ft-lb)
  T1: 13 — 23 (1.3 — 2.3, 9 — 17)
  T2: 18 — 31 (1.8 — 3.2, 13 — 23)
  T3: 28 — 38 (2.9 — 3.9, 21 — 28)
  T4: 44 — 54 (4.5 — 5.5, 33 — 40)
  T5: 47 — 67 (4.8 — 6.8, 35 — 49)
  T6: 54 — 83 (5.5 — 8.5, 40 — 61)

7923TG45

Automatic transmission crossmember and related components—2002–2004 Impreza and Legacy

temporarily tighten the engine-to-transmission mounting nuts
- Rear crossmember to the rear cushion rubber mounts. Align the rear cushion guide with the rear

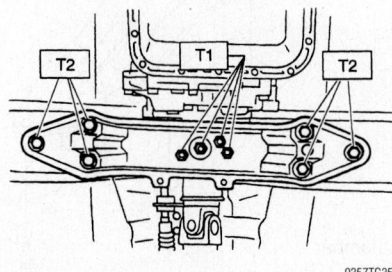

9357TG35

Location of the rear crossmember bolts—2002–2004 3.0L engine

crossmember guide hole and tighten nuts.
- Rear crossmember to the chassis. Tighten the rear crossmember bolts to 39–49 ft. lbs. (53–66 Nm) on all except the 3.0L engine. On the 3.0L engine tighten the T1 bolts to 26 ft. lbs. (35 Nm) and the T2 bolts to 55 ft. lbs. (75 Nm).
- Engine to transmission retaining nuts to 34–40 ft. lbs. (46–54 Nm)

9. Remove the transmission jack from the vehicle.

10. Remove the engine support tool and install the buffer rod.

11. Install or connect the following:

- Axle shafts to the transmission using new spring pins
- Transverse link temporarily to the front crossmember. Do not complete final torque at this point.
- Stabilizer temporarily to the transverse link
- Parking brake cable bracket to the transverse link
- Transverse link-to-front crossmember mounting bolts and transverse link-to-stabilizer mounting bolts, with the tires placed on the ground
- Transverse link to front crossmember (self-locking nuts) to 40–62 ft. lbs. (54–84 Nm) and the transverse

link to stabilizer to 18–32 ft. lbs. (24–44 Nm).
- Gearshift cable to the select lever. Be sure the lever operates smoothly all across the operating range.
- Driveshaft. Tighten the driveshaft-to-rear differential retaining bolts to 17–24 ft. lbs. (23–33 Nm) and center bearing location retaining bolts to 25–33 ft. lbs. (34–45 Nm).
- Oil cooler hoses
- Engine to transmission bolts to 34–40 ft. lbs. (46–54 Nm)
- Starter
- Pitching stopper. Be sure to tighten the bolt for the body side first. Tightening torque for the body side bolt is 35–49 ft. lbs. (47–67 Nm) on all except 3.0L engine; on the 3.0L engine tighten the bolt to 37 ft. lbs. (50 Nm). The engine or transmission side bolt is torque to 33–40 ft. lbs. (44–54 Nm)) on all except 3.0L engine; on the 3.0L engine tighten the bolt to 43 ft. lbs. (58 Nm).
- Torque converter-to-driveplate mounting bolts to 17–20 ft. lbs. (23–27 Nm)
- ATF level gauge guide
- ATF cooler hoses to the pipes of the transmission side
- Timing hole inspection plug, air intake boot and air breather hose to the pitching stopper
- O2S connector
- Transmission harness connector
- Inhibitor switch connector
- Revolution sensor connector, if equipped
- Transmission ground terminal
- Speedometer cable. Tighten the cable nut by side, then turn it approximately 30 degrees more with a tool.
- Exhaust system and exhaust brackets or hangers that attach to the transmission, as necessary
- Air cleaner case stay
- Air intake duct with air cleaner case
- Battery ground cable

12. Refill and check transmission oil level.

13. Road test the vehicle for proper operation across all operating ranges.

WRX

1. Before servicing the vehicle, refer to the Precautions Section.

2. Drain the transmission fluid.

3. Remove or disconnect the following:
- Negative battery cable
- Intercooler
- Center and rear exhaust pipes and the muffler
- Transmission harness connector
- Transmission ground terminal
- Starter
- Pitching stopper
- Torque converter service hole plug
- Bolts that hold the torque converter to the driveplate
- Automatic Transmission Fluid (ATF) level gauge

4. Support the engine assembly with special engine support tool.
- Engine-to-transmission mounting and bolt(s) on the right side
- Undercover
- Heat shield cover
- Buffer rod

5. Drain the transmission fluid.
- ATF cooler hoses from the pipes of the transmission side
- ATF level gauge guide

➡Matchmark the installed position of the driveshaft before removal.

- Driveshaft. Plug the opening at the rear of extension housing to prevent oil from flowing out.
- Gearshift cable from the transmission select lever
- Stabilizer from the transverse link
- Transverse link bolts and lower the link
- Spring pins
- Halfshafts from the transmission

➡Discard the old spring pin and always install a new pin.

6. Place a transmission jack under the transmission.
- Engine to transmission mounting nuts

➡Do not place the jack under the oil pan otherwise the oil pan may be damaged.

- Rear cushion rubber mounting nuts and the rear crossmember

7. Move the torque converter and transmission as a unit away from the engine and lower it from the vehicle.

To install:

8. Install or connect the following:
- Transmission to the engine and temporarily tighten the engine-to-transmission mounting nuts
- Rear crossmember to the rear cushion rubber mounts. Align the rear cushion guide with the rear crossmember guide hole and tighten nuts.
- Rear crossmember to the chassis. Tighten the rear crossmember T1 bolts to 26 ft. lbs. (35 Nm) and the T2 bolts to 51 ft. lbs. (70 Nm). Refer to the accompanying illustration for bolt location.
- Engine to transmission retaining nuts to 36 ft. lbs. (50 Nm)
- Starter
- Torque converter clutch plate bolts to 18 ft. lbs. (25 Nm0
- Torque converter plug

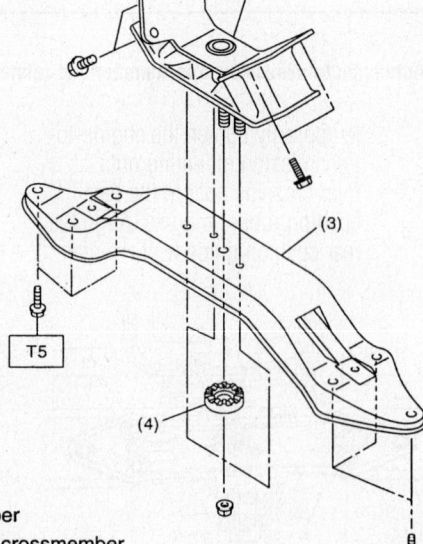

(1)	Pitching stopper
(2)	Rear cushion rubber
(3)	Transmission rear crossmember
(4)	Stopper

09490_SBCR_G0083

Automatic transmission crossmember and related components—2005–2006

- Pitching stopper. Be sure to tighten the bolt for the body side first. Tightening torque for the body side bolt is 43 ft. lbs. (58 Nm). The engine or transmission side bolt is torque to 37 ft. lbs. (50 Nm).
- Axle shafts to the transmission using new spring pins
- Transverse link temporarily to the front crossmember. Do not complete final torque at this point.
- Stabilizer temporarily to the transverse link
- Transverse link-to-front crossmember mounting bolts and transverse link-to-stabilizer mounting bolts, with the tires placed on the ground
- Transverse link to front crossmember (self-locking nuts) to 22 ft. lbs. (30 Nm) and the transverse link to stabilizer to 37 ft. lbs. (50 Nm).
- Gearshift cable to the select lever. Be sure the lever operates smoothly all across the operating range.
- ATF level gauge guide
- Oil cooler hoses
- Driveshaft
- Heat shield cover
- Center, rear exhaust pipes and the muffler
- Undercover
- ATF fluid level gauge
- Transmission harness connectors
- Transmission ground terminal
- Air cleaner case stay
- Intercooler
- Battery ground cable

9. Refill and check transmission oil level.

10. Road test the vehicle for proper operation across all operating ranges.

2005–2006

IMPREZA AND WRX

1. Before servicing the vehicle, refer to the Precautions Section.

2. Open the hood to the full open position.

3. Disconnect the negative battery cable.

4. Drain the transmission fluid.

5. On non turbocharged engine, remove the air intake chamber and intake duct. Remove the air intake chamber stay.

6. On turbocharged engine, remove the intercooler.

7. Disconnect the transmission harness connector and ground cable.

8. Remove the starter.

9. On 2006 vehicles, remove the throttle body assembly.

10. Remove the pitching stopper.

11. To separate the torque converter clutch from the drive plate, remove the service hole plug. Remove the bolts that retain the flex plate to the torque converter. When rotating the engine do so in the proper direction of rotation. Install tool ST49827720, or equivalent, to the torque converter case to hold the assembly in place.

12. Remove the transmission filler gauge and housing. Plug the opening.

13. Remove the pitching stopper bracket.

14. Position engine support tool ST41099AC000, or equivalent to hold the engine assembly in place.

15. Remove the upper side transmission to engine retaining bolts.

16. Raise and support the vehicle safely. Remove the undercover.

17. On non turbocharged engine, remove the front, center and rear exhaust pipes and muffler.

18. On turbocharged engine, remove the center and rear exhaust pipes and muffler.

19. Disconnect the transmission cooler hoses from the transmission side. Remove the oil charge pipe.

20. Remove the driveshaft.

21. Remove the shift select cable.

22. Disconnect the stabilizer link from the transverse link. Remove the bolt securing the ball joint of the transverse link to the housing.

23. Pull out the front halfshaft from the transmission. Place a cloth between the tool and the transmission to avoid damaging the side retainer of the transmission.

24. Remove the bolts that hold the clutch housing cover. Remove the nuts that retain the lower engine to transmission.

25. Position a transmission jack under the transmission.

➡ **Make sure that the support plates of the transmission jack do not contact the oil pan.**

26. Remove the transmission rear crossmember retaining bolts. Remove the crossmember from the vehicle.

27. While gradually lowering the transmission jack, fully contact the engine support, and then tilt the engine rearward.

28. Remove the transmission from the vehicle.

➡ **Move the transmission and torque converter away from the engine as an assembly.**

To install:

29. Installation is the reverse of the removal procedure.

30. Be sure to use a new differential side oil seal.

31. Be sure to fill the transmission with the proper grade and type transmission fluid.

32. Start the engine and check for leaks, correct as required.

33. Roadtest the vehicle.

LEGACY AND OUTBACK

1. Before servicing the vehicle, refer to the Precautions Section.

2. Open the hood to the full open position.

3. Disconnect the negative battery cable.

4. Drain the transmission fluid.

5. Remove the air intake duct. Remove the air intake chamber. Remove the air cleaner case assembly.

6. Disconnect the transmission harness connectors and the ground terminal.

7. Remove the starter.

8. Remove the pitching stopper.

9. To separate the torque converter clutch from the drive plate, remove the service hole plug. Remove the bolts that retain the flex plate to the torque converter. When rotating the engine do so in the proper direction of rotation. Install tool ST49827720, or equivalent, to the torque converter case to hold the assembly in place.

10. Remove the transmission filler gauge and housing. Plug the opening.

11. Remove the throttle body assembly.

12. Disconnect the engine harness and then remove the connector from the bracket. Remove the bracket.

13. Remove the pitching stopper bracket.

14. Position engine support tool ST41099AC000, or equivalent to hold the engine assembly in place.

15. Remove the upper side transmission to engine retaining bolts.

16. Raise and support the vehicle safely. Remove the undercover.

17. Remove the front, center and rear exhaust pipes and muffler. Remove the heat shield cover.

18. Disconnect the transmission cooler hoses from the transmission side. Remove the oil charge pipe.

19. Remove the driveshaft.

20. Remove the shift select cable.

21. Remove the two brackets which hold the front stabilizer. Remove the bolt securing the ball joint of the front arm to the housing.

22. Pull out the front halfshaft from the transmission. Place a cloth between the tool and the transmission to avoid damaging the side retainer of the transmission.

23. Remove the bolts that hold the clutch housing cover. Remove the nuts that retain the lower engine to transmission.

24. Position a transmission jack under the transmission.

➡**Make sure that the support plates of the transmission jack do not contact the oil pan.**

25. Remove the transmission rear crossmember retaining bolts. Remove the crossmember from the vehicle.

26. While gradually lowering the transmission jack, fully contact the engine support, and then tilt the engine rearward.

➡**Retract the support until there is 0.39 inch clearance between the crossmember and the torque converter case.**

27. Remove the transmission from the vehicle.

➡**Move the transmission and torque converter away from the engine as an assembly.**

To install:

28. Installation is the reverse of the removal procedure.

29. Be sure to use a new differential side oil seal.

30. Be sure to fill the transmission with the proper grade and type transmission fluid.

31. Start the engine and check for leaks, correct as required.

32. Roadtest the vehicle.

BAJA

1. Before servicing the vehicle, refer to the Precautions Section.

2. Open the hood to the full open position.

3. Disconnect the negative battery cable.

4. Drain the transmission fluid.

5. On non turbocharged engine, remove the air cleaner case or air intake chamber. Remove the air cleaner case stay.

6. On turbocharged engine, remove the intercooler.

7. Disconnect the transmission harness connectors and the ground terminal.

8. Remove the starter.

9. Remove the pitching stopper.

10. To separate the torque converter clutch from the drive plate, remove the service hole plug. Remove the bolts that retain the flex plate to the torque converter. When rotating the engine do so in the proper direction of rotation. Install tool ST49827720, or equivalent, to the torque converter case to hold the assembly in place.

11. Remove the transmission filler gauge and housing. Plug the opening.

12. Position engine support tool ST41099AC000, or equivalent to hold the engine assembly in place.

13. Remove the upper side transmission to engine retaining bolts.

14. Raise and support the vehicle safely. Remove the undercover.

15. On non turbocharged engine, remove the front, center and rear exhaust pipes and muffler.

16. On turbocharged engine, remove the center and rear exhaust pipes and muffler.

17. If equipped, remove the heat shield cover.

18. Disconnect the transmission cooler hoses from the transmission side. Remove the oil charge pipe.

19. Remove the driveshaft.

20. Remove the shift select cable.

21. Disconnect the stabilizer link from the transverse link. Remove the bolt securing the joint of the transverse link to the housing.

22. Remove the spring pins and separate the front halfshaft out of the transmission.

23. Remove the bolts that hold the clutch housing cover. Remove the nuts that retain the lower engine to transmission.

24. Position a transmission jack under the transmission.

➡**Make sure that the support plates of the transmission jack do not contact the oil pan.**

25. Remove the transmission rear crossmember retaining bolts. Remove the crossmember from the vehicle.

26. While gradually lowering the transmission jack, fully contact the engine support, and then tilt the engine backward.

27. Remove the transmission from the vehicle.

➡**Move the transmission and torque converter away from the engine as an assembly.**

To install:

28. Installation is the reverse of the removal procedure.

29. Be sure to use a new differential side oil seal.

30. Be sure to fill the transmission with the proper grade and type transmission fluid.

31. Start the engine and check for leaks, correct as required.

32. Roadtest the vehicle.

ADJUSTMENT

Some 2002–2004 vehicles are equipped with a mechanical clutch system that is adjustable. All other vehicles are equipped with a hydraulic system that is not adjustable.

Cable

The clutch cable can be adjusted at the cable bracket where the cable is attached to the side of the transmission housing.

1. Before servicing the vehicle, refer to the Precautions Section.

2. Remove the circlip and clamp.

3. Slide the cable end in the direction desired, then replace the circlip and clamp into the nearest gutters on the cable end.

➡**The cable should not be stretched out straight nor should it have right angle kinks in it. Any straightening should be gradual.**

4. Check the clutch for proper operation.

Pedal Height

Adjust the pedal with the return stop bolt, so that its pad is on the same level as the brake pedal pad.

Check to be sure that the stroke of the pedal is 5.12–5.31 in. (130–135mm). Check the clutch release fork stroke. It should be 0.67 in. (17mm).

Free-Play

1. Before servicing the vehicle, refer to the Precautions Section.

2. Remove the clutch release lever return spring from the lever, and loosen the locknut on the fork adjusting nut.

➡**Be careful not to twist the cable during adjustment**

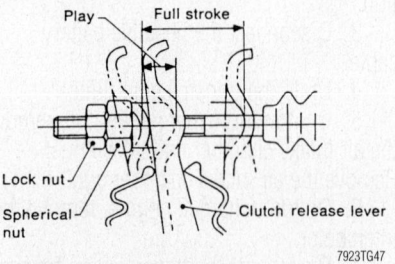

7923TG47

Be sure to tighten the locknut after making the necessary adjustments—2002–2004 vehicles with mechanical clutch and non turbocharged

3. Turn the adjusting nut (spherical nut) until a release fork free-play of 0.14–0.18 in. (3.5–4.5mm) is obtained.

4. Tighten the locknut.

5. Install the return spring on the lever. Hook the long hook side of the return spring with the lever.

6. Check the pedal free-play. It should be 0.12–0.16 in. (3.0–4.0mm).

7. Adjust the pedal free-play, as necessary, with the pedal adjusting bolt.

REMOVAL & INSTALLATION

2002–2004

✻✻ CAUTION

The clutch driven disc may contain asbestos that has been determined to be a cancer causing agent. Never clean clutch surfaces with compressed air. Avoid inhaling any dust from any clutch surface. When cleaning clutch surfaces, use a commercially available brake cleaning fluid.

1. Before servicing the vehicle, refer to the Precautions Section.

2. Remove or disconnect the following:
- Negative battery cable
- Transmission

3. Gradually unscrew the bolts which hold the pressure plate assembly on the flywheel. Loosen the bolts only 1 turn at a time, working around the pressure plate.

4. When all of the bolts have been removed, remove the clutch plate and disc.

✻✻ WARNING

Do not get oil or grease on the clutch facing.

5. Remove the 2 retaining springs and remove the throwout bearing and the release fork.

➡Do not disassemble either the clutch cover or disc. Inspect the parts for wear or damage and replace any parts as necessary. Replace the clutch disc if there is any oil or grease on the facing. Do not wash or attempt to lubricate the throwout bearing. If it requires replacement, the bearing may be removed and a new one installed in the holder by means of a press.

To install:

6. Fit the release fork boot on the front of the transmission housing.

7. Install or connect the following:
- Release fork

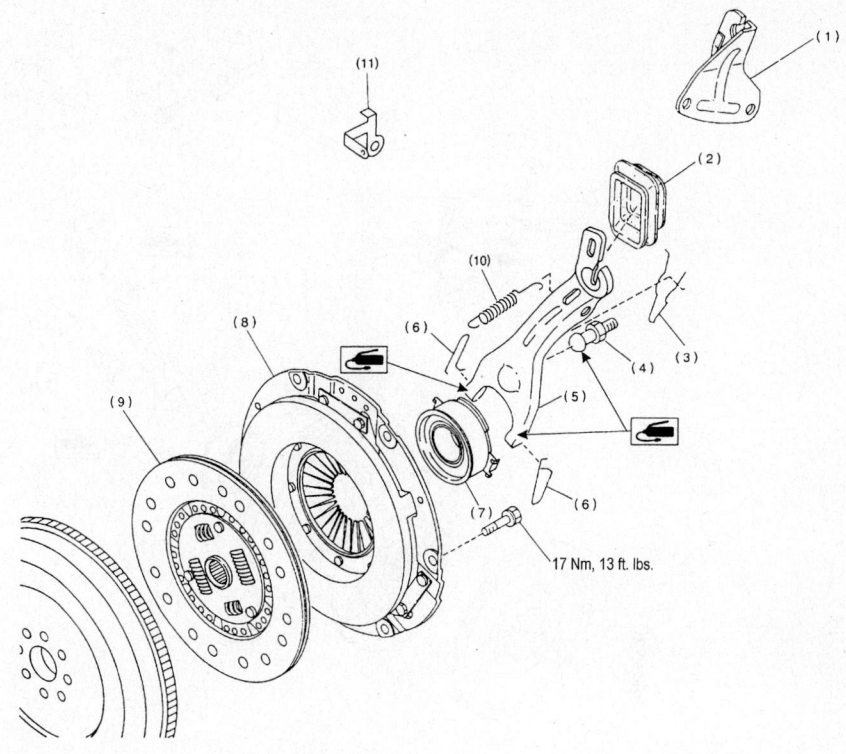

(1) Clutch cable bracket	(6) Clip	(10) Return spring (Models without
(2) Clutch release lever sealing	(7) Clutch release bearing	hill holder only)
(3) Retainer spring	(8) Clutch cover	(11) Clutch return spring bracket
(4) Pivot	(9) Clutch disc	
(5) Clutch release lever		

7923TG48

Clutch and related components—2002–2004 mechanical clutch, non-turbocharged vehicles

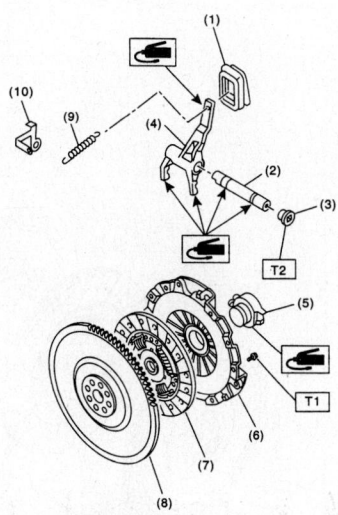

(1) Clutch release lever sealing	(6) Clutch cover	**Tightening torque: N·m (kgf-m, ft-lb)**
(2) Release lever shaft	(7) Clutch disc	**T1: 15.7 (1.6, 11.6)**
(3) Plug	(8) Flywheel	**T2: 44 (4.5, 32.5)**
(4) Release lever	(9) Spring	
(5) Release bearing	(10) Bracket	

9357TG36

Clutch and related components—2002–2004 mechanical clutch, turbocharged vehicles

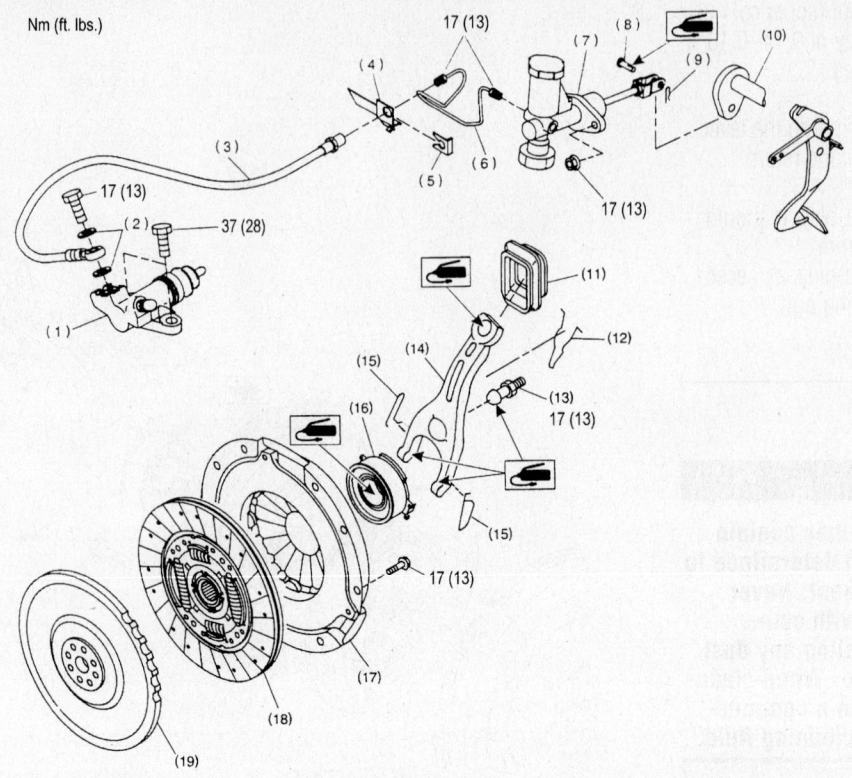

Nm (ft. lbs.)

(1) Operating cylinder	(8) Clevis pin
(2) Washer	(9) Snap pin
(3) Clutch hose	(10) Lever
(4) Bracket	(11) Clutch release lever sealing
(5) Clamp	(12) Retainer spring
(6) Pipe	(13) Pivot
(7) Master cylinder ASSY	
	(14) Release lever
	(15) Clip
	(16) Release bearing
	(17) Clutch cover
	(18) Clutch disc
	(19) Flywheel

7923TG49

Clutch and related components—2002–2004 hydraulic clutch

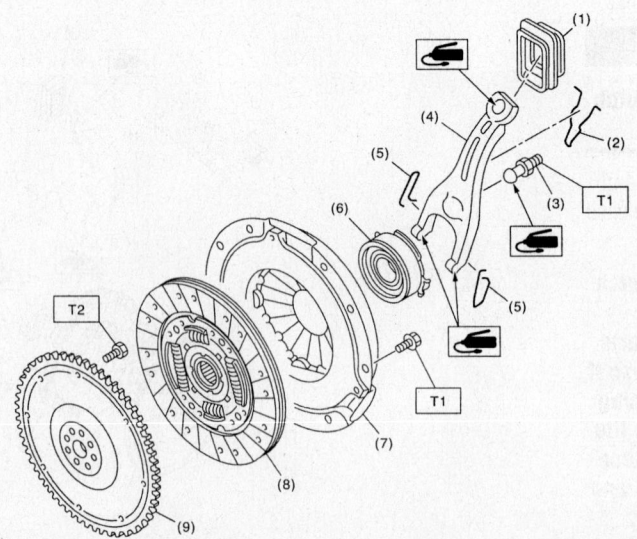

(1) Dust cover	(6) Release bearing
(2) Lever spring	(7) Clutch cover
(3) Pivot	(8) Clutch disc
(4) Release lever	(9) Flywheel
(5) Clip	

Tightening torque: N·m (kgf-m, ft-lb)	
T1:	16 (1.6, 11.8)
T2:	72 (7.3, 52.8)

09490_SBCR_G0084

Clutch and related components—2005–2006 non turbocharged vehicle

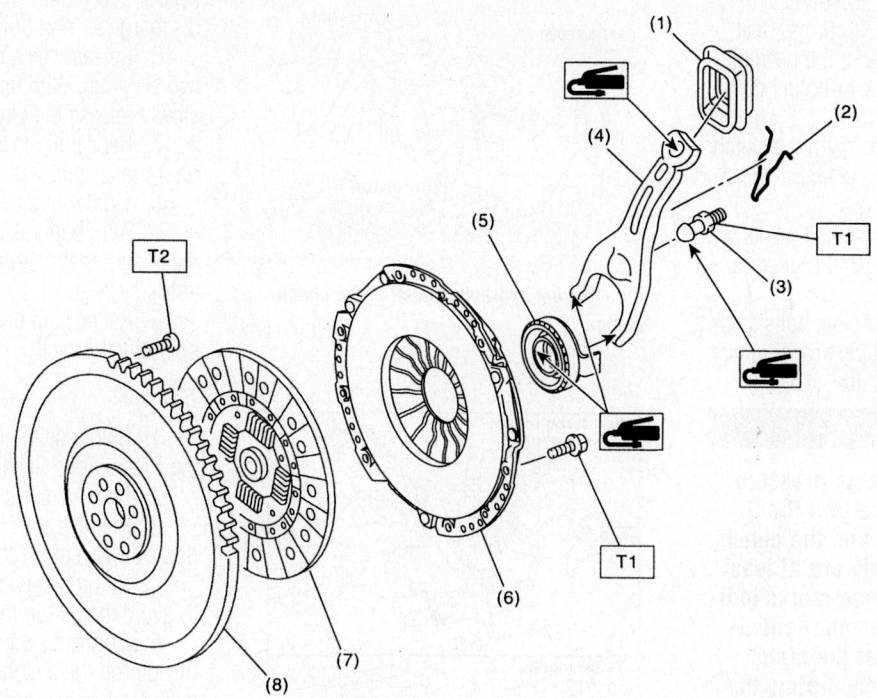

(1)	Dust cover	(5)	Release bearing
(2)	Lever spring	(6)	Clutch cover
(3)	Pivot	(7)	Clutch disc
(4)	Release lever	(8)	Flywheel

Tightening torque: N·m (kgf-m, ft-lb)	
T1:	16 (1.6, 11.8)
T2:	72 (7.3, 52.8)

09490_SBCR_G0085

Clutch and related components—2005–2006 turbocharged vehicle

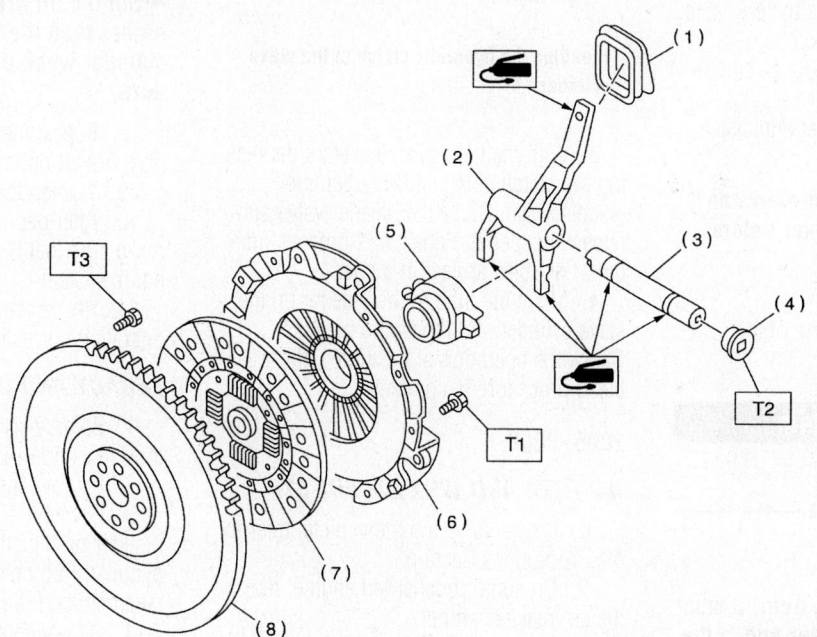

(1)	Dust cover	(5)	Release bearing
(2)	Release lever	(6)	Clutch cover
(3)	Clutch release lever shaft	(7)	Clutch disc
(4)	Plug	(8)	Flywheel

Tightening torque: N·m (kgf-m, ft-lb)	
T1:	16 (1.6, 11.8)
T2:	44 (4.5, 32.5)
T3:	75 (7.6, 55.3)

09490_SBCR_G0086

Clutch and related components—2005–2006 STI

- Throwout bearing assembly and secure it with the 2 springs. Coat the inside diameter of the bearing holder and the fork-to-holder contact points with grease.

8. Insert a pilot shaft through the clutch cover and disc, then insert the end of the pilot into the needle bearing.

9. If equipped, position the **O** marks on the clutch cover and flywheel 120 degrees apart.

10. Tighten the pressure plate bolts gradually, 1 turn at a time, until the proper torque is reached. Tighten to 11 ft. lbs. (15 Nm).

✳✳ WARNING

When installing the clutch pressure plate assembly, be sure that the O marks on the flywheel and the clutch pressure plate assembly are at least 120 degrees apart. These marks indicate the direction of residual unbalance. Also, be sure that the clutch disc is installed properly, noting the FRONT and REAR markings.

11. After installation of the transmission, perform the adjustments outlined above.

2005–2006

1. Before servicing the vehicle, refer to the Precautions Section.
2. Disconnect the negative battery cable.
3. Remove the transmission.
4. Install tool ST499747100, or equivalent on the flywheel.
5. Remove the clutch cover and clutch disc. Do not disassemble

➡**Be sure to put alignment marks on the flywheel and clutch cover before removing the clutch cover.**

To install:

6. Installation is the reverse of the removal procedure.

Hydraulic Clutch System

BLEEDING

2002–2004

➡**To properly bleed the system, it must be bled at the slave cylinder and at the damper.**

1. Before servicing the vehicle, refer to the Precautions Section.
2. Connect a vinyl tube to the air bleeder on the clutch operating (slave) cylinder. Put the other end in a jar with clean clutch fluid.

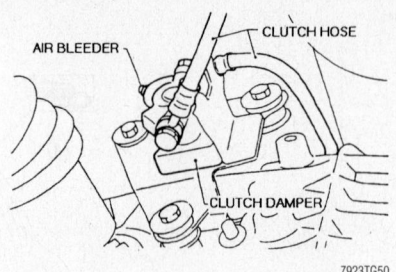

Bleeding the hydraulic clutch at the clutch damper

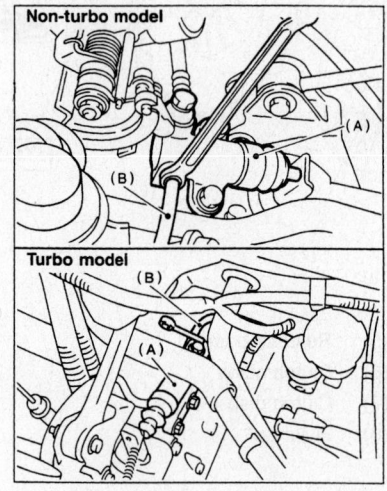

(A) Operating cylinder
(B) Vinyl tube

Bleeding the hydraulic clutch at the slave cylinder

3. With the help of an assistant depressing the clutch pedal, slowly open the bleeder valve. Close the bleeder valve and release the pedal. Repeat this process until no air bubbles appear in the jar.

4. Move the tube to the bleeder on the slave cylinder and repeat the process. Check the operation of the clutch after the bleed procedure is complete.

2005–2006

IMPREZA AND WRX EXCEPT STI

1. Before servicing the vehicle, refer to the Precautions Section.
2. On non turbocharged engine, remove the air intake chamber.
3. On turbocharged engine, remove the intercooler.
4. Connect a vinyl tube to the air bleeder on the clutch operating (slave) cylinder. Put the other end in a jar with clean clutch fluid.
5. Slowly depress the clutch pedal and keep it depressed. Open the air bleeder to discharge air and fluid.

6. Release the air bleeder for one or two seconds. With the bleeder closed, slowly release the clutch pedal.

7. Repeat the procedure until there are no more air bubbles in the vinyl tube.

8. Tighten the air bleeder.

9. After depressing the clutch pedal, make sure that there are no leaks in the entire system

10. Recheck to ensure that the clutch is operating correctly.

STI

1. Before servicing the vehicle, refer to the Precautions Section.
2. Remove the intercooler.
3. Remove the clutch operating cylinder. Do not remove the clutch hose.
4. Using a service clamp, fix the piston to avoid the piston from jumping out.
5. Connect a vinyl tube to the air bleeder on the clutch operating (slave) cylinder. Put the other end in a jar with clean clutch fluid.
6. Slowly depress the clutch pedal and keep it depressed. Open the air bleeder to discharge air and fluid.
7. Release the air bleeder for one or two seconds. With the bleeder closed, slowly release the clutch pedal.

➡**Set the air breather screw position higher than the tip of the operating cylinder when performing this procedure.**

8. Repeat the procedure until there are no more air bubbles in the vinyl tube.
9. Tighten the air bleeder.
10. After depressing the clutch pedal, make sure that there are no leaks in the entire system
11. Recheck to ensure that the clutch is operating correctly.

LEGACY AND OUTBACK

1. Before servicing the vehicle, refer to the Precautions Section.
2. Remove the air intake chamber.
3. Connect a vinyl tube to the air bleeder on the clutch operating (slave) cylinder. Put the other end in a jar with clean clutch fluid.
4. Slowly depress the clutch pedal and keep it depressed. Open the air bleeder to discharge air and fluid.
5. Release the air bleeder for one or two seconds. With the bleeder closed, slowly release the clutch pedal.
6. Repeat the procedure until there are no more air bubbles in the vinyl tube.

7. Tighten the air bleeder.

8. After depressing the clutch pedal, make sure that there are no leaks in the entire system

9. Recheck to ensure that the clutch is operating correctly.

BAJA

1. Before servicing the vehicle, refer to the Precautions Section.

2. On non turbocharged engine, remove the air intake chamber.

3. On turbocharged engine, remove the intercooler.

4. Connect a vinyl tube to the air bleeder on the master cylinder. Put the other end in a jar with clean clutch fluid.

5. Slowly depress the clutch pedal and keep it depressed. Open the air bleeder to discharge air and fluid.

6. Release the air bleeder for one or two seconds. With the bleeder closed, slowly release the clutch pedal.

7. Repeat the procedure until there are no more air bubbles in the vinyl tube.

8. Tighten the air bleeder.

9. Connect a vinyl tube to the air bleeder on the clutch operating (slave) cylinder. Put the other end in a jar with clean clutch fluid.

10. Slowly depress the clutch pedal and keep it depressed. Open the air bleeder to discharge air and fluid.

11. Release the air bleeder for one or two seconds. With the bleeder closed, slowly release the clutch pedal.

12. Repeat the procedure until there are no more air bubbles in the vinyl tube.

13. Tighten the air bleeder.

14. After depressing the clutch pedal, make sure that there are no leaks in the entire system

15. Recheck to ensure that the clutch is operating correctly.

Transfer Case Assembly

REMOVAL & INSTALLATION

The transfer case must be removed as an assembly with the transmission.

Halfshaft

REMOVAL & INSTALLATION

Front—2002–2004

1. Before servicing the vehicle, refer to the Precautions Section.

2. Remove or disconnect the following:

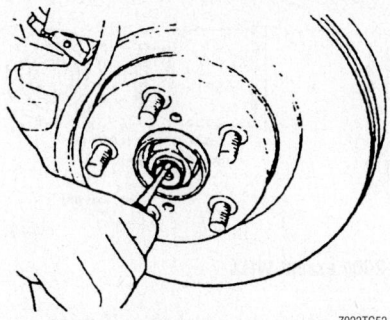

7923TG52

Unstaking the front axle nut—2002–2004

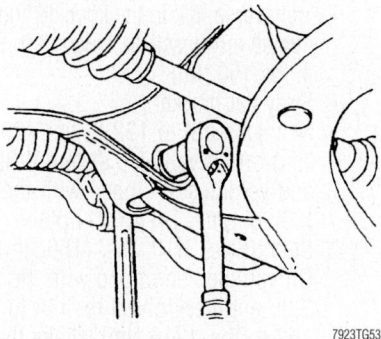

7923TG53

Remove the front transverse link arm from the crossmember—2002–2004

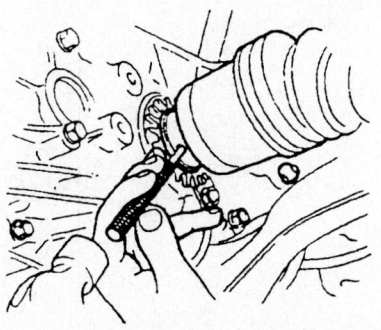

7923TG54

Drive out the front halfshaft-to-transmission roll pin—2002–2004

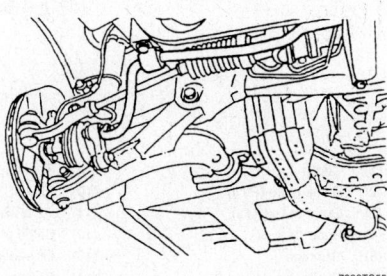

7923TG55

Remove the front sway bar bracket—2002–2004

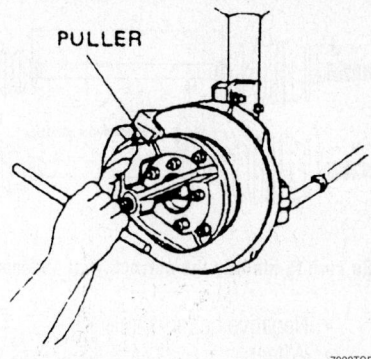

7923TG56

Using a special puller tool, press the front axle shaft from the spindle housing—2002–2004

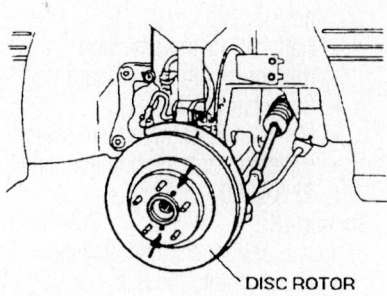

7923TG57

Use two 8mm bolts (arrows) to loosen the front rotor from the spindle housing—2002–2004

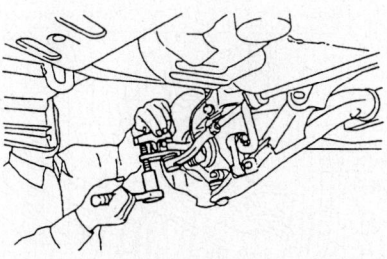

7923TG58

Using a special tool to separate the front tie rod end from the steering knuckle—2002–2004

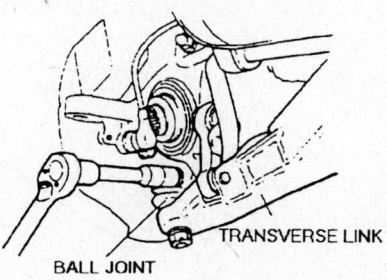

7923TG59

Remove the front transverse link arm from the spindle housing—2002–2004

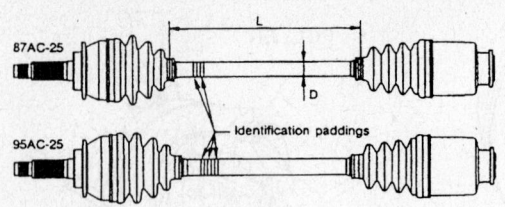

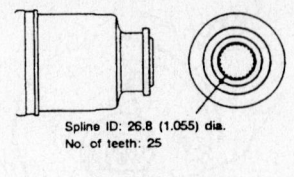

Spline ID: 26.8 (1.055) dia.
No. of teeth: 25

Unit: mm (in)

7923TG60

Be sure to identify the correct front halfshaft—2002–2004 except WRX

- Negative battery cable
- Wheel
- Axle nut, unstake the nut before attempting removal
- Stabilizer link from the transverse link
- Transverse link ball joint from the housing
- Halfshaft-to-transmission roll/spring pin and discard it
- Sway bar bracket
- Halfshaft from the transmission
- Halfshaft from the hub using puller 92707000

To install:

3. Install or connect the following:
- Halfshaft into the hub

4. Using installer 922431000 and adapter 927390000, pull the halfshaft through the hub.

5. Install or connect the following:
- Temporarily tighten a new axle nut

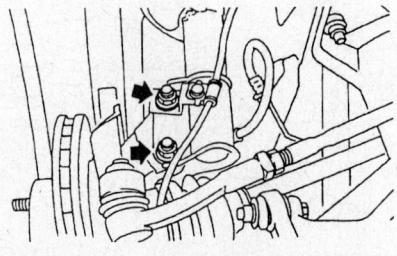

Before loosening the front strut-to-housing bolts (arrows), matchmark the camber adjustment bolt and strut—2002–2004

7923TG61

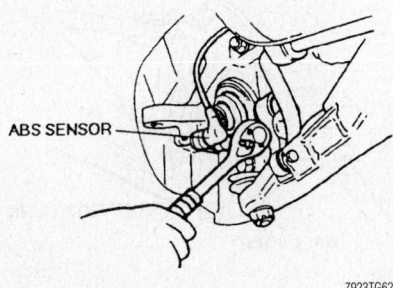

ABS SENSOR

7923TG62

Removing the front ABS sensor—2002–2004

- Align the halfshaft roll/spring pin hole
- Halfshaft onto the transmission
- New pin.
- Transverse link to the knuckle and tighten a new self-locking nut to 36 ft. lbs. (50 Nm)
- Sway bar bracket
- New axle nut to 152 ft. lbs. (206 Nm) on all vehicles except WRX and vehicles equipped with the 3.0L engine. On WRX, tighten the nut to 137 ft. lbs. (186 Nm). On vehicles equipped with the 3.0L engine, tighten the nut to 159 ft. lbs. (216 Nm). Stake the nut

- Wheel
- Negative battery cable

Front—2005–2006

IMPREZA AND WRX

1. Before servicing the vehicle, refer to the Precautions Section.
2. Raise and support the vehicle safely.
3. Remove the wheels and tires.
4. Raise the caulking portion of the axle nut. Depress the brake pedal and remove the axle nut.

➡ **Be sure to loosen the axle nut after removing the tire and wheel from the vehicle. Failure to do this may damage the wheel bearings.**

5. Remove the stabilizer link from the transverse link. Disconnect the transverse link from the housing.
6. Remove the front halfshaft assembly.

➡ **Use axle shaft puller tools ST92647000 and ST28099PA110, or equivalent, if it is difficult to remove the halfshaft.**

7. Remove the front halfshaft from the

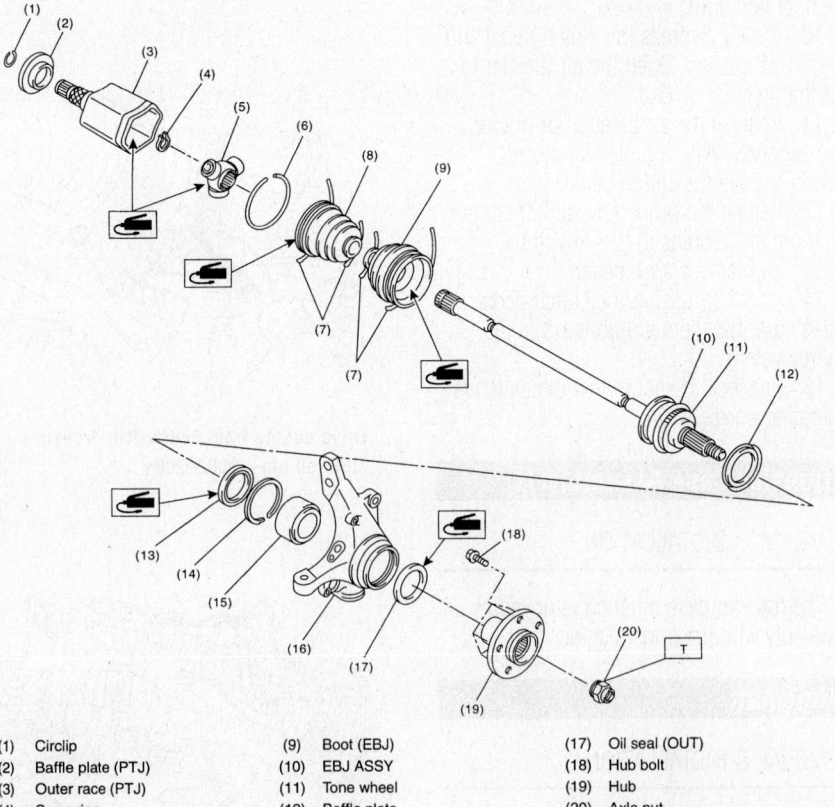

(1)	Circlip	(9)	Boot (EBJ)	(17)	Oil seal (OUT)
(2)	Baffle plate (PTJ)	(10)	EBJ ASSY	(18)	Hub bolt
(3)	Outer race (PTJ)	(11)	Tone wheel	(19)	Hub
(4)	Snap ring	(12)	Baffle plate	(20)	Axle nut
(5)	Trunnion	(13)	Oil seal (IN)		
(6)	Circlip	(14)	Snap ring		
(7)	Boot band	(15)	Bearing		
(8)	Boot (PTJ)	(16)	Housing		

Tightening torque: N·m (kgf-m, ft-lb)
T: 220 (22.4, 162)

09490_SBCR_G0087

Front halfshaft (EBJ+PTJ type) and related components—2005–2006 Impreza and WRX

transmission, using a pry bar. Be careful not to damage the holder portion.

To install:

8. Installation is the reverse of the removal procedure.

9. Be sure to replace the differential side seal with a new one.

10. Be sure to use new locknuts.

11. Always tighten the axle shaft nut before installing the tire and wheel assembly. Tighten the nut to 162 ft. lbs.

LEGACY AND OUTBACK

1. Before servicing the vehicle, refer to the Precautions Section.

2. Raise and support the vehicle safely.

3. Remove the wheels and tires.

4. If equipped with manual transmission, drain the gear oil.

5. If equipped with automatic transmission, drain the differential gear oil.

6. Raise the caulking portion of the axle nut. Depress the brake pedal and remove the axle nut.

➡**Be sure to loosen the axle nut after removing the tire and wheel from the vehicle. Failure to do this may damage the wheel bearings.**

7. Remove the stabilizer link from the front arm. Disconnect the front ball arm from the housing.

8. Remove the front halfshaft assembly.

➡**Use axle shaft puller tools ST92647000 and ST28099PA110, or equivalent, if it is difficult to remove the halfshaft.**

9. Remove the front halfshaft from the transmission, using a pry bar. Be careful not to damage the holder portion.

To install:

10. Installation is the reverse of the removal procedure.

11. Be sure to replace the differential side seal with a new one.

12. Be sure to use new locknuts.

13. Always tighten the axle shaft nut before installing the tire and wheel assembly. Tighten the nut to 162 ft. lbs.

BAJA (2.5L SOHC)

1. Before servicing the vehicle, refer to the Precautions Section.

2. Disconnect the negative battery cable.

3. Raise and support the vehicle safely.

4. Remove the wheels and tires.

5. Raise the caulking portion of the axle nut. Depress the brake pedal and remove the axle nut.

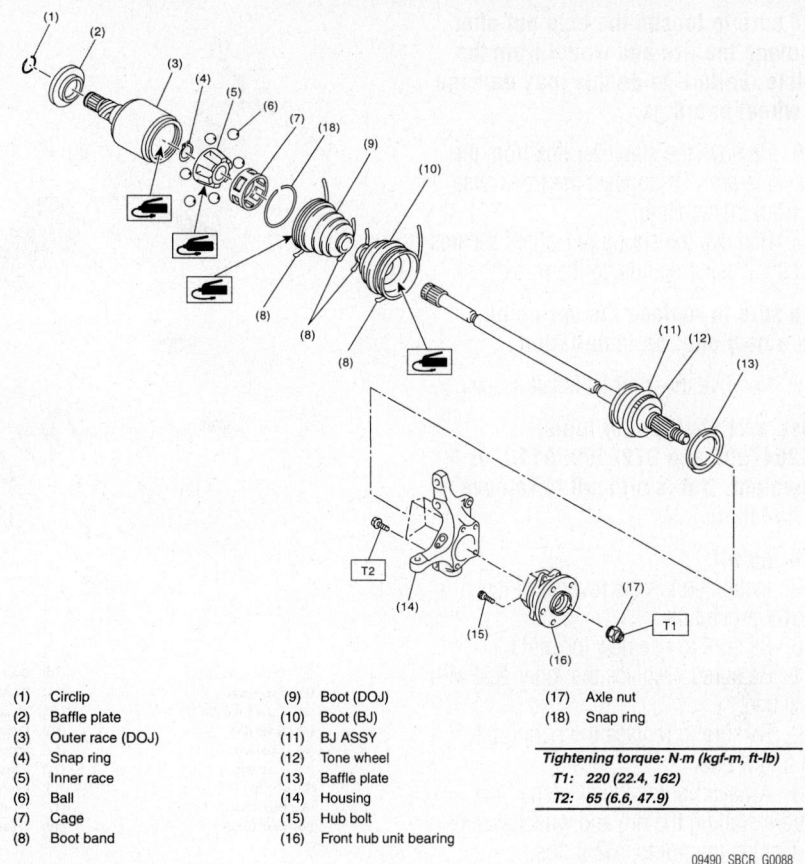

(1)	Circlip	(9)	Boot (DOJ)	(17)	Axle nut
(2)	Baffle plate	(10)	Boot (BJ)	(18)	Snap ring
(3)	Outer race (DOJ)	(11)	BJ ASSY		
(4)	Snap ring	(12)	Tone wheel		
(5)	Inner race	(13)	Baffle plate		
(6)	Ball	(14)	Housing		
(7)	Cage	(15)	Hub bolt		
(8)	Boot band	(16)	Front hub unit bearing		

Tightening torque: N·m (kgf-m, ft-lb)
T1: 220 (22.4, 162)
T2: 65 (6.6, 47.9)

09490_SBCR_G0088

Front halfshaft (BJ+DOJ type) and related components—2005–2006 Impreza and WRX

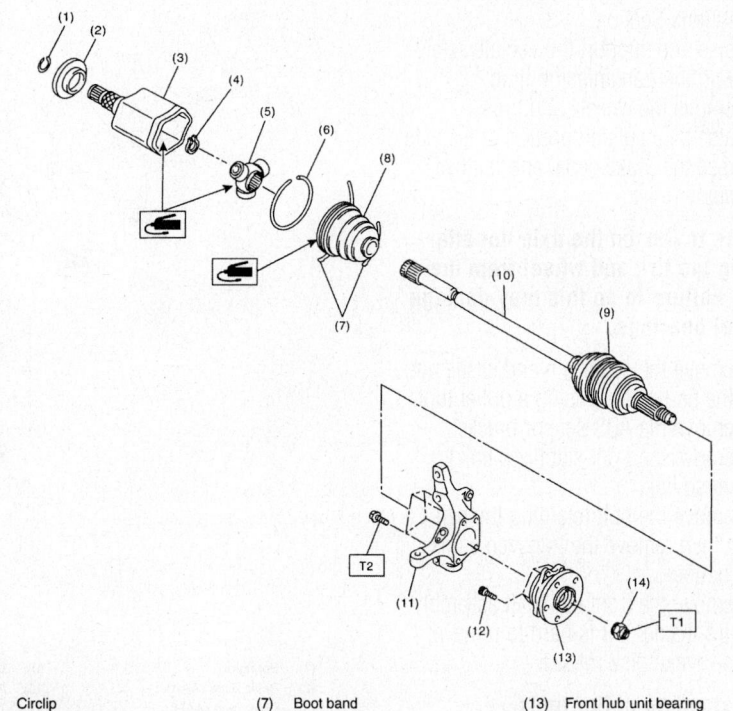

(1)	Circlip	(7)	Boot band	(13)	Front hub unit bearing
(2)	Baffle plate	(8)	Boot (PTJ)	(14)	Axle nut
(3)	Outer race (PTJ)	(9)	Boot (EBJ)		
(4)	Snap ring	(10)	EBJ shaft ASSY		
(5)	Trunnion	(11)	Housing		
(6)	Snap ring	(12)	Hub bolt		

Tightening torque: N·m (kgf-m, ft-lb)
T1: 220 (22.4, 162)
T2: 65 (6.6, 47.9)

09490_SBCR_G0089

Front halfshaft and related components—2005–2006 Legacy and Outback

➡**Be sure to loosen the axle nut after removing the tire and wheel from the vehicle. Failure to do this may damage the wheel bearings.**

6. Remove the stabilizer link from the transverse arm. Disconnect the transverse arm from the housing.

7. Remove the spring pin which secures the transmission spindle to the inner joint.

➡**Be sure to replace the spring pin with a new one, on installation.**

8. Remove the front halfshaft assembly.

➡**Use axle shaft puller tools ST92647000 and ST28099PA110, or equivalent, if it is difficult to remove the halfshaft.**

To install:

9. Installation is the reverse of the removal procedure.

10. Be sure to use new locknuts.

11. Be sure to replace the inner seal with a new one.

12. Be sure to replace the spring pin with a new one.

13. Always tighten the axle shaft nut before installing the tire and wheel assembly. Tighten the nut to 162 ft. lbs.

BAJA (2.5L DOHC)

1. Before servicing the vehicle, refer to the Precautions Section.

2. Raise and support the vehicle safely.

3. Drain the transmission fluid.

4. Remove the wheels and tires.

5. Raise the caulking portion of the axle nut. Depress the brake pedal and remove the axle nut.

➡**Be sure to loosen the axle nut after removing the tire and wheel from the vehicle. Failure to do this may damage the wheel bearings.**

6. Remove the cotter pin and castle nut. Remove the tie rod end, using a puller tool.

7. Remove the ABS sensor bracket.

8. Remove the front stabilizer link for the transverse link.

9. Remove the bolt retaining the ball joint, and then remove the transverse link form the housing.

10. Remove the front halfshaft assembly from the front axle. If it is hard to remove, remove the brake disk rotor.

➡**Use axle shaft puller tools ST92647000 and ST28099PA110, or equivalent, if it is difficult to remove the halfshaft. Do not hammer the halfshaft to remove it. Be careful not to damage the oil seal or wheel tone.**

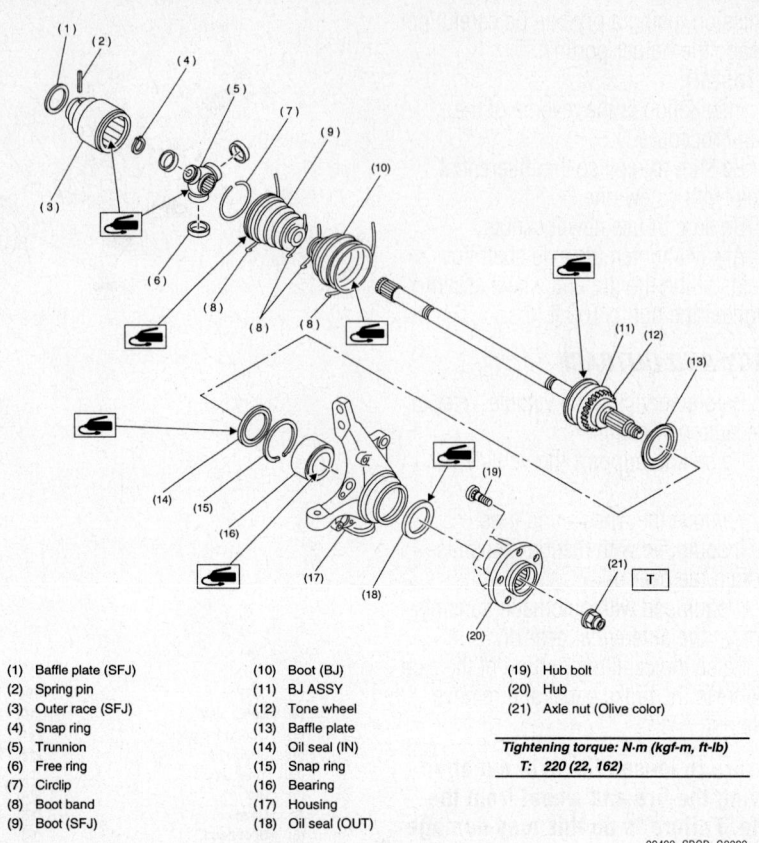

(1)	Baffle plate (SFJ)	(10)	Boot (BJ)	(19)	Hub bolt
(2)	Spring pin	(11)	BJ ASSY	(20)	Hub
(3)	Outer race (SFJ)	(12)	Tone wheel	(21)	Axle nut (Olive color)
(4)	Snap ring	(13)	Baffle plate		
(5)	Trunnion	(14)	Oil seal (IN)		
(6)	Free ring	(15)	Snap ring		
(7)	Circlip	(16)	Bearing		
(8)	Boot band	(17)	Housing		
(9)	Boot (SFJ)	(18)	Oil seal (OUT)		

Tightening torque: N·m (kgf-m, ft-lb)
T: 220 (22, 162)

09490_SBCR_G0090

Front halfshaft and related components—2005–2006 Baja 2.5L SOHC

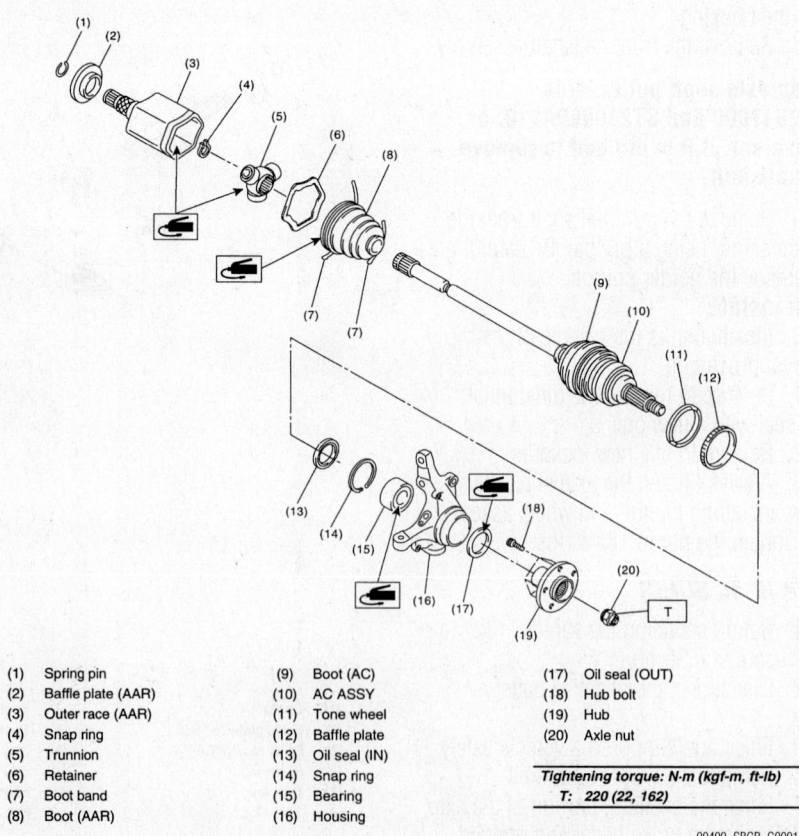

(1)	Spring pin	(9)	Boot (AC)	(17)	Oil seal (OUT)
(2)	Baffle plate (AAR)	(10)	AC ASSY	(18)	Hub bolt
(3)	Outer race (AAR)	(11)	Tone wheel	(19)	Hub
(4)	Snap ring	(12)	Baffle plate	(20)	Axle nut
(5)	Trunnion	(13)	Oil seal (IN)		
(6)	Retainer	(14)	Snap ring		
(7)	Boot band	(15)	Bearing		
(8)	Boot (AAR)	(16)	Housing		

Tightening torque: N·m (kgf-m, ft-lb)
T: 220 (22, 162)

09490_SBCR_G0091

Front halfshaft and related components—2005–2006 Baja 2.5L DOHC

11. Remove the front halfshaft from the transmission, using a pry bar. Be careful not to damage the holder portion.

To install:

12. Installation is the reverse of the removal procedure.

13. Be sure to use new locknuts.

14. Be sure to replace the inner seal with a new one.

15. Be sure to replace the spring pin with a new one.

16. Always tighten the axle shaft nut before installing the tire and wheel assembly. Tighten the nut to 162 ft. lbs.

Rear—2002–2004

EXCEPT IMPREZA

1. Before servicing the vehicle, refer to the Precautions Section.

2. Remove or disconnect the following:
   - Negative battery cable
   - Wheel
   - Axle nut, unstake the nut before attempting removal

3. Remove the rear differential on vehicles equipped with a T–type as follows:

   a. Place the gear shifter into the **Neutral** position.

   b. Release the parking brake.

   c. Remove the rear exhaust pipe and muffler.

   d. Remove the heat shield cover.

   e. Remove the driveshaft (propeller).

   f. Remove the rear differential protector, if equipped.

   g. Place a transmission jack under the differential and loose the nuts that attach the differential to the rear crossmember.

   h. Remove the driveshaft joint from the differential using a suitable shaft removal tool such as ST 28099PA100.

   i. Remove the protector nut.

   j. Remove the differential front member and support the differential assembly with the transmission jack making sure to securely attach the differential to the jack with a chain or strap.

   k. Remove the nuts attaching the differential to the crossmember.

   l. Remove the differential stud bolt from the rear crossmember bushing. You may have to carefully adjust the angle and position of the jack to facilitate stud removal.

   m. Once the stud bolt has been removed, lower the jack, making sure the rear drive shaft does not strike the lateral link bolt.

   n. Pull the driveshaft out of the differential and remove the differential.

4. Remove the rear differential on vehicles equipped with a VA–type as follows:

   a. Place the gear shifter into the **Neutral** position.

   b. Release the parking brake.

   c. Remove the rear exhaust pipe and muffler.

   d. Remove the heat shield cover.

   e. Remove the driveshaft (propeller).

   f. Place a transmission jack under the differential and loosen the nuts that attach the differential to the rear crossmember.

   g. Remove the driveshaft joint from the differential using a suitable shaft removal tool such as ST 28099PA100.

   h. Remove the protector nut.

   i. Remove the nuts with attach the differential to the front member. Support the differential assembly with the transmission jack.

   j. Remove the differential front member and securely attach the differential to the jack with a chain or strap.

   k. Remove the nuts attaching the differential to the crossmember.

   l. Remove the differential stud bolt from the rear crossmember bushing. You may have to carefully adjust the angle and position of the jack to facilitate stud removal.

   m. Once the stud bolt has been removed, lower the jack, making sure the rear drive shaft does not strike the lateral link bolt.

   n. Pull the driveshaft out of the differential and remove the differential.

5. Remove the axle nut and using a suitable puller, remove the halfshaft from the being careful not to damage the tone wheel using puller ST 926470000 and plate ST 927140000.

To install:

6. Insert the axle into the hub splines being careful not to damage the tone wheel using adapter installer ST1 922431000 and adapter ST2 927390000. Tighten the axle nut until snug.

7. Install the rear differential on vehicles equipped with a VA–type as follows:

   a. Place the differential on a transmission jack and fasten securely with a chain or band.

   b. Place a seal protector tool such as ST28099PA090 on the differential.

   c. Insert the splined shaft of the halfshaft until the spline portion is inside the oil seal.

   d. Remove the seal protector.

   e. Insert the driveshaft into the differential until it is fully seated.

   f. Adjust the transmission jack as

necessary so the stud bolt is inserted correctly into the crossmember bushing.

   g. Once the stud bolt is inserted, raise the transmission jack until the differential is level.

   h. Temporarily tighten the rear crossmember nuts.

   i. Remove the band or chain securing the differential to the transmission jack and raise the differential enough to move the jack away. Install the differential member and two inner bolts to 81 ft. lbs. (110 Nm) and the outer bolt to 48 ft. lbs. (65 Nm).

   j. Tighten the crossmember self locking nut to 51 ft. lbs. (70 Nm) and the protector nut to 47 ft. lbs. (64 Nm).

   k. Remove the jack and install the driveshaft (propeller).

   l. Install the heat shield cover.

   m. Install the rear exhaust pipe and muffler.

   n. Apply the parking brake.

   o. Place the gear shifter into the **Park** position.

8. Remove the filler plug and fill the differential to the level. Tighten the filler plug to 25 ft. lbs. (34 Nm).

9. Install the rear differential on vehicles equipped with a T–type as follows:

   a. Place the differential on a transmission jack and fasten securely with a chain or band.

   b. Place seal protector tool ST 28099PA090 over the differential side oil seal and insert the shaft until the spline portion is inside the seal. Remove the seal protector.

   c. Insert the axle shaft completely into the differential

   d. Adjust the transmission jack as necessary so the stud bolt is inserted correctly into the crossmember bushing.

   e. Once the stud bolt is inserted, raise the transmission jack until the differential is level.

   f. Temporarily tighten the rear crossmember nuts .

   g. Remove the band or chain securing the differential to the transmission jack and raise the differential enough to move the jack away. Install the differential member and tighten the T1 bolts to 48 ft. lbs. (65 Nm) and the T2 bolt to 81 ft. lbs. (110 Nm).

   h. Tighten the rear crossmember self locking nut to 51 ft. lbs. (70 Nm) and the protector nut to 47 ft. lbs. (64 Nm).

   i. Remove the jack and install the driveshaft (propeller).

   j. Install the heat shield cover.

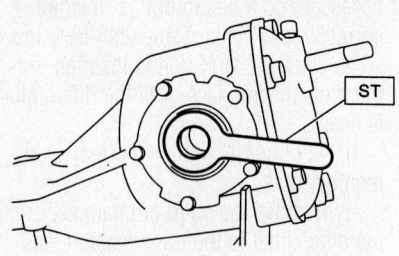

Place seal protector tool ST 28099PA090 over the rear differential side oil seal—2002–2004 Legacy, Outback and Baja

42356-SBCR-G46

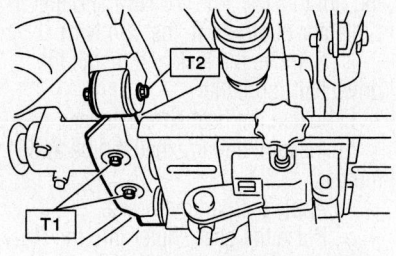

Tighten the rear differential front member and tighten the T1 and T2 bolts to the specifications—2002–2004 Legacy, Outback and Baja

42356-SBCR-G47

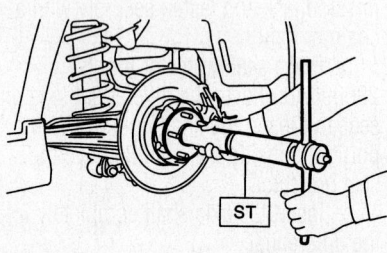

Install the rear axle into the hub splines using adapter installer ST1 922431000 and adapter ST2 927390000—2002–2004 Legacy, Outback and Baja

42356-SBCR-G48

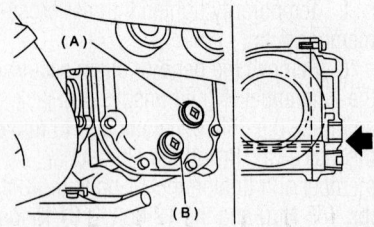

(A) Filler plug
(B) Drain plug

42356-SBCR-G49

Location of the rear differential drain and fill plugs—2002–2004 Legacy, Outback and Baja

k. Install the rear exhaust pipe and muffler.

l. Apply the parking brake.

m. Place the gear shifter into the **Park** position.

10. Remove the filler plug and fill the differential to the level. Tighten the filler plug to 36 ft. lbs. (49 Nm).

11. Install or connect the following:
- Axle nut and tighten to 174 ft. lbs. (235 Nm). Stake the nut.
- Wheels
- Negative battery cable

12. Check all fluids and road test the vehicle.

IMPREZA

1. Before servicing the vehicle, refer to the Precautions Section.

2. Remove or disconnect the following:

- Negative battery cable
- Axle nut
- Wheels
- Halfshaft from the differential using remover tool ST 28099PA100

➡The side spline shaft circlip comes out with the shaft.

✳✳ CAUTION

Be careful not to damage the side bearing retainer use the bolt at the 5 O'clock position as a supporting point when using the removal tool.

- Axle shaft from the hub using puller ST 9266470000 and puller plate 927140000

To install:

3. Install the axle shaft into the hub. Use installer ST 922431000 and adapter ST 927390000 to pull the drive shaft into place and side tighten the axle nut.

4. Place a seal protector tool such as ST28099PA090 on the differential.

5. Insert the splined shaft of the half-shaft until the spline portion is inside the oil seal.

6. Remove the seal protector.

7. Insert the driveshaft into the differential until it is fully seated.

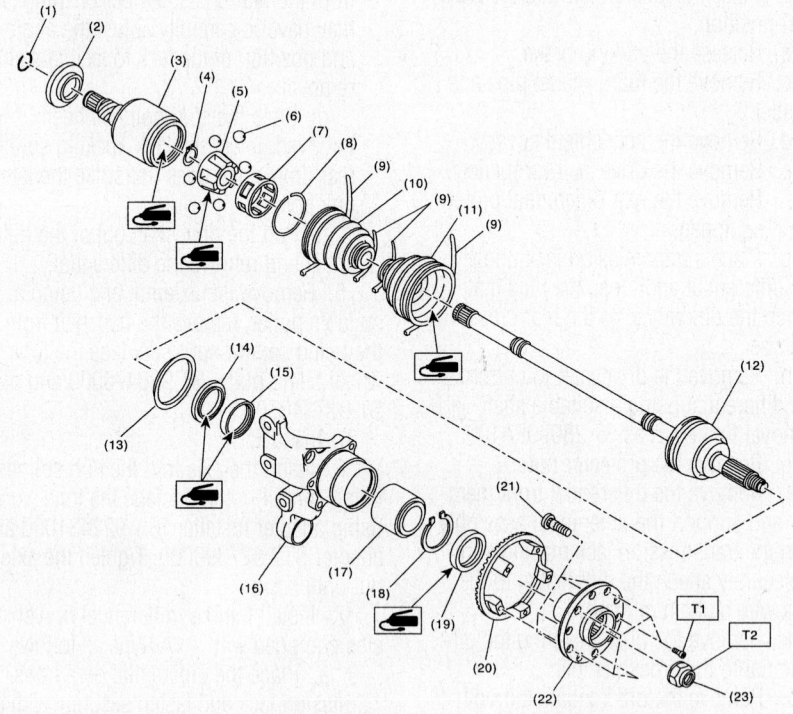

(1) Circlip	(9) Boot band	(19) Oil seal (OUT)
(2) Baffle plate (DOJ)	(10) Boot (DOJ)	(20) Tone wheel
(3) Outer race DOJ: Except STI model	(11) Boot	(21) Hub bolt
Outer race EDJ: STI model	(12) EBJ ASSY	(22) Hub
(4) Snap ring	(13) Baffle plate	(23) Axle nut
(5) Inner race	(14) Oil seal (IN. No. 2)	
(6) Ball	(15) Oil seal (IN.)	
(7) Cage	(16) Housing	
(8) Snap ring	(17) Bearing	
	(18) Snap ring	

Tightening torque: N·m (kgf-m, ft-lb)	
T1:	13 (1.3, 9.4)
T2:	190 (19.4, 140)

09490_SBCR_G0092

Rear halfshaft and related components—2005–2006 Impreza and WRX

8. Insert the splined shaft of the half-shaft until the spline portion is inside the oil seal.

9. Torque the new axle nut to 137 ft. lbs. (186 Nm) on 2002 vehicles or 174 ft. lbs. (235 Nm) on 2002–2003 vehicles and stake the nut.
- Wheel
- Negative battery cable

Rear—2005–2006

IMPREZA AND WRX

1. Before servicing the vehicle, refer to the Precautions Section.
2. Disconnect the negative battery cable.
3. Raise and support the vehicle safely.
4. Remove the wheels and tires.
5. Unlock the axle nut. Depress the parking brake. Remove the axle nut.

➡**Remove the axle nut with vehicle weight NOT applied to the axle.**

6. Disconnect the stabilizer link. Remove the bolt that retains the trailing link to the housing.
7. Remove the bolts which secure the front lateral link and rear lateral link to the housing.
8. Remove the ABS wheel speed sensor from the back plate.
9. Remove the halfshaft from the rear axle. If it is hard to remove, remove the brake disk rotor.

➡**Use axle shaft puller tools ST92647000 and ST28099PA110, or equivalent, if it is difficult to remove the halfshaft. Do not hammer the halfshaft to remove it. Be careful not to damage the oil seal or wheel tone.**

10. Remove the halfshaft from the differential using tool ST28099PA100.

➡**Using the tool will prevent damage to the side bearing retainer.**

To install:
11. Installation is the reverse of the removal procedure.
12. Tighten the axle nut to 140 ft. lbs. with the parking brake applied.

LEGACY AND OUTBACK

1. Before servicing the vehicle, refer to the Precautions Section.
2. Disconnect the negative battery cable.
3. Raise and support the vehicle safely.
4. Remove the wheels and tires.
5. Unlock the axle nut. Depress the parking brake. Remove the axle nut.

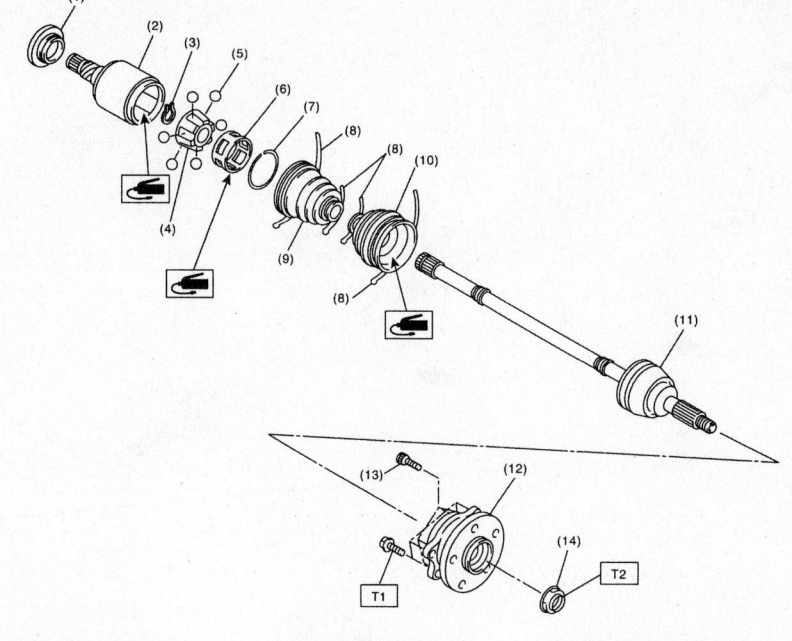

(1)	Baffle plate (DOJ)	(8)	Boot band	(13)	Hub bolt
(2)	Outer race (DOJ)	(9)	Boot (DOJ)	(14)	Axle nut (olive color)
(3)	Snap ring	(10)	Boot (BJ)		
(4)	Inner race	(11)	BJ shaft ASSY (2.5 i AT model)		
(5)	Ball		EBJ shaft ASSY (Except for 2.5 i AT model)		
(6)	Cage				
(7)	Snap ring	(12)	Rear hub unit bearing		

Tightening torque: N·m (kgf-m, ft-lb)	
T1:	65 (6.6, 47.9)
T2:	240 (24.5, 177)

09490_SBCR_G0093

Rear halfshaft and related components—2005–2006 Legacy and Outback

➡**Be sure to loosen the axle nut after removing the tire and wheel from the vehicle. Failure to do this may damage the wheel bearings.**

6. Remove the rear differential assembly.
7. Remove the axle nut and rear half-shaft.

➡**Use axle shaft puller tools ST92647000 and ST28099PA110, or equivalent, if it is difficult to remove the halfshaft. Do not hammer the halfshaft to remove it. Be careful not to damage the oil seal or magnetic encoder.**

To install:
8. Installation is the reverse of the removal procedure.
9. Tighten the axle nut to 177 ft. lbs. with the parking brake applied.

➡**Install the tire and wheel assembly after installation of the axle nut. Failure to follow this order, may cause damage to the wheel bearings.**

BAJA

1. Before servicing the vehicle, refer to the Precautions Section.
2. Disconnect the negative battery cable.

3. Raise and support the vehicle safely.
4. Remove the wheels and tires.
5. Unlock the axle nut. Depress the parking brake. Remove the axle nut.

➡**Be sure to loosen and retighten the axle nut after removing the tire and wheel assembly from the vehicle. Failure to follow this rule may damage the wheel bearings.**

6. Remove the rear differential assembly.
7. Remove the axle nut and rear half-shaft.

➡**Use axle shaft puller tools ST92647000 and ST28099PA110, or equivalent, if it is difficult to remove the halfshaft. Do not hammer the halfshaft to remove it. Be careful not to damage the oil seal or magnetic encoder.**

To install:
8. Installation is the reverse of the removal procedure.
9. Tighten the axle nut to 177 ft. lbs. with the parking brake applied.

➡**Install the tire and wheel assembly after installation of the axle nut. Failure to follow this order, may cause damage to the wheel bearings.**

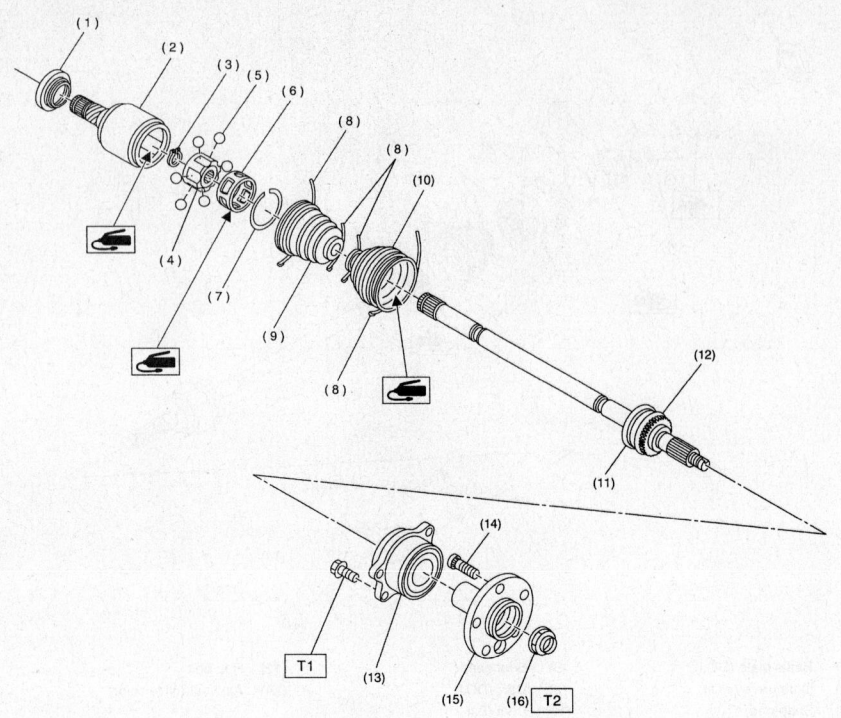

(1) Baffle plate (DOJ)	(8) Boot band	(15) Hub
(2) Outer race (DOJ)	(9) Boot (DOJ)	(16) Axle nut (Olive color)
(3) Snap ring	(10) Boot (BJ)	
(4) Inner race	(11) BJ ASSY	
(5) Ball	(12) Tone wheel	
(6) Cage	(13) Hub unit bearing	
(7) Circlip	(14) Hub bolt	

Tightening torque: N·m (kgf-m, ft-lb)
T1: 66 (6.7, 48.5)
T2: 240 (24, 177)

09490_SBCR_G0094

Rear halfshaft and related components—2005–2006 Baja

CV-Joints

OVERHAUL

Front

1. Before servicing the vehicle, refer to the Precautions Section.
2. Place alignment marks on the shaft and outer race.

3. Remove the inner boot band and boot.
4. Remove the circlip from the inner joint outer race using a suitable prytool.
5. Outer race from the shaft assembly and wipe off the grease.
6. Place alignment marks on the free ring and trunnion as shown in the illustration.
7. Remove the free ring from the trunnion.
8. Place alignment marks on the trunnion and shaft as shown in the illustration.
9. Remove the snapring and trunnion.
10. Place the shaft in a vise between wooden blocks.

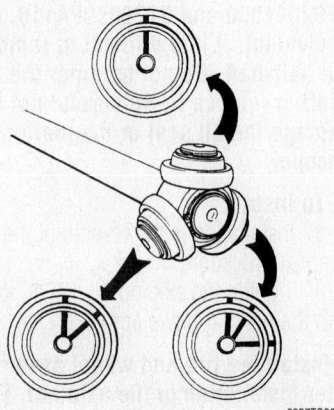

9357TG38

Place alignment marks on the free ring and trunnion as shown—Front halfshaft

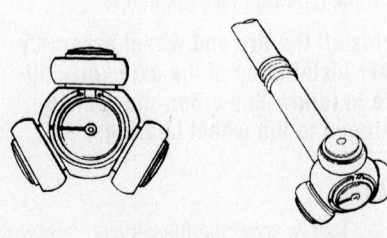

9357TG39

Place alignment marks on the trunnion and shaft as shown—Front halfshaft

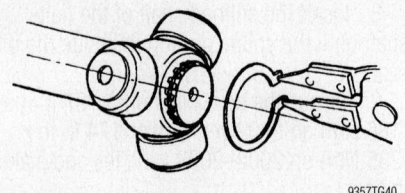

9357TG40

Remove the snapring and trunnion—Front halfshaft

11. Using a suitable prytool, raise the outer boot band claws.
12. Cut and remove the boot.
13. Only the boot can be replaced, the joint is not serviceable and must be replaced if damaged.

To install:
14. Place the half shaft in a vise.
15. Place the outer boot and small band on the shaft.
16. Apply 2.12–2.47 oz. (60–70g) of supplied grease to the joint.
17. Apply 0.71–1.06 oz. (20–30g) of supplied grease to the whole inner surface of the boot, and apply some grease to the shaft.
18. Install the boot to the joint groove, and attach the large boot band as shown.
19. Install the boot to the shaft groove, and attach the small boot band as shown.
20. Using boot band plier tool 28099A000 to tighten the large band to 116 ft. lbs. (157 Nm) and the small band to 98 ft. lbs. (133 Nm).
21. Place the inner boot on the center of the shaft.
22. Align the alignment marks from earlier and install the trunnion and snapring. Make sure the snapring is fully engaged.
23. Apply 3.53–3.88 oz. (100–110g) of supplied grease to the joint outer race.
24. Apply a coat of supplied grease to the free ring and trunnion.

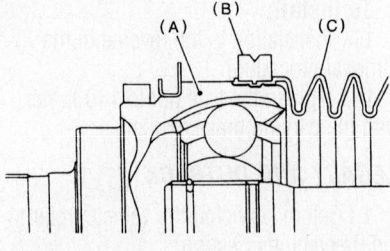

(A) EBJ
(B) Lorge boot band
(C) Boot

9357TG41

Position the boot to the joint groove, and attach the large boot band as shown— Front halfshaft

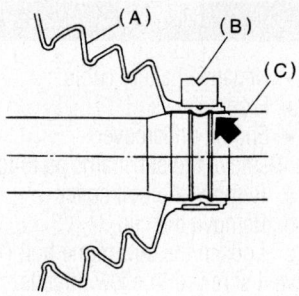

(A) Boot
(B) Small boot band
(C) Shaft

9357TG42

Position the boot to the shaft groove, and attach the small boot band as shown— Front halfshaft

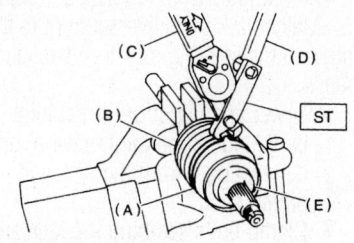

(A) Large boot band
(B) Boot
(C) Torque wrench
(D) Socket flex handle
(E) BJ

9357TG43

Using boot band plier tool 28099A000 tighten the boot bands—Front halfshaft

25. Align the marks on the free ring and trunnion and install the free ring.

26. Align the marks on the shaft and outer race and install the outer race.

27. Pull on the shaft lightly to ensure the circlip is completely engaged.

28. Apply an even coat 1.06–1.41 oz. (30–40g) of the supplied grease to the entire inner surface of the boot.

29. Install the boot and band.

30. Once the band is properly tightened, cut off any excess to leave only 0.39 inch (10mm) and bend it over.

31. Install the shaft.

Rear

The Double Offset Joint (DOJ), is the only part of the assembly that can be replaced, if any of the other components are defective then the shaft should be replaced.

1. Before servicing the vehicle, refer to the Precautions Section.

2. Straighten the bent claw of the large clamp at the Double Offset Joint (DOJ) end of the boot.

3. Loosen the band using pliers being careful not to damage the bolt.

4. Remove the small boot band from the DOJ using the same technique.

5. Remove the boot from the large end of the DOJ outer race.

6. Remove the round circlip using a suitable prytool from the neck of the joint outer race.

7. Remove the joint outer race.

❋❋ CAUTION

The grease used is for CV–Joints, do not replace with another type of grease

8. Clean the grease and remove the balls. Be careful not to loose any of the 6 balls.

9. Turn the cage by a half pitch to the track groove of the inner race and shift the cage.

10. Remove the snap ring that secures the inner race to the shaft and remove the inner race.

11. Take the cage from the shaft and remove the boot.

12. The other boots may be removed in the same manner as the DOJ boot.

13. Wrap the shaft splines with tape to prevent damage.

To install:

14. The following grease must be used during assembly:

a. BJ side (Non- turbocharged engine): Molylex No. 2 #723223010.

b. EBJ side (Turbocharged engine): NTG2218 # 28093AA000.

15. DOJ side: VU–3A702 (Yellow) # 23223GA050.

16. Install the BJ or EBJ boots and fill it with 2.12–2.47 oz. (60–70g) of grease.

17. Place the DOJ boot on the center of the shaft.

18. Insert the DOJ cage onto the shaft. Make sure to insert the cage with the cut–out portion facing shaft end.

19. Install the inner race on the shaft and fasten with the snap ring. Make sure the snap ring is firmly engaged.

20. Install the cage (previously fitted) to the inner race on the shaft. Fit the cage with the protruded part aligned with the track on the inner race and then turn a half pitch.

21. Fill the DOJ inner race with 2.82–3.17 oz. (80–90g) of grease.

22. Apply a coat of grease to the to the cage pocket and 6 balls.

23. Insert the 6 balls.

24. Align the outer race track and ball positions and place where the shaft, inner race, cage and balls were located prior to removal and then install the outer race.

25. Install the circlip into the groove on the outer race.

➡**Make sure the balls, cage and inner race are fully seated. make sure not to place the matched position of the circlip in the ball groove of the outer race. Pull the shaft lightly to make sure the circlip is fully engaged.**

26. Apply an even coat 0.71–1.06 oz. (20–30g) of grease to the entire inner surface of the boot and to the shaft.

27. Make sure the boot is free from any dirt or foreign materials prior to installation.

28. Place the outer race of the boot at the center of its travel.

29. Put a band through the boot clip and wind twice in alignment with the groove on the boot.

30. Pinch the end of the band using tool ST 9250910000 and tighten securely until it cannot be moved by hand. Make sure there is appropriate air inside the boot.

31. Tap on the clip with a punch to lock it making sure not to damage the damaged while tapping. Cut any excess off the band leaving 0.39 inch (10mm) and bend the remaining portion over the clip. Make sure the end of the band is close to the clip.

32. Install the remaining boot clamps in the same manner.

STEERING

Air Bag

✷✷ CAUTION

These vehicles are equipped with an air bag system. The system must be disabled before performing service on or around system components, steering column, instrument panel components, wiring and sensors. Failure to follow safety and disabling procedures could result in accidental air bag deployment, possible personal injury and unnecessary system repairs.

PRECAUTIONS

Several precautions must be observed when handling the inflator module to avoid accidental deployment and possible personal injury.

• Never carry the inflator module by the wires or connector on the underside of the module.

• When carrying a live inflator module, hold it securely with both hands, and ensure that the bag and trim cover are pointed away.

• Place the inflator module on a bench or other surface with the bag and trim cover facing up.

• With the inflator module on the bench, never place anything on or close to the module that may be thrown in the event of an accidental deployment.

DISARMING

1. Before servicing the vehicle, refer to the Precautions Section.
2. Be sure to position the front wheels in the straight ahead position.
3. Disconnect the negative battery cable. Tape the battery cable for added protection.
4. Wait more than 20 seconds before starting work.

Power Steering Gear

REMOVAL & INSTALLATION

2002–2004

IMPREZA

1. Before servicing the vehicle, refer to the Precautions Section.
2. Remove or disconnect the following:
   • Negative battery cable

• Front wheels
• Under cover
• Bolt, clip on the sub-frame
• Sub-frame bolts. Leave bolt (1) connected by a few threads and remove the bolts in the sequence illustrated. Once the other bolts are removed, remove bolt (1) and the sub-frame. Refer to the illustration for bolt location.
• Front Y-pipe
• Tie rod end cotter pin and nut
• Tie rod ends from the steering knuckle using a puller
• Jack-up plate and front sway bar
• Fluid lines from the rack and pinion

3. Matchmark the universal joint to the serration in the steering rack for installation reference.

   • Lower and upper universal joint bolts and lift the joint upward, disconnecting it from the rack and pinion shaft
   • Clamp bolts securing the rack and pinion to the crossmember
   • Rack and pinion

To install:

4. Install the rack and pinion. Tighten the clamp bolts to 43 ft. lbs. (59 Nm).

   • Fluid pipes and tighten to 9 ft. lbs. (13 Nm)

5. Align the steering rack to the universal joint. Push the long yoke of the joint all the way into the serrated position of the steering shaft, setting the bolt hole in the cut-out. Pull the short yoke all the way out of the serrated portion of the rack and pinion, setting the bolt hole in the cut-out. Insert the bolt through the short yoke. Pull the yoke and ensure the bolt is properly engaged in the cut-out. Fasten the short yoke side with the spring washer and bolt, then fasten the yoke side. Tighten the bolts to 17 ft. lbs. (24 Nm).

6. Install or connect the following:
   • Tie rod ends to the steering knuckle
   • Sway bar and jack-up plate
   • Y-pipe with new gaskets and nuts
   • Sub-frame. Tighten the T1 bolts to 25 ft. lbs. (34 Nm), the T2 bolts to 41 ft. lbs. (55 Nm) and the T3 bolts to 52 ft. lbs. (71 Nm).
   • Wheels

7. Fill and bleed the steering system.

WRX

1. Before servicing the vehicle, refer to the Precautions Section.
2. Remove or disconnect the following:

• Negative battery cable
• Front wheels
• Engine undercover

3. Remove the sub-frame as follows:
   a. Remove the bolt cover.
   b. Remove the clip.
   c. Loosen the sub-frame bolt (1) and leave it screwed in a few threads. Remove the remaining bolts in the following order: 2, 3, 4, 5 and 6. See illustration for bolt location.

4. Remove or disconnect the following:
   • Front exhaust pipe
   • Tie rod end cotter pin and nut
   • Tie rod ends from the steering knuckle using a puller
   • Jack-up plate and front sway bar
   • Fluid lines from the rack and pinion

5. Matchmark the universal joint to the serration in the steering rack for installation reference.

   • Lower and upper universal joint bolts and lift the joint upward, disconnecting it from the rack and pinion shaft
   • Clamp bolts securing the rack and pinion to the crossmember
   • Rack and pinion

To install:

6. Install the rack and pinion. Tighten the clamp bolts to 43 ft. lbs. (59 Nm).

7. Align the steering rack to the universal joint. Push the long yoke of the joint all the way into the serrated position of the steering shaft, setting the bolt hole in the cut-out. Pull the short yoke all the way out of the serrated portion of the rack and pinion, setting the bolt hole in the cut-out. Insert the bolt through the short yoke. Pull the yoke and ensure the bolt is properly engaged in the cut-out. Fasten the short yoke side with the spring washer and bolt, then fasten the yoke side. Tighten the bolts to 19 ft. lbs. (27 Nm).

8. Install or connect the following:
   • Tie rod ends to the steering knuckle
   • Sway bar and jack-up plate

9. Install the sub-frame and bolts. Tighten the bolts as follows, referring to the illustration for location:
   a. T1: 25 ft. lbs. (34 Nm).
   b. T2: 41 ft. lbs. (55 Nm).
   c. T3: 52 ft. lbs. (71 Nm).
   • Bolt and clip
   • Exhaust pipe with new gaskets and nuts
   • Engine undercover
   • Wheels

10. Fill and bleed the steering system.

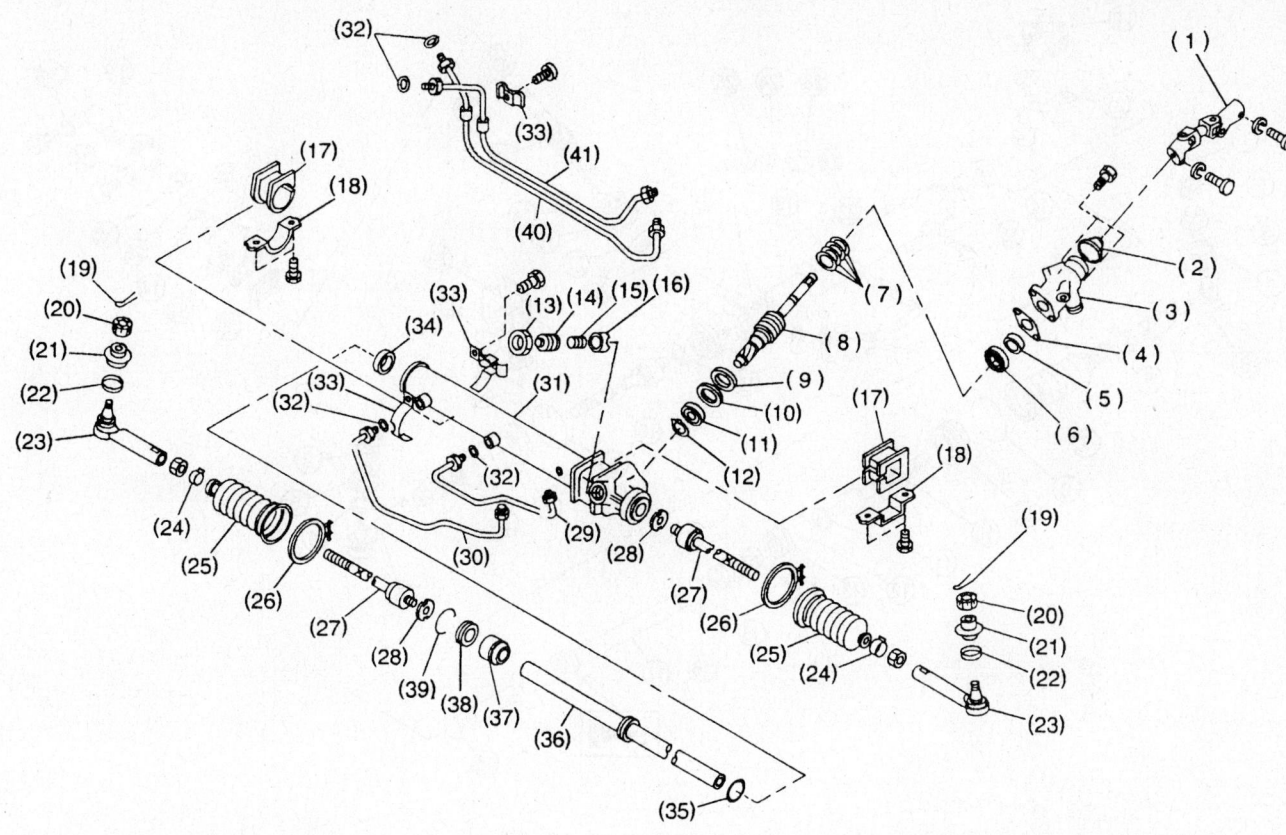

(1)	Universal joint	(15)	Spring	(29)	Pipe B
(2)	Dust cover	(16)	Sleeve	(30)	Pipe A
(3)	Valve housing	(17)	Adapter	(31)	Steering body
(4)	Gasket	(18)	Clamp	(32)	O-ring
(5)	Oil seal	(19)	Cotter pin	(33)	Clamp
(6)	Special bearing	(20)	Castle nut	(34)	Oil seal
(7)	Seal ring	(21)	Dust cover	(35)	Piston ring
(8)	Pinion and valve ASSY	(22)	Clip	(36)	Rack
(9)	Oil seal	(23)	Tie-rod end	(37)	Rack bushing
(10)	Back-up washer	(24)	Clip	(38)	Rack stopper
(11)	Ball bearing	(25)	Boot	(39)	Circlip
(12)	Snap ring	(26)	Band	(40)	Pipe E
(13)	Lock nut	(27)	Tie-rod	(41)	Pipe F
(14)	Adjusting screw	(28)	Lock washer		

7923TG64

Typical steering rack and related components—Impreza

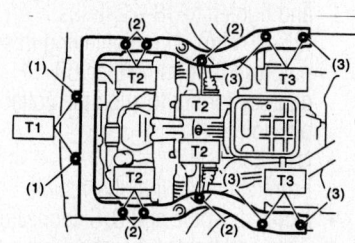

(1) M8 bolt
(2) M12 bolt
(3) M10 bolt

42356-SBCR-G61

**Sub frame bolt tightening sequence—
2002–2004 Impreza and WRX**

LEGACY, OUTBACK AND BAJA

1. Before servicing the vehicle, refer to the Precautions Section.
2. Remove or disconnect the following:
   • Negative battery cable
   • Air intake duct
   • Front axle nut, loosen only at this time
   • Front tire and wheel assemblies
   • Electrical connector from the Oxygen Sensor (O2S)
   • Front exhaust pipe assembly
   • Tie rod end cotter pin and loosen the castle nut
   • Tie rod ends from the steering knuckle arm using a ball joint puller
   • Jack up plate and the front stabilizer bar
3. From the power steering rack, remove the center pressure pipe, connect a vinyl hose to the pipe and joint, then turn the steering wheel to discharge the fluid into a container.

➡ **When discharging the power steering fluid (line A and B), turn the steering wheel fully, left and right. Be sure to disconnect the other pipe and drain the fluid in the same manner.**

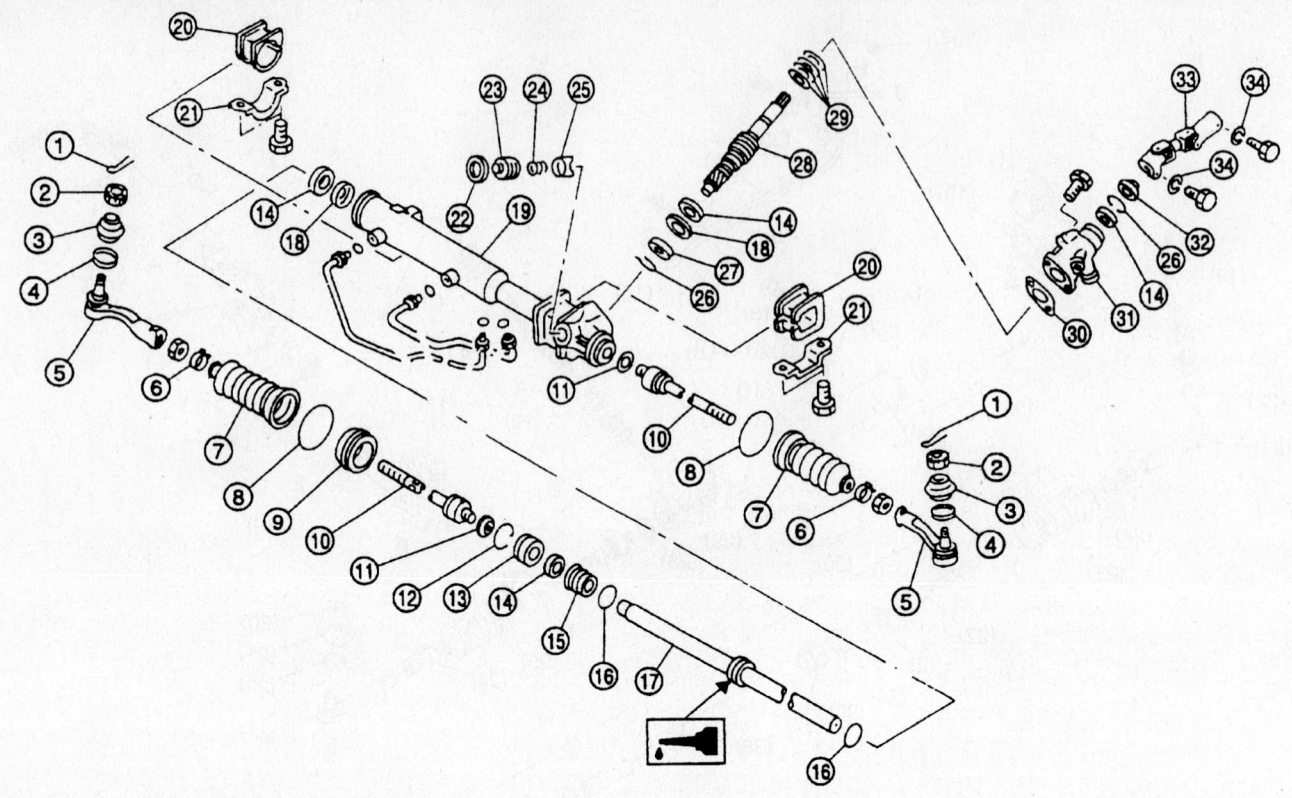

①	Cotter pin	⑬	Rack stopper	㉔	Spring
②	Castle nut	⑭	Oil seal	㉕	Sleeve
③	Dust cover	⑮	Rack bushing	㉖	C-ring
④	Clip	⑯	O-ring	㉗	Ball bearing
⑤	Tie-rod end	⑰	Rack	㉘	Valve
⑥	Clip	⑱	Back-up washer	㉙	Seal ring
⑦	Boot	⑲	Rack housing	㉚	Packing
⑧	Clip	⑳	Adapter	㉛	Valve housing
⑨	Spacer	㉑	Clamp	㉜	Dust seal
⑩	Tie-rod	㉒	Lock nut	㉝	Universal joint
⑪	Lock washer	㉓	Adjusting screw	㉞	Spring washer
⑫	Circlip				

7923TG63

Typical steering rack and related components—Legacy

4. From the control valve of the gear-box assembly, remove the power steering **C** and **D** pressure pipes. Remove pipe **D** first and pipe **C** second.

5. If not disconnected when draining the fluid from the control valve of the gear-box assembly, remove the power steering **A** and **B** pressure pipes. Remove pipe **A** first and pipe **B** second.

6. Remove or disconnect the following:
• Universal joint assembly. Match-mark the assembly before removal.
• Power steering gearbox-to-cross-member assembly bolts
• Gearbox assembly from the vehicle

To install:
7. Install or connect the following:
• Power steering rack and tighten the rack to crossmember bolts to 35–52 ft. lbs. (47–70 Nm)
• Universal joint assembly making sure to align the matchmarks
• Power steering pressure pipes and tighten to 7–12 ft. lbs. (10–16 Nm)
• Universal joint assembly-to-power steering gearbox bolts and tighten to 16–19 ft. lbs. (22–24 Nm) and the universal joint assembly-to-steering shaft bolts 16–19 ft. lbs. (22–24 Nm)

• Tie rod end to steering knuckle nut and tighten to 18–22 ft. lbs. (25–29 Nm). After tightening this nut, turn it no more than 60 degrees further to align the cotter pin hole.
• New cotter pin
• Front stabilizer into the vehicle
• Exhaust Y-pipe and O2S connector
• Tires and tighten the wheel lug nuts to specification

8. Partially lower the vehicle, then refill and bleed the power steering system.

9. Check for fluid leaks and the fluid level, then install the jack up plate.

10. Check and adjust the toe-in and the steering angle.

2005–2006

IMPREZA AND WRX

1. Before servicing the vehicle, refer to the Precautions Section.
2. Disconnect the negative battery cable.
3. Loosen the front wheel nut.
4. Raise and support the vehicle safely.
5. Remove the tires and wheels.
6. Remove the under cover.
7. Remove the sub frame. Leave bolt (1) connected by a few threads and remove the bolts in the sequence illustrated. Once the other bolts are removed, remove bolt (1) and the sub frame.
8. Remove the front exhaust pipe.
9. Remove the cotter pin and castle nut. Using a puller, remove the tie rod end from the knuckle arm.
10. Remove the jack up plate. Remove the front stabilizer.
11. Disconnect the power steering fluid pipe at the center of the gearbox and attach a vinyl hose. Discharge the fluid into a suitable container by turning the steering wheel fully clockwise and counterclockwise. Disconnect the other fluid line, and repeat the discharge procedure.
12. Remove the steering wheel. Make a matchmark on the universal joint. Remove the universal joint bolts and remove the joint from the vehicle.
13. Disconnect the fluid lines from the steering gear, pressure hose first.
14. Remove the steering gear retaining bolts and clamps securing the steering gear to the crossmember. Remove the steering gear from the vehicle.

To install:

15. Insert the steering gear into the crossmember. Be careful not to damage the gearbox boot.
16. Tighten the steering gear to the crossmember bracket to 44.3 ft. lbs.

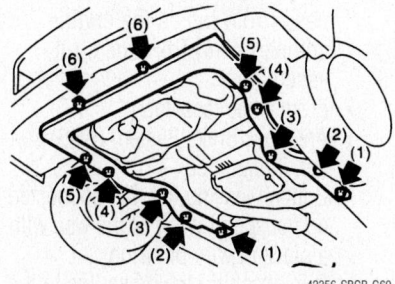

Sub frame bolt removal sequence— Impreza and WRX

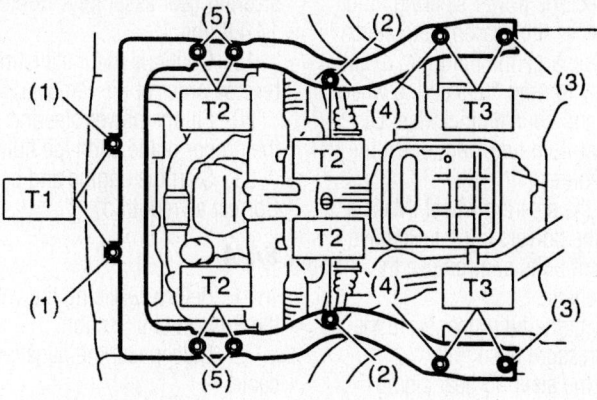

(1)	M8 bolt	(4)	M10 bolt
(2)	M12 bolt	(5)	M12 bolt
(3)	M10 bolt		

Tightening torque:
T1: 34 N·m (3.5 kgf-m, 25 ft-lb)
T2: 55 N·m (5.6 kgf-m, 41 ft-lb)
T3: 70 N·m (7.1 kgf-m, 52 ft-lb)

09490_SBCR_G0095

Sub frame bolt tightening sequence—2005–2006 Impreza and WRX

17. Connect the fluid lines.
18. Align the cutout at the serrated section of the column shaft and yoke. Insert the universal joint into the column shaft.
19. Align the mating marks and insert the universal joint to serrated section of the steering gear assembly. Tighten the bolt to 17.4 ft. lbs.
20. Continue the installation in the reverse order of the removal procedure.
21. When installing the sub frame be sure to torque the bolts to specification and in the proper sequence.
22. Fill the power steering system with the proper grade and type fluid.
23. Start the engine and check for leaks. Correct as required.

LEGACY AND OUTBACK

1. Before servicing the vehicle, refer to the Precautions Section.
2. Disconnect the negative battery cable.
3. Loosen the front wheel nut.
4. Raise and support the vehicle safely.
5. Remove the tires and wheels.
6. Remove the under cover.
7. Remove the front exhaust pipe.
8. Remove the cotter pin and castle nut. Using a puller, remove the tie rod end from the knuckle arm.
9. Remove the front crossmember support plate. Remove the jack up plate. Remove the front stabilizer.

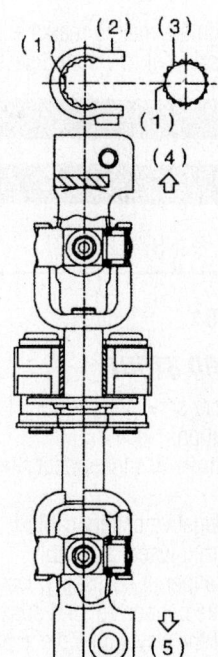

(1)	Cutout portion
(2)	Yoke
(3)	Column shaft
(4)	Column shaft side
(5)	Gearbox side

09490_SBCR_G0013
Steering column joint alignment— 2005–2006

10. Disconnect the power steering fluid pipe at the center of the gearbox and attach a vinyl hose. Discharge the fluid into a suitable container by turning the steering wheel fully clockwise and counterclockwise. Disconnect the other fluid line, and repeat the discharge procedure.

11. Remove the steering wheel. Make a matchmark on the universal joint. Remove the universal joint bolts and remove the joint from the vehicle.

12. Disconnect the fluid lines from the steering gear, pressure hose first.

13. Remove the steering gear clamp bolts securing the steering gear to the crossmember and remove the clamp. Remove the bolts that retain the steering gear bracket. Remove the steering gear and bracket from the vehicle.

To install:

14. Insert the steering gear into the crossmember. Be careful not to damage the gearbox boot.

15. Tighten the steering gear to the crossmember bracket to 44.1 ft. lbs.

16. Connect the fluid lines.

17. Align the cutout at the serrated section of the column shaft and yoke. Insert the universal joint into the column shaft.

18. Align the mating marks and insert the universal joint to serrated section of the steering gear assembly. Tighten the bolt to 17.4 ft. lbs.

19. Continue the installation in the reverse order of the removal procedure.

20. Fill the power steering system with the proper grade and type fluid.

21. Start the engine and check for leaks. Correct as required.

BAJA

1. Before servicing the vehicle, refer to the Precautions Section.

2. Disconnect the negative battery cable.

3. Loosen the front wheel nut.

4. Raise and support the vehicle safely.

5. Remove the tires and wheels.

6. Remove the front exhaust pipe assembly.

7. Remove the cotter pin and castle nut. Using a puller, remove the tie rod end from the knuckle arm.

8. Remove the jack up plate. Remove the front stabilizer.

9. Disconnect the power steering fluid pipe at the center of the gearbox and attach a vinyl hose. Discharge the fluid into a suitable container by turning the steering wheel fully clockwise and counterclockwise. Disconnect the other fluid line, and repeat the discharge procedure.

10. Remove the steering wheel. Make a matchmark on the universal joint. Remove the universal joint bolts and remove the joint from the vehicle.

11. Disconnect the fluid lines from the steering gear, return hose first.

12. Remove the steering gear clamp bolts securing the steering gear to the crossmember. Remove the steering gear from the vehicle.

To install:

13. Insert the steering gear into the crossmember. Be careful not to damage the gearbox boot.

14. Tighten the steering gear to the crossmember bracket to 43.0 ft. lbs.

15. Connect the fluid lines.

16. Align the cutout at the serrated section of the column shaft and yoke. Insert the universal joint into the column shaft.

17. Align the mating marks and insert the universal joint to serrated section of the steering gear assembly. Tighten the bolt to 17.4 ft. lbs.

18. Continue the installation in the reverse order of the removal procedure.

19. Fill the power steering system with the proper grade and type fluid.

20. Start the engine and check for leaks. Correct as required.

FRONT SUSPENSION

Strut

REMOVAL & INSTALLATION

2002–2004

STANDARD STRUT

1. Before servicing the vehicle, refer to the Precautions Section.

2. Remove or disconnect the following:

- Negative battery cable
- Front wheel assembly
- Caliper, if necessary, leaving the line connected, and suspend it out of the way with a piece of wire or string
- Clip or bolt attaching the brake line to the strut housing
- Bolt securing the Anti-lock Brake System (ABS) sensor harness, if equipped

➡**Scribe a matchmark on the camber adjusting bolt which secures the strut to the housing.**

- 2 bolts and nuts securing the strut to the steering knuckle. Notice that the shaft of the top bolt is not round. This bolt is used for camber adjustment, and must always be installed in the top hole.
- 3 nuts securing the strut to the body in the engine compartment
- Strut and coil spring assembly from the vehicle

To install:

3. Install or connect the following:

- Strut assembly into the vehicle
- Upper strut retainer nuts and tighten the nuts to 15 ft. lbs. (20 Nm)
- Lower strut nuts and bolts. Be sure the alignment adjustment bolt is installed in the top mounting hole. Tighten the nuts to 112 ft. lbs. (152 Nm) for Impreza and Outback, or to 130 ft. lbs. (177 Nm) for Legacy and WRX.
- ABS sensor harness, if equipped and tighten the bolt to 24 ft. lbs. (32 Nm)

- Caliper, if removed
- Brake line to the strut and install the clip or bolt
- Front wheel
- Negative battery cable

4. Check the front end alignment and adjust as necessary.

PNEUMATIC STRUT

1. Before servicing the vehicle, refer to the Precautions Section.

2. Remove or disconnect the following:

- Negative battery cable
- Air line and height sensor harness (from inside the engine compartment) from the strut assembly
- Front wheel assembly
- Anti-lock Brake System (ABS) sensor, if equipped
- Caliper, leaving the line connected, and suspend it out of the way with a piece of wire or string
- Clip attaching the brake line to the strut housing

3. Matchmark the camber adjustment

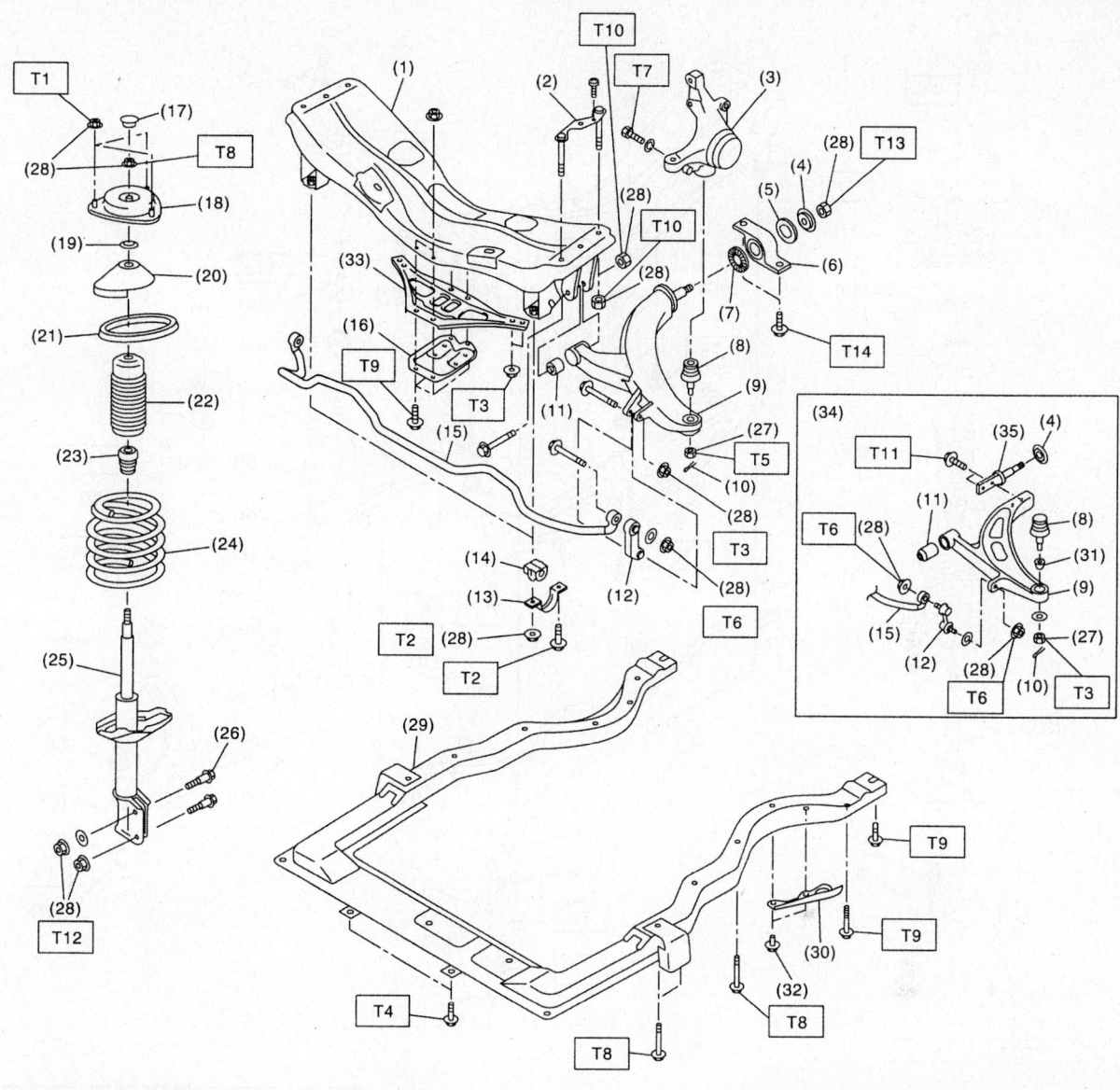

(1)	Front crossmember	(19)	Spacer
(2)	Bolt ASSY	(20)	Upper spring seat
(3)	Housing	(21)	Rubber seat
(4)	Washer	(22)	Dust cover
(5)	Stopper rubber (Rear)	(23)	Helper
(6)	Rear bushing	(24)	Coil spring
(7)	Stopper rubber (Front)	(25)	Damper strut
(8)	Ball joint	(26)	Adjusting bolt
(9)	Transverse link	(27)	Castle nut
(10)	Cotter pin	(28)	Self-locking nut
(11)	Front bushing	(29)	Sub frame
(12)	Stabilizer link	(30)	Cover
(13)	Clamp	(31)	Boss
(14)	Bushing	(32)	Clip
(15)	Stabilizer	(33)	Jack-up plate (Turbo model and STI model)
(16)	Jack-up plate (Non-turbo model)	(34)	Sedan turbo model and STI model
(17)	Dust seal	(35)	Fitting
(18)	Strut mount		

Tightening torque: N·m (kgf-m, ft-lb)
T1: 20 (2.0, 14.5)
T2: 25 (2.5, 18.1)
T3: 30 (3.1, 22)
T4: 34 (3.5, 25)
T5: 40 (4.1, 30)
T6: 45 (4.6, 33)
T7: 50 (5.1, 37)
T8: 55 (5.6, 41)
T9: 70 (7.1, 52)
T10: 95 (9.7, 70.1)
T11: 155 (15.8, 114)
T12: 175 (17.8, 129)
T13: 190 (19.4, 140)
T14: 250 (25.5, 184)

09490_SBCR_G0096

Front suspension and related components—2005–2006 Impreza and WRX

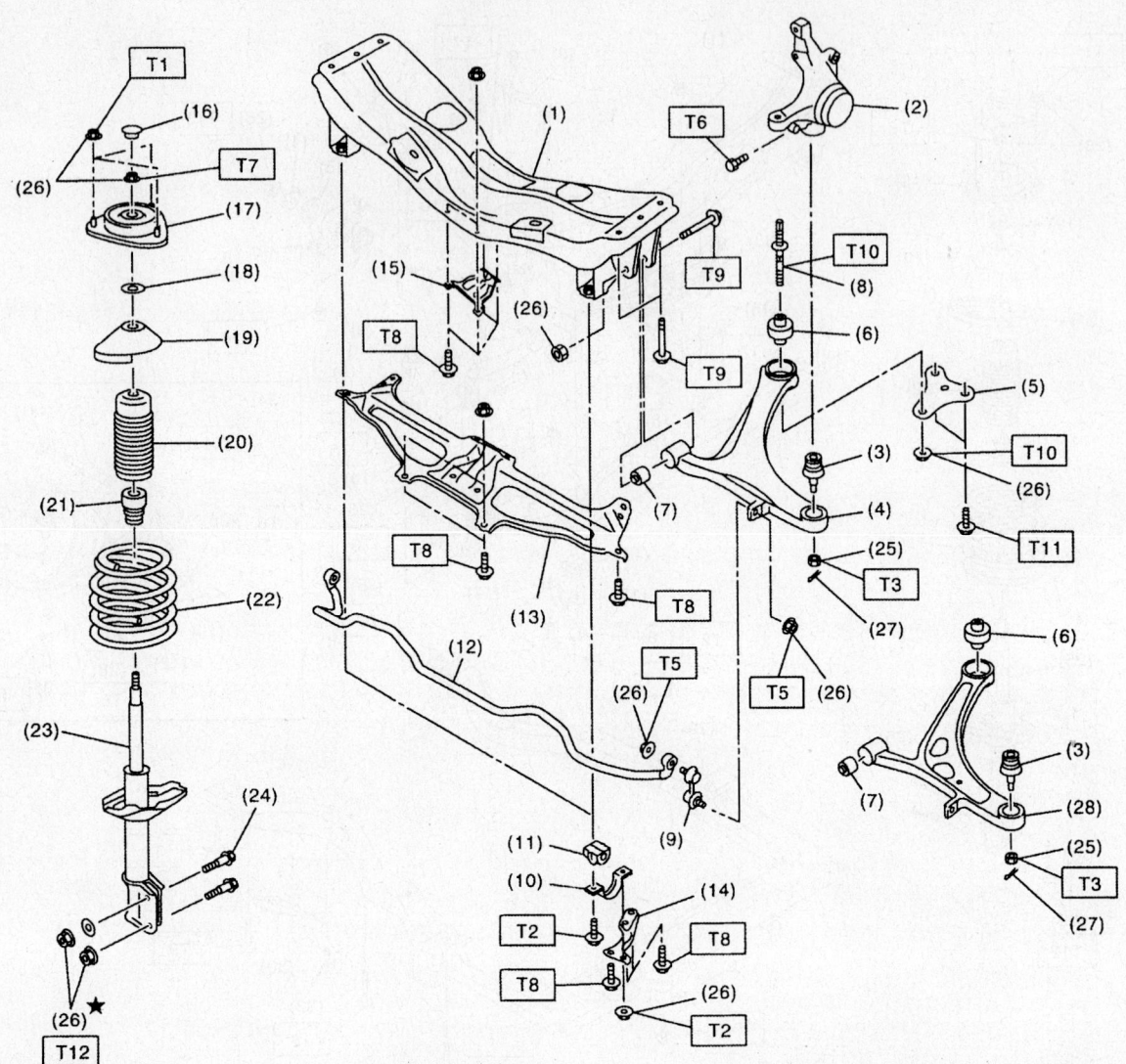

(1)	Front crossmember	(15)	Jack-up plate
(2)	Housing	(16)	Dust seal
(3)	Ball joint	(17)	Strut mount
(4)	Front arm (steel type)	(18)	Spacer
(5)	Support plate	(19)	Upper spring seat
(6)	Rear bushing	(20)	Dust cover
(7)	Front bushing	(21)	Helper
(8)	Stud bolt	(22)	Coil spring
(9)	Stabilizer link	(23)	Damper strut
(10)	Bracket	(24)	Adjusting bolt
(11)	Bushing	(25)	Castle nut
(12)	Stabilizer	(26)	Self-locking nut
(13)	Crossmember support plate (Large type)	(27)	Cotter pin
(14)	Crossmember support plate (Small type)	(28)	Front arm (aluminium type)

Tightening torque: N·m (kgf-m, ft-lb)

T1:	**20 (2.0, 14.5)**
T2:	**25 (2.5, 18.1)**
T3:	**30 (3.1, 22)**
T4:	**39 (4.0, 28.8)**
T5:	**45 (4.6, 33.2)**
T6:	**50 (5.1, 36.9)**
T7:	**55 (5.6, 41)**
T8:	**60 (6.1, 44.3)**
T9:	**95 (9.7, 70.1)**
T10:	**110 (11.2, 81.1)**
T11:	**150 (15.3, 110.6)**
T12:	**152 (15.5, 112.1)**

09490_SBCR_G0097

Front suspension and related components—2005–2006 Legacy and Outback

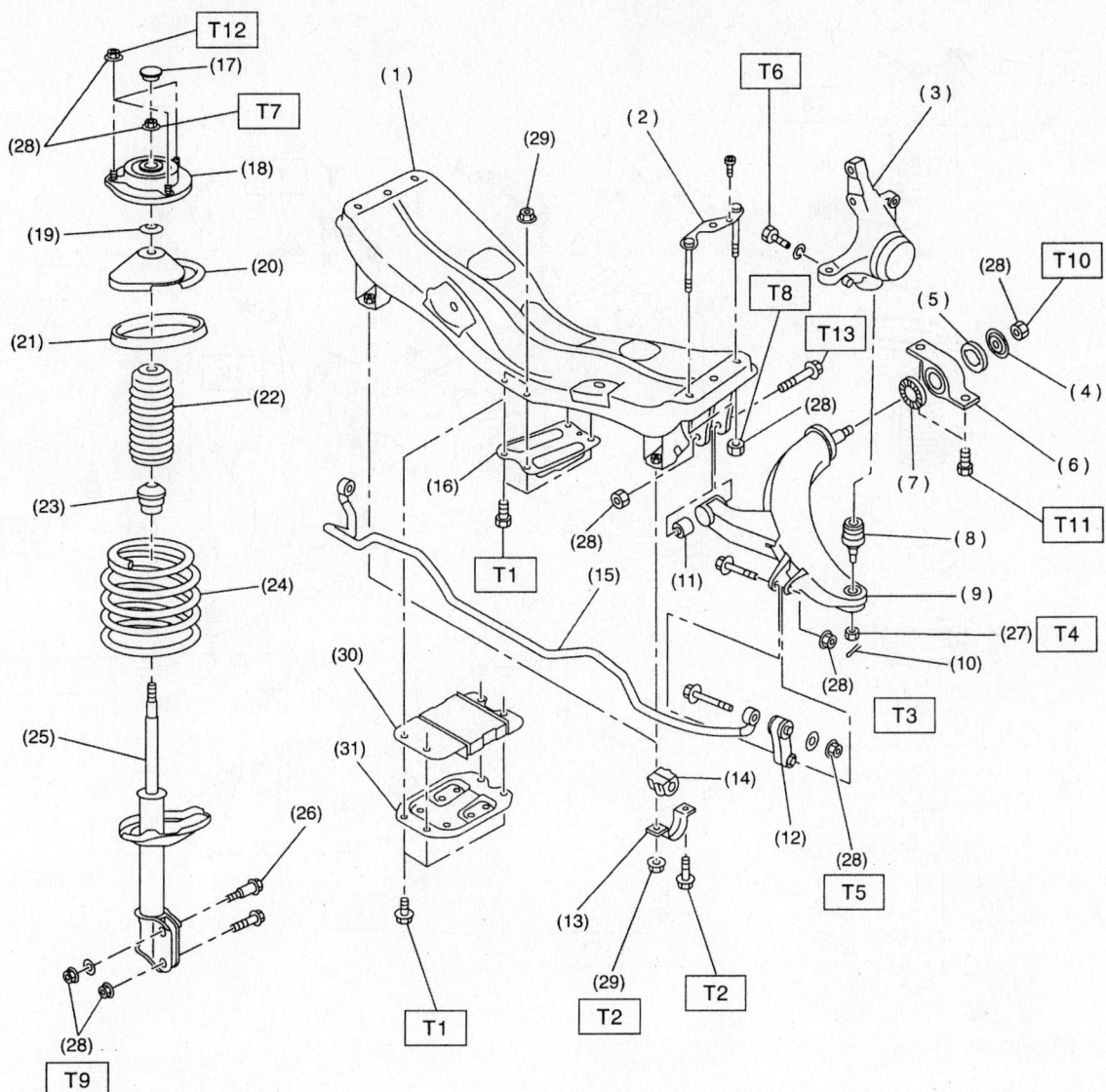

(1) Front crossmember
(2) Bolt ASSY
(3) Housing
(4) Washer
(5) Stopper rubber (Rear)
(6) Rear bushing
(7) Stopper rubber (Front)
(8) Ball joint
(9) Transverse link
(10) Cotter pin
(11) Front bushing
(12) Stabilizer link
(13) Clamp
(14) Bushing
(15) Stabilizer
(16) Jack-up plate

(17) Dust seal
(18) Strut mount
(19) Spacer
(20) Upper spring seat
(21) Rubber seat
(22) Dust cover
(23) Helper
(24) Coil spring
(25) Damper strut
(26) Adjusting bolt
(27) Castle nut
(28) Self-locking nut
(29) Flange nut
(30) Dynamic damper (Non-TURBO, MT model)

(31) Dynamic plate (Non-TURBO, MT model)

Tightening torque: N·m (kgf-m, ft-lb)
T1: 70 (7.1, 52)
T2: 25 (2.5, 18.1)
T3: 30 (3.1, 22)
T4: 39 (4, 29)
T5: 45 (4.6, 33)
T6: 50 (5.1, 37)
T7: 55 (5.6, 41)
T8: 100 (10.2, 74)
T9: 152 (16, 116)
T10: 186 (19.0, 137)
T11: 245 (25.0, 181)
T12: 20 (2.0, 14.5)
T13: 95 (9.7, 71)

09490_SBCR_G0098

Front suspension and related components—2005–2006 Baja

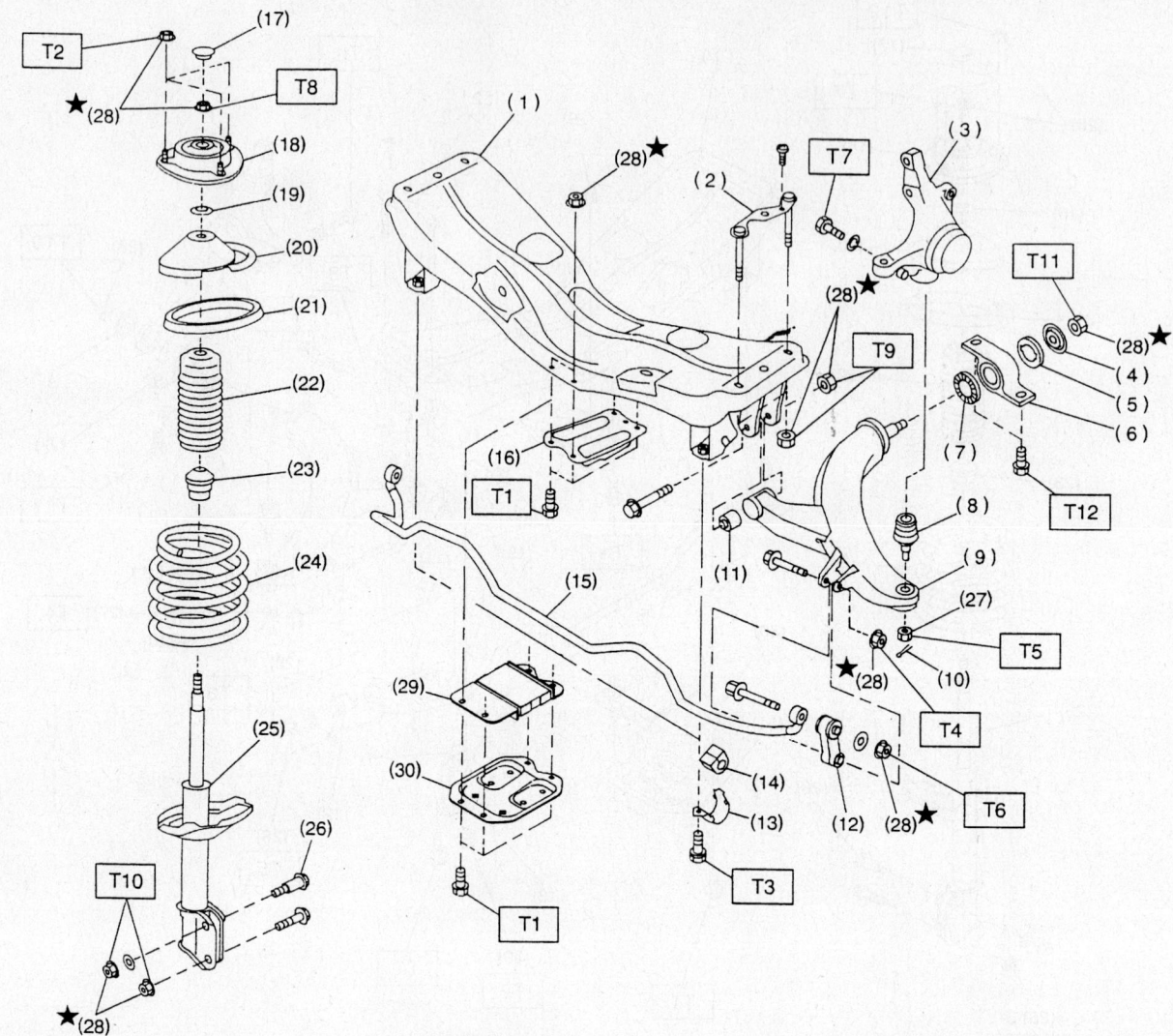

(1) Crossmember
(2) Bolt ASSY
(3) Housing
(4) Washer
(5) Stop rubber (Rear)
(6) Rear bushing
(7) Stop rubber (Front)
(8) Ball joint
(9) Transverse link
(10) Cotter pin
(11) Front bushing
(12) Stabilizer link
(13) Clamp
(14) Bushing
(15) Stabilizer

(16) Jack-up plate (Except 2500 cc MT model)
(17) Dust seal
(18) Strut mount
(19) Spacer
(20) Upper spring seat
(21) Rubber seat
(22) Dust cover
(23) Helper
(24) Coil spring
(25) Damper strut
(26) Adjusting bolt
(27) Castle nut
(28) Self-locking nut
(29) Dynamic damper (2500 cc MT model)

(30) Jack-up plate (2500 cc MT model)

Tightening torque: N·m (kg-m, ft-lb)
T1: 18±5 (1.8±0.5, 13.0±3.6)
T2: 20±6 (2.0±0.6, 14.5±4.3)
T3: 25±4 (2.5±0.4, 18.1±2.9)
T4: 29±5 (3.0±0.5, 21.7±3.6)
T5: 39 (4, 29)
T6: 44±6 (4.5±0.6, 32.5±4.3)
T7: 49±10 (5.0±1.0, 36±7)
T8: 54±5 (5.5±0.5, 39.8±3.6)
T9: 98±15 (10.0±1.5, 72±11)
T10: 152±20 (15.5±2.0, 112±14)
T11: 186±10 (19.0±1.0, 137±7)
T12: 245±49 (25.0±5.0, 181±36)

Rear

Front

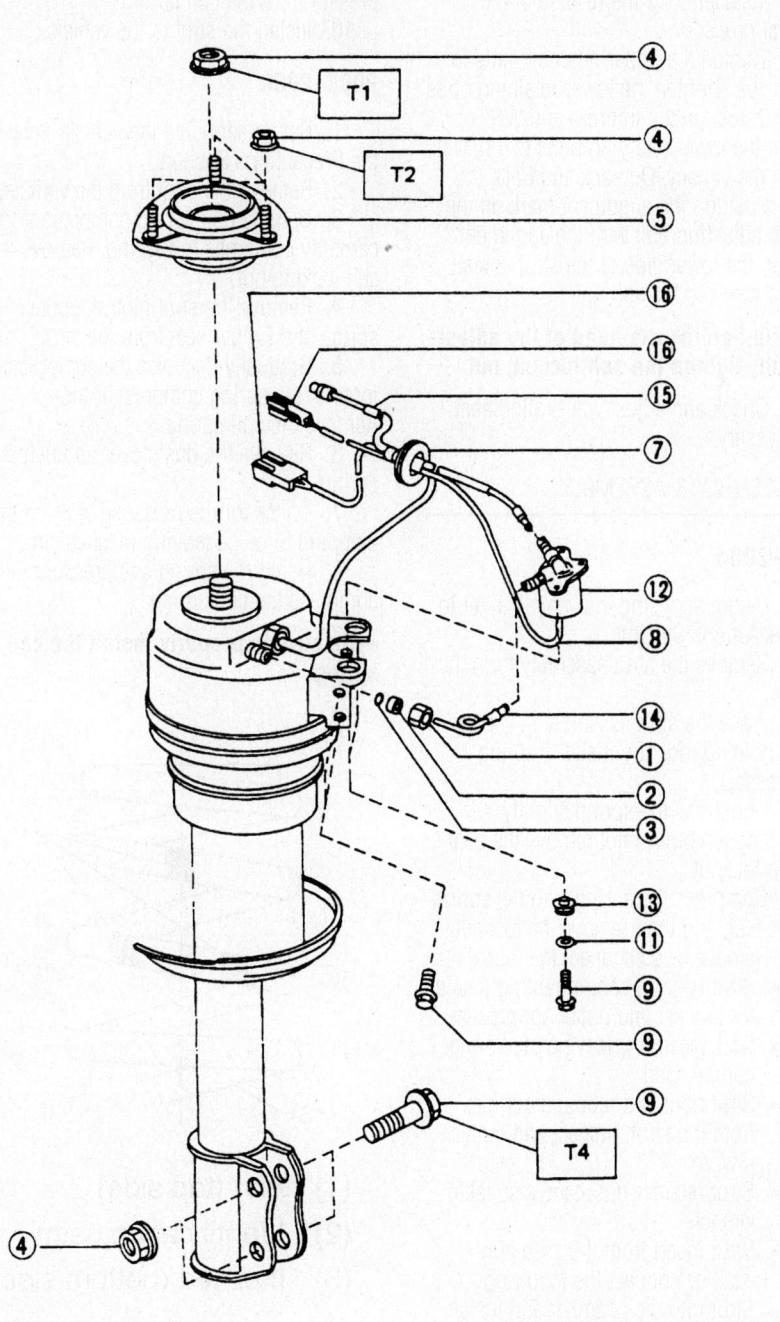
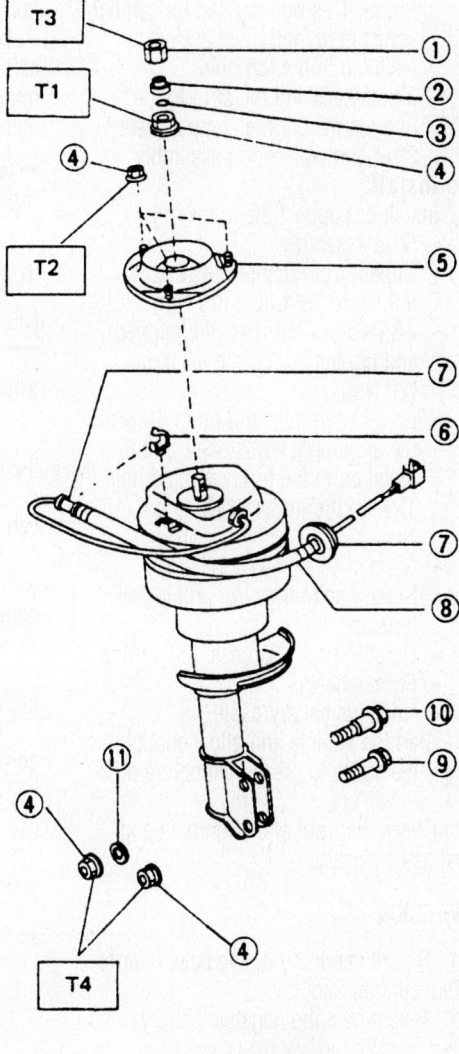

Tightening torque: N·m (kg-m, ft-lb)
T1: 49 — 69 (5 — 7, 36 — 51)
T2: 14 — 25 (1.4 — 2.6, 10 — 19)
T3: 7 — 17 (0.7 — 1.7, 5.1 —12.3)
T4: 186 — 235 (19 — 24, 137 — 174)

1   Cap
2   Air bushing
3   O-ring
4   Self lock nut
5   Strut mount
6   Clip
7   Grommet
8   Corrugate tube
9   Flange bolt
10 Adjusting bolt
11 Washer
12 Solenoid valve
13 Insulator
14 Air pipe for solenoid valve
15 Air pipe
16 Connector

7923TG69

Front and rear pneumatic suspension and related components—2002–2004 Legacy and Impreza

bolt to the strut housing as reference for installation.

- Bolt securing the ABS sensor harness, if equipped
- 2 bolts and nuts securing the strut to the steering knuckle. Notice that the shaft of the top bolt is not round. This bolt is used for camber adjustment, and must always be installed in the top hole.
- 3 nuts securing the strut to the body in the engine compartment
- Strut and coil spring assembly

To install:

4. Install or connect the following:
- Strut assembly
- Upper strut retainer nuts and tighten to 15 ft. lbs. (20 Nm)
- ABS sensor harness (if equipped), and tighten the bolt to 15 ft. lbs. (20 Nm)
- Lower strut nuts and bolts. Be sure the alignment adjustment bolt is installed in the top mounting hole. Tighten the nuts to 112 ft. lbs. (152 Nm).
- Caliper
- Brake line to the strut and install the clip
- Height sensor harness and air line
- Front wheel
- Negative battery cable

5. Start the vehicle and allow enough time for the struts to pressurize before driving.

6. Check the front end alignment and adjust as necessary.

2005–2006

1. Before servicing the vehicle, refer to the Precautions Section.

2. Disconnect the negative battery cable.

3. Raise and support the vehicle safely.

4. Remove the tire and wheel.

5. Remove the bolt retaining the brake hose to the strut.

6. Make and alignment mark on the camber adjusting bolt which secures the strut to the housing.

7. Remove the bolt retaining the ABS wheel speed sensor harness.

8. Remove the two bolts retaining the strut to the housing.

➡**While holding the head of the adjusting bolt, loosen the self locking nut.**

9. Remove the three upper strut retaining nuts.

10. Remove the strut from the vehicle.

To install:

11. Installation is the reverse of the removal procedure.

12. Tighten the upper retaining nuts to 14.5 ft. lbs. Tighten the lower retaining bolts to 129 ft. lbs. on the Impreza and WRX. Tighten the lower retaining bolts to 112.1 ft. lbs. on the Legacy, Outback and Baja.

13. Position the alignment mark on the camber adjusting bolt with the alignment mark on the lower side of the strut. Install using a new self locking nut.

➡**While holding the head of the adjusting bolt, tighten the self locking nut.**

14. Check and adjust wheel alignment, as necessary.

DISASSEMBLY & ASSEMBLY

2002–2004

1. Before servicing the vehicle, refer to the Precautions Section.

2. Remove the strut assembly from the vehicle.

3. Place the strut assembly in a vise with a holding tool and install a spring compressor.

4. Compress the spring slightly.

5. Loosen but do not remove the bearing cap locknut.

6. Compress the spring with the spring compressor, and then remove the locknut.

7. Remove or disconnect the following:
- Strut bearing cap, mounting insulator bracket and upper spring seat
- Coil assembly, leaving the spring compressed
- Strut boot and rebound bumper from the strut. Inspect and replace if worn.
- Strut retainer nut using a suitable wrench
- Strut insert from the assembly

8. Install or connect the following:
- Strut into the chamber and install the retainer nut. Tighten the nut snugly.
- Rebound bumper and the boot to the strut piston rod
- Coil spring on the strut assembly. Be sure the spring is properly positioned on the lower bracket.
- Upper spring seat, mounting insulator and bearing cap. Be sure the upper spring seat is facing the proper direction.
- Locknut and tighten to 36–43 ft. lbs. (49–59 Nm) on all except Legacy and Outback or to 41 ft. lbs. (55Nm) on Legacy and Outback.

9. Loosen and remove the spring compressor from the coil spring.

10. Install the strut to the vehicle.

2005–2006

1. Before servicing the vehicle, refer to the Precautions Section.

2. Remove the strut from the vehicle.

3. Using a coil spring compressor tool, carefully compress the spring. Remove the self locking nut.

4. Remove the strut mount, upper spring and rubber seat from the strut.

5. Gradually decrease the compression force of the spring compressor tool. Remove the coil spring.

6. Remove the dust cover and helper spring.

7. Check for the presence of air in the damping force generating mechanism.

8. Using the spring compression tool, compress the coil spring.

➡**Be sure to properly install the coil spring.**

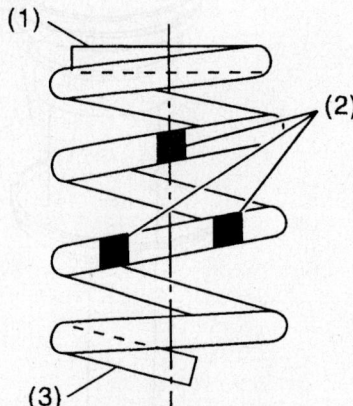

(1) Flat (top side)

(2) Identification paint

(3) Inclined (bottom side)

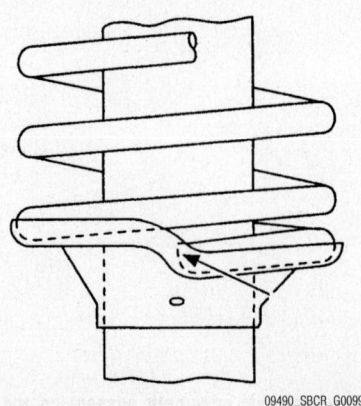

09490_SBCR_G0099

Front strut spring alignment—2005–2006

9. Position the coil spring so that its end face fits good into the spring seat.

10. Install the helper spring and dust cover to the piston rod.

11. Pull the piston rod fully upward, and install the rubber seat and spring seat.

12. Install the strut mount to the piston rod, and then tighten the self locking nut, temporarily. Be sure to use a new self locking nut.

13. Use a hexagon wrench to prevent the strut rod from turning. Tighten the self locking nut to 41 ft. lbs.

14. Carefully loosen the coil spring.

Stabilizer Bar

REMOVAL & INSTALLATION

2002–2004

1. Before servicing the vehicle, refer to the Precautions Section.

2. Raise and support the vehicle safely.

3. Remove the jack up plate from the lower part of the crossmember.

4. On Outback Sport, remove the sub frame.

5. Remove the bolt and nut which secures the stabilizer to the crossmember.

6. Remove the bolts which secure the stabilizer link to the front transverse link.

7. Remove the stabilizer bar from the vehicle.

To install:

8. Installation is the reverse of the removal procedure.

9. Install the rubber bushing, on the front crossmember side, while aligning it with the paint mark on the stabilizer bar.

10. Be sure that the bushings and the stabilizer have the same identification colors.

11. Always fully tighten the rubber bushings when the wheels are in full contact with the ground and the vehicle is at curb height.

2005–2006

IMPREZA AND WRX

1. Before servicing the vehicle, refer to the Precautions Section.

2. Raise and support the vehicle safely.

3. Remove the jack up plate from the lower part of the crossmember.

4. Remove the sub frame.

5. Remove the nut which secures the stabilizer link to the front transverse link.

6. Remove the bolts that secure the stabilizer bar to the crossmember.

7. Remove the stabilizer bar from the vehicle.

To install:

8. Installation is the reverse of the removal procedure.

9. Install the rubber bushing, on the front crossmember side, while aligning it with the paint mark on the stabilizer bar.

10. Be sure that the bushings and the stabilizer have the same identification colors.

11. Always fully tighten the rubber bushings when the wheels are in full contact with the ground and the vehicle is at curb height.

LEGACY AND OUTBACK

1. Before servicing the vehicle, refer to the Precautions Section.

2. Raise and support the vehicle safely.

3. Remove the engine under cover.

4. Remove the front crossmember support plate.

5. Remove the stabilizer bar link.

6. Remove the stabilizer bar bracket bolts and bushings.

7. Remove the stabilizer bar from the vehicle.

To install:

8. Installation is the reverse of the removal procedure.

9. Be sure to use new self locking nuts, as required.

10. Install the rubber bushing, on the front crossmember side, while aligning it with the paint mark on the stabilizer bar.

11. Be sure that the bushings and the stabilizer have the same identification colors.

12. The stabilizer bracket has a set orientation. Install it with the arrow mark facing the upper side of the vehicle.

13. Always fully tighten the rubber bushings when the wheels are in full contact with the ground and the vehicle is at curb height.

BAJA

1. Before servicing the vehicle, refer to the Precautions Section.

2. Raise and support the vehicle safely.

3. Remove the jack up plate from the lower part of the crossmember.

4. Remove the nut which secures the stabilizer link to the front transverse link.

5. Remove the bolts that secure the stabilizer bar to the crossmember.

6. Remove the stabilizer bar from the vehicle.

To install:

7. Installation is the reverse of the removal procedure.

8. Install the rubber bushing, on the front crossmember side, while aligning it with the paint mark on the stabilizer bar.

9. Be sure that the bushings and the stabilizer have the same identification colors.

10. Always fully tighten the rubber bushings when the wheels are in full contact with the ground and the vehicle is at curb height.

Lower Ball Joint

REMOVAL & INSTALLATION

2002–2004

1. Before servicing the vehicle, refer to the Precautions Section.

2. Remove or disconnect the following:
   - Negative battery cable
   - Front wheel and tire assembly
   - Ball joint castle nut cotter pin and discard the cotter pin
   - Castle nut
   - Ball joint from the lower control arm using a suitable puller or pry tool
   - Bolt securing the ball joint to the steering knuckle
   - Ball joint using a suitable wedge to expand the steering knuckle connection point

To install:

3. Install or connect the following:
   - Ball joint to the steering knuckle
   - Retaining bolt and tighten to 37 ft. lbs. (50 Nm)
   - Ball joint to the lower control arm and tighten the castle nut on all except WRX to 29 ft. lbs. (39 Nm). On WRX sedan, tighten to 22 ft. lbs. (30 Nm) and 33 ft. lbs. (45 Nm) on all other WRX. Then, tighten the castle nut an additional 60 degrees until the slot in the castle nut is aligned with the cotter pin hole in the ball joint.
   - New cotter pin
   - Wheel
   - Negative battery cable

2005–2006

IMPREZA AND WRX

1. Before servicing the vehicle, refer to the Precautions Section.

2. Disconnect the negative battery cable.

3. Raise and support the vehicle safely.

4. Remove the tire and wheel.

5. Remove the cotter pin from the ball stud. Remove the castle nut. Extract the ball stud from the transverse link.

6. Remove the bolt securing the ball joint to the housing. Extract the ball joint from the housing.

To install:

7. Installation is the reverse of the removal procedure.

8. Install the ball joint to the transverse link arm and tighten the castle nut to 22 ft. lbs. (30 Nm) on sedan with DOHC and STI. On all other vehicles, tighten to 30 ft. lbs. (40 Nm). Tighten the castle nut an additional 60 degrees until the slot in the castle nut is aligned with the cotter pin hole in the ball joint.

9. Check and adjust alignment, as required.

2005–2006

LEGACY AND OUTBACK

1. Before servicing the vehicle, refer to the Precautions Section.

2. Disconnect the negative battery cable.

3. Raise and support the vehicle safely.

4. Remove the tire and wheel.

5. Remove the stabilizer bar brackets and bushings (both sides).

6. Remove the cotter pin from the ball stud. Remove the castle nut. Extract the ball stud from the transverse link.

7. Remove the bolt securing the ball joint to the housing. Extract the ball joint from the housing.

To install:

8. Installation is the reverse of the removal procedure.

9. Install the ball joint to the transverse link arm and tighten the castle nut to 22 ft. lbs. (30 Nm), if the front arm is aluminum and 28.8 ft. lbs. (39 Nm), if the front arm is steel. Tighten the castle nut an additional 60 degrees until the slot in the castle nut is aligned with the cotter pin hole in the ball joint.

10. The stabilizer bracket has a set orientation. Install it with the arrow mark facing the upper side of the vehicle.

11. Always fully tighten the rubber bushings when the wheels are in full contact with the ground and the vehicle is at curb height.

12. Check and adjust alignment, as required.

BAJA

1. Before servicing the vehicle, refer to the Precautions Section.

2. Disconnect the negative battery cable.

3. Raise and support the vehicle safely.

4. Remove the tire and wheel.

5. Remove the cotter pin from the ball stud. Remove the castle nut. Extract the ball stud from the transverse link.

6. Remove the bolt securing the ball

joint to the housing. Extract the ball joint from the housing.

To install:

7. Installation is the reverse of the removal procedure.

8. Install the ball joint to the transverse link arm and tighten the castle nut to 29 ft. lbs. (39 Nm). Tighten the castle nut an additional 60 degrees until the slot in the castle nut is aligned with the cotter pin hole in the ball joint.

9. Check and adjust alignment, as required.

Front Lower Control Arm

REMOVAL & INSTALLATION

2002–2004

EXCEPT WRX

1. Before servicing the vehicle, refer to the Precautions Section.

2. Remove or disconnect the following:
   • Tire and wheel assembly

3. On Impreza, remove the sub-frame as follows:
   a. Remove the bolt cover.
   b. Remove the clip.
   c. Loosen the sub-frame bolt (1) and leave it screwed in a few threads. Remove the remaining bolts in the following order: 2, 3, 4, 5 and 6. See illustration for bolt location.
   • Sway link from the lower control arm

• Bolt securing the ball joint to the steering knuckle
• Nuts (NOT the bolts) securing the lower control arm to the cross-member
• 2 bolts holding the bushing bracket of the control arm to the body
• Ball joint from the steering knuckle
• Bolts securing the lower control arm to the crossmember, then the lower control

To install:

4. Install or connect the following:
   • Lower control arm; temporarily tighten the 2 bolts used to secure the rear bushing of the lower control arm to the body

➡ **These bolts should be tightened so they can still move back and forth in the oblong shaped hole in the bracket that holds the bushing.**

5. On Impreza , install the sub frame as follows:

6. Install the sub-frame and bolts. Tighten the bolts as follows, referring to the illustration for location:
   a. T1: 25 ft. lbs. (34 Nm).
   b. T2: 41 ft. lbs. (55 Nm).
   c. T3: 52 ft. lbs. (71 Nm).

7. Install or connect the following:
   • Bolts used to secure the lower control arm to the crossmember and temporarily tighten the nuts
   • Ball joint into the steering knuckle and secure with the retaining bolt

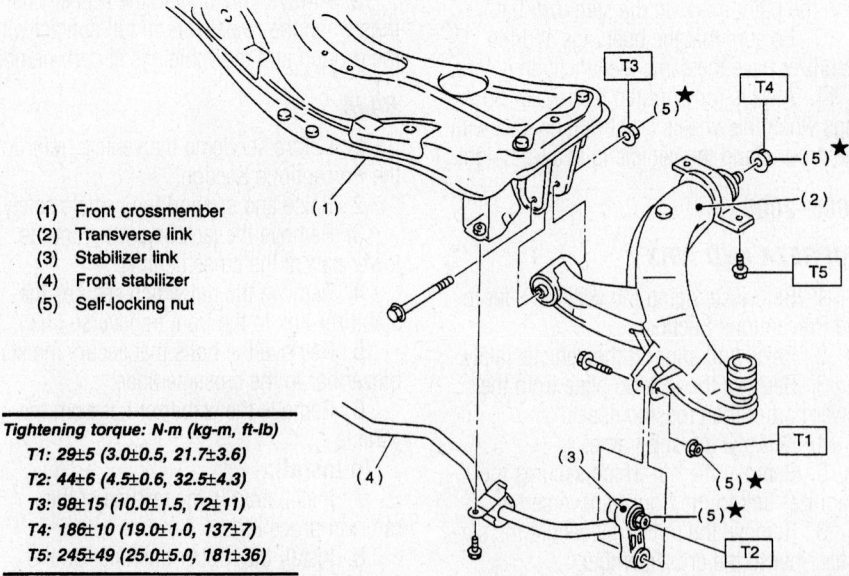

(1) Front crossmember
(2) Transverse link
(3) Stabilizer link
(4) Front stabilizer
(5) Self-locking nut

Tightening torque: N·m (kg-m, ft-lb)
T1: 29±5 (3.0±0.5, 21.7±3.6)
T2: 44±6 (4.5±0.6, 32.5±4.3)
T3: 98±15 (10.0±1.5, 72±11)
T4: 186±10 (19.0±1.0, 137±7)
T5: 245±49 (25.0±5.0, 181±36)

9307TG07

Lower control arm (transverse link) and related components—2002–2004 Impreza and Legacy

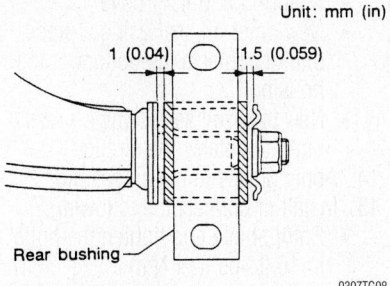

Unit: mm (in)

1 (0.04)   1.5 (0.059)

Rear bushing

9307TG08

Proper control arm to rear bushing clearance specifications—2002–2004

- Sway link to the control arm and temporarily tighten the bolts

※ WARNING

Discard loosened self-locking nut and replace with a new one.

8. Lower the vehicle, and then tighten the bolts to the following specifications:
   a. Lower control arm-to-sway bar: 22 ft. lbs. (30 Nm). On Impreza tighten the retainers to on vehicles except sedan turbocharged to 22 ft. lbs. (30 Nm) and 33 ft. lbs. on sedan turbocharged.

➡**Move the rear bushing back and forth until the control arm-to-rear bushing clearance if established. Refer to the illustration for specifications.**

   b. Lower control arm-to-crossmember: 74 ft. lbs. (100 Nm).
   c. Lower control arm-to-rear link bushing-to-body: 184 ft. lbs. (250 Nm).
9. Check the wheel alignment and adjust if necessary.

WRX

1. Before servicing the vehicle, refer to the Precautions Section.
2. Remove or disconnect the following:
   - Tire and wheel assembly
3. Remove the sub-frame as follows:
   a. Remove the bolt cover.
   b. Remove the clip.
   c. Loosen the sub-frame bolt (1) and leave it screwed in a few threads. Remove the remaining bolts in the following order: 2, 3, 4, 5 and 6. See illustration for bolt location.
   - Sway link from the lower control arm
   - Bolt securing the ball joint to the steering knuckle
   - Nuts (NOT the bolts) securing the lower control arm to the crossmember

- 2 bolts holding the bushing bracket of the control arm to the body
- Ball joint from the steering knuckle
- Bolts securing the lower control arm to the crossmember, then the lower control

To install:
4. Install or connect the following:
   - Lower control arm; temporarily tighten the 2 bolts used to secure the rear bushing of the lower control arm to the body

➡**These bolts should be tightened so they can still move back and forth in the oblong shaped hole in the bracket that holds the bushing.**

   - Bolts used to secure the lower control arm to the crossmember and temporarily tighten the nuts
   - Ball joint into the steering knuckle and secure with the retaining bolt
   - Sway link to the control arm and temporarily tighten the bolts

※ WARNING

Discard loosened self-locking nut and replace with a new one.

5. Lower the vehicle, and then tighten the bolts to the following specifications:
   a. Lower control arm-to-sway bar: 33 ft. lbs. (45 Nm) on sedan or 22 ft. lbs. (30 Nm) on all except sedan.
   b. Lower control arm and crossmember to 74 ft. lbs. (100 Nm).
   c. Lower control arm rear bushing and body to 184 ft. lbs. (250 Nm).

➡**Move the rear bushing back and forth until the control arm-to-rear bushing clearance if established. Refer to the illustration for specifications.**

6. Install the sub-frame and bolts. Tighten the bolts in the proper sequence to specification:
   a. T1: 25 ft. lbs. (34 Nm).
   b. T2: 41 ft. lbs. (55 Nm).
   c. T3: 52 ft. lbs. (71 Nm).
7. Check the wheel alignment and adjust if necessary.

2005–2006

IMPREZA, WRX AND BAJA

1. Before servicing the vehicle, refer to the Precautions Section.
2. Disconnect the negative battery cable.
3. Raise and support the vehicle safely.
4. Remove the tire and wheel.
5. Remove the sub frame, on Impreza and WRX.

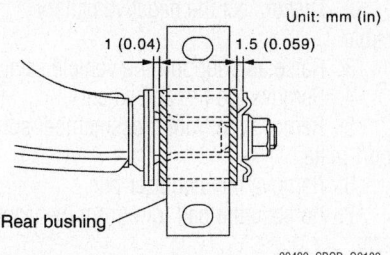

Unit: mm (in)

1 (0.04)   1.5 (0.059)

Rear bushing

09490_SBCR_G0100

Transverse link rear bushing alignment location and specification—2005–2006 Impreza, WRX and Baja

6. Disconnect the stabilizer link from the transverse link.
7. Remove the bolt securing the ball joint of the transverse link to housing.
8. Remove the nut (do not remove the bolt) securing the transverse link to the crossmember.
9. Remove the two bolts securing the bushing bracket of the transverse link to the vehicle body at the rear bushing.
10. Remove the ball joint from the housing.
11. Remove the bolt securing the transverse link to the crossmember. Remove the transverse link from the crossmember.

To install:
12. Install the transverse link to its mounting.
13. Temporarily tighten the two bolts used to secure the rear bushing of the transverse link.

➡**These bolts should be tightened so that they can still move back and forth in the oblong shaped hole in the bracket, which holds the bushing.**

14. Continue the installation in the reverse order of the removal procedure.
15. Always fully tighten the rubber bushings when the wheels are in full contact with the ground and the vehicle is at curb height.
16. Tighten the transverse link rear bushing to body to 184 ft. lbs. (250 Nm) on Impreza and WRX. Tighten the transverse link rear bushing to body to 181 ft. lbs. (250 Nm) on Baja.

➡**Move the rear bushing back and forth until the transverse link to rear bushing clearance is established, before tightening.**

17. Check and adjust alignment, as required.

LEGACY AND OUTBACK

1. Before servicing the vehicle, refer to the Precautions Section.

2. Disconnect the negative battery cable.

3. Raise and support the vehicle safely.

4. Remove the tire and wheel.

5. Remove the front crossmember support plate.

6. Remove the stabilizer bar.

7. Remove the ball joint from the front arm.

8. Remove the nut securing the front arm to the crossmember.

➡️**Do not remove the bolt.**

9. Remove the front arm support plate.

10. Remove the bolt securing the front arm to the crossmember and pull the front arm out of the crossmember.

11. To remove the stud bolt, use tool ST20299AG020.

➡️**Do not remove the stud bolt unless it is necessary. Always replace the removed parts with new ones.**

To install:

12. Installation is the reverse of the removal procedure.

13. Tighten the stud bolt to 81.1 ft. lbs. (110 Nm), if removed.

14. Tighten the support plate to front arm bolts to 81.1 ft. lbs. (110 Nm).

15. Tighten the support plate to body bolts to 110.6 ft. lbs. (150 Nm).

16. Always fully tighten the rubber bushings when the wheels are in full contact with the ground and the vehicle is at curb height.

17. Check and adjust alignment, as required.

CONTROL ARM BUSHING REPLACEMENT

1. Remove the control arm from the vehicle.

2. Mount the control arm in a soft jawed vise.

3. Use either a press or a control arm bushing fixture (C-clamp like tool) along with a slotted washer and a piece of pipe (slightly larger than the bushing) and press out the old bushing.

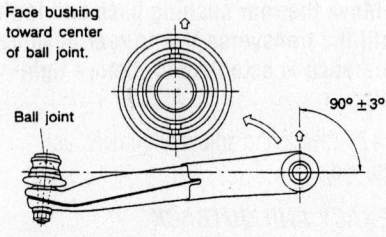

Face bushing toward center of ball joint.

Ball joint

90° ± 3°

9307TG09

The front control arm bushing must be installed in the proper direction

4. Clean the inside bushing contact surfaces of rust and old rubber.

To install:

5. Apply a light coating of grease to both the replacement busing and bushing contact surfaces on the control arm.

6. Align the bushing.

7. Install the bushing using the press tool. A bushing install clamp can also be used to compress the bushing into the control arm.

8. Install the control arm on the vehicle.

Wheel Bearings

ADJUSTMENT

The wheel bearings are not adjustable.

REMOVAL & INSTALLATION

2002–2004

1. Before servicing the vehicle, refer to the Precautions Section.

2. Remove or disconnect the following:
- Steering knuckle assembly from the vehicle

3. Position the steering knuckle in a soft-jawed vise.

4. Press the hub from the steering knuckle. If the inner bearing race remains in the hub, press it out.

5. Remove or disconnect the following:
- Rotor shield
- Inner and outer seals
- Snapring from the steering knuckle

6. Press the inner bearing race to remove the outer bearing.

7. Remove or disconnect the following:
- Tone ring, if equipped with Anti-lock Brake System (ABS)
- Wheel lugs from the hub using a suitable press

➡️**To prevent deforming the hub, do not hammer the lugs out.**

To install:

8. Install or connect the following:
- Wheel lugs into the hub using a suitable press

9. If equipped with ABS, clean all foreign material from the hub and tone ring.

10. Install or connect the following:
- Tone ring

11. Clean the inside of the steering knuckle.

12. Remove the plastic lock from the inner race and press a new greased bearing into the hub by pressing the outer race.

13. Install or connect the following:

- Snapring into its groove
- New outer oil seal using a press, until it contacts the bottom of the housing
- New inner oil seal using a press, until it contacts the circlip

14. Apply grease to the oil seal lips.

15. Install or connect the following:
- Rotor shield and tighten the bolts to 10 ft. lbs. (14 Nm)
- Hub to the steering knuckle

16. Press a new bearing into the hub by driving the inner race.

17. Install the steering knuckle on the vehicle.

2005–2006

IMPREZA, WRX, LEGACY AND OUTBACK

➡️**It may be necessary to remove the front halfshaft from the vehicle.**

1. Before servicing the vehicle, refer to the Precautions Section.

2. Disconnect the negative battery cable.

3. Raise and support the vehicle safely.

4. Remove the tire and wheel.

5. Remove the crimped section of the axle nut. Depress the brake pedal and remove the axle nut.

➡️**Be sure to loosen the axle nut after removing the tire and wheel from the vehicle. Failure to do this may damage the wheel bearings.**

6. Remove the disc brake caliper from its mounting as suspend it to the side, with wire. Do not disconnect the brake line.

7. Remove the rotor.

➡️**If the rotor is seized within the hub, remove the rotor by installing an 8mm bolt in the screw hole on the rotor.**

8. Remove the ABS wheel speed sensor assembly and harness.

9. Remove the four bolts from the housing.

10. Remove the front hub unit bearing. Use tools ST92647000 and ST28099PA100 if necessary.

To install:

11. Tighten the front hub unit bearing to housing bolts to 47.9 ft. lbs. (65 Nm).

12. Be sure to use a new axle nut. Tighten the axle nut, temporarily.

13. Install the rotor. Install the caliper.

14. While depressing the brake pedal, tighten the axle nut (olive color) to 162 ft. lbs. (220 Nm). Lock it securely in place.

➡️**Install the tire and wheel after instal-**

lation of the axle nut. Failure to do this may result in wheel bearing damage. Do not over tighten, as this too could cause wheel bearing damage.

15. Continue the installation in the reverse order of the removal procedure.

BAJA

1. Before servicing the vehicle, refer to the Precautions Section.
2. Disconnect the negative battery cable.
3. Raise and support the vehicle safely.
4. Remove the tire and wheel.
5. Remove the crimped section of the axle nut. Depress the brake pedal and remove the axle nut.

➡**Be sure to loosen the axle nut after removing the tire and wheel from the vehicle. Failure to do this may damage the wheel bearings.**

6. Remove the stabilizer link.
7. Remove the disc brake caliper from its mounting as suspend it to the side, with wire. Do not disconnect the brake line.
8. Remove the rotor.

➡**If the rotor is seized within the hub, remove the rotor by installing an 8mm bolt in the screw hole on the rotor.**

9. Remove the cotter pin and castle nut which secure the tie rod end to the housing knuckle arm.
10. Using a puller, remove the tie rod ball joint from the knuckle arm.
11. Remove the ABS wheel speed sensor assembly and harness. Remove the bolt that retains the sensor harness to the strut.
12. Remove the transverse link ball joint from the housing.
13. Remove the inner joint from the transmission spindle.

➡**Do not pull the inner joint when removing the front halfshaft.**

14. Remove the front halfshaft from the hub. Use tools ST92647000 and ST28099PA100 if necessary.

➡**Be careful not to damage the oil seal lip and tone wheel when removing the halfshaft. If the front halfshaft is removed, replace the inner seal with a new one.**

15. After scribing an alignment mark on the camber adjusting bolt head, remove the bolts which connect the housing and the strut. Disconnect the housing from the strut.
To install:
16. While aligning the alignment mark on the camber adjusting bolt head, connect the housing and strut. Tighten the bolt to 130 ft. lbs. (177 Nm).

➡**When the self locking nut is removed, replace it with a new one.**

17. Install the halfshaft.
18. Continue the installation in the reverse order of the removal procedure.
19. Be sure to use a new axle nut. While depressing the brake pedal, tighten the axle nut to 162 ft. lbs. (220 Nm). Lock it securely in place.

➡**Install the tire and wheel after installation of the axle nut. Failure to do this may result in wheel bearing damage. Do not over tighten, as this too could cause wheel bearing damage.**

20. Continue the installation in the reverse order of the removal procedure.

REAR SUSPENSION

Strut

REMOVAL & INSTALLATION

2002–2004

STANDARD STRUT

1. Before servicing the vehicle, refer to the Precautions Section.
2. Remove or disconnect the following:
   - Rear seat assembly, on Sedan only
   - Rear speaker grille and service hole cap, on Wagon only
   - Strut mount cap
   - Wheel and tire assembly
   - Brake hose clip
   - Union bolt from the brake caliper. Move the brake hose out of the way.
   - Lower nuts and bolts securing the strut to the rear wheel housing
   - Retainer nuts securing the strut bearing cap to the strut tower, from inside the vehicle
   - Strut from the vehicle
To install:
3. Install or connect the following:
   - Strut onto the vehicle, making sure to position the strut properly in the upper strut tower mounts. Refer to the illustration.

   - Strut retainer nuts and tighten to 14 ft. lbs. (20 Nm) on all except Legacy and Outback. On Legacy and Outback, tighten the nut to 22 ft. lbs. (30 Nm).
   - Strut to the rear wheel knuckle assembly using the retainer nuts and bolts, and tighten the bolts to 162 ft. lbs. (220 Nm) on WRX and 145 ft. lbs. (196 Nm) on all others
   - Brake union bolt and tighten to 13 ft. lbs. (18 Nm)
   - Brake hose clip
4. Bleed the brakes.
   - Wheel
   - Strut mount cap
   - Rear seat, on Sedan
   - Speaker grille, on Wagon

PNEUMATIC STRUT

1. Before servicing the vehicle, refer to the Precautions Section.
2. Remove or disconnect the following:

   - Negative battery cable
   - Rear seat assembly, on the Sedan
   - Rear speaker grille and service hole cap, on the Wagon
   - Strut mount cap
   - Air line from the top of the strut assembly
   - Height sensor and solenoid valve

   wiring harnesses from the strut assembly
   - Wheel and tire assembly
   - Brake hose clip
   - Union bolt from the brake caliper. Move the brake hose out of the way.
   - Lower nuts and bolts securing the strut to the rear wheel housing
   - Retainer nuts securing the strut bearing cap to the strut tower (from inside the vehicle)
   - Strut from the vehicle
To install:
3. Install or connect the following:
   - Strut on to the vehicle, making sure to position the strut properly in the upper strut tower mounts. Refer to the illustration if needed. Install the retainer nuts, and tighten to 11 ft. lbs. (15 Nm).
   - Strut to the rear wheel knuckle assembly, using the retainer nuts and bolts, and tighten the bolts to 145 ft. lbs. (196 Nm)
   - Brake union bolt and tighten to 13 ft. lbs. (18 Nm)
   - Brake hose clip
4. Bleed the brakes.
   - Wheel
   - Height sensor and solenoid valve wiring harnesses to the strut

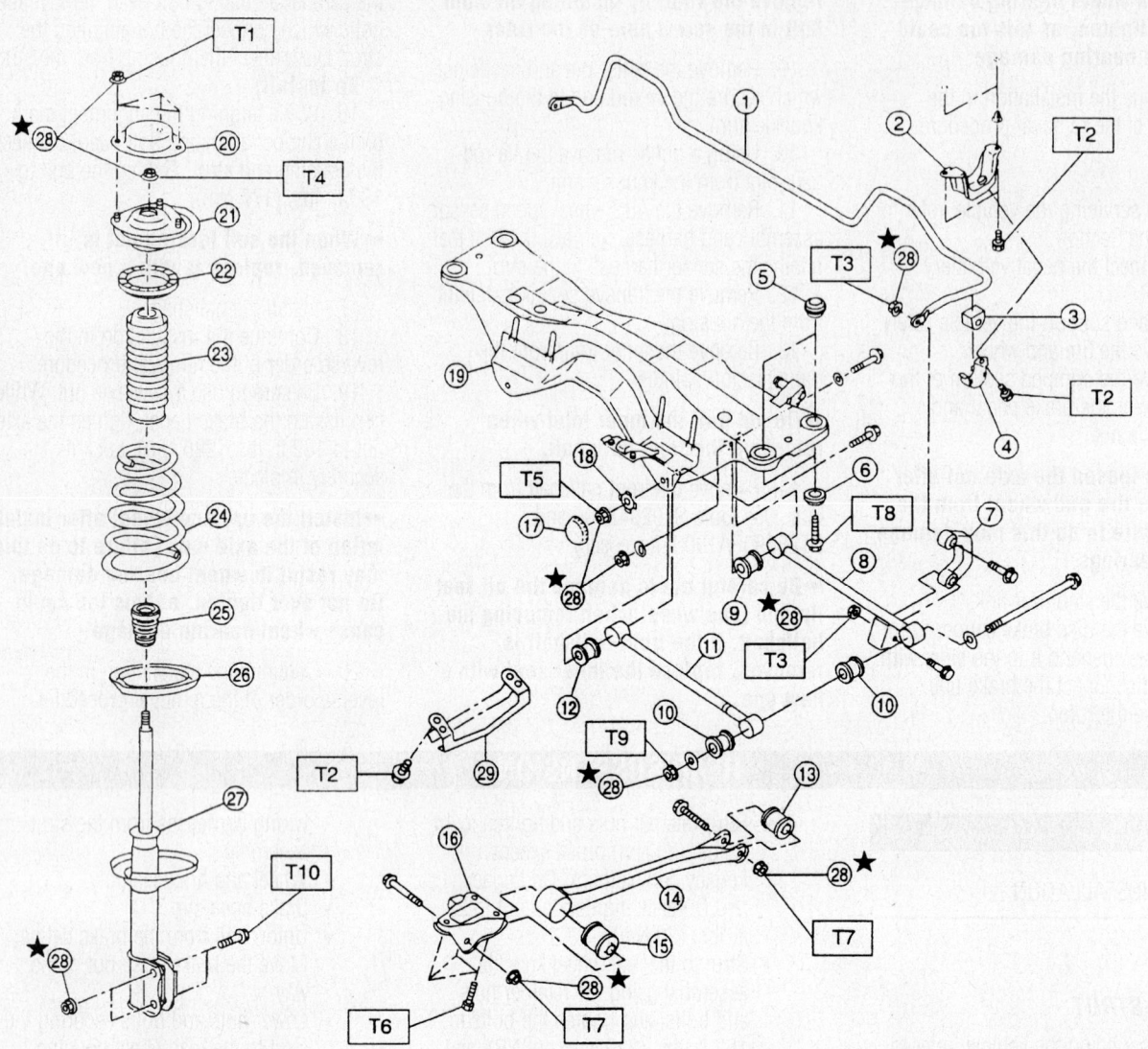

1. Stabilizer
2. Stabilizer bracket
3. Stabilizer bushing
4. Clamp
5. Floating bushing
6. Stopper
7. Stabilizer link
8. Rear lateral link
9. Bushing (C)
10. Bushing (A)
11. Front lateral link
12. Bushing (B)
13. Trailing link rear bushing
14. Trailing link
15. Trailing link front bushing
16. Trailing link bracket
17. Cap (Protection)
18. Washer
19. Crossmember
20. Strut mount cap
21. Strut mount
22. Rubber seat upper
23. Dust cover
24. Coil spring
25. Helper
26. Rubber seat lower
27. Damper strut
28. Self-locking nut
29. Crossmember reinforcement lower (Sedan model only)

Tightening torque: N·m (kg-m, ft-lb)
T1: 20 ± 6 (2.0 ± 0.6, 14.5 ± 4.3)
T2: 25 ± 7 (2.5 ± 0.7, 18.1 ± 5.1)
T3: 44 ± 6 (4.5 ± 0.6, 32.5 ± 4.3)
T4: 59 ± 10 (6.0 ± 1.0, 43 ± 7)
T5: 98 ± 15 (10.0 ± 1.5, 72 ± 11)
T6: 98 ± 20 (10.0 ± 2.0, 72 ± 14)
T7: 113 ± 15 (11.5 ± 1.5, 83 ± 11)
T8: 127 ± 20 (13.0 ± 2.0, 94 ± 14)
T9: 137 ± 20 (14.0 ± 2.0, 101 ± 14)
T10: 196^{+39}_{-10} ($20.0^{+4.0}_{-1.0}$, 145^{+29}_{-7})

Rear suspension and related components—2002–2004 Legacy and Impreza

7923TG71

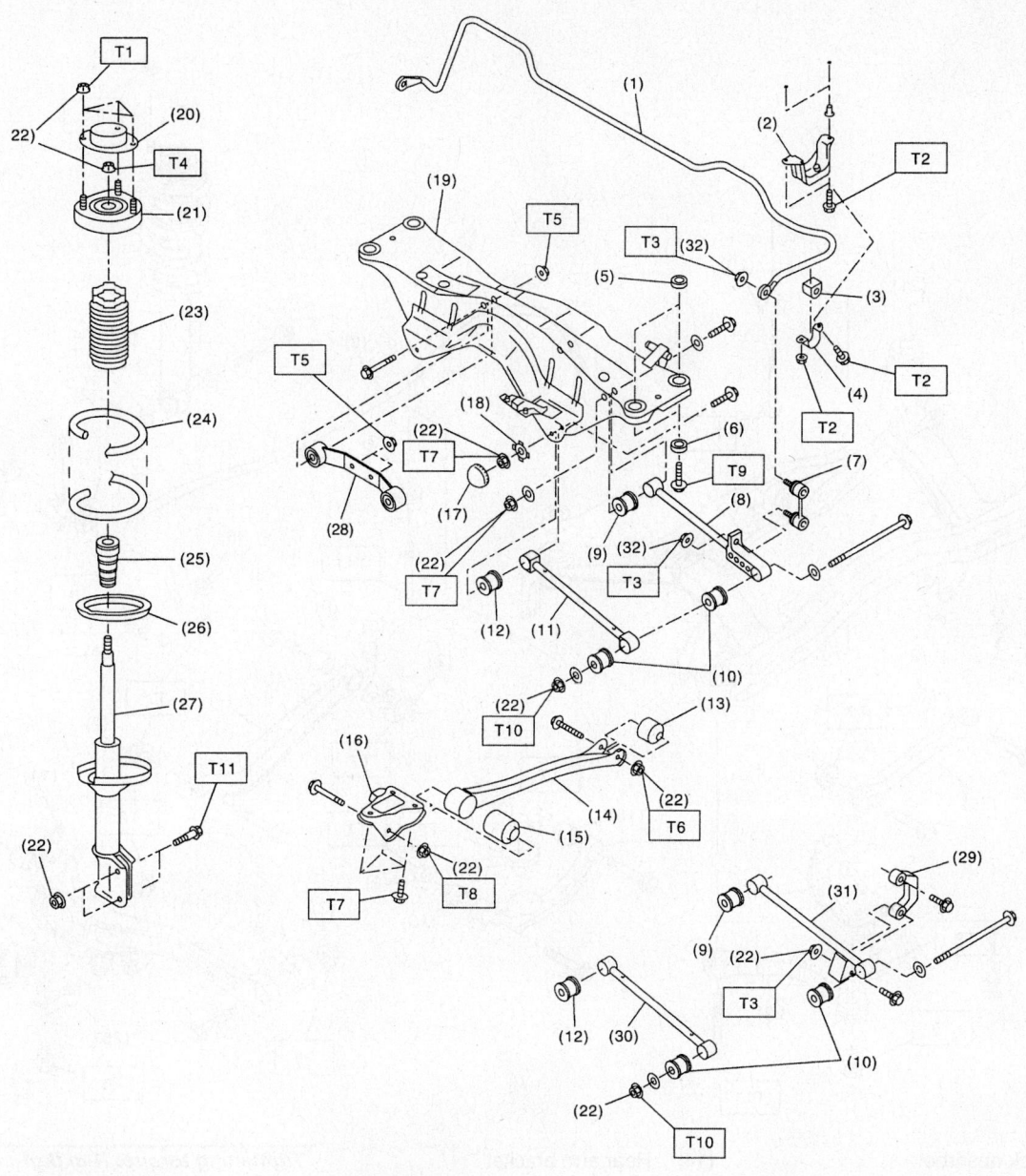

(1)	Stabilizer
(2)	Stabilizer bracket
(3)	Stabilizer bushing
(4)	Clamp
(5)	Floating bushing
(6)	Stopper
(7)	Stabilizer link (STI model)
(8)	Rear lateral link (STI model)
(9)	Bushing (C)
(10)	Bushing (A)
(11)	Front lateral link (Turbo model and STI model)
(12)	Bushing (B)
(13)	Trailing link rear bushing
(14)	Trailing link
(15)	Trailing link front bushing

(16)	Trailing link bracket
(17)	Cap (Protection)
(18)	Washer
(19)	Rear crossmember
(20)	Strut mount cap
(21)	Strut mount
(22)	Self-locking nut
(23)	Dust cover
(24)	Coil spring
(25)	Helper
(26)	Lower rubber seat
(27)	Damper strut
(28)	Differential rear member
(29)	Stabilizer link (Except STI model)
(30)	Front lateral link (Non-turbo model)

(31)	Rear lateral link (Except STI model)
(32)	Flange nut

Tightening torque: N·m (kgf-m, ft-lb)

T1:	**20 (2.0, 14.5)**
T2:	**25 (2.5, 18.1)**
T3:	**45 (4.6, 33.2)**
T4:	**55 (5.6, 40.6)**
T5:	**70 (7.1, 52)**
T6:	**90 (9.2, 66)**
T7:	**100 (10.2, 74)**
T8:	**115 (11.7, 85)**
T9:	**130 (13.3, 96)**
T10:	**140 (14.3 103)**
T11:	**200 (20.0, 145)**

09490_SBCR_G0101

Rear suspension and related components—2005–2006 Impreza and WRX

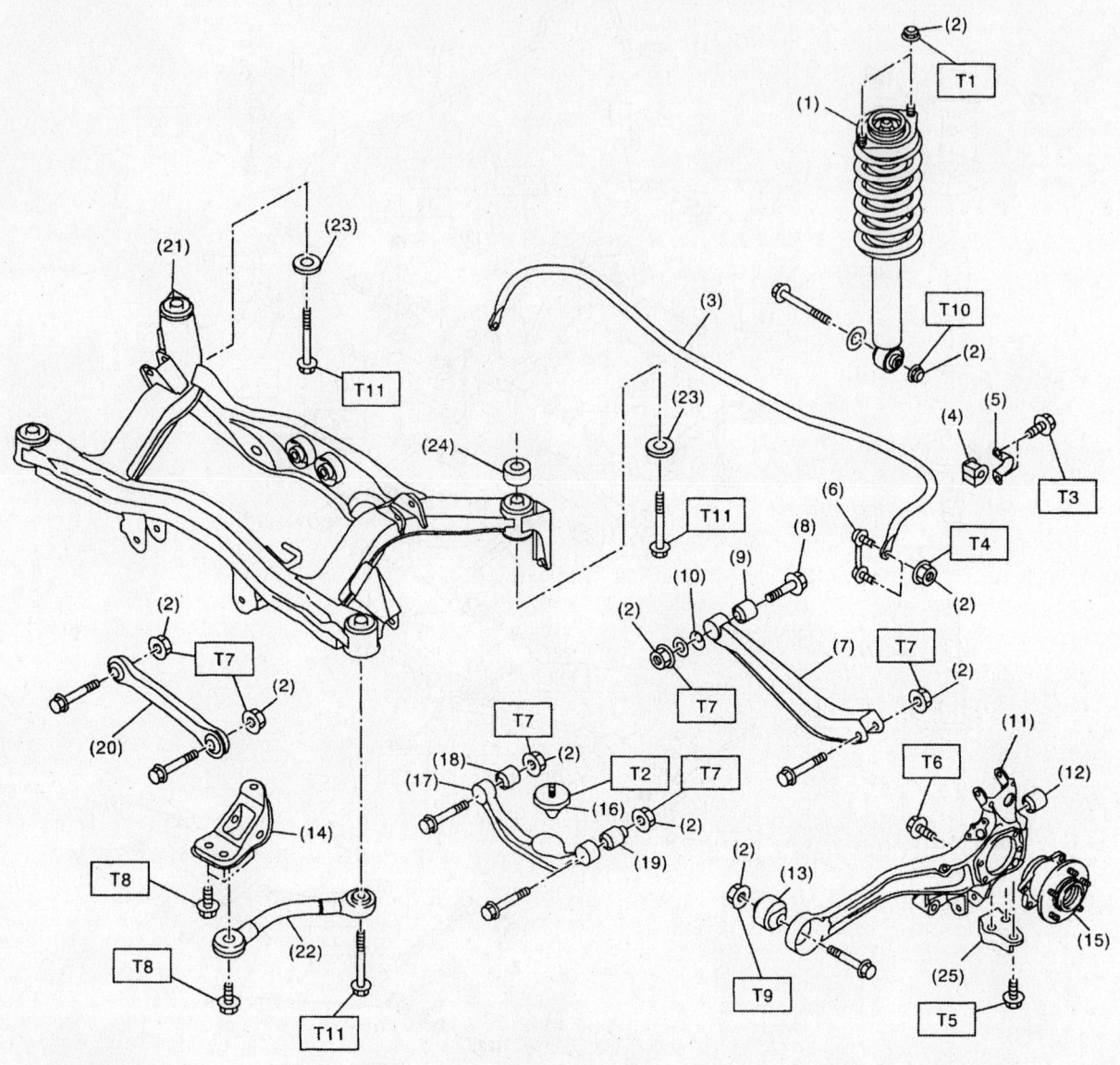

(1)	Shock absorber	(14)	Rear arm bracket
(2)	Self-locking nut	(15)	Hub bearing unit
(3)	Stabilizer	(16)	Helper
(4)	Stabilizer bushing	(17)	Upper link
(5)	Bracket	(18)	Upper link bushing (inner side)
(6)	Stabilizer link	(19)	Upper link bushing (outer side)
(7)	Rear link	(20)	Front link
(8)	Adjusting bolt	(21)	Rear sub frame
(9)	Rear link bushing	(22)	Sub frame support arm
(10)	Adjusting washer	(23)	Sub frame lower stopper
(11)	Rear arm	(24)	Spacer
(12)	Rear arm rear bushing	(25)	Stabilizer link bracket (model with aluminum rear arm)
(13)	Rear arm front bushing		

Tightening torque: N·m (kgf-m, ft-lb)

T1:	30 (3.1, 22.4)
T2:	32 (3.3, 24)
T3:	40 (4.1, 30)
T4:	44 (4.5, 32.5)
T5:	60 (6.1, 44)
T6:	65 (6.6, 48)
T7:	120 (12.2, 89)
T8:	125 (12.7, 92)
T9:	150 (15.3, 111)
T10:	160 (16.3, 118)
T11:	175 (17.8, 129)

09490_SBCR_G0102

Rear suspension and related components—2005–2006 Legacy and Outback

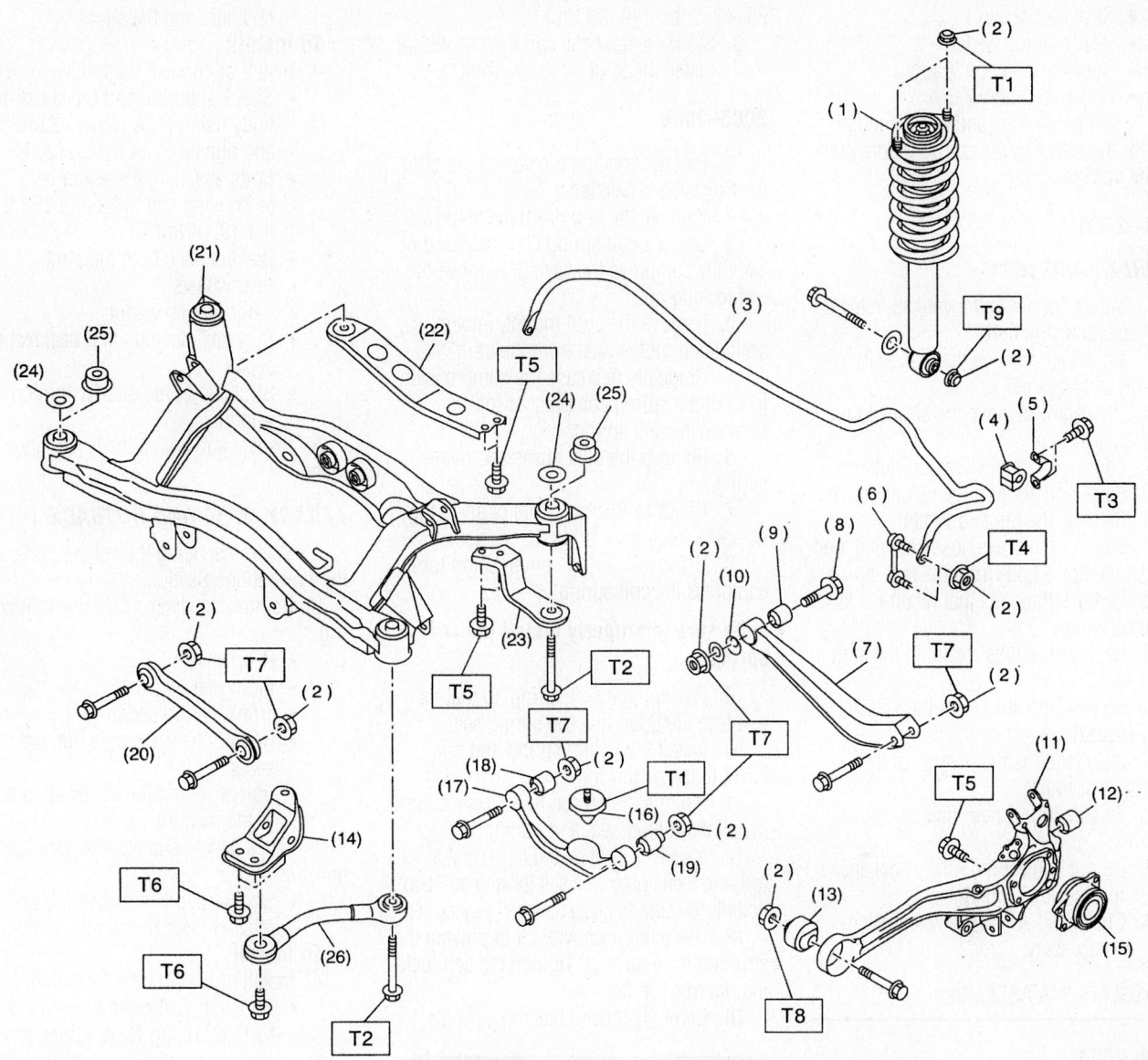

(1)	Shock absorber	(15)	Hub bearing unit
(2)	Self-locking nut	(16)	Helper
(3)	Stabilizer	(17)	Link upper
(4)	Stabilizer bushing	(18)	Link upper bushing (Inside)
(5)	Clamp	(19)	Link upper bushing (Outside)
(6)	Stabilizer link	(20)	Link front
(7)	Link rear	(21)	Rear sub frame
(8)	Adjusting bolt	(22)	Support sub frame (RH)
(9)	Link rear bushing	(23)	Support sub frame (LH)
(10)	Adjusting washer	(24)	Stopper upper (Except OUTBACK model)
(11)	Rear arm	(25)	Stopper upper (OUTBACK MODEL)
(12)	Rear arm rear bushing	(26)	Support sub frame front
(13)	Rear arm front bushing		
(14)	Rear arm bracket		

Tightening torque: N·m (kgf-m, ft-lb)

T1:	30 (3.1, 22.4)
T2:	175 (17.8, 129)
T3:	40 (4.1, 30)
T4:	44 (4.5, 32.5)
T5:	65 (6.6, 48)
T6:	125 (12.8, 92)
T7:	120 (12.2, 88)
T8:	150 (15.3, 111)
T9:	160 (16.3, 118)

09490_SBCR_G0103

Rear suspension and related components—2005–2006 Baja

- Air line to the top of the strut
- Strut mount cap
- Rear seat, on Sedan
- Speaker grille, on Wagon
- Negative battery cable

5. Start the vehicle, and allow enough time for the shock to pressurize before driving the vehicle.

2005–2006

IMPREZA AND WRX

1. Before servicing the vehicle, refer to the Precautions Section.

2. On sedan, remove the rear seat cushion and backrest.

3. On wagon, remove the strut cap on the quarter trim.

4. Loosen the rear wheel lug nuts.

5. Raise and support the vehicle safely.

6. Remove the tire and wheel.

7. Remove the brake hose clip, and then remove the brake hose from the rear strut.

8. Remove the bolts that retain the strut to the housing.

9. Remove the nuts retaining the strut to the body.

10. Remove the strut from the vehicle.

To install:

11. Installation is the reverse of the removal procedure.

12. Be sure to use new locknuts, as required.

13. Do not subject the ABS wheel speed sensor to excessive tension.

14. Check and adjust the wheel alignment, as necessary.

DISASSEMBLY & ASSEMBLY

2002–2004

1. Before servicing the vehicle, refer to the Precautions Section.

2. Remove the strut assembly from the vehicle and secure in a soft jawed vise.

3. Compress the coil spring with a spring compressor until the upper spring seat can be turned by hand.

4. Remove the self-locking nut on the top of the strut assembly, and then remove the upper spring seat.

5. Remove the coil spring and compressor. If the spring is being replaced, slowly release the spring from the compressor and compress the new coil spring.

To install:

6. Place the proper end of the coil spring on the lower spring seat on the strut.

7. Install the insulator, upper spring seat and strut mount on the strut piston. Install a new self-locking nut. Tighten the nut to 36–43 ft. lbs. (49–59 Nm).

8. Slowly release the spring compressor.

9. Install the strut on to the vehicle.

2005–2006

1. Before servicing the vehicle, refer to the Precautions Section.

2. Remove the strut from the vehicle.

3. Using a coil spring compressor tool, carefully compress the spring. Remove the self locking nut.

4. Remove the strut mount, upper spring and rubber seat from the strut.

5. Gradually decrease the compression force of the spring compressor tool. Remove the coil spring.

6. Remove the dust cover and helper spring.

7. Check for the presence of air in the damping force generating mechanism.

8. Using the spring compression tool, compress the coil spring.

➡**Be sure to properly install the coil spring.**

9. Position the coil spring so that its end face fits good into the spring seat.

10. Install the helper spring and dust cover to the piston rod.

11. Pull the piston rod fully upward, and install the rubber seat and spring seat.

12. Install the strut mount to the piston rod, and then tighten the self locking nut, temporarily. Be sure to use a new self locking nut.

13. Use a hexagon wrench to prevent the strut rod from turning. Tighten the self locking nut to 41 ft. lbs.

14. Carefully loosen the coil spring.

Shock Absorber

REMOVAL & INSTALLATION

2002–2004

IMPREZA

1. Before servicing the vehicle, refer to the Precautions Section.

2. Remove or disconnect the following:

- Rear seat cushion and backrest on sedan
- Strut cap from the quarter trim on wagon
- Rear wheels
- Brake hose clip from the strut
- Bolts attaching the shock absorber to housing

3. Use a jack to support the rear suspension.

- Shock absorber-to-body upper retainers and the shock

To install:

4. Install or connect the following:

- Shock absorber and the shock-to-body using NEW upper retainers and tighten to 14 ft. lbs. (20 Nm).
- Bolts attaching the shock absorber to housing and tighten to 162 ft. lbs. (220 Nm)
- Brake hose clip to the strut
- Rear wheels
- Floor mat on wagon
- Rear seat cushion and backrest on sedan
- Strut cap to the quarter trim on wagon

5. Inspect and adjust the wheel alignment.

LEGACY, BAJA AND OUTBACK

1. Before servicing the vehicle, refer to the Precautions Section.

2. Remove or disconnect the following:

- Rear wheels
- Floor mat on wagon
- Trunk mat on sedan
- Roll up the trunk side trim on sedan
- Bolt attaching the shock absorber to the rear arm

3. Use a jack to support the rear suspension.

- Shock absorber-to-body upper retainers and the shock

To install:

4. Install or connect the following:

- Shock absorber and the shock-to-body using NEW upper retainers and tighten to 22 ft. lbs. (30 Nm).

5. Place the vehicle jack (the one supplied with the vehicle) upside down and place it between the link rear and the subframe. Place a cloth between the jack and areas it is touching to prevent damage to the link rear or sub-frame Adjust the jack so the shock is aligned with the rear are at the correct holes and install the lower shock bolts.

6. Support the shock/rear rear arm horizontally with a jack and tighten the nuts and bolts to 118 ft. lbs. (160 Nm).

7. Install or connect the following:

- Floor mat on wagon
- Trunk side trim on sedan
- Trunk mat on sedan
- Rear wheels

8. Inspect and adjust the wheel alignment.

2005–2006

LEGACY, OUTBACK AND BAJA

1. Before servicing the vehicle, refer to the Precautions Section.

2. On sedan, roll up the trunk side trim.

3. On wagon, remove the luggage floor mat.

4. On Baja, remove the rear pillar lower trim.

5. Loosen the rear wheel lug nuts.

6. Raise and support the vehicle safely.

7. Remove the tire and wheel.

8. Remove the bolts that retain the shock absorber to the rear arm.

9. Using a jack, support the shock absorber.

10. Remove the nuts that retain the shock absorber mount to the vehicle.

11. Remove the shock absorber from the vehicle.

To install:

12. Installation is the reverse of the removal procedure.

13. Be sure to use new bolts and nuts, as required.

14. Check and adjust the wheel alignment, as necessary.

Stabilizer Bar

REMOVAL & INSTALLATION

1. Before servicing the vehicle, refer to the Precautions Section.

2. Loosen the rear wheel lug nuts.

3. Raise and support the vehicle safely.

4. Remove the tire and wheel.

5. Remove the bolts that secure the stabilizer link to the rear arm.

6. Remove the bolts which secure the stabilizer bar to the sub frame.

7. Remove the stabilizer bar from the vehicle.

To install:

8. Installation is the reverse of the removal procedure.

9. Be sure that the stabilizer bar and the bushings have the same identification markings and/or colors.

10. Be sure to use new bolts and nuts, as required.

11. Always fully tighten the rubber bushings when the wheels are in full contact with the ground and the vehicle is at curb height.

12. Check and adjust the wheel alignment, as necessary.

Rear Lower Control Arm

REMOVAL & INSTALLATION

Trailing Link

2002–2004 IMPREZA

1. Before servicing the vehicle, refer to the Precautions Section.

2. Remove or disconnect the following:
- Tire and wheel assembly
- Rear parking bracket clamp and Anti-lock Brake System (ABS) sensor harness, if equipped
- Bolts that secure the trailing link to the bracket
- Bolt that secures the trailing link to the rear housing
- Trailing link from the vehicle

To install:

3. Install or connect the following:
- Trailing link and the through-bolts and nuts. DO NOT tighten the nuts and bolts at this time.
- ABS sensor bracket, if equipped, and parking brake cable to the trailing link.
- Tire and wheel assembly

4. Tighten the trailing link-to-bracket bolt to 72 ft. lbs. (98 Nm) and the nut to 83 ft. lbs. (113 Nm).

5. Tighten the trailing link-to-rear housing to 83 ft. lbs. (113 Nm).

6. Check the wheel alignment and adjust if necessary.

2005–2006 IMPREZA AND WRX

1. Before servicing the vehicle, refer to the Precautions Section.

2. Loosen the wheel nuts.

3. Raise and support the vehicle safely.

4. Remove the tire and wheel.

5. Remove both the rear parking brake clamp and the ABS wheel speed sensor harness.

6. Remove the bolt which secures the trailing link to the trailing link bracket.

7. Remove the bolt which secures the trailing link to the rear housing.

8. Remove the trailing link from the vehicle.

To install:

9. Installation is the reverse of the removal procedure.

10. Be sure to use new bolts and nuts, as required.

11. Always fully tighten the rubber bushings when the wheels are in full contact with the ground and the vehicle is at curb height.

12. Check and adjust the wheel alignment, as necessary.

Lateral Link

2002–2004 IMPREZA

1. Before servicing the vehicle, refer to the Precautions Section.

2. Remove or disconnect the following:
- Tire and wheel assembly
- Stabilizers on turbocharged engine
- Anti-lock Brake System (ABS) sensor harness from the trailing link, if equipped.
- Bolts that secure the lateral link to the rear housing.

➡Discard the old self-locking nuts and replace with new ones during installation.

- Bolts which secure the trailing link to the rear housing
- Halfshaft from the rear differential using a suitable tool.

➡On all except 2.2L engine, do not remove the circlip attached to the inside of the differential. On 2.2L engine, the side spline circlip comes out together with the shaft. Be careful not to damage the side bearing retainer.

3. Scribe an alignment mark on the rear lateral link adjusting bolt and crossmember.

4. Remove or disconnect the following:
- Outer lateral link bolt securing the lateral link to the housing
- Bolts securing the front and rear lateral links to the crossmember
- Lateral links from the vehicle

To install:

5. Install or connect the following:
- Bolts securing the front and rear lateral links to the crossmember and hand-tighten
- Outer lateral link bolt securing the lateral link to the housing and hand-tighten
- Halfshaft to the rear differential
- Bolts that secure the lateral link to the rear housing
- Bolts which secure the trailing link to the rear housing
- ABS sensor harness to the trailing link, if equipped
- Tire and wheel assembly

6. Tighten the lateral link bolts as shown in the illustration.

7. Check the wheel alignment and adjust if necessary.

2005–2006 IMPREZA AND WRX

1. Before servicing the vehicle, refer to the Precautions Section.

2. Loosen the wheel nuts.

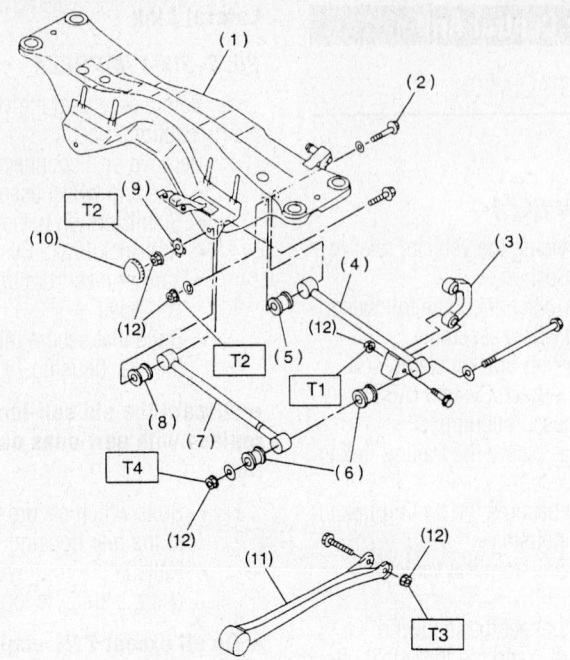

(1) Crossmember	(8) Bushing (B)
(2) Adjusting bolt	(9) Washer
(3) Stabilizer link	(10) Cap
(4) Rear lateral link	(11) Trailing link
(5) Bushing (C)	(12) Self-locking nut
(6) Bushing (A)	
(7) Front lateral link	

Tightening torque: N·m (kg-m, ft-lb)
T1: 44±6 (4.5±0.6, 32.5±4.3)
T2: 98±15 (10.0±1.5, 72±11)
T3: 113±15 (11.5±1.5, 83±11)
T4: 137±20 (14.0±2.0, 101±14)

9307TG10

Lateral link mounting and tightening specifications—2002–2004 Impreza

3. Raise and support the vehicle safely.
4. Remove the tire and wheel.
5. Remove the stabilizers.
6. Remove the ABS wheel speed sensor harness from the trailing arm.
7. Remove the bolt securing the trailing link to the rear housing.
8. Remove the bolts which secure the lateral link assembly to the rear housing.
9. Remove the DOJ from the rear differential, using tool ST28099PA100 or equivalent.

➡**The side spline snapring comes out together with the shaft. Be careful not to damage the side bearing retainer when using the special tool.**

10. Scribe an alignment mark on the rear lateral link adjusting bolt and crossmember.
11. Remove the bolts securing the front and rear lateral links to the crossmember, detach the lateral links.

➡**To loosen the adjusting bolt, always loosen the nut while holding the head of the adjusting bolt.**

To install:
12. Installation is the reverse of the removal procedure.

13. Be sure to use new bolts and nuts, as required.
14. Always fully tighten the rubber bushings when the wheels are in full contact with the ground and the vehicle is at curb height.
15. Check and adjust the wheel alignment, as necessary.

Rear Arm

2002–2004 LEGACY AND BAJA

1. Before servicing the vehicle, refer to the Precautions Section.
2. Remove or disconnect the following:

- Tire and wheel assembly
- Wheel bearing assembly, refer to the procedure in this section
- Bolt holding the parking brake cable to the control arm
- Bolt securing the brake hose to the rear arm
- Bolt securing the Anti-lock Brake Sensor (ABS) sensor to the rear arm
- Brake line from the wheel cylinder with a flare nut wrench, if equipped with drum brakes. Plug the line to avoid contaminating the system.

Suspend the brake backing plate from the sub-frame.
- Nut securing the stabilizer link to the rear arm
- Bolt holding the shock absorber to the rear arm

3. Use a suitable transmission jack to support the rear arm horizontally.
4. Remove or disconnect the following:
- Bolt securing the rear arm to the body
- Nut securing the front link to the rear arm, loosen
- Nut securing the rear link to the rear arm, loosen
- Bolts holding the rear arm to the links
- Rear arm

To install:
5. Use a transmission jack to support the rear arm.
6. Install or connect the following:
- Rear arm and temporarily tighten the bolts securing the rear arm to the link
- Wheel bearing unit
- Bolt securing the ABS sensor to the rear arm
- Brake hose-to-rear arm bolt
- Parking brake cable clamp-to-rear arm bolt

➡**Place a rag or cloth between the jack and its mating area to avoid scratching the rear link and sub-frame.**

7. Place the tire changing jack (supplied with the car) upside down and position between the rear link and sub-frame. Adjust the jack position so the rear shock absorber is aligned with the rear arm at their corresponding holes. Install the lower shock absorber bolts.
8. Using the transmission jack, support the rear arm horizontally, then tighten the nuts and bolts holding the rear arm, front and rear links, upper link and shock absorber. Refer to the specifications in the illustration.
9. Install the tire and wheel assembly.
10. Check and adjust the alignment, if necessary.

2005–2006 LEGACY AND OUTBACK

1. Before servicing the vehicle, refer to the Precautions Section.
2. Loosen the wheel nuts.
3. Raise and support the vehicle safely.
4. Remove the tire and wheel.
5. Remove the sub frame support arm.
6. Remove the bearing unit.
7. Hang the backing plate from the sub frame.

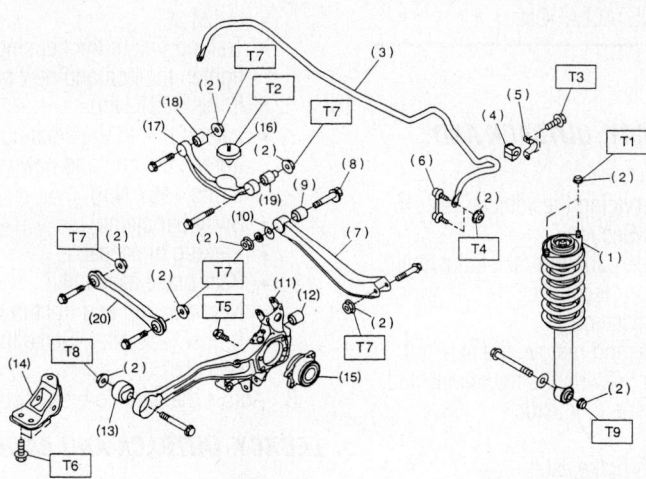

(1)	Shock absorber	(12)	Rear arm rear bushing
(2)	Self-locking nut	(13)	Rear arm front bushing
(3)	Stabilizer	(14)	Rear arm bracket
(4)	Stabilizer bushing	(15)	Hub bearing unit
(5)	Clamp	(16)	Helper
(6)	Stabilizer link	(17)	Link upper
(7)	Link rear	(18)	Link upper bushing (Inside)
(8)	Adjusting bolt	(19)	Link upper bushing (Outside)
(9)	Link rear bushing	(20)	Link front
(10)	Adjusting washer		
(11)	Rear arm		

Tightening torque: N·m (kg-m, ft-lb)
T1: 30±7 (3.1±0.7, 22.4±5.1)
T2: 32±10 (3.3±1.0, 23.9±7.2)
T3: 39±7 (4.0±0.7, 28.9±5.1)
T4: 44±6 (4.5±0.6, 32.5±4.3)
T5: 66±10 (6.7±1.0, 48.5±7.2)
T6: 108±15 (11±1.5, 80±11)
T7: 123±15 (12.5±1.5, 90±11)
T8: 147±20 (15±2, 108±14)
T9: 157±20 (16±2, 116±14)

9307TG16

Rear control arm mounting and tightening specifications—2002–2004 Legacy

8. Remove the bolt that secures the parking brake cable clamp to the rear arm bracket.

9. Remove the bolt which holds the brake hose bracket and the ABS wheel speed sensor bracket to the rear arm.

10. Remove the bolts which secure the brake hose bracket to the rear arm. Remove the bolts which secure the ABS wheel speed sensor to the rear arm.

11. Remove the stabilizer link from the rear arm. Remove the shock absorber from the rear arm.

12. Support the rear arm assembly, horizontally using a transmission jack.

13. Remove the nuts that retain the rear arm to the bracket. Remove the rear arm bracket.

14. Loosen the nut which holds the front link to the rear arm.

15. Loosen the nut which holds the rear link to the rear arm.

16. Loosen the nut which holds the upper link to the rear arm.

17. Remove the rear arm from the vehicle.

To install:

18. Installation is the reverse of the removal procedure.

19. Be sure to use new bolts and nuts, as required.

20. Always fully tighten the rubber bushings when the wheels are in full contact with the ground and the vehicle is at curb height.

21. Check and adjust the wheel alignment, as necessary.

2005–2006 BAJA

1. Before servicing the vehicle, refer to the Precautions Section.

2. Loosen the wheel nuts.

3. Raise and support the vehicle safely.

4. Remove the tire and wheel.

5. Remove the sub frame support arm.

6. Remove the bearing unit.

7. Remove the bolt securing the parking brake cable clamp to the rear arm.

8. Remove the bolt securing the brake hose to the rear arm.

9. Remove the bolt securing the ABS sensor to the rear arm.

10. Hang the backing plate from the sub frame.

11. Remove the nut securing the stabilizer link from the rear arm. Remove the shock absorber retaining bolt from the rear arm.

12. Support the rear arm assembly, horizontally using a transmission jack.

13. Remove the bolt securing the rear arm to the body.

14. Loosen the nut which holds the front link to the rear arm.

15. Loosen the nut which holds the rear link to the rear arm.

16. Loosen the nut which holds the upper link to the rear arm.

17. Remove the bolts securing the rear arm to the links. Remove the rear arm from the vehicle.

To install:

18. Installation is the reverse of the removal procedure.

19. Be sure to use new bolts and nuts, as required.

20. Always fully tighten the rubber bushings when the wheels are in full contact with the ground and the vehicle is at curb height.

21. Check and adjust the wheel alignment, as necessary.

CONTROL ARM BUSHING REPLACEMENT

1. Remove the control arm from the vehicle.

2. Scribe a matchmark on the control arm and rear bushing.

3. Loosen the nut and remove the rear bushing. Discard the nut.

To install:

4. Install the rear bushing to the control arm, making sure to align the marks made during removal.

5. Install a new nut.

Upper Link

2005–2006 LEGACY, OUTBACK AND BAJA

1. Before servicing the vehicle, refer to the Precautions Section.

2. Loosen the wheel nuts.

3. Raise and support the vehicle safely.

4. Remove the tire and wheel.

5. Support the rear arm horizontally using a transmission jack.

6. Remove the bolt which secures the upper link to the sub frame.

7. Remove the bolts which secure the upper link to the rear arm. Remove the upper link.

To install:

8. Installation is the reverse of the removal procedure.

9. Be sure to use new bolts and nuts, as required.

10. Always fully tighten the rubber bushings when the wheels are in full contact with the ground and the vehicle is at curb height.

11. Check and adjust the wheel alignment, as necessary.

Front Link

2005–2006 LEGACY, OUTBACK AND BAJA

1. Before servicing the vehicle, refer to the Precautions Section.

2. Loosen the wheel nuts.

3. Raise and support the vehicle safely.

4. Remove the tire and wheel.

5. Support the rear arm horizontally using a transmission jack.

6. Remove the bolt which secures the front link to the sub frame.

7. Remove the bolts which secure the front link to the rear arm. Remove the front link.

To install:

8. Installation is the reverse of the removal procedure.

9. Be sure to use new bolts and nuts, as required.

10. Always fully tighten the rubber bushings when the wheels are in full contact with the ground and the vehicle is at curb height.

11. Check and adjust the wheel alignment, as necessary.

Rear Link

2005–2006 LEGACY, OUTBACK AND BAJA

1. Before servicing the vehicle, refer to the Precautions Section.

2. Loosen the wheel nuts.

3. Raise and support the vehicle safely.

4. Remove the tire and wheel.

5. Remove the stabilizer bar.

6. Support the rear arm horizontally using a transmission jack.

7. Remove the bolts which secure the rear link to the rear arm.

8. Place alignment marks on the rear link adjusting bolt and sub frame.

9. Remove the bolt which secures the rear link to the sub frame. Remove the rear link.

➡**When loosening the adjusting bolt, make sure to hold the bolt head in place when loosening the nut.**

To install:

10. Installation is the reverse of the removal procedure.

11. Be sure to use new bolts and nuts, as required.

12. Always fully tighten the rubber bushings when the wheels are in full contact with the ground and the vehicle is at curb height.

13. Check and adjust the wheel alignment, as necessary.

Wheel Bearings

ADJUSTMENT

The wheel bearings are not adjustable.

REMOVAL & INSTALLATION

2002–2004

EXCEPT LEGACY, OUTBACK AND BAJA

1. Before servicing the vehicle, refer to the Precautions Section.

2. Loosen the parking brake adjustment.

3. Remove or disconnect the following:
   - Wheel assembly
   - Unstake and remove the axle nut
   - Caliper, leaving the line connected, and suspend it aside
   - Rotor
   - Parking brake cable
   - Sway bar clamp
   - Bolt securing the lateral link to the housing
   - Bolts securing the trailing link to the housing
   - Halfshaft
   - Bolts securing the strut to the housing
   - Speed sensor from the backing plate, if equipped with Anti-lock Brake System (ABS)
   - Housing assembly
   - Hub from the rear housing using Hub Stand 92708000 and Puller 927420000
   - Backing plate from the housing
   - Outer, inner and sub oil seals
   - Snapring
   - Bearing by pressing the inner race

To install:

4. Clean the housing thoroughly.

➡**Do not remove the plastic lock from the inner race when installing the bearing.**

5. Install or connect the following:
   - New bearing into the housing by pressing the outer race and pack the bearing with grease
   - Snapring and ensure that it fits properly
   - New outer seal until it contacts the snapring using a press
   - New inner seal until it contacts the bottom using a press
   - New sub oil seal and apply grease to the oil seal lip
   - Backing plate and tighten the bolts to 43 ft. lbs. (58 Nm)
   - Hub into the housing using installer 927450000 to press it into position
   - Housing to the strut and tighten the bolts to 119 ft. lbs. (162 Nm)
   - Speed sensor, if equipped with ABS

   - Halfshaft
   - Trailing link to the housing and tighten the bolt and new nut to 94 ft. lbs. (127 Nm)
   - Lateral link to the housing and tighten the bolt and new nut to 116 ft. lbs. (157 Nm)
   - Sway bar clamp
   - Parking brake cable
   - Rear brake assembly
   - New axle nut and tighten it to 152 ft. lbs. (206 Nm). Stake the nut.
   - Wheel

6. Adjust the parking brake cable.

LEGACY, OUTBACK AND BAJA

1. Before servicing the vehicle, refer to the Precautions Section.

2. Loosen the parking brake adjustment.

3. Remove or disconnect the following:
   - Wheel assembly
   - Unstake and remove the axle nut
   - Parking brake lever
   - ABS sensor
   - Caliper, leaving the line connected, and suspend it aside
   - Rotor
   - Four bolts from the rear arm and the hub and bearing assembly

4. Disassemble the hub/bearing assembly as follows:
   a. Place the hub/bearing in a press.
   b. Using dummy collar tool ST 398507703 to press the bearing from the hub.
   c. Place the hub assembly on hub stand ST 927080000 and using a common puller and tool ST 399520105, remove the inner race of the bearing.

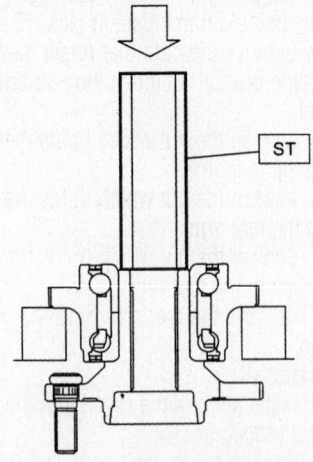

42356-SBCR-G50

Use dummy collar tool ST 398507703 to press the bearing from the hub— 2002–2004 Legacy, Outback and Baja

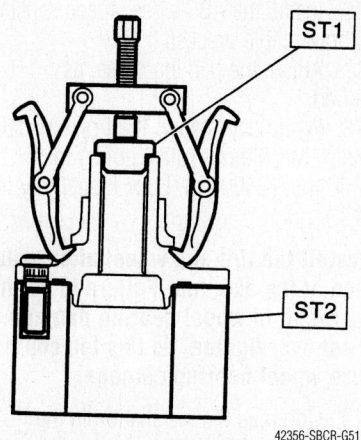

42356-SBCR-G51

Place the hub assembly on stand ST 927080000 and use a common puller and tool ST 399520105, remove the bearing inner race—2002–2004 Legacy, Outback and Baja

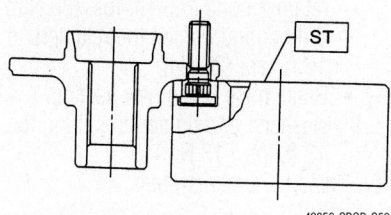

42356-SBCR-G52

Using the hub stand tool, press out the hub bolt—2002–2004 Legacy, Outback and Baja

Discard the bearing as it should not be reused.

   d. Using the hub stand tool, press out the hub bolt.

To install:

5. Install the bearing assembly to the hub as follows:

   a. Press a new hub bolt into place using the hub stand. Make sure the bolt closely contacts the hub. use a 0.47 inch (12mm) hole in the hub stand to prevent the bolt from tilting while installing.

   b. Using the hub stand, hub installer ST 927450000 and spacer ST 28499AE000, press the NEW bearing assembly into the hub. Make sure to always press on the inner race while installing.

6. Install or connect the following:

- Hub assembly with the mounting holes on the backing plate and install the hub assembly and backing plate. Temporarily tighten the axle nuts. be careful not to damage the tone ring.

ST1 927080000 HUB STAND
ST2 927450000 HUB INSTALLER
ST3 28499AE000 SPACER

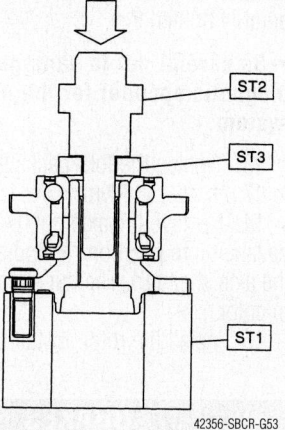

42356-SBCR-G53

Installing the bearing into the hub—2002–2004 Legacy, Outback and Baja

- Four bolts and tighten to 48 ft. lbs. (66 Nm).

7. Remove the axle nut, use axle shaft installer tool ST 922431000 and adapter 927390000 to pull the axle shaft into position and hand tighten the axle nut

- Rotor
- Caliper
- ABS sensor
- Parking brake lever
- Tighten the axle nut to 174 ft. lbs. (235 Nm) and stake the nut.
- Wheel

8. Adjust the parking brake cable.

2005–2006

IMPREZA AND WRX

1. Before servicing the vehicle, refer to the Precautions Section.
2. Loosen the wheel nuts.
3. Disconnect the negative battery cable.
4. Raise and support the vehicle safely.
5. Remove the tire and wheel.
6. Remove the crimped section of the axle nut. Depress the parking brake and remove the axle nut.

➡**Be sure to loosen the axle nut after removing the tire and wheel from the vehicle. Failure to do this may damage the wheel bearings.**

7. Release the parking brake lever and loosen the locknut.
8. Remove the brake caliper from its mounting. Wire it to the side; do not allow it to hang. Do not disconnect the brake fluid line.

9. Remove the rotor.

➡**If the rotor is seized within the hub, remove the rotor by installing an 8mm bolt in the screw hole on the rotor.**

10. Disconnect the parking brake cable end.
11. Disconnect the rear stabilizer from the rear lateral link.
12. Remove the bolts that retain the trailing link assembly to the rear housing.
13. Remove the bolts that retain the lateral assembly to the rear housing.
14. Remove the rear ABS wheel speed sensor from the backing plate.
15. Disengage the BJ from the housing splines, and remove the rear halfshaft assembly. If necessary, use tool ST92647000 and ST28099PA110.

➡**Be careful not to damage the oil seal lip when removing the rear halfshaft. When the rear halfshaft is replaced also replace the inner oil seal.**

16. Remove the bolts that secure the rear housing to the strut. Separate the two.

To install:

17. Installation is the reverse of the removal procedure.
18. Be sure to use new bolts and nuts, as required.
19. Always fully tighten the rubber bushings when the wheels are in full contact with the ground and the vehicle is at curb height.
20. Check and adjust the wheel alignment, as necessary.
21. Adjust the rear parking brake, as required.

LEGACY, OUTBACK AND BAJA

1. Before servicing the vehicle, refer to the Precautions Section.
2. Loosen the wheel nuts.
3. Disconnect the negative battery cable.
4. Raise and support the vehicle safely.
5. Remove the tire and wheel.
6. Remove the crimped section of the axle nut. Depress the parking brake and remove the axle nut.

➡**Be sure to loosen the axle nut after removing the tire and wheel from the vehicle. Failure to do this may damage the wheel bearings.**

7. Release the parking brake lever.
8. Remove the rear ABS wheel speed sensor.

9. Remove the brake caliper from its mounting. Wire it to the side; do not allow it to hang. Do not disconnect the brake fluid line.

10. Remove the rotor. Matchmark the rotor and hub and bearing to aid in reassembly.

➡ **If the rotor is seized within the hub, remove the rotor by installing an 8mm bolt in the screw hole on the rotor.**

11. Remove the hub and bearing unit. If it is hard to remove, use tools ST926470000 and ST927140000.

➡ **Be careful not to damage the magnetic encoder for the ABS system.**

To install:

12. Align the hub and bearing unit to the mounting hole of the backing plate. Install the assembly. Using a new axle nut temporarily tighten it.

➡ **Be careful not to damage the magnetic encoder for the ABS system.**

13. Tighten the four backing plate bolts to 47.9 ft. lbs. (65 Nm).

14. Remove the axle nut. Draw the rear halfshaft into position. Temporarily tighten the axle shaft nut. The nut should be olive in color.

15. Install the rotor. Install the caliper.

16. Install the ABS wheel speed sensor and brake cable bracket.

17. Adjust the parking brake, as required.

18. While depressing the brake pedal, tighten the axle nut (olive color) to 177 ft. lbs. (240 Nm). Lock it securely in place.

➡ **Install the tire and wheel after installation of the axle nut. Failure to do this may result in wheel bearing damage. Do not over tighten, as this too could cause wheel bearing damage.**

19. Continue the installation in the reverse order of the removal procedure.

BRAKES

Brake Caliper

REMOVAL & INSTALLATION

2002–2004

FRONT—WRX

1. Before servicing the vehicle, refer to the Precautions Section.

2. Remove or disconnect the following:
   - Front wheels
   - Brake hose from the caliper body
   - Caliper retainer bolts and the caliper
   - Caliper bracket, if necessary

To install:

3. Compress the piston assembly into the cylinder bore.

4. Install or connect the following:

   - Caliper bracket to the spindle assembly, and secure in place with the retainer bolts. Tighten the retainer bolts to 59 ft. lbs. (80 Nm).
   - Caliper and tighten the retainers to 19 ft. lbs. (26 Nm)
   - Brake hose using new sealing washers, and tighten the fitting to 13 ft. lbs. (18 Nm)

5. Bleed the brake system.

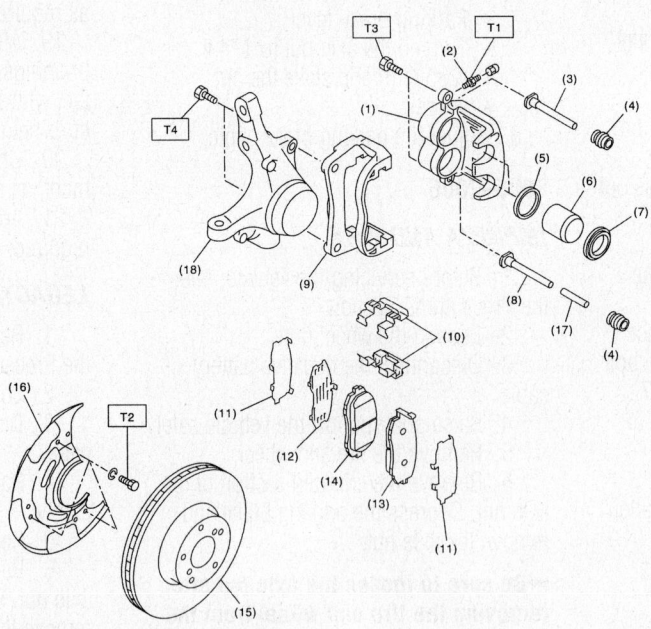

(1)	Caliper body	(9)	Support	(17)	Bushing
(2)	Air bleeder screw	(10)	Pad clip	(18)	Housing
(3)	Guide pin (Green)	(11)	Outer shim		
(4)	Pin boot	(12)	Inner shim		
(5)	Piston seal	(13)	Pad (Outside)		
(6)	Piston	(14)	Pad (Inside)		
(7)	Piston boot	(15)	Disc rotor		
(8)	Lock pin (Yellow)	(16)	Disc cover		

Tightening torque: N·m (kgf-m, ft-lb)

T1:	8 (0.8, 5.8)
T2:	18 (1.8, 13.0)
T3:	26.5 (2.7, 19.5)
T4:	80 (8.2, 59)

09490_SBCR_G0104

Front disc brakes and related components 15 inch type—2005–2006 Impreza and WRX

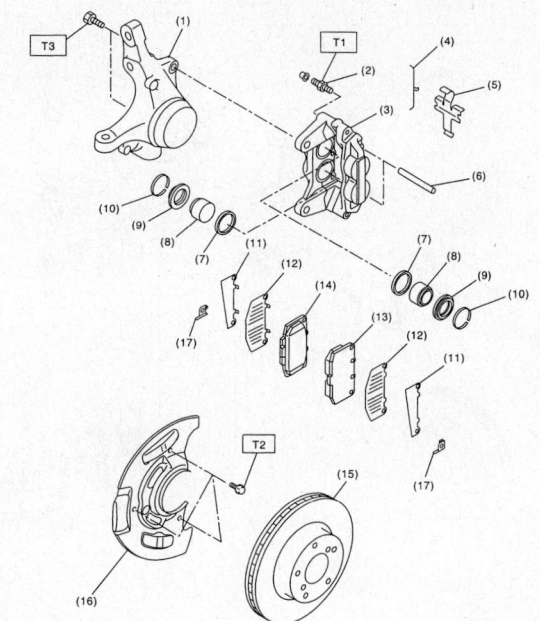

(1)	Housing	(10)	Boot ring	
(2)	Air bleeder screw	(11)	Outer shim	
(3)	Caliper body	(12)	Inner shim	
(4)	M clip	(13)	Pad (Outside)	
(5)	Cross spring	(14)	Pad (Inside)	
(6)	Pad pin	(15)	Disc rotor	
(7)	Piston seal	(16)	Disc cover	
(8)	Piston	(17)	Spacer	
(9)	Piston boot			

Tightening torque: N·m (kgf-m, ft-lb)
T1: 8 (0.8, 5.8)
T2: 18 (1.8, 13.0)
T3: 80 (8.2, 59)

09490_SBCR_G0105

Front disc brakes and related components 16 inch type—2005–2006 Impreza and WRX

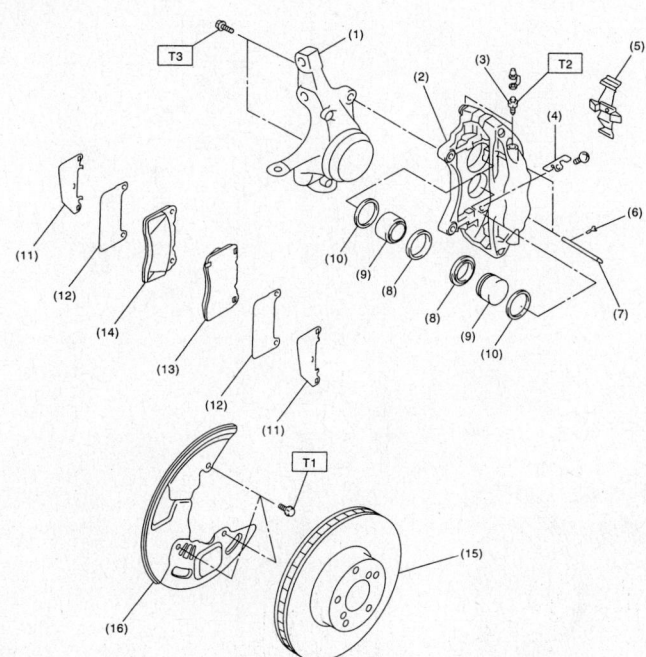

(1)	Housing	(8)	Piston boot	(15)	Disc rotor
(2)	Caliper body	(9)	Piston	(16)	Disc cover
(3)	Air bleeder screw	(10)	Piston seal		
(4)	Guide plate	(11)	Pad shim (Outside)		
(5)	Cross spring	(12)	Pad shim (Inside)		
(6)	Clip	(13)	Pad (Outside)		
(7)	Pad pin	(14)	Pad (Inside)		

Tightening torque: N·m (kgf-m, ft-lb)
T1: 18 (1.8, 13.0)
T2: 20 (2.0, 14.5)
T3: 155 (15.8, 114.3)

09490_SBCR_G0106

Front disc brakes and related components 17 inch type—2005–2006 Impreza and WRX

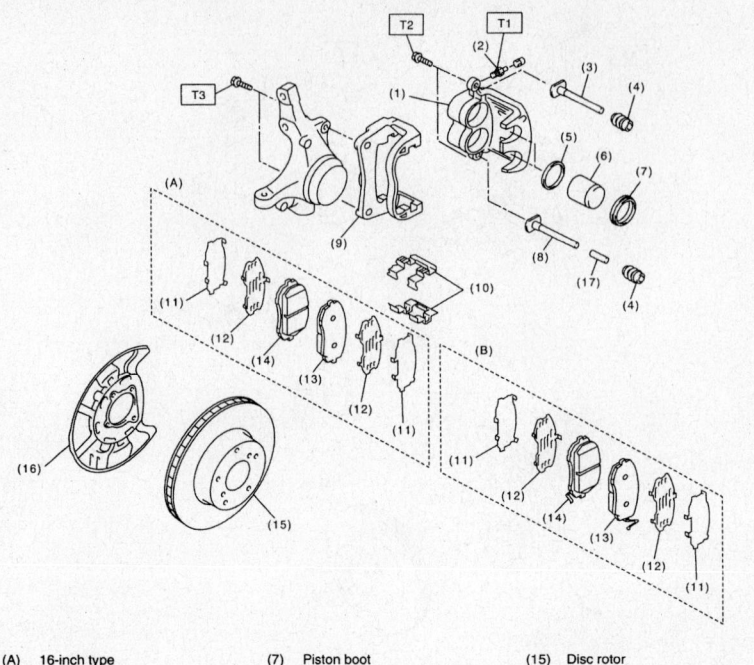

(A)	16-inch type	(7)	Piston boot	(15)	Disc rotor	
(B)	17-inch type	(8)	Lock pin (Yellow)	(16)	Disc cover	
		(9)	Support	(17)	Bushing	
(1)	Caliper body	(10)	Pad clip			
(2)	Air bleeder screw	(11)	Outer shim			
(3)	Guide pin (Green)	(12)	Inner shim			
(4)	Pin boot	(13)	Pad (Outside)			
(5)	Piston seal	(14)	Pad (Inside)			
(6)	Piston					

Tightening torque: N·m (kgf-m, ft-lb)
T1: 8 (0.8, 5.8)
T2: 27 (2.8, 19.9)
T3: 80 (8.2, 59)

09490_SBCR_G0107

Front disc brakes and related components—2005–2006 Outback and Legacy

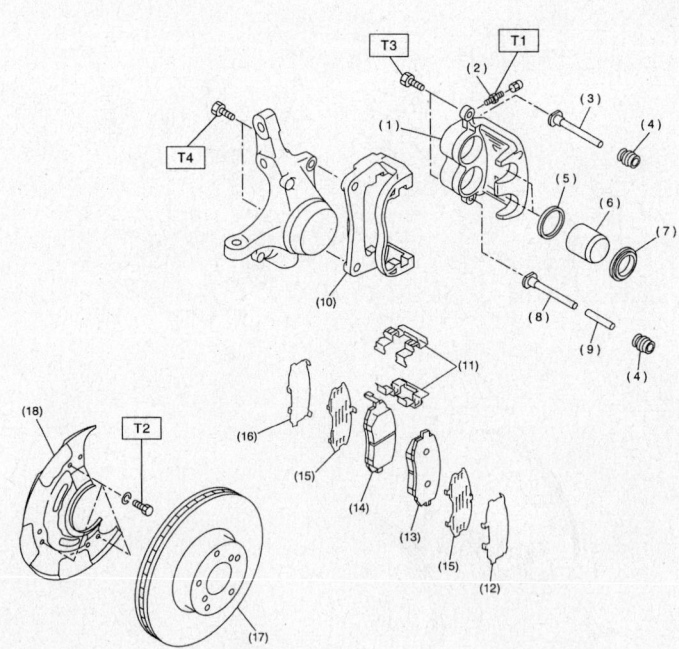

(1)	Caliper body	(9)	Bushing	(17)	Disc rotor	
(2)	Air bleeder screw	(10)	Support	(18)	Disc cover	
(3)	Guide pin (Green)	(11)	Pad clip			
(4)	Pin boot	(12)	Outer shim			
(5)	Piston seal	(13)	Pad (Outside)			
(6)	Piston	(14)	Pad (Inside)			
(7)	Piston boot	(15)	Rubber coated shim			
(8)	Lock pin (Yellow)	(16)	Inner shim			

Tightening torque: N·m (kgf-m, ft-lb)
T1: 8 (0.8, 5.8)
T2: 18 (1.8, 13.0)
T3: 37 (3.8, 27.5)
T4: 80 (8.2, 59)

09490_SBCR_G0108

Front disc brakes and related components—2005–2006 Baja

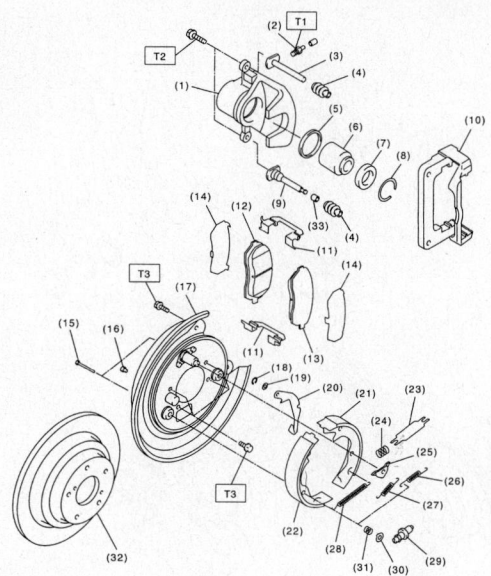

(1)	Caliper body	(14)	Shim	(27)	Primary shoe return spring
(2)	Air bleeder screw	(15)	Shoe hold-down pin	(28)	Adjusting spring
(3)	Guide pin (Green)	(16)	Cover	(29)	Adjuster
(4)	Pin boot	(17)	Back plate	(30)	Shoe hold-down cup
(5)	Piston seal	(18)	Retainer	(31)	Shoe hold-down spring
(6)	Piston	(19)	Spring washer	(32)	Disc rotor
(7)	Piston boot	(20)	Parking brake lever	(33)	Bushing
(8)	Boot ring	(21)	Parking brake shoe (Secondary)		
(9)	Lock pin (Yellow)	(22)	Parking brake shoe (Primary)		
(10)	Support	(23)	Strut		
(11)	Pad clip	(24)	Strut shoe spring		
(12)	Inner pad	(25)	Shoe guide plate		
(13)	Outer pad	(26)	Secondary shoe return spring		

Tightening torque: N·m (kgf-m, ft-lb)
T1: *8 (0.8, 5.8)*
T2: *37 (3.8, 27.5)*
T3: *53 (5.4, 39.1)*

09490_SBCR_G0109

Rear disc brakes and related components 14 inch type—2005–2006 Impreza and WRX

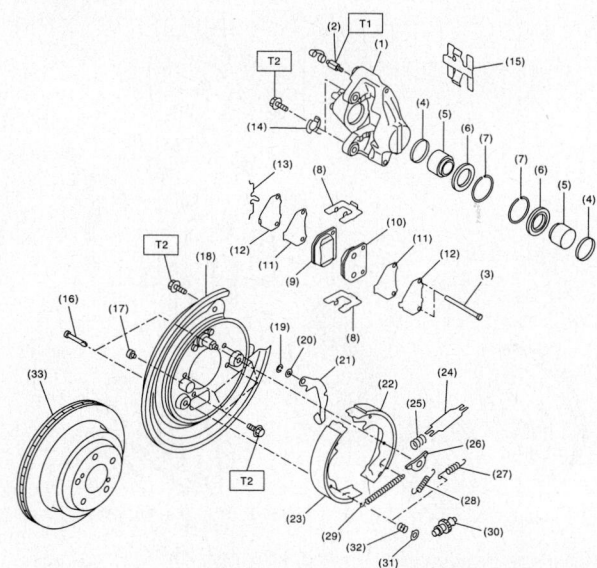

(1)	Caliper body	(14)	Washer	(27)	Secondary shoe return spring
(2)	Air bleeder screw	(15)	Cross spring	(28)	Primary shoe return spring
(3)	Pad pin	(16)	Shoe hold-down pin	(29)	Adjusting spring
(4)	Piston seal	(17)	Cover	(30)	Adjuster
(5)	Piston	(18)	Back plate	(31)	Shoe hold-down cup
(6)	Piston boot	(19)	Retainer	(32)	Shoe hold-down spring
(7)	Boot ring	(20)	Spring washer	(33)	Disc rotor
(8)	Pad clip	(21)	Parking brake lever		
(9)	Inner pad	(22)	Parking brake shoe (Secondary)		
(10)	Outer pad	(23)	Parking brake shoe (Primary)		
(11)	Inner shim	(24)	Strut		
(12)	Outer shim	(25)	Strut shoe spring		
(13)	M clip	(26)	Shoe guide plate		

Tightening torque: N·m (kgf-m, ft-lb)
T1: *8 (0.8, 5.8)*
T2: *53 (5.4, 39.1)*

09490_SBCR_G0110

Rear disc brakes and related components 15 inch type—2005–2006 Impreza and WRX

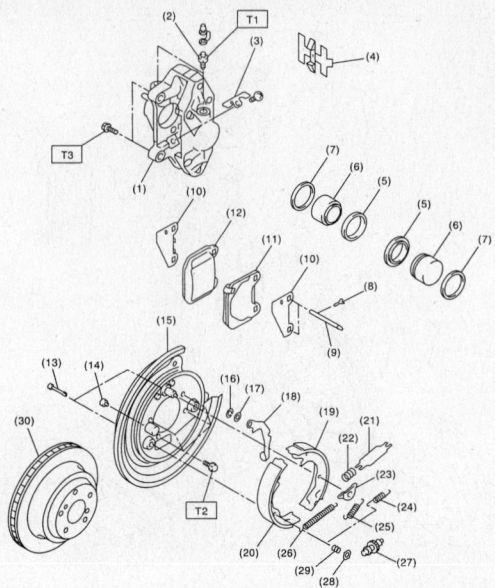

(1)	Caliper body	(13)	Shoe hold-down pin	(25)	Primary shoe return spring
(2)	Air bleeder screw	(14)	Cover	(26)	Adjusting spring
(3)	Guide plate	(15)	Back plate	(27)	Adjuster
(4)	Cross spring	(16)	Retainer	(28)	Shoe hold-down cup
(5)	Piston boot	(17)	Spring washer	(29)	Shoe hold-down spring
(6)	Piston	(18)	Parking brake lever	(30)	Disc rotor
(7)	Piston seal	(19)	Parking brake shoe (Secondary)		
(8)	Clip	(20)	Parking brake shoe (Primary)		
(9)	Pad pin	(21)	Strut		
(10)	Pad shim	(22)	Strut shoe spring		
(11)	Pad (Outside)	(23)	Shoe guide plate		
(12)	Pad (Inside)	(24)	Secondary shoe return spring		

Tightening torque: N·m (kgf-m, ft-lb)
T1: 20 (2.0, 14.5)
T2: 53 (5.4, 39.1)
T3: 65 (6.6, 47.9)

09490_SBCR_G0111

Rear disc brakes and related components 17 inch type—2005–2006 Impreza and WRX

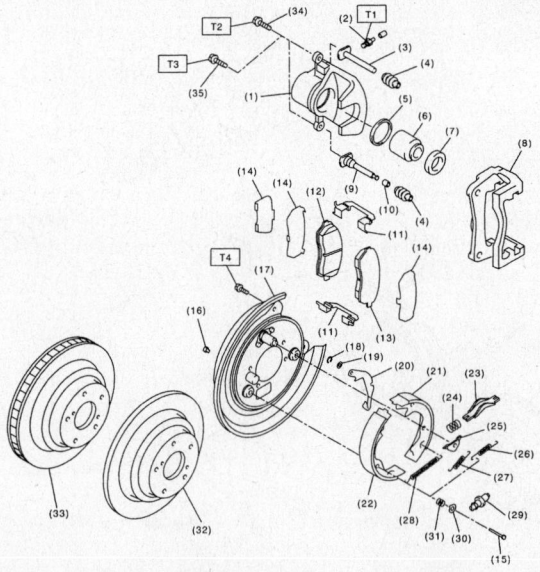

(1)	Caliper body	(15)	Shoe hold-down pin	(29)	Adjuster
(2)	Air bleeder screw	(16)	Cover	(30)	Shoe hold-down cup
(3)	Guide pin (Green)	(17)	Back plate	(31)	Shoe hold-down spring
(4)	Pin boot	(18)	Retainer	(32)	Disc rotor (Solid type)
(5)	Piston seal	(19)	Spring washer	(33)	Disc rotor (Ventilated type)
(6)	Piston	(20)	Parking brake lever	(34)	Bolt (For solid disc brake)
(7)	Piston boot	(21)	Parking brake shoe (Secondary)	(35)	Bolt (For ventilated disc brake)
(8)	Support	(22)	Parking brake shoe (Primary)		
(9)	Lock pin (Yellow)	(23)	Strut		
(10)	Bushing	(24)	Strut shoe spring		
(11)	Pad clip	(25)	Shoe guide plate		
(12)	Inner pad	(26)	Secondary shoe return spring		
(13)	Outer pad	(27)	Primary shoe return spring		
(14)	Shim	(28)	Adjusting spring		

Tightening torque: N·m (kgf-m, ft-lb)
T1: 8 (0.8, 5.8)
T2: 27 (2.8, 19.9)
T3: 37 (3.7, 27.2)
T4: 53 (5.4, 39.1)

09490_SBCR_G0112

Rear disc brakes and related components—2005–2006 Outback and Legacy

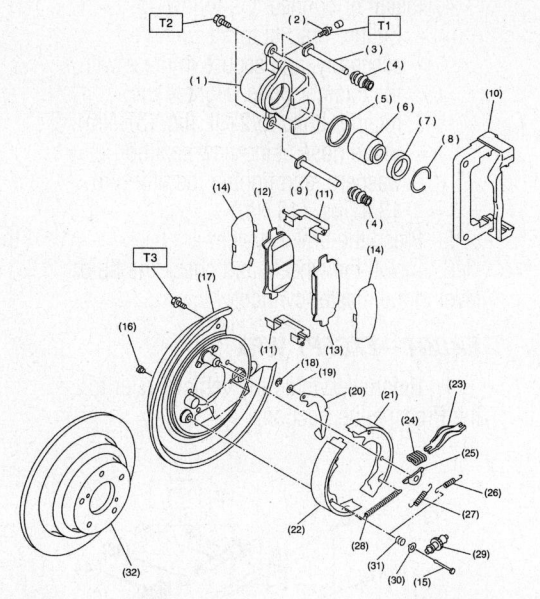

(1)	Caliper body	(14)	Shim	(27)	Primary shoe return spring
(2)	Air bleeder screw	(15)	Shoe hold-down pin	(28)	Adjusting spring
(3)	Guide pin (Green)	(16)	Cover	(29)	Adjuster
(4)	Pin boot	(17)	Back plate	(30)	Shoe hold-down cup
(5)	Piston seal	(18)	Retainer	(31)	Shoe hold-down spring
(6)	Piston	(19)	Spring washer	(32)	Disc rotor
(7)	Piston boot	(20)	Parking brake lever		
(8)	Boot ring	(21)	Parking brake shoe (Secondary)		
(9)	Lock pin (Yellow)	(22)	Parking brake shoe (Primary)		
(10)	Support	(23)	Strut		
(11)	Pad clip	(24)	Strut shoe spring		
(12)	Inner pad	(25)	Shoe guide plate		
(13)	Outer pad	(26)	Secondary shoe return spring		

Tightening torque: N·m (kgf-m, ft-lb)
T1: 8 (0.8, 5.8)
T2: 39 (4.0, 28.9)
T3: 52 (5.3, 38.3)

09490_SBCR_G0113

Rear disc brakes and related components—2005–2006 Baja

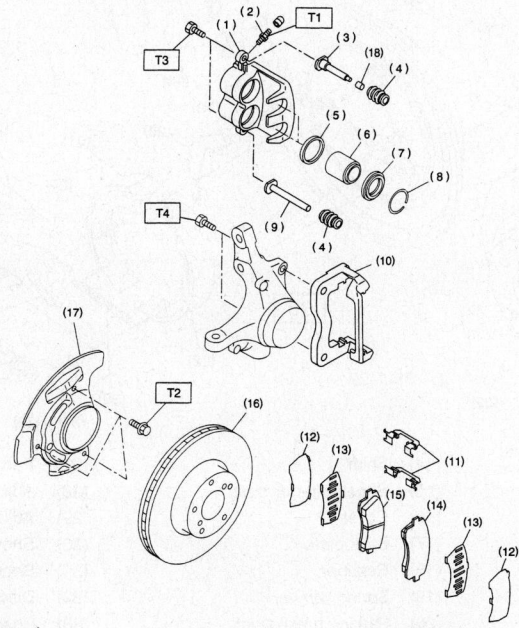

(1)	Caliper body	(9)	Lock pin (Yellow)	(17)	Disc cover
(2)	Air bleeder screw	(10)	Support	(18)	Bush
(3)	Guide pin (Green)	(11)	Pad clip		
(4)	Pin boot	(12)	Outer shim		
(5)	Piston seal	(13)	Inner shim		
(6)	Piston	(14)	Pad (Outside)		
(7)	Piston boot	(15)	Pad (Inside)		
(8)	Boot ring	(16)	Disc rotor		

Tightening torque: N·m (kgf-m, ft-lb)
T1: 8 (0.8, 5.8)
T2: 18 (1.8, 13.0)
T3: 37 (3.8, 27.5)
T4: 80 (8.2, 59)

9357TG65

Front disc brakes and related components—WRX

6. Install the wheels and check the fluid level in the master cylinder.

REAR—WRX

1. Before servicing the vehicle, refer to the Precautions Section.
2. Remove or disconnect the following:
   • Rear wheels
   • Brake hose from the caliper body
   • Caliper bracket retainer bolts
   • Caliper and bracket assembly off the rotor

To install:

3. Compress the piston assembly into the cylinder bore.

4. Install or connect the following:
   • Caliper bracket to the spindle assembly, and secure in place with the retainer bolts. Tighten the retainer bolts to 27 ft. lbs. (37 Nm).
   • Brake hose using new sealing washers, and tighten the fitting to 13 ft. lbs. (18 Nm)
5. Bleed the brake system.
6. Install the wheels and check the fluid level in the master cylinder.

FRONT—EXCEPT WRX

1. Before servicing the vehicle, refer to the Precautions Section.

2. Remove or disconnect the following:
   • Front wheels
   • Brake hose from the caliper body
   • Caliper retainer bolts and the caliper
   • Caliper bracket, if necessary

To install:

3. Compress the piston assembly into the cylinder bore.

4. Install or connect the following:
   • Caliper bracket to the spindle assembly, and secure in place with the retainer bolts. Tighten the retainer bolts to 58 ft. lbs. (78 Nm).

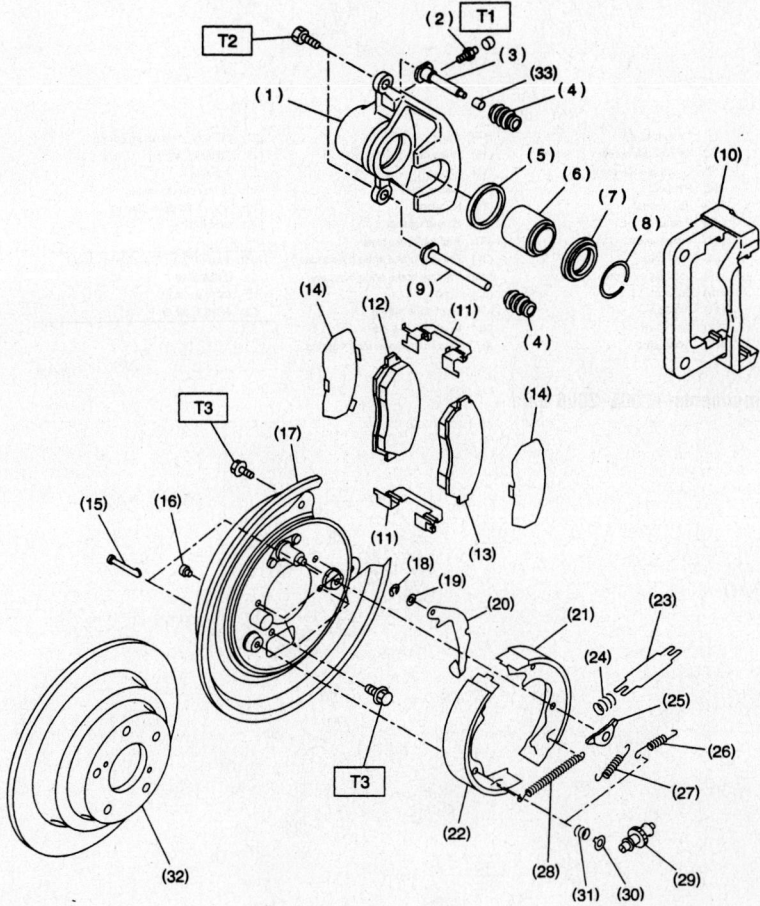

(1)	Caliper body	(14)	Shim	(27)	Primary shoe return spring
(2)	Air bleeder screw	(15)	Shoe hold-down pin	(28)	Adjusting spring
(3)	Guide pin (Green)	(16)	Cover	(29)	Adjuster
(4)	Pin boot	(17)	Back plate	(30)	Shoe hold-down cup
(5)	Piston seal	(18)	Retainer	(31)	Shoe hold-down spring
(6)	Piston	(19)	Spring washer	(32)	Disc rotor
(7)	Piston boot	(20)	Parking brake lever	(33)	Bush
(8)	Boot ring	(21)	Parking brake shoe (Secondary)		
(9)	Lock pin (Yellow)	(22)	Parking brake shoe (Primary)		
(10)	Support	(23)	Strut		
(11)	Pad clip	(24)	Strut shoe spring		
(12)	Inner pad	(25)	Shoe guide plate		
(13)	Outer pad	(26)	Secondary shoe return spring		

Tightening torque: N·m (kgf-m, ft-lb)
T1: 8 (0.8, 5.8)
T2: 37 (3.8, 27.5)
T3: 53 (5.4, 39.1)

9357TG66

Rear disc brakes and related components—WRX

- Caliper and tighten the retainers to 29 ft. lbs. (40 Nm) on 2002 vehicles and Outback 19 ft. lbs. (26 Nm)
- Brake hose using new sealing washers, and tighten the fitting to 13 ft. lbs. (18 Nm)

5. Bleed the brake system.

6. Install the wheels and check the fluid level in the master cylinder.

7. Pump the brake pedal several times to seat the brakes before attempting to move the vehicle and road test the vehicle.

REAR—EXCEPT WRX

1. Before servicing the vehicle, refer to the Precautions Section.

2. Install or connect the following:
- Rear wheels
- Brake hose from the caliper body
- Caliper bracket retainer bolts
- Caliper and bracket assembly off the rotor

To install:

3. Compress the piston assembly into the cylinder bore.

4. Install or connect the following:
- Caliper bracket to the spindle assembly, and secure in place with the retainer bolts, torque the bolts to 58 ft. lbs. (78 Nm)
- Caliper and tighten the retainers to 29 ft. lbs. (40 Nm) on 2002 vehicles except Outback and 19 ft. lbs. (26 Nm) on Outback
- Brake hose using new sealing washers, and tighten the fitting to 13 ft. lbs. (18 Nm)

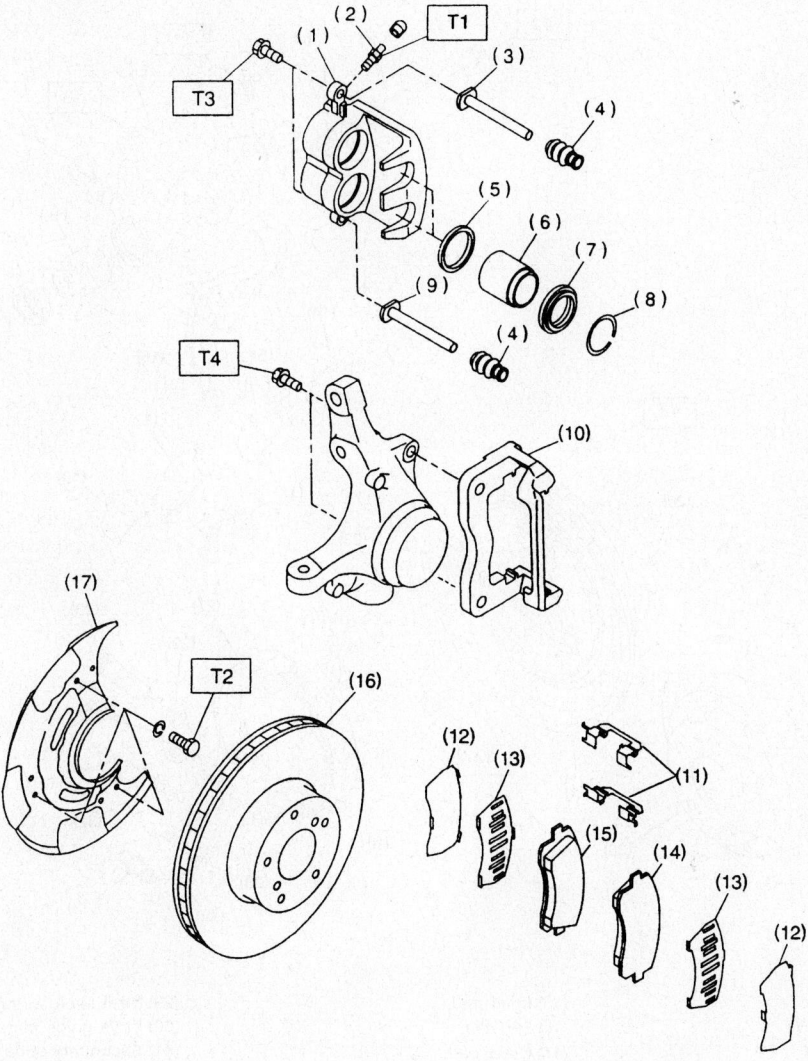

(1)	Caliper body	(9)	Lock pin (Yellow)	(17)	Disc cover
(2)	Air bleeder screw	(10)	Support		
(3)	Guide pin (Green)	(11)	Pad clip		
(4)	Pin boot	(12)	Outer shim		
(5)	Piston seal	(13)	Inner shim		
(6)	Piston	(14)	Pad (Outside)		
(7)	Piston boot	(15)	Pad (Inside)		
(8)	Boot ring	(16)	Disc rotor		

Tightening torque: N·m (kg-m, ft-lb)
T1: 8±1 (0.8±0.1, 5.8±0.7)
T2: 18±5 (1.8±0.5, 13.0±3.6)
T3: 37±5 (3.8±0.5, 27.5±3.6)
T4: 78±10 (8.0±1.0, 58±7)

93016G60

Front disc brakes and related components—except WRX

5. Bleed the brake system. Install the wheels. Check the fluid level in the master cylinder.

6. Pump the brake pedal several times to seat the brakes before attempting to move the vehicle and road test the vehicle.

2005–2006

FRONT—IMPREZA AND WRX)

1. Before servicing the vehicle, refer to the Precautions Section.

2. Loosen the wheel nuts.
3. Disconnect the negative battery cable.
4. Raise and support the vehicle safely.
5. Remove the tire and wheel.
6. Remove the union bolt. Disconnect the brake line from the brake caliper. Be sure to properly catch the fluid to avoid damage to painted surfaces and improper disposal.
7. Remove the caliper from its mounting.

To install:

8. Installation is the reverse of the removal procedure.

9. Tighten the caliper retaining bolts to specification.

10. Check the brake fluid level, correct as required.

11. Bleed the hydraulic system, as required.

REAR—IMPREZA AND WRX

1. Before servicing the vehicle, refer to the Precautions Section.
2. Loosen the wheel nuts.
3. Disconnect the negative battery cable.

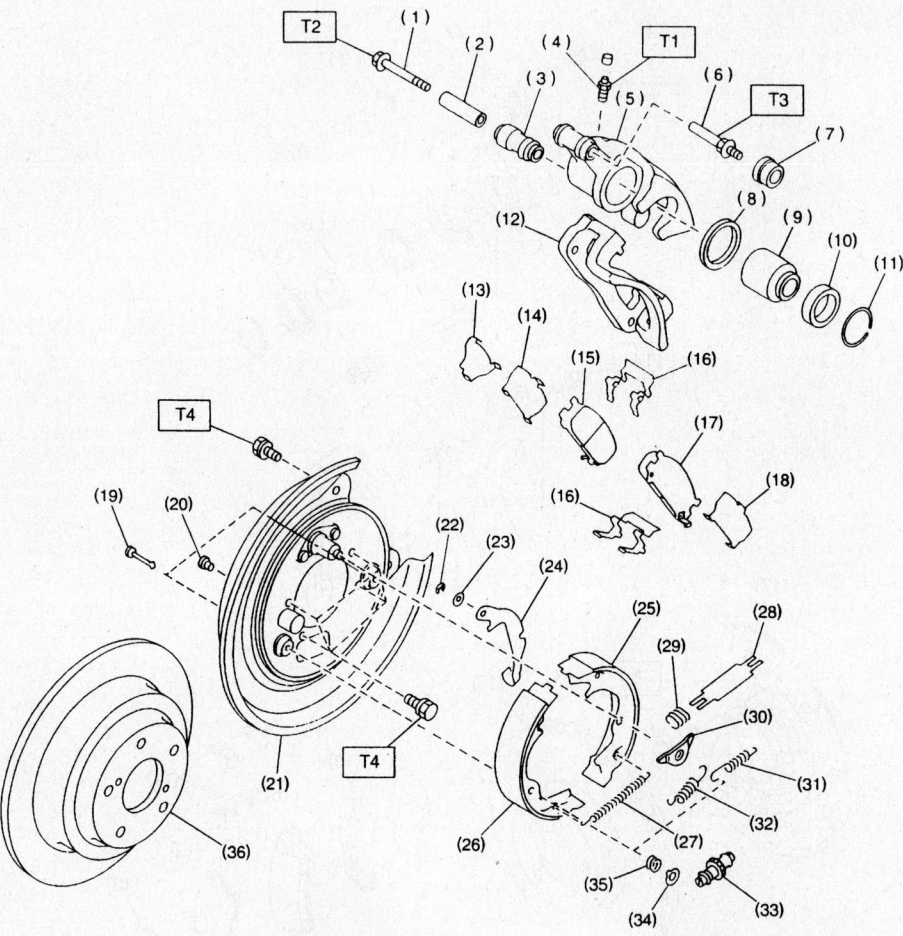

(1) Lock pin	(15) Inner pad	(29) Strut shoe spring
(2) Lock pin sleeve	(16) Pad clip	(30) Shoe guide plate
(3) Lock pin boot	(17) Outer pad	(31) Secondary shoe return spring
(4) Air bleeder screw	(18) Outer shim	(32) Primary shoe return spring
(5) Caliper body	(19) Shoe hold-down pin	(33) Adjuster
(6) Guide pin	(20) Cover	(34) Shoe hold-down cup
(7) Guide pin boot	(21) Back plate	(35) Shoe hold-down spring
(8) Piston seal	(22) Retainer	(36) Disc rotor
(9) Piston	(23) Spring washer	
(10) Piston boot	(24) Parking brake lever	
(11) Boot ring	(25) Parking brake shoe (Secondary)	
(12) Support	(26) Parking brake shoe (Primary)	
(13) Shim	(27) Adjusting spring	
(14) Inner shim	(28) Strut	

Tightening torque: N·m (kg-m, ft-lb)
T1: 8 ± 1 (0.8 ± 0.1, 5.8 ± 0.7)
T2: 20 ± 4 (2.0 ± 0.4, 14.5 ± 2.9)
T3: 26 ± 5 (2.7 ± 0.5, 19.5 ± 3.6)
T4: 52 ± 6 (5.3 ± 0.6, 38.3 ± 4.3)

93016G61

Rear disc brakes and related components—except WRX

4. Raise and support the vehicle safely.

5. Remove the tire and wheel.

6. Remove the union bolt. Disconnect the brake line from the brake caliper. Be sure to properly catch the fluid to avoid damage to painted surfaces and improper disposal.

7. Remove the caliper from its mounting.

To install:

8. Installation is the reverse of the removal procedure.

9. Tighten the caliper retaining bolts to specification.

10. Check the brake fluid level, correct as required.

11. Bleed the hydraulic system, as required.

FRONT—LEGACY, OUTBACK AND BAJA

1. Before servicing the vehicle, refer to the Precautions Section.

2. Loosen the wheel nuts.

3. Disconnect the negative battery cable.

4. Raise and support the vehicle safely.

5. Remove the tire and wheel.

6. Remove the union bolt. Disconnect the brake line from the brake caliper. Be sure to properly catch the fluid to avoid damage to painted surfaces and improper disposal.

7. Remove the caliper from its mounting.

To install:

8. Installation is the reverse of the removal procedure.

9. Tighten the caliper retaining bolts to specification.

10. Check the brake fluid level, correct as required.

11. Bleed the hydraulic system, as required.

REAR—LEGACY, OUTBACK AND BAJA

1. Before servicing the vehicle, refer to the Precautions Section.

2. Loosen the wheel nuts.

3. Disconnect the negative battery cable.

4. Raise and support the vehicle safely.

5. Remove the tire and wheel.

6. Remove the union bolt. Disconnect the brake line from the brake caliper. Be sure to properly catch the fluid to avoid damage to painted surfaces and improper disposal.

7. Remove the caliper from its mounting.

To install:

8. Installation is the reverse of the removal procedure.

9. Tighten the caliper retaining bolts to specification.

10. Check the brake fluid level, correct as required.

11. Bleed the hydraulic system, as required.

Brake Pads

REMOVAL & INSTALLATION

2002–2004

FRONT—WRX

1. Before servicing the vehicle, refer to the Precautions Section.

2. Remove or disconnect the following:
   - Wheels
   - Lock pin bolts from the lower portion of the caliper
   - Caliper upward to access the pads
   - Disc brake pads

To install:

3. Compress the caliper piston.

4. Install or connect the following:
   - New pads into the caliper brackets, being sure all shims and clips are in their original positions
   - Calipers down into position and install the lock pin bolts. Tighten the lock pin bolts to 19 ft. lbs. (26 Nm).
   - Wheels

5. Check the fluid level in the master cylinder, pump the brake pedal several times to seat the brakes before attempting to move the vehicle and road test the vehicle.

REAR—WRX

1. Before servicing the vehicle, refer to the Precautions Section.

2. Remove or disconnect the following:
   - Portion of brake fluid from the master cylinder reservoir
   - Wheels
   - Lock pin bolts from the lower portion of the caliper
   - Caliper upward to access the pads
   - Disc brake pads

To install:

3. Compress the caliper piston.

4. Install or connect the following:
   - New pads into the caliper bracket,

being sure all shims and clips are in their original positions
   - Caliper down into position and install the lock pin bolts. Tighten the lock pin bolt to 27 ft. lbs. (37 Nm).
   - Wheels

5. Check the fluid level in the master cylinder, pump the brake pedal several times to seat the brakes before attempting to move the vehicle and road test the vehicle.

FRONT—EXCEPT WRX

1. Before servicing the vehicle, refer to the Precautions Section.

2. Remove or disconnect the following:
   - Wheels
   - Lock pin bolts from the lower portion of the caliper
   - Caliper by swinging it upward to access the pads
   - Disc brake pads

To install:

3. Compress the caliper pistons.

4. Install or connect the following:
   - New pads into the caliper brackets, being sure all shims and clips are in their original positions
   - Caliper and tighten the retainers to 29 ft. lbs. (40 Nm) on 2002 vehicles except Outback and 19 ft. lbs. (26 Nm) on Outback
   - Wheels

5. Check the fluid level in the master cylinder, pump the brake pedal several times to seat the brakes before attempting to move the vehicle and road test the vehicle.

REAR—EXCEPT WRX

1. Before servicing the vehicle, refer to the Precautions Section.

2. Remove or disconnect the following:
   - Wheels
   - Small portion of brake fluid from the master cylinder reservoir
   - Parking brake cable from the caliper lever, if equipped
   - Lock pin bolts from the lower portion of the caliper
   - Caliper by swinging it upward to access the pads
   - Disc brake pads

To install:

3. Compress the caliper pistons.

4. Install or connect the following:
   - New pads into the caliper brackets, being sure all shims and clips are in their original positions
   - Caliper down into position and

install the lock pin bolts. Tighten the lock pin bolts to 29 ft. lbs. (40 Nm).

- Parking brake cable
- Wheels

5. Check the fluid level in the master cylinder, pump the brake pedal several times to seat the brakes before attempting to move the vehicle and road test the vehicle.

2005–2006

FRONT—IMPREZA AND WRX

1. Before servicing the vehicle, refer to the Precautions Section.
2. Loosen the wheel nuts.
3. Disconnect the negative battery cable.
4. Raise and support the vehicle safely.
5. Remove the tire and wheel.
6. On 15 inch type, remove the lower caliper bolt. Raise the caliper body upward and support it. Do not disconnect the fluid line.
7. On 16 inch type, remove the "M" clip. Remove the pad pins and cross spring. Expand the pads and then push the piston back.
8. On 17 inch type, remove the clip. Remove the pad pins and cross spring. Expand the pads and then push the piston back.
9. Remove the disc brake pads.

To install:

10. Installation is the reverse of the removal procedure.
11. On 15 inch wheel, apply a thin coat of Molykote AS-880N (part number K0779YA010) or equivalent to the frictional portion between the pad and pad clip, and the pad and pad inner shim.
12. On 16 inch type, apply a thin coat of Molykote AS-880N (part number K0779YA010) or equivalent to the frictional portion between the pad and pad inner shim.
13. On 17 inch type, apply a thin coat of Molykote AS-880N (part number K0779YA010) or equivalent to the frictional portion between the pad and pad shim.
14. Check the brake fluid level, correct as required.
15. Bleed the hydraulic system, as required.

REAR—IMPREZA AND WRX

1. Before servicing the vehicle, refer to the Precautions Section.
2. Loosen the wheel nuts.
3. Disconnect the negative battery cable.

4. Raise and support the vehicle safely.
5. Remove the tire and wheel.
6. On 14 inch type, remove the lower caliper bolt. Raise the caliper body upward and support it. Do not disconnect the fluid line.
7. On 15 inch type, remove the "M" clip. Remove the pad pins and cross spring. Expand the pads and then push the piston back.
8. On 17 inch type, remove the clip. Remove the pad pins and cross spring. Expand the pads and then push the piston back.
9. Remove the disc brake pads. On 15 inch type also remove the shims.

To install:

10. Installation is the reverse of the removal procedure.
11. On 14 inch wheel, apply a thin coat of Molykote AS-880N (part number K0779YA010) or equivalent to the frictional portion between the pad and pad clip.
12. On 15 and 17 inch types, apply a thin coat of Molykote AS-880N (part number K0779YA010) or equivalent to the frictional portion between the pad and pad inner shim.
13. Check the brake fluid level, correct as required.
14. Bleed the hydraulic system, as required.

FRONT—LEGACY, OUTBACK AND BAJA

1. Before servicing the vehicle, refer to the Precautions Section.
2. Loosen the wheel nuts.
3. Disconnect the negative battery cable.
4. Raise and support the vehicle safely.
5. Remove the tire and wheel.
6. Remove the lower caliper bolt. Raise the caliper body upward and support it. Do not disconnect the fluid line.
7. Remove the disc brake pads.

To install:

8. Installation is the reverse of the removal procedure.
9. Apply a thin coat of Molykote M7439 (part number 003602001) or equivalent to the pad clip.
10. Apply a thin coat of Molykote AS-880N (part number K0779YA010) or equivalent to the contact surface between the pad and pad inner shim.
11. On Legacy and Outback, apply a thin coat of Molykote AS-880N (part number K0779YA010) or equivalent to the three contact surfaces between the inner shim and outer shim of the outer pads.

12. Check the brake fluid level, correct as required.
13. Bleed the hydraulic system, as required.

REAR—LEGACY, OUTBACK AND BAJA

1. Before servicing the vehicle, refer to the Precautions Section.
2. Loosen the wheel nuts.
3. Disconnect the negative battery cable.
4. Raise and support the vehicle safely.
5. Remove the tire and wheel.
6. Remove the caliper bolt. Do not disconnect the fluid line.
7. Remove the disc brake pads.

To install:

8. Installation is the reverse of the removal procedure.
9. Apply a thin coat of Molykote M7439 (part number 003602001) or equivalent to the pad clip.
10. Apply a thin coat of Molykote AS-880N (part number K0779YA010) or equivalent to the contact surface between the pad and shim.
11. Check the brake fluid level, correct as required.
12. Bleed the hydraulic system, as required.

Brake Drums

REMOVAL & INSTALLATION

WRX

1. Before servicing the vehicle, refer to the Precautions Section.
2. Remove the rear wheels.
3. Release the parking brake.
4. If necessary, remove the adjusting hole cover from the backing plate and using a suitable tool, back off the shoe adjuster.
5. If the drum is difficult to remove, insert an 8mm bolt into the hole on the drum to push it off.
6. Remove the drum.

To install:

7. Install the drum.
8. Adjust the brake shoes.
9. Install the rear wheels.

Except WRX

1. Before servicing the vehicle, refer to the Precautions Section.
2. Remove or disconnect the following:
   - Rear wheels
   - Center cap by prying it off
   - Cotter pin, castle nut, and center retainer washer

- Drum

To install:

3. Install or connect the following:
- Drum
- Center retainer washer and castle nut. Tighten the castle nut to 108 ft. lbs. (147 Nm). Install a new cotter pin.
- Center cap

4. Adjust the brake shoes.
5. Install the rear wheels.

Brake Shoes

REMOVAL & INSTALLATION

1. Before servicing the vehicle, refer to the Precautions Section.
2. Remove or disconnect the following:
- Wheels
- Brake drum
- Both return springs

- Both retaining clips
- Brake shoes from the adjuster side first, then the wheel cylinder side, and pull them off the backing plate
- Parking brake cable from the parking lever on the trailing brake shoe, if equipped with rear drum parking brakes

To install:

3. Apply brake grease to the backing plate where the brake shoes contact it.

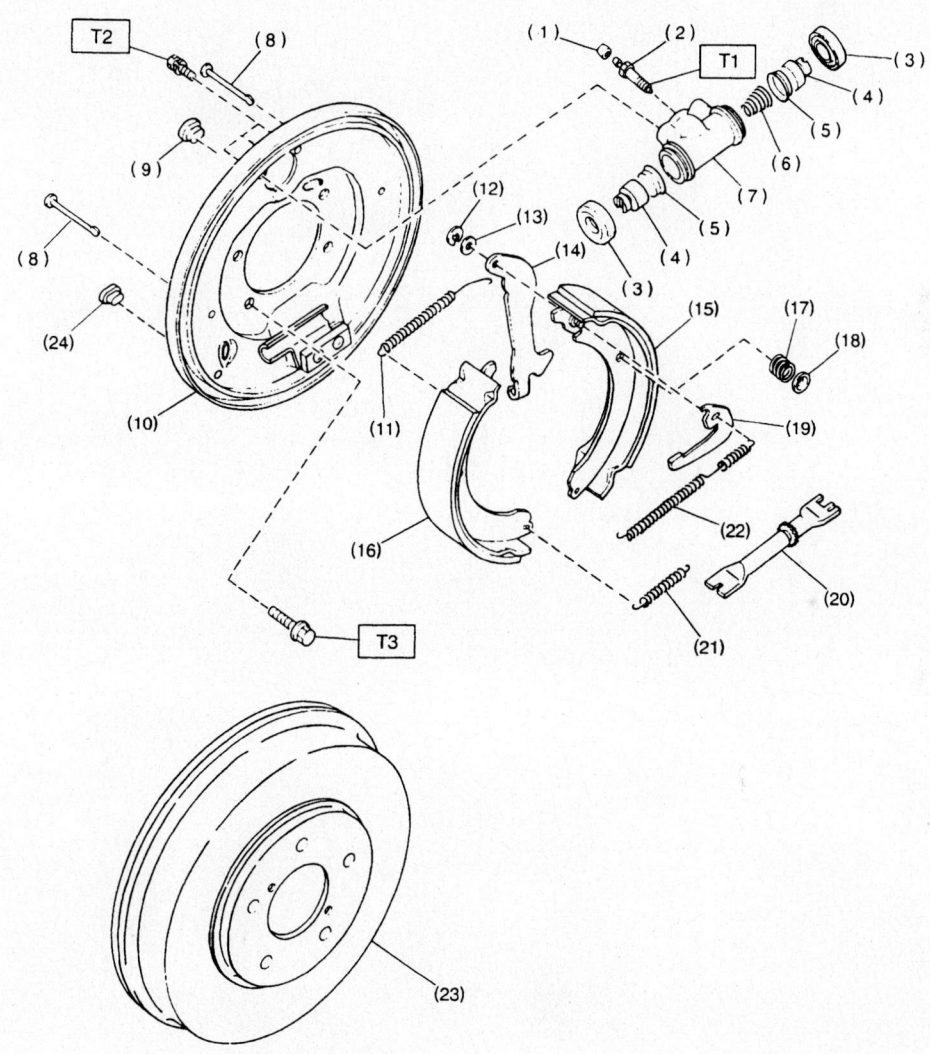

(1) Air bleeder cap
(2) Air bleeder screw
(3) Boot
(4) Piston
(5) Cup
(6) Spring
(7) Wheel cylinder body
(8) Pin
(9) Plug
(10) Back plate

(11) Upper shoe return spring
(12) Retainer
(13) Washer
(14) Parking brake lever
(15) Brake shoe (Trailing)
(16) Brake shoe (Leading)
(17) Shoe hold-down spring
(18) Cup
(19) Adjusting lever
(20) Adjuster

(21) Lower shoe return spring
(22) Adjusting spring
(23) Drum
(24) Plug

Tightening torque: N·m (kg-m, ft-lb)
T1: 8±1 (0.8±0.1, 5.8±0.7)
T2: 10±2 (1.0±0.2, 7.2±1.4)
T3: 52±6 (5.3±0.6, 38.3±4.3)

93016G62

Typical rear drum brake assembly and related components

4. Install or connect the following:
- Parking brake cable to the parking lever on the trailing brake shoe, if equipped with rear drum parking brakes
- Brake shoes to the wheel cylinder, then to the adjuster. Secure in place with the 2 pins and retaining clips.
- Return springs. The upper spring is thinner.
- Drum and adjust the brake shoes
- Wheels

5. Adjust the parking brake.

6. Check the fluid level in the master cylinder, pump the brake pedal several times to seat the brakes before attempting to move the vehicle and road test the vehicle.

SUBARU

Forester

BRAKES 7-80
DRIVE TRAIN 7-62
ENGINE REPAIR 7-17
FRONT SUSPENSION 7-72
FUEL SYSTEM 7-59
REAR SUSPENSION 7-76
SPECIFICATIONS AND
MAINTENANCE CHARTS 7-3
Engine and Vehicle Identification
 Chart...7-3
General Engine Specifications 7-3
Engine Tune-Up Specifications.......7-4
Firing Order 7-4
Accessory Drive Belt Routing 7-5
Capacities 7-5
Valve Specifications.......................7-6
Camshaft Specifications 7-7
Crankshaft and Connecting Rod
 Specifications 7-8
Piston and Ring
 Specifications 7-8
Torque Specifications 7-9
Wheel Alignment 7-14
Tire, Wheel and Ball Joint
 Specifications 7-14
Brake Specifications 7-15
Scheduled Maintenance
 Intervals...................................7-16
STEERING 7-70
A
Air Bag.......................................7-70
 Disarming 7-70
 Precautions 7-70
Alternator 7-17
 Removal & Installation..............7-17
Automatic Transmission 7-63
 Removal & Installation..............7-63
B
Brake Caliper 7-80
 Removal & Installation..............7-80
Brake Drums...............................7-86
 Removal & Installation..............7-86
Brake Shoes................................7-86
 Removal & Installation..............7-86

C
Camshaft and Valve Lifters 7-39
 Inspection 7-45
 Removal & Installation..............7-39
Clutch 7-64
 Adjustment..............................7-64
 Removal & Installation..............7-64
CV-Joints....................................7-67
 Overhaul 7-67
Cylinder Head 7-29
 Removal & Installation..............7-29
D
Disc Brake Pads...........................7-86
 Removal & Installation..............7-86
Distributor...................................7-17
E
Engine Assembly 7-17
 Removal & Installation..............7-17
Exhaust Manifold 7-38
 Removal & Installation..............7-38
F
Front Crankshaft Seal 7-39
 Removal & Installation..............7-39
Fuel Filter 7-59
 Removal & Installation..............7-59
Fuel Injector...............................7-60
 Removal & Installation..............7-60
Fuel Pump 7-59
 Removal & Installation..............7-59
Fuel System Pressure
 Relieving 7-59
Fuel System Service
 Precautions..............................7-59
H
Halfshafts...................................7-65
 Removal & Installation..............7-65
Heater Core................................7-23
 Removal & Installation..............7-23
Hydraulic Clutch System 7-65
 Bleeding..................................7-65
I
Ignition Timing 7-17
 Adjustment..............................7-17
Intake Manifold 7-33
 Removal & Installation..............7-33

Intercooler..................................7-32
 Removal & Installation..............7-32
L
Lower Ball Joint...........................7-74
 Removal & Installation..............7-74
Lower Control Arm (Front) 7-74
 Control Arm Bushing
  Replacement..........................7-75
 Removal & Installation..............7-74
Lower Control Arm (Rear).............7-78
 Control Arm Bushing
  Replacement..........................7-78
 Removal & Installation..............7-78
M
Manual Transmission 7-62
 Removal & Installation..............7-62
O
Oil Pan.......................................7-48
 Removal & Installation..............7-48
Oil Pump 7-50
 Removal & Installation..............7-50
P
Piston and Ring 7-58
 Positioning 7-58
Power Steering Gear 7-70
 Removal & Installation..............7-70
R
Rear Main Seal 7-51
 Removal & Installation..............7-51
Rocker Arms/Shafts 7-32
 Removal & Installation..............7-32
S
Stabilizer Bar (Front)....................7-74
 Removal & Installation..............7-74
Stabilizer Bar (Rear).....................7-77
 Removal & Installation..............7-77
Starter.......................................7-47
 Removal & Installation..............7-47
Strut (Front) 7-72
 Disassembly & Assembly 7-73
 Removal & Installation..............7-72
Strut (Rear) 7-76
 Disassembly & Assembly 7-76
 Removal & Installation..............7-76

T

Timing Belt, Cover and
Crankshaft seal 7-51
Removal & Installation..............7-51
Transfer Case Assembly.................7-65
Removal & Installation..............7-65
Turbocharger...................................7-32
Removal & Installation..............7-32

V

Valve Lash 7-45
Adjustment................................7-45

W

Water Pump 7-20
Removal & Installation..............7-20
Wheel Bearings (Front).................7-75
Adjustment................................7-75
Removal & Installation..............7-75

Wheel Bearings (Rear) 7-78
Adjustment................................7-78
Removal & Installation.............7-78

SPECIFICATIONS AND MAINTENANCE CHARTS

ENGINE AND VEHICLE IDENTIFICATION

				Engine Code				Model Year	
Code ①	Liters (cc)	Cu. In.	Cyl.	Fuel Sys.	Type	Eng. Mfg.		Code ②	Year
6 ③	2.5 (2457)	150	4	MFI	DOHC	Subaru		2	2002
6	2.5 (2457)	150	4	MFI	SOHC	Subaru		3	2003

MFI: Multiport Fuel Injection

DOHC: Double Overhead Camshafts

SOHC: Single Overhead Camshaft

① 6th digit of the VIN

② 10th digit of the VIN

③ Turbocharged

Code ②	Year
4	2004
5	2005
6	2006

09490_FORE_C0001

GENERAL ENGINE SPECIFICATIONS

Year	Model	Engine Displacement Liters (VIN)	Net Horsepower @ rpm	Net Torque @ rpm (ft. lbs.)	Bore x Stroke (in.)	Compression Ratio	Oil Pressure psi @ rpm
2002	Forester	2.5 (6) ①	165@5600	162@4000	3.92x3.11	10.1:1	14@600
2003	Forester	2.5 (6) ①	165@5600	162@4000	3.92x3.11	10.1:1	14@600
2004	Forester	2.5 (6) ①	165@5600	162@4000	3.92x3.11	10.0:1	14@600
		2.5 (6) ②	210@5600	235@4600	3.92x3.11	8.2:1	14@600
2005	Forester	2.5 (6) ①	173@5600	166@4400	3.92x3.11	10.0:1	14@600
		2.5 (6) ②	230@5600	235@3600	3.92x3.11	8.2:1	14@600
2006	Forester	2.5 (6) ①	173@5600	166@4400	3.92x3.11	10.0:1	14@600
		2.5 (6) ②	230@5600	235@3600	3.92x3.11	8.4:1	14@600

① SOHC

② DOHC

09490_FORE_C0002

ENGINE TUNE-UP SPECIFICATIONS

Year	Engine Displacement Liters (VIN)	Spark Plugs Gap (in.)	Ignition Timing (deg.) ① ② MT	AT	Fuel Pump (psi)	Idle Speed (rpm) MT	AT	Valve Clearance ③ In.	Ex.
2002	2.5 (6) ④	0.039-0.043	⑤	⑥	34-38	650	700	0.0071-0.0087	0.0090-0.0106
2003	2.5 (6) ④	0.039-0.043	⑤	⑥	34-38	600-800	550-750	0.0071-0.0087	0.0090-0.0106
2004	2.5 (6) ④	0.039-0.043	⑤	⑥	⑦	550-750	600-800	0.0071-0.0087	0.0090-0.0106
	2.5 (6) ⑧	0.028-0.031	⑨	⑨	34-38	600-800	600-800	0.0071-0.0087	0.0090-0.0106
2005	2.5 (6) ④	0.039-0.043	⑤	⑥	⑦	550-750	600-800	0.0063-0.0095	0.0082-0.0114
	2.5 (6) ⑧	0.028-0.031	⑨	⑨	⑩	600-800	600-800	0.0055-0.0095	0.0118-0.0158
2006	2.5 (6) ④	0.039-0.043	⑤	⑥	⑦	550-750	600-800	0.0063-0.0095	0.0082-0.0114
	2.5 (6) ⑧	0.028-0.031	⑨	⑨	⑩	600-800	600-800	0.0071-0.0087	0.0130-0.0146

① At idle
② BTDC: before top dead center
③ With engine cold
④ SOHC
⑤ 10 degrees +/-8 at 650rpm
⑥ 15 degrees +/-8 at 700rpm
⑦ 41-46 while disconnecting pressure regulator vacuum hose from intake manifold
   33-38 after connecting pressure regulator vacuum hose
⑧ DOHC
⑨ 17 degrees +/-10 at 700rpm
⑩ 48-53 while disconnecting pressure regulator vacuum hose from intake manifold
   40-45 after connecting pressure regulator vacuum hose

09490_FORE_C0003

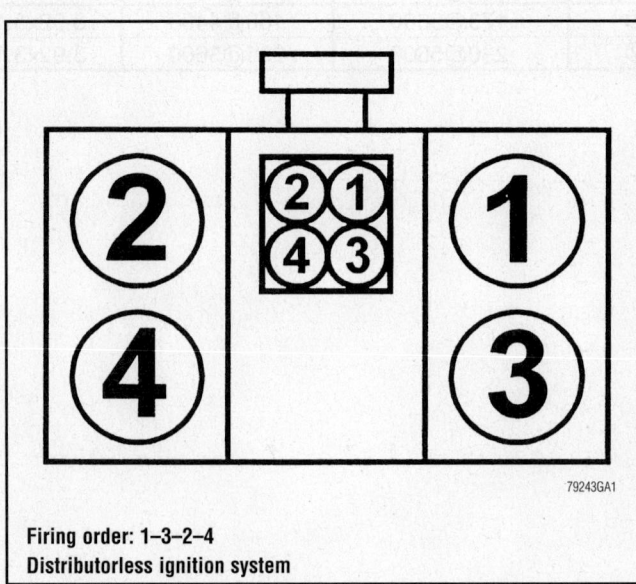

Firing order: 1-3-2-4
Distributorless ignition system

79243GA1

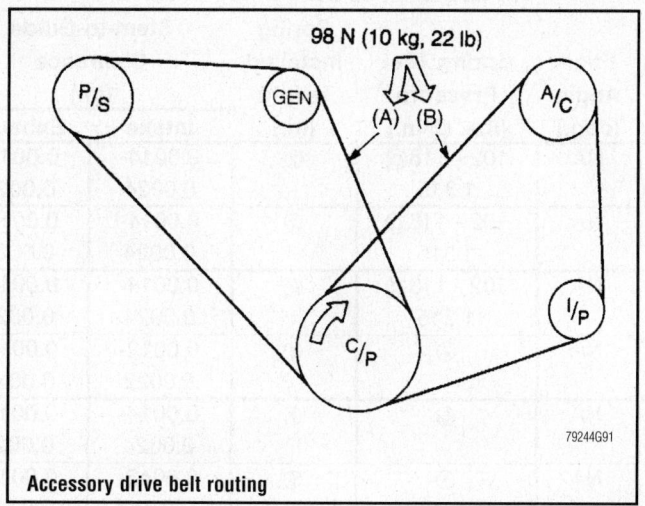

98 N (10 kg, 22 lb)

Accessory drive belt routing

79244G91

CAPACITIES

Year	Model	Engine Displacement Liters (VIN)	Engine Oil with Filter (qts.)	Transmission (pts.)			Transfer Case (pts.)	Drive Axle		Fuel Tank (gal.)	Cooling System (qts.)
				4-Spd	5-Spd	Auto.		Front (pts.)	Rear (pts.)		
2002	Forester	2.5 (6)	4.2	—	7.4	19.6	—	2.6 ①	1.6	15.9	6.3
2003	Forester	2.5 (6)	4.2	—	7.4	19.6	—	2.6 ①	1.6	15.9	②
2004	Forester	2.5 (6)	4.2	—	7.4	19.6	—	2.6 ①	1.6	15.9	③
2005	Forester	2.5 (6)	4.2	—	7.4	19.6	—	2.6 ①	1.6	15.9	③
2006	Forester	2.5 (6)	④	—	7.4	19.6	—	2.6 ①	1.6	15.9	③

① automatic transmission differential

② manual transmission: 7.3
automatic transmission: 7.2

③ SOHC: manual transmission: 7.3. Automatic transmission: 7.2
DOHC: manual transmission: 7.8. Automatic transmission: 7.7

④ SOHC: 4.4
DOHC: 4.5

09490_ FORE_C0004

VALVE SPECIFICATIONS

Year	Engine Displacement Liters (VIN)	Seat Angle (deg.)	Face Angle (deg.)	Spring Test Pressure (lbs. @ in.)	Spring Installed Height (in.)	Stem-to-Guide Clearance (in.)		Stem Diameter (in.)	
						Intake	Exhaust	Intake	Exhaust
2002	2.5 (6) ①	90	NA	102 - 118@ 1.315	②	0.0014- 0.0024	0.0016- 0.0026	0.2343- 0.2348	0.2341- 0.2346
2003	2.5 (6) ①	90	NA	102 - 118@ 1.315	②	0.0014- 0.0024	0.0016- 0.0026	0.2343- 0.2348	0.2341- 0.2346
2004	2.5 (6) ①	90	NA	102 - 118@ 1.315	②	0.0014- 0.0024	0.0016- 0.0026	0.2343- 0.2348	0.2341- 0.2346
	2.5 (6) ③	90	NA	④	⑤	0.0012- 0.0022	0.0016- 0.0026	0.2344- 0.2350	0.2341- 0.2346
2005	2.5 (6) ①	90	NA	⑥	⑦	0.0014- 0.0024	0.0016- 0.0026	0.2343- 0.2348	0.2341- 0.2346
	2.5 (6) ③	90	NA	④	⑤	0.0012- 0.0022	0.0016- 0.0026	0.2344- 0.2350	0.2341- 0.2346
2006	2.5 (6) ①	90	NA	⑨	⑩	0.0014- 0.0024	0.0016- 0.0026	0.2343- 0.2348	0.2341- 0.2346
	2.5 (6) ③	90	NA	④	⑤	0.0012- 0.0022	0.0016- 0.0026	0.2344- 0.2350	0.2341- 0.2346

NA: Not Available
① SOHC
② Free length: 1.8913 in.
③ DOHC
④ Set: 46.1-52.8@1.417
  Lift: 95.8-110.0@1.043
⑤ Free length: 1.863 in.
⑥ Set: 48.0-55.0@1.772
  Lift: 119.0-130.0@1.366
⑦ Free length: 2.1378 in.
⑧ Set: 52.9-60.8@1.772
  Lift: 130.3-143.9@1.366
⑨ Free length: 2.173 in.
⑩ Free length: 2.173 in.

09490_FORE_C0005

CAMSHAFT SPECIFICATIONS
All measurements in inches unless noted

Year	Engine Displacement Liters (VIN)	Journal Dia.	Brg. Oil Clearance	Shaft End-play ①	Runout	Lobe Height	
						Intake	Exhaust
2002	2.5 (6) ②	1.2570-1.2577	0.0022-0.0035	0.0012-0.0035	NA	1.5545-1.5585	1.5455-1.5495
2003	2.5 (6) ②	1.2570-1.2577	0.0022-0.0035	0.0012-0.0035	NA	1.5545-1.5585	1.5455-1.5495
2004	2.5 (6) ②	1.2570-1.2577	0.0022-0.0035	0.0012-0.0035	NA	1.5545-1.5585	③
	2.5 (6) ④	⑤	0.0015-0.0028	0.0027-0.0047	NA	1.8330-1.8370	1.8410-1.8440
2005	2.5 (6) ②	1.2570-1.2577	0.0022-0.0035	0.0012-0.0035	NA	1.5545-1.5585	⑥
	2.5 (6) ④	⑤	0.0015-0.0028	0.0027-0.0047	NA	1.8330-1.8370	1.8410-1.8440
2006	2.5 (6) ②	1.2570-1.2577	0.0022-0.0035	0.0012-0.0035	NA	⑦	1.5783-1.5822
	2.5 (6) ④	⑤	0.0015-0.0028	0.0027-0.0047	NA	1.8330-1.8370	1.8410-1.8440

NA: Not Available

① Side clearance

② SOHC

③ Exhaust #1: 1.5455-1.5495
   Exhaust #1: 1.5686-1.5726

④ DOHC

⑤ Front: 1.4939-1.4946
   Except front: 1.1790-1.1796

⑥ California: 1.5686-1.5726
   Except California: 1.5638-1.5677

⑦ Constant: 1.5778-1.5817
   Low speed: 1.3851-1.3891
   High speed: 1.5872-1.5911

09490_ FORE_C0006

CRANKSHAFT AND CONNECTING ROD SPECIFICATIONS

All measurements are given in inches.

Year	Engine Displacement Liters (VIN)	Crankshaft				Connecting Rod		
		Main Brg. Journal Dia.	Main Brg. Oil Clearance	Shaft End-play	Thrust on No.	Journal Diameter	Oil Clearance	Side Clearance
2002	2.5 (6) ①	2.5194-2.5200	②	0.0012-0.0098	3	1.8891-1.8898	0.0005-0.0015	0.0028-0.0130
2003	2.5 (6) ①	2.3619-2.3625	②	0.0012-0.0098	3	1.8891-1.8898	0.0005-0.0015	0.0028-0.0130
2004	2.5 (6) ①	2.3619-2.3625	②	0.0012-0.0098	3	1.8891-1.8898	0.0005-0.0015	0.0028-0.0130
	2.5 (6) ③	2.3619-2.3625	0.0004-0.0012	0.0012-0.0045	NA	NA	0.0007-0.0018	0.0028-0.0130
2005	2.5 (6) ①	2.3619-2.3625	0.0004-0.0012	0.0012-0.0045	NA	NA	0.0006-0.0017	0.0028-0.0130
	2.5 (6) ③	2.3619-2.3625	0.0004-0.0012	0.0012-0.0045	NA	NA	0.0007-0.0018	0.0028-0.0130
2006	2.5 (6) ①	2.3619-2.3625	0.0004-0.0012	0.0012-0.0045	NA	NA	0.0006-0.0017	0.0028-0.0130
	2.5 (6) ③	2.3619-2.3625	0.0004-0.0012	0.0012-0.0045	NA	NA	0.0007-0.0018	0.0028-0.0130

NA: Not Available

① SOHC

② Journals 1 and 5: 0.0001 - 0.0016 in.
  Journals 2 and 4: 0.0004 - 0.0018 in.
  Journal 3: 0.0004 - 0.0016 in.

③ DOHC

09490_ FORE_C0007

PISTON AND RING SPECIFICATIONS

All measurements are given in inches.

Year	Engine Displacement Liters (VIN)	Piston Clearance	Ring Gap			Ring Side Clearance		
			Top Compression	Bottom Compression	Oil Control	Top Compression	Bottom Compression	Oil Control
2002	2.5 (6) ①	0.0004-0.0012	0.0079-0.0138	0.0146-0.0250	0.0079-0.0197	0.0016-0.0031	0.0012-0.0028	NA
2003	2.5 (6) ①	0.0004-0.0012	0.0079-0.0138	0.0146-0.0250	0.0079-0.0197	0.0016-0.0031	0.0012-0.0028	NA
2004	2.5 (6) ①	0.0004-0.0012	0.0079-0.0138	0.0146-0.0250	0.0079-0.0197	0.0016-0.0031	0.0012-0.0028	NA
	2.5 (6) ②	NA	0.0079-0.0098	0.0150-0.0203	0.0079-0.0197	0.0016-0.0031	0.0012-0.0028	NA
2005	2.5 (6) ①	NA	0.0079-0.0138	0.0144-0.0203	0.0079-0.0197	0.0016-0.0031	0.0012-0.0028	NA
	2.5 (6) ②	NA	0.0079-0.0098	0.0150-0.0203	0.0079-0.0197	0.0016-0.0031	0.0012-0.0028	NA
2006	2.5 (6) ①	NA	0.0079-0.0138	0.0144-0.0203	0.0079-0.0197	0.0016-0.0031	0.0012-0.0028	NA
	2.5 (6) ②	NA	0.0079-0.0098	0.0150-0.0203	0.0079-0.0197	0.0016-0.0031	0.0012-0.0028	NA

NA: Not Available
① SOHC
② DOHC

09490_ FORE_C0008

TORQUE SPECIFICATIONS
All readings in ft. lbs.

Year	Engine Displacement Liters (VIN)	Cylinder Head Bolts	Main ① Bearing Bolts	Rod Bearing Bolts	Crankshaft Damper Bolts	Flywheel Bolts	Manifold Intake	Manifold Exhaust	Spark Plugs	Oil Pan Drain Plug
2002	2.5 (6) ②	③	④	31 - 44	123 - 137	51 - 55	17 - 20	19 - 26	13 - 17	33
2003	2.5 (6) ②	③	④	31 - 44	123 - 137	51 - 55	17 - 20	19 - 26	13 - 17	33
2004	2.5 (6) ②	③	④	31 - 44	123 - 137	51 - 55	17 - 20	19 - 26	13 - 17	33
	2.5 (6) ⑤	⑥	⑦	38.4	132.7	⑧	18	⑨	13-17	33
2005	2.5 (6) ②	⑩	⑦	33.3	132.7	52.8	18	19-26	13-17	33
	2.5 (6) ⑤	⑥	⑦	38.4	132.7	⑧	18	⑨	13-17	33
2006	2.5 (6) ②	⑩	⑦	33.3	132.7	52.8	18	19-26	13-17	33
	2.5 (6) ⑤	⑥	⑦	38.4	132.7	⑧	18	⑨	13-17	33

① Engine block connecting bolts

② SOHC

③ Step 1: Tighten all bolts to 22 ft. lbs.
  Step 2: Tighten all bolts to 51 ft. lbs.
  Step 3: Loosen all bolts 180 degrees.
  Step 4: Repeat Step 3.
  Step 5: Tighten bolts 1 and 2 to 25 ft. lbs.
  Step 6: Tighten bolts 3, 4, 5 and 6 to 11 ft. lbs.
  Step 7: Tighten all bolts 80 to 90 degrees.
  Step 8: Repeat Step 7. Do not exceed 180 degrees total tightening.

④ Split engine case connecting bolts:
  Short bolts: 17-20 ft. lbs.
  Long bolts: 33-37 ft. lbs.
  Smaller short bolts (if used) 5 ft. lbs.

⑤ DOHC

⑥ Step 1: Tighten all bolts to 21.4 ft. lbs.
  Step 2: Tighten all bolts to 51 ft. lbs.
  Step 3: Loosen all bolts 180 degrees, and again 180 degrees
  Step 4: Tighten all bolts to 36 ft. lbs.
  Step 5: + 80-90 degrees
  Step 6: + 40-45 degrees

⑦ Step 1: Left side 10mm bolts (A-D) 7.2 ft. lbs.
  Step 2: Right side 10mm bolts (E-J) 7.2 ft. lbs.
  Step 3: Left side 10mm bolts (A-D) 13 ft. lbs.
  Step 4: Right side 10mm bolts (E-J) 13 ft. lbs.
  Step 5: Left side bolts (A and C) +90 degrees, bolts (B and D) 29.5 ft. lbs.
  Step 6: Right side 10mm bolts (E-J) + 90 degrees
  Step 7: Left side cylinder block connecting bolts (A-G) 18.1 ft. lbs., bolt (H) 4.7 ft. lbs.

⑧ Automatic: 53.1 ft. lbs.
  Manual: 38.4 ft. lbs.

⑨ Exhaust pipe to manifold 26 ft. lbs.
  Right upper cover 5.5 ft. lbs.
  Front exhaust pipe 13.7 ft. lbs.
  Right manifold to turbocharger joint pipe 26-28 ft. lbs.
  Lower covers 13.7 ft. lbs.

⑩ Step 1: Tighten all bolts to 22 ft. lbs.
  Step 2: Tighten all bolts to 51 ft. lbs.
  Step 3: Loosen all bolts 180 degrees, and again 180 degrees
  Step 4: Tighten all bolts to 36 ft. lbs.
  Step 5: + 80-90 degrees
  Step 6: + 40-45 degrees

09490_ FORE_C0009

CAUTION:
Remove oil in the mating surface of bearing and cylinder block before installation. Also apply a coat of engine oil to crankshaft pins.

1) Position the crankshaft on the #2 and #4 cylinder block.

2) Apply fluid packing to the mating surface of #1 and #3 cylinder block, and position it on #2 and #4 cylinder block.

CAUTION:
Do not allow fluid packing to jut into O-ring grooves, oil passages, bearing grooves, etc.

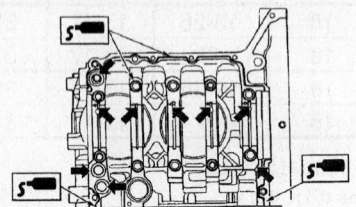

3) Temporarily tighten the 10 mm cylinder block connecting bolts in alphabetical sequence shown in the figure.

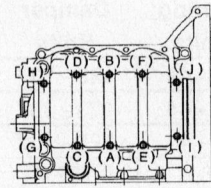

4) Tighten the 10 mm cylinder block connecting bolts in alphabetical sequence.

Tightening torque:
47±3 N·m (4.8±0.3 kgf-m, 34.7±2.2 ft-lb)

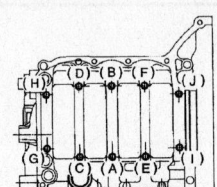

5) Tighten the 8 mm and 6 mm cylinder block connecting bolts in alphabetical sequence shown in the figure.

Tightening torque:
(A) — (G): 25 N·m (2.5 kgf-m, 18.1 ft-lb)
(H): 6.4 N·m (0.65 kgf-m, 4.7 ft-lb)

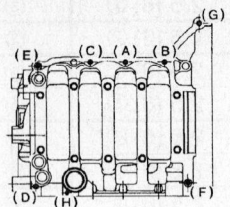

09490_SBCR_G0002

Main bearing torque sequence—2002 SOHC engine

CAUTION:
Remove oil in the mating surface of bearing and cylinder block before installation. Also apply a coat of engine oil to crankshaft pins.

1) Position the crankshaft on the #2 and #4 cylinder block.

2) Apply fluid packing to the mating surface of #1 and #3 cylinder block, and position it on #2 and #4 cylinder block.

NOTE:
Do not allow fluid packing to jut into O-ring grooves, oil passages, bearing grooves, etc.

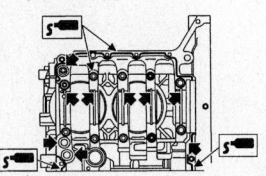

3) Tighten 10 mm cylinder block connecting bolts in alphabetical sequence shown in figure. (LH side)

Tightening torque:
15 N·m (1.5 kgf-m, 11 ft-lb)

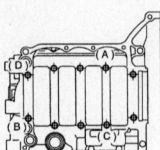

4) Tighten 10 mm cylinder block connecting bolts in alphabetical sequence shown in figure. (RH side)

Tightening torque:
15 N·m (1.5 kgf-m, 11 ft-lb)

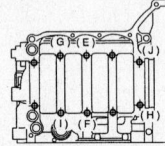

5) Tighten bolts (A to D) on left side of cylinder block more 90° in alphabetical sequence.

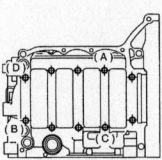

6) Tighten bolts (E to J) on right side of cylinder block more 90° in alphabetical sequence.

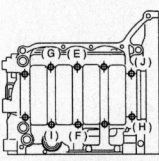

7) Tighten 8 mm and 6 mm cylinder block connecting bolts in alphabetical sequence shown in figure.

Tightening torque:
(A) — (G): 25 N·m (2.5 kgf-m, 18.1 ft-lb)
(H): 6.4 N·m (0.65 kgf-m, 4.7 ft-lb)

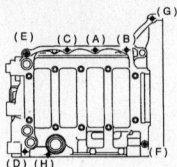

09490_SBCR_G0003

Main bearing torque sequence—2003–2004 SOHC engine

NOTE:
Remove oil on the mating surface of bearing and cylinder block before installation. Apply engine oil to crankshaft pins.

1) Position the crankshaft and O-rings on the #1 and #3 cylinder block.

2) Apply liquid gasket to the mating surface of #1 and #3 cylinder block, and position #2 and #4 cylinder block.

Liquid gasket:
Three bond 1215 (Part No. 004403007) or equivalent

NOTE:
Do not allow liquid gasket to jut into O-ring grooves, oil passages, bearing grooves, etc.

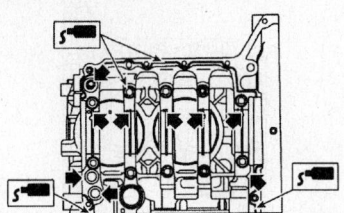

3) Apply a coat of engine oil to the washer and bolt thread.

4) Tighten the 10 mm cylinder block connecting bolts on LH side (A — D) in alphabetical sequence.

Tightening torque:
9.75 N·m (1.0 kgf-m, 7.2 ft-lb)

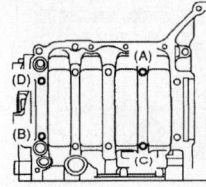

5) Tighten the 10 mm cylinder block connecting bolts on RH side (E — J) in alphabetical sequence.

Tightening torque:
9.75 N·m (1.0 kgf-m, 7.2 ft-lb)

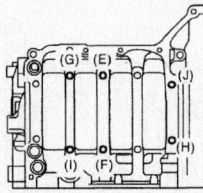

6) Further tighten the LH side bolts (A — D) in alphabetical sequence.

Tightening torque:
18 N·m (1.8 kgf-m, 13 ft-lb)

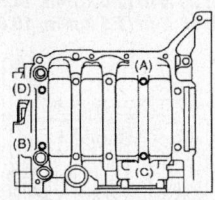

7) Further tighten the RH side bolts (E — J) in alphabetical sequence.

Tightening torque:
18 N·m (1.8 kgf-m, 13 ft-lb)

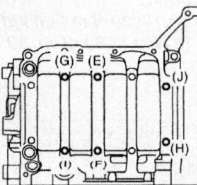

8) Further tighten the LH side bolts (A — D) in alphabetical sequence.
• (A), (C): Angle tightening

Tightening angle:
90°
• (B), (D): Torque tightening

Tightening torque:
40 N·m (4.1 kgf-m, 29.6 ft-lb)

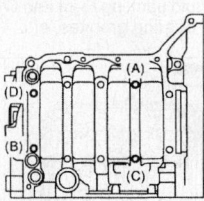

9) Tighten the RH side bolts (E — J) 90° further in alphabetical sequence.

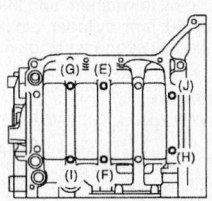

10) Tighten the 8 mm and 6 mm cylinder block connecting bolts on LH side (A — H) in alphabetical sequence.

Tightening torque:
(A) — (G): 25 N·m (2.5 kgf-m, 18.1 ft-lb)
(H): 6.4 N·m (0.65 kgf-m, 4.7 ft-lb)

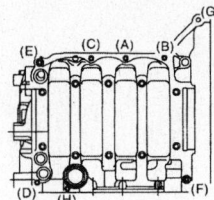

09490_SBCR_G0004

Main bearing torque sequence—2005–2006 SOHC engine

1) Remove oil in the mating surface of bearing and cylinder block before installation. Also apply a coat of engine oil to crankshaft pins.

2) Position the crankshaft on #2 and #4 cylinder block.

3) Apply fluid packing to the mating surface of #1 and #3 cylinder block, and position it on #2 and #4 cylinder block.

NOTE:
Do not allow fluid packing to jut into O-ring grooves, oil passages, bearing grooves, etc.

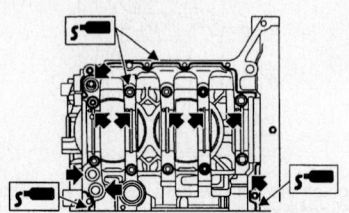

4) Apply engine oil to washers and thread of bolts.

5) Tighten the 10 mm cylinder block connecting bolts in alphabetical sequence shown in the figure. (LH side)

Tightening torque:
10 N·m (1.0 kgf-m, 7.4 ft-lb)

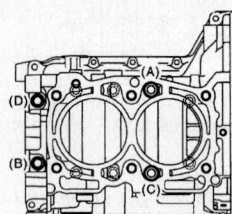

6) Tighten the 10 mm cylinder block connecting bolts in alphabetical sequence shown in the figure. (RH side)

Tightening torque:
10 N·m (1.0 kgf-m, 7.4 ft-lb)

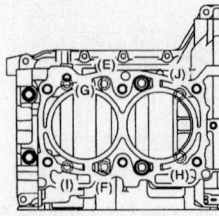

7) Further tighten the LH side bolts (A — D) in alphabetical sequence.

Tightening torque:
(A), (C): 20 N·m (2.0 kgf-m, 14.8 ft-lb)
(B), (D): 15 N·m (1.5 kgf-m, 10.8 ft-lb)

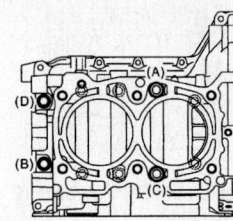

8) Further tighten the RH side bolts (E — J) in alphabetical sequence.

Tightening torque:
(E), (F), (G), (I): 20 N·m (2.0 kgf-m, 14.8 ft-lb)
(H), (J): 18 N·m (1.8 kgf-m, 13.3 ft-lb)

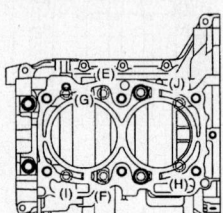

9) Further tighten the LH side bolts (A — D) by 90° in alphabetical sequence.

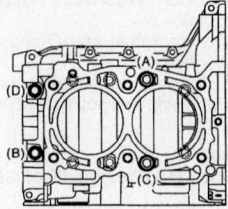

10) Further tighten the RH side bolts (E — J) by 90° in alphabetical sequence.

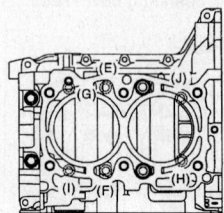

11) Tighten the 8 mm and 6 mm cylinder block connecting bolts in alphabetical sequence shown in the figure.

Tightening torque:
(A) — (G): 25 N·m (2.5 kgf-m, 18.1 ft-lb)
(H): 6.4 N·m (0.65 kgf-m, 4.7 ft-lb)

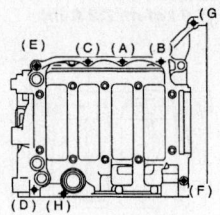

09490_SBCR_G0006

Main bearing torque sequence—2004 DOHC engine

1) Remove oil on the mating surface of bearing and cylinder block before installation. Apply engine oil to crankshaft pins.

2) Position the crankshaft and O-rings on the #1 and #3 cylinder block.

3) Apply liquid gasket to the mating surface of #1 and #3 cylinder blocks, and position cylinder block #2 and #4.

NOTE:
Do not allow liquid gasket to run over to O-ring grooves, oil passages, bearing grooves, etc.

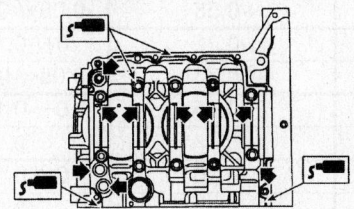

4) Apply a coat of engine oil to the washer and bolt thread.

5) Tighten the 10 mm cylinder block connecting bolts on LH side (A — D) in alphabetical sequence.

Tightening torque:
10 N·m (1.0 kgf-m, 7.2 ft-lb)

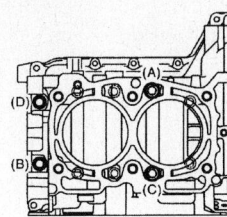

6) Tighten the 10 mm cylinder block connecting bolts on RH side (E — J) in alphabetical sequence.

Tightening torque:
10 N·m (1.0 kgf-m, 7.2 ft-lb)

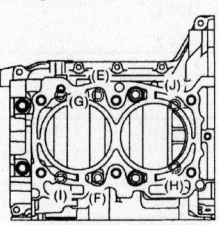

7) Further tighten the LH side bolts (A — D) in alphabetical sequence.

Tightening torque:
18 N·m (1.8 kgf-m, 13.0 ft-lb)

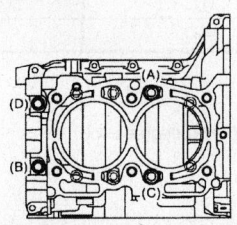

8) Further tighten the RH side bolts (E — J) in alphabetical sequence.

Tightening torque:
18 N·m (1.8 kgf-m, 13.0 ft-lb)

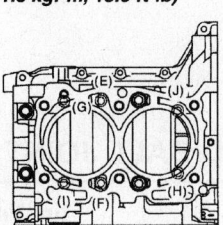

9) Further tighten the LH side bolts (A — D) in alphabetical sequence.

(A), (C): 90°
(B), (D): 40 N·m (4.1 kgf-m, 29.5 ft-lb)

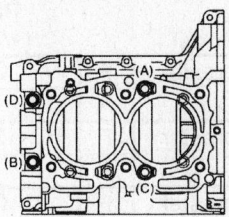

10) Tighten the RH side bolts (E — J) 90° further in alphabetical sequence.

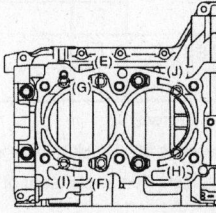

11) Tighten the 8 mm and 6 mm cylinder block connecting bolts on LH side (A — H) in alphabetical sequence.

Tightening torque:
(A) — (G): 25 N·m (2.5 kgf-m, 18.1 ft-lb)
(H): 6.4 N·m (0.65 kgf-m, 4.7 ft-lb)

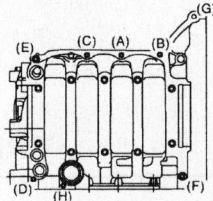

09490_SBCR_G0007

Main bearing torque sequence—2005–2006 DOHC engine

WHEEL ALIGNMENT

Year	Model		Caster Range (+/-Deg.)	Caster Preferred Setting (Deg.)	Camber Range (+/-Deg.)	Camber Preferred Setting (Deg.)	Toe-in (in.)
2002	Forester	F	0.75	+2.58	0.50	-0.25	0+/-0.12
		R	—	—	0.75	-0.58	0.08+/-0.04
2003	Forester	F	0.75	+2.58	0.50	-0.25	0+/-0.12
		R	—	—	0.75	-0.58	0.08+/-0.04
2004	Forester	F	0.75	+2.58	0.50	-0.25	0+/-0.12
		R	—	—	0.75	-0.58	0.08+/-0.04
2005	Forester	F	—	①	0.50	-0.25	0+/-0.12
		R	—	—	—	②	③
2006	Forester	F	—	①	0.50	-0.25	0+/-0.12
		R	—	—	—	②	③

① Reference: 3 degrees 03'

② SOHC: -0 degrees 50'. DOHC: -0 degrees 55'

③ 0.079 +/-0.12 inch. Toe angle (sum of both wheels) 0 degrees 10' +/-0 degrees 15'

09490_ FORE_C0010

TIRE, WHEEL AND BALL JOINT SPECIFICATIONS

Year	Model	OEM Tires Standard	OEM Tires Optional	Tire Pressures (psi) Front	Tire Pressures (psi) Rear	Wheel Size	Ball Joint Inspection	Lug Nut
2002	Forester	P205/70R15 95S	P215/60R16 94H	①	①	②	0.012 in. ③	58-72
2003	Forester	P215/60R16 94H	—	①	①	16X6.5 JJ	0.012 in. ③	68
2004	Forester	P215/60R16 94H	—	①	①	16X6.5 JJ	0.012 in. ③	68
2005	Forester	P215/60R16 94H	—	①	①	16X6.5 JJ	0.012 in. ③	66
2006	Forester	P215/60R16 94H	P215/55R17	①	①	④	0.012 in. ③	66

OEM: Original Equipment Manufacturer

PSI: Pounds Per Square Inch

STD: Standard

OPT: Optional

① See drivers door placard

② With standard tires: 6.0JJ

   With optional tires: 6.5JJ

③ Apply 154 lbs. vertical force

④ Standard: 16x6.5JJ

   Optional: 17x7.0JJ

09490_ FORE_C0011

BRAKE SPECIFICATIONS

All measurements in inches unless noted

Year	Model		Brake Disc			Brake Drum Diameter			Minimum Lining Thickness		Brake Caliper	
			Original Thickness	Minimum Thickness	Maximum Runout	Original Inside Diameter	Max. Wear Limit	Maximum Machine Diameter	Front	Rear	Bracket Bolts (ft. lbs.)	Mounting Bolts (ft. lbs.)
2002	Forester	F	0.945	0.660	0.0030	—	—	—	0.059	0.059	59	28
		R	0.390	0.335	0.0028	9.00 ①	9.08 ②	NA	—	0.059	38	28
2003	Forester	F	0.940	0.870	0.0030	—	—	—	0.059	0.059	59	28
		R	0.390	0.335	0.0028	9.00 ①	9.08 ②	NA	—	0.059	38	28
2004	Forester	F	0.940	0.870	0.0030	—	—	—	0.059	0.059	59	28
		R	0.390	0.335	0.0028	9.00 ①	9.08 ②	NA	—	0.059	38	28
2005	Forester	F	0.940	0.870	0.0030	—	—	—	0.059	0.059	59	19.2
		R	0.390	0.335	0.0028	9.00 ①	9.08 ②	NA	—	0.059	38	28
2006	Forester	F	0.940	0.870	0.0030	—	—	—	0.059	0.059	59	19.2
		R	0.390	0.335	0.0028	9.00 ①	9.08 ②	NA	—	0.059	38	28

NA: Not Available

① Parking brake drum on vehicles with rear disc brakes: 6.69

② Parking brake drum on vehicles with rear disc brakes: 6.73

09490_ FORE_C0012

SCHEDULED MAINTENANCE INTERVALS
SUBARU—FORESTER

TO BE SERVICED	TYPE OF SERVICE	VEHICLE MILEAGE INTERVAL (x1000)																
		3	7.5	15	22.5	30	37.5	45	52.5	60	67.5	75	82.5	90	97.5	105	112.5	120
Accessory drive belts	R									✓								✓
Accessory drive belts	S/I					✓				✓						✓		
Air cleaner filter	R					✓				✓				✓				✓
Automatic transmission fluid	S/I					✓				✓				✓				✓
Axle shaft joints	S/I			✓		✓		✓		✓		✓		✓		✓		✓
Brake fluid	R					✓				✓				✓				✓
Brake system lines	S/I			✓		✓		✓		✓		✓		✓		✓		✓
Clutch operation	S/I			✓		✓		✓		✓		✓		✓		✓		✓
Disc brake pads & rotors	S/I			✓		✓		✓		✓		✓		✓		✓		✓
Drums brake linings & drums	S/I					✓				✓				✓				✓
Engine coolant	R					✓				✓				✓				✓
Engine cooling system, hoses & connections	S/I					✓				✓				✓				✓
Engine oil & filter	R	✓	✓	✓	✓	✓	✓	✓	✓	✓	✓	✓	✓	✓	✓	✓	✓	✓
Front & rear axle boots	S/I			✓		✓		✓		✓		✓		✓		✓		✓
Front & rear wheel bearings	S/I			✓		✓		✓		✓		✓		✓		✓		✓
Fuel filter (2002-2004)	R					✓				✓				✓				✓
Fuel filter (2005-2006)	R									✓								✓
Parking & service brake systems' operation	S/I			✓		✓		✓		✓		✓		✓		✓		✓
Spark plugs (non turbo)	R									✓								✓
Spark plugs (turbo)	R					✓				✓				✓				✓
Steering & suspension	S/I			✓		✓		✓		✓		✓		✓		✓		✓
Timing belt	R															✓		
Timing belt	S/I					✓				✓				✓				
Tire rotation	S/I		✓	✓	✓	✓	✓	✓	✓	✓	✓	✓	✓	✓	✓	✓	✓	✓
Transmission & differential fluid levels	S/I					✓				✓				✓				✓
Valve clearance	S/I															✓		

R: Replace   S/I: Inspect and service, if needed   L: Lubricate

FREQUENT OPERATION MAINTENANCE (SEVERE SERVICE)

If a vehicle is operated under any of the following conditions it is considered severe service:

- Towing a trailer or using a camper or car-top carrier.

- Repeated short trips of less than 5 miles in temperatures below freezing, or trips of less than 10 miles in any temperature.

- Extensive idling or low-speed driving for long distances as in heavy commercial use, such as delivery, taxi or police cars.

- Operating on rough, muddy or salt-covered roads, or extensive mountain driving.

- Operating on unpaved or dusty roads.

- Driving in extremely hot (over 90°) conditions.

Engine oil and filter: replace every 3000 miles or 3 months, whichever occurs first.

Fuel filter: replace every 7500 miles or 7.5 months, whichever occurs first.

Fuel system, hoses & connections: inspect every 7500 miles or 7.5 months, whichever occurs first.

Transmission & differential fluid: replace every 15,000 miles.

Automatic transmission fluid: replace every 15,000 miles.

Brake fluid: replace every 15,000 miles.

Disc brake pads & rotors: inspect every 7500 miles or 7.5 months, whichever occurs first.

Front & rear axle boots: inspect every 7500 miles or 7.5 months, whichever occurs first.

Axle shaft boots: inspect every 7500 miles or 7.5 months, whichever occurs first.

Drum brake linings & drums: inspect every 7500 miles or 7.5 months, whichever occurs first.

Brake lines: inspect every 7500 miles or 7.5 months, whichever occurs first.

Parking & service brake system operation: inspect every 7500 miles or 7.5 months, whichever occurs first.

Clutch operation: inspect every 7500 miles or 7.5 months, whichever occurs first.

ENGINE REPAIR

➡**Disconnecting the negative battery cable on this vehicle may interfere with the functions of the on board computer systems and may require the computer to undergo a relearning process, once the negative battery cable is reconnected.**

Distributor

The Forester is equipped with a distributorless ignition system.

Alternator

REMOVAL & INSTALLATION

1. Before servicing the vehicle, refer to the Precautions Section.
2. Disconnect the negative battery cable.
3. On DOHC engine, Remove the collector cover.
4. To remove the front side V belt, remove the belt covers. Loosen the lock bolt. Loosen the slider bolt. Remove the front side belt.
5. To remove the rear side V belt, remove the belt covers. Loosen the lock bolt. Loosen the slider bolt. Remove the rear side belt. Remove the belt tensioner.
6. Disconnect the electrical connectors from the alternator.
7. Remove the alternator bracket retaining bolts. Remove the alternator from the engine.

To install:
8. Installation is the reverse of the removal procedure.
9. Check and adjust the drive belt tension, as required.

Ignition Timing

ADJUSTMENT

The ignition timing is controlled by the engine control computer and is not adjustable. To check the ignition timing proceed as follows:
1. Before servicing the vehicle, refer to the Precautions Section.
2. Warm up the engine, then turn the ignition **OFF**.
3. Connect a timing light to the No. 1 spark plug wire according to the manufactures directions.
4. Start the engine. Check the timing.
5. If the timing is not correct, there

could be a problem in the ignition control system.

Engine Assembly

REMOVAL & INSTALLATION

SOHC Engine

2002–2004

✳✳ CAUTION

The fuel injection system remains under pressure after the engine has been turned OFF. Properly relieve fuel pressure before disconnecting any fuel lines. Failure to do so may result in fire or personal injury.

1. Before servicing the vehicle, refer to the Precautions Section.
2. Relieve the fuel system pressure.
3. Drain the engine oil and coolant.
4. Discharge and recover the air conditioning system.
5. Remove or disconnect the following:
   - Negative battery cables and battery
   - Air cleaner assembly
   - Engine undercover
   - Radiator hoses and fan motor harness
   - Radiator
   - Air conditioning compressor and cap the lines
   - Air cleaner case stay
   - Front Oxygen (O_2S) sensor
6. If equipped with California emissions specifications, disconnect the rear O_2S sensor.
7. Remove or disconnect the following:
   - Engine ground terminal
   - Crankshaft Position (CKP) sensor connector
   - Camshaft Position (CMP) sensor connector
   - Knock Sensor (KS) connector
   - Alternator connector and terminal
   - Air conditioning compressor connectors, if equipped
   - Accelerator cable
   - Cruise control cable, if equipped
   - Pressure Switch
   - Brake booster hose
   - Heater inlet and outlet hoses
   - Resonator chamber
   - Front side V–belt
   - Pipe with bracket from the intake manifold

   - Power steering pump, leaving the lines connected and position it aside
   - Exhaust Y-pipe
   - Lower starter nuts
   - Lower engine-to-transmission nuts
   - Front engine mount-to-crossmember nuts
8. If equipped with an automatic transmission, perform the following:
   a. Remove the torque converter service hole plug.
   b. Matchmark the torque converter-to-driveplate.
   c. Rotate the engine to remove the torque converter-to-driveplate bolts as they become accessible.
9. Remove or disconnect the following:
   - Flywheel cover, if equipped with a manual transmission
   - Pitching stopper
   - Fuel delivery, return and evaporation hoses
10. Support the engine with a suitable lifting device attached to the engine lifting eyes.
11. Slightly raise the engine.
12. Raise the transmission with a floor jack.
    - Starter
    - Upper engine-to-transmission bolts
13. If equipped with a manual transmission, pull the engine forward then up and out of the vehicle to clear the transmission mainshaft.
14. If equipped with an automatic transmission, pull the engine forward then up and out of the vehicle.

To install:
15. If equipped with a manual transmission, apply a small amount of grease to the splines of the mainshaft.
16. Position the engine in the engine compartment and align it with the transmission.
17. Install the engine. Torque the upper bolts to 37 ft. lbs. (50 Nm).
18. Remove the lifting device and floor jack.
19. Install or connect the following:
    - Pitching stopper. Torque the bolts to 43 ft. lbs. (58 Nm) on the body side and 37 ft. lbs. (50 Nm) on the bracket side.
    - Starter
    - Flywheel cover, if equipped with a manual transmission
20. If equipped with an automatic transmission, perform the following:

a. Align the matchmarks, install the torque converter-to-driveplate bolts while rotating the engine and tighten to 18 ft. lbs. (25 Nm).

b. Install the service hole cover.

21. Install or connect the following:
- Power steering pump. Tighten the retainer bolts to 14 ft. lbs. (20 Nm).
- Power steering switch connector
- Accessory drive belt
- Resonator chamber and tighten the bolts to 24 ft. lbs. (33 Nm)
- Lower engine-to-transmission nuts. Tighten them to 36 ft. lbs. (50 Nm).
- Lower engine mounting nuts. Tighten them to 63 ft. lbs. (85 Nm) in the inner most elliptical hole in the front crossmember so the clearance is 0.16–0.24 in. (4–6mm).
- Exhaust Y-pipe with new gaskets and nuts
- Brake booster hose
- Heater inlet and outlet hoses
- Accelerator cable
- Cruise control cable, if equipped
- Engine harness connectors
- Engine ground terminal
- CKP sensor connector
- CMP sensor connector
- Knock sensor connector
- Alternator connector and terminal
- Air conditioning compressor connectors, if equipped
- Front O_2S sensor, and if removed, the rear O_2S sensor.
- Air cleaner element and cover

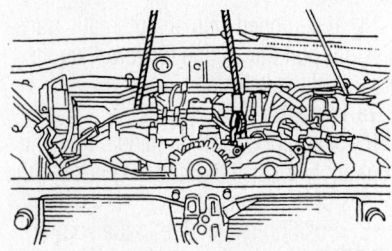

Attach a suitable lift device to the engine

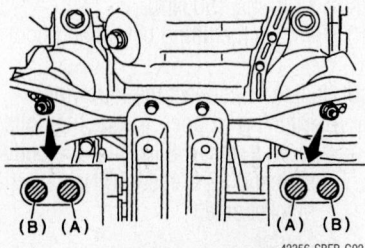

Be sure to tighten the front cushion rubber mounting bolts in the innermost elliptical hole in the front crossmember— 2002–2004

- Air conditioning lines with new O-rings, if equipped. Torque the bolts to 18 ft. lbs. (25 Nm).
- Radiator
- Engine undercover
- Negative battery cable

22. Fill the crankcase to the proper level with clean engine oil.

23. Fill and bleed the cooling system.

24. Charge the air conditioning system using an approved recovery/recycling machine.

25. If equipped, check the automatic transmission fluid level and add Dexron®II if necessary.

26. Start the engine and allow it to reach normal operating temperature. Check for leaks.

2005–2006

1. Before servicing the vehicle, refer to the Precautions Section.

2. Open the hood to the full open position and support it with the hood stay.

3. Properly relieve the fuel system pressure. Remove the fuel cap.

4. Properly disarm the SRS system.

5. Discharge the air conditioning system.

6. Disconnect the negative battery cable.

7. Drain the engine oil. Drain the cooling system.

8. Remove the air intake duct, air cleaner case and air intake chamber.

9. Remove the under cover. Remove the radiator.

10. Disconnect the air conditioning pressure hoses from the compressor.

11. Disconnect the front and rear oxygen sensor electrical connectors, the engine ground cable, the engine harness connector, the alternator connector and terminal, the air conditioning compressor electrical connector and the power steering switch connector.

12. Disconnect the brake booster hose and the heater hoses.

13. Remove the front side V-belt. Remove the power steering pipe and bracket. Remove the reservoir tank. Remove the power steering pump and position it aside.

14. Remove the front and center exhaust pipes.

15. Remove the lower transmission to engine retaining nuts.

16. Remove the front cushion rubber onto front crossmember retaining nuts.

17. If equipped with automatic transmission, remove the service plug hole cap. Remove the torque converter to drive plate retaining bolts.

18. Remove the pitching stopper. Disconnect the fuel hoses from the fuel pipes.

19. Remove the clip and disconnect the evaporator hose from the pipe.

20. Position the engine lifting fixture in place. Support the transmission using a suitable jack.

➡**Before separating the engine from the transmission, check to be sure nothing has been overlooked that will stop the engine from being removed. This is very important in order to facilitate reinstallation and because the transmission lowers under its own weight.**

21. Separate the engine from the transmission.

22. Remove the starter.

23. If equipped with automatic transmission, install tool ST498277200, with will hold the torque converter in place.

24. Remove the upper right side bolts that retain the transmission to the engine.

25. Slightly raise the engine. Raise the transmission, using the suitable jack. Move the engine horizontally until the mainshaft is withdrawn from the clutch cover. Carefully remove the engine from the vehicle.

To install:

26. Installation is the reverse of the removal procedure.

27. Torque the upper transmission to engine retaining bolts to 36.9 ft. lbs.

28. Torque the lower transmission to engine retaining bolts to 36.9 ft. lbs.

29. If equipped, tighten the torque converter to drive plate bolts to 18.1 ft. lbs.

30. Fill the engine with the proper grade and type engine oil.

31. Fill and bleed the cooling system.

32. Charge the air conditioning system.

33. If equipped. Adjust the clutch cable as required.

34. If equipped, check the automatic transmission fluid level, correct as required using the proper grade and type transmission fluid.

35. Start the engine and allow it to reach normal operating temperature. Check for leaks and correct as required.

DOHC Engine

2004

> ✳✳ **CAUTION**
>
> **The fuel injection system remains under pressure after the engine has been turned OFF. Properly relieve fuel pressure before disconnecting any fuel lines. Failure to do so may result in fire or personal injury.**

1. Before servicing the vehicle, refer to the Precautions Section.

2. Relieve the fuel system pressure.

3. Drain the engine oil and coolant.

4. Discharge and recover the air conditioning system.

5. Disconnect the negative battery cable.

6. Remove the radiator.

7. Remove the coolant filler tank.

8. Disconnect the air conditioning hoses from the compressor and cap the lines.

9. Remove the intercooler as follows:

a. Disconnect the Positive Crankcase Ventilation (PCV) hose from the pipe.

b. Remove the PCV pipe from the intercooler.

c. Disconnect the air by–pass valve hose.

d. Remove the intercooler–to–bracket bolts.

e. Loosen the clamps that connect the intercooler to the turbocharger and throttle body.

f. Disconnect the intercooler duct from the turbocharger and remove the intercooler.

10. Disconnect the following:
- Engine harness connector
- Engine ground terminal
- Left and right engine ground cables
- Alternator connector and terminal
- Air conditioning compressor connectors, if equipped
- Accelerator cable, on vehicles equipped with a manual transmission
- Clutch release spring, if equipped with a manual transmission
- Brake booster hose
- Heater inlet and outlet hoses

11. Loosen the power steering pump lock bolt and slider bolt. Remove the power steering belt.

12. Disconnect the power steering switch connector.

13. Remove the pipe with the bracket from the intake manifold and remove the power steering pump.

14. Remove the power steering tank from its bracket by pulling it up.

15. Place the power steering pump on the right wheel apron.

16. Remove the transmission cooler lines from the frame, if equipped with an automatic transmission.

17. Remove the center exhaust pipe.

18. Remove the lower transmission–to–engine nuts.

19. Remove the nuts which attach the front mount rubber to the crossmember.

20. On vehicles equipped with a manual transmission, perform the following:

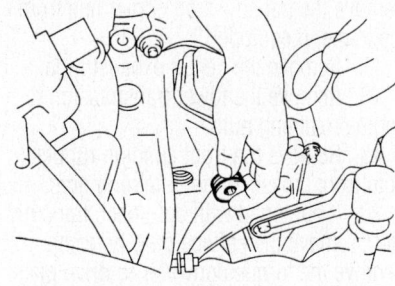

Remove the plug using a 10mm wrench— 2004 DOHC engine

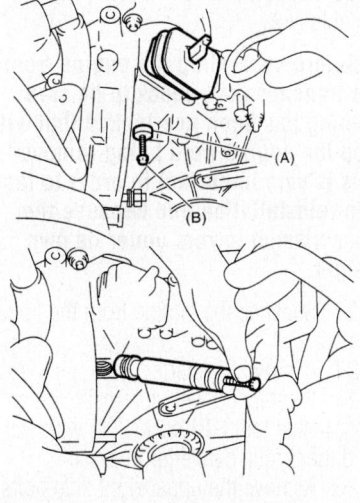

(A) Shaft
(B) Bolt

67170-SBFR-G02

Screw the 6mm bolt into the release fork and remove it—2004 DOHC engine

a. Remove the clutch operating cylinder from the transmission.

b. Remove the plug using a 10mm wrench

c. Screw the 6mm bolt into the release fork and remove it.

d. Raise the release fork and unfasten the release bearing tabs to free the fork.

21. If equipped with an automatic transmission, perform the following:

a. Remove the torque converter service hole plug.

b. Matchmark the torque converter-to-driveplate.

c. Rotate the engine to remove the torque converter-to-driveplate bolts as they become accessible.

22. Remove the pitching stopper.

23. Disconnect the fuel delivery, return and evaporation hoses

24. Remove the fuel filter and bracket.

25. Support the engine with a suitable

lifting device attached to the engine lifting eyes.

26. Support the transmission with a floor jack.

27. Remove the starter.

28. Install tool ST 498277200 to the converter housing to prevent the converter from falling.

29. Remove the upper engine-to-transmission bolts.

30. If equipped with a manual transmission, pull the engine forward then up and out of the vehicle to clear the transmission mainshaft.

31. If equipped with an automatic transmission, pull the engine forward then up and out of the vehicle.

To install:

32. If equipped with a manual transmission, perform the following:

a. Remove the release bearing from the clutch cover using a flat bladed tool.

b. Install the release bearing onto the transmission.

c. Install the release fork into the release bearing tab.

d. Apply a small amount of grease to the splines of the mainshaft.

e. Insert the release fork shaft into the release fork. Make sure the cutout portion of the release fork shaft contacts the spring pin.

f. Install the plug and tighten to 32 ft. lbs. (44 Nm).

33. Install the front mount rubber to the engine and tighten to 26 ft. lbs. (35 Nm).

34. Position the engine in the engine compartment and align it with the transmission.

35. Install the engine. Torque the upper bolts to 37 ft. lbs. (50 Nm).

36. Remove the lifting device and floor jack.

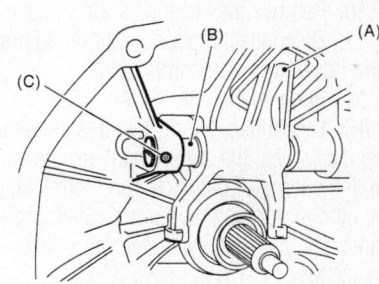

(A) Release fork
(B) Release shaft
(C) Spring pin

67170-SBFR-G03

Release fork and related components— 2004 DOHC engine

37. Install the pitching stopper. Torque the bolts to 43 ft. lbs. (58 Nm) on the body side and 37 ft. lbs. (50 Nm) on the bracket side.

38. Installation is the reverse of removal, please keep in mind the following torque specifications:

    a. Torque converter-to-driveplate bolts while rotating the engine and tighten to 18 ft. lbs. (25 Nm).

    b. Power steering pump. Tighten the retainer bolts to 14 ft. lbs. (20 Nm).

    c. Lower engine-to-transmission nuts. Tighten them to 37 ft. lbs. (50 Nm).

    d. Engine-to-crossmember nuts to 63 ft. lbs. (85 Nm).

39. Fill the crankcase to the proper level with clean engine oil.

40. Fill and bleed the cooling system.

41. Charge the air conditioning system using an approved recovery/recycling machine.

42. If equipped, check the automatic transmission fluid level and add Dexron®II if necessary.

43. Start the engine and allow it to reach normal operating temperature. Check for leaks.

2005–2006

1. Before servicing the vehicle, refer to the Precautions Section.

2. Open the hood to the full open position and support it with the hood stay.

3. Remove the collector cover.

4. Properly relieve the fuel system pressure. Remove the fuel cap.

5. Properly disarm the SRS system.

6. Discharge the air conditioning system.

7. Disconnect the negative battery cable.

8. Drain the engine oil. Drain the cooling system.

9. Remove the radiator. Remove the coolant filler tank.

10. Remove the secondary air pump.

11. Disconnect the air conditioning pressure hoses from the compressor.

12. Remove the intercooler.

13. Disconnect the right and left engine ground cables, the engine harness connector, the alternator connector and terminal, the air conditioning compressor electrical connector and the power steering switch connector.

14. Disconnect the brake booster hose and the heater hoses.

15. Remove the front side V-belt. Remove the power steering pipe and bracket. Remove the reservoir tank. Remove the power steering pump and position it aside.

16. Raise and support the vehicle safely. Remove the transmission cooler line from the frame, if equipped.

17. Remove the center exhaust pipe.

18. Remove the lower transmission to engine retaining nuts.

19. Remove the front cushion rubber onto front crossmember retaining nuts.

20. If equipped with automatic transmission, remove the service plug hole cap. Remove the torque converter to drive plate retaining bolts.

21. Remove the pitching stopper. Disconnect the fuel hoses from the fuel pipes.

22. Position the engine lifting fixture in place. Support the transmission using a suitable jack.

➡ **Before separating the engine from the transmission, check to be sure nothing has been overlooked that will stop the engine from being removed. This is very important in order to facilitate reinstallation and because the transmission lowers under its own weight.**

23. Separate the engine from the transmission.

24. Remove the starter.

25. If equipped with automatic transmission, install tool ST498277200, with will hold the torque converter in place.

26. Remove the upper right side bolts that retain the transmission to the engine.

27. Slightly raise the engine. Raise the transmission, using the suitable jack. Move the engine horizontally until the mainshaft is withdrawn from the clutch cover. Carefully remove the engine from the vehicle.

To install:

28. Installation is the reverse of the removal procedure.

29. Torque the upper transmission to engine retaining bolts to 36.9ft. lbs.

30. Torque the lower transmission to engine retaining bolts to 36.9ft. lbs.

31. If equipped, tighten the torque converter to drive plate bolts to 18.1 ft. lbs.

32. Fill the engine with the proper grade and type engine oil.

33. Fill and bleed the cooling system.

34. Charge the air conditioning system.

35. If equipped. Adjust the clutch cable as required.

36. If equipped, check the automatic transmission fluid level, correct as required using the proper grade and type transmission fluid.

37. Start the engine and allow it to reach normal operating temperature. Check for leaks and correct as required.

Water Pump

REMOVAL & INSTALLATION

SOHC Engine

2002–2004

1. Before servicing the vehicle, refer to the Precautions Section.

2. Drain the coolant into a suitable container.

3. Remove or disconnect the following:

- Negative battery cable
- Engine undercover
- Electrical connections from the radiator fan motor and sub motor(s)
- Bolt that retains the water by-pass pipe of the oil cooler onto the oil pump on vehicles equipped with an automatic transmission
- Radiator outlet hose
- Radiator fan motor assembly
- Accessory drive belts
- Timing belt and tensioner
- Belt idler number 2
- Camshaft Position (CMP) sensor
- Left side camshaft pulleys and left side rear timing belt cover
- Tensioner bracket
- Radiator hose and heater hose from the water pump

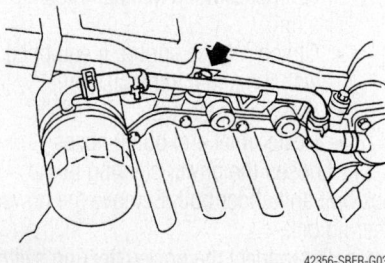

42356-SBFR-G03

Remove the bolt retaining the water by-pass pipe of the oil cooler onto the oil pump on vehicles equipped with automatic transmission—2002–2004 SOHC engine

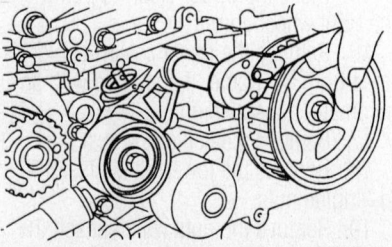

42356-SBFR-G04

Automatic belt tensioner location—SOHC engine

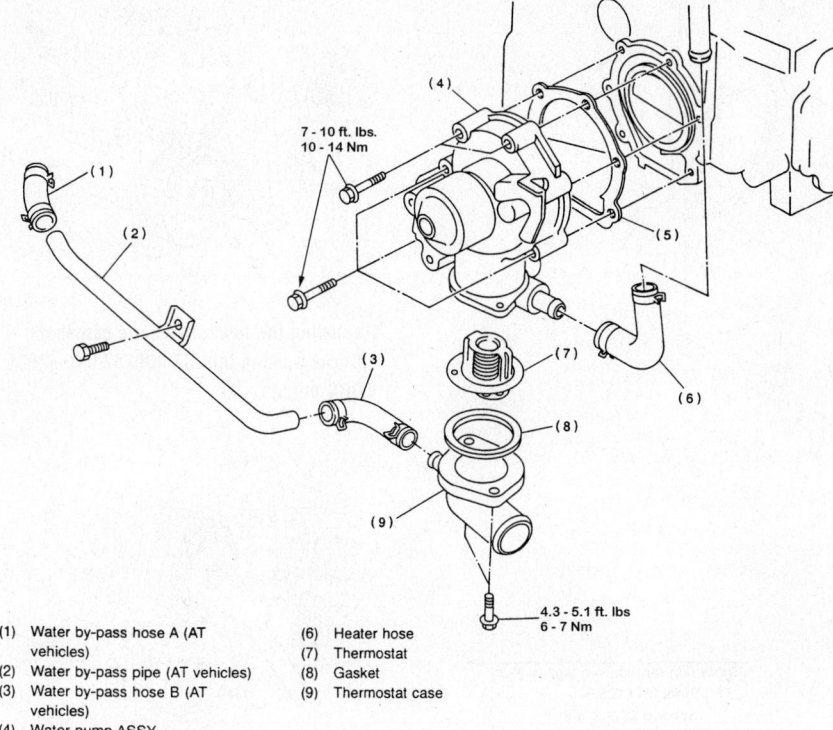

(1) Water by-pass hose A (AT vehicles)
(2) Water by-pass pipe (AT vehicles)
(3) Water by-pass hose B (AT vehicles)
(4) Water pump ASSY
(5) Gasket
(6) Heater hose
(7) Thermostat
(8) Gasket
(9) Thermostat case

7924XG01

Water pump mounting and related components—2002–2004 SOHC engine

- Water pump retainer bolts
- Water pump

To install:

4. Clean the gasket mating surfaces thoroughly. Always use new gaskets during installation.

5. Install or connect the following:
- Water pump. Torque the bolts, in sequence, to 9 ft. lbs. 12 Nm). After tightening the bolts once, retighten to the same specification again.
- Radiator hose and heater hose to the water pump
- Left side rear timing belt cover, left side camshaft pulleys and tensioner bracket
- CMP sensor
- Timing belt and tensioner

- Accessory drive belts
- Water pipe bypass pipe retaining bolt
- Radiator fan motor assembly
- Radiator outlet hose

42356-SBFR-G05

Belt idler number two location—SOHC engine

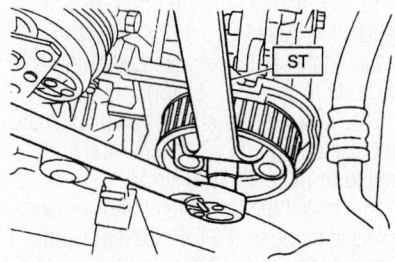

42356-SBFR-G06

Remove the left side camshaft pulleys—SOHC engine

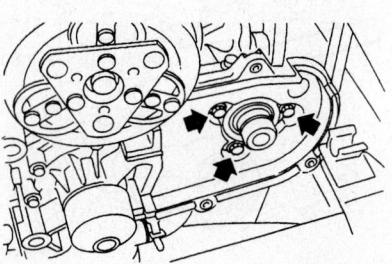

42356-SBFR-G07

Remove the left hand belt cover number two—SOHC engine

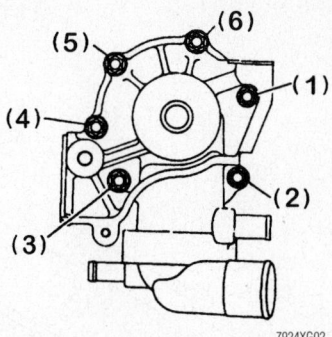

7924XG02

Water pump bolt tightening sequence—2002–2004 SOHC and 2004 DOHC engines

- Engine undercover
- Negative battery cable
6. Fill the system with coolant.
7. Start the engine and allow it to reach operating temperature.
8. Check for leaks.

2005–2006

1. Before servicing the vehicle, refer to the Precautions Section.
2. Disconnect the negative battery cable.
3. Drain the cooling system. Remove the radiator.
4. To remove the front side V belt, remove the belt covers. Loosen the lock bolt. Loosen the slider bolt. Remove the front side belt.
5. To remove the rear side V belt, remove the belt covers. Loosen the lock bolt. Loosen the slider bolt. Remove the rear side belt. Remove the belt tensioner.
6. Remove the crankshaft pulley bolt.
7. Lock the crankshaft in place using tool ST499977100, or equivalent.
8. Remove the crankshaft pulley.
9. Remove the timing belt.
10. Remove the automatic belt tensioner adjuster. Remove the belt idler number 2.
11. Using tool ST18231AA010, remove the left side cam sprocket. Remove the left side belt cover number 2. Remove the tensioner bracket.
12. Disconnect the hose from the water pump.
13. Remove the water pump retaining bolts. Remove the water pump from the engine.

To install:

14. Clean the gasket mating surfaces thoroughly. Always use new gaskets during installation.
15. Installation is the reverse of the removal procedure.
16. Tighten the water pump retaining bolts in two stages and in the proper

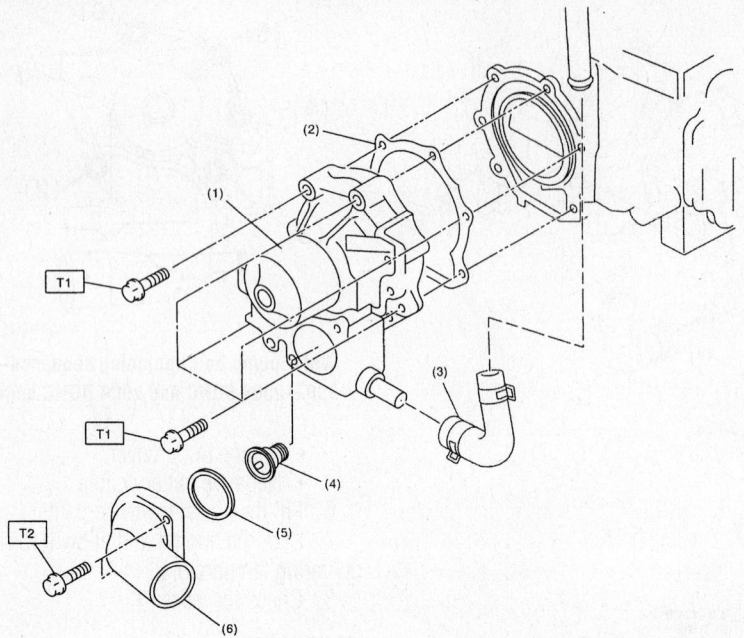

(1)	Water pump ASSY	(5)	Gasket
(2)	Gasket	(6)	Thermostat cover
(3)	Heater by-pass hose		
(4)	Thermostat		

Tightening torque: N·m (kgf-m, ft-lb)	
T1:	First 12 (1.2, 8.9)
	Second 12 (1.2, 8.9)
T2:	12 (1.2, 8.9)

09490_FORE_G0001

Water pump and related components—2005–2006 SOHC engine

sequence to 8.9 ft. lbs. and than again to 8.9 ft. lbs.

17. Be sure to fill the cooling system with the proper grade and type coolant.

18. Start the engine and check for leaks, correct as required.

DOHC Engine

2002–2004

1. Before servicing the vehicle, refer to the Precautions Section.

2. Drain the coolant into a suitable container.

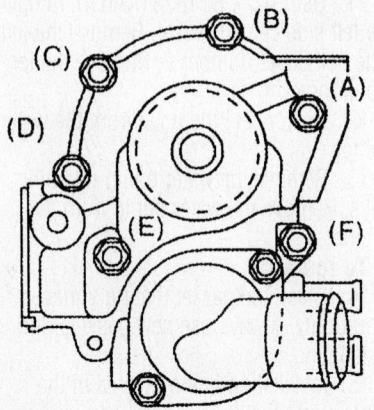

09490_FORE_G0003

Water pump bolt tightening sequence— 2005–2006 SOHC and 2004 DOHC engines

3. Disconnect the negative battery cable.

4. Remove the radiator.

5. Remove the drive belts.

6. Remove the crankshaft pulley and the timing belt.

7. Remove the automatic belt tension adjuster **A**. Refer to the illustration for component location.

8. Remove the belt idler **B**. Refer to the illustration for component location.

9. Remove the belt idle No. 2 **C**. Refer to the illustration for component location.

10. Remove the Camshaft Position (CMP) sensor.

11. Remove the left hand camshaft

67170-SBFR-G04

Location of the belt tension adjuster (A), belt idler (B) and belt idler No. 2 (C)— DOHC engine

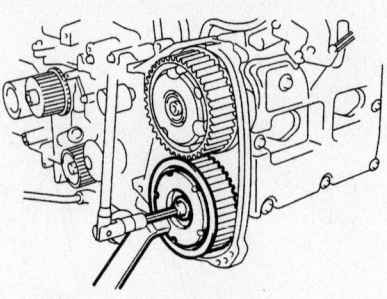

67170-SBFR-G05

Removing the lower left hand camshaft sprockets using tool ST 499207400—2004 DOHC engine

67170-SBFR-G06

Removing the upper left hand camshaft sprockets using tool ST 499977500—2004 DOHC engine

sprockets using tools ST 499207400 and ST 499977500 as illustrated.

12. Remove belt cover No.2 from the left hand side.

13. Remove the tensioner bracket.

14. Disconnect the hose from the water pump, remove the pump bolts and the pump.

To install:

15. Clean the gasket mating surfaces thoroughly. Always use new gaskets during installation.

16. Install the water pump. Torque the bolts, in sequence, to 9 ft. lbs. (12 Nm). After tightening the bolts once, retighten to the same specification again.

17. Attach the hose to the water pump.

18. Install the tensioner bracket and tighten the bolts to 18 ft. lbs. (25 Nm).

19. Install belt cover No.2 and tighten the bolts to 3 ft. lbs. (5 Nm).

20. Install the lower left hand camshaft sprocket using tool ST 499207400 and tighten to 72 ft. lbs. (98 Nm)

21. Install the upper left hand camshaft sprocket using tool ST 499977500 and tighten to 22 ft. lbs. (29 Nm) plus an additional 45 degrees.

22. Install the CMP.

23. Install belt idler No.2 and the belt idler.

24. Install the automatic belt tension adjuster and tighten to 29 ft. lbs. (39 Nm).

25. Install the timing belt.

26. Install the crankshaft pulley and tighten the bolt to 132 ft. lbs. (180 Nm).

27. Install the drive belts.

28. Install the radiator.

29. Fill the system with coolant.

30. Start the engine and allow it to reach operating temperature.

31. Check for leaks.

2005–2006

1. Before servicing the vehicle, refer to the Precautions Section.

2. Disconnect the negative battery cable.

3. Drain the cooling system. Remove the radiator.

4. Remove the collector cover.

5. To remove the front side V belt, remove the belt covers. Loosen the lock bolt. Loosen the slider bolt. Remove the front side belt.

6. To remove the rear side V belt, remove the belt covers. Loosen the lock bolt. Loosen the slider bolt. Remove the rear side belt. Remove the belt tensioner.

7. Remove the crankshaft pulley bolt.

8. Lock the crankshaft in place using tool ST499977100, or equivalent.

9. Remove the crankshaft pulley.

10. Remove the timing belt.

11. Remove the automatic belt tensioner adjuster. Remove the idler belt. Remove the belt idler number 2.

12. Remove the camshaft position sensor.

13. Using tool ST499977500, remove the left side cam sprocket. Remove the left side belt cover number 2. Remove the tensioner bracket.

14. Disconnect the hose from the water pump.

15. Remove the water pump retaining bolts. Remove the water pump from the engine.

To install:

16. Clean the gasket mating surfaces thoroughly. Always use new gaskets during installation.

17. Installation is the reverse of the removal procedure.

18. Tighten the water pump retaining bolts in two stages and in the proper sequence to 8.9 ft. lbs. and than again to 8.9 ft. lbs.

19. Be sure to fill the cooling system with the proper grade and type coolant.

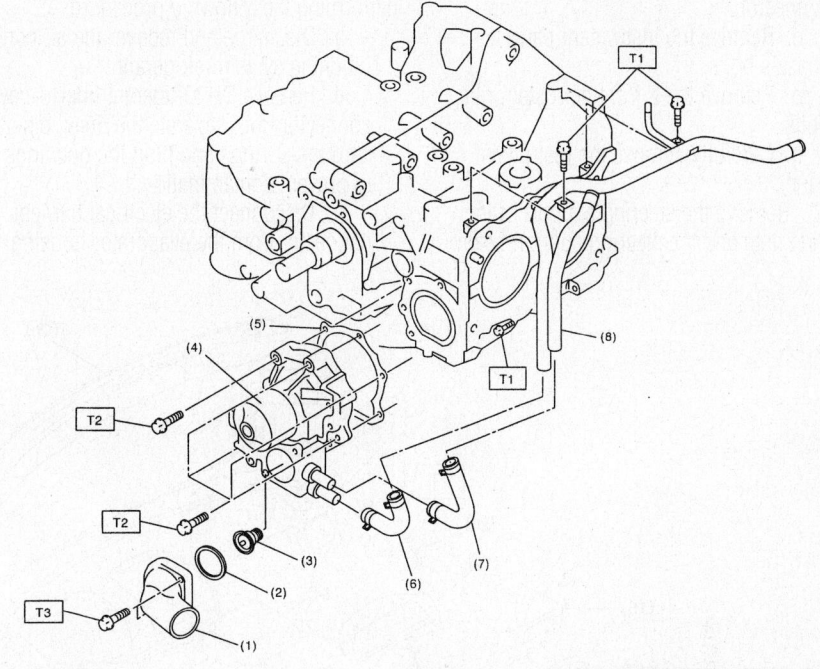

(1)	Thermostat cover	(6)	Heater by-pass hose
(2)	Gasket	(7)	Coolant filler by-pass hose
(3)	Thermostat	(8)	Water by-pass pipe
(4)	Water pump ASSY		
(5)	Gasket		

Tightening torque: N·m (kgf-m, ft-lb)	
T1:	6.4 (0.65, 4.7)
T2:	First 12 (1.2, 8.9)
	Second 12 (1.2, 8.9)
T3:	12 (1.2, 8.9)

09490_FORE_G0002

Water pump and related components—2005–2006 DOHC engine

20. Start the engine and check for leaks, correct as required.

Heater Core

REMOVAL & INSTALLATION

2002

1. Before servicing the vehicle, refer to the Precautions Section.

2. Disconnect the negative battery cable.

✳✳ CAUTION

After disconnecting the negative battery cable, wait for at least 20 seconds for the air bag module to deplete its energy.

3. Drain the engine coolant into a clean container for reuse.

4. Disconnect the heater hoses from the heater core.

5. Remove the instrument panel by performing the following procedure:

a. If equipped with a manual transmission, remove the shift knob.

b. Remove both the front and rear console covers.

c. Remove the console box-to-chassis screws and the console box.

d. Remove the 3 lower left side cover assembly screws, disengage the 3 upper clips, and remove the cover assembly.

e. Using a screwdriver, disconnect the data link connector from the lower cover.

f. Remove the knee panel.

g. At the glove box, remove the right side cover screw, the clip and the side cover.

h. Remove the glove box screws and remove the glove box.

i. Remove the center panel bezel.

j. Remove the audio assembly screws, disconnect the electrical connectors and remove the audio assembly.

k. Remove the 2 steering column-to-instrument panel bolts and lower the steering column.

l. Move the temperature control switch to FULL HOT, the mode selector switch to DEF and the recirculation switch to FRESH positions.

m. Disconnect the temperature control cable and the mode control cable from the heater housing.; then, the recirculation control cable from the intake housing.

n. Disconnect the electrical harness connectors.

o. Remove the instrument panel-to-chassis bolts.

p. Remove the 2 front defroster grille bolts.

q. Carefully, remove the instrument panel.

6. Remove the steering support beam bracket nuts and the steering support beam.

7. Remove the evaporator housing by performing the following procedure:

a. Discharge and recover the air conditioning system refrigerant.

b. Remove the refrigerant line-to-cowl connector bolt, separate the lines, discard the O-rings and plug the openings to prevent contamination.

c. Disconnect the electrical harness connector from the evaporator housing.

d. Disconnect the drain hose.

e. Remove the evaporator housing nut/bolts and the evaporator housing.

8. Remove the heater housing-to-chassis bolts and the heater housing.

9. Remove the heater core from the heater housing.

To install:

10. Install the heater core to the heater housing.

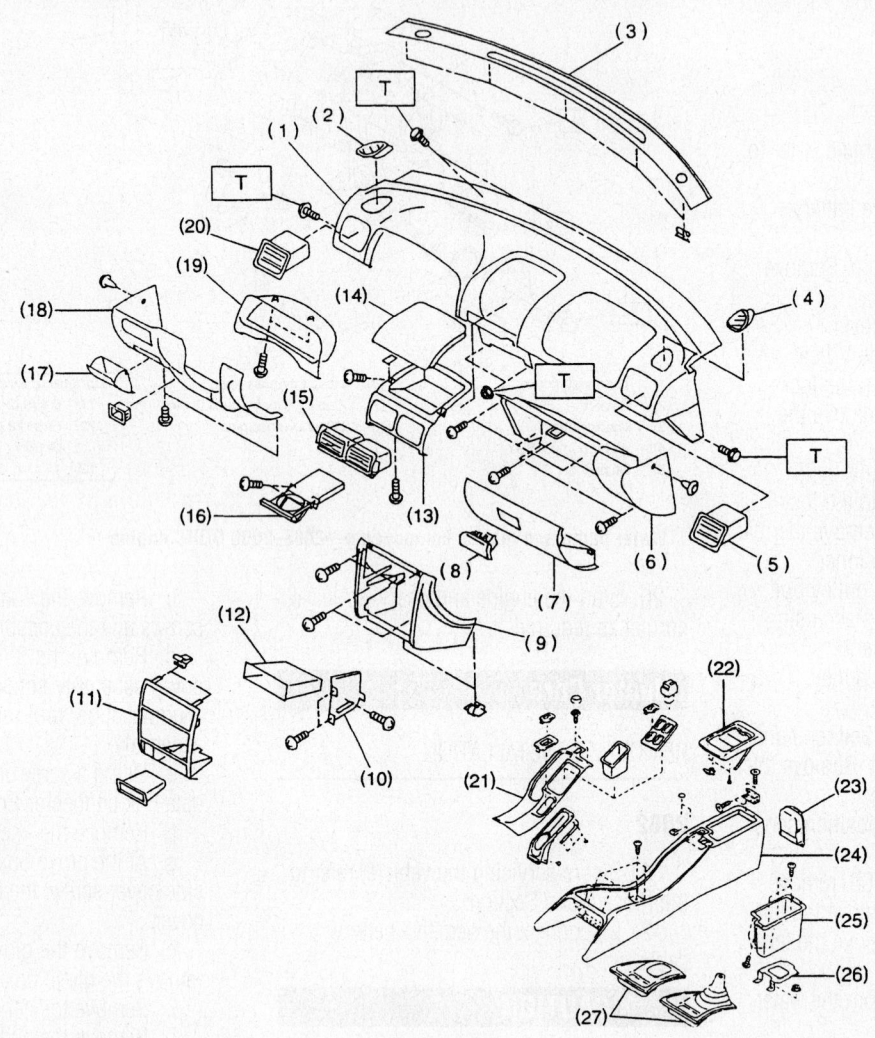

1	Pad & frame	12	Pocket	23	Rear cup holder
2	Grille side (D)	13	Panel center	24	Console box
3	Front def. grille	14	Center pocket lid	25	Console pocket
4	Grille side (P)	15	Grille center	26	Rear console BRKT
5	Grille vent (P)	16	Cup holder	27	Front cover
6	Glove box panel	17	Side pocket		
7	Glove box lid	18	Lower cover ASSY		
8	Knob	19	Meter visor		
9	Instrument panel center console	20	Grille vent (D)		
10	BRKT (Radio)	21	Console cover		
11	Center console cover	22	Console lid		

Tightening torque: N·m (kg-m, ft-lb)
T: 7±1 (0.7±0.1, 5.1±0.7)

93113GI8

Instrument panel and related components—2002–2004

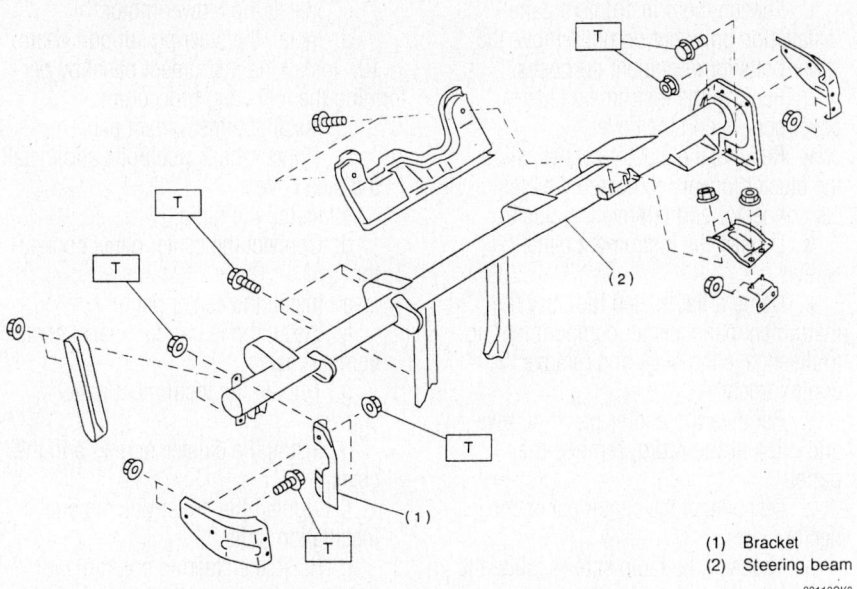

(1) Bracket
(2) Steering beam

93113GK8

Steering support beam assembly and related components

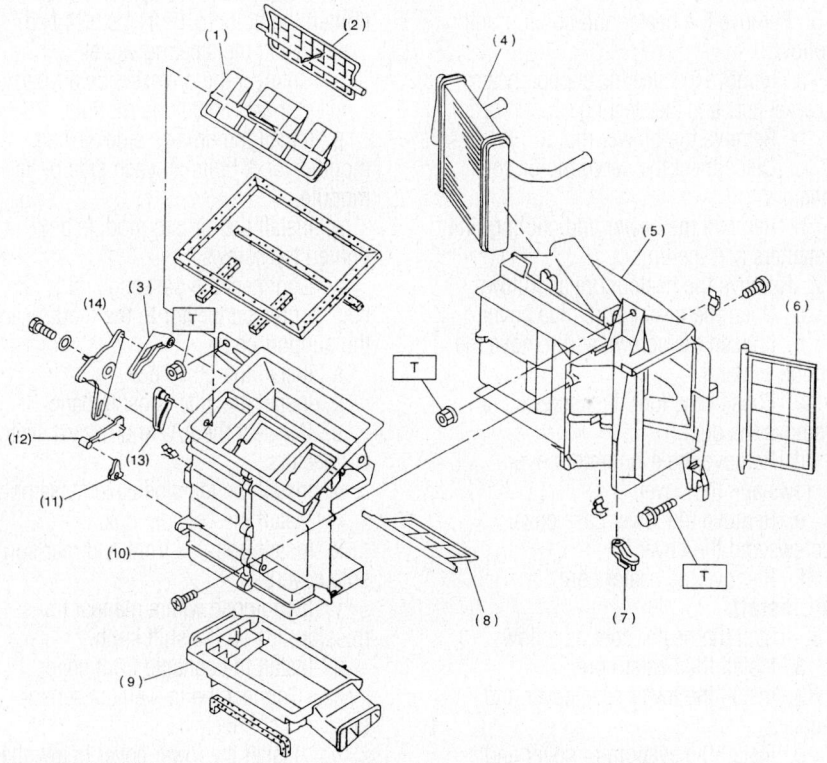

1	Vent door	7	Mix lever	13	Vent lever
2	DEF door	8	Foot door	14	Side link
3	DEF lever	9	Foot duct		
4	Heater core	10	Heater case REAR		
5	Heater case FRONT	11	Foot lever lower		
6	Mix door	12	Foot lever upper		

Tightening torque: N·m (kg-m, ft-lb)
T: 7.35±1.96
(0.750±0.200, 5.421±1.446)

93113GI7

Heater core and related components—2002–2004

11. Install the heater housing and the heater housing-to-chassis bolts.

12. Install the steering support beam and the steering support beam bracket nuts.

13. Install the evaporator housing by performing the following procedure:

 a. Install the evaporator housing and the evaporator housing nut/bolts.

 b. Connect the drain hose.

 c. Connect the electrical harness connector to the evaporator housing.

 d. Using new O-rings, assemble the refrigerant lines and install the refrigerant line-to-cowl connector bolt.

14. Install the instrument panel by performing the following procedure:

 a. Carefully, install the instrument panel.

 b. Install the 2 front defroster grille bolts.

 c. Install the instrument panel-to-chassis bolts.

 d. Connect the electrical harness connectors.

 e. Connect the temperature control cable and the mode control cable to the heater housing. Then, the recirculation control cable to the intake housing.

 f. Install the steering column and lower the 2 steering column-to-instrument panel bolts and torque to 14–21 ft. lbs. (20–30 Nm).

 g. Install the audio assembly, connect the electrical connectors and install the audio assembly screws.

 h. Install the center panel bezel.

 i. Install the glove box and the glove box screws.

 j. At the glove box, install the right side cover, the clip and the side cover screw.

 k. Install the knee panel.

 l. Connect the data link connector to the lower cover.

 m. Install the lower left side cover assembly, engage the 3 upper clips and install the cover assembly screws.

 n. Install the console box and the console box-to-chassis screws.

 o. Install both the front and rear console covers.

 p. If equipped with a manual transmission, install the shift knob.

15. Connect the heater hoses to the heater core.

16. Refill the cooling system.

17. Connect the negative battery cable.

18. Evacuate and charge the air conditioning system refrigerant.

19. Run the engine to normal operating

temperatures; then, check the climate control operation and check for leaks.

2003–2004

1. Before servicing the vehicle, refer to the Precautions Section.

2. Disconnect the negative battery cable.

> **✳✳ CAUTION**
>
> **After disconnecting the negative battery cable, wait for at least 20 seconds for the air bag module to deplete its energy.**

3. Drain the engine coolant into a clean container for reuse.

4. Disconnect the heater hoses from the heater core.

5. Remove the instrument panel by performing the following procedure:

   a. Remove the lower cover from below the steering wheel.

   b. Disconnect the in–vehicle sensor hose and connector.

   c. Remove the console front cover.

   d. If equipped with a manual transmission, remove the shift knob.

   e. Remove both the front and rear console covers.

   f. Remove the console box-to-chassis screws and the console box.

   g. Remove the 3 lower left side cover assembly screws, disengage the 3 upper clips, and remove the cover assembly.

   h. Disconnect the A/C and hazard switch connectors.

   i. Remove the side console panel.

   j. Remove the glove box screws and remove the glove box.

   k. Disconnect the passenger side air bag module connector from the support beam and the module.

   l. Unfasten the air bag module screws and remove the module.

   m. Remove the driver's side air bag module Torx® bolts from each side of the module.

   n. Slide the module forward and disconnect the air bag connector and remove the module.

   o. Remove the steering wheel.

   p. Matchmark the universal joint and remove the bolts and the joint.

   q. Remove the trim panel from under the instrument panel.

   r. Remove the knee guard panel and steering column lower covers.

   s. Remove the 2 steering column-to-instrument panel bolts and lower the steering column.

   t. Loosen the 4 instrument panel installation bolts but do not remove the lower bolts for alignment purposes.

   u. Remove the instrument cluster cover screws and the cover.

   v. Remove the cluster screws, slide the cluster forward to detach the electrical connector and remove the cluster.

   w. Loosen the instrument panel screws.

   x. Using a flat bladed tool, pry the instrument panel center compartment up to disengage the clips and remove the compartment.

   y. Remove the center panel screws and clips at the radio, remove the panel.

   z. Disconnect the center panel connector.

   aa. Loosen the radio screws, slide the radio forward and detach the antenna and electrical connectors and remove the radio.

   bb. Remove the side covers from the instrument panel and loosen the two bolts.

   cc. Remove the instrument panel.

6. Remove the heater and cooling unit as follows:

   a. Remove the steering support beam bracket nuts and the steering support beam.

   b. Remove the blower motor.

   c. Disconnect the servo motor connectors.

   d. Remove the heater and cooling unit retainers and the unit.

7. Remove the heater core as follows:

   a. Open the heater core pipe cover.

   b. Loosen the screws and remove the mode actuator.

   c. Loosen the foot duct screws and remove the duct.

   d. Remove the evaporator cover screws and the cover.

   e. Remove the lower case cover screws and the cover

   f. Remove the heater core.

To install:

8. Install the heater core as follows:

   a. Install the heater core.

   b. Install the lower case cover and screws

   c. Install the evaporator cover and screws.

   d. Install the foot duct and screws.

   e. Install the mode actuator.

   f. Close the heater core pipe cover.

9. Install the heater and cooling unit as follows:

   a. Install the heater and cooling unit and retainers.

   b. Connect the servo motor connectors.

   c. Install the blower motor.

   d. Install the steering support beam.

10. Install the instrument panel by performing the following procedure:

   a. Install the instrument panel.

   b. Tighten the 2 side bolts and install the side covers.

   c. Install the radio.

   d. Connect the center panel connector.

   e. Install the center panel.

   f. Install the instrument panel center compartment.

   g. Tighten the instrument panel screws.

   h. Install the cluster, screws and the cover.

   i. Tighten the 4 instrument panel installation bolts.

   j. Raise the steering column and tighten the bolts to 18 ft. lbs. 925 Nm).

   k. Install the knee guard panel and steering column lower covers.

   l. Install the trim panel under the instrument panel.

   m. Install the universal joint and tighten the bolts to 17 ft. lbs. (24 Nm).

   n. Install the steering wheel.

   o. Connect the driver's side air bag connector and install the module.

   p. Install the driver's side air bag module Torx® bolts on each side of the module.

   q. Install the air bag module and tighten the screws.

   r. Connnect the passenger side air bag module connector to the module and the support beam.

   s. Install the glove box.

   t. Install the side console panel.

   u. Connect the A/C and hazard switch connectors.

   v. Install the left side cover assembly.

   w. Install the console box.

   x. Install both the front and rear console covers.

   y. If equipped with a manual transmission, install the shift knob.

   z. Install the console front cover.

   aa. Connect the in–vehicle sensor hose and connector.

   bb. Install the lower cover below the steering wheel.

11. Connect the heater hoses to the heater core.

12. Refill the cooling system.

13. Connect the negative battery cable.

14. Evacuate and charge the air conditioning system refrigerant.

15. Run the engine to normal operating temperatures; then, check the climate control operation and check for leaks.

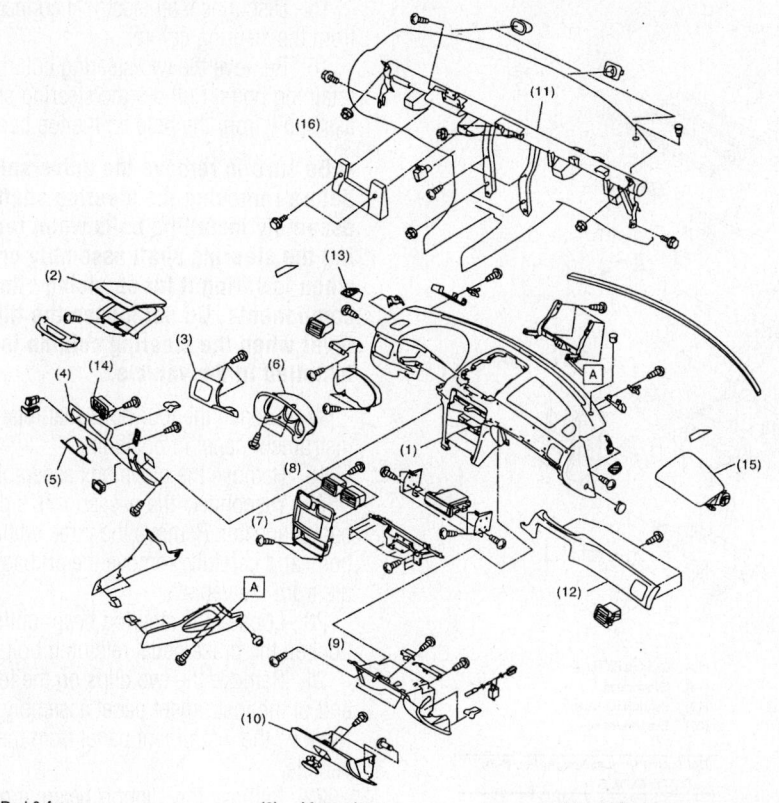

(1) Pad & frame
(2) Center compartment (Model without navigation system)
(3) Grille cover
(4) Lower cover
(5) Coin box
(6) Meter visor
(7) Center panel
(8) Air vent grille
(9) Glove box panel
(10) Glove box lid
(11) Steering beam
(12) Grille cover
(13) Air vent grille (Defroster)
(14) Switch panel
(15) Passenger's airbag module
(16) Knee guard panel

09490_FORE_G0004

Instrument panel and related components—2005–2006

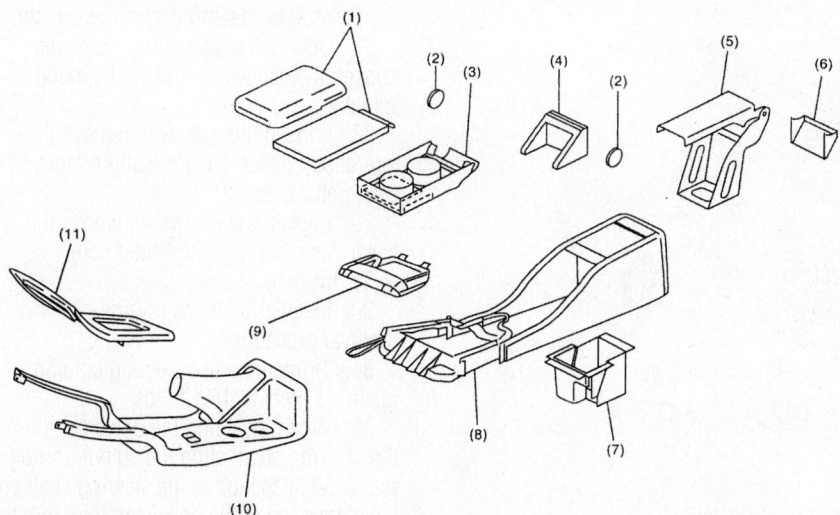

(1) Arm cover
(2) Hinge cover
(3) Cup holder
(4) Hinge front cover
(5) Armrest frame
(6) Hinge rear cover
(7) Console pocket
(8) Console box
(9) Console lid and tray
(10) Console cover
(11) Front cover

09490_FORE_G0005

Console and related components—2005–2006

2005–2006

1. Before servicing the vehicle, refer to the Precautions Section.

2. Position the front tires and wheels in the straight ahead position.

3. Disarm the SRS system. Wait for at least 20 seconds for the air bag module to deplete its energy before starting any work.

➡ **The SRS system wiring harnesses and connectors are yellow in color. Do not use electrical test equipment on these circuits. Be careful not to damage the SRS system harness when servicing.**

4. Disconnect the negative battery cable.

5. Drain the engine coolant. Properly discharge the air conditioning system.

6. Remove the bolts retaining the expansion valve. Disconnect the heater hoses from the heater core.

7. Loosen the screws and clips and remove the lower instrument panel assembly.

8. Remove the console front cover and console cover.

9. Loosen the screws and clip and remove the side console panel.

10. Remove the two torx head bolts on the side of the steering wheel. Disconnect the horn harness.

11. Disconnect the air bag connector on the back of the air bag module. Remove the air bag module from the steering wheel.

12. Matchmark the steering wheel and the steering column shaft. Remove the steering wheel retaining nut. Using, a steering gear puller tool, remove the steering wheel.

13. At the base of the steering column matchmark the universal joint. Remove the joint bolt. Remove the joint.

14. Remove the trim panel, under the instrument panel, knee guard panel and lower steering column cover.

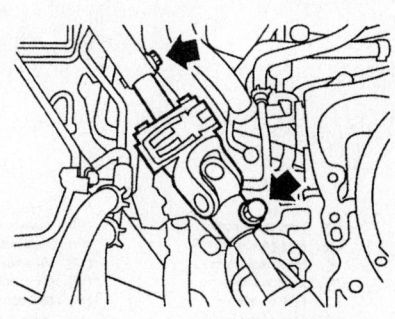

09490_TRIB_G0008

Steering column universal joint and retaining bolt location

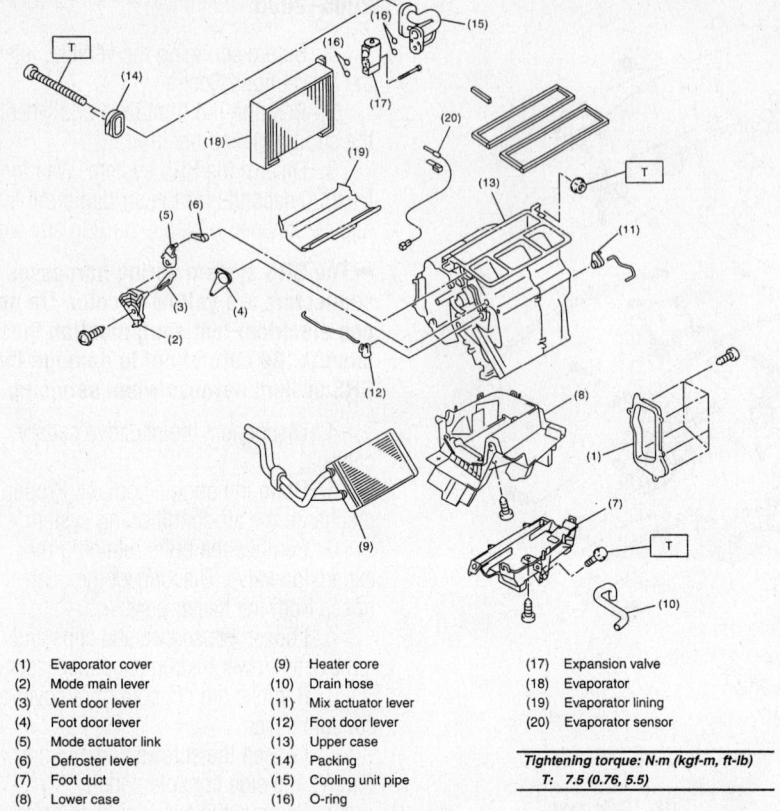

(1)	Evaporator cover	(9)	Heater core	(17)	Expansion valve
(2)	Mode main lever	(10)	Drain hose	(18)	Evaporator
(3)	Vent door lever	(11)	Mix actuator lever	(19)	Evaporator lining
(4)	Foot door lever	(12)	Foot door lever	(20)	Evaporator sensor
(5)	Mode actuator link	(13)	Upper case		
(6)	Defroster lever	(14)	Packing		
(7)	Foot duct	(15)	Cooling unit pipe		
(8)	Lower case	(16)	O-ring		

Tightening torque: N·m (kgf-m, ft-lb)
T: 7.5 (0.76, 5.5)

09490_FORE_G0006

Manual air conditioning heater core and related components—2005–2006

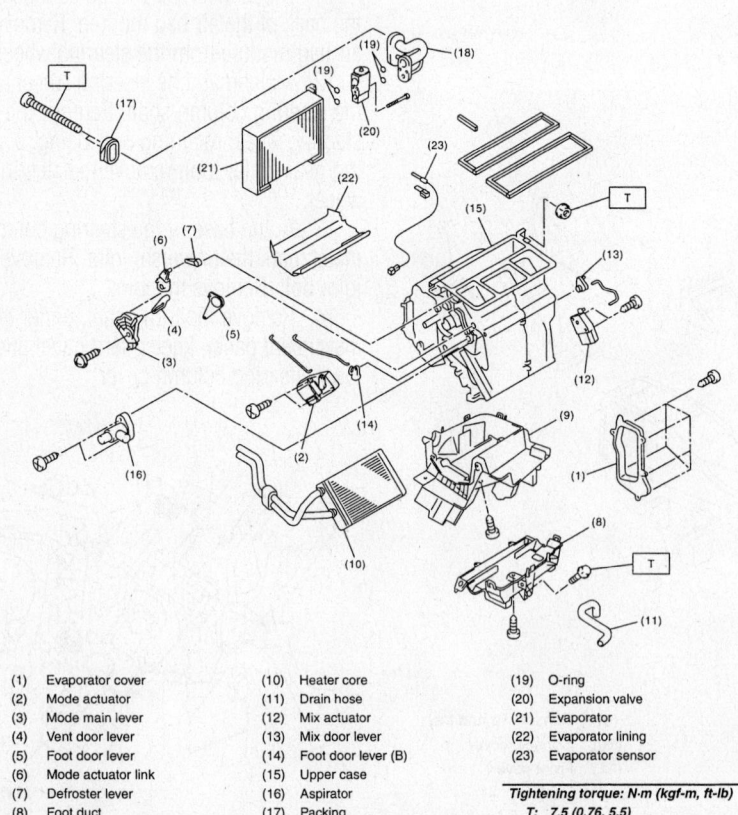

(1)	Evaporator cover	(10)	Heater core	(19)	O-ring
(2)	Mode actuator	(11)	Drain hose	(20)	Expansion valve
(3)	Mode main lever	(12)	Mix actuator	(21)	Evaporator
(4)	Vent door lever	(13)	Mix door lever	(22)	Evaporator lining
(5)	Foot door lever	(14)	Foot door lever (B)	(23)	Evaporator sensor
(6)	Mode actuator link	(15)	Upper case		
(7)	Defroster lever	(16)	Aspirator		
(8)	Foot duct	(17)	Packing		
(9)	Lower case	(18)	Cooling unit pipe		

Tightening torque: N·m (kgf-m, ft-lb)
T: 7.5 (0.76, 5.5)

09490_FORE_G0007

Automatic air conditioning heater core and related components—2005–2006

15. Disconnect all electrical connectors from the steering column.

16. Remove the two steering column retaining bolts. Pull out the steering shaft assembly from the hole on the toe board.

➡**Be sure to remove the universal joint before removing the steering shaft assembly installing bolts when removing the steering shaft assembly or when lowering it for servicing other components. Do not loosen the tilt lever when the steering column is not installed in the vehicle.**

17. Loosen the four bolts that retain the instrument panel in position.

18. Remove the glove box assembly.

19. Disconnect the passenger's side air bag connector. Remove the three retaining bolts and carefully remove the air bag module from the vehicle.

20. Loosen the steering beam bolts. Loosen the brake pedal retaining bolts.

21. Remove the two clips on the leading end of the instrument panel assembly. Remove the instrument panel from the vehicle.

22. Remove the support beam, if not removed with the instrument panel.

23. Disconnect the connectors from the blower motor unit assembly and the intake actuator register.

24. Loosen the bolt and nut and remove the blower motor unit from the vehicle.

25. Disconnect the servo motor connector.

26. Loosen the nuts and remove the heater/cooling assembly from the vehicle.

27. Open the heater core pipe cover. Loosen the screws to remove the mode actuator.

28. Loosen the screws to remove the foot duct. Loosen the screws to remove the evaporator cover.

29. Loosen the screws to remove the lower case. Remove the heater core.

To install:

30. Installation is the reverse of the removal procedure.

31. Tighten the two steering column retaining bolts to 18.1 ft. lbs.

32. When installing the universal joint to the steering shaft, align the cutout portion at the serrated section of the steering shaft and yoke, then insert the universal; joint into the steering shaft.

33. After installing the roll connector, be sure to check for proper adjustment:

- Check that the front wheels are in the straight ahead position
- Turn the roll connector pin (A) clockwise until it stops

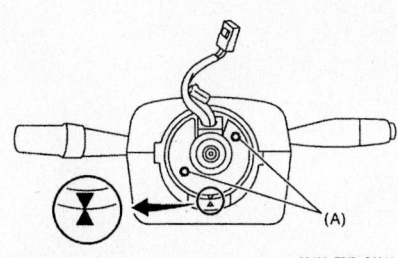

Roll pin alignment

09490_TRIB_G0011

- Turn the roll connector pins (A) approximately 3.25 turns until the triangle marks are aligned
34. Tighten the steering wheel retaining nut to 33.2 ft. lbs. (45 Nm).
35. Refill the cooling system with the proper grade and type coolant.
36. Evacuate, charge and leak test the air conditioning system.
37. Start the engine and check for leaks, correct as required.

Cylinder Head

REMOVAL & INSTALLATION

2002–2004

1. Before servicing the vehicle, refer to the Precautions Section.
2. Remove or disconnect the following:
- Negative battery cable
- Accessory drive belts

N·m (kg-m, ft-lb)

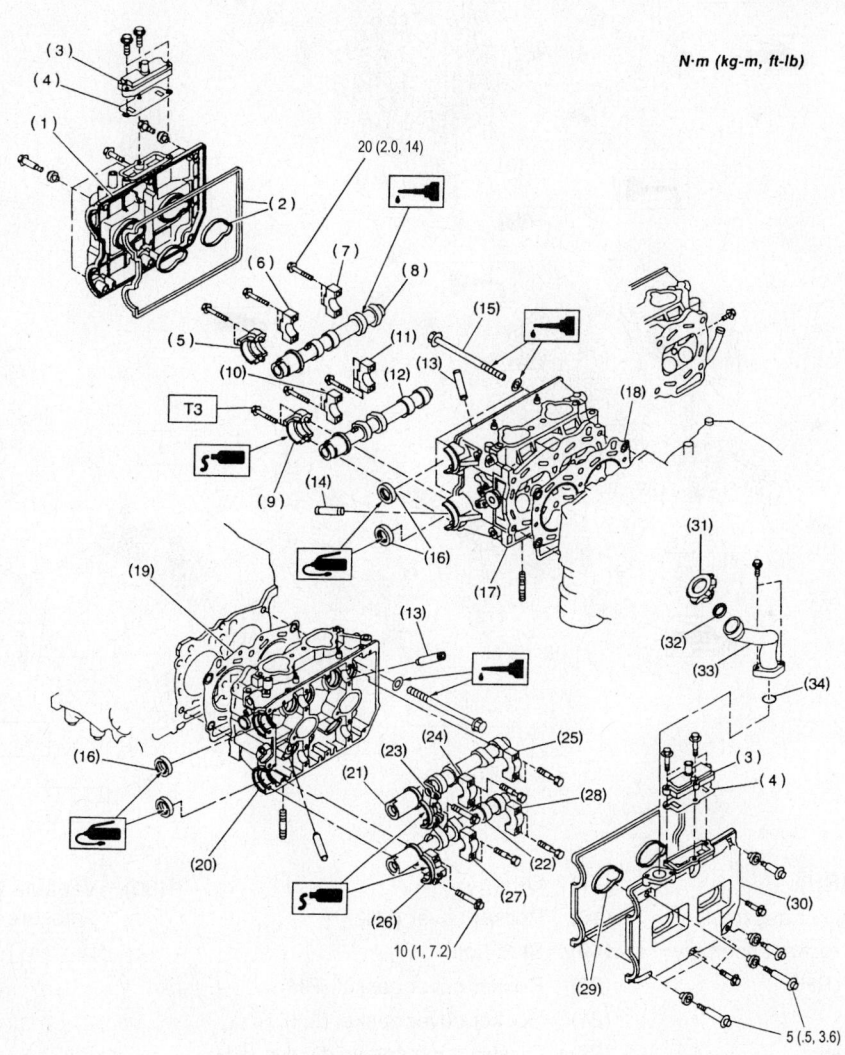

(1) Rocker cover (RH)
(2) Rocker cover gasket (RH)
(3) Oil separator cover
(4) Gasket
(5) Intake camshaft cap (Front RH)
(6) Intake camshaft cap (Center RH)
(7) Intake camshaft cap (Rear RH)
(8) Intake camshaft (RH)
(9) Exhaust camshaft cap (Front RH)
(10) Exhaust camshaft cap (Center RH)
(11) Exhaust camshaft cap (Rear RH)
(12) Exhaust camshaft (RH)
(13) Intake valve guide
(14) Exhaust valve guide

(15) Cylinder head bolt
(16) Oil seal
(17) Cylinder head (RH)
(18) Cylinder head gasket (RH)
(19) Cylinder head gasket (LH)
(20) Cylinder head (LH)
(21) Intake camshaft (LH)
(22) Exhaust camshaft (LH)
(23) Intake camshaft cap (Front LH)
(24) Intake camshaft cap (Center LH)
(25) Intake camshaft cap (Rear LH)
(26) Exhaust camshaft (Front LH)
(27) Exhaust camshaft cap (Center LH)
(28) Exhaust camshaft cap (Rear LH)

(29) Rocker cover gasket (LH)
(30) Rocker cover (LH)
(31) Oil filler cap
(32) Gasket
(33) Oil filler duct
(34) O-ring

Cylinder head and related components—DOHC engine

7923TG07

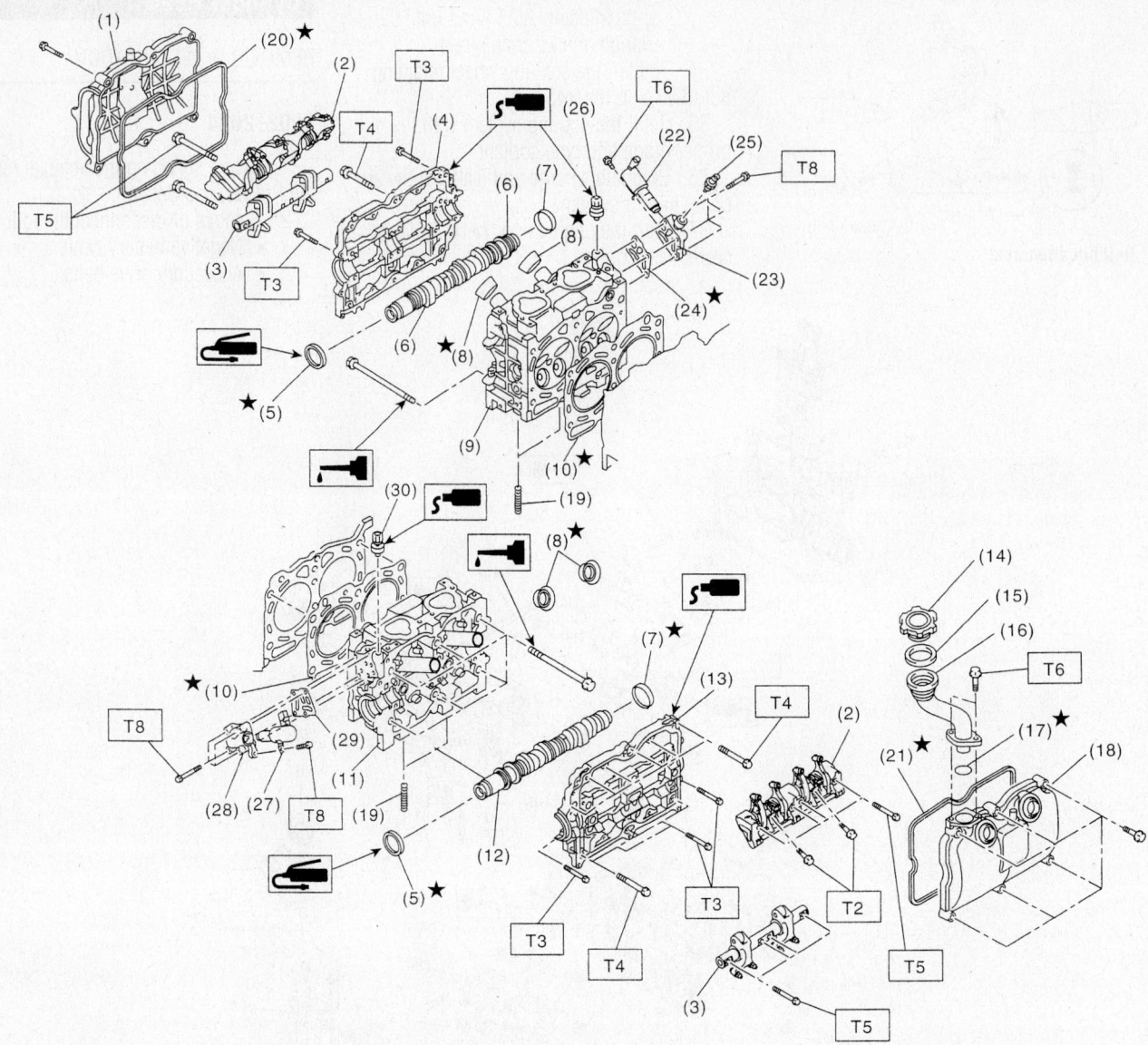

(1)	Rocker cover (RH)
(2)	Intake valve rocker assembly
(3)	Exhaust valve rocker assembly
(4)	Camshaft cap (RH)
(5)	Oil seal
(6)	Camshaft (RH)
(7)	Plug
(8)	Spark plug pipe gasket
(9)	Cylinder head (RH)
(10)	Cylinder head gasket
(11)	Cylinder head (LH)
(12)	Camshaft (LH)
(13)	Camshaft cap (LH)
(14)	Oil filler cap
(15)	Gasket
(16)	Oil filler duct
(17)	O-ring
(18)	Rocker cover (LH)
(19)	Stud bolt
(20)	Rocker cover gasket (RH)
(21)	Rocker cover gasket (LH)
(22)	Oil switching solenoid valve (RH)
(23)	Oil switching solenoid valve holder (RH)
(24)	Gasket
(25)	Oil temperature sensor
(26)	Variable valve lift diagnosis oil pressure switch (RH)
(27)	Oil switching solenoid valve (LH)
(28)	Oil switching solenoid valve holder (LH)
(29)	Gasket
(30)	Variable valve lift diagnosis oil pressure switch (LH)

Tightening torque: N·m (kgf-m, ft-lb)

T3:	9.75 (1.0, 7.2)
T4:	18 (1.8, 13.0)
T5:	25 (2.5, 18.1)
T6:	6.4 (0.65, 4.7)
T7:	8 (0.8, 5.9)
T8:	10 (1.0, 7.4)

Cylinder head and related components—SOHC engine

- Power steering pump
- Alternator and bracket
- Valve rocker cover
- Connector bracket attaching bolt
- Crankshaft Position (CKP) and Camshaft Position (CMP) sensors
- Coolant filler tank

3. Relieve the fuel system pressure and disconnect the fuel pipes.

- Intake manifold and gasket
- Water pipe
- Timing belt, camshaft sprocket and related components
- Oil level gauge guide attaching bolt on the left cylinder head
- Valve covers
- Camshafts
- Cylinder head bolts, in the proper sequence. On the DOHC engine leave bolts A and D installed loosely to prevent the cylinder head from falling. On the SOHC engine leave bolts A and C installed loosely to prevent the cylinder head from falling.
- Cylinder head from the block using a plastic-faced hammer, if needed
- Remove bolts A and D on the DOHC engine and bolts A and C on the SOHC engine.
- Cylinder head and gasket

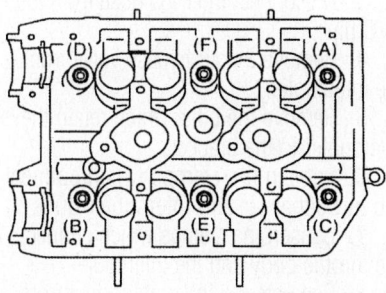

09490_SBCR_G0017

Cylinder head bolt loosening sequence—DOHC engine

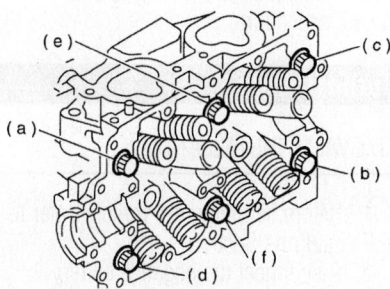

09490_SBCR_G0018

Cylinder head bolt loosening sequence—SOHC engine

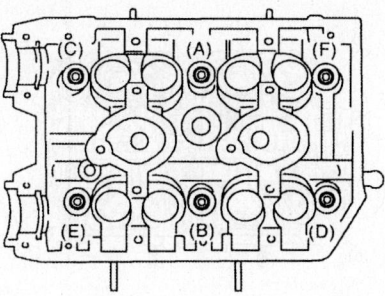

09490_SBCR_G0019

Cylinder head bolt tightening sequence—DOHC engine

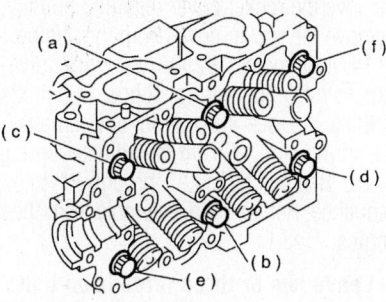

09490_SBCR_G0020

Cylinder head bolt tightening sequence—SOHC engine

4. Clean all gasket material from both mating surfaces.

To install:

5. Inspect the cylinder head for warpage. Warpage should not exceed 0.0020 in. (0.05mm).

6. Install a new head gasket and the cylinder head.

7. Secure the head in place with the mounting bolts. Coat each bolts with clean engine oil, and hand-tighten. Tighten the cylinder head bolts to specification.

8. Install or connect the following:

- Camshafts
- Valve covers
- Oil level gauge guide attaching bolt on the left cylinder head
- Timing belt, camshaft sprockets and related components
- Water pipe
- Intake manifold and tighten the bolts to specification
- Fuel delivery pipes
- Blow-by hose
- Knock sensor
- CKP and CMP sensors
- Connector bracket attaching bolt
- Spark plug wires
- Valve rocker cover and tighten the bolts to 48 inch lbs. (9 Nm)
- Alternator

- Power steering pump
- Accessory drive belt
- Negative battery cable

9. Start the engine and allow it to reach operating temperature. Check for leaks.

2005–2006

1. Before servicing the vehicle, refer to the Precautions Section.

2. Disconnect the negative battery cable.

3. If equipped with DOHC engine, remove the collector cover.

4. To remove the front side V belt, remove the belt covers. Loosen the lock bolt. Loosen the slider bolt. Remove the front side belt.

5. To remove the rear side V belt, remove the belt covers. Loosen the lock bolt. Loosen the slider bolt. Remove the rear side belt. Remove the belt tensioner.

6. Remove the crankshaft pulley bolt.

7. Lock the crankshaft in place using tool ST499977100, or equivalent.

8. Remove the crankshaft pulley.

9. Remove the left side timing belt cover.

10. If equipped with DOHC engine, remove the right side timing belt cover.

11. Remove the front timing belt cover.

12. Remove the timing belt.

13. Remove the camshaft position sensor.

14. Remove the camshaft sprockets.

➡**Be sure to lock the camshaft in place using tool ST18231AA010 (SOHC engine) and ST499207400 (DOHC engine), or equivalent.**

15. Remove the intake manifold.

16. If equipped with SOHC engine, remove the bolt that retains the air conditioning compressor bracket to the cylinder head.

17. Remove the rocker cover retaining bolts. Remove the rocker cover.

18. If equipped with SOHC engine, remove the rocker arm assembly.

19. If equipped with SOHC engine remove the camshaft. If equipped with DOHC engine, remove the camshafts.

20. Remove the cylinder head, in the proper sequence. On the DOHC engine leave bolts A and D installed loosely to prevent the cylinder head from falling. On the SOHC engine leave bolts A and C installed loosely to prevent the cylinder head from falling.

21. Loosen the cylinder head from the block using a plastic-faced hammer, if needed.

22. Remove bolts A and D on the DOHC engine and bolts A and C on the SOHC

engine. Remove the cylinder head from the engine. Discard the gasket

23. Clean all gasket material from both mating surfaces.

To install:

24. Installation is the reverse of the removal procedure.

25. Apply a thin coat of clean engine oil to the washers and cylinder head bolts.

26. Tighten the cylinder head retaining bolts to specification and in the proper sequence.

27. Start the engine and allow it to reach operating temperature.

28. Check for leaks, correct as required.

Rocker Arms/Shafts

REMOVAL & INSTALLATION

SOHC Engine

1. Before servicing the vehicle, refer to the Precautions Section.

2. Disconnect the negative battery cable.

3. To remove the front side V belt, remove the belt covers. Loosen the lock bolt. Loosen the slider bolt. Remove the front side belt.

4. To remove the rear side V belt, remove the belt covers. Loosen the lock bolt. Loosen the slider bolt. Remove the rear side belt. Remove the belt tensioner.

5. Remove the crankshaft pulley bolt.

6. Lock the crankshaft in place using tool ST499977100, or equivalent.

7. Remove the crankshaft pulley.

8. Remove the left side timing belt cover.

9. Remove the front timing belt cover.

10. Remove the timing belt.

11. Remove the camshaft position sensor.

12. Remove the camshaft sprockets, No.1 and No.2.

➡**Be sure to lock the camshaft in place using tool ST18231AA010, or equivalent.**

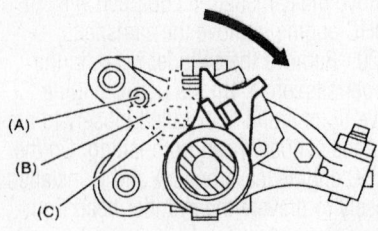

(A) Adjuster pin
(B) Spring stopper
(C) Spring

09490_SBCR_G0030

Rocker arm spring stopper removal—
SOHC engine

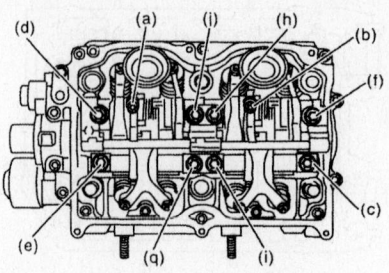

09490_SBCR_G0031

Rocker arm assembly bolt loosening
sequence—SOHC engine

13. Disconnect the PCV valve hose. Remove the rocker cover retaining bolts. Remove the rocker cover from the engine.

14. Use tool St18258AA000, or equivalent, to rotate the spring stopper in direction of the arrow, see illustration, and remove the spring stopper from the adjuster spring.

15. Remove bolts "A" through "J" in sequence. Remove the assembly from the engine.

➡**Leave two or three threads on bolts "I" and "J" engaged in order to retain the rocker arm assembly.**

16. Position tool ST18354AA000 on the rocker arm assembly. Remove the assembly from the engine.

To install:

17. Install the rocker arm assembly on its mounting.

18. Temporarily tighten the bolts equally in the proper sequence.

➡**Do not temporarily tighten bolts "I" and "J". Position tool ST18354AA000 to the rocker arm assembly.**

19. Tighten bolts "A" through "H" to specification (18.1 ft. lbs.).

20. Tighten bolts "I" through "J" to specification (5.9 ft. lbs.).

21. Use tool ST18258AA000 to rotate the spring stopper in the opposite direction of the arrow, see illustration, to fasten the adjuster pin.

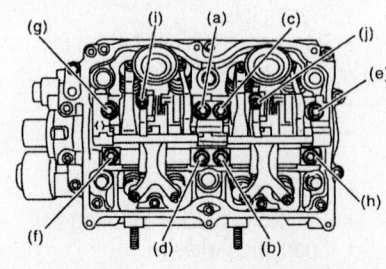

09490_SBCR_G0032

Rocker arm assembly bolt tightening
sequence—SOHC engine

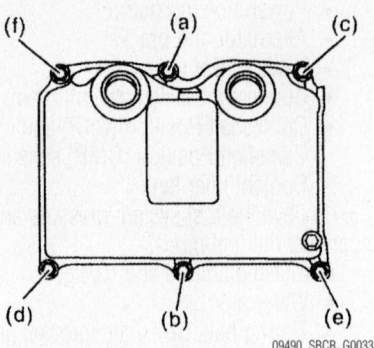

09490_SBCR_G0033

Rocker arm cover bolt tightening
sequence—SOHC engine

22. Adjust the valve clearance.

23. Use a new gasket and install the rocker arm cover. Torque the retaining bolts to 4.7 ft. lbs and in the proper sequence. Recheck and retorque to 4.7 ft. lbs in the proper sequence.

24. Continue the installation in the reverse order of the removal procedure.

Intercooler

REMOVAL & INSTALLATION

1. Before servicing the vehicle, refer to the Precautions Section.

2. Disconnect the negative battery cable.

3. Disconnect the PVC hose from the PVC pipe.

4. Remove the air bypass valve from the intercooler.

5. Remove the bolts which retain the intercooler to the bracket.

6. Loosen the clamps which connect the turbocharger and intercooler duct.

7. Loosen the clamps which connect the throttle body and the intercooler.

8. Separate the intercooler duct from the turbocharger.

9. Remove the intercooler sensor from the throttle body.

To install:

10. Installation is the reverse of the removal procedure.

Turbocharger

REMOVAL & INSTALLATION

1. Before servicing the vehicle, refer to the Precautions Section.

2. Disconnect the negative battery cable.

3. Raise and support the vehicle safely.

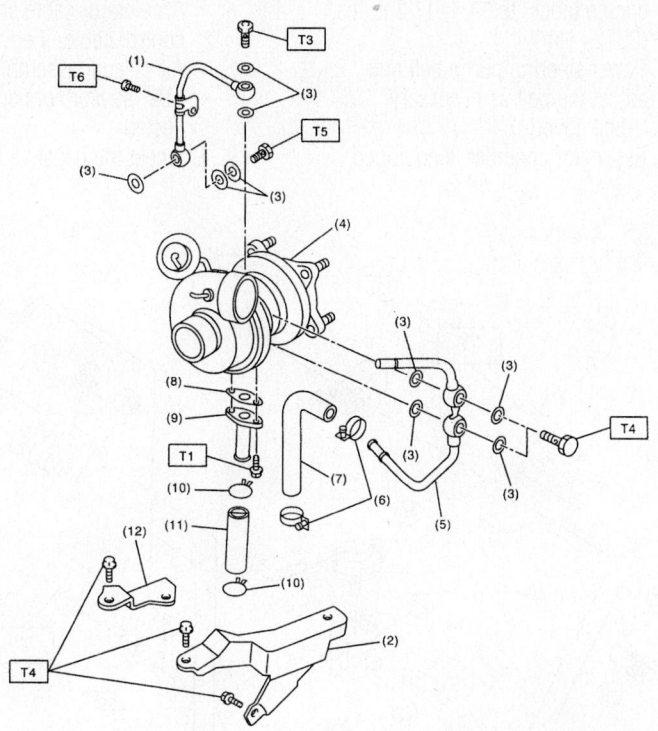

(1)	Oil inlet pipe A	(8)	Gasket
(2)	Turbocharger bracket LH	(9)	Oil outlet pipe
(3)	Metal gasket	(10)	Clip
(4)	Turbocharger	(11)	Oil outlet hose
(5)	Water pipe	(12)	Turbocharger bracket RH
(6)	Clamp		
(7)	Engine coolant hose		

Tightening torque: N·m (kgf-m, ft-lb)	
T1:	4.4 (0.45, 3.3)
T2:	20 (2.0, 14.8)
T3:	16 (1.6, 11.6)
T4:	33 (3.4, 24.6)
T5:	29 (3.0, 21.7)
T6:	4.9 (0.50, 3.6)

09490_FORE_G0008

Turbocharger and related components

4. Remove the collector cover, if equipped.
5. Remove the center exhaust pipe.
6. Lower the vehicle.
7. Separate the turbocharger joint pipe from the turbocharger.
8. Disconnect the engine coolant hose from the coolant filler tank.
9. Loosen the clamp that attaches the turbocharger to the intake duct.
10. Remove the oil pipe–to–turbocharger bolt and remove the oil pipe from the turbocharger.
11. Remove the turbocharger bracket and disconnect the oil outlet hose from the pipe.
12. Remove the turbocharger.
To install:
13. Installation is the reverse of removal, please note the following torque specs:
    a. Turbocharger–to–intake duct: 2 ft. lbs. (3 Nm).
    b. Oil pipe–to–turbocharger: 11 ft. lbs. (16 Nm).
    c. Joint pipe–to–turbocharger using a new gasket: 25 ft. lbs. (35 Nm).
    d. Turbocharger bracket: 24 ft. lbs. (33 Nm).

Intake Manifold

REMOVAL & INSTALLATION

2002–2004

1. Before servicing the vehicle, refer to the Precautions Section.
2. Disconnect the negative battery cable.
3. Drain the cooling system into a suitable container.
4. Remove or disconnect the following:
   • Mass Air Flow (MAF) sensor connector, if necessary
   • Air intake duct, air cleaner upper cover and the air cleaner element
5. Properly release the fuel pressure.
   • Accelerator cable and the cruise control cable, if equipped
   • Resonator chamber, if equipped
   • V-belt cover(s)
6. Loosen the lock bolt and slider bolt, then remove the power steering belt
7. Remove or disconnect the following:
   • Power steering pipe bracket-to-manifold bolts

• Bolts holding the power steering pump to the bracket
• Connector from the power steering pump switch, if equipped
• Power steering pump, and place on the right side wheel apron. Do NOT disconnect the fluid lines.
• Spark plug wires from the spark plugs
• Positive Crankcase Ventilation (PCV) hose and vacuum hose from the intake manifold
• Engine coolant hoses from the throttle body
• Brake booster hose
• Air cleaner case stay (right side) and engine harness bracket, if necessary
• Engine harness connectors form the bulkhead harness connectors
• Engine Coolant Temperature (ECT) sensor connector
• Knock sensor, Camshaft Position (CMP) sensor and Crankshaft Position (CKP) sensor electrical connectors
• Oil pressure switch
• Fuel hoses from the fuel pipes
• Intake manifold mounting bolts
• Intake manifold and discard the gaskets

➡ **The intake manifold sits on pins that protrude from the cylinder heads. Be sure the pins remain in the cylinder heads.**

To install:
8. Install or connect the following:
   • New gaskets
   • Intake manifold to the engine. Tighten the mounting bolts to 16.7–19.5 ft. lbs. (23–27 Nm).
   • Fuel hoses to the fuel pipes, be sure to secure the hoses with new clamps
   • Knock sensor, CMP sensor, CKP sensor, oil pressure switch and ECT sensor wiring
   • Engine harness bracket and engine harness connectors to bulkhead connectors
   • EGR pipe, if removed
   • Brake booster hose
   • Engine coolant hose and air bypass hose to idle air control solenoid valve, if removed
   • Engine coolant hoses to the throttle body
   • PCV hoses to intake manifold
   • Spark plug wires to ignition coil or plugs, as applicable
   • Power steering pump to bracket.

Tighten the bolts to 13–16.6 ft. lbs. (17.6–22.6 Nm).
- Power steering pipe brackets on right side of the intake manifold
- Bolt, which installs the power steering pump stiffener on the

engine block, to 14.4–17.3 ft. lbs. (20–24 Nm)
- Power steering pump belt and adjust the belt as necessary
- V-belt cover(s)
- Resonator chamber, if equipped

- Accelerator cable and the cruise control cable, if equipped
- Air cleaner assembly
- MAF sensor connector, if disconnected
9. Connect the negative battery cable

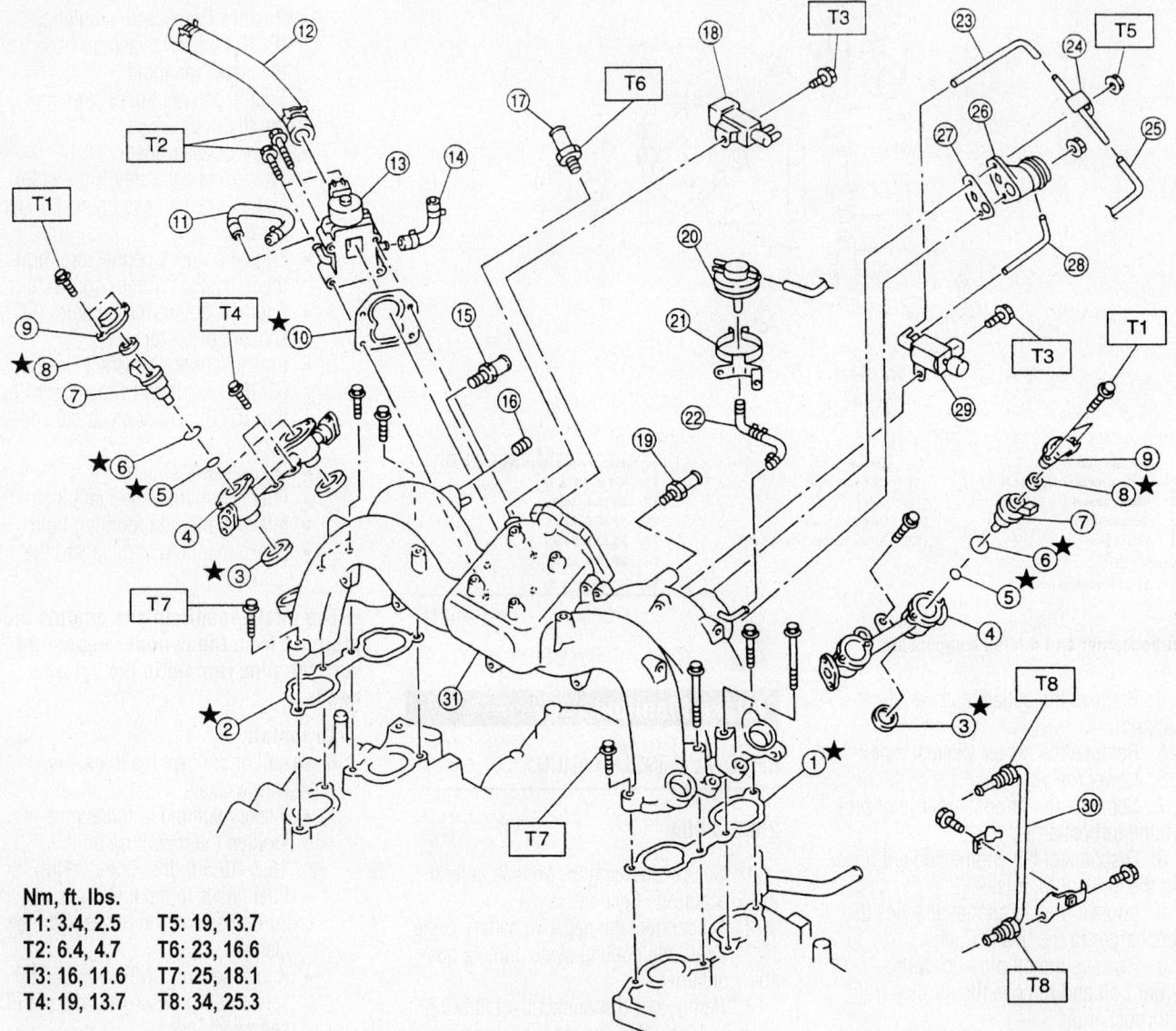

Nm, ft. lbs.

T1: 3.4, 2.5	T5: 19, 13.7
T2: 6.4, 4.7	T6: 23, 16.6
T3: 16, 11.6	T7: 25, 18.1
T4: 19, 13.7	T8: 34, 25.3

① Intake manifold gasket RH	⑫ Air by-pass hose
② Intake manifold gasket LH	⑬ Idle air control solenoid valve
③ Fuel injector pipe insulator	⑭ Engine coolant hose A
④ Fuel injector pipe	⑮ Nipple (AT model)
⑤ O-ring A	⑯ Plug
⑥ O-ring B	⑰ PCV valve
⑦ Fuel injector	⑱ Purge control solenoid valve
⑧ Insulator	⑲ Nipple
⑨ Fuel injector cap	⑳ BPT
⑩ Gasket	㉑ BPT holder bracket
⑪ Engine coolant hose B	

㉒ Back pressure hose
㉓ EGR vacuum hose A
㉔ EGR vacuum pipe
㉕ EGR vacuum hose C
㉖ EGR valve
㉗ Gasket
㉘ EGR vacuum hose B
㉙ EGR solenoid valve
㉚ EGR pipe
㉛ Intake manifold

7923TG17

Intake manifold and related components—2002–2004 SOHC engine

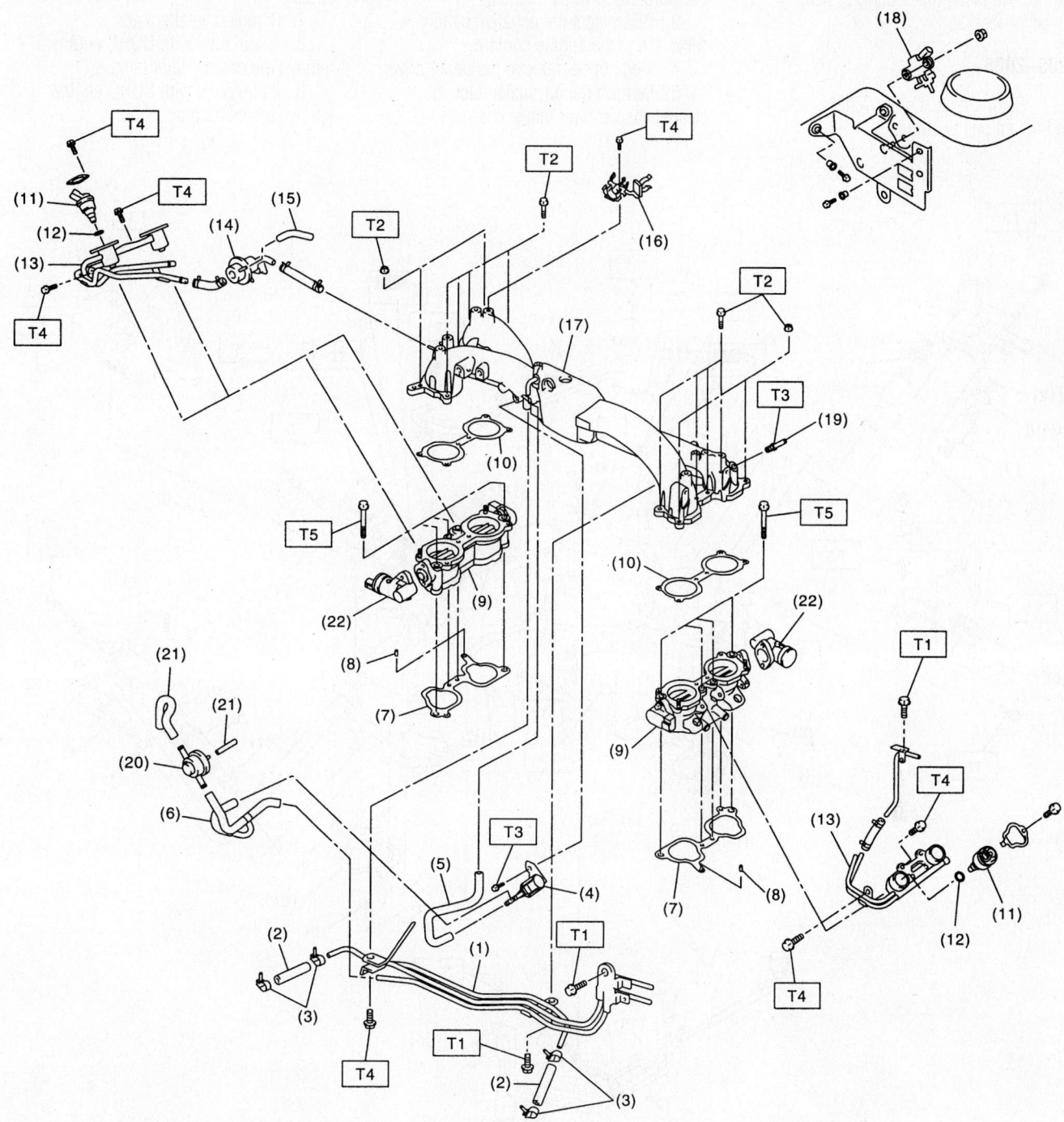

(1) Fuel pipe ASSY
(2) Fuel hose
(3) Clip
(4) Purge control solenoid valve
(5) Vacuum hose
(6) Vacuum control hose
(7) Intake manifold gasket
(8) Guide pin

(9) Tumble generator valve ASSY
(10) Tumble generator valve gasket
(11) Fuel injector
(12) O-ring
(13) Fuel injector pipe
(14) Pressure regulator
(15) Pressure regulator hose
(16) Blow-by hose stay

(17) Intake manifold
(18) Wastegate control solenoid valve ASSY
(19) Nipple
(20) Purge valve
(21) Purge hose
(22) Tumble generator valve actuator

Intake manifold and related components—2002–2004 DOHC engine

09490_SBCR_G0026

and refill the cooling system. Start the engine, and bleed the cooling system. Check for leaks.

2005–2006

1. Before servicing the vehicle, refer to the Precautions Section.

2. Properly relieve the fuel system pressure. Remove the fuel cap.

3. Disconnect the negative battery cable. Drain the engine coolant.

4. If equipped, remove the under cover.

5. Remove the air intake duct, air cleaner case and air intake chamber.

6. If equipped, remove the inter-cooler.

7. Remove the alternator.

8. If equipped with DOHC engine, remove the coolant filler tank.

9. If equipped with SOHC engine, remove the spark plug wires.

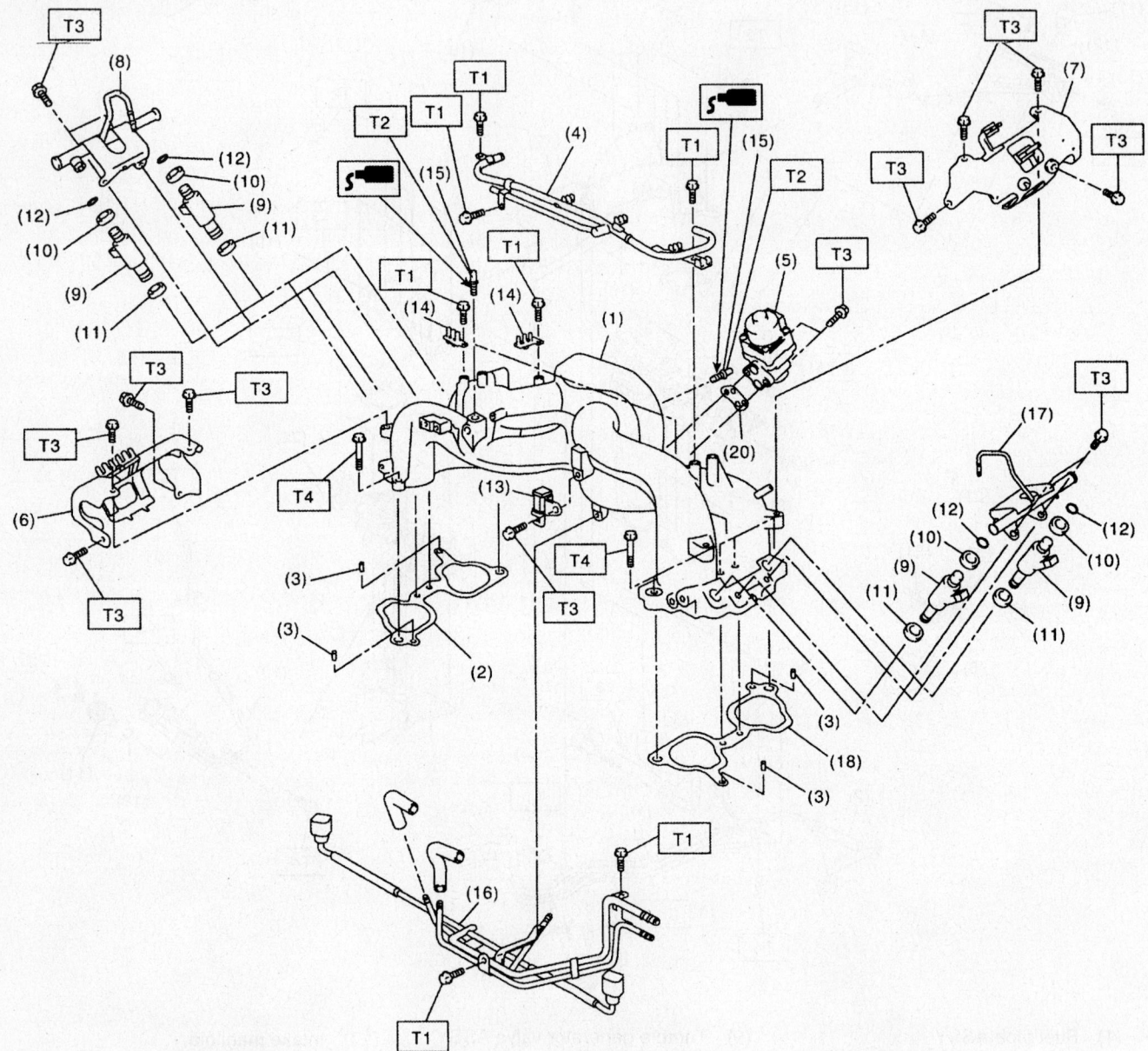

(1)	Intake manifold	(7)	Fuel pipe protector LH	(13)	Purge control solenoid valve
(2)	Gasket (RH)	(8)	Fuel injector pipe RH	(14)	Plug cord holder
(3)	Guide pin	(9)	Fuel injector	(15)	Nipple
(4)	PCV pipe	(10)	O-ring	(16)	Fuel pipe
(5)	EGR valve	(11)	O-ring	(17)	Fuel injector pipe LH
(6)	Fuel pipe protector RH	(12)	O-ring	(18)	Gasket (LH)

Intake manifold and related components—2005–2006 SOHC engine

10. Disconnect the engine coolant hoses from the throttle body. Disconnect the brake booster hose.

11. Disconnect the PCV hose from the intake manifold. Disconnect the engine harness electrical connectors from the bulkhead harness connectors.

12. Disconnect the engine coolant temperature sensor electrical connector, knock sensor electrical connector and crankshaft position sensor connector.

13. Disconnect the power steering pump switch electrical connector, oil pressure switch connector and camshaft sensor connector.

14. If equipped with DOHC engine, disconnect the oil flow solenoid valve electrical connector and the ignition coil connector.

15. If equipped with SOHC engine, remove the EGR pipe from the intake manifold and disconnect the fuel lines from the fuel pipe.

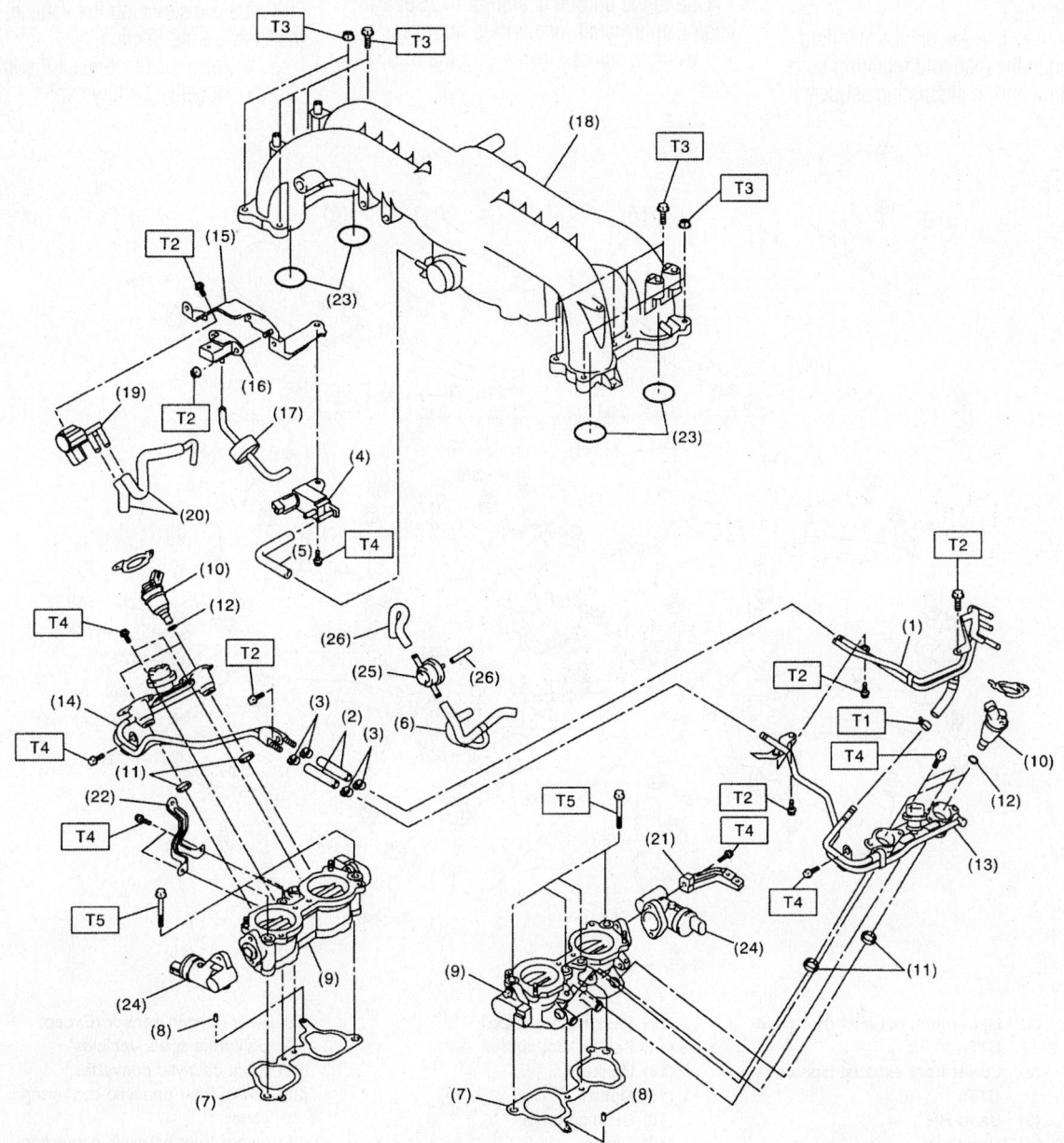

(1)	Fuel pipe ASSY	(10)	Fuel injector	(19)	Wastegate control solenoid valve ASSY
(2)	Fuel hose	(11)	Seal ring		
(3)	Clip	(12)	O-ring	(20)	Vacuum hose
(4)	Purge control solenoid valve	(13)	Fuel injector pipe LH	(21)	Ground stay
(5)	Vacuum hose	(14)	Fuel injector pipe RH	(22)	Coolant filler tank stay
(6)	Vacuum control hose	(15)	Solenoid valve bracket	(23)	O-ring
(7)	Intake manifold gasket	(16)	Manifold absolute pressure sensor	(24)	Tumble generator valve actuator
(8)	Guide pin	(17)	Filter	(25)	Purge valve
(9)	Intake manifold (lower)	(18)	Intake manifold	(26)	Purge hose

Intake manifold and related components—2005–2006 DOHC engine

16. If equipped with DOHC engine, disconnect the fuel delivery hose, return hose and evaporation hose.

17. Remove the intake manifold retaining bolts. Remove the intake manifold from the engine.

To install:

18. Installation is the reverse of the removal procedure.

19. Be sure to use new intake manifold gaskets. Torque the manifold retaining bolts to specification and in alternating sequence.

20. Be sure to fill the cooling system with the proper grade and type engine coolant.

21. Start the engine and check for leaks, correct as required.

Exhaust Manifold

Due to the unique design of the Subaru engine an exhaust manifold is not used. The exhaust enters directly into the front Y-pipe.

REMOVAL & INSTALLATION

❊❊ **CAUTION**

The exhaust pipe may be hot; DO NOT perform any work until the system has completely cooled.

1. Before servicing the vehicle, refer to the Precautions Section.
2. Remove or disconnect the following:
   • Negative battery cable

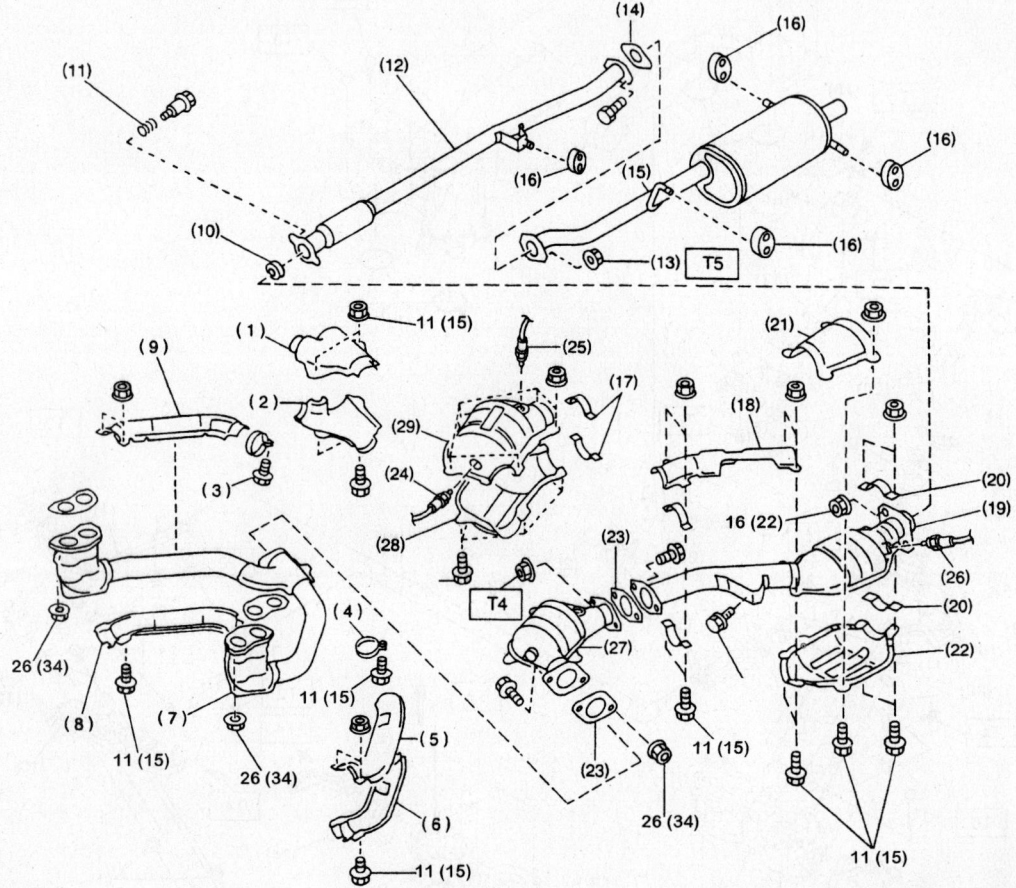

(1) Upper front exhaust pipe cover CTR
(2) Lower front exhaust pipe cover CTR
(3) Band RH
(4) Band LH
(5) Upper front exhaust pipe cover LH
(6) Lower front exhaust pipe cover LH
(7) Front exhaust pipe
(8) Lower front exhaust pipe cover RH
(9) Upper front exhaust pipe cover RH
(10) Gasket
(11) Spring
(12) Rear exhaust pipe
(13) Self-locking nut
(14) Gasket
(15) Muffler
(16) Cushion rubber
(17) Clamp
(18) Upper center exhaust pipe cover
(19) Center exhaust pipe
(20) Clamp B
(21) Upper rear catalytic converter cover
(22) Lower rear catalytic converter cover
(23) Gasket
(24) Front oxygen sensor
(25) Rear oxygen sensor (California spec. vehicles)
(26) Rear oxygen sensor (Except California spec. vehicles)
(27) Front catalytic converter
(28) Lower front catalytic converter cover
(29) Upper front catalytic converter cover

Exhaust system and related components—typical

7924XG07A

- Front Oxygen (O₂) sensor electrical connectors
- Rear Oxygen (O₂) sensor electrical connectors
- Center exhaust pipe from front exhaust pipe
- Nuts that secure the exhaust pipe to the cylinder head
- Front pipe-to-front catalytic converter mounting nuts

3. Discard the gaskets.

To install:

4. Clean all gasket surfaces completely.
5. Install or connect the following:
- Catalytic converter to front exhaust pipe using new gasket. Torque the bolts to 22 ft. lbs. (30 Nm).
- Exhaust pipe to the cylinder head using new gaskets. Torque the mounting nuts to 22 ft. lbs. (30 Nm).
- Exhaust pipe to the center pipe using new gaskets. Torque the mounting nuts to 26 ft. lbs. (35 Nm).
- Rear O₂sensors electrical connectors
- Front O₂sensors electrical connectors
- Negative battery cable

6. Start the engine and check for exhaust leaks.

Front Crankshaft Seal

REMOVAL & INSTALLATION

2002–2004

1. Before servicing the vehicle, refer to the Precautions Section.
2. Remove or disconnect the following:
- Negative battery cable
- Radiator electric fan motor wiring connectors
- Coolant reservoir tank
- 4 bolts that secure the radiator shroud, and then remove the fan assembly

3. Position the No. 1 piston to Top Dead Center (TDC) of its compression stroke.
- Accessory drive belt cover
- Air conditioning compressor drive belt and tensioner

4. Secure the crankshaft pulley with tool No. ST499977000.
- Crankshaft pulley bolt and pulley
- Left timing belt cover mounting bolts and the left cover
- Right timing belt cover mounting bolts and the right cover

- Center timing belt cover mounting bolts and the center cover
- Timing belt
- Timing belt crankshaft sprocket
- Crankshaft seal from the oil pump housing

To install:

5. Install or connect the following:
- New crankshaft seal, using a suitable seal driver
- Timing belt crankshaft sprocket and the timing belt
- Center, right, then the left timing belt covers. Tighten the bolts to 44 inch lbs. (5 Nm).
- Crankshaft pulley and tighten the bolt to 94 ft. lbs. (127 Nm)
- Air conditioning compressor drive belt tensioner and the drive belts
- Fan shroud and fan motor assembly
- Accessory drive belt cover
- Negative battery cable

Camshaft and Valve Lifters

REMOVAL & INSTALLATION

SOHC Engine

2002–2003

1. Before servicing the vehicle, refer to the Precautions Section.
2. Remove or disconnect the following:
- Negative battery cable
- Timing belt covers
- Timing belt
- Camshaft sprockets
- Spark plug wires
- Oil level gauge guide and Camshaft Position (CMP) sensor support
- Positive Crankcase Ventilation (PCV) hose
- Valve cover
- Rocker arm assembly

3. Remove the camshaft cap as follows:
a. Bolts "A" through "B" in alphabetical sequence

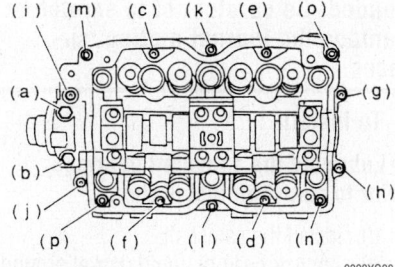

Camshaft cap removal sequence— 2002–2003 SOHC engine

b. Loosen bolts "C" through "J" equally all the way in alphabetical sequence
c. Bolts "K" through "P" in alphabetical sequence using tool 499497000

4. Remove or disconnect the following:
- Camshaft
- Oil seal, if necessary
- Plug from the rear side of the camshaft

To install:

➡**Lubricate the camshaft bearings prior to camshaft installation.**

5. Install or connect the following:
- Camshaft
- Camshaft cap. Apply liquid gasket on the edge of the cam cap mating surface 0.12 inch (3mm) thick

6. Temporarily tighten bolts "G" through "J" in alphabetical sequence
7. Install the valve rocker assembly. Torque the bolts "A" through "H" to 18 ft. lbs. (25 Nm).
8. Torque bolts "I" through "N" to 13 ft. lbs. (25 Nm) in alphabetical sequence using tool 499497000
9. Torque bolts "O" through "X" to 7.2 ft. lbs. (10 Nm) in alphabetical sequence
10. Install or connect the following:
- Oil seal to the camshaft using tool 499597000 oil seal guide and 49958700 oil seal installer
- Plug to the rear side of the camshaft using tool 499587700 oil seal installer
- Valve cover
- Oil level gauge guide and CMP sensor support
- PCV house
- Spark plug wires
- Camshaft sprockets. Torque the bolts to 58 ft. lbs. (78 Nm).
- Timing belt
- Timing belt covers. Torque the bolts to 3.6 ft. lbs. (5 Nm).
- Negative battery cable

2004

1. Before servicing the vehicle, refer to the Precautions Section.
2. Disconnect the negative battery cable.
3. Remove the drive belts.
4. Remove the crankshaft pulley.
5. Remove the timing belt covers.
6. Remove the timing belt.
7. Remove the camshaft sprocket.
8. Remove the crankshaft sprocket.
9. Remove timing belt cover No.2 from the left and right hand sides.
10. Remove the tensioner bracket.

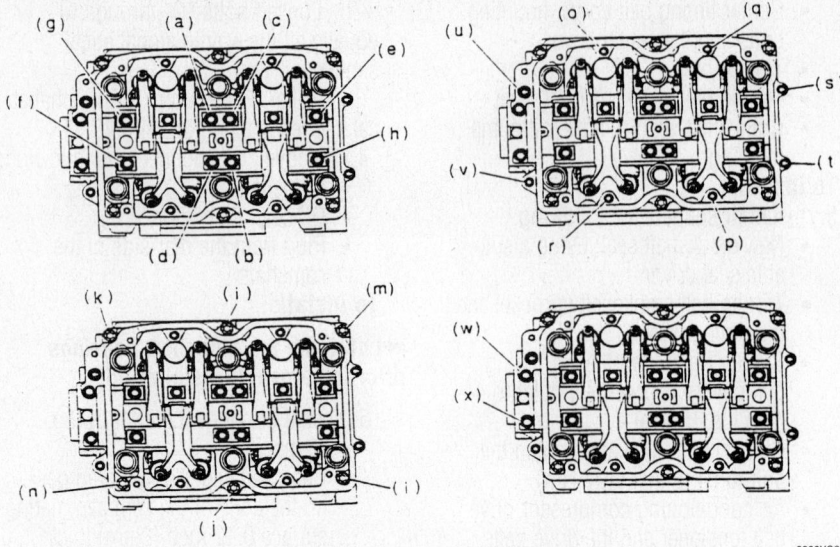

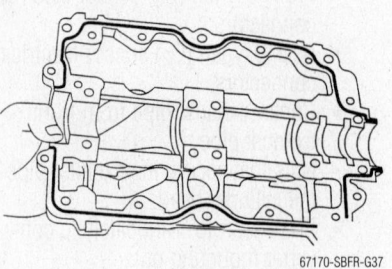

Apply a bead of liquid gasket around the camshaft cap—2004 SOHC engine

Camshaft cap tightening sequence—2002–2003 SOHC engine

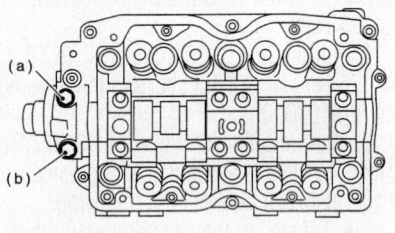

Remove camshaft cap bolts A and B using several passes—2004 SOHC engine

11. Remove the Camshaft Position (CMP) sensor support from the left hand side.

12. Remove the oil level gauge guide from the left hand side.

13. Remove the spark plug wires.

14. Remove the valve cover and gasket.

15. Remove the valve rocker assembly.

➡**Before removing the camshaft cap bolts mark and note their location so they may be reinstalled in their original positions.**

16. Loosen the camshaft cap bolts using several passes in the sequences illustrated in the following order:

a. Remove bolts A and B.

b. Loosen bolts C through J in sequence.

c. Using a suitable size torque bit, remove bolts K through P in sequence.

17. Remove the camshaft caps, camshaft and oil seal.

18. Remove the plug from the rear right hand side of the camshaft.

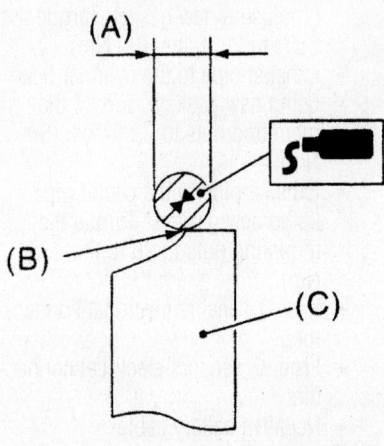

Apply liquid gasket on the edge (B) of the cam cap (C) mating surface 0.12 inch (3mm) thick (A)—2004 SOHC engine

✱✱ CAUTION

Do not remove the oil seal unless needed. Be careful not to scratch or damage the journal mating surfaces.

To install:

➡**Lubricate the camshaft bearings prior to camshaft installation.**

19. Install the camshaft.

20. Apply a bead of liquid gasket around the camshaft cap as illustrated.

21. Install the camshaft cap. Apply liquid

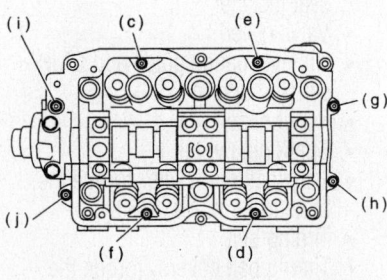

Remove camshaft cap bolts C through J using several passes—2004 SOHC engine

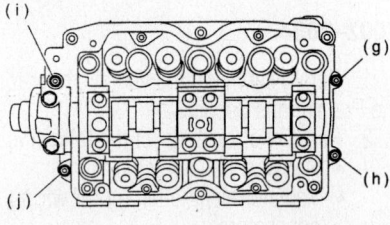

Temporarily tighten bolts "G" through "J" in sequence—2004 SOHC engine

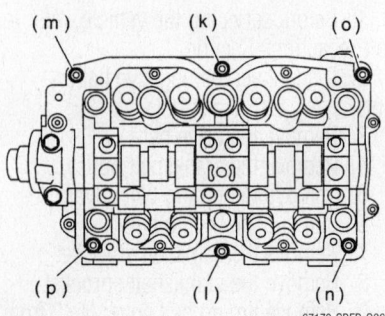

Remove camshaft cap bolts K through P using several passes—2004 SOHC engine

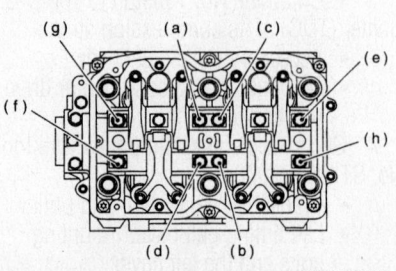

Tighten bolts "A" through "H" in sequence—2004 SOHC engine

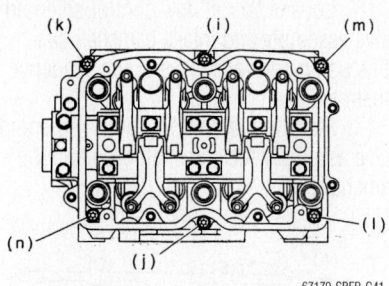

Tighten bolts "I" through "N" in sequence—2004 SOHC engine

67170-SBFR-G41

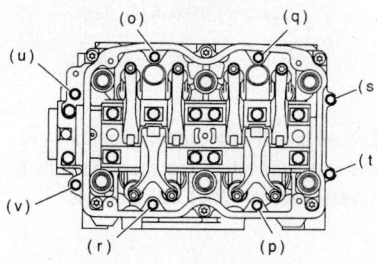

Tighten bolts "O" through "V" in sequence—2004 SOHC engine

67170-SBFR-G42

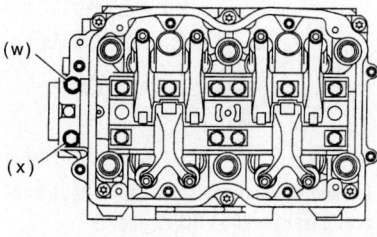

Tighten bolts "W" through "X" in sequence—2004 SOHC engine

67170-SBFR-G43

gasket on the edge (B) of the cam cap (C) mating surface 0.12 inch (3mm) thick (A).

22. Temporarily tighten bolts "G" through "J" in alphabetical sequence.

23. Install the valve rocker assembly. Torque the bolts "A" through "H" to 18 ft. lbs. (25 Nm).

24. Torque bolts "I" through "N" to 13 ft. lbs. (25 Nm) in alphabetical sequence using Torx tool 499497000.

25. Torque bolts "O" through "V" to 7.2 ft. lbs. (10 Nm) in alphabetical sequence.

26. Torque bolts "W" through "X" to 7.2 ft. lbs. (10 Nm) in alphabetical sequence.

27. Install the oil seal to the camshaft using tool 499597000 oil seal guide and 499587500 oil seal installer.

28. Install the plug to the rear side of the camshaft using tool 499587700 oil seal installer.

29. Adjust the valve clearance.

30. Install the valve cover.

31. Connect the spark plug wires.

32. Connect the PVC house.

33. Install the oil level gauge guide and CMP sensor support

34. Install the tensioner bracket and tighten to 18 ft. lbs. (25 Nm).

35. Install the right and left hand timing belt cover No.2 and tighten the bolts to 3 ft. lbs. (5 Nm).

36. Install the crankshaft and camshaft sprockets.

37. Install the timing belt and cover.

38. Install the crankshaft pulley and tighten the bolt to 130 ft. lbs. (177 Nm).

39. Install the drive belts.

40. Connect the negative battery cable.

2005–2006

1. Before servicing the vehicle, refer to the Precautions Section.

2. Disconnect the negative battery cable.

3. To remove the front side V belt, remove the belt covers. Loosen the lock bolt. Loosen the slider bolt. Remove the front side belt.

4. To remove the rear side V belt, remove the belt covers. Loosen the lock bolt. Loosen the slider bolt. Remove the rear side belt. Remove the belt tensioner.

5. Remove the crankshaft pulley bolt.

6. Lock the crankshaft in place using tool ST499977100, or equivalent.

7. Remove the crankshaft pulley.

8. Remove the left side timing belt cover.

9. Remove the front timing belt cover.

10. Remove the timing belt.

11. Remove the camshaft position sensor.

12. Remove the camshaft sprockets.

➡**Be sure to lock the camshaft in place using tool ST18231AA010, or equivalent.**

13. Remove the left and right timing belt N02 covers.

➡**Do not damage or loose the rubber seal when removing the covers.**

14. Remove the tensioner bracket. Remove the camshaft position sensor support, on the left side.

15. Remove the oil level gauge guide, on the left side.

16. Remove the valve rocker arm assembly.

17. Remove camshaft cap retaining bolts "A" and "B" in the proper sequence.

18. Loosen camshaft cap bolts "C"

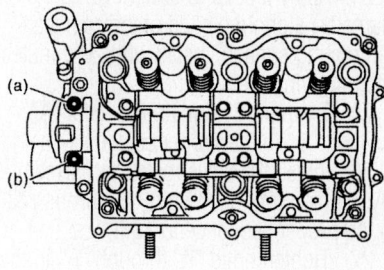

Camshaft bolt loosening sequence—2005–2006 SOHC engine

09490_SBCR_G0034

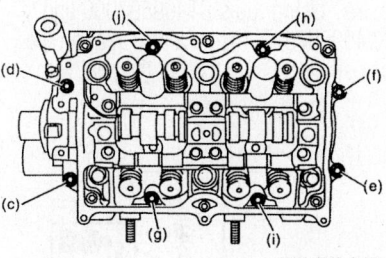

Camshaft bolt loosening sequence—2005–2006 SOHC engine

09490_SBCR_G0035

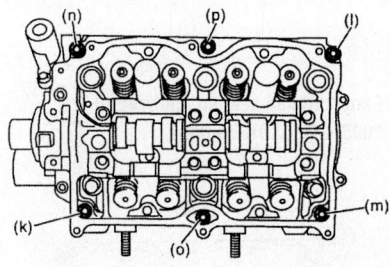

Camshaft bolt loosening sequence—2005–2006 SOHC engine

09490_SBCR_G0036

through "J" all the way in the proper sequence.

19. Remove camshaft cap bolts "K" through "P" in the proper sequence using a torx head bit.

20. Remove the camshaft caps.

21. Remove the camshaft. Remove the oil seal. Remove the plug from the rear side of the camshaft.

➡**Do not remove the oil seal unless necessary.**

To install:

➡**Lubricate the camshaft journals with clean engine oil prior to installation.**

22. Install the camshaft into the cylinder head.

23. Apply liquid gasket to the mating surfaces of the camshaft cap.

24. Apply a bead of sealant (0.12 inch in diameter) along the edge of the camshaft cap mating surface. Install with 20 minutes after applying the sealant.

25. Temporarily tighten the bolts "A" through "D" in the proper sequence.

26. Install the valve rocker arm assembly. Tighten torx head bolts "E" through "J" in the proper sequence to 13 ft. lbs.

27. Tighten bolts "K" through "R" in the proper sequence to 7.2 ft. lbs.

28. Tighten bolts "S" through "T" in the proper sequence to 7.2 ft. lbs.

➡ **Be sure to use a new seal washer.**

29. Using tools ST499597000 and ST499587500, install a new seal on the camshaft. Be sure to coat the seal with clean engine oil before installation. Use tool ST499587700 and install the plug.

30. Adjust the valve clearance.

31. Continue the installation in the reverse order of the removal procedure.

DOHC Engine

2004

1. Before servicing the vehicle, refer to the Precautions Section.

2. Disconnect the negative battery cable.

3. Remove the drive belts.

4. Remove the crankshaft pulley.

5. Remove the timing belt covers.

6. Remove the timing belt.

7. Remove the camshaft sprockets.

8. Remove the crankshaft sprocket

9. Disconnect the oil flow control solenoid valve connector.

10. Remove the tensioner bracket.

11. Remove timing belt cover No.2 from the left hand side and the timing belt cover from the right hand side.

12. Remove the spark plug wires.

13. Remove the oil level gauge guide from the left hand side.

14. Remove the valve cover and gasket.

15. Remove the oil pipe.

➡ **Before removing the camshaft cap bolts mark and note their location so they may be reinstalled in their original positions.**

16. Loosen the oil flow control solenoid valve assembly and intake camshaft cap bolts using several passes in the sequence illustrated.

17. Remove the oil flow control solenoid valve assembly the camshaft caps and the camshaft.

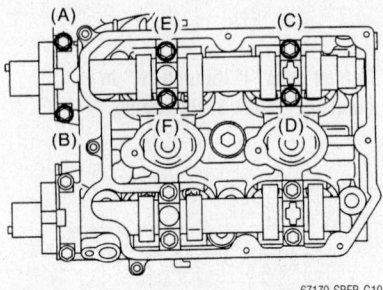

67170-SBFR-G10

Loosen the oil flow control solenoid valve assembly and intake camshaft cap bolts in several passes—2004 DOHC engine

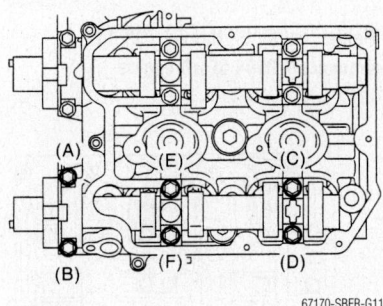

67170-SBFR-G11

Loosen exhaust camshaft cap bolts in several passes—2004 DOHC engine

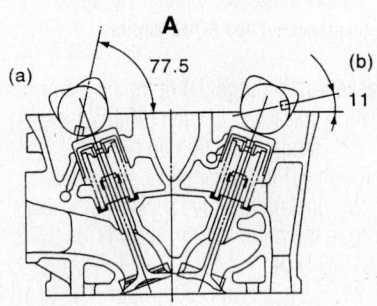

67170-SBFR-G12

Camshaft alignment—2004 DOHC engine

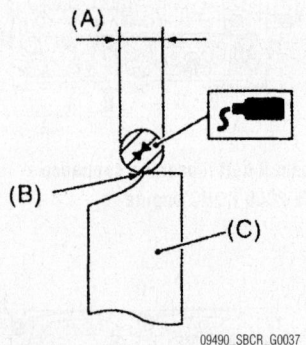

09490_SBCR_G0037

Camshaft cap sealant application— 2005–2006 SOHC engine

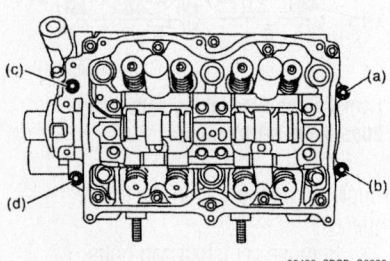

09490_SBCR_G0038

Camshaft bolt tightening sequence— 2005–2006 SOHC engine

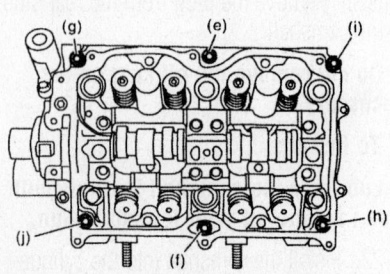

09490_SBCR_G0039

Camshaft bolt tightening sequence— 2005–2006 SOHC engine

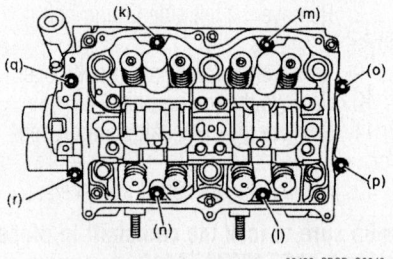

09490_SBCR_G0040

Camshaft bolt tightening sequence— 2005–2006 SOHC engine

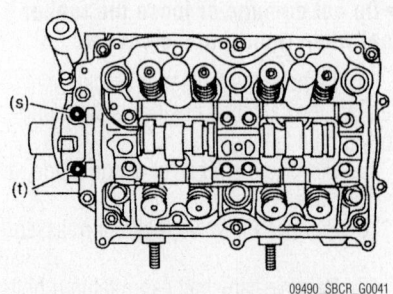

09490_SBCR_G0041

Camshaft bolt tightening sequence— 2005–2006 SOHC engine

18. Loosen the exhaust camshaft cap bolts using several passes in the sequence illustrated. Remove the caps and camshaft.

To install:

19. Lubricate the camshaft bearings prior to camshaft installation. Install the camshaft so that each valve is close to or in contact with the base circle of the cam lobe.

➡ **When the camshafts are positioned as shown in the accompanying illustration, the camshafts need to be rotated at a minimum to align with the timing belt during installation. The right hand camshaft will not need to be rotated when set at the position show in the accompanying illustration. The left hand intake camshaft should be rotated 80 degrees clockwise and the left hand exhaust camshaft should be rotated 45 degrees counterclockwise.**

20. Install the intake camshaft caps and oil flow control solenoid valve assembly as follows:

a. Apply liquid gasket sparingly to the cap mating surfaces as illustrated. Do not use an excessive amount of gasket maker as this can cause excess packing to come out and block the oil seal which will cause an oil leak.

21. Apply clean engine oil to the cap bearing surface and install the cap on the

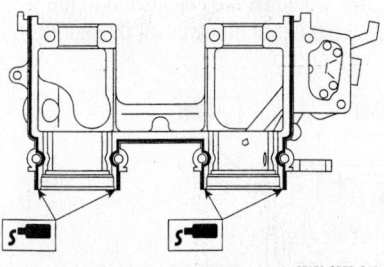

Apply liquid gasket sparingly to the cap mating surfaces—2004 DOHC engine

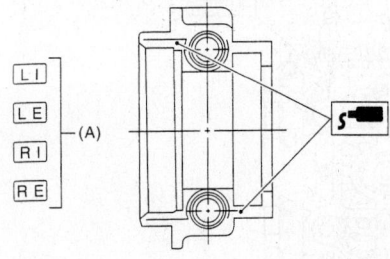

Apply clean engine oil to the cap bearing surface and install the cap on the camshaft indicated by the letter A—2004 DOHC engine

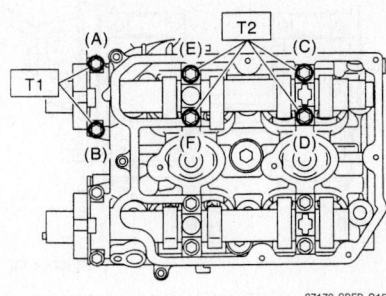

Intake camshaft and oil control valve bearing cap torque sequence—2004 DOHC engine

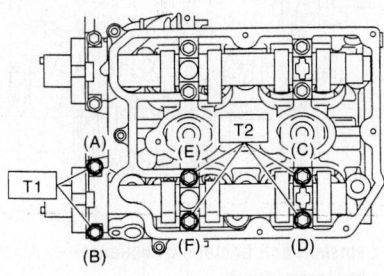

Exhaust camshaft bearing cap torque sequence—2004 DOHC engine

camshaft indicated by the letter **A** in the accompanying illustration.

a. Tighten the camshaft cap and oil control valve assembly bolts in two to three passes. Apply clean engine oil to the cap bearing surface and install the cap on the camshaft indicated by the letter **A** n the sequence illustrated. Tighten bolts C, D E and F to 7 ft. lbs. (10 Nm) and bolts A and B to 14 ft. lbs. (20 Nm).

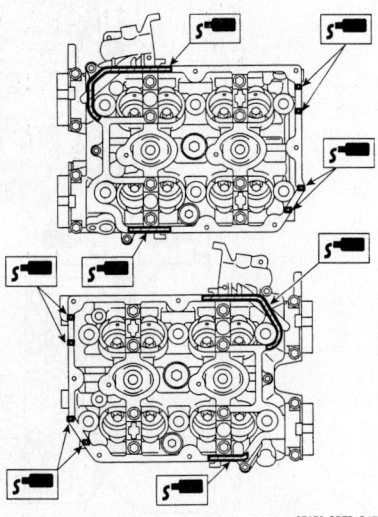

Apply liquid gasket maker to the areas shown before installing the valve cover—2004 DOHC engine

22. Install the exhaust caps and tighten the bolts using two or three passes. Tighten bolts C, D E and F to 7 ft. lbs. (10 Nm) and bolts A and B to 14 ft. lbs. (20 Nm).

23. Install the oil seal to the camshaft using tool 499587600 oil seal guide and 499597200 oil seal installer

24. Install a new gasket on the valve cover, also install the peripheral and ignition coil gaskets.

25. Apply liquid gasket maker to the points shown in the accompanying illustration.

26. Install the valve cover and make sure the gasket is positioned correctly. Tighten the valve cover bolts in the sequence illustrated to 5 ft. lbs. (7 Nm).

27. Install the oil pipe and tighten to 21 ft. lbs. (29 Nm).

28. Connect the oil flow control solenoid valve connector.

29. Connect the spark plug wires.

30. Install the right and left hand timing belt cover No.2 and tighten the bolts to 3 ft. lbs. (5 Nm).

31. Install the tensioner bracket and tighten to 18 ft. lbs. (25 Nm).

32. Install the crankshaft and camshaft sprockets.

33. Install the timing belt and cover.

34. Install the crankshaft pulley and tighten the bolt to 132 ft. lbs. (180 Nm).

35. Install the drive belts.

36. Connect the negative battery cable.

2005–2006

1. Before servicing the vehicle, refer to the Precautions Section.

2. Disconnect the negative battery cable.

3. Remove the collector cover.

4. To remove the front side V belt, remove the belt covers. Loosen the lock bolt. Loosen the slider bolt. Remove the front side belt.

5. To remove the rear side V belt, remove the belt covers. Loosen the lock bolt. Loosen the slider bolt. Remove the rear side belt. Remove the belt tensioner.

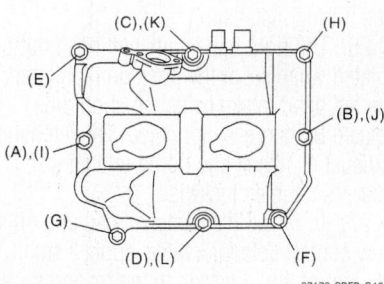

Tighten the valve cover bolts in this sequence—2004 DOHC engine

6. Remove the crankshaft pulley bolt.

7. Lock the crankshaft in place using tool ST499977100, or equivalent.

8. Remove the crankshaft pulley.

9. Remove the left side timing belt cover.

10. Remove the right side timing belt cover.

11. Remove the front timing belt cover.

12. Remove the timing belt.

13. Remove the camshaft position sensor.

14. Remove the camshaft sprockets.

➡ **Be sure to lock the camshaft in place using tool ST499207400, or equivalent.**

15. Lock the crankshaft in place using tool ST499977100, or equivalent.

16. Remove the crankshaft pulley.

17. Remove the tensioner bracket. Remove the right and left timing belt No.2 covers.

18. Remove the spark plug wires. Remove the oil level gauge, on the left side.

19. Remove the rocker cover and gasket. Remove the oil pipe.

20. Loosen the oil flow control solenoid valve assembly and the intake camshaft cap bolts equally and in the proper sequence.

21. Loosen the exhaust camshaft cap bolts equally and in the proper sequence.

22. Remove the oil flow control solenoid valve assembly, intake camshaft cap and camshaft.

23. Remove the exhaust camshaft caps and camshaft.

➡ **Arrange the camshafts caps so that they can be installed in their original positions.**

To install:

➡ **Lubricate the camshaft journals with clean engine oil prior to installation.**

24. Install the camshaft so that the valves are closed or in contact with the "base circle" of the cam lobe.

25. If the camshafts are positioned as shown in the, the camshafts need to be rotated at a minimum to align the timing belt during installation.

26. The right hand camshaft need not be rotated when set at the position illustrated, the left hand intake camshaft should be rotated 80 degrees clockwise. The left hand exhaust camshaft should be rotated 45 degrees counterclockwise.

27. To install the camshaft cap and oil flow control solenoid valve, apply a small amount of liquid gasket to the mating surface of the cap.

Do not apply an excessive amount of

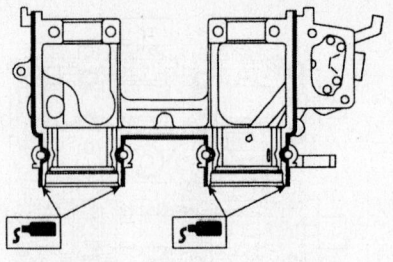

Camshaft cap liquid gasket application points—2005–2006 DOHC engine

09490_SBCR_G0042

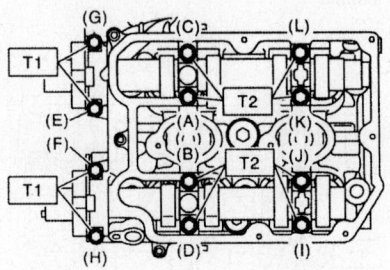

09490_SBCR_G0043

Camshaft bolt tightening sequence—2005–2006 DOHC engine

sealant, as it will squish out and flow toward the seal resulting in an oil leak.

28. Apply a thin coat of engine oil to the cap bearing surface and install the cap according to the cap identification mark. Gradually tighten the cap bolts in two stages, first to 7.2 ft. lbs. and than to 14.5 ft. lbs. in the proper sequence.

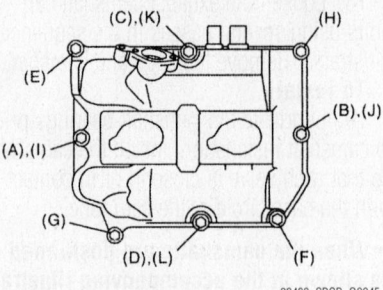

09490_SBCR_G0045

Rocker cover tightening sequence—2005–2006 DOHC engine

➡ **After tightening the camshaft cap, ensure that the camshaft rotates slightly while holding it at base circle.**

29. Using tools ST49587600 and ST499597200, install a new seal on the camshaft. Be sure to coat the seal with clean engine oil before installation. Use tool ST499587700 and install the plug.

30. Install a new gasket on the rocker cover. Apply liquid gasket to the cylinder head (see illustration).

➡ **Apply an extra amount of liquid gasket around the semicircular plugs, 5mm or more.**

31. Temporarily tighten the rocker cover retaining bolts, in the proper sequence, and then tighten to 4.7 ft. lbs. Install the oil pipe.

32. Continue the installation in the reverse order of the removal procedure.

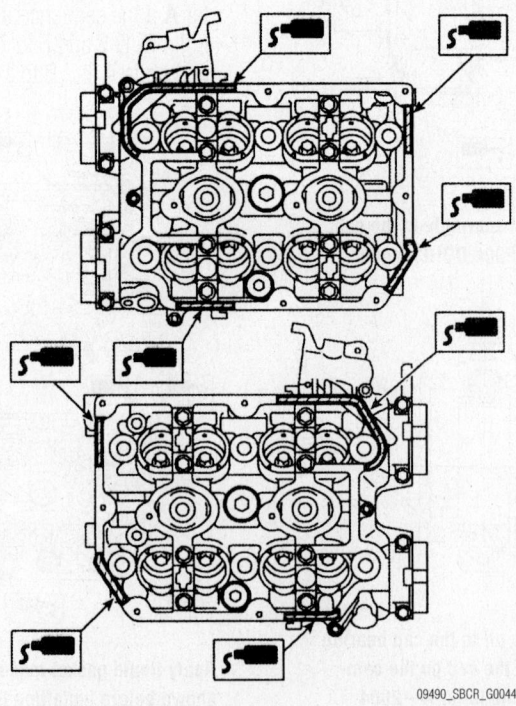

09490_SBCR_G0044

Rocker cover liquid gasket application points—2005–2006 DOHC engine

INSPECTION

1. Before servicing the vehicle, refer to the Precautions Section.

2. Remove the camshaft from the engine.

3. Check the camshaft bearing journals for damage and binding.

4. If the journals are binding, check the cylinder head for damage.

5. Check the cylinder head for clogged oil holes.

6. Check the camshaft surface for abnormal wear and damage. Replace the camshaft, as required.

7. Measure the camshaft lobe surface and replace the camshaft if not within specification.

8. Measure the camshaft journal diameter and replace the camshaft if not within specification.

9. Measure the camshaft run out and replace the camshaft if not within specification.

Valve Lash

ADJUSTMENT

SOHC Engine

2002–2004

➡**The valve adjustment should be performed while the engine is cold. A Shim Replace Kit 498187100 will be needed to perform the valve adjustment.**

1. Before servicing the vehicle, refer to the precautions in the beginning of this section.

2. Adjustment should be performed when engine is cold.

3. Remove or disconnect the following:
- Negative battery cable
- Engine coolant reservoir tank
- Timing belt cover on the left hand side

4. When inspecting the No. 1 and 3 cylinders remove the following:
- Air intake duct as a unit
- Resonator chamber
- Spark plug wires from the No. 1 and 3 cylinders
- Blow-by house from valve cover
- Engine undercover
- Timing belt cover on the right hand side
- Valve cover on the right hand side

5. When inspecting the No. 2 and 4 cylinders remove the following:
- Battery and battery tray

- Window washer motor connectors front and rear
- Rear gate glass washer hose from the washer motor
- Washer tank mounting bolts and secure out of the way
- Spark plug wires from the No. 2 and 4 cylinders
- Blow by house from valve cover
- Timing belt cover on the right hand side
- Valve cover on the left hand side

6. Set No. 1 cylinder to Top Dead Center (TDC).

➡**When arrow mark on the left hand side comes exactly to the top, No. 1 cylinder piston is brought to TDC of the compression stroke.**

7. Check the valve clearance:

8. If any valve needs adjustment, perform the following:

a. Loosen the valve rocker nut and screw.

b. Place a thickness gauge in at as horizontal a direction as possible with respect to the valve stem and face.

c. Adjust the screw until proper clearance is obtained.

d. Tighten the rocker nut after adjusted.

9. Install or connect the following:
- Valve covers left and right
- Timing belt covers
- Blow-by houses to valve covers
- Spark plug wires
- Washer tank
- Rear gate glass washer hose to the washer motor
- Washer motor connectors
- Battery and battery tray
- Engine undercover
- Resonator chamber
- Air intake duct unit
- Engine coolant reservoir tank

2005–2006

➡**The valve adjustment should be performed while the engine is cold.**

1. Before servicing the vehicle, refer to the precautions in the beginning of this section.

2. Raise and support the vehicle safely.

3. Remove the under cover.

4. Lower the vehicle.

5. Disconnect the negative battery cable.

6. To remove the front side V belt, remove the belt covers. Loosen the lock bolt. Loosen the slider bolt. Remove the front side belt.

7. To remove the rear side V belt, remove the belt covers. Loosen the lock bolt. Loosen the slider bolt. Remove the rear side belt. Remove the belt tensioner.

8. Remove the crankshaft pulley bolt.

9. Lock the crankshaft in place using tool ST499977100, or equivalent.

10. Remove the crankshaft pulley.

11. Remove the left side timing belt cover.

12. Remove the fuel injector.

13. Remove the rocker cover.

14. Position the number one piston at TDC of the compression stroke.

➡**When the arrow (see illustration) on the camshaft sprocket (left side) comes exactly to the top, number one cylinder piston is at TDC of the compression stroke.**

15. Measure the valve clearance, using a feeler gauge.

16. If adjustment is needed, loosen the valve rocker nut and screw. Position the feeler gauge.

➡**Insert the feeler gauge in a horizontally as possible with respect to the valve stem end face. Adjust the exhaust valve clearance while lifting up the vehicle.**

17. While noting the valve clearance, tighten the rocker adjusting screw.

18. When the proper valve clearance is obtained, tighten the valve rocker nut to 7.2 ft. lbs.

19. Adjust the valve clearance on the remaining cylinders, following the above procedure.

➡**Be sure to position the pistons to their respective TDC positions on the compression stroke, before checking and adjusting the valves. By rotating the crankshaft pulley clockwise every 180 degrees from the state that number one piston is on TDC of the compression stroke, the remaining pistons come to TDC of the compression stroke in the following order, #3, #2, and #4.**

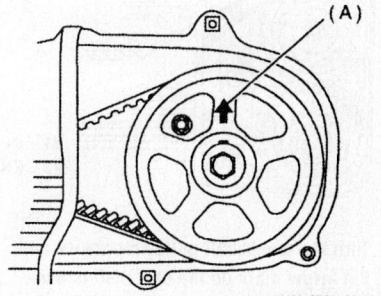

09490_SBCR_G0050

TDC alignment—2005–2006 SOHC engine

20. After adjustment, replace any removed components.

21. Be sure to use new gaskets and seals, as required.

DOHC Engine

➡ **The valve adjustment should be performed while the engine is cold.**

1. Before servicing the vehicle, refer to the precautions in the beginning of this section.

2. Raise and support the vehicle safely.

3. Remove the under cover.

4. Lower the vehicle.

5. Remove the collector cover.

6. Disconnect the negative battery cable.

7. Remove the air intake duct.

8. Remove the bolt that retains the right side timing belt cover. Remove the remaining bolts and remove the right side timing belt cover.

9. Disconnect the ignition coil electrical connector. Remove the ignition coil.

10. Position a suitable container under the vehicle.

11. Disconnect the PCV hose from the rocker cover. Remove the rocker cover retaining bolts. Remove the rocker cover from the vehicle.

12. Position the number one piston at TDC of the compression stroke.

13. Using a feeler gauge, measure and record the clearance of the number one cylinder intake and the number three cylinder exhaust valves.

➡ **Insert the feeler gauge in a horizontally as possible with respect to the valve lifter. Measure and record the exhaust valve clearance while lifting up the vehicle.**

14. Rotate the crankshaft pulley clockwise until the arrow mark on the camshaft is positioned as shown to measure and record

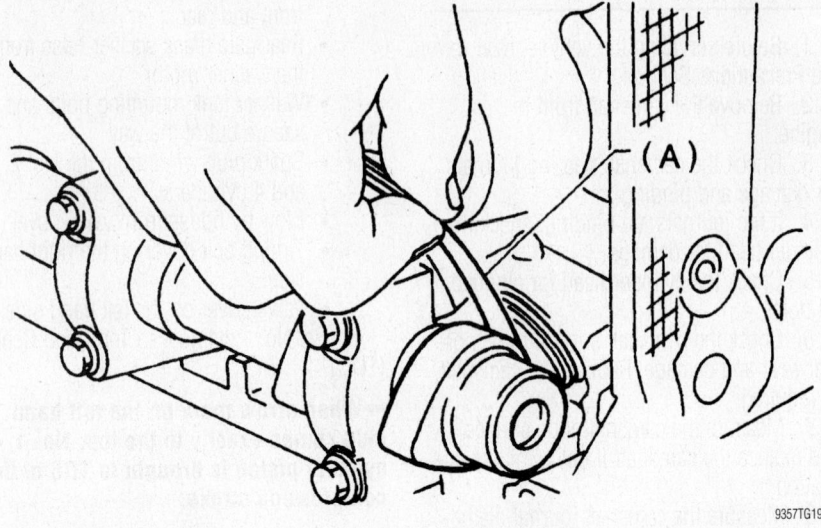

Use a feeler gauge to inspect the valve clearance—DOHC engine

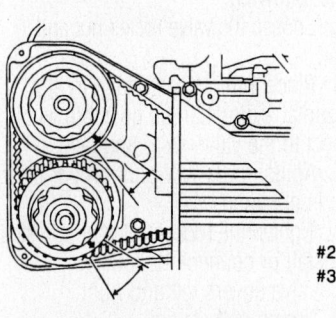

#2 EX.
#3 IN.

9357TG20

Turn the crankshaft pulley clockwise until the arrow mark on the camshaft is positioned as shown to measure the # 2 exhaust and # 3 intake valves—DOHC engine

the clearance on the number two exhaust and number three intake valves.

15. Rotate the crankshaft pulley clockwise until the arrow mark on the camshaft is positioned as shown to measure and record the number two intake and number four exhaust valves.

16. Rotate the crankshaft pulley clockwise until the arrow mark on the camshaft is positioned as shown to measure and record

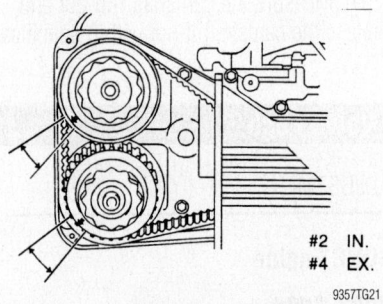

#2 IN.
#4 EX.

9357TG21

Turn the crankshaft pulley clockwise until the arrow mark on the camshaft is positioned as shown to measure the # 2 intake and # 4 exhaust valves—DOHC engine

the number one exhaust and number four intake valves.

17. If adjustment is required, remove the camshafts.

18. Remove and measure the thickness of the valve lifter. Select a suitable shim, using the shim selection chart.

19. Install the replacement shim to the lifter.

20. After all shims have been adjusted, inspect the valve clearances again.

21. After completion, install all removed components.

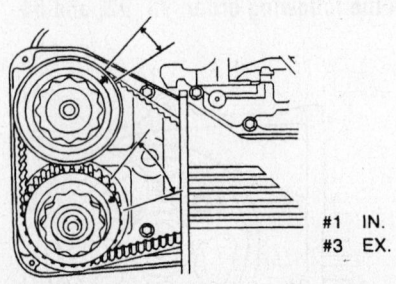

#1 IN.
#3 EX.

9357TG18

Turn the crankshaft pulley clockwise until the arrow mark on the camshaft is positioned as shown to measure the # 1 intake and # 3 exhaust valves—DOHC engine

Unit: (mm)
Intake valve: $S = (V + T) - 0.20$
Exhaust valve: $S = (V + T) - 0.35$
S: Valve lifter thickness required
V: Measured valve clearance
T: Valve lifter thickness to be used

09490_SBCR_G0051

Use this table to help you select a suitable shim—DOHC engine

Part No.	Thickness mm (in)
13228 AB102	4.68 (0.1843)
13228 AB112	4.69 (0.1846)
13228 AB122	4.70 (0.1850)
13228 AB132	4.71 (0.1854)
13228 AB142	4.72 (0.1858)
13228 AB152	4.73 (0.1862)
13228 AB162	4.74 (0.1866)
13228 AB172	4.75 (0.1870)
13228 AB182	4.76 (0.1874)
13228 AB192	4.77 (0.1878)
13228 AB202	4.78 (0.1882)
13228 AB212	4.79 (0.1886)
13228 AB222	4.80 (0.1890)
13228 AB232	4.81 (0.1894)
13228 AB242	4.82 (0.1898)
13228 AB252	4.83 (0.1902)
13228 AB262	4.84 (0.1906)
13228 AB272	4.85 (0.1909)
13228 AB282	4.86 (0.1913)
13228 AB292	4.87 (0.1917)
13228 AB302	4.88 (0.1921)
13228 AB312	4.89 (0.1925)
13228 AB322	4.90 (0.1929)
13228 AB332	4.91 (0.1933)
13228 AB342	4.92 (0.1937)
13228 AB352	4.93 (0.1941)
13228 AB362	4.94 (0.1945)
13228 AB372	4.95 (0.1949)
13228 AB382	4.96 (0.1953)
13228 AB392	4.97 (0.1957)
13228 AB402	4.98 (0.1961)
13228 AB412	4.99 (0.1965)
13228 AB422	5.00 (0.1969)
13228 AB432	5.01 (0.1972)
13228 AB442	5.02 (0.1976)
13228 AB452	5.03 (0.1980)
13228 AB462	5.04 (0.1984)
13228 AB472	5.05 (0.1988)
13228 AB482	5.06 (0.1992)
13228 AB492	5.07 (0.1996)
13228 AB502	5.08 (0.2000)
13228 AB512	5.09 (0.2004)
13228 AB522	5.10 (0.2008)
13228 AB532	5.11 (0.2012)
13228 AB542	5.12 (0.2016)
13228 AB552	5.13 (0.2020)
13228 AB562	5.14 (0.2024)
13228 AB572	5.15 (0.2028)
13228 AB582	5.16 (0.2031)
13228 AB592	5.17 (0.2035)
13228 AB602	5.18 (0.2039)
13228 AB612	5.19 (0.2043)

Part No.	Thickness mm (in)
13228 AB622	5.20 (0.2047)
13228 AB632	5.21 (0.2051)
13228 AB642	5.22 (0.2055)
13228 AB652	5.23 (0.2059)
13228 AB662	5.24 (0.2063)
13228 AB672	5.25 (0.2067)
13228 AB682	5.26 (0.2071)
13228 AB692	5.27 (0.2075)
13228 AB702	4.38 (0.1724)
13228 AB712	4.40 (0.1732)
13228 AB722	4.42 (0.1740)
13228 AB732	4.44 (0.1748)
13228 AB742	4.46 (0.1756)
13228 AB752	4.48 (0.1764)
13228 AB762	4.50 (0.1771)
13228 AB772	4.52 (0.1780)
13228 AB782	4.54 (0.1787)
13228 AB792	4.56 (0.1795)
13228 AB802	4.58 (0.1803)
13228 AB812	4.60 (0.1811)
13228 AB822	4.62 (0.1819)
13228 AB832	4.64 (0.1827)
13228 AB842	4.66 (0.1835)
13228 AB852	5.29 (0.2083)
13228 AB862	5.31 (0.2091)
13228 AB872	5.33 (0.2098)
13228 AB882	5.35 (0.2106)
13228 AB892	5.37 (0.2114)
13228 AB902	5.39 (0.2122)
13228 AB912	5.41 (0.2123)
13228 AB922	5.43 (0.2138)
13228 AB932	5.45 (0.2146)
13228 AB942	5.47 (0.2154)
13228 AB952	5.49 (0.2161)
13228 AB962	5.51 (0.2169)
13228 AB972	5.53 (0.2177)
13228 AB982	5.55 (0.2185)
13228 AB992	5.57 (0.2193)
13228 AC002	5.59 (0.2201)
13228 AC012	5.61 (0.2209)
13228 AC022	5.63 (0.2217)
13228 AC032	5.65 (0.2224)

09490_SBCR_G0052

Valve adjusting shim chart—DOHC engine

Starter

REMOVAL & INSTALLATION

2002–2004

1. Before servicing the vehicle, refer to the Precautions Section.
2. Remove or disconnect the following:
   - Negative battery
   - Air intake duct and assembly
   - Wires
   - Starter

To install:

3. Install or connect the following:
   - Starter. Torque the mounting bolts to 37 ft. lbs. (50 Nm).
   - Wires
   - Air intake duct and assembly
   - Negative battery

2005–2006

1. Before servicing the vehicle, refer to the Precautions Section.
2. Disconnect the negative battery cable.
3. Remove the air intake chamber, on non turbocharged engine.
4. Remove the intercooler, on turbocharged engine.
5. Remove the air intake chamber stay, on non turbocharged engine.
6. Disconnect the electrical connectors from the starter.
7. Remove the starter retaining bolts. Remove the starter from the vehicle.

To install:

8. Installation is the reverse of the removal procedure.

9. Torque the starter retaining bolts to 37 ft. lbs.

Oil Pan

REMOVAL & INSTALLATION

2002–2004

1. Before servicing the vehicle, refer to the Precautions Section.

2. Remove or disconnect the following:
- Negative battery cable
- Air intake duct
- Mass Air Flow (MAF) sensor on turbocharged engine
- Air intake boot and air cleaner upper cover on turbocharged engine
- Intercooler on turbocharged engine
- Front Oxygen Sensor (O_2S) electrical connector
- Pitching stopper
- Upper radiator brackets

3. Support the engine with a suitable lifting device.

- Front wheel and tire assemblies

4. Lift up the engine slightly.
- Engine undercover

5. Drain the oil from the engine into a suitable container.

6. Install the drain plug with a new gasket and tighten it to 33–36 ft. lbs. (43–47 Nm).

7. Remove or disconnect the following:
- Rear O_2S electrical connector
- Exhaust Y-pipe
- Nuts that secure the front engine mounts to the front crossmember
- Oil pan mounting bolts

8. While supporting the oil pan, use a

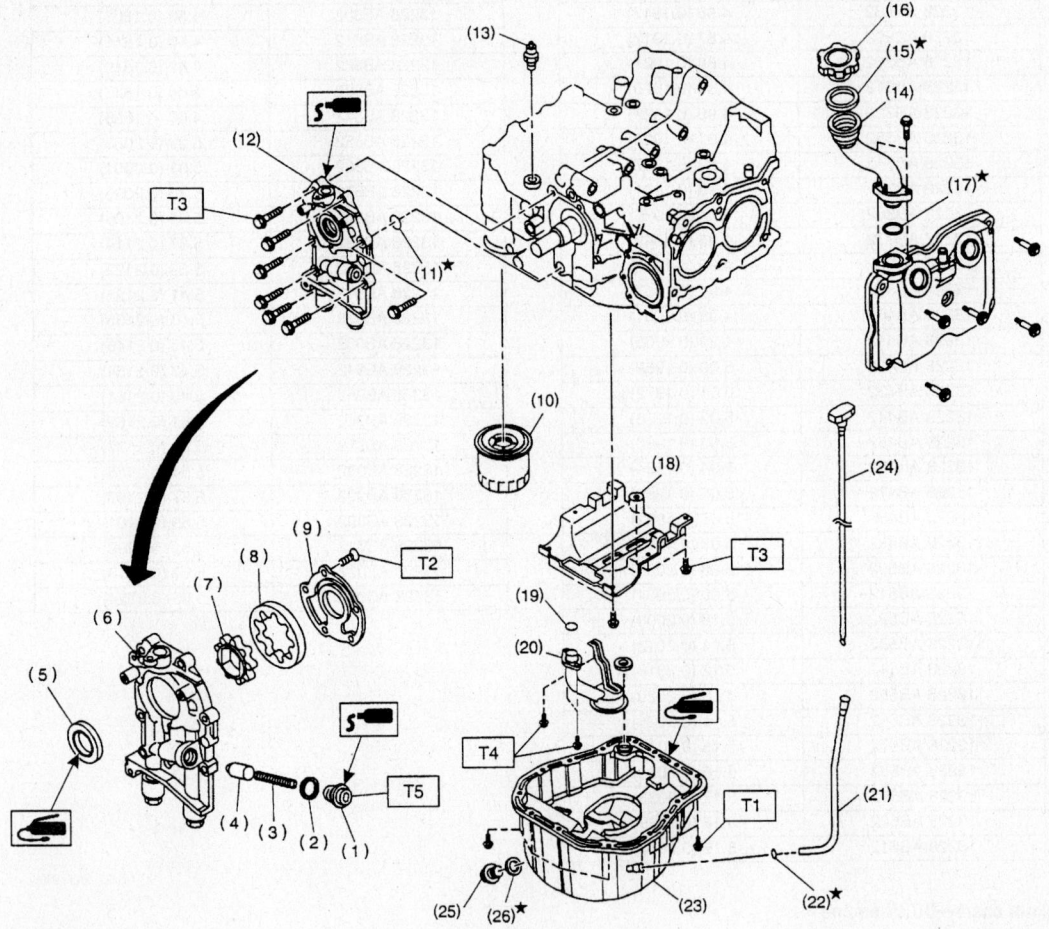

1) Plug	(13) Oil pressure switch	(25) Drain plug
2) Washer	(14) Oil filler duct	(26) Metal gasket
3) Relief valve spring	(15) O-ring	
4) Relief valve	(16) Oil filler cap	
5) Oil seal	(17) O-ring	
6) Oil pump case	(18) Baffle plate	
7) Inner rotor	(19) O-ring	
8) Outer rotor	(20) Oil strainer	
9) Oil pump cover	(21) Oil level gauge guide	
10) Oil filter	(22) O-ring	
11) O-ring	(23) Oil pan	
12) Oil pump ASSY	(24) Oil level gauge	

Tightening torque: N·m (kg-m, ft-lb)
T1: 5 (0.5, 3.6)
T2: 5^{+1}_{-0} $(0.5^{+0.1}_{-0}, 3.6^{+0.7}_{-0})$
T3: 6.4 (0.65, 4.7)
T4: 10 (1.0, 7.2)
T5: 44.1±3.4 (4.5±0.35, 32.5±2.5)

9307TG03

Oil pan and related components

rubber mallet and tap the oil pan to free it from the engine.

9. Clean all gasket material from both mating surfaces.

To install:

10. Apply a continuous bead of sealer to a new oil pan gasket.

11. Install the oil pan assembly. Tighten the bolts to specification and in alternating sequence.

12. Lower the engine onto the front crossmember.

13. Install or connect the following:
- Front engine mount nuts
- Y-pipe with new gaskets. Tighten the pipe-to-engine nuts to 23 ft. lbs. (30 Nm)
- Rear O₂S electrical connector
- Engine undercover
- Front wheel and tire assemblies

14. Remove the engine lifting device.
- Front O₂S sensor electrical connector
- Pitching stopper. Tighten the front bolt to 40 ft. lbs. (54 Nm) and the rear bolt to 49 ft. lbs. (67 Nm).
- Upper radiator brackets
- MAF sensor on turbocharged engine
- Air intake boot and air cleaner upper cover on turbocharged engine
- Intercooler on turbocharged engine
- Air intake duct
- Negative battery cable

15. Fill the engine to the proper level with the recommended oil and run the engine. Check for leaks.

2005–2006

SOHC ENGINE

1. Before servicing the vehicle, refer to the Precautions Section.

2. Disconnect the negative battery cable.

3. Raise and support the vehicle safely.

4. Remove the front tires and wheels.

5. Lower the vehicle.

6. Remove the air intake duct and the air cleaner case. Remove the air intake chamber.

7. Remove the pitching stopper.

8. Remove the hood stay holder and the radiator upper brackets.

9. Properly support the engine with a lifting device and wire ropes.

10. Lift the vehicle and support it safely.

➡**When lifting the vehicle, raise the wire ropes at the same time.**

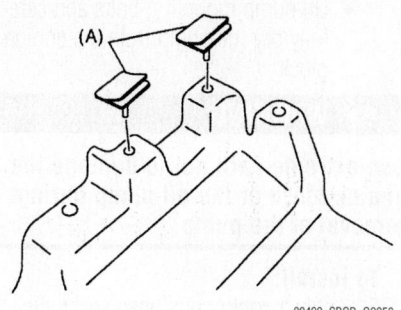

09490_SBCR_G0058

Oil pan baffle plate seal location and positioning—2005–2006

11. Remove the under cover.

12. Drain the engine oil.

13. Remove the front and center exhaust pipes.

14. Remove the nuts which retain the front cushion rubber onto the front crossmember.

15. Remove the bolts that retain the oil pan to the cylinder block, with the engine in the raised position.

16. Insert an oil pan gasket cutter tool into the gap between the cylinder block and the oil pan. Remove the oil pan from the engine.

➡**Do not use a screwdriver or similar tool in place of the cutter tool.**

17. Remove the oil strainer, if required. Remove the baffle plate, if required.

To install:

18. Be sure to clean the old gasketing material from the mating surfaces.

19. Apply a continuous bead of sealer to a new oil pan gasket.

20. Make sure that the seals (A) are installed securely on the baffle plate and in the direction shown in the illustration. Install the baffle plate; tighten the retaining bolts to 4.7 ft. lbs.

21. Replace the O-ring and install the oil strainer. Tighten the bolt to 7.2 ft. lbs.

22. Apply liquid gasket, part number 004403012 or equivalent, to the oil pan mating surface. Install the oil pan. Torque the retaining bolts to specification.

23. Continue the installation in the reverse order of the removal procedure.

24. Torque the front cushion mounting bolts to 63 ft. lbs.

25. Be sure to fill the engine with the correct grade and type engine oil.

26. Start the engine and check for leaks. Correct as required.

DOHC ENGINE

1. Before servicing the vehicle, refer to the Precautions Section.

2. Disconnect the negative battery cable.

3. Raise and support the vehicle safely.

4. Remove the front tires and wheels.

5. Remove the collector cover.

6. Lower the vehicle.

7. Disconnect the connector from the MAF sensor.

8. Remove the air intake boot and air cleaner upper cover.

9. Remove the intercooler.

10. Remove the pitching stopper.

11. Remove the radiator upper brackets.

12. Properly support the engine with a lifting device and wire ropes.

13. Lift the vehicle and support it safely.

➡**When lifting the vehicle, raise the wire ropes at the same time.**

14. Remove the under cover.

15. Drain the engine oil.

16. Remove the front exhaust pipe.

17. Remove the nuts which retain the front cushion rubber onto the front crossmember.

18. Remove the bolts that retain the oil pan to the cylinder block, with the engine in the raised position.

19. Insert an oil pan gasket cutter tool into the gap between the cylinder block and the oil pan. Remove the oil pan from the engine.

➡**Do not use a screwdriver or similar tool in place of the cutter tool.**

20. Remove the oil strainer, if required. Remove the baffle plate, if required.

To install:

21. Be sure to clean the old gasketing material from the mating surfaces.

22. Apply a continuous bead of sealer to a new oil pan gasket.

23. Make sure that the seals (A) are installed securely on the baffle plate and in the direction shown in the illustration. Install the baffle plate; tighten the retaining bolts to 4.7 ft. lbs.

24. Replace the O-ring and install the oil strainer. Tighten the bolt to 7.2 ft. lbs.

25. Apply liquid gasket, part number 004403012 or equivalent, to the oil pan mating surface. Install the oil pan. Torque the retaining bolts to specification.

26. Continue the installation in the reverse order of the removal procedure.

27. Torque the front cushion mounting bolts to 61 ft. lbs.

28. Be sure to fill the engine with the correct grade and type engine oil.

29. Start the engine and check for leaks. Correct as required.

Oil Pump

REMOVAL & INSTALLATION

2002–2004

1. Before servicing the vehicle, refer to the Precautions Section.
2. Drain the cooling system.
3. Drain the engine oil into a separate container.
4. Remove or disconnect the following:
   - Negative battery cable
   - Engine undercover
   - Water pipe and hose between oil cooler and water pump
   - Radiator
   - Crankcase Position (CKP) Sensor
   - Camshaft Position (CMP) Sensor
   - Belt(s) and rear tensioner
   - Crankshaft pulley
   - Water pump
   - Timing belt, and belt guide, if equipped
   - Crankshaft sprocket

- Oil pump mounting bolts and carefully pry the pump from the engine block

✵✵ WARNING

Use extreme care not to damage the engine block or the oil pump during removal of the pump.

To install:

5. Apply a continuous bead sealant to the mating surfaces of the oil pump.
6. Install or connect the following:
   - New front seal to the oil pump coat the inside of the seal with engine oil
   - New O-ring to the oil pump
   - Oil pump. Torque the bolts to 56 inch lbs. (6.4 Nm).
   - Crankshaft sprocket
   - Timing belt, and belt guide, if equipped
   - Water pump
   - Crankshaft pulley
   - Belt(s) and rear tensioner

- CMP Sensor
- CKP Sensor
- Radiator
- Engine coolant pipe
- Engine undercover
- Negative battery cable

7. Refill the cooling system.
8. Refill the engine to the proper level with the recommended oil.

2005–2006

1. Before servicing the vehicle, refer to the Precautions Section.
2. Disconnect the negative battery cable. Drain the cooling system.
3. On the DOHC engine, remove the collector cover.
4. Raise and support the vehicle safely.
5. Remove the under cover.
6. On the DOHC engine, remove the bolts which retain the water pipe of the oil cooler to the oil pump. Remove the water pipe and hoses between the oil cooler and the water pump.
7. Lower the vehicle. Remove the radiator.
8. To remove the front side V belt, remove the belt covers. Loosen the lock bolt. Loosen the slider bolt. Remove the front side belt.
9. To remove the rear side V belt, remove the belt covers. Loosen the lock bolt. Loosen the slider bolt. Remove the rear side belt.
10. On the SOHC engine remove the belt tensioner.
11. On the DOHC engine remove the rear side V belt tensioner
12. Remove the crankshaft position sensor.
13. Remove the crankshaft pulley bolt.
14. Lock the crankshaft in place using tool ST499977100 for the SOHC engine and tool ST499207400 for the DOHC engine, or equivalent.
15. Remove the crankshaft pulley.
16. Remove the water pump.
17. If equipped, remove the timing belt guide. Remove the crankshaft sprocket.
18. Remove the oil pump retaining bolts.

➡**When disassembling and checking the oil pump, loosen the relief valve plug before removing the oil pump from its mounting.**

19. Using a flat tip tool remove the oil pump from the engine.

To install:

20. Be sure all mating surfaces are clean and free of dirt.
21. Apply liquid gasket part number

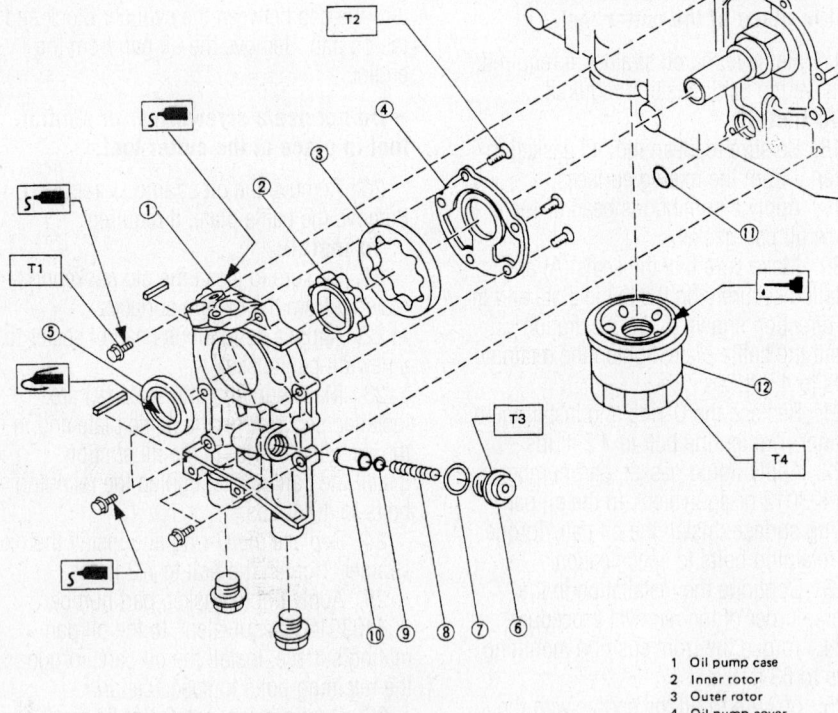

Tightening torque: N·m (kg-m, ft-lb)
T1: 6 – 7 (0.6 – 0.7, 4.3 – 5.1)
T2: 4 – 7 (0.4 – 0.7, 2.9 – 5.1)
T3: 40 – 48 (4.1 – 4.9, 30 – 35)
T4: 12 – 16 (1.2 – 1.6, 9 – 12)

1 Oil pump case
2 Inner rotor
3 Outer rotor
4 Oil pump cover
5 Front oil seal
6 Plug
7 Washer
8 Relief spring
9 Washer
10 Relief valve
11 O-ring
12 Oil filter

7923TG37

Oil pump and related components

004403007, or equivalent to the mating surfaces of the oil pump.

22. Be sure to replace the O-ring with a new one.

23. Apply a thin coat of clean engine oil to the inside of the oil seal.

24. Position the oil pump to its mounting, aligning the notched area with the crankshaft and push the pump straight.

➡**Be sure that the oil seal lip is not folded.**

25. Install the oil pump. Apply liquid gasket part number 004403042, or equivalent to the three retaining bolt threads. Install the bolts and tighten to 4.7 ft. lbs.

26. Continue the installation in the reverse order of the removal procedure.

27. Be sure to fill the cooling system with the proper grade and type coolant.

28. Start the engine and check for leaks. Correct, as required.

Rear Main Seal

REMOVAL & INSTALLATION

1. Before servicing the vehicle, refer to the Precautions Section.

2. Remove or disconnect the following:
- Engine from the vehicle
- Clutch assembly/flywheel, if equipped with manual transmission
- Torque converter flexplate from the crankshaft, if equipped with an automatic transmission

3. Using a seal removal tool, pry the oil seal from the housing.

To install:

4. Utilizing the appropriate seal installer, install or connect the following:
- New oil seal and press it into the housing using oil seal guide ST 499597100 and installer ST 499587200.
- Clutch assembly/flywheel, if equipped

- Flywheel/flexplate and tighten the bolts to 53 ft. lbs. (72 Nm), if equipped
- Engine into the vehicle

Timing Belt, Cover and Crankshaft seal

REMOVAL & INSTALLATION

SOHC Engine

2002–2004

1. Before servicing the vehicle, refer to the Precautions Section.

When servicing the timing belt, note the following:

a. The intake and exhaust camshafts can be rotated independently when the timing belt is removed. If the intake and exhaust valves are lifted off of their seats simultaneously, their heads will contact each other, possibly causing damage.

b. When the timing belt is removed, the camshafts are positioned so that none of the valves are lifted off of their seats, resulting in a "zero-lift" position.

c. The left-hand cylinder head camshaft must be rotated from the "zero-

Remove the timing belt guide on vehicles equipped with manual transmission—2002–2004 SOHC engine

lift" position as little as possible when orienting it for timing belt installation, otherwise possible valve head interference may occur.

d. Never allow the camshafts to rotate in the direction shown in the accompanying illustration, which would cause both the intake and exhaust valves to lift simultaneously, causing interference.

2. Remove all necessary components to gain access to the timing belt.

3. If equipped with manual transmissions, loosen the 2 timing belt guide mounting bolts, then separate the guide from the engine block.

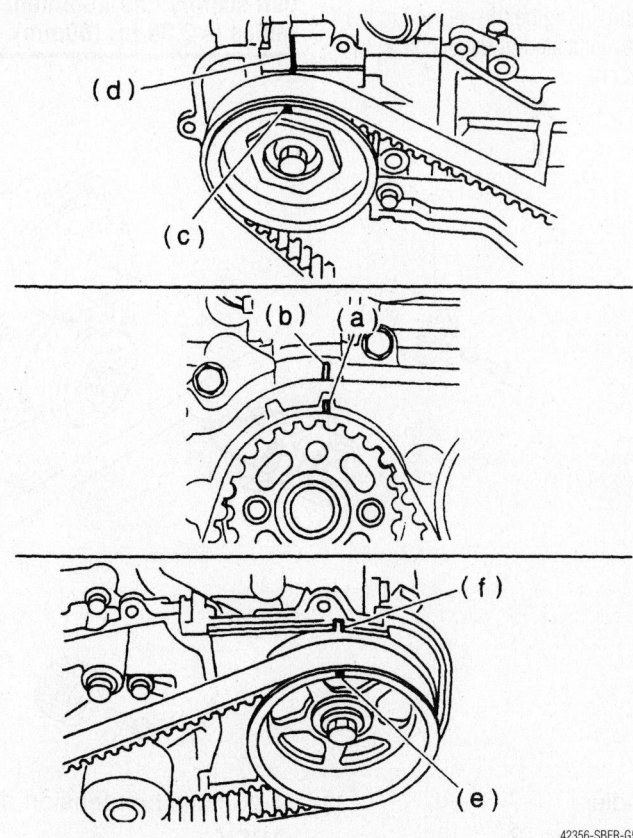

Before removing the timing belt, turn the crankshaft sprocket until all of the alignment marks are aligned as indicated—2002–2004 SOHC engine

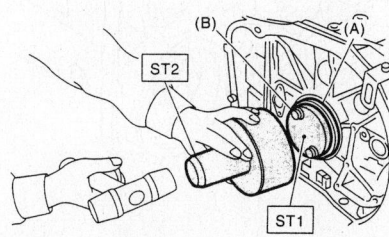

(A) Rear oil seal
(B) Flywheel attaching bolt

67170-SBFR-G27

Installing the rear main seal

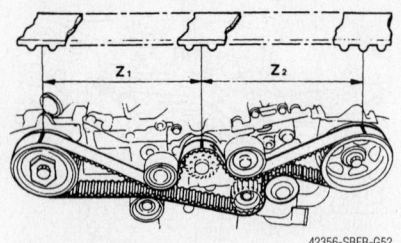

42356-SBFR-G52

If the original marks on the timing belt are worn or faded, make new alignment marks in the positions indicated—2002–2004 SOHC engine

4. If the directional arrow and alignment marks on the timing belt are faded, and the belt is to be reused, remark the belt with white paint or a grease pencil as follows:

a. Using a Subaru tool No. ST-499987500 Crankshaft Socket, or equivalent, installed on the crankshaft sprocket, rotate the crankshaft until the crankshaft sprocket, left-hand exhaust camshaft sprocket, left-hand intake camshaft sprocket, right-hand intake camshaft sprocket and right-hand exhaust camshaft sprocket timing mark notches are aligned with the respective marks on the belt cover and engine block.

b. Make alignment and/or arrow marks on the timing belt in relation to the sprockets as indicated in the accompanying illustration.

- Z1: 46.8 tooth length
- Z2: 43.7 tooth length

5. Loosen the center bolt from the timing belt idler pulley, then remove the idler pulley from the engine block.

✳✳ WARNING

After removing the timing belt, DO NOT rotate the camshafts. Damage to the valves may occur.

6. Carefully remove the timing belt from all of the sprockets.

7. Remove the automatic belt tension adjuster assembly as follows:

a. Remove the 2 timing belt idler pulleys, as indicated in the accompanying illustration.

b. Loosen the automatic tension adjuster assembly mounting bolts, then separate the adjuster assembly from the engine block.

To install:

✳✳ WARNING

Do not allow oil, grease, or coolant to come in contact with the timing belt. If this occurs, quickly and thoroughly remove all traces of the compound. Also, never bend the timing belt sharply; the minimum bending radius is 2.36 in. (60mm).

8. Inspect the camshaft and crankshaft sprocket teeth for abnormal or excessive wear or scratches. Ensure there is no freeplay between the sprocket and the key. Inspect the crankshaft sprocket sensor notch for damage or contamination with debris or dirt.

➡**When preparing the automatic tension adjuster assembly for installation, adhere to the following points:**

- Always use a vertical press, rather than a horizontal press or vise, to depress the adjuster assembly rod
- Depress the adjuster rod in a vertical position ONLY
- Depress the adjuster rod slowly (taking more than 3 minutes) with a force of 66 lbs. (30 kg)
- Do not allow the press force to exceed 2205 lbs. (1000 kg)
- Press the adjuster rod in as far as the end surface of the cylinder — do not press the rod into the cylinder, which may cause damage to the assembly
- Do not release the press force from the rod until the stopper pin is completely inserted in the cylinder

9. Prepare the automatic timing belt tension adjuster assembly for installation as follows:

a. Position the adjuster assembly in a vertical press.

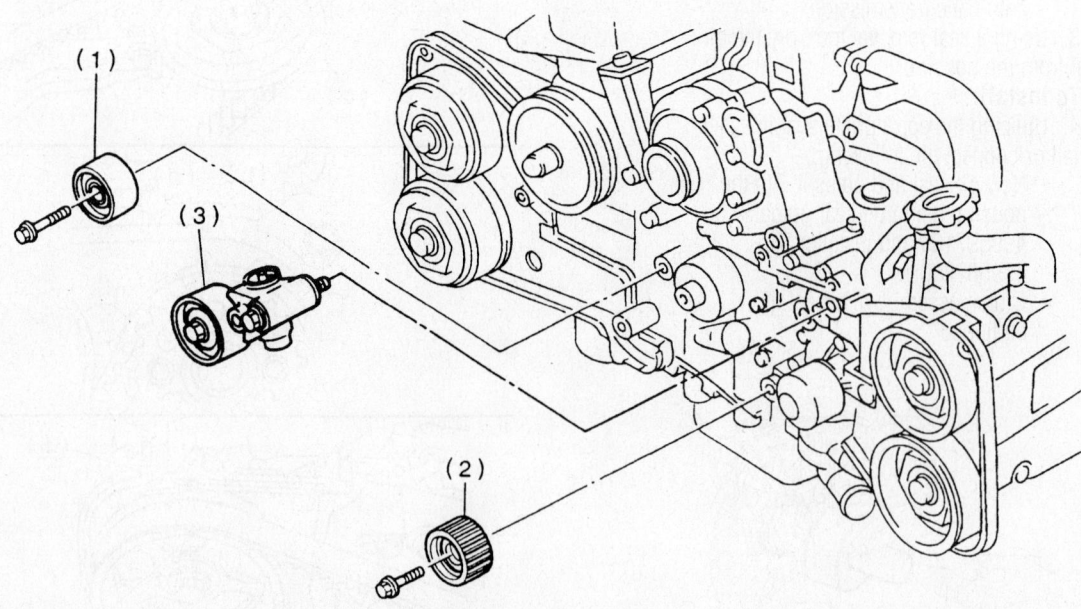

(1) Belt idler
(2) Belt idler No. 2
(3) Automatic belt tension adjuster ASSY

79245G52

It is necessary to remove the automatic adjuster assembly and reset the pushrod for timing belt installation—2002–2004 SOHC engine

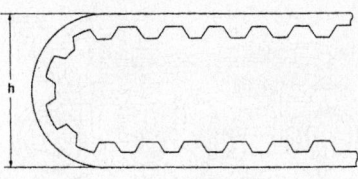

Never bend the timing belt into a radius tighter than 2.36 in./60mm (h), otherwise it will be damaged beyond—2002–2004 SOHC engine

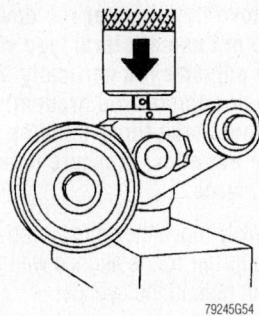

Use a vertical press to push the adjuster rod into its housing until it is flush with the assembly's outer surface . . . —2002–2004 SOHC engine

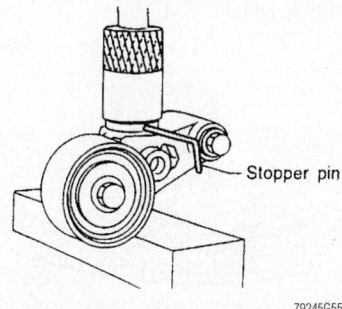

. . . then insert a 0.08 in. (2mm) diameter pin or Allen wrench into the housing and rod holes to hold it in position—2002–2004 SOHC engine

b. Slowly depress the adjuster rod with a force of 66 lbs. (30 kg) until the hole in the rod is aligned with the hole in the adjuster cylinder housing.

c. Insert a 0.08 in. (2mm) diameter stopper pin or Allen wrench through the hole in the cylinder housing and rod, then slowly release the press force from the adjuster rod.

10. Install the adjuster assembly onto the engine block.

11. Install timing belt idler pulley No. 2 on the engine block. Tighten the bolts to 28 ft. lbs. (39 Nm).

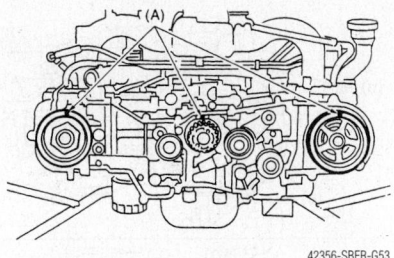

Make sure the sprockets are aligned as shown—2002–2004 SOHC engine

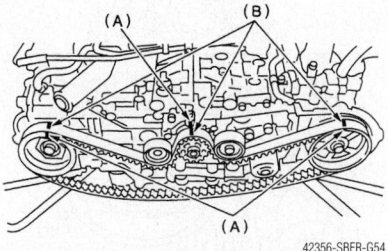

After installing the timing belt, the alignment marks should be positioned as shown. If not, remove the belt and align the sprockets and reinstall the belt—2002–2004 SOHC engine

12. Install the timing belt idler pulley No. 1 on the engine block. Tighten the bolts to 28 ft. lbs. (39 Nm).

13. If the camshaft and crankshaft timing marks are no longer aligned, perform the following:

a. Position the crankshaft sprocket so that its mark is aligned with the mark on the oil pump cover on the engine block.

b. Align the single line mark on the right-hand exhaust camshaft sprocket with the notch on the belt cover.

c. Rotate the right-hand intake camshaft so that the single line mark is aligned with the notch on the belt cover.

→At this point, the double line marks on both right-hand camshaft sprockets should be aligned.

d. Turn the left-hand exhaust (lower) camshaft counterclockwise (as viewed from the front of the engine) until the single line mark is aligned with the notch on the belt cover.

e. Position the single line mark on the left-hand intake camshaft sprocket so that it is aligned with the notch on the belt cover. When rotating the camshaft, do so only in a clockwise direction (as viewed from the front of the engine).

→At this point, the double line marks on both left-hand camshaft sprockets should be aligned.

f. Ensure the timing marks are aligned as shown in the accompanying illustration. If they are not, repeat Substeps 12a through 12e until they are properly aligned.

14. Install the timing belt around the camshaft, crankshaft and idler pulleys so that the positioning marks on the timing belt are aligned with the marks on the sprockets as follows:

a. Position the timing belt on the crankshaft sprocket so that the marks are aligned.

b. Route the belt down and under the left-hand, upper idler pulley, then up and around the left-hand intake camshaft sprocket, ensuring the camshaft sprocket mark is aligned with the mark on the belt.

c. Route the belt down and around the left-hand exhaust camshaft sprocket, making sure the marks are properly aligned, then up and over the first lower idler pulley and down and around the second lower idler pulley.

d. While holding the timing belt on the inner, left-hand, lower idler pulley, route the other side of the timing belt (from the crankshaft sprocket) down and under the right-hand upper idler pulley.

e. Route the timing belt up and around the right-hand intake camshaft sprocket so that the belt and sprocket marks are aligned.

f. Position the belt down and around the right-hand exhaust camshaft sprocket, ensuring the positioning marks are aligned.

15. Install the right-hand lower idler pulley so that the timing belt is routed over the top side of it.

→Once the belt is completely installed on all of the pulleys and sprockets, ensure that the positioning marks are still all aligned.

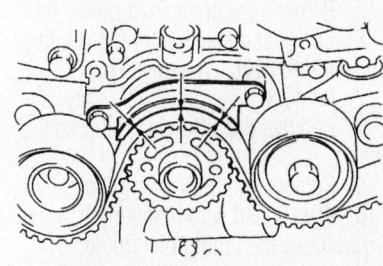

On vehicles equipped with manual transmissions, ensure the timing belt-to-guide clearance (arrows) is correct before tightening the mounting bolts—2002–2004 SOHC engine

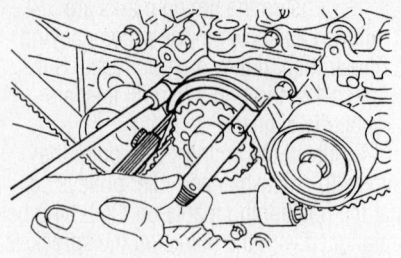

42356-SBFR-G55

On vehicles equipped with manual transmission, use a feeler gauges to adjust the clearance between the timing belt and the belt guide—2002–2004 SOHC engine

16. After ensuring all of the marks are still aligned, use a pair of pliers to withdraw the stopper pin or Allen wrench from the adjuster assembly housing.

17. On vehicles with manual transmissions, perform the following:

    a. Install the timing belt guide by temporarily tightening the mounting bolts.

    b. Position the timing belt guide so that there is 0.019–0.059 in. (0.5–1.5mm) clearance between the timing belt and the belt guide.

    c. Tighten the guide mounting bolts securely, then double check the guide clearance.

18. Install the timing belt covers and all remaining engine components.

2005–2006

1. Before servicing the vehicle, refer to the Precautions Section.

2. Disconnect the negative battery cable.

3. To remove the front side V belt, remove the belt covers. Loosen the lock bolt. Loosen the slider bolt. Remove the front side belt.

4. To remove the rear side V belt, remove the belt covers. Loosen the lock bolt. Loosen the slider bolt. Remove the rear side belt. Remove the belt tensioner.

5. Remove the crankshaft pulley bolt.

6. Lock the crankshaft in place using tool ST499977100, or equivalent.

7. Remove the crankshaft pulley.

8. Remove the left side timing belt cover.

9. Remove the front timing belt cover.

10. If equipped with manual transmission, remove the timing belt guide.

➡**If the belt is going to be reused and the alignment mark on the belt is not readable, put a new mark on the belt to indicate the direction of rotation. Using tool ST499987500, turn the crankshaft to**

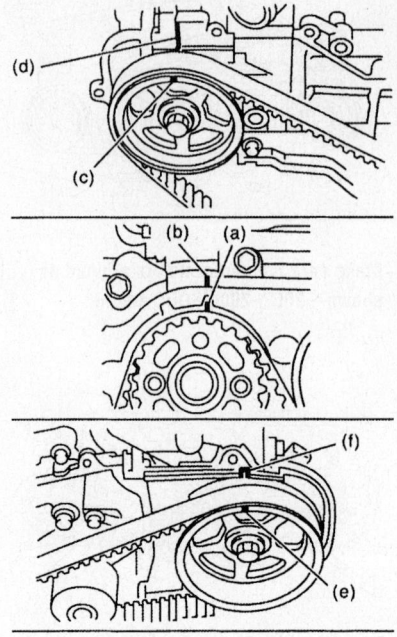

09490_SBCR_G0060

Timing belt alignment—2005–2006 SOHC engine

align the mark of the sprocket "A" to the cylinder mark notch "B". Ensure that the right side cam sprocket mark "C", cam cap and cylinder head matching surface "D" or left side cam sprocket mark "E", timing belt cover notch "F" are properly aligned. Paint an alignment mark on the belt in relation to the crankshaft sprocket and camshaft sprockets. Z1 measurement is 46.8 teeth. Z2 measurement is 43.7 teeth.

11. Remove both the number two belt idlers. Remove the timing belt from the engine.

12. Remove the number one belt idler. Remove the automatic belt tension adjuster assembly.

To install:

13. Attach the automatic belt tension adjuster assembly to a vertical pressing tool.

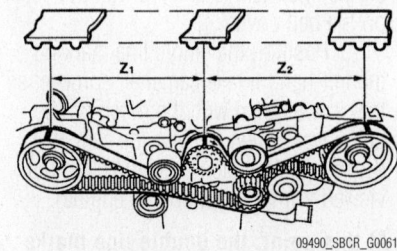

09490_SBCR_G0061

Timing belt Z1 and Z2 teeth measurement—2005–2006 SOHC engine

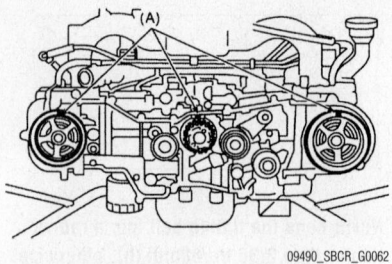

09490_SBCR_G0062

Timing mark alignment position A—2005–2006 SOHC engine

➡**Always use a vertical type pressing tool to move the adjuster rod downward. Do not use a lateral type vise. Push the adjuster rod vertically. Press in the push adjuster rod gradually, which should take three minutes or more. Do not allow pressure to exceed 2,205 lb. force.**

14. Slowly move the adjuster rod down until the adjuster rod is aligned with the stopper pin hole in the cylinder.

➡**Press the adjuster rod as far as the end surface of the cylinder. Do not press the adjuster rod into the cylinder. Doing so may damage the cylinder.**

15. Using a 0.08 inch stopper pin, insert it into the stopper pin hole in the cylinder. Secure the adjuster rod.

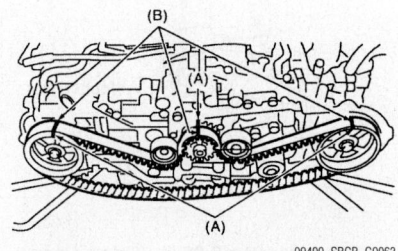

09490_SBCR_G0063

Timing mark alignment position B—2005–2006 SOHC engine

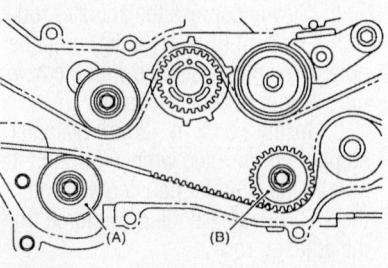

09490_SBCR_G0064

(A) Belt idler (No. 2)
(B) Belt idler No. 2

Belt idler number two locations—2005–2006 SOHC engine

→**Do not release the press pressure
until the stopper pin is completely
inserted in the hole.**

16. Install the automatic belt tensioner
assembly. Tighten the retaining bolt to 28.9
ft. lbs.

17. Install the belt idler number one.
Tighten the retaining bolt to 28.9 ft. lbs.

18. Turn the number one and number
two camshaft sprockets, using tool
ST499207100 or tool ST18231AA010 and
position the alignment marks "A" on each at
the highest position.

19. While aligning the alignment mark
"B" on the timing belt with mark "A" on the
sprockets, position the timing belt properly.

20. Install both belt idler number two's.
Tighten the retaining bolt to 28.9 ft. lbs.

21. After checking to be sure the marks
on the timing belt and the camshaft sprockets are aligned remove the stopper pin from
the belt tension adjuster.

22. Install the timing belt guide, if
equipped with manual transmission. Temporarily tighten the bolts. Check and adjust
the clearance between the belt and the
guide. It should be 0.039 +/- 0.020 inch.
Tighten the bolts to 7.2 ft. lbs.

23. Continue the installation in the
reverse order of the removal procedure.

DOHC Engine

1. Before servicing the vehicle, refer to
the Precautions Section.

2. Disconnect the negative battery
cable.

3. Remove the collector cover.

4. To remove the front side V belt,
remove the belt covers. Loosen the lock
bolt. Loosen the slider bolt. Remove the
front side belt.

5. To remove the rear side V belt,
remove the belt covers. Loosen the lock
bolt. Loosen the slider bolt. Remove the rear
side belt. Remove the belt tensioner.

6. Remove the crankshaft pulley bolt.

7. Lock the crankshaft in place using
tool ST499977100, or equivalent.

8. Remove the crankshaft pulley.

9. Remove the left side timing belt
cover.

10. Remove the right side timing belt
cover.

11. Remove the front timing belt cover.

12. Remove the timing belt guide, if
equipped.

→**If the belt is going to be reused and
the alignment mark on the belt is not
readable, put a new mark on the belt to
indicate the direction of rotation. Using**

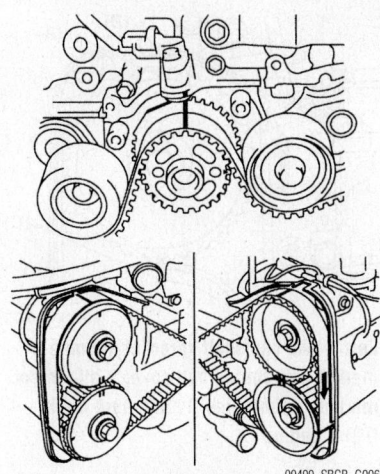

Timing belt alignment—DOHC engine

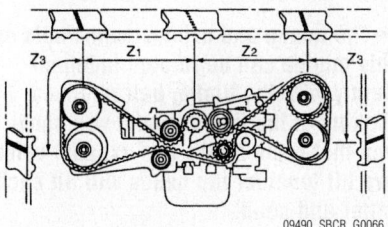

Timing belt Z1, Z2 and Z3 teeth measurement—DOHC engine

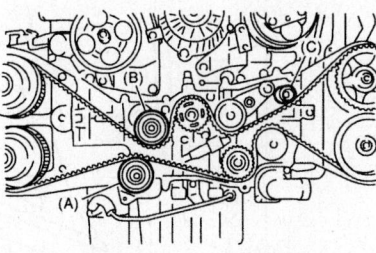

Belt idler identification and location—
DOHC engine

tool ST499987500, turn the crankshaft
to align the mark on the crankshaft
sprocket, intake camshaft sprocket
(left), exhaust camshaft sprocket (left),
intake camshaft sprocket (right),
exhaust camshaft sprocket (right) with
the notches of the timing belt cover and
cylinder block. Paint an alignment
mark on the belts in relation to the
camshaft sprockets. Z1 measurement
is 54.4 teeth. Z2 measurement is 51.0
teeth and Z3 measurement is 28.0
teeth.

13. Remove the belt idler belt "A".
Remove the timing belt.

14. Remove the belt idlers belt "B" and
"C".

15. Remove the belt idler number two.
Remove the automatic belt tension adjuster
assembly.

To install:

16. Attach the automatic belt tension
adjuster assembly to a vertical pressing
tool.

→**Always use a vertical type pressing
tool to move the adjuster rod downward. Do not use a lateral type vise.
Push the adjuster rod vertically. Press
in the push adjuster rod gradually,
which should take three minutes or
more. Do not allow pressure to exceed
2,205 lb. force.**

17. Slowly move the adjuster rod down
until the adjuster rod is aligned with the
stopper pin hole in the cylinder.

→**Press the adjuster rod as far as the
end surface of the cylinder. Do not
press the adjuster rod into the cylinder. Doing so may damage the cylinder.**

18. Using a 0.08 inch stopper pin, insert
it into the stopper pin hole in the cylinder.
Secure the adjuster rod.

→**Do not release the press pressure
until the stopper pin is completely
inserted in the hole.**

19. Install the automatic belt tensioner
assembly. Tighten the retaining bolt to 28.9
ft. lbs.

20. Install the belt idler number two.
Tighten the retaining bolt to 28.9 ft. lbs.

21. Install the belt idlers. Tighten the
retaining bolts to 28.9 ft. lbs.

22. Align the mark "A" on the crankshaft
sprocket with the mark on the oil pump at
the cylinder block.

23. Align the single line mark "A" on the
right exhaust camshaft sprocket with the
notch "B" on the timing belt cover.

24. Align single line mark "A" on the
right intake camshaft sprocket with the
notch "B" on the timing cover. Ensure that
the double lines "C" on the intake and
exhaust camshaft sprockets are aligned.

25. Align single line mark "A" on the left
exhaust camshaft sprocket with the notch
"B" on the timing cover by turning the
sprocket counterclockwise as viewed from
the front of the engine.

26. Align single line mark "A" on the left
intake camshaft sprocket with the notch "B"
on the timing cover, by turning the sprocket
clockwise as viewed from the front of the
engine. Ensure that the double lines "C" on
the intake and exhaust camshaft sprockets
are aligned.

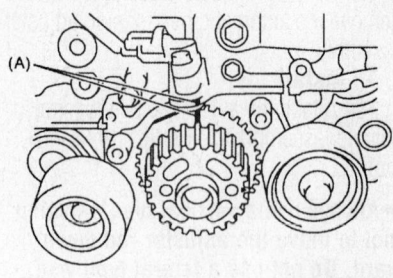

Crankshaft sprocket mark A to oil pump cover alignment—DOHC engine

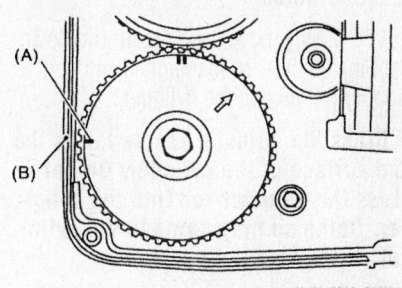

Right exhaust camshaft sprocket alignment mark A with timing belt cover B alignment mark—DOHC engine

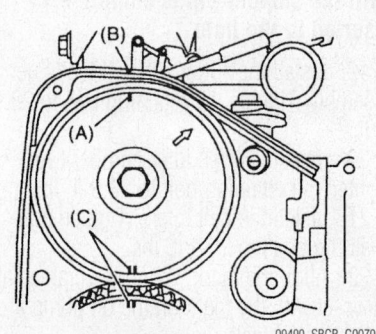

Right intake camshaft sprocket alignment mark A with timing belt cover B alignment mark an double line C alignment mark—DOHC engine

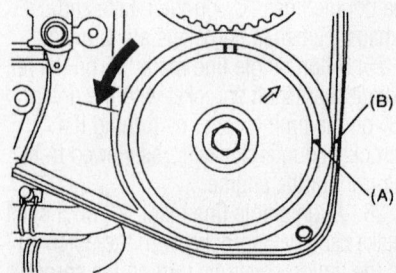

Left exhaust camshaft sprocket alignment mark A with timing belt cover B alignment mark—2005–DOHC engine

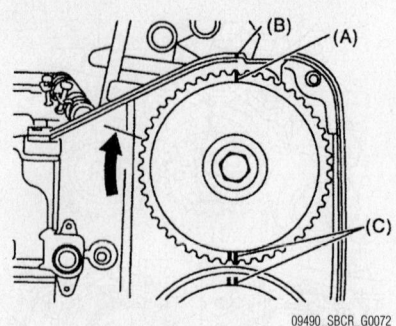

Left intake camshaft sprocket alignment mark A with timing belt cover B alignment mark an double line C alignment mark—DOHC engine

27. Make sure that the camshaft and crankshaft sprockets are positioned properly.

➡The intake and exhaust camshafts on this engine can be rotated independently with the timing belt removed. By looking at the illustration it will show you that if the intake and exhaust valve are lift together the heads will hit each other and bend.

➡When the timing belts are not installed, 4 camshafts are held at "zero lift" position, where all cams on the camshafts do not push the intake and exhaust valves down (under this condition all valves remain unlifted). When the camshafts are rotated to install the timing belts, # 2 intake and # 4 exhaust cam of the left hand camshafts are held to push their corresponding valves down. Under this condition these valves are held lifted. The right side camshafts are held in so that their cams do not push the valves down. The left hand camshafts must be rotated from the "zero lift" position to the position where the timing belt is to be installed at as small an angle as possible, in order to prevent mutual interference of intake and exhaust valve heads. Do not allow the camshafts to rotate in the direction illustrated as this causes both the intake and exhaust valves to lift off at the same time with will cause valve damage.

28. When installing the belt, make sure to align the marks made during removal or

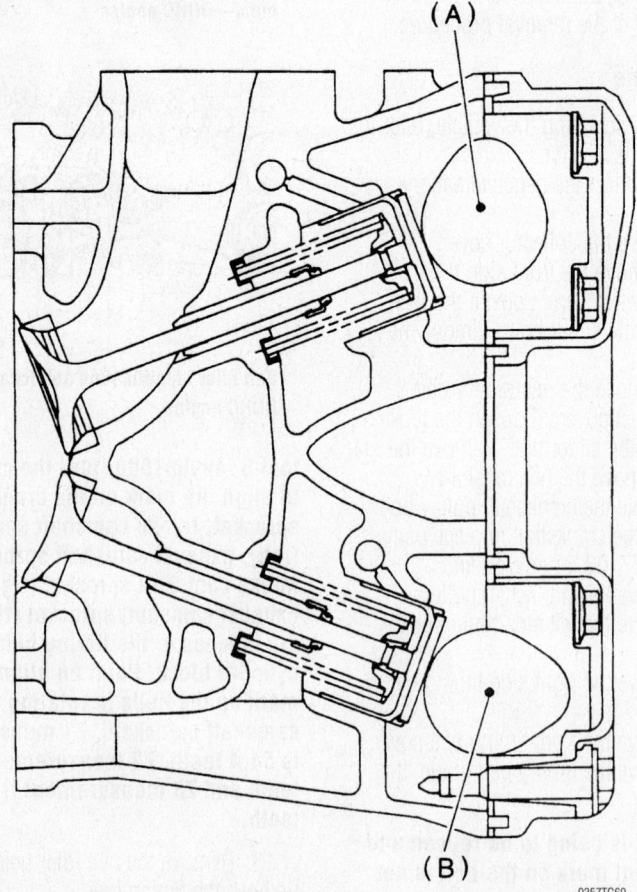

If the intake and exhaust valve are lifted together the heads will hit each other and bend — DOHC engine

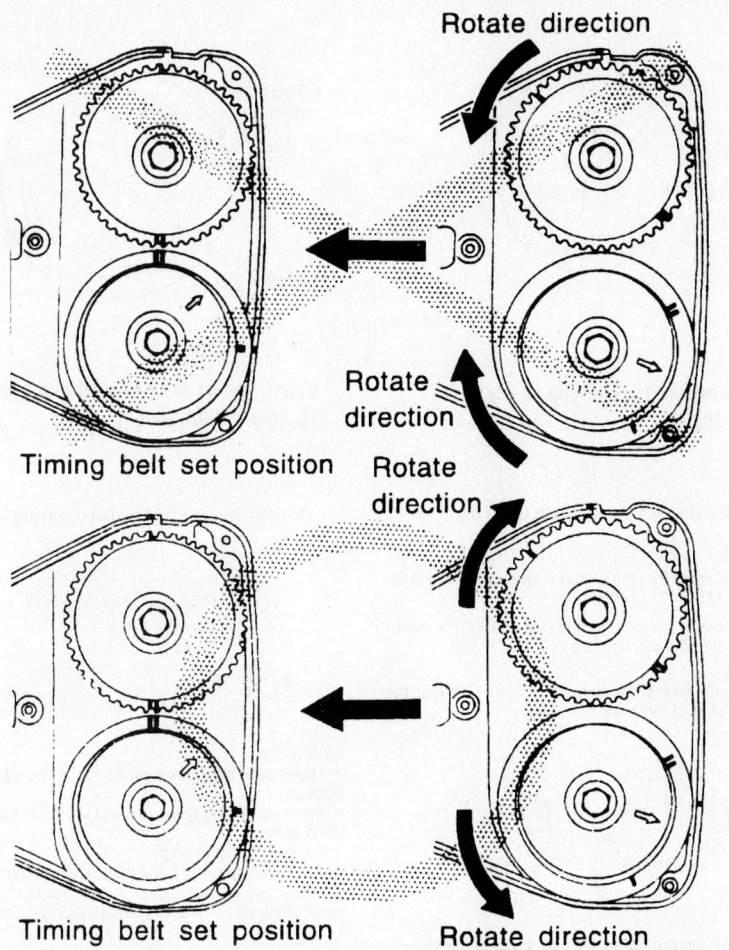

Rotate direction

Timing belt set position

Rotate direction

Rotate direction

Timing belt set position

Rotate direction

9357TG61

Do not allow the camshafts to rotate in the direction shown as this causes both the intake and exhaust valves to lift off at the same time with will cause valve damage—DOHC engine

if using a new belt, align the in alphabetical order as shown in the illustration.

✳✳ WARNING

Disengagement of more than 3 timing belt teeth may result in contact between the valve and piston. Always make sure the belts rotation is correct.

29. Install the timing belt.

➡ Align the alignment mark on the timing belt with marks on the sprocket in the order shown in the illustration.

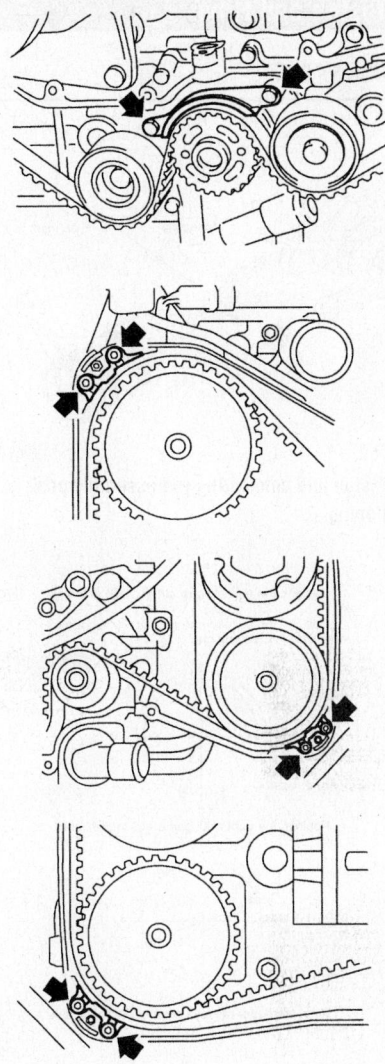

09490_SBCR_G0074

Timing belt guide bolt location and tightening specification—DOHC engine

While aligning the timing marks, position the timing belt properly.

30. Install the belt idlers. Tighten the retaining bolts to 28.9 ft. lbs.

➡ **Make sure that the marks on the timing belt and sprockets are aligned.**

31. After checking to be sure the marks on the timing belt and the camshaft sprockets are aligned remove the stopper pin from the belt tension adjuster.

32. Install the timing belt guide, if equipped with manual transmission. Temporarily tighten the bolts. Check and adjust the clearance between the belt and the guide. It should be 0.039 +/- 0.020 inch. Tighten the guide bolts to specification, see illustration.

33. Continue the installation in the reverse order of the removal procedure.

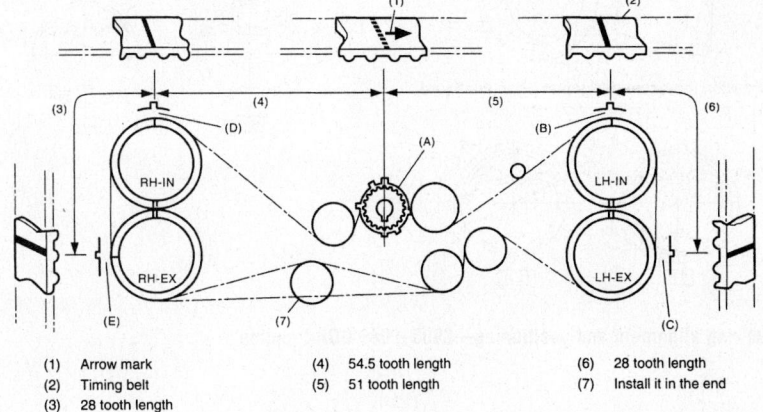

(1) Arrow mark
(2) Timing belt
(3) 28 tooth length
(4) 54.5 tooth length
(5) 51 tooth length
(6) 28 tooth length
(7) Install it in the end

RH-IN   RH-EX   LH-IN   LH-EX

(D) (A) (B) (E) (C)

09490_SBCR_G0073

Timing belt alignment and installation sequence—DOHC engine

Piston and Ring

POSITIONING

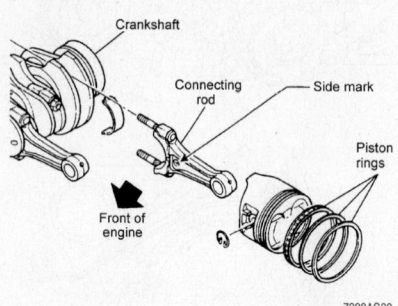

7923AG83

Piston and connecting rod assembly positioning

Position the top ring gap at (A) or (B) in the figure.
Position the second ring gap at 180° on the reverse side the top ring gap.

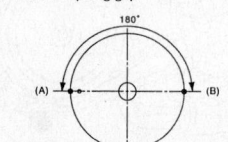

Position the upper rail gap at (C) in the figure.

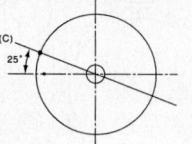

Align the upper rail spin stopper (D) to the side hole (E) on the piston.

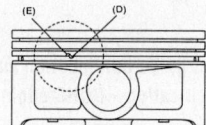

Position the expander gap at (F) in the figure on the 180° opposite direction of (C).

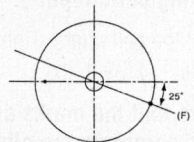

Position the lower rail gap at (G) in the figure.

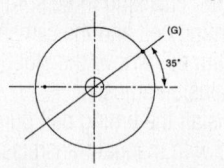

09490_SBCR_G0079

Piston ring alignment and positioning— 2005–2006 SOHC engine

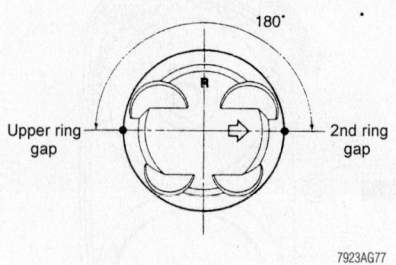

7923AG77

Compression ring end-gap spacing— 2002–2004

Position the top ring gap at (A) or (B) in the figure.

NOTE:
Assemble so that the piston ring mark "R" faces the upper side of the piston.

Position the second ring gap at 180° on the reverse side the top ring gap.

NOTE:
Assemble so that the piston ring mark "R" faces the upper side of the piston.

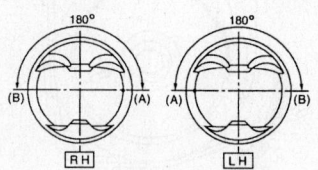

Position the upper rail gap at (C) in the figure.

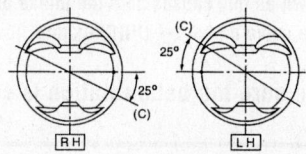

Align the upper rail spin stopper (E) to the side hole (D) on the piston.

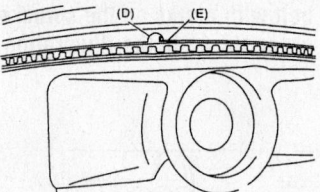

Position the expander gap at (F) in the figure.

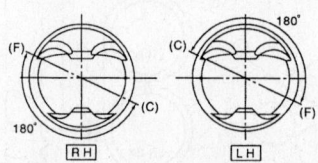

09490_SBCR_G0080

Piston ring alignment and positioning—2005–2006 DOHC engine

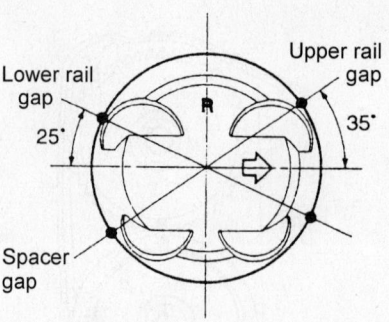

7923AG78

Upper, spacer and lower oil ring end-gap spacing—2002–2004

Position the lower rail gap at (G) in the figure.

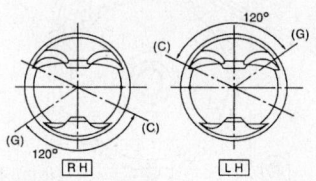

CAUTION:
• **Make sure ring gaps do not face the same direction.**
• **Make sure ring gaps are not within the piston skirt area.**
Install the snap ring.
Install the snap rings in the piston holes located opposite to the service holes in cylinder block when positioning all pistons in corresponding cylinders.

NOTE:
Use new snap rings.

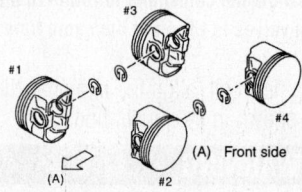

CAUTION:
Piston front mark faces towards the front of engine.

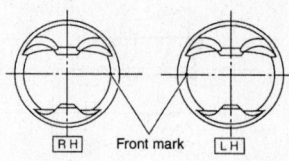

FUEL SYSTEM

Fuel System Service Precautions

Safety is the most important factor when performing not only fuel system maintenance but any type of maintenance. Failure to conduct maintenance and repairs in a safe manner may result in serious personal injury or death. Maintenance and testing of the vehicle's fuel system components can be accomplished safely and effectively by adhering to the following rules and guidelines.

• To avoid the possibility of fire and personal injury, always disconnect the negative battery cable unless the repair or test procedure requires that battery voltage be applied.

• Always relieve the fuel system pressure prior to disconnecting any fuel system component (injector, fuel rail, pressure regulator, etc.), fitting or fuel line connection. Exercise extreme caution whenever relieving fuel system pressure, to avoid exposing skin, face and eyes to fuel spray. Please be advised that fuel under pressure may penetrate the skin or any part of the body that it contacts.

• Always place a shop towel or cloth around the fitting or connection prior to loosening to absorb any excess fuel due to spillage. Ensure that all fuel spillage (should it occur) is quickly removed from engine surfaces. Ensure that all fuel soaked cloths or towels are deposited into a suitable waste container.

• Always keep a dry chemical (Class B) fire extinguisher near the work area.

• Do not allow fuel spray or fuel vapors to come into contact with a spark or open flame.

• Always use a backup wrench when loosening and tightening fuel line connection fittings. This will prevent unnecessary stress and torsion to fuel line piping. Always follow the proper torque specifications.

• Always replace worn fuel fitting O-rings with new. Do not substitute fuel hose or equivalent, where fuel pipe is installed.

Fuel System Pressure

RELIEVING

➡**This procedure must be performed prior to servicing any component of the fuel injection system.**

1. Before servicing the vehicle, refer to the Precautions Section.

2. Disconnect the connector from the fuel pump relay.

3. Crank the engine for 5 seconds or more to relieve the fuel pressure. If the engine starts during this time, allow it to run until it stalls.

4. Connect the fuel pump relay connector after repairs are completed.

Fuel Filter

REMOVAL & INSTALLATION

2002–2004

1. Before servicing the vehicle, refer to the Precautions Section.

2. Properly relieve the fuel system pressure.

3. Remove or disconnect the following:
• Negative battery cable
• Fuel delivery hoses from the fuel filter
• Fuel filter from its holder

To install:

4. Install or connect the following:
• Fuel filter into its mounting bracket
• Fuel delivery hoses and tighten the hose clamps
• Negative battery cable

2005–2006

1. Before servicing the vehicle, refer to the Precautions Section.

2. Properly relieve the fuel system pressure. Disconnect the negative battery cable.

3. Remove the fuel pump assembly.

4. Separate the fuel filter from the fuel pump. Turn the filter holder around and remove the fuel filter.

To install:

Fuel Pump

REMOVAL & INSTALLATION

2002–2004

1. Before servicing the vehicle, refer to the Precautions Section.

2. Properly relieve the fuel system pressure.

3. Drain the fuel tank by removing the drain plugs from the tank and draining into an approved container. Once the fuel has been drained, replace the plugs and tighten to 19 ft. lbs. (26 Nm).

4. Remove the rear seat cushion and access panel.

5. Disconnect the negative battery cable.

6. Clean any debris away from the fuel pump mounting to prevent it from entering the tank.
• Fuel pump electrical connector
• Fuel delivery and return hoses
• Fuel pump mounting nuts
• Fuel pump out of the fuel tank

To install:

7. Replace the sealing gaskets for the fuel pump.

8. Install or connect the following:
• Fuel pump into the tank. Torque the mounting nuts in sequence to 39 inch lbs. (4.4 Nm).
• Fuel delivery and return hoses
• Fuel pump electrical connector
• Fuel filler cap
• Negative battery cable

9. Start the vehicle and check for leaks.

10. Install the fuel pump access cover and rear seat cushion.

2005–2006

1. Before servicing the vehicle, refer to the Precautions Section.

2. Properly relieve the fuel system pressure.

3. Disconnect the negative battery cable.

4. Remove the fuel filler cap.

5. Raise and support the vehicle safely.

6. Drain the fuel tank.

7. Remove the luggage floor mat.

8. Remove the fuel pump access cover retaining screws. Remove the access cover.

9. Disconnect the electrical connector from the fuel pump.

10. Disconnect and plug the fuel line hoses at the fuel pump.

11. Remove the clips and disconnect the jet pump hose.

12. Remove the fuel pump assembly retaining nuts. Remove the fuel pump from the vehicle.

To install:

13. Installation is the reverse of the removal procedure.

14. Be sure to use a new gasket and retainer.

15. Tighten the fuel pump retaining nuts to 3.3 ft. lbs. in an alternating sequence pattern.

16. Continue the installation in the reverse order of the removal procedure.

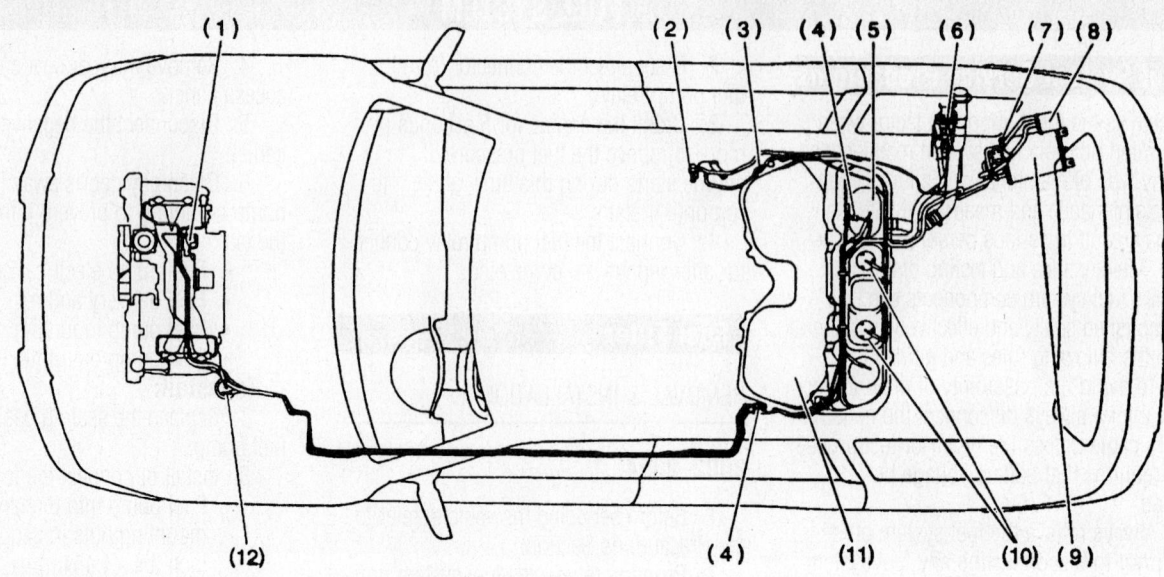

(1) Purge control solenoid valve
(2) Roll over valve
(3) Pressure control solenoid
(4) Quick connector

(5) Fuel pump
(6) Fuel tank pressure sensor
(7) Vent control solenoid valve
(8) Air filter

(9) Canister
(10) Fuel cut valve
(11) Fuel tank
(12) Fuel filter

7924XG12

Fuel system and related components—typical

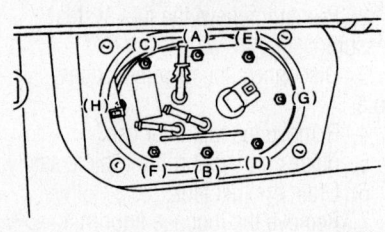

7924XG13

Fuel pump mounting bolt tightening sequence

17. Start the engine and check for leaks, correct as required.

Fuel Injector

REMOVAL & INSTALLATION

SOHC Engine

2002–2004

1. Before servicing the vehicle, refer to the Precautions Section.
2. Properly relieve the fuel system pressure.
3. Disconnect the negative battery cable.

4. To remove the right side injectors, remove or disconnect the following:
- Air cleaner ducts and resonator chamber, on California vehicles
- Mass Air Flow (MAF) sensor connector and air intake duct and air cleaner upper cover as a unit, on non-California vehicles
- Air cleaner element
- Spark plug wires from the right side spark plugs
- V-belt covers and power steering pump belt
- Power steering pump brackets-to-intake manifold bolts
- Power steering pump-to-bracket bolts, then position the pump on the right side wheel apron

5. To remove the injectors on the left side, remove or disconnect the following:
- Windshield washer motor electrical connector
- Electrical connector from the rear window washer, if necessary
- Rear window washer hose from the washer motor and plug or cap the line
- Two bolts that secure the washer tank to the body

- Washer tank and secure it out of the way
- Spark plug wires from the left side spark plugs
- Fuel pipe protector

6. Remove or disconnect the following (for either side):
- Band that secures the engine harness to the fuel injector pipe, if equipped
- Intake manifold protector, if equipped
- Fuel injector electrical connector(s)
- Bolts that hold the fuel injector pipe (fuel rail) to the intake manifold, if applicable

7. Pull up on the injector pipe (fuel rail), then remove the fuel injector(s) from the intake manifold. Remove and discard the injector O-rings.

To install:
- Install or connect the following:
- New injector O-rings
- Fuel injector(s) into the intake manifold
- Retaining clips, if applicable
- Injector pipe (fuel rail) and secure with the retaining bolts. Tighten the bolts to 14 ft. lbs. (19 Nm).
- Fuel injector electrical connector(s)

- Intake manifold protector, if equipped
- Band that secures the engine harness to the fuel injector pipe, if equipped

8. To install the injectors on the left side, install or connect the following:
- Fuel pipe protector
- Spark plug wires to the left side spark plugs
- Washer tank and secure with the two mounting bolts
- Rear window washer hose to the washer motor
- Electrical connector to the rear window washer, if necessary
- Windshield washer motor electrical connector

9. To install the right side injectors, install or connect the following:
- Power steering pump into position
- Pump-to-bracket bolts
- Power steering pump brackets-to-intake manifold bolts
- Power steering pump belt and V-belt covers
- Spark plug wires to the right side spark plugs
- Air cleaner element
- Air intake duct and upper cover and the MAF sensor connector
- Air cleaner ducts and resonator chamber, on California vehicles

2005–2006

1. Before servicing the vehicle, refer to the Precautions Section.
2. Properly relieve the fuel system pressure. Remove the fuel cap.
3. Disconnect the negative battery cable.
4. If removing the right side injectors, remove the air intake duct and air cleaner

case. Remove the number one and three spark plug wires.

5. If removing the right side injectors, remove the front side V belt, remove the belt covers. Loosen the lock bolt. Loosen the slider bolt. Remove the front side belt. Remove the bolts that retain the power steering hoses to the intake manifold protector. Do not disconnect the power steering hoses. Remove the bolts which retain the power steering pump to the bracket. Disconnect the power steering pump switch connector. Remove the reservoir tank from the bracket, by pulling upward. Place the power steering pump and tank assembly to the side.

6. If removing the left side injectors, remove the battery. Remove the number two and four spark plug wires.
7. Remove the fuel line protector cover(s).
8. Disconnect the electrical connector from the fuel injector.
9. Remove the bolts that hold the fuel injector line to the intake manifold.
10. Remove the fuel injector retaining clip. Remove the fuel injector while lifting up the fuel injector line.

To install:
11. Installation is the reverse of the removal procedure.
12. Be sure to use new O-rings.
13. Start the engine and check for leaks, correct as required.

DOHC Engine

1. Before servicing the vehicle, refer to the Precautions Section.
2. Properly relieve the fuel system pressure. Remove the fuel cap.
3. Disconnect the negative battery cable. Remove the collector cover. Drain the radiator, as required.

4. If removing the right side injectors, remove the air cleaner upper cover and air intake boot. Remove the air cleaner element. Remove the coolant filler tank.

5. If removing the right side injectors, remove the front side V belt, remove the belt covers. Loosen the lock bolt. Loosen the slider bolt. Remove the front side belt.

6. If removing the right side injectors, disconnect the power steering switch connector. Remove the bolts that retain the power steering hoses to the intake manifold protector. Do not disconnect the power steering hoses. Remove the bolts which retain the power steering pump to the bracket. Remove the reservoir tank from the bracket, by pulling upward. Place the power steering pump and tank assembly to the side.

7. If removing the left side injectors, remove the intake manifold.
8. Remove the fuel line protector cover(s).
9. Disconnect the electrical connector from the fuel injector.
10. If removing the left side injectors, disconnect the connector from the purge control solenoid valve and remove the valve.
11. Remove the harness band that holds the engine harness to the fuel injector line.
12. Remove the bolts that hold the fuel injector line to the intake manifold.
13. Remove the fuel injector while lifting up the fuel injector line.

To install:
14. Installation is the reverse of the removal procedure.
15. Be sure to use new O-rings and insulators.
16. Start the engine and check for leaks, correct as required.

DRIVE TRAIN

Manual Transmission

REMOVAL & INSTALLATION

1. Before servicing the vehicle, refer to the Precautions Section.

2. Open the hood to the full open position.

3. Disconnect the negative battery cable.

4. Drain the transmission gear oil.

5. On non turbocharged engine, remove the air intake duct and cleaner case.

6. On turbocharged engine, remove the intercooler assembly.

7. Disconnect the neutral position switch connector and the back up light switch connector. Disconnect the VSS sensor electrical connector.

8. On 2002–2003 vehicles, disconnect the ground cable.

9. Remove the starter. Remove the clutch operating cylinder from the transmission, and suspend it with wire.

10. Remove the pitching stopper. Position engine support tool ST41099AA010, or equivalent to hold the engine assembly in place.

11. Remove the bolts which holds the upper side of the transmission to the engine.

12. On non turbocharged engine, remove the front and center exhaust pipes.

13. On turbocharged engine, remove the center exhaust pipe.

14. Remove the rear exhaust pipe and muffler. If equipped, remove the heat shield cover.

15. Remove the hanger bracket from the right side of the transmission.

16. Remove the driveshaft.

17. Remove the gear shift rod and the stay from the transmission.

18. Separate the stabilizer link from the transverse link. Remove the bolt securing the ball joint of the transverse link to housing. Separate the transverse link and the housing.

19. On 2002–2003 vehicles, remove the spring pins and separate the front halfshafts from each side of the transmission. Remove the halfshafts.

20. On 2004–2006 vehicles, using tool ST28399SA000, remove the halfshaft from the transmission side. Hold the transmission side joint (AARi) of the front halfshaft by hand and extract the housing from the transmission by pressing it out-

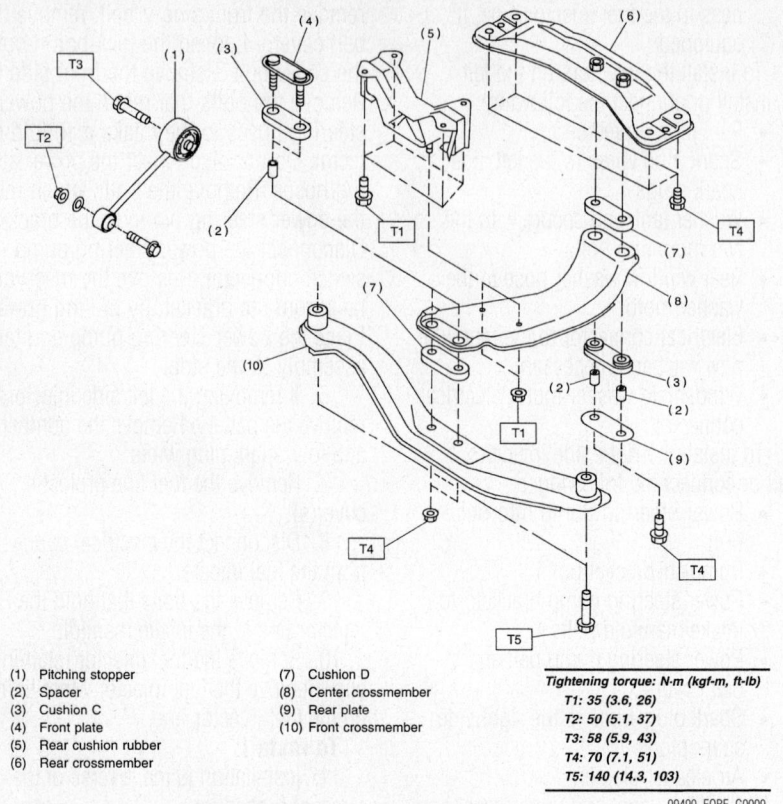

(1)	Pitching stopper	(7)	Cushion D	
(2)	Spacer	(8)	Center crossmember	
(3)	Cushion C	(9)	Rear plate	
(4)	Front plate	(10)	Front crossmember	
(5)	Rear cushion rubber			
(6)	Rear crossmember			

Tightening torque: N·m (kgf-m, ft-lb)
T1: 35 (3.6, 26)
T2: 50 (5.1, 37)
T3: 58 (5.9, 43)
T4: 70 (7.1, 51)
T5: 140 (14.3, 103)

09490_FORE_G0009

Manual transmission crossmember and related components—2002–2003

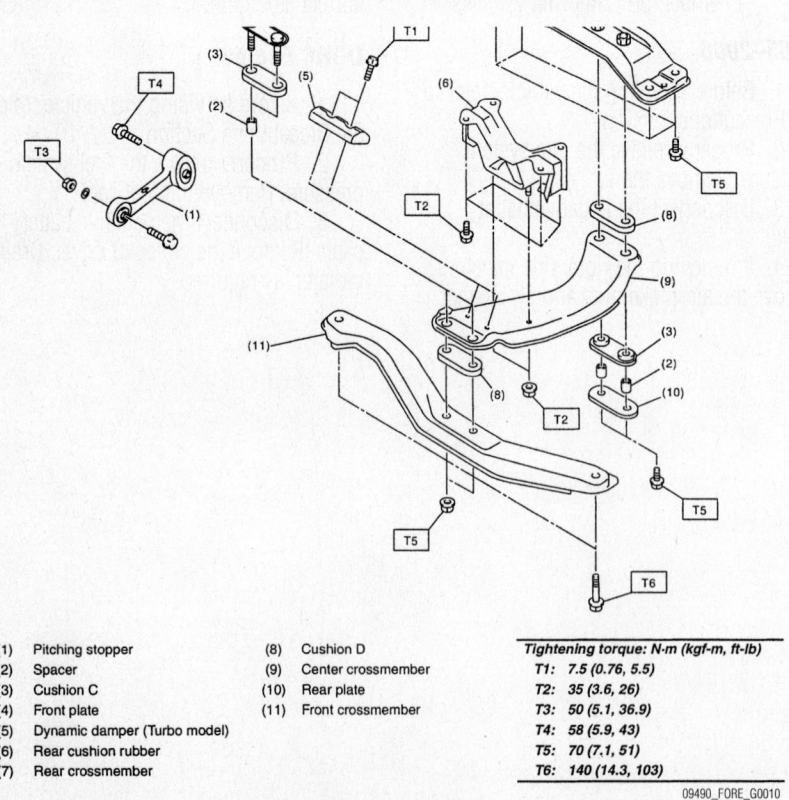

(1)	Pitching stopper	(8)	Cushion D	
(2)	Spacer	(9)	Center crossmember	
(3)	Cushion C	(10)	Rear plate	
(4)	Front plate	(11)	Front crossmember	
(5)	Dynamic damper (Turbo model)			
(6)	Rear cushion rubber			
(7)	Rear crossmember			

Tightening torque: N·m (kgf-m, ft-lb)
T1: 7.5 (0.76, 5.5)
T2: 35 (3.6, 26)
T3: 50 (5.1, 36.9)
T4: 58 (5.9, 43)
T5: 70 (7.1, 51)
T6: 140 (14.3, 103)

09490_FORE_G0010

Manual transmission crossmember and related components—2004–2006

side so as not to stretch the boot on the AARi side.

➡**Face letters "MT" on the handle of the tool to the transmission side. Use an angle so that the protrusion of the tool touches the transmission case during the operation.**

21. Remove the nuts that hold the lower side of the transmission to the engine.

22. Position a transmission jack under the transmission assembly.

23. Remove the transmission rear cross-member retaining bolts. Remove the rear crossmember from the vehicle.

24. Carefully remove the transmission from the vehicle.

➡**Move the transmission jack rearward until the main shaft is withdrawn from the clutch cover.**

25. Separate the transmission assembly and rear cushion rubber.

To install:

26. Installation is the reverse of the removal procedure.

27. Tighten the engine to transmission retaining bolts to 36.9 ft. lbs.

28. Be sure to fill the transmission with the proper grade and type transmission fluid.

29. Start the engine and check for leaks, correct as required.

30. Roadtest the vehicle.

Automatic Transmission

REMOVAL & INSTALLATION

1. Before servicing the vehicle, refer to the Precautions Section.

2. Open the hood to the full open position.

3. Disconnect the negative battery cable.

4. On non turbocharged engine, remove the air intake chamber and intake duct. Remove the air intake chamber stay.

5. On non turbocharged 2004–2006 vehicles, remove the resonator chamber.

6. On turbocharged engine, remove the intercooler.

7. Disconnect the transmission harness connector and ground cable.

8. Remove the starter. Remove the pitching stopper.

9. To separate the torque converter clutch from the drive plate, remove the service hole plug. Remove the bolts that retain the flex plate to the torque converter. When rotating the engine do so in the proper

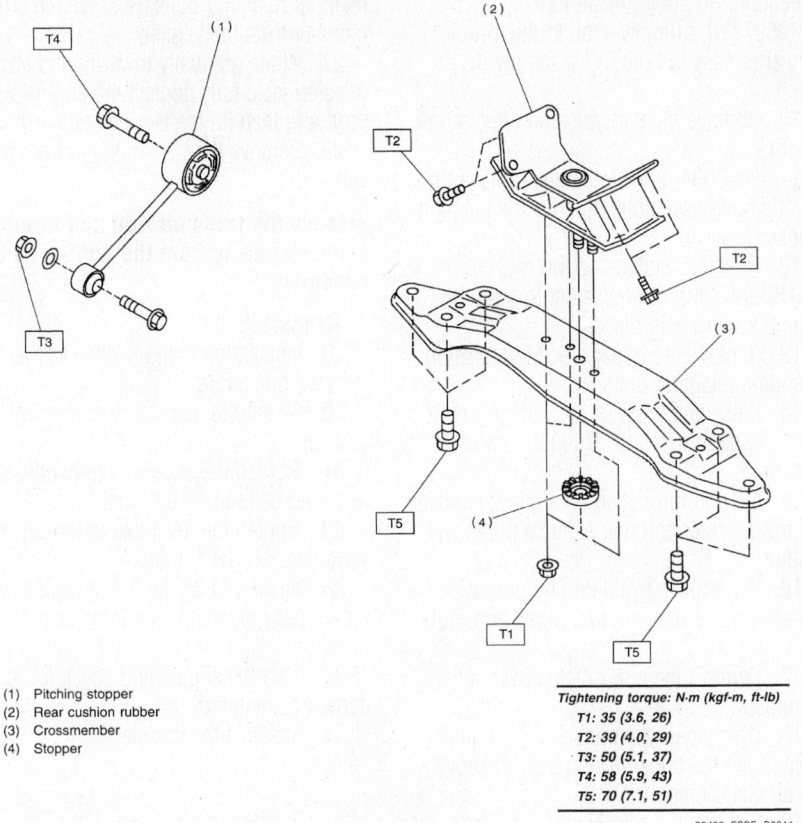

(1) Pitching stopper
(2) Rear cushion rubber
(3) Crossmember
(4) Stopper

Tightening torque: N·m (kgf-m, ft-lb)
T1: 35 (3.6, 26)
T2: 39 (4.0, 29)
T3: 50 (5.1, 37)
T4: 58 (5.9, 43)
T5: 70 (7.1, 51)

09490_FORE_G0011

Automatic transmission crossmember and related components—2002–2003

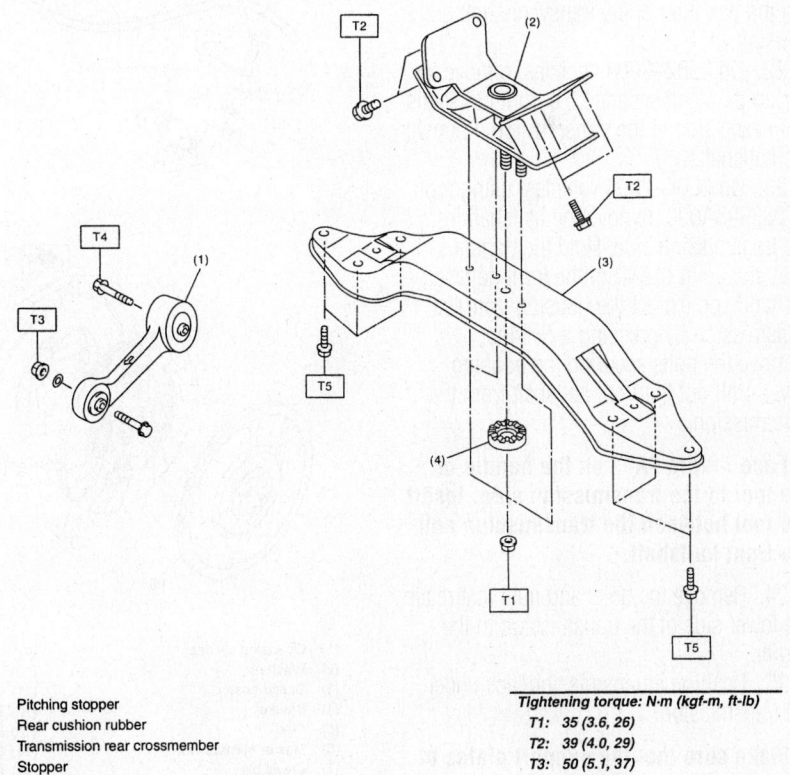

(1) Pitching stopper
(2) Rear cushion rubber
(3) Transmission rear crossmember
(4) Stopper

Tightening torque: N·m (kgf-m, ft-lb)
T1: 35 (3.6, 26)
T2: 39 (4.0, 29)
T3: 50 (5.1, 37)
T4: 58 (5.9, 43)
T5: 70 (7.1, 51)

09490_FORE_G0012

Automatic transmission crossmember and related components—2004–2006

direction of rotation. Install tool ST49827720, or equivalent, to the torque converter case to hold the assembly in place.

10. Remove the transmission filler gauge dipstick.

11. On 2004–2006 vehicles, remove the throttle body assembly. Remove the pitching stopper bracket.

12. Position engine support tool ST41099AC000, or equivalent to hold the engine assembly in place.

13. Remove the upper side transmission to engine retaining bolts.

14. Raise and support the vehicle safely. Remove the undercover. Drain the transmission fluid.

15. On non turbocharged engine, remove the front, center and rear exhaust pipes and muffler.

16. On turbocharged engine, remove the center and rear exhaust pipes and muffler.

17. Remove the heat shield cover, if equipped.

18. Disconnect the transmission cooler hoses from the transmission side. Remove the oil charge pipe.

19. Remove the driveshaft.

20. Remove the shift select cable.

21. Disconnect the stabilizer link from the transverse link. Remove the bolt securing the ball joint of the transverse link to the housing.

22. On 2002–2003 vehicles, remove the spring pins and separate the front halfshafts from each side of the transmission. Remove the halfshafts.

23. On 2004–2006 vehicles, using tool ST28399SA000, remove the halfshaft from the transmission side. Hold the transmission side joint (AARi) of the front halfshaft by hand and extract the housing from the transmission by pressing it outside. Remove the bolts securing the housing cover. Pull out the front halfshaft from the transmission.

➡ Face letters "AT" on the handle of the tool to the transmission side. Insert the tool between the transmission and the front halfshaft.

24. Remove the bolts and nuts that retain the lower side of the transmission to the engine.

25. Position a transmission jack under the transmission.

➡ Make sure that the support plates of the transmission jack do not contact the oil pan.

26. Remove the transmission rear cross-member retaining bolts. Remove the cross-member from the vehicle.

27. While gradually lowering the transmission jack, fully contact the engine support, and then tilt the engine rearward.

28. Remove the transmission from the vehicle.

➡ Move the transmission and torque converter away from the engine as an assembly.

To install:

29. Installation is the reverse of the removal procedure.

30. Be sure to use a new differential side oil seal.

31. Tighten the engine to transmission retaining bolts to 36.9 ft. lbs.

32. Tighten the flex plate to torque converter bolts to 18.1 ft. lbs.

33. Be sure to fill the transmission with the proper grade and type transmission fluid.

34. Start the engine and check for leaks, correct as required.

35. Roadtest the vehicle.

Clutch

ADJUSTMENT

This vehicle is equipped with a hydraulic clutch that is self-adjusting, therefore no adjustment is possible or necessary.

REMOVAL & INSTALLATION

✳✳ CAUTION

The clutch driven disc may contain asbestos, which has been determined to be a cancer-causing agent. Never clean clutch surfaces with compressed air. Avoid inhaling any dust from any clutch surface. When cleaning clutch surfaces, use a commercially available brake cleaning fluid.

1. Before servicing the vehicle, refer to the Precautions Section.

2. Remove or disconnect the following:

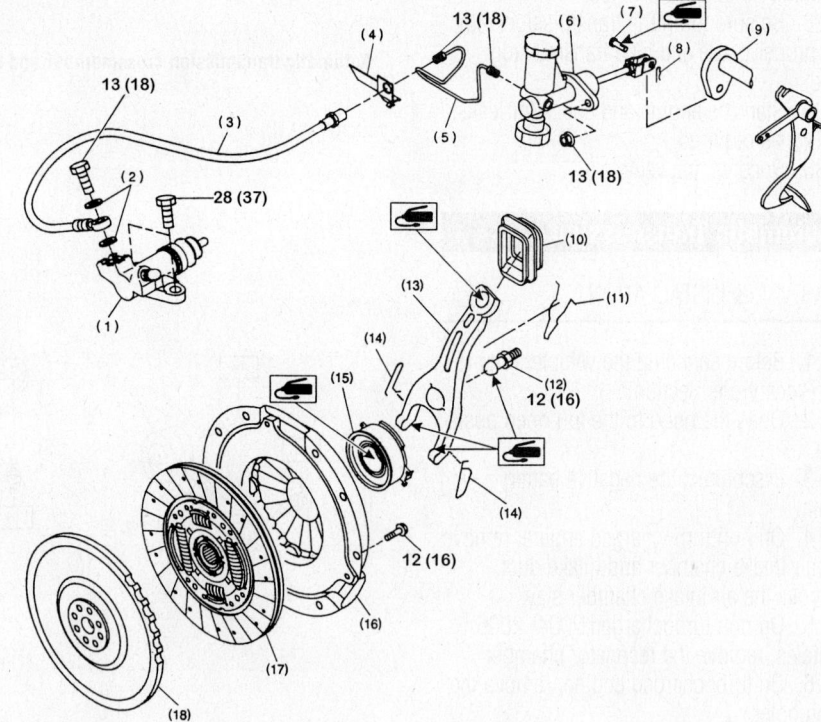

(1) Operating cylinder	(9) Lever
(2) Washer	(10) Clutch release lever sealing
(3) Clutch hose	(11) Retainer spring
(4) Bracket	(12) Pivot
(5) Pipe	(13) Release lever
(6) Master cylinder ASSY	(14) Clip
(7) Clevis pin	(15) Release bearing
(8) Snap pin	(16) Clutch cover
	(17) Clutch disc
	(18) Flywheel

Clutch and related components

- Negative battery cable
- Transmission

※※ WARNING

Removing the bolts on one side of the pressure plate will warp the pressure plate, rendering it useless.

3. Gradually unscrew the six 6mm bolts that hold the pressure plate assembly on the flywheel. Loosen the bolts only 1 turn at a time, working around the pressure plate. Do not unscrew all the bolts on one side at one time.

4. Remove or disconnect the following:
- Clutch plate and disc
- 2 retaining springs, the throwout bearing and the release fork

→Do not disassemble either the clutch cover or disc. Inspect the parts for wear or damage and replace any parts as necessary. Replace the clutch disc if there is any oil or grease on the facing. Do not wash or attempt to lubricate the throwout bearing, because it is sealed and permanently lubricated. If it requires replacement, the bearing may be removed and a new one installed in the holder by means of a press.

To install:

5. Fit the release fork boot on the front of the transmission housing.

6. Install or connect the following:
- Release fork

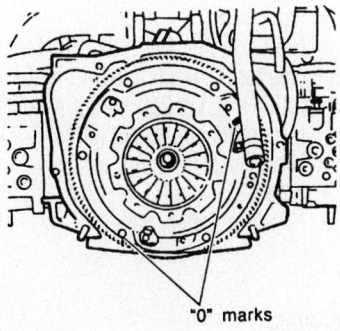

"0" marks

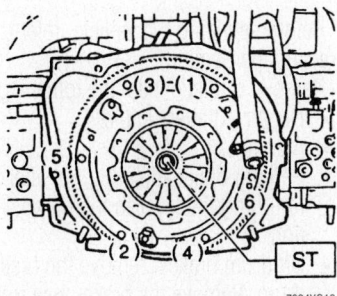

7924XG16

Clutch cover alignment and tightening sequence

- Throwout bearing assembly and secure it with the 2 springs

→Coat the inside diameter of the throwout bearing and the release lever contact points with grease.

- Clutch alignment tool through the clutch cover and disc, then insert the end of the tool into the needle bearing

7. Tighten the pressure plate bolts following the illustrated sequence, 1 turn at a time, until the proper torque is reached. Tighten to 12 ft. lbs. (16 Nm).

※※ WARNING

When installing the clutch pressure plate assembly, be sure that the O marks on the flywheel and the clutch pressure plate assembly are at least 120 degrees apart. These marks indicate the direction of residual unbalance. Also, be sure that the clutch disc is installed properly, noting the FRONT and REAR markings.

8. Install the transmission.

Hydraulic Clutch System

BLEEDING

2002–2004

→To properly bleed the system, it must be bled at the slave cylinder and at the damper. Each of these has an air bleeder on it.

1. Before servicing the vehicle, refer to the Precautions Section.

2. Remove any components necessary to access the slave cylinder.

3. On turbocharged engines, remove the cylinder without disconnecting the clutch hose.

4. Connect a vinyl tube to the air bleeder on the damper and put the other end in a jar with clean clutch fluid.

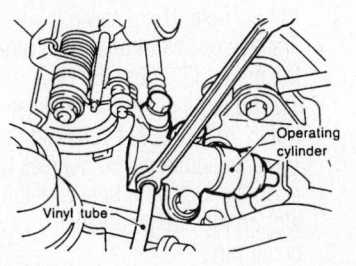

7924XG18

Bleeding the hydraulic clutch at the slave cylinder

→Do not let the fluid level fall too low in the master cylinder. Do not release the pedal with the bleeder open.

5. With the help of an assistant depressing the clutch pedal, slowly open the bleeder valve. Close the bleeder valve and release the pedal. Repeat this process until no air bubbles appear in the jar.

6. Move the tube to the bleeder on the slave cylinder and repeat the process. Check the operation of the clutch after the bleed procedure is complete.

7. When complete, reinstall all removed components. On turbocharged engines tighten the slave cylinder bolts to 27 ft. lbs. (37 Nm).

2005–2006

1. Before servicing the vehicle, refer to the Precautions Section.

2. On non turbocharged engine, remove the air intake chamber.

3. On turbocharged engine, remove the intercooler.

4. Connect a vinyl tube to the air bleeder on the clutch operating (slave) cylinder. Put the other end in a jar with clean clutch fluid.

5. Slowly depress the clutch pedal and keep it depressed. Open the air bleeder to discharge air and fluid.

6. Release the air bleeder for one or two seconds. With the bleeder closed, slowly release the clutch pedal.

7. Repeat the procedure until there are no more air bubbles in the vinyl tube.

8. Tighten the air bleeder.

9. After depressing the clutch pedal, make sure that there are no leaks in the entire system

10. Recheck to ensure that the clutch is operating correctly.

Transfer Case Assembly

REMOVAL & INSTALLATION

The transfer case is an integral part of the transmission.

Halfshafts

REMOVAL & INSTALLATION

Front

1. Before servicing the vehicle, refer to the Precautions Section.

2. Remove or disconnect the following:
- Negative battery cable

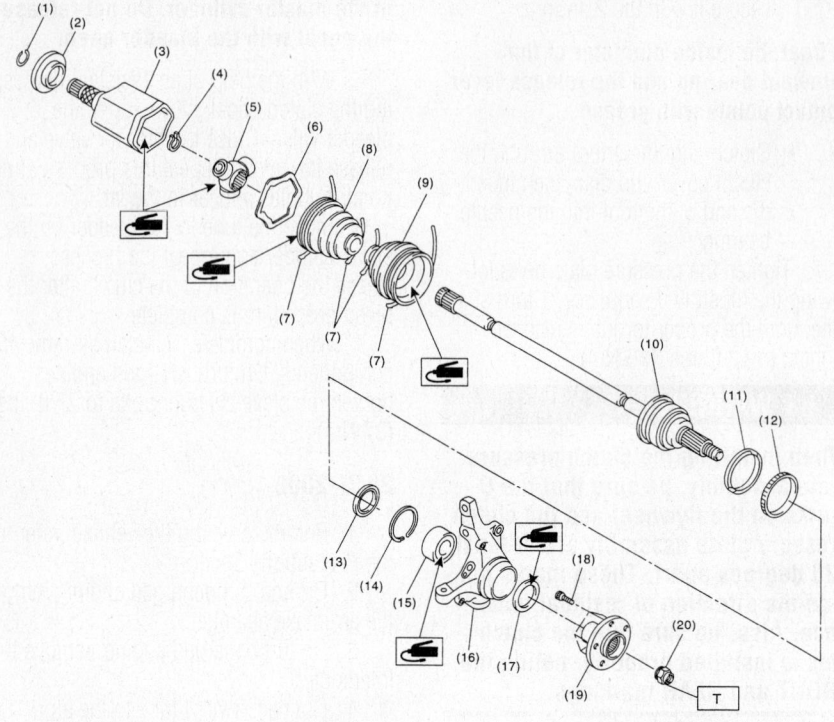

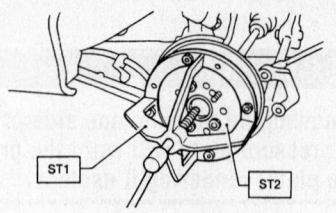

| ST1 | 926470000 | AXLE SHAFT PULLER |
| ST2 | 927140000 | PLATE |

7924XG29

Be sure not to damage the threads when removing the front or rear halfshafts

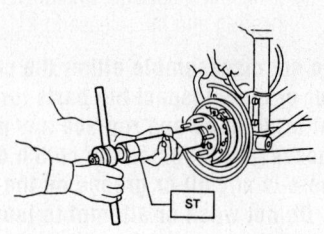

| ST1 | 922431000 | AXLE SHAFT INSTALLER |
| ST2 | 927390000 | ADAPTER |

7924XG30

To avoid using a hammer when installing the halfshafts, use the proper tools

(1)	Spring pin	(9)	Boot (AC)	(17)	Oil seal (OUT)
(2)	Baffle plate (SFJ)	(10)	AC ASSY	(18)	Hub bolt
(3)	Outer race (SFJ)	(11)	Tone wheel	(19)	Hub
(4)	Snap ring	(12)	Baffle plate	(20)	Axle nut
(5)	Trunnion	(13)	Oil seal (IN)		
(6)	Retainer	(14)	Snap ring		
(7)	Boot band	(15)	Bearing		
(8)	Boot (AARi)	(16)	Housing		

67170-SBFR-G30A

Front halfshaft and related components

- Drain the transmission fluid
- Wheel and tire
- Axle nut. Loosen the axle nut after removing the tire and wheel from the vehicle. Failure to do this may damage the wheel bearings.
- Stabilizer link from transverse link
- Brake caliper and suspend from the housing using wire
- Brake rotor
- Tie rod from the knuckle
- Anti-Lock Brake (ABS) sensor and harness
- Transverse link ball joint from the knuckle
- Halfshaft-to-transmission roll pin and discard it
- Halfshaft from the transmission

3. Using Axle Shaft Puller 926470000 and Plate 927140000, remove the halfshaft from the hub.

➡**For automatic transmission use the tool with the "AT" stamped side facing the transmission side. For manual**

transmission use the tool with the "MT" facing the transmission side.

To install:

4. Be sure to replace the differential side retainer oil seal.

5. Install the halfshaft into the hub.

6. Using halfshaft Installer 922431000 and Adapter 927390000, pull the halfshaft through the hub.

7. Install and temporarily tighten a new axle nut.

8. Install or connect the following:
- Halfshaft onto the transmission
- On 2002 vehicles align the halfshaft roll pin hole. Use a new roll pin
- Transverse link to housing. Torque the nut to 37 ft. lbs. (50 Nm).
- Tie rod and tighten the castellated nut to 20 ft. lbs. (27 Nm) and then up to an additional 60 degrees to align the cotter pin hole with the slot on the nut, then install a new cotter pin
- On 2003–2006 vehicles, check AARi retainer position

- Stabilizer link
- Brake rotor
- Brake caliper
- ABS sensor and harness, if equipped
- New axle nut. On 2002–2004 vehicles torque the nut to 140 ft. lbs. (190 Nm). On 2005–2006 vehicles torque the nut to 162 ft. lbs. (220 Nm). Stake the nut.
- Wheel
- Negative battery cable

Rear

2002–2004

1. Before servicing the vehicle, refer to the Precautions Section.

2. Remove or disconnect the following:
- Negative battery cable
- Axle nut
- Parking brake adjusting nut after returning the lever to the off position
- On drum brakes, remove the brake drum. Remove the brake hose from the wheel cylinder and plug the line and wheel cylinder.

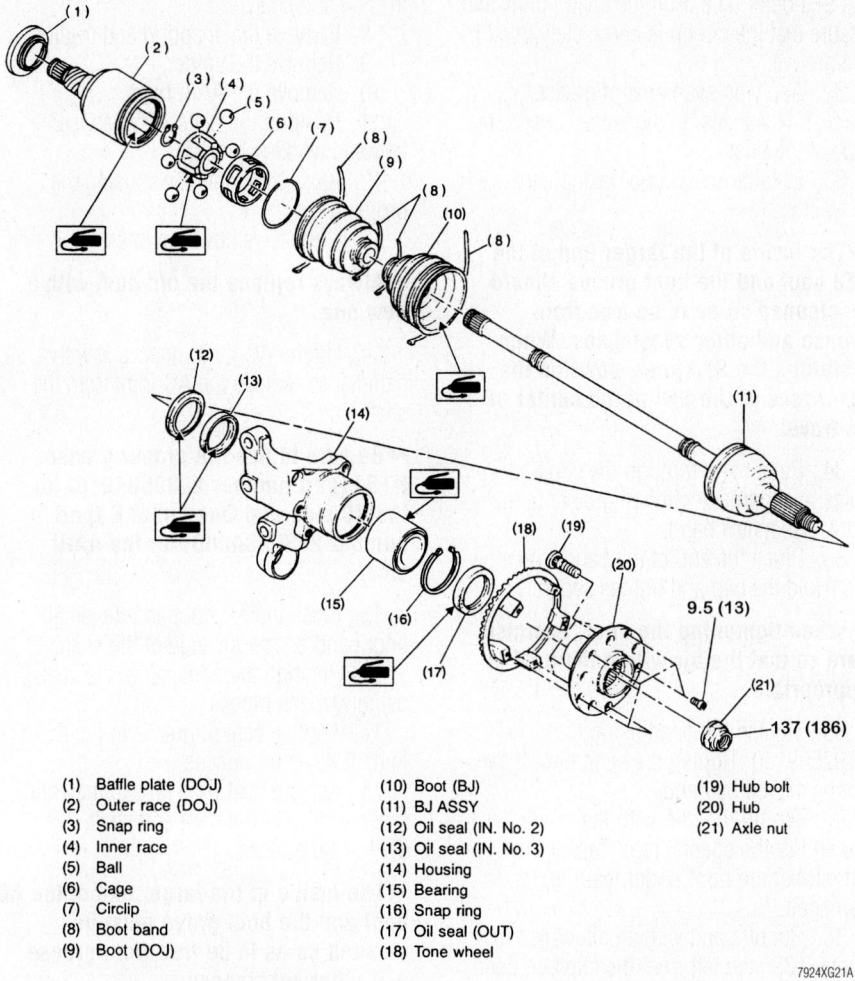

(1)	Baffle plate (DOJ)	
(2)	Outer race (DOJ)	
(3)	Snap ring	
(4)	Inner race	
(5)	Ball	
(6)	Cage	
(7)	Circlip	
(8)	Boot band	
(9)	Boot (DOJ)	
(10)	Boot (BJ)	
(11)	BJ ASSY	
(12)	Oil seal (IN. No. 2)	
(13)	Oil seal (IN. No. 3)	
(14)	Housing	
(15)	Bearing	
(16)	Snap ring	
(17)	Oil seal (OUT)	
(18)	Tone wheel	
(19)	Hub bolt	
(20)	Hub	
(21)	Axle nut	

7924XG21A

Rear halfshaft and related components

- On disc brakes, remove the brake caliper and rotor
- Parking brake cable from the lever
- Lateral link assembly from rear housing
- ABS sensor from the backing plate
- Halfshaft from the differential

3. Using Axle Shaft Puller 926470000 and Plate 927140000, remove the halfshaft from the hub.

To install:

4. Install the halfshaft into the rear housing.

5. Using halfshaft Installer 922431000 and Adapter 927390000, pull the halfshaft into place.

6. Install and temporarily tighten a new axle nut.

7. Install or connect the following:
- Halfshaft-to-differential align roll pin holes and slide the halfshaft onto the splines
- New roll pin

- Trailing link assembly-to-rear housing bolt and nut.
- Rear housing assembly and strut assembly with a new nut and torque to 145 ft. lbs. (196 Nm).
- Rear stabilizer and rear lateral link. Torque bolt and new nut to 32 ft. lbs. (44 Nm).
- Parking brake cable to the lever
- On drum brakes, brake hose to the wheel cylinder. Brake drum.
- On disc brakes, brake rotor and caliper. Adjust the parking brake.

8. Bleed the brake system and adjust the parking brake.
- New axle nut. Torque the nut to 137 ft. lbs. (186 Nm).
- ABS sensor to the backing plate
- Wheel
- Negative battery cable

2005–2006

1. Before servicing the vehicle, refer to the Precautions Section.

2. Disconnect the negative battery cable.

3. Raise and support the vehicle safely.

4. Remove the wheels and tires.

5. Unlock the axle nut. Depress the parking brake. Remove the axle nut.

➡**Be sure to loosen the axle nut after removing the tire and wheel from the vehicle. Failure to do this may damage the wheel bearings.**

6. Remove the rear stabilizer link. Remove the bolts that retain the trailing link to the housing.

7. Remove the bolts that retain the front and rear lateral link to the housing.

8. Remove the rear ABS wheel speed sensor from the backing plate.

9. Remove the rear halfshaft.

➡**Use axle shaft puller tools ST92647000 and ST28099PA110, or equivalent, if it is difficult to remove the halfshaft. Do not hammer the halfshaft to remove it. Be careful not to damage the oil seal or magnetic encoder.**

To install:

10. Installation is the reverse of the removal procedure.

11. Tighten the axle nut to 140 ft. lbs. (190 Nm), with the parking brake applied.

➡**Install the tire and wheel assembly after installation of the axle nut. Failure to follow this order, may cause damage to the wheel bearings.**

CV-Joints

OVERHAUL

Front

2002

1. Before servicing the vehicle, refer to the Precautions Section.

2. Remove the halfshaft from the vehicle.

3. Place alignment marks on the shaft and outer race.

4. Remove the SFJ boot band and boot.

5. Remove the circlip from the SFJ outer race.

6. Remove the SFJ outer race from the shaft assembly. Wipe off the grease.

➡**This is special grease. Do not confuse it with other greases.**

7. Place an alignment mark on the free ring and trunion.

8. Remove the free ring from the

trunion. Be careful with the free ring position.

9. Place an alignment mark on the trunion and shaft.

10. Remove the snapring and trunion.

11. Remove the SFJ boot.

12. Position the halfshaft in a vise, between wooden blocks.

13. Raise the boot claws, using the proper tools.

14. Cut and remove the boot.

➡**Always replace the old boot with a new one.**

15. Disassembly is complete, the BJ cannot be disassembled.

➡**Be sure to use the proper grease, NTG2218 (part number 28093AA000) for the BJ side and SSG6003 (part number 28093TA000) for the SFJ side.**

16. Place the BJ boot and the small boot band on the BJ side of the shaft.

17. Position the halfshaft in the vise, using wooden blocks.

18. Apply a coat of grease to the BJ joint. 2.12–2.47 ounces.

19. Apply a coat of grease to the entire inner surface of the boot and shaft. 0.71–1.06 ounces.

➡**The inside of the larger end of the BJ boot and the boot grove must be cleaned so as to be free from grease and other substances.**

20. Install the boot projecting portion to the BJ groove.

21. Set the large boot band in place.

22. Install the boot projecting portion to the shaft groove.

23. Tighten the boot bands using tool ST28099AC100, a torque wrench and flex handle. Tighten the large boot band to 116 ft. lbs. Tighten the small boot band to 98 ft. lbs.

24. Place the SFJ boot at the center of the shaft.

25. Align the alignment marks and install the trunion on the shaft.

26. Install the snapring to the shaft. Be sure that the snapring is completely fitted in the shaft groove.

27. Fill 3.53–3.88 ounces of grease into the interior of the SFJ outer race.

28. Apply a coat of grease to the free ring and trunion.

29. Align the alignment marks on the free ring and trunion. Install the free ring. Be careful with the free ring position.

30. Align the alignment marks on the shaft and outer race, and install the outer race.

31. Install the circlip in the groove on the SFJ outer race. Pull the shaft lightly and assure that the circlip is completely fitted in the groove.

32. Apply an even coat of grease, 1.06–1.41 ounces, to the entire inner surface of the boot.

33. Install the SFJ boot taking care not to twist it.

➡**The inside of the larger end of the SFJ boot and the boot groove should be cleaned so as to be free from grease and other substances. When installing the SFJ boot, position the outer race of the SFJ at the center of its travel.**

34. Put a band through the clip and wind twice in alignment with the groove of the boot. Use a new band.

35. Pinch the end of the band with a pliers. Hold the clip and tighten securely.

➡**When tightening the boot, exercise care so that the air within the boot is appropriate.**

36. Tighten the band using tool ST92509100. Tighten the band until it cannot be moved by hand.

37. Tap on the clip with the punch, at the end of the special tool. Tap to the extent that the boot underneath is not damaged.

38. Cut off band with an allowance of about 0.39 inch left from the clip and bend the allowance over the clip.

➡**Be careful so that the end of the band is close contact with the clip.**

39. Fix up the boot on the BJ in the same manner.

➡**Extend and retract the SFJ to provide equal grease coating.**

2003–2004

1. Before servicing the vehicle, refer to the Precautions Section.

2. Remove the halfshaft from the vehicle.

3. Place alignment marks on the shaft and outer race.

4. Remove the AARi boot band and boot.

5. Remove the retainer from the AARi outer race.

6. Remove the AARi outer race from the shaft assembly. Wipe off the grease.

➡**This is special grease. Do not confuse it with other greases.**

7. Place an alignment mark on the trunion and shaft.

8. Remove the snapring and trunion.

9. Remove the spider.

10. Remove the AARi boot.

11. Position the halfshaft in a vise, between wooden blocks.

12. Raise the boot claws, using the proper tools.

13. Cut and remove the boot.

➡**Always replace the old boot with a new one.**

14. Hit the AC joint inner race with a hammer to remove the AC joint from the shaft.

➡**Be sure to use the proper grease, NTBJ (part number 28395SA010) for the AC side and One Luber C (part number 28395SA000) for the AARi side.**

15. Place the AC boot and the small boot band on the AC side of the shaft.

16. Position the halfshaft in the vise, using wooden blocks.

17. Apply a coat of grease to the AC joint. 2.82–3.00 ounces.

18. Apply a coat of grease to the entire inner surface of the boot and shaft. 0.71–1.06 ounces.

➡**The inside of the larger end of the AC boot and the boot grove must be cleaned so as to be free from grease and other substances.**

19. Install the boot projecting portion to the AC groove.

20. Set the large boot band in place.

21. Install the boot projecting portion to the shaft groove.

22. Tighten the boot bands using tool ST28099AC000, a torque wrench and flex handle. Caulked portion clearance of the large boot band should be 0.051 inch, or more. Caulked portion clearance of the small boot band should be 0.051 inch, or more.

23. Place the AARi boot and retainer to the shaft and position it at the center of the shaft.

24. Align the alignment marks and install the trunion on the shaft.

25. Fill 3.53–3.88 ounces of grease into the interior of the AARi outer race.

26. Apply a coat of grease to the free ring and trunion.

27. Align the alignment marks on the free ring and trunion. Install the free ring. Be careful with the free ring position.

28. Align the alignment marks on the

shaft and outer race, and install the outer race.

29. Apply an even coat of grease, 1.06–1.41 ounces, to the entire inner surface of the boot.

30. Install the AARi boot taking care not to twist it.

→**The inside of the larger end of the AARi boot and the boot groove should be cleaned so as to be free from grease and other substances. When installing the AARi boot, position the outer race of the AARi at the center of its travel.**

31. Install a new large boot band and small boot band in the specified position.

32. Tighten the band using tool ST280099AC000. Caulked portion clearance of the large boot band should be 0.04 inch, or less. Caulked portion clearance of the small boot band should be 0.04 inch, or less.

→**Extend and retract the AARi to provide equal grease coating.**

2005–2006

1. Before servicing the vehicle, refer to the Precautions Section.

2. Remove the halfshaft from the vehicle.

3. Place alignment marks on the shaft and outer race.

4. Remove the AARi boot band and boot.

5. Remove the retainer from the AARi outer race.

6. Remove the AARi outer race from the shaft assembly. Wipe off the grease.

→**This is special grease. Do not confuse it with other greases.**

7. Place an alignment mark on the trunion and shaft.

8. Remove the snapring and trunion.

9. Remove the spider.

10. Remove the AARi boot.

→**Be sure to use the proper grease, One Louver C (part number 28395SA000).**

11. Place the AARi boot and retainer to the shaft and position it at the center of the shaft.

12. Align the alignment marks and install the trunion on the shaft.

13. Fill 3.53–3.88 ounces of grease into the interior of the AARi outer race.

14. Apply a coat of grease to the free ring and trunion.

15. Align the alignment marks on the shaft and outer race. Install the outer race.

16. Apply an even coat of grease, 1.06–1.41 ounces, to the entire inner surface of the boot.

17. Install the AARi boot taking care not to twist it.

→**The inside of the larger end of the AARi boot and the boot groove should be cleaned so as to be free from grease and other substances. When installing the AARi boot, position the outer race of the AARi at the center of its travel.**

18. Check the position of the retainer.

19. Install a new large boot band and small boot band in the specified position.

20. Tighten the band using tool ST280099AC000. Clearance at the crimped section of the large boot band should be 0.04 inch, or less. Clearance at the crimped section of the small boot band should be 0.04 inch, or less.

→**Extend and retract the AARi to provide equal grease coating.**

Rear

The Double Offset Joint (DOJ), is the only part of the assembly that can be replaced, if any of the other components are defective then the shaft should be replaced.

1. Before servicing the vehicle, refer to the Precautions Section.

2. Remove the halfshaft from the vehicle.

3. Straighten the bent claw of the large clamp at the Double Offset Joint (DOJ) end of the boot.

4. Loosen the band using pliers being careful not to damage the bolt.

5. Remove the small boot band from the DOJ using the same technique.

6. Remove the boot from the large end of the DOJ outer race.

7. Remove the round circlip using a suitable prytool from the neck of the joint outer race.

8. Remove the joint outer race.

✳✳ CAUTION

The grease used is for CV–Joints, do not replace with another type of grease

9. Clean the grease and remove the balls. Be careful not to loose any of the 6 balls.

10. Turn the cage by a half pitch to the track groove of the inner race and shift the cage.

11. Remove the snapring that secures the inner race to the shaft and remove the inner race.

12. Take the cage from the shaft and remove the boot.

13. The other boots may be removed in the same manner as the DOJ boot.

14. Wrap the shaft splines with tape to prevent damage.

→**Be sure to use the proper grease. On 2002–2004 vehicles, Molylex No. 2 (part number 723223010) for the BJ side and VU-3A702 (yellow) (part number 23223GA050) for the DOJ side. On 2005–2006 vehicles, NTG2218-M (part number 28395AG010) for the EBJ side and NKG205 (part number 28495AG000) for the DOJ side.**

15. Install the BJ or EBJ boots and fill it with 2.12–2.47 ounces of grease.

16. Place the DOJ boot on the center of the shaft.

17. Insert the DOJ cage onto the shaft. Make sure to insert the cage with the cut–out portion facing shaft end.

18. Install the inner race on the shaft and fasten with the snap ring. Make sure the snap ring is firmly engaged.

19. Install the cage (previously fitted) to the inner race on the shaft. Fit the cage with the protruded part aligned with the track on the inner race and then turn a half pitch.

20. Fill the DOJ inner race with 2.82–3.17 ounces of grease.

21. Apply a coat of grease to the cage pocket and 6 balls.

22. Insert the 6 balls.

23. Align the outer race track and ball positions and place where the shaft, inner race, cage and balls were located prior to removal and then install the outer race.

24. Install the circlip into the groove on the outer race.

→**Make sure the balls, cage and inner race are fully seated. Make sure not to place the matched position of the circlip in the ball groove of the outer race. Pull the shaft lightly to make sure the circlip is fully engaged.**

25. Apply an even coat 0.71–1.06 ounces of grease to the entire inner surface of the boot and to the shaft.

26. Make sure the boot is free from any dirt or foreign materials prior to installation.

27. Place the outer race of the boot at the center of its travel.

28. Put a band through the boot clip and wind twice in alignment with the groove on the boot.

29. Pinch the end of the band using

tool ST 9250910000 and tighten securely until it cannot be moved by hand. Make sure there is appropriate air inside the boot.

30. Tap on the clip with a punch to lock it making sure not to damage the damaged

while tapping. Cut any excess off the band leaving 0.39 inch and bend the remaining portion over the clip. Make sure the end of the band is close to the clip.

31. Install the remaining boot clamps in the same manner.

STEERING

Air Bag

❋❋ CAUTION

These vehicles are equipped with an air bag system. The system must be disabled before performing service on or around system components, steering column, instrument panel components, wiring and sensors. Failure to follow safety and disabling procedures could result in accidental air bag deployment, possible personal injury and unnecessary system repairs.

PRECAUTIONS

Several precautions must be observed when handling the inflator module to avoid

accidental deployment and possible personal injury.

• Never carry the inflator module by the wires or connector on the underside of the module.

• When carrying a live inflator module, hold it securely with both hands, and ensure that the bag and trim cover are pointed away.

• Place the inflator module on a bench or other surface with the bag and trim cover facing up.

• With the inflator module on the bench, never place anything on or close to the module that may be thrown in the event of an accidental deployment.

DISARMING

1. Before servicing the vehicle, refer to the Precautions Section.

2. Be sure to position the front wheels in the straight ahead position.

3. Disconnect the negative battery cable. Tape the battery cable for added protection.

4. Wait more than 20 seconds before starting work.

Power Steering Gear

REMOVAL & INSTALLATION

2002–2004

1. Before servicing the vehicle, refer to the Precautions Section.

2. Remove or disconnect the following:
• Negative battery cable
• Front wheels
• Engine undercover

3. Remove the sub frame while refer-

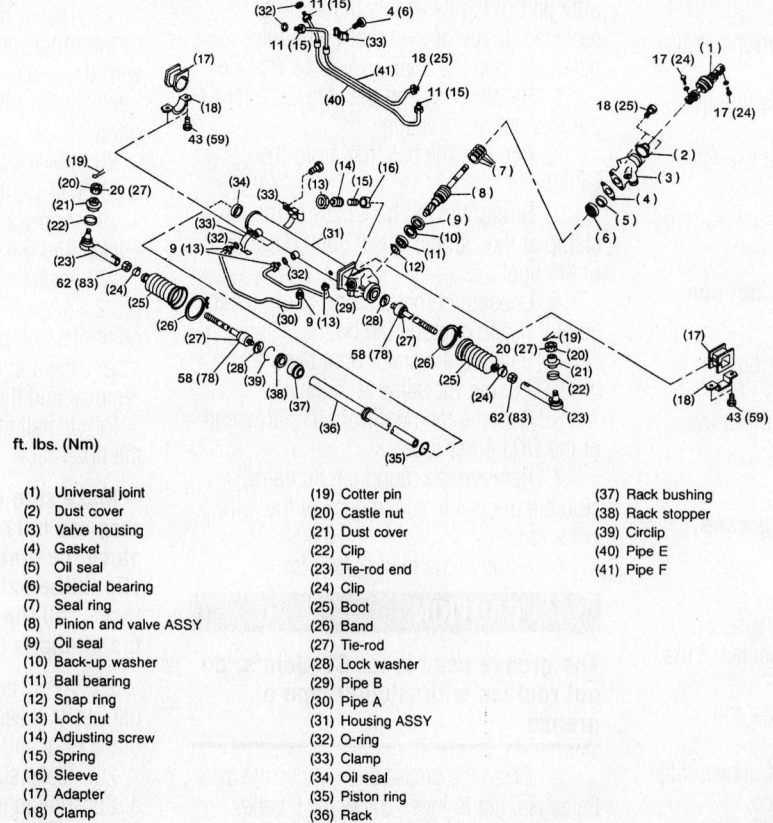

ft. lbs. (Nm)

(1) Universal joint
(2) Dust cover
(3) Valve housing
(4) Gasket
(5) Oil seal
(6) Special bearing
(7) Seal ring
(8) Pinion and valve ASSY
(9) Oil seal
(10) Back-up washer
(11) Ball bearing
(12) Snap ring
(13) Lock nut
(14) Adjusting screw
(15) Spring
(16) Sleeve
(17) Adapter
(18) Clamp

(19) Cotter pin
(20) Castle nut
(21) Dust cover
(22) Clip
(23) Tie-rod end
(24) Clip
(25) Boot
(26) Band
(27) Tie-rod
(28) Lock washer
(29) Pipe B
(30) Pipe A
(31) Housing ASSY
(32) O-ring
(33) Clamp
(34) Oil seal
(35) Piston ring
(36) Rack

(37) Rack bushing
(38) Rack stopper
(39) Circlip
(40) Pipe E
(41) Pipe F

Power steering gear and related components

7924XG22

ring to the accompanying illustrations for bolt locations as follows:

    a. Loosen bolt (1) but leave it connected by a few threads.

    b. Remove bolts 2, 3, 4, 5 and 6 (in that order) and remove the subframe.

- Y-pipe
- Tie rod end cotter pin and nut
- Jack-up plate and front sway bar
- Fluid lines from the rack and pinion

4. Matchmark the universal joint to the serration in the steering rack for installation reference.

5. Remove or disconnect the following:

- Universal joint bolts and lift the joint upward disconnecting it from the rack and pinion shaft.
- Rack and pinion

To install:

6. Install the rack and pinion. Torque the clamp bolts to 43 ft. lbs. (59 Nm).

7. Align the steering rack to the universal joint. Push the long yoke of the joint all the way into the serrated position of the steering shaft, setting the bolt hole in the cut-out. Pull the short yoke all the way out of the serrated portion of the rack and pinion, setting the bolt hole in the cut-out. Insert the bolt through the short yoke. Pull the yoke and ensure the bolt is properly engaged in the cut-out. Fasten the short yoke side with the spring washer and bolt, then fasten the yoke side. Tighten the bolts to 17 ft. lbs. (24 Nm).

8. Install or connect the following:

- Tie rod ends to the steering knuckle
- Sway bar and jack-up plate
- Y-pipe with new gaskets and nuts

9. Install the sub frame while referring to the accompanying illustrations for bolt locations as follows:

    a. Replace any M12 bolts with new ones. Tighten the bolts marked in the illustration as (T1) to 41 ft. lbs. (55 Nm) and the bolts marked (T2) to 52 ft. lbs. (71 Nm).

    b. Inspect all the bolts and make sure they are torqued to the proper specification.

10. Install or connect the following:

- Engine undercover
- Wheels

11. Fill and bleed the steering system.

2005–2006

1. Before servicing the vehicle, refer to the Precautions Section.

2. Disconnect the negative battery cable.

3. Loosen the front wheel nut.

4. Raise and support the vehicle safely.

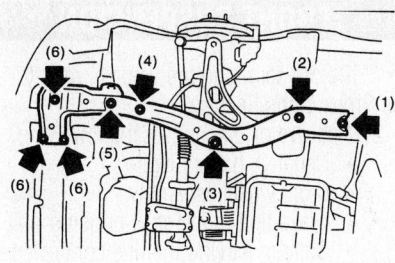

Sub frame bolt removal sequence

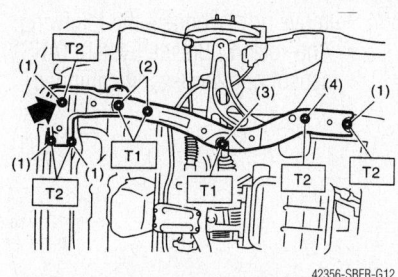

Sub frame bolt tightening specification location

5. Remove the tires and wheels.

6. Remove the undercover.

7. Remove the sub frame. Leave bolt (1) connected by a few threads and remove the bolts in the sequence illustrated. Once the other bolts are removed, remove bolt (1) and the sub frame.

8. Remove the front exhaust pipe.

9. Remove the cotter pin and castle nut. Using a puller, remove the tie rod end from the knuckle arm.

10. Remove the jack up plate. Remove the front stabilizer.

11. Disconnect the power steering fluid pipe at the center of the gearbox and attach a vinyl hose. Discharge the fluid into a suitable container by turning the steering wheel fully clockwise and counterclockwise. Disconnect the other fluid line, and repeat the discharge procedure.

12. Remove the steering wheel. Make a matchmark on the universal joint. Remove the universal joint bolts and remove the joint from the vehicle.

13. Disconnect the fluid lines from the steering gear, pressure hose first.

14. Remove the steering gear retaining bolts and clamps securing the steering gear to the crossmember. Remove the steering gear from the vehicle.

To install:

15. Insert the steering gear into the crossmember. Be careful not to damage the gearbox boot.

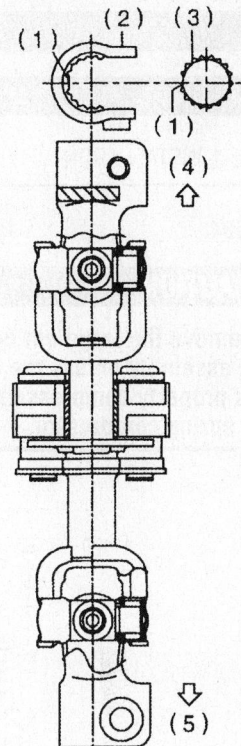

(1)   Cutout portion
(2)   Yoke
(3)   Column shaft
(4)   Column shaft side
(5)   Gearbox side

Steering column joint alignment— 2005–2006

16. Tighten the steering gear to the crossmember bracket to 44.1 ft. lbs.

17. Connect the fluid lines.

18. Align the cutout at the serrated section of the column shaft and yoke. Insert the universal joint into the column shaft.

19. Align the mating marks and insert the universal joint to serrated section of the steering gear assembly. Tighten the bolt to 17.4 ft. lbs.

20. Continue the installation in the reverse order of the removal procedure.

21. When installing the sub frame be sure to torque the bolts to specification and in the proper sequence. Tighten bolts marked (T1) to 41 ft. lbs. (55 Nm) and the bolts marked (T2) to 52 ft. lbs. (71 Nm).

22. Fill the power steering system with the proper grade and type fluid.

23. Start the engine and check for leaks. Correct as required.

FRONT SUSPENSION

Strut

REMOVAL & INSTALLATION

2002–2004

✳✳ CAUTION

Do not remove the large nut on top of the strut assembly unless the coil spring is properly compressed with a suitable spring compressor.

1. Before servicing the vehicle, refer to the Precautions Section.
2. Remove or disconnect the following:
   - Negative battery cable
   - Front wheel assembly
   - Brake line to the strut housing bolt
   - Caliper, leaving the line connected and suspend it out of the way
3. Matchmark the camber adjustment bolt to the strut housing as reference for installation.
4. Remove or disconnect the following:
   - Anti-locking Brakes System (ABS) sensor and harness, if equipped
   - Strut from the steering knuckle. Note that the shaft of the top bolt is not round.
   - Strut from the body in the engine compartment
   - Strut and coil spring assembly

To install:

5. Install the strut and coil assembly. Torque the upper strut retainer nuts to 15 ft. lbs. (20 Nm).
6. Align matchmark on camber adjustment bolt and strut housing.
7. Install or connect the following:
   - Lower strut nuts and bolts. Tighten

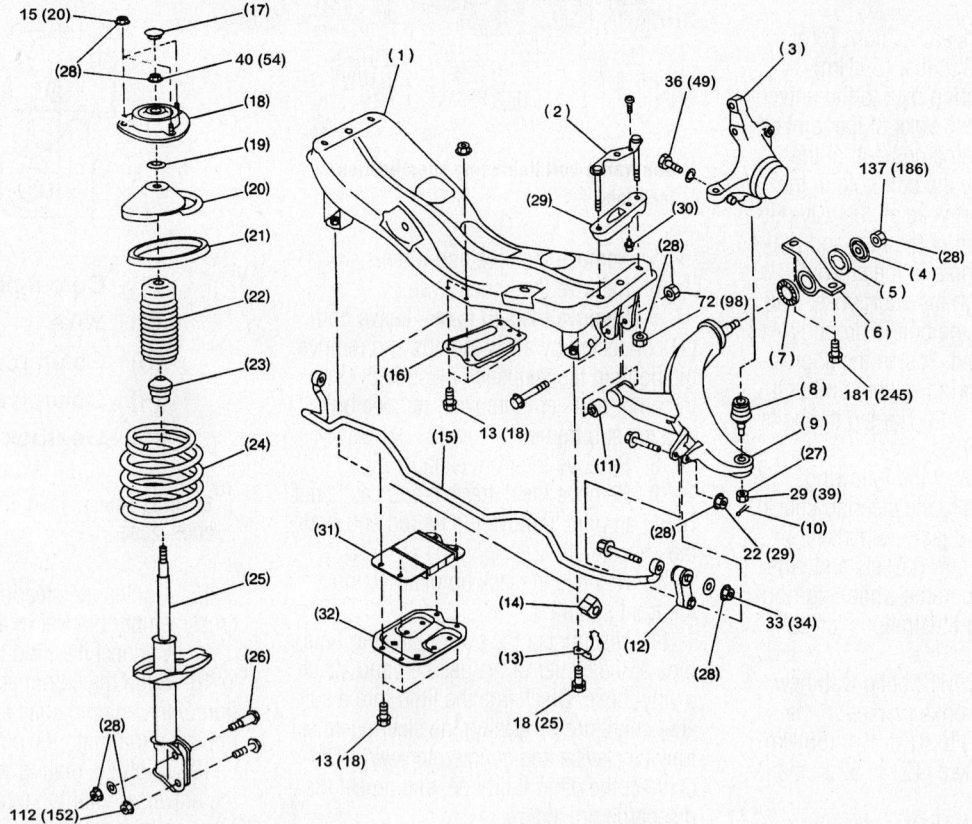

(1) Front crossmember	(17) Dust seal
(2) Bolt ASSY	(18) Strut mount
(3) Housing	(19) Spacer
(4) Washer	(20) Upper spring seat
(5) Stopper rubber (Rear)	(21) Rubber seat
(6) Rear bushing	(22) Dust cover
(7) Stopper rubber (Front)	(23) Helper
(8) Ball joint	(24) Coil spring
(9) Transverse link	(25) Damper strut
(10) Cotter pin	(26) Adjusting bolt
(11) Front bushing	(27) Castle nut
(12) Stabilizer link	(28) Self-locking nut
(13) Clamp	(29) Adapter front crossmember
(14) Bushing	(30) Clip
(15) Stabilizer	(31) Dynamic damper (MT model)
(16) Jack-up plate (Except MT model)	(32) Jack-up plate (MT model)

7924XG23A

Front suspension and related components—2002–2004

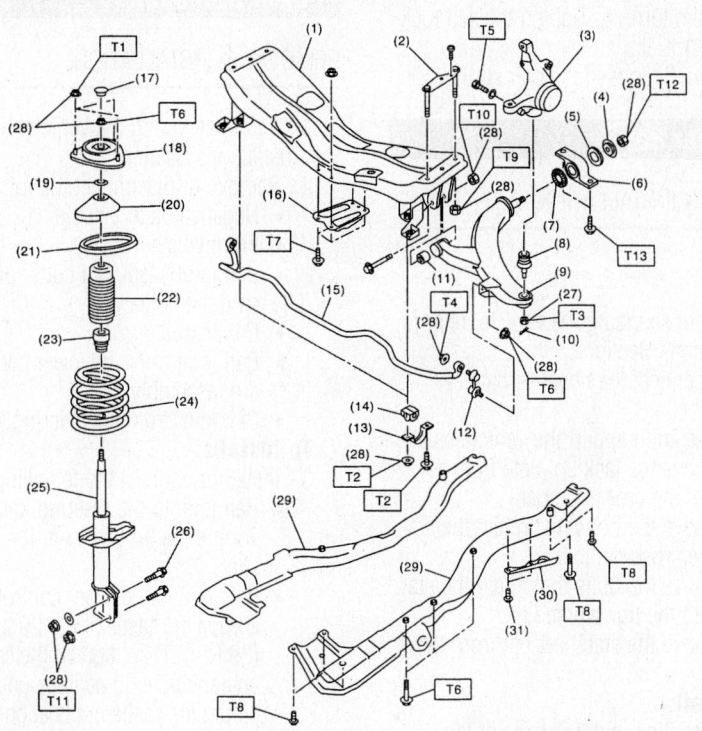

(1)	Front crossmember	(17)	Dust seal
(2)	Bolt ASSY	(18)	Strut mount
(3)	Housing	(19)	Spacer
(4)	Washer	(20)	Upper spring seat
(5)	Stopper rubber (Rear)	(21)	Rubber seat
(6)	Rear bushing	(22)	Dust cover
(7)	Stopper rubber (Front)	(23)	Helper
(8)	Ball joint	(24)	Coil spring
(9)	Transverse link	(25)	Damper strut
(10)	Cotter pin	(26)	Adjusting bolt
(11)	Front bushing	(27)	Castle nut
(12)	Stabilizer link	(28)	Self-locking nut
(13)	Clamp	(29)	Sub frame
(14)	Bushing	(30)	Cover
(15)	Stabilizer	(31)	Clip
(16)	Jack-up plate		

Tightening torque: N·m (kgf-m, ft-lb)
- T1: 20 (2.0, 14.5)
- T2: 25 (2.5, 18.1)
- T3: 40 (4.1, 30) (Tighten an additional 60°)
- T4: 45 (4.6, 33)
- T5: 50 (5.1, 37)
- T6: 55 (5.6, 41)
- T7: 70 (7.1, 52)
- T8: 71 (7.2, 52)
- T9: 100 (10.2, 74)
- T10: 125 (12.7, 92.3)
- T11: 175 (17.8, 129)
- T12: 190 (19.4, 140)
- T13: 250 (25.5, 184)

09490_FORE_G0013

Front suspension and related components—2003–2006

the nuts, while securing the bolts to 129 ft. lbs. (175 Nm) on 2002–2004 vehicles.

- ABS sensor and harness. Torque the bolt to 24 ft. lbs. (32 Nm), if equipped.
- Brake line to the strut bolt. Torque the bolt to 24 ft. lbs. (32 Nm).
- Caliper
- Front wheel
- Negative battery cable

8. Check and adjust the front end alignment.

2005–2006

1. Before servicing the vehicle, refer to the Precautions Section.
2. Raise and support the vehicle safely.
3. Remove the tire and wheel.

4. Remove the bolt retaining the brake hose to the strut.

5. Make and alignment mark on the camber adjusting bolt which secures the strut to the housing.

6. Remove the bolt retaining the ABS wheel speed sensor harness.

7. Remove the two bolts retaining the strut to the housing.

➡**While holding the head of the adjusting bolt, loosen the self locking nut.**

8. Remove the three upper strut retaining nuts.

9. Remove the strut from the vehicle.

To install:

10. Installation is the reverse of the removal procedure.

11. Tighten the upper retaining nuts to

14.5 ft. lbs. Tighten the lower retaining bolts to 129 ft. lbs.

12. Position the alignment mark on the camber adjusting bolt with the alignment mark on the lower side of the strut. Install using a new self locking nut.

➡**While holding the head of the adjusting bolt, tighten the self locking nut.**

13. Check and adjust wheel alignment, as necessary.

DISASSEMBLY & ASSEMBLY

2002–2004

1. Before servicing the vehicle, refer to the Precautions Section.
2. Remove the strut assembly from the vehicle.
3. Place the strut assembly in a vise with a holding tool and install a spring compressor.
4. Compress the spring slightly.
5. Loosen but do not remove the bearing cap locknut.
6. Compress the spring with the spring compressor, and then remove the locknut.
7. Remove or disconnect the following:
- Strut bearing cap, mounting insulator bracket and upper spring seat
- Coil assembly, leaving the spring compressed
- Strut boot and rebound bumper from the strut. Inspect and replace if worn.
- Strut retainer nut using a suitable wrench
- Strut insert from the assembly
8. Install or connect the following:
- Strut into the chamber and install the retainer nut. Tighten the nut snugly.
- Rebound bumper and the boot to the strut piston rod
- Coil spring on the strut assembly. Be sure the spring is properly positioned on the lower bracket.
- Upper spring seat, mounting insulator and bearing cap. Be sure the upper spring seat is facing the proper direction.
- Locknut and tighten to 41 ft. lbs. (55Nm)
9. Loosen and remove the spring compressor from the coil spring.
10. Install the strut to the vehicle.

2005–2006

1. Before servicing the vehicle, refer to the Precautions Section.

2. Remove the strut from the vehicle.

3. Using a coil spring compressor tool, carefully compress the spring. Remove the self locking nut.

4. Remove the strut mount, upper spring and rubber seat from the strut.

5. Gradually decrease the compression force of the spring compressor tool. Remove the coil spring.

6. Remove the dust cover and helper spring.

7. Check for the presence of air in the damping force generating mechanism.

8. Using the spring compression tool, compress the coil spring.

➡**Be sure to properly install the coil spring.**

9. Position the coil spring so that its end face fits good into the spring seat.

10. Install the helper spring and dust cover to the piston rod.

11. Pull the piston rod fully upward, and install the rubber seat and spring seat.

12. Install the strut mount to the piston rod, and then tighten the self locking nut, temporarily. Be sure to use a new self locking nut.

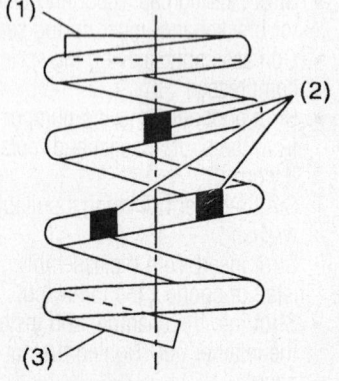

(1) **Flat (top side)**

(2) **Identification paint**

(3) **Inclined (bottom side)**

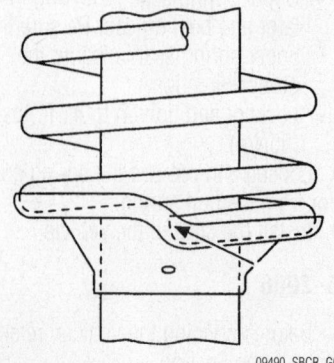

09490_SBCR_G0099

Strut spring alignment—2005–2006

13. Use a hexagon wrench to prevent the strut rod from turning. Tighten the self locking nut to 41 ft. lbs.

14. Carefully loosen the coil spring.

Stabilizer Bar

REMOVAL & INSTALLATION

2002

1. Before servicing the vehicle, refer to the Precautions Section.

2. Disconnect the negative battery cable.

3. Raise and support the vehicle safely.

4. Remove the jack up plate from the lower part of the crossmember.

5. Remove the bolts that retain the stabilizer to the crossmember.

6. Remove the bolts that retain the stabilizer link to the transverse link.

7. Remove the stabilizer bar from the vehicle.

To install:

8. Installation is the reverse of the removal procedure.

9. Install the rubber bushing, on the front crossmember side, while aligning it with the paint mark on the stabilizer bar.

10. Be sure that the bushings and the stabilizer have the same identification colors.

11. Always fully tighten the rubber bushings when the wheels are in full contact with the ground and the vehicle is at curb height.

2003–2006

1. Before servicing the vehicle, refer to the Precautions Section.

2. Raise and support the vehicle safely.

3. Remove the jack up plate from the lower part of the crossmember.

4. Remove the sub frame.

5. Remove the nut which secures the stabilizer link to the front transverse link.

6. Remove the bolts that secure the stabilizer bar to the crossmember.

7. Remove the stabilizer bar from the vehicle.

To install:

8. Installation is the reverse of the removal procedure.

9. Install the rubber bushing, on the front crossmember side, while aligning it with the paint mark on the stabilizer bar.

10. Be sure that the bushings and the stabilizer have the same identification colors.

11. Always fully tighten the rubber bushings when the wheels are in full contact with the ground and the vehicle is at curb height.

Lower Ball Joint

REMOVAL & INSTALLATION

1. Before servicing the vehicle, refer to the Precautions Section.

2. Remove or disconnect the following:
- Negative battery cable
- Front wheel
- Ball joint castle nut cotter pin, discard the cotter pin
- Castle nut
- Ball joint from the lower control arm assembly
- Ball joint from the steering knuckle

To install:

3. Install or connect the following:
- Ball joint to the steering knuckle. Torque the bolt to 36 ft. lbs. (49 Nm).
- Ball joint to the lower control arm. Torque the castle nut to 29 ft. lbs. (39 Nm). Then, tighten the castle nut an additional 60 degrees until the slot in the castle nut is aligned with the cotter pin hole in the ball joint.
- New cotter pin
- Wheel
- Negative battery cable

Lower Control Arm

REMOVAL & INSTALLATION

2002–2004

1. Before servicing the vehicle, refer to the Precautions Section.

2. Remove or disconnect the following:
- Wheel assembly
- Stabilizer link
- Ball joint from housing
- Mounting bolts
- Control arm

To install:

3. Install or connect the following:
- Control arm to stabilizer. Torque the nut/bolt to 22 ft. lbs. (29 Nm).
- Control arm to crossmember. Torque the nut/bolt to 72 ft. lbs. (98 Nm).
- Control arm to rear mount. Torque the bolts to 181 ft. lbs. (245 Nm).
- Ball joint to housing, Torque the nut to 36 ft. lbs. (49 Nm).
- Wheel assembly

2005–2006

1. Before servicing the vehicle, refer to the Precautions Section.

2. Disconnect the negative battery cable.

3. Raise and support the vehicle safely.

4. Remove the tire and wheel.

5. Remove the sub frame.

6. Disconnect the stabilizer link from the transverse link.

7. Remove the bolt securing the ball joint of the transverse link to housing.

8. Remove the nut (do not remove the bolt) securing the transverse link to the crossmember.

9. Remove the two bolts securing the bushing bracket of the transverse link to the vehicle body at the rear bushing.

10. Remove the ball joint from the housing.

11. Remove the bolt securing the transverse link to the crossmember. Remove the transverse link from the crossmember.

To install:

12. Install the transverse link to its mounting.

13. Temporarily tighten the two bolts used to secure the rear bushing of the transverse link.

➡ **These bolts should be tightened so that they can still move back and forth in the oblong shaped hole in the bracket, which holds the bushing.**

14. Continue the installation in the reverse order of the removal procedure.

15. Always fully tighten the rubber bushings when the wheels are in full contact with the ground and the vehicle is at curb height.

16. Tighten the transverse link rear bushing to body to 184 ft. lbs. (250 Nm)

17. Check and adjust alignment, as required.

CONTROL ARM BUSHING REPLACEMENT

1. Before servicing the vehicle, refer to the Precautions Section.

2. Remove the control arm from the vehicle.

3. Mount the control arm in a soft jawed vise.

4. To remove the front bushing, using the proper removal tool press out the old bushing.

5. To remove the rear bushing, scribe an aligning mark on the transverse link and the rear bushing. Loosen the nut and remove the rear bushing.

6. Clean the inside bushing contact surfaces of rust and old rubber.

7. Align the bushing.

8. To install the front bushing, use the press tool.

9. To install the rear bushing, align the

mating marks and tighten using a new lock-nut to 140 ft. lbs.

➡ **While holding the rear bushing so as not to change the position of the alignment marks, tighten the locknut.**

10. Install the control arm on the vehicle.

Wheel Bearings

ADJUSTMENT

The wheel bearings are not adjustable.

REMOVAL & INSTALLATION

2002–2004

1. Before servicing the vehicle, refer to the Precautions Section.

2. Remove the steering knuckle assembly.

3. Position the steering knuckle in a soft-jawed vise.

4. Press the hub from the steering knuckle. If the inner bearing race remains in the hub, press it out.

5. Remove or disconnect the following:
- Rotor shield
- Inner and outer seals
- Snapring from the steering knuckle

6. Press the inner bearing race to remove the outer bearing.

7. Remove the Anti-lock Brakes (ABS) tone ring, if equipped

8. Press the wheel lugs from the hub.

➡ **To prevent deforming the hub, do not hammer the lugs out.**

To install:

9. Press new wheel lugs into the hub.

10. If equipped, clean all foreign material from the hub and tone ring. Install the tone ring.

11. Clean the inside of the steering knuckle.

12. Remove the plastic lock from the inner race and press a new greased bearing into the hub by pressing the outer race.

13. Install the snapring into its groove.

14. Press a new outer oil seal until it contacts the bottom of the housing.

15. Press a new inner oil seal until it contacts the circlip.

16. Apply grease to the oil seal lips.

17. Install the rotor shield and tighten the bolts to 10 ft. lbs. (14 Nm).

18. Attach the hub to the steering knuckle.

19. Press a new bearing into the hub by driving the inner race.

20. Install the steering knuckle on the vehicle.

2005–2006

1. Before servicing the vehicle, refer to the Precautions Section.

2. Disconnect the negative battery cable.

3. Raise and support the vehicle safely.

4. Remove the tire and wheel.

5. Remove the crimped section of the axle nut. Depress the brake pedal and remove the axle nut.

➡ **Be sure to loosen the axle nut after removing the tire and wheel from the vehicle. Failure to do this may damage the wheel bearings.**

6. Removed the stabilizer link.

7. Remove the disc brake caliper from its mounting as suspend it to the side, with wire. Do not disconnect the brake line.

8. Remove the rotor.

➡ **If the rotor is seized within the hub, remove the rotor by installing an 8mm bolt in the screw hole on the rotor.**

9. Remove the cotter pin and castle nut securing the tie rod end to the housing knuckle arm.

10. Using a puller, remove the tie rod ball joint from the knuckle arm.

11. Remove the ABS wheel speed sensor assembly and harness.

12. Remove the transverse link ball joint from the housing.

13. Remove the front halfshaft. Use tools ST92647000 and ST28099PA100 if necessary. Hang the halfshaft to the side using wire.

14. Scribe an alignment mark on the camber adjusting bolt head. Remove the bolts connecting the housing and strut.

15. Disconnect the housing from the strut.

To install:

16. Installation is the reverse of the removal procedure.

17. Be sure to replace the differential side retainer oil seal.

18. Be sure to use new locknuts, as required.

19. While depressing the brake pedal, tighten the NEW axle nut to 162 ft. lbs. (220 Nm). Lock it securely in place.

➡ **Install the tire and wheel after installation of the axle nut. Failure to do this may result in wheel bearing damage. Do not over tighten, as this too could cause wheel bearing damage.**

20. Continue the installation in the reverse order of the removal procedure.

REAR SUSPENSION

Strut

REMOVAL & INSTALLATION

2002–2004

✳✳ CAUTION

Do not remove the large nut on top of the strut assembly unless the coil spring is properly retained with a spring compressor.

1. Before servicing the vehicle, refer to the Precautions Section.
2. Remove or disconnect the following:
   - Strut mount cap located at the rear interior quarter trim
   - Wheel
   - Brake hose clip
   - Union bolt from the brake caliper, if equipped with disc brakes and move the brake hose out of the way
   - Brake hose and pipe from strut and drum, if equipped with drum brakes
3. Support rear with jack.
4. Remove or disconnect the following:
5. Remove or disconnect the following:
   - Retainer nuts securing the strut bearing cap to the strut tower, from inside the vehicle
   - Lower nuts and bolts securing the strut to the rear wheel housing
   - Strut

To install:
6. Install or connect the following:
   - Strut on to the vehicle, making sure to position the strut with the "4WD" mark on the strut mount facing the outside of the vehicle as shown in the illustration. Torque the retaining nuts to 15 ft. lbs. (20 Nm).
   - Strut and mount cap. Torque the strut mount cap bolts to 14.5 ft. lbs. (20 Nm).
   - Strut to the rear wheel knuckle assembly. Torque the retainer nuts/bolts to 148 ft. lbs. (200 Nm) on 2002–2004 vehicles.
   - Union bolt, if equipped with disc brakes. Torque the bolt and to 13 ft. lbs. (18 Nm).
   - Brake hose to brake pipe, if equipped with drum brakes. Torque to 10 ft. lbs. (15 Nm).
   - Brake hose clip
   - Wheel
   - Strut mount cap
7. Bleed the brakes.

2005–2006

1. Before servicing the vehicle, refer to the Precautions Section.
2. Loosen the rear wheel nuts.
3. Raise and support the vehicle safely.
4. Remove the tire and wheel.
5. Remove the brake hose clip and remove the brake hose from the rear strut.
6. Remove the bolts that retain the rear strut to the housing.

➡**Do not apply excessive tension to the brake hose and ABS wheel speed sensor harness.**

7. Remove the nuts which secure the strut mount to the vehicle body.

To install:
8. Installation is the reverse of the removal procedure.
9. Be sure to use new self locking locknuts.
10. Tighten the strut mount to body bolts to 14.5 ft. lbs.
11. Tighten the rear strut to housing bolts to 148 ft. lbs.
12. Check and adjust the rear wheel alignment, as required.

DISASSEMBLY & ASSEMBLY

2002–2004

1. Before servicing the vehicle, refer to the Precautions Section.

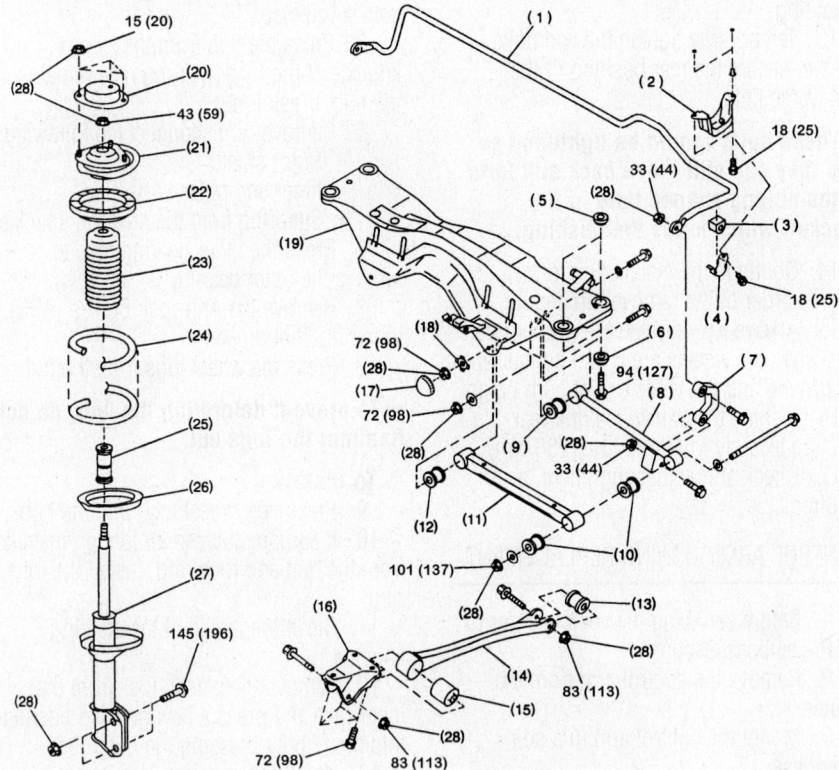

(1)	Stabilizer	(15)	Trailing link front bushing
(2)	Stabilizer bracket	(16)	Trailing link bracket
(3)	Stabilizer bushing	(17)	Cap (Protection)
(4)	Clamp	(18)	Washer
(5)	Floating bushing	(19)	Rear crossmember
(6)	Stopper	(20)	Strut mount cap
(7)	Stabilizer link	(21)	Strut mount
(8)	Rear lateral link	(22)	Rubber seat upper
(9)	Bushing (C)	(23)	Dust cover
(10)	Bushing (A)	(24)	Coil spring
(11)	Front lateral link	(25)	Helper
(12)	Bushing (B)	(26)	Rubber seat lower
(13)	Trailing link rear bushing	(27)	Damper strut
(14)	Trailing link	(28)	Self-locking nut

Rear suspension and related components—2002–2004

7924XG24A

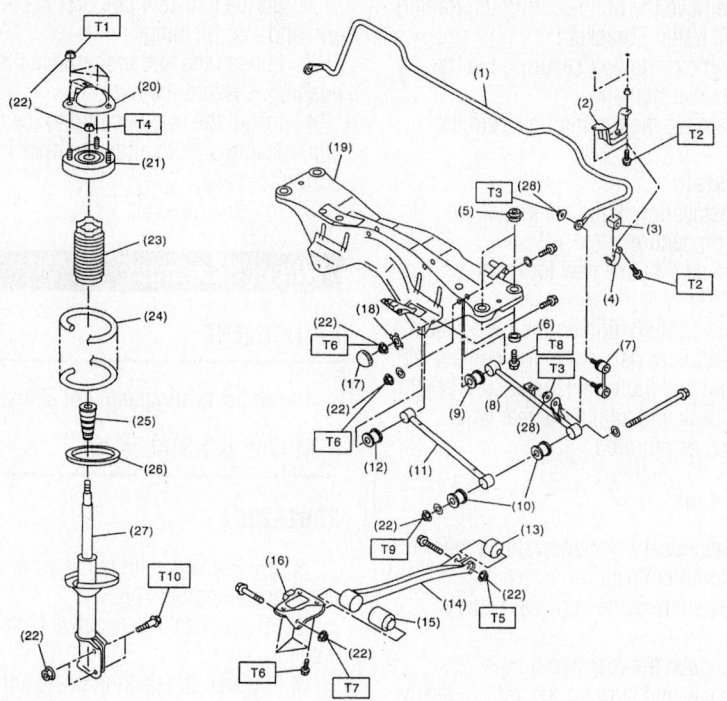

(1)	Stabilizer	(15)	Trailing link front bushing	(28)	Flange nut	
(2)	Stabilizer bracket	(16)	Trailing link bracket			
(3)	Stabilizer bushing	(17)	Cap (Protection)	**Tightening torque: N·m (kgf-m, ft-lb)**		
(4)	Clamp	(18)	Washer	T1:	20 (2.0, 14.5)	
(5)	Floating bushing	(19)	Rear crossmember	T2:	25 (2.5, 18.1)	
(6)	Stopper	(20)	Strut mount cap	T3:	45 (4.6, 33.2)	
(7)	Stabilizer link	(21)	Strut mount	T4:	60 (6.1, 44)	
(8)	Rear lateral link	(22)	Self-locking nut	T5:	90 (9.2, 66)	
(9)	Bushing	(23)	Dust cover	T6:	100 (10.2, 74)	
(10)	Bushing	(24)	Coil spring	T7:	115 (11.7, 85)	
(11)	Front lateral link	(25)	Helper	T8:	130 (13.3, 96)	
(12)	Bushing	(26)	Rubber seat lower	T9:	140 (14.3, 103)	
(13)	Trailing link rear bushing	(27)	Damper strut	T10:	200 (20.4, 148)	
(14)	Trailing link					

09490_FORE_G0014

Rear suspension and related components—2005–2006

2. Remove the strut assembly from the vehicle.

3. Place the strut assembly in a vise with a holding tool and install a spring compressor.

4. Compress the spring slightly.

5. Loosen but do not remove the bearing cap locknut.

6. Compress the spring with the spring compressor, and then remove the locknut.

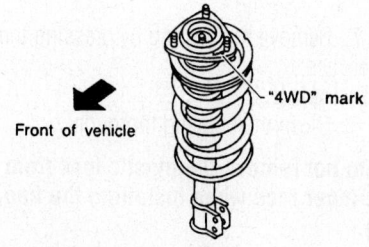

"4WD" mark

Front of vehicle

7924XG31

Rear strut upper bearing positioning— 2002–2004

7. Remove or disconnect the following:

- Strut bearing cap, mounting insulator bracket and upper spring seat
- Coil assembly, leaving the spring compressed
- Strut boot and rebound bumper from the strut. Inspect and replace if worn.
- Strut retainer nut using a suitable wrench
- Strut insert from the assembly

8. Install or connect the following:

- Strut into the chamber and install the retainer nut. Tighten the nut snugly.
- Rebound bumper and the boot to the strut piston rod
- Coil spring on the strut assembly. Be sure the spring is properly positioned on the lower bracket.
- Upper spring seat, mounting insulator and bearing cap. Be sure the

upper spring seat is facing the proper direction.

- Locknut and tighten to 41 ft. lbs. (55Nm)

9. Loosen and remove the spring compressor from the coil spring.

10. Install the strut to the vehicle.

2005–2006

1. Before servicing the vehicle, refer to the Precautions Section.

2. Remove the strut from the vehicle.

3. Using a coil spring compressor tool, carefully compress the spring. Remove the self locking nut.

4. Remove the strut mount, upper spring and rubber seat from the strut.

5. Gradually decrease the compression force of the spring compressor tool. Remove the coil spring.

6. Remove the dust cover and helper spring.

7. Check for the presence of air in the damping force generating mechanism.

8. Using the spring compression tool, compress the coil spring.

➡**Be sure to properly install the coil spring.**

9. Position the coil spring so that its end face fits good into the spring seat.

10. Install the helper spring and dust cover to the piston rod.

11. Pull the piston rod fully upward, and install the rubber seat and spring seat.

12. Install the strut mount to the piston rod, and then tighten the self locking nut, temporarily. Be sure to use a new self locking nut.

13. Use a hexagon wrench to prevent the strut rod from turning. Tighten the self locking nut to 41 ft. lbs.

14. Carefully loosen the coil spring.

Stabilizer Bar

REMOVAL & INSTALLATION

2002–2003

1. Before servicing the vehicle, refer to the Precautions Section.

2. Disconnect the negative battery cable.

3. Raise and support the vehicle safely.

4. Remove the bolts that retain the stabilizer link to the rear lateral link.

5. Remove the bolts that retain the stabilizer bracket.

6. Separate the rear stabilizer and stabilizer link.

To install:

7. Installation is the reverse of the removal procedure.

8. Install the rubber bushing, while aligning it with the paint mark on the stabilizer bar.

9. Be sure that the bushings and the stabilizer have the same identification colors.

10. Always fully tighten the rubber bushings when the wheels are in full contact with the ground and the vehicle is at curb height.

11. Tighten the stabilizer to stabilizer link bolts to 33 ft. lbs. Tighten the stabilizer link to rear lateral link bolts to 33 ft. lbs. Tighten the stabilizer to stabilizer bracket bolts to 18.1 ft. lbs.

2004–2006

1. Before servicing the vehicle, refer to the Precautions Section.

2. Raise and support the vehicle safely.

3. Remove the stabilizer link.

4. Remove the bolt which retains the stabilizer to the stabilizer bracket.

5. Remove the stabilizer bar from the vehicle.

To install:

6. Installation is the reverse of the removal procedure.

7. Be sure to use new locknuts, as required.

8. Install the rubber bushing, while aligning it with the paint mark on the stabilizer bar.

9. Be sure that the bushings and the stabilizer have the same identification colors.

10. Always fully tighten the rubber bushings when the wheels are in full contact with the ground and the vehicle is at curb height.

11. Tighten the stabilizer link to rear lateral link bolts to 33 ft. lbs. Tighten the stabilizer to stabilizer bracket bolts to 18.1 ft. lbs.

Lower Control Arm

REMOVAL & INSTALLATION

Trailing Link

1. Before servicing the vehicle, refer to the Precautions Section.

2. Disconnect the negative battery cable.

3. Loosen the rear wheel nuts.

4. Raise and support the vehicle safely.

5. Remove the tire and wheel assembly.

6. Remove the rear parking brake clamp and the ABS sensor harness.

7. Remove the bolt securing the trailing link to the trailing bracket.

8. Remove the bolt securing the trailing link to the rear housing.

9. Remove the trailing link from the vehicle.

To install:

10. Installation is the reverse of the removal procedure.

11. Be sure to use new locknuts, as required.

12. Always fully tighten the rubber bushings when the wheels are in full contact with the ground and the vehicle is at curb height.

13. Check and adjust the rear wheel alignment, as required.

Lateral Link

1. Before servicing the vehicle, refer to the Precautions Section.

2. Disconnect the negative battery cable.

3. Loosen the rear wheel nuts.

4. Raise and support the vehicle safely.

5. Remove the tire and wheel assembly.

6. Remove the stabilizer.

7. Remove the ABS sensor harness from the trailing link.

8. Remove the bolt securing the trailing link to the housing.

9. Remove the bolts that secure the lateral link assembly to the rear housing.

10. Remove the DOJ from the rear differential using tool St28099PA100.

➡**Be careful not to damage the side bearing retainer, with the tool.**

11. Scribe an alignment mark on the rear lateral link adjusting bolt and crossmember.

12. Remove the bolts securing the front and rear lateral links to the crossmember, detach the lateral links.

➡**To loosen the adjusting bolt, always loosen the nut while holding the head of the adjusting bolt.**

To install:

13. Installation is the reverse of the removal procedure.

14. Be sure to use new bolts and nuts, as required.

15. Always fully tighten the rubber bushings when the wheels are in full contact with the ground and the vehicle is at curb height.

16. Check and adjust the wheel alignment, as necessary.

CONTROL ARM BUSHING REPLACEMENT

1. Remove the control arm from the vehicle.

2. Scribe a matchmark on the control arm and rear bushing.

3. Loosen the nut and remove the rear bushing. Discard the nut.

4. Install the rear bushing to the control arm, making sure to align the marks made during removal.

5. Install a new nut.

Wheel Bearings

ADJUSTMENT

The wheel bearings are not adjustable.

REMOVAL & INSTALLATION

2002–2004

1. Before servicing the vehicle, refer to the Precautions Section.

2. Disconnect the negative battery cable.

3. Loosen the parking brake adjustment.

4. Remove or disconnect the following:
   - Wheel
   - Axle nut
   - Caliper, leaving the line connected if equipped with disc brakes and suspend it aside, then remove the rotor
   - Drum and brake line, if equipped with drum brakes
   - Parking brake cable
   - Rear stabilizer from lateral link
   - Trailing link to the housing
   - Lateral link to the housing
   - Halfshaft
   - Anti-lock Brakes (ABS), speed sensor from the backing plate, if equipped
   - Strut from the housing
   - Housing assembly

5. Using Hub Stand 92708000 and Hub Remover 927420000

6. Remove or disconnect the following:
   - Hub from the rear housing
   - Backing plate from the housing
   - Outer, inner and sub oil seals.
   - Snapring

7. Remove the bearing by pressing the inner race.

To install:

8. Clean the housing thoroughly.

➡**Do not remove the plastic lock from the inner race when installing the bearing.**

9. Install the new bearing into the housing by pressing the outer race.

ST1 927080000 HUB STAND
ST2 927420000 HUB REMOVER

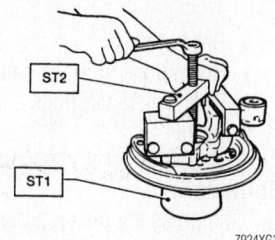

7924XG32

Use the proper tools to separate the hub from the housing to prevent damage

10. Pack the bearing with grease.

11. Install the snapring.

12. Using Installer 927460000 seal driver, press in a new outer seal until it comes in contact with the snapring.

13. Using Installer 927450000 seal driver, press in a new inner seal until it contacts the bottom.

14. Install or connect the following:
- New sub oil seal, apply grease to the oil seal lip
- Backing plate. Torque the bolts to 38 ft. lbs. (52 Nm).

15. Using Installer 927450000 bearing driver, press in the hub into the housing.

16. Install or connect the following:
- Housing to the strut. Torque the bolts to 108 ft. lbs. (147 Nm).
- Halfshaft
- Lateral link to the housing. Torque the bolt and new nut to 101 ft. lbs. (137 Nm).
- Trailing link to the housing. Torque the bolt and new nut to 94 ft. lbs. (127 Nm).
- Stabilizer to rear lateral link
- Parking brake cable and brake
- Brake line, if equipped with drum brakes
- Rotor and caliper, if equipped with disc brakes
- ABS speed sensor, if equipped
- New axle nut and tighten it to 137 ft. lbs. (186 Nm). Stake the nut.
- Wheel
- Negative battery cable

17. Adjust the parking brake cable.

2005–2006

DISC BRAKE

1. Before servicing the vehicle, refer to the Precautions Section.

2. Loosen the wheel nuts.

3. Disconnect the negative battery cable.

4. Raise and support the vehicle safely.

5. Remove the tire and wheel.

6. Remove the crimped section of the axle nut. Depress the parking brake and remove the axle nut.

➡**Be sure to loosen the axle nut after removing the tire and wheel from the vehicle. Failure to do this may damage the wheel bearings.**

7. Release the parking brake lever and loosen the locknut.

8. Remove the brake caliper from its mounting. Wire it to the side; do not allow it to hang. Do not disconnect the brake fluid line.

9. Remove the rotor.

➡**If the rotor is seized within the hub, remove the rotor by installing an 8mm bolt in the screw hole on the rotor.**

10. Disconnect the parking brake cable end.

11. Disconnect the rear stabilizer from the rear lateral link.

12. Remove the bolts that retain the trailing link assembly to the rear housing.

13. Remove the bolts that retain the lateral assembly to the rear housing.

14. Remove the rear ABS wheel speed sensor from the backing plate.

15. Disengage the BJ from the housing splines, and remove the rear halfshaft assembly. If necessary, use tool ST92647000 and ST927140000.

➡**Be careful not to damage the oil seal lip when removing the rear halfshaft. When the rear halfshaft is replaced also replace the inner oil seal.**

16. Remove the bolts that secure the rear housing to the strut. Separate the two.

To install:

17. Installation is the reverse of the removal procedure.

18. Be sure to use new bolts and nuts, as required.

19. Always fully tighten the rubber bushings when the wheels are in full contact with the ground and the vehicle is at curb height.

20. Check and adjust the wheel alignment, as necessary.

21. Adjust the rear parking brake, as required.

DRUM BRAKE

1. Before servicing the vehicle, refer to the Precautions Section.

2. Loosen the wheel nuts.

3. Disconnect the negative battery cable.

4. Raise and support the vehicle safely.

5. Remove the tire and wheel.

6. Remove the crimped section of the axle nut. Depress the parking brake and remove the axle nut.

➡**Be sure to loosen the axle nut after removing the tire and wheel from the vehicle. Failure to do this may damage the wheel bearings.**

7. Release the parking brake lever and loosen the locknut.

8. Remove the brake drum.

9. Disconnect the brake hose from the wheel cylinder.

10. Disconnect the parking brake cable end from the parking brake lever.

11. Disconnect the rear stabilizer from the rear lateral link.

12. Remove the bolts that retain the trailing link assembly to the rear housing.

13. Remove the bolts that retain the lateral assembly to the rear housing.

14. Remove the rear ABS wheel speed sensor from the backing plate.

15. Disengage the BJ from the housing splines, and remove the rear halfshaft assembly. If necessary, use tool ST92647000 and ST927140000.

➡**Be careful not to damage the oil seal lip when removing the rear halfshaft. When the rear halfshaft is replaced also replace the inner oil seal.**

16. Remove the bolts that secure the rear housing to the strut. Separate the two.

To install:

17. Installation is the reverse of the removal procedure.

18. Be sure to use new bolts and nuts, as required.

19. Always fully tighten the rubber bushings when the wheels are in full contact with the ground and the vehicle is at curb height.

20. Check and adjust the wheel alignment, as necessary.

21. Adjust the rear parking brake, as required.

BRAKES

Brake Caliper

REMOVAL & INSTALLATION

2002–2004

FRONT

1. Before servicing the vehicle, refer to the Precautions Section.
2. Remove or disconnect the following:
   - Front wheels
   - Brake hose from the caliper body
   - Caliper retainer bolts and the caliper
   - Cliper bracket, if necessary

To install:

3. Compress the piston assembly into the cylinder bore.

4. Install or connect the following:
   - Caliper bracket to the spindle assembly, if removed and secure in place with the retainer bolts. Tighten the retainer bolts to 59 ft. lbs. (80 Nm).
   - Caliper and tighten the retainers to 28 ft. lbs. (37 Nm).
   - Brake hose using new sealing washers, and tighten the fitting to 13 ft. lbs. (18 Nm)
5. Bleed the brake system.
6. Install the wheels and check the fluid level in the master cylinder.

REAR

1. Before servicing the vehicle, refer to the Precautions Section.

2. Remove or disconnect the following:
   - Rear wheels
   - Brake hose from the caliper body
   - Caliper retainer bolts and the caliper
   - Caliper bracket, if necessary

To install:

3. Compress the piston assembly into the cylinder bore.
4. Install or connect the following:
   - Caliper bracket, if removed and secure in place with the retainer bolts. Tighten the retainer bolts to 38 ft. lbs. (52 Nm).
   - Caliper and tighten the retainers to 28 ft. lbs. (37 Nm).
   - Brake hose using new sealing

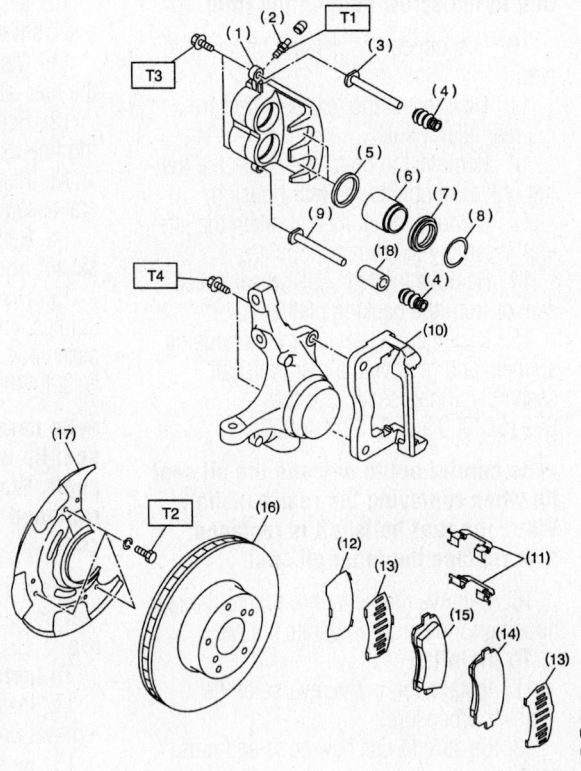

(1) Caliper body	(10) Support
(2) Air bleeder screw	(11) Pad clip
(3) Guide pin (Green)	(12) Outer shim (For Australia)
(4) Pin boot	(13) Inner shim (For Australia)
(5) Piston seal	(14) Pad (Outside)
(6) Piston	(15) Pad (Inside)
(7) Piston boot	(16) Disc rotor
(8) Boot ring	(17) Disc cover
(9) Lock pin (Yellow)	(18) Bushing

Tightening torque: N·m (kgf-m, ft-lb)
T1: 8 (0.8, 5.8)
T2: 18 (1.8, 13.0)
T3: 38 (3.9, 28)
T4: 80 (8.2, 59)

Front disc brake and related components—2002

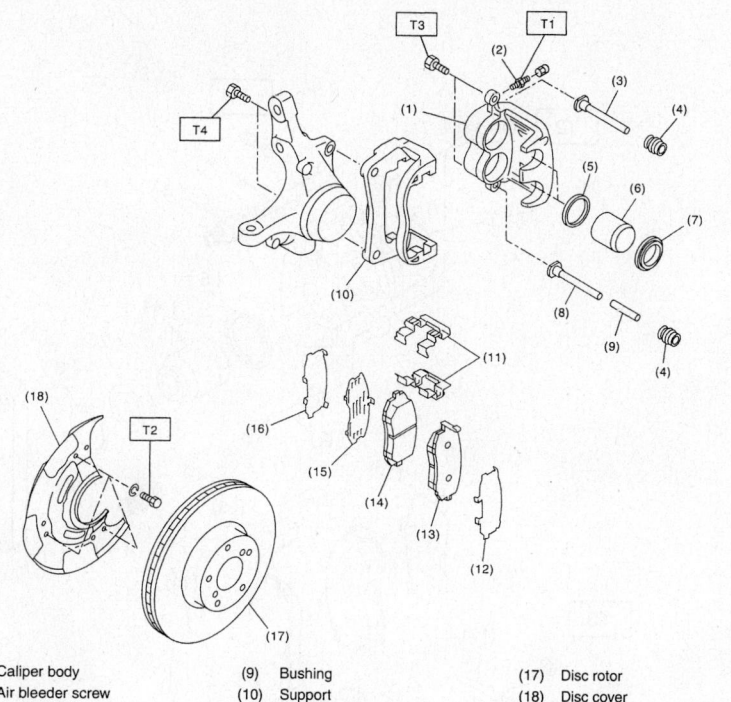

(1)	Caliper body	(9)	Bushing	(17)	Disc rotor
(2)	Air bleeder screw	(10)	Support	(18)	Disc cover
(3)	Guide pin (Green)	(11)	Pad clip		
(4)	Pin boot	(12)	Outer shim		
(5)	Piston seal	(13)	Pad (Outside)		
(6)	Piston	(14)	Pad (Inside)		
(7)	Piston boot	(15)	Rubber coated shim		
(8)	Lock pin (Yellow)	(16)	Inner shim		

Tightening torque: N·m (kgf-m, ft-lb)

T1:	8 (0.8, 5.8)
T2:	18 (1.8, 13.0)
T3:	37 (3.8, 27.5)
T4:	80 (8.2, 59)

09490_FORE_G0016

Front disc brake and related components—2003

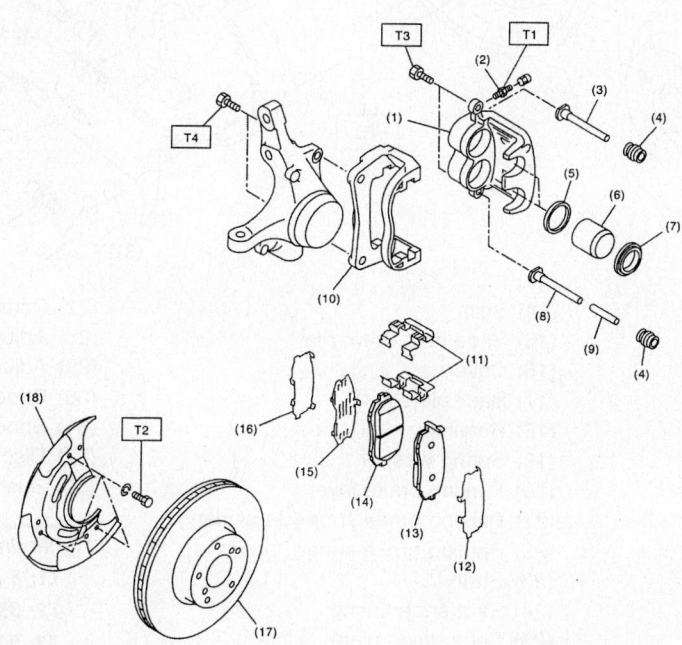

(1)	Caliper body	(9)	Bushing	(17)	Disc rotor
(2)	Air bleeder screw	(10)	Support	(18)	Disc cover
(3)	Guide pin (Green)	(11)	Pad clip		
(4)	Pin boot	(12)	Outer shim		
(5)	Piston seal	(13)	Outer pad		
(6)	Piston	(14)	Inner pad		
(7)	Piston boot	(15)	Rubber coat shim		
(8)	Lock pin (Yellow)	(16)	Inner shim		

Tightening torque: N·m (kgf-m, ft-lb)

T1:	8 (0.8, 5.8)
T2:	18 (1.8, 13.0)
T3:	26 (2.7, 19.2)
T4:	80 (8.2, 59)

09490_FORE_G0017

Front disc brake and related components—2004–2006

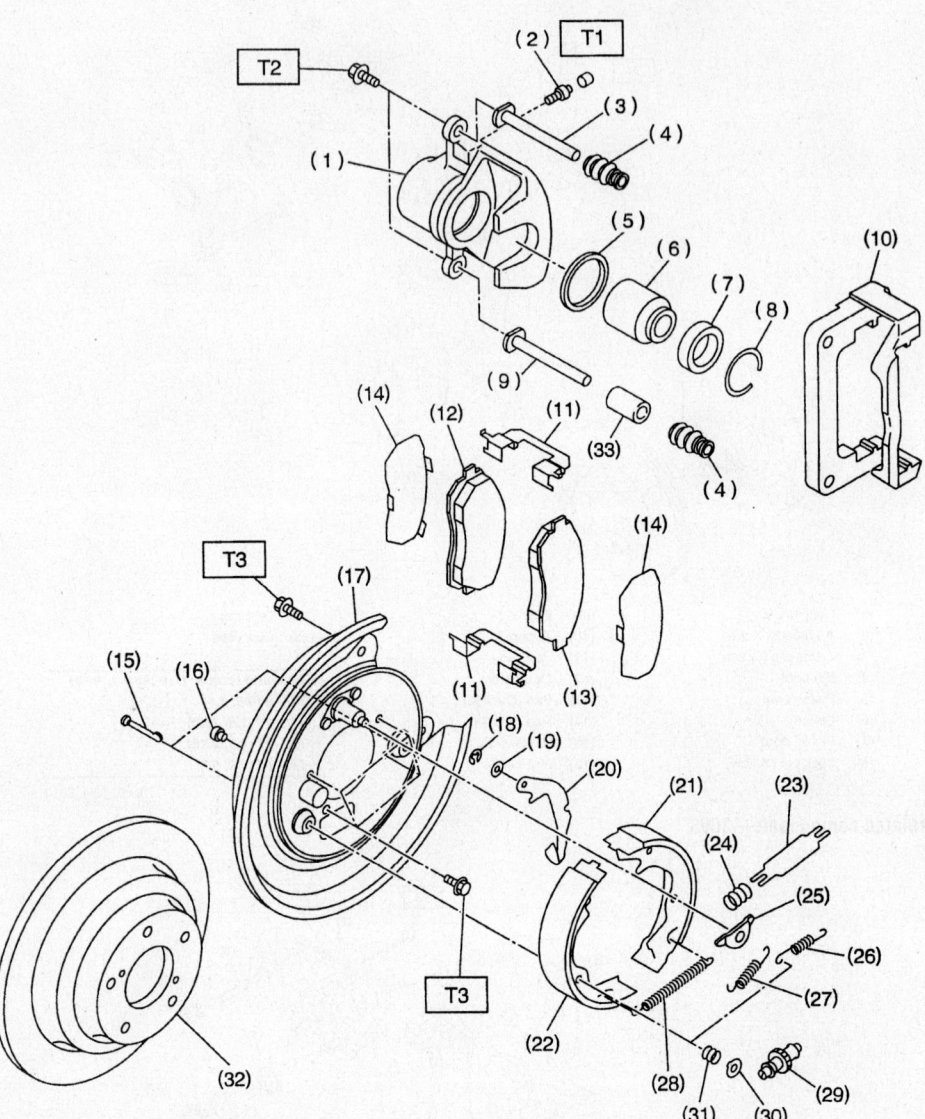

(1) Caliper body
(2) Air bleeder screw
(3) Guide pin (Green)
(4) Pin boot
(5) Piston seal
(6) Piston
(7) Piston boot
(8) Boot ring
(9) Lock pin (Yellow)
(10) Support
(11) Pad clip
(12) Inner pad
(13) Outer pad

(14) Shim
(15) Shoe hold-down pin
(16) Cover
(17) Back plate
(18) Retainer
(19) Spring washer
(20) Parking brake lever
(21) Parking brake shoe (Secondary)
(22) Parking brake shoe (Primary)
(23) Strut
(24) Strut shoe spring
(25) Shoe guide plate
(26) Secondary shoe return spring

(27) Primary shoe return spring
(28) Adjusting spring
(29) Adjuster
(30) Shoe hold-down cup
(31) Shoe hold-down spring
(32) Disc rotor
(33) Bushing

Tightening torque: N·m (kgf-m, ft-lb)
T1: 8 (0.8, 5.8)
T2: 38 (3.9, 28)
T3: 53 (5.4, 39)

09490_FORE_G0018

Rear disc brake and related components—2002

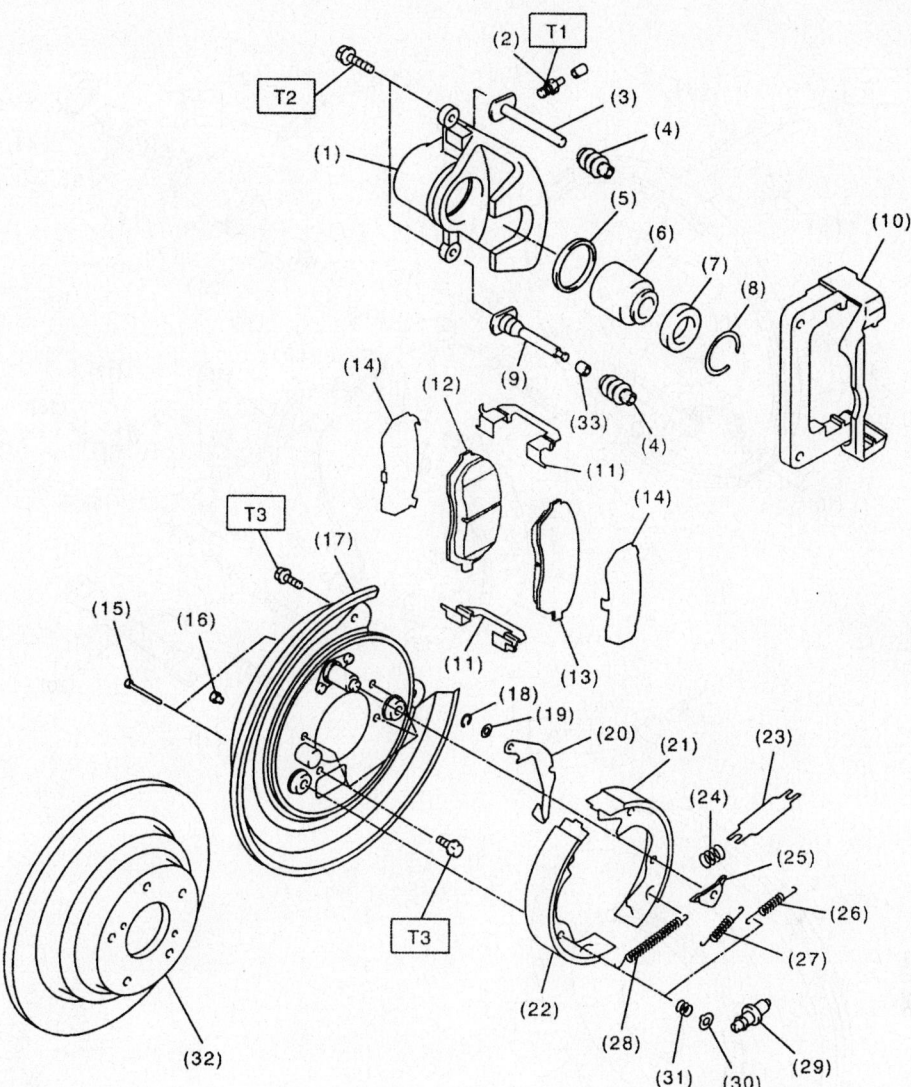

(1)	Caliper body	(14)	Shim	(27)	Primary shoe return spring
(2)	Air bleeder screw	(15)	Shoe hold-down pin	(28)	Adjusting spring
(3)	Guide pin (Green)	(16)	Cover	(29)	Adjuster
(4)	Pin boot	(17)	Back plate	(30)	Shoe hold-down cup
(5)	Piston seal	(18)	Retainer	(31)	Shoe hold-down spring
(6)	Piston	(19)	Spring washer	(32)	Disc rotor
(7)	Piston boot	(20)	Parking brake lever	(33)	Bushing
(8)	Boot ring	(21)	Parking brake shoe (Secondary)		
(9)	Lock pin (Yellow)	(22)	Parking brake shoe (Primary)		
(10)	Support	(23)	Strut		
(11)	Pad clip	(24)	Strut shoe spring		
(12)	Inner pad	(25)	Shoe guide plate		
(13)	Outer pad	(26)	Secondary shoe return spring		

Tightening torque: N·m (kgf-m, ft-lb)
T1: 8 (0.8, 5.8)
T2: 37 (3.8, 27.5)
T3: 52 (5.3, 38.3)

09490_FORE_G0019

Rear disc brake and related components—2003–2006

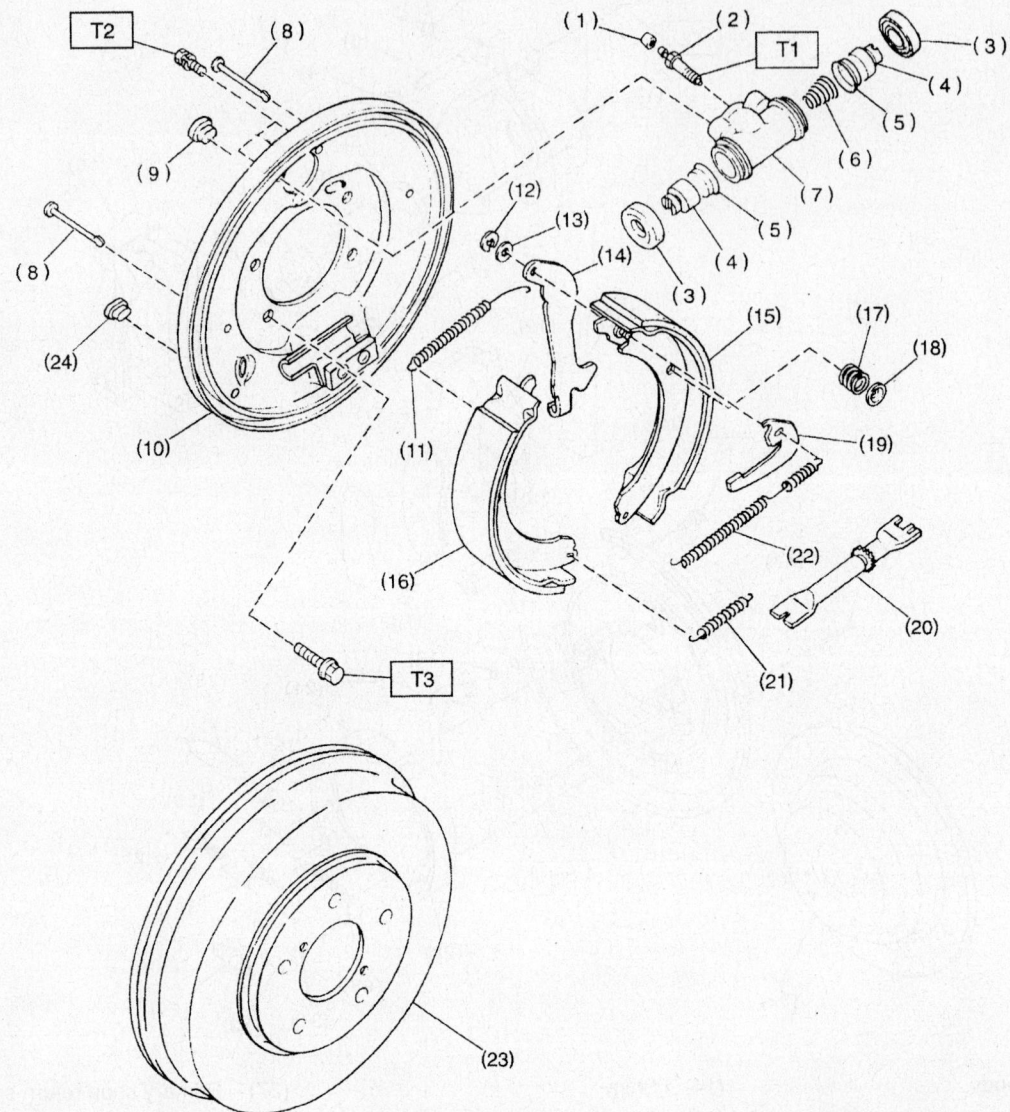

(1) Air bleeder cap
(2) Air bleeder screw
(3) Boot
(4) Piston
(5) Cup
(6) Spring
(7) Wheel cylinder body
(8) Pin
(9) Plug
(10) Back plate

(11) Upper shoe return spring
(12) Retainer
(13) Washer
(14) Parking brake lever
(15) Brake shoe (Trailing)
(16) Brake shoe (Leading)
(17) Shoe hold-down spring
(18) Cup
(19) Adjusting lever
(20) Adjuster

(21) Lower shoe return spring
(22) Adjusting spring
(23) Drum
(24) Plug

Tightening torque: N·m (kgf-m, ft-lb)
T1: 8 (0.8, 5.8)
T2: 10 (1.0, 7.2)
T3: 53 (5.4, 39)

09490_FORE_G0020

Drum brake and related components—2002

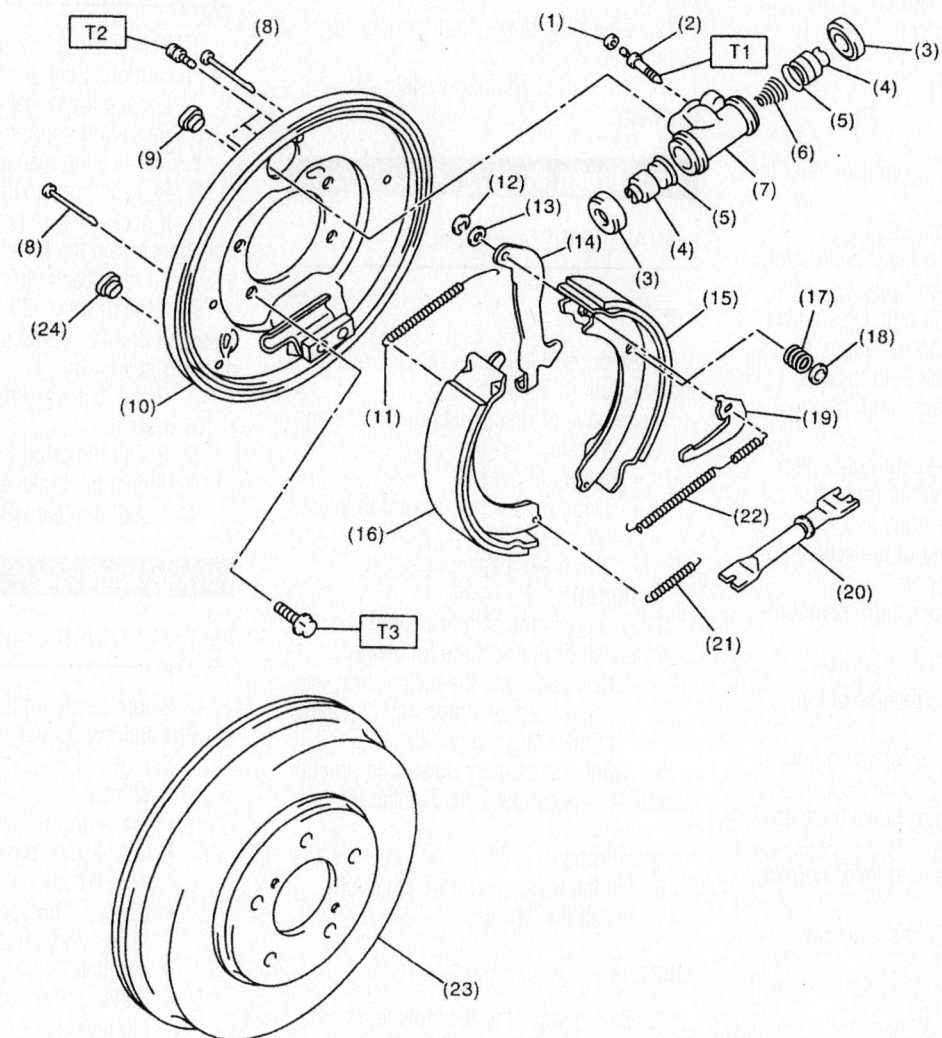

(1)	Air bleeder cap	(11)	Upper shoe return spring	(21)	Lower shoe return spring
(2)	Air bleeder screw	(12)	Retainer	(22)	Adjusting spring
(3)	Dust boots	(13)	Washer	(23)	Brake drum
(4)	Piston	(14)	Parking brake lever	(24)	Plug
(5)	Cup	(15)	Brake shoe (Trailing)		
(6)	Spring	(16)	Brake shoe (Leading)		
(7)	Wheel cylinder body	(17)	Shoe hold-down spring		
(8)	Hold-down pin	(18)	Hold-down cup		
(9)	Plug	(19)	Adjusting lever		
(10)	Back plate	(20)	Adjuster ASSY		

Tightening torque: N·m (kgf-m, ft-lb)

T1: 8 (0.8, 5.8)
T2: 10 (1.0, 7.2)
T3: 52 (5.3, 38.3)

09490_FORE_G0021

Drum brake and related components—2003–2006

washers, and tighten the fitting to 13 ft. lbs. (18 Nm)

5. Bleed the brake system.

6. Install the wheels and check the fluid level in the master cylinder.

2005–2006

FRONT

1. Before servicing the vehicle, refer to the Precautions Section.

2. Loosen the wheel nuts.

3. Raise and support the vehicle safely.

4. Remove the tire and wheel.

5. Remove the union bolt. Disconnect the brake line from the brake caliper. Be sure to properly catch the fluid to avoid damage to painted surfaces and improper disposal.

6. Remove the bolt securing the lock pin (yellow) to caliper body assembly.

7. Raise the caliper body assembly and move it toward the center of the vehicle to separate it from the support.

8. Remove the caliper from its mounting.

To install:

9. Installation is the reverse of the removal procedure.

10. Tighten the caliper retaining bolts to specification.

11. Be sure to use new brake hose gaskets.

12. Check the brake fluid level, correct as required.

13. Bleed the hydraulic system, as required.

REAR

1. Before servicing the vehicle, refer to the Precautions Section.

2. Loosen the wheel nuts.

3. Raise and support the vehicle safely.

4. Remove the tire and wheel.

5. Disconnect the brake hose from the brake caliper. Be sure to properly catch the fluid to avoid damage to painted surfaces and improper disposal.

6. Remove the bolt securing the lock pin (yellow) to caliper body assembly.

7. Raise the caliper body assembly and move it toward the center of the vehicle to separate it from the support.

8. Remove the caliper from its mounting.

To install:

9. Installation is the reverse of the removal procedure.

10. Tighten the caliper retaining bolts to specification.

11. Be sure to use new brake hose gaskets.

12. Check the brake fluid level, correct as required.

13. Bleed the hydraulic system, as required.

Disc Brake Pads

REMOVAL & INSTALLATION

Front

1. Before servicing the vehicle, refer to the Precautions Section.

2. Remove or disconnect the following:
   - Wheels
   - Lower caliper bolt
   - Swing the caliper upward to access the pads
   - Disc brake pads

To install:

3. Compress the caliper piston.

4. Install or connect the following:
   - New pads into the caliper brackets, being sure all shims and clips are in their original positions.

5. Swing the calipers down into position and install the caliper bolt. Tighten the bolt to specification.

6. Fill the master cylinder reservoir.

7. Install the wheels.

Rear

1. Before servicing the vehicle, refer to the Precautions Section.

2. Remove or disconnect the following:
   - Wheels
   - Lower caliper bolt
   - Swing the caliper upward to access the pads
   - Disc brake pads

To install:

3. Compress the caliper piston.

4. Install or connect the following:
   - New pads into the caliper brackets, being sure all shims and clips are in their original positions.

5. Swing the calipers down into position and install the caliper bolt. Tighten the bolt to specification.

6. Fill the master cylinder reservoir.

7. Install the wheels.

Brake Drums

REMOVAL & INSTALLATION

1. Before servicing the vehicle, refer to the Precautions Section.

2. Loosen the wheel nuts.

3. Raise and support the vehicle safely.

4. Remove the rear wheels.

5. Release the parking brake.

6. If necessary, remove the adjusting hole cover from the backing plate and using a suitable tool, back off the shoe adjuster.

7. If the drum is difficult to remove, insert an 8mm bolt into the hole on the drum to push it off.

8. Remove the drum.

To install:

9. Install the drum.

10. Adjust the brake shoes.

11. Install the rear wheels.

Brake Shoes

REMOVAL & INSTALLATION

1. Before servicing the vehicle, refer to the Precautions Section.

2. Remove or disconnect the following:
   - Wheels
   - Brake drum
   - Hold–down pins, springs and cups from the shoes
   - Lower return spring from both shoes
   - Shoes and adjuster from the backing plate
   - Parking brake cable from the parking lever on the trailing brake shoe
   - Upper shoe return spring and adjusting spring from the shoe

To install:

3. Apply brake grease to the backing plate where the brake shoes contact it.

4. Install or connect the following:
   - Upper return spring to the shoes
   - Parking brake cable to the lever
   - Shoes on the backing plate
   - Hold–down pins, springs and cups to the shoes
   - Lower return spring to both shoes

5. Set the outside diameter of the shoes less than 0.020–0.031 in. (0.5–0.8mm) compared to the inside diameter of the drum.

6. Install the drum and adjust the brake shoes.

7. Install the wheels.

SUBARU

B9 Tribeca

8

BRAKES 8-33
DRIVE TRAIN 8-24
ENGINE REPAIR................8-9
FRONT SUSPENSION.........8-29
FUEL SYSTEM 8-23
REAR SUSPENSION 8-31
SPECIFICATIONS AND
MAINTENANCE CHARTS 8-2
Engine and Vehicle Identification
  Chart...................................8-2
General Engine Specifications 8-2
Engine Tune-Up Specifications.......8-2
Accessory Drive Belt Routing 8-3
Capacities 8-3
Valve Specifications......................8-3
Camshaft Specifications 8-4
Crankshaft and Connecting Rod
  Specifications 8-4
Piston and Ring Specifications.......8-4
Torque Specifications 8-4
Wheel Alignment 8-7
Tire, Wheel and Ball Joint
  Specifications 8-7
Brake Specifications.....................8-7
Scheduled Maintenance Intervals 8-8
STEERING 8-28

A
Air Bag......................................8-28
  Disarming 8-28
  Precautions 8-28
Alternator 8-9
  Removal & Installation...............8-9
Automatic Transmission 8-24
  Removal & Installation...............8-24

B
Brake Caliper 8-33
  Removal & Installation 8-33
Brake Pads.................................8-34
  Removal & Installation...............8-34

C
Camshaft and Valve Lifters 8-15
  Inspection 8-17
  Removal & Installation...............8-15
CV-Joints...................................8-26
  Overhaul 8-26

Cylinder Head...........................8-13
  Removal & Installation..............8-13

E
Engine Assembly 8-9
  Removal & Installation...............8-9
Exhaust Manifold 8-15
  Removal & Installation...............8-15

F
Front Crankshaft Seal 8-15
  Removal & Installation...............8-15
Front Lateral Link........................8-32
  Removal & Installation...............8-32
Fuel Filter 8-23
  Removal & Installation...............8-23
Fuel Injector 8-24
  Removal & Installation...............8-24
Fuel Pump 8-23
  Removal & Installation...............8-23
Fuel System Pressure 8-23
  Relieving 8-23
Fuel System Service
  Precautions..............................8-23

H
Halfshaft....................................8-25
  Removal & Installation...............8-25
Heater Core................................8-10
  Removal & Installation...............8-10

I
Ignition Timing 8-9
  Adjustment..............................8-9
Intake Manifold 8-14
  Removal & Installation...............8-14

L
Lower Ball Joint..........................8-30
  Removal & Installation...............8-30
Lower Control Arm (Front) 8-30
  Control Arm Bushing
    Replacement 8-30
  Removal & Installation...............8-30
Lower Control Arm (Rear)............8-32
  Lower Control Arm Bushing
    Replacement 8-32
  Removal & Installation...............8-32

O
Oil Pan 8-19
  Removal & Installation...............8-19

Oil Pump 8-20
  Removal & Installation..............8-20

P
Piston and Ring 8-23
  Positioning 8-23
Power Steering Gear 8-28
  Removal & Installation...............8-28

R
Rear Lateral Link.........................8-32
  Removal & Installation...............8-32
Rear Main Seal 8-20
  Removal & Installation...............8-20

S
Shock Absorber 8-31
  Removal & Installation...............8-31
Stabilizer Bar (Front)...................8-30
  Removal & Installation...............8-30
Stabilizer Bar (Rear)....................8-31
  Removal & Installation...............8-31
Starter Motor 8-19
  Removal & Installation...............8-19
Strut...8-29
  Disassembly & Assembly 8-29
  Removal & Installation...............8-29

T
Timing Chain, Sprockets, Front
  Cover and Seal 8-20
  Removal & Installation...............8-20
Transfer Case Assembly................8-25
  Removal & Installation...............8-25

U
Upper Control Arm 8-32
  Removal & Installation...............8-32

V
Valve Lash 8-17
  Adjustment..............................8-17

W
Water Pump 8-10
  Removal & Installation...............8-10
Wheel Bearings (Front).................8-30
  Adjustment..............................8-30
  Removal & Installation...............8-30
Wheel Bearings (Rear)..................8-32
  Adjustment..............................8-32
  Removal & Installation...............8-32

SPECIFICATIONS AND MAINTENANCE CHARTS

ENGINE AND VEHICLE IDENTIFICATION CHART

| Engine Code | | | | | | | Model Year | |
Code ①	Liters (cc)	Cu. In.	Cyl.	Fuel Sys.	Type	Eng. Mfg.	Code ②	Year
8	3.0 (3000)	183	6	MFI	DOHC	Subaru	6	2006

MFI: Multiport Fuel Injection

DOHC: Double Overhead Camshaft

① 6th digit of the VIN

② 10th digit of the VIN

09490_TRIB_C0001

GENERAL ENGINE SPECIFICATIONS

Year	Model	Engine Displacement Liters (VIN)	Net Horsepower @ rpm	Net Torque @ rpm (ft. lbs.)	Bore x Stroke (in.)	Com-pression Ratio	Oil Pressure @ rpm
2006	B9 Tribeca	3.0 (8)	250@6600	219@4200	3.51x3.15	10.7:1	20@600

09490_TRIB_C0002

ENGINE TUNE-UP SPECIFICATIONS

Year	Liters (VIN)	Engine Displacement Liters (VIN)	Spark Plug Gap (in.)	Ignition Timing (deg.) ①		Fuel Pump (psi)	Idle Speed (rpm)		Valve Clearance ②	
				MT	AT		MT	AT	In.	Ex.
2006	B9 Tribeca	3.0 (8)	0.028-0.031	–	15	③	–	④	⑤	⑥

Note: The Vehicle Emission Control Information label often reflects specification changes made during production.

The lable figures must be used if they differ from those in this chart.

① Before Top Dead Center. At idle. +/- 8 degrees.

② Measured with engine cold

③ 49-50.5 at operating temperature

④ 600-700 in N. 720-820 in N with AC on.

⑤ 0.0079 +0.0016/-0.0024

⑥ 0.0138 +/-0.0020

09490_TRIB_C0003

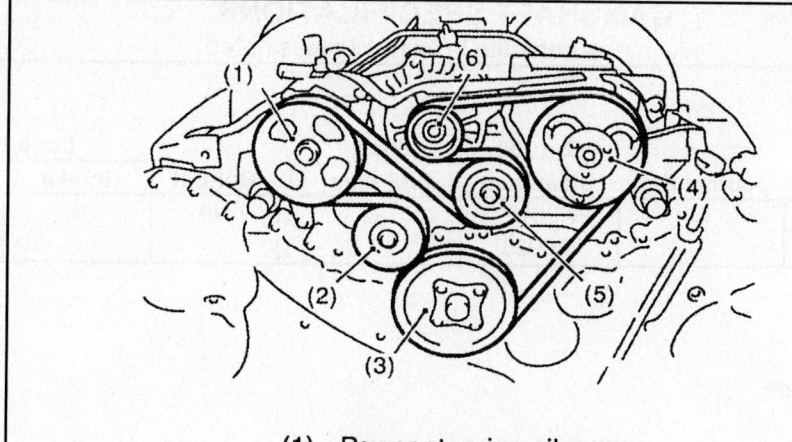

(1) Power steering oil pump
(2) Belt tension adjuster
(3) Crank pulley
(4) A/C compressor
(5) Belt idler
(6) Generator

09490_TRIB_G0001

Accessory drive belt routing

CAPACITIES

Year	Liters (VIN)	Engine Displacement Liters (VIN)	Engine Oil with Filter (qts.)	Transmission (pts.) 5-Spd	Transmission (pts.) Auto.	Transfer Case (pts.)	Drive Axle Front ① (pts.)	Drive Axle Rear (pts.)	Fuel Tank (gal.)	Cooling System (qts.)
2006	B9 Tribeca	3.0 (8)	5.8	–	17.2	–	2.4	1.4	14.1	6.5

Note: All capacities are approximate. Add fluid gradually and check to be sure a proper fluid level is obtained.

① A/T differential

09490_TRIB_C0004

VALVE SPECIFICATIONS

Year	Engine Displacement Liters (VIN)	Seat Angle (deg.)	Face Angle (deg.)	Spring Test Pressure (lbs. @ in.)	Spring Installed Height (in.)	Stem-to-Guide Clearance (in.) Intake	Stem-to-Guide Clearance (in.) Exhaust	Stem Diameter (in.) Intake	Stem Diameter (in.) Exhaust
2006	3.0 (8)	90	NA	NA	①	0.0012-0.0022	0.0016-0.0026	002148-0.2154	0.2148-0.2150

NA: Not Available

① Free length: Intake inner 1.557 in., outer 1.621 in. exhaust 1.824 in.

09490_TRIB_C0005

CAMSHAFT SPECIFICATIONS
All measurements in inches unless noted

Year	Engine Displacement Liters (VIN)	Journal Dia.	Brg. Oil Clearance	Shaft End-play ①	Runout	Lobe Height Intake	Lobe Height Exhaust
2006	3.0 (8)	②	0.0015-0.0028	③	NA	④	1.6398-1.6437

NA: Not Available

① Side clearance

② Front: 1.4939-1.4946
   Except front: 1.0215-1.0222

③ Intake: 0.0030-0.0053
   Exhaust: 0.0012-0.0035

④ High: 1.6571-1.6610
   Low 1: 1.5016-1.5055
   Low 2: 1.3756-1.3795

09490_TRIB_C0006

CRANKSHAFT AND CONNECTING ROD SPECIFICATIONS
All measurements are given in inches.

Year	Engine Displacement Liters (VIN)	Crankshaft Main Brg. Journal Dia.	Crankshaft Main Brg. Oil Clearance	Crankshaft Shaft End-play	Crankshaft Thrust on No.	Connecting Rod Journal Diameter	Connecting Rod Oil Clearance	Connecting Rod Side Clearance
2006	3.0 (8)	2.5194-2.5200	0.0004-0.0012	0.0012-0.0045	NA	NA	0.0006-0.0017	0.0028-0.0130

NA: Not Available

09490_TRIB_C0007

PISTON AND RING SPECIFICATIONS
All measurements are given in inches.

Year	Engine Displacement Liters (VIN)	Piston Clearance	Ring Gap Top Compression	Ring Gap Bottom Compression	Ring Gap Oil Control	Ring Side Clearance Top Compression	Ring Side Clearance Bottom Compression	Ring Side Clearance Oil Control
2006	3.0 (8)	NA	0.0079-0.0138	0.0138-0.0197	0.0079-0.0236	0.0016-0.0031	0.0012-0.0028	0.0018-0.0049

NA: Not Available

09490_TRIB_C0008

TORQUE SPECIFICATIONS
All readings in ft. lbs.

Year	Engine Displacement Liters (VIN)	Cylinder Head Bolts	Main Bearing Bolts	Rod Bearing Bolts	Crankshaft Damper Bolts	Flywheel Bolts	Manifold		Spark Plugs	Oil Pan Drain Plug
							Intake	Exhaust		
2006	3.0 (8)	①	②	39	131	60	18	22	15	33

① Step 1: Tighten all bolts to 14 ft. lbs.

Step 2: Tighten all bolts to 37 ft. lbs.

Step 3: Loosen all bolts 180 degrees, and again 180 degrees

Step 4: Tighten all bolts to 14 ft. lbs.

Step 5: Bolts (1-4) 35.4 ft. lbs.

Step 6: Bolts (5-8) 33 ft. lbs.

Step 7: + 90 degrees

Step 8: Bolts (1-4) 45 degrees

② Step 1: Bolts (1-11 and 13) 18 ft. lbs.

Bolts (12 and 14) 14. ft. lbs.

Step 2: Retighten bolts (1-11 and 13) 18. ft. lbs.

Retighten bolts (12 and 14) 14. ft. lbs.

Step 3: + 90 degrees

Step 4: Upper bolt to cylinder block 18 ft. lbs.

Step 7: + 90 degrees

Step 8: Bolts (1-4) 45 degrees

09490_TRIB_C0009

1) After setting the cylinder block to ST, install the crankshaft bearing.

NOTE:
Remove oil on the mating surface of bearing and cylinder block before installation. Apply a coat of engine oil to crankshaft pins.
2) Position the crankshaft and connecting rod on #2, #4 and #6 cylinder block.
3) Apply liquid gasket to the mating surface of the #1, #3 and #5 cylinder blocks, and position it on #2, #4 and #6 cylinder blocks.

NOTE:
Do not allow liquid gasket to run over to oil passages, bearing grooves, etc.

Applying liquid gasket diameter:
1.0±0.2 mm (0.039±0.008 in)

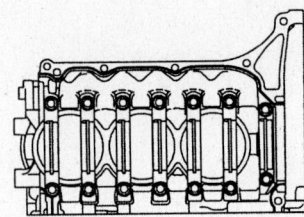

4) Apply a coat of engine oil to the washer and bolt thread.
5) Tighten all bolts in the numerical order as shown in the figure.

Tightening torque:
(1) — (11), (13): 25 N·m (2.5 kgf-m, 18 ft-lb)
(12), (14): 20 N·m (2.0 kgf-m, 14 ft-lb)

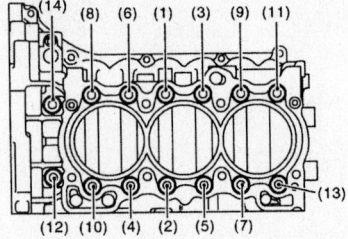

6) Retighten all bolts in the numerical order as shown in the figure.

Tightening torque:
(1) — (11), (13): 25 N·m (2.5 kgf-m, 18.4 ft-lb)
(12), (14): 20 N·m (2.0 kgf-m, 14 ft-lb)

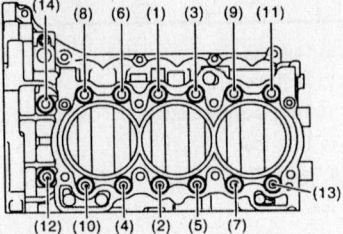

7) Tighten all bolts 90° — 110° in the numerical order as shown in the figure.

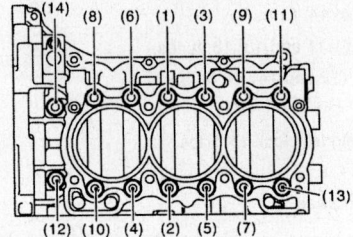

8) Install the upper bolt to cylinder block.

Tightening torque:
25 N·m (2.5 kgf-m, 18 ft-lb)

NOTE:
Remove the liquid gasket which is running over to sealing surface between cylinder block and rear chain cover, cylinder block and oil pan upper, after tightening the bolts which combine the cylinder block.

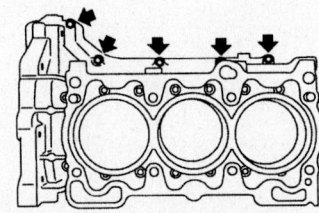

09490_SBCR_G0009

Main bearing torque sequence

WHEEL ALIGNMENT

Year	Model		Caster		Camber		Toe-in (in.)
			Range (+/-Deg.)	Preferred Setting (Deg.)	Range (+/-Deg.)	Preferred Setting (Deg.)	
2006	B9 Tribeca	F	—	①	—	②	③
		R	—	—	—	④	⑤

① Referential value: 4 degrees 04'

② Difference between right and left is 45' or less: 0 degrees 00'

③ 0 +/-0.12: toe angle (sum of both wheels) 0 degrees +/-0 degrees 14'

④ Difference between right and left is 45' or less: -0 degrees 31'

⑤ 0.08 +/-0.08: toe angle (sum of both wheels) 0 degrees +/-0 degrees 14'

09490_TRIB_C0010

TIRE, WHEEL AND BALL JOINT SPECIFICATIONS

Year	Model	OEM Tires		Tire Pressures (psi)		Wheel Size	Ball Joint Inspection	Lug Nut
		Standard	Optional	Front	Rear			
2006	Legacy	P255/55 R18	None	33	32	8JJ	—	81

OEM: Original Equipment Manufacturer

PSI: Pounds Per Square Inch

09490_TRIB_C0011

BRAKE SPECIFICATIONS
All measurements in inches unless noted

Year	Model		Brake Disc			Brake Drum Diameter			Minimum Lining Thickness		Brake Caliper	
			Original Thickness	Minimum Thickness	Maximum Runout	Original Inside Diameter	Max. Wear Limit	Maximum Machine Diameter	Front	Rear	Bracket Bolts (ft. lbs.)	Mounting Bolts (ft. lbs.)
2006	B9 Tribeca	F	1.180	1.100	0.0020	NA	NA	NA	0.059	—	88.5	19.9
		R	0.710	0.630	0.0020	NA	NA	NA	—	0.059	27.2	19.9

NA: Not Available

09490_TRIB_C0012

SCHEDULED MAINTENANCE INTERVALS

SUBARU—B9 TRIBECA

TO BE SERVICED	TYPE OF SERVICE	VEHICLE MILEAGE INTERVAL (x1000)												
		7.5	15	22.5	30	37.5	45	52.5	60	67.5	75	82.5	90	97.5
Engine oil & filter ①	R	✓	✓	✓	✓	✓	✓	✓	✓	✓	✓	✓	✓	✓
Brake lines	S/I		✓		✓		✓		✓		✓		✓	
Disc brake pads & discs, front & rear axle boots & axle shaft joint portions	S/I		✓		✓		✓		✓		✓		✓	
Parking brake	S/I		✓		✓		✓		✓		✓		✓	
Steering & suspension	S/I		✓		✓		✓		✓		✓		✓	
Air filter element	R				✓				✓				✓	
Engine coolant	R				✓				✓				✓	
Spark plugs	R				✓				✓				✓	
Automatic transmission fluid & filter	S/I				✓				✓				✓	
Brake fluid	R				✓				✓				✓	
Brake linings & drums	S/I				✓				✓				✓	
Camshaft drive belt	S/I				✓				✓				✓	
Coolant level, hoses & clamps	S/I				✓				✓				✓	
Drive belts	S/I				✓				✓				✓	
Fuel system, hoses & connections	S/I				✓				✓				✓	
Transmission and/or differential gear fluid	S/I				✓				✓				✓	
Tires (rotate)	S/I	✓	✓	✓	✓	✓	✓	✓	✓	✓	✓	✓	✓	✓
Front & rear wheel bearing	I								✓					

R: Replace   S/I: Service or Inspect

① first oil change 3,000

FREQUENT OPERATION MAINTENANCE (SEVERE SERVICE)

If a vehicle is operated under any of the following conditions it is considered severe service:

- Extremely dusty areas.

- 50% or more of the vehicle operation is in 32°C (90°F) or higher temperatures, or constant operation in temperatures below 0°C (32°F).

- Prolonged idling (vehicle operation in stop and go traffic).

- Frequent short running periods (engine does not warm to normal operating temperatures).

- Police, taxi, delivery usage or trailer towing usage.

Oil & oil filter change: change every 3750 miles.

Air filter element: service or inspect every 15,000 miles.

Automatic transmission fluid: service or inspect every 15,000 miles.

Brake linings & drums: service or inspect every 15,000 miles.

Coolant level, hoses & clamps: service or inspect every 15,000 miles.

Drive belts: service or inspect every 15,000 miles.

Transmission/differential gear oil: service or inspect every 15,000 miles.

Front & rear wheel bearing: service or inspect every 30,000 miles.

09490_TRIB_C0013

ENGINE REPAIR

➡**Disconnecting the negative battery cable may interfere with the functions of the on board computer systems and may require the computer to undergo a relearning process, once the negative battery cable is reconnected.**

Alternator

REMOVAL & INSTALLATION

1. Before servicing the vehicle, refer to the Precautions Section.
2. Disconnect the negative battery cable.
3. Remove the collector cover.
4. Disconnect the electrical connector and the terminal from the alternator.
5. Position a suitable tool on the belt tension assembly mounting bolt. Rotate the tool clockwise and loosen the drive belt. Remove the drive belt. Remove the drive belt cover.

➡**Upon installation make sure that the automatic belt tensioner indicator indicates proper belt alignment.**

6. Remove the alternator retaining bolts. Remove the alternator from the vehicle.
   To install:
7. Installation is the reverse of the removal procedure.
8. Tighten the retaining bolts to 18 ft. lbs. (25 Nm).

Ignition Timing

ADJUSTMENT

This vehicle is equipped with Distributorless Ignition System (DIS). The ignition timing is controlled by the Powertrain Control Module (PCM) and is not adjustable.

Engine Assembly

✳✳ CAUTION

Whenever working near any of the Supplemental Restraint System (SRS) components, such as the impact sensors, the air bag module, steering column and instrument panel, properly disable the SRS.

REMOVAL & INSTALLATION

1. Before servicing the vehicle, refer to the Precautions Section.

2. Open the hood to the full open position.

➡**Change the bolt mounting position of the hood hinge from the top mounting to the bottom.**

3. Remove the collector cover.
4. Properly relieve the fuel system pressure. Remove the fuel cap.
5. Properly disarm the SRS system.
6. Discharge the air conditioning system.
7. Disconnect the negative battery cable. Remove the battery.
8. Drain the engine oil. Drain the cooling system.
9. Remove the air intake duct, air cleaner case and air intake chamber.
10. Remove the engine front cover.
11. Remove the radiator.

➡**Be sure to protect the condenser from damage.**

12. Remove the fuel hose bracket.
13. Position a suitable tool on the belt tension assembly mounting bolt. Rotate the tool clockwise and loosen the drive belt. Remove the drive belt. Remove the drive belt cover.

➡**Upon installation make sure that the automatic belt tensioner indicator, indicates proper belt alignment.**

14. Disconnect the air conditioning pressure hoses from the compressor.
15. Disconnect the engine harness connector, ground cable, power steering switch connector, alternator connector and terminal and air conditioning compressor electrical connector.
16. Disconnect the brake booster hose and the heater hoses.
17. Remove the power steering pump with the bracket.

➡**Do not disconnect the hoses from the pump body. Position the pump assembly aside, using wire.**

18. Remove the reservoir tank with the bracket.

➡**Do not disconnect the hoses from the tank body.**

19. Remove the upper bolts which hold the vacuum pump bracket to the engine and transmission.
20. Raise and support the vehicle safely.
21. Remove the center exhaust pipe.

➡**Do not let the front exhaust pipe interfere with the coolant pipes on the engine side. Remove the ground cable.**

22. Remove the lower bolts that hold the vacuum pump bracket to the engine and transmission. Remove the vacuum pump with the bracket.
23. Remove the lower transmission to engine retaining nuts.
24. Remove the front cushion rubber onto front crossmember retaining nuts.
25. Lower the vehicle.
26. Remove the service plug hole cap. Remove the torque converter to drive plate retaining bolts.
27. Remove the pitching stopper. Disconnect the fuel hoses from the fuel pipes.
28. Position the engine lifting fixture in place. Support the transmission using a suitable jack.

➡**Before separating the engine from the transmission, check to be sure nothing has been overlooked that will stop the engine from being removed. This is very important in order to facilitate reinstallation and because the transmission lowers under its own weight.**

29. Separate the engine from the transmission.
30. Remove the starter.
31. Install tool ST498277200, with will hold the torque converter in place.
32. Remove the upper side bolts that retain the transmission to the engine.
33. Slightly raise the engine. Raise the transmission, using the suitable jack. Move the engine horizontally until the mainshaft is withdrawn from the clutch cover. Carefully remove the engine from the vehicle.

➡**Be careful not to damage adjacent body parts or panels with the crank pulley, oil level gauge etc.**

34. Remove the front cushion rubber mounts.
    To install:
35. Installation is the reverse of the removal procedure.
36. Torque the upper transmission to engine retaining bolts to 37 ft. lbs. (50 Nm).
37. Torque the lower transmission to engine retaining bolts to 37 ft. lbs. (50 Nm).
38. Tighten the torque converter to drive plate bolts to 18 ft. lbs. (25 Nm).
39. Fill the engine with the proper grade and type engine oil.
40. Fill and bleed the cooling system.

41. Charge the air conditioning system.

42. Check the automatic transmission fluid level, correct as required using the proper grade and type transmission fluid.

43. Start the engine and allow it to reach normal operating temperature. Check for leaks and correct as required.

Water Pump

REMOVAL & INSTALLATION

1. Before servicing the vehicle, refer to the Precautions Section.

2. Disconnect the negative battery cable.

3. Drain the cooling system. Remove the radiator.

4. Position a suitable tool on the belt tension assembly mounting bolt. Rotate the tool clockwise and loosen the drive belt. Remove the drive belt. Remove the drive belt cover.

➡ **Upon installation make sure that the automatic belt tensioner indicator, indicates proper belt alignment.**

5. Remove the front timing chain cover. Remove the timing chain assembly.

6. Remove the water pump retaining bolts. Remove the water pump from the engine.

➡ **If the pump cannot be removed easily; screw in a bolt (A) to the threaded end and remove the water pump.**

To install:

7. Installation is the reverse of the removal procedure.

8. Be sure to use a new O ring. Apply engine coolant to the O ring, prior to installation.

9. Torque the water pump retaining bolts to 4.7 ft. lbs. (6.4 Nm).

10. Fill the radiator with the proper grade and type engine coolant.

11. Start the engine and check for leaks. Correct as required.

Heater Core

REMOVAL & INSTALLATION

➡ **Position the front wheels in the straight ahead position.**

1. Before servicing the vehicle, refer to the Precautions Section.

2. Disarm the SRS system. Wait at least 20 seconds before starting any repair work.

➡ **Air bag connectors are colored yellow. Do not use electrical equipment on these circuits. Be careful not to damage the airbag system harness when servicing.**

3. Disconnect the negative battery cable.

4. Drain the engine coolant. Discharge the air conditioning system.

5. Remove the bolts securing the expansion valve and pipe.

6. Disconnect and plug the heater hoses.

7. To remove the instrument panel upper:

- Remove the front door scuff plate and pillar lower trim
- Remove the left side instrument panel side cover
- Remove the clips, and remove the left side console lower panel
- Remove the clips, and remove the instrument panel lower under cover
- Remove the clips and hooks, disconnect the connectors, and then remove the instrument panel lower cover
- Remove the glove box
- Remove the front console panel
- Remove the instrument panel ornament and the driver's and passenger's side inner panels
- Remove the screws, and pull out the control panel. Disconnect the electrical harness and remove the control panel.

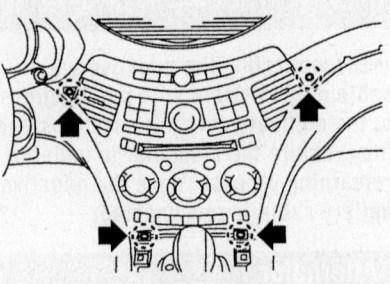

09490_TRIB_G0004

Front control panel retaining clip locations

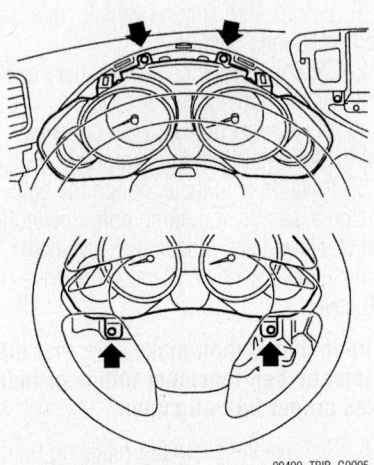

09490_TRIB_G0005

Instrument cluster retaining screw locations

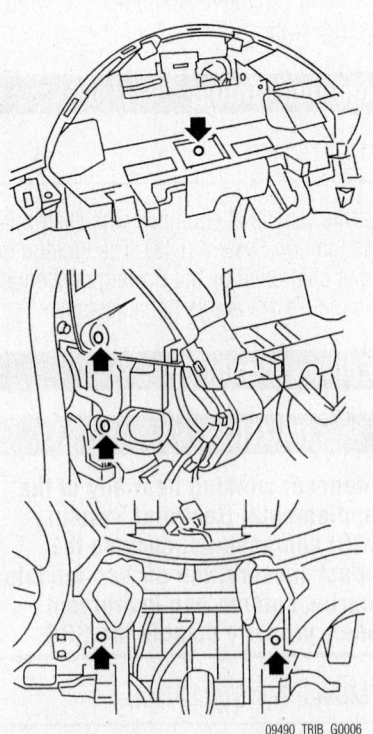

09490_TRIB_G0006

Upper Instrument panel retaining screw locations

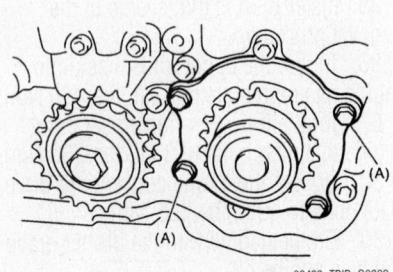

09490_TRIB_G0002

Water pump location and removal

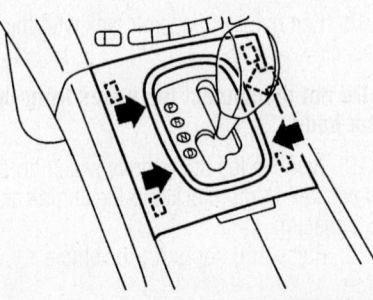

09490_TRIB_G0003

Indicator ring retainer removal location

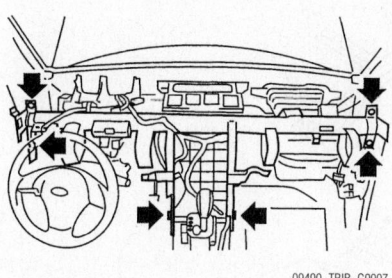

09490_TRIB_G0007

Steering support beam retaining bolt locations

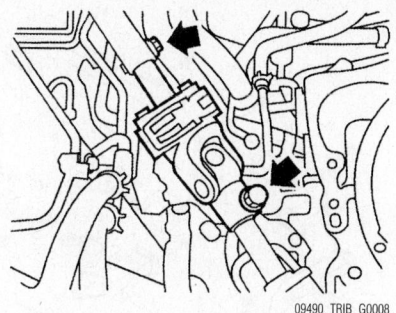

09490_TRIB_G0008

Steering column universal joint and retaining bolt location

- Remove the instrument panel lower cover
- Remove the console ring indicator and front console panel cover. The ring indicator can be removed by inserting the clip remover tool or equivalent into the positions indicated by the arrows in the illustration. Remove the screws. Remove the bolts inside the upper pocket. Pull out the console lower pocket and remove the console box.
- Remove the screws and clips and

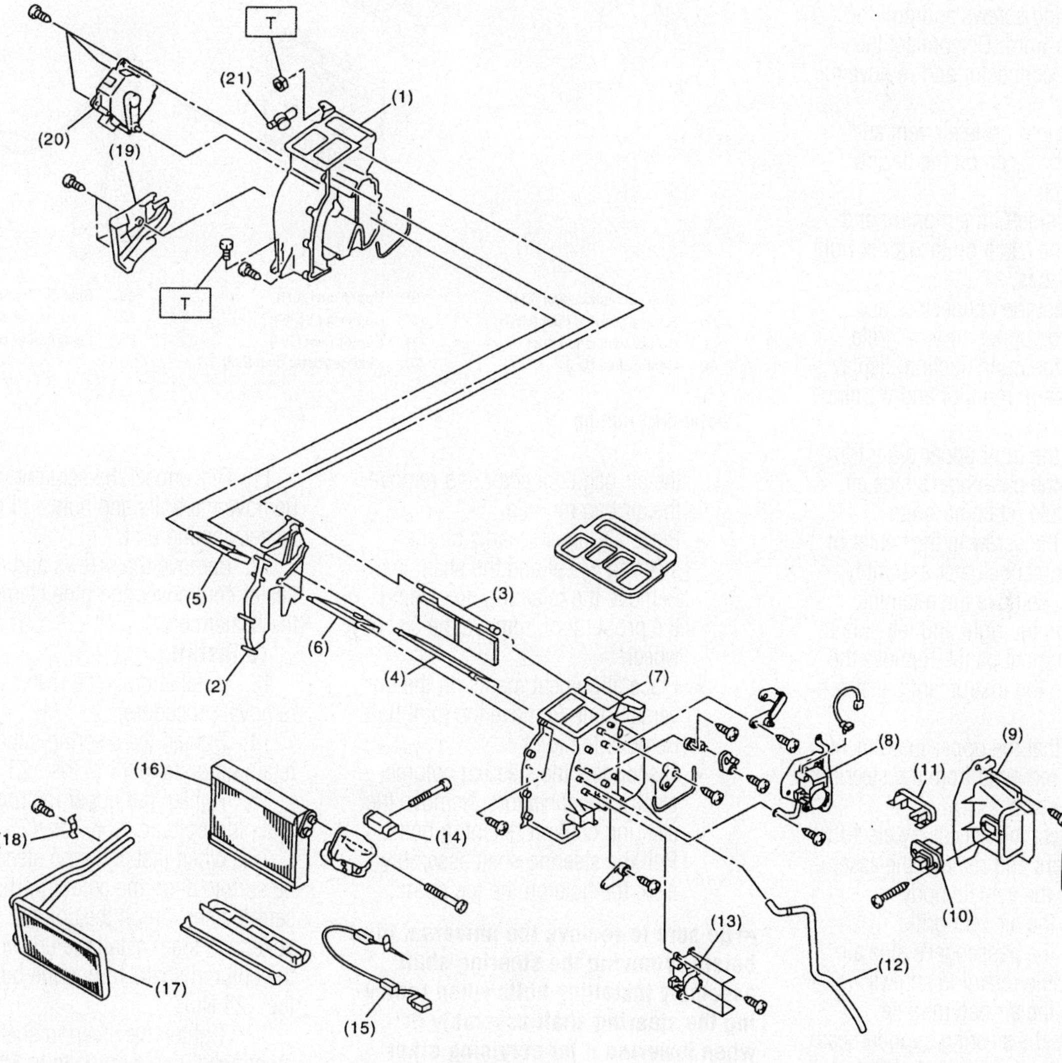

(1)	Heater unit case LH	(9)	Evaporator cover	(17)	Heater Core
(2)	Separator	(10)	Power transistor	(18)	Heater pipe clamp
(3)	Mode door RR	(11)	Pipe cover	(19)	Heater core cover
(4)	Mode door FR	(12)	Drain hose	(20)	Air mix door actuator LH
(5)	Air mix door LH	(13)	Air mix door actuator RH	(21)	Aspirator
(6)	Air mix door RH	(14)	Expansion valve		
(7)	Heater unit case RH	(15)	Evaporator sensor		
(8)	Mode door actuator	(16)	Evaporator		

Tightening torque: N·m (kgf-m, ft-lb)
   T: 7.5 (0.76, 5.5)

09490_TRIB_G0009

Heating and cooling unit and related components

remove the console side upper panel

- Remove the screws and pull out the front control panel
- Disconnect the harness connectors and remove the front control panel
- Remove the audio system retaining screws and partially pull the assembly forward. Disconnect the electrical connectors and antenna feed cable. Remove the radio assembly.
- Position the tilt steering in the lowest detent
- Remove the instrument cluster hood. Remove the instrument cluster retaining screws and pull the cluster forward. Disconnect the electrical connector and remove the assembly
- Lift the upper grille air vent and remove the catch on the hazard side switch
- Hold the side of the monitor and remove the catch on the upper right and left sides
- Disconnect the connectors and remove the upper air vent grille
- Remove the multifunction display or navigation monitor and warning box
- Remove the front upper pillar trim
- Remove the passenger's side air bag module retaining bolts
- Remove the screw in the center of the instrument cluster assembly housing. Remove the retaining screws on the right and left side of the instrument panel. Remove the screws in the instrument panel center
- Be sure that the upper instrument panel is removed from the steering support beam
- Disconnect the necessary electrical connectors and remove the assembly from the vehicle body
- Remove the air vent grille
- Remove the passenger's side air bag module retaining screws. Remove the air bag module.

8. To remove the steering support beam:
- Remove the knee guard plate retaining bolts, remove the knee guard plate

➡**The steering column may have to be removed from the vehicle for this procedure. Be sure the front wheels are in the straight ahead position.**

- Remove the screws on the side of the steering wheel. Disconnect the horn from the harness. Disconnect

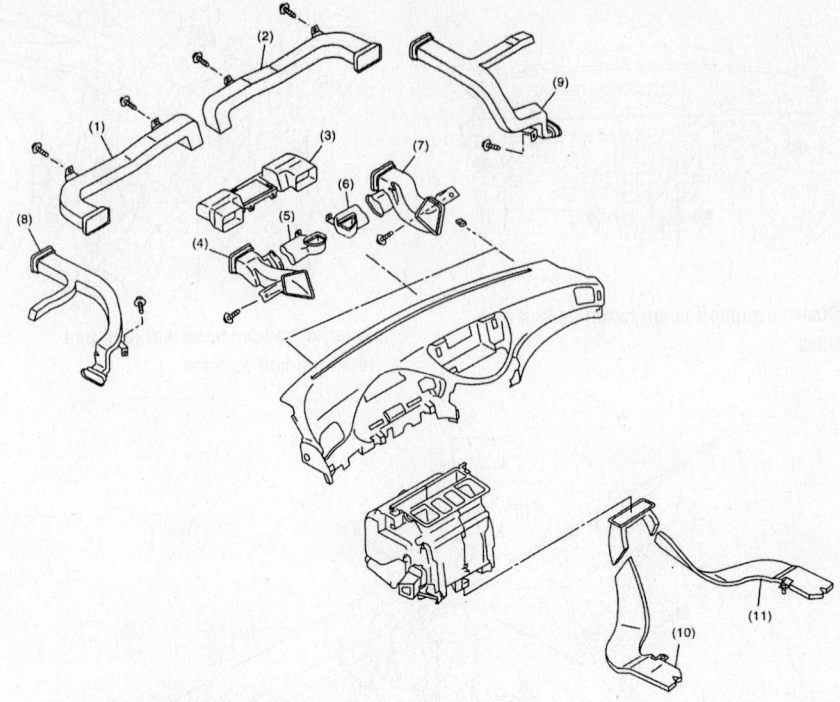

(1)	Side ventilation duct (LH)	(5)	Upper duct (LH)	(9)	Side defroster duct (RH)
(2)	Side ventilation duct (RH)	(6)	Upper duct (RH)	(10)	Rear heater duct (LH)
(3)	Center ventilation duct	(7)	Center duct (RH)	(11)	Rear heater duct (RH)
(4)	Center duct (LH)	(8)	Side defroster duct (LH)		

09490_TRIB_G0010

Heater duct routing

the air bag connector and remove the air bag module.
- Place alignment marks on the steering wheel and the shaft. Remove the retaining nut. Using the proper tool, remove the steering wheel.
- Place alignment marks on the universal joint. Remove the joint bolt. Remove the joint.
- Disconnect the steering column electrical connectors. Remove the steering column retaining bolts. Pull the steering shaft assembly from the hole on the toe board.

➡**Be sure to remove the universal joint before removing the steering shaft assembly installing bolts when removing the steering shaft assembly or when lowering it for servicing other components.**

- Remove the bolts and remove the steering support beam

9. Disconnect the electrical connectors from the air conditioning control module, intake door actuator, blower motor, power transistor and blower resistor.

10. Lift up the floor mat. Loosen the bolt and nut and remove the blower motor assembly.

11. Disconnect the actuator connector. Remove the bolts and nuts and remove the heater/cooling unit.

12. Remove the screws and remove the heater core cover and pipe clamp. Remove the heater core.

To install:

13. Installation is the reverse of the removal procedure.

14. Tighten the steering support beam retaining bolts to 18 ft. lbs. (25 Nm).

15. Tighten the upper instrument panel retaining bolts to 18 ft. lbs. (25 Nm).

16. When installing the steering column be sure to align the cutout portion at the serrated section on the column shaft and yoke, and then install the universal joint into the column shaft. Torque the bolt to 17.4 ft. lbs. (24 Nm).

17. Tighten the steering shaft to instrument panel retaining bolts to 18 ft. lbs. (25 Nm).

18. After installing the roll connector, be sure to check for proper adjustment:
- Check that the front wheels are in the straight ahead position
- Turn the roll connector pin (A) clockwise until it stops
- Turn the roll connector pins (A) approximately 3.25 turns until the triangle marks are aligned

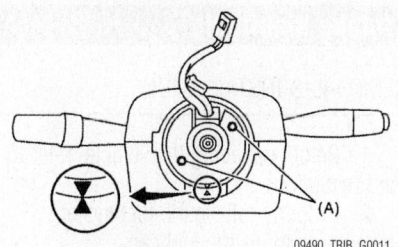

Roll pin alignment

19. Tighten the steering wheel retaining nut to 33.2 ft. lbs. (45 Nm).

20. Refill the cooling system with the proper grade and type coolant.

21. Evacuate, charge and leak test the air conditioning system.

22. Start the engine and check for leaks, correct as required.

Cylinder Head

REMOVAL & INSTALLATION

1. Before servicing the vehicle, refer to the Precautions Section.

2. Disconnect the negative battery cable.

3. Remove the crankshaft pulley cover.

4. Remove the crankshaft pulley bolt.

5. Lock the crankshaft in place using tool ST499977100, or equivalent.

6. Remove the crankshaft pulley.

7. Remove the front timing chain cover.

➡**Bolts are three different sizes. Be careful to install the correct bolt in the correct hole.**

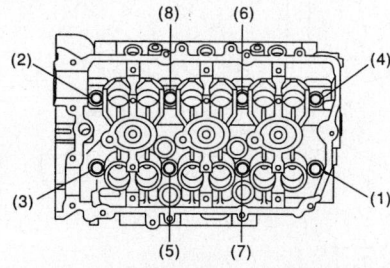

Cylinder head bolt loosening sequence

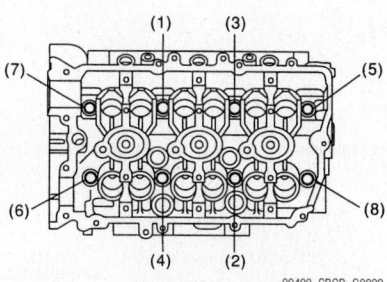

Cylinder head bolt tightening sequence

8. Remove the timing chain.

9. Remove the camshaft sprocket.

➡**Be sure to lock the camshaft in place using tool ST499977500, or equivalent.**

10. Remove the crankshaft sprocket.

11. Remove the oil pump.

12. Remove the water pump.

13. Remove the rear timing chain cover retaining bolts. Remove the rear timing chain cover.

➡**There are seven different size bolts. Be sure not to confuse them on installation.**

Fluid gasket application diameter:
(A) 1.0±0.5 mm (0.039±0.020 in)
(B) 3.0±1.0 mm (0.118±0.039 in)

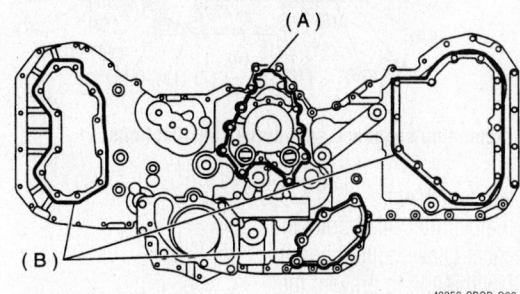

Rear timing chain cover sealant application

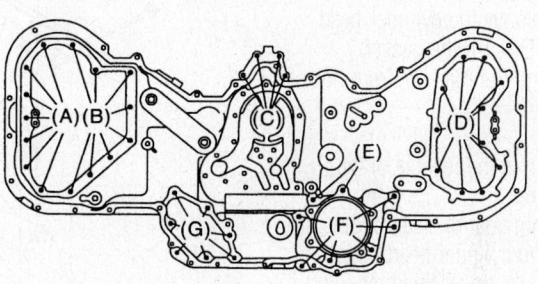

(A)	M6 × 14	(E) M8 × 40
(B)	M6 × 18 (Silver)	(F) M8 × 30
(C)	M6 × 30	(G) M6 × 22
(D)	M6 × 18	

Rear timing chain bolt sizes and locations

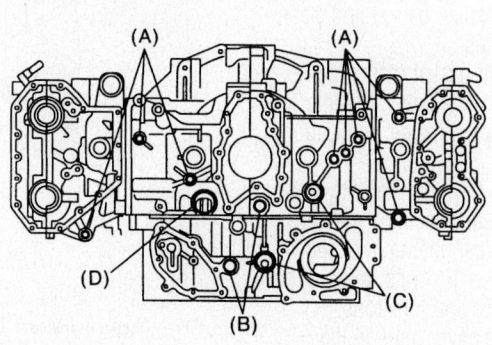

(A)	14.2 × 1.9	(C) 25 × 2
(B)	19.2 × 2.4	(D) 31.2 × 1.9

Rear cover O-ring sizes and locations

(1) — (11)	9 N·m (0.9 kgf-m, 6.5 ft-lb)
(12) — (19)	20 N·m (2.0 kgf-m, 14 ft-lb)
(20) — (30)	9 N·m (0.9 kgf-m, 6.5 ft-lb)
(31) — (38)	12 N·m (1.2 kgf-m, 8.7 ft-lb)
(39) — (45)	9 N·m (0.9 kgf-m, 6.5 ft-lb)

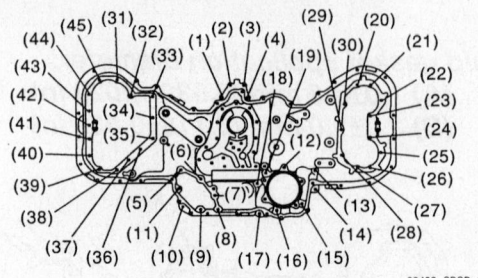

09490_SBCR_G0025

Rear cover bolt tightening sequence and torque specifications

Intake Manifold

REMOVAL & INSTALLATION

1. Before servicing the vehicle, refer to the Precautions Section.
2. Properly relieve the fuel system pressure. Remove the fuel cap.
3. Disconnect the negative battery cable. Drain the engine coolant.
4. Remove the air cleaner case and air intake chamber.
5. Remove the alternator.
6. Disconnect the electrical connector from the throttle body. Disconnect the engine coolant hoses from the throttle body.
7. Disconnect the engine harness connector. Disconnect the PCV hose. Discon-

14. Remove the camshafts.
15. Remove the cylinder head bolts in the proper sequence. Leave bolts 2 and 4 connected by a few threads to prevent the head from falling. Tap the head with a plastic mallet to separate it from the block.
16. Remove bolts 2 and 4 from the cylinder head. Remove the cylinder head from the engine. Discard the gasket.
17. Clean all gasket material from both mating surfaces.

To install:

18. Installation is the reverse of the removal procedure.
19. Apply a thin coat of clean engine oil to the washers and cylinder head bolts.
20. Tighten the cylinder head retaining bolts to specification and in the proper sequence.
21. Install the rear chain cover as follows:

➡ **There are several size bolts used, refer to the illustration for size and locations.**

a. Rear chain cover gasket and clean the mating surfaces.
b. Apply liquid gasket maker to the mating surfaces of the cover. Refer to the illustration for gasket maker application and diameter.
c. Install new O-rings. Refer to the illustration for O-ring location and size.
d. Install the rear chain cover and temporarily tighten the bolts, refer to the illustration for size and locations.
e. Tighten the cover bolts in the sequence illustrated to the specifications shown in the illustration.
22. Start the engine and allow it to reach operating temperature.
23. Check for leaks, correct as required.

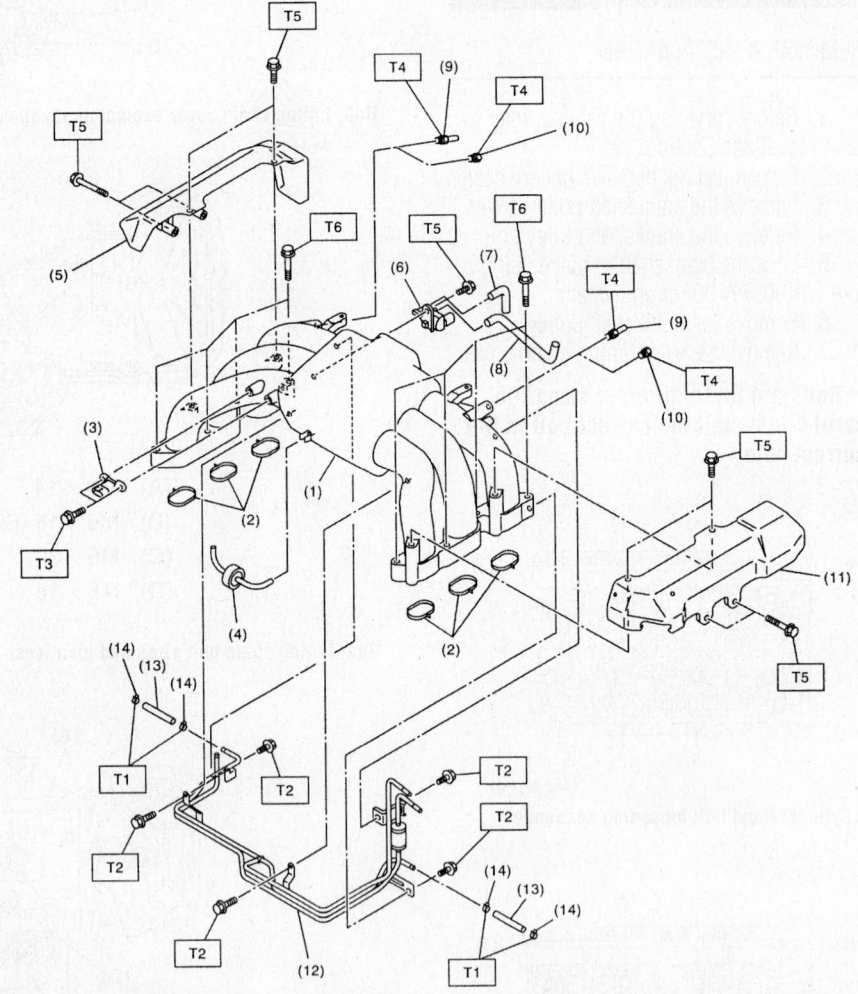

(1)	Intake manifold	(6)	Purge control solenoid valve	(11)	Fuel pipe protector LH
(2)	O-ring	(7)	Hose	(12)	Fuel pipe ASSY
(3)	Manifold absolute pressure sensor	(8)	Hose	(13)	Hose
(4)	Filter	(9)	Nipple	(14)	Clamp
(5)	Fuel pipe protector RH	(10)	Plug		

Intake manifold and related components

09490_SBCR_G0029

nect the brake booster hose. Disconnect the fuel hoses from the fuel pipe.

8. Remove the left side fuel line protector. Remove the engine harness from the left side fuel injector pipe. Remove the bolts which hold the fuel injector pipe to the left cylinder head.

9. Remove the right side fuel line protector. Remove the engine harness from the right side fuel injector pipe. Remove the bolts which hold the fuel injector pipe to the right cylinder head.

10. Remove the right and left intake manifold retaining bolts. Remove the intake manifold from the engine.

To install:

11. Installation is the reverse of the removal procedure.

12. Be sure to use new intake manifold O-rings. Torque the manifold retaining bolts to specification and in alternating sequence.

13. Be sure to fill the cooling system with the proper grade and type engine coolant.

14. Start the engine and check for leaks, correct as required.

Exhaust Manifold

Due to the unique design of the Subaru engine, an exhaust manifold is not used. The exhaust enters directly into the front Y-pipe.

REMOVAL & INSTALLATION

1. Before servicing the vehicle, refer to the Precautions Section.

2. Remove or disconnect the following:
- Negative battery cable
- Air cleaner case, if necessary
- Front Oxygen Sensor (O₂S)
- Front undercover
- Rear O₂S electrical connector
- Y-pipe-to-rear pipe mounting nuts and separate the Y-pipe from the rear pipe
- Bolts that secure the front Y-pipe to the cylinder head
- Y-pipe from the hanger bracket
- Front exhaust pipe from the catalytic converter and discard the gaskets

To install:

3. Clean all gasket surfaces completely.

4. Install or connect the following:
- New gaskets
- Front catalytic converter to front exhaust pipe.
- Y-pipe. Temporarily tighten the bolt that holds the center exhaust pipe to the hanger bracket.

- Y-pipe, to the cylinder head.
- Y-pipe to the rear pipe. Tighten the retainers.

5. Tighten the center exhaust pipe to hanger bracket
- Rear O₂S electrical connector
- Front O₂S electrical connector
- Front undercover, if equipped
- Air cleaner case, if removed
- Negative battery cable

6. Start the engine and check for exhaust leaks.

Front Crankshaft Seal

REMOVAL & INSTALLATION

1. Before servicing the vehicle, refer to the Precautions Section.

2. Remove or disconnect the following:
- Negative battery cable
- Drive belt

3. Secure the crankshaft pulley with tool No. 499977100.

4. Remove or disconnect the following:
- Crankshaft pulley bolt and pulley
- Front chain cover
- Timing chain
- Crankshaft seal

To install:

5. Using a suitable seal driver, install a new crankshaft seal.

6. Install or connect the following:
- Timing chain
- Front chain cover. Refer to the timing chain procedure in this section.
- Crankshaft pulley and tighten the bolt to 131 ft. lbs. (178 Nm)
- Drive belt
- Negative battery cable

Camshaft and Valve Lifters

REMOVAL & INSTALLATION

1. Before servicing the vehicle, refer to the Precautions Section.

2. Disconnect the negative battery cable.

3. Remove the crankshaft pulley cover.

4. Remove the crankshaft pulley bolt.

5. Lock the crankshaft in place using tool ST499977100, or equivalent.

6. Remove the crankshaft pulley.

7. Remove the front timing chain cover.

➡**Bolts are three different sizes. Be careful to install the correct bolt in the correct hole.**

8. Remove the timing chain.

9. Remove the camshaft sprocket.

➡**Be sure to lock the camshaft in place using tool ST499977500, or equivalent.**

10. Remove the crankshaft sprocket.

11. Remove the oil pump.

12. Remove the water pump.

13. Remove the rear timing chain cover retaining bolts. Remove the rear timing chain cover.

➡**There are seven different size bolts. Be sure not to confuse them on installation.**

14. Disconnect the oil pipe.

15. Remove the rocker cover retaining bolts. Remove the rocker cover. Discard the gasket.

16. Remove the plugs, see illustration for location.

17. Loosen the camshaft cap bolts in the proper sequence. Remove the camshaft caps and remove the camshaft.

To install:

18. Apply a coat of engine oil to the journals on the camshafts and place the camshafts into position.

19. To install the camshaft cap, apply a small amount of liquid gasket to the mating surface of the cap. Do not apply an excessive amount of sealant, as it will squish out and flow toward the cam journal resulting in engine seizure.

20. Apply a thin coat of engine oil to the cap bearing surface and install the cap.

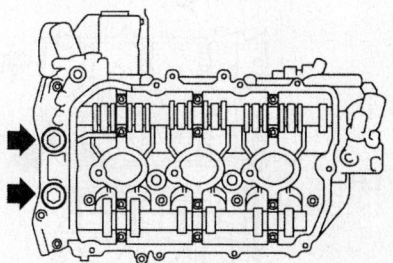

09490_SBCR_G0046

Camshaft plug bolt location

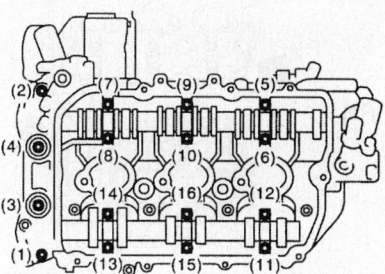

09490_SBCR_G0047

Camshaft bolt loosening sequence

Tighten cap bolts 1 through 12 to 12 ft. lbs. and bolts 13 through 16 to 7.2 ft. lbs. in the proper sequence. Install the plugs and torque to 44 ft. lbs.

21. Apply fluid gasket maker of the of the cylinder heads and valve covers.

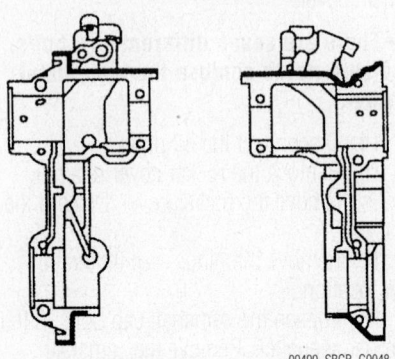

Camshaft cap liquid gasket application

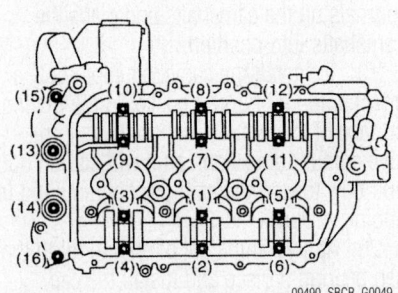

Camshaft bolt tightening sequence

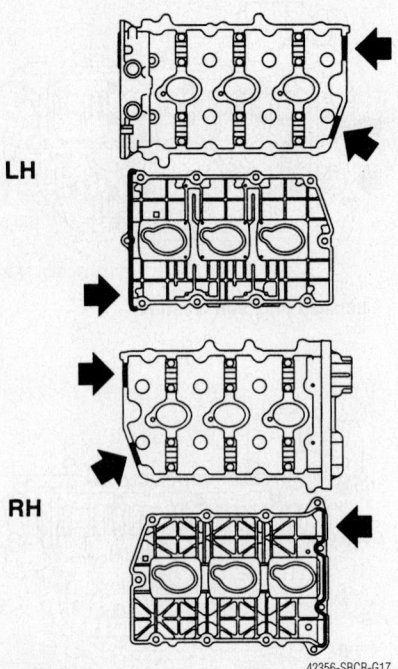

Apply fluid gasket maker of the of the cylinder heads and valve covers as shown

✳✳ CAUTION

Do not apply too much gasket maker. This may cause excess fluid gasket maker to come out and flow towards the camshaft journal resulting in engine damage.

22. Tighten the valve cover bolts to 4.7 ft. lbs. and in the proper sequence.

23. Install the rear chain cover as follows:

➡ **There are several size bolts used, refer to the illustration for size and locations.**

a. Rear chain cover gasket and clean the mating surfaces.

b. Apply liquid gasket maker to the mating surfaces of the cover. Refer to the illustration for gasket maker application and diameter.

c. Install new O-rings. Refer to the illustration for O-ring location and size

d. Install the rear chain cover and temporarily tighten the bolts, refer to the illustration for size and locations.

e. Tighten the cover bolts in the sequence illustrated to the specifications shown in the illustration.

24. Continue the installation in the reverse order of the removal procedure.

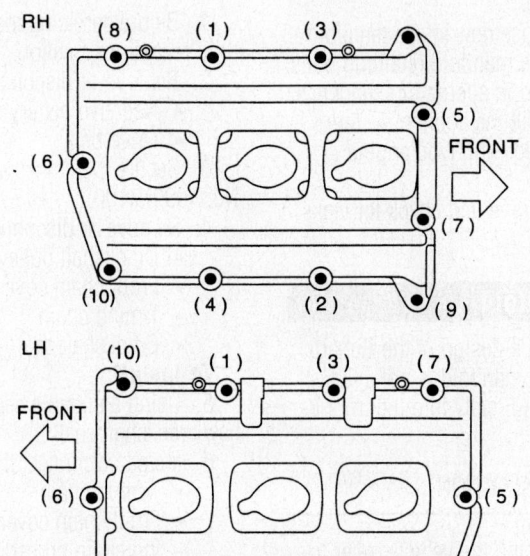

Valve cover bolt tightening sequence

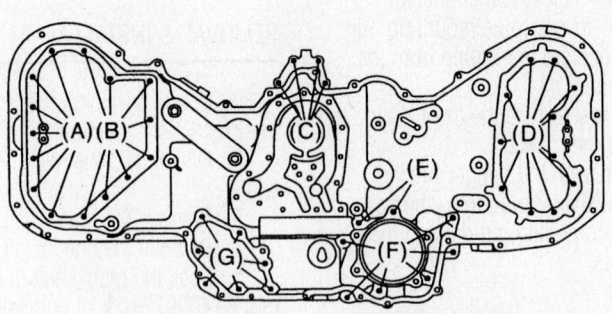

(A)	M6 × 14	(E) M8 × 40
(B)	M6 × 18 (Silver)	(F) M8 × 30
(C)	M6 × 30	(G) M6 × 22
(D)	M6 × 18	

Rear timing chain bolt sizes and locations

Fluid gasket application diameter:
(A) 1.0±0.5 mm (0.039±0.020 in)
(B) 3.0±1.0 mm (0.118±0.039 in)

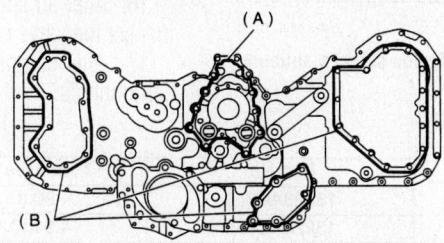

42356-SBCR-G06

Rear timing chain cover sealant application

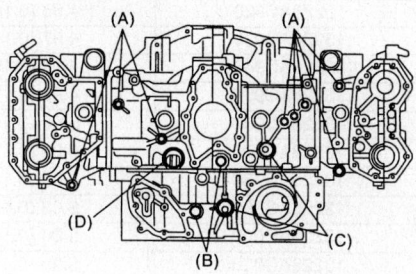

| (A) | 14.2 × 1.9 | (C) | 25 × 2 |
| (B) | 19.2 × 2.4 | (D) | 31.2 × 1.9 |

09490_SBCR_G0024

Rear cover O-ring sizes and locations

(1) — (11)	9 N·m (0.9 kgf-m, 6.5 ft-lb)
(12) — (19)	20 N·m (2.0 kgf-m, 14 ft-lb)
(20) — (30)	9 N·m (0.9 kgf-m, 6.5 ft-lb)
(31) — (38)	12 N·m (1.2 kgf-m, 8.7 ft-lb)
(39) — (45)	9 N·m (0.9 kgf-m, 6.5 ft-lb)

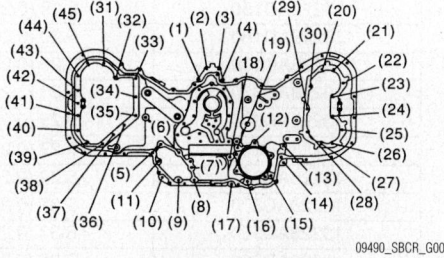

09490_SBCR_G0025

Rear cover bolt tightening sequence and torque specifications

INSPECTION

1. Before servicing the vehicle, refer to the Precautions Section.
2. Remove the camshaft from the engine.
3. Check the camshaft bearing journals for damage and binding.
4. If the journals are binding, check the cylinder head for damage.
5. Check the cylinder head for clogged oil holes.

6. Check the camshaft surface for abnormal wear and damage. Replace the camshaft, as required.
7. Measure the camshaft lobe surface and replace the camshaft if not within specification.
8. Measure the camshaft journal diameter and replace the camshaft if not within specification.
9. Measure the camshaft run out and replace the camshaft if not within specification.

Valve Lash

ADJUSTMENT

➡ **The valve adjustment should be performed while the engine is cold.**

1. Before servicing the vehicle, refer to the precautions in the beginning of this section.
2. Raise and support the vehicle safely.
3. Remove the under cover.
4. Lower the vehicle.
5. Remove the collector cover.
6. Disconnect the negative battery cable.
7. On the right side, remove the air intake duct and air cleaner case. Remove the fuel tank protector. Disconnect the oil pressure switch electrical connector. Remove the ignition coil.
8. On the left side, remove the battery and battery carrier. Disconnect the PCV hose from the rocker cover. Remove the ignition coil.
9. Remove the rocker cover retaining bolts. Remove the rocker covers from the engine.
10. Rotate the crankshaft clockwise until the cam is set in position, see illustration.
11. Using a feeler gauge measure and record the clearance of the intake and exhaust valve.

➡ **Measure it within the range of +/- 30 degrees from the specified position, shown in the illustration. Measure it in the low cam for the intake side. Insert the feeler gauge in a horizontally as possible with respect to the valve lifter.**

12. Further turn the crankshaft pulley clockwise and then measure and record the valve clearance again.
13. If adjustment is required, remove the camshafts.
14. Remove and measure the thickness

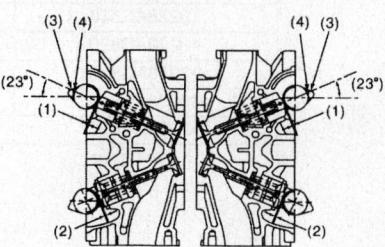

(1) Valve clearance (Intake side)
(2) Valve clearance (Exhaust side)
(3) High lift cam
(4) Low lift cam

09490_SBCR_G0053

Valve adjustment crankshaft positioning

Unit: (mm)

S = (V + T) − 0.35	
S: Valve lifter thickness required	
V: Measured valve clearance	
T: Valve lifter thickness to be used	

09490_SBCR_G0054

Use this table to help you select a suitable exhaust valve shim

Unit: (mm)

S = (V + T) − 0.20	
S: Required shim thickness	
V: Measured valve clearance	
T: Shim thickness to be used	

09490_SBCR_G0055

Use this table to help you select a suitable intake valve shim

of the valve lifter. Select a suitable shim, using the shim selection chart.

15. Install the replacement shim to the lifter.

16. After all shims have been adjusted, inspect the valve clearances again.

17. After completion, install all removed components.

Part No.	Thickness mm (in)
13228AD180	4.32 (0.1701)
13228AD190	4.34 (0.1709)
13228AD200	4.36 (0.1717)
13228AD210	4.38 (0.1724)
13228AD220	4.40 (0.1732)
13228AD230	4.42 (0.1740)
13228AD240	4.44 (0.1748)
13228AD250	4.46 (0.1756)
13228AD260	4.48 (0.1764)
13228AD270	4.50 (0.1772)
13228AD280	4.52 (0.1780)
13228AD290	4.54 (0.1787)
13228AD300	4.56 (0.1795)
13228AD310	4.58 (0.1803)
13228AD320	4.60 (0.1811)
13228AC580	4.62 (0.1819)
13228AC590	4.63 (0.1823)
13228AC600	4.64 (0.1827)
13228AC610	4.65 (0.1831)
13228AC620	4.66 (0.1835)
13228AC630	4.67 (0.1839)
13228AC640	4.68 (0.1843)
13228AC650	4.69 (0.1846)
13228AC660	4.70 (0.1850)
13228AC670	4.71 (0.1854)
13228AC680	4.72 (0.1858)
13228AC690	4.73 (0.1862)
13228AC700	4.74 (0.1866)
13228AC710	4.75 (0.1870)
13228AC720	4.76 (0.1874)
13228AC730	4.77 (0.1878)
13228AC740	4.78 (0.1882)
13228AC750	4.79 (0.1886)
13228AC760	4.80 (0.1890)
13228AC770	4.81 (0.1894)
13228AC780	4.82 (0.1898)
13228AC790	4.83 (0.1902)
13228AC800	4.84 (0.1906)
13228AC810	4.85 (0.1909)
13228AC820	4.86 (0.1913)
13228AC830	4.87 (0.1917)
13228AC840	4.88 (0.1921)
13228AC850	4.89 (0.1925)

Part No.	Thickness mm (in)
13228AC860	4.90 (0.1929)
13228AC870	4.91 (0.1933)
13228AC880	4.92 (0.1937)
13228AC890	4.93 (0.1941)
13228AC900	4.94 (0.1945)
13228AC910	4.95 (0.1949)
13228AC920	4.96 (0.1953)
13228AC930	4.97 (0.1957)
13228AC940	4.98 (0.1961)
13228AC950	4.99 (0.1965)
13228AC960	5.00 (0.1969)
13228AC970	5.01 (0.1972)
13228AC980	5.02 (0.1976)
13228AC990	5.03 (0.1980)
13228AD000	5.04 (0.1984)
13228AD010	5.05 (0.1988)
13228AD020	5.06 (0.1992)
13228AD030	5.07 (0.1996)
13228AD040	5.08 (0.2000)
13228AD050	5.09 (0.2004)
13228AD060	5.10 (0.2008)
13228AD070	5.11 (0.2012)
13228AD080	5.12 (0.2016)
13228AD090	5.13 (0.2020)
13228AD100	5.14 (0.2024)
13228AD110	5.15 (0.2028)
13228AD120	5.16 (0.2032)
13228AD130	5.17 (0.2035)
13228AD140	5.18 (0.2039)
13228AD150	5.19 (0.2043)
13228AD160	5.20 (0.2047)
13228AD170	5.21 (0.2051)
13228AD330	5.23 (0.2059)
13228AD340	5.25 (0.2067)
13228AD350	5.27 (0.2075)
13228AD360	5.29 (0.2083)
13228AD370	5.31 (0.2091)
13228AD380	5.33 (0.2098)
13228AD390	5.35 (0.2106)
13228AD400	5.37 (0.2114)
13228AD410	5.39 (0.2122)
13228AD420	5.41 (0.2130)
13228AD430	5.43 (0.2138)
13228AD440	5.45 (0.2146)
13228AD450	5.47 (0.2154)
13228AD460	5.49 (0.2161)
13228AD470	5.51 (0.2169)
13228AD480	5.53 (0.2177)
13228AD490	5.55 (0.2185)
13228AD500	5.57 (0.2193)
13228AD510	5.59 (0.2201)

09490_SBCR_G0056

Exhaust valve adjusting shim chart

Part No.	Thickness mm (in)
13218AK890	1.92 (0.0756)
13218AK900	1.94 (0.0764)
13218AK910	1.96 (0.0772)
13218AK920	1.98 (0.0780)
13218AK930	2.00 (0.0787)
13218AK940	2.02 (0.0795)
13218AK950	2.04 (0.0803)
13218AK960	2.06 (0.0811)
13218AK970	2.07 (0.0815)
13218AK980	2.08 (0.0819)
13218AK990	2.09 (0.0823)
13218AL000	2.10 (0.0827)
13218AL010	2.11 (0.0831)
13218AL020	2.12 (0.0835)
13218AL030	2.13 (0.0839)
13218AL040	2.14 (0.0843)
13218AL050	2.15 (0.0846)
13218AL060	2.16 (0.0850)
13218AL070	2.17 (0.0854)
13218AL080	2.18 (0.0858)
13218AL090	2.19 (0.0862)
13218AL100	2.20 (0.0866)
13218AL110	2.21 (0.0870)
13218AL120	2.22 (0.0874)
13218AL130	2.23 (0.0878)
13218AL140	2.24 (0.0882)
13218AL150	2.25 (0.0886)
13218AL160	2.26 (0.0890)
13218AL170	2.27 (0.0894)
13218AL180	2.28 (0.0898)
13218AL190	2.29 (0.0902)
13218AL200	2.30 (0.0906)
13218AL210	2.31 (0.0909)
13218AL220	2.32 (0.0913)
13218AL230	2.33 (0.0917)
13218AL240	2.34 (0.0921)
13218AL250	2.35 (0.0925)
13218AL260	2.36 (0.0929)
13218AL270	2.37 (0.0933)
13218AL280	2.38 (0.0937)
13218AL290	2.39 (0.0941)
13218AL300	2.40 (0.0945)
13218AL310	2.41 (0.0949)

Part No.	Thickness mm (in)
13218AL320	2.42 (0.0953)
13218AL330	2.43 (0.0957)
13218AL340	2.44 (0.0961)
13218AL350	2.45 (0.0965)
13218AL360	2.46 (0.0969)
13218AL370	2.47 (0.0972)
13218AL380	2.48 (0.0976)
13218AL390	2.49 (0.0980)
13218AL400	2.50 (0.0984)
13218AL410	2.51 (0.0988)
13218AL420	2.52 (0.0992)
13218AL430	2.53 (0.0996)
13218AL440	2.54 (0.1000)
13218AL450	2.55 (0.1004)
13218AL460	2.56 (0.1008)
13218AL470	2.57 (0.1012)
13218AL480	2.58 (0.1016)
13218AL490	2.59 (0.1020)
13218AL500	2.60 (0.1024)
13218AL510	2.61 (0.1028)
13218AL520	2.62 (0.1032)
13218AL530	2.64 (0.1039)
13218AL540	2.66 (0.1047)
13218AL550	2.68 (0.1055)
13218AL560	2.70 (0.1063)
13218AL570	2.72 (0.1071)
13218AL580	2.74 (0.1079)
13218AL590	2.76 (0.1087)

09490_SBCR_G0057

Intake valve adjusting shim chart

Starter Motor

REMOVAL & INSTALLATION

1. Before servicing the vehicle, refer to the Precautions Section.
2. Disconnect the negative battery cable.
3. Remove the collector cover.
4. Remove the air intake chamber.
5. Disconnect the electrical connectors from the starter.
6. Remove the starter retaining bolts. Remove the starter from the vehicle.

To install:
7. Installation is the reverse of the removal procedure.
8. Torque the starter retaining bolts to 37 ft. lbs.

Oil Pan

REMOVAL & INSTALLATION

➡ **If removing the upper oil pan, the engine must first be removed from the vehicle.**

1. Before servicing the vehicle, refer to the Precautions Section.
2. Disconnect the negative battery cable.
3. Raise and support the vehicle safely.
4. Remove the under cover. Drain the engine oil.
5. Remove the lower oil pan retaining bolts.
6. Insert an oil pan gasket cutter tool into the gap between the upper oil pan and the lower oil pan. Remove the lower oil pan from the engine.

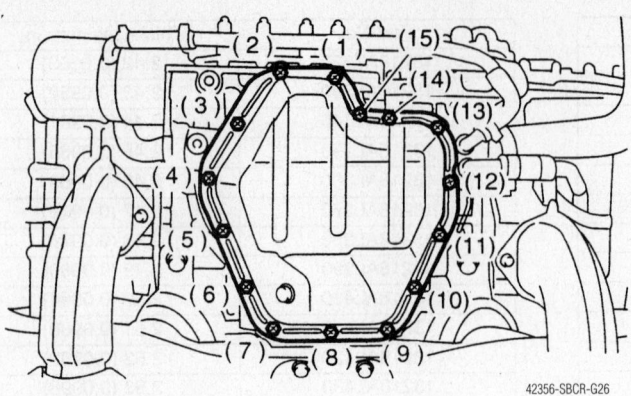

Oil pan bolt torque sequence

42356-SBCR-G26

➡**Do not use a screwdriver or similar tool in place of the cutter tool.**

7. Remove the oil strainer, if required.

To install:

8. Be sure to clean the old gasketing material from the mating surfaces.

9. Replace the O-ring and install the oil strainer. Tighten the bolt to 4.7 ft. lbs.

10. Apply a continuous bead (0.039 inch thick) of liquid gasket, part number K0877YA018 or equivalent, to the mating surfaces. Install the oil pan. Torque the retaining bolts to specification and in the proper sequence.

11. Continue the installation in the reverse order of the removal procedure.

12. Be sure to fill the engine with the correct grade and type engine oil.

13. Start the engine and check for leaks. Correct as required.

Oil Pump

REMOVAL & INSTALLATION

1. Before servicing the vehicle, refer to the Precautions Section.

2. Disconnect the negative battery cable. Drain the cooling system.

3. Remove the collector cover.

4. Raise and support the vehicle safely.

5. Remove the under cover.

6. Lower the vehicle. Remove the radiator.

7. To remove the front side V belt, remove the belt covers. Loosen the lock bolt. Loosen the slider bolt. Remove the front side belt.

8. To remove the rear side V belt, remove the belt covers. Loosen the lock bolt. Loosen the slider bolt. Remove the rear side belt.

9. Remove the crankshaft pulley cover.

10. Remove the crankshaft pulley bolt.

11. Lock the crankshaft in place using tool ST499977100, or equivalent.

12. Remove the crankshaft pulley.

13. Remove the front timing chain cover.

➡**Bolts are three different sizes. Be careful to install the correct bolt in the correct hole.**

14. Remove the timing chain.

15. Remove the crankshaft sprocket.

16. Remove the oil pump cover retaining bolts. Remove the oil pump cover.

17. Remove the inner and outer rotors.

To install:

18. Be sure all mating surfaces are clean and free of dirt.

19. Apply a thin coat of clean engine oil to the complete area of the inner and outer rotors.

20. Position the inner rotor in place. Position the outer rotor in place.

21. Install the pump cover. Tighten the retaining bolts to 4.7 ft. lbs. and in the proper sequence.

➡**Make sure that the bolts are installed in the correct positions.**

22. Continue the installation in the reverse order of the removal procedure.

23. Be sure to fill the cooling system with the proper grade and type coolant.

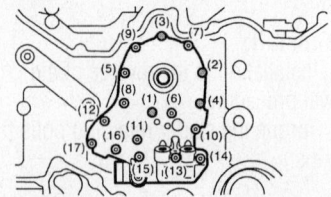

Bolt installing position	Bolt dimension
(1) and (3)	6 x 14 x 14
(2) and (4)	6 x 35 x 18
(5), (6), (7), (8), (9), (10) and (11)	6 x 35 x 15
(12), (15), (16) and (17)	6 x 16 x 16
(13) and (14)	6 x 26 x 15

09490_SBCR_G0059

Oil pump tightening sequence and bolt location

24. Start the engine and check for leaks. Correct, as required.

Rear Main Seal

REMOVAL & INSTALLATION

1. Before servicing the vehicle, refer to the Precautions Section.

2. Remove or disconnect the following:
- Engine from the vehicle
- Clutch assembly/flywheel using the Clutch Disc Guide tool 499747000, if equipped with a manual transmission
- Torque converter flexplate from the crankshaft, if equipped with an automatic transmission
- Oil seal from the cylinder block using a small prybar

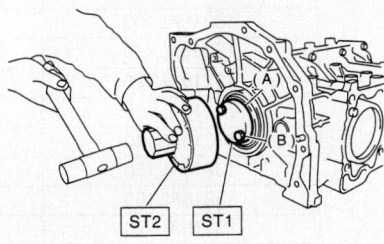

(A) Rear oil seal
(B) Drive plate attaching bolt

42356-SBCR-G28

Installing the rear main seal using oil seal guide ST1 499597100 and ST2 499598200

To install:

3. Install or connect the following:
- New oil seal by pressing it into the cylinder block using the appropriate driver and hammer
- Flywheel housing using new gaskets and sealant where necessary.
- Flywheel and tighten the bolts to specification.
- Engine

Timing Chain, Sprockets, Front Cover and Seal

REMOVAL & INSTALLATION

1. Before servicing the vehicle, refer to the Precautions Section.

2. Disconnect the negative battery cable.

3. Remove the crankshaft pulley cover.

4. Remove the crankshaft pulley bolt.

5. Lock the crankshaft in place using tool ST499977100, or equivalent.

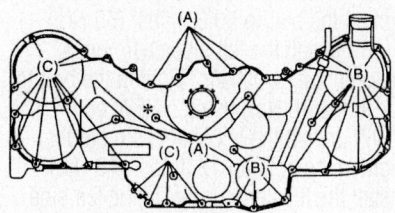

(A) M6 × 16
(B) M6 × 30
(C) M6 × 45
*: Sealing washer

09490_SBCR_G0075

Front cover bolt sizes and locations

6. Remove the crankshaft pulley.
7. Remove the front timing chain cover.

➡**Bolts are three different sizes. Be careful to install the correct bolt in the correct hole.**

8. Remove the right side chain tensioner.

➡**Be careful the plunger does not come out.**

9. Remove the right chain guide, between the right side cams. Remove the right side chain guide. Remove the right side chain tensioner lever. Remove the right timing chain.
10. Remove the left side chain tensioner.

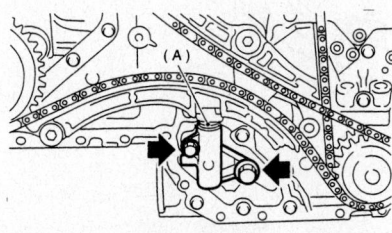

42356-SBCR-G30

The chain tensioner plunger A does not come out—right hand side

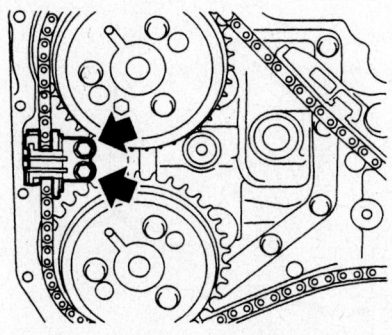

42356-SBCR-G31

Location of the chain guide between the cams—right hand side

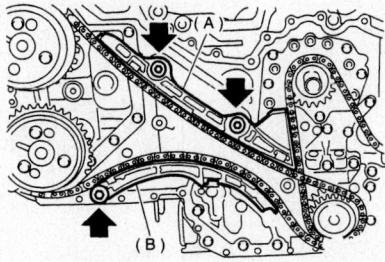

42356-SBCR-G32

Location of the chain guide—right hand side

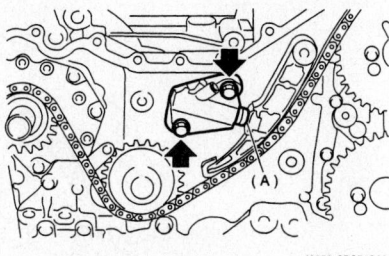

42356-SBCR-G33

The chain tensioner plunger A does not come out—left hand side

➡**Be careful the plunger does not come out.**

11. Remove the left side chain tensioner lever. Remove the left chain guide, between the left side cams. Remove the chain guide.
12. Remove the center chain guide. Remove the upper idler sprocket. Remove the left timing chain.
13. Remove the lower idler sprocket.

To install:

14. Make sure all components are clean. Apply oil to the chain guide, tensioner lever and idler sprockets.
15. Place the screw, spring, pin and tension rod into the tensioner body.
16. While pressing the tensioner onto a rubber mat, twist it to the left and right to shorten the rod. Place a thin pin into the

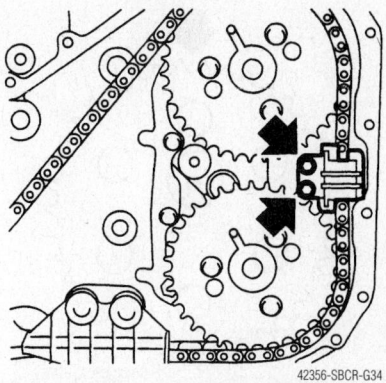

42356-SBCR-G34

Location of the chain guide between the cams—left hand side

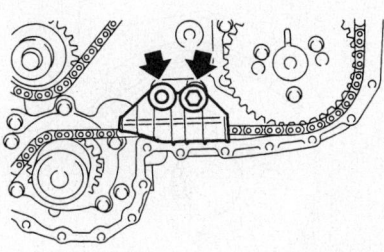

42356-SBCR-G35

Location of the chain guide—left hand side

holes between the rod and body to hold it in place. Always perform this task on a rubber mat.

17. Using the crankshaft socket tool, align the **TOP MARK** on the crankshaft sprocket to the 9 O'clock position, as shown in the illustration.
18. Align the key groove on the exhaust camshaft sprocket to the 12 O'clock position, as shown in the illustration.
19. Align the intake camshaft sprocket, as shown in the illustration.
20. Turn the crankshaft sprocket clockwise; align the **TOP MARK** to the 12 O'clock position. Piston number one is now at Top Dead Center (TDC).

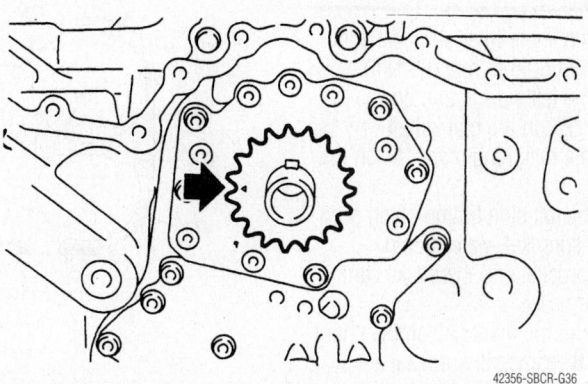

42356-SBCR-G36

Align the TOP MARK on the crankshaft sprocket to the 9 O'clock position

Align the **TOP MARK** on the camshaft sprocket to the 12 O'clock position

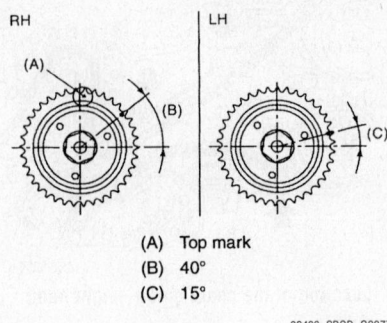

(A) Top mark
(B) 40°
(C) 15°

Intake camshaft alignment

25. Install the upper chain idler and tighten the bolt to 50.6 ft. lbs. (69 Nm).

26. Install the left side chain guide, between the camshafts. Tighten the bolt to 4.7 ft. lbs. (6 Nm) using a NEW bolt.

27. Install the left side chain guide. Tighten the bolts to 12 ft. lbs. (16 Nm). Install the tensioner lever on the left side and tighten the bolt to 12 ft. lbs. (16 Nm). Install the chain tensioner on the left side and tighten the bolts to 12 ft. lbs. (16 Nm).

28. Install the right side timing chain. Align the marks of the left and right timing chains on the lower idler sprocket. Install

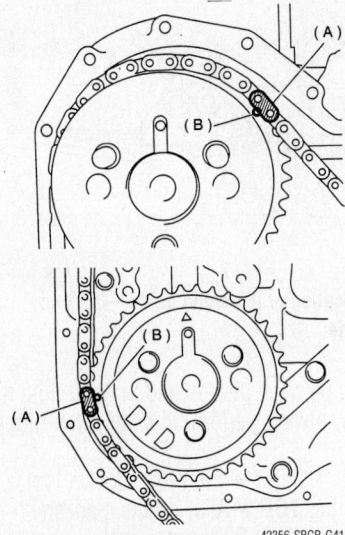

Make sure the mark A on the chain and the mark B camshaft sprocket are aligned the same way as the one on the crankshaft sprocket–right hand side

✳✳ CAUTION

Do not rotate the camshaft or crankshaft sprockets until the chain is completely routed or damage will occur.

21. Install the lower idler sprocket and tighten the bolt to 50.6 ft. lbs. (69 Nm).

22. Install the left side timing chain, align the mark "B" on the crankshaft sprocket with the matching mark "A" on the timing chain.

23. Route the left side timing chain onto the lower idler sprocket, water pump, exhaust cam sprocket and the intake cam sprocket in that order.

24. Make sure the mark "A" on the chain and the mark "B" camshaft sprocket are aligned the same way as the one on the crankshaft sprocket or damage will occur.

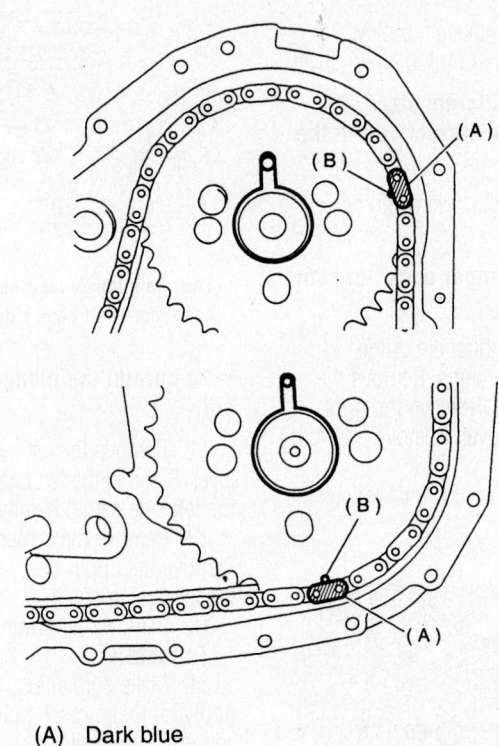

(A) Dark blue
(B) Mark

Make sure the mark A on the chain and the mark B camshaft sprocket are aligned the same way as the one on the crankshaft sprocket–left hand side

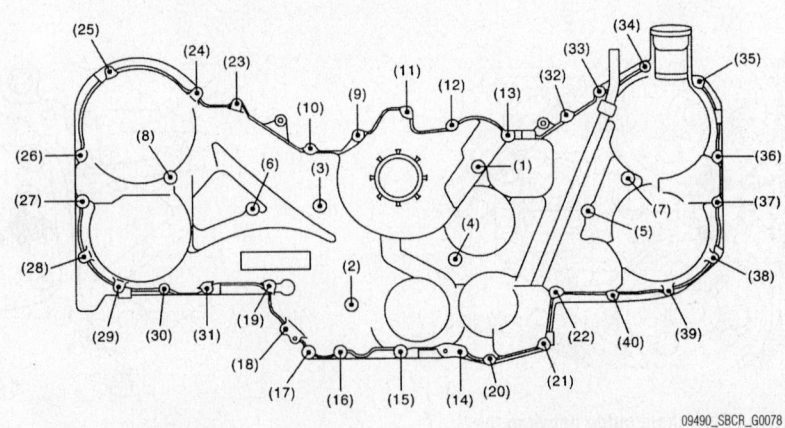

Front timing cover bolt tightening sequence

the right side timing chain to the right side intake camshaft sprocket and the right side exhaust camshaft sprocket in this order.

29. Make sure the mark "A" on the chain and the mark "B" camshaft sprocket are aligned the same way as the one on the crankshaft sprocket or damage will occur.

30. Install the right side chain guide. Install the right side chain tensioner lever and tighten the bolts to 12 ft. lbs. (16 Nm). Install the right side chain guide and tighten the NEW bolt to 4.7 ft. lbs. (6 Nm). Install the right side chain tensioner and tighten the bolts to 12 ft. lbs. (16 Nm).

31. Adjust the clearance between the chain guide on the right side and the center chain guide so that there is range between 0.331–0.339 inch (8.4–8.6mm).

32. Install the center chain guide and tighten the NEW bolt to 5.8 ft. lbs. (8 Nm).

33. Check the match marks on each sprocket and corresponding timing chain are correct, remove the stopper from the tensioner.

34. Clean the mating surfaces on the front timing cover. Apply a bead of liquid gasket 0.020 inch in diameter to the mating surface of the front timing chain cover. Install the timing chain cover. Torque the retaining bolts to 4.8 ft. lbs. and in the proper sequence.

35. Continue the installation in the reverse order of the removal procedure.

Piston and Ring

POSITIONING

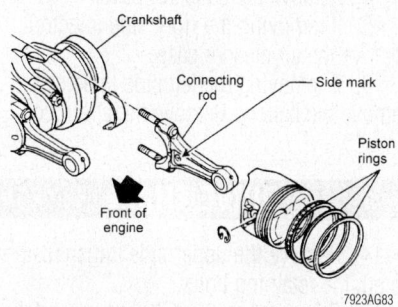

Piston and connecting rod assembly positioning

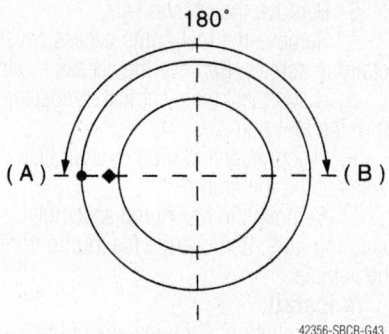

Top ring end-gap (A), second ring gap (B)

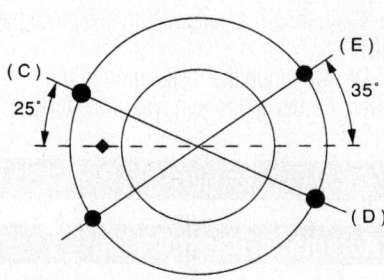

Upper rail end-gap (C), expander gap (D) and lower rail gap (E)

FUEL SYSTEM

Fuel System Service Precautions

Safety is the most important factor when performing not only fuel system maintenance, but any type of maintenance. Failure to conduct maintenance and repairs in a safe manner may result in serious personal injury or death. Maintenance and testing of the vehicle's fuel system components can be accomplished safely and effectively by adhering to the following rules and guidelines.

• To avoid the possibility of fire and personal injury, always disconnect the negative battery cable unless the repair or test procedure requires that battery voltage be applied.

• Always relieve the fuel system pressure prior to disconnecting any fuel system component (injector, fuel rail, pressure regulator, etc.), fitting or fuel line connection. Exercise extreme caution whenever relieving fuel system pressure, to avoid exposing skin, face and eyes to fuel spray. Please be advised that fuel under pressure may penetrate the skin or any part of the body that it contacts.

• Always place a shop towel or rag around the fitting or connection prior to loosening to absorb any excess fuel due to spillage. Ensure that all fuel spillage (should it occur) is quickly removed from engine surfaces. Ensure that all fuel soaked cloths or towels are deposited into a suitable waste container.

• Always keep a dry chemical (Class B) fire extinguisher near the work area.

• Do not allow fuel spray or fuel vapors to come into contact with a spark or open flame.

• Always use a back-up wrench when loosening and tightening fuel line connection fittings. This will prevent unnecessary stress and torsion to fuel line piping.

• Always replace worn fuel fitting O-rings with new. Do not substitute fuel hose or equivalent, where fuel pipe is installed.

Fuel System Pressure

RELIEVING

➡This procedure must be performed prior to servicing any component of the fuel injection system.

1. Before servicing the vehicle, refer to the Precautions Section.

2. Remove the fuel pump fuse from the main fuse box.

3. Start the engine and run until it stalls.

4. Crank the engine for 5 seconds or more to ensure the fuel pressure is properly relieved. If the engine starts during this time, allow it to run until it stalls.

5. Turn the ignition switch to the OFF position. Remove the key.

6. Disconnect the negative battery cable.

7. Remove the fuel cap.

Fuel Filter

REMOVAL & INSTALLATION

➡The fuel filter is an integral part of the fuel pump, no filter replacement procedures are given by the manufacturer.

Fuel Pump

REMOVAL & INSTALLATION

1. Before servicing the vehicle, refer to the Precautions Section.

2. Properly relieve the fuel system pressure.

3. Disconnect the negative battery cable.

4. Remove the fuel filler cap.

5. Drain the fuel from the fuel tank into a suitable container.

6. Remove the second seat.

7. Remove the fuel pump access cover retaining screws. Remove the access cover.

8. Disconnect the electrical connector from the fuel pump.

9. Disconnect and plug the fuel line hoses at the fuel pump.

10. Remove the fuel pump assembly retaining nuts. Remove the fuel pump from the vehicle.

To install:

11. Installation is the reverse of the removal procedure.

12. Be sure to use a new gasket.

13. Tighten the fuel pump retaining nuts to 3.2 ft. lbs. in an alternating sequence pattern.

14. Continue the installation in the reverse order of the removal procedure.

15. Start the engine and check for leaks, correct as required.

Fuel Injector

REMOVAL & INSTALLATION

1. Before servicing the vehicle, refer to the Precautions Section.

2. Properly relieve the fuel system pressure. Remove the fuel cap.

3. Disconnect the negative battery cable.

4. Remove the collector cover.

5. If removing the right side injectors, remove the air cleaner case.

6. If removing the left side injectors, remove the battery. Remove the alternator harness from the left side fuel line protector cover.

7. Remove the fuel line protector cover(s).

8. Disconnect the electrical connector from the fuel injector.

9. Remove the engine harness from the fuel injector line.

10. Remove the bolt that hold the fuel injector line to the intake manifold.

11. Remove the fuel injector while lifting up the fuel injector line.

To install:

12. Installation is the reverse of the removal procedure.

13. Be sure to use new O-rings and insulators.

14. Start the engine and check for leaks, correct as required.

DRIVE TRAIN

Automatic Transmission

REMOVAL & INSTALLATION

1. Before servicing the vehicle, refer to the Precautions Section.

2. Open the hood to the full open position.

3. Disconnect the negative battery cable.

4. Remove the collector cover.

5. Remove the air intake chamber. Remove the air breather hose.

6. Remove the starter.

7. Disconnect the front oxygen sensor and the transmission harness connector.

8. Disconnect the engine harness connectors and remove the rear engine hanger.

9. Remove the vacuum pipe and hose assembly. Remove the brake vacuum pump.

10. Remove the throttle body retaining bolts and slide the throttle body over.

11. To separate the torque converter clutch from the drive plate, remove the service hole plug. Remove the bolts that retain the flex plate to the torque converter. When rotating the engine do so in the proper direction of rotation. Install tool ST49827720, or equivalent, to the torque converter case to hold the assembly in place.

12. Remove the pitching stopper. Remove the pitching stopper bracket.

13. Position engine support tool ST41099AC010 and ST41099AC020 or equivalent to hold the engine assembly in place.

14. Remove the upper side transmission to engine retaining bolts.

15. Raise and support the vehicle safely. Remove the undercover.

16. Remove the front, and rear exhaust pipe and muffler. Remove the heat shield cover.

17. Drain the transmission. Remove the oil charge pipe.

18. Disconnect the connector from the

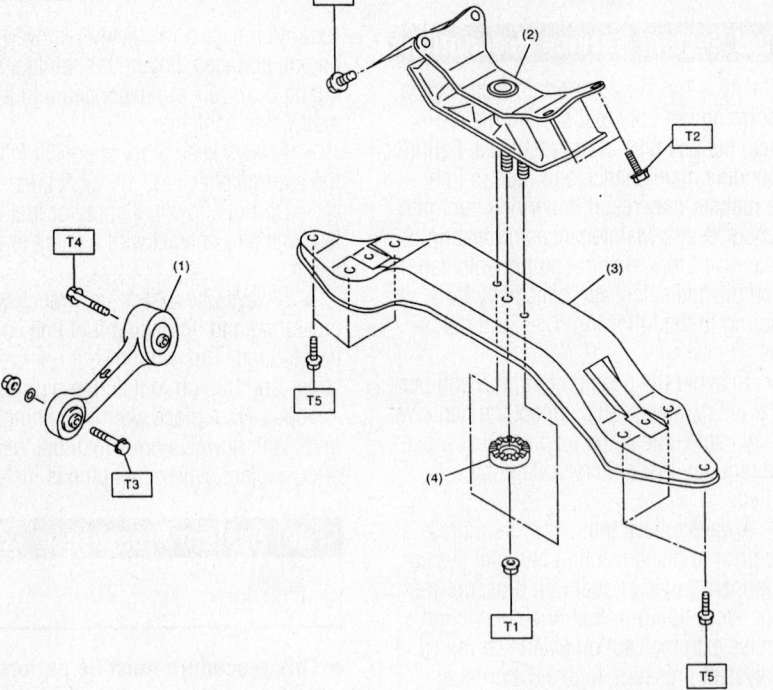

(1)	Pitching stopper	
(2)	Rear cushion rubber	
(3)	Crossmember	
(4)	Stopper	

Tightening torque: N·m (kgf-m, ft-lb)
T1: 35 (3.6, 26)
T2: 40 (4.1, 29.5)
T3: 50 (5.1, 36.9)
T4: 58 (5.9, 42.8)
T5: 75 (7.6, 55.3)

09490_TRIB_G0012

Automatic transmission crossmember and related components

number one turbine speed sensor. Remove the number one turbine speed sensor from the transmission body. Be sure to cover the hole to prevent dirt from entering.

➡ **Failure to follow this procedure may cause interference between the vehicle body and the sensor when removing the transmission.**

19. Remove the driveshaft.
20. Remove the shift select cable.
21. Disconnect the hose from the transmission inlet and outlet fluid lines.
22. Remove the front crossmember support plate.
23. Remove the two clutch housing cover retaining bolts. Remove the front stabilizer clamp.
24. Remove the bolts that retain the front ball joint to the housing.
25. Pull out the front halfshaft from the transmission. Place a cloth between the tool and the transmission to avoid damaging the side retainer of the transmission.
26. Position a transmission jack under the transmission.

➡ **Make sure that the support plates of the transmission jack do not contact the oil pan.**

27. Remove the transmission rear crossmember retaining bolts. Remove the crossmember from the vehicle.
28. Remove the transmission to engine lower retaining bolts.
29. Remove the transmission from the vehicle. Turn the engine support fixture assembly to the left and lower the rear of the engine for easy removal.

➡ **Move the transmission and torque converter away from the engine as an assembly.**

To install:

30. Installation is the reverse of the removal procedure.
31. Be sure to use a new differential side oil seal.
32. Be sure to fill the transmission with the proper grade and type transmission fluid.
33. Start the engine and check for leaks, correct as required.
34. Roadtest the vehicle.

Transfer Case Assembly

REMOVAL & INSTALLATION

The transfer case must be removed as an assembly with the transmission.

Halfshaft

REMOVAL & INSTALLATION

Front

1. Before servicing the vehicle, refer to the Precautions Section.
2. Raise and support the vehicle safely.
3. Remove the wheels and tires.
4. Raise the crimped portion of the axle nut. Depress the brake pedal and remove the axle nut.

➡ **Be sure to loosen the axle nut after removing the tire and wheel from the vehicle. Failure to do this may damage the wheel bearings.**

5. Remove the stabilizer link from the front arm. Disconnect the front arm ball joint from the front housing.
6. Remove the front halfshaft assembly.

➡ **Use axle shaft puller tools ST92647000 and ST28099PA110, or equivalent, if it is difficult to remove the halfshaft.**

7. Remove the front halfshaft from the transmission, using a pry bar. Be careful not to damage the holder portion.

To install:

8. Installation is the reverse of the removal procedure.
9. Be sure to replace the differential side seal with a new one.
10. Be sure to use new locknuts.
11. While depressing the brake pedal, tighten a new axle shaft nut (olive color) to 177 ft. lbs. (240 Nm).

➡ **Always tighten the axle shaft nut before installing the tire and wheel assembly**

Rear

1. Before servicing the vehicle, refer to the Precautions Section.
2. Disconnect the negative battery cable.
3. Raise and support the vehicle safely.
4. Remove the wheels and tires.
5. Unlock the axle nut. Depress the parking brake. Remove the axle nut.

➡ **Be sure to loosen the axle nut after removing the tire and wheel from the vehicle. Failure to do this may damage the wheel bearings.**

6. Remove the rear differential assembly.
7. Remove the axle nut and rear halfshaft.

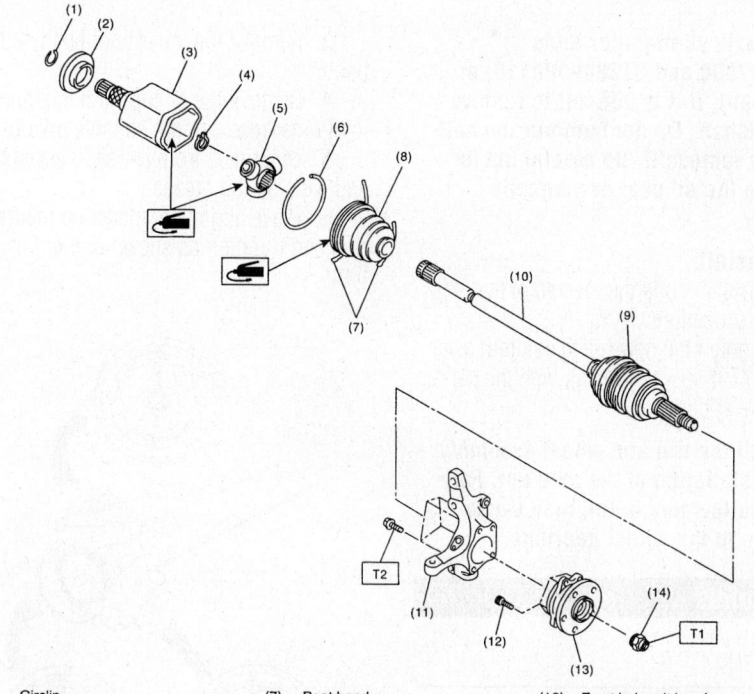

(1)	Circlip	(7)	Boot band	(13)	Front hub unit bearing
(2)	Baffle plate	(8)	Boot (PTJ)	(14)	Axle nut
(3)	Outer race (PTJ)	(9)	Boot (EBJ)		
(4)	Snap ring	(10)	EBJ shaft ASSY		
(5)	Trunnion	(11)	Front housing		
(6)	Snap ring	(12)	Hub bolt		

Tightening torque: N·m (kgf-m, ft-lb)
T1: 240 (24.5, 177)
T2: 65 (6.6, 47.9)

09490_TRIB_G0013

Front halfshaft and related components

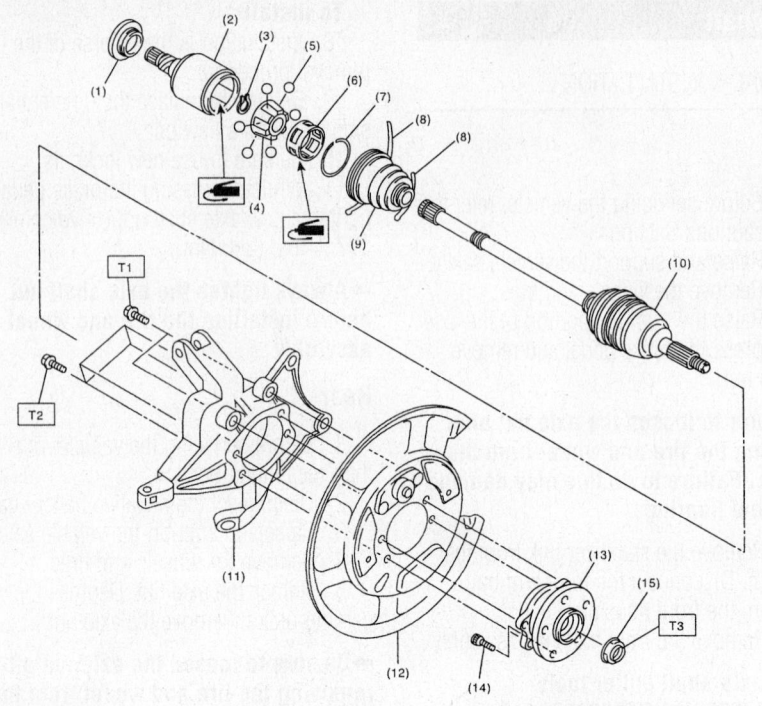

(1)	Baffle plate (DOJ)	(8)	Boot band	(15) Axle nut (olive color)
(2)	Outer race (DOJ)	(9)	Boot (DOJ)	
(3)	Snap ring	(10)	EBJ shaft ASSY	**Tightening torque: N·m (kgf-m, ft-lb)**
(4)	Inner race	(11)	Rear housing	T1: 75 (7.6, 55.3)
(5)	Ball	(12)	Back plate	T2: 65 (6.6, 47.9)
(6)	Cage	(13)	Rear hub unit bearing	T3: 240 (24.5, 177)
(7)	Snap ring	(14)	Hub bolt	

09490_TRIB_G0014

Rear halfshaft and related components

➡**Use axle shaft puller tools ST92647000 and ST28099PA110, or equivalent, if it is difficult to remove the halfshaft. Do not hammer the halfshaft to remove it. Be careful not to damage the oil seal or magnetic encoder.**

To install:

8. Installation is the reverse of the removal procedure.

9. Tighten the new (olive colored) axle nut to 177 ft. lbs. (240 Nm), with the parking brake applied.

➡**Install the tire and wheel assembly after installation of the axle nut. Failure to follow this order, may cause damage to the wheel bearings.**

CV-Joints

OVERHAUL

Front

1. Before servicing the vehicle, refer to the Precautions Section.

2. Place alignment marks on the shaft and outer race.

3. Remove the inner boot band and boot.

4. Remove the circlip from the inner joint outer race using a suitable prytool.

5. Outer race from the shaft assembly and wipe off the grease.

6. Place alignment marks on the free ring and trunnion as shown in the illustration.

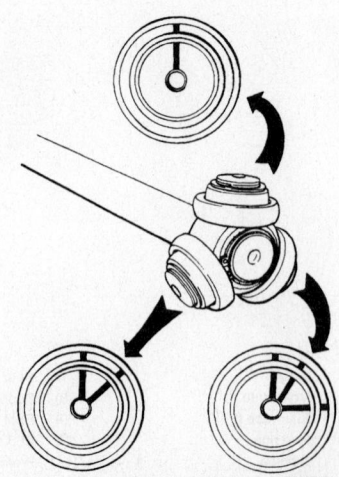

9357TG38

Place alignment marks on the free ring and trunnion as shown—Front halfshaft

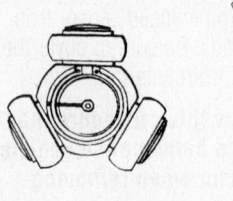

9357TG39

Place alignment marks on the trunnion and shaft as shown—Front halfshaft

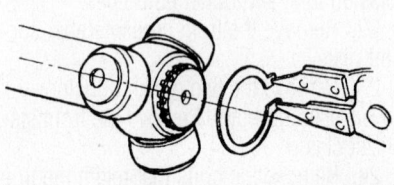

9357TG40

Remove the snapring and trunnion—Front halfshaft

7. Remove the free ring from the trunnion.

8. Place alignment marks on the trunnion and shaft as shown in the illustration.

9. Remove the snapring and trunnion.

10. Place the shaft in a vise between wooden blocks.

11. Using a suitable prytool, raise the outer boot band claws.

12. Cut and remove the boot.

13. Only the boot can be replaced, the joint is not serviceable and must be replaced if damaged.

To install:

14. Place the half shaft in a vise.

15. Place the outer boot and small band on the shaft.

16. Apply 3.53–3.88 oz. (100–110g) of supplied grease to the joint.

17. Apply 1.06–1.41 oz. (30–40g) of supplied grease to the whole inner surface

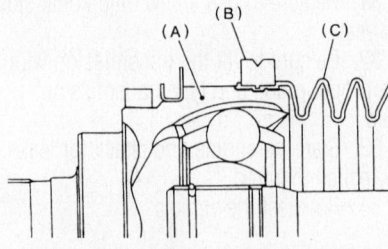

(A) EBJ
(B) Lorge boot band
(C) Boot

9357TG41

Position the boot to the joint groove, and attach the large boot band as shown—Front halfshaft

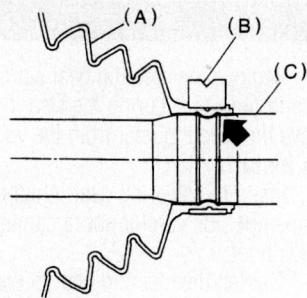

(A) Boot
(B) Small boot band
(C) Shaft

9357TG42

Position the boot to the shaft groove, and attach the small boot band as shown— Front halfshaft

of the boot, and apply some grease to the shaft.

18. Install the boot to the joint groove, and attach the large boot band as shown.

19. Install the boot to the shaft groove, and attach the small boot band as shown.

20. Use boot band plier tool ST925091000 to tighten the large and small bands

21. Place the inner boot on the center of the shaft.

22. Align the alignment marks from earlier and install the trunnion and snapring. Make sure the snapring is fully engaged.

23. Apply 3.53–3.88 oz. (100–110g) of supplied grease to the joint outer race.

24. Apply a coat of supplied grease to the free ring and trunnion.

25. Align the marks on the free ring and trunnion and install the free ring.

26. Align the marks on the shaft and outer race and install the outer race.

27. Pull on the shaft lightly to ensure the circlip is completely engaged.

28. Apply an even coat 1.06–1.41 oz. (30–40g) of the supplied grease to the entire inner surface of the boot.

29. Install the boot and band.

30. Once the band is properly tightened, cut off any excess to leave only 0.39 inch (10mm) and bend it over.

31. Install the shaft.

Rear

The Double Offset Joint (DOJ), is the only part of the assembly that can be replaced, if any of the other components are defective then the shaft should be replaced.

1. Before servicing the vehicle, refer to the Precautions Section.

2. Straighten the bent claw of the large clamp at the Double Offset Joint (DOJ) end of the boot.

3. Loosen the band using pliers being careful not to damage the bolt.

4. Remove the small boot band from the DOJ using the same technique.

5. Remove the boot from the large end of the DOJ outer race.

6. Remove the round circlip using a suitable prytool from the neck of the joint outer race.

7. Remove the joint outer race.

✳✳ CAUTION

The grease used is for CV–Joints, do not replace with another type of grease

8. Clean the grease and remove the balls. Be careful not to loose any of the 6 balls.

9. Turn the cage by a half pitch to the track groove of the inner race and shift the cage.

10. Remove the snapring that secures the inner race to the shaft and remove the inner race.

11. Take the cage from the shaft and remove the boot.

12. The other boots may be removed in the same manner as the DOJ boot.

13. Wrap the shaft splines with tape to prevent damage.

To install:

14. The following grease must be used during assembly:

15. DOJ side: NKG205 (part number 28495AG00A).

16. Install the BJ or EBJ boots and fill it with 2.12–2.47 oz. (60–70g) of grease.

17. Place the DOJ boot on the center of the shaft.

18. Insert the DOJ cage onto the shaft. Make sure to insert the cage with the cut–out portion facing shaft end.

19. Install the inner race on the shaft and fasten with the snapring. Make sure the snapring is firmly engaged.

20. Install the cage (previously fitted) to the inner race on the shaft. Fit the cage with the protruded part aligned with the track on the inner race and then turn a half pitch.

21. Fill the DOJ inner race with 2.82–3.17 oz. (80–90g) of grease.

22. Apply a coat of grease to the to the cage pocket and 6 balls.

23. Insert the 6 balls.

24. Align the outer race track and ball positions and place where the shaft, inner race, cage and balls were located prior to removal and then install the outer race.

25. Install the circlip into the groove on the outer race.

➡**Make sure the balls, cage and inner race are fully seated. make sure not to place the matched position of the circlip in the ball groove of the outer race. Pull the shaft lightly to make sure the circlip is fully engaged.**

26. Apply an even coat 0.71–1.06 oz. (20–30g) of grease to the entire inner surface of the boot and to the shaft.

27. Make sure the boot is free from any dirt or foreign materials prior to installation.

28. Place the outer race of the boot at the center of its travel.

29. Put a band through the boot clip and wind twice in alignment with the groove on the boot.

30. Pinch the end of the band using tool ST 9250910000 and tighten securely until it cannot be moved by hand. Make sure there is appropriate air inside the boot.

31. Tap on the clip with a punch to lock it making sure not to damage the damaged while tapping. Cut any excess off the band leaving 0.39 inch (10mm) and bend the remaining portion over the clip. Make sure the end of the band is close to the clip.

32. Install the remaining boot clamps in the same manner.

STEERING

Air Bag

❋❋ CAUTION

This vehicle is equipped with an air bag system. The system must be disabled before performing service on or around system components, steering column, instrument panel components, wiring and sensors. Failure to follow safety and disabling procedures could result in accidental air bag deployment, possible personal injury and unnecessary system repairs.

PRECAUTIONS

Several precautions must be observed when handling the inflator module to avoid accidental deployment and possible personal injury.

• Never carry the inflator module by the wires or connector on the underside of the module.

• When carrying a live inflator module, hold it securely with both hands, and ensure that the bag and trim cover are pointed away.

• Place the inflator module on a bench or other surface with the bag and trim cover facing up.

• With the inflator module on the bench, never place anything on or close to the module that may be thrown in the event of an accidental deployment.

DISARMING

1. Before servicing the vehicle, refer to the Precautions Section.
2. Be sure to position the front wheels in the straight ahead position.
3. Disconnect the negative battery cable. Tape the battery cable for added protection.
4. Wait more than 20 seconds before starting work.

Power Steering Gear

REMOVAL & INSTALLATION

1. Before servicing the vehicle, refer to the Precautions Section.
2. Disconnect the negative battery cable.
3. Loosen the front wheel nut.
4. Raise and support the vehicle safely.

5. Remove the tires and wheels.
6. Remove the under cover.
7. Remove the front exhaust pipe assembly.
8. Remove the cotter pin and castle nut. Using a puller, remove the tie rod end from the knuckle arm.
9. Remove the front crossmember support plate. Remove the jack up plate. Remove the front stabilizer.
10. Disconnect the power steering fluid pipe at the center of the gearbox and attach a vinyl hose. Discharge the fluid into a suitable container by turning the steering wheel fully clockwise and counterclockwise. Disconnect the other fluid lines, and repeat the discharge procedure.
11. Remove the steering wheel. Make a matchmark on the universal joint. Remove the universal joint bolts and remove the joint from the vehicle.
12. Disconnect the fluid lines from the steering gear, pressure hose first.

13. Remove the steering gear clamp bolts and bracket securing the steering gear. Remove the steering gear from the vehicle.

To install:

14. Insert the steering gear into the crossmember. Be careful not to damage the gearbox boot.
15. Tighten the steering gear to the crossmember bracket to 44.1 ft. lbs.
16. Connect the fluid lines.
17. Align the cutout at the serrated section of the column shaft and yoke. Insert the universal joint into the column shaft.
18. Align the mating marks and insert the universal joint to serrated section of the steering gear assembly. Tighten the bolt to 17.4 ft. lbs.
19. Continue the installation in the reverse order of the removal procedure.
20. Fill the power steering system with the proper grade and type fluid.
21. Start the engine and check for leaks. Correct as required.

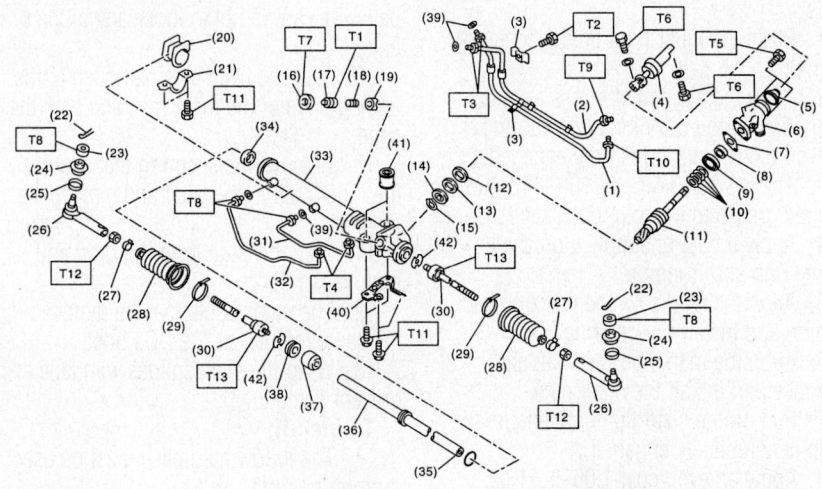

(1)	Pipe C	(20)	Adapter	(39)	O-ring
(2)	Pipe D	(21)	Clamp	(40)	Bracket
(3)	Clamp plate	(22)	Cotter pin	(41)	Bushing
(4)	Universal joint	(23)	Castle nut	(42)	Lock washer
(5)	Dust seal	(24)	Dust cover		
(6)	Valve housing	(25)	Clip		
(7)	Gasket	(26)	Tie-rod end		
(8)	Oil seal	(27)	Clip		
(9)	Bushing	(28)	Boot		
(10)	Seal ring	(29)	Band		
(11)	Pinion & valve ASSY	(30)	Tie-rod		
(12)	Oil seal	(31)	Pipe B		
(13)	Back-up washer	(32)	Pipe A		
(14)	Ball bearing	(33)	Steering body		
(15)	Snap ring	(34)	Oil seal		
(16)	Lock nut	(35)	Piston ring		
(17)	Adjusting screw	(36)	Rack		
(18)	Spring	(37)	Rack bushing		
(19)	Sleeve	(38)	Holder		

Tightening torque: N·m (kgf-m, ft-lb)	
T1:	3.9 (0.4, 2.9)
T2:	10 (1.02, 7.4)
T3:	15 (1.5, 10.8)
T4:	17 (1.7, 12.5)
T5:	20 (2.0, 14.8)
T6:	24 (2.4, 17.4)
T7:	25 (2.5, 18.1)
T8:	27 (2.75, 19.9)
T9:	29 (3.0, 21.4)
T10:	37 (3.8, 27.3)
T11:	60 (6.1, 44.1)
T12:	85 (8.7, 62.7)
T13:	130 (13.3, 95.9)

09490_TRIB_G0015

Power steering gear and related components

FRONT SUSPENSION

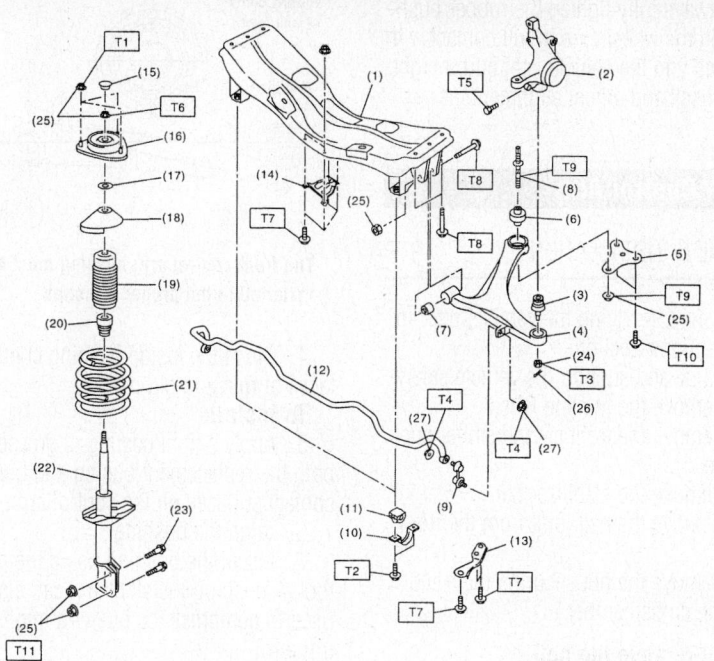

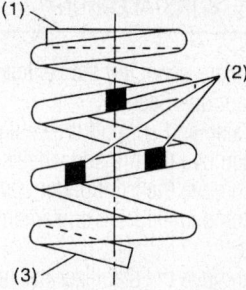

(1) Flat (top side)
(2) Identification paint
(3) Inclined (bottom side)

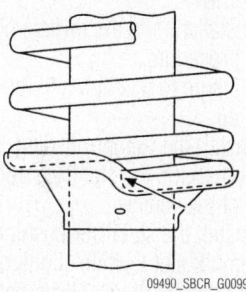

09490_SBCR_G0099

Front strut spring alignment

(1)	Front crossmember	(15)	Dust seal	
(2)	Housing	(16)	Strut mount	
(3)	Ball joint	(17)	Spacer	
(4)	Front arm	(18)	Upper spring seat	
(5)	Arm support plate	(19)	Dust cover	
(6)	Rear bushing	(20)	Front helper	
(7)	Front bushing	(21)	Front coil spring	
(8)	Stud bolt	(22)	Front strut	
(9)	Stabilizer link	(23)	Adjusting bolt	
(10)	Stabilizer clamp	(24)	Castle nut	
(11)	Stabilizer bushing	(25)	Self–locking nut	
(12)	Front stabilizer	(26)	Cotter pin	
(13)	Crossmember support plate	(27)	Flange nut (with WAX)	
(14)	Jack–up plate			

Tightening torque: N·m (kgf-m, ft-lb)
T1: 20 (2.0, 14.5)
T2: 25 (2.5, 18.1)
T3: 45 (4.6, 33.2)
T4: 60 (6.1, 44.3)
T5: 50 (5.1, 36.9)
T6: 55 (5.6, 40.6)
T7: 60 (6.1, 44.3)
T8: 95 (9.7, 70.1)
T9: 145 (14.8, 106.9)
T10: 150 (15.3, 110.6)
T11: 175 (17.8, 129.1)

09490_TRIB_G0016

Front suspension and related components

Strut

REMOVAL & INSTALLATION

1. Before servicing the vehicle, refer to the Precautions Section.
2. Disconnect the negative battery cable.
3. Raise and support the vehicle safely.
4. Remove the tire and wheel.
5. Remove the bolt retaining the brake hose to the strut.
6. Make and alignment mark on the camber adjusting bolt which secures the strut to the housing.
7. Remove the clip retaining the ABS wheel speed sensor harness.
8. Remove the two bolts retaining the strut to the housing.

➡**While holding the head of the adjusting bolt, loosen the self locking nut.**

9. Remove the three upper strut retaining nuts.
10. Remove the strut from the vehicle.

To install:
11. Installation is the reverse of the removal procedure.
12. Tighten the upper retaining nuts to 14.5 ft. lbs. Tighten the lower retaining bolts to 129 ft. lbs.
13. Position the alignment mark on the camber adjusting bolt with the alignment mark on the lower side of the strut. Install using a new self locking nut.

➡**While holding the head of the adjusting bolt, tighten the self locking nut.**

14. Check and adjust wheel alignment, as necessary.

DISASSEMBLY & ASSEMBLY

1. Before servicing the vehicle, refer to the Precautions Section.

2. Remove the strut from the vehicle.
3. Using a coil spring compressor tool, carefully compress the spring. Remove the self locking nut.
4. Remove the strut mount, upper spring and rubber seat from the strut.
5. Gradually decrease the compression force of the spring compressor tool. Remove the coil spring.
6. Remove the dust cover and helper spring.
7. Check for the presence of air in the damping force generating mechanism.
8. Using the spring compression tool, compress the coil spring.

➡**Be sure to properly install the coil spring.**

9. Position the coil spring so that its end face fits good into the spring seat.
10. Install the helper spring and dust cover to the piston rod.
11. Pull the piston rod fully upward, and install the rubber seat and spring seat.
12. Install the strut mount to the piston rod, and then tighten the self locking nut, temporarily. Be sure to use a new self locking nut.
13. Use a hexagon wrench to prevent the strut rod from turning. Tighten the self locking nut to 41 ft. lbs.
14. Carefully loosen the coil spring.

Stabilizer Bar

REMOVAL & INSTALLATION

1. Before servicing the vehicle, refer to the Precautions Section.
2. Raise and support the vehicle safely.
3. Remove the tire and wheels.
4. Remove the front under cover.
5. Remove the front crossmember support plate.
6. Remove the stabilizer bar link.
7. Remove the stabilizer bar bracket bolts and bushings.
8. Remove the stabilizer bar from the vehicle.

To install:

9. Installation is the reverse of the removal procedure.
10. Be sure to use new self locking nuts, as required.
11. Install the rubber bushing, so that the paint mark on the stabilizer bar is on the left side of the vehicle.
12. Install the stabilizer bushing (front crossmember side), while aligning it with the paint mark on the stabilizer bar.
13. Tighten the stabilizer link bolts to 44.3 ft. lbs. (60 Nm). Tighten the stabilizer bar clamp to 18.1 ft. lbs. (25 Nm).
14. Always fully tighten the rubber bushings when the wheels are in full contact with the ground and the vehicle is at curb height.

Lower Ball Joint

REMOVAL & INSTALLATION

1. Before servicing the vehicle, refer to the Precautions Section.
2. Raise and support the vehicle safely.
3. Remove the tire and wheel.
4. Remove the stabilizer bar brackets and bushings (both sides).
5. Remove the cotter pin from the ball stud. Remove the castle nut. Extract the ball stud from the front arm.
6. Remove the bolt securing the ball joint to the housing. Extract the ball joint from the housing.

To install:

7. Installation is the reverse of the removal procedure.
8. Install the ball joint to the front arm and tighten the castle nut to 33.2 ft. lbs. (45 Nm). Tighten the castle nut an additional 60 degrees until the slot in the castle nut is aligned with the cotter pin hole in the ball joint.

9. Install the stabilizer bracket. Tighten to 18.1 ft. lbs. (25 Nm).
10. Always fully tighten the rubber bushings when the wheels are in full contact with the ground and the vehicle is at curb height.
11. Check and adjust alignment, as required.

Lower Control Arm

REMOVAL & INSTALLATION

1. Before servicing the vehicle, refer to the Precautions Section.
2. Raise and support the vehicle safely.
3. Remove the tire and wheel.
4. Remove the front crossmember support plate.
5. Remove the stabilizer bar.
6. Remove the ball joint from the front arm.
7. Remove the nut securing the front arm to the crossmember.

➡**Do not remove the bolt.**

8. Remove the front arm support plate.
9. Remove the bolt securing the front arm to the crossmember and pull the front arm out of the crossmember.
10. To remove the stud bolt, use tool ST20299AG020.

➡**Do not remove the stud bolt unless it is necessary. Always replace the removed parts with new ones.**

To install:

11. Installation is the reverse of the removal procedure.
12. Tighten the stud bolt to 81.1 ft. lbs. (110 Nm), if removed.
13. Tighten the support plate to front arm bolts to 106.9 ft. lbs. (145 Nm).
14. Tighten the support plate to body bolts to 110.6 ft. lbs. (150 Nm).
15. Always fully tighten the rubber bushings when the wheels are in full contact with the ground and the vehicle is at curb height.
16. Check and adjust alignment, as required.

CONTROL ARM BUSHING REPLACEMENT

1. Remove the control arm from the vehicle.
2. Mount the control arm in a soft jawed vise.
3. Use either a press or a control arm bushing fixture (C-clamp like tool) along with a slotted washer and a piece of pipe (slightly larger than the bushing) and press out the old bushing.

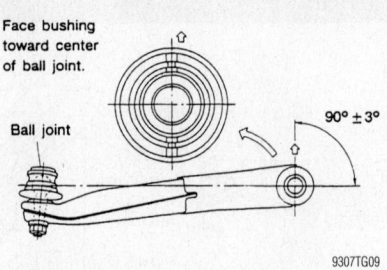

Face bushing toward center of ball joint.

Ball joint

90° ± 3°

9307TG09

The front control arm bushing must be installed in the proper direction

4. Clean the inside bushing contact surfaces of rust and old rubber.

To install:

5. Apply a light coating of grease to both the replacement busing and bushing contact surfaces on the control arm.
6. Align the bushing.
7. Install the bushing using the press tool. A bushing install clamp can also be used to compress the bushing into the control arm.
8. Install the control arm on the vehicle.

Wheel Bearings

ADJUSTMENT

The wheel bearings are not adjustable.

REMOVAL & INSTALLATION

1. Before servicing the vehicle, refer to the Precautions Section.
2. Disconnect the negative battery cable.
3. Raise and support the vehicle safely.
4. Remove the tire and wheel.
5. Remove the crimped section of the axle nut. Depress the brake pedal and remove the axle nut.

➡**Be sure to loosen the axle nut after removing the tire and wheel from the vehicle. Failure to do this may damage the wheel bearings.**

6. Remove the disc brake caliper from its mounting as suspend it to the side, with wire. Do not disconnect the brake line.
7. Remove the rotor.

➡**If the rotor is seized within the hub, remove the rotor by installing an 8mm bolt in the screw hole on the rotor.**

8. As required, remove the front halfshaft.
9. Remove the four bolts from the front housing.

➡**Be careful not to damage the magnetic encoder of the ABS sensor. Use the proper tools when working around this component.**

10. Remove the front halfshaft from the hub. Use tools ST92647000 and ST28099PA100 if necessary.

To install:

11. Installation in the reverse order of the removal procedure.

12. Tighten the four front hub retaining bolts to 47.9 ft. lbs. (15 Nm).

13. Be sure to use a new axle nut (olive colored). While depressing the brake pedal, tighten the axle nut to 177 ft. lbs. (240 Nm). Lock it securely in place.

➡**Install the tire and wheel after installation of the axle nut. Failure to do this may result in wheel bearing damage. Do not over tighten, as this too could cause wheel bearing damage.**

14. Continue the installation in the reverse order of the removal procedure.

REAR SUSPENSION

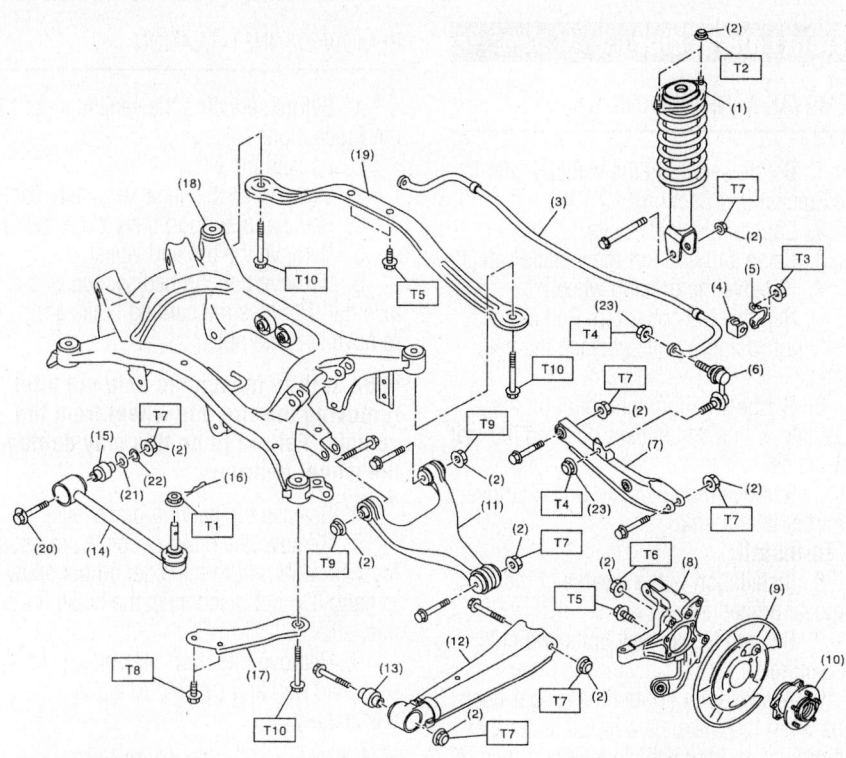

9. Detach the rear lateral link.

10. Remove the nuts that retain the shock absorber mount to the vehicle.

11. Remove the shock absorber from the vehicle.

To install:

12. Installation is the reverse of the removal procedure.

13. Be sure to use new bolts and nuts, as required.

14. Always fully tighten the rubber bushings when the wheels are in full contact with the ground and the vehicle is at curb height.

15. Check and adjust the wheel alignment, as necessary.

(1)	Shock absorber	(13)	Trailing link bushing		
(2)	Self–locking nut	(14)	Front lateral link		
(3)	Rear Stabilizer	(15)	Front lateral link bushing		
(4)	Stabilizer bushing	(16)	Snap pin		
(5)	Stabilizer clamp	(17)	Front sub frame support plate		
(6)	Stabilizer link	(18)	Rear sub frame		
(7)	Rear lateral link	(19)	Rear sub frame support plate		
(8)	Rear housing	(20)	Adjusting bolt		
(9)	Back plate	(21)	Adjusting washer		
(10)	Hub unit bearing	(22)	Washer		
(11)	Upper arm	(23)	Flange nut (WAX)		
(12)	Trailing link				

Tightening torque: N·m (kgf-m, ft-lb)
T1: 27 (2.8, 19.9)
T2: 30 (3.1, 22.4)
T3: 35 (3.6, 25.8)
T4: 60 (6.1, 44)
T5: 65 (6.6, 48)
T6: 75 (7.6, 89)
T7: 120 (12.2, 89)
T8: 125 (12.7, 92)
T9: 150 (15.3, 111)
T10: 200 (20.4, 148)

09490_TRIB_G0017

Rear suspension and related components

Shock Absorber

REMOVAL & INSTALLATION

1. Before servicing the vehicle, refer to the Precautions Section.

2. Remove the strut cap (of the quarter trim).

3. Loosen the rear wheel lug nuts.

4. Raise and support the vehicle safely.

5. Remove the tire and wheel.

6. Remove the nut and detach the rear stabilizer link.

7. Using a jack, support the shock absorber.

8. Remove the bolts on the bottom of the shock absorber.

Stabilizer Bar

REMOVAL & INSTALLATION

1. Before servicing the vehicle, refer to the Precautions Section.

2. Loosen the rear wheel lug nuts.

3. Raise and support the vehicle safely.

4. Remove the tire and wheel.

5. Remove the stabilizer link.

6. Remove the clamp and bushing bolts which secure the stabilizer bar. Remove the clamps and bushings.

7. Remove the stabilizer bar from the vehicle.

To install:

8. Installation is the reverse of the removal procedure.

9. Be sure that the stabilizer bar and the bushings have the same identification markings and/or colors.

10. Be sure to use new bolts and nuts, as required.

11. Tighten the stabilizer link retaining bolts to 44 ft. lbs (60 Nm).

12. Tighten the stabilizer clamp retaining bolts to 25.8 ft. lbs (35 Nm).

13. Always fully tighten the rubber bushings when the wheels are in full contact with the ground and the vehicle is at curb height.

14. Check and adjust the wheel alignment, as necessary.

Lower Control Arm

REMOVAL & INSTALLATION

1. Before servicing the vehicle, refer to the Precautions Section.
2. Loosen the wheel nuts.
3. Raise and support the vehicle safely.
4. Remove the tire and wheel.
5. Remove the bracket and remove the parking brake cable from the guide.
6. Remove the ABS wheel speed sensor harness from the trailing link.
7. As required, support the assembly before removing the trailing link.
8. Remove the trailing link from the vehicle.

To install:
9. Installation is the reverse of the removal procedure.
10. Be sure to use new bolts and nuts, as required.
11. Tighten the trailing link retaining bolts to 89 ft. lbs (120 Nm).
12. Tighten the parking brake cable bracket to 24 ft. lbs (33 Nm).
13. Always fully tighten the rubber bushings when the wheels are in full contact with the ground and the vehicle is at curb height.
14. Check and adjust the wheel alignment, as necessary.

LOWER CONTROL ARM BUSHING REPLACEMENT

1. Remove the control arm from the vehicle.
2. Scribe a matchmark on the control arm and rear bushing.
3. Press out the old bushing.
4. Install the rear bushing to the control arm, making sure to align the marks made during removal.

Upper Control Arm

REMOVAL & INSTALLATION

1. Before servicing the vehicle, refer to the Precautions Section.
2. Loosen the wheel nuts.
3. Raise and support the vehicle safely.
4. Remove the tire and wheel.
5. As required, support the assembly before removing the upper control arm.
6. Remove the bolts. Remove the upper control arm from the vehicle.

To install:
7. Installation is the reverse of the removal procedure.

8. Be sure to use new bolts and nuts, as required.
9. Tighten the upper arm to rear sub frame retaining bolts to 111 ft. lbs (150 Nm).
10. Tighten the upper arm to rear housing retaining bolts to 89 ft. lbs (120 Nm).
11. Always fully tighten the rubber bushings when the wheels are in full contact with the ground and the vehicle is at curb height.
12. Check and adjust the wheel alignment, as necessary.

Front Lateral Link

REMOVAL & INSTALLATION

1. Before servicing the vehicle, refer to the Precautions Section.
2. Loosen the wheel nuts.
3. Raise and support the vehicle safely.
4. Remove the tire and wheel.
5. Remove the cotter pin. Remove the castle nut. Using a puller, detach the ball joint.
6. Scribe an alignment mark on the front lateral link adjustment bolt and the rear sub frame.
7. Remove the adjusting bolt. Remove the front lateral link.

To install:
8. Installation is the reverse of the removal procedure.
9. Be sure to use new bolts and nuts, as required.
10. Always fully tighten the rubber bushings when the wheels are in full contact with the ground and the vehicle is at curb height.
11. Check and adjust the wheel alignment, as necessary.

Rear Lateral Link

REMOVAL & INSTALLATION

1. Before servicing the vehicle, refer to the Precautions Section.
2. Loosen the wheel nuts.
3. Raise and support the vehicle safely.
4. Remove the tire and wheel.
5. Remove the nut and detach the stabilizer link.
6. Remove the bolts on the bottom side of the shock absorber.
7. Remove the bolts and remove the rear lateral link.

To install:
8. Installation is the reverse of the removal procedure.

9. Be sure to use new bolts and nuts, as required.
10. Always fully tighten the rubber bushings when the wheels are in full contact with the ground and the vehicle is at curb height.
11. Check and adjust the wheel alignment, as necessary.

Wheel Bearings

ADJUSTMENT

The wheel bearings are not adjustable.

REMOVAL & INSTALLATION

1. Before servicing the vehicle, refer to the Precautions Section.
2. Loosen the wheel nuts.
3. Disconnect the negative battery cable.
4. Raise and support the vehicle safely.
5. Remove the tire and wheel.
6. Remove the crimped section of the axle nut. Depress the parking brake and remove the axle nut.

➡**Be sure to loosen the axle nut after removing the tire and wheel from the vehicle. Failure to do this may damage the wheel bearings.**

7. Release the parking brake lever.
8. Remove the brake caliper from its mounting. Wire it to the side; do not allow it to hang. Do not disconnect the brake fluid line.
9. Remove the rotor. Matchmark the rotor and hub and bearing to aid in reassembly.

➡**If the rotor is seized within the hub, remove the rotor by installing an 8mm bolt in the screw hole on the rotor.**

10. Remove the four hub unit bolts from the rear arm.
11. Remove the hub and bearing unit. If it is hard to remove, use tools ST926470000 and ST927140000.

➡**Be careful not to damage the magnetic encoder for the ABS system.**

To install:
12. Align the hub and bearing unit to the mounting hole of the backing plate. Install the assembly. Using a new axle nut temporarily tighten it.

➡**Be careful not to damage the magnetic encoder for the ABS system.**

13. Tighten the four backing plate bolts to 47.9 ft. lbs. (65 Nm).
14. Remove the axle nut. Draw the rear halfshaft into position. Temporarily tighten

the axle shaft nut. The nut should be olive in color.

15. Install the rotor. Install the caliper.

16. Adjust the parking brake, as required.

17. While depressing the brake pedal, tighten the axle nut (olive color) to 177 ft. lbs. (240 Nm). Lock it securely in place.

➡️**Install the tire and wheel after installation of the axle nut. Failure to do this may result in wheel bearing damage.**

Do not over tighten, as this too could cause wheel bearing damage.

18. Continue the installation in the reverse order of the removal procedure.

BRAKES

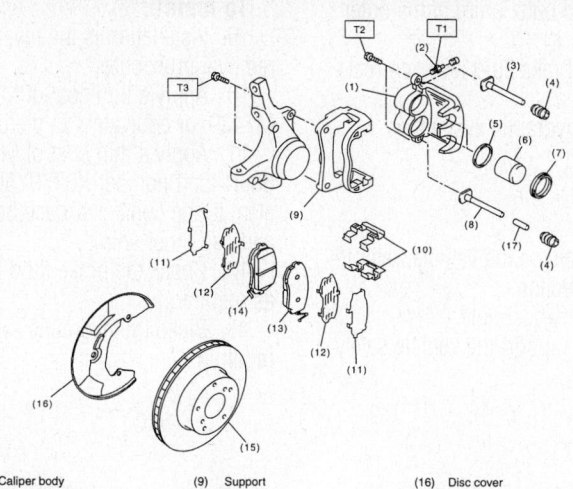

Front disc brake and related components

(1)	Caliper body	(9)	Support	(16)	Disc cover
(2)	Air bleeder screw	(10)	Pad clip	(17)	Bushing
(3)	Guide pin (Green)	(11)	Outer shim		
(4)	Pin boot	(12)	Inner shim		
(5)	Piston seal	(13)	Pad (Outside)		
(6)	Piston	(14)	Pad (Inside)		
(7)	Piston boot	(15)	Disc rotor		
(8)	Lock pin (Yellow)				

Tightening torque: N·m (kgf-m, ft-lb)
T1: 8 (0.8, 5.8)
T2: 27 (2.8, 19.9)
T3: 120 (12.2, 88.5)

09490_TRIB_G0018

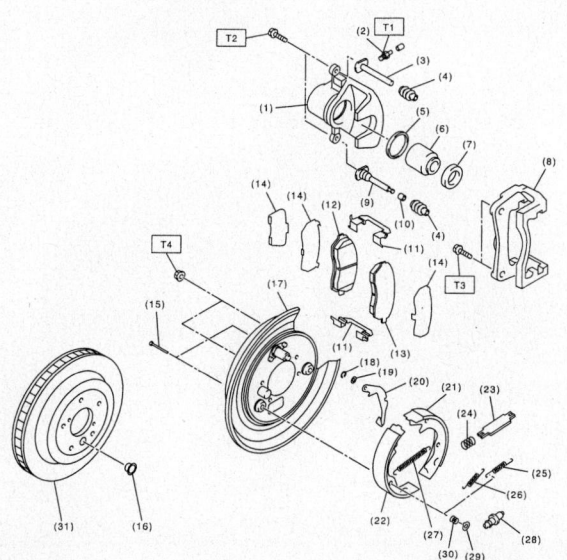

Rear disc brake and related components

(1)	Caliper body	(13)	Outer pad	(25)	Secondary shoe return spring
(2)	Air bleeder screw	(14)	Shim	(26)	Primary shoe return spring
(3)	Guide pin (Green)	(15)	Shoe hold pin	(27)	Adjusting spring
(4)	Pin boot	(16)	Cover	(28)	Adjuster
(5)	Piston seal	(17)	Back plate	(29)	Brake shoe cup
(6)	Piston	(18)	Retainer	(30)	Brake shoe spring
(7)	Piston boot	(19)	Spring washer		
(8)	Support	(20)	Parking brake lever		
(9)	Lock pin (Yellow)	(21)	Parking brake shoe (Secondary)		
(10)	Bushing	(22)	Parking brake shoe (Primary)		
(11)	Pad clip	(23)	Strut		
(12)	Inner pad	(24)	Strut shoe spring		

Tightening torque: N·m (kgf-m, ft-lb)
T1: 8 (0.8, 5.8)
T2: 27 (2.8, 19.9)
T3: 37 (3.7, 27.2)
T4: 66 (6.7, 48.7)

09490_TRIB_G0019

Brake Caliper

REMOVAL & INSTALLATION

Front

1. Before servicing the vehicle, refer to the Precautions Section.

2. Loosen the wheel nuts.

3. Raise and support the vehicle safely.

4. Remove the tire and wheel.

5. Remove the union bolt. Disconnect the brake line from the brake caliper. Be sure to properly catch the fluid to avoid damage to painted surfaces and improper disposal.

6. Remove the bolt retaining the lock pin to the caliper body.

7. Remove the caliper from its mounting.

To install:

8. Installation is the reverse of the removal procedure.

9. Tighten the caliper retaining bolts to 88.5 ft. lbs. (120 Nm)..

10. Check the brake fluid level, correct as required.

11. Bleed the hydraulic system, as required.

Rear

1. Before servicing the vehicle, refer to the Precautions Section.

2. Loosen the wheel nuts.

3. Raise and support the vehicle safely.

4. Remove the tire and wheel.

5. Disconnect the brake line from the brake caliper. Be sure to properly catch the fluid to avoid damage to painted surfaces and improper disposal.

6. Remove the caliper retaining bolts. Remove the caliper from its mounting.

To install:

7. Installation is the reverse of the removal procedure.

8. Tighten the caliper retaining bolts to specification.

9. Check the brake fluid level, correct as required.

10. Bleed the hydraulic system, as required.

Brake Pads

REMOVAL & INSTALLATION

Front

1. Before servicing the vehicle, refer to the Precautions Section.
2. Loosen the wheel nuts.
3. Raise and support the vehicle safely.
4. Remove the tire and wheel.
5. Remove the caliper bolt.
6. Raise the caliper body and properly support it.

➡**Do not disconnect the brake fluid line from the caliper.**

7. Remove the pads.

To install:

8. Installation is the reverse of the removal procedure.

9. Apply a thin coat of Molykote M7439, or equivalent to the pad clip.
10. Apply a thin coat of Molykote AS-880N (part number K0779YA010) or equivalent to the contact surface between the pad and pad inner shim.
11. Apply a thin coat of Molykote AS-880N (part number K0779YA010) or equivalent to the three contact surfaces between the inner shim and outer shim of the outer pads.
12. Check the brake fluid level, correct as required.
13. Bleed the hydraulic system, as required.

Rear

1. Before servicing the vehicle, refer to the Precautions Section.
2. Loosen the wheel nuts.
3. Raise and support the vehicle safely.
4. Remove the tire and wheel.
5. Remove the brake hose bracket.
6. Remove the caliper bolt.
7. Raise the caliper body and properly support it.

➡**Do not disconnect the brake fluid line from the caliper.**

8. Remove the pads.

To install:

9. Installation is the reverse of the removal procedure.

10. Apply a thin coat of Molykote M7439, or equivalent to the pad clip.
11. Apply a thin coat of Molykote AS-880N (part number K0779YA010) or equivalent to the contact surface between the pad and pad inner shim.
12. Check the brake fluid level, correct as required.
13. Bleed the hydraulic system, as required.

SUZUKI

Aerio • Verona

DRIVE TRAIN 9-38
ENGINE REPAIR 9-10
FRONT BRAKES 9-54
FRONT SUSPENSION 9-47
FUEL SYSTEM 9-35
REAR BRAKES 9-56
REAR SUSPENSION 9-51
SPECIFICATIONS AND
MAINTENANCE CHARTS 9-2
Engine and Vehicle Identification 9-2
General Engine Specifications 9-2
Engine Tune-Up Specifications........9-2
Firing Order 9-3
Accessory Drive Belt Routing 9-3
Capacities 9-4
Valve Specifications....................9-4
Crankshaft and Connecting
 Rod Specifications....................9-5
Piston and Ring Specifications 9-5
Torque Specifications 9-6
Wheel Alignment 9-6
Tire, Wheel and Ball Joint
 Specifications 9-7
Brake Specifications 9-7
Scheduled Maintenance
 Intervals................................9-8
STEERING 9-45
A
Air Bag 9-45
 Disarming 9-45
 Precautions............................9-45
 Rearming 9-45
Alternator 9-10
 Installation 9-10
 Removal 9-10
B
Brake Caliper (Front) 9-54
 Removal & Installation...............9-54
Brake Caliper (Rear)....................9-56
 Removal & Installation...............9-56
Brake Drums.............................9-56
 Removal & Installation...............9-56
Brake Shoes 9-57
 Removal & Installation...............9-57
C
Camshaft and Valve Lifters 9-24
 Removal & Installation...............9-24

Clutch 9-39
 Bleeding................................9-39
 Removal & Installation...............9-39
Coil Spring (Front) 9-48
 Removal & Installation...............9-48
Coil Spring (Rear).......................9-52
 Removal & Installation...............9-52
CV-Joints.................................9-42
 Overhaul 9-42
Cylinder Head 9-17
 Removal & Installation...............9-17
D
Disc Brake Pads (Front)................9-54
 Removal & Installation...............9-54
Disc Brake Pads (Rear) 9-56
 Removal & Installation...............9-56
Distributor................................9-10
 Removal & Installation...............9-10
E
Engine Assembly 9-10
 Removal & Installation...............9-10
Exhaust Manifold 9-21
 Removal & Installation...............9-21
F
Front Crankshaft Seal 9-22
 Removal & Installation...............9-22
Fuel Filter 9-35
 Removal & Installation...............9-35
Fuel Injector 9-37
 Removal & Installation...............9-37
Fuel Pump 9-35
 Removal & Installation...............9-35
Fuel System Pressure 9-35
 Relieving 9-35
Fuel System Service
 Precautions............................9-35
H
Halfshaft..................................9-40
 Removal & Installation...............9-40
Heater Core..............................9-14
 Removal & Installation...............9-14
I
Ignition Timing 9-10
 Adjustment.............................9-10
Intake Manifold 9-19
 Removal & Installation...............9-19
L
Lower Ball Joint 9-48
 Removal & Installation...............9-48

Lower Control Arm (Front) 9-48
 Control Arm Bushing
  Replacement.........................9-49
 Removal & Installation...............9-48
Lower Control Arm (Rear).............9-52
 Removal & Installation...............9-52
O
Oil Pan....................................9-26
 Removal & Installation...............9-26
Oil Pump 9-28
 Removal & Installation...............9-28
P
Piston and Ring 9-34
 Positioning 9-34
R
Rack and Pinion Steering Gear 9-45
 Removal & Installation...............9-45
Rear Main Seal 9-29
 Removal & Installation...............9-29
Rocker Arms/Shafts 9-19
S
Starter Motor 9-26
 Removal & Installation...............9-26
Strut (Front)..............................9-47
 Removal & Installation...............9-47
Strut (Rear) 9-51
 Removal & Installation...............9-51
T
Timing Chain 9-29
 Removal & Installation...............9-29
Transaxle Assembly 9-38
 Removal & Installation...............9-38
Transfer Case 9-40
 Removal & Installation...............9-40
V
Valve Lash 9-26
 Adjustment.............................9-26
W
Water Pump 9-13
 Removal & Installation...............9-13
Wheel Bearings (Front).................9-49
 Adjustment.............................9-49
 Removal & Installation...............9-49
Wheel Bearings (Rear) 9-53
 Adjustment.............................9-53
 Removal & Installation...............9-53

SPECIFICATIONS AND MAINTENANCE CHARTS

VEHICLE AND ENGINE IDENTIFICATION

		Engine							Model Year	
Code	Liters (cc)	Cu. in.	Cyl.	Fuel Sys.	Engine Type	Eng. Mfg.			Code	Year
4	2.0 (1955)	121.7	4	MFI	DOHC	Suzuki			2	2002
6	2.3 (2290)	140	4	MFI	DOHC	Suzuki			3	2003
L	2.5 (2492)	152	6	MFI	DOHC	Suzuki			4	2004

MFI: Multiport Fuel Injection

DOHC: Dual Overhead Camshaft

5	2005
6	2006

09490-SUZC-C0001

GENERAL ENGINE SPECIFICATIONS

Year	Engine ID/VIN	Engine Displacement Liters (cc)	Fuel System Type	Net Horsepower @ rpm	Net Torque @ rpm (ft. lbs.)	Bore x Stroke (in.)	Com-pression Ratio	Oil Pressure @ rpm
2002	4	2.0 (1955)	MFI	145@5700	136@3000	3.31x3.54	9.3:1	55.5-66.8@4000
2003	4	2.0 (1955)	MFI	145@5700	136@3000	3.31x3.54	9.3:1	55.5-66.8@4000
2004	6	2.3 (2290)	MFI	155@5400	152@3000	3.54x3.54	9.3:1	55.5@4000
	L	2.5 (2492)	MFI	155@5800	177@4000	3.00x3.50	9.8:1	12.3-21@750
2005	6	2.3 (2290)	MFI	155@5400	152@3000	3.54x3.54	9.3:1	55.5@4000
	L	2.5 (2492)	MFI	155@5800	177@4000	3.00x3.50	9.8:1	12.3-21@750
2006	6	2.3 (2290)	MFI	155@5400	152@3000	3.54x3.54	9.3:1	55.5@4000
	L	2.5 (2492)	MFI	155@5800	177@4000	3.00x3.50	9.8:1	12.3-21@750

MFI: Multiport Fuel Injection

09490-SUZC-C0002

GASOLINE ENGINE TUNE-UP SPECIFICATIONS

Year	Engine Displacement Liters (cc)	Engine ID/VIN	Spark Plugs Gap (in.)	Ignition Timing (deg.) MT	AT	Fuel Pump (psi)		Idle Speed (rpm) MT	AT	Valve Clearance In.	Ex.
2002	2.0 (1955)	4	0.040-0.043	5B	5B	31.3-36.9	①	700-800	700-800	HYD	HYD
2003	2.0 (1955)	4	0.040-0.043	5B	5B	31.3-36.9	①	700-800	700-800	HYD	HYD
2004	2.3 (2290)	6	0.040-0.043	5B	5B	31.3-36.9	①	700-800	700-800	HYD	HYD
	2.5 (2492)	L	0.039-0.043	—	5B	45	①	—	700-800	HYD	HYD
2005	2.3 (2290)	6	0.040-0.043	5B	5B	31.3-36.9	①	700-800	700-800	HYD	HYD
	2.5 (2492)	L	0.039-0.043	—	5B	45	①	—	700-800	HYD	HYD
2006	2.3 (2290)	6	0.040-0.043	5B	5B	31.3-36.9	①	700-800	700-800	HYD	HYD
	2.5 (2492)	L	0.039-0.043	—	5B	45	①	—	700-800	HYD	HYD

Note: The Vehicle Emission Control Information label often reflects specification changes made during production.

The label figures must be used if they differ from those in this chart.

HYD: Hydraulic

B: Before top dead center

① At idle

09490-SUZC-C0003

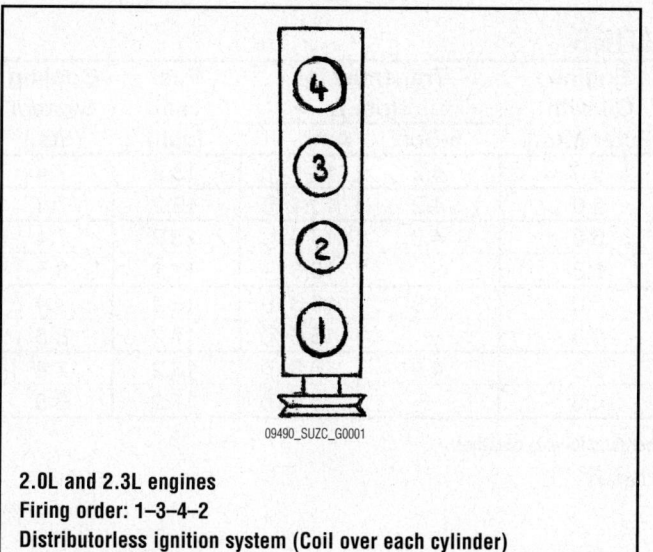

2.0L and 2.3L engines
Firing order: 1–3–4–2
Distributorless ignition system (Coil over each cylinder)

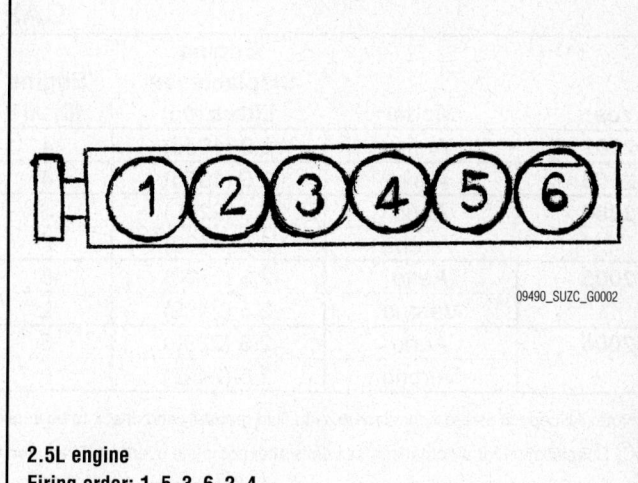

2.5L engine
Firing order: 1–5–3–6–2–4
Distributorless ignition system (Coil over each cylinder)

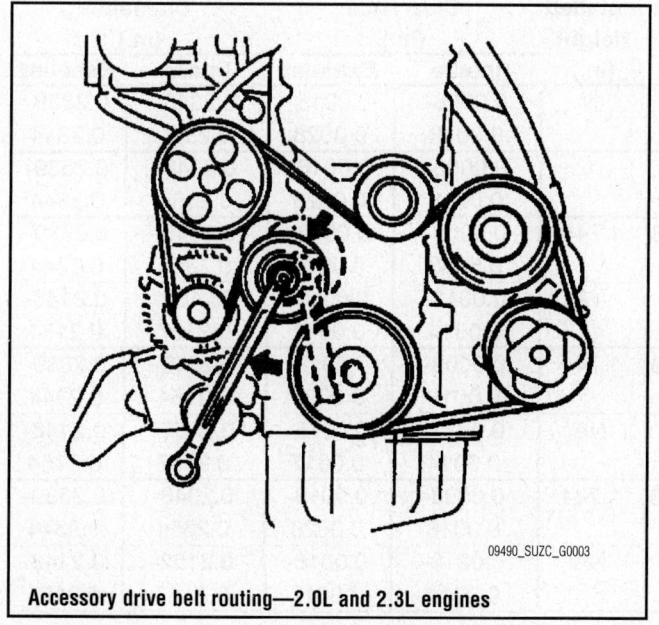

Accessory drive belt routing—2.0L and 2.3L engines

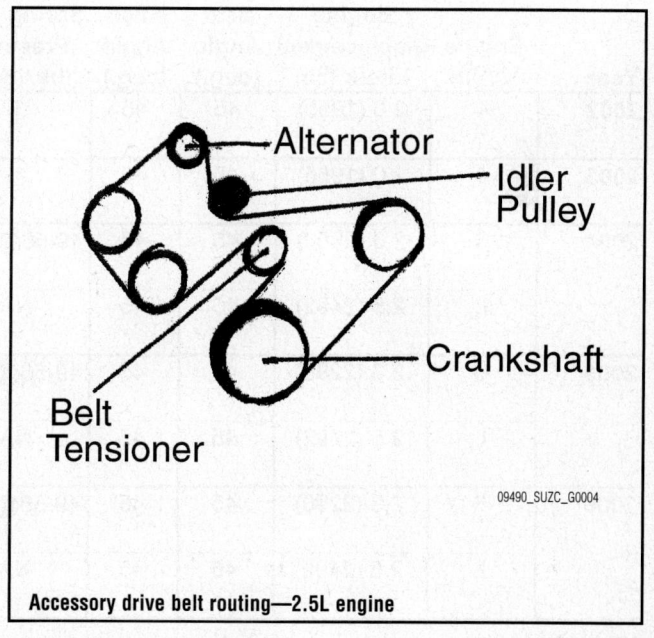

Accessory drive belt routing—2.5L engine

CAPACITIES

Year	Model	Engine Displacement Liters (cc)	Engine ID/VIN	Engine Oil with Filter (qts.)	Transmission (pts.) 5-Spd	Auto.	Fuel Tank (gal.)	Cooling System (qts.)
2002	Aerio	2.0 (1955)	4	5.0	4.2	13.1 ①	13.2	7.4
2003	Aerio	2.0 (1955)	4	5.0	4.2	13.1 ①	13.2	7.4
2004	Aerio	2.3 (2290)	6	5.0	4.9	16.3 ①	13.2	7.4
	Verona	2.5 (2492)	L	6.8	—	15.6 ①	17.2	8.5
2005	Aerio	2.3 (2290)	6	5.0	4.9	16.3 ①	13.2	7.4
	Verona	2.5 (2492)	L	6.8	—	15.6 ①	17.2	8.5
2006	Aerio	2.3 (2290)	6	5.0	4.9	16.3 ①	13.2	7.4
	Verona	2.5 (2492)	L	6.8	—	15.6 ①	17.2	8.5

Note: All capacities are approximate. Add fluid gradualy and check to be sure a proper fluid level is obtained.

① Specification for automatic transaxle is after complete overhaul. Drain and fill will be less

09490-SUZC-C0004

VALVE SPECIFICATIONS

Year	Engine ID/VIN	Engine Displacement Liters (cc)	Seat Angle (deg.)	Face Angle (deg.)	Spring Test Pressure (lbs. @ in.)	Spring Installed Height (in.)	Stem-to-Guide Clearance (in.) Intake	Exhaust	Stem Diameter (in.) Intake	Exhaust
2002	4	2.0 (1955)	45	45	①	②	0.0008-0.0018	0.0018-0.0028	0.2348-0.2354	0.2339-0.2344
2003	4	2.0 (1955)	45	45	①	②	0.0008-0.0018	0.0018-0.0028	0.2348-0.2354	0.2339-0.2344
2004	6	2.3 (2290)	45	45	49-56@1.33	1.744	0.0008-0.0018	0.0018-0.0028	0.2348-0.2354	0.2339-0.2344
	L	2.5 (2492)	45	45	NA	NA	0.0013-0.0014	0.0016-0.0017	0.2152-0.2157	0.2148-0.2154
2005	6	2.3 (2290)	45	45	49-56@1.33	1.744	0.0008-0.0018	0.0018-0.0028	0.2348-0.2354	0.2339-0.2344
	L	2.5 (2492)	45	45	NA	NA	0.0013-0.0014	0.0016-0.0017	0.2152-0.2157	0.2148-0.2154
2006	6	2.3 (2290)	45	45	49-56@1.33	1.744	0.0008-0.0018	0.0018-0.0028	0.2348-0.2354	0.2339-0.2344
	L	2.5 (2492)	45	45	NA	NA	0.0013-0.0014	0.0016-0.0017	0.2152-0.2157	0.2148-0.2154

NA: Not Available

① Inner: 15.2-17.4@1.08
   Outer: 33.9-39.2@1.25

② Inner: 1.4204 inches
   Outer: 1.5921 inches

09490-SUZC-C0005

CRANKSHAFT AND CONNECTING ROD SPECIFICATIONS

All measurements are given in inches.

Year	Engine ID/VIN	Engine Displacement Liters (cc)	Crankshaft				Connecting Rod		
			Main Brg. Journal Dia.	Main Brg. Oil Clearance	Shaft End-play	Thrust on No.	Journal Diameter	Oil Clearance	Side Clearance
2002	4	2.0 (1955)	①	②	0.0039-0.0138	3	1.9768-1.9685	0.0018-0.0024	0.0099-0.0150
2003	4	2.0 (1955)	①	②	0.0039-0.0138	3	1.9768-1.9685	0.0018-0.0024	0.0099-0.0150
2004	6	2.3 (2290)	①	0.0013-0.0019	0.0039-0.0138	3	1.9768-1.9685	0.0018-0.0025	0.0099-0.0150
	L	2.5 (2492)	2.2827-2.2835	0.0008-0.0019	0.0020-0.0100	3	1.8583-1.8587	0.0012-0.0025	0.0032-0.0102
2005	6	2.3 (2290)	①	0.0013-0.0019	0.0039-0.0138	3	1.9768-1.9685	0.0018-0.0025	0.0099-0.0150
	L	2.5 (2492)	2.2827-2.2835	0.0008-0.0019	0.0020-0.0100	3	1.8583-1.8587	0.0012-0.0025	0.0032-0.0102
2006	6	2.3 (2290)	①	0.0013-0.0019	0.0039-0.0138	3	1.9768-1.9685	0.0018-0.0025	0.0099-0.0150
	L	2.5 (2492)	2.2827-2.2835	0.0008-0.0019	0.0020-0.0100	3	1.8583-1.8587	0.0012-0.0025	0.0032-0.0102

① No. 1: 2.2835-2.2837
　No. 2: 2.2832-2.2835
　No. 3: 2.2830-2.2832

② For crankshaft stamped "1" through "3": 0.0010-0.0018 inches
　For crankshaft stamped "4" through "9": 0.0011-0.0017 inches

09490-SUZC-C0006

PISTON AND RING SPECIFICATIONS

All measurements are given in inches.

Year	Engine ID/VIN	Engine Displacement Liters (cc)	Piston Clearance	Ring Gap			Ring Side Clearance		
				Top Compression	Bottom Compression	Oil Control	Top Compression	Bottom Compression	Oil Control
2002	4	2.0 (1955)	0.0008-0.0015	0.0079-0.0137	0.0138-0.0196	0.0079-0.0275	0.0012-0.0027	0.0008-0.0023	snug
2003	4	2.0 (1955)	0.0008-0.0015	0.0079-0.0137	0.0138-0.0196	0.0079-0.0275	0.0012-0.0027	0.0008-0.0023	snug
2004	6	2.3 (2290)	0.0008-0.0015	0.0079-0.0125	0.0126-0.0185	0.0079-0.0196	0.0012-0.0027	0.0008-0.0023	snug
	L	2.5 (2492)	0.0006-0.0011	0.0079-0.0137	0.0138-0.0196	0.0079-0.0275	0.0016-0.0031	0.0004-0.0019	0.0023-0.0059
2005	6	2.3 (2290)	0.0008-0.0015	0.0079-0.0125	0.0126-0.0185	0.0079-0.0196	0.0012-0.0027	0.0008-0.0023	snug
	L	2.5 (2492)	0.0006-0.0011	0.0079-0.0137	0.0138-0.0196	0.0079-0.0275	0.0016-0.0031	0.0004-0.0019	0.0023-0.0059
2006	6	2.3 (2290)	0.0008-0.0015	0.0079-0.0125	0.0126-0.0185	0.0079-0.0196	0.0012-0.0027	0.0008-0.0023	snug
	L	2.5 (2492)	0.0006-0.0011	0.0079-0.0137	0.0138-0.0196	0.0079-0.0275	0.0016-0.0031	0.0004-0.0019	0.0023-0.0059

09490-SUZC-C0007

TORQUE SPECIFICATIONS
All readings in ft. lbs.

Year	Engine ID/VIN	Engine Displacement Liters (cc)	Cylinder Head Bolts	Main Bearing Bolts	Rod Bearing Bolts	Crankshaft Damper Bolts	Flywheel Bolts	Manifold Intake	Manifold Exhaust	Spark Plugs	Lug Nut
2002	4	2.0 (1955)	①	②	33	108.5	51	18	40	18	61.5
2003	4	2.0 (1955)	①	②	33	108.5	51	18	40	18	61.5
2004	6	2.3 (2290)	①	③	④	108.5	51	18	40	18	61.5
	L	2.5 (2492)	⑤	⑥	⑦	369-443	52-59	7-12	20-22	15-22	74
2005	6	2.3 (2290)	①	③	④	108.5	51	18	40	18	61.5
	L	2.5 (2492)	⑤	⑥	⑦	369-443	52-59	7-12	20-22	15-22	74
2006	6	2.3 (2290)	①	③	④	108.5	51	18	40	18	61.5
	L	2.5 (2492)	⑤	⑥	⑦	369-443	52-59	7-12	20-22	15-22	74

① Step 1: 38.5 ft. lbs. (53 Nm)
  Step 2: 61 ft. lbs. (84 Nm)
  Step 3: loosen bolts 1-10
  Step 4: 38.5 ft. lbs. (53 Nm)
  Step 5: 76 ft. lbs. (105 Nm)
  Step 6 - (M6 bolt) 8 ft. lbs. (11 Nm)

② Step 1 - bolts 1-10: 29 ft. lbs. (40 Nm)
  Step 2 - loosen bolts 1-10
  Step 3 - bolts 1-10: 29 ft. lbs. (40 Nm)
  Step 4 - bolts 11-22: (10mm) 42.5 ft. lbs. (59 Nm);
  (8mm) 19.5 ft. lbs. (27 Nm)

③ Step 1 - bolts 1-10: 29 ft. lbs. (40 Nm)
  Step 2 - loosen bolts 1-10
  Step 3 - bolts 1-10: 29 ft. lbs. (40 Nm)
  Step 4 - bolts 1-10: (10mm) 42 ft. lbs. (58 Nm);
  bolts 11-22: (8mm) 18 ft. lbs. (25 Nm)

④ Step 1: 11 ft. lbs. (15 Nm)
  Step 2: turn head bolts 45 degrees
  Step 3: turn head bolts an additional 45 degrees

⑤ Step 1: 15 ft. lbs. (20 Nm)
  Step 2: 18 ft. lbs. (25 Nm)
  Step 3: turn head bolts 70 degrees
  Step 4: turn head bolts an additional 70 degrees

⑥ Step 1 - bed plate inner bolts: 24 ft. lbs. (33 Nm)
  Step 2 - bed plate inner bolts an additional 135 degrees
  Step 3 - bed plate outer bolts: 15-19 ft. lbs. (20-26 Nm)

⑦ Step 1: 15 ft. lbs. (20 Nm)
  Step 2: turn head bolts 45 degrees
  Step 3: turn head bolts an additional 90 degrees

09490-SUZC-C0008

WHEEL ALIGNMENT

Year	Model		Caster Range (+/-Deg.)	Caster Preferred Setting (Deg.)	Camber Range (+/-Deg.)	Camber Preferred Setting (Deg.)	Toe-in (in.)	Steering Axis Inclination (Deg.)
2002	Aerio	F	0	+2.11	1.00	+0.10	0+/-0.08	—
		R	—	—	—	—	—	—
2003	Aerio	F	1.10	2.00	1.00	+0.10	0+/-0.08	—
		R	—	—	—	—	—	—
2004	Aerio	F	0	+2.11	1.00	+0.10	0.04+/-0.08	—
		R	—	—	—	—	—	—
	Verona	F	1.00	-3.00	1.00	-0.30	6.0+/-10.0	—
		R	—	—	1.00	-1.30	9.0+/-10.0	—
2005	Aerio	F	0	+2.11	1.00	+0.10	0.04+/-0.08	—
		R	—	—	—	—	—	—
	Verona	F	1.00	-3.00	1.00	-0.30	6.0+/-10.0	—
		R	—	—	1.00	-1.30	9.0+/-10.0	—
2006	Aerio	F	0	+2.11	1.00	+0.10	0.04+/-0.08	—
		R	—	—	—	—	—	—
	Verona	F	1.00	-3.00	1.00	-0.30	6.0+/-10.0	—
		R	—	—	1.00	-1.30	9.0+/-10.0	—

09490-SUZC-C0009

TIRE, WHEEL AND BALL JOINT SPECIFICATIONS

| Year | Model | OEM Tires | | Tire Pressures (psi) | | Wheel | Ball Joint |
		Standard	Optional	Front	Rear	Size	Inspection
2002	Aerio	P185/65R14	None	30	30	5.5JJ	①
	Aerio LX & SX	P195/55R15	None	30	30	5.5JJ	①
2003	Aerio	P185/65R14	None	30	30	5.5JJ	①
	Aerio LX & SX	P195/55R15	None	30	30	5.5JJ	①
2004	Aerio	P185/65R14	None	30	30	5.5JJ	①
	Aerio LX & SX	P195/55R15	None	30	30	5.5JJ	①
	Verona	P205/65R15	None	29	29	6.0-J	①
	Verona LX	P205/60R16	None	29	29	6.0-J	①
2005	Aerio	P185/65R14	None	30	30	5.5JJ	①
	Aerio LX & SX	P195/55R15	None	30	30	5.5JJ	①
	Verona	P205/65R15	None	29	29	6.0-J	①
	Verona LX	P205/60R16	None	29	29	6.0-J	①
2006	Aerio	P185/65R14	None	30	30	5.5JJ	①
	Aerio LX & SX	P195/55R15	None	30	30	5.5JJ	①
	Verona	P205/65R15	None	29	29	6.0-J	①
	Verona LX	P205/60R16	None	29	29	6.0-J	①

OEM: Original Equipment Manufacturer

PSI: Pounds Per Square Inch

STD: Standard

OPT: Optional

① Replace if any measurable movement is found.

09490-SUZC-C0010

BRAKE SPECIFICATIONS
All measurements in inches unless noted

| Year | Model | Brake Disc | | | Brake Drum Diameter | | | Minimum Lining Thickness | | Brake Caliper | |
		Original Thickness	Minimum Thickness	Maximum Runout	Original Inside Diameter	Max. Wear Limit	Maximum Machine Diameter	Front	Rear	Bracket bolts (ft. lbs.)	Mounting bolts (ft. lbs.)
2002	Aerio	0.790	0.710	0.004	7.87	7.95	7.95	0.080 ①	0.100 ①	62	26
2003	Aerio	0.790	0.710	0.004	7.87	7.95	7.95	0.080 ①	0.100 ①	62	26
2004	Aerio	0.790	0.710	0.004	7.87	7.95	7.95	0.080 ①	0.100 ①	62	26
	Verona	NA	0.860	0.002	—	—	—	0.310 ①	0.080 ①	70	②
2005	Aerio	0.790	0.710	0.004	7.87	7.95	7.95	0.080 ①	0.100 ①	62	26
	Verona	NA	0.860	0.002	—	—	—	0.310 ①	0.080 ①	70	②
2006	Aerio	0.790	0.710	0.004	7.87	7.95	7.95	0.080 ①	0.100 ①	62	16
	Verona	NA	0.860	0.002	—	—	—	0.310 ①	0.080 ①	70	②

NA: Not Available

① Measurement is for lining and backing together.

② Front: 19 ft. lbs.

Rear: 33 ft. lbs.

09490-SUZC-C0011

SCHEDULED MAINTENANCE INTERVALS
SUZUKI—AERIO & VERONA

TO BE SERVICED	TYPE OF SERVICE	VEHICLE MILEAGE INTERVAL (x1000)												
		7.5	15	22.5	30	37.5	45	52.5	60	67.5	75	82.5	90	97.5
Engine oil & filter	R	✓	✓	✓	✓	✓	✓	✓	✓	✓	✓	✓	✓	✓
Automatic transmission fluid & filter ①	S/I	✓	✓	✓	✓	✓	✓	✓	✓	✓	✓	✓	✓	✓
Clutch pedal free travel	S/I	✓	✓	✓	✓	✓	✓	✓	✓	✓	✓	✓	✓	✓
Drive axle boots	S/I	✓	✓	✓	✓	✓	✓	✓	✓	✓	✓	✓	✓	✓
Gear shift control lever/shift operation	S/I	✓	✓	✓	✓	✓	✓	✓	✓	✓	✓	✓	✓	✓
Inspect & rotate tires	S/I	✓	✓	✓	✓	✓	✓	✓	✓	✓	✓	✓	✓	✓
Manual transmission oil ②	S/I	✓	✓	✓	✓	✓	✓	✓	✓	✓	✓	✓	✓	✓
Power steering system	S/I	✓	✓	✓	✓	✓	✓	✓	✓	✓	✓	✓	✓	✓
Suspension system	S/I	✓	✓	✓	✓	✓	✓	✓	✓	✓	✓	✓	✓	✓
Brake discs, pads, drums & shoes	S/I	✓		✓		✓		✓		✓		✓		✓
Brake hoses, pipes, brake lever & cable	S/I	✓		✓		✓		✓		✓		✓		✓
Brake fluid ③	S/I		✓		✓		✓		✓		✓		✓	
Brake pedal	S/I		✓		✓		✓		✓		✓		✓	
Cooling system, hoses & connections	S/I		✓		✓		✓		✓		✓		✓	
Fuel tank, cap & lines	S/I		✓		✓		✓		✓		✓		✓	
Transfer case oil	S/I	✓			✓		✓		✓		✓			
Air cleaner filter element	R				✓				✓				✓	
Engine coolant	R				✓				✓				✓	
Spark plugs	R				✓				✓				✓	
Drive belts	S/I				✓				✓				✓	
Exhaust system	S/I				✓				✓				✓	

09490-SUZC-C0012

SCHEDULED MAINTENANCE INTERVALS
SUZUKI—AERIO & VERONA

TO BE SERVICED	TYPE OF SERVICE	VEHICLE MILEAGE INTERVAL (x1000)												
		7.5	15	22.5	30	37.5	45	52.5	60	67.5	75	82.5	90	97.5
Automatic transmission fluid hose	R						✓							
Ignition coils	S/I												✓	

R: Replace   S/I: Service or Inspect

① Replace every 100,000 miles.

② Replace every 15,000 miles.

③ Replace every 60,000 miles.

FREQUENT OPERATION MAINTENANCE (SEVERE SERVICE)

If a vehicle is operated under any of the following conditions it is considered severe service:

- **Extremely dusty areas.**

- **50% or more of the vehicle operation is in 32°C (90°F) or higher temperatures, or constant operation in temperatures below 0°C (32°F).**

- **Prolonged idling (vehicle operation in stop and go traffic).**

- **Frequent short running periods (engine does not warm to normal operating temperatures).**

- **Police, taxi, delivery usage or trailer towing usage.**

Oil & oil filter: change every 3000 miles.

Brake discs, pads, drums & shoes: service or inspect initially at 3000 miles, 6000 miles, & every 12,000 miles thereafter.

Brake hoses & pipes: service or inspect initially at 3000 miles, 6000 miles & every 12,000 miles thereafter.

Air cleaner filter element: service or inspect ever 3000 miles & replace every 30,000 miles (if not replaced previously).

Automatic transmission fluid & filter: service or inspect every 6000 miles & replace every 15,000 miles (if not replaced previously).

Clutch pedal free travel: service or inspect every 6000 miles.

Inspect & rotate tires: service or inspect every 6000 miles.

Manual transmission oil: service or inspect every 6000 miles & replace every 12,000 miles (if not replaced previously).

Power steering system: service or inspect every 6000 miles.

Steering system: service or inspect every 6000 miles.

Suspension system: service or inspect every 6000 miles.

Drive belts: service or inspect every 15,000 miles.

Exhaust system: service or inspect every 15,000 miles.

09490-SUZC-C0013

ENGINE REPAIR

➡ **Disconnecting the negative battery cable on some vehicles may interfere with the functions of the on board computer systems and may require the computer to undergo a relearning process, once the negative battery cable is reconnected.**

Distributor

REMOVAL & INSTALLATION

These models utilize a Distributorless Ignition System (DIS). With this system, the Electronic Control Module (ECM) determines proper ignition timing and time for the primary ignition coil circuit to turn **ON** and **OFF**.

Alternator

REMOVAL

Aerio

1. Before servicing the vehicle, refer to the Precautions Section.
2. Remove or disconnect the following:
   - Negative battery cable
   - Air cleaner outlet hose and air cleaner assembly
   - Power steering return pipe clamp bolt
   - Right side drive shaft
   - Right engine under cover
   - Serpentine drive belt
   - Power steering suction hose
   - Union bolt
   - Power steering high pressure pipe and gasket
   - Power steering pressure switch connector
   - Power steering pump bolts
   - Power steering pump
   - **B** terminal wire and connector from the alternator
   - Alternator cover
   - Alternator mounting bolts
   - Alternator

Verona

1. Before servicing the vehicle, refer to the Precautions Section.
2. Remove or disconnect the following:
   - Negative battery cable
   - Serpentine accessory drive belt by turning the automatic tensioner

roller bolt clockwise to relieve tension on the belt
   - Harness connector from the back of the alternator
   - Alternator lead to the battery
   - Alternator bolts and alternator

INSTALLATION

Aerio

1. Install or connect the following:
   - Alternator
   - Alternator mounting bolts. Tighten to upper bolt to 16 ft. lbs. (23 Nm); and lower bolt to 37 ft. lbs. (50 Nm).
   - Alternator cover
   - **B** terminal wire and connector to the alternator
   - Power steering pump and bolts. Tighten the bolts to 18 ft. lbs. (25 Nm).
   - Power steering pressure switch connector
   - Power steering high pressure pipe and gasket
   - Union bolt and tighten to 26 ft. lbs. (55 Nm)
   - Power steering suction hose
   - Serpentine drive belt
   - Right engine under cover
   - Right side drive shaft
   - Power steering return pipe clamp bolt
   - Air cleaner outlet hose and air cleaner assembly
   - Negative battery cable

Verona

1. Install or connect the following:
   - Alternator
   - Alternator mounting bolts. Tighten to upper bolt to 26 ft. lbs. (35 Nm); and lower bolt to 30 ft. lbs. (40 Nm).

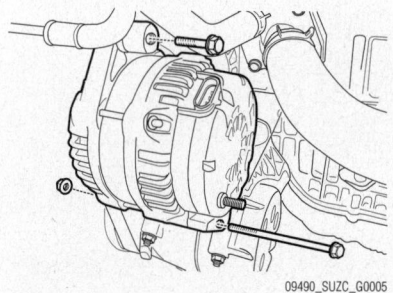

Alternator mounting—Verona

09490_SUZC_G0005

   - Alternator lead to the battery
   - Harness connector to the back of the alternator
   - Serpentine accessory drive belt. Relieve tension on the belt by first applying downward pressure on the automatic tension roller bolt and releasing pressure once the belt is in place.
   - Negative battery cable

Ignition Timing

ADJUSTMENT

Ignition timing is controlled by the Electronic Control Module (ECM). The ECM receives signals from various sensors mounted on the engine. No ignition timing adjustment is possible.

Engine Assembly

REMOVAL & INSTALLATION

2.0L and 2.3L Engines

✳✳ CAUTION

The fuel injection system remains under pressure after the engine has been turned OFF. Properly relieve fuel pressure before disconnecting any fuel lines. Failure to do so may result in fire or personal injury.

1. Before servicing the vehicle, refer to the Precautions Section.
2. Properly relieve the fuel system pressure.
3. Disconnect the negative battery cable.
4. Remove engine hood after disconnecting windshield washer hose.
5. Drain cooling system.
6. Remove radiator with cooling fan.
7. Remove top engine cover (2.3L)
8. Remove air cleaner outlet hose.
9. Remove air cleaner case assembly.
10. Disconnect the following cables:
    a. Accelerator cable from throttle body.
    b. Shift and select cable from transmission (M/T).
    c. Gear select cable from transmission (A/T).
11. Remove accelerator cable with its bracket from intake manifold.

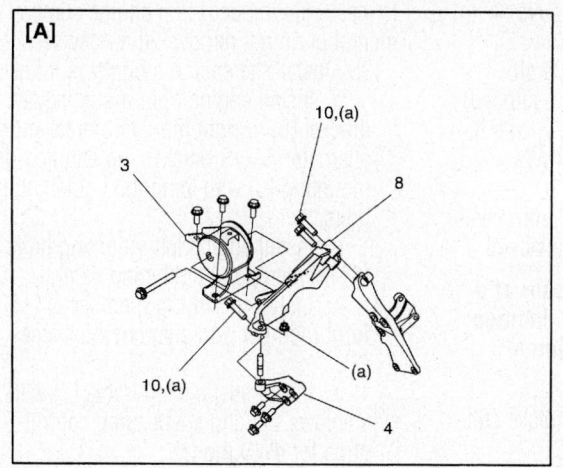

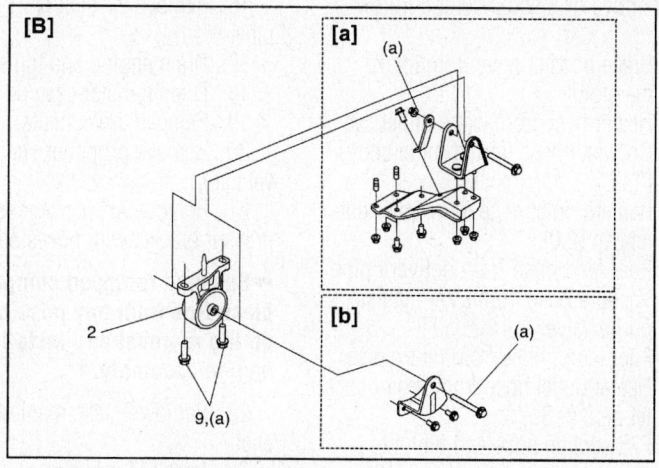

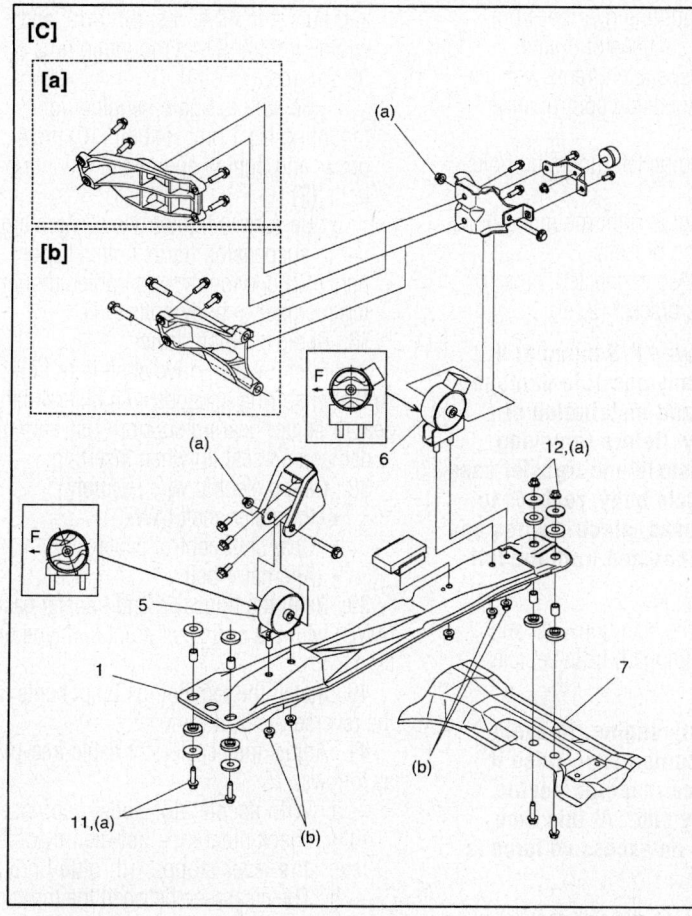

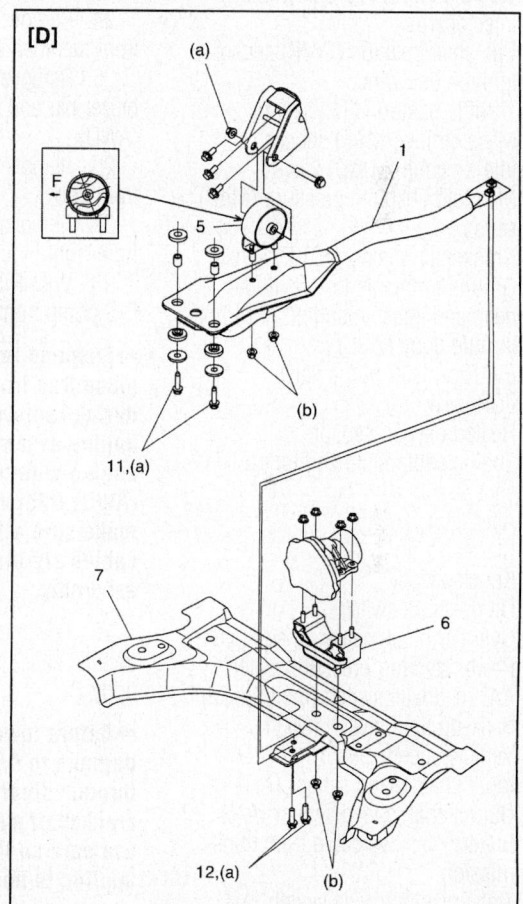

A. Engine right mounting assembly
B. Engine left mounting assembly
C. Engine front and rear mounting assemblies for 2WD model
D. Engine front and rear mounting assemblies for AWD model
a. For M/T model
b. For A/T model

1. Mounting member
2. Engine left mounting
3. Engine right mounting
4. Right mounting bracket
5. Engine front mounting
6. Engine rear mounting
7. Front member and suspension frame

8. Engine right mounting no. 1 bracket
9. Engine left mounting bolts
10. Engine right mounting no. 1 bracket bolt
11. Mounting member-to-body bolts
12. Mounting member-to-suspension frame bolts and nuts

Engine mounting components—2.0L and 2.3L engines

09490_SUZC_G0006

12. Remove or disconnect the following:

- Brake booster hose from intake manifold
- Heater hose from water outlet cap
- Coolant hoses from throttle body (2.0L)
- Radiator inlet hose from water outlet cap (2.0L)
- Fuel feed hose from delivery pipe
- Vacuum hose from EVAP canister purge valve
- Fuel return hose from return pipe
- Heater outlet hose from heater outlet pipe (2.3L)
- P/S suction hose and high pressure pipe from P/S pump (2.3L)
- Injector wire
- Camshaft position (CMP) sensor
- Ignition coil wire
- Throttle position (TP) sensor
- Mass airflow (MAF) sensor
- Idle air control (IAC) valve
- Manifold absolute pressure (MAP) sensor
- Crankshaft position (CKP) sensor
- Ground terminals from cylinder head and intake manifold (2.0L) or throttle body (2.3L)
- EVAP canister purge valve
- EGR valve
- Heated oxygen sensor
- Engine coolant temperature (ECT) sensor
- Alternator
- Starter
- Knock sensor
- Oil pressure switch
- Cylinder block heater (if equipped)
- Power steering pressure switch
- A/C magnetic switch (if equipped)
- Back-up light switch (M/T)
- Vehicle speed sensor
- Input shaft speed sensor (A/T)
- Output shaft speed sensor (A/T)
- Battery negative cable from transmission
- Transmission range switch (A/T)
- Valve body and trans fluid temperature sensor connector (A/T)

13. Remove clutch operating cylinder from transmission with hose still attached (M/T).

➡ **Suspend removed clutch operating cylinder at a place free from any possible damage during removal and installation of engine assembly.**

14. Remove right and left engine under covers.

15. Remove alternator belt.

16. Remove exhaust No.1 and No. 2 pipe.
17. Drain engine and transaxle oil.
18. Drain transfer case oil (if equipped).
19. Remove drive shafts.
20. Remove propeller shaft (AWD vehicle).
21. Remove A/C compressor from compressor bracket with hoses still attached.

➡ **Suspend removed compressor at a place free from any possible damage during removal and installation of engine assembly.**

22. Remove intake manifold intake stiffener.
23. Install lifting device.
24. Remove mounting member from front member and suspension frame.
25. Remove suspension frame with stabilizer bar and suspension control arms (AWD).
26. Remove engine left mounting bolts from body.
27. Remove engine right mounting bolts from right mounting bracket.
28. With P/S hose connected, detach P/S pump from its bracket (2.0L).

➡ **Suspend removed P/S pump at a place free from any possible damage during removal and installation of engine assembly. Before removing engine with transaxle and transfer case (AWD) from vehicle body, recheck to make sure all hoses, electric wires and cables are disconnected from the full assembly.**

29. Lower engine with transaxle and transfer case (if equipped) from vehicle body.

➡ **Before lowering engine, to avoid damage to A/C compressor, raise it through clearance made on engine crankshaft pulley side. At this time, use care so that no excessive force is applied to hoses.**

30. Disconnect the transfer case from the transaxle assembly (AWD vehicles).
31. Disconnect transaxle from engine.
32. Remove clutch cover and clutch disc.

To install:
33. Install or connect the following:
- Clutch cover and clutch disc. Torque the clutch cover bolts to 17 ft. lbs. (23 Nm).
- Transaxle to engine assembly
- Transfer case assembly to transaxle
34. Lift engine with transaxle and trans-

fer case (if equipped) into engine compartment, but do not remove lifting device.

35. Install the engine mounts as follows:
   a. Install engine right mounting No. 1 bracket (8) to right mounting bracket (4), alternator & P/S bracket and engine right mounting (3) with temporal tightening its bolts.

   b. Install engine left mounting bolts (9) with temporal tightening its nut.

   c. Install mounting member (1) to front member and suspension frame (7).

   d. Install suspension frame (7) with stabilizer bar and suspension control arms for 4WD model.

   e. Tighten the engine mounting bolts and nuts A to 40 ft. lbs. (55 Nm); and engine front and rear mounting nuts B to 32.5 ft. lbs. (45 Nm).

   f. Be sure to tighten engine right mounting No.1 bracket bolt (10) previously, and tighten engine left mounting bolts (9).

   g. Be sure to tighten mounting member to suspension frame bolts (12) or nuts (12) previously, and tighten mounting member to body bolts (11).

36. Remove lifting device.
37. Push in each drive shaft joint fully so that snap ring engages with differential gear or center bearing support. Use care not to damage oil seal lip when inserting.

38. Clamp electric wire securely.
- Propeller shaft (AWD)
- Gear shift control cable
- Alternator belt

39. Refill the transaxle and transfer case (AWD) with the correct amount and type of fluid.

40. Install the remaining components in the reverse order of removal.

41. Adjust the accelerator cable free-play as follows:
   a. With accelerator pedal depressed fully, check clearance between throttle lever and lever stopper (throttle body).

   b. The clearance between the throttle lever and lever stopper (throttle body) should be within 0.02–0.07 inches (0.5–2.0mm).

   c. If measured value is out of specification, adjust it to specification with cable adjusting nut.

42. Fill the engine with engine oil and the cooling system with coolant.

43. Fill the power steering reservoir and bleed the power steering system.

44. Run the engine and verify that there are no fuel, coolant, transaxle or exhaust leaks.

2.5L Engine

※※ CAUTION

The fuel injection system remains under pressure after the engine has been turned OFF. Properly relieve fuel pressure before disconnecting any fuel lines. Failure to do so may result in fire or personal injury.

1. Before servicing the vehicle, refer to the Precautions Section.
2. Remove the fuel pump fuse.
3. Start the engine and repeat cranking until the remaining fuel in the fuel line is all consumed.
4. Disconnect the negative battery cable.
5. Drain the engine coolant.
6. Drain the engine oil.
7. Drain the transaxle oil.
8. Drain the power steering oil.
9. Recover the refrigerant.
10. Remove the engine cover.
11. Remove the air filter snorkel and housing assembly.
12. Remove the breather hose.
13. Disconnect the coolant reservoir hoses and coolant reservoir.
14. Disconnect the power steering pump outlet pipe.
15. Disconnect the fuel rail hose and fuel feeding/return pipe.
16. Disconnect the connectors on the O2S, EGR, charcoal canister purge solenoid valve, MAP sensor, Throttle body, engine oil pressure sensor, CMP sensor, knock sensor, VIS and the engine coolant temperature sensor.
17. Detach wiring harness from the wiring harness bracket on the cylinder block.
18. Disconnect the engine block ground by removing the engine block ground bolts.
19. Remove the battery.
20. Disconnect the automatic transaxle cable.
21. Disconnect the inhibit switch connector.
22. Disconnect the connectors on the transaxle.
23. Disconnect the alternator connector and wiring harness.
24. Remove the upper and lower radiator hoses.
25. Remove the front wheels.
26. Remove the engine under covers.
27. Remove the front axle shaft.
28. Remove the front muffler.

29. Support the engine assembly using an engine assembly lift support.
30. Remove the impact beam.
31. Remove the front engine mount bracket-to-damper bush bolt/nut and the rear transaxle mount bracket-to-damper bush bolt/nut.
32. Remove the center front suspension member bolts and the center front suspension member.
33. Disconnect the connector on the compressor.
34. Disconnect the refrigerant suction and discharge pipe block from the compressor.
35. Remove the oil filter.
36. Support the engine and transaxle with the power pack stand DW010-010.
37. Remove the engine mount bracket (RH).
38. Remove the engine mount bracket (LH).
39. Separate the engine and transaxle assembly from the engine compartment.
40. Remove the starter
41. Remove three transaxle torque converter bolts.
42. Remove the transaxle-to-engine block bolts.
43. Remove the rear transaxle mount bracket.
44. Remove the front damper bush bracket.
45. Remove the engine block-to-transaxle bolts.
46. Remove the engine assembly from the transaxle.

To install:
47. Raise the engine and transaxle into the engine compartment.
48. Installation should follow the removal procedure in the reverse order.
49. Tighten the engine block-to-transaxle bolts near the rear transaxle bracket to 44–63 ft. lbs. (60–85 Nm).
50. Tighten the engine block-to-transaxle bolts near the front engine mount bracket to 44–63 ft. lbs. (60–85 Nm).
51. Tighten the transaxle-to-engine block bolts near the rear support bracket to 44–63 ft. lbs. (60–85 Nm).
52. Tighten the 3 flexible plate-to-torque converter retaining bolts to 30–37 ft. lbs. (40–50 Nm).
53. Install the rear transaxle mount bracket and tighten the rear transaxle mount bracket bolts to 26–55 ft. lbs. (55–75 Nm).
54. Install the front damper bush bracket and tighten the front damper bush bracket bolts to 26–55 ft. lbs. (55–75 Nm).

55. Install the remaining components in the reverse order of removal
56. Fill the cooling system, engine, transaxle and power steering pump.
57. Adjust all cables and check all connections.
58. Connect the negative battery cable to the battery.
59. Start the vehicle and check for leaks.

Water Pump

REMOVAL & INSTALLATION

2.0L and 2.3L Engines

1. Before servicing the vehicle, refer to the Precautions Section.
2. Disconnect the negative battery cable.
3. Drain the cooling system into a suitable container and tighten the drain plug.
4. Remove or disconnect the following:
   - Condenser cooling fan (if equipped with A/C)
   - Radiator outlet hose from thermostat cap
   - Heater outlet pipe bolt
   - Serpentine belt
   - Compressor from its bracket with hoses still attached (if equipped with A/C)
   - Water pump bolts and the water pump

➡ **Do not lose dowel pins when removing water pump.**

To install:
5. Install or connect the following:
   - New O-ring to water pump.

➡ **Do not forget to install dowel pins on water pump side before mounting water pump to cylinder block.**

   - Heater outlet pipe to water pump

※※ WARNING

Use NEW bolts to install water pump to cylinder block. Failure to do so may result water leakage.

   - Water pump with new bolts to cylinder block and tighten to 18 ft. lbs. (25 Nm)
   - Compressor to its bracket (if equipped with A/C)
   - Heater outlet pipe bolt
   - Radiator outlet hose to thermostat cap
   - Serpentine belt
6. Fill the cooling system.
7. Connect the negative battery cable.

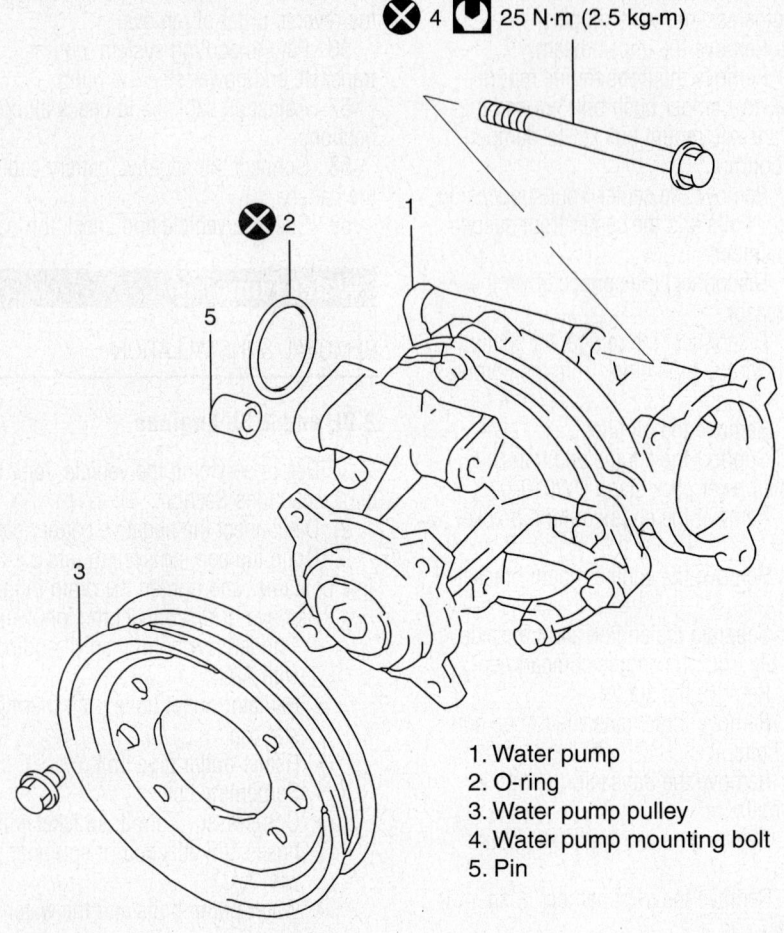

⊗ 4 ⬒ 25 N·m (2.5 kg-m)

1. Water pump
2. O-ring
3. Water pump pulley
4. Water pump mounting bolt
5. Pin

09490_SUZC_G0007

Water pump and related components—2.0L and 2.3L engines

8. Start the engine and top off the coolant as necessary.
9. Check the cooling system for leaks.

2.5L Engine

1. Before servicing the vehicle, refer to the Precautions Section.
2. Disconnect the negative battery cable.
3. Drain the cooling system and tighten the drain plug.
4. Remove or disconnect the following:
   - Serpentine belt
   - Coolant hose from the thermostat housing and the water pump
   - Power steering pump mounting bolts
   - Alternator
   - Compressor
   - Bracket bolts
   - Auto tensioner mounting bolt and auto tensioner
   - Water pump mounting bolts and water pump
   - Water pump gasket

5. Inspect the water pump body for cracks and leaks.
6. Clean the mating surfaces of the water pump and the engine block.

To install:

7. Clean the water pump mounting surface of old gasket material.
8. Install or connect the following:
   - New water pump gasket to the water pump
   - Water pump onto the engine block and tighten the mounting bolts to 17 ft. lbs. (23 Nm)
   - Auto tensioner with the mounting bolt. Tighten the auto tensioner bolt to 33 ft. lbs. (45 Nm).
   - Bracket and bracket bolts. Tighten the bracket retaining bolts to 33 ft. lbs. (45 Nm).
   - Power steering pump and retaining bolts. Tighten the power steering pump retaining bolts to 18 ft. lbs. (25 Nm).
   - Compressor
   - Alternator
   - Serpentine belt

9. Fill the cooling system.
10. Connect the negative battery cable.
11. Start the engine and top off the coolant as necessary.
12. Check the cooling system for leaks.

Heater Core

REMOVAL & INSTALLATION

Aerio

1. Before servicing the vehicle, refer to the Precautions Section.
2. Disconnect the negative battery cable.
3. Recover refrigerant from A/C system.

➡**The amount of removed compressor oil must be measured for replenishing compressor oil.**

4. Remove suction pipe mounting bolts.
5. Disconnect suction pipe and condenser outlet pipe using special tools 09991-15410 and 09991-15420.

❋❋ WARNING

Keep internal parts of air conditioning system free from moisture and dirt. When disconnecting any line from system, install a blind plug or cap to the fitting immediately.

6. Drain engine coolant and disconnect heater hoses from heater unit.
7. Remove heater unit mounting nuts.
8. Remove instrument panel as follows:
   a. Disable air bag system.
   b. Remove steering column hole cover.
   c. Detach steering lower shaft.
   d. Remove glove box and hood latch release lever.
   e. Remove console box.
   f. Remove instrument panel center lower covers.
   g. Remove front pillar trims, front pillar lower garnishes, front side sill scuffs and dash side trims.
   h. Disconnect instrument panel harness connectors and antenna cable which need to be disconnected for removal for instrument panel.
   i. Remove instrument panel ground wire.
   j. Remove instrument panel mounting bolts.
   k. Remove instrument panel with steering column, steering support member and instrument panel harness.

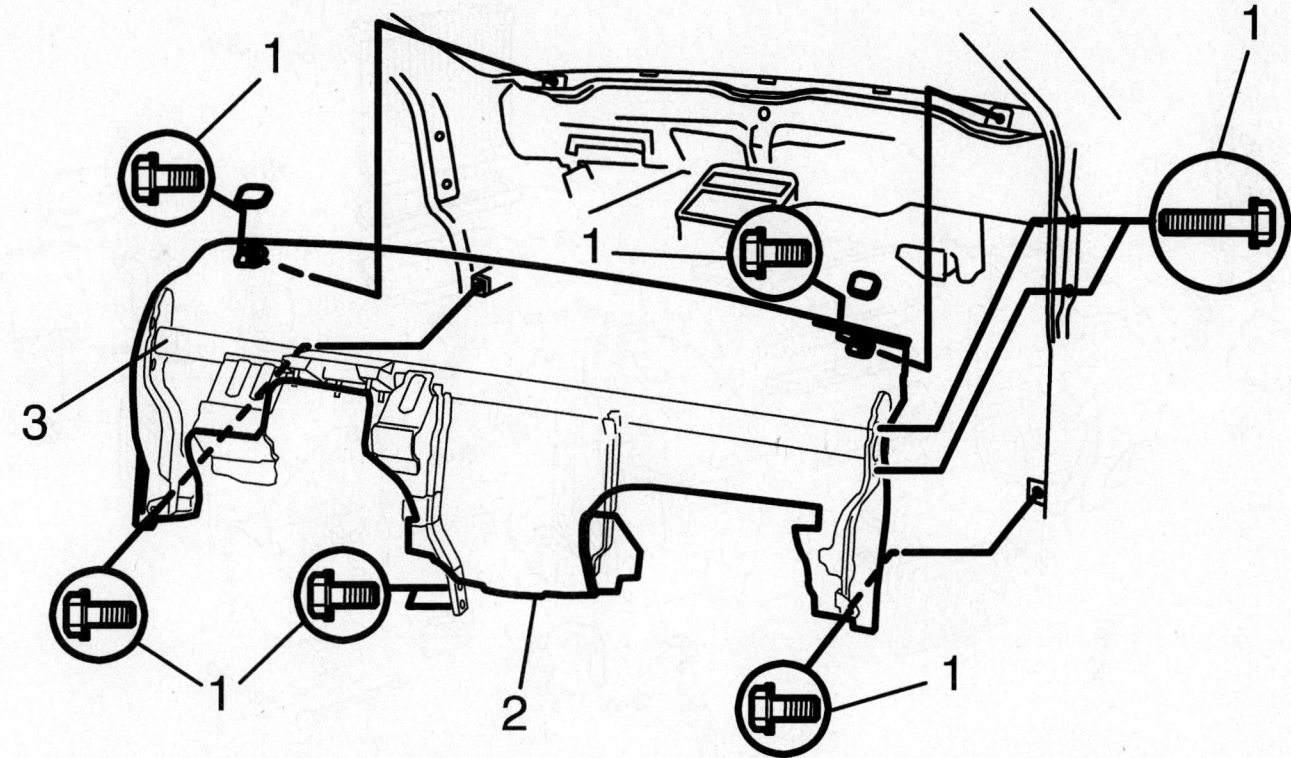

1. Instrument panel mounting bolts
2. Instrument panel
3. Steering support member

09490_SUZC_G0008

Instrument panel and mounting bolt locations—Aerio

9. Disconnect main harness connector and clamp, air intake control actuator connector and blower motor connector.

10. Remove blower motor relay from blower unit.

11. Remove blower unit from vehicle body.

12. Disconnect couplers from the following parts:

- Temperature control actuator
- Air flow control actuator
- A/C evaporator inlet air temperature sensor
- A/C evaporator outlet air temperature sensor
- Blower motor controller

13. Remove heater and cooling unit drain hose .

14. Remove heater and cooling unit from vehicle body.

15. Remove the heater core.

To install:

16. Reverse removal procedure to install heater and cooling unit noting the following instructions:

- Heater core to the heater and cooling unit
- Heater and cooling unit into the vehicle

- Blower unit
- Blower motor relay

➡**When installing each part, be careful not to catch any cable or wiring harness.**

17. Replenish specified amount of compressor oil to compressor suction side.

18. Evacuate and charge refrigerant.

19. Sufficiently apply compressor oil to fitting surface of O-ring and pipe.

20. Install the instrument panel.

21. Refill the cooling system.

22. Connect the negative battery cable.

23. Operate the engine to normal operating temperatures; then, check the climate control operation and check for leaks.

Verona

1. Before servicing the vehicle, refer to the Precautions Section.

2. Drain the cooling system.

3. Recover the refrigerant.

4. Disconnect the negative battery cable.

5. Remove the instrument panel assembly as follows:

a. Remove the floor console.

b. Remove the sun sensor by gently prying up on it and disengaging the connector.

c. Remove the tweeter speakers by gently prying up on it and disengaging the connectors.

d. Remove the stereo cassette AM/FM radio by first removing the audio system trim plate, mounting screws and then disengaging the audio system electrical connectors.

e. Remove the instrument cluster dimmer switch assembly.

f. Remove the instrument cluster trim panel.

g. Remove the automatic temperature controls assembly.

h. Remove the instrument cluster.

i. Remove the kick panels.

j. Remove the glove box.

k. Remove the screws and the glove box housing.

l. Disconnect the glove box housing electrical connectors.

m. Remove the knee bolster.

n. Remove the screws and the instrument panel side covers.

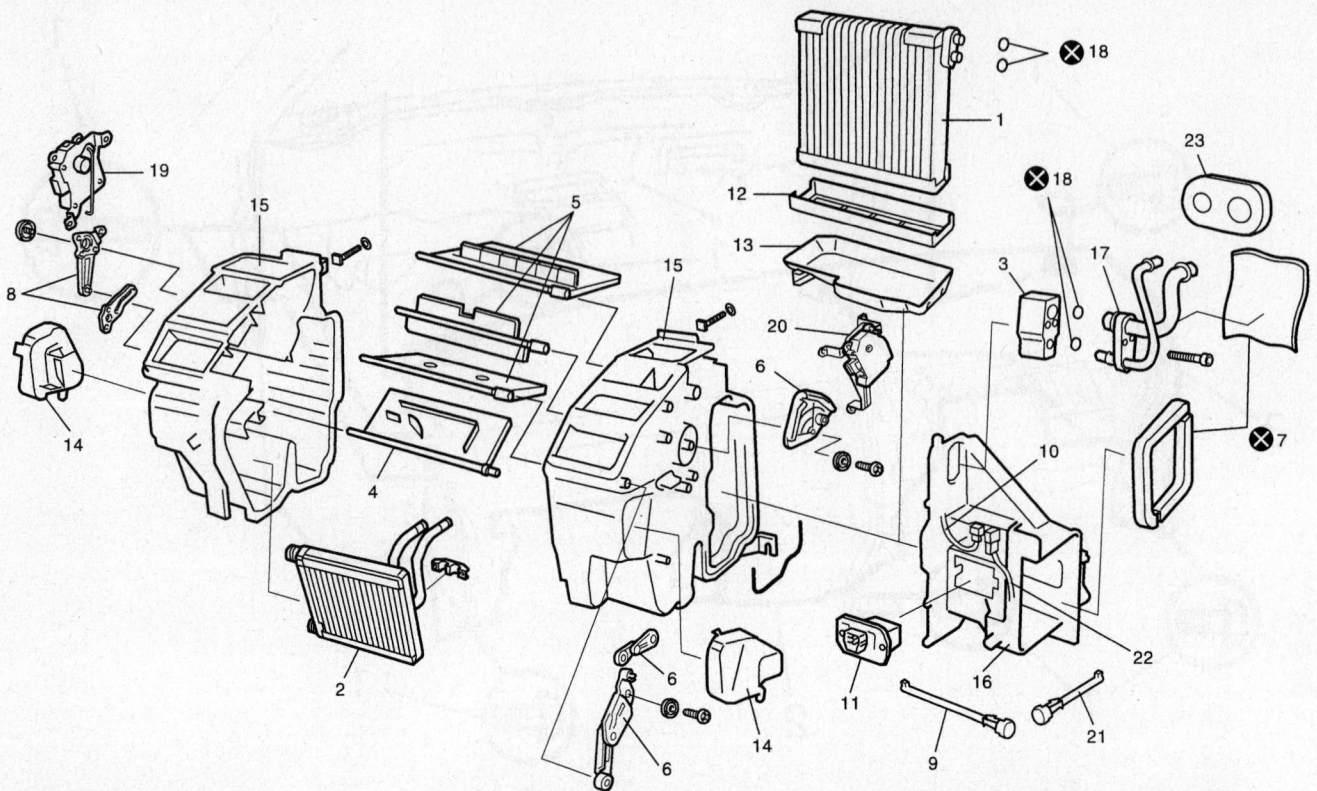

1. A/C evaporator
2. Heater core
3. Expansion valve
4. Temperature selector door
5. Air outlet selector door
6. Air outlet selector link
7. Packing
8. Temperature selector link
9. A/C evaporator outlet
   air temperature sensor clamp
10. A/C evaporator outlet air
    temperature sensor harness
11. Blower motor controller
12. A/C evaporator undercover
13. Lower packing
14. Foot air nozzle
15. Heater and cooling case
16. Air joint duct
17. Expansion valve pipe
18. O-ring
19. Temperature control actuator
20. Air flow control actuator
21. A/C evaporator inlet air
    temperature sensor clamp
22. A/C evaporator inlet air
    temperature sensor harness
23. Grommet

09490_SUZC_G0009

Exploded view of the heater core, heater housing and related components—Aerio

o. Remove the screw and instrument panel fuse block.

p. Remove the nuts and the bolts securing the steering column.

q. Disconnect the steering column electrical connector and lower the steering column.

r. Remove the instrument panel bolts below the windshield.

s. Remove the fixing screws securing the instrument panel.

t. Remove the nuts securing the sides of the instrument panel to the body.

u. Disconnect the instrument panel electrical connectors.

v. Remove the instrument panel.

6. Remove the retaining nuts that secure the suction hose and the liquid evaporator pipe blocks at the fire wall.

7. Loosen the clamp bolts from the suction hose and the liquid evaporator pipe to allow movement of the hose and the pipe.

8. Remove the evaporator drain hose.

9. Compress the heater hose clamps at the fire wall and slide the clamps toward the engine.

10. Remove the 2 heater hoses from the core lines at the fire wall.

11. Remove the screws that secure the A/C module to the fire wall on either side of the heater hoses.

12. Remove the A/C module screw, that is located above the fuel filter, from the engine compartment side of the fire wall.

13. Have an assistant support the A/C module from inside the vehicle. Remove the A/C module screws from the evaporator flange on the engine compartment side of the fire wall. The A/C module will start to drop.

14. Pull the case straight away from the fire wall.

15. Remove the A/C module from the vehicle.

16. Remove the heater core from the A/C module as follows:

a. Remove the wiring harness from the A/C module.

b. Slide the cable eyelet off the post on the temperature door lever.

c. Remove the mode motor ventilation link.

d. Remove the air inlet case with the screws.

e. Remove the evaporator core case cover.

f. Remove the heater core cover with the screws.

g. Remove the heater core clip and heater core from the case.

To install:

17. Install the heater core to the case.

18. Position the A/C module in the vehicle.

19. Slowly raise the A/C module into position and hold it against the fire wall while the screws are installed and tightened from the engine side of the fire wall.

20. Align and install the A/C module screws above the fuel filter and at the evaporator flange. Tighten the A/C module screws to 71 inch lbs. (8 Nm).

21. Install the A/C module screws adjacent to the heater hoses and tighten the A/C module screws to 71 inch lbs. (8 Nm).

22. Install the evaporator drain hose.

23. Install the 2 heater hoses.

24. Slide the heater hose clamps into position.

25. Install the instrument panel assembly.

26. Install new O-rings on the suction hose and the liquid evaporator pipe at the fire wall and put the pipes back in place.

27. Install the retaining nuts that secure the suction hose and the liquid evaporator pipe blocks at the fire wall. Tighten the suction hose and the liquid evaporator pipe retaining nuts to 89 inch lbs. (10 Nm).

28. Refill the cooling system.

29. Recharge the A/C system.

30. Connect the negative battery cable.

31. Operate the engine to normal operating temperatures; then, check the climate control operation and check for leaks.

Cylinder Head

REMOVAL & INSTALLATION

2.0L and 2.3L Engines

✳✳ CAUTION

The fuel injection system remains under pressure after the engine has been turned OFF. Properly relieve fuel pressure before disconnecting any fuel lines. Failure to do so may result in fire or personal injury.

1. Before servicing the vehicle, refer to the Precautions Section.

2. Disconnect the negative battery cable.

3. Relieve the fuel system pressure.

4. Drain the cooling system.

5. Remove or disconnect the following:
- Engine assembly
- Timing chains
- Camshafts and valve lash adjusters
- Intake manifold rear stiffener
- Coolant pipe from intake manifold
- Right engine mounting bracket
- Power steering pump bracket

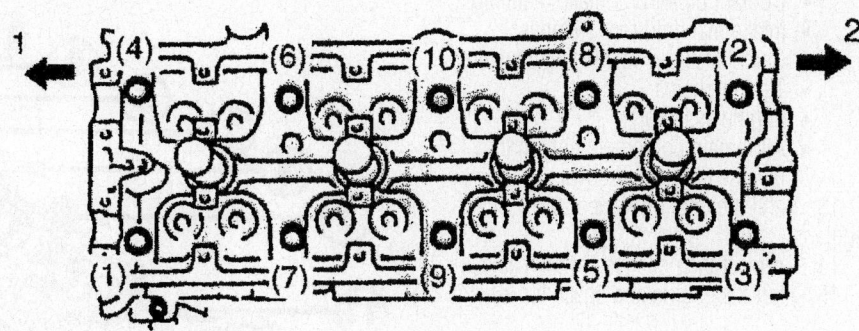

1. Crankshaft pulley side
2. Flywheel side

09490_SUZC_G0010

Cylinder head bolt loosening sequence—2.0L and 2.3L engines

6. Loosen the cylinder head bolts in reverse order of tightening including the M6 bolt, located on the bottom left portion of the cylinder head. Once each bolt is loose, remove the bolts from the cylinder head.

7. Check to be sure all components are removed or disconnected before removing the cylinder head.

8. Remove the cylinder head with the intake manifold, exhaust manifold and water outlet cap as an assembly.

9. Clean the cylinder block and cylinder head mating surfaces of any old gasket material, oil or dust and clean any engine coolant from the cylinders.

To install:

10. Match the matchmark on the crank timing sprocket and the mating surface of the cylinder block and lower crankcase.

11. Install or connect the following:

- Knock pins to cylinder block
- New cylinder head gasket with the top mark facing up and toward the crankshaft pulley
- Cylinder head to the engine block

12. Apply engine oil to the bolt threads and tighten the cylinder head bolts, in sequence, using the following 6 Steps:

a. Step 1: 38.5 ft. lbs. (53 Nm)

b. Step 2: 61 ft. lbs. (84 Nm)

c. Step 3: Loosen all of the cylinder head bolts

d. Step 4: 38.5 ft. lbs. (53 Nm)

e. Step 5: 76 ft. lbs. (105 Nm)

f. Step 6: Bolt M6 (bottom left of cylinder head)—8 ft. lbs. (11 Nm)

- Right engine mounting bracket with new mounting bolts. Tighten to 40 ft. lbs. (55 Nm).
- Power steering pump bracket

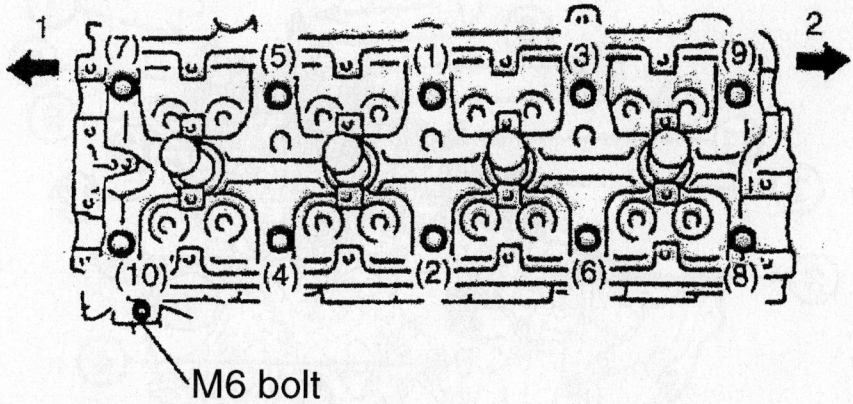

M6 bolt

1. Crankshaft pulley side
2. Flywheel side

09490_SUZC_G0011

Cylinder head bolt torque sequence—2.0L and 2.3L engines

- Coolant pipe from intake manifold
- Intake manifold rear stiffener
- Camshafts and valve lash adjusters
- Timing chains
- Engine assembly
- All remaining components in the reverse order of removal

13. Refill the cooling system with coolant.

14. Connect the negative battery cable.

15. Start the engine and check for leaks.

2.5L Engine

> **✳✳ CAUTION**
>
> **The fuel injection system remains under pressure after the engine has been turned OFF. Properly relieve fuel pressure before disconnecting any fuel lines. Failure to do so may result in fire or personal injury.**

1. Before servicing the vehicle, refer to the Precautions Section.

2. Remove the fuel pump fuse.

3. Start the engine and repeat cranking until the remaining fuel in the fuel line is consumed.

4. Disconnect the negative battery cable.

5. Drain the cooling system.

6. Drain the engine oil and transaxle oil.

7. Drain the power steering fluid.

8. Recover the A/C refrigerant.

9. Remove or disconnect the following:
   - Engine assembly
   - Timing chain
   - Intake manifold support bracket
   - Power steering pump

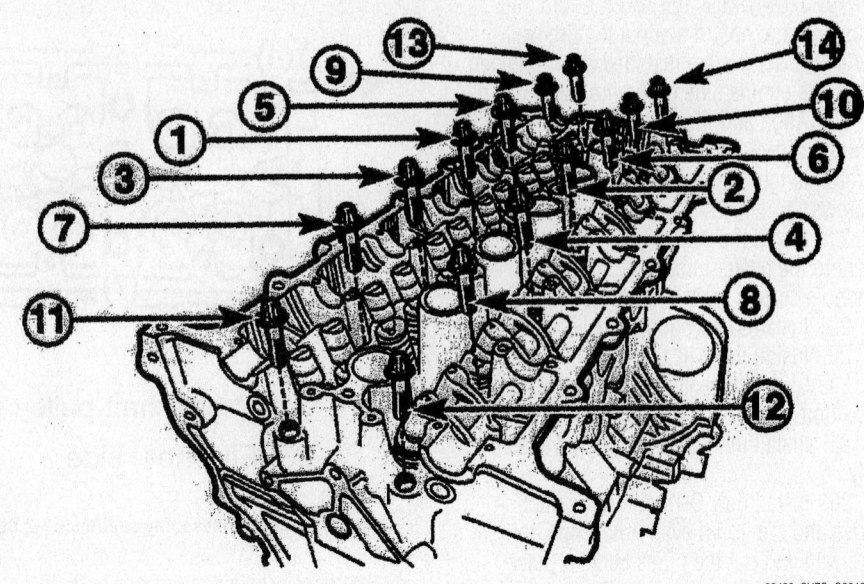

09490_SUZC_G0013

Cylinder head bolt tightening sequence—2.5L engine

- EGR valve, pipes and adapter
- Heater pipe
- Exhaust manifold and gasket
- Camshaft position (CMP) sensor
- Engine coolant temperature (ECT) sensor
- Coolant outlet port

10. Loosen the cylinder head bolts in reverse order of the tightening sequence. Once each bolt is loose, remove the bolts from the cylinder head.

11. Check to be sure all components are removed or disconnected before removing the cylinder head.

12. Remove the cylinder head and gasket, as well as any liquid gasket material.

13. Clean the cylinder block and cylinder head mating surfaces of any old gasket material, oil or dust and clean any engine coolant from the cylinders.

To install:

14. Install or connect the following:
   - Dowel pins on top of cylinder block
   - Liquid gasket on top of cylinder block
   - New cylinder head gasket
   - Liquid gasket on the upper surface of the new cylinder head gasket
   - Cylinder head to the engine block

15. Install the cylinder head to the engine block and and tighten the cylinder head bolts, in sequence, using the following 4 Steps:
   a. Step 1: 15 ft. lbs. (20 Nm)
   b. Step 2: 18 ft. lbs. (25 Nm)
   c. Step 3: Turn the cylinder head bolts an additional 70 degrees
   d. Step 4: Turn the cylinder head bolts 70 degrees again
   - Coolant outlet port. Tighten the bolts to 71–106 inch lbs. (8–12 Nm).
   - ECT sensor
   - CMP sensor
   - Exhaust manifold and new gasket
   - Heater pipe
   - EGR adapter and tighten the retaining bolts to 13–16 ft. lbs. (18–22 Nm).
   - EGR pipes and valve
   - Power steering pump. Tighten the retaining bolts to 15–22 ft. lbs. (20–30 Nm).
   - Intake manifold support bracket

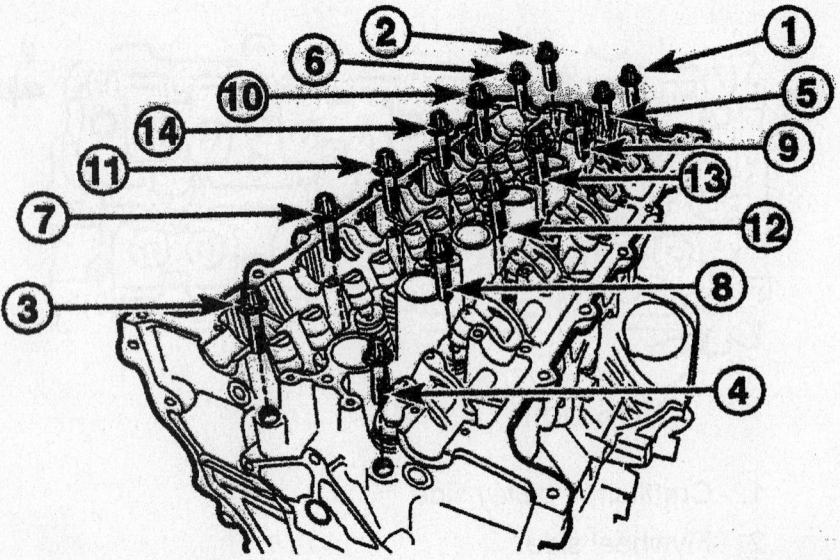

09490_SUZC_G0012

Cylinder head bolt loosening sequence—2.5L engine

Tighten the bracket bolt to 15–22 ft. lbs. (20–30 Nm).
- Timing chain and cover
- Engine assembly
- All remaining components in the reverse order of removal

16. Refill the engine and transaxle with correct type and amount of oil.

17. Refill the cooling system with coolant.

18. Refill the power steering system with correct type and amount of power steering fluid.

19. Connect the negative battery cable.

20. Charge the A/C system.

21. Install the fuel pump fuse.

22. Start the engine and check for leaks.

Rocker Arms/Shafts

The 2.0L and 2.3L engines do not utilize rocker arms/shafts. On the 2.0L and 2.3L engines, the camshaft directly actuates the valves.

2.5L Engine

1. Before servicing the vehicle, refer to the Precautions Section.

2. Relieve the fuel system pressure.

3. Remove or disconnect the following:
- Cylinder head
- Camshafts
- Rocker arm shaft plugs at both ends of cylinder block
- Front and rear rocker arm shafts using rocker arm shaft installer/remover DW110-160

✳ WARNING

To prevent the oil from leaking through the pressure chamber, do not flip the cylinder head over when removing or installing the rocker arm.

- Rocker arms

To install:

4. Installation should follow the removal procedure in the reverse order.

5. Install or connect the following:
- Rocker arms
- Rocker arm shafts using the rocker arm shaft installer/remover DW110-160

➡ **Front rocker arm shaft which will be installed to the timing chain side can be identified by the hole for the rocker arm spring.**

- Sealant to the rocker arm shaft plugs

- Rocker arm shaft plugs. Tighten the rocker arm shaft plug to 26–33 ft. lbs. (35–45 Nm).
- Camshafts
- Cylinder head
- Negative battery cable

6. Start the engine and check for any water or oil leaks when finished.

7. Check ignition timing as necessary.

Intake Manifold

REMOVAL & INSTALLATION

2.0L and 2.3L Engines

✳✳ CAUTION

The fuel system pressure must be relieved before disconnecting any fuel lines. Failure to do so may result in personal injury.

1. Before servicing the vehicle, refer to the Precautions Section.

2. Properly relieve the fuel system pressure.

3. Disconnect the negative battery cable.

4. Drain the coolant from the vehicle.

✳✳ CAUTION

To help avoid the danger of being burned, do not remove the drain plug and the radiator cap while the engine is still hot. Scalding fluid and steam can be blown out under pressure if the plug and cap are taken off too soon.

5. Remove the engine cover.

6. Remove the air cleaner outlet hose.

7. Disconnect the accelerator cable from throttle valve lever.

8. Remove the accelerator cable, with cable bracket from intake manifold.

9. Disconnect the following electric lead wires:
- IAC valve connector
- TP sensor connector
- EGR valve connector
- EVAP canister purge valve connector
- MAP sensor connector
- Ground terminal from throttle body

10. Disconnect the following hoses:
- Brake booster hose from intake manifold
- PCV hose from PCV valve
- Fuel pressure regulator vacuum hose from intake manifold
- Coolant hoses from throttle body

- Breather hose from throttle body
- Vacuum hose from EVAP canister purge valve

11. Remove intake manifold front stiffener.

12. Remove rear stiffener, and disconnect coolant pipe from intake manifold.

13. Remove intake manifold with throttle body from cylinder head, and the gasket.

To install:

14. Before installing the gasket, make sure that the mating surfaces of the intake manifold and the cylinder head are clean and undamaged.

15. Install a new intake manifold gasket onto the cylinder head.

16. Install the intake manifold and throttle body onto the cylinder head.

17. Install the intake manifold mounting nuts and bolts. Tighten the nuts and bolts to 18 ft. lbs. (25 Nm).

18. Connect coolant pipe to intake manifold and install rear stiffener. Tighten bolts to 18 ft. lbs. (25 Nm).

19. Install intake manifold front stiffener. Tighten bolts to 40 ft. lbs. (55 Nm).

20. Connect the following hoses:
- Brake booster hose to intake manifold
- PCV hose to PCV valve
- Fuel pressure regulator vacuum hose to intake manifold
- Coolant hoses to throttle body
- Breather hose to throttle body
- Vacuum hose to EVAP canister purge valve

21. Connect the following electric lead wires:
- IAC valve connector
- TP sensor connector
- EGR valve connector
- EVAP canister purge valve connector
- MAP sensor connector
- Ground terminal to throttle body

22. Connect accelerator cable to throttle valve lever and install its bracket to the intake manifold.

23. Adjust accelerator cable play as follows:

a. With accelerator pedal depressed fully, check clearance between throttle lever (2) and lever stopper (1) (throttle body) which should be within specification.

b. Clearance specification between throttle lever and lever stopper (with pedal depressed fully) "b" should measure 0.02–0.07 inches (0.5–2.0mm).

c. If measured value is out of specification, adjust it to specification with cable adjusting nut (4).

24. Install air cleaner outlet hose.

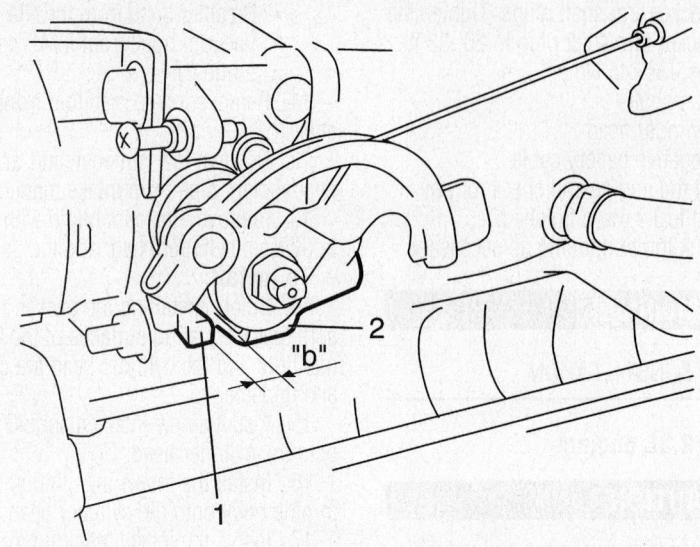

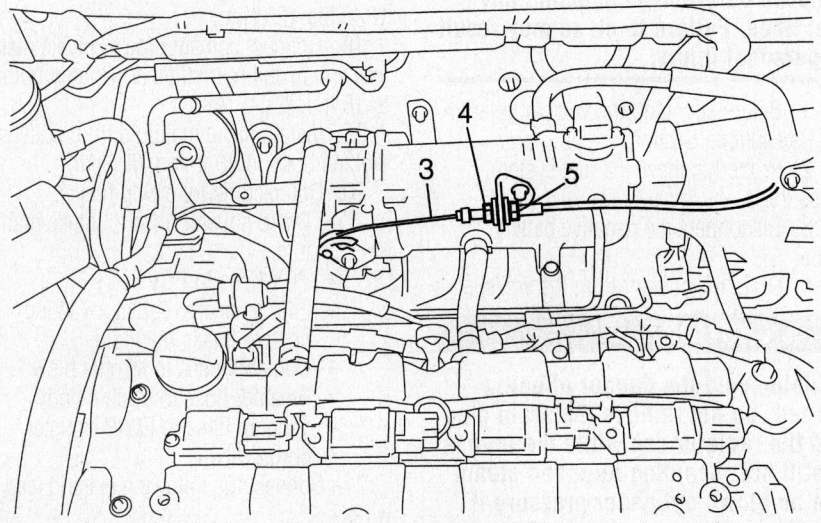

1. Lever stopper
2. Throttle lever
3. Accelerator cable
4. Cable adjusting nut
5. Lock nut
"b" Clearance between throttle lever and lever stopper

09490_SUZC_G0014

Proper adjustment of the accelerator cable—2.0L and 2.3L engines

25. Install engine cover.
26. Check to ensure that all removed parts are back in place. Reinstall any necessary parts which have not been reinstalled.
27. Refill the cooling system.
28. Connect the negative battery cable.
29. Start the engine and check for fuel and cooling system leaks.

2.5L Engine

❋❋ **CAUTION**

The fuel system pressure must be relieved before disconnecting any fuel lines. Failure to do so may result in personal injury.

1. Before servicing the vehicle, refer to the Precautions Section.
2. Remove the fuel pump fuse.
3. Start the engine and repeat cranking until the remaining fuel in the fuel line is all consumed.
4. Disconnect the negative battery cable.
5. Drain the engine coolant.
6. Remove the engine cover.
7. Remove the air filter snorkel.
8. Remove the VIS vacuum tank.
9. Disconnect the fuel pressure regulator vacuum hose.
10. Disconnect the fuel line.
11. Remove the breather hose.
12. Remove the PCV vacuum hose.

13. Remove the air filter outlet hose.
14. Remove the throttle body assembly.
15. Disconnect the charcoal canister purge solenoid connector and hose.
16. Disconnect the Oxygen Sensor (O2S) connector.
17. Disconnect the Manifold Absolute Pressure (MAP) sensor connector.
18. Disconnect the Crankshaft Position (CKP) sensor connector.
19. Disconnect the Exhaust Gas Recirculation (EGR) solenoid connector.
20. Disconnect the Camshaft Position (CMP) sensor connector.
21. Remove the Intake Air Temperature (IAT) sensor.
22. Disconnect the ignition coil connector.
23. Remove the vacuum hose on the brake booster.
24. Remove the EGR pipe from the intake manifold.
25. Remove the engine coolant temperature (ECT) sensor.
26. Remove the wiring harness from the support bracket on the cylinder block.
27. Remove the intake manifold retaining nuts/bolts and the intake manifold.
28. Remove the bolts on the intake manifold support bracket.
29. Remove the intake manifold gaskets.
30. Clean the contact surfaces between intake manifold and the cylinder head.

➡**Remove the fuel rail after removing the intake manifold.**

To install:
31. Installation should follow the removal procedure in the reverse order.

➡**Install the intake manifold after installing the fuel rail.**

32. Install new intake manifold gaskets and intake manifold.
33. Install the intake manifold mounting nuts and bolts. Tighten the nuts and bolts to 89–124 inch lbs. (10–14 Nm).

❋❋ **WARNING**

Genuine intake manifold retaining bolts within the specifications should be used to prevent damaged intake manifold from impairing the airflow to the cylinder.

34. Tighten the 2 bolts on the intake manifold support bracket to 89–124 inch lbs. (10–14 Nm).
35. Tighten the intake EGR pipe retaining bolts on the intake manifold to 71–106 inch lbs. (8–12 Nm).
36. Install the IAT (Intake Air Tempera-

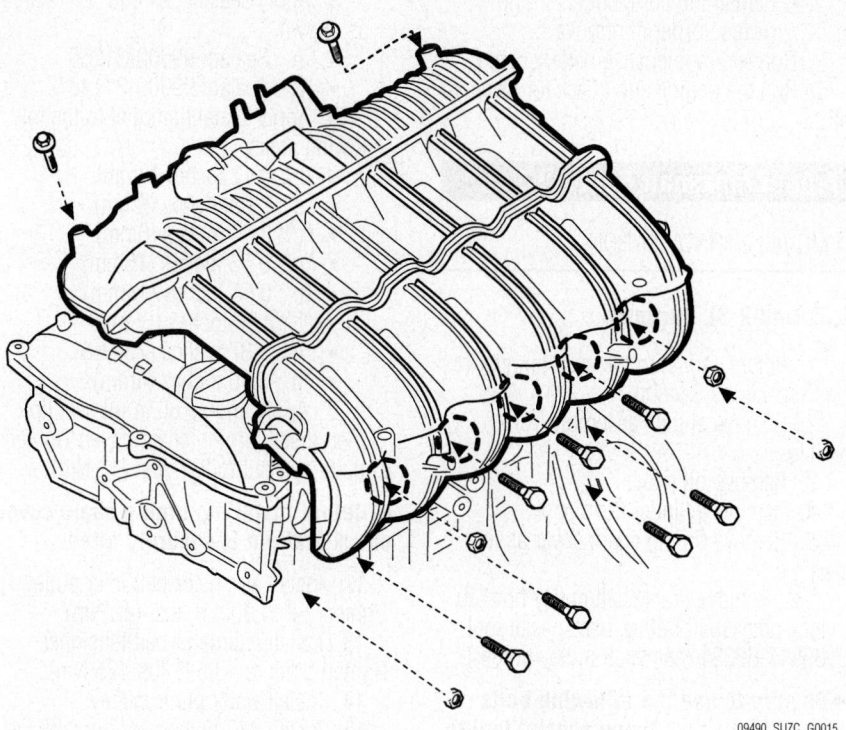

Intake manifold, nuts and bolts—2.5L engine

09490_SUZC_G0015

ture) sensor. Tighten the IAT sensor to 15–22 ft. lbs. (20–30 Nm).

37. Install the throttle body assembly. Tighten the retaining bolts to 6–13 ft. lbs. (8–18 Nm).

38. Connect the fuel pressure regulator vacuum hose.

39. Connect the fuel line.

40. Install the VIS vacuum tank. Tighten the retaining bolts to 62–80 inch lbs. (7–9 Nm).

41. Install the air filter snorkel.

42. Install the engine cover. Tighten the retaining bolts to 71–106 inch lbs. (8–12 Nm).

43. Check to ensure that all removed parts are back in place. Reinstall any necessary parts which have not been reinstalled.

44. Refill the cooling system.

45. Connect the negative battery cable.

46. Start the engine and check for fuel and cooling system leaks.

Exhaust Manifold

REMOVAL & INSTALLATION

2.0L and 2.3L Engines

✳ CAUTION

To avoid the danger of being burned, do not service the exhaust system while it is hot. Service should be performed only after the system cools down.

1. Before servicing the vehicle, refer to the Precautions Section.

2. Remove or disconnect the following:
- Negative battery cable
- Heated Oxygen (HO2S) sensor electrical connector and A/C magnet clutch connector
- Exhaust manifold cover
- 2 bolts attaching the exhaust pipe to the exhaust manifold
- Engine under cover (right side)

- Exhaust pipe stiffener
- Exhaust No.1 pipe
- Air cleaner outlet hose and air cleaner case assembly
- Wire harness clamp

3. Support engine with engine support jack and remove engine right mounting No.1 bracket.
- Engine right mounting bracket
- Exhaust manifold mounting nuts and bolts
- Exhaust manifold and the gasket

➡ **Be careful not to damage fins of radiator.**

To install:

4. Install or connect the following:
- New gasket to the cylinder head
- Exhaust manifold. Tighten manifold bolts and nuts in sequence to 40 ft. lbs. (55 Nm).
- Engine right mounting bracket and new bracket bolts. Tighten the bracket bolts to 40 ft. lbs. (55 Nm).

➡ **Use new engine right mounting bracket bolt. Reusing bolt may result exhaust leakage.**

- Engine right mounting No.1 bracket. Tighten the bolts and nut to 40 ft. lbs. (55 Nm).
- Air cleaner case assembly and air cleaner outlet hose
- New pipe gasket and exhaust No.1 pipe. Tighten the exhaust manifold-to-exhaust pipe nut to 40 ft. lbs. (55 Nm); and the No.1-to-No. 2 exhaust pipe bolt to 31.5 ft. lbs. (43 Nm).
- Exhaust pipe stiffener and tighten exhaust pipe stiffener bolts at the exhaust pipe first, and at the engine next to 40 ft. lbs. (55 Nm)

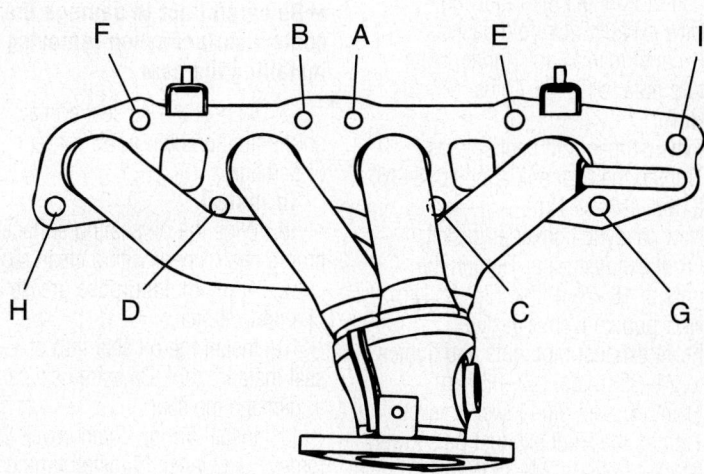

Exhaust manifold tightening sequence—2.0L and 2.3L engines

09490_SUZC_G0016

- Engine under cover (right side)
- Exhaust manifold cover
- Heated Oxygen (HO2S) sensor electrical connector and A/C magnet clutch connector
- Remaining components in the reverse order of removal

5. Connect the negative battery cable.
6. Run the engine and check for exhaust leaks.

2.5L Engine

❈❈ CAUTION

To avoid the danger of being burned, do not service the exhaust system while it is hot. Service should be performed only after the system cools down.

1. Before servicing the vehicle, refer to the Precautions Section.
2. Remove or disconnect the following:
- Negative battery cable
- Front exhaust pipe lower bracket bolts and front exhaust pipe lower bracket
- Front exhaust pipe nuts
- Pup converter gaskets
- Front exhaust pipe-to-front muffler nuts
- Catalytic converter gasket
- Front exhaust pipe assembly from the rubber hanger
- Oxygen Sensor (O2S) connectors
- Exhaust manifold heat shield
- Exhaust manifold retaining nuts and the exhaust manifolds
- Exhaust manifold gaskets

To install:

3. Clean and inspect the sealing surfaces of the exhaust manifold and the cylinder head.
4. Install or connect the following:
- New exhaust manifold gaskets
- Exhaust manifolds. Tighten manifold nuts to 20–22 ft. lbs. (27–30 Nm).
- Exhaust manifold heat shields. Tighten the heat shield bolts 13–16 ft. lbs. (18–22 Nm).
- New catalytic converter gasket
- Front exhaust pipe. Tighten the nuts to 18–26 ft. lbs. (25–35 Nm).
- New pup converter gasket
- Front exhaust pipe nuts and tighten to 24–35 ft. lbs. (32–48 Nm)
- Front exhaust pipe lower bracket. Tighten the front exhaust pipe lower bracket bolts to 24–35 ft. lbs. (32–48 Nm).

- Remaining components in the reverse order of removal

5. Connect the negative battery cable.
6. Run the engine and check for exhaust leaks.

Front Crankshaft Seal

REMOVAL & INSTALLATION

2.0L and 2.3L Engines

1. Before servicing the vehicle, refer to the Precautions Section.
2. Remove engine assembly from vehicle.
3. Remove oil pan.
4. Remove cylinder head cover.
5. Remove timing chain cover as follows:

a. Remove crankshaft pulley bolt. To lock crankshaft pulley, use special tool 09917-68221 (camshaft pulley holder).

➡**Be sure to use the following bolts instead of pins for fixing special tool to crankshaft pulley. Bolt size: M8, P1.25 L = 25 mm (0.98 in.) Strength: 7T**

b. Remove crankshaft pulley. To remove crankshaft pulley, use special tools 09944-36011 and 09926-58010 (steering wheel remover and bearing puller attachment).

c. Remove A/C compressor bracket.

d. Remove alternator belt idler pulley, water pump pulley and alternator belt tensioner.

e. Remove timing chain cover bolts and nut.

6. Tape the end of a flat-bladed tool to avoid damaging the timing chain cover. Pry out the oil seal using the taped end of the tool.

➡**Be careful not to damage the oil seal contact surface when removing or installing the seal.**

7. Inspect the oil seal contact surface on the timing chain cover for signs of wear or damage.

To install:

8. Wipe the oil sealing surface on the timing chain cover with a clean rag.
9. Apply multipurpose grease to the lip of a new oil seal.
10. Install the oil seal into place using a seal installer tool. Be extremely careful not to damage the seal.
11. Install timing chain cover. Reverse removal sequence to install timing chain cover noting the following points:

a. Apply sealant "A" and "B" to area as shown.
- "A": Sealant 99000-31250
- "B": Sealant 99000-31140

b. Apply sealant amount to the following areas.
- "a": 0.12 inches (3mm)
- "b": 0.08 inches (2mm)
- "c": 0.24 inches (6mm)
- "d": 0.63 inches (16mm)
- "e": 0.55 inches (14mm)
- "f": 2.56 inches (65mm)
- "g": 2.87 inches (73mm)
- "h": 0.16 inches (4mm)

c. Apply engine oil to oil seal lip, then install timing chain cover. Tighten bolts and nut to 8 ft. lbs. (11 Nm).

➡**Before installing timing chain cover, check that pin is securely fitted.**

12. Install alternator belt idler pulley. Tighten nut to 30.5 ft. lbs. (42 Nm).
13. Install alternator belt tensioner. Tighten bolts to 18.5 ft. lbs. (25 Nm).
14. Install water pump pulley.
15. Install A/C compressor bracket (if equipped). Tighten bracket bolts to 40 ft. lbs. (55 Nm).
16. Install cylinder head cover with new gaskets and seal washers and tighten nuts to 8 ft. lbs. (11 Nm).
17. Install oil pan.
18. Install crankshaft pulley using pulley holder tool. Tighten the crankshaft pulley bolt to 108.5 ft. lbs. (150 Nm).
19. Install engine assembly.
20. Install all remaining components in the reverse order of the removal procedure.

2.5L Engine

1. Before servicing the vehicle, refer to the Precautions Section.
2. Disconnect the negative battery cable.
3. Remove the engine.
4. Remove the flex plate bolts and the flex plate.
5. Support the engine assembly with an engine stand.
6. Remove the cylinder head cover.
7. Remove the drive belt.
8. Remove the oil pan.
9. Install the crankshaft pulley holder DW110-130-01 to the crankshaft pulley bolt.

➡**Very high torque is required to remove or to install the crankshaft pulley. Never remove the crankshaft pulley when the vehicle is hung on the lift.**

10. Remove the crankshaft pulley bolt

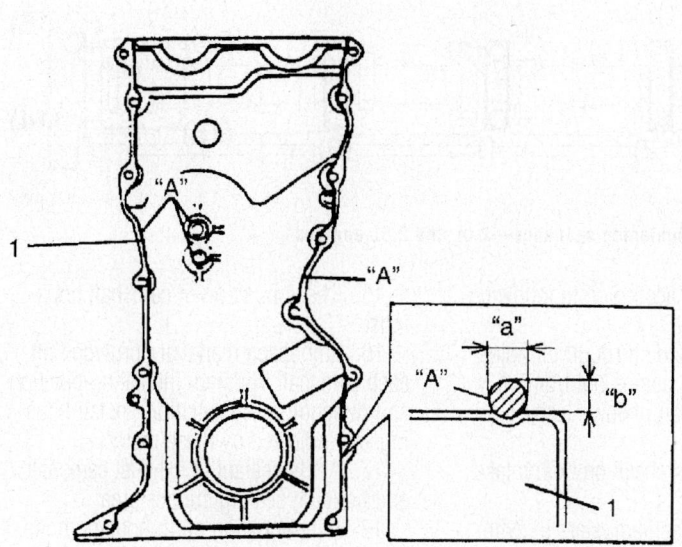
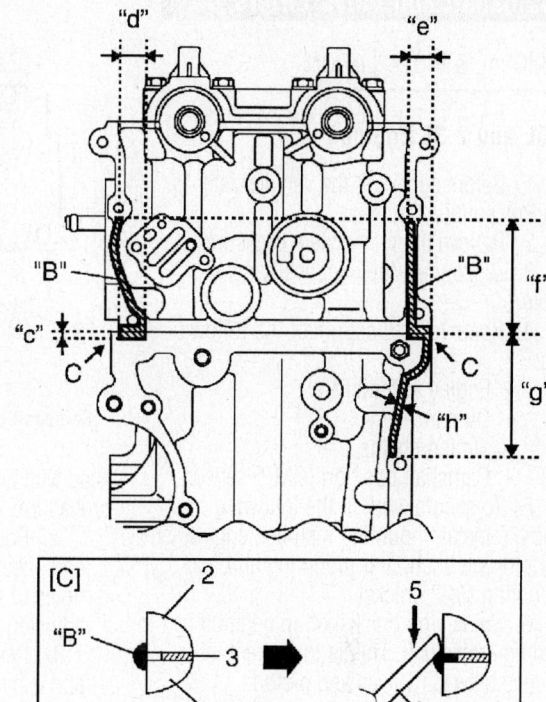

1. Timing chain cover
2. Cylinder head
3. Cylinder head gasket
4. Cylinder block
5. Rub into
6. Jig

09490_SUZC_G0017

Sealant application on timing belt cover—2.0L and 2.3L engines

using the crankshaft pulley installer/remover tool DW110-130-02.

11. Remove the engine mount bracket support.

12. Remove the water pump pulley.

13. Remove the drive belt automatic tensioner retaining bolts and the tensioner.

14. Remove the timing chain cover.

15. Tape the end of a flat-bladed tool to avoid damaging the timing chain cover. Pry out the oil seal using the taped end of the tool.

➡**Be careful not to damage the oil seal contact surface when removing or installing the seal.**

16. Inspect the oil seal contact surface on the timing chain cover for signs of wear or damage.

To install:

17. Wipe the oil sealing surface on the timing chain cover with a clean rag.

18. Apply multipurpose grease to the lip of a new oil seal.

19. Install the oil seal into place using seal installer tool DW110-180-01. Be extremely careful not to damage the seal.

20. Apply the liquid gasket (Loctite® 5900-M8585) on the timing chain cover.

Install the timing chain cover install guide pin DW110-140 and position the timing chain cover.

➡**Be sure the applied liquid gasket is not damaged when installing the timing chain cover.**

21. Install the timing chain cover. Tighten the timing chain cover bolt (center) to 13–16 ft. lbs. (18–22 Nm); and tighten the remaining timing chain cover bolts to 53–89 inch lbs. (6–10 Nm).

22. Install the drive belt automatic tensioner and tighten the drive belt automatic tensioner retaining bolt to 15–22 ft. lbs. (20–30 Nm).

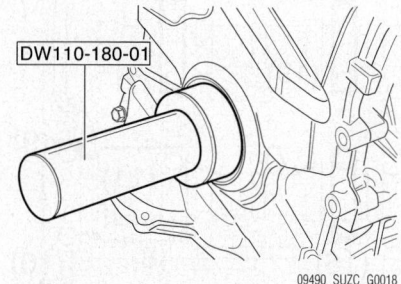

09490_SUZC_G0018

Install the oil seal into place using seal installer tool DW110-180-01—2.5L engine

23. Install the water pump pulley and tighten the water pump pulley bolts to 71–106 inch lbs. (8–12 Nm).

24. Install the engine mount bracket support and tighten the engine mount bracket support bolts to 18–23 ft. lbs. (25–31 Nm).

25. Install the crankshaft pulley. Tighten the crankshaft pulley bolt using the crankshaft pulley holder DW110-130-01 and the crankshaft pulley installer/remover tool DW110-130-02. Tighten the crankshaft pulley bolt to 369–443 ft. lbs. (500–600 Nm).

26. Install the oil pan.

27. Install the cylinder head cover along with a new gasket. Tighten the cylinder head cover bolts to 106–124 inch lbs. (12–14 Nm).

28. Install the flex plate and tighten the flexible plate bolts to 52–59 ft. lbs. (70–80 Nm).

29. Install the transaxle to the engine assembly.

30. Install the engine and transaxle assembly.

31. Install all remaining components in the reverse order of the removal procedure.

32. Connect the negative battery cable.

33. Start the engine and check for leaks.

Camshaft and Valve Lifters

REMOVAL & INSTALLATION

2.0L and 2.3L Engines

1. Before servicing the vehicle, refer to the Precautions Section.
2. Relieve the fuel system pressure.
3. Disconnect the negative battery cable.
4. Remove or disconnect the following:

- Engine assembly
- Oil pan
- Timing chains
- Camshaft position (CMP) sensor

5. To secure work in the following steps, reinstall mounting member, engine front torque bush, rear mounting and rear mounting No.2 bracket.
6. Set key on crankshaft in position by turning crankshaft. This is to prevent interference between valves and piston.
7. Loosen camshaft housing bolts in such order as indicated in figure and remove them.

- Camshaft housings
- Camshafts
- Valve lash adjusters

➡**Never disassemble hydraulic valve lash adjuster. Don't apply force (1) to body of adjuster, oil in high pressure chamber in adjuster will leak.**

8. Immerse removed adjuster in clean engine oil and keep it there until reinstalling it so as to prevent oil leakage. If it is left in air, place it with its bucket body facing down. Don't place on its side or with bucket body facing up.

To install:

9. Before installing valve lash adjuster to cylinder head, fill oil passage of cylinder

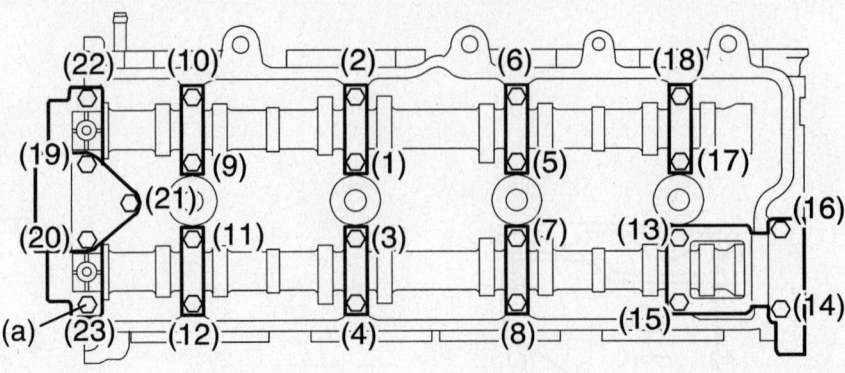

Camshaft cap bolt tightening sequence—2.0L and 2.3L engines

09490_SUZC_G0020

head with engine oil according to following procedure:

a. Pour engine oil through oil holes and check that oil comes out from oil holes in sliding part of valve lash adjuster.

b. Perform this check on both intake and exhaust sides.

10. Install valve lash adjusters to cylinder head. Apply engine oil around valve lash adjuster and then install it to cylinder head.

11. Match matchmark on crank timing sprocket and mating surface of cylinder block and lower crankcase.

12. Lubricate the lobes and journals of the camshaft with clean engine oil.

13. Install or connect the following:

- Camshafts

14. Apply oil to sliding surface of each camshaft and camshaft journal then install them by aligning match marks on cylinder head and camshafts.

➡**Install camshaft in such direction that its end with groove for CMP sensor installation comes to exhaust side.**

- Camshaft housing pins

15. Check position of camshaft housings.

16. Embossed marks are provided on each camshaft housing, indicating position and direction for installation. Install housings as indicated by these marks.

17. Apply sealant to exhaust camshaft end housing sealing surface area.

18. After applying oil to housing bolts, tighten them temporarily first. Then tighten them by following numerical order in figure.

19. Tighten a little at a time and evenly among bolts and repeat tightening sequence two or three times before they are tightened to 8 ft. lbs. (11 Nm).

- CMP sensor
- Timing chains
- Oil pan
- Engine assembly
- Negative battery cable

20. Start the engine and check for any water or oil leaks when finished.

21. Check ignition timing as necessary.

➡**Don't turn camshafts or start engine (i.e., valves should not be operated) for about half an hour after reinstalling hydraulic valve lash adjusters and camshafts. As it takes time for valves to settle in place, operating engine within half an hour after their installation may cause interference to occur between valves themselves or valves and piston.**

2.5L Engine

1. Before servicing the vehicle, refer to the Precautions Section.
2. Relieve the fuel system pressure.
3. Disconnect the negative battery cable.
4. Remove or disconnect the following:

- Cylinder head
- Intake cam sprocket and exhaust cam sprocket

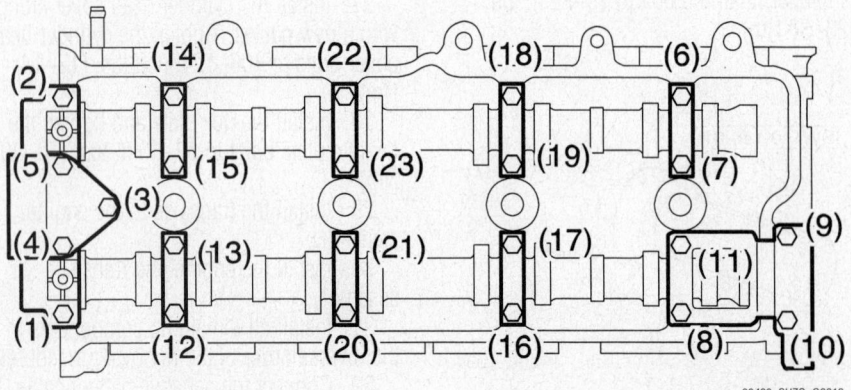

Loosen the camshaft bearing caps in the sequence shown—2.0L and 2.3L engine

09490_SUZC_G0019

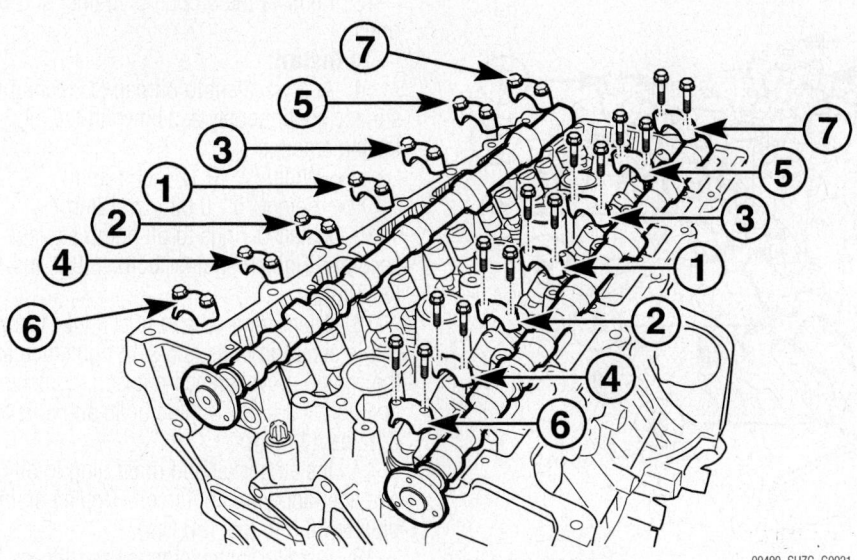

Loosen the camshaft bearing caps in the sequence shown—2.5L engine

09490_SUZC_G0021

- Camshaft cap bolts gradually and in sequence

➡**Make sure the engraved position numbers on the camshaft caps are not mixed up. Camshaft caps should follow the numeric order when installing, intake camshaft caps 1-3-5-7-9-11-13 and exhaust camshaft caps 2-4-6-8-10-12-14.**

- Intake camshaft and exhaust camshaft

To install:

5. Installation should follow the removal procedure in the reverse order.

6. Lubricate the lobes and journals of the camshafts with clean engine oil.

7. Install or connect the following:
- Intake and exhaust camshafts

➡**The intake camshaft can be identified by the groove near the cam sprocket flange. The exhaust camshaft has no groove.**

- Camshaft caps in sequence

➡**Make sure the engraved position numbers on the camshaft caps are not mixed up. Camshaft caps should follow the numeric order when installing, intake camshaft caps 1-3-5-7-9-11-13 and exhaust camshaft caps 2-4-6-8-10-12-14.**

8. Tighten the camshaft cap bolts in sequence to 89–124 inch lbs. (10–14 Nm).
- Cam sprockets. Tighten the cam sprocket bolts to 13–16 ft. lbs. (18–22 Nm).

➡**Intake and exhaust cam sprockets are identified by the engraved marks, "IN" for intake and "EX" for exhaust respectively.**

- Cylinder head
- Negative battery cable

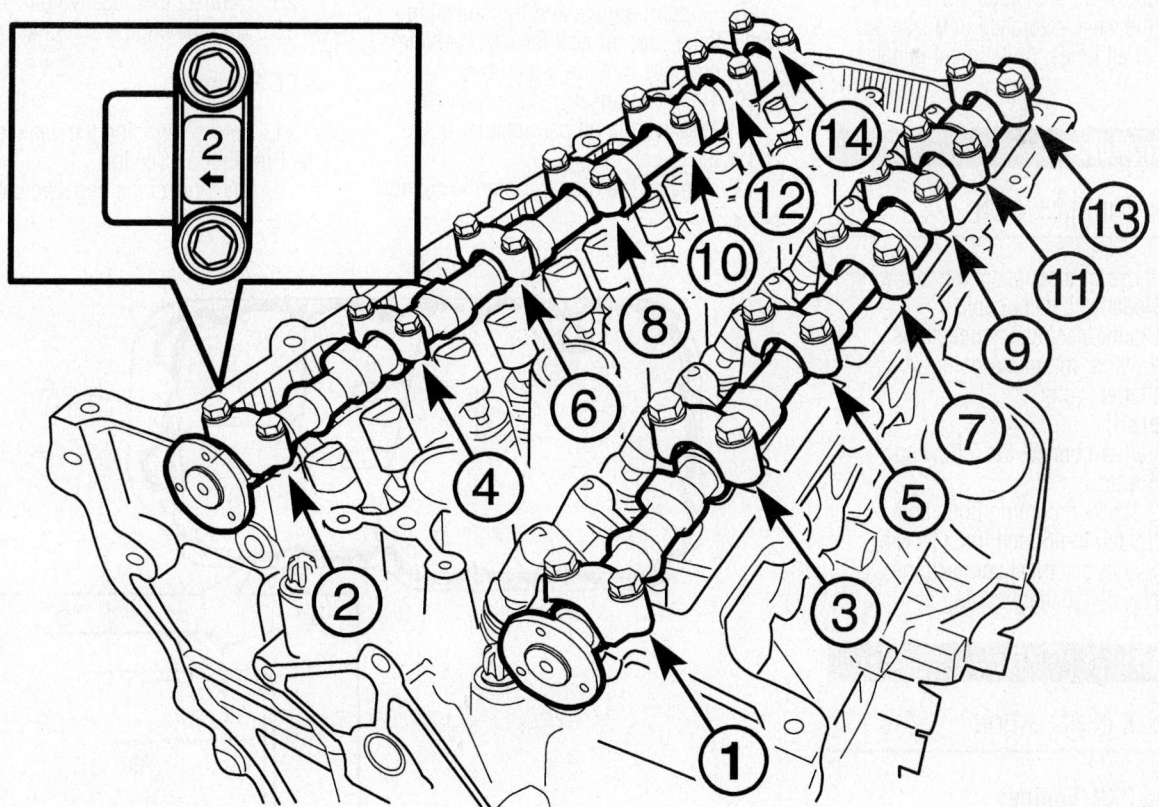

09490_SUZC_G0022

Camshaft cap installation sequence—2.5L engine

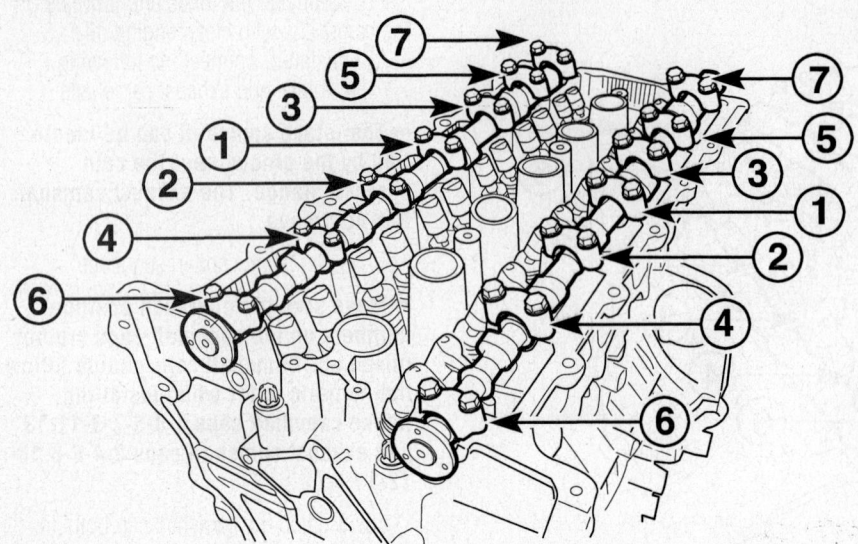

Camshaft cap bolt tightening sequence—2.5L engine

9. Start the engine and check for any water or oil leaks when finished.

10. Check ignition timing as necessary.

Valve Lash

ADJUSTMENT

Hydraulic valve lash adjusters are used to adjust the valve clearance to **0** lash automatically at all times. Adjustment is not required.

Starter Motor

REMOVAL & INSTALLATION

1. Remove or disconnect the following:
   - Negative battery cable
   - Starter electrical connections
   - 2 starter mounting bolts
   - Starter motor

To install:
2. Install or connect the following:
   - Starter
   - 2 starter mounting bolts and tighten to 96 inch lbs. (10 Nm)
   - Starter electrical connections
   - Negative battery cable

Oil Pan

REMOVAL & INSTALLATION

2.0L and 2.3L Engines

1. Before servicing the vehicle, refer to the Precautions Section.

2. Disconnect the negative battery cable.

3. Remove the oil level gauge.

4. Raise and safely support the front of the vehicle.

5. Drain engine oil by removing drain plug.

6. Remove engine under cover(s).

7. Remove exhaust No.1 pipe.

8. Support engine and transmission with engine support jack for 2WD vehicle.

9. Remove mounting member.

10. Remove transaxle lower stiffener.

11. Remove the oil pan retainer bolts and nuts.

12. Remove the oil pan from the cylinder block.

13. Remove the oil pump strainer and O-rings.

To install:
14. Apply sealant to oil pan (1) mating surface continuously as shown in the following amount:
   - Width "a": 0.12 inch (3mm)
   - Height "b": 0.08 inch (2mm)

15. Install O-rings to oil pump strainer securely. Tighten strainer bolts to 8 ft. lbs. (11 Nm).

16. After fitting oil pan to cylinder block, run in securing bolts and start tightening at the center: move wrench outward, tightening one bolt at a time. Tighten bolts and nuts to 8 ft. lbs. (11 Nm).

17. Install gasket and drain plug to oil pan after applying engine oil. Tighten drain plug to 36.5 ft. lbs. (50 Nm).

18. Install transmission lower stiffener.

19. Install mounting member for 2WD vehicle. Tighten the engine mounting bolt and nut to 40 ft. lbs. (55 Nm); and tighten the engine front and rear mounting nut to 32.5 ft. lbs. (45 Nm).

20. Remove engine support jack for 2WD vehicle.

21. Install exhaust No.1 pipe.

22. Install engine under cover(s).

23. Install oil level gauge.

24. Refill the engine with oil.

25. Connect the negative battery cable.

26. Start the engine and check for leaks.

2.5L Engine

1. Before servicing the vehicle, refer to the Precautions Section.

2. Disconnect the negative battery cable.

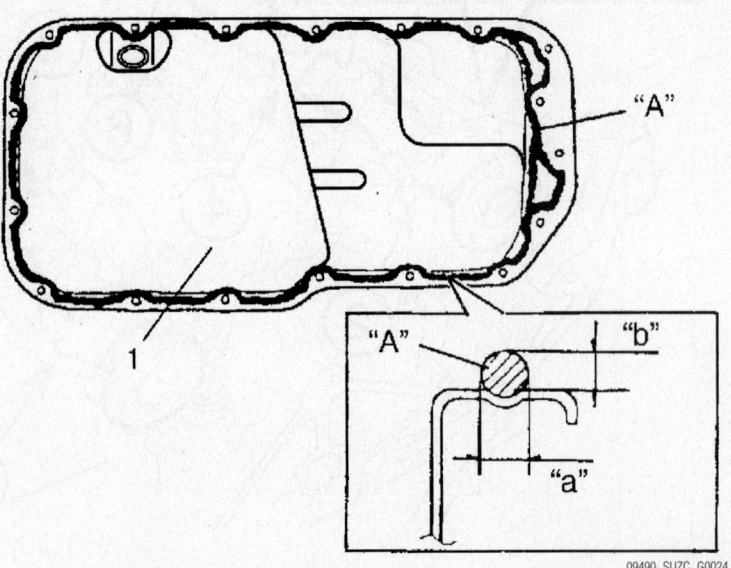

Oil pan seal thickness—2.0L and 2.3L engines

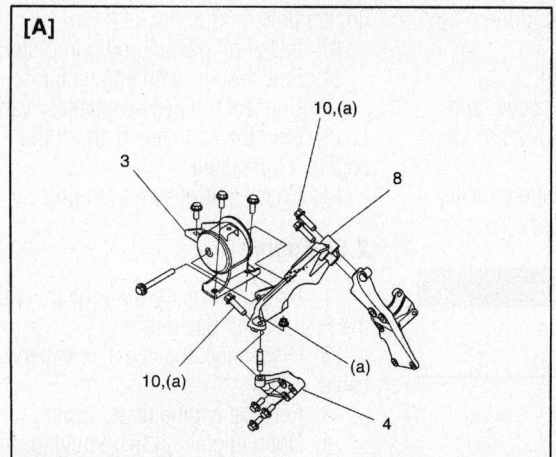

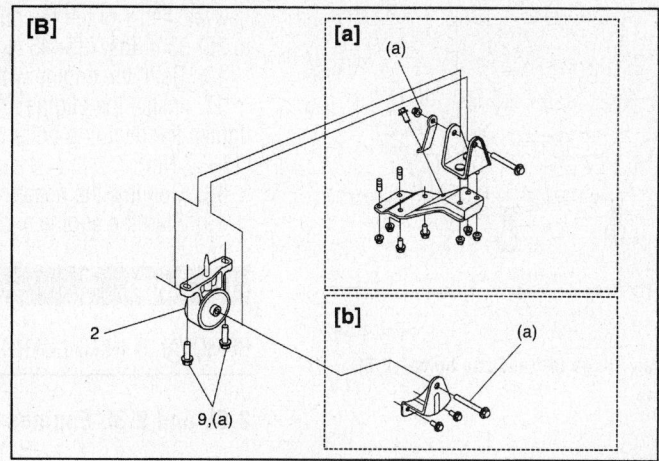

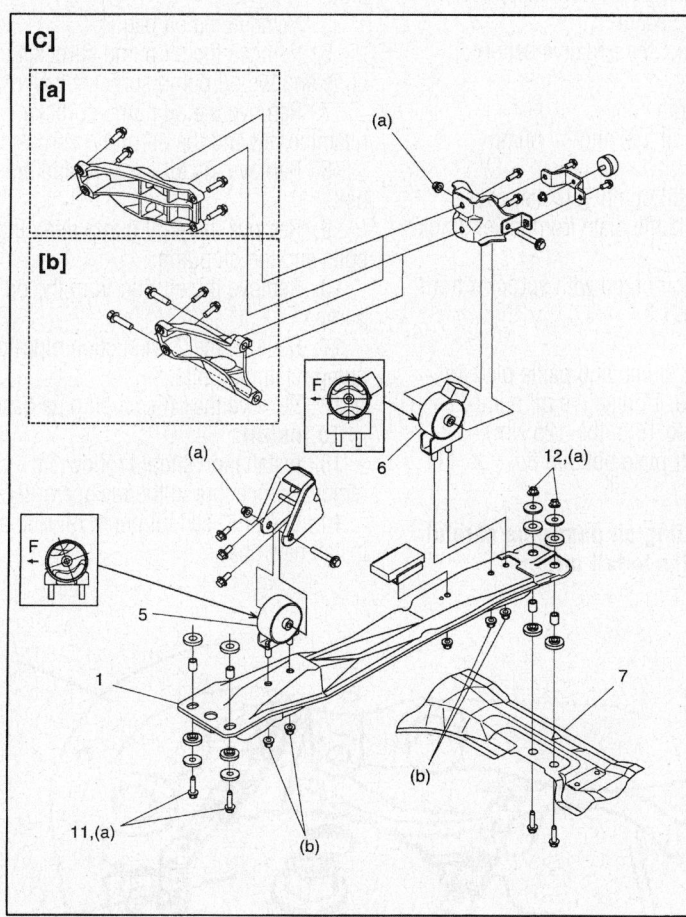

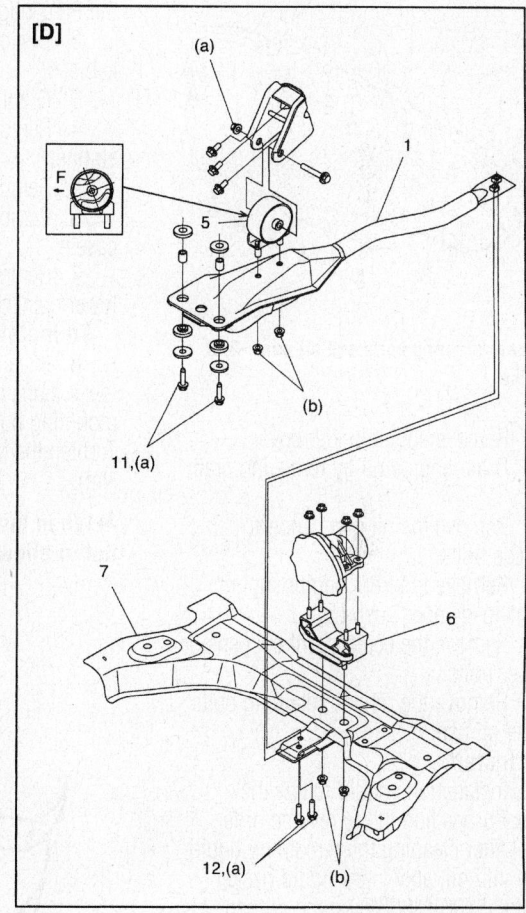

A. Engine right mounting assembly
B. Engine left mounting assembly
C. Engine front and rear mounting
   assemblies for 2WD model
D. Engine front and rear mounting
   assemblies for AWD model
a. For M/T model
b. For A/T model

1. Mounting member
2. Engine left mounting
3. Engine right mounting
4. Right mounting bracket
5. Engine front mounting
6. Engine rear mounting
7. Front member and
   suspension frame

8. Engine right mounting
   no. 1 bracket
9. Engine left mounting bolts
10. Engine right mounting
    no. 1 bracket bolt
11. Mounting member-to-
    body bolts
12. Mounting member-to-
    suspension frame bolts
    and nuts

09490_SUZC_G0006

Engine mounting components—2.0L and 2.3L engines

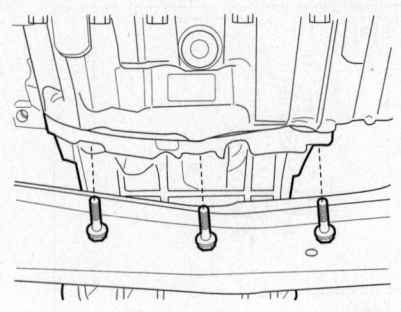

Oil pan flange-to-transaxle bolts—2.5L engine

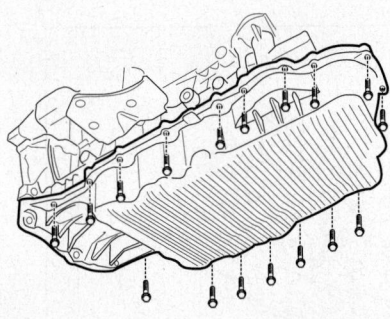

Oil pan retaining bolts and oil pan—2.5L engine

3. Remove engine under cover.

4. Drain engine oil by removing drain plug.

5. Remove the oil pan flange-to-transaxle bolts.

6. Remove the front engine mount bracket-to-damper bush bolt.

7. Remove the center front suspension member bolts.

8. Remove the oil pan retaining bolts and the oil pan.

To install:

9. Installation should follow the removal procedure in the reverse order.

10. After cleaning the remaining liquid gasket and oil, apply new liquid gasket (Loctite® 5900-M8585) on the oil pan.

11. After cleaning the remaining liquid gasket and oil on the bed plate flange, install the oil pan and tighten the oil pan retaining bolts to 71–106 inch lbs. (8–12 Nm).

12. Tighten the center front suspension member bolts to 59–74 ft. lbs. (80–100 Nm).

13. Tighten the front engine mount bracket-to-damper bush bolt to 52–66 ft. lbs. (70–90 Nm).

14. Tighten the oil pan flange-to-transaxle bolts to 20–26 ft. lbs. (27–35 Nm).

15. Install the oil pan drain plug and a new washer. Tighten the oil pan drain plug to 26–33 ft. lbs. (35–45 Nm).

16. Refill the engine with oil.

17. Install the engine under cover and tighten the retaining bolts to 13–20 ft. lbs. (17–27 Nm).

18. Connect the negative battery cable.

19. Start the engine and check for leaks.

Oil Pump

REMOVAL & INSTALLATION

2.0L and 2.3L Engines

1. Before servicing the vehicle, refer to the Precautions Section.

2. Disconnect the negative battery cable.

3. Drain engine oil.

4. Remove oil pan and oil pump strainer.

5. Remove oil pump sprocket cover.

6. Remove baffle plate from lower crank case.

7. Remove oil pump with sprocket from lower crankcase.

To install:

8. Install oil pump and baffle plate to lower crank case. Tighten the oil pump mounting bolts to 18 ft. lbs. (25 Nm). Tighten the bafflr plate bolts to 8 ft. lbs. (11 Nm).

➥ **When installing oil pump, be careful not to allow pins to fall off.**

9. Install oil pump sprocket cover and tighten bolts to 8 ft. lbs. (11 Nm).

10. Install oil pan and oil pump strainer.

11. Refill engine with engine oil.

12. Connect the negative battery cable.

13. Start the engine and check the engine oil pressure.

14. Check that no leaks are present.

2.5L Engine

1. Before servicing the vehicle, refer to the Precautions Section.

2. Disconnect the negative battery cable.

3. Remove engine under cover.

4. Drain engine oil by removing drain plug.

5. Remove the oil pan.

6. Remove the oil pump sprocket cover bolts and the oil pump sprocket cover.

7. Remove the oil pump sprocket retaining nut and the oil pump sprocket.

8. Remove the oil suction pipe bracket bolt.

9. Remove the 4 oil pump retaining bolts and the oil pump.

10. Remove the oil ring from the oil pump.

11. Remove the 2 oil suction pipe-to-oil pump retaining bolts.

12. Remove the oil suction pipe gasket.

To install:

13. Installation should follow the removal procedure in the reverse order.

14. Discard used oil pump oil ring and install new one.

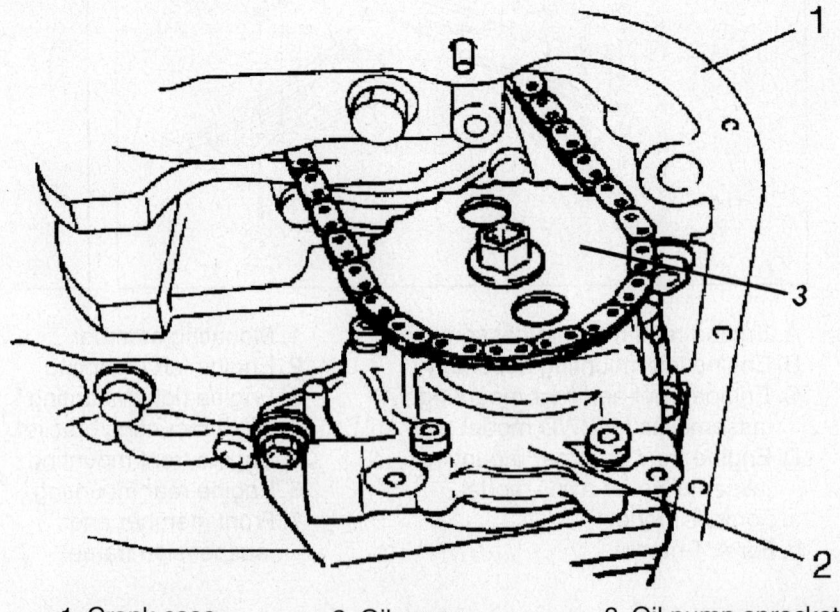

1. Crank case   2. Oil pump   3. Oil pump sprocket

Crank case (1), oil pump (2) and oil pump sprocket (3)—2.0L and 2.3L engine

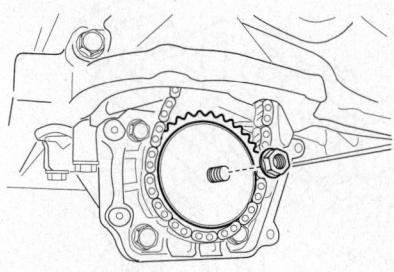

Oil pump sprocket and retaining nut—2.5L engine

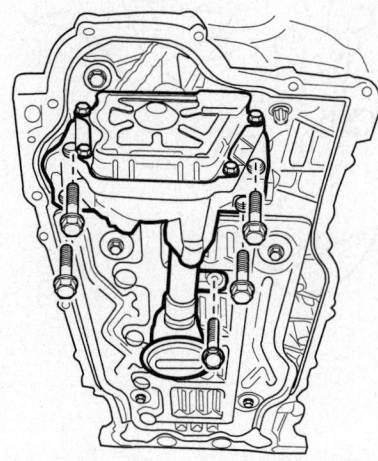

Oil pump and 4 retaining bolts—2.5L engine

15. Tighten the oil suction pipe-to-oil pump retaining bolts, oil suction pipe bracket bolt, and the oil pump retaining bolt to the following torques:
- Oil suction pipe-to-oil pump retaining bolts to 106–124 inch lbs. (8–12 Nm).
- Oil suction pipe bracket bolt to 106–124 inch lbs. (8–12 Nm).
- Oil pump retaining bolts to 13–16 ft. lbs. (18–22 Nm).

16. Install the oil pump sprocket. Tighten the oil pump sprocket retaining nut to 13–16 ft. lbs. (18–22 Nm).

➡**Make sure that the dotted surface on the sprocket faces toward the oil pump.**

17. Install the oil pump sprocket cover. Tighten the oil pump sprocket cover retaining bolts to 44–62 inch lbs. (5–7 Nm).
18. Install the oil pan.
19. Refill the engine with oil.
20. Install the engine under cover and tighten the retaining bolts to 13–20 ft. lbs. (17–27 Nm).
21. Connect the negative battery cable.
22. Start the engine and check for leaks.

Rear Main Seal

REMOVAL & INSTALLATION

1. Before servicing the vehicle, refer to the Precautions Section.
2. Remove or disconnect the following:
- Transaxle assembly
- Flexplate/flywheel from the crankshaft
3. Carefully pry the oil seal out of the retainer without scratching the sealing surface of the crankshaft.

To install:
4. Apply engine oil the lip of the new seal.
5. Install or connect the following:
- Seal in the retainer using a suitable seal driver
- Flexplate/flywheel
- Transaxle assembly

Timing Chain

REMOVAL & INSTALLATION

2.0L and 2.3L Engines

1. Before servicing the vehicle, refer to the Precautions Section.
2. Remove engine assembly from vehicle.
3. Remove oil pan.
4. Remove cylinder head cover.
5. Remove timing chain cover as follows:
   a. Remove crankshaft pulley bolt. To lock crankshaft pulley, use special tool 09917-68221 (camshaft pulley holder).

➡**Be sure to use the following bolts instead of pins for fixing special tool to crankshaft pulley. Bolt size: M8, P1.25 L = 25 mm (0.98 in.) Strength: 7T**

   b. Remove crankshaft pulley. To remove crankshaft pulley, use special tools 09944-36011 and 09926-58010 (steering wheel remover and bearing puller attachment).
   c. Remove A/C compressor bracket.
   d. Remove alternator belt idler pulley, water pump pulley and alternator belt tensioner.
   e. Remove timing chain cover bolts and nut.
6. For reinstallation of timing chain, turn crankshaft so that timing marks on cylinder head and lower crankcase match with those on sprockets as shown in figure.
7. Remove second timing chain as follows:

   a. Turn crankshaft to meet following conditions.
- Key (I) on crankshaft is positioned as shown.
- Arrow mark on idler sprocket (II) points upward.
- Marks on sprockets (III) match with marks on cylinder head. Note that this step must be followed for reinstallation of timing chain.
   b. Remove timing chain tensioner adjuster No.2 and gasket. To remove them, slacken second timing chain by turning intake camshaft counterclockwise a little while pushing back pad.
   c. Remove intake and exhaust camshaft timing sprocket bolts. To remove them, fit a spanner to hexagonal part at the center of camshaft to hold it stationary.
   d. Remove camshaft timing sprockets and second timing chain.

✳✳ WARNING

After second timing chain is removed, never turn intake camshaft, exhaust camshaft and crankshaft independently more than such an extent as shown. If turned, interference may occur between piston and valves and valves themselves, and parts related to piston and valves may be damaged.

8. Remove timing chain guide No.1.
9. Remove timing chain tensioner adjuster No.1.
10. Remove timing chain tensioner.
11. Remove idler sprocket and first timing chain.
12. Remove crankshaft timing sprocket.

To install:
13. Check that match mark on crankshaft timing sprocket is in match with timing mark on lower crankcase.
14. Install crankshaft timing sprocket as shown in figure.
15. Apply oil to bush of idler sprocket and install idler sprocket and sprocket shaft.
16. Install first timing chain by aligning dark blue plate of first timing chain and match mark on idler sprocket.
17. Bring yellow plate of first timing chain into match with match mark on crankshaft timing sprocket.
18. Apply engine oil to sliding surface of timing chain tensioner and then install it as shown in figure. Tighten tensioner nut to 18 ft. lbs. (25 Nm).
19. With latch of tensioner adjuster No.1 returned and plunger pushed back into

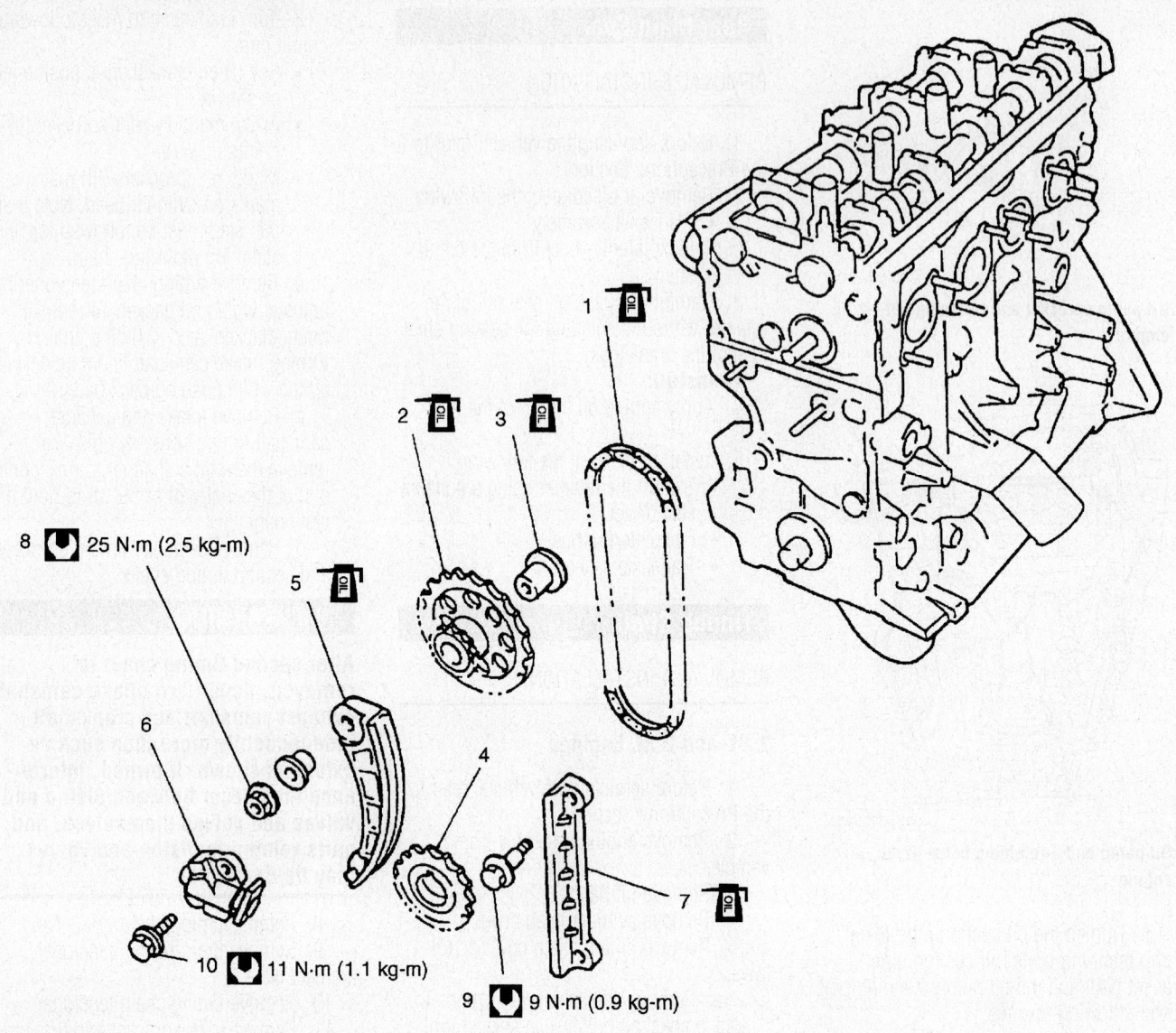

8 🔧 25 N·m (2.5 kg-m)

10 🔧 11 N·m (1.1 kg-m)

9 🔧 9 N·m (0.9 kg-m)

1. First timing chain
2. Idler sprocket
3. Idler sprocket shaft
4. Crankshaft timing sprocket
5. Timing chain tensioner
6. Timing chain tensioner adjuster No. 1
7. Timing chain guide No. 1
8. Timing chain tensioner nut
9. Timing chain guide No. 1 bolt
10. Timing chain tensioner adjuster No. 1 bolt

09490_SUZC_G0031

Exploded view of first timing chain and related components—2.0L and 2.3L engines

body, insert stopper into latch and body. After inserting it, check to make sure that plunger will not come out.

20. Install timing chain tensioner adjuster No.1. Tighten the timing chain tensioner adjuster No. 1 bolt to 8 ft. lbs. (11 Nm).

21. Pull out stopper from adjuster No.1.

22. Apply engine oil to sliding surface of timing chain guide No.1 and then install it. Tighten guide bolts to 6.5 ft. lbs. (9 Nm).

23. Check that dark blue and yellow plates of first timing chain are in match with match marks on sprockets respectively.

24. Install second timing chain as follows:

a. Check that match mark on crank timing sprocket is in match with timing mark on lower crankcase.

b. Check that arrow mark on idler sprocket faces upward.

c. Check that knock pins of intake and exhaust camshafts are aligned with timing marks on cylinder head.

d. Install second timing chain by aligning yellow plate of second timing chain and match marks on idler sprocket.

e. Install sprockets to intake and exhaust camshafts by aligning dark blue plate of second timing chain, match marks on intake sprocket and exhaust sprocket respectively.

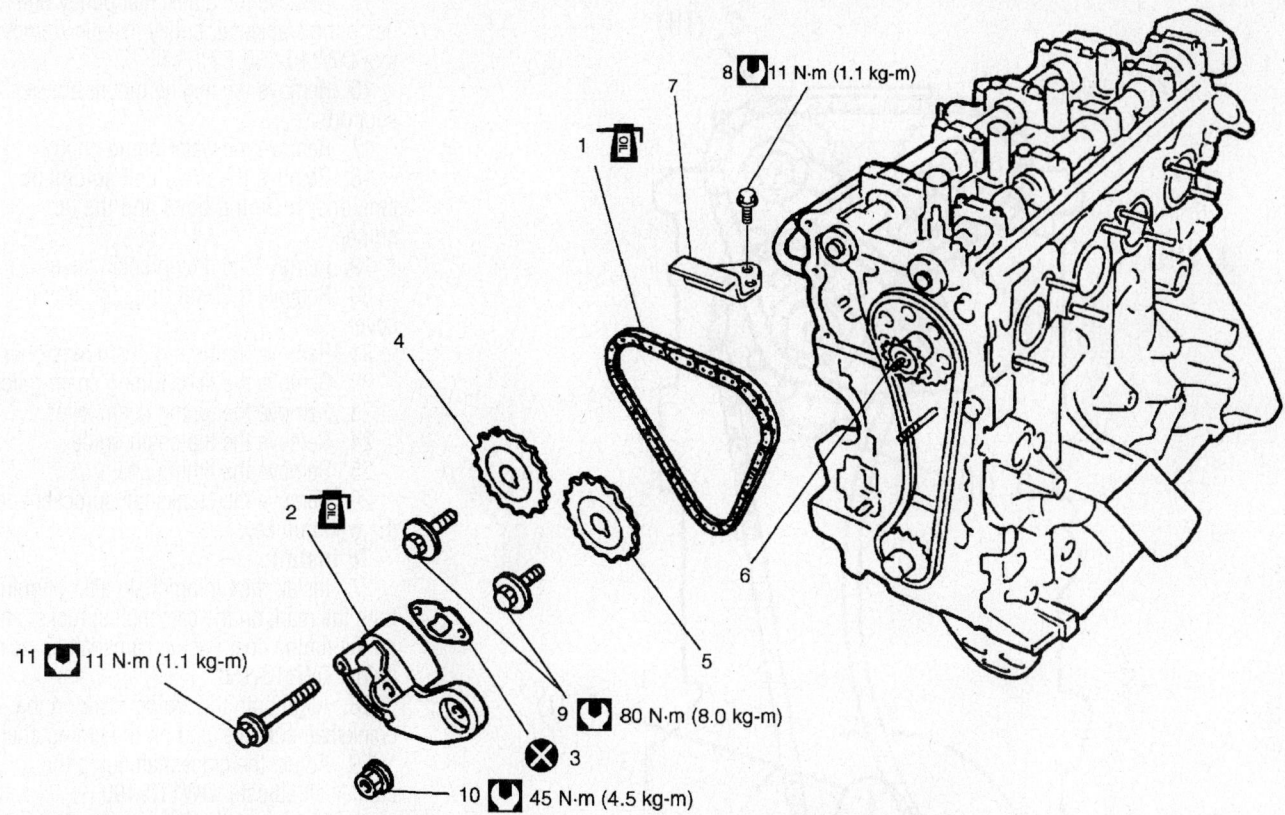

1. Second timing chain (apply engine oil)
2. Timing chain tensioner adjuster no. 2
   (apply engine oil to sliding surface)
3. Tensioner adjuster no.2 gasket
4. Intake camshaft timing sprocket
5. Exhaust camshaft timing sprocket
6. Idler sprocket
7. Timing chain guide no. 2
8. Camshaft housing bolt
9. Camshaft timing sprocket bolt
10. Timing chain tensioner adjuster
    no. 2 nut
11. Timing chain tensioner adjuster
    no. 2 bolt

09490_SUZC_G0030

Exploded view of second timing chain and related components—2.0L and 2.3L engines

➥**As an arrow mark is provided on both sides, camshaft timing sprocket has no specific installation direction.**

   f. Install intake and exhaust camshaft timing sprocket bolts. To install it, fit a spanner to hexagonal part at the center of camshaft to hold it stationary. Tighten the camshaft timing sprocket bolt to 57.5 ft. lbs. (80 Nm).

   g. Push back plunger into tensioner body and hold it at the position by inserting stopper into body.

   h. Install timing chain tensioner adjuster No.2 with new gasket. Tighten the timing chain tensioner adjuster No. 2 bolts to 8 ft. lbs. (11 Nm) and No. 2 nut to 33 ft. lbs. (45 Nm).

   i. Pull out stopper from timing chain tensioner adjuster No.2.

   j. Turn crankshaft two rotations clockwise then align timing mark on crankshaft and timing mark on cylinder block.

   k. Check that timing marks of cylinder head and cylinder block are in match with match marks on sprockets respectively.

   l. Apply oil to timing chains, tensioner, tensioner adjusters, sprockets, and guides.

25. Install timing chain cover. Reverse removal sequence to install timing chain cover noting the following points:
   a. Apply sealant "A" and "B" to area as shown.
   • "A": Sealant 99000-31250
   • "B": Sealant 99000-31140
   b. Apply sealant amount to the following areas.
   • "a": 0.12 inches (3mm)
   • "b": 0.08 inches (2mm)
   • "c": 0.24 inches (6mm)
   • "d": 0.63 inches (16mm)
   • "e": 0.55 inches (14mm)
   • "f": 2.56 inches (65mm)
   • "g": 2.87 inches (73mm)
   • "h": 0.16 inches (4mm)
   c. Apply engine oil to oil seal lip, then install timing chain cover. Tighten bolts and nut to 8 ft. lbs. (11 Nm).

➥**Before installing timing chain cover, check that pin is securely fitted.**

26. Install alternator belt idler pulley. Tighten nut to 30.5 ft. lbs. (42 Nm).

27. Install alternator belt tensioner. Tighten bolts to 18.5 ft. lbs. (25 Nm).

28. Install water pump pulley.

29. Install A/C compressor bracket (if equipped). Tighten bracket bolts to 40 ft. lbs. (55 Nm).

30. Install cylinder head cover with new gaskets and seal washers and tighten nuts to 8 ft. lbs. (11 Nm).

31. Install oil pan.

32. Install crankshaft pulley using pulley

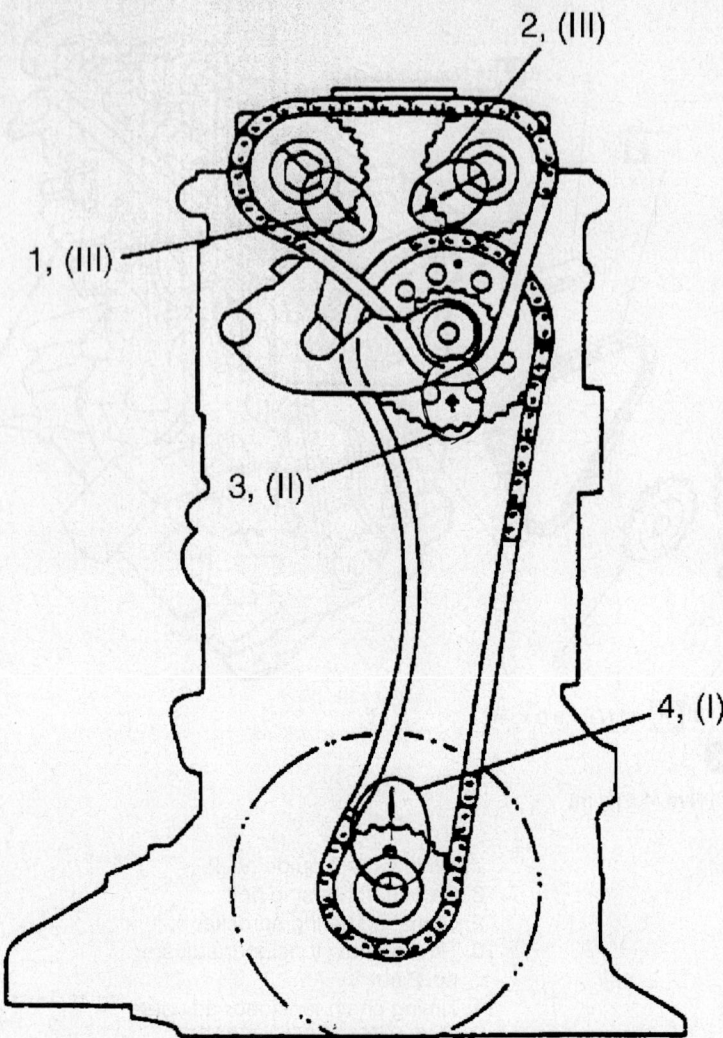

2, (III)

1, (III)

3, (II)

4, (I)

1. Timing marks of intake camshaft timing sprocket	2. Timing marks of exhaust camshaft timing sprocket	3. Arrow mark on idler sprocket
		4. Key on crankshaft

09490_SUZC_G0032

Turn crankshaft so that timing marks on cylinder head and lower crankcase match with those on sprockets as shown—2.0L and 2.3L engines

holder tool. Tighten the crankshaft pulley bolt to 108.5 ft. lbs. (150 Nm).

33. Install engine assembly.

34. Install all remaining components in the reverse order of the removal procedure.

2.5L Engine

1. Before servicing the vehicle, refer to the Precautions Section.

2. Remove the fuel pump fuse.

3. Start the engine and repeat cranking until the remaining fuel in the fuel line is all consumed.

4. Disconnect the negative battery cable.

5. Drain the engine coolant, engine oil, transaxle oil and power steering oil.

6. Recover the refrigerant.

7. Remove the engine.

8. Remove the flex plate bolts and the flex plate.

9. Support the engine assembly with an engine stand.

10. Remove the intake manifold.

11. Remove the cylinder head cover.

12. Remove the drive belt.

13. Remove the oil pan.

14. Install the crankshaft pulley holder DW110-130-01 to the crankshaft pulley bolt.

➡**Very high torque is required to remove or to install the crankshaft pulley. Never remove the crankshaft pulley when the vehicle is hung on the lift.**

15. Remove the crankshaft pulley bolt using the crankshaft pulley installer/remover tool DW110-130-02.

16. Remove the engine mount bracket support.

17. Remove the water pump pulley.

18. Remove the drive belt automatic tensioner retaining bolts and the tensioner.

19. Remove the timing chain cover.

20. Remove 2 dowel rings for timing cover.

21. Remove the timing chain tensioner.

22. Remove the fixed timing chain guide.

23. Remove the timing chain lever.

24. Remove the top chain guide.

25. Remove the timing chain.

26. Remove the crankshaft sprocket and the woodruff key.

To install:

27. Install the timing chain after aligning both the mark on the camshaft sprocket and on the timing chain using camshaft sprocket holder DW110-150.

28. Align both the timing mark on the crankshaft sprocket and on the timing chain.

29. Adjust the crankshaft using the crankshaft adjuster DW110-190.

30. Make sure the 3 timing marks on the timing chain are exactly aligned with the marks on the exhaust camshaft sprocket, intake camshaft sprocket, and crankshaft sprocket respectively.

31. Install the fixed timing chain guide and tighten the fixed timing chain guide bolts to 13–16 ft. lbs. (18–22 Nm).

32. Install the timing chain lever and tighten the timing chain lever bolt to 13–16 ft. lbs. (18–22 Nm).

33. Install the timing chain tensioner and tighten the timing chain tensioner bolt to 13–16 ft. lbs. (18–22 Nm).

34. Install the top chain guide and tighten the top chain guide bolt to 80–97 inch lbs. (9–11 Nm).

35. Install 2 dowel rings.

36. Apply the liquid gasket (Loctite® 5900-M8585) on the timing chain cover.

Install the timing chain cover install guide pin DW110-140 and position the timing chain cover.

➡**Be sure the applied liquid gasket is not damaged when installing the timing chain cover.**

37. Install the timing chain cover. Tighten the timing chain cover bolt (center) to 13–16 ft. lbs. (18–22 Nm); and tighten the remaining timing chain cover bolts to 53–89 inch lbs. (6–10 Nm).

38. Install the drive belt automatic tensioner and tighten the drive belt automatic

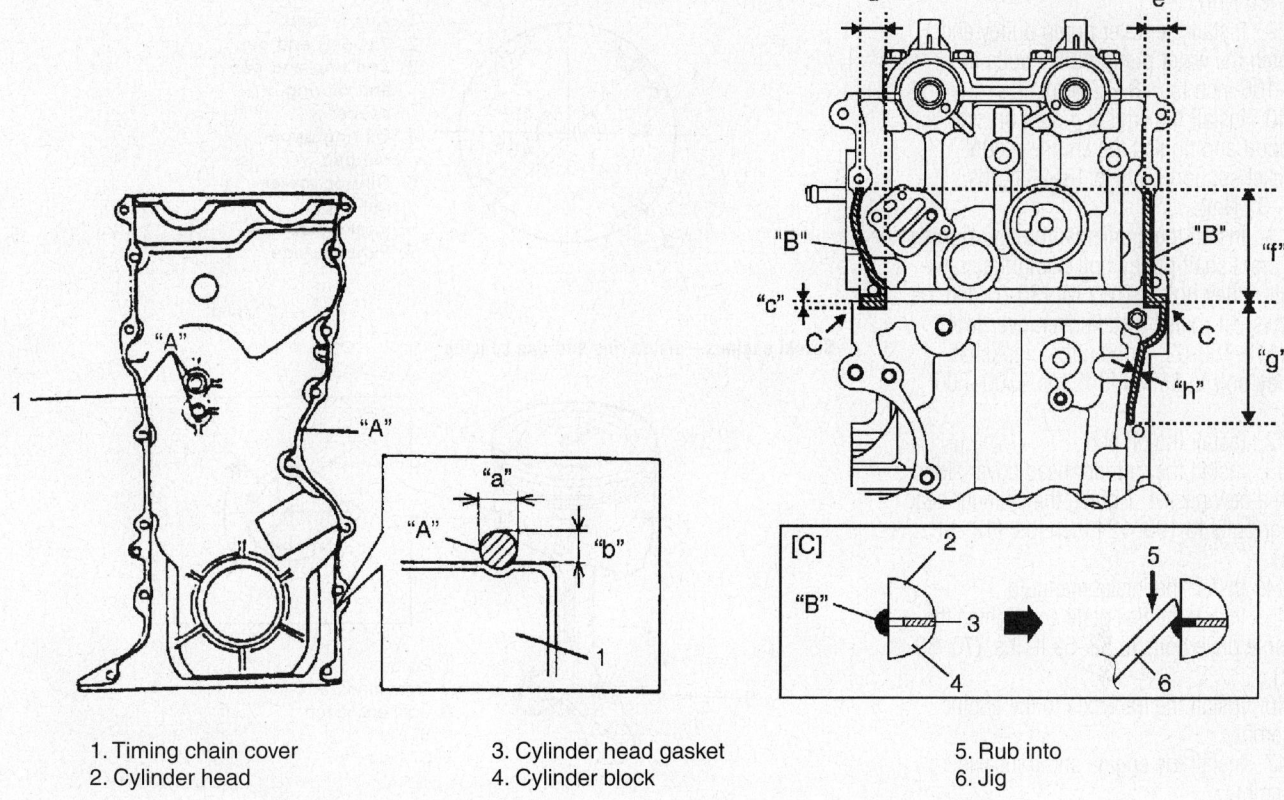

1. Timing chain cover
2. Cylinder head

3. Cylinder head gasket
4. Cylinder block

5. Rub into
6. Jig

09490_SUZC_G0017

Sealant application on timing belt cover—2.0L and 2.3L engines

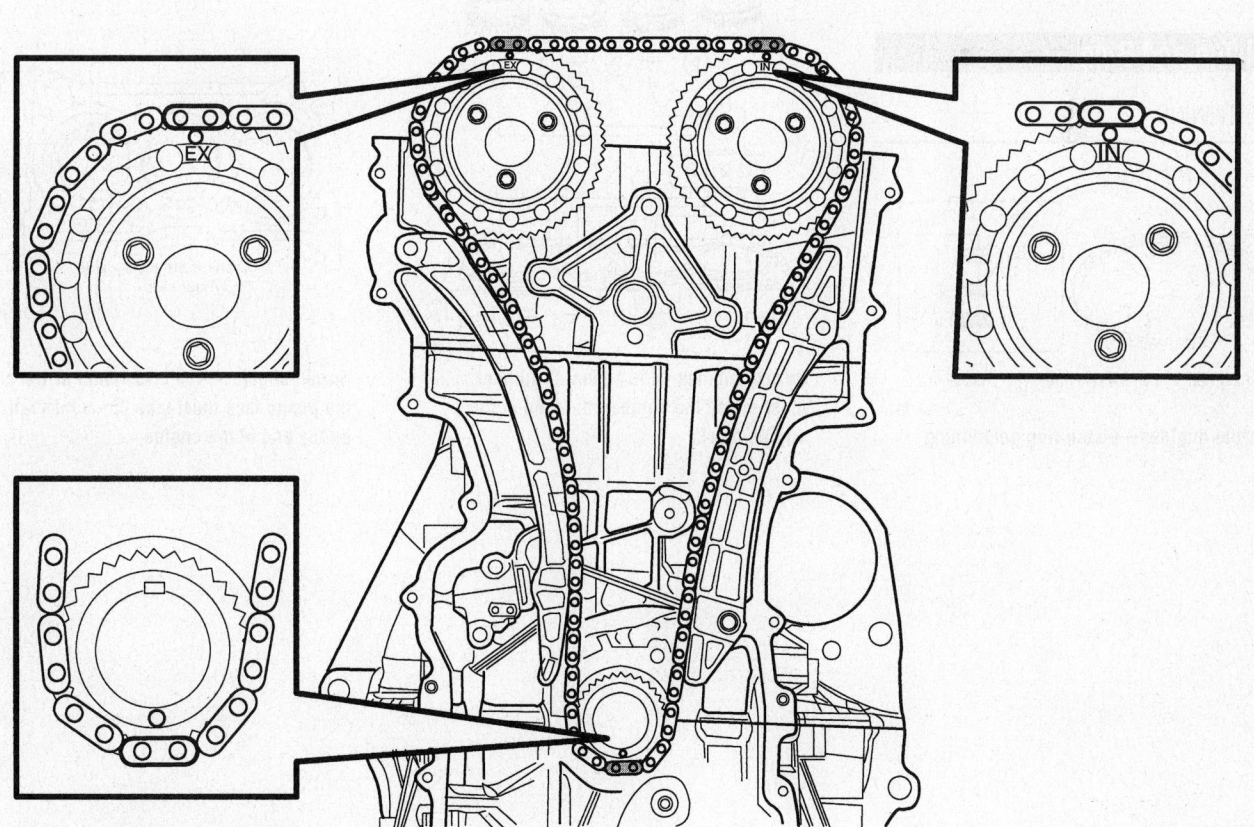

09490_SUZC_G0033

View of the timing chain and timing marks—2.5L engine

tensioner retaining bolt to 15–22 ft. lbs. (20–30 Nm).

39. Install the water pump pulley and tighten the water pump pulley bolts to 71–106 inch lbs. (8–12 Nm).

40. Install the engine mount bracket support and tighten the engine mount bracket support bolts to 18–23 ft. lbs. (25–31 Nm).

41. Install the crankshaft pulley. Tighten the crankshaft pulley bolt using the crankshaft pulley holder DW110-130-01 and the crankshaft pulley installer/remover tool DW110-130-02. Tighten the crankshaft pulley bolt to 369–443 ft. lbs. (500–600 Nm).

42. Install the oil pan.

43. Install the cylinder head cover along with a new gasket. Tighten the cylinder head cover bolts to 106–124 inch lbs. (12–14 Nm).

44. Install the intake manifold.

45. Install the flex plate and tighten the flexible plate bolts to 52–59 ft. lbs. (70–80 Nm).

46. Install the transaxle to the engine assembly.

47. Install the engine and transaxle assembly.

48. Install all remaining components in the reverse order of the removal procedure.

Piston and Ring

POSITIONING

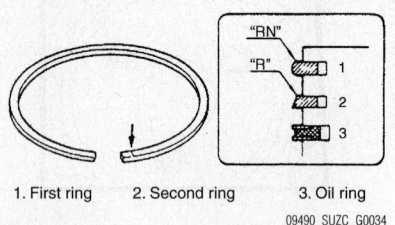

1. First ring   2. Second ring   3. Oil ring

09490_SUZC_G0034

Suzuki engines—piston ring positioning

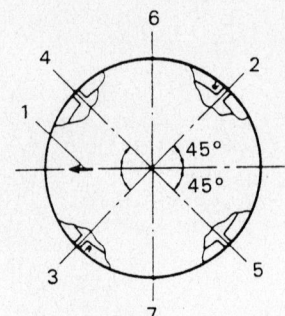

1. Arrow mark
2. 1st ring end gap
3. 2nd ring end gap and oil ring specer
4. Oil ring upper rail gap
5. Oil ring lower rail gap
6. Intake side
7. Exhaust side

7923AG85

Suzuki engines—piston ring end-gap spacing

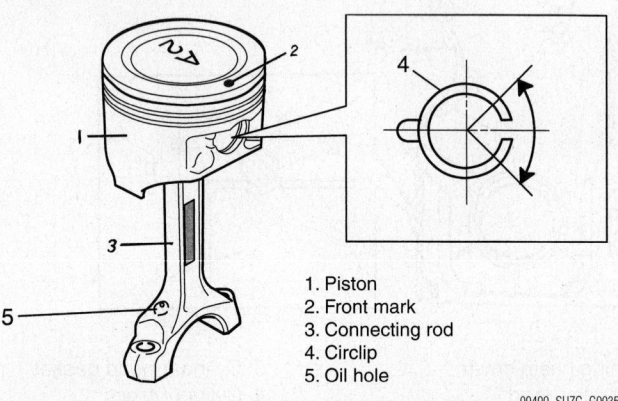

1. Piston
2. Front mark
3. Connecting rod
4. Circlip
5. Oil hole

09490_SUZC_G0035

Suzuki engines—piston/connecting rod assembly-to-engine positioning

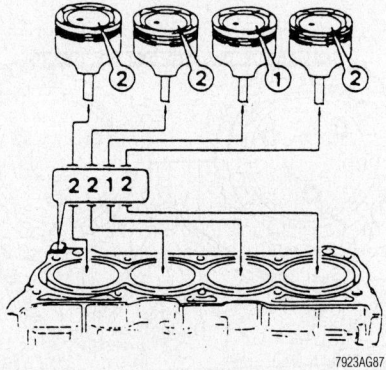

7923AG87

Suzuki engines—the piston ID number must match the number stamped in the engine block

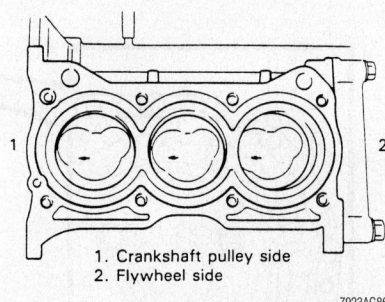

1. Crankshaft pulley side
2. Flywheel side

7923AG86

Suzuki engines—the directional arrow on the piston face must face the crankshaft pulley end of the engine

FUEL SYSTEM

Fuel System Service Precautions

Safety is the most important factor when performing not only fuel system maintenance, but any type of maintenance. Failure to conduct maintenance and repairs in a safe manner may result in serious personal injury or death. Maintenance and testing of the vehicle's fuel system components can be accomplished safely and effectively by adhering to the following rules and guidelines.

- To avoid the possibility of fire and personal injury, always disconnect the negative battery cable unless the repair or test procedure requires that battery voltage be applied.

- Always relieve the fuel system pressure prior to disconnecting any fuel system component (injector, fuel rail, pressure regulator, etc.), fitting or fuel line connection. Exercise extreme caution whenever relieving the fuel system pressure, to avoid exposing your skin, face and eyes to fuel spray. Please be advised that fuel under pressure may penetrate the skin or any part of the body that it contacts.

- Always place a shop towel or cloth around the fitting or connection prior to loosening to absorb any excess fuel due to spillage. Ensure that all fuel spillage (should it occur) is quickly removed from the engine surfaces. Ensure that all fuel soaked cloths or towels are deposited into a suitable waste container.

- Always keep a dry chemical (Class B) fire extinguisher near the work area.

- Do not allow fuel spray or fuel vapors to come into contact with a spark or open flame.

- Always use a back-up wrench when loosening and tightening fuel line connection fittings. This will prevent unnecessary stress and torsion on fuel line piping. Always follow the proper torque specifications.

- Always replace worn fuel fitting O-rings with new. Do not substitute fuel hose or equivalent, where fuel pipe is installed.

Fuel System Pressure

RELIEVING

✻✻ CAUTION

Care should be used when working around the fuel system. DO NOT smoke or expose the fuel system to any open flames. Keep a fire extinguisher handy.

1. Before servicing the vehicle, refer to the precautions in the beginning of this section.
2. Disconnect the negative battery cable from the battery.
3. Place the vehicle in **PARK** for automatic transmission or **NEUTRAL** for manual transmission.
4. Remove the relay box cover.
5. Disconnect the fuel pump relay from the relay box.
6. Remove the fuel filler cap from the filler neck to release the fuel vapor pressure in the fuel tank.
7. Start the vehicle and allow the engine to run until it stalls.
8. Crank the engine 3 more revolutions to eliminate any remaining pressure in the fuel lines.
9. Disconnect the negative battery cable.
10. Connect the fuel pump relay to the relay box.
11. After servicing the fuel system, connect the negative battery cable.
12. Start the engine and check for leaks in the system.

Fuel Filter

REMOVAL & INSTALLATION

The fuel filter on the Verona is located under the vehicle along the fuel line.

The fuel filter for Aerio models is an integral component of the in-tank fuel pump assembly. Refer to the Fuel Pump Removal procedure later in this section.

Verona

✻✻ CAUTION

The fuel system pressure must be relieved before disconnecting any fuel lines. Failure to do so may result in personal injury.

1. Before servicing the vehicle, refer to the Precautions Section.
2. Properly relieve the fuel system pressure.
3. Disconnect the negative battery cable.
4. Place a container under the fuel filter.
5. Remove or disconnect the following:
   - Fuel filter protector
   - Rear fuel feeding tube from the fuel filter

✻✻ CAUTION

A small amount of fuel may be released after the fuel hose is disconnected. Cover the hose and pipe with a shop towel.

   - Rear fuel filter tube from the fuel filter
   - Fuel filter mounting bracket bolt and remove the fuel filter from the vehicle
   - Fuel filter from the mounting bracket

To install:

6. Install or connect the following:
   - Fuel filter on the mounting bracket
   - Fuel filter in position on the underside of the vehicle. Secure with mounting bolt.
   - Rear fuel filter tube to the fuel filter
   - Rear fuel feeding tube to the fuel filter
   - Fuel filter protector
   - Negative battery cable
7. With the ignition **ON** and the engine **OFF** check for leaks.

Fuel Pump

REMOVAL & INSTALLATION

Aerio

1. Before servicing the vehicle, refer to the Precautions Section.
2. Relieve the pressure from the fuel system.
3. Disconnect the negative battery cable.
4. Drain the fuel from the tank by pumping the fuel out through the fuel tank filler.

✻✻ CAUTION

Use a gasoline safe hand operated pump device to drain the fuel tank.

5. Remove or disconnect the following:
   - Fuel filler cap
6. Raise and safely support the vehicle.
   - Muffler
   - Propeller shaft No. 2
   - EVAP canister and bracket
   - Fuel filler hose and breather hose
   - Filter inlet hose from canister air suction filter
7. Due to absence of fuel tank drain plug, drain fuel tank by pumping fuel out

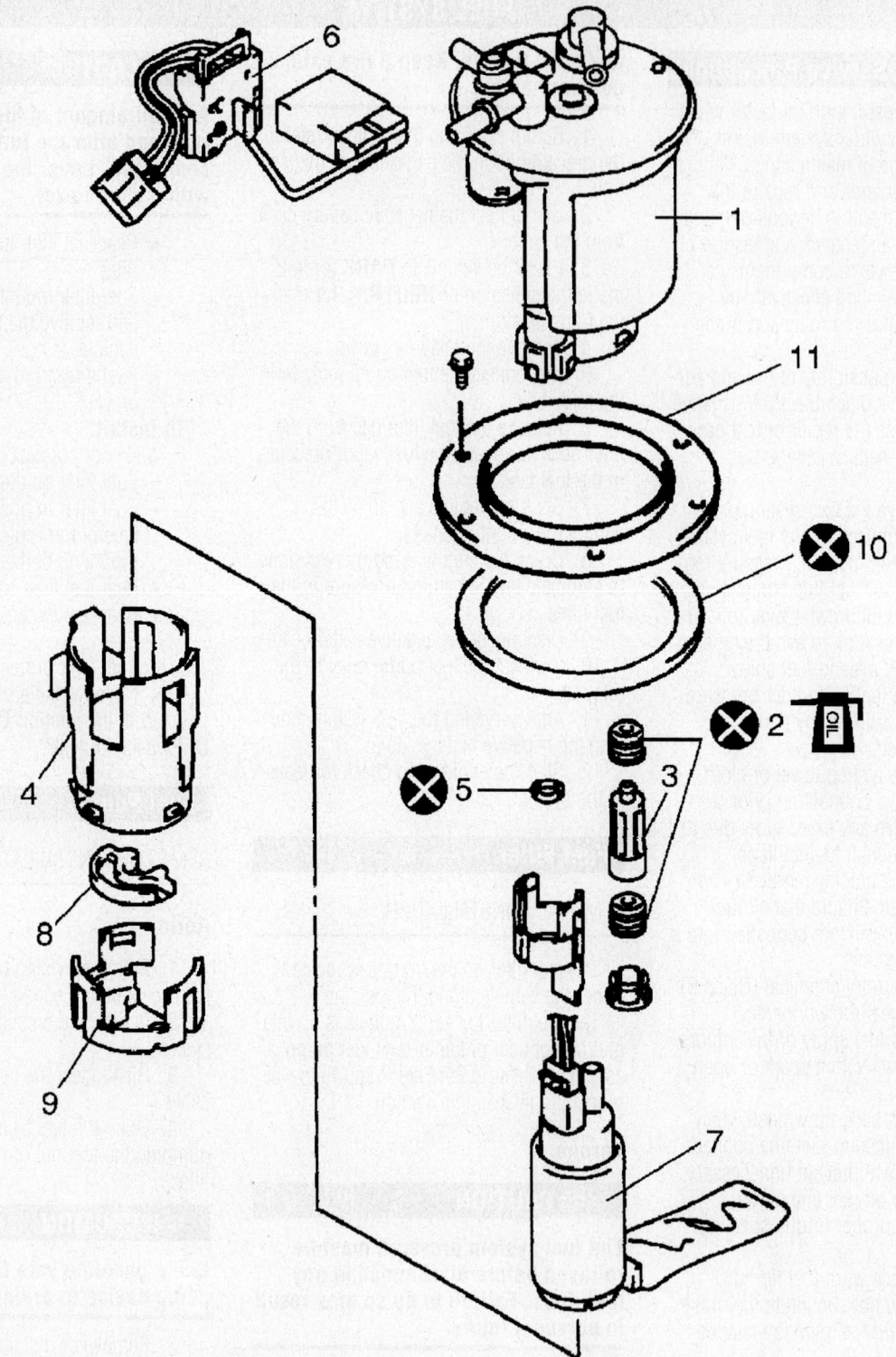

1. Fuel filter assembly
2. Grommet
3. Tube
4. Housing
5. O-ring
6. Fuel level sensor (fuel sender gauge)
7. Fuel pump
8. Cushion
9. Bracket
10. Gasket
11. Fuel pump plate

09490_SUZC_G0036

Fuel pump assembly components—Aerio

through fuel tank filler. Use hand operated pump device to drain fuel tank.

> ❊ **CAUTION**
>
> **Do not force pump hose into fuel tank, or pump hose may damage fuel tank inlet valve. Never drain or store fuel in an open container due to possibility of fire or explosion.**

8. Disconnect fuel pipe joints from fuel pipes. For quick joint, disconnect it as follows:
  a. Remove mud, dust and/or foreign material between pipe and joint by blowing compressed air.
  b. Unlock joint lock by inserting special tool 09919-47020 between pipe and joint.
  c. Disconnect joint from pipe.

> ❊ **CAUTION**
>
> **A small amount of fuel may be released after the fuel hose is disconnected. In order to reduce the chance of personal injury, cover the hose and pipe to be disconnected with a shop cloth. Be sure to put that cloth in an approved container when disconnection is completed.**

9. Support the fuel tank using a transmission jack.
  • Two bands that secure the fuel tank
10. Slowly and carefully lower the fuel tank enough to disconnect the fuel pump connector, tank pressure and temperature sensor connector and ground wire harness.
  • Fuel tank from vehicle
  • Fuel cut valve hose, fuel feed line and fuel return pipe from fuel pump assembly
  • Fuel pump assembly from fuel tank
To install:
11. Clean mating surfaces of fuel pump assembly and fuel tank.
12. Put plate on fuel pump assembly by matching the protrusion of fuel pump assembly to plate hole.
13. Install new gasket and fuel pump assembly with plate to fuel tank and tighten the assembly bolts to 7.5 ft. lbs. (10 Nm).
14. Install or connect the following:
  • Fuel cut valve hose, fuel feed line and fuel return pipe to the fuel pump assembly

➡ **When connecting joint, clean outside surface of pipe where joint is to be inserted, push joint into pipe till joint lock clicks and check to ensure that pipes are connected securely, or fuel leak may occur.**

15. Raise fuel tank with jack and connect fuel pump connector, tank pressure and temperature sensor connector and ground wire harness and then clamp wire harness.
  • Fuel tank to the vehicle. Tighten the fuel tank strap bolts to 36.5 ft. lbs. (50 Nm).
  • Fuel filler hose, breather hose and filter inlet hose. Clamp them securely and tighten to 2.5 ft. lbs. (3.5 Nm).
  • Fuel feed pipe, fuel return pipe and purge pipe and clamp securely

➡ **When connecting joint, clean outside surfaces of pipe where joint is to be inserted, push joint into pipe till joint lock clicks and check to ensure that pipes are connected securely, or fuel leak may occur.**

  • EVAP canister and bracket
  • Propeller shaft No. 2
  • Muffler
  • Lower the vehicle
  • Fuel filler cap
  • Negative battery cable
16. Fill the fuel tank enough to check for fuel leaks.
17. Turn the ignition switch to the **ON** position, but leave the engine **OFF** and check for fuel leaks.

Verona

1. Before servicing the vehicle, refer to the Precautions Section.
2. Relieve the pressure from the fuel system.
3. Disconnect the negative battery cable.
4. Drain the fuel from the tank by pumping the fuel out through the fuel tank filler.

> ❊ **CAUTION**
>
> **Use a gasoline safe hand operated pump device to drain the fuel tank.**

5. Remove or disconnect the following:
  • Exhaust front muffler
  • Fuel tank shield
6. Support the fuel tank using a transmission jack.
  • Two bands that secure the fuel tank
7. Slowly and carefully lower the fuel tank until enough space for the fuel tank filler tube removal is achieved.
8. Pull the fuel tank filler tube release fully backward and disconnect the fuel tank filler tube.
  • Fuel inlet line
  • Fuel return line

> ❊ **CAUTION**
>
> **A small amount of fuel may be released after the fuel hose is disconnected. Cover the hose and pipe to be disconnected with a shop cloth.**

  • EVAP vent hose
  • Fuel pump harness connectors
  • Fuel tank
  • Fuel pump and sending unit from fuel tank
To install:
9. Install or connect the following:
  • Fuel strainer on the fuel pump

➡ **Always install a new fuel pump strainer when replacing the fuel pump.**

  • Fuel pump and sending unit into the fuel tank
  • Electrical connectors to the fuel pump
  • Remaining components in the reverse order of removal
  • Negative battery cable
10. Fill the fuel tank enough to check for fuel leaks.
11. Turn the ignition switch to the **ON** position, but leave the engine **OFF** and check for fuel leaks.

Fuel Injector

REMOVAL & INSTALLATION

Aerio

1. Before servicing the vehicle, refer to the Precautions Section.
2. Relieve the pressure from the fuel system.
3. Disconnect the negative battery cable.
4. Remove or disconnect the following:
  • Engine cover
  • Vacuum hose from intake manifold
  • Fuel injector electrical connections
  • Fuel rail from cylinder head
  • Fuel injector(s)
  • Injector O-ring and discard
To install:
5. Install or connect the following:
  • Grommet to the injector
  • New injector O-ring
  • Check if fuel rail insulators for damage or scoring and replace if necessary
  • Insulators and cushions to fuel injector and fuel rail
6. Coat the injector O-rings with a thin coat of gasoline.

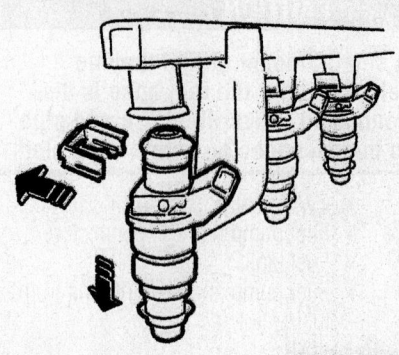

09490_SUZC_G0037

After removing the retaining clip, pull the injector down and out—2.5L engine

- Fuel injector into cylinder head and fuel rail
- Fuel rail mounting bolts and tighten to 18 ft. lbs. (25 Nm). Make sure injectors can rotate smoothly.

- Fuel injector electrical connections
- Vacuum hose to the intake manifold
- Engine cover

7. Connect the negative battery cable.

8. With the ignition **ON** and the engine **OFF** check for leaks.

Verona

1. Before servicing the vehicle, refer to the Precautions Section.

2. Relieve the pressure from the fuel system.

3. Disconnect the negative battery cable.

4. Remove or disconnect the following:

- Fuel feeding pipe from fuel rail
- Intake manifold
- Fuel injector wiring harness connector
- Fuel rail assembly

- Fuel injector retaining clips
- Fuel injectors by pulling down and out

To install:

➡ **Different injectors are calibrated for different flow rates. When ordering new fuel injectors, be certain to order the identical part number that is inscribed on the old injector.**

5. Install or connect the following:

- Injector to the fuel rail and secure with retaining clip
- Injector and fuel rail assembly
- Fuel injector wiring harness connector
- Intake manifold
- Fuel feeding pipe to the fuel rail

6. Connect the negative battery cable.

7. With the ignition **ON** and the engine **OFF** check for leaks.

DRIVE TRAIN

Transaxle Assembly

REMOVAL & INSTALLATION

Manual

AERIO

1. Before servicing the vehicle, refer to the Precautions Section.

2. Disconnect the negative battery cable.

3. Drain the transaxle oil.

4. Remove or disconnect the following:

- Clutch operating cylinder with hose still attached
- Gear control cables
- All the wiring harness clamps and connectors involved with the transaxle removal, tag if necessary for location to aid during installation
- Ground cable at the transaxle
- Starter
- Exhaust No. 1 pipe bolts

5. Support engine with lifting device.

- Engine under covers
- Exhaust No. 1 pipe and exhaust No. 2 pipe
- Lower stiffener
- Left and right ball joints from steering knuckles
- Drive shaft joints
- Center shaft support and center shaft
- Dynamic damper
- Engine rear mounting

- Engine rear mounting No. 1 bracket with No. 2 bracket
- Transaxle to engine bolts and nut

6. Support transaxle with a transmission jack.

- Left engine mounting with bracket

7. Remove any remaining attached parts from the transaxle.

8. Pull transaxle out so as to disconnect the input shaft from the clutch disc and then remove it.

To install:

9. Install or connect the following:

- Transaxle assembly. Use care when inserting the input shaft into the clutch assembly. If the spline on the input shaft does not align with the clutch assembly spline, turn the crankshaft slightly to aid in spline alignment.
- Left engine mounting with bracket
- Transaxle-to-engine bolts and nut. Tighten the nut and bolts to 44 ft. lbs. (61 Nm).
- Engine rear mounting No. 1 bracket with No. 2 bracket
- Engine rear mounting. Tighten the nuts to 32.5 ft. lbs. (45 Nm).
- Dynamic damper. Tighten the bolts to 18 ft. lbs. (25 Nm).
- Center shaft support and center shaft
- Drive shaft joints
- Left and right ball joints from steering knuckles
- Lower stiffener. Tighten the bolts to 7 ft. lbs. (10 Nm).

- Exhaust No. 1 pipe and exhaust No. 2 pipe
- Engine under covers

10. Remove engine lifting device.

- Starter
- All the wiring harness clamps and connectors involved with the transaxle removal
- Gear control cables
- Clutch operating cylinder and hose. Tighten the bolts to 16 ft. lbs. (23 Nm).
- Negative battery cable and the ground cable on the transaxle

11. Refill the transaxle with the recommended lubricant.

12. Check the function of the engine, clutch and transaxle.

Automatic

AERIO

1. Before servicing the vehicle, refer to the Precautions Section.

2. Disconnect the negative battery cable.

3. Drain the cooling system, engine oil, transaxle fluid and transfer case fluid (AWD).

4. Remove the engine/transaxle assembly from the vehicle. Refer to the engine procedure earlier in this chapter.

5. Remove or disconnect the following:

- Engine rear mounting No. 1 bracket and engine rear mounting No. 2 bracket with stiffener (2WD)
- Transfer case assembly (AWD)
- Center bearing support bolts and

center bearing support with center shaft from differential side gear (2WD)
- Lower stiffener
- Torque converter bolts
- Starter
- Transaxle with the torque converter from the engine compartment

➡**When removing the transaxle from the engine, move it parallel with the crankshaft and use care so not to apply excessive force to the drive plate and torque converter.**

✳✳ CAUTION

Be sure to keep the transaxle with the torque converter horizontal or facing up throughout the work. Should it be tilted with converter down, the converter may fall off and cause personal injury.

To install:
6. Install or connect the following:
   - Transaxle to the engine assembly
   - Transaxle attaching bolts and nut and tighten to 61.5 ft. lbs. (85 Nm).
   - Torque converter bolts and tighten the bolts to 18 ft. lbs. (25 Nm)
   - Lower stiffener. Tighten the lower stiffener bolts to the transaxle first and the lower stiffener bolt to the engine second. Tighten the bolts to 36.5 ft. lbs. (50 Nm).
   - Starter
   - Center shaft to differential side gear (2WD). Tighten the center bearing support bolts to 40 ft. lbs. (55 Nm).
   - Engine rear mounting No. 1 bracket and engine rear mounting No. 2 bracket with stiffener (2WD)
   - Transfer case assembly (AWD)
7. Install the engine/transaxle assembly into the vehicle. Refer to the engine procedure earlier in this chapter.
8. Fill the cooling system, engine, transaxle and transfer case (AWD).
9. Connect the negative battery cable.

VERONA

1. Before servicing the vehicle, refer to the Precautions Section.
2. Disconnect the negative battery cable.
3. Drain the transaxle fluid.
4. Remove or disconnect the following:
   - Battery
   - Air filter housing
5. Install an engine support fixture

- Transaxle wiring harness from transaxle
- Park/neutral switch connector
- Shift control cable from the transaxle
- Upper transaxle-to-engine mounting bolts
- Left transaxle mounting bracket
6. Raise and safely support the front of the vehicle securely on jackstands.
   - Centermember
   - Impact bar
   - Oil cooler pipes from transaxle
   - Drive shafts
   - Starter
   - Flywheel-to-torque converter bolts
   - Rear transaxle mounting bracket bolts
7. Support the transaxle assembly using transaxle support fixture DW260-010.
   - Oil pan flange-to-transaxle bolts
   - Left lower engine-to-transaxle mounting bolts from rear transaxle mounting bracket side
   - Right lower engine-to-transaxle mounting bolts
   - Transaxle assembly from vehicle

To install:
8. Install or connect the following:
   - Transaxle to the engine assembly
9. Support the transaxle assembly using transaxle support fixture DW260-010.
   - Right lower engine-to-transaxle mounting bolts and tighten to 54 ft. lbs. (73 Nm)
   - Left lower engine-to-transaxle mounting bolts and tighten to 54 ft. lbs. (73 Nm)
   - Oil pan flange-to-transaxle bolts and tighten to 23 ft. lbs. (31 Nm)
   - Rear transaxle mounting bracket bolts and tighten to 48 ft. lbs. (65 Nm)
   - Flywheel-to-torque converter bolts and tighten to 33 ft. lbs. (45 Nm)
   - Starter
   - Drive shafts
   - Oil cooler pipes to the transaxle
   - Impact bar
   - Centermember
10. Lower the vehicle.
    - Left transaxle mounting bracket
    - Upper transaxle-to-engine mounting bolts and tighten to 54 ft. lbs. (73 Nm)
    - Clip to shift control cable and shift cable to the transaxle
    - Park/neutral switch connector
    - Transaxle wiring harness to the transaxle
11. Remove the engine support fixture
    - Air filter housing

- Battery
12. Fill the transaxle with the correct amount and type of fluid.
13. Connect the negative battery cable.

Clutch

REMOVAL & INSTALLATION

1. Before servicing the vehicle, refer to the Precautions Section.
2. Remove the transaxle.
3. Hold the flywheel stationary.
4. Matchmark the pressure plate and flywheel for installation reference.
5. Loosen the pressure plate attaching bolts 1 turn at a time (evenly) until the spring pressure is released.
6. Remove the clutch disc and pressure plate.

To install:
7. Clean the flywheel mating surfaces of all oil, grease and metal deposits. Inspect flywheel for cracks, heat checking or other defects and replace or resurface as necessary.
8. Check the wear on the facings of the clutch disc by measuring the depth of each rivet head depression. Replace clutch disc when rivet heads are 0.02 in. (0.5mm) below the surface of clutch surface.
9. Check the diaphragm spring and pressure plate for wear or damage. If the spring or plate is excessively worn, replace the pressure plate assembly.
10. Check the pilot bearing for smooth operation. If the bearing does not spin freely, replace it.
11. Position the clutch disc and pressure plate with the matchmarks aligned and install a clutch alignment tool.
12. Install the pressure plate bolts. Tighten the mounting bolts evenly and in a crisscross pattern to 17 ft. lbs. (23 Nm). Remove the alignment tool and the flywheel holding tool.
13. Lightly lubricate the transaxle input shaft splines, pilot bearing surface of the input shaft, and the release bearing with grease.
14. Install the transaxle.

BLEEDING

1. Remove bleeder plug cap.
2. Connect a vinyl tube to the bleeder plug.
3. Depress clutch pedal several times, and then loosen the bleeder plug with the pedal depressed.
4. When fluid no longer comes out,

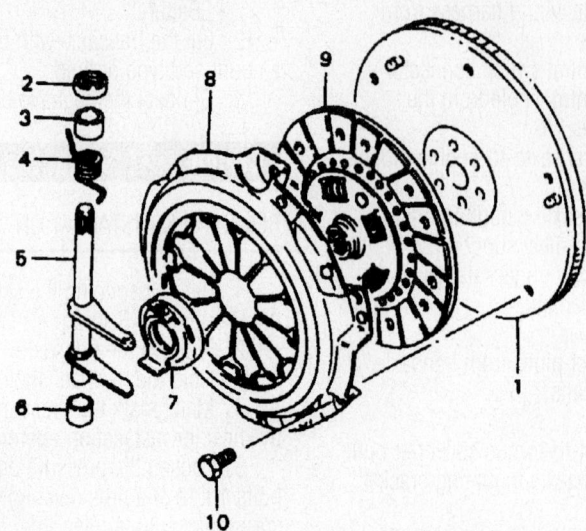

1. Flywheel
2. Release shaft seal
3. No. 2 bush
4. Return spring
5. Release shaft
6. No. 1 bush
7. Release bearing
8. Clutch cover
9. Clutch disc
10. Clutch cover bolt

7923UG27

Clutch component identification

tighten the bleeder plug, and then release the clutch pedal.

5. Repeat the previous 2 steps until all the air in the fluid is completely bled.

6. Tighten the bleeder plug.

7. Install the bleeder plug cap.

8. Check that all of the air has been bled from the clutch line.

9. Check the fluid level and add as necessary.

Transfer Case

REMOVAL & INSTALLATION

Aerio with AWD

1. Before servicing the vehicle, refer to the Precautions Section.

2. Disconnect the negative battery cable.

3. Remove or disconnect the following:

- Transfer breather hose from clamp on intake manifold
- Exhaust cover and exhaust No. 1 pipe nuts
- Heated oxygen sensor

4. Support engine using engine hanger.

- Front wheels
- Transaxle fluid and transfer case fluid
- Ball joints from steering knuckles
- Stabilizer joint

- Engine rear mounting nuts
- Exhaust No. 2 pipe hanger bush from suspension frame
- Engine front mounting nuts and mounting member to body bolts

5. Support the suspension frame using a transmission jack.

6. Remove 6 suspension frame bolts and lower the suspension frame with suspension arms and stabilizer.

- Front of propeller shaft from transfer output flange
- Exhaust No. 1 pipe stay
- Exhaust No. 1 pipe
- Exhaust No. 2 pipe
- Right side drive shaft
- Transfer case to transaxle stiffener
- Transfer case to engine stiffener

7. Support the transfer case assembly using a transmission jack.

- Transfer case mounting bolts
- Transfer case from transaxle

To install:

8. Remount the transfer case assembly by reversing the removal procedure.

9. Tighten the bolts and nuts to the specified torque as follows:

- Engine rear mounting bracket nut: 14.5 ft. lbs. (20 Nm)
- Transfer case mounting bolts: 36.5 ft. lbs. (50 Nm)
- Transfer case to engine stiffener bolts: 36.5 ft. lbs. (50 Nm)
- Transfer case to transaxle stiffener bolts: 36.5 ft. lbs. (50 Nm)

- Propeller shaft No. 1 bolts: 17 ft. lbs. (23 Nm)
- Exhaust No. 2 pipe to muffler bolts: 31.5 ft. lbs. (43 Nm)
- Suspension frame bolts: 65 ft. lbs. (90 Nm)
- Engine rear mounting nuts: 40 ft. lbs. (55 Nm)
- Ball joint lock nuts: 43.5 ft. lbs. (60 Nm)
- Exhaust No. 2 pipe to No. 1 pipe bolts: 31.5 ft. lbs. (43 Nm)
- Exhaust No. 1 pipe stay bolts: 36.5 ft. lbs. (50 Nm)
- Engine front mounting nuts: 32.5 ft. lbs. (45 Nm)
- Mounting member to body bolts: 39.5 ft. lbs. (55 Nm)
- Wheel lug nuts: 61.5 ft. lbs. (85 Nm)
- Remaining components in the reverse order of removal

10. Fill the transaxle and the transfer case with the correct type and amount of fluid.

11. Connect the negative battery cable.

Halfshaft

REMOVAL & INSTALLATION

Aerio

FRONT

1. Before servicing the vehicle, refer to the Precautions Section.

2. Disconnect the negative battery cable.

3. Remove drive shaft nut and washer.

4. Raise and safely support the front of the vehicle securely on jackstands.

5. Remove wheel.

6. Drain transaxle oil.

7. Remove tie rod end cotter pin and castle nut.

8. Disconnect tie rod end from steering knuckle by using special tool 09913-65210.

9. Using plastic hammer, drive out drive shaft joint so as to release snap ring fitting of joint spline at center shaft.

10. Disconnect front suspension control arm ball stud from steering knuckle after removing ball stud bolt and nut.

11. Remove drive shaft assembly.

➡**To prevent damage of boots, be careful not to contact them with other parts when removing drive shaft assembly.**

12. Remove center shaft support bolts and remove center shaft support with center

- Transaxle oil filler/level and drain plug (for M/T): 15.5 ft. lbs. (21 Nm)
- Transaxle fluid drain plug (for A/T): 29 ft. lbs. (40 Nm)
- Ball stud bolt: 43.5 ft. lbs. (60 Nm)
- Tie rod end castle nut: 25.5–40 ft. lbs. (35–55 Nm)
- New drive shaft nut: 127 ft. lbs. (175 Nm)
- Wheel nut: 61.5 ft. lbs. (85 Nm)
- Center shaft support bolt: 40 ft. lbs. (55 Nm)

14. Apply sealant to drain plug and filler/level plug for manual transaxle.

15. Fill transaxle with the correct type and amount of oil.

16. Connect the negative battery cable.

REAR

1. Before servicing the vehicle, refer to the Precautions Section.

2. Disconnect the negative battery cable.

3. Remove drive shaft nut and washer.

4. Raise and safely support the rear of the vehicle securely on jackstands.

5. Remove wheel.

6. Drain rear differential oil.

7. Remove parking brake wire mounting bolt (1), rear control rod outer bolts (2) and rear trailing rod rear bolt (3).

8. Install used clamp (2) to differential side joint (1) and pull out drive shaft from rear differential gear case by using tire lever (3).

9. Remove drive shaft.

To install:

❄❄ WARNING

Protect oil seals and boots from any damage, preventing them form unnecessary contact while installing drive shaft. Do not hit joint boot with hammer. Inserting joint only by hands is allowed. Make sure that differential side joint is inserted fully and its snap ring is seated as it was.

10. Clean rear wheel bearing oil seal and then apply grease. Replace it if required.

11. Install drive shaft assembly by reversing removal procedure noting the following points:

   a. Tighten each bolt and nut to the specified torque.

- Rear control rod outer bolt: 68.5 ft. lbs. (95 Nm)
- Rear trailing rod rear bolt: 68.5 ft. lbs. (95 Nm)

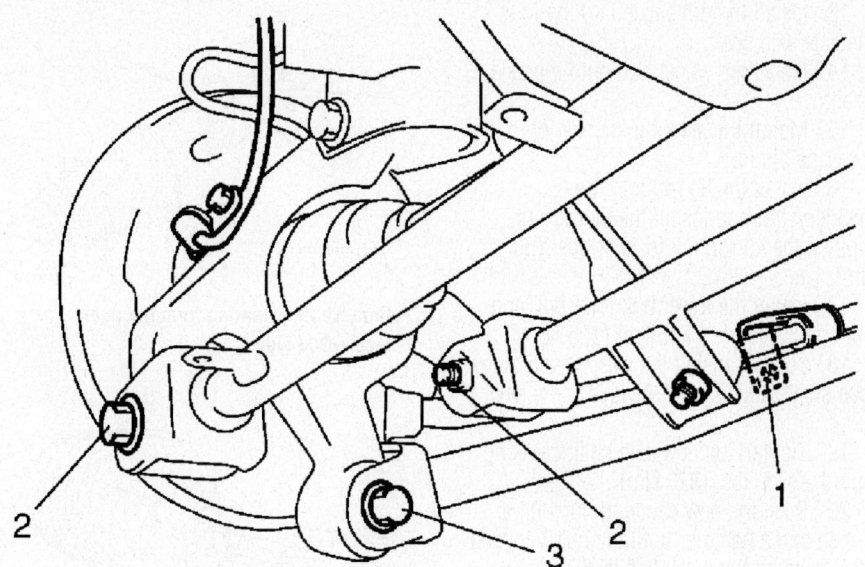

1. Parking brake wire mounting bolt
2. Rear control rod outer bolts
3. Rear trailing rod rear bolt

09490_SUZC_G0038

Parking brake wire mounting bolt (1), rear control rod outer bolts (2) and rear trailing rod rear bolt (3)—Aerio with AWD

shaft from differential side gear shaft (if equipped).

To install:

13. Install drive shaft assembly by reversing removal procedure noting the following points:

   a. Tighten each bolt and nut to the specified torque.

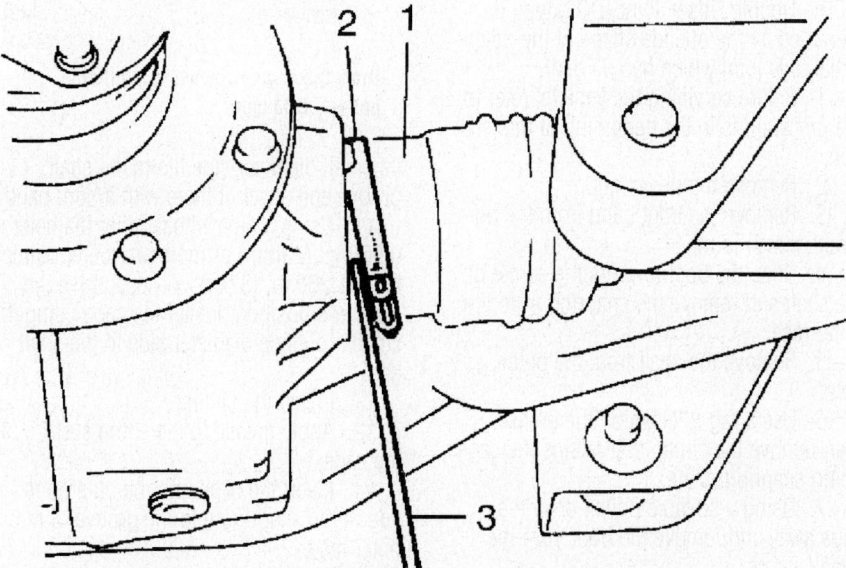

1. Differential side joint
2. Used clamp
3. Tire lever

09490_SUZC_G0039

Pull out drive shaft from rear differential gear case by using tire lever with a used clamp—Aerio with AWD

- New drive shaft nut: 127 ft. lbs. (175 Nm)

12. Clean rear wheel bearing oil seal and then apply grease. Replace it if required.

13. Apply sealant to drain plug for rear differential gear case.

14. Fill rear differential gear case with the correct type and amount of oil.

15. Connect the negative battery cable.

Verona

1. Before servicing the vehicle, refer to the Precautions Section.

2. Disconnect the negative battery cable.

3. Raise and safely support the front of the vehicle securely on jackstands.

4. Remove wheel.

5. Remove drive shaft nut and washer.

6. Disconnect front suspension control arm ball stud from steering knuckle after removing ball stud bolt and nut.

➡**Use only a tool made specifically for separating the lower ball joint. Failure to use the correct tool may cause damage to the ball joint and the seal.**

7. Remove tie rod end nut.

8. Disconnect tie rod end from steering knuckle by using ball joint separator KM-507-B.

➡**Use only a tool made specifically for separating the tie rod from the steering knuckle. Failure to use the correct tool may cause damage to the knuckle/strut assembly.**

9. Remove the drive axle from the intermediate axle shaft using the axle shaft remover KM-460-A.

10. Remove the intermediate axle shaft mounting bolts and the shaft from the transaxle.

➡**Support the unfastened end of the drive axle. Do not allow the drive axle to dangle freely from the transaxle for any length of time after it has been removed from the wheel hub. Place a drain pan below the transaxle to catch the escaping fluid. Cap the transaxle drive opening after the drive axle has been removed to keep the fluid in and any contamination out.**

To install:

11. Clean the hub seal and the transaxle seal. Be careful to not damage the seals.

12. Install the intermediate axle shaft into the transaxle. Tighten the intermediate axle shaft mounting bolts to 38 ft. lbs. (52 Nm).

13. Install the drive axle into the intermediate axle shaft.

14. Install the wheel hub onto the axle shaft.

15. Mount the steering knuckle onto the lower ball joint.

16. Install the tie rod into the knuckle/strut and install the tie rod nut. Tighten the tie rod nut to 37 ft. lbs. (50 Nm).

17. Install the lower ball joint bolt and nut and tighten to 77 ft. lbs. (105 Nm).

18. Loosely install the washer and a new axle shaft caulking nut. Always use a new nut.

19. Tighten the new axle shaft caulking nut to 236 ft. lbs. (320 Nm).

20. Peen the new caulking nut with a punch and a hammer until the nut is locked into place on the axle shaft hub.

21. Install the wheels. Loosely install the nuts.

22. Lower the vehicle to the floor and tighten the wheel nuts to 74 ft. lbs. (100 Nm).

23. Connect the negative battery cable.

24. Refill the transaxle fluid to the proper level.

CV-Joints

OVERHAUL

Double Offset Joint (DOJ) Type

The Double Offset Joint (DOJ) type is identified by the outside shape of the differential side joint which has no dent.

1. Before servicing the vehicle, refer to the precautions in the beginning of this section.

2. Remove the driveshaft.

3. Remove the boot band from the differential side joint.

4. Slide the boot towards the center of the shaft and remove the snapring from the outer race.

5. Remove the shaft from the outer race.

6. Use a rag to clean off the grease, then remove the circlip that retains the cage using snapring pliers.

7. Using a suitable puller, draw the cage away and remove the boot from the shaft.

8. Inspect all components for wear and/or damage and replace as necessary.

To install:

9. Clean all components and allow them to completely dry.

10. Install the boot onto the shaft until

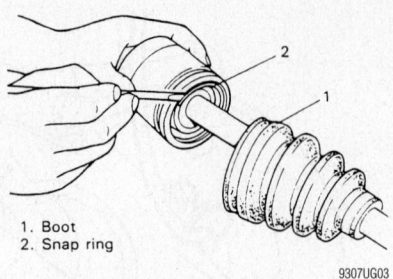

1. Boot
2. Snap ring

9307UG03

Remove the snapring from the outer race—DOJ type

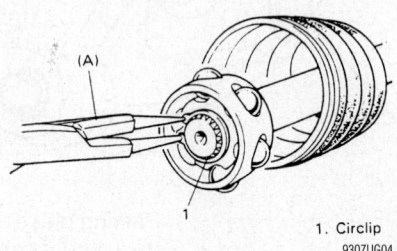

(A)

1. Circlip

9307UG04

Remove the circlip that retains the cage—DOJ type

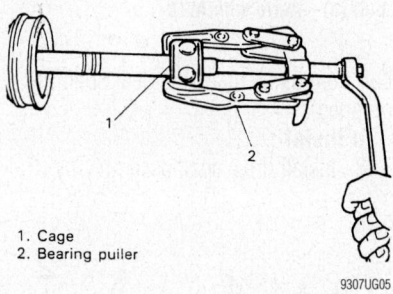

1. Cage
2. Bearing puller

9307UG05

Draw the cage away using a suitable puller—DOJ type

its small diameter side fits to the shaft groove and attach it there with a boot band.

11. Using a pipe whose inner diameter is 0.906 in. (23mm) or more and outer diameter is 1.260 in. (32mm) or less, drive the cage into position. Install the cage using the smaller outside diameter side to the shaft end.

12. Install the circlip.

13. Apply grease to the entire surface of the cage.

14. Insert the cage into the outer race and fit the snapring into the groove of the outer race.

➡**Position the opening of the snapring so that it will not be lined up with a ball.**

15. Apply grease to the inside of the outer race and fit the boot to the outer race.

16. After fitting the boot, insert a screw-

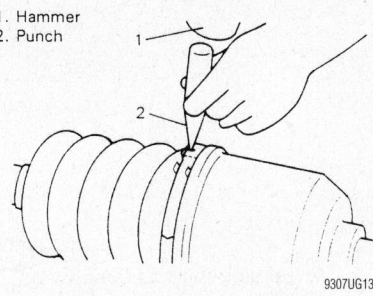

1. Hammer
2. Punch

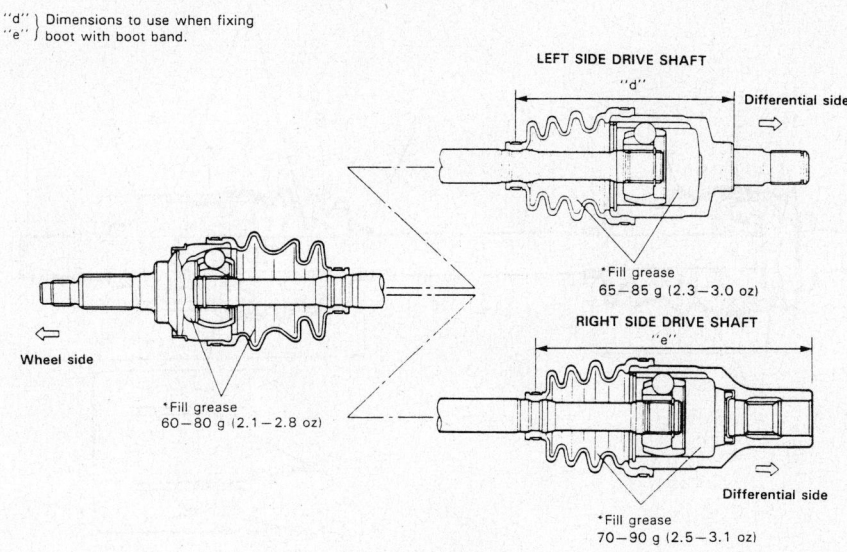

"d" } Dimensions to use when fixing
"e" } boot with boot band.

LEFT SIDE DRIVE SHAFT

"d"

Differential side ⇨

*Fill grease
65–85 g (2.3–3.0 oz)

RIGHT SIDE DRIVE SHAFT

"e"

Differential side

*Fill grease
70–90 g (2.5–3.1 oz)

⇦ Wheel side

*Fill grease
60–80 g (2.1–2.8 oz)

*Be sure to use grease supplied with spare parts.

9307UG06

Adjust the boot so that measurements D and E are as illustrated—DOJ type

9307UG13

Using a punch, caulk the center of the band folded over the fixture—Esteem models

driver into the boot on the outer race side and allow air to enter the boot so that the air pressure in the boot equals atmospheric pressure

17. Fix the boot to the outer race with a boot band. Before tightening the band adjust the boot so that measurement D which is 8.10 in. (205.8mm) and measurement E which is 7.34 in. (186.4mm) are as shown in the accompanying illustration.

18. Install the driveshaft in the vehicle.

Tripod Joint Type

The Tripod joint type can be identified by the 3 dent lines on the outside of the differential side joint.

1. Before servicing the vehicle, refer to the precautions in the beginning of this section.

2. Remove the driveshaft.

3. Remove the boot band and the Tripod joint housing.

4. Use a rag to clean off the grease, then remove the circlip using snapring pliers.

5. Remove the spider by using a suitable puller.

6. Remove the boot band and pull the differential side boot from the shaft.

7. Remove the dynamic damper band, then pull the damper through the shaft, if equipped.

8. Remove the boot bands from the wheel side joint boot and pull the boot through the shaft.

9. Inspect all components for wear and/or damage and replace as necessary.

To install:

10. Clean all components and allow them to completely dry.

11. On Verona models, perform the following:

a. Apply grease to the wheel side joint. Use black grease in the tube included with the wheel side boot kit.

b. Install the wheel side boot on the shaft.

c. Fill the inside of the boot with grease and fasten the boot with a band.

d. Install the dynamic damper on the shaft, if equipped.

e. Install the differential side boot on the shaft.

f. Apply grease to the Tripod joint. Use the yellow grease in the tube included in the differential side joint kit.

g. Using a pipe whose inner diameter is 0.906 in. (23mm) or more and outer diameter is 1.260 in. (32mm) or less,

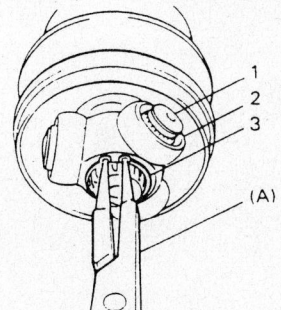

1. Spider
2. Bearing
3. Circlip

9307UG07

Remove the circlip that retains the spider assembly—Tripod type

drive the spider into position. Face the chamfered side inward and fasten it in place with the circlip.

h. Fill the boot with grease, then install the housing and joint it with the boot.

i. Fasten the boot bands.

12. On Aerio models, perform the following:

a. Apply grease to the wheel side joint. Use black grease in the tube included with the wheel side boot kit.

b. Install the wheel side boot on the shaft.

c. Fill the inside of the boot with grease.

d. Fit the boot band into the groove in the boot.

e. Tighten the boot band until its outer diameter is 3.05 in. (77.5mm).

f. Fold the boot band over the metal fixture.

g. Using a punch, caulk the center of the band folded over the fixture.

h. Cut the band about 0.28 in. (7mm) from the clip.

i. Clamp the band with the clip.

j. Fix the small diameter side of the boot band into the groove in the boot.

k. Tighten the boot band until its outer diameter is 1.14 in. (29mm).

l. Fold the boot band over the metal fixture.

m. Using a punch, caulk the center of the band folded over the fixture.

n. Install the dynamic damper on the right side driveshaft.

o. Install the differential side boot on the shaft.

p. Apply grease to the Tripod joint. Use the yellow grease in the tube included in the differential side joint kit.

q. Install the spider into position. Face the chamfered side inward and fasten it in place with the circlip.

r. Fill the boot with grease, then

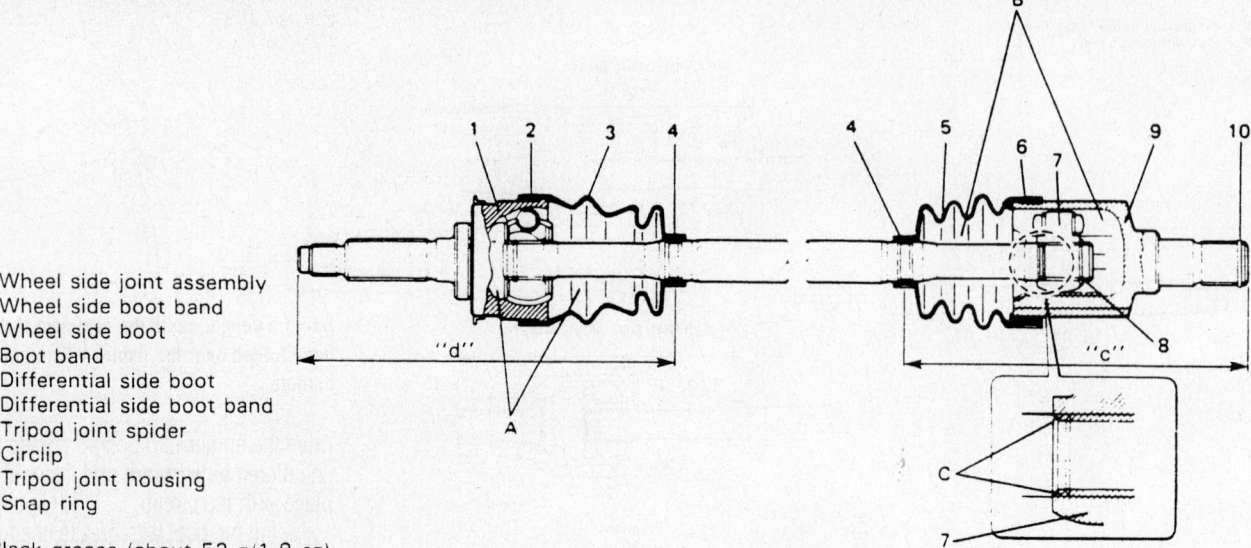

1. Wheel side joint assembly
2. Wheel side boot band
3. Wheel side boot
4. Boot band
5. Differential side boot
6. Differential side boot band
7. Tripod joint spider
8. Circlip
9. Tripod joint housing
10. Snap ring

A: Black grease (about 52 g/1.8 oz)
B: Yellow grease (about 100 g/3.5 oz)
C: Chamfered spline

9307UG14

Adjust the boot so that measurement D is correct—Esteem models equipped with a Tripod type joint

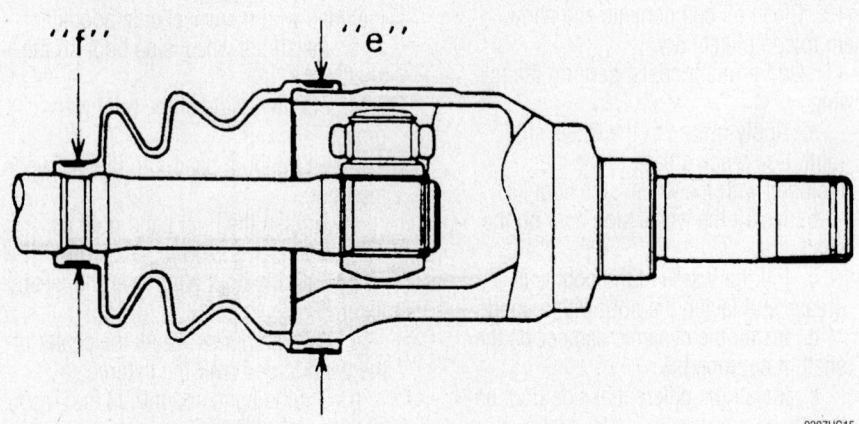

9307UG15

When tightening the bands, make sure that diameters E and F are correct—Esteem models equipped with a Tripod type joint

install the housing and joint it with the boot.

s. Adjust the boot so that the when measured from the tip of the driveshaft (wheels side), to the inner (small) band; the measurement is 8 in. (204mm).

13. Insert a screwdriver into the boot on the outer race side and allow air to enter the boot so that the air pressure in the boot equals atmospheric pressure

a. Fasten the boot band so that diameter E which measures 3.11 in. (79mm) and diameter F which measures 1.14 in. (29mm) are correct. Refer to the accompanying illustration for the diameter locations.

14. Install the driveshaft.

STEERING

Air Bag

❋❋ CAUTION

Most vehicles are equipped with an air bag system. The system must be disabled before performing service on or around system components, steering column, instrument panel components, wiring and sensors. Failure to follow safety and disabling procedures could result in accidental air bag deployment, possible personal injury and unnecessary system repairs.

PRECAUTIONS

Several precautions must be observed when handling the inflator module to avoid accidental deployment and possible personal injury.
• Never carry the inflator module by the wires or connector on the underside of the module.
• When carrying a live inflator module, hold securely with both hands, and ensure

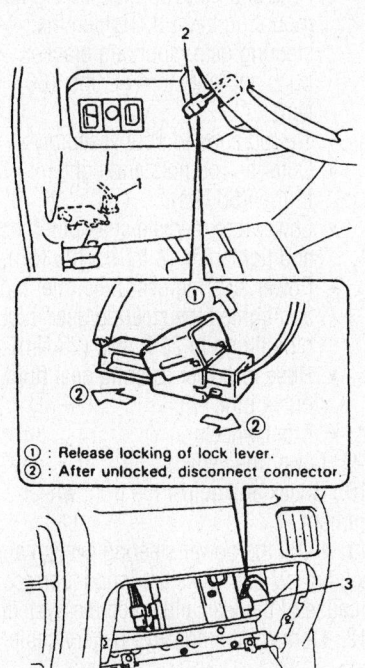

① : Release locking of lock lever.
② : After unlocked, disconnect connector.

1 Air bag fuse box
2 Yellow connector of driver air bag (inflator) module
3 Yellow connectors of passenger air bag (inflator) module
4 Glove box

7923UG29

Air bag connector locations

that the bag and trim cover are pointed away.
• Place the inflator module on a bench or other surface with the bag and trim cover facing up.
• With the inflator module on the bench, never place anything on or close to the module that may be thrown in the event of an accidental deployment.

DISARMING

❋❋ WARNING

When performing service on or around the air bag system components or wiring, disable the air bag system. Failure to follow the procedures could result in possible deployment, personal injury or unneeded system repairs.

1. Before servicing the vehicle, refer to the precautions in the beginning of this section.
2. Disconnect the negative battery cable.
3. Turn the steering wheel so the wheels are pointing straight ahead.
4. Turn the ignition switch to the **LOCK** position and remove the key.
5. Remove the **AIR BAG-IG** fuse from the air bag fuse box located near the junction/fuse box.
6. Remove the left side steering wheel side cap and disconnect the yellow connector for the driver's side air bag (inflator) module.
7. Pull out the glove box while pushing in on the stoppers from the left and the right sides. Disconnect the yellow connector for the passenger air bag (inflator) module.

REARMING

❋❋ WARNING

When performing service on or around the air bag system components or wiring, disable the air bag system. Failure to follow the procedures could result in possible deployment, personal injury or unneeded system repairs.

1. Before servicing the vehicle, refer to the precautions in the beginning of this section.
2. Connect the negative battery cable.

3. Turn the ignition switch to the **LOCK** position and remove the key.
4. Connect the yellow connector for the passenger side air bag (inflator) module and the yellow connector for the driver's side air bag (inflator) module. Be sure to lock each connector with the lock lever.
5. Install the glove box assembly.
6. Install the left side steering wheel side cover.
7. Install the **AIR BAG-IG** fuse in the air bag fuse box.
8. Turn the ignition **ON** and verify that the **AIR BAG** warning lamp flashes 7 times, then turns off. If the system does not operate as described, diagnosis and repairs to the air bag system are necessary.

Rack and Pinion Steering Gear

REMOVAL & INSTALLATION

Aerio

❋❋ WARNING

Be sure to set the front wheels straight ahead and remove the ignition key from the cylinder before starting repairs. The contact coil of the air bag system may be damaged if the key is not removed and the wheels are not straight ahead.

1. Before servicing the vehicle, refer to the Precautions Section.
2. Disconnect the negative battery cable.
3. Take out fluid in P/S fluid reservoir with syringe or such.
4. Remove or disconnect the following:
• Steering column joint cover
• Steering shaft upper joint bolt, loosen but do not remove
• Steering shaft lower joint bolt
• Lower joint from the pinion
• Front wheels
• Tie rod ends from the steering knuckles
• Exhaust pipe No. 2
• Engine rear mounting together with its bracket from engine and member
• Engine mounting member
• Transfer case (AWD)

➥**When the lines are disconnected plug the lines or place an oil pan under the vehicle.**

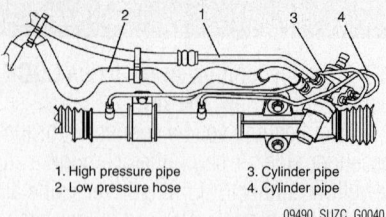

1. High pressure pipe
2. Low pressure hose
3. Cylinder pipe
4. Cylinder pipe

09490_SUZC_G0040

Fluid pipes and hose to the power steering gear—Aerio

- Cylinder pipes and from steering gear box, using flare nut wrench
- High pressure pipe and low pressure hose from steering gear box, using flare nut wrench
- Steering gear box mounting bolts and steering gear box

To install:

✳✳ WARNING

Be sure to confirm that steering wheel and front tires (wheels) are in straight position when inserting steering lower joint into steering pinion shaft.

5. Install or connect the following:
   - Steering gear, brackets and mounting bolts. Tighten bolts to 40 ft. lbs. (55 Nm).

➡ **If a plug was put to disconnected pipe when removing steering gear box, remove that plug before reconnecting pipe.**

- Cylinder lines on the rack and pinion and tighten their fittings to 18 ft. lbs. (25 Nm)
- High and low pressure lines to the steering gear. Tighten the fittings to 25 ft. lbs. (35 Nm).
- Transfer case (AWD)
- Engine mounting member
- Engine rear mounting together with its bracket to the engine and member
- Exhaust pipe No. 2
- Tie rod ends to the steering knuckles. Tighten the castle nuts to 25.5–39.5 ft. lbs. (35–55 Nm).
- Front wheels

6. Be sure the steering wheel is straight and the front wheels are pointing straight ahead.
   - Steering shaft to the rack and pinion
   - Lower steering shaft-to-rack and pinion clinch bolt and tighten both steering joint bolts (upper and lower) to 18 ft. lbs. (25 Nm)
   - Front wheels
7. Lower the vehicle.
8. Connect the negative battery cable.
9. Fill the power steering system and then bleed the power steering system.
10. Check and adjust the front wheel alignment.

Verona

✳✳ WARNING

Be sure to set the front wheels straight ahead and remove the ignition key from the cylinder before starting repairs. The contact coil of the air bag system may be damaged if the key is not removed and the wheels are not straight ahead.

1. Before servicing the vehicle, refer to the Precautions Section.
2. Disconnect the negative battery cable.
3. Raise and safely support the front of the vehicle.
4. Remove or disconnect the following:
   - Front wheels
   - Power steering gear fluid inlet pipe
   - Hose from power steering gear fluid outlet pipe. Place a drain pan under the steering gear to catch the power steering fluid.
5. Position the steering gear straight ahead by turning the steering wheel until the steering wheel spokes are vertical and pointed to the left.
6. Scribe a mark on the stub shaft housing that lines up with a mark on the intermediate shaft lower coupling.
   - Intermediate shaft pinch bolt
   - Outer tie rod nuts
   - Tie rod ends from the strut assembly using ball joint remover KM-507-B

- Nuts and bolts from steering gear mounting brackets
- Crossmember assembly for easy removal of the rack and pinion assembly

✳✳ WARNING

When removing the crossmember assembly it should be supported by Jack stands, in advance. Failure to support the crossmember properly can result in personal injury. In addition, remove the centermember bolt with caution when the exhaust pipe is hot.

- Speed sensitive power steering (SSPS) solenoid valve connector
- Rack and pinion assembly from crossmember assembly

To install:

7. Install the rack and pinion assembly from below. The steering gear must be in a straight-ahead position, and the steering wheel spokes must be vertical and pointing to the left. Align the marks on the shafts to ensure proper positioning. Seat the stub shaft into the intermediate shaft.
8. Install or connect the following:
   - SSPS solenoid valve connector.
   - Nuts and bolts on the steering gear mounting bracket. Tighten the steering gear mounting bracket bolts and nuts to 44 ft. lbs. (60 Nm).
   - Tie rod ends to strut assembly
   - Outer tie rod nuts and tighten to 37 ft. lbs. (50 Nm).
   - Lower intermediate shaft pinch bolt and tighten to 18 ft. lbs. (25 Nm).
   - Power steering gear fluid inlet pipe and tighten the steering gear inlet pipe fitting to 21 ft. lbs. (28 Nm).
   - Hose to power steering gear fluid outlet pipe
   - Front wheels
9. Lower the vehicle.
10. Check and adjust the front wheel alignment.
11. Refill the power steering system and check for leaks. If leaks are found, correct the cause of the leak and bleed the system.
12. Connect the negative battery cable.

FRONT SUSPENSION

Strut

REMOVAL & INSTALLATION

Aerio

1. Before servicing the vehicle, refer to the Precautions Section.
2. Disconnect the negative battery cable.
3. Raise and safely support the front of the vehicle.
4. Remove or disconnect the following:
   - Wheel
   - Stabilizer joint from strut bracket

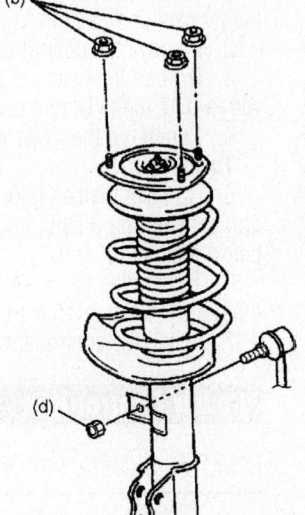

a. Strut bracket nuts
b. Strut support nuts
c. Brake hose mounting bolt
d. Stabilizer joint nut

- Brake hose mounting bolt and brake hose
- ABS wheel speed sensor harness (if equipped) from strut bracket
- Strut bracket bolts
- Strut support nuts

➡**Hold strut by hand so that it will not fall off.**

- Strut assembly

5. Using a spring compressor tool, turn special tool bolts alternately until spring tension is released. Whether it is released or not can be known by whether strut turns lightly while strut spring is held stationary.

➡**Use a commercially available spring compressor and follow the operation procedure described in the Instruction Manual supplied with that spring compressor.**

6. While keeping spring compressed, remove strut nut.
7. Disassemble strut assembly.

To install:

8. Assemble the coil spring into the strut assembly as follows:
   a. Compress spring with special tool until total length becomes about 9.8 inches (250mm).
   b. Install coil spring lower seat and compressing coil spring, and mate spring end with stepped part of lower seat.
   c. Install bump stopper and dust cover onto strut rod.
   d. Pull strut rod as far up as possible and use care not to allow it to retract into strut.
   e. Install spring seat on coil spring and then spring upper seat aligning "OUT" mark on spring upper seat and center of strut bracket.
   f. Install bearing, strut support and strut nut in this sequence. Tighten strut nut to 50.5 ft. lbs. (70 Nm).
   g. Install rubber cap.
   - Strut assembly. Tighten the strut bracket nuts to 76 ft. lbs. (105 Nm). Tighten the strut support nuts to 20.5 ft. lbs. (28 Nm).
   - Brake hose and mounting bolt. Tighten the brake hose mounting bolt to 18 ft. lbs. (25 Nm).
   - Stabilizer joint to strut bracket. Tighten the stabilizer joint nut to 36.5 ft. lbs. (50 Nm).
   - ABS wheel speed sensor harness, if equipped
   - Wheels
9. Lower the vehicle.
10. Connect the negative battery cable.
11. After installation, confirm front wheel alignment.

Verona

1. Before servicing the vehicle, refer to the Precautions Section.
2. Disconnect the negative battery cable.
3. Raise and safely support the front of the vehicle.
4. Remove or disconnect the following:
   - Cap and nuts from the strut upper area

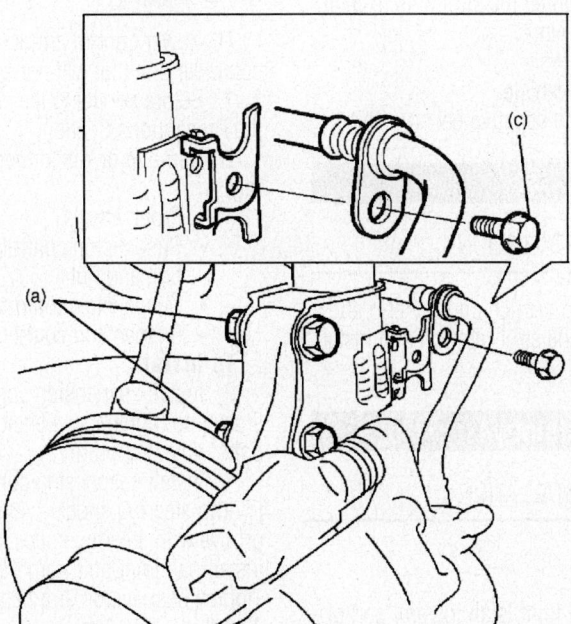

09490_SUZC_G0041

Strut assembly mounting

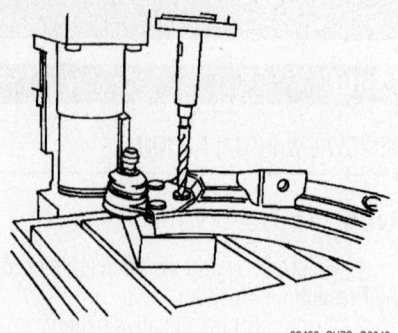

Drilling off the ball joint rivets

Exploded view of the strut assembly

- Wheels
- ABS sensor line from strut assembly (if equipped)
- E-ring and brake hose
- Strut bracket bolts and nuts
- Strut assembly

5. Remove the coil spring by disassembling the strut assembly as follows:

a. Fasten the strut assembly to the spring compressor KM-329-A. Make sure the hooks are seated on the strut spring properly.

b. Compress the front spring with the spring compressor KM-329-A.

c. Remove the dust cover from the bearing assembly.

d. Fix the strut mounting with bite and remove the piston rod nut.

e. Remove the upper strut mount, the mount bearing, the upper spring seat, the upper spring insulator, the hollow bumper, and the piston rod boot.

➡**Record the position of the front spring seat relative to the strut assembly-to-knuckle bracket. Place the front spring locator back in the same position during assembly.**

f. Release the spring.

g. Remove the spring and the lower spring insulator.

To install:

6. Install the coil spring by assembling the strut assembly as follows:

a. Install the lower spring insulator and the spring.

b. Compress the spring using the spring compressor KM-329-A.

c. Install the piston rod boot, the hollow bumper, the upper spring insulator, the upper spring seat, the upper strut mount, and the mount bearing. Be sure

the front spring seat is positioned correctly.

d. Fix the strut mounting with bite and install the piston rod nut. Tighten the piston rod nut to 52 ft. lbs. (70 Nm).

e. Install the dust cover onto the bearing assembly.

f. Release and remove the spring compressor KM-329-A.

7. Install or connect the following:

- Strut assembly, strut bracket bolts and nuts. Tighten the strut bracket bolts and nuts to 103 ft. lbs. (140 Nm).
- Brake hose and E-ring
- ABS sensor line to strut assembly (if equipped)
- Cap and nuts to the strut upper area. Tighten the strut nuts to 37 ft. lbs. (50 Nm).
- Wheels

8. Lower the vehicle.

9. Connect the negative battery cable.

Coil Spring

REMOVAL & INSTALLATION

For coil spring service on the Aerio and Verona, refer to the strut removal and installation procedure.

Lower Ball Joint

REMOVAL & INSTALLATION

Aerio

The lower ball joint is an integral part of the lower control arm assembly. If the ball joint is found to be defective the whole

lower control arm assembly must be replaced.

Verona

1. Before servicing the vehicle, refer to the Precautions Section.

2. Remove the control arm.

3. Drill off the heads of the three rivets with a 0.47 inch (12mm) drill bit.

4. Punch out the rivets with a drift.

To install:

5. Connect the ball joint to the control arm by inserting 3 ball joint bolts from below the control arm.

6. Tighten the ball joint-to-control arm nuts to 81 ft. lbs. (110 Nm).

7. Install the control arm.

Lower Control Arm

REMOVAL & INSTALLATION

Aerio

The lower control arm and ball joint are a complete unit that will not separate.

1. Before servicing the vehicle, refer to the Precautions Section.

2. Remove or disconnect the following:

- Front wheels
- Suspension control arm ball joint bolt and nut
- Suspension control arm bolts
- Suspension control arm

To install:

3. Install suspension control arm as shown but tighten suspension control arm bolts only temporarily.

4. Install suspension control arm ball joint to steering knuckle. Align ball stud groove with steering knuckle bolt hole. Then install ball joint bolt from the front direction. Tighten suspension arm ball joint nut to 43.5 ft. lbs. (60 Nm).

5. Lower hoist and vehicle in non-

loaded condition, tighten control arm mounting bolts to 65 ft. lbs. (90 Nm).

6. Check front wheel alignment.

Verona

The lower control arm and ball joint are a complete unit that will not separate.

1. Before servicing the vehicle, refer to the Precautions Section.

2. Disconnect the negative battery cable.

3. Raise and safely support the front of the vehicle.

4. Remove or disconnect the following:
- Front wheels
- Control arm joint bolt
- Ball joint from the knuckle assembly
- Control arm bracket bolts
- Stabilizer link
- Control arm bolt and control arm assembly
- Control arm bracket from the control arm by removing the nut.

To install:

❋❋ CAUTION

The weight of the vehicle must be supported by the control arms before the stabilizer link-to-control arm nuts or the stabilizer shaft-to-stabilizer link nuts are tightened. This can be done by lowering the vehicle onto jack stands under the control arms have been installed.

5. Install or connect the following:
- Control arm bracket nut and tighten to 81 ft. lbs. (110 Nm).
- Control arm bracket bolts and control arm bolt. Tighten the control arm bracket bolts to 66 ft. lbs. (90 Nm) and control arm bolt to 133 ft. lbs. (180 Nm).
- Install the stabilizer link.
- Install the control arm joint bolt and tighten to 81 ft. lbs. (110 Nm).

6. Install the wheels.

7. Check the front wheel alignment.

CONTROL ARM BUSHING REPLACEMENT

Aerio

1. Before servicing the vehicle, refer to the precautions in the beginning of this section.

2. Remove the lower control arm.

3. Cut the flange from the front bushing.

4. Use a hydraulic press to remove the front bushing.

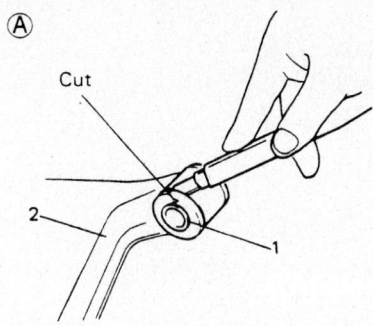

Ⓐ Cut

Ⓑ

1. Front bushing
2. Suspension arm
3. Press
4. Front bushing
5. Suspension arm

9307UG09

Cut the flange from the front bushing, then using a suitable hydraulic press; remove the rear lower control arm bushing

To install:

5. Apply a solution of soapy water to the outer diameter of the front bushing, this will aid in installation.

6. Press the front bushing into its bore using a hydraulic press until the bushing is equal on the right and left of the arm as shown in the accompanying illustration.

Verona

1. Remove the control arm.

2. Remove the control arm rear bushing using a press, the removal plate KM-307-B, and a drift.

3. Remove the control arm front bushing using a press, the remover/installer KM-508-A and the removal plate KM-307-B.

To install:

4. Coat the control arm rear shaft with a multipurpose lubricant.

5. Press the control arm front bushing into the control arm using a press, the remover/installer KM-508-A and the removal plate KM-307-B. Center the bushing.

6. Coat the outside of the control arm front bushing and the inside of the control arm with a multipurpose lubricant.

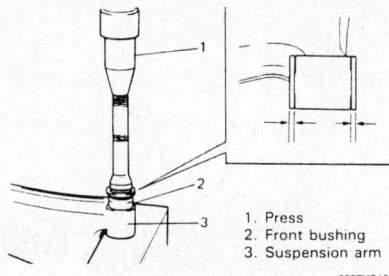

1. Press
2. Front bushing
3. Suspension arm

9307UG10

The front bushing should be positioned equally as shown after being pressed into position

7. Press the control arm front bushing into the control arm using a press, the remover/installer KM-508-A, and the removal plate KM-307-B. Center the bushing.

8. Install the control arm.

Wheel Bearings

ADJUSTMENT

The front wheel bearings are a cartridge type design and cannot be adjusted.

REMOVAL & INSTALLATION

Aerio

➡ **Always replace bearing races as a complete set.**

1. Before servicing the vehicle, refer to the precautions in the beginning of this section.

2. Remove or disconnect the following:
- Front wheel
- Brake caliper, carrier and disc from the steering knuckle
- Wheel speed sensor, if equipped with Anti-Lock Brakes (ABS)
- Tie rod from the steering knuckle
- Hub from the steering knuckle
- Steering knuckle
- Dust cover

➡ **Make note of how dust cover is positioned before removal to ensure that it is installed the exact same position.**

- Circlip
- Wheel bearing, using hydraulic press and special tool 09913-75520

➡ **When installing wheel bearing, replace it with new one.**

- Wheel bearing outside inner race

To install:

3. Face grooved rubber seal side of new wheel bearing upward as shown in figure

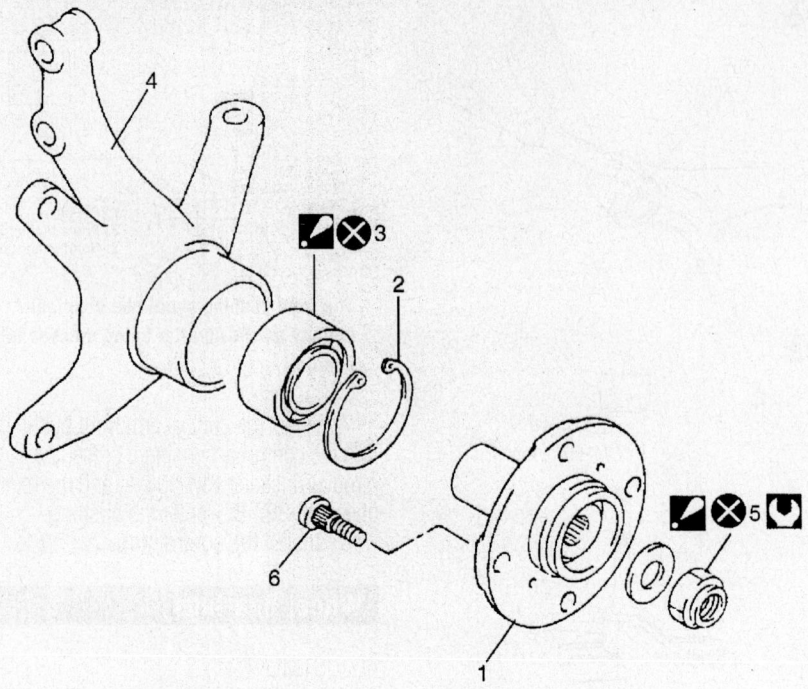

1. Front wheel hub
2. Circlip
3. Wheel bearing
4. Steering knuckle
5. Drive shaft nut
6. Hub bolt

09490_SUZC_G0044

Steering knuckle bearing components

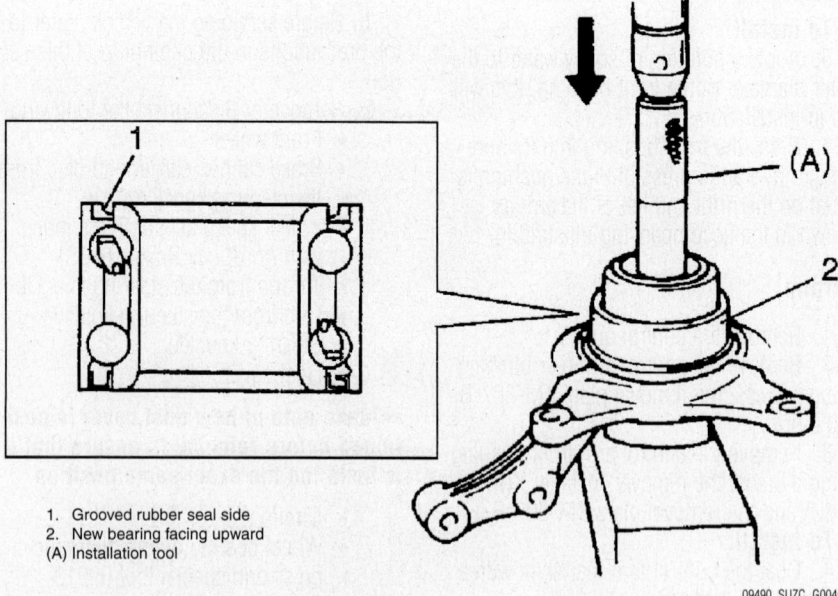

1. Grooved rubber seal side
2. New bearing facing upward
(A) Installation tool

09490_SUZC_G0045

Proper installation of wheel hub bearing

and press-fit it into knuckle using special tool 09913-75510.

4. Install or connect the following:
- Circlip
- Dust cover. Be sure that it is installed in the same position that it was removed from.

➡ **When drive in dust cover, be careful not to deform it.**

- Caulk with a punch
- Steering knuckle in the vehicle
- Tie rod end
- Brake caliper, carrier and disc on the steering knuckle
- Front wheel

Verona

➡ **Always replace bearing races as a complete set.**

1. Before servicing the vehicle, refer to the Precautions Section.
2. Remove or disconnect the following:
- Front wheel
- Caulking nut and washer
- Outer tie rod
- Control arm ball joint
- Brake caliper
- Detent screw from the brake disc
- Brake disc
- Splash shield
- Front strut bolts
- Knuckle, hub assembly
- Hub bolts from knuckle
- Hub from the knuckle

To install:

3. Install or connect the following:
- Hub to knuckle.
- Hub bolts. Tighten hub bolts to 70 ft. lbs. (95 Nm).
- Splash shield and front strut bolts. Tighten the splash shield bolts to 7 ft. lbs. (10 Nm).
- Detent screw and disc. Tighten the detent screw to 3.5 ft. lbs. (5 Nm).
- Brake caliper
- Control arm ball joint bolt and tighten to 81 ft. lbs. (110 Nm)
- Outer tie rod bolt and tighten to 37 ft. lbs. (50 Nm)
- Washer and caulking nut onto the axle shaft. Tighten the caulking nut onto the axle shaft to 236 ft. lbs. (320 Nm).
- Front wheel

REAR SUSPENSION

Strut

REMOVAL & INSTALLATION

Aerio

1. Before servicing the vehicle, refer to the Precautions Section.

→**When servicing component parts of strut assembly, loosen strut upper nut a little before removing strut assembly. This will make service work easier. Note, however, nut must not be removed at this point.**

2. Raise and safely support the front of the vehicle.

3. Remove or disconnect the following:
- Rear wheel
- Stabilizer joint from strut bracket
- E-ring securing brake hose
- ABS wheel speed sensor harness mounting bolt (if equipped)
- Strut bracket bolts and nuts. And then take brake hose off strut bracket using care not to deform brake pipe.

4. Strut support nuts. Hold strut by hand so that it will not fall off.
- Strut assembly

5. Using a spring compressor tool, turn special tool bolts alternately until spring tension is released. Whether it is released or not can be known by whether strut turns lightly while strut spring is held stationary.

→**Use a commercially available spring compressor and follow the operation procedure described in the Instruction Manual supplied with that spring compressor.**

6. While keeping spring compressed, remove strut nut.

7. Disassemble strut assembly.

To install:

8. Assemble the coil spring into the strut assembly as follows:

a. Compress spring with special tool until total length becomes about 11.42 inches (290mm).

b. Install coil spring lower seat and compressing coil spring, and mate spring end with stepped part of lower seat.

c. Install bump stopper and dust cover onto strut rod.

d. Pull strut rod as far up as possible and use care not to allow it to retract into strut.

e. Fit coil spring (rubber) seat in coil spring upper seat, making sure that their depth matches all around. No part of rubber seat should stick out higher than upper seat.

f. With "OUT" mark on spring upper seat and the center of strut bracket aligned, place upper spring seat together with spring (rubber) seat on coil spring. Put strut support on spring upper seat. Tighten strut nut to 50.5 ft. lbs. (70 Nm).

g. Loosen and remove special tool from compressing coil spring.

h. While loosening spring compressor tool, recheck that stepped part of spring seat and spring end are in place to each other.

i. Also, check to make sure that "OUT" mark on upper seat is matched with the center of strut bracket.

9. Install or connect the following:
- Strut assembly. Install the bracket bolts from the front direction and tighten the strut bracket nuts to 65 ft. lbs. (90 Nm). Tighten the strut support nuts to 20.5 ft. lbs. (28 Nm).
- E-ring securing brake hose

→**Do not twist brake hose when installing it. Install E-ring as far as it fits to bracket.**

- Stabilizer joint to strut bracket. Tighten the stabilizer joint nut to 36.5 ft. lbs. (50 Nm).
- ABS wheel speed sensor harness, if equipped
- Wheels

10. Lower the vehicle.

Verona

1. Before servicing the vehicle, refer to the Precautions Section.

a. Strut support nuts
b. Strut bracket nuts
c. Stabilizer joint nut

1. E-ring
2. Bracket

09490_SUZC_G0046

Rear strut assembly mounting

➡When servicing component parts of strut assembly, loosen strut upper nut a little before removing strut assembly. This will make service work easier. Note, however, nut must not be removed at this point.

2. Raise and safely support the front of the vehicle.

3. Remove or disconnect the following:
- Rear wheel
- Rear strut lower nut-to-knuckle
- Rear seat and rear strut upper nut-to-body.
- Rear strut assembly from the vehicle

4. Mount the rear strut assembly into the spring compressor KM-329-A. Ensure that the hooks are properly seated.

5. Compress the spring.

6. Remove the lock nut from the strut dampener rod.

7. Release the coil spring from the compressor and remove the rear strut mount, spring upper insulator, dust cover, hollow bumper, coil spring and rear spring lower insulator.

To install:

8. Install the rear spring lower insulator, coil spring, hollow bumper, dust cover, spring upper insulator and rear strut mount onto the rear strut assembly.

9. Compress the spring.

10. Install the lock nut to the strut damper rod.

11. Release the rear strut assembly from the spring compressor KM-329-A.

12. Install the rear strut with upper nut-to-body and rear seat. Tighten the rear strut upper nuts-to-body to 37 ft. lbs. (50 Nm).

13. Install the rear strut with lower nut-to-knuckle and tighten the rear strut lower nuts-to-knuckle to 66 ft. lbs. (90 Nm).

14. Install the rear wheels.

15. Lower the vehicle.

Coil Spring

REMOVAL & INSTALLATION

For coil spring service on the Aerio and Verona, refer to the strut removal and installation procedure.

Lower Control Arm

REMOVAL & INSTALLATION

Aerio

CONTROL RODS

1. Before servicing the vehicle, refer to the Precautions Section.

2. Raise and safely support the rear of the vehicle.

3. Remove the rear wheels.

4. To facilitate toe adjustment after reinstallation, put match marks on washer and on suspension frame.

5. Remove suspension frame cap (for 2WD model).

6. Remove control rod No.2 and No.1 from suspension frame and knuckle.

To install:

7. Install control rod No. 1, setting it so that its welded nut comes toward the rear.

8. Insert inner bolt and outer bolt from the vehicle front and tighten them temporarily by hand.

9. Install control rod No. 2.

10. Install control rod No. 2, setting it so that its welded nut comes toward the front.

11. Insert control rod No. 2 bolt from the vehicle front and outer bolt from the rear.

12. Install washer with its graduated part facing down.

13. With marks on washer and frame marked before removal aligned to each other, tighten bolts and nut temporarily by hand.

14. Install wheels and tighten wheel nuts to specified torque.

15. Lower hoist and bounce vehicle up and down to stabilize suspension.

➡It is the most desirable to have vehicle off hoist and in non-loaded condition when tightening them. Tighten control rod No.2 inner nut with match marks aligned.

16. Tighten control rod bolts and nuts to specified torque with vehicle weight on suspension.
- Control rod No. 1 (front side) inner nut and outer bolt: 71 ft. lbs. (98 Nm)
- Control rod No. 2 (rear side) outer bolt: 71 ft. lbs. (98 Nm)
- Control rod No. 2 (rear side) inner nut: 65 ft. lbs. (90 Nm)

17. Check rear toe and adjust it as necessary.

18. Install suspension frame cap.

TRAILING ROD

1. Before servicing the vehicle, refer to the Precautions Section.

2. Raise and safely support the rear of the vehicle.

3. Remove the rear wheels.

4. Detach parking brake cable clamp from trailing rod.

5. Remove trailing rod bolts and trailing rod.

To install:

6. Install trailing rod and tighten bolts by hand.

7. Attach parking brake cable clamp to trailing rod.

8. Lower vehicle and bounce vehicle up and down to stabilize suspension.

9. Tighten trailing rod bolts to 71 ft. lbs. (98 Nm) with vehicle weight on suspension.

Verona

FRONT PARALLEL LINK

1. Before servicing the vehicle, refer to the Precautions Section.

2. Raise and safely support the rear of the vehicle.

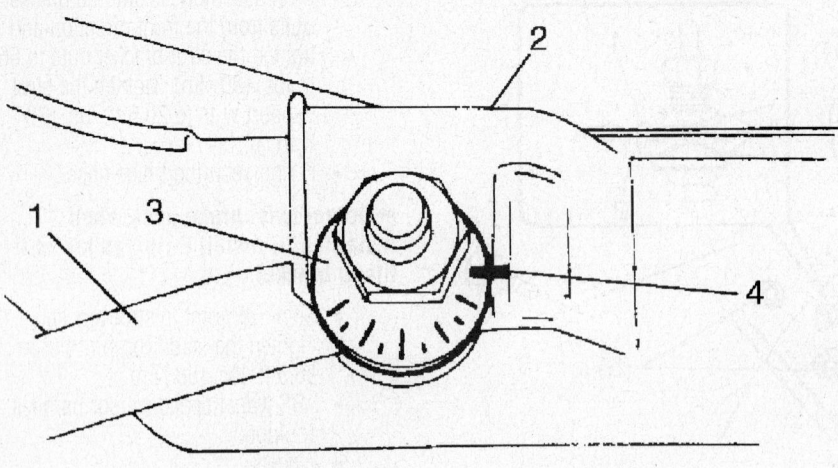

1. Control rod no. 2 (rear side)
2. Suspension frame
3. Washer
4. Match marks

09490_SUZC_G0047

Match marks on washer and on suspension frame for toe adjustment

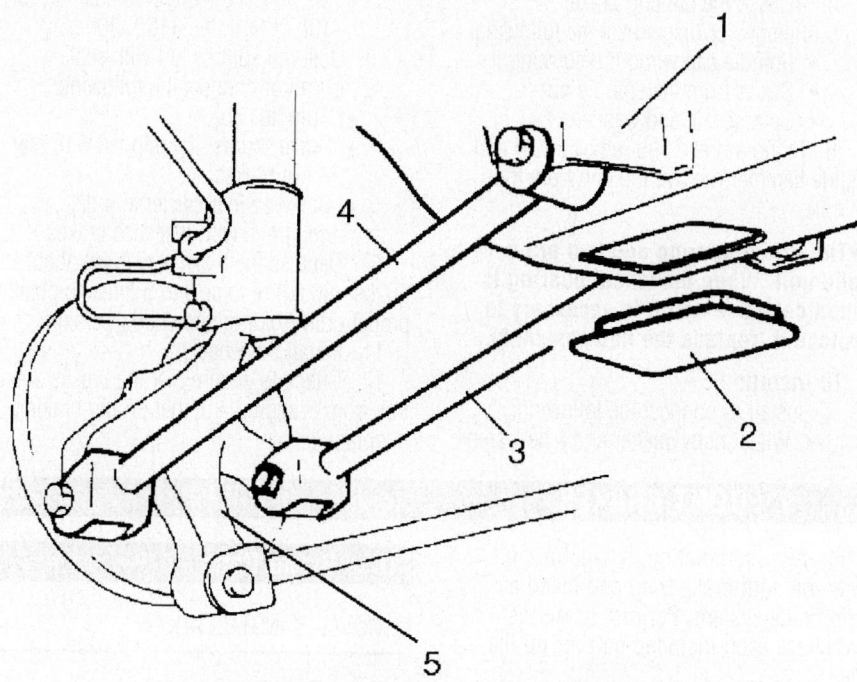

1. Suspension frame
2. Suspension frame cap (2WD)
3. Control rod no. 1
4. Control rod no. 2
5. Knuckle

09490_SUZC_G0048

View of rear suspension components—Aerio

3. Remove the rear wheels.
4. Remove the rear parallel link bolt-to-knuckle and the bolt-to-rear crossmember.
5. Remove the front parallel link.
To install:
6. Install the front parallel link bolt-to-knuckle and the cam bolt-to-rear crossmember.
7. Tighten the front parallel link bolt-to-knuckle to 81 ft. lbs. (110 Nm).
8. Adjust the rear toe, if needed.
9. Tighten the cam bolt to 59 ft. lbs. (80 Nm).
10. Install the rear wheels and lower the vehicle.

REAR PARALLEL LINK

1. Before servicing the vehicle, refer to the Precautions Section.
2. Raise and safely support the rear of the vehicle.
3. Remove the rear wheels.
4. Remove the rear parallel link.
5. Remove the stabilizer shaft-to-stabilizer link nut.
To install:
6. Install the rear parallel link with bolt-to-knuckle and the bolt-to-rear crossmember.
7. Tighten the rear parallel link bolt-to-knuckle and the bolt-to-crossmember to 81 ft. lbs. (110 Nm).

8. Install the rear wheels and lower the vehicle.

TRAILING LINK

1. Before servicing the vehicle, refer to the Precautions Section.
2. Raise and safely support the rear of the vehicle.
3. Remove the rear wheels.
4. Remove the trailing link bolt-to-knuckle and the trailing link bracket bolts.
5. Remove the trailing link.
To install:
6. Install the trailing link with bolt-to-knuckle and the bolts-to-trailing link bracket.
7. Tighten the trailing link bolt-to-knuckle to 89 ft. lbs. (120 Nm) and the trailing link bracket bolts to 66 ft. lbs. (90 Nm).
8. Install the rear wheels and lower the vehicle.

UPPER CONTROL ARM

1. Before servicing the vehicle, refer to the Precautions Section.
2. Raise and safely support the rear of the vehicle.
3. Remove the rear wheels.
4. Remove the upper arm nut-to-rear knuckle.
5. Remove the nuts upper arm-to-body.
6. Remove the upper arm.

To install:
7. Install the upper arm with nuts-to-body.
8. Tighten the upper arm nuts-to-body to 44 ft. lbs. (60 Nm).
9. Install the upper arm nut-to-rear knuckle.
10. Tighten the upper arm nut-to-rear knuckle to 81 ft. lbs. (110 Nm).
11. Install the rear wheels and lower the vehicle.

Wheel Bearings

ADJUSTMENT

The rear wheel bearings are a cartridge type design and cannot be adjusted.

REMOVAL & INSTALLATION

With Wheel Hubs

1. Before servicing the vehicle, refer to the Precautions Section.
2. Set the parking brake.
3. Remove or disconnect the following:
   • Rear wheels
   • Rear brake drums
4. Use a brass drift and knock the wheel bearings from the drum assembly.
To install:
5. Position the inner wheel bearing on the drum with the sealed side facing out. Using a rear wheel Bearing Installer 09913–76010, install the rear wheel bearing. Install the wheel bearing spacer into the drum.
6. Install or connect the following:
   • Outer wheel bearing with the sealed side facing out, using a wheel bearing installer
7. Fill the space in the brake drum in between the wheel bearings to about 40% capacity with wheel bearing grease.
8. Install or connect the following:
   • Brake drum
   • Spindle washer and a new spindle nut. Tighten the spindle nut to 58–86 ft. lbs. (80–120 Nm).
9. Coat the spindle nut and the spindle dust cap with sealer.

➡When installing the spindle cap, hammer lightly several times on the collar of the cap until the collar comes closely into contact with the brake drum. If the fitting part of the cap is deformed or damaged or if it fits loose, replace the cap with a new one.

10. Depress the brake pedal with about 66 lbs. (30 kg) of force 3 to 5 times to obtain proper drum to shoe clearance.

11. Install the wheels.

12. Check to ensure that the brake drum is free from dragging and proper braking is obtained.

Without Wheel Hubs

1. Before servicing the vehicle, refer to the Precautions Section.

2. Set the parking brake.

3. Remove or disconnect the following:
- Rear wheels
- Brake drum, if equipped with drum brakes
- Caliper, carrier and disc, if equipped with rear disc brakes

4. Release the parking brake.

5. Remove or disconnect the following:
- Spindle cap without deforming it
- Sealer from the spindle nut
- Spindle nut and washer

6. Using a brake hub removal tool and a slide hammer remove the hub from the spindle.

➡**The wheel bearing and hub are a solid unit. When the wheel bearing is found defective and it is necessary to replace it, replace the hub assembly.**

To install:

7. Install or connect the following:
- Wheel hub, washer and a new spin-

dle nut. Tighten the spindle nut to 108–144 ft. lbs. (150–200 Nm).

8. Coat the spindle nut with sealer

9. Install or connect the following:
- Spindle cap
- Brake drums, if equipped with rear drum brakes
- Brake caliper carrier and disc, if equipped with rear disc brakes

10. Depress the brake pedal with about 66 lbs. (30 kg) of force 3 to 5 times to obtain proper drum/rotor to shoe/pad clearance.

11. Install the wheels.

12. Check to ensure that the brakes are free from dragging and that proper braking is obtained.

FRONT BRAKES

Brake Caliper

REMOVAL & INSTALLATION

Aerio

1. Before servicing the vehicle, refer to the Precautions Section.

2. Remove the wheels.

3. Loosen flexible hose joint bolt a little at caliper.

➡**Be careful not to twist flexible hose while loosening the bolt.**

4. Remove caliper pin bolts.

5. Remove caliper from caliper carrier.

6. Disconnect flexible hose from caliper using care not to twist it. As this will allow brake fluid to flow out of flexible hose, have a container ready beforehand.

To install:

7. Apply grease to slide pin, then install caliper to caliper carrier.

8. Torque caliper pin bolts to 16 ft. lbs. (22 Nm).

9. Connect caliper to flexible hose.

10. Torque flexible hose joint bolt to 17 ft. lbs. (23 Nm).

❊❊ WARNING

Make sure that flexible hose is not twisted when tightening joint bolt. If it is twisted, reconnect it using care not to twist it.

11. Tighten bleeder plug.

12. Tighten wheel temporarily and lower lift.

13. Tighten wheel nuts to 61.5 ft. lbs. (85 Nm).

14. After completing installation, fill reservoir with brake fluid and bleed air from brake system. Perform brake test and check each installed part for oil leakage.

Verona

1. Before servicing the vehicle, refer to the Precautions Section.

2. Raise and safely support the rear of the vehicle.

3. To preserve wheel balance, mark the position of the front wheel relative to the wheel hub.

4. Remove the wheel.

5. Remove the caliper brake hose inlet bolt. Remove the ring seals.

6. To prevent fluid loss or contamination, plug openings at the caliper inlet and the brake hose.

7. Remove the caliper mounting bolts.

8. Remove the caliper.

To install:

9. Install the caliper

10. Install the caliper mounting bolts. Tighten the caliper mounting bolts to 19 ft. lbs. (27 Nm).

11. Install the brake hose to the caliper with the bolt and the ring seals. Tighten the caliper brake hose inlet bolt to 30 ft. lbs. (40 Nm).

12. Align the marks that were made when removing the front wheel, and install the wheel.

13. Lower the vehicle.

14. Fill the master cylinder reservoir to the proper level with clean brake fluid.

15. Bleed the air out of the brake system.

Disc Brake Pads

REMOVAL & INSTALLATION

Aerio

1. Before servicing the vehicle, refer to the Precautions Section.

2. Remove the wheels.

3. Remove the brake caliper.

4. Remove the brake pads.

To install:

5. Before installing brake pad shim, apply small amount of grease (included in spare parts) to mating surfaces of brake pad and pad shim.

6. Set brake pad springs and shim and install brake pads.

7. Install brake caliper.

8. Tighten wheel temporarily and lower lift.

9. Tighten wheel nuts to 61.5 ft. lbs. (85 Nm).

10. After completion of installation, check for brake effectiveness.

Verona

1. Before servicing the vehicle, refer to the Precautions Section.

2. Raise and safely support the rear of the vehicle.

3. To preserve wheel balance, mark the position of the front wheel relative to the wheel hub.

4. Remove the wheel.

5. Remove the lower caliper mounting bolt.

➡**Caliper removal is not necessary to service the brake pads.**

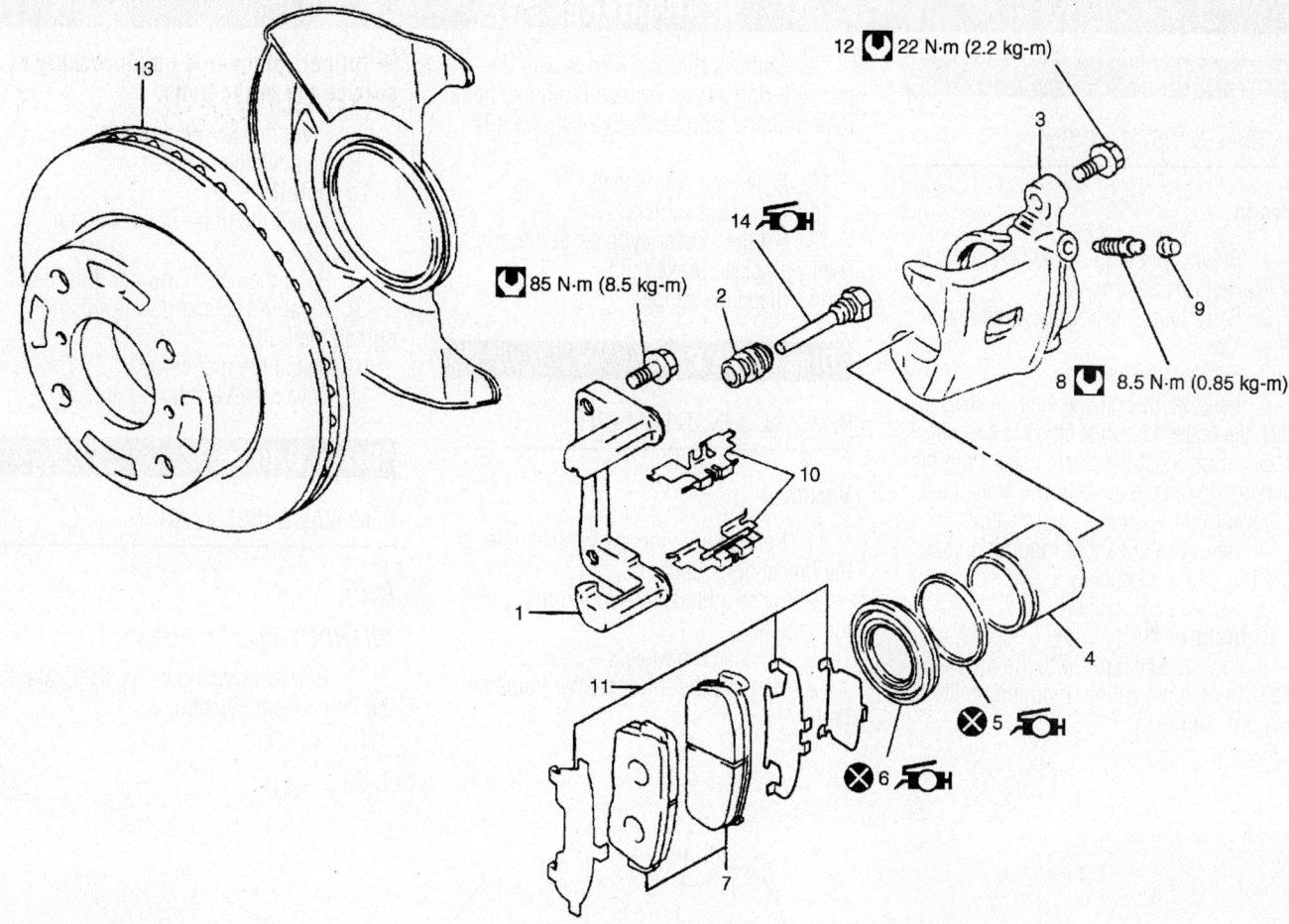

12 ⊍ 22 N·m (2.2 kg-m)

85 N·m (8.5 kg-m)

8 ⊍ 8.5 N·m (0.85 kg-m)

1. Brake caliper carrier
2. Boot
3. Caliper
4. Disc brake piston
5. Piston seal
6. Cylinder boot

7. Brake pad
8. Bleeder plug
9. Bleeder plug cap
10. Pad spring
11. Anti-noise shim
12. Caliper pin bolt

13. Brake disc
14. Slide pin
15. Retaining ring

09490_SUZC_G0049

Front caliper and related components—Aerio

6. Pivot the caliper upward.
7. Remove the brake shoes.
To install:
8. Install the brake shoes.
9. Push the caliper piston inward, if needed.

10. Pivot the caliper downward and install the mounting bolt. Tighten the caliper mounting bolt to 19 ft. lbs. (27 Nm).
11. Tighten the pushrod clevis to 13 ft. lbs. (18 Nm).

12. Align the marks that were made when removing the front wheel, and install the wheel.
13. Lower the vehicle.

REAR BRAKES

Brake Caliper

REMOVAL & INSTALLATION

Verona

1. Before servicing the vehicle, refer to the Precautions Section.
2. Raise and safely support the rear of the vehicle.
3. Remove the wheels.
4. Remove the bolt and the ring seals that attach the brake hose inlet fitting to the caliper.
5. Disconnect the brake hose. Plug the openings in the caliper and the brake hose to prevent fluid loss or contamination.
6. Remove the caliper mounting bolts from the steering knuckle.
7. Remove the caliper.

To install:

8. Install the caliper with the mounting bolts. Tighten the caliper mounting bolts to 33 ft. lbs. (45 Nm).

9. Connect the brake hose with the bolt and ring seals. Tighten the brake hose inlet bolt and ring seals to 24 ft. lbs. (32 Nm).
10. Install the rear wheels.
11. Lower the vehicle.
12. Fill the master cylinder to the proper level with clean brake fluid.
13. Bleed the caliper.

Disc Brake Pads

REMOVAL & INSTALLATION

Verona

1. Before servicing the vehicle, refer to the Precautions Section.
2. Raise and safely support the rear of the vehicle.
3. Remove the wheels.
4. Remove the lower caliper guide pin bolt.

➡**Caliper removal is not necessary to service the brake pads.**

5. Pivot the caliper upward.
6. Remove the brake shoes.

To install:

7. Install the brake shoes into the caliper.
8. Push the piston inward, if needed.
9. Pivot the caliper downward and tighten the bolt.
10. Install the rear wheels.
11. Lower the vehicle.

Brake Drums

REMOVAL & INSTALLATION

Aerio

WITHOUT WHEEL HUBS

1. Before servicing the vehicle, refer to the Precautions Section.

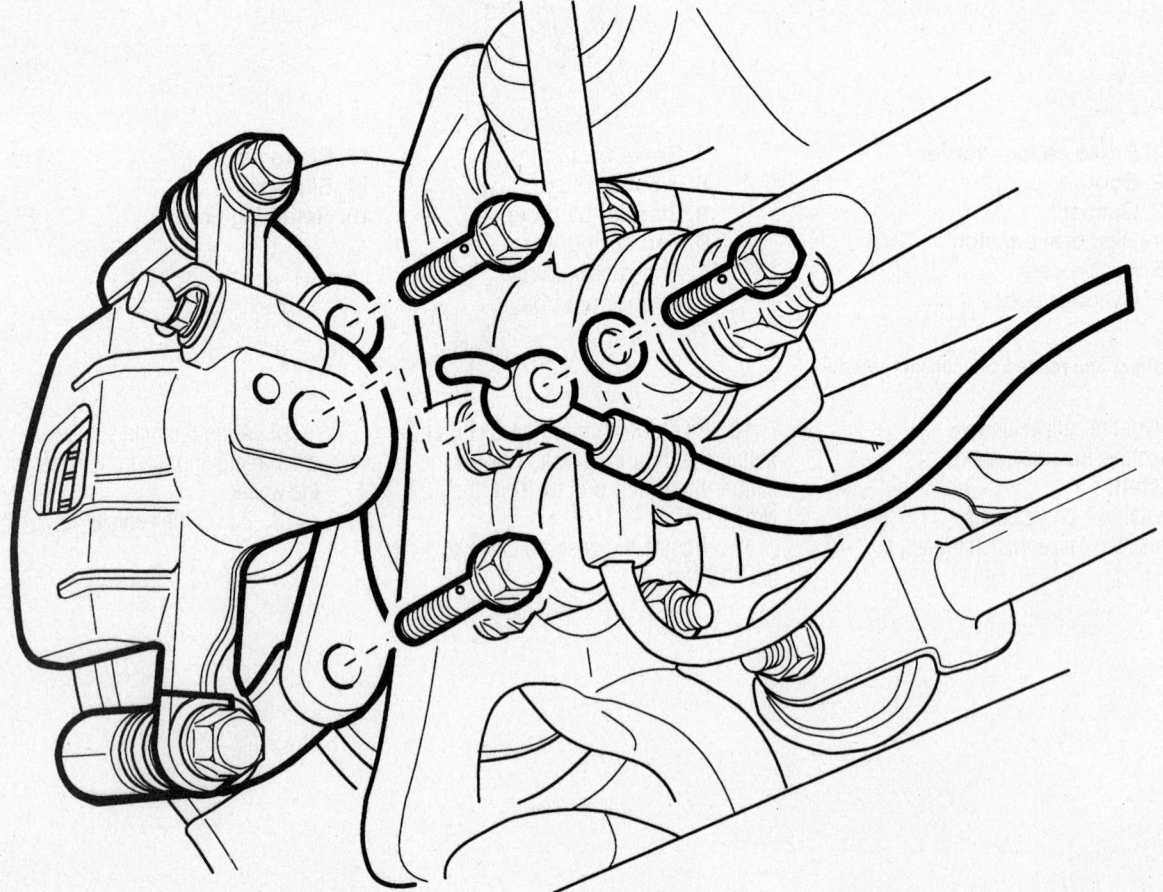

Rear brake caliper removal

09490_SUZC_G0050

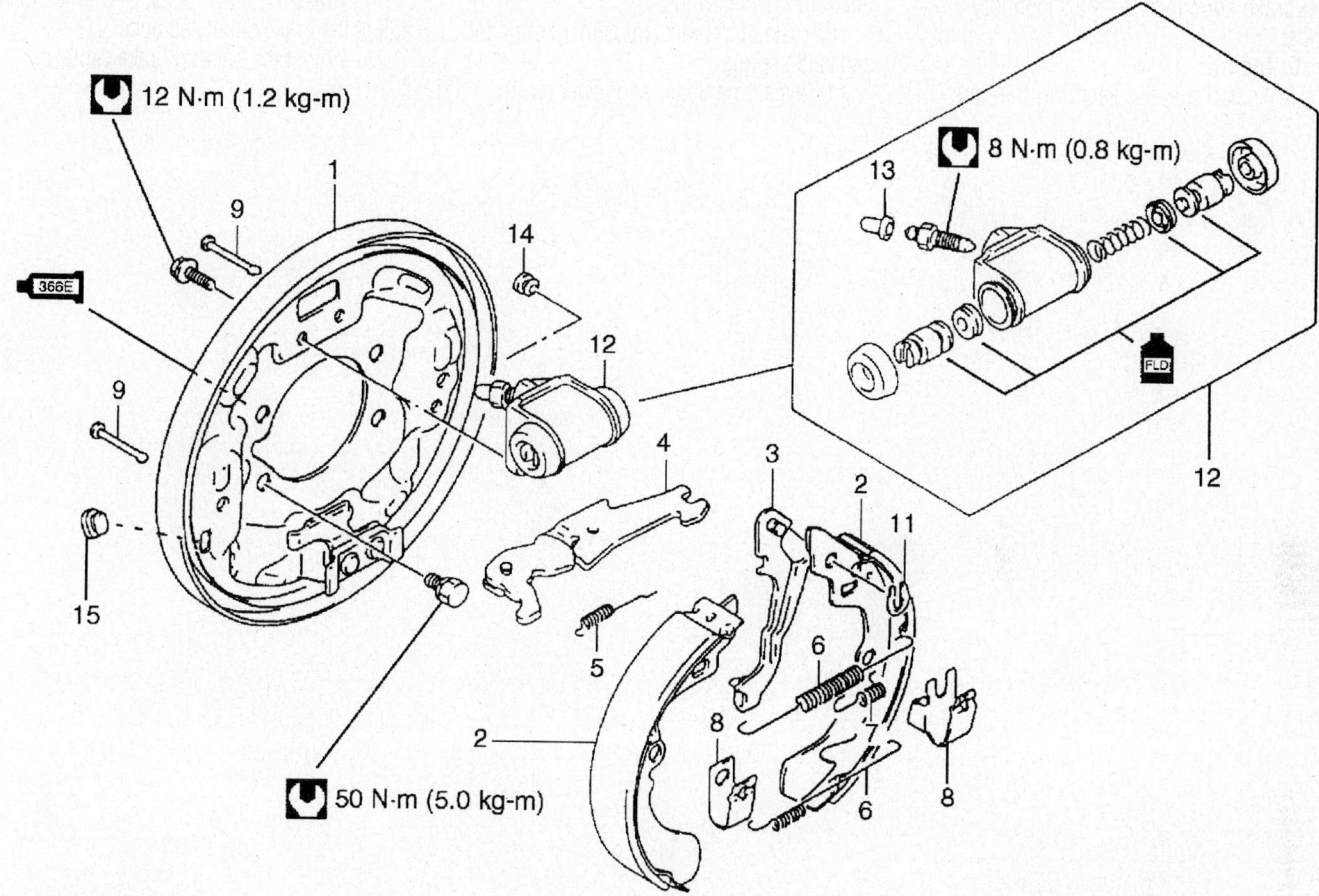

12 N·m (1.2 kg-m)

8 N·m (0.8 kg-m)

50 N·m (5.0 kg-m)

1. Brake back plate
2. Brake shoe
3. Parking brake shoe lever
4. Brake strut
5. Quadrant spring

6. Shoe return spring
7. Anti-rattle spring
8. Shoe hold-down spring
9. Shoe hold-down pin
10. Blank

11. Parking lever retainer
12. Wheel cylinder
13. Bleeder plug cap
14. Rubber plug
15. Rubber plug

09490_SUZC_G0051

Exploded view of rear drum brakes

2. Remove the rear wheels.
3. Remove the spindle dust cap.
4. Remove the sealer from the spindle nut and remove the spindle nut and washer.
5. Install the brake drum removal tool 09943-17911, to the brake drum, then attach a slide hammer to the tool and remove the brake drum.

To install:
6. Install the brake drum.
7. Install the spindle washer and a new spindle nut. Torque the spindle nut to 58–86 ft. lbs. (80–120 Nm).
8. Apply sealer to the spindle nut.

9. Install the spindle dust cap.
10. Install the wheels.

WITH WHEEL HUBS

1. Before servicing the vehicle, refer to the Precautions Section.
2. Remove the rear wheels.
3. Remove the 2 brake drum screws.
4. Pull the brake drum off using two 8mm bolts.

To install:
5. Install the brake drum.
6. Tighten the brake drum screws.
7. Install the wheels.

Brake Shoes

REMOVAL & INSTALLATION

Aerio

1. Before servicing the vehicle, refer to the Precautions Section.
2. Remove the rear wheels.
3. Remove the rear brake drums,
4. Remove the upper and lower springs from the brake shoes.
5. Remove the anti-rattle spring and brake shoe adjustment strut.

6. Remove the primary and secondary brake shoe hold-down springs and remove the shoes from vehicle.

7. Remove the clip securing the parking brake shoe lever to the secondary shoe.

To install:

8. Install the clip securing the parking brake shoe lever to the secondary shoe.

9. Install the primary and secondary brake shoes to the vehicle and secure with the hold-down springs.

10. Install the brake adjustment strut and anti-rattle spring.

11. Install the upper and lower return springs to the primary and secondary brake shoes.

12. Install the brake drum

13. Install the wheels.

14. Press the brake pedal 3–5 times to adjust the brake shoe clearance.

15. Adjust the parking brake cable.

SUZUKI

Forenza • Reno

DRIVE TRAIN 10-20
ENGINE REPAIR 10-8
FRONT BRAKES...............10-31
FRONT SUSPENSION 10-26
FUEL SYSTEM 10-18
REAR BRAKES 10-31
REAR SUSPENSION 10-29
SPECIFICATIONS AND
  MAINTENANCE CHARTS......10-2
STEERING 10-25
Engine and Vehicle Identification
  Chart.............................10-2
General Engine Specifications 10-2
Engine Tune-Up Specifications 10-2
Firing Order 10-3
Accessory Drive Belt Routing 10-3
Capacities 10-3
Valve Specifications.....................10-4
Crankshaft and Connecting Rod
  Specifications 10-4
Piston and Ring Specifications......10-4
Torque Specifications 10-5
Wheel Alignment 10-5
Tire, Wheel and Ball Joint
  Specifications....................10-5
Brake Specifications 10-6
Scheduled Maintenance
  Intervals.........................10-6

A
Air Bag..............................10-25
  Disarming 10-25
  Precautions 10-25
  Rearming 10-25
Alternator 10-8
  Removal & Installation............10-8

B
Brake Caliper (Front) 10-31
  Removal & Installation............10-31
Brake Caliper (Rear)....................10-31
  Removal & Installation............10-31

C
Camshaft and Valve Lifters 10-15
  Removal & Installation............10-15
Clutch................................10-21
  Adjustment..........................10-22
  Bleeding............................10-22
  Removal & Installation............10-21

Coil Spring (Front) 10-27
  Removal & Installation............10-27
Coil Spring (Rear).......................10-29
  Removal & Installation............10-29
CV-Joints.............................10-23
  Overhaul 10-23
Cylinder Head 10-12
  Removal & Installation............10-12

D
Disc Brake Pads (Front)...............10-31
  Removal & Installation............10-31
Disc Brake Pads (Rear)................10-32
  Removal & Installation............10-32
Distributor..............................10-8
  Removal & Installation............10-8

E
Engine Assembly 10-8
  Removal & Installation............10-8
Exhaust Manifold 10-14
  Removal & Installation............10-14

F
Front Crankshaft Seal 10-15
  Removal & Installation............10-15
Fuel Filter 10-18
  Removal & Installation............10-18
Fuel Injector...........................10-19
  Removal & Installation............10-19
Fuel Pump 10-18
  Removal & Installation............10-18
Fuel System Pressure 10-18
  Relieving 10-18
Fuel System Service
  Precautions........................10-18

H
Halfshaft.............................10-23
  Removal & Installation............10-23
Heater Core 10-10
  Removal & Installation............10-10

I
Ignition Timing 10-8
  Adjustment..........................10-8
Intake Manifold 10-13
  Removal & Installation............10-13

L
Lower Ball Joint 10-27
  Removal & Installation............10-27

Lower Control Arm (Front) 10-27
  Control Arm Bushing
    Replacement.....................10-27
  Removal & Installation............10-27
Lower Control Arm (Rear)...........10-29
  Removal & Installation............10-29

O
Oil Pan................................10-16
  Removal & Installation............10-16
Oil Pump 10-16
  Removal & Installation............10-16

P
Piston and Ring 10-17
  Positioning 10-17

R
Rack and Pinion Steering Gear 10-25
  Removal & Installation............10-25
Rear Main Seal 10-17
  Removal & Installation............10-17
Rocker Arms/Shafts 10-13

S
Starter Motor 10-16
  Removal & Installation............10-16
Strut (Front)..........................10-26
  Removal & Installation............10-26
Strut (Rear) 10-29
  Removal & Installation............10-29

T
Timing Belt 10-17
  Removal & Installation............10-17
Transaxle Assembly 10-20
  Removal & Installation............10-20

V
Valve Lash 10-16
  Adjustment..........................10-16

W
Water Pump...........................10-9
  Removal & Installation............10-9
Wheel Bearings (Front)...............10-28
  Adjustment..........................10-28
  Removal & Installation............10-28
Wheel Bearings (Rear)................10-30
  Adjustment..........................10-30
  Removal & Installation............10-31

SPECIFICATIONS AND MAINTENANCE CHARTS

VEHICLE AND ENGINE IDENTIFICATION

		Engine						Model Year	
Code	Liters (cc)	Cu. in.	Cyl.	Fuel Sys.	Engine Type	Eng. Mfg.		Code	Year
Z	2.0 (1998)	121.9	4	MFI	DOHC	Suzuki		3	2003

4	2004
5	2005
6	2006

MFI: Multi-port Fuel Injection

DOHC: Dual Overhead Camshaft

09490_RENO_C0001

GENERAL ENGINE SPECIFICATIONS

Year	Engine ID/VIN	Engine Displacement Liters (cc)	Fuel System Type	Net Horsepower @ rpm	Net Torque @ rpm (ft. lbs.)	Bore x Stroke (in.)	Compression Ratio	Oil Pressure @ rpm
2003	Z	2.0 (1998)	MFI	127@5600	131@4000	3.39x3.39	9.6:1	4.35@ ①
2004	Z	2.0 (1998)	MFI	127@5600	131@4000	3.39x3.39	9.6:1	4.35@ ①
2005	Z	2.0 (1998)	MFI	127@5600	131@4000	3.39x3.39	9.6:1	4.35@ ①
2006	Z	2.0 (1998)	MFI	127@5600	131@4000	3.39x3.39	9.6:1	4.35@ ①

MFI: Multi-port Fuel Injection

① At idle speed

09490_RENO_C0002

GASOLINE ENGINE TUNE-UP SPECIFICATIONS

Year	Engine Displacement Liters (cc)	Engine ID/VIN	Spark Plugs Gap (in.)	Ignition Timing (deg.) MT	Ignition Timing (deg.) AT	Fuel Pump (psi)	Idle Speed (rpm) MT	Idle Speed (rpm) AT	Valve Clearance In.	Valve Clearance Ex.
2003	2.0 (1998)	Z	0.039	NA	NA	55-57 ①	NA	NA	HYD	HYD
2004	2.0 (1998)	Z	0.039	NA	NA	55-57 ①	NA	NA	HYD	HYD
2005	2.0 (1998)	Z	0.039	NA	NA	55-57 ①	NA	NA	HYD	HYD
2006	2.0 (1998)	Z	0.039	NA	NA	55-57 ①	NA	NA	HYD	HYD

Note: The Vehicle Emission Control Information label often reflects specification changes made during production.

The label figures must be used if they differ from those in this chart.

NA: Not Available

HYD: Hydraulic

B: Before top dead center

① At idle

09490_RENO_C0003

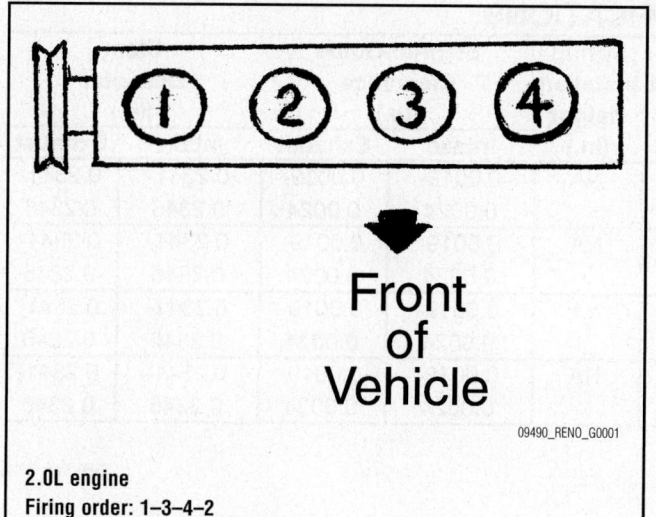

2.0L engine
Firing order: 1–3–4–2
Distributorless ignition system

09490_RENO_G0001

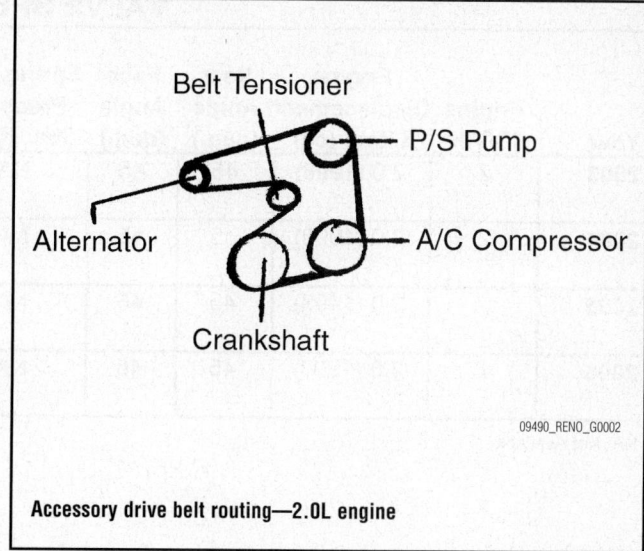

Accessory drive belt routing—2.0L engine

09490_RENO_G0002

CAPACITIES

Year	Model	Engine Displacement Liters (cc)	Engine ID/VIN	Engine Oil with Filter (qts.)	Transmission (pts.)		Fuel Tank (gal.)	Cooling System (qts.)
					5-Spd	Auto.		
2003	Forenza	2.0 (1998)	Z	4.2	3.8	14.6 ①	14.5	7.9
2004	Forenza	2.0 (1998)	Z	4.2	3.8	14.6 ①	14.5	7.9
2005	Forenza	2.0 (1998)	Z	4.2	3.8	14.6 ①	14.5	7.9
	Reno	2.0 (1998)	Z	4.2	3.8	14.6 ①	14.5	7.9
2006	Forenza	2.0 (1998)	Z	4.2	3.8	14.6 ①	14.5	7.9
	Reno	2.0 (1998)	Z	4.2	3.8	14.6 ①	14.5	7.9

Note: All capacities are approximate. Add fluid gradualy and check to be sure a proper fluid level is obtained.

① Specification for automatic transaxle is after complete overhaul. Drain and fill will be less

09490_RENO_C0004

VALVE SPECIFICATIONS

Year	Engine ID/VIN	Engine Displacement Liters (cc)	Seat Angle (deg.)	Face Angle (deg.)	Spring Test Pressure (lbs. @ in.)	Spring Installed Height (in.)	Stem-to-Guide Clearance (in.)		Stem Diameter (in.)	
							Intake	Exhaust	Intake	Exhaust
2003	Z	2.0 (1998)	45	45	NA	NA	0.0019-0.0024	0.0019-0.0024	0.2341-0.2346	0.2341-0.2346
2004	Z	2.0 (1998)	45	45	NA	NA	0.0019-0.0024	0.0019-0.0024	0.2341-0.2346	0.2341-0.2346
2005	Z	2.0 (1998)	45	45	NA	NA	0.0019-0.0024	0.0019-0.0024	0.2341-0.2346	0.2341-0.2346
2006	Z	2.0 (1998)	45	45	NA	NA	0.0019-0.0024	0.0019-0.0024	0.2341-0.2346	0.2341-0.2346

NA: Not Available

09490_RENO_C0005

CRANKSHAFT AND CONNECTING ROD SPECIFICATIONS

All measurements are given in inches.

Year	Engine ID/VIN	Engine Displacement Liters (cc)	Crankshaft				Connecting Rod		
			Main Brg. Journal Dia.	Main Brg. Oil Clearance	Shaft End-play	Thrust on No.	Journal Diameter	Oil Clearance	Side Clearance
2003	Z	2.0 (1998)	2.2824-2.2832	0.0006-0.0016	0.0027-0.0118	NA	1.9279-1.9287	NA	NA
2004	Z	2.0 (1998)	2.2824-2.2832	0.0006-0.0016	0.0027-0.0118	NA	1.9279-1.9287	NA	NA
2005	Z	2.0 (1998)	2.2824-2.2832	0.0006-0.0016	0.0027-0.0118	NA	1.9279-1.9287	NA	NA
2006	Z	2.0 (1998)	2.2824-2.2832	0.0006-0.0016	0.0027-0.0118	NA	1.9279-1.9287	NA	NA

NA: Not Available

09490_RENO_C0006

PISTON AND RING SPECIFICATIONS

All measurements are given in inches.

Year	Engine ID/VIN	Engine Displacement Liters (cc)	Piston Clearance	Ring Gap			Ring Side Clearance		
				Top Compression	Bottom Compression	Oil Control	Top Compression	Bottom Compression	Oil Control
2003	Z	2.0 (1998)	0.0004-0.0011	0.0080-0.0140	0.0110-0.0190	0.0100-0.0300	NA	NA	NA
2004	Z	2.0 (1998)	0.0004-0.0011	0.0080-0.0140	0.0110-0.0190	0.0100-0.0300	NA	NA	NA
2005	Z	2.0 (1998)	0.0004-0.0011	0.0080-0.0140	0.0110-0.0190	0.0100-0.0300	NA	NA	NA
2006	Z	2.0 (1998)	0.0004-0.0011	0.0080-0.0140	0.0110-0.0190	0.0100-0.0300	NA	NA	NA

NA: Not Available

09490_RENO_C0007

TORQUE SPECIFICATIONS
All readings in ft. lbs.

Year	Engine ID/VIN	Engine Displacement Liters (cc)	Cylinder Head Bolts	Main Bearing Bolts	Rod Bearing Bolts	Crankshaft Damper Bolts	Flywheel Bolts	Manifold Intake	Manifold Exhaust	Spark Plugs	Lug Nut
2003	Z	2.0 (1998)	①	②	③	15	④	16	16	15	81
2004	Z	2.0 (1998)	①	②	③	15	④	16	16	15	81
2005	Z	2.0 (1998)	①	②	③	15	④	16	16	15	81
2006	Z	2.0 (1998)	①	②	③	15	④	16	16	15	81

① Step 1: 18 ft. lbs. (25 Nm)
  Step 2: turn head bolts 90 degrees
  Step 3: turn head bolts 90 degrees
  Step 4: turn head bolts 90 degrees

② Step 1: 37 ft. lbs. (50 Nm)
  Step 2: turn bearing cap bolts 45 degrees
  Step 3: turn bearing cap bolts 15 degrees

③ Step 1: 26 ft. lbs. (35 Nm)
  Step 2: turn bearing cap bolts 45 degrees
  Step 3: turn bearing cap bolts 15 degrees

④ Step 1: 48 ft. lbs. (65 Nm)
  Step 2: turn bolts 30 degrees
  Step 3: turn bolts 15 degrees

09490_RENO_C0008

WHEEL ALIGNMENT

Year	Model		Caster Range (+/-Deg.)	Caster Preferred Setting (Deg.)	Camber Range (+/-Deg.)	Camber Preferred Setting (Deg.)	Toe-in (in.)
2003	Forenza	F	0.45	+4.00	0.45	-0.33	0+/-0.10
		R	—	—	0.45	-1.00	0.12+/-0.10
2004	Forenza	F	0.45	+4.00	0.45	-0.33	0+/-0.10
		R	—	—	0.45	-1.00	0.12+/-0.10
2005	Forenza	F	0.45	+4.00	0.45	-0.33	0+/-0.10
		R	—	—	0.45	-1.00	0.12+/-0.10
	Reno	F	0.45	+4.00	0.45	-0.33	0+/-0.10
		R	—	—	0.45	-1.00	0.12+/-0.10
2006	Forenza	F	0.45	+4.00	0.45	-0.33	0+/-0.10
		R	—	—	0.45	-1.00	0.12+/-0.10
	Reno	F	0.45	+4.00	0.45	-0.33	0+/-0.10
		R	—	—	0.45	-1.00	0.12+/-0.10

09490_RENO_C0009

TIRE, WHEEL AND BALL JOINT SPECIFICATIONS

Year	Model	OEM Tires Standard	OEM Tires Optional	Tire Pressures (psi) Front	Tire Pressures (psi) Rear	Wheel Size	Ball Joint Inspection
2003	Forenza	P195/55R15	None	30	30	6-J	①
2004	Forenza	P195/55R15	None	30	30	6-J	①
2005	Forenza	P195/55R15	None	30	30	6-J	①
	Reno	P195/55R15	None	30	30	6-J	①
2006	Forenza	P195/55R15	None	30	30	6-J	①
	Reno	P195/55R15	None	30	30	6-J	①

OEM: Original Equipment Manufacturer

PSI: Pounds Per Square Inch

① Replace if any measurable movement is found.

09490_RENO_C0010

BRAKE SPECIFICATIONS
All measurements in inches unless noted

| Year | Model | | | Brake Disc | | Minimum Lining Thickness | | Brake Caliper | |
			Original Thickness	Minimum Thickness	Maximum Runout	Front	Rear	Bracket bolts (ft. lbs.)	Mounting bolts (ft. lbs.)
2003	Forenza	F	NA	NA	0.001	NA	—	70	20
		R	NA	0.390	0.030	—	0.080	41	①
2004	Forenza	F	NA	NA	0.001	NA	—	70	20
		R	NA	0.390	0.030	—	0.080	41	①
2005	Forenza	F	NA	NA	0.001	NA	—	70	20
		R	NA	0.390	0.030	—	0.080	41	①
	Reno	F	NA	NA	0.001	NA	—	70	20
		R	NA	0.390	0.030	—	0.080	41	①
2006	Forenza	F	NA	NA	0.001	NA	—	70	20
		R	NA	0.390	0.030	—	0.080	41	①
	Reno	F	NA	NA	0.001	NA	—	70	20
		R	NA	0.390	0.030	—	0.080	41	①

NA: Not Available

① Caliper mounting bolt: 41 ft. lbs. (56 Nm)
Lower caliper mounting bolt: 20 ft. lbs. (27 Nm)

09490_RENO_C0011

SCHEDULED MAINTENANCE INTERVALS
SUZUKI—FORENZA & RENO

TO BE SERVICED	TYPE OF SERVICE	7.5	15	22.5	30	37.5	45	52.5	60	67.5	75	82.5	90	97.5
Engine oil & filter	R	✓	✓	✓	✓	✓	✓	✓	✓	✓	✓	✓	✓	✓
Automatic transmission fluid & filter ①	S/I	✓	✓	✓	✓	✓	✓	✓	✓	✓	✓	✓	✓	✓
Clutch pedal free travel	S/I	✓	✓	✓	✓	✓	✓	✓	✓	✓	✓	✓	✓	✓
Drive axle boots	S/I	✓	✓	✓	✓	✓	✓	✓	✓	✓	✓	✓	✓	✓
Gear shift control lever/shift operation	S/I	✓	✓	✓	✓	✓	✓	✓	✓	✓	✓	✓	✓	✓
Inspect & rotate tires	S/I	✓	✓	✓	✓	✓	✓	✓	✓	✓	✓	✓	✓	✓
Manual transmission oil ②	S/I	✓	✓	✓	✓	✓	✓	✓	✓	✓	✓	✓	✓	✓
Power steering system	S/I	✓	✓	✓	✓	✓	✓	✓	✓	✓	✓	✓	✓	✓
Suspension system	S/I	✓	✓	✓	✓	✓	✓	✓	✓	✓	✓	✓	✓	✓
Brake discs & pads	S/I	✓			✓			✓			✓		✓	✓
Brake hoses, pipes, brake lever & cable	S/I	✓		✓		✓		✓		✓		✓		✓
Brake fluid ③	S/I		✓		✓		✓		✓		✓		✓	
Brake pedal	S/I		✓		✓		✓		✓		✓		✓	
Cooling system, hoses & connections	S/I		✓		✓		✓		✓		✓		✓	
Fuel tank, cap & lines	S/I		✓		✓		✓		✓		✓		✓	
Air cleaner filter element	R				✓				✓				✓	
Engine coolant	R				✓				✓				✓	
Spark plugs	R				✓				✓				✓	
Drive belts	S/I				✓				✓				✓	
Exhaust system	S/I				✓				✓				✓	

09490_RENO_C0012

SCHEDULED MAINTENANCE INTERVALS
SUZUKI—FORENZA & RENO

TO BE SERVICED	TYPE OF SERVICE	VEHICLE MILEAGE INTERVAL (x1000)												
		7.5	15	22.5	30	37.5	45	52.5	60	67.5	75	82.5	90	97.5
Automatic transmission fluid hose	R						✓							
Ignition coils	S/I												✓	

R: Replace   S/I: Service or Inspect

① Replace every 100,000 miles.

② Replace every 15,000 miles.

③ Replace every 60,000 miles.

FREQUENT OPERATION MAINTENANCE (SEVERE SERVICE)

If a vehicle is operated under any of the following conditions it is considered severe service:

- Extremely dusty areas.

- 50% or more of the vehicle operation is in 32°C (90°F) or higher temperatures, or constant operation in temperatures below 0°C (32°F).

- Prolonged idling (vehicle operation in stop and go traffic).

- Frequent short running periods (engine does not warm to normal operating temperatures).

- Police, taxi, delivery usage or trailer towing usage.

Oil & oil filter: change every 3000 miles.

Brake discs & pads: service or inspect initially at 3000 miles, 6000 miles, & every 12,000 miles thereafter.

Brake hoses & pipes: service or inspect initially at 3000 miles, 6000 miles & every 12,000 miles thereafter.

Air cleaner filter element: service or inspect ever 3000 miles & replace every 30,000 miles (if not replaced previously).

Automatic transmission fluid & filter: service or inspect every 6000 miles & replace every 15,000 miles (if not replaced previously).

Clutch pedal free travel: service or inspect every 6000 miles.

Inspect & rotate tires: service or inspect every 6000 miles.

Manual transmission oil: service or inspect every 6000 miles & replace every 12,000 miles (if not replaced previously).

Power steering system: service or inspect every 6000 miles.

Steering system: service or inspect every 6000 miles.

Suspension system: service or inspect every 6000 miles.

Drive belts: service or inspect every 15,000 miles.

Exhaust system: service or inspect every 15,000 miles.

09490_RENO_C0013

ENGINE REPAIR

➡️Disconnecting the negative battery cable on some vehicles may interfere with the functions of the on board computer systems and may require the computer to undergo a relearning process, once the negative battery cable is reconnected.

Distributor

REMOVAL & INSTALLATION

These models utilize a Distributorless Ignition System (DIS). With this system, the Electronic Control Module (ECM) determines proper ignition timing and time for the primary ignition coil circuit to turn **ON** and **OFF**.

Alternator

REMOVAL & INSTALLATION

1. Before servicing the vehicle, refer to the Precautions Section.
2. Disconnect the negative battery cable.
3. Disconnect the manifold air temperature (MAT) sensor electrical connector and remove the air intake tube.
4. Remove all clamps from the air cleaner outlet hose, and set aside the tube.
5. Raise and safely support the front of the vehicle.
6. Disconnect the harness connector from the back of the alternator, and the alternator lead to the battery.
7. Remove the serpentine drive belt by lowering the vehicle and turning the automatic tensioner roller bolt clockwise to relieve tension on the belt.
8. Push up the power steering reservoir and set it aside.
9. Remove the alternator upper mounting bolts to the intake manifold/cylinder head support bracket, the intake manifold strap bracket, and the intake manifold-to-cylinder body strap bracket.
10. Raise and suitably support the vehicle and remove the nut and washers which hold the alternator lower bracket-to-alternator bolt. Work the bolt loose and remove the alternator.
11. Remove the alternator lower support bracket bolts.
12. Carefully remove the alternator with the lower bracket.
13. Remove the alternator lower support bracket nut, the bolt, and the washer.

To install:

14. Install the alternator to the alternator lower bracket and insert the alternator bolt. Tighten the alternator lower bracket-to-alternator nut to 18 ft. lbs. (25 Nm).
15. Install the alternator and the lower support bracket assembly to the engine block. Tighten the alternator and the lower bracket-to-engine block bolts to 27 ft. lbs. (37 Nm).
16. Install the alternator-to-intake manifold and cylinder head support bracket bolts, the alternator-to-intake manifold strap bracket bolt, and the intake manifold-to-cylinder body strap bracket bolts over the starter.
17. Tighten the alternator-to-intake manifold and cylinder head support bracket bolts to 27 ft. lbs. (37 Nm). Tighten the alternator-to-intake manifold strap bracket bolt and the intake manifold-to-cylinder body strap bracket bolts to 16 ft. lbs. (22 Nm).
18. Route the serpentine accessory drive belt.
19. Relieve tension on the belt by first applying downward pressure on the automatic tension roller bolt and releasing pressure once the belt is in place.
20. Install the power steering reservoir.
21. Install the air cleaner outlet hose and connect the MAT electrical connector.
22. Connect the negative battery cable.

Ignition Timing

ADJUSTMENT

Ignition timing is controlled by the Electronic Control Module (ECM). The ECM receives signals from various sensors mounted on the engine. No ignition timing adjustment is possible.

Engine Assembly

REMOVAL & INSTALLATION

✳✳ CAUTION

The fuel injection system remains under pressure after the engine has been turned OFF. Properly relieve fuel pressure before disconnecting any fuel lines. Failure to do so may result in fire or personal injury.

1. Before servicing the vehicle, refer to the Precautions Section.

2. Properly relieve the fuel system pressure.
3. Disconnect the negative battery cable.
4. Remove the fuel pump fuse.
5. Start the engine. After it stalls, crank the engine for 10 seconds to rid the fuel system of fuel pressure.
6. Drain the engine oil.
7. Discharge the air conditioning (A/C) system, if equipped.
8. Drain the engine coolant.
9. Remove or disconnect the following:
   - Hood
   - Manifold air temperature (MAT) sensor connector
   - Air cleaner outlet hose from throttle body and air cleaner housing
   - Breather tubes from the camshaft cover
   - Right front wheel.
   - Right front wheel well splash shield
   - Serpentine accessory drive belt.
   - Cooling system radiator and the engine cooling fans
   - Upper radiator hose from the thermostat housing
   - Power steering return hose from the power steering pump
   - Power steering pressure hose from the power steering pump
   - Electrical connector at the direct ignition system (DIS) coil and the electronic control module (ECM) ground terminal and at the starter motor
   - Oxygen (O2) sensor connector, if equipped
   - Idle air control (IAC) valve connector
   - Throttle position sensor (TPS) connector
   - Engine coolant temperature sensor (CTS) connector
   - Alternator voltage regulator connector and power lead
   - All of the necessary vacuum lines, including the brake booster vacuum hose
   - Fuel return line at the fuel rail
   - Fuel feed line at the fuel rail
   - Fuel rail and injector channel cover as an assembly
   - Throttle cable from the throttle body and the intake manifold bracket
   - Coolant hose at the throttle body
   - Heater outlet hose at the coolant pipe
   - Coolant bypass hose from the cylinder head

- Surge tank coolant hose from the coolant pipe
- Lower radiator hose from the coolant pipe
- Starter solenoid terminal wire and power lead
- A/C compressor
- Exhaust flex pipe retaining nuts from the exhaust manifold studs
- Exhaust flex pipe retaining nuts from the catalytic converter or the connecting pipe
- Exhaust flex pipe
- Crankshaft pulley bolts and crankshaft pulley
- Vacuum lines at the charcoal canister purge solenoid
- Electrical connector at the charcoal canister purge (CCP) and the exhaust gas recirculation (EGR) solenoid
- Electrical connector at the oil pressure switch
- Crankshaft position sensor (CPS) connector
- Knock sensor connector
- Lower reaction rod bracket bolts
- Lower reaction rod bracket
- Lower reaction rod mount bolt
- Lower reaction rod mount
- Rubber cover from service hole
- Transaxle torque converter bolts through the service hole (A/T)
- Transaxle bell housing bolts and the oil pan flange bolts

10. Support the transaxle with a floor jack.
11. Install the engine lifting device.
12. Disconnect the right engine mount bracket from the engine mount by removing the retaining bolt.
13. Remove the right engine mount bracket from the engine block and frame mount.
14. Separate the engine block from the transaxle. Remove the engine.

To install:

15. Install the engine into the engine compartment
16. Align the engine alignment pins to the transaxle.
17. Install the transaxle bell housing bolts and tighten to 55 ft. lbs. (75 Nm).
18. Install the oil pan flange-to-transaxle bolts and tighten to 30 ft. lbs. (40 Nm).
19. Install or connect the following:
- Right engine mount to engine block mount and frame mount
- Right engine mount bracket retaining bolts and nuts. Tighten the engine mount bracket retaining bolts and nuts to 41 ft. lbs. (55 Nm).

20. Remove the floor jack used for support of the transmission.
21. Remove the engine lifting device.
- Transaxle torque converter bolts (A/T) and tighten to 44 ft. lbs. (60 Nm)
- Rubber service hole cover
- Lower reaction rod mount
- Lower reaction rod mount bolt and tighten to 41 ft. lbs. (55 Nm)
- Lower reaction rod bracket
- Lower reaction rod bracket bolts and tighten to 49 ft. lbs. (69 Nm)
- Vacuum lines at the CCP solenoid
- Electrical connector to the CCP and the EGR solenoid
- Oil pressure switch connector
- Crankshaft pulley and crankshaft pulley bolts. Tighten the crankshaft pulley bolts to 15 ft. lbs. (20 Nm) using a torque wrench.
- CPS connector
- Exhaust flex pipe
- Exhaust flex pipe retaining nuts to the exhaust manifold studs and tighten to 26 ft. lbs. (35 Nm)
- Exhaust flex pipe retaining nuts to the catalytic converter or the connecting pipe and tighten to 26 ft. lbs. (35 Nm)
- Power steering pressure hose
- Power steering return hose
- A/C compressor, if equipped
- Serpentine drive belt
- Right front wheel well splash shield
- Right front wheel
- Fuel feed line to the fuel rail
- Fuel return line to fuel rail
- Fuel rail and injector channel cover as an assembly
- All necessary vacuum lines including brake booster vacuum hose
- O2 sensor connector, if equipped
- Starter solenoid terminal wire and power lead
- Alternator voltage regulator connector
- Engine CTS connector
- TPS connector
- IAC valve connector
- MAP sensor connector
- Knock sensor, if necessary
- Electrical connector at the DIS ignition coil and the ECM ground terminal and at the starter motor
- Air cleaner outlet hose between the throttle body and air cleaner housing
- Breather tubes to the camshaft cover
- MAT sensor connector
- Cooling system radiator and engine cooling fans

- Lower radiator hose to the coolant pipe
- Upper radiator hose to the thermostat housing
- Heater inlet hose to the cylinder head
- Heater outlet hose to coolant pipe
- Coolant surge tank hose to coolant pipe
- Coolant hose to throttle body
- Throttle cable to throttle body and intake manifold bracket

22. Install the fuel pump fuse.
23. Connect the negative battery cable.
24. Refill the engine crankcase with engine oil.
25. Refill the engine coolant system.
26. Bleed the power steering system.
27. Recharge the A/C refrigerant system, if equipped.
28. Install the hood.
29. Run the engine and verify that there are no fuel, coolant, transaxle or exhaust leaks.

Water Pump

REMOVAL & INSTALLATION

1. Before servicing the vehicle, refer to the Precautions Section.
2. Disconnect the negative battery cable.
3. Drain the cooling system into a suitable container and tighten the drain plug.
4. Remove or disconnect the following:

- Timing belt
- Timing belt tension roller retaining bolt
- Timing belt tension roller
- Water pump mounting bolts
- Water pump from engine block
- Ring seal from the water pump

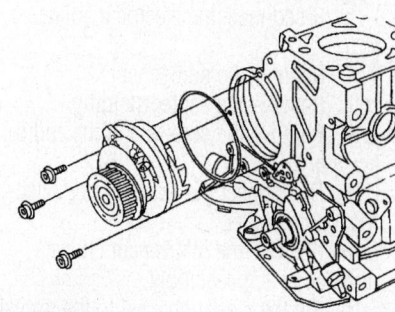

09490_RENO_G0003

Water pump, ring seal and mounting bolts—2.0L engine

To install:

5. Install or connect the following:
- New ring seal to the water pump

6. Coat the sealing surface of the ring seal with Lubriplater®.
- Water pump to engine block with flange aligned with recess of the rear timing belt cover
- Water pump mounting bolts and tighten to 18 ft. lbs. (25 Nm)
- Timing belt tension roller to oil pump with flange inserted into recess of oil pump
- Timing belt tension roller bolt. Do not tighten the bolt at this time.
- Timing belt

7. Refill the engine cooling system.
8. Connect the negative battery cable.
9. Start the engine and top off the coolant as necessary.
10. Check the cooling system for leaks.

Heater Core

REMOVAL & INSTALLATION

1. Before servicing the vehicle, refer to the Precautions Section.
2. Disconnect the negative battery cable and let the vehicle sit for 1 minute to deactivate the air bag.
3. Drain the cooling system.
4. Recover refrigerant from A/C system.
5. Remove the instrument panel as follows:

   a. Tilt the steering wheel to the lowest level.

   b. Remove the screws and the instrument cluster trim panel.

   c. Remove the steering column trim cover.

   d. Remove the A-pillar trim panels.

   e. Remove the screw from the rear portion of the floor console.

   f. Remove the screws from the front portion of the floor console.

   g. Remove the floor console.

   h. Disconnect the electrical connectors.

   i. Remove the sun sensor.

   j. Remove the center molding.

   k. Remove the screws and the audio system.

   l. Disconnect the audio system electrical connectors.

   m. Remove the instrument cluster dimmer switch assembly.

   n. Tilt the steering wheel to the lowest level.

   o. Remove the screws and the instrument cluster trim panel.

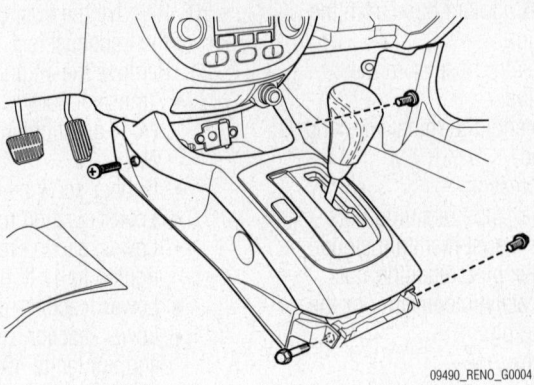

Front floor console mounting screws.

09490_RENO_G0004

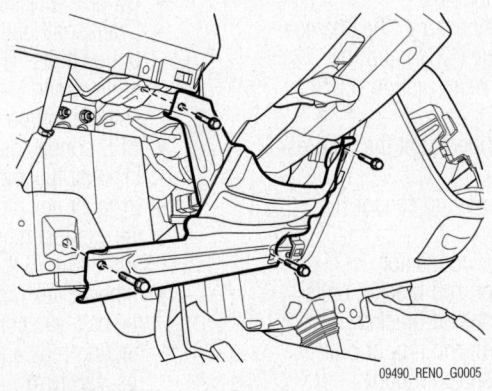

Knee bolster and mounting bolts.

09490_RENO_G0005

   p. Disconnect the electrical connectors.

   q. Remove the screws and the instrument cluster.

   r. Disconnect the electrical connectors.

   s. Remove the screws and the glove box.

   t. Remove the screws and separate the hood release cable from the hood latch release handle.

   u. Carefully pull the instrument panel under cover until the mounting clips are released and the instrument panel under cover.

   v. Remove the bolts from the driver's side knee bolster.

   w. Remove the knee bolster.

   x. Remove the screws and the instrument panel side covers.

   y. Remove the screw and the instrument panel fuse block.

   z. Remove the nuts and the bolts securing the steering column.

   aa. Disconnect the steering column electrical connector.

   bb. Lower the steering column.

   cc. Remove bolts and screws.

   dd. Remove the connecting pieces.

   ee. Remove the bolt securing the middle of the instrument panel to the body.

   ff. Remove the instrument panel screws behind the glove box brace.

   gg. Remove the bolts securing the sides of the instrument panel to body.

   hh. Disconnect the instrument panel electrical connectors.

   ii. Remove the instrument panel.

6. Remove the heater/air distributor case assembly as follows:

   a. Compress the heater hose clamps at the firewall and slide the clamps toward the engine.

   b. Remove the 2 heater hoses from the core pipes at the firewall.

   c. Remove the screws that secure the heater/air distributor case assembly to the firewall on the side of the heater core pipes.

   d. Remove the high-pressure and low-pressure pipe from the A/C expansion valve.

   e. Remove the evaporator drain hose.

   f. Have an assistant support the heater/air distributor case from inside the vehicle.

   g. Remove the screws that secure the heater/air distributor case assembly to the firewall on the side of the evaporator.

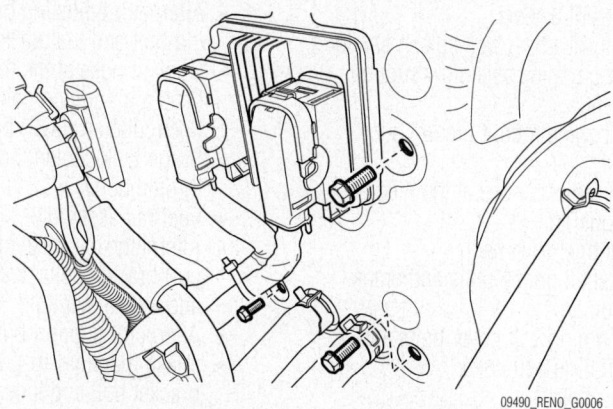

09490_RENO_G0006

Remove the screws that secure heater/air distributor case assembly to firewall on firewall side.

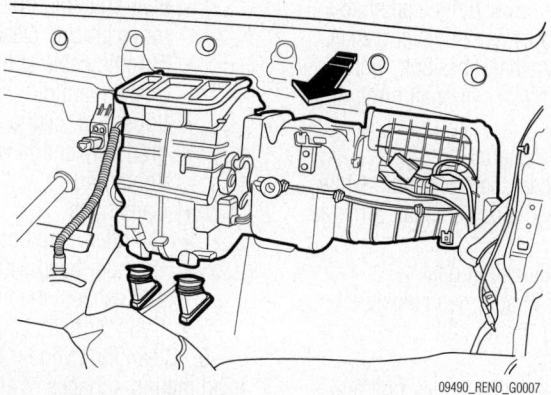

09490_RENO_G0007

Remove the heater/air distributor case assembly from the vehicle.

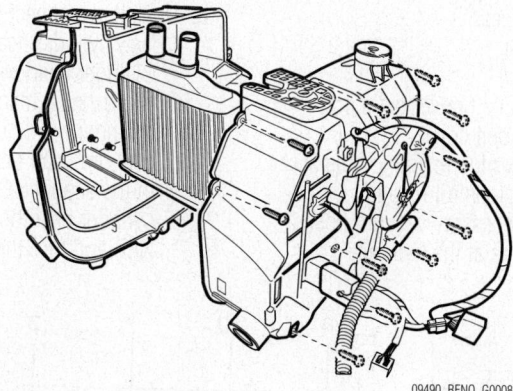

09490_RENO_G0008

Remove the heater core from the housing.

h. Disconnect the rear duct connector.

i. Remove the heater/air distributor case assembly from the vehicle.

7. Remove the wiring harness and electrical connectors from the heater/air distributor case assembly.

8. Remove the screws that connect the heater core housing and the evaporator housing.

9. Separate the sponge from heater/air distributor case assembly.

10. Remove the heater core cover screws from the heater core housing.

11. Remove the heater core.

To install:

12. Install the heater core into the case.

13. Install the screws on the heater core housing and tighten to 11 inch lbs. (1.2 Nm).

14. Install the sponge onto the heater/air distributor case assembly.

15. Install the screws that connect the heater core housing and the evaporator housing. Tighten the heater core housing-to-evaporator housing screws to 11 inch lbs. (1.2 Nm).

16. Connect the wiring harness and electrical connectors.

17. Install the heater/air distributor case assembly as follows:

a. Position the heater/air distributor case assembly in the vehicle.

b. Slowly raise the heater/air distributor case assembly into position against the firewall and hold it in position while the screws are installed and tightened from the engine side of the firewall.

c. Install the high-pressure and low-pressure pipe.

d. Install the heater/air distributor case assembly screws at the side of evaporator through the firewall from the engine compartment side. Tighten the heater/air distributor case assembly screws to 35 inch lbs. (4 Nm).

e. Install the heater/air distributor case assembly screws at the side of the heater core pipes through the firewall from the engine compartment side. Tighten the heater/air distributor case assembly screw to 35 inch lbs. (4 Nm).

f. Install the two heater hoses.

g. Slide the heater hose clamps into position.

h. Connect the rear duct connector.

18. Install the instrument panel as follows:

a. Position the instrument panel in the vehicle.

b. Connect the instrument panel electrical connectors.

c. Install the bolts securing the sides of the instrument panel to the body and tighten to 16 ft. lbs. (22 Nm).

d. Install the instrument panel screws behind the glove box brace.

e. Install the bolts securing the middle of the instrument panel to the body and tighten to 16 ft. lbs. (22 Nm).

f. Install the connecting pieces with bolts and screws. Tighten the connecting pieces bolts to 16 ft. lbs. (22 Nm).

g. Raise the steering column and connect the steering column electrical connector.

h. Install the nuts and the bolts securing the steering column and tighten to 16 ft. lbs. (22 Nm).

i. Install the instrument panel fuse block with the screw.

j. Install the instrument panel side covers with the screws.

k. Install the knee bolster.

l. Install the glove box and housing.

m. Install the instrument cluster and trim panel.

n. Install the instrument cluster dimmer switch assembly.

o. Install the audio systems and the center molding.

p. Install the sun sensor.

q. Install the floor console.

r. Install the A-pillar trim panels.

s. Install the steering column trim cover.

19. Refill the cooling system.

20. Evacuate and recharge the refrigerant.

21. Connect the negative battery cable.

22. Operate the engine to normal operating temperatures; then, check the climate control operation and check for leaks.

Cylinder Head

REMOVAL & INSTALLATION

✸✸ CAUTION

The fuel injection system remains under pressure after the engine has been turned OFF. Properly relieve fuel pressure before disconnecting any fuel lines. Failure to do so may result in fire or personal injury.

1. Before servicing the vehicle, refer to the Precautions Section.

2. Remove the fuel pump fuse.

3. Start the engine. After it stalls, crank the engine for 10 seconds to rid the fuel system of fuel pressure.

4. Disconnect the negative battery cable.

5. Disconnect the Engine Control Module (ECM) ground terminal.

6. Drain the engine coolant.

7. Remove or disconnect the following:
- Manifold Air Temperature (MAT) sensor connector
- Breather tube from the camshaft cover
- Air cleaner outlet hose from the throttle body
- Direct Ignition System (DIS) coil connector
- Oxygen (O2) sensor connector
- Electronic Throttle Control (ETC) connector
- Engine Coolant Temperature (ECT) sensor connector
- Coolant Temperature Sensor (CTS) connector
- Remove the air cleaner housing bolts and air cleaner housing

- Right front wheel
- Right front wheel well splash shield

8. Install the engine assembly support fixture DW110-060.
- Right engine mount bracket and bolts
- Upper radiator hose at the thermostat housing
- Serpentine drive belt
- Crankshaft pulley bolts and crankshaft pulley
- Front timing belt cover bolts and front timing belt cover
- Timing belt
- Breather tube at the camshaft cover
- Spark plug cover bolts and spark plug cover
- Ignition wires from the spark plugs
- Camshaft cover bolts, camshaft cover and camshaft cover gasket
- Intake camshaft gear bolt, while holding intake camshaft firmly in place
- Intake camshaft gear
- Exhaust camshaft gear bolt, while holding exhaust camshaft firmly in place
- Exhaust camshaft gear
- Timing belt automatic tensioner bolts and timing belt automatic tensioner
- Timing belt idler pulley bolt and nut and timing belt idler pulleys
- Engine mount bolts and engine mount
- Crankshaft gear
- CMP sensor
- Water pump
- Rear timing belt cover bolts and rear timing belt cover
- Exhaust flex pipe retaining nuts at the exhaust manifold studs
- All of the necessary vacuum hoses
- Fuel feed line at the fuel rail

- Alternator adjusting bracket retaining bolt and the bracket
- Coolant hose at the rear cylinder head and ignition coil Exhaust Gas Recirculation (EGR) bracket
- Surge tank coolant hose at the throttle body
- Fuel rail assembly
- Alternator-to-intake manifold support bracket bolts at cylinder head and intake manifold
- Alternator support bracket
- Intake manifold-to-generator strap bracket bolt and loosen bolt on alternator
- Move the strap clear of the intake manifold.
- Charcoal Canister Purge (CCP) and EGR solenoid bracket bolt and move bracket clear.
- Throttle cable at throttle body and intake manifold
- Loosen all cylinder head bolts gradually and in reverse order of tightening
- Camshafts
- Cylinder head bolts
- Cylinder head with intake manifold and exhaust manifold attached
- Cylinder head gasket

9. Clean the cylinder block and cylinder head mating surfaces of any old gasket material, oil or dust and clean any engine coolant from the cylinders.

To install:

10. Install or connect the following:
- New cylinder head gasket
- Cylinder head with intake manifold and exhaust manifold attached
- Cylinder head bolts. Tighten the cylinder head bolts gradually and in the sequence shown. Tighten the cylinder head bolts to 18 ft. lbs. (25 Nm) and turn the bolts another 3

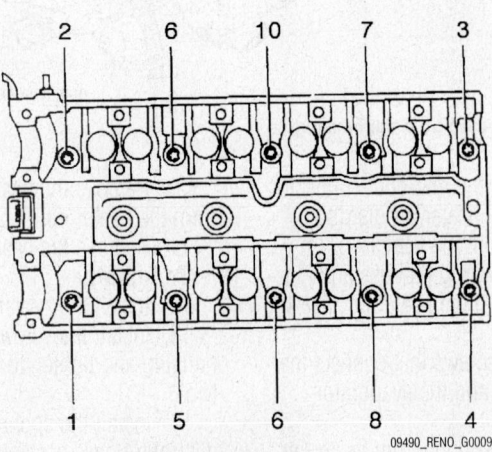

Cylinder head bolt tightening sequence—2.0L engine

09490_RENO_G0009

turns of 90 degrees using the angular torque gauge KM-470-B.
- Camshafts
- Throttle cable at throttle body and intake manifold
- Alternator-to-intake manifold support bracket
- Alternator-to-intake manifold support bracket bolts and tighten to 27 ft. lbs. (37 Nm)
- Intake manifold support bracket bolt to alternator and tighten to 16 ft. lbs. (22 Nm)
- Surge tank coolant hose at throttle body
- Coolant hose to rear cylinder head and ignition coil EGR bracket
- Fuel feed line at the fuel rail
- Fuel return line at the fuel rail
- All necessary vacuum hoses
- Fuel rail assembly
- Exhaust flex pipe retaining nuts at the exhaust manifold studs and tighten to 26 ft. lbs. (35 Nm)
- Rear timing belt cover. Tighten the rear timing belt cover bolts to 62 inch lbs. (7 Nm).
- Engine mount and retaining bolts. Tighten the engine mount retaining bolts to 33 ft. lbs. (45 Nm).
- CMP sensor
- Crankshaft gear
- Water pump
- Timing belt automatic tensioner. Tighten the timing belt automatic tensioner bolts to 18 ft. lbs. (25 Nm).
- Timing belt idler pulleys, bolt and nut. Tighten the timing belt idler pulley bolt to 18 ft. lbs. (25 Nm).
- Camshaft gears with the timing marks at the front
- Guide pin of the intake camshaft into **"IN"** bore
- Guide pin of the exhaust camshaft into **"EX"** bore
- Camshaft gears by counterholding on the hex of the camshaft with an open-ended wrench
- Intake camshaft gear with a new bolt to the camshaft. Tighten the intake camshaft gear bolt to 37 ft. lbs. (50 Nm) turn the bolt another 60 degrees and 15 degrees using the angular torque gauge.
- Exhaust camshaft gear bolt while holding the exhaust camshaft firmly in place. Tighten the exhaust camshaft gear bolt to 37 ft. lbs. (50 Nm) turn the bolt another 60 degrees and 15 degrees using the angular torque gauge.

- Small amount of gasket sealant to the corners of the front camshaft caps and to the top of the rear camshaft cover-to-cylinder head seal
- Camshaft cover, gasket and washers
- Camshaft cover bolts and tighten to 71 inch lbs. (8 Nm)
- Ignition wires to the spark plugs
- Spark plug cover and tighten the spark plug cover bolts to 71 inch lbs. (8 Nm)
- Breather tube to the camshaft cover
- Timing belt
- Front timing belt cover
- Crankshaft pulley. Tighten the crankshaft pulley bolts to 106 inch lbs. (12 Nm).
- Right engine mount bracket. Tighten the engine mount bracket retaining bolts to 41 ft. lbs. (55 Nm).

11. Remove the engine assembly lift support DW110-060.
- Serpentine drive belt
- Upper radiator hose to thermostat housing
- Right front wheel well splash shield
- Right front wheel
- Air cleaner housing
- Air cleaner outlet hose to the throttle body
- Breather tube to the camshaft cover
- MAT sensor connector
- CTS connector
- Engine CTS connector
- ETC connector
- CCP and EGR solenoid bracket bolt and tighten to 44 inch lbs. (5 Nm)
- DIS coil connector
- O2 sensor connector

- ECM ground terminal
- Fuel pump fuse

12. Refill the engine cooling system.
13. Connect the negative battery cable.
14. Start the engine and check for leaks.

Rocker Arms/Shafts

The 2.0L engine does not utilize rocker arms/shafts. On the 2.0L engine, the camshaft directly actuates the valves.

Intake Manifold

REMOVAL & INSTALLATION

✳✳ CAUTION

The fuel system pressure must be relieved before disconnecting any fuel lines. Failure to do so may result in personal injury.

1. Before servicing the vehicle, refer to the Precautions Section.
2. Remove the fuel pump fuse.
3. Start the engine. After it stalls, crank the engine for 10 seconds to rid the fuel system of fuel pressure.
4. Disconnect the negative battery cable.
5. Drain the engine coolant.
6. Remove or disconnect the following:
- Charcoal Canister Purge (CCP) and Exhaust Gas Recirculation (EGR) solenoid from the intake manifold
- Bracket bolt
- Manifold Air Temperature (MAT) sensor connector
- Air cleaner outlet hose from throttle body

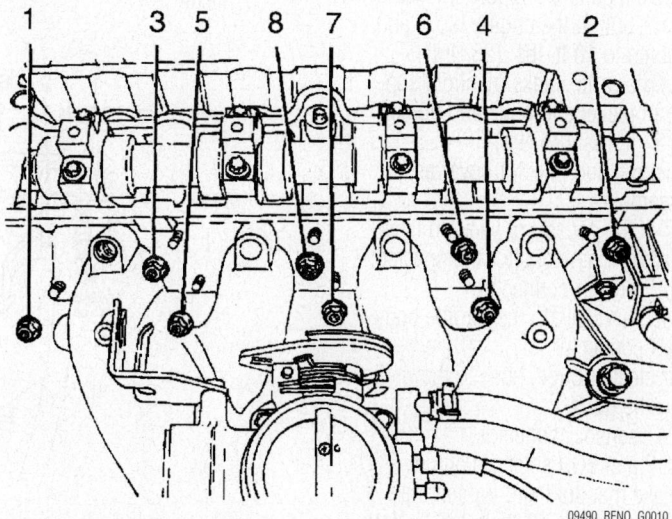

09490_RENO_G0010

Intake manifold bolt loosening sequence—2.0L engine

- Idle Air Control (IAC) valve connector
- Manifold Absolute Pressure (MAP) sensor connector
- Coolant hoses at the throttle body
- All necessary vacuum hoses, including the vacuum hose at the fuel pressure regulator and the brake booster vacuum hose at intake manifold
- Alternator-to-intake manifold strap bracket bolts and strap
- Fuel rail as an assembly
- Alternator-to-intake manifold support bracket bolts
- Alternator-to-intake manifold support bracket
- Intake manifold support bracket bolt at the engine block and the intake manifold
- Intake manifold support bracket
- Intake manifold retaining bolt and nuts in the sequence shown
- Intake manifold and intake manifold gasket

7. Clean the sealing surfaces of the intake manifold and the cylinder head.

To install:

8. Install or connect the following:
- New intake manifold gasket
- Intake manifold, intake manifold retaining bolt and nuts and tighten, in sequence, to 16 ft. lbs. (22 Nm).
- Alternator-to-intake manifold strap bracket and bolts. Tighten the alternator-to-intake manifold strap bracket bolts to 16 ft. lbs. (22 Nm).
- Intake manifold support bracket
- Intake manifold support bracket upper bolts to the intake manifold and tighten to 18 ft. lbs. (25 Nm)
- Intake manifold support bracket lower bolt to the engine block and tighten to 18 ft. lbs. (25 Nm)
- Alternator-to-intake manifold support bracket and bolts and tighten to 27 ft. lbs. (37 Nm)
- Fuel rail and injector cover as an assembly
- All of the necessary vacuum lines that were previously disconnected
- MAP sensor connector
- Coolant hoses to the throttle body
- ETC connector
- Air cleaner outlet hose to the throttle body
- MAT sensor connector
- CCP and EGR solenoid at the intake manifold and tighten the bracket bolt to 44 inch lbs. (5 Nm)
- Fuel pump fuse

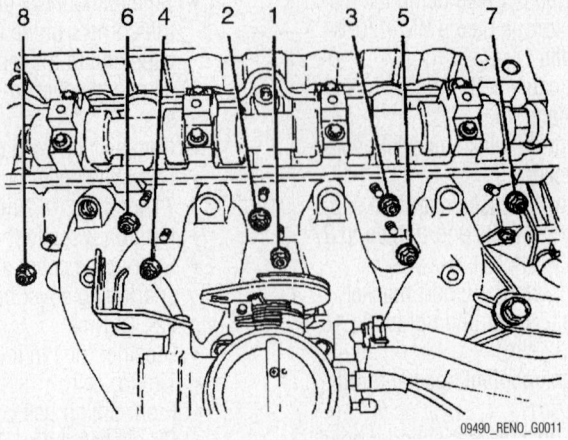

Intake manifold bolt tightening sequence—2.0L engine

9. Check to ensure that all removed parts are back in place.
10. Refill the cooling system.
11. Connect the negative battery cable.
12. Start the engine and check for fuel and cooling system leaks.

Exhaust Manifold

REMOVAL & INSTALLATION

✳✳ CAUTION

To avoid the danger of being burned, do not service the exhaust system while it is hot. Service should be performed only after the system cools down.

1. Before servicing the vehicle, refer to the Precautions Section.
2. Disconnect the negative battery cable.
3. Remove or disconnect the following:
- Heated Oxygen (HO2) sensor connector

- Exhaust manifold heat shield bolts and exhaust manifold heat shield
- Exhaust flex pipe retaining nuts from exhaust manifold studs
- Exhaust manifold retaining nuts in the sequence shown
- Exhaust manifold and gasket

4. Clean the sealing surfaces of the exhaust manifold and the cylinder head.

To install:

5. Install or connect the following:
- New exhaust manifold gasket
- Exhaust manifold and retaining nuts. Tighten the exhaust manifold retaining nuts to 16 ft. lbs. (22 Nm) in the sequence shown.
- Exhaust flex pipe retaining nuts to the exhaust manifold studs and tighten to 26 ft. lbs. (35 Nm)
- Exhaust manifold heat shield and tighten the bolts to 71 inch lbs. (8 Nm)
- O2 sensor connector

6. Connect the negative battery cable.
7. Run the engine and check for exhaust leaks.

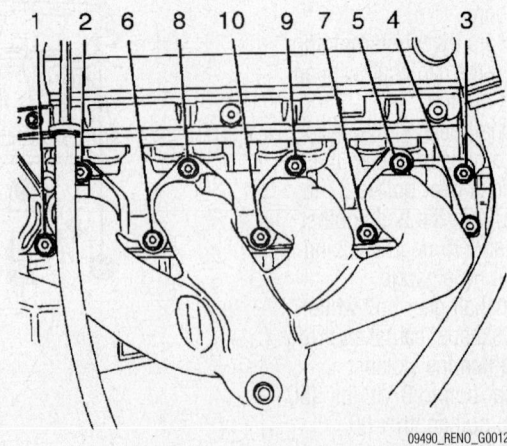

Exhaust manifold retaining nut loosening sequence—2.0L engine

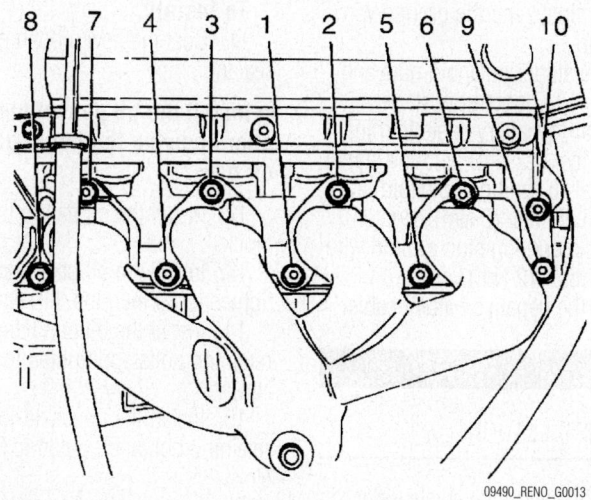

Exhaust manifold retaining nut tightening sequence—2.0L engine

Front Crankshaft Seal

REMOVAL & INSTALLATION

1. Before servicing the vehicle, refer to the Precautions Section.
2. Disconnect the negative battery cable.
3. Drain the engine oil.
4. Remove the timing belt.
5. Remove the oil pump.
6. Remove the front crankshaft seal from the oil pump housing using a suitable prying tool.

❋ CAUTION

Be careful not to damage the oil pump sealing surface when removing or installing the seal.

To install:

7. Install the oil pump.
8. Install a new oil pump-to-crankshaft seal. Coat the lip of the seal with a thin coat of grease. Use an oil seal guide tool.
9. Install the timing belt.
10. Refill engine with engine oil.
11. Connect the negative battery cable.
12. Start the engine and check the engine oil pressure.
13. Check that no leaks are present.

Camshaft and Valve Lifters

REMOVAL & INSTALLATION

1. Before servicing the vehicle, refer to the Precautions Section.
2. Relieve the fuel system pressure.
3. Disconnect the negative battery cable.

4. Remove or disconnect the following:
   • Timing belt
   • Breather tube and engine ventilation hose at the camshaft cover
   • Spark plug cover bolts and spark plug cover
   • Ignition wires from the spark plugs
   • Camshaft cover bolts and camshaft cover washers
   • Camshaft cover and the camshaft cover gasket
   • Intake camshaft gear bolt while holding the intake camshaft firmly in place
   • Intake camshaft gear
   • Exhaust camshaft gear bolt while holding the exhaust camshaft firmly in place
   • Exhaust camshaft gear
5. Loosen the camshaft bearing cap bolts in stages of one-half to one turn.
   • Camshaft bearing cap bolts from the cylinder head

   • Camshaft bearing caps. Maintain the correct positions for installation.
   • Camshafts
   • Seal ring from the camshafts
6. Check the camshaft and bearing seats for wear and replace them if necessary.
7. Remove the valve tappet adjusters. Maintain the correct positions for installation.

To install:

8. Lubricate the valve tappet adjusters with engine oil.
9. Install the valve tappet adjusters.

❋❋ WARNING

Take extreme care to prevent any scratches, nicks or damage to the camshafts.

10. Lubricate the camshaft journals and the camshaft caps with engine oil.
11. Install or connect the following:
    • Intake camshaft
    • Intake camshaft caps in their original positions
    • Intake camshaft cap bolts
    • Exhaust camshaft
    • Exhaust camshaft caps in their original positions
    • Exhaust camshaft cap bolts
    • Camshaft cap bolts gradually and in sequence shown. Tighten the camshaft bearing cap bolts to 71 inch lbs. (8 Nm).
12. Measure the intake camshaft end play and the exhaust camshaft end play. End play should not exceed 0.0015–0.0055 inch (0.040–0.14mm).
    • Intake camshaft gear
    • Intake camshaft gear bolt while

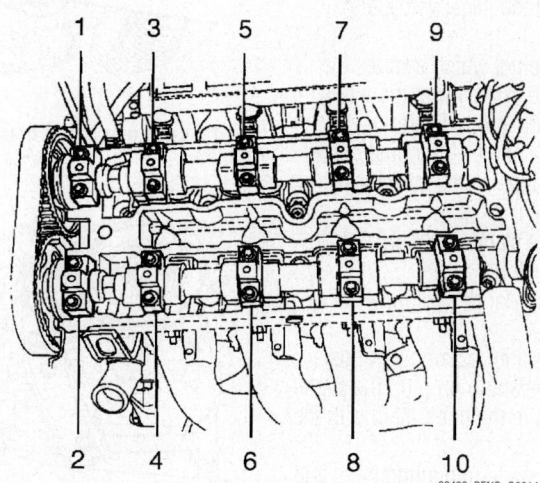

Camshaft cap bolt tightening sequence—2.0L engine

holding the intake camshaft firmly in place. Tighten the intake camshaft gear bolt to 37 ft. lbs. (50 Nm) plus 60 degrees and 15 degrees using the angular torque gauge KM-470-B.

- Exhaust camshaft gear
- Exhaust camshaft gear bolt while holding the exhaust camshaft firmly in place. Tighten the exhaust camshaft gear bolt to 37 ft. lbs. (50 Nm) plus 60 degrees and 15 degrees using the angular torque gauge KM-470-B.
- Camshaft cover and gasket
- Camshaft cover washers and bolts and tighten to 71 inch lbs. (8 Nm)
- Ignition wires to the spark plugs
- Spark plug cover and bolts and tighten to 71 inch lbs. (8 Nm)
- Breather tube and engine ventilation hose to the camshaft cover
- Timing belt
- Negative battery cable

13. Start the engine and check for any leaks when finished.

Valve Lash

ADJUSTMENT

Hydraulic valve lash adjusters are used to adjust the valve clearance to **0** lash automatically at all times. Adjustment is not required.

Starter Motor

REMOVAL & INSTALLATION

1. Before servicing the vehicle, refer to the Precautions Section.
2. Disconnect the negative battery cable.
3. Remove the nut which secures the starter ground wire to the lower mounting stud and remove the ground wire.
4. Remove the starter-to-engine block mounting bolt and the starter-to-transaxle mounting bolt.
5. Remove the starter solenoid nuts to disconnect the electrical cable.
6. Remove the starter assembly.

To install:

7. Place the starter assembly in position using an assistant to prop up the starter to aid in screwing in the upper stud with the weld nut.
8. Install the starter mounting bolts and tighten to 15 ft. lbs. (21 Nm).
9. Position the starter electrical wire on

the solenoid terminals and the ground wire on the lower stud.

10. Install the starter solenoid nuts and the ground wire nut. Tighten the starter solenoid terminal-to-battery cable terminal nut to 106 inch lbs. (12 Nm), and the starter solenoid terminal-to-ignition solenoid terminal nut to 53 inch lbs. (6 Nm). Tighten the starter lower mounting stud ground wire nut to 106 inch lbs. (12 Nm).
11. Connect the negative battery cable.

Oil Pan

REMOVAL & INSTALLATION

1. Before servicing the vehicle, refer to the Precautions Section.
2. Disconnect the negative battery cable.
3. Drain engine oil by removing drain plug.
4. Remove exhaust front pipe.
5. Remove the oil pan flange-to-transaxle retaining bolts.
6. Remove the transaxle-to-oil pan flange retaining bolt.
7. Remove transaxle rear mount bracket.
8. Remove the lower reaction rod.
9. Remove the oil pan retaining bolts.
10. Remove the oil pan from the cylinder block.

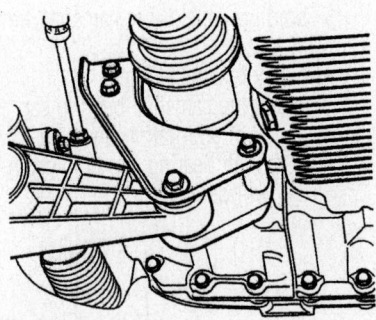

Transaxle rear mount bracket—2.0L

09490_RENO_G0015

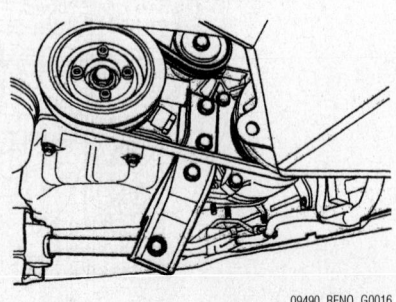

Lower reaction rod—2.0L engine

09490_RENO_G0016

To install:

11. Coat the new oil pan gasket with sealant.

➡️**Install the oil pan within 5 minutes after applying the liquid gasket to the oil pan.**

12. Install the oil pan to the cylinder block.
13. Install the oil pan retaining bolts and tighten to 89 inch lbs. (10 Nm).
14. Install the oil pan flange-to-transaxle retaining bolts and tighten to 30 ft. lbs. (40 Nm).
15. Install the transaxle-to-oil pan flange retaining bolt and tighten to 30 ft. lbs. (40 Nm).
16. Install transaxle rear mount bracket.
17. Install the lower reaction rod and bolts. Tighten the lower reaction rod bolts to 49 ft. lbs. (69 Nm).
18. Install exhaust front pipe and tighten the retaining nuts to 26 ft. lbs. (35 Nm).
19. Refill the engine with oil.
20. Connect the negative battery cable.
21. Start the engine and check for leaks.

Oil Pump

REMOVAL & INSTALLATION

1. Before servicing the vehicle, refer to the Precautions Section.
2. Disconnect the negative battery cable.
3. Drain the engine oil.
4. Remove the timing belt.
5. Remove the rear timing belt cover.
6. Disconnect the oil pressure switch connector.
7. Remove the oil pan.
8. Remove the oil suction pipe and support bracket bolts.
9. Remove the oil suction pipe.
10. Remove the oil pump retaining bolts.

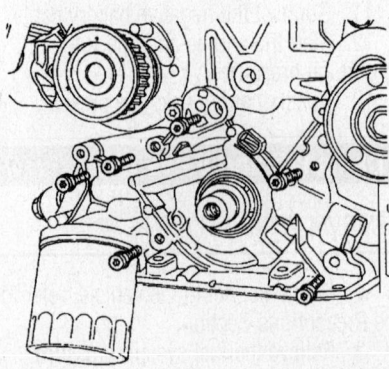

Oil pump and retaining bolts—2.0L engine

09490_RENO_G0017

11. Carefully separate the oil pump and gasket from the engine block and oil pan.

12. Remove the oil pump.

To install:

13. Apply Loctite® 242 to the oil pump bolts and Room Temperature Vulcanizing (RTV) sealant to the new oil pump gasket.

14. Install the gasket to the oil pump and install the oil pump to the engine block with the bolts and tighten to 89 inch lbs. (10 Nm).

15. Install a new oil pump-to-crankshaft seal. Coat the lip of the seal with a thin coat of grease.

16. Coat the threads of the oil suction pipe and support bracket bolts with Loctite® 242.

17. Install the oil suction pipe and the bolts. Tighten the oil suction pipe bolts to 89 inch lbs. (10 Nm) and support bracket bolts to 53 inch lbs. (6 Nm).

18. Install the oil pan.

19. Connect the oil pressure switch connector.

20. Install the rear timing belt cover.

21. Install the timing belt.

22. Refill engine with engine oil.

23. Connect the negative battery cable.

24. Start the engine and check the engine oil pressure.

25. Check that no leaks are present.

Rear Main Seal

REMOVAL & INSTALLATION

1. Before servicing the vehicle, refer to the Precautions Section.

2. Remove or disconnect the following:
   - Engine assembly
   - Flexplate/flywheel from the crankshaft

3. Carefully pry the oil seal out of the retainer without scratching the sealing surface of the crankshaft.

To install:

4. Apply engine oil the lip of the new seal.

5. Install or connect the following:
   - Seal in the retainer using a suitable seal driver
   - Flexplate/flywheel
   - Engine assembly

Timing Belt

REMOVAL & INSTALLATION

1. Before servicing the vehicle, refer to the Precautions Section.

2. Disconnect the negative battery cable.

3. Disconnect the Manifold Air Temperature (MAT) sensor connector.

4. Disconnect the air cleaner outlet hose from the throttle body.

5. Disconnect the breather tube from the camshaft cover.

6. Remove the air cleaner housing bolts.

7. Remove the air cleaner housing.

8. Remove the right front wheel.

9. Remove the right front wheel well splash shield.

10. Remove the serpentine drive belt.

11. Remove the crankshaft pulley bolts and the crankshaft pulley.

12. Remove the right engine mount bracket.

13. Remove the front timing belt cover bolts.

14. Remove the front timing belt cover.

15. Using the crankshaft gear bolt, rotate the crankshaft clockwise until the timing mark on the crankshaft gear is aligned with the notch at the bottom of the rear timing belt cover.

16. Align the camshaft gears with the notch on the camshaft cover.

17. Remove the timing belt.

18. Loosen the automatic tensioner bolt. Turn the hex-key tab to relieve belt tension.

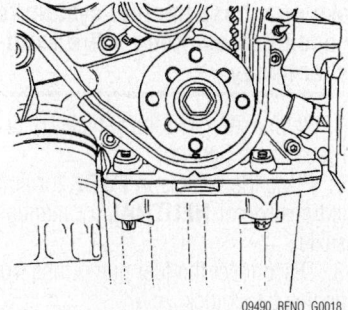

09490_RENO_G0018

Aligning the timing mark on the crankshaft gear with the notch at the bottom of the rear timing belt cover.

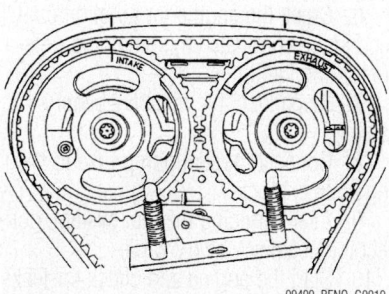

09490_RENO_G0019

Aligning the camshaft gears with the notches on the camshaft cover.

To install:

19. Align the timing mark on the crankshaft gear with the notch on the bottom of the rear timing belt cover.

20. Align the timing marks on the camshaft gears, using the intake gear mark for the intake gear and the exhaust gear mark for the exhaust gear.

21. Install the timing belt.

22. Turn the hex-key tab in a clockwise direction to tension the belt. Turn until the pointer aligns with the notch.

23. Install the automatic tensioner bolt and tighten to 18 ft. lbs. (25 Nm).

24. Rotate the crankshaft two full turns clockwise using the crankshaft pulley bolt.

25. Recheck the automatic tensioner pointer.

26. Install the front timing belt cover.

27. Install the front timing belt cover bolts and tighten to 53 inch lbs. (6 Nm).

28. Install the right engine mount bracket.

29. Install the crankshaft pulley and tighten the crankshaft pulley bolts to 15 ft. lbs. (20 Nm).

30. Install the serpentine drive belt.

31. Install the right front wheel well splash shield.

32. Install the right front wheel.

33. Install the air cleaner housing.

34. Install the air cleaner housing bolts and tighten to 89 inch lbs. (10 Nm).

35. Connect the air cleaner outlet hose to the throttle body.

36. Connect the breather tube to the camshaft cover.

37. Connect the MAT sensor connector.

38. Connect the negative battery cable.

Piston and Ring

POSITIONING

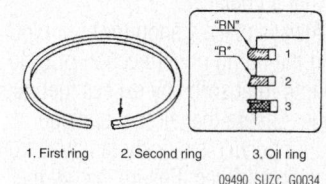

09490_SUZC_G0034

Suzuki engines—piston ring positioning

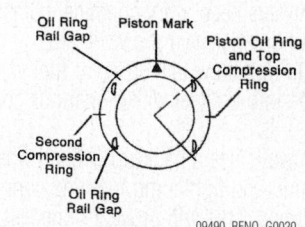

09490_RENO_G0020

Piston ring end-gap spacing—2.0L engine

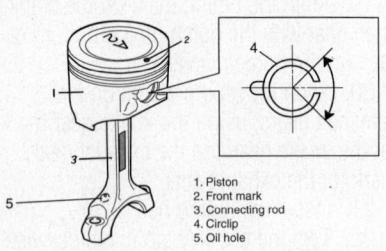

Suzuki engines—piston/connecting rod assembly-to-engine positioning

1. Piston
2. Front mark
3. Connecting rod
4. Circlip
5. Oil hole

09490_SUZC_G0035

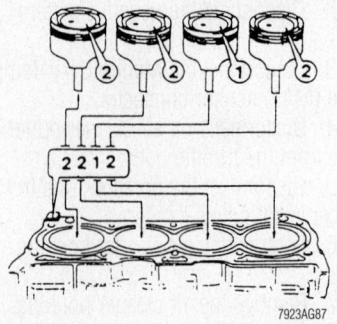

Suzuki engines—the piston ID number must match the number stamped in the engine block

7923AG87

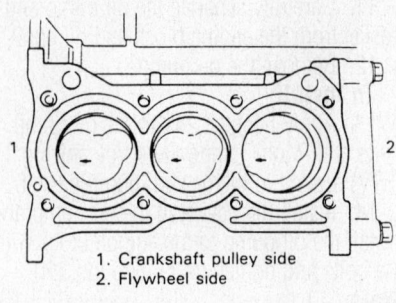

1. Crankshaft pulley side
2. Flywheel side

Suzuki engines—the directional arrow on the piston face must face the crankshaft pulley end of the engine

7923AG86

FUEL SYSTEM

Fuel System Service Precautions

Safety is the most important factor when performing not only fuel system maintenance, but any type of maintenance. Failure to conduct maintenance and repairs in a safe manner may result in serious personal injury or death. Maintenance and testing of the vehicle's fuel system components can be accomplished safely and effectively by adhering to the following rules and guidelines.

• To avoid the possibility of fire and personal injury, always disconnect the negative battery cable unless the repair or test procedure requires that battery voltage be applied.

• Always relieve the fuel system pressure prior to disconnecting any fuel system component (injector, fuel rail, pressure regulator, etc.), fitting or fuel line connection. Exercise extreme caution whenever relieving the fuel system pressure, to avoid exposing your skin, face and eyes to fuel spray. Please be advised that fuel under pressure may penetrate the skin or any part of the body that it contacts.

• Always place a shop towel or cloth around the fitting or connection prior to loosening to absorb any excess fuel due to spillage. Ensure that all fuel spillage (should it occur) is quickly removed from the engine surfaces. Ensure that all fuel soaked cloths or towels are deposited into a suitable waste container.

• Always keep a dry chemical (Class B) fire extinguisher near the work area.

• Do not allow fuel spray or fuel vapors to come into contact with a spark or open flame.

• Always use a back-up wrench when loosening and tightening fuel line connection fittings. This will prevent unnecessary stress and torsion on fuel line piping.

Always follow the proper torque specifications.

• Always replace worn fuel fitting O-rings with new. Do not substitute fuel hose or equivalent, where fuel pipe is installed.

Fuel System Pressure

RELIEVING

✳✳ CAUTION

Care should be used when working around the fuel system. DO NOT smoke or expose the fuel system to any open flames. Keep a fire extinguisher handy.

1. Before servicing the vehicle, refer to the Precautions Section.
2. Place the vehicle in **PARK** for automatic transaxle or **NEUTRAL** for manual transaxle.
3. Disconnect the fuel pump fuse from the engine fuse block.
4. Remove the fuel filler cap from the filler neck to release the fuel vapor pressure in the fuel tank.
5. Start the vehicle and allow the engine to run until it stalls.
6. Crank the engine for an additional 10 seconds to eliminate any remaining pressure in the fuel lines.
7. Disconnect the negative battery cable.
8. Connect the fuel pump fuse to the engine fuse block.
9. After servicing the fuel system, connect the negative battery cable.
10. Start the engine and check for leaks in the system.

Fuel Filter

REMOVAL & INSTALLATION

The fuel filter for Forenza and Reno models is an integral component of the in-tank fuel pump assembly. Refer to the Fuel Pump Removal procedure later in this section.

Fuel Pump

REMOVAL & INSTALLATION

1. Before servicing the vehicle, refer to the Precautions Section.
2. Relieve the pressure from the fuel system.
3. Disconnect the negative battery cable.
4. Remove the rear seat cushion from the floor by lifting it off of the retaining brackets and sliding it forward.
5. Remove or disconnect the following:
   • Fuel pump access cover
   • Electrical connector at fuel pump assembly
   • Fuel line
   • Fuel pump assembly clip
   • Fuel pump assembly from fuel tank

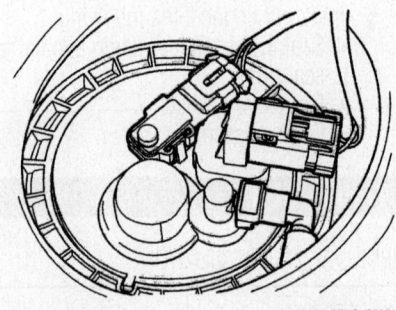

09490_RENO_G0021

Fuel pump assembly location under rear seat—Forenza/Reno

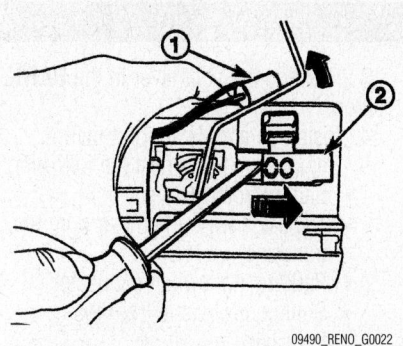

Fuel level sensor (1) and sender housing (2)—Forenza/Reno

6. Remove the fuel sender unit from the fuel pump housing as follows:

a. Disconnect the insulator connector.

b. Push the terminal wedge in the insulator connector.

c. Push the wedge outside and then pull the wires to disconnect from insulator.

d. Remove the fuel-level sensor from the sender housing.

e. Remove the sender housing.

f. Remove the fuel sender assembly.

To install:

✳✳ WARNING

The components of the fuel sender unit must be installed in the same position as removed or it will perform inaccurately.

7. Install the fuel sender unit onto the fuel pump housing as follows:

a. Wind the wires on the sender assembly.

b. Install the sender assembly to the fuel pump assembly.

c. Install the fuel level sensor onto sender housing.

d. Connect the wire into the insulator connector.

e. Connect the insulator connector.

8. Clean the gasket mating surface on the fuel tank.

9. Install or connect the following:
- New fuel pump mounting gasket
- Fuel pump assembly into fuel tank
- Fuel pump assembly clip
- Electrical connector to fuel pump assembly
- Fuel line
- Fuel pump access cover

10. Connect the negative battery cable.

11. Perform operational check of the fuel pump.

12. Install the rear seat cushion by inserting the metal loops into the rear retaining brackets and pressing the front of the seat cushion down.

Fuel Injector

REMOVAL & INSTALLATION

1. Before servicing the vehicle, refer to the Precautions Section.

2. Relieve the pressure from the fuel system.

3. Disconnect the negative battery cable.

4. Remove or disconnect the following:
- Intake Air Temperature (IAT) sensor connector
- Breather hose from cylinder head cover
- Positive Crankcase Ventilation (PCV) hose from the cylinder head cover
- Throttle cable from throttle body and bracket
- Fuel injector electrical connections
- Fuel feed line at the fuel rail
- Fuel rail retaining bolts
- Fuel rail with injectors attached
- Fuel injector retaining clips
- Fuel injectors from rail by pulling down and out
- Injector O-ring and discard

To install:

➡**Different injectors are calibrated for different flow rates. When ordering new fuel injectors, be certain to order**

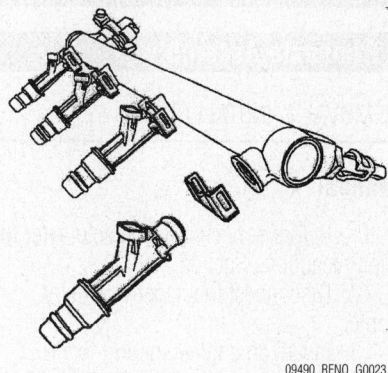

After removing the retaining clip, pull the injector down and out—2.0L engine

the identical part number that is inscribed on the old injector.

5. Lubricate the new injector O-rings with engine oil.

6. Install or connect the following:
- New injector O-ring
- Injector to the fuel rail with terminals facing outward, and secure with retaining clip. Be sure that the clips are parallel to the injector harness connector.
- Fuel injector and fuel rail assembly into the cylinder head
- Fuel rail retaining bolts and tighten to 18 ft. lbs. (25 Nm)
- Fuel feed line to the fuel rail
- Fuel injector electrical connections. Rotate each injector as required.
- Throttle cable to the throttle body and bracket
- PCV hose to the cylinder head cover
- Breather hose to the cylinder head cover
- IAT sensor connector

7. Connect the negative battery cable.

8. With the ignition **ON** and the engine **OFF** check for leaks.

DRIVE TRAIN

Transaxle Assembly

REMOVAL & INSTALLATION

Manual

1. Before servicing the vehicle, refer to the Precautions Section.
2. Disconnect the negative battery cable.
3. Install an engine support fixture.
4. Remove or disconnect the following:
   - Battery from the vehicle
   - Battery tray
   - Shift linkage
   - Drive shaft
   - Backup lamp switch electrical connector
   - Speedometer speed sensor electrical connector
   - Clutch release cylinder pipe by releasing clip
   - Damping block connection nut and bolt
   - Three rear mounting bracket bolts
   - Rear mounting bracket from the transaxle
   - Two rear damping block retaining bolts
   - Rear damping block from the front cross member
   - Two cage retaining bolts
   - Three transaxle upper mounting bracket bolts
   - Upper mounting bracket and cage
   - Three transaxle upper retaining bolts
5. Support transaxle with a transmission jack.
   - Seven transaxle lower retaining bolts
   - Transaxle assembly from vehicle
   To install:
6. Support transaxle with a transmission jack.

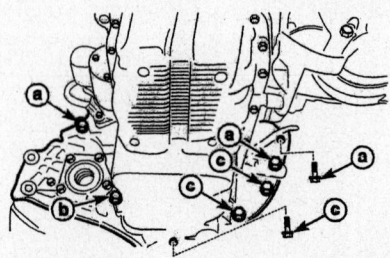

09490_RENO_G0024

Manual transaxle lower retaining bolts—Forenza/Reno

7. Install the transaxle assembly. Use care when inserting the input shaft into the clutch assembly. If the spline on the input shaft does not align with the clutch assembly spline, turn the crankshaft slightly to aid in spline alignment.
8. Install the seven transaxle lower retaining bolts and tighten to the following torque values:
   - Lower retaining bolts "**a**": 54 ft. lbs. (73 Nm)
   - Lower retaining bolts "**b**": 23 ft. lbs. (31 Nm)
   - Lower retaining bolts "**c**": 15 ft. lbs. (21 Nm)
9. Install or connect the following:
   - Three transaxle upper retaining bolts and tighten to 54 ft. lbs. (73 Nm)
   - Cage retaining bolt and cage
   - Three transaxle upper mounting bracket bolts and bracket. Tighten the transaxle upper mounting bracket bolts to 35 ft. lbs. (48 Nm).
   - Two rear damping block retaining bolts and tighten to 50 ft. lbs. (68 Nm).
   - Rear damping block to the front cross member
   - Three rear mounting bracket bolts and the bracket. Tighten the rear mounting bracket bolts to 66 ft. lbs. (90 Nm).
   - Damping block connection nut and bolt and tighten to 50 ft. lbs. (68 Nm).
   - Clutch release cylinder pipe and secure with clip
   - Speedometer speed sensor electrical connector
   - Backup lamp switch electrical connector
   - Drive shaft
   - Shift linkage
   - Battery tray
   - Battery
10. Remove the engine support fixture.
11. Connect the positive battery cable first.
12. Connect the negative battery cable second.
13. Inspect the fluid level in the transaxle and add, if necessary.
14. Check the function of the engine, clutch and transaxle.

Automatic

1. Before servicing the vehicle, refer to the Precautions Section.
2. Disconnect the negative battery cable.

3. Position the shift lever in the **PARK** position.
4. Install an engine support fixture.
5. Remove or disconnect the following:
   - Battery from the vehicle
   - Transaxle wiring harness from the transaxle
   - Park/neutral switch connector
   - Shift control cable from the transaxle
   - Upper transaxle-to-engine bolts
   - Left transaxle mounting bracket
6. Raise and safely support the vehicle.
7. Drain the transaxle fluid.
   - Oil cooler pipes from transaxle
   - Drive shafts
   - Starter
   - Flywheel-to-torque converter bolts
8. Support the transaxle assembly using transaxle support fixture DW260-010.
   - Rear transaxle mounting bracket bolts and damping block connection bolt and nut
   - Lower engine-to-transaxle bolts
   - Transaxle assembly from vehicle
   To install:
9. Install the transaxle assembly into the vehicle.
10. Support the transaxle assembly using transaxle support fixture DW260-010.
11. Install the lower engine-to-transaxle bolts and tighten to the following torque values:
   - Lower engine-to-transaxle bolts "**a**": 55 ft. lbs. (75 Nm)
   - Lower engine-to-transaxle bolt "**b**": 15 ft. lbs. (21 Nm)
   - Lower engine-to-transaxle bolts "**c**": 23 ft. lbs. (31 Nm)
12. Install or connect the following:
   - Rear transaxle mounting bracket bolts and tighten to 45 ft. lbs. (62 Nm)
   - Damping block connection bolt and nut and tighten to 50 ft. lbs. (68 Nm)
   - Torque converter bolts and tighten the bolts to 33 ft. lbs. (45 Nm)

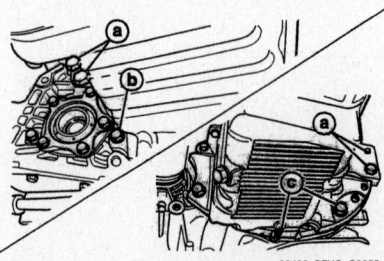

09490_RENO_G0025

Lower engine-to-transaxle bolts—Forenza/Reno

a
b
09490_RENO_G0026

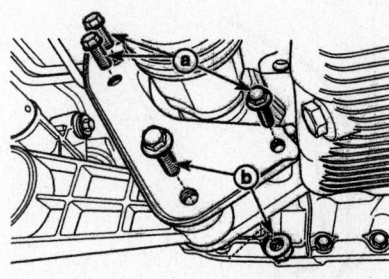

Rear transaxle mounting bracket bolts (a) and damping block connection bolt and nut (b)—Forenza/Reno

- Starter
- Drive shafts
- Oil cooler pipes to the transaxle

13. Lower the vehicle.
    - Left transaxle mounting bracket and tighten the bolts to 35 ft. lbs. (48 Nm)
    - Upper transaxle-to-engine bolts and tighten to 55 ft. lbs. (75 Nm)
    - Clip to shift control cable and shift cable to the transaxle
    - Park/neutral switch connector
    - Transaxle wiring harness to the transaxle

14. Remove the engine support fixture
15. Install the battery and connect the positive battery cable first.
16. Fill the transaxle with the correct amount and type of fluid.
17. Connect the negative battery cable.

Clutch

REMOVAL & INSTALLATION

1. Before servicing the vehicle, refer to the Precautions Section.

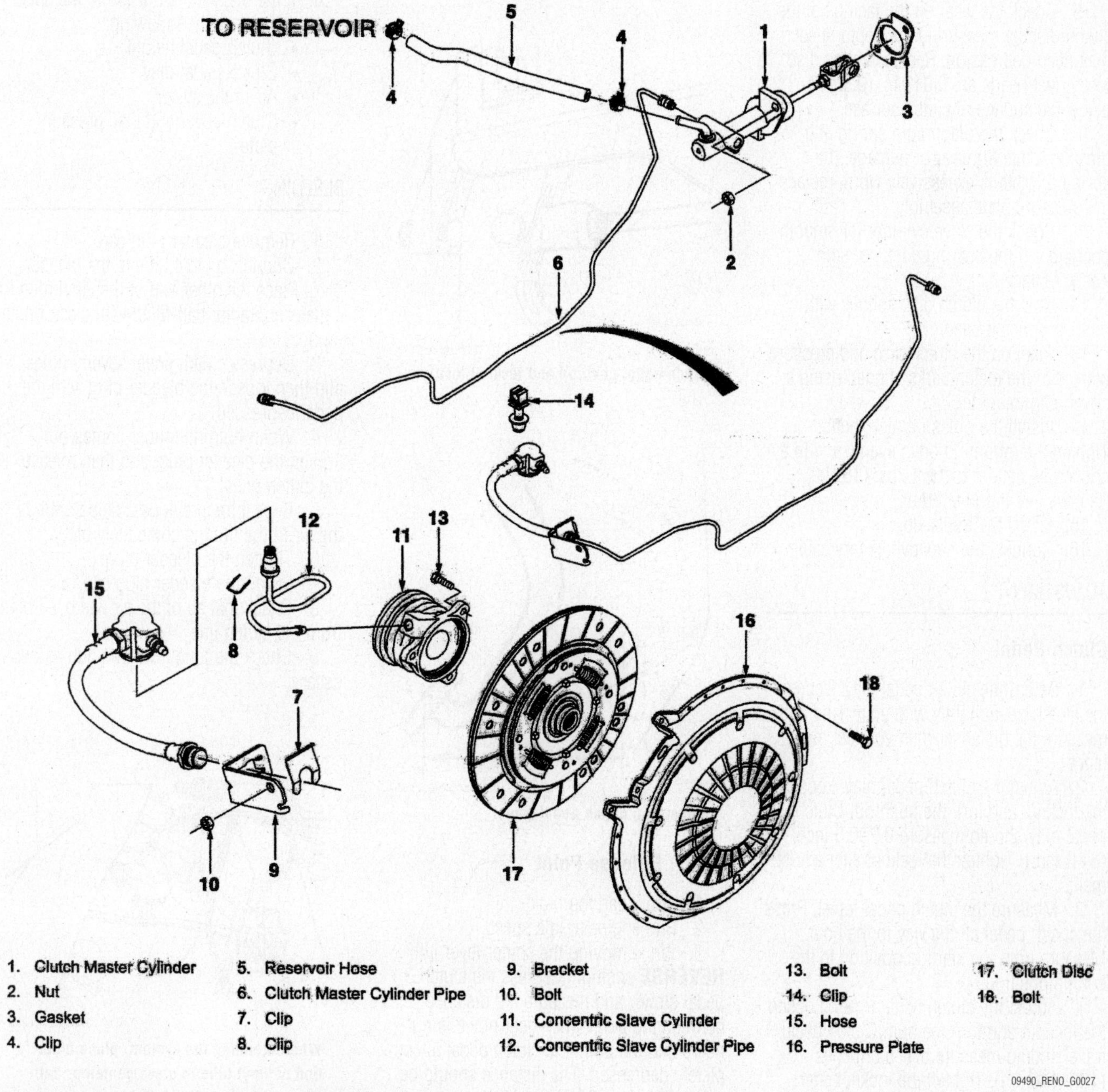

TO RESERVOIR
1
2
3
4
4
5
6
7
8
9
10
11
12
13
14
15
16
17
18
1. Clutch Master Cylinder
2. Nut
3. Gasket
4. Clip
5. Reservoir Hose
6. Clutch Master Cylinder Pipe
7. Clip
8. Clip
9. Bracket
10. Bolt
11. Concentric Slave Cylinder
12. Concentric Slave Cylinder Pipe
13. Bolt
14. Clip
15. Hose
16. Pressure Plate
17. Clutch Disc
18. Bolt
09490_RENO_G0027

Exploded view of clutch components—Forenza/Reno

2. Disconnect the negative battery cable.

3. Remove the transaxle.

4. Hold the flywheel stationary.

5. Matchmark the pressure plate and flywheel for installation reference.

6. Loosen the pressure plate attaching bolts 1 turn at a time (evenly) until the spring pressure is released.

7. Remove the clutch disc and pressure plate.

To install:

8. Clean the flywheel mating surfaces of all oil, grease and metal deposits. Inspect flywheel for cracks, heat checking or other defects and replace or resurface as necessary.

9. Check the wear on the facings of the clutch disc by measuring the depth of each rivet head depression. Replace clutch disc when rivet heads are 0.012 in. (0.3mm) below the surface of clutch surface.

10. Check the diaphragm spring and pressure plate for wear or damage. If the spring or plate is excessively worn, replace the pressure plate assembly.

11. Check the slave cylinder for smooth operation. If the bearing does not spin freely, replace it.

12. Coat the clutch disc splines with multi-purpose grease.

13. Position the clutch disc and pressure plate with the matchmarks aligned using a clutch alignment tool.

14. Install the pressure plate bolts. Tighten the mounting bolts evenly and in a crisscross pattern to 11 ft. lbs. (15 Nm). Remove the alignment tool.

15. Install the transaxle.

16. Connect the negative battery cable.

ADJUSTMENT

Clutch Pedal

1. Determine clutch pedal play. Depress the clutch pedal lightly with your hand and measure the distance when you feel resistance

2. Adjust the clutch pedal play. Loosen the locknut and turn the pushrod. Clutch pedal play should measure 0.2–0.5 inch (6–12mm). Tighten the locknut after adjustment.

3. Measure the clutch pedal travel. Press the clutch pedal all the way to the floor. Measure from the starting position to the ending position.

4. Adjust the clutch pedal travel. Loosen the locknut and turn the bolt. Clutch pedal travel should measure 5.1–5.5 inches (130–140mm). Tighten the locknut after adjustment.

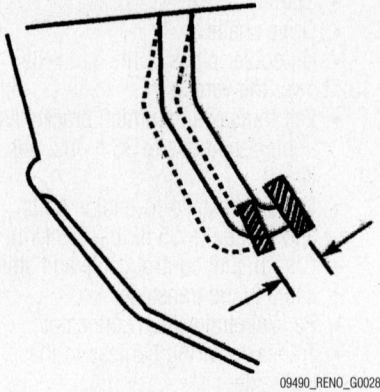

09490_RENO_G0028

Measuring clutch pedal play

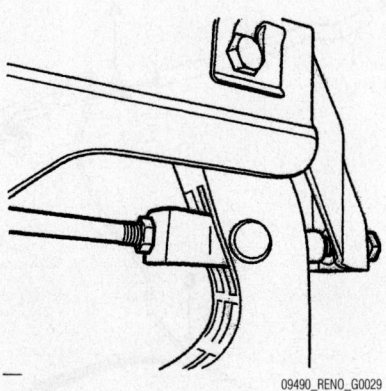

09490_RENO_G0029

Clutch pedal pushrod and locknut location

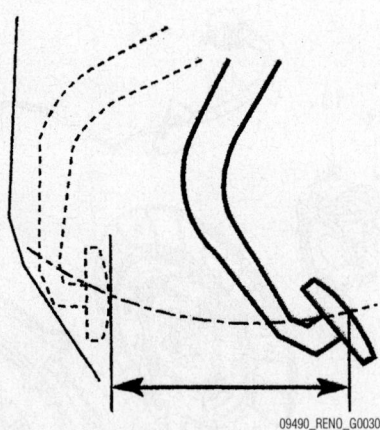

09490_RENO_G0030

Measuring clutch pedal travel

Clutch Release Point

1. Apply parking brake.
2. Run engine at idle speed.
3. While moving the shifter lever into **REVERSE** position, depress the clutch pedal slowly and measure the distance between the point when gear noise is not heard and the point the clutch pedal is completely depressed. The distance should be 1.2–1.6 inches (30–40mm).

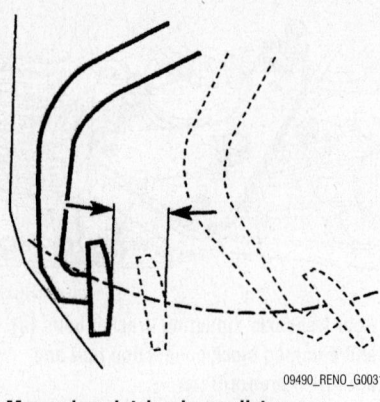

09490_RENO_G0031

Measuring clutch release distance

4. If the distance is not within the specified value, check the following:
- Clutch pedal height
- Clutch pedal play
- Air in the system
- Clutch cover and disc pressure plate

BLEEDING

1. Remove bleeder plug cap.
2. Connect a vinyl tube to the bleeder plug. Place the other end of the vinyl tube in a glass container half-filled with clean brake fluid.
3. Depress clutch pedal several times, and then loosen the bleeder plug with the pedal depressed.
4. When fluid no longer comes out, tighten the bleeder plug, and then release the clutch pedal.
5. Repeat the previous 2 steps until all the air in the fluid is completely bled.
6. Tighten the bleeder plug.
7. Install the bleeder plug cap.
8. Check that all of the air has been bled from the clutch line.
9. Check the fluid level and add as necessary.

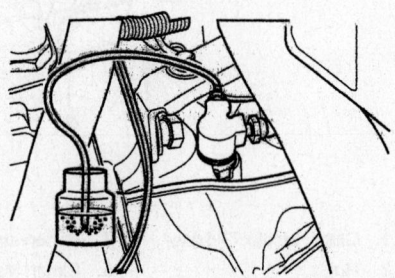

09490_RENO_G0032

When bleeding the system, place other end of vinyl tube in glass container half-filled with clean brake fluid.

Halfshaft

REMOVAL & INSTALLATION

1. Before servicing the vehicle, refer to the Precautions Section.
2. Disconnect the negative battery cable.
3. Raise and safely support the vehicle.
4. Remove wheel.
5. Remove drive shaft nut and washer. Discard nut.
6. Disconnect front suspension control arm ball stud from steering knuckle after removing pinch bolt and nut.

➡**Use only a tool made specifically for separating the lower ball joint. Failure to use the correct tool may cause damage to the ball joint and the seal.**

7. Remove tie rod end nut.
8. Disconnect tie rod end from steering knuckle by using ball joint separator KM-507-B.

➡**Use only a tool made specifically for separating the tie rod from the steering knuckle. Failure to use the correct tool may cause damage to the knuckle/strut assembly.**

9. If equipped with automatic transaxle, perform the following:
    a. Remove the damping block connection nut and bolt.
    b. Remove the rear mounting bracket bolts and the bracket.
10. Push the drive shaft from the wheel hub.
11. Remove drive shaft assembly from the transaxle using axle shaft removal tool DW340-110 (A/T), or KM-460-A (M/T).

➡**Support the unfastened end of the drive axle. Do not allow the drive axle to dangle freely from the transaxle for any length of time after it has been removed from the wheel hub. Place a drain pan below the transaxle to catch the escaping fluid. Cap the transaxle drive opening after the drive axle has been removed to keep the fluid in and any contamination out.**

To install:
12. Clean the hub seal and the transaxle seal. Be careful to not damage the seals.
13. Install the drive shaft into the transaxle.
14. Install the wheel hub onto the axle shaft.
15. If equipped with automatic transaxle, perform the following:

a. Install the rear mounting bracket bolts and the bracket and tighten to 45 ft. lbs. (62 Nm).
    b. Install the damping block connection nut and bolt and tighten to 50 ft. lbs. (68 Nm).
16. Install the tie rod into the knuckle/strut and install the tie rod nut. Tighten the tie rod nut to 41 ft. lbs. (55 Nm).
17. Install the lower ball joint pinch bolt and nut and tighten to 44 ft. lbs. (60 Nm).
18. Loosely install the washer and a new axle shaft caulking nut. Always use a new nut.
19. Install the wheels. Loosely install the nuts.
20. Lower the vehicle to the floor and tighten the wheel nuts to 74 ft. lbs. (100 Nm).
21. Tighten the new axle shaft caulking nut to 221 ft. lbs. (300 Nm).
22. Peen the new caulking nut with a punch and a hammer until the nut is locked into place on the axle shaft hub.
23. Refill the transaxle fluid to the proper level.
24. Connect the negative battery cable.

CV-Joints

OVERHAUL

Double Offset Joint (DOJ) Type

The Double Offset Joint (DOJ) type is identified by the outside shape of the differential side joint which has no dent.

1. Before servicing the vehicle, refer to the Precautions Section.
2. Remove the drive shaft.
3. Remove the large seal retaining clamp. Discard the clamp.
4. Remove the small seal retaining clamp. Discard the clamp.
5. Degrease the joint.
6. Spread the snap ring using snap ring pliers and remove the outer joint of the axle shaft.
7. Remove the seal from the joint assembly.
8. Inspect all components for wear and/or damage and replace as necessary.

To install:
9. Clean all components and allow them to completely dry.
10. Install the seal onto the axle shaft.

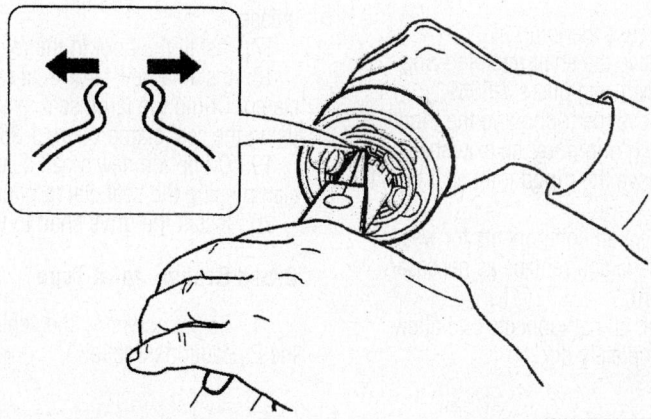

09490_RENO_G0033

Spread the snap ring using snap ring pliers —DOJ type

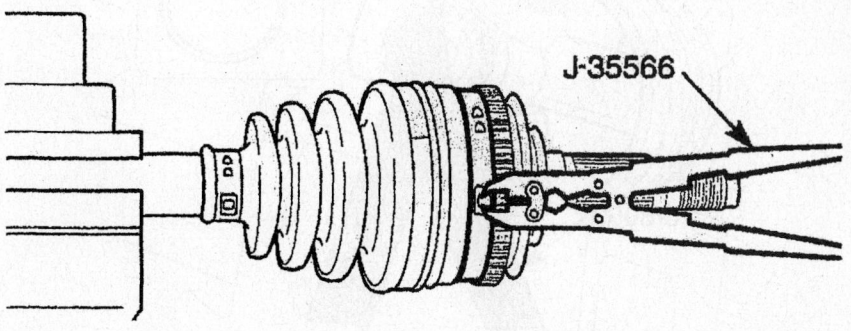

09490_RENO_G0034

Crimp the new large seal retaining clamp using seal clamp pliers—DOJ type

11. Spread the snap ring using snap ring pliers and install the outer joint the axle shaft.

12. Fill the joint seal with 4.2–4.9 ounces (120–140g) of the recommended grease.

13. Repack the joint with 4.2–4.9 ounces (120–140g) of the recommended grease.

14. Install a new large seal retaining clamp and a new small seal retaining clamp.

15. Crimp the new small seal retaining clamp and the new large seal retaining clamp using the seal clamp pliers J-35566.

16. Install the drive axle shaft to the vehicle.

Tripod Joint Type

The Tripod joint type can be identified by the 3 dent lines on the outside of the differential side joint.

1. Before servicing the vehicle, refer to the Precautions Section.

2. Remove the drive shaft.

3. Remove the large seal retaining clamp. Discard the clamp.

4. Remove the small seal retaining clamp. Discard the clamp.

5. Separate the joint housing from the boot.

6. Degrease the joint.

7. Remove the shaft retaining ring using the snap ring pliers J-8059.

8. Remove the tripod and the tripod joint retaining ring from the axle shaft.

9. Remove the tripod joint seal from the axle shaft.

10. Inspect all components for wear and/or damage and replace as necessary.

To install:

11. Clean all components and allow them to completely dry.

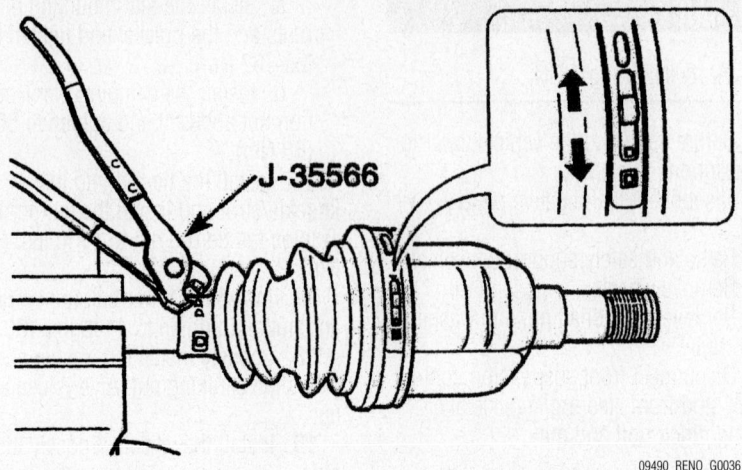

Crimp the new seal retaining clamps using seal clamp pliers—Tripod joint

09490_RENO_G0036

12. Install a new small seal retaining clamp onto the seal.

13. Install the seal onto the axle shaft.

14. Install the shaft retaining ring onto the axle shaft using the snap ring pliers J-8059.

15. Fill the tripod housing with 6.9–7.6 ounces (195–215g) of the recommended grease.

16. Repack the tripod with 6.9–7.6 ounces (195–215g) of the recommended grease.

17. Install the boot to the joint housing.

18. Install a new large seal retaining clamp. Crimp the large seal retaining clamp using the seal clamp pliers J-35566.

19. Crimp the new small seal retaining clamp using the seal clamp pliers J-35566.

20. Install the drive shaft to the vehicle.

Cross Groove Joint Type

1. Before servicing the vehicle, refer to the Precautions Section.

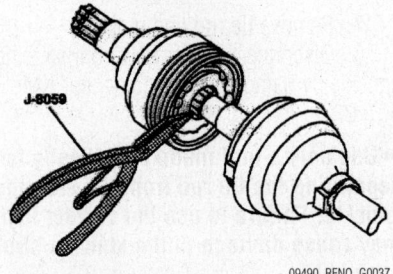

Spread the snap ring using snap ring pliers —Cross groove joint

09490_RENO_G0037

Remove/install the axle shaft from the joint assembly—Cross groove joint

09490_RENO_G0038

2. Remove the drive shaft.

3. Remove the large seal retaining clamp. Discard the clamp.

4. Remove the small seal retaining clamp. Discard the clamp.

5. Degrease the joint.

6. Remove the shaft retaining ring using the snap ring pliers J-8059.

7. Remove the axle shaft from the joint assembly.

8. Remove the seal from the joint assembly.

To install:

9. Clean all components and allow them to completely dry.

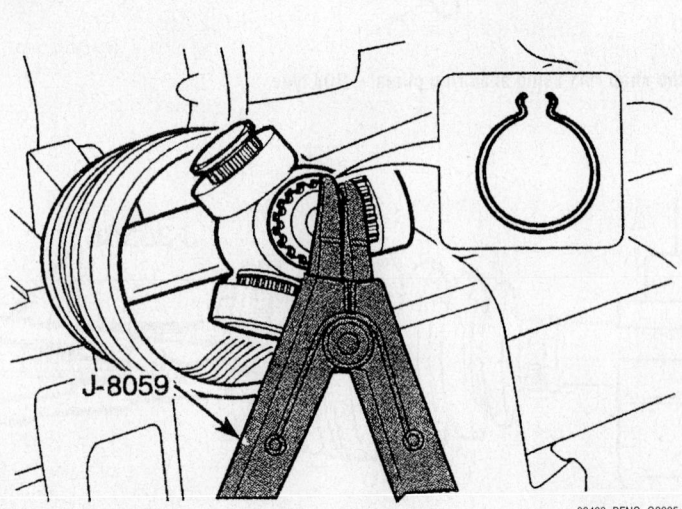

Spread the snap ring using snap ring pliers —Tripod joint

09490_RENO_G0035

10. Install a new small seal retaining clamp onto the seal. Do not crimp.

11. Install the seal onto the axle shaft.

12. Install the joint assembly onto the axle shaft.

13. Install the shaft retaining ring using the snap ring pliers J-8059.

14. Fill the joint assembly with 4.2–4.9 ounces (120–140g) of the recommended grease.

15. Repack the joint with 4.2–4.9 ounces (120–140g) of the recommended grease.

16. Install the a new large seal retaining clamp.

17. Crimp the new large seal retaining clamp using the seal clamp pliers J-35566.

18. Crimp the new small retaining clamp using the seal clamp pliers J-35566.

STEERING

Air Bag

❋ CAUTION

Most vehicles are equipped with an air bag system. The system must be disabled before performing service on or around system components, steering column, instrument panel components, wiring and sensors. Failure to follow safety and disabling procedures could result in accidental air bag deployment, possible personal injury and unnecessary system repairs.

PRECAUTIONS

Several precautions must be observed when handling the inflator module to avoid accidental deployment and possible personal injury.

• Never carry the inflator module by the wires or connector on the underside of the module.

• When carrying a live inflator module, hold securely with both hands, and ensure that the bag and trim cover are pointed away.

• Place the inflator module on a bench or other surface with the bag and trim cover facing up.

• With the inflator module on the bench, never place anything on or close to the module that may be thrown in the event of an accidental deployment.

DISARMING

❋ WARNING

When performing service on or around the air bag system components or wiring, disable the air bag system. Failure to follow the procedures could result in possible deployment, personal injury or unneeded system repairs.

1. Before servicing the vehicle, refer to the Precautions Section.

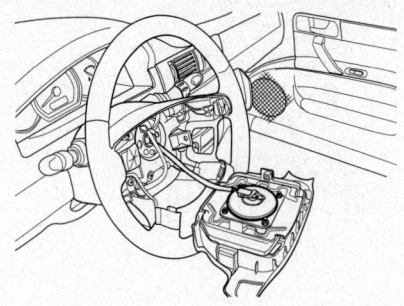

09490_RENO_G0039

Air bag connector location behind the steering wheel air bag module

2. Disconnect the negative battery cable.

3. Turn the steering wheel so the wheels are pointing straight ahead.

4. Turn the ignition switch to the **LOCK** position and remove the key.

5. Remove the **AIR BAG-F1** fuse from the instrument panel (I/P) fuse block located at the end of the instrument panel on the driver side.

6. Wait more than 1 minute for SIR capacitor to discharge.

7. Remove the mounting bolts, steering wheel air bag module and disconnect the yellow connector for the driver's side air bag (inflator) module from behind the module.

8. Remove the glove box. Disconnect the yellow connector for the passenger air bag (inflator) module.

REARMING

❋ WARNING

When performing service on or around the air bag system components or wiring, disable the air bag system. Failure to follow the procedures could result in possible deployment, personal injury or unneeded system repairs.

1. Before servicing the vehicle, refer to the Precautions Section.

2. Connect the negative battery cable.

3. Turn the ignition switch to the **LOCK** position and remove the key.

4. Connect the yellow connector for the passenger side air bag (inflator) module and the yellow connector for the driver's side air bag (inflator) module.

5. Install the glove box assembly.

6. Install the passenger side air bag module and tighten the mounting bolts to 97 inch lbs. (11 Nm).

7. Install the steering wheel air bag module and tighten the mounting bolts to 11 ft. lbs. (15 Nm).

8. Install the **AIR BAG-F1** fuse into the instrument panel fuse block.

9. Turn the ignition **ON** and verify that the **AIR BAG** indicator lamp flashes 7 times, then turns off. If the system does not operate as described, diagnosis and repairs to the air bag system are necessary.

Rack and Pinion Steering Gear

REMOVAL & INSTALLATION

1. Before servicing the vehicle, refer to the Precautions Section.

2. Disconnect the negative battery cable.

3. Raise and safely support the vehicle.

4. Remove or disconnect the following:

• Front wheels

• Power steering gear fluid outlet pipe. Place a drain pan under the steering gear to catch the power steering fluid.

• Power steering gear fluid inlet pipe

5. Position the steering gear straight ahead by turning the steering wheel until the steering wheel spokes are vertical and pointed to the left.

6. Scribe a mark on the stub shaft housing that lines up with a mark on the intermediate shaft lower coupling.

• Intermediate shaft pinch bolt

• Outer tie rod nuts

• Tie rod ends from strut assembly using ball joint remover KM-507-B

• Crossmember assembly

• Transaxle center bracket (automatic transaxle)

• Bolts securing transaxle center bracket to transaxle and engine and move center bracket out of the way (manual transaxle)

- Nuts and bolts from the steering gear mounting bracket
- Return line from the clip on the crossmember
- Rack and pinion assembly from crossmember assembly

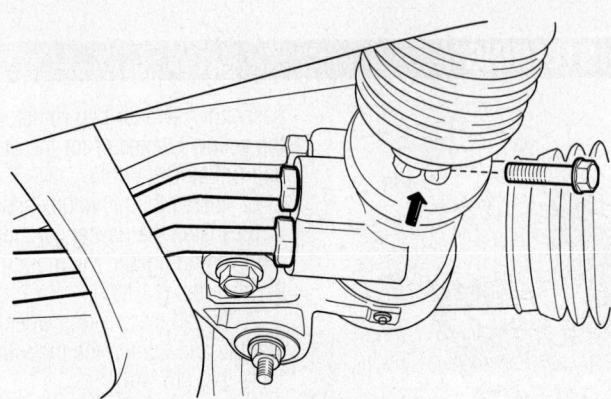

Location of intermediate shaft pinch bolt and power steering gear fluid inlet and outlet pipes—Forenza/Reno

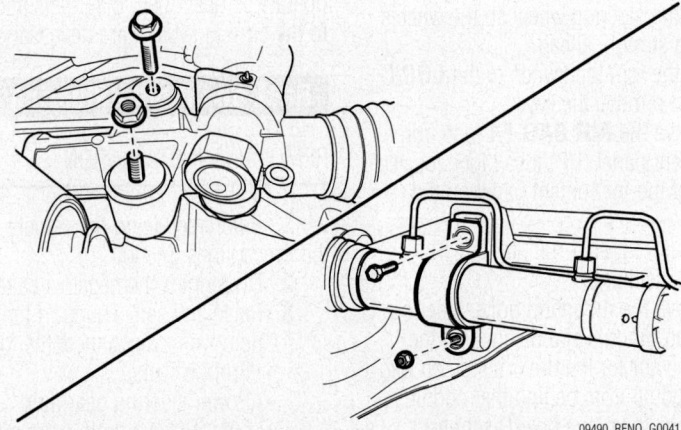

Power steering gear mounting bracket nuts and bolts—Forenza/Reno

To install:

7. Install the rack and pinion assembly onto the crossmember. The steering gear must be in a straight-ahead position, and the steering wheel spokes must be vertical and pointing to the left. Align the marks on the shafts to ensure proper positioning. Seat the stub shaft into the intermediate shaft.

8. Install or connect the following:
- Bolts and nuts on the steering gear mounting bracket. Tighten the steering gear mounting bracket bolts and nuts to 44 ft. lbs. (60 Nm).
- Return line into clip on the crossmember and tighten to 71 inch lbs. (8 Nm)

9. On vehicles equipped with a manual transaxle, position the transaxle center bracket in place and install the bolts securing the bracket to the engine and the transaxle. Tighten the transaxle center bracket-to-transaxle bolts and the transaxle center bracket-to-engine bolt to 59 ft. lbs. (80 Nm).

10. On vehicles equipped with an automatic transaxle, install the transaxle center bracket.
- Crossmember. Tighten the rear crossmember-to-body bolts to 145 ft. lbs. (196 Nm) and the front crossmember-to-body bolts to 96 ft. lbs. (130 Nm).
- Tie rod ends to strut assembly
- Outer tie rod nuts and tighten to 37 ft. lbs. (50 Nm)
- Lower intermediate shaft pinch bolt and tighten to 18 ft. lbs. (25 Nm)
- Power steering gear fluid inlet and outlet pipes and tighten the pipe fittings to 21 ft. lbs. (28 Nm)
- Front wheels

11. Lower the vehicle.

12. Check and adjust the front wheel alignment.

13. Refill the power steering system and check for leaks. If leaks are found, correct the cause of the leak and bleed the system.

14. Connect the negative battery cable.

FRONT SUSPENSION

Strut

REMOVAL & INSTALLATION

1. Before servicing the vehicle, refer to the Precautions Section.
2. Disconnect the negative battery cable.
3. Remove the strut upper cap and nut.
4. Raise and safely support the front of the vehicle.
5. Remove or disconnect the following:
- Wheels
- ABS sensor line from strut assembly (if equipped)
- Brake line from securing bracket on the strut assembly
- Stabilizer shaft link by removing the stabilizer link-to-strut assembly nut
- Steering knuckle by removing the steering knuckle-to-strut assembly nuts and bolts
- Strut assembly

6. Remove the coil spring by disassembling the strut assembly as follows:
  a. Fasten the strut assembly to the spring compressor DW320-010 or KM-329-A. Make sure the hooks are seated on the strut spring properly.
  b. Compress the front spring with the spring compressor.
  c. Use an open end wrench to hold the threaded piston rod while removing

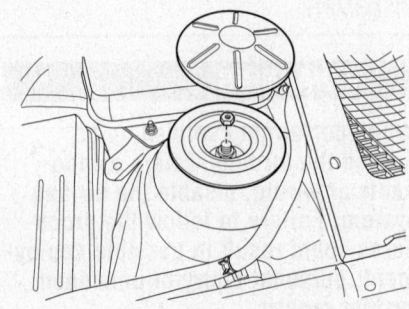

Strut upper cap and nut

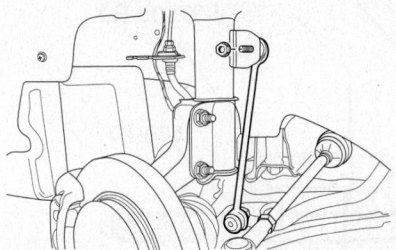

Stabilizer link-to-strut assembly and retaining nut

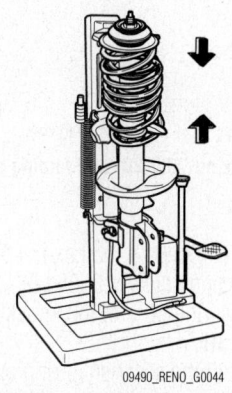

Strut assembly mounted in spring compressor tool

the piston rod nut and the washer with a commercially available double ring spanner, sharply offset.

    d. Remove the upper strut mount, the mount bearing, the upper spring seat, the upper spring insulator and the hollow bumper.

    e. Release the spring.

    f. Remove the spring and the lower spring insulator.

To install:

    7. Install the coil spring by assembling the strut assembly as follows:

    a. Install the lower spring insulator and the spring.

    b. Compress the spring using the spring compressor KM-329-A.

    c. Install the hollow bumper, the upper spring insulator, the upper spring seat, the upper strut mount, and the mount bearing. Be sure the upper spring seat is clipped to the front spring locator.

    d. Install the piston rod nut. Tighten the piston rod nut to 55 ft. lbs. (75 Nm).

    e. Release and remove the spring compressor KM-329-A.

    8. Install or connect the following:
- Strut assembly
- Strut assembly to steering knuckle and tighten the steering knuckle-to-strut assembly nuts and the bolts to 89 ft. lbs. (120 Nm)

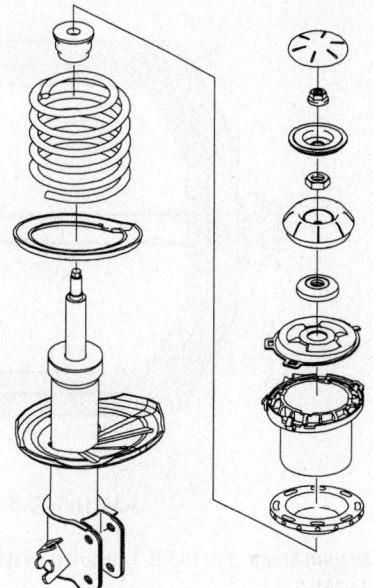

Exploded view of strut assembly

- Stabilizer shaft link to strut assembly and tighten the stabilizer link-to-strut assembly nut to 35 ft. lbs. (47 Nm)
- Brake line to securing bracket on the strut assembly
- ABS sensor line to strut assembly (if equipped)
- Wheels

    9. Lower the vehicle.

    10. Install the nuts securing the strut assembly to the body of the vehicle and tighten to 48 ft. lbs. (65 Nm).

    11. Connect the negative battery cable.

Coil Spring

REMOVAL & INSTALLATION

    For coil spring service on the Forenza and Reno, refer to the strut removal and installation procedure.

Lower Ball Joint

REMOVAL & INSTALLATION

    1. Before servicing the vehicle, refer to the Precautions Section.

    2. Remove the control arm.

    3. Drill off the heads of the three rivets with a 0.47 inch (12mm) drill bit.

    4. Punch out the rivets with a drift.

To install:

    5. Connect the ball joint to the control arm by inserting 3 ball joint bolts from below the control arm.

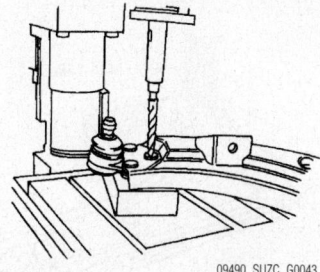

Drilling off the ball joint rivets

    6. Tighten the ball joint-to-control arm nuts to 74 ft. lbs. (100 Nm).

    7. Install the control arm.

Lower Control Arm

REMOVAL & INSTALLATION

    1. Before servicing the vehicle, refer to the Precautions Section.

    2. Disconnect the negative battery cable.

    3. Raise and safely support the front of the vehicle. Let the control arms hang free.

    4. Remove or disconnect the following:
- Front wheels
- Pinch bolt and nut from the ball joint
- Ball joint from knuckle assembly using ball joint remover KM-507-B
- Control arm-to-crossmember bolts
- Control arm from vehicle

To install:

    5. Install or connect the following:
- Control arm
- Control arm-to-crossmember bolts. Do not tighten the bolts.
- Ball joint to steering knuckle
- Ball joint pinch bolt and the nut. Tighten ball joint pinch bolt and nut to 44 ft. lbs. (60 Nm).
- Control arm-to-crossmember bolts. Tighten the front control arm-to-crossmember bolt to 92 ft. lbs. (125 Nm) and the rear control arm-to-crossmember bolt to 81 ft. lbs. (110 Nm).

    6. Install the wheels.

    7. Lower the vehicle.

    8. Connect the negative battery cable.

    9. Check the front wheel alignment.

CONTROL ARM BUSHING REPLACEMENT

    1. Before servicing the vehicle, refer to the Precautions Section.

    2. Remove the control arm.

    3. Remove the split sleeves from the rear control arm bushing.

    4. Press off the control arm rear damp-

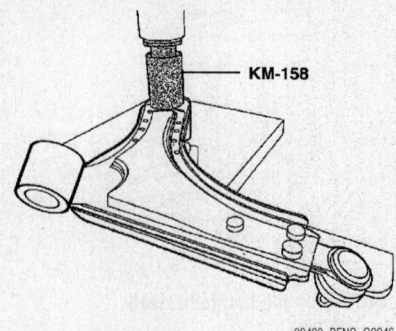

09490_RENO_G0046

Press off/on control arm bushing using a press and remover/installer tool

ing bushing using a press and the remover/installer KM-158.

To install:

5. Press the control arm rear damping bushing into the control arm using a press and the remover/installer KM-158.

6. Install the split sleeves into the rear control arm bushing.

7. Install the control arm.

Wheel Bearings

ADJUSTMENT

The front wheel bearings are a cartridge type design and cannot be adjusted.

REMOVAL & INSTALLATION

1. Before servicing the vehicle, refer to the Precautions Section.

2. Remove the drive shaft from the front wheel hub.

3. Remove the inner snap ring.

4. Remove the wheel hub with the support bridge J-37105-B-1, the bearing adapter J-37105-B-2, the hex nut 500-20, and the forcing screw J-36661-2.

5. Remove the brake shield.

6. Remove the outer snap ring.

7. Remove the wheel bearing with the support bridge J-37105-B-1, the bearing

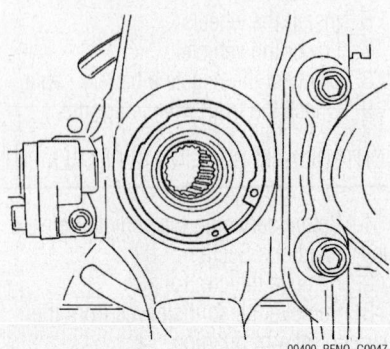

09490_RENO_G0047

Wheel hub/bearing inner snap ring

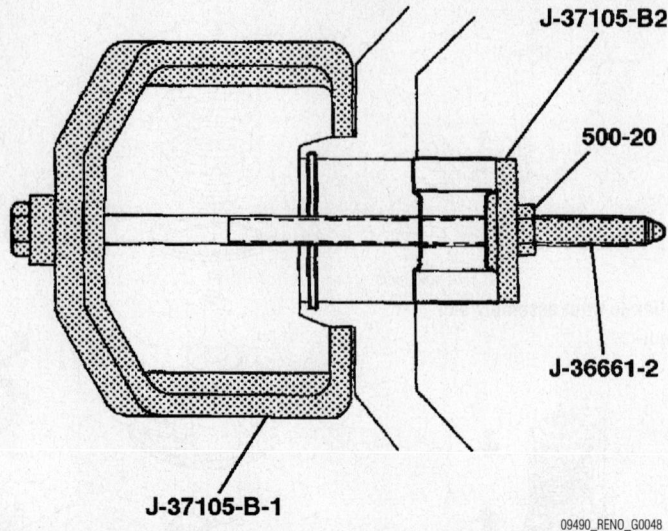

09490_RENO_G0048

Support bridge J-37105-B-1, bearing adapter J-37105-B-2, hex nut 500-20, and forcing screw J-6661-2

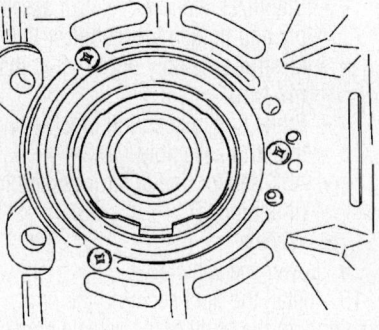

09490_RENO_G0049

Wheel hub/bearing outer snap ring

adapter J-37105-B-2, the hex nut 500-20, and the forcing screw J-36661-2.

8. Clean the bore of the knuckle.

To install:

9. Install the outer snap ring and push the wheel bearing into place with the support bridge J-37105-B-1, the bearing adapter J-37105-B-2, the hex nut 500-20, and the forcing screw J-36661-2.

10. Install the brake shield.

11. Install the inner snap ring.

12. Push the wheel hub into place with the hub adapter J-37105-B-3, the bearing adapter J-37105-B-2, the hex nut 500-20, and the forcing screw J-36661-2.

13. Install the drive axle into the front wheel hub.

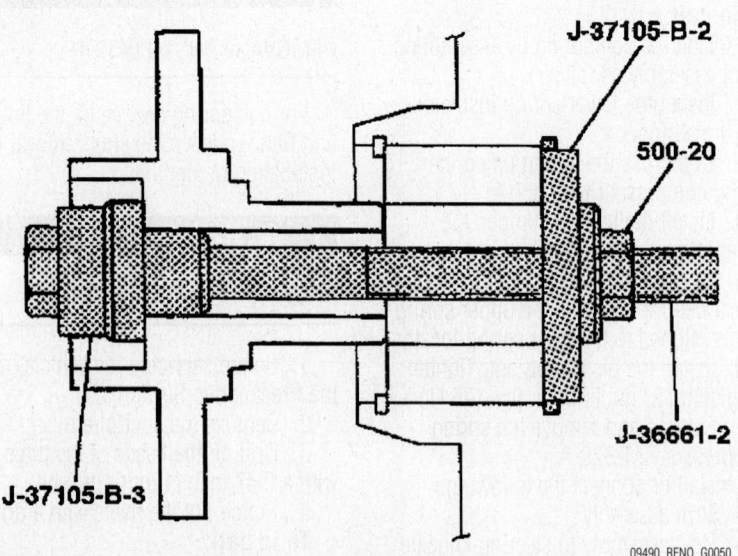

09490_RENO_G0050

Hub adapter J-37105-B-3, bearing adapter J-37105-B-2, hex nut 500-20, and forcing screw J-36661-2

REAR SUSPENSION

Strut

REMOVAL & INSTALLATION

1. Before servicing the vehicle, refer to the Precautions Section.
2. Remove the trunk carpeting that covers the rear strut mounting nuts. For station wagons, remove the panels that cover the luggage compartment wheelhouse trim panel.
3. Remove the rear strut mounting nuts.
4. Raise and suitably support the vehicle.
5. Remove the wheel.
6. Disconnect the parking brake.
7. Remove the clip that holds the brake hose to the strut assembly.
8. Remove the stabilizer link-to-strut assembly nut and disconnect the stabilizer link from the strut assembly.

9. Remove the knuckle-to-strut assembly nuts and the bolts.
10. Remove the rear strut assembly from the vehicle.
11. Remove the coil spring by disassembling the strut assembly as follows:
  a. Mount the rear strut assembly into the spring compressor (KM-329-A or DW320-010). Ensure that the hooks are properly seated.
  b. Compress the spring.
  c. Remove the lock nut from the strut dampener rod by using SST, J-42468.
  d. Remove the rear strut mount.
  e. Remove the rear spring upper seat, the dust cover, and the hollow bumper.
  f. Release the coil spring.
  g. Remove the rear spring and the rear spring lower seat.

To install:
12. Install the coil spring by assembling the strut assembly as follows:

  a. Install the rear spring lower seat and the rear spring.
  b. Compress the spring using the spring compressor tool.
  c. Install the hollow bumper, the dust cover, and the rear spring upper seat.
  d. Install the rear strut mount.
  e. Install the lock nut onto the strut dampener rod by using SST, J-42468.
  f. Tighten the ball joint-to-control arm nuts to 74 ft. lbs. (100 Nm) and the strut dampener-to-strut mount nut to 55 ft. lbs. (75 Nm).
  g. Release the spring.
  h. Remove the strut assembly from the spring compressor.
13. Install the rear strut assembly into the vehicle.
14. Secure the strut assembly by loosely attaching the strut mount-to-body nuts.
15. Install the knuckle-to-strut assembly nuts and bolts. Do not tighten.
16. Install the clip holding the brake hose to the strut assembly.
17. Tighten the knuckle-to-strut assembly nuts to 74 ft. lbs. (100 Nm).
18. Connect the stabilizer link to the strut assembly and install the stabilizer link-to-strut assembly nut. Tighten the stabilizer link-to-strut assembly nut to 35 ft. lbs. (47 Nm).
19. Connect the parking brake.
20. Install the rear wheels.
21. Lower the vehicle.
22. Tighten the strut mount-to-body nuts to 22 ft. lbs. (30 Nm).
23. Install the trunk carpeting over the rear strut mounting nuts. For station wagons, remove the panels that cover the luggage compartment wheelhouse trim panel.

Coil Spring

REMOVAL & INSTALLATION

For coil spring service on the Forenza and Reno, refer to the strut removal and installation procedure.

Lower Control Arm

REMOVAL & INSTALLATION

Front Parallel Link

1. Before servicing the vehicle, refer to the Precautions Section.
2. Raise and safely support the rear of the vehicle.

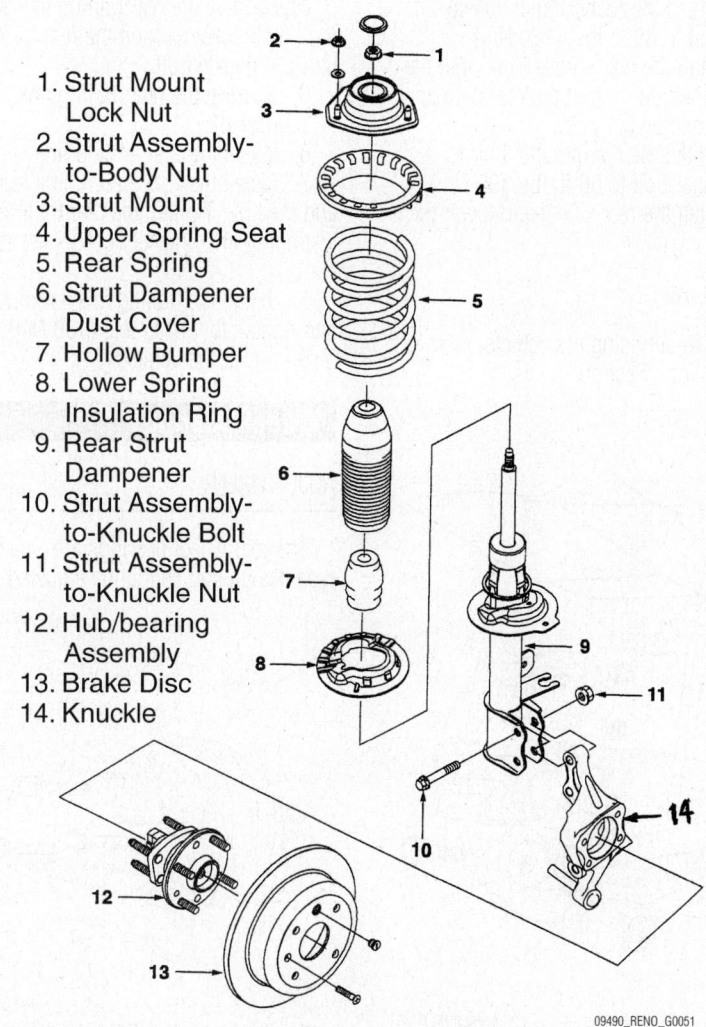

1. Strut Mount Lock Nut
2. Strut Assembly-to-Body Nut
3. Strut Mount
4. Upper Spring Seat
5. Rear Spring
6. Strut Dampener Dust Cover
7. Hollow Bumper
8. Lower Spring Insulation Ring
9. Rear Strut Dampener
10. Strut Assembly-to-Knuckle Bolt
11. Strut Assembly-to-Knuckle Nut
12. Hub/bearing Assembly
13. Brake Disc
14. Knuckle

09490_RENO_G0051

Exploded view of the rear strut assembly and related components

3. Remove the rear wheels.

4. For vehicles equipped with the antilock braking system, remove the ABS sensor from the knuckle and the ABS housing assembly from the front parallel link.

5. Remove the front parallel link bolt from the rear crossmember.

6. Remove the front parallel link bolt from the rear knuckle.

7. Remove the front parallel link.

To install:

8. Install the front parallel link.

9. Install the front parallel link onto the rear knuckle with the bolt.

10. Tighten the front parallel link-to-knuckle bolt to 89 ft. lbs. (120 Nm).

11. Install the front parallel link onto the rear cross-member with the bolt. Do not tighten.

12. For vehicles equipped with the antilock braking system, install the ABS housing assembly onto the front parallel link and the ABS sensor line into the knuckle.

13. Install the rear wheels and lower the vehicle.

14. Perform a rear toe adjustment.

Rear Parallel Link

1. Before servicing the vehicle, refer to the Precautions Section.

2. Raise and safely support the rear of the vehicle.

3. Remove the rear wheel.

4. Remove the rear parallel link bolt from the rear crossmember.

5. Remove the rear parallel link bolt from the rear knuckle.

6. Remove the rear parallel link.

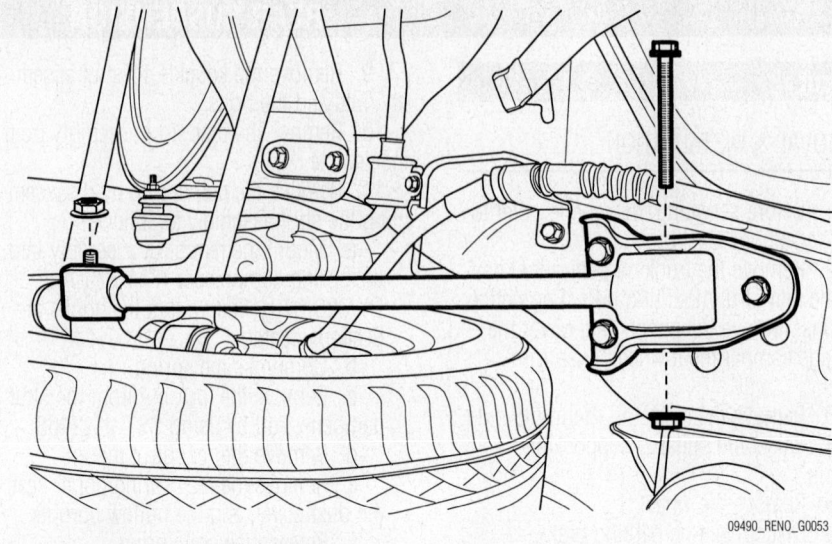

09490_RENO_G0053

Trailing link and trailing link bracket

To install:

7. Install the rear parallel link.

8. Install the rear parallel link onto the rear knuckle with the bolt.

9. Tighten the rear parallel link-to-knuckle bolt to 89 ft. lbs. (120 Nm).

10. Install the rear parallel link onto the rear crossmember. Install the rear parallel link-to-crossmember bolt.

11. Tighten the rear parallel link-to-crossmember bolt to 66 ft. lbs. (90 Nm).

12. Install the rear wheel and lower the vehicle.

Trailing Link

1. Before servicing the vehicle, refer to the Precautions Section.

2. Raise and safely support the rear of the vehicle.

3. Remove the rear trailing link-to-rear knuckle nut.

4. Remove the rear trailing link-to-trailing link bracket nut and the rear trailing link-to-knuckle bolt.

5. Remove the rear trailing link.

To install:

6. Install the rear trailing link.

7. Install the rear trailing link bracket nut and the bolt. Tighten the rear trailing link-to-trailing link bracket nut to 74 ft. lbs. (100 Nm).

8. Install the trailing link-to-knuckle nut and tighten to 110 ft. lbs. (150 Nm).

9. Lower the vehicle.

Wheel Bearings

ADJUSTMENT

The rear wheel bearings are a cartridge type design and cannot be adjusted.

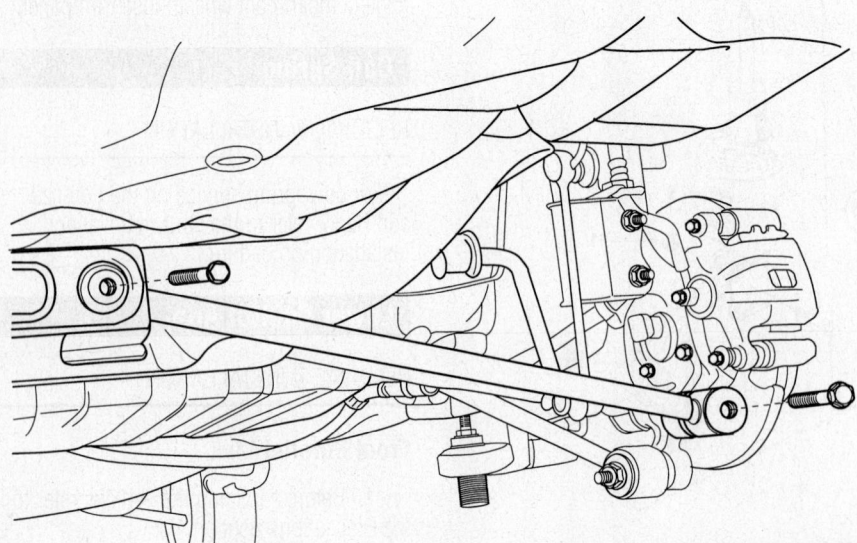

09490_RENO_G0052

Rear parallel link and retaining bolts

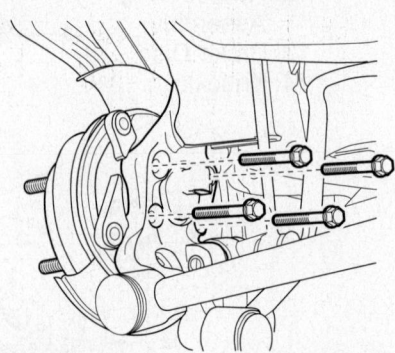

09490_RENO_G0054

Remove/install 4 hub/bearing assembly mounting bolts

REMOVAL & INSTALLATION

1. Before servicing the vehicle, refer to the Precautions Section.
2. Raise and safely support the vehicle.
3. Remove the wheel.

4. Remove the rear brake caliper and rear brake disc.
5. Remove the hub bolts and hub assembly.
   To install:
6. Install the hub assembly.
7. Tighten the hub assembly bolts to 48 ft. lbs. (65 Nm).

8. Install the hub nut.
9. Install the rear brake disc and rear brake caliper.
10. Install the wheel.
11. Lower the vehicle.
12. Check to ensure that the brakes are free from dragging and that proper braking is obtained.

FRONT BRAKES

Brake Caliper

REMOVAL & INSTALLATION

1. Before servicing the vehicle, refer to the Precautions Section.
2. Raise and safely support the front of the vehicle.
3. To preserve wheel balance, mark the position of the front wheel relative to the wheel hub.
4. Remove the wheel.
5. Remove the bolt and the washers attaching the brake hose to the caliper.
6. Disconnect the brake hose, and plug the openings in the caliper and the brake hose to prevent fluid loss and contamination.
7. Remove the caliper mounting bolts from the steering knuckle, and remove the caliper assembly.
   To install:
8. Install the caliper assembly with the mounting bolts.
9. Tighten the caliper-to-steering knuckle mounting bolts to 70 ft. lbs. (95 Nm).
10. Connect the brake hose to the caliper

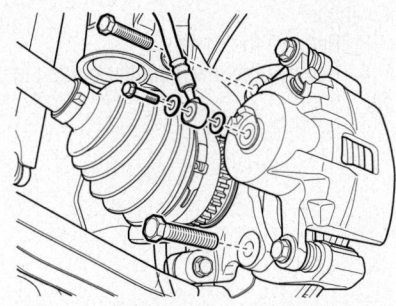

09490_RENO_G0055

Front disc brake caliper removal/installation

with bolt and washer. Tighten the caliper brake hose inlet bolt to 30 ft. lbs. (40 Nm).
11. Align the marks that were made when removing the front wheel, and install the wheel.
12. Lower the vehicle.
13. Fill the master cylinder to the proper level with clean brake fluid.
14. Bleed the air out of the brake system.
15. Recheck the fluid level.
16. Repeatedly press the brake pedal to bring the pads in contact with the rotor.

Disc Brake Pads

REMOVAL & INSTALLATION

1. Before servicing the vehicle, refer to the Precautions Section.
2. Raise and safely support the front of the vehicle.
3. Remove the wheels.
4. Remove the lower caliper mounting bolt.

➡**Caliper removal is not necessary to service the brake pads.**

5. Pivot the caliper upward.
6. Remove the brake pads.
   To install:
7. Measure the minimum brake shoe lining thickness.
8. Install the brake pads.
9. Push the caliper piston inward, if needed.
10. Pivot the caliper downward and install the lower mounting bolt. Tighten the lower caliper mounting bolt to 20 ft. lbs. (27 Nm).
11. Install the wheels.
12. Lower the vehicle.

REAR BRAKES

Brake Caliper

REMOVAL & INSTALLATION

1. Before servicing the vehicle, refer to the Precautions Section.
2. Raise and safely support the rear of the vehicle.
3. To preserve wheel balance, mark the position of the front wheel relative to the wheel hub.
4. Remove the wheels.
5. Remove the bolt and the ring seals that attach the brake hose inlet fitting to the caliper.
6. Disconnect the brake hose. Plug the openings in the caliper and the brake hose to prevent fluid loss or contamination.

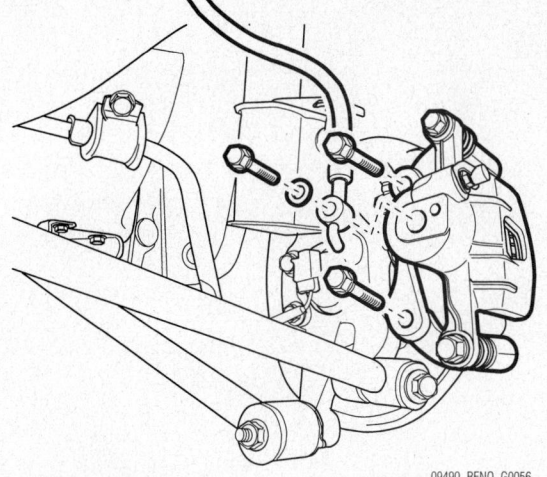

09490_RENO_G0056

Rear disc brake caliper removal/installation

7. Remove the caliper mounting bolts from the knuckle.

8. Remove the caliper.

To install:

9. Install the caliper assembly with the mounting bolts.

10. Tighten the caliper mounting bolts to 41 ft. lbs. (56 Nm).

11. Connect the brake hose to the caliper with bolt and ring seals. Tighten the caliper brake hose inlet bolt to 24 ft. lbs. (32 Nm).

12. Align the marks that were made when removing the front wheel, and install the wheel.

13. Lower the vehicle.

14. Fill the master cylinder to the proper level with clean brake fluid.

15. Bleed the air out of the brake system.

16. Recheck the fluid level.

17. Repeatedly press the brake pedal to bring the pads in contact with the rotor.

Disc Brake Pads

REMOVAL & INSTALLATION

1. Before servicing the vehicle, refer to the Precautions Section.

2. Raise and safely support the rear of the vehicle.

3. Remove the wheels.

4. Remove the lower caliper guide pin bolt.

➡**Caliper removal is not necessary to service the brake pads.**

5. Pivot the caliper upward.

6. Remove the brake pads.

To install:

7. Measure the minimum brake shoe lining thickness.

8. Install the brake pads into the caliper.

9. Push the piston inward, if needed.

10. Pivot the caliper downward and tighten the bolt to 20 ft. lbs. (27 Nm).

11. Install the rear wheels.

12. Lower the vehicle.

DRIVE TRAIN 11-42
ENGINE REPAIR 11-15
FRONT BRAKES 11-53
FRONT SUSPENSION 11-48
FUEL SYSTEM 11-41
REAR BRAKES 11-55
REAR SUSPENSION 11-50
**SPECIFICATIONS AND
  MAINTENANCE CHARTS** 11-3
STEERING 11-46
Engine and Vehicle Identification
  Chart 11-3
General Engine Specifications 11-3
Engine Tune-Up Specifications 11-4
Firing Order 11-4
Accessory Drive Belt Routing 11-5
Capacities 11-6
Valve Specifications 11-7
Crankshaft and Connecting Rod
  Specifications 11-8
Piston and Ring Specifications 11-9
Torque Specifications 11-10
Wheel Alignment 11-11
Tire, Wheel and Ball Joint
  Specifications 11-11
Brake Specifications 11-12
Scheduled Maintenance
  Intervals 11-13

A
Air Bag 11-46
  Arming 11-46
  Disarming 11-46
  Precautions 11-46
Alternator 11-15
  Installation 11-15
  Removal 11-15
Automatic Locking Hubs 11-45
  Removal & Installation 11-45
Axle Housing Assembly 11-46
  Removal & Installation 11-46
Axle Shaft, Bearing and Seal 11-45
  Removal & Installation 11-45

B
Brake Caliper 11-53
  Removal & Installation 11-53

Brake Drums 11-55
  Removal & Installation 11-55
Brake Shoes 11-56
  Removal & Installation 11-56

C
Camshaft and Valve Lifters 11-29
  Removal & Installation 11-29
Clutch 11-43
  Adjustments 11-43
  Removal & Installation 11-43
Coil Spring Front Suspension 11-49
  Removal & Installation 11-49
Coil Spring Rear Suspension 11-50
  Removal & Installation 11-50
Control Arms 11-50
  Removal & Installation 11-50
CV-Joints 11-44
  Overhaul 11-44
Cylinder Head 11-23
  Removal & Installation 11-23

D
Disc Brake Pads 11-53
  Removal & Installation 11-53
Distributor 11-15
  Removal 11-15

E
Engine Assembly 11-15
  Removal & Installation 11-15
Exhaust Manifold 11-28
  Removal & Installation 11-28

F
Front Crankshaft Seal 11-29
  Removal & Installation 11-29
Fuel Filter 11-41
  Removal & Installation 11-41
Fuel Injector 11-41
  Removal & Installation 11-41
Fuel Pump 11-41
  Removal & Installation 11-41
Fuel System Pressure 11-41
  Relieving 11-41
Fuel System Service
  Precautions 11-41

H
Halfshaft 11-43
  Removal & Installation 11-43

Heater Core 11-19
  Removal & Installation 11-19
Hydraulic Clutch System 11-43
  Bleeding 11-43

I
Ignition Timing 11-15
  Adjustment 11-15
Intake Manifold 11-26
  Removal & Installation 11-26

L
Lower Ball Joint 11-49
  Removal & Installation 11-49
Lower Control Arm 11-49
  Control Arm Bushing
    Replacement 11-50
  Removal & Installation 11-49

M
Manual Locking Hubs 11-44
  Removal & Installation 11-44

O
Oil Pan 11-32
  Removal & Installation 11-32
Oil Pump 11-33
  Removal & Installation 11-33

P
Pinion Seal 11-45
  Removal & Installation 11-45
Piston and Ring 11-40
  Positioning 11-40
Power Rack and Pinion Steering
  Gear 11-47
  Removal & Installation 11-47

R
Rear Main Seal 11-34
  Removal & Installation 11-34
Recirculating Ball Steering
  Gear 11-47
  Removal & Installation 11-47
Rocker Arms/Shafts 11-26
  Removal & Installation 11-26

S
Shock Absorber 11-50
  Removal & Installation 11-50
Spindle Bearings 11-45
  Removal, Packing &
    Installation 11-45

Starter Motor 11-32
   Removal & Installation...........11-32
Strut...11-48
   Removal & Installation...........11-48
T
Timing Belt 11-38
   Removal & Installation...........11-38
Timing Chain, Sprockets, Front
  Cover and Seal 11-34
   Removal & Installation...........11-34

Transfer Case Assembly...............11-43
   Removal & Installation...........11-43
Transmission Assembly 11-42
   Removal & Installation...........11-42
V
Valve Lash 11-31
   Adjustment........................11-31
W
Water Pump 11-18
   Removal & Installation...........11-18

Wheel Bearings Front
  Suspension.................................11-50
   Adjustment........................11-50
   Removal & Installation...........11-50
Wheel Bearings Rear
  Suspension.................................11-53
   Adjustment........................11-53
   Removal & Installation...........11-53

SPECIFICATIONS AND MAINTENANCE CHARTS

ENGINE AND VEHICLE IDENTIFICATION CHART

Engine Code								Model Year	
Code	Liters (cc)	Cu. In.	Cyl.	Fuel Sys.	Engine Type	Eng. Mfg.		Code	Year
0	1.6 (1590)	97	4	MFI	SOHC	Suzuki		2	2002
5	2.0 (1995)	121.7	4	MFI	DOHC	Suzuki		3	2003
6	2.5 (2493)	152	6	MFI	DOHC	Suzuki		4	2004
9	2.7 (2736)	167	6	MFI	DOHC	Suzuki		5	2005
								6	2006

MFI: Multiport Fuel Injection

DOHC: Dual Overhead Cam

SOHC: Single Overhead Cam

09490_SUZT_C0001

GENERAL ENGINE SPECIFICATIONS

Year	Model	Engine Displacement Liters (cc)	Engine ID/VIN	Fuel System Type	Net Horsepower @ rpm	Net Torque @ rpm (ft. lbs.)	Bore x Stroke (in.)	Compression Ratio	Oil Pressure @ rpm
2002	Vitara	1.6 (1590)	0	MFI	95@5600	98@4000	2.95x3.54	9.5:1	47-61@4000
		2.0 (1995)	5	MFI	127@6000	134@3000	3.31x3.54	9.3:1	55-67@4000
	Grand Vitara	2.5 (2493)	6	MFI	165@6500	162@4000	3.31x2.95	9.5:1	55-67@4000
	XL-7	2.7 (2736)	9	MFI	185@6000	184@4000	3.46x2.95	9.5:1	55-67@4000
2003	Vitara	2.0 (1995)	5	MFI	127@6000	134@3000	3.31x3.54	9.3:1	55-67@4000
	Grand Vitara	2.5 (2493)	6	MFI	165@6500	162@4000	3.31x2.95	9.5:1	55-67@4000
	XL-7	2.7 (2736)	9	MFI	185@6000	184@4000	3.46x2.95	9.5:1	55-67@4000
2004	Vitara	2.0 (1995)	5	MFI	127@6000	134@3000	3.31x3.54	9.3:1	55-67@4000
	Grand Vitara	2.5 (2493)	6	MFI	165@6500	162@4000	3.31x2.95	9.5:1	55-67@4000
	XL-7	2.7 (2736)	9	MFI	185@6000	184@4000	3.46x2.95	9.5:1	55-67@4000
2005	Grand Vitara	2.5 (2493)	6	MFI	165@6500	162@4000	3.31x2.95	9.5:1	55-67@4000
	XL-7	2.7 (2736)	9	MFI	185@6000	184@4000	3.46x2.95	9.5:1	55-67@4000
2006	Grand Vitara	2.7 (2736)	9	MFI	185@6000	184@4500	3.46x2.95	9.5:1	36.25@3000
	XL-7	2.7 (2736)	9	MFI	185@6000	184@4000	3.46x2.95	9.5:1	55-67@4000

MFI: Multi-port Fuel Injection

09490_SUZT_C0002

ENGINE TUNE-UP SPECIFICATIONS

Year	Engine Displacement Liters (cc)	Engine ID/VIN	Spark Plugs Gap (in.)	Ignition Timing (deg.) MT	Ignition Timing (deg.) AT	Fuel Pump (psi)	Idle Speed (rpm) MT	Idle Speed (rpm) AT	Valve Clearance In.	Valve Clearance Ex.
2002	1.6 (1590)	0	0.040	5B	5B	30-37	700-800	700-800	0.0050-0.0070	0.0050-0.0070
	2.0 (1995)	5	0.040	5B	5B	30-37	700-800	700-800	HYD	HYD
	2.5 (2493)	6	0.039-0.043	5B	5B	38-44	700-800	700-800	HYD	HYD
	2.7 (2736)	9	0.039-0.043	5B	5B	38-44	640-740	640-740	HYD	HYD
2003	2.0 (1995)	5	0.040	5B	5B	30-37	700-800	700-800	HYD	HYD
	2.5 (2493)	6	0.039-0.043	5B	5B	38-44	700-800	700-800	HYD	HYD
	2.7 (2736)	9	0.039-0.043	5B	5B	38-44	640-740	640-740	HYD	HYD
2004	2.0 (1995)	5	0.040	5B	5B	30-37	700-800	700-800	HYD	HYD
	2.5 (2493)	6	0.039-0.043	5B	5B	38-44	700-800	700-800	HYD	HYD
	2.7 (2736)	9	0.039-0.043	5B	5B	38-44	600-700	600-700	HYD	HYD
2005	2.5 (2493)	6	0.039-0.043	5B	5B	38-44	700-800	700-800	HYD	HYD
	2.7 (2736)	9	0.039-0.043	5B	5B	38-44	600-700	600-700	HYD	HYD
2006	2.7 (2736)	9	0.039-0.043	5B	5B	38-44	600-700	600-700	HYD	HYD

HYD: Hydraulic

09490_SUZT_C0003

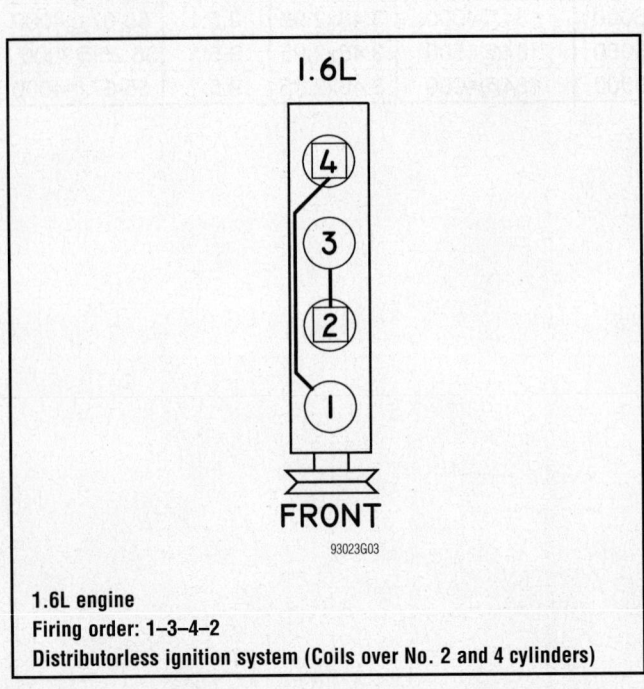

1.6L

FRONT

93023G03

1.6L engine
Firing order: 1–3–4–2
Distributorless ignition system (Coils over No. 2 and 4 cylinders)

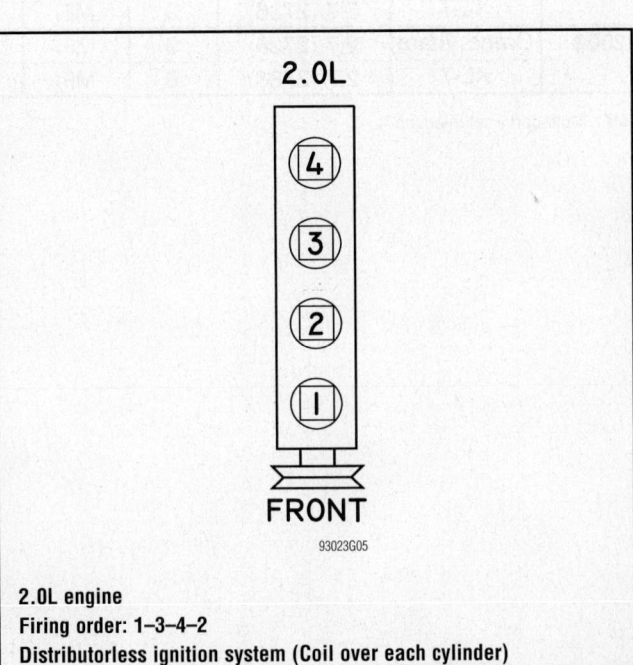

2.0L

FRONT

93023G05

2.0L engine
Firing order: 1–3–4–2
Distributorless ignition system (Coil over each cylinder)

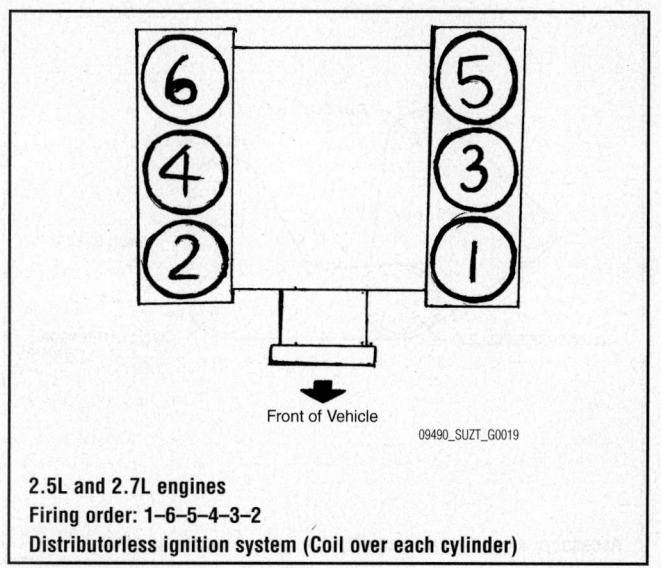

2.5L and 2.7L engines
Firing order: 1–6–5–4–3–2
Distributorless ignition system (Coil over each cylinder)

09490_SUZT_G0019

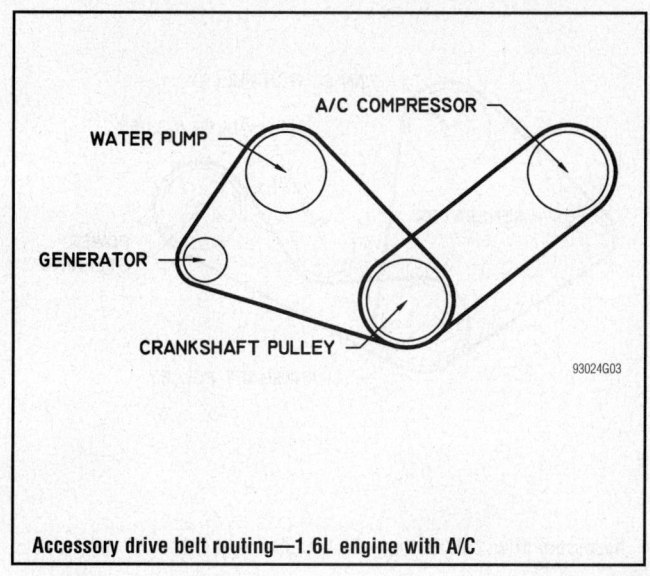

Accessory drive belt routing—1.6L engine with A/C

93024G03

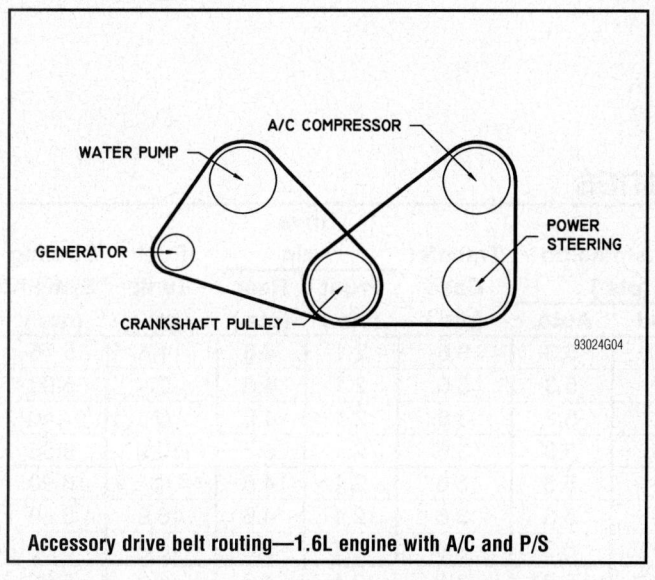

Accessory drive belt routing—1.6L engine with A/C and P/S

93024G04

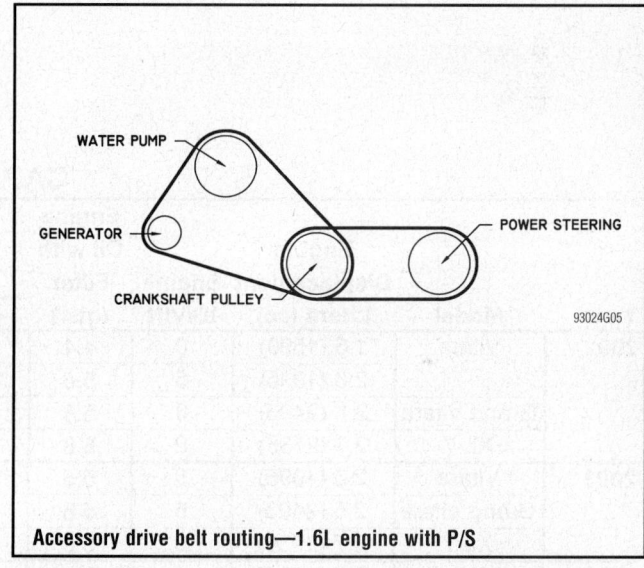

Accessory drive belt routing—1.6L engine with P/S

93024G05

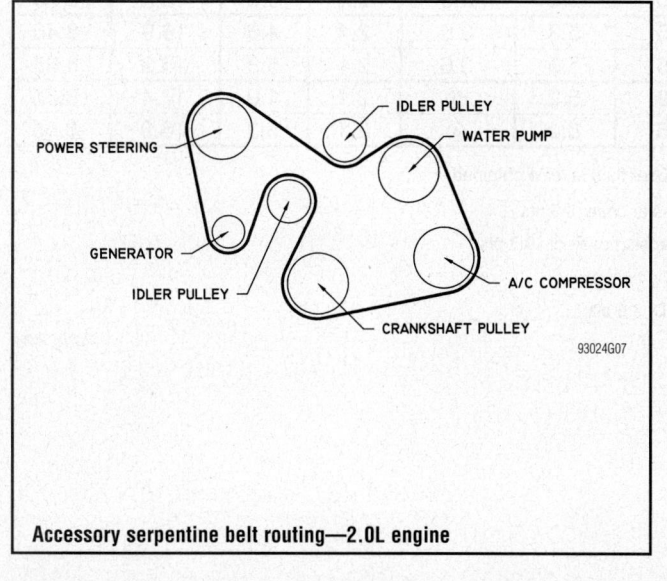

Accessory serpentine belt routing—2.0L engine

93024G07

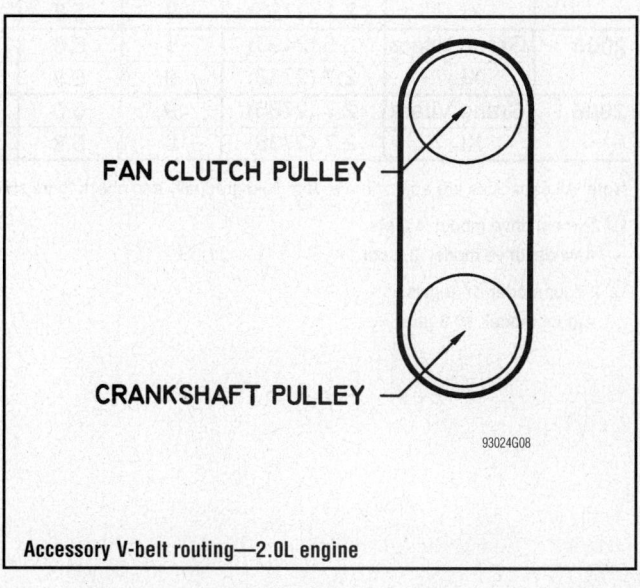

Accessory V-belt routing—2.0L engine

93024G08

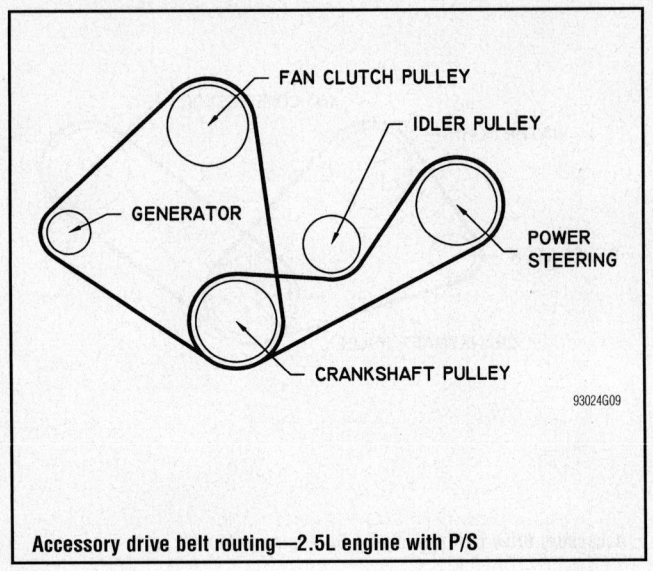

Accessory drive belt routing—2.5L engine with P/S

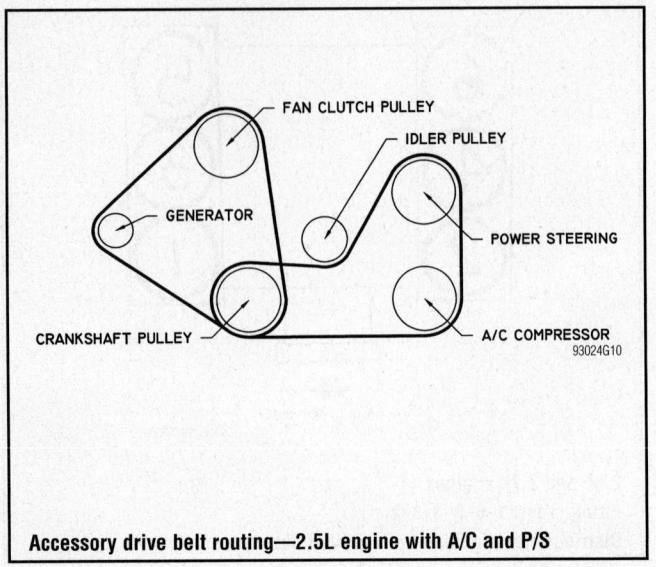

Accessory drive belt routing—2.5L engine with A/C and P/S

CAPACITIES

Year	Model	Engine Displacement Liters (cc)	Engine ID/VIN	Engine Oil with Filter (qts.)	Transmission (pts.) 5-Spd	Transmission (pts.) Auto.	Transfer Case (pts.)	Drive Axle Front (pts.)	Drive Axle Rear (pts.)	Fuel Tank (gal.)	Cooling System (qts.)
2002	Vitara	1.6 (1590)	0	4.4	①	5.3	3.6	2.1	4.6	14.8	5.75
		2.0 (1995)	5	5.5	①	5.3	3.6	2.1	4.6	②	6.90
	Grand Vitara	2.5 (2493)	6	5.8	①	5.3	3.6	2.1	4.6	②	7.40
	XL-7	2.7 (2736)	9	5.8	5.5	5.3	3.6	2.1	5.3	16.9	8.65
2003	Vitara	2.0 (1995)	5	5.5	①	5.3	3.6	2.1	4.6	②	6.90
	Grand Vitara	2.5 (2493)	6	5.8	5.5	5.3	3.6	2.1	4.6	16.9	8.40
	XL-7	2.7 (2736)	9	5.8	5.5	5.3	3.6	2.1	5.3	16.9	8.65
2004	Vitara	2.0 (1995)	5	5.5	①	5.3	3.6	2.1	4.6	②	6.90
	Grand Vitara	2.5 (2493)	6	5.8	5.5	5.3	3.6	2.1	4.6	16.9	8.40
	XL-7	2.7 (2736)	9	5.8	5.5	5.3	3.6	2.1	5.3	16.9	8.65
2005	Grand Vitara	2.5 (2493)	6	5.8	5.5	5.3	3.6	2.1	4.6	16.9	8.40
	XL-7	2.7 (2736)	9	5.8	5.5	5.3	3.6	2.1	5.3	16.9	8.65
2006	Grand Vitara	2.7 (2736)	9	5.0	4.0	5.2	③	2.1	1.9	17.4	8.65
	XL-7	2.7 (2736)	9	5.8	5.5	5.3	④	2.1	5.3	16.9	8.65

Note: All capacities are approximate. Add fluid gradually and check to be sure a proper fluid level is obtained.

① 2-wheel drive model: 4.0 pts.
4-wheel drive model: 3.2 pts.

② 2-door model: 14.8 gals.
4-door model: 16.9 gals.

③ Transfer case: 3.3 pts
Extension case oil: 0.8 pts.

④ 2WD: 4.6 pts
4WD: 3.6 pts.

09490_SUZT_C0004

VALVE SPECIFICATIONS

Year	Engine Displacement Liters (cc)	Engine ID/VIN	Seat Angle (deg.)	Face Angle (deg.)	Spring Test Pressure (lbs. @ in.)	Spring Installed Height (in.)	Stem-to-Guide Clearance (in.)		Stem Diameter (in.)	
							Intake	Exhaust	Intake	Exhaust
2002	1.6 (1590)	0	45	45	24-28@1.24	1.245	0.0008-0.0018	0.0018-0.0028	0.2152-0.2157	0.2142-0.2148
	2.0 (1995)	5	45	45	①	1.250	0.0008-0.0027	0.0018 0.0035	0.2348-0.2354	0.2339-0.2344
	2.5 (2493)	6	45	45	②	③	0.0008-0.0028	0.0018-0.0035	0.2348-0.2354	0.2339-0.2344
	2.7 (2736)	9	45	45	②	③	0.0008-0.0028	0.0018-0.0035	0.2348-0.2354	0.2339-0.2344
2003	2.0 (1995)	5	45	45	①	1.250	0.0008-0.0027	0.0018 0.0035	0.2348-0.2354	0.2339-0.2344
	2.5 (2493)	6	45	45	②	③	0.0008-0.0028	0.0018-0.0035	0.2348-0.2354	0.2339-0.2344
	2.7 (2736)	9	45	45	②	③	0.0008-0.0028	0.0018-0.0035	0.2348-0.2354	0.2339-0.2344
2004	2.0 (1995)	5	45	45	①	1.250	0.0008-0.0027	0.0018 0.0035	0.2348-0.2354	0.2339-0.2344
	2.5 (2493)	6	45	45	②	③	0.0008-0.0028	0.0018-0.0035	0.2348-0.2354	0.2339-0.2344
	2.7 (2736)	9	45	45	②	③	0.0008-0.0028	0.0018-0.0035	0.2348-0.2354	0.2339-0.2344
2005	2.5 (2493)	6	45	45	②	③	0.0008-0.0028	0.0018-0.0035	0.2348-0.2354	0.2339-0.2344
	2.7 (2736)	9	45	45	②	③	0.0008-0.0028	0.0018-0.0035	0.2348-0.2354	0.2339-0.2344
2006	2.7 (2736)	9	45	45	②	③	0.0008-0.0028	0.0018-0.0035	0.2348-0.2354	0.2339-0.2344

① Inner: 13.6-17.4@1.08
  Outer: 30.4-39.2@1.25

② Inner: 13.0-17.4@1.08
  Outer: 29.3-39.2@1.25

③ Inner: 1.3780-1.4205 inches
  Outer: 1.5441-1.5921 inches

09490_SUZT_C0005

CRANKSHAFT AND CONNECTING ROD SPECIFICATIONS

All measurements are given in inches.

Year	Engine Displacement Liters (cc)	Engine ID/VIN	Crankshaft				Connecting Rod		
			Main Brg. Journal Dia.	Main Brg. Oil Clearance	Shaft End-play	Thrust on No.	Journal Diameter	Oil Clearance	Side Clearance
2002	1.6 (1590)	0	2.0465-2.0472	0.0006-0.0023	0.0044-0.0149	3	1.7316-1.7322	0.0008-0.0031	NA
	2.0 (1995)	5	2.2828-2.2834	0.0008-0.0023	0.0039-0.0165	3	1.9660-1.9709	0.0008-0.0031	NA
	2.5 (2493)	6	①	0.0010-0.0023	0.0039-0.0150	2	1.9678-1.9685	0.0018-0.0031	0.0098-0.0177
	2.7 (2736)	9	①	0.0010-0.0023	0.0039-0.0150	2	1.9678-1.9685	0.0018-0.0031	0.0098-0.0177
2003	2.0 (1995)	5	2.2828-2.2834	0.0008-0.0023	0.0039-0.0165	3	1.9660-1.9709	0.0008-0.0031	NA
	2.5 (2493)	6	①	0.0010-0.0023	0.0039-0.0150	2	1.9678-1.9685	0.0018-0.0031	0.0098-0.0177
	2.7 (2736)	9	①	0.0010-0.0023	0.0039-0.0150	2	1.9678-1.9685	0.0018-0.0031	0.0098-0.0177
2004	2.0 (1995)	5	2.2828-2.2834	0.0008-0.0023	0.0039-0.0165	3	1.9660-1.9709	0.0008-0.0031	NA
	2.5 (2493)	6	①	0.0010-0.0023	0.0039-0.0150	2	1.9678-1.9685	0.0018-0.0031	0.0098-0.0177
	2.7 (2736)	9	①	0.0010-0.0023	0.0039-0.0150	2	1.9678-1.9685	0.0018-0.0031	0.0098-0.0177
2005	2.5 (2493)	6	①	0.0010-0.0023	0.0039-0.0150	2	1.9678-1.9685	0.0018-0.0031	0.0098-0.0177
	2.7 (2736)	9	①	0.0010-0.0023	0.0039-0.0150	2	1.9678-1.9685	0.0018-0.0031	0.0098-0.0177
2006	2.7 (2736)	9	①	0.0010-0.0023	0.0039-0.0150	2	1.9678-1.9685	0.0018-0.0031	0.0098-0.0177

NA: Not Available

① Journal 1: 2.5591-2.5593 inches
Journal 2: 2.5588-2.5591 inches
Journal 3: 2.5586-2.5588 inches

09490_SUZT_C0006

PISTON AND RING SPECIFICATIONS

All measurements are given in inches.

Year	Engine Displacement Liters (cc)	Engine ID/VIN	Piston Clearance	Ring Gap			Ring Side Clearance		
				Top Compression	Bottom Compression	Oil Control	Top Compression	Bottom Compression	Oil Control
2002	1.6 (1590)	0	0.0008-0.0015	0.0079-0.0275	0.0138-0.0275	0.0039-0.0669	0.0012-0.0027	0.0008-0.0023	NA
	2.0 (1995)	5	0.0008-0.0015	0.0079-0.0276	0.0138-0.0276	0.0079-0.0709	0.0012-0.0027	0.0008-0.0023	NA
	2.5 (2493)	6	0.0008-0.0015	0.0079-0.0276	0.0138-0.0276	0.0079-0.0709	0.0012-0.0027	0.0008-0.0023	NA
	2.7 (2736)	9	0.0008-0.0015	0.0079-0.0276	0.0138-0.0276	0.0079-0.0709	0.0012-0.0027	0.0008-0.0023	NA
2003	2.0 (1995)	5	0.0008-0.0015	0.0079-0.0276	0.0138-0.0276	0.0079-0.0709	0.0012-0.0027	0.0008-0.0023	NA
	2.5 (2493)	6	0.0008-0.0015	0.0079-0.0276	0.0138-0.0276	0.0079-0.0709	0.0012-0.0027	0.0008-0.0023	NA
	2.7 (2736)	9	0.0008-0.0015	0.0079-0.0276	0.0138-0.0276	0.0079-0.0709	0.0012-0.0027	0.0008-0.0023	NA
2004	2.0 (1995)	5	0.0008-0.0015	0.0079-0.0276	0.0138-0.0276	0.0079-0.0709	0.0012-0.0027	0.0008-0.0023	NA
	2.5 (2493)	6	0.0008-0.0015	0.0079-0.0276	0.0138-0.0276	0.0079-0.0709	0.0012-0.0027	0.0008-0.0023	NA
	2.7 (2736)	9	0.0008-0.0015	0.0079-0.0276	0.0138-0.0276	0.0079-0.0709	0.0012-0.0027	0.0008-0.0023	NA
2005	2.5 (2493)	6	0.0008-0.0015	0.0079-0.0276	0.0138-0.0276	0.0079-0.0709	0.0012-0.0027	0.0008-0.0023	NA
	2.7 (2736)	9	0.0008-0.0015	0.0079-0.0276	0.0138-0.0276	0.0079-0.0709	0.0012-0.0027	0.0008-0.0023	NA
2006	2.7 (2736)	9	0.0008-0.0015	0.0079-0.0276	0.0138-0.0276	0.0079-0.0709	0.0012-0.0027	0.0008-0.0023	NA

NA: Not Available

09490_SUZT_C0007

TORQUE SPECIFICATIONS
All readings in ft. lbs.

Year	Engine Displacement Liters (cc)	Engine ID/VIN	Cylinder Head Bolts	Main Bearing Bolts	Rod Bearing Bolts	Crankshaft Damper Bolts	Flywheel Bolts	Manifold Intake	Manifold Exhaust	Spark Plugs	Lug Nut
2002	1.6 (1590)	0	①	36-41	24-26	94 ②	58	13-20	13-20	14-21	58-80
	2.0 (1995)	5	③	④	33	109	51	13-20	13-20	14-21	58-80
	2.5 (2493)	6	③	⑤	32.5	109	51	16.0	21	18	72.5
	2.7 (2736)	9	③	⑤	32.5	109	51	16.0	21	18	72.5
2003	2.0 (1995)	5	③	④	33	109	51	13-20	13-20	14-21	58-80
	2.5 (2493)	6	③	⑤	32.5	109	51	16.0	21	18	72.5
	2.7 (2736)	9	③	⑤	32.5	109	51	16.0	21	18	72.5
2004	2.0 (1995)	5	③	④	33	109	51	13-20	13-20	14-21	58-80
	2.5 (2493)	6	③	⑤	32.5	109	51	16.0	21	18	72.5
	2.7 (2736)	9	③	⑤	32.5	109	51	16.0	21	18	72.5
2005	2.5 (2493)	6	③	⑤	32.5	109	51	16.0	21	18	72.5
	2.7 (2736)	9	③	⑤	32.5	109	51	16.0	21	18	72.5
2006	2.7 (2736)	9	③	⑤	32.5	109	51	16.0	21	18	72.5

① Step 1: 26 ft. lbs.

  Step 2: 41 ft. lbs.

  Step 3: Loosen in reverse order to 0 ft. lbs.

  Step 4: 26 ft. lbs.

  Step 5: 52 ft. lbs.

② Value shown is for crankshaft timing belt sprocket

③ Step 1: 38 ft. lbs.

  Step 2: 61 ft. lbs.

  Step 3: Loosen in reverse order to 0 ft. lbs.

  Step 4: 38 ft. lbs.

  Step 5: 76 ft. lbs.

  Step 6: Tighten 6mm bolt to 8 ft. lbs.

④ 10mm: 43.5 ft. lbs.

  8mm: 19.5 ft. lbs.

⑤ Step 1: 30.5 ft. lbs.

  Step 2: Loosen in reverse order to 0 ft. lbs.

  Step 3: 30.5 ft. lbs.

  Step 4: 43.5 ft. lbs.

  Step 5: Tighten 8mm bolt to 19.5 ft. lbs.

09490_SUZT_C0008

WHEEL ALIGNMENT

Year	Model	Caster Range (+/-Deg.)	Caster Preferred Setting (Deg.)	Camber Range (+/-Deg.)	Camber Preferred Setting (Deg.)	Toe-in (in.)	Steering Axis Inclination (Deg.)
2002	Vitara	0.00	+2.40	0.00	0.00	0+/-0.08	—
	Grand Vitara	0.00	+2.40	0.00	0.00	0+/-0.08	—
	XL-7	0.00	+2.30	0.00	0.00	0.04+/-0.08	—
2003	Vitara	0.00	+2.40	0.00	0.00	0+/-0.08	—
	Grand Vitara	0.00	+2.40	0.00	0.00	0+/-0.08	—
	XL-7	0.00	+2.30	0.00	0.00	0.04+/-0.08	—
2004	Vitara	0.00	+2.40	0.00	0.00	0+/-0.08	—
	Grand Vitara	0.00	+2.40	0.00	0.00	0+/-0.08	—
	XL-7	0.00	+2.30	0.00	0.00	0.04+/-0.08	—
2005	Grand Vitara	0.00	+2.40	0.00	0.00	0+/-0.08	—
	XL-7	0.00	+2.30	0.00	0.00	0.04+/-0.08	—
2006	Grand Vitara	0.00	+2.30	①	①	②	—
	XL-7	0.00	+2.30	0.00	0.00	0.04+/-0.08	—

① Front: 0.00 degrees
  Rear: -1.15+/-0.15 degrees

② Front: 0+/-0.08 inch
  Rear: 0.24+/-0.06 inch

09490_SUZT_C0009

TIRE, WHEEL AND BALL JOINT SPECIFICATIONS

Year	Model	OEM Tires Standard	OEM Tires Optional	Tire Pressures (psi) Front	Tire Pressures (psi) Rear	Wheel Size	Ball Joint Inspection
2002	Vitara 1.6L 2wd	P195/75R15	none	26	26	5.5-JJ	①
	Vitara 1.6L 4wd	P205/75R15	none	26	26	6.5-JJ	①
	Vitara 2.0L	P215/85R16	none	26	26	6.5-JJ	①
	Grand Vitara	P235/60R16	none	26	26	7-JJ	①
	XL-7	P235/60R16	none	26	26	7-JJ	①
2003	Vitara	P215/85R16	none	26	26	6.5-JJ	①
	Grand Vitara	P235/60R16	none	26	26	7-JJ	①
	XL-7	P235/60R16	none	26	26	7-JJ	①
2004	Vitara	P215/85R16	none	26	26	6.5-JJ	①
	Grand Vitara	P235/60R16	none	26	26	7-JJ	①
	XL-7	P235/60R16	none	26	26	7-JJ	①
2005	Grand Vitara	P235/60R16	none	26	26	7-JJ	①
	XL-7	P235/60R16	none	26	26	7-JJ	①
2006	Grand Vitara	P225/70R16	P225/65R17	26	26	6.5-J	①
	XL-7	P235/60R16	none	26	26	7-JJ	①

OEM: Original Equipment Manufacturer

PSI: Pounds Per Square Inch

① Replace if any measurable movement is found.

09490_SUZT_C0010

BRAKE SPECIFICATIONS
All measurements in inches unless noted

| Year | Model | Brake Disc | | | Brake Drum Diameter | | | Minimum Lining Thickness | | Brake Caliper | |
		Original Thickness	Minimum Thickness	Max. Runout	Original Inside Diameter	Max. Wear Limit	Max. Machine Diameter	Front	Rear	Bracket Bolts (ft. lbs.)	Mounting Bolts (ft. lbs.)
2002	Vitara	0.670	0.590	0.006	8.66	8.74	8.74	0.08	0.04	61.5	20
	Grand Vitara	0.866	0.787	0.006	8.66	8.74	8.74	0.08	0.04	61.5	①
	XL-7	0.866	0.787	0.006	10.00	10.07	10.07	0.08	0.04	61.5	①
2003	Vitara	0.670	0.590	0.006	8.66	8.74	8.74	0.08	0.04	61.5	20
	Grand Vitara	0.866	0.787	0.006	8.66	8.74	8.74	0.08	0.04	61.5	①
	XL-7	0.866	0.787	0.006	10.00	10.07	10.07	0.08	0.04	61.5	①
2004	Vitara	0.670	0.590	0.006	8.66	8.74	8.74	0.08	0.04	61.5	20
	Grand Vitara	0.866	0.787	0.006	8.66	8.74	8.74	0.08	0.04	61.5	①
	XL-7	0.866	0.787	0.006	10.00	10.07	10.07	0.08	0.04	61.5	①
2005	Grand Vitara	0.866	0.787	0.006	8.66	8.74	8.74	0.08	0.04	61.5	①
	XL-7	0.866	0.787	0.006	10.00	10.07	10.07	0.08	0.04	61.5	①
2006	Grand Vitara	0.984	0.905	0.004	10.00	10.07	10.07	0.08	0.04	61.5	26
	XL-7	0.866	0.787	0.006	10.00	10.07	10.07	0.08	0.04	61.5	①

① 10mm: 37 ft. lbs.
12mm: 62 ft. lbs.

09490_SUZT_C0011

SCHEDULED MAINTENANCE INTERVALS
SUZUKI—VITARA, GRAND VITARA AND XL-7

TO BE SERVICED	TYPE OF SERVICE	VEHICLE MILEAGE INTERVAL (x1000)												
		7.5	15	22.5	30	37.5	45	52.5	60	67.5	75	82.5	90	97.5
Engine oil & filter	R	✓	✓	✓	✓	✓	✓	✓	✓	✓	✓	✓	✓	✓
Automatic transmission fluid ①	S/I	✓	✓	✓	✓	✓	✓	✓	✓	✓	✓	✓	✓	✓
Manual transmission oil ②	S/I	✓	✓	✓	✓	✓	✓	✓	✓	✓	✓	✓	✓	✓
Steering system	S/I	✓	✓	✓	✓	✓	✓	✓	✓	✓	✓	✓	✓	✓
Transfer & differential oil ②	S/I	✓	✓	✓	✓	✓	✓	✓	✓	✓	✓	✓	✓	✓
Wheel discs & free wheeling hubs	S/I	✓	✓	✓	✓	✓	✓	✓	✓	✓	✓	✓	✓	✓
Suspension system	S/I	✓	✓	✓	✓	✓	✓	✓	✓	✓	✓	✓	✓	✓
Brake discs & pads	S/I		✓		✓		✓		✓		✓		✓	
Brake drums & shoes	S/I		✓		✓		✓		✓		✓		✓	
Brake fluid ③	S/I		✓		✓		✓		✓		✓		✓	
Brake hoses & pipes	S/I		✓		✓		✓		✓		✓		✓	
Brake pedal	S/I		✓		✓		✓		✓		✓		✓	
Brake lever & cable	S/I		✓		✓		✓		✓		✓		✓	
Clutch	S/I		✓		✓		✓		✓		✓		✓	
Idle speed	S/I		✓		✓		✓		✓		✓		✓	
Propeller shafts	S/I		✓		✓		✓		✓		✓		✓	
Valve lash (clearance)	S/I		✓		✓		✓		✓		✓		✓	
Wheel bearings	S/I		✓		✓		✓		✓		✓		✓	
Air cleaner filter element	R				✓				✓				✓	
Engine coolant	R				✓				✓				✓	
Fuel filter	R				✓				✓				✓	
Spark plugs	R				✓				✓				✓	
Cooling system hoses	S/I				✓				✓				✓	
Drive belt(s)	S/I				✓				✓				✓	
Exhaust pipes & mountings	S/I				✓				✓				✓	
Fuel lines & connections	S/I				✓				✓				✓	
Camshaft timing belt	R								✓				✓	
Distributor cap & rotor	S/I								✓					
Emission-related hoses & tubes	S/I								✓					
Oxygen sensor	S/I											✓		
EVAP canister	R	every 100,000 miles												
PCV valve	R							✓						
EGR system	S/I							✓						

09490_SUZT_C0012

SCHEDULED MAINTENANCE INTERVALS
SUZUKI—VITARA, GRAND VITARA AND XL-7

TO BE SERVICED	TYPE OF SERVICE	VEHICLE MILEAGE INTERVAL (x1000)												
		7.5	15	22.5	30	37.5	45	52.5	60	67.5	75	82.5	90	97.5
Fuel Injectors	S/I	every 100,000 miles												
TWC converter	S/I	every 100,000 miles												

R: Replace   S/I: Service or Inspect

① Replace at 100,000 miles.

② Replace oil every 30,000 miles.

③ Replace every 60,000 miles.

FREQUENT OPERATION MAINTENANCE (SEVERE SERVICE)

If a vehicle is operated under any of the following conditions it is considered severe service:

- Extremely dusty areas.

- 50% or more of the vehicle operation is in 32°C (90°F) or higher temperatures, or constant operation in temperatures below 0°C (32°F).

- Prolonged idling (vehicle operation in stop and go traffic).

- Frequent short running periods (engine does not warm to normal operating temperatures).

- Police, taxi, delivery usage or trailer towing usage.

Oil & oil filter: replace every 3000 miles.

Air cleaner filter element: service or inspect every 3000 miles & replace every 15,000 miles.

Steering wheel free play, gear box oil & linkage: service or inspect every 3000 miles.

Brake & nuts on chassis: tighten every 6000 miles.

Brake discs & pads (front): service or inspect every 6000 miles.

Brake drums & shoes (rear): service or inspect every 6000 miles.

Exhaust pipes & mountings: tighten every 6000 miles.

Propeller shafts: service or inspect every 6000 miles.

Automatic transmission fluid & filter: replace every 15,000 miles.

Distributor cap & ignition wires: service or inspect every 15,000 miles.

Drive belt(s): service or inspect every 15,000 miles.

Manual transmission oil: replace every 15,000 miles.

Transfer & differential oil: replace every 15,000 miles.

09490_SUZT_C0013

ENGINE REPAIR

➡**Disconnecting the negative battery cable on some vehicles may interfere with the functions of the on board computer system. The computer may undergo a relearning process once the negative battery cable is reconnected.**

Distributor

REMOVAL

All engines are equipped with a Distributorless Ignition System (DIS).

Alternator

REMOVAL

1. Before servicing the vehicle, refer to the Precautions Section.
2. Remove or disconnect the following:
   • Negative battery cable
   • Air cleaner assembly and hose, if necessary
   • Evaporative Emission (EVAP) canister, if necessary
   • Accessory drive belt(s)
   • Alternator harness connectors
   • Alternator mounting bracket
   • Alternator

INSTALLATION

Install or connect the following:
• Alternator
1. Install the alternator mounting bracket. Tighten the bolts as follows:
   • Vitara, XL-7 and 2002–05 Grand Vitara: 20 ft. lbs. (27 Nm)
   • 2006 Grand Vitara: 32.5 ft. lbs. (45 Nm)
2. Install the alternator harness connectors.
3. Install the accessory drive belt(s).
4. Tighten the alternator mounting bolts as follows:
   • Vitara, XL-7 and 2002–05 Grand Vitara: 24 ft. lbs. (33 Nm)
   • 2006 Grand Vitara: upper bolt—18 ft. lbs. (25 Nm)
   • 2006 Grand Vitara: lower bolt—32.5 ft. lbs. (45 Nm)
Install or connect the following:
   • EVAP canister, if necessary
   • Air cleaner assembly and hose, if necessary
   • Negative battery cable

Ignition Timing

ADJUSTMENT

1.6L and 2.7L Engines (2006 Grand Vitara)

This engine is equipped with a Distributorless Ignition System (DIS). All timing functions are controlled by the Powertrain Control Module (PCM). No adjustment is possible.

2.0L, 2.5L and 2.7L Engines (except 2006 Grand Vitara)

➡**The 2.0L, 2.5L and 2.7L engines (except 2006 Grand Vitara) use a Camshaft Position (CMP) sensor that is rotated to set base timing.**

➡**Check and adjust the ignition timing with the engine at normal operating temperature, all electrical accessories OFF and transmission in P, N for automatic transmission or neutral for manual transmission.**

1. Before servicing the vehicle, refer to the Precautions Section.
2. With the engine **OFF**, connect a jumper wire between terminals **D** and **E** of the Data Link Connector (DLC).
3. Connect a timing light to the No. 1 spark plug wire and start the engine.
4. Ignition timing at idle should be 4–6 degrees Before Top Dead Center (BTDC).
5. Adjust the timing as necessary, then turn the engine **OFF**.
6. Remove the jumper wire from the DLC and remove the timing light.

Engine Assembly

REMOVAL & INSTALLATION

1.6L Engine

1. Before servicing the vehicle, refer to the Precautions Section.
2. Relieve the fuel system pressure.
3. Drain the cooling system and engine oil.
4. Remove or disconnect the following:
   • Negative battery cable
   • Hood
   • Strut tower bar, if equipped
   • Cooling fan and shroud
   • Heater hoses
   • Radiator hoses
   • Bypass hose
   • Radiator
   • Air intake tube
   • Accelerator cable
   • Transmission cable, if equipped
   • Positive Crankcase Ventilation (PCV) valve and hose
   • Exhaust Gas Recirculation (EGR) valve and temperature sensor
   • EGR vacuum valve connector
   • EGR bypass valve connector
   • Idle Air Control (IAC) valve hoses and connector
   • Fuel lines
   • Main engine control wiring harness connectors at the firewall
   • Throttle Position (TP) sensor connector
   • Heated Oxygen (HO2S) sensor connectors

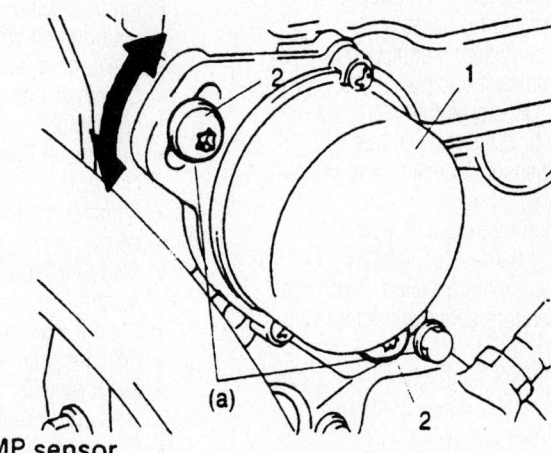

1. CMP sensor
2. Bolt

Camshaft position sensor

7924HG82

- Engine Coolant Temperature (ECT) sensor connector
- Temperature gauge sender connector
- A/C temperature switch, if equipped
- Injector harness connectors
- Alternator wiring connectors
- Manifold Absolute Pressure (MAP) sensor connector and vacuum line
- Brake booster vacuum line
- Evaporative Emission (EVAP) canister and hoses
- Distributor, if equipped
- Front skidplate, if equipped
- Power steering hoses
- A/C compressor, if equipped
- Flywheel access cover
- Torque converter, if equipped
- Clutch cable, if equipped
- Transmission cooler lines, if equipped
- Exhaust front pipe
- Starter motor
- Transmission flange fasteners and support the transmission
- Left and right engine mounts
- Engine

To install:

5. Install or connect the following:
- Engine. Tighten the mount bolts to 40 ft. lbs. (54 Nm).
- Transmission flange fasteners. Tighten them to 62 ft. lbs. (85 Nm).
- Starter motor
- Exhaust front pipe
- Transmission cooler lines, if equipped
- Clutch cable, if equipped
- Torque converter, if equipped. Tighten the bolts to 40 ft. lbs. (54 Nm).
- Flywheel access cover
- A/C compressor, if equipped
- Power steering hoses
- Front skidplate, if equipped
- Distributor, if equipped
- EVAP canister and hoses
- Brake booster vacuum line
- MAP sensor connector and vacuum line
- Alternator wiring connectors
- Injector harness connectors
- A/C temperature switch, if equipped
- Temperature gauge sender connector
- ECT sensor connector
- HO2S sensor connectors
- TP sensor connector
- Main engine control wiring harness connectors at the firewall
- Fuel lines
- IAC valve hoses and connector

- EGR bypass valve connector
- EGR vacuum valve connector
- EGR valve and temperature sensor
- PCV valve and hose
- Transmission cable, if equipped
- Accelerator cable
- Air intake tube
- Radiator
- Bypass hose
- Radiator hoses
- Heater hoses
- Cooling fan and shroud
- Strut tower bar, if equipped
- Hood
- Negative battery cable

6. Fill the crankcase to the correct level.
7. Fill the cooling system.
8. Start the engine and check for leaks.

2.0L, 2.5L and 2.7L Engines (except 2006 Grand Vitara)

1. Before servicing the vehicle, refer to the Precautions Section.
2. Relieve the fuel system pressure.
3. Drain the cooling system.
4. Drain the engine oil.
5. Remove or disconnect the following:
- Negative battery cable
- Hood
- Heater hoses
- Radiator hoses
- Cooling fan and shroud
- Radiator overflow tank
- Radiator
- Accelerator cable
- Transmission cable, if equipped
- Strut tower bar, if equipped
- Air intake assembly
- Engine oil dipstick tube
- Transmission oil dipstick tube, if equipped
- Ignition coil covers
- Ignition coil connectors
- Injector connectors
- Camshaft Position (CMP) sensor connector
- Crankshaft Position (CKP) sensor connector
- Throttle Position (TP) sensor connector
- Mass Air Flow (MAF) sensor connector
- Idle Air Control (IAC) valve
- Intake manifold ground cable
- Evaporative Emission (EVAP) canister purge valve
- Exhaust Gas Recirculation (EGR) valve connector
- Heated Oxygen (HO2S) sensor connectors

- Engine Coolant Temperature (ECT) sensor connector
- Alternator wiring connectors
- Oil pressure gauge sender connector
- Power Steering Pressure (PSP) switch connector
- Alternator bracket ground cable
- Brake booster vacuum line
- Tank pressure control vacuum valve hose
- Fuel lines
- EVAP canister
- Power steering pump
- A/C compressor
- Steering shaft lower assembly
- Front differential housing, if equipped
- Exhaust front pipe and bracket
- Transmission oil cooler lines, if equipped
- Transmission stiffener brackets, if equipped
- Flywheel access cover
- Torque converter, if equipped
- Starter motor
- Transmission flange fasteners and support the transmission
- Left and right engine mounts
- Engine

To install:

6. Install or connect the following:
- Engine
- Left and right engine mounts. Tighten the nuts to 36 ft. lbs. (50 Nm).
- Transmission flange fasteners. Tighten them to 58 ft. lbs. (80 Nm).
- Starter motor
- Torque converter. Tighten the bolts to 47 ft. lbs. (65 Nm).
- Flywheel access cover
- Transmission stiffener brackets. Tighten the bolts to 36 ft. lbs. (50 Nm).
- Transmission oil cooler lines, if equipped
- Exhaust front pipe and bracket
- Front differential housing, if equipped
- Steering shaft lower assembly
- A/C compressor
- Power steering pump
- EVAP canister
- Fuel lines
- Tank pressure control vacuum valve hose
- Brake booster vacuum line
- Alternator bracket ground cable
- PSP switch connector
- Oil pressure gauge sender connector

- Alternator wiring connectors
- ECT sensor connector
- HO$_2$S sensor connectors
- EGR valve connector
- EVAP canister purge valve
- Intake manifold ground cable
- IAC valve
- MAF sensor connector
- TP sensor connector
- CKP sensor connector
- CMP sensor connector
- Injector connectors
- Ignition coil connectors
- Ignition coil covers
- Transmission oil dipstick tube, if equipped
- Engine oil dipstick tube
- Air intake assembly
- Strut tower bar, if equipped
- Transmission cable, if equipped
- Accelerator cable
- Radiator
- Radiator overflow tank
- Cooling fan and shroud
- Radiator hoses
- Heater hoses
- Hood
- Negative battery cable

7. Fill the crankcase to the correct level.
8. Fill the cooling system.
9. Start the engine and check for leaks.

2.7L Engine (2006 Grand Vitara)

1. Before servicing the vehicle, refer to the Precautions Section.
2. Relieve the fuel system pressure.
3. Drain the cooling system.
4. Drain the engine oil.
5. Drain cooling system.
6. Drain transmission oil or A/T fluid.
7. Remove or disconnect the following:
- Negative battery cable
- Hood
- Battery and battery tray
- Surge tank cover
- Air cleaner assembly and air cleaner outlet hose
- MAP sensor
- Injector harness
- Electric throttle body
- EVAP canister purge valve
- Ignition coil assemblies
- CMP sensor
- Alternator
- Starter motor
- Pressure switch of P/S pump
- Magnet clutch switch of A/C compressor, if equipped
- Engine oil pressure switch
- Ground terminals
- Knock sensor

- CKP sensor
- A/F sensor (No.1 and No.2)
- HO$_2$S sensor (No.1 and No.2)
- EGR valve
- ECT sensor
- Back up light switch (M/T)
- Input shaft speed sensor (A/T)
- Output shaft speed sensor (A/T)
- Transmission range sensor (A/T)
- Transmission wire connector (A/T)
- Transfer actuator, if equipped
- Center differential switch, if equipped
- 4 L/N switch, if equipped
- Each wire harness clamps
- Brake booster hose from intake manifold
- EVAP canister purge hose from fuel No.2 pipe
- Vacuum tank hose from intake manifold
- IMT hose from IMT valve
- Radiator inlet hose and outlet hose from each pipe
- Heater inlet hose and outlet hose from each pipe
- Fuel feed hose and return hose from each pipe
- Reservoir hose from water outlet cap
- Clutch oil pipe from transmission front case (M/T)
- A/T fluid cooler hose from radiator (A/T)
- High pressure pipe and suction hose from P/S pump
- Suction hose and discharge hose from A/C compressor
- Shift control lever (M/T)
- A/T select cable from transmission (A/T)
- Exhaust No.1 and No.2 pipes
- Front propeller shafts, if equipped
- Rear propeller shafts

8. Support front suspension frame and engine rear mounting member using jack.

9. Lower engine with front suspension frame as follows:

    a. Hoist vehicle and remove front wheels.

    b. Remove engine under cover.

    c. Remove suspension control arm.

    d. Remove right side and left side front drive shaft assembly, if equipped.

    e. Remove stabilizer joints. When loosening joint nut, hold stud with hexagon wrench.

    f. Disconnect steering lower shaft from pinion shaft.

    g. Detach low pressure return hose from low pressure return pipe and then disconnect pipe bracket.

    h. Remove gear box union bolt.

    i. Fix radiator to body with rope to avoid the radiator from falling off when front suspension frame lowered.

    j. Support engine assembly by using a chain hoist.

    k. Support suspension frame at the specified positions.

- Front suspension frame mounting bolts
- Engine rear mounting member bolts

10. Before lowering engine, recheck to make sure all hoses, electric wires and cables are disconnected from engine.

11. Lower engine with transmission, transfer, front suspension frame and engine rear mounting member from engine compartment.

- Transmission from engine, if necessary
- Engine with engine front mounting bracket from engine front mounting, if necessary
- Clutch cover and clutch disk, if necessary

To install:

12. Install clutch cover and clutch disc, if removed.

13. Install engine with engine front

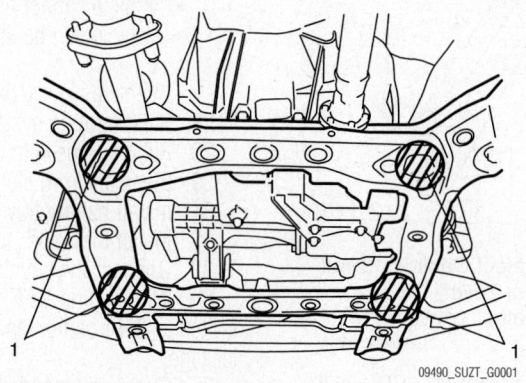

09490_SUZT_G0001

Support the suspension frame at the designated mounting points shown—2006 Grand Vitara

mounting bracket to engine front mounting. Tighten the engine front mounting bracket bolts to 40 ft. lbs. (55 Nm).

➡ **Be sure to align dowel pin of engine front mounting with dowel hole of engine front mounting bracket.**

14. Connect transmission to engine, if removed.

15. Lift engine with transmission, front suspension frame and engine rear mounting member into engine compartment with jack.

16. Tighten engine rear mounting member bolts to 40 ft. lbs. (55 Nm).

17. Lift engine with front suspension frame to install in vehicle as follows:

a. Tighten suspension frame mounting bolts to 98 ft. lbs. (135 Nm) and engine front body side mounting nuts to 40 ft. lbs. (55 Nm).

➡ **If reuse suspension frame mounting bolt, apply engine oil to thread, bearing and trunk surface.**

b. Remove chain hoist from engine and rope from the radiator.

c. Install front propeller shaft, if equipped.

d. Tighten pipe bracket bolts to 8 ft. lbs. (11 Nm) and then insert low pres-

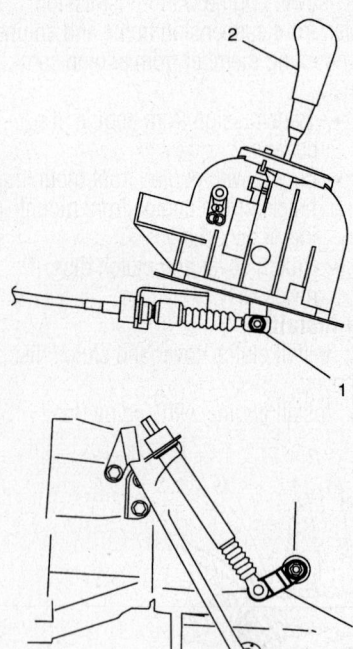

1. Manual Select Cable End Nut
2. Shift Select Lever
3. Manual Select Lever

09490_SUZT_G0002

Shift select cable adjustment (A/T)—2006 Grand Vitara

sure return hose to low pressure return pipe.

e. Tighten union gear box bolt to 25.5 ft. lbs. (35 Nm).

f. Connect steering lower shaft to pinion shaft steering.

g. Install stabilizer joints, and tighten nuts to 43.5 ft. lbs. (60 Nm). When tightening, hold stud with hexagon wrench.

h. Install right side and left side front drive shaft assembly, if equipped.

i. Install suspension control arm.

j. Install engine under cover.

k. Install front wheels and lower hoist.

l. After installation, be sure to fill specified power steering fluid and bleed air.

18. Remove engine jack.

19. Install rear propeller shafts.

20. Install front propeller shafts, if equipped.

21. Install exhaust No.1 and No.2.

22. Connect suction hose and discharge hose to A/C compressor.

23. Connect high pressure pipe and suction hose to the P/S pump.

24. Return disconnected hoses, cables and electric wire.

25. Check all removed parts are back in place.

26. For A/T model, adjust select cable as follows:

a. Remove manual selector, with select cable connected.

b. Loosen manual select cable end nut.

c. Shift select lever and manual select lever to **N**.

d. Tighten manual select cable end nut to 9.5 ft. lbs. (13 Nm)

27. After select cable was installed, check for the following:

- Push vehicle with selector lever shifted to **P**. Vehicle should not move.
- Vehicle cannot be driven in **N**.
- Vehicle can be driven in **D**, **4** and **L**.
- Vehicle can be backed in **R**.

28. Refill cooling system.

29. Refill engine with engine oil.

30. Bleed air from clutch system.

31. Install battery tray and battery.

32. Install the hood.

33. Connect the negative battery cable.

34. With engine **OFF**, turn ignition switch to **ON** position and check for fuel leakage.

35. Start engine and check coolant, oil and exhaust gas leakage at each connection.

Water Pump

REMOVAL & INSTALLATION

1.6L Engines

1. Before servicing the vehicle, refer to the Precautions Section.

2. Drain the cooling system.

3. Remove or disconnect the following:

- Negative battery cable
- Accessory drive belts
- Cooling fan and shroud
- Front cover
- Timing belt
- Oil dipstick tube
- Alternator bracket
- Timing belt tensioner
- Water pump

To install:

4. Install or connect the following:

- Water pump with a new gasket. Tighten the bolts to 106 inch lbs. (12 Nm).
- Timing belt tensioner
- Alternator bracket
- Oil dipstick tube. Tighten the bolt to 97 inch lbs. (11 Nm).
- Timing belt
- Front cover
- Cooling fan and shroud
- Accessory drive belts
- Negative battery cable

5. Fill the cooling system.

6. Start the engine and check for leaks.

2.0L Engines

1. Before servicing the vehicle, refer to the Precautions Section.

2. Drain the cooling system.

3. Remove or disconnect the following:

- Negative battery cable
- Radiator hose at the thermostat housing
- Heater outlet pipe bolt

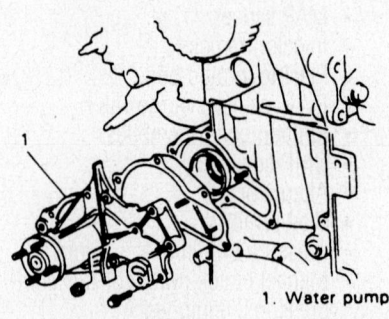

1. Water pump

7924HG04

Exploded view of the water pump mounting—1.6L engines

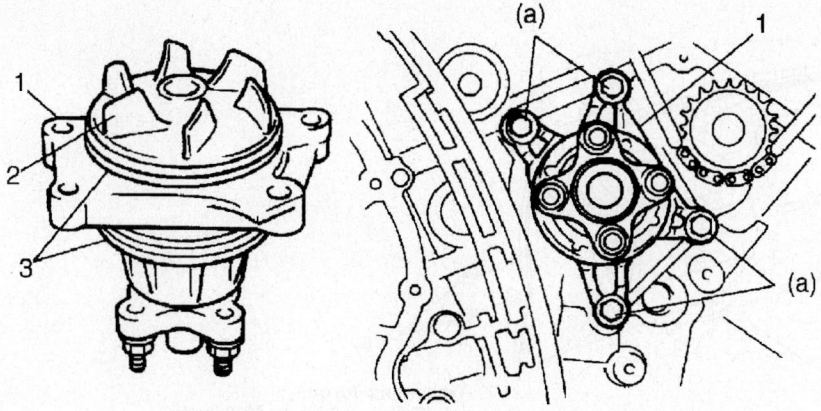

1. Water Pump
2. Impeller
3. O-ring Seals
(a) Water Pump Mounting Bolts

09490_SUZT_G0003

Water pump and related components—2.5L and 2.7L engines

- Alternator belt
- Water pump

To install:

➡**Use new water pump bolts for assembly.**

4. Install or connect the following:
- Water pump with a new O-ring seal. Tighten the bolts to 19 ft. lbs. (27 Nm).
- Alternator belt
- Heater outlet pipe bolt
- Radiator hose at the thermostat housing
- Negative battery cable
5. Fill the cooling system.
6. Start the engine and check for leaks.

2.5L and 2.7L Engines

1. Before servicing the vehicle, refer to the Precautions Section.
2. Drain engine oil.
3. Drain the cooling system.
4. Remove or disconnect the following:
- Negative battery cable
- Accessory drive belts
- Front cover
- Water pump

To install:

5. Install or connect the following:
- Water pump with a new O-ring seal. Tighten the bolts to 19 ft. lbs. (27 Nm).
- Front cover
- Accessory drive belts
- Negative battery cable
6. Fill the cooling system.
7. Fill engine with engine oil.
8. Start the engine and check for leaks.

Heater Core

REMOVAL & INSTALLATION

All Models (except 2006 Grand Vitara)

1. Before servicing the vehicle, refer to the Precautions Section.
2. Disconnect the negative battery cable.
3. Disable the air bag system.
4. Drain the cooling system into a clean container for reuse.
5. Disconnect the heater hoses from the heater core.
6. Remove the instrument panel by performing the following procedure:
   a. Remove the console.
   b. Remove the glove box and the column hole cover.
   c. Disconnect the electrical connector and cables from the heater housing and blower motor assembly.
   d. Remove the steering column.
   e. Disconnect the speedometer connector and the speedometer assembly.
   f. Remove the hood opener.
   g. Disconnect the instrument panel electrical connectors.
   h. Remove the instrument panel-to-chassis screws and bolts.
   i. Using an assistant, remove the instrument panel.
7. If equipped with air conditioning, perform the following procedure:
   a. Discharge and recover the air conditioning refrigerant.
   b. If equipped with a G16 or a J20 engine, disconnect the suction pipe and liquid pipe from the air conditioning housing. Plug the openings to prevent contamination.
   c. If equipped with an H25 or H27 engine, disconnect the compressor suction pipe and receiver/drier outlet pipe from the air conditioning housing. Plug the openings to prevent contamination.
   d. Remove the blower motor assembly.
   e. Disconnect the thermistor wire coupler.
   f. Remove the air conditioning housing.
8. Disconnect the rear duct from the heater housing.
9. Disconnect the mode actuator electrical connectors.
10. If equipped, remove the air conditioning controller.
11. If equipped, disconnect and remove the Sensing and Diagnostic Module (SDM) or air bag controller module.
12. Remove the heater housing.
13. Remove the heater core pipe clamps and grommet.
14. Remove the heater core from the heater housing.

To install:

15. Install the heater core to the heater housing.
16. Install the heater core pipe clamps and grommet.
17. Install the heater housing.
18. If equipped, connect and install the Sensing and Diagnostic Module (SDM) or air bag controller module.
19. If equipped, install the air conditioning controller.
20. Connect the mode actuator electrical connectors.
21. Connect the rear duct from the heater housing.
22. If equipped with air conditioning, perform the following procedure:
   a. Install the air conditioning housing.
   b. Connect the thermistor wire coupler.
   c. Install the blower motor assembly.
   d. If equipped with an H25 or H27 engine, connect the compressor suction pipe and receiver/drier outlet pipe to the air conditioning housing.
   e. If equipped with a G16 or a J20 engine, connect the suction pipe and liquid pipe to the air conditioning housing.
   f. Evacuate and charge the air conditioning system.
23. Install the instrument panel by performing the following procedure:
   a. Using an assistant, install the instrument panel.

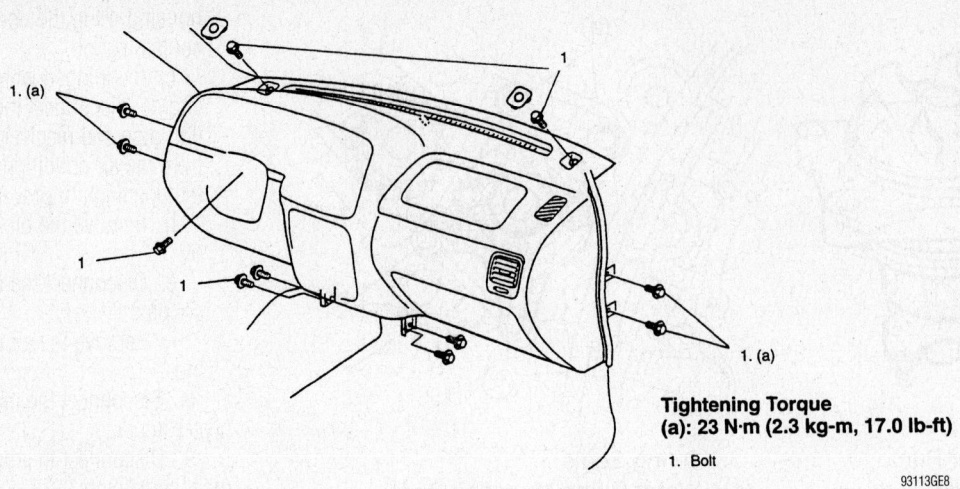

Tightening Torque
(a): 23 N·m (2.3 kg-m, 17.0 lb-ft)

1. Bolt

93113GE8

View of the instrument panel and fasteners—Suzuki Vitara

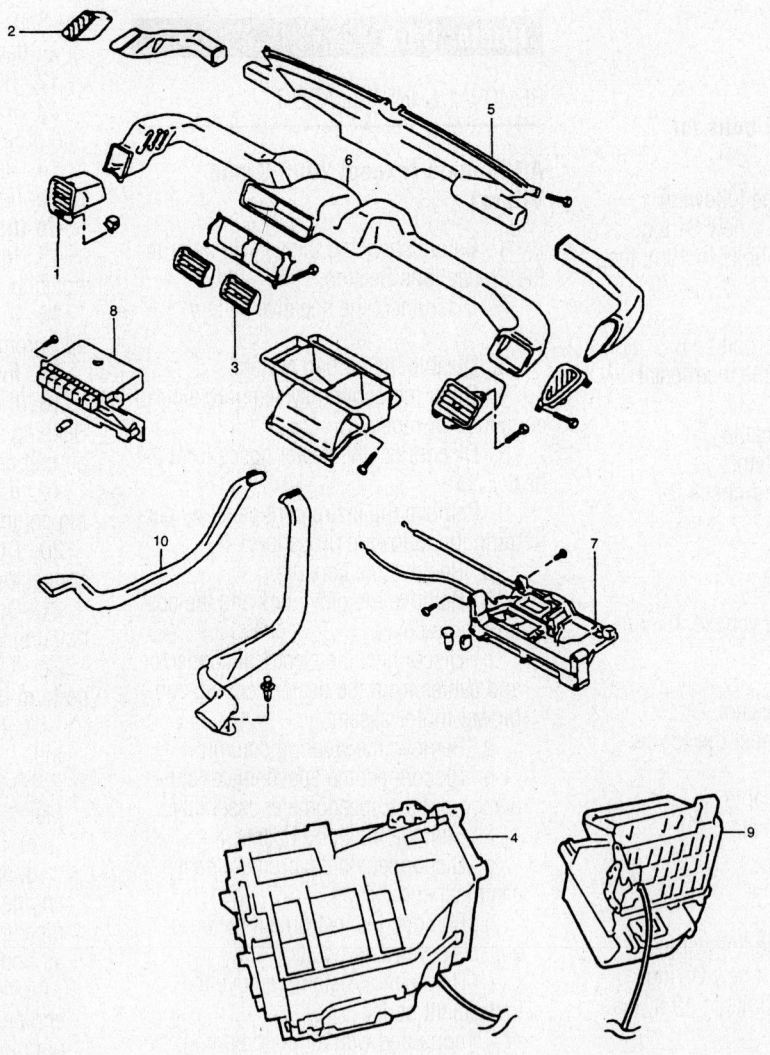

1. Side ventilator outlet
2. Side defroster outlet
3. Center ventilatior outlet
4. Heater unit
5. Defroster duct
6. Ventilator duct
7. Control lever
8. Mode control switch
9. Blower unit
10. Rear duct

93113GE9

Exploded view of the heater housing and ventilation ducts—Suzuki Vitara

b. Install the instrument panel-to-chassis screws and bolts.

c. Connect the instrument panel electrical connectors.

d. Install the hood opener.

e. Connect the speedometer connector and the speedometer assembly.

f. Install the steering column.

g. Connect the electrical connector and cables to the heater housing and blower motor assembly.

h. Install the glove box and the column hole cover.

i. Install the console.

24. Connect the heater hoses to the heater core.

25. Refill the cooling system.

26. To enable the air bag system, perform the following procedure:

a. Push inward on the glove box while pushing the stopper located at both sides and connect the passenger's side air bag module yellow connector and lock it.

b. Under the steering column, connect the contact coil/combination switch assembly's yellow connector.

c. In the fuse box, install the AIR BAG fuse.

d. Turn the ignition switch ON and verify that the AIR BAG warning light flashes 7 times and turns OFF; if the system does not operate as described, per-

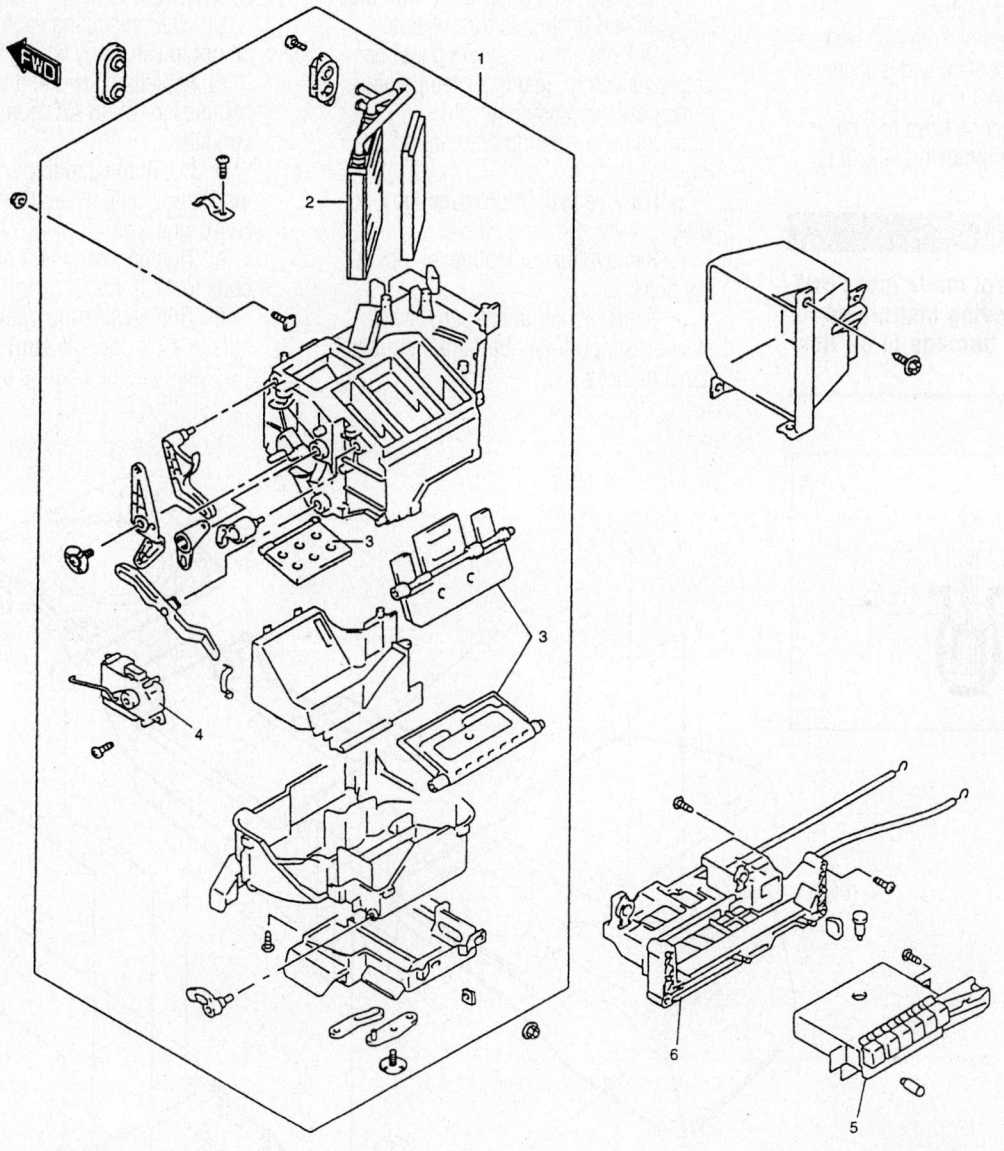

1. Heater assembly
2. Heater core
3. Damper
4. Mode actuator
5. Mode control switch
6. Control lever assembly

93113GE0

Exploded view of the heater core, heater housing and related components—Suzuki Vitara

form the Air Bad Diagnostic System Check.

27. Connect the negative battery cable.

28. Run the engine to normal operating temperatures; then, check the climate control operation and check for leaks.

2006 Grand Vitara

1. Before servicing the vehicle, refer to the Precautions Section.

2. Disconnect the negative battery cable.

3. Disable air bag system.

4. Recover refrigerant from A/C system.

5. Drain engine coolant and disconnect heater hoses from HVAC unit.

6. Disconnect suction hose and condenser outlet hose by removing attaching bolt.

❄❄ CAUTION

Position heat control mode into FOOT MODE before removing instrument panel to avoid the damage to air flow control door.

7. Remove instrument panel as follows:

a. Remove driver side instrument panel under cover and passenger side instrument panel under cover.

b. Remove steering column hole cover.

c. Turn steering wheel to remove steering column cover screws.

d. Remove steering column covers.

e. Remove glove box.

f. Remove hood latch release lever.

g. Remove console box.

h. Remove front pillar trim, front side sill scuff and dash side trim panels.

i. Disconnect instrument panel harness connectors, inside air temperature sensor duct and antenna cable.

j. Remove steering column mounting.

k. Remove instrument panel ground wire.

l. Remove instrument panel mounting bolts.

m. Remove instrument panel with steering support member and instrument panel harness.

8. Disconnect rear duct from HVAC unit.

9. Detach wiring, connectors and clamps from HVAC unit.

10. Remove HVAC unit.

11. Remove heater core pipe clamp and then, pull out heater core from HVAC unit.

To install:

12. Install heater core by reversing removal procedure, noting the following items:

a. When installing heater core, be careful not to damage fins of heater core.

b. When installing each part, be careful not to catch any wiring harness.

c. Replenish specified amount of A/C compressor oil to A/C compressor suction side.

d. Install the padding uniformly to the installation hole while installing the HVAC unit.

e. Tighten instrument panel mounting bolts to 17 ft. lbs. (23 Nm)

f. Tighten steering column mounting nuts to 18 ft. lbs. (25 Nm)

13. Refill engine cooling system.

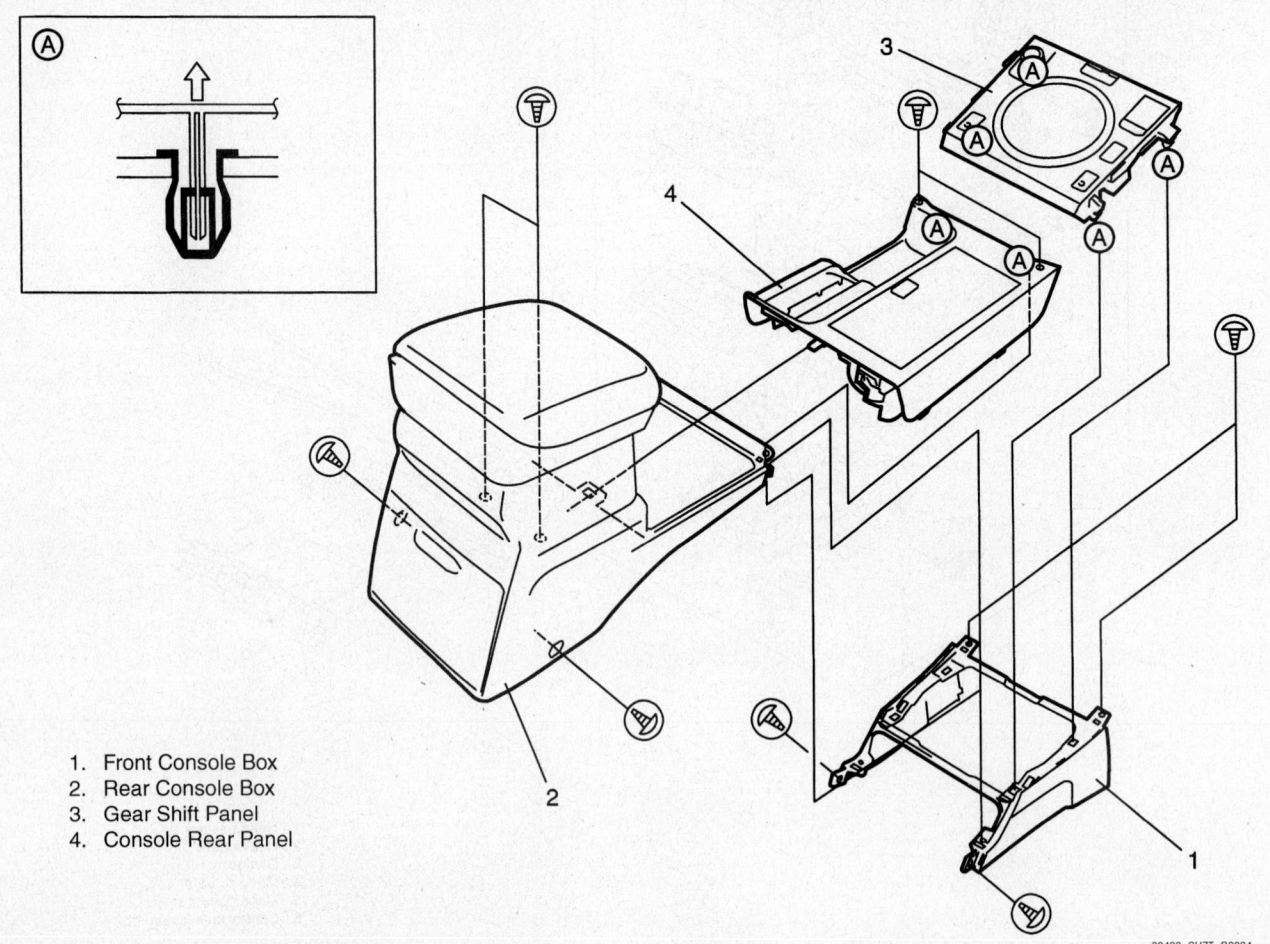

1. Front Console Box
2. Rear Console Box
3. Gear Shift Panel
4. Console Rear Panel

Exploded view of console components—2006 Grand Vitara

09490_SUZT_G0004

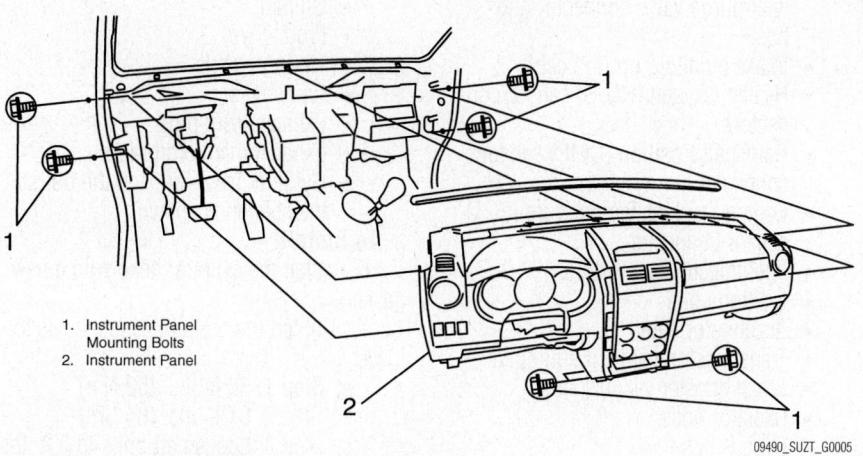

1. Instrument Panel
   Mounting Bolts
2. Instrument Panel

09490_SUZT_G0005

Instrument panel assembly removal—2006 Grand Vitara

- Heated Oxygen (HO2S) sensor connectors
- Exhaust manifold heat shield
- Exhaust manifold bracket
- Exhaust front pipe
- Exhaust manifold
- A/C compressor
- Power steering pump
- Front cover
- Timing belt
- Camshaft timing sprocket
- Rear timing belt cover
- Distributor and spark plug wires, if equipped
- Ignition coils and wiring, if equipped with Distributorless Ignition System (DIS)

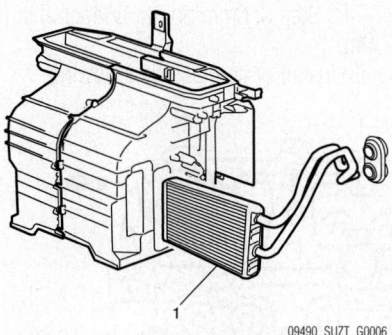

09490_SUZT_G0006

Heater core removal from HVAC unit— 2006 Grand Vitara

14. Evacuate and charge A/C system.
15. Enable air bag system.
16. Connect the negative battery cable.
17. Run the engine to normal operating temperatures; then, check the climate control operation and check for leaks.

Cylinder Head

REMOVAL & INSTALLATION

1.6L Engine

1. Before servicing the vehicle, refer to the Precautions Section.
2. Relieve the fuel system pressure.
3. Drain the cooling system.
4. Remove or disconnect the following:
   - Negative battery cable
   - Accessory drive belts
   - Air intake pipe
   - Fuel lines
   - Upper radiator hose
   - Coolant bypass hose
   - Alternator bracket
   - Intake manifold brackets
   - Intake manifold

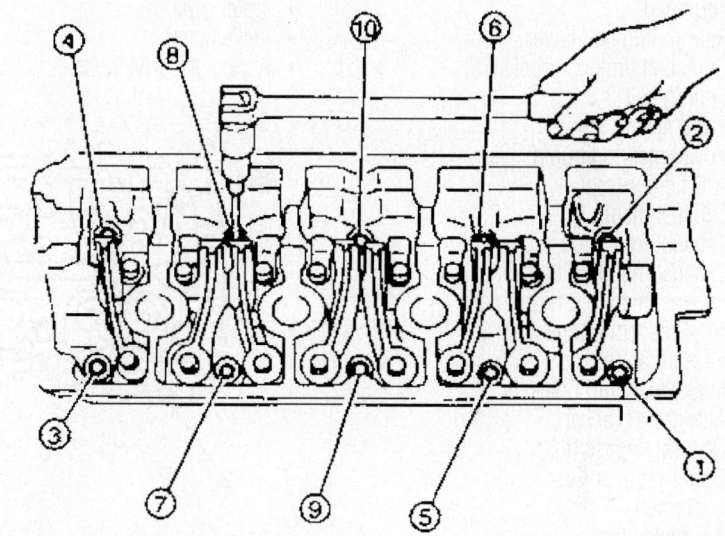

9308HG01

Cylinder head loosening sequence—1.6L engine

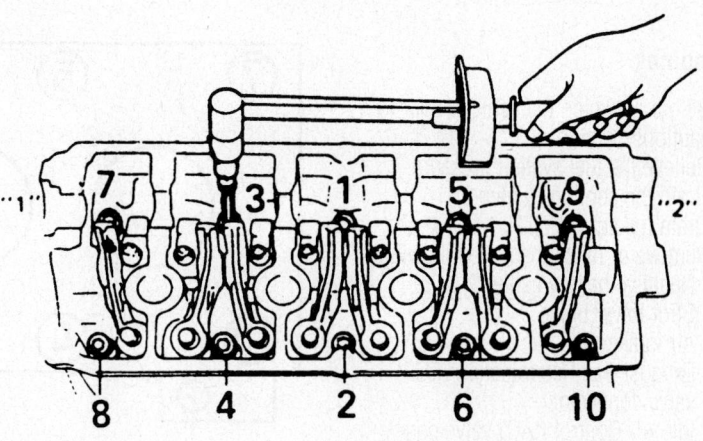

''1'': Camshaft pulley side
''2'': Distributor side

7924HG08

Cylinder head torque sequence—1.6L engine

- Valve cover
- Cylinder head. Loosen the bolts in the sequence shown.

To install:

5. Install the cylinder head with a new gasket.

6. Tighten the bolts in sequence as follows:

    a. Step 1: 26 ft. lbs. (35 Nm)

    b. Step 2: 41 ft. lbs. (55 Nm)

    c. Step 3: Loosen all bolts to 0 ft. lbs. (0 Nm)

    d. Step 4: 26 ft. lbs. (35 Nm)

    e. Step 5: 52 ft. lbs. (70 Nm)

7. Install or connect the following:

- Valve cover
- Ignition coils and wiring, if equipped with DIS
- Distributor and spark plug wires, if equipped
- Rear timing belt cover
- Camshaft timing sprocket
- Timing belt
- Front cover
- Power steering pump
- A/C compressor
- Exhaust manifold
- Exhaust front pipe
- Exhaust manifold bracket
- Exhaust manifold heat shield
- HO2S sensor connectors
- Intake manifold
- Intake manifold brackets
- Alternator bracket
- Coolant bypass hose
- Upper radiator hose
- Fuel lines
- Air intake pipe
- Accessory drive belts
- Negative battery cable

8. Fill the cooling system.

9. Start the engine and check for leaks.

2.0L Engines

1. Before servicing the vehicle, refer to the Precautions Section.

2. Relieve the fuel system pressure.

3. Drain the cooling system.

4. Drain the engine oil.

5. Remove or disconnect the following:

- Negative battery cable
- Strut tower brace
- Air intake tube
- Exhaust Gas Recirculation (EGR) valve connector
- Idle Air Control (IAC) valve connector
- Throttle Position (TP) sensor connector
- Evaporative Emission (EVAP) can-

ister purge valve connector and hose
- Intake manifold ground cable
- Heated Oxygen (HO2S) sensor connectors
- Camshaft Position (CMP) sensor connector
- Engine Coolant Temperature (ECT) sensor connector
- Fuel injector connectors
- Ignition coils
- Accelerator cable
- Transmission cable, if equipped
- Brake booster vacuum hose
- Radiator hose
- Bypass hose
- Heater hose
- Fuel lines
- Intake manifold bracket
- Water pipe
- Valve cover
- Accessory drive belts

- Oil pan
- Front cover
- Timing chains
- Camshafts
- Exhaust front pipe
- Exhaust manifold bracket
- Cylinder head. Loosen the bolts in the sequence shown.

To install:

6. Install the cylinder head with a new gasket.

7. Tighten the bolts in sequence as follows:

    a. Step 1: 38 ft. lbs. (52 Nm)

    b. Step 2: 61 ft. lbs. (84 Nm)

    c. Step 3: Loosen all bolts to 0 ft. lbs. (0 Nm)

    d. Step 4: 38 ft. lbs. (52 Nm)

    e. Step 5: 76 ft. lbs. (105 Nm)

    f. Step 6: 6mm bolt to 96 inch lbs. (8 Nm)

8. Install or connect the following:

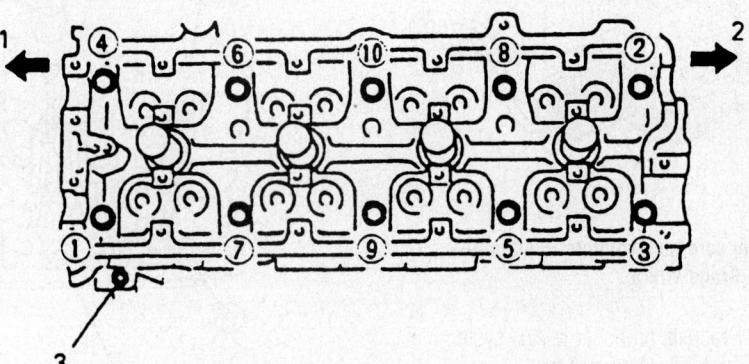

1.   **Crankshaft pulley side**
2.   **Flywheel side**
3.   **Bolt (M6)**

7924HG09

Cylinder head loosening sequence—2.0L engines

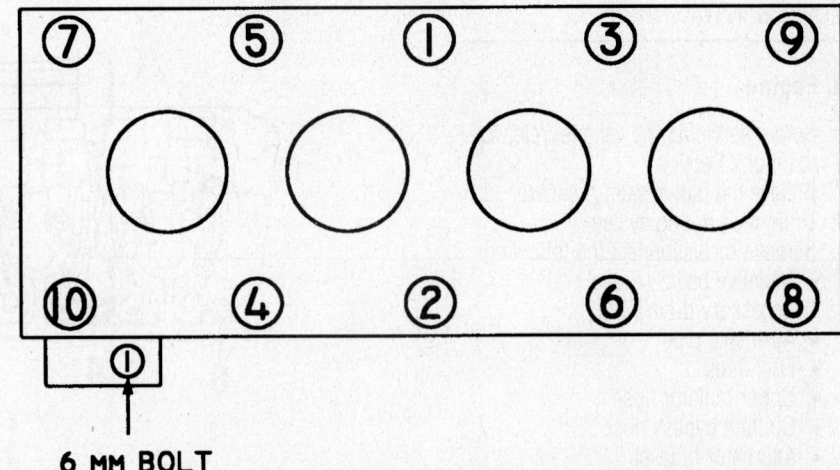

6 MM BOLT

9308HG07

Cylinder head torque sequence—2.0L engines

- Exhaust manifold bracket
- Exhaust front pipe
- Camshafts
- Timing chains
- Front cover
- Oil pan
- Accessory drive belts
- Valve cover
- Water pipe
- Intake manifold bracket
- Fuel lines
- Heater hose
- Bypass hose
- Radiator hose
- Brake booster vacuum hose
- Transmission cable, if equipped
- Accelerator cable
- Ignition coils
- Fuel injector connectors
- ECT sensor connector
- CMP sensor connector
- HO2S sensor connectors
- Intake manifold ground cable
- EVAP canister purge valve connector and hose
- TP sensor connector
- IAC valve connector
- EGR valve connector
- Air intake tube
- Strut tower brace
- Negative battery cable

9. Fill the crankcase to the correct level.
10. Fill the cooling system.
11. Start the engine and check for leaks.

2.5L and 2.7L Engines

1. Before servicing the vehicle, refer to the Precautions Section.
2. Relieve the fuel system pressure.
3. Drain the cooling system and engine oil.
4. Remove or disconnect the following:
   - Negative battery cable
   - Intake manifold
   - Ignition coil covers and ignition coils
   - Valve covers
   - Oil pan
   - Timing chain cover and timing chains
   - Camshaft Position (CMP) sensor
   - Camshafts
   - Exhaust manifolds
   - Water outlet caps
   - Cylinder heads. Loosen the bolts in the sequence shown.

To install:

5. Install the cylinder heads with new gaskets. Tighten the bolts in sequence as follows:
   a. Step 1: 38 ft. lbs. (52 Nm)

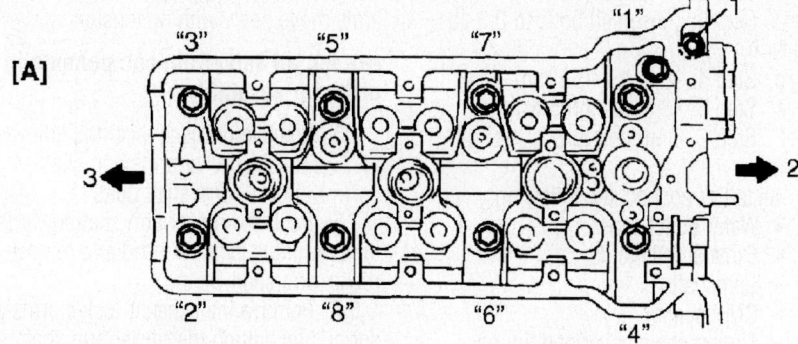

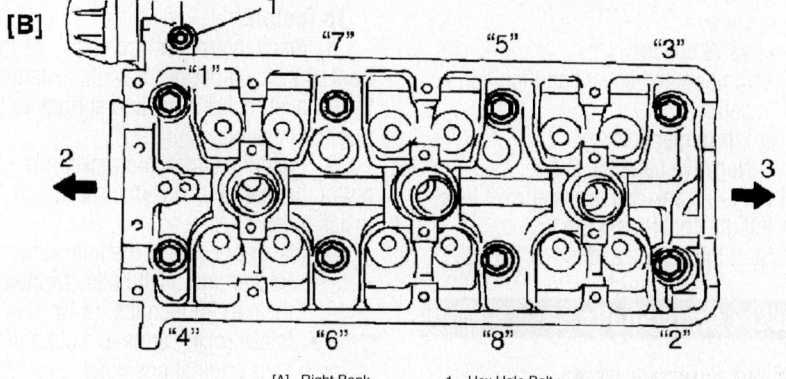

[A] Right Bank   1. Hex Hole Bolt
[B] Left Bank   2. Timing Chain Side
                3. Flywheel Side

09490_SUZT_G0007

Cylinder head loosening sequence—2.5L and 2.7L engines

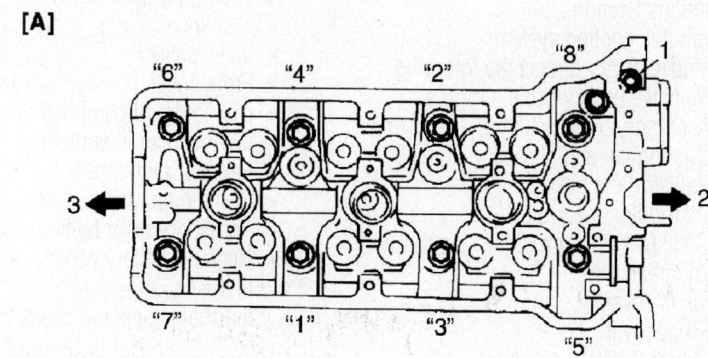

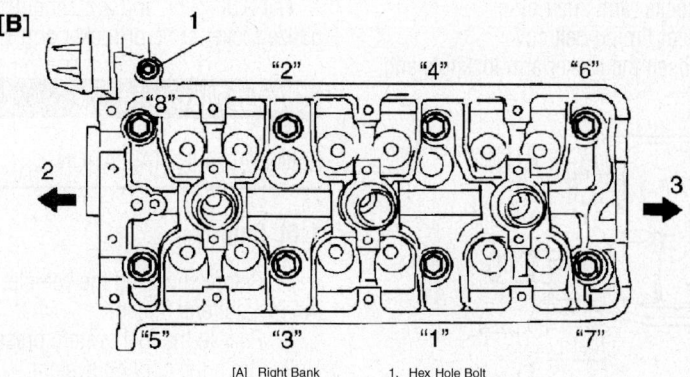

[A] Right Bank   1. Hex Hole Bolt
[B] Left Bank   2. Timing Chain Side
                3. Flywheel Side

09490_SUZT_G0008

Cylinder head torque sequence—2.5L and 2.7L engines

b. Step 2: 61 ft. lbs. (84 Nm)

c. Step 3: Loosen all bolts to 0 ft. lbs. (0 Nm)

d. Step 4: 38 ft. lbs. (52 Nm)

e. Step 5: 76 ft. lbs. (105 Nm)

f. Step 6: 6mm bolt to 8 ft. lbs. (11 Nm)

6. Install or connect the following:
- Water outlet caps
- Exhaust manifolds
- Camshafts
- CMP sensor
- Timing chain cover and timing chains
- Oil pan
- Valve covers
- Ignition coil covers and ignition coils
- Intake manifold
- Negative battery cable

7. Fill the crankcase to the correct level
8. Fill the cooling system.
9. Start the engine and check for leaks.

Rocker Arms/Shafts

REMOVAL & INSTALLATION

1.6L Engine

1. Before servicing the vehicle, refer to the Precautions Section.
2. Drain the cooling system.
3. Remove or disconnect the following:
- Negative battery cable
- Accessory drive belts
- Cooling fan and shroud
- Radiator and hoses
- Distributor, if equipped
- Camshaft Position (CMP) sensor, if equipped
- Valve cover
- Front cover
- Timing belt
- Camshaft
- Rocker arm shaft plug
- Rear timing belt cover

4. Loosen the rocker arm locknuts and

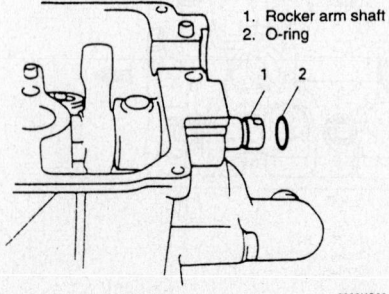

1. Rocker arm shaft
2. O-ring

9308HG02

Rocker arm shaft and O-ring—1.6L engine

back the valve adjusters off until all rocker arms move freely with no tension.

➡**Keep all valvetrain components in order for assembly.**

5. Remove or disconnect the following:
- Intake rocker arms and clips
- Rocker arm shaft bolts

6. Push the rocker arm shaft towards the rear of the cylinder head and remove the rocker arm shaft O-ring.

7. Remove the exhaust rocker arms and springs by pulling the rocker arm shaft out of the front of the cylinder head.

To install:

8. Insert the rocker arm shaft into the front of the cylinder head, while installing the exhaust rocker arms and springs in their original positions.

9. Push the end of the rocker arm shaft out of the rear of the cylinder head and install a new O-ring.

10. Install or connect the following:
- Rocker arm shaft bolts. Tighten them to 96 inch lbs. (8 Nm).
- Intake rocker arms and clips in their original positions
- Rear timing belt cover
- Rocker arm shaft plug. Tighten it to 24 ft. lbs. (33 Nm).
- Camshaft
- Timing belt and adjust the valve clearance
- Valve cover
- Front cover
- Distributor, if equipped
- CMP sensor, if equipped
- Radiator and hoses
- Cooling fan and shroud
- Accessory drive belts
- Negative battery cable

11. Fill the cooling system.
12. Start the engine and check for leaks.

2.0L, 2.5L and 2.7L Engines

The 2.0L, 2.5L and 2.7L engines do not utilize rocker arms or rocker arm shafts.

Intake Manifold

REMOVAL & INSTALLATION

1.6L Engine

1. Before servicing the vehicle, refer to the Precautions Section.
2. Relieve the fuel system pressure.
3. Drain the cooling system.
4. Remove or disconnect the following:
- Negative battery cable
- Air intake pipe

- Accelerator cable and bracket
- Transmission cable, if equipped
- Throttle Position (TP) sensor connector
- Idle Air Control (IAC) valve connector and hoses
- Engine Coolant Temperature (ECT) sensor connector
- Coolant temperature gauge sender connector
- A/C coolant temperature switch connector, if equipped
- Evaporative Emission (EVAP) canister purge valve connector and vacuum line
- Exhaust Gas Recirculation (EGR) temperature sensor connector
- EGR vacuum valve connector
- EGR bypass valve connector
- EGR vacuum lines
- Fuel injector connectors
- Intake manifold ground cable
- Transmission vacuum line, if equipped
- Brake booster vacuum line
- Manifold Absolute Pressure (MAP) sensor vacuum line
- Fuel lines
- Upper radiator hose
- Coolant bypass hose
- Alternator bracket
- Intake manifold brackets
- Intake manifold

To install:

5. Install or connect the following:
- Intake manifold with a new gasket. Tighten the nuts to 17 ft. lbs. (23 Nm).
- Intake manifold brackets. Tighten the fasteners to 36 ft. lbs. (50 Nm).
- Alternator bracket. Tighten the fasteners to 36 ft. lbs. (50 Nm).
- Coolant bypass hose
- Upper radiator hose
- Fuel lines
- MAP sensor vacuum line
- Brake booster vacuum line
- Transmission vacuum line, if equipped
- Intake manifold ground cable
- Fuel injector connectors
- EGR vacuum lines
- EGR bypass valve connector
- EGR vacuum valve connector
- EGR temperature sensor connector
- EVAP canister purge valve connector and vacuum line
- A/C coolant temperature switch connector, if equipped
- Coolant temperature gauge sender connector
- ECT sensor connector

- IAC valve connector and hoses
- TP sensor connector
- Transmission cable, if equipped
- Accelerator cable and bracket
- Air intake pipe
- Negative battery cable

6. Fill the cooling system.
7. Start the engine and check for leaks.

2.0L Engine

1. Before servicing the vehicle, refer to the Precautions Section.
2. Relieve the fuel system pressure.
3. Drain the cooling system.
4. Remove or disconnect the following:
   - Negative battery cable
   - Air intake tube
   - Exhaust Gas Recirculation (EGR) valve connector
   - Idle Air Control (IAC) valve connector
   - Throttle Position (TP) sensor connector
   - Evaporative Emissions (EVAP) canister purge valve connector and hose
   - Intake manifold ground cable
   - Manifold Absolute Pressure (MAP) sensor connector
   - Accelerator cable
   - Transmission cable, if equipped
   - Brake booster vacuum hose
   - Positive Crankcase Ventilation (PCV) valve and hose
   - Fuel pressure regulator vacuum hose
   - Intake manifold vacuum hose
   - Throttle body coolant hoses
   - Water bypass pipe
   - Fuel lines
   - Fuel supply manifold with injectors attached
   - Intake manifold support brackets
   - Intake manifold water pipe
   - Intake manifold

To install:

5. Install or connect the following:
   - Intake manifold with a new gasket. Tighten the fasteners to 17 ft. lbs. (23 Nm).
   - Intake manifold water pipe
   - Intake manifold front support bracket. Tighten the bolts to 36 ft. lbs. (50 Nm).
   - Intake manifold rear support bracket. Tighten the bolts to 18 ft. lbs. (25 Nm).
   - Fuel supply manifold with injectors attached
   - Fuel lines
   - Water bypass pipe

- Throttle body coolant hoses
- Intake manifold vacuum hose
- Fuel pressure regulator vacuum hose
- PCV valve and hose
- Brake booster vacuum hose
- Transmission cable, if equipped
- Accelerator cable
- MAP sensor connector
- Intake manifold ground cable
- EVAP canister purge valve connector and hose
- TP sensor connector
- IAC valve connector
- EGR valve connector
- Air intake tube
- Negative battery cable

6. Fill the cooling system.
7. Start the engine and check for leaks.

2.5L and 2.7L Engine (except 2006 Grand Vitara)

1. Before servicing the vehicle, refer to the Precautions Section.
2. Relieve the fuel system pressure.
3. Drain the cooling system.
4. Remove or disconnect the following:
   - Negative battery cable
   - Strut tower bar
   - Intake Air Temperature (IAT) sensor connector
   - Fuse/relay box, if necessary
   - Intake control valve, if equipped
   - Manifold Absolute Pressure (MAP) sensor connector
   - Surge tank cover
   - Air intake assembly
   - Accelerator cable
   - Transmission cable, if equipped
   - Throttle body coolant hoses
   - Fuel injector connectors
   - Throttle Position (TP) sensor connector
   - Mass Air Flow (MAF) sensor connector
   - Idle Air Control (IAC) valve connector
   - Intake manifold ground cables
   - Brake booster vacuum hose
   - Evaporative Emissions (EVAP) canister purge valve connector and hoses
   - Exhaust Gas Recirculation (EGR) valve connector
   - Positive Crankcase Ventilation (PCV) valve and hose
   - Heater hoses
   - EGR pipe
   - Fuel lines
   - Throttle body and intake collector
   - Intake manifold

To install:

5. Install or connect the following:
   - Intake manifold with new gaskets. Tighten the fasteners to 16 ft. lbs. (23 Nm).
   - Throttle body and intake collector with new gaskets. Tighten the fasteners to 102 inch lbs. (12 Nm).
   - Fuel lines
   - EGR pipe
   - Heater hoses
   - PCV valve and hose
   - EGR valve connector
   - EVAP canister purge valve connector and hoses
   - Brake booster vacuum hose
   - Intake manifold ground cables
   - IAC valve connector
   - MAF sensor connector
   - TP sensor connector
   - Fuel injector connectors
   - Throttle body coolant hoses
   - Transmission cable, if equipped
   - Accelerator cable
   - Air intake assembly
   - Surge tank cover
   - MAP sensor connector
   - Intake control valve, if equipped
   - Fuse/relay box, if necessary
   - IAT sensor connector
   - Strut tower bar
   - Negative battery cable

6. Fill the cooling system.
7. Start the engine and check for leaks.

2.7L Engine (2006 Grand Vitara)

1. Before servicing the vehicle, refer to the Precautions Section.
2. Relieve the fuel system pressure.
3. Drain the cooling system.
4. Remove or disconnect the following:
   - Negative battery cable
   - Throttle body assembly
   - Water hose from heater outlet pipe
   - Heater outlet pipe bolts
   - MAP sensor connector and fuel injector harness connector
   - Harness clamp bracket (right side) from intake collector
   - Vacuum hose of EVAP canister purge valve from intake collector
   - EVAP canister purge valve connector and ground terminals
   - Harness clamp bracket (left side) from intake collector
   - EGR valve assembly connector, EGR pipe and PCV hose
   - Intake collector from intake manifold, and then water hose
   - EGR valve assembly and PCV pipe from intake collector, if necessary

- Brake booster hose and vacuum tank hose from intake manifold then EVAP canister purge valve hose from fuel No.2 pipe
- Fuel feed and return hoses
- Fuel No.2 pipe bolts
- Intake manifold

To install:
5. Install or connect the following:
- Intake manifold with new gaskets. Tighten the fasteners to 16 ft. lbs. (23 Nm).
- Fuel No.2 pipe bolts
- Fuel feed and return hoses
- Brake booster hose and vacuum tank hose to the intake manifold then EVAP canister purge valve hose to the fuel No.2 pipe
- EGR valve assembly with a new gasket and PCV pipe to the intake collector, if necessary
- Intake collector and a new gasket to the intake manifold, and then water hose
- EGR valve assembly connector, EGR pipe with a new gasket and PCV hose
- Harness clamp bracket (left side) to the intake collector
- EVAP canister purge valve connector and ground terminals
- Vacuum hose of EVAP canister purge valve to the intake collector
- Harness clamp bracket (right side) to the intake collector
- MAP sensor connector and fuel injector harness connector
- Heater outlet pipe bolts
- Water hose to the heater outlet pipe
- Throttle body assembly
- Negative battery cable
6. Refill cooling system.
7. Upon completion of installation, verify that there is no fuel leakage at each connection.

Exhaust Manifold

REMOVAL & INSTALLATION

1.6L and 2.0L Engines

1. Before servicing the vehicle, refer to the Precautions Section.
2. Remove or disconnect the following:
- Negative battery cable
- Strut tower bar, if equipped
- Air intake assembly and bracket
- Heated Oxygen (HO2S) sensor connector
- Exhaust front pipe

- Exhaust manifold heat shield
- Exhaust manifold bracket, if equipped
- Exhaust manifold

To install:
3. Install or connect the following:
- Exhaust manifold with a new gasket. Tighten the fasteners to 13–20 ft. lbs. (18–28 Nm).
- Exhaust manifold bracket, if equipped. Tighten the bolts to 36–43 ft. lbs. (50–60 Nm).
- Exhaust manifold heat shield
- Exhaust front pipe. Tighten the fasteners to 29–43 ft. lbs. (40–60 Nm).
- HO2S sensor connector
- Air intake assembly and bracket
- Strut tower bar, if equipped. Tighten the fasteners to 66 ft. lbs. (90 Nm).
- Negative battery cable
4. Start the engine and check for leaks.

2.5L and 2.7L Engines (except 2006 Grand Vitara)

1. Before servicing the vehicle, refer to the Precautions Section.
2. Remove or disconnect the following:
- Negative battery cable
- Strut tower bar
- Air intake assembly
- Heated Oxygen (HO2S) sensor connectors
- Oil dipstick tube
- Exhaust Gas Recirculation (EGR) pipe
- Exhaust manifold heat shields
- Evaporative Emissions (EVAP) canister
- Front driveshaft, if equipped
- Exhaust front pipe
- Exhaust manifold brace
- Exhaust manifolds

To install:
3. Install or connect the following:
- Exhaust manifolds with new gaskets. Tighten the nuts to 21 ft. lbs. (30 Nm).
- Exhaust manifold brace
- Exhaust front pipe. Tighten the fasteners to 37 ft. lbs. (50 Nm).
- Front driveshaft, if equipped
- EVAP canister
- Exhaust manifold heat shields
- EGR pipe
- Oil dipstick tube
- HO2S sensor connectors
- Air intake assembly
- Strut tower bar
- Negative battery cable
4. Start the engine and check for leaks.

2.7L Engine (2006 Grand Vitara)

1. Before servicing the vehicle, refer to the Precautions Section.
2. Disconnect the negative battery cable.
3. Properly relieve the fuel system pressure.
4. Remove or disconnect the following:
- Battery and battery tray, if necessary
- Air cleaner assembly
- A/F sensor connector
- Heated Oxygen (HO2S) sensor
- Exhaust Gas Recirculation (EGR) pipe, if necessary
- Fuel feed and return hoses, if necessary
- Brake booster hose and EVAP canister purge valve hose from intake manifold and fuel No. 2 pipe, if necessary
- Vacuum tank assembly, if necessary
- Steering lower shaft assembly, if necessary
- Suction and discharge hoses from A/C compressor, if necessary
- Oil filter, if necessary
- Engine hook, if necessary
- Oil dipstick tube, if necessary
- Exhaust manifold heat shield
- Exhaust front pipe
- Exhaust manifold brace
- Exhaust manifold and gasket

To install:
5. Install or connect the following:
- Exhaust manifold with new gasket. Tighten the nuts to 21 ft. lbs. (30 Nm).
- Exhaust manifold brace. Tighten the bolts to 21 ft. lbs. (30 Nm).
- Exhaust front pipe and new gasket. Tighten the bolt and nut to 37 ft. lbs. (50 Nm).
- Exhaust manifold heat shield
- Oil dipstick tube, if necessary
- Engine hook, if necessary
- Oil filter, if necessary
- Suction and discharge hoses to the A/C compressor, necessary
- Steering lower shaft assembly, if necessary
- Vacuum tank assembly, if necessary
- Brake booster hose and EVAP canister purge valve hose to the intake manifold and fuel No. 2 pipe, if necessary
- Fuel feed and return hoses, if necessary
- EGR pipe and new gasket, if necessary

- Heated Oxygen (HO$_2$S) sensor and tighten to 32 ft. lbs. (45 Nm)
- A/F sensor connector
- Air cleaner assembly
- Battery and battery tray, if necessary. Be sure to connect the positive battery cable first, then the negative battery cable second.

6. Connect the negative battery cable.

7. Turn the ignition switch to the **ON** position and check for fuel leaks.

8. Start the engine and check for exhaust leaks.

Front Crankshaft Seal

REMOVAL & INSTALLATION

1.6L Engine

1. Before servicing the vehicle, refer to the Precautions Section.
2. Drain the cooling system.
3. Remove or disconnect the following:
- Negative battery cable
- Accessory drive belts
- Cooling fan and shroud
- Water pump pulley
- Crankshaft pulley
- Front cover
- Timing belt
- Crankshaft timing sprocket
- Front crankshaft seal

To install:

4. Install or connect the following:
- Front crankshaft seal flush with the oil pump housing
- Crankshaft timing sprocket. Tighten the bolt to 94 ft. lbs. (128 Nm).
- Timing belt
- Front cover
- Crankshaft pulley. Tighten the bolts to 12 ft. lbs. (16 Nm).
- Water pump pulley
- Cooling fan and shroud
- Accessory drive belts
- Negative battery cable

5. Start the engine and check for leaks.

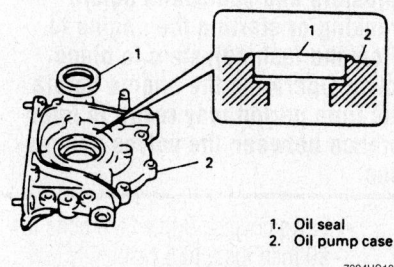

1. Oil seal
2. Oil pump case

7924HG13

Install the new oil pump seal flush with the oil pump housing—1.6L engine

Camshaft and Valve Lifters

REMOVAL & INSTALLATION

1.6L Engine

1. Before servicing the vehicle, refer to the Precautions Section.
2. Drain the cooling system.
3. Remove or disconnect the following:
- Negative battery cable
- Radiator
- Accessory drive belts
- Crankshaft pulley
- Front cover
- Timing belt

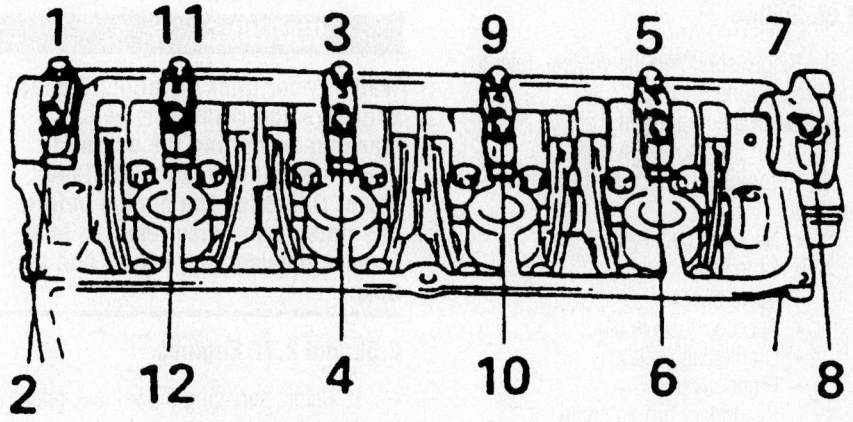

Camshaft housing torque sequence—1.6L engine

7924HG25

- Camshaft sprocket
- Valve cover
- Distributor and case, if equipped
- Camshaft Position (CMP) sensor and case, if equipped

4. Loosen the rocker arm locknuts and back the valve adjusters off until all rocker arms move freely with no tension.

5. Remove or disconnect the following:
- Camshaft bearing caps. Loosen the bolts in reverse of the tightening sequence.
- Camshaft

To install:

6. Install or connect the following:
- Camshaft

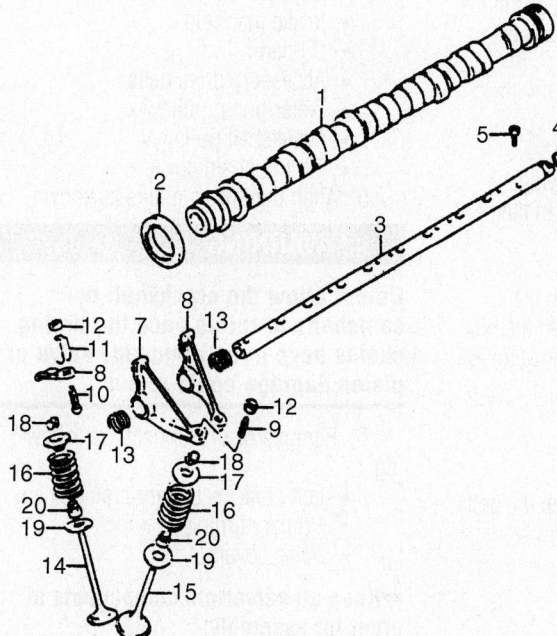

1. Camshaft
2. Camshaft oil seal
3. Rocker arm shaft
4. O ring
5. Rocker shaft bolt
6. Rocker arm (IN)
7. Rocker arm No. 1 (EX)
8. Rocker arm No. 2 (EX)
9. Valve adjusting screw
10. Valve adjusting screw
11. Clip
12. Lock nut
13. Rocker arm spring
14. Intake valve
15. Exhaust valve
16. Valve spring
17. Valve spring retainer
18. Valve cotter
19. Valve spring seat
20. Valve stem seal

7924HG24

Exploded view of the valve train components—1.6L engine

- Camshaft bearing caps. Tighten the bolts in sequence to 96 inch lbs. (11 Nm).
- CMP sensor and case, if equipped
- Distributor and case, if equipped
- Camshaft sprocket. Tighten the bolt to 44 ft. lbs. (60 Nm).
- Timing belt and adjust the valve clearance
- Valve cover
- Front cover
- Crankshaft pulley. Tighten the bolts to 12 ft. lbs. (16 Nm).
- Accessory drive belts
- Radiator
- Negative battery cable

7. Fill the cooling system.
8. Start the engine and check for leaks.

2.0L Engine

1. Before servicing the vehicle, refer to the Precautions Section.
2. Drain the engine oil.
3. Drain the cooling system.
4. Remove or disconnect the following:

- Negative battery cable
- Oil pan
- Valve cover
- Accessory drive belts
- Crankshaft pulley
- Front cover
- Secondary timing chain
- Camshaft Position (CMP) sensor

➡Keep all valvetrain components in order for installation.

- Camshaft bearing caps. Loosen the bolts in several steps in reverse of the tightening sequence.
- Camshafts
- Hydraulic lash adjusters

To install:

5. Install or connect the following:
- Hydraulic lash adjusters in their original positions
- Camshafts
- Camshaft bearing caps in their original positions. Tighten the bolts in several steps in sequence to 96 inch lbs. (11 Nm).
- CMP sensor
- Secondary timing chain
- Front cover
- Crankshaft pulley. Tighten the bolt to 109 ft. lbs. (148 Nm).
- Accessory drive belts
- Valve cover
- Oil pan
- Negative battery cable

6. Fill the crankcase to the correct level.

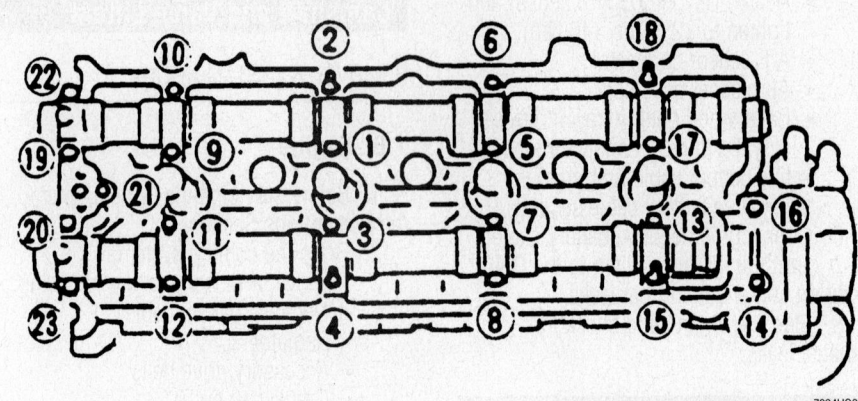

Camshaft housing torque sequence—2.0L engines

7. Fill the cooling system.
8. Start the engine and check for leaks.

※ WARNING

Wait ½ hour after installing the lash adjusters and camshafts before cranking or starting the engine to allow the lash adjusters to bleed down. Operating the engine before this time period may result in interference between the valves and pistons.

2.5L and 2.7L Engines

1. Before servicing the vehicle, refer to the Precautions Section.
2. Drain the engine oil.
3. Drain the cooling system.
4. Remove or disconnect the following:
- Negative battery cable
- Intake manifold
- Oil pan
- Accessory drive belts
- Water pump pulley
- Crankshaft pulley
- Timing chain cover

5. Align the timing marks as shown.

※ WARNING

Do not allow the crankshaft or camshafts to rotate once the timing chains have been removed. Valve or piston damage could result.

6. Remove or disconnect the following:

- Left bank secondary timing chain
- Primary timing chain
- Valve covers

➡Keep all valvetrain components in order for assembly.

- Right bank camshaft bearing caps. Loosen the bolts in several steps

and in reverse of the tightening sequence shown.
- Right bank secondary timing chain, exhaust and intake camshafts as an assembly
- Camshaft Position (CMP) sensor
- Left bank camshaft bearing caps. Loosen the bolts in several steps and in reverse of the tightening sequence shown.
- Left bank camshafts
- Hydraulic lash adjusters

To install:

7. Install or connect the following:
- Hydraulic lash adjusters in their original positions
- Left bank camshafts
- Left bank camshaft bearing caps. Tighten the bolts in several steps, in the sequence shown, to 102 inch lbs. (12 Nm).
- CMP sensor
- Right bank secondary timing chain, exhaust and intake camshafts as an assembly
- Right bank camshaft bearing caps. Tighten the bolts in several steps, in the sequence shown, to 102 inch lbs. (12 Nm).

※ WARNING

Wait ½ hour after installing the lash adjusters and camshafts before cranking or starting the engine to allow the lash adjusters to bleed down. Operating the engine before this time period may result in interference between the valves and pistons.

- Valve covers. Tighten the bolts to 90 inch lbs. (10.5 Nm).
- Primary timing chain
- Left bank secondary timing chain

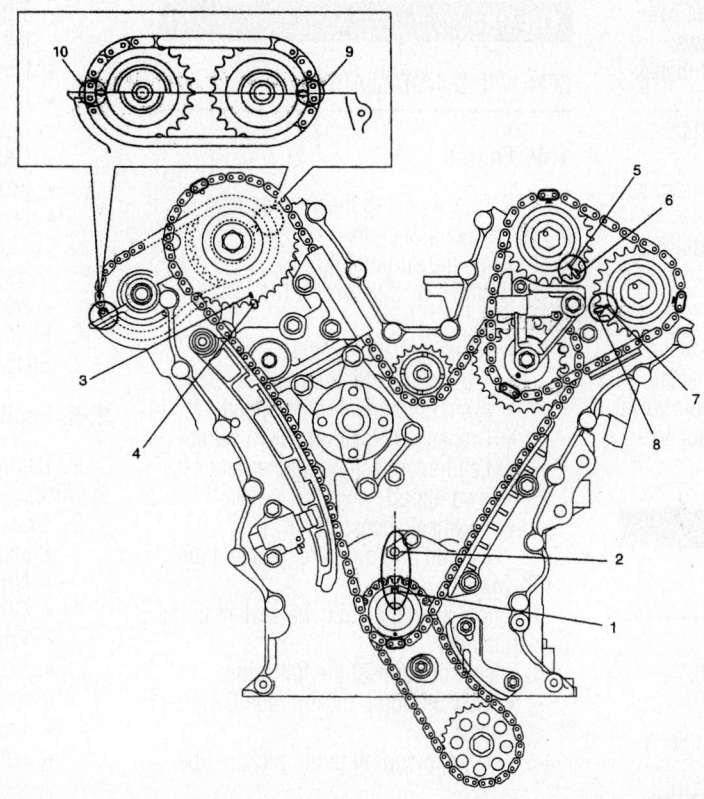

3. Timing mark of RH (No.2) bank 1st timing chain sprocket	6. Timing mark of LH (No.1) bank 2nd timing chain	9. Timing mark of RH (No.2) bank 2nd timing chain intake sprocket
4. Timing mark of RH (No.2) bank 1st timing chain	7. Timing mark of LH (No.1) bank 2nd timing chain exhaust sprocket	10. Timing mark of RH (No.2) bank 2nd timing chain exhaust sprocket
5. Timing mark of LH (No.1) bank 2nd timing chain intake sprocket	8. Timing mark of LH (No.1) bank 2nd timing chain	

09490_SUZT_G0009

Timing chain alignment marks—2.5L and 2.7L engines

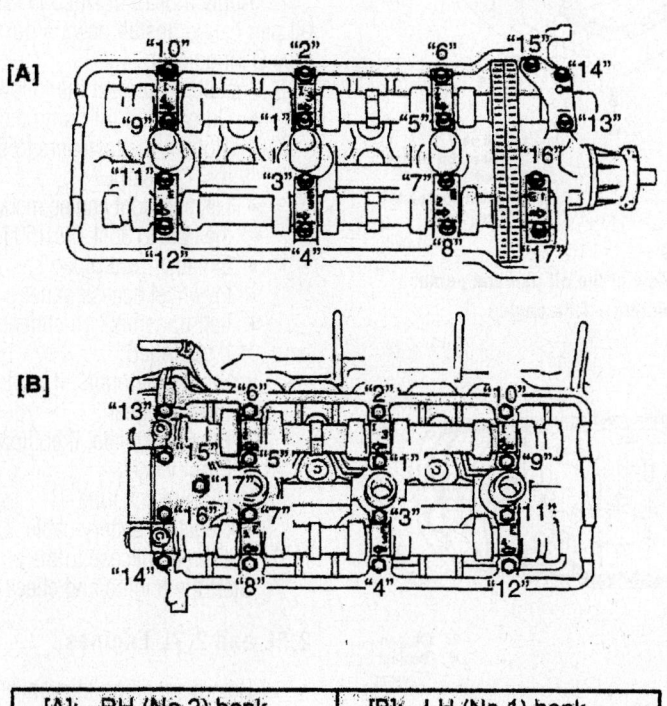

| [A]: RH (No.2) bank | [B]: LH (No.1) bank |

09490_SUZT_G0010

Left and right bank camshaft housing tightening sequence—2.5L and 2.7L engines

- Timing chain cover
- Crankshaft pulley. Tighten the bolt to 109 ft. lbs. (148 Nm).
- Water pump pulley
- Accessory drive belts
- Oil pan
- Intake manifold
- Negative battery cable
8. Fill the crankcase to the correct level.
9. Fill the cooling system.
10. Start the engine and check for leaks.

Valve Lash

ADJUSTMENT

1.6L Engine

➡ **Measure valve clearance with the engine cold.**

1. Before servicing the vehicle, refer to the Precautions Section.
2. Remove the valve cover.
3. Set the engine to Top Dead Center (TDC) of the compression stroke for the cylinder to be adjusted.

4. Check the valve clearance. The valve clearance specifications are as follows:
- Intake valves: 0.005–0.007 inches (0.13–0.17mm)
- Exhaust valves: 0.005–0.007 inches (0.13–0.17mm)

5. After adjustment, tighten the locknuts to 11–14 ft. lbs. (15–19 Nm).

6. Repeat for each valve to be adjusted.

2.0L, 2.5L and 2.7L Engines

2.0L, 2.5L and 2.7L engines utilize automatic hydraulic lash adjusters to maintain proper valve lash at all times. Periodic valve lash inspection and adjustment is not necessary or possible.

Starter Motor

REMOVAL & INSTALLATION

All Models (except 2006 Grand Vitara)

1. Before servicing the vehicle, refer to the Precautions Section.

2. Remove or disconnect the following:
- Negative battery cable
- Starter motor wiring connectors
- Starter motor

To install:

3. Install or connect the following:
- Starter motor. Tighten the bolts to 22 ft. lbs. (30 Nm).
- Starter motor wiring connectors. Tighten the solenoid nut to 11 ft. lbs. (15 Nm).
- Negative battery cable

2006 Grand Vitara

1. Before servicing the vehicle, refer to the Precautions Section.

2. Remove or disconnect the following:
- Negative battery cable
- Alternator
- Magnetic switch lead wire and battery cable from starter motor terminals
- Starter motor

To install:

3. Install or connect the following:
- Starter motor. Tighten the mounting bolts to 40 ft. lbs. (55 Nm).
- Magnetic switch lead wire and battery cable to the starter motor terminals. Tighten the battery cable nut to 8 ft. lbs. (11 Nm).
- Alternator
- Negative battery cable

Oil Pan

REMOVAL & INSTALLATION

1.6L Engine

1. Before servicing the vehicle, refer to the Precautions Section.

2. Drain the engine oil.

3. Remove or disconnect the following:
- Negative battery cable
- Front skidplate, if equipped
- Front differential, if equipped
- Crankshaft Position (CKP) sensor
- Left transmission stiffener bracket, if equipped
- Flywheel access panel
- Oil pan and oil pump pickup tube

To install:

4. Apply a bead of silicone sealant to the oil pan flange.

5. Install or connect the following:
- New oil pump pickup tube O-ring seal
- Oil pan and oil pump pickup tube.

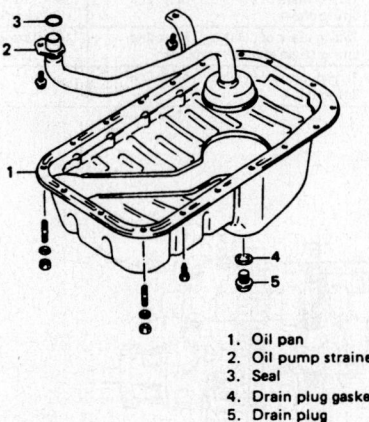

1. Oil pan
2. Oil pump strainer
3. Seal
4. Drain plug gasket
5. Drain plug

7924HG11

Exploded view of the oil pan and pump pickup mounting—1.6L engine

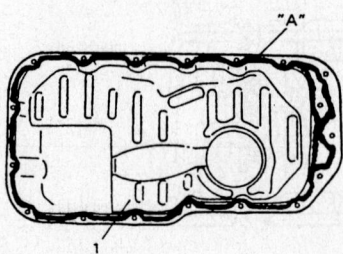

1. Oil pan
A. Sealant

7924HG79

Before installing the oil pan, apply a continuous bead of silicone sealant to the oil pan mating flange—all engines

Tighten the fasteners to 97 inch lbs. (11 Nm).
- Flywheel access panel
- Left transmission stiffener bracket, if equipped
- CKP sensor
- Front differential, if equipped
- Front skidplate, if equipped. Tighten the bolts to 40 ft. lbs. (55 Nm).
- Negative battery cable

6. Fill the crankcase to the correct level.

7. Start the engine and check for leaks.

2.0L Engine

1. Before servicing the vehicle, refer to the Precautions Section.

2. Drain the engine oil.

3. Remove or disconnect the following:
- Negative battery cable
- Oil dipstick tube
- Front wheels
- Front skidplate, if equipped
- Steering gear
- Front differential, if equipped
- Left transmission stiffener bracket, if equipped
- Flywheel access panel
- Exhaust front pipe
- Left and right motor mounts and raise the engine about 1 inch (25mm) for clearance
- Oil pan and oil pump pickup tube

To install:

4. Apply a bead of silicone sealant to the oil pan flange. Install new oil pump pickup tube O-ring seals.

5. Install or connect the following:
- Oil pan and oil pump pickup tube. Tighten the fasteners to 97 inch lbs. (11 Nm).
- Left and right engine mounts. Tighten the nuts to 36 ft. lbs. (50 Nm).
- Exhaust front pipe
- Flywheel access panel
- Left transmission stiffener bracket, if equipped
- Front differential, if equipped
- Steering gear
- Front skidplate, if equipped
- Front wheels
- Oil dipstick tube
- Negative battery cable

6. Fill the crankcase to the correct level.

7. Start the engine and check for leaks.

2.5L and 2.7L Engines

1. Before servicing the vehicle, refer to the Precautions Section.

2. Drain the engine oil.

3. Remove or disconnect the following:

1. O-ring

9302HG05

Lower crankcase O-ring seal

- Negative battery cable
- Oil dipstick tube
- Front wheels
- Front skidplate, if equipped
- Steering gear
- Front differential, if equipped
- Lower oil pan
- Oil pickup tube bracket
- Radiator outlet pipe
- Upper oil pan and oil pickup tube

To install:

4. Install a new O-ring to the lower crankcase.

5. Apply a bead of silicone sealant to the upper oil pan flange.

6. Install or connect the following:
- New oil pump pickup tube O-ring seals
- Upper oil pan and oil pump pickup tube and tighten the M6 bolts to 8 ft. lbs. (11 Nm), M8 bolts to 20 ft. lbs. (25 Nm) and the oil pan nuts to 8 ft. lbs. (11 Nm)
- Radiator outlet pipe
- Oil pickup tube bracket

- Lower oil pan. Tighten the bolts to 97 inch lbs. (11 Nm).
- Front differential, if equipped
- Steering gear
- Front skidplate, if equipped
- Front wheels
- Oil dipstick tube
- Negative battery cable

7. Fill the crankcase to the correct level.

8. Start the engine and check for leaks.

Oil Pump

REMOVAL & INSTALLATION

1.6L Engine

1. Before servicing the vehicle, refer to the Precautions Section.
2. Drain the engine oil.
3. Drain the cooling system.
4. Remove or disconnect the following:
- Negative battery cable
- Accessory drive belts
- Crankshaft pulley
- Front cover
- Timing belt
- Alternator and bracket
- A/C compressor and bracket, if equipped
- Oil pan and oil pump pickup tube
- Crankshaft timing sprocket
- Oil pump

➡**The oil pump bolts are different lengths. Note their location for assembly.**

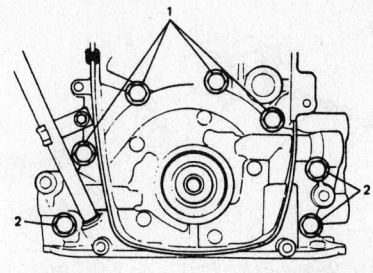

1. No. 1 bolts (short)
2. No. 2 bolts (long)

7924HG14

Oil pump housing short (1) and long bolt (2) locations—1.6L engines

To install:

5. Install or connect the following:
- Oil pump with a new gasket. Tighten the bolts to 97 inch lbs. (11 Nm).
- Crankshaft timing sprocket. Tighten the bolt to 94 ft. lbs. (130 Nm).
- Oil pan and oil pump pickup tube
- A/C compressor and bracket, if equipped
- Alternator and bracket
- Timing belt
- Front cover
- Crankshaft pulley. Tighten the bolts to 12 ft. lbs. (16 Nm).
- Accessory drive belts
- Negative battery cable

6. Fill the crankcase to the correct level.

7. Start the engine and check for leaks.

2.0L Engine

1. Before servicing the vehicle, refer to the Precautions Section.

(a) M6 bolts
(b) M8 bolts
(c) Nuts

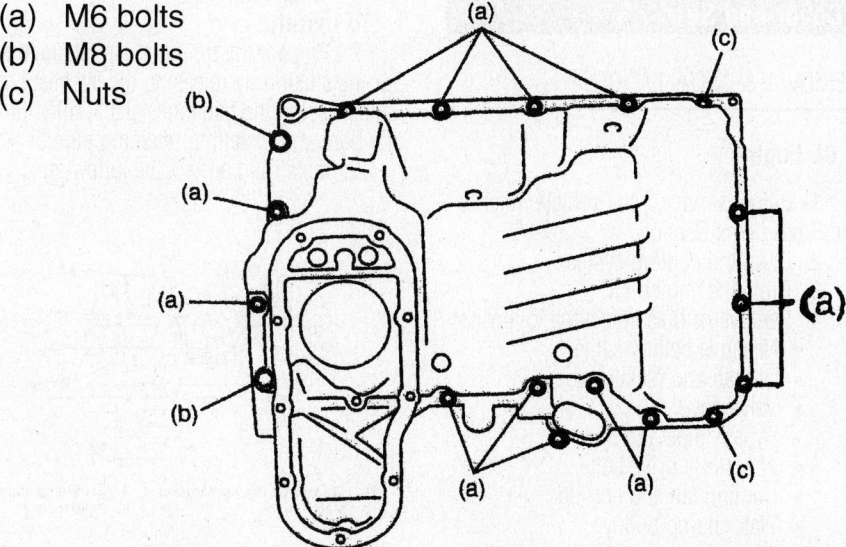

09490_SUZT_G0011

Upper oil pan fastener identifications—2.5L and 2.7L engines

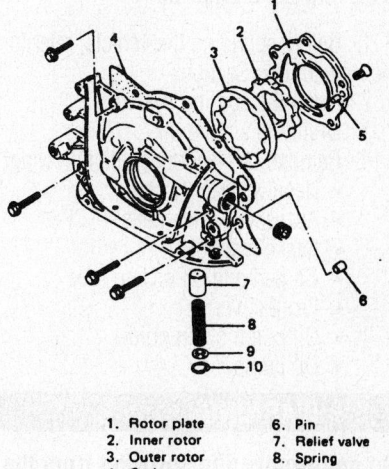

1. Rotor plate	6. Pin
2. Inner rotor	7. Relief valve
3. Outer rotor	8. Spring
4. Gasket	9. Retainer
5. Pin	10. Retainer ring

7924HG12

Exploded view of the oil pump housing—1.6L engines

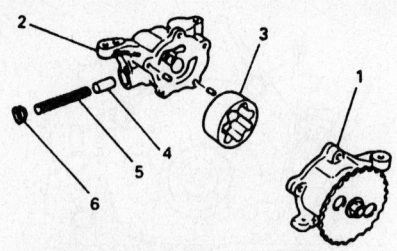

1. Oil pump case No.1   4. Relief valve
2. Oil pump case No.2   5. Relief spring
3. Outer rotor   6. Retainer

7924HG15

Exploded view of oil pump—2.0L and 2.5L engines

2. Drain the engine oil.
3. Remove or disconnect the following:
   - Negative battery cable
   - Oil pan and pickup tube
   - Oil pump sprocket cover
   - Oil pump

❋❋ WARNING

Do not remove the sprocket from the oil pump. Damage to the oil pump center shaft and abnormal pump operation may result.

To install:
4. Install or connect the following:
   - Oil pump. Tighten the bolts to 20 ft. lbs. (27 Nm).
   - Oil pump sprocket cover. Tighten the bolts to 108 inch lbs. (12 Nm).
   - Oil pan and pickup tube
   - Negative battery cable
5. Fill the crankcase to the correct level.
6. Start the engine and check for leaks.

2.5L and 2.7L Engines

1. Before servicing the vehicle, refer to the Precautions Section.
2. Drain the cooling system.
3. Drain the engine oil.
4. Remove or disconnect the following:
   - Negative battery cable
   - Accessory drive belts
   - Intake manifold
   - Oil pan and oil pickup tube
   - Front cover
   - Oil pump chain guide
   - Oil pump

❋❋ WARNING

Do not remove the sprocket from the oil pump. Damage to the oil pump center shaft and abnormal pump operation may result.

To install:
5. Install or connect the following:
   - Oil pump. Tighten the bolts to 20 ft. lbs. (27 Nm).
   - Oil pump chain guide. Tighten the bolts to 97 inch lbs. (11 Nm).
   - Front cover
   - Oil pan and oil pickup tube
   - Intake manifold
   - Accessory drive belts
   - Negative battery cable
6. Fill the crankcase to the correct level.
7. Fill the cooling system.
8. Start the engine and check for leaks.

Rear Main Seal

REMOVAL & INSTALLATION

1. Before servicing the vehicle, refer to the Precautions Section.
2. Remove or disconnect the following:
   - Negative battery cable
   - Transmission
   - Clutch assembly, if equipped
   - Flywheel
   - Rear main seal
To install:
3. Install or connect the following:
   - Rear main seal flush with the cylinder block
   - Flywheel. Tighten the bolts in a crossing pattern to 58 ft. lbs. (79 Nm) for 1.6L engines or to 51 ft. lbs. (70 Nm) for all other engines.
   - Clutch assembly, if equipped
   - Transmission
   - Negative battery cable

Timing Chain, Sprockets, Front Cover and Seal

REMOVAL & INSTALLATION

2.0L Engine

1. Before servicing the vehicle, refer to the Precautions Section.
2. Drain the cooling system.
3. Drain the engine oil.
4. Remove or disconnect the following:
   - Negative battery cable
   - Oil pan and pickup tube
   - Valve cover
   - Bypass pipe and hose
   - Accessory drive belts
   - Cooling fan and shroud
   - Water pump pulley
   - Alternator belt tensioner and idler pulleys
   - Upper radiator hose

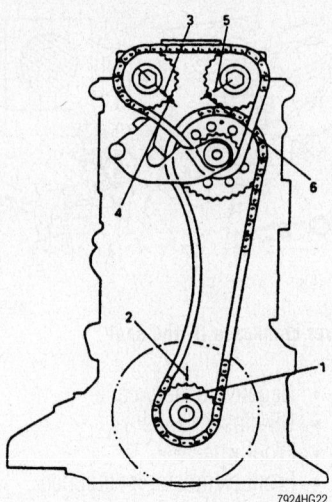

7924HG22

Timing mark alignment—2.0L engines

   - A/C compressor and bracket, if equipped
   - Crankshaft pulley
   - Front crankshaft seal
   - Front cover
5. Rotate the crankshaft to align the timing marks as shown.

❋❋ WARNING

Do not allow the crankshaft or camshafts to rotate once the timing chains have been removed. Valve or piston damage could result.

6. Remove or disconnect the following:
   - Second timing chain tensioner
   - Camshaft sprockets and second timing chain
   - First timing chain tensioner
   - Timing chain idler sprocket and first timing chain
To install:
7. Prepare the timing chain tensioners for installation by releasing the latches, compressing the tensioner piston fully into the bore and installing retaining pins.
8. Install or connect the following:

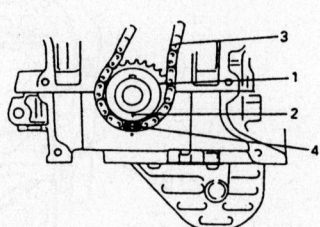

1. Crankshaft timing sprocket   3. 1st timing chain
2. Match mark   4. Yellow plate

7924HG17

Crankshaft and first timing chain alignment—2.0L engines

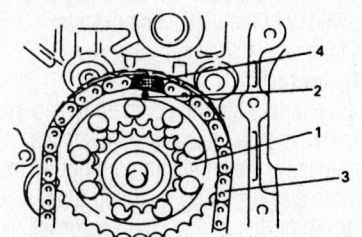

1. Idler sprocket
2. Match mark on idler sprocket
3. 1st timing chain
4. Dark blue plate

7924HG16

Idler sprocket and first timing chain alignment—2.0L engines

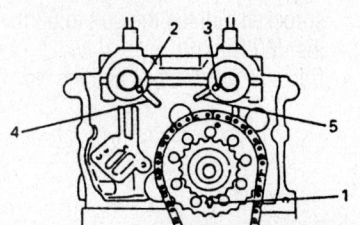

1. Arrow mark on idler sprocket
2. Knock pin of intake camshaft
3. Knock pin of exhaust camsaft
4. Timing mark of intake side
5. Timing mark of exhaust side

7924HG19

Idler sprocket and camshaft alignment—2.0L engines

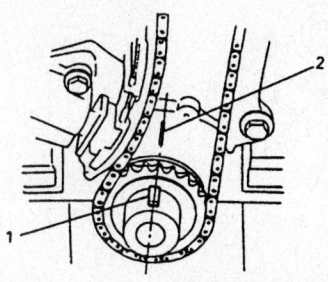

1. Crank timing sprocket key
2. Timing mark

7924HG18

Crankshaft sprocket alignment—2.0L engines

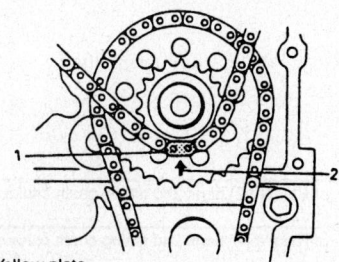

1. Yellow plate
2. Match mark of 2nd timing chain (Arrow mark)

7924HG20

Idler sprocket and second timing chain alignment—2.0L engines

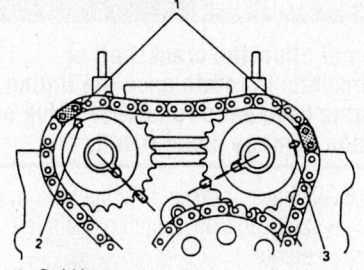

1. Dark blue
2. Arrow mark on intake camshaft timing sprocket
3. Arrow mark on exhaust camshaft timing sprocket

7924HG21

Camshaft sprocket and second timing chain alignment—2.0L engines

- Timing chain idler sprocket and first timing chain with the matchmarks and colored links aligned as shown
- First timing chain tensioner. Tighten the bolts to 97 inch lbs. (11 Nm).
- Camshaft sprockets and second timing chain with the matchmarks and colored links aligned as shown
- Second timing chain tensioner. Tighten the bolts to 97 inch lbs. (11 Nm) and the nut to 33 ft. lbs. (45 Nm).

9. Tighten the camshaft sprocket bolts to 59 ft. lbs. (80 Nm).

10. Remove the timing chain tensioner retaining pins.

11. Rotate the crankshaft two complete turns and check that the timing marks align.

12. Install or connect the following:
- Front cover. Apply sealant as shown.
- Front crankshaft seal
- Crankshaft pulley. Tighten the bolt to 109 ft. lbs. (130 Nm).
- A/C compressor and bracket, if equipped
- Upper radiator hose
- Alternator belt tensioner and idler pulleys
- Water pump pulley
- Cooling fan and shroud
- Accessory drive belts
- Bypass pipe and hose
- Valve cover
- Oil pan and pickup tube
- Negative battery cable

13. Fill the crankcase to the correct level.

14. Fill the cooling system.

15. Start the engine and check for leaks.

2.5L and 2.7L Engines

1. Before servicing the vehicle, refer to the Precautions Section.

2. Drain the cooling system.

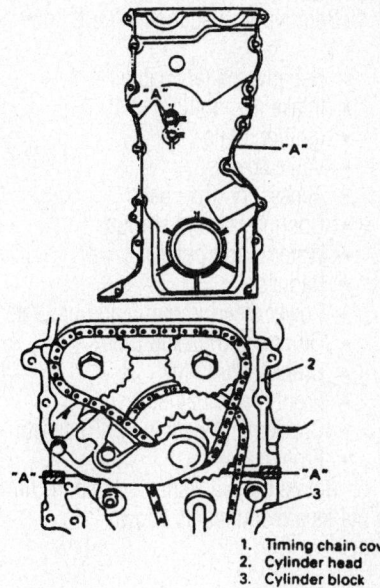

1. Timing chain cover
2. Cylinder head
3. Cylinder block

7924HG10

Prior to installing the timing chain cover on the engine block and cylinder head, apply silicone sealant to the cover as indicated (areas marked A)—2.0L and 2.5L engines

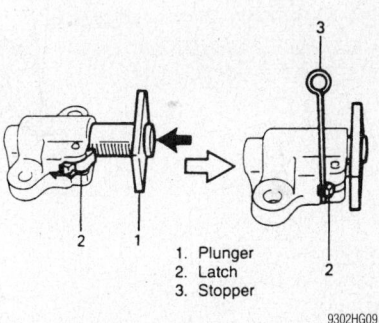

1. Plunger
2. Latch
3. Stopper

9302HG09

Preparing the No. 1 timing chain tensioner adjuster for installation—2.0L and 2.5L engines

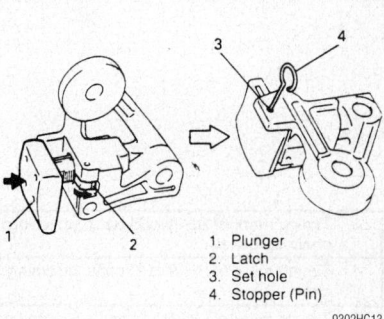

1. Plunger
2. Latch
3. Set hole
4. Stopper (Pin)

9302HG13

No. 2 timing chain tensioner—left bank tensioner shown—2.0L and 2.5L engines

3. Drain the engine oil.

4. Remove or disconnect the following:

- Negative battery cable
- Intake manifold
- Ignition coils
- Valve covers
- Accessory drive belts
- Cooling fan and shroud
- Water pump pulley
- Radiator
- Power steering pump and brackets
- Oil pan and pickup tube
- Crankshaft pulley
- Front crankshaft seal
- Crankshaft Position (CKP) sensor
- Front cover

5. Rotate the crankshaft so that the timing marks are aligned as shown.

✴✴ WARNING

Do not allow the crankshaft or camshafts to rotate once the timing chains have been removed. Valve or piston damage could result.

6. Remove or disconnect the following:
- Left bank No. 2 timing chain tensioner
- Left bank intake and exhaust camshaft sprockets with the No. 2 timing chain
- No. 1 timing chain guides
- No. 1 timing chain tensioner
- Center idler sprocket and the No. 1 timing chain
- Right bank No. 1 timing chain sprocket

7. The right bank No. 2 timing chain is removed with the intake and exhaust camshafts.

To install:

8. Prepare the timing chain tensioners for installation by releasing the latches, compressing the tensioner piston fully into the bore and installing retaining pins.

9. Align the timing chain sprocket matchmarks and colored chain links as shown during assembly.

10. Install or connect the following:
- Right bank intake and exhaust camshafts with the No. 2 timing chain
- Right bank No. 1 timing chain sprocket. Tighten the bolt to 58 ft. lbs. (80 Nm).
- Center idler sprocket and the No. 1

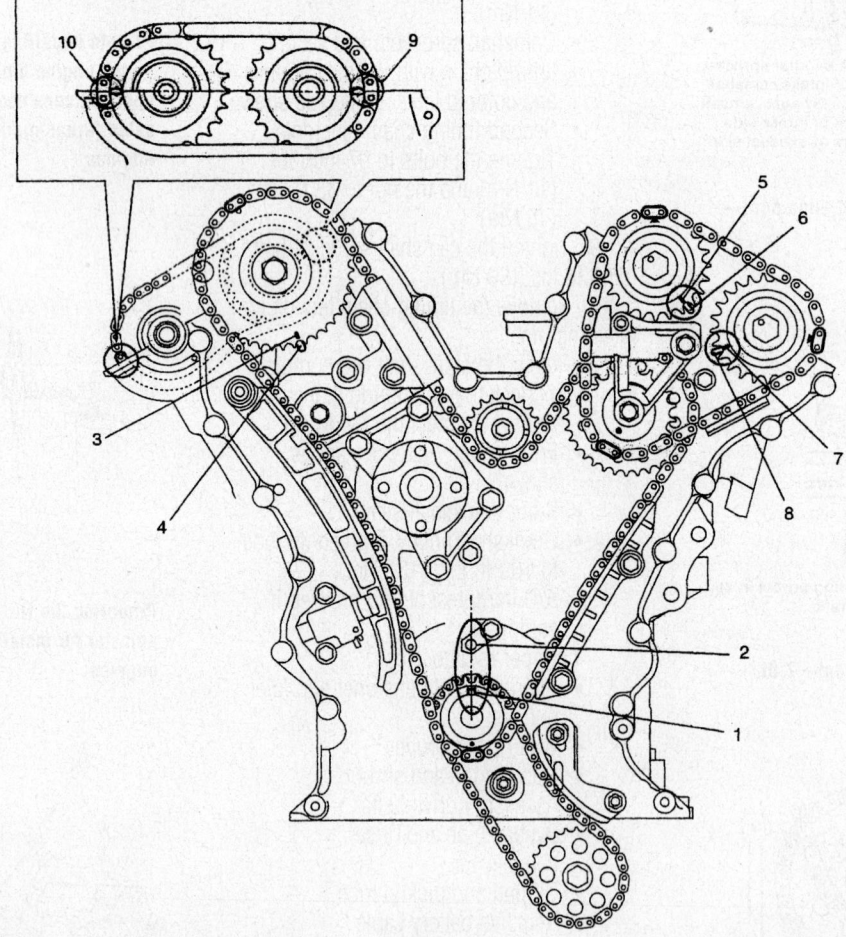

3.	Timing mark of RH (No.2) bank 1st timing chain sprocket	6.	Timing mark of LH (No.1) bank 2nd timing chain	9.	Timing mark of RH (No.2) bank 2nd timing chain intake sprocket
4.	Timing mark of RH (No.2) bank 1st timing chain	7.	Timing mark of LH (No.1) bank 2nd timing chain exhaust sprocket	10.	Timing mark of RH (No.2) bank 2nd timing chain exhaust sprocket
5.	Timing mark of LH (No.1) bank 2nd timing chain intake sprocket	8.	Timing mark of LH (No.1) bank 2nd timing chain		

09490_SUZT_G0009

Timing chain alignment marks—2.5L and 2.7L engines

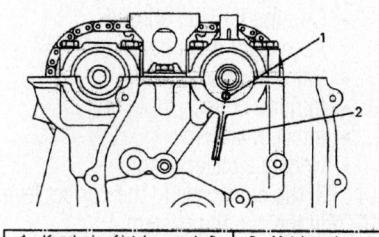

1.	Knock pin of intake camshaft	2.	Match mark

09490_SUZT_G0012

Right bank camshaft timing marks—2.5L and 2.7L engines

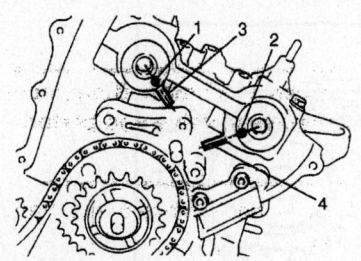

1.	Knock pin of LH (No.1) bank intake camshaft	3.	Match mark of intake side
2.	Knock pin of LH (No.1) bank exhaust camshaft	4.	Match mark of exhaust side

09490_SUZT_G0014

Left bank camshaft alignment—2.5L and 2.7L engines

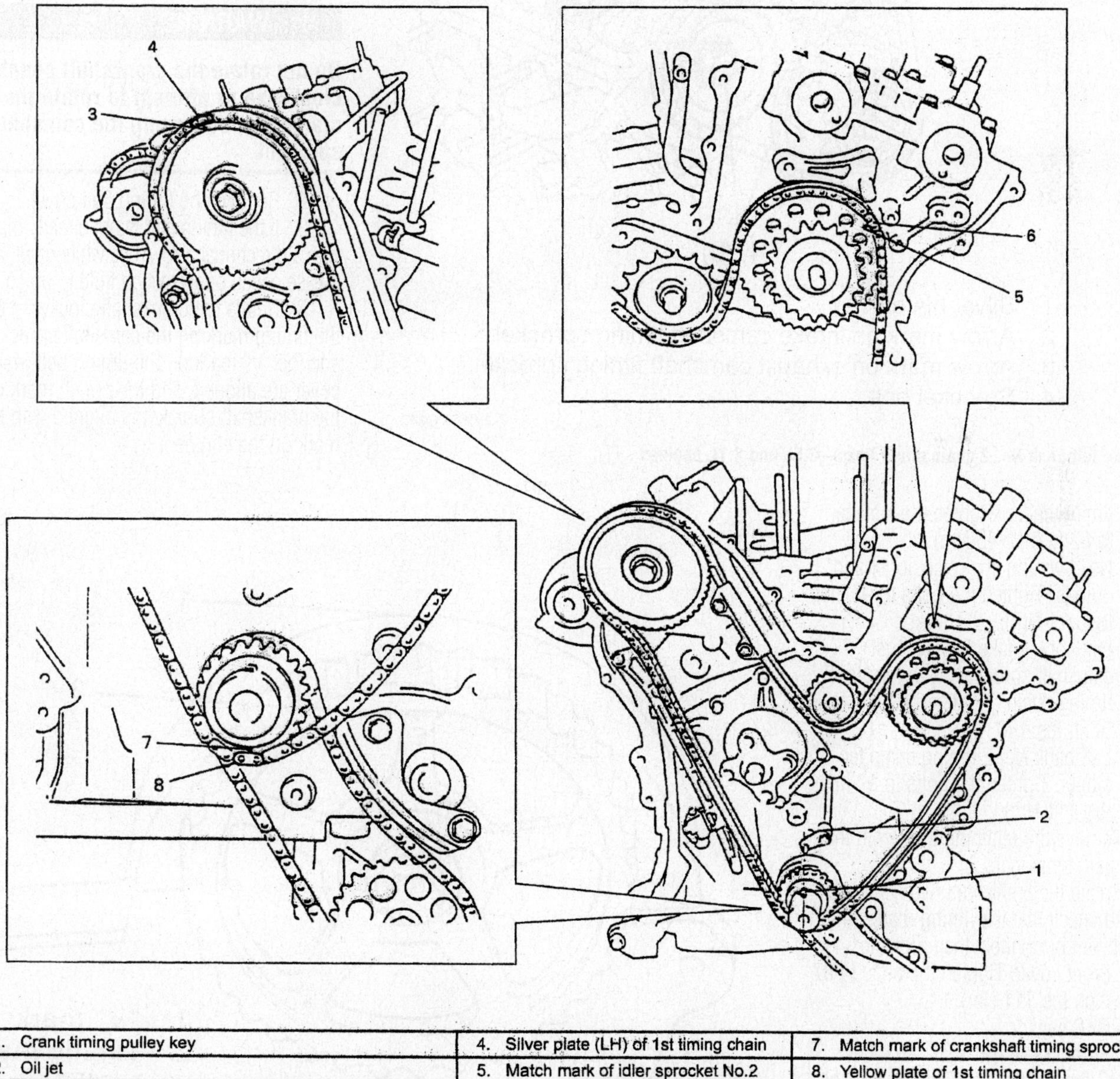

1.	Crank timing pulley key	4.	Silver plate (LH) of 1st timing chain	7.	Match mark of crankshaft timing sprocket
2.	Oil jet	5.	Match mark of idler sprocket No.2	8.	Yellow plate of 1st timing chain
3.	Match mark of RH (No.2) bank 1st timing chain sprocket	6.	Silver plate (LH) of 1st timing chain		

09490_SUZT_G0013

No. 1 timing chain alignment—2.5L and 2.7L engines

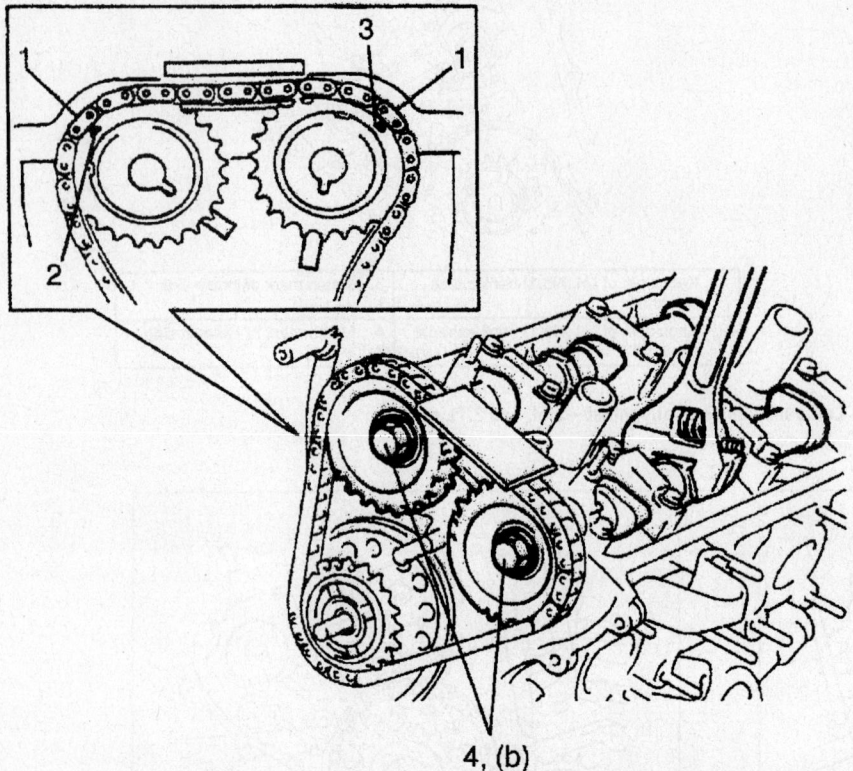

1. Silver plate
2. Arrow mark on intake camshaft timing sprocket
3. Arrow mark on exhaust camshaft timing sprocket
4. Sprocket bolt

09490_SUZT_G0015

Align the left bank No. 2 chain silver links—2.5L and 2.7L engines

timing chain. Tighten the fastener to 32 ft. lbs. (45 Nm).
- No. 1 timing chain tensioner and guides. Tighten the bolts to 97 inch lbs. (11 Nm).
- Left bank intake and exhaust camshaft sprockets with the No. 2 timing chain. Tighten the bolts to 57 ft. lbs. (80 Nm).
- Left bank No. 2 timing chain tensioner. Tighten the bolts to 97 inch lbs. (11 Nm).

11. Remove the retaining pins from the timing chain tensioners.

12. Rotate the crankshaft two complete turns and check that the timing marks align.

13. Install or connect the following:
- Front cover. Tighten the bolts to 97 inch lbs. (11 Nm).
- CKP sensor
- Front crankshaft seal
- Crankshaft pulley. Tighten the bolt to 109 ft. lbs. (148 Nm).
- Oil pan and pickup tube
- Power steering pump and brackets
- Radiator

- Water pump pulley
- Cooling fan and shroud
- Accessory drive belts
- Valve covers
- Ignition coils
- Intake manifold
- Negative battery cable

14. Fill the crankcase to the correct level.
15. Fill the cooling system.
16. Start the engine and check for leaks.

Timing Belt

REMOVAL & INSTALLATION

1.6L Engine

❉❉ WARNING

Do not rotate the crankshaft counterclockwise or attempt to rotate the crankshaft by turning the camshaft sprocket.

1. Remove the timing belt cover.
2. If the timing belt is not already marked with a directional arrow, use white paint, a grease pencil or correction fluid to do so.
3. Rotate the crankshaft clockwise until the timing mark on the camshaft sprocket and the "V" mark on the timing belt inside cover are aligned, and the punch mark on the crankshaft sprocket is aligned with the mark on the engine.

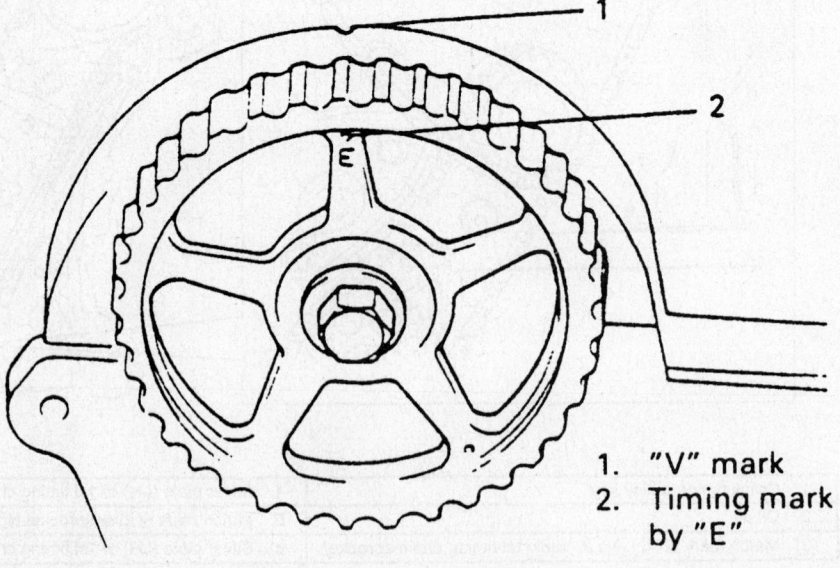

1. "V" mark
2. Timing mark by "E"

Camshaft timing marks — Suzuki Vitara 1.6L 16-valve engine

79245G22

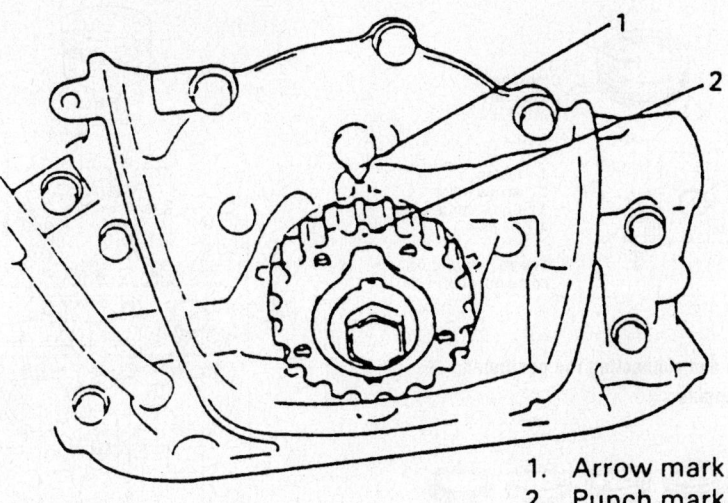

1. Arrow mark
2. Punch mark

79245G23

Align the punch mark with the arrow for proper timing belt installation — Suzuki Vitara 1.6L 16-valve engine

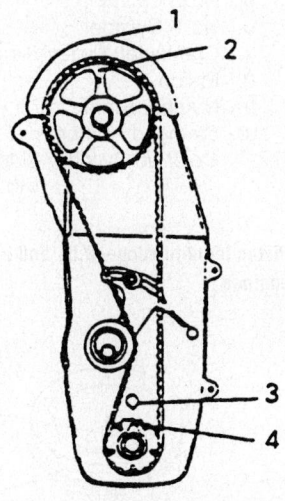

1. "V" mark on cylinder head cover
2. Timing mark by "E" on camshaft timing belt pulley
3. Arrow mark on oil pump case
4. Punch mark on crankshaft timing belt pulley

79245G47

Rotate the crankshaft clockwise until the camshaft and crankshaft timing marks are aligned — Suzuki Vitara 1.6L 16-valve engine

✳✳ WARNING

Do not rotate the crankshaft or camshaft once the timing belt is removed, because the valves and pistons can come into contact, which may cause internal engine damage.

4. Disconnect one end of the tensioner spring. Loosen the timing belt tensioner bolt and stud, then, using your finger, press the tensioner plate up and remove the timing belt from the crankshaft and camshaft sprockets.

5. Remove the timing belt tensioner, tensioner plate and spring from the engine.

6. Install Suzuki tool 09917-68220, or equivalent, onto the camshaft sprocket to hold the camshaft from rotating. Loosen the camshaft sprocket retaining bolt, then pull the camshaft sprocket off of the end of the camshaft.

7. Remove the crankshaft timing belt sprocket by loosening the center bolt, while preventing the crankshaft from rotating. To hold the crankshaft from turning, use Suzuki tool 09927-56010, or equivalent, or a large prybar inserted in the transmission housing slot and the flywheel teeth. Pull the sprocket off of the end of the crankshaft. Be sure to retain the crankshaft sprocket key and belt guide for assembly.

8. If necessary, remove the timing belt inside cover from the cylinder head.

To install:

9. If necessary, install the timing belt inside cover.

10. Slide the timing belt guide on the crankshaft so that the concave side faces the oil pump, then install the sprocket key in the groove in the crankshaft.

11. Slide the pulley onto the crankshaft, and install the center retaining bolt. Tighten the center bolt to 80 ft. lbs. (110 Nm). To hold the crankshaft from turning, use Suzuki tool 09927-56010, or equivalent, or a large prybar inserted in the transmission housing slot and the flywheel teeth.

12. Install the timing belt camshaft sprocket, ensuring that the slot in the sprocket engages the camshaft (pulley) pin; this ensures that the sprocket is properly positioned on the end of the camshaft. Secure the camshaft with the holding tool used during removal, then tighten the sprocket bolt to 44 ft. lbs. (60 Nm).

13. Assemble the timing belt tensioner plate and the tensioner, making sure that the lug of the tensioner plate engages the tensioner.

✳✳ WARNING

If any binding is felt when adjusting the timing belt tension by turning the crankshaft, STOP turning the engine, because the pistons may be hitting the valves.

14. Install the timing belt tensioner, tensioner plate and spring on the engine. Tighten the mounting bolt and stud only finger-tight at this time. Ensure that when the tensioner is moved in a counterclockwise direction, the tensioner moves in the same direction. If the tensioner does not move, remove it and the tensioner plate to reassemble them properly.

15. Loosen all rocker arm valve lash locknuts and adjusting screws. This will permit movement of the camshaft without any rocker arm associated drag, which is essential for proper timing belt tensioning. If the camshaft does not rotate freely (free of rocker arm drag), the belt will not be properly tensioned.

16. Rotate the camshaft sprocket clockwise until the timing mark on the sprocket and the "V" mark on the timing belt inside cover are aligned.

17. Using a wrench, or socket and breaker bar, on the crankshaft sprocket center bolt, turn the crankshaft clockwise until the punch mark on the sprocket is

aligned with the arrow mark on the oil pump.

18. With the camshaft and crankshaft marks properly aligned, push the tensioner up with your finger and install the timing belt on the 2 sprockets, ensuring that the drive side of the belt is free of all slack. Release your finger from the tensioner. Be sure to install the timing belt so that the directional arrow is pointing in the appropriate direction.

➡️**In this position, the No. 4 cylinder is at Top Dead Center (TDC) on the compression stroke.**

19. Rotate the crankshaft clockwise 2 full revolutions, then tighten the tensioner stud to 97 inch lbs. (11 Nm). Then, tighten the tensioner bolt to 18 ft. lbs. (24 Nm).

20. Ensure that all 4 timing marks are still aligned as before; if they are not, remove the timing belt, and install and tension it again.

21. Install the timing belt cover and all related components.

Piston and Ring

POSITIONING

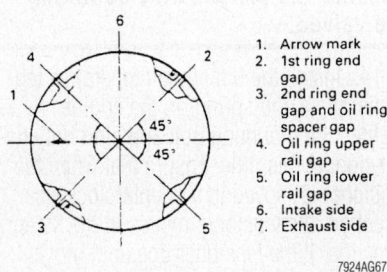

1. Arrow mark
2. 1st ring end gap
3. 2nd ring end gap and oil ring spacer gap
4. Oil ring upper rail gap
5. Oil ring lower rail gap
6. Intake side
7. Exhaust side

7924AG67

Piston ring end-gap spacing—All engines

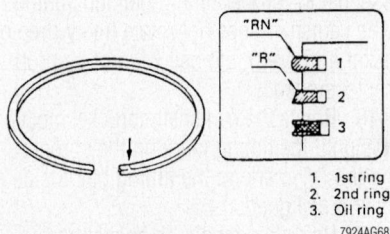

1. 1st ring
2. 2nd ring
3. Oil ring

7924AG68

Compression ring identification marks— All engines

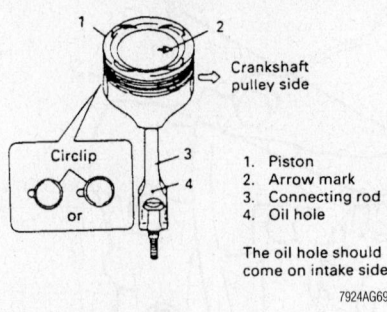

1. Piston
2. Arrow mark
3. Connecting rod
4. Oil hole

The oil hole should come on intake side

7924AG69

Piston and connecting rod positioning— 1.6L engine

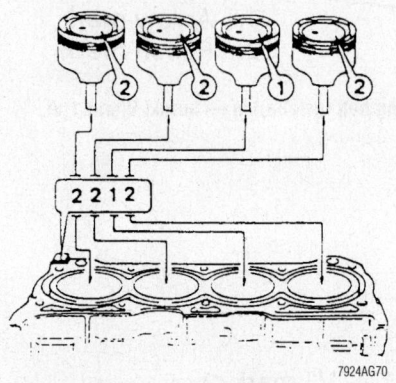

2 2 1 2

7924AG70

Piston installation—1.6L engine

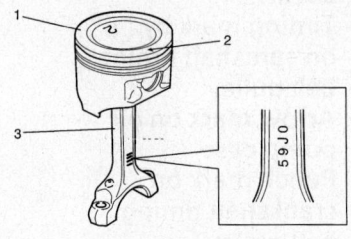

1. Piston
2. Arrow Mark
3. Connecting Rod

09490_SUZT_G0016

Piston and connecting rod positioning— 2.0L, 2.5L and 2.7L engines

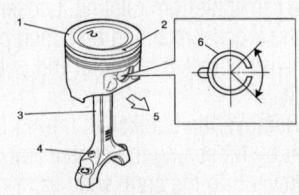

1. Piston
2. Arrow Mark
3. Connecting Rod
4. Oil Hole
5. Crankshaft Pulley Side
6. Circlip

09490_SUZT_G0017

Piston pin circlip installation—2.0L, 2.5L and 2.7L engines

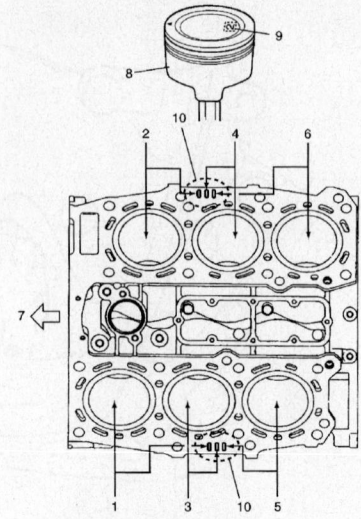

1. No. 1 Cylinder
2. No. 2 Cylinder
3. No. 3 Cylinder
4. No. 4 Cylinder
5. No. 5 Cylinder
6. No. 6 Cylinder
7. Crankshaft Pulley Side
8. Piston
9. Number Stamp
10. Stamped Number or Painted Color on Cylinder Block

09490_SUZT_G0018

Piston identification—2.5L and 2.7L engines

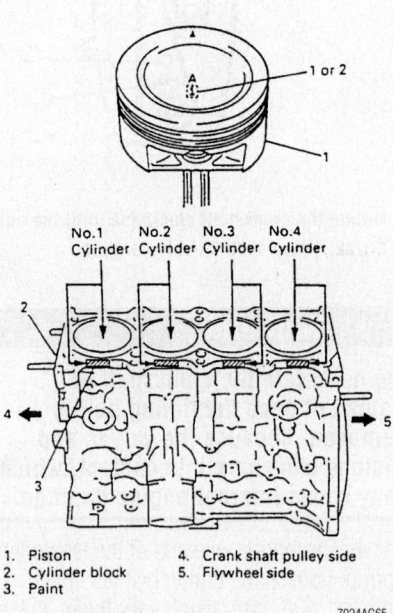

1. Piston
2. Cylinder block
3. Paint
4. Crank shaft pulley side
5. Flywheel side

7924AG65

Piston identification—2.0L engines
Match pistons with "1" indicators to red cylinder paint marks

Match pistons with "2" indicators to blue cylinder paint marks

FUEL SYSTEM

Fuel System Service Precautions

Safety is the most important factor when performing not only fuel system maintenance but any type of maintenance. Failure to conduct maintenance and repairs in a safe manner may result in serious personal injury or death. Maintenance and testing of the vehicle fuel system components can be accomplished safely and effectively by adhering to the following rules and guidelines.

- To avoid the possibility of fire and personal injury, always disconnect the negative battery cable unless the repair or test procedure requires that battery voltage be applied.
- Always relieve the fuel system pressure prior to disconnecting any fuel system component (injector, fuel rail, pressure regulator, etc.), fitting or fuel line connection. Exercise extreme caution whenever relieving fuel system pressure to avoid exposing skin, face and eyes to fuel spray. Please be advised that fuel under pressure may penetrate the skin or any part of the body that it contacts.
- Always place a shop towel or cloth around the fitting or connection prior to loosening to absorb any excess fuel due to spillage. Ensure that all fuel spillage (should it occur) is quickly removed from engine surfaces. Ensure that all fuel soaked cloths or towels are deposited into a suitable waste container.
- Always keep a dry chemical (Class B) fire extinguisher near the work area.
- Do not allow fuel spray or fuel vapors to come into contact with a spark or open flame.
- Always use a backup wrench when loosening and tightening fuel line connection fittings. This will prevent unnecessary stress and torsion to fuel line piping.
- Always replace worn fuel fitting O-rings with new. Do not substitute fuel hose or equivalent, where fuel pipe is installed.

Fuel System Pressure

RELIEVING

✳✳ CAUTION

Care should be used when working around the fuel system. DO NOT smoke or expose the fuel system to any open flames. Keep a fire extinguisher handy.

1. Before servicing the vehicle, refer to the Precautions Section.
2. Place the vehicle in **PARK** for automatic transmission or **NEUTRAL** for manual transmission, set parking brake and block drive wheels.
3. Disconnect the fuel pump relay.
4. Remove the fuel filler cap from the filler neck to release the fuel vapor pressure in the fuel tank.
5. Start the engine and run it until it stops from lack of fuel. Crank the engine 2–3 times for a 3 second period. The fuel lines should now be depressurized.
6. After servicing, connect the fuel pump relay.

Fuel Filter

REMOVAL & INSTALLATION

All Models (except 2006 Grand Vitara)

1. Before servicing the vehicle, refer to the Precautions Section.
2. Relieve fuel system pressure.
3. Remove or disconnect the following:
   - Negative battery cable
   - Fuel lines from the fuel filter
   - Fuel filter

To install:

4. Install or connect the following:
   - Fuel filter and tighten the bracket bolt. Note the fuel flow directional arrow.
   - Fuel lines to the fuel filter.
   - Negative battery cable
5. Start the engine and inspect the fuel filter connections for leaks.

2006 Grand Vitara

The fuel filter for the 2006 Grand Vitara is an integral component of the in-tank fuel pump assembly. Refer to the Fuel Pump Removal procedure later in this section.

Fuel Pump

REMOVAL & INSTALLATION

1. Before servicing the vehicle, refer to the Precautions Section.
2. Relieve the fuel system pressure.
3. Remove or disconnect the following:
   - Negative battery cable
   - Exhaust center pipe and rear propeller shaft, if necessary
   - Fuel filler hose and vent hose

- Fuel tank inlet valve and drain the fuel tank
- Fuel filter inlet hose
- Evaporative Emissions (EVAP) vapor hose
- Fuel return line
- Fuel tank skidplate
- Fuel pump connector
- Fuel tank pressure sensor connector
- Fuel tank
- Fuel pump module

To install:

4. Install or connect the following:
   - Fuel pump module with a new seal. Tighten the bolts to 44 inch lbs. (5 Nm) for all models except 2006 Grand Vitara; and 8 ft. lbs. (11 Nm) for 2006 Grand Vitara
   - Fuel tank. Tighten the strap bolts to 37 ft. lbs. (50 Nm).
   - Fuel tank pressure sensor connector
   - Fuel pump connector
   - Fuel tank skidplate
   - Fuel return line
   - EVAP vapor hose
   - Fuel filter inlet hose
   - Fuel tank inlet valve and drain the fuel tank
   - Fuel filler hose and vent hose
   - Exhaust center pipe and rear propeller shaft, if necessary
   - Negative battery cable
5. Fill the fuel tank.
6. Start the engine and check for leaks.

Fuel Injector

REMOVAL & INSTALLATION

1.6L and 2.0L Engines

1. Before servicing the vehicle, refer to the Precautions Section.
2. Relieve the fuel system pressure.
3. Remove or disconnect the following:
   - Negative battery cable
   - Front intake manifold bracket, if equipped
   - Positive Crankcase Ventilation (PCV) valve and hose
   - Fuel injector harness connectors
   - Fuel line bracket
   - Fuel supply manifold
   - Fuel injectors

To install:

4. Install or connect the following:
   - Fuel injectors with new O-ring seals

- Fuel supply manifold. Tighten the bolts to 17 ft. lbs. (23 Nm).
- Fuel line bracket
- Fuel injector harness connectors
- PCV valve and hose
- Front intake manifold bracket, if equipped
- Negative battery cable

5. Start the engine and check for leaks.

2.5L and 2.7L Engines

1. Before servicing the vehicle, refer to the Precautions Section.
2. Relieve the fuel system pressure.

3. Remove or disconnect the following:
- Negative battery cable
- Air intake tube
- Throttle body intake collector
- Fuel lines
- Fuel pressure regulator vacuum line
- Fuel injector harness connectors
- Fuel supply manifold connect pipe
- Fuel supply manifolds
- Fuel injectors

To install:

4. Install or connect the following:
- Fuel injectors with new O-ring seals

- Fuel supply manifolds. Tighten the bolts to 18 ft. lbs. (25 Nm).
- Fuel supply manifold connect pipe. Tighten the bolts to 22 ft. lbs. (30 Nm).
- Fuel injector harness connectors
- Fuel pressure regulator vacuum line
- Fuel lines
- Throttle body intake collector
- Air intake tube
- Negative battery cable

5. Start the engine and check for leaks.

DRIVE TRAIN

Transmission Assembly

REMOVAL & INSTALLATION

Manual Transmission

1. Before servicing the vehicle, refer to the Precautions Section.
2. Drain the transmission fluid.
3. Drain the transfer case fluid, if equipped.
4. Remove or disconnect the following:
- Negative battery cable
- Shift lever boots
- Gear shift lever
- Transfer case shift lever, if equipped
- 4WD switch connector, if equipped
- Reverse light switch connector
- Starter motor
- Front driveshaft, if equipped
- Rear driveshaft
- Speedometer cable, if equipped
- Vehicle Speed (VSS) sensor, if equipped

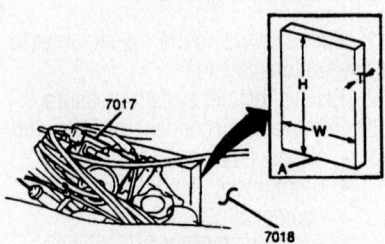

A   WOOD BLOCK
H   200 mm (8.0")
T   45 mm (1.8")
W   100-150 mm (4.0-6.0")
7017 DISTRIBUTOR CAP
7018 BULKHEAD

7924HG37

Support the engine with a wooden block between the cylinder head and the firewall—All models

- Clutch slave cylinder or cable
- Flywheel access cover
- Transmission flange bolts and nuts
- Transmission braces, if equipped

5. Support the transmission with a jack and remove the transmission mount and crossmember.
6. Place a wooden block at the rear of the cylinder head as shown to support the engine when the transmission is removed.
7. Lower the transmission away from the vehicle.

To install:

8. Install or connect the following:
- Transmission. Tighten the flange fasteners to 62 ft. lbs. (85 Nm) on the 1.6L engine, and 58 ft. lbs. (80 Nm) on the 2.0L, 2.5L and 2.7L engines.
- Transmission mount and crossmember. Tighten the fasteners to 29–43 ft. lbs. (40–60 Nm).
- Transmission braces, if equipped. Tighten the bolts to 62–72 ft. lbs. (85–98 Nm).
- Flywheel access cover
- Clutch slave cylinder or cable
- VSS sensor, if equipped
- Speedometer cable, if equipped
- Rear driveshaft
- Front driveshaft, if equipped
- Starter motor
- Reverse light switch connector
- 4WD switch connector, if equipped
- Transfer case shift lever, if equipped
- Gear shift lever
- Shift lever boots
- Negative battery cable

9. Fill the transmission to the correct level.
10. Fill the transfer case, if equipped.

Automatic Transmission

1. Before servicing the vehicle, refer to the Precautions Section.
2. Drain the transfer case oil, if equipped.
3. Remove or disconnect the following:
- Negative battery cable
- Center console and transfer case shift lever, if equipped
- Transmission dipstick tube
- Transmission wiring harness connectors
- Starter motor
- Front driveshaft, if equipped
- Rear driveshaft
- Gear select cable and bracket
- Throttle Valve (TV) cable, if equipped
- Exhaust front pipe
- Transmission oil cooler lines
- Transmission brace
- Flywheel access cover
- Torque converter
- Speedometer cable, if equipped
- Vehicle Speed (VSS) sensor connector, if equipped
- Transmission flange bolts and nuts
- Transmission braces, if equipped

4. Support the transmission with a jack and remove the transmission mount and crossmember.
5. Place a wooden block at the rear of the cylinder head as shown to support the engine when the transmission is removed.
6. Lower the transmission away from the vehicle.

To install:

7. Install or connect the following:
- Transmission. Tighten the flange fasteners to 62 ft. lbs. (85 Nm) on the 1.6L engine, and 58 ft. lbs. (80 Nm) on the 2.0L, 2.5L and 2.7L engines.
- Transmission mount and cross-

member. Tighten the fasteners to 29–43 ft. lbs. (40–60 Nm).

- Transmission braces, if equipped. Tighten the bolts to 62–72 ft. lbs. (85–98 Nm).
- VSS sensor connector, if equipped
- Speedometer cable, if equipped
- Torque converter. Tighten the bolts to 47 ft. lbs. (65 Nm).
- Flywheel access cover
- Transmission brace
- Transmission oil cooler lines
- Exhaust front pipe
- TV cable, if equipped
- Gear select cable and bracket
- Rear driveshaft
- Front driveshaft, if equipped
- Starter motor
- Transmission wiring harness connectors
- Transmission dipstick tube
- Center console and transfer case shift lever, if equipped
- Negative battery cable

8. Fill the transmission to the correct level.
9. Fill the transfer case, if equipped.

Clutch

ADJUSTMENTS

These vehicles are equipped with a hydraulic clutch system. No adjustment is necessary.

REMOVAL & INSTALLATION

1. Before servicing the vehicle, refer to the Precautions Section.
2. Remove the transmission.
3. Loosen the pressure plate mounting bolts in a 2-step crisscross sequence until the spring tension is relieved.
4. Remove the pressure plate and the clutch disc.

To install:

5. Using a clutch alignment tool, assemble the clutch disc and pressure plate onto the flywheel.
6. Tighten the pressure plate bolts in multiple passes to 17 ft. lbs. (23 Nm).
7. Install the transmission.
8. Check for proper clutch operation.

Hydraulic Clutch System

BLEEDING

1. Before servicing the vehicle, refer to the Precautions Section.

2. Fill the master cylinder reservoir to the MAX line with clean brake fluid and keep it at least half full throughout the bleeding procedure.
3. From beneath the vehicle, remove the bleeder plug cap, then attach a clear vinyl tube to the slave cylinder bleeder plug. Insert the open end of the hose into a container.
4. Have an assistant depress the clutch pedal. Open the bleeder after the pedal is depressed.
5. Close the bleeder before releasing the clutch pedal.
6. Repeat until all air bubbles are gone from the hydraulic fluid.
7. Install the bleeder plug cap.
8. Fill the clutch master cylinder fluid reservoir to the specified full level.

Transfer Case Assembly

REMOVAL & INSTALLATION

1. Before servicing the vehicle, refer to the Precautions Section.
2. Drain the transfer case oil.
3. Remove or disconnect the following:

- Negative battery cable
- Camshaft Position (CMP) sensor
- Center console
- Transmission shift lever and case, if equipped with a manual transmission
- Transfer case shift lever
- Front and rear driveshafts
- Exhaust center pipe
- Speedometer cable or Vehicle Speed (VSS) sensor, as equipped
- Vent hose
- 4WD switch connector

4. Support the transmission with a jack and remove the transmission mount and crossmember.
5. Place a wooden block at the rear of the cylinder head as shown to support the engine when the transfer case is removed.
6. Lower the transfer case away from the vehicle.

To install:
7. Install or connect the following:

- Transfer case. Tighten the bolts to 30 ft. lbs. (41 Nm).
- 4WD switch connector
- Vent hose
- Speedometer cable or VSS sensor, as equipped
- Exhaust center pipe

- Front and rear driveshafts. Tighten the bolts to 36 ft. lbs. (50 Nm).
- Transfer case shift lever
- Transmission shift lever and case, if equipped with a manual transmission
- Center console
- CMP sensor
- Negative battery cable

8. Fill the transfer case.

Halfshaft

REMOVAL & INSTALLATION

Front

LEFT

1. Before servicing the vehicle, refer to the Precautions Section.
2. Remove or disconnect the following:

- Front wheel
- Hub drive flange or locking hub, as equipped
- Snapring
- Thrust washer
- Halfshaft flange fasteners
- Halfshaft

To install:
3. Install or connect the following:

- Halfshaft. Tighten the flange bolts to 37 ft. lbs. (50 Nm).
- Thrust washer
- Snapring
- Locking hub, if equipped. Tighten the bolts to 24 ft. lbs. (33 Nm).
- Hub drive flange, if equipped. Tighten the bolts to 35 ft. lbs. (48 Nm).
- Front wheel

RIGHT

1. Before servicing the vehicle, refer to the Precautions Section.
2. Remove or disconnect the following:

- Front wheel
- Hub drive flange or locking hub, as equipped
- Snapring
- Thrust washer
- Brake caliper
- Wheel speed sensor, if equipped
- Brake rotor
- Stabilizer bar link
- Outer tie rod end
- Lower ball joint
- Strut bracket bolts
- Steering knuckle and wheel hub

3. Pry the inboard joint out of the differential and remove the halfshaft.

1. Drive shaft oil seal
2. Double off-set joint (DOJ)
3. Joint circlip
4. DOJ boot
5. Ball joint boot
6. Ball joint assembly (RH side)
7. Drive shaft assembly (LH side)
8. Left drive shaft
9. Drive shaft bearing circlip
10. Drive shaft bearing

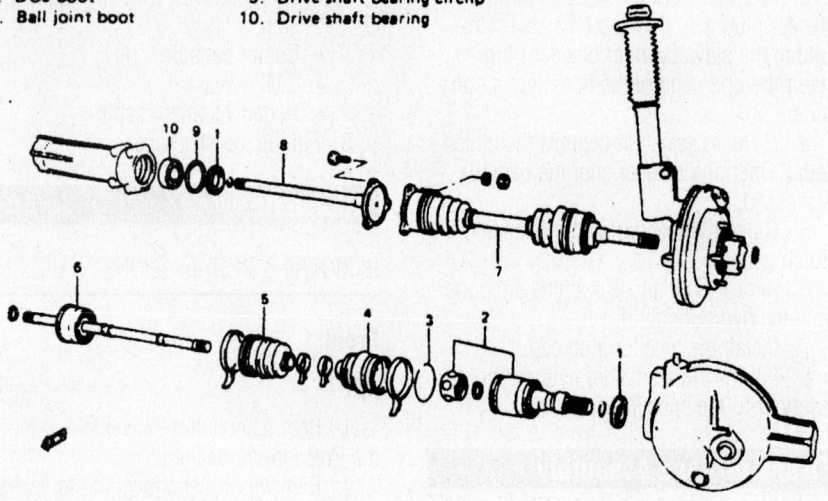

7924HG31

Exploded view of the left- and right-hand halfshaft assemblies

To install:

4. Insert the inboard joint into the differential until the circlip is felt to seat.

5. Install or connect the following:
- Steering knuckle and wheel hub
- Strut bracket bolts. Tighten them to 70 ft. lbs. (95 Nm).
- Lower ball joint. Tighten the nut to 40 ft. lbs. (55 Nm).
- Outer tie rod end. Tighten the nut to 35 ft. lbs. (48 Nm).
- Stabilizer bar link. Tighten the nut to 21 ft. lbs. (29 Nm).
- Brake rotor
- Wheel speed sensor, if equipped
- Brake caliper
- Thrust washer
- Snapring

Rear drive shaft flange nuts (1), match mark (2), rear drive shaft flange (3), and rear drive shaft (4)—2006 Grand Vitara

09490_SUZT_G0020

- Locking hub, if equipped. Tighten the bolts to 24 ft. lbs. (33 Nm).
- Hub drive flange, if equipped. Tighten the bolts to 35 ft. lbs. (48 Nm).
- Front wheel

6. Check the wheel alignment and adjust as necessary.

REAR (2006 GRAND VITARA)

1. Before servicing the vehicle, refer to the Precautions Section.
2. Disconnect the negative battery cable.
3. Remove drive shaft nut and washer.
4. Raise and safely support the rear of the vehicle securely on jackstands.
5. Remove wheel.
6. Match mark the rear drive shaft flange and rear drive shaft.
7. Remove the flange nuts and remove drive shaft.

To install:

✳✳ WARNING

Protect oil seals and boots from any damage, preventing them form unnecessary contact while installing drive shaft. Do not hit joint boot with hammer. Inserting joint only by hands is allowed.

8. Install drive shaft assembly by reversing removal procedure noting the following points:

a. Align the match marks of the rear drive shaft to the drive shaft flange.

b. Tighten each bolt and nut to the specified torque.
- Rear drive shaft flange nuts: 58 ft. lbs. (80 Nm)
- New drive shaft nut: 145 ft. lbs. (200 Nm)

9. Connect the negative battery cable.

CV-Joints

OVERHAUL

Outer CV-Joint

The outer CV-joint is serviced with the axle shaft as an assembly. The outer CV-joint boot can be serviced by removing the inner CV-joint.

Inner CV-Joint

1. Before servicing the vehicle, refer to the Precautions Section.
2. Remove or disconnect the following:
- Halfshaft from the vehicle
- Grease boot clamps
- Outer race snapring
- Outer race
- Shaft snapring
- Inner race, cage and balls

To install:

3. Install or connect the following:
- Inner race, cage and balls
- Shaft snapring
- Outer race
- Outer race snapring

4. Fill the outer race and the grease boot with CV-joint grease and tighten the boot clamps.

5. Install the axle halfshaft.

Manual Locking Hubs

REMOVAL & INSTALLATION

1. Before servicing the vehicle, refer to the Precautions Section.
2. Set the selector knob to the **FREE** position.
3. Remove or disconnect the following:
- Hub cover assembly
- Hub body assembly

To install:

4. Install the hub body. Tighten the bolts to 18 ft. lbs. (25 Nm).
5. Align the hub cover stopper nail with the groove in the hub body and install the hub cover. Tighten the bolts to 90 inch lbs. (10 Nm).
6. Check for proper hub operation.

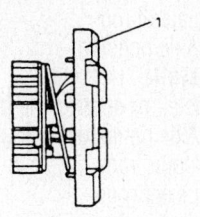

1. Cover

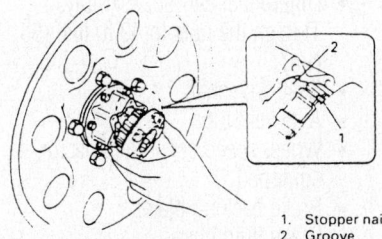

1. Stopper nail
2. Groove
9308HG06

Manual hub alignment—All models

Automatic Locking Hubs

REMOVAL & INSTALLATION

1. Before servicing the vehicle, refer to the Precautions Section.
2. Unlock the hub by setting the transfer case in the 2H position and driving backwards at least 6.5 feet (2 meters).
3. Remove or disconnect the following:
   - Hub sub assembly
   - Hub brake assembly

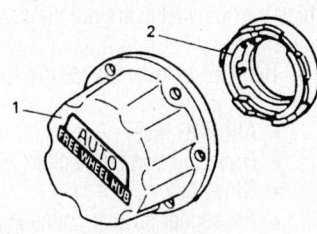

1. Free wheeling hub sub assembly
2. Free wheeling hub brake assembly

9308HG04

Automatic hub—All models

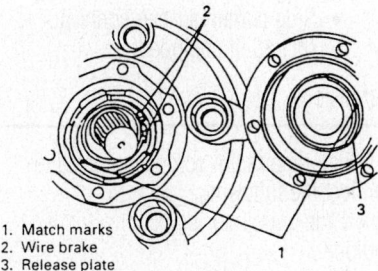

1. Match marks
2. Wire brake
3. Release plate

9308HG05

Automatic hub matchmarks—All models

To install:

4. Align the brake assembly key with the slot in the spindle and install the brake assembly.
5. Align the matchmark on the sub assembly with the mark on the brake assembly and install the sub assembly. Tighten the hub bolts to 24 ft. lbs. (33 Nm).
6. Check for proper hub operation.

Spindle Bearings

REMOVAL, PACKING & INSTALLATION

1. Before servicing the vehicle, refer to the Precautions Section.
2. Support the control arm with a stand or floor jack.
3. Remove or disconnect the following:
   - Front wheel
   - Locking hub or drive flange, as equipped
   - Brake caliper and rotor
   - Wheel hub and bearing assembly
   - Outer tie rod end
   - Lower ball joint
   - Strut bracket bolts
   - Wheel spindle and steering knuckle assembly
   - Inner oil seal
   - Spindle bearing

To install:

4. Fill the recess in the wheel spindle with lithium grease.
5. Coat the spindle bearing and wheel spindle mating surfaces with sealant.
6. Press or drive the spindle bearing into the wheel spindle.
7. Install or connect the following:
   - Inner oil seal
   - Wheel spindle and steering knuckle assembly
   - Strut bracket bolts
   - Lower ball joint
   - Outer tie rod end
   - Wheel hub and bearing assembly
   - Brake caliper and rotor
   - Locking hub or drive flange, as equipped
   - Front wheel

Axle Shaft, Bearing and Seal

REMOVAL & INSTALLATION

All Models (except 2006 Grand Vitara)

1. Before servicing the vehicle, refer to the Precautions Section.

2. Loosen the parking brake cable for clearance.
3. Remove or disconnect the following:
   - Rear wheel
   - Brake drum
   - Wheel speed sensor, if equipped
   - Bearing retainer nuts
   - Axle shaft and bearing
   - Axle shaft inner oil seal
4. If equipped with ABS, grind a flat spot on the wheel speed sensor tone ring, then split the ring with a chisel.
5. Grind flat spots on the bearing retainer and split it with a chisel.
6. Press the wheel bearing off the axle shaft.
7. Remove the bearing retainer and the outer oil seal.

To install:

8. Install or connect the following:
   - Outer oil seal to the bearing retainer
   - Bearing retainer to the axle shaft
   - Bearing and retainer ring pressed onto the axle shaft
   - Wheel speed sensor tone ring pressed onto the axle shaft, if equipped
   - Axle shaft inner oil seal
   - Axle shaft and bearing
   - Bearing retainer nuts. Tighten them to 17 ft. lbs. (23 Nm).
   - Wheel speed sensor, if equipped
   - Brake drum
   - Rear wheel
9. Fill the rear differential to the correct level.

Pinion Seal

REMOVAL & INSTALLATION

1. Before servicing the vehicle, refer to the Precautions Section.
2. Remove or disconnect the following:
   - Driveshaft
   - Wheels
   - Brake calipers and pads or brake drum

➡**The brake calipers and pads or brake drum must be removed so that there is no additional drag when measuring pinion bearing preload.**

3. Use an inch lb. torque wrench and measure and record the amount of torque required to maintain pinion rotation through several revolutions.
4. Remove or disconnect the following:
   - Pinion flange
   - Pinion seal

- Pinion bearing
- Collapsible spacer

To install:

➡ **Use a new collapsible spacer and flange nut for assembly.**

5. Install or connect the following:
   - Collapsible spacer
   - Pinion bearing
   - Pinion seal
   - Pinion flange
6. Rotate the pinion flange occasionally while tightening the flange nut to make sure the pinion bearings seat correctly.
7. Take frequent bearing preload torque readings. Tighten the flange nut to achieve the preload torque readings originally recorded.

> ## ⚙ CAUTION
>
> **Never loosen the pinion nut to reduce bearing preload. If it is necessary to reduce bearing preload, install a new collapsible spacer and pinion nut.**

8. Install or connect the following:
   - Driveshaft
   - Brake calipers and pads or brake drum
   - Wheels
9. Fill the differential with gear lubricant and check for leaks.

Axle Housing Assembly

REMOVAL & INSTALLATION

All Models (except 2006 Grand Vitara)

1. Before servicing the vehicle, refer to the Precautions Section.
2. Drain the gear oil.
3. Support the vehicle at the frame with a hoist or jackstands.
4. Support the rear axle with a floor jack.
5. Remove or disconnect the following:
   - Rear wheels
   - Rear brake drums
   - Rear axle shafts
   - Load sensing proportioning valve linkage, if equipped
   - Brake fluid hose
   - Brake backing plates
   - Wheel speed sensor connector, if equipped
   - Axle vent tube
   - Rear driveshaft
   - Differential carrier assembly
   - Shock absorber lower bolts
   - Coil springs
   - Upper rods
   - Lower rods

- Lateral rod
- Axle housing

To install:

6. Install or connect the following:
   - Axle housing
   - Upper rods
   - Lower rods
   - Coil springs
   - Lateral rod
   - Shock absorber lower bolts
   - Differential carrier assembly. Tighten the nuts to 40 ft. lbs. (55 Nm).
   - Rear driveshaft
   - Axle vent tube
   - Wheel speed sensor connector, if equipped
   - Brake backing plates
   - Brake fluid hose
   - Load sensing proportioning valve linkage, if equipped
   - Rear axle shafts
   - Rear brake drums
   - Rear wheels
7. Fill the rear axle to the correct level.
8. Lower the vehicle so that the rear suspension is at curb height.
9. Tighten the upper, lower and lateral rod fasteners to 65 ft. lbs. (90 Nm).
10. Tighten the lower shock absorber fasteners to 62 ft. lbs. (85 Nm).

STEERING

Air Bag

> ## ⚙ CAUTION
>
> **Some vehicles are equipped with an air bag system. The system must be disarmed before performing service on, or around, system components, the steering column, instrument panel components, wiring and sensors. Failure to follow the safety precautions and the disarming procedure could result in accidental air bag deployment, possible injury and unnecessary system repairs.**

PRECAUTIONS

Several precautions must be observed when handling the inflator module to avoid accidental deployment and possible personal injury.

- Never carry the inflator module by the wires or connector on the underside of the module.

- When carrying a live inflator module, hold securely with both hands and ensure that the bag/trim cover are pointed away.
- Place the inflator module on a bench or other surface with the bag and trim cover facing up.
- With the inflator module on the bench, never place anything on or close to the module which may be thrown in the event of an accidental deployment.
- Never use air bag component parts from another vehicle.
- If there is a chance of electrical shock to any of the air bag components, remove the air bag module before servicing the vehicle.

DISARMING

1. Before servicing the vehicle, refer to the Precautions Section.
2. Disconnect the negative battery cable.
3. Turn ignition switch to **"LOCK"** position and remove key.
4. Turn steering wheel so that the

vehicle's front wheels are pointing straight ahead.

5. Remove or disconnect the following:
   - AIR BAG fuse
   - Driver air bag connector (yellow)
   - Glove box
   - Passenger air bag connector (yellow)
   - Side air bag connector (yellow), if equipped (located under front seat cushion)
   - Rear quarter inner trim panel (if equipped with side curtain air bag)
   - Side curtain air bag connector (black), if equipped

ARMING

When repairs are complete, install or connect the following:

- Side curtain air bag connector, if equipped
- Rear quarter inner trim panel, if necessary

1. Yellow connector of driver air bag (inflator) module
2. Connector stay
3. Air bag fuse box
4. Yellow connector of passenger air bag (inflator) module
5. Glove box

7924HG36

Air bag component location and identification—All models

- Side air bag connector, if equipped
- Passenger air bag connector
- Glove box
- Driver air bag connector
- AIR BAG fuse
- Negative battery cable

1. Turn ignition switch to **"ON"** position and verify that **"AIR BAG"** warning lamp flashes 6 times and then turns off.

Recirculating Ball Steering Gear

REMOVAL & INSTALLATION

1. Before servicing the vehicle, refer to the Precautions Section.

2. Remove or disconnect the following:

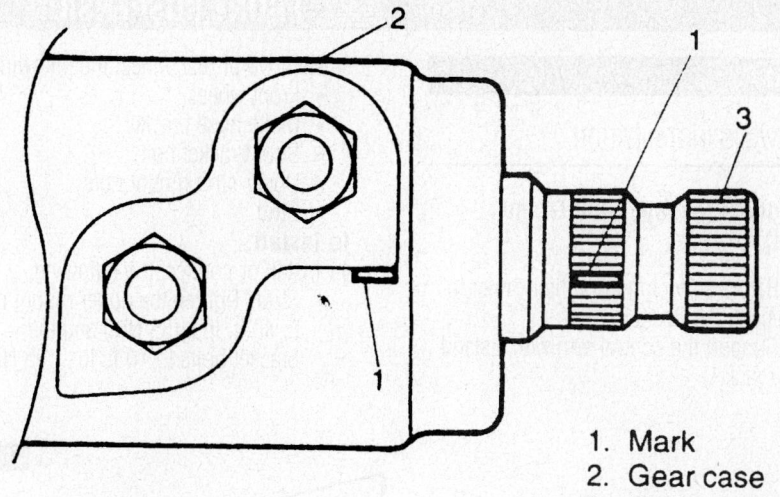

1. Mark
2. Gear case
3. Pinion shaft

9302HG14

Steering gear centering marks

- Skidplate, if equipped
- Coolant overflow tank
- Intermediate shaft pinch bolt
- Power steering hoses
- Pitman arm center link joint
- Steering gearbox
- Pitman arm

To install:

3. Install or connect the following:
- Pitman arm. Tighten the nut to 102 ft. lbs. (140 Nm).
- Steering gearbox. Tighten the bolts to 62 ft. lbs. (85 Nm).
- Pitman arm center link joint. Tighten the nut to 37 ft. lbs. (50 Nm).
- Power steering hoses
- Intermediate shaft pinch bolt. Tighten it to 18 ft. lbs. (25 Nm).
- Coolant overflow tank
- Skidplate, if equipped

4. Fill the power steering system.

5. Start the engine and check for leaks.

6. Check the wheel alignment and adjust as necessary.

Power Rack and Pinion Steering Gear

REMOVAL & INSTALLATION

1. Before servicing the vehicle, refer to the Precautions Section.

2. Remove or disconnect the following:
- Power steering hoses
- Intermediate shaft pinch bolt
- Front wheels
- Outer tie rod ends
- Steering gear

To install:

3. Install or connect the following:
- Steering gear. Tighten the bolts to 40 ft. lbs. (55 Nm).
- Outer tie rod ends. Tighten the nuts to 32 ft. lbs. (43 Nm).
- Front wheels
- Intermediate shaft pinch bolt. Tighten it to 18 ft. lbs. (25 Nm).
- Power steering hoses

4. Fill the power steering system.

5. Start the engine and check for leaks.

6. Check the wheel alignment and adjust as necessary.

FRONT SUSPENSION

Strut

REMOVAL & INSTALLATION

All Models (except 2006 Grand Vitara)

1. Before servicing the vehicle, refer to the Precautions Section.
2. Support the control arm with a stand or floor jack.

3. Remove or disconnect the following:
   • Front wheel
   • Brake hose bracket
   • Strut bracket bolts
   • Upper strut mount nuts
   • Strut

To install:

4. Install or connect the following:
   • Strut. Tighten the upper mount nuts to 40 ft. lbs. (55 Nm) and the bracket bolts to 70 ft. lbs. (95 Nm).

   • Brake hose bracket
   • Front wheel
5. Check the wheel alignment and adjust as necessary.

2006 Grand Vitara

1. Before servicing the vehicle, refer to the Precautions Section.
2. Disconnect the negative battery cable.
3. Remove the strut upper cap and loosen the strut nut.

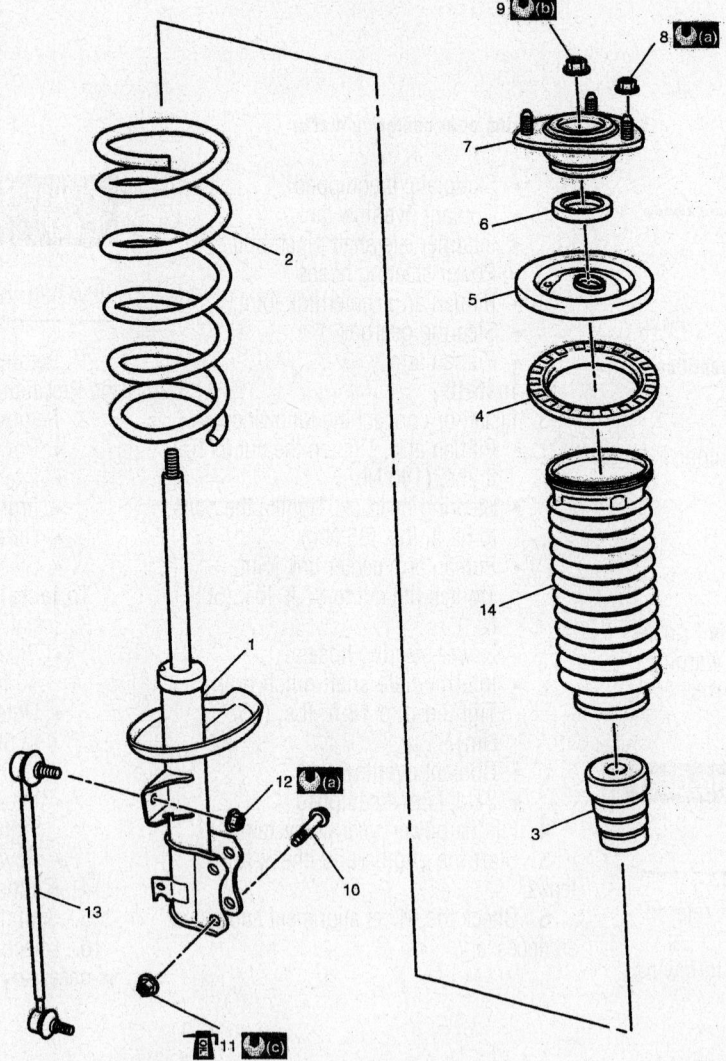

1.	Strut assembly	7.	Strut support	13.	Stabilizer joint	
2.	Coil spring	8.	Strut support nut	14.	Strut dust cover	
3.	Bump stopper	9.	Strut nut	(a) :	50 N·m (5.0 kgf-m, 36.5 lb-ft)	
4.	Coil spring seat	10.	Strut bracket bolt : Insert from vehicle front side.	(b) :	70 N·m (7.0 kgf-m, 51.0 lb-ft)	
5.	Coil spring upper seat	11.	Strut bracket nut : if reuse nut, apply engine oil to thread and bearing.	(c) :	135 N·m (13.5 kgf-m, 98.0 lb-ft)	
6.	Strut bearing	12.	Stabilizer joint nut	(d) :	60 N·m (6.0 kgf-m, 43.5 lb-ft)	

09490_SUZT_G0021

Exploded view of strut and coil spring assembly

4. Raise and safely support the front of the vehicle, allowing the front suspension to hang free.

5. Remove or disconnect the following:
- Wheels
- Stabilizer joint from strut bracket
- Brake line from securing bracket on the strut assembly
- ABS sensor line from strut assembly
- Strut bracket bolts and nuts
- Strut support nuts
- Strut assembly

6. Remove the coil spring by disassembling the strut assembly as follows:

a. Fasten the strut assembly to the spring compressor 09943-25010. Make sure the hooks are seated on the strut spring properly.

b. Compress the front spring with the spring compressor.

c. While keeping the coil spring compressed, remove the strut nut.

d. Disassemble the strut assembly.

To install:

7. Install the coil spring by assembling the strut assembly as follows:

a. Compress the coil spring until the total length becomes about 12.2 inches (310mm)

b. Install the bump stopper onto strut rod.

c. Install the compressed coil spring onto the strut , and place coil spring end onto the spring lower seat. Be sure that the end of the coil spring does not interfere with the step of the spring lower seat.

d. Pull strut rod up as far as possible.

e. Install coil spring upper seat with strut dust cover on coil spring and then spring upper seat aligning **"OUT"** mark on spring upper seat and center of strut bracket.

f. Install the bearing and strut support.

g. Install the strut nut. Tighten the strut nut to 51 ft. lbs. (70 Nm).

h. Release and remove the spring compressor.

8. Install strut assembly by reversing the removal procedure, noting the following instructions:
- Strut assembly
- Strut assembly to steering knuckle and tighten the strut bracket nut and bolt to 98 ft. lbs. (135 Nm)
- Stabilizer joint to strut assembly and tighten the stabilizer joint nut to 36.5 ft. lbs. (50 Nm)
- Brake line to securing bracket on the strut assembly. Tighten mounting bolt to 18 ft. lbs. (25 Nm)
- ABS sensor line to strut assembly.

Tighten clamp bolt to 7.5 ft. lbs. (10 Nm)
- Wheels

9. Lower the vehicle.

10. Install the nuts securing the strut assembly to the body of the vehicle and tighten to 36.5 ft. lbs. (50 Nm).

11. Connect the negative battery cable.

Coil Spring

REMOVAL & INSTALLATION

All Models (except 2006 Grand Vitara)

1. Before servicing the vehicle, refer to the Precautions Section.

2. Support the vehicle at the frame with a hoist or jackstand.

3. Support the control arm with a floor jack.

4. Remove or disconnect the following:
- Front wheel
- Brake caliper and rotor
- Locking hub or drive flange, if equipped
- Axle shaft snapring and thrust washer, if equipped
- Wheel speed sensor, if equipped
- Stabilizer bar link
- Lower ball joint
- Strut bracket bolts

5. Lower the floor jack and remove the coil spring.

To install:

➡**The bottom of the spring has a larger diameter than the top.**

6. Install the coil spring onto the control arm and raise the floor jack.

7. Install or connect the following:
- Strut bracket bolts. Tighten them to 70 ft. lbs. (95 Nm).
- Lower ball joint. Tighten the nut to 40 ft. lbs. (55 Nm).
- Stabilizer bar link. Tighten the nut to 21 ft. lbs. (29 Nm).
- Wheel speed sensor, if equipped
- Axle shaft snapring and thrust washer, if equipped
- Locking hub or drive flange, if equipped
- Brake caliper and rotor
- Front wheel

8. Check the wheel alignment and adjust as necessary.

2006 Grand Vitara

For coil spring service on the 2006 Grand Vitara, refer to the strut removal and installation procedure.

Lower Ball Joint

REMOVAL & INSTALLATION

The lower ball joint is serviced with the lower control arm as an assembly.

Lower Control Arm

REMOVAL & INSTALLATION

1. Before servicing the vehicle, refer to the Precautions Section.

2. Support the vehicle at the frame with a hoist or jackstand.

3. Support the control arm with a floor jack.

4. Remove or disconnect the following:
- Front wheel
- Brake caliper and rotor
- Locking hub or drive flange, if equipped
- Axle shaft snapring and thrust washer, if equipped
- Wheel speed sensor, if equipped
- Stabilizer bar link
- Lower ball joint
- Strut bracket bolts

5. Lower the floor jack and remove the coil spring.

6. Remove the inner control arm bolts and remove the control arm.

To install:

7. Install the inner control arm bolts.

8. Install the coil spring onto the control arm and raise the floor jack.

9. Install or connect the following:
- Strut bracket bolts. Tighten them to 70 ft. lbs. (95 Nm).
- Lower ball joint. Tighten the nut to 40 ft. lbs. (55 Nm).
- Stabilizer bar link. Tighten the nut to 21 ft. lbs. (29 Nm).
- Wheel speed sensor, if equipped
- Axle shaft snapring and thrust washer, if equipped
- Locking hub or drive flange, if equipped
- Brake caliper and rotor
- Front wheel

10. Lower the vehicle so that the front suspension is at curb height.

11. Tighten the front inner bolt to 62 ft. lbs. (85 Nm) and the rear inner bolt to 92 ft. lbs. (127 Nm).

12. Check the wheel alignment and adjust as necessary.

CONTROL ARM BUSHING REPLACEMENT

1. Before servicing the vehicle, refer to the Precautions Section.
2. Remove the control arm from the vehicle.
3. Remove the control arm bushings with a hydraulic press.

To install:

4. Lubricate the control arm bushings with liquid soap.
5. Press the bushings into the control arm until the bushing flange contacts the housing edge of the control arm.
6. Install the control arm to the vehicle.
7. Check the wheel alignment and adjust as necessary.

Wheel Bearings

ADJUSTMENT

The wheel bearings are not adjustable.

REMOVAL & INSTALLATION

All Models (except 2006 Grand Vitara)

1. Before servicing the vehicle, refer to the Precautions Section.
2. Remove or disconnect the following:
   - Front wheel
   - Brake caliper and rotor
   - Locking hub or hub drive flange, if equipped
   - Hub grease cap, if equipped
   - Wheel speed sensor, if equipped
   - Wheel bearing lockwasher
   - Wheel bearing locknut and inner washer
   - Wheel hub and bearing assembly
   - Wheel hub oil seal
   - Wheel bearing oil seal
   - Snapring
3. Press the wheel bearing and race out of the hub.

To install:

4. Press the wheel bearing and race into the hub so that the race is fully seated in the hub bore.
5. Install or connect the following:
   - Snapring
   - Wheel bearing oil seal
   - Wheel hub oil seal
   - Wheel hub and bearing assembly
   - Wheel bearing locknut and inner washer. Tighten the nut to 157 ft. lbs. (216 Nm).
   - Wheel bearing lockwasher. Tighten the retaining screws to 13 inch lbs. (1.5 Nm).
   - Wheel speed sensor, if equipped
   - Hub grease cap, if equipped
   - Locking hub or hub drive flange, if equipped
   - Brake caliper and rotor
   - Front wheel

2006 Grand Vitara

1. Before servicing the vehicle, refer to the Precautions Section.
2. Remove or disconnect the following:
   - Front wheel
   - Caulking nut and washer
   - Front wheel spindle, if equipped
   - Brake caliper and carrier
   - Brake disc
   - Hub housing bolts from knuckle
   - Hub housing from the vehicle

To install

➡ **Apply grease to the end face of inner ring of hub assembly before installation.**

3. Install or connect the following:
   - Hub housing to knuckle.
   - Hub housing bolts. Tighten hub housing bolts to 36.5 ft. lbs. (50 Nm).
   - Brake disc
   - Brake caliper and carrier. Tighten the carrier mounting bolts to 61.5 ft. lbs. (85 Nm).
   - Front wheel spindle, if equipped
   - Washer and caulking nut onto the axle shaft. Tighten the caulking nut onto the axle shaft to 145 ft. lbs. (200 Nm).
   - Caulk the axle shaft nut
   - Front wheel

REAR SUSPENSION

Shock Absorber

REMOVAL & INSTALLATION

1. Before servicing the vehicle, refer to the Precautions Section.
2. Support the rear axle housing with a hydraulic jack or stand.
3. Remove or disconnect the following:
   - Shock absorber upper locknut and retaining nut
   - Lower shock absorber mounting nut and bolt
   - Rear shock absorber

To install:

4. Install or connect the following:
   - Rear shock absorber
   - Lower mounting nut and bolt
   - Upper retaining nut and locknut
5. Torque the upper mounting nuts to 21 ft. lbs. (22–35 Nm) and the lower mounting nut/bolt to 74 ft. lbs. (100 Nm).

6. Remove the jack or stand from the rear axle assembly.

Coil Spring

REMOVAL & INSTALLATION

1. Before servicing the vehicle, refer to the Precautions Section.
2. Support the vehicle at the frame with a hoist or jackstand.
3. Support the rear axle housing with a floor jack.
4. Remove or disconnect the following:
   - Rear wheels
   - Parking brake cable hanger
   - Shock absorber lower mounting bolts
   - Wheel speed sensor harness clamps, if equipped
   - Brake pipe E-ring
   - Axle vent hose
5. Lower the floor jack and remove the coil springs.

To install:

6. Install the coil springs onto the axle spring seats and raise the floor jack.
7. Install or connect the following:
   - Axle vent hose
   - Brake pipe E-ring
   - Wheel speed sensor harness clamps, if equipped
   - Shock absorber lower mounting bolts. Tighten them to 74 ft. lbs. (100 Nm).
   - Parking brake cable hanger
   - Rear wheels

Control Arms

REMOVAL & INSTALLATION

2006 Grand Vitara

CONTROL ROD

1. Before servicing the vehicle, refer to the Precautions Section.

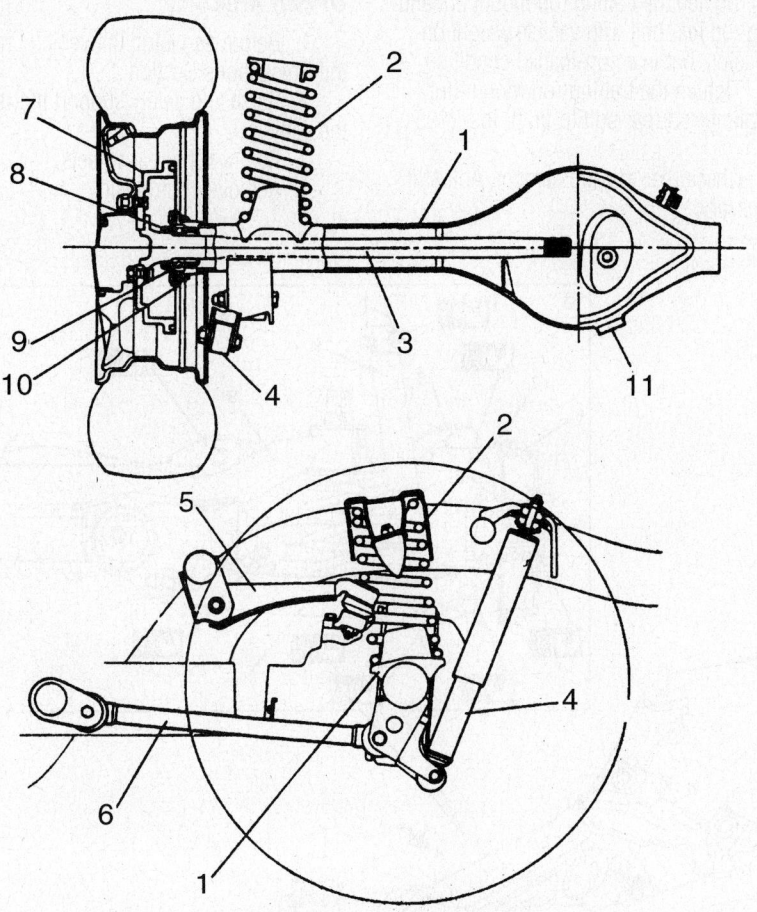

1. Rear axle housing
2. Coil spring
3. Axle shaft
4. Shock absorber
5. Upper arm
6. Trailing rod
7. Brake drum
8. Wheel bearing retainer
9. Rear wheel bearing
10. Brake back plate
11. Oil drain plug

7924HG32

Rear suspension component identification

2. Raise and safely support the rear of the vehicle.

3. Support the lower arm with a jack.

4. Remove the rear wheels.

5. Remove the parking cable hanger bolt.

6. Match mark control rod washer to suspension frame to install the bolts in correct position.

7. Remove control rod inner bolt, outer bolt, and then remove the control rod.

To install:

8. Install control rod in position and install the inner and outer bolts from the vehicle frontward. Make sure that the control rod washer is installed with the graduated part facing up and aligned with the match mark made during the removal procedure.

9. Install the parking cable hanger bolt. Tighten the bolt to 7.5 ft. lbs. (10 Nm).

10. Remove the jack from under the lower arm.

11. Install the rear wheels and lower the vehicle.

12. Tighten the control rod mount nut and control rod outer bolt (with match marks aligned) with vehicle weight on suspension, but in a non-loaded condition.

13. Tighten the control rod mount nut and control rod outer bolt to 98 ft. lbs. (135 Nm).

14. Check rear toe and camber. Adjust it if necessary.

LOWER ARM

1. Before servicing the vehicle, refer to the Precautions Section.

2. Raise and safely support the rear of the vehicle.

3. Remove the rear wheels.

4. Match mark lower arm washer to suspension frame to install the bolts in correct position.

5. Loosen lower arm mount nut.

6. Remove rear coil spring.

7. Remove suspension rod mount bolt and then remove lower arm.

To install:

8. Install lower arm to rear suspension frame and install the lower arm inner bolt from the vehicle rearward. Make sure that the lower arm washer is installed with the graduated part facing up and aligned with the match mark made during the removal procedure. Tighten the bolt and nut temporarily by hand.

9. Install the rear coil spring.

10. Install the rear wheels and lower the vehicle.

11. Tighten the lower arm outer bolt and lower arm mount nut (with match marks aligned), shock absorber bolts with vehicle weight on suspension, but in a non-loaded condition.

12. Tighten the lower arm outer bolt and mount nut to 98 ft. lbs. (135 Nm).

13. Tighten the shock absorber upper and lower bolts.

14. Check rear toe and camber. Adjust it if necessary.

TRAILING ROD

1. Before servicing the vehicle, refer to the Precautions Section.

2. Raise and safely support the rear of the vehicle.

3. Remove the rear wheels.

4. Support the lower arm with a jack.

5. Remove the air suction pipe bolts (5-door model only)

6. Remove the trailing rod front bolt and rear bolt and then remove trailing rod.

7. Remove the trailing rod mount bracket.

To install:

8. Install the trailing rod mount bracket. Tighten the new mount bracket bolts to 76 ft. lbs. (105 Nm).

9. Install the trailing rod into position and install the front and rear bolts posi-

tioned from the body inside. Tighten the bolts temporarily by hand.

10. Install and tighten the air suction pipe bolts (5-door model only)

11. Remove the jack from under the lower arm.

12. Install the rear wheels and lower the vehicle.

13. Tighten the trailing rod mount nut and trailing rod rear bolt with vehicle weight on suspension, but in a non-loaded condition.

14. Tighten the trailing rod mount nut and trailing rod rear bolt to 98 ft. lbs. (135 Nm).

15. Check rear toe and camber. Adjust it if necessary.

UPPER ARM

1. Before servicing the vehicle, refer to the Precautions Section.

2. Raise and safely support the rear of the vehicle.

3. Remove the rear wheels.

4. Remove control rod.

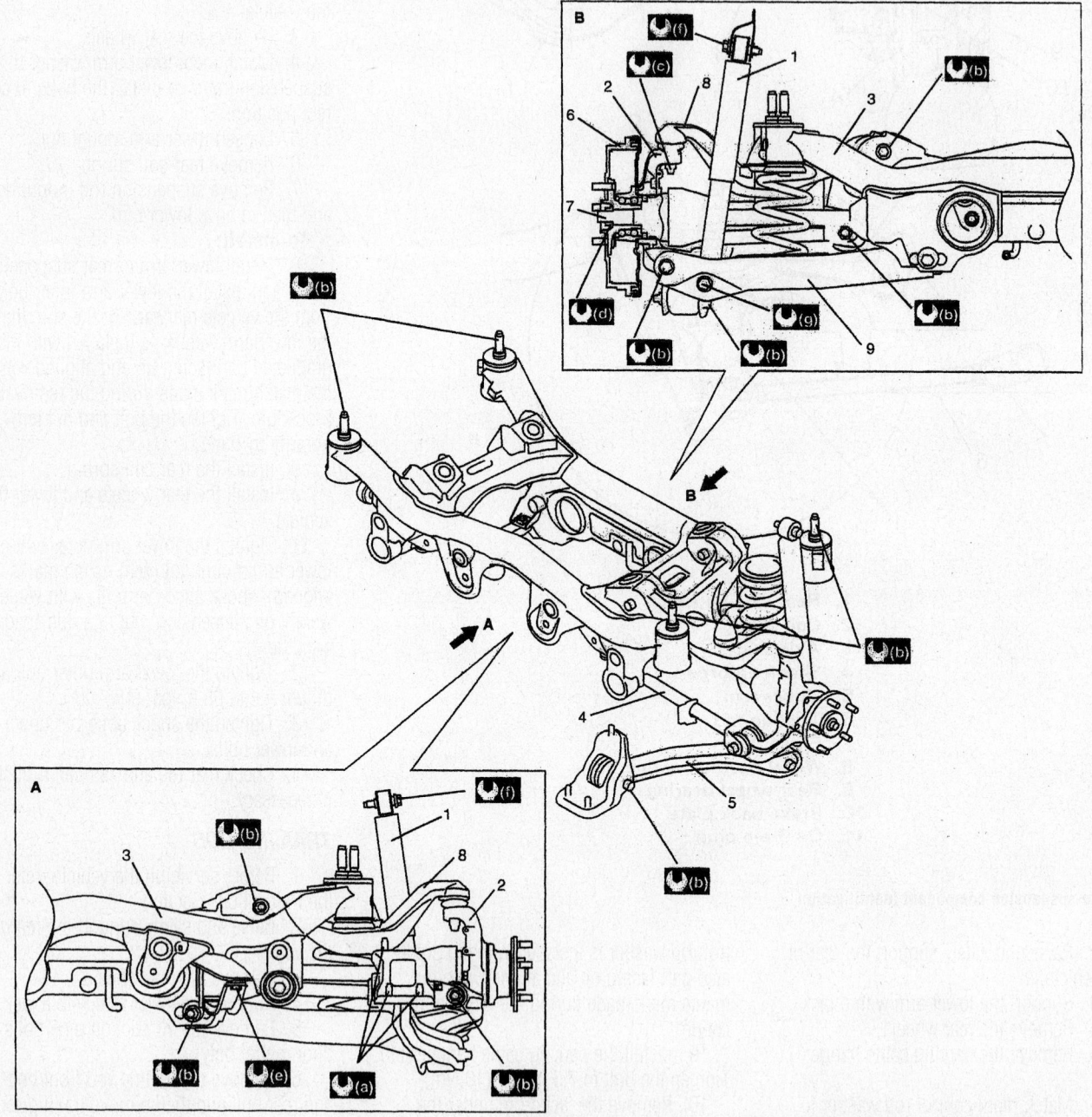

[A]: View A	4. Control rod	9. Lower Arm	🔧(e) : 50 N·m (5.0 kgf-m, 36.5 lb-ft)
[B]: View B	5. Trailing rod	🔧(a) : 105 N·m (10.5 kgf-m, 76.0 lb-ft)	🔧(f) : 60 N·m (6.0 kgf-m, 43.5 lb-ft)
1. Rear shock absorber	6. Rear brake drum	🔧(b) : 135 N·m (13.5 kgf-m, 98.0 lb-ft)	🔧(g) : 90 N·m (9.0 kgf-m, 65.0 lb-ft)
2. Rear suspension knuckle	7. Rear drive shaft	🔧(c) : 55 N·m (5.5 kgf-m, 40.0 lb-ft)	
3. Rear suspension frame	8. Upper Arm	🔧(d) : 200 N·m (20.0 kgf-m, 145.0 lb-ft)	

Exploded view of rear suspension—2006 Grand Vitara

5. Remove trailing rod.

6. Remove lower arm.

7. Remove the rear suspension knuckle.

8. Remove wheel sensor bolts from upper arm.

9. Remove the upper arm bolts and then remove the upper arm.

To install:

10. Install the upper arm into position to the rear suspension frame.

11. Install the upper arm bolts from the upper arm inside and tighten temporarily by hand.

12. Install and tighten the wheel sensor bolts to 8 ft. lbs. (11 Nm).

13. Install the rear suspension knuckle.

14. Install the trailing rod.

15. Install the control rod.

16. Install the lower arm.

17. Install the rear wheels and lower the vehicle.

18. Tighten all of the bolts and nuts with vehicle weight on suspension, but in a non-loaded condition.

19. Tighten all of the bolts and nuts to 98 ft. lbs. (135 Nm).

20. Check rear toe and camber. Adjust it if necessary.

Wheel Bearings

ADJUSTMENT

The rear wheel bearings on the 2006 Grand Vitara are a cartridge type design and cannot be adjusted.

REMOVAL & INSTALLATION

2006 Grand Vitara

1. Before servicing the vehicle, refer to the Precautions Section.

2. Set the parking brake.

3. Remove or disconnect the following:

- Rear wheels
- Brake drum

4. Release the parking brake.

5. Remove or disconnect the following:

- Rear axle nut and washer
- Rear brake shoe, if necessary
- Hub housing bolts from knuckle
- Hub housing from the vehicle

➡**The wheel bearing and hub are a solid unit. When the wheel bearing is**

found defective and it is necessary to replace it, replace the hub assembly.

To install:

6. Install or connect the following:

➡**Apply grease to the end face of inner ring of hub assembly before installation.**

7. Install or connect the following:

- Hub housing to knuckle.
- Hub housing bolts. Tighten hub housing bolts to 36.5 ft. lbs. (50 Nm).
- Rear brake shoe, If necessary
- New rear axle nut. Tighten the axle nut to 145 ft. lbs. (200 Nm). Caulk the axle nut.

8. Depress the brake pedal with about 66 lbs. (30 kg) of force 3 to 5 times to obtain proper drum/rotor to shoe/pad clearance.

9. Install the wheels.

10. Check to ensure that the brakes are free from dragging and that proper braking is obtained.

FRONT BRAKES

Brake Caliper

REMOVAL & INSTALLATION

1. Raise and safely support the vehicle.

2. Remove the wheels.

3. Disconnect and plug the brake line.

4. Remove the caliper mounting bolts (guide pins) and remove the caliper from the vehicle.

To install:

5. Install the caliper on the vehicle. Tighten the mounting bolts as follows:

- Vitara: 20 ft. lbs. (27 Nm)
- XL-7 and 2002–05 Grand Vitara: 10mm bolt to 37 ft. lbs. (50 Nm) and 12mm bolt to 62 ft. lbs. (84Nm)
- 2006 Grand Vitara: 26 ft. lbs. (36 Nm)

6. Connect the hydraulic brake line,

using 2 new washers. Torque the union bolt to 17 ft. lbs. (23 Nm).

7. Replace the front wheels.

8. Lower the vehicle.

9. Fill the brake reservoir and bleed the hydraulic brake system.

Disc Brake Pads

REMOVAL & INSTALLATION

1. Siphon about ⅔ of the fluid out of the master cylinder.

2. Raise and safely support the vehicle.

3. Remove the wheels.

4. Remove the brake caliper mounting bolts and remove the caliper from the mounting bracket.

5. Support the caliper with a wire.

6. Using a large pair of plies or a C-clamp compress the caliper piston back into the bore.

7. Remove the disc brake pads and any shims from the caliper mounting bracket.

To install:

8. Install the brake pads and any shims removed from the caliper mounting bracket.

9. Install the caliper on the mounting bracket and install the mounting bolts.

10. Install the front wheels and lower the vehicle.

✷✷ CAUTION

Do not attempt to drive the vehicle until after the following step is performed.

11. Depress the brake pedal repeatedly until a firm pedal is obtained. Do not attempt to drive the vehicle unless a firm pedal is obtained.

12. Check the fluid level in the master cylinder. Add fresh brake fluid, as necessary.

13. Road-test the vehicle.

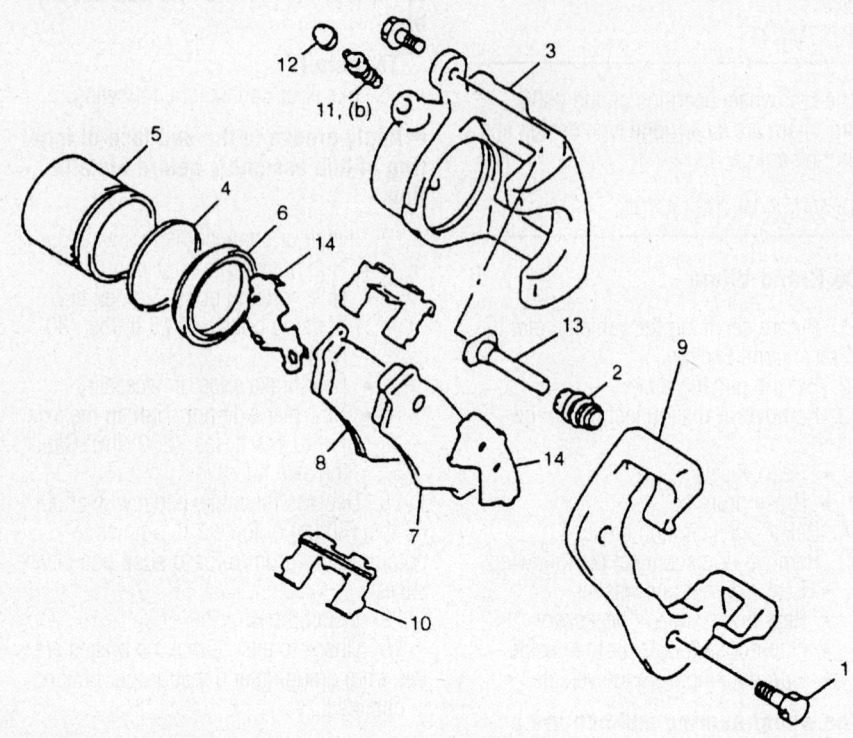

1. Caliper (slide) pin bolt
2. Boot
3. Disc brake caliper
   (disc brake cylinder)
4. Piston seal
5. Disc brake piston
6. Cylinder boot
7. Disc brake inner pad
8. Disc brake outer pad
9. Brake caliper carrier
10. Pad spring
11. Bleeder plug
12. Bleeder plug cap
13. Caliper pin
14. Anti noise shim
15. Inner shim

Tightening torque
(a): 8.0 N·m (0.80 kg-m, 6.0 lb-ft)
(b): 8.5 N·m (0.85 kg-m, 6.5 lb-ft)

93026G41

Front disc brake components—Vitara

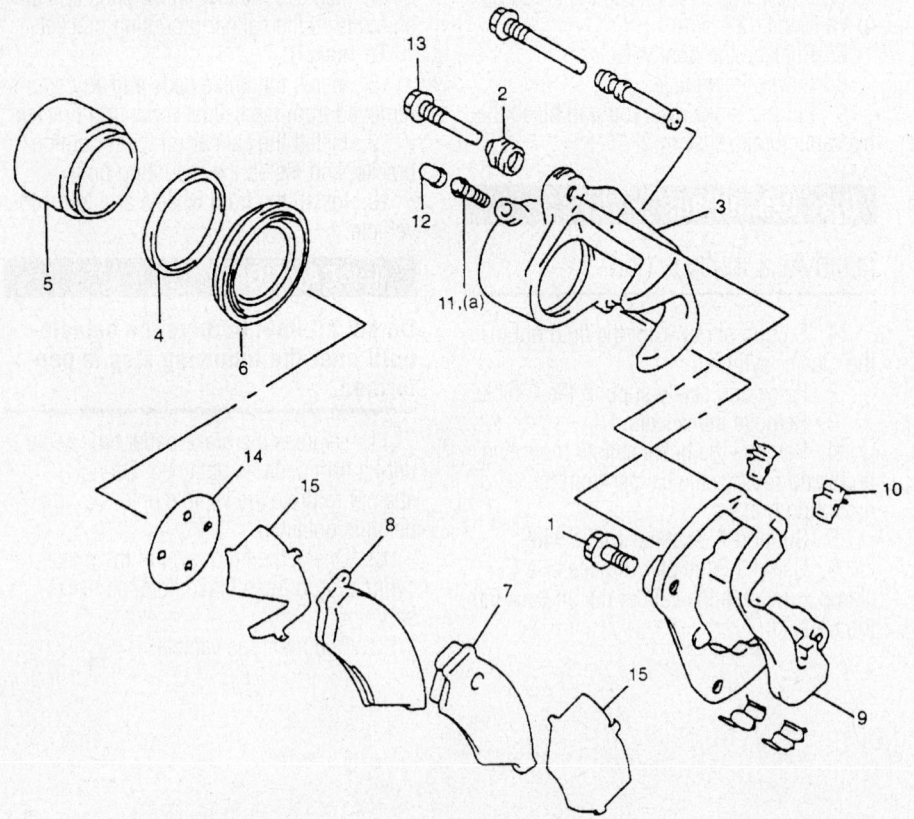

1. Caliper (slide) pin bolt
2. Boot
3. Disc brake caliper
   (disc brake cylinder)
4. Piston seal
5. Disc brake piston
6. Cylinder boot
7. Disc brake inner pad
8. Disc brake outer pad
9. Brake caliper carrier
10. Pad spring
11. Bleeder plug
12. Bleeder plug cap
13. Caliper pin
14. Anti noise shim
15. Inner shim

Tightening torque
(a): 8.0 N·m (0.80 kg-m, 6.0 lb-ft)

93026G42

Front disc brake components—Grand Vitara

REAR BRAKES

Brake Drums

REMOVAL & INSTALLATION

1. Raise and safely support the vehicle.
2. Remove the rear wheel(s).
3. Release the parking brake.

4. Remove the parking brake lever cover screws and loosen the brake cable locking nut.

5. Install 2, 8mm bolts into the brake drum holes and uniformly tighten each bolt. Tighten each bolt until the brake drum is removed from the vehicle. If there is difficulty in removing the drum, insert a small tool

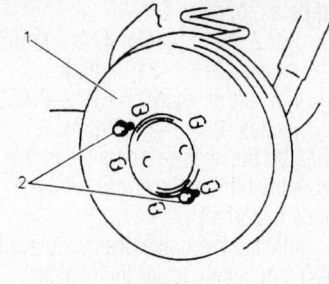

1 DRUM
2 TWO 8mm BOLTS

93026G39

Removing the brake drum with the two 8mm bolts

through the hole in the rear of the backing plate, and hold the automatic adjusting lever away from the adjuster. Using another narrow, flat tool at the same time, reduce the brake shoe adjuster by turning the adjusting wheel.

To install:

6. Install the brake drum and pull the parking brake lever all the way up until a clicking sound can no longer be heard.

7. Verify that the rear wheels will not turn. If the rear wheels turn, adjust the parking brake cable as necessary.

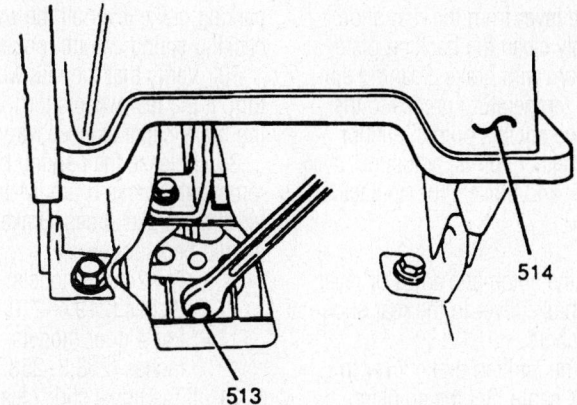

513 PARKING BRAKE CABLE LOCKNUT

514 PARKING BRAKE LEVER COVER

93026G38

Reducing the adjuster to remove the brake drum

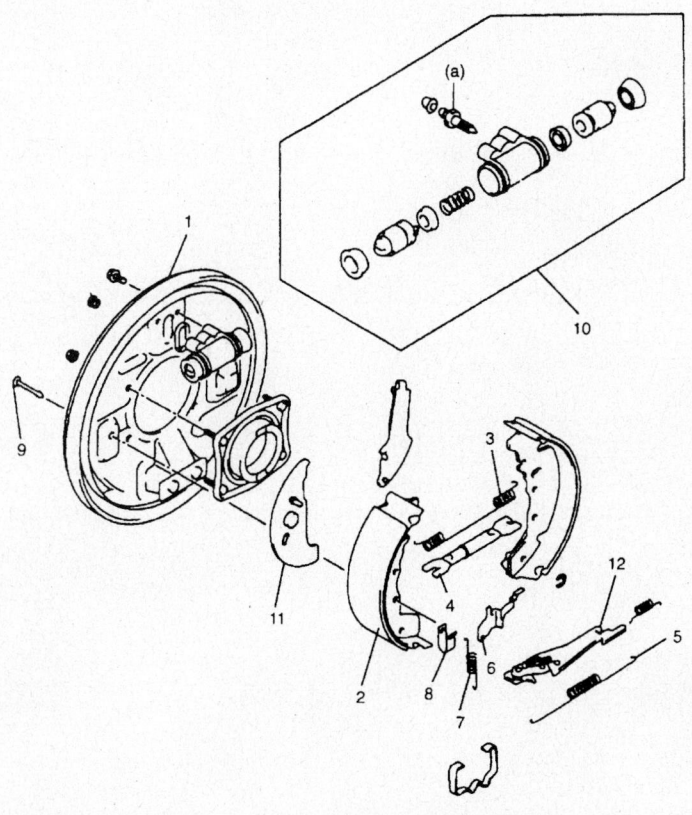

1. Brake back plate
2. Brake shoe
3. Shoe return upper spring
4. Adjuster
5. Shoe return lower spring
6. Adjuster lever
7. Adjuster spring
8. Shoe hold down spring
9. Shoe hold down pin
10. Wheel cylinder
11. Link
12. Brake strut

Tightening torque
(a): 7.5 N·m (0.75 kg-m, 5.5 lb-ft)

93026G43

Exploded view of the rear brake components—Vitara and Grand Vitara

8. Release the parking brake and remove the brake drum. Measure the diameter of the brake shoes. Outer diameter should be as follows:
- For 2 door models: 8.638–8.650 inches (219.4–219.7mm)
- For 4 door models: 9.972–9.988 inches (253.3–253.7mm)

9. If the brake shoe clearance is not correct, adjust the brake shoes until the clearance is correct.

10. Reinstall the brake drum, replace the wheel(s), and safely lower the vehicle.

11. Adjust the parking brake and install the cover with the 2 screws.

12. Road-test the vehicle for proper brake operation.

Brake Shoes

REMOVAL & INSTALLATION

1. Raise and safely support the vehicle.
2. Remove the rear wheel(s).
3. Remove the brake drum.
4. Using a suitable tool, remove the brake shoe return spring.
5. Using a brake spring hold-down tool, disengage the hold-down spring and retainers from the front shoe. Remove the hold-down retainer pinch
6. Disconnect the anchor spring from the front shoe and remove the front shoe.

7. Remove the anchor spring from the rear shoe. Using a brake spring hold-down tool, disengage the hold-down spring and retainers from the rear shoe. Remove the hold-down pinch

8. Disengage the parking brake lever from the parking brake cable and remove the rear shoe.

9. Remove the C-washer, the automatic adjuster lever and spring, the C-washer, and the parking brake lever from the rear shoe.

10. Thoroughly clean the backing plate and brake hardware with brake cleaning solvent. Apply high temperature grease to the backing plate shoe contact points, anchor plate and shoe contact points, adjusting bolt, and adjuster and brake shoe contact points.

To install:

11. Reinstall the automatic adjuster lever and the parking brake lever to the rear shoe using new C-washers.

12. Connect the parking brake lever to the parking brake cable. Set the adjuster and spring to the rear shoe.

13. Set the rear brake shoe in place, install the hold-down pin and install the hold-down spring and retainers. Make sure that the shoe is inserted in the wheel cylinder and that the other end is in the anchor plate.

14. Install the anchor spring to the rear shoe.

15. Install the front shoe to the other end of the anchor spring and set the front shoe in place. Make sure that the front shoe engages the wheel cylinder, adjuster mechanism and spring, and the anchor plate.

16. Reinstall the front brake shoe hold-down pin and secure with the hold-down spring and retainers using a suitable tool.

17. Install the return spring.

18. Install the brake drum and pull the parking brake lever all the way up until a clicking sound can no longer be heard.

19. Verify that the rear wheels will not turn. If the rear wheels turn, adjust the parking brake cable as necessary.

20. Release the parking brake and remove the brake drum. Measure the diameter of the brake shoes. Brake diameter should be as follows:
- For 2 door models: 8.638–8.650 inches (219.4–219.7mm)
- For 4 door models: 9.972–9.988 inches (253.3–253.7mm)

21. If the brake shoe clearance is not correct, adjust the brake shoes until the clearance is correct.

22. Reinstall the brake drum, replace the wheel(s), and safely lower the vehicle.

23. Road-test the vehicle for proper brake operation.

BRAKES **12-141**
DRIVE TRAIN **12-108**
ENGINE REPAIR..............**12-11**
FUEL SYSTEM **12-103**
SPECIFICATIONS AND
MAINTENANCE CHARTS......**12-2**
Engine and Vehicle
  Identification 12-2
General Engine Specifications 12-2
Engine Tune-Up Specifications 12-2
Firing Order 12-3
Accessory Drive Belt Routing 12-3
Capacities 12-4
Valve Specifications......................12-4
Crankshaft and Connecting Rod
  Specifications 12-5
Camshaft and Bearing
  Specifications Chart 12-6
Piston and Ring Specifications 12-7
Torque Specifications 12-7
Wheel Alignment 12-8
Tire, Wheel and Ball Joint
  Specifications 12-9
Brake Specifications 12-9
Scheduled Maintenance
  Intervals.............................12-10
STEERING AND
SUSPENSION **12-123**
A
Air Bag....................................12-123
  Disarming 12-123
Alternator 12-11
  Removal 12-11
Automatic Transmission 12-108
  Removal & Installation..........12-108
B
Brake Drums............................12-144
  Removal & Installation..........12-144
Brake Shoes............................12-145
  Removal & Installation..........12-145
C
Camshaft and Valve Lifters 12-59
  Inspection 12-73
  Removal & Installation..........12-59
Coil Spring 12-129
  Removal & Installation..........12-129
CV-Joints................................12-119
  Overhaul 12-119

Cylinder Head............................12-23
  Removal & Installation............12-23
D
Distributor..................................12-11
E
Engine Assembly 12-12
  Removal & Installation............12-12
Exhaust Manifold 12-57
  Removal & Installation............12-57
F
Front Brake Caliper....................12-141
  Removal & Installation..........12-141
Front Crankshaft Seal 12-58
  Removal & Installation............12-58
Front Disc Brake Pads 12-144
  Removal & Installation..........12-144
Front Wheel Bearing 12-138
  Adjustment.........................12-138
  Removal & Installation..........12-138
Fuel Injector.............................12-105
  Removal & Installation..........12-105
Fuel Pump 12-103
  Removal & Installation..........12-103
Fuel System Pressure 12-103
  Relieving............................12-103
H
Halfshaft...................................12-117
  Removal & Installation..........12-117
Heater Core...............................12-14
  Removal & Installation............12-14
I
Ignition Timing 12-12
  Adjustment...........................12-12
Intake Manifold...........................12-51
  Removal & Installation............12-51
L
Lower Ball Joint.........................12-133
  Removal & Installation..........12-133
Lower Control Arm 12-134
  Control Arm Bushing
    Replacement.....................12-138
  Removal & Installation..........12-134
O
Oil Cooler 12-89
  Removal & Installation............12-89
Oil Pan.......................................12-82
  Removal & Installation............12-82

Oil Pump 12-86
  Removal & Installation............12-86
P
Pinion Seal 12-122
  Removal & Installation..........12-122
Piston and Ring 12-102
  Positioning 12-102
Power Rack And Pinion Steering
  Gear...................................12-123
  Removal & Installation..........12-123
R
Rear Air Spring 12-130
  Removal & Installation..........12-130
Rear Axle Shaft, Bearing and
  Seal....................................12-119
  Removal & Installation..........12-119
Rear Brake Caliper....................12-143
  Removal & Installation..........12-143
Rear Disc Brake Pads 12-144
  Removal & Installation..........12-144
Rear Main Seal 12-90
  Removal & Installation............12-90
S
Shock Absorber 12-125
  Removal & Installation..........12-125
Starter Motor 12-82
  Removal & Installation............12-82
T
Timing Belt 12-90
  Removal & Installation............12-90
Timing Chain, Sprockets, Front
  Cover and Seal 12-99
  Removal & Installation............12-99
Transfer Case Assembly.............12-116
  Removal & Installation..........12-116
U
Upper Ball Joint........................12-131
  Removal & Installation..........12-131
Upper Control Arm 12-133
  Control Arm Bushing
    Replacement.....................12-134
  Removal & Installation..........12-133
V
Valve Lash 12-73
  Adjustment...........................12-73
W
Water Pump 12-21
  Removal & Installation............12-21

SPECIFICATION AND MAINTENANCE CHARTS

ENGINE AND VEHICLE IDENTIFICATION

		Engine						Model Year	
Code ①	Liters (cc)	Cu. In.	Cyl.	Fuel Sys.	Engine Type	Eng. Mfg.		Code ②	Year
5VZ-FE	3.4 (3378)	206	6	MFI	DOHC	Toyota		2	2002
1GR-FE	4.0 (3956)	241	6	SFI	DOHC	Toyota		3	2003
2UZ-FE	4.7 (4664)	285	8	SFI	DOHC	Toyota		4	2004
								5	2005
								6	2006

① Stamped on the left side of the engine block

② 10th digit of the Vehicle Identification Number (VIN)

SFI: Sequential Fuel Injection

MFI: Multi-port Fuel Injection

DOHC: Double Overhead Camshaft

09490_4RUN_C0001

GENERAL ENGINE SPECIFICATIONS

Year	Model	Engine Displacement Liters	Engine Series ID	Net Horsepower @ rpm	Net Torque @ rpm (ft. lbs.)	Bore x Stroke (in.)	Compression Ratio	Oil Pressure @ rpm
2002	4Runner	3.4	5VZ-FE	183@4800	217@3600	3.68x3.23	9.6:1	36-75@3000
2003	4Runner	4.0	1GR-FE	245@5200	283@3400	3.70x3.74	10.0:1	43-85@3000
		4.7	2UZ-FE	235@4800	320@3400	3.70x3.31	9.6:1	43-85@3000
2004	4Runner	4.0	1GR-FE	245@5200	282@3400	3.70x3.74	10.0:1	43-85@3000
		4.7	2UZ-FE	235@4800	315@3400	3.70x3.31	9.6:1	43-85@3000
2005	4Runner	4.0	1GR-FE	245@5200	282@3800	3.70x3.74	10.0:1	43-85@3000
		4.7	2UZ-FE	270@5400	320@3400	3.70x3.31	9.6:1	43-85@3000
2006	4Runner	4.0	1GR-FE	236@5200	266@4000	3.70x3.74	10.0:1	43-85@3000
		4.7	2UZ-FE	260@5400	306@3400	3.70x3.31	10.0:1	43-85@3000

09490_4RUN_C0002

ENGINE TUNE-UP SPECIFICATIONS

Year	Engine Displacement Liters	Engine ID	Spark Plug Gap (in.)	Ignition Timing (deg.)*	Fuel Pump (psi)	Idle Speed (rpm)	Valve Clearance Intake	Valve Clearance Exhaust
2002	3.4	5VZ-FE	0.039-0.043	8-12	38-44	650-750	0.006-0.009	0.011-0.014
2003	4.0	1GR-FE	0.043	8-12	41-42	650-750	0.006-0.010	0.011-0.015
	4.7	2UZ-FE	0.043	8-12	38-44	650-750	0.006-0.010	0.010-0.014
2004	4.0	1GR-FE	0.043	8-12	41-42	650-750	0.006-0.010	0.011-0.015
	4.7	2UZ-FE	0.043	8-12	38-44	650-750	0.006-0.010	0.010-0.014
2005	4.0	1GR-FE	0.043	8-12	41-42	650-750	0.006-0.010	0.011-0.015
	4.7	2UZ-FE	0.043	5-15	38-44	650-750	0.006-0.010	0.010-0.014
2006	4.0	1GR-FE	0.043	8-12	41-42	650-750	0.006-0.010	0.011-0.015
	4.7	2UZ-FE	0.043	5-15	38-44	650-750	0.006-0.010	0.010-0.014

NOTE: The Vehicle Emission Control Information label often reflects specification changes made during production.

The label figures must be used if they differ from those in this chart.

* With terminals TC and CG connected to DLC3

09490_4RUN_C0003

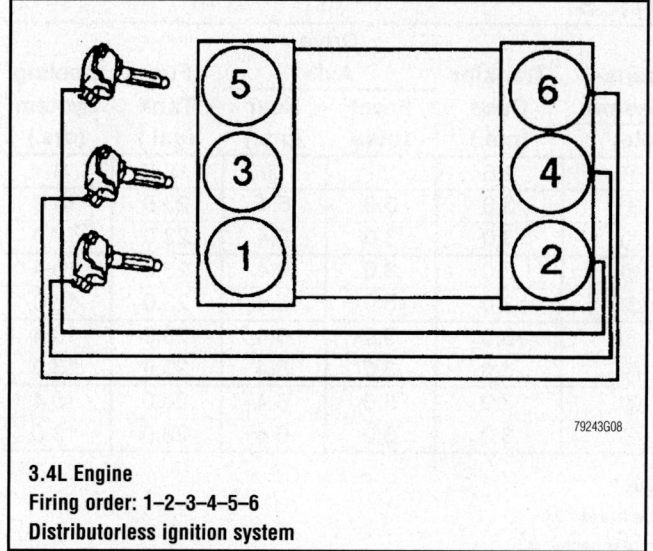

**3.4L Engine
Firing order: 1–2–3–4–5–6
Distributorless ignition system**

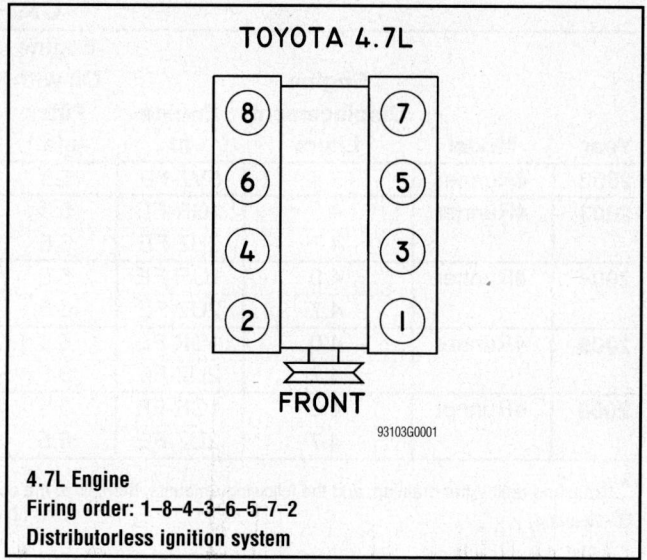

**4.7L Engine
Firing order: 1–8–4–3–6–5–7–2
Distributorless ignition system**

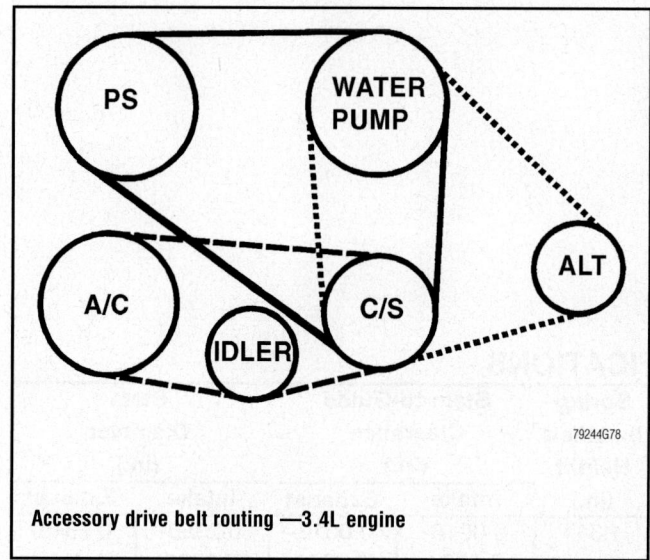

Accessory drive belt routing —3.4L engine

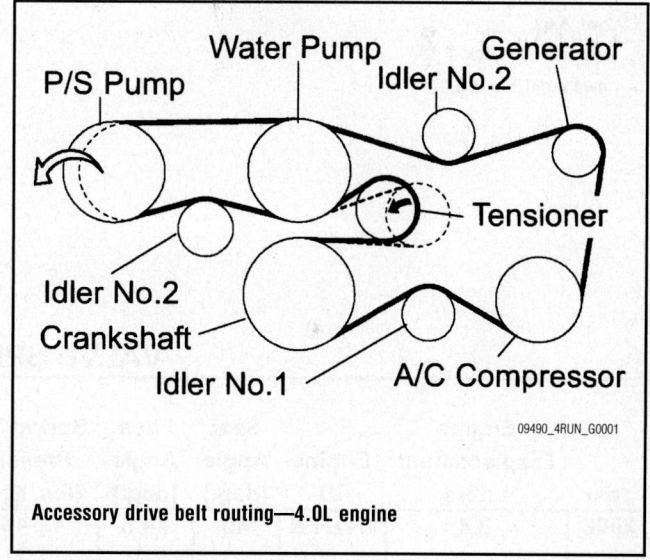

Accessory drive belt routing—4.0L engine

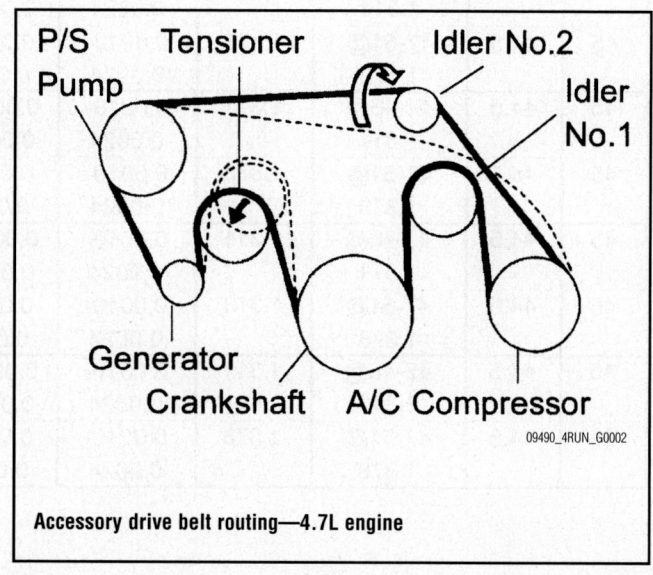

Accessory drive belt routing—4.7L engine

CAPACITIES

Year	Model	Engine Displacement Liters	Engine ID	Engine Oil with Filter (qts.)	Trans- mission (pts.)*	Transfer Case (pts.)	Drive Axle Front (pts.)	Rear (pts.)	Fuel Tank (gal.)	Cooling System (qts.)
2002	4Runner	3.4	5VZ-FE	5.5	①	2.6	②	③	18.5	④
2003	4Runner	4.0	1GR-FE	5.5	⑤	3.0	3.0	6.4	23.0	10.4
		4.7	2UZ-FE	6.5	⑤	3.0	3.0	6.4	23.0	13.0
2004	4Runner	4.0	1GR-FE	5.5	⑤	3.0	3.0	6.4	23.0	10.4
		4.7	2UZ-FE	6.5	⑤	3.0	3.0	6.4	23.0	13.0
2005	4Runner	4.0	1GR-FE	5.5	⑤	3.0	3.0	6.4	23.0	10.4
		4.7	2UZ-FE	6.5	⑤	3.0	3.0	6.4	23.0	13.0
2006	4Runner	4.0	1GR-FE	5.5	⑤	3.0	3.0	6.4	23.0	10.4
		4.7	2UZ-FE	6.5	⑤	3.0	3.0	6.4	23.0	13.0

* Drain and refill. After draining, add the following amounts, then, fill to the cold full line.

① 2wd: 3.4
   4wd: 4.2

② Without ADD: 2.32
   With ADD: 2.44

③ 2wd: 5.82
   4wd w/o diff. Lock: 5.18
   4wd w/diff. Lock: 5.82

④ With rear heater: 9.5
   Without rear heater: 8.5

⑤ A340E, A340F: 4.2
   A750E, A750F: 6.4

09490_4RUN_C0004

VALVE SPECIFICATIONS

Year	Engine Displacement Liters	Engine ID	Seat Angle (deg.)	Face Angle (deg.)	Spring Test Pressure (lbs. @ in.)	Spring Installed Height (in.)	Stem-to-Guide Clearance (in.) Intake	Exhaust	Stem Diameter (in.) Intake	Exhaust
2002	3.4	5VZ-FE	45	44.5	42-46@ 1.311	1.311	0.0010- 0.0024	0.0012- 0.0026	0.2350- 0.2356	0.2348- 0.2354
2003	4.0	1GR-FE	45	44.5	42-46@ 1.311	1.311	0.0010- 0.0024	0.0012- 0.0026	0.2154- 0.2159	0.2152- 0.2158
	4.7	2UZ-FE	45	44.5	47-51@ 1.378	1.378	0.0010- 0.0024	0.0012- 0.0026	0.2154- 0.2159	0.2152- 0.2157
2004	4.0	1GR-FE	45	44.5	42-46@ 1.311	1.311	0.0010- 0.0024	0.0012- 0.0026	0.2154- 0.2159	0.2152- 0.2158
	4.7	2UZ-FE	45	44.5	47-51@ 1.378	1.378	0.0010- 0.0024	0.0012- 0.0026	0.2154- 0.2159	0.2152- 0.2157
2005	4.0	1GR-FE	45	44.5	42-46@ 1.311	1.311	0.0010- 0.0024	0.0012- 0.0026	0.2154- 0.2159	0.2152- 0.2158
	4.7	2UZ-FE	45	44.5	47-51@ 1.378	1.378	0.0010- 0.0024	0.0012- 0.0026	0.2154- 0.2159	0.2152- 0.2157
2006	4.0	1GR-FE	45	44.5	42-46@ 1.311	1.311	0.0010- 0.0024	0.0012- 0.0026	0.2154- 0.2159	0.2152- 0.2158
	4.7	2UZ-FE	45	44.5	47-51@ 1.378	1.378	0.0010- 0.0024	0.0012- 0.0026	0.2154- 0.2159	0.2152- 0.2157

09490_4RUN_C0005

CRANKSHAFT AND CONNECTING ROD SPECIFICATIONS

All measurements are given in inches.

Year	Engine Displacement Liters	Engine ID	Crankshaft				Connecting Rod		
			Main Brg. Journal Dia.	Main Brg. Oil Clearance	Shaft End-play	Thrust on No.	Journal Diameter	Oil Clearance	Side Clearance
2002	3.4	5VZ-FE	2.5191-2.5197	①	0.0008-0.0087	2	2.1648-2.1654	0.0009-0.0021	0.0059-0.0130
2003	4.0	1GR-FE	2.8342-2.8346	0.0007-0.0012	0.0016-0.0094	3	2.2044-2.2047	0.0010-0.0018	0.0059-0.0118
	4.7	2UZ-FE	2.6373-2.6378	②	0.0008-0.0087	3	2.0465-2.0472	0.0011-0.0021	0.0063-0.0138
2004	4.0	1GR-FE	2.8342-2.8346	0.0007-0.0012	0.0016-0.0094	3	2.2044-2.2047	0.0010-0.0018	0.0059-0.0118
	4.7	2UZ-FE	2.6373-2.6378	②	0.0008-0.0087	3	2.0465-2.0472	0.0011-0.0021	0.0063-0.0138
2005	4.0	1GR-FE	2.8342-2.8346	0.0007-0.0012	0.0016-0.0094	3	2.2044-2.2047	0.0010-0.0018	0.0059-0.0118
	4.7	2UZ-FE	2.6373-2.6378	②	0.0008-0.0087	3	2.0465-2.0472	0.0011-0.0021	0.0063-0.0138
2006	4.0	1GR-FE	2.8342-2.8346	0.0007-0.0012	0.0016-0.0094	3	2.2044-2.2047	0.0010-0.0018	0.0059-0.0118
	4.7	2UZ-FE	2.6373-2.6378	②	0.0008-0.0087	3	2.0465-2.0472	0.0011-0.0021	0.0063-0.0138

① No. 1: 0.0008-0.0015 in.
   All others: 0.0009-0.0017 in.

② No. 1 and No. 5: 0.0011-0.0018 in.
   All others: 0.0016-0.0023 in.

09490_4RUN_C0006

CAMSHAFT AND BEARING SPECIFICATIONS CHART

All measurements are given in inches.

Year	Engine Displ. Liters	Engine ID/VIN	Journal Dia.	Brg. Oil Clearance	Shaft End-play	Runout	Lobe Height	
							Intake	Exhaust
2002	3.4	5VZ-FE	1.0610-1.0616	0.0014-0.0028	0.0013-0.0031	0.0024	1.6657-1.6697	1.6520-1.6559
2003	4.0	1GR-FE	①	②	0.0016-0.0035	0.0024	1.7389-1.7428	1.7551-1.7591
	4.7	2UZ-FE	1.0612-1.0618	0.0012-0.0028	③	0.0031	1.6512-1.6551	1.6520-1.6559
2004	4.0	1GR-FE	①	②	0.0016-0.0035	0.0024	1.7389-1.7428	1.7551-1.7591
	4.7	2UZ-FE	1.0612-1.0618	0.0012-0.0028	③	0.0031	1.6512-1.6551	1.6520-1.6559
2005	4.0	1GR-FE	①	②	0.0016-0.0035	0.0024	1.7389-1.7428	1.7551-1.7591
	4.7	2UZ-FE	1.0612-1.0618	0.0012-0.0028	③	0.0031	1.6512-1.6551	1.6520-1.6559
2006	4.0	1GR-FE	①	②	0.0016-0.0035	0.0024	1.7389-1.7428	1.7551-1.7591
	4.7	2UZ-FE	1.0612-1.0618	0.0012-0.0028	③	0.0031	1.6512-1.6551	1.6520-1.6559

① No. 1 Journal: 1.4162-1.4167 in.
   All Others: 0.9039-0.9045 in.

② No. 1 Cam, No. 1 Journal: 0.0003-0.0015 in.
   All Other Cams, No. 1 Journal: 0.0016-0.0031 in.
   All Other Cams, All Other Journals: 0.0010-0.0024 in.

③ Intake: 0.0000-0.0016 in.
   Exhaust: 0.0012-0.0028 in.

09490_4RUN_C0007

PISTON AND RING SPECIFICATIONS

All measurements are given in inches.

Year	Engine Displ. Liters	Engine ID	Piston Clearance	Ring Gap			Ring Side Clearance		
				Top Comp.	Bottom Comp.	Oil Control	Top Comp.	Bottom Comp.	Oil Control
2002	3.4	5VZ-FE	0.0053-0.0060	0.0118-0.0197	0.0157-0.0236	0.0059-0.0217	0.0016-0.0031	0.0012-0.0028	SNUG
2003	4.0	1GR-FE	0.0031-0.0040	0.0118-0.0157	0.0157-0.0197	0.0039-0.0157	0.0008-0.0028	0.0008-0.0024	0.0028-0.0060
	4.7	2UZ-FE	0.0035-0.0044	0.0118-0.0197	0.0157-0.0256	0.0051-0.0189	0.0012-0.0031	0.0012-0.0028	SNUG
2004	4.0	1GR-FE	0.0031-0.0040	0.0118-0.0157	0.0157-0.0197	0.0039-0.0157	0.0008-0.0028	0.0008-0.0024	0.0028-0.0060
	4.7	2UZ-FE	0.0035-0.0044	0.0118-0.0197	0.0157-0.0256	0.0051-0.0189	0.0012-0.0031	0.0012-0.0028	SNUG
2005	4.0	1GR-FE	0.0031-0.0040	0.0118-0.0157	0.0157-0.0197	0.0039-0.0157	0.0008-0.0028	0.0008-0.0024	0.0028-0.0060
	4.7	2UZ-FE	0.0035-0.0044	0.0118-0.0158	0.0157-0.0217	0.0051-0.0150	0.0012-0.0032	0.0008-0.0024	SNUG
2006	4.0	1GR-FE	0.0031-0.0040	0.0118-0.0157	0.0157-0.0197	0.0039-0.0157	0.0008-0.0028	0.0008-0.0024	0.0028-0.0060
	4.7	2UZ-FE	0.0035-0.0044	0.0118-0.0197	0.0157-0.0256	0.0051-0.0189	0.0012-0.0031	0.0012-0.0028	SNUG

09490_4RUN_C0008

TORQUE SPECIFICATIONS

All readings in ft. lbs.

Year	Engine Displacement Liters	Engine ID	Cylinder Head Bolts	Main Bearing Bolts	Rod Bearing Bolts	Crankshaft Damper Bolts	Flywheel Bolts	Manifold		Spark Plugs	Oil Pan Drain Plug
								Intake	Exhaust		
2002	3.4	5VZ-FE	①	②	③	213	61	13	30	13	28
2003	4.0	1GR-FE	④	⑤	③	184	61	19	22	15	30
	4.7	2UZ-FE	⑥	⑦	③	181	⑧	13	33	13	29
2004	4.0	1GR-FE	④	⑤	③	184	61	19	22	15	30
	4.7	2UZ-FE	⑥	⑦	③	181	⑧	13	33	13	29
2005	4.0	1GR-FE	④	⑤	③	184	61	19	22	15	30
	4.7	2UZ-FE	⑨	⑦	③	181	⑧	13	33	13	29
2006	4.0	1GR-FE	④	⑤	③	184	61	19	22	15	30
	4.7	2UZ-FE	⑨	⑦	③	181	⑧	13	33	13	29

① 12-pointed bolts
    Step 1: 25 ft. lbs.
    Step 2: +90 degrees
    Step 3: +90 degrees
    Recessed heads: 13 ft. lbs.

② Step 1: 45 ft. lbs.
    Step 2: Plus 90 degrees

③ Step 1: 18 ft. lbs.
    Step 2: Plus 90 degrees

④ Right side: step 1, 27 ft. lbs.; step 2, plus 180 degrees
    Left side: step 1, 27 ft. lbs.; step 2, plus 180 degrees; 14mm cap screw, 22 ft. lbs.

⑤ 12-point bolt heads: Step 1: 45 ft. lbs.
    Step 2: Plus 90 degrees; 12mm capscrew: 18 ft. lbs.

⑥ Step 1: 24 ft. lbs.
    Step 2: plus 90 degrees
    Step 3: plus 90 degrees

⑦ Step 1: 20 ft. lbs.
    Step 2: plus 90 degrees

⑧ Step 1: 22 ft. lbs.
    Step 2: plus 90 degrees

⑨ Step 1: 36 ft. lbs.
    Step 2: plus 90 degrees

09490_4RUN_C0009

WHEEL ALIGNMENT

Year	Model	Caster Range (+/-Deg.)	Caster Preferred Setting (Deg.)	Camber Range (+/-Deg.)	Camber Preferred Setting (Deg.)	Toe-in (in.)	Steering Axis Inclination (Deg.)
2002	2WD	0.75	+3.25	0.75	-0.25	0.08+/-0.08	11+/-0.75
	4WD	0.75	+3.06	0.75	-0.25	0.08+/-0.08	11+/-0.75
2003	2WD w/o air suspension	0.75	+3.38	0.75	-0.47	0.04+/-0.08	12.97+/-0.75
	2WD w/air suspension	0.75	+3.55	0.75	-0.50	0.04+/-0.08	13.00+/-0.75
	4WD w/o air suspension	0.75	+3.22	0.75	-0.15	0.04+/-0.08	12.65+/-0/75
	4WD w/air suspension	0.75	+3.37	0.75	-0.17	0.04+/-0.08	12.67+/-0.75
2004	2WD w/o air suspension	0.75	+3.38	0.75	-0.47	0.04+/-0.08	12.97+/-0.75
	2WD w/air suspension	0.75	+3.55	0.75	-0.50	0.04+/-0.08	13.00+/-0.75
	4WD w/o air suspension	0.75	+3.22	0.75	-0.15	0.04+/-0.08	12.65+/-0/75
	4WD w/air suspension	0.75	+3.37	0.75	-0.17	0.04+/-0.08	12.67+/-0.75
2005	2WD w/o air suspension	0.75	+3.38	0.75	-0.47	0.04+/-0.08	12.97+/-0.75
	2WD w/air suspension	0.75	+3.55	0.75	-0.50	0.04+/-0.08	13.00+/-0.75
	4WD w/o air suspension	0.75	+3.22	0.75	-0.15	0.04+/-0.08	12.65+/-0/75
	4WD w/air suspension	0.75	+3.37	0.75	-0.17	0.04+/-0.08	12.67+/-0.75
2006	2WD w/o air suspension	0.75	+3.38	0.75	-0.47	0.04+/-0.08	12.97+/-0.75
	2WD w/air suspension	0.75	+3.55	0.75	-0.50	0.04+/-0.08	13.00+/-0.75
	4WD w/o air suspension	0.75	+3.22	0.75	-0.15	0.04+/-0.08	12.65+/-0/75
	4WD w/air suspension	0.75	+3.37	0.75	-0.17	0.04+/-0.08	12.67+/-0.75

09490_4RUN_C0010

TIRE, WHEEL AND BALL JOINT SPECIFICATIONS

Year	Model	OEM Tires		Tire Pressures (psi)		Wheel Size	Ball Joint Inspection	Lugnut Torque (ft. lbs.)
		Standard	Optional	Front	Rear			
2002	4Runner SR5	P225/75R15	P265/70R16	Std: 29 Opt: 32	Std: 29 Opt: 32	7-JJ	①	83
	4Runner Limited	P265/70R16	None	32	32	7-JJ	①	83
2003	Sport / SR5	P265/70R16	P265/65R17	32	32	7-JJ	①	83
	Limited	P265/70R16	None	32	32	7-JJ	①	83
2004	SR5	P265/70R16	P265/65R17	32	32	②	③	83
	Sport	P265/65R17	None	32	32	7.5JJ	③	83
	Limited	P265/65R17	None	32	32	7.5JJ	③	83
2005	SR5	P265/70R16	P265/65R17	32	32	②	③	83
	Sport	P265/65R17	None	32	32	7.5JJ	③	83
	Limited	P265/65R17	None	32	32	7.5JJ	③	83
2006	SR5	P265/70R16	P265/65R17	32	32	②	③	83
	Sport	P265/65R17	None	32	32	7.5JJ	③	83
	Limited	P265/60R18	None	32	32	NA	③	83

OEM: Original Equipment Manufacturer

PSI: Pounds Per Square Inch

STD: Standard

OPT: Optional

NA: Not Available

① Turning torque: upper 6-39 in. lbs.; lower 0.8-21.7 in. lbs.

② 16 in. steel: 7J

   16 inch aluminum: 7JJ

   17 inch: 7.5JJ

③ Turning torque: upper less than 40 inch lbs.; lower, less than 27 inch lbs.

09490_4RUN_C0011

BRAKE SPECIFICATIONS
All measurements in inches unless noted

Year	Model		Brake Disc			Brake Drum Diameter			Minimum Lining Thickness	Brake Caliper	
			Original Thickness	Minimum Thickness	Maximum Runout	Original Inside Diameter	Max. Wear Limit	Maximum Machine Diameter		Bracket Bolts (ft. lbs.)	Mounting Bolts (ft. lbs.)
2002	4Runner		0.866	0.787	0.0028	11.61	—	11.69	0.039	—	90
2003	4Runner	F	1.102	1.024	0.0020	—	—	—	0.039	—	91
		R	0.709	0.630	0.0079	—	—	—	0.039	77	65
2004	4Runner	F	1.102	1.024	0.0020	—	—	—	0.039	—	91
		R	0.709	0.630	0.0079	—	—	—	0.039	77	65
2005	4Runner	F	1.102	1.024	0.0020	—	—	—	0.039	—	91
		R	0.709	0.630	0.0079	—	—	—	0.039	77	65
2006	4Runner	F	1.102	1.024	0.0020	—	—	—	0.039	—	91
		R	0.709	0.630	0.0079	—	—	—	0.039	77	65

09490_4RUN_C0012

SCHEDULED MAINTENANCE INTERVALS

TOYOTA—4RUNNER

TO BE SERVICED	TYPE OF SERVICE	VEHICLE MILEAGE INTERVAL (x1000)																		
		5	10	15	20	25	30	35	40	45	50	55	60	65	70	75	80	85	90	95
Automatic transmission and differential fluid	S/I			✓			✓			✓			✓			✓			✓	
Ball joints and boots	S/I			✓			✓			✓			✓			✓			✓	
Brake system	S/I			✓			✓			✓			✓			✓			✓	
Charcoal canister	S/I												✓							
Drive belts	S/I						✓						✓						✓	
Driveshaft bushing	L						✓						✓						✓	
Engine coolant	R						✓						✓						✓	
Engine oil & filter	R	✓	✓	✓	✓	✓	✓	✓	✓	✓	✓	✓	✓	✓	✓	✓	✓	✓	✓	✓
Exhaust system	S/I			✓			✓			✓			✓			✓			✓	
Fuel tank cap gasket	S/I						✓						✓						✓	
Halfshaft boots & flange bolts	S/I			✓			✓			✓			✓			✓			✓	
Limited slip differential fluid	R						✓						✓						✓	
Manual transmission and differential fluid	S/I						✓						✓						✓	
Platinum spark plugs	R												✓							
Propeller shaft (4WD)	L			✓			✓			✓			✓			✓			✓	
Propeller shaft bolts	S/I			✓			✓			✓			✓			✓			✓	
Rack and pinion assembly	S/I			✓			✓			✓			✓			✓			✓	
Rear wheel bearing	L						✓						✓						✓	
Steering linkage	S/I			✓			✓			✓			✓			✓			✓	
Valves	S/I												✓							

R: Replace   S/I: Service or Inspect   L: Lubricate

FREQUENT OPERATION MAINTENANCE (SEVERE SERVICE)

If a vehicle is operated under any of the following conditions it is considered severe service:

- Towing a trailer or using a camper or car-top carrier.

- Repeated short trips of less than 5 miles in temperatures below freezing.

- Excessive idling or low-speed driving for long distances as in heavy commercial use, such as delivery, taxi or police cars.

- Operating on rough, muddy or salt-covered roads.

- Operating on unpaved or dusty roads.

Oil filter: service or inspect every 5000 miles or 4 months, whichever occurs first.

Brake linings and discs or drums: service or inspect every 5000 miles or 4 months, whichever occurs first.

Steering linkage: service or inspect every 5000 miles or 4 months, whichever occurs first.

Ball joints and boots: service or inspect every 5000 miles or 4 months, whichever occurs first.

Brake discs & pads (front): service or inspect every 6000 miles.

Halfshaft boots: service or inspect every 5000 miles or 4 months. Retighten the flange bolts, whichever occurs first.

Body chassis bolts and nuts: service or inspect every 5000 miles or 4 months, whichever occurs first.

Transmission and differential fluid: replace every 15,000 miles or 12 months, whichever occurs first.

Transfer case and differential fluid: replace every 15,000 miles or 12 months, whichever occurs first.

Timing belt: replace every 60,000 miles or 48 months, whichever occurs first.

ENGINE REPAIR

➡ Disconnecting the negative battery cable on some vehicles may interfere with the functions of the on board computer system. The computer may undergo a relearning process once the negative battery cable is reconnected.

Distributor

All models are equipped with a distibutorless ignition system.

Alternator

REMOVAL

3.4L Engine

1. Before servicing the vehicle, refer to the Precautions Section.
2. Remove or disconnect the following:
   - Negative battery cable
   - Alternator wiring
   - Alternator locknut, pivot bolt, nut and adjusting bolt
   - Drive belt
   - Alternator

To install:
Install or connect the following:
- Alternator
- Drive belt. Tighten the locknut 25 ft. lbs. (33 Nm) and the pivot bolt 38 ft. lbs. (51 Nm).
- Alternator wiring
- Negative battery cable

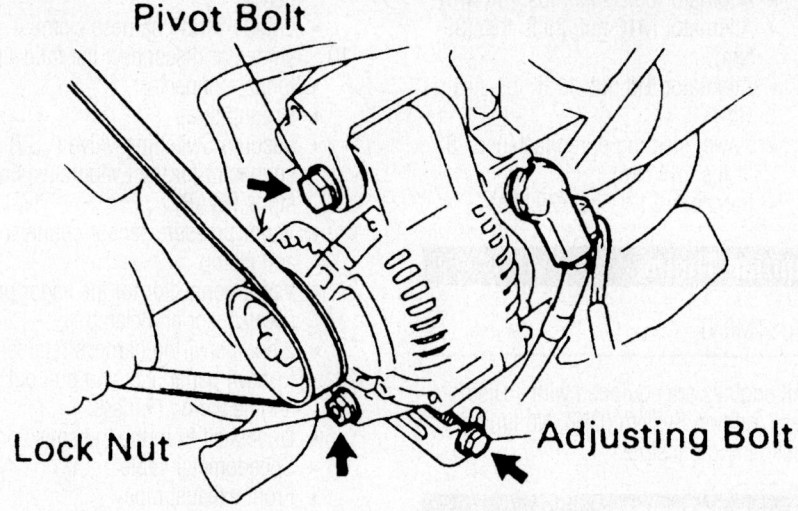

Locations of the adjusting and pivot bolts and the locknut—3.4L engine

4.0L Engine

1. Before servicing the vehicle, refer to the Precautions Section.
2. Remove the V-bank cover.
3. Remove the engine under-cover.
4. Remove the accessory drive belt.
5. Remove the battery.
6. Disconnect the wiring harness from the alternator.
7. Remove the 2 mounting bolts and lift out the alternator.
8. Installation is the reverse of removal. Torque the mounting bolts to 32 ft. lbs. (43 Nm).

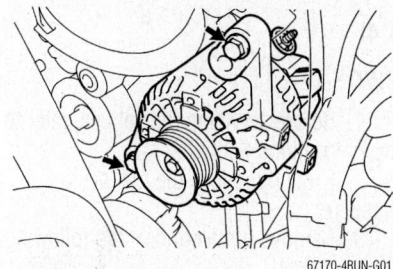

Alternator mounting—4.0L engine

4.7L Engine

1. Before servicing the vehicle, refer to the Precautions Section.
2. Disconnect the battery ground cable.
3. Remove the radiator upper seal (11 clips).
4. Remove the fan and shroud.
5. Remove the power steering pump and position it out of the way without disconnecting the hoses.
6. Disconnect the alternator wiring harness.
7. Remove the bolt and 2 nuts, and lift out the alternator.

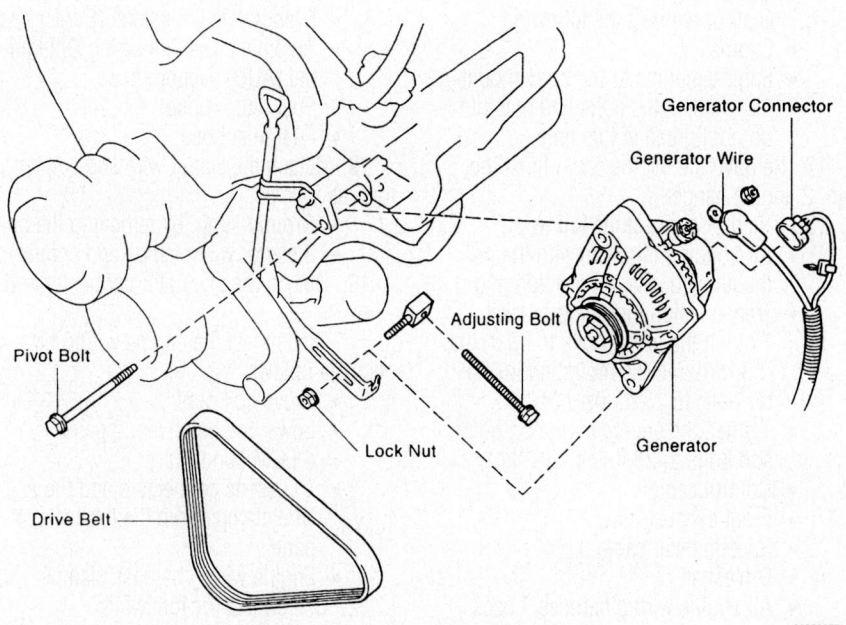

Exploded view of the alternator and drive belt—3.4L engine

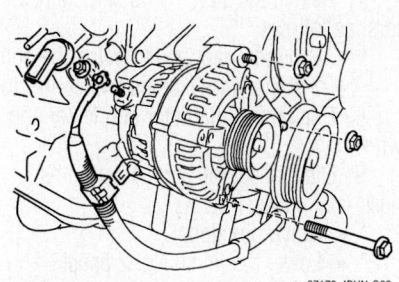

Alternator mounting—4.7L engine

8. Installation is the reverse of removal. Observe the following torques:
- Alternator bolt: 29 ft. lbs. (39 Nm)
- Alternator M10 nut: 29 ft. lbs. (39 Nm)
- Alternator M8 nut: 11 ft. lbs. (16 Nm)
- Power steering pump fasteners: 32 ft. lbs. (43 Nm)
- Fan nuts: 21 ft. lbs. (29 Nm)

Ignition Timing

ADJUSTMENT

All engines are equipped with a Distributorless Ignition System (DIS). No timing adjustment is possible.

Engine Assembly

REMOVAL & INSTALLATION

3.4L

2WD

1. Before servicing the vehicle, refer to the Precautions Section.
2. Properly relieve the fuel system pressure.
3. Remove or disconnect the following:
- Hood
- Battery
- Engine under covers
4. Drain the cooling system.
5. Drain the engine oil.
- Radiator
- Fan with the fluid coupling and fan pulleys
- Air cleaner cap
- Mass Air Flow (MAF) meter and the resonator
- Air cleaner case and filter
6. Disconnect the following hoses:
- Heater hoses
- Brake booster vacuum hose
- Evaporative Emissions (EVAP) hose
- Fuel return hose
- Fuel inlet hose
7. Detach the starter wire and connectors, as follows:
- Ground strap, by removing the bolt
- 3 starter wire clamps and connector
8. Detach the alternator connector and wire.
9. Disconnect the engine wiring harness, as follows:
- Glove box door
- Lower the finish No. 2 panel
- 4 ECM connectors

- 2 cassette connectors and the 2 wire clamps from the lower finish panel
- Engine wiring harness clamp
10. Remove or disconnect the following:
- Igniter connector
- Ground strap
- Vacuum Switching Valve (VSV) connector for the Evaporative Emissions (EVAP)
- Vapor pressure sensor connector and clamp
- Vapor connector for the vapor pressure sensor and clamp
- 2 engine wiring harness retainer-to-cowl panel nuts and pull out the engine wiring harness
- Driveshaft from the transmission
- Speedometer cable
- Front exhaust pipe
- Nut and the control cable
11. Support the transmission with a suitable jack.
12. Remove or disconnect the following:
- Transmission rear mounting bracket by removing the 8 bolts
- Bolt and the air conditioning compressor wire clamp, if equipped with air conditioning
13. If necessary, install a No. 2 engine hanger with 2 bolts. Tighten the 2 bolts to 30 ft. lbs. (40 Nm).
14. Attach the engine hoist chain to the 2 engine hangers.
15. Remove or disconnect the following:
- 4 engine front mounting insulators-to-frame bolts and nuts
- Engine and transmission

To install:
16. Install or connect the following:
- Engine
- Engine mounts to the body mountings. Install the bolts and nuts but do not tighten at this time.
17. Remove the engine chain hoist the No. 2 engine hanger.
18. Install or connect the following:
- Air conditioning wire with the bolt, if equipped with air conditioning
- Transmission mounting bracket. Tighten the frame bolts to 43 ft. lbs. (58 Nm) and the mounting insulator bolts to 13 ft. lbs. (18 Nm).
- Tighten the engine mounting nuts and bolts to 28 ft. lbs. (38 Nm).
- Control cable
- Front exhaust pipe
- Speedometer cable
- Driveshaft
- All engine wiring harness, hoses and cables
- Fan with the fluid coupling and fan

pulleys. Tighten the nuts to 48 inch lbs. (5.4 Nm).
- Air cleaner case and air filter
- MAF meter, resonator and the air cleaner cap
- Radiator
19. Fill the engine with oil.
20. Fill the engine and radiator with coolant.
21. Install or connect the following:
- Engine undercover
- Battery
- Hood
22. Start the engine and check for leaks.
23. Make any necessary adjustments and road test the vehicle.

4WD

1. Before servicing the vehicle, refer to the Precautions Section.
2. Remove or disconnect the following:
- Transmission
- Hood
3. Release the fuel system pressure.
4. Remove or disconnect the following:
- Battery
- Engine undercovers
5. Drain the cooling system.
6. Drain the engine oil.
7. Remove or disconnect the following:
- Radiator
- Fan with the fluid coupling and fan pulleys
- Air cleaner cap
- Mass Air Flow (MAF) meter and the resonator
8. Disconnect the following hoses:
- Heater hoses
- Brake booster vacuum hose
- Evaporative Emissions (EVAP) hose
- Automatic Disconnecting Differential (ADD) vacuum hose
- Fuel return hose
- Fuel inlet hose
9. Detach the starter wire and connectors, as follows:
- Ground strap, by removing the bolt
- 3 starter wire clamps and connector
10. Detach the alternator connector and wire.
11. Disconnect the engine wiring harness, as follows:
- Glove box door
- Lower the finish No. 2 panel
- 4 ECM connectors
- 2 cassette connectors and the 2 wire clamps from the lower finish panel
- Engine wiring harness clamp
12. Disconnect the following:
- Igniter connector
- Ground strap

- Vacuum Switching Valve (VSV) connector for the EVAP
- Vapor pressure sensor connector and clamp
- Vapor connector for the vapor pressure sensor and clamp

13. Remove or disconnect the following:
- 2 engine wiring harness retainer-to-cowl panel nuts and wiring harness
- Air conditioning compressor wire clamp, if equipped with air conditioning

14. If necessary, install a No. 2 engine hanger with 2 bolts. Tighten the 2 bolts to 30 ft. lbs. (40 Nm).

15. Attach the engine hoist chain to the 2 engine hangers.

16. Remove or disconnect the following:
- 4 engine front mounting insulators-to-frame bolts and nuts
- Engine

To install:

17. Install or connect the following:
- Engine
- Engine mounts-to-body mountings. Install the bolts and nuts but do not tighten at this time.

18. Remove the engine chain hoist the No. 2 engine hanger.

19. Install or connect the following:
- Air conditioning wire with the bolt, if equipped with air conditioning
- Tighten the engine mounting nuts and bolts to 28 ft. lbs. (38 Nm).
- Engine wiring harness
- Engine wiring harness clamp
- All wires, hoses and cables
- Fan with the fluid coupling and fan pulleys. Tighten the nuts to 48 inch lbs. (5.4 Nm).
- Air cleaner case and air filter
- MAF meter, resonator and the air cleaner cap
- Radiator

20. Fill the engine with oil.

21. Fill the engine and radiator with coolant.

22. Install or connect the following:
- Transmission and refill it with transmission oil
- Engine undercover
- Battery
- Hood

23. Start the engine, make any necessary adjustments and check for leaks.

4.0L Engine

1. Before servicing the vehicle, refer to the Precautions Section.
2. Relieve fuel system pressure

3. Remove the transmission.
4. Drain the cooling system.
5. Drain the engine oil.
6. Remove the battery.
7. Remove the hood.
8. Remove the V-bank cover.
9. Remove the air cleaner assembly.
10. Remove the radiator support upper seal (11 clips).
11. Remove the radiator.
12. Remove the fan shroud.
13. Remove the accessory drive belt.
14. Remove the fan.
15. Remove the fan pulley.
16. Remove the power steering pump.
17. Remove the alternator.
18. Remove the A/C compressor. Don't disconnect the lines.
19. Disconnect the fuel lines.
20. Disconnect the heater hoses.
21. Remove the upper intake manifold (air surge tank).
22. Remove the right front door scuff plate.
23. Remove the right cowl side trim plate.
24. Remove the glove compartment door.
25. Disconnect the connectors for the ECM, 4wd ECU, and instrument panel wiring. Pull the harness into the engine compartment.
26. Disconnect the front differential connector.
27. Remove the engine ground cable.
28. Attach lifting plates as shown.
29. Take up the weight of the engine with a crane.
30. Remove the engine mounting bracket bolts.
31. Remove the engine.
32. Installation is the reverse of removal. Observe the following torques:
- Engine mount brackets: 28 ft. lbs. (38 Nm)
- Hood: 10 ft. lbs. (13 Nm)
- V-bank cover: 66 inch lbs. (7.5 Nm)

4.7L Engine

1. Before servicing the vehicle, refer to the Precautions Section.
2. Relieve fuel system pressure.
3. Remove the transmission.
4. Remove the hood.
5. Remove the throttle body cover.
6. Remove the air cleaner assembly.
7. Remove the engine under-covers.
8. Drain the coolant and engine oil.
9. Disconnect the fuel lines.
10. Remove the fuel vapor line.
11. Remove the accessory drive belt.
12. Remove the power steering pump.
13. Remove the alternator.
14. Remove the A/C compressor. Don't disconnect the lines.
15. Remove the fan.
16. Remove the transmission filler tube.
17. Remove the oil level sending unit.
18. Remove the exhaust manifolds.
19. Disconnect the heater hoses.
20. Remove the right front door scuff plate.
21. Remove the right cowl side trim plate.
22. Remove the glove compartment door.
23. Disconnect the connectors for the ECM, 4wd ECU, and instrument panel wiring. Pull the harness into the engine compartment.
24. Disconnect the front differential connector.
25. Remove the engine ground cables.
26. Attach a crane to the lifting plates.
27. Take up the weight of the engine with a crane.
28. Remove the engine mounting bracket bolts.
29. Remove the engine.
30. Installation is the reverse of removal. Observe the following torques:
- Engine mount brackets: 28 ft. lbs. (38 Nm)
- Exhaust manifolds: 33 ft. lbs. (44 Nm)
- Oil level sending unit: 11 ft. lbs. (15 Nm)

Engine lifting bracket positions—4.0L engine

67170-4RUN-G03

Engine mounts and related parts—4.7L engine

67170-4RUN-G03

- Fan: 21 ft. lbs. (29 Nm)
- A/C compressor: bolt, 34 ft. lbs. (47 Nm); nut, 18 ft. lbs. (25 Nm)
- Power steering pump: 32 ft. lbs. (43 Nm)
- Hood: 10 ft. lbs. (13 Nm)
- V-bank cover: 66 inch lbs. (7.5 Nm)

Heater Core

REMOVAL & INSTALLATION

2002 Models

FRONT HEATER

1. Before servicing the vehicle, refer to the Precautions Section.
2. Disconnect the negative battery cable.

⚡ CAUTION

After the negative battery cable has been disconnected, wait at least 1½ minutes for the air bag module to deplete its energy.

3. Drain the cooling system into a clean container for reuse.
4. Disconnect the heater hoses from the heater core.
5. Remove the steering wheel by performing the following procedure:
   a. Position the front wheels in the straight-ahead position.
   b. At both sides of the steering wheel, remove the side covers.
   c. Using a Torx® wrench, loosen the steering wheel screws until the screw's circumference ring catches on the screw case.
   d. Carefully, lift the air bag module, disconnect the electrical connector and remove the air bag.

⚡ CAUTION

Place the air bag module in a safe location with the front facing upward.

   e. Remove the steering wheel nut.
   f. Using a steering wheel puller, press the steering wheel from the steering column.
6. Remove the instrument panel and reinforcement by performing the following procedure:
   a. Remove both front door scuff plates.
   b. Remove both cowl side trims.
   c. Remove the 2 hood lock release

lever screws and the hood lock release lever.
   d. Remove the 2 fuel lid release lever screws and the fuel lid release lever.
   e. Remove the 4 lower finish panel bolts and the panel.
   f. Remove the No. 1 and No. 2 heater-to-register duct screw and the ducts.
   g. Pry out the starter switch bezel.
   h. Remove the steering column cover screws and the covers.
   i. Remove the combination switch-to-steering column screws, disconnect the electrical connector and the combination switch.
   j. Remove the steering column-to-instrument panel nuts/bolts and the lower steering column bolt; then carefully, remove the steering column.
   k. Remove the 4 cluster finish panel screws and the panel.
   l. Remove the 4 combination meter screws, disconnect the electrical connectors and remove the combination meter.
   m. Pry out the parking brake hole cover.
   n. Pry out the upper console panel.
   o. Disengage the 7 center cluster finish panel clips and remove the panel.

➡**Remove the center cluster finish panel clips by starting at the bottom and working toward the top.**

   p. Remove the heater control knobs.
   q. Remove the 2 rear console box bolts/screws and the rear console box.
   r. Remove the upper console panel garnish.
   s. Remove the 2 glove compartment door screws and the door.
   t. Disconnect the passenger's side air bag module electrical connector.
   u. Remove the glove box light.
   v. Remove the 3 lower No. 2 finish panel bolts and the panel.
   w. Remove the 3 glove compartment door reinforcement bolts and the reinforcement.
   x. Remove the No. 4 heater-to-register duct.
   y. Remove the radio assembly.
   z. Remove the side bracket bolt and the bracket.
   aa. If equipped with manual air conditioning, remove the heater control assembly.
   bb. If equipped with automatic air conditioning, remove the air conditioning control assembly.
   cc. Remove the instrument panel-to-

chassis nut and 2 bolts; then, remove the instrument panel.
   dd. Remove the instrument panel reinforcement-to-chassis nuts/bolts and the reinforcement.
7. Remove the defroster nozzle and heater-to-register duct.
8. Remove the evaporator housing by performing the following procedure:
   a. Discharge and recover the air conditioning system refrigerant.
   b. Disconnect the refrigerant lines from the evaporator core. Discard the O-rings and plug the openings to prevent contamination.
   c. Disconnect the electrical connectors.
   d. Remove the 3 evaporator housing-to-chassis screws and the housing.
9. Disconnect the mode control servomotor connector.
10. Disconnect the aspirator hose from the room temperature sensor.
11. Disconnect the heater valve control cable.
12. Remove the heater housing-to-chassis nuts and the heater housing.
13. Remove the 3 heater core-to-heater housing screws, the 2 clips and clamp.
14. Remove the heater core from the heater housing.

To install:

15. Install the 3 heater core-to-heater housing screws, the 2 clips and clamp.
16. Install the heater housing and the heater housing-to-chassis nuts.
17. Connect the heater valve control cable.
18. Connect the aspirator hose to the room temperature sensor.
19. Connect the mode control servomotor connector.
20. Install the defroster nozzle and heater-to-register duct.
21. Install the evaporator housing by performing the following procedure:
   a. Install the evaporator housing and the 3 housing-to-chassis screws.
   b. Connect the electrical connectors.
   c. Using new O-rings, connect the refrigerant lines to the evaporator core.
22. Install the instrument panel and reinforcement by performing the following procedure:
   a. Install the instrument panel reinforcement and the reinforcement-to-chassis nuts/bolts.
   b. Install the instrument panel and the instrument panel-to-chassis nut and 2 bolts.
   c. If equipped with automatic air conditioning, install the air conditioning control assembly.

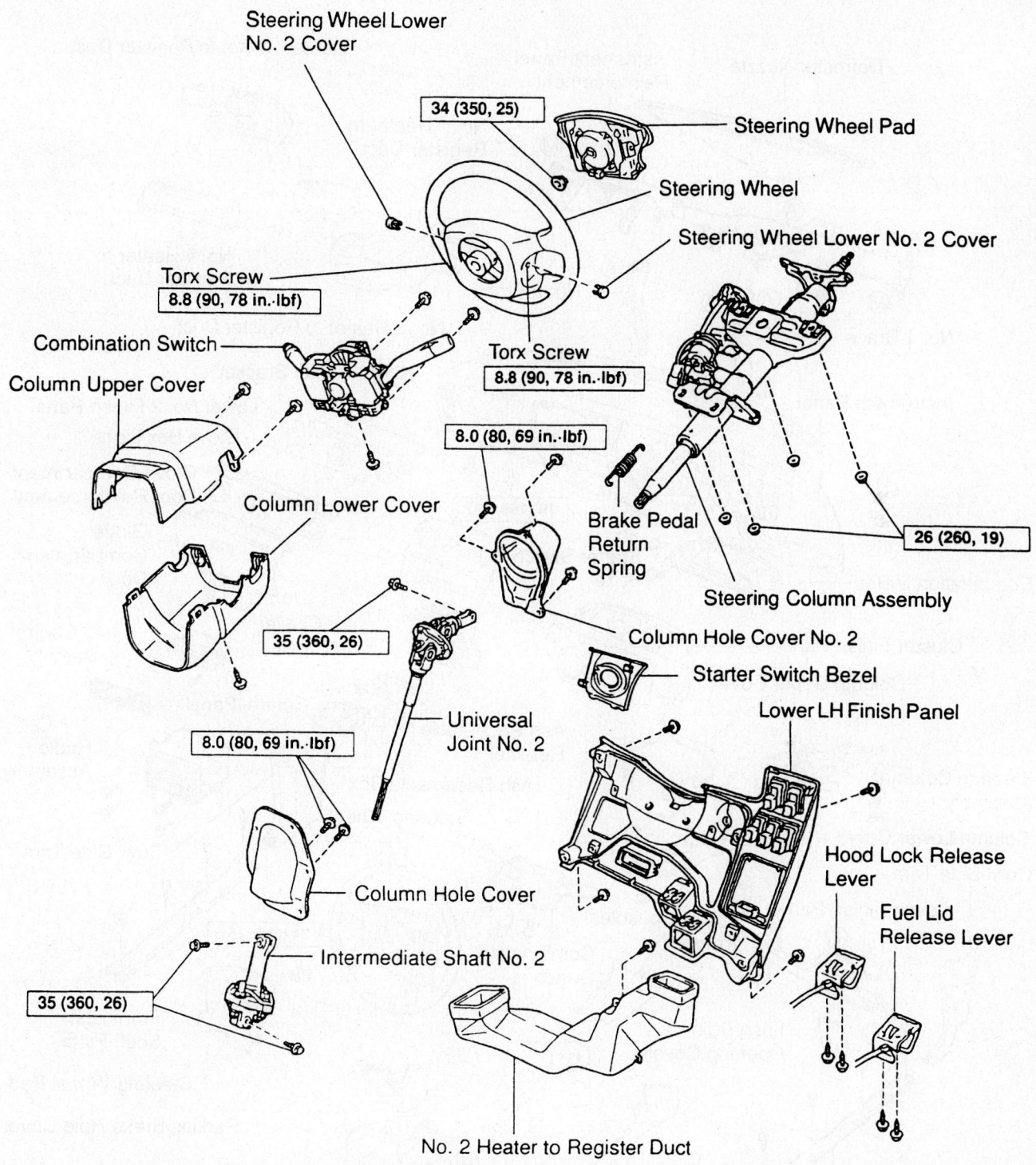

Steering Wheel Lower No. 2 Cover

34 (350, 25)

Steering Wheel Pad

Steering Wheel

Steering Wheel Lower No. 2 Cover

Torx Screw
8.8 (90, 78 in.·lbf)

Combination Switch

Column Upper Cover

Torx Screw
8.8 (90, 78 in.·lbf)

8.0 (80, 69 in.·lbf)

Column Lower Cover

Brake Pedal Return Spring

26 (260, 19)

35 (360, 26)

Steering Column Assembly

Column Hole Cover No. 2

Starter Switch Bezel

Lower LH Finish Panel

8.0 (80, 69 in.·lbf)

Universal Joint No. 2

Column Hole Cover

Intermediate Shaft No. 2

Hood Lock Release Lever

Fuel Lid Release Lever

35 (360, 26)

No. 2 Heater to Register Duct

N·m (kgf·cm, ft·lbf) : Specified torque

93113GJ9

Exploded view of the steering wheel, air bag module, steering column and related components—2002

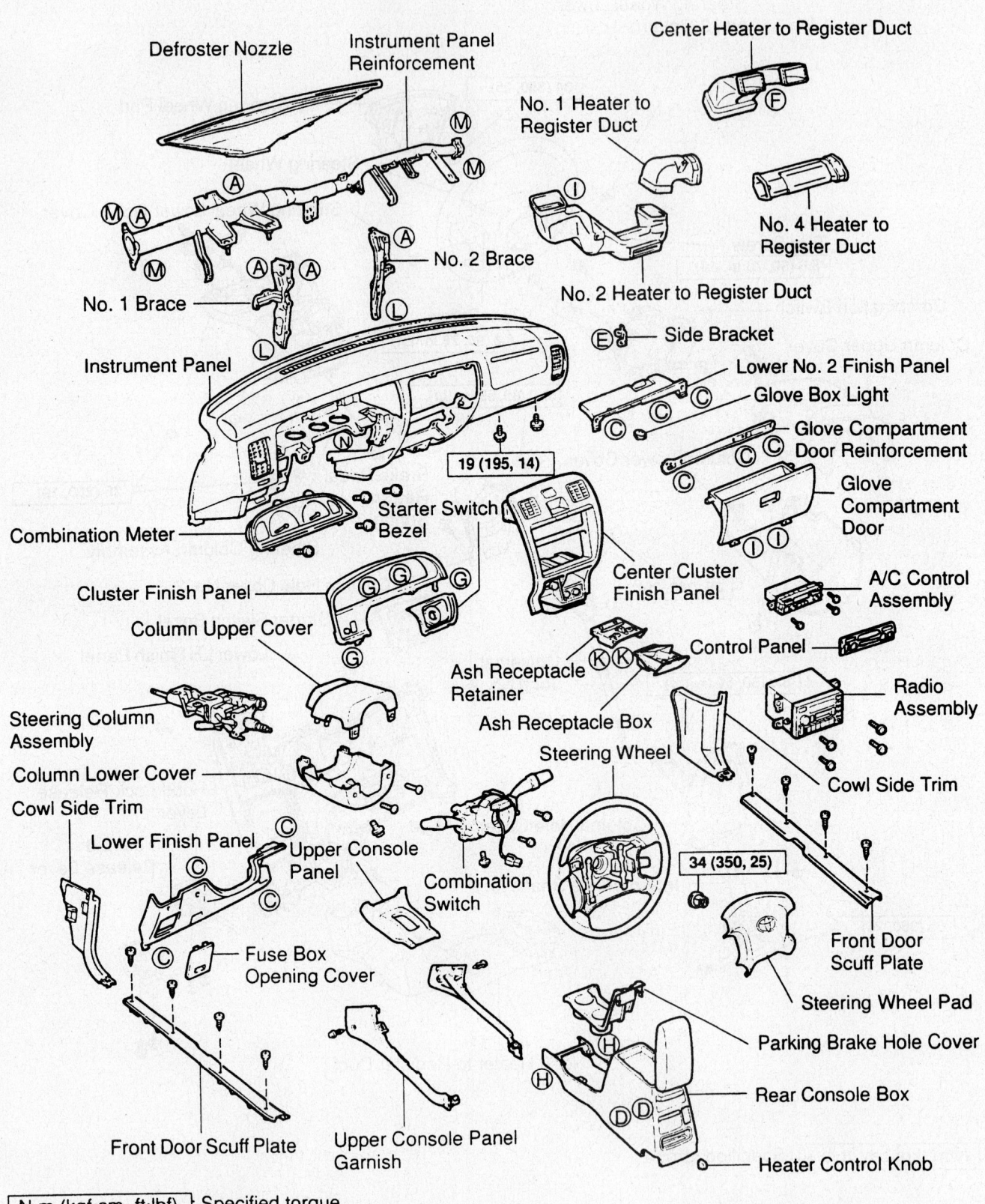

Defroster Nozzle

Instrument Panel Reinforcement

Center Heater to Register Duct

No. 1 Heater to Register Duct

No. 4 Heater to Register Duct

No. 2 Brace

No. 1 Brace

No. 2 Heater to Register Duct

Instrument Panel

Side Bracket

Lower No. 2 Finish Panel

Glove Box Light

Glove Compartment Door Reinforcement

Glove Compartment Door

19 (195, 14)

Combination Meter

Starter Switch Bezel

Cluster Finish Panel

Center Cluster Finish Panel

A/C Control Assembly

Control Panel

Column Upper Cover

Radio Assembly

Steering Column Assembly

Ash Receptacle Retainer

Ash Receptacle Box

Column Lower Cover

Steering Wheel

Cowl Side Trim

Cowl Side Trim

Lower Finish Panel

Upper Console Panel

Combination Switch

34 (350, 25)

Fuse Box Opening Cover

Front Door Scuff Plate

Steering Wheel Pad

Parking Brake Hole Cover

Rear Console Box

Front Door Scuff Plate

Upper Console Panel Garnish

Heater Control Knob

N·m (kgf·cm, ft·lbf) : Specified torque

93113GJ0

Exploded view of the instrument panel and related components—2002

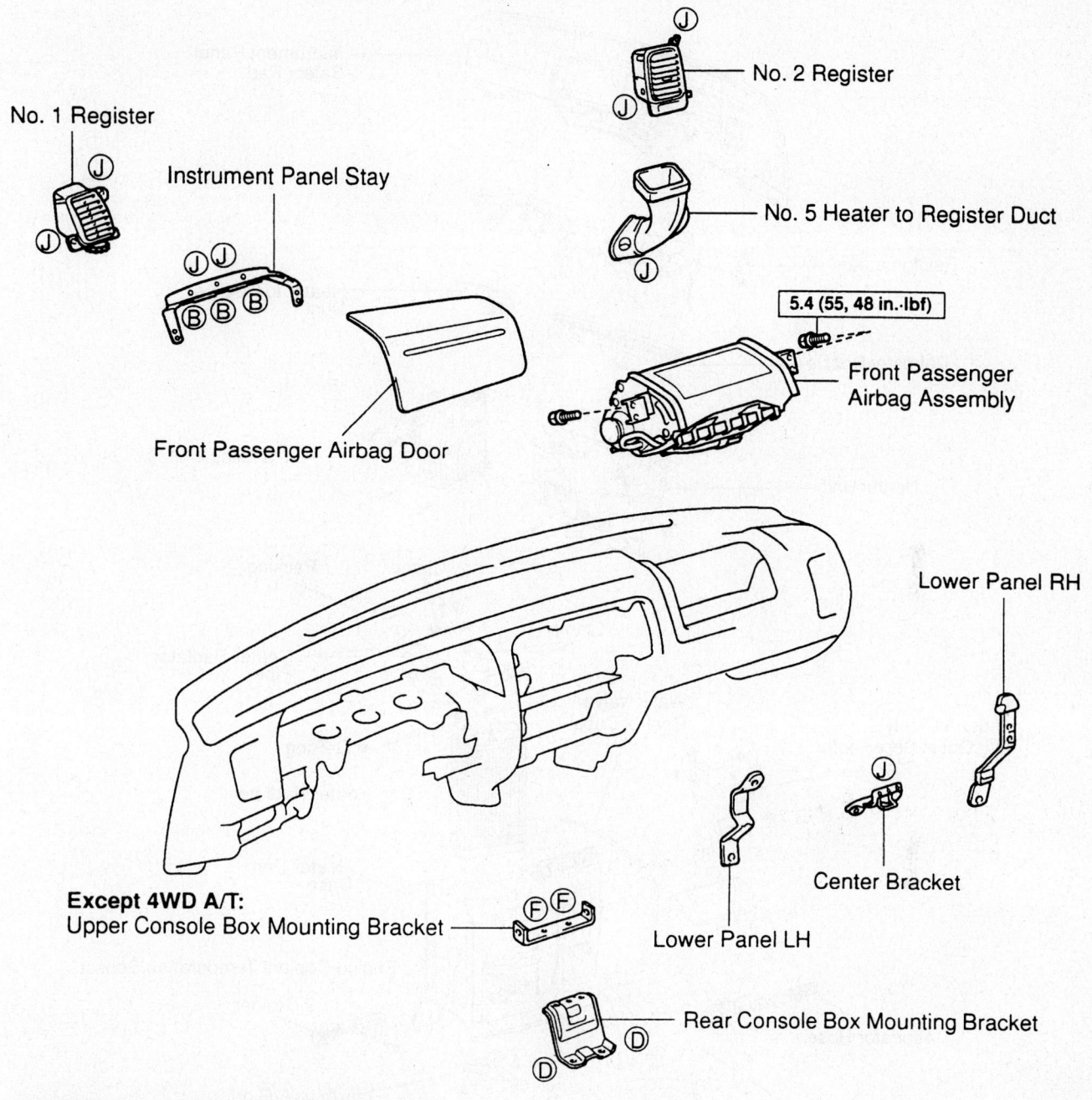

No. 1 Register

Instrument Panel Stay

No. 2 Register

No. 5 Heater to Register Duct

5.4 (55, 48 in.·lbf)

Front Passenger Airbag Door

Front Passenger Airbag Assembly

Lower Panel RH

Except 4WD A/T:
Upper Console Box Mounting Bracket

Center Bracket

Lower Panel LH

Rear Console Box Mounting Bracket

N·m (kgf·cm, ft·lbf) : Specified torque

93113GK1

Exploded view of the instrument panel air bag module, ventilation components and brackets—2002

d. If equipped with manual air conditioning, install the heater control assembly.

e. Install the side bracket and the bracket bolt.

f. Install the radio assembly.

g. Install the No. 4 heater-to-register duct.

h. Install the glove compartment door reinforcement and the 3 reinforcement bolts.

i. Install the lower No. 2 finish panel and the 3 panel bolts.

j. Install the glove box light.

k. Connect the passenger's side air bag module electrical connector.

l. Install the glove compartment door and the 2 door screws.

m. Install the upper console panel garnish.

n. Install the rear console box and the 2 rear console box bolts/screws.

o. Install the heater control knobs.

p. Install the center cluster finish panel and engage the 7 panel clips.

q. Install the upper console panel.

r. Install the parking brake hole cover.

s. Install the combination meter, connect the electrical connectors and install the 4 combination meter screws.

t. Install the cluster finish panel and the 4 panel screws.

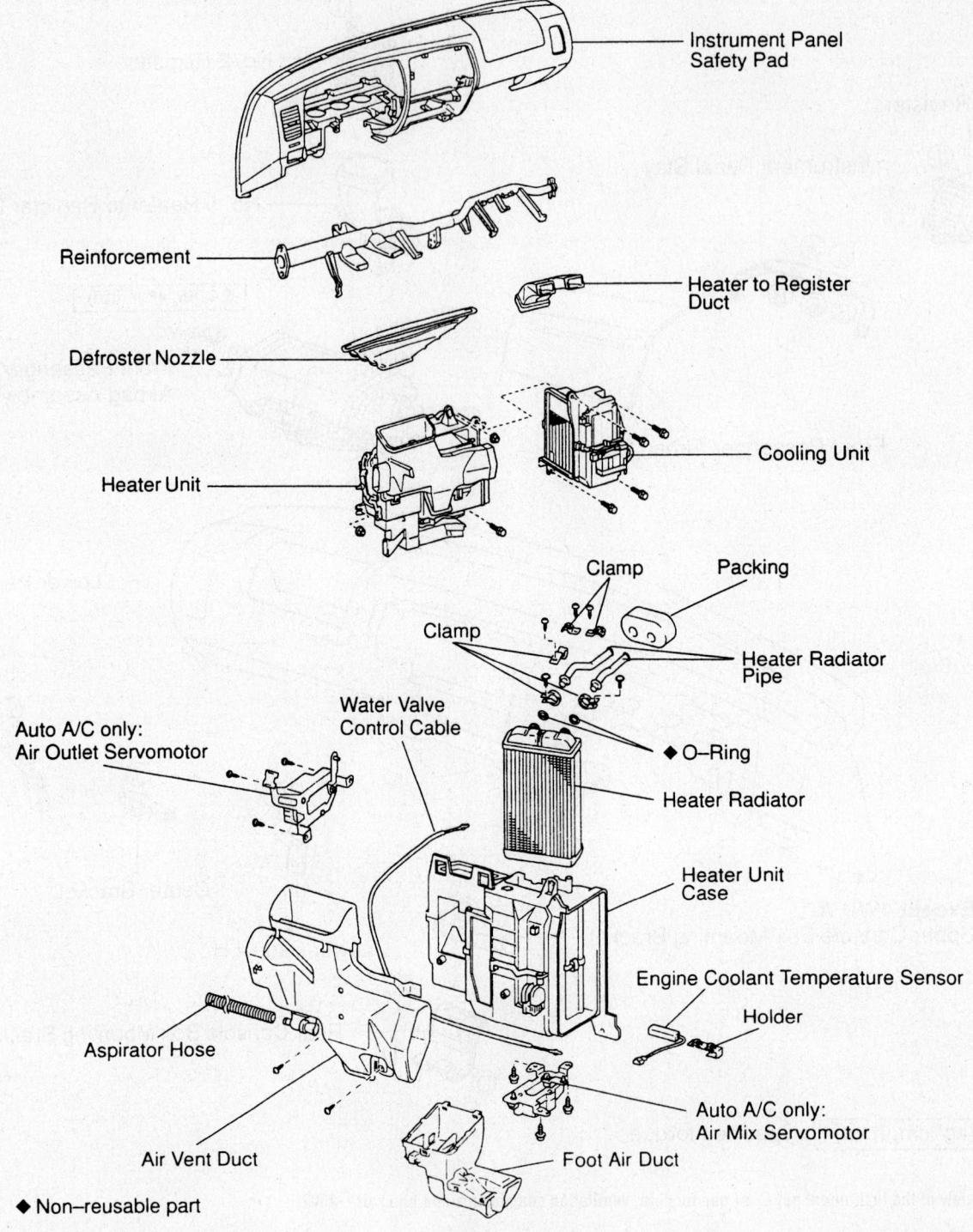

Instrument Panel Safety Pad

Reinforcement

Heater to Register Duct

Defroster Nozzle

Heater Unit

Cooling Unit

Clamp

Packing

Clamp

Heater Radiator Pipe

Auto A/C only: Air Outlet Servomotor

Water Valve Control Cable

◆ O–Ring

Heater Radiator

Heater Unit Case

Engine Coolant Temperature Sensor

Holder

Aspirator Hose

Auto A/C only: Air Mix Servomotor

Air Vent Duct

Foot Air Duct

◆ Non–reusable part

93113GK2

Exploded view of the front heater core, heater housing, evaporator housing and related components—2002

u. Install the steering column and torque the steering column-to-instrument panel nuts to 19 ft. lbs. (26 Nm) and the lower steering column bolt to 26 ft. lbs. (35 Nm).

v. Install the combination switch-to-steering column, connect the electrical connector and the combination switch screws.

w. Install the steering column cover and the cover screws.

x. Pry out the starter switch bezel.

y. Install the No. 1 and No. 2 heater-to-register duct and the duct screws.

z. Install the lower finish panel and the 4 panel bolts.

aa. Install the fuel lid release lever and the 2 fuel lid release lever screws.

bb. Install the hood lock release lever and the 2 hood lock release lever screws.

cc. Install both cowl side trims.

dd. Install both front door scuff plates.

23. Install the steering wheel by performing the following procedure:

a. Install the steering wheel to the steering column.

b. Install the steering wheel nut and torque the nut to 25 ft. lbs. (34 Nm).

c. Carefully, install the air bag module and connect the electrical connector.

d. Using a Torx® wrench, torque the steering wheel screws to 78 inch lbs. (8.8 Nm).

e. At both sides of the steering wheel, install the side covers.

24. Connect the heater hoses to the heater core.

25. Refill the cooling system to the correct level.

26. Connect the negative battery cable.

27. Evacuate and charge the air conditioning system.

28. Run the engine to normal operating temperatures; then, check the climate control operation and check for leaks.

REAR AUXILIARY HEATER

1. Before servicing the vehicle, refer to the Precautions Section.

2. Disconnect the negative battery cable.

3. Drain the cooling system into a clean container for reuse.

4. Remove the front seats.

5. Remove the center console box.

6. Move the floor carpet backward.

7. Disconnect the rear heater hoses from the rear heater core.

8. Remove the rear heater duct bolt, screw and duct.

9. Remove the rear heater control assembly.

10. Disconnect the electrical connectors.

11. Remove the 3 rear heater housing-to-chassis screws and the housing.

12. Remove the blower resistor, the rear

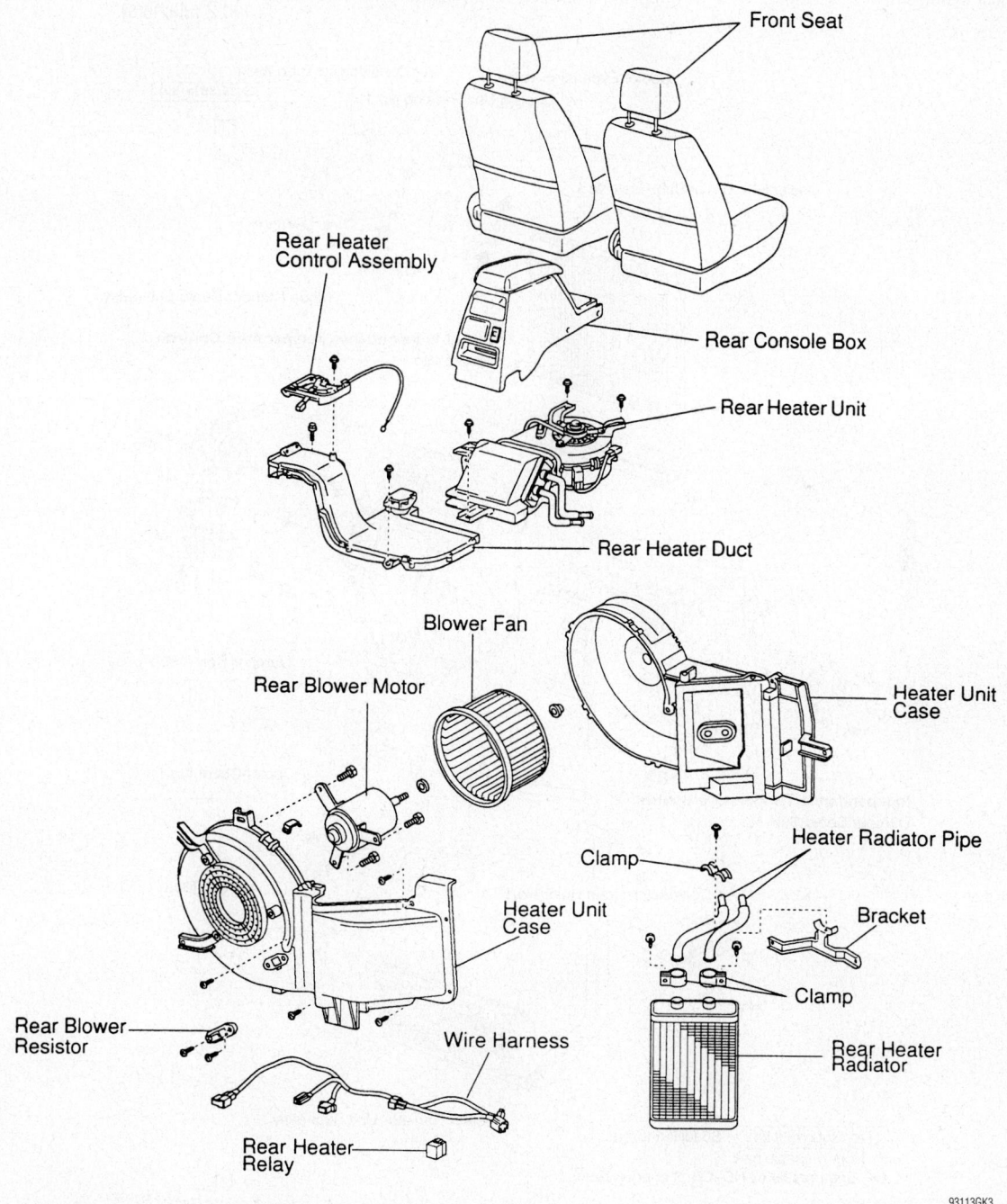

Exploded view of the rear heater housing and heater core—2002

heater relay and with wiring harness from the rear heater housing.

13. Remove the rear heater housing case screws and separate the cases.

14. Remove the rear heater core.

To install:

15. Install the rear heater core.

16. Assemble the rear heater housing case and install the case screws.

17. Install the blower resistor, the rear heater relay and wiring harness to the rear heater housing.

18. Install the rear heater housing and the 3 housing-to-chassis screws.

19. Connect the electrical connectors.

20. Install the rear heater control assembly.

21. Install the rear heater duct, bolt and screw.

22. Connect the rear heater hoses to the rear heater core.

23. Move the floor carpet foreword.

24. Install the center console box.

25. Install the front seats.

26. Refill the cooling system to the correct level.

27. Connect the negative battery cable.

28. Run the engine to normal operating temperatures; then, check the climate control operation and check for leaks.

2003–06 Models

1. Before servicing the vehicle, refer to the Precautions Section.

2. Properly discharge and recover the refrigerant.

3. Disconnect the refrigerant lines.

4. Disconnect the heater hoses.

5. Remove the instrument panel pad:

   a. Disable the air bag system.

   b. Remove the lower finish panel (2 bolts and 4 clips).

   c. Remove the console upper rear panel (5 clips and 2 retainers)

   d. Remove the upper panel (6 clips and 2 retainers).

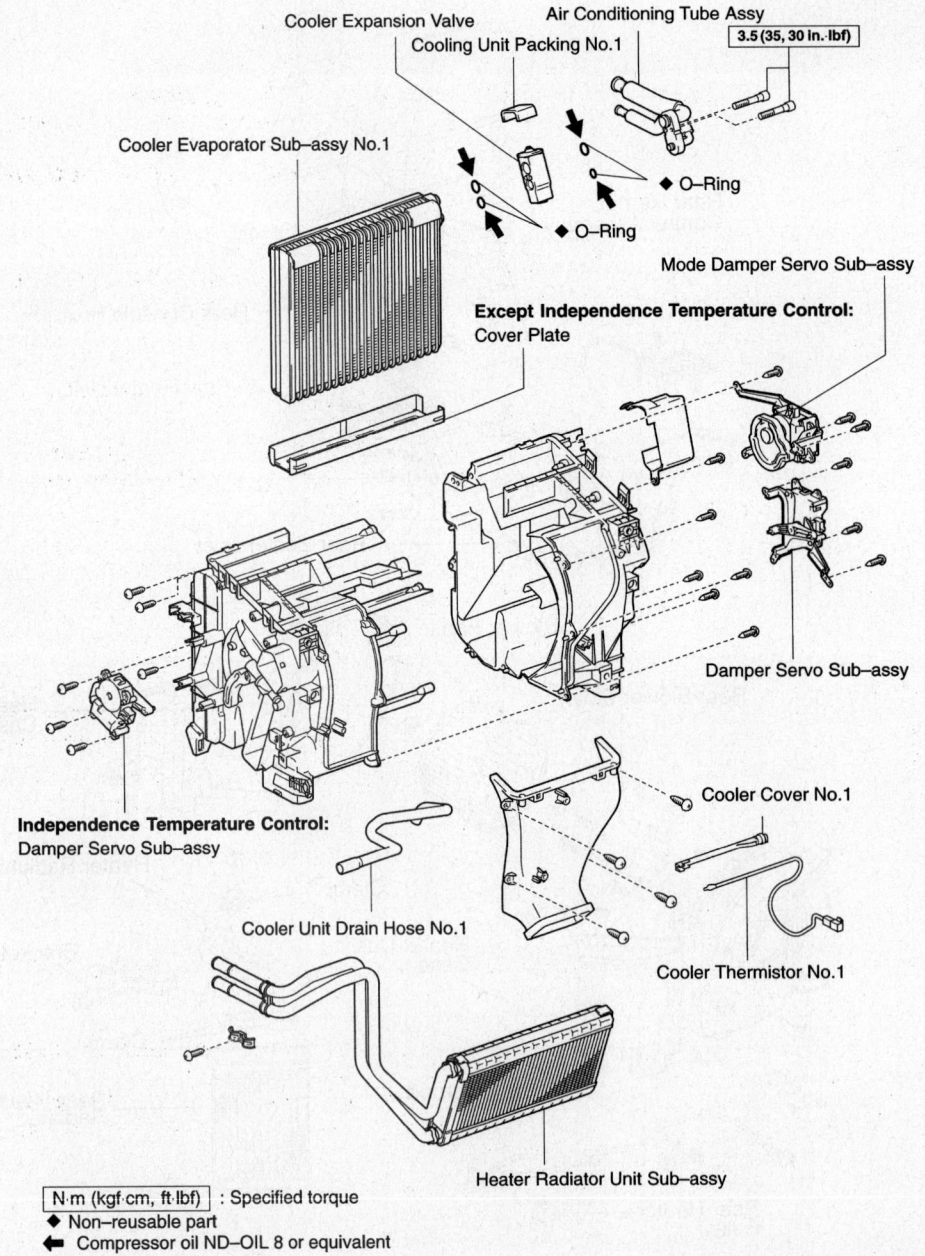

Cooler Expansion Valve
Cooling Unit Packing No.1
Air Conditioning Tube Assy
3.5 (35, 30 in.·lbf)
Cooler Evaporator Sub–assy No.1
◆ O–Ring
◆ O–Ring
Mode Damper Servo Sub–assy
Except Independence Temperature Control: Cover Plate
Damper Servo Sub–assy
Independence Temperature Control: Damper Servo Sub–assy
Cooler Cover No.1
Cooler Unit Drain Hose No.1
Cooler Thermistor No.1
Heater Radiator Unit Sub–assy

N·m (kgf·cm, ft·lbf) : Specified torque
◆ Non–reusable part
◀ Compressor oil ND–OIL 8 or equivalent

Heater core and related components—2003–06 models

67170-4RUN-G04

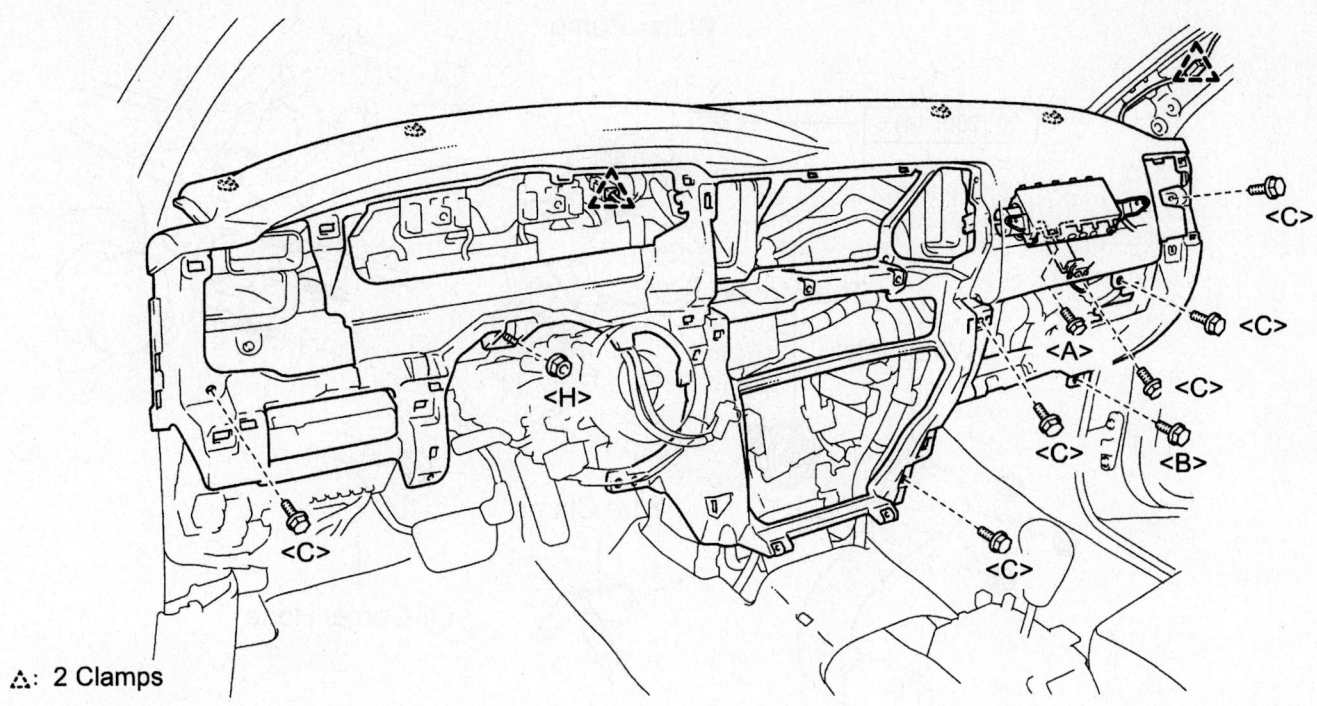

⚠: 2 Clamps

09490_4RUN_G0003

Fastener location of the instrument panel pad—2003–06 models

e. Remove the console upper panel garnishes.

f. Remove the A/C control unit (1 screw and 4 clips).

g. Remove the cluster finish center panel (3 bolts and 8 clips).

h. Remove the radio.

i. Remove the steering wheel.

j. Remove the steering column covers.

k. Remove the turn signal switch.

l. Remove the lower left instrument panel (2 bolts and 2 retainers).

m. Remove the instrument cluster finish panel (7 clips).

n. Remove the instrument cluster (3 screws and 2 bolts).

o. Remove the console box rear assembly (6 bolts).

p. Remove the right front door scuff plate.

q. Remove the right cowl trim board.

r. Remove the instrument panel under-cover.

s. Remove the glove compartment door.

t. Disconnect the passenger air bag connector.

u. Remove the instrument panel lower finish panel.

v. Remove the instrument panel register.

w. Remove the assist grip.

x. Remove the A-pillar trim panels.

y. Remove the instrument panel safety pad and passenger air bag.

z. Remove the speakers.

aa. Remove the defroster nozzle.

bb. Remove the heater ducts.

cc. Remove the meter hood retainer.

dd. Remove the instrument panel bracket.

ee. Remove the automatic light control sensor.

ff. Remove the thermistor.

gg. Disconnect the instrument panel wiring.

hh. Remove the antenna wire.

ii. Remove the navigation antenna assembly.

jj. Remove the instrument panel left door air bag.

kk. Remove the instrument panel safety pad.

ll. Remove the instrument panel passenger door air bag.

6. Remove the ECM.

7. Remove the A/C amplifier.

8. Remove the side defroster ducts.

9. Remove the heater ducts.

10. Remove the rear air ducts.

11. Remove the console duct.

12. Remove the air ducts.

13. Remove the instrument panel brace bracket.

14. Remove the steering column.

15. Remove the instrument panel reinforcement (18 bolts, 8 nuts and 27 clamps).

16. Remove the lower defroster nozzle.

17. Remove the A/C case.

18. Remove the mode damper servo.

19. Remove the damper servo.

20. Remove the heater core.

21. Installation is the reverse of removal. Evacuate, charge and leak test the system. Adjust the spiral cable.

Water Pump

REMOVAL & INSTALLATION

3.4L

1. Before servicing the vehicle, refer to the Precautions Section.

2. Disconnect the negative battery cable.

3. Drain the cooling system.

4. Remove or disconnect the following:
- Timing belt
- Thermostat
- No. 2 oil cooler hose from the water pump
- Water pump

5. Thoroughly clean the mating surfaces.

To install:

6. Apply sealant (PN 08826-00100) to the water pump.

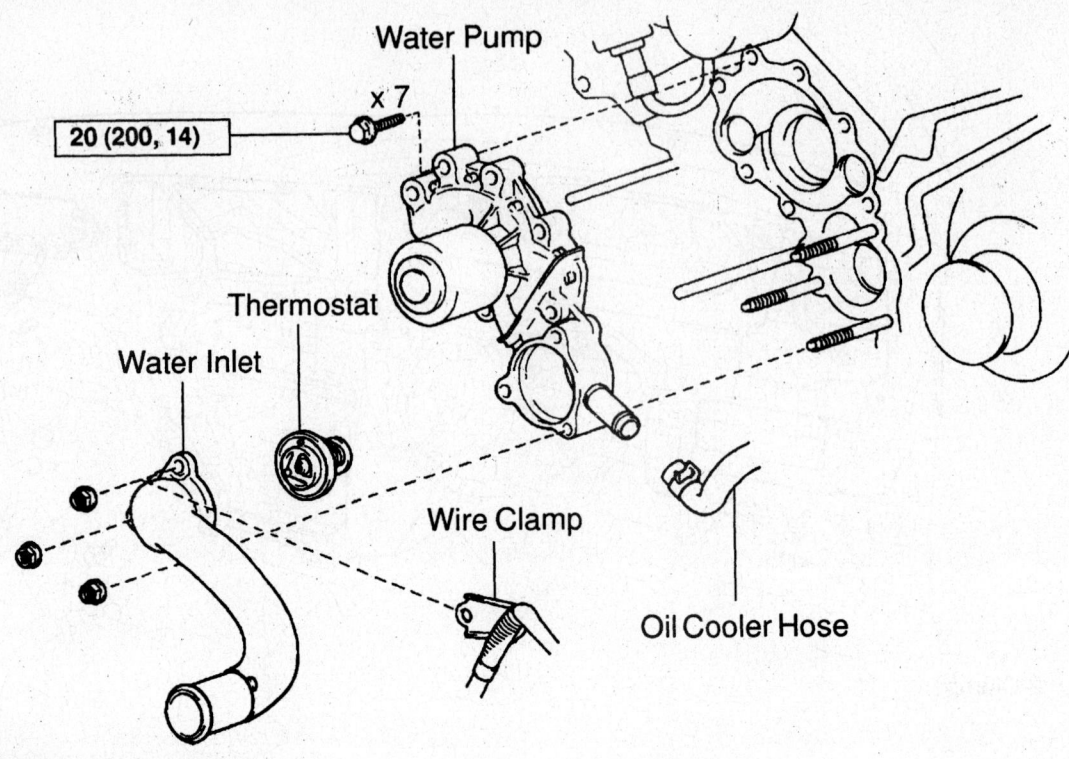

Water Pump

x 7

20 (200, 14)

Thermostat

Water Inlet

Wire Clamp

Oil Cooler Hose

N·m(kgf·cm, ft·lbf) : Specified torque

◆ Non–reusable part

67170-4RUN-G132

Water pump installation—3.4L engine

※※ WARNING

Parts must be assembled within 5 minutes of application. Otherwise the material must be removed and reapplied.

7. Install or connect the following:
- Water pump. Tighten the bolts to 14 ft. lbs. (20 Nm).
- No. 2 oil cooler hose
- Thermostat
- Timing belt
- Negative battery cable

8. Refill the cooling system to the correct level.

9. Start the engine and check for leaks.

4.0L Engine

1. Before servicing the vehicle, refer to the Precautions Section.
2. Remove the engine under-cover.
3. Drain the cooling system.
4. Remove the V-bank cover.
5. Remove the radiator support upper seal.
6. Remove the accessory drive belt.
7. Remove the fan.
8. Remove the air cleaner assembly.

9. Remove the water inlet.
10. Remove the idler pulley.
11. Remove the alternator.
12. Remove the A/C compressor. Don't disconnect the lines.
13. Remove the belt tensioner.
14. Remove the 17 bolts and the water pump. Discard the gasket.

To install:

15. Install the water pump with a new gasket. Torque the 10mm bolts to 80 inch lbs. (9 Nm); the 12mm bolts to 17 ft. lbs. (23 Nm).
16. Install the tensioner.
17. Install the compressor.
18. Install the alternator.

67170-4RUN-G05

Water pump mounting—4.0L engine

19. Install the idler pulleys. Torque to 29 ft. lbs. (39 Nm).
20. Install the water inlet. Use a new O-ring coated with soapy water. Torque the 5 bolts to 80 inch lbs. (9 Nm).
21. Install the air cleaner.
22. Install the fan with the bolts loose.
23. Install the drive belt. Then tighten the fan bolts to 15 ft. lbs. (21 Nm).
24. Install the upper seal.
25. Refill the cooling system to the correct level.
26. Install the V-bank cover. Torque to 66 inch lbs. (7.5 Nm).
27. Install the under-cover. Torque to 21 ft. lbs. (29 Nm).

4.7L Engine

1. Before servicing the vehicle, refer to the Precautions Section.
2. Remove the timing belt.
3. Remove the water inlet housing. Remove the silicone gasket material.
4. Remove the 5 bolts, 2 studs and the nut. Remove the water pump. Discard gasket.
5. Installation is the reverse of removal. Use soapy water on the new bypass pipe O-ring. Use a new water pump gasket and

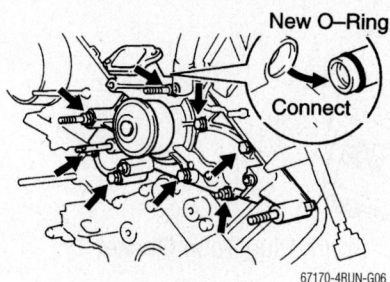

Water pump installation—4.7L engine

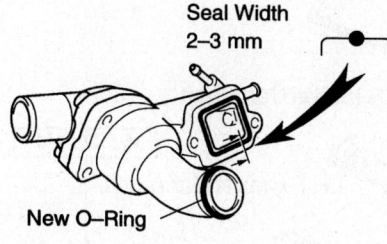

Seal Width
2–3 mm

New O–Ring

Water inlet installation—4.7L engine

torque the bolts to 16 ft. lbs. (21 Nm); the studs and the nut to 13 ft. lbs. (18 Nm).

6. Apply a 2–3 mm wide bead of silicone gasket material to the inlet housing and install it within 5 minutes. Torque to 13 ft. lbs. (18 Nm).

Cylinder Head

REMOVAL & INSTALLATION

3.4L

1. Before servicing the vehicle, refer to the Precautions Section.
2. Disconnect the negative battery cable.
3. Relieve the fuel system pressure.
4. Remove the engine undercover.
5. Drain the cooling system.
6. Remove or disconnect the following:
   - Front exhaust pipe
   - Air cleaner cap
   - Mass Air Flow (MAF) meter and the resonator
7. Disconnect the following cables:
   - Actuator cable with the bracket, if equipped with cruise control
   - Accelerator cable
   - Throttle cable, if equipped with an automatic transmission
8. Disconnect the following hoses:
   - Heater hose
   - Brake booster vacuum hose
   - Evaporative Emissions (EVAP) hose
   - Automatic Disconnecting Differential (ADD) vacuum hose, for 4-wheel drive

   - Fuel inlet and fuel return hose
9. Remove or disconnect the following:
   - Spark plug wires with the ignition coils
   - Spark plugs
   - Intake chamber stay
   - No. 2 timing belt cover
   - Air intake chamber assembly
10. Remove the following connectors and hoses:
    - Throttle Position (TP) sensor connector
    - Idle Air Control (IAC) valve connector
    - Positive Crankcase Ventilation (PCV) hoses
    - Water bypass hoses
    - Air assist hose from the throttle body
    - Intake air connector
11. Disconnect the engine wiring harness protector, as follows:
    - 6 injector connectors
    - Engine Coolant Temperature (ECT) sensor and sender gauge connectors
    - Engine wiring harness protector from the cylinder head
12. Remove or disconnect the following:
    - Fuel pressure regulator
    - Intake manifold assembly
13. Set the No. 1 cylinder at Top Dead Center (TDC) of the compression stroke, as follows:
    a. Turn the crankshaft pulley and align its groove with the timing mark **0** of the No. 1 timing belt cover.
    b. Check that the timing marks of the camshaft timing pulleys and the No. 3 timing belt cover are aligned. If not, turn the crankshaft pulley 1 revolution (360 degrees).
14. Remove or disconnect the following:
    - Timing belt tensioner, by alternately loosening the 2 bolts
    - Timing belt
15. Remove the camshaft timing pulleys, as follows:
    a. Using Variable Pin Wrench Set 09960-10010, remove the pulley bolt, the timing pulley and the knock pin.
    b. Remove the 2 timing pulleys with the timing belt.
16. Remove or disconnect the following:
    - Bolt and the No. 2 idler pulley
    - Camshaft Position (CMP) sensor
    - No. 3 timing belt cover
    - Alternator from the engine
    - Alternator bracket
    - Power steering pump and move it aside without disconnecting the pump lines

    - Exhaust crossover pipe and gaskets, by removing the 6 nuts
    - Left-hand exhaust manifold, by removing the heat insulator and 6 nuts
    - Right-hand exhaust manifold, by removing the heat insulator and 6 nuts
    - 8 bolts, seal washers, cylinder head cover and gasket.

➡**Remove both cylinder head covers**

   - Semi-circular plugs
   - Right exhaust and intake camshafts
   - Left exhaust and intake camshafts
   - Valve lifters and shims from the cylinder head; arrange the valve lifters and shims in correct order

17. Remove the cylinder heads, as follows:
    - Bolt and ground strap
    - Cylinder head (recessed head) bolt on the cylinder head, using an 8mm hexagon wrench; then, repeat the procedure for the other side.
    - 8 cylinder head (12-pointed head) bolts, on each cylinder head.

➡**Loosen the bolts in several passes, in the sequence shown.**

   - 16 cylinder head bolts and plate washers
   - Cylinder head

   To install:

18. Clean all surfaces.
19. Install or connect the following:
    - New cylinder head gaskets
    - Cylinder heads
20. Apply a light coat of engine oil on the threads and under the heads of the cylinder head bolts.
21. Tighten the cylinder head bolts using several passes, in sequence, as follows:
    - Step 1: 25 ft. lbs. (34 Nm)
    - Step 2: Mark the front of the cylinder head bolt with paint
    - Step 3: Turn 90 degrees
    - Step 4: Check that the painted mark is now at a 90 degrees angle to the front
22. Install the recessed head cylinder head bolts, as follows:
    - Step 1: Apply a light coat of engine oil on the threads and under the heads of the cylinder head bolts
    - Step 2: Tighten the cylinder head bolts, using a 8mm hexagon wrench, to 13 ft. lbs. (18 Nm).
    - Bolt and ground strap
23. Install or connect the following:
    - Valve lifters and shims

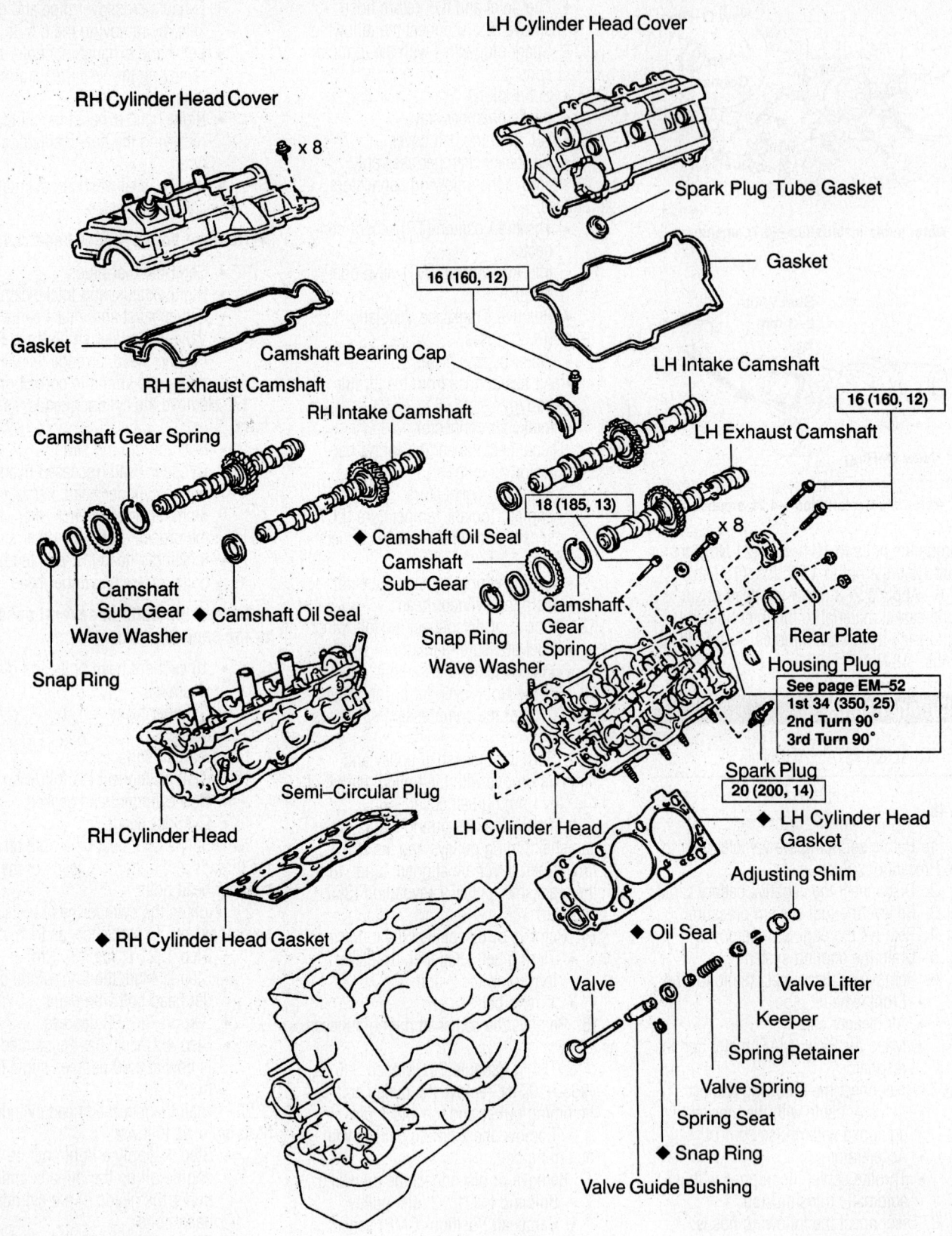

LH Cylinder Head Cover

RH Cylinder Head Cover

x 8

Spark Plug Tube Gasket

Gasket

16 (160, 12)

Gasket

Camshaft Bearing Cap

LH Intake Camshaft

RH Exhaust Camshaft

RH Intake Camshaft

16 (160, 12)

Camshaft Gear Spring

LH Exhaust Camshaft

LH Intake Camshaft

18 (185, 13)

◆ Camshaft Oil Seal

Camshaft Sub–Gear

Camshaft Sub–Gear

◆ Camshaft Oil Seal

Camshaft Gear Spring

Wave Washer

Snap Ring

Rear Plate

Snap Ring

Wave Washer

Housing Plug

See page EM–52
1st 34 (350, 25)
2nd Turn 90°
3rd Turn 90°

Semi–Circular Plug

Spark Plug
20 (200, 14)

RH Cylinder Head

LH Cylinder Head

◆ LH Cylinder Head Gasket

Adjusting Shim

◆ RH Cylinder Head Gasket

◆ Oil Seal

Valve Lifter

Valve

Keeper

Spring Retainer

Valve Spring

Spring Seat

◆ Snap Ring

Valve Guide Bushing

N·m (kgf·cm, ft·lbf) : Specified torque
◆ Non–reusable part

67170-4RUN-G133

Cylinder heads and related parts—3.4L engine

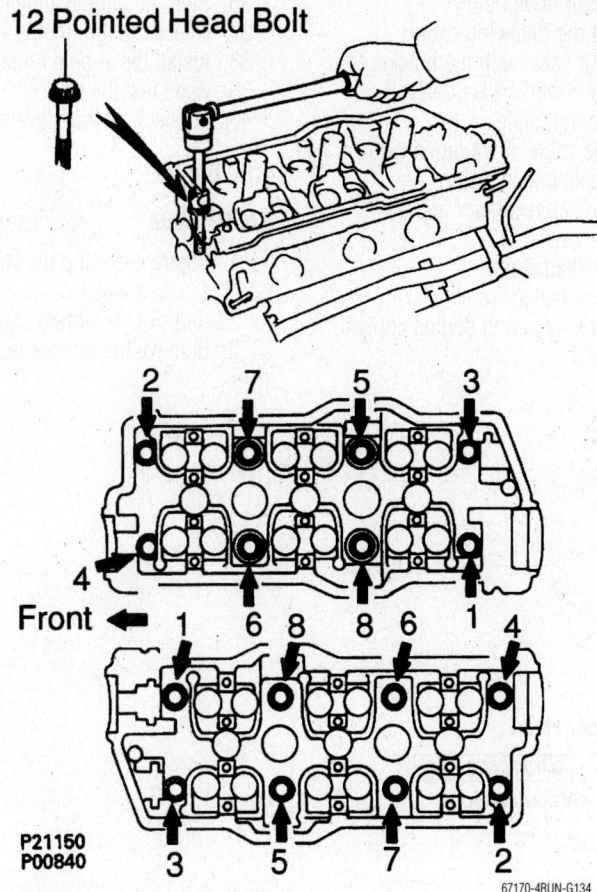

12 Pointed Head Bolt

Front ←

P21150
P00840

67170-4RUN-G134

Head bolt loosening sequence—3.4L engine

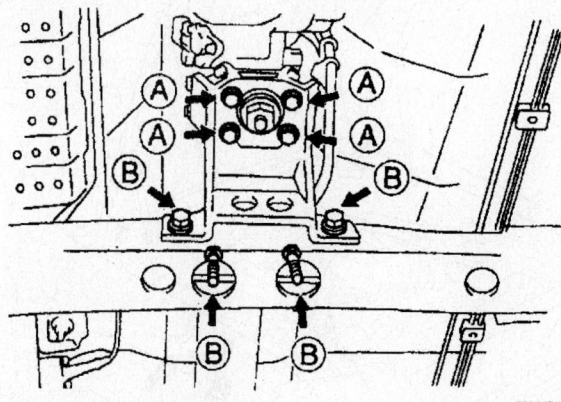

7924YG04

Cylinder head torque sequence—3.4L engine

➡**Check that the valve lifter rotates smoothly by hand.**

- Right intake and exhaust camshafts
- Left intake and exhaust camshafts

24. Check and adjust the valve clearance.

25. Install or connect the following:
- Semi-circular plugs
- Cylinder head covers. Uniformly,

tighten the bolts, in several passes, to 53 inch lbs. (6 Nm).
- Exhaust manifolds with new gaskets. Tighten the nuts to 30 ft. lbs. (40 Nm).
- Exhaust manifold heat insulators. Tighten the nuts to 71 inch lbs. (8 Nm).
- Exhaust crossover pipe. Tighten the nuts to 33 ft. lbs. (45 Nm).
- Power steering pump

- Alternator bracket. Tighten the fasteners to 14 ft. lbs. (18 Nm).
- Alternator
- No. 3 timing belt cover. Tighten the bolts to 80 inch lbs. (9 Nm).
- CMP sensor. Tighten it to 71 inch lbs. (8 Nm).
- Timing belt
- No. 2 timing belt idler bolt. Tighten the bolt to 30 ft. lbs. (40 Nm).

➡**Check that the pulley bracket moves smoothly.**

- Left camshaft timing pulley

26. Set the No. 1 cylinder to TDC of the compression stroke, as follows:
   a. Connect the timing belt to the left camshaft timing pulley.
   b. Check that the installation mark on the timing belt is aligned with the end of the No. 1 timing belt cover.
   c. Install the right camshaft timing pulley and the timing belt.
   d. Set the timing belt tensioner. Alternately, tighten the bolts to 20 ft. lbs. (28 Nm).

27. Using pliers, remove the 1.5mm hexagon wrench from the belt tensioner.

28. Check the valve timing.

29. Install or connect the following:
- New gaskets and the intake manifold assembly. Tighten the bolts and nuts to 13 ft. lbs. (18 Nm).
- Intake manifold stay. Tighten the bolts to 14 ft. lbs. (18 Nm).
- Fuel pressure regulator

30. Connect the engine wiring harness to the intake manifold, as follows:
- Engine wiring harness to the cylinder head
- 3 engine wiring harness clamps

31. Install or connect the following:
- 6 injector connectors
- ECT sender gauge connector
- ECT sensor connector
- Intake air connector
- Air intake chamber assembly. Tighten the bolts and nuts to 13 ft. lbs. (18 Nm).
- Intake chamber stay
- No. 2 timing belt cover. Tighten the bolts to 80 inch lbs. (9 Nm).
- PCV hoses
- Water bypass hoses
- Air assist hose to the throttle body
- IAC valve connector
- TP sensor connector
- CMP sensor connector to the No. 2 timing belt cover
- 3 spark plug wire clamps

32. Connect the following hoses:
- Brake booster vacuum hose

- EVAP hose
- Automatic Disconnecting Differential (ADD) vacuum hose, for 4-wheel drive
- Fuel inlet and fuel return hose
- Heater hose

33. Install or connect the following:
- Oil dipstick and guide, using a new O-ring
- Spark plugs
- Spark plug wires, with the ignition coils

- Alternator drive belt

34. Connect the following cables:
- Actuator cable with the bracket, if equipped with cruise control
- Accelerator cable
- Throttle cable, if equipped with an automatic transmission
- MAF meter, resonator and air cleaner cap
- Front exhaust pipe
- Negative battery cable

35. Fill the radiator with engine coolant.

36. Start the engine and check for leaks.
37. Check the ignition timing.
38. Install the engine undercover.
39. Road test the vehicle.
40. Recheck all fluid levels.

4.0L Engine

RIGHT SIDE

1. Before servicing the vehicle, refer to the Precautions Section.
2. Remove the timing chain.
3. Remove the air cleaner.

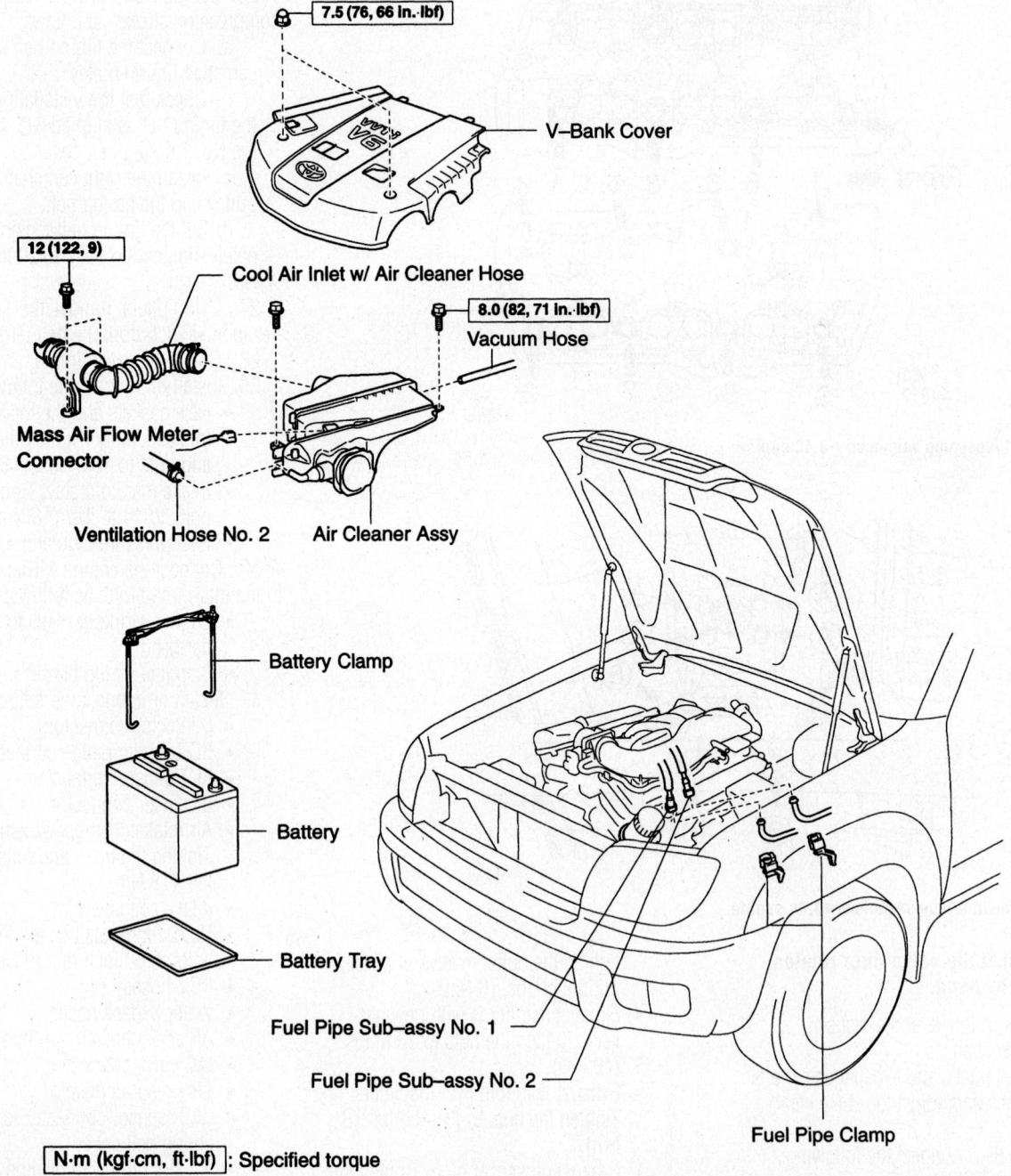

7.5 (76, 66 in.·lbf) — V–Bank Cover

12 (122, 9)

Cool Air Inlet w/ Air Cleaner Hose

8.0 (82, 71 in.·lbf)
Vacuum Hose

Mass Air Flow Meter Connector

Ventilation Hose No. 2   Air Cleaner Assy

Battery Clamp

Battery

Battery Tray

Fuel Pipe Sub–assy No. 1

Fuel Pipe Sub–assy No. 2

Fuel Pipe Clamp

N·m (kgf·cm, ft·lbf) : Specified torque

67170-4RUN-G15

For right cylinder head removal, remove these parts—Part 1

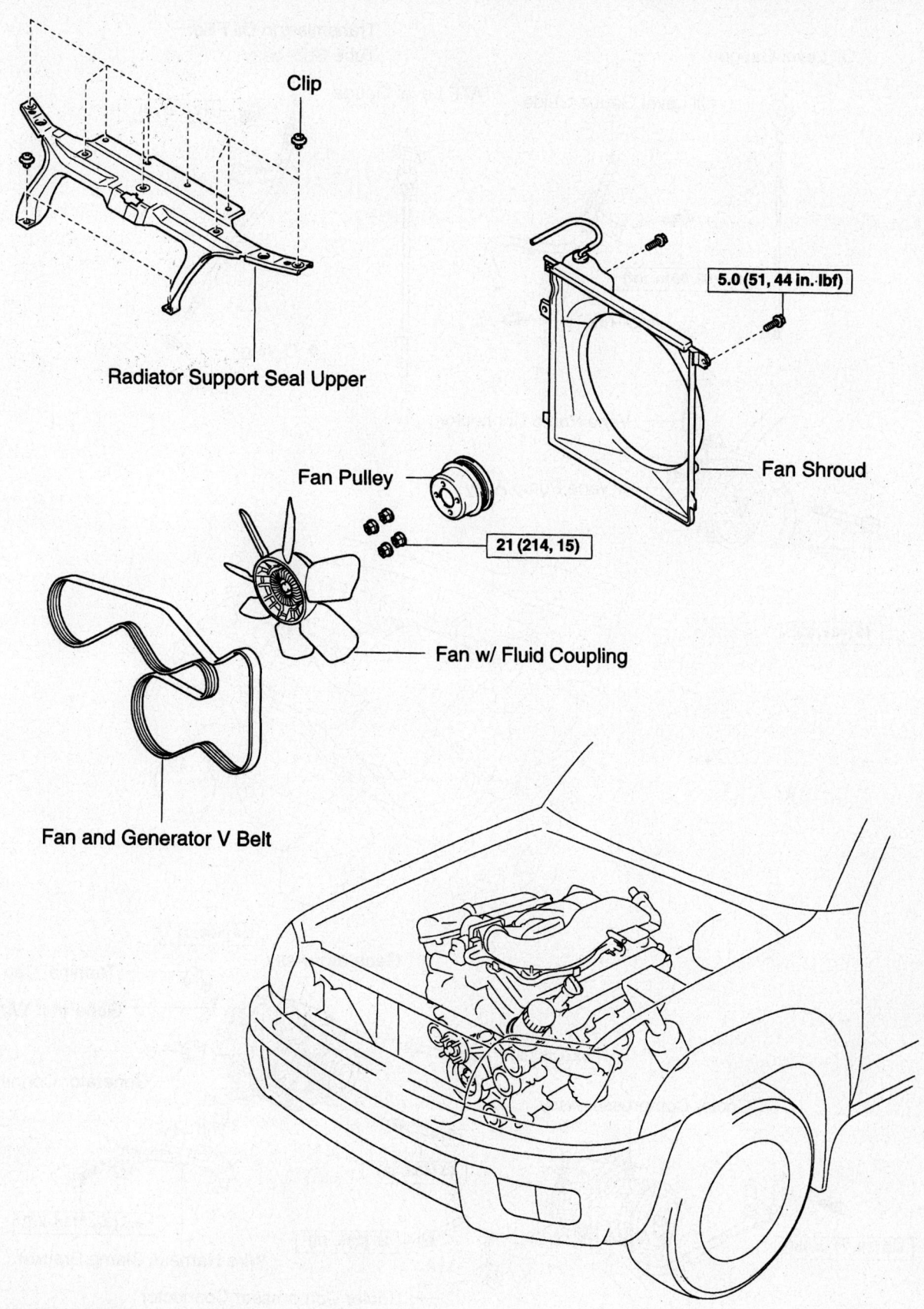

Clip

Radiator Support Seal Upper

5.0 (51, 44 in.·lbf)

Fan Shroud

Fan Pulley

21 (214, 15)

Fan w/ Fluid Coupling

Fan and Generator V Belt

N·m (kgf·cm, ft·lbf) : Specified torque

For right cylinder head removal, remove these parts—Part 2

67170-4RUN-G08

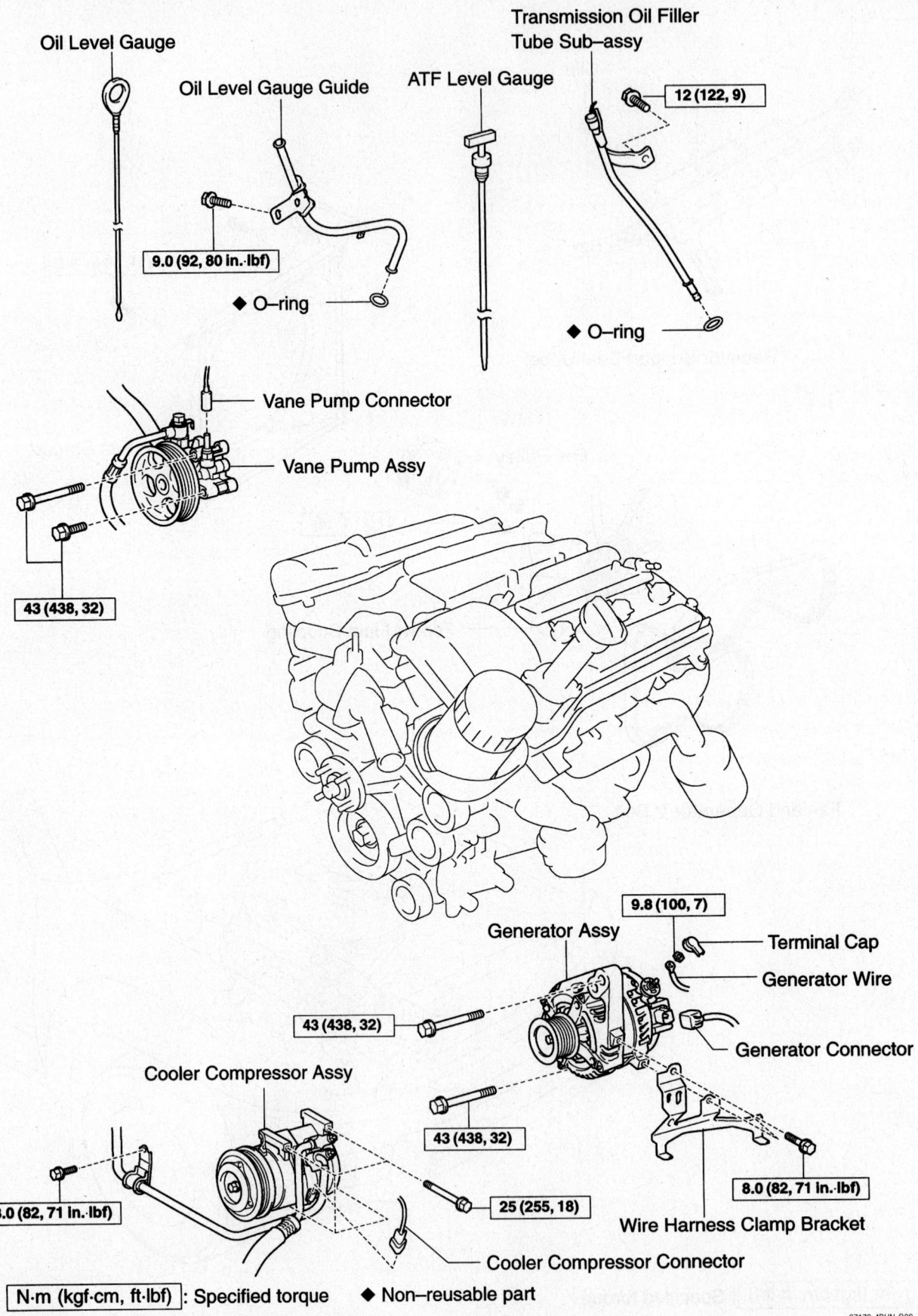

Oil Level Gauge

Oil Level Gauge Guide

ATF Level Gauge

Transmission Oil Filler
Tube Sub–assy

12 (122, 9)

9.0 (92, 80 in.·lbf)

◆ O–ring

◆ O–ring

Vane Pump Connector

Vane Pump Assy

43 (438, 32)

9.8 (100, 7)

Generator Assy

Terminal Cap

Generator Wire

43 (438, 32)

Generator Connector

Cooler Compressor Assy

43 (438, 32)

8.0 (82, 71 in.·lbf)

25 (255, 18)

8.0 (82, 71 in.·lbf)

Wire Harness Clamp Bracket

Cooler Compressor Connector

N·m (kgf·cm, ft·lbf) : Specified torque   ◆ Non–reusable part

For right cylinder head removal, remove these parts—Part 3

67170-4RUN-G09

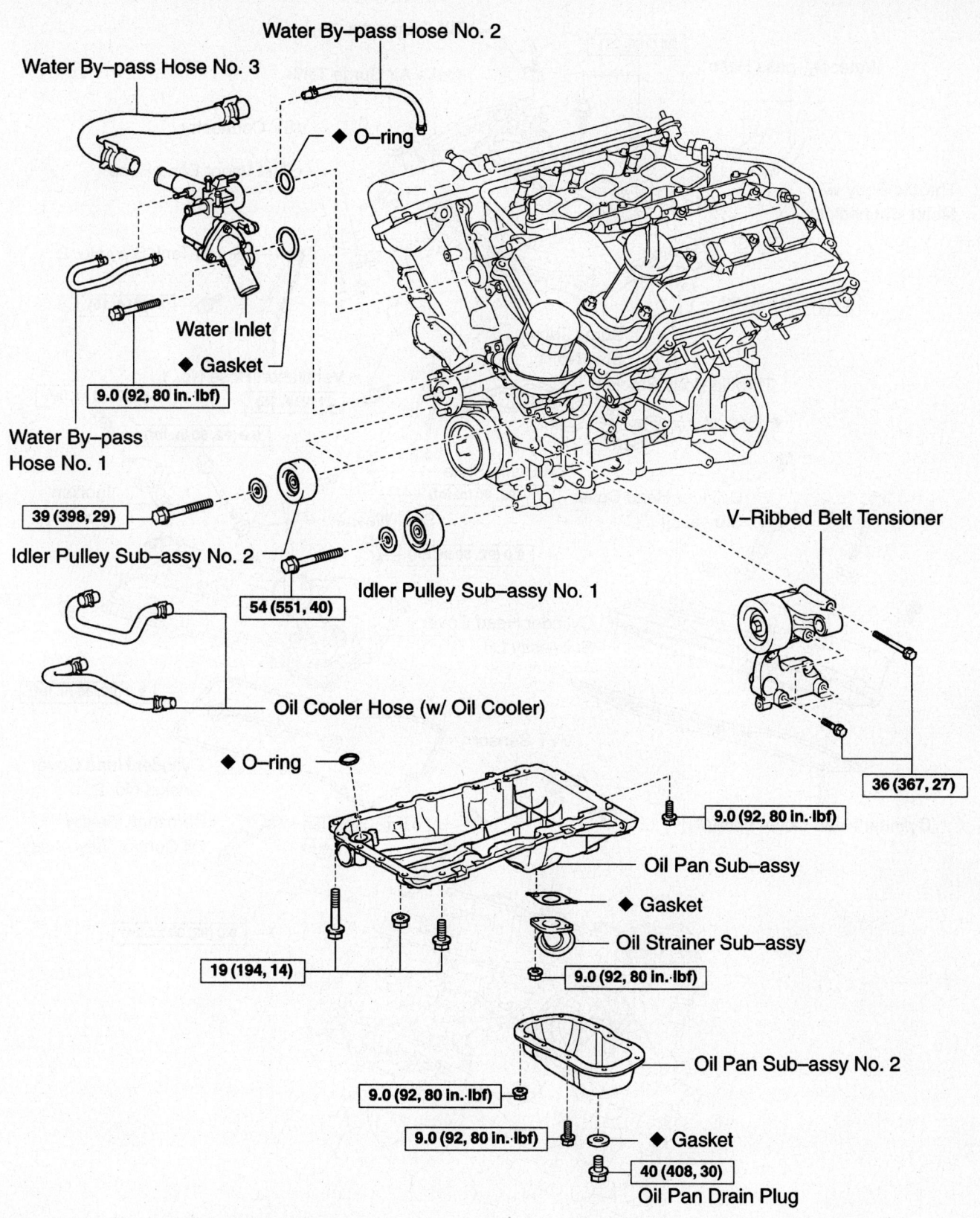

Water By-pass Hose No. 3

Water By-pass Hose No. 2

◆ O-ring

Water Inlet

◆ Gasket

9.0 (92, 80 in.·lbf)

Water By-pass Hose No. 1

39 (398, 29)

Idler Pulley Sub-assy No. 2

54 (551, 40)

Idler Pulley Sub-assy No. 1

V-Ribbed Belt Tensioner

Oil Cooler Hose (w/ Oil Cooler)

36 (367, 27)

◆ O-ring

9.0 (92, 80 in.·lbf)

Oil Pan Sub-assy

◆ Gasket

Oil Strainer Sub-assy

19 (194, 14)

9.0 (92, 80 in.·lbf)

Oil Pan Sub-assy No. 2

9.0 (92, 80 in.·lbf)

9.0 (92, 80 in.·lbf)

◆ Gasket

40 (408, 30)

Oil Pan Drain Plug

N·m (kgf·cm, ft·lbf) : Specified torque

◆ Non-reusable part

67170-4RUN-G10

For right cylinder head removal, remove these parts—Part 4

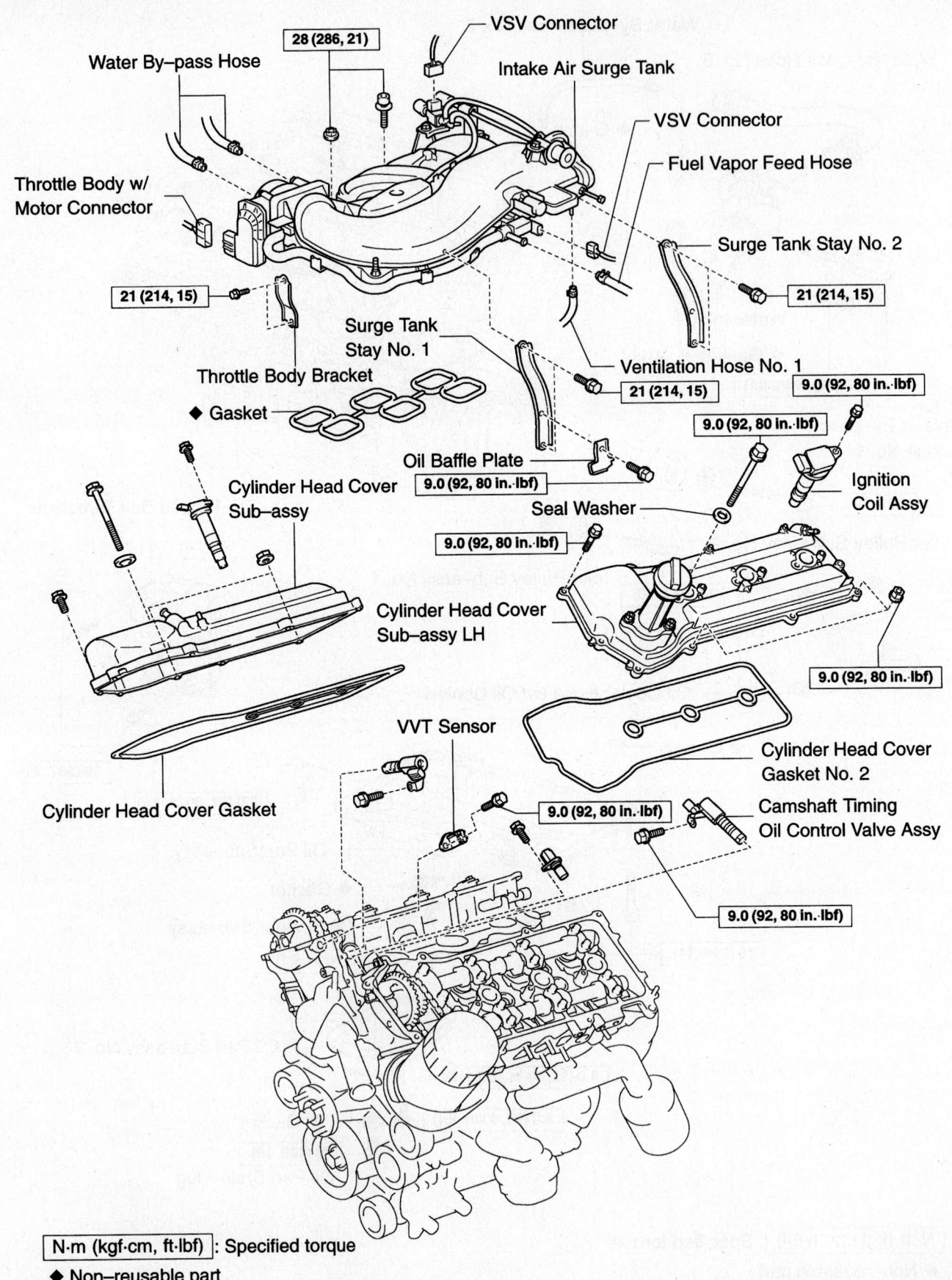

28 (286, 21)

VSV Connector

Water By-pass Hose

Intake Air Surge Tank

VSV Connector

Fuel Vapor Feed Hose

Throttle Body w/ Motor Connector

Surge Tank Stay No. 2

21 (214, 15)

21 (214, 15)

Surge Tank Stay No. 1

Ventilation Hose No. 1

21 (214, 15)

9.0 (92, 80 in.·lbf)

Throttle Body Bracket

9.0 (92, 80 in.·lbf)

◆ Gasket

Oil Baffle Plate

9.0 (92, 80 in.·lbf)

Ignition Coil Assy

Cylinder Head Cover Sub-assy

Seal Washer

9.0 (92, 80 in.·lbf)

Cylinder Head Cover Sub-assy LH

9.0 (92, 80 in.·lbf)

VVT Sensor

Cylinder Head Cover Gasket No. 2

Cylinder Head Cover Gasket

9.0 (92, 80 in.·lbf)

Camshaft Timing Oil Control Valve Assy

9.0 (92, 80 in.·lbf)

N·m (kgf·cm, ft·lbf) : Specified torque

◆ Non-reusable part

67170-4RUN-G11

For right cylinder head removal, remove these parts—Part 5

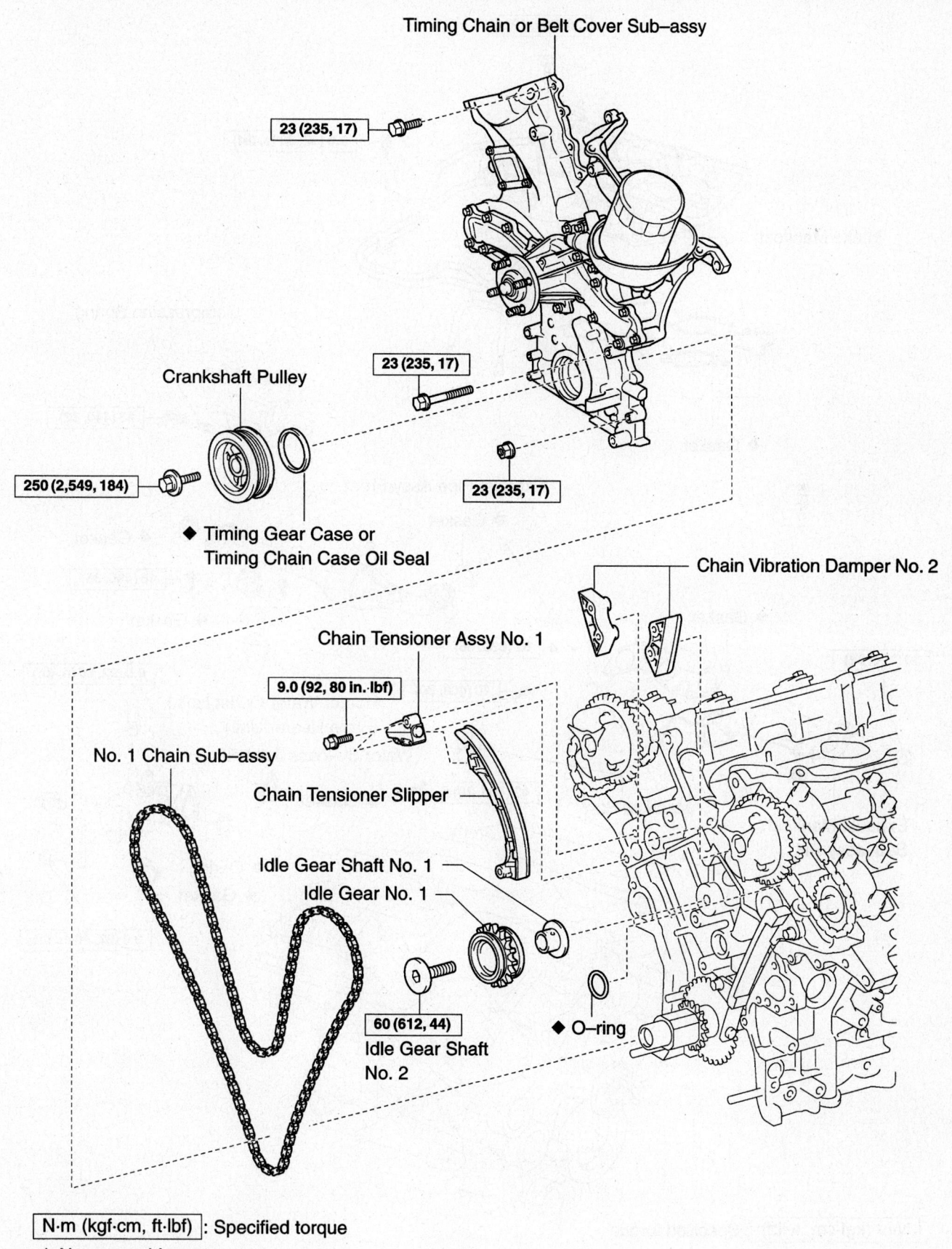

Timing Chain or Belt Cover Sub–assy

23 (235, 17)

Crankshaft Pulley

23 (235, 17)

250 (2,549, 184)

23 (235, 17)

◆ Timing Gear Case or
Timing Chain Case Oil Seal

Chain Vibration Damper No. 2

Chain Tensioner Assy No. 1

9.0 (92, 80 in.·lbf)

No. 1 Chain Sub–assy

Chain Tensioner Slipper

Idle Gear Shaft No. 1

Idle Gear No. 1

60 (612, 44)
Idle Gear Shaft
No. 2

◆ O–ring

N·m (kgf·cm, ft·lbf) : Specified torque

◆ Non–reusable part

67170-4RUN-G12

For right cylinder head removal, remove these parts—Part 6

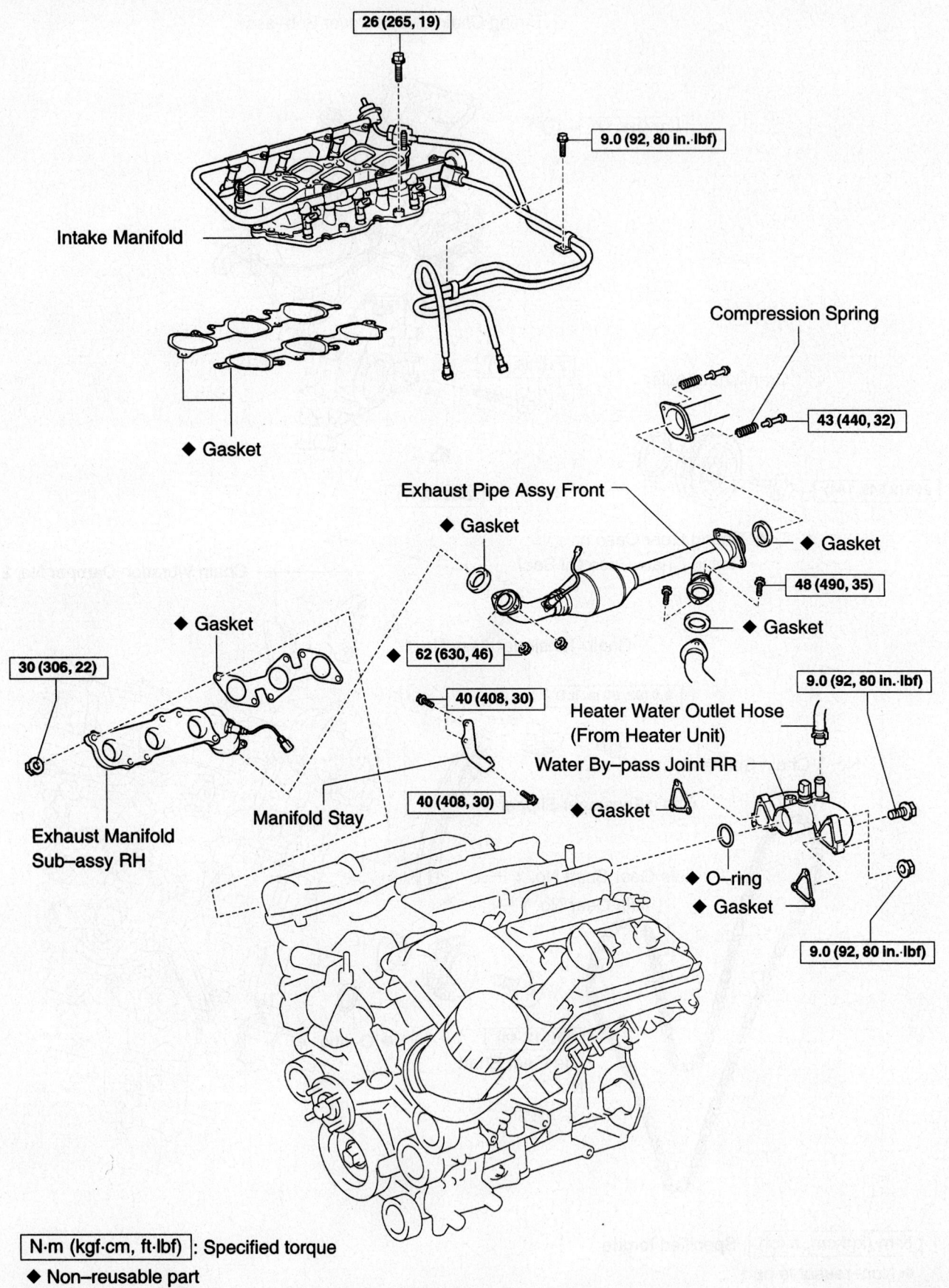

26 (265, 19)

9.0 (92, 80 in.·lbf)

Intake Manifold

Compression Spring

43 (440, 32)

◆ Gasket

Exhaust Pipe Assy Front

◆ Gasket

◆ Gasket

◆ Gasket

48 (490, 35)

◆ Gasket

◆ Gasket

30 (306, 22)

◆ 62 (630, 46)

9.0 (92, 80 in.·lbf)

40 (408, 30)

Heater Water Outlet Hose
(From Heater Unit)

Water By-pass Joint RR

40 (408, 30)

◆ Gasket

Exhaust Manifold
Sub-assy RH

Manifold Stay

◆ O-ring

◆ Gasket

9.0 (92, 80 in.·lbf)

N·m (kgf·cm, ft·lbf) : Specified torque

◆ Non-reusable part

67170-4RUN-G13

For right cylinder head removal, remove these parts—Part 7

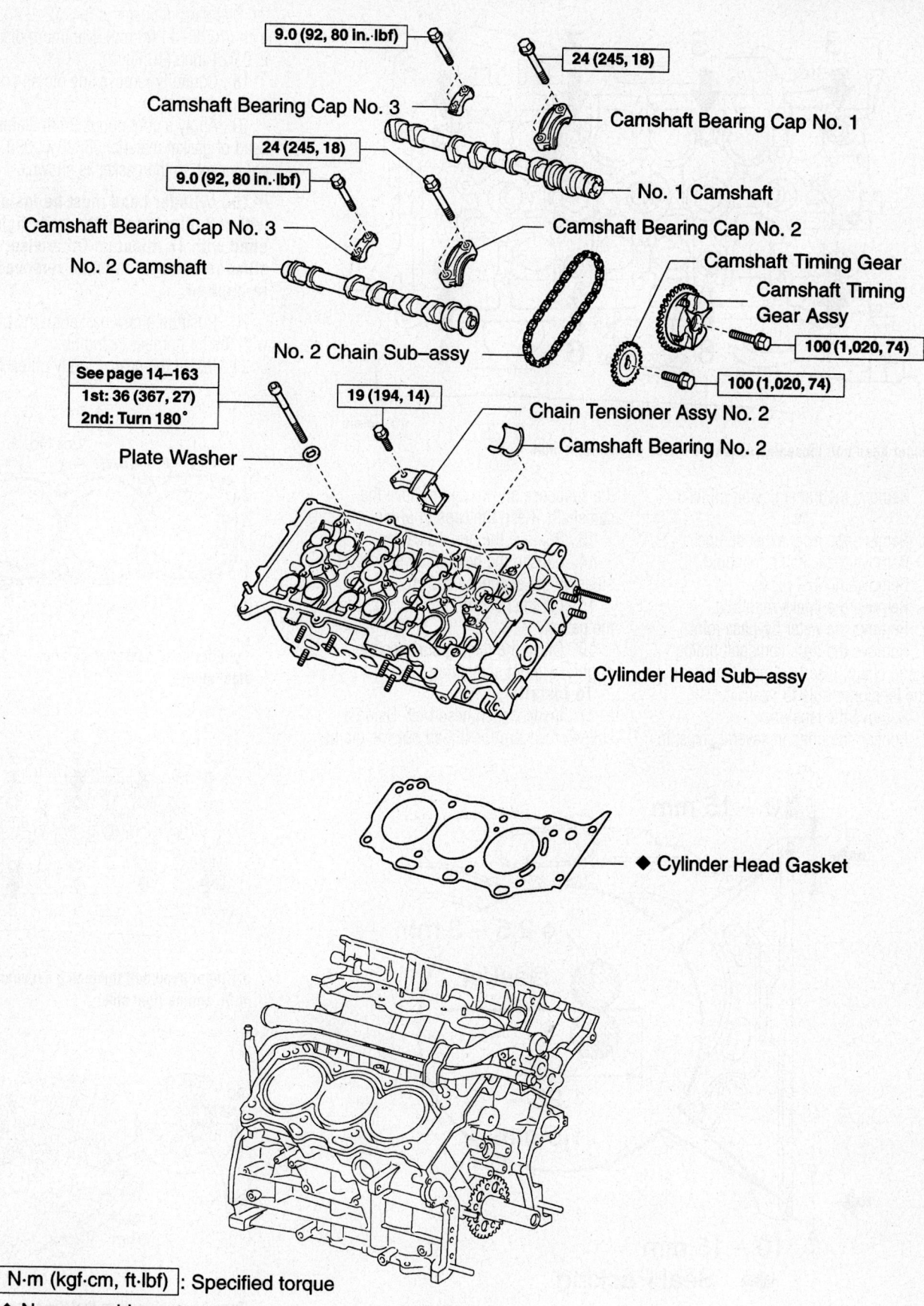

9.0 (92, 80 in.·lbf)

24 (245, 18)

Camshaft Bearing Cap No. 3

Camshaft Bearing Cap No. 1

24 (245, 18)

9.0 (92, 80 in.·lbf)

No. 1 Camshaft

Camshaft Bearing Cap No. 3

Camshaft Bearing Cap No. 2

No. 2 Camshaft

Camshaft Timing Gear

Camshaft Timing Gear Assy

No. 2 Chain Sub–assy

100 (1,020, 74)

See page 14–163
1st: 36 (367, 27)
2nd: Turn 180°

19 (194, 14)

100 (1,020, 74)

Chain Tensioner Assy No. 2

Camshaft Bearing No. 2

Plate Washer

Cylinder Head Sub–assy

◆ Cylinder Head Gasket

N·m (kgf·cm, ft·lbf) : Specified torque

◆ Non–reusable part

67170-4RUN-G14

For right cylinder head removal, remove these parts—Part 8

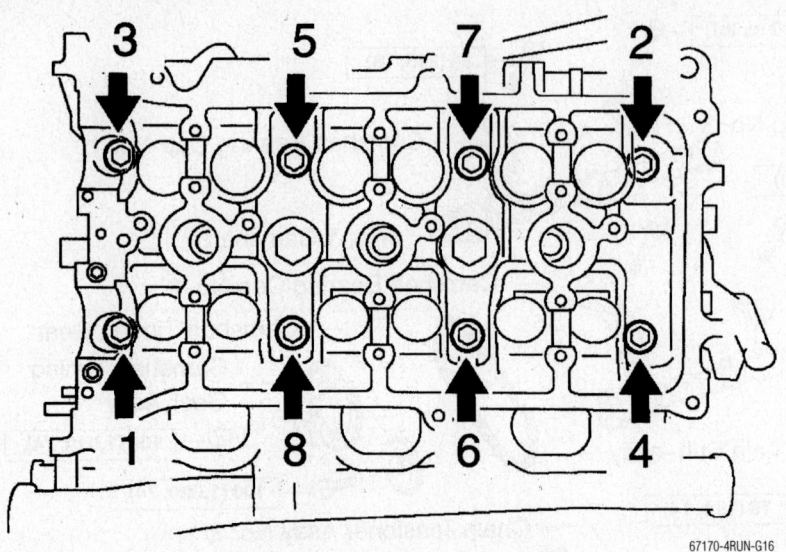

Cylinder head bolt loosening sequence—4.0L engine right side

4. Remove the transmission oil filler tube.

5. Remove the front exhaust pipe.

6. Remove the exhaust manifold.

7. Remove the fuel pipes.

8. Remove the intake manifold.

9. Remove the water by-pass joint.

10. Remove the right camshaft timing gears and chain. Insert a 1mm diameter pin into the tensioner hole to retain it.

11. Remove the tensioner.

12. Loosen the bolts in several steps, in the sequence shown and remove the camshafts. Keep the caps in order.

13. Remove the ground cables.

14. Loosen the head bolts in several steps, in the sequence shown.

15. Lift the head from the block. Discard the gasket.

16. Clean the mating surfaces thoroughly, without scratching.

To install:

17. Inspect each head bolt. Using a caliper, measure the thread outside diame-

ter. Standard diameter is 0.4272–0.4331 inch (10.85–11.00mm). Minimum diameter is 0.421 inch (10.7mm).

18. Carefully remove any old silicone gasket material.

19. Apply a continuous 3mm diameter bead of gasket material 08826-00080, or equivalent, to the gasket as shown.

➡**The cylinder head must be installed within 3 minutes and the bolts tightened with 15 minutes. Otherwise, the silicone material must be removed and re-applied.**

20. Position a new gasket on the block with the lot number facing up.

21. Install the head. Lightly oil each

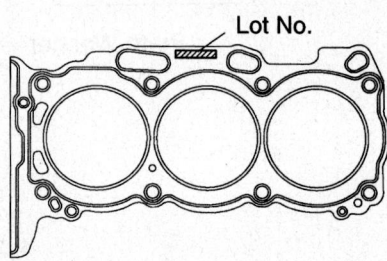

Cylinder head gasket placement—4.0L right side

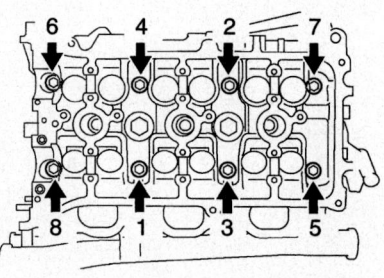

Cylinder head bolt tightening sequence—4.0L engine right side

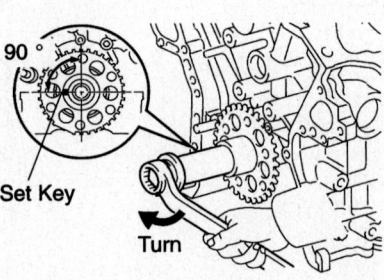

Turn the crankshaft so that the crankshaft key is at the 9:00 o'clock position—4.0L engine right side

Apply a continuous 3mm diameter bead of gasket material 08826-00080, or equivalent, to the gasket as shown—4.0L engine right side

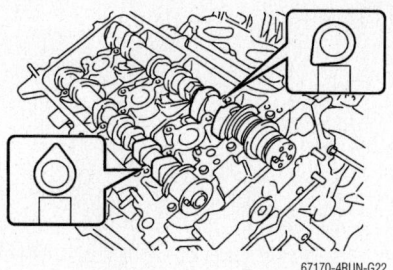

Place the camshafts on the head as shown—4.0L engine right side

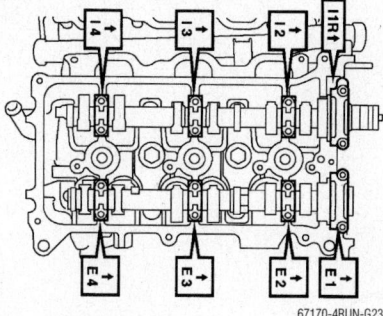

Install the bearing caps—4.0L engine right side

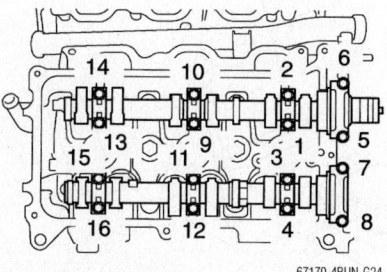

Tighten the cap bolts, in several even passes, in the sequence shown—4.0L

head bolt with clean engine oil. Tighten the bolts, in several even passes, to 27 ft. lbs. (36 Nm).

22. Using a torque angle gauge, tighten each bolt, in sequence, an additional 180 degrees. If an angle gauge is not available, make a paint mark on each bolt head and the cylinder head, and tighten each bolt 180 degrees.

23. Install the camshafts.

➡The camshaft thrust clearance is very small. The camshaft must be kept level during installation. If the camshaft is not kept level, the portion of the head which receives the thrust may crack or be damaged causing the camshaft to seize or break. To avoid this, the following steps must be taken:

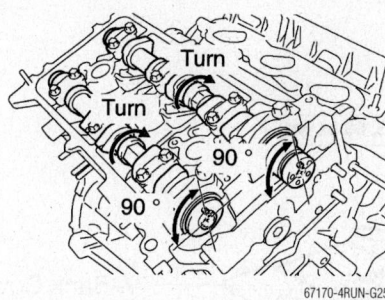

Turn the camshafts clockwise until the knock pin is in a position 90 degrees to the head—4.0L engine right side

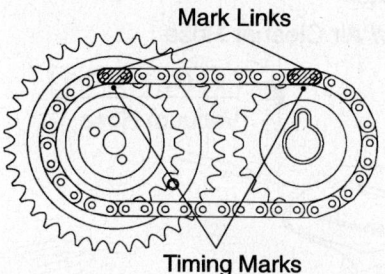

Install the chain aligning the yellow links with the timing marks on the gears—4.0L engine right side

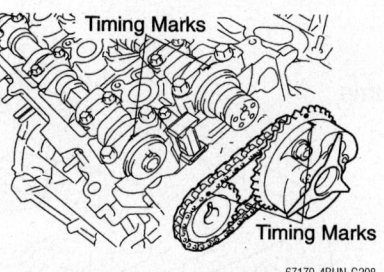

Align the timing marks on the timing gears with the marks on the bearing caps, and install the gear/chain assembly

a. Turn the crankshaft so that the crankshaft key is at the 9 o'clock position.

b. Place the camshafts on the head as shown.

24. Install the bearing caps.

a. Tighten the cap bolts, in several even passes, in the sequence shown, to 80 inch lbs (9 Nm) for 10mm bolts; 18 ft. lbs. (24 Nm) for 12mm bolts.

b. Turn the camshafts clockwise until the knock pin is in a position 90 degrees to the head.

25. Install the tensioner. Torque to 14 ft. lbs. (19 Nm).

26. Install the chain aligning the yellow links with the timing marks on the gears.

27. Align the timing marks on the timing gears with the marks on the bearing caps, and install the gear/chain assembly. Torque the bolts to 74 ft. lbs. (100 Nm).

28. Install the water bypass joint. Apply soapy water to the new O-ring and using 2 new gaskets. Torque the 2 bolts and 4 nuts to 80 inch lbs. (9 Nm).

29. Install the intake manifold, using new gaskets. Torque the 10 bolts, in several even passes, to 19 ft. lbs. (26 Nm).

30. Connect the fuel lines.

31. Install the exhaust manifold. Torque to 22 ft. lbs. (30 Nm). Torque the manifold stay to 30 ft. lbs. (40 Nm).

32. The remainder of installation is the reverse of removal.

LEFT SIDE

1. Before servicing the vehicle, refer to the Precautions Section.

2. Remove the timing chain.

3. Remove the air cleaner.

4. Remove the front exhaust pipe.

5. Remove the exhaust manifold.

6. Remove the fuel pipes.

7. Remove the intake manifold.

8. Remove the water by-pass joint.

9. Remove the chain vibration damper.

10. Remove the left camshaft timing gears and chain. Insert a 1mm diameter pin into the tensioner hole to retain it.

11. Remove the tensioner.

12. Loosen the bolts in several steps, in the sequence shown and remove the camshafts. Keep the caps in order.

13. Remove the ground cable.

14. Loosen the 2 head bolts in several steps, in the sequence shown.

15. Loosen the remaining 8 head bolts in several steps, in the sequence shown.

16. Lift the head from the block. Discard the gasket.

17. Clean the mating surfaces thoroughly, without scratching.

To install:

18. Inspect each head bolt. Using a caliper, measure the thread outside diameter. Standard diameter is 0.4272–0.4331 inch (10.85–11.00mm). Minimum diameter is 0.421 inch (10.7mm).

19. Carefully remove any old silicone gasket material.

20. Apply a continuous 3mm diameter bead of gasket material 08826-00080, or equivalent, to the gasket as shown.

➡The cylinder head must be installed within 3 minutes and the bolts tightened with 15 minutes. Otherwise, the silicone material must be removed and re-applied.

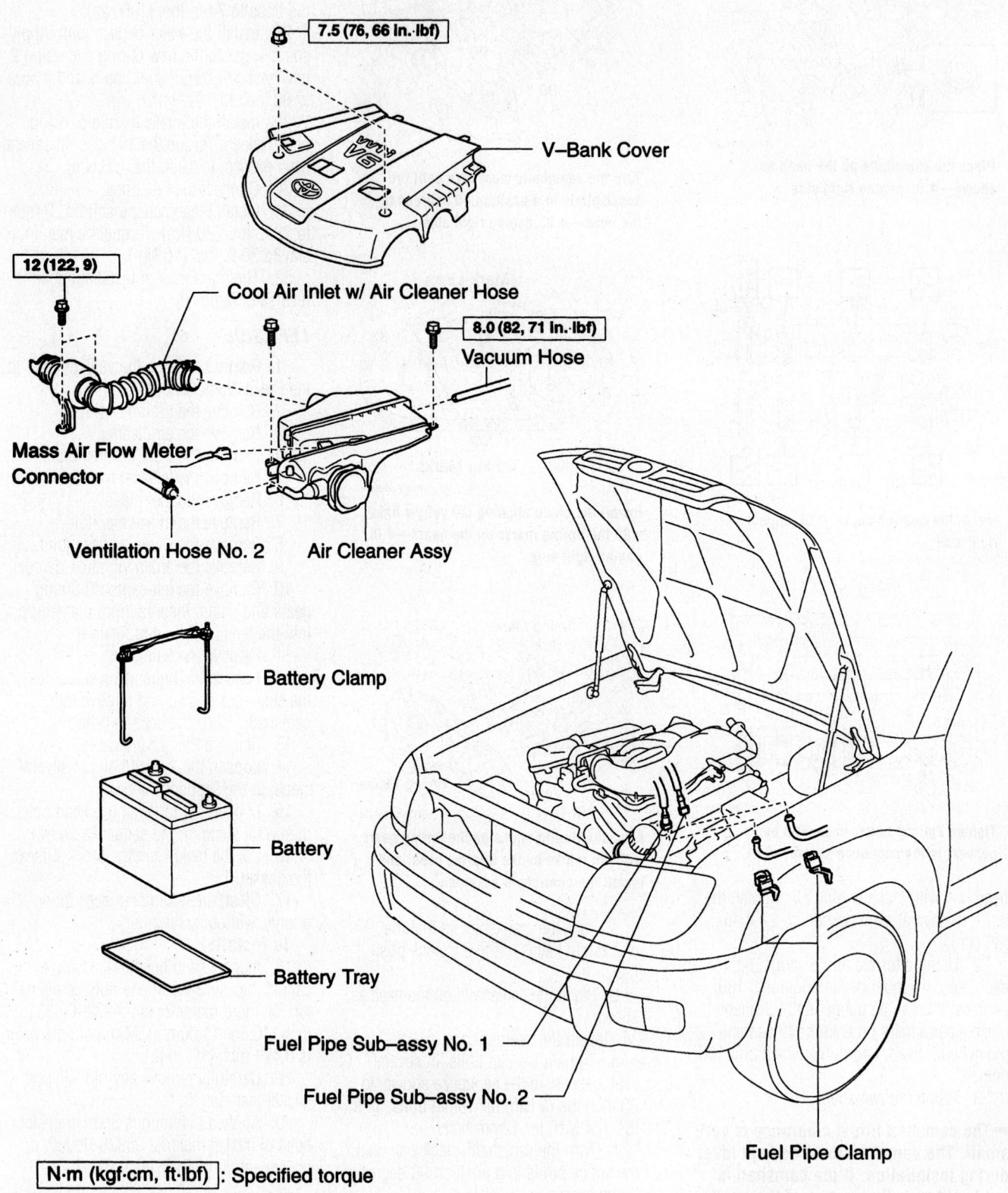

7.5 (76, 66 in.·lbf)

V–Bank Cover

12 (122, 9)

Cool Air Inlet w/ Air Cleaner Hose

8.0 (82, 71 in.·lbf)

Vacuum Hose

Mass Air Flow Meter
Connector

Ventilation Hose No. 2   Air Cleaner Assy

Battery Clamp

Battery

Battery Tray

Fuel Pipe Sub–assy No. 1

Fuel Pipe Sub–assy No. 2

Fuel Pipe Clamp

N·m (kgf·cm, ft·lbf) : Specified torque

67170-4RUN-G15

For left cylinder head removal, remove these parts—Part 1

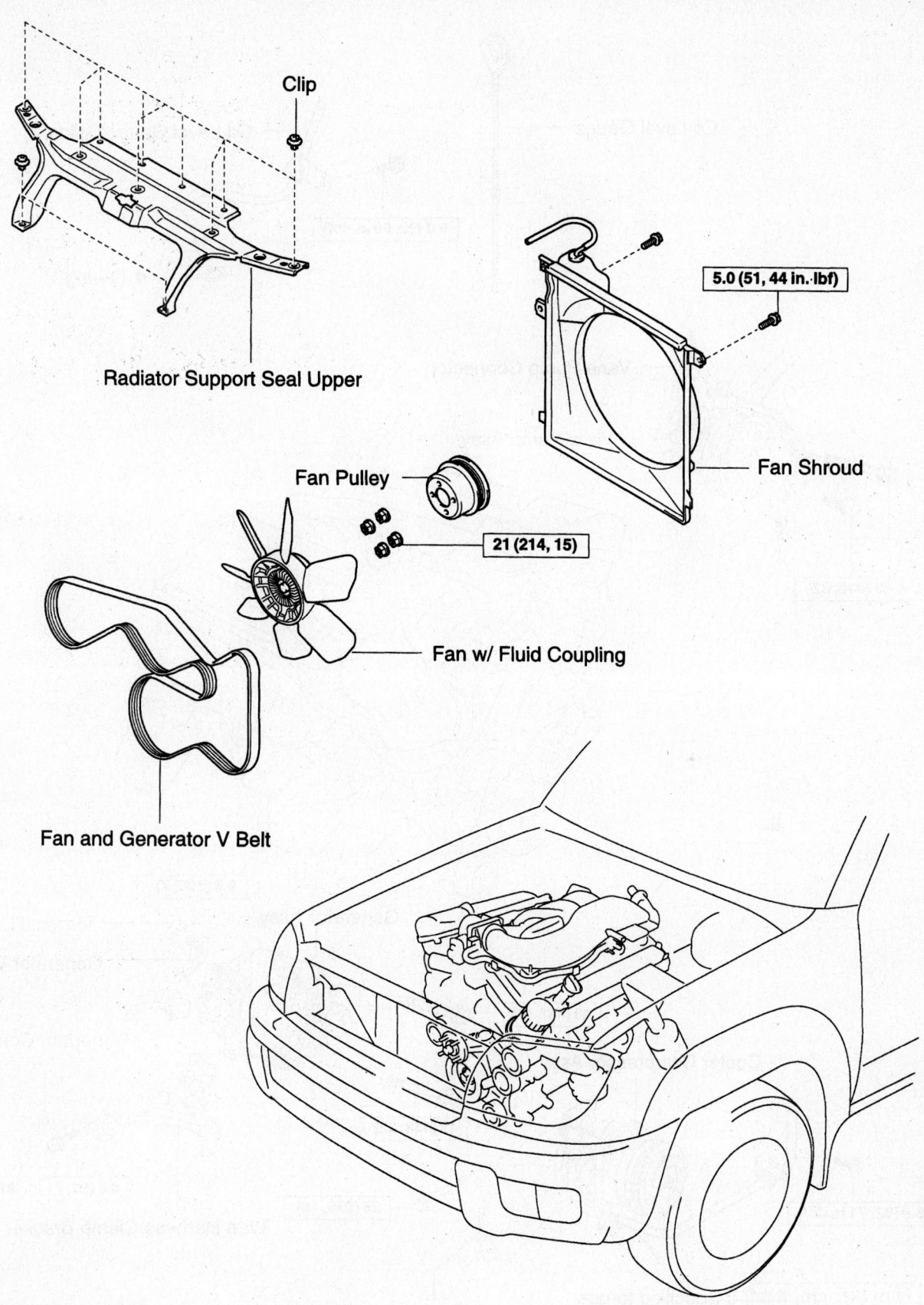

Clip

Radiator Support Seal Upper

5.0 (51, 44 in.·lbf)

Fan Shroud

Fan Pulley

21 (214, 15)

Fan w/ Fluid Coupling

Fan and Generator V Belt

N·m (kgf·cm, ft·lbf) : Specified torque

For left cylinder head removal, remove these parts—Part 2

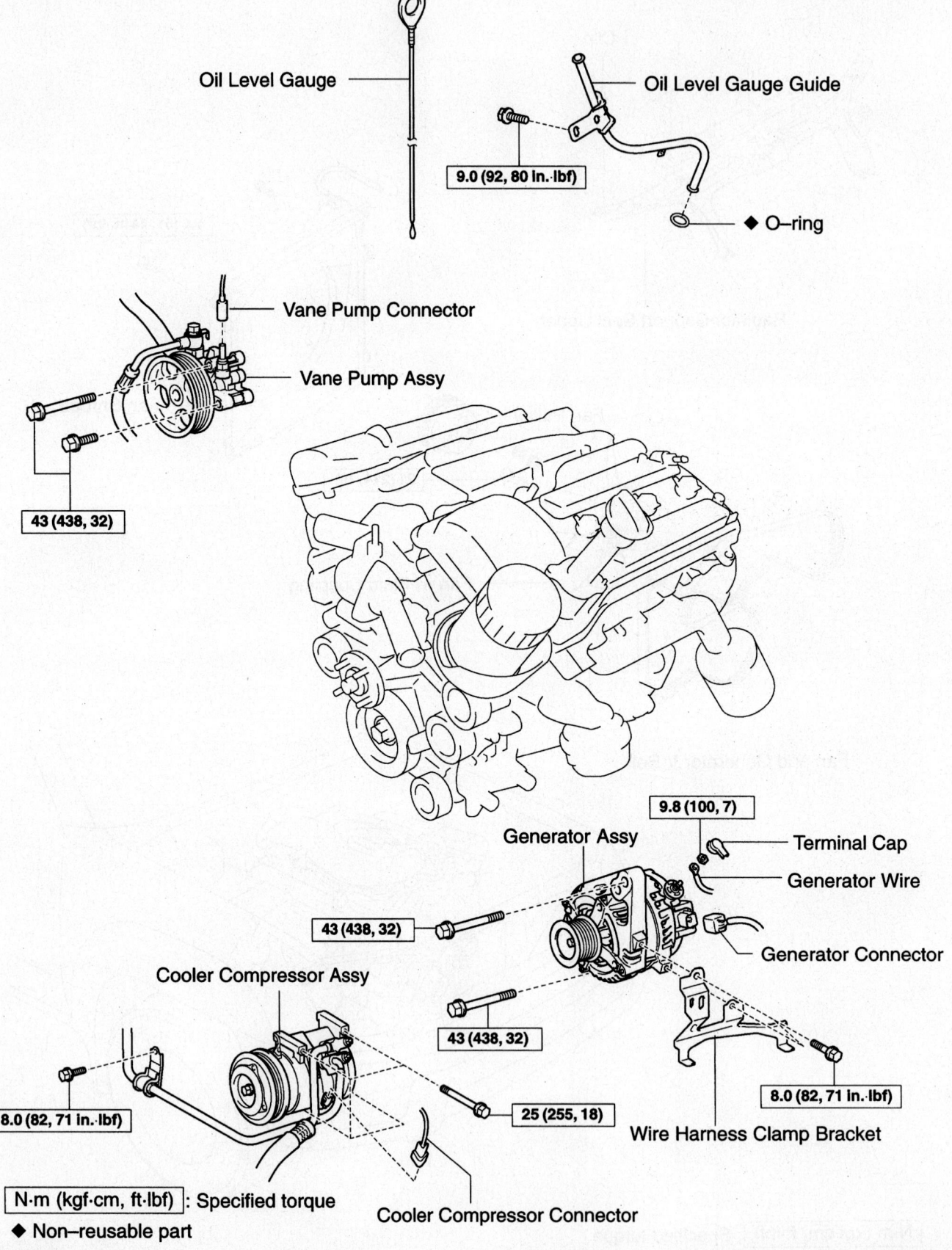

Oil Level Gauge

Oil Level Gauge Guide

9.0 (92, 80 in.·lbf)

◆ O–ring

Vane Pump Connector

Vane Pump Assy

43 (438, 32)

9.8 (100, 7)

Generator Assy

Terminal Cap

Generator Wire

43 (438, 32)

Generator Connector

Cooler Compressor Assy

43 (438, 32)

8.0 (82, 71 in.·lbf)

8.0 (82, 71 in.·lbf)

25 (255, 18)

Wire Harness Clamp Bracket

N·m (kgf·cm, ft·lbf) : Specified torque

Cooler Compressor Connector

◆ Non–reusable part

67170-4RUN-G27

For left cylinder head removal, remove these parts—Part 3

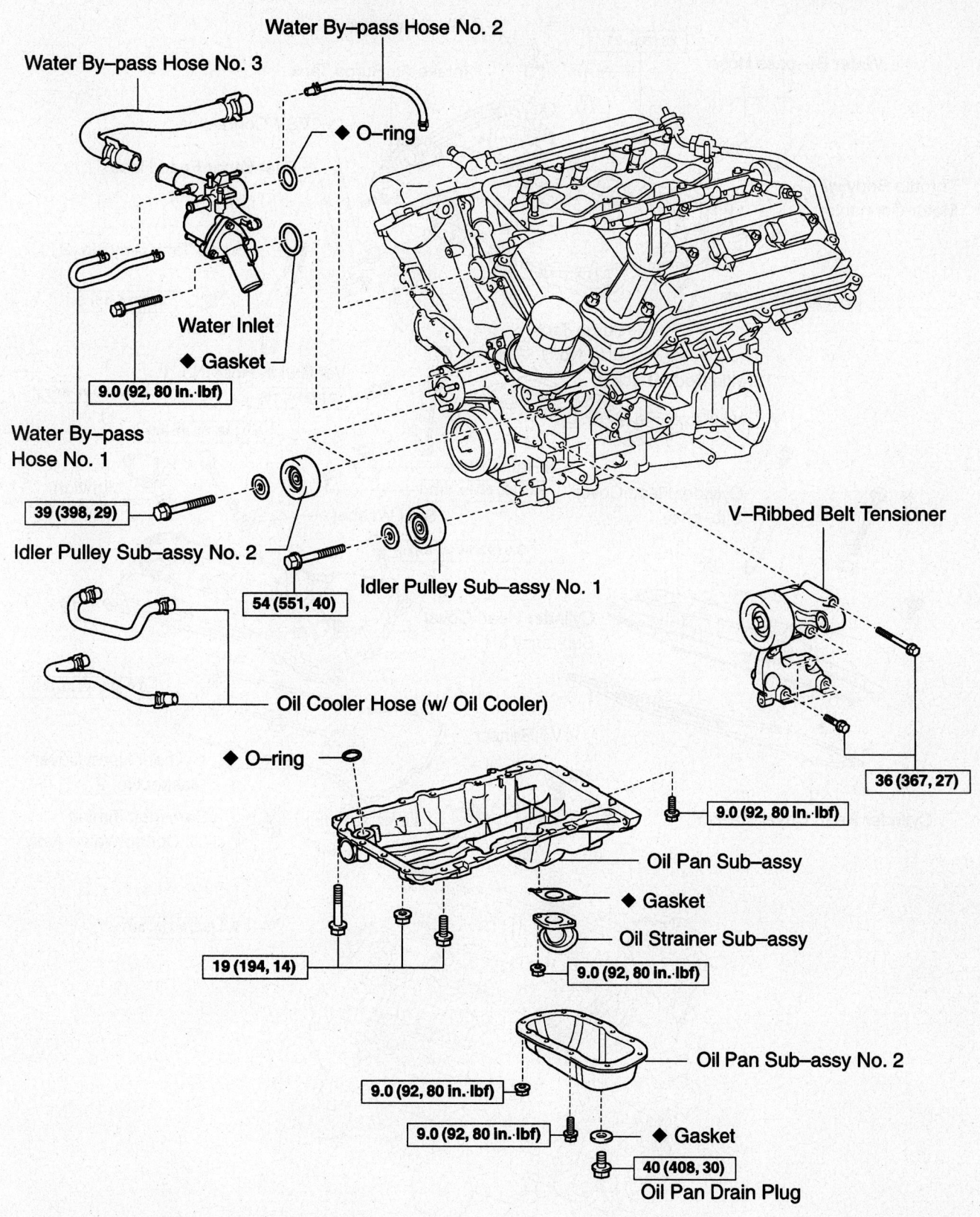

Water By-pass Hose No. 2

Water By-pass Hose No. 3

◆ O-ring

Water Inlet

◆ Gasket

9.0 (92, 80 in.·lbf)

Water By-pass Hose No. 1

39 (398, 29)

Idler Pulley Sub-assy No. 2

54 (551, 40)

Idler Pulley Sub-assy No. 1

Oil Cooler Hose (w/ Oil Cooler)

V-Ribbed Belt Tensioner

36 (367, 27)

◆ O-ring

9.0 (92, 80 in.·lbf)

Oil Pan Sub-assy

◆ Gasket

Oil Strainer Sub-assy

19 (194, 14)

9.0 (92, 80 in.·lbf)

Oil Pan Sub-assy No. 2

9.0 (92, 80 in.·lbf)

9.0 (92, 80 in.·lbf)

◆ Gasket

40 (408, 30)

Oil Pan Drain Plug

N·m (kgf·cm, ft·lbf) : Specified torque

◆ Non-reusable part

67170-4RUN-G10

For left cylinder head removal, remove these parts—Part 4

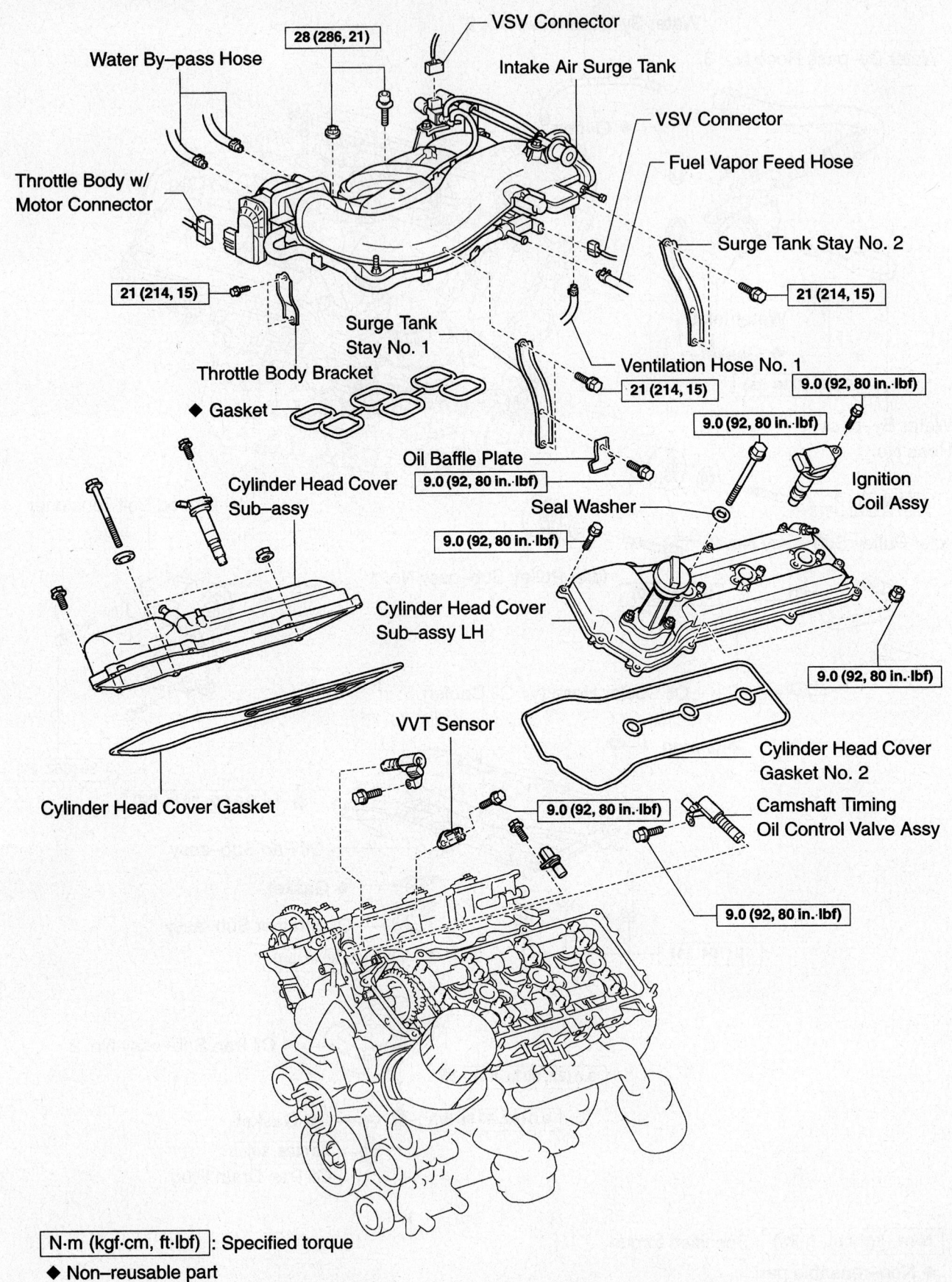

Water By–pass Hose

28 (286, 21)

VSV Connector

Intake Air Surge Tank

VSV Connector

Fuel Vapor Feed Hose

Throttle Body w/ Motor Connector

Surge Tank Stay No. 2

21 (214, 15)

21 (214, 15)

Surge Tank Stay No. 1

Throttle Body Bracket

Ventilation Hose No. 1

21 (214, 15)

9.0 (92, 80 in.·lbf)

◆ Gasket

9.0 (92, 80 in.·lbf)

Oil Baffle Plate

9.0 (92, 80 in.·lbf)

Ignition Coil Assy

Cylinder Head Cover Sub–assy

Seal Washer

9.0 (92, 80 in.·lbf)

Cylinder Head Cover Sub–assy LH

9.0 (92, 80 in.·lbf)

VVT Sensor

Cylinder Head Cover Gasket No. 2

Cylinder Head Cover Gasket

9.0 (92, 80 in.·lbf)

Camshaft Timing Oil Control Valve Assy

9.0 (92, 80 in.·lbf)

N·m (kgf·cm, ft·lbf) : Specified torque

◆ Non–reusable part

67170-4RUN-G11

For left cylinder head removal, remove these parts—Part 5

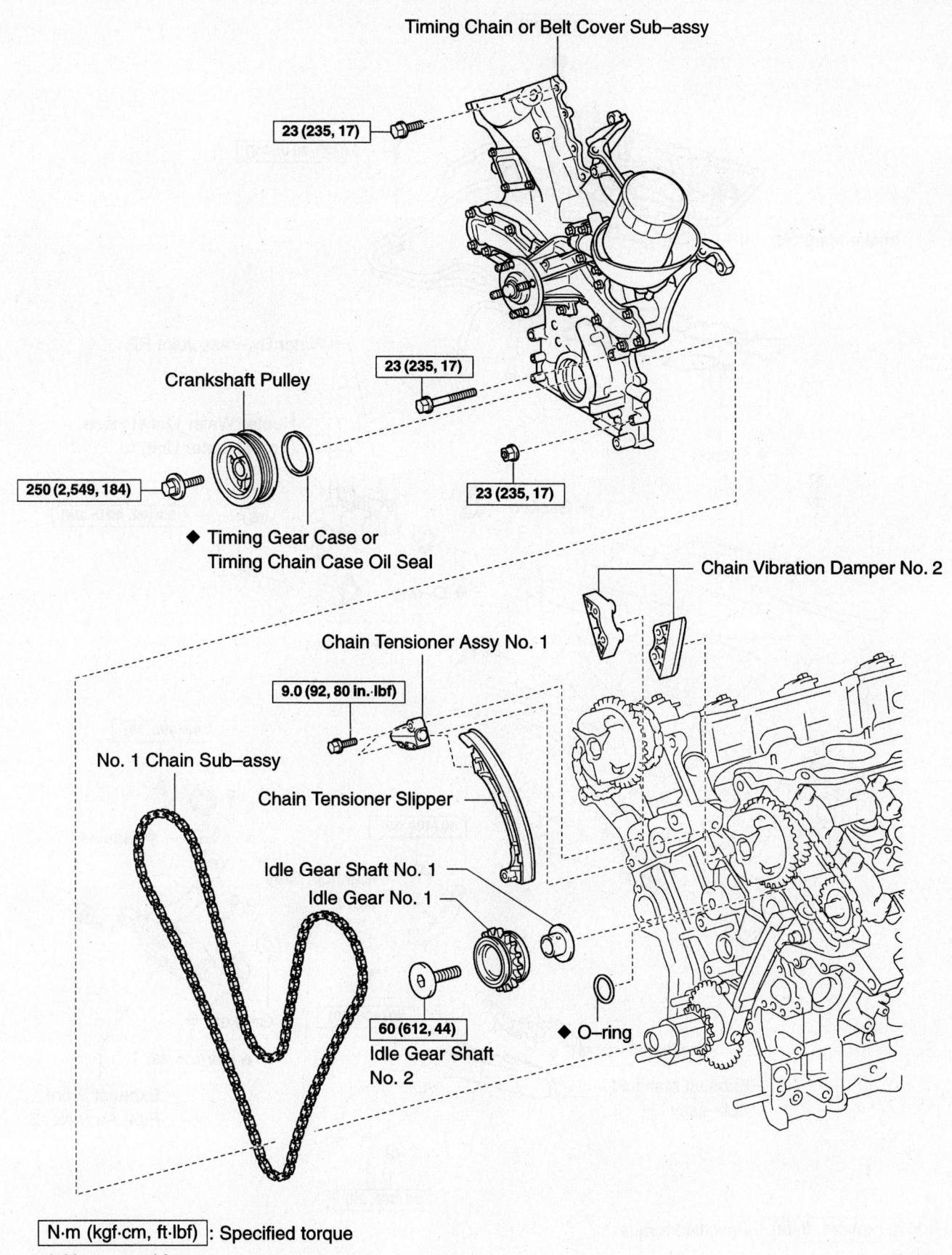

Timing Chain or Belt Cover Sub–assy

23 (235, 17)

Crankshaft Pulley

23 (235, 17)

250 (2,549, 184)

23 (235, 17)

◆ Timing Gear Case or
Timing Chain Case Oil Seal

Chain Vibration Damper No. 2

Chain Tensioner Assy No. 1

9.0 (92, 80 in.·lbf)

No. 1 Chain Sub–assy

Chain Tensioner Slipper

Idle Gear Shaft No. 1

Idle Gear No. 1

60 (612, 44)

Idle Gear Shaft
No. 2

◆ O–ring

N·m (kgf·cm, ft·lbf) : Specified torque

◆ Non–reusable part

67170-4RUN-G12

For left cylinder head removal, remove these parts—Part 6

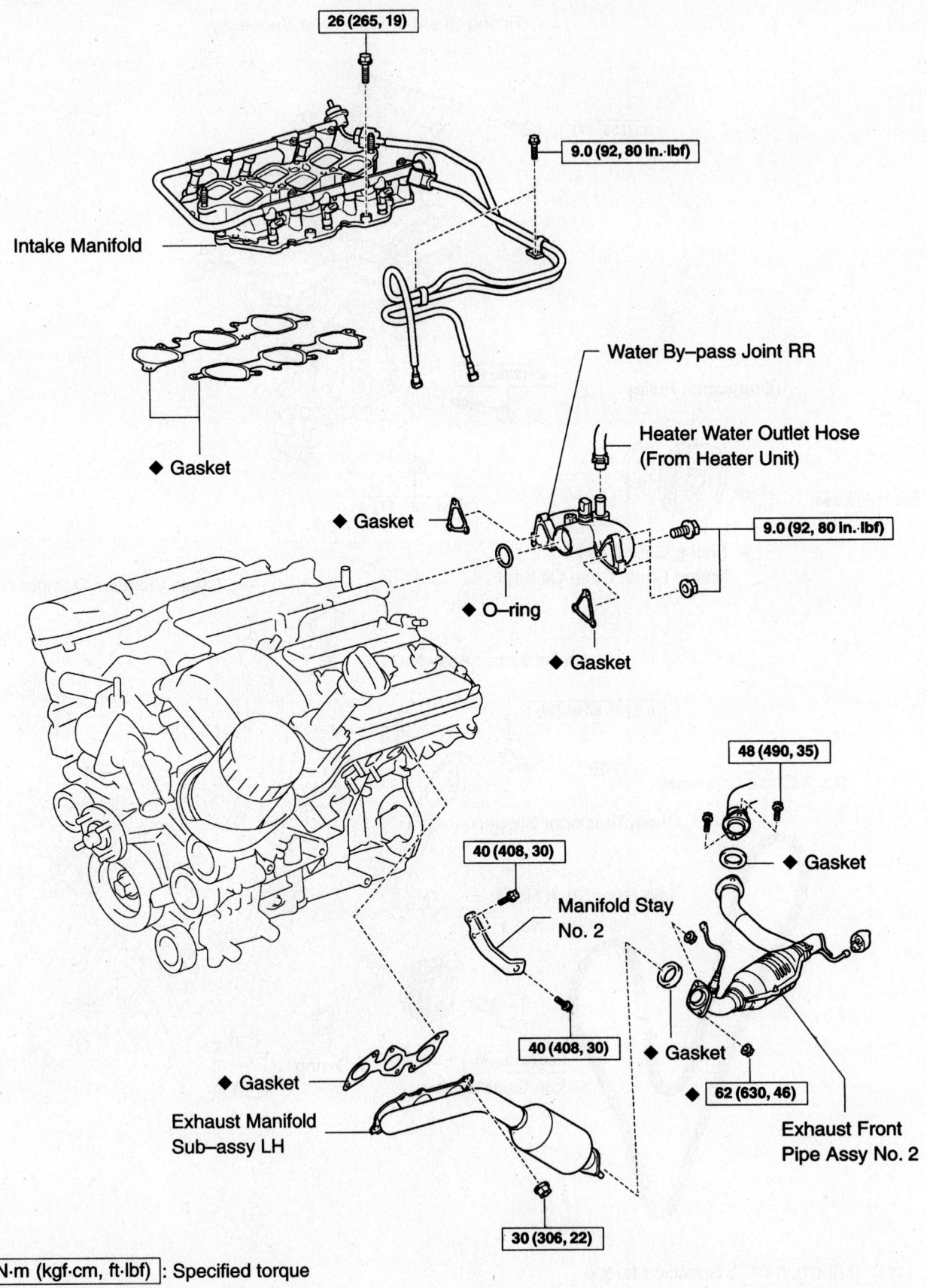

26 (265, 19)

9.0 (92, 80 in.·lbf)

Intake Manifold

◆ Gasket

Water By–pass Joint RR

Heater Water Outlet Hose
(From Heater Unit)

◆ Gasket

9.0 (92, 80 in.·lbf)

◆ O–ring

◆ Gasket

48 (490, 35)

◆ Gasket

40 (408, 30)

Manifold Stay
No. 2

40 (408, 30)

◆ Gasket

62 (630, 46)

◆ Gasket

Exhaust Manifold
Sub–assy LH

Exhaust Front
Pipe Assy No. 2

30 (306, 22)

N·m (kgf·cm, ft·lbf) : Specified torque

◆ Non–reusable part

67170-4RUN-G28

For left cylinder head removal, remove these parts—Part 7

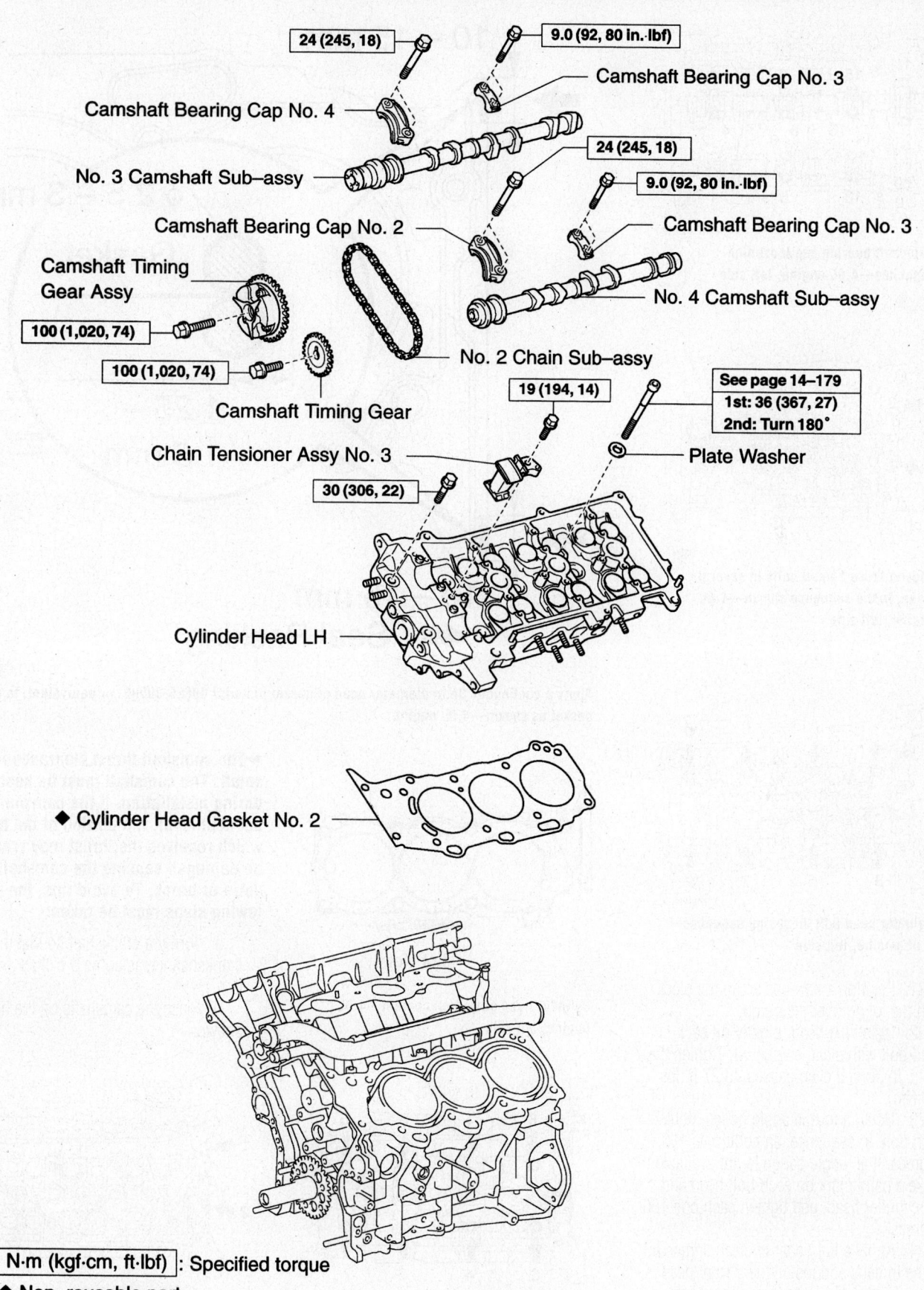

24 (245, 18)

9.0 (92, 80 in.·lbf)

Camshaft Bearing Cap No. 4

Camshaft Bearing Cap No. 3

No. 3 Camshaft Sub–assy

24 (245, 18)

9.0 (92, 80 in.·lbf)

Camshaft Bearing Cap No. 2

Camshaft Bearing Cap No. 3

Camshaft Timing Gear Assy

No. 4 Camshaft Sub–assy

100 (1,020, 74)

100 (1,020, 74)

Camshaft Timing Gear

No. 2 Chain Sub–assy

See page 14–179
1st: 36 (367, 27)
2nd: Turn 180°

19 (194, 14)

Chain Tensioner Assy No. 3

Plate Washer

30 (306, 22)

Cylinder Head LH

◆ Cylinder Head Gasket No. 2

N·m (kgf·cm, ft·lbf) : Specified torque

◆ Non–reusable part

67170-4RUN-G29

For left cylinder head removal, remove these parts—Part 8

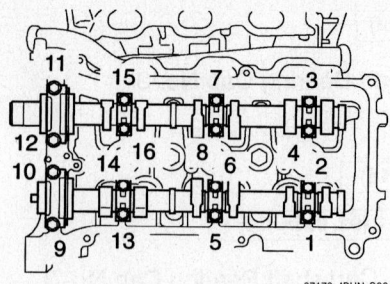

Camshaft bearing cap loosening sequence—4.0L engine, left side

67170-4RUN-G30

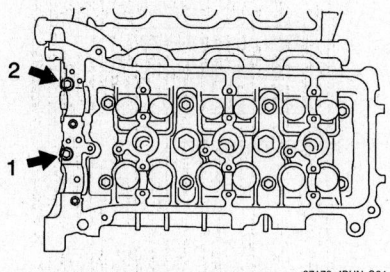

Loosen these 2 head bolts in several steps, in the sequence shown—4.0L engine, left side

67170-4RUN-G31

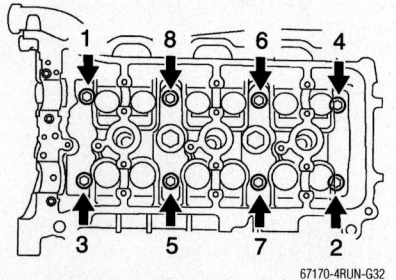

Cylinder head bolt loosening sequence—4.0L engine, left side

67170-4RUN-G32

21. Position a new gasket on the block with the lot number facing up.

22. Install the head. Lightly oil each head bolt with clean engine oil. Tighten the bolts, in several even passes, to 27 ft. lbs. (36 Nm).

23. Using a torque angle gauge, tighten each bolt, in sequence, an additional 180 degrees. If an angle gauge is not available, make a paint mark on each bolt head and the cylinder head, and tighten each bolt 180 degrees.

24. Apply a light coat of clean engine oil to the threads and install the 2 front head bolts. Tighten the bolts in the sequence shown to 22 ft. lbs. (30 Nm).

25. Install the camshafts.

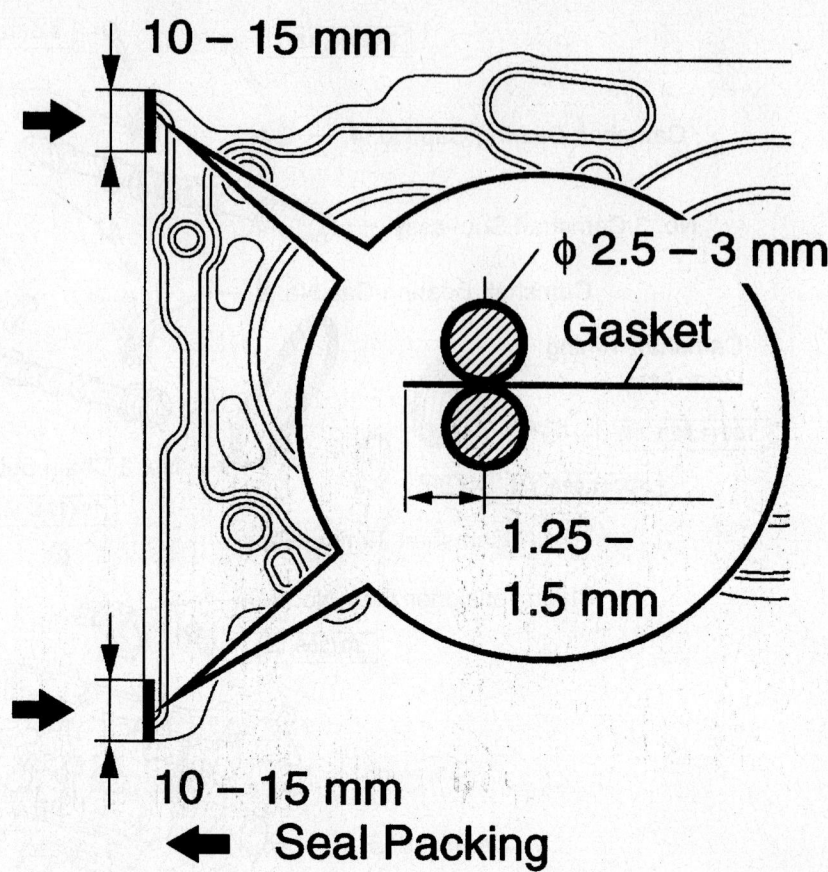

10 – 15 mm

φ 2.5 – 3 mm

Gasket

1.25 – 1.5 mm

10 – 15 mm

← Seal Packing

67170-4RUN-G18

Apply a continuous 3mm diameter bead of gasket material 08826-00080, or equivalent, to the gasket as shown—4.0L engine

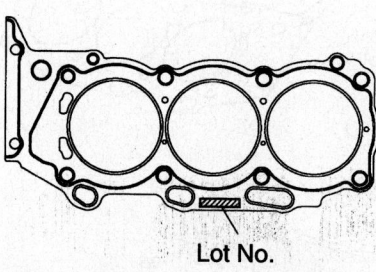

Lot No.

67170-4RUN-G33

Cylinder head gasket placement—4.0L engine, left side

➡The camshaft thrust clearance is very small. The camshaft must be kept level during installation. If the camshaft is not kept level, the portion of the head which receives the thrust may crack or be damaged causing the camshaft to seize or break. To avoid this, the following steps must be taken:

   a. Turn the crankshaft so that the crankshaft key is at the 9 o'clock position.

   b. Place the camshafts on the head as shown.

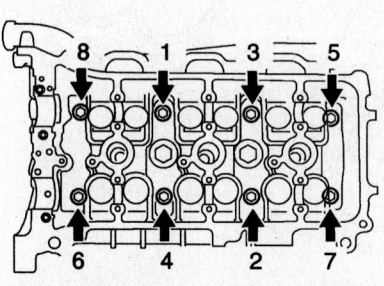

67170-4RUN-G34

Cylinder head bolt tightening sequence—4.0L engine

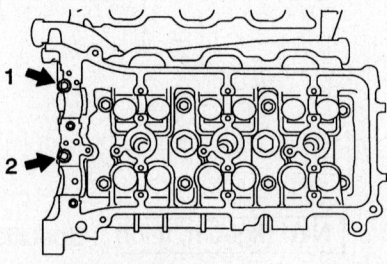

67170-4RUN-G35

Front head bolt torque sequence—4.0L engine, left side

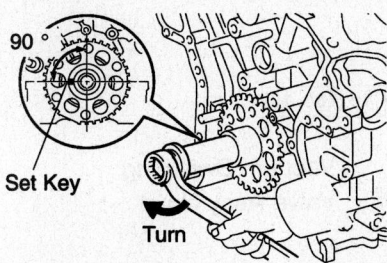

Turn the crankshaft so that the crankshaft key is at the 9 o'clock position—4.0L engine, left side

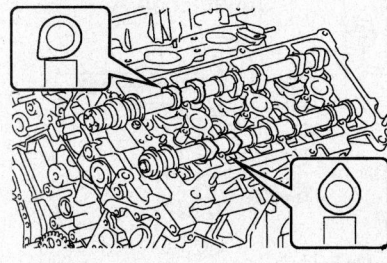

Place the camshafts on the head as shown—4.0L engine, left side

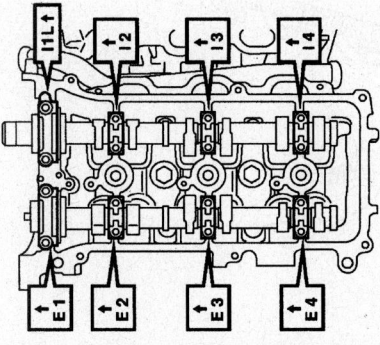

Install the bearing caps—4.0L engine, left side

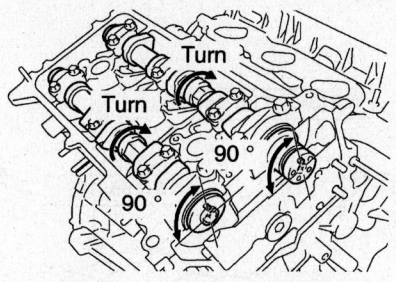

Turn the camshafts clockwise until the knock pin is in a position 90 degrees to the head

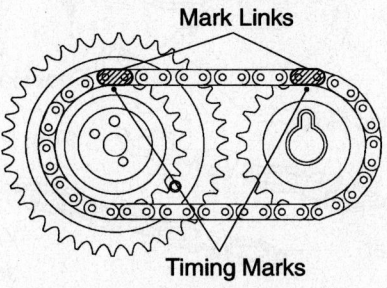

Install the chain aligning the yellow links with the timing marks on the gears

26. Install the bearing caps.

a. Tighten the cap bolts, in several even passes, in the sequence shown, to 80 inch lbs (9 Nm) for 10mm bolts; 18 ft. lbs. (24 Nm) for 12mm bolts.

b. Turn the camshafts clockwise until the knock pin is in a position 90 degrees to the head.

27. Install the tensioner. Torque to 14 ft. lbs. (19 Nm).

28. Install the chain aligning the yellow links with the timing marks on the gears.

29. Align the timing marks on the timing

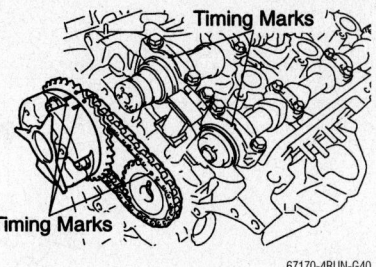

Align the timing marks on the timing gears with the marks on the bearing caps

gears with the marks on the bearing caps, and install the gear/chain assembly. Torque the bolts to 74 ft. lbs. (100 Nm).

30. Install the water bypass joint. Apply soapy water to the new O-ring and using 2 new gaskets. Torque the 2 bolts and 4 nuts to 80 inch lbs. (9 Nm).

31. Install the intake manifold, using new gaskets. Torque the 10 bolts, in several even passes, to 19 ft. lbs. (26 Nm).

32. Connect the fuel lines.

33. Install the exhaust manifold. Torque to 22 ft. lbs. (30 Nm). Torque the manifold stay to 30 ft. lbs. (40 Nm).

34. The remainder of installation is the reverse of removal.

4.7L Engine

RIGHT SIDE

1. Before servicing the vehicle, refer to the Precautions Section.

2. Remove the timing belt.

3. Remove the camshaft.

4. Remove the transmission oil filler tube.

5. Remove the engine oil dipstick tube.

6. Disconnect the fuel lines.

7. Disconnect the vapor hose.

8. Disconnect the injector harness.

9. Disconnect the water bypass hoses.

10. Remove the intake manifold.

11. Remove the water inlet housing.

12. Remove the front water bypass joint.

13. Remove the water bypass assembly.

14. Remove the rear water bypass joint.

15. Remove the ground cable.

16. Remove the front exhaust pipe.

17. Loosen the 10 head bolts, evenly and in several passes, in the sequence shown. Remove the bolts and washers.

18. Clean the mating surfaces thoroughly, without scratching.

To install

19. Check the thread diameter of each head bolt. Outer diameter should be 0.3862–0.3921 inch (9.810–0.960mm). Minimum should be 0.3819 inch (9.700mm).

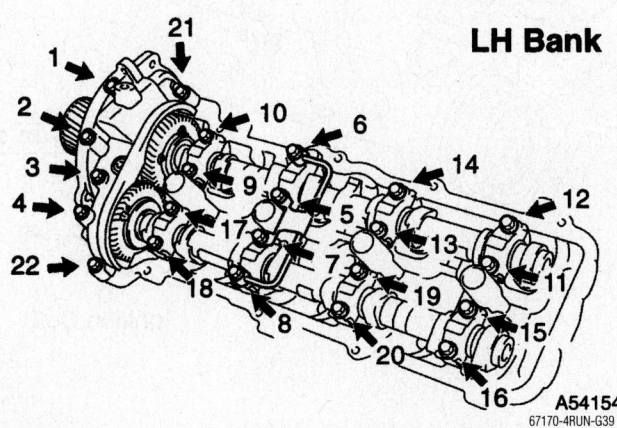

Tighten the cap bolts, in several even passes, in the sequence shown—4.0L engine, left side

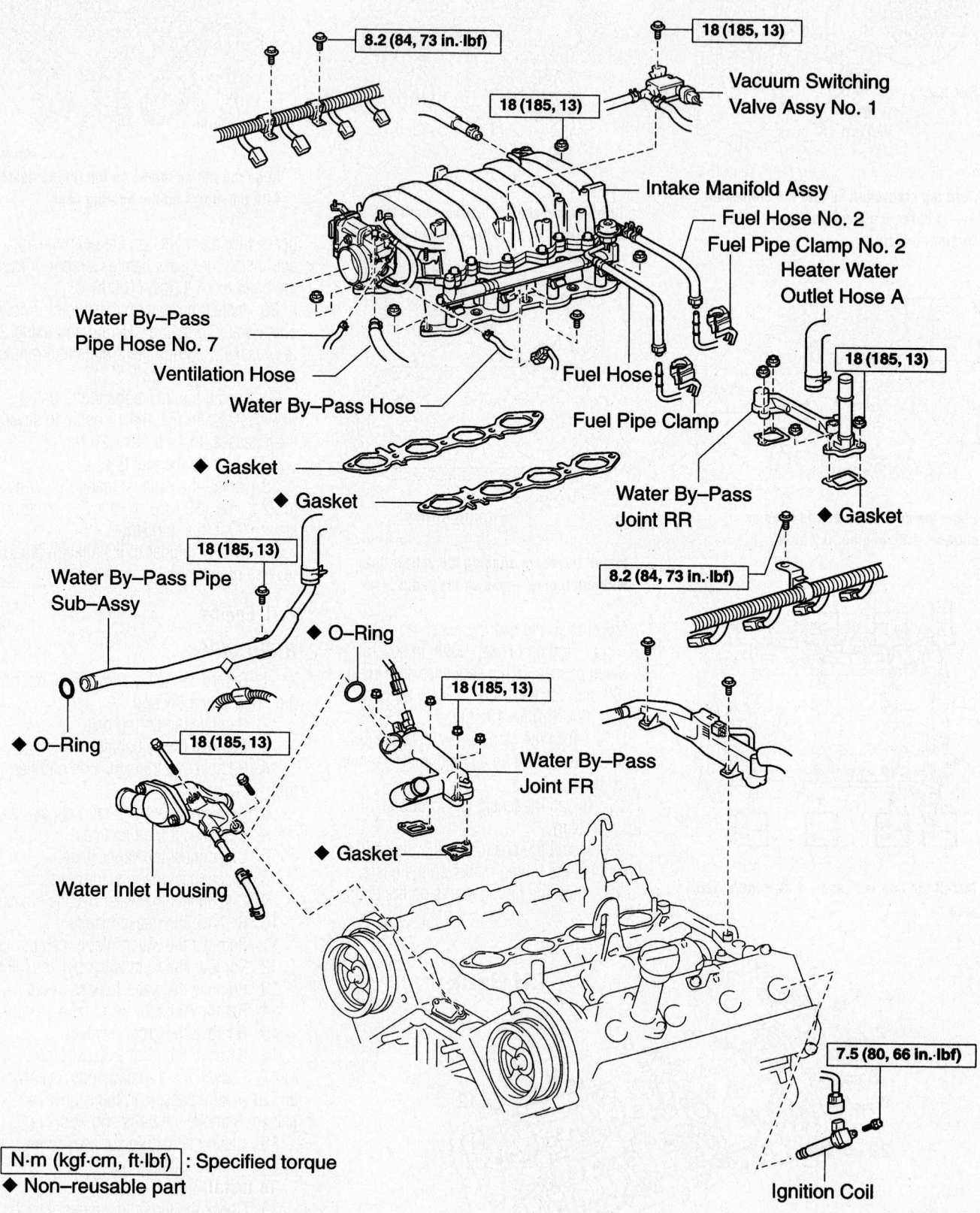

8.2 (84, 73 in.·lbf)

18 (185, 13)

Vacuum Switching
Valve Assy No. 1

18 (185, 13)

Intake Manifold Assy

Fuel Hose No. 2

Fuel Pipe Clamp No. 2

Heater Water
Outlet Hose A

Water By–Pass
Pipe Hose No. 7

18 (185, 13)

Ventilation Hose

Water By–Pass Hose

Fuel Hose

Fuel Pipe Clamp

◆ Gasket

◆ Gasket

Water By–Pass
Joint RR

◆ Gasket

18 (185, 13)

8.2 (84, 73 in.·lbf)

Water By–Pass Pipe
Sub–Assy

◆ O–Ring

◆ O–Ring

18 (185, 13)

18 (185, 13)

Water By–Pass
Joint FR

Water Inlet Housing

◆ Gasket

7.5 (80, 66 in.·lbf)

Ignition Coil

N·m (kgf·cm, ft·lbf) : Specified torque
◆ Non–reusable part

67170-4RUN-G41

For 4.7L engine right cylinder head removal, remove these parts—Part 1

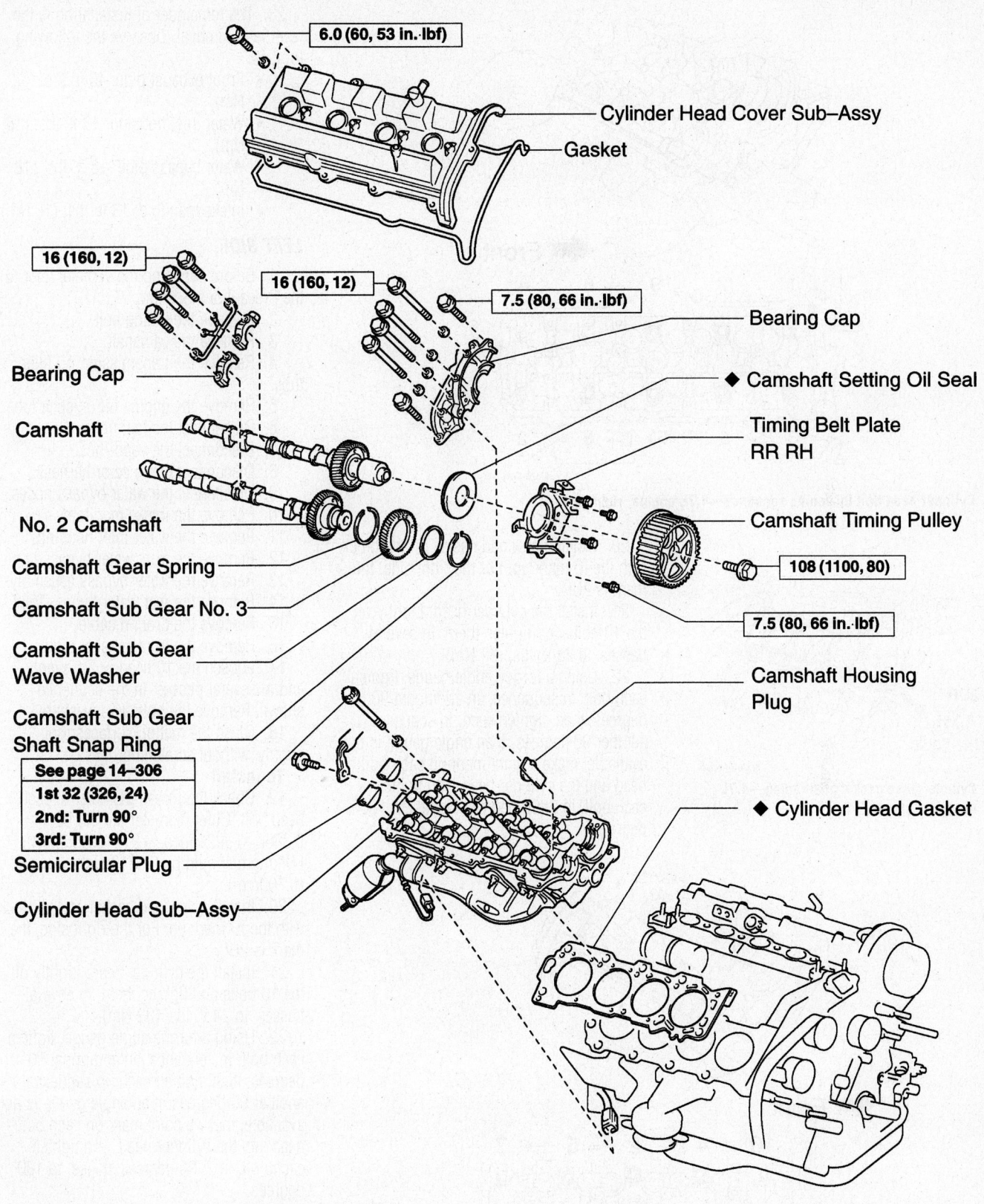

6.0 (60, 53 in.·lbf)

Cylinder Head Cover Sub–Assy

Gasket

16 (160, 12)

16 (160, 12)

7.5 (80, 66 in.·lbf)

Bearing Cap

Bearing Cap

◆ Camshaft Setting Oil Seal

Camshaft

Timing Belt Plate RR RH

No. 2 Camshaft

Camshaft Timing Pulley

Camshaft Gear Spring

108 (1100, 80)

Camshaft Sub Gear No. 3

Camshaft Sub Gear
Wave Washer

7.5 (80, 66 in.·lbf)

Camshaft Sub Gear
Shaft Snap Ring

Camshaft Housing
Plug

See page 14–306
1st 32 (326, 24)
2nd: Turn 90°
3rd: Turn 90°

◆ Cylinder Head Gasket

Semicircular Plug

Cylinder Head Sub–Assy

N·m (kgf·cm, ft·lbf) : Specified torque

◆ Non–reusable part

For 4.7L engine right cylinder head removal, remove these parts—Part 2

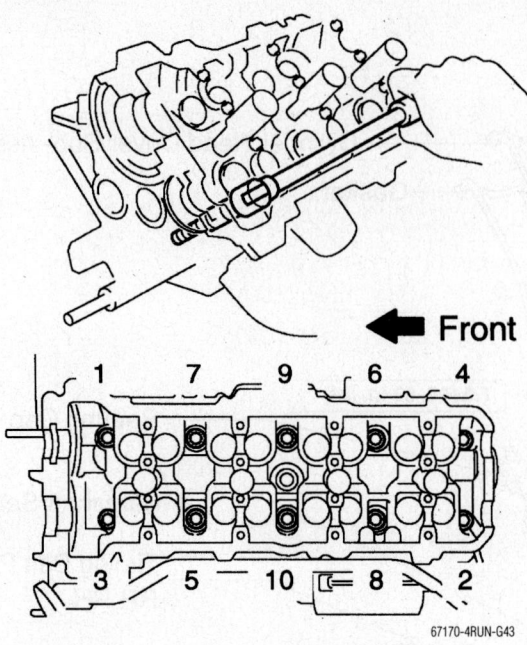

← Front

Cylinder head bolt loosening sequence—4.7L engine, right side

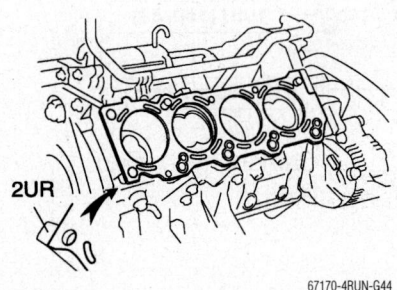

Cylinder head gasket positioning—4.7L engine, right side

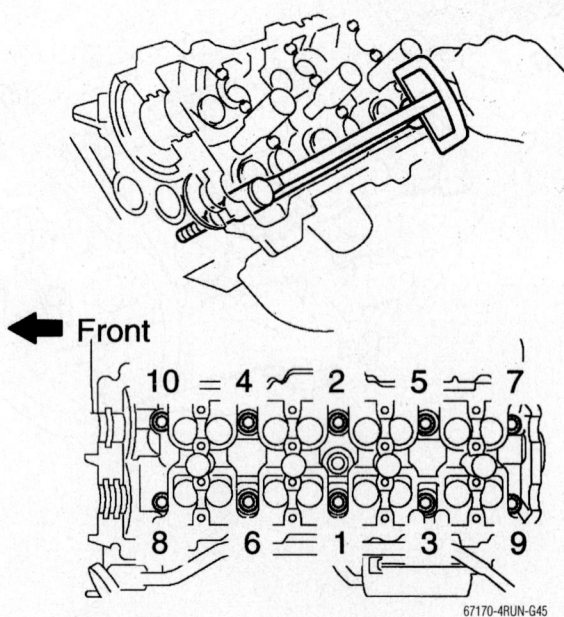

← Front

Cylinder head bolt torque sequence—4.7L engine, right side

20. Install a new had gasket on the block with the ID mark up. For the right side, the mark is 2UR.

21. Install the cylinder head. Lightly oil the 10 bolts and tighten them, in several passes, to 24 ft. lbs. (32 Nm).

22. Using a torque angle gauge, tighten each bolt, in sequence, an additional 90 degrees, then, tighten each, in sequence another 90 degrees. If an angle gauge is not available, make a paint mark on each bolt head and the cylinder head, and tighten each bolt, in 2, 90 degrees stages, to 180 degrees.

23. The remainder of installation is the reverse of removal. Observe the following torques:

- Front exhaust pipe: 46 ft. lbs. (62 Nm)
- Water inlet housing: 13 ft. lbs. (18 Nm)
- Water bypass pipe: 13 ft. lbs. (18 Nm)
- Intake manifold: 13 ft. lbs. (18 Nm)

LEFT SIDE

1. Before servicing the vehicle, refer to the Precautions Section.
2. Remove the timing belt.
3. Remove the camshaft.
4. Remove the transmission oil filler tube.
5. Remove the engine oil dipstick tube.
6. Disconnect the fuel lines.
7. Disconnect the vapor hose.
8. Disconnect the injector harness.
9. Disconnect the water bypass hoses.
10. Remove the intake manifold.
11. Remove the water inlet housing.
12. Remove the front water bypass joint.
13. Remove the water bypass assembly.
14. Remove the rear water bypass joint.
15. Remove the ground cable.
16. Remove the front exhaust pipe.
17. Loosen the 10 head bolts, evenly and in several passes, in the sequence shown. Remove the bolts and washers.
18. Clean the mating surfaces thoroughly, without scratching.

To install

19. Check the thread diameter of each head bolt. Outer diameter should be 0.3862–0.3921 inch (9.810–0.960mm). Minimum should be 0.3819 inch (9.700mm).

20. Install a new had gasket on the block with the ID mark up. For the right side, the mark is 2UL.

21. Install the cylinder head. Lightly oil the 10 bolts and tighten them, in several passes, to 24 ft. lbs. (32 Nm).

22. Using a torque angle gauge, tighten each bolt, in sequence, an additional 90 degrees, then, tighten each, in sequence another 90 degrees. If an angle gauge is not available, make a paint mark on each bolt head and the cylinder head, and tighten each bolt, in 2, 90 degrees stages, to 180 degrees.

23. The remainder of installation is the reverse of removal. Observe the following torques:

- Front exhaust pipe: 46 ft. lbs. (62 Nm)
- Water inlet housing: 13 ft. lbs. (18 Nm)

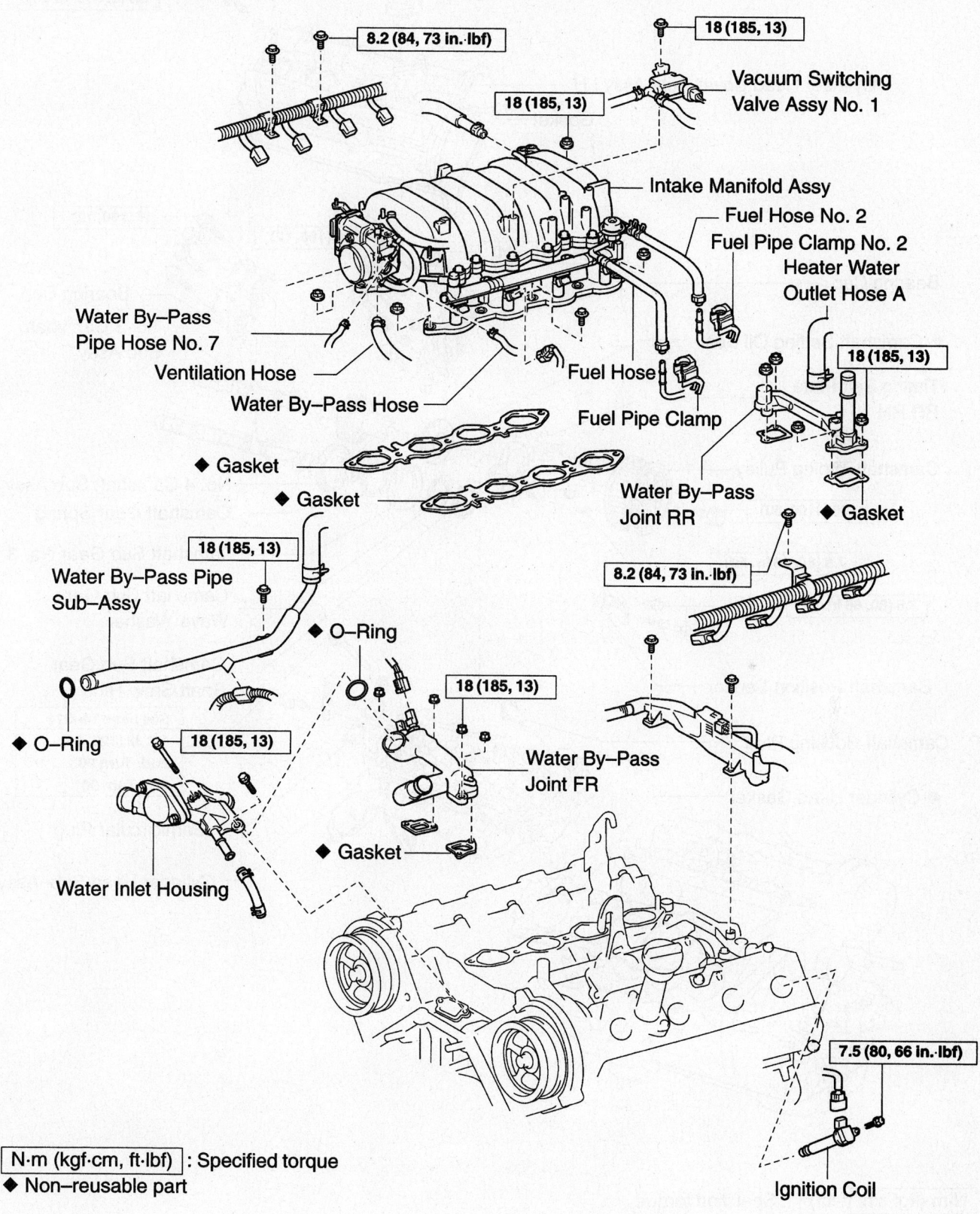

8.2 (84, 73 in.·lbf)

18 (185, 13)

18 (185, 13)

Vacuum Switching
Valve Assy No. 1

Intake Manifold Assy

Fuel Hose No. 2
Fuel Pipe Clamp No. 2
Heater Water
Outlet Hose A

Water By–Pass
Pipe Hose No. 7

Ventilation Hose

Water By–Pass Hose

Fuel Hose

Fuel Pipe Clamp

18 (185, 13)

◆ Gasket

◆ Gasket

Water By–Pass
Joint RR

◆ Gasket

18 (185, 13)

Water By–Pass Pipe
Sub–Assy

◆ O–Ring

8.2 (84, 73 in.·lbf)

◆ O–Ring

18 (185, 13)

18 (185, 13)

Water By–Pass
Joint FR

◆ Gasket

Water Inlet Housing

7.5 (80, 66 in.·lbf)

Ignition Coil

N·m (kgf·cm, ft·lbf) : Specified torque
◆ Non–reusable part

67170-4RUN-G41

For 4.7L engine left cylinder head removal, remove these parts—Part 1

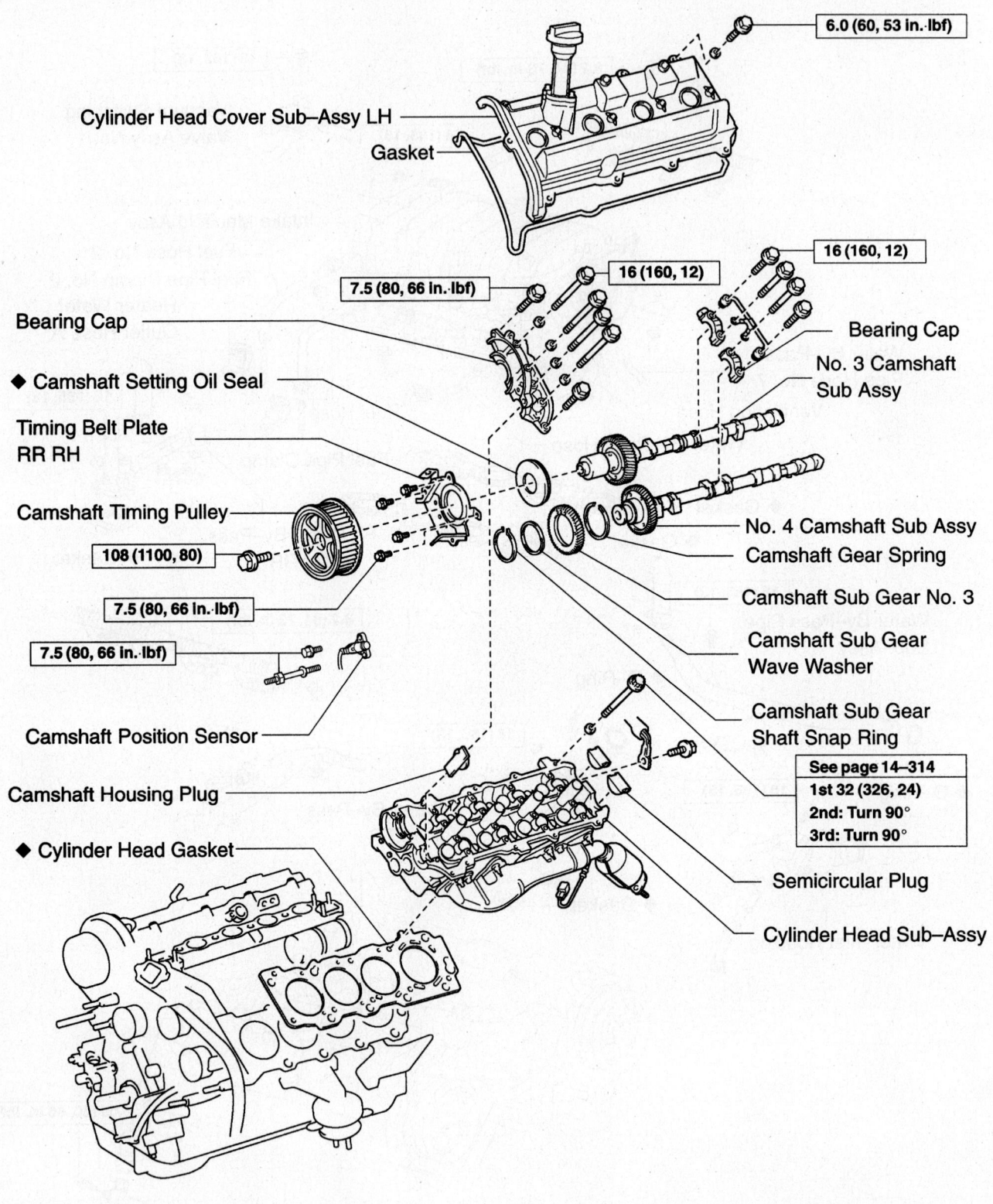

Cylinder Head Cover Sub–Assy LH

Gasket

6.0 (60, 53 in.·lbf)

16 (160, 12)

16 (160, 12)

7.5 (80, 66 in.·lbf)

Bearing Cap

◆ Camshaft Setting Oil Seal

Timing Belt Plate RR RH

Camshaft Timing Pulley

108 (1100, 80)

7.5 (80, 66 in.·lbf)

7.5 (80, 66 in.·lbf)

Camshaft Position Sensor

Camshaft Housing Plug

◆ Cylinder Head Gasket

Bearing Cap

No. 3 Camshaft Sub Assy

No. 4 Camshaft Sub Assy

Camshaft Gear Spring

Camshaft Sub Gear No. 3

Camshaft Sub Gear Wave Washer

Camshaft Sub Gear Shaft Snap Ring

See page 14–314
1st 32 (326, 24)
2nd: Turn 90°
3rd: Turn 90°

Semicircular Plug

Cylinder Head Sub–Assy

N·m (kgf·cm, ft·lbf) : Specified torque

◆ Non–reusable part

67170-4RUN-G46

For 4.7L engine left cylinder head removal, remove these parts—Part 2

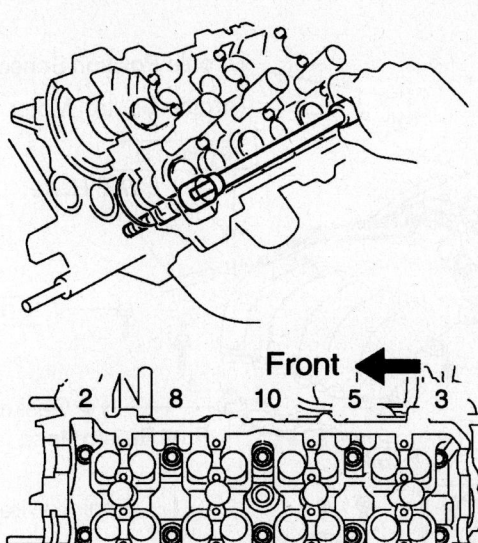

Cylinder head bolt loosening sequence—4.7L engine, left side

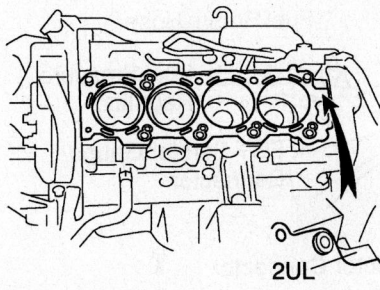

Cylinder head gasket positioning—4.7L engine, left side

- Water bypass pipe: 13 ft. lbs. (18 Nm)
- Intake manifold: 13 ft. lbs. (18 Nm)

Intake Manifold

REMOVAL & INSTALLATION

3.4L

1. Before servicing the vehicle, refer to the Precautions Section.
2. Disconnect the negative battery cable.

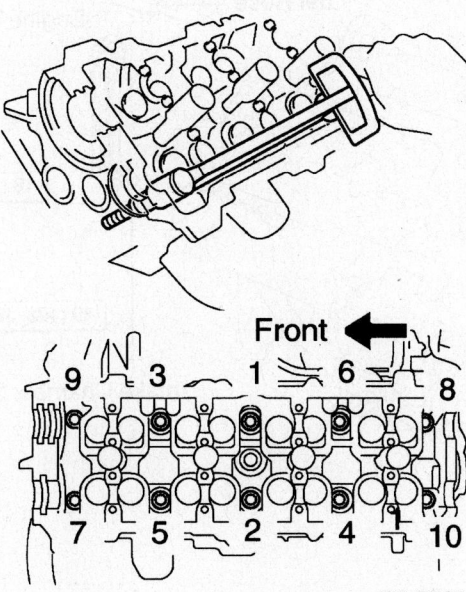

Cylinder head bolt torque sequence—4.7L engine, left side

3. Relieve the fuel system pressure.
4. Remove the engine undercover.
5. Drain the cooling system.
6. Remove or disconnect the following:
   - Air cleaner cap
   - Mass Air Flow (MAF) meter and the resonator
   - Actuator cable with the bracket, if equipped with cruise control
   - Accelerator cable
   - Throttle cable, if equipped with automatic transmission
7. Disconnect the following hoses:
   - Heater hose
   - Brake booster vacuum hose
   - Evaporative Emissions (EVAP) hose
   - Automatic Disconnecting Differential (ADD) vacuum hose, for 4-Wheel drive
   - Fuel inlet and fuel return hose
8. Remove or disconnect the following:
   - Spark plug wires, with the ignition coils
   - Intake chamber stay
   - No. 2 timing belt cover
   - Air intake chamber assembly
   - Throttle Position (TP) sensor connector
   - Idle Air Control (IAC) valve connector
   - Positive Crankcase Ventilation (PCV) hoses
   - Water bypass hoses
   - Air assist hose from the throttle body
   - Intake air connector
   - Engine wiring harness
   - Fuel return hose
   - Vacuum hose, from the fuel pressure regulator
   - Ground strap, from the intake air connector
   - Data Link Connector 1 (DLC1), from the bracket
   - 6 injector connectors
   - Engine Coolant Temperature (ECT) sensor and sender gauge connectors
   - Engine wiring harness protector from the cylinder head
   - Fuel pressure regulator
   - Intake manifold assembly
   - Intake manifold stay
   - Intake manifold, delivery pipes and the injectors assembly with the gaskets

To install:

9. Install or connect the following:
   - New gaskets
   - Intake manifold assembly. Tighten the bolts and nuts to 13 ft. lbs. (18 Nm).

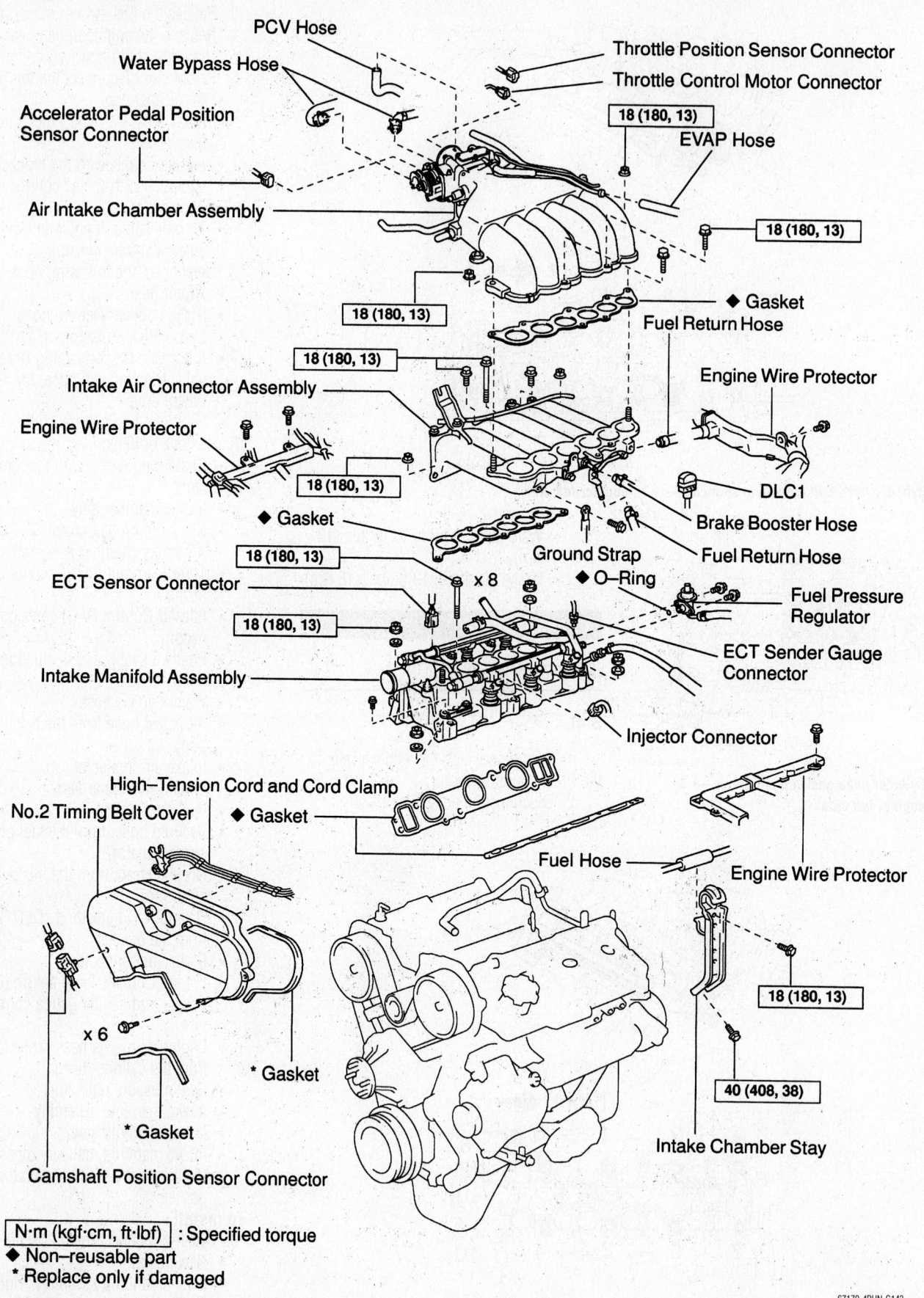

PCV Hose

Water Bypass Hose

Accelerator Pedal Position
Sensor Connector

Air Intake Chamber Assembly

Throttle Position Sensor Connector

Throttle Control Motor Connector

18 (180, 13)

EVAP Hose

18 (180, 13)

18 (180, 13)

◆ Gasket

Fuel Return Hose

Intake Air Connector Assembly

Engine Wire Protector

18 (180, 13)

18 (180, 13)

Engine Wire Protector

DLC1

Brake Booster Hose

Fuel Return Hose

◆ Gasket

18 (180, 13)

Ground Strap

◆ O–Ring

ECT Sensor Connector

x 8

Fuel Pressure
Regulator

18 (180, 13)

ECT Sender Gauge
Connector

Intake Manifold Assembly

Injector Connector

High–Tension Cord and Cord Clamp

No.2 Timing Belt Cover

◆ Gasket

Fuel Hose

Engine Wire Protector

18 (180, 13)

x 6

* Gasket

40 (408, 38)

* Gasket

Intake Chamber Stay

Camshaft Position Sensor Connector

N·m (kgf·cm, ft·lbf) : Specified torque
◆ Non–reusable part
* Replace only if damaged

67170-4RUN-G143

Upper and lower intake manifold and related parts—3.4L engine

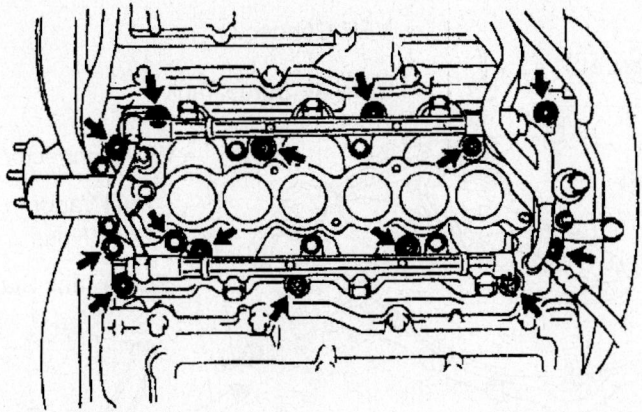

7924YG38

Intake manifold bolts and nuts—3.4L engine

- Intake manifold stay. Tighten the bolts to 14 ft. lbs. (18 Nm).
- Fuel pressure regulator
- Engine wiring harness to the cylinder head, by installing the 3 bolts
- 3 engine wiring harness clamps
- 6 injector connectors
- ECT sender gauge connector
- ECT sensor connector
- Intake manifold. Tighten the bolts and nuts to 14 ft. lbs. (18.5 Nm).
- DLC1 to the bracket on the intake manifold
- Ground strap to the intake manifold, by installing the bolt
- Brake booster vacuum hose, to the intake air connector
- 2 fuel return hoses
- Engine wiring harness to the intake manifold
- Air intake chamber assembly to the engine. Tighten the bolts and nuts to 14 ft. lbs. (18.5 Nm).
- Intake chamber stay. Tighten the bolts to 30 ft. lbs. (40 Nm).
- New O-ring to the oil filler tube
- Oil filler tube end into the tube hole in the oil pan
- Oil filler tube and No. 1 throttle cable clamp
- No. 2 timing belt cover. Tighten the bolts to 80 inch lbs. (9 Nm).
- PCV hoses
- Water bypass hoses
- Air assist hose to the throttle body
- IAC valve connector
- TP sensor connector
- Brake booster vacuum hose
- EVAP hose
- Automatic Disconnecting Differential (ADD) vacuum hose, for 4-Wheel drive
- Fuel inlet and fuel return hose

- 3 spark plug wire clamps to the No. 2 timing belt cover
- CMP connector to the No. 2 timing belt cover
- Spark plug wires with the ignition coils
- Heater hose
- Actuator cable with the bracket, if equipped with cruise control
- Accelerator cable
- Throttle cable, if equipped with automatic transmission
- MAF meter, resonator and the air cleaner cap
- Engine undercover
- Negative battery cable

10. Refill the cooling system to the correct level.

11. Start the engine and check for leaks.

4.0L Engine

1. Before servicing the vehicle, refer to the Precautions Section.
2. Drain the cooling system.
3. Remove the V-bank cover.
4. Remove the air cleaner assembly.
5. Remove the alternator.
6. Remove the power steering pump. Don't disconnect the lines.
7. Remove the A/C compressor. Don't disconnect the lines.
8. Disconnect the fuel lines.
9. Remove the water inlet and outlet hoses.
10. Remove the water bypass hoses.
11. Remove the fuel vapor hose.
12. Remove the vent hose.
13. Disconnect the 2 VSV connectors.
14. Disconnect the wiring at the throttle body.
15. Remove the throttle body bracket.
16. Remove the oil baffle plate.
17. Remove the upper manifold supports (stays).

18. Remove the 2 nuts and 4 bolts and lift off the upper manifold. Discard the gasket.
19. Disconnect the fuel injectors.
20. Disconnect the fuel supply lines.
21. Remove the intake manifold brace (stay).
22. Remove the 10 bolts and lift off the intake manifold. Discard the gaskets.
23. Installation is the reverse of removal. There is no torque sequence for the intake manifold bolts. Torque them evenly and in several passes, to 19 ft. lbs. (26 Nm). Torque the intake manifold brace to 30 ft. lbs. (40 Nm).
24. Observe the following torques:
   - Upper intake manifold (surge tank) nuts and bolts: 21 ft. lbs. (28 Nm)
   - Throttle body: 15 ft. lbs. (21 Nm)
   - V-bank cover: 66 inch lbs. (7.5 Nm)
25. Refill the cooling system to the correct level.
26. Start the engine and check for leaks.

4.7L Engine

1. Before servicing the vehicle, refer to the Precautions Section.
2. Drain the cooling system.
3. Relieve the fuel system pressure.
4. Remove or disconnect the following:
   - Negative battery cable
   - Accelerator cable
   - Throttle Position (TP) sensor connector
   - Accelerator pedal position sensor
   - Throttle motor connector
   - Evaporative Emissions (EVAP) vacuum switching valve connector
   - Fuel injector connectors
   - Engine Coolant Temperature (ECT) sensor connector
   - ETC gauge sender connector
   - Heated Oxygen (HO$_2$S) sensor connectors
   - Fuel pressure regulator vacuum hose
   - Positive Crankcase Ventilation (PCV) valve and hose
   - EVAP hoses
   - Power steering vacuum hoses
   - Water bypass hose
   - Engine control wiring harness clamps
   - Cylinder head ground cables
   - Intake manifold wire harness protector
   - EVAP pipe
   - Engine appearance cover brackets
   - Intake manifold

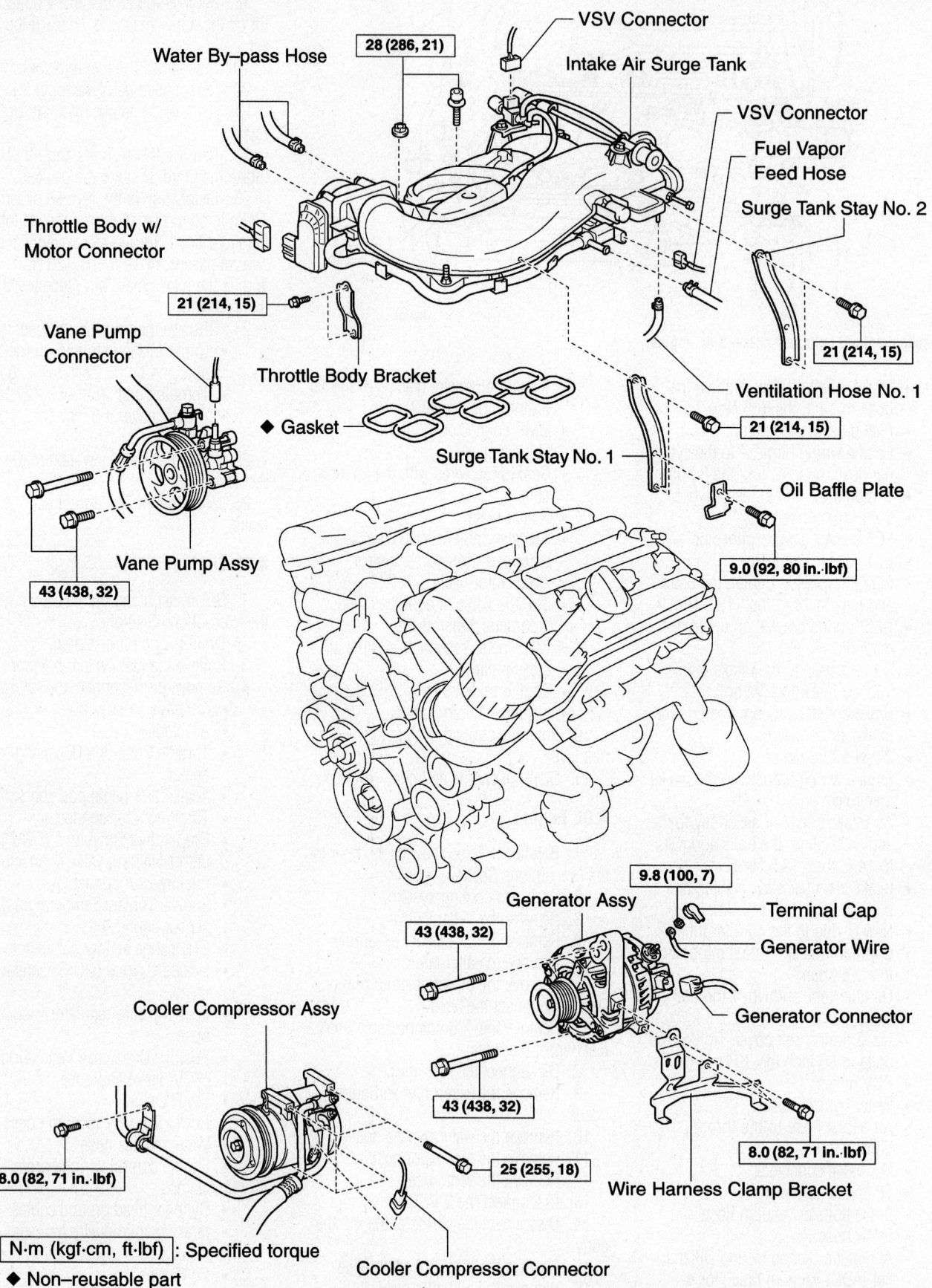

VSV Connector

Water By–pass Hose

`28 (286, 21)`

Intake Air Surge Tank

VSV Connector

Fuel Vapor Feed Hose

Surge Tank Stay No. 2

Throttle Body w/ Motor Connector

`21 (214, 15)`

Vane Pump Connector

Throttle Body Bracket

`21 (214, 15)`

Ventilation Hose No. 1

◆ Gasket

`21 (214, 15)`

Surge Tank Stay No. 1

Oil Baffle Plate

Vane Pump Assy

`9.0 (92, 80 in.·lbf)`

`43 (438, 32)`

`9.8 (100, 7)`

Generator Assy

Terminal Cap

`43 (438, 32)`

Generator Wire

Cooler Compressor Assy

Generator Connector

`43 (438, 32)`

`8.0 (82, 71 in.·lbf)`

`25 (255, 18)`

`8.0 (82, 71 in.·lbf)`

Wire Harness Clamp Bracket

`N·m (kgf·cm, ft·lbf)` : Specified torque

◆ Non–reusable part

Cooler Compressor Connector

67170-4RUN-G50

Upper intake manifold (intake air surge tank) and related parts—4.0L engine

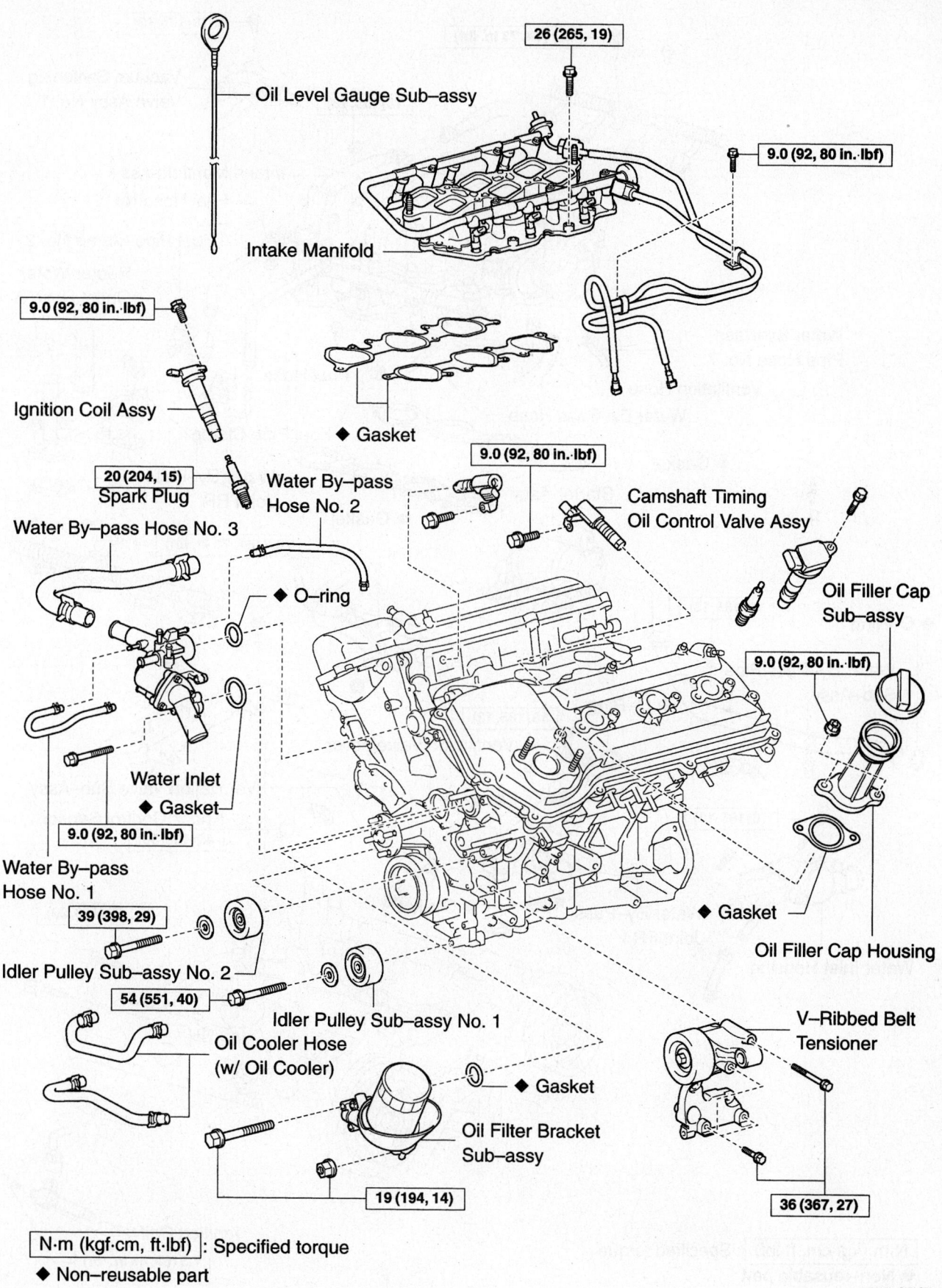

26 (265, 19)

Oil Level Gauge Sub–assy

9.0 (92, 80 in.·lbf)

Intake Manifold

9.0 (92, 80 in.·lbf)

Ignition Coil Assy

◆ Gasket

9.0 (92, 80 in.·lbf)

20 (204, 15)
Spark Plug

Water By–pass
Hose No. 2

Camshaft Timing
Oil Control Valve Assy

Water By–pass Hose No. 3

Oil Filler Cap
Sub–assy

◆ O–ring

9.0 (92, 80 in.·lbf)

Water Inlet

◆ Gasket

9.0 (92, 80 in.·lbf)

Water By–pass
Hose No. 1

39 (398, 29)

◆ Gasket

Oil Filler Cap Housing

Idler Pulley Sub–assy No. 2

V–Ribbed Belt
Tensioner

54 (551, 40)

Idler Pulley Sub–assy No. 1

Oil Cooler Hose
(w/ Oil Cooler)

◆ Gasket

Oil Filter Bracket
Sub–assy

19 (194, 14)

36 (367, 27)

N·m (kgf·cm, ft·lbf) : Specified torque

◆ Non–reusable part

Lower intake manifold assembly and related parts—4.0L engine

67170-4RUN-G51

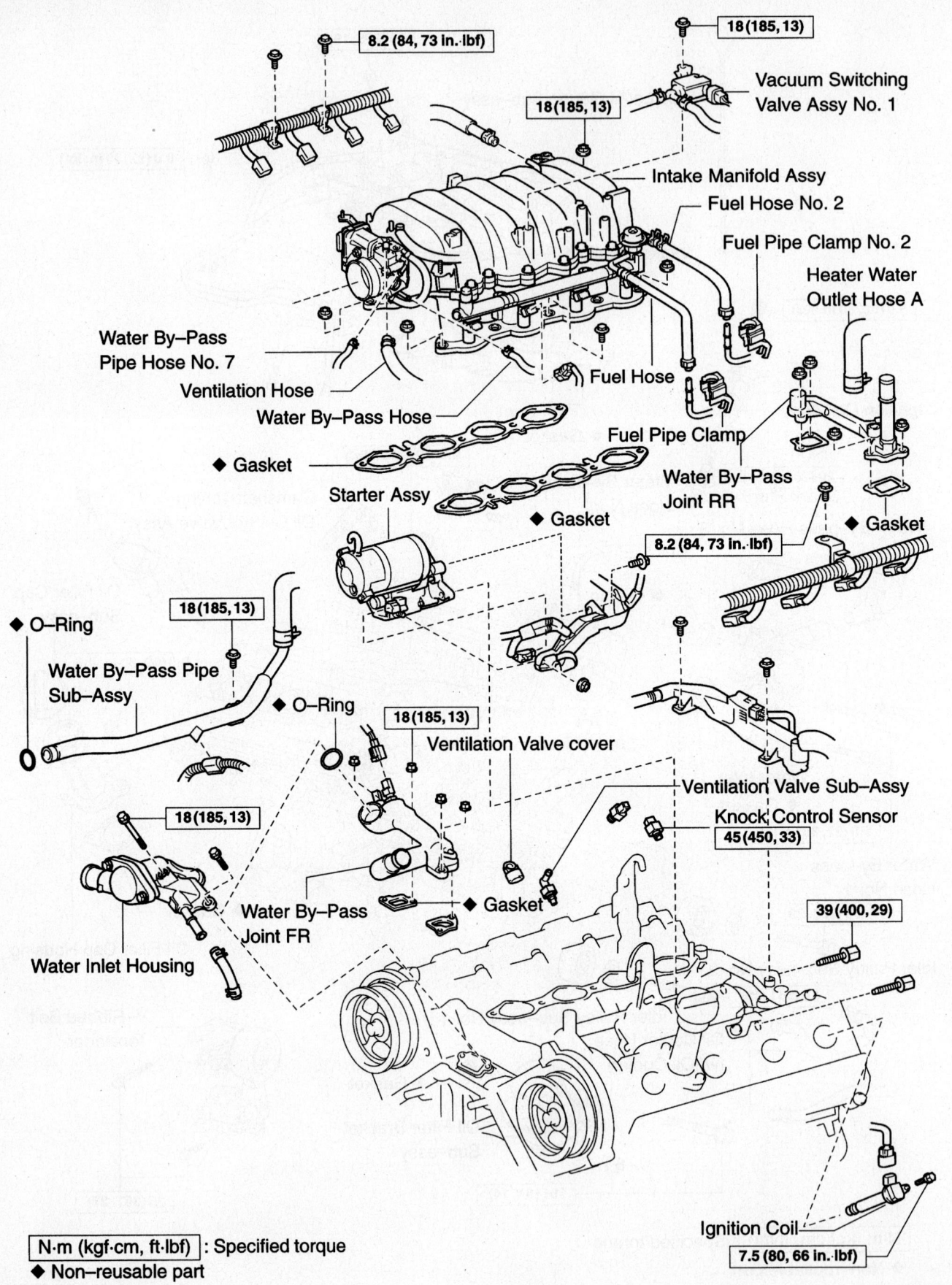

8.2 (84, 73 in.·lbf)

18 (185, 13)

Vacuum Switching Valve Assy No. 1

18 (185, 13)

Intake Manifold Assy

Fuel Hose No. 2

Fuel Pipe Clamp No. 2

Heater Water Outlet Hose A

Water By–Pass Pipe Hose No. 7

Ventilation Hose

Water By–Pass Hose

Fuel Hose

◆ Gasket

Starter Assy

◆ Gasket

Fuel Pipe Clamp

Water By–Pass Joint RR

◆ Gasket

8.2 (84, 73 in.·lbf)

◆ O–Ring

18 (185, 13)

Water By–Pass Pipe Sub–Assy

◆ O–Ring

18 (185, 13)

Ventilation Valve cover

Ventilation Valve Sub–Assy

Knock Control Sensor

45 (450, 33)

39 (400, 29)

18 (185, 13)

Water Inlet Housing

Water By–Pass Joint FR

◆ Gasket

Ignition Coil

7.5 (80, 66 in.·lbf)

N·m (kgf·cm, ft·lbf) : Specified torque

◆ Non–reusable part

Intake manifold and related parts—4.7L engine

67170-4RUN-G52

To install:

5. Install or connect the following:
   - Intake manifold. Tighten the fasteners to 13 ft. lbs. (18 Nm).
   - Engine appearance cover brackets
   - EVAP pipe
   - Intake manifold wire harness protector
   - Cylinder head ground cables
   - Engine control wiring harness clamps
   - Water bypass hose
   - Power steering vacuum hoses
   - EVAP hoses

   - PCV valve and hose
   - Fuel pressure regulator vacuum hose
   - HO$_2$S sensor connectors
   - ETC gauge sender connector
   - ECT sensor connector
   - Fuel injector connectors
   - EVAP vacuum switching valve connector
   - Throttle motor connector
   - Accelerator pedal position sensor
   - TP sensor connector
   - Accelerator cable
   - Negative battery cable

6. Refill the cooling system to the correct level.
7. Start the engine and check for leaks.

Exhaust Manifold

REMOVAL & INSTALLATION

3.4L Engine

1. Before servicing the vehicle, refer to the Precautions Section.

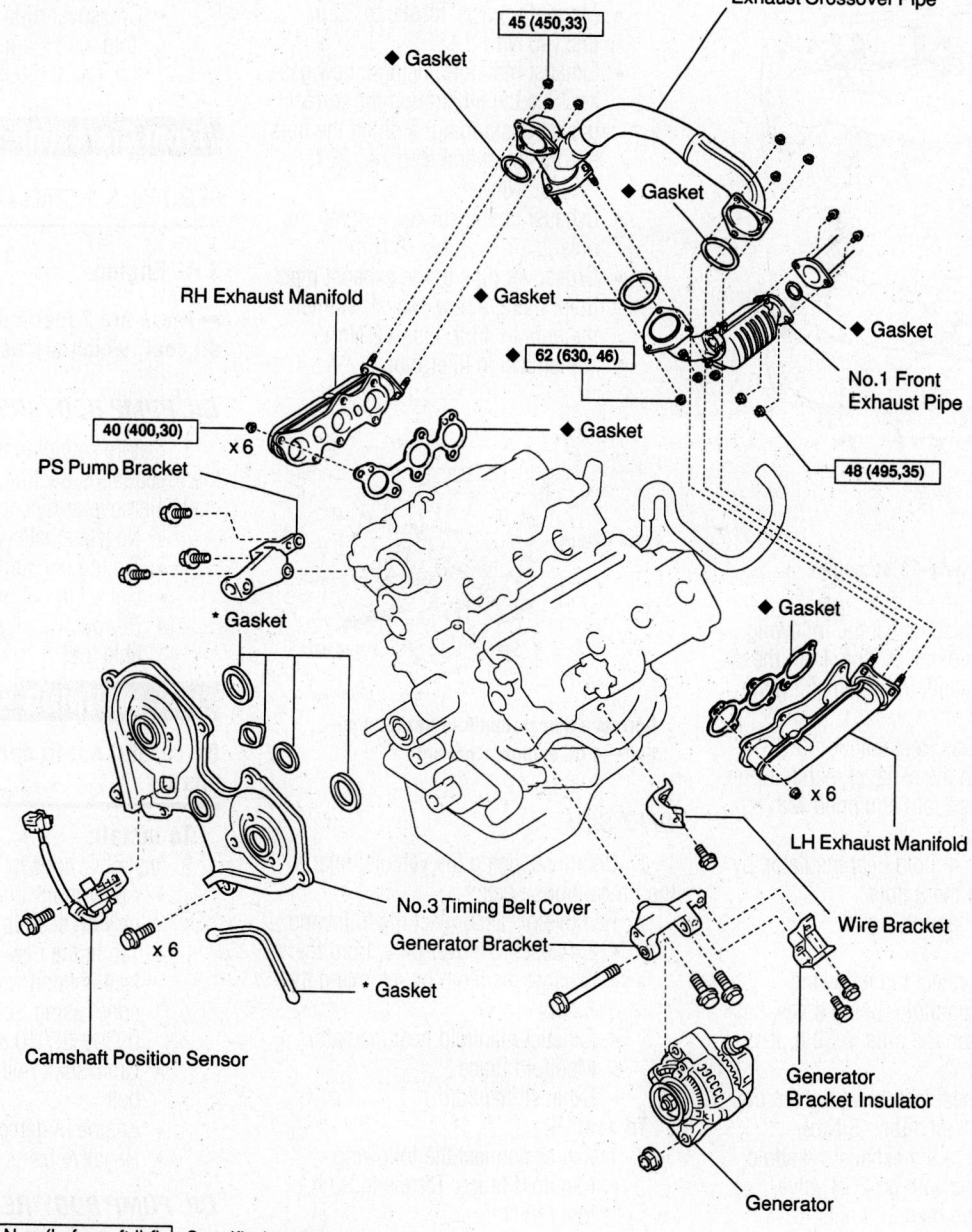

Exhaust Crossover Pipe

45 (450,33)

◆ Gasket

◆ Gasket

RH Exhaust Manifold

◆ Gasket

◆ Gasket

62 (630, 46)

No.1 Front Exhaust Pipe

40 (400,30) x 6

◆ Gasket

48 (495,35)

PS Pump Bracket

* Gasket

◆ Gasket

x 6

LH Exhaust Manifold

No.3 Timing Belt Cover

Generator Bracket

Wire Bracket

x 6

* Gasket

Camshaft Position Sensor

Generator Bracket Insulator

Generator

N·m (kgf·cm, ft·lbf) :Specified torque
◆ Non–reusable part
* Replace only if damaged

67170-4RUN-G144

Exhaust manifolds and related parts—3.4L engine

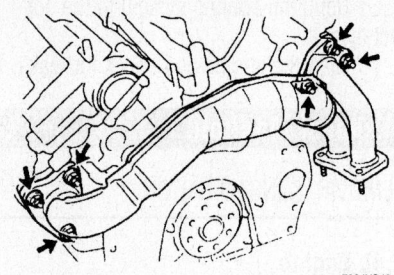

Exhaust crossover pipe mounting nut locations—3.4L engine

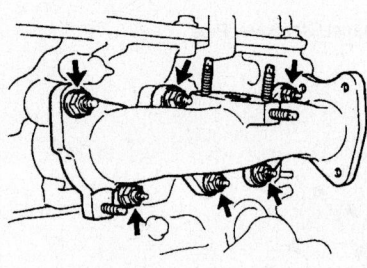

Exhaust manifold nuts—3.4L engine

2. Remove or disconnect the following:
- Exhaust crossover pipe, from the exhaust manifold by removing the 3 nuts
- Exhaust Gas Recirculation (EGR) pipe, from the exhaust manifold, on the left manifold equipped with an EGR valve
- Exhaust manifold heat insulator, by removing the 3 nuts
- Exhaust manifold

To install:

3. Install or connect the following:
- Exhaust manifold, using a new gasket. Tighten the nuts to 30 ft. lbs. (40 Nm).
- Exhaust heat insulator. Tighten the nuts to 71 inch lbs. (8 Nm).
- EGR pipe to the exhaust manifold, if equipped with an EGR valve. Tighten the manifold nuts to 14 ft. lbs. (18 Nm) and the clamp nuts to 71 inch lbs. (8 Nm).
- Crossover pipe to the exhaust manifold, using a new gasket. Tighten the nuts to 33 ft. lbs. (45 Nm).

4.0L Engine

RIGHT SIDE

1. Before servicing the vehicle, refer to the Precautions Section.
2. Remove or disconnect the following:
- Transmission filler tube
- Exhaust crossover pipe, from the exhaust manifold by removing the 3 nuts
- Exhaust manifold heat insulator
- Manifold brace
- Exhaust manifold

To install:

3. Install or connect the following:
- Manifold brace. Torque to 30 ft. lbs. (40 Nm)
- Exhaust manifold, using a new gasket. See the illustration for correct gasket positioning. Tighten the nuts evenly, in several steps, to 22 ft. lbs. (30 Nm).
- Exhaust heat insulator. Tighten the nuts to 71 inch lbs. (8 Nm).
- Crossover pipe to the exhaust manifold, using a new gasket. Tighten the nuts to 46 ft. lbs. (62 Nm).
- Transmission filler tube

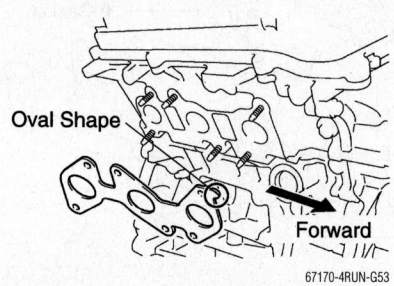

Correct exhaust manifold gasket position—4.0L engine, right side

LEFT SIDE

1. Before servicing the vehicle, refer to the Precautions Section.
2. Remove or disconnect the following:
- Exhaust crossover pipe, from the exhaust manifold by removing the 3 nuts
- Exhaust manifold heat insulator
- Manifold brace
- Exhaust manifold

To install:

3. Install or connect the following:
- Manifold brace. Torque to 30 ft. lbs. (40 Nm)
- Exhaust manifold, using a new gasket. See the illustration for correct gasket positioning. Tighten the nuts evenly, in several steps, to 22 ft. lbs. (30 Nm).

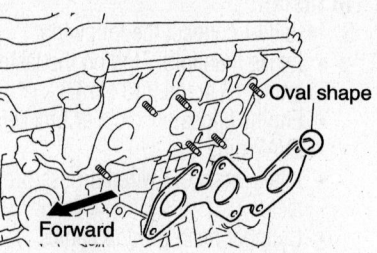

Correct exhaust manifold gasket position—4.0L engine, left side

- Exhaust heat insulator. Tighten the nuts to 71 inch lbs. (8 Nm).
- Crossover pipe to the exhaust manifold, using a new gasket. Tighten the nuts to 46 ft. lbs. (62 Nm).

Front Crankshaft Seal

REMOVAL & INSTALLATION

3.4L Engine

➡There are 2 methods to replace the oil seal, which are as follows:

OIL PUMP BODY INSTALLED

1. Before servicing the vehicle, refer to the Precautions Section.
2. Remove or disconnect the following:
- Negative battery cable
- Timing belt and crankshaft pulley
- Cut off the oil seal lip, using a knife
- Pry out the oil seal, using a suitable tool

❊❊ WARNING

Be careful not to damage the crankshaft.

To install:

3. Install or connect the following:
- Apply multi-purpose grease to the new oil seal lip
- Tap in the new oil seal until its surface is flush with the oil pump case edge, using Seal Driver tool 09309-37010 and a mallet
- Crankshaft pulley and the timing belt
- Engine undercover, if removed
- Negative battery cable

OIL PUMP BODY REMOVED

1. Before servicing the vehicle, refer to the Precautions Section.
2. Carefully pry out the seal using a suitable tool.

3. Apply multi-purpose grease to the new oil seal lip.

4. Using Seal Driver tool 09309-37010, drive the new seal into place.

4.0L Engine

1. Before servicing the vehicle, refer to the Precautions Section.

2. Remove the V-bank cover.

3. Remove the radiator support upper seal.

4. Remove the engine under-cover.

5. Remove the accessory drive belt.

6. Remove the fan.

7. Remove the crankshaft pulley.

8. Remove the seal, using a small pry-bar. Take car to avoid scratching the crankshaft.

To install:

9. Coat the new oil seal lip with MP grease.

10. Using a seal driver, install the seal flush with the timing case.

11. Install the crank pulley. Use and installer, such as 09213-54015, or equivalent. Torque the bolt to 184 ft. lbs. (250 Nm).

12. Install the fan. Torque to 15 ft. lbs. (21 Nm).

13. The remainder of installation is the reverse of removal.

67170-4RUN-G55

Seal removal—4.0L engine

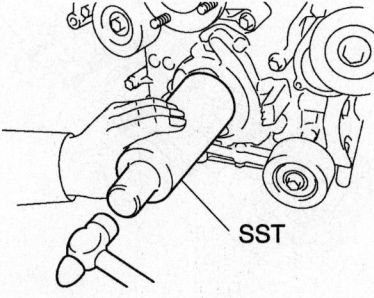

SST

67170-4RUN-G56

Seal installation—4.0L engine

4.7L Engine

For seal replacement, see the Front Cover, Timing Belt and Seal procedure, later in this chapter.

Camshaft and Valve Lifters

REMOVAL & INSTALLATION

3.4L Engine

1. Before servicing the vehicle, refer to the Precautions Section.

2. Release the fuel pressure.

3. Remove or disconnect the following:
   - Negative battery cable
   - Engine undercover

4. Drain the cooling system.

5. Remove or disconnect the following:
   - Air cleaner cap
   - Mass Air Flow (MAF) meter and the resonator
   - Actuator cable with the bracket, if equipped with cruise control
   - Accelerator cable
   - Throttle cable, if equipped with an automatic transmission
   - Heater hose
   - Brake booster vacuum hose
   - Evaporative Emission (EVAP) hose
   - Automatic Disconnecting Differential (ADD) vacuum hose, for 4-wheel drive
   - Fuel inlet and fuel return hose
   - Spark plug wires, with the ignition coils
   - Intake chamber stay
   - Camshaft Position (CMP) sensor connector from the No. 2 timing belt cover
   - 3 spark plug wire clamps from the No. 2 timing belt cover
   - 6 bolts and the No. 2 timing belt cover
   - Air intake chamber assembly
   - Throttle Position (TP) sensor connector
   - Idle Air Control (IAC) valve connector
   - Positive Crankcase Ventilation (PCV) hoses
   - Water bypass hoses
   - Air assist hose from the throttle body
   - Intake air connector
   - 6 injector connectors
   - Engine Coolant Temperature (ECT) sensor and sender gauge connectors
   - Engine wiring harness protector, from the cylinder head

6. Set the No. 1 cylinder at Top Dead Center (TDC) of the compression stroke, as follows:

   a. Turn the crankshaft pulley and align its groove with the timing mark **0** on the No. 1 timing belt cover.

   b. Check that the timing marks of the camshaft timing pulleys and the No. 3 timing belt cover are aligned. If not, turn the crankshaft pulley 1 revolution (360 degrees).

7. Remove or disconnect the following:
   - Timing belt tensioner, by alternately loosening the 2 bolts
   - Timing belt
   - Camshaft timing pulley bolt, the timing pulley and the knock pin, using Variable Pin Wrench Set 09960-10010
   - Both timing pulleys
   - Bolt and the No. 2 idler pulley
   - Camshaft Position (CMP) sensor
   - Timing belt cover
   - 8 bolts, seal washers, cylinder head covers and gasket
   - Semi-circular plugs

8. Remove the right exhaust camshafts, as follows:

   a. Bring the service bolt hole of the driven sub-gear upward by turning the hexagon head portion of the exhaust camshaft with a wrench.

   b. Align the timing mark (2 dot marks) of the camshaft drive and driven gears by turning the camshaft with a wrench.

   c. Secure the exhaust camshaft sub-gear to the driven gear with a service bolt (6mm diameter, 16–20mm bolt length and 1.0mm in thread pitch).

➡**When removing the camshaft, be sure that the torsional spring force of the sub-gear has been eliminated by the above operation.**

   d. Uniformly loosen and remove the bearing cap bolts in several passes, in the sequence shown.

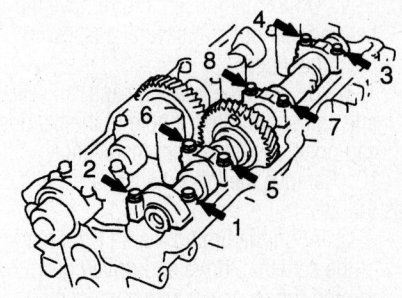

67170-4RUN-G135

Intake camshaft bearing cap loosening sequence—3.4L engine, left side

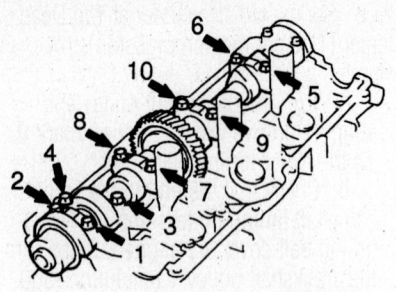

Exhaust camshaft bearing cap loosening sequence—3.4L engine, right side

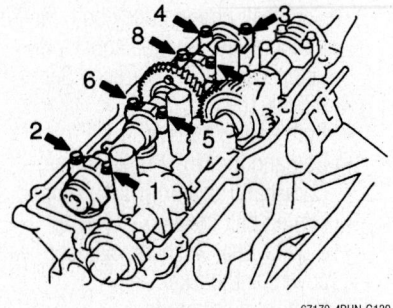

Intake camshaft bearing cap loosening sequence—3.4L engine, right side

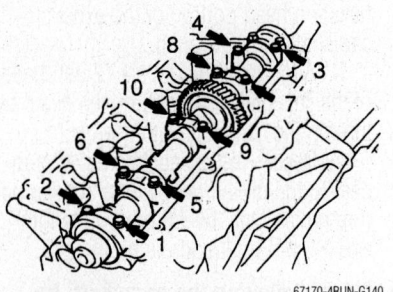

Exhaust camshaft bearing cap loosening sequence—3.4L engine, right side

e. Remove the bearing caps and camshaft. Make a note of the bearing cap positions for proper installation.

9. Remove the right-hand intake camshaft, as follows:

a. Uniformly loosen and remove the bearing cap bolts in several passes, in the sequence.

b. Remove the bearing caps, oil seal and camshaft. Make a note of the bearing cap positions for proper installation.

10. Remove the left exhaust camshafts, as follows:

a. Align the timing mark (1 dot mark) of the camshaft drive and driven gears by turning the camshaft with a wrench.

b. Secure the exhaust camshaft sub-gear to the driven gear with a service bolt

(6mm diameter, 16–20mm bolt length and 1.0mm in thread pitch).

➡**When removing the camshaft, be sure the torsional spring force of the sub-gear has been eliminated by the above operation.**

c. Uniformly loosen and remove the bearing cap bolts in several passes, in the sequence.

d. Remove the bearing caps and camshaft. Make a note of the bearing cap positions for proper installation.

➡**Do not pry on or attempt to force the camshaft with a tool or other object.**

11. Remove the left-hand intake camshaft, as follows:

a. Uniformly loosen and remove the bearing cap bolts in several passes, in the sequence.

b. Remove the bearing caps, oil seal and camshaft.

➡**Make a note of the bearing cap positions for proper installation.**

12. Remove the valve lifters and shims from the cylinder head. Arrange the valve lifters and shims in correct order.

To install:

13. Install the valve lifters and shims. Check that the valve lifter rotates smoothly by hand.

14. Install the right intake camshaft, as follows:

a. Apply engine oil to the thrust portion of the intake camshaft.

b. Position the intake camshaft at 90 degrees angle of the timing mark (2 dot marks) on the cylinder head.

c. Install the bearing caps in their proper locations. Apply a light coat of engine oil to the threads and install the cap bolts.

d. Apply a light coat of engine oil on the threads and under the heads of the bearing cap bolts.

e. Uniformly tighten the cap bolts in

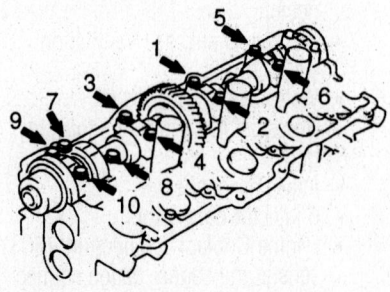

Intake camshaft bearing cap torque sequence—3.4L engine, left side

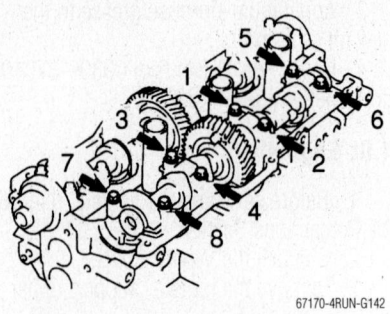

Exhaust camshaft bearing cap torque sequence—3.4L engine, left side

the sequence shown to 12 ft. lbs. (16 Nm).

15. Install the right exhaust camshaft, as follows:

a. Apply engine oil to the thrust portion of the intake camshaft.

b. Align the timing marks (2 dot marks) of the camshaft drive and driven gears.

c. Roll down the exhaust camshaft onto the bearing journals while engaging the gears with each other. Install the bearing caps in their proper locations.

d. Apply a light coat of engine oil to the threads and install the cap bolts.

e. Apply a light coat of engine oil on the threads and under the heads of the bearing cap bolts.

f. Uniformly tighten the cap bolts in the sequence to 12 ft. lbs. (16 Nm).

g. Remove the service bolt from the driven sub-gear. Check that the intake and exhaust camshafts turn smoothly.

h. Align the timing marks (2 dot marks) of the camshaft drive and driven gears by turning the camshaft with a wrench.

16. Install the left intake camshaft, as follows:

a. Apply engine oil to the thrust portion of the intake camshaft.

b. Position the intake camshaft at 90 degrees angle of the timing mark (1 dot mark) on the cylinder head.

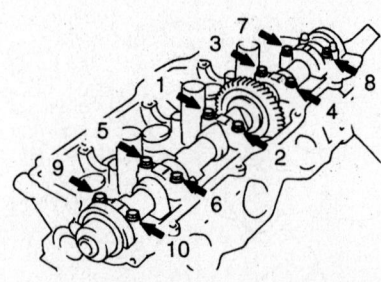

Intake camshaft bearing cap torque sequence—3.4L engine, right side

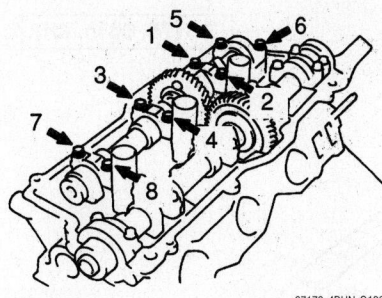

Exhaust camshaft bearing cap torque sequence—3.4L engine, right side

c. Install the bearing caps in their proper locations. Apply a light coat of engine oil to the threads and install the cap bolts.

d. Apply a light coat of engine oil on the threads and under the heads of the bearing cap bolts.

e. Uniformly tighten the cap bolts in the sequence to 12 ft. lbs. (16 Nm).

17. Install the left exhaust camshaft, as follows:

a. Apply engine oil to the thrust portion of the intake camshaft.

b. Align the timing marks (1 dot mark) of the camshaft drive and driven gears.

c. Roll down the exhaust camshaft onto the bearing journals while engaging the gears with each other. Install the bearing caps in their proper locations.

d. Apply a light coat of engine oil to the threads and install the cap bolts.

e. Apply a light coat of engine oil on the threads and under the heads of the bearing cap bolts.

f. Uniformly tighten the cap bolts in the sequence to 12 ft. lbs. (16 Nm).

g. Remove the service bolt.

18. Check and adjust the valve clearance.

19. Install or connect the following:
- Semi-circular plugs
- Cylinder head covers. Tighten the bolts, in several passes, to 53 inch lbs. (6 Nm).
- No. 3 timing belt cover. Tighten the 6 bolts to 80 inch lbs. (9 Nm).
- CMP sensor. Tighten it to 71 inch lbs. (8 Nm).
- No. 2 timing belt idler. Tighten the bolt to 30 ft. lbs. (40 Nm).

➡**Check that the pulley bracket moves smoothly.**

20. Install the left camshaft timing pulley, as follows:

a. Install the knock pin to the camshaft.

b. Align the knock pin hose of the camshaft with the knock pin groove of the timing pulley.

c. Slide the timing pulley on the camshaft with the flange side facing outward. Tighten the pulley bolt to 81 ft. lbs. (110 Nm).

21. Set the No. 1 cylinder to Top Dead Center (TDC) of the compression stroke, as follows:

a. Turn the crankshaft pulley, and align its groove with the timing mark **0** on the No. 1 timing belt cover.

b. Turn the camshaft, align the knock pin hole of the camshaft with the timing mark of the No. 3 timing belt cover.

c. Turn the camshaft timing pulley, align the timing marks of the camshaft timing pulley and the No. 3 timing belt cover.

22. Install or connect the following:
- Timing belt to the left camshaft timing pulley. Check that the installation mark on the timing belt is aligned with the end of the No. 1 timing belt cover.
- Right camshaft timing pulley
- Timing belt

23. Set the timing belt tensioner, as follows:

a. Using a press, slowly press in the pushrod using 220–2,205 lbs. (981–9,807 N) of force.

b. Align the holes of the pushrod and housing, pass a 1.5mm hexagon wrench through the holes to keep the setting position of the pushrod.

c. Release the press and install the dust boot to the tensioner.

24. Install the timing belt tensioner and alternately tighten the bolts to 20 ft. lbs. (28 Nm). Using pliers, remove the 1.5mm hexagon wrench from the belt tensioner.

25. Check the valve timing, as follows:

a. Slowly turn the crankshaft pulley 2 revolutions from the TDC-to-TDC. Always turn the crankshaft pulley clockwise.

b. Check that each pulley aligns with the timing marks. If the timing marks do not align, remove the timing belt and reinstall it.

26. Install or connect the following:
- Engine wiring harness to the cylinder head
- 3 engine wiring harness clamps.
- 6 injector connectors
- ECT sender gauge connector
- ECT sensor connector
- Intake air connector
- Air intake chamber assembly. Tighten the 4 bolts and 2 nuts to 13 ft. lbs. (18 Nm).
- Intake chamber stay. Tighten the 2 bolts to 30 ft. lbs. (40 Nm).
- New O-ring to the oil filler tube
- Oil filler tube end into the oil pan tube hole
- Oil filler tube and No. 1 throttle cable clamp
- No. 2 timing belt cover. Tighten the bolts to 80 inch lbs. (9 Nm).
- Remaining components
- Negative battery cable

27. Refill the cooling system to the correct level..

28. Start the engine and check for leaks.

29. Check the ignition timing.

30. Install the engine undercover.

31. Road test the vehicle.

32. Recheck all fluid levels.

4.0L Engine

RIGHT SIDE

1. Before servicing the vehicle, refer to the Precautions Section.

2. Drain the cooling system.

3. Remove the V-bank cover.

4. Remove the air cleaner assembly.

5. Remove the water bypass hoses.

6. Remove the fuel vapor hose.

7. Remove the vent hose.

8. Disconnect the 2 VSV connectors.

9. Disconnect the wiring at the throttle body.

10. Remove the throttle body bracket.

11. Remove the oil baffle plate

12. Remove the upper manifold supports (stays).

13. Remove the 2 nuts and 4 bolts and lift off the upper manifold. Discard the gasket.

14. Remove the cylinder head cover.

15. Turn the crankshaft pulley to align the notch with the timing mark "0" on the chain cover.

16. Check that the camshaft and chain timing marks are aligned as shown. If they're not, rotate the crankshaft one full turn (360 degrees) clockwise.

17. Using paint, matchmark the chain links that correspond to the timing marks on the camshaft sprockets.

18. Remove the 4 bolts and the timing chain cover plate and gasket.

19. While turning the stopper plate of the tensioner upward, push in on the plunger. Then, turn the stopper plate downward and insert a 3.5mm diameter metal pin into the holes in the stopper plate and tensioner.

20. Remove the 2 bolts and the tensioner.

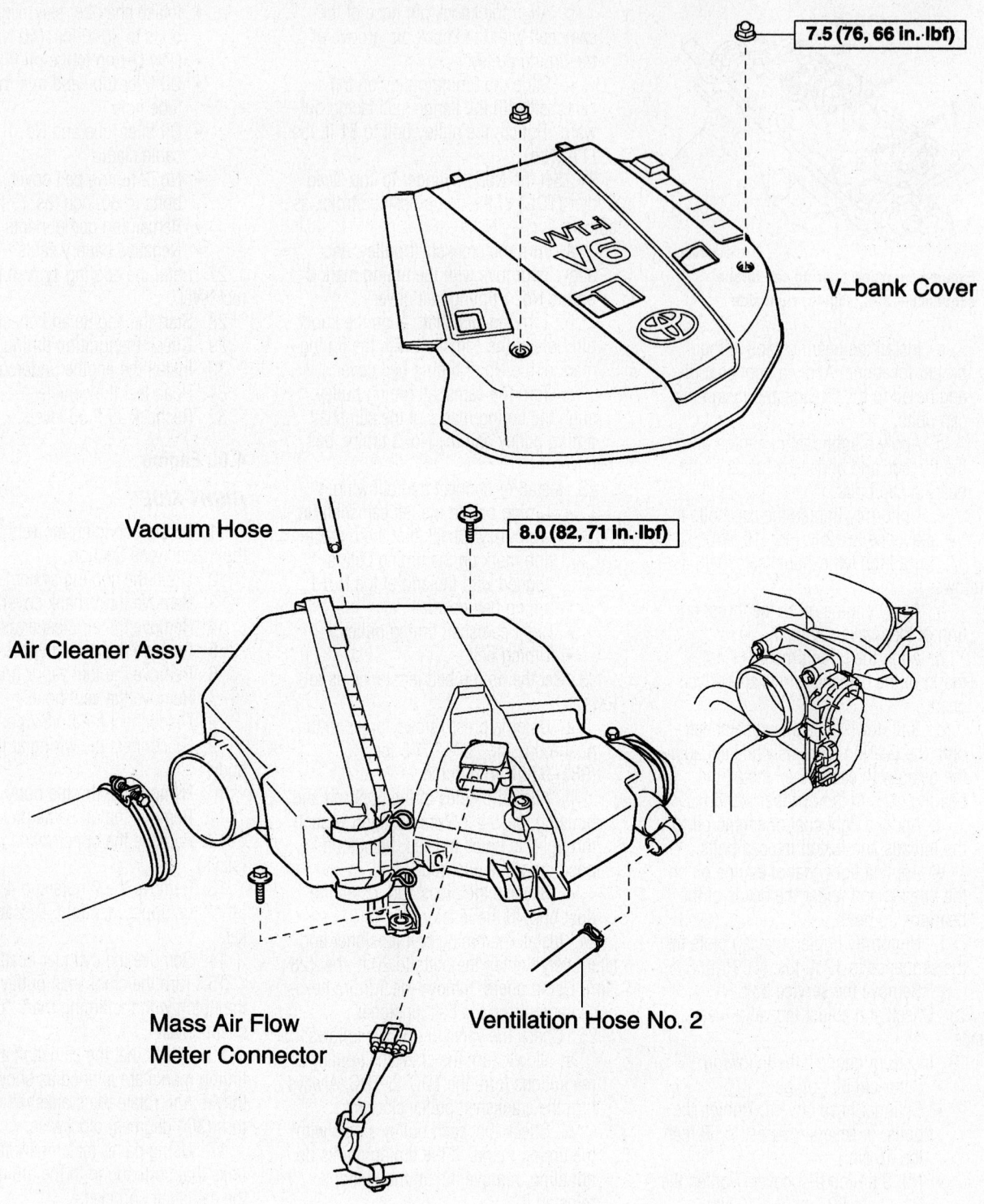

7.5 (76, 66 in.·lbf)

V–bank Cover

Vacuum Hose

8.0 (82, 71 in.·lbf)

Air Cleaner Assy

Mass Air Flow Meter Connector

Ventilation Hose No. 2

N·m (kgf·cm, ft·lbf) : Specified torque

V-bank cover and air cleaner assembly—4.0L engine

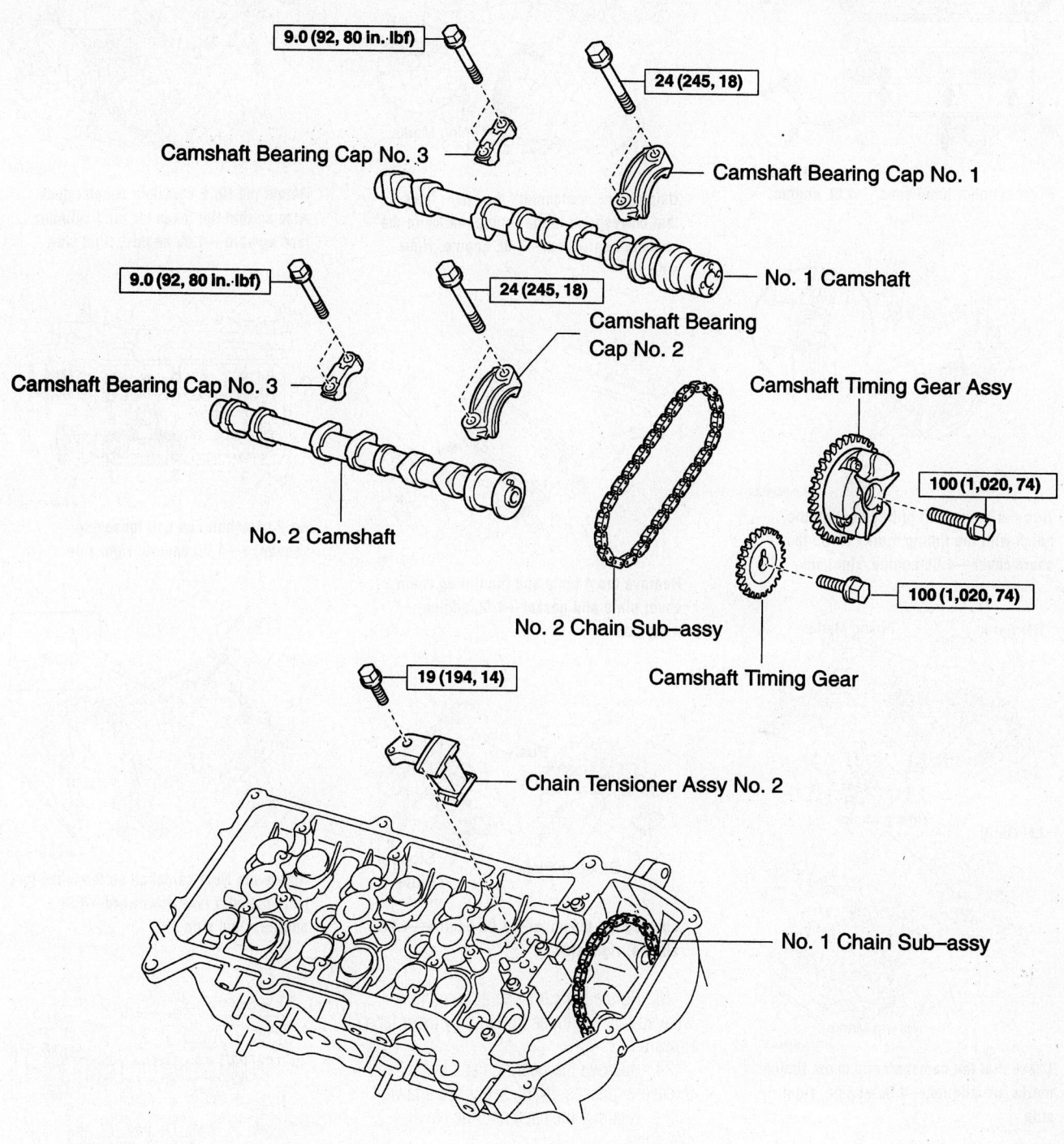

9.0 (92, 80 in.·lbf)

Camshaft Bearing Cap No. 3

24 (245, 18)

Camshaft Bearing Cap No. 1

No. 1 Camshaft

9.0 (92, 80 in.·lbf)

24 (245, 18)

Camshaft Bearing Cap No. 2

Camshaft Bearing Cap No. 3

Camshaft Timing Gear Assy

No. 2 Camshaft

100 (1,020, 74)

No. 2 Chain Sub–assy

Camshaft Timing Gear

100 (1,020, 74)

19 (194, 14)

Chain Tensioner Assy No. 2

No. 1 Chain Sub–assy

N·m (kgf·cm, ft·lbf) : Specified torque

Right side camshaft and related parts—4.0L engine

67170-4RUN-G58

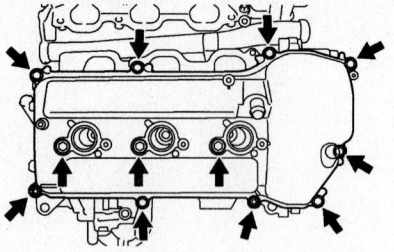

Right cylinder head cover—4.0L engine

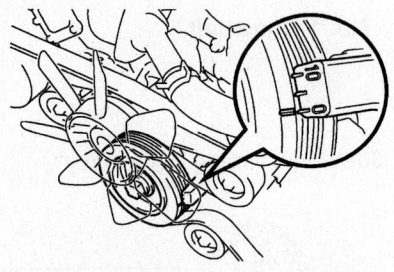

Turn the crankshaft pulley to align the notch with the timing mark "0" on the chain cover—4.0L engine, right side

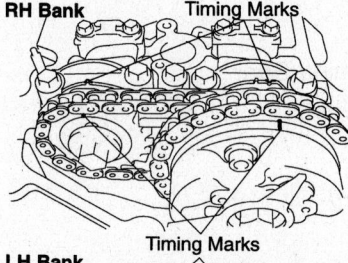

Check that the camshaft and chain timing marks are aligned—4.0L engine, right side

☀☀ WARNING

The camshaft thrust clearances are very small. Keep the camshaft level while removing it to avoid damage to the cylinder head.

21. While lifting the No.2 tensioner, insert a 1mm dia. pin to hold it.
22. Remove the No.2 (outer) camshaft timing gear bolt. Remove the gear.

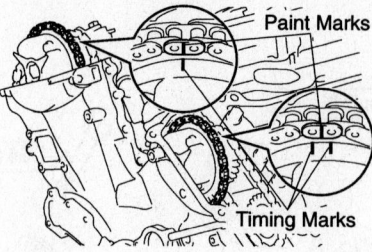

Using paint, matchmark the chain links that correspond to the timing marks on the camshaft sprockets—4.0L engine, right side

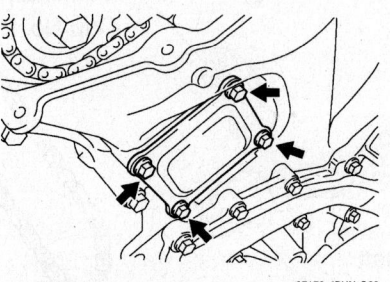

Remove the 4 bolts and the timing chain cover plate and gasket—4.0L engine, right side

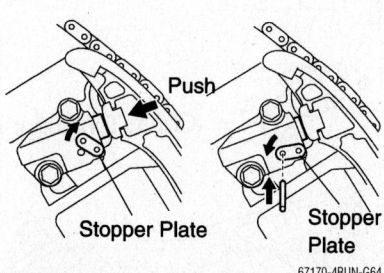

Locking the tensioner—4.0L engine—4.0L engine, right side

23. Rotate the camshaft counterclockwise so that the lobes for No.1 cylinder face upward.
24. Remove the bearing caps evenly and in several passes, in the sequence shown.
25. Remove the caps and the No.2 camshaft.
26. Remove the No.1 camshaft sprocket bolt and remove the sprocket. Lift the chain from the sprocket and secure it.
27. Rotate the No.1 camshaft so the lobes for No.1 cylinder face downward, as shown.
28. Remove the bearing caps evenly and in several passes, in the sequence shown.
29. Remove the caps and the No.1 camshaft. Keep the caps in order.
30. Tie the chain so that it doesn't fall into the case.

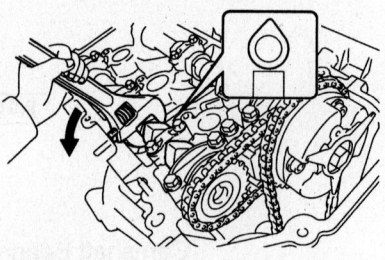

Rotate the No.2 camshaft counterclockwise so that the lobes for No.1 cylinder face upward—4.0L engine, right side

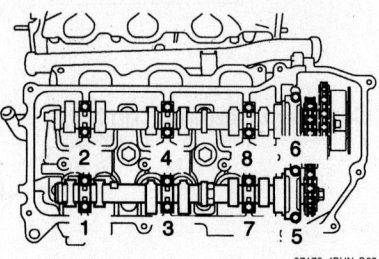

No.2 camshaft cap bolt loosening sequence—4.0L engine, right side

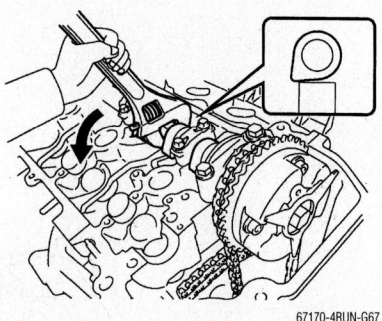

Rotate the No.1 camshaft so the lobes for No.1 cylinder face downward—4.0L engine, right side

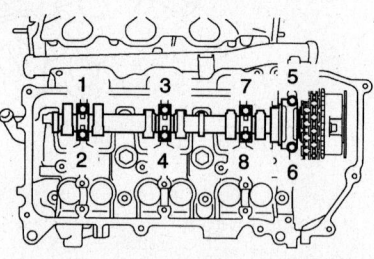

Remove the no.1 camshaft bearing caps evenly and in several passes, in the sequence shown—4.0L engine, right side

To install:
31. Align the painted link with the timing mark on the sprocket.
32. Install the gear and chains on the No.1 camshaft.

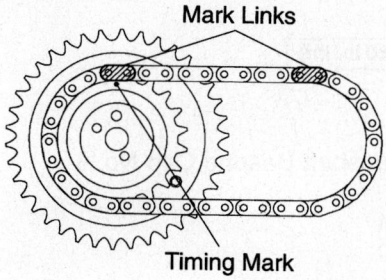

Painted link and timing mark alignment—4.0L engine, right side

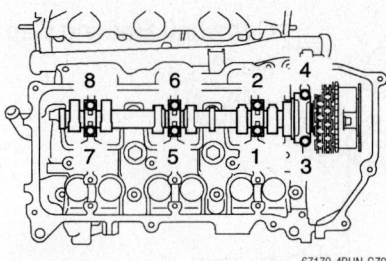

No.1 camshaft bolt torque sequence—4.0L engine, right side

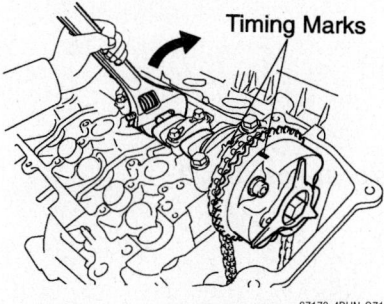

Rotate the No.1 camshaft clockwise so that the timing mark on the camshaft sprocket is aligned with the timing mark on the bearing cap—4.0L engine, right side

33. Install the bolt hand-tight.

34. Place the camshaft on the head with the No.1 cylinder lobes facing downward as before.

35. Install the caps. Lightly oil the bolts and install them.

36. Torque the bolts evenly, and in several passes, in the sequence shown, to 10mm bolts, 80 inch lbs. (9 Nm); 12m bolts (18 ft. lbs. (24 Nm).

37. Rotate the No.1 camshaft clockwise so that the timing mark on the camshaft sprocket is aligned with the timing mark on the bearing cap.

38. Align the paint mark on the chain with the timing mark on the sprocket.

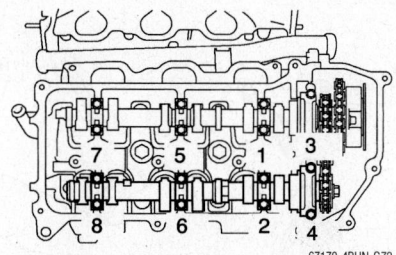

No.2 camshaft bolt torque sequence—4.0L engine, right side

39. Torque the sprocket bolt to 74 ft. lbs. (100 Nm).

40. Install the No.2 chain tensioner.

41. Temporarily, install the No.2 camshaft sprocket, aligning the timing marks. Torque the tensioner bolt to 14 ft. lbs. (19 Nm).

42. Place the No.2 camshaft on the head with the No.1 cylinder lobes facing upwards, as before.

43. Install the caps. Lightly oil the bolts and install them.

44. Torque the bolts evenly, and in several passes, in the sequence shown, to 10mm bolts, 80 inch lbs. (9 Nm); 12m bolts 18 ft. lbs. (24 Nm).

45. Rotate the No.2 camshaft clockwise so that the alignment pin is aligned with the hole in the sprocket.

46. Torque the sprocket bolt to 74 ft. lbs. (100 Nm).

47. Remove the pin from the No.2 tensioner.

48. Install the No.1 tensioner. Torque the bolts to 80 inch lbs. (9 Nm). Remove the pin.

49. Install the timing chain cover plate, using a new gasket. Torque the bolts to 80 inch lbs. (9 Nm).

50. Turn the crankshaft pulley two complete revolutions clockwise and align the "0" mark with the notch on the pulley. Check that the camshaft timing marks are aligned as shown.

51. Check the valve clearance as described below.

52. Clean all old gasket material from the head and head cover. Apply a 2–3mm dia. Bead of silicone gasket material to the head mating surface. Place the cover in position within 3 minutes.

53. Install the washers and bolts. Torque the bolts to 80 inch lbs. (9 Nm) within 15 minutes.

54. The remainder of installation is the reverse of removal.

LEFT SIDE

1. Before servicing the vehicle, refer to the Precautions Section.
2. Drain the cooling system.
3. Remove the V-bank cover.
4. Remove the air cleaner assembly.
5. Remove the water bypass hoses.

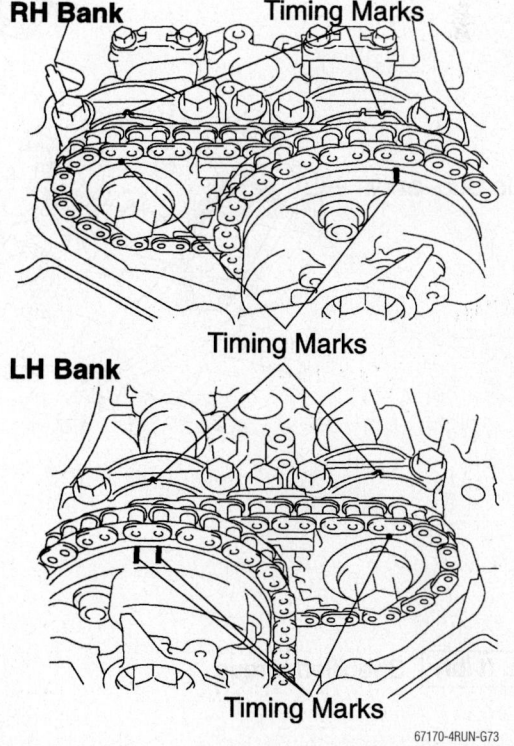

Check that the camshaft timing marks are aligned as shown—4.0L engine, right side

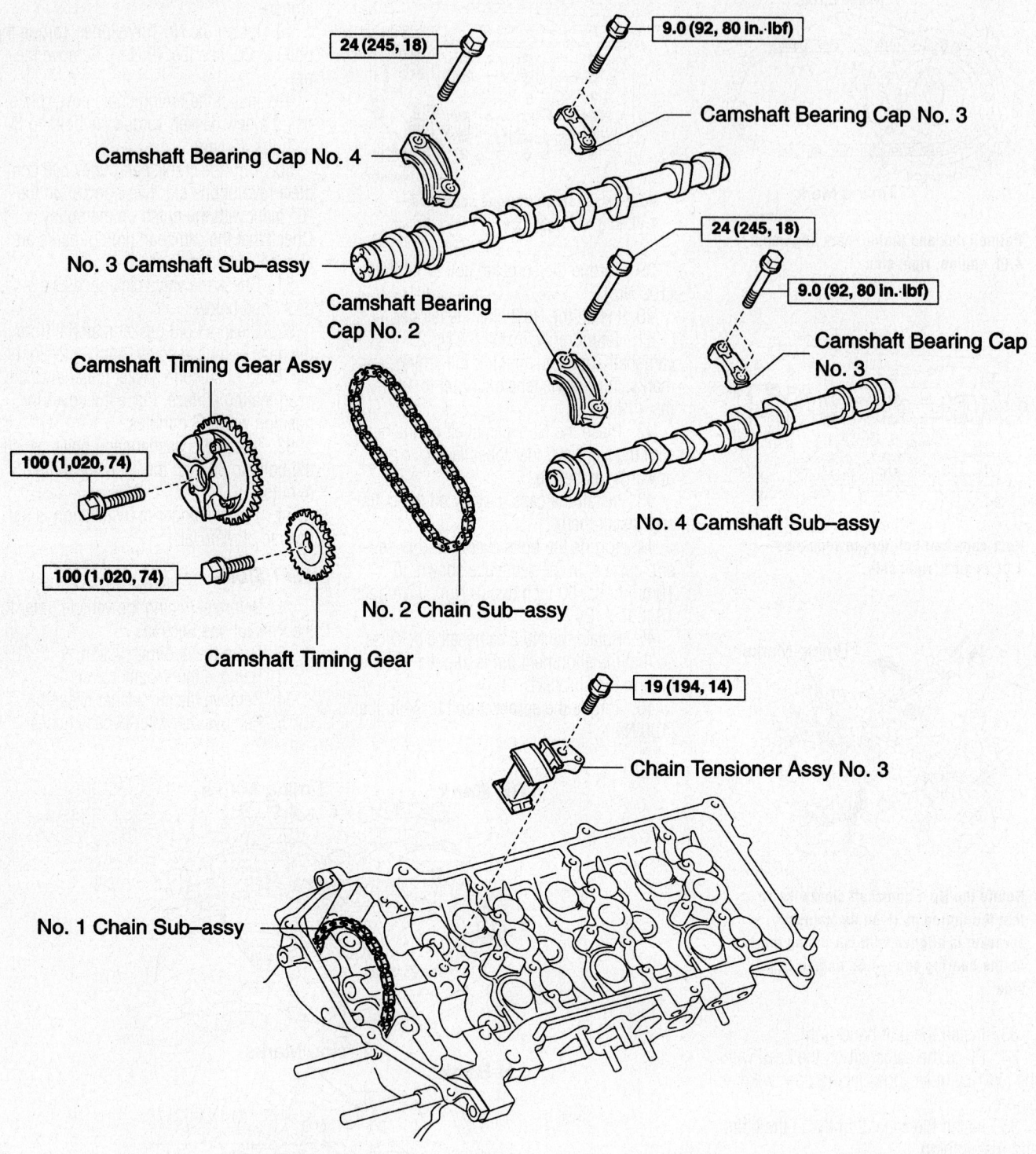

Camshaft Bearing Cap No. 4

24 (245, 18)

No. 3 Camshaft Sub–assy

9.0 (92, 80 in.·lbf)

Camshaft Bearing Cap No. 3

Camshaft Bearing
Cap No. 2

24 (245, 18)

9.0 (92, 80 in.·lbf)

Camshaft Bearing Cap
No. 3

Camshaft Timing Gear Assy

100 (1,020, 74)

100 (1,020, 74)

No. 4 Camshaft Sub–assy

Camshaft Timing Gear

No. 2 Chain Sub–assy

19 (194, 14)

Chain Tensioner Assy No. 3

No. 1 Chain Sub–assy

N·m (kgf·cm, ft·lbf) : Specified torque

67170-4RUN-G74

Left side camshaft and related parts—4.0L engine

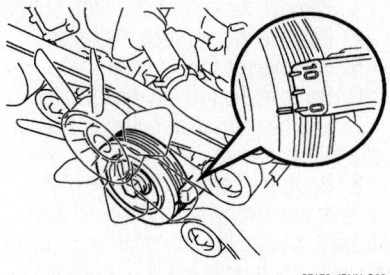

Turn the crankshaft pulley to align the notch with the timing mark "0" on the chain cover—4.0L engine

6. Remove the fuel vapor hose.
7. Remove the vent hose.
8. Disconnect the 2 VSV connectors.
9. Disconnect the wiring at the throttle body.
10. Remove the throttle body bracket.
11. Remove the oil baffle plate
12. Remove the upper manifold supports (stays).
13. Remove the 2 nuts and 4 bolts and lift off the upper manifold. Discard the gasket.
14. Remove the ignition coil assembly.
15. Remove the cylinder head cover.
16. Turn the crankshaft pulley to align the notch with the timing mark "0" on the chain cover.
17. Check that the camshaft and chain timing marks are aligned as shown. If they're not, rotate the crankshaft one full turn (360 degrees) clockwise.
18. Using paint, matchmark the chain links that correspond to the timing marks on the camshaft sprockets.

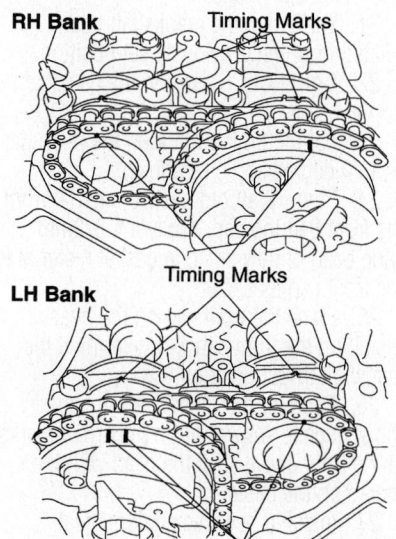

RH Bank   **Timing Marks**

Timing Marks

LH Bank

Timing Marks

Check that the camshaft and chain timing marks are aligned—4.0L engine

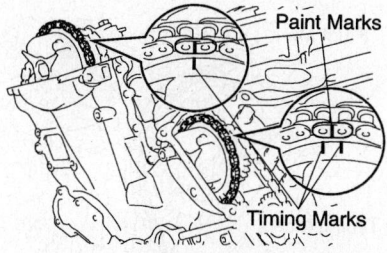

Paint Marks

Timing Marks

Using paint, matchmark the chain links that correspond to the timing marks on the camshaft sprockets—4.0L engine

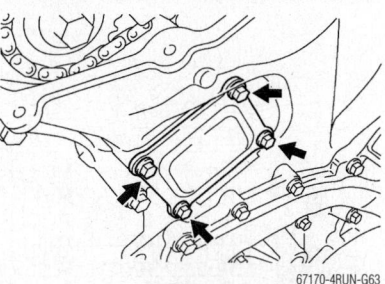

Remove the 4 bolts and the timing chain cover plate and gasket—4.0L engine

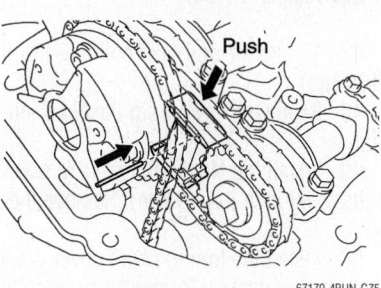

Push

Locking the tensioner—4.0L engine left side

19. Remove the 4 bolts and the timing chain cover plate and gasket.
20. While turning the stopper plate of the tensioner upward, push in on the plunger. Then, turn the stopper plate downward and insert a 3.5mm diameter metal pin into the holes in the stopper plate and tensioner.
21. Remove the 2 bolts and the tensioner.

✷✷ WARNING

The camshaft thrust clearances are very small. Keep the camshaft level while removing it to avoid damage to the cylinder head.

22. While lifting the No.4 tensioner, insert a 1mm dia. pin to hold it.
23. Remove the No.4 (outer) camshaft timing gear bolt. Remove the gear.

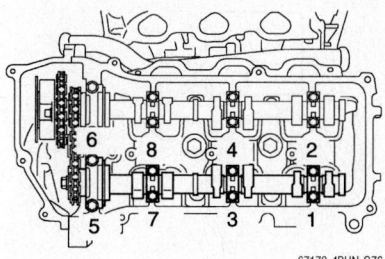

No.4 camshaft bearing cap bolt loosening sequence—4.0L engine, left side

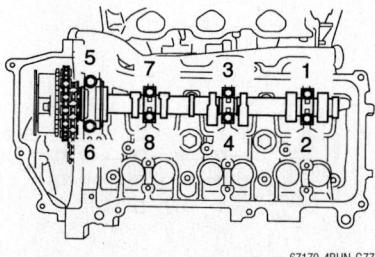

Remove the no.3 camshaft bearing caps evenly and in several passes, in the sequence shown—4.0L engine, left side

24. Remove the bearing caps evenly and in several passes, in the sequence shown.
25. Remove the caps and the No.4 camshaft.
26. Remove the No.3 tensioner.
27. Release the chain tension by turning the crank pulley slightly counterclockwise.
28. Remove the No.3 camshaft sprocket bolt and remove the sprocket. Lift the chain from the sprocket and secure it.
29. Remove the bearing caps evenly and in several passes, in the sequence shown.
30. Remove the caps and the No.3 camshaft. Keep the caps in order.
31. Tie the chain so that it doesn't fall into the case.

To install:
32. Align the painted link with the timing mark on the sprocket.
33. Install the gear and chains on the No.3 camshaft.

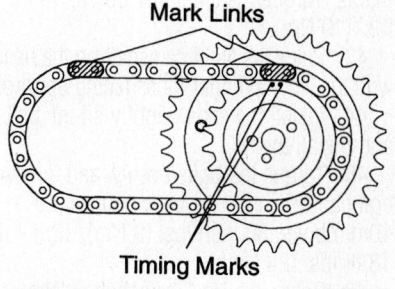

Mark Links

Timing Marks

Painted link and timing mark alignment—4.0L engine, left side

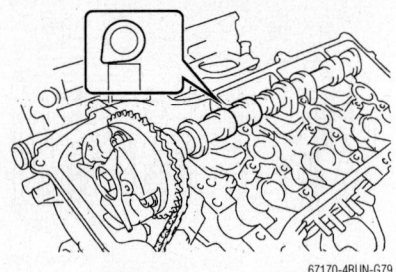

Place the camshaft on the head with the No.3 cylinder lobes facing downward as shown

67170-4RUN-G79

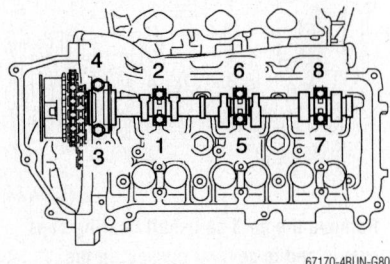

No.3 camshaft bolt torque sequence—4.0L engine, left side

67170-4RUN-G80

34. Install the bolt hand-tight.

35. Place the camshaft on the head with the No.3 cylinder lobes facing downward as shown.

36. Install the caps. Lightly oil the bolts and install them.

37. Torque the bolts evenly, and in several passes, in the sequence shown, to 10mm bolts, 80 inch lbs. (9 Nm); 12m bolts, 18 ft. lbs. (24 Nm).

38. Rotate the No.3 camshaft clockwise so that the timing mark on the camshaft sprocket is aligned with the timing mark on the bearing cap.

39. Align the paint mark on the chain with the timing mark on the sprocket.

40. Torque the sprocket bolt to 74 ft. lbs. (100 Nm).

41. Install the No.4 chain tensioner.

42. Temporarily, install the No.4 camshaft sprocket, aligning the timing marks. Torque the tensioner bolt to 14 ft. lbs. (19 Nm).

43. Place the No.4 camshaft on the head with the No.1 cylinder lobes facing upwards.

44. Install the caps. Lightly oil the bolts and install them.

45. Torque the bolts evenly, and in several passes, in the sequence shown, to 10mm bolts, 80 inch lbs. (9 Nm); 12m bolts 18 ft. lbs. (24 Nm).

46. Rotate the No.2 camshaft clockwise so that the alignment pin is aligned with the hole in the sprocket.

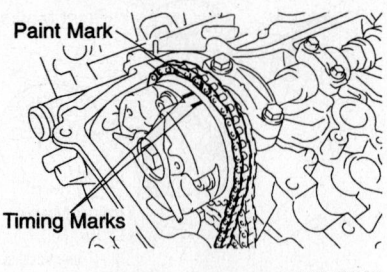

Rotate the No.3 camshaft clockwise so that the timing mark on the camshaft sprocket is aligned with the timing mark on the bearing cap—4.0L engine, left side

67170-4RUN-G81

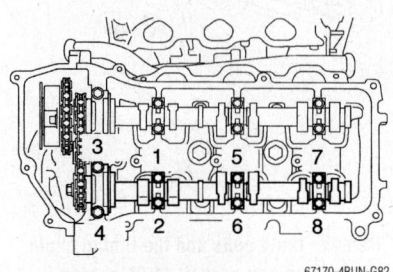

No.4 camshaft bolt torque sequence—4.0L engine, left side

67170-4RUN-G82

47. Torque the sprocket bolt to 74 ft. lbs. (100 Nm).

48. Remove the pin from the No.3 tensioner.

49. Install the No.4 tensioner. Torque the bolts to 80 inch lbs. (9 Nm). Remove the pin.

50. Install the timing chain cover plate, using a new gasket. Torque the bolts to 80 inch lbs. (9 Nm).

51. Turn the crankshaft pulley two complete revolutions clockwise and align the "0" mark with the notch on the pulley. Check that the camshaft timing marks are aligned.

52. Check the valve clearance as described below.

53. Clean all old gasket material from the head and head cover. Apply a 2–3mm dia. Bead of silicone gasket material to the head mating surface. Place the cover in position within 3 minutes.

54. Install the washers and bolts. Torque the bolts to 80 inch lbs. (9 Nm) within 15 minutes.

55. The remainder of installation is the reverse of removal.

4.7L Engine

RIGHT SIDE

1. Before servicing the vehicle, refer to the Precautions Section.

2. Remove the timing belt.

3. Remove the camshaft timing pulley.

4. Remove the right rear belt plate.

5. Remove the ignition coil.

6. Remove the cylinder head cover.

7. Set the crankshaft pulley at the correct angle as shown.

8. Bring the service bolt hole of the sub-gear upward by turning the exhaust camshaft. Secure the sub-gear to the main gear with a bolt 6mm x 1.0 x 16-20mm long.

9. Set the timing mark (1 dot) of the camshaft main gear at approximately 10 degrees by turning the exhaust camshaft.

10. Loosen the 22 camshaft bearing cap bolts, evenly and in several passes, in the sequence shown.

11. Remove the oil feed pipe and bearing caps. Keep the caps in order.

12. Place the camshaft in a soft-jawed vise as shown.

13. Turn the sub-gear clockwise and remove the service bolt.

14. Remove the snapring, wave washer, sub-gear and camshaft gear spring.

15. Remove the camshaft housing plug.

16. Remove the semi-circular plug.

To install:

17. Remove all old silicone gasket material from the camshaft housing plug. Apply new gasket material as shown, and install the plug.

18. Assemble the camshaft gear spring, sub-gear and wave washer. Install the snapring.

19. Align the holes in the main and sub-gears by turning the sub-gear counterclockwise. Install the service bolt.

20. Apply MP grease to the thrust portion of the camshafts.

21. Verify that the crankshaft hasn't rotated from the previously set position.

22. CAREFULLY position the camshafts on the head.

23. Set the timing mark on the main gear at a 10 degree angle as shown.

24. Remove all old gasket material from the front bearing cap. Apply a 1.5-2mm wide bead of new silicone gasket material to the front cap as shown.

25. Place the front cap on the head. Installing the front cap will determine the camshaft thrust position.

26. Position the other bearing caps in the sequence shown. Align the arrow marks at the front and rear of the head with the marks on the caps.

27. Install a new camshaft oil seal.

28. Apply a light coating of clean engine oil to the bolt threads and install the bolts. Note the length of the bolts in the accompanying illustration.

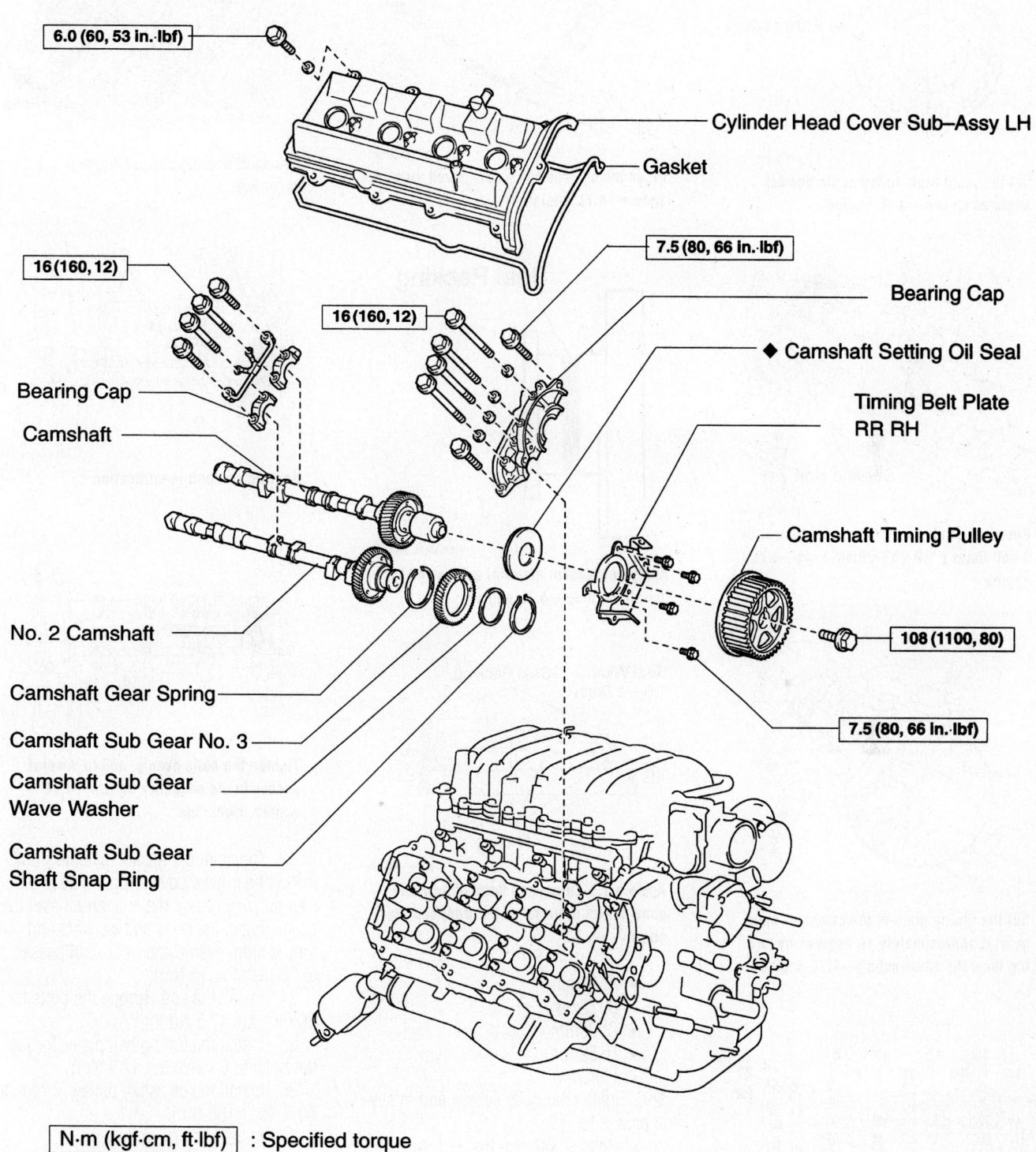

6.0 (60, 53 in.·lbf)

Cylinder Head Cover Sub-Assy LH

Gasket

16 (160, 12)

7.5 (80, 66 in.·lbf)

Bearing Cap

16 (160, 12)

◆ Camshaft Setting Oil Seal

Bearing Cap

Timing Belt Plate
RR RH

Camshaft

Camshaft Timing Pulley

No. 2 Camshaft

Camshaft Gear Spring

108 (1100, 80)

Camshaft Sub Gear No. 3

Camshaft Sub Gear
Wave Washer

7.5 (80, 66 in.·lbf)

Camshaft Sub Gear
Shaft Snap Ring

N·m (kgf·cm, ft·lbf) : Specified torque
◆ Non-reusable part

67170-4RUN-G83

Right camshaft and related parts—4.7L engine

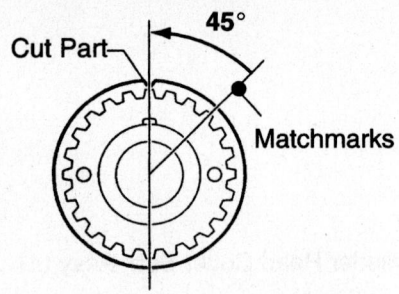

Set the crankshaft pulley at the correct angle as shown—4.7L engine

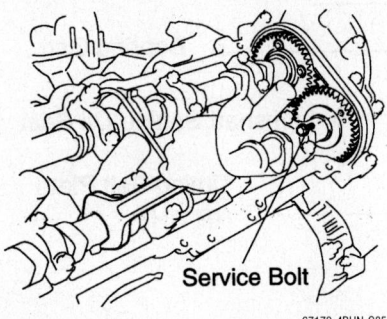

Secure the sub-gear to the main gear with a bolt 6mm x 1.0 x 16-20mm long—4.7L engine

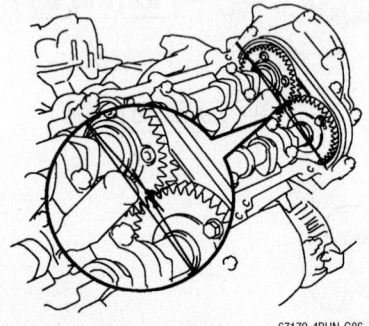

Set the timing mark of the camshaft main gear at approximately 10 degrees by turning the exhaust camshaft—4.7L engine

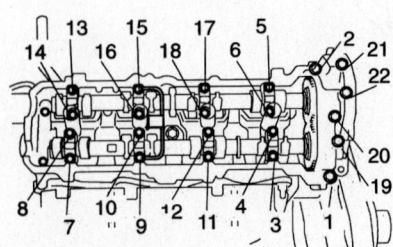

Loosen the 22 camshaft bearing cap bolts, evenly and in several passes, in the sequence shown—4.7L engine

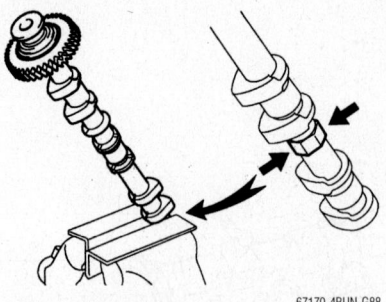

Place the camshaft in a soft-jawed vise as shown—4.7L engine

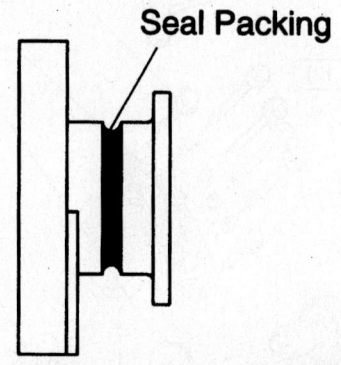

Apply new gasket material as shown, and install the plug—4.7L engine

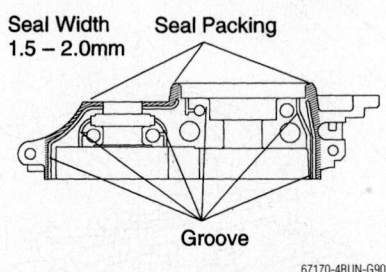

Apply a 1.5-2mm wide bead of new silicone gasket material to the front cap as shown

- A: 94mm
- B: 72mm
- C: 25mm
- D: 52mm
- E: 38mm

29. Tighten the bolts evenly, and in several passes to
- Bolts C: 66 inch lbs. (7.5 Nm)
- All other bolts: 12 ft. lbs. (16 Nm)

30. Turn the camshafts to access the service bolt and remove it.

31. Check the valve clearance.

32. Clean the semi-circular plugs, apply new silicone gasket material and install them.

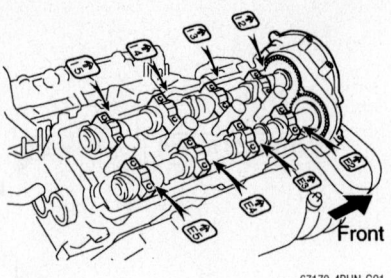

Camshaft bearing cap positioning sequence

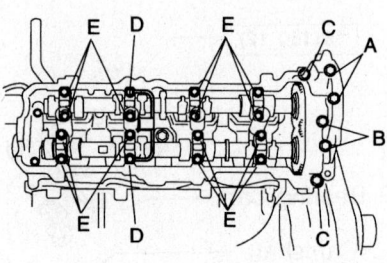

Bearing cap bolt identification

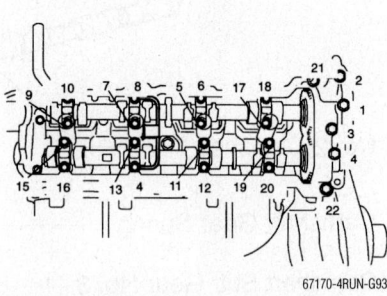

Tighten the bolts evenly, and in several passes in the sequence shown—4.7L engine, right side

33. Clean all old gasket material from the cylinder head cover. Apply a bead of new silicone gasket material and install the cover. Install the bolts and washers and torque them evenly and in several passes, to 53 inch lbs. (6 Nm).

34. Install the coil. Torque the bolts to 66 inch lbs. (7.5 Nm).

35. Install the timing belt plate. Torque the bolts to 66 inch lbs. (7.5 Nm).

36. Install the camshaft pulley. Torque to 80 ft. lbs. (108 Nm).

37. Install the timing belt.

LEFT SIDE

1. Before servicing the vehicle, refer to the Precautions Section.

2. Remove the timing belt.

3. Remove the camshaft timing pulley.

4. Remove the camshaft position sensor.

5. Remove the timing belt rear plate.

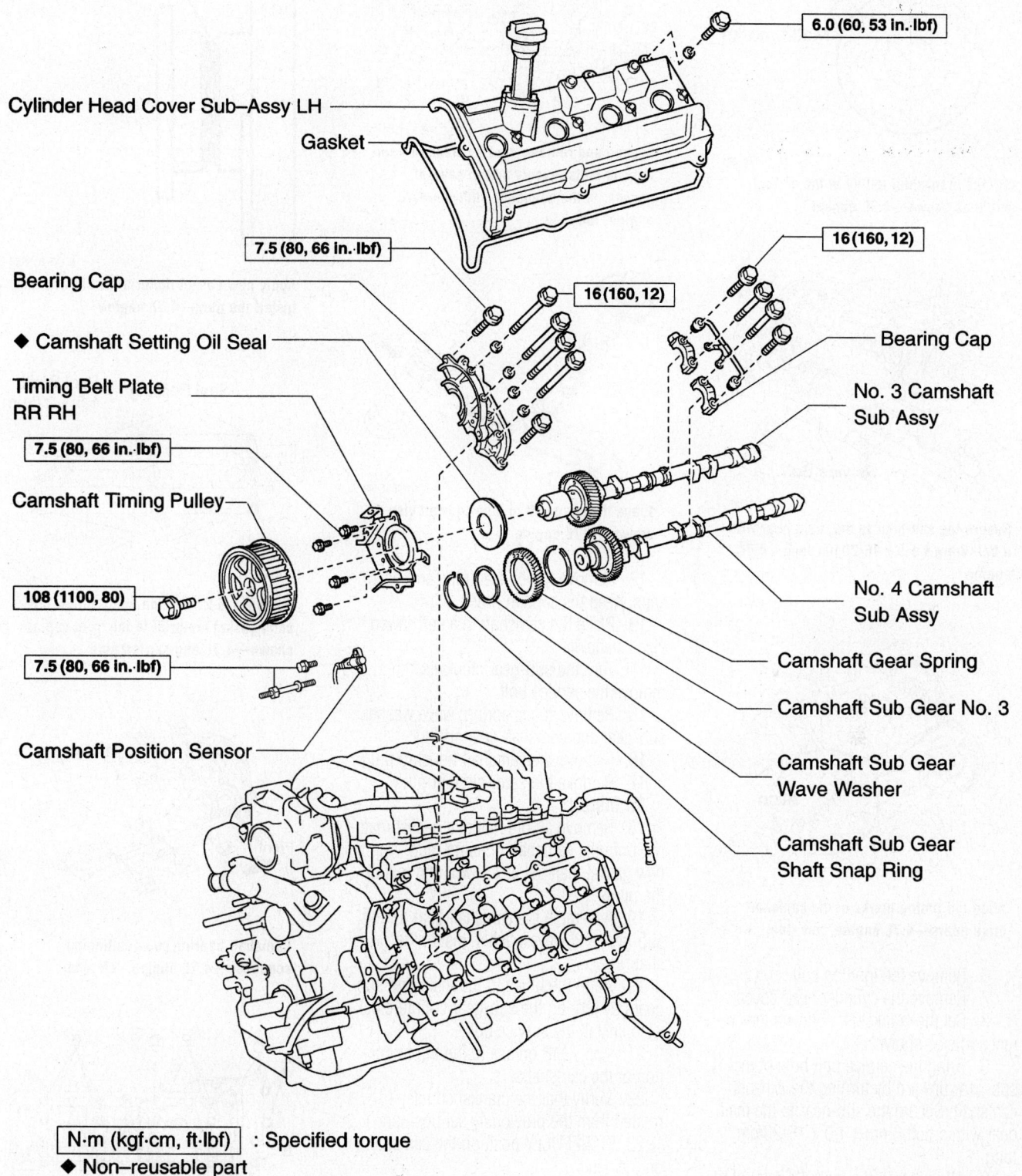

Cylinder Head Cover Sub–Assy LH

Gasket

6.0 (60, 53 in.·lbf)

Bearing Cap

7.5 (80, 66 in.·lbf)

◆ Camshaft Setting Oil Seal

Timing Belt Plate
RR RH

7.5 (80, 66 in.·lbf)

Camshaft Timing Pulley

108 (1100, 80)

7.5 (80, 66 in.·lbf)

Camshaft Position Sensor

16 (160, 12)

16 (160, 12)

Bearing Cap

No. 3 Camshaft
Sub Assy

No. 4 Camshaft
Sub Assy

Camshaft Gear Spring

Camshaft Sub Gear No. 3

Camshaft Sub Gear
Wave Washer

Camshaft Sub Gear
Shaft Snap Ring

N·m (kgf·cm, ft·lbf) : Specified torque
◆ Non–reusable part

67170-4RUN-G94

Left camshaft and related parts—4.7L engine

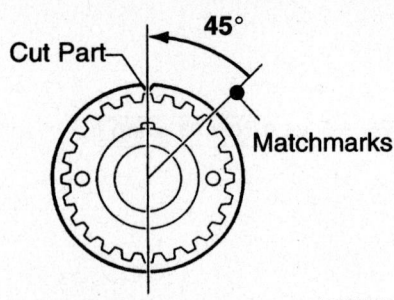

Set the crankshaft pulley at the correct angle as shown—4.7L engine

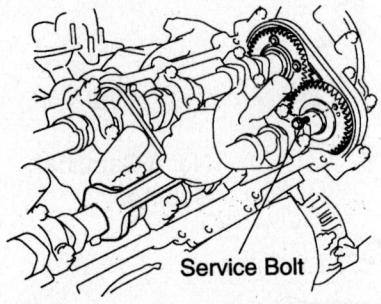

Secure the sub-gear to the main gear with a bolt 6mm x 1.0 x 16-20mm long—4.7L engine

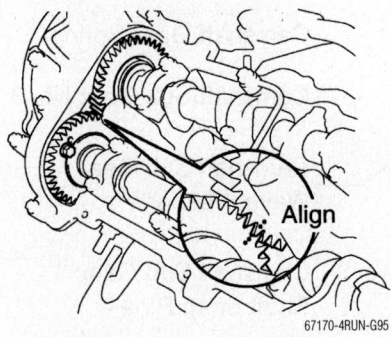

Align the timing marks of the camshaft drive gears—4.7L engine, left side

6. Remove the ignition coil.
7. Remove the cylinder head cover.
8. Set the crankshaft pulley at the correct angle as shown.
9. Bring the service bolt hole of the sub-gear upward by turning the exhaust camshaft. Secure the sub-gear to the main gear with a bolt 6mm x 1.0 x 16-20mm long.
10. Align the timing mark (2 dots) of the camshaft drive gears by turning the exhaust camshaft as shown.
11. Loosen the 22 camshaft bearing cap bolts, evenly and in several passes, in the sequence shown.

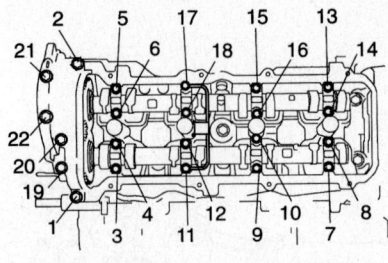

Loosen and remove the 22 camshaft bearing cap bolts, evenly and in several passes, in the sequence shown—4.7L engine, left side

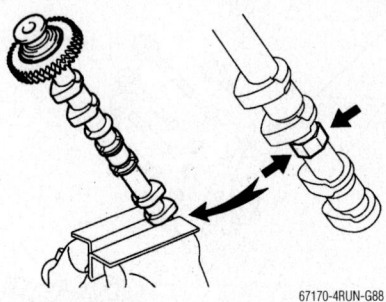

Place the camshaft in a soft-jawed vise as shown—4.7L engine

12. Remove the oil feed pipe and bearing caps. Keep the caps in order.
13. Place the camshaft in a soft-jawed vise as shown.
14. Turn the sub-gear clockwise and remove the service bolt.
15. Remove the snapring, wave washer, sub-gear and camshaft gear spring.
16. Remove the camshaft housing plug.
17. Remove the semi-circular plug.

To install:
18. Remove all old silicone gasket material from the camshaft housing plug. Apply new gasket material as shown, and install the plug.
19. Assemble the camshaft gear spring, sub-gear and wave washer. Install the snapring.
20. Align the holes in the main and sub-gears by turning the sub-gear counterclockwise. Install the service bolt.
21. Apply MP grease to the thrust portion of the camshafts.
22. Verify that the crankshaft hasn't rotated from the previously set position.
23. CAREFULLY position the camshafts on the head.
24. Set the timing mark on the main gear at a 10 degree angle as shown.
25. Remove all old gasket material from the front bearing cap. Apply a 1.5-2mm wide bead of new silicone gasket material to the front cap as shown.

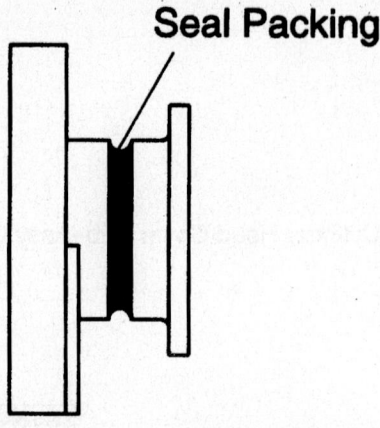

Apply new gasket material as shown, and install the plug—4.7L engine

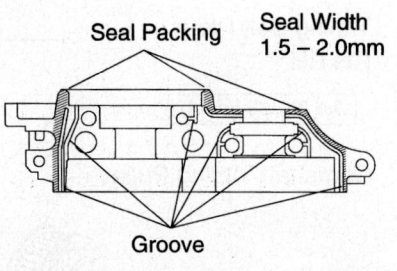

Apply a 1.5-2mm wide bead of new silicone gasket material to the front cap as shown—4.7L engine, left side

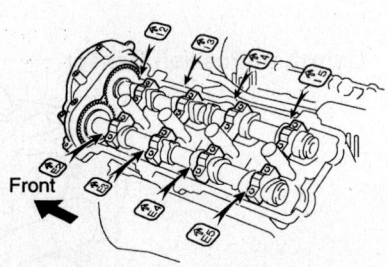

Camshaft bearing cap positioning sequence—4.7L engine, left side

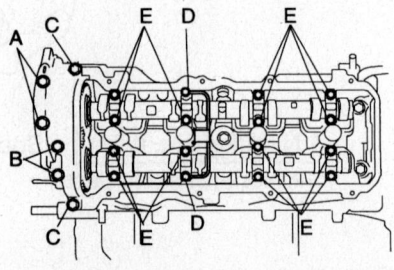

Bearing cap bolt identification—4.7L engine, left side

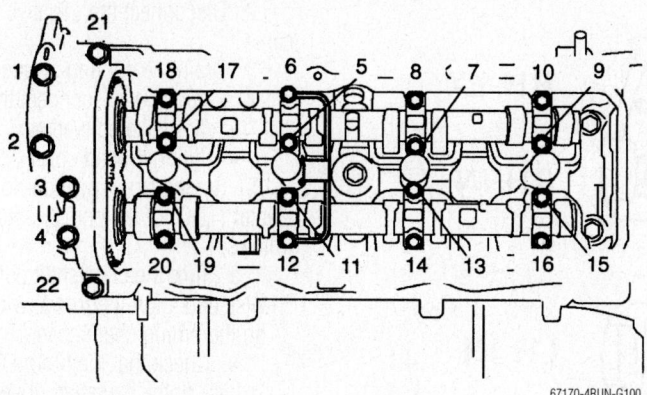

67170-4RUN-G100

Tighten the bolts evenly, and in several passes in the sequence shown—4.7L engine, left side

26. Place the front cap on the head. Installing the front cap will determine the camshaft thrust position.

27. Position the other bearing caps in the sequence shown. Align the arrow marks at the front and rear of the head with the marks on the caps.

28. Install a new camshaft oil seal.

29. Apply a light coating of clean engine oil to the bolt threads and install the bolts. Note the length of the bolts in the accompanying illustration.
- A: 94mm
- B: 72mm
- C: 25mm
- D: 52mm
- E: 38mm

30. Tighten the bolts evenly, and in several passes in the sequence shown, to:
- Bolts C: 66 inch lbs. (7.5 Nm)
- All other bolts: 12 ft. lbs. (16 Nm)

31. Turn the camshafts to access the service bolt and remove it.

32. Check the valve clearance.

33. Clean the semi-circular plugs, apply new silicone gasket material and install them.

34. Clean all old gasket material from the cylinder head cover. Apply a bead of new silicone gasket material and install the cover. Install the bolts and washers and torque them evenly and in several passes, to 53 inch lbs. (6 Nm).

35. Install the coil. Torque the bolts to 66 inch lbs. (7.5 Nm).

36. Install the timing belt plate. Torque the bolts to 66 inch lbs. (7.5 Nm).

37. Install the camshaft pulley. Torque to 80 ft. lbs. (108 Nm).

38. Install the timing belt.

INSPECTION

Runout

1. Before servicing the vehicle, refer to the Precautions Section.

2. Remove the camshafts.

3. Place the camshaft on a V-block, on a precise flat table.

4. Set the dial indicator to center journal.

5. Turn the camshaft to one direction by hand and measure the camshaft runout.

6. Runout should measure less than follows:
   a. 3.4L and 4.0L Engine: 0.0024 in. (0.06 mm).
   b. 4.7L Engine: 0.0031 in. (0.08 mm).

7. Camshaft should be replaced if it exceeds the limit.

Cam Height

1. Before servicing the vehicle, refer to the Precautions Section.

2. Remove the camshafts.

3. Measure the cam height with a micrometer.

4. The intake camshaft should measure as follows:
   a. 3.4L Engine: between 1.6657–1.6697 in. (42.310–42.410 mm) and not less than 1.6598 in. (42.160 mm).
   b. 4.0L Engine: between 1.7389–1.7428 in. (44.168–44.268 mm) and not less than 1.7330 in. (44.018 mm).
   c. 4.7L Engine: between 1.6512–

1.6551 in. (44.940–42.040 mm) and not less than 1.8262 in. (46.385 mm).

5. The exhaust camshaft should measure as follows:
   a. 3.4L Engine: between 1.6520–1.6559 in. (41.960–42.060 mm) and not less than 1.6461 in. (41.810 mm).
   b. 4.0L Engine: between 1.7551–1.7591 inches (44.580–44.680 mm) and not less than 1.6453 inches (41.790 mm).
   c. 4.7L Engine: between 1.6520–1.6559 inches (44.960–42.060 mm) and not less than 1.6461 inches (41.810 mm).

6. Camshaft should be replaced if it exceeds the limit.

Camshaft Journal Diameter

1. Before servicing the vehicle, refer to the Precautions Section.

2. Remove the camshafts from the vehicle.

3. Using a micrometer, measure the journal diameter.

4. Journal diameter should measure as follows:
   a. 3.4L Engine: All journals between 1.0610–1.0616 in. (26.949–26.965 mm).
   b. 4.0L Engine: No. 1 journal between 1.4162–1.4167 in. (35.971–35.985 mm). All other journals between 0.9039–0.9045 in. (22.959–22.975 mm).
   c. 4.7L Engine: All journals between 1.0612–1.0618 in. (26.954–26.970 mm).

5. The camshaft should be replaced if it exceeds the limit.

Valve Lash

ADJUSTMENT

3.4L Engine

1. Before servicing the vehicle, refer to the Precautions Section.

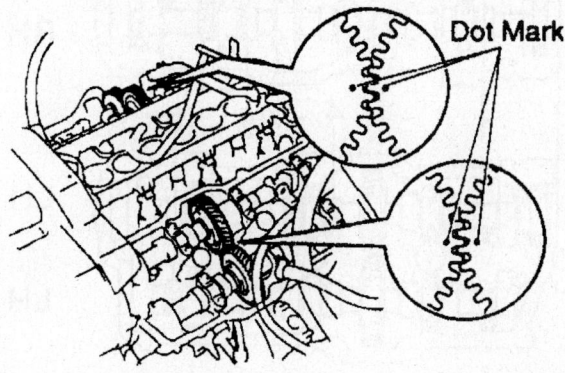

Aligning the timing marks—3.4L engine

7924YG81

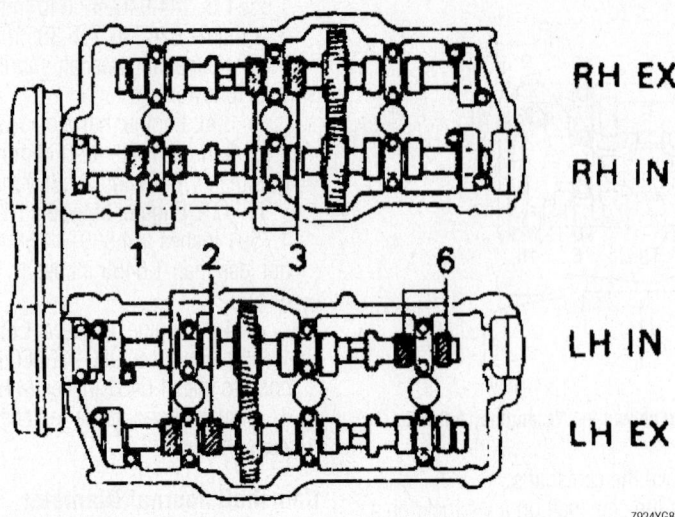

First valve adjustment—3.4L engine

RH EX

RH IN

1 2 3 6

LH IN

LH EX

7924YG82

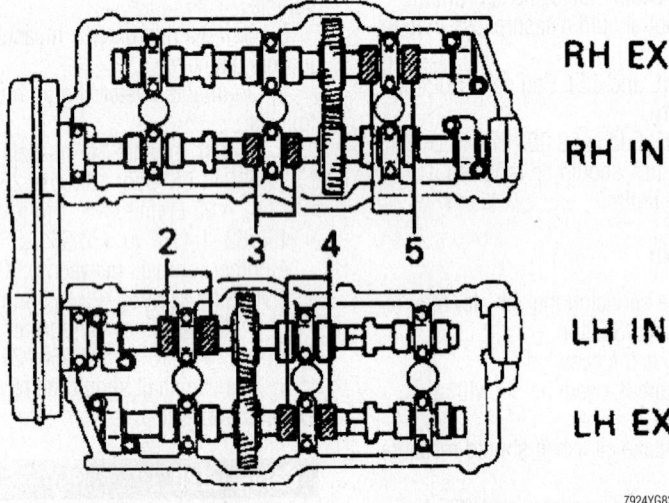

Second valve adjustment—3.4L engine

RH EX

RH IN

2 3 4 5

LH IN

LH EX

7924YG83

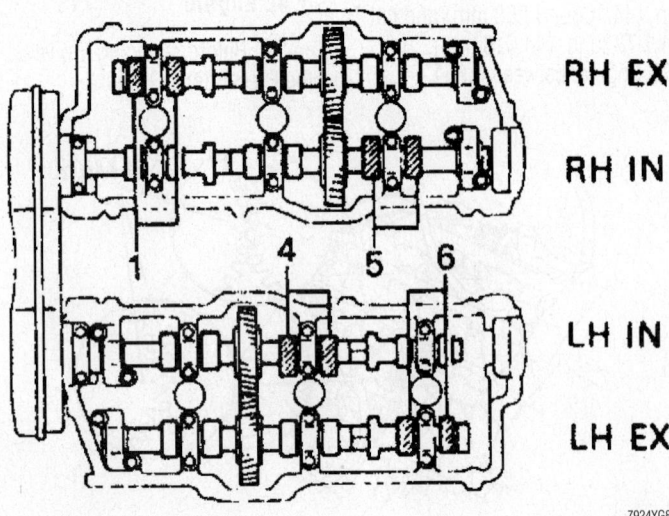

Third valve adjustment—3.4L engine

RH EX

RH IN

1 4 5 6

LH IN

LH EX

7924YG84

2. Disconnect the negative battery cable.

3. Drain the cooling system.

4. Remove or disconnect the following:
- Air intake connector
- Cylinder head cover

5. Set the No. 1 cylinder to Top Dead Center (TDC) of the compression stroke, as follows:

   a. Turn the crankshaft pulley clockwise and align its groove with the **0** mark on the timing chain cover.

   b. Check that the timing marks (1 and 2 dots) of the camshaft drive and driven gears are in a straight line on the cylinder head surface. If not, turn the crankshaft 1 revolution (360 degrees) and align the marks.

6. Inspect the valve clearance, as follows:

   a. Measure the clearance between the valve lifter and the camshaft. Measure the 1st intake and the 3rd exhaust valves on the right head and the 6th intake and the 2nd exhaust valves on the left head.

   b. Turn the crankshaft ⅔ of a revolution (240 degrees) and adjust the 3rd intake and the 5th exhaust valves on the right head and the 2nd intake and the 4th exhaust valves on the left head.

   c. Turn the crankshaft ⅔ of a revolution (240 degrees) and adjust the 5th intake and the 1st exhaust valves on the right head and the 4th intake and the 6th exhaust valves on the left head.

7. Valve clearance cold should be:
- Intake: 0.006–0.009 in. (0.13–0.23mm)
- Exhaust: 0.011–0.014 in. (0.27–0.37mm)

Front of No.1 and Rear of No.6 Cylinders

SST (B)   SST (A)

Others

SST (B)   SST (A)

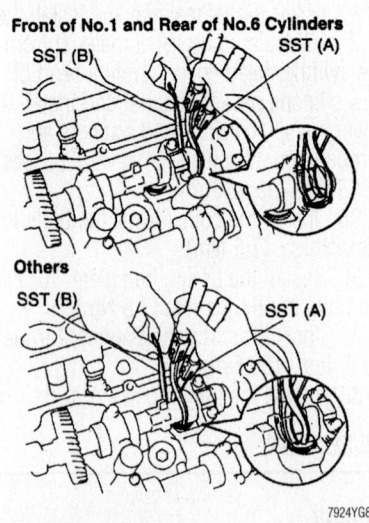

7924YG85

Removing the adjusting shim—3.4L engine

8. Adjust the valve clearance by using adjusting shims, as follows:

a. Turn the equipment camshaft so that the cam lobe for the valve to be adjusted faces up.

b. Turn the valve lifter so that the notches are perpendicular to the camshaft.

c. Using SST 09248-55040, press down the valve lifter and place SST 09248-05420, between the camshaft and the valve lifter. Remove SST 09248-55040.

d. Remove the adjusting shim with a small flat prying tool and a magnetic finger.

e. Determine the replacement adjusting shim size according to the following formula or use the adjusting shim charts.

f. Using a micrometer, measure the thickness of the removed shim. Calculate the thickness of a new shim so that the valve clearance comes within the specified value.

- T: Thickness of the removed shim
- A: Measured valve clearance
- N: Thickness of the new shim

g. Intake: $N = T + (A - 0.007$ in. $(0.18$mm$))$

h. Exhaust: $N = T + (A - 0.013$ in. $(0.32$mm$))$

i. Install a new adjusting shim. Place it on the valve lifter. Using the SST 09248-55040, press down the valve lifter and remove SST 09248-05420.

j. Recheck the valve clearance.

9. Install or connect the following:
- Cylinder head cover
- Intake air connector
- Negative battery cable

10. Refill with engine coolant.

11. Start the engine and check for leaks.

4.0L Engine

1. Before servicing the vehicle, refer to the Precautions Section.

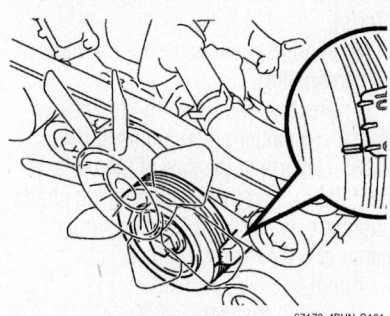

Turn the crankshaft clockwise and align its groove with the "0" mark on the timing cover—4.0L engine

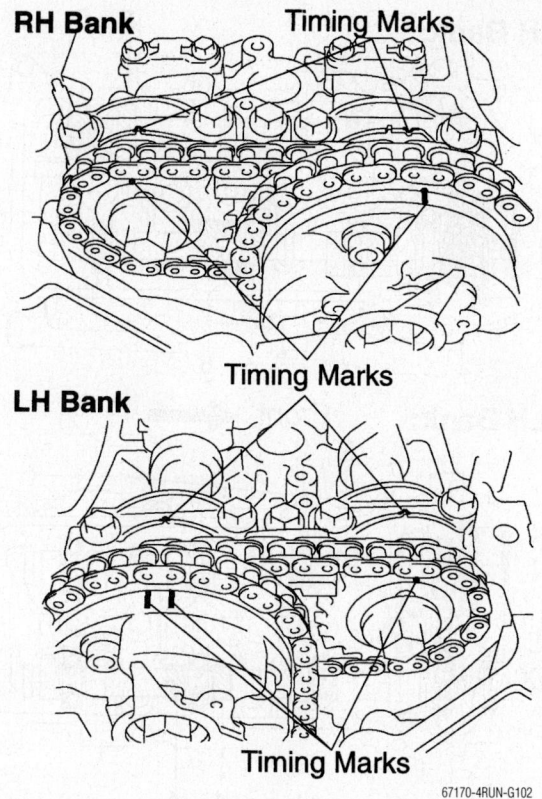

Check that the timing marks on the camshaft sprockets are aligned with the timing marks on the bearings caps—4.0L engine

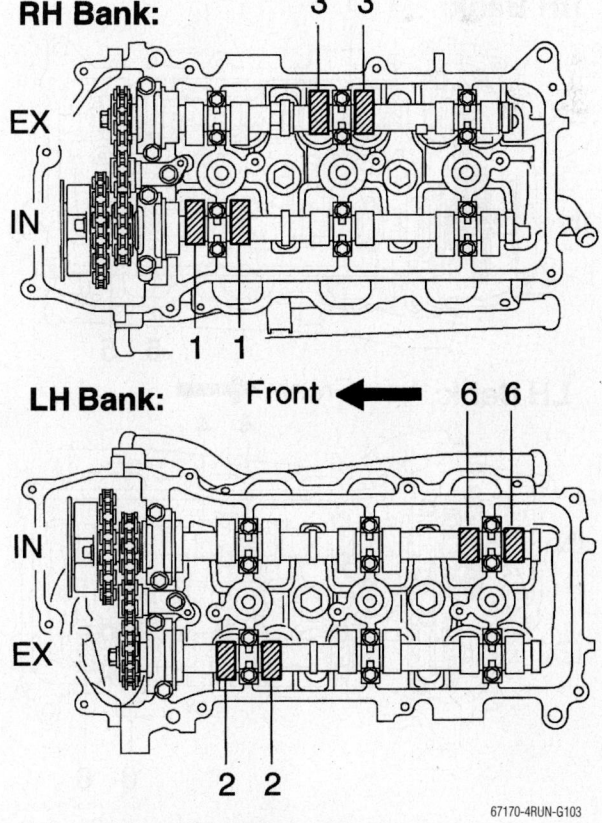

Valve clearance check, Figure A—4.0L engine

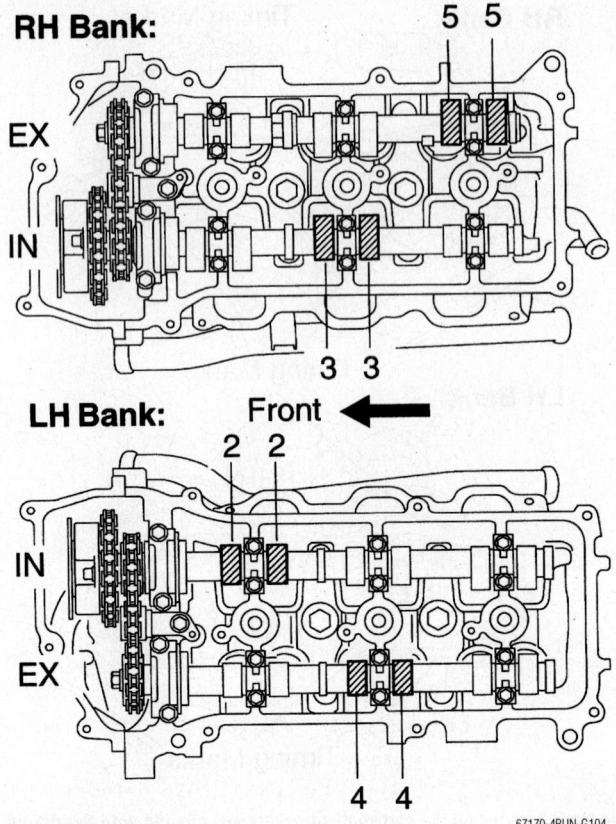

RH Bank:

EX

IN

5 5

3 3

LH Bank: Front ←

2 2

IN

EX

4 4

Valve clearance check, Figure B—4.0L engine

67170-4RUN-G104

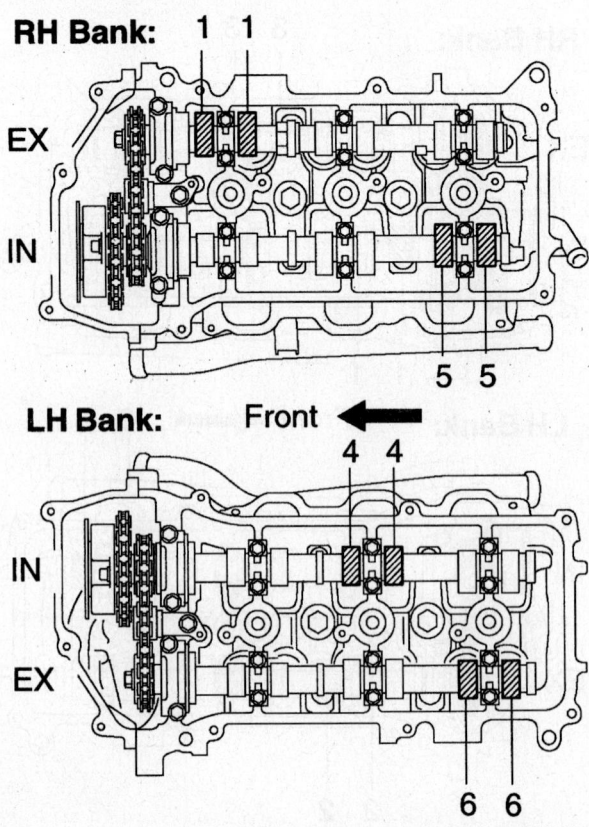

RH Bank: 1 1

EX

IN

5 5

LH Bank: Front ←

4 4

IN

EX

6 6

Valve clearance check, Figure C—4.0L engine

67170-4RUN-G105

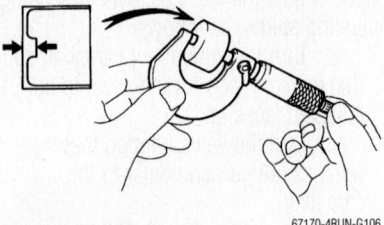

67170-4RUN-G106

Measure the lifter at these points—4.0L engine

2. Drain the cooling system.

3. Remove the V-bank cover.

4. Remove the air cleaner assembly.

5. Remove the upper intake manifold (air surge tank).

6. Remove the ignition coil.

7. Remove the cylinder head cover.

8. Turn the crankshaft clockwise and align its groove with the "0" mark on the timing cover.

9. Check that the timing marks on the camshaft sprockets are aligned with the timing marks on the bearings caps, as shown. If not, rotate the crankshaft 360 degrees clockwise.

10. Check the valve clearance for the valves shown in Figure A. Clearance should be:

- Intake: 0.006–0.010 in. (0.15–0.25mm)
- Exhaust: 0.011–0.015 inch (0.29–0.39mm)

11. If any are not correct, record how far off they are. This will determine replacement lifter(s).

12. Turn the crankshaft ⅔ turn (240 degrees) clockwise. Check the valves shown in Figure B.

13. If any are not correct, record how far off they are. This will determine replacement lifter(s).

14. Turn the crankshaft ⅔ turn (240 degrees) clockwise. Check the valves shown in Figure C.

15. If any are not correct, record how far off they are. This will determine replacement lifter(s).

16. If adjustment is necessary, remove the camshaft(s).

17. Remove the lifters to be replaced. Measure the lifter at the points shown.

18. Determine the size of the replacement lifter using the accompanying charts. Lifters are available in 35 sizes, in increments of 0.020mm, from 5.060mm to 5.740mm.

19. Install the camshaft(s).

20. The remainder of installation is the reverse of removal.

Valve Lifter Selection Chart (Intake)

New Lifter Thickness mm (in.)

No.	Thickness	No.	Thickness	No.	Thickness
06	5.060 (0.1992)	30	5.300 (0.2087)	54	5.540 (0.2181)
08	5.080 (0.2000)	32	5.320 (0.2094)	56	5.560 (0.2189)
10	5.100 (0.2008)	34	5.340 (0.2102)	58	5.580 (0.2197)
12	5.120 (0.2016)	36	5.360 (0.2110)	60	5.600 (0.2205)
14	5.140 (0.2024)	38	5.380 (0.2118)	62	5.620 (0.2213)
16	5.160 (0.2031)	40	5.400 (0.2126)	64	5.640 (0.2220)
18	5.180 (0.2039)	42	5.420 (0.2134)	66	5.660 (0.2228)
20	5.200 (0.2047)	44	5.440 (0.2142)	68	5.680 (0.2236)
22	5.220 (0.2055)	46	5.460 (0.2150)	70	5.700 (0.2244)
24	5.240 (0.2063)	48	5.480 (0.2157)	72	5.720 (0.2252)
26	5.260 (0.2071)	50	5.500 (0.2165)	74	5.740 (0.2260)
28	5.280 (0.2079)	52	5.520 (0.2173)		

Intake valve clearance (Cold):
0.15 – 0.25 mm (0.0059 – 0.0098 in.)

EXAMPLE:
The 5.250 mm (0.2067 in.) lifter is installed, and the measured clearance is 0.400 mm (0.0158 in.).
Replace the 5.250 mm (0.2067 in.) shim with a new No. 46 lifter.

Intake valve lifter selection chart—4.0L engine

67170-4RUN-G107

Valve Lifter Selection Chart (Exhaust)

Exhaust valve clearance (Cold):
0.29 – 0.39 mm (0.0114 – 0.0154 in.)

EXAMPLE:

The 5.340 mm (0.2102 in.) lifter is installed, and the measured clearance is 0.480 mm (0.0189 in.).
Replace the 5.340 mm (0.2102 in.) shim with a new No. 48 lifter.

New Lifter Thickness mm (in.)

No.	Thickness	No.	Thickness	No.	Thickness
06	5.060 (0.1992)	30	5.300 (0.2087)	54	5.540 (0.2181)
08	5.080 (0.2000)	32	5.320 (0.2094)	56	5.560 (0.2189)
10	5.100 (0.2008)	34	5.340 (0.2102)	58	5.580 (0.2197)
12	5.120 (0.2016)	36	5.360 (0.2110)	60	5.600 (0.2205)
14	5.140 (0.2024)	38	5.380 (0.2118)	62	5.620 (0.2213)
16	5.160 (0.2031)	40	5.400 (0.2126)	64	5.640 (0.2220)
18	5.180 (0.2039)	42	5.420 (0.2134)	66	5.660 (0.2228)
20	5.200 (0.2047)	44	5.440 (0.2142)	68	5.680 (0.2236)
22	5.220 (0.2055)	46	5.460 (0.2150)	70	5.700 (0.2244)
24	5.240 (0.2063)	48	5.480 (0.2157)	72	5.720 (0.2252)
26	5.260 (0.2071)	50	5.500 (0.2165)	74	5.740 (0.2260)
28	5.280 (0.2079)	52	5.520 (0.2173)		

Installed lifter thickness mm (in.) column headings (top of chart):

5.060 (0.1992), 5.080 (0.2000), 5.100 (0.2008), 5.120 (0.2016), 5.140 (0.2024), 5.160 (0.2031), 5.180 (0.2039), 5.200 (0.2047), 5.210 (0.2051), 5.220 (0.2055), 5.240 (0.2063), 5.250 (0.2067), 5.260 (0.2071), 5.270 (0.2075), 5.280 (0.2079), 5.300 (0.2087), 5.310 (0.2091), 5.320 (0.2094), 5.330 (0.2098), 5.340 (0.2102), 5.350 (0.2106), 5.360 (0.2110), 5.370 (0.2114), 5.380 (0.2118), 5.390 (0.2122), 5.400 (0.2126), 5.410 (0.2130), 5.420 (0.2134), 5.430 (0.2138), 5.440 (0.2142), 5.450 (0.2146), 5.460 (0.2150), 5.470 (0.2154), 5.480 (0.2157), 5.490 (0.2161), 5.500 (0.2165), 5.510 (0.2169), 5.520 (0.2173), 5.530 (0.2177), 5.540 (0.2181), 5.550 (0.2185), 5.560 (0.2189), 5.570 (0.2193), 5.580 (0.2197), 5.590 (0.2201), 5.600 (0.2205), 5.620 (0.2213), 5.640 (0.2220), 5.660 (0.2228), 5.680 (0.2236), 5.700 (0.2244), 5.720 (0.2252), 5.740 (0.2260)

Measured clearance mm (in.) row headings (left of chart):

0.000 - 0.020 (0.0000 - 0.0008), 0.021 - 0.040 (0.0008 - 0.0016), 0.041 - 0.060 (0.0016 - 0.0024), 0.061 - 0.080 (0.0024 - 0.0031), 0.081 - 0.100 (0.0032 - 0.0039), 0.101 - 0.120 (0.0040 - 0.0047), 0.121 - 0.140 (0.0048 - 0.0055), 0.141 - 0.160 (0.0056 - 0.0063), 0.161 - 0.180 (0.0063 - 0.0071), 0.181 - 0.200 (0.0071 - 0.0079), 0.201 - 0.220 (0.0079 - 0.0087), 0.221 - 0.240 (0.0087 - 0.0094), 0.241 - 0.260 (0.0095 - 0.0102), 0.261 - 0.280 (0.0103 - 0.0110), 0.281 - 0.289 (0.0111 - 0.0114), 0.391 - 0.410 (0.0154 - 0.0161), 0.411 - 0.430 (0.0162 - 0.0169), 0.431 - 0.450 (0.0170 - 0.0177), 0.451 - 0.470 (0.0178 - 0.0185), 0.471 - 0.490 (0.0185 - 0.0193), 0.491 - 0.510 (0.0193 - 0.0201), 0.511 - 0.530 (0.0201 - 0.0209), 0.531 - 0.550 (0.0209 - 0.0217), 0.551 - 0.570 (0.0217 - 0.0224), 0.571 - 0.590 (0.0225 - 0.0232), 0.591 - 0.610 (0.0233 - 0.0240), 0.611 - 0.630 (0.0241 - 0.0248), 0.631 - 0.650 (0.0248 - 0.0256), 0.651 - 0.670 (0.0256 - 0.0264), 0.671 - 0.690 (0.0264 - 0.0272), 0.691 - 0.710 (0.0272 - 0.0280), 0.711 - 0.730 (0.0280 - 0.0287), 0.731 - 0.750 (0.0288 - 0.0295), 0.751 - 0.770 (0.0296 - 0.0303), 0.771 - 0.790 (0.0304 - 0.0311), 0.791 - 0.810 (0.0311 - 0.0319), 0.811 - 0.830 (0.0319 - 0.0327), 0.831 - 0.850 (0.0327 - 0.0335), 0.851 - 0.870 (0.0335 - 0.0343), 0.871 - 0.890 (0.0343 - 0.0350), 0.891 - 0.910 (0.0351 - 0.0358), 0.911 - 0.930 (0.0359 - 0.0366), 0.931 - 0.950 (0.0367 - 0.0374), 0.951 - 0.970 (0.0374 - 0.0382), 0.971 - 0.990 (0.0382 - 0.0390), 0.991 - 1.010 (0.0390 - 0.0398)

6717D-4RUN-G108

Exhaust valve lifter selection chart—4.0L engine

4.7L Engine

1. Before servicing the vehicle, refer to the Precautions Section.
2. Drain the cooling system.
3. Remove the battery ground cable.
4. Remove the throttle body cover.
5. Remove the air cleaner.
6. Remove the radiator upper support seal.
7. Remove the accessory drive belt.
8. Remove the oil cooler pipe.
9. Remove the timing belt covers.
10. Remove the cylinder head covers.
11. Turn the crankshaft clockwise and align its groove with the "0" mark on the timing cover.
12. Check that the timing marks on the camshaft sprockets are aligned with the timing marks on the bearings caps, as shown. If not, rotate the crankshaft 360 degrees clockwise.
13. Check the valve clearance on the valves indicated in Figure A. Valve clearance should be:

- Intake: 0.006–0.010 inch (0.15–0.25mm)
- Exhaust: 0.010–0.014 inch (0.25–0.35mm)

14. If any valve is not within specifications, record how far out it is. This will determine the replacement shim.

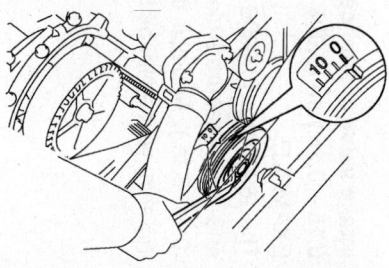

Turn the crankshaft clockwise and align its groove with the "0" mark on the timing cover—4.7L engine

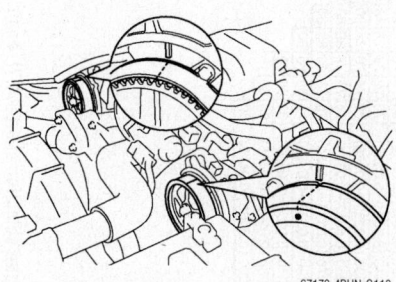

Check that the timing marks on the camshaft sprockets are aligned with the timing marks on the bearings caps—4.7L engine

RH Cylinder Head

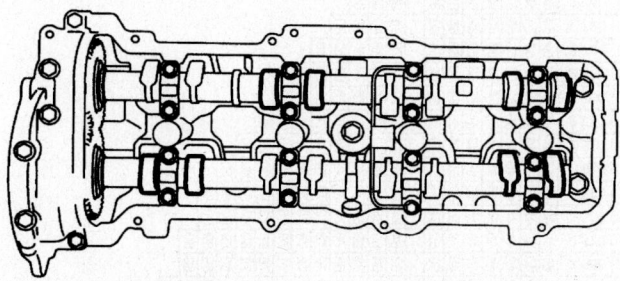

LH Cylinder Head   Front ⬅

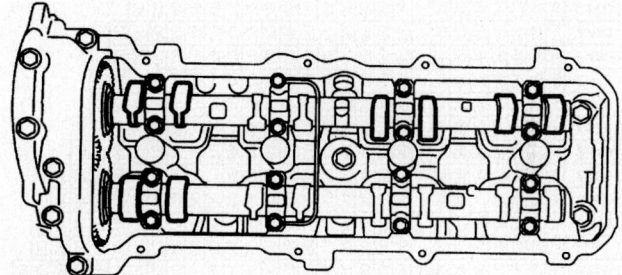

67170-4RUN-G111

Valve clearance check, Figure A—4.7L engine

RH Cylinder Head

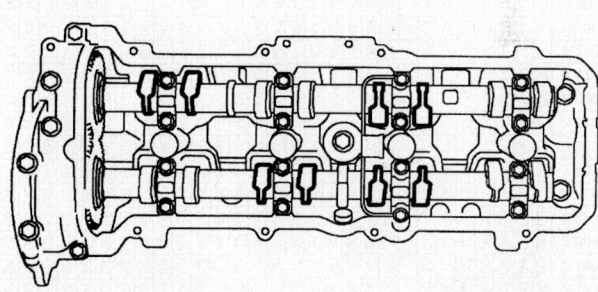

LH Cylinder Head   Front ⬅

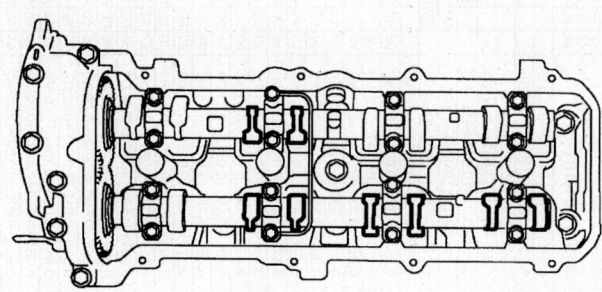

67170-4RUN-G112

Valve clearance check, Figure B—4.7L engine

Adjusting Shim Selection Chart (Intake)

The intake valve shim selection chart is a matrix with installed shim thickness across the top and measured clearance down the side.

Installed shim thickness — mm (in.):

2.000 (0.0787), 2.020 (0.0795), 2.040 (0.0803), 2.060 (0.0811), 2.080 (0.0819), 2.100 (0.0827), 2.120 (0.0835), 2.140 (0.0843), 2.160 (0.0850), 2.180 (0.0858), 2.200 (0.0866), 2.210 (0.0870), 2.220 (0.0874), 2.230 (0.0878), 2.240 (0.0882), 2.250 (0.0886), 2.260 (0.0890), 2.270 (0.0894), 2.280 (0.0898), 2.290 (0.0902), 2.300 (0.0906), 2.310 (0.0909), 2.320 (0.0913), 2.330 (0.0917), 2.340 (0.0921), 2.350 (0.0925), 2.360 (0.0929), 2.370 (0.0933), 2.380 (0.0937), 2.390 (0.0941), 2.400 (0.0945), 2.410 (0.0949), 2.420 (0.0953), 2.430 (0.0957), 2.440 (0.0961), 2.450 (0.0965), 2.460 (0.0969), 2.470 (0.0972), 2.480 (0.0976), 2.490 (0.0980), 2.500 (0.0984), 2.510 (0.0988), 2.520 (0.0992), 2.530 (0.0996), 2.540 (0.1000), 2.550 (0.1004), 2.560 (0.1008), 2.570 (0.1012), 2.580 (0.1016), 2.590 (0.1020), 2.600 (0.1024), 2.620 (0.1031), 2.640 (0.1039), 2.660 (0.1047), 2.680 (0.1055), 2.700 (0.1063), 2.720 (0.1071), 2.740 (0.1079), 2.760 (0.1087), 2.780 (0.1094), 2.800 (0.1102)

Measured clearance — mm (in.):

0.000–0.030 (0.0000–0.0012), 0.031–0.050 (0.0012–0.0020), 0.051–0.070 (0.0020–0.0028), 0.071–0.090 (0.0028–0.0035), 0.091–0.110 (0.0036–0.0043), 0.111–0.130 (0.0044–0.0051), 0.131–0.149 (0.0052–0.0059), 0.150–0.250 (0.0059–0.0098), 0.251–0.270 (0.0099–0.0106), 0.271–0.290 (0.0107–0.0114), 0.291–0.310 (0.0115–0.0122), 0.311–0.330 (0.0122–0.0130), 0.331–0.350 (0.0130–0.0138), 0.351–0.370 (0.0138–0.0146), 0.371–0.390 (0.0146–0.0154), 0.391–0.410 (0.0154–0.0161), 0.411–0.430 (0.0162–0.0169), 0.431–0.450 (0.0170–0.0177), 0.451–0.470 (0.0178–0.0185), 0.471–0.490 (0.0185–0.0193), 0.491–0.510 (0.0193–0.0201), 0.511–0.530 (0.0201–0.0209), 0.531–0.550 (0.0209–0.0217), 0.551–0.570 (0.0217–0.0224), 0.571–0.590 (0.0225–0.0232), 0.591–0.610 (0.0233–0.0240), 0.611–0.630 (0.0241–0.0248), 0.631–0.650 (0.0248–0.0256), 0.651–0.670 (0.0256–0.0264), 0.671–0.690 (0.0264–0.0272), 0.691–0.710 (0.0272–0.0280), 0.711–0.730 (0.0280–0.0287), 0.731–0.750 (0.0288–0.0295), 0.751–0.770 (0.0296–0.0303), 0.771–0.790 (0.0304–0.0311), 0.791–0.810 (0.0311–0.0319), 0.811–0.830 (0.0319–0.0327), 0.831–0.850 (0.0327–0.0335), 0.851–0.870 (0.0335–0.0343), 0.871–0.890 (0.0343–0.0350), 0.891–0.910 (0.0351–0.0358), 0.911–0.930 (0.0359–0.0366), 0.931–0.950 (0.0367–0.0374), 0.951–0.970 (0.0374–0.0382), 0.971–0.990 (0.0382–0.0390), 0.991–1.010 (0.0390–0.0398), 1.011–1.030 (0.0398–0.0406), 1.031–1.050 (0.0406–0.0413)

Shim number / thickness reference

Shim No.	Thickness	Shim No.	Thickness	Shim No.	Thickness
00	2.000 (0.0787)	28	2.280 (0.0898)	56	2.560 (0.1008)
02	2.020 (0.0795)	30	2.300 (0.0906)	58	2.580 (0.1016)
04	2.040 (0.0803)	32	2.320 (0.0913)	60	2.600 (0.1024)
06	2.060 (0.0811)	34	2.340 (0.0921)	62	2.620 (0.1031)
08	2.080 (0.0819)	36	2.360 (0.0929)	64	2.640 (0.1039)
10	2.100 (0.0827)	38	2.380 (0.0937)	66	2.660 (0.1047)
12	2.120 (0.0835)	40	2.400 (0.0945)	68	2.680 (0.1055)
14	2.140 (0.0843)	42	2.420 (0.0953)	70	2.700 (0.1063)
16	2.160 (0.0850)	44	2.440 (0.0961)	72	2.720 (0.1071)
18	2.180 (0.0858)	46	2.460 (0.0969)	74	2.740 (0.1079)
20	2.200 (0.0866)	48	2.480 (0.0976)	76	2.760 (0.1087)
22	2.220 (0.0874)	50	2.500 (0.0984)	78	2.780 (0.1094)
24	2.240 (0.0882)	52	2.520 (0.0992)	80	2.800 (0.1102)
26	2.260 (0.0890)	54	2.540 (0.1000)		

Intake valve clearance (Cold):
0.15 – 0.25 mm (0.006 – 0.010 in.)

EXAMPLE:

The 2.300mm (0.0906 in.) shim is installed, and the measured clearance is 0.440 mm (0.0173 in.). Replace the 2.300 mm (0.0906 in.) shim with a No. 54 shim.

Intake valve shim selection chart—4.7L engine

67170-4RUN-G113

Adjusting Shim Selection Chart (Exhaust)

Measured clearance mm (in.) / Installed shim thickness mm (in.)	2.000 (0.0787)	2.020 (0.0795)	2.040 (0.0803)	2.060 (0.0811)	2.080 (0.0819)	2.100 (0.0827)	2.120 (0.0835)	2.140 (0.0843)	2.160 (0.0850)	2.180 (0.0858)	2.200 (0.0866)	2.210 (0.0870)	2.220 (0.0874)	2.230 (0.0878)	2.240 (0.0882)	2.250 (0.0886)	2.260 (0.0890)	2.270 (0.0894)	2.280 (0.0898)	2.290 (0.0902)	2.300 (0.0906)	2.310 (0.0909)	2.320 (0.0913)	2.330 (0.0917)	2.340 (0.0921)	2.350 (0.0925)	2.360 (0.0929)	2.370 (0.0933)	2.380 (0.0937)	2.390 (0.0941)	2.400 (0.0945)	2.410 (0.0949)	2.420 (0.0953)	2.430 (0.0957)	2.440 (0.0961)	2.450 (0.0965)	2.460 (0.0969)	2.470 (0.0972)	2.480 (0.0976)	2.490 (0.0980)	2.500 (0.0984)	2.510 (0.0988)	2.520 (0.0992)	2.530 (0.0996)	2.540 (0.1000)	2.550 (0.1004)	2.560 (0.1008)	2.570 (0.1012)	2.580 (0.1016)	2.590 (0.1020)	2.600 (0.1024)	2.620 (0.1031)	2.640 (0.1039)	2.660 (0.1047)	2.680 (0.1055)	2.700 (0.1063)	2.720 (0.1071)	2.740 (0.1079)	2.760 (0.1087)	2.780 (0.1094)	2.800 (0.1102)
0.000–0.030 (0.0000–0.0012)												00	00	00	00	00	00	00	00	00	02	02	04	04	06	06	08	08	10	10	12	12	14	14	16	16	18	18	20	20	22	22	24	24	26	26	28	28	30	30	32	34	36	38	40	42	44	46	48	50	52
0.031–0.050 (0.0012–0.0020)											00	00	00	00	00	00	00	02	02	04	04	06	06	08	08	10	10	12	12	14	14	16	16	18	18	20	20	22	22	24	24	26	26	28	28	30	30	32	32	34	34	36	38	40	42	44	46	48	50	52	54
0.051–0.070 (0.0020–0.0028)										00	00	00	00	00	00	02	02	04	04	06	06	08	08	10	10	12	12	14	14	16	16	18	18	20	20	22	22	24	24	26	26	28	28	30	30	32	32	34	34	36	36	38	40	42	44	46	48	50	52	54	56
0.071–0.090 (0.0028–0.0035)									00	00	00	00	00	02	02	04	04	06	06	08	08	10	10	12	12	14	14	16	16	18	18	20	20	22	22	24	24	26	26	28	28	30	30	32	32	34	34	36	36	38	38	40	42	44	46	48	50	52	54	56	58
0.091–0.110 (0.0036–0.0043)								00	00	00	00	02	02	04	04	06	06	08	08	10	10	12	12	14	14	16	16	18	18	20	20	22	22	24	24	26	26	28	28	30	30	32	32	34	34	36	36	38	38	40	40	42	44	46	48	50	52	54	56	58	60
0.111–0.130 (0.0044–0.0051)							00	00	00	00	02	04	04	06	06	08	08	10	10	12	12	14	14	16	16	18	18	20	20	22	22	24	24	26	26	28	28	30	30	32	32	34	34	36	36	38	38	40	40	42	42	44	46	48	50	52	54	56	58	60	62
0.131–0.150 (0.0052–0.0059)						00	00	00	00	02	04	06	06	08	08	10	10	12	12	14	14	16	16	18	18	20	20	22	22	24	24	26	26	28	28	30	30	32	32	34	34	36	36	38	38	40	40	42	42	44	44	46	48	50	52	54	56	58	60	62	64
0.151–0.170 (0.0059–0.0067)					00	00	00	00	02	04	06	08	08	10	10	12	12	14	14	16	16	18	18	20	20	22	22	24	24	26	26	28	28	30	30	32	32	34	34	36	36	38	38	40	40	42	42	44	44	46	46	48	50	52	54	56	58	60	62	64	66
0.171–0.190 (0.0067–0.0075)				00	00	00	00	02	04	06	08	10	10	12	12	14	14	16	16	18	18	20	20	22	22	24	24	26	26	28	28	30	30	32	32	34	34	36	36	38	38	40	40	42	42	44	44	46	46	48	48	50	52	54	56	58	60	62	64	66	68
0.191–0.210 (0.0075–0.0083)			00	00	00	00	02	04	06	08	10	12	12	14	14	16	16	18	18	20	20	22	22	24	24	26	26	28	28	30	30	32	32	34	34	36	36	38	38	40	40	42	42	44	44	46	46	48	48	50	50	52	54	56	58	60	62	64	66	68	70
0.211–0.230 (0.0083–0.0091)		00	00	00	00	02	04	06	08	10	12	14	14	16	16	18	18	20	20	22	22	24	24	26	26	28	28	30	30	32	32	34	34	36	36	38	38	40	40	42	42	44	44	46	46	48	48	50	50	52	52	54	56	58	60	62	64	66	68	70	72
0.231–0.249 (0.0091–0.0098)		00	00	00	02	04	06	08	10	12	14	16	16	18	18	20	20	22	22	24	24	26	26	28	28	30	30	32	32	34	34	36	36	38	38	40	40	42	42	44	44	46	46	48	48	50	50	52	52	54	54	56	58	60	62	64	66	68	70	72	74
0.250–0.350 (0.0098–0.0138)																																																													
0.351–0.370 (0.0138–0.0146)	06	08	10	12	14	16	18	20	22	24	26	28	28	30	30	32	32	34	34	36	36	38	38	40	40	42	42	44	44	46	46	48	48	50	50	52	52	54	54	56	56	58	58	60	60	62	62	64	64	66	66	68	70	72	74	76	78	80	80	80	
0.371–0.390 (0.0146–0.0154)	08	10	12	14	16	18	20	22	24	26	28	30	30	32	32	34	34	36	36	38	38	40	40	42	42	44	44	46	46	48	48	50	50	52	52	54	54	56	56	58	58	60	60	62	62	64	64	66	66	68	68	70	72	74	76	78	80	80	80		
0.391–0.410 (0.0154–0.0161)	10	12	14	16	18	20	22	24	26	28	30	32	32	34	34	36	36	38	38	40	40	42	42	44	44	46	46	48	48	50	50	52	52	54	54	56	56	58	58	60	60	62	62	64	64	66	66	68	68	70	70	72	74	76	78	80	80	80			
0.411–0.430 (0.0162–0.0169)	12	14	16	18	20	22	24	26	28	30	32	34	34	36	36	38	38	40	40	42	42	44	44	46	46	48	48	50	50	52	52	54	54	56	56	58	58	60	60	62	62	64	64	66	66	68	68	70	70	72	72	74	76	78	80	80	80				
0.431–0.450 (0.0170–0.0177)	14	16	18	20	22	24	26	28	30	32	34	36	36	38	38	40	40	42	42	44	44	46	46	48	48	50	50	52	52	54	54	56	56	58	58	60	60	62	62	64	64	66	66	68	68	70	70	72	72	74	74	76	78	80	80	80					
0.451–0.470 (0.0178–0.0185)	16	18	20	22	24	26	28	30	32	34	36	38	38	40	40	42	42	44	44	46	46	48	48	50	50	52	52	54	54	56	56	58	58	60	60	62	62	64	64	66	66	68	68	70	70	72	72	74	74	76	76	78	80	80	80						
0.471–0.490 (0.0185–0.0193)	18	20	22	24	26	28	30	32	34	36	38	40	40	42	42	44	44	46	46	48	48	50	50	52	52	54	54	56	56	58	58	60	60	62	62	64	64	66	66	68	68	70	70	72	72	74	74	76	76	78	78	80	80	80							
0.491–0.510 (0.0193–0.0201)	20	22	24	26	28	30	32	34	36	38	40	42	42	44	44	46	46	48	48	50	50	52	52	54	54	56	56	58	58	60	60	62	62	64	64	66	66	68	68	70	70	72	72	74	74	76	76	78	78	80	80	80	80								
0.511–0.530 (0.0201–0.0209)	22	24	26	28	30	32	34	36	38	40	42	44	44	46	46	48	48	50	50	52	52	54	54	56	56	58	58	60	60	62	62	64	64	66	66	68	68	70	70	72	72	74	74	76	76	78	78	80	80	80	80	80									
0.531–0.550 (0.0209–0.0217)	24	26	28	30	32	34	36	38	40	42	44	46	46	48	48	50	50	52	52	54	54	56	56	58	58	60	60	62	62	64	64	66	66	68	68	70	70	72	72	74	74	76	76	78	78	80	80	80	80	80	80										
0.551–0.570 (0.0217–0.0224)	26	28	30	32	34	36	38	40	42	44	46	48	48	50	50	52	52	54	54	56	56	58	58	60	60	62	62	64	64	66	66	68	68	70	70	72	72	74	74	76	76	78	78	80	80	80	80	80	80	80											
0.571–0.590 (0.0225–0.0232)	28	30	32	34	36	38	40	42	44	46	48	50	50	52	52	54	54	56	56	58	58	60	60	62	62	64	64	66	66	68	68	70	70	72	72	74	74	76	76	78	78	80	80	80	80	80	80	80													
0.591–0.610 (0.0233–0.0240)	30	32	34	36	38	40	42	44	46	48	50	52	52	54	54	56	56	58	58	60	60	62	62	64	64	66	66	68	68	70	70	72	72	74	74	76	76	78	78	80	80	80	80	80	80	80															
0.611–0.630 (0.0241–0.0248)	32	34	36	38	40	42	44	46	48	50	52	54	54	56	56	58	58	60	60	62	62	64	64	66	66	68	68	70	70	72	72	74	74	76	76	78	78	80	80	80	80	80	80	80																	
0.631–0.650 (0.0248–0.0256)	34	36	38	40	42	44	46	48	50	52	54	56	56	58	58	60	60	62	62	64	64	66	66	68	68	70	70	72	72	74	74	76	76	78	78	80	80	80	80	80	80	80																			
0.651–0.670 (0.0256–0.0264)	36	38	40	42	44	46	48	50	52	54	56	58	58	60	60	62	62	64	64	66	66	68	68	70	70	72	72	74	74	76	76	78	78	80	80	80	80	80	80	80																					
0.671–0.690 (0.0264–0.0272)	38	40	42	44	46	48	50	52	54	56	58	60	60	62	62	64	64	66	66	68	68	70	70	72	72	74	74	76	76	78	78	80	80	80	80	80	80	80																							
0.691–0.710 (0.0272–0.0280)	40	42	44	46	48	50	52	54	56	58	60	62	62	64	64	66	66	68	68	70	70	72	72	74	74	76	76	78	78	80	80	80	80	80	80	80																									
0.711–0.730 (0.0280–0.0287)	42	44	46	48	50	52	54	56	58	60	62	64	64	66	66	68	68	70	70	72	72	74	74	76	76	78	78	80	80	80	80	80	80	80																											
0.731–0.750 (0.0288–0.0295)	44	46	48	50	52	54	56	58	60	62	64	66	66	68	68	70	70	72	72	74	74	76	76	78	78	80	80	80	80	80	80	80																													
0.751–0.770 (0.0296–0.0303)	46	48	50	52	54	56	58	60	62	64	66	68	68	70	70	72	72	74	74	76	76	78	78	80	80	80	80	80	80	80																															
0.771–0.790 (0.0304–0.0311)	48	50	52	54	56	58	60	62	64	66	68	70	70	72	72	74	74	76	76	78	78	80	80	80	80	80	80	80																																	
0.791–0.810 (0.0311–0.0319)	50	52	54	56	58	60	62	64	66	68	70	72	72	74	74	76	76	78	78	80	80	80	80	80	80	80																																			
0.811–0.830 (0.0319–0.0327)	52	54	56	58	60	62	64	66	68	70	72	74	74	76	76	78	78	80	80	80	80	80	80	80																																					
0.831–0.850 (0.0327–0.0335)	54	56	58	60	62	64	66	68	70	72	74	76	76	78	78	80	80	80	80	80	80	80																																							
0.851–0.870 (0.0335–0.0343)	56	58	60	62	64	66	68	70	72	74	76	78	78	80	80	80	80	80	80	80																																									
0.871–0.890 (0.0343–0.0350)	58	60	62	64	66	68	70	72	74	76	78	80	80	80	80	80	80	80																																											
0.891–0.910 (0.0351–0.0358)	60	62	64	66	68	70	72	74	76	78	80	80	80	80	80	80																																													
0.911–0.930 (0.0359–0.0366)	62	64	66	68	70	72	74	76	78	80	80	80	80	80																																															
0.931–0.950 (0.0367–0.0374)	64	66	68	70	72	74	76	78	80	80	80	80																																																	
0.951–0.970 (0.0374–0.0382)	66	68	70	72	74	76	78	80	80	80																																																			
0.971–0.990 (0.0382–0.0390)	68	70	72	74	76	78	80	80	80																																																				
0.991–1.010 (0.0390–0.0398)	70	72	74	76	78	80	80	80																																																					
1.011–1.030 (0.0398–0.0406)	72	74	76	78	80	80	80																																																						
1.031–1.050 (0.0406–0.0413)	74	76	78	80	80	80																																																								
1.051–1.070 (0.0414–0.0421)	76	78	80	80	80																																																									
1.071–1.090 (0.0422–0.0429)	78	80	80	80																																																										
1.091–1.110 (0.0430–0.0437)	80	80	80																																																											
1.111–1.130 (0.0437–0.0445)	80	80																																																												
1.131–1.150 (0.0445–0.0453)	80																																																													

Exhaust valve clearance (Cold):
0.25 – 0.35 mm (0.010 – 0.014 in.)

EXAMPLE:
The 2.300mm (0.0906 in.) shim is installed, and the measured clearance is 0.440 mm (0.0173 in.). Replace the 2.300 mm (0.0906 in.) shim with a No. 44 shim.

Shim No.	Thickness	Shim No.	Thickness	Shim No.	Thickness
00	2.000 (0.0787)	28	2.280 (0.0898)	56	2.560 (0.1008)
02	2.020 (0.0795)	30	2.300 (0.0906)	58	2.580 (0.1016)
04	2.040 (0.0803)	32	2.320 (0.0913)	60	2.600 (0.1024)
06	2.060 (0.0811)	34	2.340 (0.0921)	62	2.620 (0.1031)
08	2.080 (0.0819)	36	2.360 (0.0929)	64	2.640 (0.1039)
10	2.100 (0.0827)	38	2.380 (0.0937)	66	2.660 (0.1047)
12	2.120 (0.0835)	40	2.400 (0.0945)	68	2.680 (0.1055)
14	2.140 (0.0843)	42	2.420 (0.0953)	70	2.700 (0.1063)
16	2.160 (0.0850)	44	2.440 (0.0961)	72	2.720 (0.1071)
18	2.180 (0.0858)	46	2.460 (0.0969)	74	2.740 (0.1079)
20	2.200 (0.0866)	48	2.480 (0.0976)	76	2.760 (0.1087)
22	2.220 (0.0874)	50	2.500 (0.0984)	78	2.780 (0.1094)
24	2.240 (0.0882)	52	2.520 (0.0992)	80	2.800 (0.1102)
26	2.260 (0.0890)	54	2.540 (0.1000)		

67170-4RUN-G114

Exhaust valve shim selection chart—4.7L engine

15. Turn the crankshaft 1 full turn (360 degrees) clockwise.

16. Check the valve clearance on the valves indicated in Figure B. Valve clearance should be:
- Intake: 0.006–0.010 inch (0.15–0.25mm)
- Exhaust: 0.010–0.014 inch (0.25–0.35mm)

17. If any valve is not within specifications, record how far out it is. This will determine the replacement shim.

18. If any valve requires adjustment, remove the camshaft(s).

19. Remove the lifter(s) and shims to be replaced. Using a micrometer, measure the shim thickness. Use the accompanying charts to determine the size of the new shim. Shims are available in 41 sizes in increments of 0.020mm from 2.00mm to 2.80mm.

20. Once all out of specification shims are replaced, install the camshafts.

21. The remainder of installation is the reverse of removal.

Starter Motor

REMOVAL & INSTALLATION

3.4L Engine

1. Before servicing the vehicle, refer to the Precautions Section.
2. Remove or disconnect the following:
- Negative battery cable

- Starter electrical connectors
- Starter

To install:

3. Install or connect the following:
- Starter. Tighten the fasteners to 29 ft. lbs. (39 Nm).
- Electrical connections
- Negative battery cable

4.0L Engine

WITH 2WD

1. Before servicing the vehicle, refer to the Precautions Section.
2. Remove the rear under-cover.
3. Remove the No.2 manifold support.
4. Disconnect the starter wiring.
5. Remove the 2 bolts and lower the starter.
6. Installation is the reverse of removal. Torque the starter mounting bolts to 27 ft. lbs. (37 Nm). Torque the manifold brace to 30 ft. lbs. (40 Nm). Torque the under-cover to 21 ft. lbs. (29 Nm).

WITH 4WD

1. Before servicing the vehicle, refer to the Precautions Section.
2. Remove the rear under-cover.
3. Remove the front exhaust pipe.
4. Remove the fender splash shield.
5. Disconnect the steering intermediate shaft.
6. Disconnect the starter wiring.
7. Remove the 2 bolts and lower the starter.
8. Installation is the reverse of removal.

Torque the starter mounting bolts to 27 ft. lbs. (37 Nm). Torque the intermediate shaft bolts to 27 ft. lbs. (36 Nm). Torque the under-cover to 21 ft. lbs. (29 Nm).

4.7L Engine

1. Before servicing the vehicle, refer to the Precautions Section.
2. Drain the cooling system.
3. Remove or disconnect the following:
- Negative battery cable
- Engine appearance cover
- Air intake assembly
- Fuel supply hose
- Fuel injector connectors
- Throttle position sensor connector
- Coolant temperature sensor connector
- Coolant hoses
- Intake manifold
- Coolant by-pass hoses
- Air pump assembly, if equipped
- Starter wiring
- Starter
4. Installation is reverse order of removal. Tighten the starter mounting bolts to 29 ft. lbs. (39 Nm).

Oil Pan

REMOVAL & INSTALLATION

3.4L Engine

1. Before servicing the vehicle, refer to the Precautions Section.

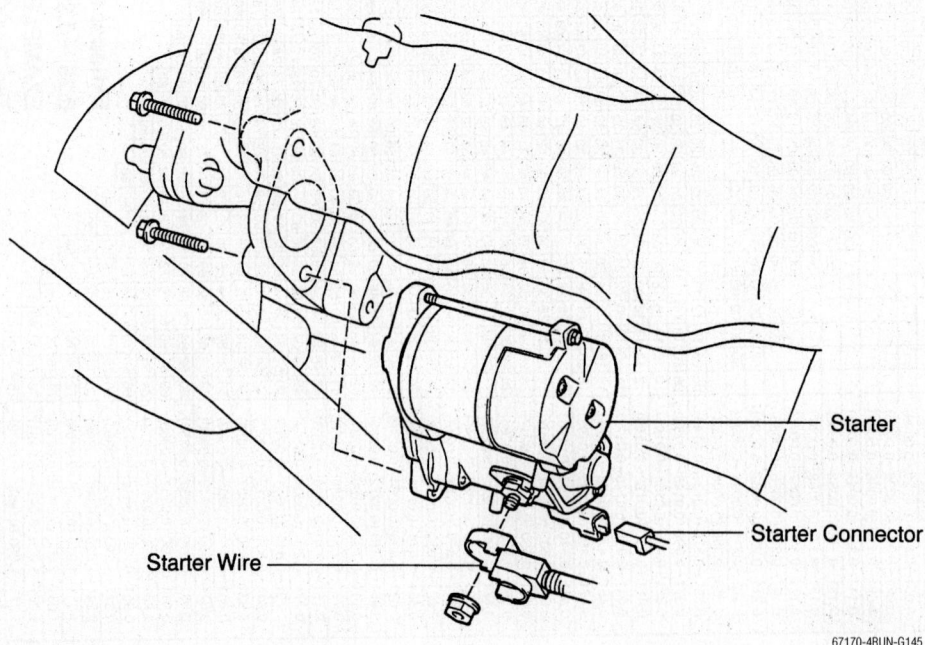

Starter mounting—3.4L engine

Starter

Starter Connector

Starter Wire

67170-4RUN-G145

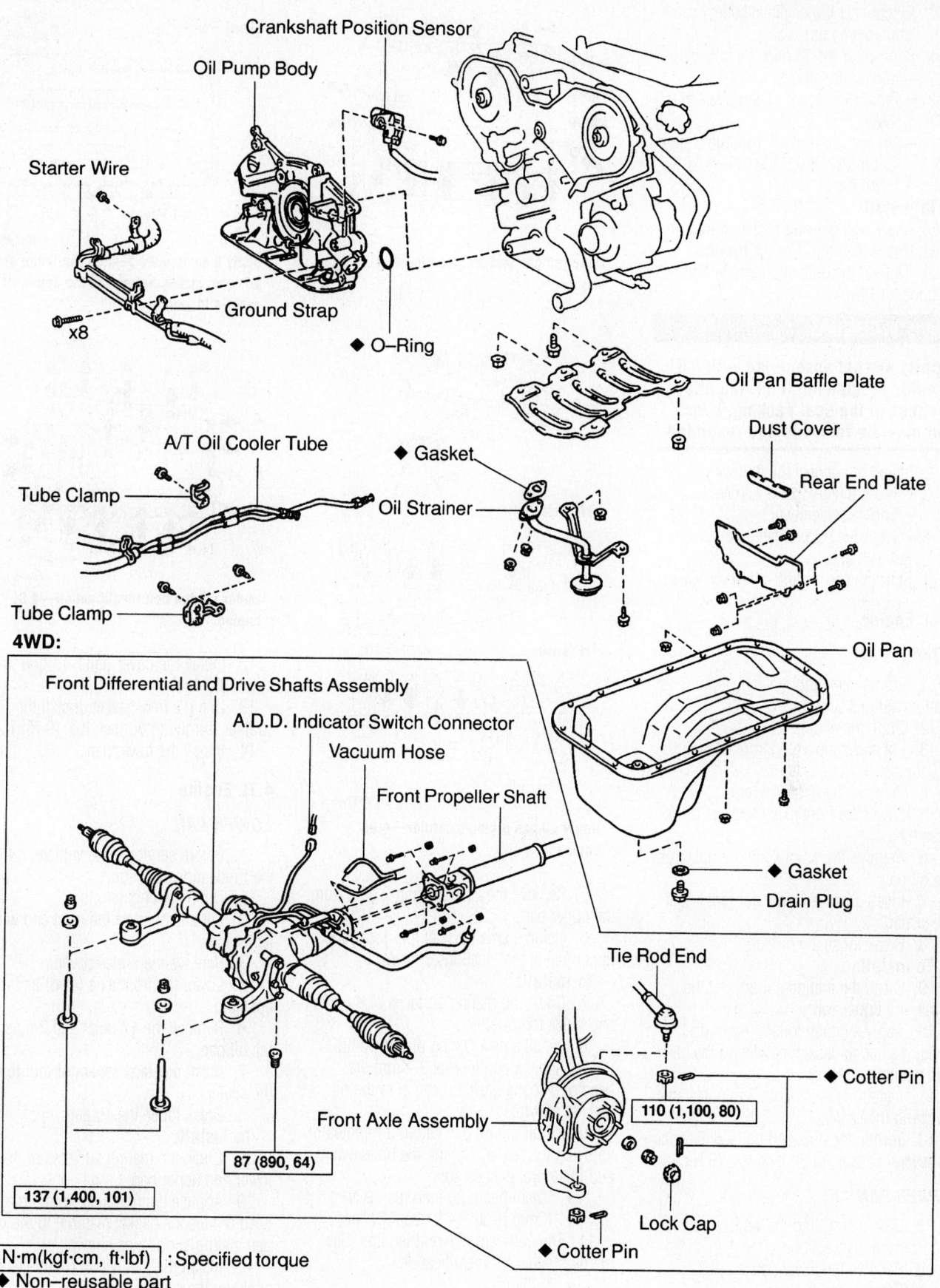

Crankshaft Position Sensor

Oil Pump Body

Starter Wire

Ground Strap

x8

◆ O–Ring

Oil Pan Baffle Plate

Dust Cover

Rear End Plate

A/T Oil Cooler Tube

Tube Clamp

◆ Gasket

Oil Strainer

Tube Clamp

Oil Pan

4WD:

Front Differential and Drive Shafts Assembly

A.D.D. Indicator Switch Connector

Vacuum Hose

Front Propeller Shaft

◆ Gasket

Drain Plug

Tie Rod End

◆ Cotter Pin

110 (1,100, 80)

Front Axle Assembly

Lock Cap

87 (890, 64)

◆ Cotter Pin

137 (1,400, 101)

N·m(kgf·cm, ft·lbf) : Specified torque
◆ Non–reusable part

67170-4RUN-G146

Oil pan and related parts—3.4L engine

2. Disconnect the negative battery cable.
3. Drain the engine oil.
4. Remove or disconnect the following:
   • Engine undercover
   • Front differential, if equipped with 4WD
   • Oil pan, separate it from the engine using SST 09032-00100 and a brass bar

To install:

5. Apply seal packing to the oil pan.
6. Install the oil pan to the cylinder block. Tighten the nuts and bolts to 108 inch lbs. (13 Nm)

❄❄ WARNING

If parts are not assembled within 5 minutes of applying time, the effectiveness of the seal packing is lost and must be removed and reapplied.

7. Install or connect the following:
   • Front differential, if removed
   • Engine undercover
   • Negative battery cable
8. Fill with engine oil.
9. Start the engine and check for leaks.

4.0L Engine

LOWER PAN

1. Before servicing the vehicle, refer to the Precautions Section.
2. Drain the engine oil.
3. Lift and support the front end with jackstands.
4. Remove the under-covers.
5. Lower the front axle (4wd) as far as possible.
6. Remove the 10 bolts and 2 nuts from the oil pan.
7. Insert a gasket separator tool to cut the sealer.
8. Break loose the oil pan.

To install:

9. Clean the mating surfaces of the lower and upper pan.
10. Apply a continuous 3–4mm dia. bead of silicone gasket material to the oil pan mating surface.
11. Install the pan within 3 minutes after applying the sealer.
12. Torque the nuts and bolts evenly and in several passes, to 80 inch lbs. (9 Nm).

UPPER PAN

1. Before servicing the vehicle, refer to the Precautions Section.
2. Remove the lower pan.
3. Remove the strainer.
4. Remove the 4 bellhousing cover bolts and remove the cover.

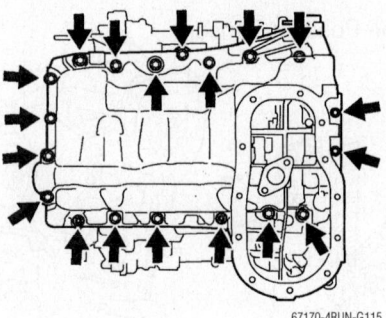

Upper oil pan fastener locations—4.0L engine
67170-4RUN-G115

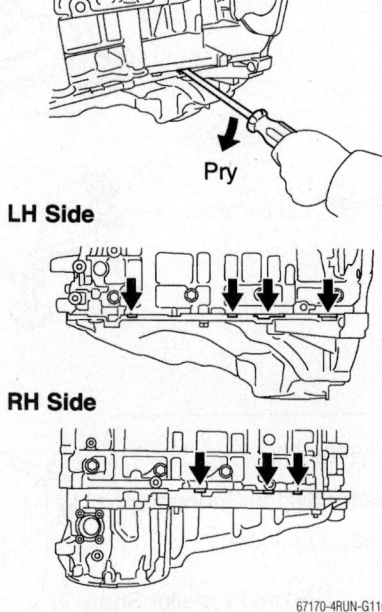

LH Side

RH Side

Upper oil pan prying locations—4.0L engine
67170-4RUN-G116

5. Remove the 17 bolts and 2 nuts from the upper pan.
6. Using a small prybar, pry loose the upper pan at the locations shown.

To install:

7. Clean the mating surfaces of the pan and block thoroughly.
8. Install a new O-ring on the pump.
9. Apply a continuous 3–4mm dia. bead of silicone gasket material to the oil pan, as shown.
10. Install the oil pan within 3 minutes of applying the sealer. Tighten the bolts evenly and in several passes, to:
    • 10mm head: 80 inch lbs. (9 Nm)
    • 12mm head: 14 ft. lbs. (19 Nm)
11. The bolts are different lengths. See the illustration for identification.
    • A: 25mm
    • B: 40mm
    • C: 14mm

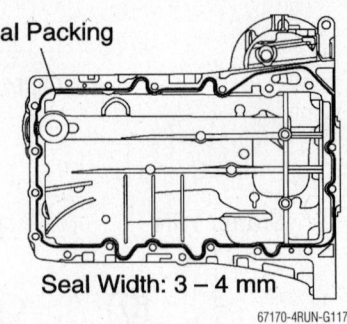

Seal Packing

Seal Width: 3 – 4 mm
67170-4RUN-G117

Apply a continuous 3–4mm dia. bead of silicone gasket material to the upper oil pan—4.0L engine

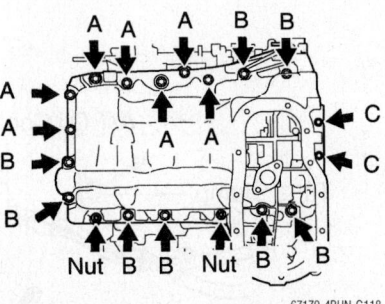

Upper oil pan bolt identification—4.0L engine
67170-4RUN-G118

12. Install the cover bolts. Torque to 27 ft. lbs. (37 Nm).
13. Using a new gasket, install the strainer. Torque to 80 inch lbs. (9 Nm).
14. Install the lower pan.

4.7L Engine

LOWER PAN

1. Before servicing the vehicle, refer to the Precautions Section.
2. Drain the engine oil.
3. Lift and support the front end with jackstands.
4. Remove the under-covers.
5. Lower the front axle (4wd) as far as possible.
6. Remove the 17 bolts and 2 nuts from the oil pan.
7. Insert a gasket separator tool to cut the sealer.
8. Break loose the oil pan.

To install:

9. Clean the mating surfaces of the lower and upper pan.
10. Apply a continuous 3–5mm dia. bead of silicone gasket material to the oil pan mating surface, as shown.
11. Install the pan within 5 minutes after applying the sealer.
12. Torque the nuts and bolts evenly and in several passes, to 66 inch lbs. (7.5 Nm).

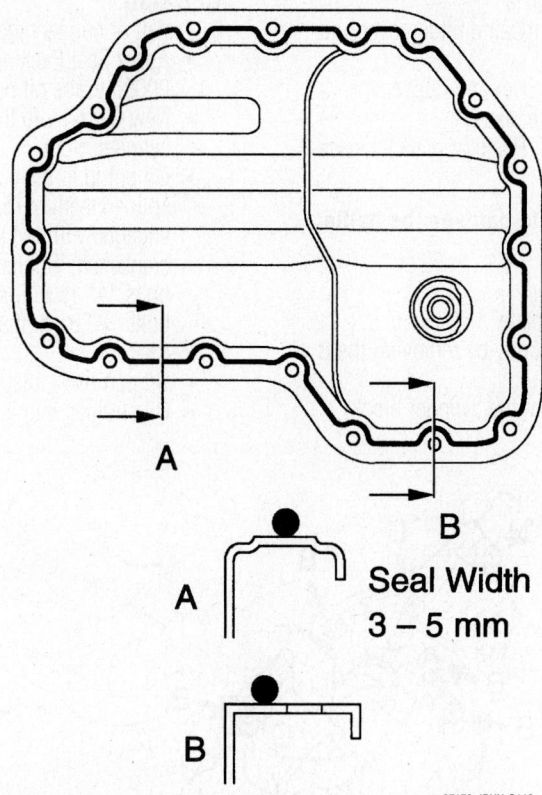

67170-4RUN-G119

Apply a continuous 3–5mm dia. bead of silicone gasket material to the oil pan mating surface

10. Install the oil pan within 5 minutes of applying the sealer. Tighten the bolts and nuts evenly and in several passes, to:
- Bolts A and D: 66 inch lbs. (7.5 Nm)

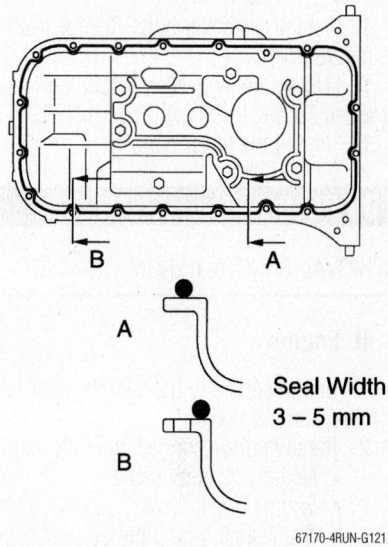

67170-4RUN-G121

Apply a continuous 3–5mm dia. bead of silicone gasket material to the upper oil pan—4.7L engine

UPPER PAN

1. Before servicing the vehicle, refer to the Precautions Section.
2. Remove the lower pan.
3. Remove the baffle.
4. Remove the 4 bellhousing cover bolts and remove the cover.
5. Remove the 18 bolts and 2 nuts from the upper pan.
6. Using a small prybar, pry loose the upper pan.

To install:

7. Clean the mating surfaces of the pan and block thoroughly.
8. Install a new O-ring on the pump.
9. Apply a continuous 3–5mm dia. bead of silicone gasket material to the oil pan, as shown.

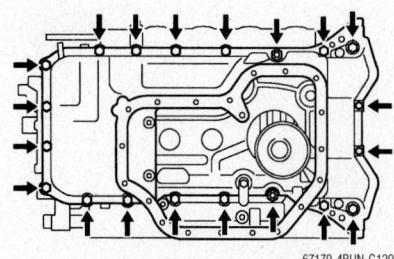

67170-4RUN-G120

Upper oil pan fastener locations—4.7L engine

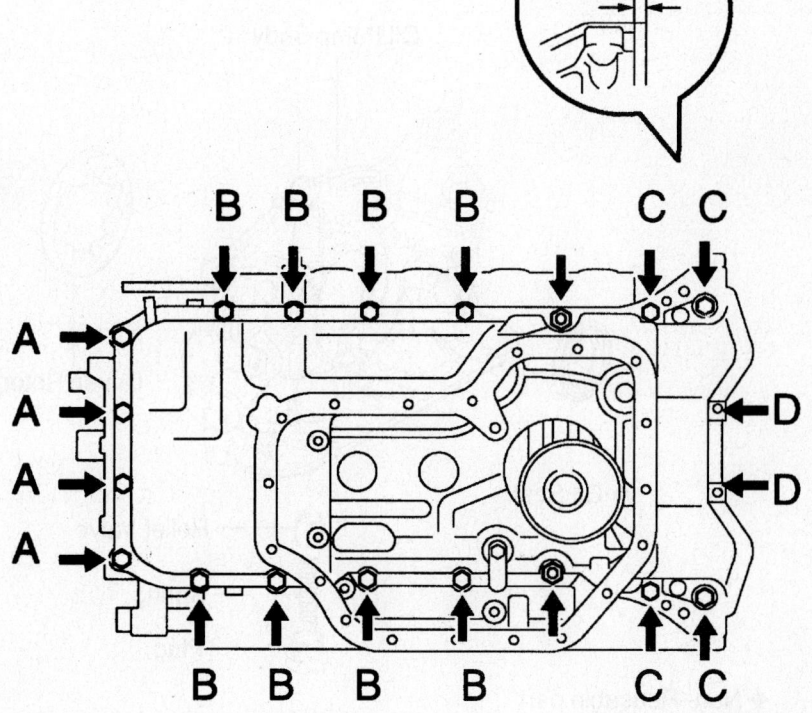

67170-4RUN-G122

Upper oil pan bolt identification—4.7L engine

- Bolts B, C and the nuts: 21 ft. lbs. (28 Nm)

11. The bolts are different lengths. See the illustration for identification.
   - A: 20mm
   - B: 25mm
   - C: 60mm
   - D: 35mm

12. Install the cover bolts. Torque to 27 ft. lbs. (37 Nm).

13. Using a new gasket, install the strainer. Torque to 80 inch lbs. (9 Nm).

14. Install the lower pan.

Oil Pump

REMOVAL & INSTALLATION

3.4L Engine

1. Before servicing the vehicle, refer to the Precautions Section.

2. Remove or disconnect the following:
   - Negative battery cable
   - Engine undercover
   - Crankshaft timing pulley
   - Front differential, if equipped with 4WD

3. Drain the engine oil from the engine.

4. Remove or disconnect the following:
   - Timing belt and crankshaft gear
   - Oil cooler tube and clamp, if equipped with automatic transmission

- Stiffener plate
- Flywheel housing undercover and dust cover
- Rear end cover and dust cover
- Starter wire clamp
- Crankshaft Position (CKP) sensor
- Oil pan

➡**Be careful not to damage the baffle plate flange.**

- Oil strainer
- Oil baffle plate
- Oil pump body by removing the 8 bolts.
- O-ring from the cylinder block

To install:

5. Install or connect the following:
   - Apply Seal Packing PN 08826-00080 to the oil pump
   - New O-ring into the groove of the cylinder block
   - Oil pump to the crankshaft with the splined teeth of the drive rotor engaged with the large teeth of the crankshaft. Tighten the oil pump bolts "A" 15 ft. lbs. (20 Nm) and bolts "B" 31 ft. lbs. (42 Nm)
   - CKP
   - Oil pan baffle plate
   - Oil strainer with a new gasket.

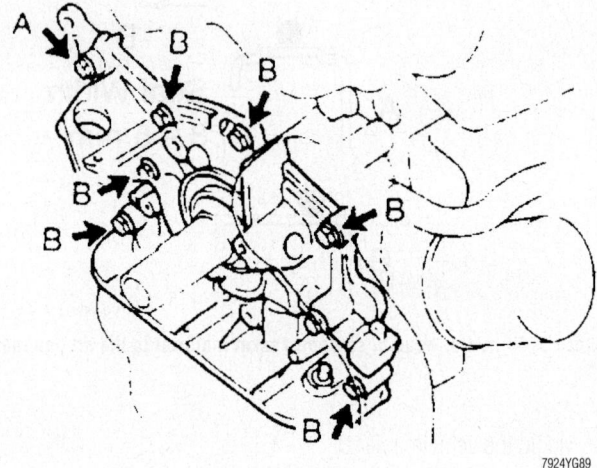

Oil pump bolt identification—3.4L engine

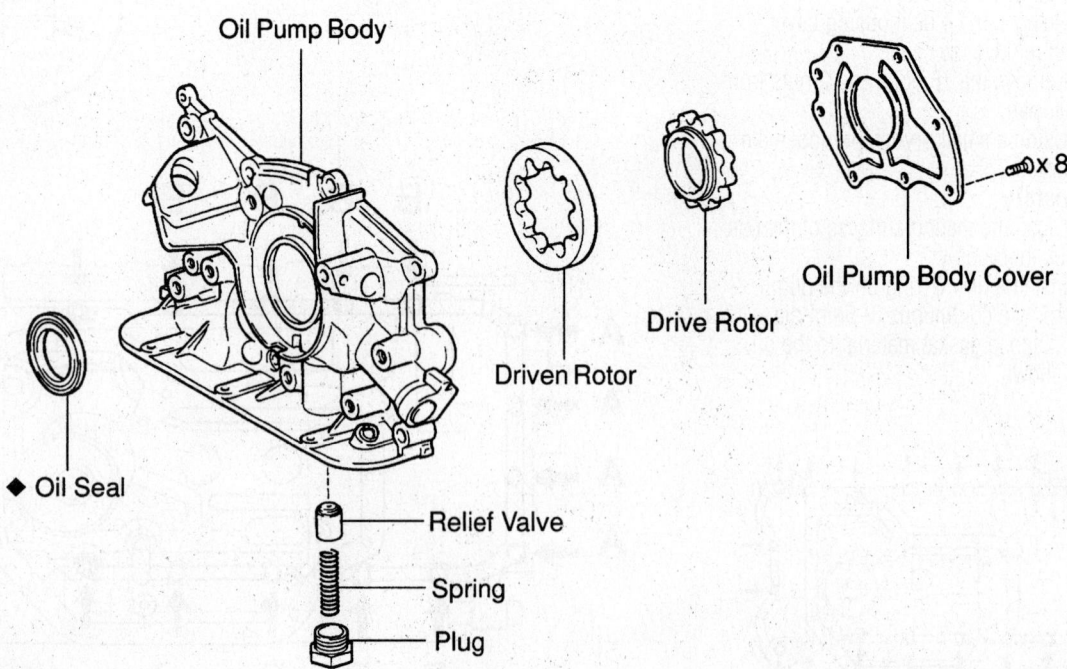

◆ Oil Seal

◆ Non–Reusable part

Oil pump—3.4L engine

Tighten the bolts to 13 ft. lbs. (18 Nm).
- Remaining components
- Negative battery cable
6. Fill with engine oil.
7. Start the engine and check for leaks.

4.0L Engine

1. Before servicing the vehicle, refer to the Precautions Section.
2. Remove the timing chain cover.
3. Remove the oil pipe, 3 bolts.
4. Remove the 2 O-rings.

5. Remove the 7 bolts, oil pump cover drive and driven rotors.
6. Check that the relief valve falls smoothly into the valve hole under its own weight.
7. Place the rotors into the timing chain cover with the marks facing upward.

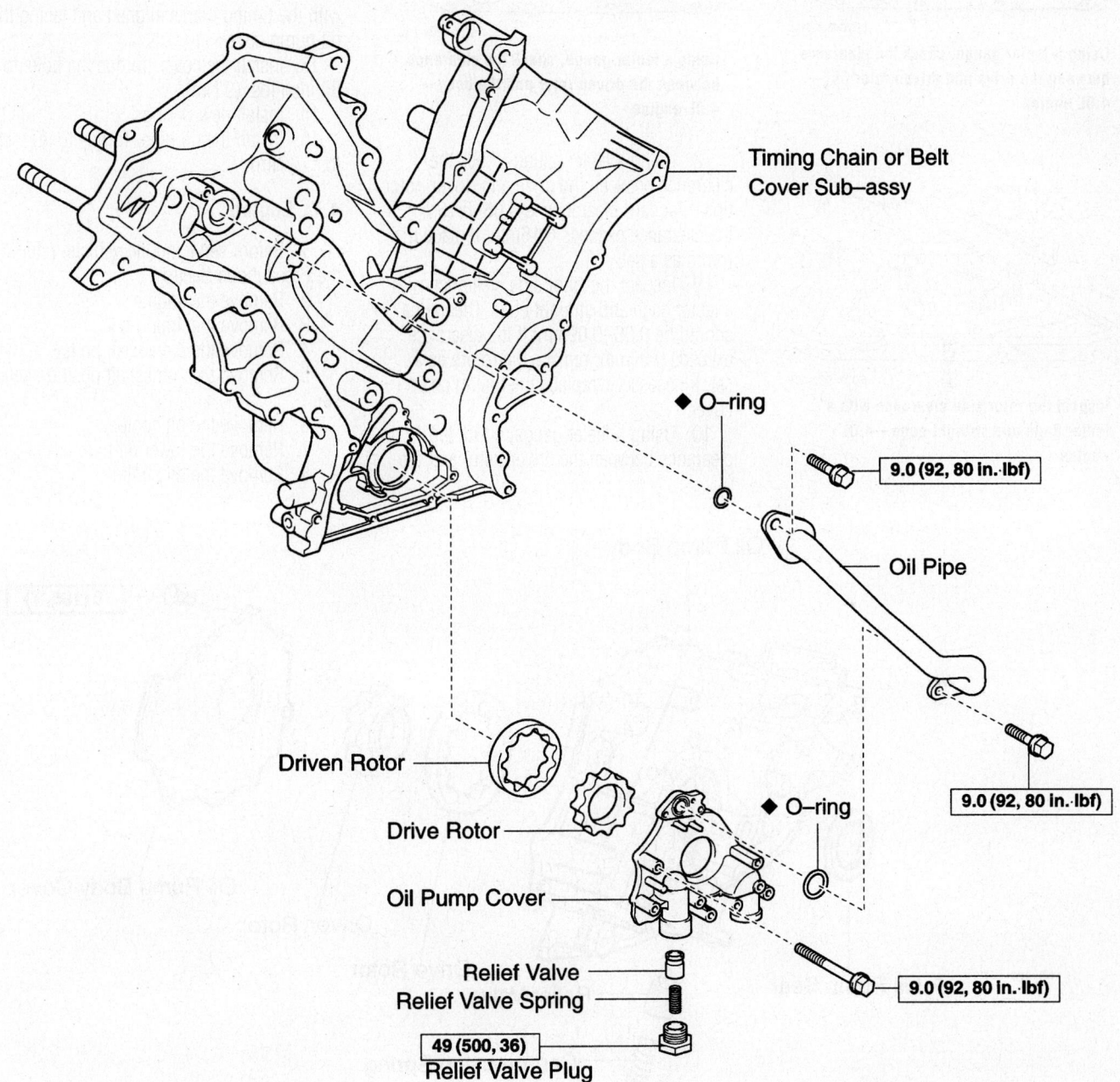

Timing Chain or Belt Cover Sub–assy

◆ O–ring

9.0 (92, 80 in.·lbf)

Oil Pipe

9.0 (92, 80 in.·lbf)

Driven Rotor

Drive Rotor

◆ O–ring

Oil Pump Cover

9.0 (92, 80 in.·lbf)

Relief Valve

Relief Valve Spring

49 (500, 36)

Relief Valve Plug

N·m (kgf·cm, ft·lbf) : Specified torque

◆ Non–reusable part

67170-4RUN-G123

Oil pump and related parts—4.0L engine

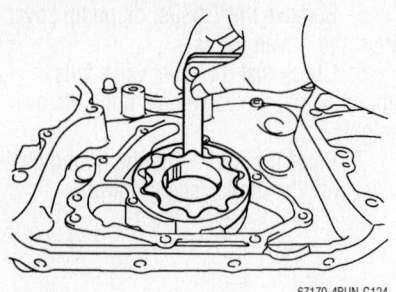

67170-4RUN-G124

Using a feeler gauge, check the clearance between the drive and driven rotor tips—4.0L engine

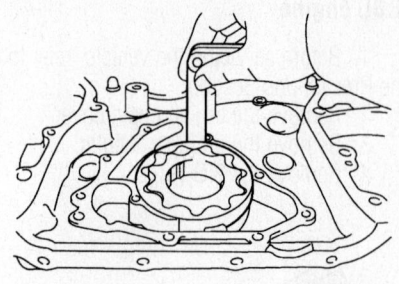

67170-4RUN-G126

Using a feeler gauge, check the clearance between the driven rotor and the body—4.0L engine

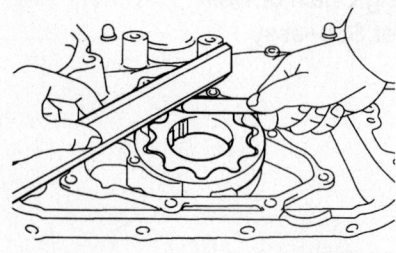

67170-4RUN-G125

Inspect the rotor side clearance with a feeler gage and straight edge—4.0L engine

8. Using a feeler gauge, check the clearance between the drive and driven rotor tips. Standard clearance is 0.06–0.16mm. If the clearance exceeds 0.16mm, replace the rotors as a set.

9. Inspect the rotor side clearance with a feeler gage and straight edge. Clearance should be 0.03–0.09mm. If the clearance exceeds 0.09mm, replace the rotors as a set. If necessary, replace the timing chain cover.

10. Using a feeler gauge, check the clearance between the driven rotor and the body. Clearance should be 0.250–0.325mm. If clearance exceeds 0.325mm, replace the rotors as a set. If necessary, replace the timing chain cover.

11. Coat the relief valve with clean engine oil and install it. Torque to 36 ft. lbs. (49 Nm).

12. Coat the rotors with clean engine oil. Place the rotors in the timing chain cover with the timing marks aligned and facing the oil pump cover.

13. Install the cover. Torque the bolts to 80 inch lbs. (9 Nm).

14. Install new O-rings.

15. Install the oil pipe. Torque to 80 inch lbs. (9 Nm).

4.7L Engine

1. Before servicing the vehicle, refer to the Precautions Section.

2. Remove the engine.

3. Remove the timing belt.

4. Remove the crankshaft pulley.

5. Remove the crankshaft position sensor.

6. Remove the oil cooler.

7. Remove the lower oil pan.

8. Remove the oil strainer.

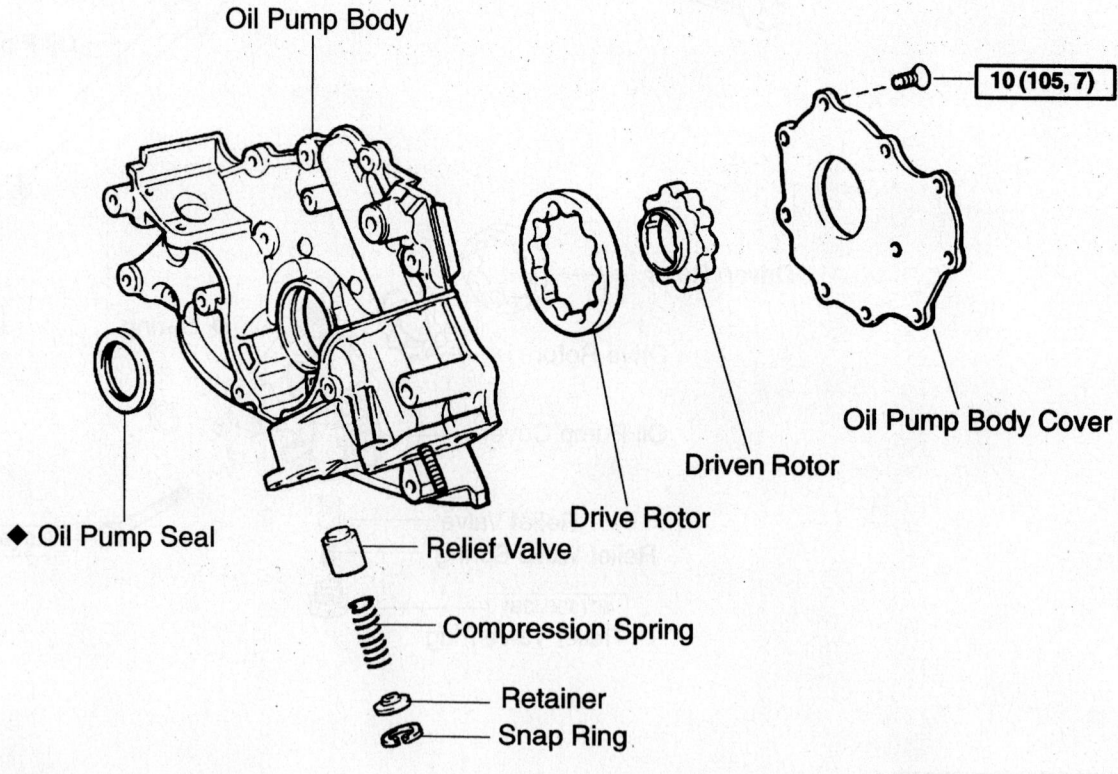

Oil Pump Body

10 (105, 7)

Oil Pump Body Cover

Driven Rotor

Drive Rotor

◆ Oil Pump Seal

Relief Valve

Compression Spring

Retainer

Snap Ring

| N·m (kgf·cm, ft·lbf) | : Specified torque

◆ Non–reusable part

Oil pump and related parts—4.7L engine

67170-4RUN-G127

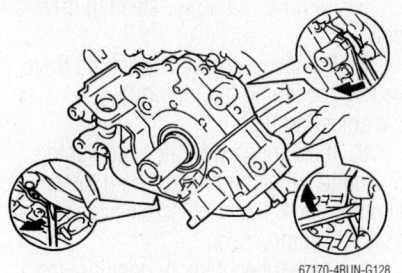

Oil pump prying positions—4.7L engine

9. Remove the baffle plate.
10. Remove the upper oil pan.
11. Remove the 8 bolts and, prying at the positions shown, pry off the oil pump.
12. Remove the O-ring from the block.
13. Remove the timing belt cover seal.

To install:

14. Thoroughly clean the mating surfaces of the pump and block.
15. Install a new front cover seal.
16. Install a new O-ring on the block.
17. Apply new formed-in-place silicone gasket material to the oil pump, as shown. The pump must be installed within 5 minutes of sealer application.
18. Install the bolts. Torque the bolts evenly and in several passes, to:

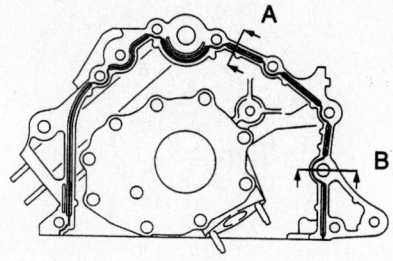

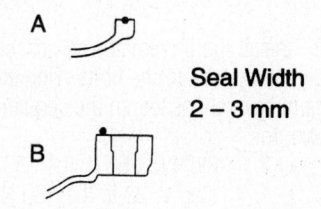

Seal Width 2 – 3 mm

Gasket material application—4.7L engine oil pump

- Bolts with 6mm and 12mm heads: 11 ft. lbs. (16 Nm)
- Bolts with 14mm heads: 23 ft. lbs. (31 Nm)

19. The remainder of installation is the reverse of removal.

Oil Cooler

REMOVAL & INSTALLATION

3.4L Engine

1. Before servicing the vehicle, refer to the Precautions Section.
2. Drain the cooling system.
3. Disconnect the 2 hoses from the cooler.
4. Remove the bolt, gaskets, cooler and O-ring.
5. Installation is the reverse of removal. Use a new O-ring. Torque the cooler bolt to 43 ft. lbs. (59 Nm).

4.0L Engine

1. Before servicing the vehicle, refer to the Precautions Section.
2. Drain the cooling system.
3. Remove the oil filter.
4. Remove the V-bank cover.
5. Disconnect the 2 hoses from the cooler.
6. Remove the bolt, washer and cooler.
7. Installation is the reverse of removal. Use new O-rings. Torque the cooler bolt to

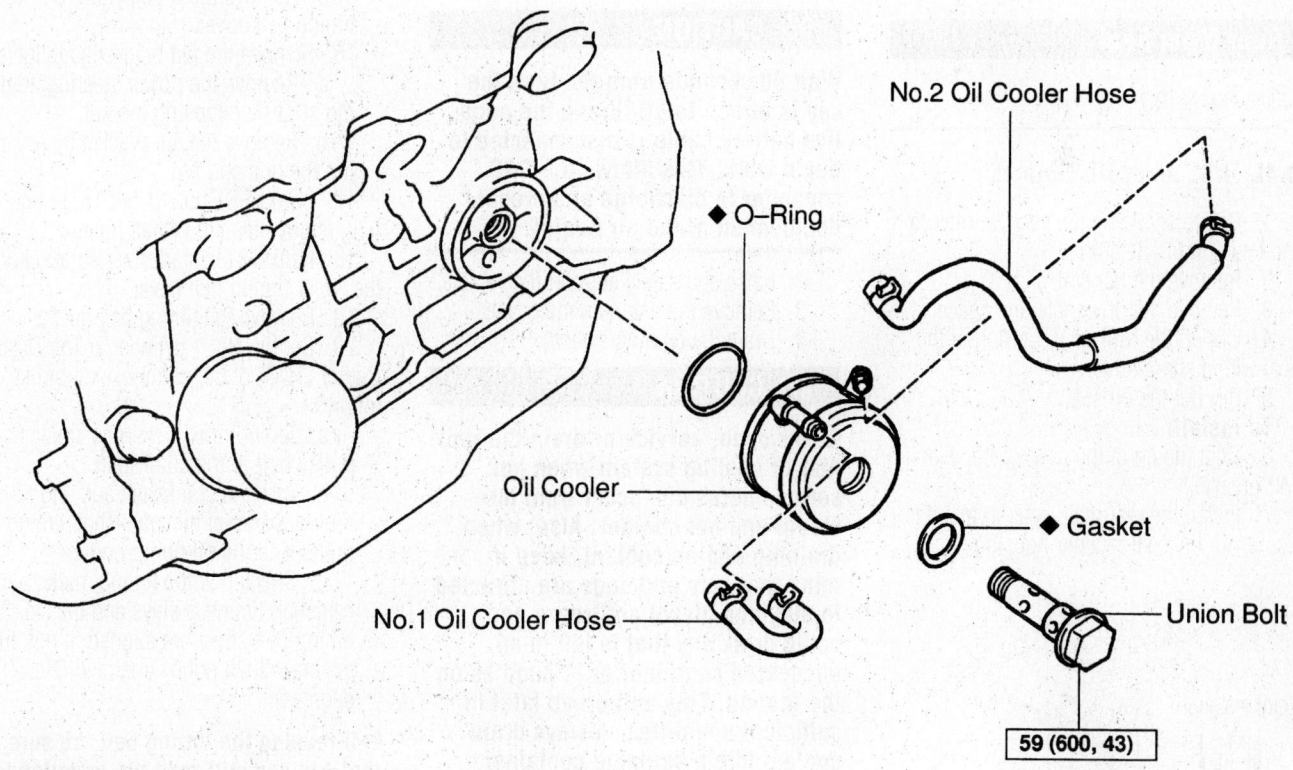

No.2 Oil Cooler Hose

◆ O–Ring

Oil Cooler

◆ Gasket

No.1 Oil Cooler Hose

Union Bolt

59 (600, 43)

N·m(kgf·cm, ft·lbf) : Specified torque
◆ Non–reusable part

Oil Cooler—3.4L engine

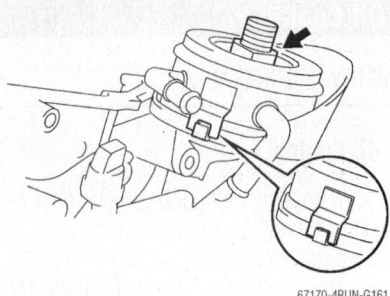

Oil Cooler—4.0L engine

67170-4RUN-G161

50 ft. lbs. (68 Nm). Torque the V-bank cover to 66 inch lbs. (7.5 Nm).

4.7L Engine

1. Before servicing the vehicle, refer to the Precautions Section.
2. Remove the engine under-cover.
3. Drain the cooling system.
4. Drain the engine oil.
5. Remove the oil filter.
6. Disconnect the 2 hoses from the cooler.
7. Remove the bolt, washer and cooler. Discard the O-ring.
8. Installation is the reverse of removal. Use new O-rings. Torque the cooler bolt to 51 ft. lbs. (69 Nm).

Rear Main Seal

REMOVAL & INSTALLATION

3.4L, 4.0L and 4.7L Engines

1. Before servicing the vehicle, refer to the Precautions Section.
2. Remove the transmission
3. Remove the driveplate and spacers.
4. Cut off the rubber lip portion of the seal with a sharp knife.
5. Pry out the oil seal.

To install:

6. Coat the lip of the new spacer with MP grease.
7. Install the rear main seal so that it is flush with the seal retainer housing.

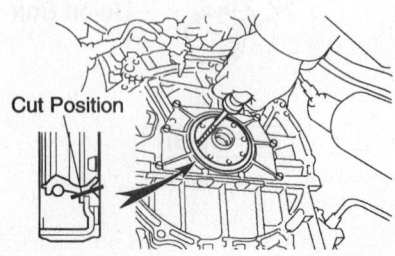

Cut Position

67170-4RUN-G130

Cutting the seal—3.4L and 4.0L engines

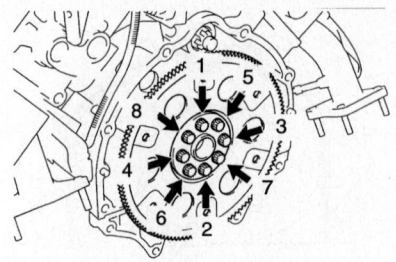

67170-4RUN-G131

Driveplate bolt torque sequence

8. Install the driveplate and spacers. Apply thread lock to the bolts. Tighten the bolts in several passes, in the sequence shown, to:
   - 3.4L and 4.0L: 61 ft. lbs. (83 Nm)
   - 4.7L: step 1, 22 ft. lbs. (30 Nm); step 2, plus 90 degrees; step 3, plus an additional 90 degrees
9. Install the transmission.

Timing Belt

REMOVAL & INSTALLATION

3.4L Engine

1. Disconnect the negative battery cable.

✳✳ CAUTION

Wait 90 seconds from the time the key is turned to LOCK and the negative battery cable is disconnected to begin work. This allows the SRS capacitor to discharge and prevent deployment of the air bag(s).

2. Raise and safely support the vehicle.
3. Remove the engine undercover.
4. Drain the cooling system.

✳✳ CAUTION

Never open, service or drain the radiator or cooling system when hot; serious burns can occur from the steam and hot coolant. Also, when draining engine coolant, keep in mind that cats and dogs are attracted to ethylene glycol antifreeze and could drink any that is left in an uncovered container or in puddles on the ground. This will prove fatal in sufficient quantities. Always drain coolant into a sealable container. Coolant should be reused unless it is contaminated or is several years old.

5. Disconnect the upper radiator hose from the engine.

6. Remove the power steering drive belt.
7. Remove the air conditioning drive belt by loosening the idler pulley nut and the adjusting bolt.
8. If equipped with air conditioning, disconnect the compressor from the engine and set aside. Do not disconnect the lines from the compressor.
9. If equipped with air conditioning, disconnect the air conditioning bracket.
10. Remove the fan with the fluid coupling and fan pulleys.
11. Loosen the lockbolt, pivot bolt, and the adjusting bolt and the alternator drive belt.
12. Remove the No. 2 fan shroud by removing the 2 clips.
13. Disconnect the power steering pump from the engine and set aside. Do not disconnect the lines from the pump.
14. Remove the oil dipstick and the guide.
15. Remove the No. 2 timing belt cover as follows:
    a. Detach the camshaft position sensor connector from the No. 2 timing belt cover.
    b. Disconnect the 4 spark plug wire clamps from the No. 2 timing belt cover.
    c. Remove the 6 bolts and remove the timing belt cover.
16. Remove the fan bracket as follows:
    a. Remove the power steering adjusting strut by removing the nut.
    b. Remove the fan bracket by removing the bolt and nut.
17. Using SST 09213-54015, or equivalent, remove the crankshaft pulley.
18. Remove the starter wire bracket and the No. 1 timing belt cover.
19. Remove the timing belt guide.
20. Set the No. 1 cylinder at Top Dead Center (TDC) of the compression stroke, as follows:
    a. Temporarily install the crankshaft pulley bolt to the crankshaft.
    b. Turn the crankshaft and align the timing marks of the crankshaft timing pulley and the oil pump body.
    c. Check that the timing marks of the camshaft timing pulleys and the No. 3 timing belt cover are aligned. If not, turn the crankshaft pulley one revolution (360 degrees).

➡️If reusing the timing belt, be sure that you can still read the installation marks. If not, place new installation marks on the timing belt to match the timing marks of the camshaft timing pulleys.

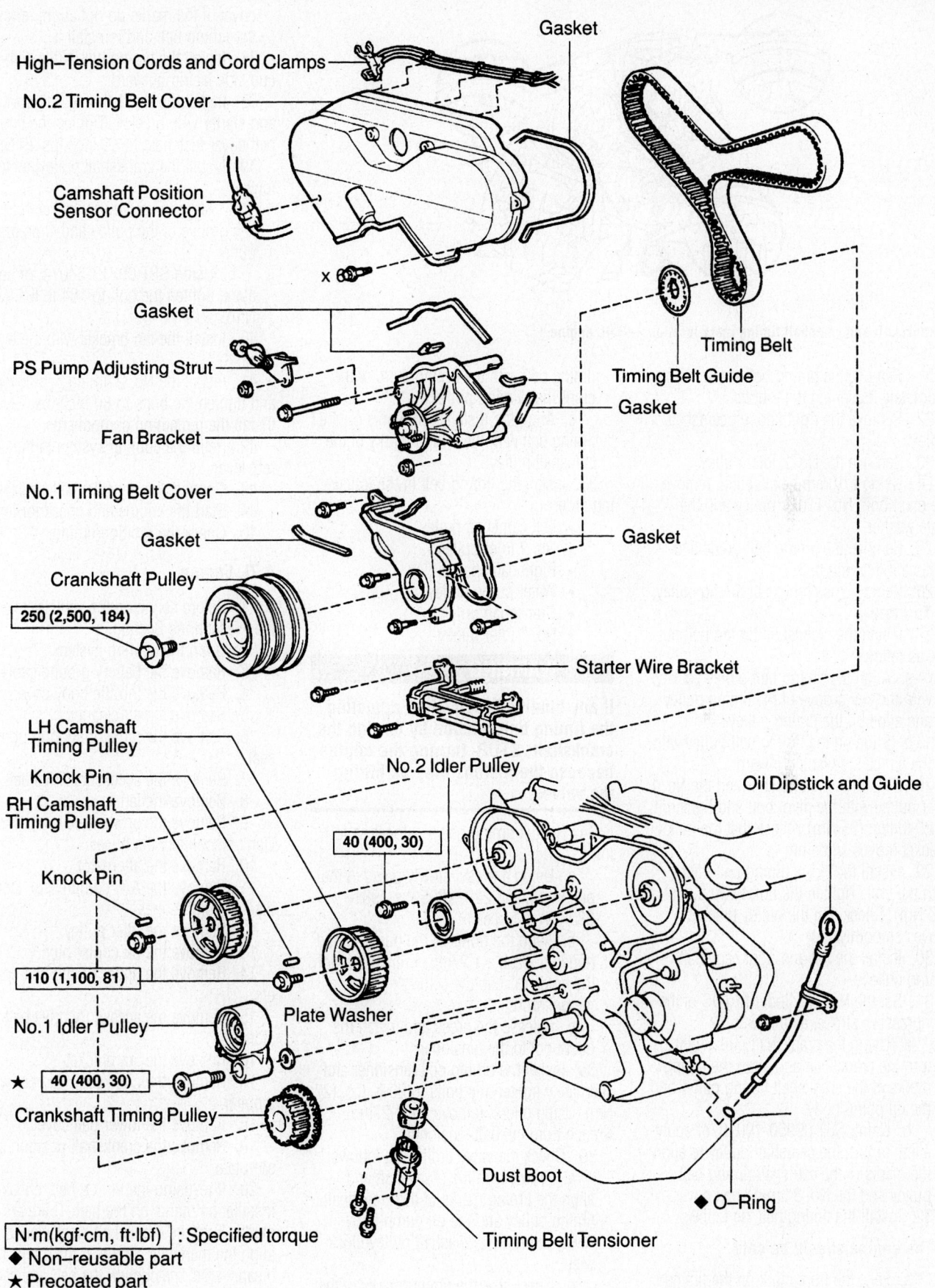

High–Tension Cords and Cord Clamps

No.2 Timing Belt Cover

Camshaft Position Sensor Connector

x

Gasket

Gasket

PS Pump Adjusting Strut

Fan Bracket

Timing Belt

Timing Belt Guide

Gasket

No.1 Timing Belt Cover

Gasket

Crankshaft Pulley

250 (2,500, 184)

Gasket

Starter Wire Bracket

LH Camshaft Timing Pulley

Knock Pin

RH Camshaft Timing Pulley

Knock Pin

No.2 Idler Pulley

40 (400, 30)

Oil Dipstick and Guide

110 (1,100, 81)

Plate Washer

No.1 Idler Pulley

★ 40 (400, 30)

Crankshaft Timing Pulley

Dust Boot

◆ O–Ring

Timing Belt Tensioner

N·m(kgf·cm, ft·lbf) : Specified torque
◆ Non–reusable part
★ Precoated part

Timing belt and related parts—3.4L engine

67170-4RUN-G148

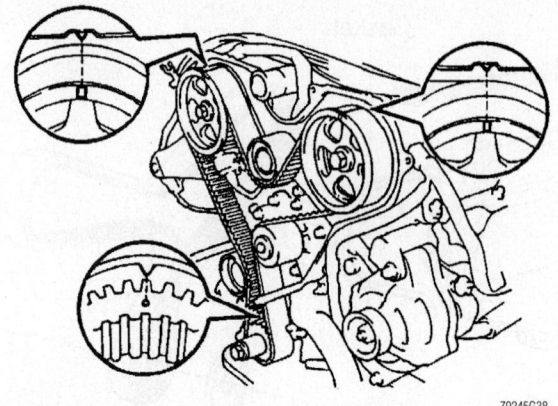

Crankshaft and camshaft timing mark locations—3.4L engine

21. Remove the timing belt tensioner by alternately loosening the 2 bolts.

22. Remove the right and left camshaft pulleys.

23. Remove the No. 2 idler pulley.

24. Using a 10mm hex wrench, remove the pivot bolt, No. 1 idler pulley and the plate washer.

25. Remove the timing belt guide and remove the timing belt.

26. Remove the crankshaft timing pulley.

To install:

27. Install the crankshaft timing belt pulley, as follows:

   a. Align the timing belt pulley set key with the key groove of the timing pulley and slide on the timing pulley.

   b. Slide on the timing belt pulley with the flange side facing inward.

28. Install the plate washer and the No. 1 idler pulley with the pivot bolt and tighten it to 26 ft. lbs. (35 Nm). Check that the pulley bracket moves smoothly.

29. Install the No. 2 timing belt idler with the bolt. Tighten the bolt to 30 ft. lbs. (40 Nm). Check that the pulley bracket moves smoothly.

30. Install the left and right camshaft timing pulleys.

31. Set the No. 1 cylinder to TDC of the compression stroke, as follows:

   a. Using the crankshaft pulley bolt, turn the crankshaft and align the timing marks of the crankshaft timing pulley and the oil pump body.

   b. Using SST 09960-10010, or equivalent, to turn the camshaft pulley to align the marks of the camshaft timing belt pulley and the No. 3 timing belt cover.

32. Install the timing belt, as follows:

➡The engine should be cold.

   a. Face the front mark on the timing belt forward.

   b. Align the installation mark on the timing belt with the timing mark of the crankshaft timing pulley.

   c. Align the installation marks on the timing belt with the timing marks of the camshaft pulleys.

33. Install the timing belt in the following order:

- Left camshaft pulley
- No. 2 idler pulley
- Right camshaft pulley
- Water pump pulley
- Crankshaft pulley
- No. 1 idler pulley

❋❋ WARNING

If any binding is felt when adjusting the timing belt tension by turning the crankshaft, STOP turning the engine, because the pistons may be hitting the valves.

34. Set the timing belt tensioner as follows:

   a. Using a press, slowly press in the pushrod using 220–2205 lbs. (981–9807 N) of force.

   b. Align the holes of the pushrod and housing, pass a 1.27mm wrench through the holes to keep the setting position of the pushrod.

   c. Release the press and install the dust boot to the tensioner.

35. Install the timing belt tensioner and alternately tighten the bolts to 20 ft. lbs. (27 Nm). Using pliers, remove the 1.27mm wrench from the belt tensioner.

36. Check the valve timing, as follows:

   a. Slowly turn the crankshaft and align the timing marks of the crankshaft timing pulley and the oil pump body. Always turn the crankshaft pulley clockwise.

   b. Check that the timing marks of the right and left timing pulleys align with the timing marks of the No. 3 timing belt cover. If the marks do not align, remove the timing belt and reinstall it.

37. Install the timing belt guide with the cup side facing outward.

38. Install the No. 1 timing belt cover and starter wire bracket. Tighten the timing belt cover fasteners to 80 inch lbs. (9 Nm).

39. Install the crankshaft pulley, as follows:

   a. Align the pulley set key with the key groove of the pulley and slide the pulley.

   b. Using SST 09213-54014, or equivalent, tighten the bolt to 184 ft. lbs. (250 Nm).

40. Install the fan bracket with the bolt and nut.

41. Install the No. 2 timing belt cover, and tighten the bolts to 80 inch lbs. (9 Nm). Install the remaining components.

42. Refill the cooling system to the correct level.

43. Connect the negative battery cable.

44. Start the engine and check for leaks.

45. Check the ignition timing.

4.7L Engine

1. Before servicing the vehicle, refer to the Precautions Section.

2. Drain the cooling system.

3. Remove the battery ground cable.

4. Remove the throttle body cover.

5. Remove the air cleaner.

6. Remove the radiator support upper seal.

7. Remove the accessory drive belt.

8. Remove the fan.

9. Remove the power steering pump. Don't disconnect the hoses.

10. Remove the alternator.

11. Remove the A/C compressor. Don't disconnect the lines.

12. Remove the idler pulley.

13. Remove the oil cooler pipe.

14. Remove the upper timing belt covers.

15. Remove the accessory drive belt tensioner.

16. Remove the fan bracket.

17. Remove the crankshaft damper with a puller.

18. Remove the lower belt cover.

19. Remove the crankshaft position sensor plate.

20. If re-using the timing belt, check the installation marks on the belt. There are 3 marks. Turn the crankshaft clockwise to align the marks as shown. If the marks have disappeared, matchmark the belt and each pulley.

21. Using the Crankshaft Pulley Holding

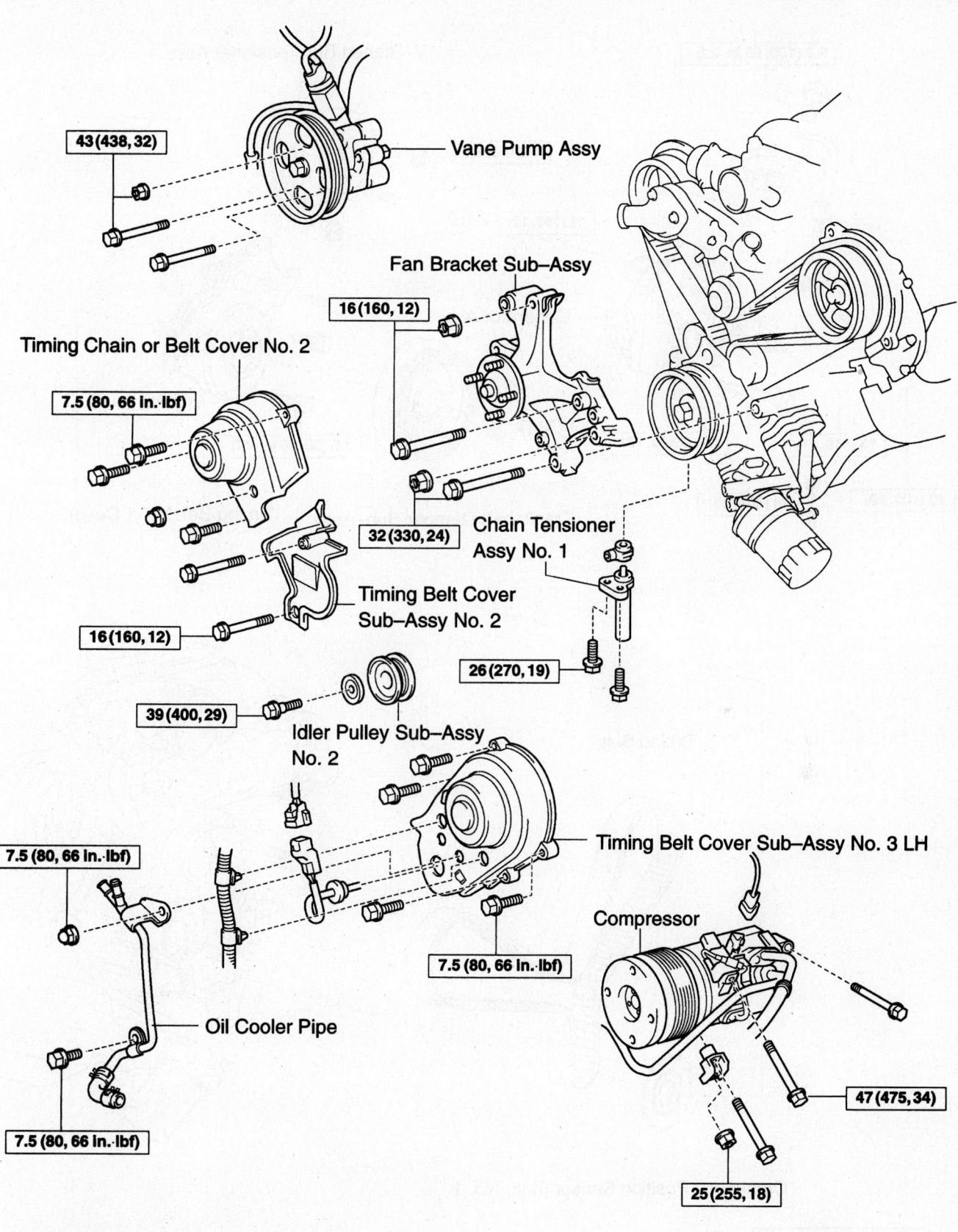

43 (438, 32) — Vane Pump Assy

Fan Bracket Sub–Assy

16 (160, 12)

Timing Chain or Belt Cover No. 2

7.5 (80, 66 in.·lbf)

Chain Tensioner Assy No. 1

32 (330, 24)

Timing Belt Cover Sub–Assy No. 2

16 (160, 12)

26 (270, 19)

39 (400, 29)

Idler Pulley Sub–Assy No. 2

Timing Belt Cover Sub–Assy No. 3 LH

7.5 (80, 66 in.·lbf)

Compressor

7.5 (80, 66 in.·lbf)

Oil Cooler Pipe

47 (475, 34)

7.5 (80, 66 in.·lbf)

25 (255, 18)

N·m (kgf·cm, ft·lbf) : Specified torque

67170-4RUN-G149

These parts must be removed prior to timing belt removal—4.7L engine

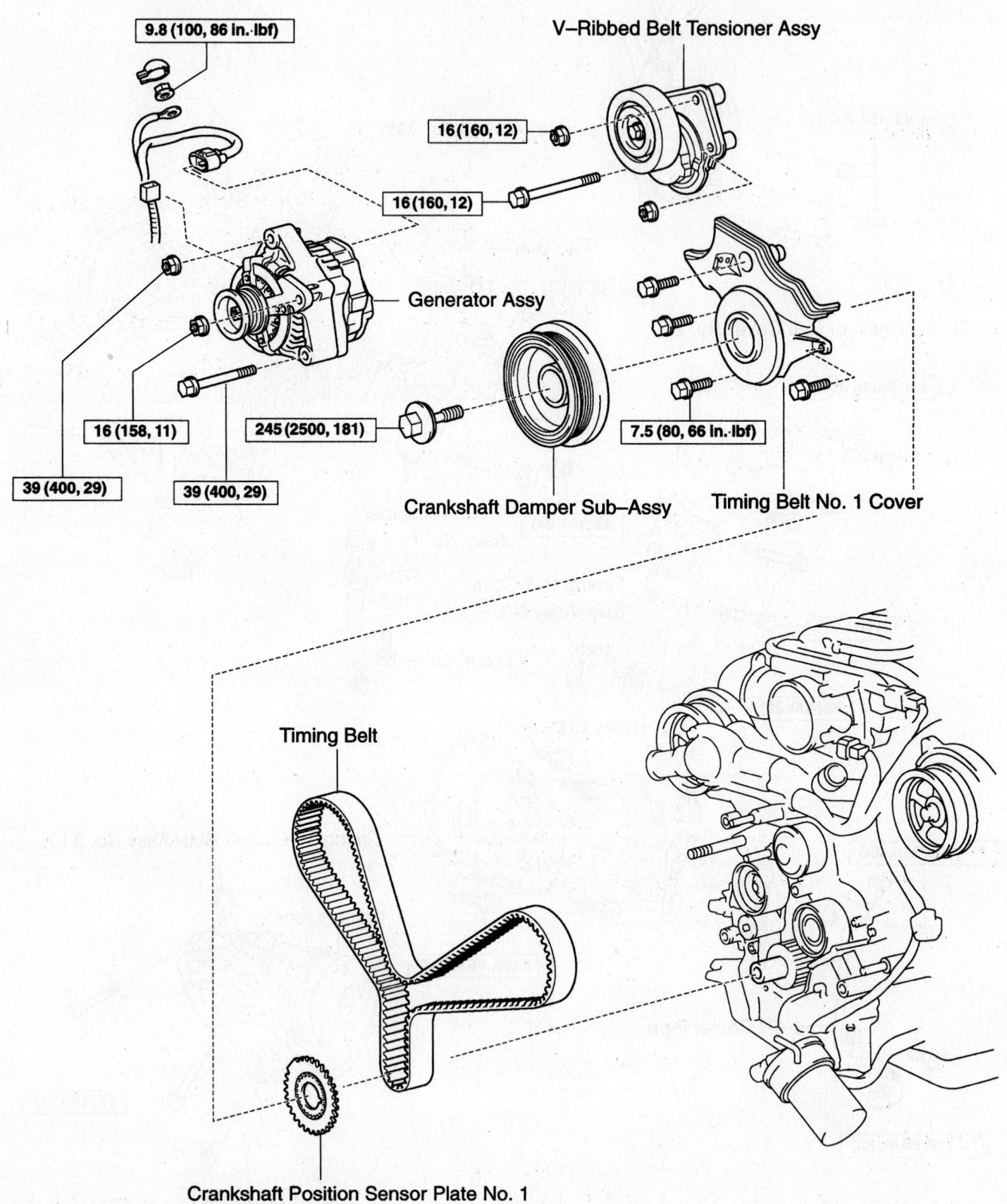

V-Ribbed Belt Tensioner Assy

9.8 (100, 86 in.·lbf)

16 (160, 12)

16 (160, 12)

Generator Assy

16 (158, 11)

245 (2500, 181)

7.5 (80, 66 in.·lbf)

39 (400, 29)

39 (400, 29)

Crankshaft Damper Sub–Assy

Timing Belt No. 1 Cover

Timing Belt

Crankshaft Position Sensor Plate No. 1

N·m (kgf·cm, ft·lbf) : Specified torque

Timing belt and related parts—4.7L engine

67170-4RUN-G150

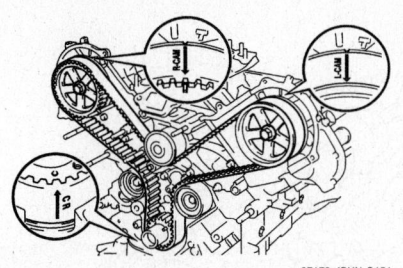

67170-4RUN-G151

Turn the crankshaft clockwise to align the marks

tool 09213-70010, Bolt tool 90105-08076 and Companion Flange Holding tool 09330-00021, or equivalent, loosen the crankshaft pulley bolt.

22. Position the No. 1 cylinder to approximately 50 degrees After Top Dead Center (ATDC) of the compression stroke by performing the following procedures:

a. Rotate the crankshaft pulley (CLOCKWISE) to align its groove with the timing mark "0" on the lower (No. 1) timing belt cover.

b. Check that the camshaft sprocket timing marks are aligned with the rear timing belt plate marks; if not, rotate the crankshaft 1 revolution (360 degrees).

c. Rotate the crankshaft pulley approximately 50 degrees (CLOCKWISE) and align the crankshaft pulley timing mark between the centers of the crankshaft pulley bolt and the idler pulley bolt.

✲✲ WARNING

If the timing belt is disengaged, having the crankshaft pulley in the wrong angle can cause the valve to come into contact with the piston when removing the camshaft pulley.

23. Remove the crankshaft pulley bolt.

➡ **If reusing the timing belt and the installation marks have disappeared, place new installation marks on the timing belt to match the camshaft timing sprocket marks.**

➡ **To avoid meshing the timing sprocket and the timing belt, secure one with a string; then, place matchmarks on the timing belt and the right-side camshaft timing sprocket.**

24. Remove the timing belt tensioner bolts and the tensioner.

25. Using the Camshaft Holding tool 09960-10010, or equivalent, slightly turn the left-side camshaft sprocket clockwise to loosen the tension spring. Then, disconnect the timing belt from the camshaft sprockets.

26. Remove the alternator by performing the following procedures:

a. Disconnect the electrical connector from the alternator.

b. Remove the rubber cap/nut and disconnect the battery wire from the alternator.

c. Disconnect the wire clamp from the alternator cord clip.

d. Remove the alternator-to-engine nuts/bolts and the alternator.

27. Remove the serpentine drive belt tensioner nuts/bolts and the tensioner.

28. Using the Crankshaft Puller Assembly tool 09950-50012, or equivalent, press the crankshaft pulley from the crankshaft.

✲✲ WARNING

DO NOT rotate the crankshaft pulley.

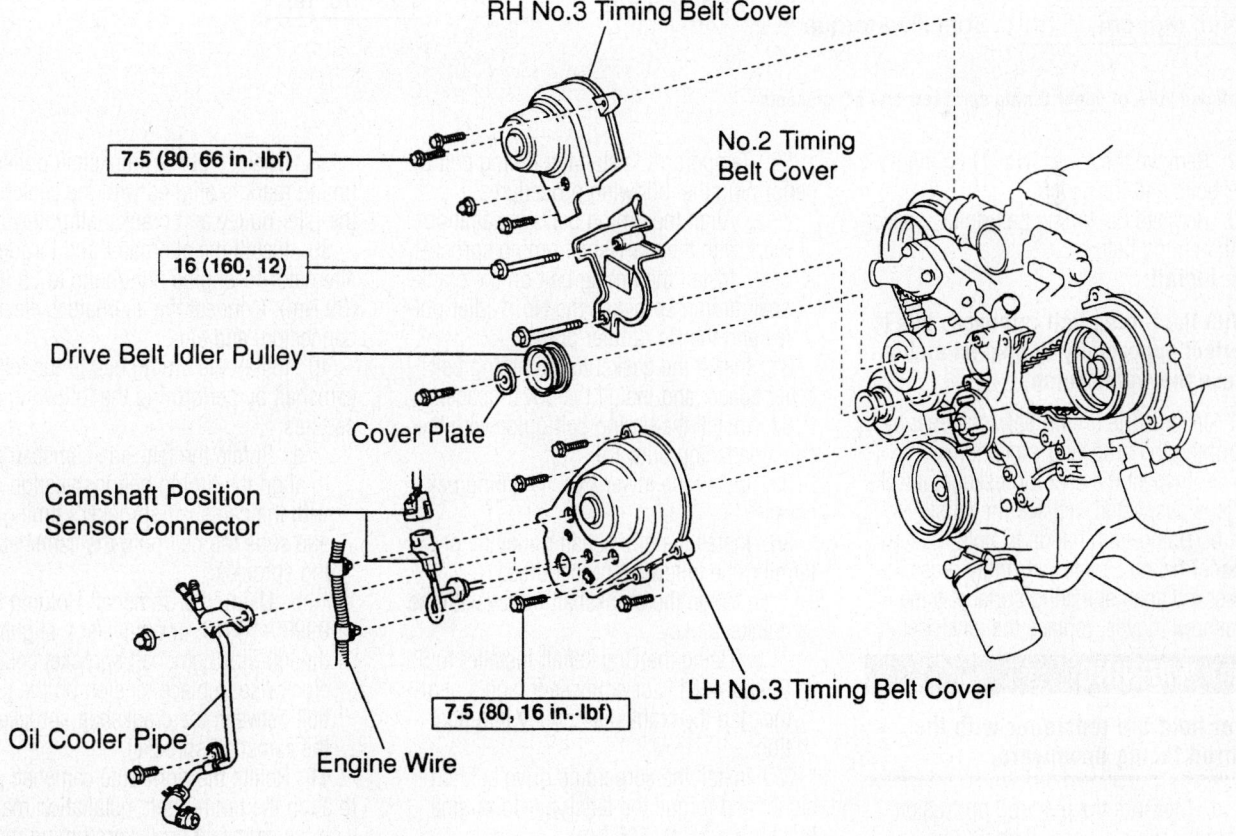

RH No.3 Timing Belt Cover

No.2 Timing Belt Cover

7.5 (80, 66 in.·lbf)

16 (160, 12)

Drive Belt Idler Pulley

Cover Plate

Camshaft Position Sensor Connector

7.5 (80, 16 in.·lbf)

LH No.3 Timing Belt Cover

Oil Cooler Pipe

Engine Wire

N·m (kgf·cm, ft·lbf) : Specified torque

93025G25

Exploded view of upper timing belt covers

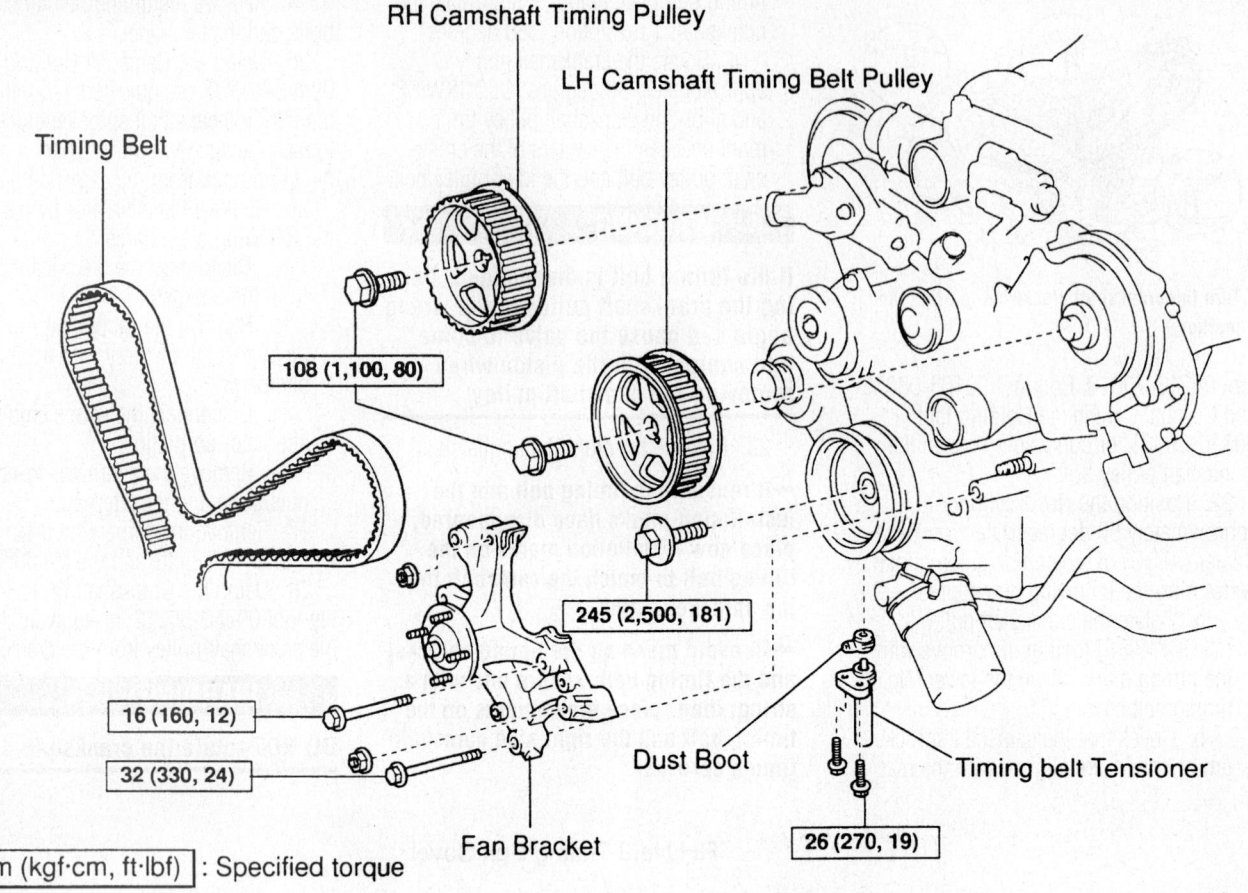

RH Camshaft Timing Pulley

LH Camshaft Timing Belt Pulley

Timing Belt

108 (1,100, 80)

245 (2,500, 181)

16 (160, 12)

32 (330, 24)

Dust Boot

Timing belt Tensioner

Fan Bracket

26 (270, 19)

N·m (kgf·cm, ft·lbf) : Specified torque

93025G26

Exploded view of upper timing sprockets and components

29. Remove the lower (No. 1) timing belt cover bolts and the cover.

30. Remove the timing belt guide, spacer and the timing belt.

To install:

➡ **With the timing belt removed, this is a perfect opportunity to inspect and/or replace the water pump.**

31. Inspect the timing belt tensioner by performing the following procedures:

a. Inspect the seal for leakage; if leakage is suspected, replace the tensioner.

b. Using both hands to hold the tensioner facing upward, strongly press the pushrod against a solid surface. If the pushrod moves, replace the tensioner.

✱✱ WARNING

Never hold the tensioner with the pushrod facing downward.

c. Measure the pushrod protrusion from the housing end, it should be 0.413–0.453 in. (10.5–11.5mm). If the protrusion is not as specified, replace the tensioner.

32. Temporarily install the timing belt by performing the following procedures:

a. Align the timing belt's installation mark with the crankshaft timing sprocket.

b. Install the timing belt on the crankshaft timing sprocket, the No. 1 idler pulley and the No. 2 idler pulley.

33. Install the gasket to the timing belt cover spacer and install the cover spacer.

34. Install the timing belt guide with the cup side facing outward.

35. Install the lower (No. 1) timing belt cover.

36. Install the crankshaft pulley by performing the following procedures:

a. Align the crankshaft pulley with the crankshaft key.

b. Using the Crankshaft Installer tool 09223-46011, or equivalent, and a hammer, tap the crankshaft pulley into position.

37. Install the serpentine drive belt tensioner and torque the tensioner-to-engine bolts to 12 ft. lbs. (16 Nm).

➡ **To install the serpentine drive belt tensioner, use a bolt 4.18 in. (106mm) in length.**

38. Check that the crankshaft pulley's timing mark is aligned with the centers of the idler pulley and crankshaft pulley bolts.

39. Install the alternator and torque the alternator-to-engine nuts/bolts to 29 ft. lbs. (39 Nm). Connect the alternator's electrical connectors and clip.

40. Install the timing belt to the left-side camshaft by performing the following procedures:

a. Rotate the left-side camshaft pulley to align the timing belt installation mark with the camshaft sprocket's timing mark and slide the belt onto the camshaft timing sprocket.

b. Using the Camshaft Holding tool 09960-10010, or equivalent, slightly turn the left-side camshaft sprocket counterclockwise to place tension on the timing belt between the crankshaft sprocket and the camshaft sprocket.

41. Rotate the right-side camshaft pulley to align the timing belt installation mark with the camshaft sprocket's timing mark and slide the belt onto the camshaft timing sprocket.

42. Using a vertical press, slowly press

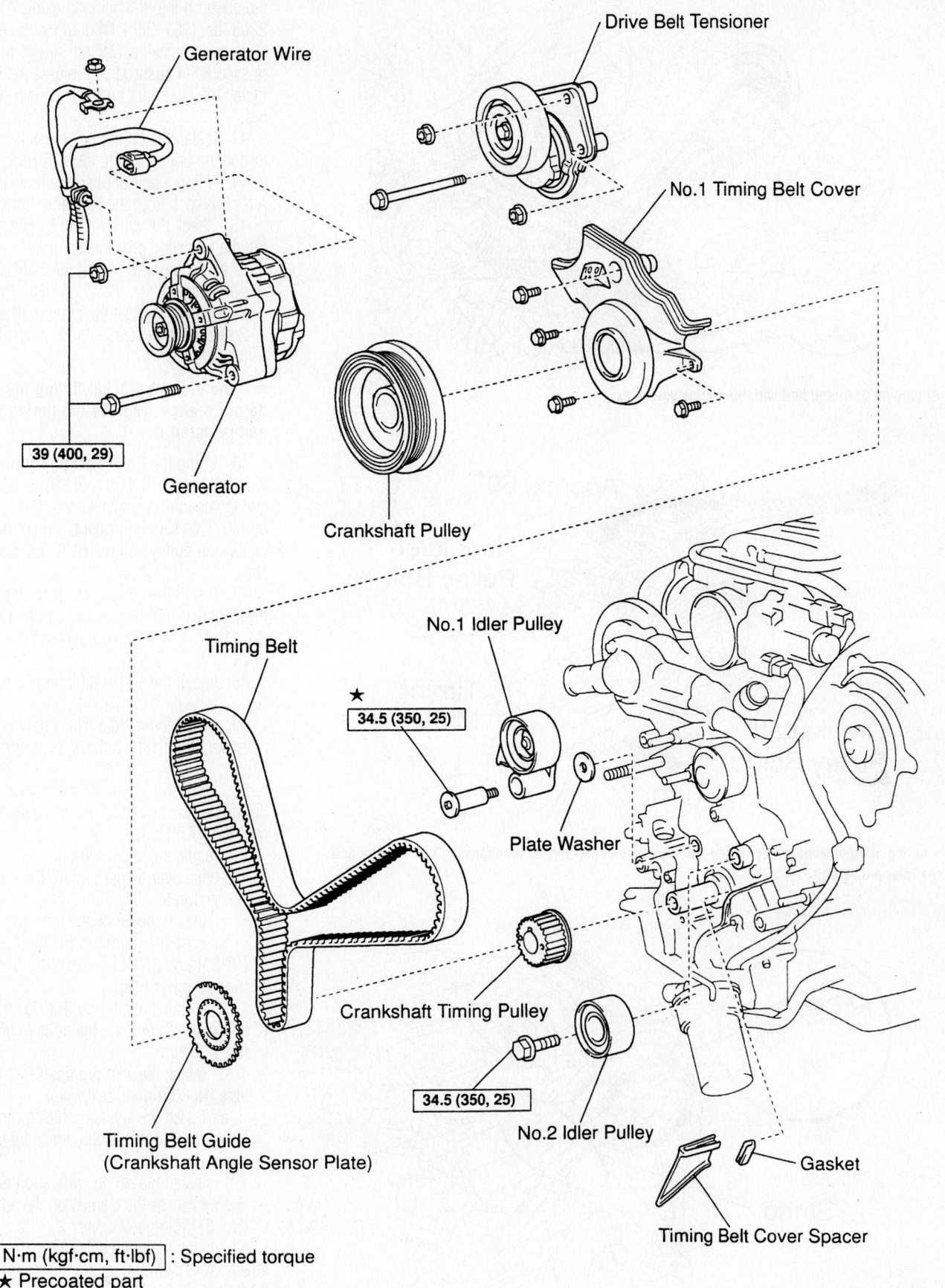

Generator Wire

Drive Belt Tensioner

No.1 Timing Belt Cover

39 (400, 29)

Generator

Crankshaft Pulley

Timing Belt

No.1 Idler Pulley

★ **34.5 (350, 25)**

Plate Washer

Crankshaft Timing Pulley

34.5 (350, 25)

No.2 Idler Pulley

Timing Belt Guide
(Crankshaft Angle Sensor Plate)

Gasket

Timing Belt Cover Spacer

N·m (kgf·cm, ft·lbf) : Specified torque
★ Precoated part

93025G27

Exploded view of lower timing belt cover, sprockets and components

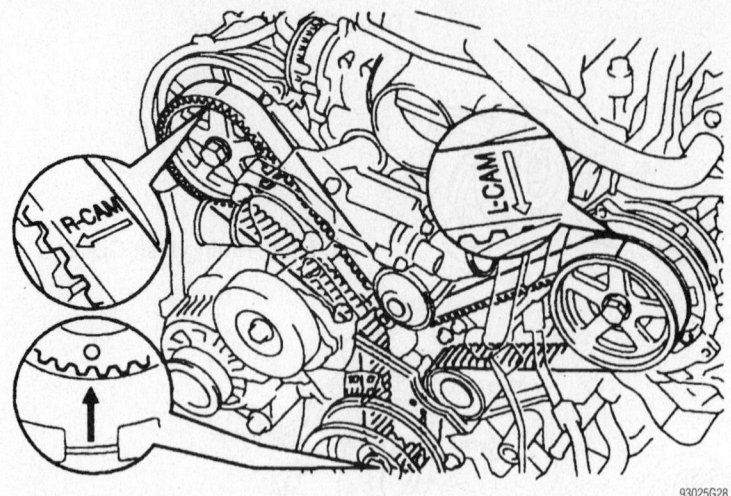

Alignment of timing belt with the timing sprockets

93025G28

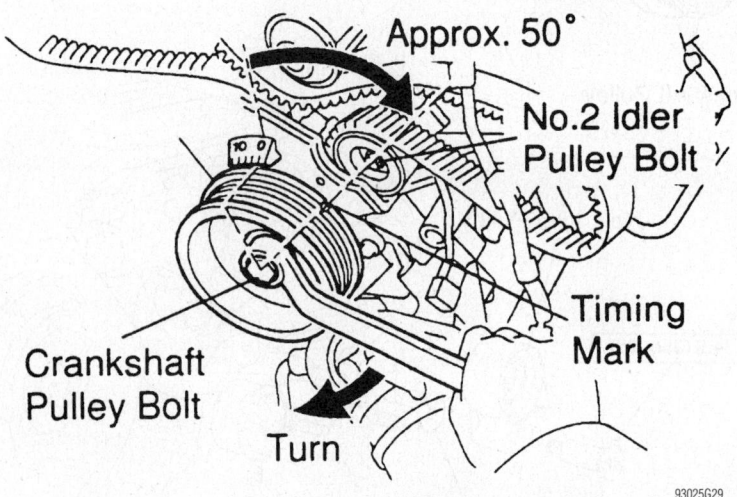

Aligning of crankshaft pulley timing mark with the center line of the crankshaft pulley bolt and the idler pulley bolt

93025G29

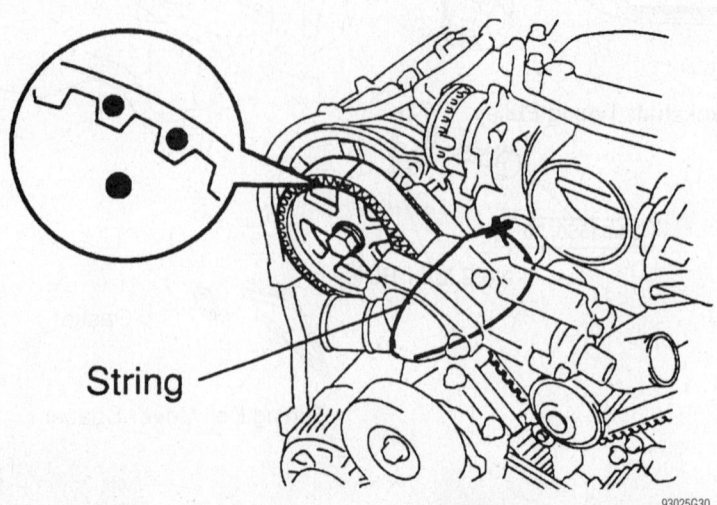

Securing the timing belt with string and matchmarking the camshaft with the timing belt

93025G30

the pushrod into the housing using 200–2205 lbs. (981–9807 N) until the holes align, then, install a 1.27mm Allen® wrench to secure the pushrod and release the press. Install the dust boot on the tensioner housing.

43. Install the timing belt tensioner and torque the bolts to 19 ft. lbs. (26 Nm).

44. Using a pair of pliers, remove the Allen® wrench from the tensioner housing.

45. Check the valve timing by performing the following procedure:

    a. Temporarily install the crankshaft pulley bolt.

    b. Slowly, rotate the crankshaft pulley 2 revolutions (CLOCKWISE) and realign the TDC marks.

➡ If the pulley/sprocket timing marks do not realign, remove the timing belt and reinstall it.

46. Using the Crankshaft Pulley Holding tool 09213-70010, Bolt tool 90105-08076 and Companion Flange Holding tool 09330-00021, or equivalent, torque the crankshaft pulley bolt to 181 ft. lbs. (245 Nm).

47. Install the cooling fan bracket and torque the 12mm (head size) bolt to 12 ft. lbs. (16 Nm) and the 14mm (head size) bolt to 24 ft. lbs. (32 Nm).

48. Install the air conditioning compressor.

49. Install the middle (No. 2) timing belt cover and torque the bolts to 12 ft. lbs. (16 Nm).

50. Install the upper right-side (No. 3) timing belt cover and torque the bolts to 66 inch lbs. (7.5 Nm).

51. Install the upper left-side (No. 3) timing belt cover by performing the following procedures:

    a. Install the oil cooler tube and bolt.

    b. Feed the Camshaft Position Sensor (CPS) through the left-side (No. 3) timing belt cover hole.

    c. Install the left-side (No. 3) timing belt cover and torque the bolts to 66 inch lbs. (7.5 Nm).

    d. Install the wire grommet to the left-side (No. 3) timing belt cover.

    e. Install the sensor connector to the connector bracket and connect the sensor connector.

    f. Install the sensor wire and the engine wire to the clamps on the left-side (No. 3) timing belt cover.

52. Install the drive belt idler pulley and cover plate; then, torque the pulley bolt to 27 ft. lbs. (37 Nm).

53. To complete the installation, reverse the removal procedures.

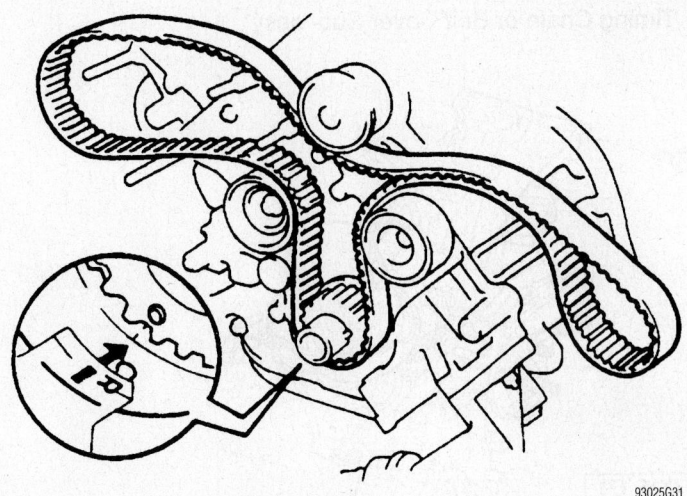

Installing the timing belt on the crankshaft sprocket

93025G31

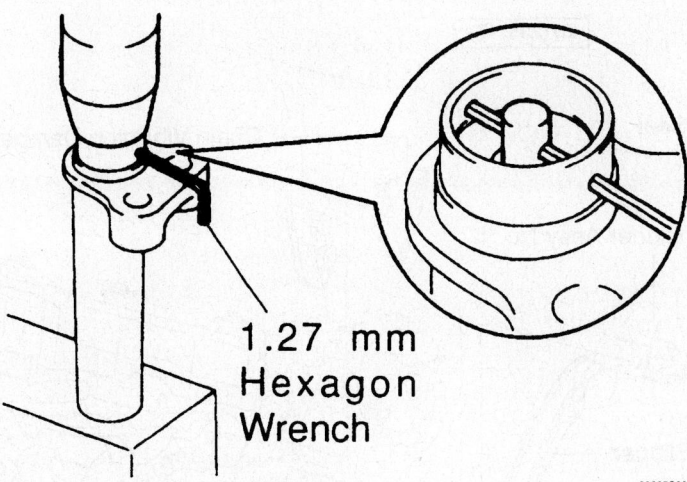

1.27 mm
Hexagon
Wrench

93025G32

Securing the timing belt tensioner pushrod

93025G33

Checking the TDC alignment marks after rotating the crankshaft 2 revolutions

54. Refill the cooling system and connect the negative battery cable.

Timing Chain, Sprockets, Front Cover and Seal

REMOVAL & INSTALLATION

4.0L Engine

1. Before servicing the vehicle, refer to the Precautions Section.
2. Remove the steering gear.
3. Remove the front differential carrier.
4. Drain the cooling system.
5. Remove the battery.
6. Drain the engine oil.
7. Remove the V-bank cover.
8. Remove the radiator upper support seal.
9. Remove the accessory drive belt.
10. Remove the fan.
11. Remove the vent hose.
12. Remove the air cleaner.
13. Remove the oil dipstick tube.
14. Remove the water inlet.
15. Remove the power steering pump. Don't disconnect the lines.
16. Remove the alternator. Don't disconnect the lines.
17. Remove the A/C compressor. Don't disconnect the lines.
18. Remove the accessory drive belt tensioner.
19. Remove the idler pulleys.
20. Remove the crankshaft pulley.
21. Remove the lower oil pan.
22. Remove the oil strainer.
23. Remove the flywheel housing cover.
24. Remove the upper oil pan.
25. Remove the upper intake manifold (air surge tank).
26. Remove the ignition coil.
27. Remove the cylinder head covers.
28. Remove the camshaft timing control valve.
29. Remove the VVT sensor.
30. Remove the oil filter bracket.
31. Remove the timing chain cover (24 bolts, 2 nuts).
32. Remove the timing gear case oil seal.
33. Turn the crankshaft clockwise to align the key with the timing mark on the block. This sets No.1 cylinder at TDC compression.
34. Check that the timing marks on the camshaft sprockets are aligned with the timing marks on the bearing caps as shown. If not, rotate the crankshaft 1 full turn (360 degrees) clockwise and re-check.
35. While turning the chain tensioner

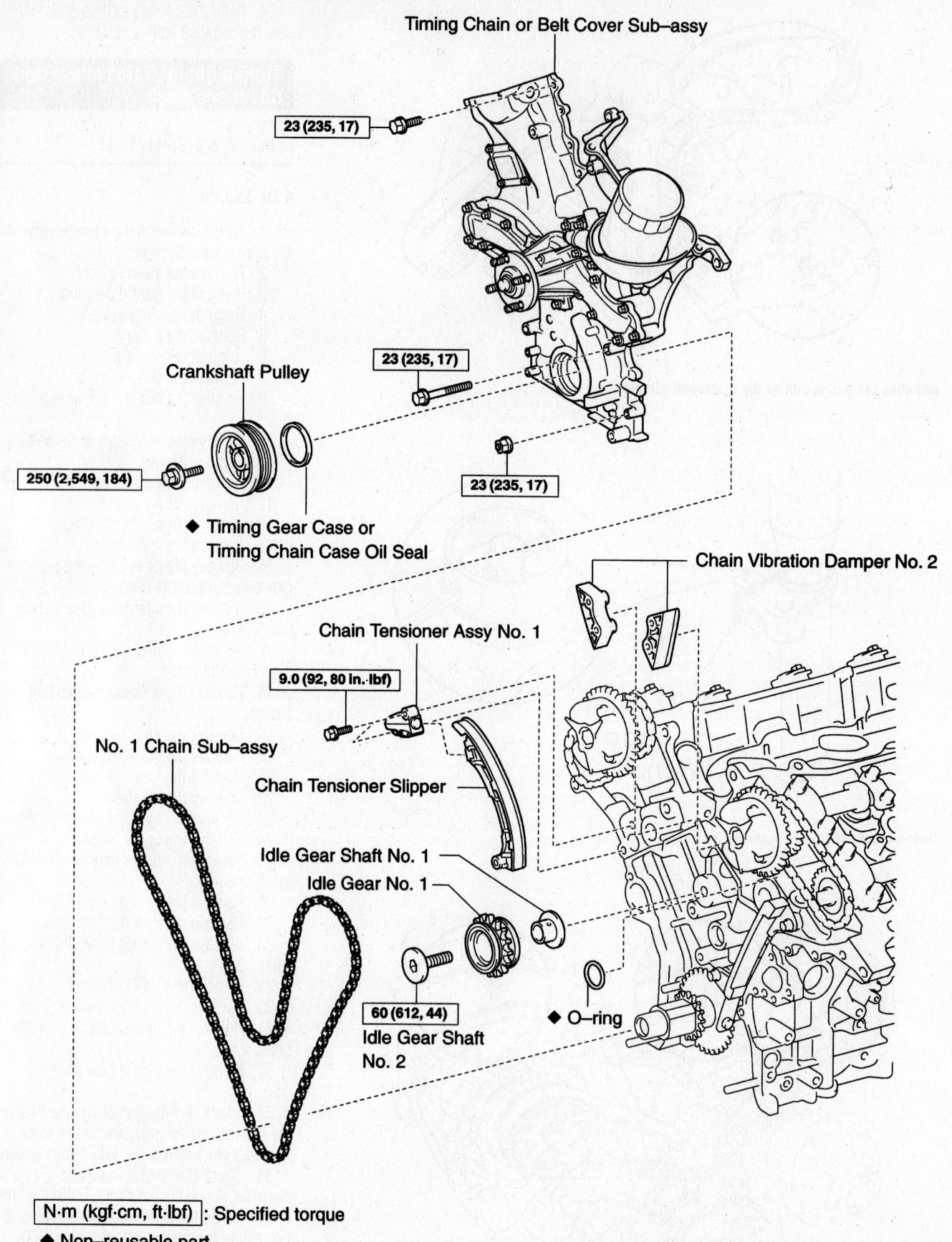

Timing Chain or Belt Cover Sub–assy

23 (235, 17)

23 (235, 17)

Crankshaft Pulley

250 (2,549, 184)

23 (235, 17)

◆ **Timing Gear Case or Timing Chain Case Oil Seal**

Chain Vibration Damper No. 2

Chain Tensioner Assy No. 1

9.0 (92, 80 in.·lbf)

No. 1 Chain Sub–assy

Chain Tensioner Slipper

Idle Gear Shaft No. 1

Idle Gear No. 1

60 (612, 44)

Idle Gear Shaft No. 2

◆ **O–ring**

N·m (kgf·cm, ft·lbf) : Specified torque

◆ Non–reusable part

Timing chain and related parts—4.0L engine

67170-4RUN-G152

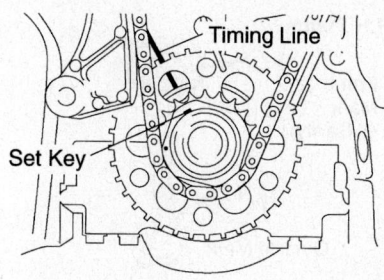

Align the key with the timing mark on the block—4.0L engine

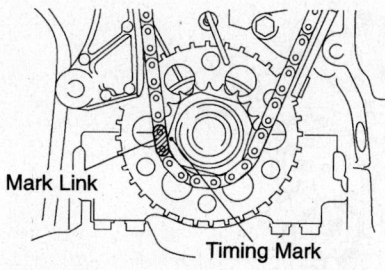

Place the chain on the crankshaft sprocket, aligning the yellow link with the timing mark on the sprocket—4.0L engine

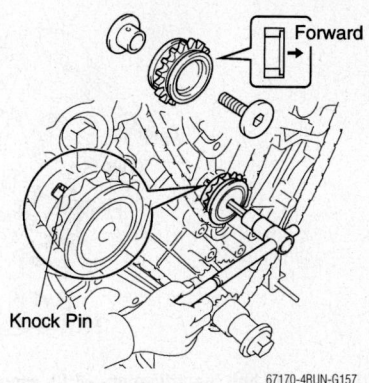

Idler gear installation—4.0L engine

stopper plate upward, push in the plunger. Turn the stopper plate downward and install a 3.5mm pin into the holes to hold the stopper plate.

36. Remove the tensioner slipper.
37. Remove the idle gear.
38. Remove the vibration damper.
39. Remove the chain.

To install:

40. Install the tensioner slipper.
41. Install the tensioner. Torque to 80 inch lbs. (9 Nm).
42. Verify that all timing marks are still aligned.
43. Place the chain on the crankshaft

Install the chain on the camshafts, aligning the orange links with the timing marks—4.0L engine

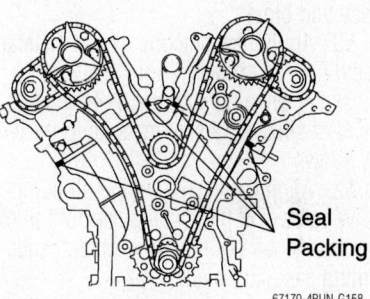

Gasket material application points on the block—4.0L engine

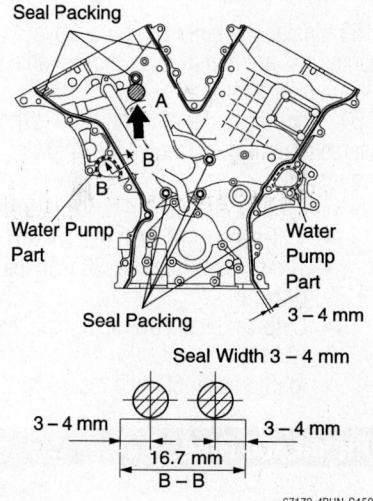

Apply a continuous 3–4mm dia. bead of silicone gasket material to the case—4.0L engine

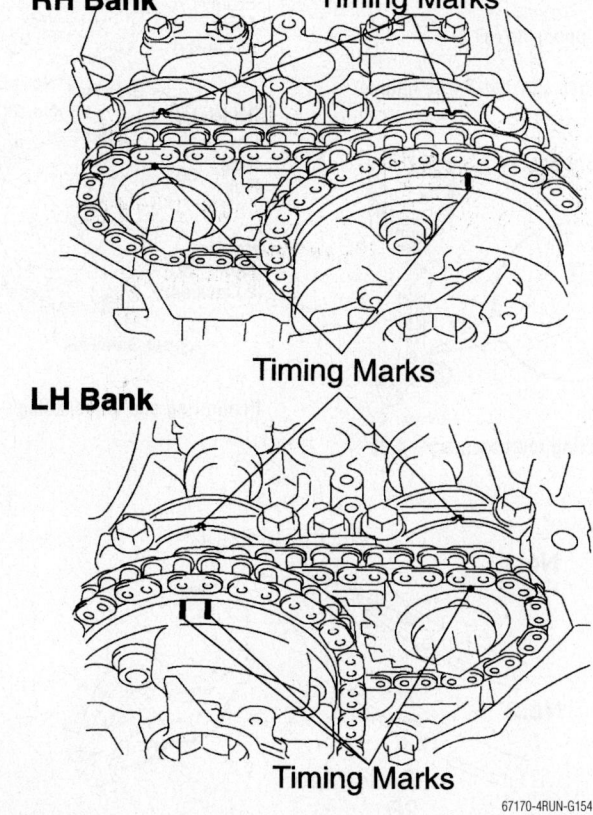

Check that the timing marks on the camshaft sprockets are aligned with the timing marks on the bearing caps—4.0L engine

sprocket, aligning the yellow link with the timing mark on the sprocket.

44. Install the chain on the camshafts, aligning the orange links with the timing marks.
45. Install the vibration damper.
46. Install the idler gear. Torque to 44 ft. lbs. (60 Nm).
47. Remove the pin from the tensioner.
48. Install the case oil seal.

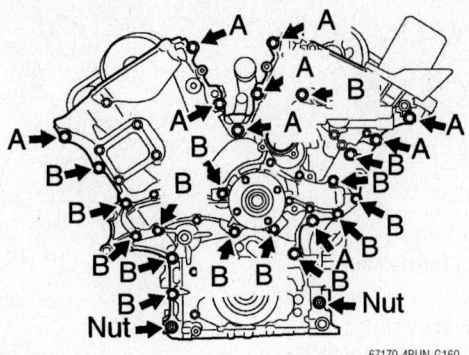

Timing case bolt identification—4.0L engine

49. Clean all gasket material from the case and block.

50. Apply new silicone gasket material to the block, as shown

51. Apply a continuous 3–4mm dia. bead of silicone gasket material to the case as shown.

52. Align the keyway of the oil pump drive rotor with the rectangular portion of the crankshaft timing sprocket, and slide the timing case into place.

➡ **Install the timing case with 3 minutes of applying the gasket material. Do not apply gasket material to point "A" in the illustration.**

53. Install the bolts and nuts. Bolts marked "A" are 25mm long; those marked "B" are 55mm long.

54. The remainder of installation is the reverse of removal. Observe the following torques:

- Oil filter bracket: 14 ft. lbs. (19 Nm)
- VVT sensor: 71 inch lbs. (8 Nm)
- Timing control valve: 80 inch lbs. (9 Nm)
- Cylinder head covers: 80 inch lbs. (9 Nm)
- ignition coil: 80 inch lbs. (9 Nm)

Piston and Ring

POSITIONING

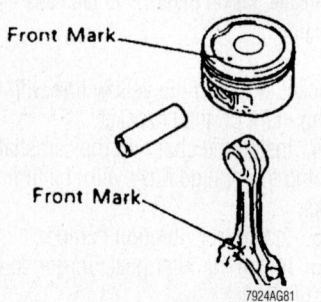

Piston to connecting rod assembly—3.4L engines

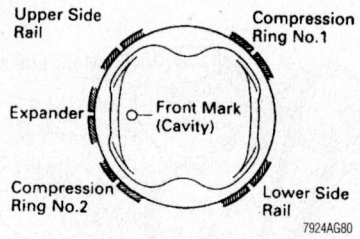

Piston ring end-gap spacing—3.4L engine

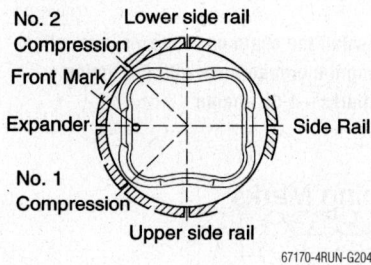

Piston ring end-gap spacing—4.0L engine

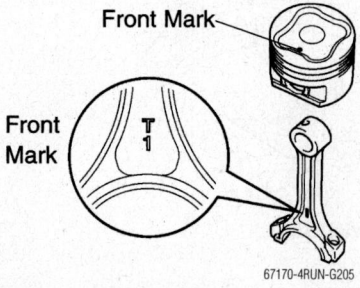

Piston to connecting rod assembly—4.0L engines

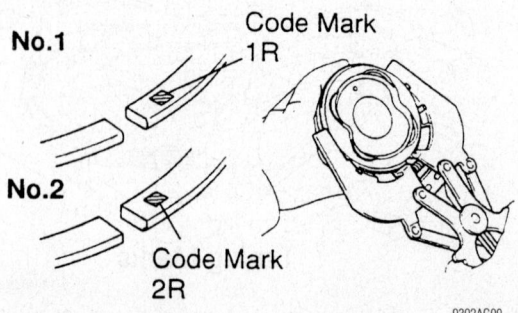

Piston ring identification—4.7L engine

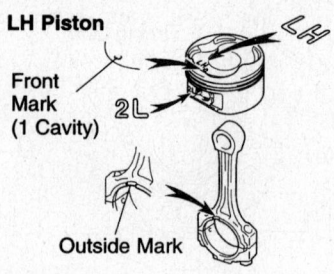

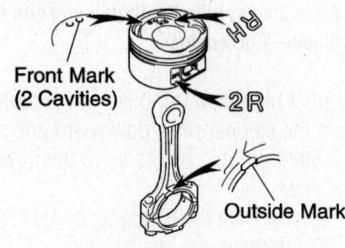

Piston to connecting rod assembly—4.7L engines

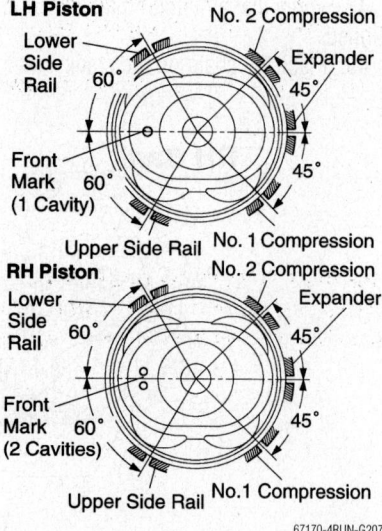

Piston ring end-gap spacing—4.7L engine

FUEL SYSTEM

Fuel System Pressure

RELIEVING

2002

1. Before servicing the vehicle, refer to the Precautions Section.
2. Disconnect the negative battery terminal.
3. Place a catch-pan under the joint to be disconnected. A large quantity of fuel may be released when the joint is opened.
4. Wear eye or full face protection.
5. Place a shop towel over the area and slowly loosen the joint using a wrench of the correct size. Use a back-up wrench if needed.
6. Allow the fuel left in the line to bleed off slowly before fully disconnecting the joint.
7. Plug the opened lines immediately to prevent fuel spillage or the entry of dirt.
8. Dispose of the released fuel properly.
9. After connecting fuel lines, connect the negative battery cable and start the engine.
10. Check for leaks and repair as needed.

2003–06

1. Before servicing the vehicle, refer to the Precautions Section.
2. Remove the fuel pump relay from the engine compartment relay block. Start the engine and allow it to run out of fuel. Turn the ignition off. Try to restart the engine to make sure there is no fuel pressure. Turn the ignition off, disconnect the battery ground and install the relay. Remove the fuel tank cap.
3. Place a catch-pan under the joint to be disconnected. A large quantity of fuel may be released when the joint is opened.
4. Wear eye or full face protection.
5. Place a shop towel over the area and slowly loosen the joint using a wrench of the correct size. Use a back-up wrench if needed.
6. Allow the fuel left in the line to bleed off slowly before fully disconnecting the joint.
7. Plug the opened lines immediately to prevent fuel spillage or the entry of dirt.

Fuel Pump

REMOVAL & INSTALLATION

1. Before servicing the vehicle, refer to the Precautions Section.

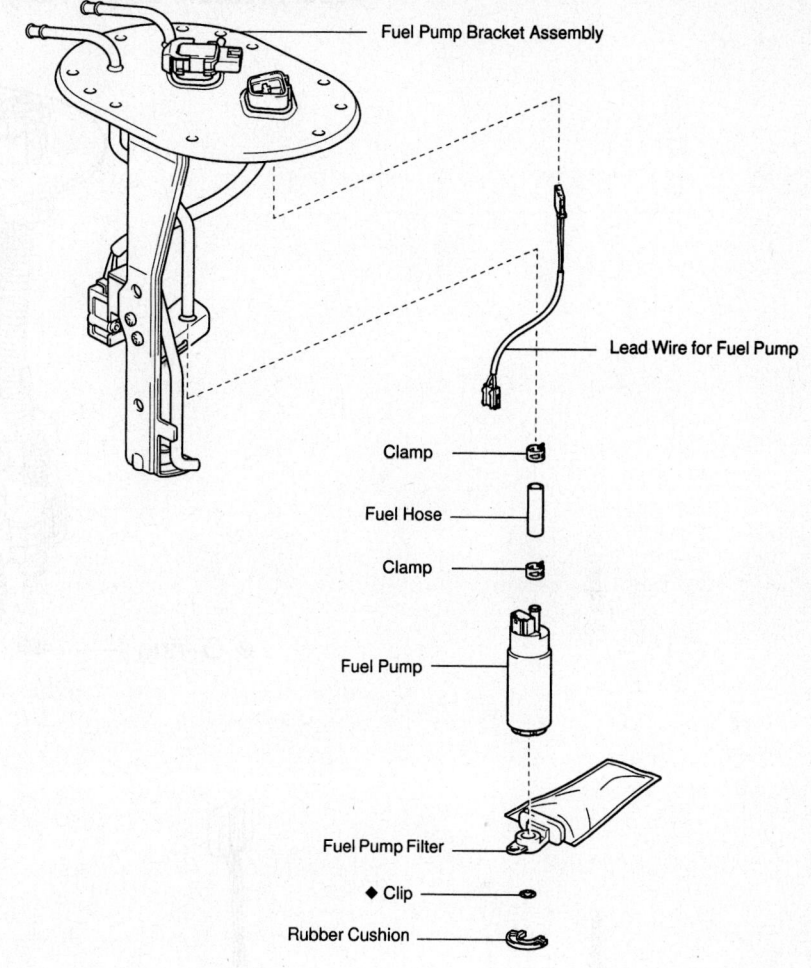

◆ Non–reusable part

67170-4RUN-G162

Fuel pump and related parts—3.4L engine

2. Relieve the fuel pressure.
3. Disconnect the negative battery cable from the battery.
4. Drain the fuel from the fuel tank.
5. Remove or disconnect the following:

- Fuel tank
- Fuel pump connector from the clamp
- Access plate bolts, then pull out the fuel pump assembly from the fuel tank
- Gasket(s) from the pump bracket
- Fuel pump connector
- Bracket from the lower side of the fuel pump
- Fuel pump from the fuel hose
- Rubber cushion, the clip and the fuel filter at the bottom of the fuel pump

To install:

6. Install or connect the following:

- Fuel pump filter to the fuel pump with a new clip
- Fuel pump to the fuel pump bracket
- Fuel hose to the outlet port of the fuel pump
- Fuel pump connector
- Fuel pump assembly with a new gasket(s). Tighten the bolts to 31 inch lbs. (4 Nm).
- Fuel pump connector to the clamp
- Fuel tank
- All electrical and fuel connections
- Negative battery cable

7. Refill the fuel tank and check for leaks.

4.0L and 4.7L Engines

1. Before servicing the vehicle, refer to the Precautions Section.

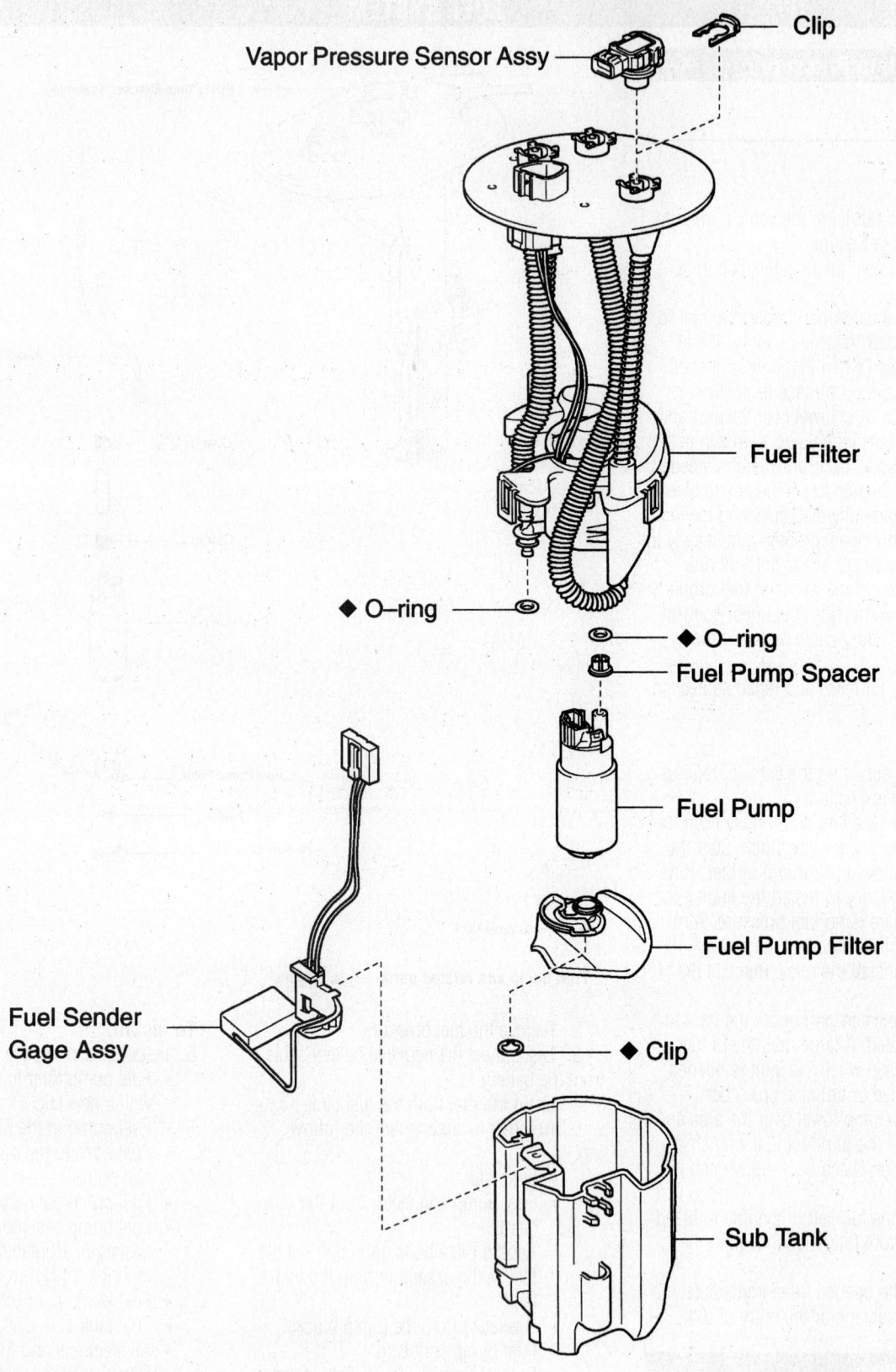

Vapor Pressure Sensor Assy

Clip

Fuel Filter

◆ O–ring

◆ O–ring

Fuel Pump Spacer

Fuel Pump

Fuel Pump Filter

Fuel Sender
Gage Assy

◆ Clip

Sub Tank

◆ Non–reusable part

67170-4RUN-G163

Fuel pump and related parts—4.0L and 4.7L engines

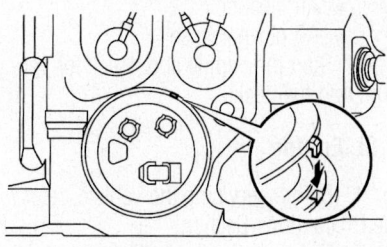

When installing the module, make sure that the tab on the plate fits into the slot in the tank

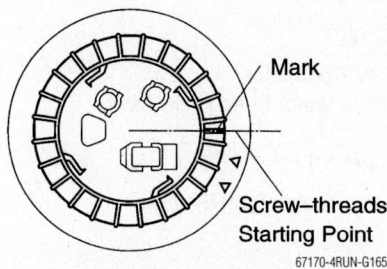

Align the mark on the retainer with a starting point on the tank

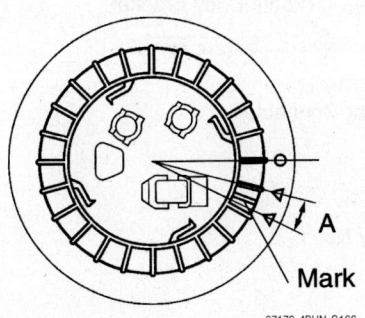

Using the tool, turn the retainer 1 full turn more and position the mark on the retainer into a range "A" on the tank

2. Relieve the fuel pressure.
3. Disconnect the negative battery cable from the battery.
4. Drain the fuel from the fuel tank.
5. Remove the rear floor mat.
6. Remove the floor access cover.
7. Disconnect the wiring from the pump.
8. Disconnect the filler hose.
9. Remove the skid plate.
10. Disconnect the supply and return lines.
11. Disconnect the breather hose.
12. Disconnect the vent hose.
13. Support the tank with a suitable jack. Remove the bolts and tank bands and lower the tank from the vehicle.
14. Disconnect the supply and return lines from the pump and tank.

15. Using a special tool, such as 09808-14020, or equivalent, unscrew the retainer. Pull out the module.
16. Installation is the reverse of removal. Use a new gasket on the module. When installing the module, make sure that the tab on the plate fits into the slot in the tank. Use a new retainer ring. Align the mark on the retainer with a starting point on the tank. Hold the module and turn the retainer 1 full turn, by hand.
17. Using the tool, turn the retainer 1 full turn more and position the mark on the retainer into a range "A" on the tank, as shown.
18. The remainder of installation is the reverse of removal. Torque the band bolts to 30 ft. lbs. (40 Nm); the skid plate bolts to 15 ft. lbs. (20 Nm).

Fuel Injector

REMOVAL & INSTALLATION

3.4L Engine

1. Before servicing the vehicle, refer to the Precautions Section.
2. Depressurize the fuel system.

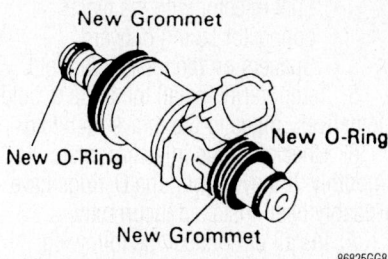

Install new O-rings and grommets on each injector—3.4L engine

3. Remove or disconnect the following:
   - Air cleaner hose
   - Upper half of the intake manifold
   - Fuel pressure regulator
   - Fuel inlet pipe
   - Fuel injector electrical connections
   - Fuel rail with the injectors
   - Spacers from the intake manifold
   - Injectors from the delivery pipes
   - O-rings and grommets, discard them

To install:
4. Install or connect the following:
   - New grommets and O-rings on each injector, lubricated with a light coat of gasoline

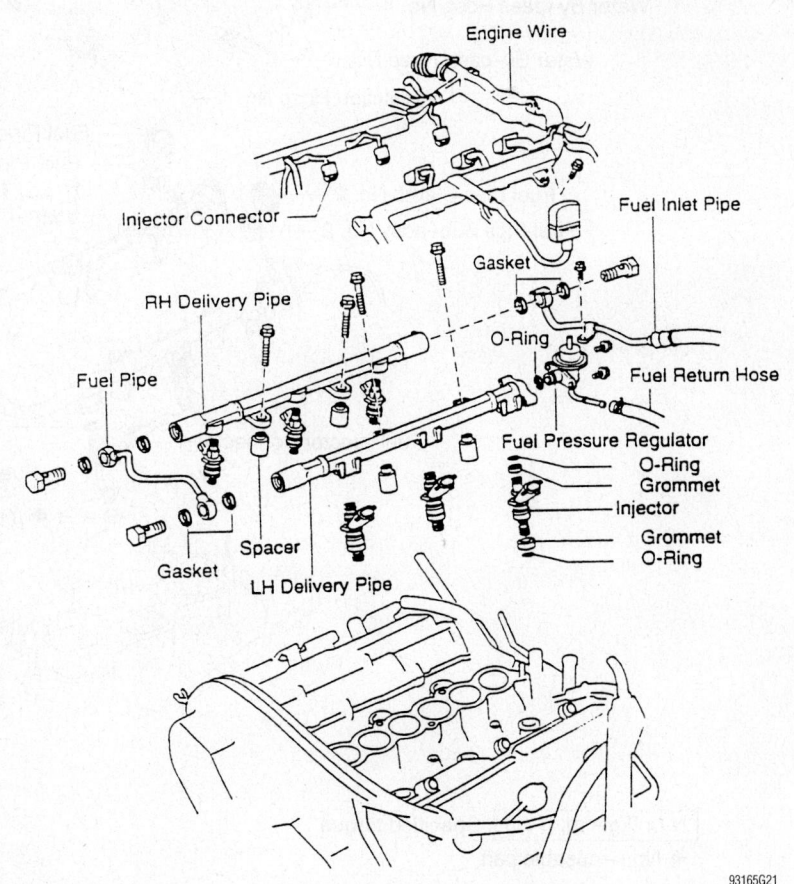

Fuel injector arrangement and related components—3.4L engine

- Fuel injector with the electrical connector facing outward
- Spacers on the intake manifold

5. Temporarily install the bolts to hold the delivery pipes to the intake manifold.

6. Check that the injectors rotate smoothly. If they do not, the O-rings have probably been installed incorrectly.

7. Install or connect the following:
- Fuel injector electrical connectors
- Fuel pipe with new gaskets. Tighten

the bolts to 25 ft. lbs. (34 Nm) and the delivery pipes-to-intake manifold bolts to 10 ft. lbs. (13 Nm).
- Fuel pipe union with new gaskets. Tighten the clamp bolt to 71 inch lbs. (8 Nm).
- Fuel pressure regulator

8. Inspect the vacuum lines and connections. Look for any loose connections, sharp bends or damage.

9. Install or connect the following:

- Air cleaner
- Air cleaner hose

10. Start the engine and check for vacuum and fuel leaks.

4.0L Engine

1. Before servicing the vehicle, refer to the Precautions Section.

2. Depressurize the fuel system.

3. Drain the cooling system.

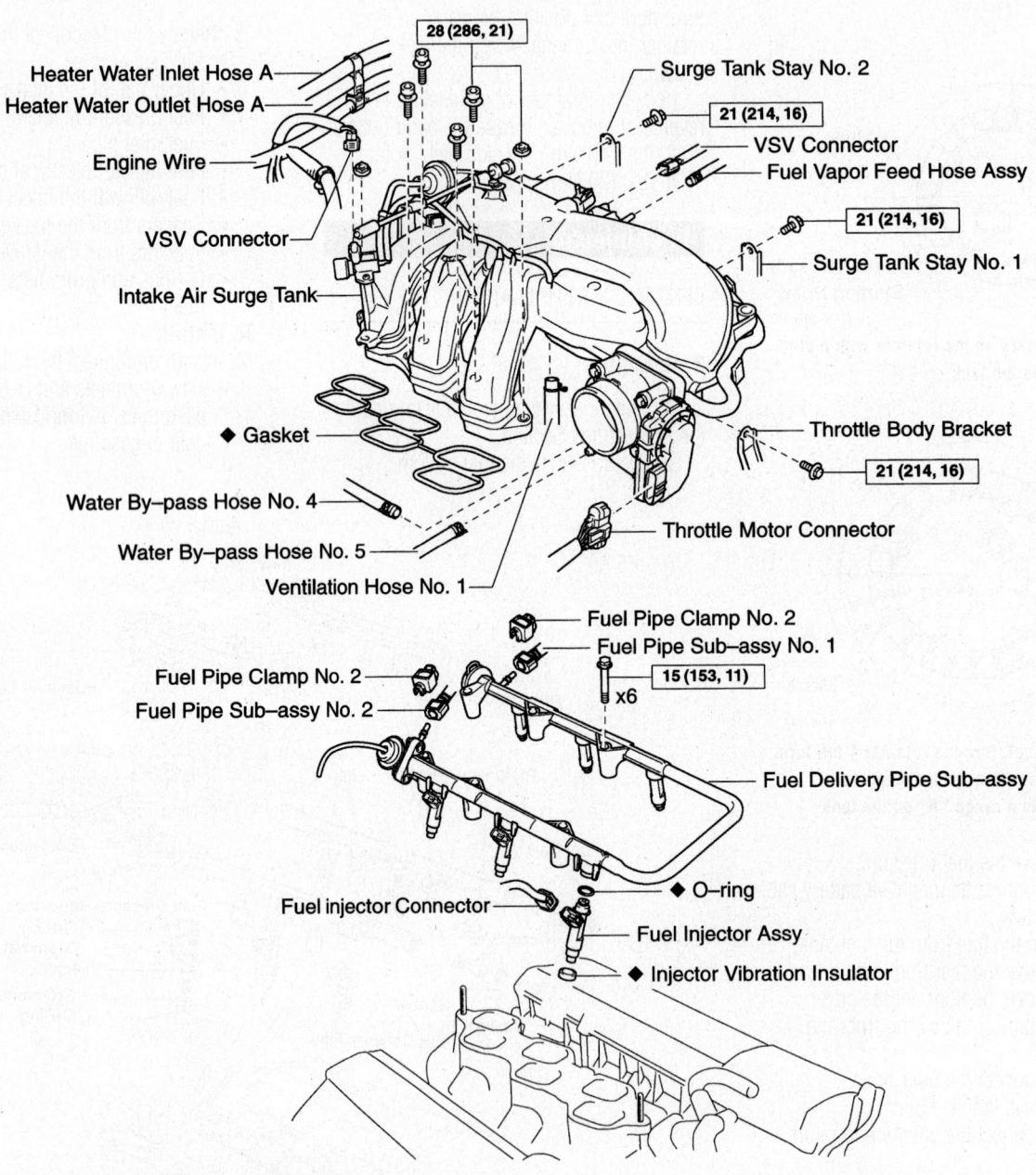

28 (286, 21)

Heater Water Inlet Hose A
Heater Water Outlet Hose A
Engine Wire
VSV Connector
Intake Air Surge Tank
◆ Gasket
Water By-pass Hose No. 4
Water By-pass Hose No. 5
Ventilation Hose No. 1

Surge Tank Stay No. 2
21 (214, 16)
VSV Connector
Fuel Vapor Feed Hose Assy
21 (214, 16)
Surge Tank Stay No. 1
Throttle Body Bracket
21 (214, 16)
Throttle Motor Connector

Fuel Pipe Clamp No. 2
Fuel Pipe Sub-assy No. 1
15 (153, 11)
x6
Fuel Pipe Clamp No. 2
Fuel Pipe Sub-assy No. 2
Fuel Delivery Pipe Sub-assy
◆ O-ring
Fuel injector Connector
Fuel Injector Assy
◆ Injector Vibration Insulator

N·m (kgf·cm, ft·lbf) : Specified torque
◆ Non-reusable part

67170-4RUN-G167

Fuel injectors and related parts—4.0L engine

4. Remove the V-bank cover.

5. Remove the air cleaner.

6. Disconnect the water bypass hoses.

7. Disconnect the fuel vapor hose.

8. Disconnect the vent hose.

9. Remove the upper intake manifold (air surge tank).

10. Disconnect the fuel supply lines.

11. Disconnect the injector wires.

12. Remove the injector bolts and pull the injector/fuel rail assembly from the heads.

13. Remove the injectors from the fuel rail.

To install:

14. Install a new insulator on each injector.

15. Coat new O-rings with gasoline and install them on the injectors.

16. Install the injectors on the fuel rail with a twisting motion. Position the connector outward.

17. Install the assembly onto the heads. Install the bolts finger-tight. Make sure each

injector twists freely If any doesn't, replace the O-ring.

18. Tighten the bolts to 11 ft. lbs. (15 Nm).

19. Connect the wiring.

20. The remainder of installation is the reverse of removal.

4.7L Engine

1. Before servicing the vehicle, refer to the Precautions Section.

2. Depressurize the fuel system.

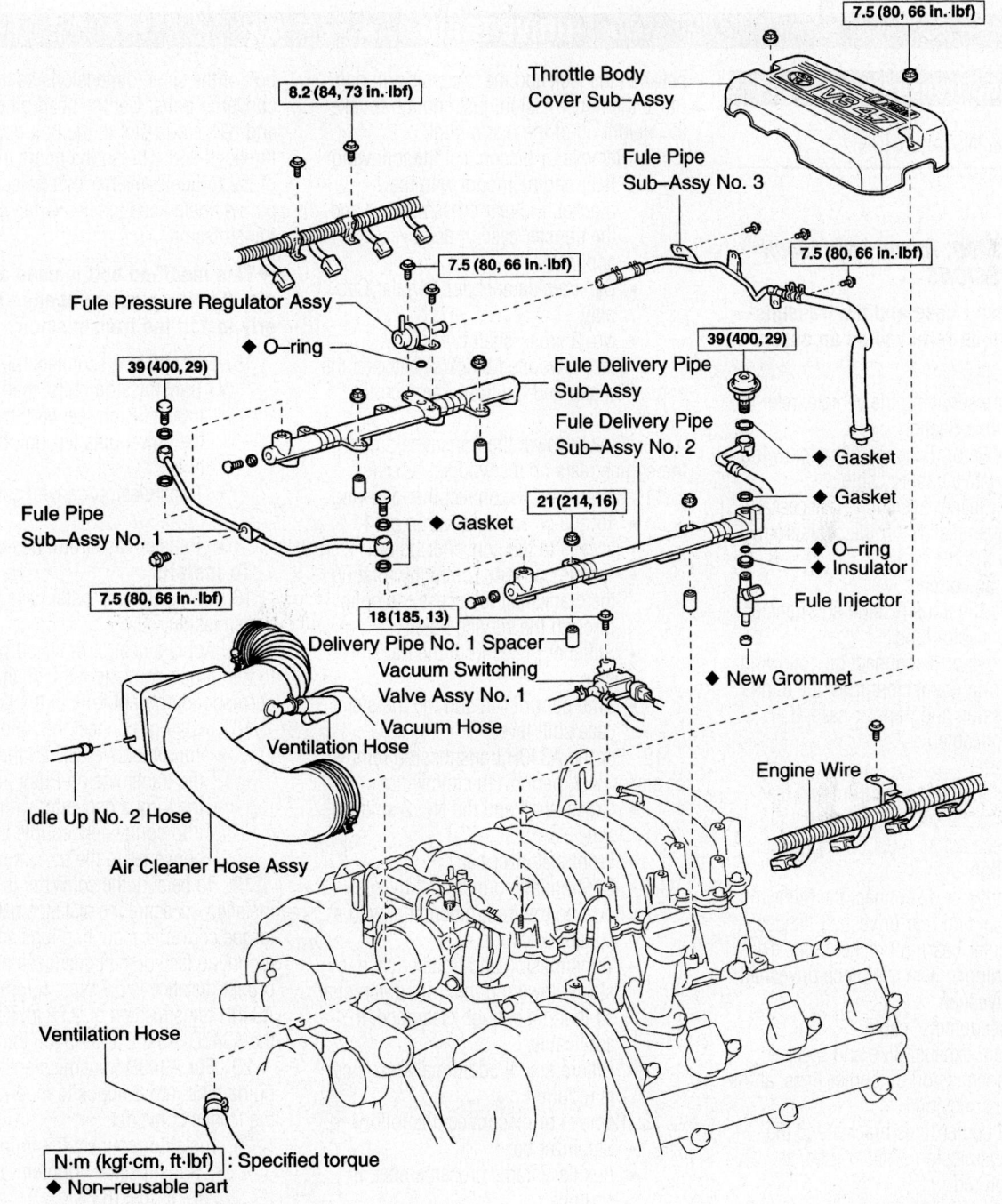

Fuel injectors and related parts—4.7L engine

3. Remove the air cleaner.

4. Disconnect the fuel pipes.

5. Disconnect the fuel hose from the pulsation damper.

6. Disconnect the vacuum switching valve.

7. Disconnect the wiring from the injectors.

8. Remove the nuts and remove the left and right fuel rails with the injectors.

9. Remove the injectors from the rails. Discard the O-rings, grommets and insulators from each injector.

To install:

10. Install a new grommet and insulator on each injector.

11. Coat each new O-ring with clean gasoline and install them on the injectors.

12. Install the injectors on the rails with a twisting motion. Make sure that they can rotate freely.

13. Install the rails/injectors on the intake manifold. Hand-tighten the nuts. Check that each injector can rotate freely. If not, replace the O-ring. Tighten the nuts to 16 ft. lbs. (21 Nm).

14. Using new gaskets, connect the fuel line to the pulsation damper. Torque to 29 ft. lbs. (39 Nm).

15. Install the vacuum switching valve. Torque to 13 ft. lbs. (18 Nm).

16. Connect fuel pipe No.3. Torque to 66 inch lbs.

17. Connect fuel pipe No.1 using new gaskets. Torque to 29 ft. lbs. (39 Nm).

18. Install the air cleaner.

19. Install the throttle body cover.

DRIVE TRAIN

Automatic Transmission

REMOVAL & INSTALLATION

2002

MODEL A340D, A340E AND A340H TRANSMISSIONS

➡The transfer case and the transmission should be removed as an assembly.

1. Before servicing the vehicle, refer to the Precautions Section.

2. Remove or disconnect the following:
   - Negative battery cable
   - Air cleaner assembly, if necessary
   - Transmission throttle cable from the throttle body
   - Engine undercover

3. Drain the transmission and transfer case (if applicable) fluid.

4. Remove or disconnect the following:
   - Wiring connectors from the transmission and transfer case, if applicable.
   - Starter

5. Matchmarks on the front and rear driveshaft flanges and the differential pinion flanges. These marks must be aligned during installation.

6. Remove or disconnect the following:
   - Front and rear driveshaft flanges.
   - Center bearing bracket bolts, if equipped with a 2-piece driveshaft
   - Driveshaft
   - Speedometer cable
   - Front exhaust pipe and bracket
   - Transmission oil cooler lines, at the transmission
   - Oil cooler lines bracket and the transmission oil filler tube, as required

7. Support the transmission, using a suitable jack with a wooden block placed between the jack and the transmission pan. Raise the transmission, just enough to take the weight off of the rear mount.

8. Remove or disconnect the following:
   - Rear engine mount with the bracket, the rear crossmember and the transfer case undercover, if applicable
   - Dynamic damper, for Regular Cab only
   - No. 2 cross-shaft bracket

9. Place a wooden block(s) between the engine oil pan and the front frame crossmember.

10. Slowly, lower the transmission until the engine rests on the wooden block(s).

11. Remove or disconnect the following:
   - Torque converter cover to gain access to the converter bolts
   - Torque converter bolts, by rotating the crankshaft to access the bolts through the service holes
   - Stiffener plates from the transmission
   - Shift control rod and the transfer case shift lever

12. For the A340H transmission remove or disconnect, perform the following:
   - Cross-shaft and the No. 2 shifting rod
   - Front stabilizer bar
   - Differential mount bolts, by supporting the front differential with a suitable jack
   - Transmission and transfer case, by slowly lowering the front differential so there is enough clearance, if applicable
   - Differential, if additional clearance is required

13. Remove or disconnect the following:
   - Stabilizer bar
   - Auxiliary frame crossmember, if equipped

14. For A340D transmissions, obtain a bolt of the same dimensions as the torque converter bolts. Cut the head off of the bolt and hacksaw a slot in the bolt opposite the threaded end. Thread the guide pin into one of the torque converter bolt holes. The guide pin will help keep the converter with the transmission.

➡This modified bolt is used as a guide pin. 2 guide pins are needed to properly install the transmission.

15. Remove or disconnect the following:
   - Transmission bolts, then move the transmission rearward by prying on the dowel pins through the service hole
   - Transmission/transfer case assembly
   - Transfer case from the transmission

To install:

16. Connect the transfer case to the transmission.

17. Apply a coat of multi-purpose grease to the torque converter stub shaft and the corresponding pilot hole in the flexplate.

18. Install or connect the following:
   - Torque converter into the front of the transmission Push inward on the torque converter while rotating it to completely couple the torque converter to the transmission.

19. To be sure the converter is properly installed, measure the distance between the torque converter mounting lugs and the front mounting face of the transmission. The proper distance is 0.71 in. (18mm) for the A340H transmission or 0.79 in. (20mm) for the A340D, A340E and A340F transmissions.

20. For A340D transmissions, install guide pins into 2 opposite mounting lugs of the torque converter.

21. Install or connect the following:
   - Transmission. Tighten the bolts to 47 ft. lbs. (63 Nm).
   - Torque converter bolts, by rotating the crankshaft. Tighten the bolts

evenly to 30 ft. lbs. (41 Nm) for the A340H, A3430D and A340E transmissions or to 20 ft. lbs. (27 Nm) for the A340F transmission.

- Torque converter access cover

22. Remove the wood block(s) from under the engine oil pan.

23. Install or connect the following:
- Transmission crossmember. Tighten the bolts to 70 ft. lbs. (95 Nm).
- Rear mount and bracket. Tighten the bracket bolts to 43 ft. lbs. (58 Nm) and the bracket-to-rear mount bolts to 108 inch lbs. (13 Nm).
- Transmission onto the crossmember. Tighten the transmission-to-mount bolts to 18 ft. lbs. (25 Nm).

24. Remove the wooden blocks from between the frame and the engine and the support from under the transmission.

25. Install or connect the following:
- Front differential, for the A340H transmission. Tighten the 2 rear mount bolts to 123 ft. lbs. (167 Nm) and the front mount through-bolt to 108 ft. lbs. (147 Nm).

➡**If the differential oil was drained, refill it at this time.**

- Shift control rod and the transfer case shift lever
- Front stabilizer bar, if applicable
- Cross-shaft and the No. 2 shifting rod, if applicable
- Stiffener plates. Tighten the bolts to 27 ft. lbs. (37 Nm).
- Transfer case undercover and the dynamic damper, if equipped. Tighten the dynamic damper mount bolts to 27 ft. lbs. (37 Nm).
- No. 2 cross-shaft bracket
- Oil filler tube and the oil cooler pipe bracket
- Oil cooler lines to the transmission. Tighten the fittings to 25 ft. lbs. (34 Nm).
- Front exhaust pipe and the support bracket
- Speedometer cable
- Front and rear driveshaft flanges with the differential pinion flanges, by aligning the matchmarks. Tighten the bolts to 54 ft. lbs. (74 Nm).
- Starter
- Wiring connectors to the transmission and the transfer case, if applicable
- Engine undercover
- Transmission throttle cable, by adjusting it
- Air cleaner assembly, if removed
- Negative battery cable

26. Refill the transmission and the transfer case, if applicable.
27. Start the engine and check for leaks.
28. Road test the vehicle for proper operation.
29. Recheck all fluid levels.

MODEL A340F TRANSMISSION

1. Before servicing the vehicle, refer to the Precautions Section.
2. Remove or disconnect the following:
- Negative battery cable
- Throttle cable, from the engine compartment
- Automatic Transmission Fluid (ATF) dipstick
- Oil filler pipe

3. Remove the transmission shift lever assembly and transfer shift lever, as follows:
- Rear console upper panel, by disconnecting the connectors
- Heater control knobs
- Center cluster finish panel, by disconnecting the connectors
- Transfer shift lever knob, without the 2–4 selector
- Bolt and the transfer shift lever knob, with the 2–4 selector
- Front console upper panel
- 2–4 selector connector, if equipped
- Transfer shift lever knob
- Shift control rod
- Transmission shift lever assembly connector and the 8 screws
- Shift lever snapring, using pliers and pull out it from the transfer case
- Engine undercover
- Front and rear driveshafts
- Exhaust pipe

4. Disconnect the following connectors from the transmission:
- No. 2 Vehicle Speed Sensor (VSS) connector
- Solenoid connector
- Automatic Transmission Fluid (ATF) temperature sensor connector
- Park/neutral position switch connector

5. Detach the following connectors from the transfer case:
- Transfer neutral position switch connector
- Transfer L4 position switch connector
- Transfer 4WD position switch connector
- Actuator connector (2–4 selector only)

6. Remove or disconnect the following:
- Wiring harness from the transmission and the transfer case

- Both oil cooler pipes
- Rear end-plate and torque converter clutch mounting bolt

7. Support the transmission with a suitable jack.

8. Remove or disconnect the following:
- Engine rear mount bolts
- 4 bolts and the crossmember
- Starter
- Transmission

To install:

9. Install or connect the following:
- Transmission. Tighten the bolts to 53 ft. lbs. (71 Nm).
- Starter. Tighten the bolts to 29 ft. lbs. (39 Nm).
- Crossmember. Tighten the 4 bolts to 48 ft. lbs. (65 Nm).
- Engine rear mount. Tighten the 4 bolts to 14 ft. lbs. (19 Nm).
- Clutch converter bolts, by installing the green colored bolt before the other 5. Tighten the bolts to 30 ft. lbs. (41 Nm).
- Rear end-plate. Tighten the bolts to 13 ft. lbs. (18 Nm).
- Both oil cooler pipes. Tighten to 25 ft. lbs. (34 Nm).
- Oil cooler pipe clamps. Tighten the 10mm head bolt to 48 inch lbs. (5 Nm) and the 12mm head bolt to 108 inch lbs. (13 Nm).
- Wiring harness to the transmission and the transfer case
- Remaining components

10. Fill the transmission and transfer case with transmission fluid.
- Throttle cable
- Negative battery cable

2003–06

A340E

1. Before servicing the vehicle, refer to the Precautions Section.
2. Remove the engine under-covers.
3. Remove the exhaust pipe.
4. Remove the driveshaft.
5. Drain the transmission.
6. Remove the front crossmember brackets.
7. Remove the filler tube.
8. Remove the transmission cooler lines.
9. Remove the transmission control cable and bracket.
10. Support the transmission with a suitable transmission jack.
11. Remove the crossmember.
12. Tilt the transmission downward, slightly and disconnect the wiring.
13. Remove the starter.
14. Remove the manifold stays.

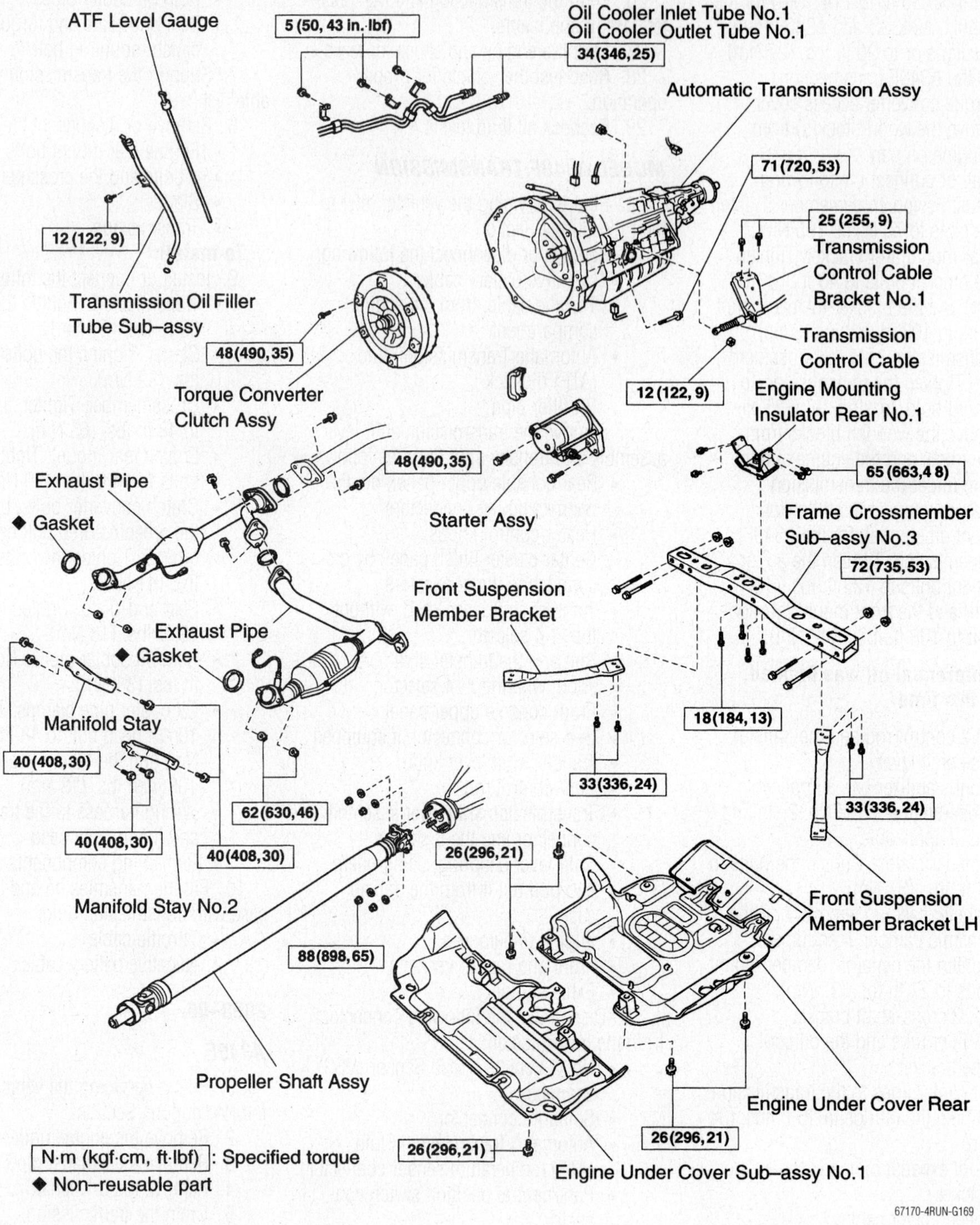

ATF Level Gauge

5 (50, 43 in. lbf)

Oil Cooler Inlet Tube No.1
Oil Cooler Outlet Tube No.1

34 (346, 25)

Automatic Transmission Assy

71 (720, 53)

12 (122, 9)

25 (255, 9)

Transmission Oil Filler
Tube Sub–assy

48 (490, 35)

Torque Converter
Clutch Assy

Transmission
Control Cable
Bracket No.1

12 (122, 9)

Transmission
Control Cable

Engine Mounting
Insulator Rear No.1

65 (663, 4 8)

Exhaust Pipe

◆ Gasket

48 (490, 35)

Starter Assy

Frame Crossmember
Sub–assy No.3

72 (735, 53)

Exhaust Pipe

◆ Gasket

Front Suspension
Member Bracket

Manifold Stay

40 (408, 30)

18 (184, 13)

33 (336, 24)

33 (336, 24)

62 (630, 46)

40 (408, 30)

Manifold Stay No.2

40 (408, 30)

26 (296, 21)

Front Suspension
Member Bracket LH

88 (898, 65)

Propeller Shaft Assy

Engine Under Cover Rear

N·m (kgf·cm, ft·lbf): Specified torque

26 (296, 21)

26 (296, 21)

◆ Non–reusable part

26 (296, 21)

Engine Under Cover Sub–assy No.1

67170-4RUN-G169

2003–06 A340E and related parts

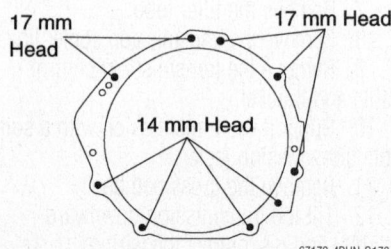

Transmission-to-engine bolt identification—2003–06 A340E

67170-4RUN-G170

15. Remove the bellhousing cover plate
and turn the crankshaft to access and
remove the 6 torque converter bolts.

16. Remove the 9 transmission-to-
engine bolts and remove the transmission.

17. Remove the rear insulator.

18. Installation is the reverse of removal.
Observe the following torques:

- Rear insulator: 48 ft. lbs. (65 Nm)
- Transmission-to-engine bolts (see
illustration): 14mm head, 27 ft. lbs.

(37 Nm); 17mm head, 53 ft. lbs.
(71 Nm)

- Torque converter bolts (install black
bolt first): 35 ft. lbs. (48 Nm)
- Manifold stays: 30 ft. lbs. (40 Nm)
- Crossmember-to-frame: 53 ft. lbs.
(72 Nm)
- Crossmember-to-insulator: 13 ft.
lbs. (18 Nm)
- Crossmember brackets: 24 ft. lbs.
(33 Nm)

- Bellhousing cover: 13 ft. lbs. (18 Nm)
- Engine under-covers: 21 ft. lbs. (26 Nm)
- Transmission drain plug: 15 ft. lbs. (20 Nm)

A750E

1. Before servicing the vehicle, refer to the Precautions Section.
2. Remove the engine under-covers.
3. Remove the exhaust pipe.

4. Remove the driveshaft.
5. Drain the transmission.
6. Remove the front crossmember brackets.
7. Remove the transmission cooler lines.

2WD:

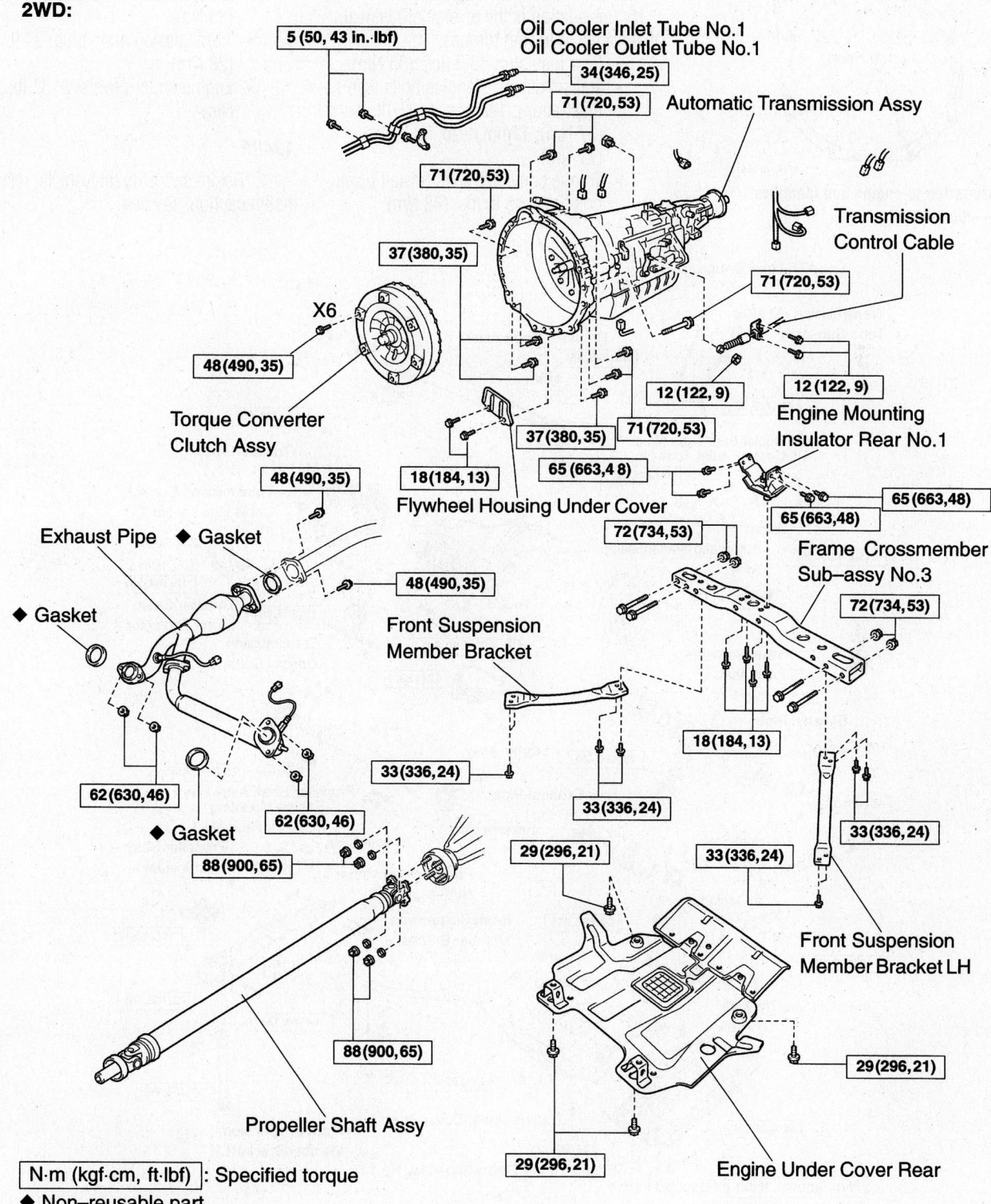

5 (50, 43 in.·lbf)
Oil Cooler Inlet Tube No.1
Oil Cooler Outlet Tube No.1
34(346,25)
71(720,53)
Automatic Transmission Assy
71(720,53)
37(380,35)
Transmission Control Cable
71(720,53)
X6
48(490,35)
Torque Converter Clutch Assy
48(490,35)
18(184,13)
37(380,35)
71(720,53)
65(663,48)
Flywheel Housing Under Cover
12(122,9)
12(122,9)
Engine Mounting Insulator Rear No.1
65(663,48)
72(734,53)
65(663,48)
Exhaust Pipe   ◆ Gasket
Frame Crossmember Sub–assy No.3
72(734,53)
48(490,35)
◆ Gasket
Front Suspension Member Bracket
62(630,46)
62(630,46)
18(184,13)
◆ Gasket
33(336,24)
33(336,24)
88(900,65)
33(336,24)
33(336,24)
29(296,21)
33(336,24)
Front Suspension Member Bracket LH
29(296,21)
88(900,65)
Propeller Shaft Assy
29(296,21)
Engine Under Cover Rear

N·m (kgf·cm, ft·lbf) : Specified torque
◆ Non–reusable part

67170-4RUN-G171

2003–06 A750E and related parts

8. Remove the transmission control cable and bracket.

9. Support the transmission with a suitable transmission jack.

10. Remove the crossmember.

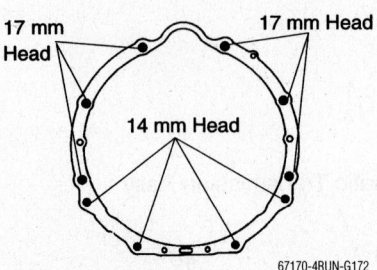

Transmission-to-engine bolt identification—2003–A750E

17 mm Head

17 mm Head

14 mm Head

67170-4RUN-G172

11. Tilt the transmission downward, slightly and disconnect the wiring.

12. Remove the bellhousing cover plate and turn the crankshaft to access and remove the 6 torque converter bolts.

13. Remove the 10 transmission-to-engine bolts and remove the transmission.

14. Remove the rear insulator.

15. Installation is the reverse of removal. Observe the following torques:

- Rear insulator: 48 ft. lbs. (65 Nm)
- Transmission-to-engine bolts (see illustration): 14mm head, 27 ft. lbs. (37 Nm); 17mm head, 53 ft. lbs. (71 Nm)
- Torque converter bolts (install black bolt first): 35 ft. lbs. (48 Nm)

- Manifold stays: 30 ft. lbs. (40 Nm)
- Bellhousing cover: 13 ft. lbs. (18 Nm)
- Crossmember-to-frame: 53 ft. lbs. (72 Nm)
- Crossmember-to-insulator: 13 ft. lbs. (18 Nm)
- Crossmember brackets: 24 ft. lbs. (33 Nm)
- Transmission drain plug: 21 ft. lbs. (28 Nm)
- Engine under-covers: 21 ft. lbs. (26 Nm)

A340F

1. Before servicing the vehicle, refer to the Precautions Section.

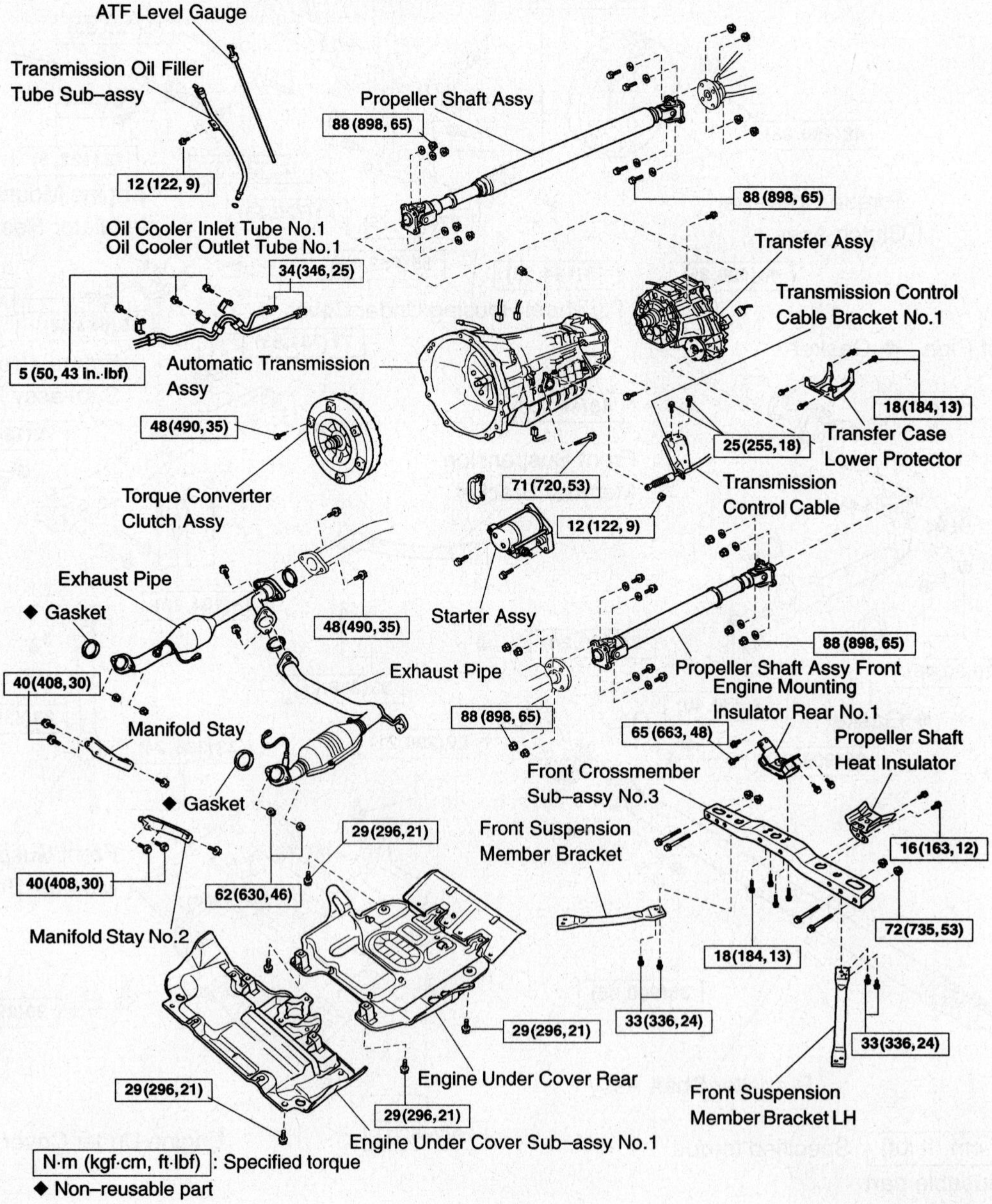

ATF Level Gauge

Transmission Oil Filler Tube Sub-assy

Propeller Shaft Assy

88 (898, 65)

88 (898, 65)

12 (122, 9)

Oil Cooler Inlet Tube No.1
Oil Cooler Outlet Tube No.1

34 (346, 25)

Transfer Assy

Transmission Control Cable Bracket No.1

5 (50, 43 in.·lbf)

Automatic Transmission Assy

48 (490, 35)

18 (184, 13)

Transfer Case Lower Protector

25 (255, 18)

Transmission Control Cable

Torque Converter Clutch Assy

71 (720, 53)

12 (122, 9)

Exhaust Pipe

◆ Gasket

40 (408, 30)

48 (490, 35)

Starter Assy

Exhaust Pipe

88 (898, 65)

88 (898, 65)

Propeller Shaft Assy Front Engine Mounting Insulator Rear No.1

Manifold Stay

65 (663, 48)

Propeller Shaft Heat Insulator

◆ Gasket

Front Crossmember Sub-assy No.3

16 (163, 12)

40 (408, 30)

29 (296, 21)

Front Suspension Member Bracket

62 (630, 46)

72 (735, 53)

Manifold Stay No.2

18 (184, 13)

33 (336, 24)

33 (336, 24)

29 (296, 21)

29 (296, 21)

Engine Under Cover Rear

Front Suspension Member Bracket LH

29 (296, 21)

29 (296, 21)

Engine Under Cover Sub-assy No.1

N·m (kgf·cm, ft·lbf) : Specified torque
◆ Non-reusable part

67170-4RUN-G173

2003–06 A340F and related parts

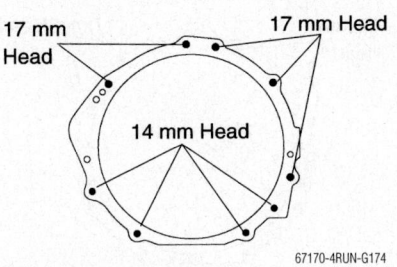

17 mm Head

17 mm Head

14 mm Head

67170-4RUN-G174

Transmission-to-engine bolt identification—2003–06 A340F

2. Remove the engine under-covers.
3. Remove the exhaust pipe.
4. Remove the front and rear drive-shafts.
5. Drain the transmission.
6. Remove the driveshaft heat insulator.
7. Remove the front crossmember brackets.
8. Remove the filler tube.
9. Remove the transmission cooler lines.

10. Remove the transmission control cable and bracket.
11. Support the transmission with a suitable transmission jack.
12. Remove the crossmember.
13. Remove the skid plate.
14. Tilt the transmission downward, slightly and disconnect the wiring.
15. Remove the starter.
16. Remove the manifold stays.
17. Remove the bellhousing cover plate

4WD:

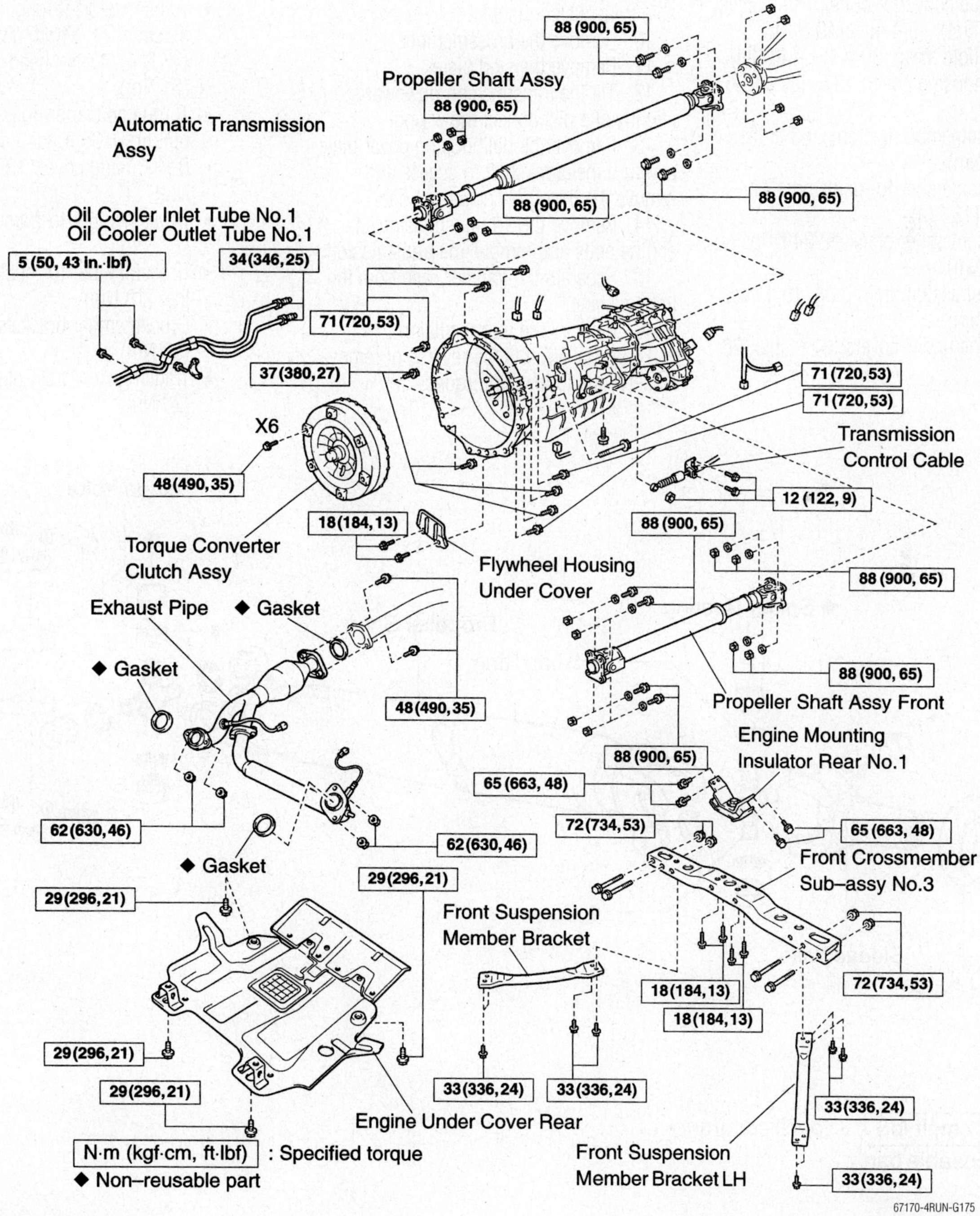

Propeller Shaft Assy

88 (900, 65)

88 (900, 65)

88 (900, 65)

Automatic Transmission Assy

Oil Cooler Inlet Tube No.1
Oil Cooler Outlet Tube No.1

5 (50, 43 in.·lbf)

34 (346, 25)

71 (720, 53)

37 (380, 27)

71 (720, 53)
71 (720, 53)

Transmission Control Cable

12 (122, 9)

X6

48 (490, 35)

Torque Converter Clutch Assy

18 (184, 13)

88 (900, 65)

88 (900, 65)

Flywheel Housing Under Cover

Exhaust Pipe   ◆ Gasket

88 (900, 65)

88 (900, 65)

Propeller Shaft Assy Front

◆ Gasket

48 (490, 35)

Engine Mounting Insulator Rear No.1

88 (900, 65)

65 (663, 48)

65 (663, 48)

62 (630, 46)

72 (734, 53)

Front Crossmember Sub–assy No.3

62 (630, 46)

◆ Gasket

29 (296, 21)

29 (296, 21)

Front Suspension Member Bracket

72 (734, 53)

29 (296, 21)

18 (184, 13)

18 (184, 13)

29 (296, 21)

29 (296, 21)

33 (336, 24)

33 (336, 24)

33 (336, 24)

N·m (kgf·cm, ft·lbf) : Specified torque
◆ Non–reusable part

Engine Under Cover Rear

Front Suspension Member Bracket LH

33 (336, 24)

67170-4RUN-G175

2003–06 A750F and related parts

and turn the crankshaft to access and remove the 6 torque converter bolts.

18. Remove the 9 transmission-to-engine bolts and remove the transmission.

19. Separate the transfer case from the transmission.

20. Remove the rear insulator.

21. Installation is the reverse of removal. Observe the following torques:

- Rear insulator: 48 ft. lbs. (65 Nm)
- Transmission-to-engine bolts (see illustration): 14mm head, 27 ft. lbs. (37 Nm); 17mm head, 53 ft. lbs. (71 Nm)
- Torque converter bolts (install black bolt first): 35 ft. lbs. (48 Nm)
- Manifold stays: 30 ft. lbs. (40 Nm)
- Bellhousing cover: 13 ft. lbs. (18 Nm)
- Crossmember-to-frame: 53 ft. lbs. (72 Nm)
- Crossmember-to-insulator: 13 ft. lbs. (18 Nm)
- Crossmember brackets: 24 ft. lbs. (33 Nm)
- Transmission drain plug: 15 ft. lbs. (20 Nm)
- Engine under-covers: 21 ft. lbs. (26 Nm)

A750F

1. Before servicing the vehicle, refer to the Precautions Section.

2. Remove the engine under-covers.

3. Remove the exhaust pipe.

4. Remove the front and rear driveshafts.

5. Drain the transmission.

6. Remove the front crossmember brackets.

7. Remove the transmission cooler lines.

8. Remove the transmission control cable and bracket.

9. Support the transmission with a suitable transmission jack.

10. Remove the crossmember.

11. Remove the skid plate.

12. Tilt the transmission downward, slightly and disconnect the wiring.

13. Remove the bellhousing cover plate and turn the crankshaft to access and remove the 6 torque converter bolts.

14. Remove the 10 transmission-to-engine bolts and remove the transmission.

15. Separate the transfer case from the transmission.

16. Remove the rear insulator.

17. Installation is the reverse of removal. Observe the following torques:

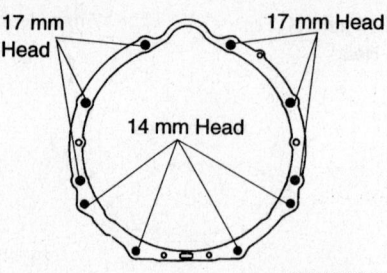

Transmission-to-engine bolt identification—2003–06 A750F

- Rear insulator: 48 ft. lbs. (65 Nm)
- Transmission-to-engine bolts (see illustration): 14mm head, 27 ft. lbs. (37 Nm); 17mm head, 53 ft. lbs. (71 Nm)
- Torque converter bolts (install black bolt first): 35 ft. lbs. (48 Nm)
- Bellhousing cover: 13 ft. lbs. (18 Nm)
- Crossmember-to-frame: 53 ft. lbs. (72 Nm)
- Crossmember-to-insulator: 13 ft. lbs. (18 Nm)
- Crossmember brackets: 24 ft. lbs. (33 Nm)
- Transmission drain plug: 15 ft. lbs. (20 Nm)

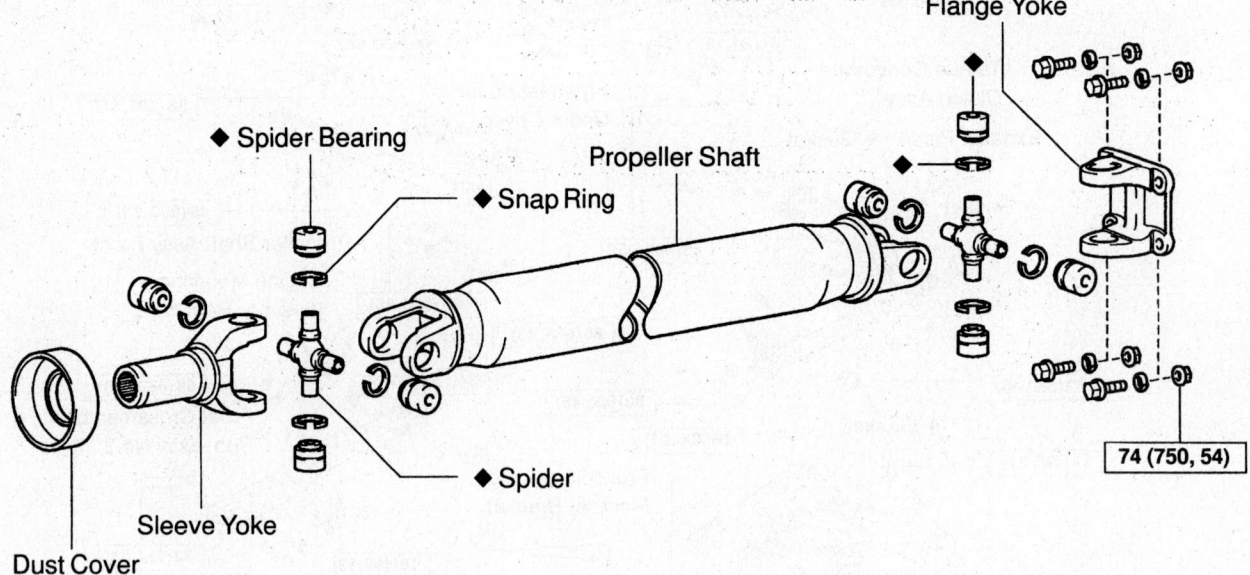

N·m (kgf·cm, ft·lbf) : Specified torque

◆ Non–reusable part

Rear driveshaft with 2WD—2002

Front Propeller Shaft

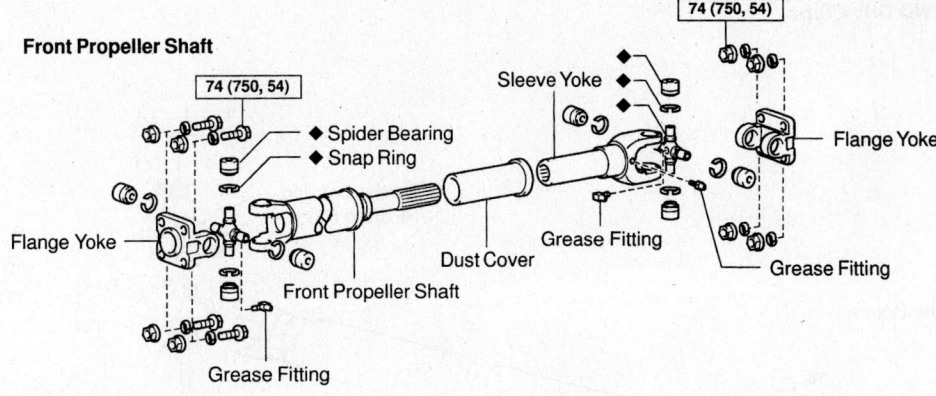

74 (750, 54)

Sleeve Yoke
◆ Spider Bearing
◆ Snap Ring
Flange Yoke
Flange Yoke
Grease Fitting
Dust Cover
Grease Fitting
Front Propeller Shaft
Grease Fitting

Rear Propeller Shaft

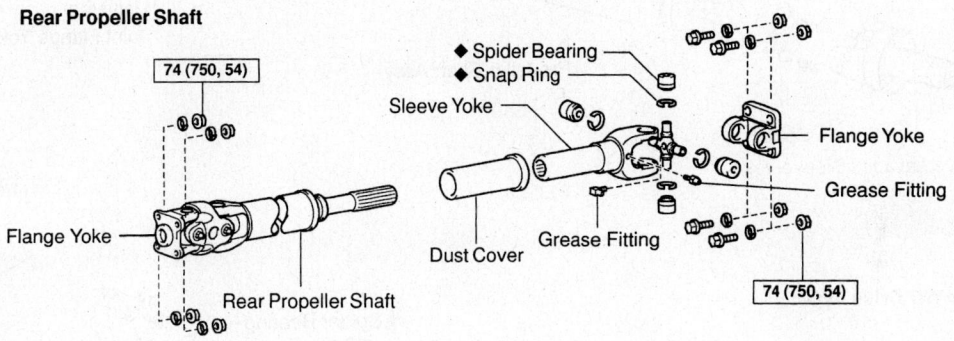

74 (750, 54)

◆ Spider Bearing
◆ Snap Ring
Sleeve Yoke
Flange Yoke
Grease Fitting
Flange Yoke
Dust Cover
Grease Fitting
Rear Propeller Shaft
74 (750, 54)

N·m (kgf·cm, ft·lbf) : Specified torque
◆ Non–reusable part

67170-4RUN-G180

Front and rear driveshafts with 4WD—2002

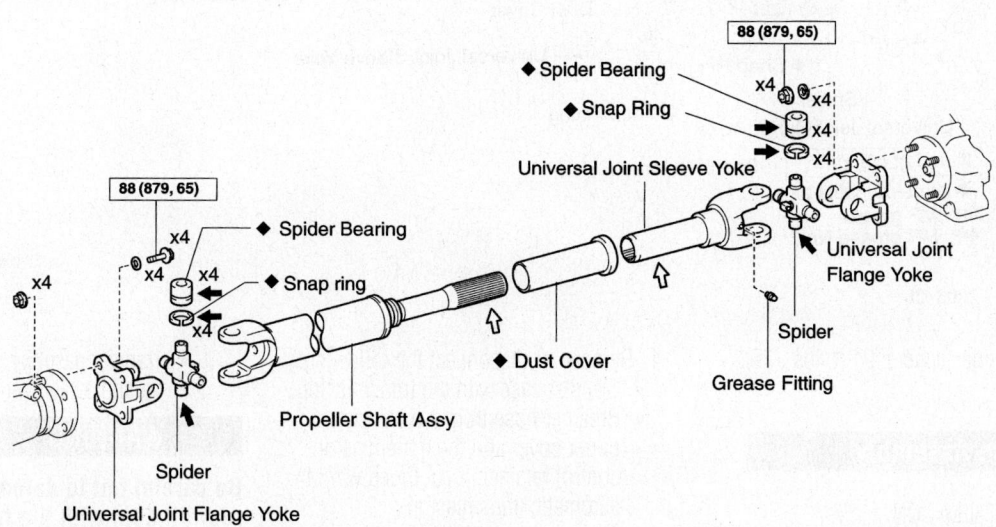

88 (879, 65)

◆ Spider Bearing
◆ Snap Ring
Universal Joint Sleeve Yoke
88 (879, 65)
◆ Spider Bearing
Universal Joint Flange Yoke
◆ Snap ring
Spider
◆ Dust Cover
Grease Fitting
Propeller Shaft Assy
Spider
Universal Joint Flange Yoke

N·m (kgf·cm, ft·lbf) : Specified torque
◆ Non–reusable part
⇐ MP grease
⇐ MP grease No.2

67170-4RUN-G177

Front driveshaft—2003–06

2WD Drive Type:

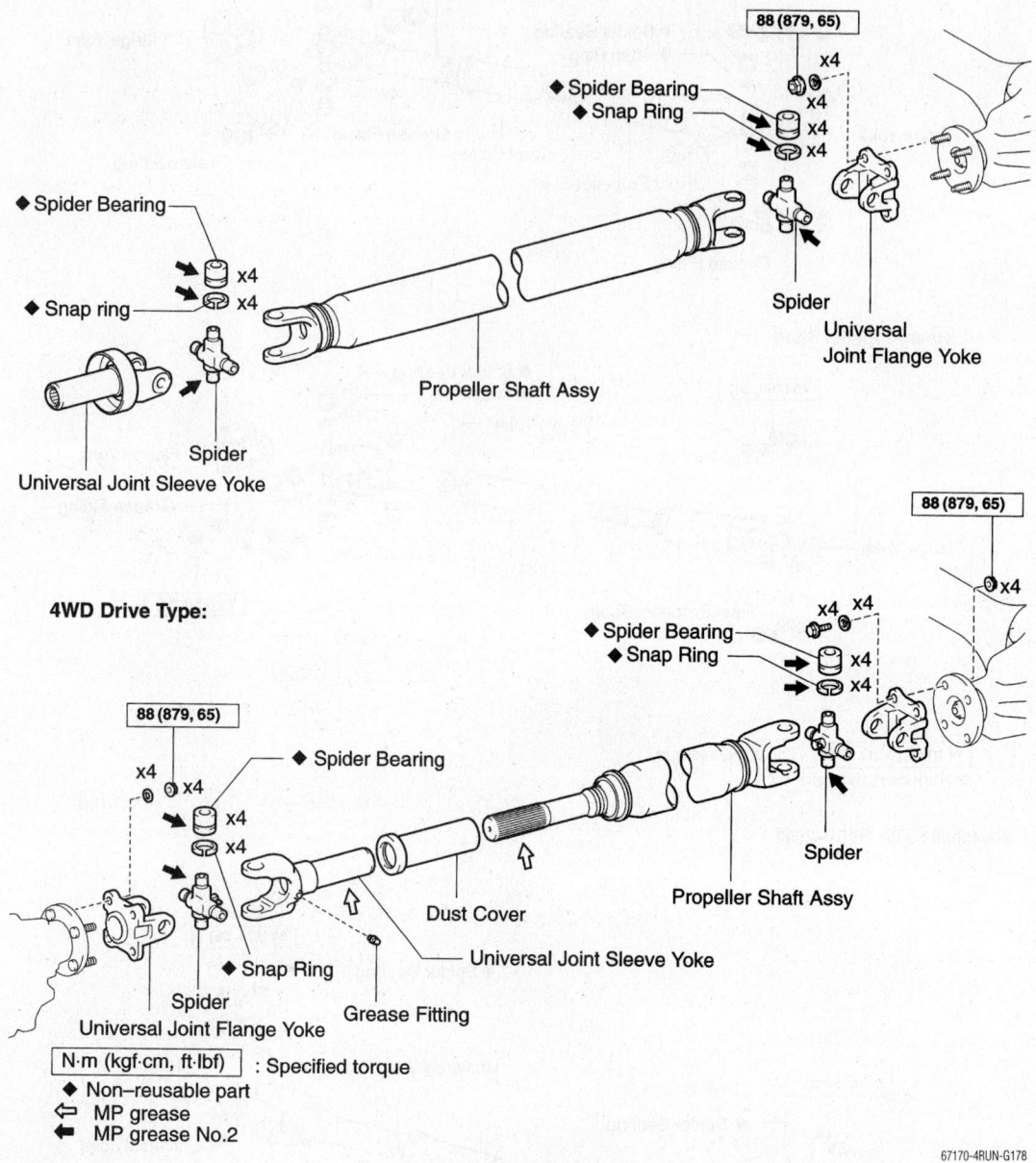

88 (879, 65)

◆ Spider Bearing
◆ Snap Ring
x4
x4
x4
x4

Spider

Universal
Joint Flange Yoke

◆ Spider Bearing

◆ Snap ring
x4
x4

Spider

Universal Joint Sleeve Yoke

Propeller Shaft Assy

4WD Drive Type:

88 (879, 65)
x4

◆ Spider Bearing
◆ Snap Ring
x4 x4
x4
x4

Spider

Universal Joint Flange Yoke

88 (879, 65)
x4
x4
x4
x4

◆ Spider Bearing

Spider

Propeller Shaft Assy

Dust Cover

Universal Joint Sleeve Yoke

◆ Snap Ring

Spider

Universal Joint Flange Yoke

Grease Fitting

N·m (kgf·cm, ft·lbf) : Specified torque

◆ Non–reusable part
⇦ MP grease
⬅ MP grease No.2

67170-4RUN-G178

Rear driveshafts—2003–06

- Engine under-covers: 21 ft. lbs. (26 Nm)

Transfer Case Assembly

REMOVAL & INSTALLATION

2002

1. Before servicing the vehicle, refer to the Precautions Section.
2. Disconnect the negative battery cable.
3. Drain the transmission and the transfer case.

4. Remove or disconnect the following:
- Transfer case with the transmission
- Breather hose from the transfer upper cover and the transmission control retainer, if equipped with an automatic transmission
- Rear engine mounting
- Dynamic damper
5. Remove the driveshaft upper dust cover and the transfer from the transmissions, as follows:
- Dust cover bolt from the bracket
- Transfer case adapter rear mounting bolts

- Transfer case, by pulling it straight up and away from the transmission.

✱✱ WARNING

Be careful not to damage the adapter rear oil seal with the transfer input gear spline.

To install:

6. Install the transfer case and the driveshaft upper dust cover to the transmission with a new gasket, as follows:
- Shift the 2 shift fork shafts to the high 4 position

- Apply MP grease to the adapter oil seal
- New gasket to the transfer adapter
- Transfer case to the transmission.

✳✳ WARNING

Take care not to damage the oil seal by the input gear spline.

- Transfer case adapter. Tighten the rear bolts to 27 ft. lbs. (37 Nm).
- Dust cover to the bracket. Tighten the bolt to 17 ft. lbs. (23 Nm).
7. Install or connect the following:
- Engine rear mount. Tighten the bolts to 19 ft. lbs. (25 Nm).

- Dynamic damper. Tighten the bolts to 27 ft. lbs. (37 Nm).
- Breather hose, if equipped with an automatic transmission
- Transfer case with the transmission to the engine
8. Fill the transmission and the transfer case with oil.
9. Test drive the vehicle and check the abnormal noise and smooth operation.
10. Recheck the fluid levels.

2003–06

1. Before servicing the vehicle, refer to the Precautions Section.

2. Remove the transmission and transfer case as an assembly.
3. Remove the 8 bolts and 2 clamps, and separate the transfer case from the transmission.
4. Installation is the reverse of removal. Torque the bolts to 17 ft. lbs. (24 Nm).

Halfshaft

REMOVAL & INSTALLATION

2002

1. Before servicing the vehicle, refer to the Precautions Section.

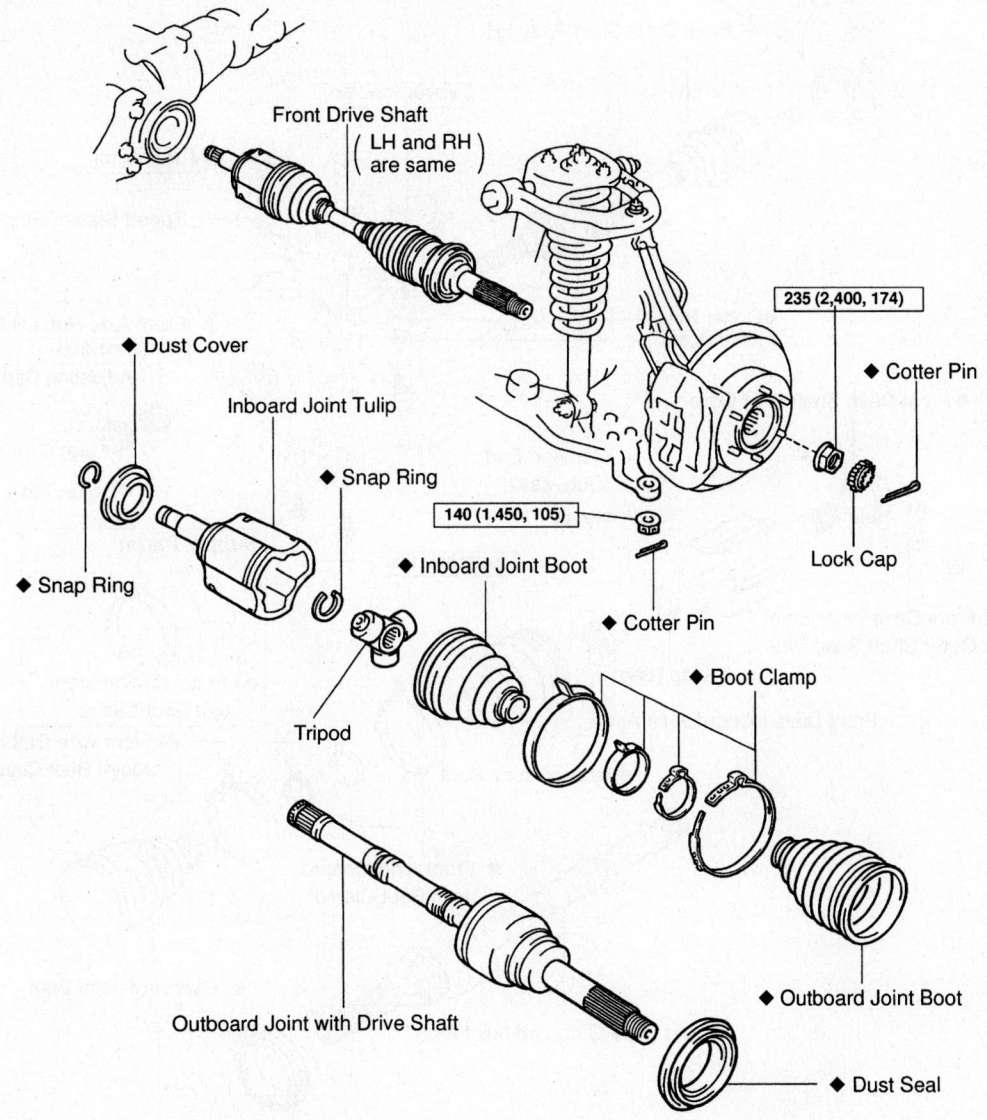

Front Drive Shaft
(LH and RH
are same)

235 (2,400, 174)

◆ Dust Cover

Inboard Joint Tulip

◆ Snap Ring

◆ Snap Ring

140 (1,450, 105)

Tripod

◆ Inboard Joint Boot

◆ Cotter Pin

◆ Cotter Pin

Lock Cap

◆ Boot Clamp

Outboard Joint with Drive Shaft

◆ Outboard Joint Boot

◆ Dust Seal

N·m (kgf·cm, ft·lbf) : Specified torque
◆ Non–reusable part

Front halfshaft and related parts—2002

67170-4RUN-G181

2. Remove the front wheel.

3. Drain the differential oil from the differential.

4. Remove the halfshaft locknut, as follows:
- Grease cap
- Cotter pin and the lockcap
- Locknut, while applying the brakes

5. Remove or disconnect the following:
- Halfshaft, using a brass bar and a hammer
- Lower control arm
- Halfshaft, by pushing the steering knuckle outward
- Snapring from the inboard shaft

To install:

6. Install or connect the following:
- Snapring to the inboard shaft
- Halfshaft
- Steering knuckle
- Lower control arm. Tighten the nut to 105 ft. lbs. (142 Nm).

7. Connect the halfshaft, as follows:
- Set the snapring opening side facing downward
- Strike the inboard joint into the differential, using SST 09631-10030 and a hammer
- Check that the halfshaft cannot be pulled out by hand

8. Install or connect the following:
- Locknut, while applying the brakes. Tighten it to 174 ft. lbs. (235 Nm).
- Adjusting cap and new cotter pin
- Grease cap
- Front wheel

9. Fill the differential with oil.

2003–06

1. Before servicing the vehicle, refer to the Precautions Section.

2. Remove the front wheel.

3. Drain the differential oil from the differential.

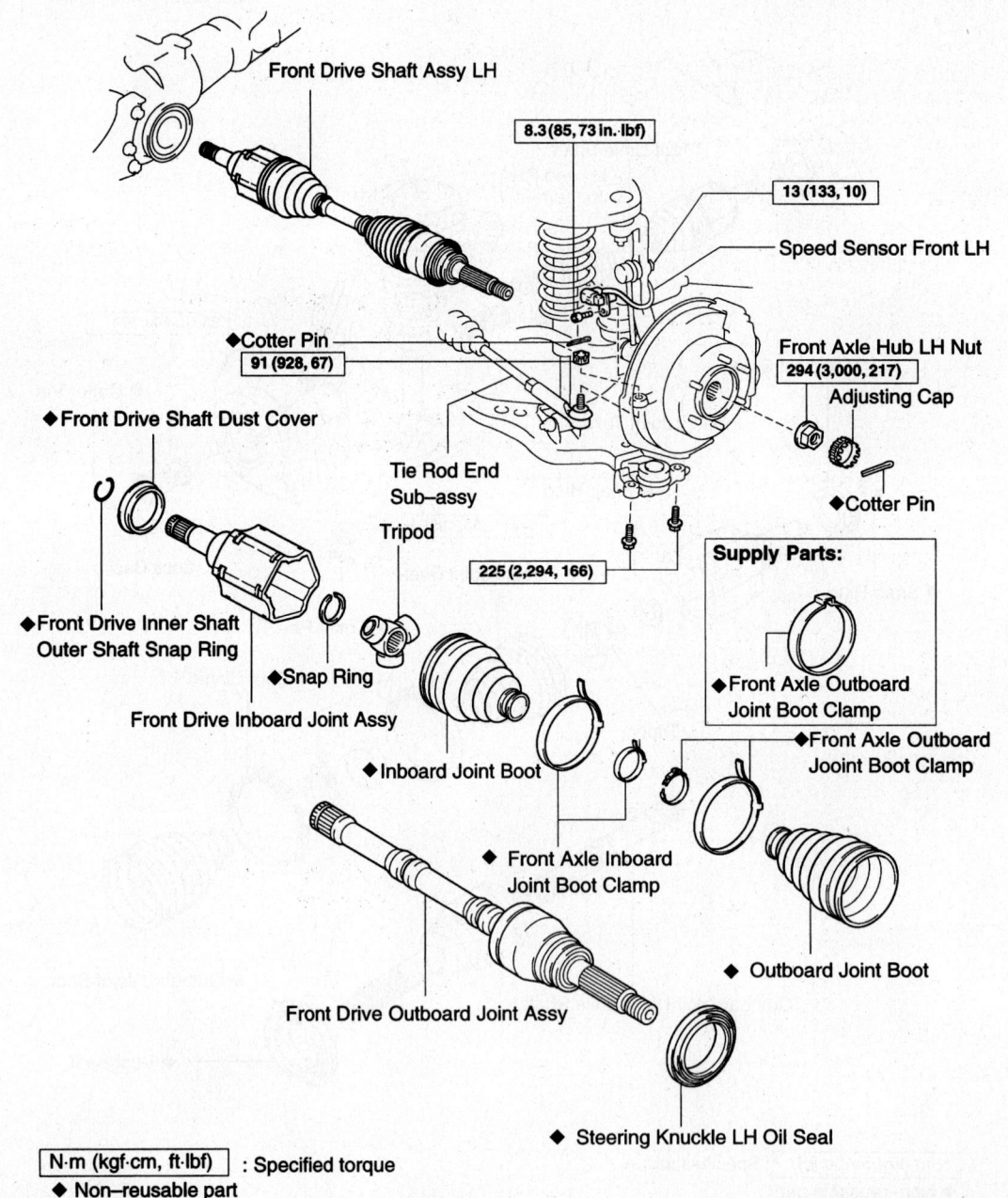

Front halfshaft and related parts, left side shown—2003–06

4. Remove the speed sensor harness.

5. Remove the halfshaft locknut, as follows:

- Grease cap
- Cotter pin and the lockcap
- Locknut, while applying the brakes

6. Remove or disconnect the following:

- Tie rod end from the knuckle
- Lower control arm from the ball joint
- Halfshaft from the steering knuckle
- Halfshaft from the differential with a slide hammer

To install:

7. Install or connect the following:

- Halfshaft
- Steering knuckle
- Lower control arm. Tighten the nut to 166 ft. lbs. (225 Nm).

8. Connect the halfshaft, as follows:

- Strike the inboard joint into the differential, using SST 09631-10030 and a hammer
- Check that the halfshaft cannot be pulled out by hand

9. Install or connect the following:

- Tie rod end. Torque to 67 ft. lbs. (91 Nm)
- Speed sensor harness
- Locknut, while applying the brakes. Tighten it to 217 ft. lbs. (294 Nm).
- Adjusting cap and new cotter pin
- Grease cap
- Front wheel

10. Fill the differential with oil.

CV-Joints

OVERHAUL

Outboard Joint

The outboard joint is replaced with halfshaft; no overhaul is possible or necessary.

Inboard (Tri-Pot) Joint

1. Before servicing the vehicle, refer to the Precautions Section.

2. Remove the halfshaft from the vehicle.

3. Remove the large clamp from the inboard joint.

4. Remove the small clamp, using side cutters, from the inboard joint.

5. Slide the inboard joint boot toward the outboard joint.

6. Matchmark the inboard joint to the halfshaft.

7. Remove the inboard joint housing from the halfshaft.

8. Remove the snapring from the end of the halfshaft.

9. Matchmark the halfshaft to the tri-pot joint.

10. Remove the tri-pot from the halfshaft, using a brass bar and a hammer.

✳✳ WARNING

Do not tap on the tri-pot joint.

11. Remove the inboard and outboard boots from the halfshaft.

✳✳ WARNING

Do not disassemble the outboard joint.

To assemble:

12. Wrap vinyl tape around the halfshaft splines to prevent damaging the boots.

13. Install the outboard and inboard boots to the halfshaft with the small end clamps.

14. Assemble the tri-pot joint to the halfshaft with the beveled side facing the outboard joint and align the matchmarks.

15. Install the tri-pot joint, using a brass bar and a hammer.

✳✳ WARNING

Do not tap on the roller.

16. Install the snapring.

17. Lubricate the outboard joint with ½ of the grease supplied with the kit.

18. Assemble the boot to the outboard joint

19. Assemble the inboard joint housing to the halfshaft by aligning the matchmarks.

20. Temporarily install the boot onto the tri-pot housing.

21. Make sure the boots are positioned in the shaft grooves.

22. Install a new inboard joint clamp.

23. Crimp the large clamp with tool 09521-24010 so that the crimp clearance is 0.039–0.059 in. (1.0–1.5mm).

24. Install the halfshaft.

Rear Axle Shaft, Bearing and Seal

REMOVAL & INSTALLATION

2002

1. Before servicing the vehicle, refer to the Precautions Section.

2. Remove or disconnect the following:

- Rear wheel
- Brake drum

3. Check the bearing backlash and axle shaft deviation, as follows:

a. Using a dial indicator, check that

the backlash in the bearing shaft direction. The maximum is 0.027 in. (0.7mm).

b. If the backlash exceeds the maximum, replace the bearing.

c. Using a dial indicator, check the deviation at the surface of the axle shaft outside the hub bolt. Maximum is 0.0039 in. (0.1mm).

d. If the deviation exceeds the maximum, replace the axle shaft.

4. Remove or disconnect the following:

- Anti-lock Brake System (ABS) speed sensor from the axle housing, if equipped
- Axle shaft assembly by removing the 4 nuts from the backing plate
- O-ring from the axle housing
- Bearing and retainer (differential side) and ABS speed sensor rotor, if equipped
- Snapring from the axle shaft
- Axle shaft

➡ **Inspect the axle shaft and flange runouts. The axle shaft run-out should be 0.079 in. (2.0mm) and the flange run-out should be 0.004 in. (0.1mm).**

- Outer seal
- Bearing from the axle shaft

To install:

5. Install or connect the following:

- Bearing from the axle shaft
- Outer seal
- Axle shaft
- Snapring to the axle shaft
- Bearing and retainer (differential side) and ABS speed sensor rotor, if equipped
- O-ring to the axle housing
- Axle shaft assembly. Tighten the 4 backing plate nuts to 48 ft. lbs. (66 Nm).
- Anti-lock Brake System (ABS) speed sensor to the axle housing, if equipped
- Brake drum
- Rear wheel

2003–06

1. Before servicing the vehicle, refer to the Precautions Section.

2. Remove or disconnect the following:

- Rear wheel
- Speed sensor
- Brake caliper
- Brake rotor
- Parking brake assembly
- Axle shaft assembly by removing the 4 nuts from the backing plate
- O-ring from the axle housing
- Bearing and retainer (differential

side) and ABS speed sensor rotor, if equipped
- Snapring from the axle shaft
- Axle shaft

➡**Inspect the axle shaft and flange run-outs. The axle shaft run-out should be 0.079 in. (2.0mm) and the flange run-out should be 0.004 in. (0.1mm).**

- Outer seal
- Bearing from the axle shaft, using a press

To install:
3. Install or connect the following:
- Bearing from the axle shaft
- Outer seal
- Axle shaft
- Snapring to the axle shaft
- Bearing and retainer (differential side) and ABS speed sensor rotor, if equipped
- O-ring to the axle housing
- Axle shaft assembly. Tighten the 4

backing plate nuts to 91 ft. lbs. (123 Nm).
- Anti-lock Brake System (ABS) speed sensor to the axle housing, if equipped
- Brake rotor
- Brake caliper
- Rear wheel

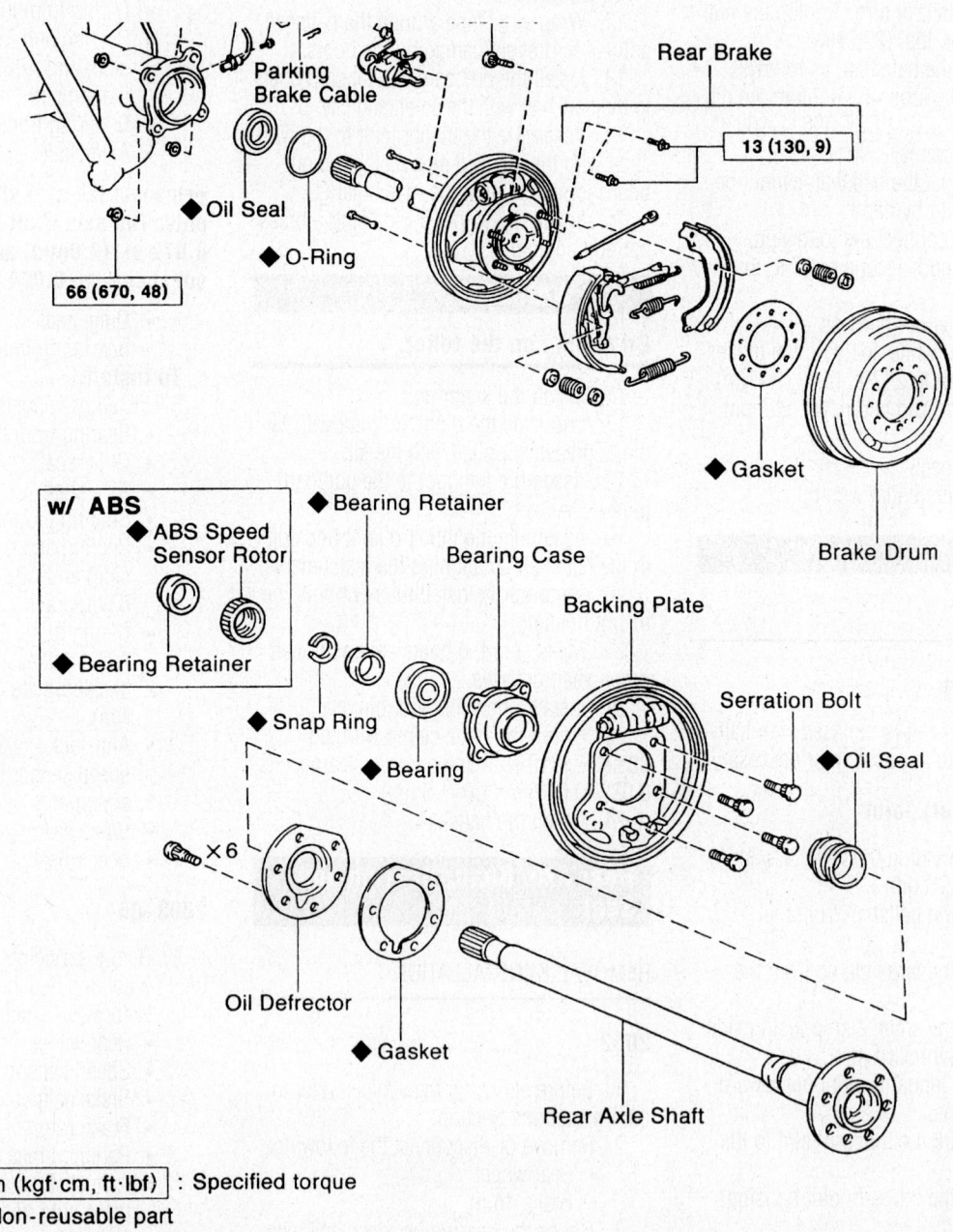

Exploded view of the rear axle shaft and components—2002

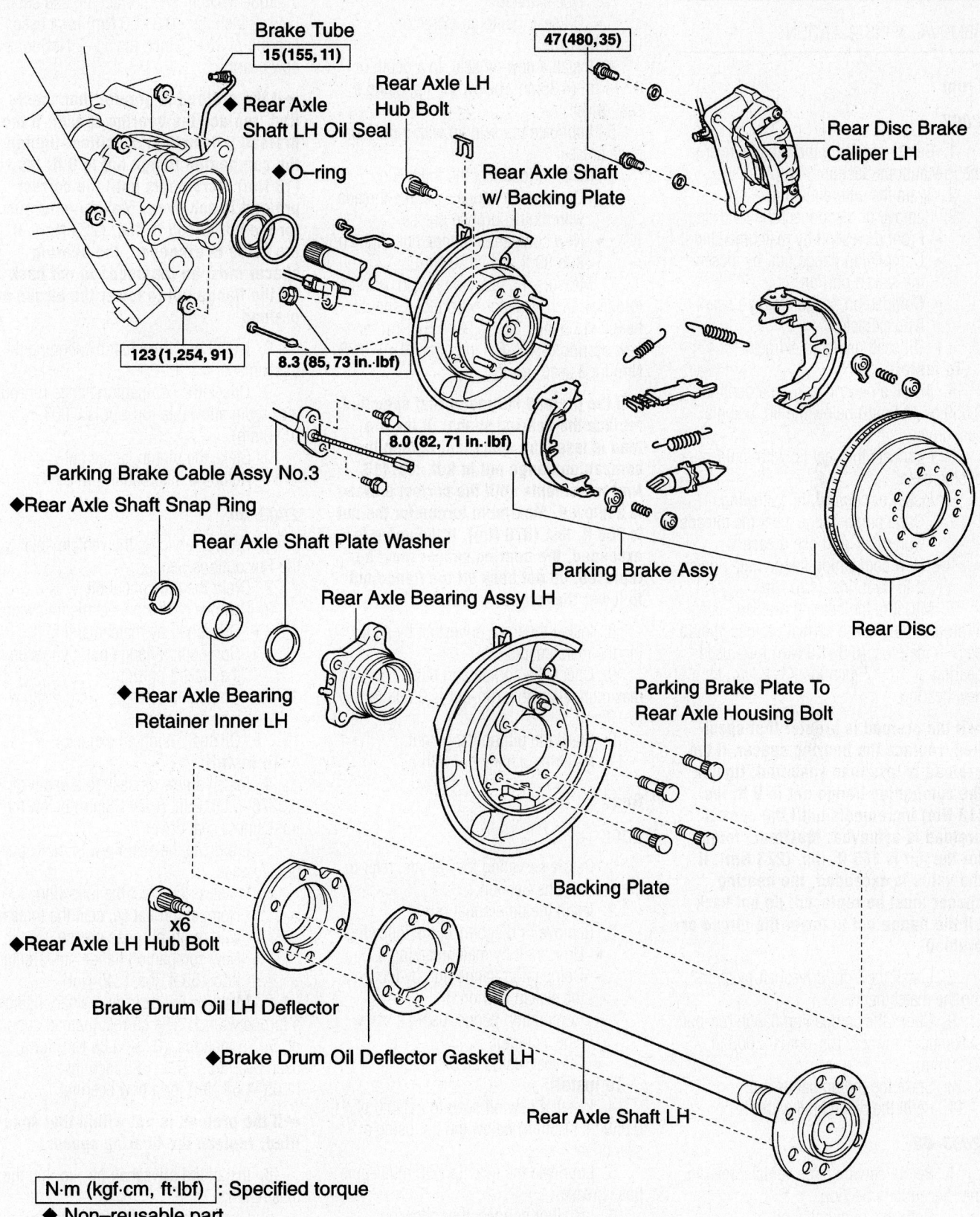

Brake Tube
15 (155, 11)

Rear Axle Shaft LH Oil Seal

O-ring

47 (480, 35)

Rear Axle LH Hub Bolt

Rear Disc Brake Caliper LH

Rear Axle Shaft w/ Backing Plate

123 (1,254, 91)

8.3 (85, 73 in.·lbf)

8.0 (82, 71 in.·lbf)

Parking Brake Cable Assy No.3

◆Rear Axle Shaft Snap Ring

Parking Brake Assy

Rear Axle Shaft Plate Washer

Rear Disc

Rear Axle Bearing Assy LH

◆Rear Axle Bearing Retainer Inner LH

Parking Brake Plate To Rear Axle Housing Bolt

Backing Plate

x6

◆Rear Axle LH Hub Bolt

Brake Drum Oil LH Deflector

◆Brake Drum Oil Deflector Gasket LH

Rear Axle Shaft LH

N·m (kgf·cm, ft·lbf) : Specified torque

◆ Non-reusable part

67170-4RUN-G183

Exploded view of the rear axle shaft and components—2003–06

Pinion Seal

REMOVAL & INSTALLATION

Front

2002

1. Before servicing the vehicle, refer to the Precautions Section.
2. Drain the differential oil.
3. Remove or disconnect the following:
   - Front driveshaft by matchmarking it
   - Companion flange nut, by loosen the staked portion
   - Companion flange, using a screw-type extractor
   - Oil seal, using an extractor

To install:

4. Install a new oil seal, to a depth of 0.059 in. (1.5mm) below the lip, using a seal driver.
5. Lubricate the seal lip with multi-purpose grease.
6. Install or connect the following:
   - Companion flange, coat the threads with multi-purpose grease
   - New companion flange nut. Tighten it to 89 ft. lbs. (120 Nm).
7. Measure the bearing preload, using a torque wrench. The correct preload should be 5–9 inch lbs. (0.6–1.0 Nm) for a used bearing or 10–17 inch lbs. (1–2 Nm) for a new bearing.

➡**If the preload is greater that specified, replace the bearing spacer. If the preload is less than specified, tighten the companion flange nut in 9 ft. lbs. (13 Nm) increments until the correct preload is achieved. Maximum torque for the nut is 165 ft. lbs. (223 Nm). If the value is exceeded, the bearing spacer must be replaced; do not back off the flange nut to lower the torque or preload.**

8. Install the front driveshaft by aligning the matchmarks.
9. Check the companion flange run-out; maximum allowable run-out is 0.003 in. (0.10mm).
10. Stake the pinion flange nut.
11. Refill the differential with oil.

2003–06

1. Before servicing the vehicle, refer to the Precautions Section.
2. Drain the differential oil.
3. Remove or disconnect the following:
   - Front driveshaft by matchmarking it
   - Companion flange nut, by loosen the staked portion

   - Companion flange, using a screw-type extractor
   - Oil seal, using an extractor

To install:

4. Install a new oil seal, to a depth of 0.171 in. (4.35mm) below the lip, using a seal driver.
5. Lubricate the seal lip with multi-purpose grease.
6. Install or connect the following:
   - Companion flange, coat the threads with multi-purpose grease
   - New companion flange nut. Tighten it to 89 ft. lbs. (120 Nm).
7. Measure the bearing preload, using a torque wrench. The correct preload should be 8.7–13.9 inch lbs. (0.98–1.57 Nm) for a new bearing or 4.3–6.9 inch lbs. (0.49–0.78 Nm) for a used bearing.

➡**If the preload is greater that specified, replace the bearing spacer. If the preload is less than specified, tighten the companion flange nut in 9 ft. lbs. (13 Nm) increments until the correct preload is achieved. Maximum torque for the nut is 268 ft. lbs. (370 Nm). If the value is exceeded, the bearing spacer must be replaced; do not back off the flange nut to lower the torque or preload.**

8. Install the front driveshaft by aligning the matchmarks.
9. Check the companion flange run-out; maximum allowable run-out is 0.003 in. (0.10mm).
10. Stake the pinion flange nut.
11. Refill the differential with oil.

Rear

2002

1. Before servicing the vehicle, refer to the Precautions Section.
2. Drain the differential oil.
3. Remove or disconnect the following:
   - Driveshaft by matchmarking it
   - Companion flange nut, by loosen the staked portion
   - Companion flange, using a screw-type extractor
   - Oil seal, using an extractor

To install:

4. Install a new oil seal, to a depth of 0.059 in. (1.5mm) below the lip, using a seal driver.
5. Lubricate the seal lip with multi-purpose grease.
6. Install or connect the following:
   - Companion flange, coat the threads with multi-purpose grease
   - New companion flange nut. Tighten it to 89 ft. lbs. (120 Nm).

7. Measure the bearing preload, using a torque wrench. The correct preload should be 5–9 inch lbs. (0.6–1.0 Nm) for a used bearing or 10–17 inch lbs. (1–2 Nm) for a new bearing.

➡**If the preload is greater that specified, replace the bearing spacer. If the preload is less than specified, tighten the companion flange nut in 9 ft. lbs. (13 Nm) increments until the correct preload is achieved. Maximum torque for the nut is 165 ft. lbs. (223 Nm). If the value is exceeded, the bearing spacer must be replaced; do not back off the flange nut to lower the torque or preload.**

8. Install the driveshaft by aligning the matchmarks.
9. Check the companion flange run-out; maximum allowable run-out is 0.003 in. (0.10mm).
10. Stake the pinion flange nut.
11. Refill the differential with oil.

2003–06

1. Before servicing the vehicle, refer to the Precautions Section.
2. Drain the differential oil.
3. Remove or disconnect the following:
   - Driveshaft by matchmarking it
   - Companion flange nut, by loosen the staked portion
   - Companion flange, using a screw-type extractor
   - Oil seal, using an extractor

To install:

4. Install a new oil seal, to a depth of 0.0276–0.0512 in. (0.7–1.3mm) below the lip, using a seal driver.
5. Lubricate the seal lip with multi-purpose grease.
6. Install or connect the following:
   - Companion flange, coat the threads with multi-purpose grease
   - New companion flange nut. Tighten it to 159 ft. lbs. (215 Nm).
7. Measure the bearing preload, using a torque wrench. The correct preload should be 5–7.5 inch lbs. (0.56–0.85 Nm) for a used bearing or 9.3–14.5 inch lbs. (1.05–1.64 Nm) for a new bearing.

➡**If the preload is not within that specified, replace the bearing spacer.**

8. Install the driveshaft by aligning the matchmarks.
9. Check the companion flange run-out; maximum allowable run-out is 0.003 in. (0.10mm).
10. Stake the pinion flange nut.
11. Refill the differential with oil.

STEERING AND SUSPENSION

Air Bag

✳✳ CAUTION

Some vehicles are equipped with an air bag system. The system must be disarmed before performing service on, or around, system components, the steering column, instrument panel components, wiring and sensors. Failure to follow the safety precautions and the disarming procedure could result in accidental air bag deployment, possible injury and unnecessary system repairs.

DISARMING

To avoid personal injury when working on vehicles equipped with an air bag, the negative battery cable must be disconnected and at least 90 seconds must elapse before working on the system. Failure to do so may result in deployment of the air bag.

Power Rack And Pinion Steering Gear

REMOVAL & INSTALLATION

2002

1. Before servicing the vehicle, refer to the Precautions Section.
2. Disarm the airbag system.
3. Place the wheels in the straight-ahead position.

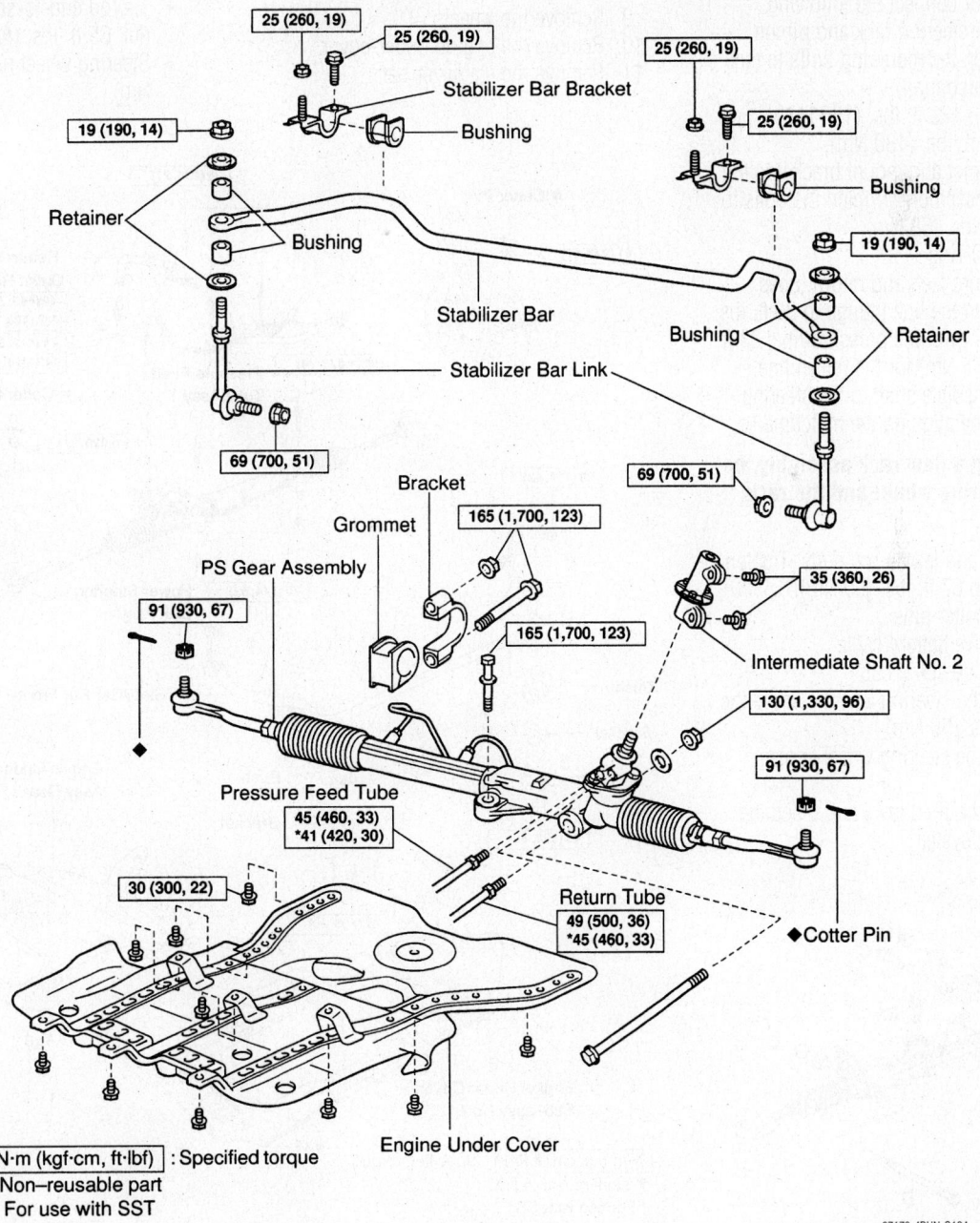

25 (260, 19)
25 (260, 19)
25 (260, 19)
Stabilizer Bar Bracket
Bushing
25 (260, 19)
19 (190, 14)
Bushing
Retainer
Bushing
19 (190, 14)
Stabilizer Bar
Bushing
Retainer
Stabilizer Bar Link
69 (700, 51)
Bracket
69 (700, 51)
Grommet
165 (1,700, 123)
PS Gear Assembly
35 (360, 26)
91 (930, 67)
165 (1,700, 123)
Intermediate Shaft No. 2
130 (1,330, 96)
91 (930, 67)
Pressure Feed Tube
45 (460, 33)
*41 (420, 30)
30 (300, 22)
Return Tube
49 (500, 36)
*45 (460, 33)
◆Cotter Pin
Engine Under Cover

N·m (kgf·cm, ft·lbf) : Specified torque
◆Non–reusable part
* For use with SST

Steering gear and related parts—2002

67170-4RUN-G184

4. Remove or disconnect the following:
- Negative battery cable
- Steering wheel
- Engine under-cover
- Stabilizer bar
- Right and left tie rod ends from the knuckle
- Intermediate shaft from the steering rack, by matchmarking it
- Pressure feed and the return tubes, using SST 09631-22020
- Mount bracket and the grommet, from the power steering rack assembly
- Power steering rack and pinion

To install:

5. Install or connect the following:
- Power steering rack and pinion. Tighten the mounting bolts to (see the illustration):
- A & C: 123 ft. lbs. (165 Nm)
- B: 96 ft. lbs. (130 Nm).
- Grommet and mount bracket to the gear assembly. Tighten the bolts to 65 ft. lbs. (88 Nm).
- New O-ring
- Pressure feed and return tubes. Tighten the line fittings to 30 ft. lbs. (41 Nm) for the pressure line; 33 ft. lbs. (45 Nm) for the return line.
- Intermediate shaft to the steering rack, by aligning the matchmarks

➡️If installing a new rack assembly, be sure the steering wheel and the rack are centered.

- Right and left tie rod ends. Tighten nuts to 67 ft. lbs. (90 Nm).
- New cotter pins
- Negative battery cable
6. Center the spiral cable.
7. Install the steering wheel. Torque the nut to 37 ft. lbs. (50 Nm)
8. Check the steering wheel center point.
9. Check the fluid level and bleed the power steering system.

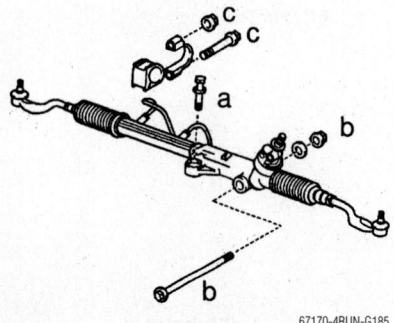

Steering gear bolt identification—2002

67170-4RUN-G185

10. Install the under-cover. Torque to 22 ft. lbs. (30 Nm).
11. Check the front wheel alignment.

2003–06

1. Before servicing the vehicle, refer to the Precautions Section.
2. Disarm the airbag system.
3. Remove the negative battery cable.
4. Place the wheels in the straight-ahead position.
5. Remove the steering wheel.
6. Remove the lower steering column cover.
7. Remove the turn signal switch.
8. Remove the spiral cable.
9. Remove the wheels.
10. Remove the engine under-cover.
11. Remove the stabilizer bar.

12. Remove the tie rod ends from the knuckles.
13. Disconnect the pressure and return lines from the gear.
14. Remove the 2 bolts and nuts. The nuts have detents. Never turn the nuts, just the bolts.
15. Remove the gear.
16. Installation is the reverse of removal. Center the spiral cable. Observe the following torques"
- Pressure and return lines: 30 ft. lbs. (40 Nm)
- Steering gear: 74 ft. lbs. (100 Nm)
- Tie rod end-to-knuckle: 67 ft. lbs. (91 Nm)
- Tie rod end-to-steering gear lock-nut: 65 ft. lbs. (88 Nm)
- Steering wheel nut: 37 ft. lbs. (50 Nm)

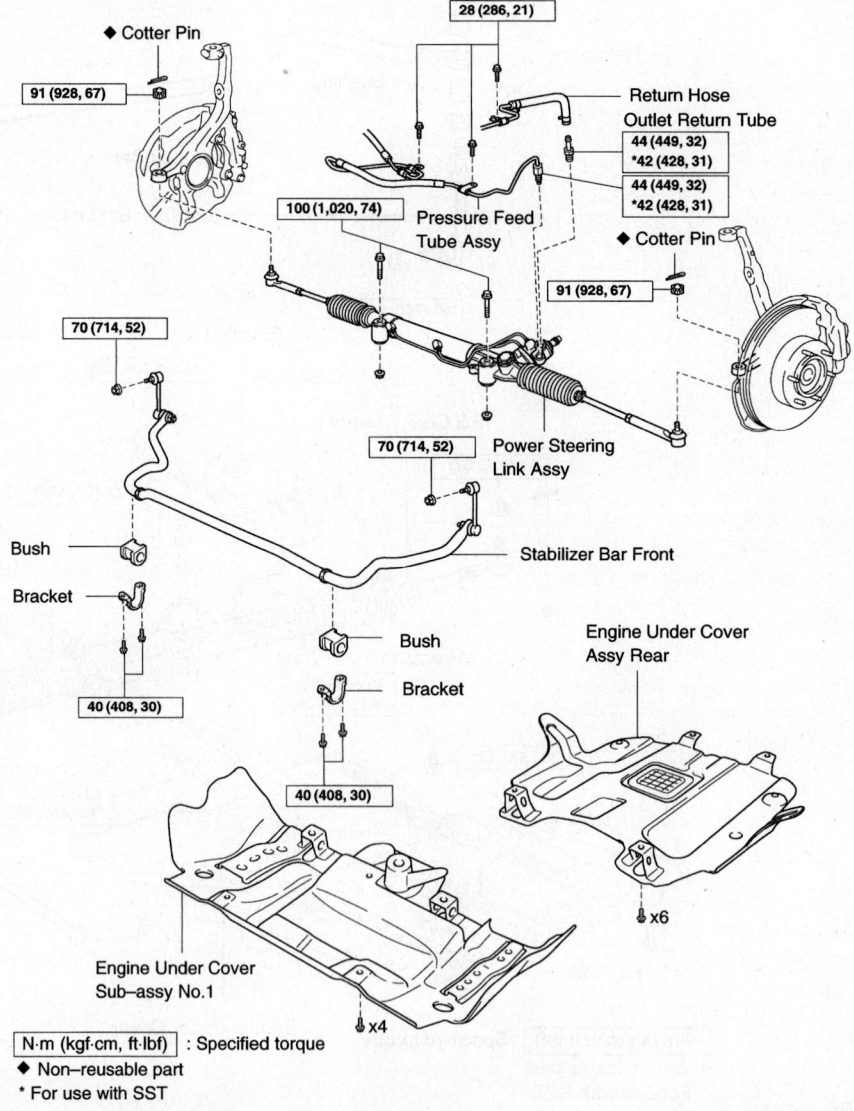

Steering gear and related parts—2003–06

67170-4RUN-G186

Shock Absorber

REMOVAL & INSTALLATION

Front

2002

1. Before servicing the vehicle, refer to the Precautions Section.
2. Remove or disconnect the following:
   - Front wheel

- Shock from the lower control arm
- 3 upper nuts
- Shock absorber

To install:
3. Install or connect the following:
   - Shock absorber. Tighten the 3 nuts to 47 ft. lbs. (64 Nm).
   - Lower shock-to-lower control arm. Tighten the bolt to 101 ft. lbs. (135 Nm).
   - Wheels
4. Check the vehicle alignment.

2003–06

1. Before servicing the vehicle, refer to the Precautions Section.
2. Remove or disconnect the following:
   - Front wheel
   - Stabilizer bar
3. With air shocks, lower the suspension as far as possible and disconnect the air line.
   - Shock from the lower control arm
   - 3 upper nuts
   - Shock absorber

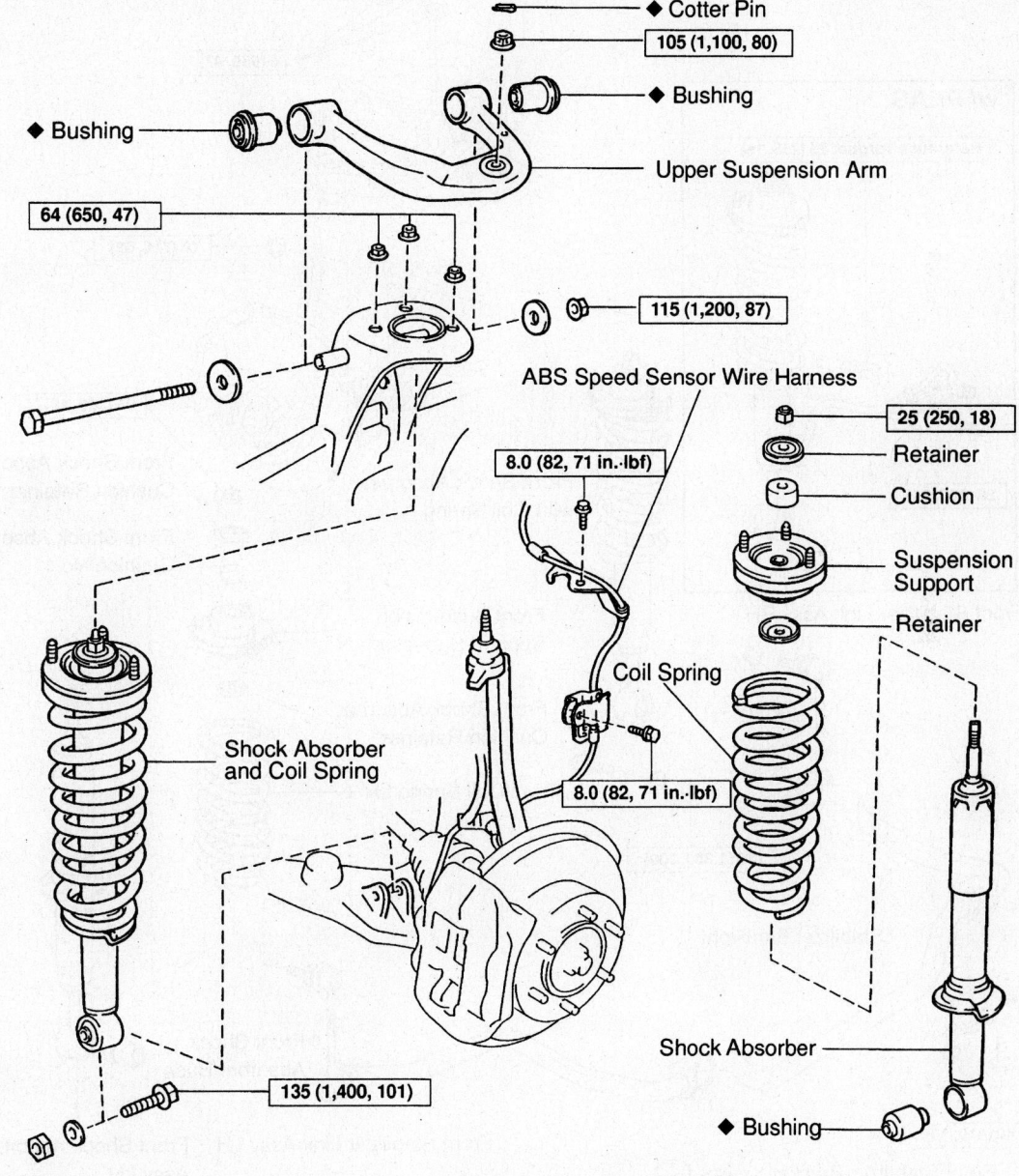

- ◆ Cotter Pin
- 105 (1,100, 80)
- ◆ Bushing
- ◆ Bushing
- Upper Suspension Arm
- 64 (650, 47)
- 115 (1,200, 87)
- ABS Speed Sensor Wire Harness
- 25 (250, 18)
- Retainer
- Cushion
- Suspension Support
- Retainer
- 8.0 (82, 71 in. lbf)
- Coil Spring
- Shock Absorber and Coil Spring
- 8.0 (82, 71 in. lbf)
- 135 (1,400, 101)
- Shock Absorber
- ◆ Bushing

N·m (kgf·cm, ft·lbf) : Specified torque

◆ Non–reusable part

Front shock absorber and related parts—2002

67170-4RUN-G187

To install:

4. Install or connect the following:
- Shock absorber. Tighten the 3 nuts to 47 ft. lbs. (64 Nm).
- Lower shock-to-lower control arm. Temporarily tighten the lower bolt. When the weight of the vehicle is on the suspension, tighten the bolt to 100 ft. lbs. (135 Nm).

5. Connect the air line. Torque to 18 ft. lbs. (25 Nm)

6. Install the wheels.
7. Check the vehicle alignment.

Rear

2002

1. Before servicing the vehicle, refer to the Precautions Section.
2. Remove the wheel.
3. Lower the floor jack to take tension off of the spring.

4. Remove or disconnect the following:
- Shock absorber from the rear axle housing
- Nut, retainers and the cushions holding the shock absorber to the frame
- Shock absorber with the washers and bushings

To install:

5. Install the shock absorber to the frame with the washers and bushings.

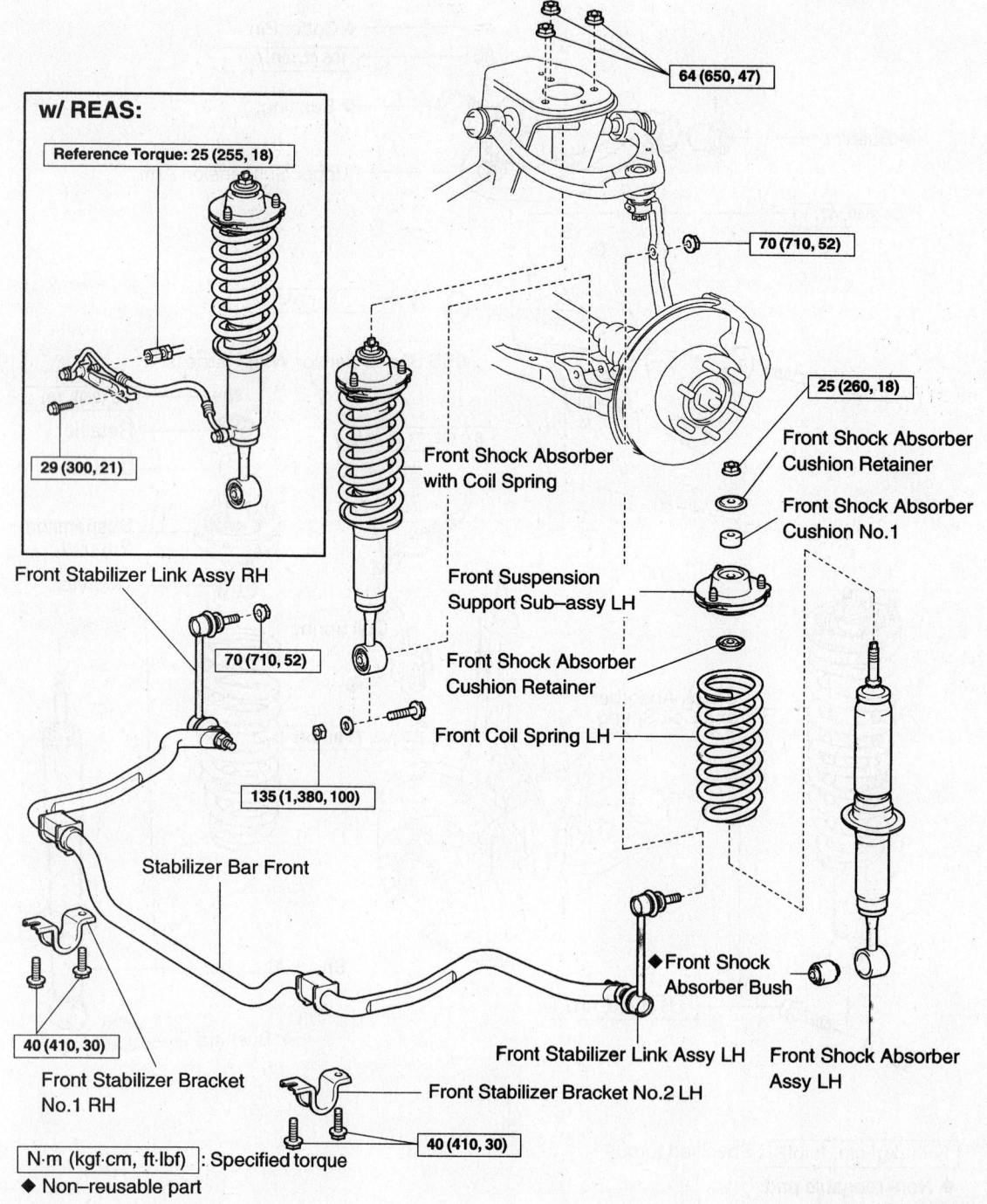

w/ REAS:

Reference Torque: 25 (255, 18)

29 (300, 21)

Front Stabilizer Link Assy RH

70 (710, 52)

Stabilizer Bar Front

135 (1,380, 100)

40 (410, 30)

Front Stabilizer Bracket No.1 RH

64 (650, 47)

70 (710, 52)

Front Shock Absorber with Coil Spring

Front Suspension Support Sub–assy LH

Front Shock Absorber Cushion Retainer

Front Coil Spring LH

25 (260, 18)

Front Shock Absorber Cushion Retainer

Front Shock Absorber Cushion No.1

Front Shock Absorber Bush

Front Stabilizer Link Assy LH

Front Stabilizer Bracket No.2 LH

40 (410, 30)

Front Shock Absorber Assy LH

N·m (kgf·cm, ft·lbf) : Specified torque
◆ Non–reusable part

67170-4RUN-G188

Front shock absorber and related parts—2003–06

6. Tighten the shock absorber-to-frame nut to 14 ft. lbs. (20 Nm)

7. Connect the shock absorber to the rear axle housing. Tighten the bolt to 47 ft. lbs. (64 Nm)

8. Install the wheels.

2003–06 W/O AIR SUSPENSION

1. Before servicing the vehicle, refer to the Precautions Section.

2. Remove the wheel.

3. Support the rear axle with a suitable jack.

4. Remove the lower shock bolt.

5. Remove the upper nut, washers, bushings and retainer.

6. Installation is the reverse of removal. Torque the upper nut to 18 ft. lbs. (25 Nm); the lower bolt to 72 ft. lbs. (98 Nm).

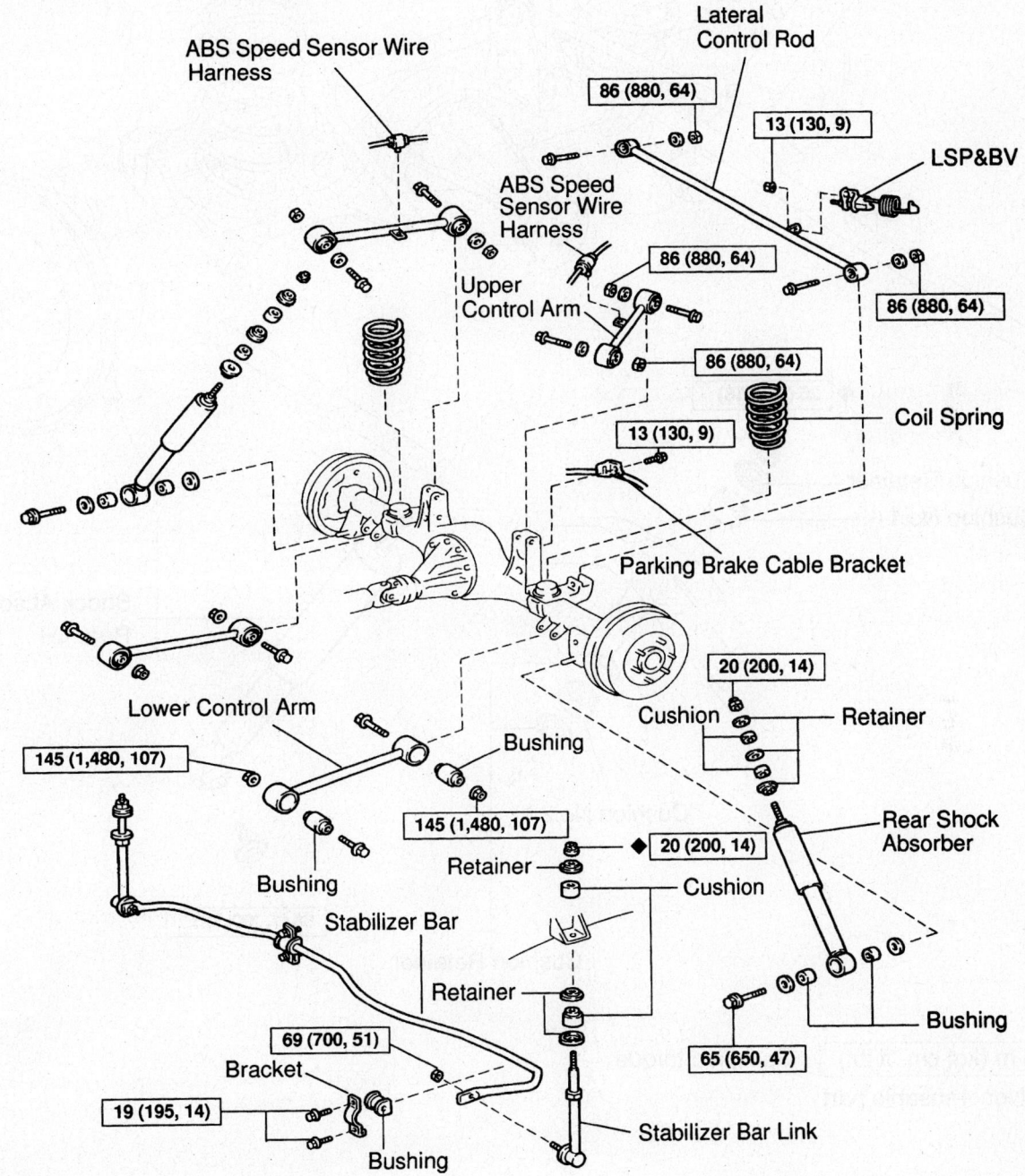

ABS Speed Sensor Wire Harness

Lateral Control Rod

86 (880, 64)

13 (130, 9)

LSP&BV

ABS Speed Sensor Wire Harness

86 (880, 64)

86 (880, 64)

Upper Control Arm

86 (880, 64)

Coil Spring

13 (130, 9)

Parking Brake Cable Bracket

20 (200, 14)

Cushion — Retainer

Lower Control Arm

Bushing

145 (1,480, 107)

145 (1,480, 107)

Retainer

◆ 20 (200, 14)

Cushion

Rear Shock Absorber

Bushing

Bushing

Stabilizer Bar

Retainer

Bushing

69 (700, 51)

Bracket

65 (650, 47)

19 (195, 14)

Stabilizer Bar Link

Bushing

N·m (kgf·cm, ft·lbf) : Specified torque

◆ Non–reusable part

Rear suspension components—2002

67170-4RUN-G190

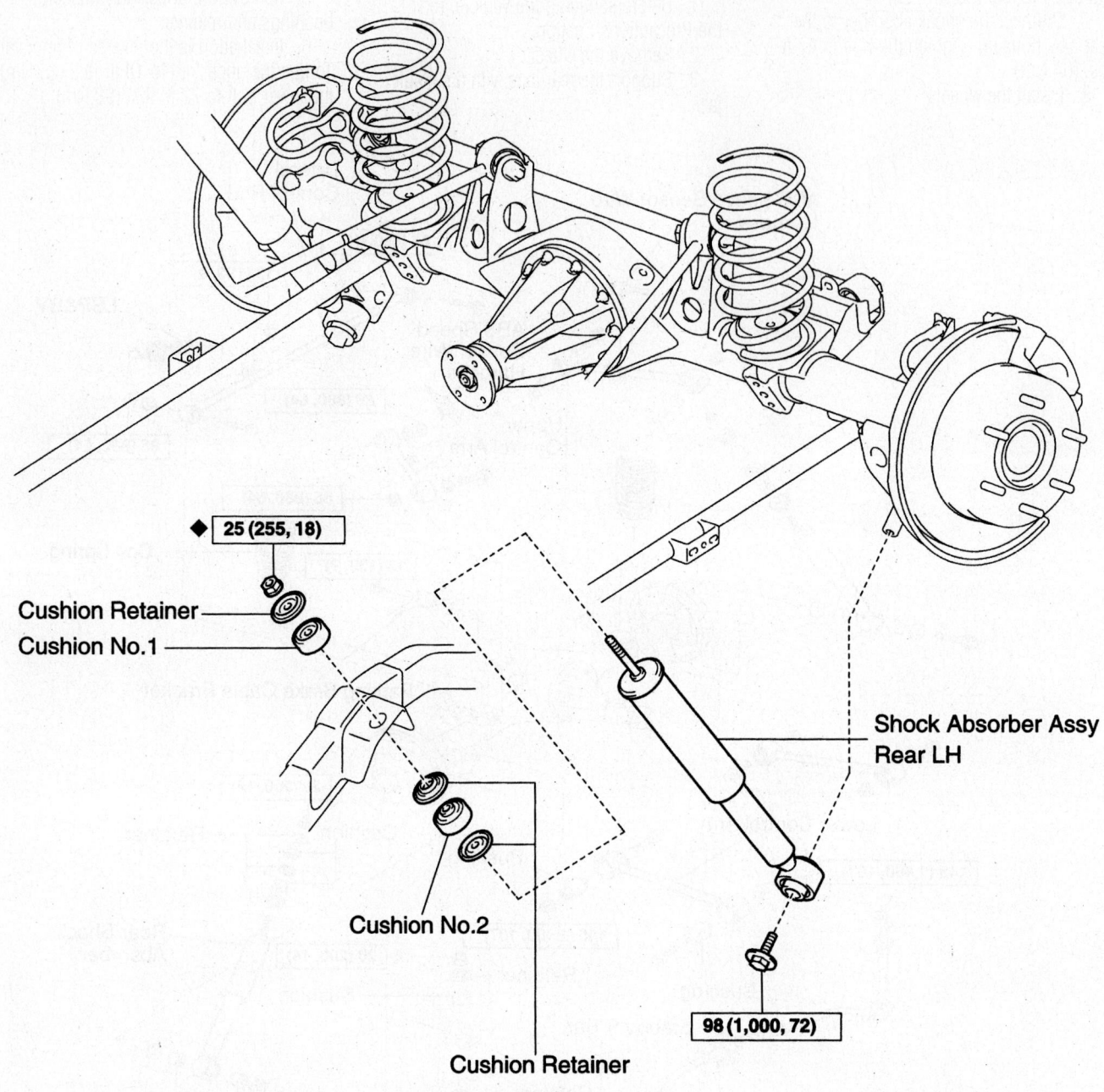

◆ 25 (255, 18)

Cushion Retainer

Cushion No.1

Shock Absorber Assy
Rear LH

Cushion No.2

98 (1,000, 72)

Cushion Retainer

N·m (kgf·cm, ft·lbf) : Specified torque

◆ Non–reusable part

67170-4RUN-G191

Rear suspension—2003–06 without air suspension

2003–06 WITH AIR SUSPENSION

1. Before servicing the vehicle, refer to the Precautions Section.
2. Remove the wheel.
3. Remove the stabilizer bar link.
4. Remove the air line bracket nut.

5. Lower the axle to fully extend the shocks.
6. Disconnect the air line coupling.
7. Remove the lower shock bolt.
8. Remove the upper nut, washers, bushings and retainer.

9. Installation is the reverse of removal. Torque the upper nut to 18 ft. lbs. (25 Nm); the lower bolt to 72 ft. lbs. (98 Nm). Torque the bracket bolt to 221 ft. lbs. (29 Nm); the coupling to 18 ft. lbs. (25 Nm).

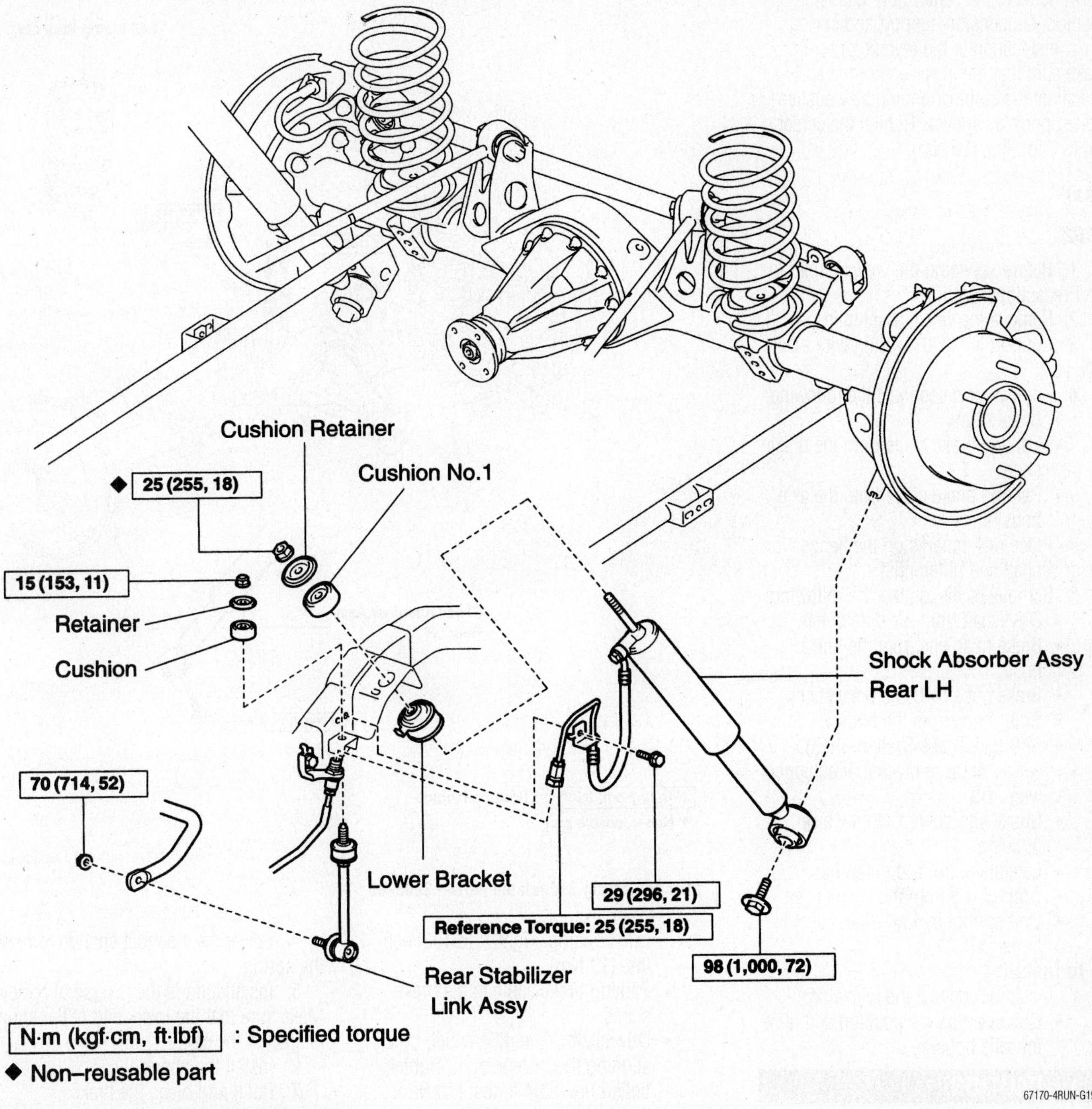

Cushion Retainer

Cushion No.1

◆ 25 (255, 18)

15 (153, 11)

Retainer

Cushion

Shock Absorber Assy
Rear LH

70 (714, 52)

Lower Bracket

29 (296, 21)

Reference Torque: 25 (255, 18)

Rear Stabilizer
Link Assy

98 (1,000, 72)

N·m (kgf·cm, ft·lbf) : Specified torque

◆ Non–reusable part

67170-4RUN-G192

Rear suspension—2003–06 with air suspension

Coil Spring

REMOVAL & INSTALLATION

Front

1. Before servicing the vehicle, refer to the Precautions Section.
2. Remove the front shock absorber.
3. Place the strut in a compressor.
4. Set the compressor to span an 8 coil set.
5. Compress the spring just enough for disassembly.

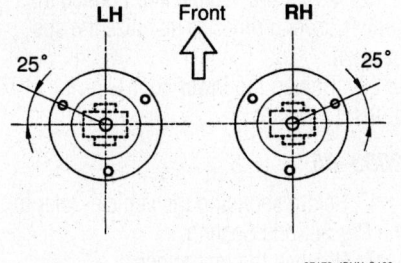

67170-4RUN-G193

Suspension support orientation—2002

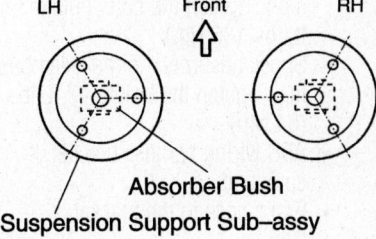

Absorber Bush

Suspension Support Sub–assy

67170-4RUN-G194

Suspension support orientation—2003–06

6. Remove the center nut, retainers, cushion, suspension support and spring.

7. Installation is the reverse of removal. Make sure that the lower end of the spring is against the stop. Orient the new suspension support as shown. Tighten the center nut to 18 ft. lbs. (25 Nm).

Rear

2002

1. Before servicing the vehicle, refer to the Precautions Section.
2. Remove the wheel assemblies.
3. Support the axle housing with a suitable jack.
4. Remove or disconnect the following:
   - Brake drum
   - Parking brake cable from the brake shoe
   - Parking brake cable from the axle housing.
5. Place matchmarks on the flanges for the driveshaft and differential.
6. Remove or disconnect the following:
   - Driveshaft from the differential
   - Brake hose line from the brake hose
   - Brake hose-to-brake bracket clip
   - Brake hose from the body
   - Anti-lock Brake System (ABS) wiring harness bracket, if equipped with ABS
   - Shock absorbers from the axle housing
   - Lateral control rod nuts/bolts
   - Control rod from the suspension
   - Coil spring, by lower the rear axle housing

To install:

7. Install or connect the following:
   - Coil springs into position and raise the axle housing.

> ✸✸ **CAUTION**
>
> **Be sure to fit the lower end of the coil spring into the gap of the spring seat on the lower control arm.**

   - Lateral control rod to the suspension. Tighten the bolts/nuts to 64 ft. lbs. (86 Nm).
   - Shock absorbers to the axle housing. Tighten the bolt to 47 ft. lbs. (64 Nm).
   - ABS wiring harness bracket, if equipped
   - Brake hose to the bracket
   - Clip
   - Brake line to the brake hose. Tighten the tube.
   - Parking brake cable bracket to the

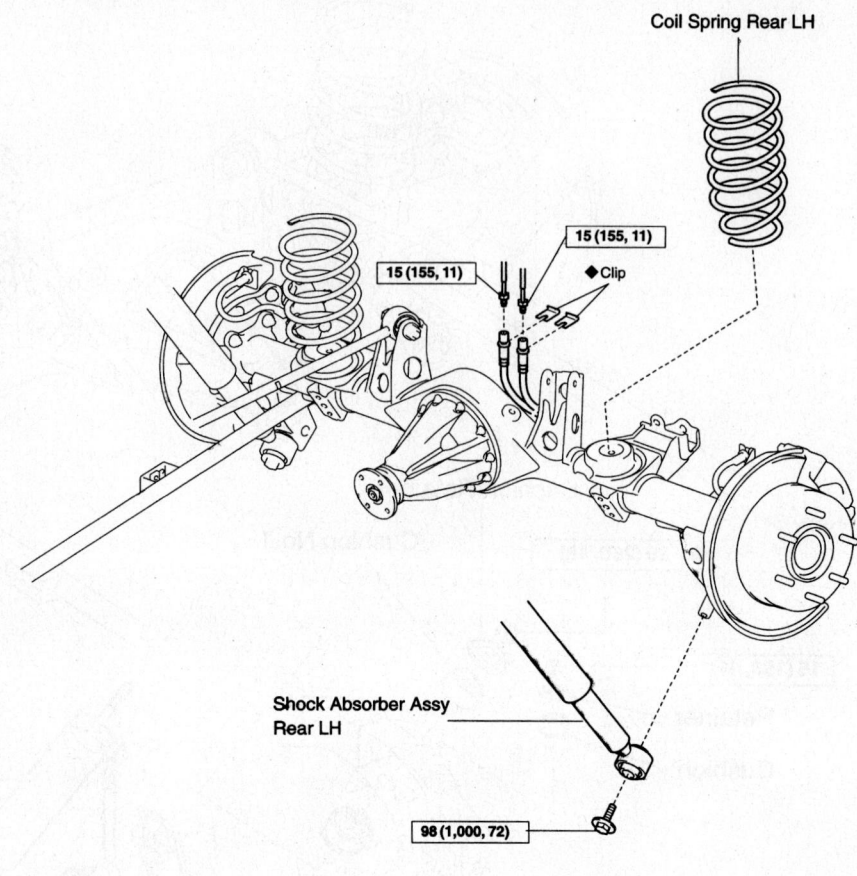

Coil Spring Rear LH

15 (155, 11)

15 (155, 11)

◆ Clip

Shock Absorber Assy Rear LH

98 (1,000, 72)

| N·m (kgf·cm, ft·lbf) | : Specified torque
◆ Non-reusable part

67170-4RUN-G195

Rear coil spring and related parts—2003–06

axle housing. Tighten to 108 inch lbs. (13 Nm).
   - Parking brake cable to the brake shoes
   - Driveshaft to the differential, by aligning the matchmarks. Tighten bolts/nuts to 54 ft. lbs. (73 Nm).
8. Fill the differential with the proper amount and type of oil.
9. Install the brake drum.
10. Bleed the brake system.
11. Install the wheel assemblies.
12. Lower the vehicle and bounce the vehicle several times to stabilize the suspension.
13. Tighten the lower control arm to 107 ft. lbs. (145 Nm).

2003–06

1. Before servicing the vehicle, refer to the Precautions Section.
2. Remove the rear shock.
3. Disconnect the brake pipes from the hoses at the bracket.

4. Lower the axle just enough to remove the spring.
5. Installation is the reverse of removal. Make sure that the lower end of the spring is against the stop.
6. Install the shock.
7. Refill and bleed the brakes.

Rear Air Spring

REMOVAL & INSTALLATION

1. Before servicing the vehicle, refer to the Precautions Section.
2. Remove the wheels.
3. Support the frame with a suitable jack.
4. Lower the axle as far as it will go.
5. Disconnect the height control tube.
6. Remove the clip at the top of the pneumatic cylinder.
7. Discharge the air to collapse the cylinder.

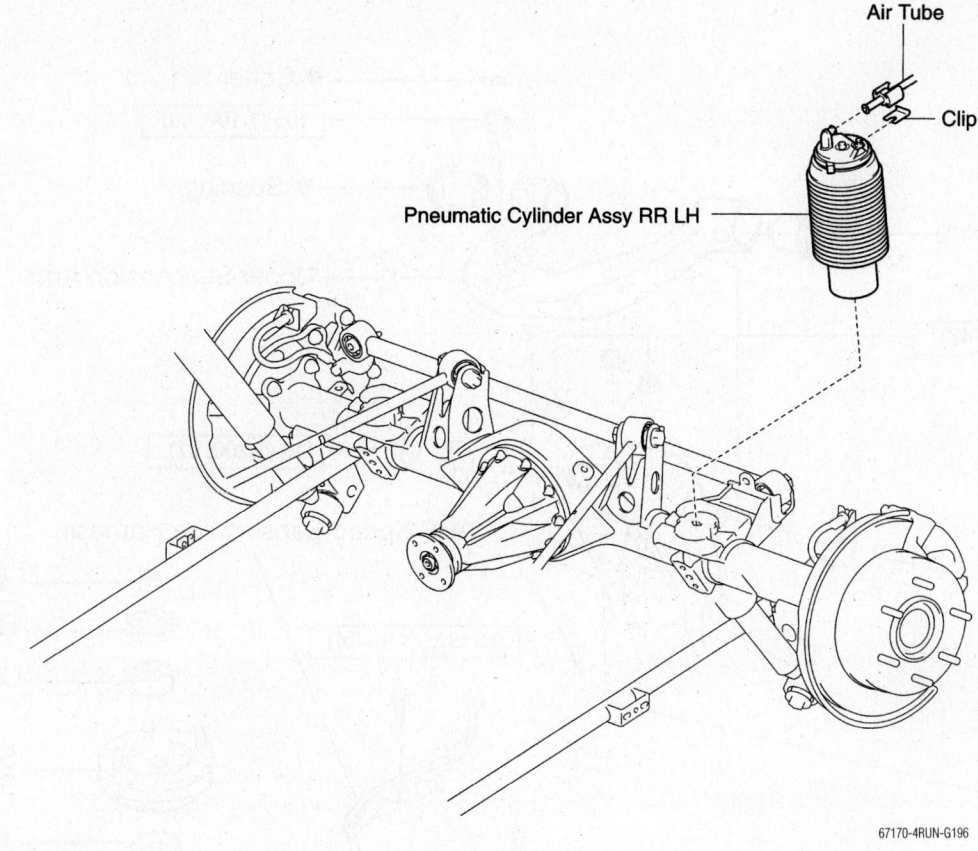

Air Tube

Clip

Pneumatic Cylinder Assy RR LH

67170-4RUN-G196

Rear air spring

8. Turn the cylinder 90 degrees and remove it from the axle.

To install

➡**Never extend the cylinder. If the cylinder is to be re-used, new O-rings, plate and height control plug must be used.**

9. Install the cylinder with the clip.
10. Connect the height control tube.
11. Raise the axle until it just contacts the bottom of the spring and install the clip at the lower end into the keyhole in the axle.

➡**Don't extend the cylinder until the wheels are on the ground.**

12. Once the vehicle is on the ground, start the engine and check for leaks.

Upper Ball Joint

REMOVAL & INSTALLATION

2002

1. Before servicing the vehicle, refer to the Precautions Section.
2. Remove or disconnect the following:
   • Front wheels

   • Strut assembly
   • Grease cap
3. If equipped with 4WD, disconnect the halfshaft, as follows:
   • Cotter pin and lockcap
   • Locknut, while applying the brakes
4. Remove or disconnect the following:
   • Anti-lock Brake System (ABS) speed sensor and wiring harness clamp from the steering knuckle, if equipped with ABS
   • Brake line bracket from the steering knuckle
   • Front brake caliper and the rotor
   • Lower ball joint
5. Remove the steering knuckle with the axle hub, as follows:
   • Cotter pin and loosen the nut
   • Steering knuckle from the upper control arm, using SST 09950-40010
   • Steering knuckle
   • Upper ball joint
   • Wire and the boot
   • Snapring
   • Upper ball joint, using SST 09950-40010 and a deep socket wrench.

To install:
6. Install the upper ball joint, as follows:

   • New ball joint with a new snapring
   • New boot secured with a new wire
7. Install or connect the following:
   • Steering knuckle with the axle hub to the upper control arm. Tighten the nut to 80 ft. lbs. (108 Nm).
   • New cotter pin
   • Lower ball joint. Tighten the 4 bolts to 59 ft. lbs. (80 Nm).
   • Rotor and caliper. Tighten the caliper bolts to 90 ft. lbs. (123 Nm).
   • Brake line bracket to the steering knuckle. Tighten it to 21 ft. lbs. (28 Nm).
   • ABS speed sensor and wiring harness clamp to the steering knuckle, if equipped. Tighten the bolts to 72 inch lbs. (8 Nm).
   • Halfshaft, if disconnected. Tighten the nut to 174 ft. lbs. (235 Nm).
   • Grease cap
   • Strut
   • Front wheel
8. Check the alignment

2003–06

The upper ball joint is not replaceable. If defective, the arm must be replaced.

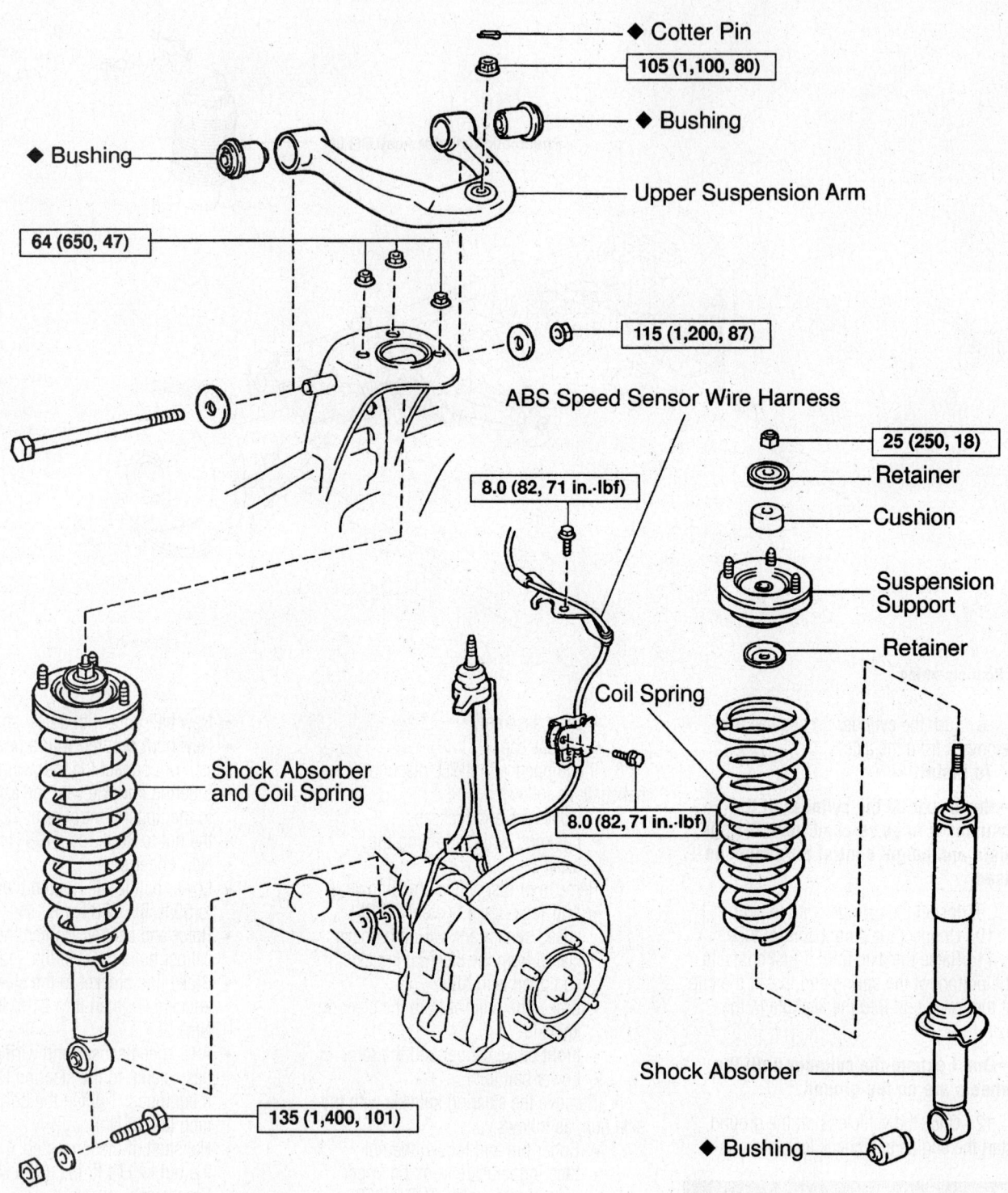

◆ Cotter Pin

105 (1,100, 80)

◆ Bushing

◆ Bushing

Upper Suspension Arm

64 (650, 47)

115 (1,200, 87)

ABS Speed Sensor Wire Harness

25 (250, 18)

Retainer

Cushion

Suspension Support

8.0 (82, 71 in.·lbf)

Coil Spring

Retainer

Shock Absorber and Coil Spring

8.0 (82, 71 in.·lbf)

Shock Absorber

◆ Bushing

135 (1,400, 101)

N·m (kgf·cm, ft·lbf) : Specified torque

◆ Non–reusable part

67170-4RUN-G197

Upper control arm and related parts—2002

Lower Ball Joint

REMOVAL & INSTALLATION

1. Before servicing the vehicle, refer to the Precautions Section.
2. Remove the wheel.
3. Disconnect the tie rod end, as follows:

- Loosen the 4 bolts
- Cotter pin and nut from the tie rod end
- Tie rod end from the steering knuckle, using SST 09610-20012

4. Remove the lower ball joint, as follows:

- Cotter pin and the nut from the lower ball joint
- Lower ball joint from the lower suspension arm
- Lower ball joint

To install:

5. Install or connect the following:

- Lower ball joint to the lower control arm. Tighten the 4 bolts to 59 ft. lbs. (80 Nm).
- Nut. Tighten it to 105 ft. lbs. (142 Nm).
- Tie rod end to the steering knuckle. Tighten the nut to 66 ft. lbs. (90 Nm).
- Wheel

2003–06

The lower ball joint is not replaceable. If defective, the arm must be replaced.

Upper Control Arm

REMOVAL & INSTALLATION

2002

1. Before servicing the vehicle, refer to the Precautions Section.
2. Remove or disconnect the following:

- Shock and coil spring assembly

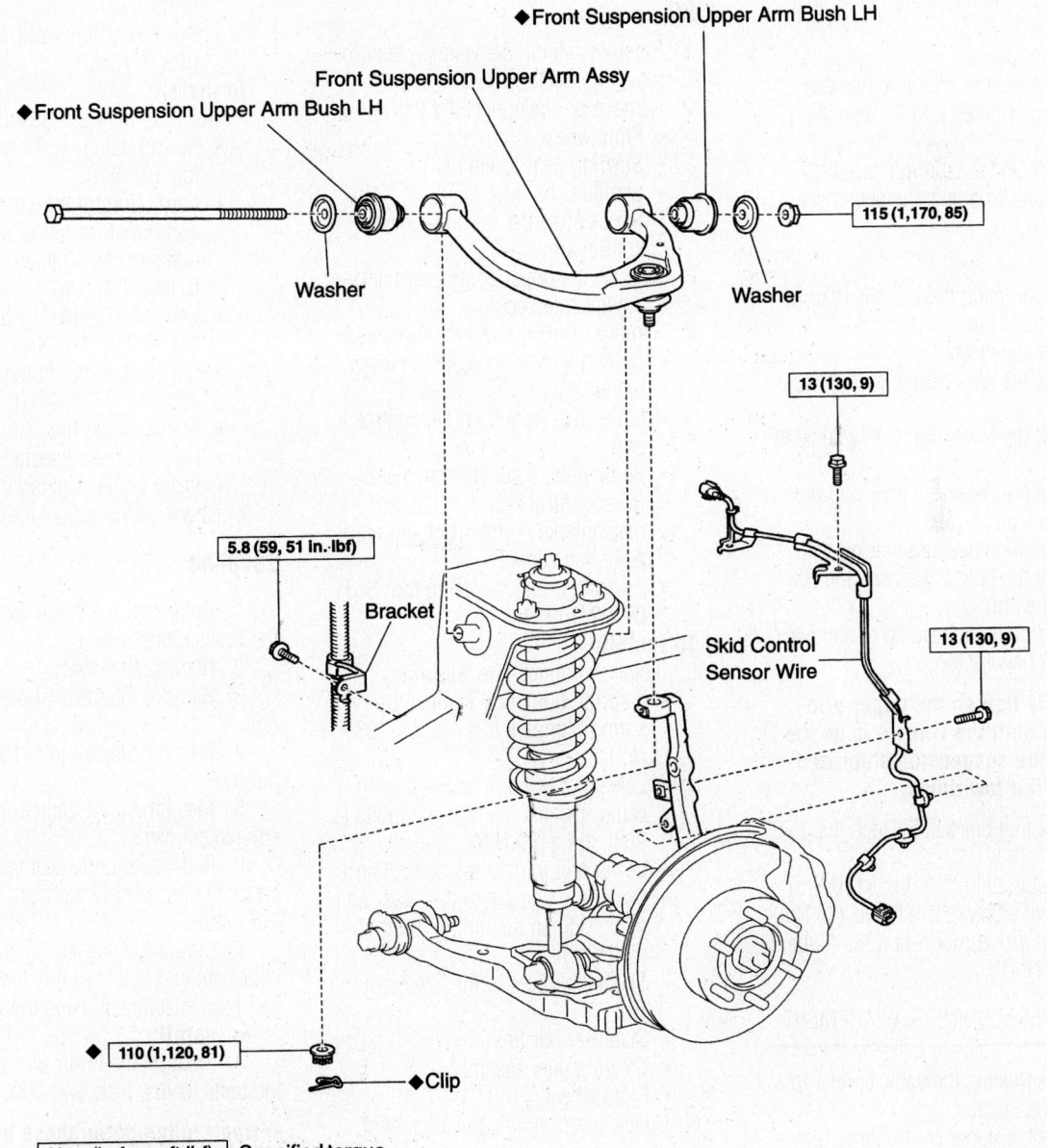

◆Front Suspension Upper Arm Bush LH

Front Suspension Upper Arm Assy

◆Front Suspension Upper Arm Bush LH

Washer

Washer

115 (1,170, 85)

13 (130, 9)

5.8 (59, 51 in.·lbf)

Bracket

Skid Control Sensor Wire

13 (130, 9)

◆ 110 (1,120, 81)

◆Clip

N·m (kgf·cm, ft·lbf) : Specified torque
◆ Non–reusable part

67170-4RUN-G198

Upper control arm and related parts—2003–06

- Anti-lock Brake System (ABS) speed sensor wire harness clamp

3. Disconnect the upper ball joint, as follows:

- Cotter pins and loosen the nut
- Upper ball joint from the control arm, using a ball joint separator
- Support the steering knuckle
- Nut

4. Detach the control arm, by removing the nut, bolt, washers and lowering the arm.

To install:

5. Install or connect the following:

- Upper control arm with the washer, bolt and nut. Tighten the nut to 87 ft. lbs. (115 Nm).
- Upper ball joint to the control arm. Tighten the mounting nut to 80 ft. lbs. (105 Nm).
- New cotter pin
- ABS speed sensor wire harness clamp. Tighten it to 71 inch lbs. (8 Nm).
- Shock and coil spring assembly

6. Check and/or adjust the alignment.

2003–06

1. Before servicing the vehicle, refer to the Precautions Section.

2. Remove the wheel.

3. Remove the ABS wiring from the arm and strut.

4. Support the lower arm with a suitable jack.

5. Remove the cotter pin and nut from the upper ball stud.

6. Remove the brake line bracket bolt.

7. Remove the bolt, 2 washers and nut and remove the arm.

8. Installation is the reverse of removal. Observe the following torques:

➡ **Do not fully tighten the upper arm through-bolt until the vehicle is on the ground and the suspension stabilized by jouncing it a few times.**

- Brake line bracket: 51 inch lbs. (5.8 Nm)
- Ball stud nut: 81 ft. lbs. (110 Nm)
- ABS wire brackets: 9 ft. lbs. (13 Nm)
- Upper arm through-bolt: 85 ft. lbs. (115 Nm)

CONTROL ARM BUSHING REPLACEMENT

1. Before servicing the vehicle, refer to the Precautions Section.

2. Remove the upper control arm from the vehicle.

3. Pry up the bushing flange, using a chisel and a hammer.

4. Using tools 09613-26010, 09613-20060 and 09950-00020 and a shop press, remove the bushing.

To install:

5. Using tools 09223-00010, 09506-35010 and a shop press, press the new bushing into the upper control arm.

6. Install the upper control arm to the chassis.

7. Check and/or adjust the alignment.

Lower Control Arm

REMOVAL & INSTALLATION

2002

2WD

1. Before servicing the vehicle, refer to the Precautions Section.

2. Remove or disconnect the following:

- Front wheel
- Steering gear assembly
- Stabilizer bar link
- Shock absorber from the lower control arm

3. Support the upper control arm and the steering knuckle securely.

4. Remove or disconnect the following:

- Cotter pin and nut from the lower ball joint
- Lower ball joint from the control arm
- Bolts, nuts, adjusting cams and lower control arm, by placing matchmarks on the front and rear adjusting cams
- 2 spring bumpers, using tool SST 09922-10010

To install:

5. Install or connect the following:

- 2 spring bumpers. Tighten the spring bumpers to 17 ft. lbs. (23 Nm).
- Lower control arm and adjusting cams. Tighten the nuts and bolts to 96 ft. lbs. (130 Nm),
- Lower ball joint to the control arm
- Cotter pin and nut to the lower ball joint. Tighten the nut to 105 ft. lbs. (142 Nm).
- Shock absorber to the lower control arm
- Stabilizer bar link
- Steering gear assembly
- Front wheel

6. Check and/or adjust the alignment.

4WD

1. Before servicing the vehicle, refer to the Precautions Section.

2. Remove or disconnect the following:

- Front wheel
- Steering gear assembly
- Stabilizer bar link
- Shock absorber from lower control arm

3. Support the upper control and steering knuckle securely.

4. Remove or disconnect the following:

- Cotter pin and nut from the lower ball joint
- Lower ball joint from the lower control arm

5. Place matchmarks on the front and rear adjusting cams.

6. Remove or disconnect the following:

- 2 bolts, nuts, adjusting cams and lower control arm
- Spring bumpers with a special tool 09922-10010.

To install:

7. Install or connect the following:

- Spring bumpers. Tighten to 17 ft. lbs. (23 Nm).
- Lower control arm, placing it in the appropriate position with the matchmarks. Tighten the arm to 96 ft. lbs. (130 Nm).
- Lower ball joint. Tighten the nut to 105 ft. lbs. (142 Nm).
- Shock absorber to the lower control arm
- Stabilizer bar link
- Steering gear assembly
- Front wheel. Tighten the lug nuts.

8. Check and/or adjust the alignment.

2003–06

1. Before servicing the vehicle, refer to the Precautions Section.

2. Remove the wheel.

3. Remove the lower shock absorber bolt.

4. Remove the ball joint attachment bolts.

5. Matchmark the camber and toe adjustment cams.

6. Remove the nuts, camber adjustment cams, toe adjustment cams, toe plate and lower arm.

7. Mount the arm in a vise. Remove the cotter pin and ball stud nut, then remove the ball joint attachment using a puller.

To install:

8. Position the lower arm and install the bolts, cams, plate and nuts.

➡ **Don't fully tighten these bolts until the vehicle is on the ground and the suspension stabilized by jouncing a few times.**

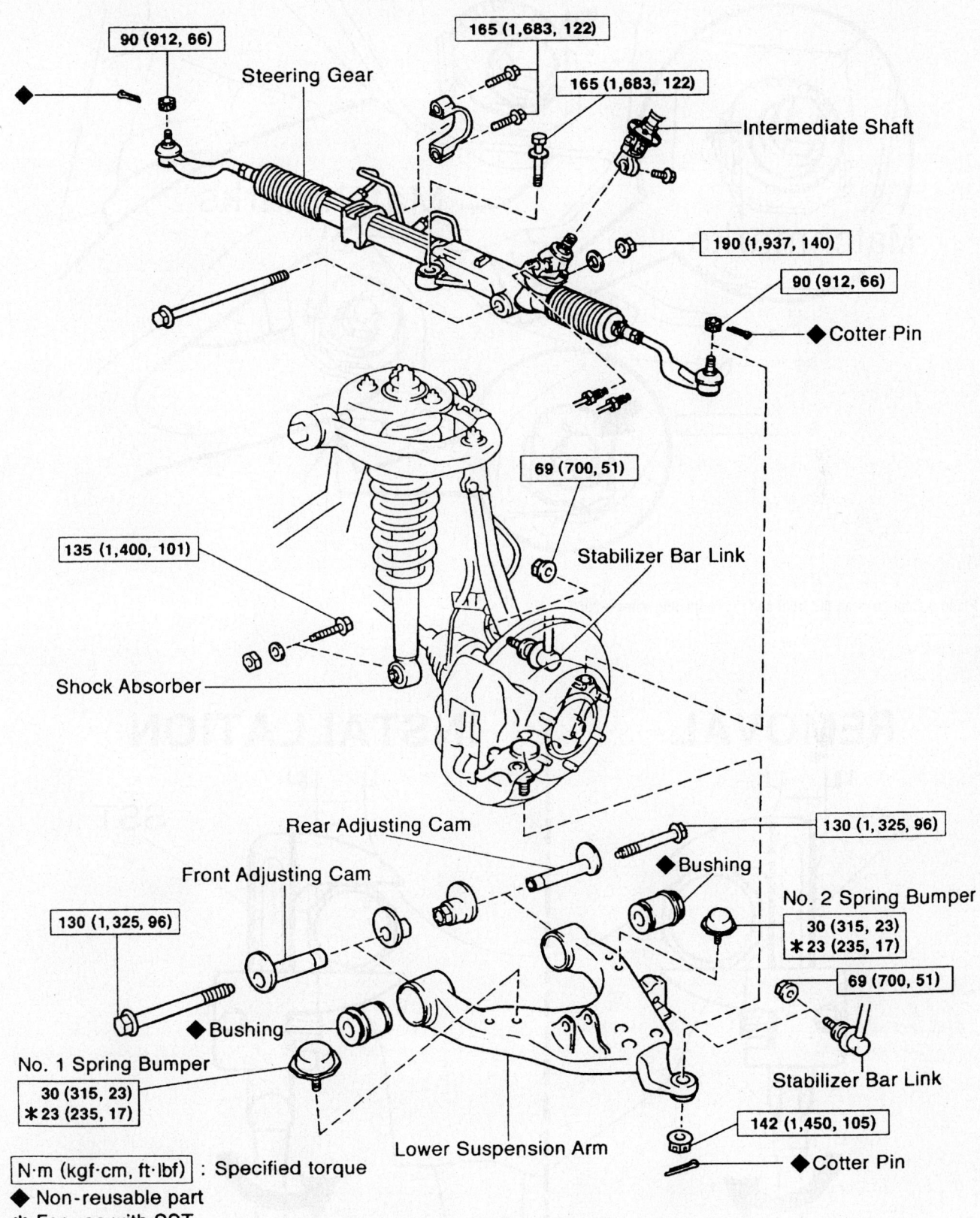

90 (912, 66)

Steering Gear

165 (1,683, 122)

165 (1,683, 122)

Intermediate Shaft

190 (1,937, 140)

90 (912, 66)

◆ Cotter Pin

69 (700, 51)

Stabilizer Bar Link

135 (1,400, 101)

Shock Absorber

Rear Adjusting Cam

130 (1,325, 96)

◆ Bushing

No. 2 Spring Bumper

Front Adjusting Cam

30 (315, 23)
✳ 23 (235, 17)

130 (1,325, 96)

69 (700, 51)

◆ Bushing

Stabilizer Bar Link

No. 1 Spring Bumper

30 (315, 23)
✳ 23 (235, 17)

142 (1,450, 105)

Lower Suspension Arm

◆ Cotter Pin

N·m (kgf·cm, ft·lbf) : Specified torque
◆ Non-reusable part
✳ For use with SST

86828G20

Exploded view of the lower control arm and associated components—2002

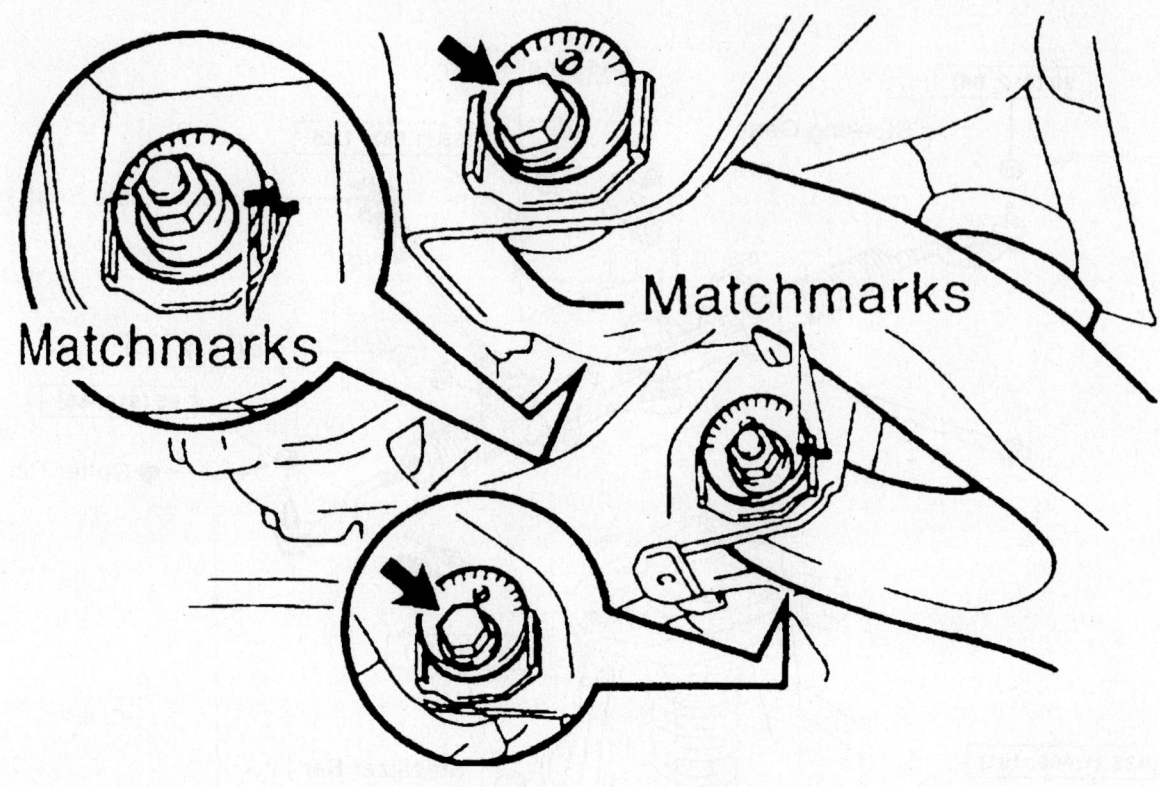

Place matchmarks on the front and rear adjusting cams—2002

86828G21

REMOVAL

INSTALLATION

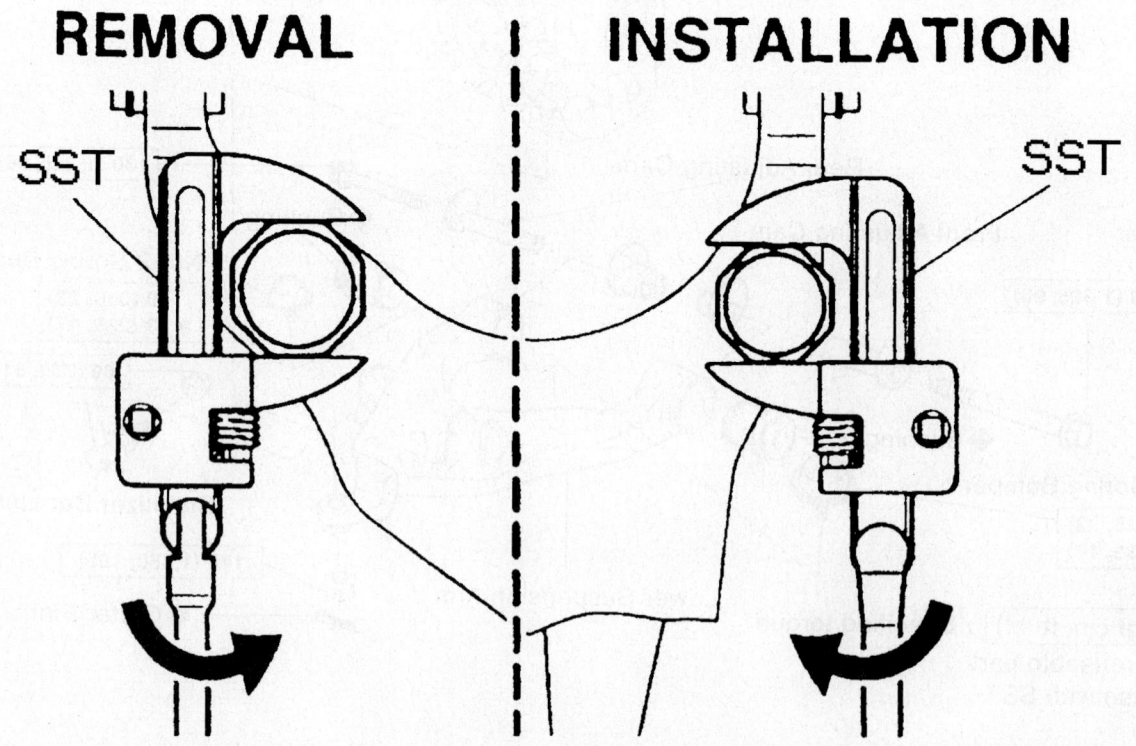

SST

SST

86828G22

Remove the spring bumpers with special tool 09922-10010—2002

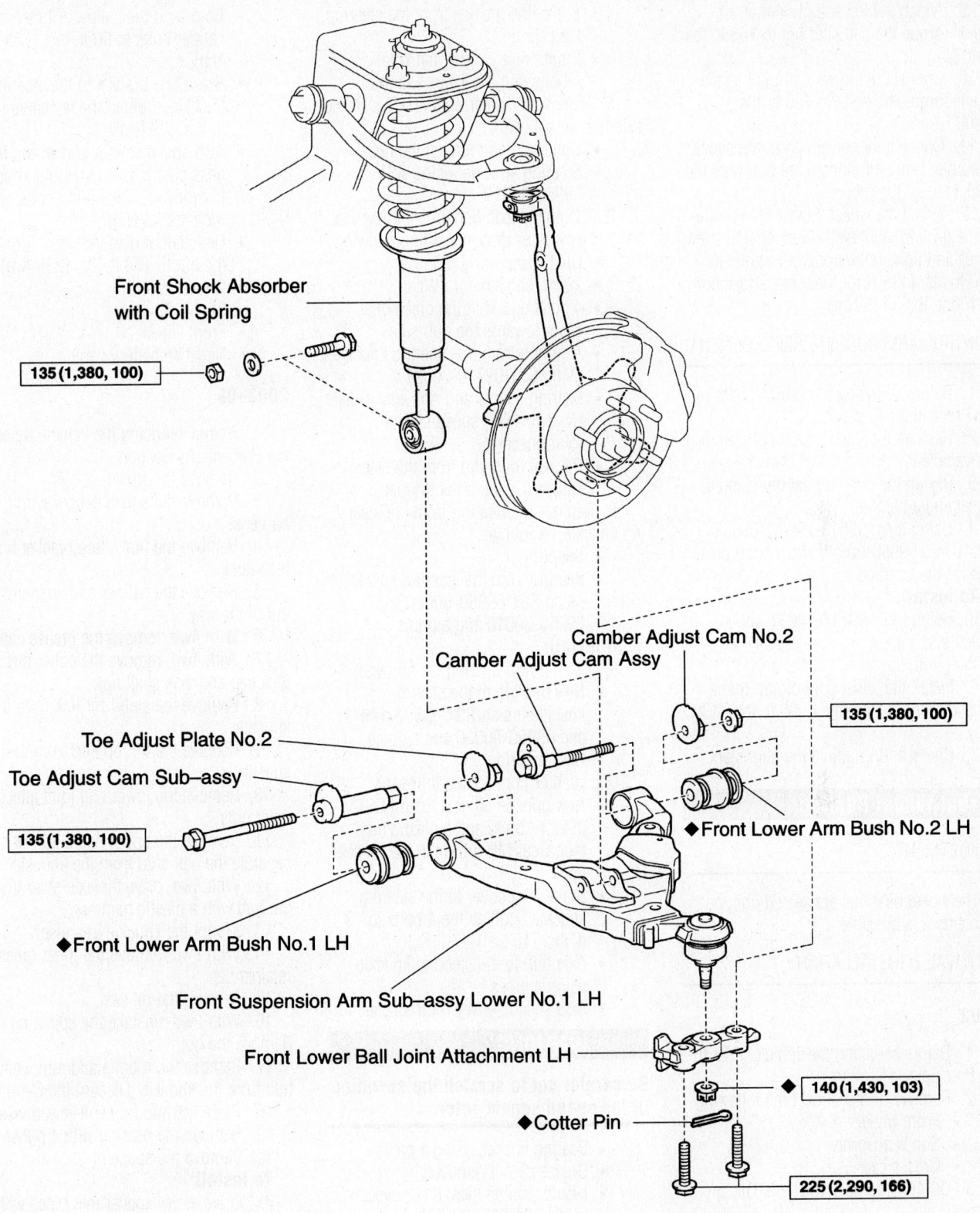

Front Shock Absorber with Coil Spring

135 (1,380, 100)

Camber Adjust Cam No.2
Camber Adjust Cam Assy

135 (1,380, 100)

Toe Adjust Plate No.2

Toe Adjust Cam Sub–assy

135 (1,380, 100)

◆Front Lower Arm Bush No.2 LH

◆Front Lower Arm Bush No.1 LH

Front Suspension Arm Sub–assy Lower No.1 LH

Front Lower Ball Joint Attachment LH

140 (1,430, 103)

◆Cotter Pin

225 (2,290, 166)

N·m (kgf·cm, ft·lbf) : Specified torque

◆ Non–reusable part

67170-4RUN-G199

Lower control arm and related parts—2003–06

9. Attach the lower ball joint attachment. Torque the ball stud nut to 103 ft. lbs. (140 Nm).

10. Connect the lower ball joint attachment. Torque the bolts to 166 ft. lbs. (225 Nm).

11. Connect the lower end of the shock absorber. Don't full tighten the bolt until the vehicle is on the ground.

12. Install the wheel. Lower the vehicle to the ground. Jounce the suspension a few times and tighten the upper arm bolts to 100 ft. lbs. (135 Nm); the lower shock bolt to 100 ft. lbs. (135 Nm).

CONTROL ARM BUSHING REPLACEMENT

1. Before servicing the vehicle, refer to the Precautions Section.

2. Remove the lower control arm from the vehicle.

3. Pry up the bushing flange, using a chisel and a hammer.

4. Using tools 09613-26010, 09632-36010 and 09950-00020 and a shop press, remove the bushing.

To install:

5. Using tools 09316-20011, 09710-30021 and a shop press, press the new bushing into the lower control arm.

6. Install the lower control arm to the chassis. Torque the bolts to 96 ft. lbs. (130 Nm).

7. Check and/or adjust the alignment.

Front Wheel Bearing

ADJUSTMENT

The wheel bearings are sealed unit; no adjustment is possible.

REMOVAL & INSTALLATION

2002

1. Before servicing the vehicle, refer to the Precautions Section.

2. Remove or disconnect the following:
- Front wheels
- Shock absorber
- Grease cap

3. On 4WD, remove the halfshaft, as follows:
- Cotter pin and lockcap
- Locknut, while applying the brakes

4. Remove or disconnect the following:
- Anti-lock Brake System (ABS) speed sensor and wiring harness clamp from the steering knuckle, if equipped with ABS

- Brake line bracket from the steering knuckle
- Front brake caliper and rotor
- 4 bolts and the lower ball joint

5. Remove the steering knuckle with the axle hub, as follows:
- Cotter pin and loosen the nut
- Steering knuckle, using SST 09950-40010

6. Clamp the axle hub in a soft jaw vise.

7. Remove or disconnect the following:
- Grease cap, for 2WD
- Inside oil seal, for 4WD
- 4 bolts and shift the brake dust cover towards the hub side
- Axle hub from the steering knuckle, using SST 09710-30021
- Bearing spacer and Anti-lock Brake System (ABS) speed sensor rotor/spacer
- Oil seal (outside) from the steering knuckle, using a flat pry bar

8. Remove the bearing from the steering knuckle, as follows:
- Snapring
- Bearing from the steering knuckle, using SST 09950-60020 and 09950-70010 and a press

To install:

9. Install a new bearing, as follows:
- New bearing to the steering knuckle, using SST 09527-17011 and 09950-60020 and a press
- New snapring

10. Install or connect the following:
- New outside oil seal, using SST 09223-15030 and a plastic hammer. Coat MP grease to the oil seal lip.
- Brake dust cover to the steering knuckle. Tighten the 4 bolts to 13 ft. lbs. (18 Nm).
- Axle hub to the steering knuckle, using a press
- ABS speed sensor rotor/spacer.

✳✳ WARNING

Be careful not to scratch the serration of the speed sensor rotor.

- Bearing spacer, using a press
- Grease cap, if removed
- New inside oil seal, if removed, using SST 09527-17011 and a plastic hammer
- Steering knuckle with the axle hub. Tighten the nut to 80 ft. lbs. (108 Nm).
- New cotter pin
- Lower ball joint. Tighten the 4 bolts to 59 ft. lbs. (80 Nm).

- Rotor and the caliper. Tighten the caliper bolts to 90 ft. lbs. (123 Nm).
- Brake line bracket to the steering knuckle. Tighten the fasteners to 21 ft. lbs. (28 Nm).
- ABS speed sensor and wiring harness clamp to the steering knuckle, if removed. Tighten the bolts to 72 inch lbs. (8 Nm).
- Halfshaft, if disconnected. Tighten the nut to 174 ft. lbs. (235 Nm).
- Grease cap
- Shock absorber
- Front wheel
- Negative battery cable

2003–06

1. Before servicing the vehicle, refer to the Precautions Section.

2. Remove the wheel.

3. Remove the speed sensor wiring harness.

4. Remove the brake line bracket from the knuckle.

5. Remove the caliper and suspend it out of the way.

6. With 4wd, remove the grease cap.

7. With 4wd, remove the cotter pin, lock cap and axle shaft nut.

8. Remove the stabilizer link from the knuckle.

9. Remove the tie rod end from the knuckle.

10. Remove the lower ball joint attachment nuts.

11. Remove the upper ball stud nut and separate the ball stud from the knuckle.

12. With 4wd, drive the axle shaft from the hub with a plastic hammer.

13. Mount the knuckle in a vise.

14. With 2wd, remove the inner grease retainer cap.

15. Remove the oil seal.

16. With 2wd, unstake the adjusting nut. Remove the nut.

17. Remove the 4 bolts and remove the hub from the knuckle. Discard the O-ring.

18. Place the hub in a soft-jawed vise.

19. Remove the bearing with a puller set.

20. Remove the spacer.

To install:

21. Drive a new spacer into place with a hammer and brass drift.

22. Using a press install the new bearing.

23. Coat a new O-ring with MP grease and install it.

24. Install the dust cover and attach the hub to the knuckle. Torque the bolts to 59 ft. lbs. (80 Nm).

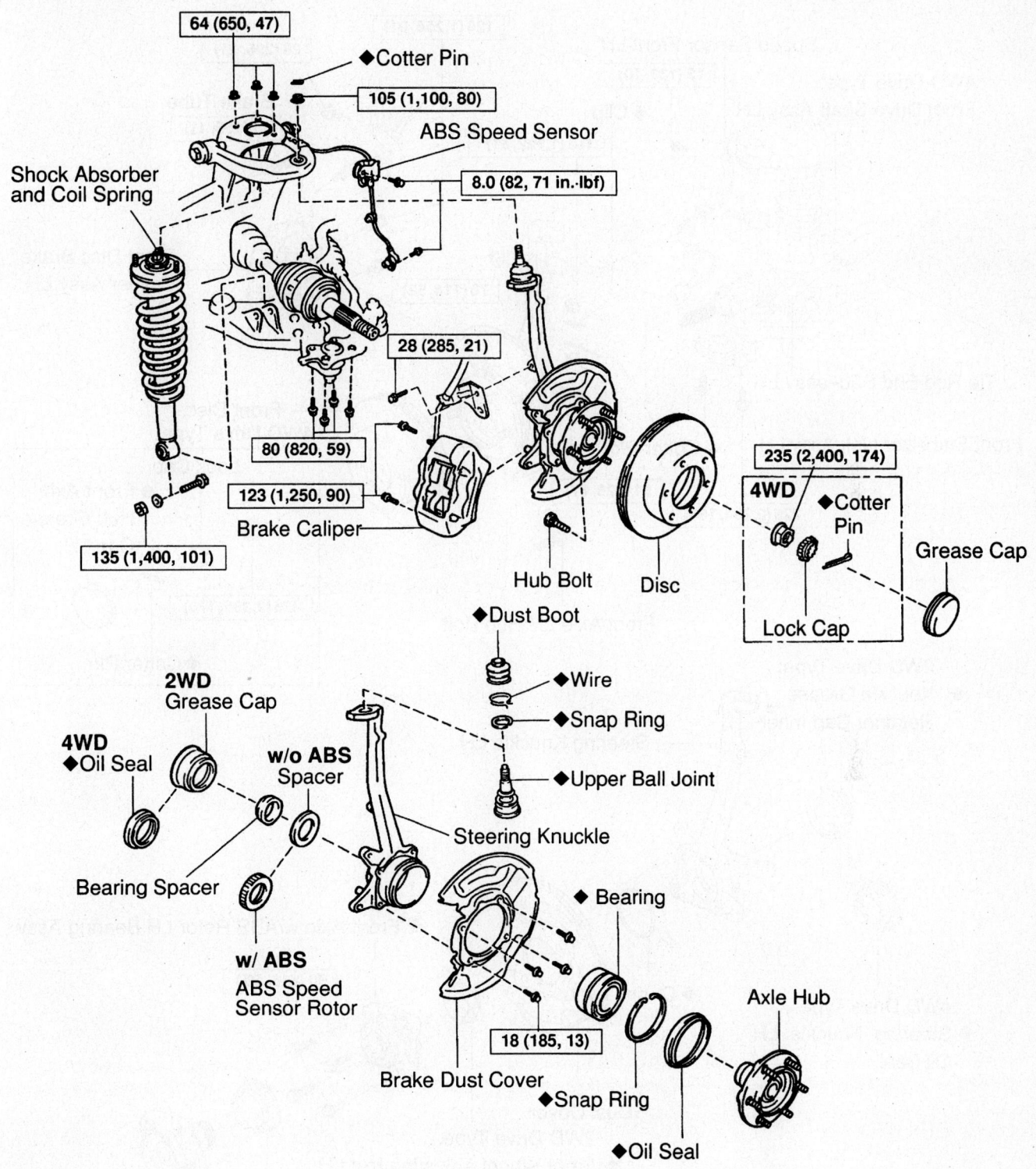

64 (650, 47)

◆Cotter Pin

105 (1,100, 80)

ABS Speed Sensor

Shock Absorber
and Coil Spring

8.0 (82, 71 in.·lbf)

28 (285, 21)

235 (2,400, 174)

4WD

◆Cotter
Pin

Grease Cap

Lock Cap

80 (820, 59)

123 (1,250, 90)

Brake Caliper

Hub Bolt

Disc

135 (1,400, 101)

◆Dust Boot

◆Wire

2WD
Grease Cap

4WD
◆Oil Seal

w/o ABS
Spacer

◆Snap Ring

◆Upper Ball Joint

Bearing Spacer

Steering Knuckle

w/ ABS

ABS Speed
Sensor Rotor

◆ Bearing

Axle Hub

18 (185, 13)

Brake Dust Cover

◆Snap Ring

◆Oil Seal

N·m (kgf·cm, ft·lbf) : Specified torque

◆ Non–reusable part

Front hub and related parts—2002

67170-4RUN-G200

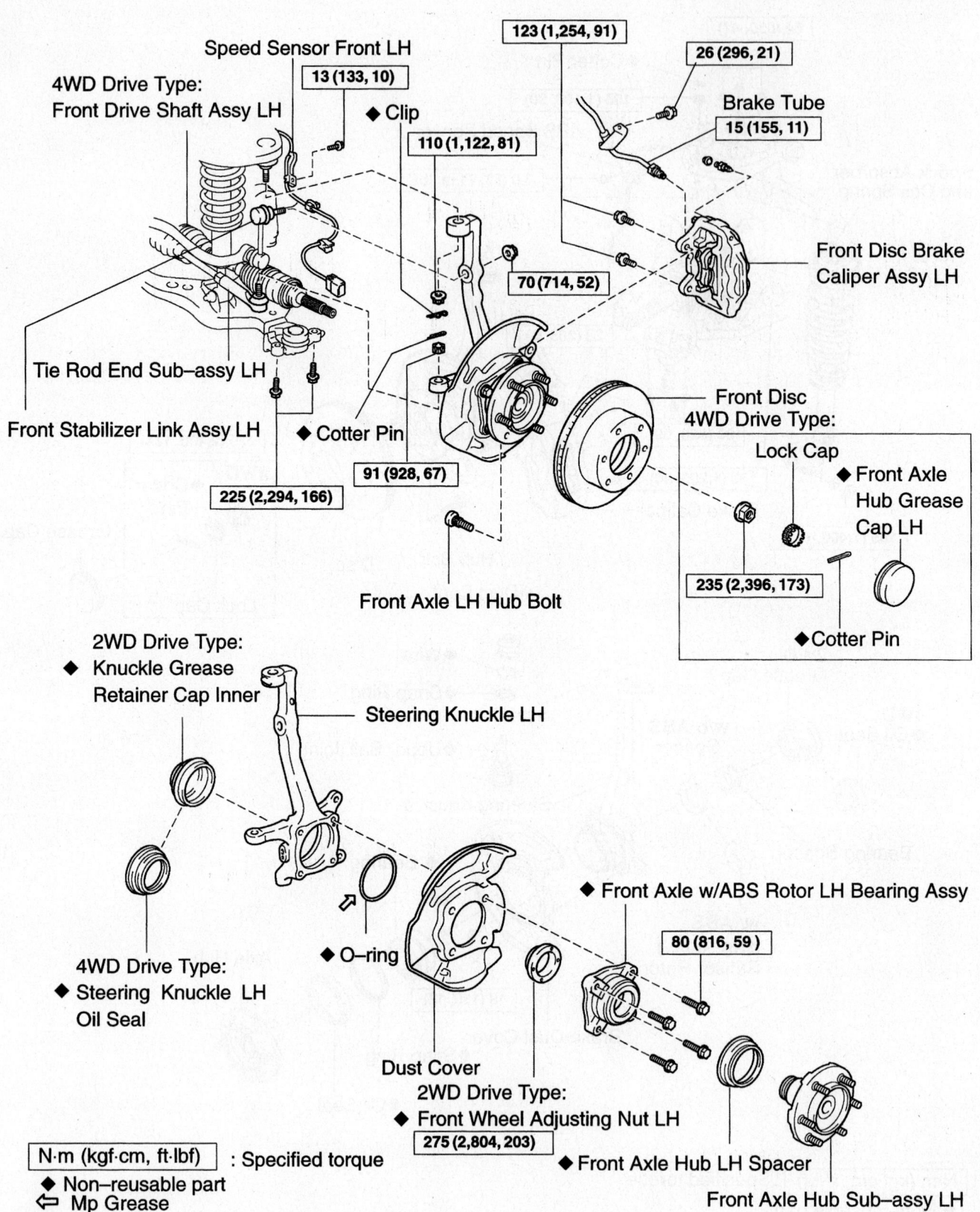

Speed Sensor Front LH

123 (1,254, 91)

26 (296, 21)

13 (133, 10)

◆ Clip

4WD Drive Type:
Front Drive Shaft Assy LH

Brake Tube

15 (155, 11)

110 (1,122, 81)

Front Disc Brake
Caliper Assy LH

70 (714, 52)

Tie Rod End Sub–assy LH

Front Stabilizer Link Assy LH

◆ Cotter Pin

Front Disc
4WD Drive Type:

Lock Cap

◆ Front Axle
Hub Grease
Cap LH

91 (928, 67)

225 (2,294, 166)

235 (2,396, 173)

◆ Cotter Pin

Front Axle LH Hub Bolt

2WD Drive Type:
◆ Knuckle Grease
Retainer Cap Inner

Steering Knuckle LH

4WD Drive Type:
◆ Steering Knuckle LH
Oil Seal

◆ Front Axle w/ABS Rotor LH Bearing Assy

80 (816, 59)

◆ O–ring

Dust Cover
2WD Drive Type:
◆ Front Wheel Adjusting Nut LH

275 (2,804, 203)

◆ Front Axle Hub LH Spacer

Front Axle Hub Sub–assy LH

N·m (kgf·cm, ft·lbf) : Specified torque
◆ Non–reusable part
⇐ Mp Grease

25. With 2wd, install a new adjusting nut. Torque to 203 ft. lbs. (275 Nm).

26. With 4wd, press a new oil seal into place.

27. With 2wd, install a new grease cap, using a hammer and brass drift.

28. The remainder of installation is the reverse of removal Use new cotter pins. Observe the following torques:

- Upper ball stud nut: 81 ft. lbs. (110 Nm)
- Lower ball joint attachment: 166 ft. lbs. (225 Nm)
- Tie rod end nut: 67 ft. lbs. (91 Nm)

- Stabilizer link: 52 ft. lbs. (70 Nm)
- Axle shaft nut: 173 ft. lbs. (235 Nm)
- Caliper: 91 ft. lbs. (123 Nm)
- Brake line bracket: 21 ft. lbs. (29 Nm)

BRAKES

Front Brake Caliper

REMOVAL & INSTALLATION

2002

1. Before servicing the vehicle, refer to the Precautions Section.

2. Raise and support the vehicle safely.

3. Remove the wheels.

4. Disconnect the brake hose from the caliper by removing the union bolt and 2 gaskets. Plug the end of the hose to prevent loss of fluid.

5. Remove the bolts that attach the caliper to the torque plate.

6. Lift the bottom of the caliper up and remove the caliper assembly.

To install:

7. Grease the caliper slides and bolts with lithium grease or equivalent. Install the caliper and secure with the bolts. Torque the bolts to 90 ft. lbs. (123 Nm).

8. Connect the brake hose to the caliper, using 2 new washers. Make sure the flexible hose lock is securely in the lock hole of the caliper. Torque the union bolt to 11 ft. lbs. (15 Nm).

9. Fill the brake system to the proper level and bleed the brake system.

10. Install the tire and wheel assembly.

11. Top off the brake fluid level in the master cylinder. Check for leaks and proper brake operation.

2003–06

1. Before servicing the vehicle, refer to the Precautions Section.

2. Remove the wheel.

3. Remove the clips and remove the anti-rattle pins.

4. Remove the anti-rattle spring.

5. Remove the pads and shims.

6. Disconnect the brake line. Plug the opening to prevent fluid loss.

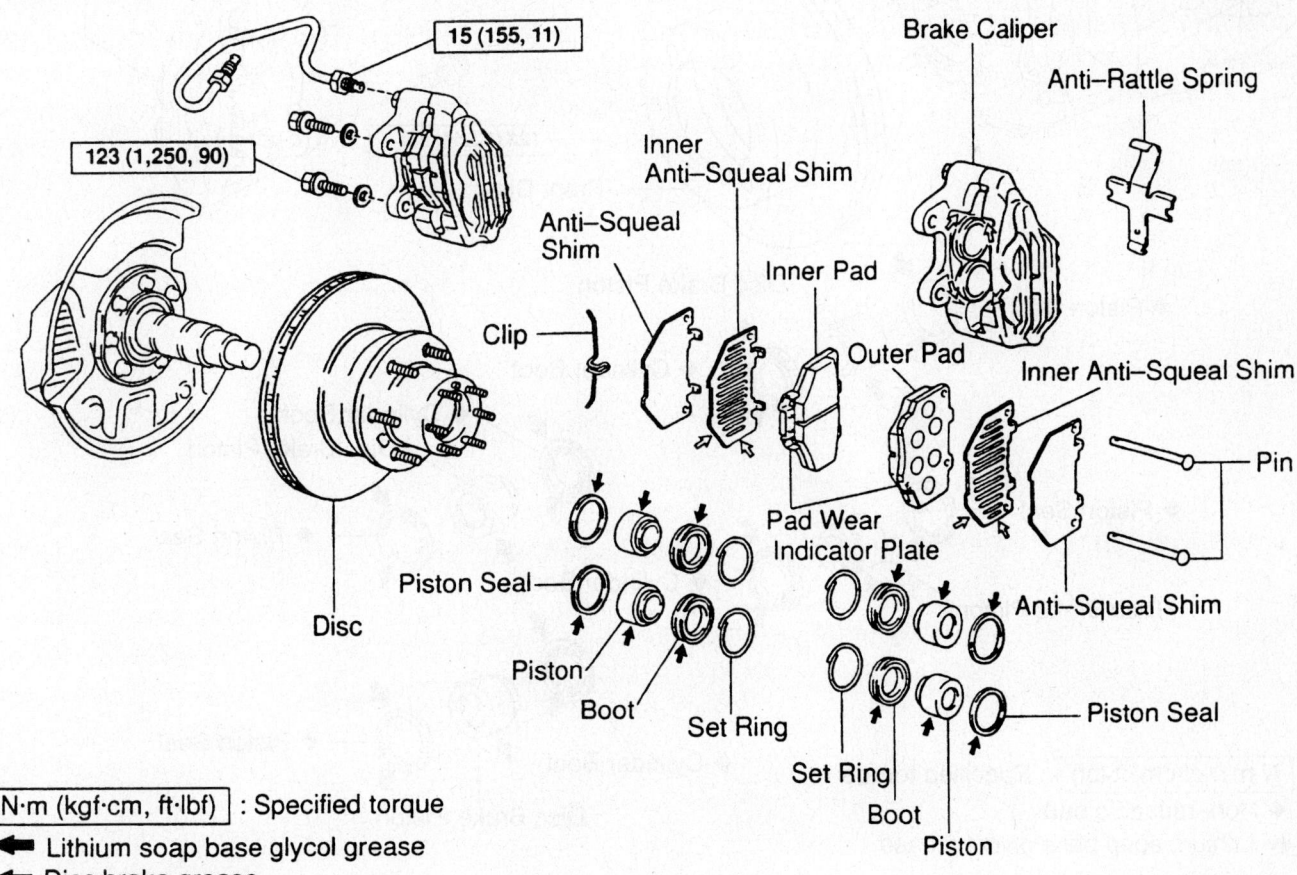

N·m (kgf·cm, ft·lbf) : Specified torque
◄ Lithium soap base glycol grease
⇐ Disc brake grease

Caliper assembly—2002

93026G71

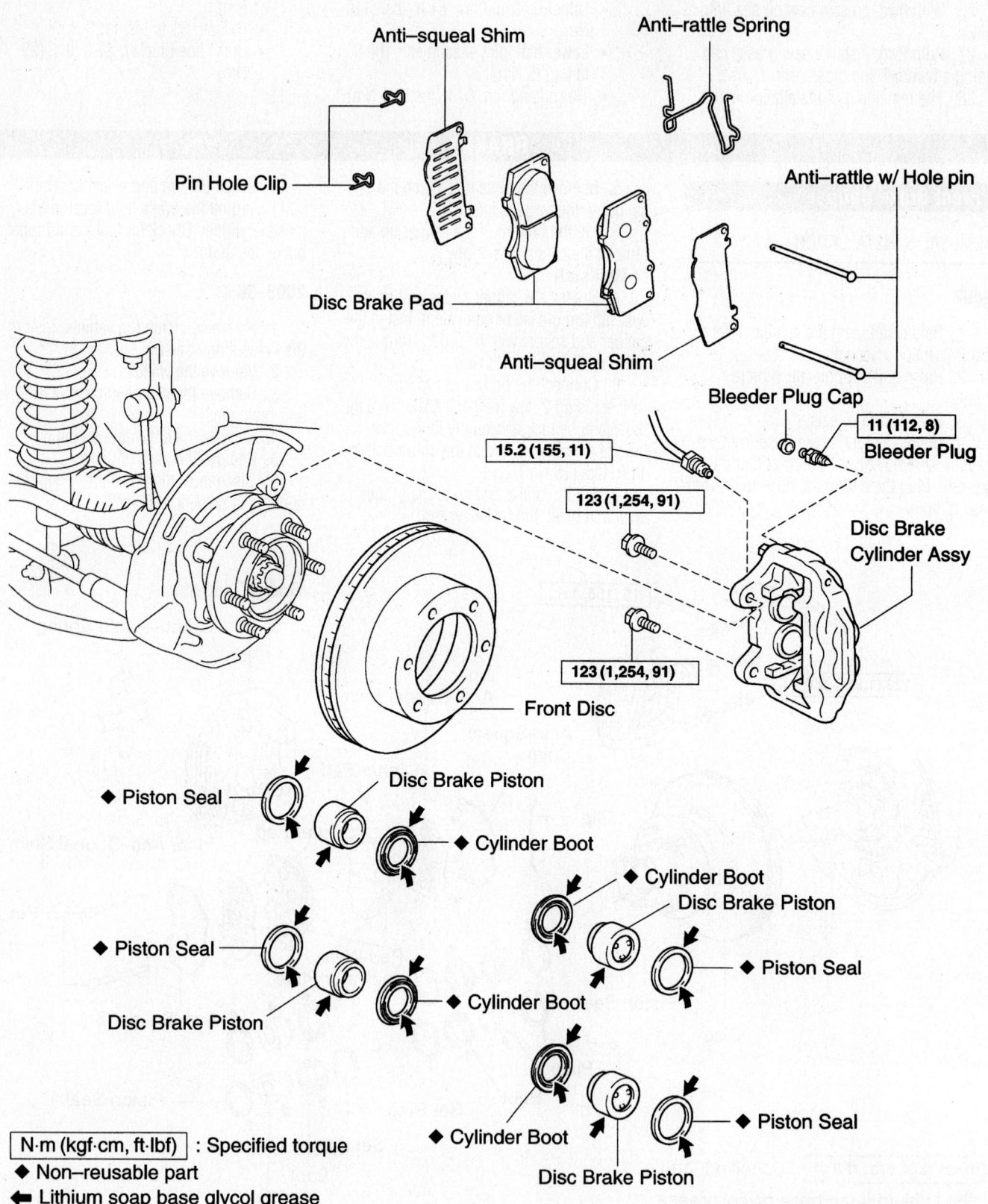

Anti–squeal Shim

Anti–rattle Spring

Pin Hole Clip

Anti–rattle w/ Hole pin

Disc Brake Pad

Anti–squeal Shim

Bleeder Plug Cap

15.2 (155, 11)

11 (112, 8)
Bleeder Plug

123 (1,254, 91)

Disc Brake
Cylinder Assy

123 (1,254, 91)

Front Disc

◆ Piston Seal

Disc Brake Piston

◆ Cylinder Boot

◆ Cylinder Boot
Disc Brake Piston

◆ Piston Seal

◆ Piston Seal

◆ Cylinder Boot

◆ Piston Seal

Disc Brake Piston

◆ Cylinder Boot

N·m (kgf·cm, ft·lbf) : Specified torque
◆ Non–reusable part
◀ Lithium soap base glycol grease

◆ Cylinder Boot

Disc Brake Piston

67170-4RUN-G202

Front brake caliper—2003–06

7. Remove the caliper mounting bolts and lift off the caliper.

To install:

8. Install the caliper and torque the bolts to 91 ft. lbs. (123 Nm).

9. Connect the brake line. Torque to 11 ft. lbs. (15 Nm).

10. Install the shims and pads.

11. Install the anti-rattle pins, spring and new clips. Make sure that the clip with the handle faces the inside.

12. Refill and bleed the system.

13. Depress the brake pedal a few times to seat the pads.

Rear Brake Caliper

REMOVAL & INSTALLATION

1. Before servicing the vehicle, refer to the Precautions Section.

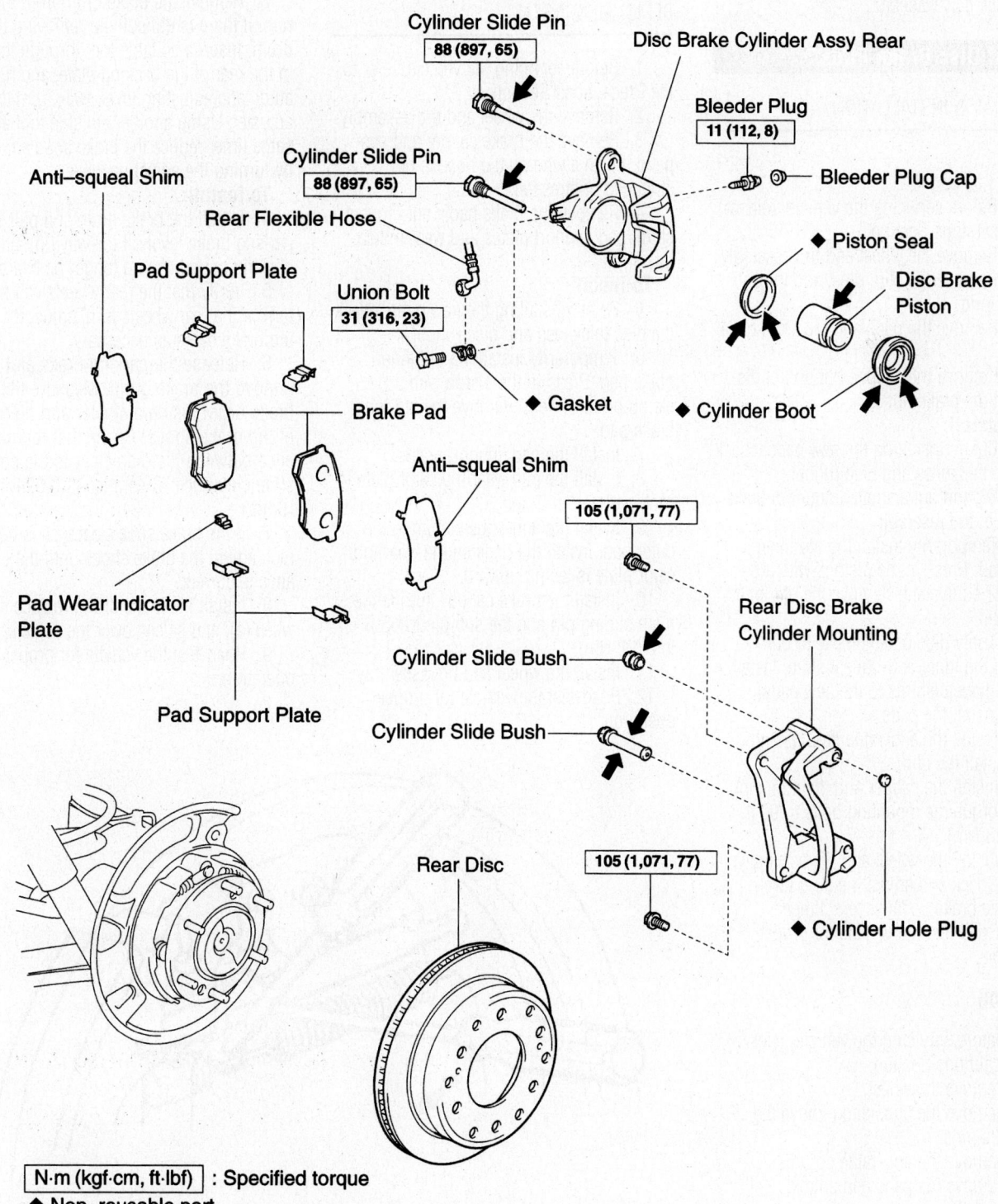

Cylinder Slide Pin
88 (897, 65)

Disc Brake Cylinder Assy Rear

Bleeder Plug
11 (112, 8)

Bleeder Plug Cap

Cylinder Slide Pin
88 (897, 65)

Anti–squeal Shim

Rear Flexible Hose

Pad Support Plate

Union Bolt
31 (316, 23)

◆ Piston Seal

Disc Brake Piston

◆ Gasket

◆ Cylinder Boot

Brake Pad

Anti–squeal Shim

105 (1,071, 77)

Pad Wear Indicator Plate

Rear Disc Brake Cylinder Mounting

Cylinder Slide Bush

Cylinder Slide Bush

Pad Support Plate

Rear Disc

105 (1,071, 77)

◆ Cylinder Hole Plug

N·m (kgf·cm, ft·lbf) : Specified torque
◆ Non–reusable part
◀ Lithium soap base glycol grease

67170-4RUN-G203

Rear brake caliper

2. Remove the wheel.

3. Remove the brake hose from the caliper. Plug the hose to prevent fluid loss.

4. Remove the caliper pins and lift off the caliper.

5. Installation is the reverse of removal. Torque the caliper pins to 65 ft. lbs. (88 Nm). Torque the hose bolt to 23 ft. lbs. (31 Nm). Use new washers.

Front Disc Brake Pads

REMOVAL & INSTALLATION

2002

1. Before servicing the vehicle, refer to the Precautions Section.

2. Remove the wheel and tire assembly.

3. Remove the clip, pins, and the anti-rattle spring.

4. Remove the pads and the anti-squeal shims.

5. Remove the caliper, but do not disconnect the brake hose.

To install:

6. Before installing the new pads, check the disc thickness and disc runout.

7. Siphon out a small amount of brake fluid from the reservoir.

8. Temporarily install the old inner brake pad. Press in the pistons with a C-clamp or equivalent. Remove the old inner brake pad.

9. Apply disc brake grease to both sides of the inner anti-squeal shim. Install the anti-squeal shims to the new pads.

10. Install the pads.

11. Install the anti-rattle springs and pins. Install the clip.

12. Install the caliper and the mounting bolts. Torque the mounting bolts to 90 ft. lbs. (123 Nm).

13. Install the wheel and tire assembly.

14. Check and adjust the fluid level. Apply the brake pedal several times.

15. Road-test the vehicle for proper operation.

2003–06

1. Before servicing the vehicle, refer to the Precautions Section.

2. Remove the wheel.

3. Remove the clips and remove the anti-rattle pins.

4. Remove the anti-rattle spring.

5. Remove the pads and shims.

To install:

6. Compress the piston with a C-clamp or large pliers.

7. Install the shims and pads.

8. Install the anti-rattle pins, spring and new clips. Make sure that the clip with the handle faces the inside.

9. Depress the brake pedal a few times to seat the pads.

Rear Disc Brake Pads

REMOVAL & INSTALLATION

1. Before servicing the vehicle, refer to the Precautions Section.

2. Remove the wheel and tire assembly.

3. Remove the brake caliper and suspend it with a wire so the hose is not stretched or stressed.

4. Remove the brake pads, anti-squeal shim, pad support plates and wear indicators.

To install:

5. Before installing the new pads, check the disc thickness and disc runout.

6. Temporarily install the old inner brake pad. Press in the piston with a C-clamp or equivalent. Remove the old inner brake pad.

7. Install the pad support plates.

8. Install the pad wear indicator plate to each pads.

9. Install the anti-squeal shim to the outer pad. Install the pads so the wear indicator plate is facing upward.

10. Install the brake caliper. Torque the main sliding pin and the sub pin to 65 ft. lbs. (88 Nm).

11. Install the wheel and tire assembly.

12. Road-test the vehicle for proper operation.

Brake Drums

REMOVAL & INSTALLATION

1. Before servicing the vehicle, refer to the Precautions Section.

2. Remove the rear wheel(s).

3. Remove the brake drum from the axle hub. If there is difficulty in removing the drum, insert a suitable tool through the hole in the rear of the backing plate, and hold the automatic adjusting lever away from the adjuster. Using another suitable tool at the same time, reduce the brake shoe adjuster by turning the adjusting wheel.

To install:

4. Install the brake drum and pull the parking brake lever all the way up until a clicking sound can no longer be heard.

5. Verify that the rear wheels will not turn. If the rear wheels turn, adjust the parking brake cable as necessary.

6. Release the parking brake and remove the brake drum. Measure the brake drum inside diameter and diameter of the brake shoes. Check that the difference between the diameters is the correct shoe clearance. Clearance is 0.024 in. (6mm).

7. If the brake shoe clearance is not correct, adjust the brake shoes until the clearance is correct.

8. Install the brake drum, replace the wheel(s), and safely lower the vehicle.

9. Road-test the vehicle for proper brake operation.

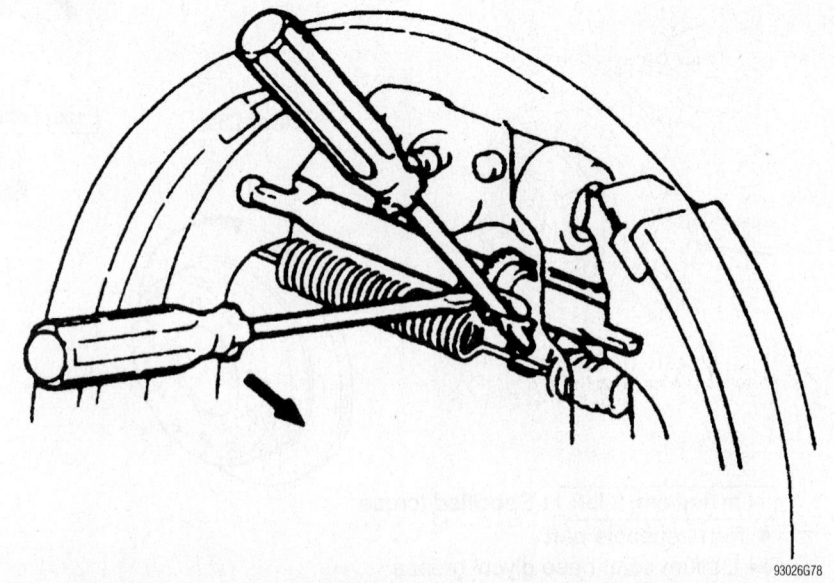

93026G78

Use a brake adjusting tool (brake spoon) and a prytool to adjust the brake shoes through the adjusting hole

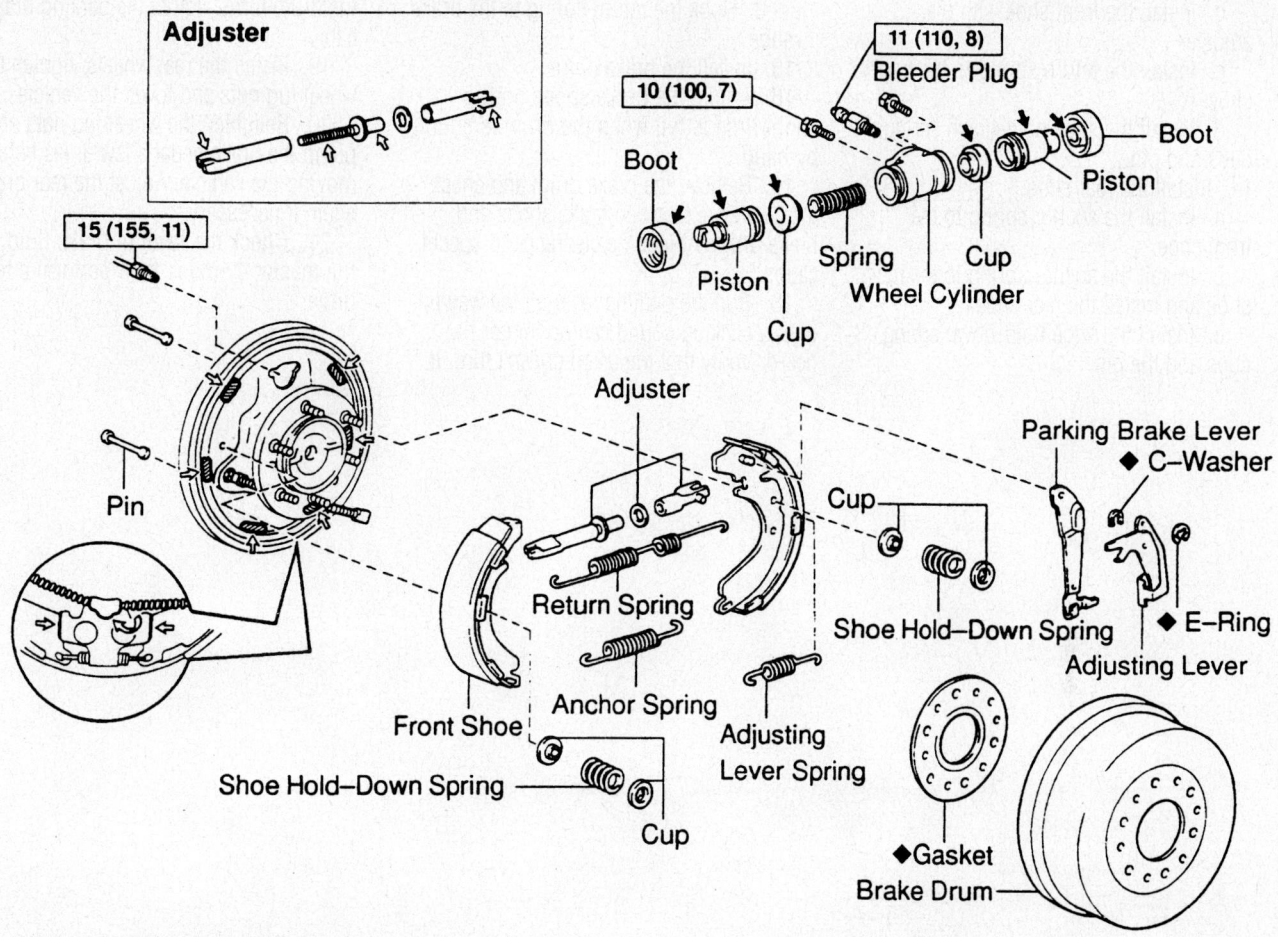

Adjuster

11 (110, 8)
Bleeder Plug

10 (100, 7)

Boot
Boot

Piston

Piston
Spring   Cup
Cup   Wheel Cylinder

15 (155, 11)

Adjuster

Parking Brake Lever
◆ C–Washer

Pin
Cup

◆ E–Ring
Return Spring
Shoe Hold–Down Spring
Adjusting Lever

Front Shoe
Anchor Spring
Adjusting Lever Spring

Shoe Hold–Down Spring

Cup

◆Gasket
Brake Drum

| N·m (kgf·cm, ft·lbf) | : Specified torque
◆ Non–reusable part
➡ Lithium soap base glycol grease
➡ High temperature grease

Exploded view of the rear brake drums components

93026G77

Brake Shoes

REMOVAL & INSTALLATION

1. Before servicing the vehicle, refer to the Precautions Section.
2. Loosen the rear wheel lug nuts slightly.
3. Block the front wheels, raise the rear of the vehicle, and safely support it with a suitable jack.
4. Remove the wheel lug nuts and the wheel.
5. Remove the brake drum.
6. If the drum is difficult to remove, perform the following:
   a. Insert a flat prying tool through the hole in the brake drum and hold the automatic adjusting lever away from the adjuster.

b. Reduce the brake shoe adjustment by turning the adjuster bolt with a brake tool.
   c. The drum should now be loose enough to remove without much effort.
7. Remove the rear shoe.
   a. Carefully unhook the return spring from the brake shoe.
   b. Remove the shoe hold-down spring, cups and the pin.
   c. Disconnect the anchor spring from the rear shoe and remove the rear shoe.
   d. Disconnect the anchor spring from the front shoe.
8. Remove the front shoe.
   a. Remove the shoe hold-down spring, cups and pin.
   b. Remove the return spring from the front shoe.

c. Remove the front shoe with the adjuster.
   d. Disconnect the parking brake cable from the front shoe.
To install:
9. Inspect the shoes for signs of unusual wear or scoring.
10. Check the wheel cylinder for any sign of fluid seepage or frozen pistons.
11. Clean and inspect the brake backing plate and all other components. Check that the brake drum inner diameter is within specified limits. Lubricate the backing plate at the positions the brakes come in contact with the backing plate. Also lubricate the anchor plate.
12. Mount the automatic adjuster assembly onto a new rear brake shoe.
13. Install the front shoe.
   a. Install the parking brake cable to the front shoe.

b. Install the front shoe with the adjuster.

c. Install the return spring to the front shoe.

d. Install the shoe hold-down spring, cups and pin.

14. Install the rear shoe.

a. Install the anchor spring to the front shoe.

b. Install the anchor spring to the rear shoe and install the rear shoe.

c. Install the shoe hold-down spring, cups and the pin.

d. Hook the return spring to the brake shoe.

15. Install the brake drum.

16. Adjust the brake shoes until a slight drag is felt when the drum is spun by hand.

17. Remove the brake drum and check the clearance between brake shoes and brake drum. Adjust the clearance to specification.

18. Pull the parking lever all the way up until a clicking sound can no longer be heard. Verify that the drum doesn't turn. If the drum turns, adjust the parking brake cable.

19. Install the rear wheels, tighten the wheel lug nuts and lower the vehicle.

20. Retighten the wheel lug nuts and pump the brake pedal a few times before moving the vehicle. Adjust the rear brakes again if necessary.

21. Check the level of brake fluid in the master cylinder, then perform a test drive.

TOYOTA

Celica • Corolla • ECHO • MR2

13

BRAKES **13-105**
DRIVE TRAIN **13-77**
ENGINE REPAIR **13-16**
FUEL SYSTEM **13-73**
SPECIFICATIONS AND
  MAINTENANCE CHARTS......**13-2**
Engine and Vehicle
  Identification 13-2
General Engine Specifications 13-2
Engine Tune-Up Specifications 13-3
Firing Order 13-4
Accessory Drive Belt Routing 13-4
Capacities 13-5
Valve Specifications...................13-6
Crankshaft and Connecting Rod
  Specifications 13-7
Cramshaft and Bearing
  Specifications 13-8
Piston and Ring Specifications 13-9
Torque Specifications 13-10
Wheel Alignment 13-11
Tire, Wheel and Ball Joint
  Specifications 13-12
Brake Specifications 13-13
Scheduled Maintenance
  Intervals...........................13-14
STEERING AND
  SUSPENSION.................**13-93**
A
Air Bag...............................13-93
  Disarming 13-93
  Rearming 13-93
Alternator............................13-16
  Removal & Installation............13-16
Automatic Transaxle....................13-82
  Removal & Installation.............13-82
B
Brake Caliper 13-105
  Removal & Installation..........13-105
Brake Drums..........................13-110
  Removal & Installation..........13-110

Brake Shoes............................13-112
  Removal & Installation..........13-112
C
Camshaft(s) 13-54
  Inspection 13-58
  Removal & Installation.............13-54
Clutch 13-84
  Adjustments.......................13-85
  Removal & Installation.............13-84
CV-Joints..............................13-88
  Overhaul 13-88
Cylinder Head 13-45
  Removal & Installation............13-45
D
Disc Brake Pads.......................13-110
  Removal & Installation..........13-110
E
Engine Assembly 13-17
  Removal & Installation.............13-17
Exhaust Manifold......................13-53
  Removal & Installation............13-53
F
Fuel Filter 13-73
  Removal & Installation.............13-73
Fuel Injectors 13-74
  Removal & Installation.............13-74
Fuel Pump 13-73
  Removal & Installation.............13-73
Fuel System Pressure 13-73
  Relieving 13-73
H
Halfshaft.............................13-85
  Removal & Installation.............13-85
Heater Core..........................13-25
  Removal & Installation.............13-25
Hydraulic Clutch System 13-85
  Bleeding...........................13-85
I
Ignition Timing 13-17
  Adjustment........................13-17

Intake Manifold.........................13-52
  Removal & Installation............13-52
L
Lower Ball Joint........................13-100
  Removal & Installation...........13-100
M
Manual Transaxle Assembly 13-77
  Removal & Installation............13-77
O
Oil Pan................................13-66
  Removal & Installation...........13-66
Oil Pump 13-67
  Removal & Installation............13-67
P
Piston and Ring Positioning........13-72
R
Rack and Pinion Steering Gear 13-93
  Removal & Installation............13-93
Rear Main Seal 13-69
  Removal & Installation...........13-69
Rear Shock Absorber and
  Spring..............................13-100
  Removal And Installation 13-100
S
Strut and Coil Spring.................13-97
  Removal & Installation............13-97
T
Timing Chain, Sprockets,
  Front Cover and Seal..............13-69
  Removal & Installation...........13-69
V
Valve Lash 13-59
  Adjustment.........................13-59
W
Water Pump 13-24
  Removal & Installation............13-24
Wheel Bearings........................13-102
  Adjustment.........................13-102
  Removal & Installation..........13-102

SPECIFICATIONS AND MAINTENANCE CHARTS

ENGINE AND VEHICLE IDENTIFICATION

Code ①	Liters (cc)	Cu. In.	Cyl.	Fuel Sys.	Engine Type	Eng. Mfg.	Code ②	Year
1NZ-FE	1.5 (1496)	91	4	EFI	DOHC	Toyota	2	2002
1ZZ-FE	1.8 (1794)	109	4	EFI	DOHC	Toyota	3	2003
2ZZ-GE	1.8 (1796)	109	4	EFI	DOHC	Toyota	4	2004
							5	2005
							6	2006

EFI: Electronic Fuel Injection

SFI: Sequential Multi-port Fuel Injection

DOHC: Double Overhead Camshaft

① 8th digit of VIN

② 10th digit of VIN

09490_CORO_C0001

GENERAL ENGINE SPECIFICATIONS

Year	Model	Engine Displacement Liters (cc)	Engine Series (ID/VIN)	Net Horsepower @ rpm	Net Torque @ rpm (ft. lbs.)	Bore x Stroke (in.)	Com-pression Ratio	Oil Pressure psi @ idle
2002	Corolla	1.8 (1794)	1ZZ-FE	120@5600	122@4400	3.11x3.60	10.0:1	4.3
	Celica	1.8 (1794)	1ZZ-FE	140@6400	125@4200	3.11x3.60	10.0:1	4.3
	Celica GT-S	1.8 (1796)	2ZZ-GE	180@7600	130@6800	3.23x3.35	11.5:1	5.7
	Echo	1.5 (1496)	1NZ-FE	108@5999	105@3999	2.95x3.32	10.5:0	4.3
	MR 2	1.8 (1794)	1ZZ-FE	140@6400	125@4200	3.11x3.60	10.0:1	4.3
2003	Corolla	1.8 (1794)	1ZZ-FE	120@5600	122@4400	3.11x3.60	10.0:1	4.3
	Celica	1.8 (1794)	1ZZ-FE	140@6400	125@4200	3.11x3.60	10.0:1	4.3
	Celica GT-S	1.8 (1796)	2ZZ-GE	180@7600	130@6800	3.23x3.35	11.5:1	4.3
	Echo	1.5 (1496)	1NZ-FE	108@5999	105@3999	2.95x3.32	10.5:0	4.3
	MR 2	1.8 (1794)	1ZZ-FE	140@6400	125@4200	3.11x3.60	10.0:1	4.3
2004	Corolla	1.8 (1794)	1ZZ-FE	120@5600	122@4400	3.11x3.60	10.0:1	4.3
	Celica	1.8 (1794)	1ZZ-FE	140@6400	125@4200	3.11x3.60	10.0:1	4.3
	Celica GT-S	1.8 (1796)	2ZZ-GE	180@7600	130@6800	3.23x3.35	11.5:1	5.7
	Echo	1.5 (1496)	1NZ-FE	108@6000	105@4000	2.95x3.33	10.5:1	4.3
	MR 2	1.8 (1794)	1ZZ-FE	140@6400	125@4200	3.11x3.60	10.0:1	4.3
2005	Corolla	1.8 (1794)	1ZZ-FE	120@5600	122@4400	3.11x3.60	10.0:1	4.3
	Corolla	1.8 (1796)	2ZZ-GE	170@7600	127@4400	3.23x3.35	11.5:1	5.7
	Celica	1.8 (1794)	1ZZ-FE	140@6400	125@4200	3.11x3.60	10.0:1	4.3
	Celica GT-S	1.8 (1796)	2ZZ-GE	180@7600	130@6800	3.23x3.35	11.5:1	5.7
	Echo	1.5 (1496)	1NZ-FE	108@6000	105@4000	2.95x3.33	10.5:1	4.3
	MR 2	1.8 (1794)	1ZZ-FE	140@6400	125@4200	3.11x3.60	10.0:1	4.3
2006	Corolla	1.8 (1794)	1ZZ-FE	126@6000	122@4200	3.11x3.60	10.0:1	4.3
	Corolla XRS	1.8 (1796)	2ZZ-GE	164@7600	125@4400	3.23x3.35	11.5:1	5.7

EFI: Electronic Fuel Injection

09490_CORO_C0002

ENGINE TUNE-UP SPECIFICATIONS

Year	Engine Displacement Liters (cc)	Engine ID/VIN	Spark Plug Gap (in.)	Ignition Timing (deg.) ①	Fuel Pump (psi)	Idle Speed (rpm)		Valve Clearance	
						MT	AT	In.	Ex.
2002	1.5 (1496)	1NZ-FE	0.043	8-12B	44-50	700-800	700-800	0.006-0.010	0.011-0.014
	1.8 (1794)	1ZZ-FE	0.043	8-12B	44-50	650-750	700-800	0.006-0.010	0.010-0.014
	1.8 (1796)	2ZZ-GE	0.043	8-12B	44-50	750-850	700-800	0.006-0.010	0.014-0.018
2003	1.5 (1496)	1NZ-FE	0.043	8-12B	44-50	700-800	700-800	0.006-0.010	0.011-0.014
	1.8 (1794)	1ZZ-FE	0.043	8-12B	44-50	650-750	700-800	0.006-0.010	0.010-0.014
	1.8 (1796)	2ZZ-GE	0.043	8-12B	44-50	750-850	700-800	0.006-0.010	0.014-0.018
2004	1.5 (1496)	1NZ-FE	0.043	8-12B	44-50	700-800	700-800	0.006-0.010	0.011-0.014
	1.8 (1794)	1ZZ-FE	0.043	8-12B	44-50	650-750	700-800	0.006-0.010	0.010-0.014
	1.8 (1796)	2ZZ-GE	0.043	8-12B	44-50	750-850	700-800	0.006-0.010	0.014-0.018
2005	1.5 (1496)	1NZ-FE	0.043	8-12B	44-50	700-800	700-800	0.006-0.010	0.011-0.014
	1.8 (1794)	1ZZ-FE	0.043	8-12B	44-50	650-750	700-800	0.006-0.010	0.010-0.014
	1.8 (1796)	2ZZ-GE	0.043	8-12B	44-50	750-850	700-800	0.006-0.010	0.014-0.018
2006	1.8 (1794)	1ZZ-FE	0.043	8-12B	44-50	650-750	650-750	0.0059-0.0098	0.0098-0.0138
	1.8 (1796)	2ZZ-GE	0.043	8-12B	44-50	750-850	750-850	0.0031-0.0071	0.0087-0.0126

Note: The Vehicle Emission Control Information label often reflects specification changes made during production. The label figures must be used if they differ from those in this chart.

① With terminal TE1 and E1 connected of DLC1
With terminal TC and CG of DLC1 disconnected

09490_CORO_C0003

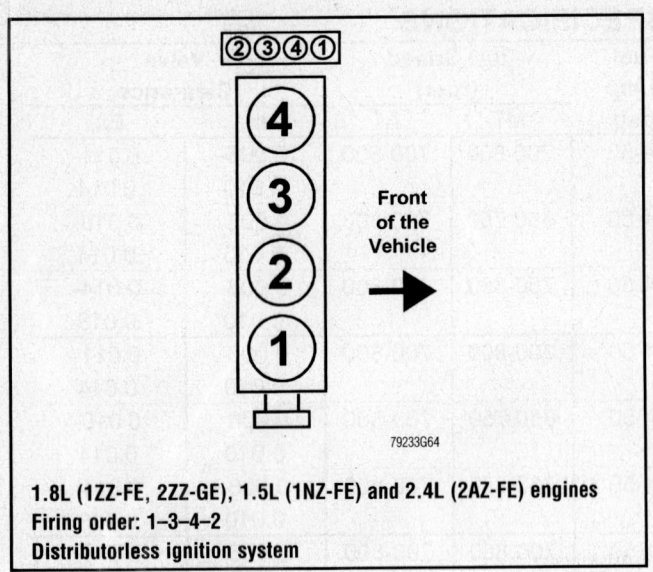

1.8L (1ZZ-FE, 2ZZ-GE), 1.5L (1NZ-FE) and 2.4L (2AZ-FE) engines
Firing order: 1–3–4–2
Distributorless ignition system

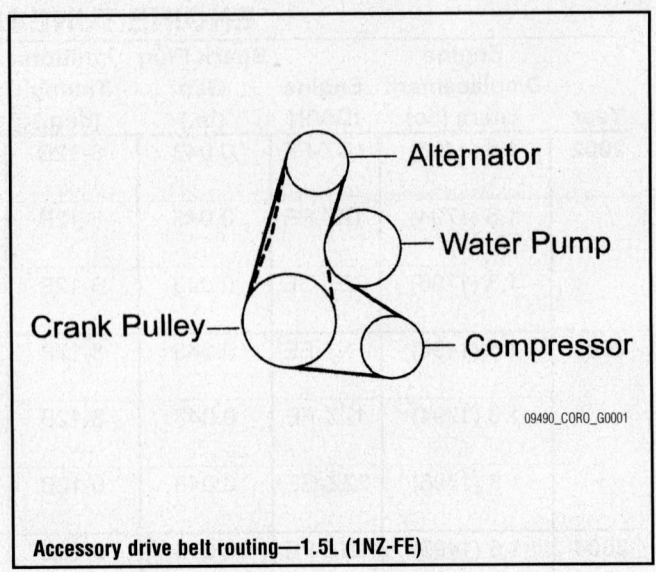

Accessory drive belt routing—1.5L (1NZ-FE)

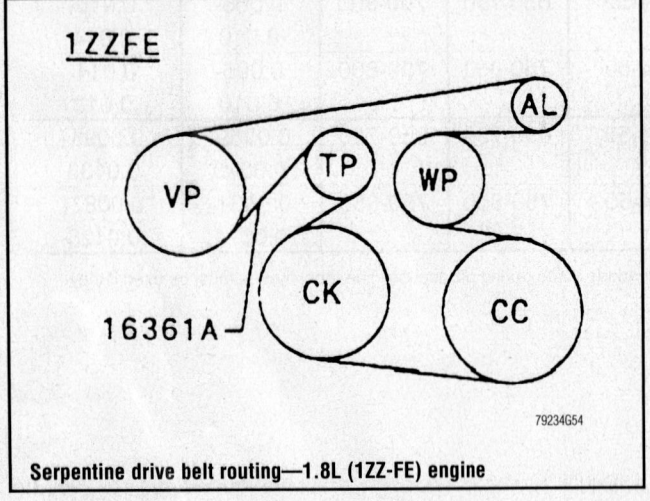

Serpentine drive belt routing—1.8L (1ZZ-FE) engine

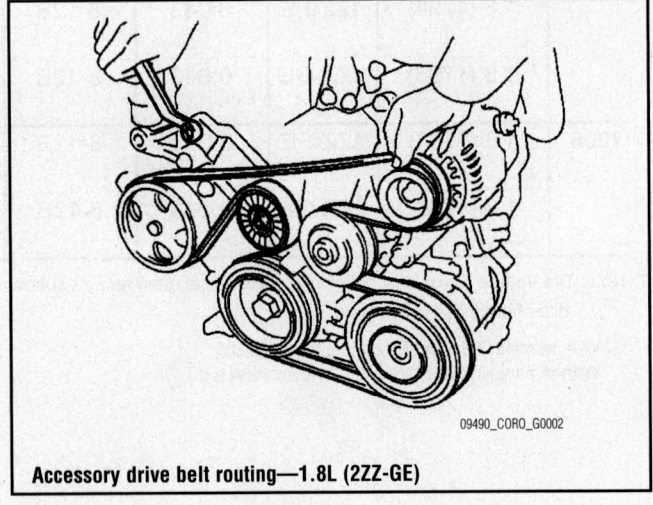

Accessory drive belt routing—1.8L (2ZZ-GE)

CAPACITIES

Year	Model	Engine Displacement Liters (cc)	Engine ID/VIN	Engine Oil with Filter	Transmission (pts.) 5-Spd	6-Spd	Auto.	Front Drive Axle (pts.)	Fuel Tank (gal.)	Cooling System (qts.)
2002	Celica	1.8 (1794)	1ZZ-FE	3.9	4.0	—	6.2	①	14.5	③
	Celica GT-S	1.8 (1796)	2ZZ-GE	4.8	—	4.8	8.4	①	14.5	④
	Corolla	1.8 (1794)	1ZZ-FE	3.9	4.0	—	②	①	13.2	⑥
	Echo	1.5 (1496)	1NZ-FE	3.9	4.0	—	6.2	①	11.9	⑤
	MR 2	1.8 (1794)	1ZZ-FE	3.9	⑨	—	—	①	12.7	11.0
2003	Celica	1.8 (1794)	1ZZ-FE	3.9	4.0	—	6.2	①	14.5	③
	Celica GT-S	1.8 (1796)	2ZZ-GE	4.8	—	4.8	7.4	①	14.5	④
	Corolla	1.8 (1794)	1ZZ-FE	3.9	4.0	—	6.6	①	13.2	6.9
	Echo	1.5 (1496)	1NZ-FE	3.9	4.0	—	6.2	①	11.9	⑤
	MR 2	1.8 (1794)	1ZZ-FE	3.9	⑨	⑩	—	①	12.7	11.0
2004	Celica	1.8 (1794)	1ZZ-FE	3.9	4.0	—	6.2	①	14.5	③
	Celica GT-S	1.8 (1796)	2ZZ-GE	4.8	—	4.8	7.4	①	14.5	④
	Corolla	1.8 (1794)	1ZZ-FE	3.9	4.0	—	6.4	①	13.2	6.9
	Echo	1.5 (1496)	1NZ-FE	3.9	4.0	—	6.2	①	11.9	⑤
	MR 2	1.8 (1794)	1ZZ-FE	3.9	⑨	⑩	—	①	12.7	11.0
2005	Celica	1.8 (1794)	1ZZ-FE	3.9	4.0	—	6.2	①	14.5	③
	Celica GT-S	1.8 (1796)	2ZZ-GE	4.8	—	4.8	7.4	①	14.5	④
	Corolla	1.8 (1794)	1ZZ-FE	3.9	4.0	—	6.4	①	13.2	6.9
	Corolla XRS	1.8 (1796)	2ZZ-GE	4.8	—	4.8	—	①	13.2	7.0
	Echo	1.5 (1496)	1NZ-FE	3.9	4.0		6.2	①	11.9	⑤
	MR 2	1.8 (1794)	1ZZ-FE	3.9	⑨	⑩	—	①	12.7	11.0
2006	Corolla	1.8 (1794)	1ZZ-FE	4.4	4.0	—	6.4	①	13.2	6.9
	Corolla XRS	1.8 (1796)	2ZZ-GE	4.7	—	4.8	—	①	13.2	7.0

Note: All capacities are approximate. Add fluid gradually and check to be sure a proper fluid level is obtained.

① Included in transaxle capacity

② 3-Spd: 5.2
   4-Spd: 6.6

③ M/T: 6.0
   A/T: 5.9

④ M/T: 6.2
   A/T: 6.1

⑤ w/MT: 4.7
   w/AT: 4.5

⑥ w/MT: 6.1
   w/AT: 6.0

⑨ w/Limited Slip: 3.8
   w/o Limited Slip: 4.0

⑩ w/Limited Slip: 3.8
   w/o Limited Slip: 4.0

09490_CORO_C0004

VALVE SPECIFICATIONS

Year	Engine Displacement Liters (cc)	Engine ID/VIN	Seat Angle (deg.)	Face Angle (deg.)	Spring Test Pressure (lbs. @ in.)	Spring Installed Height (in.)	Stem-to-Guide Clearance (in.)		Stem Diameter (in.)	
							Intake	Exhaust	Intake	Exhaust
2002	1.5 (1496)	1NZ-FE	45	44.5	33.5-37@ 1.280	1.280	0.0010- 0.0024	0.0012- 0.0026	0.1957- 0.1963	0.1955- 0.1961
	1.8 (1794)	1ZZ-FE	45	44.5	31.3-34.8@ 1.252	1.323	0.0010- 0.0024	0.0012- 0.0025	0.2154- 0.2159	0.2152- 0.2158
	1.8 (1796)	2ZZ-GE	45	44.5	①	1.516	0.0010- 0.0023	0.0012- 0.0025	0.2150- 0.2156	0.2144- 0.2154
2003	1.5 (1496)	1NZ-FE	45	44.5	33.5-37@ 1.280	1.280	0.0010- 0.0024	0.0012- 0.0026	0.1957- 0.1963	0.1955- 0.1961
	1.8 (1794)	1ZZ-FE	45	44.5	31.3-34.8@ 1.252	1.323	0.0010- 0.0024	0.0012- 0.0025	0.2154- 0.2159	0.2152- 0.2158
	1.8 (1796)	2ZZ-GE	45	44.5	①	1.516	0.0010- 0.0023	0.0012- 0.0025	0.2150- 0.2156	0.2144- 0.2154
2004	1.5 (1496)	1NZ-FE	45	44.5	33.5-37@ 1.280	1.280	0.0010- 0.0024	0.0012- 0.0026	0.1957- 0.1963	0.1955- 0.1961
	1.8 (1794)	1ZZ-FE	45	44.5	31.3-34.8@ 1.252	1.323	0.0010- 0.0024	0.0012- 0.0026	0.2154- 0.2159	0.2152- 0.2157
	1.8 (1796)	2ZZ-GE	45	44.5	①	1.516	0.0010- 0.0023	0.0012- 0.0025	0.2150- 0.2156	0.2144- 0.2154
2005	1.5 (1496)	1NZ-FE	45	44.5	33.5-37.1@ 1.280	1.280	0.0010- 0.0024	0.0012- 0.0026	0.1957- 0.1963	0.1955- 0.1961
	1.8 (1794)	1ZZ-FE	45	44.5	75.4-83.3@ 0.949	1.323	0.0010- 0.0024	0.0012- 0.0026	0.2154- 0.2159	0.2152- 0.2158
	1.8 (1796)	2ZZ-GE	45	44.5	NA	③	0.0010- 0.0023	0.0012- 0.0025	0.2145- 0.2156	0.2144- 0.2154
2006	1.8 (1794)	1ZZ-FE	45	44.5	75.4-83.3@ 0.949	1.323	0.0010- 0.0024	0.0012- 0.0026	0.2154- 0.2159	0.2152- 0.2158
	1.8 (1796)	2ZZ-GE	45	44.5	NA	③	0.0010- 0.0023	0.0012- 0.0025	0.2145- 0.2156	0.2144- 0.2154

① Intake: 49.6-55.5@1.516
  Exhaust: 47.6-52.6@1.516

② Inner spring free length: 1.799 in. (45.7 mm)

③ Intake spring free length: 1.827 in.
  Exhaust spring free length: 1.831 in.

09490_CORO_C0005

CRANKSHAFT AND CONNECTING ROD SPECIFICATIONS

All measurements are given in inches.

Year	Engine Displacement Liters (cc)	Engine ID/VIN	Crankshaft				Connecting Rod		
			Main Brg. Journal Dia.	Main Brg. Oil Clearance	Shaft End-play	Thrust on No.	Journal Diameter	Oil Clearance	Side Clearance
2002	1.5 (1496)	1NZ-FE	①	0.0004-0.0009	0.0035-0.0075	3	1.5745-1.5748	0.0006-0.0016	0.0063-0.0142
	1.8 (1762)	1ZZ-FE	②	0.0006-0.0013	0.0016-0.0094	3	1.7320-1.7323	0.0011-0.0024	0.0063-0.0135
	1.8 (1796)	2ZZ-GE	②	0.0006-0.0013	0.0016-0.0094	3	1.7713-1.7717	0.0011-0.0020	0.0063-0.0135
2003	1.5 (1496)	1NZ-FE	①	0.0004-0.0009	0.0035-0.0075	3	1.5745-1.5748	0.0006-0.0016	0.0063-0.0142
	1.8 (1762)	1ZZ-FE	②	0.0006-0.0013	0.0016-0.0094	3	1.7320-1.7323	0.0011-0.0024	0.0063-0.0135
	1.8 (1796)	2ZZ-GE	②	0.0006-0.0013	0.0016-0.0094	3	1.7713-1.7717	0.0011-0.0020	0.0063-0.0135
2004	1.5 (1496)	1NZ-FE	①	0.0004-0.0009	0.0035-0.0075	3	1.5745-1.5748	0.0006-0.0016	0.0063-0.0142
	1.8 (1762)	1ZZ-FE	③	0.0006-0.0013	0.0016-0.0094	3	1.7320-1.7323	0.0011-0.0024	0.0063-0.0135
	1.8 (1796)	2ZZ-GE	②	0.0006-0.0013	0.0016-0.0094	3	1.7713-1.7717	0.0011-0.0020	0.0063-0.0135
2005	1.5 (1496)	1NZ-FE	①	0.0004-0.0009	0.0035-0.0075	3	1.5745-1.5748	0.0006-0.0016	0.0063-0.0142
	1.8 (1762)	1ZZ-FE	③	0.0006-0.0013	0.0016-0.0094	3	1.7320-1.7323	0.0011-0.0024	0.0063-0.0135
	1.8 (1796)	2ZZ-GE	④	0.0006-0.0013	0.0016-0.0094	3	1.7713-1.7717	0.0011-0.0020	0.0063-0.0135
2006	1.8 (1762)	1ZZ-FE	③	0.0006-0.0013	0.0016-0.0094	3	1.7320-1.7323	0.0011-0.0024	0.0063-0.0135
	1.8 (1796)	2ZZ-GE	④	0.0006-0.0013	0.0016-0.0094	3	1.7713-1.7717	0.0011-0.0020	0.0063-0.0135

① Reference mark:
  0: 1.81102-1.81110
  1: 1.81110-1.81118
  2: 1.81118-1.81126
  3: 1.81126-1.81133
  4: 1.81133-1.81141
  5: 1.81141-1.81149

② Reference mark:
  0: 1.8897-1.8898
  1: 1.8896-1.8897
  2: 1.8895-1.8896
  3: 1.8894-1.8895
  4. 1.8893-1.8894
  5: 1.8892-1.8893

③ Reference mark:
  0: 1.8897-1.8898
  1: 1.8896-1.8897
  2: 1.8896
  3: 1.8895
  4: 1.8894
  5: 1.8893-1.8894

④ Reference mark:
  0: 1.8897-1.8898
  1: 1.8896-1.8897
  2: 1.8896
  3: 1.8895
  4: 1.8894
  5: 1.8892-1.8893

09490_CORO_C0006

CAMSHAFT AND BEARING SPECIFICATIONS CHART
All measurements are given in inches.

Year	Engine Displacement Liters (cc)	Engine ID/VIN	Journal Dia.	Brg. Oil Clearance	Shaft End-play	Runout	Lobe Height	
							Intake	Exhaust
2002	1.5 (1496)	1NZ-FE	①	0.0016-0.0037	0.0016-0.0037	0.0012	1.7566-1.7605	1.7585-1.7624
	1.8 (1762)	1ZZ-FE	①	0.0014-0.0028	0.0016-0.0037	0.0012	1.7454-1.7493	1.7229-1.7268
	1.8 (1796)	2ZZ-GE	②	③	0.0016-0.0055	0.0012	④	⑤
2003	1.5 (1496)	1NZ-FE	①	0.0016-0.0037	0.0016-0.0037	0.0012	1.7566-1.7605	1.7585-1.7624
	1.8 (1762)	1ZZ-FE	①	0.0014-0.0028	0.0016-0.0037	0.0012	1.7454-1.7493	1.7229-1.7268
	1.8 (1796)	2ZZ-GE	②	③	0.0016-0.0055	0.0012	④	⑤
2004	1.5 (1496)	1NZ-FE	①	0.0016-0.0037	0.0016-0.0037	0.0012	1.7566-1.7605	1.7585-1.7624
	1.8 (1762)	1ZZ-FE	①	0.0014-0.0028	0.0016-0.0037	0.0012	1.7454-1.7493	1.7229-1.7268
	1.8 (1796)	2ZZ-GE	②	③	0.0016-0.0055	0.0012	④	⑤
2005	1.5 (1496)	1NZ-FE	①	0.0016-0.0037	0.0016-0.0037	0.0012	1.7566-1.7605	1.7585-1.7624
	1.8 (1762)	1ZZ-FE	①	0.0014-0.0028	0.0016-0.0037	0.0012	1.7454-1.7493	1.7229-1.7268
	1.8 (1796)	2ZZ-GE	②	③	0.0016-0.0055	0.0012	④	⑤
2006	1.8 (1762)	1ZZ-FE	①	0.0014-0.0028	0.0016-0.0037	0.0012	1.7454-1.7493	1.7229-1.7268
	1.8 (1796)	2ZZ-GE	②	③	0.0016-0.0055	0.0012	④	⑤

① No. 1 Journal: 1.3563-1.3569 in.
  All Others: 0.9035-0.9041 in.

② No. 1 Journal: 1.3563-1.3569 in.
  All Others: 1.1004-1.1010 in.

③ No. 1 Journal: 0.0014-0.0030 in.
  All Others: 0.0014-0.0028 in.

④ No. 1: 1.5959-1.5998 in.
  No. 2: 1.5236-1.5276 in.

⑤ No. 1: 1.5728-1.5767 in.
  No. 2: 1.5273-1.5313 in.

09490_CORO_C0007

PISTON AND RING SPECIFICATIONS

All measurements are given in inches.

Year	Engine Displacement Liters (cc)	Engine ID/VIN	Piston Clearance	Ring Gap			Ring Side Clearance		
				Top Compression	Bottom Compression	Oil Control	Top Compression	Bottom Compression	Oil Control
2002	1.5 (1496)	1NZ-FE	0.0018-0.0027	0.0098-0.0138	0.0138-0.0197	0.0039-0.0138	0.0012-0.0028	0.0012-0.0028	SNUG
	1.8 (1762)	1ZZ-FE	0.0033-0.0042	0.0098-0.0138	0.0138-0.0197	0.0059-0.0157	0.0012-0.0028	0.0012-0.0028	SNUG
	1.8 (1796)	2ZZ-GE	0.0003-0.0015	0.0098-0.0138	0.0138-0.0197	0.0059-0.0157	0.0012-0.0028	0.0012-0.0028	SNUG
2003	1.5 (1496)	1NZ-FE	0.0018-0.0027	0.0098-0.0138	0.0138-0.0197	0.0039-0.0138	0.0012-0.0028	0.0012-0.0028	SNUG
	1.8 (1762)	1ZZ-FE	0.0026-0.0035	0.0098-0.0138	0.0138-0.0197	0.0059-0.0157	0.0008-0.0028	0.0012-0.0028	0.0012-0.0043
	1.8 (1796)	2ZZ-GE	0.0003-0.0015	0.0098-0.0138	0.0138-0.0197	0.0059-0.0157	0.0012-0.0028	0.0012-0.0028	SNUG
2004	1.5 (1496)	1NZ-FE	0.0018-0.0027	0.0098-0.0138	0.0138-0.0197	0.0039-0.0138	0.0012-0.0028	0.0012-0.0028	SNUG
	1.8 (1762)	1ZZ-FE	0.0026-0.0035	0.0098-0.0138	0.0138-0.0197	0.0059-0.0157	0.0008-0.0028	0.0012-0.0028	0.0012-0.0043
	1.8 (1796)	2ZZ-GE	0.0003-0.0015	0.0098-0.0138	0.0138-0.0197	0.0059-0.0157	0.0012-0.0028	0.0012-0.0028	SNUG
2005	1.5 (1496)	1NZ-FE	0.0022-0.0031	0.0098-0.0138	0.0138-0.0197	0.0039-0.0138	0.0012-0.0028	0.0008-0.0024	0.0008-0.0024
	1.8 (1762)	1ZZ-FE	0.0029-0.0038	0.0098-0.0138	0.0138-0.0197	0.0059-0.0157	0.0008-0.0028	0.0012-0.0028	0.0012-0.0043
	1.8 (1796)	2ZZ-GE	0.0003-0.0015	0.0098-0.0138	0.0138-0.0197	0.0059-0.0157	0.0009-0.0028	0.0012-0.0028	SNUG
2006	1.8 (1762)	1ZZ-FE	0.0029-0.0038	0.0098-0.0138	0.0138-0.0197	0.0059-0.0157	0.0008-0.0028	0.0012-0.0028	0.0012-0.0043
	1.8 (1796)	2ZZ-GE	0.0003-0.0015	0.0098-0.0138	0.0138-0.0197	0.0059-0.0157	0.0009-0.0028	0.0012-0.0028	SNUG

09490_CORO_C0008

TORQUE SPECIFICATIONS

All readings in ft. lbs.

Year	Engine Displacement Liters (cc)	Engine ID/VIN	Cylinder Head Bolts	Main Bearing Bolts	Rod Bearing Bolts	Crankshaft Damper Bolts	Flywheel Bolts	Manifold Intake	Manifold Exhaust	Spark Plugs	Lug Nuts
2002	1.5 (1496)	1NZ-FE	①	②	③	94	④	22	20	13	76
	1.8 (1794)	1ZZ-FE	④	⑤	⑥	102	④	14	27	13	76
	1.8 (1796)	2ZZ-GE	⑦	⑤	⑧	87	④	⑨	37	13	76
2003	1.5 (1496)	1NZ-FE	①	②	③	94	④	22	20	13	76
	1.8 (1794)	1ZZ-FE	④	⑤	⑥	102	④	14	27	13	76
	1.8 (1796)	2ZZ-GE	⑦	⑤	⑧	87	④	⑨	37	13	76
2004	1.5 (1496)	1NZ-FE	①	②	③	94	④	22	20	13	76
	1.8 (1794)	1ZZ-FE	④	⑤	⑥	102	④	22	27	13	76
	1.8 (1796)	2ZZ-GE	⑦	⑤	⑧	87	④	⑨	37	13	76
2005	1.5 (1496)	1NZ-FE	①	②	③	94	④	22	20	13	76
	1.8 (1794)	1ZZ-FE	④	⑤	⑥	102	④	14	27	13	76
	1.8 (1796)	2ZZ-GE	⑦	⑤	⑧	87	④	⑨	37	13	76
2006	1.8 (1794)	1ZZ-FE	④	⑤	⑥	102	④	14	27	13	76
	1.8 (1796)	2ZZ-GE	⑦	⑤	⑧	87	④	⑨	37	13	76

① Step 1: 22 ft. lbs.
  Step 2: 90 degree turn
  Step 3: 90 degree turn

② Step 1: 16 ft. lbs.
  Step 2: 90 degree turn

③ Step 1: 11 ft. lbs.
  Step 2: 90 degree turn

④ Step 1: 36 ft. lbs.
  Step 2: 90 degree turn

⑤ 12 pointed bolts:
  Step 1: 16 ft. lbs.
  Step 2: 32 ft. lbs.
  Step 3: 45 degree turn
  Step 4: 45 degree turn
  Hex head bolts: 14 ft. lbs.

⑥ Step 1: 15 ft. lbs.
  Step 2: 90 degree turn

⑦ Step 1: 26 ft. lbs.
  Step 2: 180 degree turn

⑧ Step 1: 22 ft. lbs.
  Step 2: 90 degree turn
  Recessed head bolt: 13 ft. lbs.

⑨ 4 upper bolts and 1 nut: 20 ft. lbs.
  1 lower bolt: 34 ft. lbs.

09490_CORO_C0009

WHEEL ALIGNMENT

Year	Model		Caster Range (+/-Deg.)	Caster Preferred Setting (Deg.)	Camber Range (+/-Deg.)	Camber Preferred Setting (Deg.)	Toe-in (in.)	Steering Axis Inclination (Deg.)
2002	Celica	F	0.75	+2.05	0.75	-0.77	0+/-0.08	14.97+/-0.75
		R	—	—	0.75	-1.17	0.14+/-0.08	—
	Corolla	F	0.75	+1.19	0.75	-0.18	0.04+/-0.08	12.38+/-0.75
		R	—	—	0.75	-0.92	0.16+/-0.08	—
	Echo	F	0.75	+1.60	0.75	-0.58	0+/-0.08	10.08+/-0.75
		R	—	—	0.75	-0.93	0.11+/-0.12	—
	MR 2	F	0.75	3.08	0.75	-0.47	0.06+/-0.08	14.52+/-0.75
		R	—	—	0.75	-1.05	0.12+/-0.08	—
2003	Celica	F	0.75	+2.05	0.75	-0.77	0+/-0.08	14.97+/-0.75
		R	—	—	0.75	-1.17	0.14+/-0.08	—
	Corolla	F	0.75	+1.19	0.75	-0.18	0.04+/-0.08	12.38+/-0.75
		R	—	—	0.75	-0.92	0.16+/-0.08	—
	Echo	F	0.75	+1.60	0.75	-0.58	0+/-0.08	10.08+/-0.75
		R	—	—	0.75	-0.93	0.11+/-0.12	—
	MR 2	F	0.75	3.08	0.75	-0.47	0.06+/-0.08	14.52+/-0.75
		R	—	—	0.75	-1.05	0.12+/-0.08	—
2004	Celica	F	0.75	+2.12	0.75	-0.47	0+/-0.08	13.15+/-0.75
		R	—	—	0.75	-1.18	0.12+/-0.08	—
	Celica GTS	F	0.75	+2.02	0.75	-0.42	0+/-0.08	13.07+/-0.75
		R	—	—	0.75	-1.18	0.12+/-0.08	—
	Corolla	F	0.75	+2.83	0.75	-0.53	0+/-0.08	11.35+/-0.75
		R	—	—	0.5	-1.45	0.10+/-0.10	—
	Echo	F	0.75	①	0.75	-0.58	0+/-0.08	10.08+/-0.75
		R	—	—	0.75	-0.93	0.13+/-0.12	—
	MR 2	F	0.75	3.08	0.75	-0.47	0.06+/-0.08	14.52+/-0.75
		R	—	—	0.75	-1.05	0.12+/-0.08	—
2005	Celica	F	0.75	+2.12	0.75	-0.47	0+/-0.08	13.15+/-0.75
		R	—	—	0.75	-1.18	0.12+/-0.08	—
	Celica GTS	F	0.75	+2.02	0.75	-0.42	0+/-0.08	13.07+/-0.75
		R	—	—	0.75	-1.18	0.12+/-0.08	—
	Corolla	F	0.75	+2.83	0.75	-0.53	0+/-0.08	11.35+/-0.75
		R	—	—	0.5	-1.45	0.10+/-0.10	—
	Corolla XRS	F	0.75	+3.08	0.75	-0.63	0+/-0.08	11.57+/-0.75
		R	—	—	0.5	-1.47	0.13+/-0.10	—
	Echo	F	0.75	①	0.75	-0.58	0+/-0.08	10.08+/-0.75
		R	—	—	0.75	-0.93	0.13+/-0.12	—
	MR 2	F	0.75	3.08	0.75	-0.47	0.06+/-0.08	14.52+/-0.75
		R	—	—	0.75	-1.05	0.12+/-0.08	—
2006	Corolla	F	0.75	+2.83	0.75	-0.53	0+/-0.08	11.35+/-0.75
		R	—	—	0.5	-1.45	0.10+/-0.10	—
	Corolla XRS	F	0.75	+3.08	0.75	-0.63	0+/-0.08	11.57+/-0.75
		R	—	—	0.5	-1.47	0.13+/-0.10	—

① Sedan M/S: 0.97

   Sedan P/S: 1.93

   Hatchback M/S: 0.88

   Hatchback P/S: 1.85

TIRE, WHEEL AND BALL JOINT SPECIFICATIONS

| Year | Model | OEM Tires | | Tire Pressures (psi) | | Wheel Size | Ball Joint Inspection |
		Standard	Optional	Front	Rear		
2002	Celica	205/55VR15	P205/55VR15	33	33	6.5-JJ	9-26 in. ①
	Corolla CE	P175/65R14	None	30	30	5.5-JJ	9-26 in. ①
	Corolla S, LE	P185/65R14	None	30	30	5.5-JJ	9-26 in. ①
	Echo	P175/65R14	None	32	32	5.5-JJ	9-26 in. ①
	MR 2	F: 205/50R16 R: 215/50R16	None	33	33	6-JJ	9-26 in. ①
2003	Celica	205/55VR15	P205/55VR15	33	33	6.5-JJ	9-26 in. ①
	Corolla CE	P185/65R15	None	30	30	5.5-JJ	9-43 in. ①
	Corolla S, LE	P195/65R15	None	30	30	5.5-JJ	9-43 in. ①
	Echo	P175/65R14	None	32	32	5.5-JJ	9-26 in. ①
	MR 2	F: 205/50R16 R: 215/50R16	None	33	33	6-JJ	9-26 in. ①
2004	Celica	P195/60R15	P205/55R15	29 ③	29 ③	6.5-JJ	9-26 in. ①
	Corolla CE	P185/65R15	None	30	30	6-JJ	9-43 in. ①
	Corolla S, LE	P195/65R15	None	30	30	6-JJ	9-43 in. ①
	Echo	P175/65R14	P185/60R15	32	32	5.5-JJ	5.2-30 in. ①
	MR 2	F: 205/50R16 R: 215/50R16	None	33	33	6-JJ	9-26 in. ①
2005	Celica	P195/60R15	P205/55R15	29 ③	29 ③	6.5-JJ	9-26 in. ①
	Corolla CE	P185/65R15	None	30	30	6-JJ	9-30 in. ①
	Corolla S, LE	P195/65R15	None	30	30	6-JJ	9-30 in. ①
	Corolla XRS	P195/55R16	None	32	32	6-JJ	9-30 in. ①
	Echo	P175/65R14	P185/60R15	32	32	5.5-JJ	5.2-30 in. ①
	MR 2	F: 205/50R16 R: 215/50R16	None	33	33	6-JJ	9-26 in. ①
2006	Corolla CE	P185/65R15	None	30	30	6-JJ	9-30 in. ①
	Corolla S, LE	P195/65R15	None	30	30	6-JJ	9-30 in. ①
	Corolla XRS	P195/55R16	None	32	32	6-JJ	9-30 in. ①

OEM: Original Equipment Manufacturer

PSI: Pounds Per Square Inch

① Torque required in inch lbs. to rotate ball joint when removed from the knuckle

② Sport and convertible models, 215/55R17

③ P205/55R15: 32

09490_CORO_C0011

BRAKE SPECIFICATIONS
All measurements in inches unless noted

Year	Model		Brake Disc Original Thickness	Brake Disc Minimum Thickness	Brake Disc Maximum Runout	Brake Drum Diameter Original Inside Diameter	Brake Drum Diameter Max. Wear Limit	Brake Drum Diameter Maximum Machine Diameter	Minimum Lining Thickness	Brake Caliper Bracket Bolts (ft. lbs.)	Brake Caliper Mounting Bolts (ft. lbs.)
2002	Celica	F	1.102	1.024	0.0020	—	—	—	0.039	25	79
		R	0.354	0.314	0.0059	7.87	—	7.91	0.039	—	34
	Celica GT-S	F	0.984	0.906	0.0020	—	—	—	0.039	25	79
		R	0.354	0.295	0.0059	—	—	—	0.039	—	34
	Corolla		0.866	0.787	0.0020	7.87	—	7.91	0.039	25	65
	Echo		0.709	0.630	0.0020	7.09	—	7.13	0.039	25	65
	MR 2	F	0.787	0.709	0.0020	—	—	—	0.039	80	25
		R	0.630	0.591	0.0039	—	—	—	0.039	44	15
2003	Celica	F	1.102	1.024	0.0020	—	—	—	0.039	25	79
		R	0.354	0.314	0.0059	7.87	—	7.91	0.039	—	34
	Celica GT-S	F	0.984	0.906	0.002	—	—	—	0.039	25	79
		R	0.354	0.295	0.006	—	—	—	0.039	—	34
	Corolla		0.866	0.787	0.0020	7.87	—	7.91	0.039	25	65
	Echo		0.709	0.630	0.0020	7.09	—	7.13	0.039	25	65
	MR 2	F	0.787	0.709	0.0020	—	—	—	0.039	80	25
		R	0.630	0.591	0.0039	—	—	—	0.039	44	15
2004	Celica	F	0.984	0.906	0.002	—	—	—	0.039	25	79
		R	0.354	0.295	0.006	7.87	—	7.91	0.039	—	34
	Corolla		0.984	0.906	0.0020	7.87	—	7.91	0.039	25	79
	Echo		0.709	0.630	0.0020	7.09	—	7.13	0.039	25	65
	MR 2	F	0.787	0.709	0.0020	—	—	—	0.039	80	25
		R	0.630	0.591	0.0039	—	—	—	0.039	44	15
2005	Celica	F	0.984	0.906	0.002	—	—	—	0.039	25	79
		R	0.354	0.295	0.006	7.87	—	7.91	0.039	—	34
	Corolla		0.984	0.906	0.0020	7.87	—	7.91	0.039	25	79
	Echo		0.709	0.630	0.0020	7.09	—	7.13	0.039	25	65
	MR 2	F	0.787	0.709	0.0020	—	—	—	0.039	80	25
		R	0.630	0.591	0.0039	—	—	—	0.039	44	15
2006	Corolla		0.984	0.906	0.0020	7.87	—	7.91	0.039	25	79

F: Front
R: Rear

09490_CORO_C0012

SCHEDULED MAINTENANCE INTERVALS
2002-2003—CELICA, COROLLA, ECHO & MR2 SPYDER

TO BE SERVICED	TYPE OF SERVIC	VEHICLE MILEAGE INTERVAL (x1000)												
		7.5	15	22.5	30	37.5	45	52.5	60	67.5	75	82.5	90	97.5
Engine oil & filter	R	✓	✓	✓	✓	✓	✓	✓	✓	✓	✓	✓	✓	✓
Drive belts	S/I								✓	✓	✓	✓	✓	✓
Automatic transaxle fluid & filter	S/I		✓		✓		✓		✓		✓		✓	
Ball joints & dust covers	S/I		✓		✓		✓		✓		✓		✓	
Bolts & nuts on body & chassis	S/I		✓		✓		✓		✓		✓		✓	
Brake line pipes & hoses	S/I		✓		✓		✓		✓		✓		✓	
Brake linings & drums	S/I		✓		✓		✓		✓		✓		✓	
Brake pads & discs (front & rear if equipped)	S/I		✓		✓		✓		✓		✓		✓	
Differential oil	S/I		✓		✓		✓		✓		✓		✓	
Drive shaft boots	S/I		✓		✓		✓		✓		✓		✓	
Manual transaxle oil	S/I		✓		✓		✓		✓		✓		✓	
Steering gear housing oil	S/I		✓		✓		✓		✓		✓		✓	
Steering linkage	S/I		✓		✓		✓		✓		✓		✓	
Air filter	R				✓				✓				✓	
Spark plugs	R				✓				✓				✓	
Spark plugs (platinum tip)	R								✓					
Exhaust system	S/I				✓				✓				✓	
Fuel lines & connections	S/I				✓				✓				✓	
Valve clearance	S/I				✓				✓				✓	
Engine coolant	R						✓				✓			
Fuel tank cap gasket	R								✓					
Charcoal canister	S/I								✓					

R: Replace   S/I: Service or Inspect

FREQUENT OPERATION MAINTENANCE (SEVERE SERVICE)

If a vehicle is operated under any of the following conditions it is considered severe service:

- **Extremely dusty areas.**

- **50% or more of the vehicle operation is in 32°C (90°F) or higher temperatures, or constant operation in temperatures below 0°C (32°F).**

- **Prolonged idling (vehicle operation in stop and go traffic).**

- **Frequent short running periods (engine does not warm to normal operating temperatures).**

- **Police, taxi, delivery usage or trailer towing usage.**

Oil & oil filter change: change every 2500 miles.

Bolts & nuts on chassis & body: tighten every 7500 miles.

Ball joints & dust covers: service or inspect every 12,000 miles.

Brake linings & drums: service or inspect ever 12,000 miles.

Brake pads & discs (front & rear if equipped): service or inspect every 12,000 miles.

Drive shaft boots: service or inspect every 12,000 miles.

Steering linkage: service or inspect every 12,000 miles.

Air filter: service or inspect every 15,000 miles.

Exhaust system: service or inspect every 15,000 miles.

Timing belt: replace every 60,000 miles.

SCHEDULED MAINTENANCE INTERVALS

2004-2005—CELICA, ECHO & MR2 SPYDER, 2004-2006—COROLLA

TO BE SERVICED	TYPE OF SERVICE	VEHICLE MILEAGE INTERVAL (x1000)												
		5	10	15	20	25	30	35	40	45	50	55	60	65
Air cleaner filter	R						✓						✓	
Transmission fluid	S/I						✓						✓	
Ball joints & dust covers	S/I			✓			✓			✓			✓	
Bolts & nuts on chassis & body	S/I													
Brake line pipes & hoses	S/I			✓			✓			✓			✓	
Brake pads & discs/linings & drums (front & rear)	S/I	✓	✓	✓	✓	✓	✓	✓	✓	✓	✓	✓	✓	✓
Drive belts	S/I												✓	
Driveshaft boots	S/I			✓			✓			✓			✓	
Engine coolant	S/I			✓			✓			✓			✓	
Engine coolant	R	Replace at 100,000 miles												
Engine oil & filter	R	✓	✓	✓	✓	✓	✓	✓	✓	✓	✓	✓	✓	✓
Exhaust pipes & mountings	S/I			✓			✓			✓			✓	
Fuel lines & connections	S/I						✓						✓	
Propeller shaft bolt	S/I			✓			✓			✓			✓	
Radiator core & condenser	S/I			✓			✓			✓			✓	
Front differential fluid	S/I						✓						✓	
Rotate Tires	S/I	✓	✓	✓	✓	✓	✓	✓	✓	✓	✓	✓	✓	✓
Spark plugs	R	Replace at 120,000 miles												
Steering linkage & gear box	S/I			✓			✓			✓			✓	

R: Replace   S/I: Service or Inspect

Drivebelts: After initial inspection at 60,000 miles, inspect every 15,000 miles thereafter.

FREQUENT OPERATION MAINTENANCE (SEVERE SERVICE)

 If a vehicle is operated under any of the following conditions it is considered severe service:

- Desert/Extremely dusty areas.

- Trailer towing usage.

- Prolonged idling (vehicle operation in stop and go traffic).

- Frequent short running periods (engine does not warm to normal operating temperatures).

- Police, taxi, delivery usage or trailer towing usage.

Air cleaner filter: service or inspect every 5000 miles

Ball joints & dust covers: service or inspect every 5000 miles.

Bolts & nuts on chassis & body: service or inspect every 5000 miles.

Driveshaft boots: service or inspect every 5000 miles.

Steering linkage: service or inspect every 5000 miles.

Transmission and Front differential fluid: replace every 30,000 miles.

Cabin air filter: service or inspect every 15,000 miles

09490_CORO_C00014

ENGINE REPAIR

Alternator

REMOVAL & INSTALLATION

Celica

1. Before servicing the vehicle, refer to the Precautions Section.

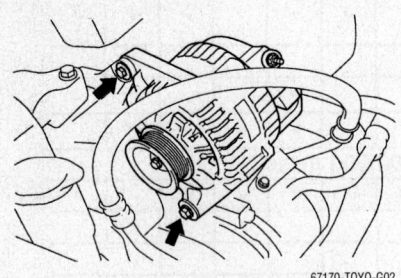

67170-TOYO-G02

Alternator with 2 mounting bolts

2. Remove or disconnect the following:

- Negative battery cable
- Accessory drive belt
- Wiring
- 2 bolts and bracket, 2ZZ-GE engines only
- Alternator

To install:

3. Install or connect the following:

- Alternator
- Torque the larger bolt—40 ft. lbs. (54 Nm); smaller bolt—18 ft. lbs. (25 Nm)
- 2ZZ-GE engines: 2 bolts and bracket. Torque to 21 ft. lbs (29 Nm).
- Accessory drive belt
- Wiring
- Negative battery cable

Corolla

2002–2004

1. Before servicing the vehicle, refer to the Precautions Section.

➡ It may be necessary to remove the gravel shield and work from underneath the car in order to gain access to the alternator retaining bolts.

2. Remove or disconnect the following:
- Negative battery cable
- Wiring from the alternator
- Accessory drive belt
- Alternator

To install:

3. Install or connect the following:
- Alternator. Torque the smaller bolt to 18 ft. lbs. (25 Nm) and the larger bolt to 40 ft. lbs. (54 Nm).

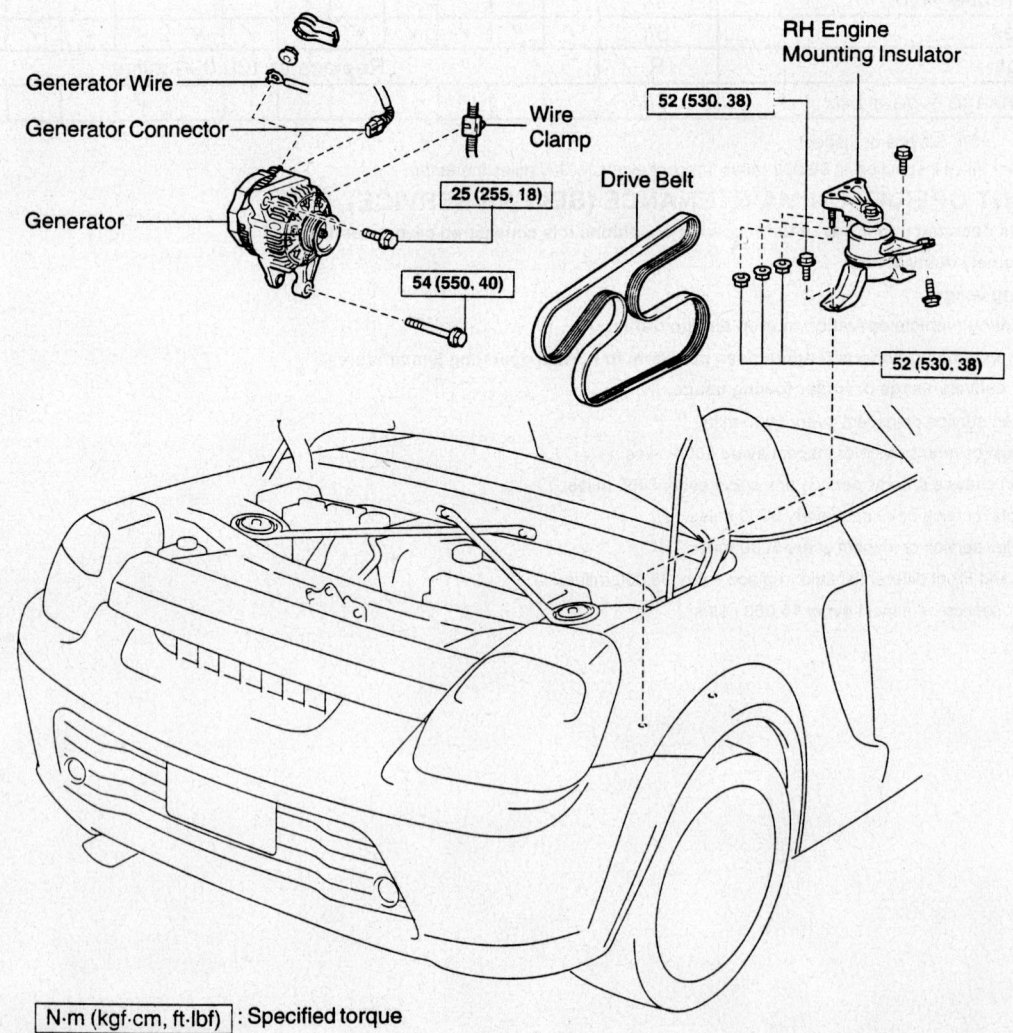

Generator Wire

Generator Connector

Generator

Wire Clamp

25 (255, 18)

54 (550, 40)

Drive Belt

RH Engine Mounting Insulator

52 (530. 38)

52 (530. 38)

N·m (kgf·cm, ft·lbf) : Specified torque

9357WG01

MR2 alternator mounting exploded view

- Alternator connector and wiring
- Accessory drive belt
- Negative battery cable

2005–2006

1. Before servicing the vehicle, refer to the Precautions Section.
2. Remove or disconnect the following:

- Negative battery cable
- Engine appearance cover, 2ZZ-GE engine only
- Air switching valve assembly, 2ZZ-GE engine only
- Engine undercover
- Accessory drive belt
- Air pump outlet hose, 2ZZ-GE engine only
- Alternator wiring
- Alternator support bracket, 2ZZ-GE engine only
- Alternator

To install:

3. Install the alternator and tighten the bolts as follows:

- Top bolt to 18 ft. lbs (25 Nm)
- Bottom bolt, 1ZZ-FE engine: 40 ft. lbs. (54 Nm)
- Bottom bolt, 2ZZ-FE engine: 43 ft. lbs. (58 Nm)

4. Install or connect the following:

- Alternator wiring
- Alternator support bracket. Tighten to 21 ft. lbs. (29 Nm).
- Air pump outlet hose. Tighten clamps to 62 inch lbs. (7 Nm).
- Accessory drive belt
- Engine undercover
- Air switching valve assembly
- Engine appearance cover
- Negative battery cable

Echo

1. Before servicing the vehicle, refer to the Precautions Section.
2. Remove or disconnect the following:

- Negative battery cable
- Accessory drive belt
- Wire clamp from the rectifier
- Alternator harness
- 2 bolts
- Alternator

To install:

3. Install or connect the following:

- Alternator and hand tighten the bolts
- Accessory drive belt and adjust if necessary. Torque the 14 mm bolt to 14 ft. lbs. (18 Nm) and the 17 mm bolt to 40 ft. lbs. (54 Nm).

- Alternator connector and clamp
- Alternator electrical wiring
- Negative battery cable

MR2

1. Before servicing the vehicle, refer to the Precautions Section.
2. Remove or disconnect the following:

- Negative battery cable
- Accessory drive belt
- Right motor mount
- Wire harness clamp
- Alternator wiring harness
- Alternator mounting bolts
- Alternator

To install:

3. Install or connect the following:

- Alternator. Tighten the upper mounting bolt to 18 ft. lbs. (25 Nm) and the lower mounting bolt to 40 ft. lbs. (54 Nm).
- Alternator wiring harness
- Wire harness clamp
- Right motor mount. Tighten the bolts to 38 ft. lbs. (52 Nm).
- Accessory drive belt
- Negative battery cable

Ignition Timing

ADJUSTMENT

➡️ **The timing on engines equipped with DIS is not adjustable.**

Engine Assembly

REMOVAL & INSTALLATION

1NZ-FE Engines

1. Before servicing the vehicle, refer to the Precautions Section.
2. Drain the cooling system.
3. Drain the engine oil.
4. Drain the transaxle fluid.
5. Remove or disconnect the following:

- Hood
- Battery and tray
- Outer front cowl top panel
- Engine under covers
- Accelerator cable
- Air cleaner cap and related parts
- Air cleaner case and related parts
- All tubes, hoses and connectors attached to the engine
- Accessory drive belts
- Alternator
- Radiator

- With A/C, position the compressor out of the way
- With MT, the clutch release cylinder
- Transaxle control cables
- Fusible link
- Power steering pump
- Center exhaust pipe
- Halfshafts
- Suspension crossmember
- Rear engine mount

6. Attach a shop crane to the engine hangers.
7. Remove all remaining engine mount bolts/nuts.
8. Remove the engine/transaxle assembly.

To install:

9. Install or connect the following:

- Engine/transaxle assembly into position
- Right engine mount insulator. Torque the bolts to 35 ft. lbs. (47 Nm).
- Left engine mount. Torque the bolts to 35 ft. lbs. (47 Nm).
- Rear engine mount bracket. Torque the bolts to 35 ft. lbs. (47 Nm).
- Rear engine mount insulator. Torque the bolt to 47 ft. lbs. (64 Nm).
- Suspension crossmember. Torque the rear mount bolts to 86 ft. lbs. (116 Nm); the front mount bolts to 52 ft. lbs. (70 Nm).
- Halfshafts
- Center exhaust pipe
- Power steering pump
- Fusible link
- Transaxle control cables
- Clutch release cylinder. Torque the 2 bolts to 10 ft. lbs. (13 Nm).
- Compressor. Torque the 4 bolts to 18 ft. lbs. (25 Nm).
- Radiator
- Alternator
- All remaining tubes, hoses and connectors
- Air cleaner case
- Air cleaner cap
- Accelerator cable
- Outer front cowl top panel
- Engine under covers
- Hood
- Battery and tray

10. Refill the cooling system to the correct level.
11. Refill with engine oil to the correct level.
12. Refill the transaxle with oil to the correct level.
13. Start the vehicle, check for leaks and repair if necessary.

1ZZ-FE Engines

2002–2004 COROLLA

1. Before servicing the vehicle, refer to the Precautions Section.
2. Relieve the fuel system pressure.
3. Drain the cooling system.
4. Drain the engine oil.
5. Drain the transaxle fluid.
6. Remove or disconnect the following:
   - Negative battery cable
   - Battery
   - Hood
   - Undercover
   - Accelerator cable
   - Throttle cable from the accelerator cable, if equipped with automatic transmission
   - Radiator and cooling fan
   - Air intake assembly
   - Coolant reservoir tank stay
   - Electrical connector, the hose, the mounting bolt, and remove the washer tank
   - Cruise control actuator
   - Manifold Absolute Pressure (MAP) sensor vacuum hose from the gas filter on the intake manifold
   - Brake booster vacuum hose from the intake manifold
   - With air conditioning: the air conditioning vacuum hose from the actuator
   - Air hose from the air pipe, if equipped with power steering
   - Air conditioning actuator connector, if equipped
7. Disconnect the following wires and connectors from the right-hand fender apron as follows:
   - Ground strap connector
   - MAP sensor connector
   - Air conditioning pressure switch, if equipped
   - Engine wiring harness from the fender apron
8. Remove or disconnect the following:
   - Data Link Connector 1 (DLC1) connector and ground strap from the left-hand fender apron
   - Engine relay box and 4 connectors
   - Charcoal canister
   - Heater hoses from water inlet housing
   - Fuel inlet and return hoses
   - With manual transmission, clutch release cylinder without disconnecting the pipe
   - Transaxle control cable(s)
9. To disconnect the engine wiring harness, disconnect or remove the following components:

- Left-hand and right-hand front door scuff plate
- Lower finish panel
- Lower panel with the glove compartment
- Radio and center cluster finish panel
- Rear console box
- On manual transmission, shift lever knob
- On automatic transmission, shifting hole bezel
- Lower center finish panel
- Floor carpet bracket
- The 3 ECM connectors and cowl wire connector

10. Remove or disconnect the remaining components:
   - Air conditioning compressor
   - Front exhaust pipe
   - Halfshafts
   - Power steering pump
   - Engine mounting center member
   - Through-bolt and nut holding the mounting insulator to the mounting bracket
   - Engine and transaxle assembly
   - Front and rear engine mounting bracket
   - Starter
   - Separate the transaxle assembly from the engine

To install:

11. Install or connect the following:
   - Engine to the transaxle
   - Starter
   - Rear engine mounting bracket bolts: 57 ft. lbs. (77 Nm).
   - Front engine mounting bracket. bolts: 57 ft. lbs. (77 Nm).
   - Engine and transaxle assembly into the vehicle
   - Engine mounting center member
   - Front engine mounting insulator through-bolt and nut. Torque the bolt to 64 ft. lbs. (87 Nm).
   - Halfshafts
   - Front exhaust pipe
   - Power steering pump. Torque the bolts to 29 ft. lbs. (39 Nm).
   - Accessory drive belt
   - Air conditioner compressor. Torque the bolts to 18 ft. lbs. (25 Nm).
12. To install and connect the engine wiring harness, perform the following:
   - Push the wire through the cowl
   - Connect the 3 ECM connectors
   - Attach the cowl wire connector
   - Floor carpet bracket
   - Center lower finish panel
   - With automatic transmission, install the shifting hole bezel, with

manual transmission, install the shift lever knob
   - Rear console box
   - Center cluster finish panel and the radio
   - Lower panel with the glove compartment door
   - Right and left-hand door scuff plates
   - Lower finish panel
13. Install or connect the following:
   - With manual transmission, clutch release cylinder
   - Transaxle control cable(s)
   - Fuel return and inlet hose. Torque the bolt to 22 ft. lbs. (29 Nm).
   - Heater hoses to the water inlet housing
   - Charcoal canister
14. Connect the following wires and connectors on the left-hand fender apron:
   - The 4 connectors to the engine relay box
   - Engine relay box
   - The DLC1 connector
   - The connector on the fender apron
   - The ground strap on the fender apron
15. Install or connect on the right-hand fender apron:
   - The ground strap connector
   - The MAP sensor connector
   - With air conditioning, the air conditioning pressure switch
   - The engine wire from the fender apron
16. Install or connect the following:
   - With A/C, the actuator connector
   - With power steering, the air hoses to the air pipe
   - The vacuum hose from the MAP sensor to the gas filter to the intake chamber
   - The brake booster vacuum hose to the air intake chamber
   - With A/C, the vacuum hose from the actuator
   - With cruise control, actuator, actuator cable and cover
   - Electrical connector and vinyl hose
   - Washer tank with the bolt
   - Coolant reservoir tank stay
   - Air cleaner
   - Radiator and cooling fan
   - With automatic transmission, connect the throttle cable.
   - Accelerator cable
   - Negative battery cable
   - Undercovers and hood
17. Refill the transaxle oil the correct level.

18. Refill the cooling system to the correct level.

19. Refill the engine with oil to the correct level.

20. Start the vehicle, check for leaks and repair if necessary.

2005–2006 COROLLA

1. Before servicing the vehicle, refer to the Precautions Section.
2. Release the fuel system pressure.
3. Drain the engine oil.
4. Drain the cooling system.
5. Drain the transaxle fluid.
6. Remove or disconnect the following:
- Negative battery cable
- Front wheels
- Engine undercovers
- Engine appearance cover
- Radiator hoses
- Transmission cooler hoses, if equipped with automatic transmission
- Radiator
- Battery and battery tray
- Battery carrier
- Air intake assembly
- Fuel supply hose
- Union to check valve hose
- Heater hoses
- Accelerator cable
- Clutch release cylinder assembly, if equipped with manual transmission
- Accessory drive belt
- Alternator
- A/C compressor
- Power steering return hose
- Glove compartment door
- Right-hand front door scuff plate
- Right-hand side cowl trim
- Engine wiring harness from the relay block
- Front floor panel brace supporting the exhaust assembly
- Front exhaust assembly
- Steering intermediate shaft

➡**Fix the steering wheel with the seatbelt to prevent rotation.**

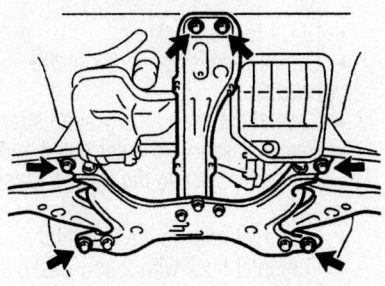

09490_CORO_G0003

Location of the suspension crossmember mounting bolts—2005–06 Corolla

- Stabilizer links
- Wheel speed sensors
- Tie rod ends
- Lower control arms
- Halfshafts

7. Secure the engine assembly with a suitable lifter.

8. Remove the bolts from the right and left motor mounts.

9. Remove the six bolts from the suspension crossmember shown in the illustration.

10. Remove the engine/transaxle assembly from the vehicle.

11. Separate the engine from the transaxle assembly.

To install:

12. Install the transaxle assembly to the engine.

13. Install the engine/transaxle assembly into the vehicle.

14. Temporarily install the suspension crossmember bolts.

15. Install the engine mount bolts. Tighten the left side to 59 ft. lbs. (80 Nm), right side to 38 ft. lbs. (52 Nm).

16. Using Special Tool 09670-00010 in the positioning holes of the crossmember, tighten the crossmember as follows:
    a. Bolt A: 116 ft. lbs. (157 Nm).
    b. Bolt B: 83 ft. lbs. (113 Nm).
    c. Front two mounting bolts: 44 ft. lbs. (60 Nm)

17. The remainder of the installation is the reverse order of removal.

18. Refill the transaxle with oil to the correct level.

19. Refill the cooling system to the correct level.

20. Refill the engine with oil to the correct level.

21. Start the engine and check for leaks.

22. Check and adjust the alignment, if needed.

CELICA

1. Before servicing the vehicle, refer to the Precautions Section.
2. Release the fuel system pressure.

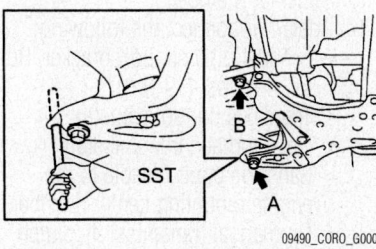

09490_CORO_G0004

Use Alignment Tool 09670-00010 to align the crossmember positioning holes— 2005–06 Corolla

3. Drain the engine oil.
4. Drain the cooling system.
5. Drain the transaxle fluid.
6. Remove or disconnect the following:
- Negative battery cable
- Battery
- Hood
- Undercover
- Accelerator cable, cable bracket, and clamps
- Air cleaner
- Cruise control actuator cable
- Radiator
- MAP sensor vacuum hose from the gas filter on the intake manifold
- Power steering air hose from the intake manifold
- Power steering hose from the air the air pipe
- Brake booster vacuum hose from the intake manifold
- Air conditioning idle-up valve
- Air conditioning idle-up valve hose from the intake manifold
- Air conditioning idle-up valve hose from the air pipe
- DLC1 from the bracket
- Engine wiring harness from the bracket
- Ground cable from the body and the ground strap from the body
- Heater hoses from the water outlet
- Heater hose from the water bypass pipe
- Fuel inlet hose from the fuel filter and the fuel return hose from the return pipe
- EVAP hose from the charcoal canister
- Engine wiring harness from the engine compartment relay box

7. To remove the engine wiring harness from the passenger's compartment, remove or disconnect the following:
- The scuff plate
- The cowl side trim
- The finish panel from the lower instrument panel
- Front side of the floor carpet
- The wiring harness from the clamp of the ECM bracket
- The 3 ECM connectors
- The circuit opening relay connector
- The 3 connectors from the connectors on the bracket
- The A/C amplifier connector
- The MAP sensor connector
- The MAP sensor wire from the clamp on the bracket
- The wire clamp from the bracket
- The 2 nuts holding the engine wiring harness to the cowl

8. Remove or disconnect the following:
- Front exhaust pipe
- Halfshafts
- Alternator drive belt
- Air conditioning drive belt, compressor connector, and compressor. Do not disconnect the air conditioning lines
- Remove the drive belt and remove the 4 bolts that secure the power steering pump. Without disconnecting the lines, securely hang the pump out of the way
- A/C relay box
- On manual transmission, clutch release cylinder from the transaxle
- Transaxle control cable(s)
- On automatic transmission, transaxle control cable from the engine mounting center member.
- Exhaust pipe support bracket.

9. To remove the engine mounting center member, remove the following components:
- The 2 dust covers from the rear side of the member
- The A/C pipe from the bracket
- The bolt and nut holding the front engine mounting bracket to the mounting insulator
- The bolt holding the rear engine mounting bracket to the insulator
- The bolt and 2 nuts holding the rear engine mounting insulator to the front suspension member
- The 2 bolts and the rear engine mounting bracket, and the center member with the rear mounting insulator

10. Remove or disconnect the remaining components:
- Attach an engine chain hoist to the engine hangers
- Left-hand engine mounting bracket from the mounting insulator
- Through-bolt and the left-hand mounting insulator
- Ground strap connector
- Right-hand engine mounting bracket from the mounting insulator
- Lift the engine and transaxle assembly from the vehicle
- Transaxle from the engine assembly

To install:

11. Install or connect the following:
- Transaxle to the engine assembly.
- Engine into the engine compartment
- Right-hand engine mounting bracket to the mounting insulator. Temporarily install the 3 nuts.
- Left-hand engine mounting insula-

tor to the body with the through-bolt
- Left-hand engine mounting bracket to the mounting insulator and install the 2 bolts and nut. Bolts and nut: 47 ft. lbs. (64 Nm).
- Left-hand engine mounting through-bolt to the body. Bolt: 54 ft. lbs. (73 Nm).
- Right-hand mounting bracket to the insulator. 12mm nut: 21 ft. lbs. (28 Nm), 14mm nut: 38 ft. lbs. (52 Nm).
- Engine ground strap connector

12. Remove the engine hoist.

13. To install the engine mounting center member, perform the following:

a. Attach the center member together with the rear engine mounting insulator to the front suspension member.

b. Temporarily install the 2 bolts and nut holding the center member to the body.

c. Install the rear engine mounting bracket. Bolts: 58 ft. lbs. (78 Nm).

d. Temporarily install the bolt and 2 nuts holding the rear engine mounting insulator to the front suspension member.

e. Temporarily install the bolt holding the rear engine mounting bracket to the insulator.

f. Temporarily install the bolt and nut holding the front engine mounting bracket to the insulator.

g. Tighten the 2 bolts holding the center member to the body to 26 ft. lbs. (35 Nm).

h. Tighten the bolt and 2 nuts holding the rear mounting insulator to the front suspension member to 59 ft. lbs. (80 Nm).

i. Tighten the bolt holding the rear engine mounting bracket to the insulator to 65 ft. lbs. (88 Nm).

j. Tighten the bolt and nut holding the front engine mounting bracket to the insulator to 65 ft. lbs. (88 Nm).

k. Install the air conditioning pipe to the bracket and install the 2 dust covers to the center member.

14. Install or connect the following:
- Exhaust pipe support bracket. Bolts to 14 ft. lbs. (19 Nm).
- Transaxle control cable(s)
- On automatic transmission, transaxle control cable to the engine mounting center member.
- On manual transmission, clutch release cylinder. Bolts: 108 inch lbs. (12 Nm), then attach the bracket with the bolt.

- A/C relay box to the body
- Power steering pump. 12mm bolts: 14 ft. lbs. (19 Nm). 14mm bolts: 29 ft. lbs. (39 Nm). Install the drive belt. Adjusting bolt: 29 ft. lbs. (39 Nm).
- A/C compressor. Bolts: 18 ft. lbs. (25 Nm).
- A/C drive belt with the adjusting bolt. Torque the locknut to 29 ft. lbs. (39 Nm). Connect the connector.
- Alternator drive belt
- Halfshafts
- Front exhaust pipe

15. To install the engine wiring harness in the passenger compartment, perform the following:

a. Push the harness through the cowl panel, install the retainer to the cowl with the 2 nuts and install the wire clamp to the bracket.

b. Connect the harness to the clamp on the ECM.

c. Connect the 3 ECM connectors and the circuit opening relay connector.

d. Connect the 3 connectors to the connectors on the bracket.

e. Connect the A/C amplifier connector.

f. Install the floor carpet, the lower instrument panel finish panel, the cowl side trim panel, and the scuff plate.

16. Install or connect the following:
- Engine wiring harness with the 2 connectors to the engine compartment relay box and install the relay box covers
- MAP sensor connector
- MAP sensor wire to the clamp on the bracket
- MAP sensor vacuum hose to the gas filter on the intake manifold
- Brake booster vacuum hose to the intake manifold
- A/C idle-up valve connector
- A/C idle-up valve hose to the intake manifold
- A/C idle-up valve hose to the air pipe
- DLC1 to the bracket
- Engine harness protector to the bracket
- Ground cable and the ground strap
- Heater hose to the water outlet and the heater hose to the water bypass pipe
- Fuel inlet hose to the fuel filter
- Fuel inlet hose with 2 new gaskets and the union bolt. Bolt: 22 ft. lbs. (30 Nm).
- Fuel return hose to the return pipe

and connect the EVAP hose to the charcoal canister
- Power steering air hoses to the intake manifold and the air pipe
- Radiator
- Actuator cable to the clamps, on models equipped with cruise control
- Accelerator cable to the throttle body, cable bracket, and the clamps
- Air cleaner
- Battery tray and battery
- Hood
- Negative battery cable

- Engine undercover
17. Refill the transaxle assembly with oil to the correct level.
18. Refill the cooling system to the correct level.
19. Refill the engine with oil to the correct level.
20. Start the vehicle, check for leaks and repair if necessary.

MR2

1. Before servicing the vehicle, refer to the Precautions Section.
2. Drain the cooling system.

3. Relieve the fuel system pressure.
4. Drain the engine oil
5. Drain the transaxle fluid
6. Remove or disconnect the following:
- Engine hood
- Rear suspension upper brace
- Engine undercovers
- Battery and tray
- Air intake assembly and intake air hose
- Accelerator cable
- Rear bumper cover
- Heated Oxygen (HO2S) sensor connectors

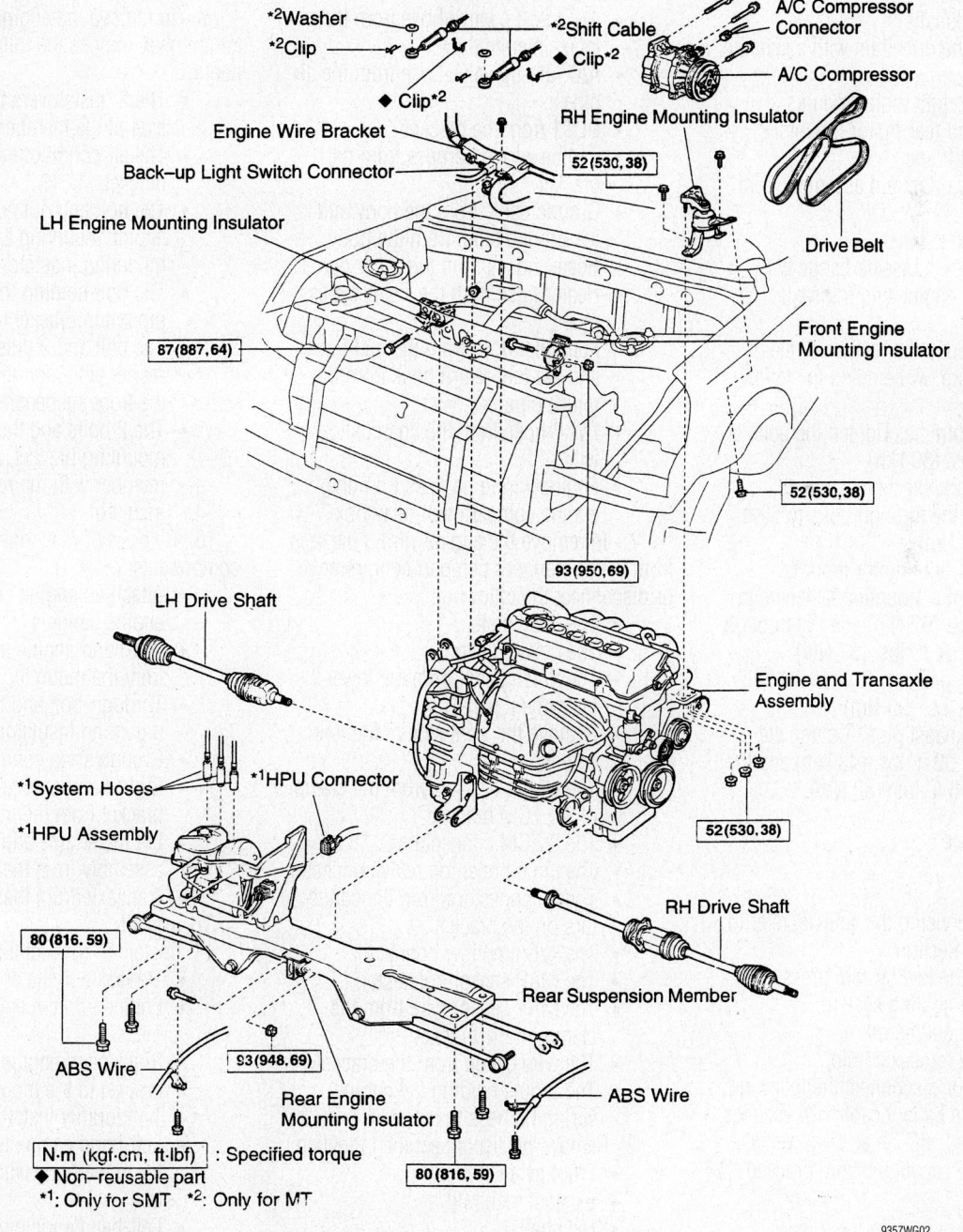

MR2 engine mounting exploded view

9357WG02

- Front exhaust pipe
- Accessory drive belt
- A/C compressor
- Engine wiring harness connectors in the luggage compartment. Pull the harness through the firewall.
- Transaxle control ECU harness connector
- Engine wire at Junction Box No. 1
- Engine ground wires
- Wire brackets
- Heater hoses
- Radiator hoses
- Fuel line
- Clutch hose
- Shift cables
- Axle halfshafts

7. Support the drivetrain with a suitable jack.
- Left and right motor mounts
- Front and rear motor mounts
- Rear subframe

8. Lower the drivetrain assembly from the vehicle.

9. Remove the starter.

10. Remove the transaxle flange bolts and separate the engine and transaxle.

To install:

11. Installation is the reverse of the removal procedure, while using the following torque values:
- Rear subframe. Tighten the bolts to 59 ft. lbs. (80 Nm).
- Front and rear motor mounts. Tighten the through bolts to 69 ft. lbs. (93 Nm).
- Left and right motor mounts. Tighten the mounting fasteners to 38 ft. lbs. (52 Nm) and the through bolts to 64 ft. lbs. (87 Nm).
- A/C compressor. Tighten the bolts to 18 ft. lbs. (25 Nm).
- Front exhaust pipe. Tighten the bolts to 32 ft. lbs. (43 Nm) and the nut to 46 ft. lbs. (62 Nm).

2ZZ-GE Engines

CELICA

1. Before servicing the vehicle, refer to the Precautions Section.
2. Release the fuel system pressure.
3. Drain the cooling system.
4. Drain the engine oil.
5. Drain the transaxle fluid.
6. Remove or disconnect the following:
- Negative battery cable. On vehicles equipped with an air bag, wait at least 90 seconds before proceeding.
- Battery
- Hood

- Undercover
- Accelerator cable, cable bracket, and clamps.
- Air cleaner
- ECM box
- Cruise control actuator cable
- Radiator
- MAP sensor vacuum hose from the gas filter on the intake manifold
- Power steering air hose from the intake manifold
- Power steering hose from the air the air pipe
- Brake booster vacuum hose from the intake manifold
- A/C idle-up valve
- A/C idle-up valve hose from the intake manifold
- A/C idle-up valve hose from the air pipe
- DLC1 from the bracket
- Engine wiring harness from the bracket
- Ground cable from the body and the ground strap from the body
- Heater hoses from the water outlet
- Heater hose from the water bypass pipe
- Fuel inlet hose from the fuel filter and the fuel return hose from the return pipe
- EVAP hose from the charcoal canister
- Engine wiring harness from the engine compartment relay box

7. To remove the engine wiring harness from the passenger's compartment, remove or disconnect the following:
- The scuff plate
- The cowl side trim
- The finish panel from the lower instrument panel
- Remove the front side of the floor carpet
- The wiring harness from the clamp of the ECM bracket
- The 3 ECM connectors
- The circuit opening relay connector
- The 3 connectors from the connectors on the bracket
- The A/C amplifier connector
- The MAP sensor connector
- The MAP sensor wire from the clamp on the bracket
- The wire clamp from the bracket
- The 2 nuts holding the engine wiring harness to the cowl

8. Remove or disconnect the following:
- Front exhaust pipe
- Exhaust manifold
- Halfshafts
- Alternator drive belt

- A/C drive belt, compressor connector, and compressor. Do not disconnect the A/C lines
- Remove the drive belt and remove the 4 bolts that secure the power steering pump. Without disconnecting the lines, securely hang the pump out of the way.
- A/C relay box
- On manual transmission, clutch release cylinder from the transaxle
- Transaxle control cable(s)
- On automatic transmission, transaxle control cable from the engine mounting center member.
- Exhaust pipe support bracket

9. To remove the engine mounting center member, remove the following components:
- The 2 dust covers from the rear side of the member
- The air conditioning pipe from the bracket
- The bolt and nut holding the front engine mounting bracket to the mounting insulator
- The bolt holding the rear engine mounting bracket to the insulator
- The bolt and 2 nuts holding the rear engine mounting insulator to the front suspension member
- The 2 bolts and the rear engine mounting bracket, and the center member with the rear mounting insulator

10. Remove or disconnect the remaining components:
- Attach an engine chain hoist to the engine hangers
- Left-hand engine mounting bracket from the mounting insulator
- Through-bolt and the left-hand mounting insulator
- Ground strap connector
- Right-hand engine mounting bracket from the mounting insulator
- Lift the engine and transaxle assembly from the vehicle
- Transaxle from the engine assembly

To install:

11. Install or connect the following:
- Transaxle to the engine assembly
- Engine into the engine compartment
- Right-hand engine mounting bracket to the mounting insulator. Temporarily install the 3 nuts.
- Left-hand engine mounting insulator to the body with the through-bolt
- Left-hand engine mounting bracket to the mounting insulator and

install the 2 bolts and nut. Bolts and nut: 47 ft. lbs. (64 Nm).
- Left-hand engine mounting through-bolt to the body. Bolt: 54 ft. lbs. (73 Nm).
- Right-hand mounting bracket to the insulator. 12mm nut: 21 ft. lbs. (28 Nm), 14mm nut: 38 ft. lbs. (52 Nm).
- Engine ground strap connector and remove the engine hoist

12. To install the engine mounting center member, perform the following:

a. Attach the center member together with the rear engine mounting insulator to the front suspension member.

b. Temporarily install the 2 bolts and nut holding the center member to the body.

c. Install the rear engine mounting bracket. Bolts: 58 ft. lbs. (78 Nm).

d. Temporarily install the bolt and 2 nuts holding the rear engine mounting insulator to the front suspension member.

e. Temporarily install the bolt holding the rear engine mounting bracket to the insulator.

f. Temporarily install the bolt and nut holding the front engine mounting bracket to the insulator.

g. Tighten the 2 bolts holding the center member to the body to 26 ft. lbs. (35 Nm).

h. Tighten the bolt and 2 nuts holding the rear mounting insulator to the front suspension member to 59 ft. lbs. (80 Nm).

i. Tighten the bolt holding the rear engine mounting bracket to the insulator to 65 ft. lbs. (88 Nm).

j. Tighten the bolt and nut holding the front engine mounting bracket to the insulator to 65 ft. lbs. (88 Nm).

k. Install the A/C pipe to the bracket and install the 2 dust covers to the center member.

13. Install or connect the following:
- Exhaust manifold
- Exhaust pipe support bracket. Bolts to 14 ft. lbs. (19 Nm).
- Transaxle control cable(s)
- On automatic transmission, transaxle control cable to the engine mounting center member.
- On manual transmission, clutch release cylinder. Bolts: 108 inch lbs. (12 Nm), then attach the bracket with the bolt
- A/C relay box to the body
- Power steering pump. 12mm bolts: 14 ft. lbs. (19 Nm). 14mm bolts: 29

ft. lbs. (39 Nm). Install the drive belt. Adjusting bolt: 29 ft. lbs. (39 Nm).
- A/C compressor. Bolts: 18 ft. lbs. (25 Nm).
- A/C drive belt with the adjusting bolt and tighten the idler pulley locknut to 29 ft. lbs. (39 Nm). Connect the connector.
- Alternator drive belt
- Halfshafts
- Front exhaust pipe

14. To install the engine wiring harness in the passenger compartment, perform the following:
- Push the harness through the cowl panel, install the retainer to the cowl with the 2 nuts and install the wire clamp to the bracket
- Connect the harness to the clamp on the ECM
- Connect the 3 ECM connectors and the circuit opening relay connector
- Connect the 3 connectors to the connectors on the bracket
- A/C amplifier connector
- Install the floor carpet, the lower instrument panel finish panel, the cowl side trim panel, and the scuff plate

15. Install or connect the following:
- Engine wiring harness with the 2 connectors to the engine compartment relay box and install the relay box covers.
- MAP sensor connector
- MAP sensor wire to the clamp on the bracket
- MAP sensor vacuum hose to the gas filter on the intake manifold
- Brake booster vacuum hose to the intake manifold
- A/C idle-up valve connector
- A/C idle-up valve hose to the intake manifold
- A/C idle-up valve hose to the air pipe
- DLC1 to the bracket
- Engine harness protector to the bracket
- Ground cable and the ground strap
- Heater hose to the water outlet and the heater hose to the water bypass pipe
- Fuel inlet hose to the fuel filter
- Fuel inlet hose with 2 new gaskets and the union bolt. Bolt: 22 ft. lbs. (30 Nm).
- Fuel return hose to the return pipe and connect the EVAP hose to the charcoal canister
- Power steering air hoses to the intake manifold and the air pipe.

- Radiator
- On models equipped with cruise control, install the actuator cable to the clamps.
- Accelerator cable to the throttle body, cable bracket, and the clamps.
- Air cleaner
- Battery tray and battery
- Hood
- Negative battery cable
- Engine undercover

16. Refill the transaxle assembly with oil to the correct level.

17. Refill the cooling system to the correct level.

18. Refill the engine with oil to the correct level.

19. Start the vehicle, check for leaks and repair if necessary.

2005–06 COROLLA

1. Before servicing the vehicle, refer to the Precautions Section.
2. Release the fuel system pressure.
3. Drain the engine oil.
4. Drain the cooling system.
5. Drain the transaxle fluid.
6. Remove or disconnect the following:
- Negative battery cable
- Front wheels
- Engine undercovers
- Engine appearance cover
- Battery and battery carrier
- Accelerator cable
- Air intake assembly
- Cruise control actuator assembly
- Union to check valve hose
- Fuel supply hose
- Heater hoses
- Radiator hoses
- Clutch release cylinder assembly, if equipped with manual transmission
- Transmission control cables
- Floor shift cable transmission control selector
- Engine wiring harness from the relay block
- Engine wiring from the ECM
- Air switching valve assembly
- Accessory drive belt
- Alternator
- A/C compressor
- Power steering return hose
- Power steering reservoir
- Shift lever knob
- Upper console panel
- Parking brake hole cover
- Console box carpet
- Rear console box
- Front exhaust assembly
- Steering intermediate shaft

➡ **Fix the steering wheel with the seatbelt to prevent rotation.**

- Stabilizer links
- Wheel speed sensors
- Tie rod ends
- Lower control arms
- Halfshafts

7. Secure the engine assembly with a suitable lifter.

8. Remove the bolts from the right and left motor mounts.

9. Remove the six bolts from the suspension crossmember shown in the illustration.

10. Remove the engine/transaxle assembly from the vehicle.

11. Separate the engine from the transaxle assembly.

To install:

12. Install the transaxle assembly to the engine.

13. Install the engine/transaxle assembly into the vehicle.

14. Temporarily install the suspension crossmember bolts.

15. Install the engine mount bolts. Tighten the left side to 59 ft. lbs. (80 Nm), right side to 38 ft. lbs. (52 Nm).

16. Using Special Tool 09670-00010 in the positioning holes of the crossmember, tighten the crossmember as follows:

    a. Bolt A: 116 ft. lbs. (157 Nm).

    b. Bolt B: 83 ft. lbs. (113 Nm).

    c. Front two mounting bolts: 44 ft. lbs. (60 Nm)

17. The remainder of the installation is the reverse order of removal.

18. Refill the transaxle with oil to the correct level.

19. Refill the cooling system to the correct level.

20. Refill the engine with oil to the correct level.

21. Start the engine and check for leaks.

22. Check and adjust the alignment, if needed.

Water Pump

REMOVAL & INSTALLATION

1NZ-FE Engines

1. Before servicing the vehicle, refer to the Precautions Section.

2. Drain the cooling system.

3. Remove or disconnect the following:

- Negative battery cable
- Accessory drive belt
- Water pump pulley, using Special Tool 09960-10010 or equivalent

- Water pump and gasket (3 bolts; 2 nuts)

To install:

4. Install or connect:

- Water pump, with a new gasket. Torque the nuts and bolts to 96 inch lbs. (11 Nm).
- Pulley. Torque the bolts to 11 ft. lbs. (15 Nm).
- Accessory drive belt
- Negative battery cable

5. Refill the cooling system to the proper level.

6. Start the vehicle, check for leaks and repair if necessary.

1ZZ-FE Engines

1. Before servicing the vehicle, refer to the Precautions Section.

2. Drain the cooling system.

3. Remove or disconnect the following:

- Negative battery cable
- Right-hand engine under cover
- Accessory drive belt
- Water pump

To install:

4. Install or connect the following:

- Water pump. Bolts marked **A** (short): 80 inch lbs. (9 Nm). Bolts marked **B** (long): 96 inch lbs. (11 Nm).
- Accessory drive belt
- Right engine under cover
- Negative battery cable

5. Refill the cooling system to the proper level.

6. Start the vehicle, check for leaks and repair if necessary.

2ZZ-GE Engines

CELICA

1. Before servicing the vehicle, refer to the Precautions Section.

2. Drain the cooling system.

3. Remove or disconnect the following:

- Negative battery cable
- Right-hand engine under cover
- Accessory drive belt
- Water pump pulley
- Water pump and O-ring

To install:

4. Install or connect the following:

- Water pump with new O-ring. Bolts A: 80 inch lbs. (9 Nm), Bolts B: 8 ft. lbs. (11 Nm).
- Water pump pulley. Bolts: 11 ft. lbs. (15 Nm).
- Accessory drive belt
- Right engine under cover
- Negative battery cable

5. Refill the cooling system to the proper level.

6. Start the vehicle, check for leaks and repair if necessary.

2005–06 COROLLA

1. Before servicing the vehicle, refer to the Precautions Section.

2. Drain the cooling system.

3. Remove or disconnect the following:

- Negative battery cable
- Right-hand engine under cover
- Engine appearance cover
- Refrigerant line
- Air switching valve assembly
- Accessory drive belt
- Alternator

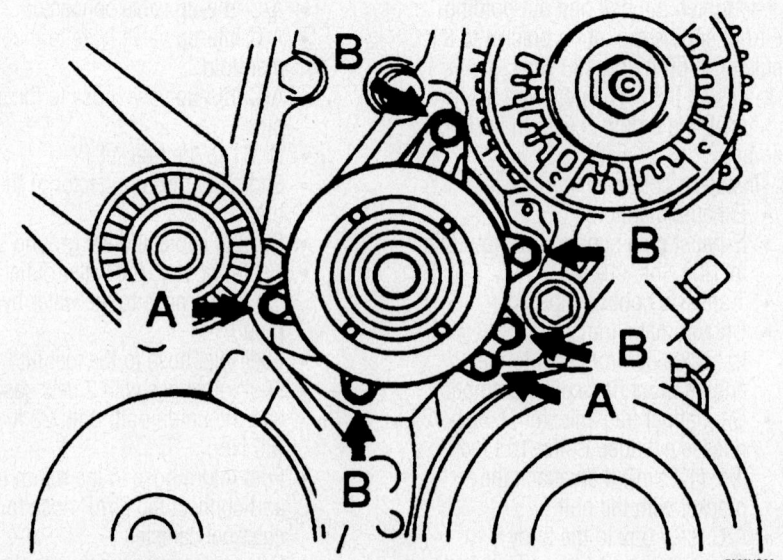

Water pump bolt identification—1.8L (1ZZ-FE) engine

7923VG06

- Water pump pulley
- Water pump

To install:

4. Install or connect the following:
   - Water pump with a new O-ring. Tighten all 6 bolts to 80 inch lbs. (9 Nm).
   - Water pump pulley. Tighten the bolts to 11 ft. lbs. (15 Nm).
   - Alternator

- Accessory drive belt
- Air switching valve assembly
- Refrigerant line
- Engine appearance cover
- Engine undercover
- Negative battery cable

5. Refill the cooling system to the correct level.
6. Start the engine and check for leaks.

Heater Core

REMOVAL & INSTALLATION

Echo

1. Before servicing the vehicle, refer to the Precautions Section.
2. Drain the cooling system.

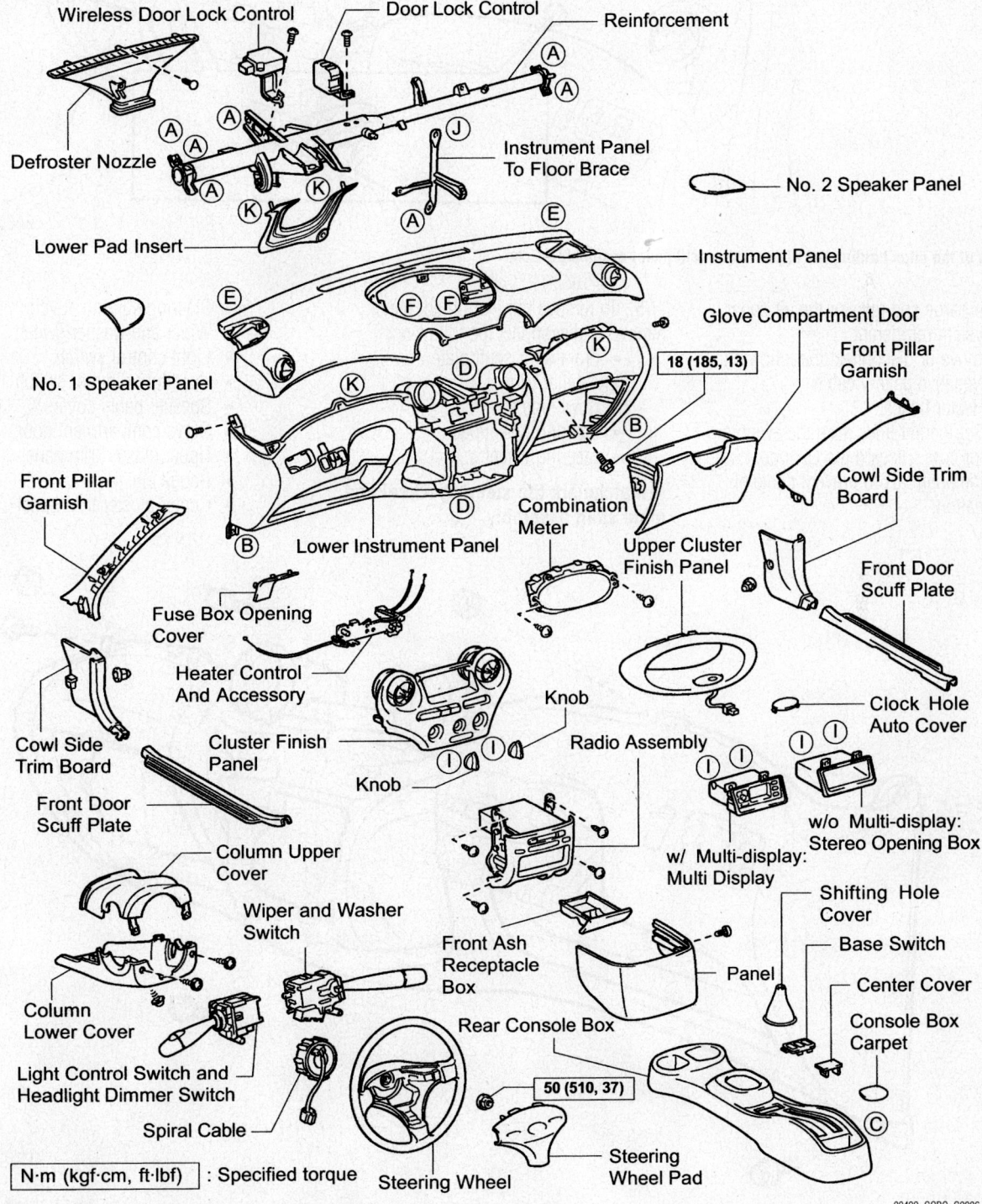

N·m (kgf·cm, ft·lbf) : Specified torque

Exploded view of the instrument panel components—Echo

09490_CORO_G0006

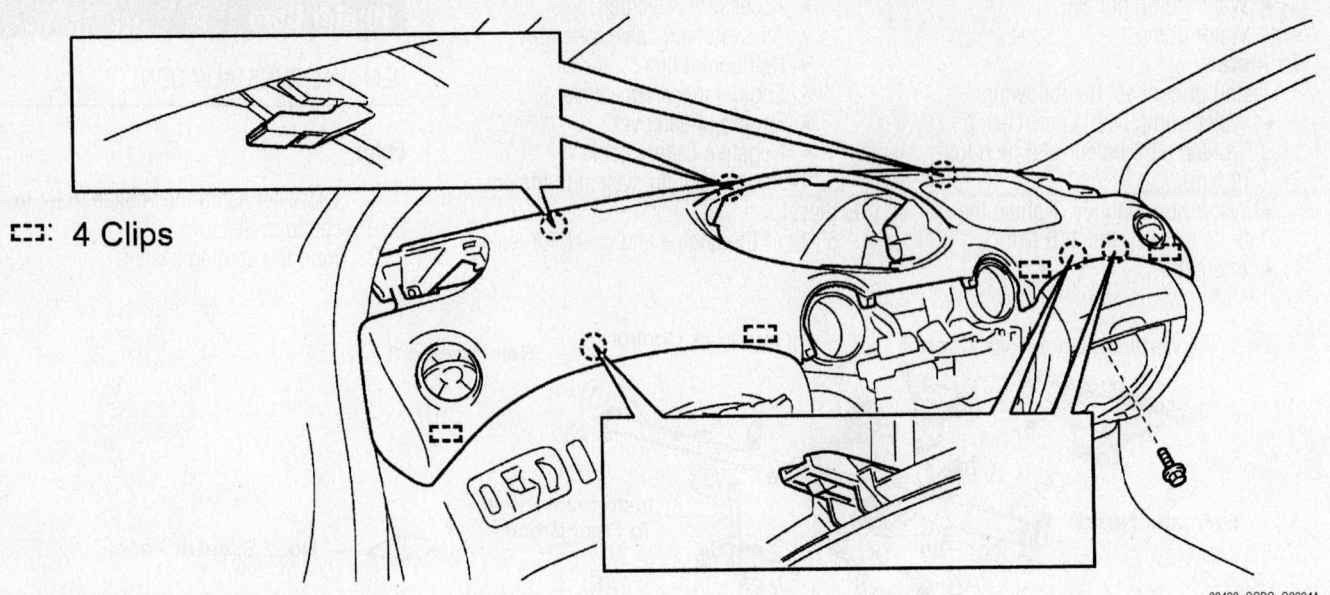

Location of the clips holding the upper instrument panel assembly—Echo

3. Discharge and recover the air conditioning system refrigerant.
4. Remove or disconnect the following:
   - Negative battery cable
   - Heater hoses
   - Refrigerant lines from the evaporator core. Discard the O-rings. Plug the openings to prevent contamination.

5. To remove the instrument panel, remove or disconnect the following:
   - Front door scuff plate
   - A-pillar trim
   - Cowl side trim
   - Steering wheel pad
   - Steering wheel

➡**Matchmark the steering wheel and main shaft assembly.**

   - Steering column covers
   - Wiper and washer switch
   - Light control switch
   - Headlight dimmer switch
   - Speaker panel covers
   - Glove compartment door
   - Upper cluster trim panel
   - Gauge cluster
   - Center cluster trim panel

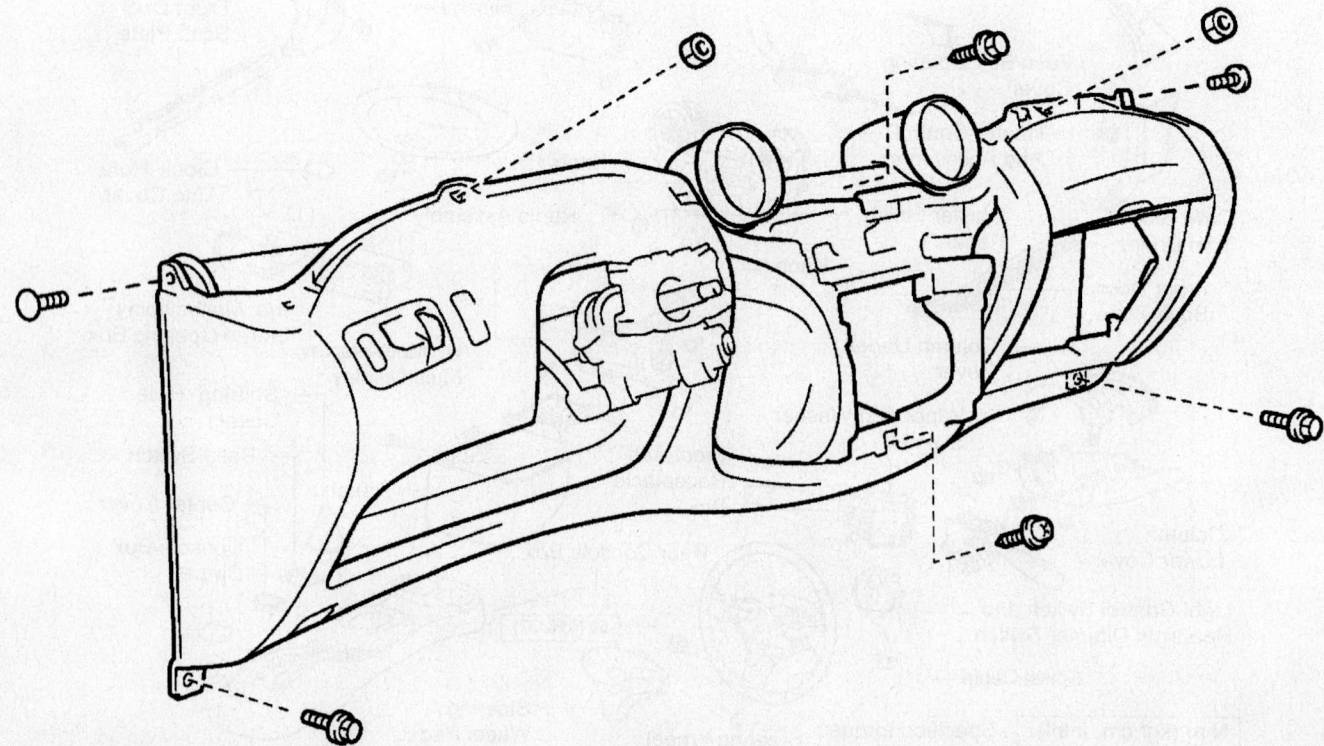

Location of the bolts holding the lower instrument panel assembly—Echo

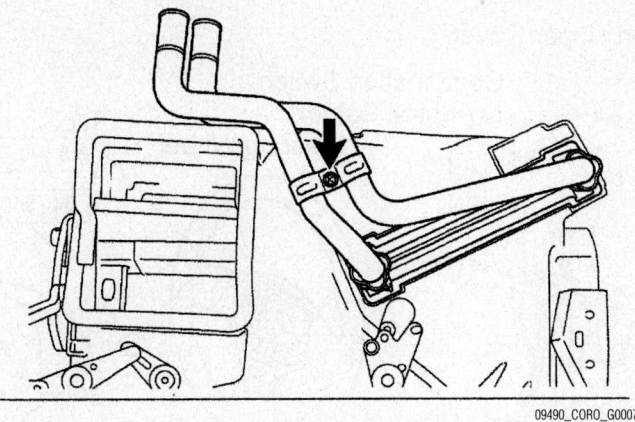

09490_CORO_G0007

Disconnect the pipe connector clamp to remove the heater core—Echo

- Bolt holding the passenger airbag assembly to the body
- Passenger airbag electrical connector
- Upper instrument panel
- Defroster nozzle
- Front ash try receptacle box
- Heater control unit
- Radio unit
- Hood release lever
- Lower instrument panel
- Lower pad insert
- Steering column
- Rear console box
- Instrument panel-to-floor brace
- Instrument panel reinforcement

6. Remove or disconnect the following:
- Engine ECU
- Air conditioning unit
- Heater unit

7. Disconnect the pipe connector clamp.

8. Remove the heater core.

9. Installation is the reverse order of removal.

Celica

1. Before servicing the vehicle, refer to the Precautions Section.

2. Disconnect the negative battery cable.

3. Drain the cooling system into a clean container for reuse.

4. Discharge and recover the air conditioning system refrigerant.

5. Remove the evaporator housing by performing the following procedure:
- Disconnect the refrigerant lines from the evaporator core. Discard the O-rings. Plug the openings to prevent contamination.
- Remove the 2 grommets and the drain pipe grommet.
- At the passenger's side, remove the lower finish panel.
- Disconnect the evaporator housing's electrical connector.
- Remove the evaporator housing-to-chassis 3 nuts and 4 bolts.
- Remove the evaporator housing.

6. Remove the heater valve-to-cowl bolt.

7. Remove the heater hoses from the heater core pipes.

8. Remove the heater pipe grommets.

9. Remove the air bag module and the steering wheel by performing the following procedure:
- Place the front wheels in the straight-ahead position.
- At both sides of the steering wheel, remove the screw covers.
- Using a Torx socket, loosen the 2 air bag module-to-steering wheel Torx screws until the circumference ring catches on the screw case.
- Carefully, remove the air bag module and disconnect the electrical connector.
- Remove the steering wheel nut.
- Using a steering wheel puller, press the steering wheel from the steering column.

10. Remove the instrument panel and reinforcement by performing the following procedure:
- Remove the front pillar lower garnishes and the pillar garnishes.
- Remove the front door scuff plates.
- Remove the cowl side trim boards.
- Remove the steering column covers.
- Pry out the upper console panel by disengaging the 4 clips.
- Remove the console box.
- Remove the No. 1 finish panel.
- Remove the finish panel and the combination switch.
- Remove the lower cluster finish panel and the cluster finish panel.
- Remove the No. 1 register and the combination meter.
- Remove the 2 center cluster finish panel screws; then, pry out the finish panel.
- Remove the air conditioning control assembly.
- Remove the lower finish panel.
- Remove the glove compartment door finish plate located inside the lower finish panel.
- Pull up the air bag electrical connector and disconnect it.
- Remove the 5 lower finish panel screws and the panel.
- Remove the 4 bolts and lower pad inserts.
- Remove the 4 lower center finish panel screws and the panel.
- Remove the No. 2 side defroster nozzle.
- Disconnect the instrument panel electrical connectors.
- Remove the instrument panel 9 bolts and 2 nuts.
- Remove the instrument panel.
- Remove the No. 1 center console bracket bolt and 2 nuts.
- Remove the No. 2 center console bracket nut and bolt.
- Remove the brake spring.
- Remove the center console bracket support 6 bolts and 2 nuts.
- Remove the center console bracket.

11. Disconnect the connector and the electrical connector from the heater housing.

12. Remove the 4 heater housing-to-chassis nuts and heater housing.

13. Remove the 2 air outlet damper control servo motor screws and the servo motor.

14. Remove the air vent duct.

15. Remove the 2 heater air duct screws, the 2 clips and the duct.

16. Remove the 3 heater pipe clamp screws and the clamps.

17. Remove the heater core from the heater housing.

To install:

18. Install the heater core to the heater housing.

19. Install the heater pipe clamp and the 3 clamp screws.

20. Install the heater air duct, the 2 clips and the 2 duct screws.

21. Install the air vent duct.

22. Install the air outlet damper control servo motor and the 2 servo motor screws.

23. Install the heater housing and 4 heater housing-to-chassis nuts.

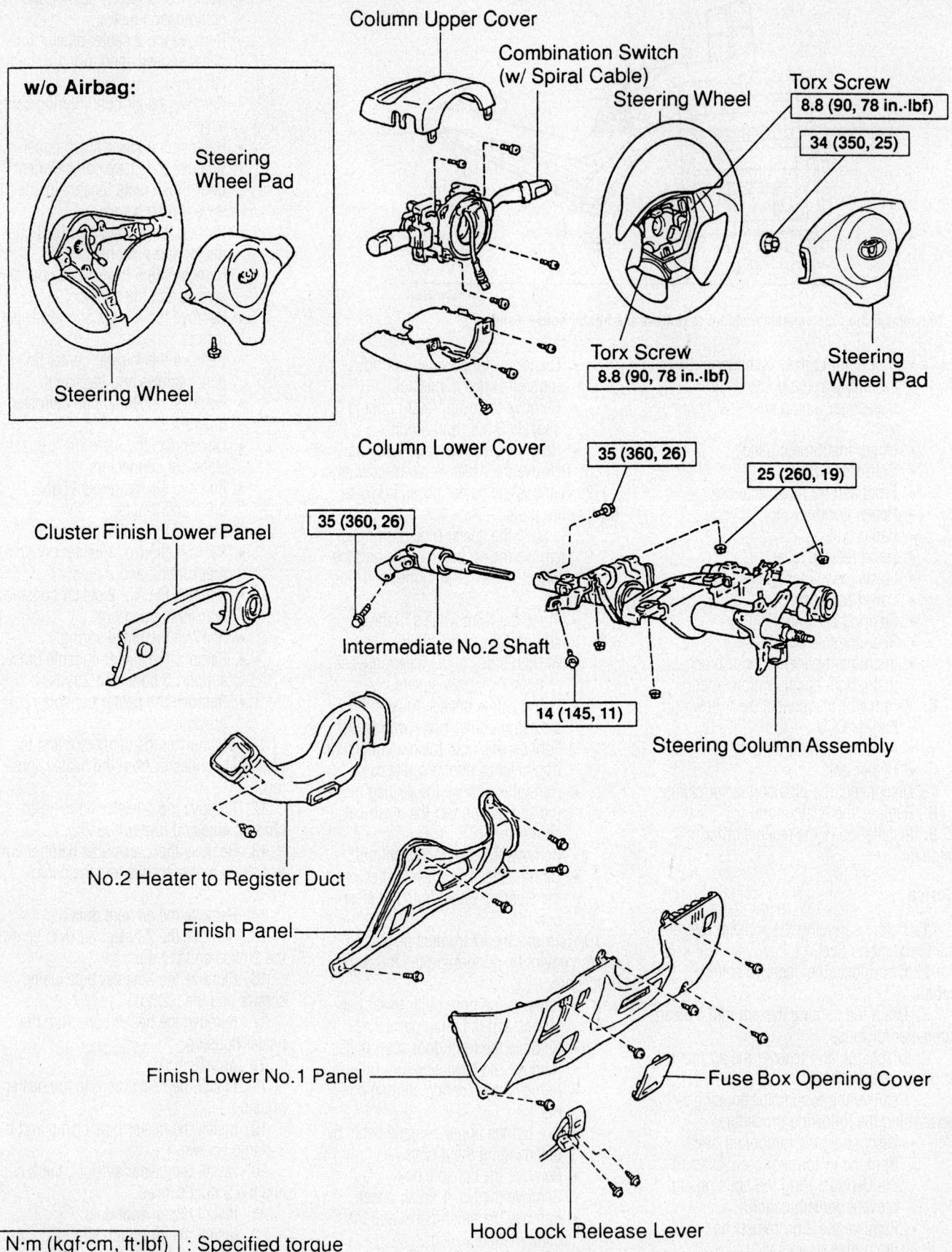

w/o Airbag:

Steering Wheel Pad

Steering Wheel

Column Upper Cover

Combination Switch (w/ Spiral Cable)

Steering Wheel

Torx Screw
8.8 (90, 78 in.·lbf)

34 (350, 25)

Torx Screw
8.8 (90, 78 in.·lbf)

Steering Wheel Pad

Column Lower Cover

Cluster Finish Lower Panel

35 (360, 26)

35 (360, 26)

25 (260, 19)

Intermediate No.2 Shaft

14 (145, 11)

Steering Column Assembly

No.2 Heater to Register Duct

Finish Panel

Finish Lower No.1 Panel

Fuse Box Opening Cover

Hood Lock Release Lever

N·m (kgf·cm, ft·lbf) : Specified torque

93112GK8

Exploded view of the air bag module, steering wheel and related components (typical)—Celica

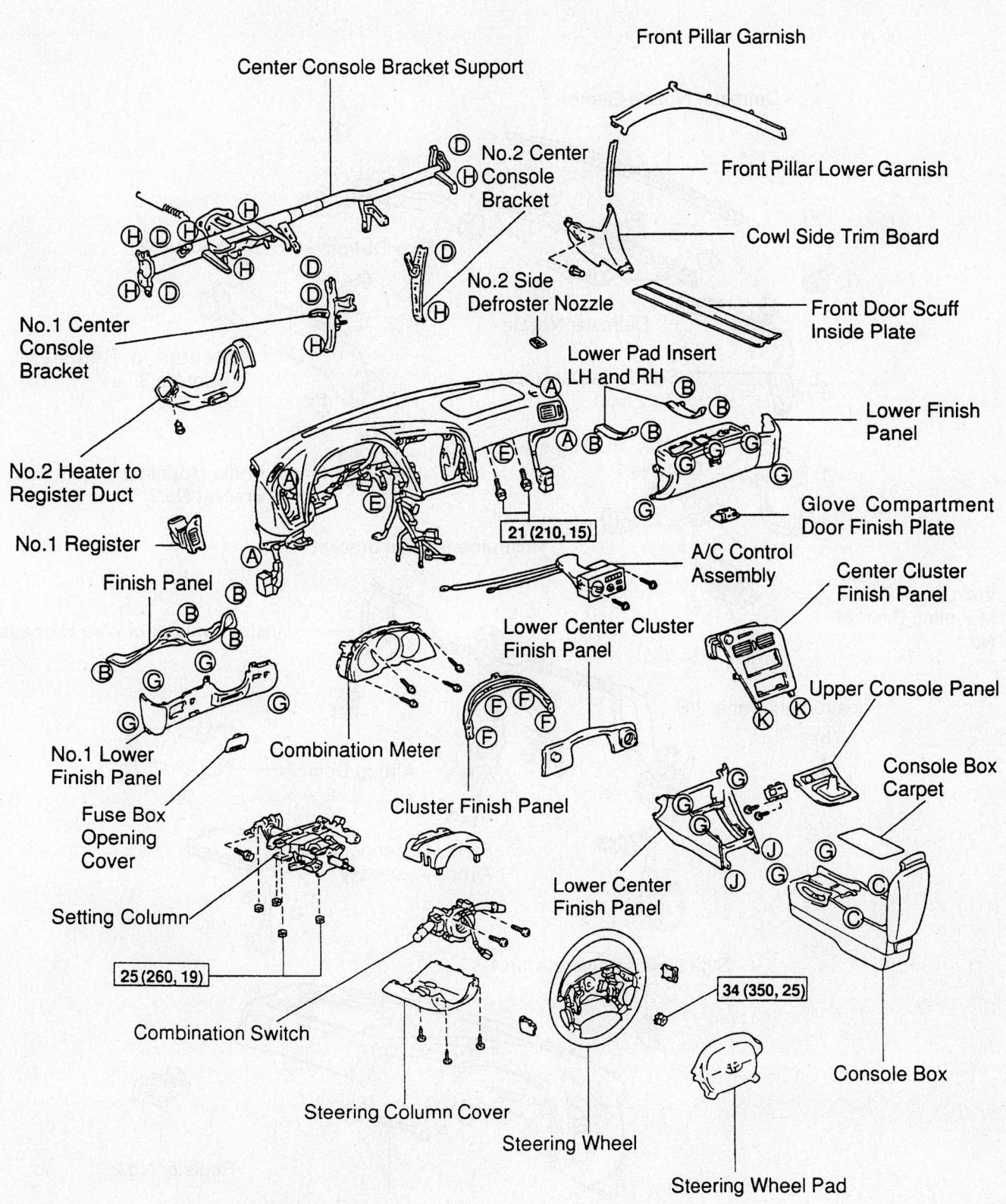

N·m (kgf·cm, ft·lbf) : Specified torque

93112GK9

Exploded view of the instrument panel and related components—Celica

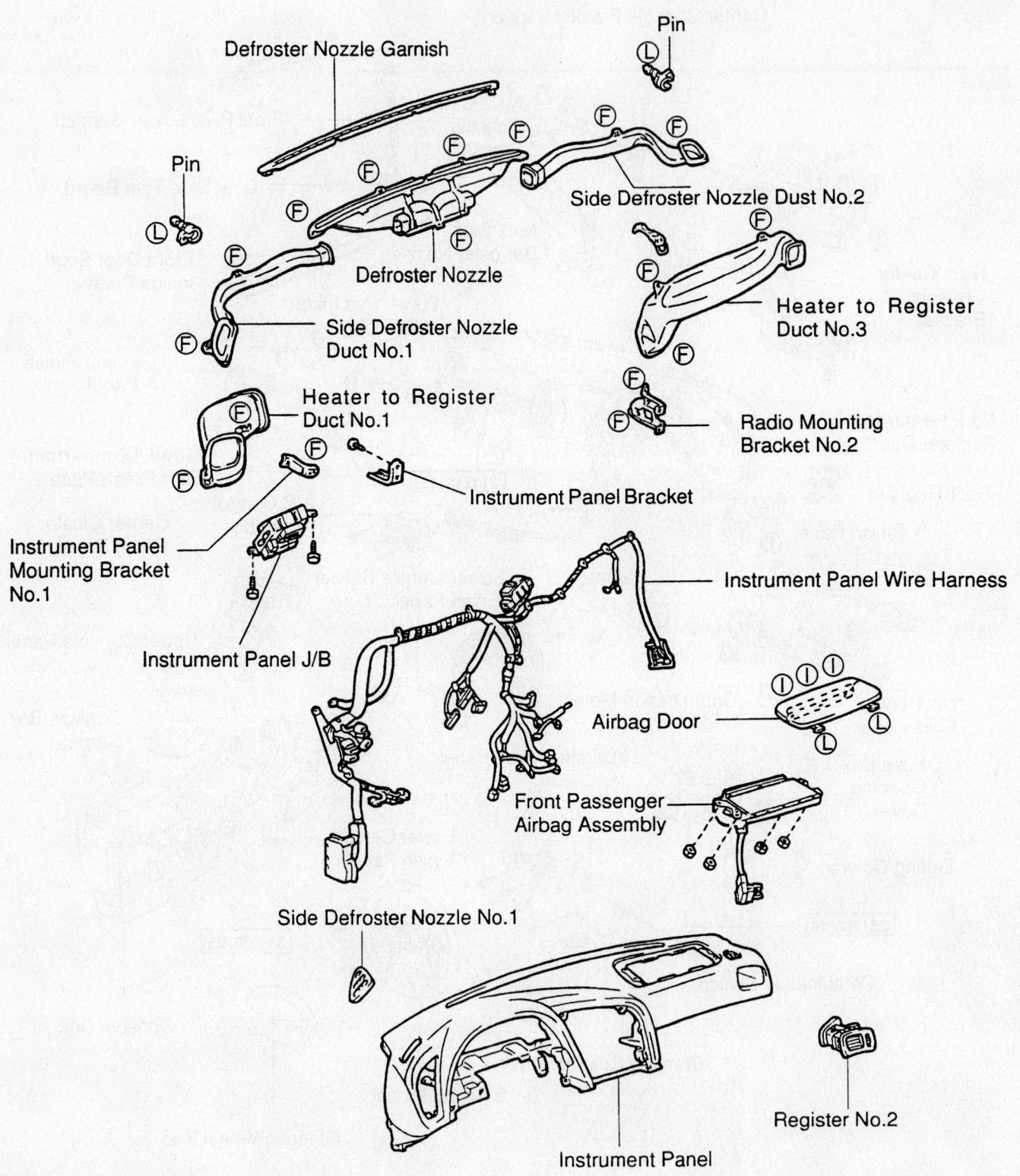

Defroster Nozzle Garnish

Pin

Pin

Side Defroster Nozzle Dust No.2

Defroster Nozzle

Side Defroster Nozzle
Duct No.1

Heater to Register
Duct No.3

Heater to Register
Duct No.1

Radio Mounting
Bracket No.2

Instrument Panel Bracket

Instrument Panel
Mounting Bracket
No.1

Instrument Panel J/B

Instrument Panel Wire Harness

Airbag Door

Front Passenger
Airbag Assembly

Side Defroster Nozzle No.1

Instrument Panel

Register No.2

93112GK0

Exploded view of the ventilation system, wiring harness and related components—Celica

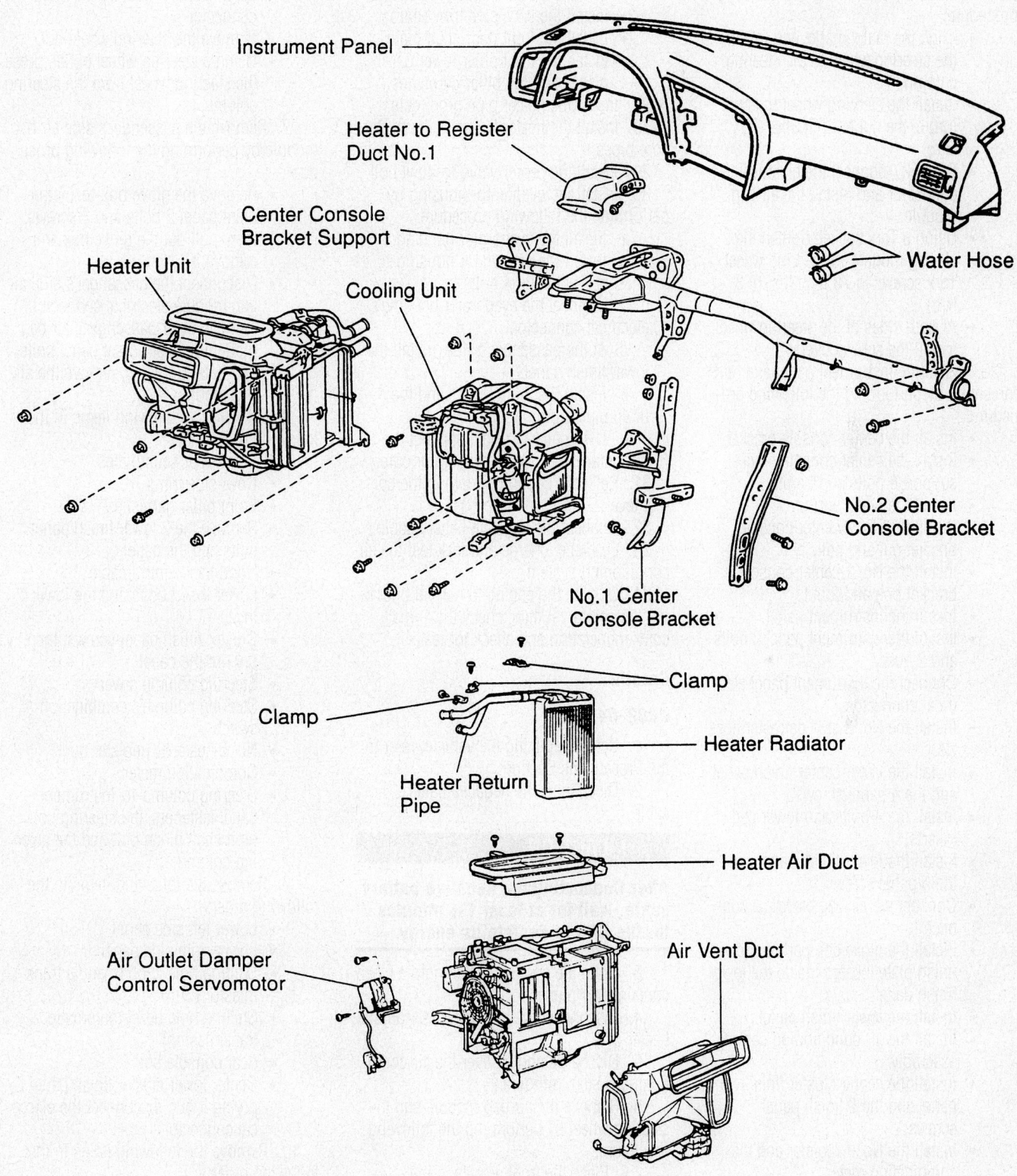

Instrument Panel

Heater to Register
Duct No.1

Center Console
Bracket Support

Heater Unit

Cooling Unit

Water Hose

No.2 Center
Console Bracket

No.1 Center
Console Bracket

Clamp

Clamp

Heater Return
Pipe

Heater Radiator

Heater Air Duct

Air Outlet Damper
Control Servomotor

Air Vent Duct

93112GL1

Exploded view of the heater core, heater housing, evaporator housing and related components—Celica

24. Connect the connector and the electrical connector to the heater housing.

25. Install the air bag module and the steering wheel by performing the following procedure:
- Align the matchmarks and install the steering wheel to the steering column.
- Install the steering wheel nut and torque the nut to 26 ft. lbs. (35 Nm).
- Carefully, connect the electrical connector and install the air bag module.
- Using a Torx socket, tighten the 2 air bag module-to-steering wheel Torx screws to 78 inch lbs. (8.8 Nm).
- At both sides of the steering wheel, install the screw covers.

26. Install the instrument panel and reinforcement by performing the following procedure:
- Install the center console bracket.
- Install the center console bracket support 6 bolts and 2 nuts.
- Install the brake spring.
- Install the No. 2 center console bracket nut and bolt.
- Install the No. 1 center console bracket bolt and 2 nuts.
- Install the instrument panel.
- Install the instrument panel 9 bolts and 2 nuts.
- Connect the instrument panel electrical connectors.
- Install the No. 2 side defroster nozzle.
- Install the lower center finish panel and the 4 panel screws.
- Install the 4 bolts and lower pad inserts.
- Install the lower finish panel and the 5 panel screws.
- Connect the air bag electrical connector.
- Install the glove compartment door finish plate located inside the lower finish panel.
- Install the lower finish panel.
- Install the air conditioning control assembly.
- Install the center cluster finish panel and the 2 finish panel screws.
- Install the No. 1 register and the combination meter.
- Install the lower cluster finish panel and the cluster finish panel.
- Install the finish panel and the combination switch.
- Install the No. 1 finish panel.

- Install the console box.
- Pry out the upper console panel by engaging the 4 clips.
- Install the steering column covers.
- Install the cowl side trim boards.
- Install the front door scuff plates.
- Install the front pillar lower garnishes and the pillar garnishes.

27. Install the heater pipe grommets.

28. Install the heater hoses to the heater core pipes.

29. Install the heater valve-to-cowl bolt.

30. Install the evaporator housing by performing the following procedure:
   a. Install the evaporator housing.
   b. Install the evaporator housing-to-chassis 3 nuts and 4 bolts.
   c. Connect the evaporator housing's electrical connector.
   d. At the passenger's side, install the lower finish panel.
   e. Install the 2 grommets and the drain pipe grommet.
   f. Using new O-rings, connect the refrigerant lines to the evaporator core.

31. Refill the cooling system to the correct level.

32. Connect the negative battery cable.

33. Evacuate, charge and leak test the air conditioning system.

34. Operate the engine to normal operating temperatures; then, check the climate control operation and check for leaks.

Corolla

2002-04

1. Before servicing the vehicle, refer to the Precautions Section.

2. Disconnect the negative battery cable.

❋❋ CAUTION

After Connecting the negative battery cable, wait for at least 1½ minutes for the SRS to deplete its energy.

3. Drain the cooling system into a clean container for reuse.

4. Disconnect the heater hoses from the heater core.

5. Discharge and recover the air conditioning system refrigerant.

6. Remove the air bag module and the steering wheel by performing the following procedure:
- Place the front wheels in the straight-ahead position.
- At both sides of the steering wheel, remove the screw covers.
- Using a Torx socket, loosen the 2 air bag module-to-steering wheel

Torx screws until the circumference ring catches on the screw case.
- Carefully, remove the air bag module and disconnect the electrical connector.
- Remove the steering wheel nut.
- Using a steering wheel puller, press the steering wheel from the steering column.

7. Remove the passenger's side air bag module by performing the following procedure:
- Remove the glove box-to-instrument panel 2 bolts and 3 screws; then, pull out the glove box and remove it.
- Disconnect the passenger's side air bag module electrical connector.
- Remove the 2 passenger's air bag module-to-instrument panel bolts and nuts. Carefully, remove the air bag module.

8. Remove the following items in the following order:
- Front door scuff plates
- Cowl side trims
- Front pillar garnishes
- Remove the 2 lower finish panel bolts and the panel
- Hood lock control cable
- Lower insert bolts and the lower insert
- Cluster finish panel screws; then, pry out the panel
- Steering column cover
- Steering column's combination switch
- No. 2 heater-to-register duct
- Combination meter
- Steering column-to-instrument panel fasteners, the steering column shaft pinch bolt and the steering column

9. Remove the following items in the following order:
- Lower left side panel
- Lower right side panel
- Shifting hole knob (manual transmission)
- Shifting hole bezel (automatic transmission)
- Rear console box
- Center lower cluster finish panel by prying it out, disconnect the electrical connector

10. Remove the following items in the following order:
- Stereo opening cover
- Lower center finish panel
- Air conditioning control panel
- Center cluster finish panel by prying it out

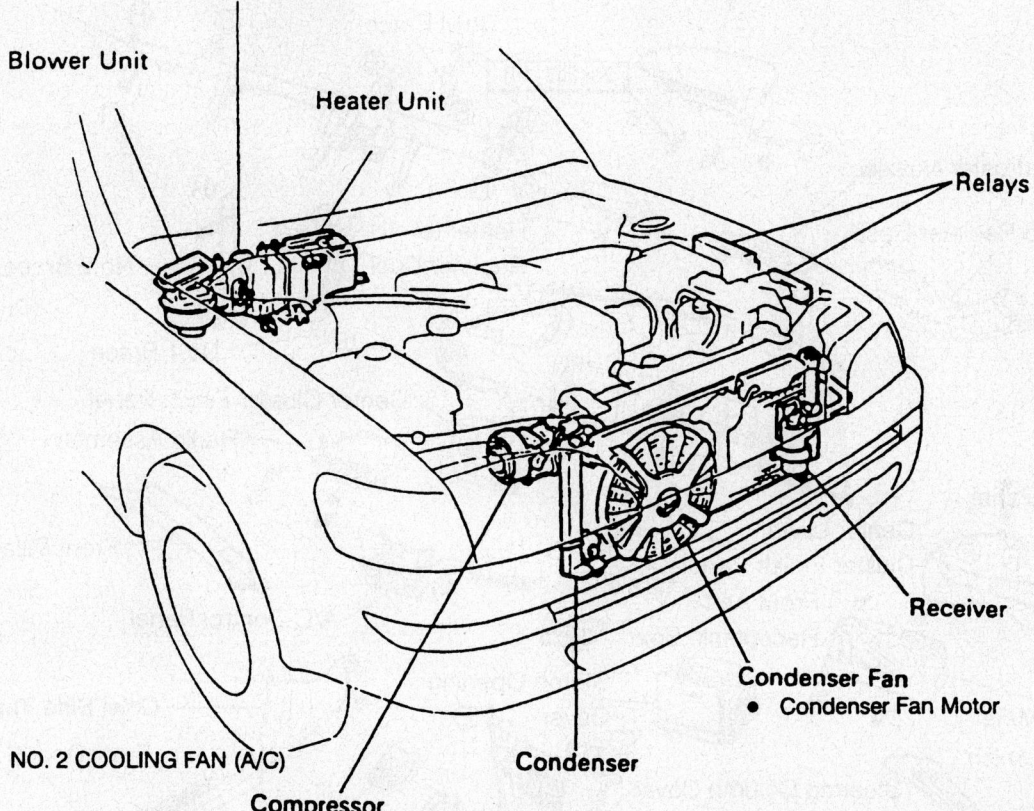

Cooling Unit
- Expansion Valve
- Evaporator
- Blower Resistor
- Thermistor

Blower Unit

Heater Unit

Relays

Receiver

Condenser Fan
- Condenser Fan Motor

NO. 2 COOLING FAN (A/C)

Condenser

Compressor
- Magnetic Clutch
- Refrigerant Temperature Switch

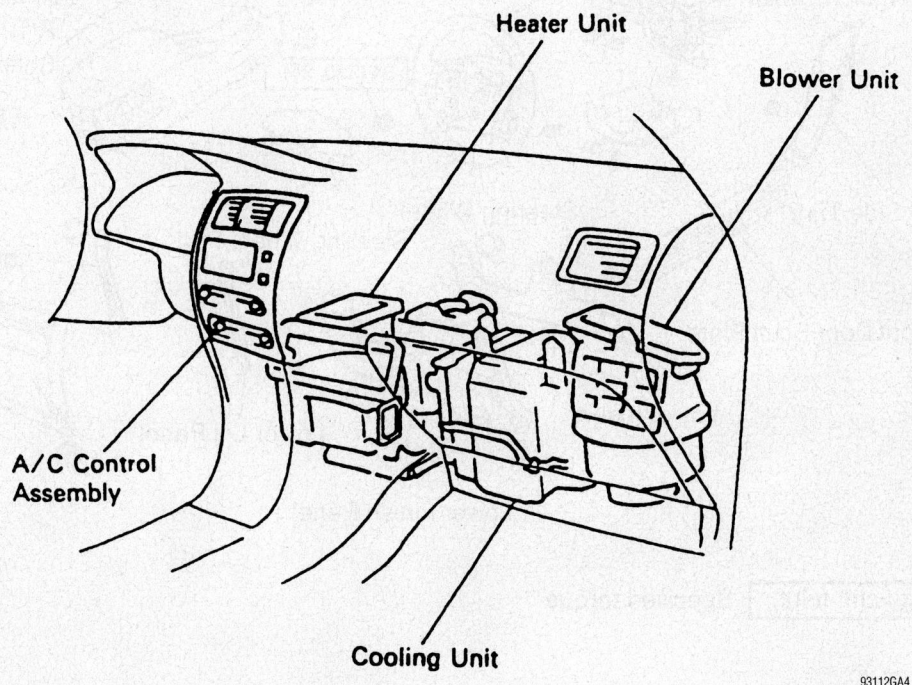

Heater Unit

Blower Unit

A/C Control
Assembly

Cooling Unit

93112GA4

View of the heater/air conditioning assembly and related components—2002–04 Corolla

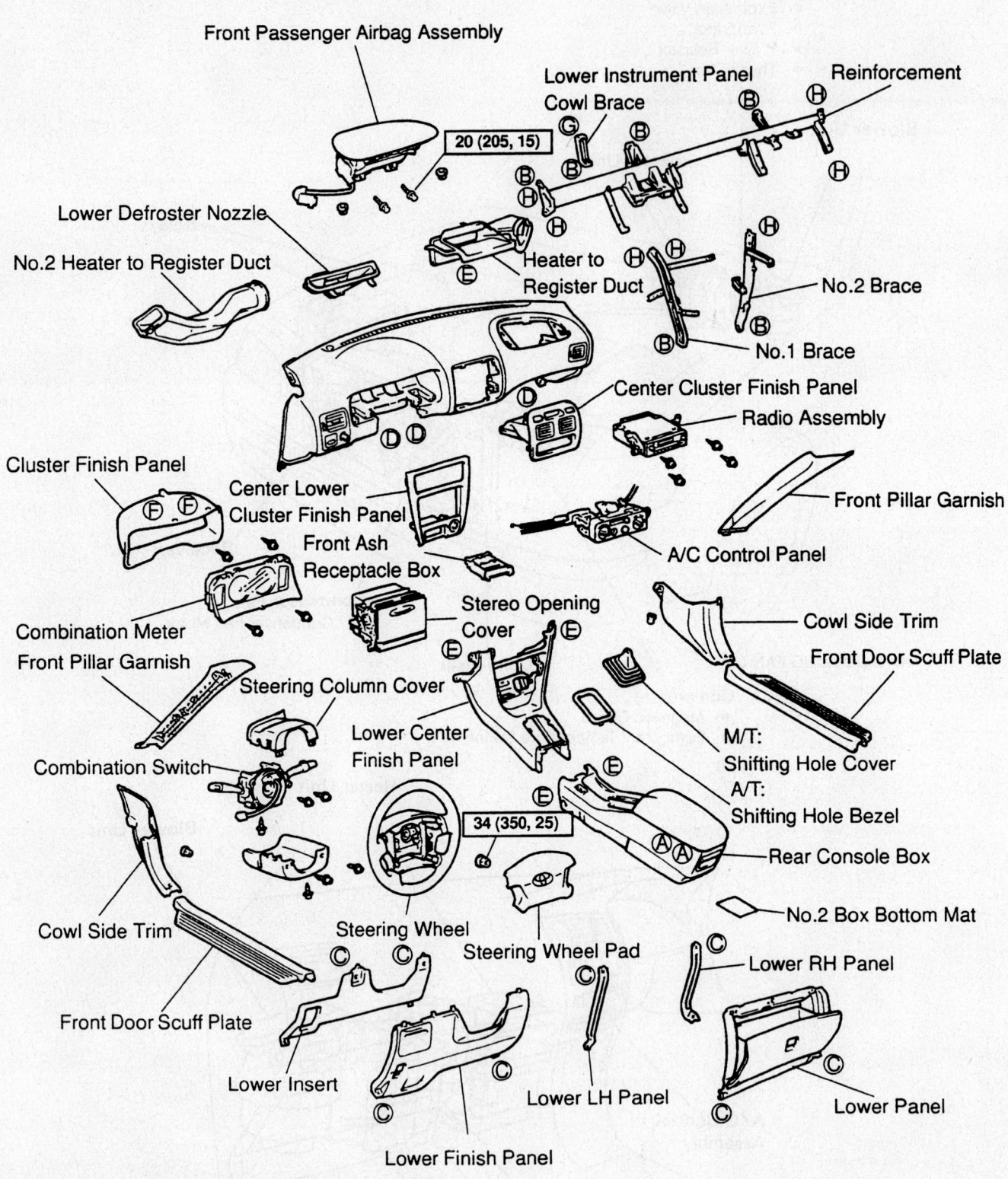

Front Passenger Airbag Assembly

Lower Instrument Panel Cowl Brace

Reinforcement

20 (205, 15)

Lower Defroster Nozzle

No.2 Heater to Register Duct

Heater to Register Duct

No.2 Brace

No.1 Brace

Center Cluster Finish Panel

Radio Assembly

Cluster Finish Panel

Center Lower Cluster Finish Panel

Front Ash Receptacle Box

Front Pillar Garnish

A/C Control Panel

Combination Meter

Front Pillar Garnish

Stereo Opening Cover

Cowl Side Trim

Front Door Scuff Plate

Steering Column Cover

Combination Switch

Lower Center Finish Panel

M/T: Shifting Hole Cover

A/T: Shifting Hole Bezel

Rear Console Box

34 (350, 25)

No.2 Box Bottom Mat

Cowl Side Trim

Steering Wheel

Steering Wheel Pad

Lower RH Panel

Front Door Scuff Plate

Lower Insert

Lower LH Panel

Lower Panel

Lower Finish Panel

N·m (kgf·cm, ft·lbf) : Specified torque

93112GM4

Exploded view of the instrument panel and related components—2002–04 Corolla

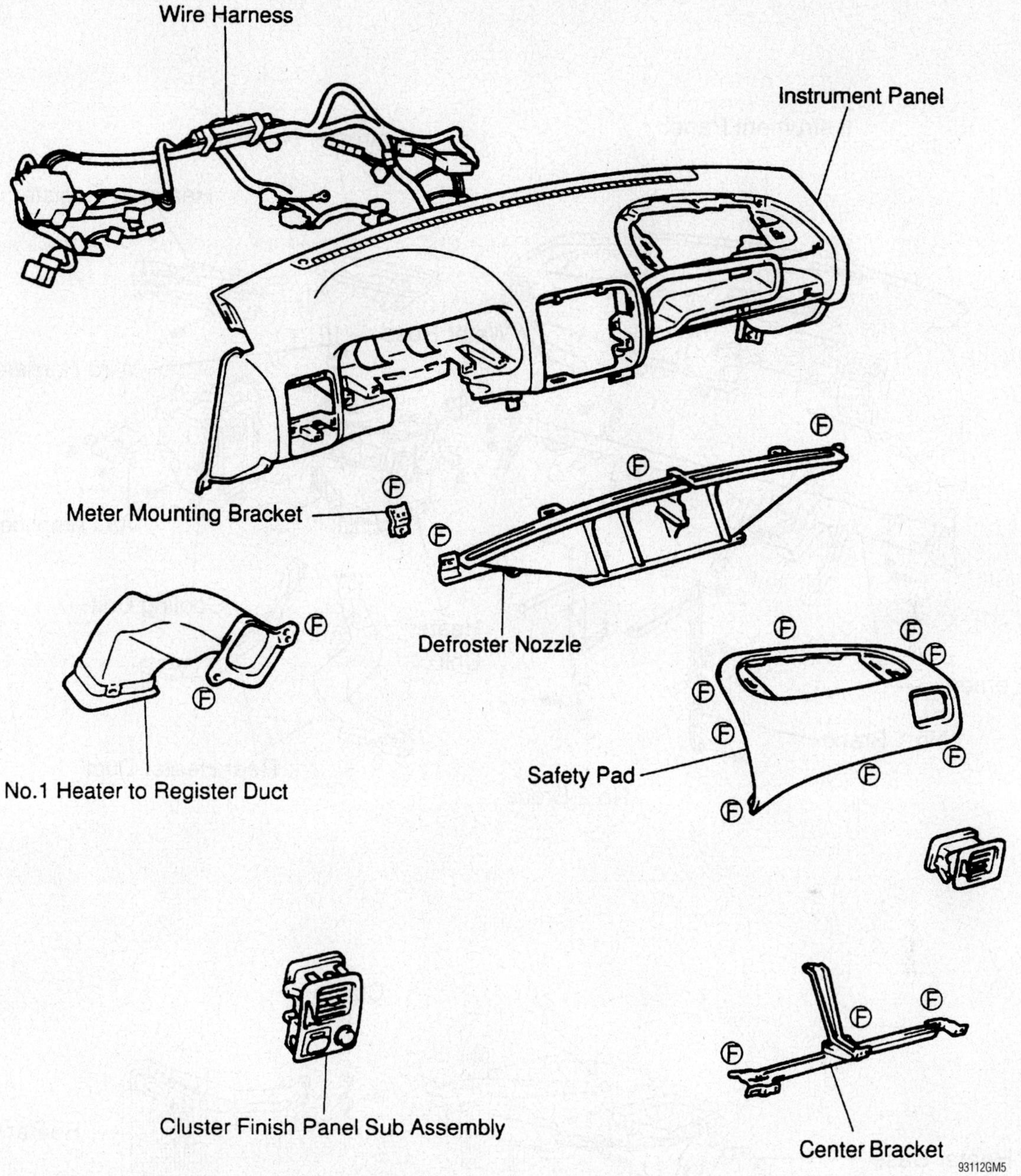

Wire Harness

Instrument Panel

Meter Mounting Bracket

Defroster Nozzle

No.1 Heater to Register Duct

Safety Pad

Cluster Finish Panel Sub Assembly

Center Bracket

93112GM5

Exploded view of the ventilation system, wiring harness and related components—2002–04 Corolla

- Radio
- Instrument panel electrical connectors
- Instrument panel screws and the instrument panel

11. Remove the following items in the following order:
- Heater-to-register duct
- Lower defroster nozzle
- No. 1 and No. 2 braces
- Reinforcement

12. Remove the evaporator housing by performing the following procedure:

a. Disconnect the cruise control actuator connector.

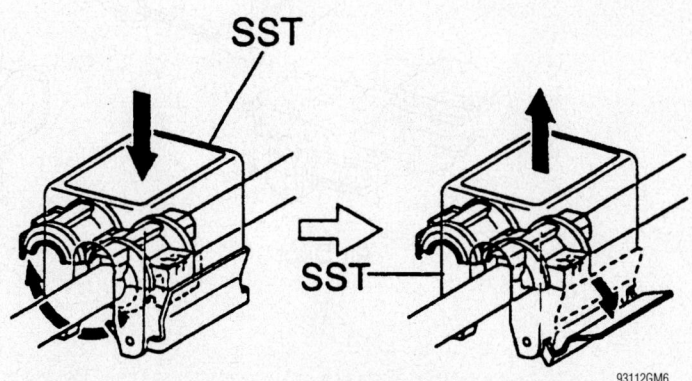

93112GM6

Using the special tool to remove the air conditioning refrigerant line clamps—All Models

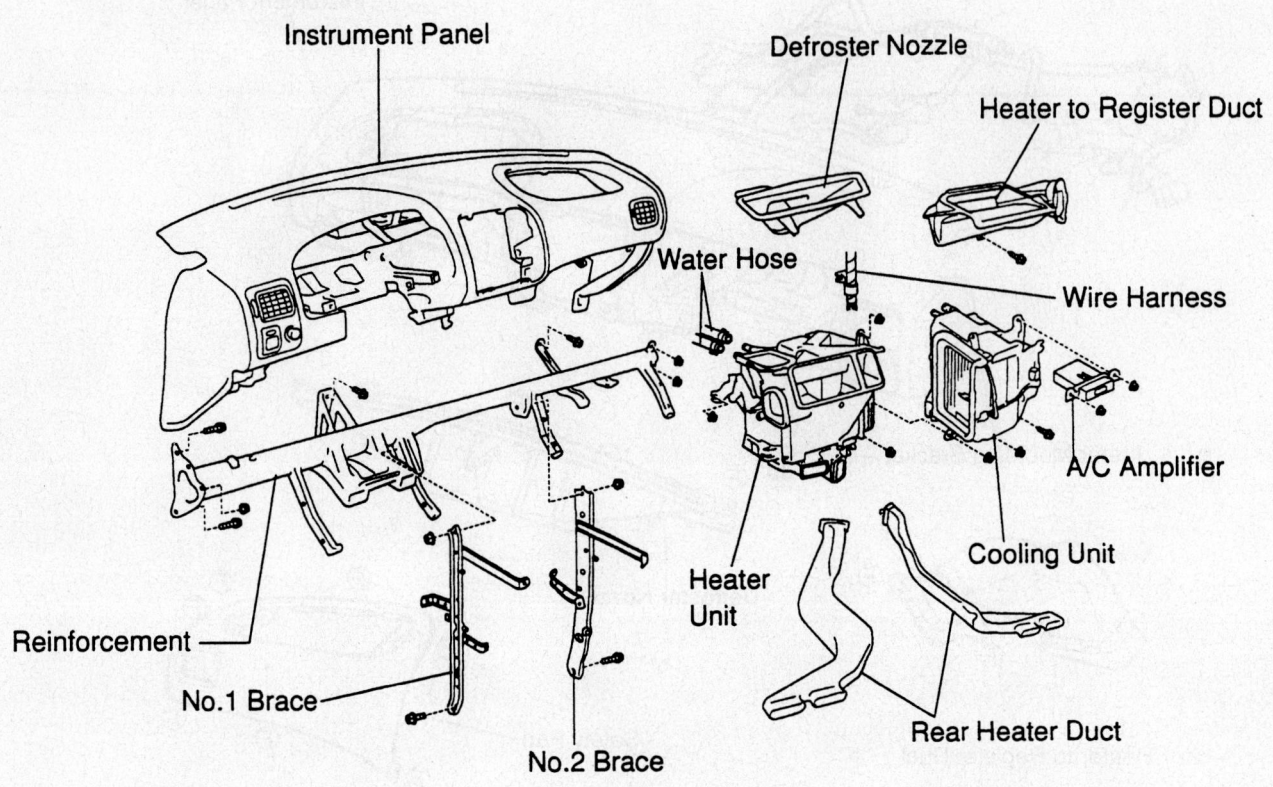

Instrument Panel

Defroster Nozzle

Heater to Register Duct

Water Hose

Wire Harness

A/C Amplifier

Cooling Unit

Heater Unit

Reinforcement

No.1 Brace

No.2 Brace

Rear Heater Duct

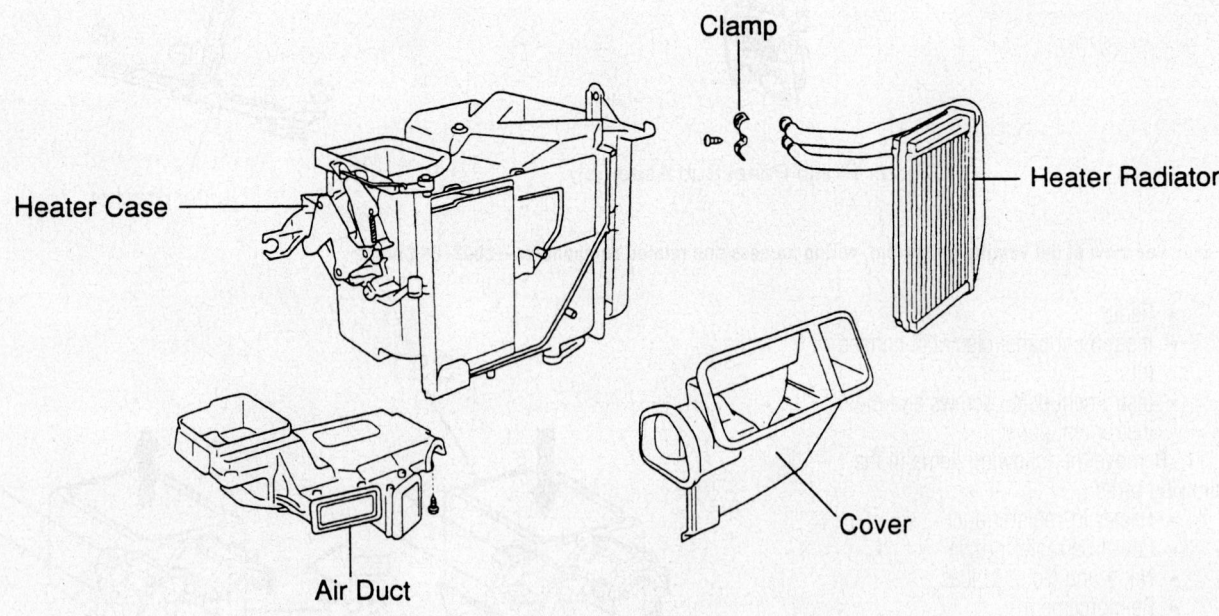

Clamp

Heater Case

Heater Radiator

Air Duct

Cover

93112GM7

Exploded view of the heater core, heater housing, evaporator housing and related components—2002–04 Corolla

b. Remove the 3 cruise control actuator set bolts.

c. Using tool No. 09870-00025 (liquid line) and/or 09870-00015 (suction line), disconnect the refrigerant line clamps. Discard the O-rings and plug the openings to prevent contamination.

d. Remove the tube grommet and the drain hose grommet.

e. Disconnect the air conditioning amplifier electrical connector.

f. Remove the 2 air conditioning amplifier nuts and the amplifier.

g. Disconnect the evaporator housing's electrical connectors.

h. Remove the evaporator housing-to-chassis 3 screws, nut and the housing.

13. Pull back the carpet and remove the rear heater ducts.

14. Disconnect the wiring harness from the heater housing.

15. Remove the 3 heater housing-to-chassis nuts and the heater housing.

16. Remove the air duct-to-heater housing screw, the 5 clips and the duct.

17. Release the 2 heater core cover clips and the cover.

18. Remove the heater core-to-heater housing screw, clamp and the heater core.

To install:

19. Install the heater core, clamp and the heater core-to-heater housing screw.

20. Install the heater core cover and the 2 cover clips.

21. Install the air duct, the 5 clips and the duct-to-heater housing screw.

22. Install the heater housing and the 3 heater housing-to-chassis nuts.

23. Connect the wiring harness to the heater housing.

24. Install the rear heater ducts and the carpet.

25. Install the evaporator housing by performing the following procedure:

a. Install the evaporator housing-to-chassis 3 screws, nut and the housing.

b. Connect the evaporator housing's electrical connectors.

c. Install the 2 air conditioning amplifier nuts and the amplifier.

d. Connect the air conditioning amplifier electrical connector.

e. Install the tube grommet and the drain hose grommet.

f. Using new O-rings, connect the refrigerant line clamps.

g. Install the 3 cruise control actuator set bolts.

h. Connect the cruise control actuator connector.

26. Install the following items in the following order:

- Reinforcement
- No. 1 and No. 2 braces
- Lower defroster nozzle
- Heater-to-register duct
- Instrument panel and the 4 instrument panel screws
- Instrument panel electrical connectors
- Radio
- Center cluster finish panel

27. Install the following items in the following order:

- Air conditioning control panel
- Lower center finish panel
- Stereo opening cover
- Electrical connector and install the center lower cluster finish panel

28. Install the following items in the following order:

- Rear console box
- Shifting hole bezel (automatic transmission)
- Shifting hole knob (manual transmission)
- Lower right side panel
- Lower left side panel
- Steering column, the steering column shaft pinch bolt and the steering column-to-instrument panel fasteners
- Combination meter
- No. 2 heater-to-register duct
- Steering column's combination switch
- Steering column cover
- Cluster finish panel and the 2 panel screws
- Lower insert and the 2 lower insert bolts
- Hood lock control cable
- Lower finish panel and the 2 panel bolts

29. Install the following items in the following order:

- Front pillar garnishes
- Cowl side trims
- Front door scuff plates

30. Install the passenger's side air bag module by performing the following procedure:

a. Carefully, Install the air bag module and the 2 passenger's air bag module-to-instrument panel bolts and nuts.

b. Connect the passenger's side air bag module electrical connector.

c. Install the glove box and the glove box-to-instrument panel 2 bolts and 3 screws.

31. Install the air bag module and the steering wheel by performing the following procedure:

a. Align the matchmarks and install the steering wheel to the steering column.

b. Install the steering wheel nut and torque the nut to 26 ft. lbs. (35 Nm).

c. Carefully, connect the electrical connector and install the air bag module.

d. Using a Torx socket, tighten the 2 air bag module-to-steering wheel Torx screws to 78 inch lbs. (8.8 Nm).

e. At both sides of the steering wheel, install the screw covers.

32. Connect the heater hoses to the heater core.

33. Refill the cooling system to the correct level.

34. Connect the negative battery cable.

35. Evacuate, charge and leak test the air conditioning system refrigerant.

2005-06

1. Before servicing the vehicle, refer to the Precautions Section.

2. Drain the cooling system.

3. Discharge and recover the air conditioning system refrigerant.

4. Remove or disconnect the following:

- Negative battery cable
- Refrigerant lines from the evaporator core. Discard the O-rings. Plug the openings to prevent contamination.
- Heater hoses

5. To remove the lower instrument panel, remove or disconnect the following:

- Horn button
- Steering wheel
- Gauge cluster
- Instrument panel heater registers
- Shift lever knob
- Shifter top console panel
- Heater control knobs
- Lower center instrument panel trim panel
- Upper center instrument cluster trim panel
- Glove compartment door
- A-pillar trim
- Passenger airbag electrical connector
- Upper instrument panel assembly
- Heater control unit
- Steering column cover
- Headlight dimmer switch
- Windshield wiper switch
- Glove compartment door stopper
- Parking brake hole cover
- Console box carpet
- Floor console
- Front door scuff plates
- Side cowl trim
- DLC3 electrical connector
- Lower instrument panel assembly

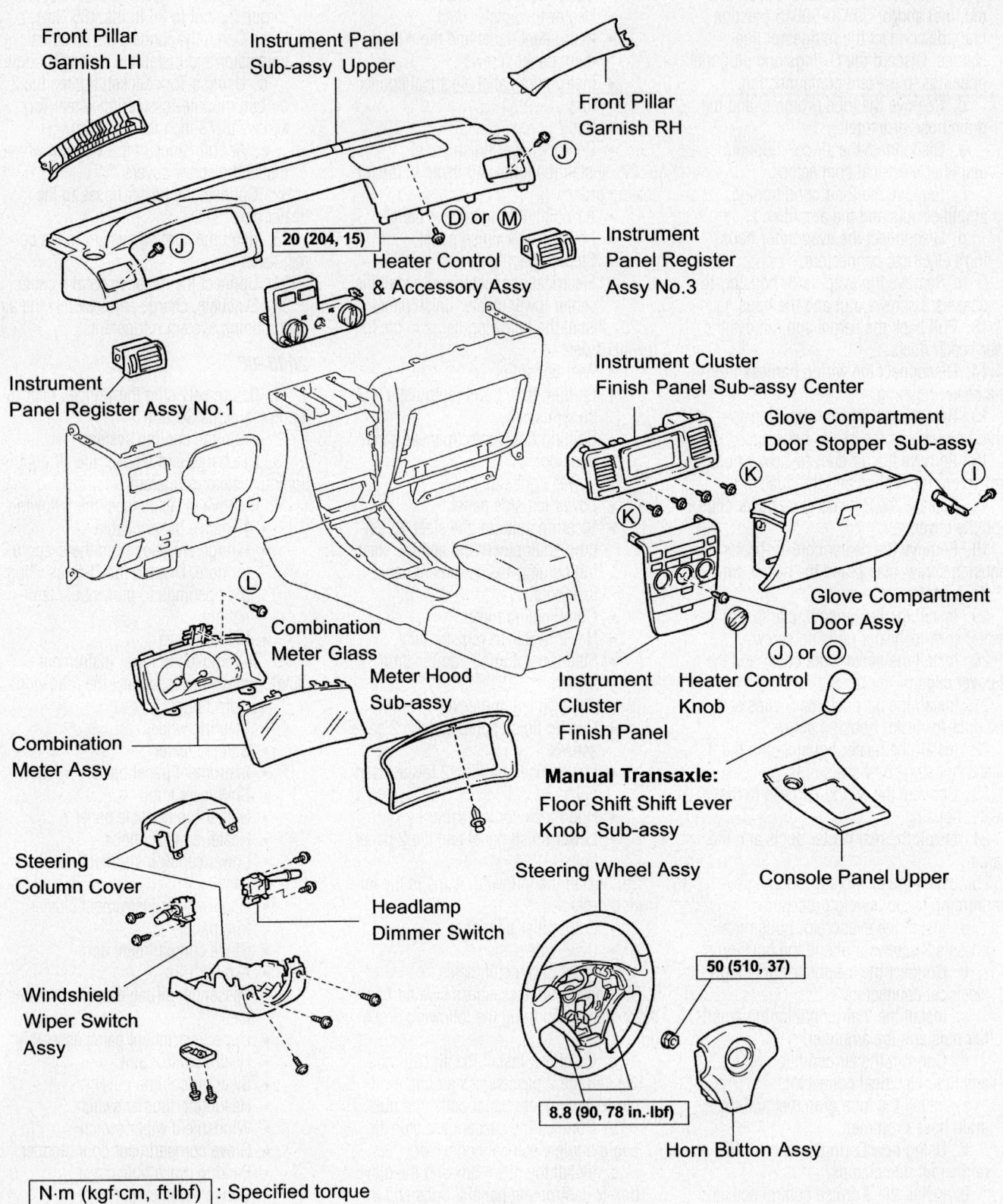

Front Pillar
Garnish LH

Instrument Panel
Sub-assy Upper

Front Pillar
Garnish RH

J

D or M

20 (204, 15)

Heater Control
& Accessory Assy

Instrument
Panel Register
Assy No.3

J

Instrument
Panel Register Assy No.1

Instrument Cluster
Finish Panel Sub-assy Center

Glove Compartment
Door Stopper Sub-assy

K

K

I

L

Combination
Meter Glass

Meter Hood
Sub-assy

Instrument
Cluster
Finish Panel

Heater Control
Knob

Glove Compartment
Door Assy

J or O

Combination
Meter Assy

Manual Transaxle:
Floor Shift Shift Lever
Knob Sub-assy

Console Panel Upper

Steering
Column Cover

Headlamp
Dimmer Switch

Steering Wheel Assy

50 (510, 37)

Windshield
Wiper Switch
Assy

8.8 (90, 78 in.·lbf)

Horn Button Assy

N·m (kgf·cm, ft·lbf) : Specified torque

09490_CORO_G0008

Exploded view of the instrument panel components—2005–06 Corolla

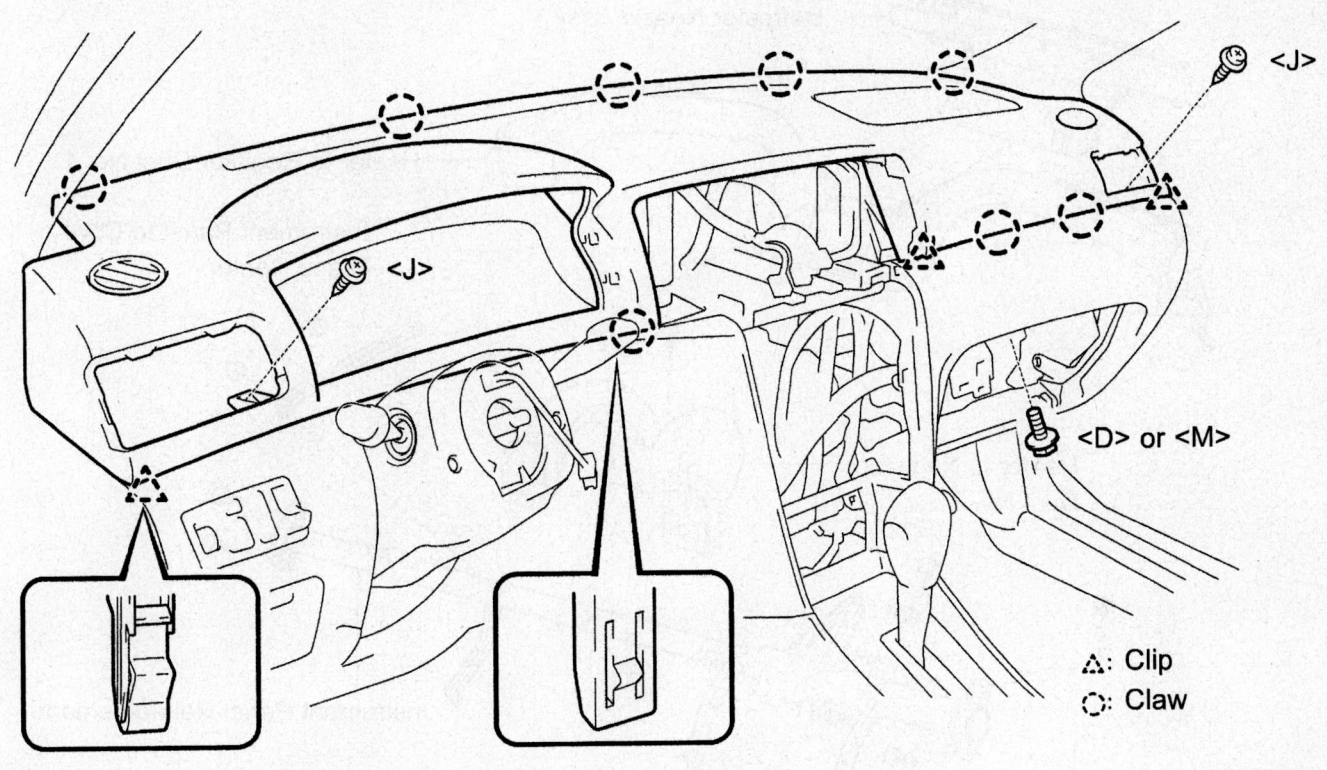

<J>

<D> or <M>

△: Clip
◯: Claw

09490_CORO_G0009

Location of the clips holding the upper instrument panel assembly—2005–06 Corolla

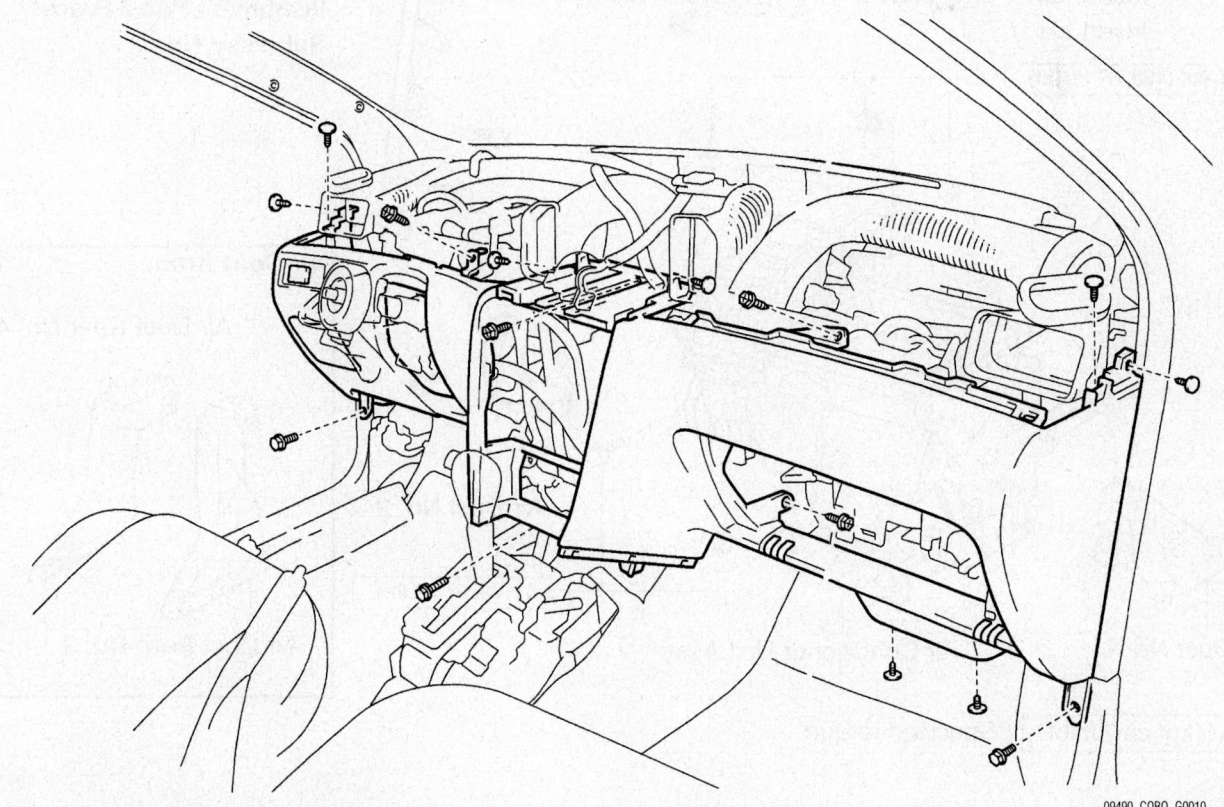

09490_CORO_G0010

Location of the bolts and screws holding the lower instrument panel assembly—2006–06 Corolla

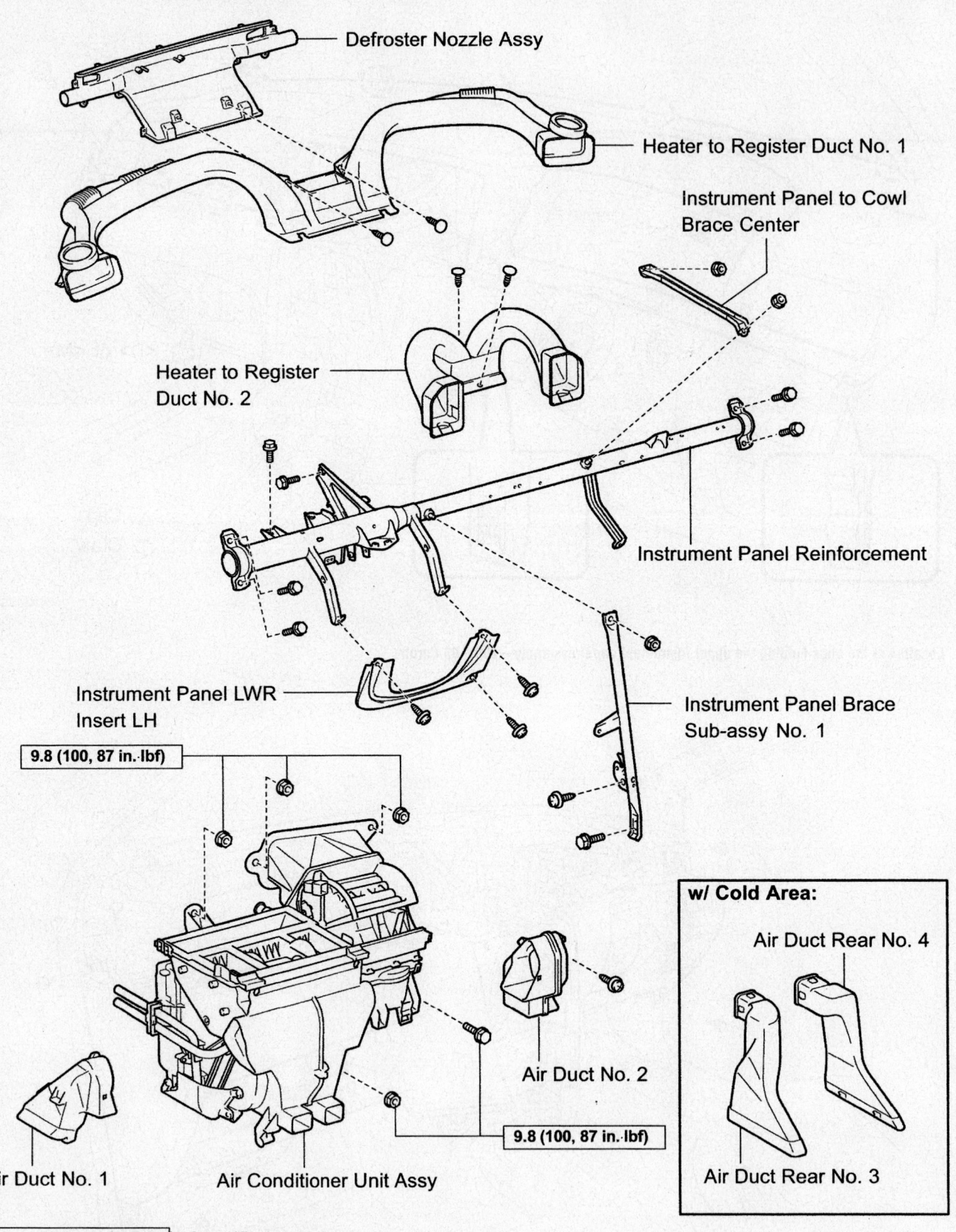

Defroster Nozzle Assy

Heater to Register Duct No. 1

Instrument Panel to Cowl Brace Center

Heater to Register Duct No. 2

Instrument Panel Reinforcement

Instrument Panel LWR Insert LH

9.8 (100, 87 in.·lbf)

Instrument Panel Brace Sub-assy No. 1

w/ Cold Area:

Air Duct Rear No. 4

Air Duct No. 2

9.8 (100, 87 in.·lbf)

Air Duct No. 1

Air Conditioner Unit Assy

Air Duct Rear No. 3

N·m (kgf·cm, ft·lbf) : Specified torque

09490_CORO_G0012

Exploded view of the ventilation system—2005–06 Corolla

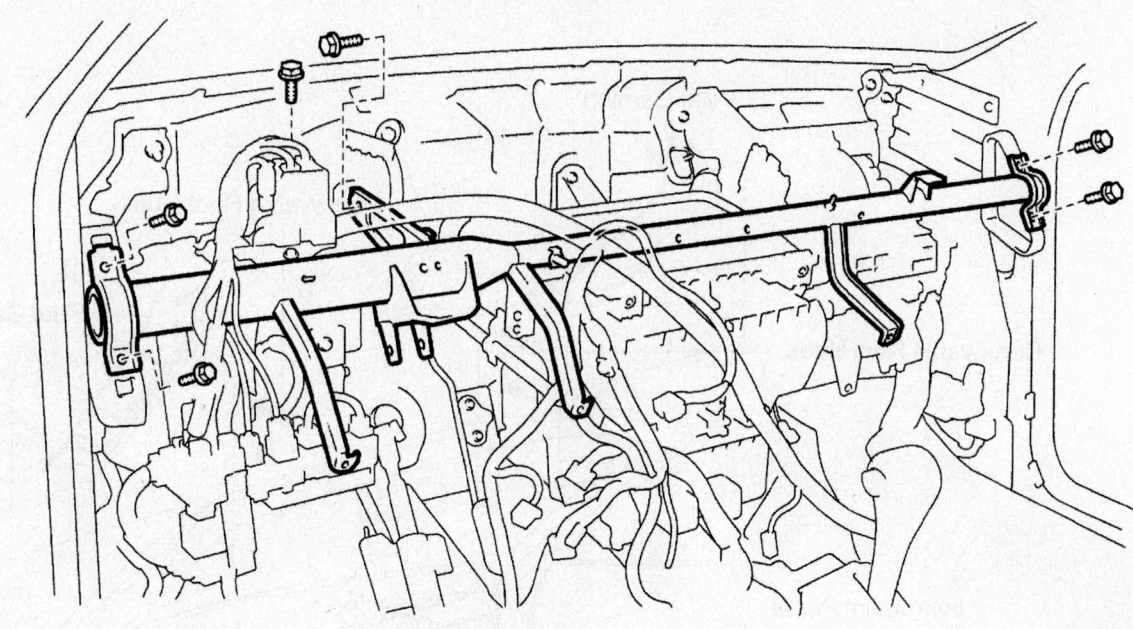

09490_CORO_G0011

Bolt locations of the instrument panel reinforcement—2005–06 Corolla

6. Remove or disconnect the following:
   - Left-hand lower instrument panel pad insert
   - Instrument panel brace
   - No. 1 air duct
   - No. 2 air duct
   - No. 3 rear air duct
   - No. 4 rear air duct
   - No. 2 heater-to-register duct
   - No. 1 heater-to-register duct
   - Defroster nozzle
   - Defroster damper control cable
   - Airmix damper control cable
   - ECM
   - Instrument panel-to-cowl brace
   - Steering column assembly
   - Instrument panel reinforcement brace
   - A/C-heater unit electrical connectors
   - AC-heater unit

7. Remove the piping cover from the heater unit.

8. Remove the heater piping clamp and remove the heater core.

9. Installation is the reverse order of removal.

10. Refill the cooling system to the correct level.

11. Start the engine and check for leaks.

MR2

1. Before servicing the vehicle, refer to the Precautions Section.

2. Drain the cooling system.

3. Discharge the A/C system.

4. Remove or disconnect the following:
   - Negative battery cable
   - Luggage compartment trim box cover
   - Heater hoses
   - Pipe grommets
   - Steering wheel
   - Door scuff plates
   - Cowl side trims
   - No. 1 lower finish panel
   - Instrument panel lower pad insert
   - Steering column covers
   - Spiral cable
   - Wiper/washer switch
   - Light control /headlight dimmer switch
   - Instrument cluster finish panel
   - Combination meter
   - Glove compartment door
   - Passenger airbag assembly
   - Passenger airbag manual switch
   - Hood release lever
   - Center instrument cluster finish panel
   - Ash tray outer case
   - Stereo finish panel
   - Instrument panel brace inner and outer covers
   - Heater control cables
   - Heater control assembly
   - Steering column
   - Front pillar garnishes
   - Instrument panel lower mount bracket
   - Center instrument panel bracket
   - Instrument panel braces
   - Instrument panel
   - Instrument panel reinforcement
   - A/C system suction tube and liquid line
   - Blower resistor connector
   - Thermistor connector
   - Cooling unit
   - Heater unit
   - Heater core

To install:
5. Install or connect the following:
   - Heater core
   - Heater unit
   - Cooling unit
   - Thermistor connector
   - Blower resistor connector
   - A/C system suction tube and liquid line
   - Instrument panel reinforcement
   - Instrument panel
   - Instrument panel braces
   - Center instrument panel bracket
   - Instrument panel lower mount bracket
   - Front pillar garnishes
   - Steering column
   - Heater control assembly
   - Heater control cables
   - Instrument panel brace inner and outer covers
   - Stereo finish panel
   - Ash tray outer case
   - Center instrument cluster finish panel
   - Hood release lever
   - Passenger airbag manual switch
   - Passenger airbag assembly
   - Glove compartment door

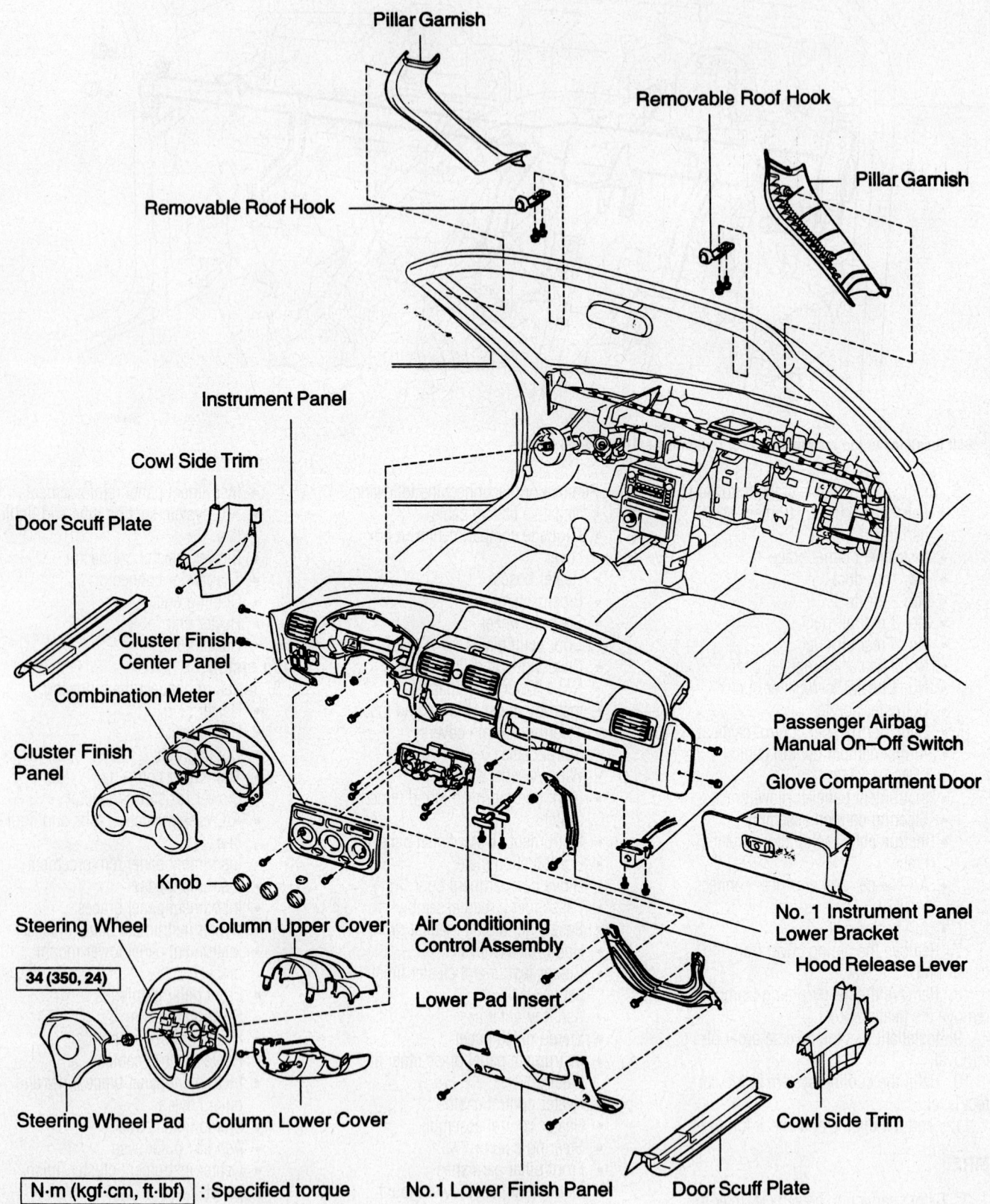

Pillar Garnish

Removable Roof Hook

Pillar Garnish

Removable Roof Hook

Instrument Panel

Cowl Side Trim

Door Scuff Plate

Cluster Finish Center Panel

Combination Meter

Cluster Finish Panel

Knob

Steering Wheel

Column Upper Cover

34 (350, 24)

Steering Wheel Pad

Column Lower Cover

Air Conditioning Control Assembly

Lower Pad Insert

Passenger Airbag Manual On–Off Switch

Glove Compartment Door

No. 1 Instrument Panel Lower Bracket

Hood Release Lever

Cowl Side Trim

N·m (kgf·cm, ft·lbf) : Specified torque

No.1 Lower Finish Panel

Door Scuff Plate

9357WG15

Instrument panel removal exploded view—MR2

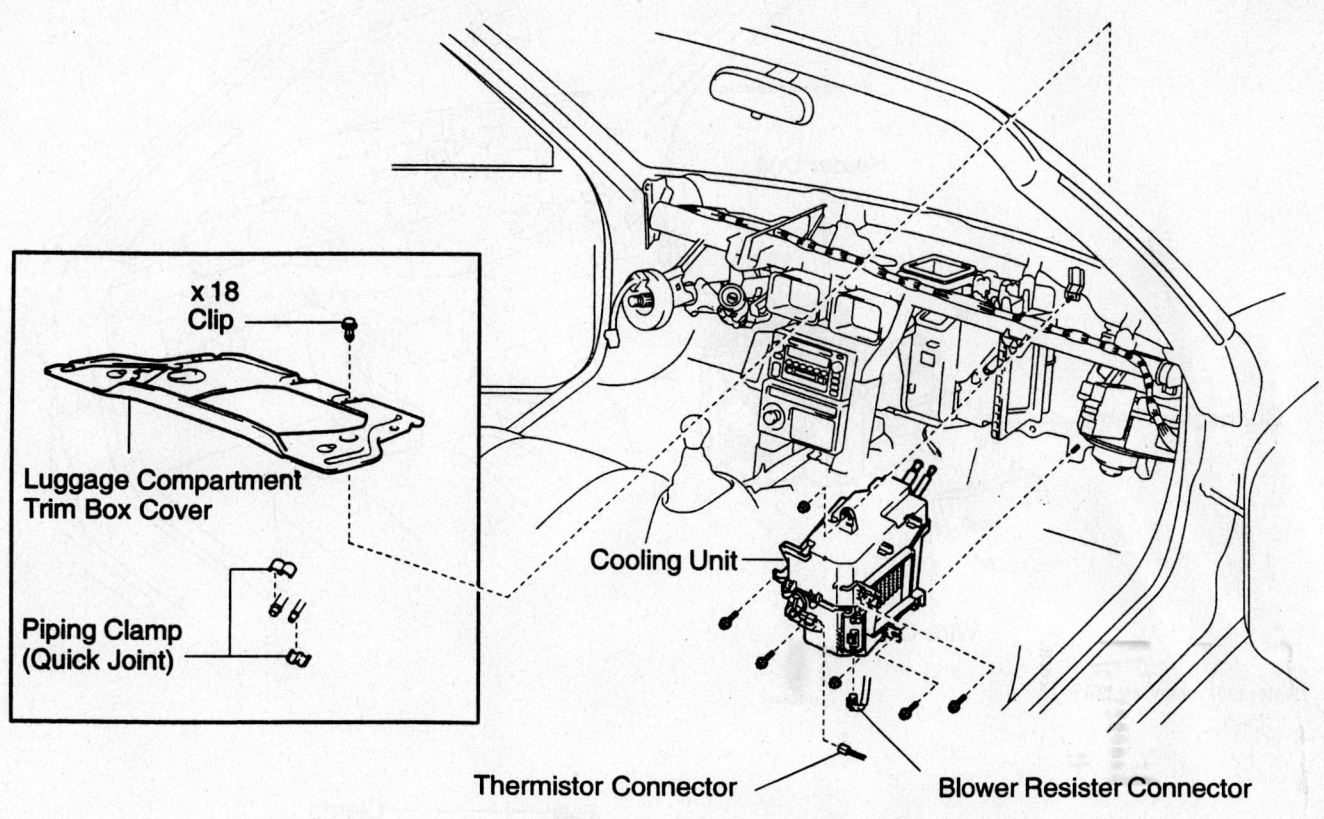

x 18 Clip

Luggage Compartment Trim Box Cover

Piping Clamp (Quick Joint)

Cooling Unit

Thermistor Connector

Blower Resister Connector

9357WG16

Evaporator case removal—MR2

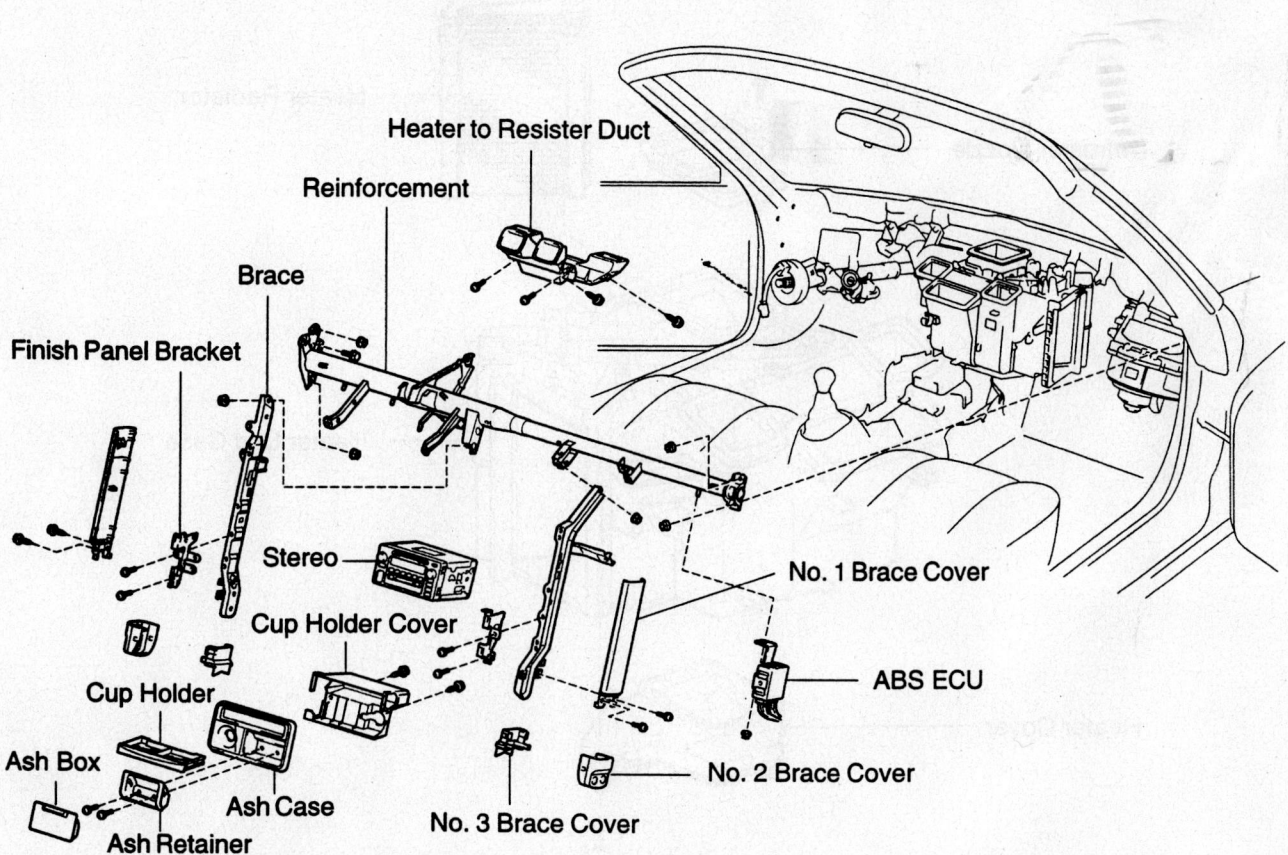

Heater to Resister Duct

Reinforcement

Brace

Finish Panel Bracket

Stereo

Cup Holder Cover

Cup Holder

Ash Box

Ash Case

Ash Retainer

No. 3 Brace Cover

No. 2 Brace Cover

No. 1 Brace Cover

ABS ECU

9357WG17

Instrument panel cross brace removal—MR2

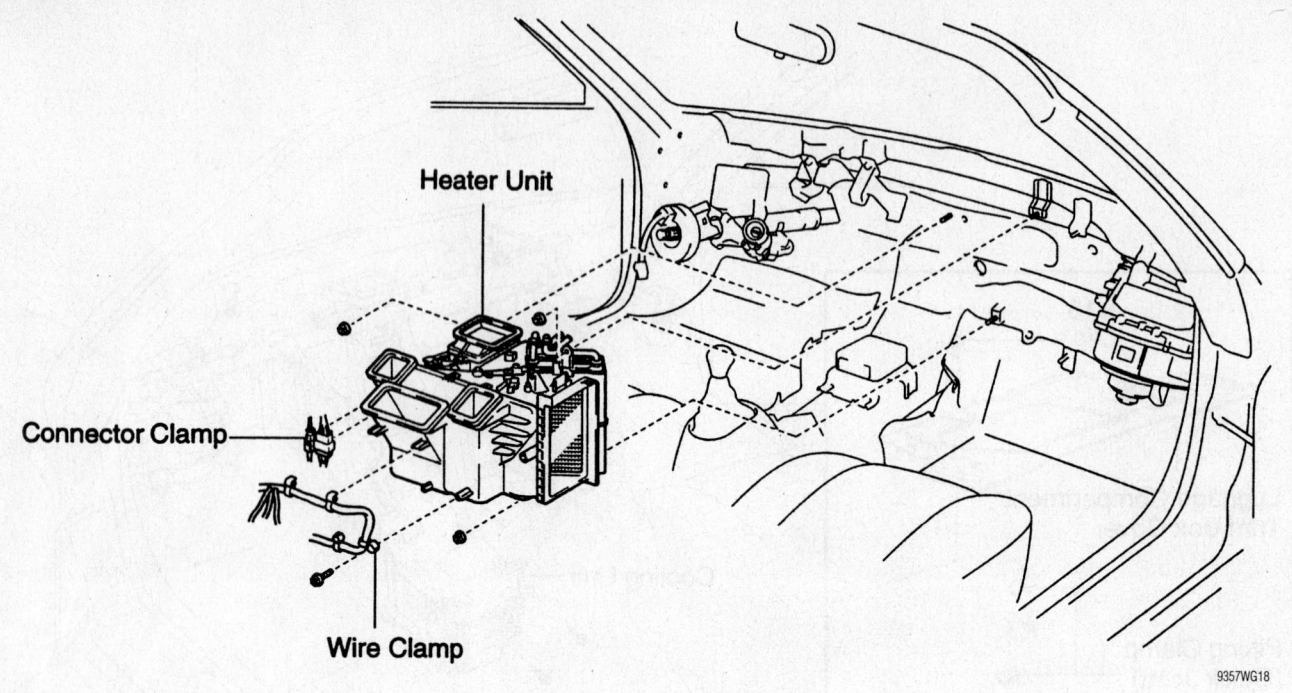

Heater Unit

Connector Clamp

Wire Clamp

Heater unit removal—MR2

9357WG18

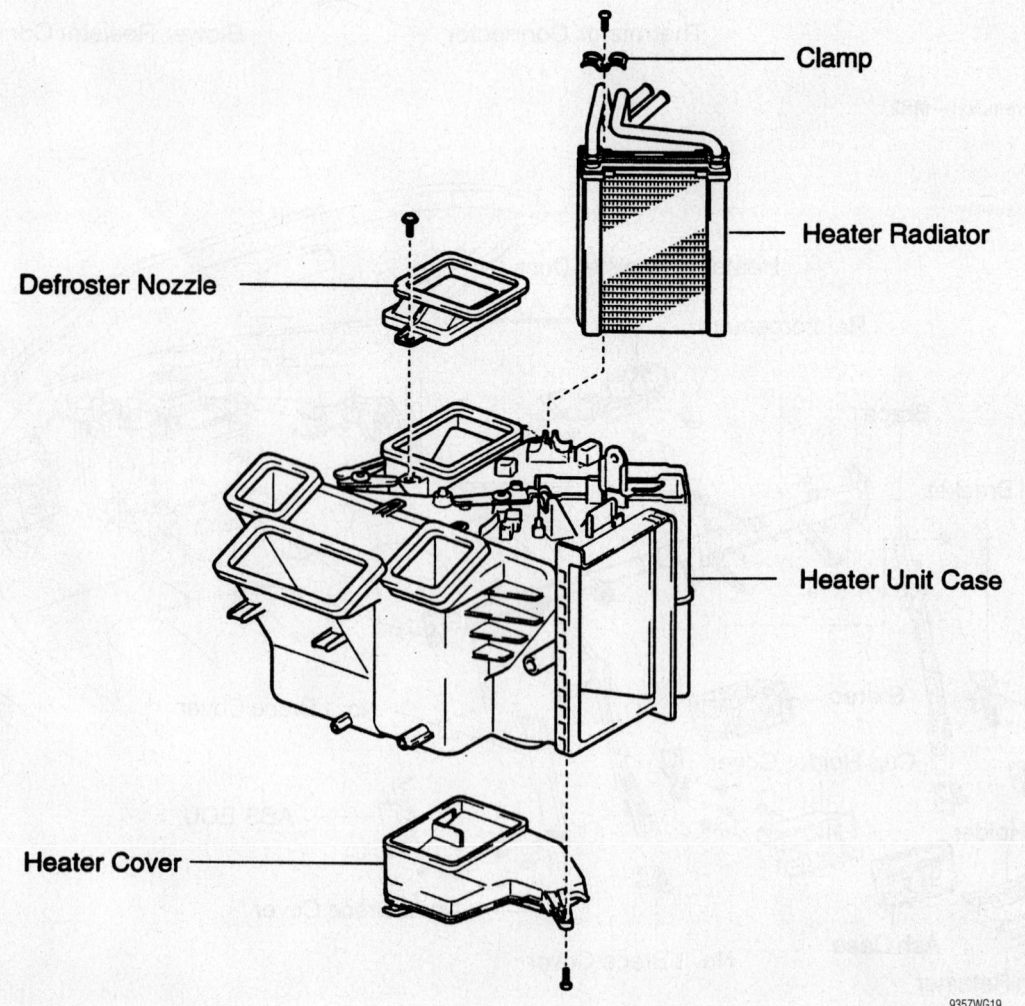

Clamp

Heater Radiator

Defroster Nozzle

Heater Unit Case

Heater Cover

9357WG19

Heater core removal—MR2

- Combination meter
- Instrument cluster finish panel
- Light control /headlight dimmer switch
- Wiper/washer switch
- Spiral cable
- Steering column covers
- Instrument panel lower pad insert
- No. 1 lower finish panel
- Cowl side trims
- Door scuff plates
- Steering wheel
- Pipe grommets
- Heater hoses
- Luggage compartment trim box cover
- Negative battery cable

6. Refill the cooling system to the correct level.

7. Recharge the A/C system.

8. Start the engine and check for leaks.

Cylinder Head

REMOVAL & INSTALLATION

1NZ-FE Engines

1. Before servicing the vehicle, refer to the Precautions Section.

2. Drain the cooling system.

3. Remove or disconnect the following:

- Negative battery cable
- Water filler
- Outer front cowl top panel
- Alternator
- Air cleaner
- Accelerator cable
- Center exhaust pipe
- Exhaust manifold support
- Exhaust manifold
- Ignition coil
- Spark plugs
- PCV hoses
- Throttle body
- Engine wiring harness at the head
- Intake manifold
- Camshaft position sensor
- ECT sensor
- Oil control valve
- PCV valve
- Oil filer cap
- Cylinder head cover
- Fuel injectors
- Timing chain cover
- Camshaft sprockets and valve timing control assembly
- Camshafts
- Cylinder head. Remove the bolts in a circular pattern, in several stages,

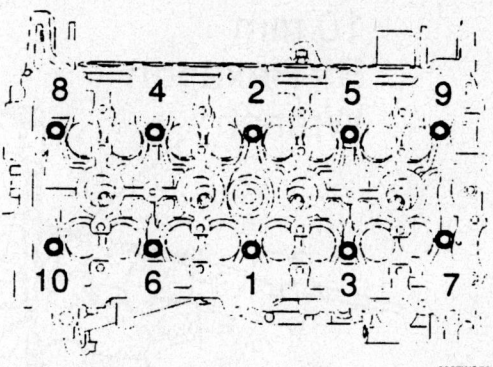

Cylinder head bolt tightening sequence—1NZ-FE engine

starting from the ends and working towards the center

To install:

4. Install or connect the following:

- Cylinder head, using a new gasket. The Lod. No. on the gasket faces UP

5. Torque the cylinder head bolts, in sequence, in 3 steps:

 a. Step 1: 22 ft. lbs. (29 Nm).

 b. Step 2: Plus a 90 degree turn.

 c. Step 3: Plus a 90 degree turn.

6. Install or connect the following:

- Water bypass pipe. Torque the bolt to 80 inch lbs. (9 Nm).
- Camshafts

7. Camshaft bearing caps in 2 stages:

 a. Step: 10 ft. lbs. (13 Nm).

 b. Step 2: 17 ft. lbs. (23 Nm).

8. Install or connect the following:

- Sprockets and valve timing controller assembly, aligning the knock pin and hole. Torque the bolts to 47 ft. lbs. (64 Nm).
- Check and adjust the valves
- Cylinder head cover
- Oil filler cap
- PCV valve
- ECT sensor
- Camshaft position sensor
- Timing chain cover
- Intake manifold
- Engine wiring harness
- Throttle body
- PCV hoses
- Spark plugs
- Ignition coils
- Exhaust manifold
- Exhaust manifold support. Torque the bolts to 27 ft. lbs. (37 Nm).
- Front exhaust pipe. Torque the nuts to 46 ft. lbs. (62 Nm).
- Accelerator cable
- Air cleaner
- Alternator
- Water filler

- Negative battery cable

9. Refill the cooling system to the correct level.

10. Start the vehicle, check for leaks and repair if necessary.

1ZZ-FE Engines

CELICA

1. Before servicing the vehicle, refer to the Precautions Section.

2. Drain the cooling system.

3. Remove or disconnect the following:

- Battery
- ECU box
- Coolant reservoir
- Air intake assembly
- Accelerator cable
- Alternator
- Exhaust pipe
- Exhaust manifold
- Coils
- Spark plugs
- PCV hoses
- Throttle body
- Injectors
- Wiring harness
- Intake manifold
- Camshaft position sensor
- ECT sensor
- PCV valve
- Oil filler cap
- Camshaft sprockets
- Camshafts
- Hoses
- Cylinder head bolts in sequence. To prevent damage to the cylinder head, loosen each bolt about ¼ of a turn during each pass until the bolts are loose.
- Cylinder head

To install:

4. Clean and degrease the surface of the cylinder head and engine block.

5. Install or connect the following:

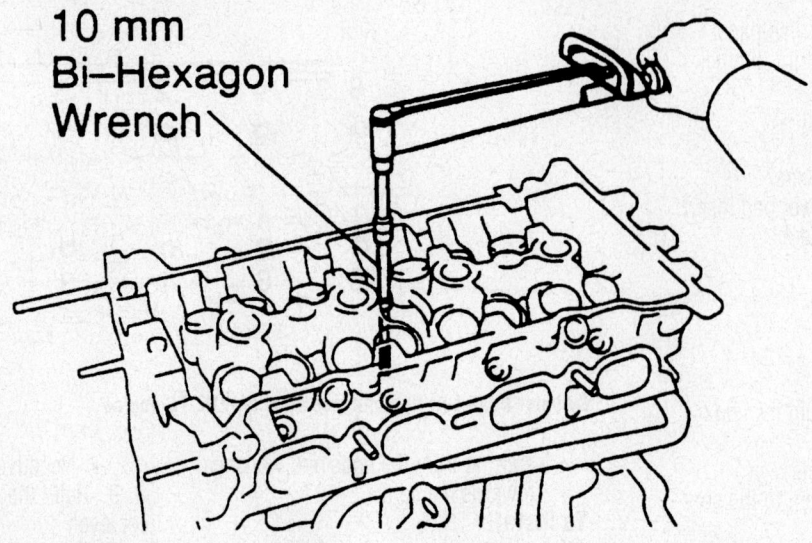

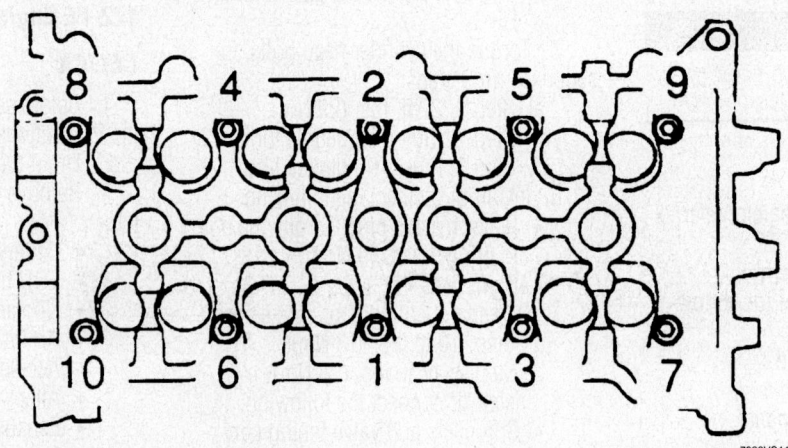

Cylinder head bolt tightening sequence—1ZZ-FE and 2ZZ-GE engines

- New gasket on the engine block with the Lod No. stamp facing up.
- Cylinder head
- Apply a light coat of oil to cylinder head bolt threads and tighten in sequence. Replace any bolt that appears deformed. Bolts: 36 ft. lbs. (49 Nm).
- Tighten each bolt in sequence an additional 90 degree turn.
- Camshafts
- Sprockets
- Oil filler cap
- PCV valve
- ECT sensor
- Intake manifold
- Wiring harness
- Exhaust manifold
- Exhaust pipe
- Alternator
- Accelerator cable
- Air cleaner

- ECM box
- Battery

6. Refill the cooling system to the correct level.

7. Start the vehicle, check for leaks and repair if necessary.

COROLLA

1. Before servicing the vehicle, refer to the Precautions Section.

2. Relieve the fuel system pressure.

3. Drain the cooling system.

4. Remove or disconnect the following:

- Negative battery cable
- Right-hand front wheel
- Right-hand engine under cover
- Engine appearance cover
- Front exhaust pipe assembly
- Manifold support
- Accessory drive belt
- Power steering pump

- Alternator
- Right-hand engine mounting insulator
- Ignition coil
- Cylinder head cover
- Accessory drive belt tensioner
- Right-hand engine mount
- Water pump
- Crankshaft position sensor

5. Set the No. 1 cylinder to Top Dead Compression (TDC).

6. Remove or disconnect the following:

- Crankshaft pulley
- Timing chain tensioner
- Front cover
- Timing chain
- Air intake assembly
- Oil dipstick tube
- Intake manifold
- Camshafts
- Fuel hose

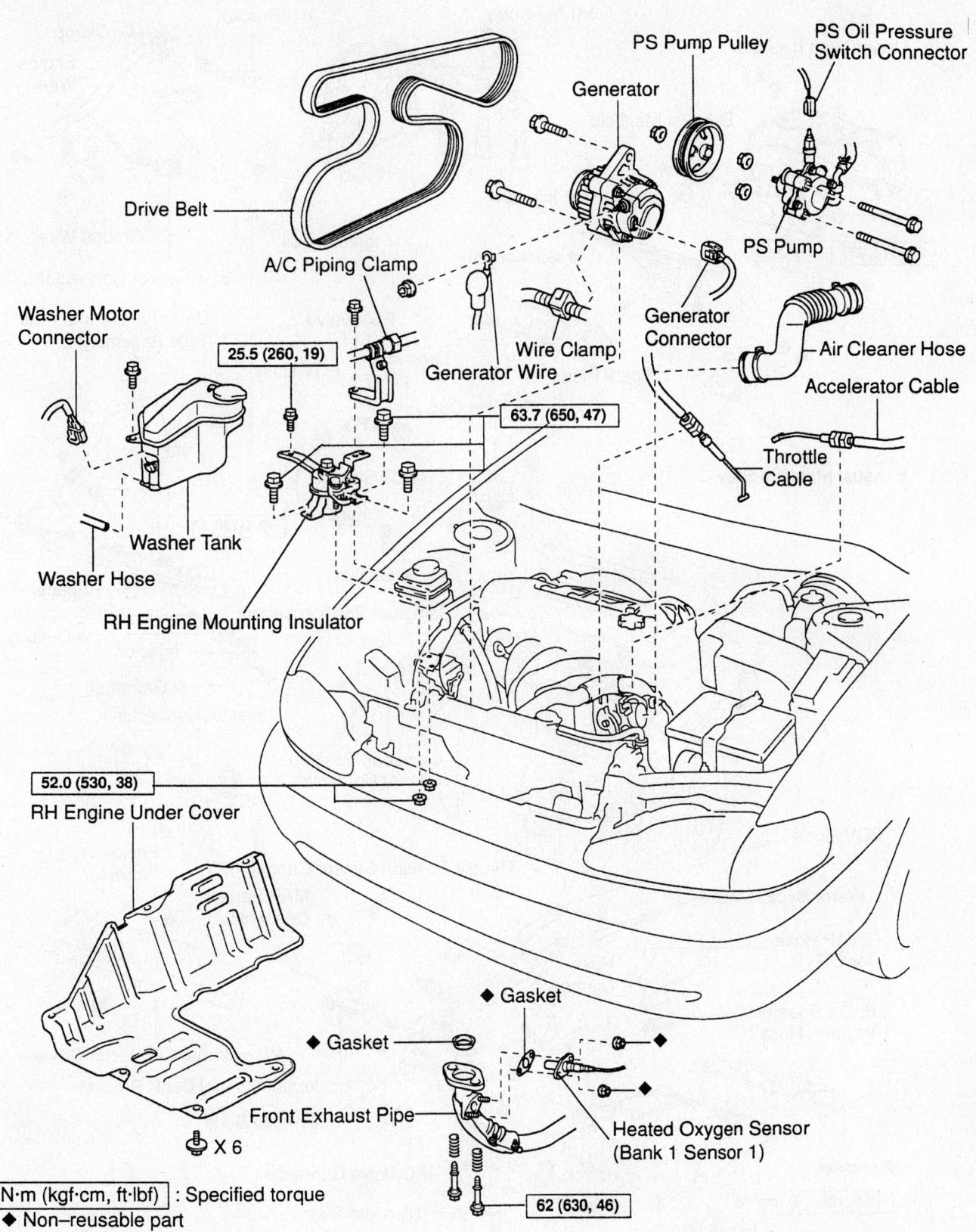

PS Pump Pulley

PS Oil Pressure
Switch Connector

Generator

Drive Belt

A/C Piping Clamp

PS Pump

Washer Motor
Connector

Generator
Connector

Wire Clamp

Generator Wire

Air Cleaner Hose

25.5 (260, 19)

Accelerator Cable

63.7 (650, 47)

Throttle
Cable

Washer Tank

Washer Hose

RH Engine Mounting Insulator

52.0 (530, 38)

RH Engine Under Cover

◆ Gasket

◆ Gasket

◆ ◆ ◆

Heated Oxygen Sensor
(Bank 1 Sensor 1)

Front Exhaust Pipe

✖ X 6

62 (630, 46)

N·m (kgf·cm, ft·lbf) : Specified torque

◆ Non–reusable part

9301WG03

Exploded view of engine accessories and right-hand engine under cover—1.8L (1ZZ-FE) engine.

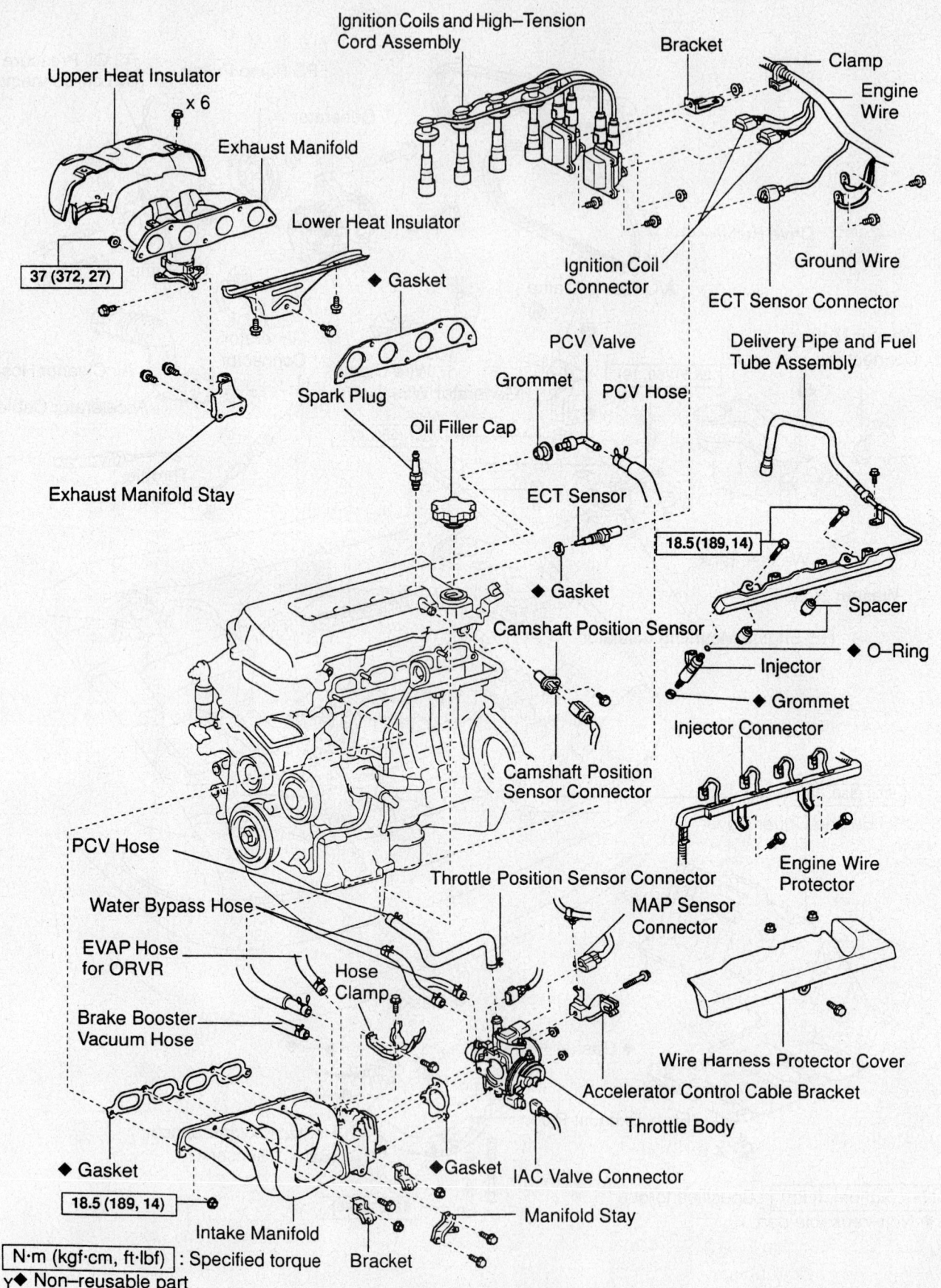

Ignition Coils and High–Tension Cord Assembly

Bracket

Clamp

Engine Wire

Upper Heat Insulator

x 6

Exhaust Manifold

Lower Heat Insulator

Ignition Coil Connector

Ground Wire

ECT Sensor Connector

37 (372, 27)

◆ Gasket

Spark Plug

Exhaust Manifold Stay

Oil Filler Cap

PCV Valve

Grommet

PCV Hose

Delivery Pipe and Fuel Tube Assembly

ECT Sensor

18.5 (189, 14)

◆ Gasket

Camshaft Position Sensor

Spacer

◆ O–Ring

Injector

◆ Grommet

Injector Connector

Camshaft Position Sensor Connector

Engine Wire Protector

PCV Hose

Throttle Position Sensor Connector

MAP Sensor Connector

Water Bypass Hose

EVAP Hose for ORVR

Hose Clamp

Brake Booster Vacuum Hose

Wire Harness Protector Cover

Accelerator Control Cable Bracket

Throttle Body

◆ Gasket

◆ Gasket

IAC Valve Connector

18.5 (189, 14)

Manifold Stay

Intake Manifold

Bracket

N·m (kgf·cm, ft·lbf) : Specified torque

γ◆ Non–reusable part

9301WG04

View of engine intake, exhaust, ignition, and fuel system location—1.8L (1ZZ-FE) engine.

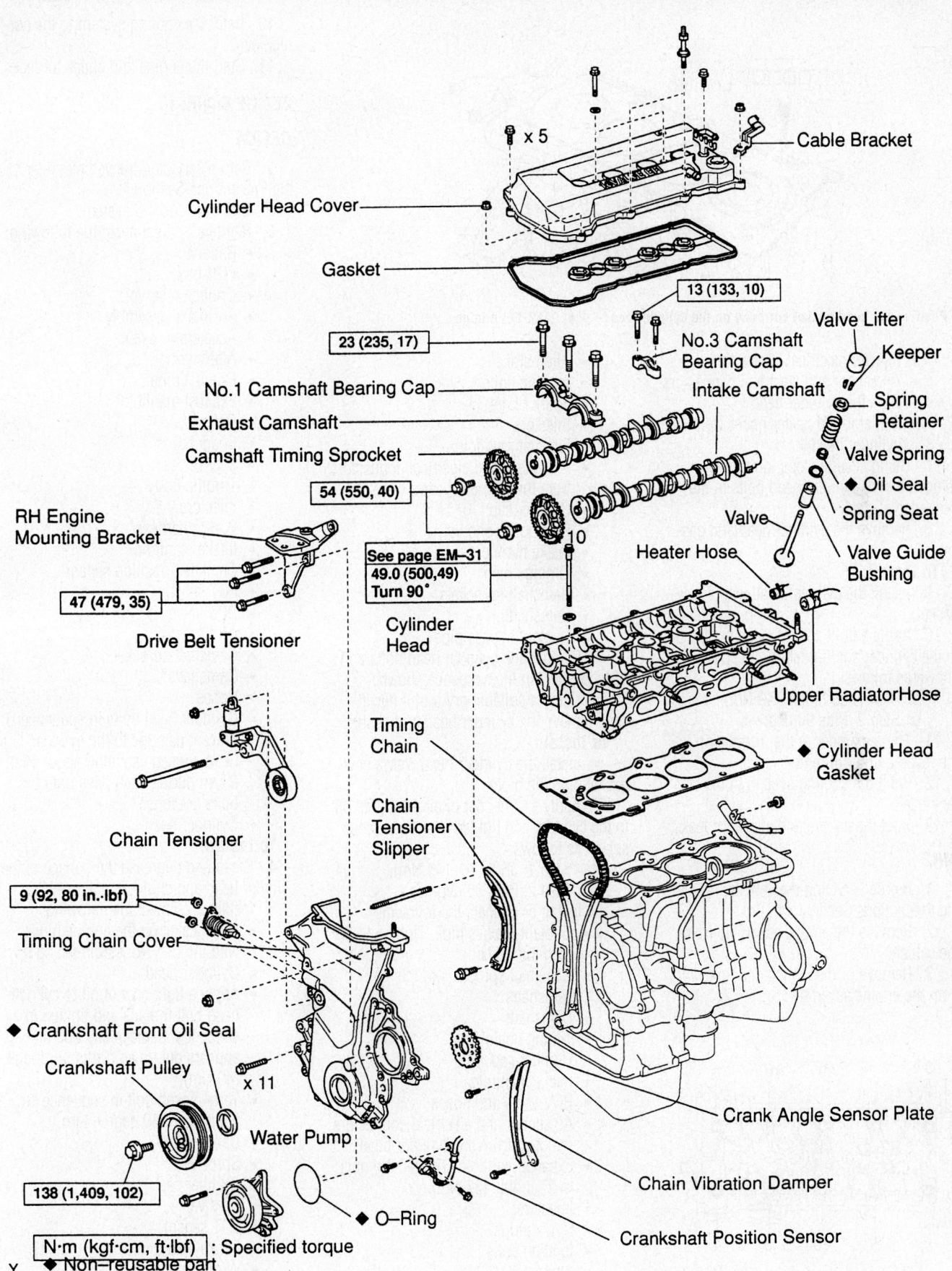

Cable Bracket

Cylinder Head Cover

Gasket

13 (133, 10)

23 (235, 17)

No.3 Camshaft Bearing Cap

Valve Lifter

Keeper

No.1 Camshaft Bearing Cap

Intake Camshaft

Spring Retainer

Exhaust Camshaft

Camshaft Timing Sprocket

Valve Spring

54 (550, 40)

◆ Oil Seal

RH Engine Mounting Bracket

Valve

Spring Seat

Heater Hose

Valve Guide Bushing

47 (479, 35)

See page EM–31
49.0 (500, 49)
Turn 90°

Drive Belt Tensioner

Cylinder Head

Upper Radiator Hose

Timing Chain

◆ Cylinder Head Gasket

Chain Tensioner Slipper

Chain Tensioner

9 (92, 80 in.·lbf)

Timing Chain Cover

◆ Crankshaft Front Oil Seal

Crankshaft Pulley

x 11

Crank Angle Sensor Plate

Water Pump

Chain Vibration Damper

138 (1,409, 102)

◆ O–Ring

Crankshaft Position Sensor

N·m (kgf·cm, ft·lbf) : Specified torque

Y   ◆ Non–reusable part

Illustration of disassembled cylinder head assembly—1.8L (1ZZ-FE) engine.

9301WG05

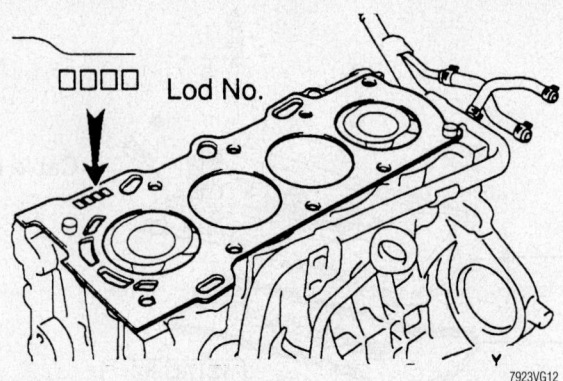

Position the head gasket correctly on the cylinder head—1.8L (1ZZ-FE) engine

- Injector electrical connectors
- All sensor and electrical connectors from the cylinder head
- Radiator and heater hoses from the cylinder head

7. Using several steps, uniformly remove the 10 cylinder head bolts in the sequence shown.

8. Remove the cylinder head and gasket.

To install:

9. Install the cylinder head with a new gasket.

10. Apply a light coat of clean engine oil to the cylinder and tighten in the sequence shown as follows:
   - a. Step 1: 36 ft. lbs. (49 Nm).
   - b. Step 2: Plus 90 degrees

11. The remainder of the installation is the reverse order of removal.

12. Refill the cooling system to the correct level.

13. Start the engine and check for leaks.

MR2

1. Before servicing the vehicle, refer to the Precautions Section.

2. Remove the engine assembly from the vehicle.

3. Remove or disconnect the following from the engine assembly:

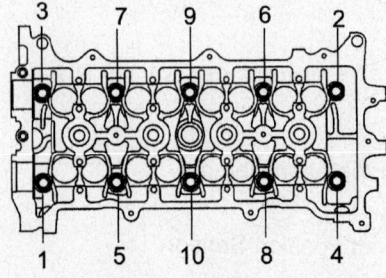

Cylinder head bolt removal sequence—1.8L (1ZZ-FE) and (2ZZ-GE) engines

- Alternator
- Ignition coils
- Spark plugs
- Injectors
- Exhaust manifold
- All sensor and electrical connectors from the cylinder head
- Oil filler cap
- PCV hoses and valve
- Intake manifold
- Timing chain
- Camshaft sprockets
- Camshafts
- Coolant bypass pipe

4. Remove the cylinder head bolts and plate washers in the sequence shown.

5. Using a suitable pry tool if necessary, remove the cylinder head and gasket.

To install:

6. Install the cylinder head with a new gasket onto the block.

7. Apply a light coat of clean engine oil to the cylinder and tighten in the sequence shown as follows:
   - a. Step 1: 36 ft. lbs. (49 Nm).
   - b. Step 2: Plus 90 degrees

8. Install or connect the following:
   - Coolant bypass pipe. Tighten to 80 inch lbs. (9 Nm).
   - Camshaft sprockets to the camshafts
   - Camshafts
   - Timing chain
   - Oil filler cap
   - Intake manifold
   - PCV valve and hoses
   - All sensor and electrical connectors removed from the cylinder head
   - Exhaust manifold. Tighten the nuts to 27 ft. lbs. (37 Nm).
   - Injectors
   - Spark plugs
   - Ignition coils
   - Alternator

9. Install the engine assembly into the vehicle.

10. Refill the cooling system to the correct level.

11. Start the engine and check for leaks.

2ZZ-GE Engines

CELICA

1. Before servicing the vehicle, refer to the Precautions Section.

2. Drain the cooling system.

3. Remove or disconnect the following:
   - Battery
   - ECU box
   - Coolant reservoir
   - Air intake assembly
   - Accelerator cable
   - Alternator
   - Exhaust pipe
   - Exhaust manifold
   - Coils
   - Spark plugs
   - PCV hoses
   - Throttle body
   - Injectors
   - Wiring harness
   - Intake manifold
   - Camshaft position sensor
   - ECT sensor
   - PCV valve
   - Oil filler cap
   - Camshaft sprockets
   - Camshafts
   - Hoses
   - Cylinder head bolts in sequence. To prevent damage to the cylinder head, loosen each bolt about ¼ of a turn during each pass until the bolts are loose.
   - Cylinder head

To install:

4. Clean and degrease the surface of the cylinder head and engine block.

5. Install or connect the following:
   - New gasket on the engine block with the Lod No. stamp facing up.
   - Cylinder head
   - Apply a light coat of oil to cylinder head bolt threads and tighten in sequence. Replace any bolt that appears deformed. Bolts: 26 ft. lbs. (49 Nm).
   - Torque each bolt in sequence an additional 180 degree turn.
   - Camshafts
   - Sprockets
   - Oil filler cap
   - PCV valve
   - ECT sensor
   - Intake manifold
   - Wiring harness
   - Exhaust manifold
   - Exhaust pipe

- Alternator
- Accelerator cable
- Air cleaner
- ECM box
- Battery

6. Refill the cooling system to the correct level.

7. Start the vehicle, check for leaks and repair if necessary.

2005–06 COROLLA

1. Before servicing the vehicle, refer to the Precautions Section.
2. Drain the cooling system.
3. Relieve the fuel system pressure.
4. Remove or disconnect the following:
   - Negative battery cable
   - Right-hand front wheel
   - Right-hand engine undercover

- Front suspension bar assembly tower
- Engine appearance cover
- Accelerator cable
- Air intake assembly
- Brake booster union to connector tube
- Fuel hose
- Radiator hose

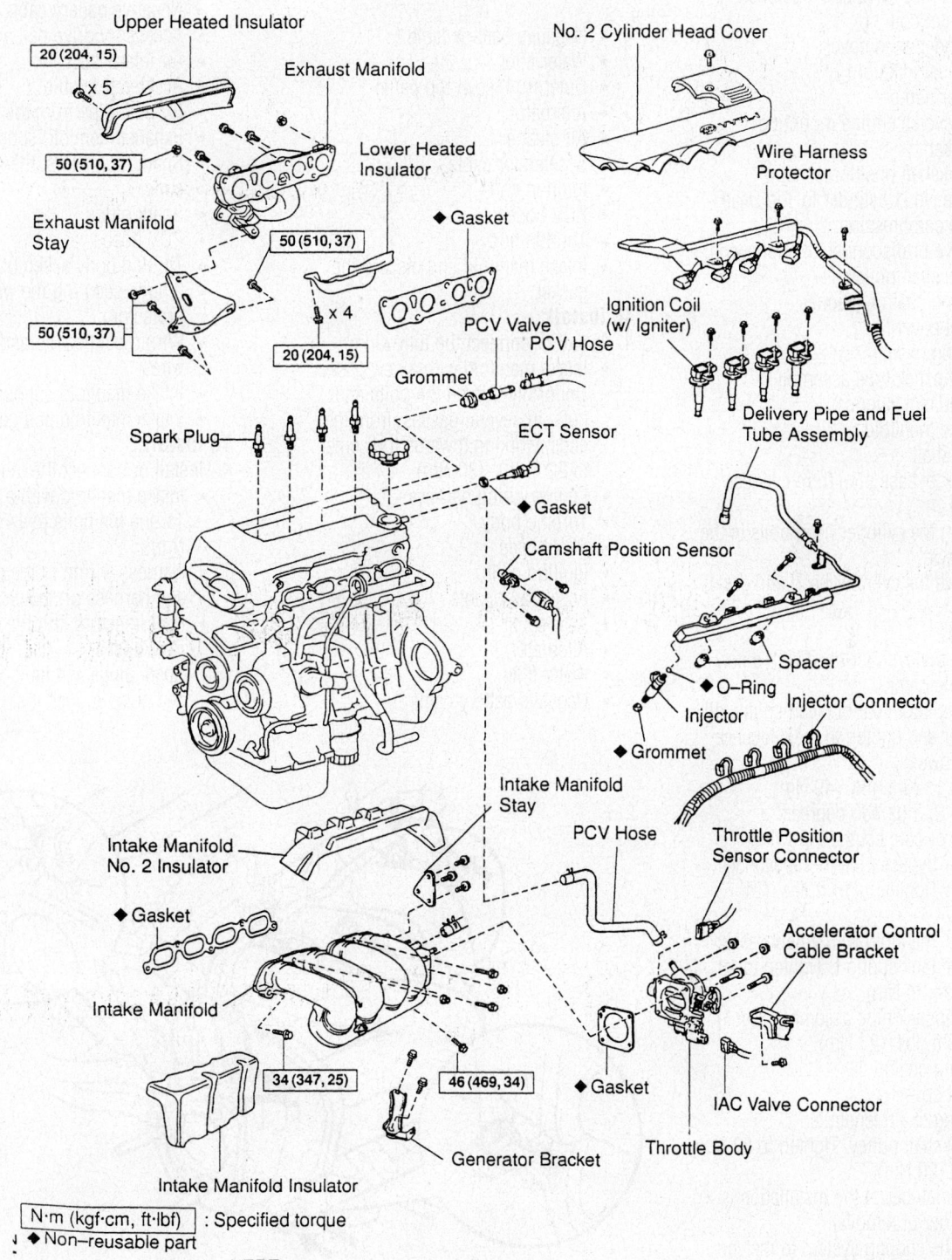

N·m (kgf·cm, ft·lbf) : Specified torque
◆ Non-reusable part

Cylinder head component exploded view—2ZZ-GE engine

9307WG91

- Heater hoses
- Coolant bypass hoses
- Refrigerant lines mounting bolts
- Air switching valve
- Accessory drive belt
- Alternator
- A/C compressor and hang securely
- Front exhaust pipe assembly
- Right-hand engine mounting insulator
- Accessory drive belt tensioner
- Ignition coil
- Cylinder head cover
- Exhaust manifold
- Water pump
- Transverse engine mounting bracket
- Crankshaft position sensor

5. Set the No. 1 cylinder to Top Dead Center (TDC) compression.

6. Remove or disconnect the following:
- Crankshaft pulley
- Timing chain tensioner
- Front cover
- Timing chain
- Oil dipstick tube assembly
- Surge tank support
- Intake manifold
- Camshaft
- Water bypass pipe from cylinder head

7. Loosen the cylinder head bolts in the sequence shown.

8. Remove the cylinder head and gasket.

To install:

9. Install the cylinder head with a new gasket onto the block.

10. Apply a light coat of clean engine oil to the cylinder and tighten in the sequence shown as follows:
   a. Step 1: 26 ft. lbs. (49 Nm).
   b. Step 2: Plus 180 degrees

11. Install or connect the following:
- Water bypass pipe to the cylinder head. Tighten to 16 ft. lbs. (21 Nm).
- Intake manifold with a new gasket
- Surge tank support. Tighten to 18 ft. lbs. (24 Nm).
- Oil dipstick tube assembly. Tighten to 18 ft. lbs. (24 Nm).
- Timing chain
- Front cover
- Timing chain tensioner
- Crankshaft pulley. Tighten to 89 ft. lbs. (120 Nm).

12. The remainder of the installation is the reverse order of removal.

13. Refill the cooling system to the correct level.

14. Start the engine and check for leaks.

Intake Manifold

REMOVAL & INSTALLATION

1NZ-FE Engines

1. Before servicing the vehicle, refer to the Precautions Section.

2. Drain the cooling system.

3. Remove or disconnect the following:
- Negative battery cable
- Water filler
- Outer front cowl top panel
- Alternator
- Air cleaner
- Accelerator cable
- Ignition coil
- PCV hoses
- Throttle body
- Intake manifold and discard the gasket

To install:

4. Install or connect the following:
- Intake manifold with a new gasket. Uniformly tighten the bolts and nuts, in several passes, from the ends, working towards the center, to 22 ft. lbs. (30 Nm).
- Engine wiring harness
- Throttle body
- PCV hoses
- Ignition coils
- Accelerator cable
- Air cleaner
- Alternator
- Water filler
- Negative battery cable

5. Refill the cooling system to the correct level.

6. Start the vehicle, check for leaks and repair if necessary.

1ZZ-FE Engines

1. Before servicing the vehicle, refer to the Precautions Section.

2. Drain the cooling system.

3. Remove or disconnect the following:
- Negative battery cable
- Accessory drive belt and alternator
- Air intake duct
- Accelerator cable
- Exhaust pipe from the manifold.
- Exhaust manifold support bracket
- Spark plug wires, then ignition coils
- Spark plugs
- PCV hoses
- Throttle body assembly
- 2 bolts securing the wiring harness protector
- Wiring connectors and ground wires
- Intake manifold support bracket
- Intake manifold and gasket

To install:

4. Install or connect the following:
- Intake manifold with a new gasket. Torque the bolts to 14 ft. lbs. (18.5 Nm).
- Harness wiring to the cylinder head and harness protector
- Fuel injectors, throttle body and the PCV hoses
- Spark plugs and ignition coils. Bolts and nuts: 80 inch lbs. (9 Nm).

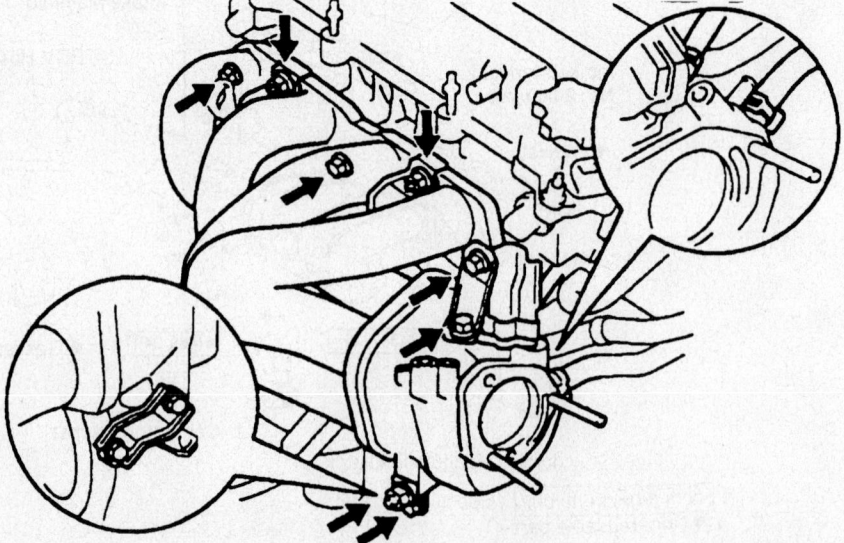

Intake manifold mounting fastener locations—1.8L (1ZZ-FE) engine

7923VG19

- Exhaust manifold and support bracket. Bolts: 37 ft. lbs. (49 Nm).
- Front exhaust pipe to the manifold. Bolts: 46 ft. lbs. (62 Nm).
- Oxygen sensor. Nuts: 14 ft. lbs. (20 Nm).
- Accelerator cable and air intake duct
- Alternator and drive belt
- Negative battery cable

5. Refill the cooling system to the correct level.

6. Start the vehicle, check for leaks and repair if necessary.

2ZZ-GE Engines

1. Before servicing the vehicle, refer to the Precautions Section.

2. Drain the cooling system.

3. Remove or disconnect the following:
- Negative battery cable
- Drive belt and alternator
- Air intake duct
- Accelerator cable
- Spark plug wires, then ignition coils
- Spark plugs
- PCV hoses
- Throttle body assembly
- Wiring harness
- Hoses and tubes connected to the head
- Intake manifold support bracket
- Intake manifold and gasket

To install:

4. Install or connect the following:
- Intake manifold with a new gasket. Bolts A: 25 ft. lbs. (34 Nm); bolt B: 34 ft. lbs. (46 Nm)
- Harness wiring to the cylinder head and harness protector
- Fuel injectors, throttle body and the PCV hoses
- Spark plugs and ignition coils. Bolts and nuts: 80 inch lbs. (9 Nm)

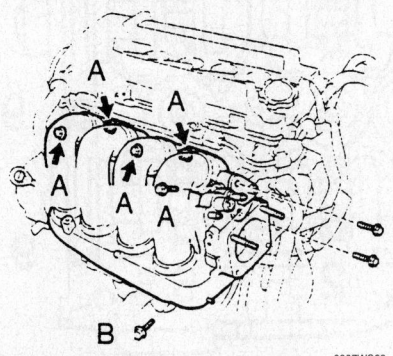

Intake manifold bolt installation—2ZZ-GE engine

9307WG93

- Oxygen sensor. Nuts: 14 ft. lbs. (20 Nm).
- Accelerator cable and air intake duct
- Alternator and drive belt
- Negative battery cable

5. Refill the cooling system to the correct level.

6. Start the vehicle, check for leaks and repair if necessary.

Exhaust Manifold

REMOVAL & INSTALLATION

1NZ-FE Engines

1. Before servicing the vehicle, refer to the Precautions Section.

2. Remove or disconnect the following:
- Negative battery cable.

➡**On vehicles equipped with an air bag, wait at least 90 seconds before proceeding after disconnecting the negative battery cable.**

- All electrical wires and vacuum hoses that interfere with removal of the exhaust manifold.
- Exhaust heat insulator
- Exhaust pipe stay
- Exhaust pipe from the manifold by removing the 2 bolts and 2 compression springs.

- Exhaust manifold and gasket

To install:

3. Clean the gasket surfaces

4. Install or connect the following:
- Exhaust manifold with a new gasket. Bolts: 20 ft. lbs. (27 Nm).
- Exhaust stay to the engine and exhaust manifold. Bolt and nuts: 29 ft. lbs. (40 Nm).
- Exhaust manifold heat insulator. Bolts: 71 inch lbs. (8 Nm).
- Exhaust pipe to the exhaust manifold with the 2 compression springs and 2 bolts. Bolts: 46 ft. lbs. (62 Nm).
- All electrical wires and vacuum hoses that were disconnected for removal of the exhaust manifold.
- Negative battery cable

1ZZ-FE Engines

1. Before servicing the vehicle, refer to the Precautions Section.

2. Drain the cooling system.

3. Remove or disconnect the following:
- Negative battery cable
- Drive belt and alternator
- Air intake duct
- Accelerator cable
- Exhaust pipe from the manifold
- Exhaust manifold support bracket
- Heat insulator from the dash panel
- Upper heat insulator
- Exhaust manifold and gasket

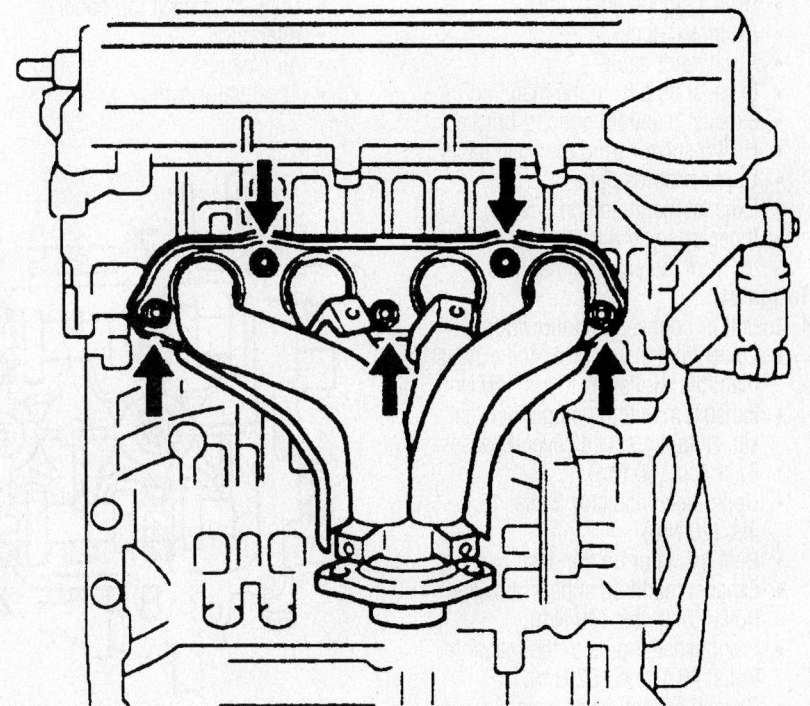

7923VG22

Exhaust manifold mounting nut locations—1.8L (1ZZ-FE) engine

- If necessary, the lower heat insulator from the exhaust manifold.

To install:

4. Install or connect the following:
- Lower heat insulator on the exhaust manifold. Bolts: 108 inch lbs. (12 Nm).
- Exhaust manifold using a new gasket. Nuts, tightened several passes: 27 ft. lbs. (37 Nm).
- Upper heat insulator. Bolts: 108 inch lbs. (12 Nm).
- Heat insulator on the dash panel
- Exhaust manifold support bracket. Bolts in an alternating pattern: 37 ft. lbs. (49 Nm).
- Front exhaust pipe to the manifold. Bolts: 46 ft. lbs. (62 Nm).
- Oxygen sensor, using new gasket and nuts. Nuts: 14 ft. lbs. (20 Nm).
- Accelerator cable and air intake duct
- Alternator and drive belt
- Negative battery cable

5. Refill the cooling system to the correct level.

6. Start the vehicle, check for leaks and repair if necessary.

2ZZ-GE Engines

1. Before servicing the vehicle, refer to the Precautions Section.
2. Drain the cooling system.
3. Remove or disconnect the following:
- Negative battery cable
- Drive belt and alternator
- Air intake duct
- Accelerator cable
- Exhaust pipe from the manifold
- Exhaust manifold support bracket
- Heat insulator from the dash panel.
- Upper heat insulator
- Exhaust manifold and gasket
- If necessary, the lower heat insulator from the exhaust manifold.

To install:

4. Install or connect the following:
- Lower heat insulator on the exhaust manifold. Bolts: 15 ft. lbs. (20 Nm).
- Exhaust manifold using a new gasket. Nuts, tightened several passes: 37 ft. lbs. (50 Nm).
- Upper heat insulator. Bolts: 15 ft. lbs. (20 Nm).
- Heat insulator on the dash panel.
- Exhaust manifold support bracket. Bolts: 37 ft. lbs. (49 Nm).
- Front exhaust pipe to the manifold. Bolts: 46 ft. lbs. (62 Nm).
- Oxygen sensor, using new gasket and nuts. Nuts: 14 ft. lbs. (20 Nm).

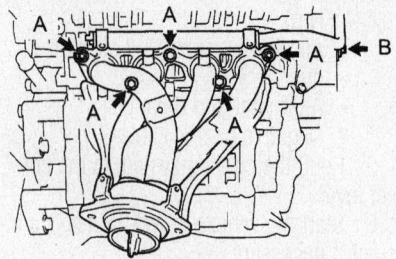

Exhaust manifold bolt locations—1.8L 2ZZ-GE engine

- Accelerator cable and air intake duct.
- Alternator and drive belt.
- Negative battery cable

5. Refill the cooling system to the correct level.

6. Start the vehicle, check for leaks and repair if necessary.

Camshaft(s)

REMOVAL & INSTALLATION

1NZ-FE Engines

1. Before servicing the vehicle, refer to the Precautions Section.
2. Drain the cooling system.
3. Remove or disconnect the following:
- Negative battery cable
- Water filler
- Outer front cowl top panel
- Alternator
- Air cleaner
- Accelerator cable

- Center exhaust pipe
- Exhaust manifold support
- Exhaust manifold
- Ignition coil
- Spark plugs
- PCV hoses
- Throttle body
- Engine wiring harness at the head
- Intake manifold
- Camshaft position sensor
- ECT sensor
- Oil control valve
- PCV valve
- Oil filer cap
- Cylinder head cover
- Fuel injectors
- Timing chain cover
- Camshaft sprockets and valve timing control assembly
- Camshafts

To install:

4. Install or connect the following:
- Camshafts. Camshaft bearing caps in 2 stages: 1st 10 ft. lbs. (13 Nm); 2nd 17 ft. lbs. (23 Nm).
- Sprockets and valve timing controller assembly, aligning the knock pin and hole. Torque the bolts to 47 ft. lbs. (64 Nm).
- Check and adjust the valves.
- Cylinder head cover
- Oil filler cap
- PCV valve
- ECT sensor
- Camshaft position sensor
- Timing chain cover
- Intake manifold
- Engine wiring harness
- Throttle body

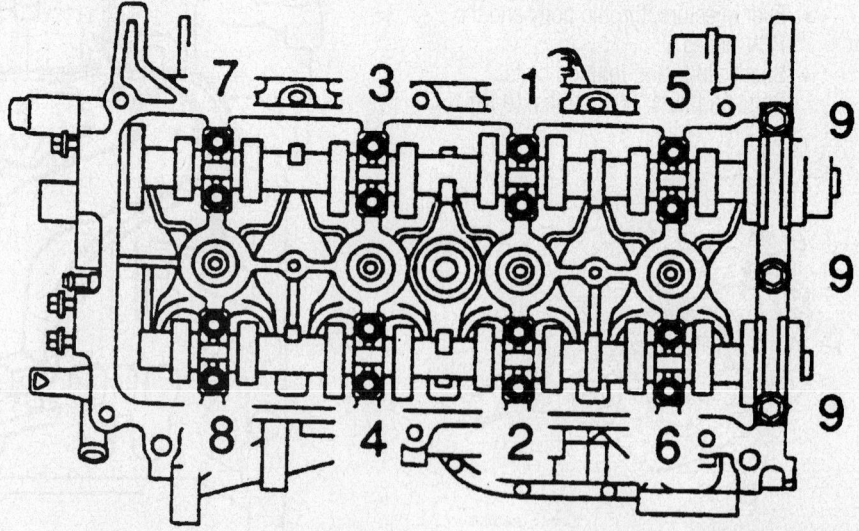

Camshaft bolt torque sequence—1NZ-FE engine

- PCV hoses
- Spark plugs
- Ignition coils
- Exhaust manifold
- Exhaust manifold support. Torque the bolts to 27 ft. lbs. (37 Nm).
- Front exhaust pipe. Torque the nuts to 46 ft. lbs. (62 Nm).
- Accelerator cable
- Air cleaner
- Alternator
- Water filler
- Negative battery cable

5. Refill the cooling system to the correct level.

6. Start the vehicle, check for leaks and repair if necessary.

1ZZ-FE Engines

1. Before servicing the vehicle, refer to the Precautions Section.

2. Remove or disconnect the following:
- Negative battery cable. On vehicles equipped with an air bag, wait at least 90 seconds before proceeding.
- Cylinder head cover

3. Turn the crankshaft so that the No. 1 piston is at TDC on the compression stroke. Check to see that the point marks on the camshaft sprockets are facing each other, if not, rotate the crankshaft 1 full revolution.

4. Tie the timing chain to each sprocket with string or wire to maintain correct valve timing.

5. Hold the camshafts with a wrench and remove the bolts securing the sprockets to the camshafts.

6. Using several passes, gradually remove the bearing cap bolts in the proper sequence. Then, remove the camshafts

To install:

7. Lubricate the camshafts with clean engine oil and place them on the cylinder head. Be sure to position the lobes for the No. 1 cylinder as shown in the illustration.

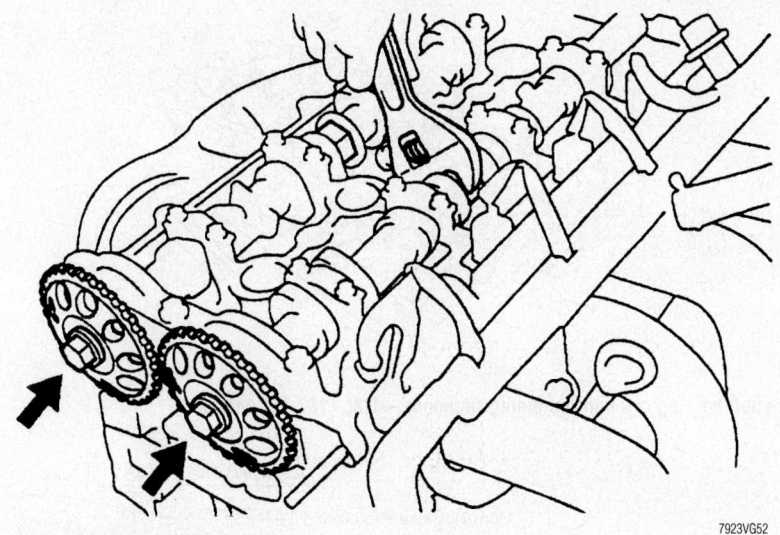

Hold the camshaft with a wrench while removing the sprocket bolt—1.8L (1ZZ-FE) engine

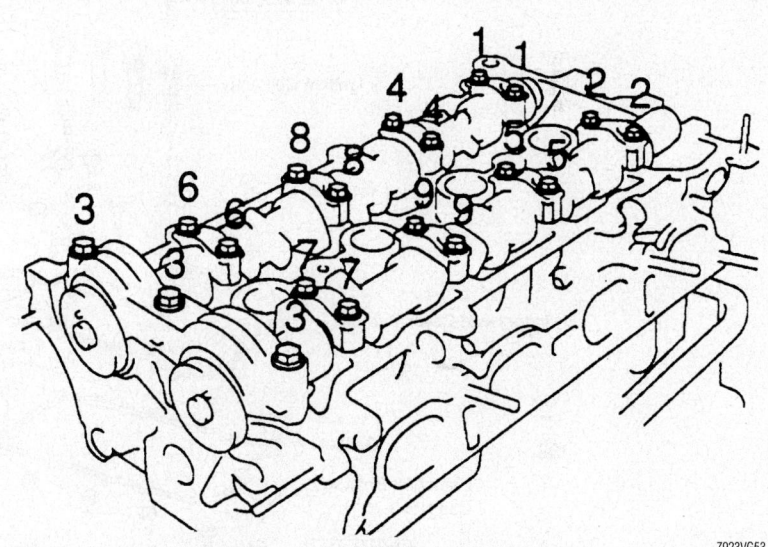

Camshaft bearing cap bolt removal sequence—1.8L (1ZZ-FE) engine

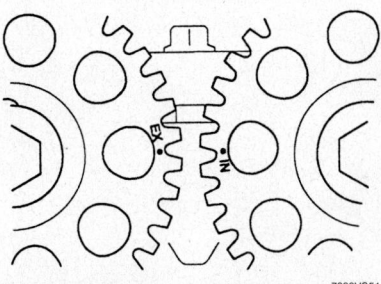

The sprocket marks will align when the No. 1 piston is at TDC on the compression stroke—1.8L (1ZZ-FE) engine

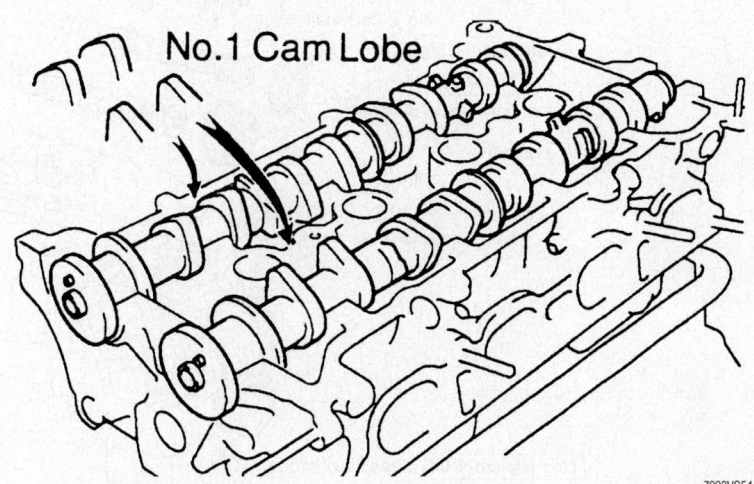

When installing the camshafts, position the lobes for the No. 1 cylinder as shown—1.8L (1ZZ-FE) engine

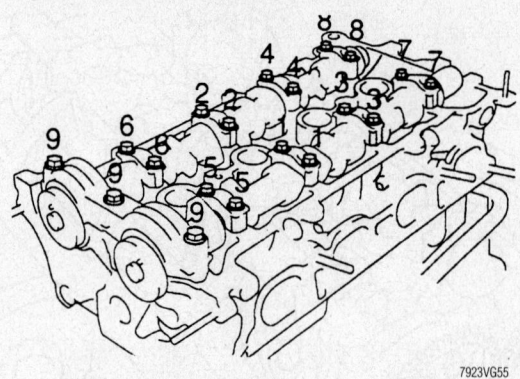

7923VG55

Camshaft bearing cap bolt tightening sequence—1.8L (1ZZ-FE) engine

8. Install the bearing caps in their original positions. Apply clean engine oil to the threads and under the heads of the bearing cap bolts. After tightening the bolts on the No. 1 bearing cap to 17 ft. lbs. (23 Nm), tighten the remaining bolts in sequence using several passes to 10 ft. lbs. (13 Nm).

9. Check the valve clearance and make adjustments as needed.

10. Install or connect the following:
- Camshaft sprockets and the chain
- Cylinder head cover
- Negative battery cable

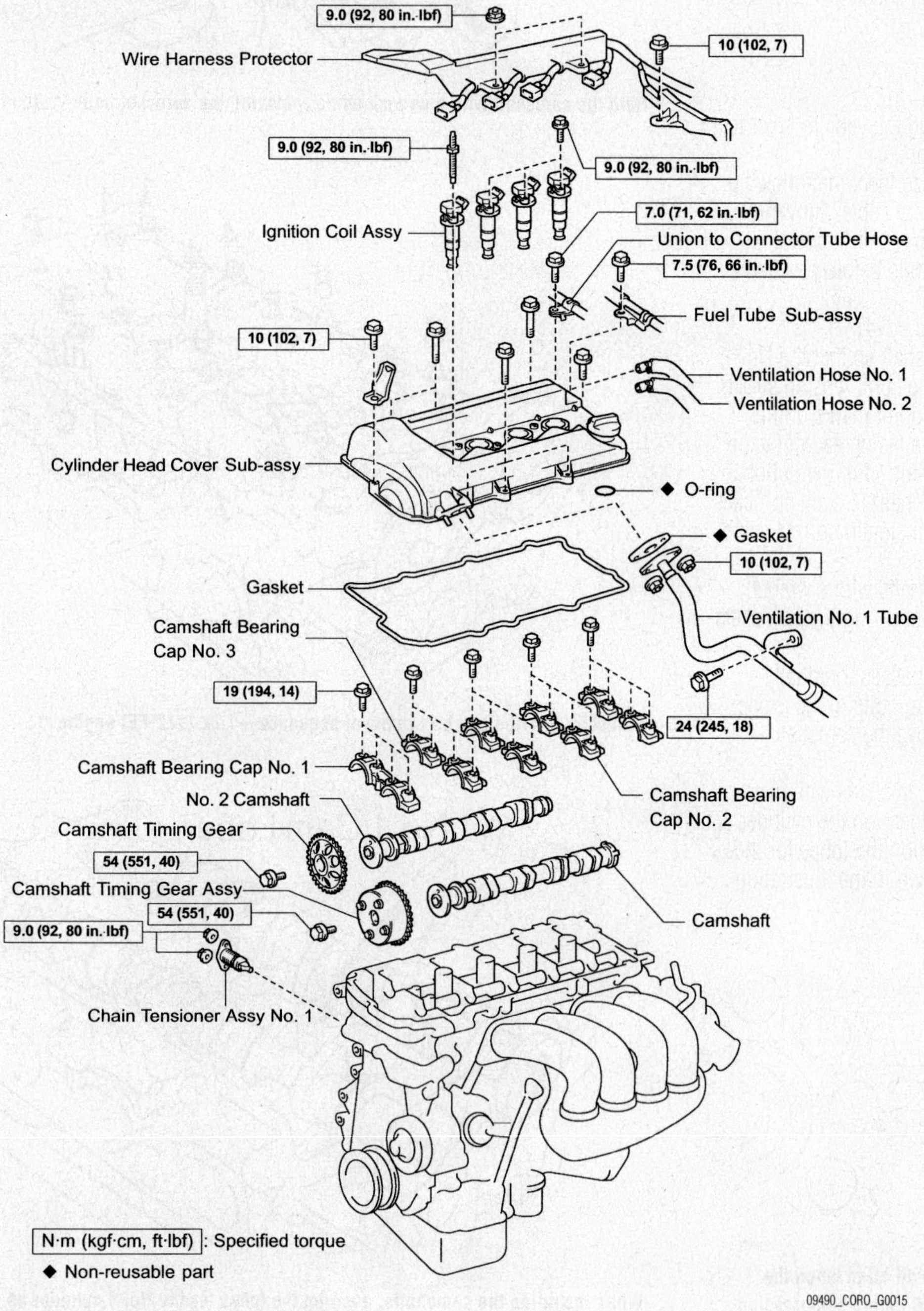

9.0 (92, 80 in.·lbf)

Wire Harness Protector

10 (102, 7)

9.0 (92, 80 in.·lbf)

9.0 (92, 80 in.·lbf)

7.0 (71, 62 in.·lbf)

Ignition Coil Assy

Union to Connector Tube Hose

7.5 (76, 66 in.·lbf)

Fuel Tube Sub-assy

10 (102, 7)

Ventilation Hose No. 1
Ventilation Hose No. 2

Cylinder Head Cover Sub-assy

◆ O-ring

◆ Gasket

10 (102, 7)

Gasket

Ventilation No. 1 Tube

Camshaft Bearing Cap No. 3

19 (194, 14)

24 (245, 18)

Camshaft Bearing Cap No. 1

No. 2 Camshaft

Camshaft Timing Gear

54 (551, 40)

Camshaft Timing Gear Assy

54 (551, 40)

Camshaft Bearing Cap No. 2

9.0 (92, 80 in.·lbf)

Camshaft

Chain Tensioner Assy No. 1

N·m (kgf·cm, ft·lbf): Specified torque

◆ Non-reusable part

09490_CORO_G0015

Exploded view of the camshaft components—2ZZ-GE engine

2ZZ-GE Engines

CELICA

1. Before servicing the vehicle, refer to the Precautions Section.

2. Remove or disconnect the following:
   - Negative battery cable. On vehicles equipped with an air bag, wait at least 90 seconds before proceeding.
   - Cylinder head cover

3. Turn the crankshaft so that the No. 1 piston is at TDC on the compression stroke. Check to see that the point marks on the camshaft sprockets are facing each other, if not, rotate the crankshaft 1 full revolution.

4. Tie the timing chain to each sprocket with string or wire to maintain correct valve timing.

5. Hold the camshafts with a wrench and remove the bolts securing the sprockets to the camshafts.

6. Using several passes, gradually remove the bearing cap bolts in the proper sequence. Then, remove the camshafts.

To install:

7. Lubricate the camshafts with clean engine oil and place them on the cylinder head. Be sure to position the lobes for the No. 1 cylinder as shown in the illustration.

8. Install the bearing caps in their original positions. Apply clean engine oil to the threads and under the heads of the bearing cap bolts. After tightening the bolts on the No. 1 bearing cap to 14 ft. lbs. (18 Nm), tighten the remaining bolts in sequence using several passes to 14 ft. lbs. (18 Nm).

9. Check the valve clearance and make adjustments as needed.

10. Install or connect the following:
    - Camshaft sprockets and the chain
    - Cylinder head cover
    - Negative battery cable

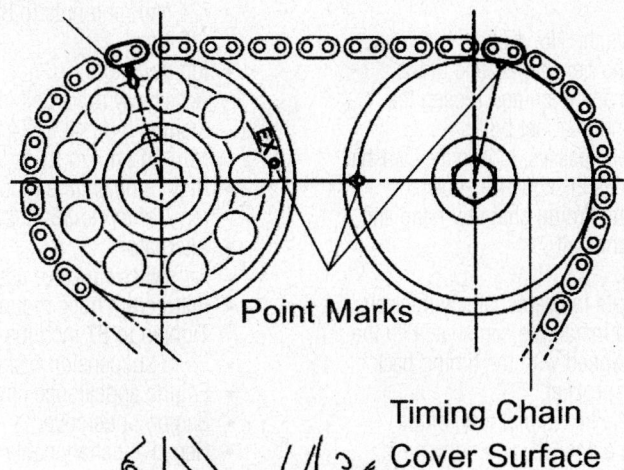

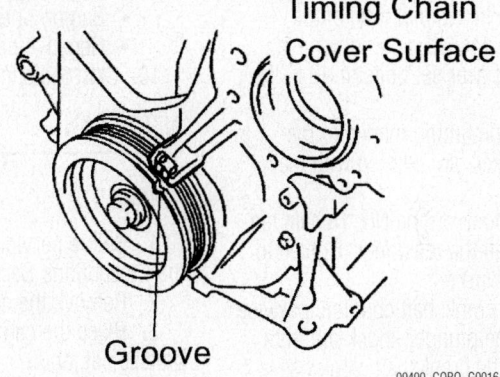

The grove in the crankshaft pulley and mark '0' on the front cover should be aligned to indicate TDC—2ZZ-GE engine

2005–06 COROLLA

1. Before servicing the vehicle, refer to the Precautions Section.

2. Remove or disconnect the following:
   - Negative battery cable
   - Right side engine undercover
   - Front suspension bar assembly
   - Engine appearance cover
   - Refrigerant lines mounting bolts
   - Air switching valve assembly
   - Accessory drive belt
   - Alternator

3. Support the engine with a suitable jack.

4. Remove or disconnect the following:
   - Right side engine mounting insulator
   - Accessory drive belt tensioner
   - Ignition coils
   - Ventilation hoses from the cylinder head cover
   - Wiring harness protector from the cylinder head cover
   - Cylinder head cover

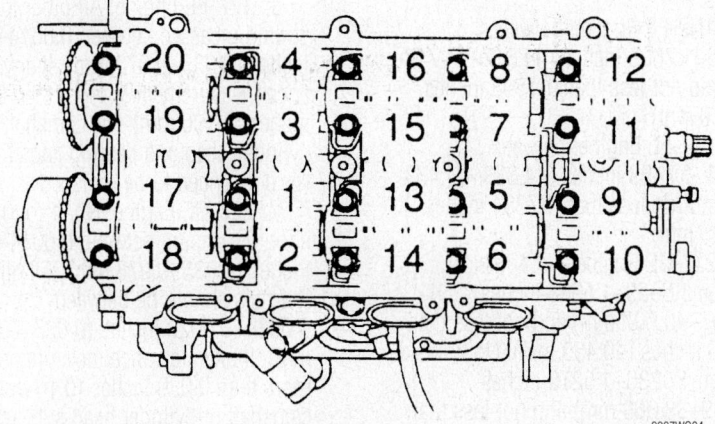

Camshaft bearing cap torque sequence—2ZZ-GE engines

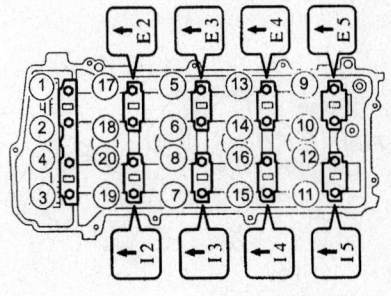

Camshaft bearing cap bolt removal sequence—2ZZ-GE engine

5. Turn the crankshaft pulley until the No. 1 cylinder is at Top Dead Center (TDC) compression.

6. Remove the No. 1 chain tensioner.

7. Hold the hexagonal lobe of the camshaft with a wrench and loosen the camshaft timing gear set bolt.

8. In several passes, loosen the camshaft bearing cap bolts in sequence shown.

9. Hold the timing chain by hand and remove the camshaft.

To install:

10. Lubricate the camshafts with clean engine oil and install the camshaft with the painted link aligned with the timing back of the camshaft sprocket.

11. Hold the hexagonal lobe of the camshaft with a wrench, and tighten the camshaft timing gear set bolts to 40 ft. lbs. (54 Nm).

12. Ensure the timing marks on the camshaft sprockets are aligned with the painted links.

13. Set the hook on the No. 1 chain tensioner and install the tensioner. Tighten to 80 inch lbs. (9 Nm).

14. Turn the crankshaft counterclockwise to disconnect the plunger knock pin from the hook. Turn the crankshaft clockwise to ensure the slipper is pushed by the plunger.

15. Install or connect the following:
- Cylinder head cover. Tighten the bolts to 88 inch lbs. (10 Nm).
- Wiring harness protector. Tighten to 80 inch lbs. (9 Nm).

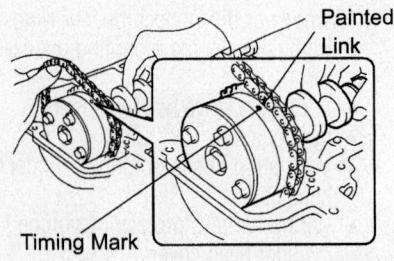

The painted link of the timing chain should align with the camshaft sprocket mark—2ZZ-GE engine

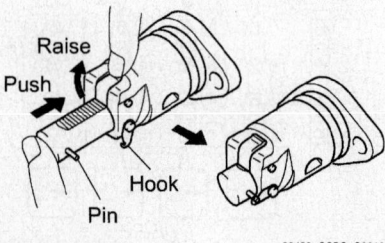

Follow the shown procedure to hook the plunger—2ZZ-GE engine

- Ventilation hoses with a new gasket. Tighten the bolts to 18 ft. lbs. (24 Nm) and nuts to 88 inch lbs. (10 Nm).
- Ignition coils
- Accessory drive belt tensioner. Tighten the bolt to 74 ft. lbs. (100 Nm) and nut to 21 ft. lbs. (29 Nm).
- Right side engine mounting insulator. Tighten to 38 ft. lbs. (52 Nm).
- Alternator
- Air switching valve assembly
- Refrigerant hose mounting bolts. Tighten to 87 inch lbs. (9.8 Nm).
- Front suspension bar assembly
- Engine appearance cover
- Engine undercover
- Negative battery cable

16. Start the engine and check for leaks.

INSPECTION

Runout

1. Before servicing the vehicle, refer to the Precautions Section.

2. Remove the camshafts.

3. Place the camshaft on a V-block, on a precise flat table.

4. Set the dial indicator to center journal.

5. Turn the camshaft to one direction by hand and measure the camshaft runout.

6. Runout should measure less than 0.0012 inches (0.03 mm).

7. Camshaft should be replaced if it exceeds the limit.

Cam Height

1. Before servicing the vehicle, refer to the Precautions Section.

2. Remove the camshafts.

3. Measure the cam height with a micrometer.

4. The intake camshaft should measure as follows:

a. 1NZ-FE Engine: between 1.7566–1.7605 inches (44.617–44.717 mm) and not less than 1.7508 inches (44.470 mm).

b. 1ZZ-FE Engine: between 1.7454–1.7493 inches (44.333–44.433 mm) and not less than 1.7394 inches (44.180 mm).

c. 2ZZ-GE Engine: No 1 intake between 1.5959–1.5998 inches (40.607–40.707 mm) and not less than 1.5925 inches (40.450 mm). No 2 intake between 1.5236–1.5276 inches (38.769–38.869 mm) and not less than 1.5201 inches (38.610 mm).

5. The exhaust camshaft should measure as follows:

a. 1NZ-FE Engine: between 1.7585–1.7624 inches (44.666–44.766 mm) and not less than 1.7528 inches (44.520 mm).

b. 1ZZ-FE Engine: between 1.7229–1.7268 inches (43.761–43.861 mm) and not less than 1.7169 inches (43.610 mm).

c. 2.4L Engine: No. 1 exhaust between 1.5728–1.5767 inches (40.019–40.119 mm) and not less than 1.5693 inches (39.860 mm). No. 2 exhaust between 1.5273–1.5313 inches (38.863–38.963 mm) and not less than 1.5240 inches (38.710 mm).

6. Camshaft should be replaced if it exceeds the limit.

Journal Oil Clearance

1. Before servicing the vehicle, refer to the Precautions Section.

2. Remove the camshafts.

3. Clean the 10 bearing caps and camshaft journals.

4. Place the 2 camshafts on the cylinder head.

5. Lay a strip of Plastigage® across each of the camshaft journals.

6. Install the 10 bearing caps.

➡ **Do not turn the camshaft.**

7. Remove the bearing caps.

8. Measure the Plastigage® at its widest point.

9. Oil clearance should measure as follows:

a. 1NZ-FE Engine: All other journals should measure between 0.0016–0.0037 inches (0.040–0.095 mm). If any clearance measurement is more than 0.0045 inches (0.115 mm), the camshaft or cylinder head and bearing caps together (or both) need to be replaced.

b. 1ZZ-FE Engine: All other journals should measure between 0.0014–0.0028 inches (0.035–0.072 mm). If any clearance measurement is more than 0.0039 inches (0.100 mm), the camshaft or cylinder head and bearing caps together (or both) need to be replaced.

c. 2ZZ-GE Engine: No. 1 journal should measure between 0.0014–0.0030 inches (0.035–0.076 mm). All other journals should be between 0.0014–0.0028 inches (0.035–0.072 mm). If any clearance measurement is more than 0.039 inches (0.10 mm), the camshaft or cylinder head sub-assembly (or both) needs to be replaced.

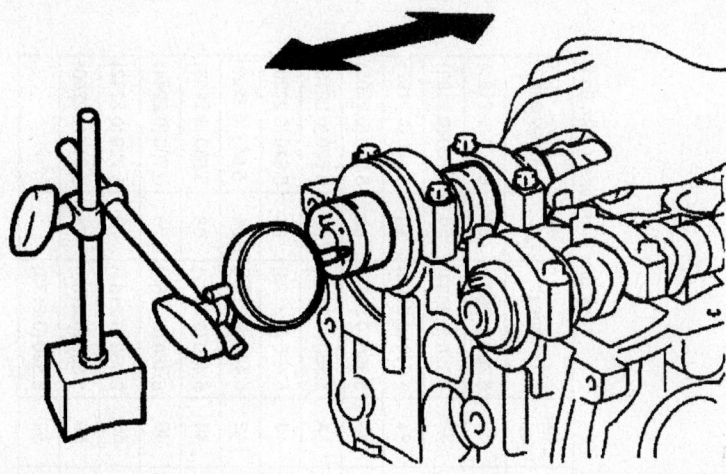

09490_SION_G0029

Measuring for camshaft endplay

End Play

1. Before servicing the vehicle, refer to the Precautions Section.

2. This procedure is performed with the camshaft installed.

3. Install a dial indicator in the thrust direction on the front end of the camshaft. Measure the end play of the dial indicator when the camshaft is moved back and forth. The dial indicator should measure as follows:

    a. 1NZ-FE Engine: between 0.0016–0.0037 inches (0.040–0.095 mm) and not exceed 0.0043 inches (0.11 mm).

    b. 1ZZ-FE Engine: Intake camshaft between 0.0016–0.0037 inches (0.040–0.095 mm) and not exceed 0.0043 inches (0.11 mm).

    c. 2ZZ-GE Engine: Intake camshaft between 0.0016–0.0055 inches (0.040–0.140 mm) and not exceed 0.0059 inches (0.15 mm).

4. Replace the cylinder head assembly if the measurement is exceeded. Replace the camshaft if damage is found on the thrust surfaces.

Valve Lash

ADJUSTMENT

1NZ-FE Engines

➡**Adjust the valve clearance when the engine is cold.**

1. Before servicing the vehicle, refer to the Precautions Section.

2. Remove or disconnect the following:
- Negative battery cable. On vehicles equipped with an air bag, wait at least 90 seconds before proceeding.
- Cylinder head cover

3. Turn the crankshaft pulley and align its groove with the timing mark **0** of the No. 1 timing cover.

4. Check that the timing marks on the camshaft sprockets and valve timing controller are facing up (12 o'clock). If not, turn the crankshaft 1 complete revolution (360 degrees).

5. Measure the clearance between the valve lifter and the camshaft. Record the measurements on the intake valves No. 1 and 2. Measure the exhaust valves at No. 1 and 3.

    a. The intake valve clearance cold is 0.006–0.010 in. (0.15–0.25mm).

    b. The exhaust valve clearance cold is 0.010–0.014 in. (0.25–0.35mm).

6. Turn the crankshaft pulley 1 revolution (360 degrees) and align the timing mark as before.

7. Measure the clearance between the valve lifter and the camshaft. Record the measurements on the intake valves No. 3 and 4. Measure the exhaust valves at No. 2 and 4.

    a. The intake valve clearance cold is 0.006–0.010 in. (0.15–0.25mm).

    b. The exhaust valve clearance cold is 0.010–0.014 in. (0.25–0.35mm).

8. To adjust the valve clearance:

    a. Set the No.1 cylinder at TDC compression. Place matchmarks on the timing chain and sprockets.

    b. Remove the 2 plugs from the timing chain cover.

    c. Turn the exhaust camshaft clockwise slightly while rotating the stopper plate on the tensioner downward. Push in on the tension plunger. When the stopper

plate cannot be easily lowered, rotate the exhaust camshaft clockwise and counter-clockwise slightly. Insert a 3mm bar into the holes in the stopper plate and tensioner to lock the tensioner. Remove the timing chain.

    d. Remove the valve timing controller assembly.

    e. Remove the lifters.

9. Determine the replacement adjusting shim size by either using the chart or the following formula:
- Intake: $N = T + A - 0.008$ in. (0.20mm)
- Exhaust: $N = T + A - 0.012$ in. (0.30mm)
- T = Thickness of removed shim
- A = Measured valve clearance
- N = Thickness of new shim

10. Install a new shim.

11. Recheck the valve clearance.

12. Install the cylinder head covers.

13. Connect the negative battery cable.

2ZZ-GE Engines

➡**Adjust the valve clearance when the engine is cold.**

1. Before servicing the vehicle, refer to the Precautions Section.

2. Remove or disconnect the following:
- Negative battery cable.
- Cylinder head covers

3. Turn the crankshaft pulley and align its groove with the timing mark **0** of the No. 1 timing cover.

4. Check that the timing marks on the camshaft sprockets are in alignment with the upper edge of the timing cover. If not, turn the crankshaft 1 complete revolution (360 degrees).

5. Measure the clearance between the valve lifter and the camshaft. Record the measurements on the intake valves No. 1 and 2. Measure the exhaust valves at No. 1 and 3.

    a. The intake valve clearance cold is 0.006–0.010 in. (0.15–0.25mm).

    b. The exhaust valve clearance cold is 0.014–0.018 in. (0.36–0.45mm).

6. Turn the crankshaft pulley 1 revolution (360 degrees) and align the groove with the timing mark **0** of the No.1 timing belt cover.

7. Measure the clearance between the valve lifter and the camshaft. Record the measurements on the intake valves No. 3 and 4. Measure the exhaust valves at No. 2 and 4.

    a. The intake valve clearance cold is 0.006–0.010 in. (0.15–0.25mm).

1ZZ–FE: Valve Lifter Selection Chart (Intake)

New lifter thickness   mm (in.)

Lifter No.	Thickness	Lifter No.	Thickness	Lifter No.	Thickness
06	5.060 (0.1992)	30	5.300 (0.2087)	54	5.540 (0.2181)
08	5.080 (0.2000)	32	5.320 (0.2094)	56	5.560 (0.2189)
10	5.100 (0.2008)	34	5.340 (0.2102)	58	5.580 (0.2197)
12	5.120 (0.2016)	36	5.360 (0.2110)	60	5.600 (0.2205)
14	5.140 (0.2024)	38	5.380 (0.2118)	62	5.620 (0.2213)
16	5.160 (0.2031)	40	5.400 (0.2126)	64	5.640 (0.2220)
18	5.180 (0.2039)	42	5.420 (0.2134)	66	5.660 (0.2228)
20	5.200 (0.2047)	44	5.440 (0.2142)	68	5.680 (0.2236)
22	5.220 (0.2055)	46	5.460 (0.2150)	70	5.700 (0.2244)
24	5.240 (0.2063)	48	5.480 (0.2157)	72	5.720 (0.2252)
26	5.260 (0.2071)	50	5.500 (0.2165)	74	5.740 (0.2260)
28	5.280 (0.2079)	52	5.520 (0.2173)		

Intake valve clearance (Cold):
0.15 – 0.25 mm (0.006 – 0.010 in.)
EXAMPLE: The 5.250 mm (0.2067 in.) lifter is installed, and
the measured clearance is 0.400 mm (0.0157 in.).
Replace the 5.250 mm (0.2067 in.) lifter with a new No. 48 lifter.

Adjusting shim chart (intake)—1ZZ-FE engine

9307WG70

1ZZ–FE: Valve Lifter Selection Chart (Exhaust)

Installed lifter thickness mm (in): 5.060 (0.1992), 5.080 (0.2000), 5.100 (0.2008), 5.120 (0.2016), 5.140 (0.2024), 5.160 (0.2031), 5.180 (0.2039), 5.200 (0.2047), 5.220 (0.2055), 5.240 (0.2063), 5.260 (0.2071), 5.280 (0.2079), 5.300 (0.2087), 5.310 (0.2091), 5.320 (0.2094), 5.330 (0.2098), 5.340 (0.2102), 5.350 (0.2106), 5.360 (0.2110), 5.370 (0.2114), 5.380 (0.2118), 5.390 (0.2122), 5.400 (0.2126), 5.410 (0.2130), 5.420 (0.2134), 5.430 (0.2138), 5.440 (0.2142), 5.450 (0.2146), 5.460 (0.2150), 5.470 (0.2154), 5.480 (0.2157), 5.490 (0.2161), 5.500 (0.2165), 5.510 (0.2169), 5.520 (0.2173), 5.530 (0.2177), 5.540 (0.2181), 5.550 (0.2185), 5.560 (0.2189), 5.570 (0.2193), 5.580 (0.2197), 5.590 (0.2201), 5.600 (0.2205), 5.620 (0.2213), 5.640 (0.2220), 5.660 (0.2228), 5.680 (0.2236), 5.700 (0.2244), 5.720 (0.2252), 5.740 (0.2260)

Measured clearance mm (in):

Measured clearance mm (in)
0.000 – 0.030 (0.0000 – 0.0012)
0.031 – 0.050 (0.0012 – 0.0020)
0.051 – 0.070 (0.0020 – 0.0028)
0.071 – 0.090 (0.0028 – 0.0035)
0.091 – 0.110 (0.0036 – 0.0043)
0.111 – 0.130 (0.0044 – 0.0051)
0.131 – 0.150 (0.0052 – 0.0059)
0.151 – 0.170 (0.0059 – 0.0067)
0.171 – 0.190 (0.0067 – 0.0075)
0.191 – 0.210 (0.0075 – 0.0083)
0.211 – 0.230 (0.0083 – 0.0091)
0.231 – 0.249 (0.0091 – 0.0098)
0.250 – 0.350 (0.0098 – 0.0138)
0.351 – 0.370 (0.0138 – 0.0146)
0.371 – 0.390 (0.0146 – 0.0154)
0.391 – 0.410 (0.0154 – 0.0161)
0.411 – 0.430 (0.0162 – 0.0169)
0.431 – 0.450 (0.0170 – 0.0177)
0.451 – 0.470 (0.0178 – 0.0185)
0.471 – 0.490 (0.0185 – 0.0193)
0.491 – 0.510 (0.0193 – 0.0201)
0.511 – 0.530 (0.0201 – 0.0209)
0.531 – 0.550 (0.0209 – 0.0217)
0.551 – 0.570 (0.0217 – 0.0225)
0.571 – 0.590 (0.0225 – 0.0232)
0.591 – 0.610 (0.0233 – 0.0240)
0.611 – 0.630 (0.0241 – 0.0248)
0.631 – 0.650 (0.0248 – 0.0256)
0.651 – 0.670 (0.0256 – 0.0264)
0.671 – 0.690 (0.0264 – 0.0272)
0.691 – 0.710 (0.0272 – 0.0280)
0.711 – 0.730 (0.0280 – 0.0287)
0.731 – 0.750 (0.0288 – 0.0295)
0.751 – 0.770 (0.0296 – 0.0303)
0.771 – 0.790 (0.0304 – 0.0311)
0.791 – 0.810 (0.0311 – 0.0319)
0.811 – 0.830 (0.0319 – 0.0327)
0.831 – 0.850 (0.0327 – 0.0335)
0.851 – 0.870 (0.0335 – 0.0343)
0.871 – 0.890 (0.0343 – 0.0350)
0.891 – 0.910 (0.0351 – 0.0358)
0.911 – 0.930 (0.0359 – 0.0366)
0.931 – 0.950 (0.0367 – 0.0374)
0.951 – 0.970 (0.0374 – 0.0382)
0.971 – 0.990 (0.0382 – 0.0390)
0.991 – 1.010 (0.0390 – 0.0398)
1.011 – 1.030 (0.0398 – 0.0406)

New lifter thickness mm (in.)

Lifter No.	Thickness	Lifter No.	Thickness	Lifter No.	Thickness
06	5.060 (0.1992)	30	5.300 (0.2087)	54	5.540 (0.2181)
08	5.080 (0.2000)	32	5.320 (0.2094)	56	5.560 (0.2189)
10	5.100 (0.2008)	34	5.340 (0.2102)	58	5.580 (0.2197)
12	5.120 (0.2016)	36	5.360 (0.2110)	60	5.600 (0.2205)
14	5.140 (0.2024)	38	5.380 (0.2118)	62	5.620 (0.2213)
16	5.160 (0.2031)	40	5.400 (0.2126)	64	5.640 (0.2220)
18	5.180 (0.2039)	42	5.420 (0.2134)	66	5.660 (0.2228)
20	5.200 (0.2047)	44	5.440 (0.2142)	68	5.680 (0.2236)
22	5.220 (0.2055)	46	5.460 (0.2150)	70	5.700 (0.2244)
24	5.240 (0.2063)	48	5.480 (0.2157)	72	5.720 (0.2252)
26	5.260 (0.2071)	50	5.500 (0.2165)	74	5.740 (0.2260)
28	5.280 (0.2079)	52	5.520 (0.2173)		

Exhaust valve clearance (Cold):
0.25 – 0.35 mm (0.010 – 0.014 in.)

EXAMPLE: The 5.340 mm (0.2102 in.) lifter is installed, and the measured clearance is 0.440 mm (0.0173 in.). Replace the 5.340 mm (0.2102 in.) lifter with a new No. 48 lifter.

Adjusting shim chart (exhaust)—1ZZ-FE engine

9307WG71

2ZZ–GE: Valve Shim Selection Chart (Intake)

The chart cross-references the **Installed lifter thickness mm (in.)** (column headings, ranging from 2.000 (0.0787) to 2.800 (0.1102)) against the **Measure clearance mm (in.)** (row headings listed below) to determine the new shim number.

Measure clearance mm (in.):

- 0.000 – 0.030 (0.0000 – 0.0012)
- 0.031 – 0.050 (0.0012 – 0.0020)
- 0.051 – 0.070 (0.0020 – 0.0028)
- 0.071 – 0.090 (0.0028 – 0.0035)
- 0.091 – 0.110 (0.0035 – 0.0043)
- 0.111 – 0.130 (0.0043 – 0.0051)
- 0.131 – 0.149 (0.0051 – 0.0059)
- 0.150 – 0.250 (0.0059 – 0.0098)
- 0.251 – 0.270 (0.0099 – 0.0106)
- 0.271 – 0.290 (0.0107 – 0.0114)
- 0.291 – 0.310 (0.0115 – 0.0122)
- 0.311 – 0.330 (0.0122 – 0.0130)
- 0.331 – 0.350 (0.0130 – 0.0138)
- 0.351 – 0.370 (0.0138 – 0.0146)
- 0.371 – 0.390 (0.0146 – 0.0154)
- 0.391 – 0.410 (0.0154 – 0.0161)
- 0.411 – 0.430 (0.0162 – 0.0169)
- 0.431 – 0.450 (0.0170 – 0.0177)
- 0.451 – 0.470 (0.0178 – 0.0185)
- 0.471 – 0.490 (0.0185 – 0.0193)
- 0.491 – 0.510 (0.0193 – 0.0201)
- 0.511 – 0.530 (0.0201 – 0.0209)
- 0.531 – 0.550 (0.0209 – 0.0217)
- 0.551 – 0.570 (0.0217 – 0.0224)
- 0.571 – 0.590 (0.0225 – 0.0232)
- 0.591 – 0.610 (0.0233 – 0.0240)
- 0.611 – 0.630 (0.0241 – 0.0248)
- 0.631 – 0.650 (0.0248 – 0.0256)
- 0.651 – 0.670 (0.0256 – 0.0264)
- 0.671 – 0.690 (0.0264 – 0.0272)
- 0.691 – 0.710 (0.0272 – 0.0280)
- 0.711 – 0.730 (0.0280 – 0.0287)
- 0.731 – 0.750 (0.0288 – 0.0295)
- 0.751 – 0.770 (0.0296 – 0.0303)
- 0.771 – 0.790 (0.0304 – 0.0311)
- 0.791 – 0.810 (0.0311 – 0.0319)
- 0.811 – 0.830 (0.0319 – 0.0327)
- 0.831 – 0.850 (0.0327 – 0.0335)
- 0.851 – 0.870 (0.0335 – 0.0343)
- 0.871 – 0.890 (0.0343 – 0.0350)
- 0.891 – 0.910 (0.0351 – 0.0358)
- 0.911 – 0.930 (0.0359 – 0.0366)
- 0.931 – 0.950 (0.0367 – 0.0374)
- 0.951 – 0.970 (0.0374 – 0.0382)
- 0.971 – 0.990 (0.0382 – 0.0390)
- 0.991 – 1.010 (0.0390 – 0.0398)
- 1.011 – 1.030 (0.0398 – 0.0406)
- 1.031 – 1.050 (0.0406 – 0.0413)

New Shim thickness mm (in.)

Shim No.	Thickness	Shim No.	Thickness	Shim No.	Thickness
00	2.000 (0.0787)	28	2.280 (0.0898)	56	2.560 (0.1008)
02	2.020 (0.0795)	30	2.300 (0.0906)	58	2.580 (0.1016)
04	2.040 (0.0803)	32	2.320 (0.0913)	60	2.600 (0.1024)
06	2.060 (0.0811)	34	2.340 (0.0921)	62	2.620 (0.1031)
08	2.080 (0.0819)	36	2.360 (0.0929)	64	2.640 (0.1039)
10	2.100 (0.0827)	38	2.380 (.0937)	66	2.660 (0.1047)
12	2.120 (0.0835)	40	2.400 (0.0945)	68	2.680 (0.1055)
14	2.140 (0.0843)	42	2.420 (0.0953)	70	2.700 (0.1063)
16	2.160 (0.0850)	44	2.440 (0.0961)	72	2.720 (0.1071)
18	2.180 (0.0858)	46	2.460 (0.0969)	74	2.740 (0.1079)
20	2.200 (0.0866)	48	2.480 (0.0976)	76	2.760 (0.1087)
22	2.220 (0.0874)	50	2.500 (0.0984)	78	2.780 (0.1094)
24	2.240 (0.0882)	52	2.520 (0.0992)	80	2.800 (0.1102)
26	2.260 (0.0890)	54	2.540 (0.1000)		

Intake valve clearance (Cold):
0.15 – 0.25 mm (0.006 – 0.010 in.)

EXAMPLE: The 2.200 mm (0.0826 in.) shim is installed, and the measured clearance is 0.400 mm (0.0157 in.).
Replace the 2.400 mm (0.0945 in.) shim with a new No. 40 shim.

Adjusting shim chart (intake)—2ZZ-GE engine

9307WG72

2ZZ–GE: Valve Shim Selection Chart (Exhaust)

New Shim thickness — mm (in.)

Shim No.	Thickness	Shim No.	Thickness	Shim No.	Thickness
00	2.000 (0.0787)	28	2.280 (0.0898)	56	2.560 (0.1008)
02	2.020 (0.0795)	30	2.300 (0.0906)	58	2.580 (0.1016)
04	2.040 (0.0803)	32	2.320 (0.0913)	60	2.600 (0.1024)
06	2.060 (0.0811)	34	2.340 (0.0921)	62	2.620 (0.1031)
08	2.080 (0.0819)	36	2.360 (0.0929)	64	2.640 (0.1039)
10	2.100 (0.0827)	38	2.380 (.0937)	66	2.660 (0.1047)
12	2.120 (0.0835)	40	2.400 (0.0945)	68	2.680 (0.1055)
14	2.140 (0.0843)	42	2.420 (0.0953)	70	2.700 (0.1063)
16	2.160 (0.0850)	44	2.440 (0.0961)	72	2.720 (0.1071)
18	2.180 (0.0858)	46	2.460 (0.0969)	74	2.740 (0.1079)
20	2.200 (0.0866)	48	2.480 (0.0976)	76	2.760 (0.1087)
22	2.220 (0.0874)	50	2.500 (0.0984)	78	2.780 (0.1094)
24	2.240 (0.0882)	52	2.520 (0.0992)	80	2.800 (0.1102)
26	2.260 (0.0890)	54	2.540 (0.1000)		

Exhaust valve clearance (Cold):
0.35 – 0.45 mm (0.014 – 0.018 in.)

EXAMPLE: The 2.200 mm (0.0862 in.) shim is installed, and the measured clearance is 0.500 mm (0.0197 in.).

Replace the 2.300 mm (0.0906 in.) shim with a new No. 30 shim.

Adjusting shim chart (exhaust)—2ZZ-GE engine

9307WG73

Valve Lifter Selection Chart (Exhaust)

The complete exhaust valve lifter selection matrix cross-references Measured clearance mm (in.) against Installed lifter thickness mm (in.) to give the new lifter number.

Measured clearance mm (in.) rows:
0.000 – 0.030 (0.0000 – 0.0012); 0.031 – 0.050 (0.0012 – 0.0020); 0.051 – 0.070 (0.0020 – 0.0028); 0.071 – 0.090 (0.0028 – 0.0035); 0.091 – 0.110 (0.0036 – 0.0043); 0.111 – 0.130 (0.0044 – 0.0051); 0.131 – 0.150 (0.0052 – 0.0059); 0.151 – 0.170 (0.0059 – 0.0067); 0.171 – 0.190 (0.0067 – 0.0075); 0.191 – 0.210 (0.0075 – 0.0083); 0.211 – 0.230 (0.0083 – 0.0091); 0.231 – 0.249 (0.0091 – 0.0098); 0.250 – 0.350 (0.0098 – 0.0138); 0.351 – 0.370 (0.0138 – 0.0146); 0.371 – 0.390 (0.0146 – 0.0154); 0.391 – 0.410 (0.0154 – 0.0161); 0.411 – 0.430 (0.0162 – 0.0169); 0.431 – 0.450 (0.0170 – 0.0177); 0.451 – 0.470 (0.0178 – 0.0185); 0.471 – 0.490 (0.0185 – 0.0193); 0.491 – 0.510 (0.0193 – 0.0201); 0.511 – 0.530 (0.0201 – 0.0209); 0.531 – 0.550 (0.0209 – 0.0217); 0.551 – 0.570 (0.0217 – 0.0224); 0.571 – 0.590 (0.0225 – 0.0232); 0.591 – 0.610 (0.0233 – 0.0240); 0.611 – 0.630 (0.0241 – 0.0248); 0.631 – 0.650 (0.0248 – 0.0256); 0.651 – 0.670 (0.0256 – 0.0264); 0.671 – 0.690 (0.0264 – 0.0272); 0.691 – 0.710 (0.0272 – 0.0280); 0.711 – 0.730 (0.0280 – 0.0287); 0.731 – 0.750 (0.0288 – 0.0295); 0.751 – 0.770 (0.0296 – 0.0303); 0.771 – 0.790 (0.0304 – 0.0311); 0.791 – 0.810 (0.0311 – 0.0319); 0.811 – 0.830 (0.0319 – 0.0327); 0.831 – 0.850 (0.0327 – 0.0335); 0.851 – 0.870 (0.0335 – 0.0343); 0.871 – 0.890 (0.0343 – 0.0350); 0.891 – 0.910 (0.0351 – 0.0358); 0.911 – 0.930 (0.0359 – 0.0366); 0.931 – 0.950 (0.0367 – 0.0374); 0.951 – 0.970 (0.0374 – 0.0382); 0.971 – 0.990 (0.0382 – 0.0390); 0.991 – 1.010 (0.0390 – 0.0398); 1.011 – 1.030 (0.0398 – 0.0406)

Installed lifter thickness mm (in.) columns range from 5.060 (0.1992) through 5.740 (0.2260).

New lifter thickness mm (in.)

Lifter No.	Thickness	Lifter No.	Thickness	Lifter No.	Thickness
06	5.060 (0.1992)	30	5.300 (0.2087)	54	5.540 (0.2181)
08	5.080 (0.2000)	32	5.320 (0.2094)	56	5.560 (0.2189)
10	5.100 (0.2008)	34	5.340 (0.2102)	58	5.580 (0.2197)
12	5.120 (0.2016)	36	5.360 (0.2110)	60	5.600 (0.2205)
14	5.140 (0.2024)	38	5.380 (0.2118)	62	5.620 (0.2213)
16	5.160 (0.2031)	40	5.400 (0.2126)	64	5.640 (0.2220)
18	5.180 (0.2039)	42	5.420 (0.2134)	66	5.660 (0.2228)
20	5.200 (0.2047)	44	5.440 (0.2142)	68	5.680 (0.2236)
22	5.220 (0.2055)	46	5.460 (0.2150)	70	5.700 (0.2244)
24	5.240 (0.2063)	48	5.480 (0.2157)	72	5.720 (0.2252)
26	5.260 (0.2071)	50	5.500 (0.2165)	74	5.740 (0.2260)
28	5.280 (0.2079)	52	5.520 (0.2173)		

Exhaust valve clearance (Cold):

0.25 - 0.35 mm (0.010 - 0.014 in.)

EXAMPLE: The 5.340 mm (0.2102 in.) lifter is installed, and the measured clearance is 0.440 mm (0.0173 in.). Replace the 5.340 mm (0.2102 in.) lifter with a new No. 48 lifter.

Adjusting shim chart (exhaust)—1NZ-FE engine

09490-CORO-G0034

Valve Lifter Selection Chart (Intake)

(Large cross-reference matrix: vertical axis = Installed lifter thickness mm (in.) from 5.060 (0.1992) to 5.740 (0.2260); horizontal axis = Measured clearance mm (in.); cell values are new lifter numbers.)

Measured clearance mm (in.):

- 0.000 - 0.030 (0.0000 - 0.0012)
- 0.031 - 0.050 (0.0012 - 0.0020)
- 0.051 - 0.070 (0.0020 - 0.0028)
- 0.071 - 0.090 (0.0028 - 0.0035)
- 0.091 - 0.110 (0.0036 - 0.0043)
- 0.111 - 0.130 (0.0044 - 0.0051)
- 0.131 - 0.149 (0.0052 - 0.0059)
- 0.150 - 0.250 (0.0059 - 0.0098)
- 0.251 - 0.270 (0.0099 - 0.0106)
- 0.271 - 0.290 (0.0107 - 0.0114)
- 0.291 - 0.310 (0.0115 - 0.0122)
- 0.311 - 0.330 (0.0122 - 0.0130)
- 0.331 - 0.350 (0.0130 - 0.0138)
- 0.351 - 0.370 (0.0138 - 0.0146)
- 0.371 - 0.390 (0.0146 - 0.0154)
- 0.391 - 0.410 (0.0154 - 0.0161)
- 0.411 - 0.430 (0.0162 - 0.0169)
- 0.431 - 0.450 (0.0170 - 0.0177)
- 0.451 - 0.470 (0.0178 - 0.0185)
- 0.471 - 0.490 (0.0185 - 0.0193)
- 0.491 - 0.510 (0.0193 - 0.0201)
- 0.511 - 0.530 (0.0201 - 0.0209)
- 0.531 - 0.550 (0.0209 - 0.0217)
- 0.551 - 0.570 (0.0217 - 0.0224)
- 0.571 - 0.590 (0.0225 - 0.0232)
- 0.591 - 0.610 (0.0233 - 0.0240)
- 0.611 - 0.630 (0.0241 - 0.0248)
- 0.631 - 0.650 (0.0248 - 0.0256)
- 0.651 - 0.670 (0.0256 - 0.0264)
- 0.671 - 0.690 (0.0264 - 0.0272)
- 0.691 - 0.710 (0.0272 - 0.0280)
- 0.711 - 0.730 (0.0280 - 0.0287)
- 0.731 - 0.750 (0.0288 - 0.0295)
- 0.751 - 0.770 (0.0296 - 0.0303)
- 0.771 - 0.790 (0.0304 - 0.0311)
- 0.791 - 0.810 (0.0311 - 0.0319)
- 0.811 - 0.830 (0.0319 - 0.0327)
- 0.831 - 0.850 (0.0327 - 0.0335)
- 0.851 - 0.870 (0.0335 - 0.0343)
- 0.871 - 0.890 (0.0343 - 0.0350)
- 0.891 - 0.910 (0.0351 - 0.0358)
- 0.911 - 0.930 (0.0359 - 0.0366)

New lifter thickness   mm (in.)

Lifter No.	Thickness	Lifter No.	Thickness	Lifter No.	Thickness
06	5.060 (0.1992)	30	5.300 (0.2087)	54	5.540 (0.2181)
08	5.080 (0.2000)	32	5.320 (0.2094)	56	5.560 (0.2189)
10	5.100 (0.2008)	34	5.340 (0.2102)	58	5.580 (0.2197)
12	5.120 (0.2016)	36	5.360 (0.2110)	60	5.600 (0.2205)
14	5.140 (0.2024)	38	5.380 (0.2118)	62	5.620 (0.2213)
16	5.160 (0.2031)	40	5.400 (0.2126)	64	5.640 (0.2220)
18	5.180 (0.2039)	42	5.420 (0.2134)	66	5.660 (0.2228)
20	5.200 (0.2047)	44	5.440 (0.2142)	68	5.680 (0.2236)
22	5.220 (0.2055)	46	5.460 (0.2150)	70	5.700 (0.2244)
24	5.240 (0.2063)	48	5.480 (0.2157)	72	5.720 (0.2252)
26	5.260 (0.2071)	50	5.500 (0.2165)	74	5.740 (0.2260)
28	5.280 (0.2079)	52	5.520 (0.2173)		

Intake valve clearance (Cold):
0.15 - 0.25 mm (0.006 - 0.010 in.)

EXAMPLE: The 5.250 mm (0.2067 in.) lifter is installed, and the measured clearance is 0.400 mm (0.0157 in.).
Replace the 5.250 mm (0.2067 in.) lifter with a new No. 48 lifter.

Adjusting shim chart (intake)—1NZ-FE engine

09490_CORO-G0035

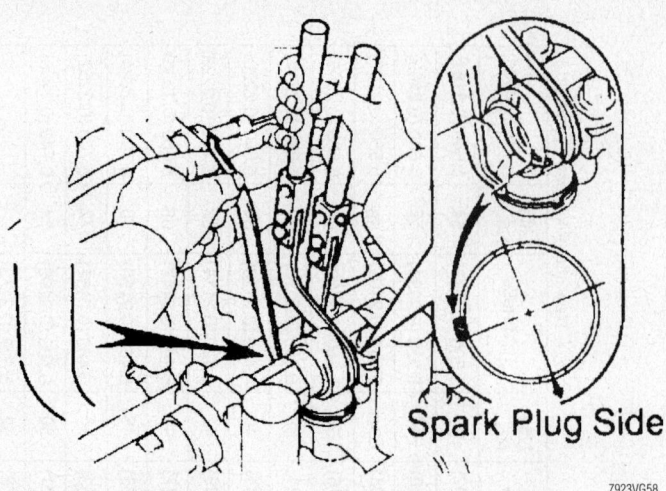

Common method of removing valve shims

7923VG58

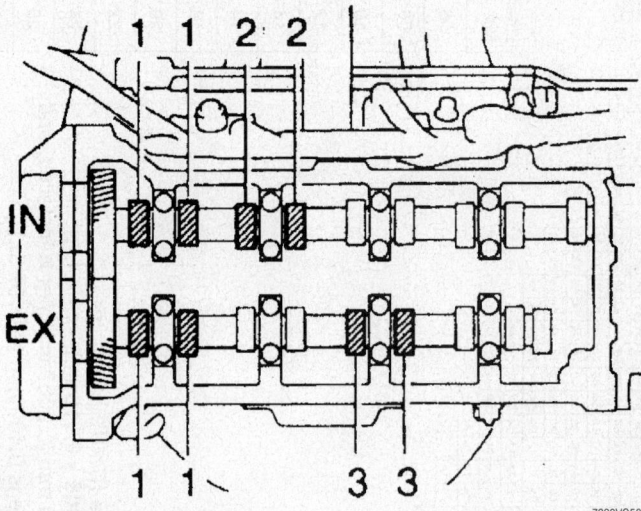

7923VG59

Intake valves (1 and 2) and exhaust valves (1 and 3)—1NZ-FE engine

b. The exhaust valve clearance cold is 0.014–0.018 in. (0.36–0.45mm).

8. To adjust the intake valve clearance:

a. Remove the intake camshaft.

b. Using a small screwdriver and a magnetic finger, remove the adjusting shim.

c. Determine the replacement adjusting shim size by either using the chart or the following formula:

- Intake: $N = T + A - 0.008$ in. (0.20mm)
- T = Thickness of removed shim
- A = Measured valve clearance
- N = Thickness of new shim

d. Install a new shim.

e. Install intake camshaft.

f. Recheck the valve clearance.

9. To adjust the exhaust valve clearance:

a. Turn the crankshaft to position the cam lobe of the camshaft on the valve to be adjusted, upward.

b. Turn the valve lifter so that the notch is perpendicular to the camshaft and facing the spark plug side.

c. Using SST 09248–55040 (valve lifter press), or equivalent, hold the camshaft in place.

d. Using SST 09248–55040 (valve lifter press), or equivalent, press down the valve lifter and place SST 09248–05420 (valve lifter stopper), or equivalent between the camshaft and valve lifter.

e. Remove the SST 09248–44040 tool.

f. Using a small screwdriver and a magnetic finger, remove the adjusting shim.

10. Determine the replacement adjusting shim size by either using the chart or the following formula:

- Exhaust: $N = T + A - 0.014$ in. (0.36mm)
- T = Thickness of removed shim
- A = Measured valve clearance
- N = Thickness of new shim

11. Install a new shim.

12. Recheck the valve clearance.

13. Install or connect the following:
- Cylinder head covers
- Negative battery cable

Oil Pan

REMOVAL & INSTALLATION

1NZ-FE Engines

1. Before servicing the vehicle, refer to the Precautions Section.

2. Drain the engine oil.

3. Remove or disconnect the following:
- Negative battery cable
- Oil filter
- Front exhaust pipe
- Engine under covers
- Oil pan bolts

4. Using a thin blade, cut the sealer holding the oil pan and lower the pan.

5. Installation is the reverse of removal. Use RTV sealer. Torque the bolts, in a crisscross pattern, to 80 inch lbs. (9 Nm).

1ZZ-FE Engines

1. Before servicing the vehicle, refer to the Precautions Section.

2. Drain the engine oil.

3. Remove or disconnect the following:
- Negative battery cable. On vehicles equipped with an air bag, wait at least 90 seconds before proceeding.
- Undercovers
- Front exhaust pipe
- Oil pan mounting bolts and nuts
- Oil pan, cutting off the applied sealer.

To install:

4. Remove any old sealant from the oil pan flange and thoroughly clean the sealing surface.

5. Install or connect the following:
- Oil pan. Tighten the bolts and nuts in several passes. Bolts and nuts: 80 inch lbs. (9 Nm).
- Front exhaust pipe
- Negative battery cable
- Undercovers

6. Refill the engine with oil to the correct level.

7. Start the vehicle, check for leaks and repair if necessary.

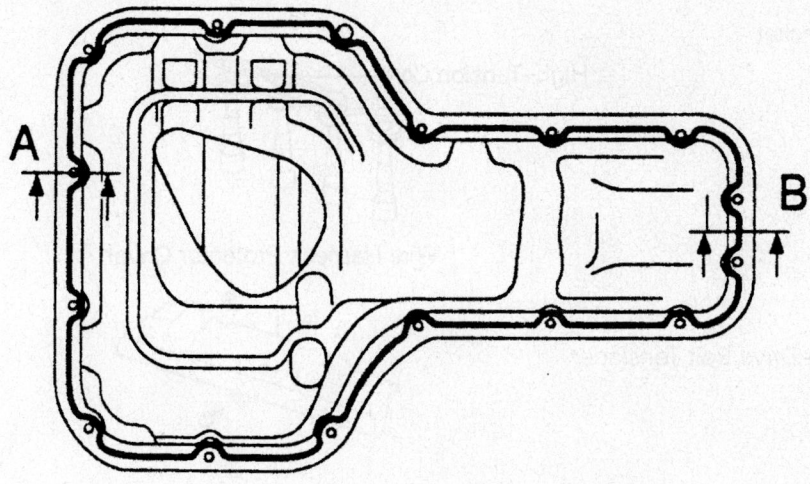

Seal Width
4 – 5 mm

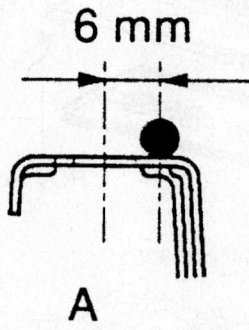

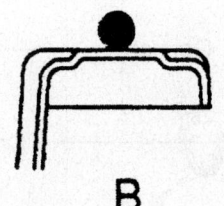

7923VG72

Apply sealant to the oil pan as shown—1.8L (1ZZ-FE) engine

2ZZ-GE Engines

1. Before servicing the vehicle, refer to the Precautions Section.
2. Drain the engine oil.
3. Remove or disconnect the following:
   - Negative battery cable. On vehicles equipped with an air bag, wait at least 90 seconds before proceeding.
   - Undercovers
   - Front exhaust pipe
   - Oil pan mounting bolts and nuts
   - Oil pan, cutting off the applied sealer.

To install:

4. Remove any old sealant from the oil pan flange and thoroughly clean the sealing surface.
5. Install or connect the following:
   - Oil pan. Tighten the bolts and nuts in several passes. Bolts and nuts: 80 inch lbs. (9 Nm).

   - Front exhaust pipe
   - Negative battery cable
   - Undercovers

6. Refill the engine with oil to the correct level.
7. Start the vehicle, check for leaks and repair if necessary.

Oil Pump

REMOVAL & INSTALLATION

1NZ-FE Engines

1. Before servicing the vehicle, refer to the Precautions Section.
2. Drain the engine oil.
3. Remove or disconnect the following:

   - Negative battery cable
   - Timing chain cover

- 2 bolts, 3 screws and the oil pump cover from the timing chain cover
- Drive and driven rotors
- Plug, spring and relief valve

4. Inspect the relief valve motion
5. Check rotor side clearance: 0.0012–0.0035 in. (0.03–0.09mm).
6. Check rotor tip clearance: 0.0024–0.0071 in. (0.06–0.18mm).
7. Check rotor-to-body clearance: 0.0098–0.0128 in. (0.250–0.325mm).

To install:

8. Install or connect the following:
   - Relief valve and spring. Torque the plug to 18 ft. lbs. (24 Nm).
   - Drive and driven rotors with the marks on the cover side
   - Cover. Torque the bolts to 80 inch lbs. (9 Nm); the screws to 96 inch lbs. (11 Nm).
   - Timing chain cover
   - Negative battery cable

9. Refill the engine with oil to the correct level.
10. Start the vehicle, check for leaks and repair if necessary.

1ZZ-FE Engines

1. Before servicing the vehicle, refer to the Precautions Section.
2. Drain the engine oil.
3. Remove or disconnect the following:
   - Negative battery cable
   - Timing chain and crankshaft sprocket
   - Oil pump and gasket

To install:

4. Clean the mounting surface.
5. Install or connect the following:
   - Oil pump, with new gasket. Bolts: 80 inch lbs. (9 Nm).
   - Crankshaft sprocket and timing chain
   - Negative battery cable

6. Refill the engine with oil to the correct level.
7. Start the vehicle, check for leaks and repair if necessary.

2ZZ-GE Engines

1. Before servicing the vehicle, refer to the Precautions Section.
2. Drain the engine oil.
3. Remove or disconnect the following:
   - Negative battery cable
   - Timing chain and crankshaft sprocket
   - Oil pump and gasket

To install:

4. Clean the mounting surface.
5. Install or connect the following:

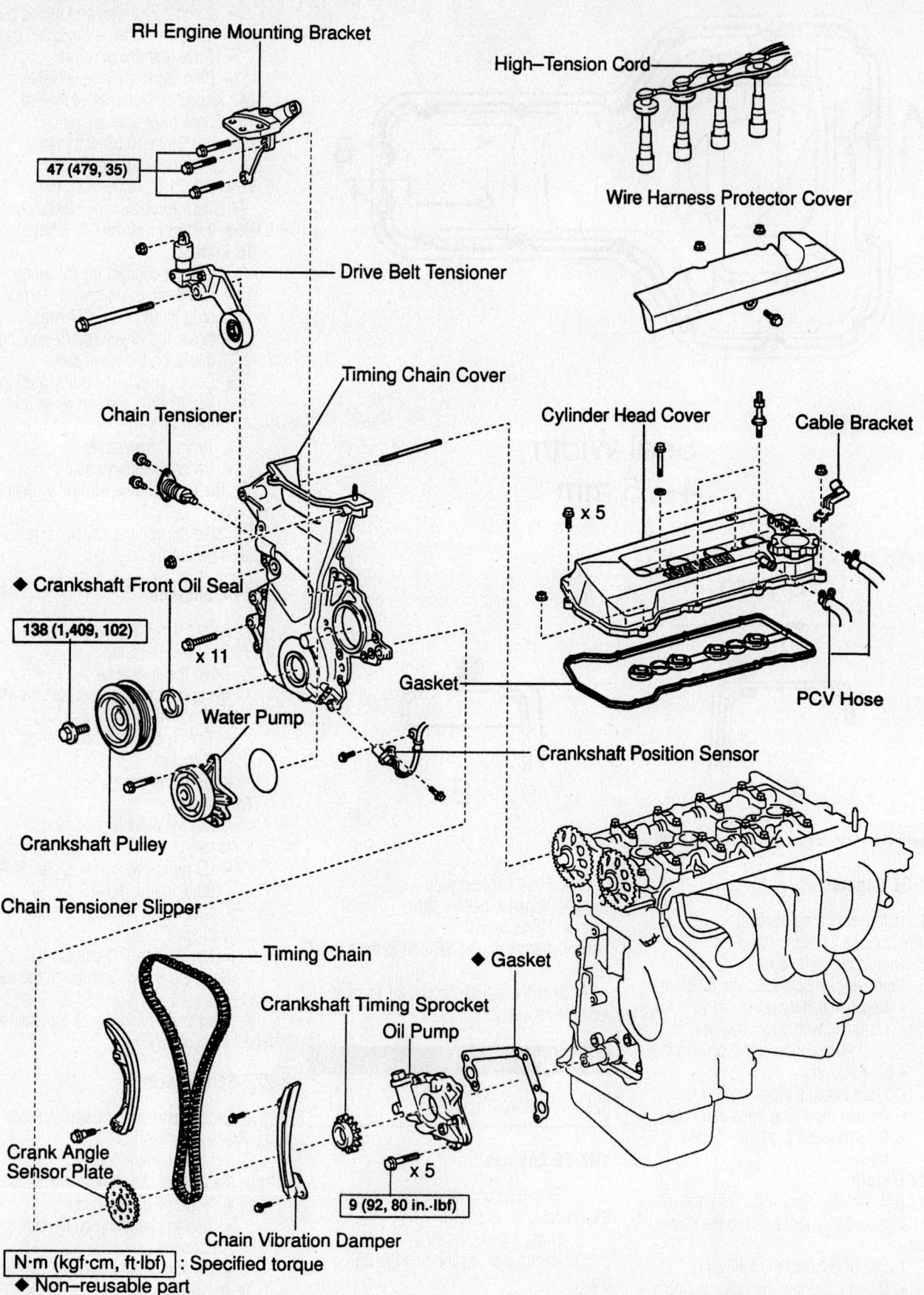

RH Engine Mounting Bracket

High–Tension Cord

47 (479, 35)

Wire Harness Protector Cover

Drive Belt Tensioner

Timing Chain Cover

Chain Tensioner

Cylinder Head Cover

Cable Bracket

x 5

◆ Crankshaft Front Oil Seal

138 (1,409, 102)

x 11

Gasket

PCV Hose

Water Pump

Crankshaft Position Sensor

Crankshaft Pulley

Chain Tensioner Slipper

Timing Chain

◆ Gasket

Crankshaft Timing Sprocket

Oil Pump

Crank Angle Sensor Plate

Chain Vibration Damper

9 (92, 80 in.·lbf)

x 5

N·m (kgf·cm, ft·lbf) : Specified torque
◆ Non–reusable part

7923VGB0

Exploded view of the oil pump mounting—1.8L (1ZZ-FE) engine

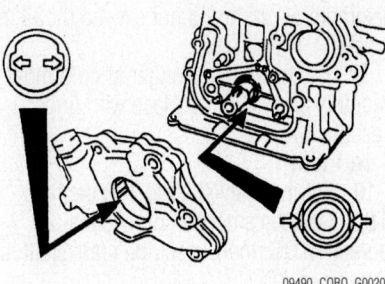

When installing the oil pump, align the spline teeth of rotor with the large teeth of the crankshaft—1ZZ-FE and 2ZZ-GE engines

09490_CORO_G0020

- Oil pump, with new gasket. Bolts: 80 inch lbs. (9 Nm).
- Crankshaft sprocket and timing chain
- Negative battery cable

6. Refill the engine with oil to the correct level.

7. Start the vehicle, check for leaks and repair if necessary.

Rear Main Seal

REMOVAL & INSTALLATION

1ZZ-FE, 2ZZ-GE and 1NZ-FE Engines

1. Remove or disconnect the following:
- Transaxle
- Clutch assembly, if equipped
- Flywheel or flexplate

2. Use a small sharp knife to cut off the lip of the oil seal. Take great care not to score any metal with the knife.

3. Use a small prytool to pry the old seal from the retaining plate. Be careful not to damage the plate. Protect the tip of the tool with tape and pad the fulcrum point with cloth.

4. Inspect the crankshaft and seal lip contact surfaces for any sign of damage.

To install:

5. Apply a light coat of multi-purpose grease to the lip of a new oil seal. Loosely fit the seal into place by hand, making sure it is not crooked.

6. Use a seal driver of the correct size to install the seal. Tap it into place until the surface of the seal is flush with the edge of the housing.

7. Install or connect the following:
- Flywheel or flexplate
- Clutch assembly, if equipped
- Transaxle

8. Start the engine and check for leaks.

Timing Chain, Sprockets, Front Cover and Seal

REMOVAL & INSTALLATION

1ZZ-FE Engines

1. Before servicing the vehicle, refer to the Precautions Section.

2. Drain the cooling system.

3. Remove or disconnect the following:
- Negative battery cable
- Right engine cover
- Accessory drive belt and generator
- Power steering pump, without disconnecting the hoses.
- Right engine mount
- Cylinder head cover
- Turn the crankshaft so the No. 1 piston is at TDC on the compression stroke.
- Crankshaft pulley
- Crankshaft position sensor from the timing chain cover.
- Accessory drive belt tensioner.
- Right engine mounting bracket
- Chain tensioner
- Water pump
- Timing chain cover
- Crankshaft angle sensor plate
- Timing chain tensioner slipper
- Timing chain and crankshaft timing sprocket.
- Timing chain vibration damper
- Valve timing control assembly and camshaft timing sprocket

4. Drive the seal from the cover.

5. Pull the chain to its full length and measure the length of any 16 consecutive links. The length should not exceed 4.827 inches (122.6mm).

6. Check the slipper and damper wear. Maximum wear should not exceed 0.039 in. (1mm).

7. The tensioner plunger should move smoothly and lock into place with finger pressure.

To install:

8. Apply engine oil from the tip of the intake camshaft, back to 16mm.

9. Align the timing mark on the valve timing controller with the knock pin and gently push the valve timing controller onto the camshaft.

10. Set the No.1 piston to TDC compression. The key on the crankshaft should be at 12 o'clock.

11. Install or connect the following:
- Sprockets. Torque the bolt to 33 ft. lbs. (45 Nm). Turn the camshafts to align the point marks on the sprockets.
- Chain damper. Bolts: 96 inch lbs. (11 Nm).
- Timing chain and crankshaft sprocket. Be sure to align the yellow chain link with the mark on the crankshaft sprocket.
- Timing chain on the camshaft sprockets. Align the yellow links with the marks on the camshaft sprockets.
- Chain tensioner slipper. Bolt: 14 ft. lbs. (18.5 Nm).
- Crankshaft angle sensor plate with the **F** mark forward
- New seal in the front cover
- Silicone sealant to the timing chain cover as illustrated
- Timing chain cover

12. Water pump. Tighten the 10mm bolts marked "C" to 80 inch lbs. (9 Nm), those marked "A" to 10 ft. lbs. (13 Nm), and the remaining 10mm bolts to 96 inch lbs. (11 Nm). Tighten the 12mm bolts to 14 ft. lbs. (18.5 Nm). Be sure to install the bolts in their original locations. Bolt lengths:
- a. A: 1.77 in. (45mm).
- b. B: 1.38 in. (35mm).
- c. C: 1.18 in. (30mm).
- d. D: 0.98 in. (25mm).

13. With a Torx wrench, tighten the stud bolt to 82 inch lbs. (9.3 Nm).

14. Install or connect the following:
- Right engine mounting bracket. Bolts, with sealant applied: 35 ft. lbs. (47 Nm).
- Accessory drive belt tensioner. Bolt: 51 ft. lbs. (69 Nm). Nut: 21 ft. lbs. (29 Nm).
- Crankshaft position sensor. Tighten to 80 inch lbs. (9 Nm).
- Crankshaft pulley. Bolt: 102 ft. lbs. (138 Nm).

15. Release the ratchet pawl and compress the chain tensioner. Place the hook on the pin to keep the tensioner compressed.

16. Install the tensioner, using a new O-ring. Torque the bolts to 80 inch lbs. (9 Nm).

17. Turn the crankshaft counterclockwise and remove the hook from the pin. Turn the crankshaft clockwise and be sure the slipper is pushed by the plunger.

18. Check the valve timing by turning the crankshaft clockwise until the mark of the pulley is aligned with the mark on the timing chain cover. The marks on the camshaft sprockets should be facing each other as shown.

19. Install or connect the following:

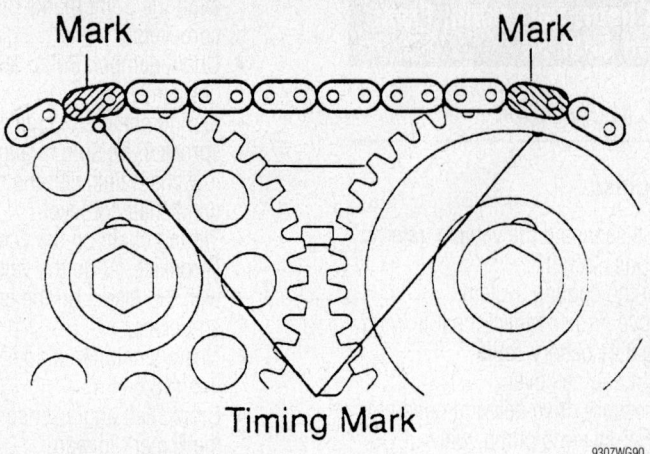

Timing chain link marks—1ZZ-FE and 2ZZ-GE engines

- Silicone sealant to the 2 areas where the timing chain cover meets the cylinder head.
- Cylinder head cover. Bolts with washers in the sequence shown: 80 inch lbs. (9 Nm). Bolts without washers: 96 inch lbs. (11 Nm).
- Right engine mount. Bolts and nuts: 38 ft. lbs. (52 Nm).
- Power steering pump
- Alternator and drive belt
- Right engine undercover
- Negative battery cable
- Washer tank

20. Refill the cooling system to the correct level.

21. Start the vehicle, check for leaks and repair if necessary.

1NZ-FE Engines

1. Before servicing the vehicle, refer to the Precautions Section.
2. Drain the cooling system.
3. Remove or disconnect the following:
   - Negative battery cable
   - Right front wheel
   - Alternator
   - Power steering pump
   - Right engine mount insulator. Use a jack and wood block for support.
   - With A/C, the bolt holding the liquid tube to the insulator
   - Ignition coils
   - Cylinder head cover

4. Place No.1 cylinder on TDC compression. Make sure that the timing marks on the camshaft sprockets and valve timing controller assembly are facing UP (12 o'clock). If not, turn the crankshaft 360 degrees to align the marks.
5. Remove or disconnect the following:
   - Crankshaft pulley bolt
   - Pulley and pin

- Crankshaft position sensor
- Right engine mount bracket
- Water pump
- Oil control valve
- 13 bolts, 1 nut and 1 stud bolt. Pry the cover off.
- 2 O-rings from the block and pan
- Chain tensioner
- Tensioner slipper
- Chain vibration damper
- Chain

6. Drive the seal from the cover.
7. Pull the chain to its full length and measure the length of any 16 consecutive links. The length should not exceed 4.85 inches (123.2mm).

8. Check the slipper and damper wear. Maximum wear should not exceed 0.039 in. (1mm).

9. The tensioner plunger should move smoothly and lock into place with finger pressure.

To install:

10. Set the crankshaft at 140 degrees ATDC. Set the camshaft sprockets at 20 degrees ATDC; then, reset the crankshaft to 20 degrees ATDC.

11. Install or connect the following:
    - New seal, driven into place until flush with the cover edge
    - Vibration damper. Torque the bolts to 80 inch lbs. (9 Nm).
    - Timing chain

➡ **A new chain will have 3 marked links to align with the 3 sprockets, as shown in the accompanying illustration.**

- Slipper
- Tensioner. Torque the bolts to 80 inch lbs. (9 Nm).
- Cover, using a 4–5mm bead of RTV sealer and new O-rings to the block and pan

12. Uniformly tighten the bolts in several passes, using the accompanying illustration, to:
    - A, C, E and G: 96 inch lbs. (11 Nm)
    - B, D and F: 18 ft. lbs. (24 Nm)

13. Bolt lengths are as follows:
    - A: 20mm

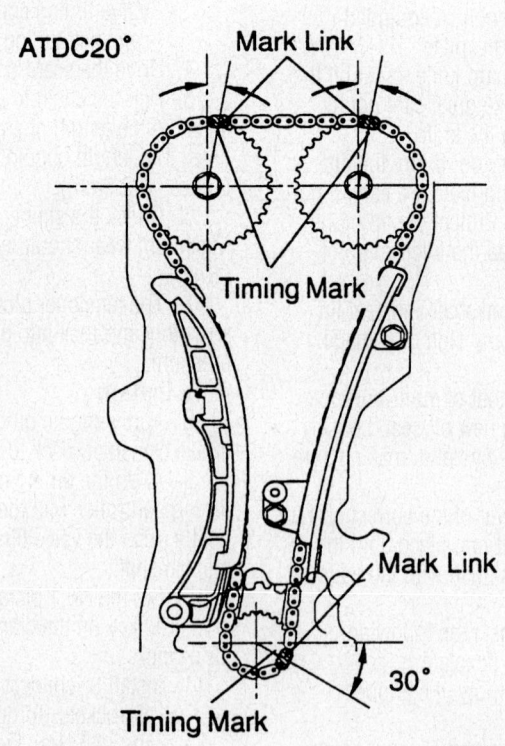

Timing chain installation—1NZ-FE engine

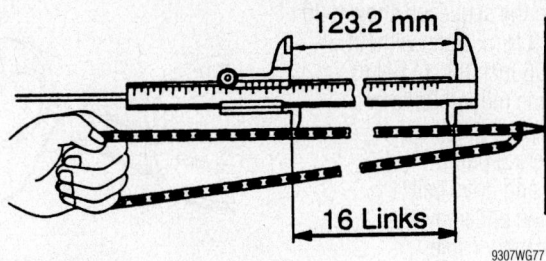

Measuring the timing chain—1NZ-FE engine

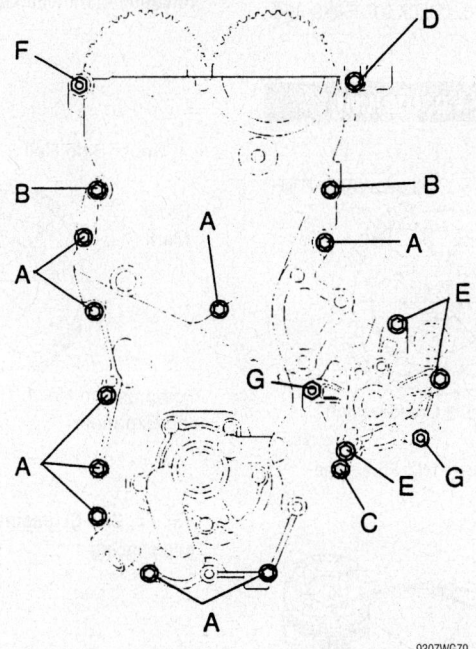

Timing cover bolt installation—1NZ-FE engine

- B: 30mm
- C: 35mm
- D: 20mm
- E: 35mm

14. Install or connect the following:
- Right engine mount bracket. Coat all but the end 2 threads of the bolt with RTV sealer. Torque the bolt to 41 ft. lbs. (55 Nm).
- Crankshaft position sensor. Torque bolt A to 66 inch lbs. (7.5 Nm); bolts B to 96 inch lbs. (11 Nm).
- Oil control valve. Torque: 71 inch lbs. (8 Nm).
- Crankshaft pulley and pin. Torque the bolt to 94 ft. lbs. (128 Nm).
- Cylinder head cover with RTV gasket material at the 2 locations shown. Uniformly tighten the bolts and nuts, in several passes, to 84 inch lbs. (10 Nm).
- PCV hoses
- Ignition coils

- Right engine mount insulator. Torque the bolts and nuts to 35 ft. lbs. (47 Nm).
- Power steering pump
- Alternator
- Right under cover
- Wheel
- Negative battery cable

15. Refill the cooling system to the correct level.

16. Start the vehicle, check for leaks and repair if necessary.

2ZZ-GE Engines

1. Before servicing the vehicle, refer to the Precautions Section.

2. Drain the cooling system.

3. Remove or disconnect the following:
- Negative battery cable
- Right engine cover
- Accessory drive belt and alternator
- Power steering pump, without disconnecting the hoses.
- Right engine mount

- Cylinder head cover
- Turn the crankshaft so the No. 1 piston is at TDC on the compression stroke.
- Crankshaft pulley
- Crankshaft position sensor from the timing chain cover.
- Accessory drive belt tensioner.
- Right engine mounting bracket
- Chain tensioner
- Water pump
- Timing chain cover
- Crankshaft angle sensor plate
- Timing chain tensioner slipper
- Timing chain and crankshaft timing sprocket.
- Timing chain vibration damper
- Valve timing control assembly and camshaft timing sprocket

4. Drive the seal from the cover.

5. Pull the chain to its full length and measure the length of any 16 consecutive links. The length should not exceed 4.827 inches (122.6mm).

6. Check the slipper and damper wear. Maximum wear should not exceed 0.039 in. (1mm).

7. The tensioner plunger should move smoothly and lock into place with finger pressure.

To install:

8. Apply engine oil from the tip of the intake camshaft, back to 16mm.

9. Align the timing mark on the valve timing controller with the knock pin and gently push the valve timing controller onto the camshaft.

10. Set the No.1 piston to TDC compression. The key on the crankshaft should be at 12 o'clock.

11. Install or connect the following:
- Sprockets. Torque the bolt to 33 ft. lbs. (45 Nm). Turn the camshafts to align the point marks on the sprockets.
- Chain damper. Bolts: 96 inch lbs. (11 Nm).
- Timing chain and crankshaft sprocket. Be sure to align the yellow chain link with the mark on the crankshaft sprocket.
- Timing chain on the camshaft sprockets. Align the yellow links with the marks on the camshaft sprockets.
- Chain tensioner slipper. Bolt: 14 ft. lbs. (18.5 Nm).
- Crankshaft angle sensor plate with the **F** mark forward
- New seal in the front cover
- Silicone sealant to the timing chain cover as illustrated

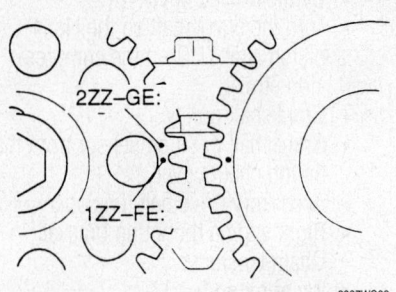

Timing mark identification at TDC compression—1ZZ-FE and 2ZZ-GE engines

- Timing chain cover

12. Install the water pump. Tighten the 10mm bolts marked "C" to 80 inch lbs. (9 Nm), those marked "A" to 10 ft. lbs. (13 Nm), and the remaining 10mm bolts to 96 inch lbs. (11 Nm). Tighten the 12mm bolts to 14 ft. lbs. (18.5 Nm). Be sure to install the bolts in their original locations. Bolt lengths:

   a. A: 1.77 in. (45mm).
   b. B: 1.38 in. (35mm).
   c. C: 1.18 in. (30mm).
   d. D: 0.98 in. (25mm).

13. With a Torx wrench, tighten the stud bolt to 82 inch lbs. (9.3 Nm).

14. Install or connect the following:
- Right engine mounting bracket. Bolts, with sealant applied: 35 ft. lbs. (47 Nm).
- Accessory drive belt tensioner. Bolt: 51 ft. lbs. (69 Nm). Nut: 21 ft. lbs. (29 Nm).
- Crankshaft position sensor. Tighten to 80 inch lbs. (9 Nm).
- Crankshaft pulley. Bolt: 102 ft. lbs. (138 Nm).

15. Release the ratchet pawl and compress the chain tensioner. Place the hook on the pin to keep the tensioner compressed.

16. Install the tensioner, using a new O-ring. Torque the bolts to 80 inch lbs. (9 Nm).

17. Turn the crankshaft counterclockwise and remove the hook from the pin. Turn the crankshaft clockwise and be sure the slipper is pushed by the plunger.

18. Check the valve timing by turning the crankshaft clockwise until the mark of the pulley is aligned with the mark on the timing chain cover. The marks on the camshaft sprockets should be facing each other as shown.

19. Install or connect the following:
- Silicone sealant to the 2 areas where the timing chain cover meets the cylinder head.
- Cylinder head cover. Bolts with

washers in the sequence shown: 80 inch lbs. (9 Nm). Bolts without washers: 96 inch lbs. (11 Nm).
- Right engine mount. Bolts and nuts: 38 ft. lbs. (52 Nm).
- Power steering pump
- Alternator and drive belt
- Right engine undercover
- Negative battery cable
- Washer tank

20. Refill the cooling system to the correct level.

21. Start the vehicle, check for leaks and repair if necessary.

Piston and Ring Positioning

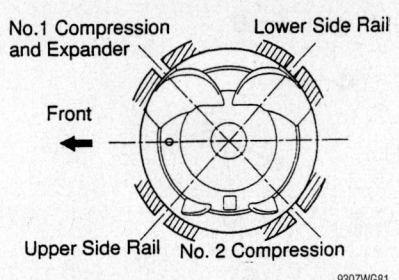

Piston ring positioning—1NZ-FE engine

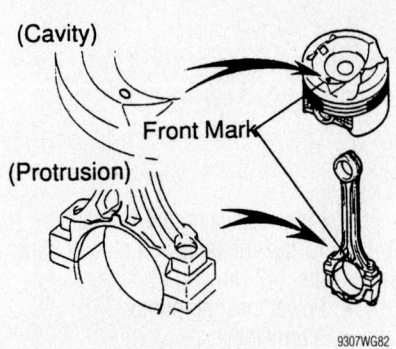

Piston and connecting rod positioning—1NZ-FE engine

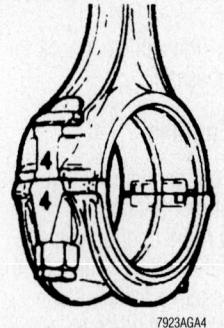

Before removing the caps from the connecting rods, be sure to matchmark them as shown

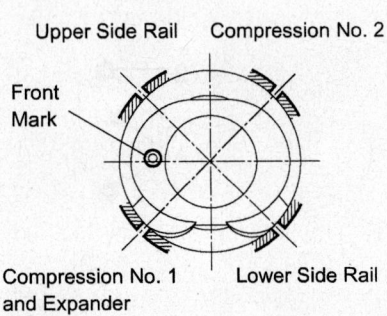

1ZZ-FE, 2ZZ-GE engine—piston ring identification mark locations

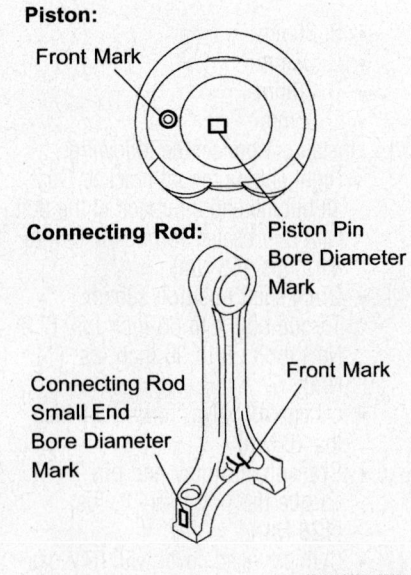

1ZZ-FE, 2ZZ-GE engine—piston ring end-gap spacing

1ZZ-FE, 2ZZ-GE engine—piston-to-connecting rod assembly

FUEL SYSTEM

Fuel System Pressure

RELIEVING

✳✳ CAUTION

Failure to relieve fuel pressure before repairs or disassembly can cause serious personal injury and/or property damage. Fuel pressure is maintained within the fuel lines, even if the engine is OFF or has not been run in a period of time. This pressure must be safely relieved before any fuel-bearing line or component is loosened or removed. On vehicles equipped with inflatable restraints or air bag systems, wait at least 90 seconds after disconnecting the battery cable before performing any other work. The back-up power will keep the restraint system energized for a period of time after the battery is disconnected.

1. Before servicing the vehicle, refer to the Precautions Section.
2. Perform the following:
   - Remove the fuse for the fuel pump
   - Start the engine until the engine stalls
   - Disconnect the negative battery cable
   - Place a catch-pan under the joint to be disconnected. A large quantity of fuel may be released when the joint is opened
   - Wear eye or full face protection
   - Place a shop towel over the area and slowly release the joint using a wrench of the correct size.
   - Allow the any fuel left in the line to bleed off slowly before fully disconnecting the joint.
   - Plug the opened lines
3. After connecting fuel lines, install the fuse for the fuel pump and start the engine.

Fuel Filter

REMOVAL & INSTALLATION

The fuel filter is in the tank as part of the fuel pump assembly.

Fuel Pump

REMOVAL & INSTALLATION

Echo

1. Before servicing the vehicle, refer to the Precautions Section.
2. Relieve the fuel system pressure.
3. Remove or disconnect the following:
   - Negative battery cable. On vehicles equipped with an air bag, wait at least 90 seconds before proceeding.
   - Rear seat cushion
   - Floor service hole cover

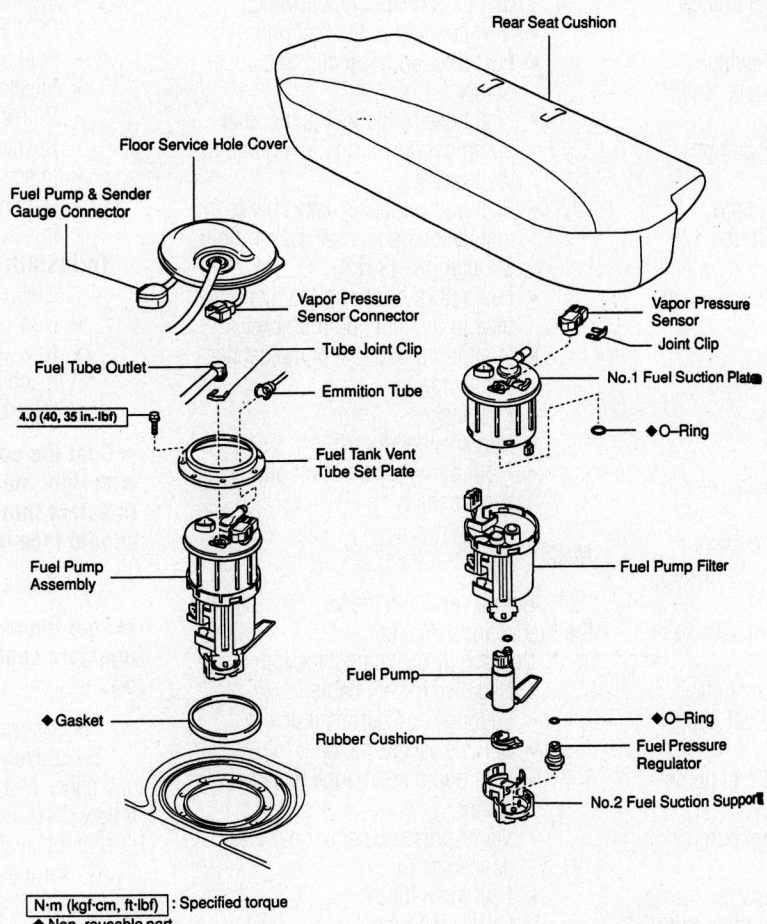

N·m (kgf·cm, ft·lbf) : Specified torque
◆ Non–reusable part

9307WG83

Fuel pump removal—Echo

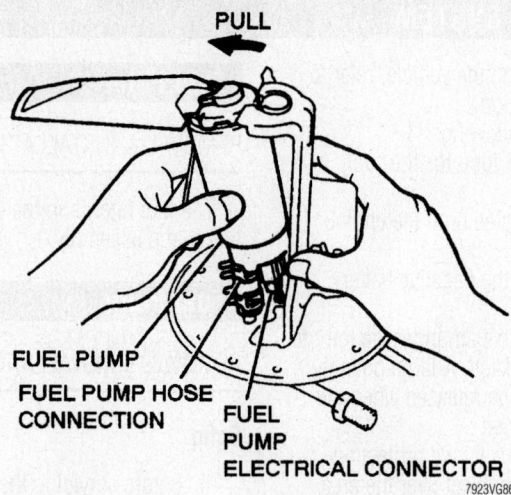

PULL

FUEL PUMP
FUEL PUMP HOSE
CONNECTION FUEL
PUMP
ELECTRICAL CONNECTOR

7923VG86

Pull the pump off the sender unit; the filter is still attached to the pump—Celica and Corolla

- Electrical connector at the fuel pump assembly
- Fuel outlet pipe from the fuel pump bracket
- Return hose from the fuel pump bracket.
- Fuel pump set plate from the fuel tank by removing the 8 bolts
- Fuel pump from the fuel bracket

To install:
4. Install or connect the following:
- Fuel pump to the fuel tank. Bolts: 35 inch lbs. (4 Nm).
- Return hose to the fuel pump bracket
- Outlet pipe to the fuel pump bracket. Tighten to 21 ft. lbs. (28 Nm).
- Service hole cover to the fuel tank
- Fuel pump connector
- Rear seat cushion
- Negative battery cable

Celica and Corolla

1. Before servicing the vehicle, refer to the Precautions Section.
2. Relieve the fuel system pressure.
3. Remove or disconnect the following:
- Negative battery cable
- Rear seat cushion and floor service hole cover
- Access plate-to-fuel tank bolts, then pull out the plate/fuel pump assembly
- Fuel pump sender and fuel pump connector
- Outlet pipe from the fuel pump bracket
- Return hose from the pump bracket
- Fuel pump bracket assembly from the fuel tank

- Lower side of the fuel pump from the pump bracket
- Fuel pump connector
- Fuel hose from the fuel pump
- Rubber cushion from the pump
- Fuel filter from the pump by removing the small clip

To install:
4. Install or connect the following:
- New cushion to the fuel pump
- Fuel filter and new clip to the fuel pump
- Fuel hose to the fuel pump, fuel pump connector and fuel pump to the bracket
- Fuel pump bracket assembly to the fuel tank using a new gasket. Bolts: 30 inch lbs. (3 Nm).
- Fuel return hose and the fuel outlet pipe to the fuel pump bracket
- Fuel pump and fuel pump sender connector
- Fuel tank
- Negative battery cable
- Floor service hole cover and rear seat cushion

MR2

1. Before servicing the vehicle, refer to the Precautions Section.
2. Remove or disconnect the following:
- Negative battery cable
- Luggage compartment box
- Service access panel
- Fuel pump and gauge harness connector
- Vapor pressure sensor harness connector
- Fuel main tube
- Emissions tube
- Fuel pump module

To install:
3. Install or connect the following:
- Fuel pump module. Tighten the bolts to 30 inch lbs. (3.4 Nm).
- Emissions tube
- Fuel main tube
- Vapor pressure sensor harness connector
- Fuel pump and gauge harness connector
- Service access panel
- Luggage compartment box
- Negative battery cable
4. Start the engine and check for leaks.

Fuel Injectors

REMOVAL & INSTALLATION

Celica and Corolla

1ZZ-FE ENGINE

1. Before servicing the vehicle, refer to the Precautions Section.
2. Properly relieve the fuel system pressure.
3. Remove or disconnect the following:
- Engine appearance cover
- PCV hose
- Fuel tube from the fuel pipe
- Injector connectors
- Delivery pipe and injectors
- Spacers from the head
- Injectors from the delivery pipe
- O-ring and grommet from each injector

To install:
4. Install or connect the following:
- New grommets
- New O-rings coated with light machine oil
- Injectors on the delivery pipe

➡**Coat the contact point on the pipe with light machine oil and twist the injectors into place. The connector should face outward.**

- Spacers

➡**Coat the seats in the head where the injectors contact, with light machine oil.**

- Delivery pipe and injectors
5. Loosely install the hold-down bolts and check that the injectors rotate smoothly. If they don't, the probable cause is incorrect O-ring installation.
6. Torque the hold-down bolts to 14 ft. lbs. (19 Nm).
7. Torque the fuel pipe bolt to 84 inch lbs. (9 Nm).

8. Connect the fuel line.
9. Install the PCV hose.
10. Install the No. 2 cover.

2ZZ-GE ENGINE

1. Before servicing the vehicle, refer to the Precautions Section.
2. Properly relieve the fuel system pressure.
3. Remove or disconnect the following:
   • Engine appearance cover
   • Fuel tube from the fuel pipe
   • Injector connectors
   • Delivery pipe and injectors
   • Spacers from the head
   • Injectors from the delivery pipe
   • O-ring and grommet from each injector

To install:

4. Install or connect the following:
   • New grommets
   • New O-rings coated with light machine oil
   • Injectors on the delivery pipe

➡**Coat the contact point on the pipe with light machine oil and twist the injectors into place. The connector should face outward.**

   • Spacers

➡**Coat the seats in the head where the injectors contact, with light machine oil.**

   • Delivery pipe and injectors
5. Loosely install the hold-down bolts and check that the injectors rotate smoothly.

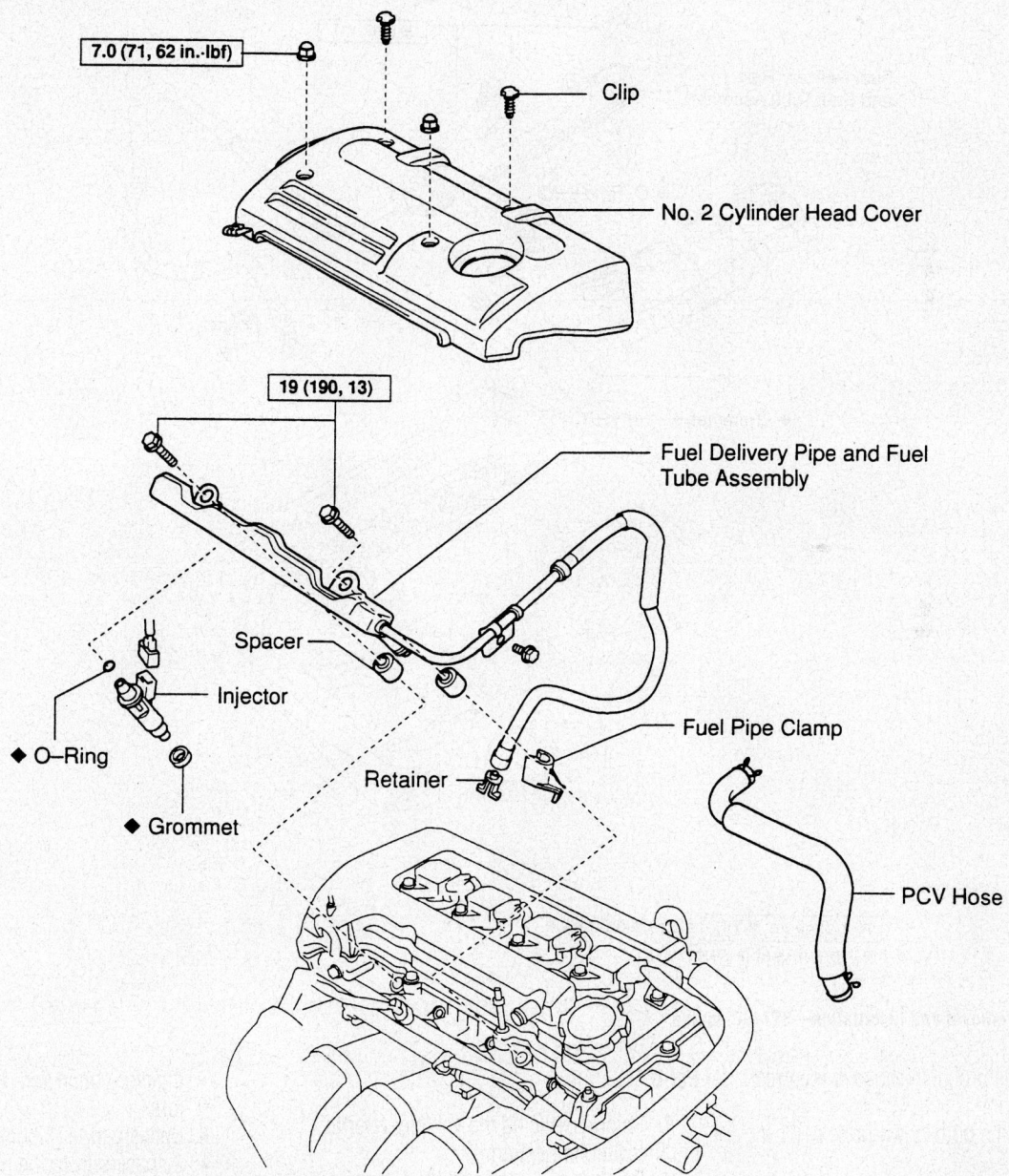

| 7.0 (71, 62 in.·lbf) |

Clip

No. 2 Cylinder Head Cover

| 19 (190, 13) |

Fuel Delivery Pipe and Fuel Tube Assembly

Spacer

Injector

◆ O–Ring

◆ Grommet

Retainer

Fuel Pipe Clamp

PCV Hose

| N·m (kgf·cm, ft·lbf) | : Specified torque

◆ Non–reusable part

9307WG95

Fuel injector removal and installation—1ZZ-FE engine

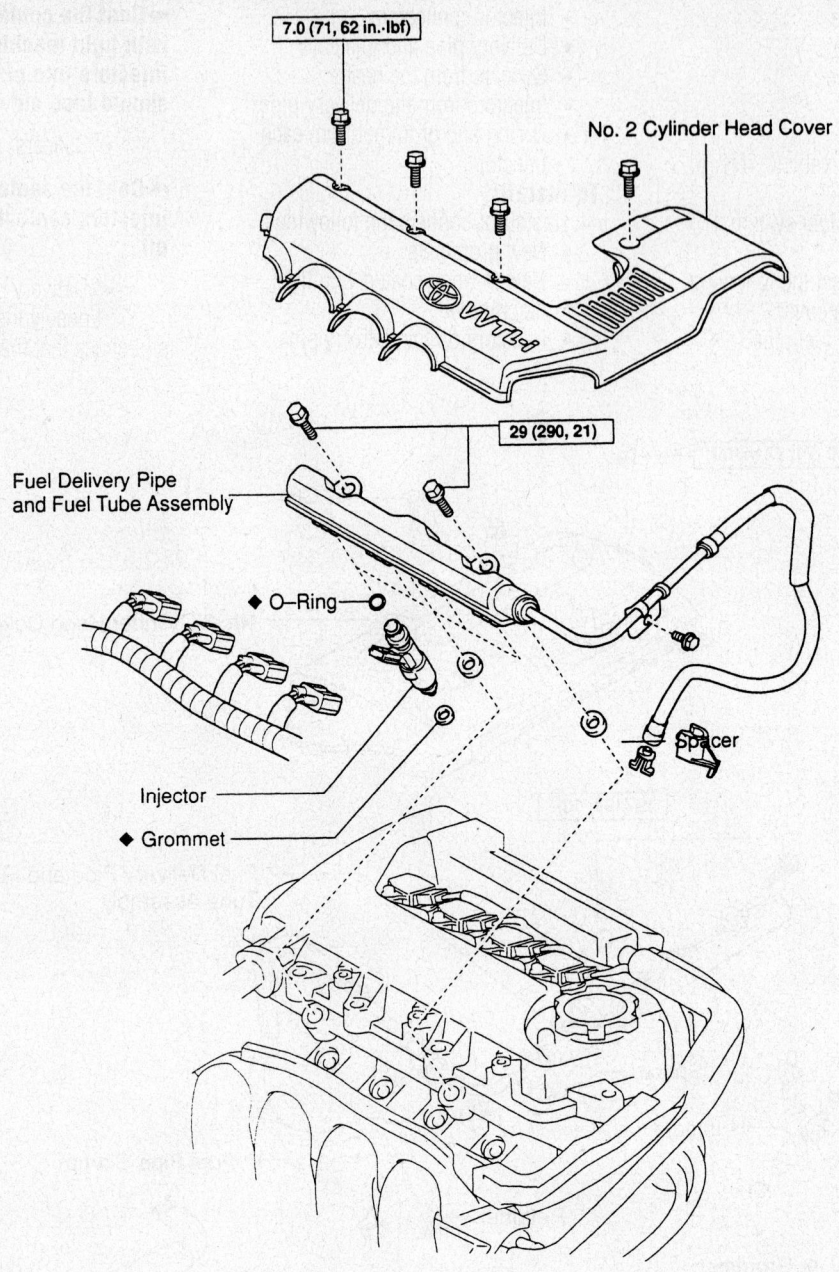

7.0 (71, 62 in.-lbf)

No. 2 Cylinder Head Cover

29 (290, 21)

Fuel Delivery Pipe
and Fuel Tube Assembly

◆ O–Ring

Spacer

Injector

◆ Grommet

N·m (kgf·cm, ft·lbf) : Specified torque
◆ Non–reusable part

9307WG96

Fuel injector removal and installation—2ZZ-GE engine

If they don't, the probable cause is incorrect O-ring installation.

6. Torque the hold-down bolts to 21 ft. lbs. (29 Nm).

7. Torque the fuel pipe bolt to 84 inch lbs. (9 Nm).

8. Connect the fuel line.

9. Install the PCV hose.

10. Install the No. 2 cover.

Echo

1. Before servicing the vehicle, refer to the Precautions Section.

2. Remove or disconnect the following:
   • Negative battery cable
   • Engine appearance cover
   • Fuel pipe clamp
   • Fuel inlet line from the fuel pipe

   • Injector connectors from the injectors
   • Delivery pipe (3 bolts) and injectors
   • 2 spacers from the head
   • Injectors from the pipe
   • O-rings and grommets

To install:

3. Install or connect the following:
   • New grommets

- New O-rings coated with clean engine oil
- Injectors to the pipe. Coat the contact area with light machine oil. The injectors twist into place. The connector should face outward
- Delivery pipe bolts. Torque: 14 ft. lbs. (19 Nm).
- Fuel pipe bolt. Torque: 80 inch lbs. (9 Nm).
- Fuel hose to fuel pipe
- Pipe clamp
- Wire harness cover
- PCV hose
- Negative battery cable

MR2

1. Before servicing the vehicle, refer to the Precautions Section.
2. Properly relieve the fuel system pressure.
3. Remove or disconnect the following:

- Negative battery cable
- Air cleaner
- Accelerator cable bracket from the throttle body
- Throttle body from the air intake chamber

- Engine hanger and air intake chamber stay
- EGR vacuum modulator if so equipped
- EGR valve and pipe if so equipped
- Air intake chamber cover and gasket
- Injector electrical connections
- Fuel inlet hose from the delivery pipe
- Fuel return hose from the fuel pressure regulator
- Fuel delivery pipe (rail)
- The 4 insulators and 2 collars from the intake manifold
- Injectors

To install:

4. Install or connect the following:

➡ **Before installing the injectors back into the fuel rail, install a NEW O-ring on each injector, coated with a light coat of gasoline (NEVER use oil of any sort).**

- Injectors

➡ **Make certain each injector can be smoothly rotated. If they do not rotate smoothly, the O-ring is not in its correct position.**

- Insulators into each injector hole
- The two spacers on the delivery pipe mounting holes in the intake manifold

5. Place the delivery pipe and injectors on the intake manifold and again check that the injectors rotate smoothly. Position the injector connector upward. Install the two bolts and tighten them to 11 ft. lbs.

6. Install or connect the following:

- Electrical connectors to each injector
- Gaskets, the inlet pipe and fuel union bolt. Bolt to 22 ft. lbs.
- Air intake chamber cover with a NEW gasket. Torque the retaining bolts in steps to 14 ft. lbs.
- All necessary hoses and electrical connections
- EGR valve and pipe if so equipped
- Engine hanger and air intake chamber stay
- EGR vacuum modulator if so equipped
- Throttle body. Torque the bolts evenly (in a X-pattern) to 16 ft. lbs.
- Accelerator cable bracket to the throttle body
- Air cleaner hose and cap
- Negative battery cable

DRIVE TRAIN

Manual Transaxle Assembly

REMOVAL & INSTALLATION

Echo

1. Before servicing the vehicle, refer to the Precautions Section.
2. Drain the transaxle fluid.
3. Remove or disconnect the following:

- Hood
- Wiper arms
- Right and left cowl top ventilator covers
- Engine appearance cover
- Battery
- Air intake assembly
- Wiring harness from the transaxle
- Transaxle control cable
- Clutch release cylinder
- Ground cable from the left engine mount
- Back-up light switch wires
- Vehicle speed sensor wiring
- 2 transaxle upper side mounting bolts
- Starter

4. At this point, attach an engine crane to support the engine.
5. Remove or disconnect the following:

- Left side engine under cover
- Both halfshafts
- 2 bolts and 1 nut securing the engine rear mount to the crossmember
- Sliding yoke
- Power steering hoses

- Support the transaxle
- Engine left mounting bracket
- Engine rear mount and bracket
- 5 transaxle lower side mount bolts
- Transaxle

To install:

6. Install or connect the following:

- Transaxle. Torque the 5 lower bolts to 25 ft. lbs. (33 Nm).
- Engine rear mount and bracket.

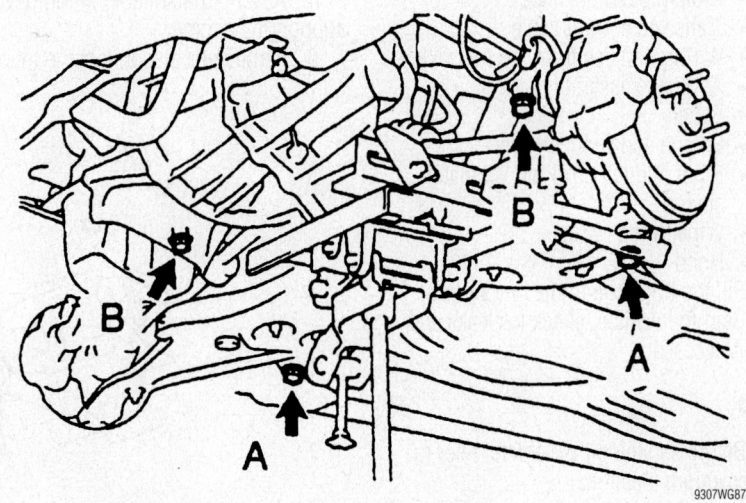

Sub-frame installation—Echo

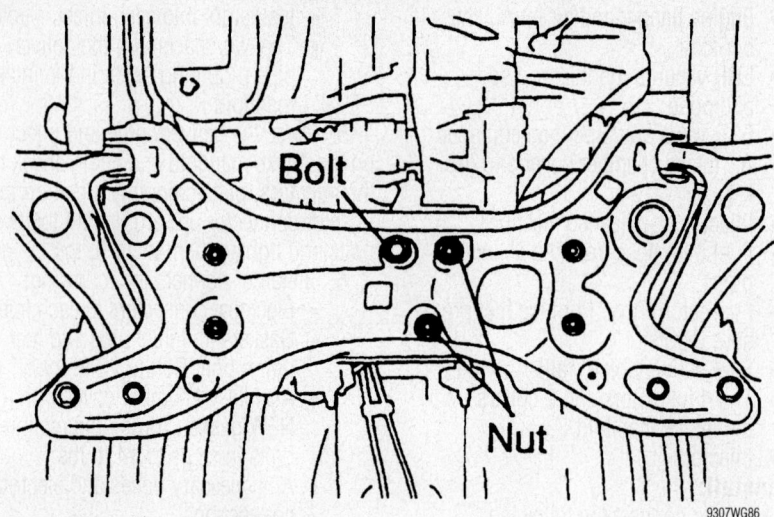

Crossmember installation—Echo

Torque the mount bolt and nut to 47 ft. lbs. (64 Nm); the bracket bolts to 36 ft. lbs. (49 Nm).
- Engine left mounting bracket. Torque the bolts to 36 ft. lbs. (49 Nm).
- Power steering hoses
- Sliding yoke
- 2 bolts and 1 nut securing the engine rear mount to the cross-member. Torque the bolts to 36 ft. lbs. (49 Nm).
- Both halfshafts
- Left side engine under cover
- Starter. Torque the bolts to 29 ft. lbs. (39 Nm).
- 2 transaxle upper side mounting bolts. Torque the bolts to 25 ft. lbs. (33 Nm).
- Vehicle speed sensor wiring
- Back-up light switch wires
- Ground cable from the left engine mount
- Clutch release cylinder
- Transaxle control cable
- Wiring harness from the transaxle
- Air intake assembly
- Battery
- Engine appearance cover
- Right and left cowl top ventilator covers
- Wiper arms
- Hood

7. Fill the transaxle to the proper level.
8. Start the vehicle, check for leaks and repair if necessary.

Celica

1. Before servicing the vehicle, refer to the Precautions Section.
2. Drain the transaxle fluid.

3. Remove or disconnect the following:
- Negative battery cable. On vehicles equipped with an air bag, wait at least 90 seconds before proceeding
- Hood
- Engine appearance cover
- Air intake assembly
- Release cylinder tube bracket
- Clutch release cylinder
- Back-up light switch connector
- Ground cable on the transaxle
- Shift cables from the transaxle
- Vehicle speed sensor connector or the speedometer cable
- Engine wire clamps
- Starter set bolt from the transaxle upper side
- Undercovers
- Halfshafts
- Front exhaust pipe and support bracket
- Starter
- Engine center support member

4. Attaching a suitable lifting device to support the engine.
5. Remove or disconnect the following:

- Engine rear mounting
- Engine front mounting bracket and insulator
- Engine left mounting bracket
- Transaxle mounting bolts from the engine rear end plate side.
- Transaxle case protector
- Engine left side and remove the 3 upper transaxle bolts.
- Transaxle

To install:

6. Position the transaxle to the engine and raise the engine right side. Align the input shaft with the clutch disc
7. Install or connect the following:
- Transaxle to the engine. 3 upper transaxle bolts: 47 ft. lbs. (64 Nm).
- Transaxle case protector. Bolts: 108 inch lbs. (13 Nm).
- 4 transaxle lower bolts. Bolt A: 17 ft. lbs. (23 Nm); Bolt B: 34 ft. lbs. (46 Nm).
- Left engine mounting bracket to the engine left mounting insulator. Bolts: 47 ft. lbs. (64 Nm).
- Engine front mounting bracket and insulator. 2 bracket bolts: 57 ft. lbs. (77 Nm). Through-bolt: 64 ft. lbs. (87 Nm).
- Engine rear mounting bracket and insulator. Bracket bolts: 57 ft. lbs. (77 Nm). Through-bolt: 64 ft. lbs. (87 Nm).
- Engine center support member
- Starter
- Front exhaust pipe and support bracket
- Halfshafts
- Transaxle oil
- Undercovers
- Engine support fixture
- Starter set bolt to the transaxle upper side. Bolt: 29 ft. lbs. (39 Nm).
- Engine wire clamps

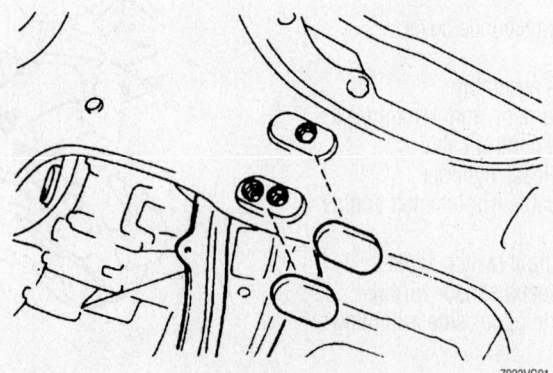

Rear mounting insulator set bolt locations—Celica

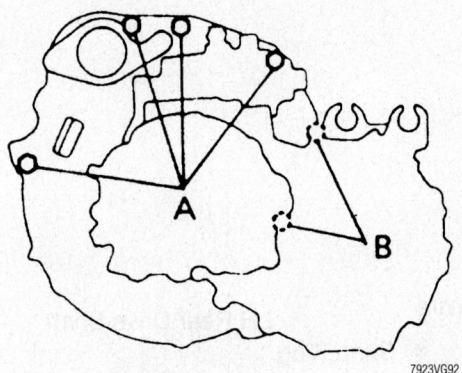

Upper transaxle mounting bolt locations—Celica

7923VG92

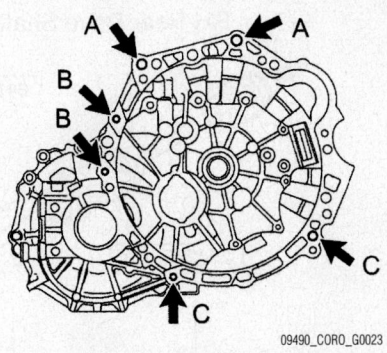

09490_CORO_G0023

Manual transaxle mounting bolt identification—Corolla

- Vehicle speed sensor connector or the speedometer cable
- Transaxle shift cables and ground cable
- Back-up light switch connector
- Release cylinder
- Air intake assembly
- Hood
- Negative battery cable

8. Fill the transaxle fluid to the proper level.

9. Start the vehicle, check for leaks and repair if necessary.

Celica GT-S

1. Before servicing the vehicle, refer to the Precautions Section.
2. Drain the transaxle fluid.
3. Remove or disconnect the following:
- Hood
- Engine appearance cover
- Radiator overflow bottle
- Battery
- Air intake assembly
- ECM box
- Wiring harness
- Battery tray
- Transaxle control cable
- Engine ground cable
- Speed sensor connector
- Back-up light switch connector
- Clutch release cylinder
- Starter
- Transaxle control cable bracket
- 2 upper transaxle mounting bolts

4. Attach an engine crane to the engine.
5. Remove or disconnect the following:
- Left engine mount bracket
- Engine under covers
- Both halfshafts
- Front exhaust pipe

6. Unbolt the steering rack from the sub-frame and suspend it.
7. Remove or disconnect the following:
- Stabilizer bar

- Wheels
- Ball joints from the lower arms
- Lower arms and sub-frame
- Engine rear mount and bracket

8. Raise the engine slightly.
9. Remove or disconnect the following:
- 4 lower transaxle side bolts
- Transaxle from the engine

To install:

10. Install or connect the following:
- Transaxle to the engine. Install the 4 lower side bolts. Torque the 2 bottom bolts to 17 ft. lbs. (23 Nm); the 2 side bolts to 35 ft. lbs. (48 Nm).

- Rear mount and bracket. Torque the bracket bolts to 47 ft. lbs. (64 Nm); the mount bolt to 64 ft. lbs. (87 Nm).
- Lower control arms-to-frame. Bolts: 101 ft. lbs. (137 Nm); tie rod ball stud nuts: 36 ft. lbs. (49 Nm).

11. Install the sub-frame. Torque the bolts as illustrated:
a. A and B: 38 ft. lbs. (52 Nm)
b. C: 116 ft. lbs. (157 Nm)
c. D: 29 ft. lbs. (39 Nm)

12. Install or connect the following:
- Steering rack. Bolts: 43 ft. lbs. (58 Nm).

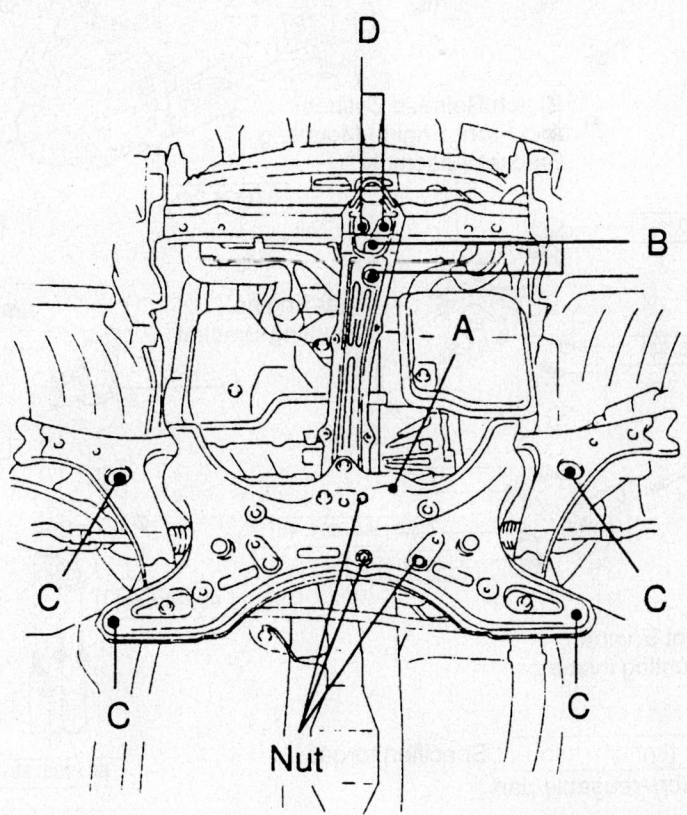

Celica sub-frame torques

9307WG99

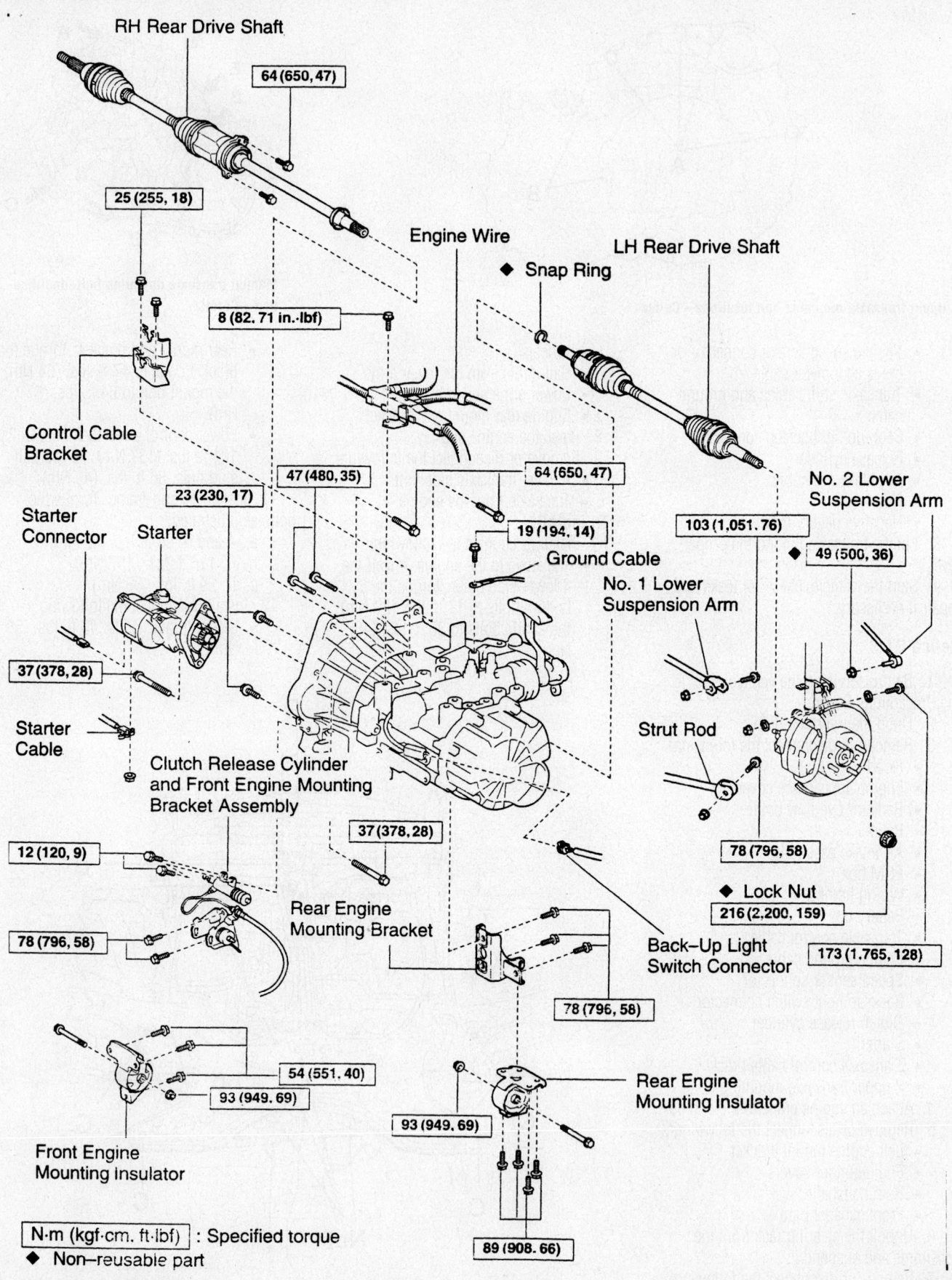

RH Rear Drive Shaft

64 (650, 47)

25 (255, 18)

Engine Wire

◆ Snap Ring

LH Rear Drive Shaft

8 (82. 71 in.·lbf)

Control Cable Bracket

47 (480, 35)

23 (230, 17)

64 (650, 47)

No. 2 Lower Suspension Arm

Starter Connector

Starter

19 (194, 14)

Ground Cable

103 (1,051. 76)

◆ 49 (500, 36)

No. 1 Lower Suspension Arm

37 (378, 28)

Starter Cable

Clutch Release Cylinder and Front Engine Mounting Bracket Assembly

Strut Rod

12 (120, 9)

37 (378, 28)

78 (796, 58)

78 (796, 58)

Rear Engine Mounting Bracket

◆ Lock Nut

216 (2,200, 159)

173 (1.765, 128)

Back-Up Light Switch Connector

78 (796, 58)

54 (551. 40)

93 (949. 69)

93 (949. 69)

Rear Engine Mounting Insulator

Front Engine Mounting Insulator

N·m (kgf·cm. ft·lbf) : Specified torque

◆ Non-reusable part

89 (908. 66)

9357WG04

Manual transaxle mounting exploded view—MR2

- Front exhaust pipe. Bolts: 32 ft. lbs. (43 Nm).
- Halfshafts
- Under covers
- Left mount bracket. Bolt: 44 ft. lbs. (60 Nm).
- Left mount. Bolts: 44 ft. lbs. (60 Nm); nut: 59 ft. lbs. (80 Nm).
- Transaxle upper side mount bolts. Torque: 47 ft. lbs. (64 Nm).
- Control cable bracket. Bolts: 18 ft. lbs. (25 Nm).
- Starter. Bolts: 28 ft. lbs. (37 Nm).
- Clutch release cylinder
- Wiring
- Control cable
- Battery tray
- ECM
- Air cleaner
- Battery
- Reservoir
- Engine appearance cover
- Hood

13. Fill the transaxle fluid to the proper level.

14. Start the vehicle, check for leaks and repair if necessary.

Corolla

1. Before servicing the vehicle, refer to the Precautions Section.
2. Drain the transaxle fluid.
3. Remove or disconnect the following:
   - Negative battery cable
   - Front wheels
   - Engine undercovers
   - Front exhaust pipe
   - Hood
   - Engine appearance cover
   - Air intake assembly
   - Battery and battery tray
   - Cruise control actuator assembly, if equipped
   - Wiring harness and ground cable
   - Back-up light switch connector
   - Vehicle speed sensor connector or the speedometer cable
   - Clutch release cylinder
   - Shift cables from the transaxle
   - Starter
   - Steering intermediate shaft
   - Halfshafts
   - Power steering lines
4. At this point, attach an engine hoist to support the engine.
5. Remove the front suspension crossmember.
6. Support the transmission with a suitable jack.
7. Remove the transverse engine mounting insulator and bracket.

8. Remove the transaxle mounting bolts and remove the transaxle from the engine assembly.

To install:

9. Align the input shaft with the clutch disc and install the transaxle to the engine. Bolts A: 47 ft. lbs. (64 Nm). Bolts B: 35 ft. lbs. (47 Nm). Bolts C: 17 ft. lbs. (23 Nm).

10. Install or connect the following:
   - Transverse engine mounting bracket. Tighten to 38 ft. lbs. (52 Nm).
   - Transverse engine mounting insulator
   - Front suspension crossmember
   - Steering intermediate shaft
   - Hole covers
   - Power steering lines
   - Halfshafts
   - Starter
   - Transaxle shift cables and ground cable
   - Clutch release cylinder
   - Back-up light switch connector
   - Vehicle speed sensor connector or the speedometer cable
   - Wiring harness and ground cable
   - Cruise control actuator assembly, if equipped
   - Battery tray and battery
   - Air intake assembly
   - Engine appearance cover
   - Hood
   - Exhaust pipe
   - Engine undercovers
   - Front wheels
   - Negative battery cable

11. Refill the transaxle with oil to the correct level.

12. Start the engine and check for leaks.

MR2

1. Before servicing the vehicle, refer to the Precautions Section.
2. Drain the transaxle oil.
3. Remove or disconnect the following:
   - Engine hood
   - Rear suspension upper brace
   - Air filter assembly
   - Battery and tray
   - Engine ground cable
   - Back-up light switch connector
   - Shift cables and bracket
   - Transaxle upper flange bolts
4. Attach an engine support fixture.
   - Left motor mount
   - Engine undercovers
   - Axle halfshafts
   - Clutch slave cylinder
   - Front motor mount and bracket
   - Starter motor

- Rear motor mount and bracket
- Transaxle lower flange bolts
- Transaxle

To install:

5. Installation is the reverse of the removal procedure, while using the following torque values:
- Upper transaxle flange bolts: 47 ft. lbs. (64 Nm)
- Lower transaxle flange bolts: **A** bolts to 35 ft. lbs. (47 Nm) and **B** bolts to 17 ft. lbs. (23 Nm). Refer to the illustration.
- Rear engine mount bracket: 58 ft. lbs. (78 Nm)
- Rear engine mount bolts: 66 ft. lbs. (89 Nm)
- Rear engine mount through bolt: 69 ft. lbs. (93 Nm)
- Starter motor: 28 ft. lbs. (37 Nm)
- Front engine mount bracket bolts: 56 ft. lbs. (78 Nm)
- Front engine mount through bolt: 69 ft. lbs. (93 Nm)

9357WG05

Upper transaxle flange bolts—MR2

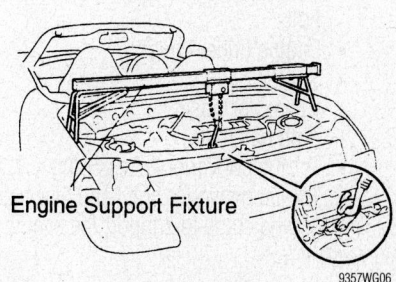

Engine Support Fixture

9357WG06

Engine support fixture—MR2

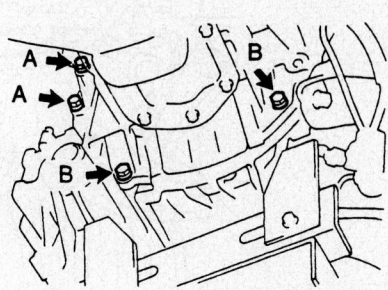

9357WG07

Lower transaxle flange bolts—MR2

- Clutch slave cylinder: 108 Inch lbs. (12 Nm)
- Left engine mount bracket: 38 ft. lbs. (52 Nm)
- Left engine mount through bolt: 64 ft. lbs. (87 Nm)
- Suspension upper brace bolts: 59 ft. lbs. (80 Nm)
- Engine hood bolts: 15 ft. lbs. (20 Nm)

6. Fill the transaxle with the correct oil.

Automatic Transaxle

REMOVAL & INSTALLATION

Echo

1. Before servicing the vehicle, refer to the Precautions Section.
2. Drain the transaxle fluid.
3. Remove or disconnect the following:
   - Hood
   - Right and left cowl top ventilator covers
   - Engine appearance cover
   - Battery
   - Air cleaner bracket
   - Wiring harness from the transaxle
   - Transaxle shift cable
   - Clutch release cylinder
   - Ground cable from the left engine mount
   - Park/Neutral switch wiring
   - Solenoid wiring
   - Direct Clutch Speed Sensor wiring
   - Filler pipe hose
   - 2 transaxle upper side mounting bolts
   - Engine under covers
   - Starter
   - Oil cooler hose
   - Both halfshafts
   - Torque converter access plug
   - Torque converter bolts and attach an engine crane to support the engine

- Front suspension subframe (lower arms, steering gear and stabilizer)
- 5 transaxle bolts
- 2 transaxle mount bolts
- Transaxle

To install:

4. Install or connect the following:
   - Transaxle. Torque the 5 bolts to 22 ft. lbs. (30 Nm).
   - 2 transaxle mount bolts. Torque the mount bolts to 36 ft. lbs. (49 Nm).
   - Torque converter bolts: 20 ft. lbs. (27 Nm).
   - Access plug
   - Subframe assembly
   - Oil cooler hoses
   - Starter. Torque the bolts to 29 ft. lbs. (39 Nm).
   - 2 transaxle upper side mounting bolts. Torque the bolts to 22 ft. lbs. (30 Nm).
   - Both halfshafts
   - Engine under covers
   - Filler hose
   - Transaxle control cable
   - Ground cable from the left engine mount
   - Wiring
   - Air cleaner bracket
   - Engine appearance cover
   - Right and left cowl top ventilator covers
   - Battery
   - Hood

5. Fill the transaxle fluid to the proper level.
6. Start the vehicle, check for leaks and repair if necessary.

Corolla

1. Before servicing the vehicle, refer to the Precautions Section.
2. Drain the transaxle fluid.
3. Remove or disconnect the following:
   - Negative battery cable
   - Hood

- Engine appearance cover
- Battery and battery tray
- Air intake assembly
- Transmission control cables and cable support
- Wiring harness clamp and throttle cable clamp
- Solenoid connector and park/neutral position switch connector. Remove the wiring harness clamps.
- Vehicle speed sensor connector
- Transaxle oil filler tube
- Oil cooler hoses
- Oxygen sensor connector

4. At this point, attach an engine hoist to support the engine.
5. Remove or disconnect the following:
   - Front exhaust pipe
   - Halfshafts
   - Transmission case protector
   - Starter
6. Support the transaxle assembly with a suitable jack.
7. Remove or disconnect the following:
   - Transverse engine mounting insulator
   - Transverse engine mounting bracket
   - Center engine support member
   - Torque convert cover
8. Turn the crankshaft to gain access and remove the torque converter bolts
9. Remove the six transaxle mounting bolts.
10. Remove the transaxle assembly.

To install:

11. Install or connect the following:
    - Transaxle. Tighten the bolts as follows: Bolt A: 47 ft. lbs. (64 Nm); Bolt B: 34 ft. lbs. (46 Nm); Bolt C: 17 ft. lbs. (23 Nm)
    - Torque converter bolts to the transaxle. Bolts: 20 ft. lbs. (28 Nm).

➡**Install the yellowish green colored bolt first.**

- Torque converter cover.
- Center engine support member.

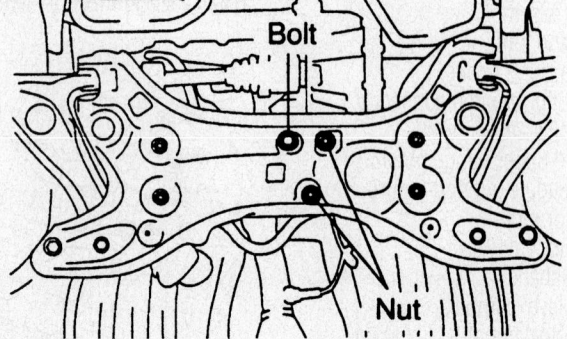

Rear mount installation—Echo

9307WG85

Transaxle mounting bolts—Corolla

67170-TOYO-G44

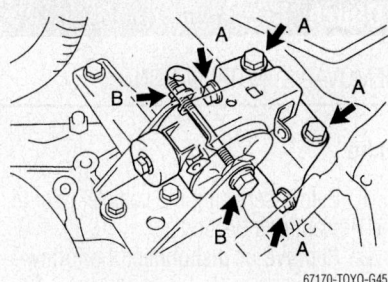

67170-TOYO-G45

Mounting insulator fastener identification—Corolla

Tighten the front bolts to 29 ft. lbs. (39 Nm) and rear bolts to 38 ft. lbs. (52 Nm).
- Transverse engine mounting brackets
- Transverse engine mounting insulators. Bolt A: 38 ft. lbs. (52 Nm); Bolt and nut B: 59 ft. lbs. (80 Nm)
- Starter
- Transmission case protector
- Halfshafts
- Front exhaust pipe
- Engine under covers
- Front wheels
- Oxygen sensor connector
- Transaxle oil filler tube
- Oil cooler hoses
- Solenoid connector and park/neutral position switch connector
- Transmission control cables and cable support
- Battery tray and battery
- Air intake assembly
- Engine appearance cover
- Hood
- Negative battery cable

12. Refill the transaxle with oil to the correct level.

13. Start the engine and check for leaks.

Celica

1. Before servicing the vehicle, refer to the Precautions Section.

2. Drain the transaxle fluid.

3. Remove or disconnect the following:
- Negative battery cable. On vehicles equipped with an air bag, wait at least 90 seconds before proceeding.
- Throttle cable from the engine
- Cruise control actuator. The cruise control actuator and bracket should be removed as an assembly.
- Air intake assembly and battery.
- Vehicle speed sensor and the transaxle ground strap.
- Engine left mounting upper side bolt
- Starter

- Park/neutral position switch connector
- Solenoid connectors
- Upper transaxle retaining bolts
- Transaxle oil cooler hoses
- Undercover
- Both halfshafts
- Shift control cable from the control shaft lever and body bracket
- Engine rear mounting through-bolt
- Front exhaust pipe
- Air conditioner pipe bracket by removing the bolt
- Shift cable from the suspension member
- The 2 power steering gear assembly set bolts and nuts
- The 3 grommets from the center crossmember
- The 13 bolts and 2 nuts holding the suspension and center crossmembers
- Crossmembers from the vehicle
- No. 1 manifold stay
- Stiffener plate
- Torque converter bolts
- Transaxle

To install:

4. Install or connect the following:
- Transaxle. 10mm bolt: 34 ft. lbs. (46 Nm); 12mm bolt: 47 ft. lbs. (64 Nm).
- Torque converter bolts, with silicone applied to threads. Bolts: 18 ft. lbs. (25 Nm).
- Stiffener plate. Bolts, alternately tightening: 12mm bolts: 15 ft. lbs. (21 Nm); 14mm bolts: 32 ft. lbs. (43 Nm).
- No. 1 manifold stay. Bolt: 15 ft. lbs. (21 Nm). Nut: 32 ft. lbs. (43 Nm).
- Raise the suspension member into position and install the 2 bolts to hold the suspension to the body. Bolts to 94 ft. lbs. (127 Nm).
- The 3 bolts to hold the rear of the lower control arms to the subframe and body. Torque the bolt that goes through the lower control arm to 123 ft. lbs. (167 Nm) and the other 2 bolts to 130 ft. lbs. (175 Nm).
- Side bolts
- Center member
- Engine rear mount and bracket. Nuts: 59 ft. lbs. (80 Nm); Bolt: 65 ft. lbs. (88 Nm).
- Engine front mount. Bolts: 59 ft. lbs. (80 Nm).
- The 2 front bolts connecting the center mount to the radiator support. Bolts: 26 ft. lbs. (35 Nm).

- Grommets to the center member
- The 2 power steering gear assembly set bolts and nuts. Nuts and bolts: 94 ft. lbs. (127 Nm).
- The 2 shift cable mounting bolts
- Air conditioner pipe bracket
- Front exhaust pipe
- Engine rear mounting bolt. Bolt: 64 ft. lbs. (88 Nm).
- Shift control cable to the control shaft lever and body bracket. Install the clips
- Left and right halfshafts
- Undercovers
- Oil cooler hoses with the 2 clips
- Upper transaxle mounting bolts. Bolts: 47 ft. lbs. (64 Nm).
- Solenoid connectors
- Park/neutral position switch connector
- Vehicle speed sensor connector and the ground strap to the transaxle
- Starter. Bolts: 29 ft. lbs. (39 Nm).
- Left mounting upper side bolt. Bolt: 47 ft. lbs. (64 Nm).
- Air intake assembly
- Cruise control actuator
- Battery and cables
- Throttle cable

5. Fill the transaxle fluid to the proper level.

6. Start the vehicle, check for leaks and repair if necessary.

Celica GT-S

1. Before servicing the vehicle, refer to the Precautions Section.

2. Drain the cooling system.

3. Drain the transaxle fluid.

4. Remove or disconnect the following:
- Hood
- Battery
- ECM
- Air intake assembly
- Engine appearance cover
- Ground cables
- Control cable bracket
- Input speed turbine sensor
- Vehicle speed sensor
- Solenoid connector
- Park/neutral switch connector
- Control cable
- Coolant reservoir
- Starter
- Engine under covers
- Halfshafts
- Stabilizer bar end links

5. Unbolt the steering rack and suspend it

- 9 bolts and 3 nuts and lower the suspension member
- Engine rear mount and attach and engine crane to support the engine.
- Upper left side engine
- Fluid cooler hoses and support the transaxle with a jack.
- Torque converter access plug
- Torque converter bolts
- Transaxle bolts
- Transaxle

To install:

6. Install or connect the following:
- Transaxle. 2 lower bolts: 17 ft. lbs. (23 Nm); 2 lower side bolts: 34 ft. lbs. (46 Nm).
- Torque converter bolts. Torque: 25 ft. lbs. (41 Nm).
- Oil cooler hoses
- Upper left side mount bolt and nut. Torque: 59 ft. lbs. (80 Nm).
- Engine rear mount insulator. Bolt: 64 ft. lbs. (87 Nm).

7. Install the suspension member. Torque the bolts as illustrated:
- a. A: 116 ft. lbs. (157 Nm).
- b. B: 38 ft. lbs. (52 Nm).
- c. C: 29 ft. lbs. (39 Nm).
- d. Nut: 38 ft. lbs. (52 Nm).

8. Install or connect the following:
- Steering rack. Bolts: 33 ft. lbs. (45 Nm).
- Stabilizer bar end links. Torque: 32 ft. lbs. (44 Nm).
- Halfshafts
- Under covers
- Starter. Torque: 28 ft. lbs. (37 Nm).
- Coolant reservoir
- Control cable
- Wiring
- Upper transaxle bolts. Torque: 47 ft. lbs. (64 Nm).
- Control cable clamp
- Engine appearance cover. Torque: 62 inch lbs. (7 Nm).
- Air cleaner
- ECM
- Battery
- Hood

9. Refill the cooling system to the correct level.

10. Fill the transaxle to the proper level.

11. Start the vehicle, check for leaks and repair if necessary.

Clutch

REMOVAL & INSTALLATION

Echo

1. Before servicing the vehicle, refer to the Precautions Section.

2. Remove or disconnect the following:

- Negative battery cable. On vehicles equipped with an air bag, wait at least 90 seconds before proceeding.
- Remove the transaxle assembly from the vehicle
- Clutch pressure plate retaining bolts
- Clutch cover
- Clutch disc
- Retaining clip and bearing from the transaxle
- Release fork and boot assembly

To install:

3. Install or connect the following:
- Clutch disc onto the flywheel

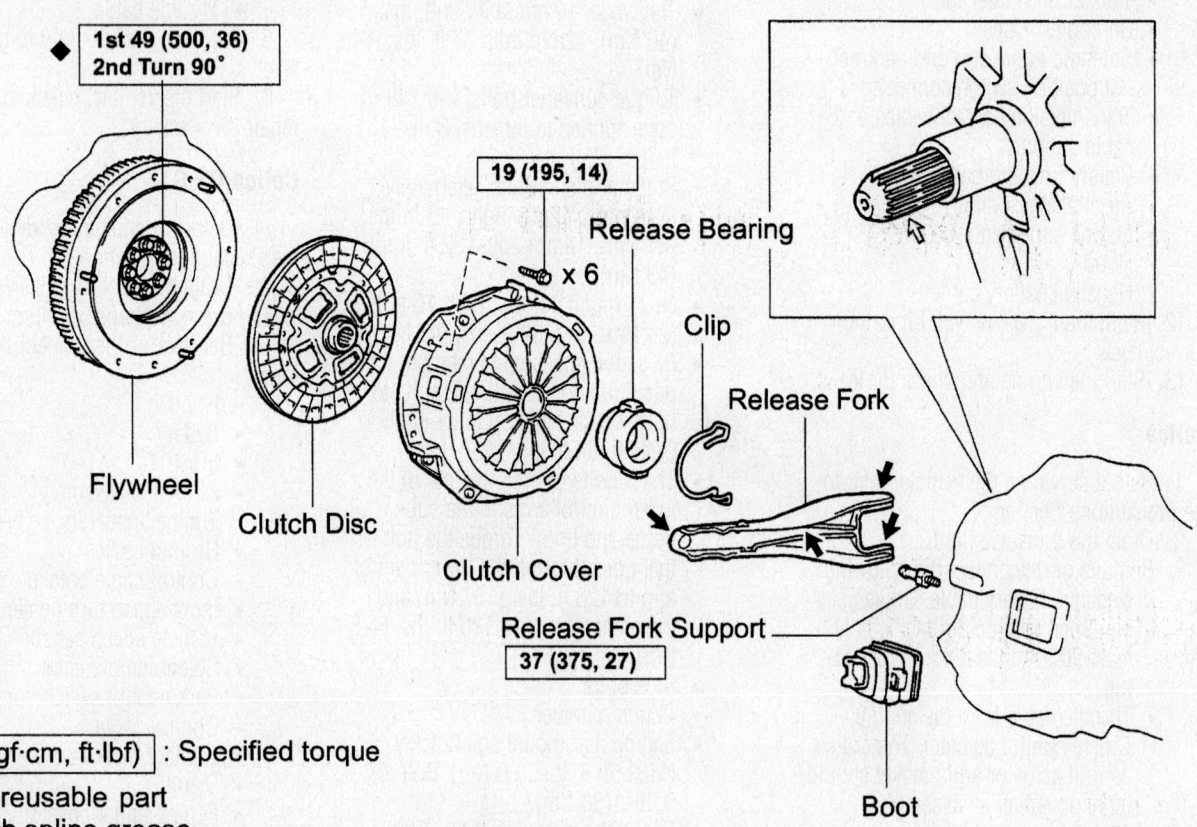

1st 49 (500, 36)
2nd Turn 90°

19 (195, 14)

Release Bearing

Clip

Release Fork

Flywheel

Clutch Disc

Clutch Cover

Release Fork Support
37 (375, 27)

Boot

N·m (kgf·cm, ft·lbf) : Specified torque

◆ Non-reusable part
⇒ Clutch spline grease
➡ Release hub grease

Clutch component assembly—Echo, others similar

09490_CORO_G0024

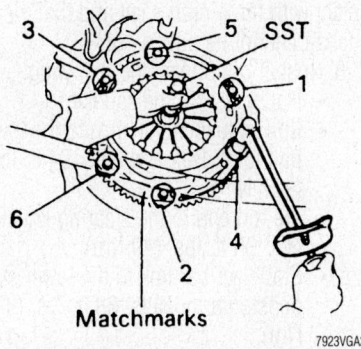

Tighten the pressure plate bolts according to the sequence shown—Echo

- Clutch cover, aligning the matchmarks
- Clutch cover retaining bolts. Bolts, tightened in a crisscross pattern: 14 ft. lbs. (19 Nm).
- Boot, release fork, hub and bearing assemblies
- Transaxle
- Negative battery cable

Corolla, Celica, and MR2

➡️**Do not allow grease or oil to get on any part of the disc, pressure plate, or flywheel surfaces.**

1. Before servicing the vehicle, refer to the Precautions Section.
2. Remove or disconnect the following:

- Negative battery cable. On vehicles equipped with an air bag, wait at least 90 seconds before proceeding
- Transaxle assembly

3. Make matchmarks on the clutch cover (pressure plate) and flywheel so that the pressure plate can be returned to its original position during installation.
4. Remove or disconnect the following:

- Release fork bearing clips
- Release bearing hub, complete with the release bearing
- Release fork and support

❈❈ CAUTION

Slowly unfasten the bolts which attach the pressure plate. Loosen each bolt 1 turn at a time until the spring tension is released. If the bolts are released improperly the clutch assembly could fly apart, causing possible injury.

- Pressure plate from the clutch cover/spring assembly

5. Inspect the disc, pressure plate and flywheel for damage and wear using a caliper to measure depth and width and a dial indicator to measure runout.

   a. The minimum clutch disc rivet head depth is 0.012 in. (0.3mm).
   b. The maximum clutch disc runout is 0.031 in. (0.8mm).
   c. The maximum pressure plate spring depth is 0.020 in. (0.5mm).
   d. The maximum pressure plate spring width is 0.236 in. (6.0mm).
   e. The maximum flywheel runout is 0.004 in. (0.1mm).
6. Replace or machine parts as necessary.

To install:

7. When reassembling, apply a thin coating of multipurpose grease to the release bearing hub and release fork contact points. Also, pack the groove inside the clutch hub with multipurpose grease and lubricate the pivot points of the release fork.
8. Install or connect the following:

- Clutch disc and pressure plate. The bolts should be tightened in 2 or 3 steps, gradually and evenly. Final bolt torque is 14 ft. lbs. (19 Nm).
- Release bearing, fork and boot
- Transaxle assembly
- Negative battery cable

ADJUSTMENTS

Hydraulic clutch actuating systems used in Toyota vehicles do not require adjustment.

Hydraulic Clutch System

BLEEDING

➡️**If any maintenance on the clutch system was performed or the system is suspected of containing air, bleed the system. Use care; brake fluid will remove the paint from any surface. If the brake fluid spills onto any painted surface, wash it off immediately with soap and water.**

1. Before servicing the vehicle, refer to the Precautions Section.
2. Fill the clutch reservoir with brake fluid. Check the reservoir level frequently and add fluid as needed.
3. Connect one end of a vinyl tube to the bleeder plug on the slave cylinder and submerge the other end into a clear container half-filled with brake fluid.
4. Slowly pump the clutch pedal several times.
5. Have an assistant hold the clutch pedal down and loosen the bleeder plug until fluid and/or air starts to run out of the bleeder plug. Close the bleeder plug while the pedal is held to the floor.

➡️**Do not allow the pedal to rise back-up while the bleeder is still open. If this happens, it will allow air to re-enter the slave cylinder and cause the clutch system not to work properly.**

6. Repeat Steps 2 and 3 until all the air bubbles are removed from the system.
7. Tighten the bleeder plug when all the air is gone.
8. Refill the master cylinder to the proper level as required.
9. Check the system for leaks.

Halfshaft

REMOVAL & INSTALLATION

Echo

1. Before servicing the vehicle, refer to the Precautions Section.
2. Drain the transaxle fluid.
3. Remove or disconnect the following:

- Negative battery cable. On vehicles equipped with an air bag, wait at least 90 seconds before proceeding.

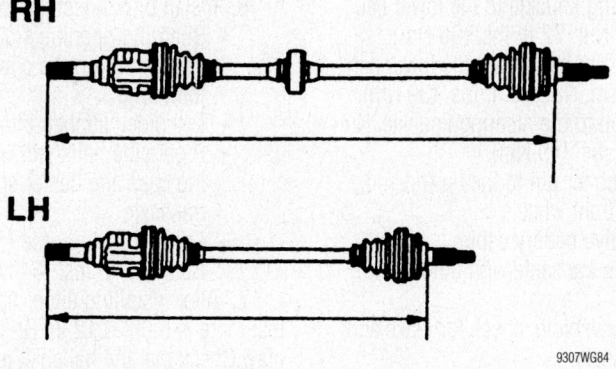

Measuring halfshaft length—Echo

- Both front wheels
- Locknut holding the halfshaft to the steering knuckle
- Tie rod end from the steering knuckle by removing the cotter pin and nut. Separate the tie rod from the steering knuckle
- Stabilizer bar link from the lower control arm. Make note of the washers and cushions positions
- Lower ball joint from the steering knuckle. Push down on the lower control arm and separate the steering knuckle from the ball joint
- Halfshaft from the steering knuckle
- Left halfshaft from the transaxle
- Snapring from the halfshaft
- Right halfshaft from the transaxle

➡ **The lockbolt is located in the center of the halfshaft, near the dampener.**

To install:

4. Install the right halfshaft to the transaxle, as follows:
   a. Coat the side gear shaft and differential case sliding surface with gear oil.
   b. Using snapring pliers, install the snapring to the halfshaft.
   c. Install the halfshaft

5. Install the left halfshaft to the transaxle, as follows:
   a. Install a new snapring to the inner spline of the halfshaft.
   b. Coat the side gear shaft and differential case sliding surface with gear oil.
   c. Install the halfshaft to the transaxle with the snapring opening facing down. The halfshaft should click into place when installing.
   d. After installation of the halfshaft, check that the halfshaft cannot be removed by hand.

6. Install or connect the following:
- Halfshaft to the steering knuckle, then install the locknut. Locknut: 159 ft. lbs. (216 Nm).
- Lock cap and new cotter pin to the halfshaft
- Steering knuckle to the lower ball joint. Nut: 72 ft. lbs. (98 Nm).
- Stabilizer bar link to the lower control arm. Nut: 29 ft. lbs. (39 Nm).
- Tie rod to the steering knuckle. Nut: 36 ft. lbs. (49 Nm).
- New cotter pin to the tie rod end
- Both front wheels
- Negative battery cable

7. Refill the transaxle with fluid to the correct level.

8. Start the vehicle, check for leaks and repair if necessary.

Celica

➡ **The hub bearing could be damaged if subjected to the full weight of the vehicle, such as if the vehicle is moved without the halfshafts. If it is absolutely necessary to place the full vehicle weight on the hub bearing, first support the bearing with SST No. 09608–16041.**

1. Before servicing the vehicle, refer to the Precautions Section.
2. Drain the transaxle fluid.
3. Remove or disconnect the following:
- Negative battery cable. On vehicles equipped with an air bag, wait at least 90 seconds before proceeding
- Both front wheels
- Cotter pin, locknut cap, and the bearing locknut
- Undercovers
- Tie rod ball joint from the steering knuckle
- Stabilizer bar link from the lower suspension arm
- Lower ball joint from the lower suspension arm
- Halfshaft from the knuckle

➡ **Be careful not to damage the inner oil seal or the ABS sensor rotor on the halfshaft.**

4. To remove the left side halfshaft, separate the halfshaft from the transaxle.
5. To remove the right side halfshaft perform the following steps:
- Remove the 2 bolts of the center bearing bracket
- Pull the halfshaft out together with the center bearing case and the center halfshaft.
- Remove the center shaft with the right-hand halfshaft from the transaxle through the bearing bracket.

➡ **Do not damage the oil seal lip.**

To install:

6. Install or connect the following:
- Snapring opening side facing downward, on the oiled inboard joint tulip
- Left side halfshaft into the transaxle
- Right side halfshaft, with the bearing case and center shaft, into the transaxle
- Center bearing case (right side). Bolts: 47 ft. lbs. (64 Nm)

7. After installing either halfshaft, check that there is 0.08–0.12 in. (2–3mm) of axial play. Check that the halfshaft is making

contact with the pinion shaft and that the halfshaft cannot be pulled out.

8. Install or connect the following:
- Halfshaft into the knuckle
- Lower suspension arm to the lower ball joint. Bolt and nuts: 94 ft. lbs. (127 Nm).
- Tie rod end to the steering knuckle. Nut: 36 ft. lbs. (49 Nm).
- Stabilizer bar link to the lower suspension arm. Nuts: 33 ft. lbs. (44 Nm).
- Front wheels
- Locknut and washer. Locknut: 159 ft. lbs. (216 Nm).
- Negative battery cable
- Locknut cap and a new cotter pin.
- Undercover

9. Fill the transaxle fluid to the proper level

10. Start the vehicle, check for leaks and repair if necessary.

Corolla

➡ **The hub bearing could be damaged if subjected to the full weight of the vehicle, such as if the vehicle is moved without the halfshafts. If it is absolutely necessary to place the full vehicle weight on the hub bearing, first support the bearing with SST No. 09608–16041.**

1. Before servicing the vehicle, refer to the Precautions Section.
2. Drain the transaxle fluid.
3. Remove or disconnect the following:
- Front wheels
- Undercovers
- Hub nut
- With ABS, speed sensor
- Stabilizer links
- Tie rod ball joint from the steering knuckle
- Lower ball joint from the lower suspension arm

4. Drive the halfshaft from the knuckle.

➡ **Most halfshafts can be separated from the knuckle using a brass or plastic hammer; some others may require the use of a puller.**

5. Remove the halfshaft from the transaxle

To install:

6. Install or connect the following:
- Snapring, opening side facing downward, to the inboard, oiled, joint tulip
- Halfshaft into the transaxle. After installing the halfshaft to the

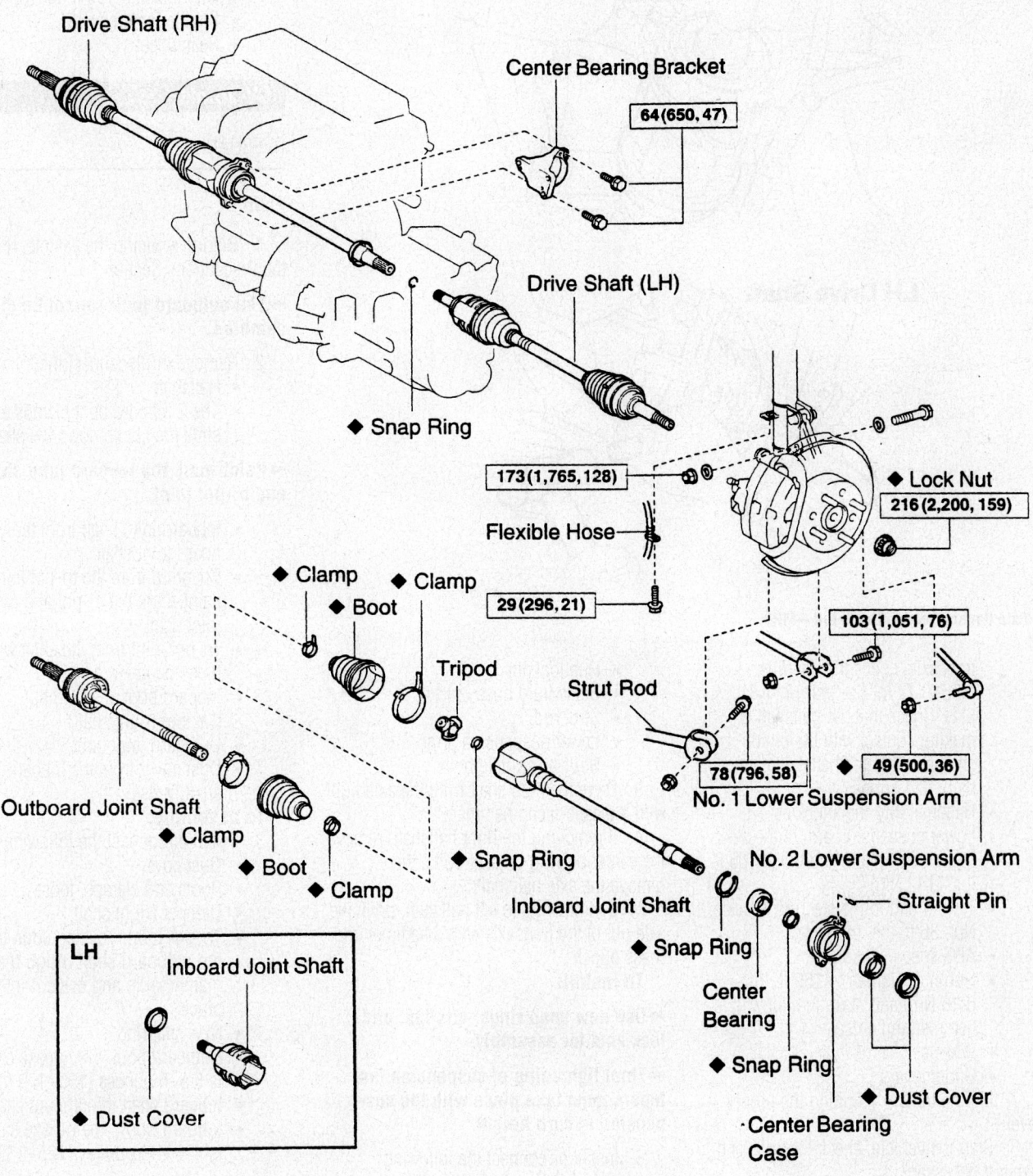

Drive Shaft (RH)

Center Bearing Bracket

64 (650, 47)

Drive Shaft (LH)

◆ Snap Ring

173 (1,765, 128)

◆ Lock Nut

216 (2,200, 159)

Flexible Hose

◆ Clamp   ◆ Clamp

◆ Boot

29 (296, 21)

103 (1,051, 76)

Tripod

Strut Rod

Outboard Joint Shaft

◆ Clamp

◆ Boot

◆ Clamp

78 (796, 58)   49 (500, 36)

No. 1 Lower Suspension Arm

◆ Snap Ring

Inboard Joint Shaft

No. 2 Lower Suspension Arm

Straight Pin

◆ Snap Ring

Center
Bearing

◆ Snap Ring

◆ Dust Cover

Center Bearing
Case

LH

Inboard Joint Shaft

◆ Dust Cover

N·m (kgf·cm, ft·lbf) : Specified torque

◆ Non–reusable part

9357WG08

Axle halfshaft mounting exploded view—MR2

RH Drive Shaft

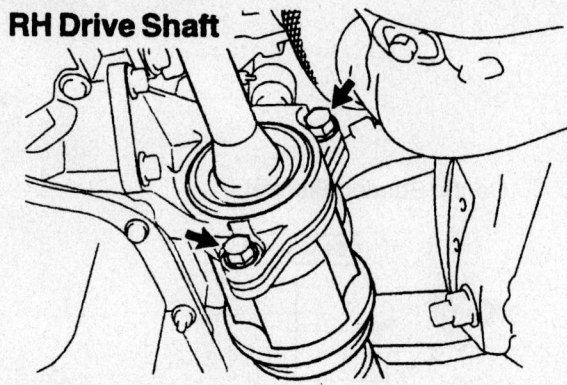

LH Drive Shaft

9357WG09

Axle halfshaft mounting detail—MR2

transaxle, check that there is 0.08–0.12 in. (2–3mm) of axial play. Check that the halfshaft is making contact with the pinion shaft and that the halfshaft cannot be pulled out

- Halfshaft into the knuckle
- Lower suspension arm to the steering knuckle. Nuts and bolts: 105 ft. lbs. (142 Nm).
- Tie rod end to the steering knuckle. Nut: 36 ft. lbs. (49 Nm).
- ABS speed sensor
- Hub nut. Tighten to 159 ft. lbs. (216 Nm) and stake the nut with a hammer and chisel.
- Wheels
- Undercovers

7. Fill the transaxle fluid to the proper level.

8. Start the vehicle, check for leaks and repair if necessary.

MR2

1. Before servicing the vehicle, refer to the Precautions Section.

2. Remove or disconnect the following:

- Rear wheel
- Engine undercovers

- Hub locknut
- Brake fluid hose bracket
- Strut rod
- Lower suspension arms
- Strut-to-spindle bolts

3. Drive the stub shaft from the axle hub with a plastic-faced hammer.

4. If removing the right halfshaft, remove the center bearing bracket bolts, then remove the axle halfshaft.

5. If removing the left halfshaft, drive the axle out of the transaxle with a hammer and brass punch.

To install:

➡**Use new snap rings, circlips, and lock nuts for assembly.**

➡**Final tightening of suspension fasteners must take place with the suspension at curb height.**

6. Install or connect the following:

- Axle halfshaft. Tighten the center bearing bracket bolts to 47 ft. lbs. (64 Nm).
- Strut-to-spindle-bolts and tighten them to 128 ft. lbs. (173 Nm).
- Lower suspension arms. Tighten the No. 1 arm bolt to 76 ft. lbs. (103 Nm) and the No. 2 bolt to 36 ft. lbs. (49 Nm).

- Strut rod. Tighten the bolt to 58 ft. lbs. (78 Nm).
- Brake fluid hose bracket
- Hub locknut: 159 ft. lbs. (216 Nm)
- Engine undercovers
- Rear wheel

CV-Joints

OVERHAUL

Echo

1. Before servicing the vehicle, refer to the Precautions Section.

➡**The outboard joint cannot be disassembled.**

2. Remove or disconnect the following:

- Halfshaft
- The 2 inboard boot clamps and slide the clamp down the shaft

➡**Paint-mark the inboard joint shaft and tri-pot joint.**

- Inboard joint shaft from the outboard joint shaft
- Snapring from the tri-pot joint, and paint-mark the tri-pot and outboard joint shaft
- Tri-pot joint from the shaft with a brass hammer
- Inboard boot and clamps
- Damper (right shaft)
- Outboard joint boot
- Dust cover from the inboard joint shaft (press)

To assemble:

3. Install or connect the following:

- Dust cover
- Boots and clamps, loosely
- Damper (right shaft)
- Tri-pot joint, beveled edge toward the outboard shaft. Align the matchmarks and drive it into place
- New snapring
- Outboard boot. The grease capacity is 5.5–6 ounces (155–170 g).
- Inboard shaft to outboard shaft
- Inboard boot. The grease capacity is 4.4–4.8 ounces (125–135 g).

➡**Toyota recommends different types of grease for the joints. OEM replacement boot kits have the grease color coded. The outboard grease is black; the inboard, yellow.**

4. Check the boots at standard halfshaft length. The right shaft should be 32.02 inches +/- 0.197 in. (813.3mm +/- 5mm);

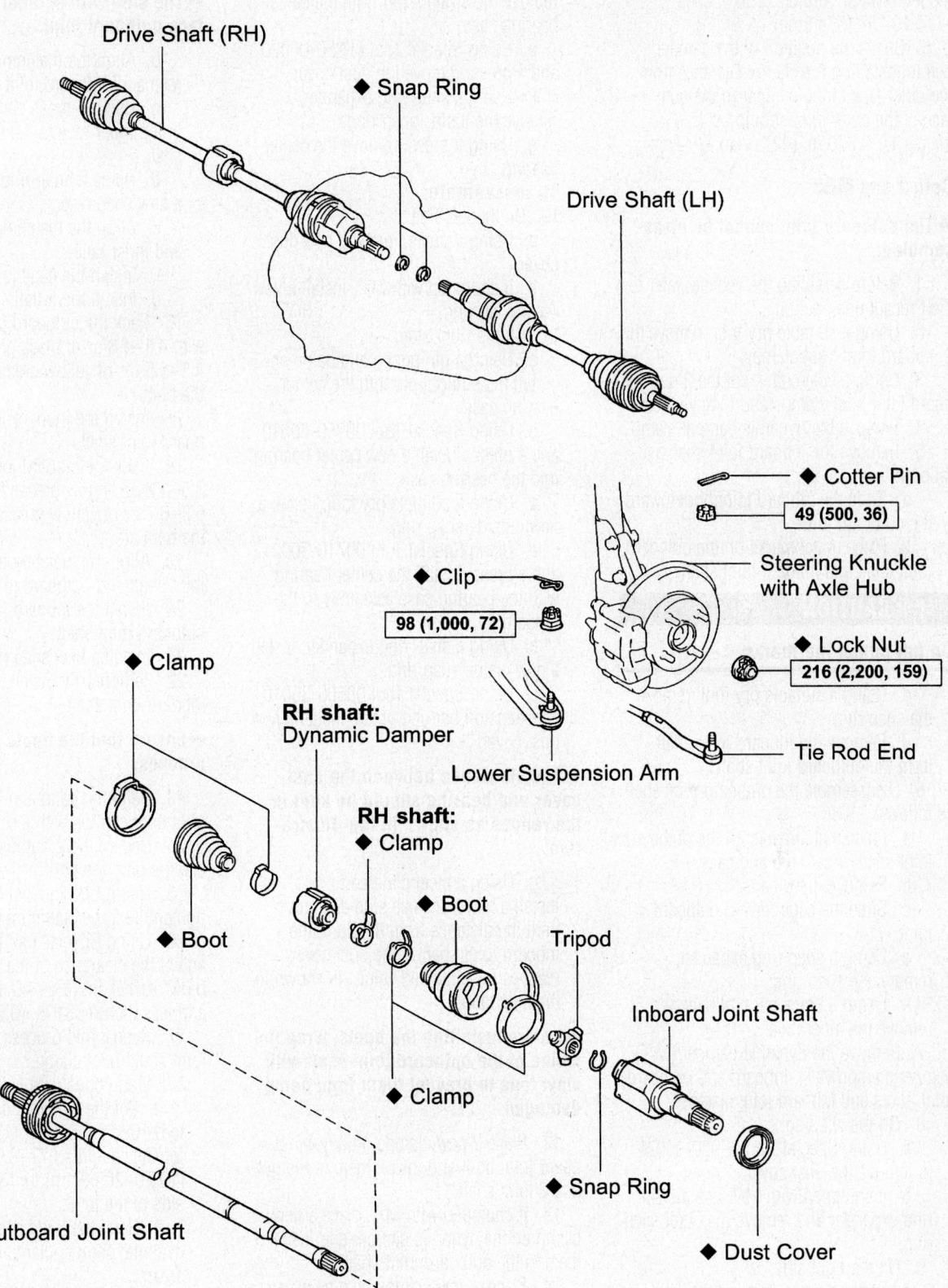

Drive Shaft (RH)

◆ Snap Ring

Drive Shaft (LH)

◆ Cotter Pin

49 (500, 36)

◆ Clip

Steering Knuckle
with Axle Hub

98 (1,000, 72)

◆ Lock Nut

216 (2,200, 159)

◆ Clamp

Tie Rod End

RH shaft:
Dynamic Damper

Lower Suspension Arm

RH shaft:
◆ Clamp

◆ Boot

◆ Boot

Tripod

Inboard Joint Shaft

◆ Clamp

◆ Clamp

◆ Snap Ring

Outboard Joint Shaft

◆ Dust Cover

N·m (kgf·cm, ft·lbf) : Specified torque

◆ Non-reusable part

09490_CORO_G0025

Exploded view of the halfshaft components—Echo

the left shaft should be 22.61 inches +/- 0.197 in. (574.3mm +/- 5mm).

5. Check the position of the damper before installing the clamp. Distance from the outer face of the damper to the outer face of the outer joint should be 16.835 inches +/- 0.079 in. (427.6mm +/- 2mm).

Celica and MR2

➡ **The outboard joint cannot be disassembled.**

1. Before servicing the vehicle, refer to the Precautions Section.
2. Using a suitable pry tool, remove the 2 inboard joint boot clamps.
3. Using a side cutter, cut the 2 outboard joint boot clamps and remove them.
4. Remove the dynamic damper clamp.
5. Remove the inboard joint shaft as follows:
   a. Slide the inboard joint boot toward the outboard joint.
   b. Place matchmarks on the outboard joint shaft and inboard joint shaft.

✳✳ CAUTION

Do not punch the marks.

   c. Using a suitable pry tool, remove the snap ring.
   d. Remove the inboard joint shaft from the outboard joint shaft.
6. Disassemble the outboard joint shaft as follows:
   a. Place matchmarks on the outboard joint shaft, inner race and cage.
   b. Remove the 6 balls.
   c. Slide the cage toward outboard joint.
   d. Using a snap ring expander, remove the snap ring.
   e. Using a brass bar and hammer, remove the inner race.
7. Remove the dynamic damper (M/T), damper clamp (M/T), inboard and outboard joint boots and inboard joint clamps.
8. On the left side:
   a. Using Special Tool 09950-00020 to remove the dust cover.
   b. If equipped with M/T, use a snap ring expander and remove the outer snap ring.
9. On the right side:
   a. Using a press, remove the transaxle side dust cover.
   b. Using a suitable pry tool, remove the outside snap ring.
   c. Using a press, remove the center bearing case.
   d. Using a pin punch and hammer,

remove the straight pin from the center bearing case.
   e. Using Special Tool 09950-00020 and a press, remove the dust cover.
   f. Using a snap ring expander, remove the inside snap ring.
   g. Using a press, remove the center bearing.

To reassemble:

10. On the left side:
   a. Using a press, install a new dust cover.
   b. If equipped with M/T, install a new outer snap ring.
11. On the right side:
   a. Using a pin punch and hammer, install the straight pin into the center bearing case.
   b. Using Special Tool 09950-60010 and a press, install a new center bearing into the bearing case.
   c. Using a suitable pry tool, install a new outside snap ring.
   d. Using Special Tool 09710-30021 and a press, install the center bearing with the bearing case assembly to the inboard joint shaft.
   e. Using a snap ring expander, install a new inside snap ring.
   f. Using Special Tool 09506-35010, an extension bar and press, install a new dust cover.

➡ **The clearance between the dust cover and bearing should be kept in the ranges as shown in the illustration.**

   g. Using a steel plate and press, install a new transaxle side dust cover until the distance from the tip of the inboard joint shaft to the dust cover reaches the specified value, as shown in the illustration.

➡ **Before installing the boots, wrap the spline of the outboard joint shaft with vinyl tape to prevent them from being damaged.**

12. Place 2 new clamps on a new outboard joint boot and install them to the outboard joint shaft.
13. If equipped with M/T, place a new clamp on the dynamic damper and install them to the outboard joint shaft.
14. Place 2 new clamps on a new inboard joint boot and install them to the outboard joint shaft.
15. Assemble the outboard joint shaft as follows:
   a. Install the cage to the outboard joint shaft.

➡ **The side with smaller diameter must face outboard joint.**

   b. Align the matchmarks on the inner race and outboard joint shaft.
   c. Using a brass bar and hammer, tap in the inner race to the outboard joint shaft.
   d. Using a snap ring expander, install a new snap ring.
   e. Align the matchmarks on the cage and inner race.
   f. Install the cage to the inner race.
   g. Install the 6 balls.
16. Pack the outboard joint and boot with 4.1–4.8 oz of black grease for Celica or 4.9–5.5 oz. of yellow grease for MR2, from the boot kit.
17. Install the inboard joint shaft to outboard joint shaft.
18. Pack the inboard joint and boot with 3.5–4.2 oz of gray grease for Celica or 6.3–6.7 oz of yellow grease for MR2, from the boot kit.
19. Align the matchmarks on the inboard joint shaft and outboard joint shaft.
20. Install the inboard joint shaft to the outboard joint shaft.
21. Install a new snap ring.
22. Temporarily install the boot to the inboard joint shaft.

➡ **Ensure that the boots are on the shaft grooves.**

23. Bend the band and lock the inboard joint boot clamps with a screwdriver.
24. Secure the 2 outboard joint boot clamps onto the boot.
25. Using Special Tool 09521-24010, tighten the large clamp on outboard joint.
26. Using Special Tool 09240-00020, adjust the clearance of the large clamp to 0.047–0.157 in. (1.2–4.0 mm) for Celica and less than 0.031 in. (0.8 mm) for MR2.
27. Repeat this process for the outboard joint small boot clamp.
28. If equipped with M/T:
   a. Set the dynamic damper distance to 6.102 ± 0.079 in. (155.0 ± 2.0 mm) for Celica and 7.677–8.071 inch (195.0–205.0mm) for MR2 from the outside of the joint.
   b. Bend the band and lock the dynamic damper clamp with a screwdriver.

Corolla

1. Before servicing the vehicle, refer to the Precautions Section.
2. Remove or disconnect the following:
   • Inboard joint boot clips

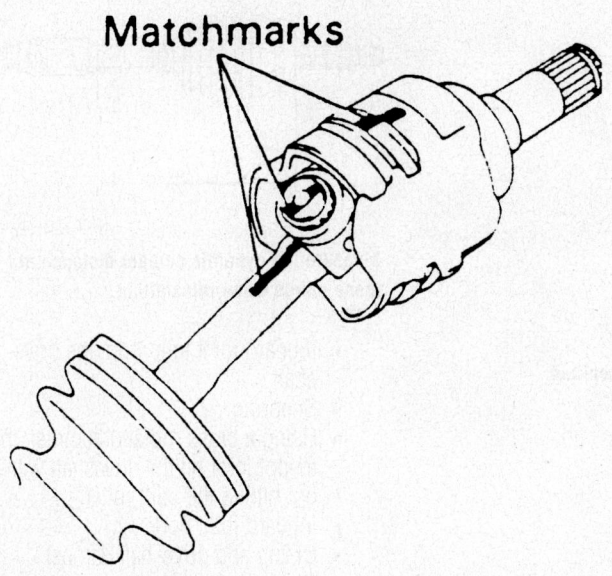

Place matchmarks on the tri-pot and outboard joints

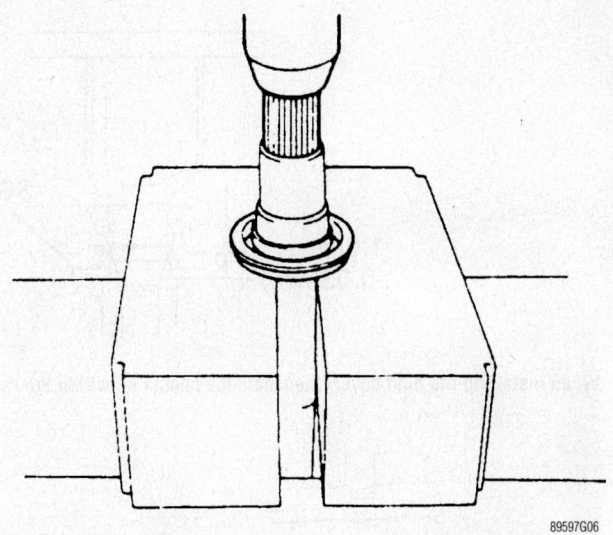

Use a press to drive the old dust cover out of the right side center driveshaft

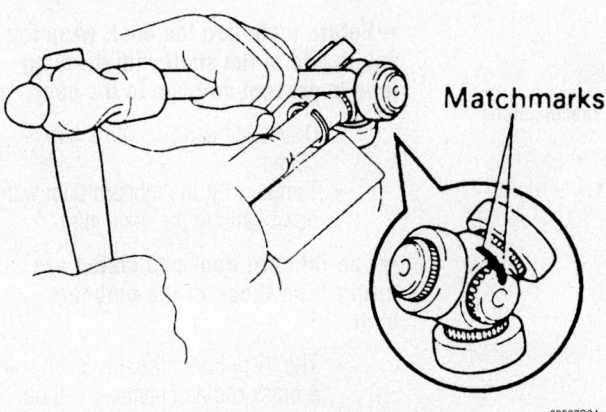

With matchmarks on the tri-pot, tap the joint for the driveshaft

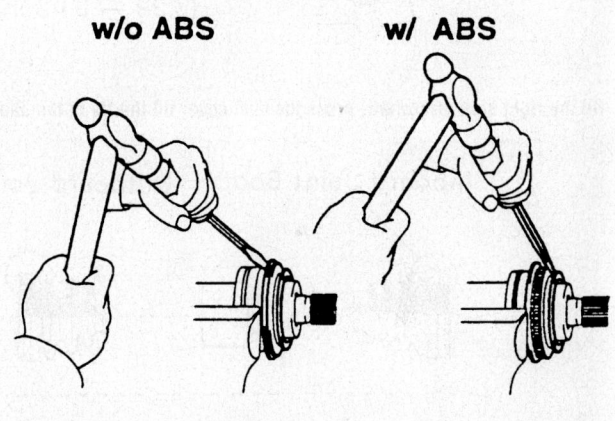

Removing the no. 2 dust deflector on models with and without ABS

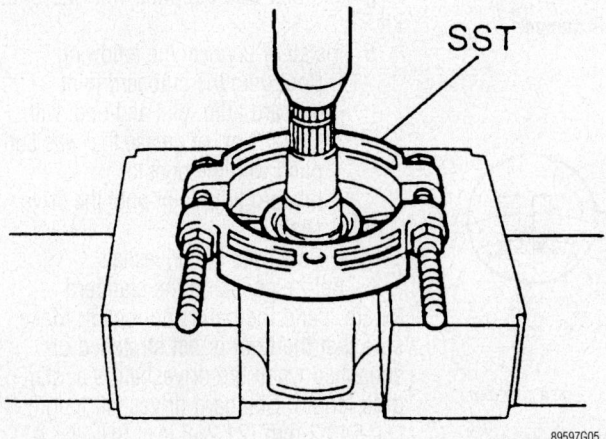

Using a press, remove the dust cover from the center driveshaft on the left side

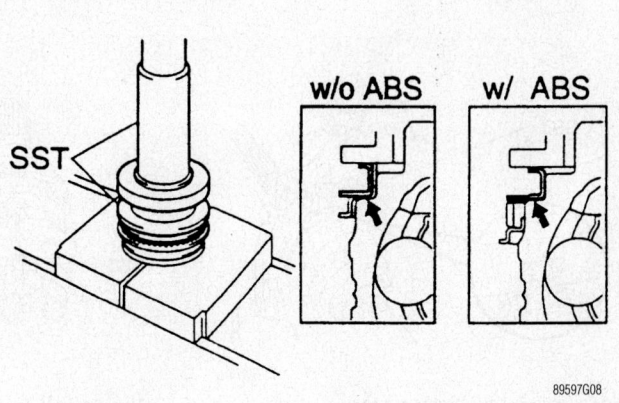

Press the new No. 2 deflector and seat it properly

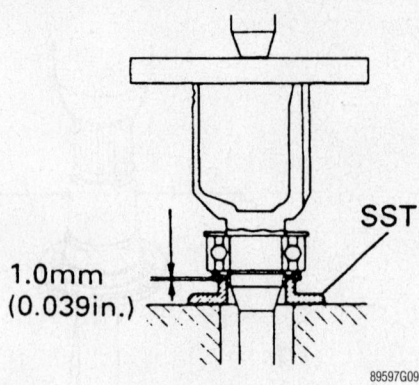

1.0mm
(0.039in.)

SST

89597G09

When installing the dust cover, the clearance should be within the ranges as specified

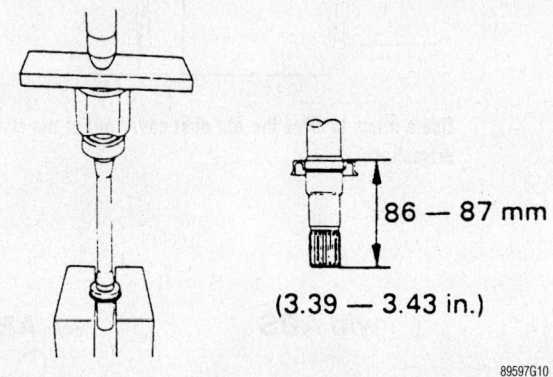

86 — 87 mm

(3.39 — 3.43 in.)

89597G10

On the right side driveshaft, press the dust cover till the tip of the shaft reaches specifications

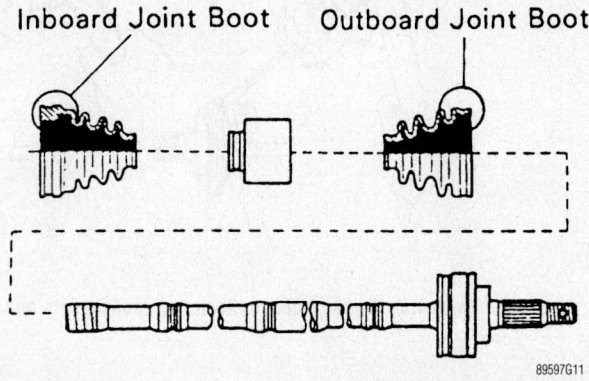

Inboard Joint Boot   Outboard Joint Boot

89597G11

Temporarily install the driveshaft boots on the inboards and outboard and dynamic damper

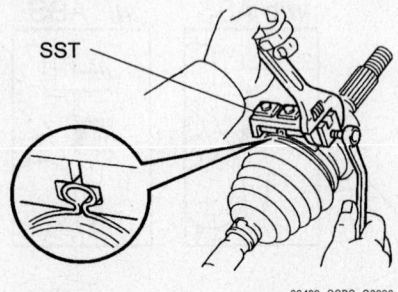

SST

09490_CORO_G0026

Use Special Tool 09521-24010 to install the boot clamp.

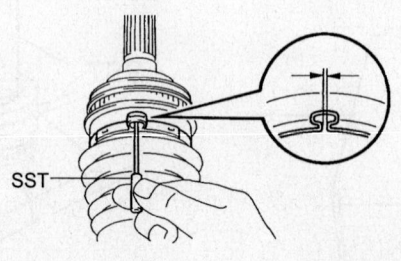

SST

09490_CORO_G0027

Use Special Tool 09240-00020 to adjust the clearance of the boot clamp.

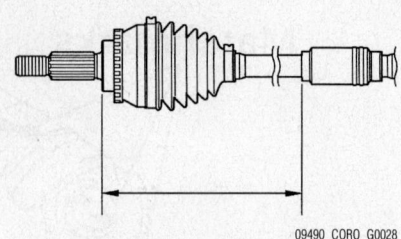

09490_CORO_G0028

Measure the dynamic damper distance at these points when reinstalling.

- Inboard joint tulip from the drive-shaft
- Snapring
- Using a brass rod and hammer, the tri-pot joint off the driveshaft without hitting the joint roller
- Inboard joint boot
- Clamp and driveshaft damper
- Clamps and the outboard drive boot. DO NOT disassemble the outboard joint.

To assemble:

3. Install or connect the following:

➡**Before installing the boot, wrap the spline end of the shaft with masking tape to prevent damage to the boot.**

- Driveshaft damper with a new clamp
- Temporarily, the inboard boot with new clamp to the drive joint

➡**The inboard boot and clamp are larger than those of the outboard boot.**

- The tri-pot onto the driveshaft with a brass rod and hammer without hitting the joint roller
- The snapring

4. Pack the outboard tulip joint and the outboard boot with about 5.4–5.7 oz. of grease that was supplied with the boot kit.

5. Install or connect the following:
- Boot onto the outboard joint
- Inboard tulip joint and boot with 5.6–6.3 oz. of grease that was supplied with the boot kit
- Inboard tulip joint onto the drive-shaft
- Boot onto the driveshaft

6. Before checking the standard length, bend the band and lock it. Make sure that the boot is not stretched or squashed when the driveshaft is at standard length. Standard driveshaft length: LH: 540.2 mm (21.268 in.); RH: 857.4 mm (33.756 in.)

STEERING AND SUSPENSION

Air Bag

DISARMING

To avoid personal injury when working on vehicles equipped with an air bag, the negative battery cable must be disconnected and at least 90 seconds must elapse before working on the system. Failure to do so may result in deployment of the air bag.

REARMING

After vehicle service is completed, reattach the battery cables (positive cable first!) to rearm the air bag system.

Rack and Pinion Steering Gear

REMOVAL & INSTALLATION

Echo

MANUAL STEERING

1. Before servicing the vehicle, refer to the Precautions Section.
2. Place the wheels in the straight-ahead position.
3. Remove or disconnect the following:
   - Steering wheel
   - Engine under covers
   - Tie rod ends
   - No. 2 column hole cover
   - Sliding yoke
   - Hood
4. Attach and engine crane to the engine for support.
5. Remove or disconnect the following:
   - Lower arms from the knuckles
   - Rear engine mount insulator
   - Front crossmember with the steering rack
   - Column hole cover sub-assembly
   - 4 bolts and nuts attaching the gear to the crossmember

➡**Because the nut has its own stopper, do not turn the nut with the bolt tight.**

To install:

6. Install or connect the following:
   - Steering rack to crossmember. Torque the nuts to 54 ft. lbs. (74 Nm).
   - Column hole cover sub-assembly
   - Front suspension assembly. Front bolts: 52 ft. lbs. (70 Nm); rear bolts: 86 ft. lbs. (116 Nm).
   - Rear mount insulator. Torque the

bolt and 2 nuts to 59 ft. lbs. (80 Nm).
   - Lower arm to knuckle. Horizontal bolt: 65 ft. lbs. (88 Nm); vertical bolt: 97 ft. lbs. (132 Nm).
   - Hood
   - Sliding yoke
   - No. 2 column hole cover
   - Tie rods
   - Under covers
   - Steering wheel. Nut: 25 ft. lbs. (34 Nm).

POWER STEERING

1. Before servicing the vehicle, refer to the Precautions Section.
2. Place the wheels in the straight-ahead position.
3. Remove or disconnect the following:
   - Steering wheel
   - Engine under covers
   - Tie rod ends
   - No. 2 column hole cover
   - Sliding yoke
   - Pressure and return lines
   - Hood
4. Attach and engine crane to the engine for support.
5. Remove or disconnect the following:
   - Lower arms from the knuckles
   - Rear engine mount insulator
   - Front crossmember with the steering rack
   - Stabilizer bar
   - Heat insulator
   - Damper
   - 4 bolts and nuts attaching the gear to the crossmember

➡**Because the nut has its own stopper, do not turn the nut with the bolt tight.**

To install:

6. Install or connect the following:
   - Steering rack to crossmember. Torque the nuts to 54 ft. lbs. (74 Nm).
   - Damper. Bolts: 13 ft. lbs. (18 Nm).
   - Heat insulator. Bolt: 26 ft. lbs. (35 Nm).
   - Stabilizer bar
   - Front suspension assembly. Front bolts: 52 ft. lbs. (70 Nm); rear bolts: 85 ft. lbs. (116 Nm).
   - Rear mount insulator. Torque the bolt and 2 nuts to 59 ft. lbs. (80 Nm).
   - Lower arm to knuckle. Horizontal bolt: 65 ft. lbs. (88 Nm); vertical bolt: 97 ft. lbs. (132 Nm).

   - Hood
   - Sliding yoke
   - No. 2 column hole cover
   - Tie rods
   - Pressure and return lines
   - Under covers
   - Steering wheel. Nut: 25 ft. lbs. (34 Nm).

Celica

1. Before servicing the vehicle, refer to the Precautions Section.
2. Remove or disconnect the following:
   - Negative battery cable. On vehicles equipped with an air bag, wait at least 90 seconds before proceeding.
   - Steering wheel using Special Tool 09950-50013

➡**Matchmark the steering wheel and main shaft assembly before removal.**

   - Right and left-hand engine undercovers
   - Left and right-hand tie rod ends. Separate the tie rod using a puller
   - Oxygen sensor

➡**Place matchmarks on the steering column intermediate shaft and the steering gear control valve shaft.**

   - Lower bolt to the intermediate shaft
   - Intermediate shaft from the control valve shaft
   - Pressure feed and return tubes from the steering rack.
   - Tube clamp bracket
   - Hood

3. Support the engine and transaxle with a support fixture.
4. Remove or disconnect the following:
   - Lower control arms from the lower ball joints
   - Through-bolts to the front and rear mounting insulators
   - Bolts holding the rear of the lower control arm to the sub-frame and body. Remove the bolts on both sides of the sub-frame.
5. Support the front sub-frame with a jack.
6. Remove or disconnect the following:
   - Bolts holding the sub-frame to the body. Lower the front sub-frame with the lower suspension arms and steering gear.
   - Power steering gear assembly by removing the set bolts and nuts from the sub-frame.

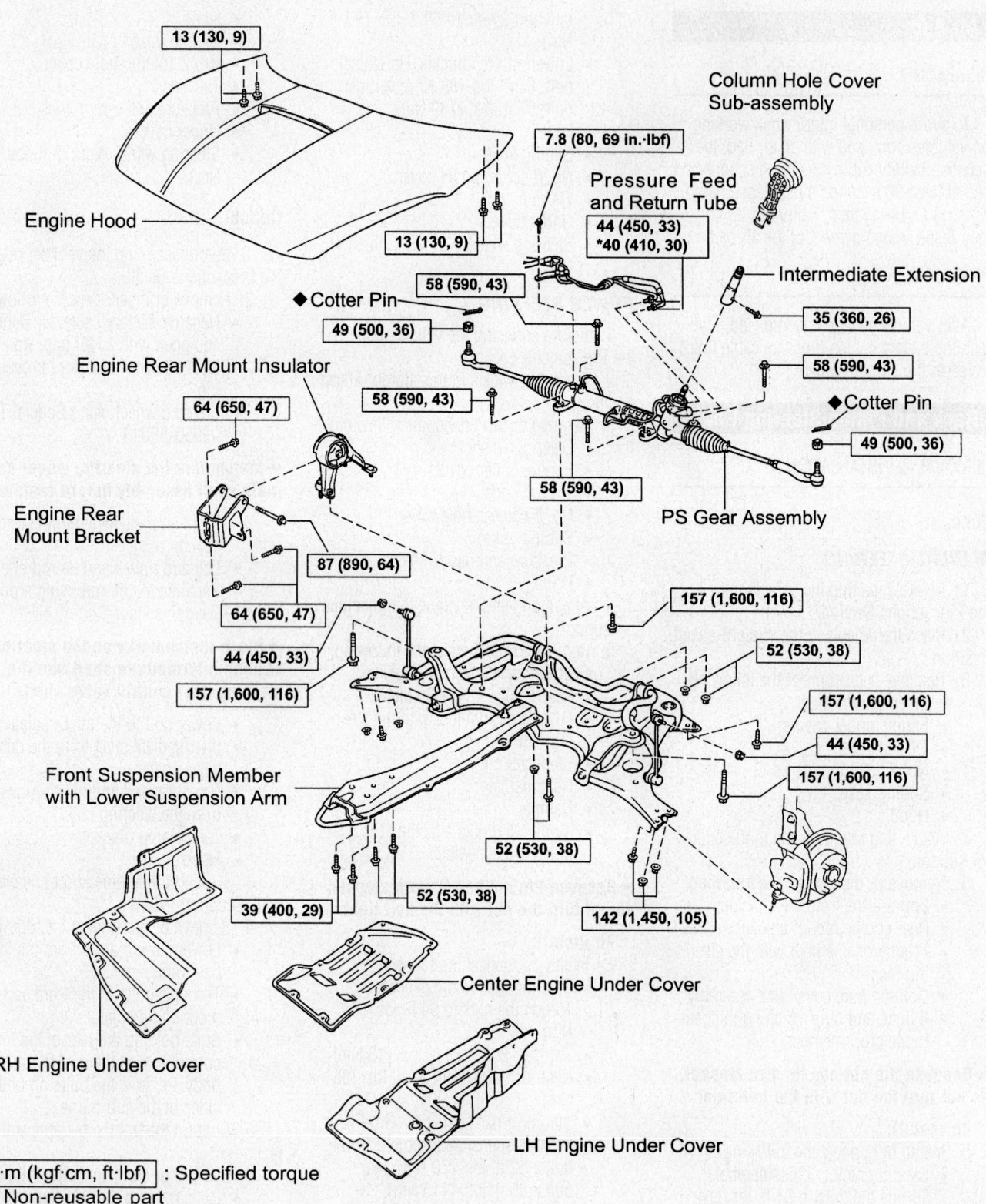

13 (130, 9)

Column Hole Cover
Sub-assembly

Engine Hood

7.8 (80, 69 in.·lbf)

Pressure Feed
and Return Tube

44 (450, 33)
*40 (410, 30)

13 (130, 9)

Intermediate Extension

58 (590, 43)

◆Cotter Pin

35 (360, 26)

49 (500, 36)

58 (590, 43)

Engine Rear Mount Insulator

◆Cotter Pin

64 (650, 47)

58 (590, 43)

49 (500, 36)

Engine Rear
Mount Bracket

87 (890, 64)

58 (590, 43)

PS Gear Assembly

64 (650, 47)

157 (1,600, 116)

44 (450, 33)

52 (530, 38)

157 (1,600, 116)

157 (1,600, 116)

44 (450, 33)

Front Suspension Member
with Lower Suspension Arm

157 (1,600, 116)

52 (530, 38)

39 (400, 29)

52 (530, 38)

142 (1,450, 105)

Center Engine Under Cover

RH Engine Under Cover

LH Engine Under Cover

N·m (kgf·cm, ft·lbf) : Specified torque
◆ Non-reusable part
* For use with SST

09490_CORO_G0030

Exploded view of the steering gear and related components—Celica, Corolla similar

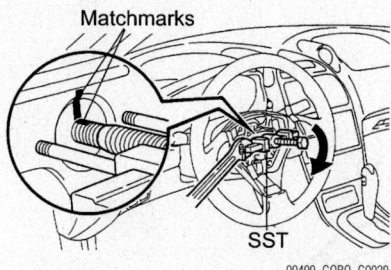

Matchmarks

SST

09490_CORO_G0029

Matchmark and remove the steering wheel—Celica

To install:

7. Install or connect the following:

- Power steering gear assembly to the sub-frame. Bolts and nuts: 43 ft. lbs. (58 Nm).
- Sub-frame to the body. Bolts: 94 ft. lbs. (127 Nm).
- Bolts to hold the rear of the lower control arms to the sub-frame and body. Bolt that goes through the lower control arm to 116 ft. lbs. (157 Nm) and the other 2 bolts to 130 ft. lbs. (175 Nm).
- Through-bolts to the front and rear mounting insulators. Bolts: 38 ft. lbs. (52 Nm).
- Lower control arms to the lower ball joints. Nuts and bolts: 105 ft. lbs. (142 Nm).
- Pressure feed and return tubes to the steering rack. Tubes: 26 ft. lbs. (36 Nm).
- Hood
- Tube clamp bracket. Bolts: 108 inch lbs. (13 Nm).

8. Align the matchmarks on the intermediate shaft and control valve shaft.

9. Install or connect the following:

- Lower bolt. Upper and lower bolts: 26 ft. lbs. (35 Nm).
- Tie rod ends to the steering knuckles
- Right and left-hand engine undercovers
- Steering wheel
- Negative battery cable

10. Bleed the power steering system.

11. Check and adjust the alignment, if needed.

Corolla

1. Before servicing the vehicle, refer to the Precautions Section.

2. Position the front wheels straight ahead.

3. Remove or disconnect the following:

- Negative battery cable

- Steering wheel, using Special Tool 09950-50013
- Front wheels
- Left and right engine undercovers
- Front exhaust assembly
- Left and right tie rod ends

➡**Matchmark the intermediate shaft before separating.**

- Steering column hole cover and loosen the upper pinch bolt on the sliding yoke
- Pressure and return power steering lines
- Stabilizer links
- Lower control arms from the ball joints
- Hood
- Engine appearance cover

4. Install an engine support and tension it to support the engine without raising it.

✳✳ CAUTION

The engine hoist is now in place and under tension. Use care when repositioning the vehicle and make necessary adjustments to the engine support.

5. Remove or disconnect the following:

6. Remove the 4 bolts to separate the center crossmember.

7. Separate the right rear mounting insulator from the crossmember.

8. Support the crossmember assembly with a suitable jack.

9. Remove the bolts from the front crossmember assembly.

➡**Matchmark the intermediate shaft and pinion shaft.**

10. Remove the pinch bolt and separate the intermediate shaft from the pinion shaft.

11. Remove the steering gear assembly from the crossmember.

12. Slide the power steering gear assembly to the right side of the vehicle.

To install:

13. Install the steering gear assembly to the crossmember. Tighten the mounting bolts to 43 ft. lbs. (59 Nm).

14. Connect the intermediate shaft to the pinion shaft. Tighten the bolts to 26 ft. lbs. (35 Nm).

15. Install or connect the following:

- Front suspension crossmember assembly. Tighten front bolts to 83 ft. lbs. (113 Nm) and rear bolts to 116 ft. lbs. (157 Nm).
- Rear right mounting insulator to the crossmember. Tighten to 38 ft. lbs. (52 Nm).

- Center crossmember bolts. Tighten to 44 ft. lbs. (60 Nm).
- Lower control arm to the lower ball joint. Bolt and nuts: 66 ft. lbs. (89 Nm).
- Stabilizer bar links to the lower control arms. Nuts: 55 ft. lbs. (74 Nm).
- Power steering lines
- Intermediate shaft to the slide yoke. Tighten upper pinch bolt to 26 ft. lbs. (35 Nm).
- Left and right-hand tie rod ends. Nuts: 36 ft. lbs. (49 Nm).
- Front exhaust assembly
- Hood
- Undercovers
- Front wheels
- Negative battery cable

16. Check and top off the power steering fluid.

17. Check and adjust the alignment, if needed.

MR2

➡**Do not allow the steering wheel to rotate when disconnected from the steering gear. Damage to the air bag spiral cable may result.**

1. Before servicing the vehicle, refer to the Precautions Section.

2. Place the wheels pointing straight ahead and lock the steering wheel in place.

3. Remove or disconnect the following:

- Luggage compartment trim box cover
- Tool box
- Power steering pressure and return lines from the power steering vane pump
- Front luggage undercover
- Outer tie rod ends
- Steering intermediate shaft
- Power steering gear mounting bolts
- Power steering gear

To install:

4. Installation is the reverse of the removal procedure, while using the following torwue values:

- Power steering gear mounting bolts: 42 ft. .lbs. (57 Nm)
- Steering intermediate shaft pinch bolt: 26 ft. lbs. (35 Nm)
- Outer tie rod ends: 36 ft. lbs. (49 Nm)

5. Fill and bleed the power steering system.

6. Check the alignment and adjust if necessary.

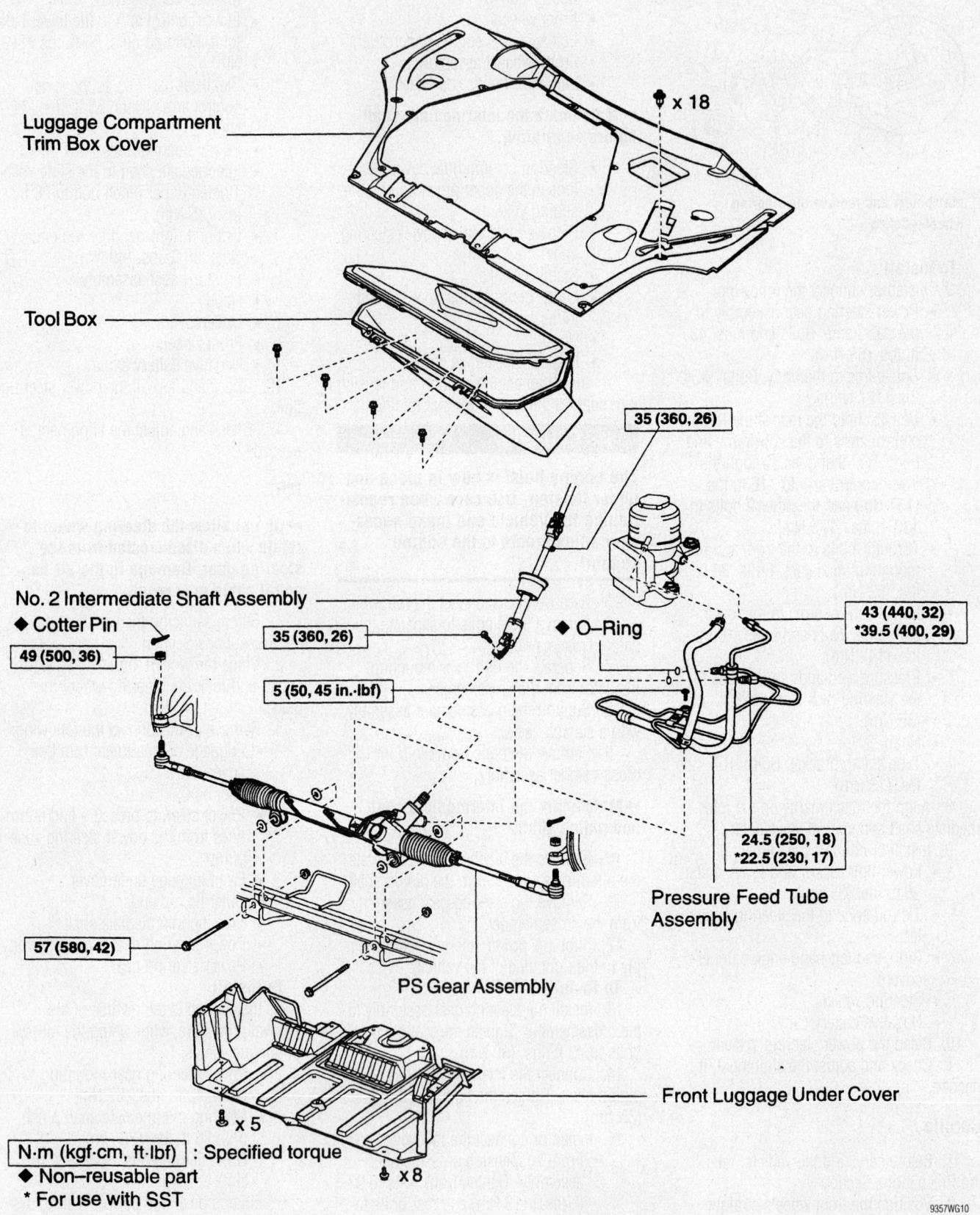

Luggage Compartment Trim Box Cover

x 18

Tool Box

35 (360, 26)

No. 2 Intermediate Shaft Assembly

◆ Cotter Pin

49 (500, 36)

35 (360, 26)

5 (50, 45 in.·lbf)

◆ O–Ring

43 (440, 32)
*39.5 (400, 29)

24.5 (250, 18)
*22.5 (230, 17)

Pressure Feed Tube Assembly

57 (580, 42)

PS Gear Assembly

Front Luggage Under Cover

x 5

x 8

N·m (kgf·cm, ft·lbf) : Specified torque
◆ Non–reusable part
* For use with SST

9357WG10

Power rack and pinion steering gear mounting exploded view—MR2

Strut and Coil Spring

REMOVAL & INSTALLATION

Front

1. Before servicing the vehicle, refer to the Precautions Section.
2. Remove or disconnect the following:
   • Negative battery cable.

✳✳ WARNING

Do not support the weight of the vehicle on the suspension arm; the arm will deform under its weight.

   • Wheel
   • Stabilizer link, except MR2
   • Bolt, and disconnect the brake hose from the strut
   • Wheel speed sensor, if equipped
   • Bolts and strut from the steering knuckle
   • Strut
3. Disassemble the strut as follows:
   a. Install a bolt and 2 nuts to the bracket at the lower portion of the strut shell and secure it in a vise.
   b. Compress the coil spring.
   c. Remove the dust cover and hold the spring seat so that it will not turn. Remove the nut on the top of the strut.
   d. Remove the suspension support, bearing, dust seal, spring seat, spring, insulators and bumper.

To install:

4. Assemble the strut as follows:
   a. Install the spring bumper to piston.
   b. Using a spring compressor, compress the spring.
   c. Install the coil spring to the strut. Fit the lower end of the coil spring into the gap of the lower seat.
   d. Install the spring seat with the insulator.
   e. Install the dust seal on the spring seat.
   f. Install the suspension support and tighten 35 ft. lbs. (47 Nm). After the nut has been tighten, release the compressor tool tension.
   g. Pack multipurpose grease into the suspension support. Install the dust cover.

➡**Do not use an impact wrench to tighten the nut. Also, check that the bearing fits into the recess in the suspension support.**

5. Install or connect the following:
   • Nuts holding the strut to the strut

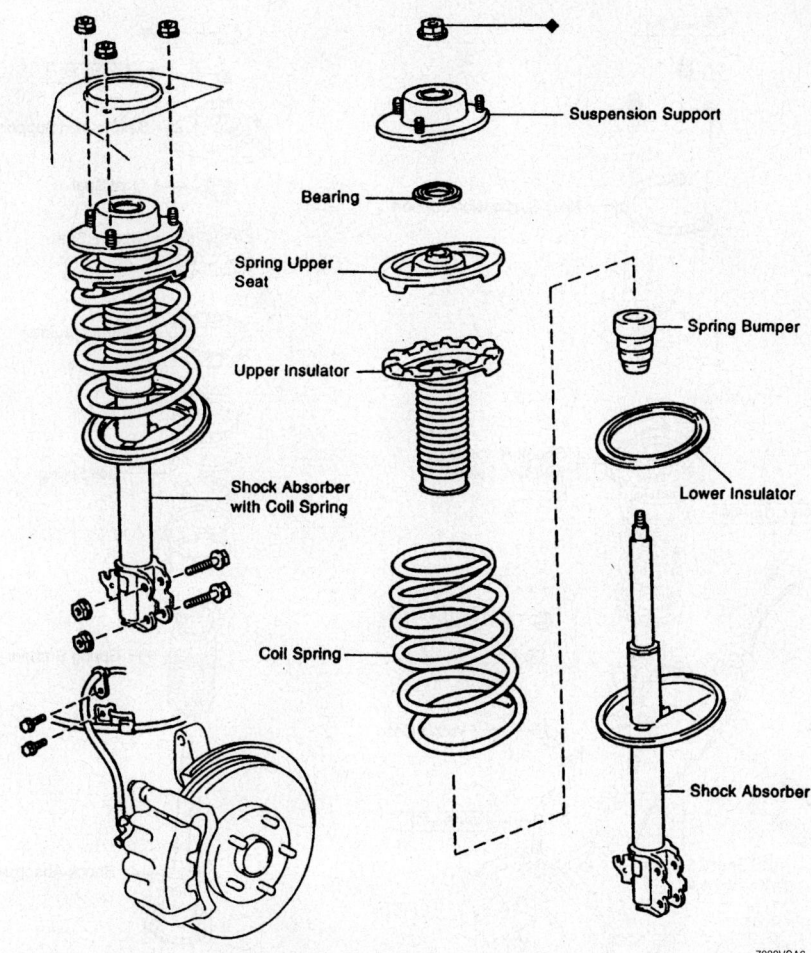

Common coil spring and strut component assembly

7923VGA6

tower. Tighten to 29 ft. lbs. (39 Nm).
   • Steering knuckle to the strut lower bracket
6. Insert the 2 bolts from the rear side and tighten the strut-to-steering knuckle arm bolts. Tighten as follows:
   a. Echo: 97 ft. lbs. (132 Nm)
   b. Corolla and Celica: 113 ft. lbs. (153 Nm)

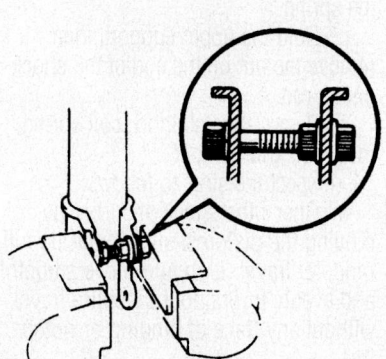

7923VGA7

Proper method of supporting the strut in a vise

   c. MR2: 103 ft. lbs. (140 Nm)
7. Install or connect the following:
   • Brake line to the steering knuckle
   • If equipped with ABS, secure the wiring harness
   • Stabilizer link, tighten to 55 ft. lbs. (74 Nm)
   • Wheel
   • Negative battery cable
8. Check and adjust the alignment, if needed.

Rear

EXCEPT MR 2

1. Before servicing the vehicle, refer to the Precautions Section.
2. Remove or disconnect the following:

   • Negative battery cable from the battery. On vehicles equipped with an air bag, wait at least 90 seconds before proceeding
   • Rear seat cushion and any trim necessary to access the strut towers

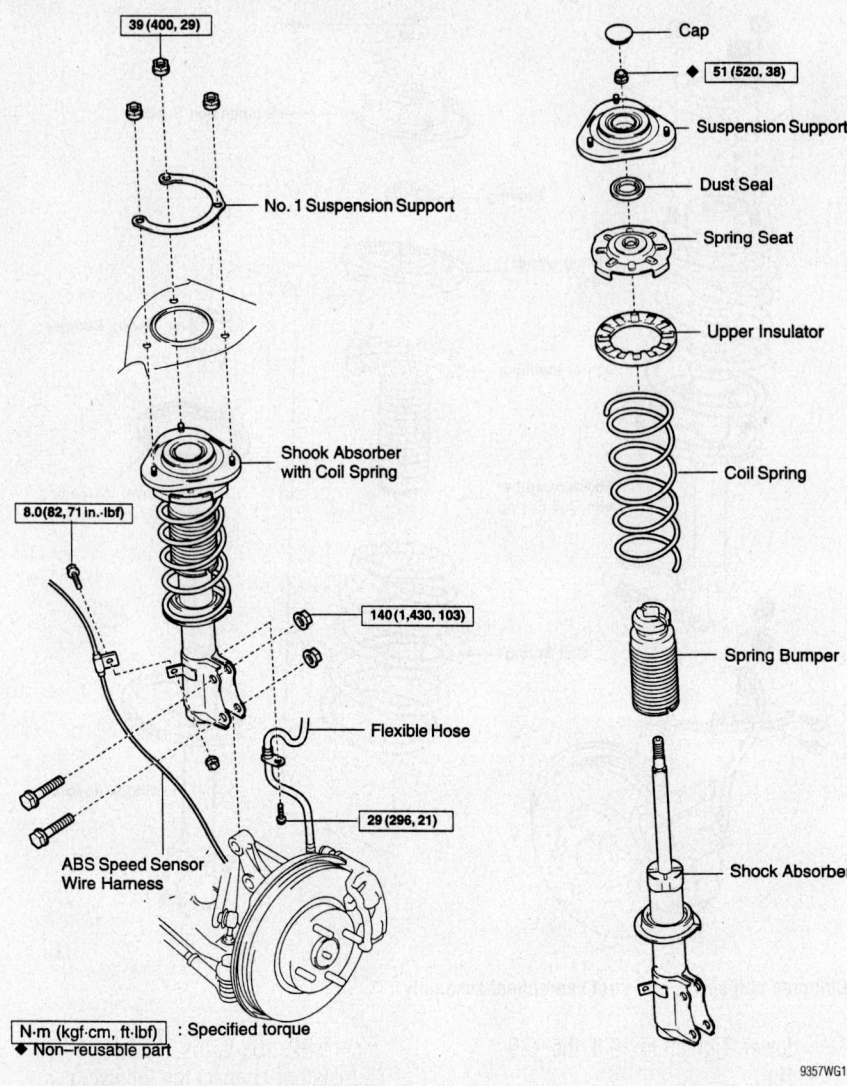

39 (400, 29)

No. 1 Suspension Support

Shock Absorber with Coil Spring

8.0 (82, 71 in.-lbf)

140 (1,430, 103)

Flexible Hose

29 (296, 21)

ABS Speed Sensor Wire Harness

N·m (kgf·cm, ft·lbf) : Specified torque
◆ Non-reusable part

Cap

◆ 51 (520, 38)

Suspension Support

Dust Seal

Spring Seat

Upper Insulator

Coil Spring

Spring Bumper

Shock Absorber

9357WG11

Front strut and coil spring mounting exploded view—MR2

3. Support the axle beam with a jack.
4. Remove or disconnect the following:
   • Wheel
   • Wheel speed sensor, if equipped
   • Stabilizer bar
5. Loosen the fasteners securing the strut to the axle carrier. Do not remove the bolts at this time.
6. Support the axle carrier with a jack.
7. Remove or disconnect the following:
   • Strut-to-strut tower nuts

❊❊ CAUTION

Do not loosen the center nut on the top of the strut piston.

   • Strut
8. Disassemble the strut as follows:
   a. Place the strut assembly in a pipe vise or strut vise.

➡**Do not attempt to clamp the strut assembly in a flat jaw vise as this will result in damage to the strut tube.**

   b. Compress the spring until the upper suspension support is free of any spring tension. Do not over-compress the spring.
   c. Hold the upper support, then remove the nut on the end of the shock piston rod.
   d. Remove the support, coil spring, insulator, and bumper.
9. Inspect the strut as follows:
   a. Check the shock absorber by moving the piston shaft through its full range of travel. It should move smoothly and evenly throughout its entire travel without any trace of binding or notching.
   b. Use a small straightedge to check

the piston shaft for any bending or deformation.
   c. Inspect the spring for any sign of deterioration or cracking. The waterproof coating on the coils should be intact to prevent rusting.
To install:

➡**Never reuse a self-locking nut. Always replace self-locking nuts and cotter pins as applicable.**

10. Assemble the strut as follows:
   a. Loosely assemble all components onto the strut assembly. Be sure the spring end aligns with the hollow in the lower seat.
   b. Align the upper suspension support with the piston rod and install the support.
   c. Align the suspension support with the strut lower bracket. This assures the spring will be properly seated top and bottom.
   d. Compress the spring to expose the strut piston rod threads.
   e. Install a new strut piston nut and tighten to 41 ft. lbs. (56 Nm).
   f. Remove the spring compressor. Be sure the paint mark on the upper support faces the outside of the strut.
11. Place the strut on the vehicle and install the nuts to hold the strut to the strut tower. Tighten the nuts to 59 ft. lbs. (80 Nm).
12. Install or connect the following:
   • Strut to the axle carrier and install the bolt and nut. Do not tighten at this time
   • Stabilizer link to the strut
   • Wheel. Bounce the vehicle up and down to stabilize the suspension
13. With the vehicle weight on the suspension, tighten the bolt holding the strut to the axle carrier as follows:
   • Corolla: 59 ft. lbs. (80 Nm)
   • Celica: 105 ft. lbs. (142 Nm)
14. Install or connect the following:
   • Rear seat cushion and any trim
   • Negative battery cable

MR2

1. Before servicing the vehicle, refer to the Precautions Section.
2. Remove or disconnect the following:
   • Rear wheel
   • Stabilizer link
   • Brake hose
   • Lower strut mounting bolts
   • Upper strut mounting nuts.
3. Disassemble the strut as follows:

a. Place the strut assembly in a pipe vise or strut vise.

b. Remove the nut, collar, suspension support, coil spring and spring bumper.

To install:

➡**Never reuse a self-locking nut. Always replace self-locking nuts and cotter pins as applicable.**

4. Assemble the strut as follows:

a. Loosely assemble all components onto the strut assembly. Be sure the spring end aligns with the hollow in the lower seat.

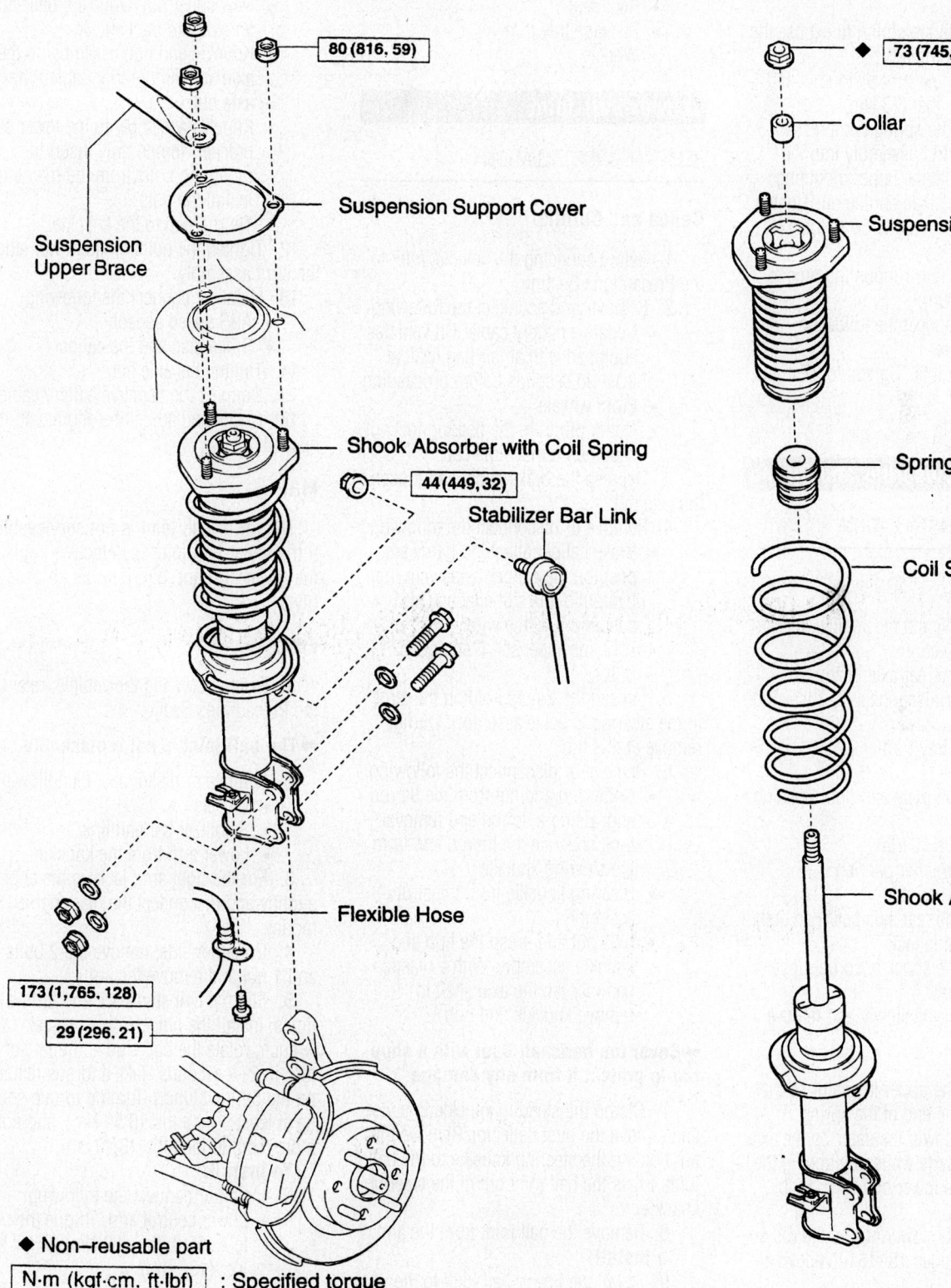

80 (816, 59)

Suspension Support Cover

Suspension Upper Brace

Shook Absorber with Coil Spring

44 (449, 32)

Stabilizer Bar Link

Flexible Hose

173 (1,765, 128)

29 (296, 21)

◆ 73 (745, 54)

Collar

Suspension Support

Spring Bumper

Coil Spring

Shook Absorber

◆ **Non-reusable part**

N·m (kgf·cm, ft·lbf) : Specified torque

9357WG12

Rear strut and coil spring assembly exploded view—MR2

b. Align the upper suspension support with the piston rod and install the support.

c. Align the suspension support with the strut lower bracket. This assures the spring will be properly seated top and bottom.

d. Compress the spring to expose the strut piston rod threads.

e. Install a new strut piston nut and tighten to 54 ft. lbs. (73 Nm).

f. Remove the spring compressor.

5. Install the strut assembly into the vehicle. Lightly coat the upper mounting bolts threads with clean engine oil. Tighten the upper mounting nuts to 59 ft. lbs. (80 Nm).

6. Tighten the lower mounting nuts to 128 ft. lbs. (173 Nm).

7. Install or connect the following:
- Brake hose
- Stabilizer link. Tighten to 32 ft. lbs. (44 Nm).
- Rear wheel

Rear Shock Absorber and Spring

REMOVAL AND INSTALLATION

Echo

1. Before servicing the vehicle, refer to the Precautions Section.

2. Support the rear axle beam.

3. Remove or disconnect the following:
- Wheels
- Package tray trim
- Rear seat
- Door scuff plate and door opening trim
- Quarter panel trim
- Roof side inner garnish
- Partition board

4. Support the rear axle beam on both sides with a suitable jack.

5. Remove the shock absorber upper nuts and lower bolt.

6. Lower the axle slowly and remove the spring.

To install:

7. Position the upper insulator so that its gap fits into the end of the spring.

8. Place the lower insulator on the axle.

9. Raise the axle while positioning the shock. Torque the lower bolt to 36 ft. lbs. (49 Nm).

10. Position the lower nut on the rod so that the rod protrudes 15–18mm above the nut.

11. Install or connect the following:

- Upper nut. Torque the nut to 18 ft. lbs. (25 Nm).
- Partition board
- Roof side inner garnish
- Quarter panel
- Door scuff plate and trim
- Rear seat
- Package tray trim
- Wheels

Lower Ball Joint

REMOVAL & INSTALLATION

Celica and Corolla

1. Before servicing the vehicle, refer to the Precautions Section.

2. Remove or disconnect the following:
- Negative battery cable. On vehicles equipped with an air bag, wait at least 90 seconds before proceeding
- Front wheels
- Cotter pin from the bearing locknut cap, then remove the cap

3. Depress the brake pedal and loosen the axle nut

4. Remove or disconnect the following:
- Brake caliper attaching hardware, position the caliper aside with the hydraulic line still attached and suspend it with a wire
- ABS speed sensor, if equipped
- Rotor

5. Loosen the 2 nuts holding the strut to the steering knuckle assembly. Do not remove at this time.

6. Remove or disconnect the following:
- Cotter pin and nut from the tie rod end. Using a tie rod end removal tool, separate the tie rod end from the steering knuckle
- Steering knuckle from the strut assembly
- Axle nut and grasp the hub and knuckle assembly. With a plastic hammer tap the axle shaft to remove knuckle and hub

➡ **Cover the halfshaft boot with a shop rag to protect it from any damage.**

7. Clamp the steering knuckle in a vise and remove the dust deflector. Remove the nut holding the steering knuckle to the ball joint. Press the ball joint out of the steering knuckle.

8. Remove the ball joint from the arm.

To install:

9. Install the Lower ball joint to the lower arm. Tighten the fasteners to:

a. Corolla: 66 ft. lbs. (89 Nm).

b. Celica: 105 ft. lbs. (142 Nm).

10. Install the ball joint to the steering knuckle. Tighten the ball joint-to-steering knuckle nut to 76 ft. lbs. (103 Nm).

11. Install or connect the following:
- New cotter pin. Drive the deflector shield onto the knuckle
- Knuckle and hub assembly to the axle and temporarily tighten the axle nut
- Knuckle assembly to the lower strut bracket. Temporarily insert the mounting bolts from the rear and install the nuts
- Tie rod end to the knuckle

12. Tighten the bolts on the lower side of the strut assembly.

13. Install or connect the following:
- ABS speed sensor
- Brake disc and the caliper

14. Tighten the axle nut.

15. Connect the negative battery cable.

16. Check and adjust the alignment, if needed.

MR2

The lower ball joint is not serviceable. If the lower ball joint is defective, replace the lower arm and ball joint as an assembly.

Echo

1. Before servicing the vehicle, refer to the Precautions Section.

➡ **The ball joint is not replaceable.**

2. Remove or disconnect the following:
- Wheels
- Stabilizer bar end link
- Lower arm from the knuckle

3. For the right arm, jack up the engine slightly and disconnect the rear engine mount.

4. On either side, remove the 2 bolts and 1 nut and remove the arm.

5. Flip the ball stud back and forth a few times. Install the nut. Using a torque wrench, rotate the ball stud at the rate of 1 turn in 2–4 seconds. Take a torque reading on the 5th revolution. Turning torque should be at least 5 inch lbs. (0.59 Nm), and not more than 30 inch lbs. (3.4 Nm).

To install:

6. Install or connect the following:
- Lower control arm. Torque the bolts to A: 65 ft. lbs. (88 Nm); B: 97 ft. lbs. (132 Nm).

➡ **DO NOT turn the nut.**

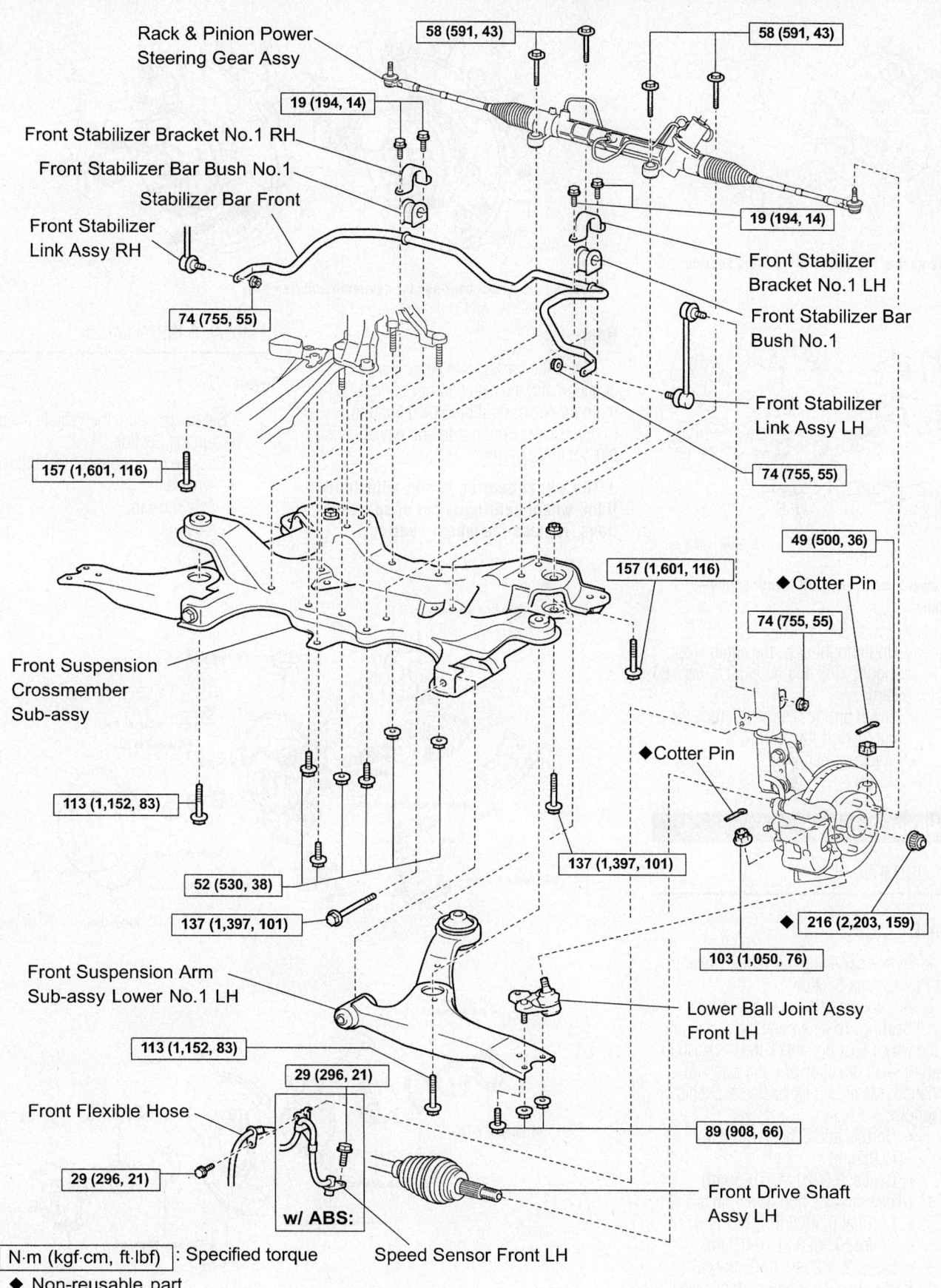

Rack & Pinion Power Steering Gear Assy

58 (591, 43)

58 (591, 43)

19 (194, 14)

Front Stabilizer Bracket No.1 RH

Front Stabilizer Bar Bush No.1

Stabilizer Bar Front

Front Stabilizer Link Assy RH

19 (194, 14)

Front Stabilizer Bracket No.1 LH

Front Stabilizer Bar Bush No.1

74 (755, 55)

Front Stabilizer Link Assy LH

157 (1,601, 116)

74 (755, 55)

49 (500, 36)

◆Cotter Pin

157 (1,601, 116)

74 (755, 55)

Front Suspension Crossmember Sub-assy

◆Cotter Pin

113 (1,152, 83)

137 (1,397, 101)

52 (530, 38)

216 (2,203, 159)

137 (1,397, 101)

103 (1,050, 76)

Front Suspension Arm Sub-assy Lower No.1 LH

Lower Ball Joint Assy Front LH

113 (1,152, 83)

29 (296, 21)

Front Flexible Hose

89 (908, 66)

29 (296, 21)

Front Drive Shaft Assy LH

w/ ABS:

N·m (kgf·cm, ft·lbf) : Specified torque

Speed Sensor Front LH

◆ Non-reusable part

09490_CORO_G0032

Exploded view of the front suspension components—Corolla, Celica similar

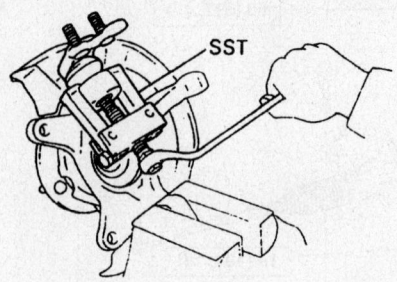

Removing the ball joint from the knuckle

7923VGA9

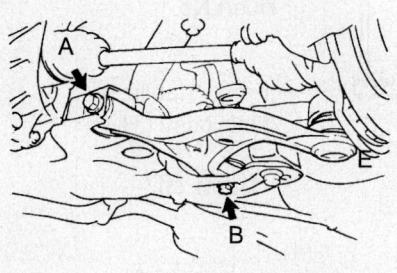

09490_CORO_G0031

**Lower control arm bolt identification—
Echo**

- On the right side, the engine rear mount. Bolt and nuts: 59 ft. lbs. (80 Nm).
- Lower arm to knuckle. Torque the nut to 72 ft. lbs. (98 Nm).
- Stabilizer bar
- Wheel

Wheel Bearings

ADJUSTMENT

Front

1. Before servicing the vehicle, refer to the Precautions Section.
2. All models use a non-adjustable wheel bearing. To determine the condition of the wheel bearing, check the backlash in bearing shaft direction and the axle hub deviation. Maximum for backlash should be as follows:
   - Corolla and Echo: 0.0020 in. (0.05mm)
   - Celica: 0.0020 in. (0.05mm)
3. Maximum axle hub deviation is:
   - Corolla: 0.0028 in. (0.07mm)
   - Celica: 0.0028 in. (0.07mm)
   - Echo: 0.0020 in. (0.05mm)
4. If the wheel bearing is out of specifications, replace the wheel bearing.

7923VGB1

Checking the wheel bearings for deviation and free-play

Rear

Check the backlash in bearing shaft direction and the axle hub deviation. Maximum for backlash should be 0.0020 in. (0.05mm). Maximum axle hub deviation is 0.0028 in. (0.07mm).

➡ **The wheel bearing is non-adjustable. If the wheel bearing is out of specifications, replace the wheel bearing.**

REMOVAL & INSTALLATION

Front

1. Before servicing the vehicle, refer to the Precautions Section.
2. Remove or disconnect the following:
   - Wheels
   - Axle nut cap
   - Axle nut

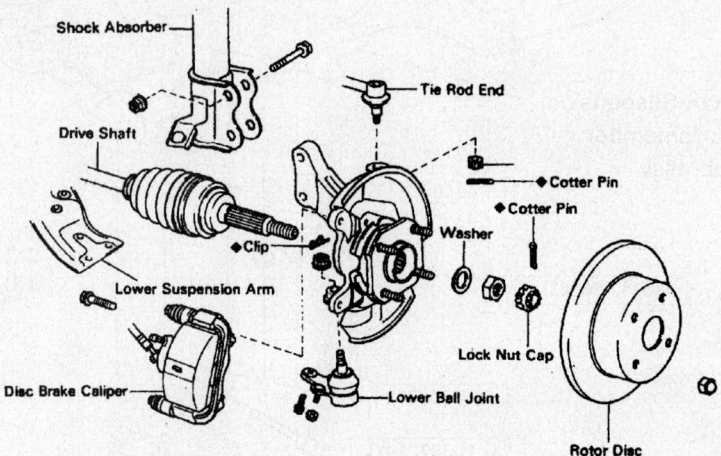

Steering knuckle and hub assembly

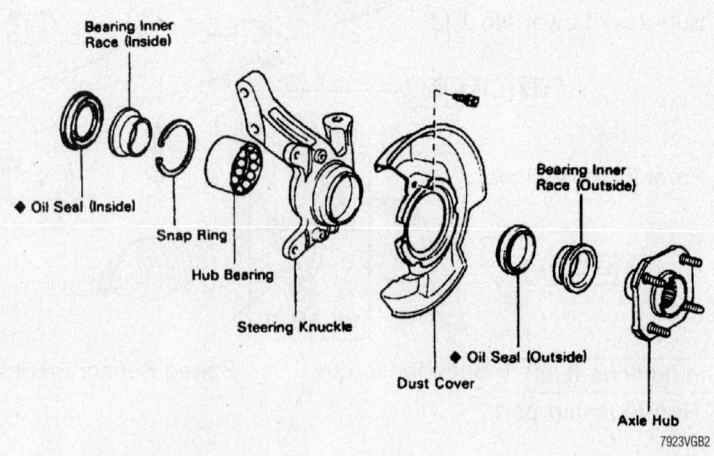

7923VGB2

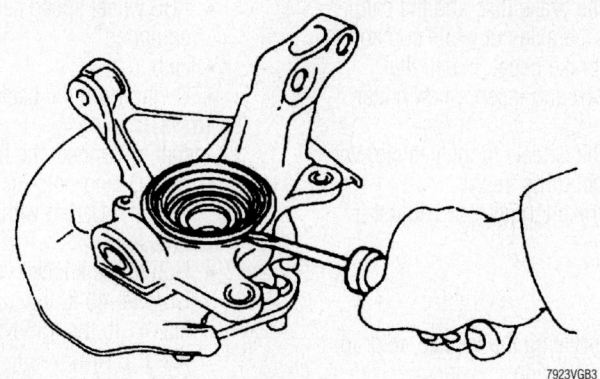

Removing the inner axle seal from the hub assembly

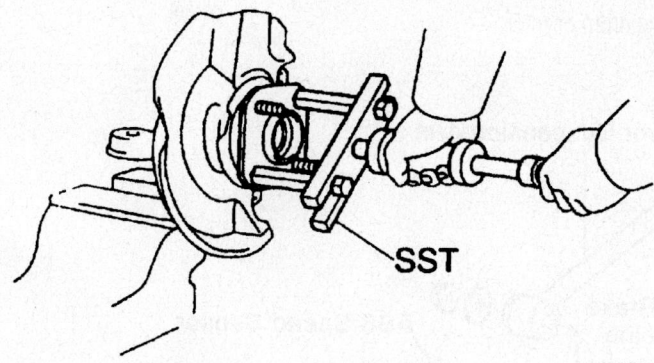

Removing the axle hub from the knuckle

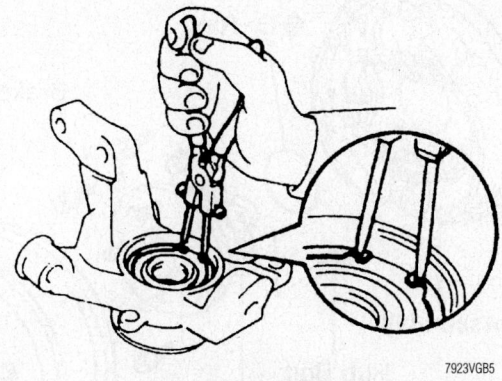

Removing the snapring from the knuckle before pressing out the bearing

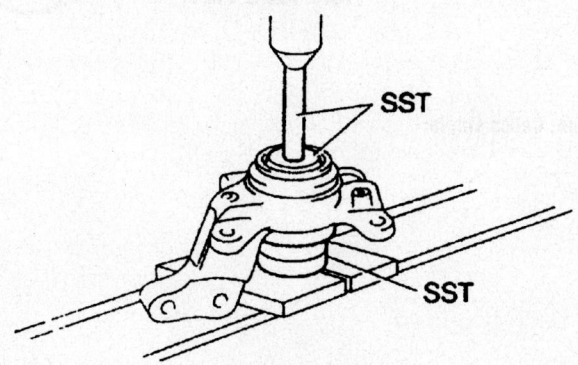

Removing the bearing from the steering knuckle using a press

- Caliper. Position the caliper aside with the hydraulic line still attached and suspend it with a wire.
- ABS speed sensor
- Rotor

3. Loosen the nuts on the lower side of the strut assembly. Do not remove at this time.

4. Remove or disconnect the following:

- Tie rod end from the steering knuckle
- Steering knuckle from the lower control arm
- Knuckle from the strut assembly
- Hub

➡**Cover the halfshaft boot with a shop rag to protect it from any damage.**

5. Clamp the steering knuckle in a vise and remove the dust deflector. Remove the nut holding the steering knuckle to the ball joint. Press the ball joint out of the steering knuckle.

6. Remove the inner axle seal.

7. Using a Torx® wrench, remove the bolts securing the dust cover.

8. Using hub puller, remove the hub and backing plate from the steering knuckle.

9. Using a proper sized driver and a press, remove the inner hub race from the axle hub.

10. Using seal removal tool, remove the outer axle seal.

11. Using snapring pliers, remove the snapring from the inner side of the steering knuckle.

12. Using a proper sized driver and a press, remove the bearing from the steering knuckle. The bearing is pressed from the front of the steering knuckle and is removed through the back of the steering knuckle.

To install:

13. Using a proper sized driver and a press, install a new bearing to the steering knuckle.

14. Install the snapring to the steering knuckle using snapring pliers.

15. Using a seal driver and a hammer, install a new outer oil seal. Apply multipurpose grease to the oil seal lip.

16. Place the dust cover on the steering knuckle. Bolts: 78 inch lbs. (9 Nm).

17. Using a press and a proper sized driver, install the axle hub to the steering knuckle.

18. Attach the ball joint to the steering knuckle. Install a new cotter pin.

19. Using a seal driver and a hammer,

install a new inner oil seal. Apply multipurpose grease to the oil seal lip.

20. Install the knuckle and hub assembly to the axle and temporarily tighten the axle nut.

21. Connect the knuckle assembly to the lower strut bracket. Temporarily insert the mounting bolts from the rear and install the nuts making sure the matchmarks made earlier are in alignment.

22. Connect the lower ball joint to lower arm.

23. Connect the tie rod end to the knuckle.

24. Tighten on the lower side of the strut assembly.

25. If equipped, install the ABS speed sensor.

26. Install the brake disc and the caliper.

27. Tighten the axle nut while someone depresses the brake pedal. Install the adjusting nut cap and insert a new cotter pin.

28. Install the wheels to the vehicle. Verify that the wheel turns freely.

29. Check the alignment and adjust if necessary.

Rear

1. Before servicing the vehicle, refer to the Precautions Section.

2. Remove or disconnect the following:

- Wheel
- Brake drum or rotor
- ABS wheel speed sensor, if equipped
- Hub
- O-ring from the backing plate

To install:

3. Install or connect the following:

- New O-ring onto the backing plate. Coat the O-ring with multipurpose grease.
- Hub to the knuckle. Bolts: Corolla—45 ft. lbs. (61 Nm); Celica—41 ft. lbs. (56 Nm); Echo, 38 ft. lbs. (52 Nm).
- ABS wheel speed sensor, if equipped
- Brake drum or rotor
- Wheel

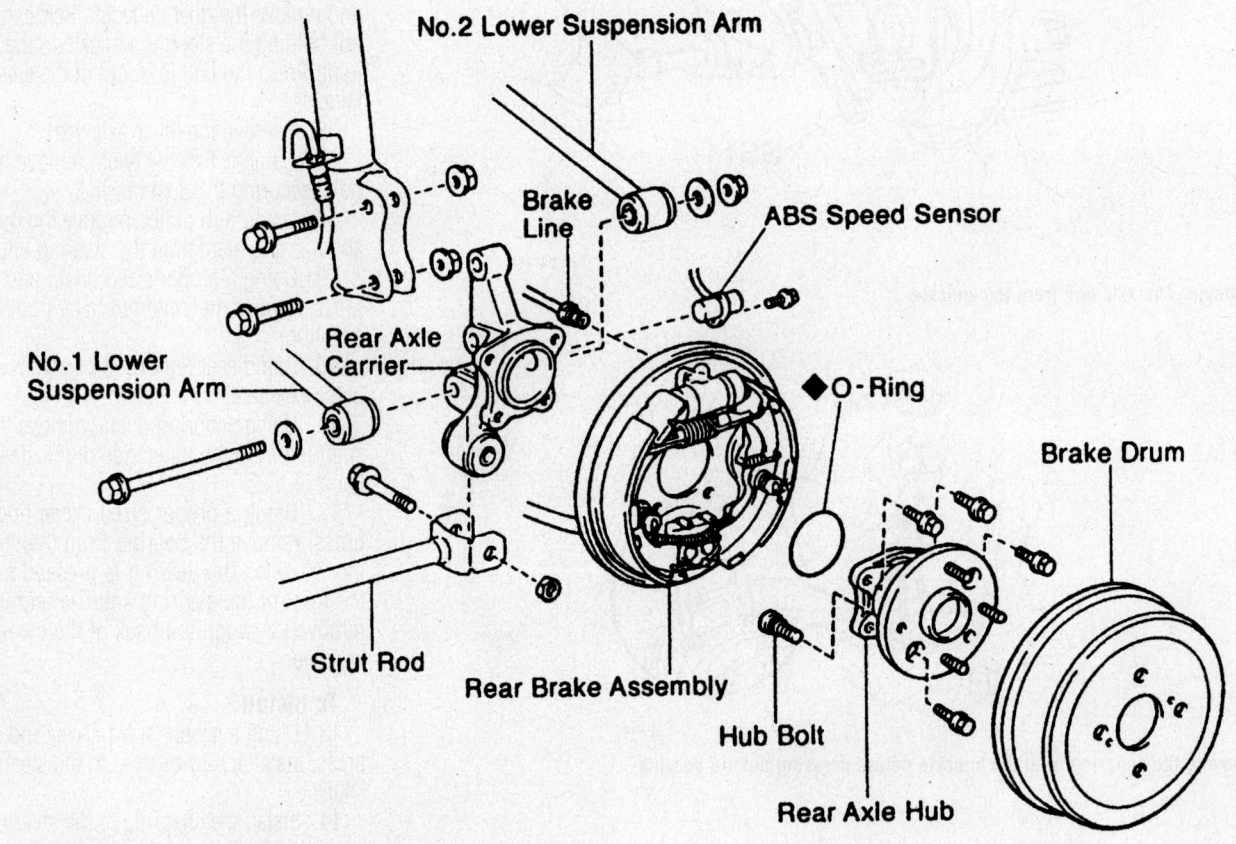

No.2 Lower Suspension Arm

Brake Line

ABS Speed Sensor

No.1 Lower Suspension Arm

Rear Axle Carrier

◆O-Ring

Brake Drum

Strut Rod

Rear Brake Assembly

Hub Bolt

Rear Axle Hub

◆ **Non-reusable part**

7923VGB9

Exploded view of the hub and wheel bearing assembly—Corolla shown, Celica similar

BRACES

Brake Caliper

REMOVAL & INSTALLATION

1. Before servicing the vehicle, refer to the Precautions Section.
2. Remove the wheels.
3. Disconnect the brake hose from the caliper.
4. Remove the bolts that attach the caliper to the torque plate. If applicable, hold the flats of the sliding pin with a wrench while loosening the caliper attaching bolts.
5. Lift up and remove the caliper assembly.

To install:

6. Install the caliper and loosely install the bolts.
7. Hold the flats of the sliding pin with a wrench, then tighten the bolts. Tighten to the front bolts to 25 ft. lbs. (34 Nm). Tighten the rear bolts to 34 ft. lbs. (47 Nm).
8. Connect the brake hose to the caliper, using 2 new washers.
9. Fill the brake system to the proper level and bleed the brake system.
10. Add brake fluid to the reservoir to fill to the correct level.
11. Lower the vehicle to the ground.

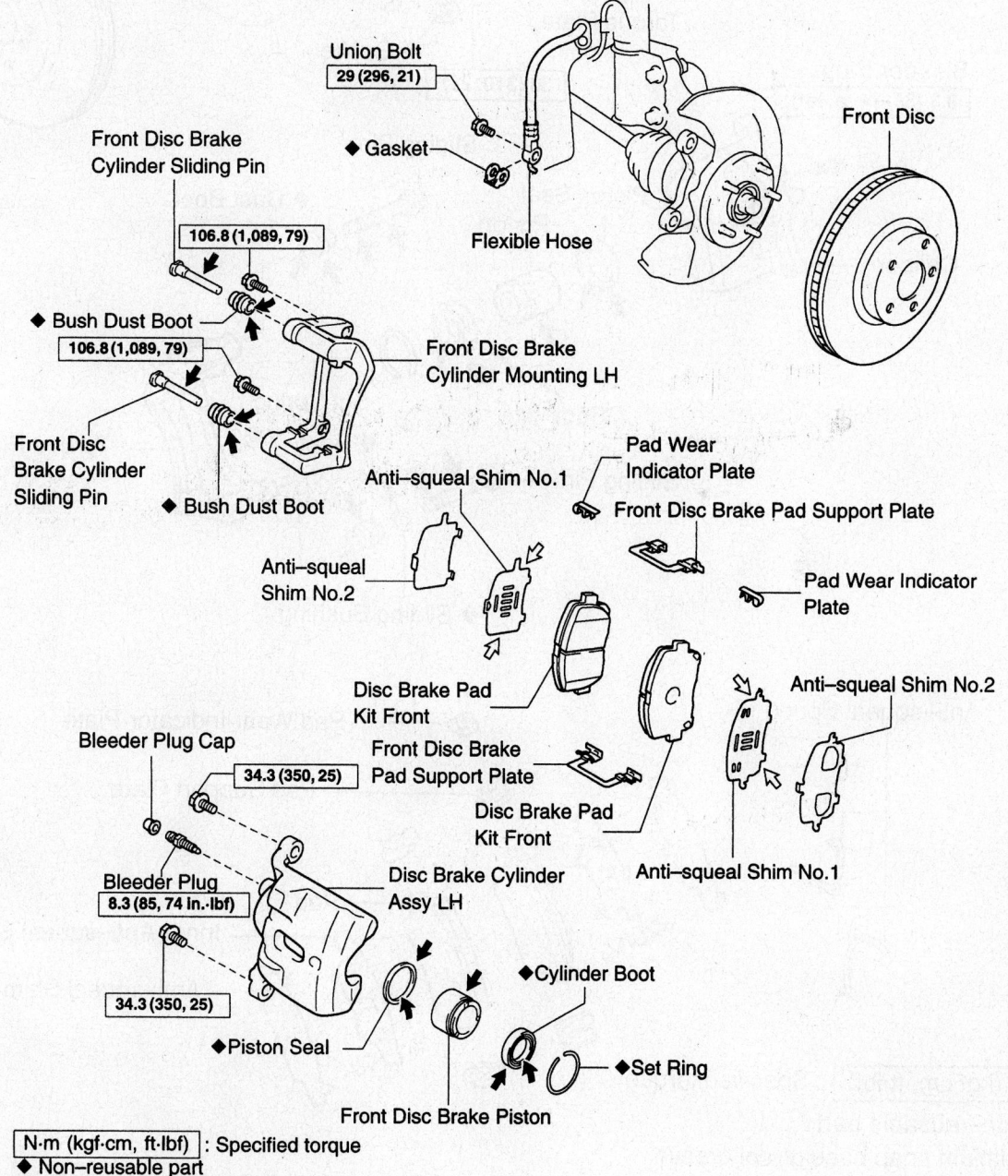

N·m (kgf·cm, ft·lbf) : Specified torque
◆ Non-reusable part
⬅ Lithium soap base glycol grease
⬅ Disc brake grease

67170-TOYO-G46

Front caliper—Corolla

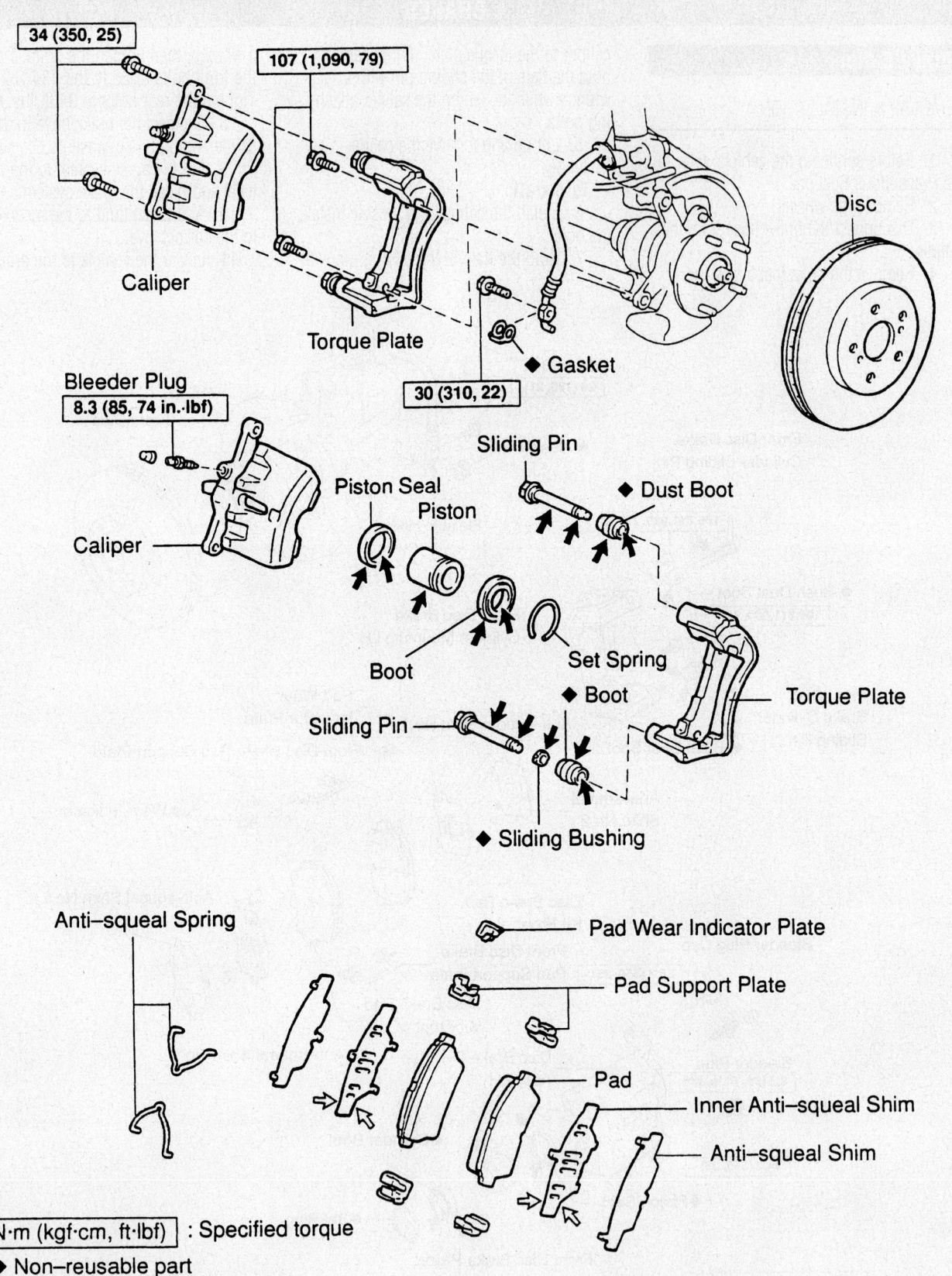

34 (350, 25)

107 (1,090, 79)

Caliper

Torque Plate

◆ Gasket

30 (310, 22)

Disc

Bleeder Plug

8.3 (85, 74 in.·lbf)

Caliper

Piston Seal

Piston

Sliding Pin

◆ Dust Boot

Boot

Set Spring

◆ Boot

Torque Plate

Sliding Pin

◆ Sliding Bushing

Anti–squeal Spring

Pad Wear Indicator Plate

Pad Support Plate

Pad

Inner Anti–squeal Shim

Anti–squeal Shim

N·m (kgf·cm, ft·lbf) : Specified torque

◆ Non–reusable part

◀ Lithium soap base glycol grease

◁ Disc brake grease

93016G67

Front caliper—Celica

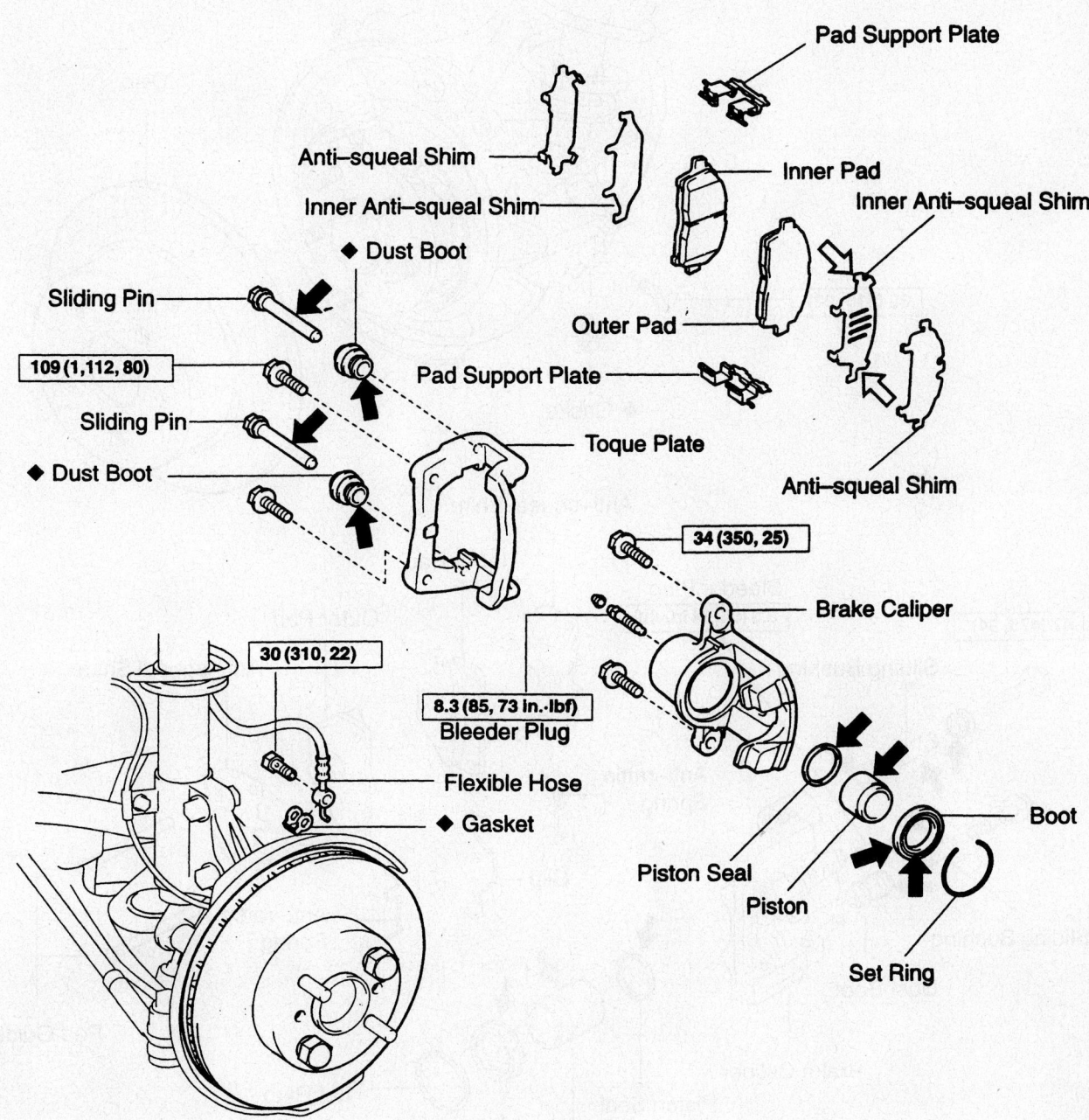

Pad Support Plate

Anti–squeal Shim

Inner Pad

Inner Anti–squeal Shim

Inner Anti–squeal Shim

Dust Boot

Sliding Pin

Outer Pad

109 (1,112, 80)

Sliding Pin

Pad Support Plate

◆ Dust Boot

Toque Plate

Anti–squeal Shim

34 (350, 25)

Brake Caliper

30 (310, 22)

8.3 (85, 73 In.·lbf)
Bleeder Plug

Flexible Hose

◆ Gasket

Piston Seal

Boot

Piston

Set Ring

N·m (kgf·cm, ft·lbf) : Specified torque
◆ Non–reusable part
➡ Lithium soap base glycol grease
⇨ Disc brake grease

9357WG13

Front caliper—MR2

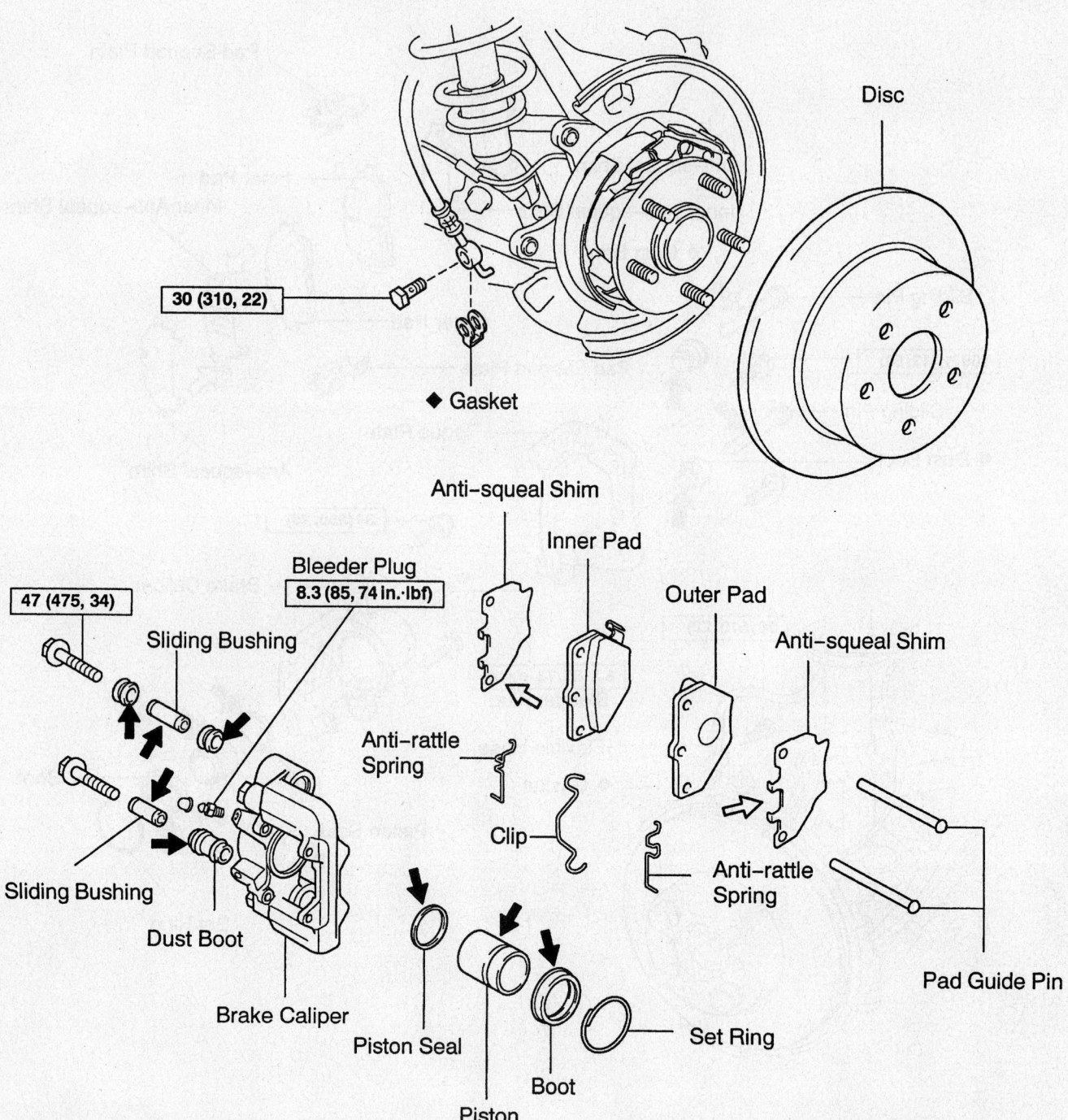

Disc

30 (310, 22)

◆ Gasket

Anti-squeal Shim

Inner Pad

Outer Pad

Anti-squeal Shim

47 (475, 34)

Bleeder Plug
8.3 (85, 74 in.·lbf)

Sliding Bushing

Anti-rattle Spring

Clip

Anti-rattle Spring

Sliding Bushing

Dust Boot

Pad Guide Pin

Brake Caliper

Piston Seal

Piston

Boot

Set Ring

N·m (kgf·cm, ft·lbf) : Specified torque

◆ Non-reusable part

◀ Lithium soap base glycol grease

◁ Disc brake grease

67170-TOYO-G47

Rear caliper—Celica

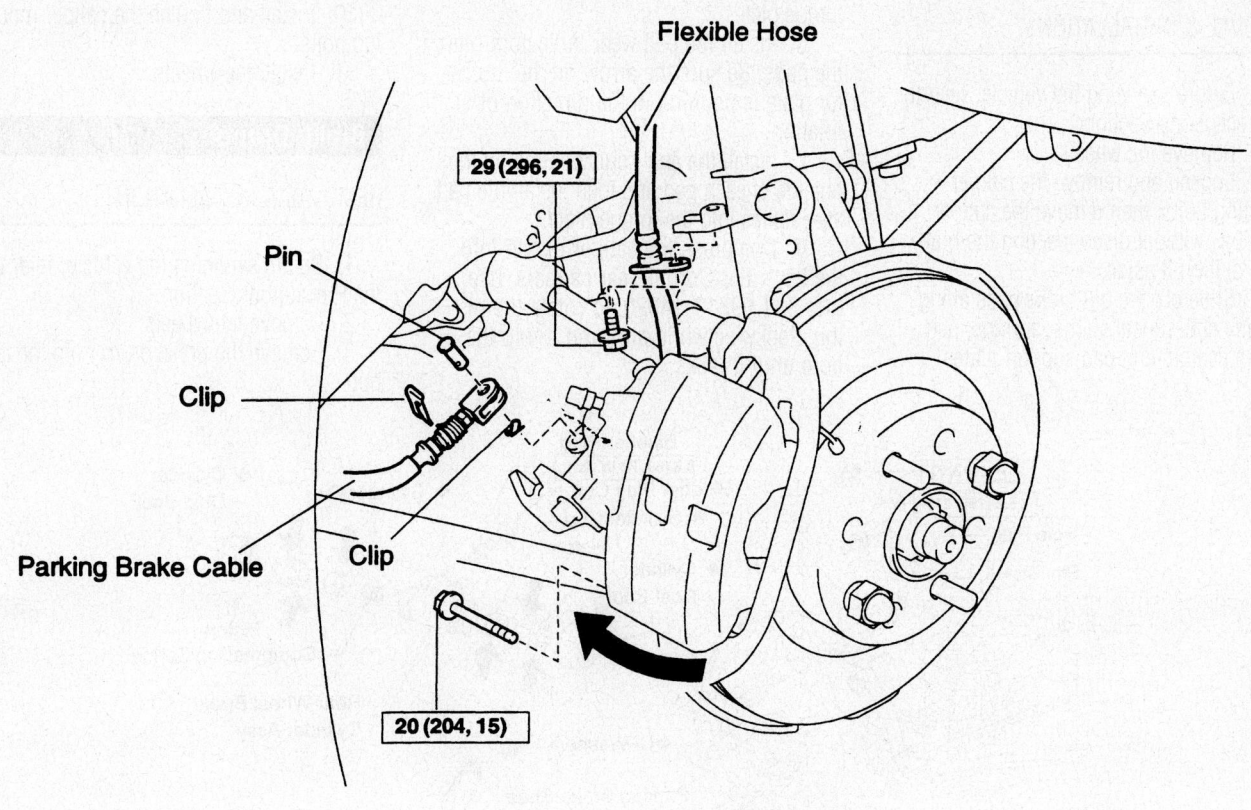

Flexible Hose

29 (296, 21)

Pin

Clip

Parking Brake Cable

Clip

20 (204, 15)

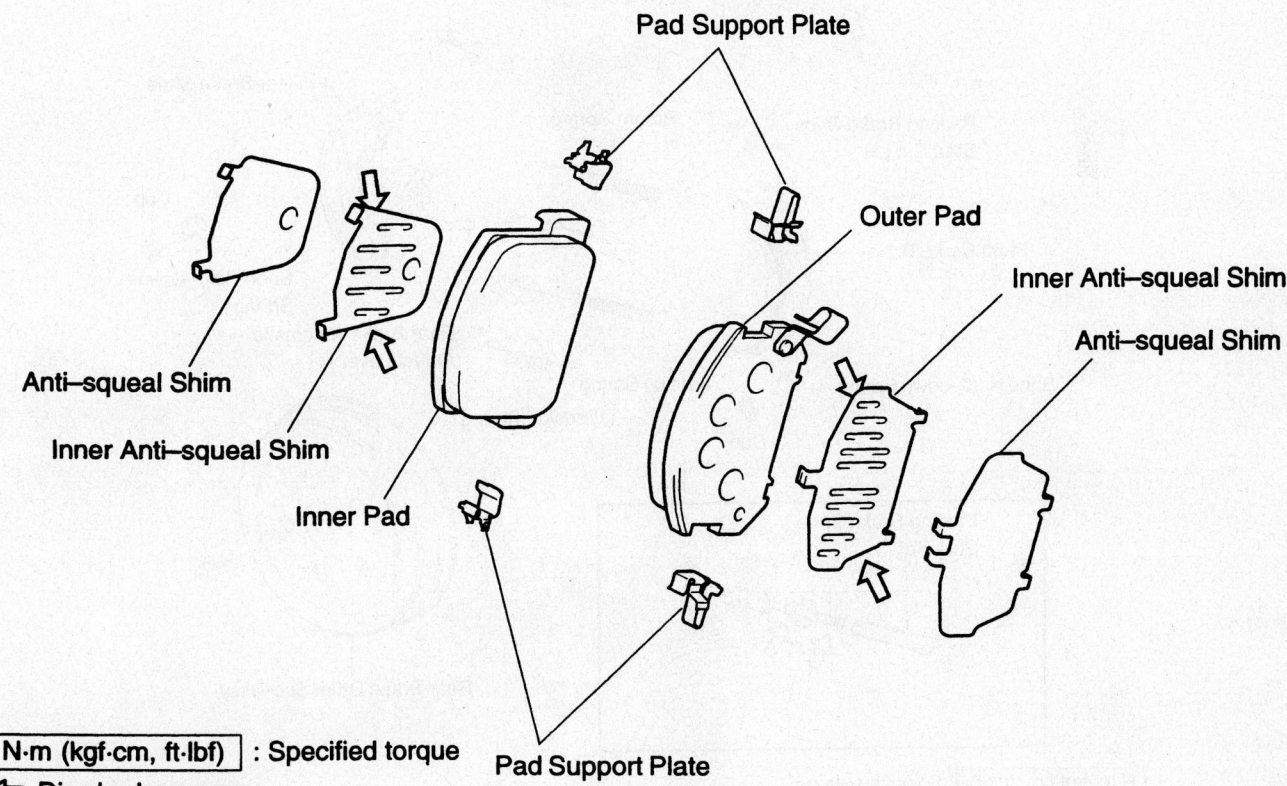

Pad Support Plate

Outer Pad

Inner Anti–squeal Shim

Anti–squeal Shim

Anti–squeal Shim

Inner Anti–squeal Shim

Inner Pad

Pad Support Plate

N·m (kgf·cm, ft·lbf) : Specified torque

⇐ Disc brake grease

9357WG14

Rear caliper—MR2

Disc Brake Pads

REMOVAL & INSTALLATION

1. Before servicing the vehicle, refer to the Precautions Section.

2. Remove the wheels.

3. Loosen and remove the caliper mounting bolts, then remove the caliper assembly, without disconnecting the brake line. Position it aside.

4. Slide out the old brake pads along with any anti-squeal shims, springs, pad wear indicators and pad support plates.

To install:

5. Install the pad support plates into the torque plate.

6. Install the pad wear indicators onto the pads. Be sure the arrow on the indicator plate is pointing in the direction of rotation.

7. Install the anti-squeal shims on the outside of each pad and then install the pad assemblies into the torque plate.

8. Compress the caliper piston into the bore. For Corolla rear calipers, use tool SST 09719-14020, to rotate the piston clockwise while pressing it into the bore until it locks.

9. Position the caliper back down over the pads.

10. Install and tighten the caliper mounting bolts.

11. Install the wheels.

Brake Drums

REMOVAL & INSTALLATION

1. Before servicing the vehicle, refer to the Precautions Section.

2. Remove the wheels.

3. Remove the brake drum from the axle hub.

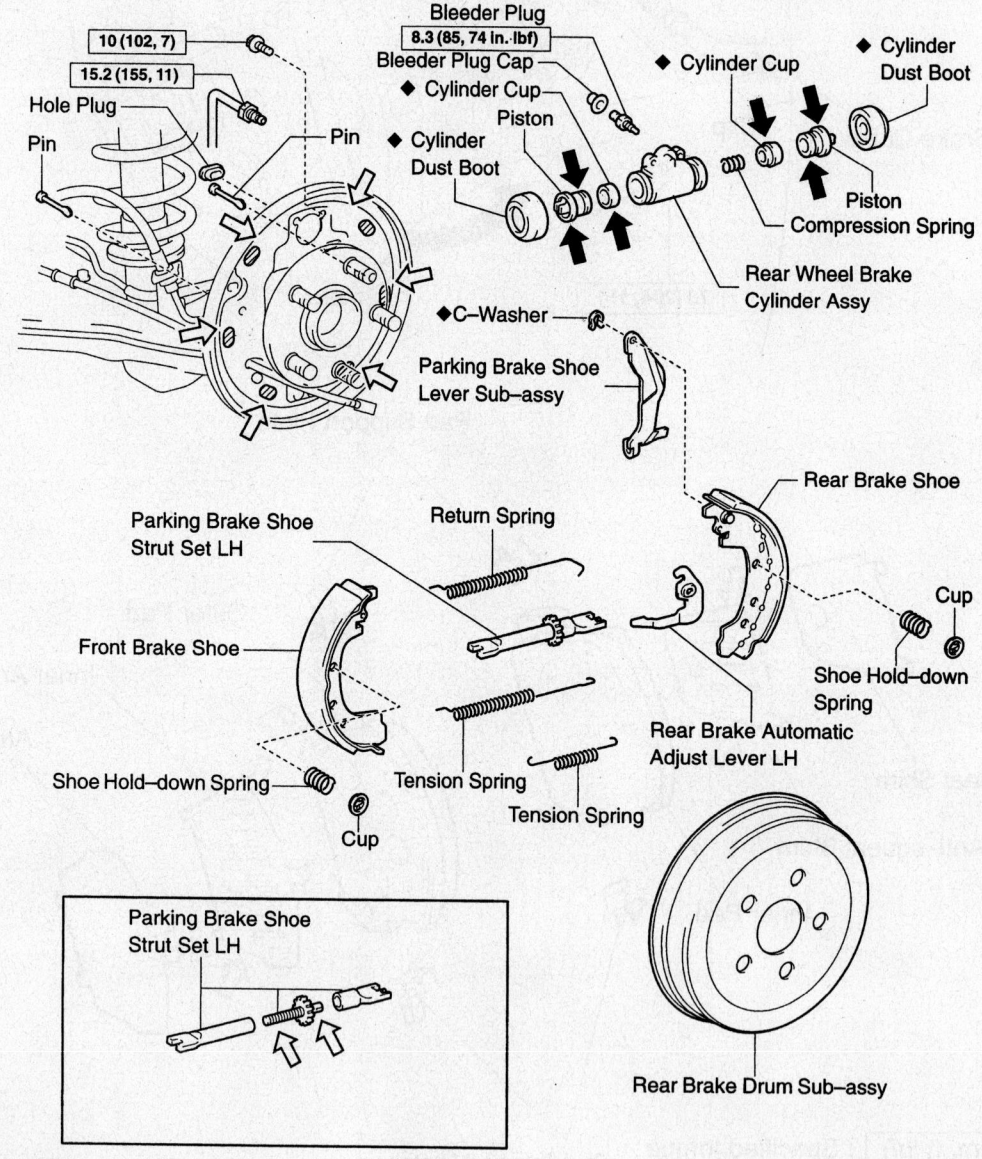

N·m (kgf·cm, ft·lbf) : Specified torque
◆ Non-reusable part
⬅ Lithium soap base glycol grease
⬅ High temperature grease

Rear drum brakes—Corolla

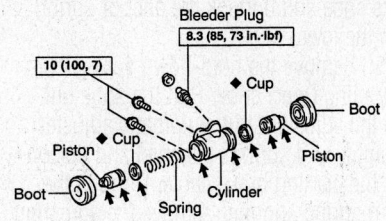

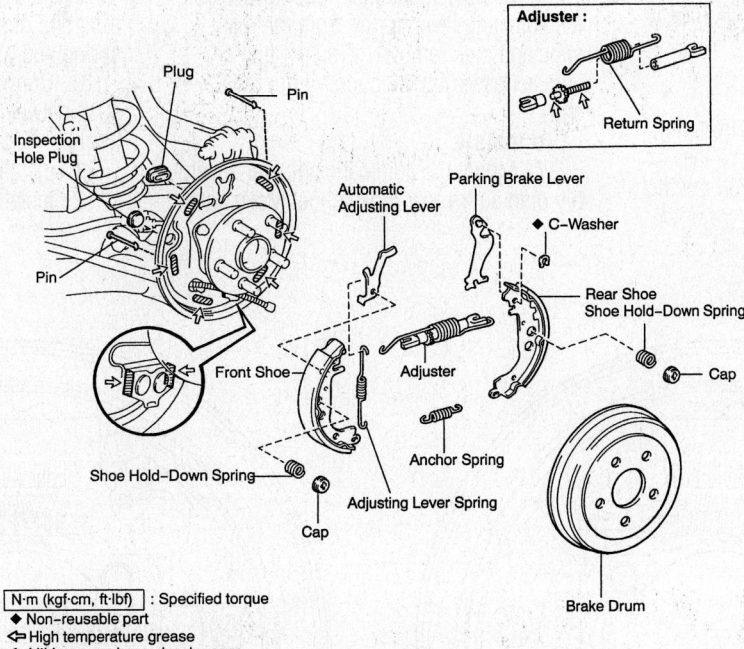

N·m (kgf·cm, ft·lbf) : Specified torque
◆ Non-reusable part
⇐ High temperature grease
⇐ Lithium soap base glycol grease

67170-TOYO-G52

Rear drum brakes—Celica

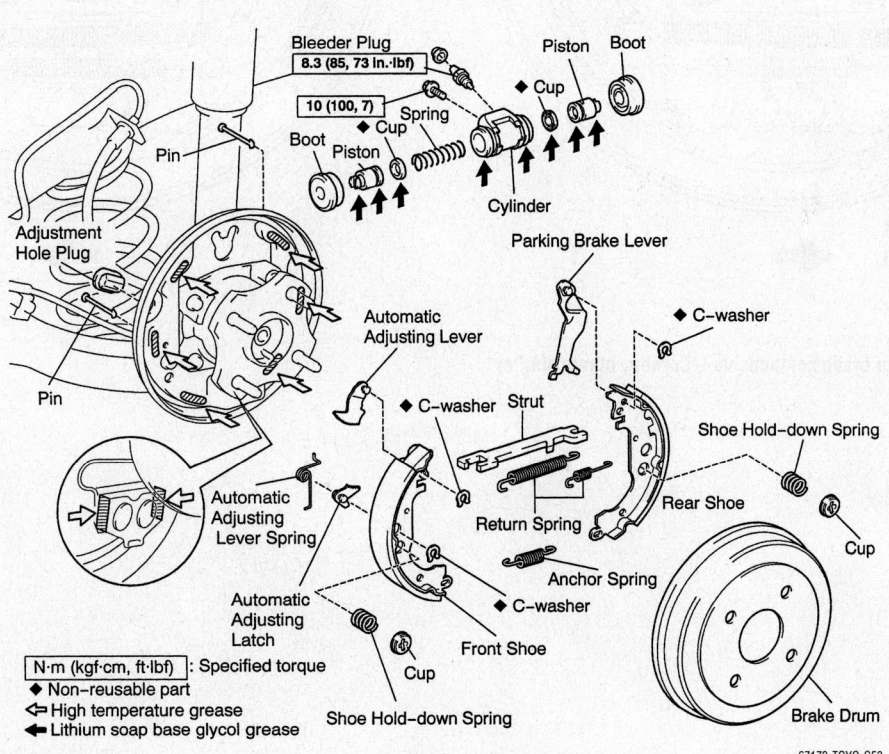

N·m (kgf·cm, ft·lbf) : Specified torque
◆ Non-reusable part
⇐ High temperature grease
⇐ Lithium soap base glycol grease

67170-TOYO-G53

Rear drum brakes—Echo

To install:

4. Install the brake drum.

5. Install the rear wheels, tighten the wheel lug nuts.

Brake Shoes

REMOVAL & INSTALLATION

1. Before servicing the vehicle, refer to the Precautions Section.

2. Remove the wheels.

3. Remove the brake drum.

4. Unhook the return spring from the leading (front) brake shoe. Remove the hold-down spring and the pin. Pull out the brake shoe and unhook the anchor spring from the lower edge.

5. Remove the hold-down spring from the trailing (rear) shoe. Pull the shoe out with the adjuster strut, automatic adjuster assembly and springs attached and disconnect the parking brake cable. Unhook the return spring and then remove the adjusting strut. Remove the anchor spring.

6. Remove the adjusting strut. Unhook the adjusting lever spring from the rear shoe and then remove the automatic adjuster assembly by popping out the C-clip.

To install:

7. Mount the automatic adjuster assembly onto a new rear brake shoe. Make sure the C-clip fits properly. Connect the adjusting strut/return spring and then install the adjusting spring.

8. Connect the parking brake cable to the rear shoe and then position the shoe so the lower end rides in the anchor plate and the upper end is against the boot in the wheel cylinder. Install the pin and the hold-down spring.

9. Install the anchor spring between the front and rear shoes. Install the hold-down spring and pin.

10. Connect the return spring/adjusting strut between the 2 shoes so it rides freely.

11. Install the drum.

12. Install the wheel.

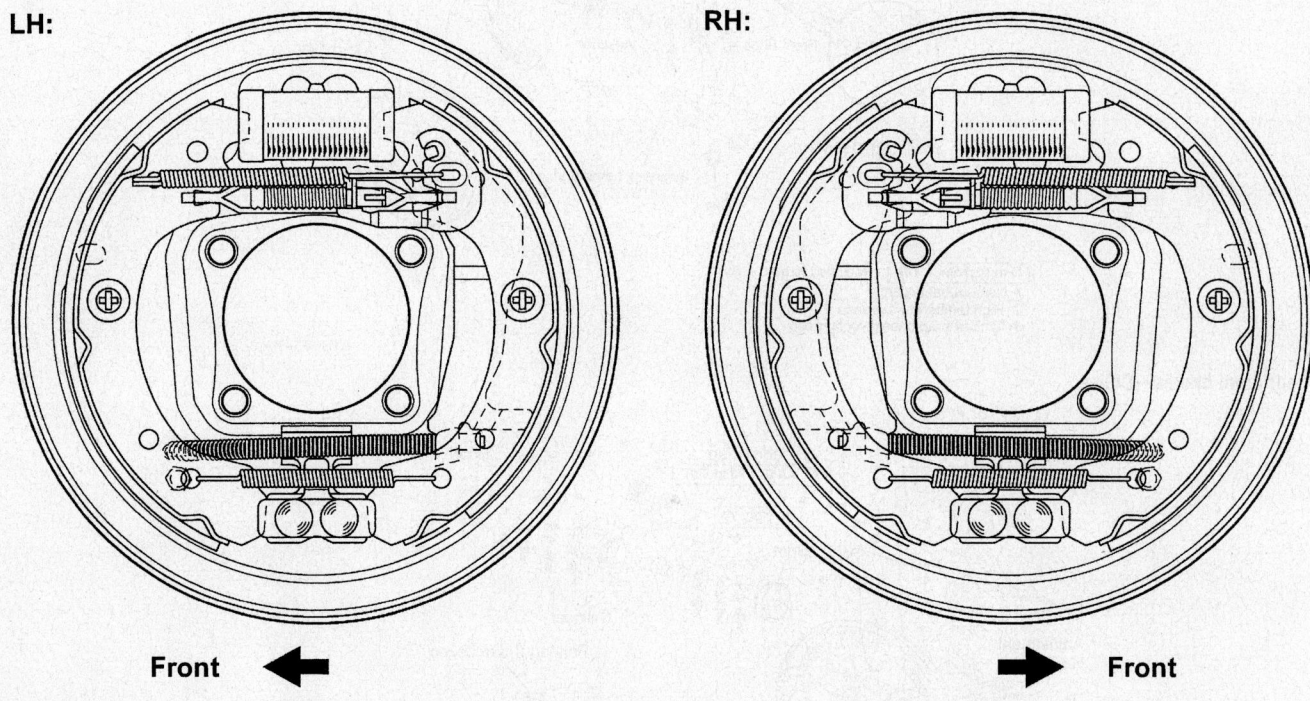

LH:

Front ←

RH:

→ **Front**

Orientation of rear drum brake components—Corolla, others similar

09490_CORO_G0033

BRAKES **14-94**
DRIVE TRAIN **14-81**
ENGINE REPAIR...............**14-16**
FUEL SYSTEM **14-74**
SPECIFICATIONS AND
  MAINTENANCE CHARTS......**14-2**
Engine and Vehicle Identification ...14-2
General Engine Specifications 14-2
Engine Tune-Up Specifications 14-3
Firing Order 14-4
Accessory Drive Belt Routing 14-4
Capacities 14-5
Valve Specifications......................14-6
Crankshaft and Connecting Rod
  Specifications 14-7
Crankshaft and Bearing
  Specifications Chart 14-8
Piston and Ring Specifications 14-9
Torque Specifications 14-10
Wheel Alignment 14-11
Tire, Wheel and Ball Joint
  Specifications 14-12
Brake Specifications 14-13
Scheduled Maintenance
  Intervals...................................14-14
STEERING AND
  SUSPENSION **14-88**
A
Air Bag......................................14-88
  Disarming 14-88
  Rearming 14-88
Alternator 14-16
  Removal & Installation............14-16
Automatic Transaxle....................14-82
  Removal & Installation............14-82
B
Brake Caliper 14-94
  Removal & Installation............14-94
Brake Drums..............................14-98
  Removal & Installation............14-98

Brake Shoes...............................14-98
  Removal & Installation............14-98
C
Camshaft(s) 14-44
  Inspection 14-50
  Removal & Installation............14-44
Clutch 14-83
  Removal & Installation............14-83
CV-Joints...................................14-85
  Overhaul 14-85
Cylinder Head.............................14-39
  Removal & Installation............14-39
D
Disc Brake Pads.........................14-98
  Removal & Installation............14-98
E
Engine Assembly 14-16
  Removal & Installation............14-16
Exhaust Manifold 14-42
  Removal & Installation............14-42
F
Front Crankshaft Seal 14-62
  Removal & Installation............14-62
Fuel Filter 14-74
  Removal & Installation............14-74
Fuel Injectors 14-76
  Removal & Installation............14-76
Fuel Pump 14-74
  Removal & Installation............14-74
Fuel System Pressure 14-74
  Relieving................................14-74
H
Halfshaft...................................14-84
  Removal & Installation............14-84
Heater Core...............................14-24
  Removal & Installation............14-24
Hydraulic Clutch System 14-84
  Bleeding................................14-84
I
Ignition Timing 14-16
  Adjustment.............................14-16

Intake Manifold 14-41
  Removal & Installation............14-41
L
Lower Ball Joint..........................14-90
  Removal & Installation............14-90
M
Manual Transaxle Assembly 14-81
  Removal & Installation............14-81
O
Oil Pan......................................14-54
  Removal & Installation............14-54
Oil Pump 14-59
  Removal & Installation............14-59
P
Piston and Ring Positioning 14-73
R
Rack and Pinion Steering Gear 14-88
  Removal & Installation............14-88
Rear Main Seal 14-61
  Removal & Installation............14-61
S
Strut and Coil Spring...................14-88
  Removal & Installation............14-88
T
Timing Belt 14-71
  Removal & Installation............14-71
Timing Chain, Sprockets, Front
  Cover and Seal 14-62
  Removal & Installation............14-62
V
Valve Lash 14-51
  Adjustment.............................14-51
W
Water Pump 14-22
  Removal & Installation............14-22
Wheel Bearings..........................14-90
  Adjustment.............................14-90
  Removal & Installation............14-91

SPECIFICATIONS AND MAINTENANCE CHARTS

ENGINE AND VEHICLE IDENTIFICATION

Code ①	Liters (cc)	Cu. In.	Cyl.	Fuel Sys.	Engine Type	Eng. Mfg.
2AZ-FE	2.4 (2398)	146	4	EFI	DOHC	Toyota
1MZ-FE	3.0 (2995)	183	6	EFI	DOHC	Toyota
3MZ-FE	3.3 (3311)	202	6	SFI	DOHC	Toyota
2GR-FE	3.5 (3456)	211	6	EFI	DOHC	Toyota

Code ②	Year
2	2002
3	2003
4	2004
5	2005
6	2006

EFI: Electronic Fuel Injection

SFI: Sequential Multi-port Fuel Injection

DOHC: Double Overhead Camshaft

① Stamped on the left side of the engine block

② 10th digit of VIN

09490_AVAL_C0001

GENERAL ENGINE SPECIFICATIONS

Year	Model	Engine Displacement Liters (cc)	Engine Series (ID/VIN)	Fuel System	Net Horsepower @ rpm	Net Torque @ rpm (ft. lbs.)	Bore x Stroke (in.)	Compression Ratio	Oil Pressure psi @ idle
2002	Avalon	3.0 (2995)	1MZ-FE	EFI	200@5200	214@5200	3.44x3.27	10.5:1	4.3
	Camry	2.4 (2398)	2AZ-FE	EFI	157@5600	162@4000	3.48x3.84	9.5:1	4.3
	Camry	3.0 (2995)	1MZ-FE	EFI	194@5200	209@4400	3.44x3.27	10.5:1	4.3
	Camry Solara	2.4 (2398)	2AZ-FE	EFI	157@5600	162@4000	3.48x3.84	9.5:1	4.3
	Camry Solara	3.0 (2995)	1MZ-FE	EFI	200@5200	214@4400	3.44x3.27	10.5:1	4.3
2003	Avalon	3.0 (2995)	1MZ-FE	EFI	200@5200	214@5200	3.44x3.27	10.5:1	4.3
	Camry	2.4 (2398)	2AZ-FE	EFI	157@5600	162@4000	3.48x3.84	9.5:1	4.3
	Camry	3.0 (2995)	1MZ-FE	EFI	194@5200	209@4400	3.44x3.27	10.5:1	4.3
	Camry Solara	2.4 (2398)	2AZ-FE	EFI	157@5600	162@4000	3.48x3.84	9.5:1	4.3
	Camry Solara	3.0 (2995)	1MZ-FE	EFI	200@5200	214@4400	3.44x3.27	10.5:1	4.3
2004	Avalon	3.0 (2995)	1MZ-FE	EFI	200@5200	214@5200	3.44x3.27	10.5:1	4.3
	Camry	2.4 (2398)	2AZ-FE	EFI	157@5600	162@4000	3.48x3.84	9.5:1	4.3
	Camry	3.0 (2995)	1MZ-FE	EFI	194@5200	209@4400	3.44x3.27	10.5:1	4.3
	Camry	3.3 (3311)	3MZ-FE	EFI	225@5600	240@3600	3.62x3.27	10.8:1	36-78@3000
	Camry Solara	2.4 (2398)	2AZ-FE	EFI	157@5600	162@4000	3.48x3.84	9.5:1	4.3
	Camry Solara	3.3 (3311)	3MZ-FE	EFI	225@5600	240@3600	3.62x3.27	10.8:1	36-78@3000
2005	Avalon	3.5 (3456)	2GR-FE	EFI	280@6200	260@4700	3.70x3.27	10.8:1	4.3
	Camry	2.4 (2362)	2AZ-FE	EFI	157@5600	162@4000	3.48x3.78	9.6:1	4.3
	Camry	3.0 (2995)	1MZ-FE	EFI	210@5800	220@4400	3.44x3.27	10.5:1	4.3
	Camry	3.3 (3311)	3MZ-FE	EFI	225@5600	240@3600	3.62x3.27	10.8:1	36-78@3000
	Camry Solara	2.4 (2398)	2AZ-FE	EFI	157@5600	162@4000	3.48x3.84	9.5:1	4.3
	Camry Solara	3.3 (3311)	3MZ-FE	EFI	225@5600	240@3600	3.62x3.27	10.8:1	36-78@3000
2006	Avalon	3.5 (3456)	2GR-FE	EFI	280@6200	260@4700	3.70x3.27	10.5:1	4.3
	Camry	2.4 (2362)	2AZ-FE	EFI	154@5700	160@4000	3.48x3.78	9.6:1	4.3
	Camry	3.0 (2995)	1MZ-FE	EFI	190@5800	197@4400	3.44x3.27	10.5:1	4.3
	Camry	3.3 (3311)	3MZ-FE	EFI	210@5600	220@3600	3.62x3.27	10.8:1	36-78@3000
	Camry Solara	2.4 (2398)	2AZ-FE	EFI	157@5600	162@4000	3.48x3.78	9.5:1	4.3
	Camry Solara	3.3 (3311)	3MZ-FE	EFI	225@5600	240@3600	3.62x3.27	10.8:1	36-78@3000

EFI: Electronic Fuel Injection

09490_AVAL_C0002

ENGINE TUNE-UP SPECIFICATIONS

Year	Engine Displacement Liters (cc)	Engine ID/VIN	Spark Plug Gap (in.)	Ignition Timing (deg.) ①	Fuel Pump (psi)	Idle Speed (rpm)		Valve Clearance	
						MT	AT	In.	Ex.
2002	2.4 (2398)	2AZ-FE	0.043	8-12 BTDC	44-50	700-800	700-800	0.007-0.011	0.011-0.015
	3.0 (2952)	1MZ-FE	0.043	8-12 BTDC	44-50	-	650-750	0.006-0.010	0.010-0.014
2003	2.4 (2398)	2AZ-FE	0.043	8-12 BTDC	44-50	700-800	700-800	0.007-0.011	0.011-0.015
	3.0 (2952)	1MZ-FE	0.043	8-12 BTDC	44-50	-	650-750	0.006-0.010	0.010-0.014
2004	2.4 (2398)	2AZ-FE	0.043	8-12 BTDC	44-50	700-800	700-800	0.007-0.011	0.011-0.015
	3.0 (2952)	1MZ-FE	0.043	8-12 BTDC	44-50	-	650-750	0.006-0.010	0.010-0.014
	3.3 (3311)	3MZ-FE	0.043	8-12 BTDC	44-50	-	650-750	0.006-0.010	0.010-0.014
2005	2.4 (2398)	2AZ-FE	0.043	8-12 BTDC	44-50	700-800	700-800	0.007-0.011	0.011-0.015
	3.0 (2952)	1MZ-FE	0.043	8-12 BTDC	44-50	650-750	650-750	0.006-0.010	0.010-0.014
	3.3 (3311)	3MZ-FE	0.043	8-12 BTDC	44-50	-	650-750	0.006-0.010	0.010-0.014
	3.5 (3456)	2GR-FE	0.043	8-12 BTDC	44-50	-	600-700	HYD	HYD
2006	2.4 (2398)	2AZ-FE	0.043	8-12 BTDC	44-50	700-800	700-800	0.007-0.011	0.011-0.015
	3.0 (2952)	1MZ-FE	0.043	8-12 BTDC	44-50	650-750	650-750	0.006-0.010	0.010-0.014
	3.3 (3311)	3MZ-FE	0.043	8-12 BTDC	44-50	-	650-750	0.006-0.010	0.010-0.014
	3.5 (3456)	2GR-FE	0.043	8-12 BTDC	44-50	-	600-700	HYD	HYD

Note: The Vehicle Emission Control Information label often reflects specification changes made during production. The label figures must be used if they differ from those in this chart.

① With terminal TE1 and E1 connected of DLC1
HYD: Hydraulic Valve Lifters

09490_AVAL_C0003

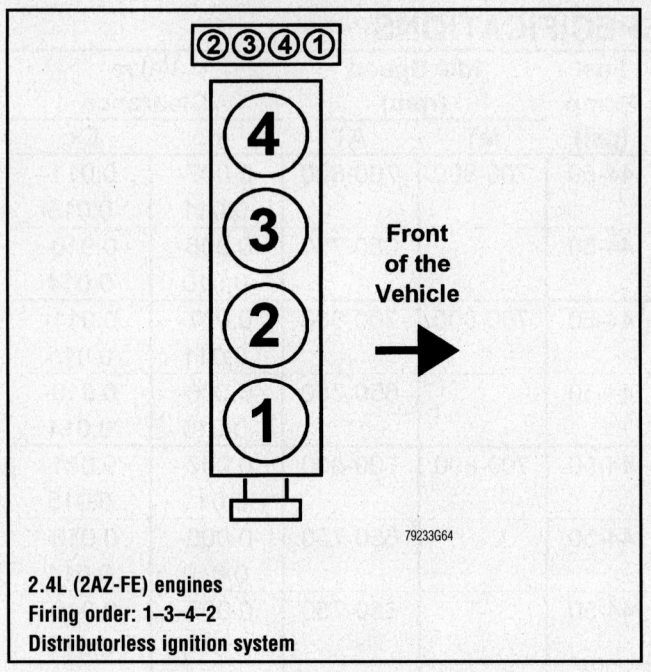

2.4L (2AZ-FE) engines
Firing order: 1–3–4–2
Distributorless ignition system

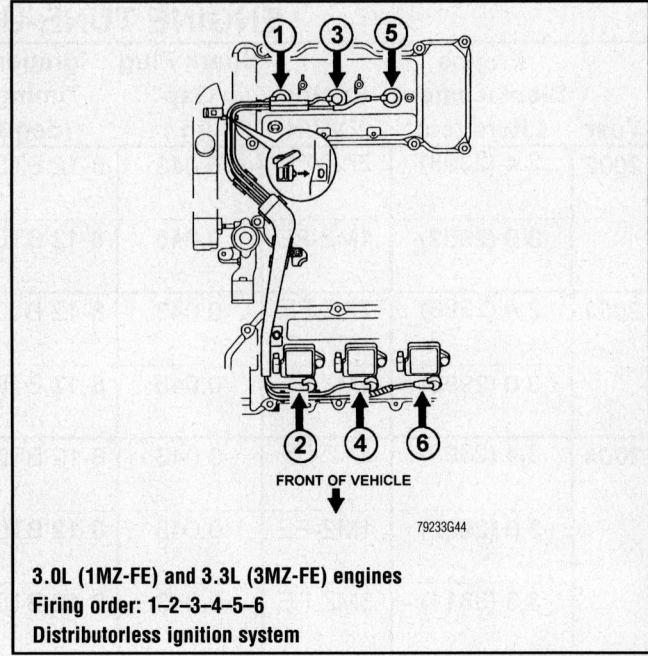

3.0L (1MZ-FE) and 3.3L (3MZ-FE) engines
Firing order: 1–2–3–4–5–6
Distributorless ignition system

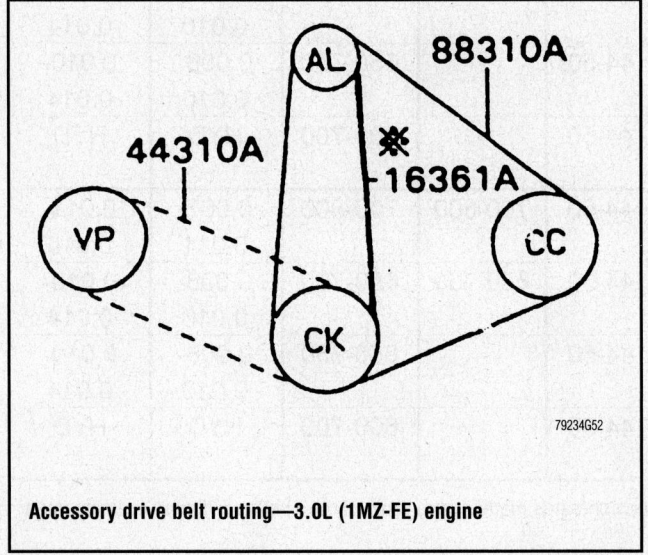

Accessory drive belt routing—3.0L (1MZ-FE) engine

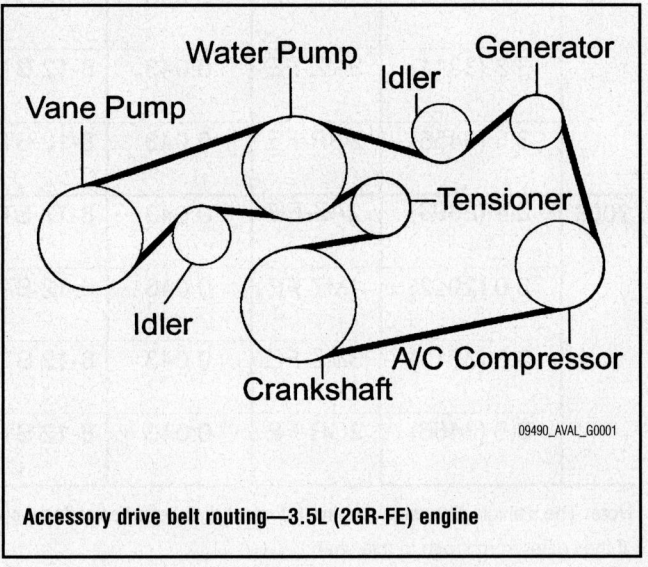

Accessory drive belt routing—3.5L (2GR-FE) engine

CAPACITIES

Year	Model	Engine Displacement Liters (cc)	Engine ID/VIN	Engine Oil with Filter	Transmission (pts.)			Fuel Tank (gal.)	Cooling System (qts.)
					4-Spd	5-Spd	Auto.		
2002	Avalon	3.0 (2995)	1MZ-FE	5.0	—	—	7.4	18.5	9.8
	Camry	2.4 (2398)	2AZ-FE	4.0	—	5.2	8.2	18.5	6.6
	Camry	3.0 (2995)	1MZ-FE	5.0	—	—	7.4	18.5	9.8
	Camry Solara	2.4 (2398)	2AZ-FE	4.0	—	5.2	8.2	18.5	6.6
	Camry Solara	3.0 (2995)	1MZ-FE	5.0	—	—	7.4	18.5	9.6
2003	Avalon	3.0 (2995)	1MZ-FE	5.0	—	—	7.4	18.5	9.8
	Camry	2.2 (2164)	2AZ-FE	4.0	—	5.2	7.4	18.5	6.6
	Camry	3.0 (2995)	1MZ-FE	5.0	—	—	7.4	18.5	9.8
	Camry Solara	2.4 (2398)	2AZ-FE	4.0	—	5.2	7.4	18.5	6.6
	Camry Solara	3.0 (2995)	1MZ-FE	5.0	—	—	7.4	18.5	9.8
2004	Avalon	3.0 (2995)	1MZ-FE	5.0	—	—	7.4	18.5	9.8
	Camry	2.4 (2398)	2AZ-FE	4.0	—	5.2	7.4	18.5	6.6
	Camry	3.0 (2995)	1MZ-FE	5.0	—	—	7.4	18.5	9.8
	Camry	3.3 (3311)	3MZ-FE	5.0	—	—	7.4	18.5	9.8
	Camry Solara	2.4 (2398)	2AZ-FE	4.0	—	5.2	7.4	18.5	6.6
	Camry Solara	3.3 (3311)	3MZ-FE	5.0	—	—	7.4	18.5	9.8
2005	Avalon	3.5 (3456)	2GR-FE	6.4	—	—	7.4	18.5	8.8
	Camry	2.4 (2398)	2AZ-FE	4.0	—	5.2	7.4	18.5	6.6
	Camry	3.0 (2995)	1MZ-FE	5.0	—	—	7.4	18.5	9.8
	Camry	3.3 (3311)	3MZ-FE	5.0	—	—	7.4	18.5	9.8
	Camry Solara	2.4 (2398)	2AZ-FE	4.0	—	5.2	7.4	18.5	6.6
	Camry Solara	3.3 (3311)	3MZ-FE	5.0	—	—	7.4	18.5	9.8
2006	Avalon	3.5 (3456)	2GR-FE	6.4	—	—	7.4	18.5	8.8
	Camry	2.4 (2398)	2AZ-FE	4.0	—	5.2	7.4	18.5	6.6
	Camry	3.0 (2995)	1MZ-FE	5.0	—	—	7.4	18.5	9.6
	Camry	3.3 (3311)	3MZ-FE	5.0	—	—	7.4	18.5	9.8
	Camry Solara	2.4 (2398)	2AZ-FE	4.0	—	5.2	7.4	18.5	6.6
	Camry Solara	3.3 (3311)	3MZ-FE	5.0	—	—	7.4	18.5	9.8

Note: All capacities are approximate. Add fluid gradually and check to be sure a proper fluid level is obtained.

09490_AVAL_C0004

VALVE SPECIFICATIONS

Year	Engine Displacement Liters (cc)	Engine ID/VIN	Seat Angle (deg.)	Face Angle (deg.)	Spring Test Pressure (lbs. @ in.)	Spring Installed Height (in.)	Stem-to-Guide Clearance (in.)		Stem Diameter (in.)	
							Intake	Exhaust	Intake	Exhaust
2002	2.4 (2398)	2AZ-FE	45	45	①	NA	0.0010-0.0024	0.0012-0.0026	0.2154-0.2159	0.2152-0.2157
	3.0 (2995)	1MZ-FE	45	44.5	41.9-46.3@ 1.331	1.331	0.0010-0.0024	0.0012-0.0026	0.2154-0.2159	0.2152-0.2157
2003	2.4 (2398)	2AZ-FE	45	45	①	NA	0.0010-0.0024	0.0012-0.0026	0.2154-0.2159	0.2152-0.2157
	3.0 (2995)	1MZ-FE	45	44.5	41.9-46.3@ 1.331	1.331	0.0010-0.0024	0.0012-0.0026	0.2154-0.2159	0.2152-0.2157
2004	2.4 (2398)	2AZ-FE	45	45	①	NA	0.0010-0.0024	0.0012-0.0026	0.2154-0.2159	0.2152-0.2157
	3.0 (2995)	1MZ-FE	45	44.5	41.9-46.3@ 1.331	1.331	0.0010-0.0024	0.0012-0.0026	0.2154-0.2159	0.2152-0.2157
	3.3 (3311)	3MZ-FE	NA	44.5	41.9-46.3@ 1.331	NA	0.0010-0.0024	0.0012-0.0026	0.2154-0.2159	0.2152-0.2157
2005	2.4 (2398)	2AZ-FE	45	45	①	NA	0.0010-0.0024	0.0012-0.0026	0.2154-0.2159	0.2152-0.2157
	3.0 (2995)	1MZ-FE	45	44.5	41.9-46.3@ 1.331	1.331	0.0010-0.0024	0.0012-0.0026	0.2154-0.2159	0.2152-0.2157
	3.3 (3311)	3MZ-FE	NA	44.5	41.9-46.3@ 1.331	NA	0.0010-0.0024	0.0012-0.0026	0.2154-0.2159	0.2152-0.2157
	3.5 (3456)	2GR-FE	NA	44.5	②	NA	0.0010-0.0024	0.0012-0.0026	0.2154-0.2159	0.2152-0.2158
2006	2.4 (2398)	2AZ-FE	45	45	①	NA	0.0010-0.0024	0.0012-0.0026	0.2154-0.2159	0.2152-0.2157
	3.0 (2995)	1MZ-FE	45	44.5	41.9-46.3@ 1.331	1.331	0.0010-0.0024	0.0012-0.0026	0.2154-0.2159	0.2152-0.2157
	3.3 (3311)	3MZ-FE	NA	44.5	41.9-46.3@ 1.331	NA	0.0010-0.0024	0.0012-0.0026	0.2154-0.2159	0.2152-0.2157
	3.5 (3456)	2GR-FE	NA	44.5	②	NA	0.0010-0.0024	0.0012-0.0026	0.2154-0.2159	0.2152-0.2158

① Inner spring free length: 1.799 in. (45.7 mm)
② Inner spring free length: 1.790 in. (45.5 mm)

09490_AVAL_C0005

CRANKSHAFT AND CONNECTING ROD SPECIFICATIONS

All measurements are given in inches.

Year	Engine Displacemen Liters (cc)	Engine ID/VIN	Crankshaft				Connecting Rod		
			Main Brg. Journal Dia.	Main Brg. Oil Clearance	Shaft End-play	Thrust on No.	Journal Diameter	Oil Clearance	Side Clearance
2002	2.4 (2398)	2AZ-FE	2.1648-2.1654	0.0007-0.0016	0.0063-0.0143	3	1.8894-1.8898	0.0009-0.0022	0.0063-0.0123
	3.0 (2995)	1MZ-FE	2.4011-2.4016	0.0010-0.0018	0.0016-0.0095	2	2.0863-2.8660	0.0015-0.0025	0.0059-0.0118
2003	2.4 (2398)	2AZ-FE	2.1648-2.1654	0.0007-0.0016	0.0063-0.0143	3	1.8894-1.8898	0.0009-0.0022	0.0063-0.0123
	3.0 (2995)	1MZ-FE	2.4011-2.4016	0.0010-0.0018	0.0016-0.0095	2	2.0863-2.8660	0.0015-0.0025	0.0059-0.0118
2004	2.4 (2398)	2AZ-FE	2.1648-2.1654	0.0007-0.0016	0.0063-0.0143	3	1.8894-1.8898	0.0009-0.0022	0.0063-0.0123
	3.0 (2995)	1MZ-FE	2.4011-2.4016	0.0010-0.0018	0.0016-0.0095	2	2.0863-2.8660	0.0015-0.0025	0.0059-0.0118
	3.3 (3311)	3MZ-FE	2.4403-2.4409	①	0.0016-0.0094	2	2.0863-2.0866	0.0015-0.0026	0.0059-0.0118
2005	2.4 (2398)	2AZ-FE	2.1648-2.1654	0.0003-0.0009	0.0016-0.0118	3	1.8894-1.8898	0.0009-0.0031	0.0063-0.0143
	3.0 (2995)	1MZ-FE	2.4011-2.4016	0.0010-0.0018	0.0016-0.0118	2	2.0863-2.8660	0.0015-0.0031	0.0059-0.0118
	3.3 (3311)	3MZ-FE	2.4403-2.4409	②	0.0016-0.0118	2	2.0863-2.0866	0.0015-0.0031	0.0059-0.0118
	3.5 (3456)	2GR-FE	2.4011-2.4016	0.0010-0.0019	0.0016-0.0094	2	NA	0.0018-0.0026	0.0059-0.0157
2006	2.4 (2398)	2AZ-FE	2.1648-2.1654	0.0003-0.0009	0.0016-0.0118	3	1.8894-1.8898	0.0009-0.0031	0.0063-0.0143
	3.0 (2995)	1MZ-FE	2.4011-2.4016	0.0010-0.0018	0.0016-0.0118	2	2.0863-2.8660	0.0015-0.0031	0.0059-0.0138
	3.3 (3311)	3MZ-FE	2.4403-2.4409	②	0.0016-0.0118	2	2.0863-2.0866	0.0015-0.0031	0.0059-0.0138
	3.5 (3456)	2GR-FE	2.4011-2.4016	0.0010-0.0019	0.0016-0.0094	2	NA	0.0018-0.0026	0.0059-0.0157

① Journal No. 1 and 4: 0.0006 - 0.0013 inch
Journal No. 2 and 3: 0.0010 - 0.0018 inch

② Journal No. 1 and 4: 0.0006 - 0.0020 in.
Journal No. 2 and 3: 0.0010 - 0.0024 in.

09490_AVAL_C0006

CAMSHAFT AND BEARING SPECIFICATIONS CHART

All measurements are given in inches.

Year	Engine Displacement Liters (cc)	Engine ID/VIN	Journal Dia.	Brg. Oil Clearance	Shaft End-play	Runout	Lobe Height Intake	Lobe Height Exhaust
2002	2.4 (2398)	2AZ-FE	①	②	③	0.0012	1.8305-1.8345	1.8104-1.8143
	3.0 (2995)	1MZ-FE	1.0610-1.0616	④	0.0016-0.0035	0.0024	1.6579-1.6618	1.6520-1.6559
2003	2.4 (2398)	2AZ-FE	①	②	③	0.0012	1.8305-1.8345	1.8104-1.8143
	3.0 (2995)	1MZ-FE	1.0610-1.0616	④	0.0016-0.0035	0.0024	1.6579-1.6618	1.6520-1.6559
2004	2.4 (2398)	2AZ-FE	1.4162-1.4167	②	⑤	0.0012	1.8346-1.8429	1.8346-1.8429
	3.0 (2995)	1MZ-FE	1.0614-1.0620	0.0010-0.0039	0.0016-0.0047	0.0024	1.6842-1.6942	1.6776-1.6876
	3.3 (3311)	3MZ-FE	1.0614-1.0620	0.0010-0.0039	0.0016-0.0047	0.0024	1.6921-1.7020	1.6874-1.6972
2005	2.4 (2398)	2AZ-FE	①	②	0.0016-0.0037	0.0012	1.8346-1.8429	1.8346-1.8429
	3.0 (2995)	1MZ-FE	1.0614-1.0620	0.0010-0.0039	0.0016-0.0037	0.0024	1.6842-1.6942	1.6776-1.6876
	3.3 (3311)	3MZ-FE	1.0614-1.0620	0.0010-0.0039	0.0016-0.0055	0.0024	1.6921-1.7020	1.6874-1.6972
	3.5 (3456)	2GR-FE	⑥	⑦	0.0031-0.0051	0.0016	1.7447-1.7487	1.7426-1.7465
2006	2.4 (2398)	2AZ-FE	①	②	0.0016-0.0037	0.0012	1.8346-1.8429	1.8346-1.8429
	3.0 (2995)	1MZ-FE	1.0614-1.0620	0.0010-0.0039	0.0016-0.0037	0.0024	1.6842-1.6942	1.6776-1.6876
	3.3 (3311)	3MZ-FE	1.0614-1.0620	0.0010-0.0039	0.0016-0.0055	0.0024	1.6921-1.7020	1.6874-1.6972
	3.5 (3456)	2GR-FE	⑥	⑦	0.0031-0.0051	0.0016	1.7447-1.7487	1.7426-1.7465

① No. 1 Journal: 1.4165-1.4167 in.
All Others: 0.9039-0.9045 in.

② Intake No. 1 Journal: 0.0003-0.0028 in.
Intake Other Journals: 0.0010-0.0039 in.
Exhaust No. 1 Journal: 0.0016-0.0039 in.
Exhaust Other Journals: 0.0010-0.0039 in.

③ Intake: 0.0016-0.0037 in.
Exhaust: 0.0032-0.0053 in.

④ Intake: 0.0014-0.0028 in.
Exhaust: 0.0010-0.0024 in.

⑤ Intake: 0.0016-0.0043 in.
Exhaust: 0.0032-0.0059 in.

⑥ No. 1 Journal: 1.4152-1.4157 in.
All Others: 1.0220-1.0226 in.

⑦ No. 1 Journal: 0.0016-0.0031 in.
All Others: 0.0010-0.0024 in.

PISTON AND RING SPECIFICATIONS
All measurements are given in inches.

Year	Engine Displacement Liters (cc)	Engine ID/VIN	Piston Clearance	Ring Gap			Ring Side Clearance		
				Top Compression	Bottom Compression	Oil Control	Top Compression	Bottom Compression	Oil Control
2002	2.4 (2398)	2AZ-FE	0.0020-0.0029	0.0087-0.0126	0.0197-0.0236	0.0039-0.0138	0.0012-0.0028	0.0012-0.0028	SNUG
	3.0 (2995)	1MZ-FE	0.0033-0.0042	0.0098-0.0138	0.0138-0.0177	0.0059-0.0157	0.0008-0.0028	0.0008-0.0024	SNUG
2003	2.4 (2398)	2AZ-FE	0.0020-0.0029	0.0087-0.0126	0.0197-0.0236	0.0039-0.0138	0.0012-0.0028	0.0012-0.0028	SNUG
	3.0 (2995)	1MZ-FE	0.0033-0.0042	0.0098-0.0138	0.0138-0.0177	0.0059-0.0157	0.0008-0.0028	0.0008-0.0024	SNUG
2004	2.4 (2398)	2AZ-FE	0.0020-0.0039	0.0087-0.0350	0.0197-0.0531	0.0039-0.0287	0.0012-0.0028	0.0012-0.0028	SNUG
	3.0 (2995)	1MZ-FE	0.0033-0.0051	0.0098-0.0374	0.0138-0.0413	0.0059-0.0394	0.0008-0.0028	0.0008-0.0024	0.0016-0.0041
	3.3 (3311)	3MZ-FE	0.0013-0.0051	0.0118-0.0374	0.0197-0.0413	0.0059-0.0394	0.0012-0.0031	0.0008-0.0024	0.0012-0.0043
2005	2.4 (2398)	2AZ-FE	0.0020-0.0039	0.0087-0.0350	0.0197-0.0531	0.0039-0.0287	0.0012-0.0028	0.0012-0.0028	SNUG
	3.0 (2995)	1MZ-FE	0.0033-0.0051	0.0098-0.0374	0.0138-0.0413	0.0059-0.0394	0.0008-0.0028	0.0008-0.0024	0.0016-0.0047
	3.3 (3311)	3MZ-FE	0.0013-0.0051	0.0118-0.0374	0.0197-0.0413	0.0059-0.0394	0.0012-0.0031	0.0008-0.0024	0.0012-0.0043
	3.5 (3456)	2GR-FE	0.0008-0.0020	0.0098-0.0138	0.0197-0.0236	0.0039-0.0157	0.0008-0.0028	0.0008-0.0024	0.0028-0.0059
2006	2.4 (2398)	2AZ-FE	0.0020-0.0039	0.0087-0.0350	0.0197-0.0531	0.0039-0.0287	0.0012-0.0028	0.0012-0.0028	SNUG
	3.0 (2995)	1MZ-FE	0.0033-0.0051	0.0098-0.0374	0.0138-0.0413	0.0059-0.0394	0.0008-0.0028	0.0008-0.0024	0.0016-0.0047
	3.3 (3311)	3MZ-FE	0.0013-0.0051	0.0118-0.0374	0.0197-0.0413	0.0059-0.0394	0.0012-0.0031	0.0008-0.0024	0.0012-0.0043
	3.5 (3456)	2GR-FE	0.0008-0.0020	0.0098-0.0138	0.0197-0.0236	0.0039-0.0157	0.0008-0.0028	0.0008-0.0024	0.0028-0.0059

09490_AVAL_C0008

TORQUE SPECIFICATIONS
All readings in ft. lbs.

Year	Engine Displacement Liters (cc)	Engine ID/VIN	Cylinder Head Bolts	Main Bearing Bolts	Rod Bearing Bolts	Crankshaft Damper Bolts	Flywheel Bolts	Manifold Intake	Manifold Exhaust	Spark Plugs	Lug Nuts
2002	2.4 (2398)	2AZ-FE	①	④	⑤	125	⑥	22	27	13	76
	3.0 (2995)	1MZ-FE	②	③	⑤	159	61	11	36	13	76
2003	2.4 (2398)	2AZ-FE	①	④	⑤	125	⑥	22	27	13	76
	3.0 (2995)	1MZ-FE	②	③	⑤	159	61	11	36	13	76
2004	2.4 (2398)	2AZ-FE	①	④	⑤	125	⑥	22	27	13	76
	3.0 (2995)	1MZ-FE	②	③	⑤	159	61	11	36	13	76
	3.3 (3311)	3MZ-FE	②	③	⑤	162	61	11	36	18	76
2005	2.4 (2398)	2AZ-FE	①	④	⑤	125	⑥	22	27	13	76
	3.0 (2995)	1MZ-FE	②	③	⑤	159	61	11	36	13	76
	3.3 (3311)	3MZ-FE	②	③	⑤	162	61	11	36	18	76
	3.5 (3456)	2GR-FE	⑦	⑧	⑤	184	61	15	15	13	76
2006	2.4 (2398)	2AZ-FE	①	④	⑤	125	⑥	22	27	13	76
	3.0 (2995)	1MZ-FE	②	③	⑤	159	61	11	36	13	76
	3.3 (3311)	3MZ-FE	②	③	⑤	162	61	11	36	18	76
	3.5 (3456)	2GR-FE	⑦	⑧	⑤	184	61	15	15	13	76

① Step 1: Several passes in sequence to 58 ft. lbs.
　Step 2: Plus 90 degrees
② Head bolt:
　Step 1: 40 ft. lbs.
　Step 2: Plus 90 degrees
　Recessed head bolt: 13 ft. lbs.
③ 6-point bolts: 20 ft. lbs.
　12-point bolts:
　Step 1: 16 ft. lbs.
　Step 2: Plus an additional 90 degrees

④ Step 1: 15 ft. lbs.
　Step 2: 29 ft. lbs.
　Step 3: Plus 90 degrees
⑤ Step 1: 18 ft. lbs.
　Step 2: Plus 90 degrees
⑥ Manual Transmission: 96 ft. lbs.
　Automatic Transmission: 72 ft. lbs.
⑦ Head bolts:
　Step 1: 27 ft. lbs.
　Step 2: Plus 90 degrees
　Step 3: Plus 90 degrees again
　14 mm head bolt: 22 ft. lbs.

⑧ 16 point bolt:
　Step 1: 45 ft. lbs.
　Step 3: Plus 90 degrees
　14 mm bolt: 38 ft. lbs.

09490_AVAL_C0009

WHEEL ALIGNMENT

Year	Model		Caster Range (+/-Deg.)	Caster Preferred Setting (Deg.)	Camber Range (+/-Deg.)	Camber Preferred Setting (Deg.)	Toe-in (in.)	Steering Axis Inclination (Deg.)
2002	Avalon	F	0.75	+2.17	0.75	-0.62	0+/-0.08	13.07+/-0.75
		R	—	—	0.75	-0.72	0.16+/-0.08	—
	Camry 4-cyl.	F	0.75	+2.18	0.75	-0.60	0+/-0.08	13.08+/-0.75
		R	—	—	0.75	-0.70	0.16+/-0.08	—
	Camry 6-cyl.	F	0.75	+2.09	0.75	-0.60	0+/-0.08	13.08+/-0.75
		R	—	—	0.75	-0.75	0.16+/-0.08	—
	Solara	F	0.75	+2.08	0.75	-0.52	0+/-0.08	12.09+/-0.75
		R	—	—	0.75	-0.65	0.16+/-0.08	—
2003	Avalon	F	0.75	+2.17	0.75	-0.62	0+/-0.08	13.07+/-0.75
		R	—	—	0.75	-0.72	0.16+/-0.08	—
	Camry 4-cyl.	F	0.75	+2.18	0.75	-0.60	0+/-0.08	11.45+/-0.75
		R	—	—	0.75	-1.27	0.16+/-0.08	—
	Camry 6-cyl.	F	0.75	+2.09	0.75	-0.60	0+/-0.08	13.08+/-0.75
		R	—	—	0.75	-0.75	0.16+/-0.08	—
	Solara	F	0.75	+2.08	0.75	-0.52	0+/-0.08	12.09+/-0.75
		R	—	—	0.75	-0.65	0.16+/-0.08	—
2004	Avalon	F	0.75	+2.17	0.75	-0.62	0+/-0.08	13.07+/-0.75
		R	—	—	0.75	-0.72	0.16+/-0.08	—
	Camry 4-cyl.	F	0.75	①	0.75	-0.72	0+/-0.08	11.45+/-0.75
		R	—	—	0.75	-1.27	0.16+/-0.08	—
	Camry 6-cyl.	F	0.75	②	0.75	-0.72	0+/-0.08	11.45+/-0.75
		R	—	—	0.75	-1.27	0.16+/-0.08	—
	Solara 4-cyl.	F	0.75	③	0.75	④	0+/-0.08	11.47+/-0.75
		R	—	—	0.75	-1.35	0.16+/-0.08	—
	Solara 6-cyl.	F	0.75	⑤	0.75	⑥	0+/-0.08	11.52+/-0.75
		R	—	—	0.75	-1.37	0.16+/-0.08	—
2005	Avalon XL	F	0.75	⑦	0.75	⑧	0+/-0.04	12.25+/-0.75
		R	—	—	0.75	⑨	0.16+/-0.08	—
	Camry 4-cyl.	F	0.75	①	0.75	-0.72	0+/-0.08	11.45+/-0.75
		R	—	—	0.75	-1.27	0.16+/-0.08	—
	Camry 6-cyl.	F	0.75	②	0.75	-0.72	0+/-0.08	11.45+/-0.75
		R	—	—	0.75	-1.27	0.16+/-0.08	—
	Solara 4-cyl.	F	0.75	③	0.75	④	0+/-0.08	11.47+/-0.75
		R	—	—	0.75	-1.35	0.16+/-0.08	—
	Solara 6-cyl.	F	0.75	⑤	0.75	⑥	0+/-0.08	11.52+/-0.75
		R	—	—	0.75	-1.37	0.16+/-0.08	—

09490_AVAL_C0010

WHEEL ALIGNMENT

Year	Model		Caster Range (+/-Deg.)	Caster Preferred Setting (Deg.)	Camber Range (+/-Deg.)	Camber Preferred Setting (Deg.)	Toe-in (in.)	Steering Axis Inclination (Deg.)
2006	Avalon XL	F	0.75	⑦	0.75	⑧	0+/-0.04	12.25+/-0.75
		R	—	—	0.75	⑨	0.16+/-0.08	—
	Camry 4-cyl.	F	0.75	①	0.75	-0.72	0+/-0.08	11.45+/-0.75
		R	—	—	0.75	-1.27	0.16+/-0.08	—
	Camry 6-cyl.	F	0.75	②	0.75	-0.72	0+/-0.08	11.45+/-0.75
		R	––	—	0.75	-1.27	0.16+/-0.08	—
	Solara 4-cyl.	F	0.75	③	0.75	④	0+/-0.08	11.47+/-0.75
		R	—	—	0.75	-1.35	0.16+/-0.08	—
	Solara 6-cyl.	F	0.75	⑤	0.75	⑥	0+/-0.08	11.52+/-0.75
		R	—	—	0.75	-1.37	0.16+/-0.08	—

① Sport: +2.72
   Except Sport: +2.65
② 3MZ-FE: +2.65
   1MZ-FE: +2.62
③ Sport: +2.93
   Except Sport: +2.90

④ Sport: -0.77
   Except Sport: -0.73
⑤ Sport: +2.88
   Except Sport: +2.83
   Convertible: +2.85

⑥ Sport: -0.77
   Except Sport: -0.75
   Convertible: -0.73

⑦ XL: +2.65
   Touring: +2.72
   XLS: +2.70
   Limited: +2.80

⑧ XL: -0.67
   Except XL: -0.72

⑨ XL: -1.15
   Touring: -1.22
   XLS: -1.22
   Limited: -1.25

09490_AVAL_C0011

TIRE, WHEEL AND BALL JOINT SPECIFICATIONS

Year	Model	OEM Tires Standard	OEM Tires Optional	Tire Pressures (psi) Front	Tire Pressures (psi) Rear	Wheel Size	Ball Joint Inspection
2002	Avalon	P205/65HR15	None	32	32	6-JJ	9-30 in. ①
	Camry, LE	P205/65HR15	None	29	29	6-JJ	9-30 in. ①
	Camry SE, XLE	P215/60HR16	None	29	29	6-JJ	9-30 in. ①
	Camry Solara	P215/60R16	P205/60R16	—	—	—	9-30 in. ①
2003	Avalon	P205/65HR15	None	32	32	6-JJ	9-30 in. ①
	Camry, LE	P205/65HR15	None	29	29	6-JJ	9-30 in. ①
	Camry SE, XLE	P215/60HR16	None	29	29	6-JJ	9-30 in. ①
	Camry Solara	P205/65R15	P205/60R16	—	—	—	9-30 in. ①
2004	Avalon	P205/65R15	P205/60R16	31	31	6-JJ	9-30 in. ①
	Camry, LE	P205/65R15	None	29	29	6-JJ	9-30 in. ①
	Camry SE, XLE	P215/60R16	None	29	29	6-JJ	9-30 in. ①
	Camry Solara	P215/60R16	P215/55R17	29	29	—	9-30 in. ①
2005	Avalon	P215/65R16	P215/55R17	29	29	6.5-JJ	9-30 in. ①
	Camry, LE	P205/65R15	None	29	29	6.5-JJ	9-30 in. ①
	Camry SE, XLE	P215/60R16	P215/55R17	29	29	6.5-JJ	9-30 in. ①
	Camry Solara	P215/60R16	P215/55R17	29	29	6.5-JJ	9-30 in. ①
2006	Avalon	P215/65R16	P215/55R17	29	29	6.5-JJ	9-30 in. ①
	Camry, LE	P205/65R15	None	29	29	6.5-JJ	9-30 in. ①
	Camry SE, XLE	P215/60R16	None	29	29	6.5-JJ	9-30 in. ①
	Camry Solara	P215/60R16	P215/55R17	29	29	6.5-JJ	9-30 in. ①

OEM: Original Equipment Manufacturer

PSI: Pounds Per Square Inch

① Torque required in inch lbs. to rotate ball joint when removed from the knuckle

② Sport and convertible models, 215/55R17

BRAKE SPECIFICATIONS
All measurements in inches unless noted

Year	Model		Brake Disc Original Thickness	Brake Disc Minimum Thickness	Brake Disc Maximum Runout	Brake Drum Diameter Original Inside Diameter	Brake Drum Diameter Max. Wear Limit	Brake Drum Diameter Maximum Machine Diameter	Minimum Lining Thickness	Brake Caliper Bracket Bolts (ft. lbs.)	Brake Caliper Mounting Bolts (ft. lbs.)
2002	Avalon	F	1.102	1.024	0.0020	—	—	—	0.039	25	79
		R	0.354	0.315	0.0059	—	—	—	0.039	25	34
	Camry	F	1.102	1.024	0.0020	—	—	—	0.039	25	79
		R	0.394	0.354	0.0059	9.00	—	9.08	0.039	14	20
	Camry Solara	F	1.102	1.024	0.0020	—	—	—	0.039	25	79
		R	0.394	0.354	0.0059	9.00	—	9.08	0.039	14	20
2003	Avalon	F	1.102	1.024	0.0020	—	—	—	0.039	25	79
		R	0.354	0.315	0.0059	—	—	—	0.039	25	34
	Camry	F	1.102	1.024	0.0020	—	—	—	0.039	25	79
		R	0.394	0.354	0.0059	9.00	—	9.08	0.039	14	20
	Camry Solara	F	1.102	1.024	0.0020	—	—	—	0.039	25	79
		R	0.394	0.354	0.0059	9.00	—	9.08	0.039	14	20
2004	Avalon	F	1.102	1.024	0.0020	—	—	—	0.039	25	79
		R	0.354	0.315	0.0059	—	—	—	0.039	25	34
	Camry	F	1.102	1.024	0.0020	—	—	—	0.039	25	79
		R	0.472	0.413	0.0059	9.00	—	9.08	0.039	29	35
	Camry Solara	F	1.102	1.024	0.0020	—	—	—	0.039	25	79
		R	0.472	0.413	0.0059	—	—	—	0.039	32	46
2005	Avalon	F	1.102	1.024	0.0020	—	—	—	0.039	25	79
		R	0.427	0.413	0.0059	—	—	—	0.039	25	34
	Camry	F	1.102	1.024	0.0020	—	—	—	0.039	25	79
		R	0.472	0.413	0.0059	8.98	—	9.08	0.039	29	35
	Camry Solara	F	1.102	1.024	0.0020	—	—	—	0.039	25	79
		R	0.472	0.413	0.0059	—	—	—	0.039	32	46
2006	Avalon	F	1.102	1.024	0.0020	—	—	—	0.039	25	79
		R	0.427	0.413	0.0059	—	—	—	0.039	25	34
	Camry	F	1.102	1.024	0.0020	—	—	—	0.039	25	79
		R	0.472	0.413	0.0059	8.98	—	9.08	0.039	29	35
	Camry Solara	F	1.102	1.024	0.0020	—	—	—	0.039	25	79
		R	0.472	0.413	0.0059	—	—	—	0.039	32	46

F: Front

R: Rear

09490_AVAL_C0013

SCHEDULED MAINTENANCE INTERVALS
2002-2003 TOYOTA—AVALON, CAMRY & SOLARA

TO BE SERVICED	TYPE OF SERVIC	VEHICLE MILEAGE INTERVAL (x1000)												
		7.5	15	22.5	30	37.5	45	52.5	60	67.5	75	82.5	90	97.5
Engine oil & filter	R	✓	✓	✓	✓	✓	✓	✓	✓	✓	✓	✓	✓	✓
Drive belts	S/I								✓	✓	✓	✓	✓	✓
Automatic transaxle fluid & filter	S/I		✓		✓		✓		✓		✓		✓	
Ball joints & dust covers	S/I		✓		✓		✓		✓		✓		✓	
Bolts & nuts on body & chassis	S/I		✓		✓		✓		✓		✓		✓	
Brake line pipes & hoses	S/I		✓		✓		✓		✓		✓		✓	
Brake linings & drums	S/I		✓		✓		✓		✓		✓		✓	
Brake pads & discs (front & rear if equipped)	S/I		✓		✓		✓		✓		✓		✓	
Differential oil	S/I		✓		✓		✓		✓		✓		✓	
Drive shaft boots	S/I		✓		✓		✓		✓		✓		✓	
Manual transaxle oil	S/I		✓		✓		✓		✓		✓		✓	
Steering gear housing oil	S/I		✓		✓		✓		✓		✓		✓	
Steering linkage	S/I		✓		✓		✓		✓		✓		✓	
Air filter	R				✓				✓				✓	
Spark plugs	R				✓				✓				✓	
Spark plugs (platinum tip)	R								✓					
Exhaust system	S/I				✓				✓				✓	
Fuel lines & connections	S/I				✓				✓				✓	
Valve clearance	S/I				✓				✓				✓	
Engine coolant	R						✓				✓			
Fuel tank cap gasket	R								✓					
Charcoal canister	S/I								✓					

R: Replace   S/I: Service or Inspect

FREQUENT OPERATION MAINTENANCE (SEVERE SERVICE)

If a vehicle is operated under any of the following conditions it is considered severe service:

- Extremely dusty areas.

- 50% or more of the vehicle operation is in 32°C (90°F) or higher temperatures, or constant operation in temperatures below 0°C (32°F).

- Prolonged idling (vehicle operation in stop and go traffic).

- Frequent short running periods (engine does not warm to normal operating temperatures).

- Police, taxi, delivery usage or trailer towing usage.

Oil & oil filter change: change every 2500 miles.

Bolts & nuts on chassis & body: tighten every 7500 miles.

Ball joints & dust covers: service or inspect every 12,000 miles.

Brake linings & drums: service or inspect ever 12,000 miles.

Brake pads & discs (front & rear if equipped): service or inspect every 12,000 miles.

Drive shaft boots: service or inspect every 12,000 miles.

Steering linkage: service or inspect every 12,000 miles.

Air filter: service or inspect every 15,000 miles.

Exhaust system: service or inspect every 15,000 miles.

Timing belt: replace every 60,000 miles.

SCHEDULED MAINTENANCE INTERVALS

2004-2006 TOYOTA—AVALON, CAMRY & SOLARA

TO BE SERVICED	TYPE OF SERVICE	VEHICLE MILEAGE INTERVAL (x1000)												
		5	10	15	20	25	30	35	40	45	50	55	60	65
Air cleaner filter	R						✓						✓	
Transmission fluid	S/I						✓						✓	
Ball joints & dust covers	S/I			✓			✓			✓			✓	
Bolts & nuts on chassis & body	S/I													
Brake line pipes & hoses	S/I			✓			✓			✓			✓	
Brake pads & discs/linings & drums (front & rear)	S/I	✓	✓	✓	✓	✓	✓	✓	✓	✓	✓	✓	✓	✓
Drive belts	S/I												✓	
Driveshaft boots	S/I			✓			✓			✓			✓	
Engine coolant	S/I			✓			✓			✓			✓	
Engine coolant	R	Replace at 100,000 miles												
Engine oil & filter	R	✓	✓	✓	✓	✓	✓	✓	✓	✓	✓	✓	✓	✓
Exhaust pipes & mountings	S/I			✓			✓			✓			✓	
Fuel lines & connections	S/I						✓						✓	
Propeller shaft bolt	S/I			✓			✓			✓			✓	
Radiator core & condenser	S/I			✓			✓			✓			✓	
Front differential fluid	S/I						✓						✓	
Rotate Tires	S/I	✓	✓	✓	✓	✓	✓	✓	✓	✓	✓	✓	✓	✓
Spark plugs	R	Replace at 120,000 miles												
Steering linkage & gear box	S/I			✓			✓			✓			✓	

R: Replace   S/I: Service or Inspect

Drivebelts: After initial inspection at 60,000 miles, inspect every 15,000 miles thereafter.

FREQUENT OPERATION MAINTENANCE (SEVERE SERVICE)

 If a vehicle is operated under any of the following conditions it is considered severe service:

- Desert/Extremely dusty areas.

- Trailer towing usage.

- Prolonged idling (vehicle operation in stop and go traffic).

- Frequent short running periods (engine does not warm to normal operating temperatures).

- Police, taxi, delivery usage or trailer towing usage.

Air cleaner filter: service or inspect every 5000 miles

Ball joints & dust covers: service or inspect every 5000 miles.

Bolts & nuts on chassis & body: service or inspect every 5000 miles.

Driveshaft boots: service or inspect every 5000 miles.

Steering linkage: service or inspect every 5000 miles.

Transmission and Front differential fluid: replace every 30,000 miles.

Cabin air filter: service or inspect every 15,000 miles

09490_AVAL_C0015

ENGINE REPAIR

Alternator

REMOVAL & INSTALLATION

1. Before servicing the vehicle, refer to the Precautions Section.
2. Remove or disconnect the following:
   - Negative battery cable
   - Air intake assembly, if necessary
   - Engine appearance cover, if necessary
   - Drive belt from the pulley
   - Harness and wire (and nut) from the alternator
   - The two bolts and engine mounting stay if necessary
   - Alternator

To install:

3. Install or connect the following:
   - Alternator
   - Drive belt. Torque the bolts. 2AZ-FE: M8 bolt—15 ft. lbs. (21 Nm); M10 bolt—38 ft. lbs. (52 Nm). 1MZ-FE/3MZ-FE: Pivot bolt—40 ft. lbs. (54 Nm); Lock bolt—14 ft. lbs. (19 Nm). 2GR-FE uses a self-adjusting tensioner.
   - Wiring
   - Air intake assembly, if removed
   - Engine appearance cover, if removed
   - Negative and starter battery cables

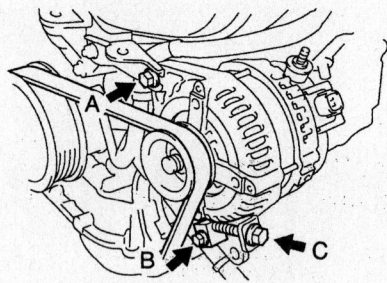

67170-TOYO-G01

(A) pivot bolt, (B) lock bolt and (C) adjusting bolt

Ignition Timing

ADJUSTMENT

➡**The timing on engines equipped with DIS is not adjustable.**

Engine Assembly

REMOVAL & INSTALLATION

2AZ-FE Engines

1. Before servicing the vehicle, refer to the Precautions Section.
2. Relieve the fuel pressure from the fuel lines.
3. Drain the engine coolant from the cooling system.
4. Drain the engine oil.
5. Drain the transmission fluid.
6. Remove or disconnect the following:
   - Front wheels
   - Engine under covers
   - Front fender apron seal
   - Engine cover sub-assembly
   - Battery cables and remove battery
   - Air cleaner assembly, brackets and inlets
   - Engine stabilizing control rod
   - Oil cooler inlet and outlet hoses
   - Engine mounting stay and bracket
   - V-belts
   - Steering gear outlet return tube
   - Union to connector tube hose
   - Transmission control cable assembly
   - Heater inlet and outlet hoses
   - Radiator inlet and outlet hoses
   - Fuel pipe sub-assembly
   - Engine wire from ECU and junction box
   - Engine harness from engine compartment junction block
   - Alternator wiring and alternator
   - A/C compressor (Do NOT disconnect hoses)
   - Front exhaust pipe support bracket
   - Front exhaust pipe assembly
   - Front stabilizer link assembly
   - Both front axle hub nuts
   - Both front speed sensors
   - Separate both outer tie rod ends
   - Separate front lower suspension arm sub assembly, both sides
   - Drive plate and torque converter clutch setting bolts (6)
   - Separate steering intermediate shaft assembly
   - Attach engine hoist
   - 4 bolts and 2 nuts from RH & LH frame side rail plate
   - 4 bolts and 2 nuts from RH & LH front suspension member

7. Carefully remove the engine assembly

To install:

8. Lower engine and transmission assembly onto the engine compartment
9. Install or connect the following:
   - RH & LH side rail plates; torque large bolt 63 ft. lbs. (85 Nm) and small bolt and nuts 24 ft. lbs. (32 Nm)
   - RH & LH front suspension member brace. Torque large bolt to 63 ft. lbs (85 Nm) and small bolt and nuts 24 ft. lbs. (32 Nm)
   - Steering intermediate shaft assembly
   - Drive plate and torque converter clutch setting bolts. Torque to 30 ft. lbs. (41 Nm).
   - LH & RH lower suspension arm sub-assemblies
   - LH & RH tie rod assemblies
   - LH & RH speed sensors
   - LH & RH front axle hub nuts. Torque to 217 ft. lbs. (294 Nm)
   - LH & RH stabilizer link assemblies
   - Front exhaust pipe assembly
   - Front exhaust support bracket
   - Fuel pipe sub-assembly
   - Transmission control cable assembly
   - A/C compressor assembly
   - Alternator belt adjusting bar and bracket
   - RH engine mounting stay
   - Engine stabilizing control rod
   - Alternator assembly
   - A/C compressor V-belt
   - Inspect drive belt deflector and tensioner
   - Air cleaner assembly, brackets and inlets
   - Verify vacuum hose connections

10. Refill the transmission fluid to the correct level.
11. Refill the engine oil to the correct level.
12. Refill the power steering fluid to the correct level.
13. Refill the cooling system to the correct level.
14. Start the engine and check for leaks.
15. Check and adjust the alignment if necessary.

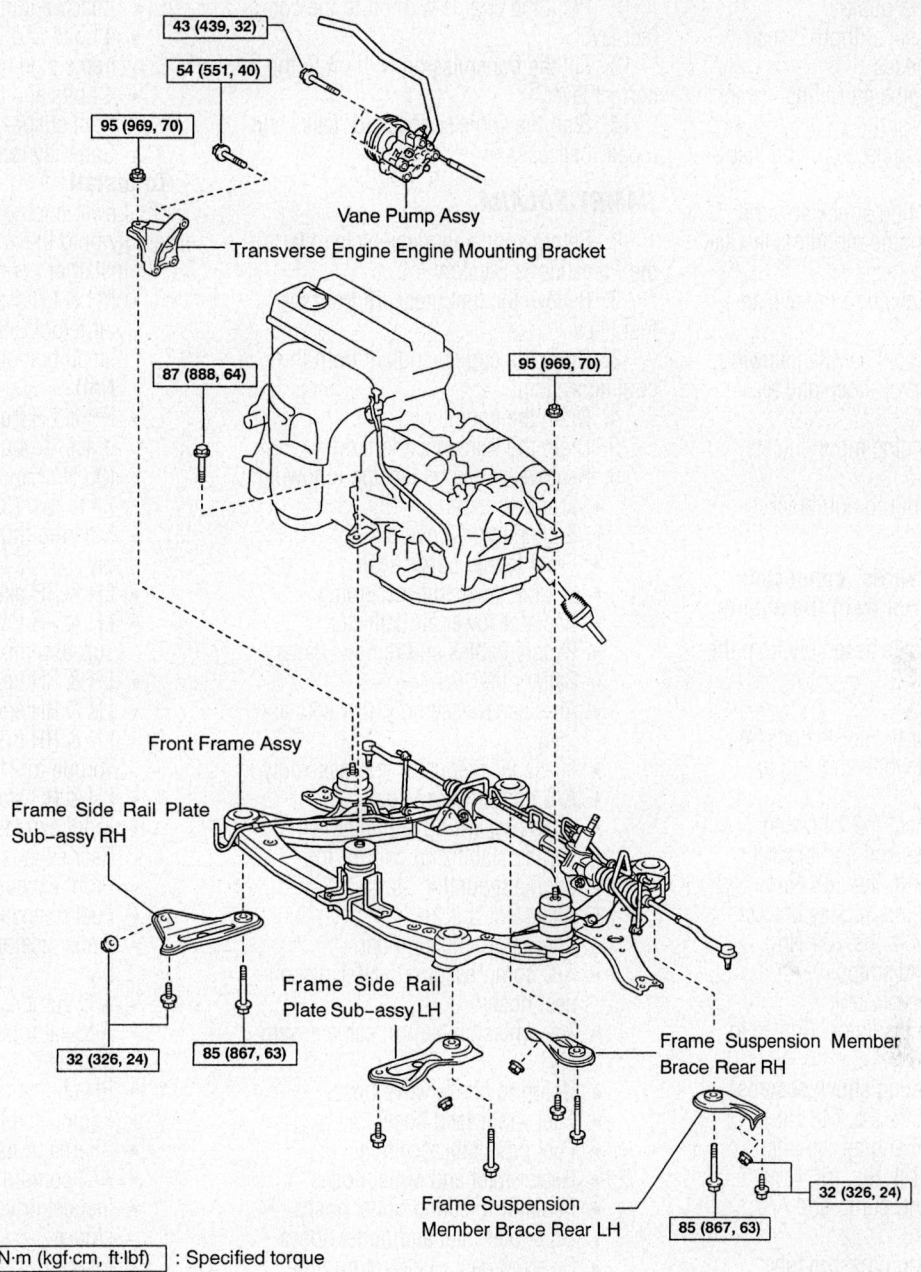

43 (439, 32)

54 (551, 40)

95 (969, 70)

Vane Pump Assy

Transverse Engine Engine Mounting Bracket

87 (888, 64)

95 (969, 70)

Front Frame Assy

Frame Side Rail Plate
Sub-assy RH

Frame Side Rail
Plate Sub-assy LH

Frame Suspension Member
Brace Rear RH

32 (326, 24)   85 (867, 63)

Frame Suspension
Member Brace Rear LH

85 (867, 63)   32 (326, 24)

N·m (kgf·cm, ft·lbf) : Specified torque

67170-TOYO-G03

2AZ-FE engine mounting

1MZ-FE Engines

AVALON

1. Before servicing the vehicle, refer to the Precautions Section.

2. Properly relieve the fuel system pressure.

3. Drain the cooling system.

4. Drain the engine oil.

5. Drain the transaxle fluid.

6. Remove or disconnect the following:

- Negative battery cable.
- Hood
- Battery and battery tray
- Accelerator and throttle cables
- Cruise control actuator, if equipped
- Air cleaner assembly, mass air flow meter and air cleaner hose
- Radiator
- Engine relay box
- 2 igniter connectors
- Noise filter connector
- Connector from the left-hand fender apron
- 2 ground straps and any other electrical connections keeping them from being removed
- Vacuum hoses from the engine.
- Fuel inlet and return hoses
- Heater hoses
- Transaxle control cable from the transaxle
- Instrument panel undercover, the lower instrument panel and glove box assembly
- 3 ECM connectors, the 5 cowl wire connectors, and the cooling fan ECM connector. Push the engine wire through the cowl panel.
- Front exhaust pipe
- Halfshafts
- Power steering pressure tube

- Power steering pump
- A/C compressor without disconnecting the hoses
- Left-hand engine mounting insulator
- Right-hand engine mounting insulator
- Engine mounting shock absorber
- Front right engine mounting insulator

7. Attach a hoist chain to the engine hangers.

8. Remove or disconnect the following:
- Coolant reservoir hose and reservoir tank
- Right-side engine mounting stay bracket
- Engine control rod and bracket assembly

➡**Make certain all wires, connectors and hoses are cleared from the engine.**

- Engine/transaxle assembly from the vehicle

To install:

9. Carefully lower the engine position. Keep the engine level while aligning the engine mounts.

10. Install or connect the following:
- Engine control rod and bracket. Tighten to 47 ft. lbs. (64 Nm).
- Right engine mount stay bracket. Tighten to 23 ft. lbs. (31 Nm).
- Engine ground straps.
- Coolant reservoir tank
- Front engine insulator. Tighten to 48 ft. lbs. (66 Nm).
- Engine mounting shock absorber. Tighten to 35 ft. lbs. (48 Nm).
- Left and right engine mounts. Tighten to 48 ft. lbs. (66 Nm).
- Power steering pump and A/C compressor
- Power steering pressure tube
- Halfshafts and front exhaust pipe
- Engine wires and connectors
- Transaxle control cable to the transaxle
- Fuel hoses and heater hoses
- All vacuum hoses, wiring and connectors
- Radiator
- Cruise control actuator
- Throttle cable and accelerator cable
- MAF meter, the air cleaner assembly, and air cleaner hose
- Hood
- Battery tray and battery
- Negative battery cable

11. Refill the cooling system to the correct level.

12. Refill the engine with oil to the correct level.

13. Fill the transmission with oil to the correct level.

14. Start the vehicle, check for leaks and repair if necessary.

CAMRY/SOLARA

1. Before servicing the vehicle, refer to the Precautions Section.

2. Relieve the fuel pressure from the fuel lines.

3. Drain the engine coolant from the cooling system.

4. Drain the engine oil.

5. Drain the transmission fluid.

6. Remove or disconnect the following:
- Front wheels
- Engine under covers
- Front fender apron seal
- V-bank cover sub-assembly
- Radiator lower air deflector
- Battery cables and remove battery
- Battery tray
- Air cleaner assembly, brackets and inlets
- Intake air resonator sub-assembly
- A/C compressor V-belt
- Alternator wiring and alternator
- Engine stabilizing control rod
- Engine mounting stay
- Alternator bracket
- Alternator adjusting bar
- A/C compressor (Do NOT disconnect hoses)
- Transmission control cable assembly
- Union to check valve hose
- Fuel vapor feed hose
- Fuel pipe sub-assembly
- Heater inlet and outlet hoses
- Radiator inlet and outlet hoses
- Oil cooler inlet and outlet hoses
- Steering gear outlet return tube
- Glove compartment door assembly
- Engine wire from ECU and junction box
- Engine harness from engine compartment junction block
- Front exhaust pipe support bracket
- Rear exhaust pipe support bracket
- Front exhaust pipe assembly
- Front stabilizer link assembly
- Rear stabilizer link assembly
- Both front axle hub nuts
- Both front speed sensors
- Separate both outer tie rod ends
- Separate front lower suspension arm sub assembly, both sides
- Left and Right drive axles
- Separate steering intermediate shaft assembly

- Attach engine hoist
- 4 bolts and 2 nuts from RH & LH frame side rail plate
- 4 bolts and 2 nuts from RH & LH front suspension member
- Carefully remove engine assembly

To install:

7. Lower engine and transmission assembly onto the engine compartment

8. Install or connect the following:
- RH & LH side rail plates; torque large bolt 63 ft. lbs. (85 Nm) and small bolt and nuts 24 ft. lbs. (32 Nm)
- RH & LH front suspension member brace. Torque large bolt to 63 ft. lbs (85 Nm) and small bolt and nuts 24 ft. lbs. (32 Nm).
- Steering intermediate shaft assembly
- LH & RH axel shaft assemblies
- LH & RH lower suspension arm sub-assemblies
- LH & RH tie rod assemblies
- LH & RH speed sensors
- LH & RH front axle hub nuts. Torque to 217 ft. lbs. (294 Nm).
- LH & RH stabilizer link assemblies
- Front exhaust pipe assembly
- Rear exhaust support bracket
- Front exhaust support bracket
- Fuel pipe sub-assembly
- Transmission control cable assembly
- A/C compressor assembly
- Alternator belt adjusting bar and bracket
- RH engine mounting stay
- Engine stabilizing control rod
- Alternator assembly
- A/C compressor V-belt
- Inspect drive belt deflector and tensioner
- Intake air resonator. Torque to 44 inch lbs. (5 Nm)
- Air cleaner assembly, brackets and inlets

9. Refill the engine with oil to the correct level.

10. Refill the transmission with fluid to the correct level.

11. Refill the cooling system to the correct level.

12. Check and adjust the alignment if necessary.

3MZ-FE Engines

1. Before servicing the vehicle, refer to the Precautions Section.

2. Relieve the fuel pressure from the fuel lines.

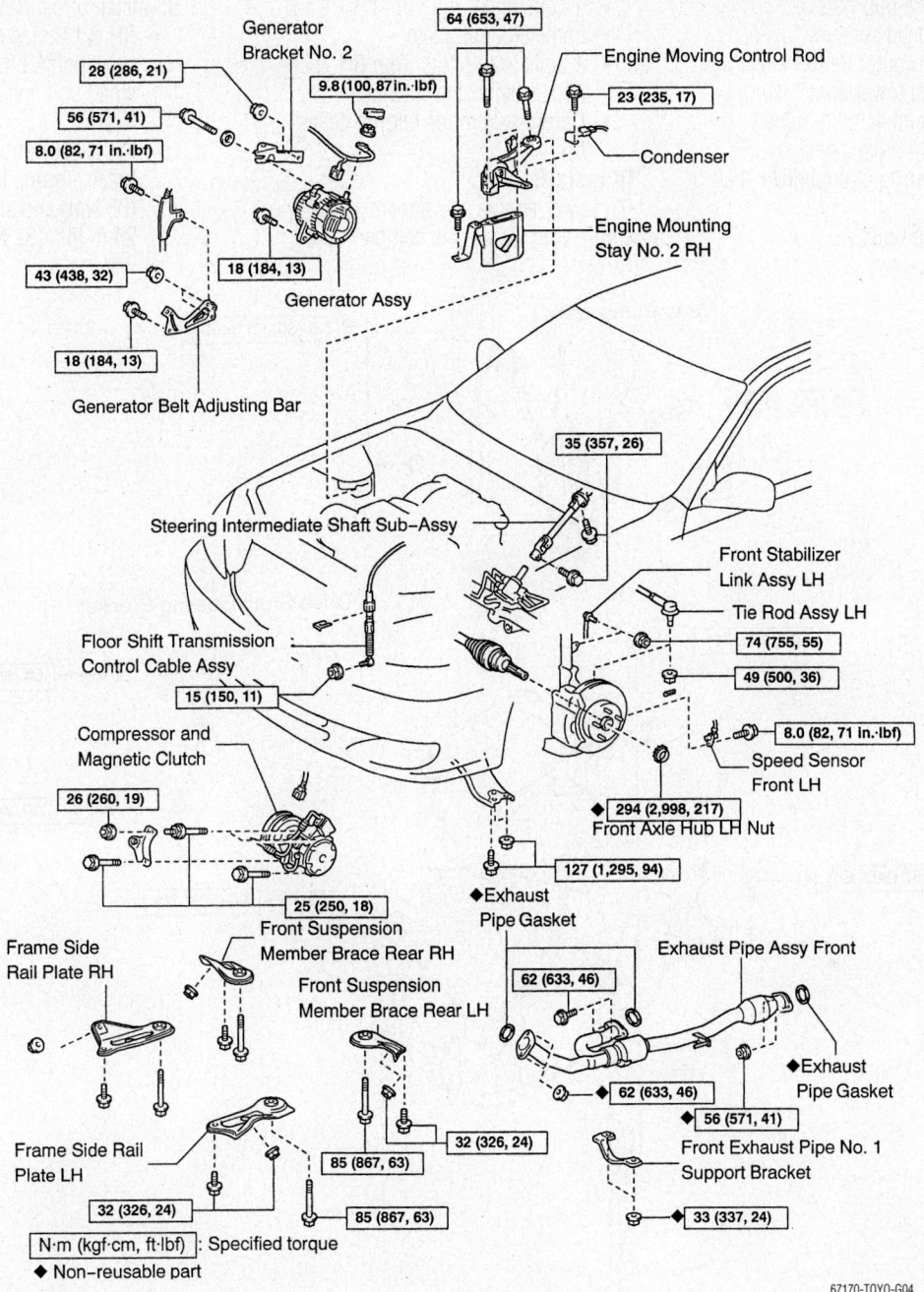

1MZ-FE and 3MZ-FE engine mounting accessories—2004

3. Drain the engine coolant from the cooling system.

4. Drain the engine oil.

5. Drain the transmission fluid.

6. Remove or disconnect the following:
- Negative battery cable
- Front wheels
- Engine under covers
- Front fender apron seal
- V-bank cover sub-assembly
- Radiator lower air deflector
- Battery and battery tray
- Air cleaner assembly, brackets and inlets
- Intake air resonator sub-assembly
- A/C compressor V-belt
- Alternator wiring and alternator
- Engine stabilizing control rod
- Engine mounting stay
- Alternator bracket
- Alternator adjusting bar
- A/C compressor (Do NOT disconnect hoses)
- Transmission control cable assembly
- Union to check valve hose
- Fuel vapor feed hose
- Fuel pipe sub-assembly
- Heater inlet and outlet hoses
- Radiator inlet and outlet hoses
- Oil cooler inlet and outlet hoses
- Steering gear outlet return tube
- Glove compartment door assembly
- Engine wire from ECU and junction box
- Engine harness from engine compartment junction block
- Front exhaust pipe support bracket
- Rear exhaust pipe support bracket
- Front exhaust pipe assembly
- Front stabilizer link assembly
- Rear stabilizer link assembly

- Both front axle hub nuts
- Both front speed sensors
- Separate both outer tie rod ends
- Separate front lower suspension arm sub assembly, both sides
- Left and Right drive axles
- Separate steering intermediate shaft assembly
- Attach engine hoist

- 4 bolts and 2 nuts from RH & LH frame side rail plate
- 4 bolts and 2 nuts from RH & LH front suspension member
- Carefully remove engine assembly

To install:

7. Lower engine and transmission assembly onto the engine compartment

8. Install or connect the following:
- RH & LH side rail plates; torque large bolt 63 ft. lbs. (85 Nm) and small bolt and nuts 24 ft. lbs. (32 Nm)
- RH & LH front suspension member brace. Torque large bolt to 63 ft. lbs (85 Nm) and small bolt and nuts 24 ft. lbs. (32 Nm).

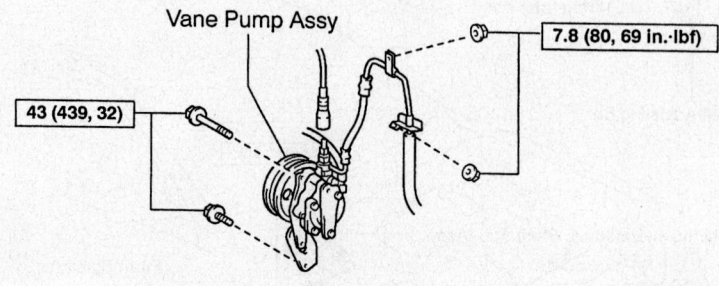

Vane Pump Assy

7.8 (80, 69 in.·lbf)

43 (439, 32)

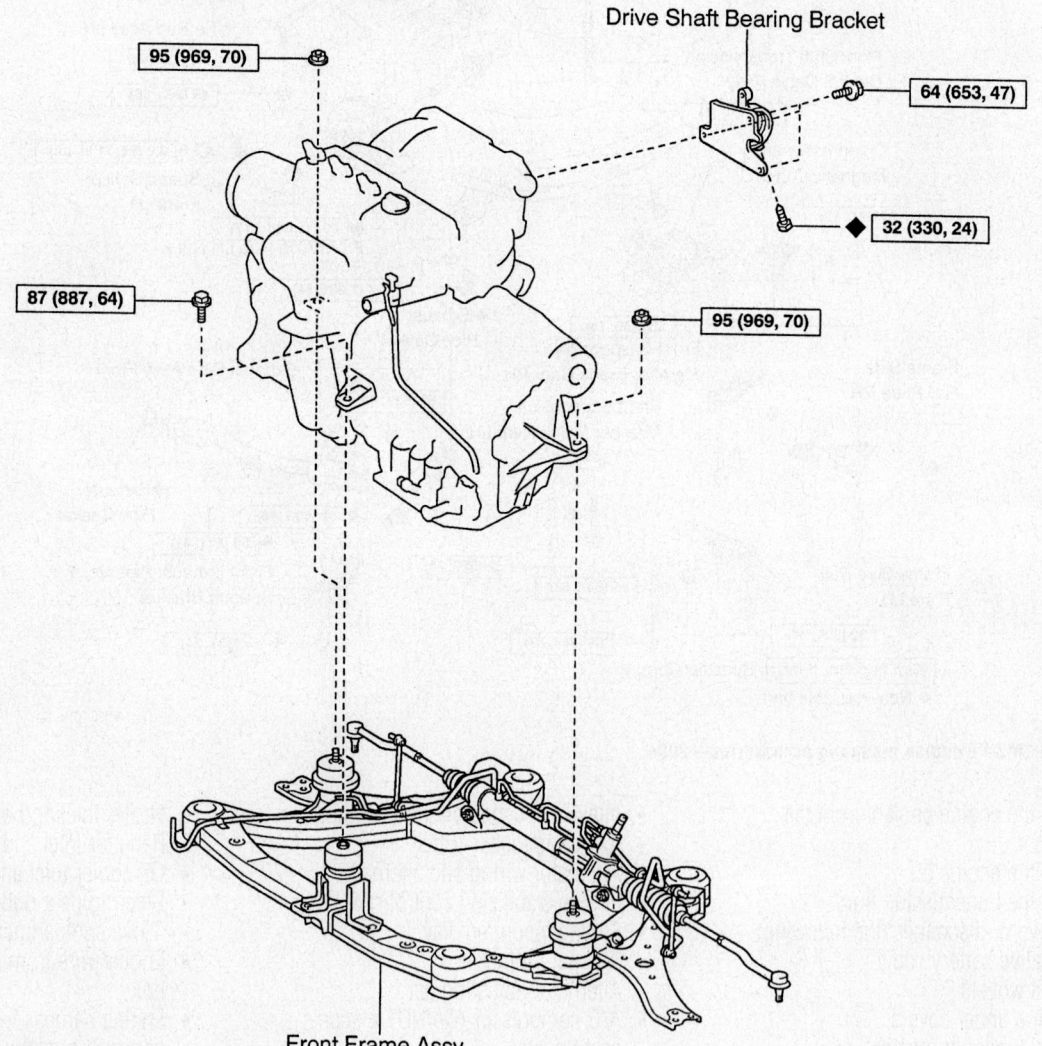

95 (969, 70)

Drive Shaft Bearing Bracket

64 (653, 47)

♦ 32 (330, 24)

87 (887, 64)

95 (969, 70)

Front Frame Assy

N·m (kgf·cm, ft·lbf) : Specified torque

♦ Non–reusable part

67170-TOYO-G05

1MZ-FE and 3MZ-FE engine mounting—2004

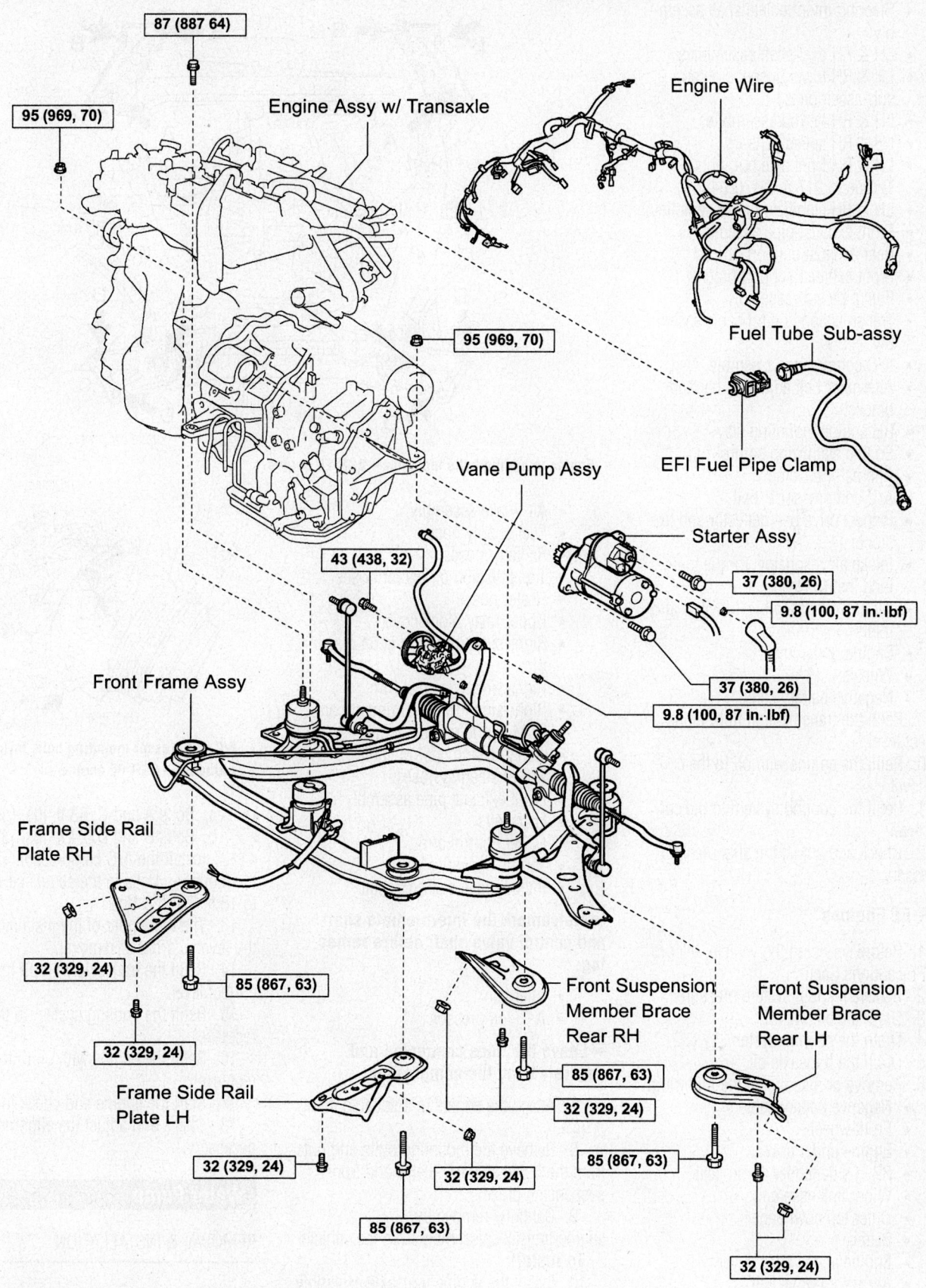

87 (887 64)

95 (969, 70)

Engine Assy w/ Transaxle

Engine Wire

Fuel Tube Sub-assy

95 (969, 70)

EFI Fuel Pipe Clamp

Vane Pump Assy

Starter Assy

43 (438, 32)

37 (380, 26)

9.8 (100, 87 in.·lbf)

37 (380, 26)

9.8 (100, 87 in.·lbf)

Front Frame Assy

Frame Side Rail Plate RH

Front Suspension Member Brace Rear RH

Front Suspension Member Brace Rear LH

32 (329, 24)

85 (867, 63)

32 (329, 24)

85 (867, 63)

Frame Side Rail Plate LH

32 (329, 24)

32 (329, 24)

85 (867, 63)

32 (329, 24)

85 (867, 63)

85 (867, 63)

32 (329, 24)

N·m (kgf·cm, ft·lbf) : Specified torque

Exploded view of the engine mounting—2GR-FE engine

09490_AVAL_G0004

- Steering intermediate shaft assembly
- LH & RH axel shaft assemblies
- LH & RH lower suspension arm sub-assemblies
- LH & RH tie rod assemblies
- LH & RH speed sensors
- LH & RH front axle hub nuts. Torque to 217 ft. lbs. (294 Nm).
- LH & RH stabilizer link assemblies
- Front exhaust pipe assembly
- Rear exhaust support bracket
- Front exhaust support bracket
- Fuel pipe sub-assembly
- Transmission control cable assembly
- A/C compressor assembly
- Alternator belt adjusting bar and bracket
- RH engine mounting stay
- Engine stabilizing control rod
- Alternator assembly
- A/C compressor V-belt
- Inspect drive belt deflector and tensioner
- Intake air resonator. Torque to 44 inch lbs. (5 Nm).
- Air cleaner assembly, brackets and inlets
- Engine undercover
- Wheels
- Negative battery cable

9. Refill the transmission fluid to the correct level.

10. Refill the engine with oil to the correct level.

11. Refill the cooling system to the correct level.

12. Check and adjust the alignment if necessary.

2GR-FE Engines

1. Before servicing the vehicle, refer to the Precautions Section.

2. Relieve the fuel system pressure.

3. Drain the engine oil.

4. Drain the cooling system.

5. Drain the transaxle oil.

6. Remove or disconnect the following:

- Negative battery cable
- Front wheel
- Engine under covers
- Right side fender apron seal
- Wiper link assembly
- Outer top cowl panel
- Battery
- Engine appearance cover
- Accessory drive belt
- Coolant overflow tank
- Right side engine moving control rod and bracket

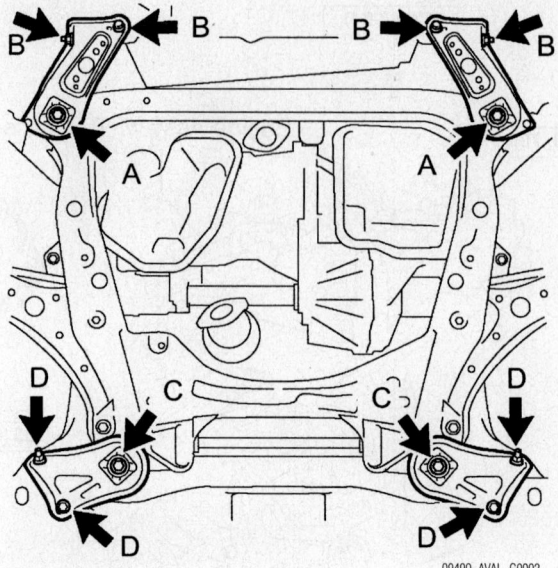

Engine mounting bolts locations—2GR-FE engine

09490_AVAL_G0002

- Air intake assembly
- Fuel supply hose
- Radiator hoses
- Transmission oil cooler hoses
- Heater hoses
- Upper relay block cover
- Right-side front door scuff plate trim
- Right-side side cowl trim
- Right-side lower instrument panel
- ECM wiring and body ground
- Transmission control cable
- Power steering hoses
- Front exhaust pipe assembly
- Halfshafts
- Lower control arms
- Torque converter bolts
- Intermediate steering shaft

➡**Matchmark the intermediate shaft and control valve shaft before removing.**

- Alternator
- A/C compressor

➡**Leave the lines connected and securely hang the compressor.**

7. Attach the engine to a suitable lifting device.

8. Remove the mounting bolts and nuts from the frame side rail plates and front suspension brace.

9. Carefully remove the engine/transaxle assembly from the vehicle.

To install:

10. Install the engine/transaxle assembly into the vehicle.

11. Install the engine mounting bolts and nuts as follows:

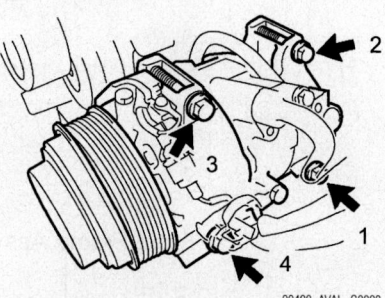

A/C compressor mounting bolts torque sequence—2GR-FE engine

09490_AVAL_G0003

a. Bolts A and C: 63 ft. lbs. (85 Nm).
b. Bolts B and D: 24 ft. lbs. (32 Nm).

12. Install the A/C compressor. Tighten the mounting bolts in the sequence shown to 18 ft. lbs. (25 Nm).

13. The remainder of the installation is the reverse order of removal.

14. Refill the transaxle with oil to the correct level.

15. Refill the cooling system to the correct level.

16. Refill the engine with oil to the correct level.

17. Start the engine and check for leaks.

18. Check and adjust the alignment if necessary.

Water Pump

REMOVAL & INSTALLATION

2AZ-FE Engines

1. Before servicing the vehicle, refer to the Precautions Section.

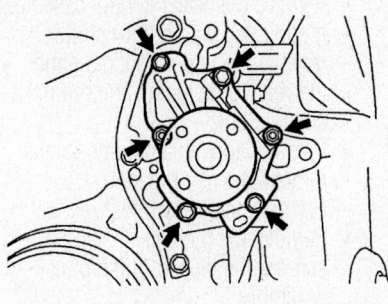

2AZ-FE water pump mounting bolts

67170-TOYO-G06

2. Disconnect the negative battery cable.
3. Drain the cooling system.
4. Remove or disconnect the following:
   - RH front wheel
   - RH fender apron seal
   - Engine stabilizer control rod
   - RH engine stay and bracket
   - Alternator
   - Water pump pulley
   - Water pump assembly

To install:

5. Install or connect the following:
   - Water pump assembly with new gasket. Torque to 80 inch lbs. (9.0 Nm)
   - Water pump pulley
   - Alternator
   - Engine mounting bracket
   - Engine stabilizing control rod
   - Engine mounting stay
   - Right front apron seal and wheel

6. Refill the cooling system to the correct level.

7. Start the engine and check for leaks.

1MZ-FE and 3MZ-FE Engines

AVALON

1. Before servicing the vehicle, refer to the Precautions Section.
2. Drain the cooling system.
3. Remove or disconnect the following:
   - Negative battery cable.

➡**On vehicles equipped with an air bag, wait at least 90 seconds before proceeding after disconnecting the negative battery cable.**

   - Timing belt
   - No. 2 idler pulley
   - 3 clamps and engine wire from the rear timing belt cover
   - Rear timing belt cover
   - Water pump

To install:

4. Install or connect the following:
   - Liquid sealer to the gasket, water pump and engine block.

- Water pump. Bolts and nuts: 53 inch lbs. (6 Nm).
- Rear timing belt cover. Bolts: 74 inch lbs. (9 Nm).
- Engine wire with the 3 clamps to the rear timing belt cover.
- No. 2 idler pulley. Bolt: 32 ft. lbs. (43 Nm).
- With the flange side **outward**, right-hand camshaft pulley. Align the knock pin hole on the camshaft pulley with the knock pin on the camshaft. Bolt: 65 ft. lbs. (88 Nm).
- With the flange side **inward**, left-hand camshaft pulley. Align the knock pin hole on the camshaft pulley with the knock pin on the camshaft. Bolt: 94 ft. lbs. (125 Nm).
- Timing belt
- Negative battery cable

5. Refill the cooling system to the correct level.

6. Start the vehicle and check for leaks.

CAMRY/SOLARA

1. Before servicing the vehicle, refer to the Precautions Section.
2. Drain the cooling system.
3. Remove or disconnect the following:
   - Negative battery cable
   - RH front wheel
   - RH fender apron seal
   - A/C drive belt
   - PS drive belt
   - Engine stabilizer control rod
   - RH engine stay
   - Alternator bracket #2
   - Crankshaft pulley
   - Both timing belt covers
   - RH engine mounting bracket
   - Timing belt cover 1 and 2
   - Timing belt, guide and idler pulley sub-assembly #1
   - Camshaft timing pulleys and idler pulley sub-assembly #2
   - Timing belt cover #3
   - Water pump assembly

To install:

4. Install or connect the following:
   - Water pump assembly with new gasket. Torque to 71 inch lbs. (8 Nm).
   - Timing belt idler #1. Torque to 25 ft. lbs (24 Nm).
   - Timing belt cover #3
   - Camshaft timing pulleys
   - Timing belt idler sub-assembly
   - Timing belt, tensioner assembly and guide
   - Engine mounting bracket
   - Upper and lower timing belts covers

1MZ-FE and 3MZ-FE water pump mounting bolts

67170-TOYO-G07

- Crankshaft pulley
- Alternator bracket
- Engine mounting stay
- Engine stabilizing control rod
- PS pump
- A/C drive belt
- Inspect drive belt tension
- Right front wheel
- Negative battery cable

5. Refill the cooling system to the correct level.

6. Start the engine and check for leaks.

2GR-FE Engine

1. Before servicing the vehicle, refer to the Precautions Section.
2. Drain the cooling system.
3. Remove or disconnect the following:
   - Negative battery cable
   - Front wheels
   - Right-side engine under cover
   - Engine appearance cover
   - Engine moving control rod and bracket
   - Left-side No. 1 engine mounting bracket
   - Accessory drive belt
   - Coolant hoses and O-rings from the water pump
   - Crankshaft pulley, using Special Tool 09213-70011 to hold the crankshaft
   - Water pump pulley, using Special Tool 09960-10010 to hold the pulley
   - Idler pulley
   - Power steering pump
   - Water pump mounting bolts
   - Water pump and gasket

To install:

4. Install the water pump with a new gasket and tighten the bolts as follows:
   a. Bolts A: 15 ft. lbs. (21 Nm).
   b. Bolts B and C: 81 inch lbs. (9.1 Nm).

➡**Bolts C must be replaced with new bolts.**

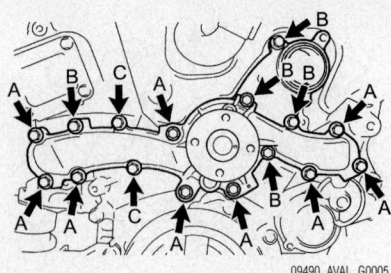

Water pump mounting bolt identification—
2GR-FE

5. Install or connect the following:
- Power steering pump
- Idler pulley. Tighten the bolt to 32 ft. lbs. (43 Nm).

➡**Idler pulley bolt is a left-hand thread.**

- Water pump pulley, using Special Tool 09960-10010. Tighten to 15 ft. lbs. (21 Nm).
- Crankshaft pulley
- Coolant oils with new O-rings
- Accessory drive belt
- Engine mounting bracket
- Engine moving control rod and bracket
- Engine appearance cover
- Engine under cover
- Front wheels
- Negative battery cable

6. Refill the cooling system to the correct level.

7. Start the engine and check for leaks.

Heater Core

REMOVAL & INSTALLATION

Avalon

2002–04

1. Before servicing the vehicle, refer to the Precautions Section.

2. Disconnect the negative battery cable.

❊❊ CAUTION

After disconnecting the negative battery cable, wait for at least 1½ minutes for the SRS to deplete its energy.

3. Drain the cooling system into a clean container for reuse.

4. Remove the air bag module and the steering wheel by performing the following procedure:
- Place the front wheels in the straight-ahead position.

- At both sides of the steering wheel, remove the screw covers.
- Using a Torx socket, loosen the 2 air bag module-to-steering wheel Torx screws until the circumference ring catches on the screw case.
- Carefully, remove the air bag module and disconnect the electrical connector.
- Remove the steering wheel nut.
- Using a steering wheel puller, press the steering wheel from the steering column.

5. Remove the instrument panel by performing the following procedure:
- Remove the front pillar garnishes and the door scuff plates.
- Remove the hood lock release lever and the cowl side trims.
- Remove the steering column covers and the combination switch.
- Remove the lower finish panel assembly and the instrument panel finish lower left side panel.
- Remove the fuse box bolt and the No. 2 heater-to-register duct.
- Remove the parking brake release lever and the No. 2 undercover.
- Remove the lower No. 2 finish panel.
- If equipped with a column shifter, disconnect the shift control cable from the shift lever housing; then, disconnect the shift control cable from the steering column cable bracket.
- Matchmark the steering column shaft and the control valve shaft.
- Remove the steering column shaft-to-intermediate shaft bolt.
- Remove the steering column-to-instrument panel nuts and remove the steering column assembly.
- Inside the glove compartment, pry out the glove compartment door finish plate.
- Pull out the air bag electrical connector and disconnect it.
- Remove the 3 glove compartment door-to-instrument panel nuts and the door.
- Remove the 4 glove compartment-to-instrument panel screws and the glove compartment; then, disconnect the glove box light connector.
- Remove the passenger's side air bag module-to-instrument panel 2 bolts and 4 nuts. Carefully, remove the air bag module from the instrument panel.
- Remove the center cluster finish panel and the radio.

- Remove the heater control assembly.
- If equipped with a floor shifter, remove the upper console panel, the rear console box and the front console box.
- If equipped with a column shifter, remove the finish panel.
- Pry out the cluster finish panel.
- Remove the 6 cluster finish panel screws, the cluster finish panel assembly.
- Remove the 4 combination meter screws and the combination meter.
- Disconnect instrument panel electrical connectors.
- Remove the instrument panel-to-chassis nuts/bolts and the remove the instrument panel.

6. Remove the instrument panels No. 2 brace.

7. Disconnect the heater hoses from the heater core.

8. Remove the 2 heater pipes-to-heater core screws and clips; then, disconnect the heater pipes from the heater core.

9. Remove the heater core O-rings and discard them.

10. Remove the heater core from the heater housing.

To install:

11. Install the heater core to the heater housing.

12. Install new heater core O-rings.

13. Install the heater pipes to the heater core; then, the 2 heater pipes-to-heater core screws and clips.

14. Install the heater hoses to the heater core.

15. Install the instrument panels No. 2 brace.

16. Install the instrument panel by performing the following procedure:
- Install the instrument panel and the instrument panel-to-chassis nuts/bolts.
- Install instrument panel electrical connectors.
- Install the combination meter and the 4 combination meter screws.
- Install the cluster finish panel, the 6 cluster finish panel assembly screws.
- Install the cluster finish panel.
- If equipped with a column shifter, install the finish panel.
- If equipped with a floor shifter, install the upper console panel, the rear console box and the front console box.
- Install the heater control assembly.
- Install the center cluster finish panel and the radio.

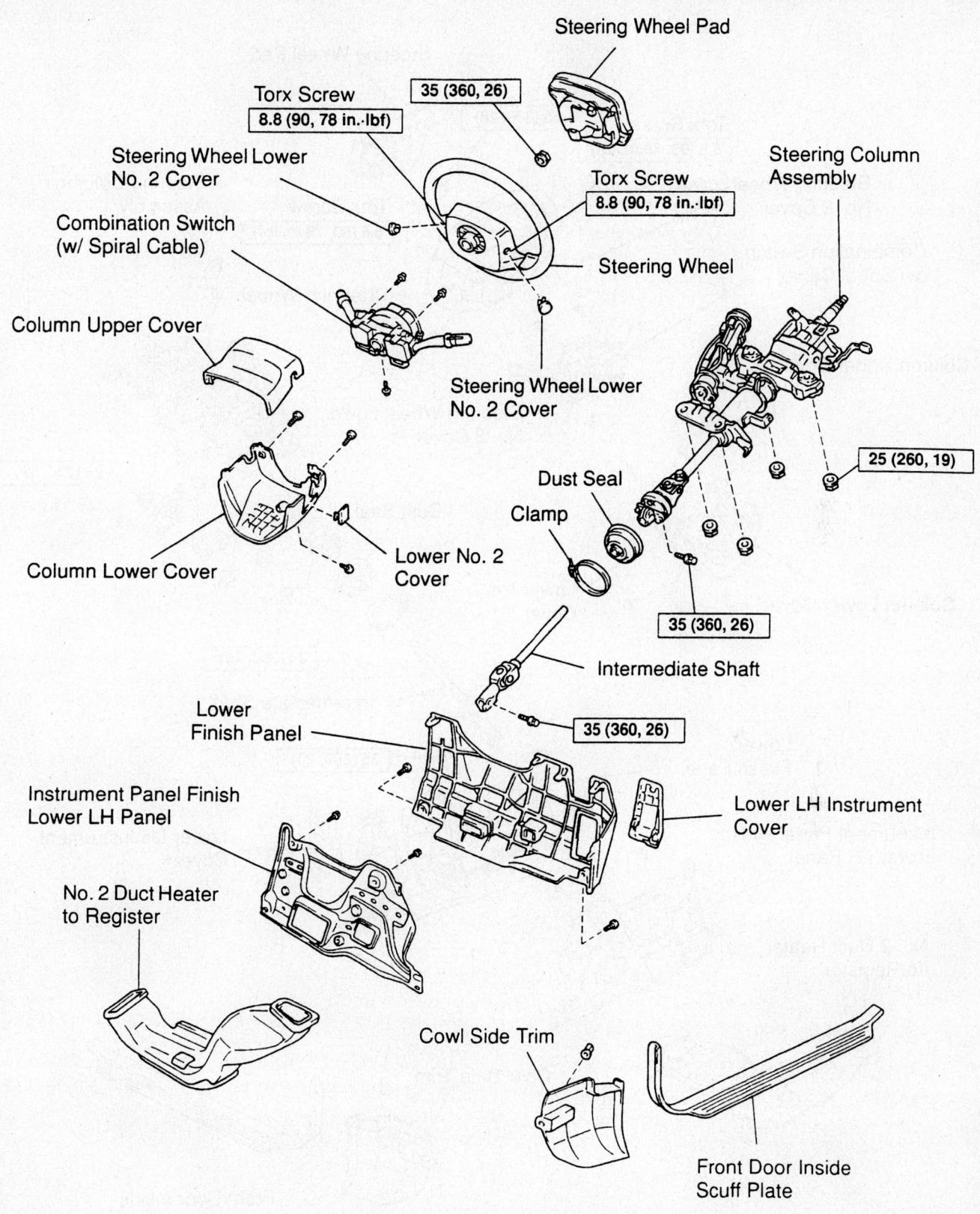

Steering Wheel Pad

Torx Screw
8.8 (90, 78 in.·lbf)

35 (360, 26)

Steering Wheel Lower No. 2 Cover

Torx Screw
8.8 (90, 78 in.·lbf)

Steering Column Assembly

Combination Switch (w/ Spiral Cable)

Steering Wheel

Column Upper Cover

Steering Wheel Lower No. 2 Cover

Dust Seal

Clamp

25 (260, 19)

Column Lower Cover

Lower No. 2 Cover

35 (360, 26)

Intermediate Shaft

Lower Finish Panel

35 (360, 26)

Instrument Panel Finish Lower LH Panel

Lower LH Instrument Cover

No. 2 Duct Heater to Register

Cowl Side Trim

Front Door Inside Scuff Plate

N·m (kgf·cm, ft·lbf) : Specified torque

93112GL2

Exploded view of the air bag module, the steering wheel, the floor shift steering column and related components—2002–04 Avalon

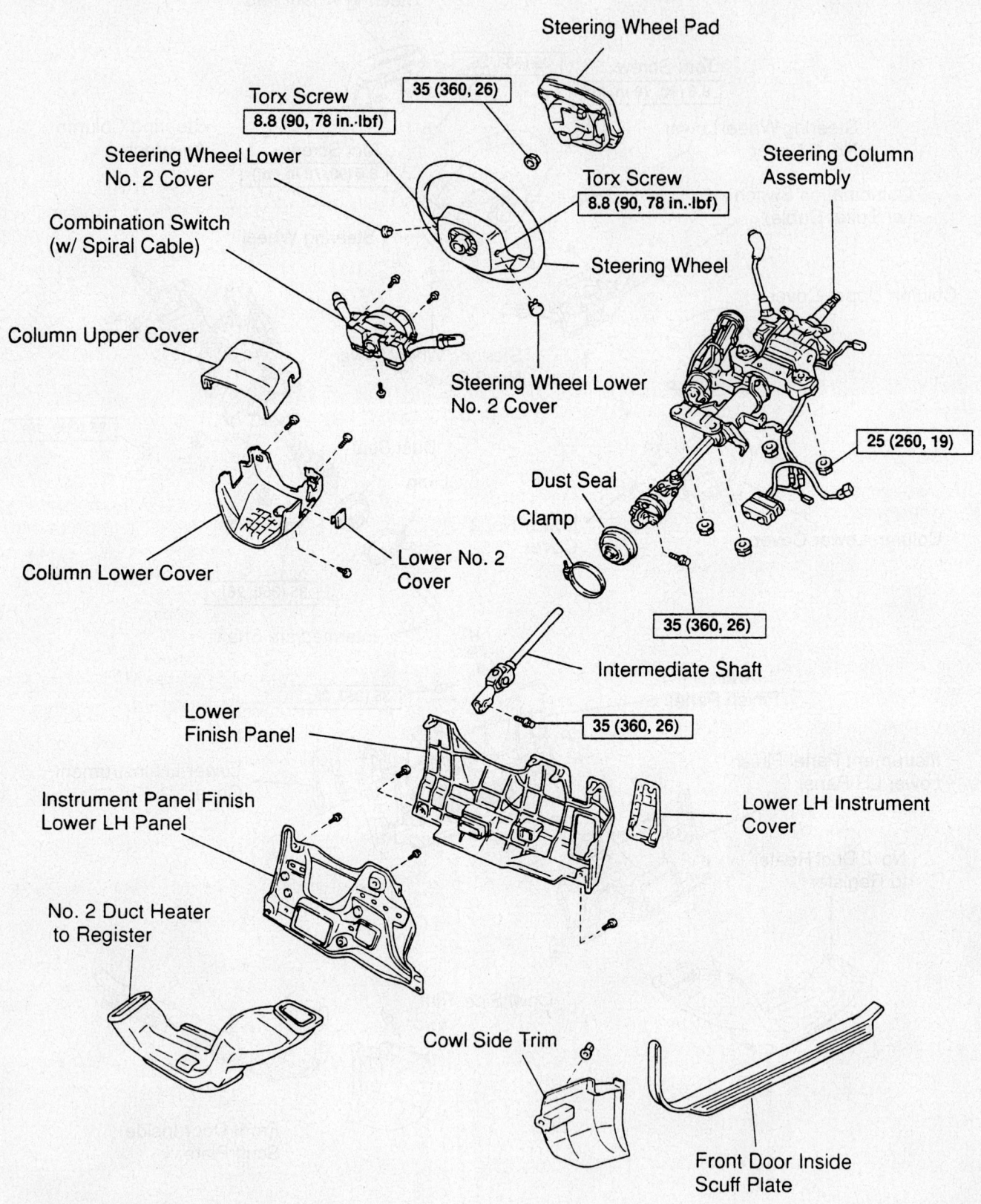

Steering Wheel Pad

35 (360, 26)

Torx Screw
8.8 (90, 78 in.·lbf)

Steering Wheel Lower
No. 2 Cover

Torx Screw
8.8 (90, 78 in.·lbf)

Steering Column
Assembly

Combination Switch
(w/ Spiral Cable)

Steering Wheel

Column Upper Cover

Steering Wheel Lower
No. 2 Cover

Dust Seal

25 (260, 19)

Clamp

Column Lower Cover

Lower No. 2
Cover

35 (360, 26)

Intermediate Shaft

Lower
Finish Panel

35 (360, 26)

Instrument Panel Finish
Lower LH Panel

Lower LH Instrument
Cover

No. 2 Duct Heater
to Register

Cowl Side Trim

Front Door Inside
Scuff Plate

N·m (kgf·cm, ft·lbf) : Specified torque

93112GL3

Exploded view of the air bag module, the steering wheel, the column shift steering column and related components—2002–04 Avalon

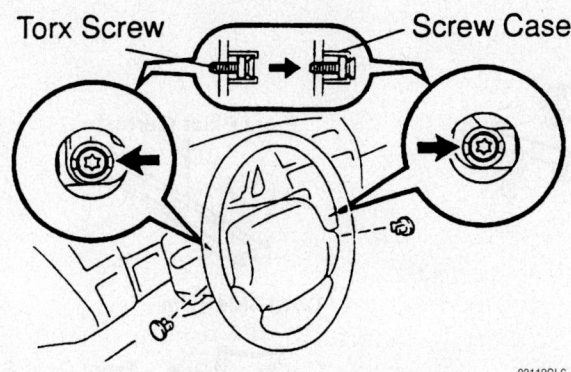

Releasing the air bag module-to-steering wheel screws—2002–04 Avalon

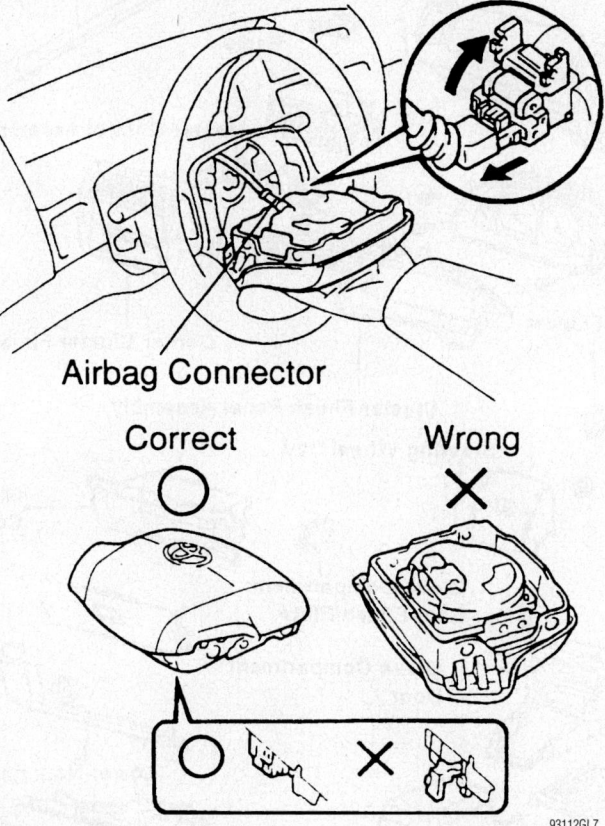

Disconnecting and positioning the air bag module—2002–04 Avalon

- Carefully, install the air bag module to the instrument panel; then, install the passenger's side air bag module-to-instrument panel 2 bolts and 4 nuts.
- Install the glove compartment and the 4 glove compartment-to-instrument panel screws; then, install the glove box light connector.
- Install the glove compartment door and the 3 door-to-instrument panel nuts.
- Connect the air bag electrical connector.
- Inside the glove compartment,

install the glove compartment door finish plate.
- Install the steering column assembly and the steering column-to-instrument panel nuts; then, torque the nuts to 19 ft. lbs. (25 Nm).
- Align the matchmarks and install the steering column shaft-to-intermediate shaft bolt.
- If equipped with a column shifter, install the shift control cable to the shift lever housing; then, install the shift control cable to the steering column cable bracket.
- Install the lower No. 2 finish panel.

- Install the parking brake release lever and the No. 2 undercover.
- Install the fuse box bolt and the No. 2 heater-to-register duct.
- Install the lower finish panel assembly and the instrument panel finish lower left side panel.
- Install the steering column covers and the combination switch.
- Install the hood lock release lever and the cowl side trims.
- Install the front pillar garnishes and the door scuff plates.

17. Install the air bag module and the steering wheel by performing the following procedure:
- Align the matchmarks and install the steering wheel to the steering column.
- Install the steering wheel nut and torque to 26 ft. lbs. (35 Nm).
- Carefully, connect the electrical connector and install the air bag module.
- Using a Torx socket, tighten the 2 air bag module-to-steering wheel Torx screws to 78 inch lbs. (8.8 Nm).
- At both sides of the steering wheel, install the screw covers.

18. Refill the cooling system.
19. Install the negative battery cable.
20. Operate the engine to normal operating temperatures; then, check the climate control operation and check for leaks.

2005–06

1. Before servicing the vehicle, refer to the Precautions Section.
2. Discharge and recover the refrigerant from the air conditioning system.
3. Drain the cooling system.
4. Remove or disconnect the following:
- Negative battery cable
- Wiper arm and blade assembly
- Right-side top cowl ventilator louver
- Wiper motor and linkage assembly
- Outside top cowl panel
- Air conditioner lines
- Heater hoses

5. To remove the instrument panel safety pad, remove or disconnect the following:

- Steering wheel covers
- Horn button
- Steering wheel assembly
- Steering column covers
- Turn signal switch assembly
- Instrument cluster finishing trim
- Instrument panel trim around the steering column

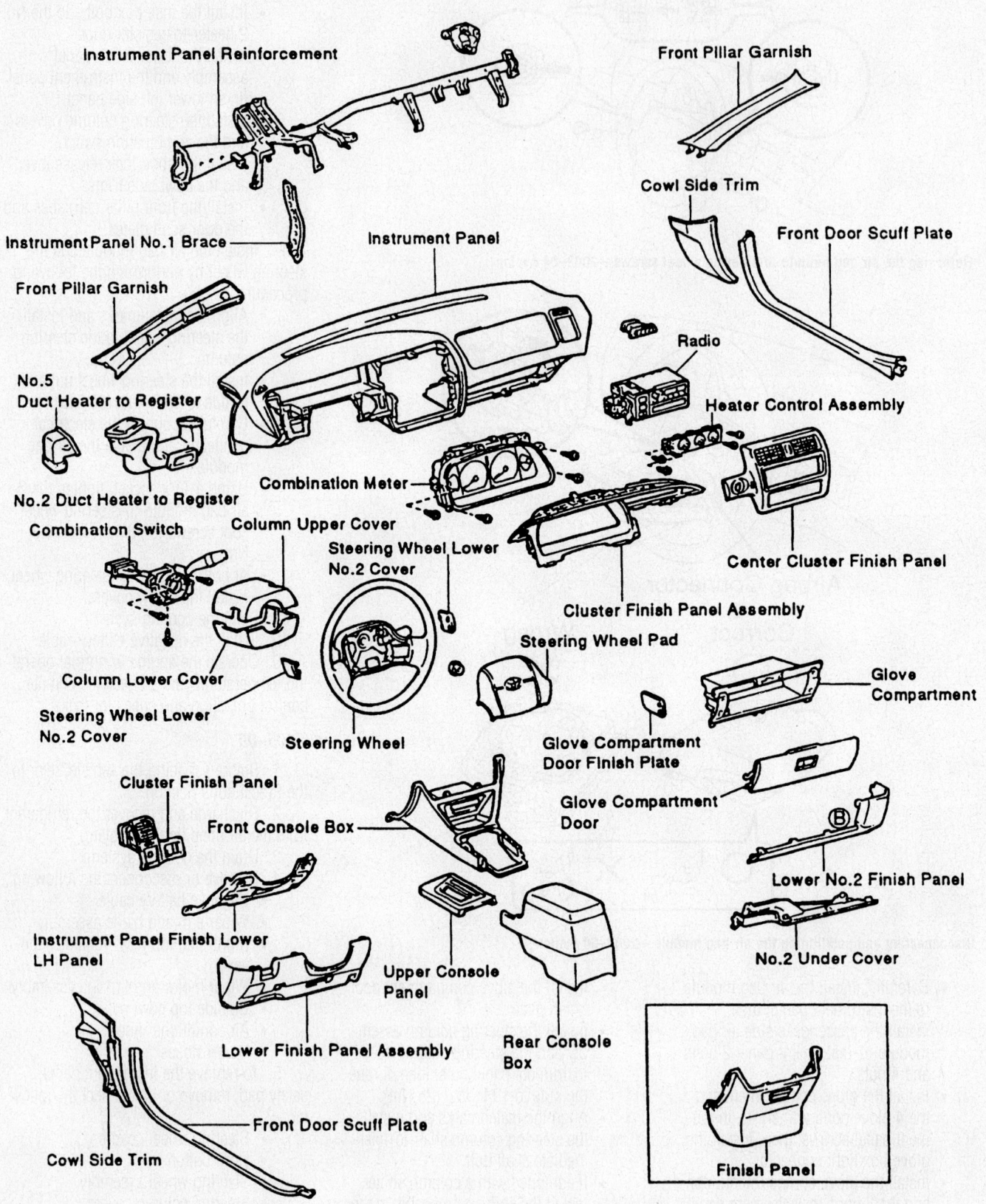

Instrument Panel Reinforcement

Front Pillar Garnish

Cowl Side Trim

Instrument Panel No.1 Brace

Instrument Panel

Front Door Scuff Plate

Front Pillar Garnish

Radio

No.5 Duct Heater to Register

Heater Control Assembly

No.2 Duct Heater to Register

Combination Meter

Combination Switch

Column Upper Cover

Steering Wheel Lower No.2 Cover

Center Cluster Finish Panel

Cluster Finish Panel Assembly

Column Lower Cover

Steering Wheel Pad

Glove Compartment

Steering Wheel Lower No.2 Cover

Steering Wheel

Glove Compartment Door Finish Plate

Cluster Finish Panel

Glove Compartment Door

Front Console Box

Lower No.2 Finish Panel

Instrument Panel Finish Lower LH Panel

Upper Console Panel

No.2 Under Cover

Lower Finish Panel Assembly

Rear Console Box

Cowl Side Trim

Front Door Scuff Plate

Finish Panel

93112GA5

Exploded view of the instrument panel and related components—2002–04 Avalon

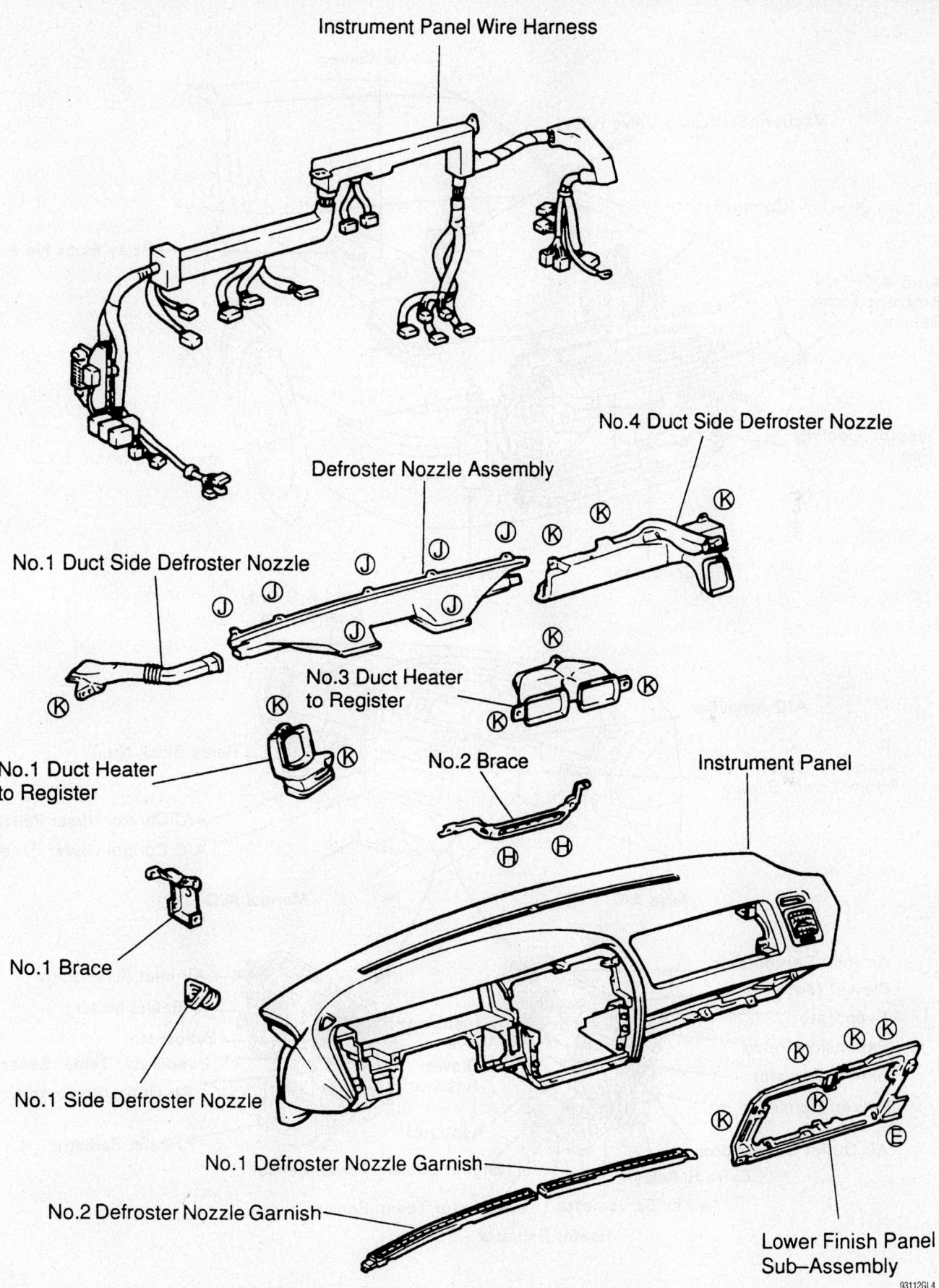

Instrument Panel Wire Harness

No.4 Duct Side Defroster Nozzle

Defroster Nozzle Assembly

No.1 Duct Side Defroster Nozzle

No.3 Duct Heater to Register

No.1 Duct Heater to Register

No.2 Brace

Instrument Panel

No.1 Brace

No.1 Side Defroster Nozzle

No.1 Defroster Nozzle Garnish

No.2 Defroster Nozzle Garnish

Lower Finish Panel Sub–Assembly

93112GL4

Exploded view of the wiring harness, ventilation system and related components—2002–04 Avalon

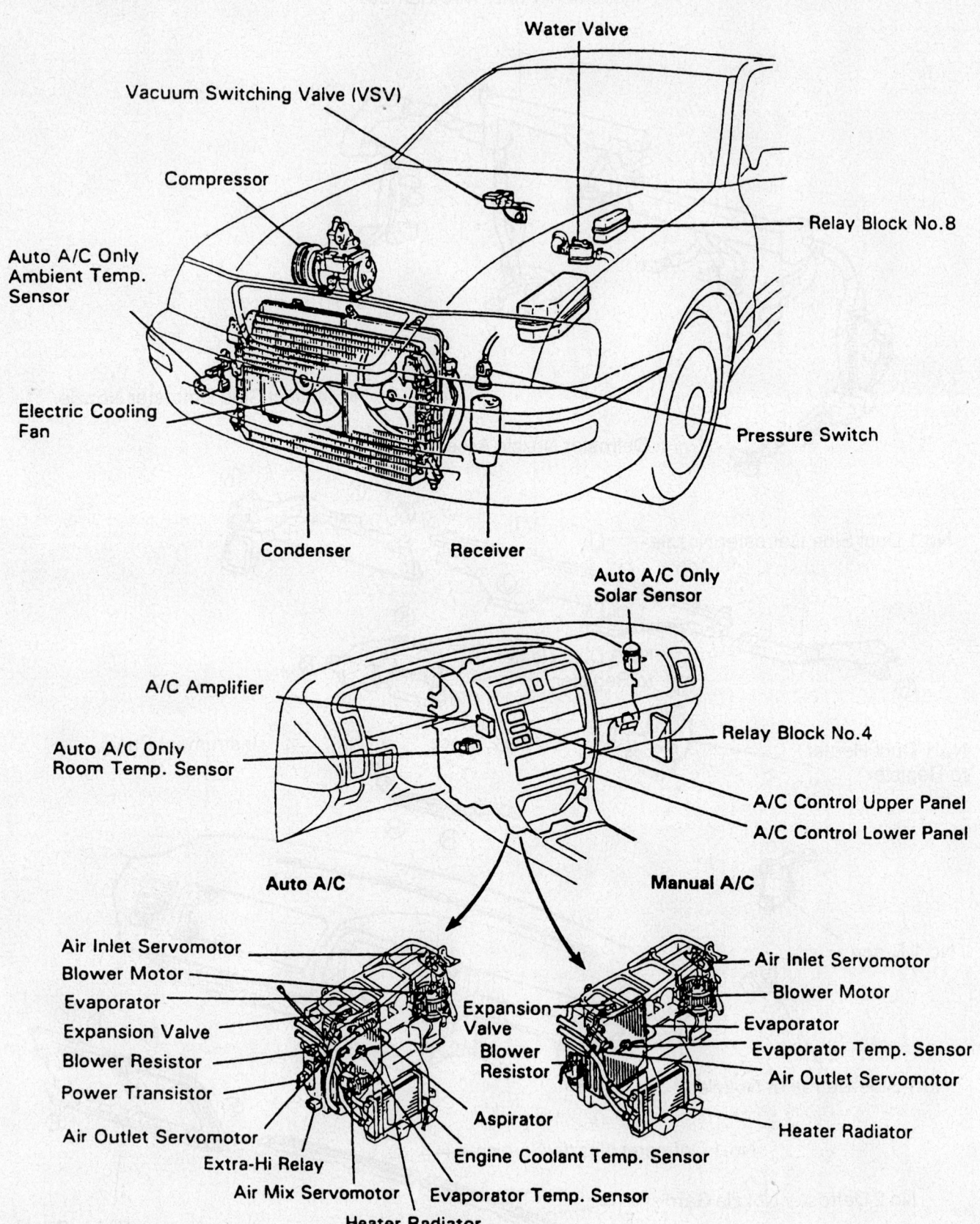

View of the heater/air conditioning assembly and related components—2002–04 Avalon

93112GA6

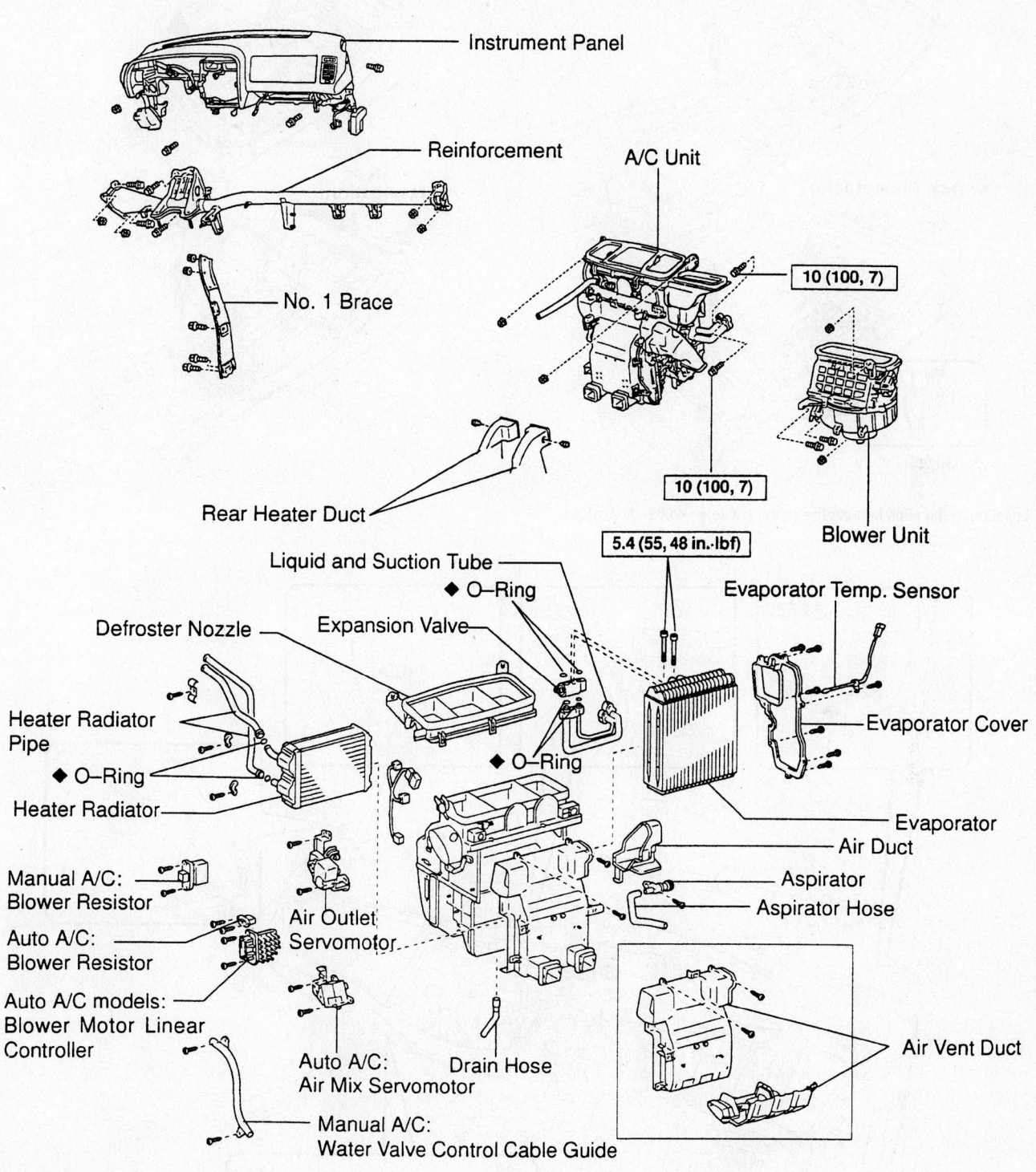

Instrument Panel

Reinforcement

A/C Unit

No. 1 Brace

10 (100, 7)

Rear Heater Duct

10 (100, 7)

Blower Unit

5.4 (55, 48 in.·lbf)

Liquid and Suction Tube

◆ O–Ring

Evaporator Temp. Sensor

Defroster Nozzle

Expansion Valve

Heater Radiator Pipe

◆ O–Ring

◆ O–Ring

Evaporator Cover

Heater Radiator

Evaporator

Air Duct

Manual A/C: Blower Resistor

Aspirator

Aspirator Hose

Auto A/C: Blower Resistor

Auto A/C models: Blower Motor Linear Controller

Air Outlet Servomotor

Auto A/C: Air Mix Servomotor

Drain Hose

Air Vent Duct

Manual A/C: Water Valve Control Cable Guide

N·m (kgf·cm, ft·lbf) : Specified torque

◆ Non–reusable part

93112GL5

Exploded view of the evaporator housing, heater housing, heater core and related components—2002–04 Avalon

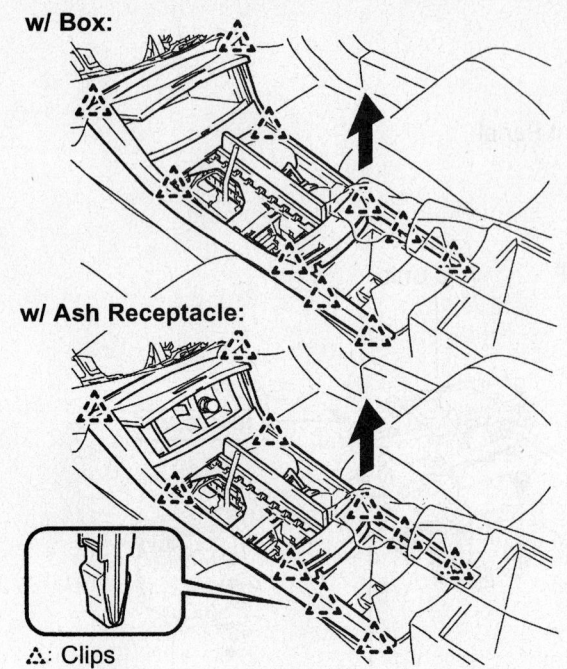

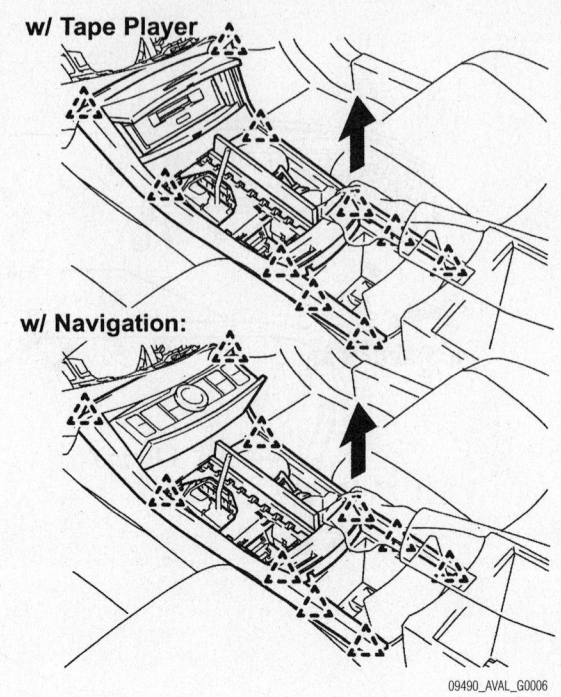

w/ Box:

w/ Tape Player:

w/ Ash Receptacle:

w/ Navigation:

△: Clips

09490_AVAL_G0006

Location of the clips holding the center console—2005–06 Avalon

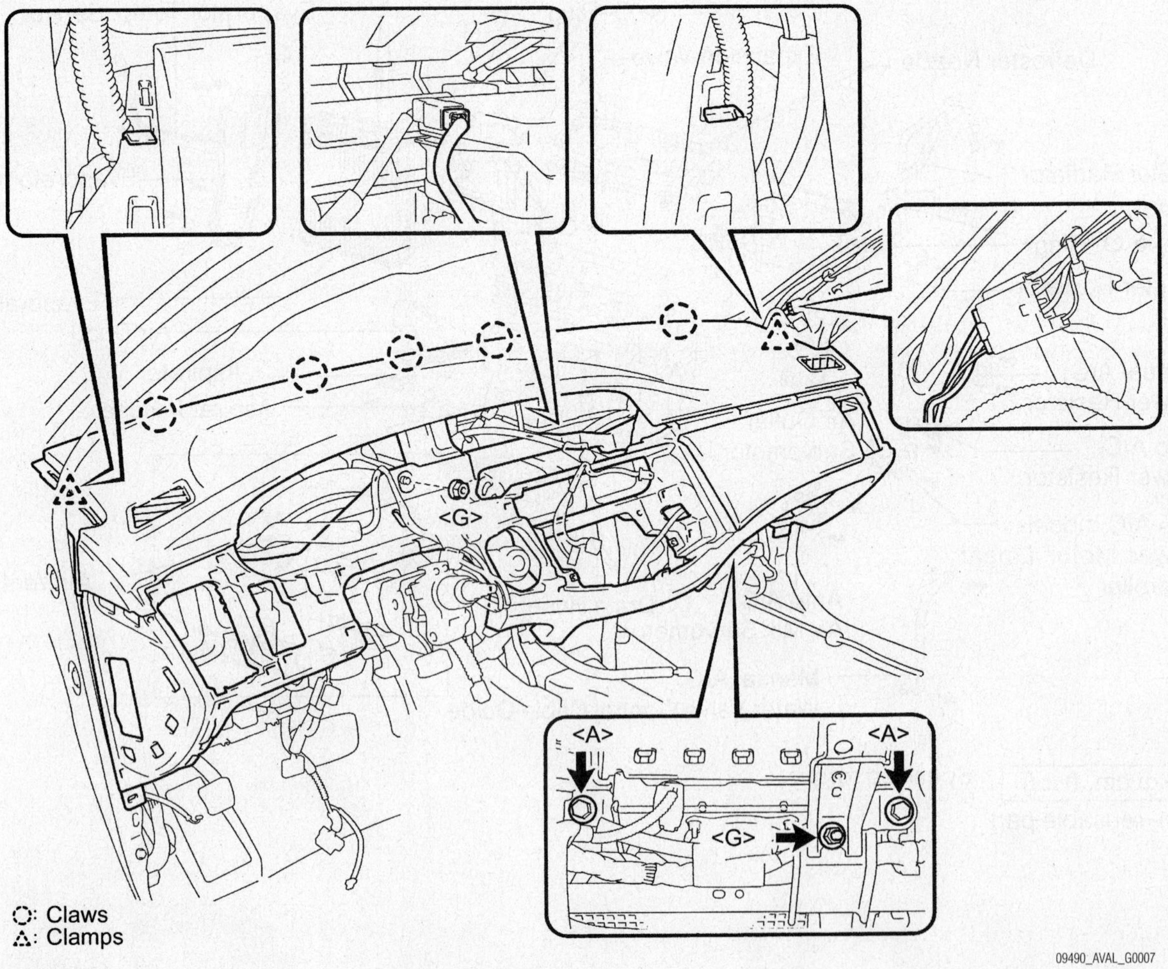

○: Claws
△: Clamps

09490_AVAL_G0007

Location of the instrument panel mounting clips and nuts—2005–06 Avalon

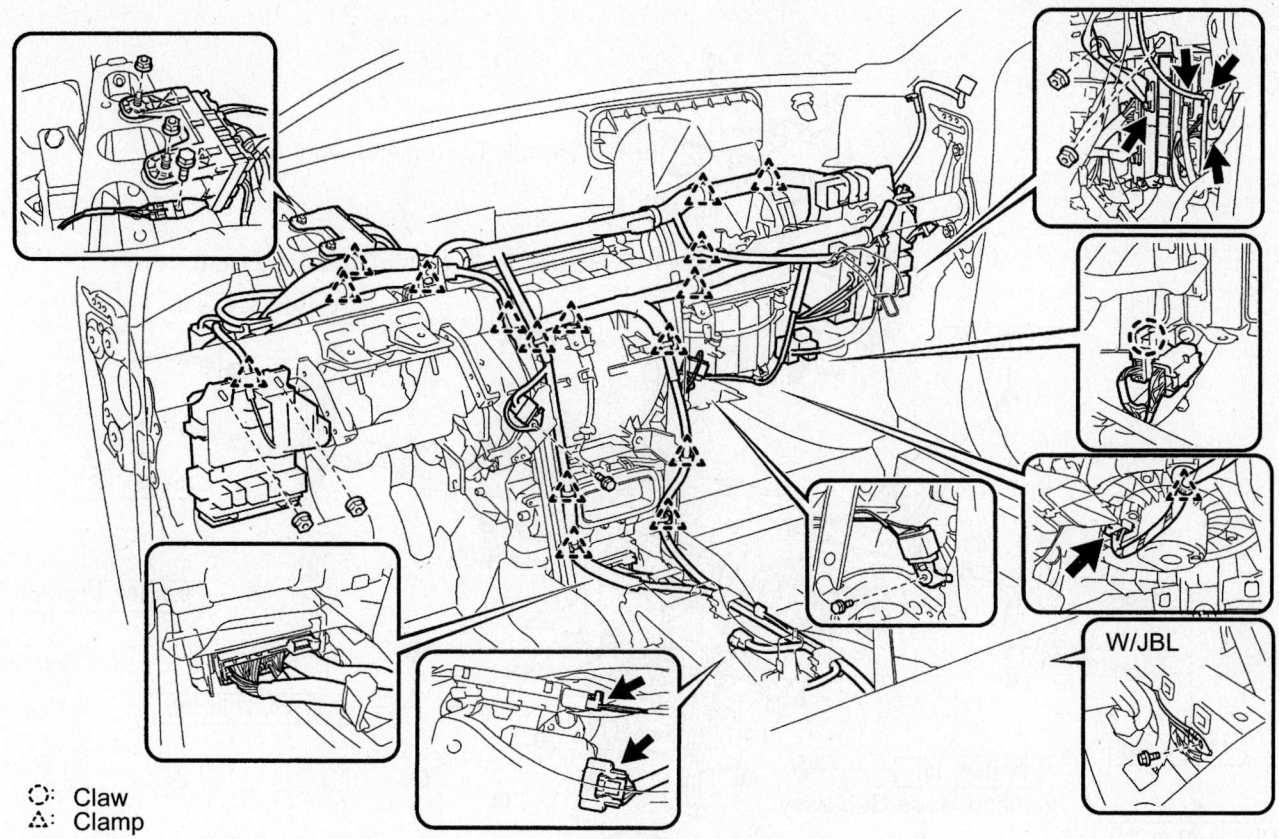

○: Claw
△: Clamp

Location of the clamps for the instrument panel reinforcement wiring harness—2005–06 Avalon

09490_AVAL_G0008

- Instrument panel center panel
- Radio
- Instrument cluster cover
- Gauge cluster assembly
- Accessory cluster assembly
- Instrument panel registers
- Speaker covers
- Front speakers
- Side cowl trim
- Front door scuff plates
- Instrument panel under cover
- Lower finishing panels
- Lower instrument panel airbag assembly
- Shift lever knob
- Shift indicator housing
- Seat heater switch, if equipped
- Center console upper panel
- Center console
- Lower center instrument panel
- Instrument side panels
- A-pillar trim
- Passenger airbag connector
- Instrument panel safety pad

6. Remove or disconnect the following:
- No. 2 heater duct
- No. 6 heater duct
- Console box duct
- Floor carpet brackets

- Rear air ducts
- No. 1 air duct
- Steering intermediate shaft
- Steering column assembly
- Instrument panel reinforcement
- Blower assembly

7. Remove or disconnect the following from the blower assembly:
- Air outlet control servomotor
- Air mix control servomotor
- Heater pipe clamps
- Heater core

8. Installation is the reverse order of removal.

9. Refill the cooling system to the correct level.

10. Start the engine and check for leaks.

Camry

➡**Removal of the heater core requires removal of the entire heater air conditioning assembly.**

1. Before servicing the vehicle, refer to the Precautions Section.

2. Drain the cooling system into a clean container for reuse.

3. Disconnect the negative battery cable. Wait 90 seconds before doing any

further work while the airbag system de-energizes.

4. Discharge and recover the air conditioning system refrigerant.

5. Disconnect A/C suction hose (No. 1) and Liquid pipe (A)
   a. Install SST to piping clamp
   b. Push down SST and release the clamp lock

✳ WARNING

Be careful not to deform the tube, when pushing the SST

➡**Cap the open fittings immediately to prevent system contamination**

6. Remove or disconnect the following:
- Heater core hoses

7. Disassemble the dash components as follows
- Lower steering column cover
- Steering wheel
- Instrument cluster finish panel sub-assembly
- Steering column cover
- Headlamp dimmer switch assembly
- Windshield wiper switch assembly
- Combination meter assembly
- Door scuff plates

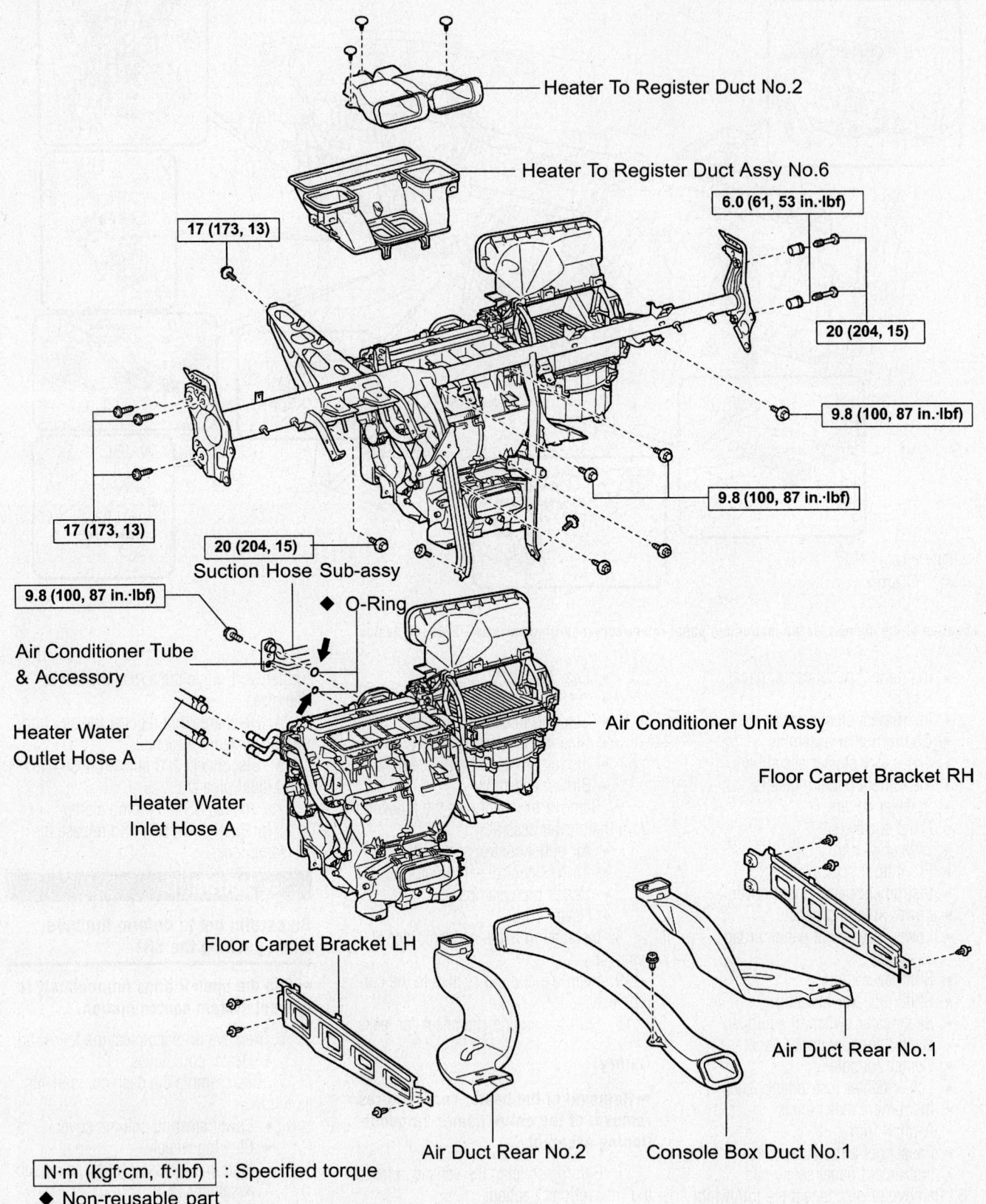

Heater To Register Duct No.2

Heater To Register Duct Assy No.6

6.0 (61, 53 in.·lbf)

17 (173, 13)

20 (204, 15)

9.8 (100, 87 in.·lbf)

9.8 (100, 87 in.·lbf)

17 (173, 13)

20 (204, 15)

Suction Hose Sub-assy

9.8 (100, 87 in.·lbf)

◆ O-Ring

Air Conditioner Tube & Accessory

Heater Water Outlet Hose A

Heater Water Inlet Hose A

Air Conditioner Unit Assy

Floor Carpet Bracket RH

Floor Carpet Bracket LH

Air Duct Rear No.1

Air Duct Rear No.2

Console Box Duct No.1

N·m (kgf·cm, ft·lbf) : Specified torque

◆ Non-reusable part

◄ Compressor Oil ND-OIL 8 or equivalent

09490_AVAL_G0009

Location of the mounting points for the instrument panel reinforcement and exploded view of the ventilation system—2005–06 Avalon

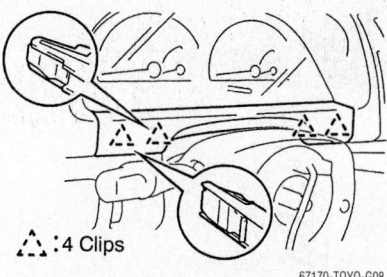

⚠ :4 Clips

67170-TOYO-G08

Instrument cluster finish panel clip locations—Camry

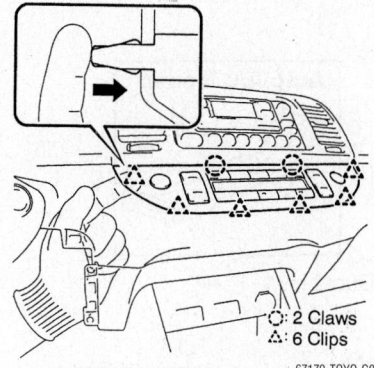

○: 2 Claws
⚠: 6 Clips

67170-TOYO-G09

Air conditioning control panel clips—Camry

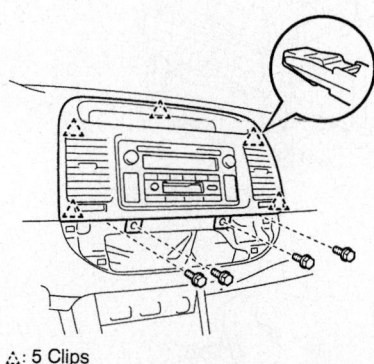

⚠: 5 Clips

67170-TOYO-G10

Radio panel clips—Camry

- LH instrument panel under cover sub-assembly
- Cowl side trim
- Coin box
- LH upper instrument panel sub assembly
- LH lower instrument panel brace
- Air conditioning control assembly
- Instrument panel center cluster finish panel
- Glove box door
- RH lower instrument panel sub-assembly
- Shift lever knob on M/T
- Remove the console panel
- Rear console box

- Cup holder and ashtray
- Upper console panel
- Front console box
- Instrument panel finish plate
- LH & RH front pillar garnish
- Instrument panel speakers
- Passenger air bag connector
- Instrument panel safety pad cap
- Instrument panel safety pad sub-assembly

❋❋ WARNING

Follow air bag removal procedures

8. Remove or disconnect the following:
- Rear air ducts
- Console box duct
- Floor shift parking lock cable assembly
- Windshield wiper relay assembly
- Instrument panel brace assemblies
- Lower instrument finish panel retainer
- Heater to foot ducts
- Steering column assembly
- Instrument panel reinforcements
- Heater blower assembly
- Lower defroster nozzles
- Air conditioning radiator assembly
- Mode damper servo sub-assembly
- Airmix damper servo sub-assembly
- Heater radiator (core) sub-assembly

To install:
Install or connect the following:
9. Heater core unit sub assembly
- Heater core into A/C assembly
- Screw and clamp
- Air conditioning radiator assembly
- Lower defroster nozzles
- Heater blower assembly
- Instrument panel reinforcements
- Steering column assembly

- Heater to foot ducts
- Lower instrument finish panel retainer
- Instrument panel brace assemblies
- Windshield wiper relay assembly and
- Floor shift parking lock cable assembly
- Console box duct
- Rear air ducts

10. Reassemble the dash components in the reverse of removal.
11. Install or connect the following:
- Heater core hoses
- A/C suction and pressure hoses, attach with bolt and plate

➡**Lubricate O-rings with compressor oil**

- Negative battery cable
12. Refill the cooling system to the correct level.
13. Evacuate and recharge A/C system
14. Start the engine and check for leaks.

Solara

1. Disconnect the negative battery cable.
2. Drain the cooling system into a clean container for reuse.
3. Disconnect the heater hoses from the heater core.
4. At the driver's side, remove the lower instrument panel.
5. Remove the left hand instrument lower panel.
6. At the heater/air conditioning housing, disengage the 3 heater protector-to-heater/air conditioning housing clips and remove the heater protector.
7. Remove the 3 heater core pipe clamp screws and the clamps.
8. Remove the 2 heater core pipe clamp

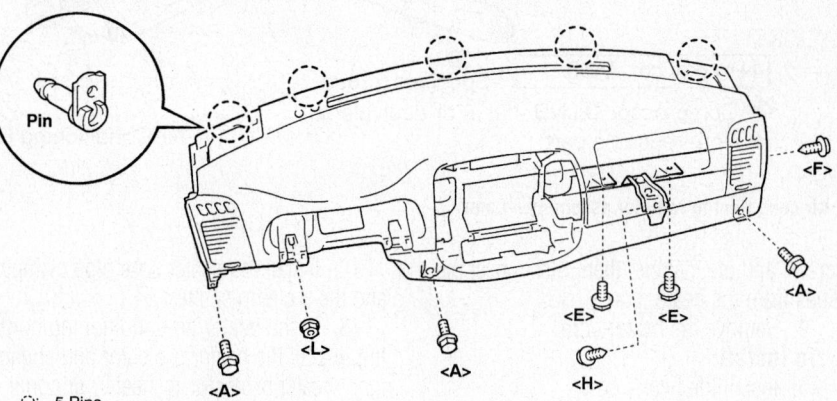

Pin

○: 5 Pins

<F>
<A>
<E>
<E>
<H>
<A>
<L>
<A>

67170-TOYO-G11

Instrument panel safety pad fastener locations—Camry

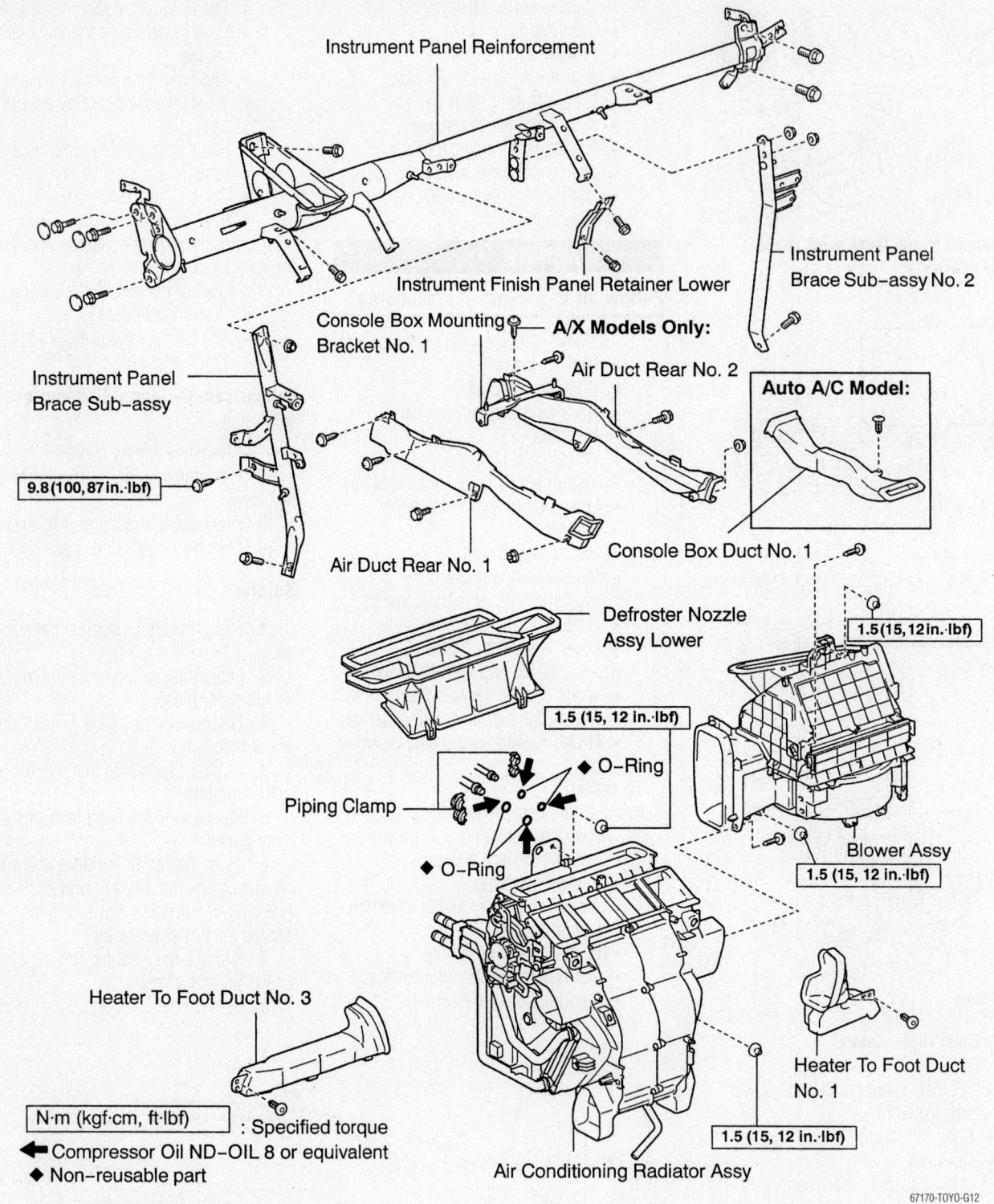

Instrument Panel Reinforcement

Instrument Finish Panel Retainer Lower

Instrument Panel Brace Sub-assy No. 2

Console Box Mounting Bracket No. 1

A/X Models Only:

Air Duct Rear No. 2

Auto A/C Model:

Instrument Panel Brace Sub-assy

9.8 (100, 87 in.·lbf)

Air Duct Rear No. 1

Console Box Duct No. 1

Defroster Nozzle Assy Lower

1.5 (15, 12 in.·lbf)

1.5 (15, 12 in.·lbf)

◆ O-Ring

Piping Clamp

◆ O-Ring

Blower Assy

1.5 (15, 12 in.·lbf)

Heater To Foot Duct No. 3

Heater To Foot Duct No. 1

N·m (kgf·cm, ft·lbf) : Specified torque

◀ Compressor Oil ND-OIL 8 or equivalent

◆ Non-reusable part

1.5 (15, 12 in.·lbf)

Air Conditioning Radiator Assy

67170-TOYO-G12

Air conditioning radiator assembly—Camry

screws and the clamps; then, disconnect the pipes from the heater core.

9. Remove the heater core.

To install:

10. Install the heater core.

11. Connect the pipes to the heater core and install the heater core pipe clamp and the 2 clamp screws.

12. Install the heater core pipe clamps and the 3 clamp screws.

13. At the heater/air conditioning housing, install the heater protector and engage the 3 heater protector-to-heater/air conditioning housing clips.

14. Install the left hand instrument lower panel.

15. At the driver's side, install the lower instrument panel.

16. Connect the negative battery cable.

17. Connect the heater hoses to the heater core.

18. Refill the cooling system to the correct level.

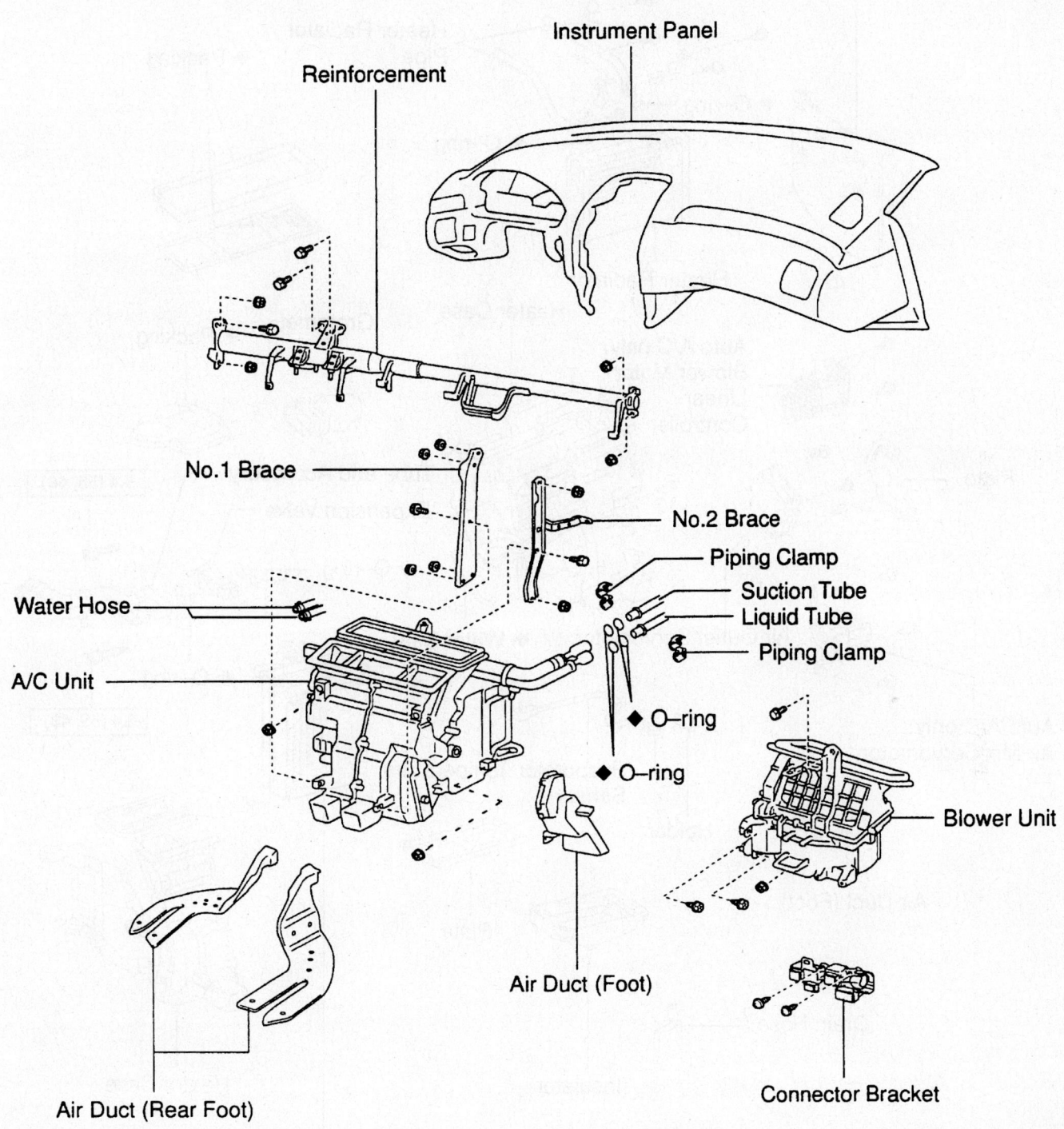

Instrument Panel

Reinforcement

No.1 Brace

No.2 Brace

Piping Clamp

Suction Tube

Liquid Tube

Piping Clamp

Water Hose

A/C Unit

◆ O–ring

◆ O–ring

Blower Unit

Air Duct (Foot)

Air Duct (Rear Foot)

Connector Bracket

◆ Non–reusable part

93112GK5

Exploded view of the instrument panel, heater/air conditioning housing and related components—Solara

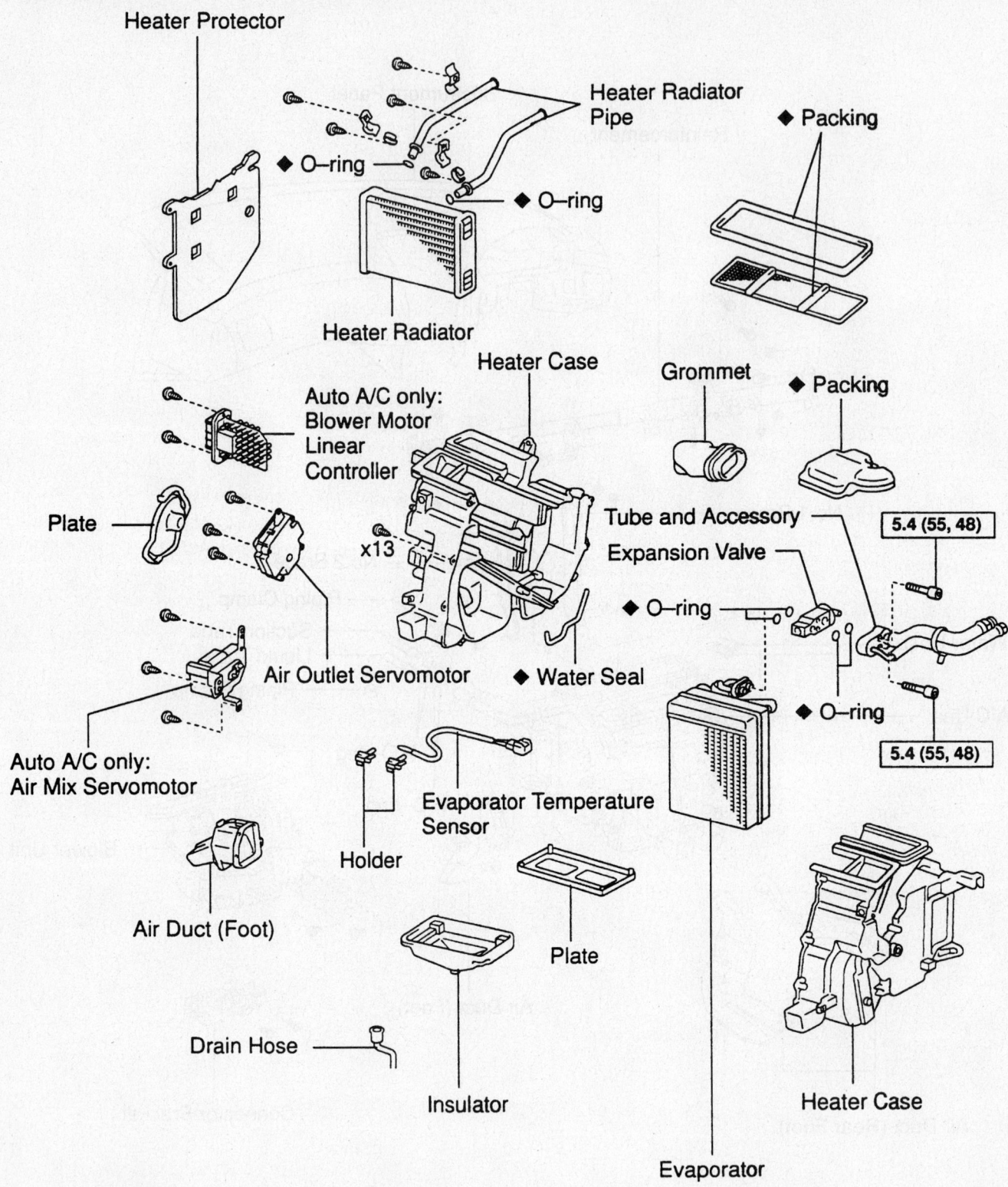

Heater Protector

Heater Radiator Pipe

◆ Packing

◆ O-ring

◆ O-ring

Heater Radiator

Heater Case

Grommet

◆ Packing

Auto A/C only:
Blower Motor
Linear
Controller

Plate

x13

Tube and Accessory

Expansion Valve

5.4 (55, 48)

◆ O-ring

Air Outlet Servomotor

◆ Water Seal

◆ O-ring

Auto A/C only:
Air Mix Servomotor

5.4 (55, 48)

Evaporator Temperature
Sensor

Air Duct (Foot)

Holder

Plate

Drain Hose

Insulator

Evaporator

Heater Case

N·m (kgf·cm, in.·lbf) : Specified torque

◆ Non-reusable part

Exploded view of the heater/air conditioning housing, heater core, evaporator core and related components—Solara

93112GK6

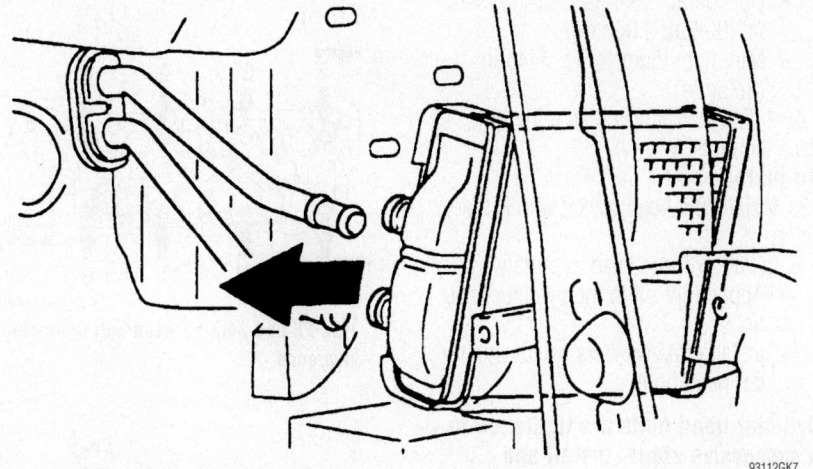

View of the heater core—Solara

93112GK7

Cylinder Head

REMOVAL & INSTALLATION

2AZ-FE Engines

1. Before servicing the vehicle, refer to the Precautions Section.
2. Drain the engine oil.
3. Drain the cooling system.
4. Remove or disconnect the following:
   - Negative battery cable.
   - Radiator hose outlet
   - Union to connector tube hose
   - Heater inlet hose
   - Fuel tube assembly
   - Intake manifold
   - Intake manifold runner valve assembly
   - Engine harness
   - Intake and exhaust manifold insulators
   - Exhaust manifold assembly
   - Timing chain
   - Camshafts
   - Camshaft bearing No. 2
   - Camshaft oil control valve assembly

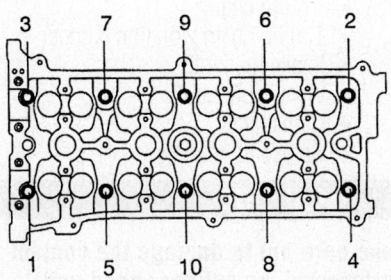

Cylinder head bolt removal sequence—2AZ-FE engine

67170-TOYO-G13

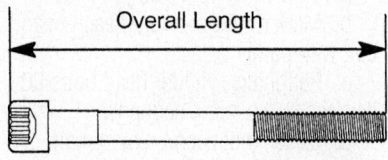

Overall Length

67170-TOYO-G14

Measuring bolt length—2AZ-FE engine

- 10 cylinder head bolts uniformly in the sequence
- Cylinder head and gasket

✽✽ WARNING

Head warpage or cracking could result from removing the bolts in an incorrect order.

5. Inspect the cylinder head set bolts. They should be 6.350–6.465 in. (161.3–164.2mm) in length. If length is greater than maximum, replace the bolts.

To install:
6. Install new head gasket
7. Install cylinder head assembly
   - Apply light oil to the cylinder head bolts
   - Install plate washers on the cylinder head bolts

➡**Cylinder head bolts are tightened in two successive steps. Install and tighten 10 cylinder head bolts in required sequence.**

   a. Tighten to 58 ft. lbs. (79 Nm)
   b. Mark the front side of each head bolt with paint.
   c. Retighten cylinder head bolts 90 degrees in the same sequence
   d. Check that each painted mark is now at a 90 degree angle to the front
8. Install or connect the following:

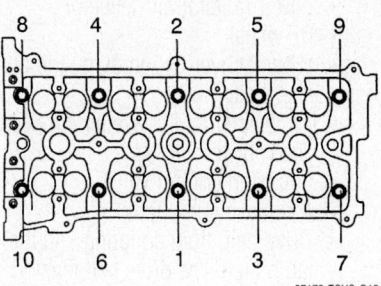

Cylinder head bolt installation sequence—2AZ-FE engine

67170-TOYO-G15

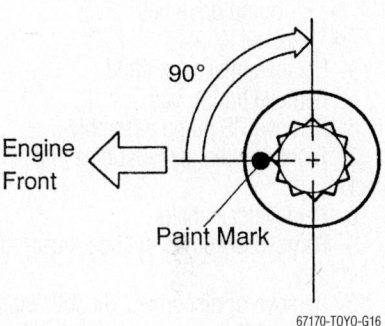

Head bolt marking procedure—2AZ-FE engine

67170-TOYO-G16

- Camshaft oil control valve assembly
- Camshaft bearing No. 2
- Camshafts
- Timing chain
- Exhaust manifold assembly
- Intake and exhaust manifold insulators
- Engine harness
- Intake manifold runner valve assembly
- Intake manifold
- Fuel tube assembly
- Heater inlet hose
- Union to connector tube hose
- Radiator hose outlet
9. Refill engine with oil to the correct level.
10. Refill the cooling system to the correct level.
11. Start the engine and check for leaks.

1MZ-FE and 3MZ-FE Engines

1. Before servicing the vehicle, refer to the Precautions Section.
2. Relieve the fuel system pressure.
3. Remove or disconnect the following:
   - Negative battery cable.
   - Oil pan protector
   - Engine undercover
   - Coolant
   - Battery clamp cover
   - Air cleaner inlet

- Lower radiator air deflector
- RF wheel
- V bank cover by removing the bolt and 2 cap nuts
- Air cleaner and intake air connector assembly
- Emission control valve
- Air intake surge tank
- Drive belt, fluid coupling and the fan pulley. The drive belt tension may be slackened by turning the tensioner counterclockwise. The pulley bolt for the drive belt tensioner has a left-handed thread.
- PS pump drive belt
- Radiator

4. Remove intake manifold
5. Remove timing belt
6. Remove PS pump assembly
7. Front and rear exhaust pipe and brackets
8. Remove camshafts
9. Remove LH or RH cylinder head assemblies
10. Remove or disconnect the following:
- The VVT Sensor connector
- Camshaft timing oil valve connector
- Engine wire harness clamp
- Remove the hex bolt
- 8 cylinder head bolts uniformly in the sequence

✸✸ WARNING

Head warpage or cracking could result from removing the bolts in an incorrect order.

11. Inspect the cylinder head set bolts. Ensure they match the following:

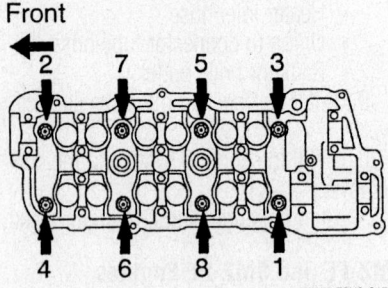

1MZ-FE and 3MZ-FE head bolt loosening sequence

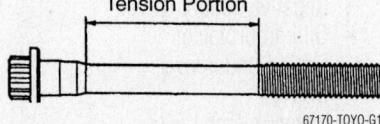

Check the diameter of the bolt in this area

- Outside diameter is .3524 to .3563 in. (8.95 to 9.05mm)
- Minimum diameter is .3445 in. (8.75mm)

12. If diameter is lees than minimum, replace the bolts.

To install:

13. Install new head gasket with R mark upward
14. Install cylinder head assembly
- Apply light oil to the cylinder head bolts
- Install plate washers on the cylinder head bolts

➥**Cylinder head bolts are tightened in two successive steps. Install and tighten 8 cylinder head bolts in required sequence.**

a. Tighten to 40 ft. lbs (54 Nm)
b. Mark the front side of each head bolt with paint.
c. Retighten cylinder head bolts 90 degrees in the same sequence
d. Check that each painted mark is now at a 90 degree angle to the front

15. Install or connect the following:
- Tighten the hex bolt to 14 ft. lbs. (19 Nm)
- Wiring harness clamp
- Camshaft timing oil valve connector
- Camshaft assemblies
- Valve cover assemblies
- Exhaust manifold assemblies and support brackets. Torque to 36 ft. lbs. (49 Nm)
- Exhaust manifold heat insulators
- PS pump assembly
- Timing belt inner cover
- Camshaft timing pulleys
- Timing belt and idler assemblies
- RH engine mounting bracket
- Timing belt covers
- Alternator bracket
- Engine mounting stay No 2
- Engine stabilizer rod
- PS drive belt
- A/C compressor drive belt
- Water outlet
- Intake manifold assembly
- Intake air surge tank
- Emission control valve set
- Air cleaner assembly
- Vacuum hoses
- V-bank cover sub-assembly
- Front suspension upper brace
- RF wheel

16. Refill the engine with oil to the correct level.
17. Refill the cooling system to the correct level.
18. Start the engine and check for leaks.

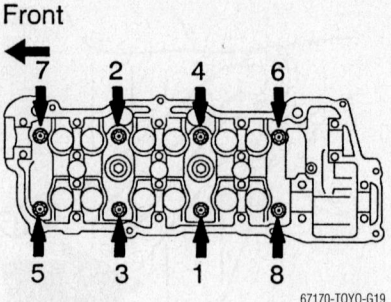

1MZ-FE and 3MZ-FE head bolt tightening sequence

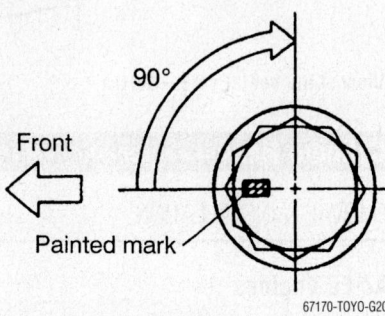

Mark the bolts as shown

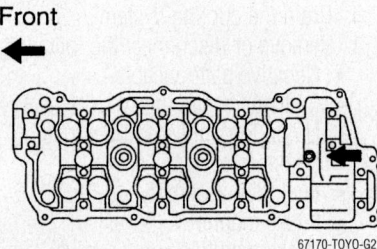

Hex head bolt location

2GR-FE Engine

1. Before servicing the vehicle, refer to the Precautions Section.
2. Drain the cooling system.
3. Drain the engine oil.
4. Relieve the fuel system pressure.
5. Remove the engine/transaxle assembly from the vehicle.
6. Remove or disconnect the following:
- Timing chain
- Timing chain vibration damper
- Intake manifold
- Water outlet pipe
- Camshafts
- Camshaft housing

✸✸ CAUTION

Take care not to damage the contact surfaces of the cylinder head and camshaft housing. Use protective tape over the pry tool.

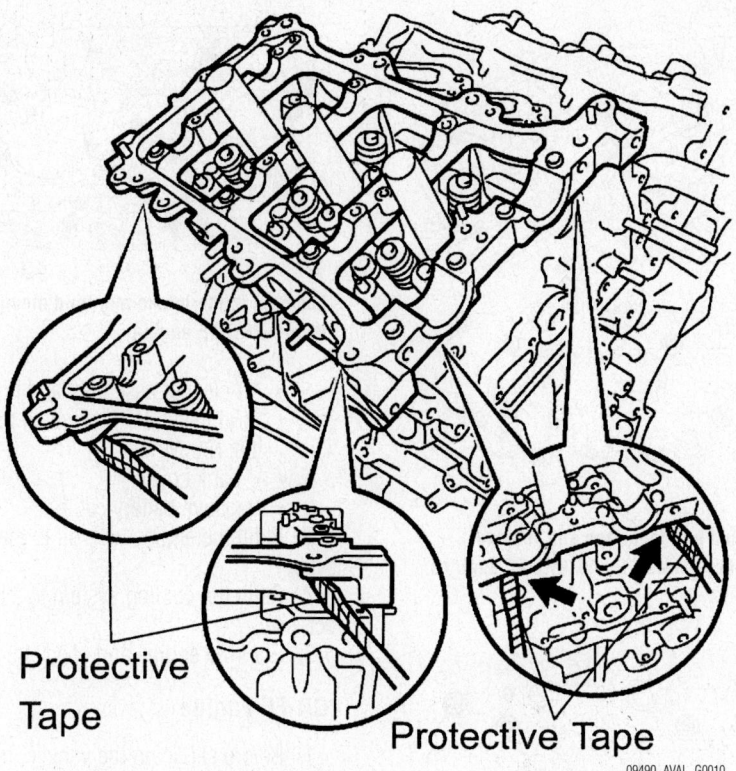

Protective Tape

Protective Tape

.09490_AVAL_G0010

Location of the prying points when removing the camshaft housing—2GR-FE engine

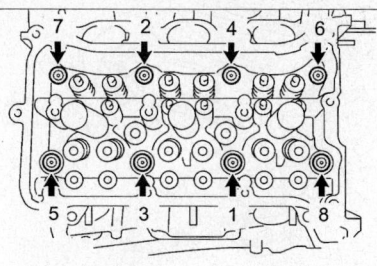

09490_AVAL_G0013

Cylinder head mounting bolt torque sequence—2GR-FE engine

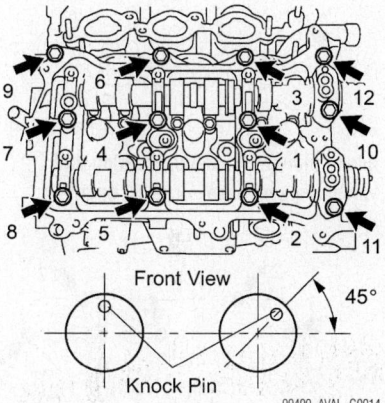

09490_AVAL_G0014

Camshaft housing torque sequence—2GR-FE engine

- Intake manifold. Tighten bolts to 15 ft. lbs. (21 Nm).
- Timing chain vibration damper
- Timing chain

15. Install the engine/transaxle assembly to the vehicle.

16. Refill the cooling system to the correct level.

17. Refill the engine with oil to the correct level.

18. Start then engine and check for leaks.

Intake Manifold

REMOVAL & INSTALLATION

2AZ-FE Engines

1. Before servicing the vehicle, refer to the Precautions Section.

2. Relieve the fuel pressure from the fuel lines.

3. Drain the engine oil and cooling system.

4. Remove or disconnect the following:
- Negative battery cable.
- Strut tower brace
- Radiator hose outlet
- Union to connector tube hose

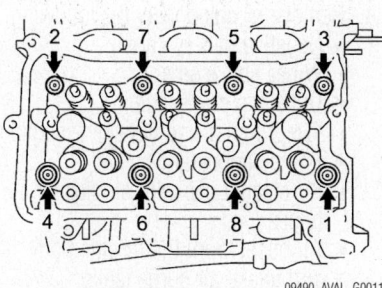

09490_AVAL_G0011

Cylinder head bolt loosening sequence—2GR-FE engine

- Rocker arms
- Valve lash adjusters
- Valve stem cap

7. Loosen the cylinder head mounting bolts in several steps in the sequence shown.

8. Remove the head bolts and plate washers.

9. Remove the cylinder head and gasket.

To install:

10. Install a new cylinder head gasket with the Lot number stamp upper side facing upward.

11. Install the cylinder head. Apply a light coat of engine oil to the threads and tighten the bolts in sequence as follows:
- a. Step 1: 27 ft. lbs. (36 Nm)
- b. Step 2: Plus 90 degrees

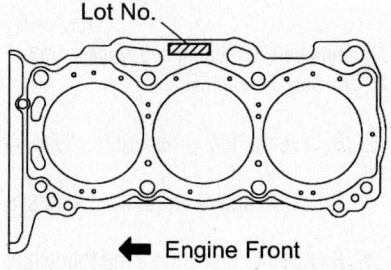

Lot No.

← Engine Front

09490_AVAL_G0012

Correct cylinder head gasket orientation—2GR-FE engine

c. Step 3: Plus an additional 90 degrees

12. Install or connect the following:
- Valve stem cap
- Valve lash adjusters
- Rocker arms

13. Install the camshaft housing as follows:

a. Apply 0.138–0.177 in. (3.5–4.5 mm) wide bead of sealant to the contact surface.

b. Install the camshaft housing and tighten the bolts in sequence to 18 ft. lbs. (25 Nm).

14. Install or connect the following:
- Camshafts
- Water outlet pipe. Tighten bolts to 7.4 ft. lbs. (10 Nm).

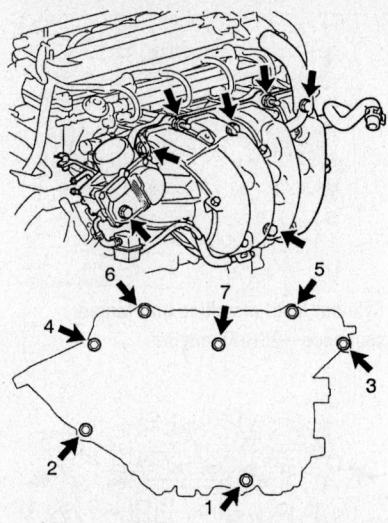

Intake manifold fastener location and loosening sequence—2AZ-FE engine

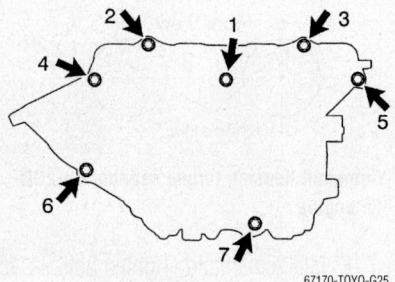

Intake manifold fastener tightening sequence—2AZ-FE engine

- Water inlet water hose
- Fuel tube assembly
- Water bypass hoses from the throttle body
- Intake manifold and gasket

To install:
- Intake manifold and gasket. Tighten to 22 ft. lbs. (30 Nm)
- Water bypass hoses from the throttle body
- Fuel tube assembly
- Union to connector tube hose
- Radiator hose outlet
- Strut tower brace. Tighten to 59 ft. lbs. (80 Nm)
- Negative battery cable

5. Refill the engine with oil to the correct level.

6. Refill the cooling system to the correct level.

7. Start the engine and check for leaks.

1MZ-FE and 3MZ-FE Engines

1. Before servicing the vehicle, refer to the Precautions Section.

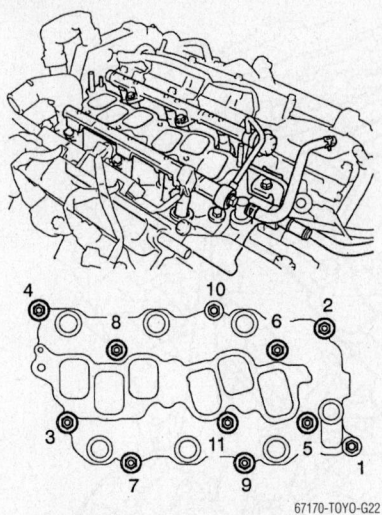

Manifold bolt locations and removal sequence—1MZ-FE and 3MZ-FE engines

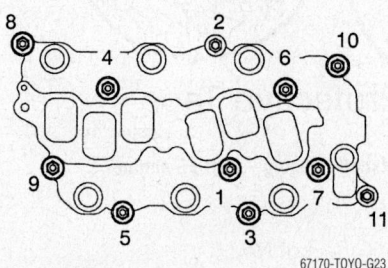

Manifold bolt tightening sequence—1MZ-FE and 3MZ-FE engines

2. Relieve the fuel pressure from the fuel lines.

3. Drain the engine oil and cooling system.

4. Remove or disconnect the following:
- Negative battery cable.
- V-bank cover
- Strut tower brace
- Air cleaner assembly and hose
- Fuel pipe assembly
- Heater inlet hose
- Manifold ground cable
- Injector plugs
- Manifold bolts and nuts in the correct sequence
- Intake manifold and gasket

To install:

5. Install or connect the following:
- Intake manifold and gasket. Tighten the bolts and nuts in the correct sequence to 11 ft. lbs. (15 Nm)
- Tighten the water outlet fasteners to 11 ft. lbs. (15 Nm)
- Injector plugs
- Manifold ground cable
- Heater inlet hose
- Fuel pipe assembly

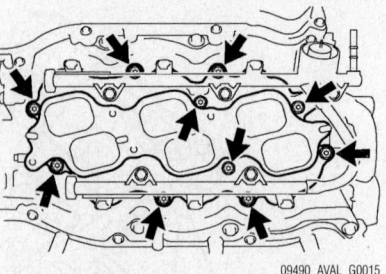

Location of the intake manifold mounting bolts—2GR-FE engine

- Air cleaner assembly and hose
- Strut tower brace. Tighten to 59 ft. lbs. (80 Nm)
- V-bank cover
- Negative battery cable

6. Refill the engine with oil to the correct level.

7. Refill the cooling system to the correct level.

8. Start the engine and check for leaks.

2GR-FE Engine

1. Before servicing the vehicle, refer to the Precautions Section.

2. Relieve the fuel system pressure.

3. Drain the cooling system.

4. Remove or disconnect the following:
- Negative battery cable
- Wiper link assembly
- Outer top cowl panel
- Wiper motor and linkage assembly
- Engine appearance cover
- Air cleaner cap with air cleaner hoses
- All hoses from the throttle body and intake air surge tanks
- Intake air surge tank
- Fuel hose from the fuel rail
- Intake manifold with gaskets.

5. Installation is the reverse order of removal. Note the following:
- Install the intake manifold with new gaskets.
- Tighten the intake manifold mounting bolts to 15 ft. lbs. (21 Nm).

Exhaust Manifold

REMOVAL & INSTALLATION

2AZ-FE Engines

1. Before servicing the vehicle, refer to the Precautions Section.

2. Remove or disconnect the following:
- Negative battery terminal
- Exhaust manifold heat insulator
- Oxygen sensor

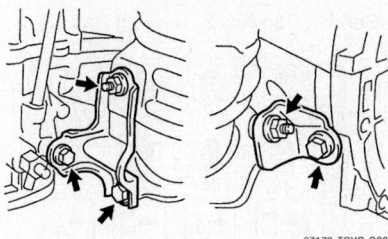

Exhaust manifold converter stays—2AZ-FE engine

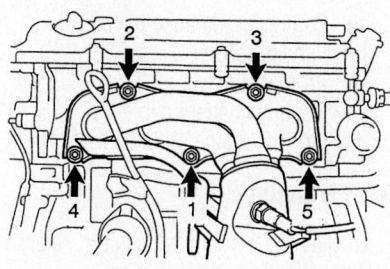

Exhaust manifold fastener tightening sequence—2AZ-FE engine

- Exhaust manifold converter stays
- Exhaust pipe-to-manifold fasteners
- Manifold nuts, manifold and gasket

To install:

3. Place a new gasket and reinstall the exhaust manifold with the nuts. Tighten to 27 ft. lbs. (33 Nm).

4. Install or connect the following:
- Exhaust manifold converter stays. Tighten to 32 ft. lbs. (44 Nm)
- Oxygen sensor
- Exhaust manifold heat insulator
- Negative battery cable

1MZ-FE Engines

1. Before servicing the vehicle, refer to the Precautions Section.

2. Remove or disconnect the following:
- Negative battery cable
- Engine undercover
- Front exhaust pipe from the exhaust manifold
- EGR pipe from the exhaust manifold
- Heated oxygen sensor connector from the right exhaust manifold
- Exhaust manifold stay
- Exhaust manifold and gasket

To install:

3. Install or connect the following:
- Exhaust manifold using new gasket. Bolts, uniformly tightened: 36 ft. lbs. (49 Nm).
- Exhaust manifold stay. Bolt and nut: 15 ft. lbs. (20 Nm).

- Heated oxygen sensor connector to the right exhaust manifold
- EGR pipe, using new gaskets, to the exhaust manifold and the engine. Nuts: 108 inch lbs. (12 Nm).
- Front exhaust pipe, using new gasket, to the exhaust manifold. Nuts: 46 ft. lbs. (62 Nm).
- Engine undercover
- Negative battery cable

3MZ-FE Engines

1. Before servicing the vehicle, refer to the Precautions Section.

2. Remove or disconnect the following:
- Negative battery cable
- Engine undercover
- Front exhaust pipes from the exhaust manifold
- Heated oxygen sensor connector from the exhaust manifold
- Heat shield insulator
- For the front manifold, the exhaust manifold stay
- Exhaust manifold fasteners in the proper sequence
- Exhaust manifold and gasket

To install:

3. Install or connect the following:
- Exhaust manifold using new gasket. Bolts, uniformly tightened in

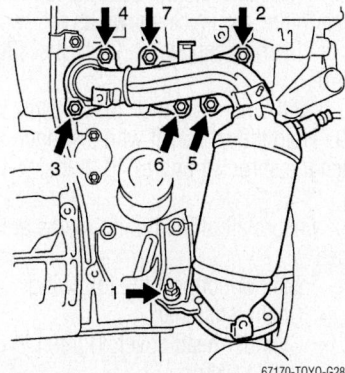

Front exhaust manifold fastener loosening sequence—3MZ-FE engine

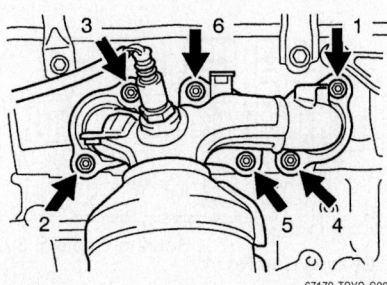

Rear exhaust manifold fastener loosening sequence—3MZ-FE engine

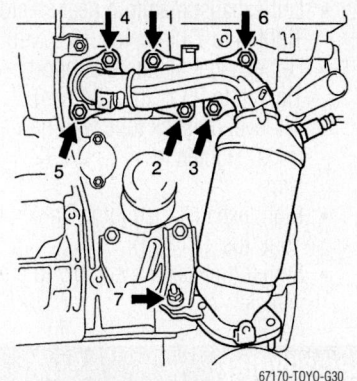

Front exhaust manifold fastener tightening sequence—3MZ-FE engine

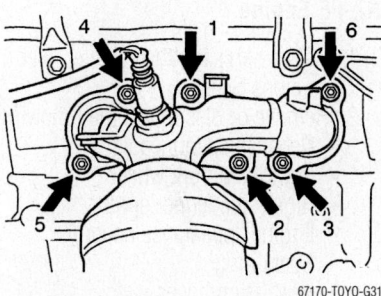

Rear exhaust manifold fastener tightening sequence—3MZ-FE engine

sequence: 36 ft. lbs. (49 Nm). Retighten bolts 1 and 2 to specification again.
- Exhaust manifold stay. Bolt and nut: 25 ft. lbs. (34 Nm).
- Heat shield insulator
- Heated oxygen sensor connector
- Front exhaust pipe, using new gasket, to the exhaust manifold. Nuts: 46 ft. lbs. (62 Nm).
- Engine undercover
- Negative battery cable

2GR-FE Engine

1. Before servicing the vehicle, refer to the Precautions Section.

2. Remove or disconnect the following:
- Engine/transaxle assembly from the vehicle
- Oxygen sensor from the right exhaust manifold
- Right exhaust manifold
- Oil level dipstick tube
- Left exhaust manifold support
- Left exhaust manifold heat shield
- Left exhaust manifold

To install:

3. Install or connect the following:
- Left exhaust manifold with a new gasket. Tighten the bolts to 15 ft. lbs. (21 Nm).

- Left exhaust manifold heat shield. Tighten to 75 inch lbs. (8.5 Nm).
- Left exhaust manifold support. Tighten to 25 ft. lbs. (34 Nm).
- Oil level dipstick tube with new O-rings. Tighten to 15 ft. lbs. (21 Nm).
- Right exhaust manifold. Tighten to 15 ft. lbs. (21 Nm).
- Engine/transaxle assembly to the vehicle.

Camshaft(s)

REMOVAL & INSTALLATION

2AZ-FE Engine

1. Before servicing the vehicle, refer to the Precautions Section.
2. Remove or disconnect the following:
   - Negative battery cable
   - Right-side front wheel
   - Right-side fender apron
   - Engine appearance cover
   - Spark plug
   - Ventilation hoses
   - Engine wire
   - Cylinder head cover
3. Turn the crankshaft clockwise and set the No. 1 cylinder at Top Dead Center (TDC) compression. Ensure the timing marks on the camshaft sprockets line up with the timing marks on the No. 1 bearing caps.
4. Place a paint mark on the timing chain to indicate TDC.
5. Remove the chain tensioner.

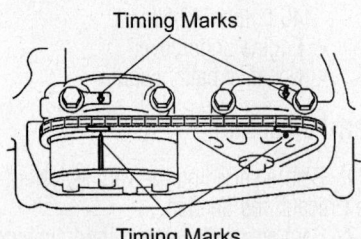

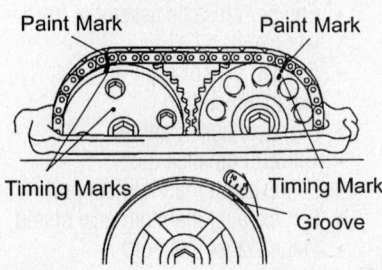

Place a paint mark on the timing chain, aligned with the timing marks already on the camshafts.—2AZ-FE engine

6. Hold the exhaust camshaft with a wrench, then loosen the sprocket bolt.
7. Uniformly remove the exhaust camshaft bearing cap bolts.
8. Remove the bearing caps.
9. Remove the exhaust camshaft sprocket from the timing chain and remove the camshaft.
10. Uniformly loosen and remove the intake camshaft bearing cap bolts.
11. Remove the bearing caps.
12. Remove the timing chain from the timing gear of the camshaft. Tie the timing chain with string so it does not fall into the front cover.
13. Remove the intake camshaft.

To install:

14. Install the timing chain on the intake camshaft timing gear, with the painted mark of the link aligned with the timing marks of the camshaft timing sprocket.
15. Install the intake bearing caps and tighten in the sequence shown as follows:
    a. Bearing cap No. 1: 22 ft. lbs. (30 Nm).
    b. All others: 80 inch lbs. (9 Nm).
16. Put the camshaft on the cylinder head with the painted mark of the link of chain aligned with the timing mark of the camshaft timing sprocket.
17. Raising the camshaft, temporarily tighten the sprocket bolt.
18. Install the exhaust bearing caps and tighten in the sequence shown as follows:
    a. Bearing cap No. 2: 22 ft. lbs. (30 Nm).
    b. All others: 80 inch lbs. (9 Nm).
19. Hold the camshaft with a wrench and tighten the sprocket bolt to 40 ft. lbs. (54 Nm).
20. Ensure all of the timing marks are aligned.
21. Install or connect the following:
    - Chain tensioner
    - Cylinder head cover. Tighten to 8 ft. lbs. (11 Nm).
    - Engine wire
    - Spark plug

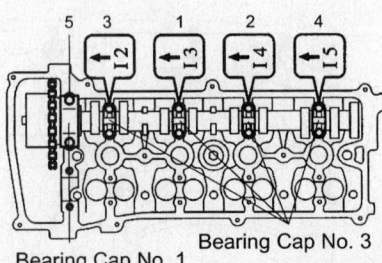

Intake camshaft bearing cap torque sequence—2AZ-FE engine

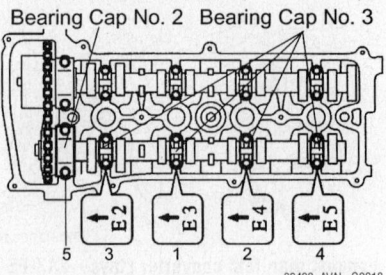

Exhaust camshaft bearing cap torque sequence—2AZ-FE engine

- Engine appearance cover
- Fender apron
- Front wheel
- Negative battery cable
22. Start the engine and check for leaks.

1MZ-FE Engines

1. Before servicing the vehicle, refer to the Precautions Section.
2. Remove or disconnect the following:
   - Timing belt and idler pulley
   - Camshaft timing pulleys
   - Cylinder head covers

➡ **The thrust clearance on both the intake and exhaust camshafts is very small; the camshafts must be kept level during removal. If the camshafts are removed without being kept level, the camshaft may be caught in the cylinder head, causing the head to break or the camshaft to seize.**

3. Remove the exhaust and intake camshafts from the right side cylinder head, as follows:
    a. Turn the camshaft with a wrench until the 2 pointed marks on the drive and driven gears are aligned. (The right camshaft gears have 2 marks apiece; the left side camshaft gears have 1 mark each.)
    b. Secure the exhaust camshaft sub-gear to the main gear using a service bolt. A bolt 0.63–0.79 in. (16–20mm) long with a 6mm thread diameter and a 1mm pitch is recommended. When removing the exhaust camshaft be sure the sub-gear is not loaded; all the force must be eliminated.
    c. Uniformly loosen and remove the exhaust camshaft bearing cap bolts in several passes and in the proper sequence. Remove the 8 bearing cap bolts and remove the caps, keeping them in the correct order.
    d. Remove the exhaust camshaft from the engine.
    e. Uniformly loosen and remove the

10 bearing cap bolts in several passes, in the proper sequence. Remove the bearing caps, keeping them in order, remove the oil seal, then, lift out the intake camshaft.

4. Remove the exhaust and intake camshafts from the left side cylinder head, as follows:

a. Turn the camshaft with a wrench until the pointed marks on the drive and driven gears are aligned. (The right camshaft gears have 2 marks apiece; the left side camshaft gears have 1 mark each.)

b. Secure the exhaust camshaft sub-gear to the main gear using a service bolt. A bolt 0.63–0.79 in. (16–20mm) long with a 6mm thread diameter and a 1mm pitch is recommended. When removing the exhaust camshaft be sure the sub-gear is not loaded; all the force must be eliminated.

c. Uniformly loosen and remove the exhaust camshaft bearing cap bolts in several passes and in the proper sequence. Remove the 8 bearing cap bolts and remove the caps. Keep the caps in the correct order.

d. Remove the exhaust camshaft from the engine.

e. Uniformly loosen and remove the 10 bearing cap bolts in several passes, in the reverse order of the installation sequence. Remove the bearing caps, keeping them in order, remove the oil seal, and then lift out the intake camshaft.

5. Remove the valve lifter shims and hydraulic lifters. If the lifters are to be reused, store them upside down in a sealed container.

To install:

6. Install the valve lifters into their original positions and shims. Check the valve clearance and replace the shims as necessary.

➡**Before installing the camshafts in either cylinder head, apply multi-purpose grease to each camshaft.**

7. Install the right camshafts, as follows:

a. Position the intake camshaft on the head so that the alignment marks are at a 90 degrees angle from vertical. The mark should be at the "3 o'clock" position.

b. Apply sealant to the No. 1 bearing cap.

c. Apply a light coat of clean engine oil to the bolt threads and under the bolt head. Install the bearing caps to their proper position. Tighten the bolts evenly and in several passes to 12 ft. lbs. (16 Nm) in the proper sequence.

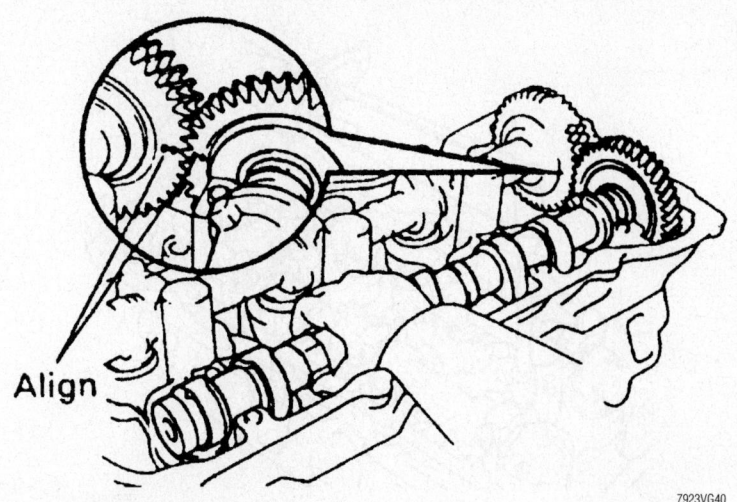

Aligning the camshaft gear timing marks for the right camshafts—3.0L (1MZ-FE) engine

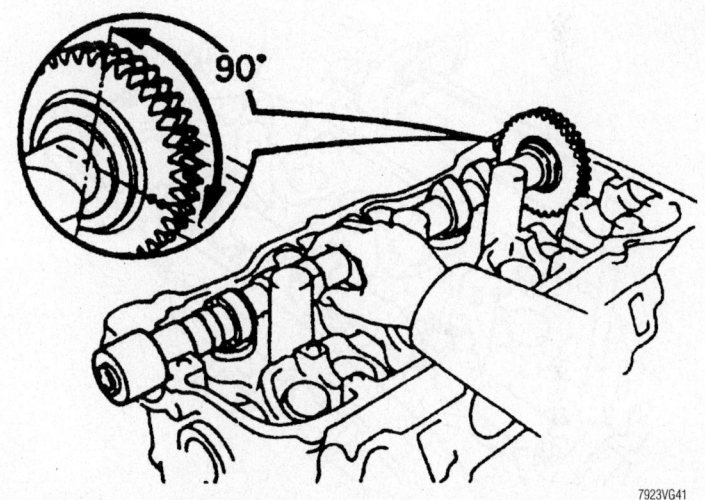

Camshaft installation for the right exhaust camshaft—3.0L (1MZ-FE) engine

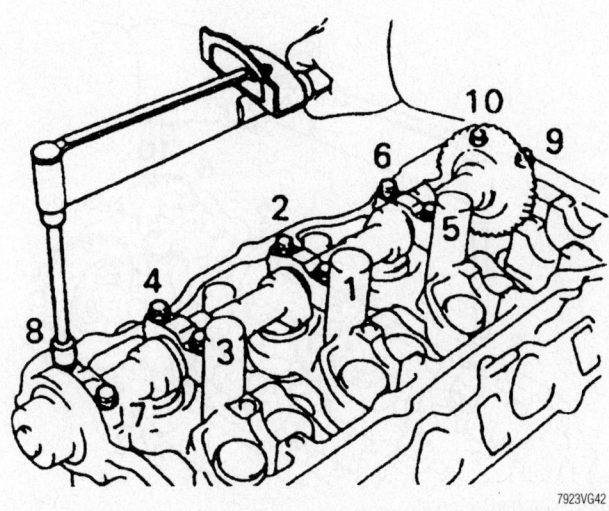

Bearing cap bolt tightening sequence for the right exhaust camshaft—3.0L (1MZ-FE) engine

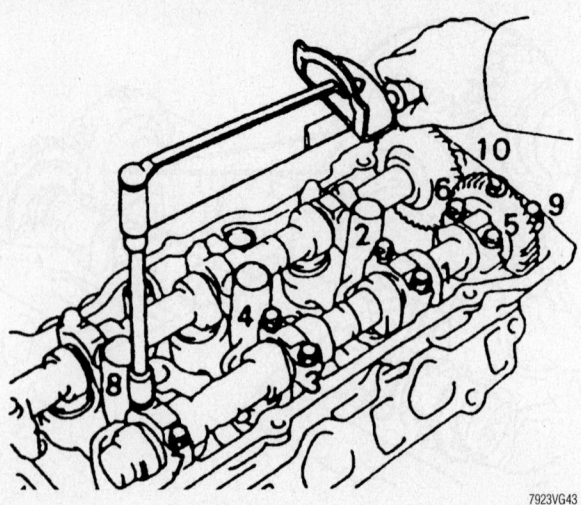

Bearing cap bolt tightening sequence for the right intake camshaft—3.0L (1MZ-FE) engine

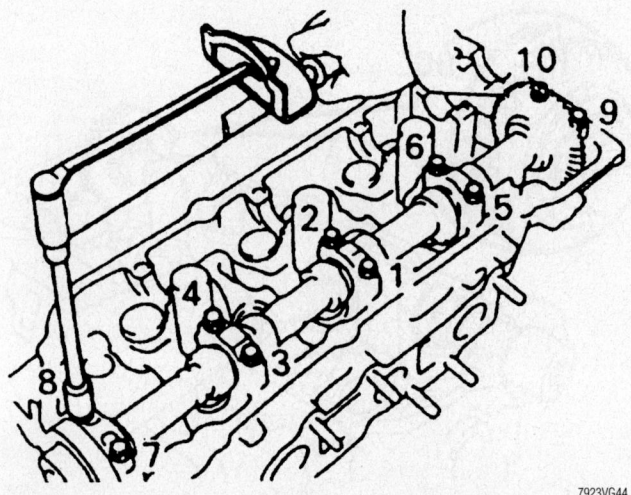

Bearing cap bolt tightening sequence for the left exhaust camshaft—3.0L (1MZ-FE) engine

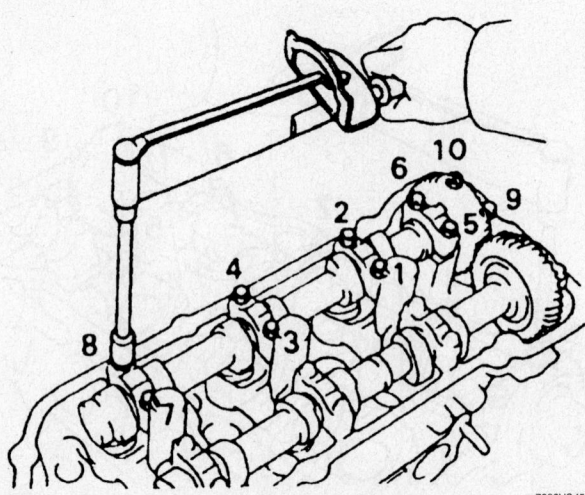

Bearing cap bolt tightening sequence for the left intake camshaft—3.0L (1MZ-FE) engine

d. Position the exhaust camshaft on the head so that the alignment marks are at a 90 degrees angle from vertical. The mark must align with the marks on the other gear.

e. Apply a light coat of clean engine oil to the bolt threads and under the bolt head. Install the bearing caps to their proper position. Tighten the bolts evenly and in several passes to 12 ft. lbs. (16 Nm) in the proper sequence.

f. Remove the service bolt.

8. Install the left camshaft, as follows:

a. Position the intake camshaft on the head so that the alignment mark is at a 90degrees angle from vertical. The mark should be at the "9 o'clock" position.

b. Apply sealant to the No. 1 bearing cap.

c. Apply a light coat of clean engine oil to the bolt threads and under the bolt head. Install the bearing caps to their proper position. Tighten the bolts evenly and in several passes to 12 ft. lbs. (16 Nm) in the proper sequence.

d. Position the exhaust camshaft on the head so that the alignment marks are at a 90 degree angle from vertical. The mark should be at the "3 o'clock" position and must align with the marks on the other gear.

e. Apply a light coat of clean engine oil to the bolt threads and under the bolt head. Install the bearing caps to their proper position. Tighten the bolts evenly and in several passes to 12 ft. lbs. (16 Nm) in the proper sequence.

f. Remove the service bolt.

9. Apply multi-purpose grease to new camshaft oil seals. Install the seals.

10. Install or connect the following:
- No. 3 (rear) timing belt cover
- Camshaft timing gears
- Idler pulley, timing belt and covers
- Cylinder head (valve) covers

3MZ-FE Engines

1. Before servicing the vehicle, refer to the Precautions Section.

2. Drain the cooling system.

3. Remove or disconnect the following:
- Negative battery cable
- RH front wheel
- Front suspension upper brace center
- V-bank cover sub-assembly
- Air cleaner assembly
- Emission control valve set
- Intake air surge tank
- Ignition coil assembly
- Cylinder head valve covers

- Front fender apron seal
- A/C Compressor and PS drive belts
- Engine stabilizing rod
- Engine mounting stay
- Alternator bracket
- Crankshaft pulley
- Timing belt covers
- Timing belt, tensioners, idlers and guides
- Camshaft timing pulleys

➡**Align the camshaft pulleys so that they can be returned to the original locations when reassembling.**

- Inner timing belt cover

❋❋ WARNING

Since the thrust clearance of the camshaft is small, the camshaft must be kept level while it is being removed. If the camshaft is not kept level, the portion of the cylinder head receiving the shaft thrust may crack or be damaged, causing the camshaft to seize or break.

4. Remove the camshaft using the following procedures
5. RH bank camshaft No.1 & LH bank camshaft No.2
   a. Align the (2 dot marks) of the camshaft drive and driven gear by turning the camshaft with a wrench.
   b. Secure the exhaust camshaft sub gear to the main gear with service bolt. Torque to 48 inch lbs. (5.4 Nm).

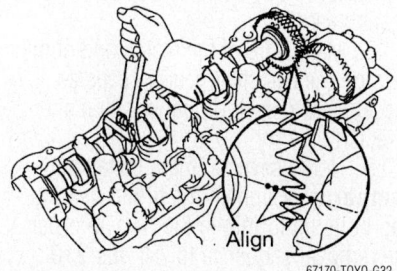

Aligning timing marks—3MZ-FE engine

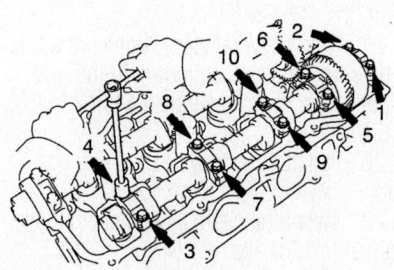

Camshaft bolt removal sequence Camshaft No.1—3MZ-FE engine

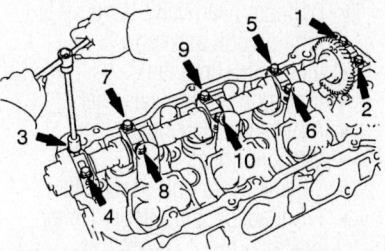

Camshaft bolt removal sequence Camshaft No.2 —3MZ-FE engine

➡**When removing the camshaft, make certain that the torsional spring force of the sub gear has been eliminated by installation of the service bolt.**

   c. Using several steps, loosen the 10 bearing cap bolts uniformly in the sequence shown in the illustration. Remove the 5 bearing caps and camshaft

❋❋ WARNING

Do not pry out camshaft

❋❋ WARNING

Do not damage contact surface of the cylinder head that receives the shaft thrust.

6. LH bank camshaft No. 3 & No. 4
   a. Using several steps, loosen the 10 bearing cap bolts uniformly in the

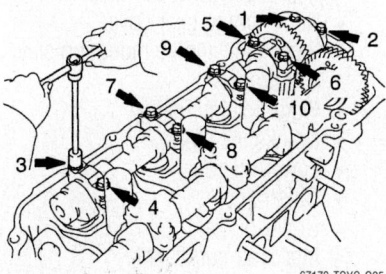

Camshaft bolt removal sequence camshaft No. 3—3MZ-FE engine

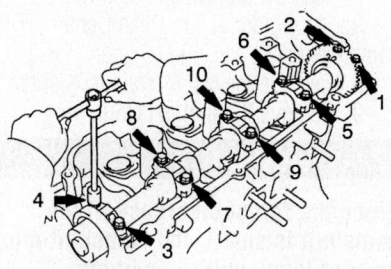

Camshaft bolt removal sequence camshaft No. 4—3MZ-FE engine

sequence shown in the illustration. Remove the 5 bearing caps and camshaft

❋❋ WARNING

Do not pry out camshaft

❋❋ WARNING

Do not damage contact surface of the cylinder head that receives the shaft thrust.

   b. Remove the oil seal from camshaft
To install:
   7. Install RH No 2 Camshaft then RH Camshaft No. 1 using same procedure

❋❋ WARNING

Since the clearance of the camshaft is small, the camshaft must be kept level while bring installed. If the camshaft is not kept level, the cylinder head or camshaft may be damaged.

   a. Apply engine oil to the thrust portion and journal of camshaft
   b. No. 2 camshaft at 90 degree angel to the timing mark (2 dot marks) on the head.
   c. Multi-purpose grease to new oil seal
   d. Oil seal to camshaft

➡**Do NOT turn over the oil seal lip**

➡**Insert oil seal until it stops**

   e. Seal packing to bearing cap No. 1

➡**Install bearing cap No. 1 within 5 minutes after applying seal packing**

➡**Do NOT expose seal packing to engine oil within 2 hours after installation**

   f. 5 bearing caps in their proper locations
   g. Apply light coat of oil to the threads of the bearing caps

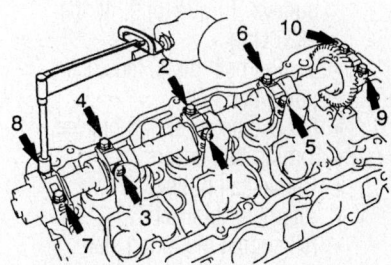

Bolt torque procedure RH camshaft No. 2—3MZ-FE engine

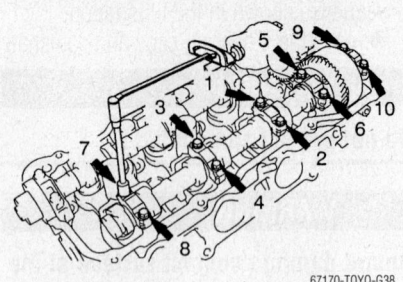

Bolt torque procedure RH camshaft No. 1—3MZ-FE engine

67170-TOYO-G38

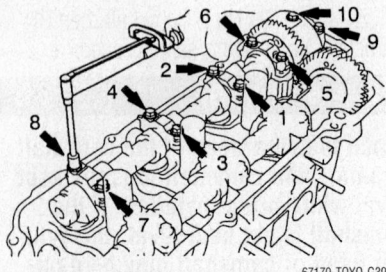

Bolt torque procedure LH camshaft No. 3—3MZ-FE engine

67170-TOYO-G39

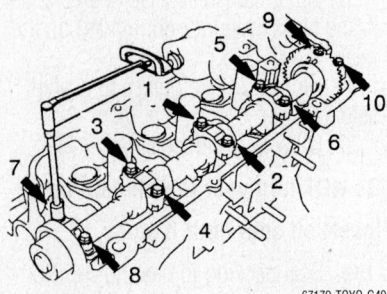

Bolt torque procedure LH camshaft No. 4—3MZ-FE engine

67170-TOYO-G40

    h. Tighten the 10 bearing cap bolts in required sequence. Torque to: 12 ft. lbs. (16 Nm)

8. Install LH camshaft No.3 and LH camshaft No.4 using the above procedure

9. Install or connect the following:
- Inner timing belt cover
- RH then LH camshaft timing pulleys. Torque to: 92 ft. lbs. (125 Nm)
- Timing belt idlers, tensioners, guide and belt
- RH engine mounting bracket
- Timing belt covers
- Crankshaft pulley
- Alternator bracket
- RH engine mounting stay
- Engine stabilizing rod
- Inspect or adjust valve lash
- A/C compressor and PS drive belts

- Cylinder head valve covers
- Ignition coil assembly
- Intake air surge tank
- Emission control valve set
- Air cleaner assembly
- Vacuum hoses
- V-bank cover sub-assembly
- Front suspension upper center brace
- RF wheel
- Negative battery cable

10. Refill the cooling system to the correct level.

11. Start the engine and check for leaks.

2GR-FE Engine

1. Before servicing the vehicle, refer to the Precautions Section.

2. Drain the cooling system.

3. Drain the engine oil.

4. Remove or disconnect the following:
- Engine/transaxle assembly
- Engine wire
- Front frame assembly for the engine
- Starter
- Transaxle assembly from the engine
- Oil dipstick tube
- Exhaust manifolds
- Drive plate

5. Secure the engine in a suitable engine stand.

6. Remove or disconnect the following:
- Idler pulley
- Right-side engine mounting bracket
- Accessory drive belt tensioner
- No. 2 timing gear cover
- Right-side engine mounting stay and bracket
- Water inlet housing
- Crankshaft pulley
- Lower oil pan
- Oil strainer
- Upper oil pan
- Intake air surge tank
- Ignition coils
- Oil pipes
- Cylinder head cover
- Front cover and seal

7. Turn the crankshaft clockwise and set the No. 1 cylinder at Top Dead Center (TDC) compression.

8. Remove the No. 1 chain tensioner.

9. Remove the timing chain.

✳✳ WARNING

Since the thrust clearance of the camshaft is small, the camshaft must be kept level while it is being removed. If the camshaft is not kept level, the portion of the cylinder head receiving the shaft thrust may crack or be damaged, causing the camshaft to seize or break.

10. While raising the chain tensioner No.2, insert a pin into the hole to fix it.

11. Holding the hexagonal portion of the exhaust camshaft with a wrench, remove the timing gear set bolt.

12. Separate the camshaft timing gears from the chain.

13. Remove the chain tensioner No. 2 and mounting bolt.

14. Remove the bearing cap gaskets.

15. Ensure the knock pin of the camshafts are positioned as shown.

16. Uniformly loosen and remove the first 8 bearing cap bolts in the sequence shown

➡**Matchmark the bearing caps for installation purposes.**

17. Using several steps, loosen and remove the remaining 12 bearing cap bolts and remove the bearing caps.

18. Remove the camshafts.

To install:

19. Apply clean engine oil to the cam lobes and cylinder head journals.

20. Install the camshafts to the camshaft housing.

➡**Before and after setting the camshaft and No.2 camshaft, check that the rocker arm is firmly set to the lash adjuster.**

21. Apply clean engine oil to the camshaft journal, camshaft housing and bearing cap.

22. Make sure of the marks and numbers on the camshaft bearing caps an place them in each proper position and direction.

23. Using several steps, install and **temporarily** tighten the first 8 bearing cap bolts uniformly in the reverse order of the removal sequence to 7 ft. lbs. (10 Nm).

24. Tighten the remaining 12 bearing cap bolts in the reverse order of removal to 18 ft. lbs. (25 Nm).

25. Using several steps, tighten the first 8 bearing cap bolts uniformly in the reverse order of the removal sequence to 12 ft. lbs. (16 Nm).

26. Install three new bearing cap gaskets.

27. Install the chain tensioner No.2 and tighten to 15 ft. lbs. (21 Nm).

28. While compressing the tensioner, insert a pin into the hole to fix it.

29. Align the mark links (yellow) with the

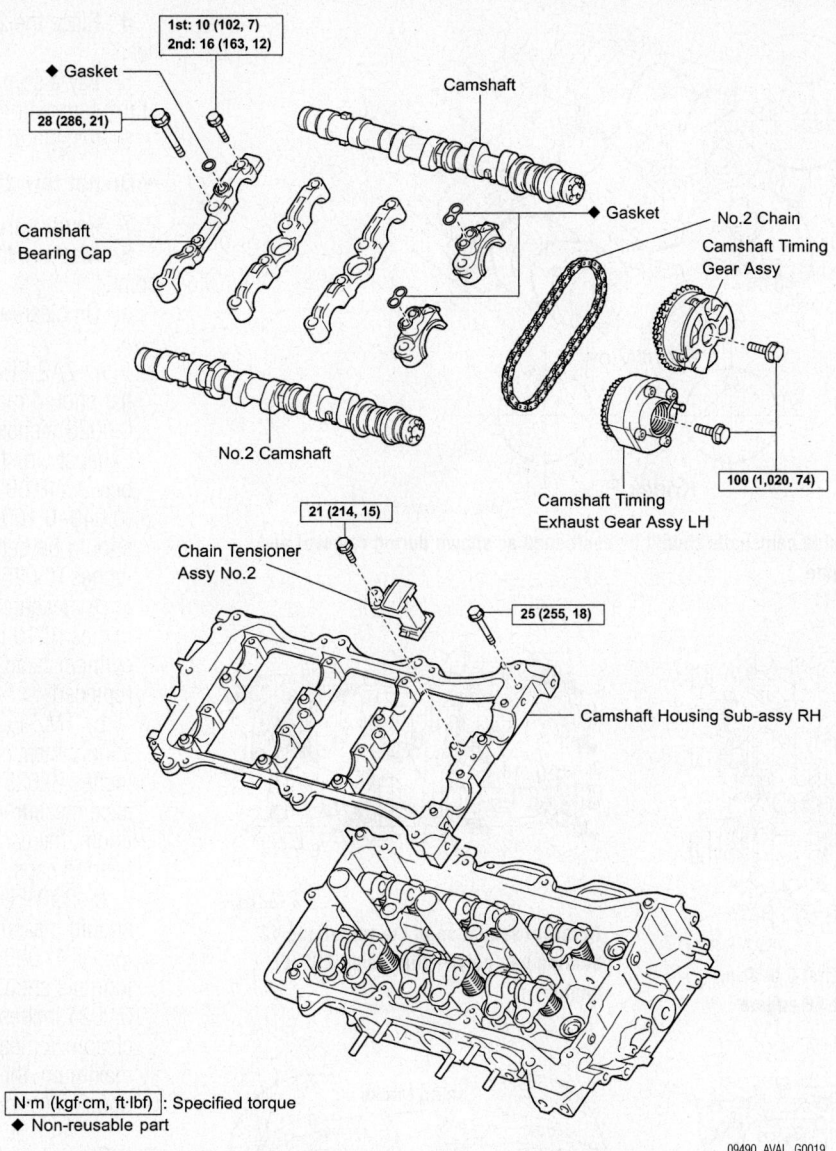

1st: 10 (102, 7)
2nd: 16 (163, 12)

◆ Gasket

Camshaft

28 (286, 21)

◆ Gasket

No.2 Chain

Camshaft
Bearing Cap

Camshaft Timing
Gear Assy

No.2 Camshaft

Camshaft Timing
Exhaust Gear Assy LH

100 (1,020, 74)

21 (214, 15)

Chain Tensioner
Assy No.2

25 (255, 18)

Camshaft Housing Sub-assy RH

N·m (kgf·cm, ft·lbf) : Specified torque
◆ Non-reusable part

09490_AVAL_G0019

Exploded view of the camshaft components—2GR-FE engine

timing marks of the camshaft timing gears as shown in the illustration.

30. Align the knock pin of the camshaft with the pin hole of the camshaft timing gear. Install the camshaft timing gear and camshaft timing exhaust gear RH with the No.2 chain installed.

31. Hold the camshaft with a wrench and tighten the sprocket bolts to 74 ft. lbs. (100 Nm).

32. Remove the pin from the tensioner.

33. The remainder of the installation is the reverse order of removal.

34. Refill the engine with oil to the correct level.

35. Refill the cooling system to the correct level.

36. Start the engine and check for leaks.

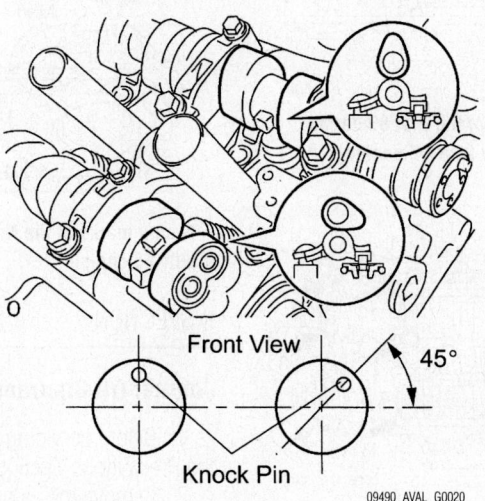

Front View

45°

Knock Pin

09490_AVAL_G0020

The knock pins of the right side camshafts should be positioned as shown during removal and installation—2GR-FE engine

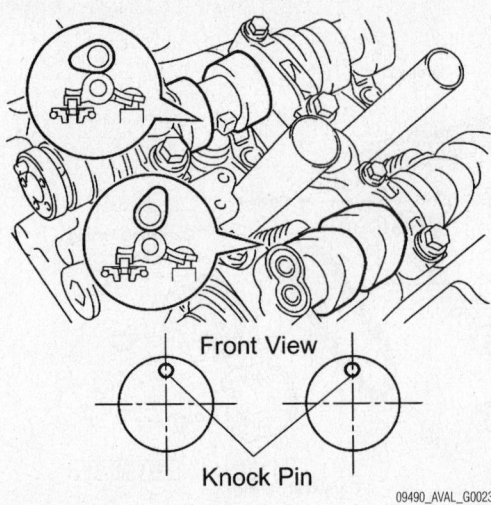

The knock pins of the left side camshafts should be positioned as shown during removal and installations—2GR-FE engine

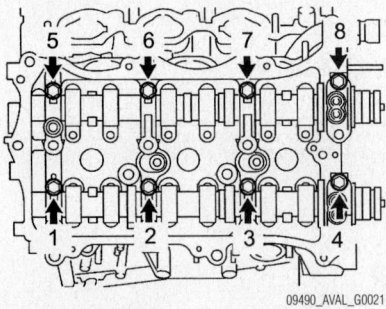

Removal sequence of the first 8 bearing cap bolts—Right side 2GR-FE engine

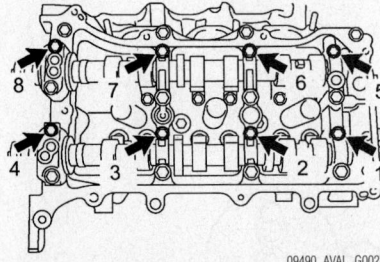

Removal sequence of the first 8 bearing cap bolts—Left side 2GR-FE engine

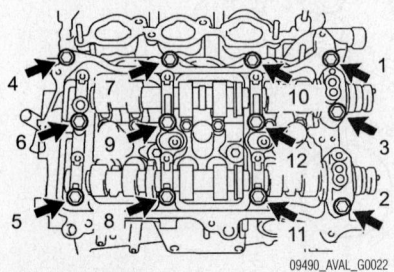

Removal sequence of the remaining 12 bearing cap bolts—Right side 2GR-FE engine

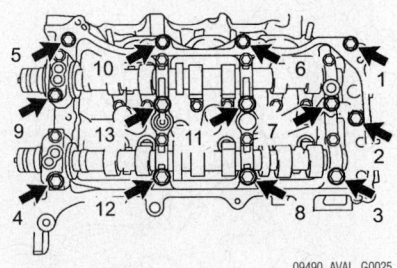

Removal sequence of the remaining 12 bearing cap bolts—Right side 2GR-FE engine

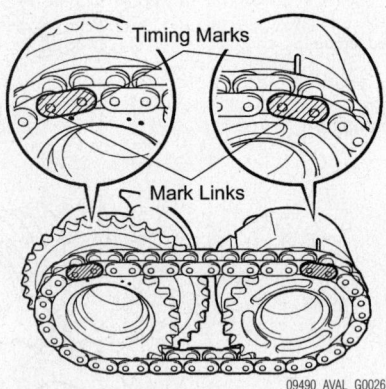

Timing marks of the No. 2 timing chain—2GR-FE engine

INSPECTION

Journal Oil Clearance

1. Before servicing the vehicle, refer to the Precautions Section.
2. Remove the camshafts.
3. Clean the 10 bearing caps and camshaft journals.

4. Place the 2 camshafts on the cylinder head.
5. Lay a strip of Plastigage® across each of the camshaft journals.
6. Install the 10 bearing caps.

→**Do not turn the camshaft.**

7. Remove the bearing caps.
8. Measure the Plastigage® at its widest point.
9. Oil clearance should measure as follows:

   a. 2AZ-FE Engine: Intake No. 1 journal should measure between 0.0003–0.0028 inches (0.007–0.070 mm). Exhaust No. 1 journal should measure between 0.0016–0.0039 inches (0.040–0.100 mm). All other journals should be between 0.0010–0.0039 inches (0.025–0.100 mm). If any clearance measurement is more than 0.039 inches (0.10 mm), the camshaft or cylinder head (or both) needs to be replaced.

   b. 1MZ-FE/3MZ-FE Engine: clearance should measure between 0.0010–0.0039 inches (0.025–0.100 mm). If any clearance measurement is more than the maximum, the camshaft or cylinder head and bearing caps, need to be replaced.

   c. 2GR-FE Engine: No. 1 journal should measure between 0.0016–0.0031 inches (0.040–0.079 mm). All other journals should be between 0.0010–0.0024 inches (0.025–0.062 mm). If any clearance measurement is more than the maximum, the camshaft or camshaft housing (or both) needs to be replaced.

End Play

1. Before servicing the vehicle, refer to the Precautions Section.
2. This procedure is performed with the camshaft installed.
3. Install a dial indicator in the thrust direction on the front end of the camshaft. Measure the end play of the dial indicator when the camshaft is moved back and forth. The dial indicator should measure as follows:

   a. 2AZ-FE Engine: Intake camshaft between 0.0016–0.0043 inches (0.040–0.110 mm). Exhaust camshaft between 0.0032–0.0059 inches (0.080–0.150 mm).

   b. 1MZ-FE/3MZ-FE Engines: between 0.0016–0.0047 inches (0.040–0.120 mm).

   c. 2GR-FE Engine: between 0.0031–0.0051 inches (0.080–0.130 mm).

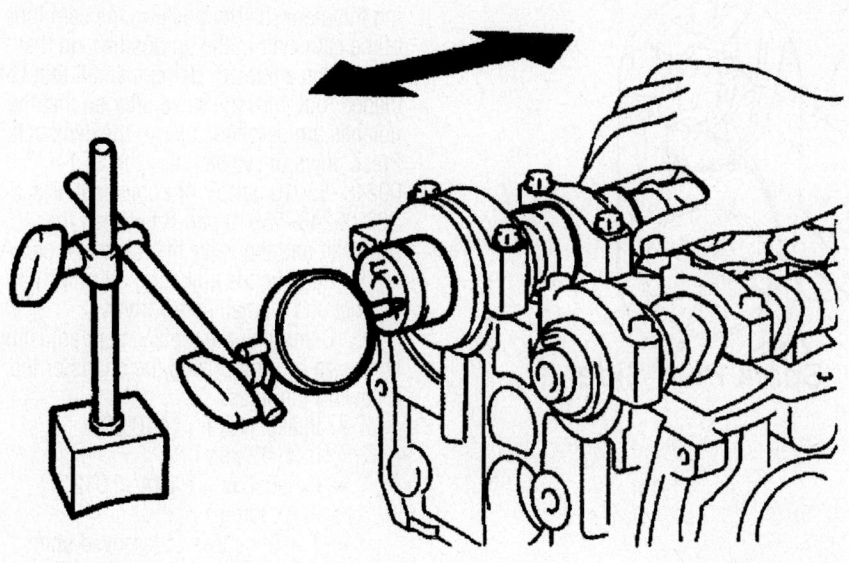

Measuring for camshaft endplay

4. Replace the cylinder head assembly if the measurement is exceeded. Replace the camshaft if damage is found on the thrust surfaces.

Valve Lash

ADJUSTMENT

1MZ-FE Engines

➡**Adjust the valve clearance when the engine is cold.**

1. Before servicing the vehicle, refer to the Precautions Section.
2. Remove or disconnect the following:
   • Negative battery cable
   • Accelerator/throttle cable from the throttle linkage
   • Air cleaner cover, air flow meter, and air duct assembly
   • V-bank cover
   • Emission control valve set
   • Air intake chamber
   • Engine harness from the injectors and the ignition coils
   • Ignition coils and keep them in order for reassembly
   • Spark plugs
   • Cylinder head covers
3. Turn the crankshaft pulley and align its groove with the timing mark **0** of the No. 1 timing cover.
4. Check that the valve lifters on the No. 1 intake are loose and the No. 1 exhaust are tight. If not, turn the crankshaft 1 complete revolution (360 degrees).

➡**All measurements should be written down. These recorded measurements will need to be used in conjunction with a mathematical formula to determine the thickness of the replacement shims.**

5. Measure the clearance between the valve lifters and the camshaft. Record the measurements on valves No. 1 and 6 intake; No. 2 and 3 exhaust.
   a. The intake valve clearance cold is 0.006–0.010 in. (0.15–0.25mm).
   b. The exhaust valve clearance cold is 0.010–0.014 in. (0.25–0.35mm).
6. Turn the crankshaft ⅔ of a revolution (240 degrees). Record the measurements on valves No. 2 and 3 intake; No. 4 and 5 exhaust.
7. Turn the crankshaft another ⅔ of a revolution. Record the measurements on valves No. 4 and 5 intake; No. 1 and 6 exhaust.

Adjusting Shim Selection Chart

New shim thickness   mm (in.)

Shim No.	Thickness	Shim No.	Thickness
1	2.500 (0.0984)	10	2.950 (0.1161)
2	2.550 (0.1004)	11	3.000 (0.1181)
3	2.600 (0.1024)	12	3.050 (0.1201)
4	2.650 (0.1043)	13	3.100 (0.1220)
5	2.700 (0.1063)	14	3.150 (0.1240)
6	2.750 (0.1083)	15	3.200 (0.1260)
7	2.800 (0.1102)	16	3.250 (0.1280)
8	2.850 (0.1122)	17	3.300 (0.1299)
9	2.900 (0.1142)		

HINT: New shims have the thickness in millimeters imprinted on the face.

Adjusting shim chart (intake and exhaust)—1MZ-FE engines

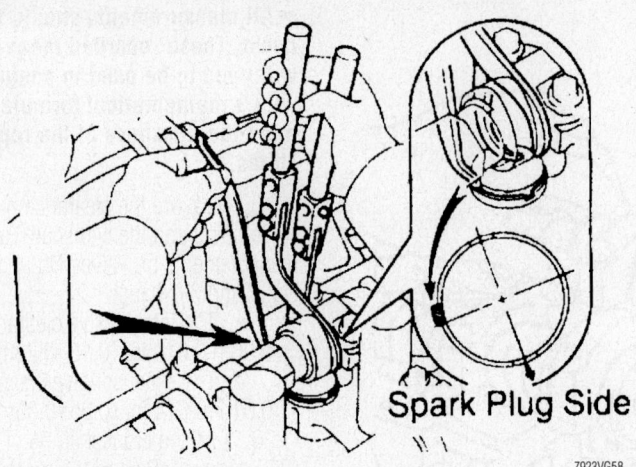

Spark Plug Side

Common method of removing valve shims

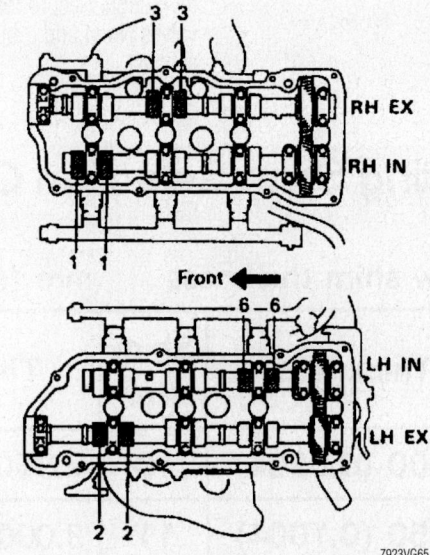

Adjust these valves during the 1st step—3.0L (1MZ-FE) engine

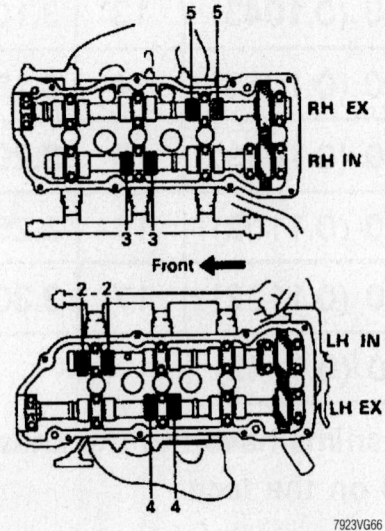

Adjust these valves during the 2nd step—3.0L (1MZ-FE) engine

8. Remove the adjusting shim by turning the crankshaft to position the cam lobe of the camshaft in the up position on the valve to be adjusted. Using a small thin flat bladed tool, turn the valve lifter so that the notches are perpendicular to the camshaft. Press down the valve lifter with SST 09248–55010 part A, or equivalent. Place SST 09248–55010 part B between the camshaft and the valve lifter; remove part A.

9. Remove the adjusting shim with a magnet and a small screwdriver.

10. Determine the replacement adjusting shim size by either using the charts or the following formulas:

- Intake: $N = T + (A–0.008$ in./0.020mm)
- Exhaust: $N = T + (A–0.012$ in./0.30mm)
- T = Thickness of removed shim
- A = Measured valve clearance
- N = Thickness of new shim

11. Select a new shim with a thickness as close as possible to the calculated value. Install the new replacement shim.

➡ **Shims are available in 17 sizes in increments of 0.0020 in. (0.050mm), from 0.0984 in. (2.500mm) to 0.1299 in. (3.300mm).**

12. Recheck the valve clearance.
13. Install or connect the following:
- Cylinder head covers
- Spark plugs and the ignition coils
- Engine wiring harness to the injectors and the coils
- Intake chamber
- Emission control valve set
- V-bank cover
- Air flow meter, air duct, and air cleaner cover
- Negative battery cable

3MZ-FE Engines

1. Before servicing the vehicle, refer to the Precautions Section.
2. Remove or disconnect the following:
- Negative battery cable
- Front suspension upper brace center
- V-bank cover sub-assembly
- Air cleaner assembly
- Emission control valve set
- Intake air surge tank
- Ignition coil assembly
- Cylinder head valve covers
3. Turn the crankshaft pulley and align its groove with the timing mark **0** of the No. 1 timing cover. Check that the timing marks of the camshaft timing pulleys and timing belt rear plates are aligned. If not, turn the

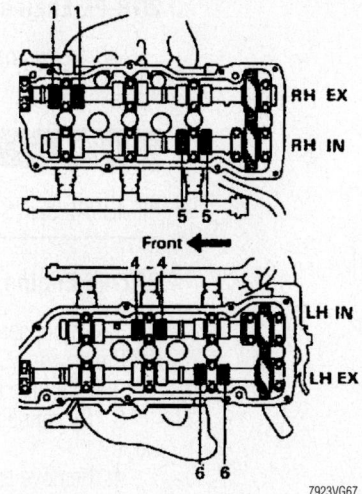

Adjust these valves during the 3rd step—3.0L (1MZ-FE) engine

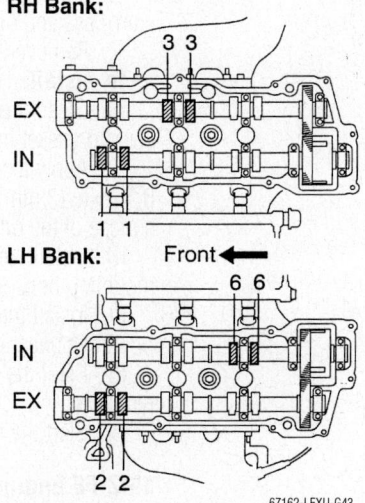

Measuring and adjust these valves first—3MZ-FE engine

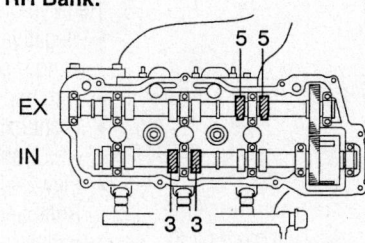

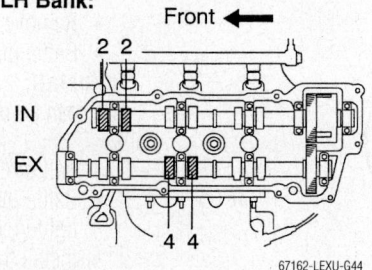

Measuring and adjust these valves second—3MZ-FE engine

crankshaft 1 revolution (360 degrees) and align the mark.

4. Measure the clearance between the valve lash adjuster and the camshaft on the valves in the first sequence and record.

   a. The intake valve clearance cold is 0.006–0.010 in. (0.15–0.25mm).

   b. The exhaust valve clearance cold is 0.010–0.014 in. (0.25–0.35mm).

5. Turn the crankshaft 2/3 revolution and (240 degrees).

6. Measure the clearance between the valve lash adjuster and the camshaft on the valves in the second sequence and record.

   a. The intake valve clearance cold is 0.006–0.010 in. (0.15–0.25mm).

   b. The exhaust valve clearance cold is 0.010–0.014 in. (0.25–0.35mm).

7. Turn the crankshaft 2/3 revolution and (240 degrees).

8. Measure the clearance between the valve lash adjuster and the camshaft on the valves in the third sequence and record.

   a. The intake valve clearance cold is 0.006–0.010 in. (0.15–0.25mm).

   b. The exhaust valve clearance cold is 0.010–0.014 in. (0.25–0.35mm).

9. Adjust the valve lash using the following procedure

10. Remove the adjusting shim and turn the crankshaft to position the cam lobe of the camshaft on the adjusting valve upward. Position the hole in the shim toward the outside of the cylinder head. Press down the valve lash adjuster with the proper tool and place the proper tool between the camshaft and the valve lash adjuster. Remove the tool.

11. Remove the adjusting shim with the proper tool.

12. Determine the thickness of the replacement shim as follows:

   a. T = Thickness of the used shim

   b. A = Measured valve clearance

   c. N = Thickness of new shim

   d. Intake: N = T + (A–0.006–0.010 in. (0.15–0.25mm))

   e. Exhaust: N = T + (A–0.010–0.014 in. (0.25–0.35mm))

➡ **Place the adjusting shim on the valve lifter with the imprinted number facing down.**

13. Reinstall the following
- Cylinder head valve covers
- Ignition coil assembly
- Intake air surge tank
- Emission control valve set
- Air cleaner assembly
- Vacuum hoses
- Front suspension upper center brace
- Negative battery cable

RH Bank:

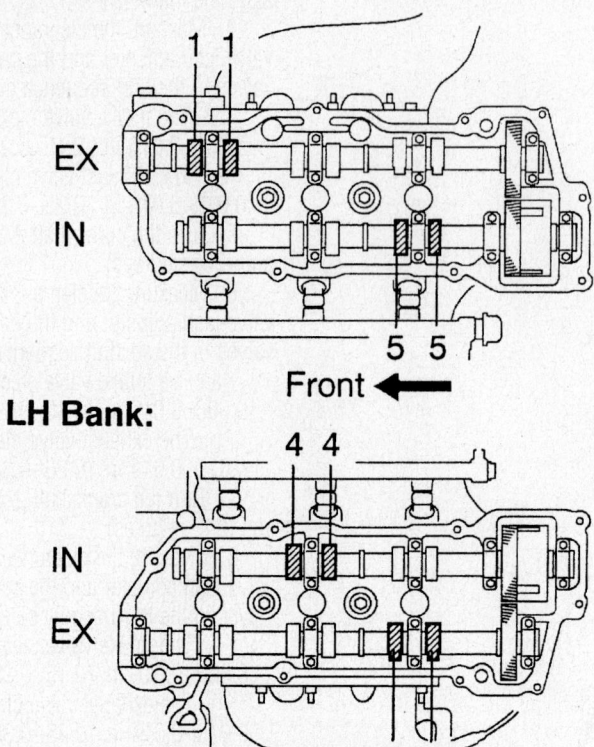

EX

IN

5 5

Front ←

LH Bank:

4 4

IN

EX

6 6

67162-LEXU-G45

Measuring and adjust these valves third—3MZ-FE engine

Front of No. 1 and No. 2 Cylinders:

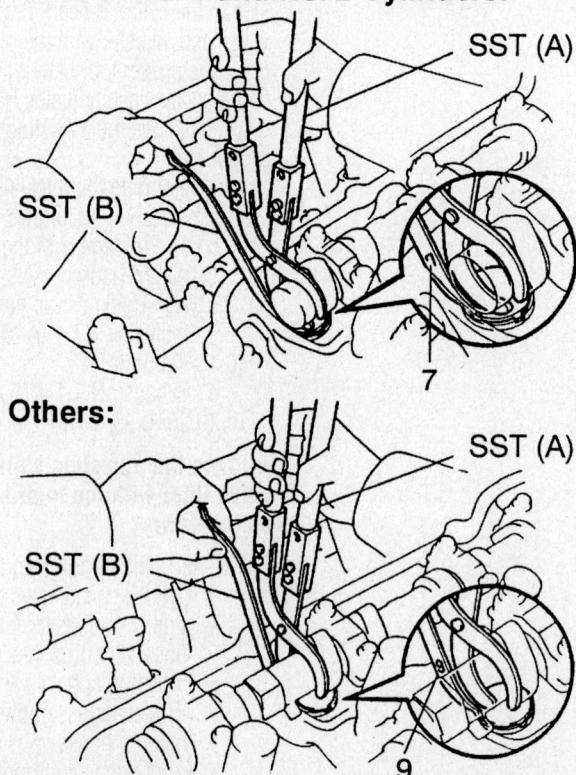

SST (A)

SST (B)

7

Others:

SST (A)

SST (B)

9

67162-LEXU-G46

Using special tool to remove the adjusting shim—3MZ-FE engine

2GR-FE Engine

2GR-FE engines use hydraulic valve lash adjusters.

Oil Pan

REMOVAL & INSTALLATION

2AZ-FE Engine

1. Before servicing the vehicle, refer to the Precautions Section.
2. Drain the engine oil.
3. Disconnect the negative battery cable.
4. Remove engine under cover and any components for access.
5. Remove the oil pan mounting bolts and nuts.
6. Insert a suitable blade between the crankcase and oil pan to cut any sealer.
7. Remove the pan.

To install:

8. Clean the pan and contact surface of any old gasket material.
9. Apply a continuous bead of sealant, 0.157 in. (3 mm) in diameter to the contact surface of the oil pan.
10. Install the oil pan and tighten the mounting bolts to 80 inch lbs. (9 Nm).
11. Install the engine undercover.
12. Connect the negative battery cable.
13. Refill the engine with oil to the correct level.
14. Start the engine and check for leaks.

1MZ-FE Engines

1. Before servicing the vehicle, refer to the Precautions Section.
2. Drain the engine oil.
3. Remove or disconnect the following:
   - Negative battery cable
   - Fender apron seal
   - Undercover
   - Front exhaust pipe bracket from the No. 1 oil pan
   - Flywheel housing undercover
   - Bolts and nuts from the No. 2 oil pan
   - Oil strainer and gasket
   - Remove the No.1 oil pan
   - Baffle plate from the No. 1 oil pan

To install:

4. Clean all mating surfaces of the oil pans.
5. Install or connect the following:
   - Baffle plate to the No. 1 oil pan and tighten: 69 inch lbs. (8 Nm).
   - Install the No. 1 oil pan. with liquid sealant. Uniformly tighten the bolts

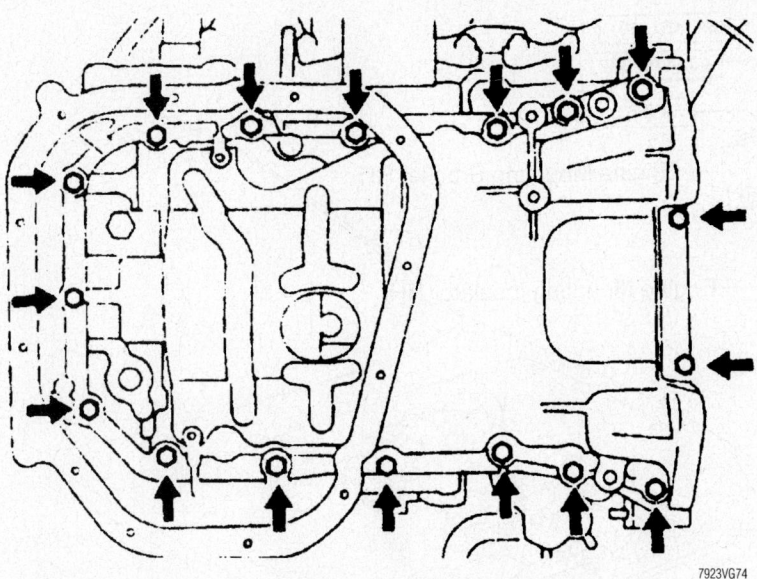

No. 1 oil pan mounting bolt locations—3.0L (1MZ-FE) engine

7923VG74

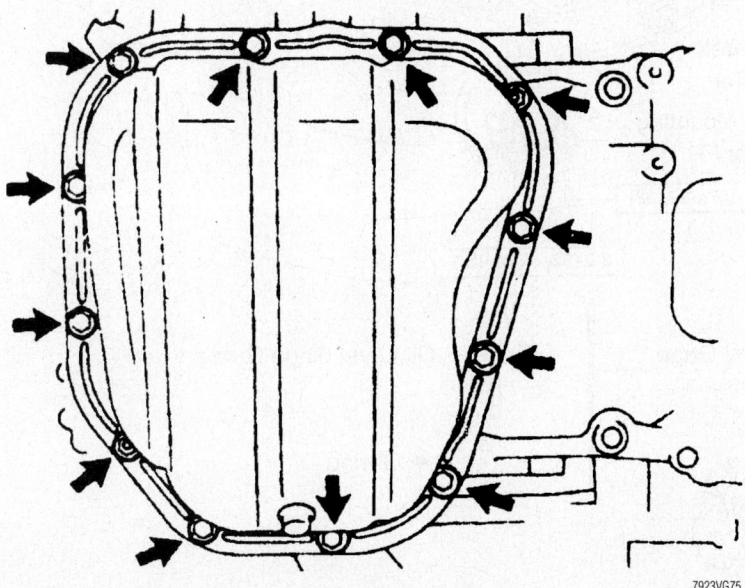

No. 2 oil pan mounting bolt locations—3.0L (1MZ-FE) engine

7923VG75

and nuts in several passes. 10mm head bolt: 69 inch lbs. (8 Nm). 12mm head bolt: 14 ft. lbs. (19.5 Nm). 14mm head bolt-27 ft. lbs. (37.2 Nm).
• Flywheel housing undercover. Tighten to 69 inch lbs. (7.8 Nm).
• Oil strainer. Tighten the nuts to 69 inch lbs. (7.8 Nm).
• No. 2 oil pan. Apply liquid sealant to the oil pan and engine block. Uniformly tighten the bolts and nuts in several passes. Bolts: 69 inch lbs. (7.8 Nm).
• Flywheel housing undercover
• Front exhaust pipe bracket to the

No. 1 oil pan. Bolts: 15 ft. lbs. (21 Nm).
• Exhaust manifolds to the front exhaust pipe nuts: 46 ft. lbs. (62 Nm). Front exhaust pipe to the center exhaust pipe. Bolts and nuts: 41 ft. lbs. (56 Nm).
• Bracket with the 2 bolts. Bolts: 14 ft. lbs. (19 Nm).
• Support stay with the 2 bolts. Bolts: 22 ft. lbs. (29 Nm).
• Undercover
• Right fender apron seal
• Negative battery cable

6. Fill the engine with oil to the correct level.

7. Start the engine and check for leaks.

3MZ-FE Engines

1. Before servicing the vehicle, refer to the Precautions Section.
2. Drain the engine oil.
3. Remove or disconnect the following:
   • Negative battery cable
   • RF wheel
   • Engine under covers
   • RH front fender apron seal
   • A/C compressor, alternator and PS drive belts
   • Engine stabilizer rod
   • Engine mounting stay No2.
   • Alternator bracket
   • Crankshaft pulley
   • Timing belt
   • Crankshaft timing pulley
   • Exhaust pipe and support brackets
   • Oil gauge guide
   • Alternator belt adjusting bar
   • A/C compressor/clutch assembly
   • A/C compressor mounting bracket
4. Separate FR engine mounting insulator

➡ **Do NOT remove the FR engine mounting at this time**

   a. Remove bolt and disconnect the power steering return hose
   b. Remove the 4 nuts
   c. Place a wooden block underneath the engine
   d. Jack up the engine and remove the engine mounting insulator

✳✳ WARNING

Be careful not to damage the oil pan

5. Remove or disconnect the following:
   • RH engine mounting bracket
   • 10 bolts and 2 nuts, gently pry off the oil pan sub assembly No.2.

✳✳ WARNING

Be careful not to damage the oil pan flange area or the contact surface of the engine block.

   • Oil strainer sub-assembly
   • Flywheel housing under cover.
   • Oil pan sub-assembly No. 1
   • Engine oil level sensor

To install:

6. Install the oil pan sub-assembly No. 1.

7. Remove any old oil sealant from contact surface. Clean the surface thoroughly.

8. Apply a continuous bead of sealant

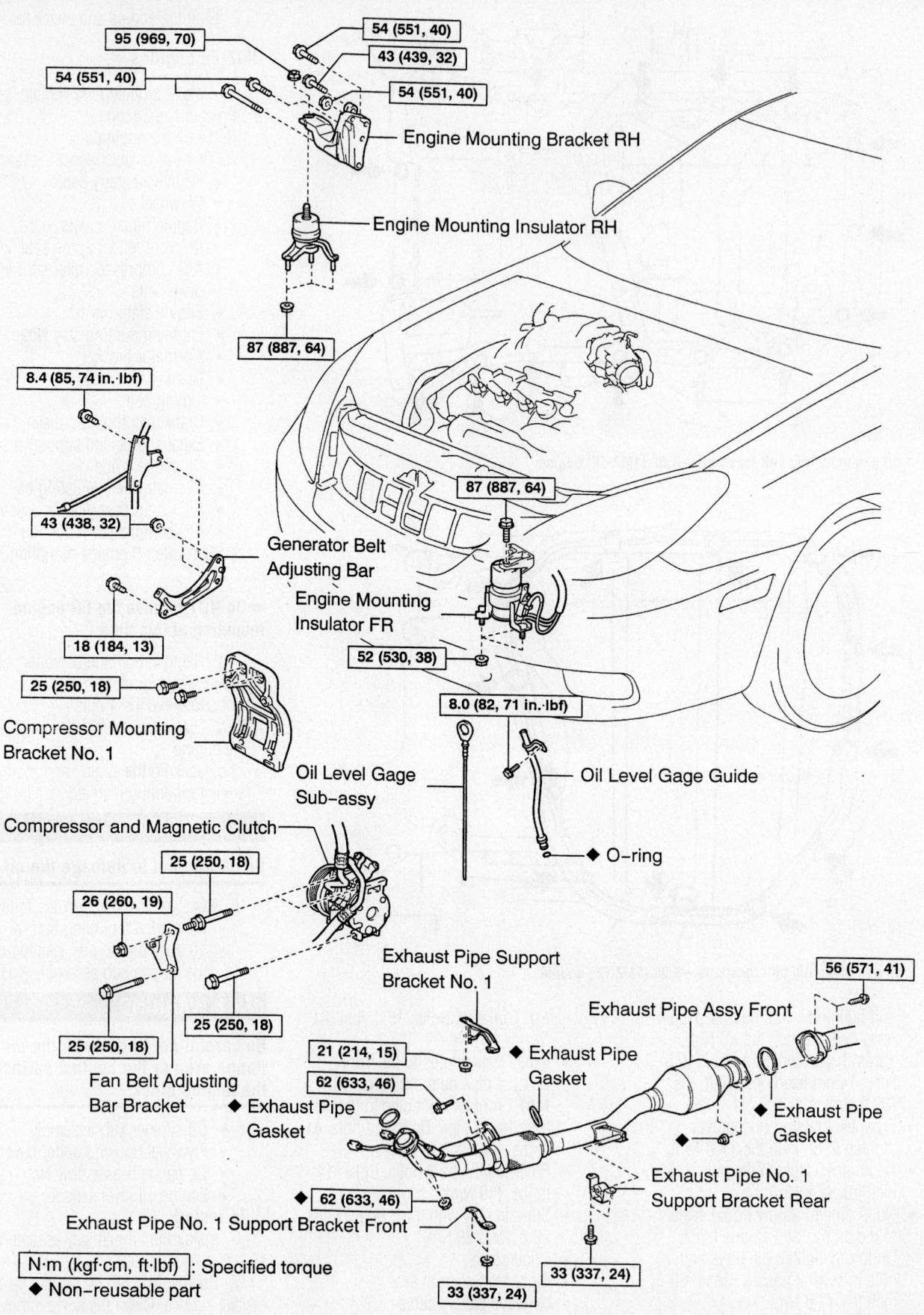

95 (969, 70)

54 (551, 40)

43 (439, 32)

54 (551, 40)

54 (551, 40)

Engine Mounting Bracket RH

Engine Mounting Insulator RH

87 (887, 64)

8.4 (85, 74 in.·lbf)

43 (438, 32)

Generator Belt Adjusting Bar

87 (887, 64)

18 (184, 13)

Engine Mounting Insulator FR

25 (250, 18)

52 (530, 38)

Compressor Mounting Bracket No. 1

8.0 (82, 71 in.·lbf)

Oil Level Gage Sub-assy

Oil Level Gage Guide

◆ O-ring

Compressor and Magnetic Clutch

25 (250, 18)

26 (260, 19)

Exhaust Pipe Support Bracket No. 1

56 (571, 41)

Exhaust Pipe Assy Front

25 (250, 18)

21 (214, 15)

◆ Exhaust Pipe Gasket

25 (250, 18)

62 (633, 46)

Fan Belt Adjusting Bar Bracket

◆ Exhaust Pipe Gasket

◆ Exhaust Pipe Gasket

◆ 62 (633, 46)

Exhaust Pipe No. 1 Support Bracket Rear

Exhaust Pipe No. 1 Support Bracket Front

33 (337, 24)

N·m (kgf·cm, ft·lbf): Specified torque

◆ Non-reusable part

33 (337, 24)

67162-LEXU-G47

Exploded view, component removal for oil pan—3MZ-FE engine

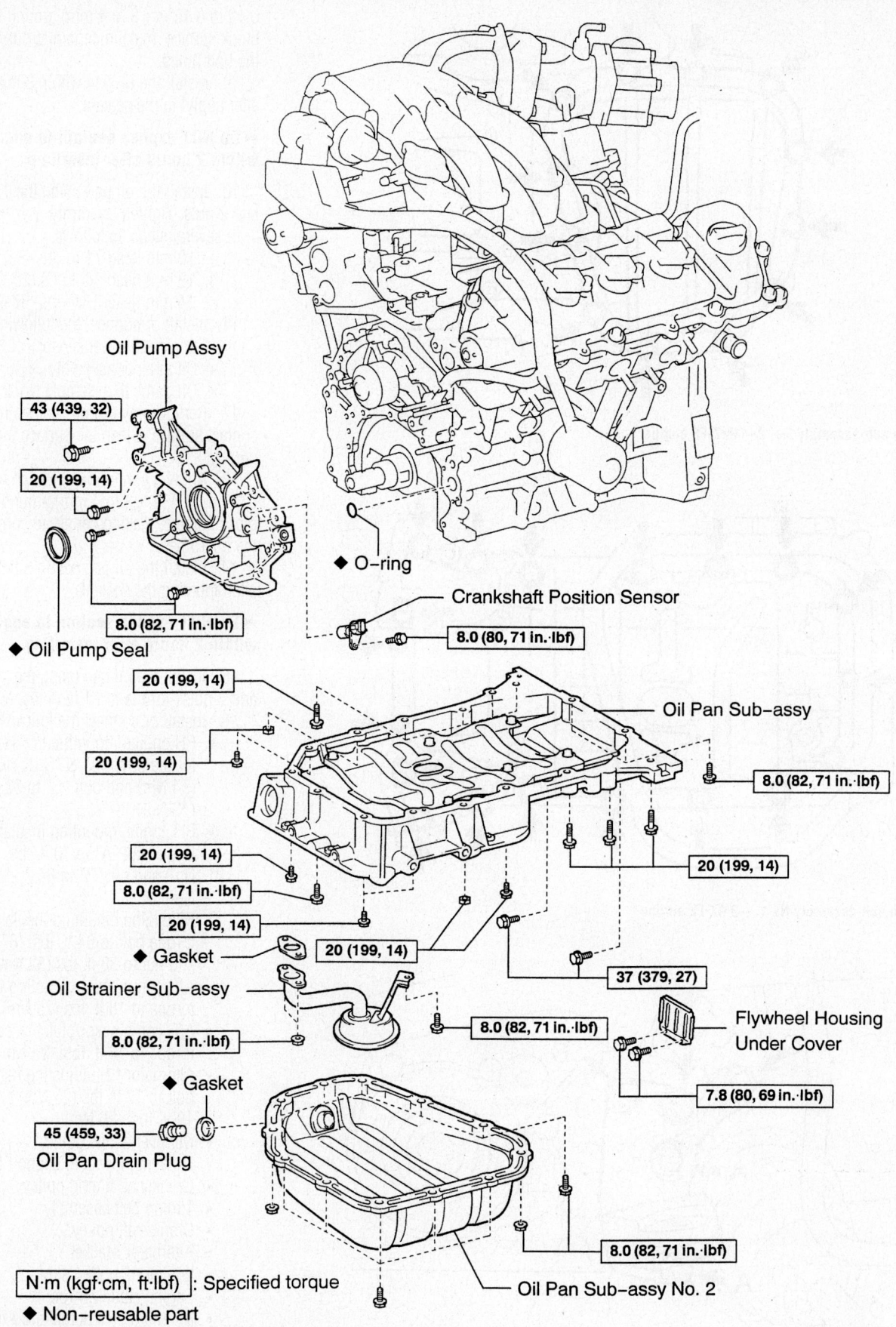

Oil Pump Assy

43 (439, 32)

20 (199, 14)

8.0 (82, 71 in.·lbf)

◆ Oil Pump Seal

◆ O-ring

Crankshaft Position Sensor

8.0 (80, 71 in.·lbf)

20 (199, 14)

Oil Pan Sub-assy

20 (199, 14)

8.0 (82, 71 in.·lbf)

20 (199, 14)
8.0 (82, 71 in.·lbf)
20 (199, 14)

◆ Gasket

20 (199, 14)

Oil Strainer Sub-assy

8.0 (82, 71 in.·lbf)

8.0 (82, 71 in.·lbf)

20 (199, 14)

37 (379, 27)

Flywheel Housing
Under Cover

7.8 (80, 69 in.·lbf)

◆ Gasket

45 (459, 33)

Oil Pan Drain Plug

8.0 (82, 71 in.·lbf)

N·m (kgf·cm, ft·lbf) : Specified torque

◆ Non-reusable part

Oil Pan Sub-assy No. 2

67162-LEXU-G48

Exploded view, removal of oil pan from engine—3MZ-FE engine

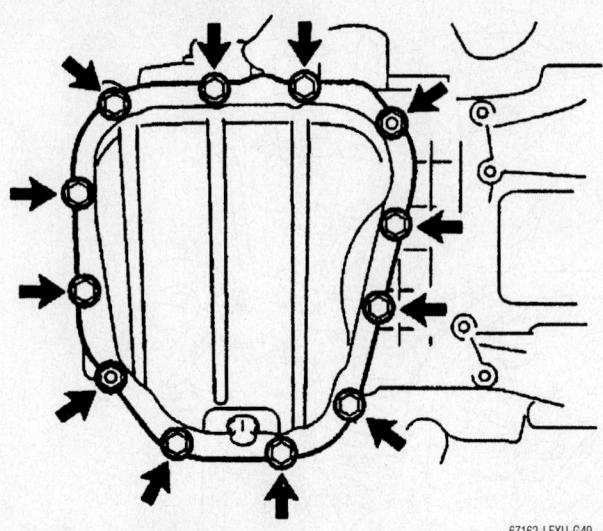

Oil pan sub-assembly No. 2—3MZ-FE engine

67162-LEXU-G49

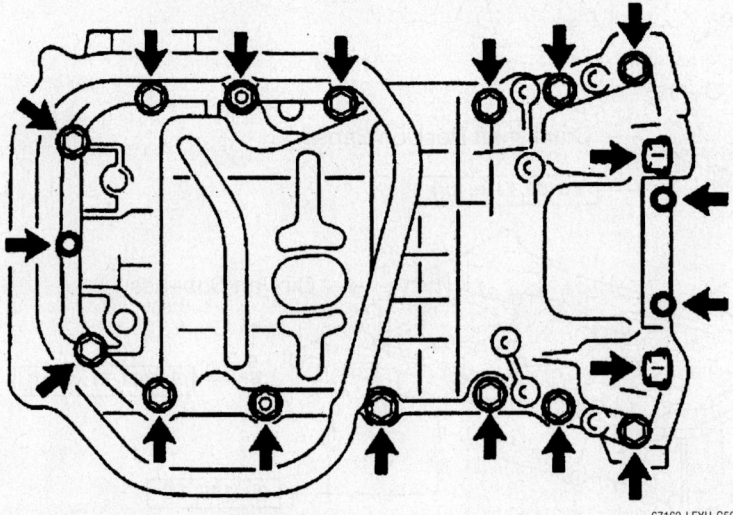

Oil pan sub-assembly No 1.—3MZ-FE engine

67162-LEXU-G50

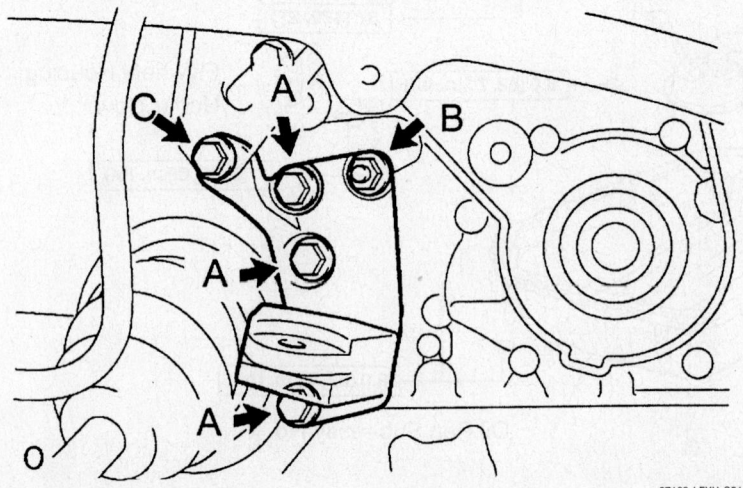

Engine mounting bracket tightening procedure—3MZ-FE engine

67162-LEXU-G51

0.12 to 0.16 in (3 to 4 mm) around the block surface, making certain to surround the bolt holes.

9. Install the oil pan within 3 minutes after applying the sealant

➡ **Do NOT expose sealant to engine oil within 2 hours after installing**

10. Install the oil pan using the 17 bolts and 2 nuts. Tighten uniformly in several steps as follows:
  a. 10 mm head 71 in. lbs (8.0 Nm).
  b. 12 mm head 14 ft. lbs.(20 Nm).
  c. 14 mm head 27 ft. lbs (37 Nm).
11. Install or connect the following:
  • Engine oil level sensor
  • Oil strainer assembly
  • Oil pan sub assembly No. 2
12. Remove any old oil sealant from contact surface. Clean the surface thoroughly.
13. Apply a continuous bead of sealant 0.16 to 0.20 in. (3 to 4 mm) around the block surface, making certain to surround the bolt holes.
14. Install the oil pan within 3 minutes after applying the sealant.

➡ **Do NOT expose sealant to engine oil within 2 hours after installing**

15. Install the oil pan using the 10 bolts and 2 nuts. Torque to 71 inch lbs. (8.0 Nm).
16. Install or connect the following:
  • RH engine mounting bracket; torque bolts "A" & "B" to 40 ft. lbs (54 Nm) and bolt "C" to 32 ft. lbs (43 Nm)
  • RH engine mounting insulator; torque nut "A" to 70 ft. lbs. (95 Nm) and nut "B" to 64 ft. lbs (87 Nm)
  • FR engine mounting insulator; torque bolt to 64 ft. lbs. (87 Nm) and nut to 38 ft. lbs (52 Nm)
  • A/C compressor mounting bracket; torque to 18 ft. lbs (25 Nm)
  • A/C compressor/clutch assembly; torque to 18 ft. lbs. (25 Nm)
  • Alternator belt adjusting bar; torque bolt to 18 ft. lbs (25 Nm) and nut to 19 ft. lbs. (26 Nm)
  • Oil level gauge guide
  • Exhaust pipes and support brackets
  • Crankshaft timing pulley
  • Timing belt assembly
  • Crankshaft pulley
  • Alternator bracket
  • Engine mounting stay
  • Engine stabilizer rod
  • Alternator, PS pump and A/C drive belts
  • RH fender apron seal

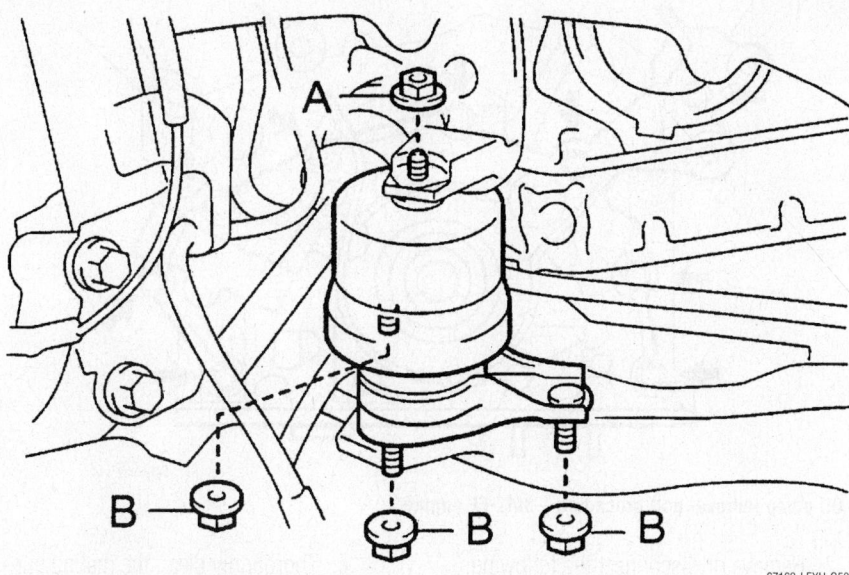

Engine mounting insulator tightening procedure—3MZ-FE engine

67162-LEXU-G52

- RF wheel
- Engine under covers
- Negative battery cable

17. Refill the engine with oil to the correct level.

18. Start the engine and check for leaks.

2GR-FE Engine

1. Before servicing the vehicle, refer to the Precautions Section.

2. Drain the engine oil.

3. Remove the engine/transaxle assembly.

4. Secure the engine in a suitable stand.

5. Remove or disconnect the following:
- Oil pipes
- Oil filter cartridge
- Lower oil pan
- Oil strainer
- Upper oil pan

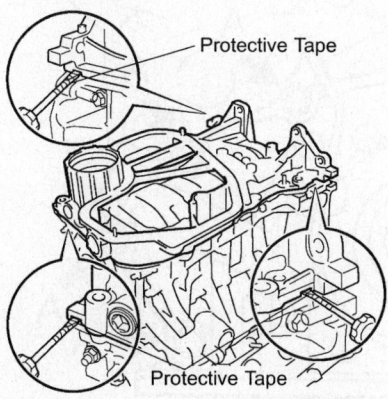

09490_AVAL_G0027

Pry locations in order to remove the upper oil pan—2GR-FE engine

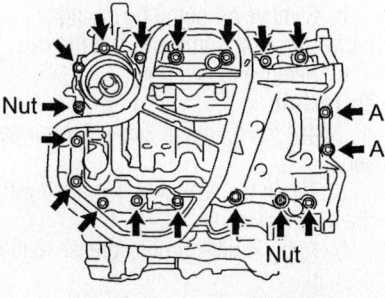

09490_AVAL_G0028

Bolt identification for the upper oil pan— 2GR-FE engine

➡ Be careful not to damage the contact surfaces of the cylinder block and oil pan. Wrap any pry tool with protective tape.

To install:

6. Apply a continuous bead of sealant 0.118–0.156 in. (3–4 mm) in diameter to the mating surface of the upper oil pan.

7. Install the upper oil pan and tighten the bolts and nuts as follows:
   a. Bolts A: 7 ft. lbs. (10 Nm)
   b. All others: 15 ft. lbs. (21 Nm).

8. Install the oil strainer. Tighten the bolts to 7 ft. lbs. (10 Nm).

9. Clean mating surface of the lower oil pan. Apply a continuous bead of sealant 0.118–0.156 in. (3–4 mm) in diameter to the mating surface of the lower oil pan.

10. Install the lower oil pan and tighten the mounting bolts to 7 ft. lbs. (10 Nm).

11. The remainder of the installation is the reverse order of removal.

Oil Pump

REMOVAL & INSTALLATION

2AZ-FE Engines

1. Before servicing the vehicle, refer to the Precautions Section.

2. Drain the engine oil.

3. Remove or disconnect the following:
- Negative battery cable
- Right-side front wheel
- Engine under covers
- Right-side fender apron seal
- Front exhaust pipe
- Engine moving control rod and bracket
- Right-side engine mounting stay and bracket
- Accessory drive belt
- Engine appearance cover
- Engine wire
- Alternator
- Power steering pump
- Ignition coils
- Ventilation hoses from the cylinder head cover
- Cylinder head cover

4. Turn the crankshaft until the No. 1 cylinder is at TDC compression.

5. Remove or disconnect the following:
- Lower oil pan
- Chain tensioner
- Accessory drive belt tensioner

6. Support the engine with a suitable lifting device.

7. Remove or disconnect the following:
- Engine mounts
- Front cover
- Timing chain
- Oil pump drive shaft gear
- Oil pump and gasket

8. The remainder of the installation is the reverse order of removal. Tighten the oil pump mounting bolts to 14 ft. lbs. (19 Nm).

1MZ-FE Engines

1. Before servicing the vehicle, refer to the Precautions Section.

2. Drain the engine oil.

3. Remove or disconnect the following:
- Negative battery cable
- Fender apron seal
- Undercover
- Front exhaust pipe
- Front exhaust pipe bracket from the No. 1 oil pan
- Alternator drive belt
- A/C compressor
- Power steering pump drive belt and adjusting strut

- Timing belt and belt pulleys
- Rear timing belt cover
- A/C compressor housing bracket
- No. 2 oil pan, oil strainer, No.1 oil pan and baffle plate
- Crankshaft position sensor
- 9 oil pump bolts. Make a note of the position of the each bolt.
- Oil pump body by prying between the oil pump and main bearing cap
- O-ring from the cylinder block
- Plug, gasket, spring, and relief valve from the oil pump body
- 9 screws, pump body cover, drive, and driven rotors

To install:

4. Install or connect the following:
   - Driven rotors, drive, pump body cover
   - 9 screws
   - Oil pump relief valve, spring, gasket, and the plug to the oil pump body
   - New O-ring on the cylinder block
   - Liquid sealant to the oil pump and engine block
   - Oil pump to the engine block
   - Bolts, uniformly tightened in several passes: 10mm head: 69 inch lbs. (8 Nm). 12mm head: 14 ft. lbs. (20 Nm)
   - Crankshaft position sensor. Bolt: 69 inch lbs. (8 Nm).
   - Baffle plate to the No. oil pan. Tighten to 69 inch lbs. (8 Nm).
   - No. 1 oil pan, oil strainer and No. 2 oil pan.
   - A/C compressor housing bracket. Bolts: 18 ft. lbs. (25 Nm).
   - Rear timing belt cover. Bolts: 74 inch lbs. (9 Nm).
   - Timing belt pulleys
   - Timing belt
   - Adjusting strut and power steering drive belt. Bolt and nut: 32 ft. lbs. (43 Nm).
   - A/C compressor
   - Alternator drive belt
   - Front exhaust pipe bracket to the No. 1 oil pan. Bolts: 15 ft. lbs. (21 Nm).
   - Front exhaust pipe
   - Undercover
   - Right fender apron seal
   - Negative battery cable

5. Fill the engine with oil to the correct level.

6. Start the vehicle and check for leaks.

3MZ-FE Engines

1. Before servicing the vehicle, refer to the Precautions Section.

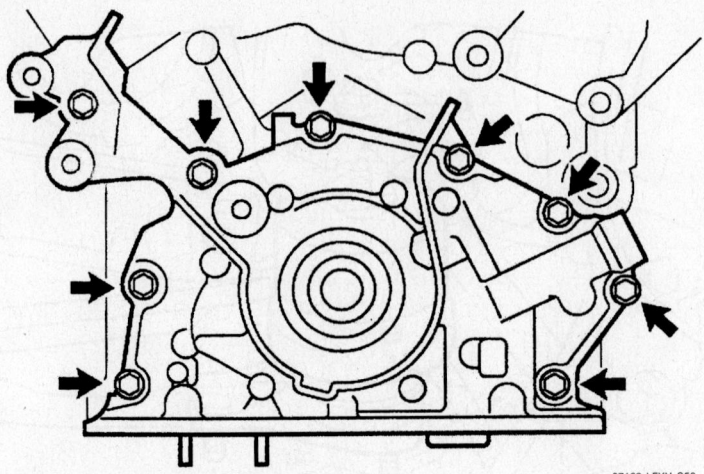

Oil pump removal bolt procedure—3MZ-FE engine

2. Remove or disconnect the following:
   - Oil pan
   - Crankshaft position sensor
3. Oil pump assembly
   a. 9 bolts
   b. Remove oil pump by prying between the oil pump and bearing cap
   c. Remove the oil ring

To install:

4. Install the oil pump seal using proper driver.
   a. Tap in the seal until it is flush with the oil pump body
   b. Apply multi-purpose grease to the seal lip.
5. Install oil pump assembly
   a. Remove any old sealant from the mating surfaces
   b. Apply a light coat of clean engine oil to the O-ring, then place it on the engine block.

   c. Thoroughly clean the mating surface of any oil or old sealant
   d. Apply a continuous bead of sealant on the oil pump body, making certain to surround the bolt holes.
   e. Install the oil pump within 3 minutes after applying the sealant.

➡**Do NOT expose the sealant to engine oil within 2 hours after installing**

   f. Align the key of the oil pump drive gear with the keyway located on the crankshaft, then slide the oil pump into place.
   g. Install the oil pump with the 9 bolts. Tighten the bolts uniformly in several steps. Torque to: Bolt A 71 in. lbs (8.0 Nm), Bolt B 14 ft lbs. (20 Nm), Bolt C 32 ft. lbs. (43 Nm)
6. Install crankshaft position sensor
7. Install oil pans using oil pan installation procedure

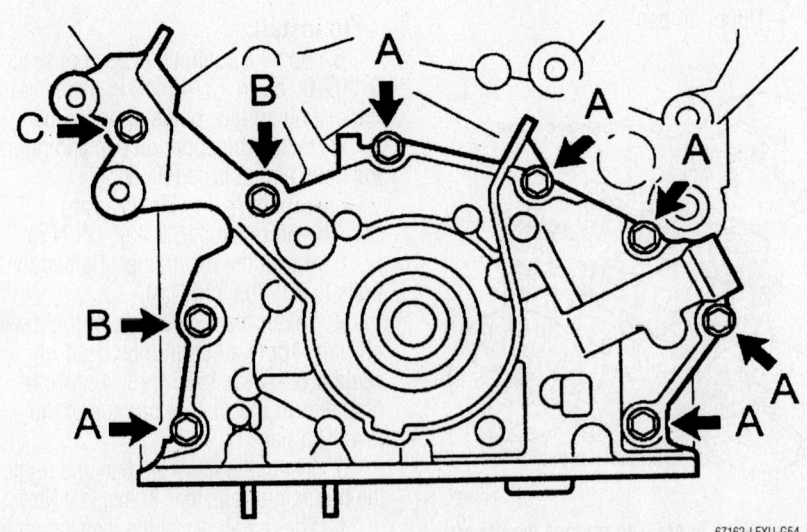

Oil pump bolt installation procedure—3MZ-FE engine

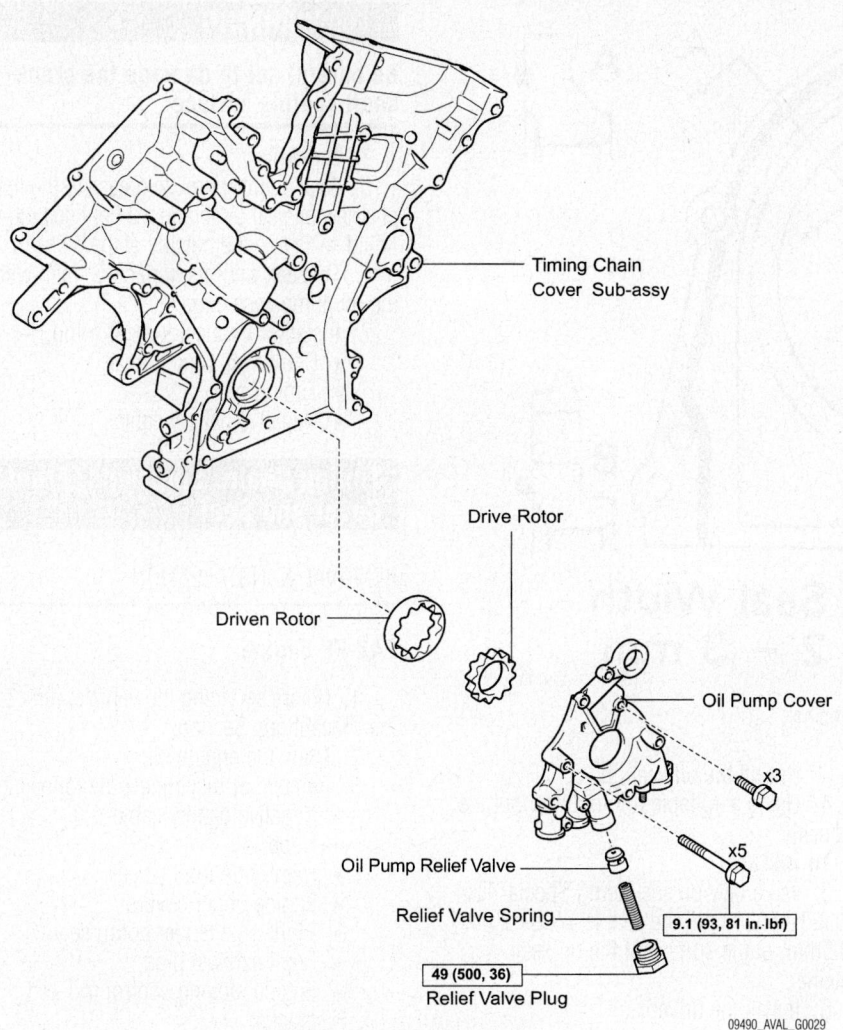

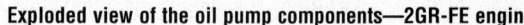

Exploded view of the oil pump components—2GR-FE engine

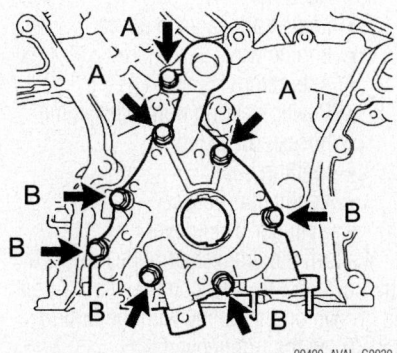

Oil pump cover bolt identification—2GR-FE engine

2GR-FE ENGINE

1. Before servicing the vehicle, refer to the Precautions Section.

2. Remove the engine/transaxle assembly from the vehicle.

3. Remove the front cover.

4. Remove the oil pump cover from the front cover and remove the oil pump.

To install:

5. Install the oil pump cover to the front cover. Bolts A are 0.87 in (22 mm) in length and Bolts B are 1.58 in. (40 mm) in length. Tighten all bolts to 81 inch lbs. (9.1 Nm).

6. Install the front cover.

7. Install the engine/transaxle assembly to the vehicle.

8. Start the engine and check for leaks.

Rear Main Seal

REMOVAL & INSTALLATION

1MZ-FE Engine

1. Remove or disconnect the following:
   - Transaxle
   - Clutch cover assembly and flywheel or driveplate
   - Remove the rear end plate.
   - Oil seal retainer and gasket. Discard the gasket or sealant.

2. Use a small prybar to pry the oil seal from the retaining plate. Be careful not to damage the plate.

To install:

3. Clean the retainer contact surfaces thoroughly and lubricate the new oil seal with multi-purpose grease.

4. Drive the oil seal into the retainer until its surface is flush with the edge of the retainer. Make sure that the seal is installed evenly in the retainer to ensure proper sealing.

5. Apply a ⅛ inch bead of sealant to the

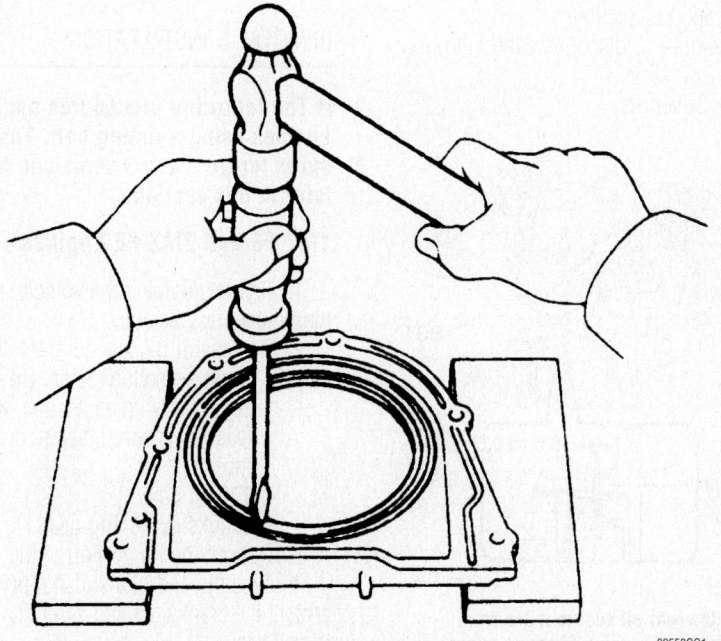

Always place the seal on blocks of wood, then tap the seal from the retainer—1MZ-FE engine

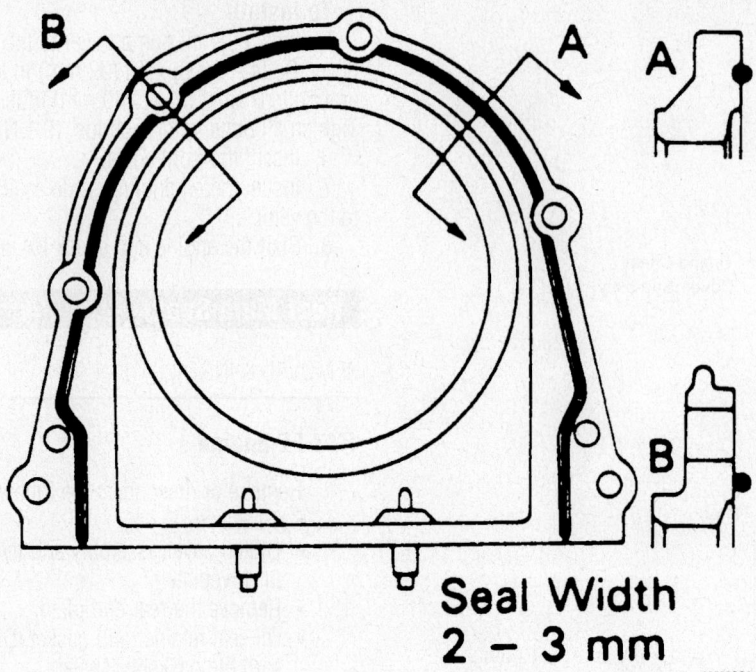

**Seal Width
2 – 3 mm**

89553GG3

Apply sealant to the rear oil seal retainer—1MZ-FE engine

oil seal retainer. Install the retainer and install the dust seal. Tighten the bolts to 69 inch lbs. (8 Nm).

6. Install the rear end plate.

7. On automatic transaxle equipped vehicles, install the driveplate.

8. On manual transaxle equipped vehicles, install the clutch disc and clutch cover.

9. Install the transaxle.

2GR-FE Engine

1. Before servicing the vehicle, refer to the Precautions Section.

2. Remove or disconnect the following:
   - Transaxle
   - Driveplate

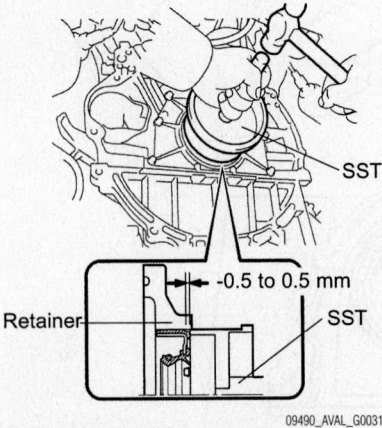

Install the rear oil seal to the correct depth using Special Tool 09223-15030—2GR-FE engine

3. Cut off the oil seal lip.

4. Using a suitable pry tool, pry out the oil seal.

To install:

5. Tap a new oil seal using Special Tool 09223-15030 until the seal is +/- 0.020 in. (0.5mm) of the surface of the oil seal retainer.

6. Install the driveplate.

7. Install the transaxle.

Front Crankshaft Seal

REMOVAL & INSTALLATION

➡**The following procedures apply to engines using a timing belt. The procedures for front cover seals can be found later in this section.**

1MZ-FE and 3MZ-FE Engines

1. Before servicing the vehicle, refer to the Precautions Section.

2. Remove or disconnect the following:
   - Negative battery cable. On vehicles equipped with an air bag, wait at least 90 seconds before proceeding.
   - Timing belt
   - Crankshaft timing gear

3. Cut out the lip portion of the oil seal.

4. Tape the end of a suitable pry bar to protect the crankshaft and carefully remove the oil seal.

✳✳ WARNING

Be careful not to damage the crankshaft sealing surface.

To install:

5. Apply multi-purpose grease to the lip of a new oil seal. Also apply a light coating of liquid sealant to the outside of the oil seal.

6. Oil seal, until its surface is flush with the oil pump case edge.

7. Install or connect the following:
   - Crankshaft timing gear
   - Timing belt
   - Negative battery cable

Timing Chain, Sprockets, Front Cover and Seal

REMOVAL & INSTALLATION

2AZ-FE Engine

1. Before servicing the vehicle, refer to the Precautions Section.

2. Drain the engine oil.

3. Remove or disconnect the following:
   - Negative battery cable
   - Hood
   - Right-side front wheel
   - Engine under covers
   - Right-side fender apron seal
   - Front exhaust pipe
   - Engine moving control rod and bracket
   - Right-side engine mounting stay and bracket
   - Accessory drive belt
   - Engine appearance cover
   - Engine wire
   - Alternator
   - Power steering pump, leave the hoses connected
   - Ignition coils
   - Ventilation hoses
   - Cylinder head cover

4. Turn the crankshaft pulley to set the No. 1 cylinder to TDC compression. Align the groove on the pulley with the timing mark '0' on the front cover.

5. Remove the crankshaft pulley as follows:

   a. TMC made: Use Special Tool 09213-54015 to hold the pulley to loosen the bolt. Use Special Tool 09950-50013 to remove the bolt and pulley.

   b. TMMK made: Use Special Tool 09960-10010 to hold the pulley to remove the bolt. Use Special Tool 09950-40011 to remove the crankshaft pulley

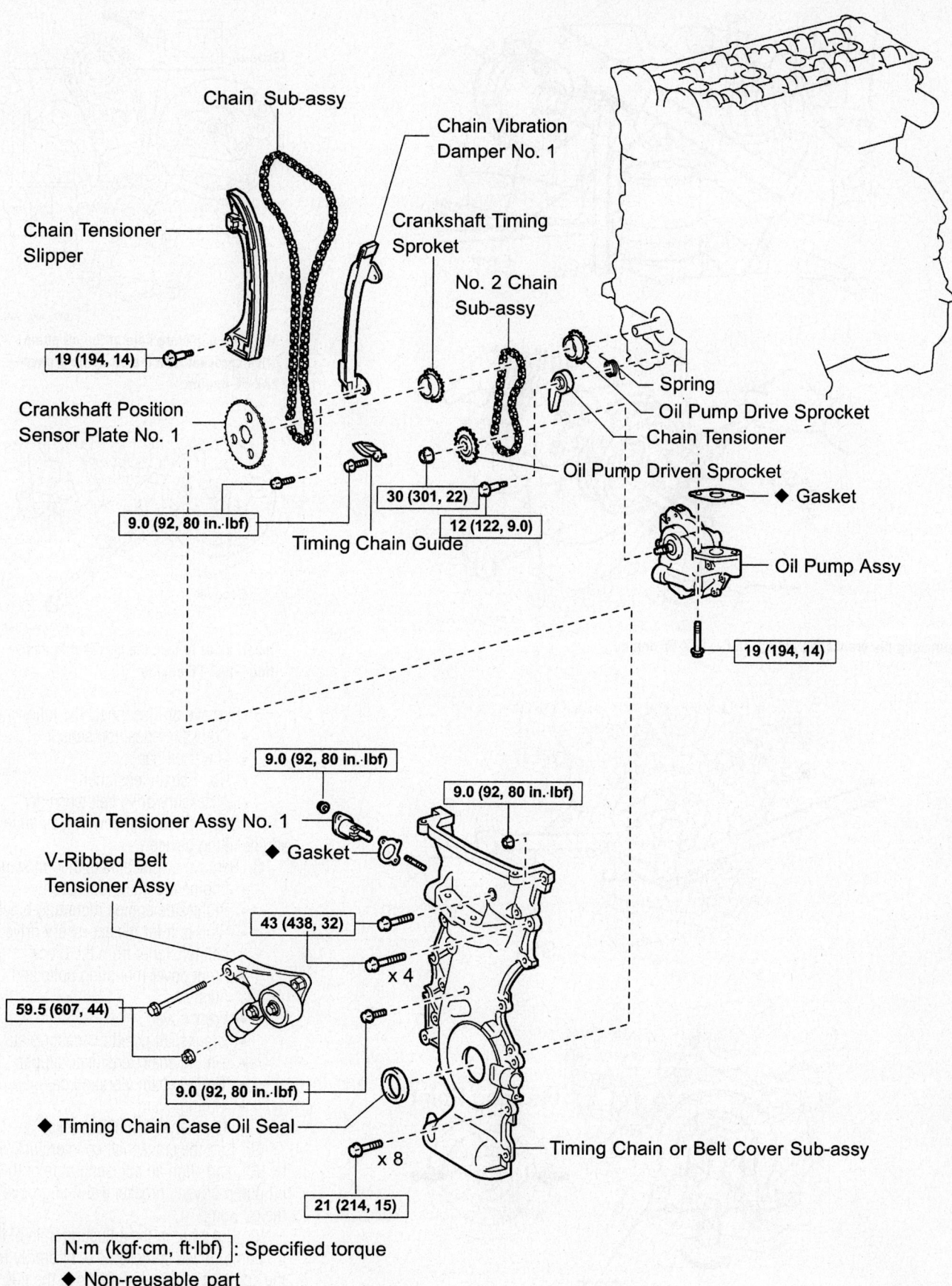

Chain Sub-assy

Chain Vibration Damper No. 1

Chain Tensioner Slipper

Crankshaft Timing Sproket

No. 2 Chain Sub-assy

19 (194, 14)

Spring

Oil Pump Drive Sprocket

Crankshaft Position Sensor Plate No. 1

Chain Tensioner

9.0 (92, 80 in.·lbf)

Oil Pump Driven Sprocket

◆ Gasket

Timing Chain Guide

30 (301, 22)

12 (122, 9.0)

Oil Pump Assy

19 (194, 14)

9.0 (92, 80 in.·lbf)

9.0 (92, 80 in.·lbf)

Chain Tensioner Assy No. 1

◆ Gasket

V-Ribbed Belt Tensioner Assy

43 (438, 32)

x 4

59.5 (607, 44)

9.0 (92, 80 in.·lbf)

◆ Timing Chain Case Oil Seal

21 (214, 15)

x 8

Timing Chain or Belt Cover Sub-assy

N·m (kgf·cm, ft·lbf) : Specified torque

◆ Non-reusable part

09490_AVAL_G0034

Exploded view of the timing chain and front cover components—2AZ-FE engine

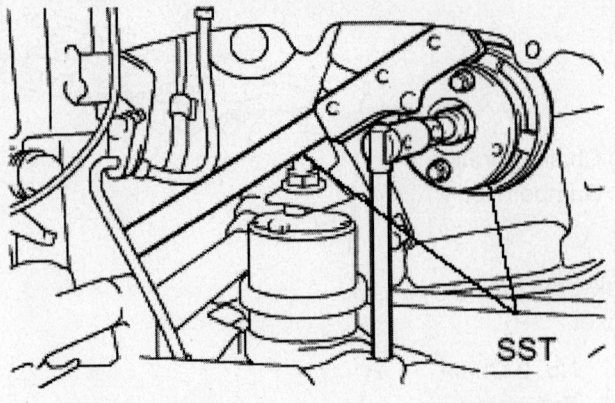

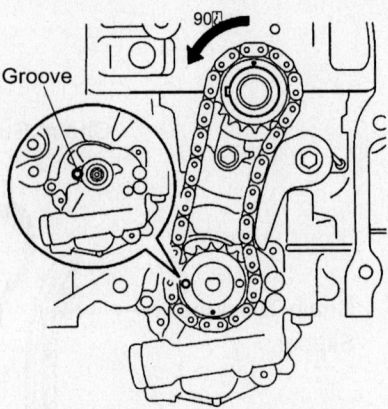

Align the adjusting hole of the oil pump drive sprocket with the oil pump groove—2AZ-FE engine

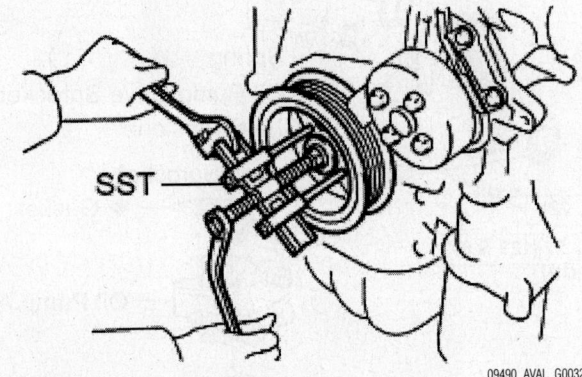

Removing the crankshaft pulley–TMC—2AZ-FE engine

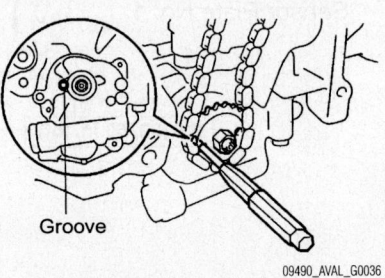

Insert a bar to lock the sprocket in position—2AZ-FE engine

6. Remove or disconnect the following:
   - Crankshaft position sensor
   - Lower oil pan
   - No. 1 chain tensioner
   - Accessory drive belt tensioner

7. Support the engine with a suitable engine lifting device.

8. Remove or disconnect the following:
   - Engine mounting insulators
   - Right-side engine mounting bracket
   - Stud bolt for the accessory drive belt tensioner from the block
   - Front cover mounting bolts and nuts
   - Front cover
   - Crankshaft position sensor plate
   - Timing chain tensioner slipper
   - Timing chain vibration damper
   - Timing chain
   - Crankshaft timing sprocket

9. Turn the crankshaft counterclockwise by 90°, and align an adjusting hole of the oil pump driven sprocket with the groove of the oil pump.

10. Put a bar in the adjusting hole of the oil pump driven sprocket to temporarily lock the sprocket in position. Remove the nut.

11. Remove the bolt, chain tensioner and spring.

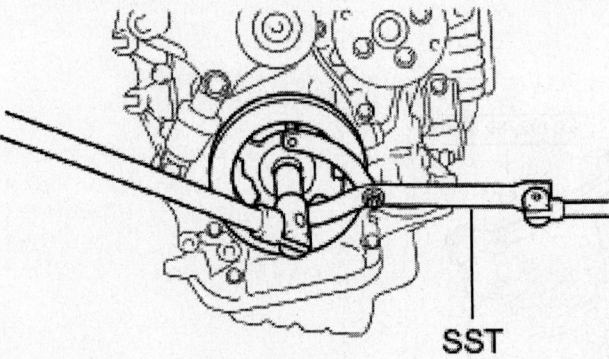

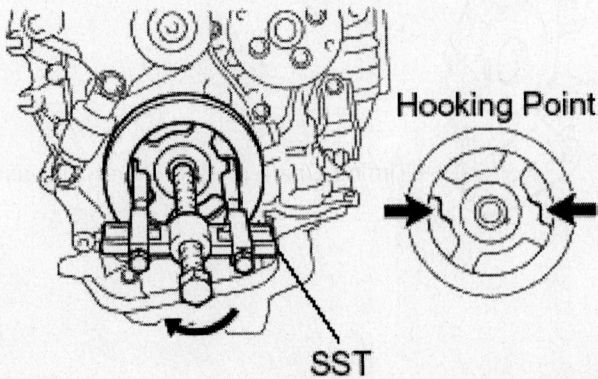

Removing the crankshaft pulley–TMMK—2AZ-FE engine

12. Remove the chain tensioner, oil pump sprocket and No. 2 timing chain.

To install:

13. Set the crankshaft key into the left horizontal position. Turn the cutout of the drive shaft to the top.

14. Align the yellow colored links with the timing marks of the sprocket as shown in the illustration.

15. Insert the sprockets with chain to the crankshaft and oil pump shaft.

16. Temporarily tighten the oil pump driven sprocket with the nut.

17. Insert the damper spring into the adjusting hole, and install the chain tensioner plate with the nut. Tighten to 9 ft. lbs. (12 Nm).

18. Align the adjusting hole of the oil pump driven sprocket with the groove of the oil pump.

19. Put a bar in the adjusting hole of the oil pump driven sprocket to temporarily lock the sprocket in position. Tighten the nut to 22 ft. lbs. (30 Nm).

20. Rotate the crankshaft counterclockwise by 90°, and align the crankshaft key to the top.

21. Install the timing chain vibration damper. Tighten to 80 inch lbs. (9 Nm).

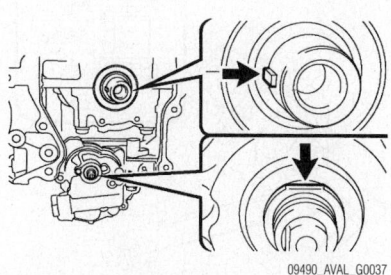

Align the crankshaft key and drive shaft cutout as shown for installation—2AZ-FE engine

09490_AVAL_G0037

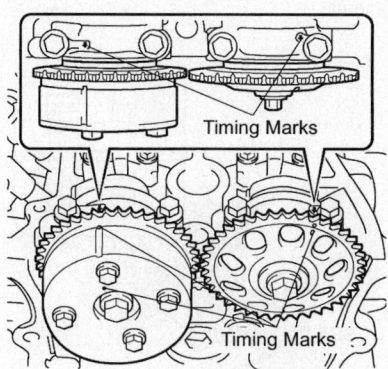

Align the marks of the sprockets and bearing caps to set the No. 1 cylinder at TDC—2AZ-FE engine

09490_AVAL_G0038

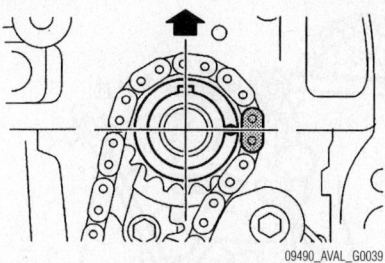

Set the crankshaft set key facing upward—2AZ-FE engine

09490_AVAL_G0039

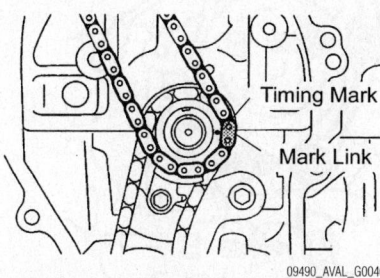

Align the marked link of the timing chain with crankshaft sprocket timing mark—2AZ-FE engine

09490_AVAL_G0040

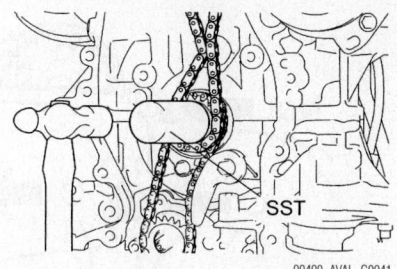

Tap the crankshaft sprocket in with Special Tool 09309-37010—2AZ-FE engine

09490_AVAL_G0041

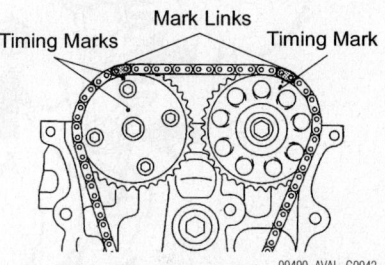

Align the marked links of the timing chain with the camshaft sprockets during installation—2AZ-FE engine

09490_AVAL_G0042

Seal Packing

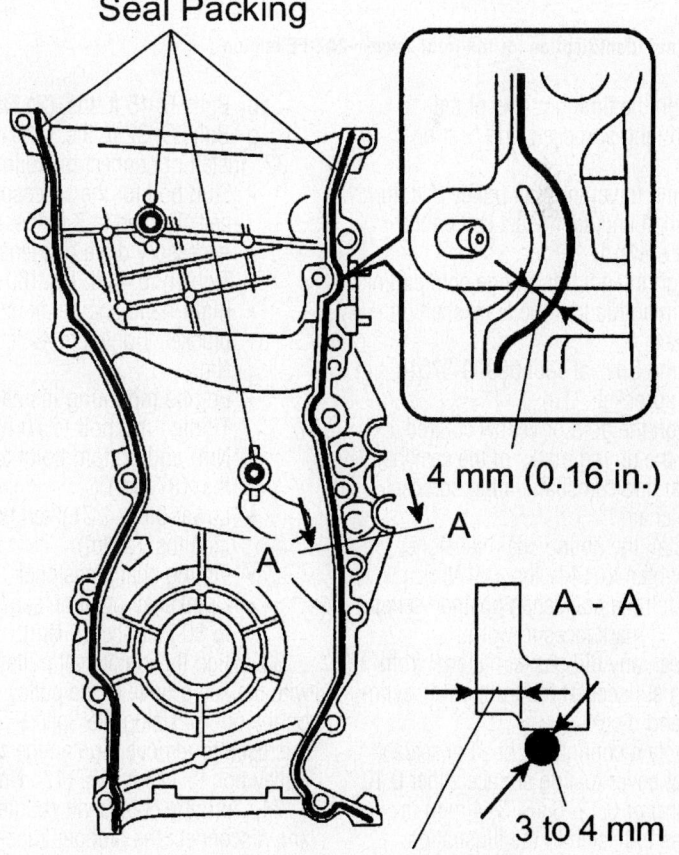

4 mm (0.16 in.)

A - A

3 to 4 mm

2.5 mm (0.098 in.)

09490_AVAL_G0043

Apply sealant to the front cover mating surface as shown—2AZ-FE engine

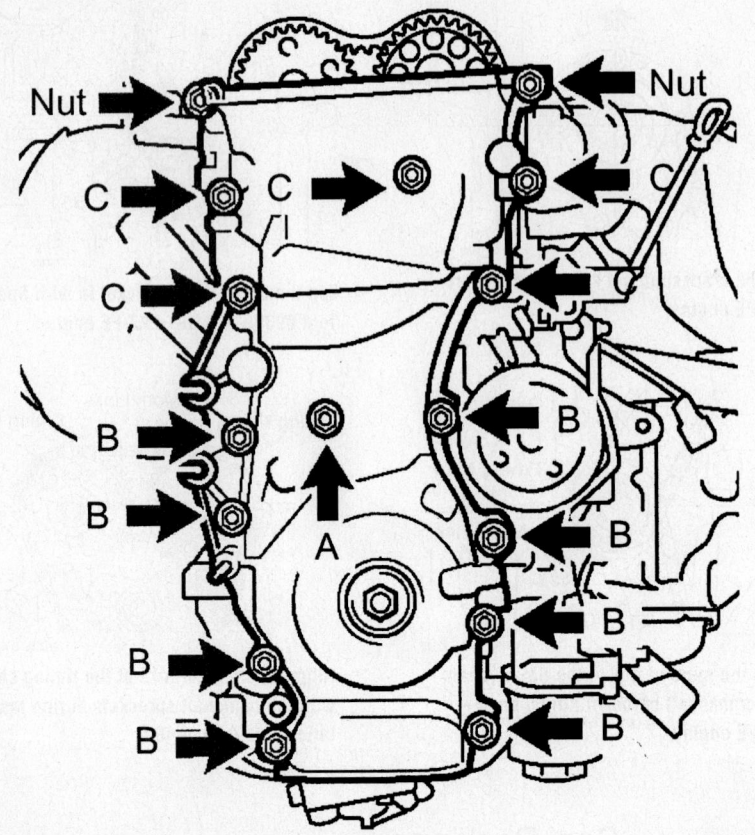

Nut ← → Nut

C ← C → C ←

C ←

B ← B → B

B ← B

B → B

A ↑

B ← B →

B ←

09490_AVAL_G0044

Bolt and nut identification for the front cover—2AZ-FE engine

22. Align the timing marks of the camshaft timing sprockets and bearing caps.

23. Using the crankshaft pulley bolt, turn the crankshaft and set the set key on the crankshaft upward.

24. Align the gold or orange colored link with the timing mark of the crankshaft timing sprocket.

25. Using Special Tool 09309-37010, tap in the sprocket.

26. Align the gold or yellow colored links with the timing marks of the camshaft timing gear and camshaft timing sprocket. Install the chain.

27. Install the timing chain tensioner slipper. Tighten to 14 ft. lbs. (19 Nm).

28. Install the crankshaft position sensor plate. The 'F' mark faces forward.

29. Clean any old gasket material from the mating surfaces of the front cover, cylinder head and block.

30. Apply a continuous bead of sealant to the front cover mating surface either 0.10 in. (2.5 mm) or 0.12–0.16 (3–4 mm) in diameter as indicated in the illustration.

31. Install the front cover and tighten as follows:

   a. Bolt A and nuts: 80 inch lbs. (9 Nm).

   b. Bolts B: 15 ft. lbs. (21 Nm).

   c. Bolts C: 32 ft. lbs. (43 Nm).

32. Install or connect the following:
- Stud bolt for the accessory drive belt. Tighten to 7 ft. lbs. (10 Nm).
- Accessory drive belt tensioner. Tighten to 44 ft. lbs. (60 Nm).
- Right-hand engine mounting bracket. Tighten to 40 ft. lbs. (54 Nm).
- Engine mounting insulator. Tighten top bolt to 70 ft. lbs. (95 Nm) and bottom bolts to 64 ft. lbs. (87 Nm).
- Lower oil pan. Tighten bolts to 80 inch lbs. (9 Nm).
- Timing chain tensioner
- Crankshaft position sensor. Tighten to 80 inch lbs. (9 Nm).

33. Align the crankshaft pulley key set with the key groove of the pulley, and side on the pulley. Using the Special Tool previously used to remove, tighten the crankshaft pulley bolt to 125 ft. lbs. (170 Nm).

34. Turn the crankshaft counterclockwise and disconnect the plunger knock pin from the hook.

35. Turn the crankshaft clockwise and check that the slipper is pushed by the plunger.

36. The remainder of the installation is the reverse order of removal.

37. Refill the engine with oil to the correct level.

38. Start the engine and check for leaks.

2GR-FE Engine

1. Before servicing the vehicle, refer to the Precautions Section.

2. Remove the engine/transaxle assembly from the vehicle.

3. Remove the transaxle.

4. Remove the oil dipstick tube.

5. Remove the driveplate.

6. Install the engine to a suitable engine stand.

7. Remove or disconnect the following:
- Idler pulley
- Right-side engine mounting bracket
- Accessory drive belt tensioner
- Water pump pulley
- No. 2 timing gear cover
- Engine mounting stay and bracket
- Water inlet housing

8. Using Special Tool 09213-70011, hold the crankshaft pulley and loosen the pulley bolt.

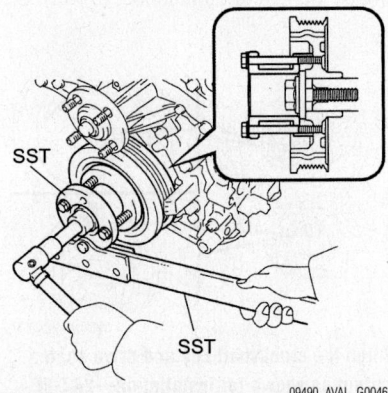

09490_AVAL_G0046

Loosen the crankshaft pulley bolt with Special Tool 09213-70011—2GR-FE engine

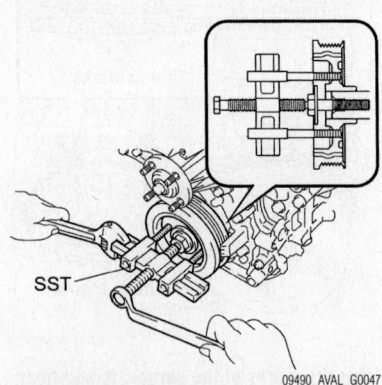

09490_AVAL_G0047

Remove the crankshaft pulley with Special Tool 09950-50013—2GR-FE engine

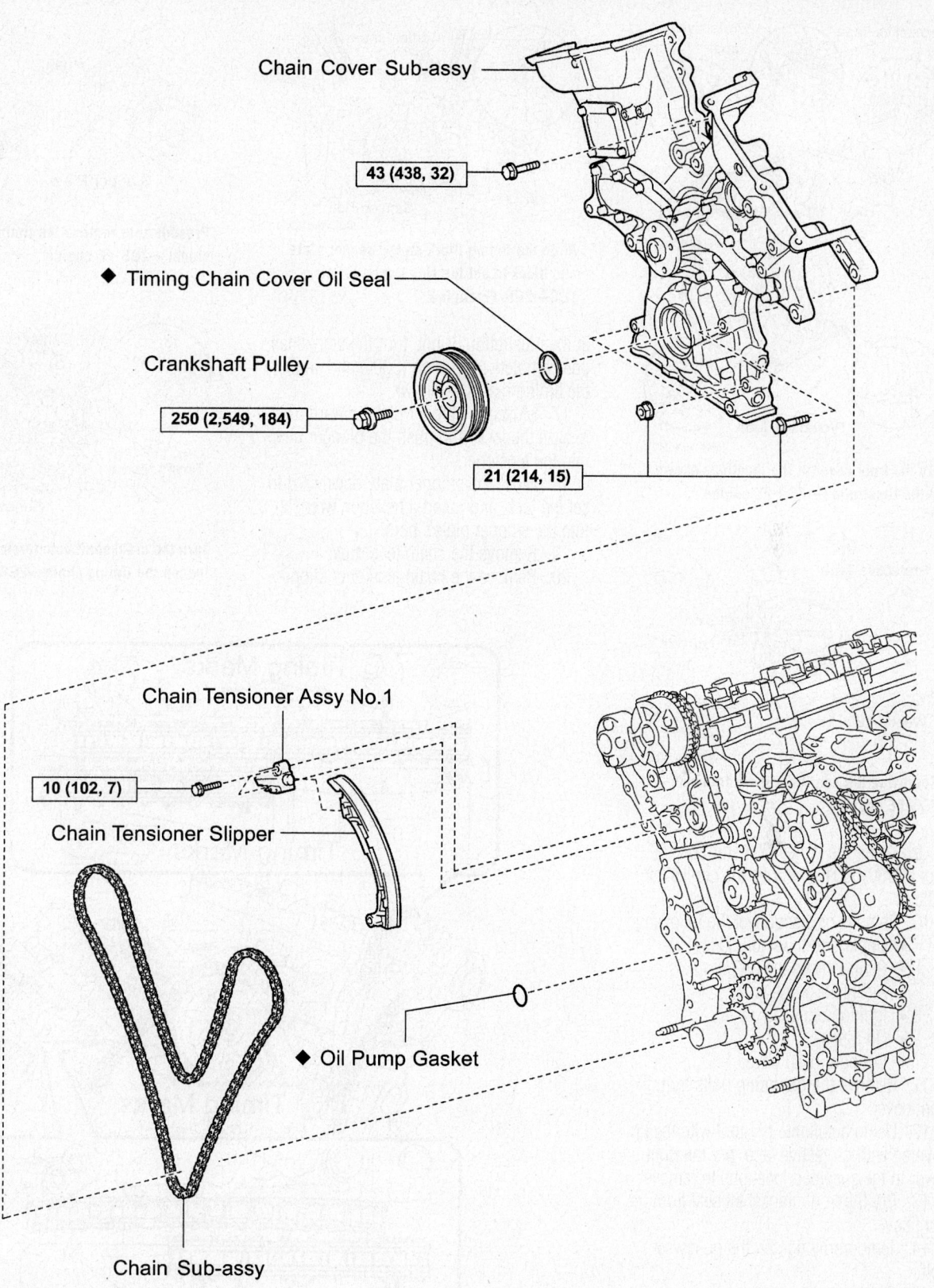

Chain Cover Sub-assy

43 (438, 32)

◆ Timing Chain Cover Oil Seal

Crankshaft Pulley

250 (2,549, 184)

21 (214, 15)

Chain Tensioner Assy No.1

10 (102, 7)

Chain Tensioner Slipper

◆ Oil Pump Gasket

Chain Sub-assy

N·m (kgf·cm, ft·lbf) : Specified torque
◆ Non-reusable part

09490_AVAL_G0045

Exploded view of the timing chain and front cover components—2GR-FE engine

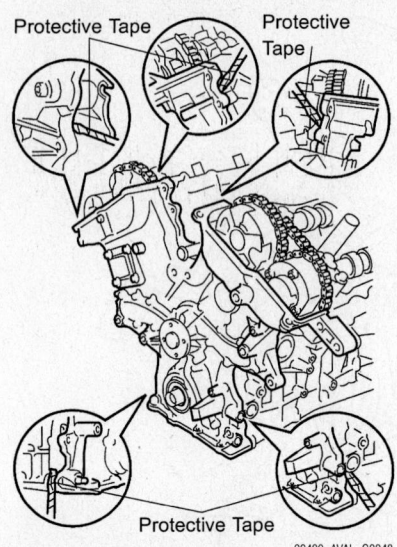

Pry the front cover at the locations shown in the illustration—2GR-FE engine

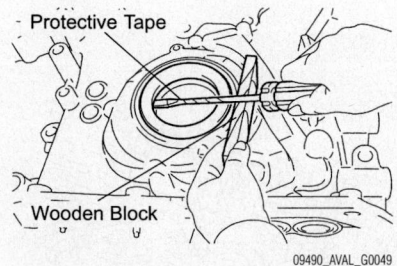

Removing the seal from the front cover— 2GR-FE engine

9. Using the pulley bolt and Special Tool 09950-50013, remove the crankshaft pulley.

10. Remove or disconnect the following:
- Upper and lower oil pans
- O-rings from the oil pump
- Air intake surge tank
- Ignition coils
- Oil pipes
- Cylinder head cover

11. Remove the mounting bolts from the front cover.

12. Using a suitable pry tool with the tip covered with protective tape, pry the front cover in the specified locations to remove.

13. Pry the front crankshaft seal from the front cover.

14. Temporarily tighten the pulley set bolt.

15. Set the timing mark on the crank angle sensor plate to the right-hand block bore center line to put the No. 1 cylinder at TDC.

16. Check that the timing marks of the camshaft timing gears are aligned with the timing marks of the bearing caps as shown

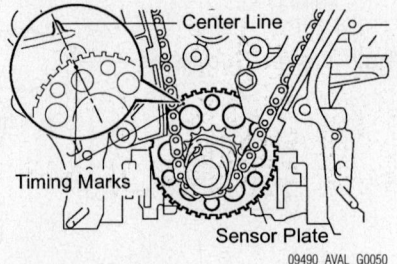

Align the timing mark on the sensor plate and block to set the No. 1 cylinder at TDC—2GR-FE engine

in the illustration. If not, turn the crankshaft one complete revolution (360°) and align the timing marks as shown.

17. Move the stopper plate upward to release the lock, and push the plunger deep into the tensioner.

18. Move the stopper plate downward to set the lock, and insert a hexagon wrench into the stopper plate's hole.

19. Remove the chain tensioner.

20. Remove the chain tensioner slipper.

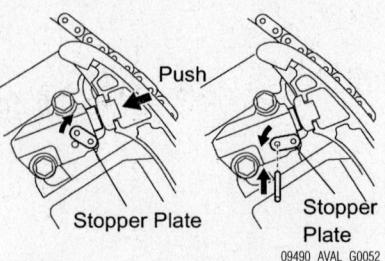

Procedure to remove the timing chain tensioner—2GR-FE engine

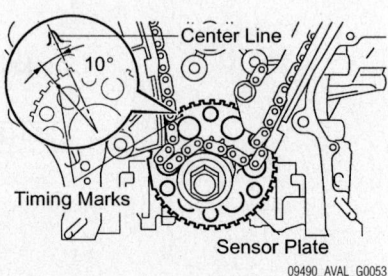

Turn the crankshaft counterclockwise to loosen the timing chain—2GR-FE engine

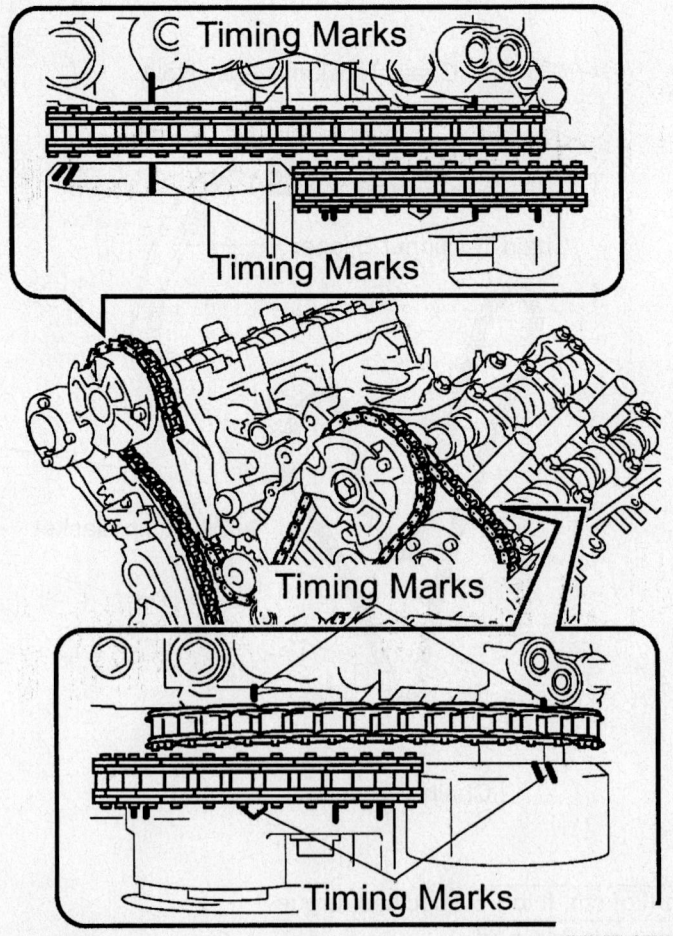

Ensure the timing marks of the camshaft timing gears are aligned with the bearing cap timing marks—2GR-FE engine

Align the timing chain marked links with the camshaft timing gear marks to install—2GR-FE engine

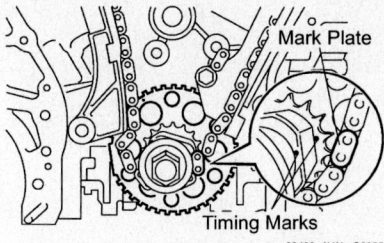

Align the timing chain marked link with the crankshaft gear to install—2GR-FE engine

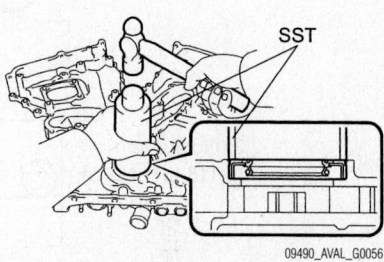

Tap a new oil seal into place in the front cover—2GR-FE engine

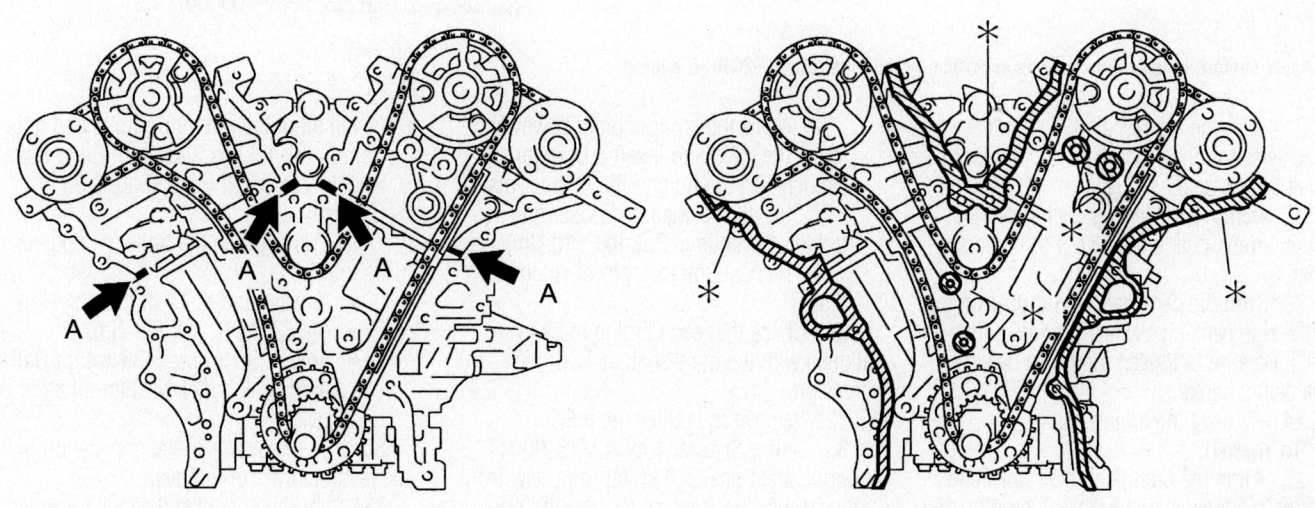

■ Seal Packing

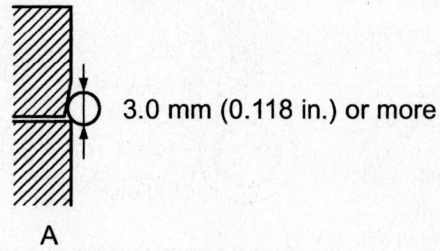

3.0 mm (0.118 in.) or more

Apply sealant to the engine block as shown when installing the front cover—2GR-FE engine

09490_AVAL_G0057

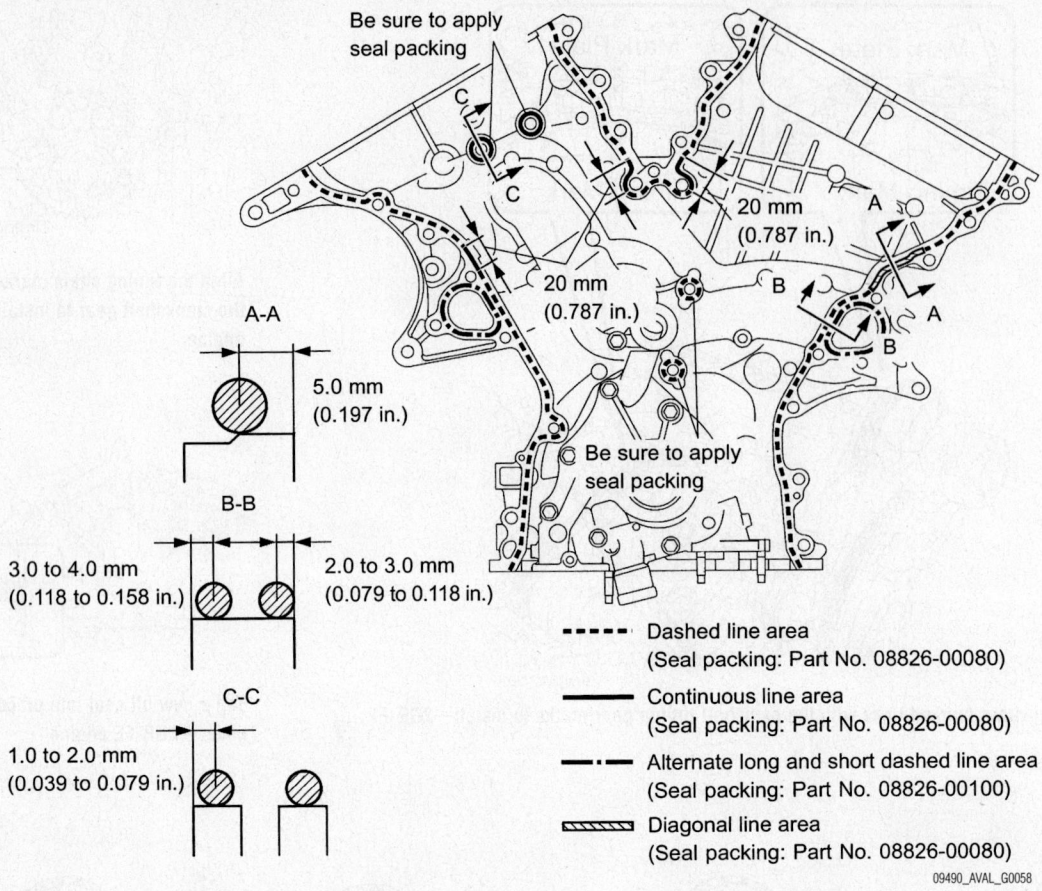

Be sure to apply seal packing

20 mm (0.787 in.)

20 mm (0.787 in.)

Be sure to apply seal packing

A-A

5.0 mm (0.197 in.)

B-B

3.0 to 4.0 mm (0.118 to 0.158 in.)

2.0 to 3.0 mm (0.079 to 0.118 in.)

C-C

1.0 to 2.0 mm (0.039 to 0.079 in.)

- - - - - Dashed line area (Seal packing: Part No. 08826-00080)

——— Continuous line area (Seal packing: Part No. 08826-00080)

—·—·— Alternate long and short dashed line area (Seal packing: Part No. 08826-00100)

▨▨▨ Diagonal line area (Seal packing: Part No. 08826-00080)

09490_AVAL_G0058

Apply sealant to the front cover as specified by the illustration—2GR-FE engine

21. Turn the crankshaft 10° counter-clockwise to loosen the chain off the crank-shaft timing gear.

22. Remove the timing chain from the crank timing gear and place it on the crank-shaft.

23. Turn the camshaft timing gear on the right-side bank clockwise (approximately 60°). Be sure to loosen the chain between the center banks.

24. Remove the timing chain.

To install:

25. Align the orange marked links and timing mark as shown in the illustration and install the timing chain.

26. Turn the camshaft timing gear on the right-side bank counterclockwise to tighten the chain between banks.

27. Align the yellow marked link and timing mark as shown in the illustration and install the chain onto the crankshaft timing gear.

28. Temporarily tighten the crankshaft pulley set bolt.

29. Install the chain tensioner slipper.

30. Install the chain tensioner as follows:

a. Move the stopper plate upward to release the lock, and push the plunger deep into the tensioner.

b. Move the stopper plate downward to set the lock, and insert a hexagon wrench into the hole of the stopper plate.

c. Install the chain tensioner and tighten the bolts to 7 ft. lbs. (10 Nm).

d. Remove the lock pin of chain ten-sioner.

31. Check that each timing mark is aligned with the crankshaft at TDC com-pression.

32. Remove the pulley set bolt.

33. Using Special Tool 09316-60011 or equivalent seal driver, tap in a new front oil seal into the front cover until its sur-face is flush with the front cover case edge. Apply multi-purpose grease to the oil seal lip.

34. Install the front cover as follows:

a. Apply a continuous bead of sealant 0.118 in. (3 mm) in diameter to the engine as shown in the illustration.

b. Apply sealant to the front cover as shown in the illustration.

c. Install a new oil pump gasket.

d. Align the oil pump's drive rotor spline and the crankshaft as shown in the illustration. Install the spline and chain cover to the crankshaft.

e. Install the front cover and loosely

install all of the mounting bolts and nuts. Bolts A are 1.57 in. (40 mm); Bolts B are 2.17 in. (55 mm); Bolts C are 0.98 in. (25 mm).

f. Fully tighten the bolts in sequence as follows:

- Areas 1 and 2: 15 ft. lbs. (21 Nm)
- Area 3: 15 ft. lbs. (21 Nm)
- Area 4: 32 ft. lbs. (43 Nm) for Bolt A; 15 ft. lbs. (21 Nm) for all other bolts

35. The remainder of the installation is the reverse order of removal.

36. Start the engine and check for leaks.

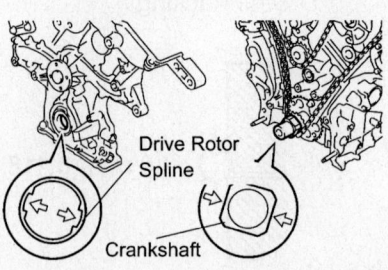

Drive Rotor Spline

Crankshaft

09490_AVAL_G0059

Correct orientation of the oil pump rotor and crankshaft during installation of the front cover—2GR-FE engine

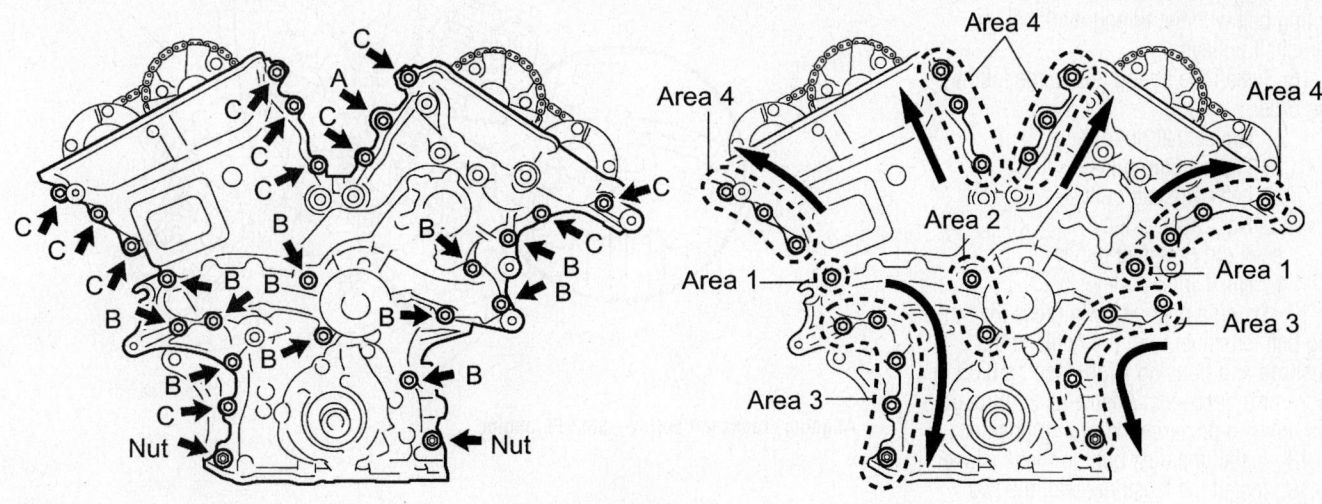

Bolt identification and torque sequence for the front cover—2GR-FE engine

Timing Belt

REMOVAL & INSTALLATION

1MZ-FE Engines

1. Before servicing the vehicle, refer to the Precautions Section.

2. Remove all necessary components for access to the timing belt covers.

> ❈❈ **CAUTION**
>
> **If equipped with an air bag, be sure to disconnect the negative battery cable and wait at least 90 seconds before proceeding.**

3. Remove the lower timing belt cover by removing the four bolts.

4. Remove the No. 2 timing belt cover as follows:

   a. Remove the bolt and disconnect the engine wire protector from the No. 3 (rear) timing belt cover.

   b. Disconnect the engine wire protector clamp from the No. 3 timing belt cover.

   c. Remove the five bolts from the No. 2 timing belt cover.

   d. Remove the No. 2 cover from the engine.

5. Remove the right engine mounting bracket by removing the nut and two bolts.

6. Remove the crankshaft timing belt guide.

7. Temporarily install the crankshaft pulley bolt.

8. Turn the crankshaft and align the crankshaft timing pulley groove with the oil pump alignment mark. Always turn the engine clockwise.

9. Ensure the timing mark of the camshaft timing pulleys and rear timing belt covers are aligned. If not, turn the engine over an additional 360 degrees (one revolution).

10. Remove the crankshaft pulley bolt.

➥**If the belt is to be reused, align the installation marks on the belt to the marks on the pulleys. If the marks have worn off, make new ones.**

11. Alternately loosen the two timing belt tensioner bolts. Remove the tensioner and dust boot.

12. Remove the timing belt.

To install:

13. Remove any oil or water from the pulleys.

14. Align the front mark of the timing belt with the dot mark of the crankshaft timing pulley.

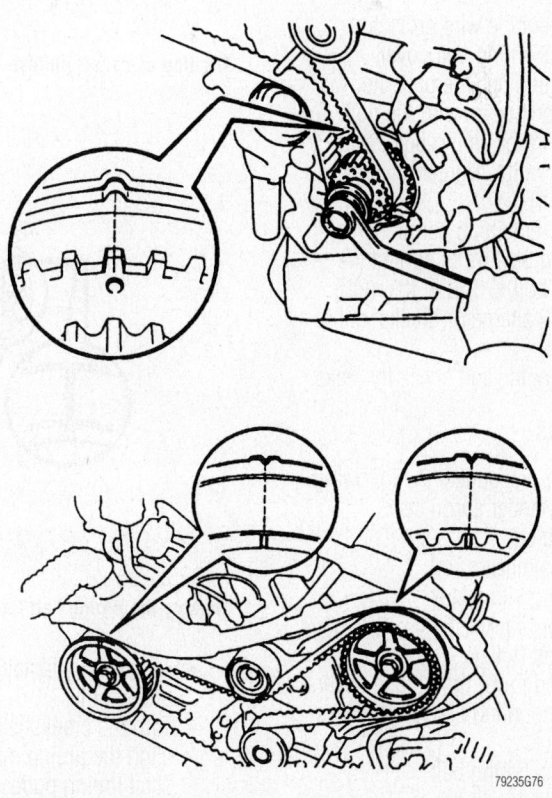

Camshaft and crankshaft timing belt sprocket alignment mark positioning for belt service—Toyota 3.0L (1MZ-FE) engine

15. Align the installation marks on the timing belt with the timing marks of the camshaft pulleys.

16. Install the timing belt in the following order:

   a. Crankshaft pulley.
   b. Water pump pulley.
   c. Left camshaft pulley.
   d. No. 2 idler pulley.
   e. Right camshaft pulley.
   f. No. 1 idler pulley.

17. Using a press, slowly press the timing belt tensioner until the holes of the pushrod and housing align. Insert a 0.05 in. (1.27mm) hexagonal Allen wrench through the holes to preserve the setting position.

18. Install the dust boot to the tensioner.

19. Install the tensioner with the two bolts. Alternately tighten the bolts to 20 ft. lbs. (27 Nm). Remove the Allen wrench.

20. Turn the crankshaft clockwise and align the crankshaft timing pulley groove with the oil pump alignment mark.

21. Ensure the camshaft timing marks align with the timing marks on the rear timing belt cover.

22. Install the timing belt guide.

23. Install the right engine mounting bracket and tighten the bolts to 21 ft. lbs. (28 Nm).

24. Install the upper timing belt cover with the five bolts. Tighten the bolts to 74 inch lbs. (8 Nm).

25. Install the engine wire protector clamp to the No. 3 timing belt cover.

26. Install the engine wire protector to the No. 3 timing belt cover with the bolt.

27. Install the lower timing belt cover by installing the four bolts. Tighten the bolts to 74 inch lbs. (8 Nm).

28. Install the remaining components. During installation be sure to tighten the crankshaft pulley bolt to 159 ft. lbs. (215 Nm) and the No. 2 alternator bracket nut to 21 ft. lbs. (28 Nm).

29. Start the engine and check for leaks.

3MZ-FE Engines

1. Remove or disconnect the following:
   - RH front fender apron seal
   - A/C drive belt
   - Engine stabilizing rod
   - Engine mounting stay
   - Alternator bracket
   - Crankshaft pulley
   - Upper and lower timing belt covers
   - Engine mounting bracket (If necessary)

2. Remove the timing belt
   - Set the No 1 cylinder to TDC/compression

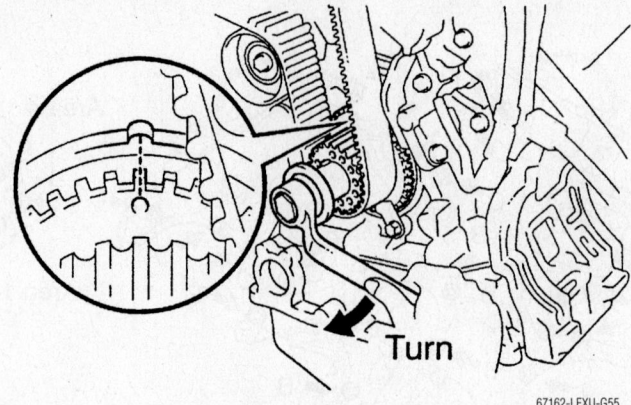

Aligning crankshaft pulley—3MZ-FE engine

67162-LEXU-G55

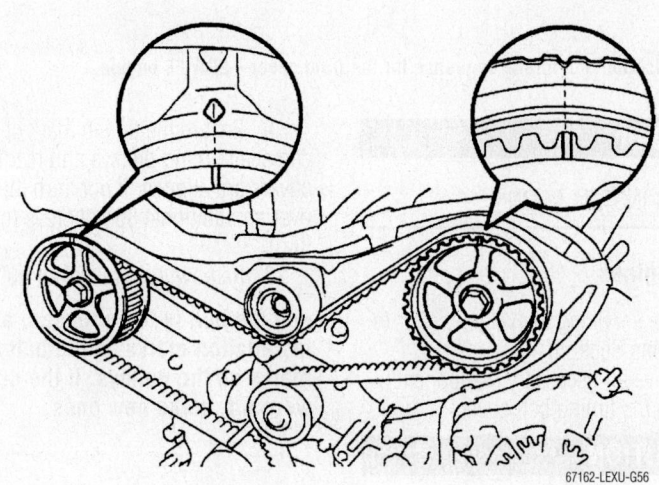

Aligning camshaft pulleys—3MZ-FE engine

67162-LEXU-G56

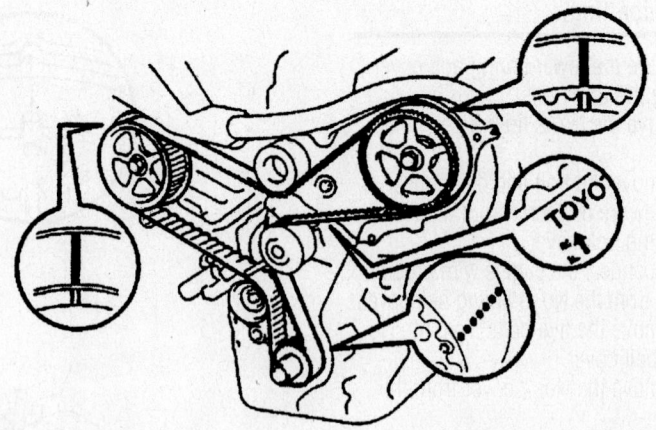

Marking the timing belt for reuse—3MZ-FE engine

67162-LEXU-G57

- Temporarily install the crankshaft pulley bolt
- Turn the crankshaft clockwise, the align the timing mark on the crankshaft timing pulley with the mark on the oil pump body
- Verify that the timing marks on the camshaft timing pulleys align with the timing marks on the inside timing belt cover. If not, turn the crankshaft on revolution (360 degrees).
- If reusing the timing belt make certain there are 3 location marks on

the timing belt corresponding with the marks on the timing gears.

- Set the No 1 cylinder to approximately 60 degrees BTDC/compression. Turn the crankshaft counterclockwise by approximately 60 degrees.

❋❋ WARNING

If the timing belt is disengaged, having the crankshaft pulley set at the wrong angle can cause contact of the piston with the valves, causing damage to the valves. Always set the crankshaft pulley at the correct angle.

- Remove timing belt tensioner

➡**Do NOT reinstall the timing belt tensioner with the plunger extended.**

3. Remove the timing belt in the following order.
- No 1 idler pulley
- RH camshaft timing pulley
- No. 2 idler pulley
- LH camshaft timing pulley
- Water pump pulley
- Crankshaft timing pulley
4. Inspect the timing belt

➡**Do not reuse the timing belt if there is evidence of fraying, oil contamination, cracking, tooth damage or wear, or belt distortion. If there is any doubt in the belt condition, replace the belt.**

To install:

5. After inspecting the pulleys for wear and checking for oil leaks install the timing belt using the following procedure.
- Temporarily install the crankshaft pulley bolt
- Make sure the crankshaft pulley is set at 60° BTDC
- Align the timing marks on the camshaft pulley with the respective marks on the inside timing cover.
- If reusing the old belt, align the marks previously made on the belt with the marks on the timing gears.
- Install the belt in the reverse order used when removing the belt
6. Install the timing belt tensioner using the following procedure

➡**Keep the tensioner in an upright position**

- Slowly depress the push rod and align the hole with the hole in the housing

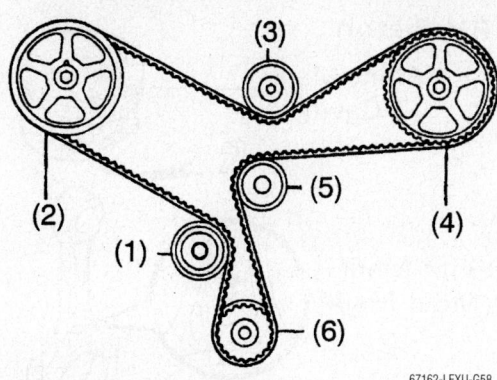

Removal of timing belt from pulleys—3MZ-FE engine

- Insert a 1.5 mm hexagon wrench through the holes to maintain the setting position of the push rod.
- Install the tensioner with the 2 bolts and torque to 20 ft. lbs (27 Nm).
- Remove the hexagon wrench
7. Slowly turn the crankshaft 2 revolutions clockwise.
8. Check to see that all timing marks are in alignment.
9. Remove the crankshaft bolt
10. Install the timing belt guide
11. The remainder of the installations is the reverse order of removal.
12. Start the engine and check for leaks.

Piston and Ring Positioning

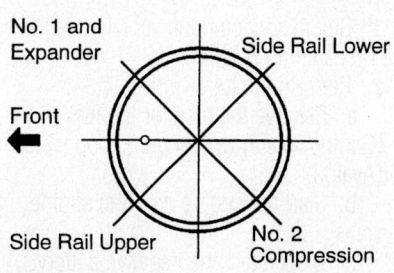

2AZ-FE engine—piston ring end-gap spacing

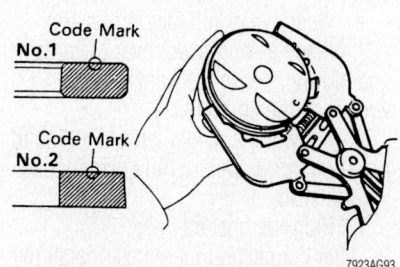

1MZ-FE engine—compression ring positioning

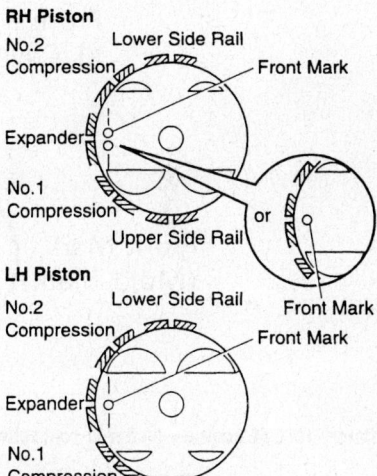

1MZ-FE engine—piston ring end-gap spacing

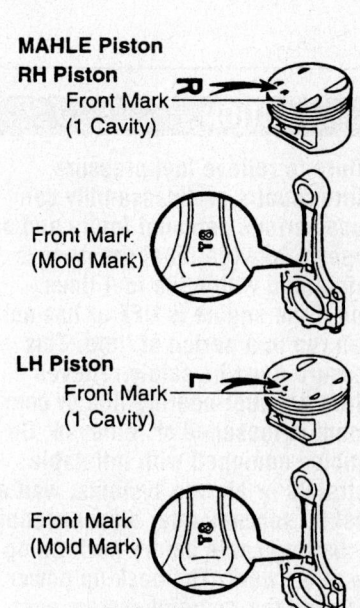

Avalon 1MZ-FE engine—piston-to-connecting rod assembly

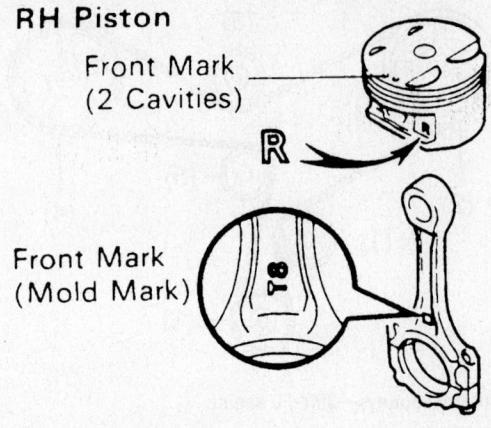

RH Piston

Front Mark
(2 Cavities)

Front Mark
(Mold Mark)

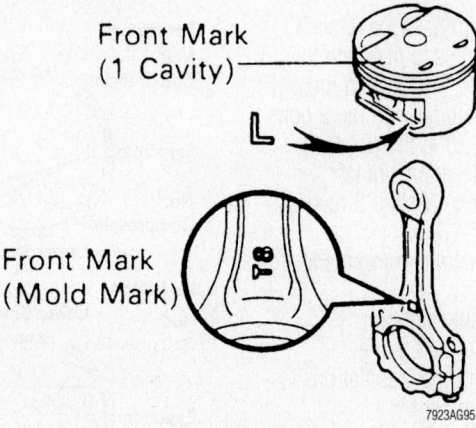

LH Piston

Front Mark
(1 Cavity)

Front Mark
(Mold Mark)

7923AG95

Camry 1MZ-FE engine—piston-to-connecting rod assembly

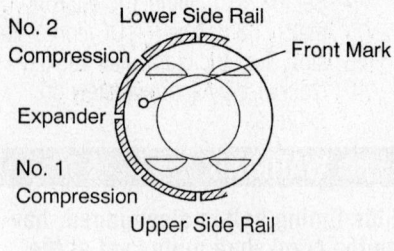

No. 2
Compression

Lower Side Rail

Front Mark

Expander

No. 1
Compression

Upper Side Rail

67170-TOYO-G42

3MZ-FE engine—piston ring end-gap spacing

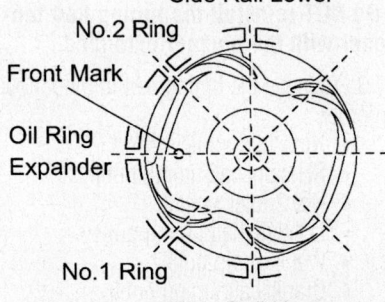

Side Rail Lower

No.2 Ring

Front Mark

Oil Ring
Expander

No.1 Ring

Side Rail Upper

09490_AVAL_G0061

2GR-FE—position ring end-gap spacing

FUEL SYSTEM

Fuel System Pressure

RELIEVING

✳✳ CAUTION

Failure to relieve fuel pressure before repairs or disassembly can cause serious personal injury and/or property damage. Fuel pressure is maintained within the fuel lines, even if the engine is OFF or has not been run in a period of time. This pressure must be safely relieved before any fuel-bearing line or component is loosened or removed. On vehicles equipped with inflatable restraints or air bag systems, wait at least 90 seconds after disconnecting the battery cable before performing any other work. The back-up power will keep the restraint system energized for a period of time after the battery is disconnected.

1. Before servicing the vehicle, refer to the Precautions Section.
2. Perform the following:
   a. Remove the fuse for the fuel pump (Camry/Solara) or fuel pump connector (Avalon).
   b. Start the engine until the engine stalls.
   c. Disconnect the negative battery cable.
   d. Place a catch-pan under the joint to be disconnected. A large quantity of fuel may be released when the joint is opened.
   e. Wear eye or full face protection.
   f. Place a shop towel over the area and slowly release the joint using a wrench of the correct size.
   g. Allow the any fuel left in the line to bleed off slowly before fully disconnecting the joint.
   h. Plug the opened lines.
3. After connecting fuel lines, install the fuse or connector for the fuel pump and start the engine.

Fuel Filter

REMOVAL & INSTALLATION

The fuel filter is in the tank as part of the fuel pump assembly.

Fuel Pump

REMOVAL & INSTALLATION

1. Before servicing the vehicle, refer to the Precautions Section.
2. Remove or disconnect the following:
   - Negative battery cable. Wait at least 90 seconds before performing any other work.
   - Rear seat bottom and seat back
   - Rear floor service hole cover
   - Fuel suction w/pump and gauge tube assembly
   - 8 bolts and fuel tank vent tube set plate
   - Vapor pressure sensor assembly

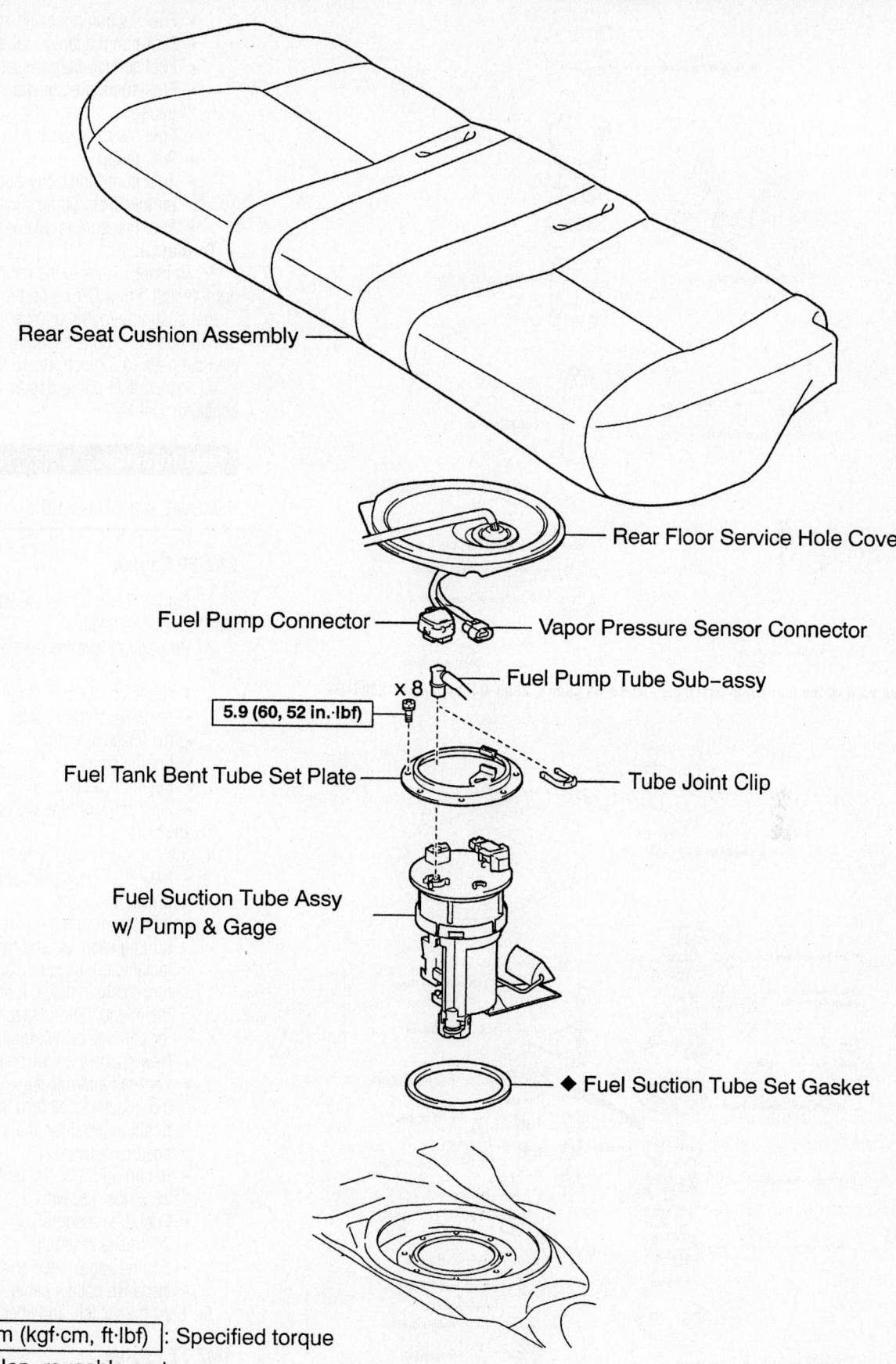

Rear Seat Cushion Assembly

Rear Floor Service Hole Cover

Fuel Pump Connector

Vapor Pressure Sensor Connector

Fuel Pump Tube Sub-assy

x 8

5.9 (60, 52 in. lbf)

Fuel Tank Bent Tube Set Plate

Tube Joint Clip

Fuel Suction Tube Assy
w/ Pump & Gage

◆ Fuel Suction Tube Set Gasket

N·m (kgf·cm, ft·lbf) : Specified torque

◆ Non-reusable part

67170-TOYO-G43

Exploded view for fuel pump removal—2002–06 Camry, 2005–04 Avalon and 2004–06 Solara

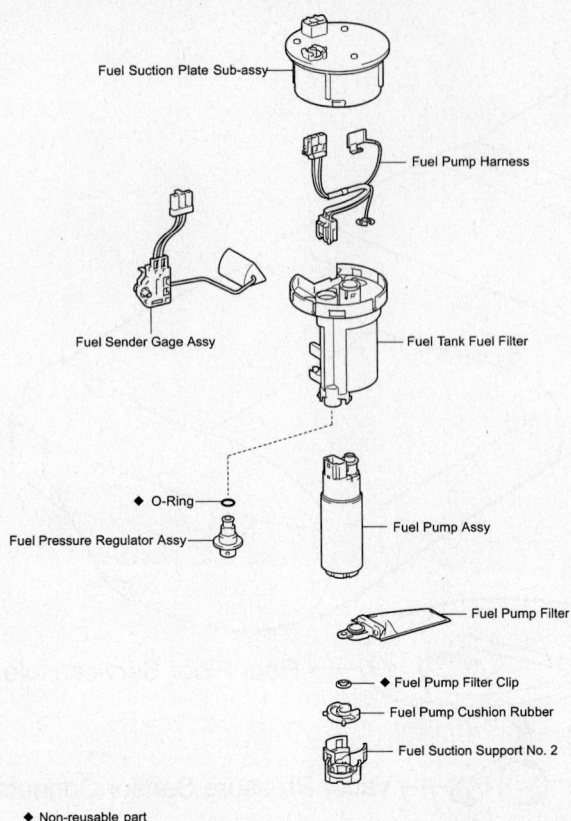

Exploded view of the fuel pump assembly—2002–06 Camry, 2005–04 Avalon and 2004–06 Solara

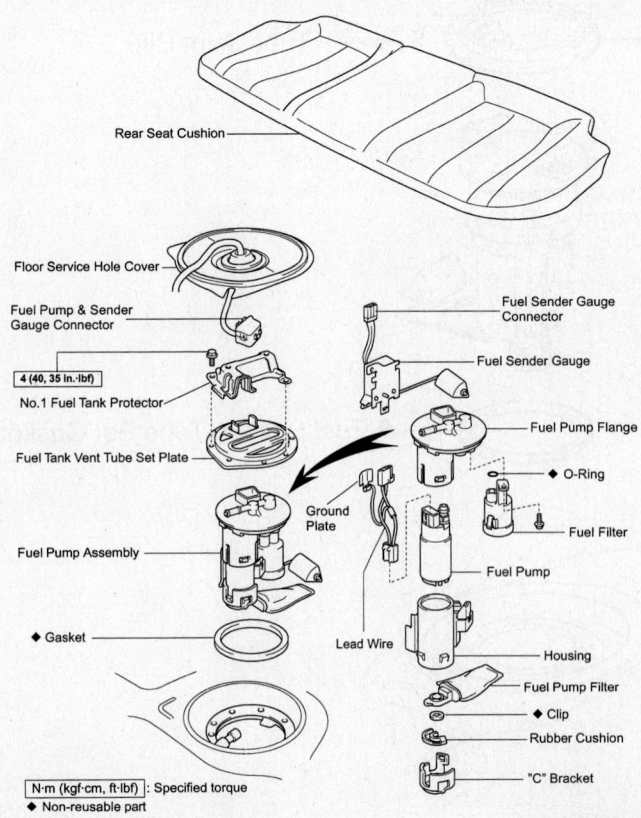

Exploded view of fuel pump removal and components—2002–04 Avalon and 2002–03 Solara

- Fuel suction hose and support
- Fuel pump cushion rubber
- Fuel sender gauge assembly
- Fuel suction plate w/sender gauge
- Fuel pump harness
- Fuel pump
- Fuel pump filter. Pry out clip and remove from pump
- Fuel pressure regulator assembly

To install:

3. To install, reverse the removal procedure. Install a new O-ring on the fuel suction w/pump and gauge assembly. Torque the 8 bolts securing the fuel tank set plate to 52 inch lbs. (6.0 Nm).

4. Inspect fuel pump operation and check for fuel leaks.

Fuel Injectors

REMOVAL & INSTALLATION

2AZ-FE Engine

1. Before servicing the vehicle, refer to the Precautions Section.
2. Properly relieve the fuel system pressure.
3. Remove or disconnect the following:
   - Negative battery cable
   - Air intake assembly
   - Engine cover
   - Fuel tube assembly
   - Fuel delivery pipe with the injectors

To install:

4. Install or connect the following:
   - New O-rings, coated with gasoline, on each injector
   - Injectors into the fuel rail while turning each left and right. After installation, check that the injectors turn freely in place; if not, remove the injector and inspect the O-ring for damage or deformation.
   - New spacers on the head
   - Fuel rail and injectors; check that the injectors still turn freely in position. Position the injector connectors outward
   - Retaining bolts, tightening them to 15 ft. lbs. (20 Nm)
   - Fuel tube assembly
   - Air intake assembly
   - Engine appearance cover.
   - Negative battery cable
5. Start the vehicle and check for leaks

1MZ-FE Engine

1. Before servicing the vehicle, refer to the Precautions Section.

2. Properly relieve the fuel system pressure.

3. Drain the cooling system.

4. Remove or disconnect the following:
- Negative battery cable. Work must be started approximately 90 seconds or longer after the negative battery cable has been disconnected, if equipped with an air bag
- Accelerator and throttle cables
- Air cleaner assembly
- V-bank cover
- Emission valve control set
- No. 2 EGR pipe
- Hydraulic motor pressure pipe from the water inlet and air inlet chamber
- Air intake chamber assembly
- Injector wiring
- Air assist pipe from the bracket on the No. 1 fuel pipe
- Air assist hoses from the intake manifold
- Fuel return hose from the No. 1 fuel pipe
- Fuel inlet hose for the fuel filter
- 2 union bolts holding the No. 2 fuel pipe to the delivery pipes
- Fuel return hose from the fuel pressure regulator
- Union bolt for the right hand delivery pipe, 2 gaskets, 2 bolts, left hand delivery pipe together with the 3 injectors and the No. 2 fuel pipe
- Union bolt for the delivery pipe and 2 gaskets from the No. 2 fuel pipe
- The 3 bolts, right hand delivery pipe together with the 3 injectors and the No. 1 fuel pipe
- The 4 spacers from the intake manifold
- The 6 injectors from the delivery pipes
- The two O-rings and two grommets from each injector

To install:

5. Install or connect the following:
- 2 new grommets to each injector
- New O-rings, with a light coat of fuel, to each injector
- Injectors
- The 4 spacers on the intake manifold
- Right hand delivery pipe and the No. 1 fuel pipe together with the 3 injectors in position on the intake manifold
- Bolt holding the right side delivery pipe, temporarily, to the intake manifold
- Left hand delivery pipe and the No. 2 fuel pipe together with the 3 injectors in position on the intake manifold

- Fuel return hose to the fuel pressure regulator

6. Temporarily install the 2 bolts holding the left hand delivery pipe to the intake manifold.

7. Temporarily install the No. 2 fuel pipe to the left side delivery pipe with the union bolt and 2 new gaskets.

8. Check that the injectors rotate smoothly. If they do not, Replace the O-rings.

9. Position the injector connector outward. Tighten the 4 bolts holding the delivery pipes to the intake manifold and tighten to 7 ft. lbs. (10 Nm). Tighten the bolt holding the No. 1 fuel pipe to the intake manifold to 14 ft. lbs. (20 Nm). Tighten the 2 union bolts holding the no. 2 fuel pipe to the delivery pipes to 24 ft. lbs. (32 Nm).

10. Install or connect the following:
- Fuel inlet and return hoses. Union bolt: 22 ft. lbs. (30 Nm).
- Fuel return hose to the No. 1 fuel pipe. Pass the fuel return hose under the heater hoses
- Air assist hoses to the intake manifold
- Air assist pipe to the bracket on the No. 1 fuel pipe
- Fuel injector wiring connectors
- Air intake chamber assembly
- Hydraulic motor pressure pipe to the intake chamber. Bolts: 69 inch lbs. (8 Nm)
- No. 2 EGR pipe with new gaskets, tighten to 9 ft. lbs. (12 Nm)
- Emission control valve set
- V-bank cover
- Air cleaner hose
- Throttle and accelerator cables
- Negative battery cable

11. Refill the cooling system to the proper level.

12. Start the vehicle and check for leaks.

3MZ-FE Engine

1. Before servicing the vehicle, refer to the Precautions Section.

2. Remove or disconnect the following:
- Negative battery cable
- Suspension upper center brace
- V-bank cover sub assembly
- Air cleaner assembly
- Emission control valve set

3. Remove intake air surge tank

4. Remove or disconnect the following:
- Throttle motor connector
- Water by-pass hoses
- Union check valve hose
- Ventilation hose
- Pressure feed hose

- Engine hangers
- Surge tank stays
- Bond cable connector
- Emission control valve bracket
- Intake air surge tank
- Gasket from intake air surge tank

5. Remove fuel pipe assembly

6. Remove or disconnect the following:
- Fuel pulsation damper and gasket
- Fuel pipe union bolt and 2 gaskets
- Bolt and separate the fuel pipe

7. Remove the fuel injector assembly

8. Remove or disconnect the following:
- 6 fuel injector connectors
- 4 bolts, then remove the fuel injector delivery pipes

☼☼ WARNING

Be careful not to drop the fuel injectors when removing the fuel delivery pipes.

- 4 delivery pipe spacers from intake manifold
- 6 insulators from the intake manifold
- Fuel injector from the fuel delivery pipes.

To install:

9. Install new fuel injector assembly

10. Install or connect the following
- A new insulator and grommet to each injector
- A light coat of gasoline to new O-rings and install them to each injector
- A light coat of gasoline on the place where a delivery pipe touches an O-ring of the injector
- Injector, while turning the clockwise and counterclockwise, into the delivery pipe
- The 6 insulators and 4 spacers in position on the intake manifold
- A light coat of gasoline on the place where an intake manifold touches an O-ring
- The delivery pipes in position on the intake manifold
- Temporarily, the 4 bolts holding the delivery pipe to the intake manifold

➡**Check that the injectors rotate smoothly. If the injectors do not rotate smoothly, the probable cause is incorrect installation of the O-rings. Replace the O-rings.**

- 4 bolts. Tighten bolts uniformly. Torque to: 7 ft. lbs. (10 Nm)
- 6 fuel injector connectors

11. Install fuel pipe assembly

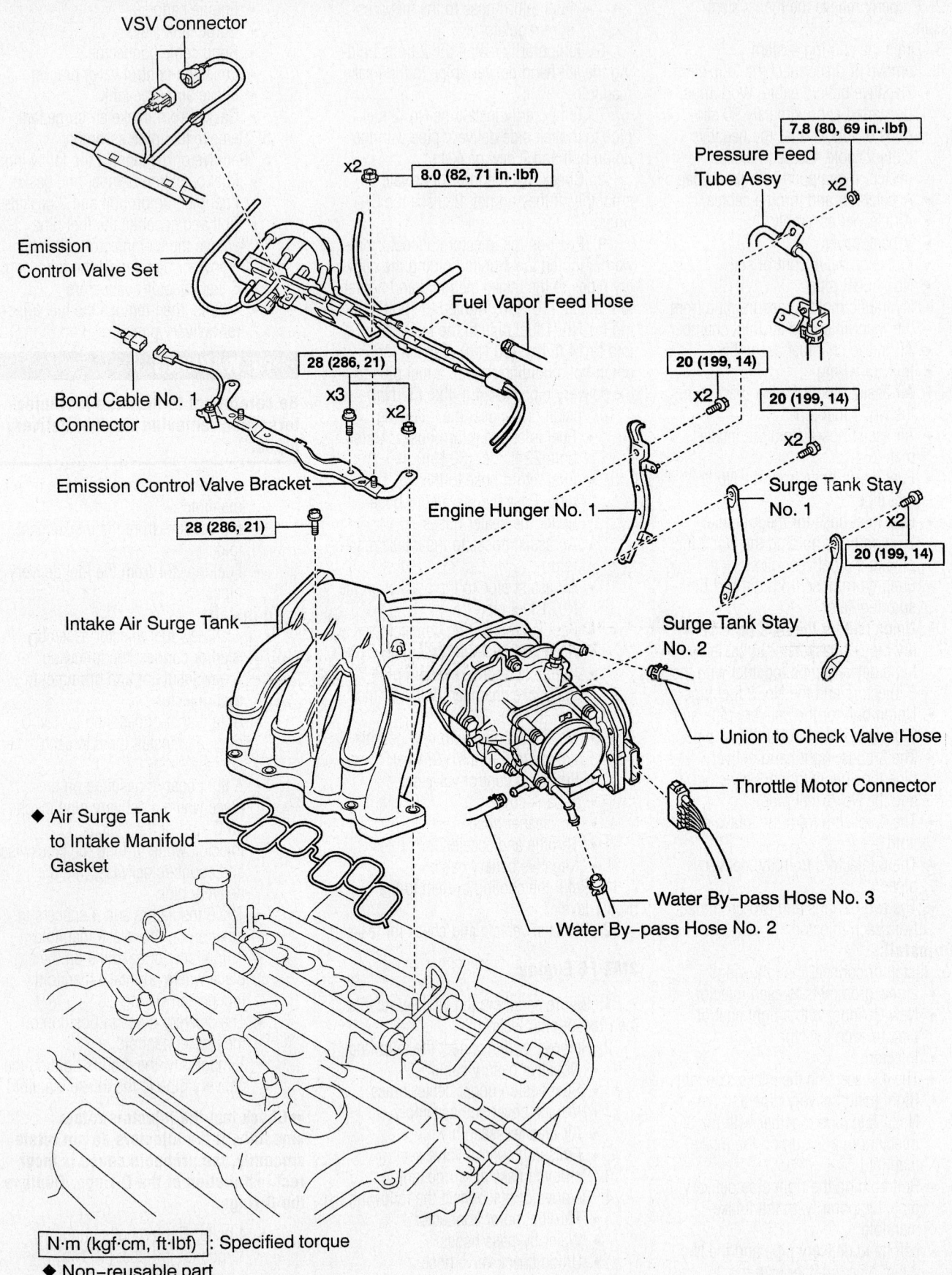

VSV Connector

Emission
Control Valve Set

Bond Cable No. 1
Connector

Emission Control Valve Bracket

Intake Air Surge Tank

◆ Air Surge Tank
to Intake Manifold
Gasket

x2 8.0 (82, 71 in.·lbf)

Fuel Vapor Feed Hose

28 (286, 21)
x3
x2

28 (286, 21)

Pressure Feed
Tube Assy

7.8 (80, 69 in.·lbf)
x2

20 (199, 14)
x2

20 (199, 14)
x2

Engine Hunger No. 1

Surge Tank Stay
No. 1
x2

Surge Tank Stay
No. 2

20 (199, 14)

Union to Check Valve Hose

Throttle Motor Connector

Water By-pass Hose No. 3

Water By-pass Hose No. 2

N·m (kgf·cm, ft·lbf) : Specified torque

◆ Non-reusable part

Exploded view of Intake Air Surge tank—3MZ-FE engine

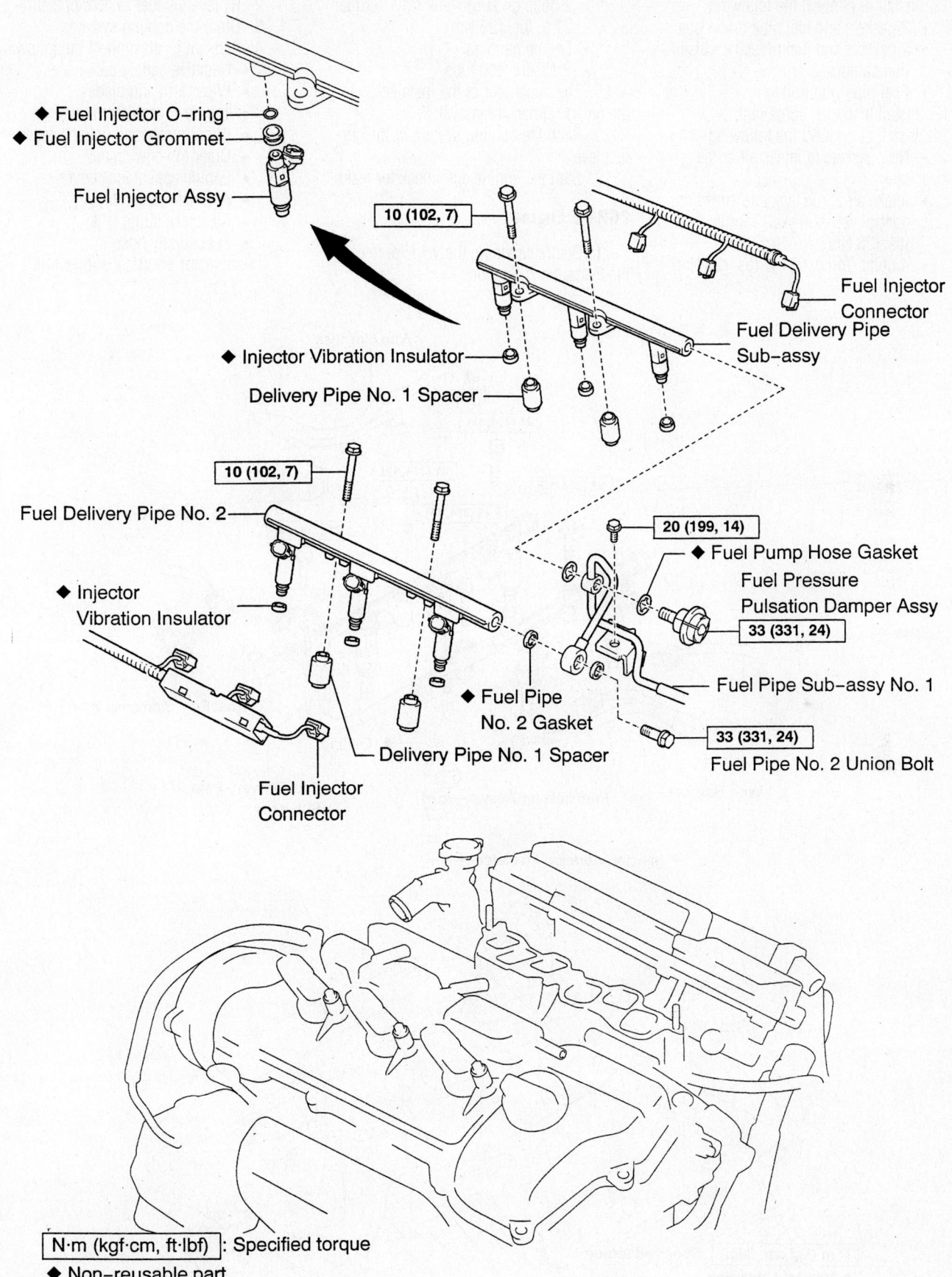

◆ Fuel Injector O-ring
◆ Fuel Injector Grommet
Fuel Injector Assy

10 (102, 7)

Fuel Injector
Connector
Fuel Delivery Pipe
Sub-assy

◆ Injector Vibration Insulator
Delivery Pipe No. 1 Spacer

10 (102, 7)

Fuel Delivery Pipe No. 2

◆ Injector
Vibration Insulator

20 (199, 14)

◆ Fuel Pump Hose Gasket
Fuel Pressure
Pulsation Damper Assy

33 (331, 24)

Fuel Pipe Sub-assy No. 1

◆ Fuel Pipe
No. 2 Gasket
Delivery Pipe No. 1 Spacer

Fuel Injector
Connector

33 (331, 24)
Fuel Pipe No. 2 Union Bolt

N·m (kgf·cm, ft·lbf) : Specified torque
◆ Non-reusable part

67162-LEXU-G62

Exploded view of fuel injector delivery pipe and injectors—3MZ-FE engine

12. Install or connect the following
- 2 gaskets and fuel pipe union bolt
- 2 gaskets and fuel pressure pulsation damper
- Fuel pipe with bolt
13. Install intake air surge tank
14. Install or connect the following
- New gaskets to intake air surge tank
- Intake air surge tank and emission control valve bracket. Torque: 21 ft. lbs. (28 Nm).
- 4 bolts. Torque: 21 ft. lbs. (28 Nm)

- 2 bolts on surge tank stay. Torque: 21 ft. lbs. (28 Nm).
- Engine hangers
- Pressure feed tube
15. The remainder of the installation is the reverse order of removal.
16. Refill the cooling system to the correct level.
17. Start the engine and check for leaks.

2GR-FE Engine

1. Before servicing the vehicle, refer to the Precautions Section.

2. Relieve the fuel system pressure.
3. Drain the cooling system.
4. Remove or disconnect the following:
- Negative battery cable
- Wiper arm and blades
- Top cowl ventilator louver
- Wiper motor and linkage assembly
- Outer top cowl panel
- Engine appearance cover
- Air intake assembly
- Air intake surge tank
- Fuel supply hose
- Injector electrical connectors

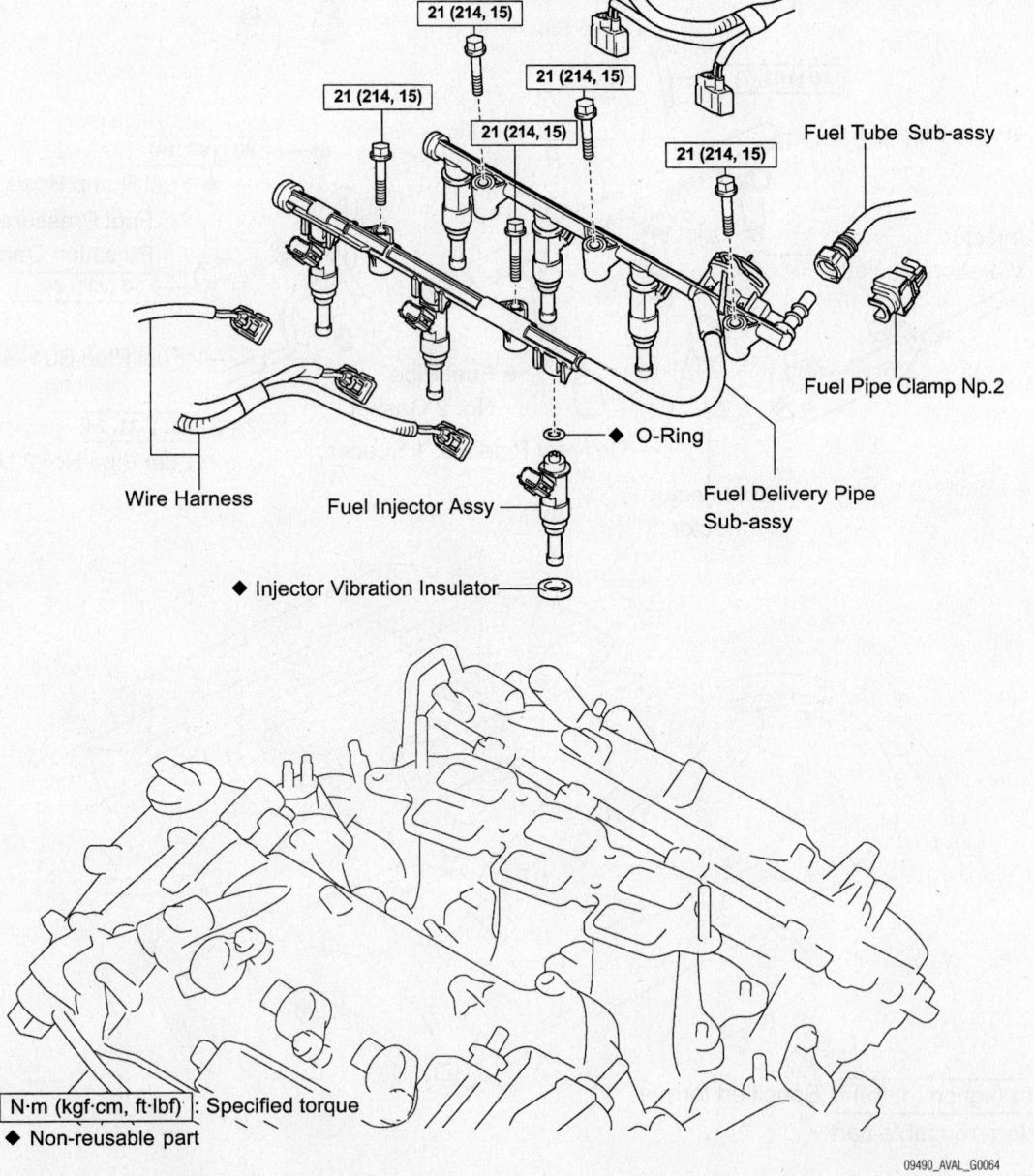

Wire Harness

21 (214, 15)

21 (214, 15)

21 (214, 15)

21 (214, 15)

21 (214, 15)

Fuel Tube Sub-assy

Fuel Pipe Clamp Np.2

◆ O-Ring

Wire Harness

Fuel Injector Assy

Fuel Delivery Pipe Sub-assy

◆ Injector Vibration Insulator

N·m (kgf·cm, ft·lbf) : Specified torque

◆ Non-reusable part

09490_AVAL_G0064

Exploded view of the fuel injector components—2GR-FE engine

- Fuel rail
- Insulators from the intake manifold
- Fuel injectors from the fuel rail

To install:

5. Apply a light coat of spindle oil or gas to new O-rings. Install the new O-rings to each injector.

6. Apply a light coat of spindle oil or gas where the fuel rail contacts the injector.

7. Push the injector, while turn the injector back and forth to install it in the fuel rail.

8. Install new insulators to the intake manifold.

9. Install the fuel rail and temporarily install the mounting bolts.

10. Check that each injectors turns smoothly. If not, reinstall the injector with a new O-ring.

11. Tighten the fuel rail mounting bolts to 15 ft. lbs. (21 Nm).

12. Install or connect the following:
- Injector electrical connectors
- Fuel supply hose
- Air intake surge tank. Tighten Bolts B to 13 ft. lbs. (18 Nm); Bolts C to 12 ft. lbs. (16 Nm); Nut to 15 ft. lbs. (21 Nm).
- Air intake assembly
- Outer top cowl panel
- Wiper motor and linkage assembly
- Top cowl ventilator louver
- Wiper arm and blades
- Negative battery cable

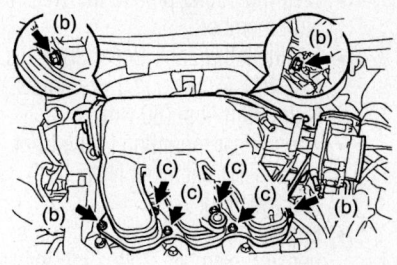

Bolt identification for the air intake surge tank installation—2GR-FE engine

13. Refill the cooling system to the correct level.

14. Start the engine and check for leaks.

DRIVE TRAIN

Manual Transaxle Assembly

REMOVAL & INSTALLATION

Camry and Solara

1. Before servicing the vehicle, refer to the Precautions Section.

2. Drain the transaxle fluid.

3. Remove or disconnect the following:
- Negative battery cable. On vehicles equipped with an air bag, wait at least 90 seconds before proceeding.
- Air cleaner
- With cruise control, cruise control actuator
- Clutch release cylinder and tube clamp
- Starter
- Back-up light switch connector and ground strap
- Wires clamp
- Clips and washers that attach the transaxle control cables to the control levers
- Transaxle control cables.
- Speed sensor connector
- Undercovers
- Left and right halfshafts
- 4 steering gear housing bolts.
- Stabilizer bar bushing bracket
- 2 set bolts and nuts
- Steering gear box from the suspension member and suspend it securely
- Exhaust pipe
- Stiffener plate
- Engine front mounting from the suspension member
- Engine rear mounting from the front suspension member
- Left engine mounting
- Steering cooler pipe from the suspension member
- 2 fender liner set screws
- The 2 bolts and 4 nuts located on the outside of the suspension member brackets
- The 4 larger bolts holding the suspension member to the vehicle body
- The 2 front lower braces, rear braces, and the front suspension member.
- Transaxle

To install:

4. Move the transaxle into position so that the input shaft spline is aligned with the clutch disc.

5. Install or connect the following:
- Transaxle into the engine and secure with the lower mounting bolts. Bolts: 10mm mounting bolts: 47 ft. lbs. (63 Nm). 12mm bolts to 34 ft. lbs. (46 Nm).
- Front suspension member and the 2 front lower braces and rear lower braces. 4 large bolts that hold the suspension member to the vehicle: 134 ft. lbs. (181 Nm); 2 outside bolts and 4 outside nuts: 24 ft. lbs. (32 Nm).
- 2 fender liner set screws

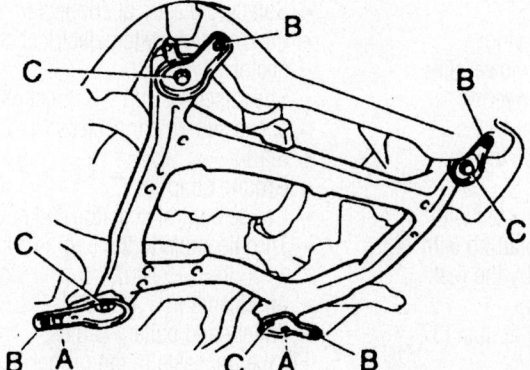

Bolt A: 32 N·m (330 kgf·cm, 24 ft·lbf)

Nut B: 36 N·m (370 kgf·cm, 27 ft·lbf)

Bolt C: 181 N·m (1,850 kgf·cm, 134 ft·lbf)

Front suspension member and fastener locations—Camry

7923VG93

- Steering cooler pipe to the suspension member
- Engine left mount. Bolts: 38 ft. lbs. (52 Nm); 2 nuts and 2 grommets. Nuts: 59 ft. lbs. (80 Nm).
- Engine rear mounting to the front suspension member. Nuts: 59 ft. lbs. (80 Nm).
- Engine front mounting to the suspension member. Bolt: 59 ft. lbs. (80 Nm).
- Stiffener plate. Bolts: 27 ft. lbs. (37 Nm).
- Exhaust pipe
- Steering gear housing to the front suspension member. Bolts and nuts: 134 ft. lbs. (181 Nm).
- Stabilizer bar bushing bracket. 4 bolts: 14 ft. lbs. (19 Nm).
- Right and left halfshafts
- Undercovers
- Vehicle speed sensor
- Control cables by installing the washers and clips
- Clamp that retains the wires to the transaxle
- Back-up light switch connector and ground cables
- Starter. Bolts: 29 ft. lbs. (39 Nm).
- Pipe clamp and clutch release cylinder to the transaxle. Bolts: 108 inch lbs. (13 Nm).
- Cruise control actuator
- Air cleaner
- Negative battery cable

6. Fill the transaxle fluid to the proper level.

7. Start the vehicle, check for leaks and repair if necessary.

Automatic Transaxle

REMOVAL & INSTALLATION

Camry, Solara and 2002–04 Avalon

1. Before servicing the vehicle, refer to the Precautions Section.
2. Drain the transaxle fluid.
3. Remove or disconnect the following:
   - Negative battery cable. On vehicles equipped with an air bag, wait at least 90 seconds before proceeding
   - Battery
   - Air cleaner assembly
   - Throttle cable from the throttle body
   - Cruise control actuator cover and connector, if equipped
   - Ground wire
   - Starter

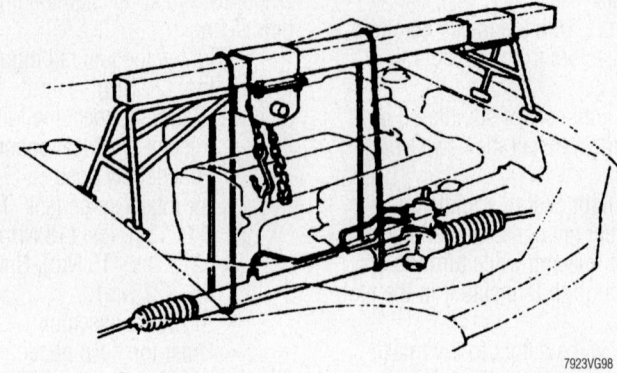

Tie the steering rack to the engine support fixture components, as shown—Avalon and Camry

- Speed sensor connectors, direct clutch speed sensor, and the park/neutral position switch connector on the transaxle
- Solenoid connector on the transaxle
- Shift control cable
- Oil cooler hoses
- Front side transaxle mounting bolts
- Front engine mounting bolts
- Oil cooler line mounting bolts from the front frame
- Upper transaxle to engine mounting bolts
- Front exhaust pipe
- Engine side covers and undercovers
- Both halfshafts
- Front side engine mounting nut
- Rear side engine mounting bolts (remove hole plugs)
- Left side transaxle mounting bolts
- Steering gear housing
- Front frame assembly
- Rear end plate mounting bolts
- Torque converter cover
- Torque converter retaining bolts
- Remaining transaxle mounting bolts
- Transaxle

To install:

4. Install or connect the following:
   - Transaxle aligning the 2 dowel pins on the block with the converter housing. 10mm bolts: 34 ft. lbs. (46 Nm); 12mm bolts: 47 ft. lbs. (64 Nm)
   - Torque converter bolts coated with sealer. Install the bolts starting with the green bolt followed by the rest. Bolts: 20 ft. lbs. (27 Nm).
   - Rear end plate. Bolts: 27 ft. lbs. (37 Nm).
   - Front frame assembly. 12mm bolts: 24 ft. lbs. (32 Nm); 19mm bolts:

134 ft. lbs. (181 Nm); Nut: 27 ft. lbs. (36 Nm).
   - Fender liner set screws
   - Steering gear to the frame. Bolts and nuts: 134 ft. lbs. (181 Nm).
   - Sway bar brackets. Bolts: 14 ft. lbs. (19 Nm).
   - Left transaxle mounting bolts. Bolts: 38 ft. lbs. (52 Nm).
   - Rear side mounting bolts and nuts. Bolts and nuts: 48 ft. lbs. (66 Nm). Install the plugs.
   - Front engine mounting nut. Nut: 59 ft. lbs. (80 Nm).
   - Halfshafts
   - Right and left engine side covers
   - Lower engine cover
   - Exhaust pipe. Nuts: 46 ft. lbs. (62 Nm).
   - Exhaust pipe to the converter. Nuts and bolts: 32 ft. lbs. (43 Nm).
   - Upper transaxle mounting bolts. Bolts: 47 ft. lbs. (64 Nm).
   - Oil cooler clamping bolts to the front frame.
   - Front side engine mounting bolts. Bolts: 59 ft. lbs. (80 Nm).
   - Front side transaxle mounting bolts. Bolts: 59 ft. lbs. (80 Nm).
   - Oil cooler hoses
   - Shift control cable
   - Solenoid electrical connector
   - Park/neutral switch electrical connector
   - Speed sensor and the direct clutch speed sensor connectors.
   - Starter
   - Ground strap
   - Cruise control actuator and cover
   - Throttle cable to the engine. Nuts: 11 ft. lbs. (15 Nm).
   - Air cleaner
   - Battery and battery cables.

5. Fill the transaxle to the proper level.
6. Start the vehicle, check for leaks and repair if necessary.

2005–06 Avalon

1. Before servicing the vehicle, refer to the Precautions Section.
2. Remove or disconnect the following:
   - Engine/transaxle assembly from the vehicle
   - Halfshafts
   - No,. 2 transmission control cable bracket
   - Wiring harness clamps
   - Starter
   - Transmission electrical connectors
   - No. 1 transmission control cable bracket
   - Transmission oil filler tube
   - Transmission oil cooler hoses
   - Transverse engine mounting bracket
   - Exhaust pipe support bracket and flywheel housing undercover
3. Turn the crankshaft to gain access to and remove the torque converter bolts.
4. Remove the transaxle mounting bolts and separate the transaxle from the engine.

To install:

5. Install the transaxle to the engine and tighten the mounting bolts as follows:
   a. Bolts A to 47 ft. lbs. (64 Nm).
   b. Bolt B to 34 ft. lbs. (46 Nm).
   c. Bolts C to 32 ft. lbs. (43 Nm).

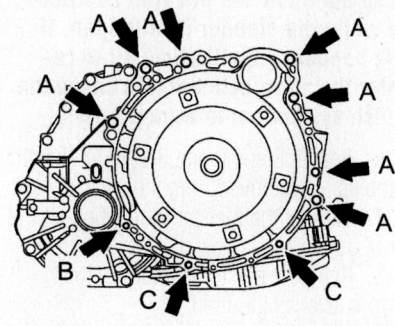

09490_AVAL_G0066

Transaxle mounting bolt identification—2005–06 Avalon

6. Install the torque converter mounting bolts. Install the green colored bolt first, then the 5 remaining bolts. Tighten to 30 ft. lbs. (41 Nm).
7. Install or connect the following:
   - Flywheel housing undercover. Tighten to 69 inch lbs. (7.8 Nm).
   - Transverse engine mounting bracket. Tighten to 47 ft. lbs. (64 Nm).
   - Oil cooler hoses. Tighten to 25 ft. lbs. (34 Nm).
   - Transmission oil filler tube. Tighten clamps to 49 inch lbs. (5.5 Nm).
   - No. 1 transmission control cable

bracket. Tighten to 9 ft. lbs. (12 Nm).
   - Transmission electrical connectors
   - Starter. Tighten mounting bolts to 27 ft. lbs. (37 Nm).
   - Wiring harness clamps
   - No. 2 transmission control cable bracket. Tighten to 9 ft. lbs. (12 Nm).
   - Halfshafts
   - Engine/transaxle assembly into the vehicle
8. Start the engine and check for leaks.

Clutch

REMOVAL & INSTALLATION

1. Before servicing the vehicle, refer to the Precautions Section.
2. Remove or disconnect the following:
   - Negative battery cable. On vehicles equipped with an air bag, wait at least 90 seconds before proceeding.
   - Remove the transaxle assembly from the vehicle
   - Clutch pressure plate retaining bolts
   - Clutch cover

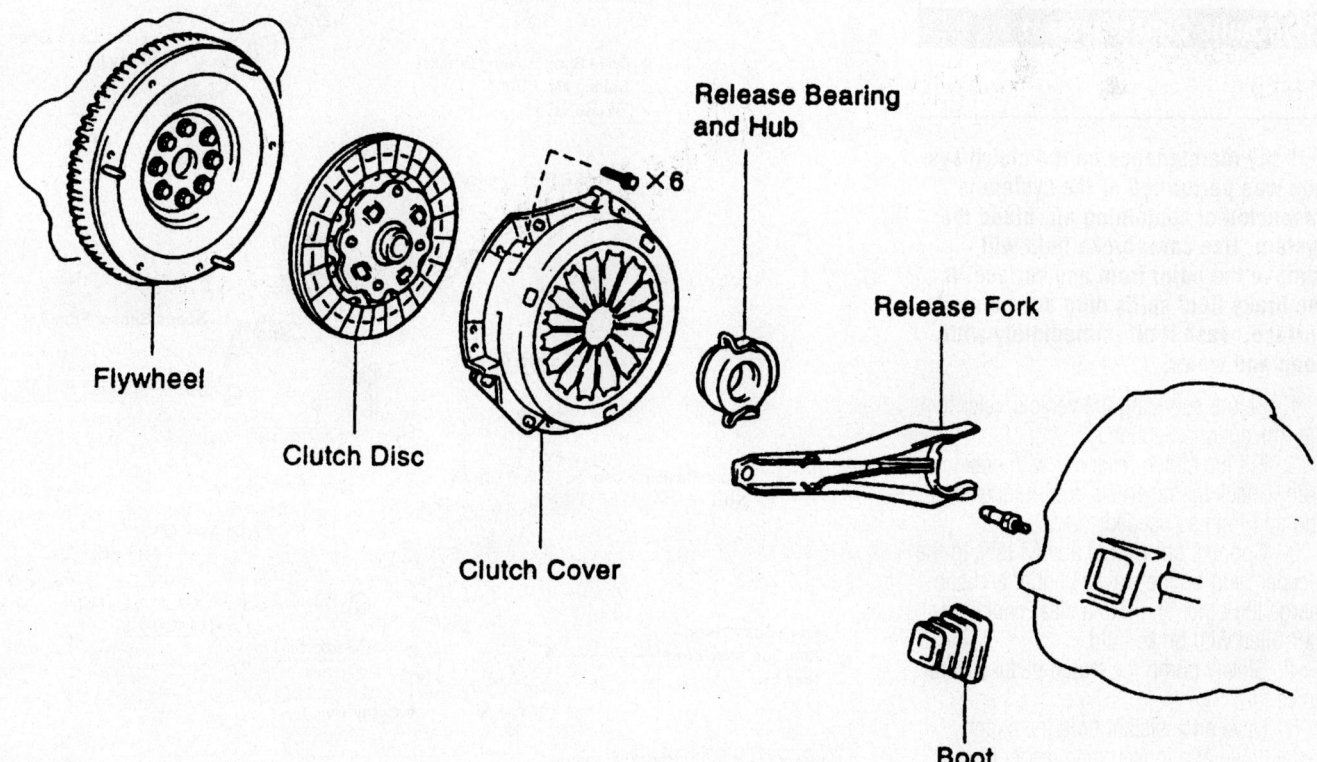

Clutch component assembly—Camry shown, Solara similar

7923VGA1

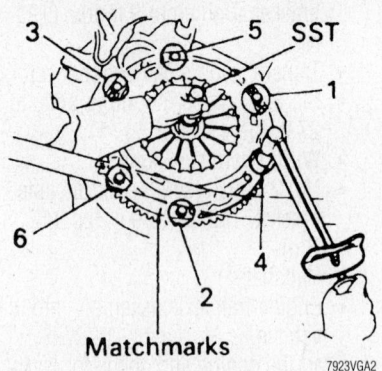

Matchmarks

7923VGA2

Tighten the pressure plate bolts according to the sequence shown—Camry

- Clutch disc
- Retaining clip and bearing from the transaxle
- Release fork and boot assembly

To install:

3. Install or connect the following:
- Clutch disc onto the flywheel
- Clutch cover, aligning the matchmarks
- Clutch cover retaining bolts. Bolts, tightened in a crisscross pattern: 14 ft. lbs. (19 Nm).
- Boot, release fork, hub and bearing assemblies
- Transaxle
- Negative battery cable

Hydraulic Clutch System

BLEEDING

➡**If any maintenance on the clutch system was performed or the system is suspected of containing air, bleed the system. Use care; brake fluid will remove the paint from any surface. If the brake fluid spills onto any painted surface, wash it off immediately with soap and water.**

1. Before servicing the vehicle, refer to the Precautions Section.
2. Fill the clutch reservoir with brake fluid. Check the reservoir level frequently and add fluid as needed.
3. Connect one end of a vinyl tube to the bleeder plug on the slave cylinder and submerge the other end into a clear container half-filled with brake fluid.
4. Slowly pump the clutch pedal several times.
5. Have an assistant hold the clutch pedal down and loosen the bleeder plug until fluid and/or air starts to run out of the bleeder plug. Close the bleeder plug while the pedal is held to the floor.

➡**Do not allow the pedal to rise back-up while the bleeder is still open. If this happens, it will allow air to re-enter the slave cylinder and cause the clutch system not to work properly.**

6. Repeat Steps 2 and 3 until all the air bubbles are removed from the system.
7. Tighten the bleeder plug when all the air is gone.
8. Refill the master cylinder to the proper level as required.
9. Check the system for leaks.

Halfshaft

REMOVAL & INSTALLATION

1. Before servicing the vehicle, refer to the Precautions Section.
2. Drain the transaxle fluid.
3. Remove or disconnect the following:

- Negative battery cable. On vehicles equipped with an air bag, wait at least 90 seconds before proceeding.
- Front fender apron seal
- Tie rod end from the steering knuckle by removing the cotter pin and nut. Separate the tie rod from the steering knuckle
- Stabilizer bar link from the lower control arm. Make note of the washers and cushions positions.
- Lower ball joint from the steering knuckle. Push down on the lower control arm and separate the steering knuckle from the ball joint
- Cotter pin, lock cap and locknut holding the halfshaft to the steering knuckle
- Halfshaft from the steering knuckle
- Left halfshaft from the transaxle
- Snapring from the halfshaft.
- Right halfshaft from the transaxle

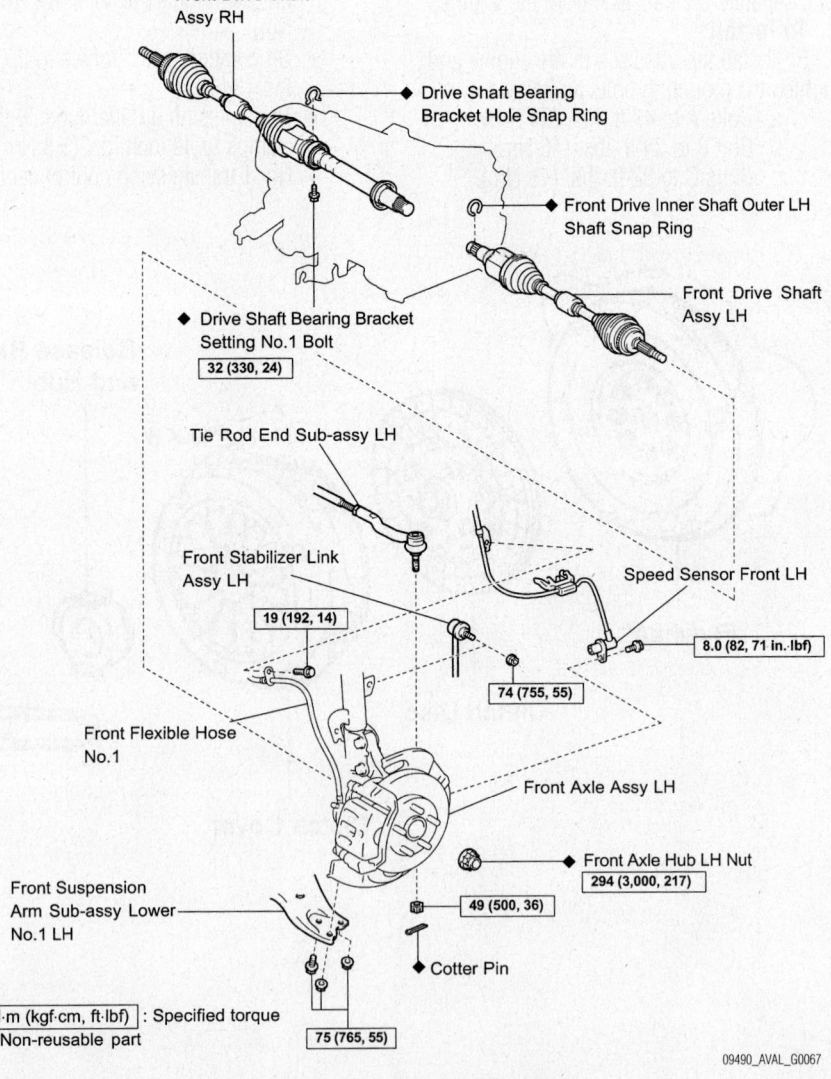

N·m (kgf·cm, ft·lbf) : Specified torque
◆ Non-reusable part

Exploded view of the halfshaft components—2005–06 Avalon shown

09490_AVAL_G0067

➥**The lockbolt is located in the center of the halfshaft, near the dampener.**

To install:

4. Install the right halfshaft to the transaxle, as follows:

a. Coat the side gear shaft and differential case sliding surface with gear oil.

b. Using snapring pliers, install the snapring to the halfshaft.

c. Install the halfshaft and the bearing lockbolt. Lockbolt: 24 ft. lbs. (32 Nm).

5. Install the left halfshaft to the transaxle, as follows:

a. Install a new snapring to the inner spline of the halfshaft.

b. Coat the side gear shaft and differential case sliding surface with gear oil.

c. Install the halfshaft to the transaxle with the snapring opening facing down. The halfshaft should click into place when installing.

d. After installation of the halfshaft, check that the halfshaft cannot be removed by hand.

6. Install or connect the following:

- Halfshaft to the steering knuckle, then install the locknut. Locknut: 217 ft. lbs. (294 Nm).
- Lock cap and new cotter pin to the halfshaft
- Steering knuckle to the lower ball joint. Nuts and bolt: 94 ft. lbs. (127 Nm).
- Stabilizer bar link to the lower control arm. Nut: 29 ft. lbs. (39 Nm).
- Tie rod to the steering knuckle. Nut: 36 ft. lbs. (49 Nm).
- New cotter pin to the tie rod end
- Front fender apron seal
- Front wheels
- Negative battery cable

7. Fill the transaxle fluid to the proper level.

8. Start the vehicle, check for leaks and repair if necessary.

CV-Joints

OVERHAUL

1. Before servicing the vehicle, refer to the Precautions Section.

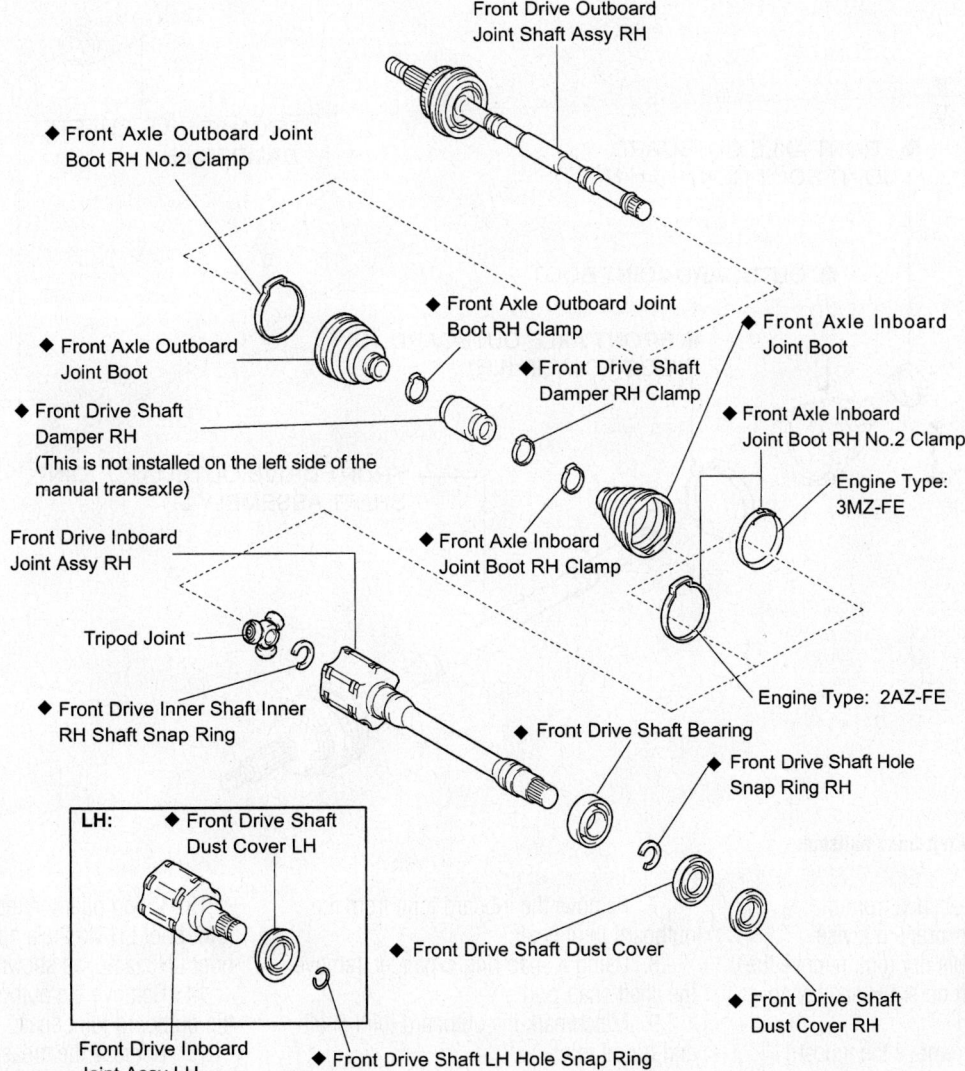

◆ Non-reusable part

Exploded view of the right-hand halfshaft

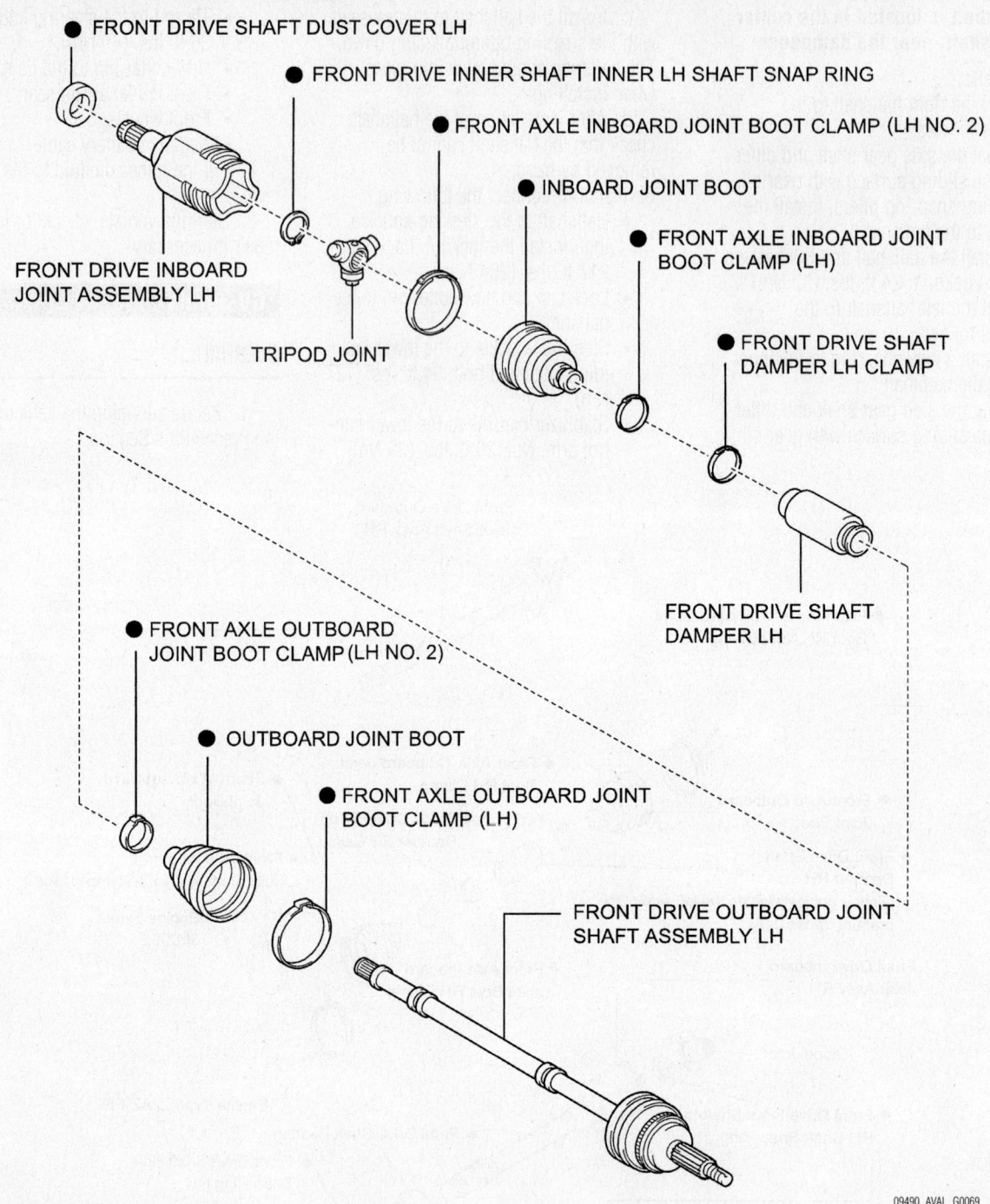

● FRONT DRIVE SHAFT DUST COVER LH

● FRONT DRIVE INNER SHAFT INNER LH SHAFT SNAP RING

● FRONT AXLE INBOARD JOINT BOOT CLAMP (LH NO. 2)

● INBOARD JOINT BOOT

● FRONT AXLE INBOARD JOINT BOOT CLAMP (LH)

FRONT DRIVE INBOARD JOINT ASSEMBLY LH

TRIPOD JOINT

● FRONT DRIVE SHAFT DAMPER LH CLAMP

FRONT DRIVE SHAFT DAMPER LH

● FRONT AXLE OUTBOARD JOINT BOOT CLAMP (LH NO. 2)

● OUTBOARD JOINT BOOT

● FRONT AXLE OUTBOARD JOINT BOOT CLAMP (LH)

FRONT DRIVE OUTBOARD JOINT SHAFT ASSEMBLY LH

09490_AVAL_G0069

Exploded view of the left-hand halfshaft

2. Remove the halfshaft from the vehicle and mount securely in a vise.

3. Using a suitable pry tool, remove the front drive inner shaft outer LH shaft snap ring.

4. Using pliers, remove the inboard joint boot No.2 clamp and inboard boot clamp, as shown in the illustration.

5. Separate the inboard joint boot from the inboard joint.

6. Matchmark the inboard joint and outboard joint shaft.

7. Remove the inboard joint from the outboard joint shaft.

8. Using a snap ring expander, remove the shaft snap ring.

9. Matchmark the outboard joint shaft and tripod joint.

10. Using a brass bar and a hammer, remove the tripod joint from the outboard joint shaft.

11. Using pliers, remove the drive shaft damper clamp, as shown in the illustration.

12. Remove the drive shaft damper.

13. Using pliers, remove the outboard joint boot LH No.2 clamp and outboard joint boot LH clamp, as shown in the illustration.

14. Remove the outboard joint boot from the outboard joint shaft.

15. Remove the grease from the outboard joint.

16. Use Special Tool 09950-0020 and a press, remove the drive shaft dust cover

17. Using a snap ring expander, remove the drive shaft hole snap ring.

18. On the right side, use Special Tool

Omega Type:

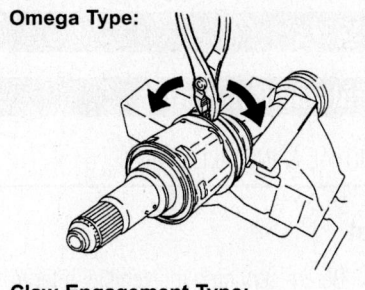

Claw Engagement Type:

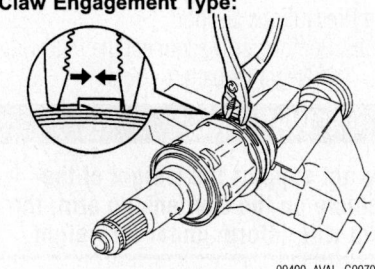

Removing the boot clamps

09490_AVAL_G0070

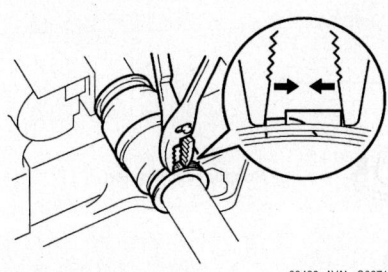

09490_AVAL_G0071

**Removing the halfshaft damper clamp—
Automatic transmission models**

09527-10011 and a press, remove the bearing.

19. Remove the bearing bracket hole snap ring.

To install:

20. Install a new bearing bracket hole snap ring to the right side halfshaft.

21. Using Special Tool 09527-30010 and a steel plate, install a new front drive shaft bearing.

22. Using a snap ring expander, install a new halfshaft hole snap ring.

23. Using Special Tool 09527-10011 and a press, install a new right side halfshaft dust cover

until the distance from the tip of the center halfshaft to the halfshaft dust cover meets the specification, as shown in the illustration.

a. 2AZ-FE Engine: 3.60+/-0.02 in. (91.5+/-0.5 mm)

b. 1MZ-FE, 3MZ-FE Engines: 4.3+/-0.02 in. (110.5+/-0.5 mm)

c. 2GR-FE Engine: 4.331–4.370 in. (110–111 mm)

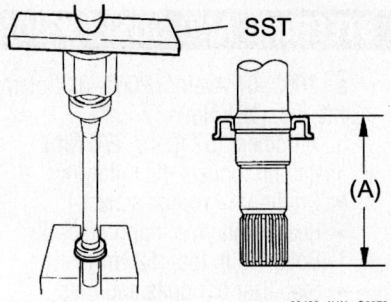

09490_AVAL_G0072

Install the right halfshaft dust cover to the specified distance

24. Using Special Tool 09527-10011 and a press, install a new left side halfshaft dust cover.

➡**Wrap the splines of the halfshaft with tape to prevent the boots from being damaged.**

25. Secure the halfshaft lightly in a soft vise.

26. Temporarily install a new outboard joint boot to the halfshaft with 2 clamps.

27. Pack the outboard joint shaft and boot with grease:

a. 2GR-FE: 4.2–4.9 oz.

b. 2AZ-FE (AT): 3.5–4.2 oz.

c. 2AZ-FE (MT): 3.7 to 4.4 oz.

d. 1MZ-FE, 3MZ-FE: 6.9–7.9 oz.

28. Using Special Tool 09521-24010, tighten the outboard joint large boot clamp.

29. Using Special Tool 09240-00020, measure the clearance of the outboard joint large boot clamp. Clearance should be as follows:

a. 2GR-FE Engine: 0.0472–0.1575 in. (1.2–4.0 mm)

b. All others: Clearance: 0.118–0.157 in. (3–4 mm)

30. Install the halfshaft damper to the halfshaft groove, if equipped with automatic transmission. Set the distance, as specified:

a. 2GR-FE: 8.051–8.209 in. (204.5 to 208.5 mm)

b. All others: 8.15+/-.08 in. (207+/-2 mm)

31. Using pliers, install a new halfshaft damper clamp.

32. Temporarily install a new inboard joint boot with 2 clamps to the halfshaft.

33. Place the beveled side of the tripod joint axial spline toward the outboard joint shaft and align the matchmarks.

34. Using a brass bar and hammer, tap in the tripod joint to the outboard joint shaft.

35. Using a snap ring expander, install a new shaft snap ring.

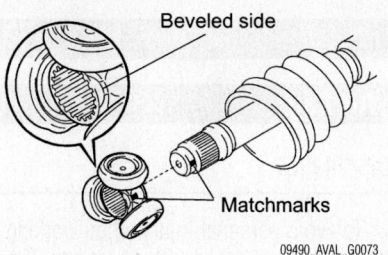

09490_AVAL_G0073

Correct orientation of the tripod joint for installation

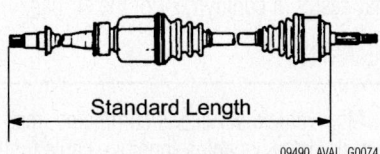

09490_AVAL_G0074

Standard length of the halfshaft should be measured as shown in the illustration.

36. Pack the outboard joint shaft and boot with grease:

a. 2GR-FE Engine: 6.7 –7.4 oz.

b. 2AZ-FE Engine: 6.0–6.7 oz.

c. 1MZ-FE, 3MZ-FE Engines: 5.5 to 6.2 oz.

37. Align the matchmarks and install the inboard joint to the outboard joint shaft.

38. Install the inboard joint boot to the inboard joint.

39. Using pliers, install the new inboard joint boot No.2 clamp and inboard joint boot clamp.

40. Install the front drive inner shaft outer LH shaft snap ring.

41. Make sure that the 2 boots are not stretched or contracted when the drive shaft is at standard length.

a. 1MZ-FE, 3MZ-FE Engine:
- Left side: 22.864+/-0.08 in. (580.76+/-2.0 mm)
- Right side: 35.354+/-0.08 in. (898.0+/-2.0 mm)

b. 2AZ-FE Engine:
- Left side: 23.276+/-0.08 in. (591.2+/-2.0 mm)
- Right side: 35.039+/-0.08 in. (890.0+/-2.0 mm)

c. 2GR-FE Engine:
- Left side: 23.150–23.307 in. (588.0–592.0 mm)
- Right side: 35.677–35.835 in. (906.2–910.2 mm)

42. Check to see that there is no play in the inboard and outboard joints and that the inboard joint slides smoothly in the thrust direction.

STEERING AND SUSPENSION

Air Bag

DISARMING

To avoid personal injury when working on vehicles equipped with an air bag, the negative battery cable must be disconnected and at least 90 seconds must elapse before working on the system. Failure to do so may result in deployment of the air bag.

REARMING

After vehicle service is completed, reattach the battery cables (positive cable first!) to rearm the air bag system.

Rack and Pinion Steering Gear

REMOVAL & INSTALLATION

1. Before servicing the vehicle, refer to the Precautions Section.
2. Position the front wheels straight ahead.
3. Disable the airbag system.
4. Remove or disconnect the following:
- Negative battery cable
- Front wheels
- Left and right front fender apron seals.
- Cotter pin and nut holding the steering knuckle to the tie rod end.
- Tie rod end from the steering knuckle.

➡ **Place matchmarks on the intermediate shaft and the control valve shaft.**

- Lower bolt holding the control valve shaft to the intermediate shaft.
- Intermediate shaft from steering rack housing.
- Tube clamp
- Return line and the pressure line from the control valve housing.
- Stabilizer bar bolts and nuts. Do not remove the bar from the vehicle.
- If necessary, rear engine mounting and bracket for additional clearance.
- On the V6 engine, oxygen sensor.
- Steering gear mounting bolts and nuts
- Steering gear

To install:

5. Install the steering gear assembly. Tighten the mounting bolts as follows:

a. 2002–04 Avalon, 2002–03 Solara: 134 ft. lbs. (181 Nm)
b. All others: 52 ft. lbs. (70 Nm)
6. Install or connect the following:
- On the V6, oxygen sensor.
- Rear engine mounting bracket. Bolts: 38 ft. lbs. (52 Nm).
- Stabilizer bar bolts and nuts.
- Tube clamp. Nut: 84 inch lbs. (10 Nm).
- Intermediate shaft to the steering rack. Bolts: 26 ft. lbs. (35 Nm).
- Tie rods to the steering knuckles with the castellated nuts.
- Front fender apron seals
- Front wheels
- Negative battery cable
- Power steering fluid

Strut and Coil Spring

REMOVAL & INSTALLATION

Front

1. Before servicing the vehicle, refer to the Precautions Section.
2. Remove or disconnect the following:
- Negative battery cable

✳✳ WARNING

Do not support the weight of the vehicle on the suspension arm; the arm will deform under its weight.

- Wheel
- Stabilizer link, if necessary

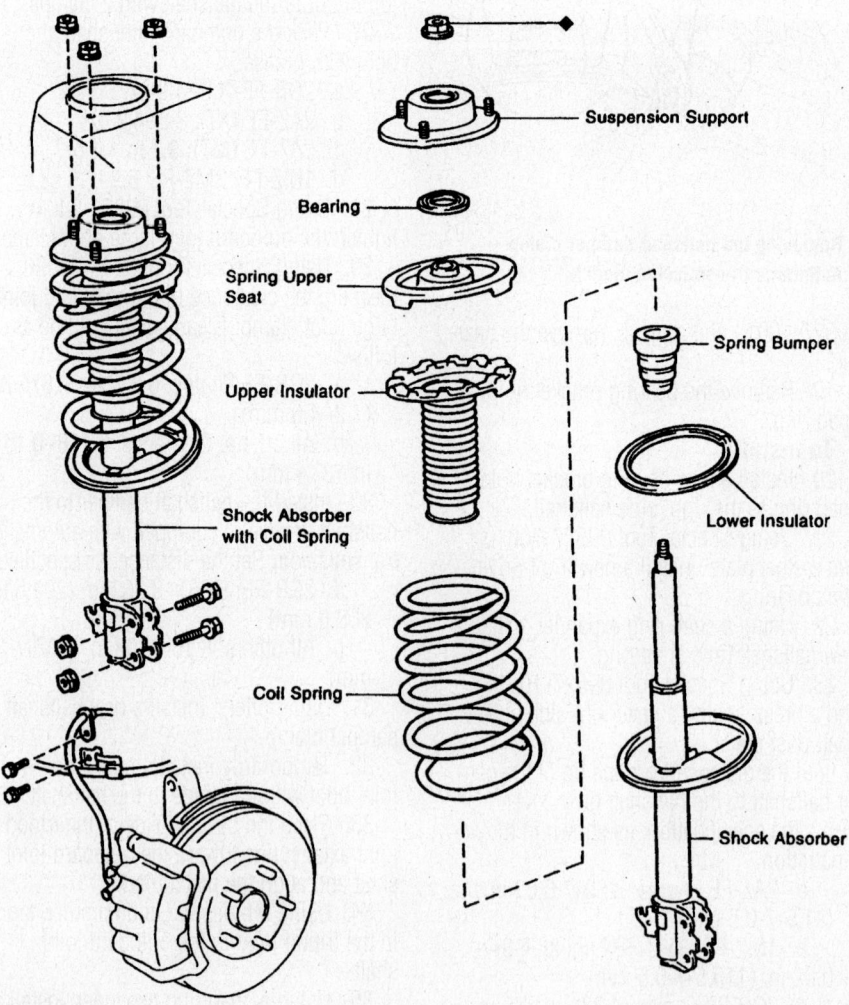

Common coil spring and strut component assembly

7923VGA6

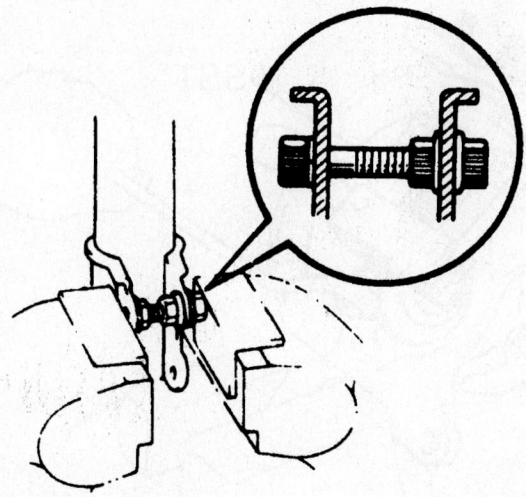

7923VGA7

Proper method of supporting the strut in a vise

- Bolt, and disconnect the brake hose from the strut
- With ABS brakes, wiring harness from the strut
- Top center locknut
- Bolts and strut from the steering knuckle
- Strut
3. To disassemble the strut:
  - Install a bolt and 2 nuts to the bracket at the lower portion of the strut shell and secure it in a vise
  - Compress the coil spring
  - Remove the dust cover and hold the spring seat so that it will not turn. Remove the nut on the top of the strut
  - Remove the suspension support, bearing, dust seal, spring seat, spring, insulators and bumper

To install:
4. To assemble the strut:
  - Install the spring bumper to piston
  - Using a spring compressor, compress the spring
  - Install the coil spring to the strut. Fit the lower end of the coil spring into the gap of the lower seat
  - Install the spring seat with the insulator
  - Install the dust seal on the spring seat
  - Install the suspension support and tighten 35 ft. lbs. (47 Nm). After the nut has been tighten, release the compressor tool tension
  - Pack multipurpose grease into the suspension support. Install the dust cover

➡**Do not use an impact wrench to tighten the nut. Also, check that the**

bearing fits into the recess in the suspension support.

5. Install the strut assembly to the vehicle.
6. Tighten the top nuts holding the strut to the strut tower as follows:
  a. 2005–06 Avalon: 63 ft. lbs. (85 Nm)
  b. All others: 59 ft. lbs. (80 Nm)
7. Install or connect the following:
  - Steering knuckle to the strut lower bracket
  - 2 bolts from the rear side and tighten the strut-to-steering knuckle arm bolts to 155 ft. lbs. (210 Nm).
  - Brake line to the steering knuckle
  - If equipped with ABS, secure the wiring harness
  - Wheel
  - Negative battery cable
8. Check and adjust the alignment, if needed.

Rear

1. Before servicing the vehicle, refer to the Precautions Section.
2. Remove or disconnect the following:
  - Negative battery cable from the battery
  - Rear seat cushion and any trim necessary to access the strut towers
3. Support the axle beam with a jack.
4. Remove or disconnect the following:
  - Wheel
  - With ABS, sensor wire from the strut
  - Stabilizer bar
5. Loosen the fasteners securing the strut to the axle carrier. Do not remove the bolts at this time.

6. Support the axle carrier with a jack.
7. Remove or disconnect the following:
  - Strut-to-strut tower nuts

✳✳ CAUTION

Do not loosen the center nut on the top of the strut piston.

- Strut
8. To disassemble the strut:
  a. Place the strut assembly in a pipe vise or strut vise.

➡**Do not attempt to clamp the strut assembly in a flat jaw vise as this will result in damage to the strut tube.**

  b. Compress the spring until the upper suspension support is free of any spring tension. Do not over-compress the spring.
  c. Hold the upper support, then remove the nut on the end of the shock piston rod.
  d. Remove the support, coil spring, insulator, and bumper.
9. Inspect the strut as follows:
  a. Check the shock absorber by moving the piston shaft through its full range of travel. It should move smoothly and evenly throughout its entire travel without any trace of binding or notching.
  b. Use a small straightedge to check the piston shaft for any bending or deformation.
  c. Inspect the spring for any sign of deterioration or cracking. The waterproof coating on the coils should be intact to prevent rusting.

To install:

➡**Never reuse a self-locking nut. Always replace self-locking nuts and cotter pins as applicable.**

10. Assemble the strut as follows:
  a. Loosely assemble all components onto the strut assembly. Be sure the spring end aligns with the hollow in the lower seat.
  b. Align the upper suspension support with the piston rod and install the support.
  c. Align the suspension support with the strut lower bracket. This assures the spring will be properly seated top and bottom.
  d. Compress the spring to expose the strut piston rod threads.
  e. Install a new strut piston nut and tighten to 36 ft. lbs. (49 Nm).
  f. Remove the spring compressor. Be sure the paint mark on the upper support faces the outside of the strut.

11. Place the strut on the vehicle and install the nuts to hold the strut to the strut tower. Tighten to 29 ft. lbs. (39 Nm).

12. Install or connect the following:
- Strut to the axle carrier and install the bolt and nut. Do not tighten at this time
- Stabilizer link to the strut
- Wheel. Bounce the vehicle up and down to stabilize the suspension

13. With the vehicle weight on the suspension, tighten the bolt holding the strut to the axle carrier to 188 ft. lbs. (255 Nm).

14. Install or connect the following:
- Rear seat cushion and any trim
- Negative battery cable

Lower Ball Joint

REMOVAL & INSTALLATION

1. Before servicing the vehicle, refer to the Precautions Section.

2. Remove or disconnect the following:
- Negative battery cable. On vehicles equipped with an air bag, wait at least 90 seconds before proceeding
- Front wheels
- Cotter pin from the bearing locknut cap, then remove the cap

3. Depress the brake pedal and loosen the axle nut

4. Remove or disconnect the following:
- Brake caliper attaching hardware, position the caliper aside with the hydraulic line still attached and suspend it with a wire
- ABS speed sensor, if equipped
- Rotor

5. Loosen the 2 nuts holding the strut to the steering knuckle assembly. Do not remove at this time.

6. Remove or disconnect the following:
- Cotter pin and nut from the tie rod end. Using a tie rod end removal tool, separate the tie rod end from the steering knuckle
- Steering knuckle from the strut assembly
- Axle nut and grasp the hub and knuckle assembly. With a plastic hammer tap the axle shaft to remove knuckle and hub

➡**Cover the halfshaft boot with a shop rag to protect it from any damage.**

7. Clamp the steering knuckle in a vise and remove the dust deflector. Remove the nut holding the steering knuckle to the ball joint. Press the ball joint out of the steering knuckle.

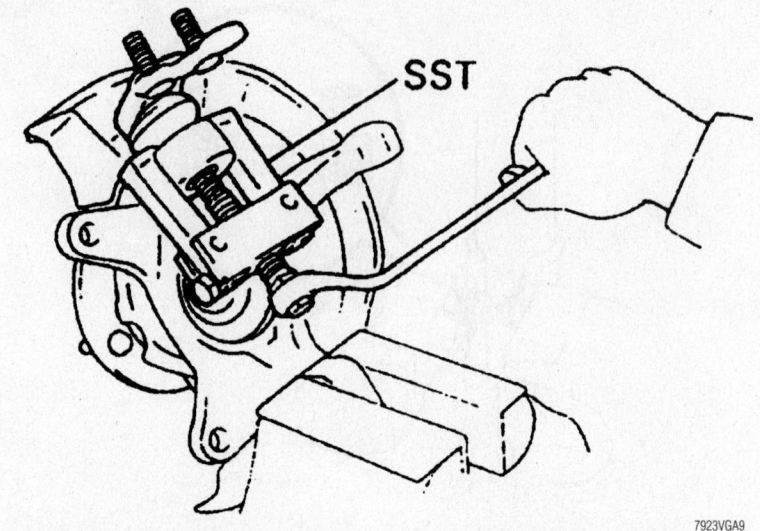

Removing the ball joint from the knuckle

8. Remove the ball joint from the arm.

To install:

9. Install the Lower ball joint to the lower arm. Tighten the fasteners to:
   a. Camry/Solara: 55 ft. lbs. (75 Nm).
   b. Avalon: 94 ft. lbs. (127 Nm).

10. Install the ball joint to the steering knuckle. Tighten the ball joint-to-steering knuckle nut to 90 ft. lbs. (123 Nm).

11. Install or connect the following:
- New cotter pin. Drive the deflector shield onto the knuckle
- Knuckle and hub assembly to the axle and temporarily tighten the axle nut
- Knuckle assembly to the lower strut bracket. Temporarily insert the mounting bolts from the rear and install the nuts
- Tie rod end to the knuckle

12. Tighten the bolts on the lower side of the strut assembly.

13. Install or connect the following:
- ABS speed sensor
- Brake disc and the caliper

14. Tighten the axle nut.

15. Connect the negative battery cable.

16. Check and adjust the alignment, if needed.

Wheel Bearings

ADJUSTMENT

Front

1. Before servicing the vehicle, refer to the Precautions Section.

2. All models use a non-adjustable wheel bearing. To determine the condition of the wheel bearing, check the backlash in bearing shaft direction and the axle hub deviation. Maximum for backlash should be 0.0020 in. (0.05mm)

3. Maximum axle hub deviation is 0.0020 in. (0.05mm)

4. If the wheel bearing is out of specifications, replace the wheel bearing.

Checking the wheel bearings for deviation and free-play

Rear

Check the backlash in bearing shaft direction and the axle hub deviation. Maximum for backlash should be 0.0020 in. (0.05mm). Maximum axle hub deviation is 0.0028 in. (0.07mm).

➡**The wheel bearing is non-adjustable. If the wheel bearing is out of specifications, replace the wheel bearing.**

REMOVAL & INSTALLATION

Front

1. Before servicing the vehicle, refer to the Precautions Section.

2. Remove or disconnect the following:
   - Negative battery cable. On vehicles equipped with an air bag, wait at least 90 seconds before proceeding
   - Wheels
   - Axle nut cap
   - Axle nut
   - Caliper. Position the caliper aside with the hydraulic line still attached and suspend it with a wire.
   - ABS speed sensor
   - Rotor
3. Loosen the nuts on the lower side of the strut assembly. Do not remove at this time.
4. Remove or disconnect the following:
   - Tie rod end from the steering knuckle

   - Steering knuckle from the lower control arm
   - Knuckle from the strut assembly
   - Hub

➡**Cover the halfshaft boot with a shop rag to protect it from any damage.**

5. Clamp the steering knuckle in a vise and remove the dust deflector. Remove the nut holding the steering knuckle to the ball joint. Press the ball joint out of the steering knuckle.
6. Remove the inner axle seal.
7. Using a Torx® wrench, remove the bolts securing the dust cover.
8. Using hub puller, remove the hub and backing plate from the steering knuckle.

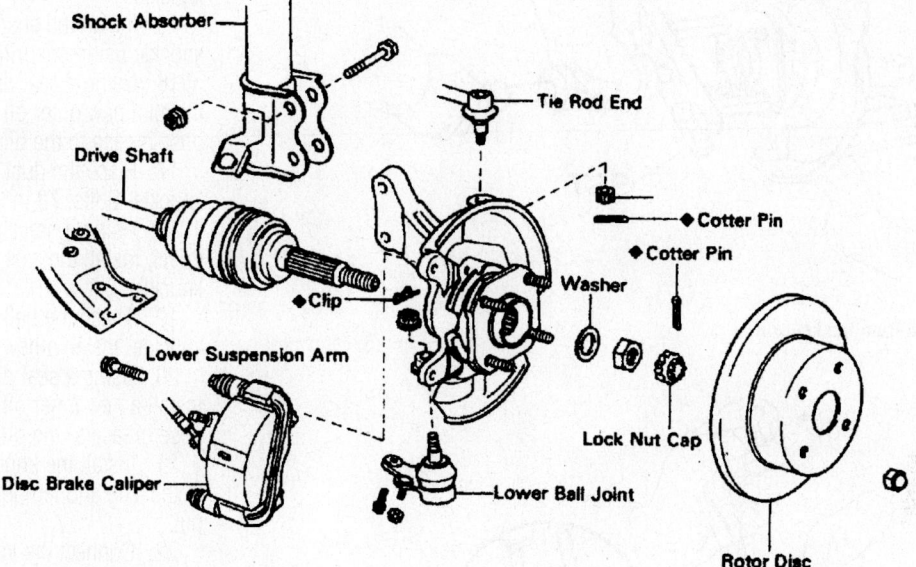

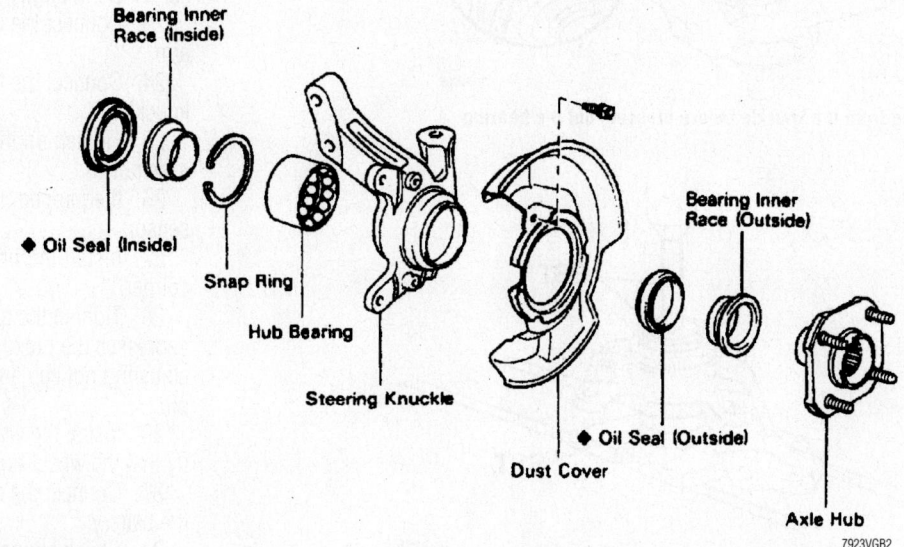

Steering knuckle and hub assembly

7923VGB2

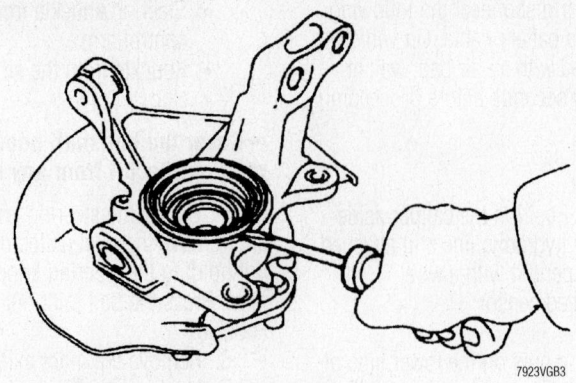

Removing the inner axle seal from the hub assembly

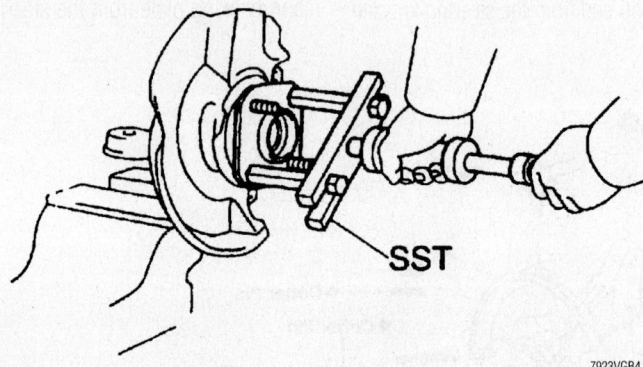

Removing the axle hub from the knuckle

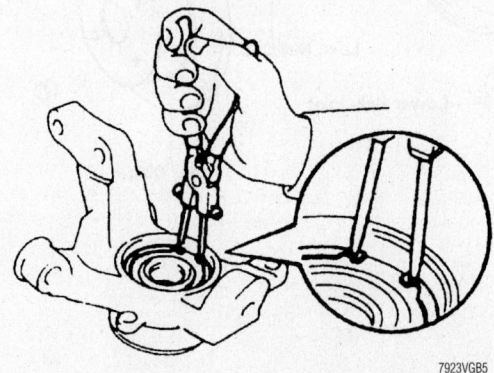

Removing the snapring from the knuckle before pressing out the bearing

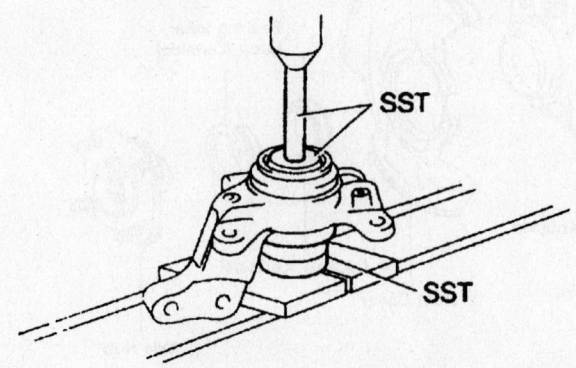

Removing the bearing from the steering knuckle using a press

9. Using a proper sized driver and a press, remove the inner hub race from the axle hub.

10. Using seal removal tool, remove the outer axle seal.

11. Using snapring pliers, remove the snapring from the inner side of the steering knuckle.

12. Using a proper sized driver and a press, remove the bearing from the steering knuckle. The bearing is pressed from the front of the steering knuckle and is removed through the back of the steering knuckle.

To install:

13. Perform the following:

14. Using a proper sized driver and a press, install a new bearing to the steering knuckle.

15. Install the snapring to the steering knuckle using snapring pliers.

16. Using a seal driver and a hammer, install a new outer oil seal. Apply multipurpose grease to the oil seal lip.

17. Place the dust cover on the steering knuckle. Bolts: 78 inch lbs. (9 Nm).

18. Using a press and a proper sized driver, install the axle hub to the steering knuckle.

19. Attach the ball joint to the steering knuckle. Install a new cotter pin.

20. Using a seal driver and a hammer, install a new inner oil seal. Apply multipurpose grease to the oil seal lip.

21. Install the knuckle and hub assembly to the axle and temporarily tighten the axle nut.

22. Connect the knuckle assembly to the lower strut bracket. Temporarily insert the mounting bolts from the rear and install the nuts making sure the matchmarks made earlier are in alignment.

23. Connect the lower ball joint to lower arm.

24. Connect the tie rod end to the knuckle.

25. Tighten on the lower side of the strut assembly.

26. If equipped, install the ABS speed sensor.

27. Install the brake disc and the caliper.

28. Tighten the axle nut while someone depresses the brake pedal. Install the adjusting nut cap and insert a new cotter pin.

29. Install the wheels to the vehicle. Verify that the wheel turns freely.

30. Connect the negative battery cable to the battery.

31. Check alignment.

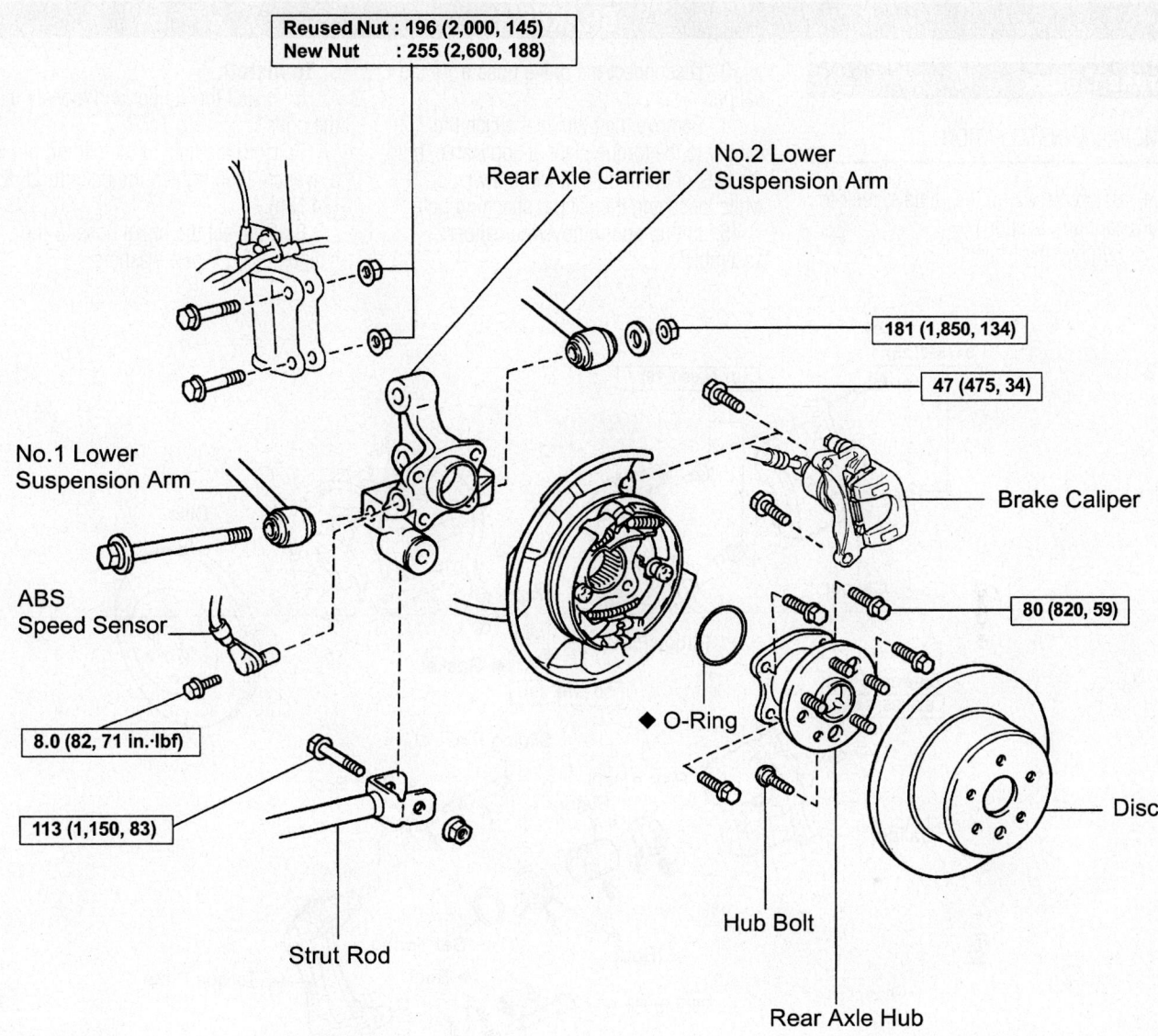

Reused Nut : 196 (2,000, 145)
New Nut : 255 (2,600, 188)

Rear Axle Carrier

No.2 Lower Suspension Arm

181 (1,850, 134)

47 (475, 34)

No.1 Lower Suspension Arm

Brake Caliper

ABS Speed Sensor

80 (820, 59)

8.0 (82, 71 in.·lbf)

◆ O-Ring

113 (1,150, 83)

Disc

Strut Rod

Hub Bolt

Rear Axle Hub

N·m (kgf·cm, ft·lbf) : Specified torque

◆ Non-reusable part

09490_AVAL_G0075

Exploded view of the hub and wheel bearing assembly—Avalon shown, others similar

Rear

1. Before servicing the vehicle, refer to the Precautions Section.
2. Remove or disconnect the following:
   - Negative battery cable
   - Wheel
   - Brake drum or rotor
   - With ABS brakes, ABS wheel speed sensor
   - Hub
   - O-ring from the backing plate

To install:

3. Install or connect the following:
   - New O-ring onto the backing plate. Coat the O-ring with multipurpose grease
   - Hub to the knuckle. Tighten to 59 ft. lbs. (80 Nm).
   - With ABS brakes, ABS wheel speed sensor
   - Brake drum or rotor
   - Wheel
   - Negative battery cable

4. Check and adjust the alignment, if needed.

BRAKES

Brake Caliper

REMOVAL & INSTALLATION

1. Before servicing the vehicle, refer to the Precautions Section.
2. Remove the wheels.

3. Disconnect the brake hose from the caliper.
4. Remove the bolts that attach the caliper to the torque plate. If applicable, hold the flats of the sliding pin with a wrench while loosening the caliper attaching bolts.
5. Lift up and remove the caliper assembly.

To install:

6. Install the caliper and loosely install the bolts.
7. Hold the flats of the sliding pin with a wrench, then tighten the bolts to 25 ft. lbs. (34 Nm).
8. Connect the brake hose to the caliper, using 2 new washers.

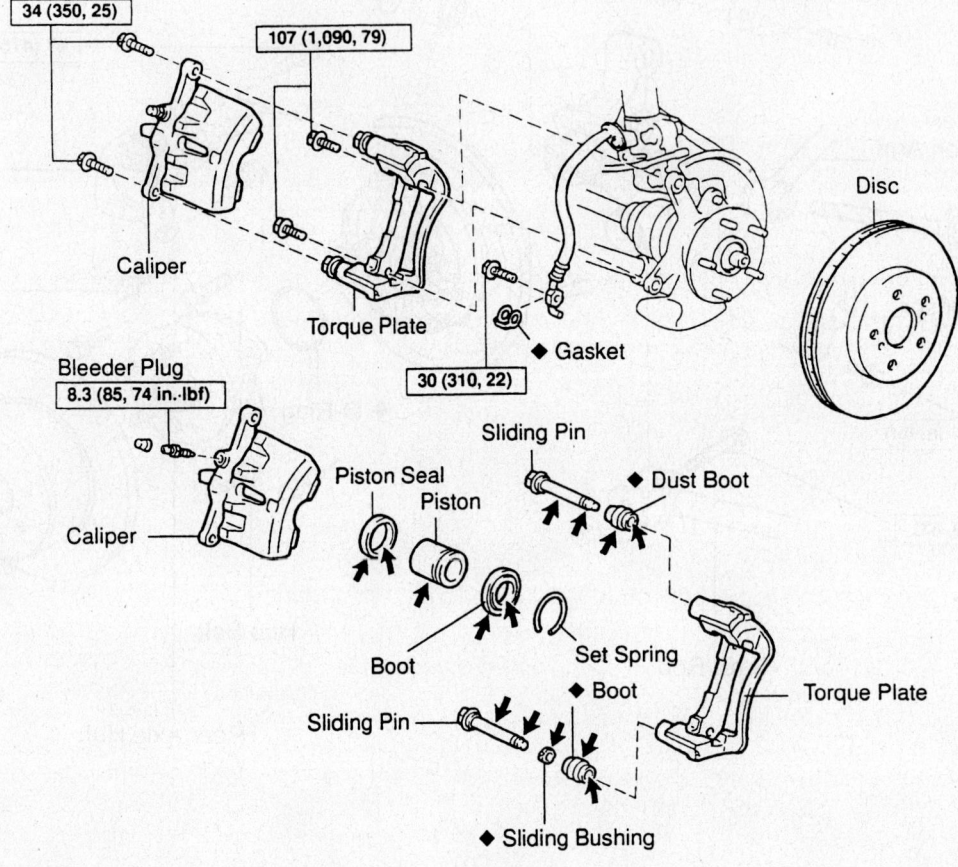

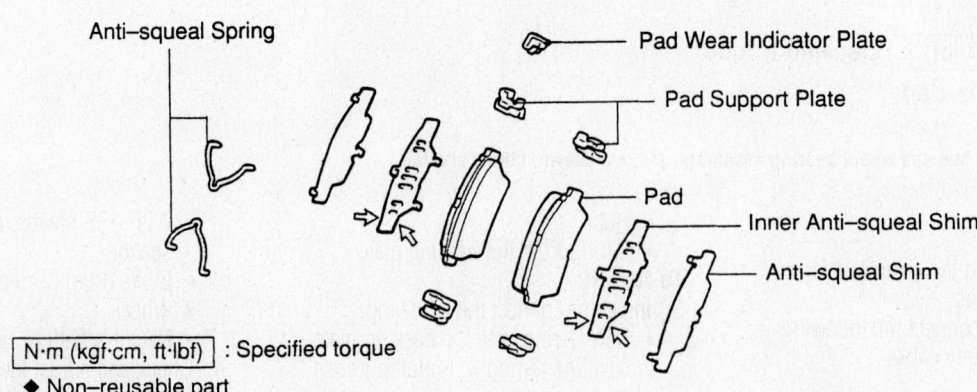

N·m (kgf·cm, ft·lbf) : Specified torque

◆ Non–reusable part

◀ Lithium soap base glycol grease

◁ Disc brake grease

93016G67

Front caliper—Camry, Avalon

TMC made:

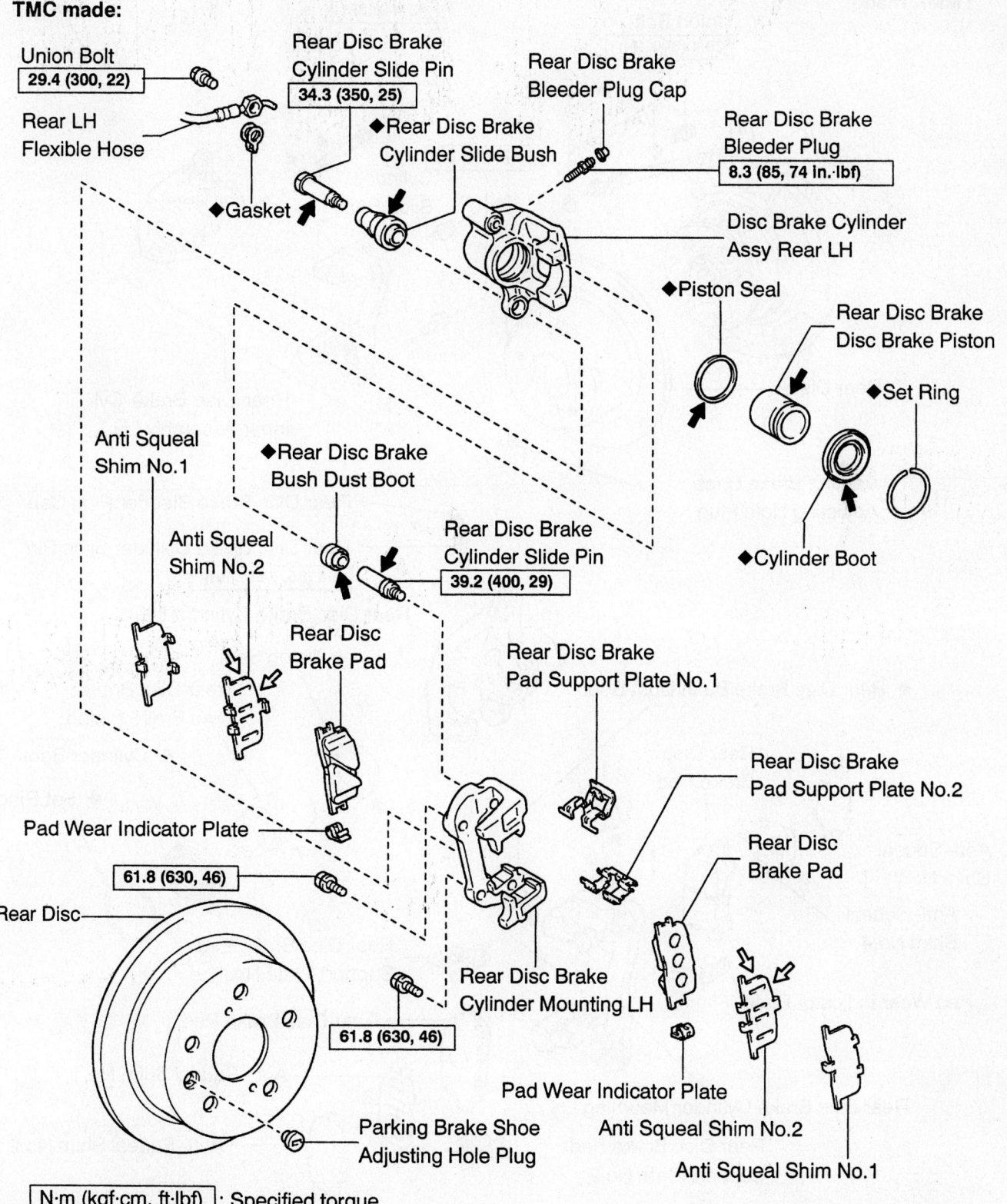

Union Bolt
29.4 (300, 22)

Rear LH
Flexible Hose

Rear Disc Brake
Cylinder Slide Pin
34.3 (350, 25)

◆Rear Disc Brake
Cylinder Slide Bush

Rear Disc Brake
Bleeder Plug Cap

Rear Disc Brake
Bleeder Plug
8.3 (85, 74 in.·lbf)

◆Gasket

Disc Brake Cylinder
Assy Rear LH

◆Piston Seal

Rear Disc Brake
Disc Brake Piston

◆Set Ring

◆Cylinder Boot

Anti Squeal
Shim No.1

◆Rear Disc Brake
Bush Dust Boot

Anti Squeal
Shim No.2

Rear Disc Brake
Cylinder Slide Pin
39.2 (400, 29)

Rear Disc
Brake Pad

Rear Disc Brake
Pad Support Plate No.1

Rear Disc Brake
Pad Support Plate No.2

Rear Disc
Brake Pad

Pad Wear Indicator Plate

61.8 (630, 46)

Rear Disc

Rear Disc Brake
Cylinder Mounting LH

61.8 (630, 46)

Pad Wear Indicator Plate

Anti Squeal Shim No.2

Anti Squeal Shim No.1

Parking Brake Shoe
Adjusting Hole Plug

N·m (kgf·cm, ft·lbf) : Specified torque
◆ Non-reusable part
◀ Lithium soap base glycol grease
◁ Disc brake grease

Rear caliper—Camry, TMC caliper

TMMK made:

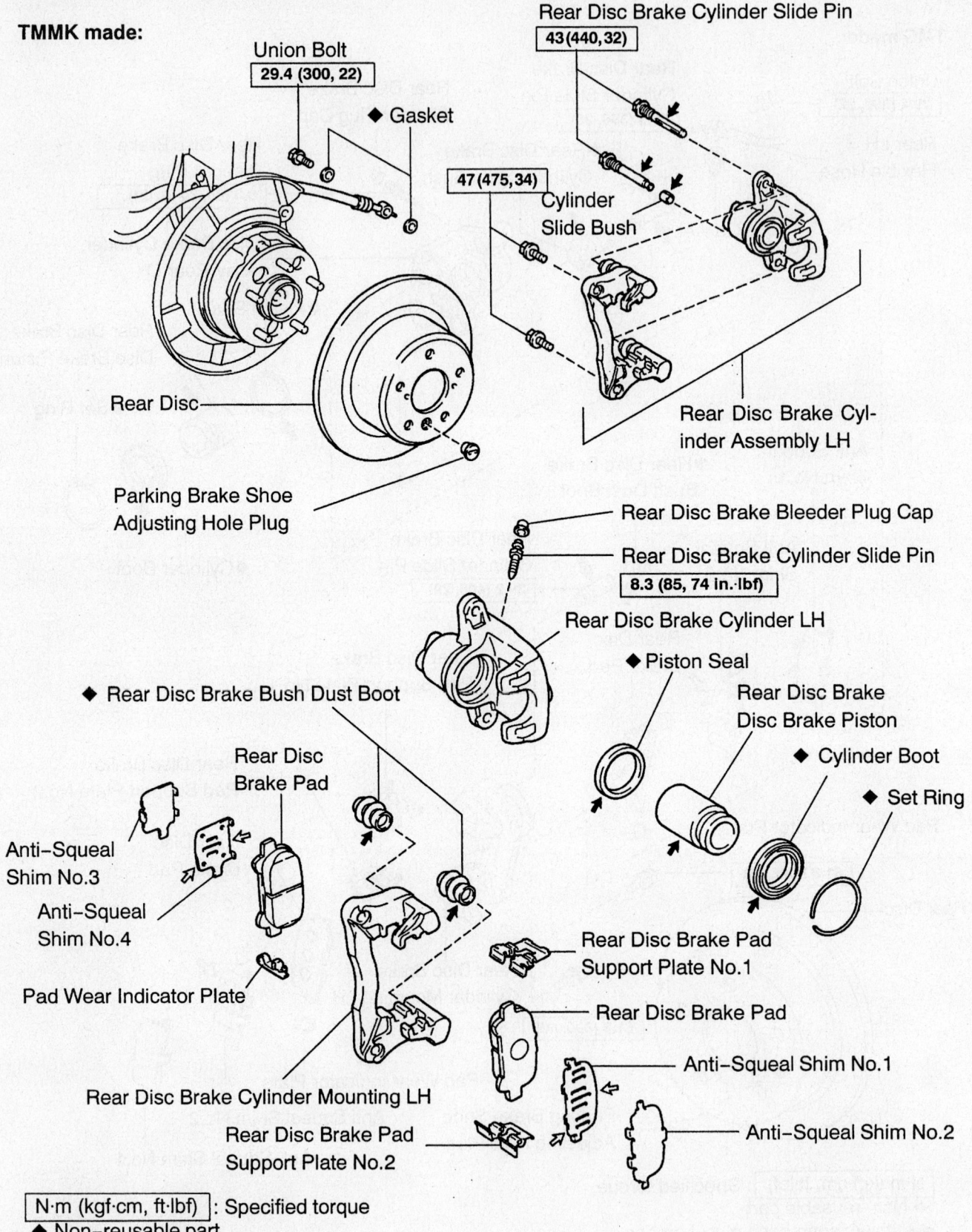

Union Bolt
29.4 (300, 22)

◆ Gasket

Rear Disc Brake Cylinder Slide Pin
43 (440, 32)

47 (475, 34)

Cylinder
Slide Bush

Rear Disc—

Rear Disc Brake Cylinder Assembly LH

Parking Brake Shoe
Adjusting Hole Plug

Rear Disc Brake Bleeder Plug Cap

Rear Disc Brake Cylinder Slide Pin
8.3 (85, 74 in.·lbf)

Rear Disc Brake Cylinder LH

◆ Piston Seal

Rear Disc Brake
Disc Brake Piston

◆ Cylinder Boot

◆ Set Ring

◆ Rear Disc Brake Bush Dust Boot

Rear Disc
Brake Pad

Anti–Squeal
Shim No.3

Anti–Squeal
Shim No.4

Pad Wear Indicator Plate

Rear Disc Brake Pad
Support Plate No.1

Rear Disc Brake Pad

Anti–Squeal Shim No.1

Rear Disc Brake Cylinder Mounting LH

Rear Disc Brake Pad
Support Plate No.2

Anti–Squeal Shim No.2

N·m (kgf·cm, ft·lbf) : Specified torque
◆ Non–reusable part
◀ Lithium soap base glycol grease
⇐ Disc brake grease

Rear caliper—Camry, TMMK caliper

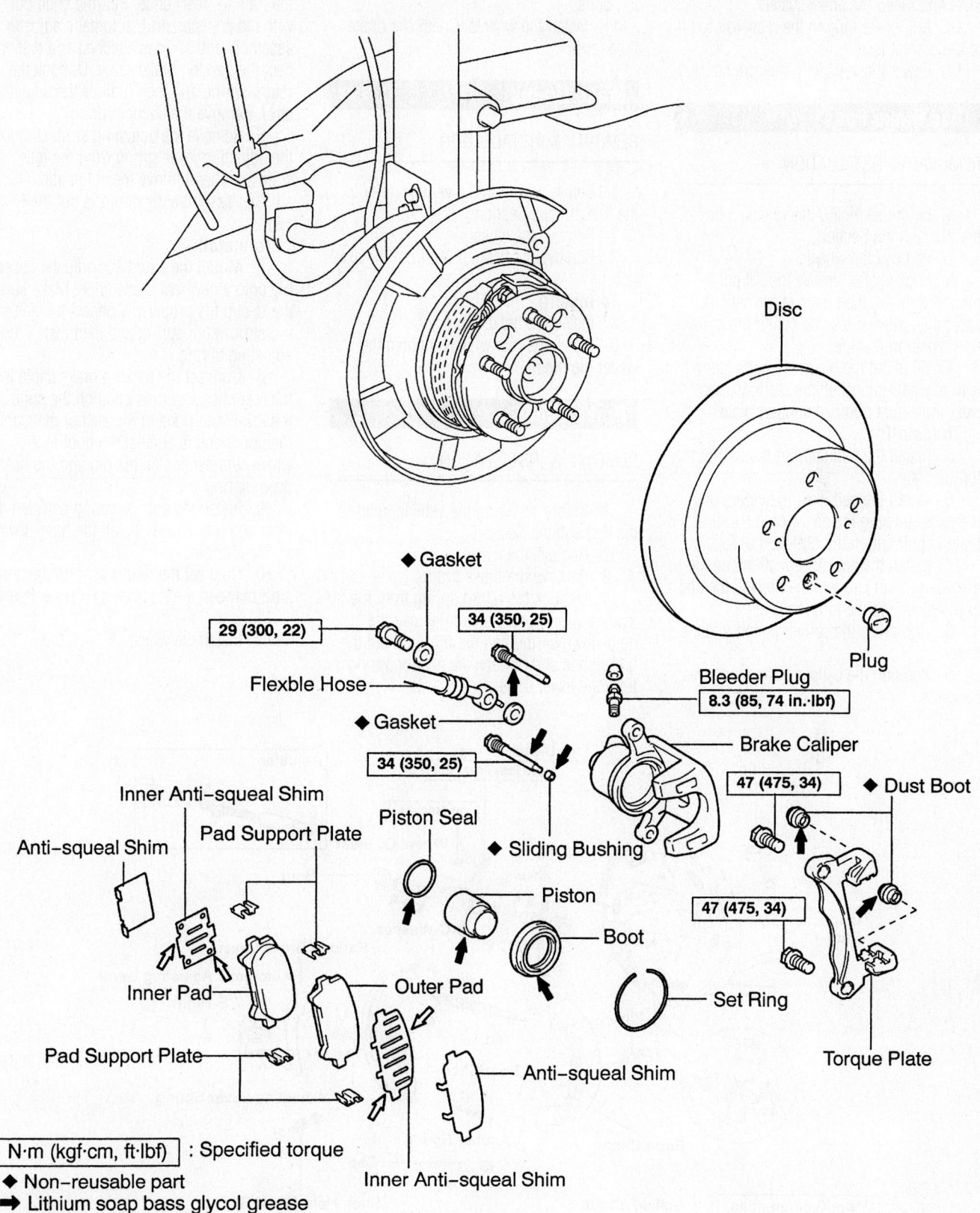

Disc

Plug

◆ Gasket

`29 (300, 22)`

Flexble Hose

◆ Gasket

`34 (350, 25)`

`34 (350, 25)`

Bleeder Plug

`8.3 (85, 74 in.·lbf)`

Brake Caliper

◆ Sliding Bushing

`47 (475, 34)`

◆ Dust Boot

`47 (475, 34)`

Inner Anti-squeal Shim

Pad Support Plate

Anti-squeal Shim

Piston Seal

Piston

Boot

Inner Pad

Outer Pad

Set Ring

Pad Support Plate

Torque Plate

Anti-squeal Shim

Inner Anti-squeal Shim

`N·m (kgf·cm, ft·lbf)` : Specified torque

◆ Non-reusable part

➡ Lithium soap bass glycol grease

⇨ Disc brake grease

67170-TOYO-G50

Rear caliper—Avalon

9. Fill the brake system to the proper level and bleed the brake system.

10. Add brake fluid to the reservoir to fill to the correct level.

11. Lower the vehicle to the ground.

Disc Brake Pads

REMOVAL & INSTALLATION

1. Before servicing the vehicle, refer to the Precautions Section.

2. Remove the wheels.

3. Loosen and remove the caliper mounting bolts, then remove the caliper assembly, without disconnecting the brake line. Position it aside.

4. Slide out the old brake pads along with any anti-squeal shims, springs, pad wear indicators and pad support plates.

To install:

5. Install the pad support plates into the torque plate.

6. Install the pad wear indicators onto the pads. Be sure the arrow on the indicator plate is pointing in the direction of rotation.

7. Install the anti-squeal shims on the outside of each pad and then install the pad assemblies into the torque plate.

8. Compress the caliper piston into the bore.

9. Position the caliper back down over the pads.

10. Install and tighten the caliper mounting bolts.

11. Install the wheels. Check the brake fluid level.

Brake Drums

REMOVAL & INSTALLATION

1. Before servicing the vehicle, refer to the Precautions Section.

2. Remove the wheels.

3. Remove the brake drum from the axle hub.

To install:

4. Install the brake drum.

5. Install the rear wheels, tighten the wheel lug nuts.

Brake Shoes

REMOVAL & INSTALLATION

1. Before servicing the vehicle, refer to the Precautions Section.

2. Remove the wheels.

3. Remove the brake drum.

4. Unhook the return spring from the leading (front) brake shoe. Remove the hold-down spring and the pin. Pull out the brake shoe and unhook the anchor spring from the lower edge.

5. Remove the hold-down spring from the trailing (rear) shoe. Pull the shoe out with the adjuster strut, automatic adjuster assembly and springs attached and disconnect the parking brake cable. Unhook the return spring and then remove the adjusting strut. Remove the anchor spring.

6. Remove the adjusting strut. Unhook the adjusting lever spring from the rear shoe and then remove the automatic adjuster assembly by popping out the C-clip.

To install:

7. Mount the automatic adjuster assembly onto a new rear brake shoe. Make sure the C-clip fits properly. Connect the adjusting strut/return spring and then install the adjusting spring.

8. Connect the parking brake cable to the rear shoe and then position the shoe so the lower end rides in the anchor plate and the upper end is against the boot in the wheel cylinder. Install the pin and the hold-down spring.

9. Install the anchor spring between the front and rear shoes. Install the hold-down spring and pin.

10. Connect the return spring/adjusting strut between the 2 shoes so it rides freely.

11. Install the drum.

12. Install the wheel.

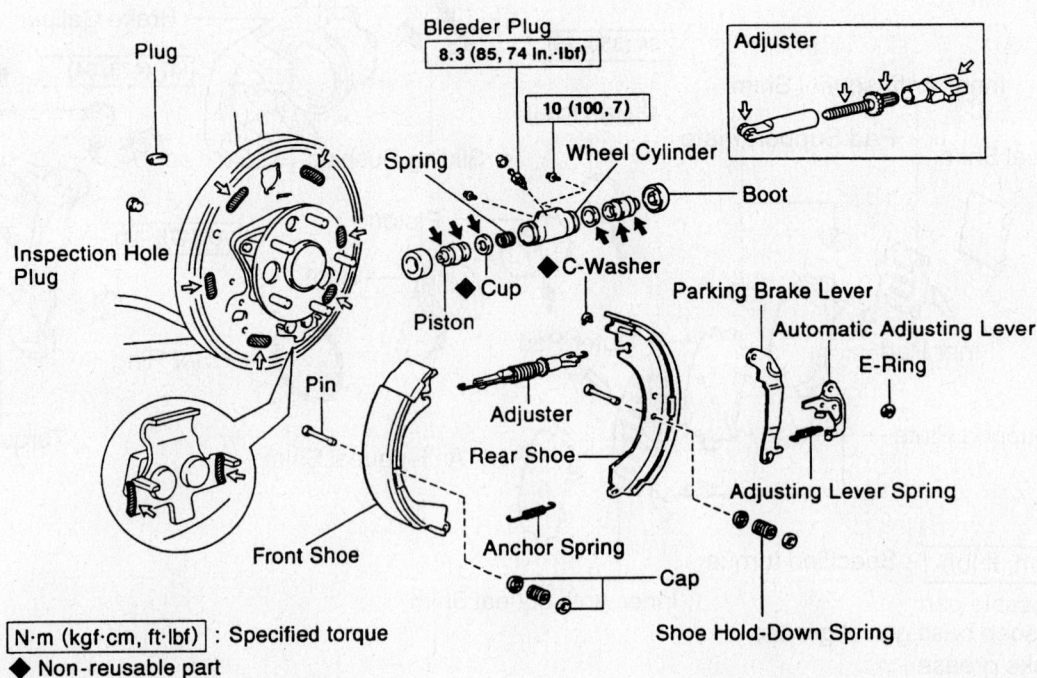

N·m (kgf·cm, ft·lbf) : Specified torque
◆ Non-reusable part
◄ Lithium soap base glycol grease
◄ High temperature grease

Rear drum brakes—Camry

93016G70

BRAKES **15-57**
DRIVE TRAIN **15-37**
ENGINE REPAIR **15-9**
FUEL SYSTEM **15-32**
SPECIFICATION AND
  MAINTENANCE CHARTS......**15-2**
Engine and Vehicle
  Identification 15-2
General Engine Specifications 15-2
Engine Tune-Up Specifications 15-3
Capacities 15-3
Valve Specifications.................15-4
Crankshaft and Connecting Rod
  Specifications 15-4
Piston and Ring Specifications 15-5
Camshaft and Bearing
  Specifications Chart 15-5
Torque Specifications 15-6
Wheel Alignment 15-6
Tire, Wheel and Ball Joint
  Specifications 15-7
Brake Specifications 15-7
Scheduled Maintenance
  Intervals............................15-8
STEERING AND
  SUSPENSION.................**15-49**

A
Air Bag...............................15-49
  Disarming 15-49
  Precautions.......................15-49
  Rearming 15-49
Alternator 15-9
  Removal & Installation.............15-9
Automatic Transaxle.................15-40
  Removal And Installation 15-40

B
Brake Caliper 15-57
  Removal And Installation 15-57
Brake Drums 15-59
  Removal And Installation 15-59
Brake Shoes.......................15-59
  Removal And Installation 15-59

C
Camshaft(s) 15-18
  Removal & Installation............15-18
Clutch 15-43
  Adjustments 15-43
  Removal & Installation 15-43
CV-Joints...........................15-47
  Overhaul 15-47
Cylinder Head 15-14
  Removal & Installation............15-14

D
Disc Brake Pads.....................15-59
  Removal And Installation 15-59

E
Engine Assembly 15-9
  Removal & Installation.............15-9
Exhaust Manifold 15-17
  Removal & Installation............15-17

F
Fuel Filter 15-33
  Removal & Installation............15-33
Fuel Injectors 15-34
  Removal & Installation............15-34
Fuel Pump 15-33
  Removal & Installation............15-33
Fuel System Pressure 15-33
  Relieving.........................15-33
Fuel System Service
  Precautions.......................15-32

H
Halfshaft............................15-45
  Removal & Installation............15-45
Heater Core.........................15-11
  Removal & Installation............15-11
Hydraulic Clutch System 15-44
  Bleeding..........................15-44

I
Ignition Timing 15-9
  Adjustment.......................15-9
Intake Manifold 15-16
  Removal & Installation............15-16

L
Lower Ball Joint.....................15-53
  Removal & Installation............15-53
Lower Control Arm 15-54
  Removal & Installation............15-54

M
Manual Transaxle Assembly 15-37
  Removal & Installation............15-37

O
Oil Pan..............................15-29
  Removal & Installation............15-29
Oil Pump 15-29
  Removal & Installation............15-29

P
Piston and Ring Positioning 15-32

R
Rack and Pinion Steering Gear 15-49
  Removal & Installation............15-49
Rear Main Seal 15-30
  Removal & Installation............15-30

S
Starter 15-29
  Removal & Installation............15-29
Strut and Coil Spring................15-51
  Removal & Installation............15-51

T
Timing Chain, Sprockets, Front
  Cover and Seal 15-30
  Removal & Installation............15-30
Transfer Case 15-44
  Removal & Installation............15-44

U
Upper Control Arm 15-54
  Removal & Installation............15-54

V
Valve Lash 15-27
  Adjustment.......................15-27

W
Water Pump 15-10
  Removal & Installation............15-10
Wheel Bearings.....................15-54
  Removal & Installation............15-54

SPECIFICATION CHARTS

ENGINE AND VEHICLE IDENTIFICATION

| Engine | | | | | | | Model Year | |
Code ①	Liters (cc)	Cu. In.	Cyl.	Fuel Sys.	Engine Type	Eng. Mfg.	Code ②	Year
1ZZ-FE	1.8 (1794)	109	4	EFI	DOHC	Toyota	3	2003
2ZZ-GE	1.8 (1796)	109.5	4	EFI	DOHC	Toyota	4	2004
							5	2005
							6	2006

EFI: Electronic Fuel Injection

DOHC: Double Overhead Camshaft

① 8th digit of VIN

② 10th digit of VIN

09490_TOY-MAT_C0001

GENERAL ENGINE SPECIFICATIONS

Year	Model	Engine Displacement Liters (VIN)	Net Horsepower @ rpm	Net Torque @ rpm (ft. lbs.)	Bore x Stroke (in.)	Com-pression Ratio	Oil Pressure @ idle
2003	Matrix	1.8 (1ZZ-FE)	①	②	3.11x3.60	10.0:1	27
	Matrix	1.8 (2ZZ-GE)	180@7600	130@6800	3.23x3.35	11.5:1	27
2004	Matrix	1.8 (1ZZ-FE)	①	②	3.11x3.60	10.0:1	27
	Matrix	1.8 (2ZZ-GE)	180@7600	130@6800	3.23x3.35	11.5:1	27
2005-06	Matrix	1.8 (1ZZ-FE)	①	②	3.11x3.60	10.0:1	27
	Matrix	1.8 (2ZZ-GE)	180@7600	130@6800	3.23x3.35	11.5:1	27

EFI: Electronic Fuel Injection

① 2WD models: 130@6000

   4WD models: 123@6000

② 2WD models: 126@4200

   4WD models: 119@4200

09490_TOY-MAT_C0002

ENGINE TUNE-UP SPECIFICATIONS

Year	Engine Displacement Liters (VIN)	Spark Plug Gap (in.)	Ignition Timing (deg.)	Fuel Pump (psi)	Idle Speed (rpm)		Valve Clearance	
					MT	AT	In.	Ex.
2003	1.8 (1ZZ-FE)	0.043	①	44-50	650-750	650-750	0.0059-0.0098	0.0098-0.0138
	1.8 (2ZZ-GE)	0.043	②	44-50	750-850	700-800	0.0031-0.0071	0.0087-0.0126
2004	1.8 (1ZZ-FE)	0.043	①	44-50	650-750	650-750	0.0059-0.0098	0.0098-0.0138
	1.8 (2ZZ-GE)	0.043	②	44-50	750-850	700-800	0.0031-0.0071	0.0087-0.0126
2005-06	1.8 (1ZZ-FE)	0.039-0.051	①	44-50	650-750	650-750	0.0059-0.0098	0.0098-0.0138
	1.8 (2ZZ-GE)	0.039-0.051	②	44-50	750-850	700-800	0.0031-0.0071	0.0087-0.0126

Note: The Vehicle Emission Control Information label often reflects specification changes made during production. The label figures must be used if they differ from those in this chart.

① With terminal TC and CG of DLC3 connected: 8-12 degrees BTDC
With terminal TC and CG of DLC3 disconnected: 10-18 degrees BTDC

② With terminal TC and CG of DLC3 connected: 8-12 degrees BTDC
With terminal TC and CG of DLC3 disconnected:
A/T: 10-18 degrees BTDC
M/T: 4-12 degrees BTDC

09490_TOY-MAT_C0003

CAPACITIES

Year	Model	Engine Displacement Liters (VIN)	Engine Oil with Filter	Transmission (pts.)			Drive Axle		Fuel Tank (gal.)	Cooling System (qts.)
				5-Spd	6-Spd	Auto.	Front (pts.)	Rear (pts.)		
2003	Matrix	1.8 (1ZZ-FE)	3.9	4.0	4.0	5.2	NA	1.04	①	6.9
	Matrix	1.8 (2ZZ-GE)	4.7	4.8	4.8	6.1	NA	—	13.2	7.1
2004	Matrix	1.8 (1ZZ-FE)	3.9	4.0	4.0	5.2	NA	1.04	①	6.9
	Matrix	1.8 (2ZZ-GE)	4.7	4.8	4.8	6.1	NA	—	13.2	7.1
2005-06	Matrix	1.8 (1ZZ-FE)	3.9	4.0	4.0	5.2	NA	1.04	①	6.9
	Matrix	1.8 (2ZZ-GE)	4.7	4.8	4.8	6.1	NA	—	13.2	7.1

Note: All capacities are approximate. Add fluid gradually and check to be sure a proper fluid level is obtained.

NA - Not available

① 2WD: 13.2 gallons
4WD: 11.9 gallons

09490_TOY-MAT_C0004

VALVE SPECIFICATIONS

Year	Engine Displacement Liters (VIN)	Seat Angle (deg.)	Face Angle (deg.)	Spring Test Pressure (lbs. @ in.)	Spring Installed Height (in.)	Stem-to-Guide Clearance (in.)		Stem Diameter (in.)	
						Intake	Exhaust	Intake	Exhaust
2003	1.8 (1ZZ-FE)	45	44.5	31.3-34.8@ 1.252	1.323	0.0010-0.0024	0.0012-0.0026	0.2154-0.2159	0.2152-0.2158
	1.8 (2ZZ-GE)	45	44.5	①	1.516	0.0010-0.0023	0.0012-0.0025	0.2145-0.2156	0.2144-0.2154
2004	1.8 (1ZZ-FE)	45	44.5	31.3-34.8@ 1.252	1.323	0.0010-0.0024	0.0012-0.0026	0.2154-0.2159	0.2152-0.2158
	1.8 (2ZZ-GE)	45	44.5	①	1.516	0.0010-0.0023	0.0012-0.0025	0.2145-0.2156	0.2144-0.2154
2005-06	1.8 (1ZZ-FE)	45	44.5	31.3-34.8@ 1.252	1.323	0.0010-0.0024	0.0012-0.0026	0.2154-0.2159	0.2152-0.2158
	1.8 (2ZZ-GE)	45	44.5	①	1.516	0.0010-0.0023	0.0012-0.0025	0.2145-0.2156	0.2144-0.2154

① Intake: 49.6-55.5@1.516
 Exhaust: 47.6-52.6@1.516

09490_TOY-MAT_C0005

CRANKSHAFT AND CONNECTING ROD SPECIFICATIONS
All measurements are given in inches.

Year	Engine Displacement Liters (cc)	Engine ID/VIN	Crankshaft				Connecting Rod		
			Main Brg. Journal Dia.	Main Brg. Oil Clearance	Shaft End-play	Thrust on No.	Journal Diameter	Oil Clearance	Side Clearance
2003	1.8 (1762)	1ZZ-FE	1.8893-1.8898	0.0006-0.0013	0.0008-0.0087	3	1.7320-1.7323	0.0011-0.0024	0.0063-0.0135
	1.8 (1796)	2ZZ-GE	1.8893-1.8898	0.0006-0.0013	0.0016-0.0094	3	1.7713-1.7717	0.0011-0.0020	0.0063-0.0135
2004	1.8 (1762)	1ZZ-FE	1.8893-1.8898	0.0006-0.0013	0.0008-0.0087	3	1.7320-1.7323	0.0011-0.0024	0.0063-0.0135
	1.8 (1796)	2ZZ-GE	1.8893-1.8898	0.0006-0.0013	0.0016-0.0094	3	1.7713-1.7717	0.0011-0.0020	0.0063-0.0135
2005-06	1.8 (1762)	1ZZ-FE	1.8893-1.8898	0.0006-0.0013	0.0008-0.0087	3	1.7320-1.7323	0.0011-0.0024	0.0063-0.0135
	1.8 (1796)	2ZZ-GE	1.8893-1.8898	0.0006-0.0013	0.0016-0.0094	3	1.7713-1.7717	0.0011-0.0020	0.0063-0.0135

09490_TOY-MAT_C0006

PISTON AND RING SPECIFICATIONS
All measurements are given in inches.

Year	Engine Displacement Liters (cc)	Engine ID/VIN	Piston Clearance	Ring Gap			Ring Side Clearance		
				Top Compression	Bottom Compression	Oil Control	Top Compression	Bottom Compression	Oil Control
2003	1.8 (1762)	1ZZ-FE	0.0026-0.0035	0.0098-0.0138	0.0138-0.0197	0.0059-0.0197	0.0008-0.0028	0.0012-0.0028	0.0012-0.0043
	1.8 (1796)	2ZZ-GE	0.0003-0.0015	0.0098-0.0138	0.0138-0.0197	NA	0.0009-0.0028	0.0012-0.0028	NA
2004	1.8 (1762)	1ZZ-FE	0.0026-0.0035	0.0098-0.0138	0.0138-0.0197	0.0059-0.0197	0.0008-0.0028	0.0012-0.0028	0.0012-0.0043
	1.8 (1796)	2ZZ-GE	0.0003-0.0015	0.0098-0.0138	0.0138-0.0197	NA	0.0009-0.0028	0.0012-0.0028	NA
2005-06	1.8 (1762)	1ZZ-FE	0.0026-0.0035	0.0098-0.0138	0.0138-0.0197	0.0059-0.0197	0.0008-0.0028	0.0012-0.0028	0.0012-0.0043
	1.8 (1796)	2ZZ-GE	0.0003-0.0015	0.0098-0.0138	0.0138-0.0197	NA	0.0009-0.0028	0.0012-0.0028	NA

NA - Not available

09490_TOY-MAT_C0007

CAMSHAFT AND BEARING SPECIFICATIONS CHART
All measurements are given in inches.

Year	Engine Displ. Liters	Engine VIN	Journal Dia.	Brg. Oil Clearance	Shaft End-play	Runout	Journal Bore	Lobe Lift	
								Intake	Exhaust
2003	1.8 (1762)	1ZZ-FE	①	0.0001-0.0024	0.0016-0.0037	0.0012	NA	1.7454-1.7493	1.7229-1.7268
	1.8 (1796)	2ZZ-GE	②	0.0001-0.0024	0.0016-0.0055	0.0012	NA	③	④
2004	1.8 (1762)	1ZZ-FE	①	0.0001-0.0024	0.0016-0.0037	0.0012	NA	1.7454-1.7493	1.7229-1.7268
	1.8 (1796)	2ZZ-GE	②	0.0001-0.0024	0.0016-0.0055	0.0012	NA	③	④
2005-06	1.8 (1762)	1ZZ-FE	①	0.0001-0.0024	0.0016-0.0037	0.0012	NA	1.7454-1.7493	1.7229-1.7268
	1.8 (1796)	2ZZ-GE	②	0.0001-0.0024	0.0016-0.0055	0.0012	NA	③	④

NA - Not available

① No 1: 1.3563-1.3569
   All others: 0.9035-0.9041

② No 1: 1.3563-1.3569
   All others: 1.1004-1.1010

③ No 1 camshaft intake: 1.6030-1.6069
   No 2 camshaft intake: 1.755-1.5795

④ No 1 camshaft exhaust: 1.5236-1.5303
   No 2 camshaft exhaust: 1.5300-1.5340

09490_TOY-MAT_C0008

TORQUE SPECIFICATIONS
All readings in ft. lbs.

Year	Engine Displacement Liters (VIN)	Cylinder Head Bolts	Main Bearing Bolts	Rod Bearing Bolts	Crankshaft Damper Bolts	Flywheel Bolts	Manifold Intake	Manifold Exhaust	Spark Plugs	Oil Pan Drain Plug
2003	1.8 (1ZZ-FE)	①	②	③	102	①	22	27	18	76
	1.8 (2ZZ-GE)	④	⑤	⑥	87	①	⑦	37	13	76
2004	1.8 (1ZZ-FE)	①	②	③	102	①	22	27	18	76
	1.8 (2ZZ-GE)	④	⑤	⑥	87	①	⑦	37	13	76
2005-06	1.8 (1ZZ-FE)	①	②	③	102	①	22	27	18	76
	1.8 (2ZZ-GE)	④	⑤	⑥	87	①	⑦	37	13	76

① Step 1: 36 ft. lbs.
　Step 2: 90 degree turn

② 12 pointed bolts:
　Step 1: 33 ft. lbs.
　Step 2: 90 degree turn
　Hex head bolts: 14 ft. lbs.

③ Step 1: 15 ft. lbs.
　Step 2: 90 degree turn

④ Step 1: 26 ft. lbs.
　Step 2: 180 degree turn

⑤ 12 pointed bolts:
　Step 1: 16 ft. lbs.
　Step 2: 32 ft. lbs.
　Step 3: 45 degree turn
　Step 4: 45 degree turn
　Hex head bolts: 13 ft. lbs.

⑥ Step 1: 22 ft. lbs.
　Step 2: 90 degree turn

⑦ Bolt A: 25 ft. lbs.
　Bolt B: 34 ft. lbs.

09490_TOY-MAT_C0009

WHEEL ALIGNMENT

Year	Model		Caster Range (+/-Deg.)	Caster Preferred Setting (Deg.)	Camber Range (+/-Deg.)	Camber Preferred Setting (Deg.)	Toe-in (in.)
2003	Matrix - 2WD	F	0.75	+2.78	0.75	-0.77	0+/-0.08
		R	—	—	0.50	-1.45	0.11+/-0.11
	Matrix - 4WD	F	0.75	+2.77	0.75	-0.48	0+/-0.08
		R	—	—	0.75	-0.73	0.08+/-0.08
2004	Matrix - 2WD	F	0.75	+2.78	0.75	-0.77	0+/-0.08
		R	—	—	0.50	-1.45	0.11+/-0.11
	Matrix - 4WD	F	0.75	+2.77	0.75	-0.48	0+/-0.08
		R	—	—	0.75	-0.73	0.08+/-0.08
2005-06	Matrix - 2WD	F	0.75	+2.78	0.75	-0.77	0+/-0.08
		R	—	—	0.50	-1.45	0.11+/-0.11
	Matrix - 4WD	F	0.75	+2.77	0.75	-0.48	0+/-0.08
		R	—	—	0.75	-0.73	0.08+/-0.08

09490_TOY-MAT_C0010

TIRE, WHEEL AND BALL JOINT SPECIFICATIONS

Year	Model	OEM Tires Standard	Optional	Tire Pressures (psi) Front	Rear	Wheel Size	Ball Joint Inspection	Lug Nuts
2003	Matrix	205/55R16	—	33	33	6.5-JJ	9-26 in. ①	76
2004	Matrix	205/55R16	—	33	33	6.5-JJ	9-26 in. ①	76
2005-06	Matrix	205/55R16	—	33	33	6.5-JJ	9-26 in. ①	76

OEM: Original Equipment Manufacturer

PSI: Pounds Per Square Inch

STD: Standard

OPT: Optional

① Torque required in inch lbs. to rotate ball joint when removed from the knuckle

09490_TOY-MAT_C0011

BRAKE SPECIFICATIONS
All measurements in inches unless noted

Year	Model		Brake Disc Original Thickness	Minimum Thickness	Maximum Runout	Brake Drum Diameter Original Inside Diameter	Max. Wear Limit	Maximum Machine Diameter	Minimum Lining Thickness	Brake Caliper Bracket Bolts (ft. lbs.)	Mounting Bolts (ft. lbs.)
2003	Matrix	F	0.984	0.906	0.0020	—	—	—	0.039	79	25
		R	0.354	0.295	0.0059	9.00	—	9.04	0.039	—	34
2004	Matrix	F	0.984	0.906	0.0020	—	—	—	0.039	79	25
		R	0.354	0.295	0.0059	9.00	—	9.04	0.039	—	34
2005-06	Matrix	F	0.984	0.906	0.0020	—	—	—	0.039	79	25
		R	0.354	0.295	0.0059	9.00	—	9.04	0.039	—	34

F: Front

R: Rear

09490_TOY-MAT_C0012

SCHEDULED MAINTENANCE INTERVALS
TOYOTA—MATRIX

TO BE SERVICED	TYPE OF SERVICE	VEHICLE MILEAGE INTERVAL (x1000)												
		7.5	15	22.5	30	37.5	45	52.5	60	67.5	75	82.5	90	97.5
Engine oil & filter	R	✓	✓	✓	✓	✓	✓	✓	✓	✓	✓	✓	✓	✓
Drive belts	S/I								✓	✓	✓	✓	✓	✓
Automatic transaxle fluid & filter	S/I		✓		✓		✓		✓		✓		✓	
Ball joints & dust covers	S/I		✓		✓		✓		✓		✓		✓	
Bolts & nuts on body & chassis	S/I		✓		✓		✓		✓		✓		✓	
Brake line pipes & hoses	S/I		✓		✓		✓		✓		✓		✓	
Brake linings & drums	S/I		✓		✓		✓		✓		✓		✓	
Brake pads & discs (front & rear if equipped)	S/I		✓		✓		✓		✓		✓		✓	
Differential oil	S/I		✓		✓		✓		✓		✓		✓	
Drive shaft boots (except Supra)	S/I		✓		✓		✓		✓		✓		✓	
Manual transaxle oil	S/I		✓		✓		✓		✓		✓		✓	
Steering gear housing oil	S/I		✓		✓		✓		✓		✓		✓	
Steering linkage	S/I		✓		✓		✓		✓		✓		✓	
Air filter	R				✓				✓				✓	
Spark plugs	R				✓				✓				✓	
Spark plugs (platinum tip)	R								✓					
Exhaust system	S/I				✓				✓				✓	
Fuel lines & connections	S/I				✓				✓				✓	
Valve clearance	S/I				✓				✓				✓	
Engine coolant	R						✓				✓			
Fuel tank cap gasket	R								✓					
Charcoal canister	S/I								✓					

R: Replace   I: Inspect   A: Adjust

FREQUENT OPERATION MAINTENANCE (SEVERE SERVICE)

If a vehicle is operated under any of the following conditions it is considered severe service:

- Extremely dusty areas.

- 50% or more of the vehicle operation is in 32°C (90°F) or higher temperatures, or constant operation in temperatures below 0°C (32°F).

- Prolonged idling (vehicle operation in stop and go traffic).

- Frequent short running periods (engine does not warm to normal operating temperatures).

- Police, taxi, delivery usage or trailer towing usage.

Oil & oil filter: change every 6000 miles.

Bolts & nuts on chassis & body: tighten every 7500 miles.

Ball joints & dust covers: service or inspect every 12,000 miles.

Brake linings & drums: service or inspect ever 12,000 miles.

Brake pads & discs (front & rear if equipped): service or inspect every 12,000 miles.

Drive shaft boots & except Supra): service or inspect every 12,000 miles.

Steering linkage: service or inspect every 12,000 miles.

Air filter: service or inspect every 15,000 miles.

Exhaust system: service or inspect every 15,000 miles.

Timing belt: replace every 60,000 miles.

ENGINE REPAIR

Alternator

REMOVAL & INSTALLATION

1. Before servicing the vehicle, refer to the precautions section.
2. Remove or disconnect the following:
   - Negative battery cable
   - Drive belt
   - Wire clamp from the clip on the rectifier end frame
   - Rubber clamp and nut
   - Alternator wiring and connector
   - Alternator

To install:

3. Install or connect the following:
   - Alternator. Torque the 12mm bolt to 18 ft. lbs. (25 Nm) and the 14mm bolt to 39 ft. lbs. (54 Nm).
   - Alternator connector and wiring
   - Rubber clamp and nut
   - Wire clamp
   - Drive belt
   - Negative battery cable

Ignition Timing

ADJUSTMENT

➡ **The timing on engines equipped with Distributorless Ignition Systems (DIS) is not adjustable.**

Engine Assembly

REMOVAL & INSTALLATION

1. Before servicing the vehicle, refer to the precautions section.
2. Relieve the fuel system pressure.
3. Drain the cooling system.
4. Drain the engine oil.
5. Drain the transaxle fluid and transfer fluid, if equipped.
6. Remove or disconnect the following:
   - Negative battery cable. Wait at least 90 seconds before proceeding.
   - Battery
   - Hood
   - Undercovers
   - Radiator inlet and outlet hoses
   - Radiator hose outlet
   - Oil cooler inlet and outlet tubes
   - Upper radiator support and radiator, if equipped with A/C
   - Battery
   - Air cleaner assembly
   - Fuel pipe clamp

- Fuel tube sub-assembly
- Accelerator control cable
- Cruise control actuator, if equipped
- Union-to-connector tube hose
- Heater inlet and outlet hoses
- Transmission shift cable(s)
- Clutch release cylinder, on manual transaxle
- Glove compartment door
- Engine relay block cover
- 3 connectors from the relay block
- 2 ground cables
- Engine wire from the Engine Control Module (ECM) and junction block
- Engine wire from the cabin
- Drive belt
- Compressor and magnetic clutch, if equipped with A/C. Unbolt and position aside, DO NOT disconnect the lines.
- Vane pump oil reservoir from the bracket
- Return tube
- Right side front door scuff plate
- Right side cowl side trim plate
- Right side rear door scuff plate, AWD
- Lower right side center pillar garnish, AWD
- Right front seat, AWD
- Column hole cover silencer sheet
- Steering intermediate shaft
- Front floor panel brace, FWD
- Center exhaust pipe, AWD
- Propeller shaft with center bearing shaft, AWD
- Front exhaust pipe
- Front hub nuts
- Tie rod ends from the steering knuckles

7. Separate the front stabilizer links and lower control arm ball joints
   - Front halfshafts
8. Remove the engine from the vehicle, as follows:

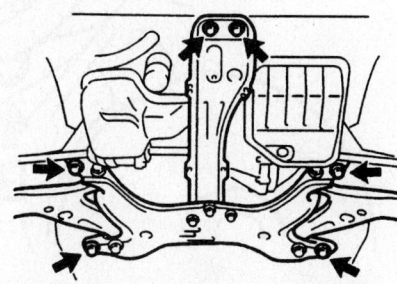

9359AB66
Remove the 6 bolts, as indicated by arrows

a. Set the engine lifter.
b. Remove the bolts and nuts, then remove the engine mounting insulator.
c. Remove the through bolt and nut, then detach the engine mounting insulator from the vehicle.
d. Remove the 6 bolts as shown.
e. Use a suitable tool to suspend the engine assembly, as shown in the figure.
f. No. 1 engine hanger: P/N 12281-15040 (1ZZ-FE), 12281-88600 (2ZZ-GE).
g. No. 2 engine hanger: P/N 12281-22021 (1ZZ-FE), 12281-88600 (2ZZ-GE).
h. Bolt: P/N 91512-B1016.
i. Torque the bolts to 28 ft. lbs. (38 Nm).

✳✳ CAUTION

Do not try to suspend the engine by hooking the chain to any other part.

j. Attach an engine chain hoist to the hangers.
k. Using the chain block and sling device, suspend the engine.
l. Remove the engine and transaxle assembly from the vehicle.

9. Remove or disconnect the following components, as necessary:
   - Vane pump
   - Steering gear
   - Crossmember
   - Manifold stay
   - Oxygen (O_2) sensor
   - Exhaust manifold
   - Starter
   - Transaxle
   - Transfer case
   - Clutch
   - Flywheel
   - Alternator
   - Ignition coil
   - Fuel delivery pipe
   - Intake manifold
   - Oil level gauge

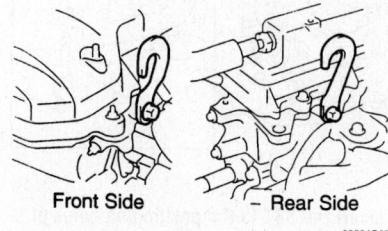

Front Side — Rear Side

9359AB67
Install the engine hangers—1ZZ-FE shown, 2ZZ-GE similar

- Water inlet and bypass pipes
- Thermostat
- Oil pressure switch
- Crankshaft Position (CKP) sensor
- Knock Sensor (KS)
- Drive belt tensioner
- Engine mounts and brackets
- Coolant Temperature Sensor (CTS)

To install:

10. Install any removed components to the engine and transaxle assembly.

11. To install the engine:

a. Place the engine and transaxle on an engine lifter.

b. Install the engine with the transaxle to the vehicle.

c. Temporarily install the crossmember and 6 bolts.

d. Install the left engine mounting insulator. Tighten the bolts to 59 ft. lbs. (80 Nm).

e. Install the right engine mounting insulator. Tighten the bolts to 38 ft. lbs. (52 Nm).

f. Insert SST 09670-00010 to the positioning holes of the right handle crossmember and on the right handle of the vehicle. Temporarily tighten bolt A, then bolt B.

g. Insert SST 09670-00010 to the positioning holes of the left handle crossmember and on the left handle of

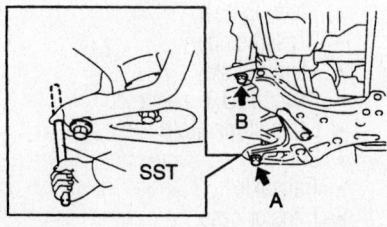

Insert the SST to the positioning holes of the right handle crossmember and on the right handle of the vehicle. Temporarily tighten bolt A, then bolt B

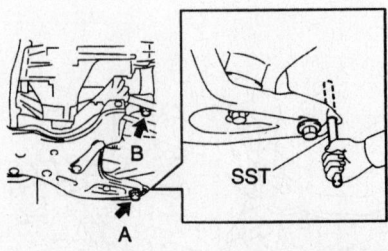

Insert the SST to the positioning holes of the left handle crossmember and on the left handle of the vehicle. Temporarily tighten bolt A, then bolt B

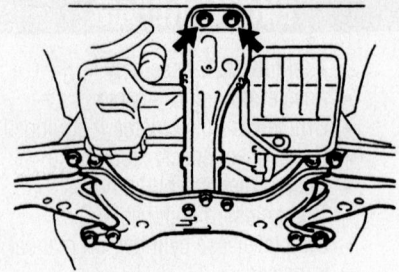

Tighten the 2 crossmember bolts, indicated by arrows

the vehicle. Temporarily tighten bolt A, then bolt B.

h. Insert the SST to the positioning holes on the right-handle crossmember and right handle. Tighten bolt A to 116 ft. lbs. (157 Nm) and bolt B to 83 ft. lbs. (113 Nm).

i. Insert the SST to the positioning holes on the left-handle crossmember and left handle. Tighten bolt A to 116 ft. lbs. (157 Nm) and bolt B to 83 ft. lbs. (113 Nm).

j. Tighten the 2 crossmember bolts, shown in the figure, to 29 ft. lbs. (39 Nm).

12. Installation of the remaining components is the reverse of the removal procedure.

13. Make sure all fluid levels are accurate, then start the engine check for leaks.

REMOVAL & INSTALLATION

1ZZ-FE Engine

1. Before servicing the vehicle, refer to the precautions section.
2. Drain the cooling system.
3. Remove or disconnect the following:
   - Negative battery cable
   - Right-hand engine under cover
   - Drive belt
   - Alternator
   - Water pump

To install:

4. Install or connect the following:
   - Water pump. Torque bolts marked **A** (short) to 80 inch lbs. (9 Nm) and bolts marked **B** (long) to 96 inch lbs. (11 Nm).
   - Alternator
   - Drive belt
   - Right engine under cover
   - Negative battery cable
5. Fill the cooling system to the proper level.
6. Start the vehicle, check for leaks and repair if necessary.

2ZZ-GE Engine

1. Before servicing the vehicle, refer to the precautions section.
2. Drain the cooling system.
3. Remove or disconnect the following:

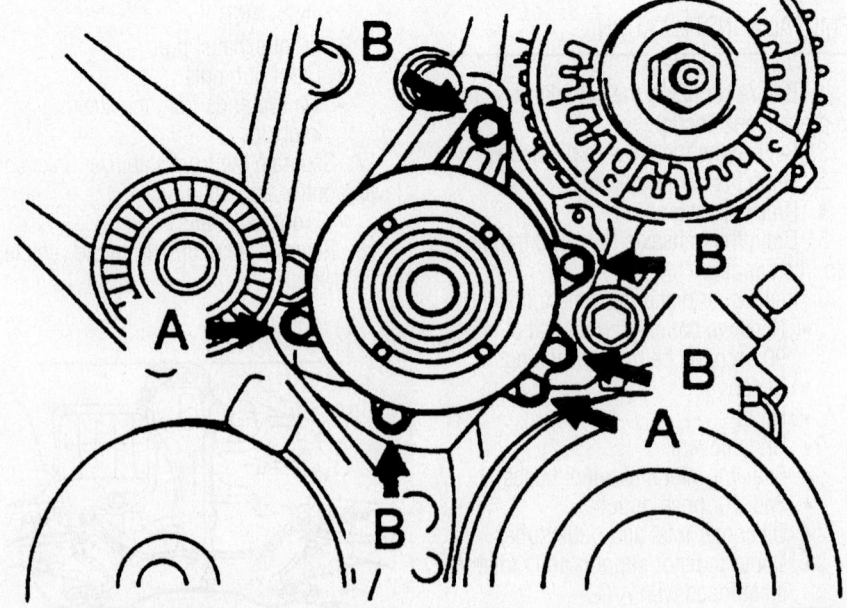

Water pump bolt identification—1.8L (1ZZ-FE) engine

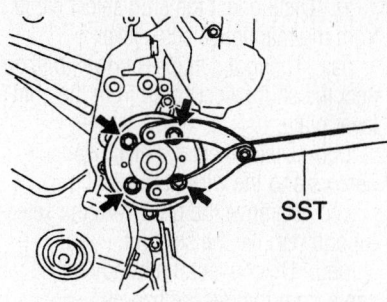

View of the special tool needed to remove and install the water pump pulley—2ZZ-GE engine

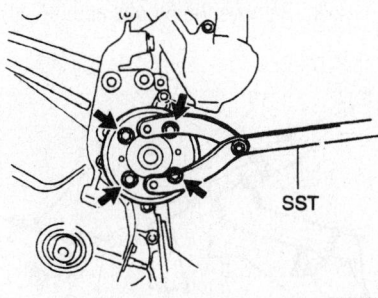

Water pump mounting and bolt locations—2ZZ-GE engine

- Negative battery cable
- Right-hand engine under cover
- Drive belt
- Alternator
- Water pump pulley, using SST 09960-10010
- Water pump and O-ring

To install:

4. Install or connect the following:

- Water pump with new O-ring. Torque the bolts to 80 inch lbs. (9 Nm).
- Water pump pulley, using SST 09960-10010. Torque the bolts to 11 ft. lbs. (15 Nm).
- Alternator
- Drive belt
- Right engine under cover
- Negative battery cable

5. Fill the cooling system to the proper level.

6. Start the vehicle, check for leaks and repair if necessary.

Heater Core

REMOVAL & INSTALLATION

1. Before servicing the vehicle, refer to the precautions section.
2. Drain the cooling system.
3. Discharge and recover the A/C system refrigerant using approved equipment.
4. Remove or disconnect the following:

- Negative battery cable
- Heater hoses from the core
- Evaporator inlet and outlet tubes from the evaporator and cap the lines to avoid system contamination

5. Remove the instrument panel as follows:

a. Disable the air bag system.

b. Using a taped flat–bladed tool, carefully pry the retaining clips attaching the center trim plate to the instrument panel.

c. Disconnect the A/C switch, hazard switch; rear defogger switch and passenger seat belt indicator switch electrical connections.

d. Remove the radio retaining screws, clamp from the radio bracket, slide the radio forward to disconnect the power and antenna connections. Remove the radio.

e. Remove the A/C switch and screw.

f. Remove the hazard switch.

g. Remove the rear defogger switch.

h. Remove the manual transmission shift knob.

i. Using a taped flat–bladed tool, carefully pry the retaining clips attaching the front floor console trim plate to the floor console assembly.

j. Disconnect the 2 cigar lighter connectors.

k. Disconnect the accessory power receptacle connectors.

l. Remove the cigar lighters and power receptacle.

m. Place both wheels in the straight ahead position.

n. Remove the bolts from the steering wheel module.

o. Release the Connector Position Assurance (CPA) from the inflator module.

p. Disconnect the steering wheel module connectors.

q. Remove the steering wheel module.

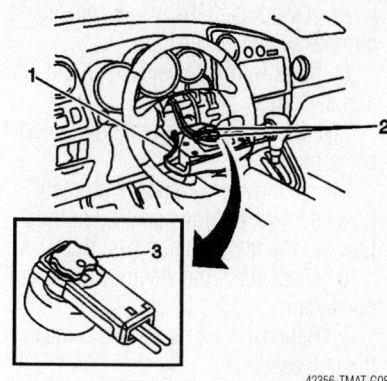

Exploded view of the CPA assembly

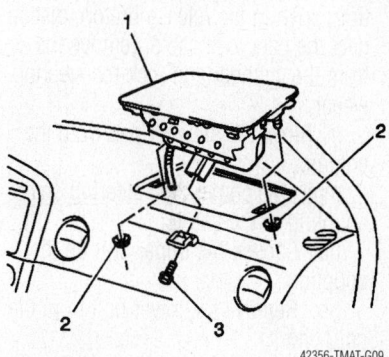

Remove the instrument panel module connectors and the passenger air bag assembly

r. Matchmark the steering wheel nut–to–shaft position, then remove the steering wheel nut and the wheel.

s. Remove the upper and lower steering column cover screws and the covers.

t. Disconnect the turn signal/headlamp assembly connectors.

u. Remove the turn signal/headlamp switch assembly

v. Remove the wiper switch by depressing the tab.

w. Remove the glove box.

x. Disconnect the instrument panel connector.

y. Remove the instrument panel module connectors and the passenger air bag assembly.

z. Remove the cluster trim plate by disengaging the clips.

aa. Remove the cluster screw and disengage the 2 lower clips.

bb. Disconnect the cluster electrical connectors and remove the cluster.

cc. Remove the windshield garnish moldings.

dd. Using a taped flat–bladed tool, carefully pry the retaining clips attaching the instrument panel left trim plate to the instrument panel.

ee. Disconnect the power mirror and dimmer switch connectors.

ff. Remove the power mirror and dimmer switches.

gg. Disconnect any remaining electrical connections.

hh. Remove the upper instrument panel screws and the panel by pulling towards the rear to disengage the tabs.

ii. Disconnect the steering wheel coil connector.

jj. Release the 3 claws and remove the coil assembly.

kk. If the vehicle is equipped with an automatic transmission, insert the key into the cylinder, turn to the ACC position, push in the release button, disconnect the park lock cable, remove the key from the cylinder and lock the steering wheel.

ll. Move the silencer pad from the column.

mm. Matchmark the steering shaft coupling to the shaft.

nn. Loosen the upper bolt on the coupling.

oo. Remove the lower bolt from the coupling.

pp. Move the coupling onto the column shaft.

qq. Disconnect the wiring harness clamps from the column.

rr. Remove the 3 bolts and the column.

ss. Remove the body hinge trim panels.

tt. Remove the sill plates.

uu. Remove the front floor console storage door.

vv. Remove the screws attaching the console to the instrument panel, pull the console rewards and up and remove the front floor console.

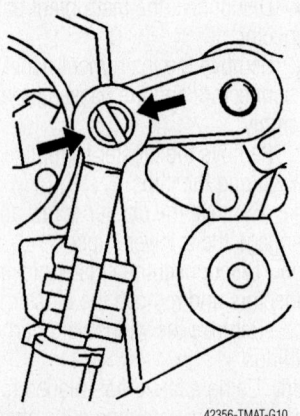

42356-TMAT-G10

Push in the clip and disconnect the cable from the manual selector shifter assembly

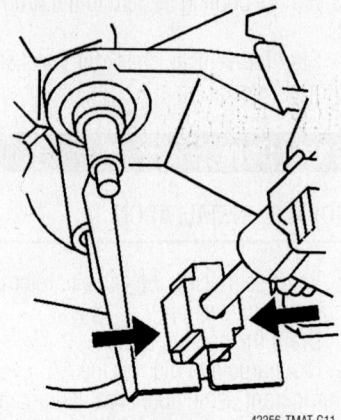

42356-TMAT-G11

Using a suitable prytool, disconnect the park lock cable from the bracket

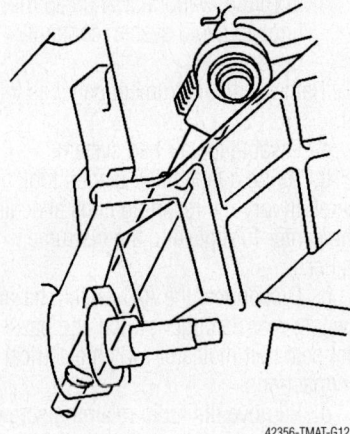

42356-TMAT-G12

Disconnect the shift select cable from the manual selector lever

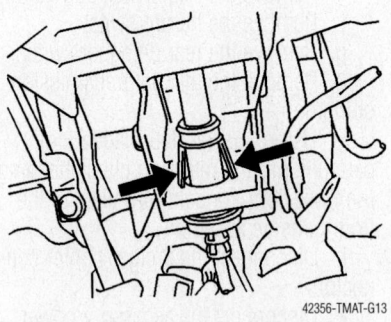

42356-TMAT-G13

Using a suitable prytool, disconnect the shift select cable from the shift lever plate

ww. Remove the HVAC retaining screw; disconnect the electrical connectors and module control, temperature control and A/C cables. Remove the unit.

xx. Push in the clip and disconnect the cable from the manual selector shifter assembly.

yy. Using a suitable prytool, disconnect the park lock cable from the bracket.

zz. Disconnect the shift select cable from the manual selector lever.

aaa. Using a suitable prytool, disconnect the shift select cable from the shift lever plate.

bbb. Disconnect the electrical connectors and the wire harness clip.

ccc. Remove the nuts from the selector and remove the selector.

ddd. Disconnect the hood release cable from the release handle.

eee. Remove the 8 bolts, 4 push retainers and the wire harness clamps from the lower instrument panel and remove the panel.

fff. Disengage the wiring harness clips, remove the bolts retaining the lower instrument panel pad and the pad.

ggg. Remove the ground cable.

hhh. Remove the connector housing bracket from the right instrument panel center support brace.

iii. Remove the left brace nut, right brace nut, left brace bolt, right brace bolt, left center support brace and right center support brace.

jjj. Remove the windshield defroster nozzle duct from the heater case.

kkk. Remove the 5 bolts and the nuts

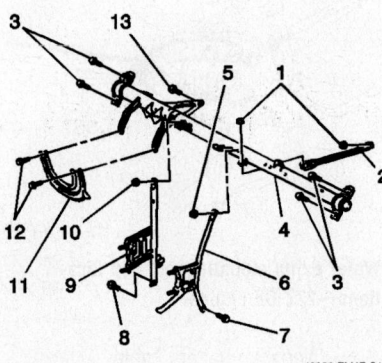

42356-TMAT-G14

Exploded view of the instrument panel reinforcement

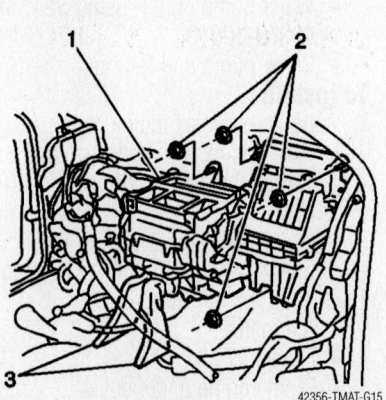

42356-TMAT-G15

Remove the HVAC module

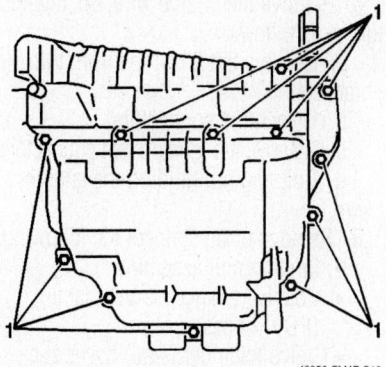

Remove the 12 bolts from the core case to access the heater core

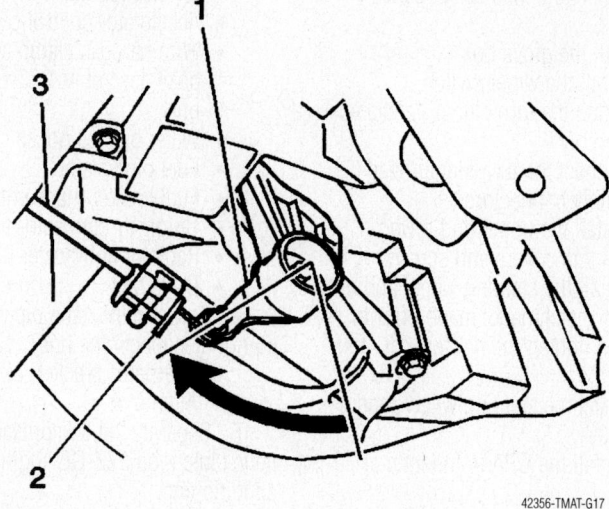

Location of the components used to adjust the temperature control cable (3) clip (2), door lever (1).

from the instrument panel reinforcement at the hinge pillars.

lll. Remove the instrument panel reinforcement.

mmm. Disconnect the blower motor connector.

nnn. Disconnect the rear ducts from the HVAC module.

ooo. Remove the HVAC module.

ppp. Remove the 12 bolts from the core case.

qqq. Remove the heater core.

To install:

a. Install the heater core.

b. Install the 12 bolts from the core case and tighten to 89 inch lbs. (10 Nm).

c. Install the HVAC module.

d. Connect the rear ducts to the HVAC module.

e. Connect the blower motor connector.

f. Install the instrument panel reinforcement.

g. Install the 5 bolts and the nuts to the instrument panel reinforcement at the hinge pillars. Tighten to 21 ft. lbs. (28 Nm).

h. Install the windshield defroster nozzle duct to the heater case.

i. Install left center support brace and right center support brace. Tighten the nuts and bolt to 15 ft. lbs. (20 Nm).

j. Install the connector housing bracket.

k. Connect the ground cable.

l. Install the lower instrument panel pad and the bolts and attach the wiring harness clips.

m. Connect the hood release cable to the release handle.

n. Install the lever and tighten the nuts to 12 ft. lbs. (18 Nm).

o. Connect the manual selector electrical connections.

p. Attach the shift cable to the shift lever plate.

q. Connect the shift select cable to the selector lever.

r. Install the park lock cable to the shift lever plate.

s. Connect the park lock cable to the manual selector lever.

t. Connect the electrical connectors and module control, temperature control and A/C cables. Install the HVAC unit.

u. Adjust the temperature control cable by setting the temperature control dial to coldest. Hold the door lever fully rearwards, clockwise. Attach the cable to the control clip.

v. Adjust the Mode linkage by setting the dial to defrost. Hold the door lever fully rearwards, clockwise. Attach the cable to the control clip.

w. Install the front floor console and tighten the screws.

x. Install the front floor console storage door.

y. Install the sill plates.

z. Install the column.

aa. Install the 3 bolts and tighten the lower bolt to 16 ft. lbs. (21 Nm) and the 2 upper bolts to 16 ft. lbs. (21 Nm).

bb. Align the matchmarks made prior to removal.

cc. Lower the coupling onto the shaft. Install the bolts and tighten to 26 ft. lbs. (35 Nm).

dd. Connect the wiring harness clamps to the column.

ee. Move the silencer pad to the column.

ff. If the vehicle is equipped with an automatic transmission, insert the key into the cylinder, turn to the ACC position, insert the park lock cable making sure the release button engages. Make

sure the key will not rotate to the lock position unless the shifter is in the park position, remove the key from the cylinder and lock the steering wheel.

gg. Install the body hinge trim panels.

hh. Make sure the turn signal switch is in the neutral position.

ii. If installing a new coil, remove the lock pin.

jj. Install the coil making sure the 3 claws engage.

kk. While holding the coil casing, turn the coil center casing counterclockwise until the coil reaches its stop.

ll. Turn the coil center casing clockwise 2 ½ turns.

mm. Align the center casing with the arrow on the outer casing.

nn. Connect the coil electrical connector.

oo. Install the upper instrument panel and screws.

pp. Connect the electrical connections.

qq. Install the power mirror and dimmer switches.

rr. Connect the power mirror and dimmer switch connectors.

ss. Install the instrument panel left trim plate to the instrument panel.

tt. Install the windshield garnish moldings.

uu. Connect the cluster electrical connectors and install the cluster.

vv. Engage the cluster lower clips and install the screw.

ww. Install the cluster trim plate.

xx. Install the passenger air bag assembly and the instrument panel module connectors.

yy. Connect the instrument panel connector.

zz. Install the glove box.

aaa. Install the wiper switch.

bbb. Install the turn signal/headlamp switch assembly

ccc. Connect the turn signal/head-lamp assembly connectors.

ddd. Install the upper and lower steering column covers and screws.

eee. Install the steering wheel and nut aligning the matchmarks made prior to removal and tighten the nut to 37 ft. lbs. (50 Nm).

fff. Connect the steering wheel module connectors.

ggg. Install the CPA to the inflator module.

hhh. Install the steering wheel module and tighten the retainers 78 inch lbs. (9 Nm)

iii. Install cigar lighters and power receptacle.

jjj. Connect the accessory power receptacle connectors.

kkk. Connect the 2 cigar lighter connectors.

lll. Install the front floor console trim plate to the floor console assembly.

mmm. Install the manual transmission shift knob.

nnn. Install the rear defogger switch.

ooo. Install the hazard switch.

ppp. Install the A/C switch and screw.

qqq. Install the radio.

rrr. Connect the A/C switch, hazard switch, rear defogger switch and passenger seat belt indicator switch electrical connections.

sss. Install the center trim plate to the instrument panel.

ttt. Connect the evaporator inlet and outlet tubes.

uuu. Connect the heater hoses to the core.

vvv. Connect the negative battery cable.

www. Recharge the A/C system and fill the cooling system.

Cylinder Head

REMOVAL & INSTALLATION

1. Before servicing the vehicle, refer to the precautions section.
2. Drain the cooling system.
3. Remove or disconnect the following:
   - Right side engine under cover
   - Right front wheel and tire
   - Cylinder head cover
   - Air cleaner assembly with hose
   - Accelerator control cable
   - Wire harness clamp and suction hose assembly, 2ZZ-GE engine only
   - Water bypass hoses
   - Fuel pipe clamp
   - Fuel tube sub-assembly
   - Union-to-connector tube hose
   - Radiator and heater inlet hoses
   - Drive belt
4. Separate the vane pipe assembly, but do not disconnect the hose, 1ZZ-FE engine.
   - Alternator bracket, 2ZZ-GE
   - Alternator
5. Separate the compressor and magnetic clutch, on 2ZZ-GE engines with air conditioning.
   - Front exhaust pipe assembly
   - Power steering pump reservoir and position it aside, 1ZZ-FE engine
6. Place a jack with a wooden block under the vehicle for support, then remove the 4 bolts and 2 nuts and remove the right side engine mount.

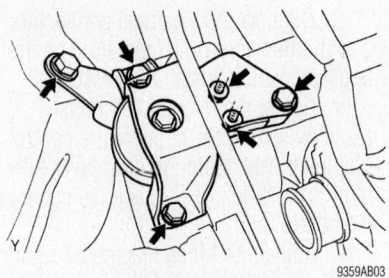

With the engine supported, remove the right side engine mount—1ZZ-FE engine shown, 2ZZ-GE similar

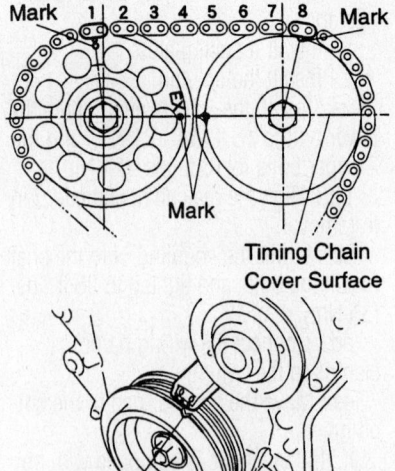

Proper timing mark alignment for TDC

7. Remove the engine wire, on 1ZZ-FE engines as follows:
   a. Remove the 5 clamps from the brackets.
   b. Detach the connectors.
   c. Remove the ignition coil connectors.
   d. Bolt and nut holding the engine wire.
8. Remove or disconnect the following:
   - Ignition coil assembly
   - Positive Crankcase Ventilation (PCV) hoses
   - Valve (cylinder head) cover sub-assembly
9. Set the No. 1 cylinder to Top Dead Center (TDC) of the compressor stroke as follows:
   a. Turn the crankshaft pulley, and align its groove with the "0" timing mark of the timing chain cover.
   b. Make sure the point marks of the camshaft timing sprockets and Variable Valve Timing (VVT) timing sprockets are in a straight line as shown. If not, turn the crankshaft 1 complete revolution (360°) and align the marks.
10. Remove or disconnect the following:
    - Crankshaft pulley, using SST 09960-10010
    - Belt tensioner
    - Exhaust manifold stay and head insulator, 2ZZ-GE engine
    - Water pump pulley and pump
    - Transverse engine mounting bracket
    - Crankshaft Position (CKP) sensor
    - No. 1 chain tensioner assembly, making sure not to revolve the crankshaft without the tensioner
    - Timing chain or belt cover
    - Timing gear cover oil seal
    - CKP sensor plate No. 1
    - Timing chain tensioner slipper
    - Timing chain vibration damper No. 1

➡ **In case you turn the camshafts with the timing chain removed, turn the crankshaft ¼ turn for the valve to avoid contact with the pistons.**

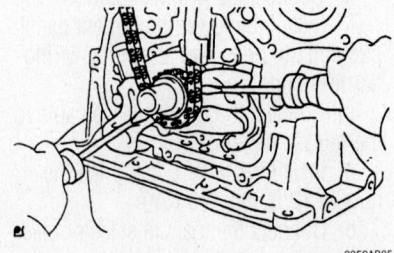

Remove the timing chain with the crankshaft gear

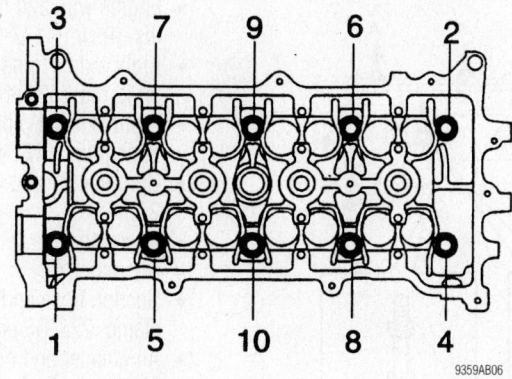

Cylinder head bolt loosening sequence

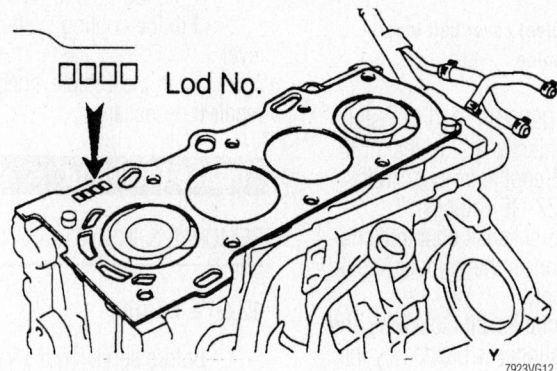

Position the head gasket correctly on the cylinder head—1.8L (1ZZ-FE) engine

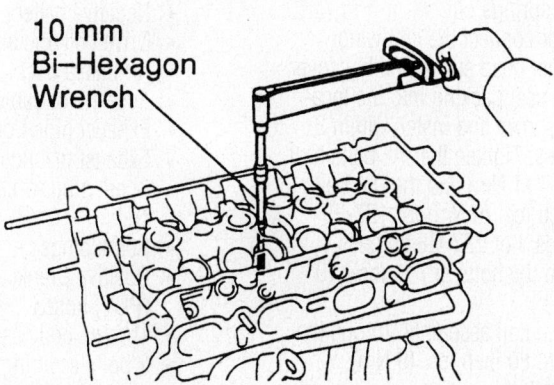

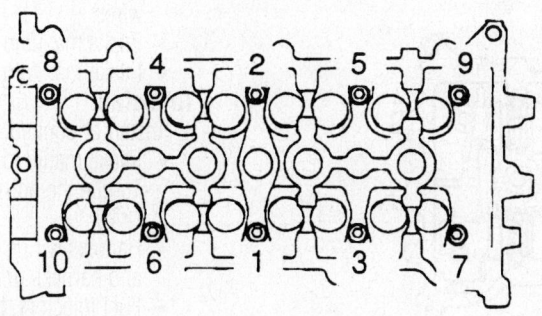

Cylinder head bolt tightening sequence—1ZZ-FE and 2ZZ-GE engines

- Timing chain sub-assembly. Remove the chain with the crankshaft gear, using screwdrivers as shown.
- Surge tank stay, 2ZZ-GE engine
- Intake manifold
- Oil level gauge
- Water bypass pipe bolts and pipe, 1ZZ-FE engine
- Camshafts
- Camshaft timing oil control valve, 1ZZ-FE engine
- Manifold stay, 1ZZ-FE engine
- Cylinder head bolts in sequence. To prevent damage to the cylinder head, loosen each bolt about ¼ of a turn during each pass until the bolts are loose.
- Cylinder head

To install:

11. Clean and degrease the surface of the cylinder head and engine block.

12. Check the length of the cylinder head bolts. They should be 5.780–5.835 in. (146.8–148.2mm) long. If they are longer than 5.846 in. (148.5mm), they must be replaced.

13. Install or connect the following:
- New gasket on the engine block with the Lot No. stamp facing up.
- Cylinder head
- Apply a light coat of oil to cylinder head bolt threads;

14. On 1ZZ-FE engines tighten the bolts as follows:
  a. Step 1: 36 ft. lbs. (49 Nm)
  b. Step 2: Tighten an additional 90 degrees.

15. On 2003–04 2ZZ-GE engines tighten the bolts as follows:
  a. Step 1: 26 ft. lbs. (35 Nm)
  b. Step 2: Tighten an additional 180 degrees.

16. On 2005–06 2ZZ-GE engines tighten the bolts as follows:
  a. Step 1: 16 ft. lbs. (25 Nm)
  b. Step 2: 36 ft. lbs. (49 Nm)
  c. Step 3: 36 ft. lbs. (49 Nm), plus an additional 90 degrees.

17. Install or connect the following:
- Manifold stay, 1ZZ-FE engine. Tighten the bolts to 36 ft. lbs. (49 Nm).
- Camshaft timing oil control valve, on 1ZZ-FE engines, and tighten to 80 inch lbs. (9 Nm)
- Camshaft
- Water by-pass pipe, on 1ZZ-FE engines, and tighten to 80 inch lbs. (9 Nm)
- Oil level gauge
- Intake manifold

- Surge tank stay, 2ZZ-GE engine. Tighten to 18 ft. lbs. (24 Nm).
- Timing chain
- Timing chain vibration damper. Tighten the bolts to 80 inch lbs. (9 Nm).
- Timing chain tensioner slipper and tighten the bolt to 14 ft. lbs. (19 Nm).
- Crankshaft position sensor plate, with the "F" mark facing forward.
- Timing gear cover oil seal
- Timing cover. For 1ZZ-FE engine, tighten the "A" bolts to 10 ft. lbs. (13 Nm), the "B" bolts to 14 ft. lbs. (19 Nm) and the stud bolt to 84 inch lbs. (9.5 Nm), using a Torx® wrench. For 2ZZ-GE engines, tighten the M8 bolts to 15 ft. lbs. (21 Nm), the M6 bolts to 8 ft. lbs. (11 Nm) and the stud bolt to 84 inch lbs. (9.5 Nm).

➡ **When installing the tensioner, make sure to set the hook again if the hook releases the plunger.**

- Timing chain tensioner. Torque the nuts to 80 inch lbs. (9 Nm).
- CKP sensor and tighten the bolts to 80 inch lbs. (9 Nm)
- Transverse engine mounting bracket. Tighten the bolts to 35 ft. lbs. (47 Nm).
- Water pump and pulley
- Exhaust manifold stay and heat insulator, 2ZZ-GE engine
- Belt tensioner. Tighten the nut to 21 ft. lbs. (29 Nm) and the bolt to 51 ft. lbs. (69 Nm) on 1ZZ-FE engines or to 74 ft. lbs. (100 Nm) on 2ZZ-GE engines.

18. Install the crankshaft pulley, as follows:

a. Align the pulley set key with the key groove of the pulley and slide on the pulley.

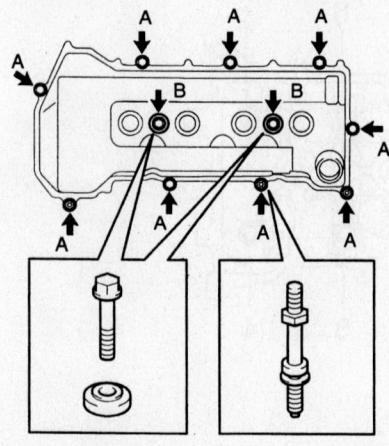

Cylinder head (valve) cover bolt locations—1ZZ-FE engine

b. Use SST 09960-11010 to install the bolt and tighten to 102 ft. lbs. (138 Nm) for 1ZZ-FE engine or to 87 ft. lbs. (118 Nm) on 2ZZ-GE engines.

c. Turn the crankshaft counterclockwise and disconnect the plunger knock pin from the hook.

d. Turn the crankshaft clockwise and check that the slipper is pushed by the plunger. If the plunger does not spring out, press the slipper into the chain tensioner with a screwdriver so that the hook is released from the knock pin and the plunger springs out.

19. Install or connect the following:

- Cylinder head sub-assembly cover. Install seal packing into the locations shown and install within 3 minutes. Tighten the "A" bolts to 8 ft. lbs. (11 Nm) and the "B" bolts to 80 inch lbs. (9 Nm) for 1ZZ-FE engines. For 2ZZ-GE engines, tighten the bolts to 7 ft. lbs. (10 Nm).
- Ignition coil assembly. Torque the bolts to 80 inch lbs. (9 Nm).

- Engine wire and tighten to 80 inch lbs. (9 Nm), 1ZZ-FE
- Right side engine mount. Tighten to 38 ft. lbs. (52 Nm).
- Front exhaust pipe
- Vane pump, 1ZZ-FE
- Compressor and magnetic clutch, 2ZZ-GE
- Alternator bracket, 2ZZ-GE engine
- Alternator
- Suction hose and wire harness clamp, 2ZZ-GE engine
- Air cleaner and hose
- Main cylinder head cover and tighten to 62 inch lbs. (7 Nm)
- Right front wheel and tire. Tighten the lug nuts to 76 ft. lbs. (103 Nm).

20. Fill the cooling system to the proper level.

21. Start the vehicle, check for leaks and repair if necessary.

Intake Manifold

REMOVAL & INSTALLATION

1ZZ-FE Engine

1. Before servicing the vehicle, refer to the precautions section.

2. Drain the cooling system.

3. Remove or disconnect the following:

- Negative battery cable
- Drive belt and alternator
- Air intake duct
- Accelerator cable
- Exhaust pipe from the manifold.
- Exhaust manifold support bracket
- Spark plug wires, then ignition coils
- Spark plugs
- Positive Crankcase Ventilation (PCV) hoses
- Throttle body assembly
- 2 bolts securing the wiring harness protector
- Wiring connectors and ground wires
- Intake manifold support bracket
- Intake manifold and gasket

To install:

4. Install or connect the following:

- Intake manifold with a new gasket. Torque the bolts to 22 ft. lbs. (30 Nm).
- Harness wiring to the cylinder head and harness protector
- Fuel injectors, throttle body and the PCV hoses
- Spark plugs and ignition coils.

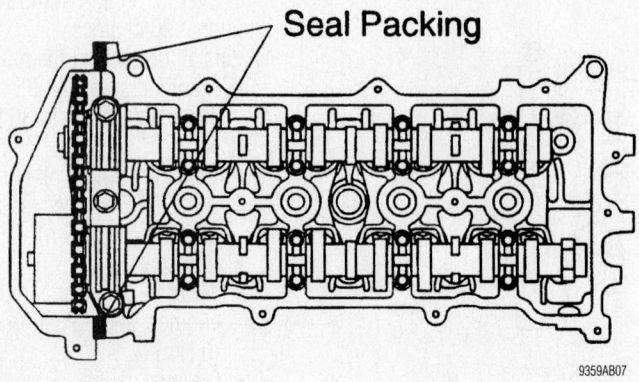

Seal packing installation locations

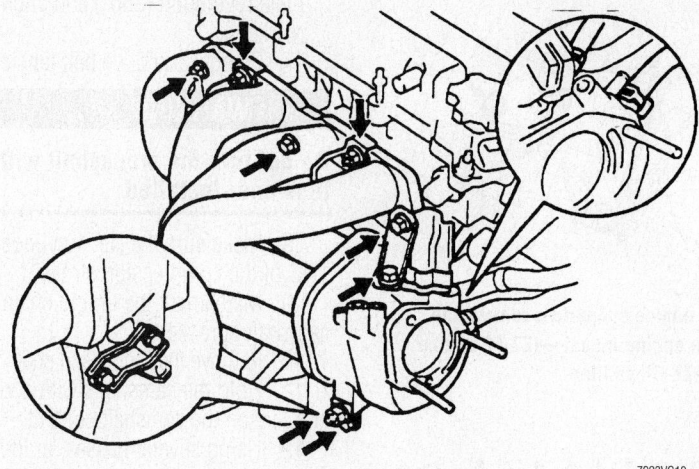

Intake manifold mounting fastener locations—1.8L (1ZZ-FE) engine

Tighten the bolts and nuts to 80 inch lbs. (9 Nm).
- Exhaust manifold and support bracket. Tighten the bolts to 37 ft. lbs. (49 Nm).
- Front exhaust pipe to the manifold. Tighten the bolts to 46 ft. lbs. (62 Nm).
- Oxygen Sensor (O_2S). Tighten the nuts to 14 ft. lbs. (20 Nm).
- Accelerator cable and air intake duct
- Alternator and drive belt
- Negative battery cable
5. Fill the cooling system.
6. Start the vehicle, check for leaks and repair if necessary.

2ZZ-GE Engine

1. Before servicing the vehicle, refer to the precautions section.
2. Drain the cooling system.
3. Remove or disconnect the following:
- Negative battery cable
- Drive belt and alternator
- Air intake duct
- Accelerator cable
- Spark plug wires, then ignition coils
- Spark plugs
- Positive Crankcase Ventilation (PCV) hoses
- Throttle body assembly
- Wiring harness
- Hoses and tubes connected to the head
- Intake manifold support bracket
- Intake manifold and gasket

To install:
4. Install or connect the following:
- Intake manifold with a new gasket. Tighten bolts A to 25 ft. lbs. (34 Nm) and bolt B to 34 ft. lbs. (46 Nm)

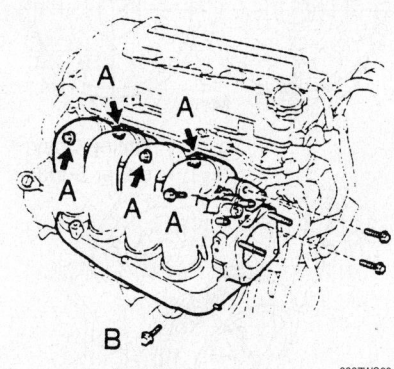

Intake manifold bolt installation—2ZZ-GE engine

- Harness wiring to the cylinder head and harness protector
- Fuel injectors, throttle body and the PCV hoses
- Spark plugs and ignition coils. Tighten the bolts and nuts to 80 inch lbs. (9 Nm).
- Oxygen Sensor (O_2S). Tighten the nuts to 14 ft. lbs. (20 Nm).
- Accelerator cable and air intake duct
- Alternator and drive belt
- Negative battery cable
5. Fill the cooling system.
6. Start the vehicle, check for leaks and repair if necessary.

Exhaust Manifold

REMOVAL & INSTALLATION

1ZZ-FE Engine

1. Before servicing the vehicle, refer to the precautions section.

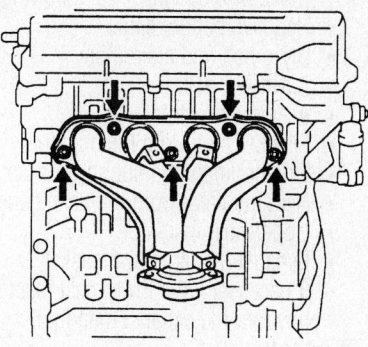

Exhaust manifold mounting nut locations—1.8L (1ZZ-FE) engine

2. Drain the cooling system.
3. Remove or disconnect the following:
- Negative battery cable
- Drive belt and alternator
- Air intake duct
- Accelerator cable
- Exhaust pipe from the manifold
- Exhaust manifold support bracket
- Heat insulator from the dash panel
- Upper heat insulator
- Exhaust manifold and gasket
- If necessary, the lower heat insulator from the exhaust manifold.

To install:
4. Install or connect the following:
- Lower heat insulator on the exhaust manifold. Tighten the bolts to 108 inch lbs. (12 Nm).
- Exhaust manifold using a new gasket. Tighten the nuts, in several passes, to 27 ft. lbs. (37 Nm).
- Upper heat insulator. Tighten the bolts to 108 inch lbs. (12 Nm).
- Heat insulator on the dash panel
- Exhaust manifold support bracket. Tighten the bolts, in an alternating pattern, to 37 ft. lbs. (49 Nm).
- Front exhaust pipe to the manifold. Tighten the bolts to 46 ft. lbs. (62 Nm).
- Oxygen Sensor (O_2S). Tighten the nuts to 14 ft. lbs. (20 Nm).
- Accelerator cable and air intake duct
- Alternator and drive belt
- Negative battery cable
5. Fill the cooling system.
6. Start the vehicle, check for leaks and repair if necessary.

2ZZ-GE Engine

1. Before servicing the vehicle, refer to the precautions section.
2. Drain the cooling system.

3. Remove or disconnect the following:
- Negative battery cable
- Drive belt and alternator
- Air intake duct
- Accelerator cable
- Exhaust pipe from the manifold
- Exhaust manifold support bracket
- Heat insulator from the dash panel.
- Upper heat insulator
- Exhaust manifold and gasket
- If necessary, the lower heat insulator from the exhaust manifold.

To install:

4. Install or connect the following:
- Lower heat insulator on the exhaust manifold. Tighten the bolts to 15 ft. lbs. (20 Nm).
- Exhaust manifold using a new gasket. Tighten the nuts, in several passes to 37 ft. lbs. (50 Nm).
- Upper heat insulator. Tighten the bolts to 15 ft. lbs. (20 Nm).
- Heat insulator on the dash panel.
- Exhaust manifold support bracket. Tighten the bolts to 37 ft. lbs. (49 Nm).
- Front exhaust pipe to the manifold. Tighten the bolts to 46 ft. lbs. (62 Nm).
- Oxygen Sensor (O2S). Tighten the nuts to 14 ft. lbs. (20 Nm).
- Accelerator cable and air intake duct.
- Alternator and drive belt.
- Negative battery cable

5. Fill the cooling system.
6. Start the vehicle, check for leaks and repair if necessary.

Camshaft(s)

REMOVAL & INSTALLATION

1ZZ-FE Engine

1. Before servicing the vehicle, refer to the precautions section.
2. Remove or disconnect the following:
- Negative battery cable
- Right side engine under cover
- Cylinder head cover
- Suction hose sub-assembly, 2ZZ-GE engine
- Drive belt
- Power steering pump reservoir and position it aside, 1ZZ-FE engine

3. Place a jack with a wooden block under the vehicle for support, then remove the 4 bolts and 2 nuts and remove the right side engine mount.
4. Remove the engine wire, on 1ZZ-FE engines:

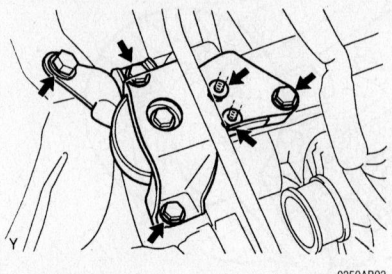

With the engine supported, remove the right side engine mount—1ZZ-FE engine shown, 2ZZ-GE similar

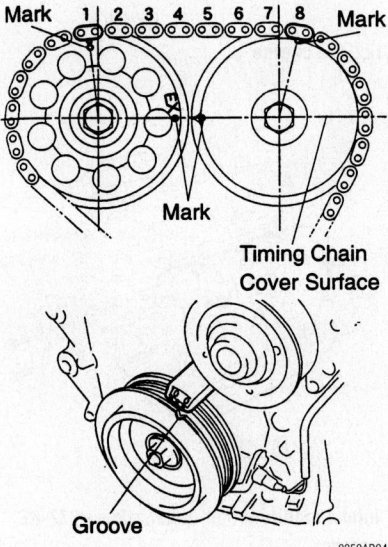

Proper timing mark alignment for TDC

a. Remove the 5 clamps from the brackets.
b. Detach the connectors.
c. Remove the ignition coil connectors.
d. Bolt and nut holding the engine wire.

5. Remove or disconnect the following:
- Ignition coil assembly
- Positive Crankcase Ventilation (PCV) hoses from the valve cover
- Valve (cylinder head) cover sub-assembly

6. Set the No. 1 cylinder to Top Dead Center (TDC) of the compressor stroke as follows:

a. Turn the crankshaft pulley, and align its groove with the "0" timing mark of the timing chain cover.
b. Make sure the point marks of the camshaft timing sprockets and VVT timing sprockets are in a straight line as shown. If not, turn the crankshaft 1 com-

plete revolution (360°) and align the marks.
7. Remove the drive belt tensioner.

8. Make sure the No. 1 cylinder is at TDC of the compression stroke.
9. Matchmark the timing chain and camshaft sprockets
10. Remove the 2 nuts and chain tensioner.
11. Hold the camshafts with a wrench and loosen the camshaft set bolt.
12. Using several passes, gradually remove the bearing cap bolts from the No. 2 camshaft, in the proper sequence.
13. Remove the camshaft and timing gear as shown.
14. Using several passes, gradually remove the bearing cap bolts from the other camshaft, in the proper sequence.
15. Remove the camshaft while holding the timing chain.

16. Tie the timing chain with a string as shown, to prevent it from dropping down into the timing chain cover.

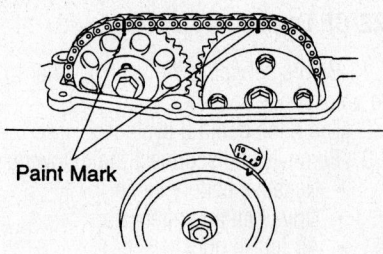

Matchmark the timing chain and cam sprockets

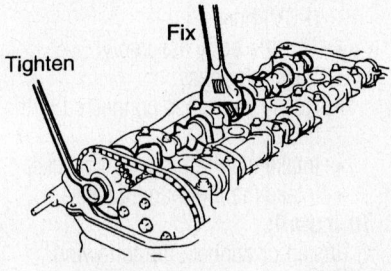

Hold the camshaft with a wrench while removing the set bolt

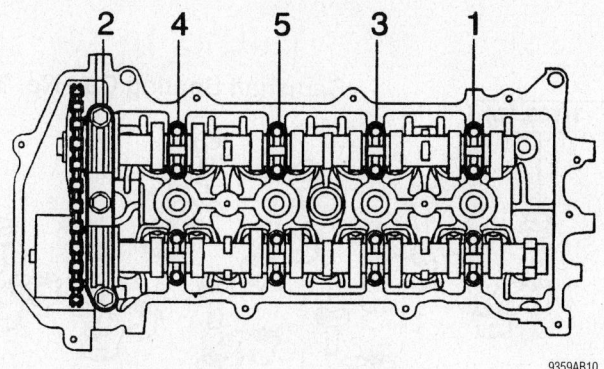

Camshaft bearing cap bolt removal sequence—1ZZ-FE engine

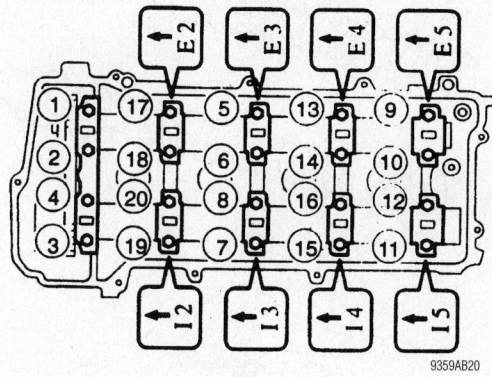

Camshaft bearing cap bolt removal sequence—2ZZ-GE engine

To install:

17. Position the camshaft on the cylinder head, then install the timing chain on the cam timing gear, with the painted links aligned with the marks on the timing gear.

18. Check the front marks and numbers and torque the camshaft cap bolts, in sequence, to 10 ft. lbs. (13 Nm) for 1ZZ-FE engine, or to 14 ft. lbs. (19 Nm) for 2ZZ-GE engines.

19. Put camshaft No. 2 on the cylinder head, with the painted links of the chain aligned with the mark on the timing gear.

20. Tighten the camshaft gear set bolt temporarily.

21. Check the front marks and numbers and torque the camshaft cap bolts, in sequence, to 10 ft. lbs. (13 Nm). Install the No. 1 bearing cap and tighten to 17 ft. lbs. (23 Nm).

22. Hold the camshaft secure with a wrench and tighten the set bolt to 40 ft. lbs. (54 Nm). Be careful not the damage the lifters.

23. Check to be sure the matchmarks on the timing chain and cam sprockets, and the alignment of the pulley groove with the timing mark on the cover are still aligned.

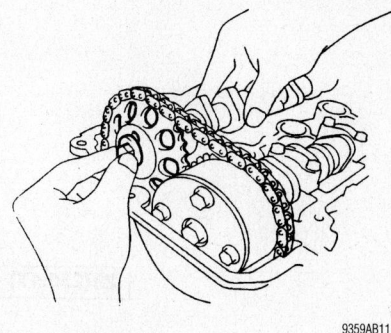

Carefully remove the cam and timing gear

24. Install the chain tensioner:
a. Make sure the O-ring is clean, then set the hook as shown.
b. Oil the tensioner, then install and tighten to 80 inch lbs. (9 Nm).

➡**When installing the tensioner, set the hook again if the hook releases the plunger.**

c. Turn the crankshaft counterclockwise, and disconnect the plunger knock pin from the hook.
d. Turn the crankshaft clockwise and check that the slipper is pushed by the plunger. If the plunger does not spring out, press the slipper into the chain ten-

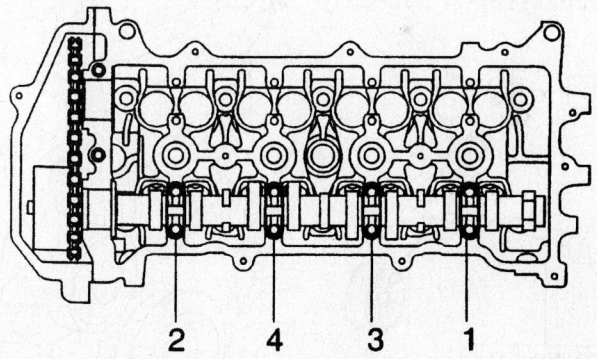

Camshaft bearing cap bolt removal sequence—1ZZ-FE engine

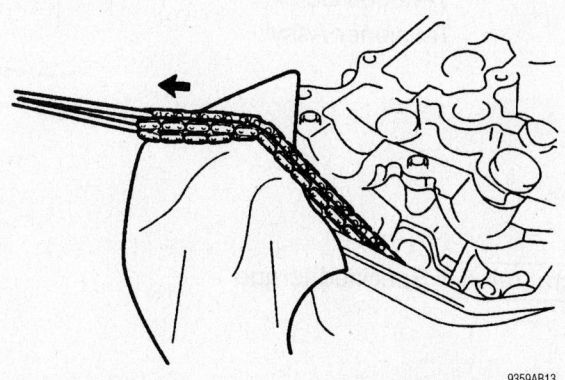

Secure the timing chain with string to prevent it from slipping down into the timing chain cover

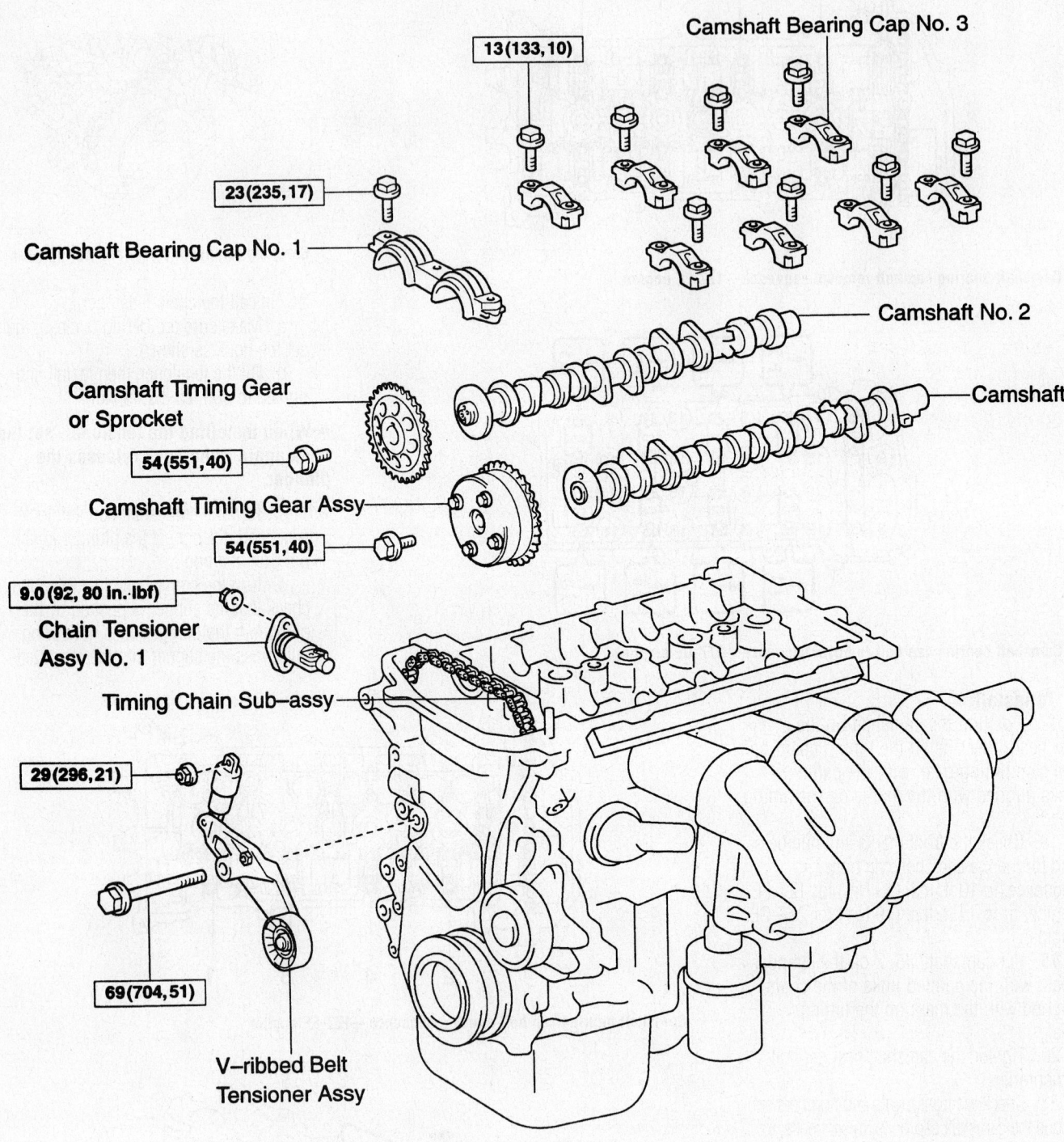

13(133,10)

Camshaft Bearing Cap No. 3

23(235,17)

Camshaft Bearing Cap No. 1

Camshaft No. 2

Camshaft Timing Gear
or Sprocket

Camshaft

54(551,40)

Camshaft Timing Gear Assy

54(551,40)

9.0 (92, 80 in.·lbf)

Chain Tensioner
Assy No. 1

Timing Chain Sub–assy

29(296,21)

69(704,51)

V–ribbed Belt
Tensioner Assy

N·m (kgf·cm, ft·lbf) : Specified torque

9359AB21

Exploded view of the camshafts and related components—1ZZ-FE engine

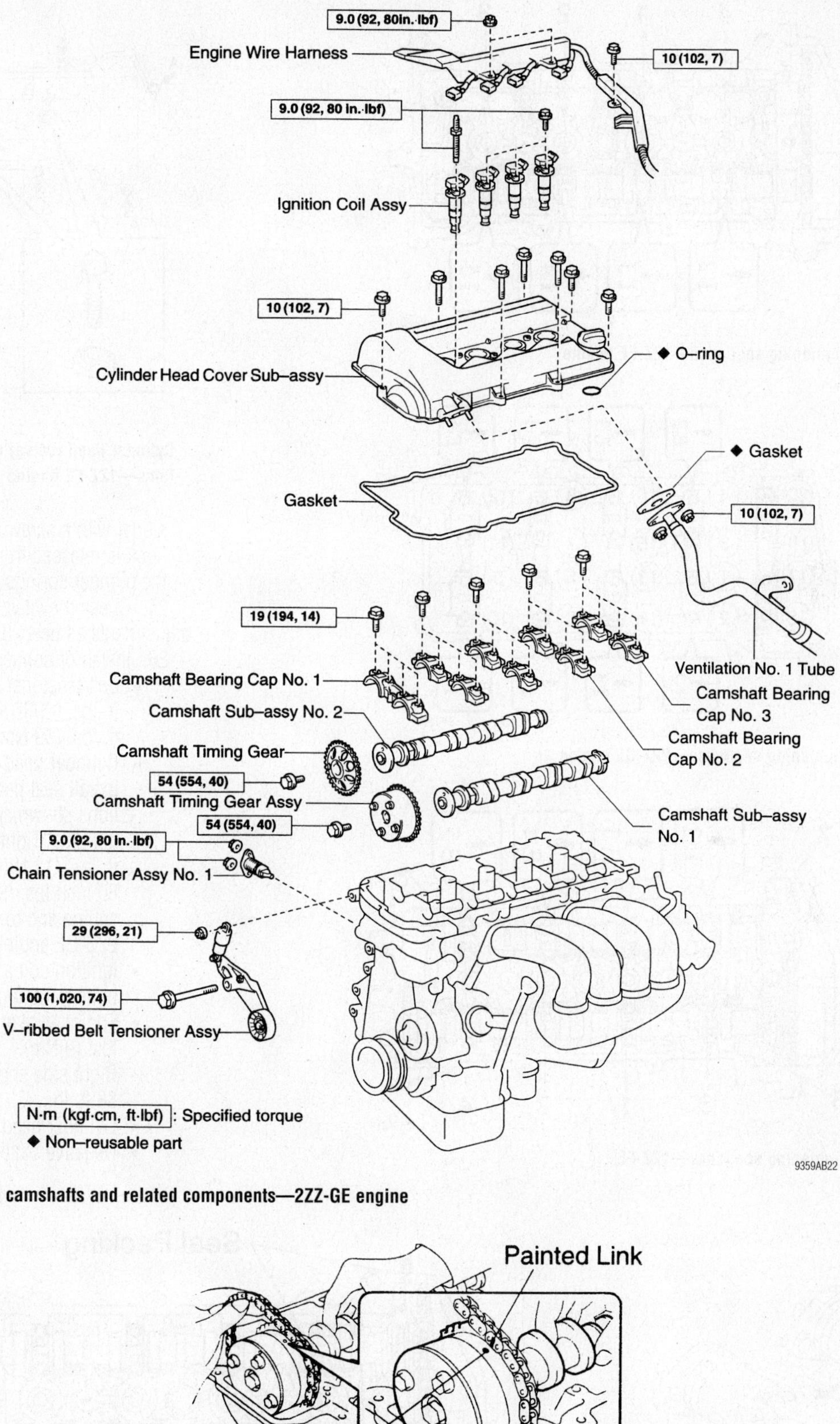

9.0 (92, 80 in.·lbf)
Engine Wire Harness

10 (102, 7)

9.0 (92, 80 in.·lbf)

Ignition Coil Assy

10 (102, 7)
Cylinder Head Cover Sub–assy

◆ O–ring

◆ Gasket

Gasket

10 (102, 7)

19 (194, 14)
Camshaft Bearing Cap No. 1
Camshaft Sub–assy No. 2
Camshaft Timing Gear

54 (554, 40)
Camshaft Timing Gear Assy

54 (554, 40)

9.0 (92, 80 in.·lbf)
Chain Tensioner Assy No. 1

29 (296, 21)

100 (1,020, 74)
V–ribbed Belt Tensioner Assy

Ventilation No. 1 Tube
Camshaft Bearing Cap No. 3
Camshaft Bearing Cap No. 2
Camshaft Sub–assy No. 1

N·m (kgf·cm, ft·lbf) : Specified torque
◆ Non–reusable part

9359AB22

Exploded view of the camshafts and related components—2ZZ-GE engine

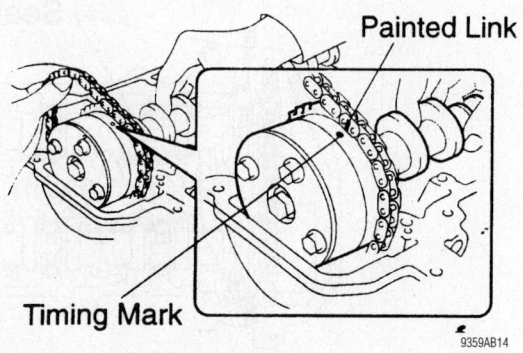

Painted Link

Timing Mark

9359AB14

Make sure the alignment marks on the timing chain and camshaft gear match up

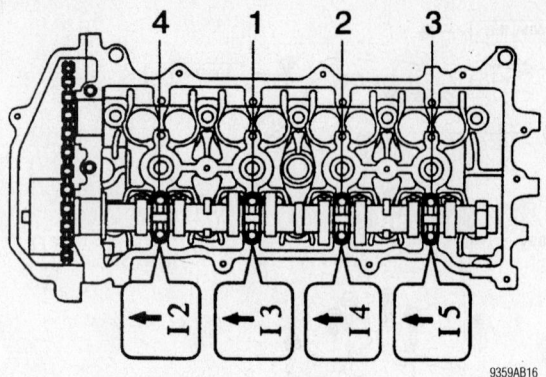

Camshaft cap bolt tightening sequence—1ZZ-FE engine

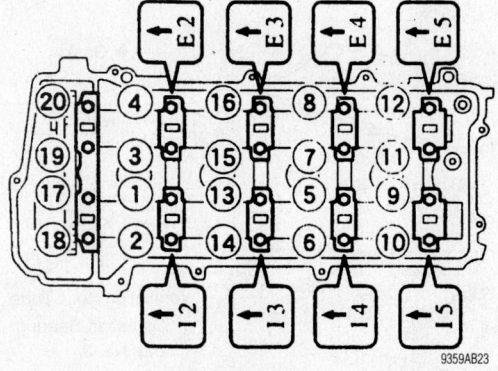

Camshaft cap bolt tightening sequence—2ZZ-GE engine

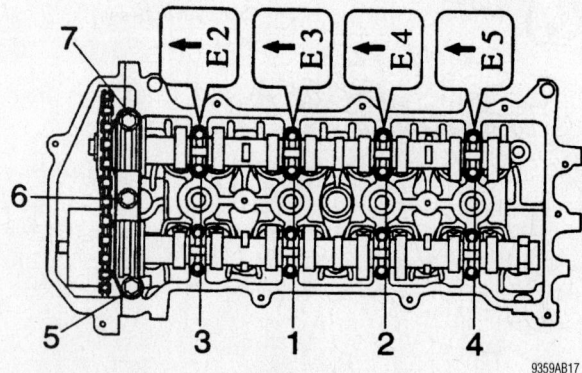

Camshaft cap bolt tightening sequence—1ZZ-FE

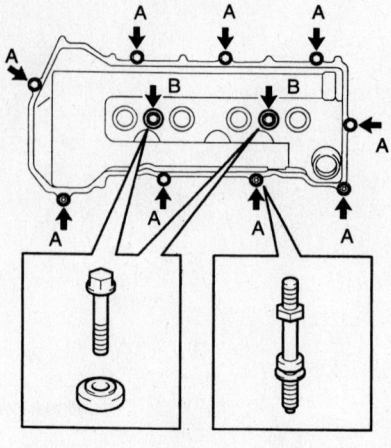

Cylinder head (valve) cover bolt locations—1ZZ-FE engine

sioner with a screwdriver so that the hook is released from the knock pin and the plunger springs out.

25. Check the valve clearance and make adjustments as needed.

26. Install or connect the following:

- Belt tensioner. Tighten the nut to 21 ft. lbs. (29 Nm) and the bolt to 51 ft. lbs. (69 Nm).
- Cylinder head sub-assembly cover. Install seal packing into the locations shown and install within 3 minutes. Tighten the "A" bolts to 8 ft. lbs. (11 Nm) and the "B" bolts to 80 inch lbs. (9 Nm) for 1ZZ-FE engine and to 7 ft. lbs. (10 Nm) for 2ZZ-GE engines.
- Ignition coil assembly. Torque the bolts to 80 inch lbs. (9 Nm).
- Engine wire and tighten to 80 inch lbs. (9 Nm)
- Right side engine mount. Tighten to 38 ft. lbs. (52 Nm).
- Cylinder head (valve) cover
- Negative battery cable

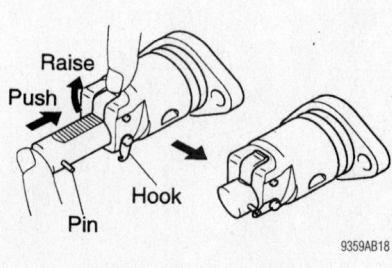

Set the timing chain tensioner hook properly

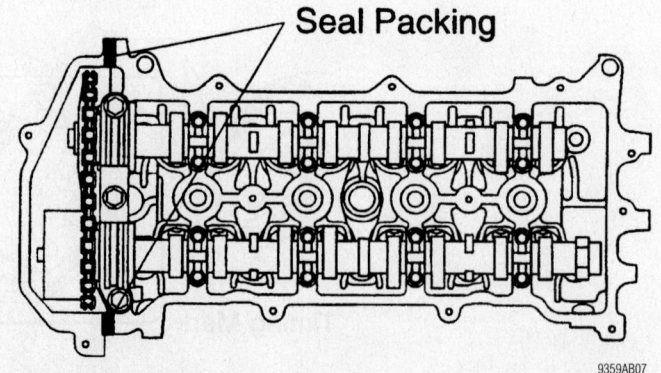

Seal packing installation locations

1ZZ-FE: Valve Lifter Selection Chart (Intake)

Valve Lifter Selection Chart (Intake) — triangular chart with "Measured clearance mm (in)" on the vertical axis and "installed lifter thickness mm (in)" on the horizontal axis. Measured clearance rows (left axis):

Measured clearance mm (in)
0.000 – 0.030 (0.0000 – 0.0012)
0.031 – 0.050 (0.0012 – 0.0020)
0.051 – 0.070 (0.0020 – 0.0028)
0.071 – 0.090 (0.0028 – 0.0035)
0.091 – 0.110 (0.0036 – 0.0043)
0.111 – 0.130 (0.0044 – 0.0051)
0.131 – 0.149 (0.0052 – 0.0059)
0.150 – 0.250 (0.0059 – 0.0098)
0.251 – 0.270 (0.0099 – 0.0106)
0.271 – 0.290 (0.0107 – 0.0114)
0.291 – 0.310 (0.0115 – 0.0122)
0.311 – 0.330 (0.0122 – 0.0130)
0.331 – 0.350 (0.0130 – 0.0138)
0.351 – 0.370 (0.0138 – 0.0146)
0.371 – 0.390 (0.0146 – 0.0154)
0.391 – 0.410 (0.0154 – 0.0161)
0.411 – 0.430 (0.0162 – 0.0169)
0.431 – 0.450 (0.0170 – 0.0177)
0.451 – 0.470 (0.0178 – 0.0185)
0.471 – 0.490 (0.0185 – 0.0193)
0.491 – 0.510 (0.0193 – 0.0201)
0.511 – 0.530 (0.0201 – 0.0209)
0.531 – 0.550 (0.0209 – 0.0217)
0.551 – 0.570 (0.0217 – 0.0224)
0.571 – 0.590 (0.0225 – 0.0232)
0.591 – 0.610 (0.0233 – 0.0240)
0.611 – 0.630 (0.0241 – 0.0248)
0.631 – 0.650 (0.0248 – 0.0256)
0.651 – 0.670 (0.0256 – 0.0264)
0.671 – 0.690 (0.0264 – 0.0272)
0.691 – 0.710 (0.0272 – 0.0280)
0.711 – 0.730 (0.0280 – 0.0287)
0.731 – 0.750 (0.0288 – 0.0295)
0.751 – 0.770 (0.0296 – 0.0303)
0.771 – 0.790 (0.0304 – 0.0311)
0.791 – 0.810 (0.0311 – 0.0319)
0.811 – 0.830 (0.0319 – 0.0327)
0.831 – 0.850 (0.0327 – 0.0335)
0.851 – 0.870 (0.0335 – 0.0343)
0.871 – 0.890 (0.0343 – 0.0350)
0.891 – 0.910 (0.0351 – 0.0358)
0.911 – 0.930 (0.0359 – 0.0366)

New lifter thickness mm (in.)

Lifter No.	Thickness	Lifter No.	Thickness	Lifter No.	Thickness
06	5.060 (0.1992)	30	5.300 (0.2087)	54	5.540 (0.2181)
08	5.080 (0.2000)	32	5.320 (0.2094)	56	5.560 (0.2189)
10	5.100 (0.2008)	34	5.340 (0.2102)	58	5.580 (0.2197)
12	5.120 (0.2016)	36	5.360 (0.2110)	60	5.600 (0.2205)
14	5.140 (0.2024)	38	5.380 (0.2118)	62	5.620 (0.2213)
16	5.160 (0.2031)	40	5.400 (0.2126)	64	5.640 (0.2220)
18	5.180 (0.2039)	42	5.420 (0.2134)	66	5.660 (0.2228)
20	5.200 (0.2047)	44	5.440 (0.2142)	68	5.680 (0.2236)
22	5.220 (0.2055)	46	5.460 (0.2150)	70	5.700 (0.2244)
24	5.240 (0.2063)	48	5.480 (0.2157)	72	5.720 (0.2252)
26	5.260 (0.2071)	50	5.500 (0.2165)	74	5.740 (0.2260)
28	5.280 (0.2079)	52	5.520 (0.2173)		

Intake valve clearance (Cold):
0.15 – 0.25 mm (0.006 – 0.010 in.)

EXAMPLE: The 5.250 mm (0.2067 in.) lifter is installed, and the measured clearance is 0.400 mm (0.0157 in.).
Replace the 5.250 mm (0.2067 in.) lifter with a new No. 48 lifter.

Adjusting shim chart (intake)—1ZZ-FE engine

9307WG70

1ZZ–FE: Valve Lifter Selection Chart (Exhaust)

New lifter thickness mm (in.)

Lifter No.	Thickness	Lifter No.	Thickness	Lifter No.	Thickness
06	5.060 (0.1992)	30	5.300 (0.2087)	54	5.540 (0.2181)
08	5.080 (0.2000)	32	5.320 (0.2094)	56	5.560 (0.2189)
10	5.100 (0.2008)	34	5.340 (0.2102)	58	5.580 (0.2197)
12	5.120 (0.2016)	36	5.360 (0.2110)	60	5.600 (0.2205)
14	5.140 (0.2024)	38	5.380 (0.2118)	62	5.620 (0.2213)
16	5.160 (0.2031)	40	5.400 (0.2126)	64	5.640 (0.2220)
18	5.180 (0.2039)	42	5.420 (0.2134)	66	5.660 (0.2228)
20	5.200 (0.2047)	44	5.440 (0.2142)	68	5.680 (0.2236)
22	5.220 (0.2055)	46	5.460 (0.2150)	70	5.700 (0.2244)
24	5.240 (0.2063)	48	5.480 (0.2157)	72	5.720 (0.2252)
26	5.260 (0.2071)	50	5.500 (0.2165)	74	5.740 (0.2260)
28	5.280 (0.2079)	52	5.520 (0.2173)		

Exhaust valve clearance (Cold):
0.25 – 0.35 mm (0.010 – 0.014 in.)

EXAMPLE: The 5.340 mm (0.2102 in.) lifter is installed, and the measured clearance is 0.440 mm (0.0173 in.).

Replace the 5.340 mm (0.2102 in.) lifter with a new No. 48 lifter.

Adjusting shim chart (exhaust)—1ZZ-FE engine

1ZZ-FE: Valve Lifter Selection Chart (Exhaust) — matrix of Installed lifter thickness mm (in.) versus Measured clearance mm (in.), giving the new lifter number.

9359AB26

Adjusting shim chart (intake)—2ZZ-GE engine

Intake valve clearance (Cold):
0.08 – 0.18 mm (0.0031 – 0.0071 in.)

EXAMPLE: The 2.200 mm (0.0826 in.) shim is installed, and the measured clearance is 0.400 mm (0.0157 in.).
Replace the 2.600 mm (0.1024 in.) shim with a new No. 60 shim.

New Shim thickness   mm (in.)

Shim No.	Thickness	Shim No.	Thickness	Shim No.	Thickness
00	2.000(0.0787)	28	2.280(0.0898)	56	2.560(0.1008)
02	2.020(0.0795)	30	2.300(0.0906)	58	2.580(0.1016)
04	2.040(0.0803)	32	2.320(0.0913)	60	2.600(0.1024)
06	2.060(0.0811)	34	2.340(0.0921)	62	2.620(0.1031)
08	2.080(0.0819)	36	2.360(0.0929)	64	2.640(0.1039)
10	2.100(0.0827)	38	2.380(0.0937)	66	2.660(0.1047)
12	2.120(0.0835)	40	2.400(0.0945)	68	2.680(0.1055)
14	2.140(0.0843)	42	2.420(0.0953)	70	2.700(0.1063)
16	2.160(0.0850)	44	2.440(0.0961)	72	2.720(0.1071)
18	2.180(0.0858)	46	2.460(0.0969)	74	2.740(0.1079)
20	2.200(0.0866)	48	2.480(0.0976)	76	2.760(0.1087)
22	2.220(0.0874)	50	2.500(0.0984)	78	2.780(0.1094)
24	2.240(0.0882)	52	2.520(0.0992)	80	2.800(0.1102)
26	2.260(0.0890)	54	2.540(0.1000)		

Adjusting shim selection chart

Columns: Installed shim thickness mm(in.) — ranging from 2.000 (0.0787) through 2.800 (0.1102) in 0.020 mm steps.

Rows: Measured clearance mm(in.):

Measure clearance mm(in.)
0.000 – 0.030 (0.0000 – 0.0012)
0.031 – 0.050 (0.0012 – 0.0020)
0.051 – 0.070 (0.0020 – 0.0028)
0.071 – 0.090 (0.0028 – 0.0035)
0.091 – 0.099 (0.0036 – 0.0039)
0.100 – 0.160 (0.0039 – 0.0063)
0.161 – 0.180 (0.0063 – 0.0071)
0.181 – 0.200 (0.0071 – 0.0079)
0.201 – 0.220 (0.0079 – 0.0087)
0.221 – 0.240 (0.0087 – 0.0094)
0.241 – 0.260 (0.0095 – 0.0102)
0.261 – 0.280 (0.0103 – 0.0110)
0.281 – 0.300 (0.0111 – 0.0118)
0.301 – 0.320 (0.0119 – 0.0126)
0.321 – 0.340 (0.0126 – 0.0134)
0.341 – 0.360 (0.0134 – 0.0142)
0.361 – 0.380 (0.0142 – 0.0150)
0.381 – 0.400 (0.0150 – 0.0157)
0.401 – 0.420 (0.0158 – 0.0165)
0.421 – 0.440 (0.0166 – 0.0173)
0.441 – 0.460 (0.0174 – 0.0181)
0.461 – 0.480 (0.0181 – 0.0189)
0.481 – 0.500 (0.0189 – 0.0197)
0.501 – 0.520 (0.0197 – 0.0205)
0.521 – 0.540 (0.0205 – 0.0213)
0.541 – 0.560 (0.0213 – 0.0220)
0.561 – 0.580 (0.0221 – 0.0228)
0.581 – 0.600 (0.0229 – 0.0236)
0.601 – 0.620 (0.0229 – 0.0244)
0.621 – 0.640 (0.0237 – 0.0244)
0.641 – 0.660 (0.0252 – 0.0260)
0.661 – 0.680 (0.0260 – 0.0268)

The body of the chart is a dense grid of two-digit shim-number values indexed by installed shim thickness (columns) against measured clearance (rows).

Adjusting shim chart (exhaust)—2ZZ-GE engine

New Shim thickness   mm (in.)

Shim No.	Thickness	Shim No.	Thickness	Shim No.	Thickness
00	2.000(0.0787)	28	2.280(0.0898)	56	2.560(0.1008)
02	2.020(0.0795)	30	2.300(0.0906)	58	2.580(0.1016)
04	2.040(0.0803)	32	2.320(0.0913)	60	2.600(0.1024)
06	2.060(0.0811)	34	2.340(0.0921)	62	2.620(0.1031)
08	2.080(0.0819)	36	2.360(0.0929)	64	2.640(0.1039)
10	2.100(0.0827)	38	2.380(0.0937)	66	2.660(0.1047)
12	2.120(0.0835)	40	2.400(0.0945)	68	2.680(0.1055)
14	2.140(0.0843)	42	2.420(0.0953)	70	2.700(0.1063)
16	2.160(0.0850)	44	2.440(0.0961)	72	2.720(0.1071)
18	2.180(0.0858)	46	2.460(0.0969)	74	2.740(0.1079)
20	2.200(0.0866)	48	2.480(0.0976)	76	2.760(0.1087)
22	2.220(0.0874)	50	2.500(0.0984)	78	2.780(0.1094)
24	2.240(0.0882)	52	2.520(0.0992)	80	2.800(0.1102)
26	2.260(0.0890)	54	2.540(0.1000)		

Exhaust valve clearance (Cold):
0.22 – 0.32 mm (0.0087 – 0.0126 in.)

EXAMPLE: The 2.200 mm (0.0862 in.) shim is installed, and the measured clearance is 0.500 mm (0.0197 in.).
Replace the 2.540 mm (0.1000 in.) shim with a new No. 54 shim.

93S9AB27

Valve Lash

ADJUSTMENT

1ZZ-FE Engine

➡ **Adjust the valve clearance when the engine is cold.**

1. Before servicing the vehicle, refer to the precautions section.
2. Remove or disconnect the following:
   - Negative battery cable.
   - Cylinder head covers
   - Engine wire
   - Ignition coil
   - Positive Crankcase Ventilation (PCV) hoses
   - Cylinder head cover sub-assembly
3. Set the No. 1 cylinder to Top Dead Center (TDC) of the compressor stroke as follows:
   a. Turn the crankshaft pulley, and align its groove with the "0" timing mark of the timing chain cover.
   b. Make sure the point marks of the camshaft timing sprockets and VVT timing sprockets are in a straight line as shown. If not, turn the crankshaft 1 complete revolution (360°) and align the marks.
4. Check the valve clearance of the first set of the valves shown:
   a. Use a feeler gauge to measure the clearance between the valve lifter and camshaft. The clearance of the intake valves should be 0.0059–0.0098 in. (0.15–0.25mm). The clearance of the exhaust valves should be 0.0098–0.0138 in. (0.25–0.35mm).
   b. Note the out-of-specification valve clearance measurements. You will need

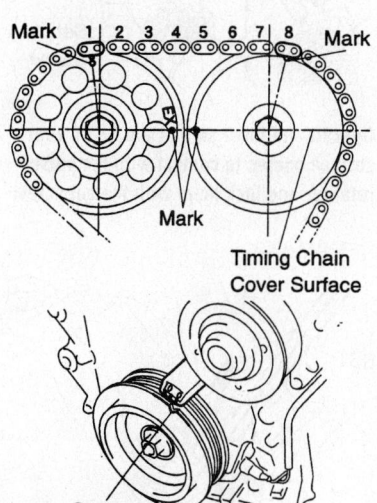

Proper timing mark alignment for TDC—1ZZ-FE and 2ZZ-GE engines

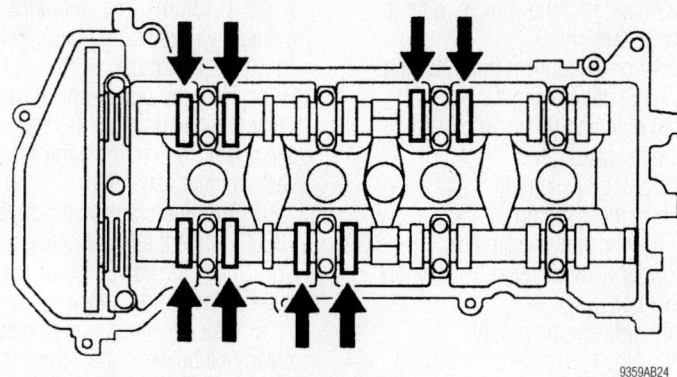

Check the clearance of the 1st set of valves–1ZZ-FE engine

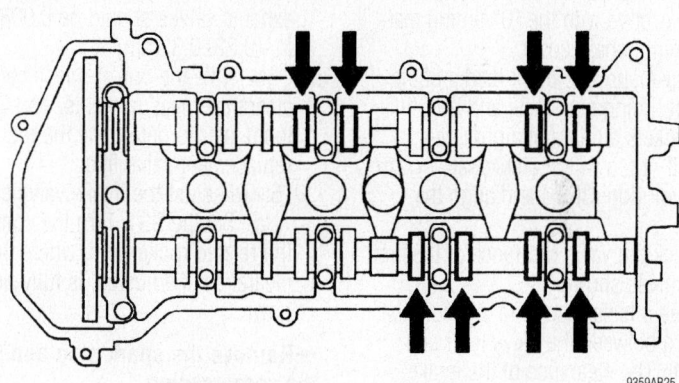

Check the clearance of the 2nd set of valves–1ZZ-FE engine

them later to determine the required replacement valve lifter.
   c. Turn the crankshaft 1 revolution (360°) to set the No. 4 cylinder to TDC.
5. Check the valve clearance of the second set of the valves shown:
   a. Use a feeler gauge to measure the clearance between the valve lifter and camshaft. The clearance of the intake valves should be 0.0059–0.0098 in. (0.15–0.25mm). The clearance of the exhaust valves should be 0.0098–0.0138 in. (0.25–0.35mm).
   b. Note the out-of-specification valve clearance measurements. You will need them later to determine the required replacement valve lifter.
6. Remove or disconnect the following:
   - Drive belt
   - Right side engine mount
   - Drive belt tensioner

✳✳ WARNING

DO NOT turn the crankshaft while the tensioner is removed!

7. Set the No. 1 cylinder to TDC of the compression stroke.
   - Camshafts
   - Valve lifters.

8. Use a micrometer to measure the thickness of the used lifter. Calculate the thickness of a new lifter. so the valve clearance comes within the specified value:
   a. A: Thickness of new lifter.
   b. B: Thickness of used lifter.
   c. C: Measured valve clearance.
   d. Intake valve clearance: A = B + (C − 0.0079 in. (0.20mm).
   e. Exhaust valve clearance: A = B + (C − 0.0118 in. (0.30mm).
   f. Select a new lifter with a thickness as close as possible to the calculated values. Lifters come in 35 sizes in increments of 0.0008 in. (0.020mm) from 0.1992–0.2260 in (5.060–5.740mm).
9. Install or connect the following:
   - Camshafts
   - Drive belt tensioner
   - Right hand engine mount
   - Cylinder head (valve) cover sub-assembly
   - Ignition coil
   - Engine wire
   - Cylinder head (valve) cover
   - Negative battery cable

2ZZ-GE Engine

➡ **Adjust the valve clearance when the engine is cold.**

1. Before servicing the vehicle, refer to the precautions section.

2. Remove or disconnect the following:
- Negative battery cable.
- Right side engine under cover
- Cylinder head cover
- Ignition coil assembly
- Wire harness clamp
- Suction hose sub-assembly
- Cylinder head cover sub-assembly
- Drive belt
- Right side engine mount

3. Set the No. 1 cylinder to Top Dead Center (TDC) of the compressor stroke as follows:

a. Turn the crankshaft pulley, and align its groove with the "0" timing mark of the timing chain cover.

b. Make sure the point marks of the camshaft timing sprockets and VVT timing sprockets are in a straight line as shown. If not, turn the crankshaft 1 complete revolution (360°) and align the marks.

4. Check the valve clearance of the first set of the valves shown:

a. Use a feeler gauge to measure the clearance between the valve lifter and camshaft. The clearance of the intake valves should be 0.0031–0.0071 in.

(0.08–0.18mm). The clearance of the exhaust valves should be 0.0087–0.0126 in. (0.22–0.32mm).

b. Note the out-of-specification valve clearance measurements. You will need them later to determine the required replacement valve lifter.

c. Turn the crankshaft 1 revolution (360°) to set the No. 4 cylinder to TDC.

5. Check the valve clearance of the second set of the valves shown:

a. Use a feeler gauge to measure the clearance between the valve lifter and camshaft. The clearance of the intake valves should be 0.0031–0.0071 in. (0.08–0.18mm). The clearance of the exhaust valves should be 0.0087–0.0126 in. (0.22–0.32mm).

b. Note the out-of-specification valve clearance measurements. You will need them later to determine the required replacement valve lifter.

6. To adjust the intake valve clearance:

a. Set the SST. Turn the crankshaft so the related rocker arm, where the valve clearance is adjusted, is fully pushed down.

➡**Remove the spark plug and take off the compression.**

b. Insert SST 09248-77010 into the plug tube. The tool cannot be inserted unless the set screw is loosened.

c. Operate the lever so that the SST's seat surface comes to contact with the valve retainer and lock them with the set screw. Clearance between the valve retainer and SST's set surface is not allowed. Be careful not to make clearance when inserting the SST, since clearance may unlock the keeper.

d. Lock the set screw on the tube side of the SST.

e. Rotate the crankshaft so that the camshaft is position as shown. During rotation, pay attention to the direction, to prevent the nose of the camshaft from interfering with the SST's shaft. Do not rotate the crankshaft excessively.

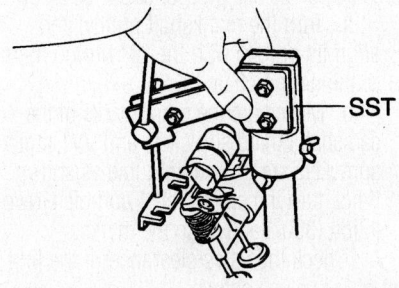

Insert the special tool into the plug tube—2ZZ-GE

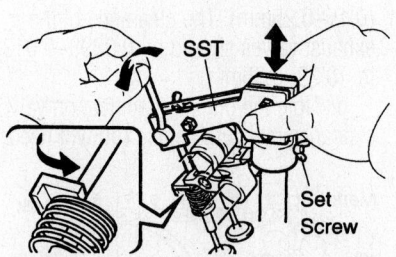

Operate the lever so that the SST's seat surface comes to contact with the valve retainer and lock them with the set screw

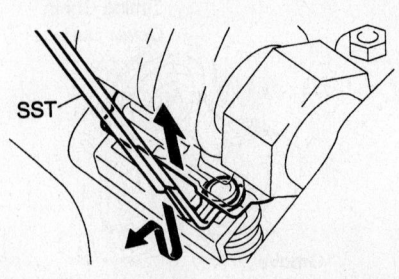

Setting the tool from the right side, makes shim removal easier—2ZZ-GE

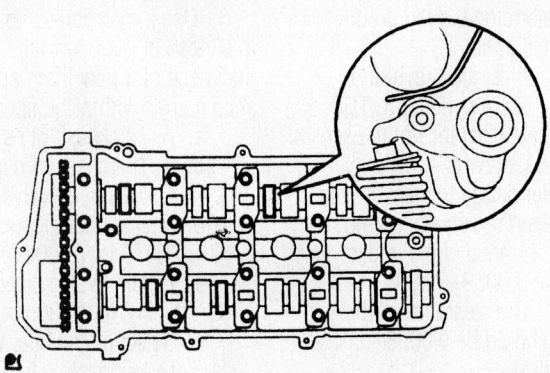

Check the clearance of the 1st set of valves–2ZZ-GE engine

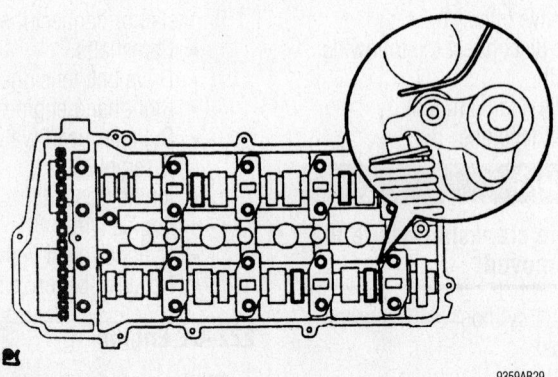

Check the clearance of the 2nd set of valves–2ZZ-GE engine

f. Lift the rocker arm to make room and remove the adjusting shim using SST 09248-77010.

7. Determine the size of the replaced shim according to the chart or the following formula:

a. Use a dial indicator to measure the thickness of the removed shim.

b. Calculate the thickness of a new shim so that the valve clearance comes within the specified value.

c. A: Thickness of new shim.

d. B: Thickness of used shim.

e. C: Measured valve clearance.

f. Intake: $A = B + (C - 0.005$ in. [0.13mm])

g. Exhaust: $A = B + (C - 0.011$ in. [0.27mm])

h. Select a new shim with a thickness as close as possible to the calculated values. Shims come in 41 sizes in increments of 0.0008 in. (0.020mm) from 0.0787–0.1102 in (2.0–2.8mm).

8. Lift the rocker arm to make room, then install the adjusting shim using the SST. To remove the tool from the shim, push down on the rocker arm.

9. Turn the crankshaft so the related rocker arm, where the valve clearance is adjusted, is fully pushed down.

10. Loosen the 2 set-screws, then remove the SST.

11. Install all components in the reverse of the removal procedure.

Starter

REMOVAL & INSTALLATION

1. Before servicing the vehicle, refer to the precautions section.

2. Remove or disconnect the following:
   - Negative battery cable
   - Right side engine undercover
   - Starter wiring
   - Starter

3. Installation is the reverse of removal. Torque the bolts to 27 ft. lbs. (37 Nm) and the nut to 7 ft. lbs. (10 Nm).

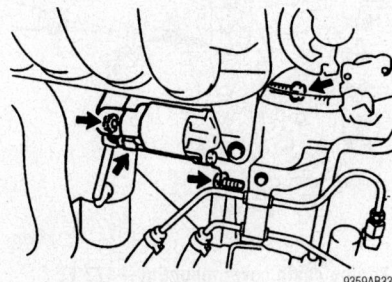

Starter mounting—Matrix

Oil Pan

REMOVAL & INSTALLATION

1ZZ-FE Engine

1. Before servicing the vehicle, refer to the precautions section.

2. Drain the engine oil.

3. Remove or disconnect the following:
   - Negative battery cable
   - Undercovers
   - Front exhaust pipe
   - Oil pan mounting bolts and nuts
   - Oil pan, cutting off the applied sealer.

To install:

4. Remove any old sealant from the oil pan flange and thoroughly clean the sealing surface.

5. Install or connect the following:
   - Oil pan. Tighten the bolts and nuts in several passes to 80 inch lbs. (9 Nm).
   - Front exhaust pipe
   - Negative battery cable
   - Undercovers

6. Fill the engine with clean oil.

7. Start the vehicle, check for leaks and repair if necessary.

2ZZ-GE Engine

1. Before servicing the vehicle, refer to the precautions section.

2. Drain the engine oil.

3. Remove or disconnect the following:
   - Negative battery cable. On vehicles equipped with an air bag, wait at

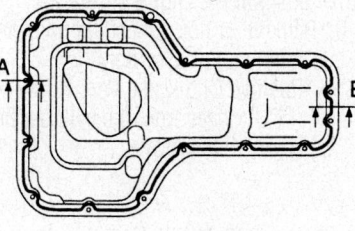

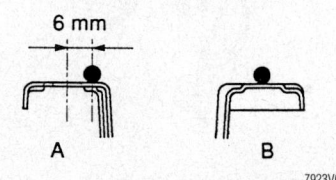

Apply sealant to the oil pan as shown— 1.8L (1ZZ-FE) engine

least 90 seconds before proceeding.
   - Undercovers
   - Front exhaust pipe
   - Oil pan mounting bolts and nuts
   - Oil pan, cutting off the applied sealer

To install:

4. Remove any old sealant from the oil pan flange and thoroughly clean the sealing surface.

5. Install or connect the following:
   - Oil pan. Tighten the bolts and nuts in several passes to 80 inch lbs. (9 Nm).
   - Front exhaust pipe
   - Negative battery cable
   - Undercovers

6. Fill the engine with clean oil.

7. Start the vehicle, check for leaks and repair if necessary.

Oil Pump

REMOVAL & INSTALLATION

1ZZ-FE Engine

1. Before servicing the vehicle, refer to the precautions section.

2. Drain the engine oil.

3. Remove or disconnect the following:
   - Negative battery cable
   - Timing chain and crankshaft sprocket
   - Timing chain vibration damper
   - Oil pump bolts, pump and gasket

To install:

4. Clean the mounting surface.

5. Install or connect the following:
   - Oil pump, with new gasket. Engage the spline teeth of the oil pump drive rotor with the larger teeth of the crankshaft, and slide the pump on.
   - Oil pump bolts and tighten to 80 inch lbs. (9 Nm)

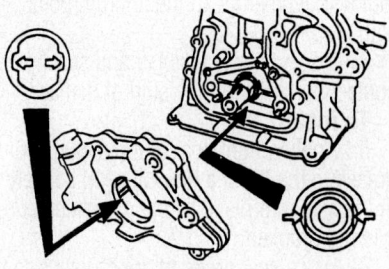

Oil pump mounting—1ZZ-FE and 2ZZ-GE engines

- Crankshaft vibration damper and tighten to 80 inch lbs. (9 Nm)
- Crankshaft sprocket and timing chain
- Negative battery cable

6. Fill the engine with clean oil.

7. Start the vehicle, check for leaks and repair if necessary.

2ZZ-GE Engine

1. Before servicing the vehicle, refer to the precautions section.

2. Drain the engine oil.

3. Remove or disconnect the following:
- Negative battery cable
- Timing chain and crankshaft sprocket
- Oil pump and gasket

To install:

4. Clean the mounting surface.

5. Install or connect the following:
- Oil pump, with new gasket. Engage the spline teeth of the oil pump drive rotor with the larger teeth of the crankshaft, and slide the pump on.
- Oil pump bolts and tighten to 80 inch lbs. (9 Nm)
- Crankshaft sprocket and timing chain
- Negative battery cable

6. Fill the engine with clean oil.

7. Start the vehicle, check for leaks and repair if necessary.

Rear Main Seal

REMOVAL & INSTALLATION

1. Remove or disconnect the following:
- Transaxle
- Clutch assembly
- Flywheel or flexplate

2. Use a small sharp knife to cut off the lip of the oil seal. Take great care not to score any metal with the knife.

3. Use a small prytool to pry the old seal from the retaining plate. Be careful not to damage the plate. Protect the tip of the tool with tape and pad the fulcrum point with cloth.

4. Inspect the crankshaft and seal lip contact surfaces for any sign of damage.

To install:

5. Apply a light coat of multi-purpose grease to the lip of a new oil seal. Loosely fit the seal into place by hand, making sure it is not crooked.

6. Use a seal driver of the correct size to install the seal. Tap it into place until the surface of the seal is flush with the edge of the housing.

Timing Chain, Sprockets, Front Cover and Seal

REMOVAL & INSTALLATION

1. Before servicing the vehicle, refer to the precautions section.

2. Drain the cooling system.

3. Remove or disconnect the following:
- Right side engine under cover
- Right front wheel and tire
- Cylinder head cover
- Wire harness clamp and suction hose assembly, 2ZZ-GE engine
- Drive belt

4. Separate the vane pipe assembly, but do not disconnect the hose, 1ZZ-FE engine.
- Alternator bracket, 2ZZ-GE
- Alternator
- Power steering pump reservoir and position it aside, 1ZZ-FE engine

5. Place a jack with a wooden block under the vehicle for support, then remove the 4 bolts and 2 nuts and remove the right side engine mount.

6. Remove the engine wire as follows, on 1ZZ-FE engines:
   a. Remove the 5 clamps from the brackets.
   b. Detach the connectors.
   c. Remove the ignition coil connectors.
   d. Bolt and nut holding the engine wire.

7. Remove the engine wire as follows, on 2ZZ-GE engines:
   a. Detach the ignition coil, oil control valve and Crankshaft Position Sensor (CKP) sensor electrical connectors.
   b. Bolt and nut for the engine ground, then position the engine wire aside

8. Remove or disconnect the following:
- Ignition coil assembly
- Positive Crankcase Ventilation (PCV) hoses from the cylinder head cover, if necessary

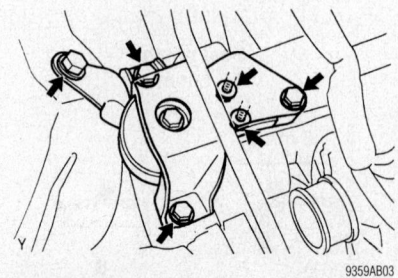

With the engine supported, remove the right side engine mount—1ZZ-FE engine shown, 2ZZ-GE similar

- Cylinder head (valve) cover sub-assembly

9. Set the No. 1 cylinder to Top Dead Center (TDC) of the compressor stroke as follows:
   a. Turn the crankshaft pulley, and align its groove with the "0" timing mark of the timing chain cover.
   b. Make sure the point marks of the camshaft timing sprockets and VVT timing sprockets are in a straight line as shown. If not, turn the crankshaft 1 complete revolution (360°) and align the marks.

- Crankshaft pulley, using SST 09960-10010
- Belt tensioner
- Water pump pulley, if equipped, and pump
- Transverse engine mounting bracket
- Crankshaft Position (CKP) sensor
- No. 1 chain tensioner assembly, making sure not to revolve the crankshaft without the tensioner

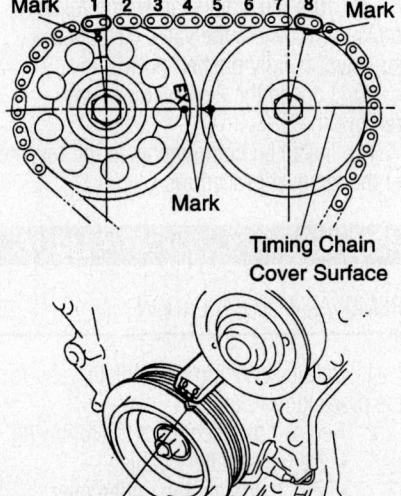

Proper timing mark alignment for TDC

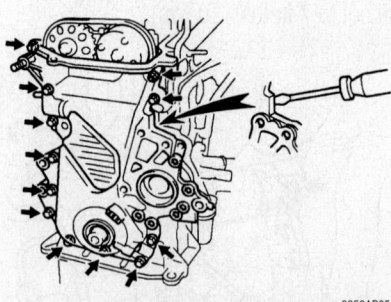

Timing chain cover mounting—1ZZ-FE engine shown, 2ZZ-GE similar

- Timing chain cover. The cover is retained with 11 bolts and nuts and a Torx® stud bolt. Pry the cover between the cylinder head and block to remove it.
- Timing gear cover oil seal
- CKP sensor plate No. 1
- Timing chain tensioner slipper

➡**In case you turn the camshafts with the timing chain removed, turn the crankshaft ¼ turn for the valve to avoid contact with the pistons.**

- Timing chain sub-assembly. Remove the chain with the crankshaft gear, using screwdrivers as shown

To install:

10. Set the No. 1 cylinder to TDC of the compression stroke:

a. Turn the hexagonal wrench head

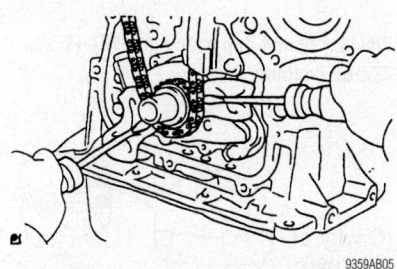

Remove the timing chain with the crankshaft gear

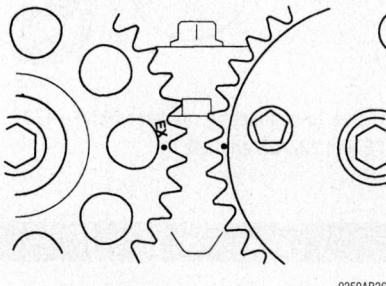

Proper alignment of the camshaft sprockets—1ZZ-FE engine

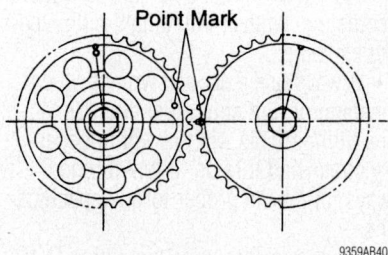

Proper alignment of the camshaft sprockets—2ZZ-GE engine

part of the camshafts, and align the point marks of the cam sprockets.

b. Using the crankshaft pulley bolt, turn the crankshaft and position the crankshaft set key upward.

11. Install or connect the following:

- Timing chain on the crank sprocket with the yellow link aligned with the mark on the crank sprocket. There are 3 yellow links on the timing chain.
- Crankshaft sprocket, using SST 09223-22010
- Timing chain on the camshaft sprockets with the yellow links aligned with the marks on the cam sprockets
- Timing chain tensioner slipper and tighten the bolt to 14 ft. lbs. (19 Nm)
- Crankshaft position sensor plate, with the "F" mark facing forward

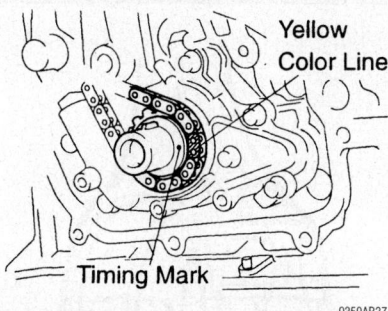

Make sure the yellow link is aligned with the crankshaft sprocket timing mark—1ZZ-FE and 2ZZ-GE engines

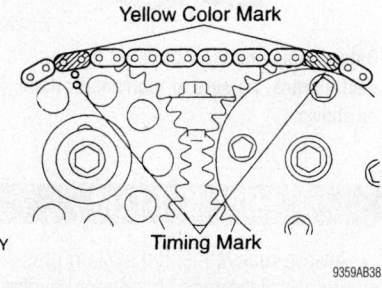

The yellow links of the timing chain must align with the camshaft sprocket timing marks—1ZZ-FE and 2ZZ-GE engines

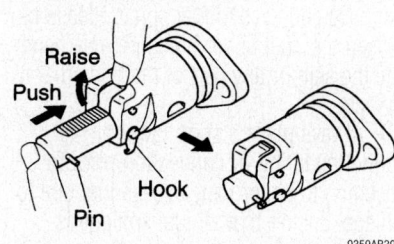

Timing chain tensioner—1ZZ-FE engine

- Timing gear cover oil seal
- Timing cover. For 1ZZ-FE engine, tighten the "A" bolts to 10 ft. lbs. (13 Nm), the "B" bolts to 14 ft. lbs. (19 Nm) and the stud bolt to 84 inch lbs. (9.5 Nm), using a Torx® wrench. For 2ZZ-GE engines, tighten the M8 bolts to 15 ft. lbs. (21 m), the M6 bolts to 8 ft. lbs. (11 Nm) and the stud bolt to 84 inch lbs. (9.5 Nm).

➡**When installing the tensioner, make sure to set the hook again if the hook releases the plunger.**

- Timing chain tensioner. Torque the nuts to 80 inch lbs. (9 Nm).
- CKP sensor and tighten the bolts to 80 inch lbs. (9 Nm)
- Transverse engine mounting bracket. Tighten the bolts to 35 ft. lbs. (47 Nm).
- Water pump and pulley
- Drive belt tensioner. Tighten the nut to 21 ft. lbs. (29 Nm) and the bolt to 51 ft. lbs. (69 Nm) on 1ZZ-FE engines or to 74 ft. lbs. (100 Nm) on 2ZZ-GE engines.

12. Install the crankshaft pulley, as follows:

a. Align the pulley set key with the key groove of the pulley and slide on the pulley.

b. Use SST 09960-11010 to install the bolt and tighten to 102 ft. lbs. (138 Nm) for 1ZZ-FE engine or to 87 ft. lbs. (118 Nm) on 2ZZ-GE engines.

c. Turn the crankshaft counterclockwise and disconnect the plunger knock pin from the hook.

d. Turn the crankshaft clockwise and check that the slipper is pushed by the plunger. If the plunger does not spring out, press the slipper into the chain tensioner with a screwdriver so that the hook is released from the knock pin and the plunger springs out.

- Cylinder head sub-assembly cover. Install seal packing into the locations shown and install within 3 minutes. Tighten the "A" bolts to 8 ft. lbs. (11 Nm) and the "B" bolts to 80 inch lbs. (9 Nm) for 1ZZ-FE engines. For 2ZZ-GE engines, tighten the bolts to 7 ft. lbs. (10 Nm).
- Ignition coil assembly. Torque the bolts to 80 inch lbs. (9 Nm).
- Engine wire and tighten to 80 inch lbs. (9 Nm)
- Right side engine mount. Tighten to 38 ft. lbs. (52 Nm).

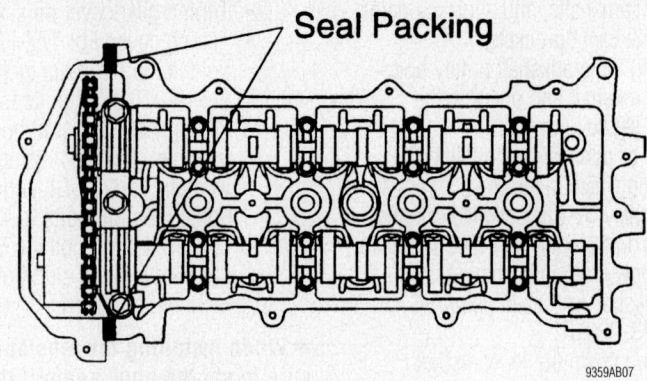

Seal Packing

Seal packing installation locations

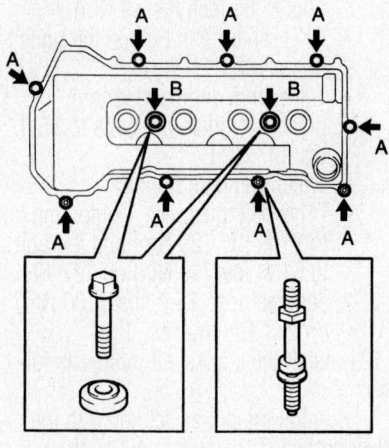

Cylinder head (valve) cover bolt locations—1ZZ-FE engine

- Alternator bracket, 2ZZ-GE engine
- Alternator
- Vane pump, 1ZZ-FE
- Main cylinder head cover and tighten to 62 inch lbs. (7 Nm)
- Right front wheel and tire. Tighten the lug nuts to 76 ft. lbs. (103 Nm).

13. Fill the cooling system to the proper level.

14. Start the vehicle, check for leaks and repair if necessary.

Piston and Ring Positioning

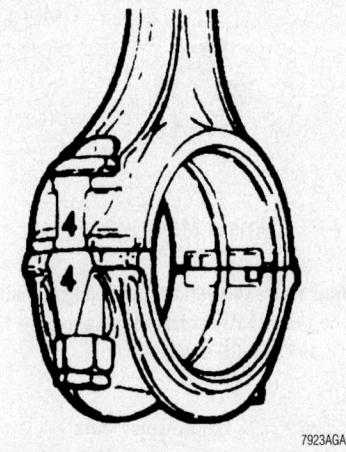

Before removing the caps from the connecting rods, be sure to matchmark them as shown

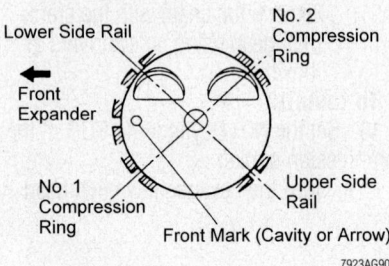

Piston ring identification mark locations—1ZZ-FE and 2ZZ-GE engines

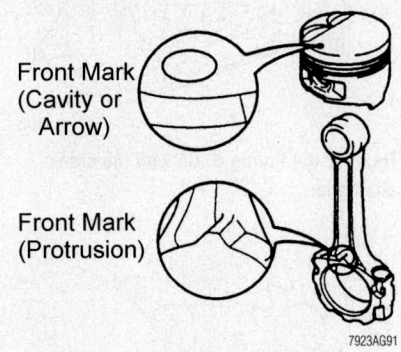

Piston ring end-gap spacing —1ZZ-FE and 2ZZ-GE engines

Piston-to-connecting rod assembly —1ZZ-FE and 2ZZ-GE engines

FUEL SYSTEM

Fuel System Service Precautions

Safety is the most important factor when performing not only fuel system maintenance, but any type of maintenance. Failure to conduct maintenance and repairs in a safe manner may result in serious personal injury or death. Work on a vehicle's fuel system components can be accomplished safely and effectively by adhering to the following rules and guidelines.

- To avoid the possibility of fire and personal injury, always disconnect the negative battery cable unless the repair or test procedure requires that battery voltage by applied.

- Always relieve the fuel system pressure prior to disconnecting any fuel system component (injector, fuel rail, pressure regulator, etc.) fitting or fuel line connection. Exercise extreme caution whenever relieving fuel system pressure, to avoid exposing skin, face and eyes to fuel spray. Please be advised that fuel under pressure may penetrate the skin or any part of the body that it contacts.

- Always place a shop towel or rag around the fitting or connection prior to loosening to absorb any excess fuel due to spillage. Ensure that all fuel spillage is quickly remove from engine surfaces. Ensure that all fuel-soaked cloths or towels

are deposited into a flame-proof waste container with a lid.

- Always keep a dry chemical (Class B) fire extinguisher near the work area.

- Do not allow fuel spray or fuel vapors to come into contact with a light bulb, spark or open flame.

- Always use a second wrench when loosening or tightening fuel line connections fittings. This will prevent unnecessary stress and torsion to fuel piping. Always follow the proper torque specifications.

- Always replace worn fuel fitting O-rings with new ones. Do not substitute fuel hose where rigid pipe is installed.

Fuel System Pressure

RELIEVING

✳✳ CAUTION

Failure to relieve fuel pressure before repairs or disassembly can cause serious personal injury and/or property damage. Fuel pressure is maintained within the fuel lines, even if the engine is OFF or has not been run in a period of time. This pressure must be safely relieved before any fuel-bearing line or component is loosened or removed. On vehicles equipped with inflatable restraints or air bag systems, wait at least 90 seconds after disconnecting the battery cable before performing any other work. The back-up power will keep the restraint system energized for a period of time after the battery is disconnected.

  1. Before servicing the vehicle, refer to the precautions section.
  2. Perform the following:
    a. Remove the rear seat cushion.
    b. Remove the rear floor service hole cover.
    c. Disconnect the fuel pump connector.
    d. Start and run the engine, until it stalls.
    e. Turn the ignition key to the **LOCK** position.
    f. Disconnect the negative battery cable.
    g. Connect the fuel pump connector.
    h. Install the service hole cover and rear seat cushion.
    i. Place a catch-pan under the joint to be disconnected. A large quantity of fuel may be released when the joint is opened.
    j. Wear eye or full face protection.
    k. Place a shop towel over the area and slowly release the joint using a wrench of the correct size.
    l. Allow the any fuel left in the line to bleed off slowly before fully disconnecting the joint.
    m. Plug the opened lines.

Fuel Filter

REMOVAL & INSTALLATION

  1. Before servicing the vehicle, refer to the precautions section.

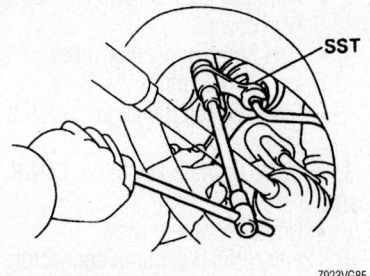

7923VG85

A line wrench with an extension may be needed to loosen the inlet line at the filter

  2. Relieve the fuel system pressure.
  3. Remove or disconnect the following:
    • Negative battery cable
    • Protective shield for the fuel filter
    • Air cleaner hose and cap, if necessary
    • Charcoal canister, if necessary
    • Slowly loosen the lower flare nut fitting until all the pressure is relieved
    • Banjo fitting and 2 metal gaskets. Discard the gaskets.

    • Fuel line with the flared nut from the filter
    • Filter from the mounting bracket
To install:
  4. Install or connect the following:
    • New fuel filter
    • Banjo fitting with a new metal gasket on each side and install the union bolt. Bolt: 22 ft. lbs. (30 Nm).
    • Flare nut to the lower connection. Nut: 22 ft. lbs. (30 Nm).
    • Charcoal canister
    • Air cleaner hose and cap
    • Protective shield
    • Negative battery cable

Fuel Pump

REMOVAL & INSTALLATION

  1. Before servicing the vehicle, refer to the precautions section.
  2. Remove or disconnect the following:
    • Negative battery cable

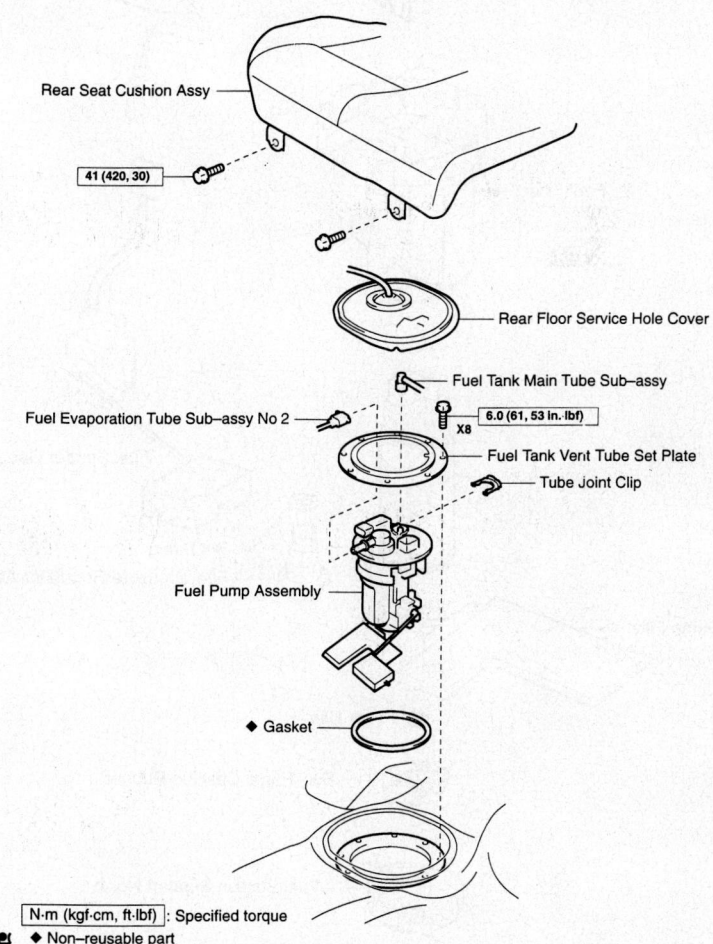

Rear Seat Cushion Assy

41 (420, 30)

Rear Floor Service Hole Cover

Fuel Tank Main Tube Sub-assy

Fuel Evaporation Tube Sub-assy No 2

6.0 (61, 53 in. lbf)
X8

Fuel Tank Vent Tube Set Plate

Tube Joint Clip

Fuel Pump Assembly

◆ Gasket

N·m (kgf·cm, ft·lbf) : Specified torque
◆ Non-reusable part

9359AB41

Exploded view of the fuel pump mounting—FWD shown, AWD similar

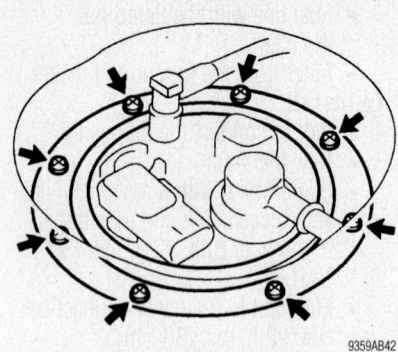

The fuel tank vent tube set plate is secured with 8 bolts on FWD vehicles

9359AB42

- Rear seat cushion and floor service hole cover
- Fuel pump and vapor pressure sensor connectors
- Start and run the engine, until it stalls

3. Turn the ignition key to the **LOCK** position.
  - Negative battery cable

4. Connect the fuel pump connector.
  - Fuel tank protector, AWD vehicles
  - Fuel tank main tube sub-assembly
  - Fuel emission tube sub-assembly No. 1, FWD vehicles
  - Fuel tank vent tube set plate. The plate is secured with 8 bolts on FWD vehicles, or 5 bolts on AWD vehicles.
  - Fuel pump assembly, being careful not to damage the filter or bend the arm of the fuel sender gauge
  - Fuel suction tube set gasket
  - Fuel suction support No. 2
  - Fuel pump rubber cushion
  - Fuel sender gauge assembly. Unplug the connector, then use a screwdriver to unlock the gauge and slide it to remove.
  - Fuel section plate sub-assembly
  - Vapor pressure sensor
  - Fuel pump harness
  - Fuel pump
  - Fuel pump filter
  - Fuel pressure regulator and O-ring

To install:

5. Install or connect the following:
  - New regulator O-ring and regulator
  - Fuel pump filter
  - Fuel pump
  - Vapor pressure sensor
  - Fuel suction tube set gasket
  - Fuel pump assembly
  - Fuel tank vent tube set plate. Tighten the bolts to 53 inch lbs. (6 Nm).
  - Connect the fuel emission tube sub-assembly
  - Fuel tank main tube sub-assembly
  - Fuel tank protector No. 2, AWD vehicles
  - Negative battery cable. Check for fuel leaks.
  - Floor service hole cover. Use butyl tape to seal the cover.
  - Rear seat cushion

Fuel Injectors

REMOVAL & INSTALLATION

1ZZ-FE Engine

1. Before servicing the vehicle, refer to the precautions section.

2. Properly relieve the fuel system pressure.

3. Remove or disconnect the following:
  - Negative battery cable.
  - No. 2 cylinder head cover
  - Positive Crankcase Ventilation (PCV) hose
  - Engine wire, unplugging the injector connectors and clamps
  - Fuel pipe clamp
  - Fuel line/tube sub-assembly

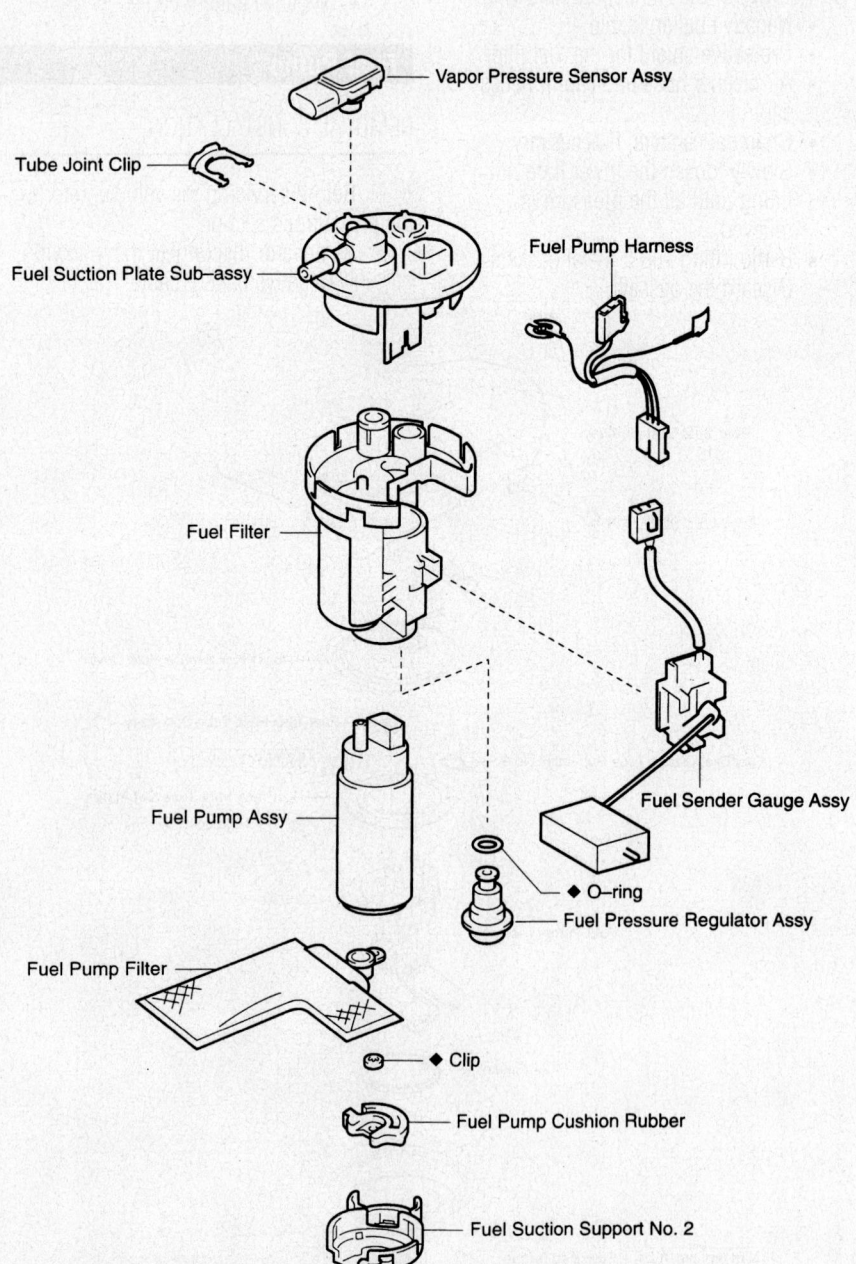

Vapor Pressure Sensor Assy
Tube Joint Clip
Fuel Suction Plate Sub–assy
Fuel Pump Harness
Fuel Filter
Fuel Pump Assy
Fuel Sender Gauge Assy
◆ O–ring
Fuel Pressure Regulator Assy
Fuel Pump Filter
◆ Clip
Fuel Pump Cushion Rubber
Fuel Suction Support No. 2

9359AB43

Fuel pump assembly components—FWD vehicles shown, AWD similar

Be careful not to drop the fuel injectors when removing the delivery pipe.

- Fuel delivery pipe sub-assembly with the injectors attached
- Delivery pipe and injectors
- Spacers from the head
- Injectors from the delivery pipe
- O-ring and grommet from each injector

To install:

4. Install or connect the following:
- New grommets
- New O-rings coated with light machine oil
- Injectors on the delivery pipe

➡**Coat the contact point on the pipe with light machine oil and twist the injectors into place. The connector should face outward.**

- Spacers

➡**Coat the seats in the head where the injectors contact, with light machine oil.**

- Delivery pipe and injectors

5. Loosely install the hold-down bolts and check that the injectors rotate smoothly. If they don't, the probable cause is incorrect O-ring installation. Torque the delivery pipe hold-down bolts to 14 ft. lbs. (19 Nm) and the fuel pipe bolt to 80 inch lbs. (9 Nm).

- Engine wire, attaching the injector connectors and clamps

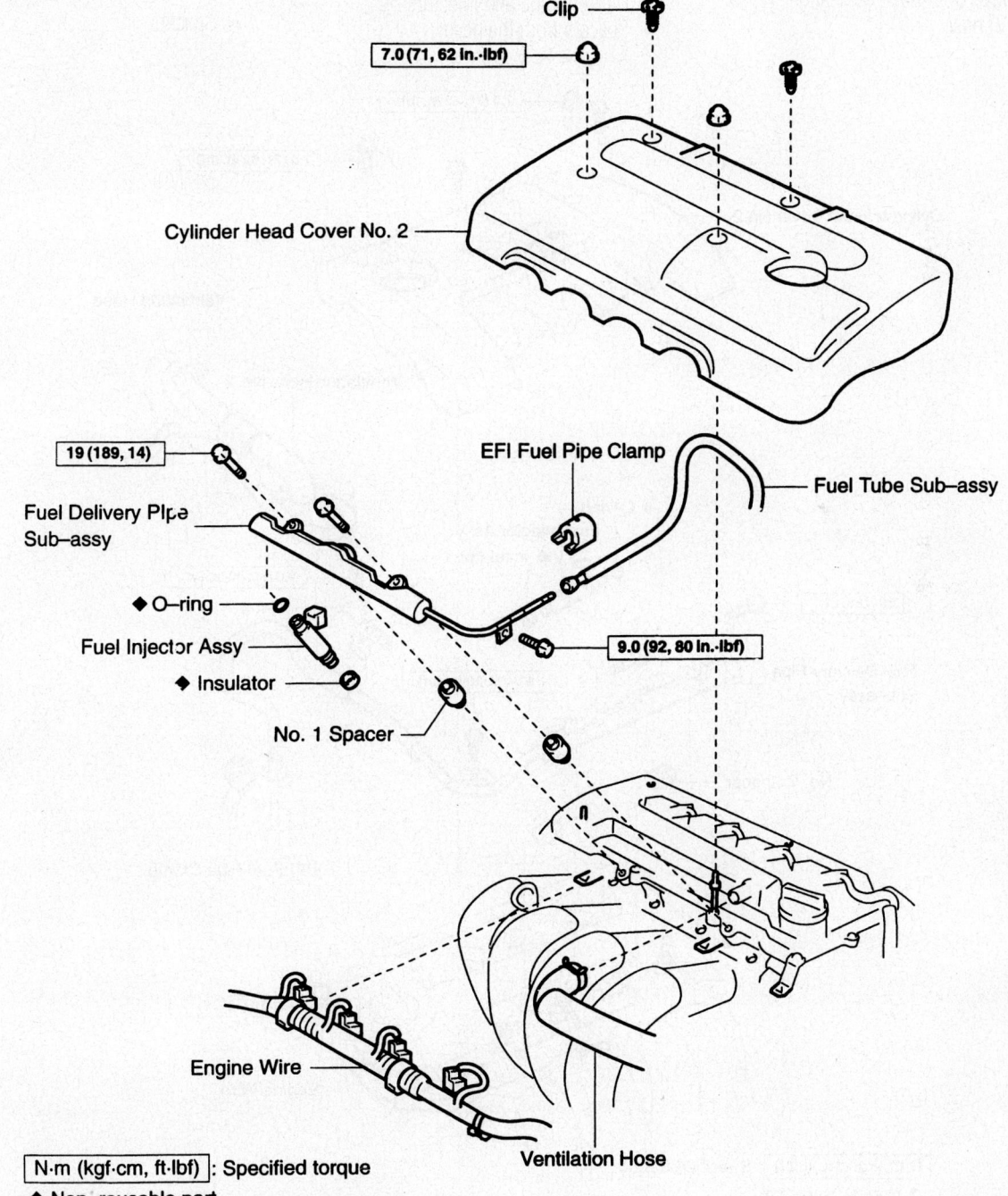

Clip

7.0 (71, 62 in.·lbf)

Cylinder Head Cover No. 2

EFI Fuel Pipe Clamp

Fuel Tube Sub–assy

19 (189, 14)

Fuel Delivery Pipe Sub–assy

◆ O–ring

Fuel Injector Assy

◆ Insulator

No. 1 Spacer

9.0 (92, 80 in.·lbf)

Engine Wire

Ventilation Hose

N·m (kgf·cm, ft·lbf) : Specified torque

◆ Non–reusable part

9359AB44

Fuel injector removal and installation—1ZZ-FE engine

- Fuel line/tube sub-assembly
- PCV hose
- No. 2 cylinder head (valve) cover

2ZZ-GE ENGINE

1. Before servicing the vehicle, refer to the precautions section.
2. Properly relieve the fuel system pressure.
3. Remove or disconnect the following:

- Negative battery cable.
- No. 2 cylinder head cover
- Positive Crankcase Ventilation (PCV) hose

- Engine wire, by removing the bolt, then unplugging the injector and Camshaft Position (CMP) sensor connectors
- Fuel pipe clamp

❋❋ WARNING

Be careful not to drop the fuel injectors when removing the delivery pipe.

- Fuel delivery pipe sub-assembly with the injectors attached
- Delivery pipe and injectors
- Spacers from the head

- Injectors from the delivery pipe
- O-ring and grommet from each injector

To install:
4. Install or connect the following:
- New grommets
- New O-rings coated with light machine oil
- Injectors on the delivery pipe

➡ **Coat the contact point on the pipe with light machine oil and twist the injectors into place. The connector should face outward.**

- Spacers

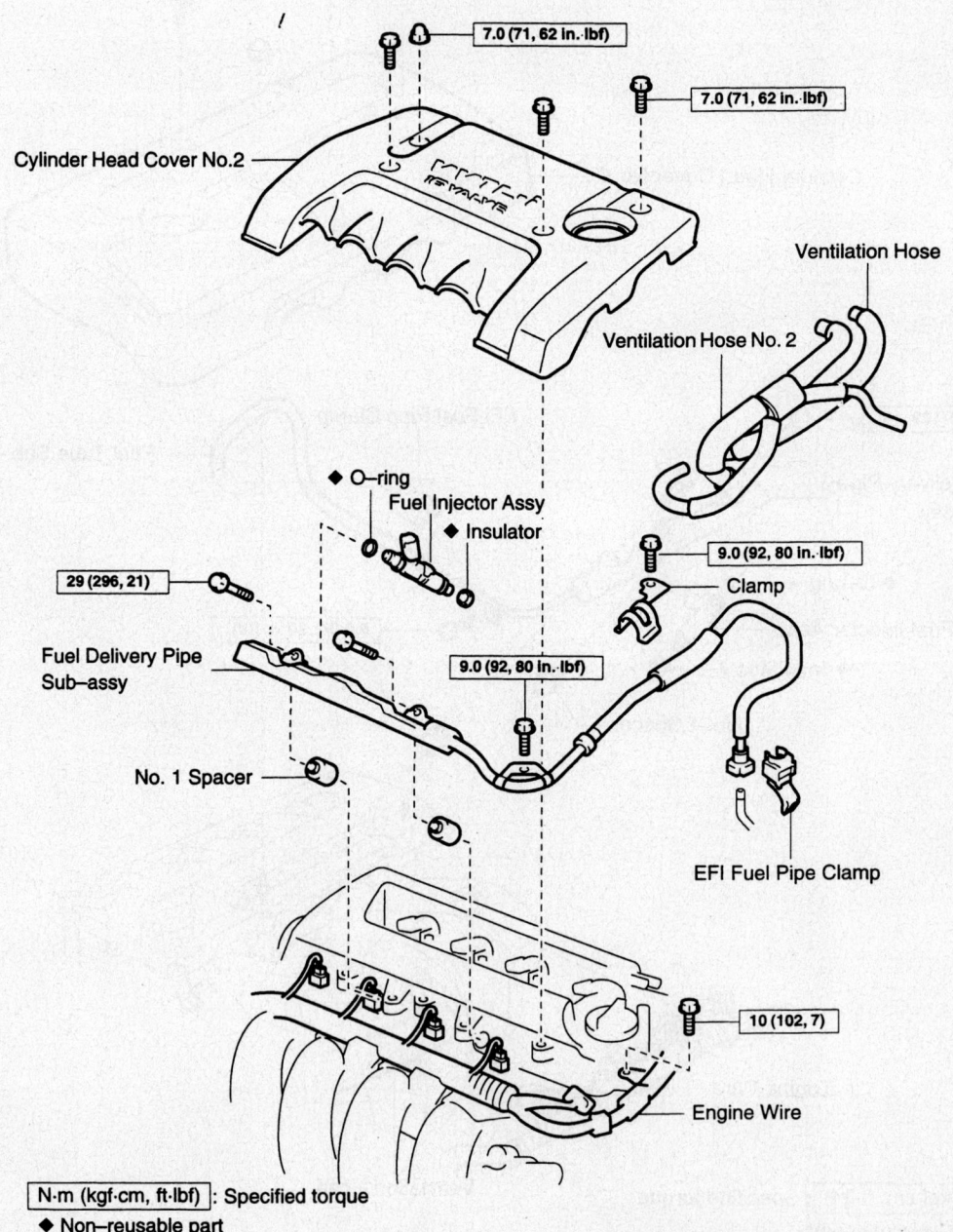

Cylinder Head Cover No.2

7.0 (71, 62 in.·lbf)
7.0 (71, 62 in.·lbf)

Ventilation Hose
Ventilation Hose No. 2

◆ O-ring
Fuel Injector Assy
◆ Insulator

9.0 (92, 80 in.·lbf)
Clamp

29 (296, 21)

Fuel Delivery Pipe Sub-assy

9.0 (92, 80 in.·lbf)

No. 1 Spacer

EFI Fuel Pipe Clamp

10 (102, 7)

Engine Wire

N·m (kgf·cm, ft·lbf): Specified torque
◆ Non-reusable part

9359AB45

Fuel injector removal and installation—2ZZ-GE engine

➡**Coat the seats in the head where the injectors contact, with light machine oil.**

• Delivery pipe and injectors

5. Loosely install the hold-down bolts and check that the injectors rotate smoothly.

If they don't, the probable cause is incorrect O-ring installation. Torque the delivery pipe hold-down bolts to 14 ft. lbs. (19 Nm) and the fuel pipe bolt to 80 inch lbs. (9 Nm).
• Fuel line/tube sub-assembly
• PCV hose

• Engine wire, by connecting the CMP sensor and injector connectors and installing the bolt. Tighten the bolt to 7 ft. lbs. (10 Nm).
• No. 2 cylinder head (valve) cover

DRIVE TRAIN

Manual Transaxle Assembly

REMOVAL & INSTALLATION

2003–04 Models

1. Before servicing the vehicle, refer to the precautions section.
2. Drain the transaxle fluid.
3. Place the front wheels in the straight-ahead position.
4. Remove or disconnect the following:
• Steering intermediate shaft
• Front wheel and tires
• Right and left side undercovers
• Exhaust pipe
• Hood
• Cylinder head (valve) cover
• Air cleaner assembly
• Battery clamp, battery, battery tray and battery carrier
• Cruise control actuator assembly, if equipped

5. Remove the wire harness as follows:
   a. Remove the wire harness clamp, 2 bolts and wire harness brackets.
   b. Remove the 2 bolts and 2 ground cables.
6. Remove or disconnect the following:
• Back-up lamp switch connector, with ABS
• Speed sensor connector, without ABS
• 5 bolts, then separate the release cylinder with the clutch pipes from the transaxle
• Shift cable clips and washer, then disconnect the cable from the transaxle and bracket
• Select cable clips and washer, then disconnect the cable from the transaxle and bracket
• Starter
• Right and left side tie rod ends
• Pressure feed tube
• Front halfshafts

7. Use a suitable tool to suspend the engine assembly, as shown in the illustration:
   a. No. 1 engine hanger: P/N 12281-22021 (5-speed M/T), 12281-88600 (6-speed M/T).
   b. No. 2 engine hanger: P/N 12281-

15040 (5-speed M/T), 12281-88600 (6-speed M/T).
   c. Bolt: P/N 91512-B1016.
   d. Torque the bolts to 28 ft. lbs. (38 Nm).

✴✴ **CAUTION**

Do not try to suspend the engine by hooking the chain to any other part.

   e. Attach an engine chain hoist to the hangers.
8. Remove or disconnect the following:
• Front suspension crossmember
9. Support the transaxle with a floor jack.
• Transverse engine mounting insulator and brackets
• Manual transaxle assembly
• Transverse engine mounting brackets from the transaxle, if necessary

To install:
• Transverse engine mounting brackets to the transaxle, if necessary
• Manual transaxle, by aligning the input shaft with the clutch disc. Torque the "A" bolts to 47 ft. lbs. (64 Nm), the "B" bolts to 35 ft. lbs.

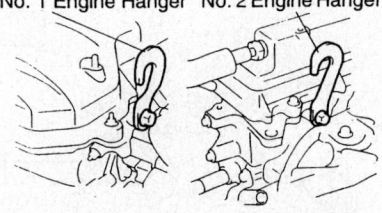

No. 1 Engine Hanger   No. 2 Engine Hanger

9359AB46

Secure the engine using the proper tools—5-speed manual transmission shown

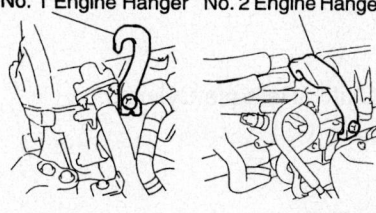

No. 1 Engine Hanger   No. 2 Engine Hanger

9359AB49

Secure the engine using the proper tools—6-speed manual transmission shown

(47 Nm) and the "C" bolts to 17 ft. lbs. (23 Nm).
• Transverse engine mounting bracket. Tighten to 38 ft. lbs. (52 Nm).
• Transverse engine mounting insulator. Tighten the "A" bolts to 38 ft. lbs. (52 Nm) and the "B" bolts to 59 ft. lbs. (80 Nm).

10. The remainder of installation is the reverse of the removal procedure, noting the following specifications:
   a. Starter mounting bolts: 27 ft. lbs. (37 Nm).
   b. Clutch release cylinder bolts: "A" bolts 19 ft. lbs. (25 Nm), "B" bolts 9 ft. lbs. (12 Nm) and "C" bolts 44 inch lbs. (5 Nm).
   c. Battery carrier bolts: 10 ft. lbs. (13 Nm).
   d. Battery clamp bolt: 44 inch lbs. (5 Nm).
   e. Battery clamp nut: 31 inch lbs. (3.5 Nm).
   f. Cylinder head cover bolts: 62 inch lbs. (7 Nm).
   g. Hood bolts: 10 ft. lbs. (13 Nm).
   h. Wheel lug nuts: 76 ft. lbs. (103 Nm).
11. Fill the transaxle fluid to the proper level.
12. Start the vehicle, check for leaks and repair if necessary.

2005 Models

1. Before servicing the vehicle, refer to the precautions section.

✴✴ **CAUTION**

Before servicing any electrical component, the ignition key must be in the OFF or LOCK position and all electrical loads must be OFF, unless instructed otherwise in these procedures. If a tool or equipment could easily come in contact with a live exposed electrical terminal, also disconnect the negative battery cable. Failure to follow these precautions may cause personal injury and/or damage to the vehicle or its components.

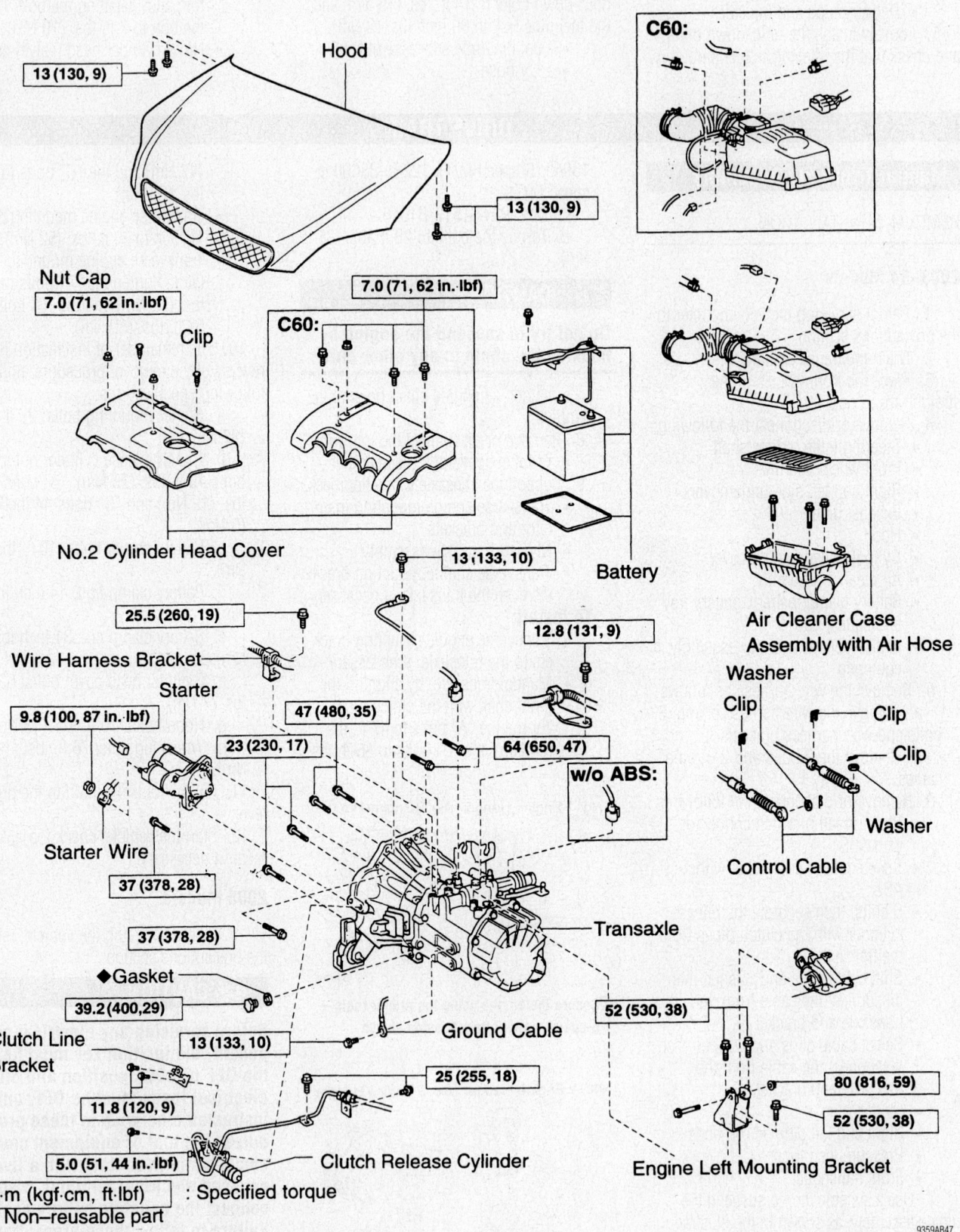

13 (130, 9)

Hood

C60:

13 (130, 9)

Nut Cap
7.0 (71, 62 in. lbf)

Clip

7.0 (71, 62 in. lbf)

C60:

No.2 Cylinder Head Cover

Battery

13 (133, 10)

Air Cleaner Case
Assembly with Air Hose

25.5 (260, 19)

Wire Harness Bracket

12.8 (131, 9)

Washer

Starter

Clip

Clip

9.8 (100, 87 in. lbf)

47 (480, 35)

Clip

23 (230, 17)

64 (650, 47)

Washer

Starter Wire

w/o ABS:

37 (378, 28)

Control Cable

37 (378, 28)

Transaxle

◆Gasket

39.2 (400, 29)

52 (530, 38)

Clutch Line
Bracket

13 (133, 10)

Ground Cable

11.8 (120, 9)

25 (255, 18)

80 (816, 59)

52 (530, 38)

5.0 (51, 44 in. lbf)

Clutch Release Cylinder

Engine Left Mounting Bracket

N·m (kgf·cm, ft·lbf) : Specified torque
◆ Non-reusable part

9359AB47

Exploded view of the manual transaxle (1 of 2)

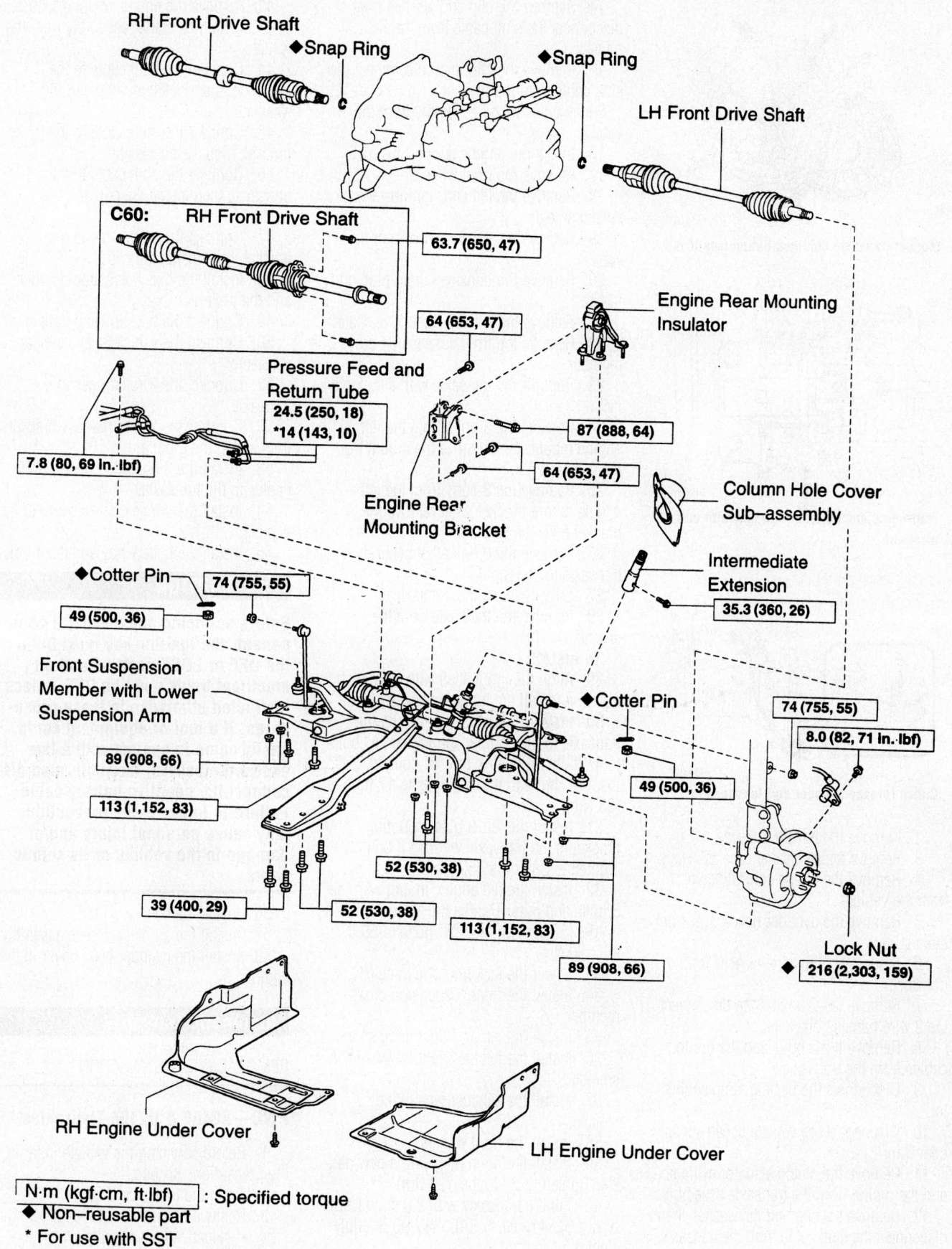

RH Front Drive Shaft

◆Snap Ring

◆Snap Ring

LH Front Drive Shaft

C60: RH Front Drive Shaft

63.7 (650, 47)

64 (653, 47)

Engine Rear Mounting Insulator

Pressure Feed and Return Tube
24.5 (250, 18)
*14 (143, 10)

87 (888, 64)

64 (653, 47)

7.8 (80, 69 in.·lbf)

Engine Rear Mounting Bracket

Column Hole Cover Sub-assembly

◆Cotter Pin

74 (755, 55)

49 (500, 36)

Intermediate Extension

35.3 (360, 26)

Front Suspension Member with Lower Suspension Arm

◆Cotter Pin

74 (755, 55)

8.0 (82, 71 in.·lbf)

89 (908, 66)

49 (500, 36)

113 (1,152, 83)

52 (530, 38)

39 (400, 29)

52 (530, 38)

113 (1,152, 83)

89 (908, 66)

Lock Nut
◆ 216 (2,303, 159)

RH Engine Under Cover

LH Engine Under Cover

N·m (kgf·cm, ft·lbf) : Specified torque
◆ Non–reusable part
* For use with SST

9359AB48

Exploded view of the manual transaxle (2 of 2)

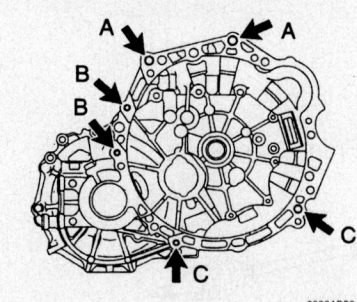

Manual transaxle bolt installation locations

9359AB50

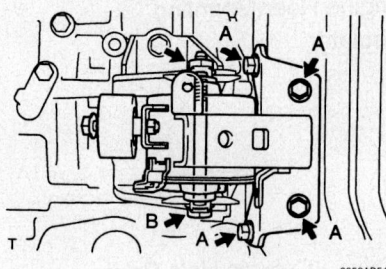

Transverse engine mounting insulator bolt locations

9359AB51

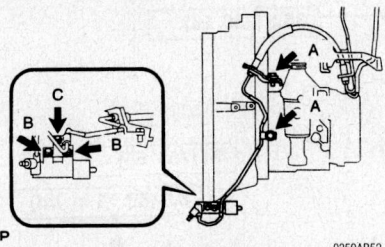

Clutch release cylinder bolt locations

9359AB52

2. Remove the battery and tray.

3. Remove the air cleaner case assembly.

4. Remove the cruise control servo from the vehicle.

5. Remove the cylinder head cover from the engine.

6. Disconnect the wire harness from the transaxle.

7. Remove the 2 bolts, then disconnect the 2 wire harness brackets.

8. Remove the 2 bolts and the ground cables from the transaxle.

9. Disconnect the backup lamp connector.

10. Disconnect the vehicle speed sensor connector.

11. Remove the clutch actuator cylinder and the piping from the transaxle assembly.

12. Remove the clip and the washer, then disconnect the shift cable from the transaxle.

13. Remove the clip, then disconnect the shift cable from the bracket.

14. Remove the clip and washer, then disconnect the shift cable from the transaxle.

15. Remove the clip, then disconnect the shift cable from the bracket.

Remove the starter assembly from the vehicle.

16. Install the engine support fixture.

17. Remove the front wheels.

18. Remove the left and right lower splash shields.

19. Remove the exhaust pipe from the vehicle.

20. Remove the transaxle drain plug and the oil.

21. Remove the left and right drive shafts.

22. Remove the front suspension cross-member.

23. Support the transaxle with a suitable jack.

24. Remove the 5 bolts from the left engine mount, then remove the mount from the vehicle.

25. Remove the 3 bolts from the left engine mount bracket, then remove the bracket from the vehicle.

26. Remove the 6 bolts that secure the transaxle to the engine.

27. Slightly lower the transaxle.

28. Remove the transaxle from the engine.

To install:

29. Align the input shaft with the clutch disc and install the transaxle to the engine.

30. Install the 6 bolts that secure the transaxle to the engine. Torque the "A" bolts to 47 ft. lbs. (64 Nm), the "B" bolts to 35 ft. lbs. (47 Nm) and the "C" bolts to 17 ft. lbs. (23 Nm).

31. Install the left engine mounting bracket to the transaxle with the 3 bolts. Tighten to 38 ft. lbs. (52 Nm).

32. Install the left engine mount with the 5 bolts and nuts. Tighten the "A" bolts to 38 ft. lbs. (52 Nm) and the "B" bolts to 59 ft. lbs. (80 Nm).

33. Lower the jack from the transaxle.

34. Install the front suspension cross-member.

35. Install the left and right drive shafts.

36. Install the left and right lower splash shields.

37. Install the exhaust pipe in the vehicle.

38. Install the front wheels.

39. Install the drain plug with a new gasket. Tighten to 29 ft. lbs. (39 Nm).

40. Fill the transaxle with 2.0 qts. (1.9L) of API GL-4 or GL-5 SAE 75W-90 or equivalent.

41. Install the fill plug with a new gasket. Tighten to 29 ft. lbs. (39 Nm).

42. Remove the engine support fixture.

43. Install the starter assembly from the vehicle.

44. Connect the shift cable to the transaxle, then install the clip and the washer.

45. Connect the shift cable to the bracket, then install the clip.

46. Connect the shift cable to the transaxle, then install the clip and the washer.

47. Connect the shift cable to the bracket, then install the clip.

48. Install the clutch actuator cylinder and the piping.

49. Connect the backup lamp connector.

50. Connect the vehicle speed sensor connector.

51. Connect the wire harness to the transaxle.

52. Connect the 2 wire harness brackets, then install the 2 bolts.

53. Install the 2 bolts and the ground cables to the transaxle.

54. Install the cruise control servo in the vehicle.

55. Install the battery tray and the 4 bolts.

❋❋ CAUTION

Before servicing any electrical component, the ignition key must be in the OFF or LOCK position and all electrical loads must be OFF, unless instructed otherwise in these procedures. If a tool or equipment could easily come in contact with a live exposed electrical terminal, also disconnect the negative battery cable. Failure to follow these precautions may cause personal injury and/or damage to the vehicle or its components.

56. Install the battery.

57. Install the air cleaner case assembly.

58. Install the cylinder head cover in the engine.

Automatic Transaxle

REMOVAL AND INSTALLATION

FWD—A246E & U240E Transaxles

1. Before servicing the vehicle, refer to the precautions section.

2. Drain the transaxle fluid.

3. Remove or disconnect the following:

- Negative battery cable
- Hood
- No. 2 cylinder head cover

- Battery and battery carrier
- Air cleaner assembly with hose
- Floor shift cable transmission control shift
- Transmission control cable support
- No. 1 transmission control cable bracket
- Wiring harness and brackets
- Transmission wire connector
- Park/neutral position switch connector, with Anti-lock Brake System (ABS)
- Speedometer sensor connector, without ABS

- Transmission revolution sensor connectors, if equipped
- Transmission fluid filler tube
- No. 1 oil cooler inlet and outlet tubes
- Foot rest
- Floor carpet
- Oxygen (O_2) sensor connector

4. Suspend the engine as follows:
   a. Disconnect the 2 Positive Crankcase Ventilation (PCV) hoses.
   b. Install the No. 1 and No. 2 engine hangers in the correct direction.
   c. No. 1 engine hanger: P/N 12281-

22021 (A246E) or 12281-88600 (U240E).
   d. No. 2 engine hanger: P/N 12281-15040 (A246E) or 12281-88600 (U240E)
   e. Bolt: P/N 91512-B1016.
   f. Torque the bolt to 28 ft. lbs. (38 Nm).
   g. Attach an engine chain hoist to the engine hangers.
- Front wheels
- Right and left engine undercovers
- Front floor panel brace, U240E transaxle
- Front exhaust pipe

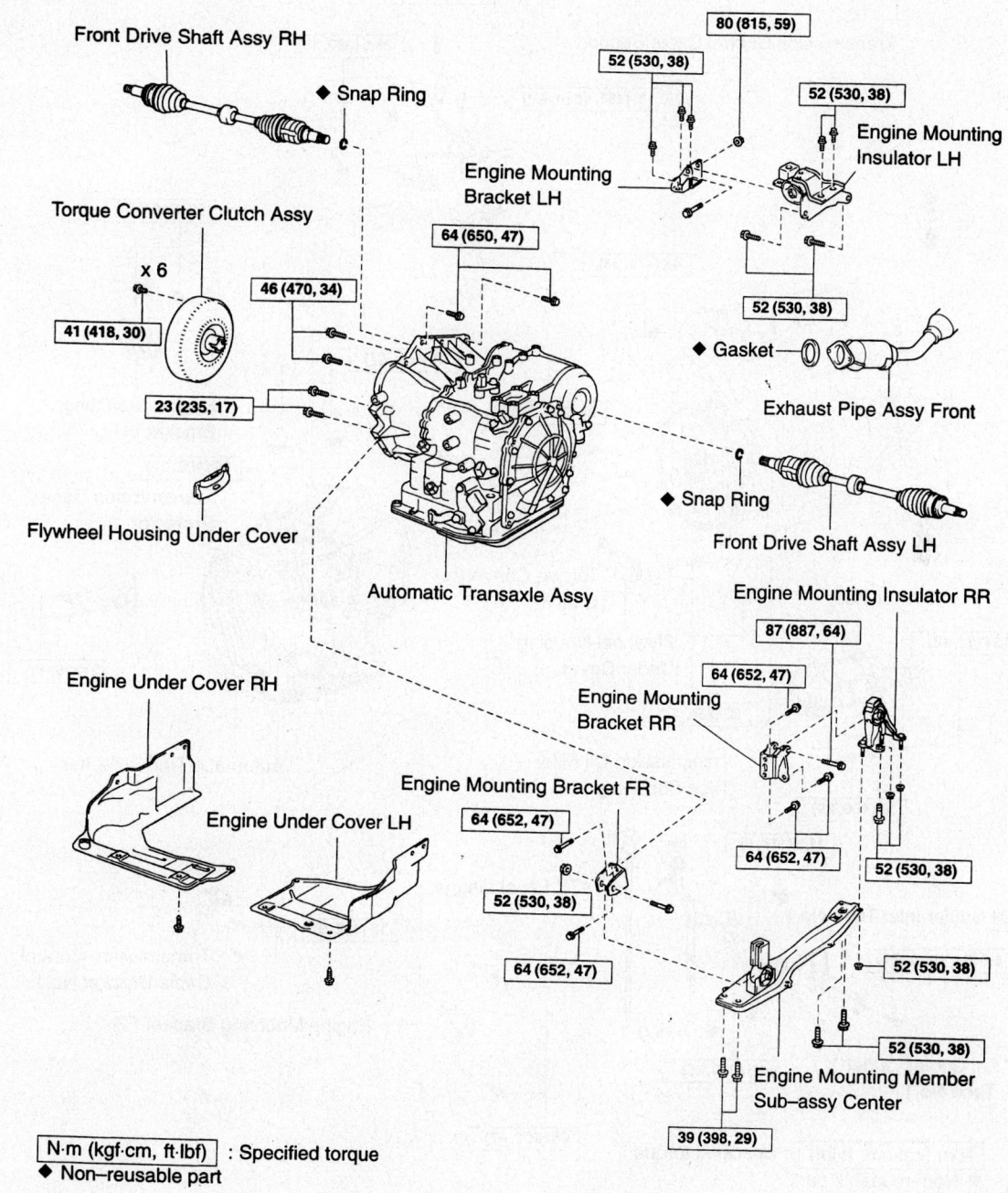

N·m (kgf·cm, ft·lbf) : Specified torque
◆ Non-reusable part

Automatic transaxle and related components—U240E transaxle shown, A246E similar

- Front halfshafts
- Automatic transmission case protector
- Starter

5. Support the transaxle with a floor jack
   - Left side transverse engine mounting insulator and bracket
   - Right side front and rear engine mount insulators
   - 4 bolts, dynamic damper and member sub-assembly
   - Front and rear right side transverse engine mounting brackets

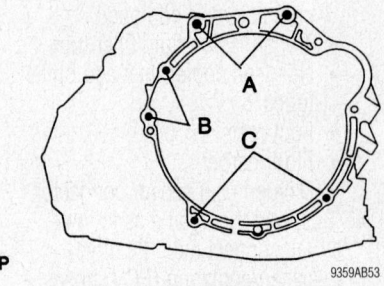

A
B
C
P
9359AB53

Automatic transaxle bolt locations

A
B
P
9359AB54

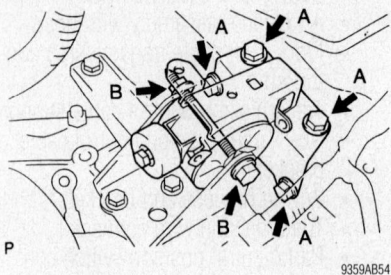

Left side engine mount insulator bolt and nut locations

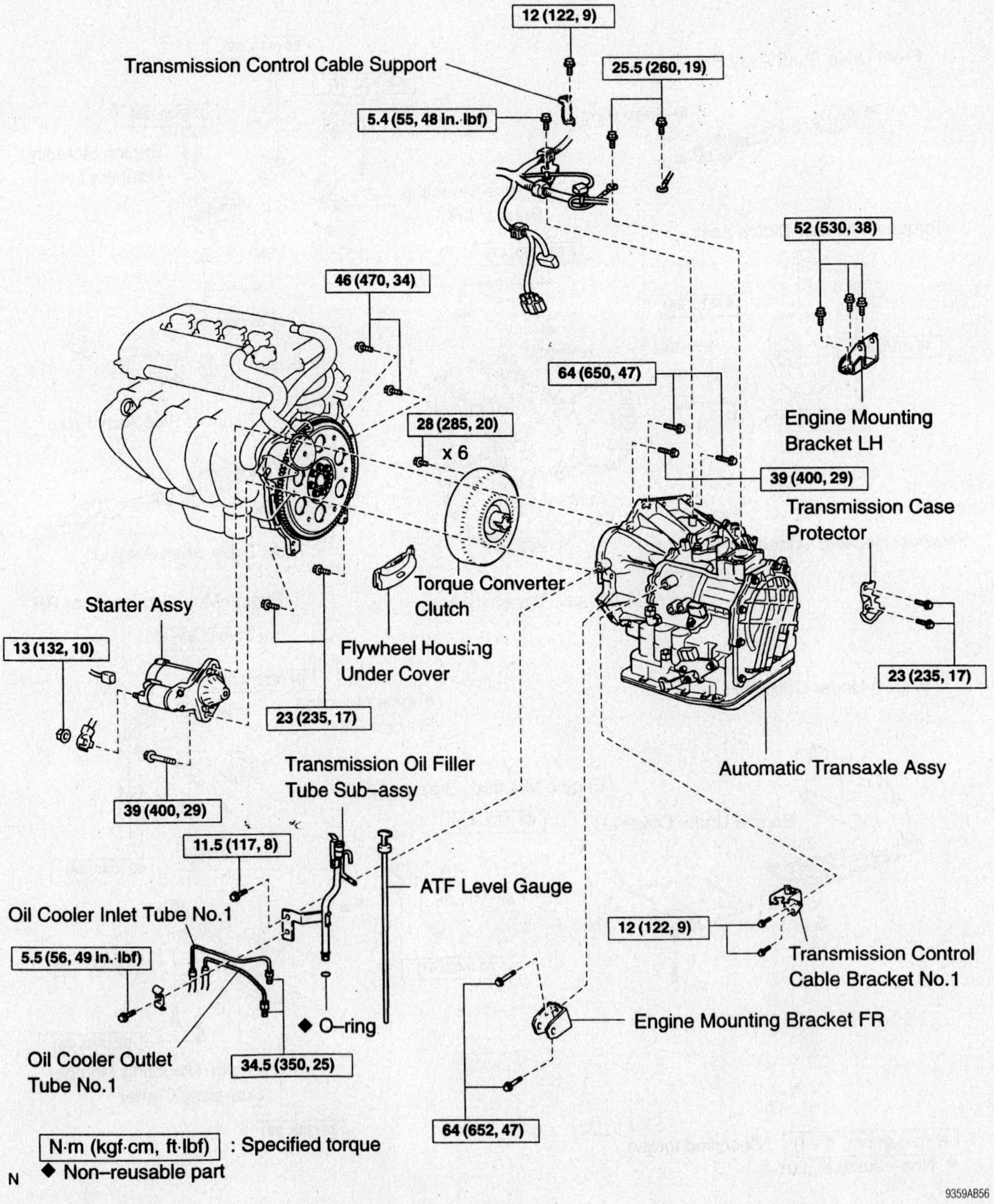

12 (122, 9)
Transmission Control Cable Support
25.5 (260, 19)
5.4 (55, 48 in.·lbf)
52 (530, 38)
46 (470, 34)
64 (650, 47)
Engine Mounting Bracket LH
28 (285, 20)
x 6
39 (400, 29)
Transmission Case Protector
Torque Converter Clutch
Starter Assy
13 (132, 10)
Flywheel Housing Under Cover
23 (235, 17)
23 (235, 17)
Transmission Oil Filler Tube Sub–assy
Automatic Transaxle Assy
39 (400, 29)
11.5 (117, 8)
ATF Level Gauge
Oil Cooler Inlet Tube No.1
12 (122, 9)
Transmission Control Cable Bracket No.1
5.5 (56, 49 in.·lbf)
◆ O–ring
Engine Mounting Bracket FR
Oil Cooler Outlet Tube No.1
34.5 (350, 25)
64 (652, 47)
N·m (kgf·cm, ft·lbf) : Specified torque
◆ Non–reusable part
N
9359AB56

Automatic transaxle and related components—U341F transaxle

- Flywheel housing undercover
- Automatic transaxle. Turn the crankshaft for access to the 6 bolts while holding the crankshaft pulley bolt with a wrench.
- Torque converter clutch

6. Installation is the reverse of the removal procedure, noting the following specifications:

   a. Automatic transaxle: Bolt "A" to 47 ft. lbs. (64 Nm), bolt "B" to 34 ft. lbs. (47 Nm) and bolt "C" to 17 ft. lbs. (23 Nm).

   b. Torque converter bolts: 20 ft. lbs. (28 Nm).

   c. Front and rear right transverse engine mounting bracket bolts: 47 ft. lbs. (64 Nm).

   d. Member sub-assembly center bolts: "A" bolts to 29 ft. lbs. (39 Nm) and "B" bolts to 38 ft. lbs. (52 Nm).

   e. Right rear engine mounting insulator-to-engine mounting bracket bolt: 64 ft. lbs. (87 Nm).

   f. Right rear engine mount insulator nuts and bolt: 38 ft. lbs. (52 Nm).

   g. Left side engine mounting bracket-to-transaxle bolts: 38 ft. lbs. (52 Nm).

   h. Left side engine mounting insulator bolts and nut: Bolt "A" to 38 ft. lbs. (52 Nm), Bolt "B" and Nut "B" to 59 ft. lbs. (80 Nm).

   i. Front right engine mount insulator-to-mounting bracket bolt and nut: 38 ft. lbs. (52 Nm).

   j. Starter bolts: 29 ft. lbs. (39 Nm).

   k. Automatic transmission case protector bolts: 14 ft. lbs. (18 Nm).

   l. Wheel lug nuts: 76 ft. lbs. (103 Nm).

   m. Oil cooler clamp bolts: 49 inch lbs. (5.5 Nm).

   n. Oil cooler inlet and outlet tubes: 25 ft. lbs. (34 Nm).

   o. Wire harness bracket bolt: 9 ft. lbs. (13 Nm).

   p. Transmission control cable bracket bolts: 9 ft. lbs. (12 Nm).

   q. Transmission control cable support: 9 ft. lbs. (12 Nm).

   r. Battery carrier: 10 ft. lbs. (13 Nm).

   s. Air cleaner assembly: 62 inch lbs. (7 Nm).

   t. Cylinder head cover bolts: 62 inch lbs. (7 Nm).

   u. Hood bolts: 10 ft. lbs. (13 Nm).

7. Fill the transaxle fluid to the proper level.

8. Start the vehicle, check for leaks and repair if necessary.

AWD—U341F Transaxle

1. Before servicing the vehicle, refer to the precautions section.

2. Drain the transaxle fluid.
3. Remove or disconnect the following:
- Negative battery cable
- Engine and transaxle assembly
- Transfer case
- Automatic transmission case protector
- Front left side halfshaft
- Transmission control cable support and bracket
- Wire harness clamp bracket, bolts and 2 wire harnesses
- Transmission wire connector
- Park/neutral position switch connector
- Transmission revolution sensor connectors, if equipped
- Transmission fluid filler tube
- Oil cooler inlet and outlet tubes
- Transverse engine mounting brackets
- Flywheel housing undercover
- Automatic transaxle. Turn the crankshaft for access to the 6 bolts while holding the crankshaft pulley bolt with a wrench.
- Torque converter clutch

4. Installation is the reverse of the removal procedure, noting the following specifications:

   a. Automatic transaxle: Bolt "A" to 47 ft. lbs. (64 Nm), bolt "B" to 34 ft. lbs. (47 Nm) and bolt "C" to 17 ft. lbs. (23 Nm).

   b. Oil cooler clamp bolts: 8 ft. lbs. (11 Nm) for the top bolt and 49 inch lbs. (5.5 Nm) for the bottom bolt

   c. Oil cooler inlet and outlet tube bolts: 25 ft. lbs. (34 Nm).

   d. Wire harness clamp bracket bolt: 48 inch lbs. (5 Nm).

   e. Transmission control cable bracket and support bolts: 9 ft. lbs. (12 Nm).

   f. Automatic transmission case protector bolts: 17 ft. lbs. (23 Nm).

5. Fill the transaxle fluid to the proper level.

6. Start the vehicle, check for leaks and repair if necessary.

Clutch

ADJUSTMENTS

Hydraulic clutch actuating systems used in Toyota vehicles do not require adjustment.

REMOVAL & INSTALLATION

1. Before servicing the vehicle, refer to the precautions section.

➡**Do not allow grease or oil to get on any part of the disc, pressure plate, or flywheel surfaces.**

2. Remove or disconnect the following:
- Negative battery cable. On vehicles equipped with an air bag, wait at least 90 seconds before proceeding
- Transaxle assembly

3. Make matchmarks on the clutch cover (pressure plate) and flywheel so that the pressure plate can be returned to its original position during installation.

4. Remove or disconnect the following:

- Release fork bearing clips
- Release bearing hub, complete with the release bearing
- Release fork and support

✳✳ CAUTION

Slowly unfasten the bolts which attach the pressure plate. Loosen each bolt 1 turn at a time until the spring tension is released. If the bolts are released improperly the clutch assembly could fly apart, causing possible injury.

- Pressure plate from the clutch cover/spring assembly

5. Inspect the disc, pressure plate and flywheel for damage and wear using a caliper to measure depth and width and a dial indicator to measure runout.

   a. The minimum clutch disc rivet head depth is 0.012 in. (0.3mm).

   b. The maximum clutch disc runout is 0.031 in. (0.8mm).

   c. The maximum pressure plate spring depth is 0.024 in. (0.6mm).

   d. The maximum pressure plate spring width is 0.197 in. (5.0mm).

   e. The maximum flywheel runout is 0.004 in. (0.1mm).

6. Replace or machine parts as necessary.

To install:

7. When reassembling, apply a thin coating of multipurpose grease to the release bearing hub and release fork contact points. Also, pack the groove inside the clutch hub with multipurpose grease and lubricate the pivot points of the release fork.

8. Install or connect the following:

- Clutch disc and pressure plate. The bolts should be tightened in 2 or 3 steps, gradually and evenly. Final bolt torque is 14 ft. lbs. (19 Nm).
- Release bearing, fork and boot
- Transaxle assembly
- Negative battery cable

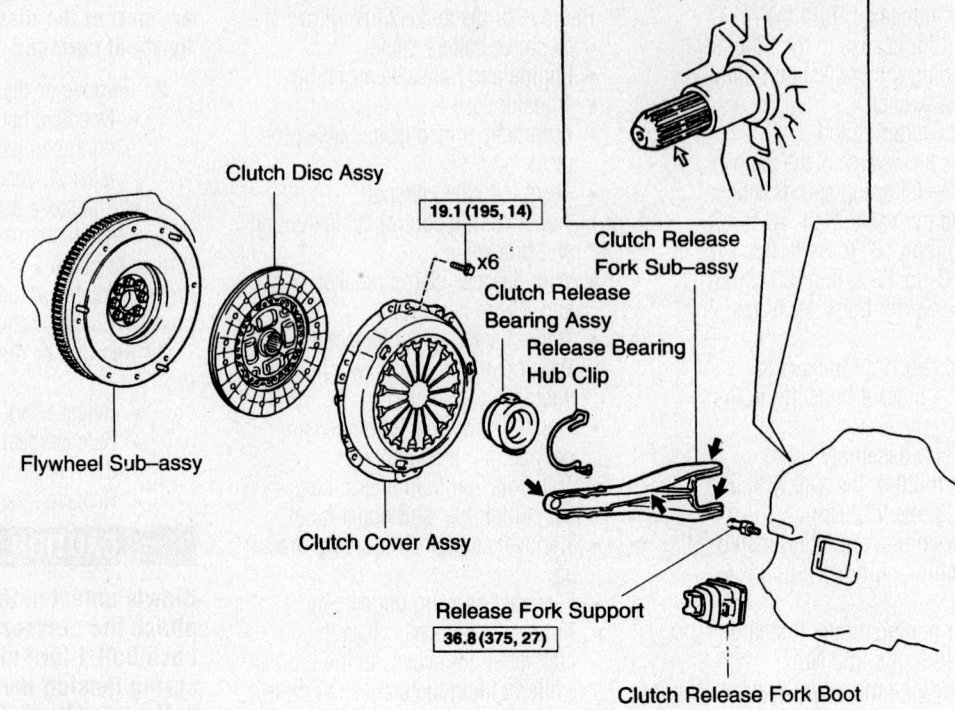

Clutch Disc Assy

19.1 (195, 14)

x6

Clutch Release Fork Sub–assy

Clutch Release Bearing Assy

Release Bearing Hub Clip

Flywheel Sub–assy

Clutch Cover Assy

Release Fork Support
36.8 (375, 27)

Clutch Release Fork Boot

N·m (kgf·cm, ft·lbf) : Specified torque
◆ Non–reusable part
⇐ Clutch spline grease
⇐ Release hub grease

9359AB57

Exploded view of the clutch components

Hydraulic Clutch System

BLEEDING

➡ If any maintenance on the clutch system was performed or the system is suspected of containing air, bleed the system. Use care; brake fluid will remove the paint from any surface. If the brake fluid spills onto any painted surface, wash it off immediately with soap and water.

1. Before servicing the vehicle, refer to the precautions section.
2. Fill the clutch reservoir with brake fluid. Check the reservoir level frequently and add fluid as needed.
3. Connect one end of a vinyl tube to the bleeder plug on the slave cylinder and submerge the other end into a clear container half-filled with brake fluid.
4. Slowly pump the clutch pedal several times.
5. Have an assistant hold the clutch pedal down and loosen the bleeder plug until fluid and/or air starts to run out of the bleeder plug. Close the bleeder plug while the pedal is held to the floor.

➡ Do not allow the pedal to rise back-up while the bleeder is still open. If this happens, it will allow air to re-enter the slave cylinder and cause the clutch system not to work properly.

6. Repeat Steps 2 and 3 until all the air bubbles are removed from the system.
7. Tighten the bleeder plug when all the air is gone.
8. Refill the master cylinder to the proper level as required.
9. Check the system for leaks.

Transfer Case

REMOVAL & INSTALLATION

1. Before servicing the vehicle, refer to the precautions section.
2. Drain the transfer case fluid.
3. Remove or disconnect the following:

- Negative battery cable. Due to the air bag system, wait at least 90 seconds before proceeding
- Engine and transaxle assembly
- Separate vane pump

- Steering gear
- Crossmember
- Manifold stay
- Oxygen (O_2) sensor
- Exhaust manifold heat shield
- Exhaust manifold
- Starter
- Right side halfshaft
- Transverse engine mounting bracket
- Center and right side transfer stiffener plates

✳ WARNING

When removing the transfer case, DO NOT touch the oil seal.

- Transfer case bolts, and transfer assembly, using a mallet to dislodge it from the transaxle
4. Installation is the reverse of the removal procedure, noting the following specifications:
   a. Transfer case stiffener case bolts: 25 ft. lbs. (34 Nm).
   b. Engine mounting bracket bolts: 47 ft. lbs. (64 Nm).
5. Add fluid to the transfer case, and check for leaks.

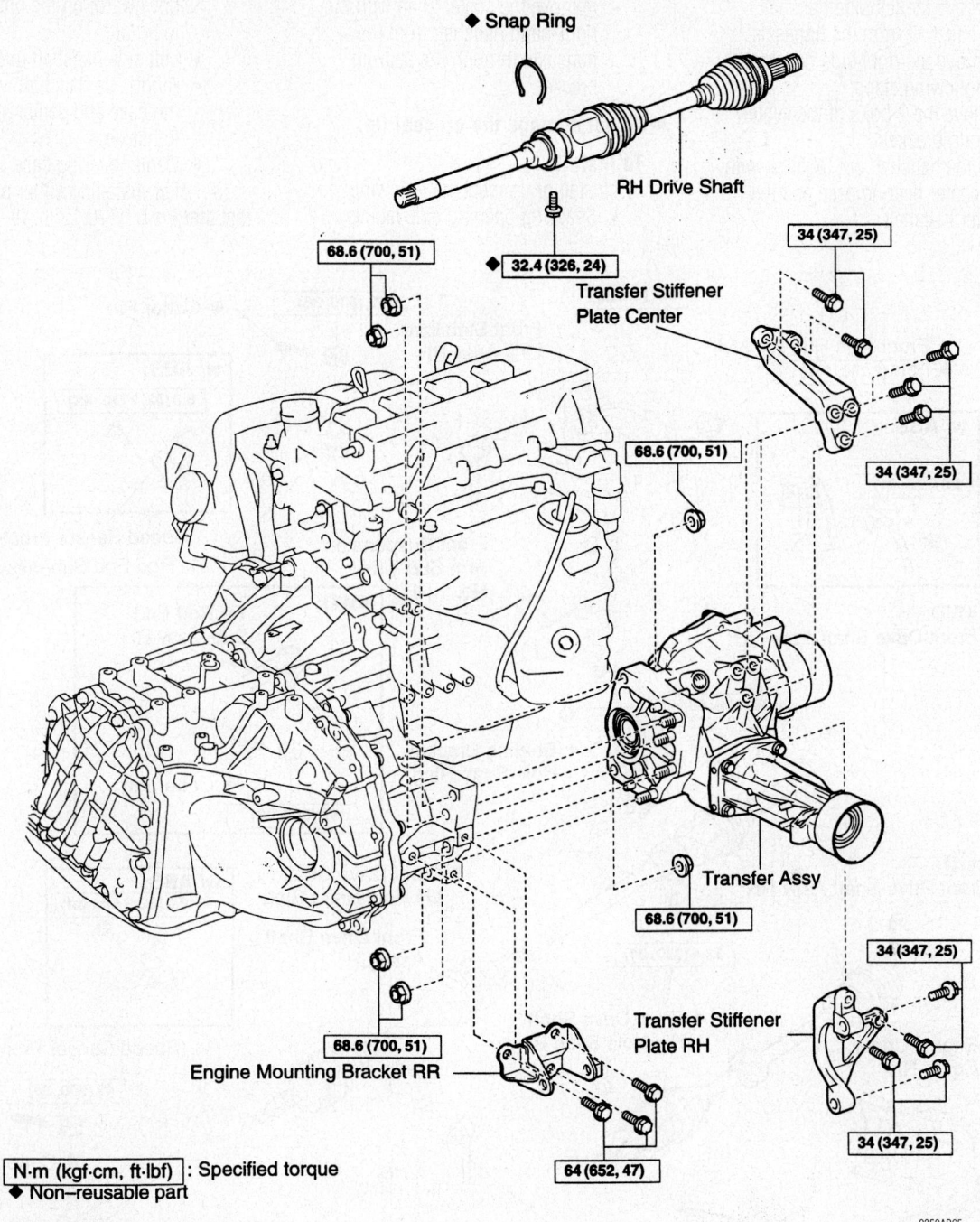

◆ Snap Ring

RH Drive Shaft

68.6 (700, 51)

◆ 32.4 (326, 24)

Transfer Stiffener Plate Center

34 (347, 25)

68.6 (700, 51)

34 (347, 25)

Transfer Assy

68.6 (700, 51)

34 (347, 25)

Transfer Stiffener Plate RH

68.6 (700, 51)

Engine Mounting Bracket RR

34 (347, 25)

64 (652, 47)

N·m (kgf·cm, ft·lbf) : Specified torque
P ◆ Non–reusable part

9359AB65

Exploded view of the transfer case mounting

Halfshaft

REMOVAL & INSTALLATION

➡ **The hub bearing could be damaged if subjected to the full weight of the vehicle, such as if the vehicle is moved without the halfshafts. If it is absolutely necessary to place the full vehicle weight on the hub bearing, first support the bearing with SST No. 09608–16041.**

1. Before servicing the vehicle, refer to the precautions section.
2. Drain the transaxle fluid.
3. Remove or disconnect the following:

- Negative battery cable. Due to the air bag system, wait at least 90 seconds before proceeding.
- Both front wheels
- Cotter pin, locknut cap, and the hub nut
- Undercovers

- Speed sensors
- Tie rod ball joint from the steering knuckle
- Stabilizer bar link from the lower suspension arm
- Lower ball joint from the lower suspension arm
- Halfshaft from the knuckle

➡ **Be careful not to damage the inner oil seal or the ABS sensor rotor on the halfshaft.**

4. To remove the left side halfshaft, separate the halfshaft from the transaxle.

5. To remove the right side halfshaft perform the following steps:

- Remove the 2 bolts of the center bearing bracket
- Pull the halfshaft out together with the center bearing case and the center halfshaft.

- Remove the center shaft with the right-hand halfshaft from the transaxle through the bearing bracket.

➡ **Do not damage the oil seal lip.**

To install:
6. Install or connect the following:
- Snapring opening side facing

downward, on the oiled inboard joint tulip
- Left side halfshaft into the transaxle
- Right side halfshaft, with the bearing case and center shaft, into the transaxle
- Center bearing case (right side).

7. After installing either halfshaft, check that there is 0.08–0.12 in. (2–3mm) of axial

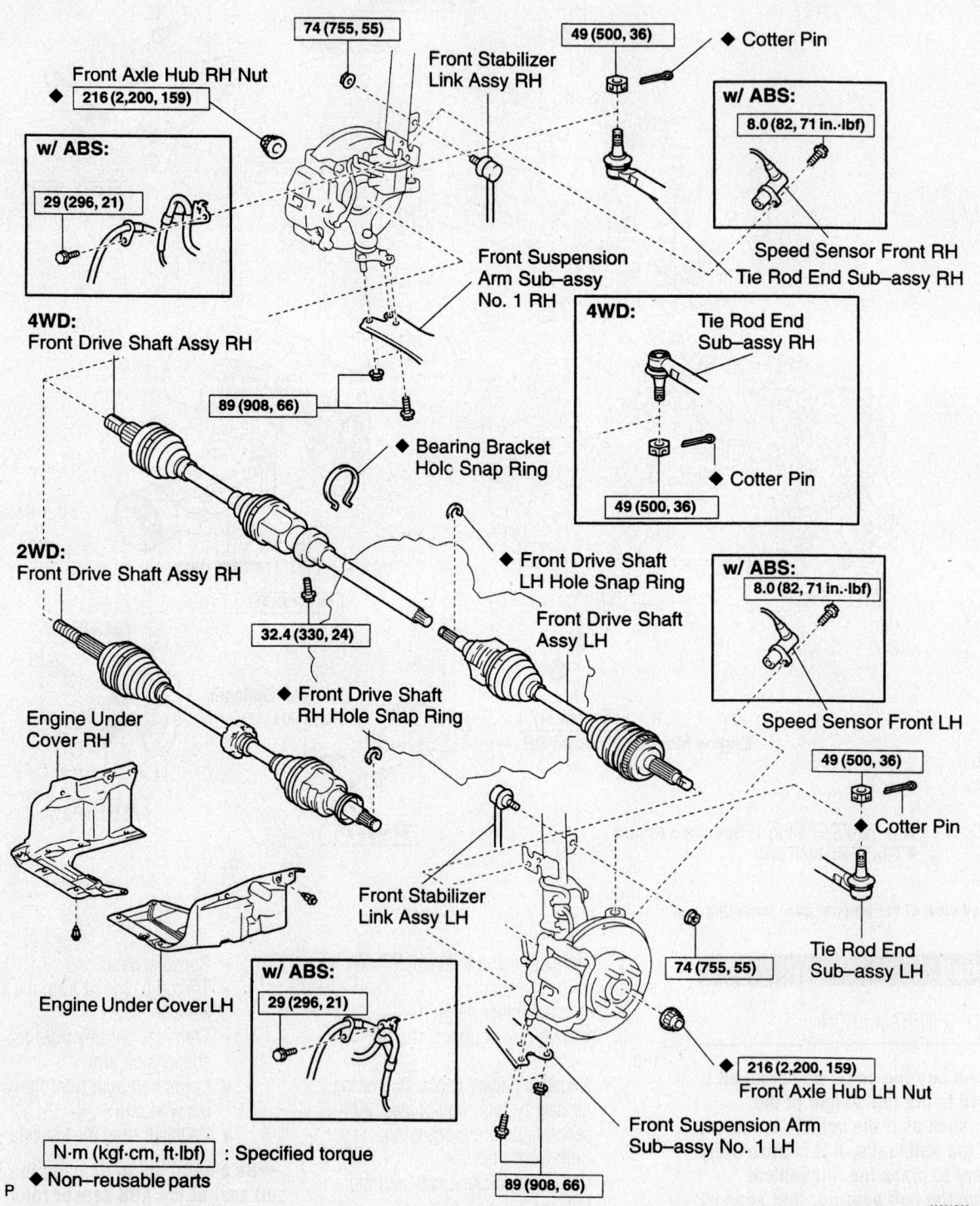

Front Axle Hub RH Nut
◆ 216 (2,200, 159)

74 (755, 55)

49 (500, 36)   ◆ Cotter Pin

w/ ABS:
8.0 (82, 71 in.·lbf)

Front Stabilizer Link Assy RH

w/ ABS:
29 (296, 21)

Front Suspension Arm Sub–assy No. 1 RH

Speed Sensor Front RH
Tie Rod End Sub–assy RH

4WD:
Front Drive Shaft Assy RH

4WD:
Tie Rod End Sub–assy RH

89 (908, 66)

◆ Bearing Bracket Holc Snap Ring

◆ Cotter Pin

49 (500, 36)

2WD:
Front Drive Shaft Assy RH

◆ Front Drive Shaft LH Hole Snap Ring

Front Drive Shaft Assy LH

w/ ABS:
8.0 (82, 71 in.·lbf)

32.4 (330, 24)

◆ Front Drive Shaft RH Hole Snap Ring

Engine Under Cover RH

Speed Sensor Front LH

49 (500, 36)

◆ Cotter Pin

Front Stabilizer Link Assy LH

74 (755, 55)

Tie Rod End Sub–assy LH

Engine Under Cover LH

w/ ABS:
29 (296, 21)

216 (2,200, 159)
Front Axle Hub LH Nut

Front Suspension Arm Sub–assy No. 1 LH

N·m (kgf·cm, ft·lbf) : Specified torque

P ◆ Non–reusable parts

89 (908, 66)

Halfshafts and related components

9359AB62

play. Check that the halfshaft is making contact with the pinion shaft and that the halfshaft cannot be pulled out.

8. Install or connect the following:
- Halfshaft into the knuckle
- Lower suspension arm to the lower ball joint. Torque the bolt and nuts to 66 ft. lbs. (89 Nm).
- Tie rod end to the steering knuckle. Tighten the nut to 36 ft. lbs. (49 Nm).
- Stabilizer bar link to the lower suspension arm. Torque the nuts to 55 ft. lbs. (74 Nm).
- Front wheels
- Hub nut and washer and tighten to 159 ft. lbs. (216 Nm)
- Negative battery cable

- Locknut cap and a new cotter pin.
- Speed sensors
- Undercover

9. Fill the transaxle fluid to the proper level

10. Start the vehicle, check for leaks and repair if necessary.

CV-Joints

OVERHAUL

1. Before servicing the vehicle, refer to the precautions section.
2. Remove or disconnect the following:
- Inboard joint boot clips
- Inboard joint tulip from the driveshaft

- Snapring
- Using a brass rod and hammer, the tri-pot joint off the driveshaft without hitting the joint roller
- Inboard joint boot
- Clamp and driveshaft damper
- Clamps and the outboard drive boot. DO NOT disassemble the outboard joint.

To assemble:

3. Install or connect the following:

➡**Before installing the boot, wrap the spline end of the shaft with masking tape to prevent damage to the boot.**

- Driveshaft damper with a new clamp

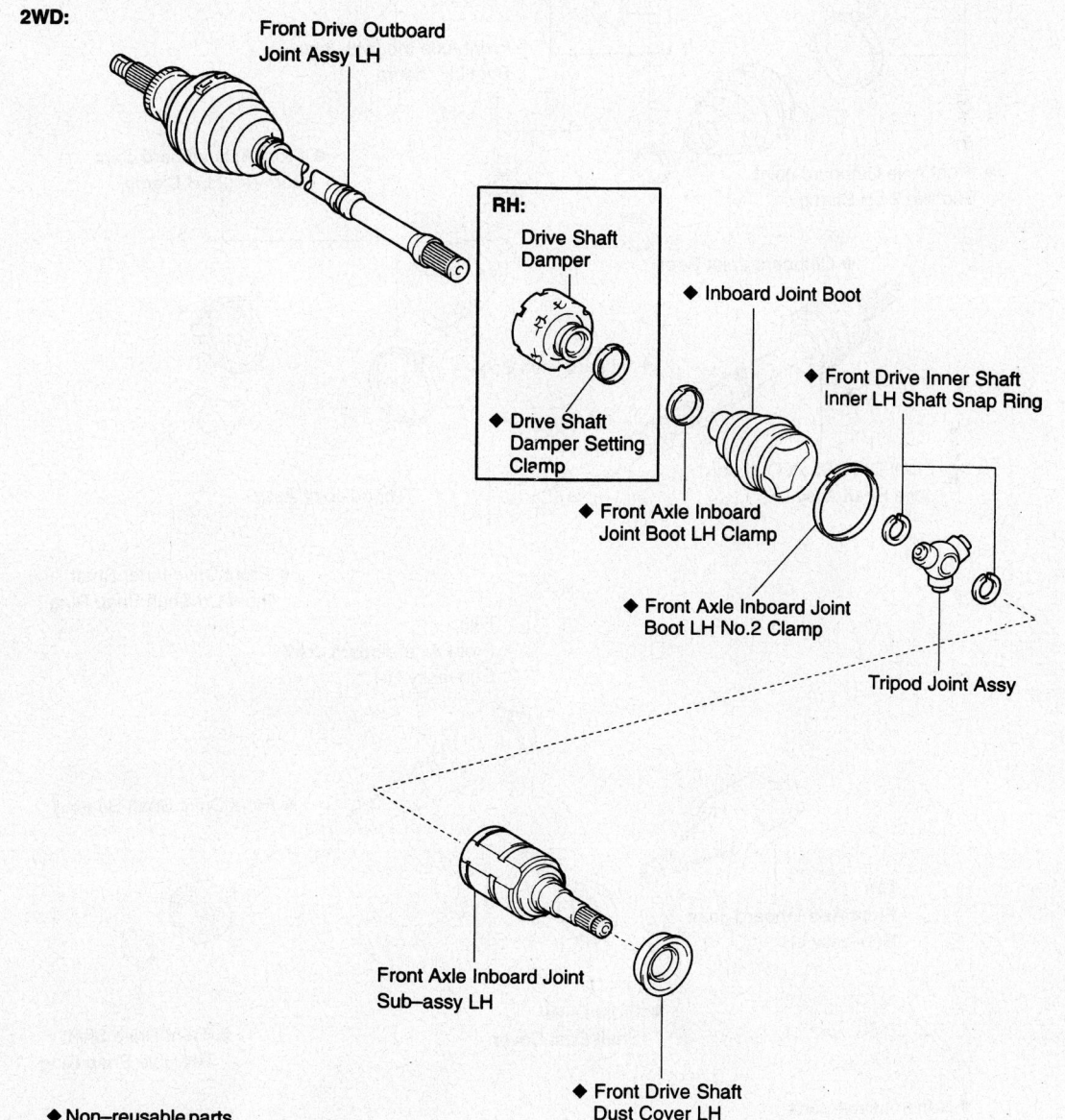

2WD:

Front Drive Outboard Joint Assy LH

RH:

Drive Shaft Damper

◆ Drive Shaft Damper Setting Clamp

◆ Inboard Joint Boot

◆ Front Drive Inner Shaft Inner LH Shaft Snap Ring

◆ Front Axle Inboard Joint Boot LH Clamp

◆ Front Axle Inboard Joint Boot LH No.2 Clamp

Tripod Joint Assy

Front Axle Inboard Joint Sub–assy LH

◆ Front Drive Shaft Dust Cover LH

◆ Non–reusable parts

P

Exploded view of the CV-joint—FWD vehicles

9359AB63

- Temporarily, the inboard boot with new clamp to the drive joint

➡The inboard boot and clamp are larger than those of the outboard boot.

- The tri-pot onto the driveshaft with a brass rod and hammer without hitting the joint roller
- The snapring

4. Pack the outboard tulip joint and the outboard boot with about 0.26–0.33 lbs. ounces of grease that was supplied with the boot kit.

5. Install or connect the following:
- Boot onto the outboard joint

6. Pack the inboard tulip joint and boot with ½ lb. of grease that was supplied with the boot kit.

- Inboard tulip joint onto the drive-shaft
- Boot onto the driveshaft

7. Before checking the standard length, bend the band and lock it. Make sure that the boot is not stretched or squashed when the driveshaft is at standard length. Standard driveshaft length: LH: 540.2 mm (21.268 in.); RH: 857.4 mm (33.756 in.)

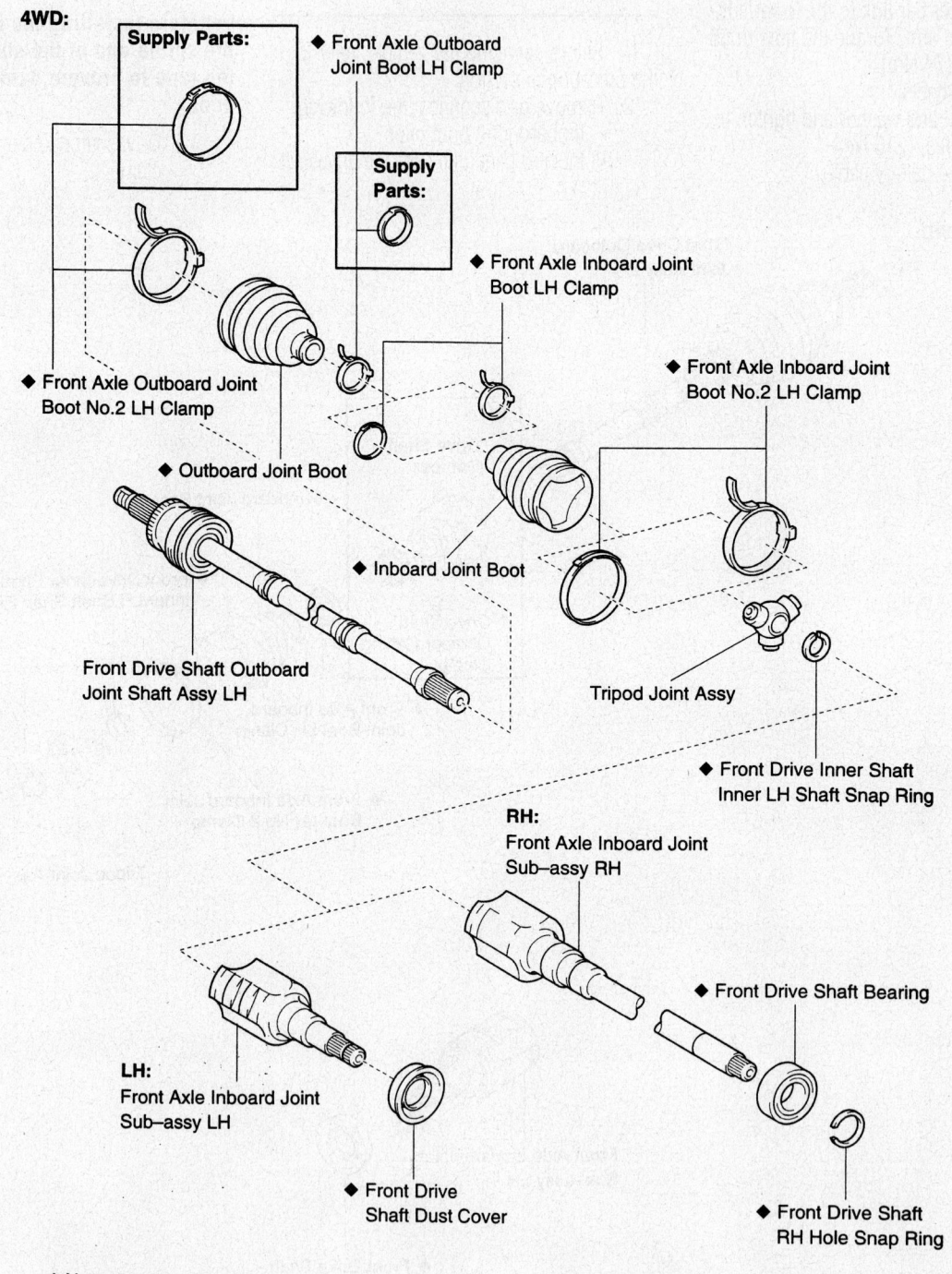

4WD:

Supply Parts:

◆ Front Axle Outboard Joint Boot LH Clamp

Supply Parts:

◆ Front Axle Inboard Joint Boot LH Clamp

◆ Front Axle Outboard Joint Boot No.2 LH Clamp

◆ Front Axle Inboard Joint Boot No.2 LH Clamp

◆ Outboard Joint Boot

◆ Inboard Joint Boot

Front Drive Shaft Outboard Joint Shaft Assy LH

Tripod Joint Assy

◆ Front Drive Inner Shaft Inner LH Shaft Snap Ring

RH:
Front Axle Inboard Joint Sub–assy RH

◆ Front Drive Shaft Bearing

LH:
Front Axle Inboard Joint Sub–assy LH

◆ Front Drive Shaft Dust Cover

◆ Front Drive Shaft RH Hole Snap Ring

◆ Non–reusable parts

P

9359AB64

Exploded view of the CV-joint—AWD vehicles

STEERING AND SUSPENSION

Air Bag

PRECAUTIONS

Several precautions must be observed when handling the inflator module to avoid accidental deployment and possible personal injury.

• Never carry the inflator module by the wires or connector on the underside of the module.

• When carrying a live inflator module, hold securely with both hands, and ensure that the bag and trim cover are pointed away.

• Place the inflator module on a bench or other surface with the bag and trim cover facing up.

• With the inflator module on the bench, never place anything on or close to the module that may be thrown in the event of an accidental deployment.

DISARMING

To avoid personal injury when working on vehicles equipped with an air bag, the negative battery cable must be disconnected and at least 90 seconds must elapse before working on the system. Failure to do so may result in deployment of the air bag.

REARMING

After vehicle service is completed, reattach the battery cables (positive cable first!) to rearm the air bag system.

Rack and Pinion Steering Gear

REMOVAL & INSTALLATION

2003–04 Models

1. Before servicing the vehicle, refer to the precautions section.
2. Position the front wheels straight ahead.
3. Remove or disconnect the following:
   • Negative battery cable. Because these vehicles are equipped with air bags, wait at least 90 seconds before proceeding.
   • Horn button
   • Steering wheel
   • Front wheels
   • Left and right engine undercovers
   • Left and right tie rod ends

• Column hose cover silencer sheet
• Steering intermediate shaft
• Pressure feed and return tubes
• Left and right side front stabilizer links
• Right and left front lower control arms from the ball joints
• Hood
• No. 2 cylinder head (valve) cover

4. Install an engine support and tension it to support the engine without raising it.
   a. No. 1 engine hanger: P/N 12281-22021 1ZZ-FE, 12281-88600 2ZZ-GE.
   b. No. 2 engine hanger: P/N 12281-15040 1ZZ-FE, 12281-88600 2ZZ-GE.
   c. Bolt: P/N 91512-B1016.
   d. Torque the bolts to 28 ft. lbs. (38 Nm).

> ❊❊ **CAUTION**
>
> Do not try to suspend the engine by hooking the chain to any other part.

   e. Attach an engine chain hoist to the hangers.

> ❊❊ **CAUTION**
>
> The engine hoist is now in place and under tension. Use care when repositioning the vehicle and make necessary adjustments to the engine support.

5. Remove or disconnect the following:
   • Bolt and nuts holding in the middle of the crossmember and support the crossmember with a jack
   • Bolts from the outer side of the suspension crossmember
   • Suspension crossmember with the steering gear assembly
   • Steering intermediate shaft, after matchmarking it
   • Rack and pinion steering gear from the crossmember

6. Installation is the reverse of the removal procedure, noting the following specifications:
   a. Steering gear bolts and nuts: 43 ft. lbs. (58 Nm) FWD, 60 ft. lbs. (82 Nm) AWD.

1ZZ-FE:

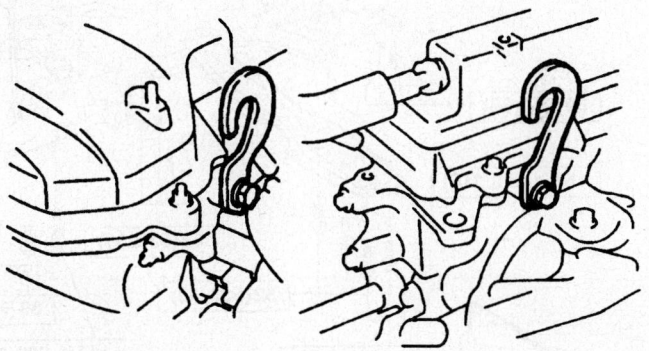

2ZZ-GE:

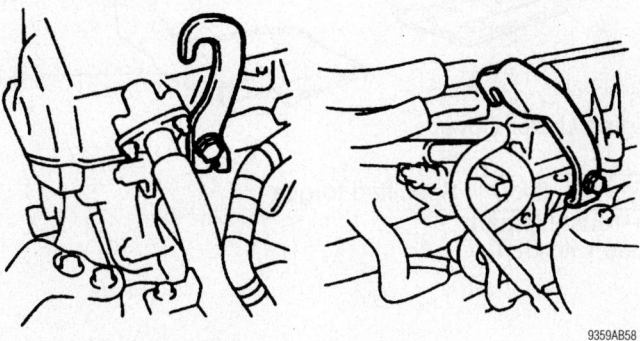

9359AB58

Proper installation of engine hangers

2WD:

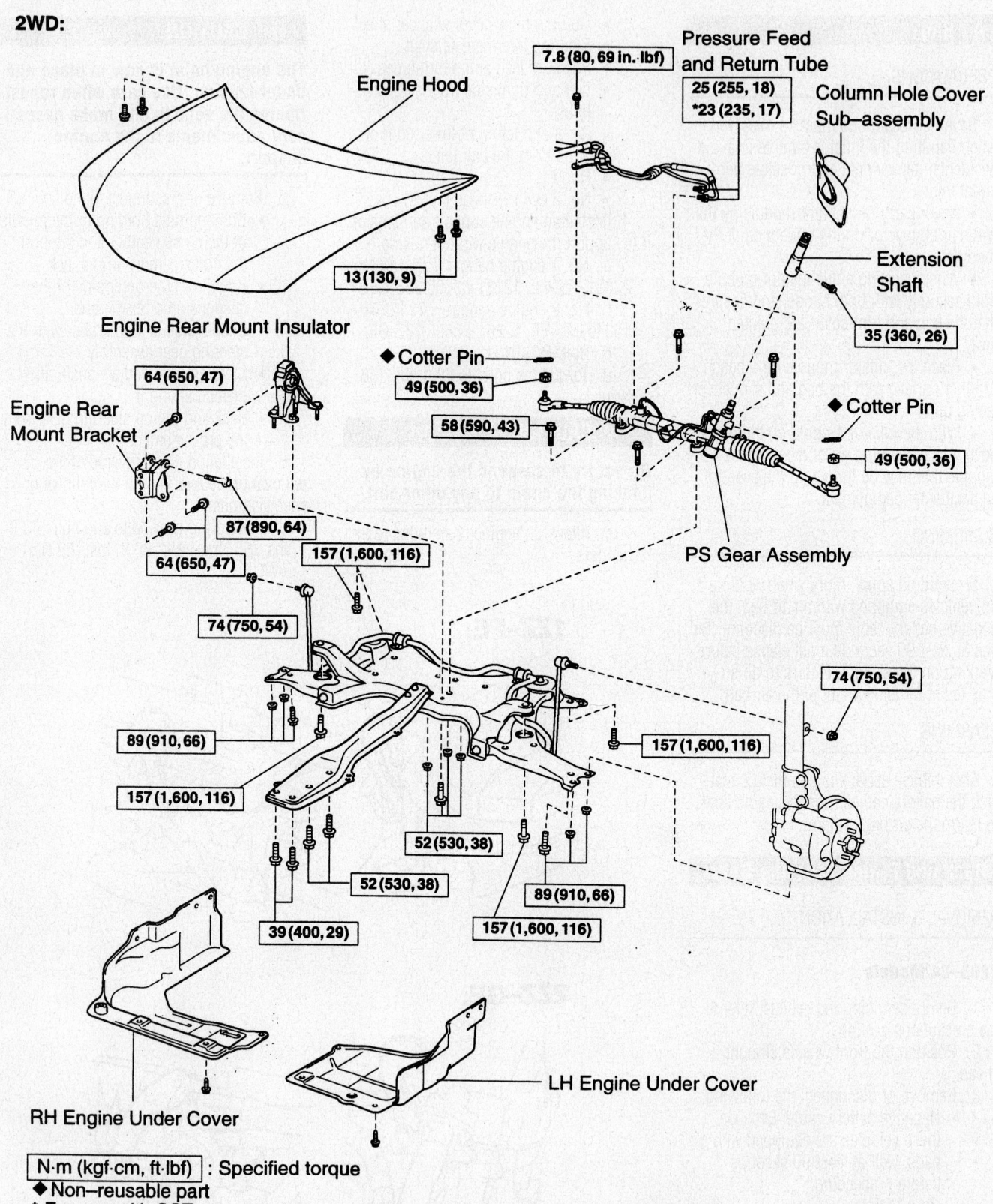

Engine Hood

7.8 (80, 69 in.·lbf)

Pressure Feed and Return Tube

25 (255, 18)
*23 (235, 17)

Column Hole Cover Sub–assembly

Extension Shaft

35 (360, 26)

Engine Rear Mount Insulator

64 (650, 47)

◆ Cotter Pin

49 (500, 36)

◆ Cotter Pin

49 (500, 36)

13 (130, 9)

Engine Rear Mount Bracket

58 (590, 43)

87 (890, 64)

64 (650, 47)

157 (1,600, 116)

PS Gear Assembly

74 (750, 54)

74 (750, 54)

89 (910, 66)

157 (1,600, 116)

157 (1,600, 116)

52 (530, 38)

52 (530, 38)

89 (910, 66)

39 (400, 29)

157 (1,600, 116)

RH Engine Under Cover

LH Engine Under Cover

N·m (kgf·cm, ft·lbf) : Specified torque
◆ Non–reusable part
* For use with SST

9359AB59

Exploded view of a typical power rack and pinion steering gear unit—FWD shown

b. Steering intermediate shaft: 26 ft. lbs. (35 Nm).

c. Suspension crossmember bolts: 116 ft. lbs. (157 Nm).

d. Engine mount insulator bolts: 38 ft. lbs. (52 Nm).

e. Center member-to-frame bolts: 29 ft. lbs. (39 Nm).

f. Stabilizer bar link-to-the lower control arms nuts: 55 ft. lbs. (74 Nm).

g. Fluid return and pressure tubes: 17 ft. lbs. (23 Nm).

h. Tie rod ends: 36 ft. lbs. (49 Nm).

i. Wheel lug nuts: 76 ft. lbs. (103 Nm).

7. Check and top off the power steering fluid.

8. Check and adjust the alignment, if needed.

2005–06 Models

1. Before servicing the vehicle, refer to the precautions section.

➡**The steering column must be in the LOCK position before disconnecting the following components:**

- Steering column
- Steering shaft coupling
- Intermediate shaft
- Lower steering shaft

➡**After disconnecting these components, do not move the front tires and wheels. Failure to follow these procedures may cause improper alignment of some components during installation and result in possible damage to the SIR coil.**

2. LOCK the steering column and verify the front wheels are in the straight ahead position.

3. Move the silencer pad away from the steering column.

4. Use paint in order to place match marks on the steering shaft coupling and on the intermediate shaft.

5. Loosen the upper coupling bolt.

6. Remove the lower coupling bolt.

7. Remove the steering column hole cover from the bulkhead.

8. Install the Engine Support Fixture.

9. Remove the front tire and wheel assemblies.

10. Remove the engine splash shields.

11. Remove the 2 outer tie rod ends.

12. Place a drain pan under the vehicle in order to collect the fluid from the power steering system.

13. Remove the pressure and return pipes from the steering gear.

14. Remove the bolt and the pipe bracket from the steering gear.

15. Remove the following components together as a unit:

- Steering gear
- Intermediate steering shaft
- Front suspension crossmember
- Trans support
- Control arms
- Front stabilizer shaft

16. Remove the bolt and the rear engine mount insulator from the crossmember.

17. Remove the 3 bolts and the rear engine mount bracket from the crossmember.

18. Use paint in order to place match marks on the intermediate shaft and on the steering gear.

19. Remove the bolt and the intermediate shaft (4).

20. Remove the 4 bolts and the steering gear from the crossmember.

To install:

21. Install the rear engine mount bracket to the crossmember.

22. Install the 3 bolts to the rear engine mount bracket. Tighten to 47 ft. lbs. (64 Nm).

23. Install the rear engine mount insulator to the crossmember.

24. Install the bolt to the rear engine mount insulator. Tighten to 64 ft. lbs. (87 Nm).

25. If you are replacing the steering gear or the intermediate shaft, copy the match marks from the old parts to the same locations on the new parts.

26. Install the steering gear and the 4 bolts to the crossmember. Tighten to 42 ft. lbs. (58 Nm).

27. Install the intermediate shaft to the steering gear. Align the match marks.

28. Install the bolt to the intermediate shaft. Tighten to 26 ft. lbs. (35 Nm).

29. Install the steering column hole cover to the bulkhead.

30. Install the following components as a unit:

- Steering gear
- Intermediate steering shaft
- Front suspension crossmember
- Trans support
- Control arms
- Front stabilizer shaft

31. Install the 2 outer tie rod ends.

32. Install the pressure and return pipes to the steering gear. Tighten the fittings to 17 ft. lbs. (23 Nm).

33. Install the pipe bracket bolt. Tighten to 69 inch lbs. (8 Nm).

34. Install the splash shields.

35. Install the front tire and wheel assemblies.

36. Remove the Engine Support Fixture.

37. Align the match marks on the intermediate shaft and on the steering shaft coupling.

38. Install the lower coupling bolt. Tighten to 26 ft. lbs. (35 Nm).

39. Tighten the upper coupling bolt. Tighten to 26 ft. lbs. (35 Nm).

40. Place the silencer pad into the correct position.

41. Fill the power steering fluid reservoir.

42. Bleed the power steering system.

43. Inspect the power steering system for leaks. Repair as necessary.

44. Measure the wheel alignment. Adjust as necessary

Strut and Coil Spring

REMOVAL & INSTALLATION

Front

1. Before servicing the vehicle, refer to the precautions section.

2. Remove or disconnect the following:
- Negative battery cable. Because of the air bag system, wait at least 90 seconds before proceeding

❊❊ WARNING

Do not support the weight of the vehicle on the suspension arm; the arm will deform under its weight.

- Wheel
- Stabilizer link from the strut
- Bolt, and disconnect the brake hose from the strut
- With ABS brakes, speed sensor wiring harness from the strut
- Lower strut bolts and nuts
- Upper strut nuts
- Strut from the steering knuckle
- Strut

3. To disassemble the strut:
- Install a bolt and 2 nuts to the bracket at the lower portion of the strut shell and secure it in a vise
- Compress the coil spring
- Dust cover and hold the spring seat so that it will not turn
- Nut on the top of the strut
- Suspension support, bearing, dust seal, spring seat, spring, insulators and bumper

To install:

4. To assemble the strut:
- Install the spring bumper to piston

5. Using a spring compressor, compress the spring.

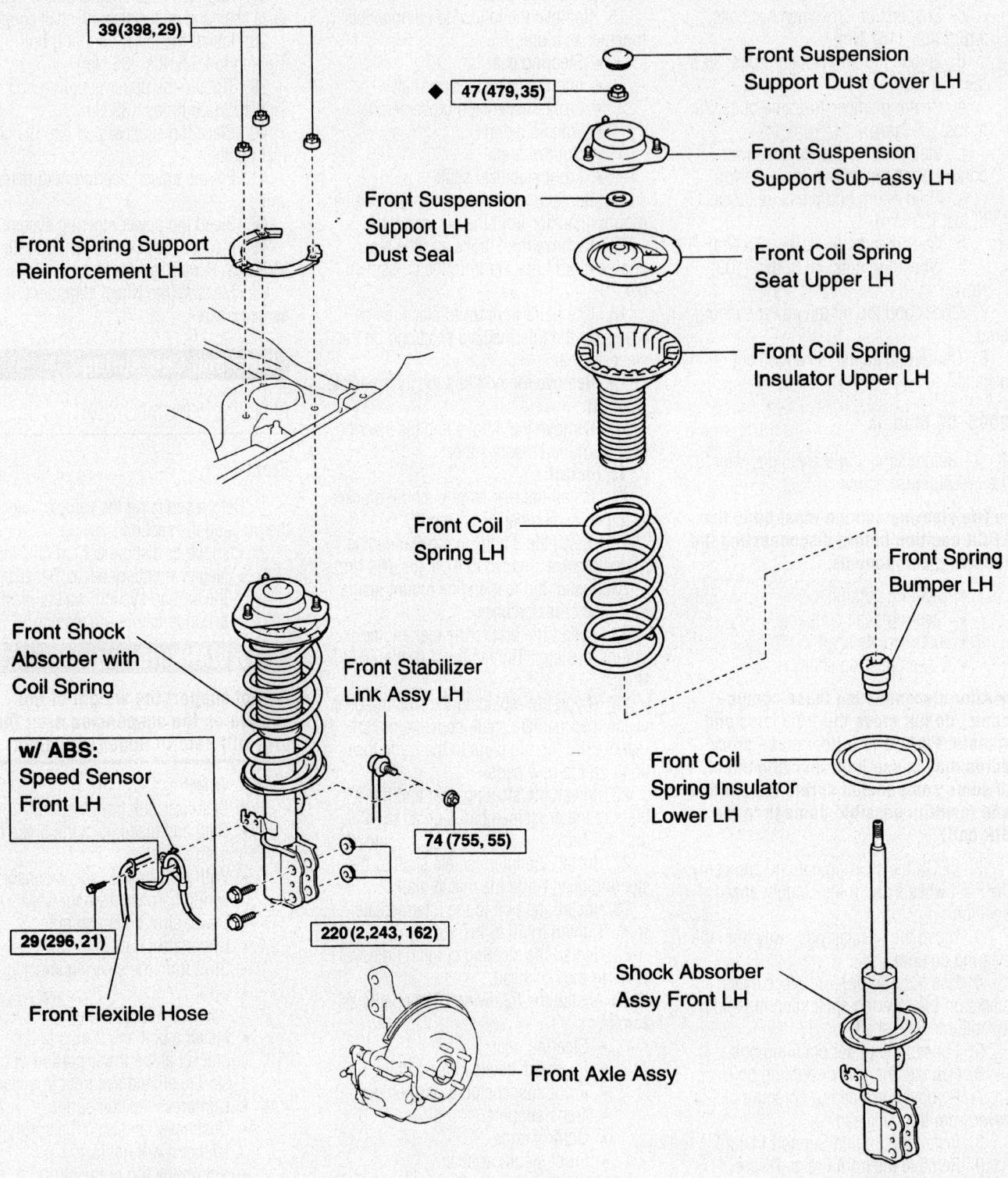

39 (398, 29)

◆ 47 (479, 35)

Front Suspension Support Dust Cover LH

Front Suspension Support Sub–assy LH

Front Suspension Support LH Dust Seal

Front Spring Support Reinforcement LH

Front Coil Spring Seat Upper LH

Front Coil Spring Insulator Upper LH

Front Coil Spring LH

Front Spring Bumper LH

Front Shock Absorber with Coil Spring

Front Stabilizer Link Assy LH

w/ ABS: Speed Sensor Front LH

Front Coil Spring Insulator Lower LH

74 (755, 55)

29 (296, 21)

220 (2,243, 162)

Front Flexible Hose

Shock Absorber Assy Front LH

Front Axle Assy

N·m (kgf·cm, ft·lbf) : Specified torque

◆ Non–reusable part

P

9359AB60

Common coil spring and strut component assembly

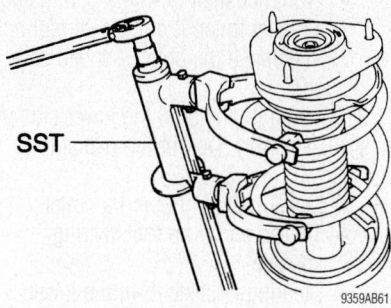

9359AB61

Proper method of supporting the strut in a vise

- Coil spring to the strut. Fit the lower end of the coil spring into the gap of the lower seat.
- Spring seat with the insulator
- Dust seal on the spring seat
- Suspension support and tighten 35 ft. lbs. (47 Nm). After the nut has been tighten, release the compressor tool tension.

6. Pack multipurpose grease into the suspension support.
- Dust cover.

➡**Do not use an impact wrench to tighten the nut. Also, check that the bearing fits into the recess in the suspension support.**

- Strut
- Nuts holding the strut to the strut tower. Tighten the nuts to 29 ft. lbs. (39 Nm).
- 2 lower strut bolts and nuts. Tighten to 162 ft. lbs. (220 Nm).
- Brake line to the steering knuckle. Tighten the line bolt to 21 ft. lbs. (29 Nm).
- Secure the wiring harness, if equipped with ABS
- Stabilizer link. Tighten the nut to 55 ft. lbs. (74 Nm).
- Wheel. Tighten the lug nuts to 76 ft. lbs. (103 Nm).
- Negative battery cable

7. Check and adjust the alignment, if needed.

Rear

1. Before servicing the vehicle, refer to the precautions section.
2. Remove or disconnect the following:
- Negative battery cable. Because of the air bag system, wait at least 90 seconds before proceeding.
- Rear wheel
- Rear deck board, luggage compartment tray and any trim necessary to access the strut towers
- Shock absorber head cover

3. On AWD vehicles, separate the rear stabilizer link.
4. For FWD vehicles:
   a. Support the axle beam with a jack.
   b. Remove the strut tower nuts and bolt.
   c. Remove the lower strut nut, cushion retainer and strut .
5. For AWD vehicles:
   a. Support the rear control arm.
   b. Remove the bolt and nut from the rear control arm.
   c. Remove the strut tower nuts.
   d. Remove the 3 rear control arm bolts.
   e. Press the rear control arm down to the outside of the vehicle, then remove the strut.
6. To disassemble the strut:
   a. Place the strut assembly in a pipe vise or strut vise.

❋❋ WARNING

Do not attempt to clamp the strut assembly in a flat jaw vise as this will result in damage to the strut tube.

   b. Compress the spring until the upper suspension support is free of any spring tension. Do not over-compress the spring.
   c. Hold the upper support, then remove the nut on the end of the shock piston rod.
   d. Remove the support, coil spring, insulator, and bumper.
7. Inspect the strut as follows:
   a. Check the shock absorber by moving the piston shaft through its full range of travel. It should move smoothly and evenly throughout its entire travel without any trace of binding or notching.
   b. Use a small straightedge to check the piston shaft for any bending or deformation.
   c. Inspect the spring for any sign of deterioration or cracking. The waterproof coating on the coils should be intact to prevent rusting.

To install:

➡**Never reuse a self-locking nut. Always replace self-locking nuts and cotter pins as applicable.**

8. Assemble the strut as follows:
   a. Loosely assemble all components onto the strut assembly. Be sure the spring end aligns with the hollow in the lower seat.
   b. Align the upper suspension support with the piston rod and install the support.
   c. Align the suspension support with the strut lower bracket. This assures the spring will be properly seated top and bottom.
   d. Compress the spring to expose the strut piston rod threads.
   e. Install a new strut piston nut and tighten to 41 ft. lbs. (56 Nm).
   f. Remove the spring compressor. Be sure the paint mark on the upper support faces the outside of the strut.
9. Install or connect the following:
- Strut on the vehicle. Tighten the strut-to-strut tower nuts to 59 ft. lbs. (80 Nm).
- Strut to the axle carrier and install the nut and cushion retainer/bolt snug. Do not fully tighten at this time.
- Strut head cover
- Rear control arm (AWD). Tighten the bolts to 48 ft. lbs. (65 Nm).
- Rear stabilizer link (AWD)
- Trunk tray, deckboard and any other trim pieces removed
- Wheel

10. With the vehicle's weight on the suspension, tighten the bolt holding the strut to the axle carrier to 59 ft. lbs. (80 Nm) for FWD vehicles, or 103 ft. lbs. (140 Nm) for AWD vehicles.
- Negative battery cable

11. Check and adjust the rear wheel alignment.

Lower Ball Joint

REMOVAL & INSTALLATION

1. Before servicing the vehicle, refer to the precautions section.
2. Remove or disconnect the following:
- Negative battery cable. Wait at least 90 seconds before proceeding.
- Front wheel

3. Depress the brake pedal and loosen the hub nut
- ABS speed sensor, if equipped
- Cotter pin and nut from the tie rod end. Using a tie rod end removal tool, separate the tie rod end from the steering knuckle.
- Lower control arm ball joint, using a suitable puller
- Separate the front halfshaft

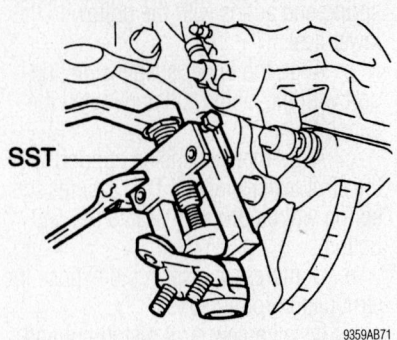

Removing the ball joint from the knuckle

- Lower ball joint cotter pin and castle nut
- Lower ball joint from the steering knuckle using a puller

To install:
4. Install or connect the following:
- Lower ball joint to the lower arm. Tighten the castle nut to 76 ft. lbs. (103 Nm).
- New cotter pin
- Front halfshaft
- Lower control arm
- Tie rod end to the knuckle
- ABS speed sensor
- Hub nut
- Wheel
- Negative battery cable
5. Check and adjust the alignment, if needed.

Upper Control Arm

REMOVAL & INSTALLATION

Rear—AWD Only

1. Before servicing the vehicle, refer to the precautions section.
2. Remove or disconnect the following:
- Negative battery cable. Wait at least 90 seconds before proceeding
- Rear wheel
- Exhaust pipe
- Propeller shaft with center bearing shaft
- Rear stabilizer links
- Rear hub nuts
- Rear brake drum
- Speed sensor
- Front brake shoe
- Parking brake shoe strut set
- Rear brake shoe
- Parking brake cables
- Rear brake hoses
- Separate the rear suspension arms
- Separate the upper control arm
- Rear drive axle assembly

- Rear strut nut and bolt
- Rear strut
- Rear suspension arm
- Rear suspension member
- Upper control arm assembly. Matchmark the camber adjust cams and rear suspension member prior to removal.
3. Installation is the reverse of the removal procedure.

Lower Control Arm

REMOVAL & INSTALLATION

1. Before servicing the vehicle, refer to the precautions section.
2. Remove or disconnect the following:
- Negative battery cable. Wait at least 90 seconds before proceeding..
- Front wheel
- Stabilizer link
- Bolt and nuts and separate the lower control arm from the lower ball joint
- Bolts and nuts, then separate the steering gear. Loosen the bolt, since the nut cannot be rotated, then suspend the steering gear.
3. Support the engine, using the engine lifting hooks and the procedure under Engine Removal & Installation.
- Crossmember
- Lower control arm from the crossmember
4. Installation is the reverse of the removal procedure.

Wheel Bearings

REMOVAL & INSTALLATION

Front

1. Before servicing the vehicle, refer to the precautions section.
2. Remove or disconnect the following:
- Negative battery cable. On vehicles equipped with an air bag, wait at least 90 seconds before proceeding.
- Wheels
- Hub nut
- Front stabilizer link
- Anti-lock Brake System (ABS) speed sensor
- Brake caliper
- Rotor
- Tie rod end from the steering knuckle
- Lower control arm ball joint

- Front halfshaft from the hub, using a mallet to tap it out. Be careful not to damage the boot or speed sensor.
3. Loosen the nuts on the lower side of the strut assembly. Do not remove at this time.
- Lower ball joint using a puller
- Tie rod end from the steering knuckle
- Steering knuckle from the lower control arm
- Knuckle from the strut assembly
- Hub

➡**Cover the halfshaft boot with a shop rag to protect it from any damage.**

4. Clamp the steering knuckle in a vise and remove the dust deflector. Remove the nut holding the steering knuckle to the ball joint. Press the ball joint out of the steering knuckle.
5. Remove the inner axle seal.
6. Using a Torx® wrench, remove the bolts securing the dust cover.
7. Using hub puller, remove the hub and backing plate from the steering knuckle.
8. Using a proper sized driver and a press, remove the inner hub race from the axle hub.
9. Using seal removal tool, remove the outer axle seal.
10. Using snapring pliers, remove the snapring from the inner side of the steering knuckle.
11. Using a proper sized driver and a press, remove the bearing from the steering knuckle. The bearing is pressed from the front of the steering knuckle and is removed through the back of the steering knuckle.

To install:
12. Perform the following:
13. Using a proper sized driver and a press, install a new bearing to the steering knuckle.
14. Install the snapring to the steering knuckle using snapring pliers.
15. Using a seal driver and a hammer, install a new outer oil seal. Apply multipurpose grease to the oil seal lip.
16. Place the dust cover on the steering knuckle. Tighten the bolts: 78 inch lbs. (9 Nm).
17. Using a press and a proper sized driver, install the axle hub to the steering knuckle.
18. Attach the ball joint to the steering knuckle. Install a new cotter pin.
19. Using a seal driver and a hammer, install a new inner oil seal. Apply multipurpose grease to the oil seal lip.

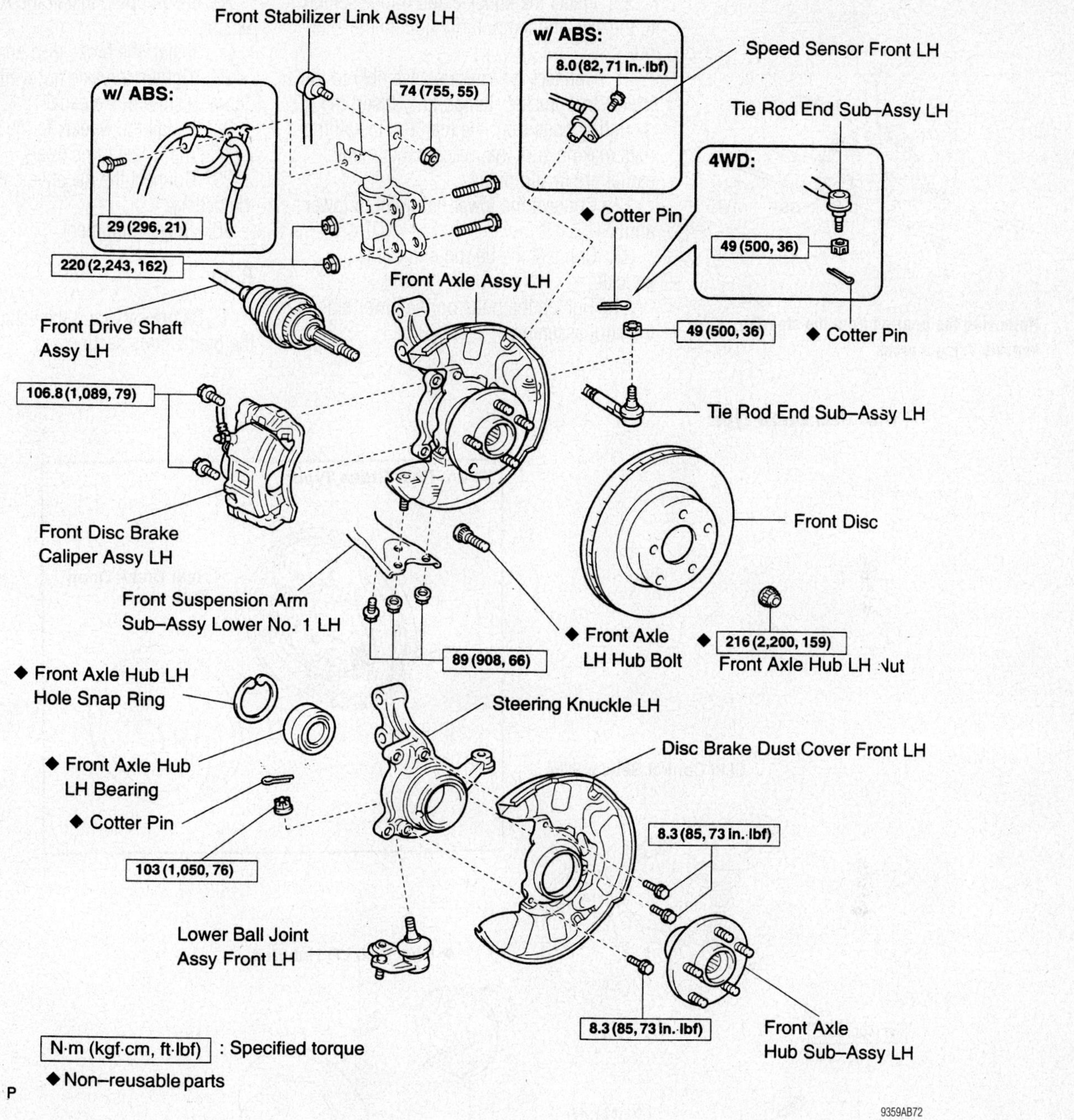

Front Stabilizer Link Assy LH

w/ ABS: 8.0 (82, 71 in.·lbf)

Speed Sensor Front LH

74 (755, 55)

Tie Rod End Sub–Assy LH

w/ ABS:

29 (296, 21)

4WD:

49 (500, 36)

♦ Cotter Pin

220 (2,243, 162)

Front Axle Assy LH

49 (500, 36)

♦ Cotter Pin

Front Drive Shaft Assy LH

106.8 (1,089, 79)

Tie Rod End Sub–Assy LH

Front Disc

Front Disc Brake Caliper Assy LH

Front Suspension Arm Sub–Assy Lower No. 1 LH

89 (908, 66)

♦ Front Axle LH Hub Bolt

216 (2,200, 159)
Front Axle Hub LH Nut

♦ Front Axle Hub LH Hole Snap Ring

Steering Knuckle LH

Disc Brake Dust Cover Front LH

♦ Front Axle Hub LH Bearing

♦ Cotter Pin

103 (1,050, 76)

8.3 (85, 73 in.·lbf)

Lower Ball Joint Assy Front LH

8.3 (85, 73 in.·lbf)

Front Axle Hub Sub–Assy LH

N·m (kgf·cm, ft·lbf) : Specified torque

P

♦ Non–reusable parts

9359AB72

Exploded view of the front hub and bearing, and related components

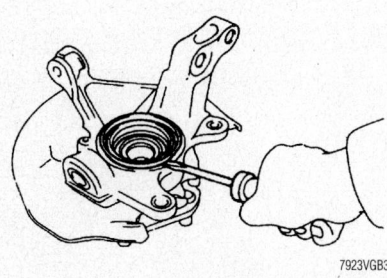

7923VGB3

Removing the inner axle seal from the hub assembly

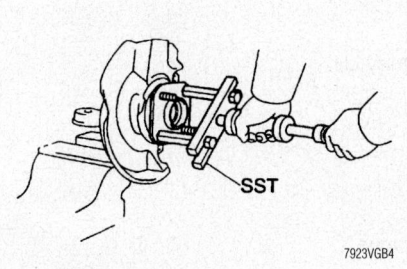

SST

7923VGB4

Removing the axle hub from the knuckle

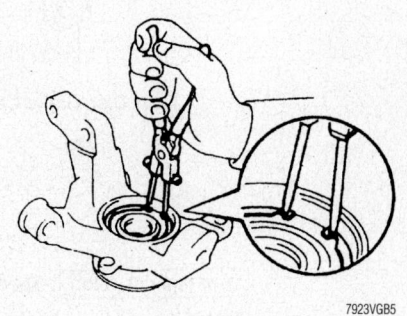

7923VGB5

Removing the snapring from the knuckle before pressing out the bearing

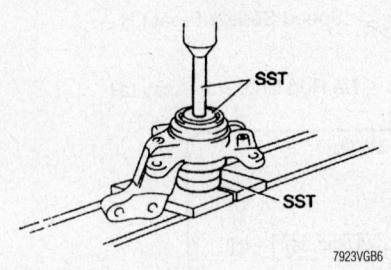

Removing the bearing from the steering knuckle using a press

20. Install the knuckle and hub assembly to the axle and temporarily tighten the axle nut.

21. Connect the knuckle assembly to the lower strut bracket. Temporarily insert the mounting bolts from the rear and install the nuts making sure the matchmarks made earlier are in alignment.

22. Connect the lower ball joint to lower arm.

23. Connect the tie rod end to the knuckle.

24. Tighten the bolts on the lower side of the strut assembly.

25. If equipped, install the ABS speed sensor.

26. Install the brake disc and the caliper.

27. Tighten the axle nut while someone depresses the brake pedal.

28. Install the wheels to the vehicle. Verify that the wheel turns freely.

29. Connect the negative battery cable to the battery.

30. Check alignment.

Rear

1. Before servicing the vehicle, refer to the precautions section.

Disc Rear Brake Type:

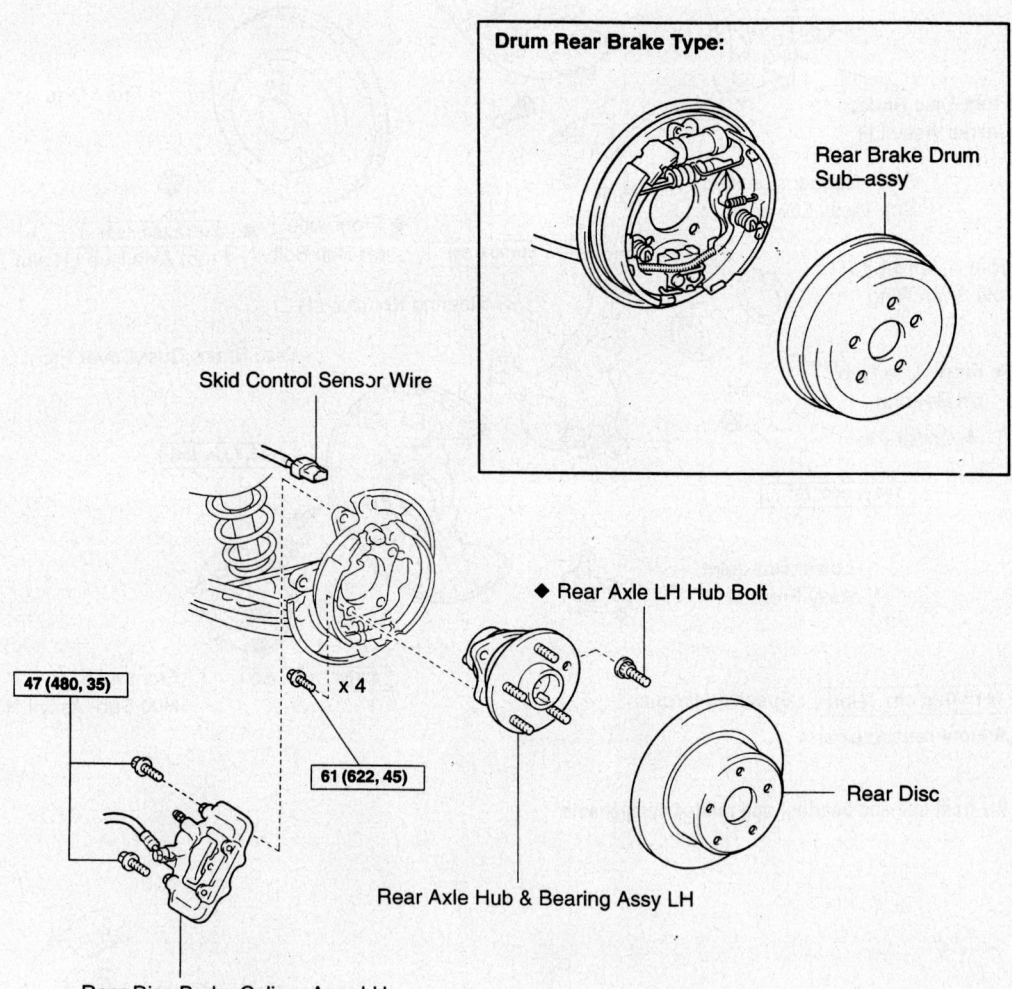

| N·m (kgf·cm, ft·lbf) | : Specified torque

◆ Non–reusable part

Exploded view of the hub and wheel bearing assembly

2. Remove or disconnect the following:
- Negative battery cable. On vehicles equipped with an air bag, wait at least 90 seconds before proceeding.
- Wheel
- Brake drum or rotor
- With ABS brakes, ABS wheel speed

sensor or skid control sensor, as applicable
- 4 hub retaining bolts
- Hub

To install:

3. Install or connect the following:
- Hub to the knuckle. Tighten the bolts to 45 ft. lbs. (61 Nm).

- ABS wheel speed or skid control sensor, if equipped
- Brake drum or rotor
- Wheel
- Negative battery cable

4. Check and adjust the alignment, if needed.

BRAKES

Brake Caliper

REMOVAL AND INSTALLATION

Front

1. Before servicing the vehicle, refer to the precautions section.

2. Remove some fluid from the reservoir with a suction pump.
3. Remove or disconnect the following:
- Front wheels
- Banjo bolt and disconnect the brake hose from the caliper. Plug the hose to prevent fluid loss and contamination.

- Mounting bolts while holding the slide pin
- Caliper

To Install:

4. Compress the caliper piston using a C–clamp or other suitable tool.
5. Install or connect the following:
- Caliper

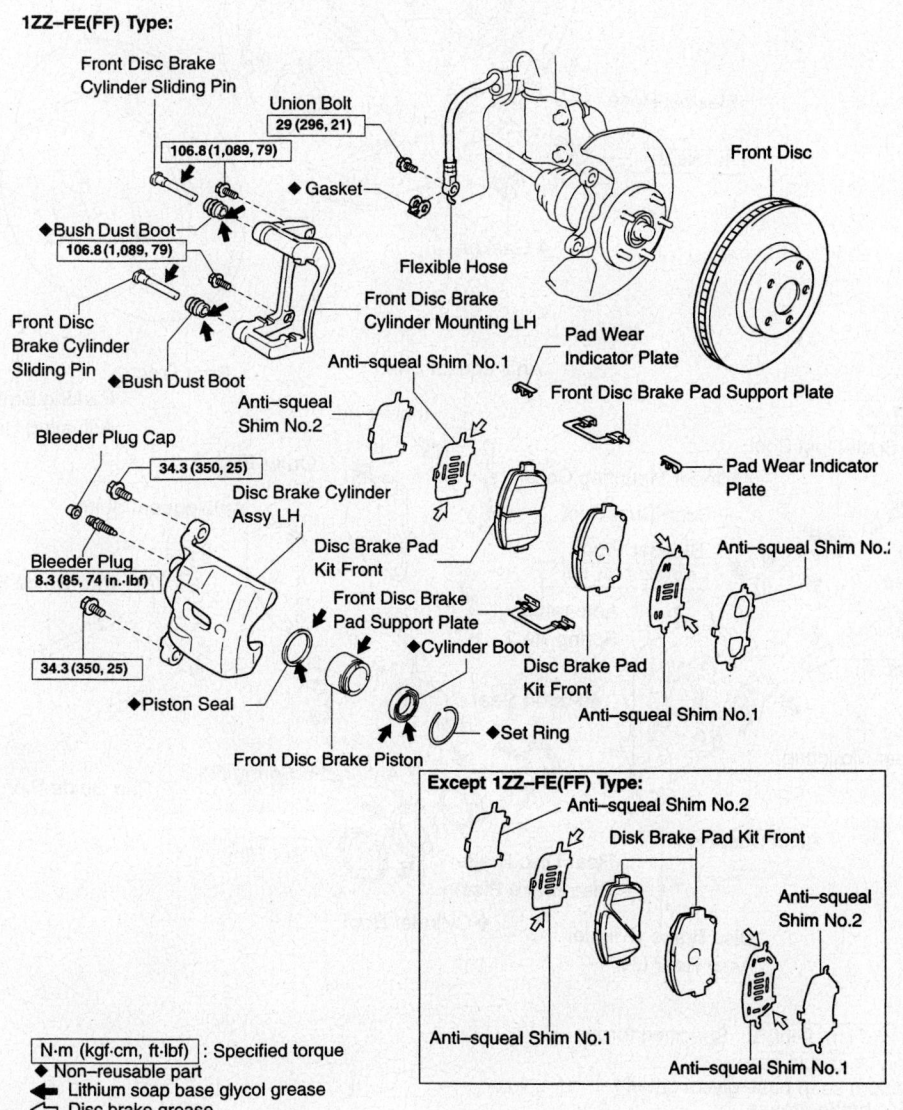

Exploded view of the front caliper components

- Mounting bolts and tighten to 25 ft. lbs. (34 Nm)
- Brake hose to the caliper using new sealing washers. Carefully torque the banjo bolt to 21 ft. lbs. (29 Nm).

6. Fill the reservoir with fluid and bleed the brakes.
- Front wheels

Rear

1. Before servicing the vehicle, refer to the precautions section.

2. Remove some fluid from the reservoir with a suction pump.

3. Remove or disconnect the following:

- Rear wheels
- Clip and both anti-rattle springs
- Two pad guide pins
- Pads with the shims
- Banjo bolt and disconnect the brake hose from the caliper. Plug the hose to prevent fluid loss and contamination.
- 2 caliper mounting bolts and the caliper from its mounting bracket

To Install:

4. Compress the caliper piston using a C-clamp or other suitable tool.

5. Install or connect the following:
- Caliper. Tighten the caliper bolts to 34 ft. lbs. (46 Nm).
- Brake hose with new sealing washers. Tighten the banjo bolt to 21 ft. lbs. (29 Nm).
- New anti-squeal shims, apply disc brake grease to the inside of the shim before installation
- Inner pad with the wear indicator facing upwards
- Outer pad
- Two pad guide pins
- Anti-rattle springs and the clip

6. Fill the reservoir with fluid and bleed

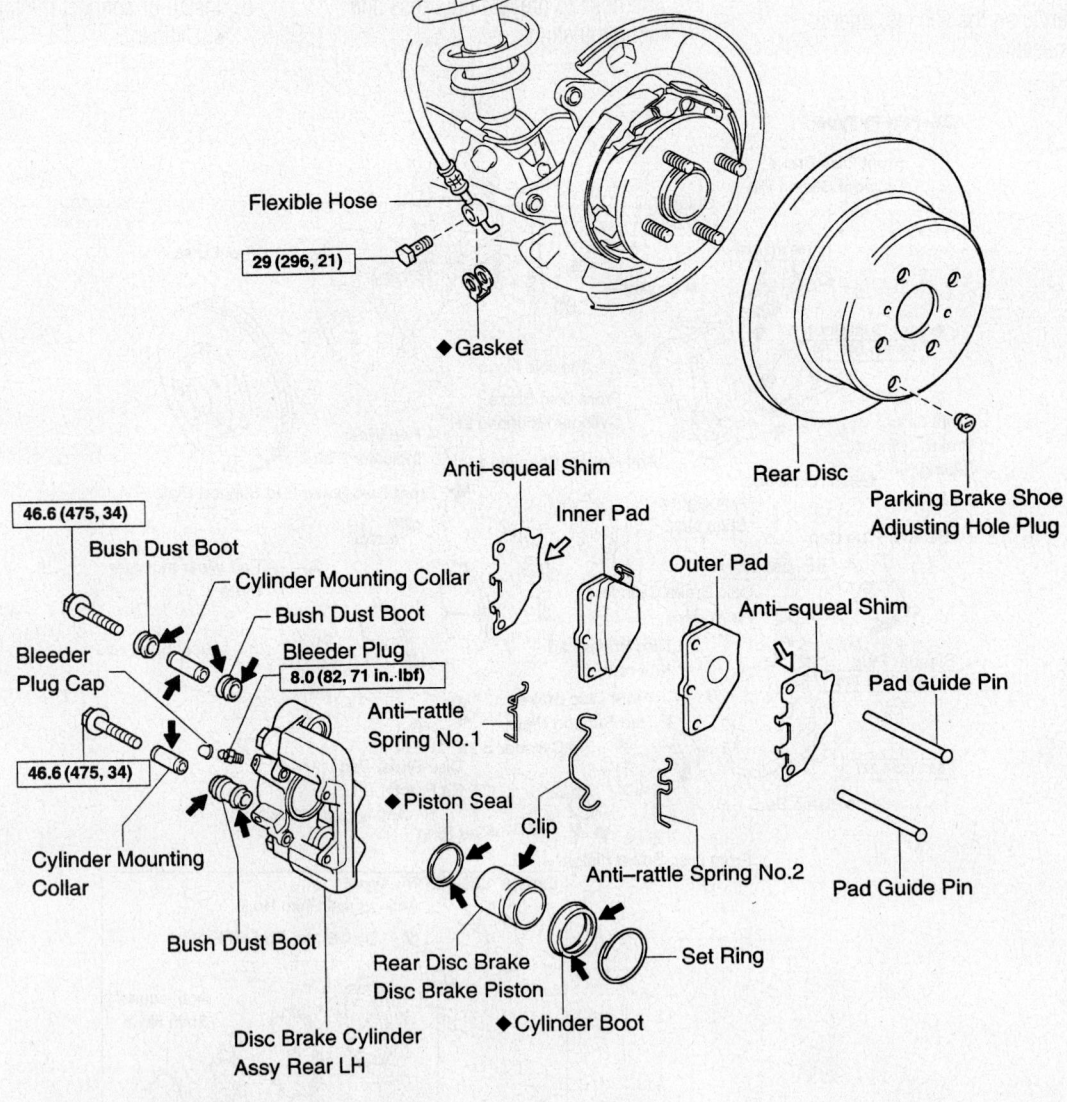

Flexible Hose

29 (296, 21)

◆Gasket

Anti-squeal Shim

Inner Pad

Rear Disc

Parking Brake Shoe Adjusting Hole Plug

Outer Pad

Anti-squeal Shim

46.6 (475, 34)

Bush Dust Boot

Cylinder Mounting Collar

Bush Dust Boot

Bleeder Plug

8.0 (82, 71 in.·lbf)

Bleeder Plug Cap

Anti-rattle Spring No.1

Pad Guide Pin

46.6 (475, 34)

◆Piston Seal

Clip

Anti-rattle Spring No.2

Pad Guide Pin

Cylinder Mounting Collar

Bush Dust Boot

Rear Disc Brake Disc Brake Piston

◆Cylinder Boot

Set Ring

Disc Brake Cylinder Assy Rear LH

N·m (kgf·cm, ft·lbf) : Specified torque
◆ Non-reusable part
◀ Lithium soap base glycol grease
◁ Disc brake grease

42356-TMAT-G02

Exploded view of the rear caliper components

the brake system. Adjust the parking brake if necessary.

- Rear wheels

Disc Brake Pads

REMOVAL AND INSTALLATION

Front

1. Before servicing the vehicle, refer to the precautions section.
2. Remove some fluid from the reservoir with a suction pump.
3. Remove or disconnect the following:
   - Front wheels
   - Mounting bolts while holding the slide pin
   - Caliper
   - Pads and shims
   - Both anti-squeal shims from the pads
   - Wear indicator plates from the pads

To Install:
4. Compress the caliper piston using a C–clamp or other suitable tool.
5. Install or connect the following:
   - Wear indicator plates
   - Both anti-squeal shims
   - Pads and shims

- Caliper
- Mounting bolts and tighten to 25 ft. lbs. (34 Nm)
6. Fill the reservoir with fluid and bleed the brakes, if necessary.
   - Front wheels

Rear

1. Before servicing the vehicle, refer to the precautions section.
2. Remove some fluid from the reservoir with a suction pump.
3. Remove or disconnect the following:
   - Rear wheels
   - Clip and both anti-rattle springs
   - Two pad guide pins
   - Pads with the shims

To Install:
4. Compress the caliper piston using a C–clamp or other suitable tool.
5. Install or connect the following:
   - New anti-squeal shims, apply disc brake grease to the inside of the shim before installation
   - Inner pad with the wear indicator facing upwards
   - Outer pad
   - Two pad guide pins
   - Anti-rattle springs and the clip
6. Fill the reservoir with fluid and bleed

the brake system. Adjust the parking brake if necessary.

- Rear wheels

Brake Drums

REMOVAL AND INSTALLATION

1. Before servicing the vehicle, refer to the precautions section.
2. Remove or disconnect the following:
   - Wheel
   - Brake drum. If the drum will not pull of the axle, back off the automatic adjuster by turning the adjusting wheel.

To install:
3. Install or connect the following:
   - Drum on the axle
   - Wheel
4. Refill the master cylinder and pump pedal to attain full brake pedal before road-testing the vehicle.

Brake Shoes

REMOVAL AND INSTALLATION

FWD Models

1. Before servicing the vehicle, refer to the precautions section.
2. Remove or disconnect the following:
   - Wheel
   - Brake drum. If the drum will not pull of the axle, back off the automatic adjuster by turning the adjusting wheel.
   - Upper side tension spring with the spacer
   - Anchor side spring using needle nosed pliers
   - Hold-down springs and pins from the front shoe
   - Upper side return spring from the front shoe
   - Front shoe
   - Left hand parking brake shoe strut set from the front shoe
   - Automatic adjuster lever from the front shoe
   - Hold-down springs and pins from the rear shoe
   - Upper side return spring from the rear shoe
   - Parking brake cable from the rear shoe using needle nosed pliers
   - Rear brake shoe
   - C–washer using a suitable pry tool from the shoe
   - Parking brake lever from the shoe

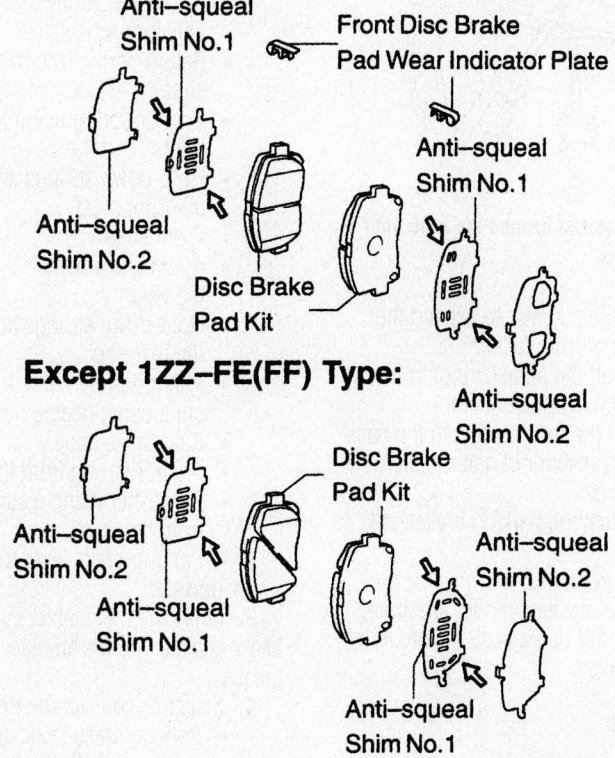

1ZZ–FE(FF) Type:

Anti–squeal Shim No.1

Front Disc Brake Pad Wear Indicator Plate

Anti–squeal Shim No.1

Anti–squeal Shim No.2

Disc Brake Pad Kit

Except 1ZZ–FE(FF) Type:

Anti–squeal Shim No.2

Disc Brake Pad Kit

Anti–squeal Shim No.2

Anti–squeal Shim No.2

Anti–squeal Shim No.1

Anti–squeal Shim No.2

Anti–squeal Shim No.1

Anti–squeal Shim No.1

42356-TMAT-G03

Exploded view of the front pads and related components

LH

RH

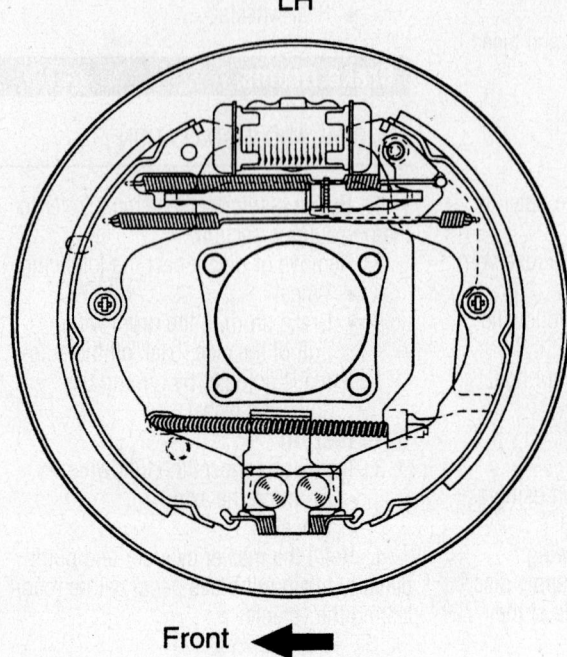

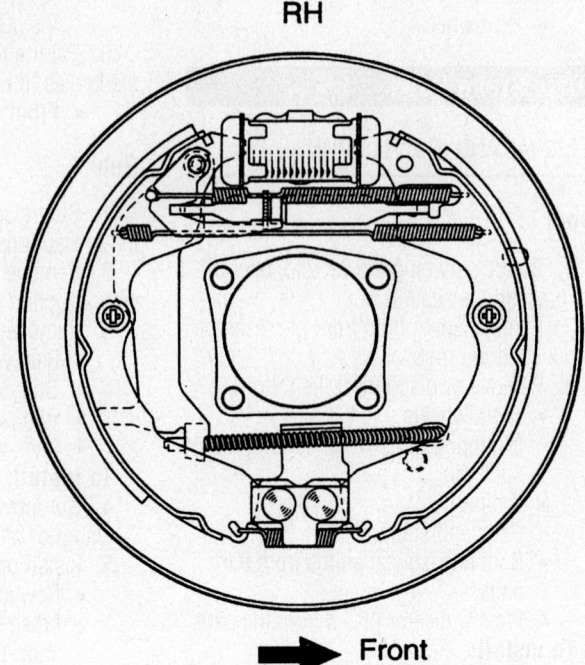

Front ◄

► Front

42356-TMAT-G04

View of the properly installed rear drum brake assembly

To install:

3. Lubricate the contact points on the backing plate and the adjuster with lithium grease.

4. Install or connect the following:
- Parking brake lever, and attach using a new C–washer
- Parking brake lever to the shoe lever
- Upper return spring to the rear shoe
- Shoe assembly onto the backing plate
- Hold-down springs and pins to retain the rear shoe
- Automatic adjuster lever

5. Apply lithium grease to the adjuster bolt.
- Left hand parking brake shoe strut set
- Upper side return spring to the front shoe
- Hold-down springs and pins to the front shoe
- Anchor side spring using needle nosed pliers
- Upper side tension spring with the spacer

6. Adjust the rear brakes as follows:

a. Temporarily install the drum and hub nuts.

b. Remove the hole plug from the backing plate.

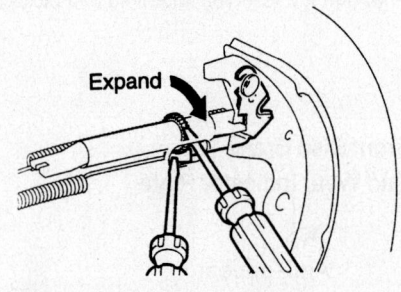

42356-TMAT-G05

Turn the adjuster to expand the shoe until the drum locks

c. Turn the adjuster to expand the shoe until the drum locks.

d. Back off the adjuster eight notches using a suitable adjustment tool.

e. Install the hole plug. into the backing plate to prevent dirt and moisture from entering.

f. Readjust the parking brake cable as necessary.

7. Install the wheels.

8. Refill the master cylinder and pump pedal to attain full brake pedal before Road-testing the vehicle.

AWD Models

1. Before servicing the vehicle, refer to the precautions section.

2. Remove or disconnect the following:
- Wheel
- Brake drum. If the drum will not pull of the axle, back off the automatic adjuster by turning the adjusting wheel.
- Return spring from the front brake shoe
- Anchor spring using needle nosed pliers
- Hold-down springs and pins from the front shoe
- Front shoe
- Automatic adjuster lever spring and the lever
- Hold-down springs and pins from the rear shoe
- Parking brake cable from the rear shoe using needle nosed pliers
- Rear brake shoe
- Anchor spring from the rear shoe
- C–washer using a suitable pry tool from the shoe
- Parking brake lever from the shoe

To install:

3. Lubricate the contact points on the backing plate and the adjuster with lithium grease.

4. Install or connect the following:
- Parking brake lever, and attach using a new C–washer
- Parking brake cable to the lever

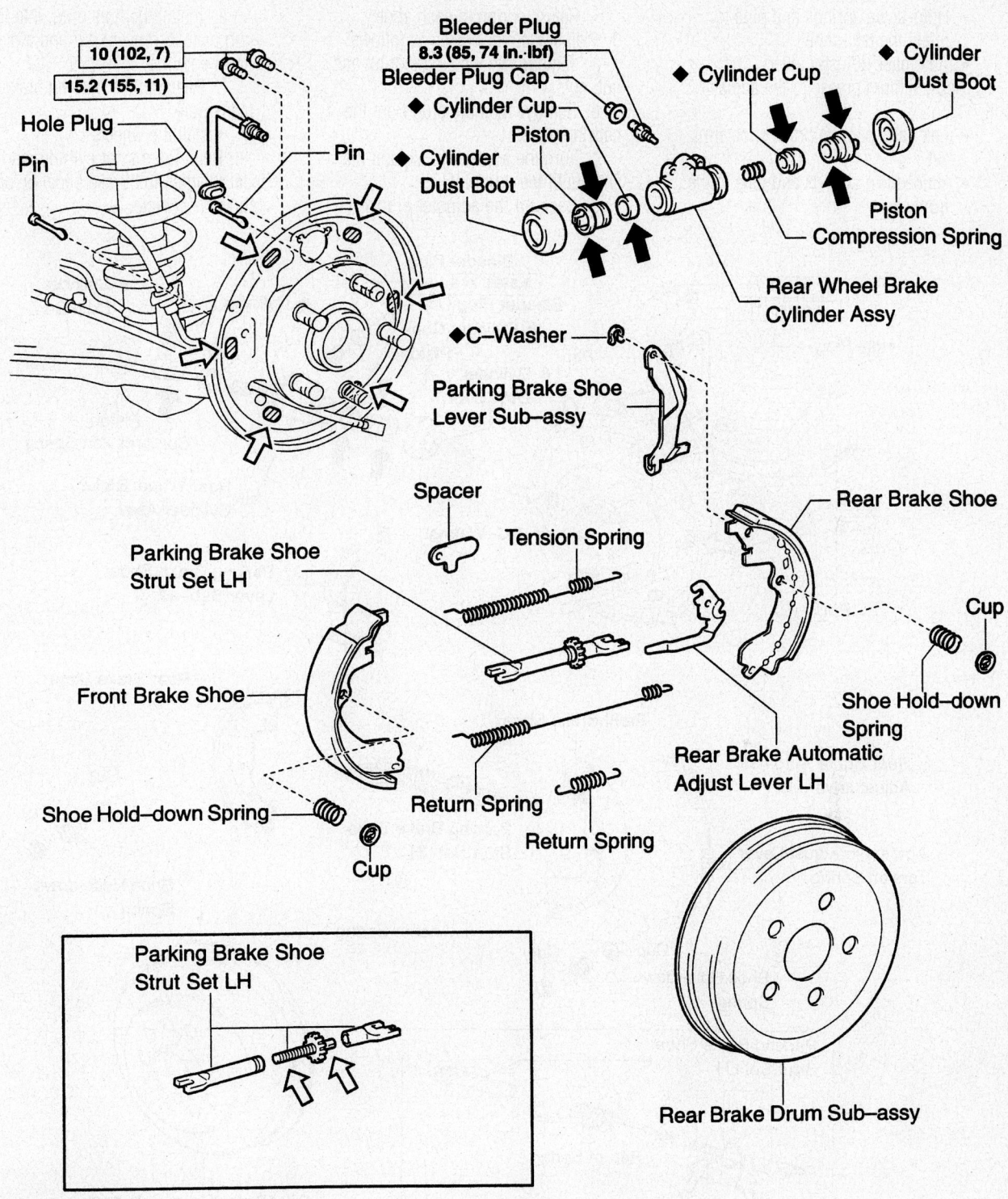

Bleeder Plug
8.3 (85, 74 In.·lbf)
Bleeder Plug Cap
◆ Cylinder Cup
◆ Cylinder
Dust Boot

10 (102, 7)
15.2 (155, 11)

Hole Plug
Pin
Pin

◆ Cylinder Cup
Piston

◆ Cylinder Cup
◆ Cylinder
Dust Boot

Piston
Compression Spring

Rear Wheel Brake
Cylinder Assy

◆C–Washer

Parking Brake Shoe
Lever Sub–assy

Spacer

Rear Brake Shoe

Parking Brake Shoe
Strut Set LH

Tension Spring

Front Brake Shoe

Cup

Shoe Hold–down
Spring

Rear Brake Automatic
Adjust Lever LH

Shoe Hold–down Spring
Return Spring

Return Spring

Cup

Parking Brake Shoe
Strut Set LH

Rear Brake Drum Sub–assy

N·m (kgf·cm, ft·lbf) : Specified torque
◆ Non–reusable part
◀ Lithium soap base glycol grease
◁ High temperature grease
P

42356-TMAT-G06

Exploded view of the drum brake components—1ZZ–FE (FWD) models

- Shoe assembly onto the backing plate
- Hold-down springs and pins to retain the rear shoe
- Automatic adjuster lever

5. Apply lithium grease to the adjuster bolt.

- Left hand parking brake shoe strut set
- Hold-down springs and pins to the front shoe

- Anchor spring to each shoe using needle nosed pliers
- Return spring to each shoe

6. Adjust the rear brakes as follows:

  a. Temporarily install the drum and hub nuts.

  b. Remove the hole plug from the backing plate.

  c. Turn the adjuster to expand the shoe until the drum locks.

  d. Back off the adjuster eight

notches using a suitable adjustment tool.

  e. Install the hole plug. into the backing plate to prevent dirt and moisture from entering.

  f. Readjust the parking brake cable as necessary.

7. Install the wheels.

8. Refill the master cylinder and pump pedal to attain full brake pedal before Road-testing the vehicle.

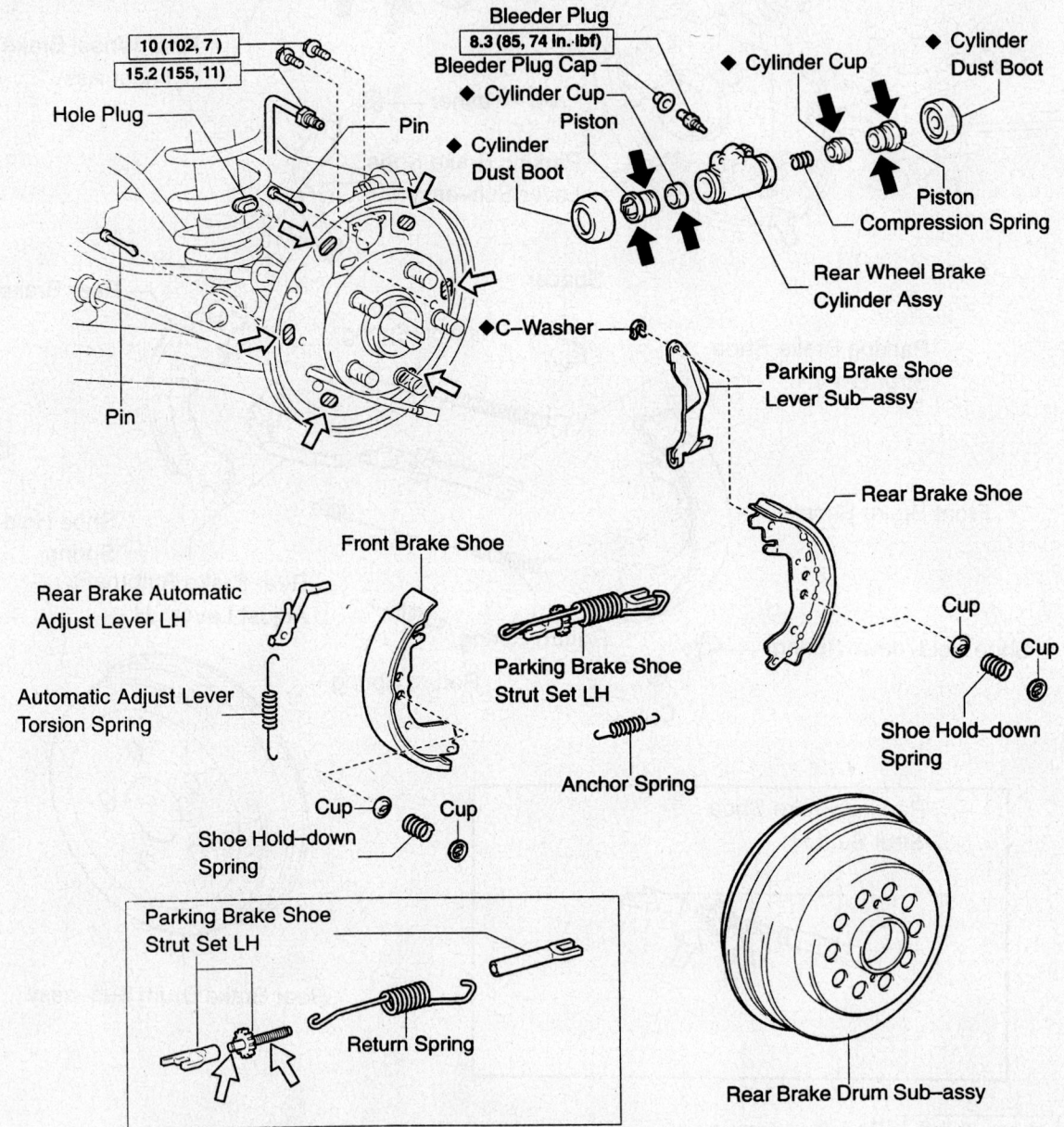

| N·m (kgf·cm, ft·lbf) | : Specified torque
◆ Non-reusable part
◀ Lithium soap base glycol grease
◁ High temperature grease

42356-TMAT-G07

Exploded view of the drum brake components—1ZZ–FE (AWD) models

TOYOTA

16

Prius

DRIVE TRAIN 16-45
ENGINE REPAIR 16-8
FRONT BRAKES 16-62
FRONT SUSPENSION 16-54
FUEL SYSTEM 16-39
REAR BRAKES 16-63
REAR SUSPENSION 16-58
SPECIFICATIONS AND
 MAINTENANCE CHARTS 16-2
STEERING 16-52
Vehicle and Engine Identification ...16-2
General Engine Specifications 16-2
Gasoline Engine Tune-Up
 Specifications 16-2
Firing Order 16-3
Accessory Drive Belt Routing 16-3
Capacities 16-4
Valve Specifications 16-4
Crankshaft and Connecting Rod
 Specifications 16-4
Piston and Ring Specifications 16-5
Torque Specifications 16-5
Wheel Alignment 16-6
Tire, Wheel and Ball Joint
 Specifications 16-6
Brake Specifications 16-6
Scheduled Maintenance
 Intervals 16-7

A
Air Bag 16-52
 Disarming 16-52
 Precautions 16-52
 Rearming 16-52
Alternator 16-8

B
Brake Caliper 16-62
 Removal & Installation 16-62
Brake Drums 16-63
 Removal & Installation 16-63
Brake Shoes 16-64
 Removal & Installation 16-64

C
Camshaft and Valve Lifters 16-29
 Removal & Installation 16-29
Coil Spring (Front) 16-55
 Removal & Installation 16-55

Coil Spring (Rear) 16-60
 Removal & Installation 16-60
CV-Joints 16-49
 Overhaul 16-49
Cylinder Head 16-26
 Removal & Installation 16-26

D
Disc Brake Pads 16-63
 Removal & Installation 16-63
Distributor 16-8
 Removal & Installation 16-8

E
Engine Assembly 16-8
 Removal & Installation 16-8
Exhaust Manifold 16-29
 Removal & Installation 16-29

F
Front Crankshaft Seal 16-29
 Removal & Installation 16-29
Fuel Filter 16-39
 Removal & Installation 16-39
Fuel Injector 16-43
 Removal & Installation 16-43
Fuel Pump 16-39
 Removal & Installation 16-39
Fuel System Pressure 16-39
 Relieving 16-39
Fuel System Service
 Precautions 16-39

H
Halfshaft 16-48
 Removal & Installation 16-48
Heater Core 16-12
 Removal & Installation 16-12

I
Ignition Timing 16-8
 Adjustment 16-8
Intake Manifold 16-28
 Removal & Installation 16-28

L
Lower Ball Joint 16-55
 Removal & Installation 16-55
Lower Control Arm 16-55
 Control Arm Bushing
  Replacement 16-57
 Removal & Installation 16-55

O
Oil Pan 16-33
 Removal & Installation 16-33
Oil Pump 16-34
 Removal & Installation 16-34

P
Piston and Ring 16-38
 Positioning 16-38

R
Rack and Pinion Steering Gear 16-52
 Removal & Installation 16-52
Rear Axle Beam 16-60
 Removal & Installation 16-60
Rear Main Seal 16-35
 Removal & Installation 16-35
Rear Stabilizer Bar 16-61
Rocker Arms/Shafts 16-28

S
Starter Motor 16-33
Strut (Shock Absorber With Coil
 Spring) (Front) 16-54
 Removal & Installation 16-54
Strut (Shock Absorber and Coil
 Spring) (Rear) 16-58
 Removal & Installation 16-58

T
Timing Chain 16-35
 Removal & Installation 16-35
Transaxle Assembly 16-45
 Removal & Installation 16-45

V
Valve Lash 16-32
 Adjustment 16-32

W
Water Pump 16-12
 Removal & Installation 16-12
Wheel Bearings (Front) 16-57
 Adjustment 16-57
 Removal & Installation 16-57
Wheel Bearings (Rear) 16-61
 Adjustment 16-61
 Removal & Installation 16-61

SPECIFICATIONS AND MAINTENANCE CHARTS

VEHICLE AND ENGINE IDENTIFICATION

		Engine						Model Year	
Code	Liters (cc)	Cu. in.	Cyl.	Fuel Sys.	Engine Type	Eng. Mfg.		Code	Year
1NZ-FXE	1.5 (1497)	91.4	4	SFI	DOHC	Toyota		2	2002

SFI: Sequential Multiport Fuel Injection

DOHC: Dual Overhead Camshaft

Code	Year
2	2002
3	2003
4	2004
5	2005
6	2006

09490_TOYP_C0001

GENERAL ENGINE SPECIFICATIONS

Year	Engine ID/VIN	Engine Displacement Liters (cc)	Fuel System Type	Net Horsepower @ rpm	Net Torque @ rpm (ft. lbs.)	Bore x Stroke (in.)	Compression Ratio	Oil Pressure @ rpm
2002	1NZ-FXE	1.5 (1497)	SFI	70@4500	82@4200	2.95x3.33	13.0:1	22-80@3000
2003	1NZ-FXE	1.5 (1497)	SFI	70@4500	82@4200	2.95x3.33	13.0:1	22-80@3000
2004	1NZ-FXE	1.5 (1497)	SFI	76@5000	82@4200	2.95x3.33	13.0:1	22-80@2500
2005	1NZ-FXE	1.5 (1497)	SFI	76@5000	82@4200	2.95x3.33	13.0:1	22-80@2500
2006	1NZ-FXE	1.5 (1497)	SFI	76@5000	82@4200	2.95x3.33	13.0:1	22-80@2500

SFI: Sequential Multiport Fuel Injection

09490_TOYP_C0002

GASOLINE ENGINE TUNE-UP SPECIFICATIONS

Year	Engine Displacement Liters (cc)	Engine ID/VIN	Spark Plugs Gap (in.)	Ignition Timing (deg.) MT	Ignition Timing (deg.) AT	Fuel Pump (psi)		Idle Speed (rpm) MT	Idle Speed (rpm) AT	Valve Clearance In.	Valve Clearance Ex.
2002	1.5 (1497)	1NZ-FXE	0.043	—	7-15B	44-50	①	—	950-1050	0.007-0.009	0.011-0.013
2003	1.5 (1497)	1NZ-FXE	0.043	—	7-15B	44-50	①	—	950-1050	0.007-0.009	0.011-0.013
2004	1.5 (1497)	1NZ-FXE	0.043	—	8-12B	44-50	①	—	950-1050	0.007-0.009	0.011-0.013
2005	1.5 (1497)	1NZ-FXE	0.043	—	8-12B	44-50	①	—	950-1050	0.007-0.009	0.011-0.013
2006	1.5 (1497)	1NZ-FXE	0.043	—	8-12B	44-50	①	—	950-1050	0.007-0.009	0.011-0.013

Note: The Vehicle Emission Control Information label often reflects specification changes made during production.

The label figures must be used if they differ from those in this chart.

B: Before top dead center

① At idle

09490_TOYP_C0003

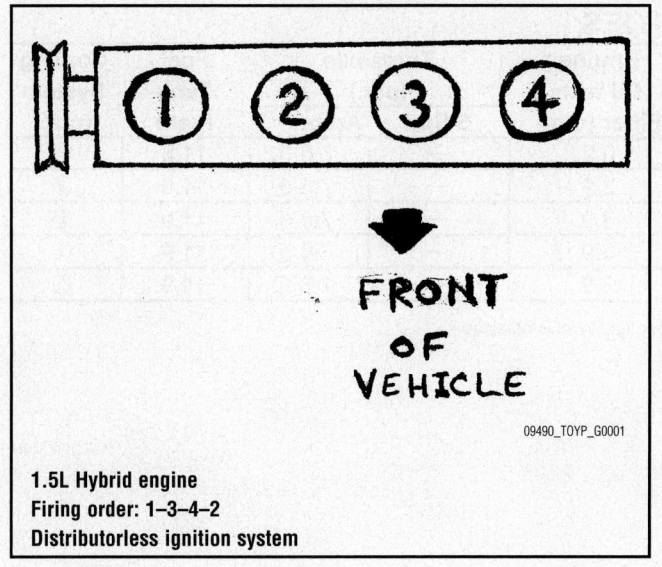

09490_TOYP_G0001

1.5L Hybrid engine
Firing order: 1–3–4–2
Distributorless ignition system

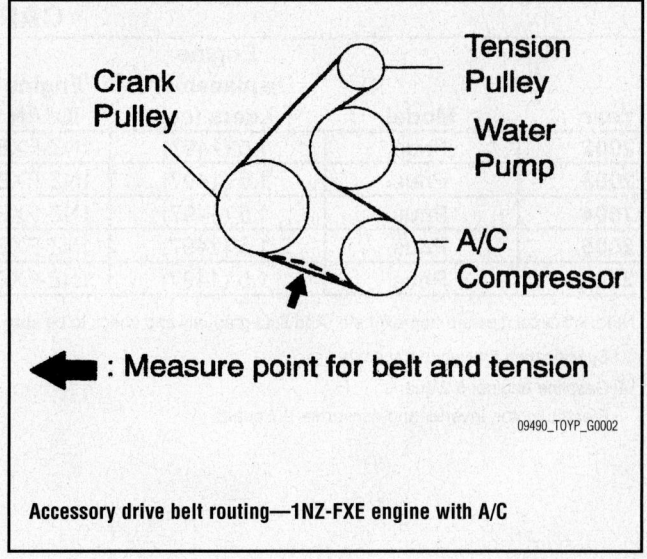

◄ : Measure point for belt and tension

09490_TOYP_G0002

Accessory drive belt routing—1NZ-FXE engine with A/C

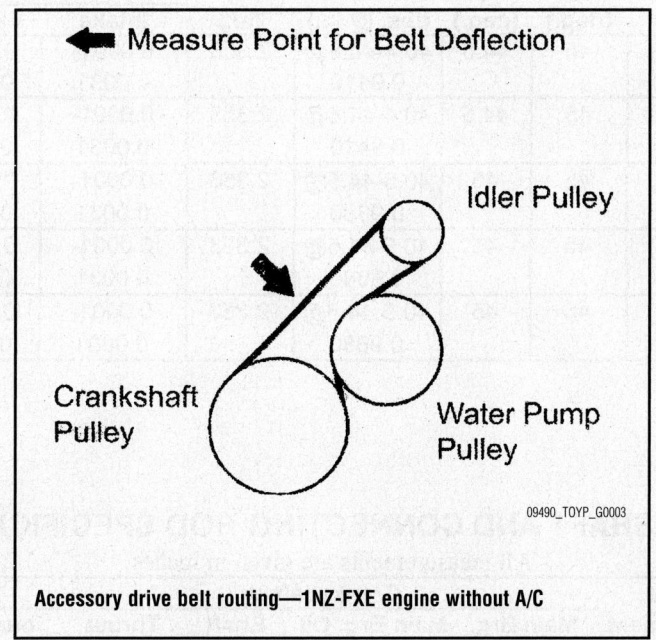

09490_TOYP_G0003

Accessory drive belt routing—1NZ-FXE engine without A/C

CAPACITIES

Year	Model	Engine Displacement Liters (cc)	Engine ID/VIN	Engine Oil with Filter (qts.)	Transaxle (pts.) 5-Spd	Transaxle (pts.) Auto.	Fuel Tank (gal.)	Cooling System (qts.)
2002	Prius	1.5 (1497)	1NZ-FXE	3.9	—	7.6 ①	11.9	②
2003	Prius	1.5 (1497)	1NZ-FXE	3.9	—	7.6 ①	11.9	②
2004	Prius	1.5 (1497)	1NZ-FXE	3.9	—	7.6 ①	11.9	②
2005	Prius	1.5 (1497)	1NZ-FXE	3.9	—	7.6 ①	11.9	②
2006	Prius	1.5 (1497)	1NZ-FXE	3.9	—	7.6 ①	11.9	②

Note: All capacities are approximate. Add fluid gradualy and check to be sure a proper fluid level is obtained.

① Specification for Hybrid transaxle.

② Gasoline engine: 5.2 quarts

Electric motor, inverter and converter: 2.7 quarts

09490_TOYP_C0004

VALVE SPECIFICATIONS

Year	Engine ID/VIN	Engine Displacement Liters (cc)	Seat Angle (deg.)	Face Angle (deg.)	Spring Test Pressure (lbs. @ in.)	Spring Installed Height (in.)	Stem-to-Guide Clearance (in.) Intake	Stem-to-Guide Clearance (in.) Exhaust	Stem Diameter (in.) Intake	Stem Diameter (in.) Exhaust
2002	1NZ-FXE	1.5 (1497)	45	44.5	40.4-44.8@ 0.9410	2.353	0.0001- 0.0031	0.0012- 0.0039	0.1957- 0.1963	0.1955- 0.1961
2003	1NZ-FXE	1.5 (1497)	45	44.5	40.4-44.8@ 0.9410	2.353	0.0001- 0.0031	0.0012- 0.0039	0.1957- 0.1963	0.1955- 0.1961
2004	1NZ-FXE	1.5 (1497)	45	45	40.5-44.5@ 0.9880	2.353	0.0001- 0.0031	0.0012- 0.0039	0.1957- 0.1963	0.1955- 0.1961
2005	1NZ-FXE	1.5 (1497)	45	45	40.5-44.5@ 0.9880	2.353	0.0001- 0.0031	0.0012- 0.0039	0.1957- 0.1963	0.1955- 0.1961
2006	1NZ-FXE	1.5 (1497)	45	45	40.5-44.5@ 0.9880	2.353	0.0001- 0.0031	0.0012- 0.0039	0.1957- 0.1963	0.1955- 0.1961

09490_TOYP_C0005

CRANKSHAFT AND CONNECTING ROD SPECIFICATIONS

All measurements are given in inches.

Year	Engine ID/VIN	Engine Displacement Liters (cc)	Crankshaft Main Brg. Journal Dia.	Crankshaft Main Brg. Oil Clearance	Crankshaft Shaft End-play	Thrust on No.	Connecting Rod Journal Diameter	Connecting Rod Oil Clearance	Connecting Rod Side Clearance
2002	1NZ-FXE	1.5 (1497)	1.8106- 1.8110	0.0004- 0.0028	0.0035- 0.0120	NA	1.5745- 1.5748	0.0006- 0.0024	0.0063- 0.0142
2003	1NZ-FXE	1.5 (1497)	1.8106- 1.8110	0.0004- 0.0028	0.0035- 0.0120	NA	1.5745- 1.5748	0.0006- 0.0024	0.0063- 0.0142
2004	1NZ-FXE	1.5 (1497)	1.8106- 1.8110	0.0004- 0.0028	0.0035- 0.0120	NA	1.5745- 1.5748	0.0006- 0.0024	0.0063- 0.0142
2005	1NZ-FXE	1.5 (1497)	1.8106- 1.8110	0.0004- 0.0028	0.0035- 0.0120	NA	1.5745- 1.5748	0.0006- 0.0024	0.0063- 0.0142
2006	1NZ-FXE	1.5 (1497)	1.8106- 1.8110	0.0004- 0.0028	0.0035- 0.0120	NA	1.5745- 1.5748	0.0006- 0.0024	0.0063- 0.0142

NA: Not Available

09490_TOYP_C0006

PISTON AND RING SPECIFICATIONS

All measurements are given in inches.

Year	Engine ID/VIN	Engine Displacement Liters (cc)	Piston Clearance	Ring Gap			Ring Side Clearance		
				Top Compression	Bottom Compression	Oil Control	Top Compression	Bottom Compression	Oil Control
2002	1NZ-FXE	1.5 (1497)	0.0018-0.0031	0.0087-0.0346	0.0126-0.0406	0.0059-0.0362	0.0012-0.0028	0.0012-0.0028	0.0012-0.0028
2003	1NZ-FXE	1.5 (1497)	0.0018-0.0031	0.0087-0.0346	0.0126-0.0406	0.0059-0.0362	0.0012-0.0028	0.0012-0.0028	0.0012-0.0028
2004	1NZ-FXE	1.5 (1497)	0.0018-0.0032	0.0079-0.0240	0.0118-0.0472	0.0039-0.0453	0.0008-0.0028	0.0008-0.0024	0.0008-0.0024
2005	1NZ-FXE	1.5 (1497)	0.0018-0.0032	0.0079-0.0240	0.0118-0.0472	0.0039-0.0453	0.0008-0.0028	0.0008-0.0024	0.0008-0.0024
2006	1NZ-FXE	1.5 (1497)	0.0018-0.0032	0.0079-0.0240	0.0118-0.0472	0.0039-0.0453	0.0008-0.0028	0.0008-0.0024	0.0008-0.0024

09490_TOYP_C0007

TORQUE SPECIFICATIONS

All readings in ft. lbs.

Year	Engine ID/VIN	Engine Displacement Liters (cc)	Cylinder Head Bolts	Main Bearing Bolts	Rod Bearing Bolts	Crankshaft Damper Bolts	Flywheel Bolts	Manifold		Spark Plugs	Lug Nut
								Intake	Exhaust		
2002	1NZ-FXE	1.5 (1497)	①	②	③	95	④	15	20	13	76
2003	1NZ-FXE	1.5 (1497)	①	②	③	95	④	15	20	13	76
2004	1NZ-FXE	1.5 (1497)	①	②	③	95	36	15	20	13	76
2005	1NZ-FXE	1.5 (1497)	①	②	③	95	36	15	20	13	76
2006	1NZ-FXE	1.5 (1497)	①	②	③	95	36	15	20	13	76

① Step 1: 21 ft. lbs. (29 Nm)
  Step 2: turn head bolts 90 degrees
  Step 3: turn head bolts 90 degrees

② Step 1: 16 ft. lbs. (22 Nm)
  Step 2: turn bearing cap bolts 90 degrees

③ Step 1: 11 ft. lbs. (15 Nm)
  Step 2: turn bearing cap bolts 90 degrees

④ Step 1: 62 ft. lbs. (84 Nm)
  Step 2: turn flywheel bolts 90 degrees

09490_TOYP_C0008

WHEEL ALIGNMENT

Year	Model		Caster Range (+/-Deg.)	Caster Preferred Setting (Deg.)	Camber Range (+/-Deg.)	Camber Preferred Setting (Deg.)	Toe-in (in.)	Steering Axis Inclination (Deg.)
2002	Prius	F	0.75	+1.03	0.75	-0.43	0.04+/-0.08	9.87+/-0.75
		R	—	—	0.75	-0.93	0.04+/-0.12	—
2003	Prius	F	0.75	+1.03	0.75	-0.43	0.04+/-0.08	9.87+/-0.75
		R	—	—	0.75	-0.93	0.04+/-0.12	—
2004	Prius	F	0.75	+3.17	0.75	-0.58	0+/-0.08	12.58+/-0.75
		R	—	—	0.50	-1.50	0.12+/-0.10	—
2005	Prius	F	0.75	+3.17	0.75	-0.58	0+/-0.08	12.58+/-0.75
		R	—	—	0.50	-1.50	0.12+/-0.10	—
2006	Prius	F	0.75	+3.17	0.75	-0.58	0+/-0.08	12.58+/-0.75
		R	—	—	0.50	-1.50	0.12+/-0.10	—

09490_TOYP_C0009

TIRE, WHEEL AND BALL JOINT SPECIFICATIONS

Year	Model	OEM Tires Standard	OEM Tires Optional	Tire Pressures (psi) Front	Tire Pressures (psi) Rear	Wheel Size	Ball Joint Inspection
2002	Prius	P175/65R14	None	35	33	5.5-JJ	①
2003	Prius	P175/65R14	None	35	33	5.5-JJ	①
2004	Prius	P185/65R15	None	35	33	6-JJ	①
2005	Prius	P185/65R15	None	35	33	6-JJ	①
2006	Prius	P185/65R15	None	35	33	6-JJ	①

OEM: Original Equipment Manufacturer

PSI: Pounds Per Square Inch

① Replace if any measurable movement is found.

09490_TOYP_C0010

BRAKE SPECIFICATIONS
All measurements in inches unless noted

Year	Model	Brake Disc Original Thickness	Brake Disc Minimum Thickness	Brake Disc Maximum Runout	Brake Drum Diameter Original Inside Diameter	Brake Drum Diameter Max. Wear Limit	Brake Drum Diameter Max. Machine Diameter	Minimum Lining Thickness Front	Minimum Lining Thickness Rear	Brake Caliper Bracket bolts (ft. lbs.)	Brake Caliper Mounting bolts (ft. lbs.)
2002	Prius	0.984	0.906	0.002	7.874	7.913	7.913	0.039	0.039	79	25
2003	Prius	0.984	0.906	0.002	7.874	7.913	7.913	0.039	0.039	79	25
2004	Prius	0.866	0.787	0.002	7.874	7.913	7.913	0.039	0.039	81	25
2005	Prius	0.866	0.787	0.002	7.874	7.913	7.913	0.039	0.039	81	25
2006	Prius	0.866	0.787	0.002	7.874	7.913	7.913	0.039	0.039	81	25

09490_TOYP_C0011

SCHEDULED MAINTENANCE INTERVALS
TOYOTA—PRIUS

TO BE SERVICED	TYPE OF SERVICE	VEHICLE MILEAGE INTERVAL (x1000)												
		7.5	15	22.5	30	37.5	45	52.5	60	67.5	75	82.5	90	97.5
Engine oil & filter	R	✓	✓	✓	✓	✓	✓	✓	✓	✓	✓	✓	✓	✓
Hybrid transaxle fluid	S/I	✓	✓	✓	✓	✓	✓	✓	✓	✓	✓	✓	✓	✓
Drive axle boots	S/I	✓	✓	✓	✓	✓	✓	✓	✓	✓	✓	✓	✓	✓
Gear shift control operation	S/I	✓	✓	✓	✓	✓	✓	✓	✓	✓	✓	✓	✓	✓
Inspect & rotate tires	S/I	✓	✓	✓	✓	✓	✓	✓	✓	✓	✓	✓	✓	✓
Power steering system	S/I	✓	✓	✓	✓	✓	✓	✓	✓	✓	✓	✓	✓	✓
Suspension system	S/I	✓	✓	✓	✓	✓	✓	✓	✓	✓	✓	✓	✓	✓
Brake discs & pads	S/I	✓		✓		✓		✓		✓		✓		✓
Brake shoes & drums	S/I	✓		✓		✓		✓		✓		✓		✓
Brake hoses & pipes	S/I	✓		✓		✓		✓		✓		✓		✓
Brake fluid	S/I		✓		✓		✓		✓		✓		✓	
Brake pedal	S/I		✓		✓		✓		✓		✓		✓	
Cooling system, hoses & connections	S/I		✓		✓		✓		✓		✓		✓	
Fuel tank, cap & lines	S/I		✓		✓		✓		✓		✓		✓	
Air cleaner filter element	R				✓				✓				✓	
Engine coolant	R				✓				✓				✓	
Spark plugs	R				✓				✓				✓	
Drive belt	S/I				✓				✓				✓	
Exhaust system	S/I				✓				✓				✓	

R: Replace   S/I: Service or Inspect

① Replace every 60,000 miles.

FREQUENT OPERATION MAINTENANCE (SEVERE SERVICE)

If a vehicle is operated under any of the following conditions it is considered severe service:

- **Extremely dusty areas.**

- **50% or more of the vehicle operation is in 32°C (90°F) or higher temperatures, or constant operation in temperatures below 0°C (32°F).**

- **Prolonged idling (vehicle operation in stop and go traffic).**

- **Frequent short running periods (engine does not warm to normal operating temperatures).**

- **Police, taxi, delivery usage or trailer towing usage.**

Oil & oil filter: change every 3000 miles.

Brake discs & pads: service or inspect initially at 3000 miles, 6000 miles, & every 12,000 miles thereafter.

Brake hoses & pipes: service or inspect initially at 3000 miles, 6000 miles & every 12,000 miles thereafter.

Air cleaner filter element: service or inspect ever 3000 miles & replace every 30,000 miles (if not replaced previously).

Hybrid transaxle fluid: service or inspect every 6000 miles & replace every 15,000 miles (if not replaced previously).

Inspect & rotate tires: service or inspect every 6000 miles.

Power steering system: service or inspect every 6000 miles.

Steering system: service or inspect every 6000 miles.

Suspension system: service or inspect every 6000 miles.

Drive belts: service or inspect every 15,000 miles.

Exhaust system: service or inspect every 15,000 miles.

ENGINE REPAIR

→Disconnecting the negative battery cable on some vehicles may interfere with the functions of the on board computer systems and may require the computer to undergo a relearning process, once the negative battery cable is reconnected.

Distributor

REMOVAL & INSTALLATION

These models utilize a Distributorless Ignition System (DIS). With this system, the Electronic Control Module (ECM) determines proper ignition timing and time for the primary ignition coil circuit to turn **ON** and **OFF**.

Alternator

The Toyota Prius, being a sophisticated, hybrid vehicle that utilizes both electric and gasoline (internal combustion) engine power for mobility does not require (or come equipped with) an alternator as a part of its charging system. The Toyota Hybrid system replaces the alternator with a pair of electrical motor-generators, a computerized shunt system to control them, a mechanical power splitter that acts as a second differential, and a battery pack that serves as an energy reservoir.

Ignition Timing

ADJUSTMENT

Ignition timing is controlled by the Electronic Control Module (ECM). The ECM receives signals from various sensors mounted on the engine. No ignition timing adjustment is possible.

Engine Assembly

REMOVAL & INSTALLATION

2002–03 Model

❊ CAUTION

The hybrid system uses high voltage circuits, so improper handling could cause electric shock or leakage. During service, be sure to follow procedures.

1. Before servicing the vehicle, refer to the Precautions Section.

2. Drain engine coolant, engine oil and transaxle fluid.
3. Properly relieve the fuel system pressure.
4. Disconnect the negative battery cable. Wait at least 90 seconds after disconnecting the negative battery cable to prevent possible airbag and seat belt pretensioner activation.
5. While wearing insulating gloves, slide up the lever of the service plug grip and remove while turning the lever to the left. Be sure to insulate the service plug with insulating tape.

❊❊ CAUTION

Do not touch the high voltage connectors and terminals for 5 minutes after removing the service plug grip.

6. Drain the HV coolant.
7. Remove the outer cowl top panel.
8. While wearing insulating gloves, remove the converter and inverter assembly as follows:
   a. Disconnect the battery power cable connector and insulate it with packaging tape.
   b. Remove the inverter terminal cover.
   c. Disconnect the circuit breaker sensor by first moving the outer cover section away toward the wire side.
   d. Remove the circuit breaker sensor.
   e. Verify that there is 0V by measuring the voltage between the terminals of the three phase connector (U-V, V-W, U-W), using a voltmeter at each terminal and body ground.
   f. Remove the mounting bolts and disconnect the MG1 and MG2 power cables. Protect the connector parts with insulating tape.
   g. Disconnect the ground cable.
   h. Disconnect the 3 water hoses from the converter and inverter assembly.
   i. Remove the 4 mounting bolts and the converter and inverter assembly.
9. Remove the heater unit water pump as follows:
   a. Disconnect the connector and wire harness clamp.
   b. Disconnect the hoses from the water pump.
   c. Remove the bolt and refrigerant line bracket.
   d. Remove the 3 bolts and remove the heater unit water pump.
10. Remove or disconnect the following:
    • Air cleaner assembly

• Engine wire clamps
• Heated oxygen sensor connector
• Power steering connectors
• Vacuum Switching Valve (VSV) connector for purge line
• VSV hose for purge line
• Ground straps from the left and right fender aprons
• Air cleaner inlet hose
• Engine coolant reservoir tank
• Upper and lower radiator hoses from radiator
• Heater hose from cylinder block
• Shift lever cable
• Brake fluid level sensor connector
• Brake fluid reservoir tank mounting bolts (2) and suspend it with rope
• Electronic Control Module (ECM) located behind glove box
• Grommet from cowl panel and pull out engine wire
• Junction box No. 1 from right fender apron
• Fuel tube from fuel pump
• Accessory drive belt
• Engine under covers
• Wiring harness from A/C compressor
• A/C compressor unit from engine. Secure the A/C compressor, still connected to the high- and low-pressure hoses, to the side of the vehicle with rope.
11. Disconnect the intermediate extension from the steering assembly as follows:
    a. Hold the steering wheel in the straight ahead position, using the seat belt looped through the steering wheel and buckled securely so the steering wheel will not turn. This will prevent an open circuit of the spiral cable.
    b. Remove the 3 clips, then remove the column hole cover.
    c. Put paint marks on the No. 2 intermediate shaft assembly and control valve shaft.
    d. Loosen the bolt on the steering gear side of the intermediate shaft.
    e. Remove the bolt on the steering column side of the intermediate shaft.
    f. Disconnect the intermediate shaft assembly from the control valve shaft.
12. Remove or disconnect the following:
    • Front exhaust pipe
    • Left and right tie rod ends
    • Front stabilizer bar links
    • Left and right ball joints from front lower control arms
    • Left and right drive shafts

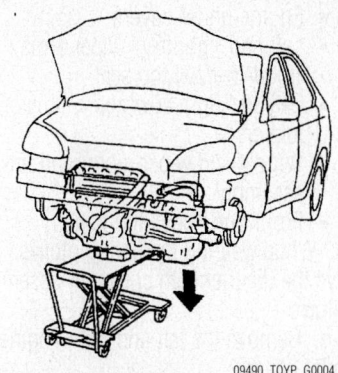

Remove/install the engine using an engine lifting device—Prius

- Torque rod
- Front suspension crossmember

13. Secure the engine/transaxle assembly onto a lifting device.

14. Remove the 2 bolts and nuts and disconnect the right engine mounting insulator from the bracket.

15. Remove the nut and disconnect the left engine mounting from the insulator.

16. While operating the engine lifter, lower the engine and transaxle assembly out of the vehicle slowly and carefully. Make sure the engine is clear of all wiring, hoses and cables.

17. Separate the engine from the transaxle unit.

To install:

18. Connect the transaxle assembly to the engine assembly.

19. Install the engine and transaxle assembly into the engine compartment using the engine lifting device. Keep the engine level and align the right and left mountings with the insulator.

20. Install or connect the following:

- Left mounting bracket to the insulator. Torque the nut to 59 ft. lbs. (80 Nm)
- Right mounting bracket to the insulator. Torque the nuts and bolts to 38 ft. lbs. (52 Nm)
- Front suspension crossmember. Torque the left and right front corner bolts to 83 ft. lbs. (113 Nm). Torque the left and right rear corner bolts to 116 ft. lbs. (157 Nm).
- Torque rod. Torque the 2 nuts and 2 bolts, at the engine, to 74 ft. lbs. (100 Nm). Torque the 2 bolts, at the body side, to 44 ft. lbs. (60 Nm).
- Left and right drive shafts
- Left and right ball joints to the front lower control arms. Torque to 105 ft. lbs. (142 Nm).
- Front stabilizer bar links. Torque to 55 ft. lbs. (74 Nm).

- Left and right tie rod ends
- Front exhaust pipe with the gaskets and O-rings. Torque the 3 exhaust pipe bolts to 46 ft. lbs. (62 Nm).

21. Connect the intermediate extension to the steering assembly as follows:

a. Align the paint marks on the No. 2 intermediate shaft assembly and control valve shaft.

b. Install the bolt on the steering column side of the intermediate shaft.

c. Tighten the bolts to 26 ft. lbs. (35 Nm).

d. Install the column hole cover with the 3 clips.

e. Remove the seat belt from the steering wheel.

22. Install or connect the following:

- A/C compressor unit to the engine. Torque the 4 mounting bolts to 18 ft. lbs. (25 Nm).
- Wiring harness to the compressor
- Accessory drive belt
- Fuel tube to the fuel pump
- Junction box No. 1 to the right fender apron
- Engine wire to cowl panel and connect grommet
- Electronic Control Module (ECM) located behind glove box
- Brake fluid reservoir tank and mounting bolts
- Brake fluid level sensor connector
- Shift lever cable to transaxle
- Heater hose to cylinder block
- Upper and lower radiator hoses to the radiator
- Engine coolant reservoir tank
- Air cleaner inlet hose
- Ground straps to the left and right fender aprons
- VSV hose for purge line
- VSV connector for purge line
- Power steering connectors
- Heated oxygen sensor connector
- Engine wire clamps
- Air cleaner assembly

23. Install the heater unit water pump as follows:

a. Install the heater unit water pump. Torque the 3 mounting bolts to 48 inch lbs. (5.4 Nm).

b. Install the bolt and refrigerant line bracket.

c. Connect the hoses to the water pump.

d. Connect the connector and wire harness clamp.

24. While wearing insulating gloves, install the converter and inverter assembly as follows:

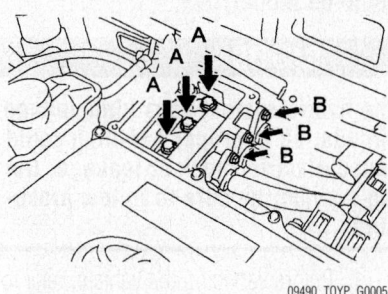

MG2 power cables and mounting bolts— 2002–03 Prius

a. Install the converter and inverter assembly and torque the 4 mounting bolts to 15 ft. lbs. (21 Nm).

b. Connect the 3 water hoses to the converter and inverter assembly.

c. Connect any remaining wiring connectors.

d. Connect the ground cable.

e. Connect the MG1 power cable and 3 mounting bolts. Torque the 3 mounting bolts to 62 inch lbs. (7 Nm).

f. Connect the 3 MG2 power cables and 6 mounting bolts. Torque the **A** mounting bolts to 14 ft. lbs. (19.5 Nm), and the **B** mounting bolts to 71 inch lbs. (8 Nm).

g. Install the circuit breaker sensor and connector cover. Torque the 2 mounting screws to 15 ft. lbs. (20 Nm).

h. Connect the circuit breaker sensor.

i. Install the inverter terminal cover and gasket. Tighten the 4 screws to 71 inch lbs. (8 Nm).

25. Install the outer cowl top panel.

26. Refill with HV coolant.

27. Refill engine cooling system.

28. Install the engine under covers.

29. Connect the negative battery cable and install the HV battery service plug.

30. Refill the engine and transaxle with oil and transaxle fluid.

31. With the converter and inverter assembly removed/installed, check that the inverter operates properly by performing the following:

a. Disconnect the auxiliary battery for 1 minute or more and then connect it again.

b. Depress accelerator pedal to a degree of 50 percent, increasing the speed up to approximately 15 km/h 3 or 4 times to check that there is no problem in the inverter operation.

32. Run the engine and verify that there are no fuel, coolant, transaxle or exhaust leaks. Also check for smooth operation.

2004–06 Model

> ✶✶ **CAUTION**
>
> **The hybrid system uses high voltage circuits, so improper handling could cause electric shock or leakage. During service, be sure to follow procedures.**

1. Before servicing the vehicle, refer to the Precautions Section.
2. Drain engine coolant, engine oil and transaxle fluid.
3. Remove the rear No. 2 floor board.
4. Remove the rear deck floor box.
5. Remove the rear No. 3 floor board.
6. Properly relieve the fuel system pressure.

7. Disconnect the negative battery cable. Wait at least 90 seconds after disconnecting the negative battery cable to prevent possible airbag and seat belt pretensioner activation.
8. While wearing insulating gloves, slide up the lever of the service plug grip and remove while turning the lever to the left. Be sure to insulate the service plug with insulating tape.

> ✶✶ **CAUTION**
>
> **Do not touch the high voltage connectors and terminals for 5 minutes after removing the service plug grip.**

9. Remove or disconnect the following:
   - Front wheels
   - Engine under covers
   - Left and right front wiper arms
   - Hood-to-cowl top seal
   - Left and right cowl top ventilator louvers
   - Windshield wiper motor and link assembly
   - Front outer cowl top panel

10. While wearing insulating gloves, remove the inverter with converter assembly as follows:
    a. Remove the left and right engine under covers.
    b. Drain the HV coolant.
    c. Remove the radiator support opening cover.
    d. Remove the inverter cover.
    e. Verify that there is 0V at the

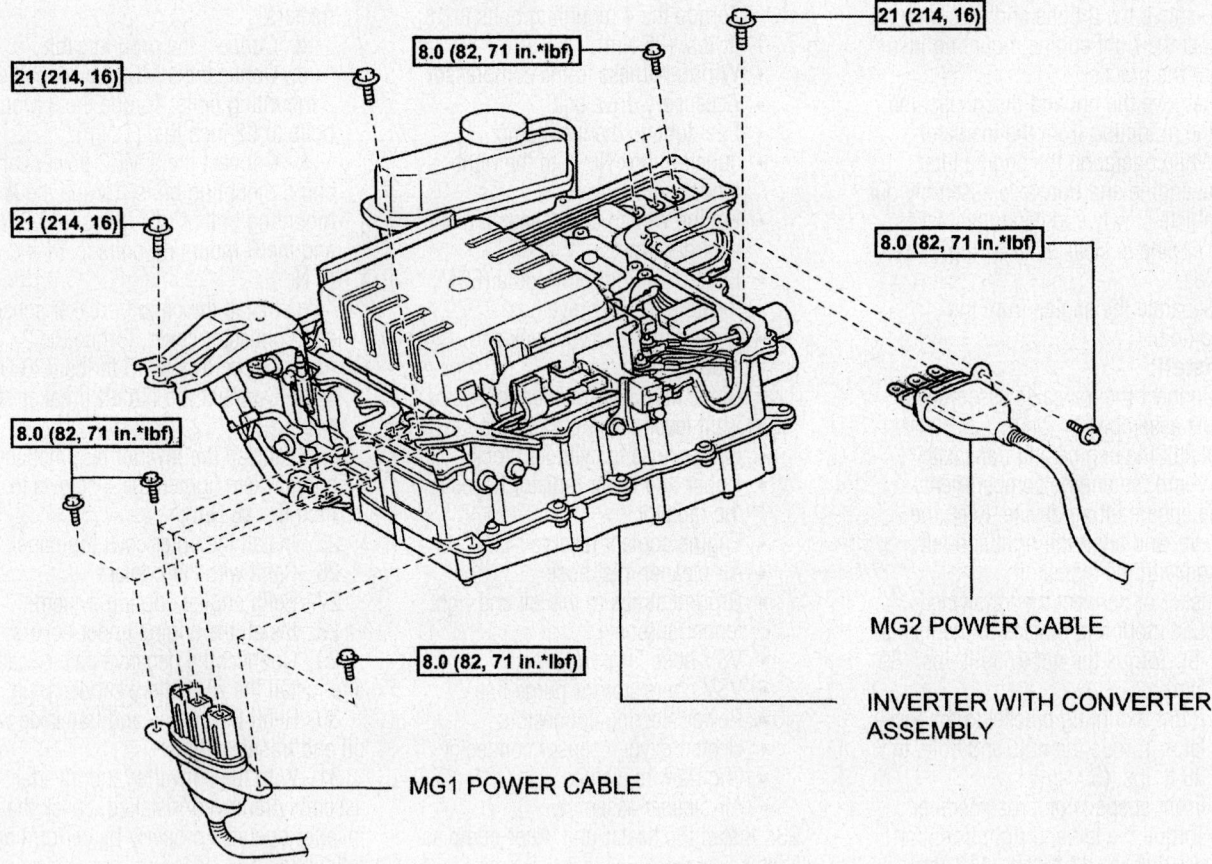

N*m (kgf*cm, ft.*lbf) : Specified torque

Inverter with converter assembly and related components—2004–06 model shown

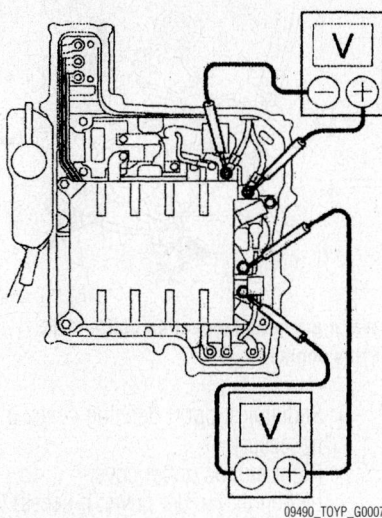

09490_TOYP_G0007

Verifying that there is 0V at the inverter with converter—Prius

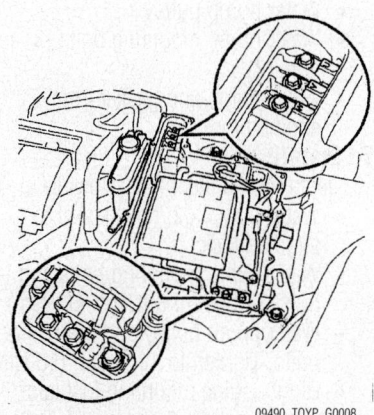

09490_TOYP_G0008

Measuring voltage between the terminals of the three phase connector—Prius

inverter with converter, using a voltmeter utilizing a measuring range of DC 400V or more.

   f. Again, verify that there is 0V by measuring the voltage between the terminals of the three phase connector (U-V, V-W, U-W), using a voltmeter utilizing a measuring range of DC 400V or more.

   g. Disconnect the No. 1, No. 2 and No. 6 inverter cooling hoses.

   h. Disconnect the No. 1 circuit breaker sensor by first moving the outer section away toward the wire side.

   i. Disconnect the 2 frame wire connectors from the inverter with converter assembly and protect the electrode and connector parts with insulating tape.

   j. Use a small screwdriver to lift up the green lock pin, then disconnect the connector for the air conditioning inverter.

   k. Disconnect any remaining wiring

connectors including the engine main wiring harness.

   l. Remove the mounting bolts and disconnect the MG1 and MG2 power cables. Protect the connector parts with insulating tape.

   m. Remove the 3 mounting bolts and the inverter with converter assembly.

11. Remove or disconnect the following:
- Radiator assembly
- No. 3 inverter cooling hose
- No. 1 heat storage water by-pass hose
- Engine ground cable
- Air cleaner inlet hose from air cleaner case
- Air cleaner assembly
- Vacuum Switching Valve (VSV) assembly
- Fuel pipe clamp
- Fuel tube from fuel delivery pipe by pinching the quick connector with your hand and pulling
- Heater water hose
- No. 3 heat storage water by-pass hose
- Connector from ECM, then pull engine wire harness toward the engine compartment side
- Harness and harness clamp from the engine room main relay block
- Ground cable
- Wiring harness from compressor
- Compressor unit from engine. Secure the compressor, still connected to the high- and low-pressure hoses, to the side of the vehicle with rope.

12. Disconnect the steering sliding yoke as follows:

   a. Hold the steering wheel in the straight ahead position, using the seat belt looped through the steering wheel and buckled securely so the steering wheel will not turn. This will prevent an open circuit of the spiral cable.

   b. Remove the 2 clips, then remove the column hole cover silencer.

   c. Loosen the bolt on the steering column side of the sliding yoke.

   d. Remove the bolt on the steering gear side of the sliding yoke.

   e. Put paint marks on the sliding yoke and intermediate shaft, then disconnect the sliding yoke.

13. Remove or disconnect the following:
- Front exhaust pipe
- Left and right front axle hub nuts
- Left front stabilizer link
- Left and right tie rod ends
- Left and right front lower control arms

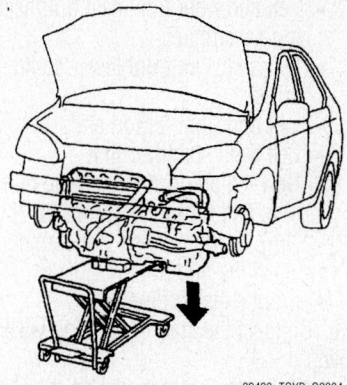

09490_TOYP_G0004

Remove/install the engine using an engine lifting device—Prius

- Left and right front axle hub/bearing assemblies
- Left and right drive shafts
- Engine torque rod
- Front suspension crossmember

14. Secure the engine/transaxle assembly onto a lifting device.

15. Remove or disconnect the following:
- Right engine mounting bracket and right engine mounting insulator
- Left engine mounting bracket and left engine mounting insulator

16. While operating the engine lifter, lower the engine and transaxle assembly out of the vehicle slowly and carefully. Make sure the engine is clear of all wiring, hoses and cables.

17. Separate the engine from the transaxle unit.

To install:

18. Connect the transaxle assembly to the engine assembly.

19. Install the engine and transaxle assembly into the engine compartment using the engine lifting device.

20. Install or connect the following:
- Right engine mounting bracket and right engine mounting insulator. Torque the nuts and bolts to 38 ft. lbs. (52 Nm)
- Left engine mounting bracket and left engine mounting insulator. Torque the nut to 59 ft. lbs. (80 Nm)
- Front suspension crossmember. Torque the left and right front corner bolts to 83 ft. lbs. (113 Nm). Torque the left and right rear corner bolts to 116 ft. lbs. (157 Nm).
- Engine torque rod. Torque the 2 nuts and 2 bolts, at the engine, to 74 ft. lbs. (100 Nm). Torque the 2 nuts and 2 bolts, at the body side, to 41 ft. lbs. (56 Nm).
- Left and right drive shafts

- Left and right front axle hub/bearing assemblies
- Left and right front lower control arms
- Left and right tie rod ends
- Left front stabilizer link
- New left and right front axle hub nuts and torque to 159 ft. lbs. (216 Nm). Using a chisel and hammer, stake the hub nut.
- Front exhaust pipe

21. Connect the steering sliding yoke as follows:

a. Align the paint marks on the sliding yoke and intermediate shaft, then connect the sliding yoke.

b. Torque the sliding yoke bolts to 26 ft. lbs. (35 Nm).

c. Install the column hole cover silencer with the 2 clips.

d. Remove the seat belt from the steering wheel.

22. Install or connect the following:
- Compressor unit to the engine. Torque the three mounting bolts to 18 ft. lbs. (25 Nm).
- Wiring harness to the compressor
- Ground cable
- Harness and harness clamp to the engine room main relay block
- Wire harness connector to the ECM
- No. 3 heat storage water by-pass hose
- Heater water hose
- Fuel tube to the fuel delivery pipe by pushing the quick connector until it makes a "click" sound. Check to make sure the connection is secure by gently, but firmly pulling on it.
- Fuel pipe clamp
- VSV assembly. Torque bolt to 5.5 ft. lbs. (7.5 Nm).
- Air cleaner assembly
- Air cleaner inlet hose to the air cleaner case
- Engine ground cable
- No. 1 heat storage water by-pass hose
- No. 3 inverter cooling hose
- Radiator assembly

23. While wearing insulating gloves, install the inverter with converter assembly as follows:

a. Install the inverter with converter assembly and tighten the three mounting bolts to 16 ft. lbs. (21 Nm).

b. Connect the MG1 and MG2 power cables and tighten the mounting bolts to 71 inch lbs. (8 Nm).

c. Connect any remaining wiring connectors including the engine main wiring harness. Be sure to insert the grommet of the engine main wiring harness into the U-shaped groove of the inverter case.

d. Engage the connector for the air conditioning inverter and secure by pushing in the lock pin.

e. Connect the 2 frame wire connectors to the inverter with converter assembly.

f. Connect the No. 1 circuit breaker sensor.

g. Connect the No. 6, No. 2 and No. 1 inverter cooling hoses.

h. Install the inverter cover and tighten the mounting fasteners to 8 ft. lbs. (11 Nm).

i. Install the radiator support opening cover.

j. Install the left and right engine under covers.

24. Install or connect the following:
- Front outer cowl top panel
- Windshield wiper motor and link assembly
- Left and right cowl top ventilator louvers
- Hood-to-cowl top seal
- Left and right front wiper arms

25. While wearing insulating gloves, joint the service plug grip with the HV battery. While pushing the service plug grip to the right, rotate the lever to the right. Slide the lever down to lock the service plug grip in place.

26. Connect the negative battery cable.

27. Install the rear No. 3 floor board.

28. Install the rear deck floor box.

29. Install the rear No. 2 floor board.

30. Refill engine coolant, engine oil and transaxle fluid.

31. Install or connect the following:
- Front wheels
- Engine under covers

32. Adjust front wheel alignment.

33. Perform initialization.

34. Run the engine and verify that there are no fuel, coolant, transaxle or exhaust leaks.

Water Pump

REMOVAL & INSTALLATION

1. Before servicing the vehicle, refer to the Precautions Section.

2. Disconnect the negative battery cable.

3. Drain the cooling system into a suitable container and tighten the drain plug.

4. Remove or disconnect the following:

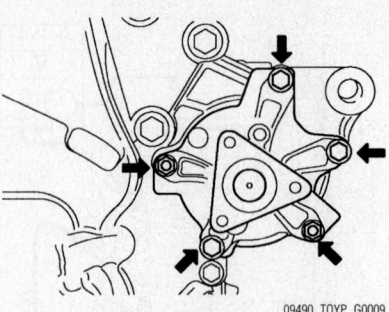

09490_TOYP_G0009

Water pump and mounting bolts—1.5L Prius engine

- Radiator support opening cover, if necessary
- Right engine under cover
- Left engine under cover, if necessary
- Accessory drive belt
- Right engine mounting insulator sub-assembly, if necessary
- Water pump pulley
- Water pump mounting bolts (3) and nuts (2)
- Water pump and gasket from engine block

To install:

5. Install or connect the following:
- Water pump and new gasket to engine block
- Water pump mounting bolts and nuts. Tighten to 8 ft. lbs. (11 Nm)
- Water pump pulley and mounting bolts. Tighten to 11 ft. lbs. (15 Nm)
- Right engine mounting insulator sub-assembly, if necessary. Torque to 38 ft. lbs. (52 Nm).
- Accessory drive belt

6. Refill the engine cooling system.

7. Connect the negative battery cable.

8. Start the engine and top off the coolant as necessary.

9. Check the cooling system for leaks.

10. Install the right engine under cover.

11. Install the left engine under cover, if necessary

12. Install the radiator support opening cover, if necessary

Heater Core

REMOVAL & INSTALLATION

2002–03 Model

1. Before servicing the vehicle, refer to the Precautions Section.

2. Disconnect the negative battery cable. Wait at least 90 seconds after disconnecting the negative battery cable to prevent

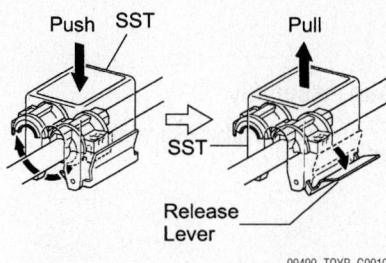

Using special pipe disconnect tool for liquid tube and suction hose

possible airbag and seat belt pretensioner activation.

3. Drain the cooling system.

4. Recover refrigerant from A/C system.

5. Remove the front wiper and outer front cowl top panel

6. Disconnect the liquid tube and suction hose as follows:

   a. Insert SST 09870-00015 (Suction tube), or SST 09870-00025 (Liquid tube) to piping clamp.

➡**Confirm the direction of the piping clamp claw and SST using the illustration showing on the caution label.**

   b. Push down SST and release the clamp lock.

➡**Be careful not to deform the tubes, when pushing SST.**

   c. Pull SST slightly and push the release lever, then remove the piping clamp with SST.

   d. Remove the piping clamp from SST.

   e. Disconnect the both tubes.

➡**Do not use tools like screwdriver to remove the tube. Cap the open fittings immediately to keep moisture or dirt out of the system.**

7. Disconnect water hoses from heater radiator (heater core) pipes.

8. Remove the instrument panel as follows:

➡**If the airbag connector is disconnected with the ignition switch at ON, DTCs will be recorded.**

   a. Place the front wheels facing straight ahead.

   b. Using a torx socket wrench, loosen the 2 torx screws until the groove along the screw circumference catches on the screw case.

   c. Pull out the wheel pad from the steering wheel and disconnect the airbag connector and disconnect.

✳✳ CAUTION

When storing the wheel pad, keep the upper surface of the pad facing upward. Never disassemble the wheel pad. When removing the wheel pad, take care not to pull the airbag wire harness.

   d. Remove the steering wheel set nut and place match marks on the steering wheel and main shaft assembly.

   e. Using SST 09950-50013, remove the steering wheel.

   f. Remove front door scuff plates.

   g. Remove cowl side trims.

   h. Remove front pillar garnishes.

   i. Remove the 2 screws and hood lock release lever.

   j. Zip the shifting hole cover open and remove the bolt and screw.

   k. Using a screwdriver, remove the lower finish panel, then disconnect the connectors and DLC3.

   l. Remove finish panel.

   m. Remove column covers.

   n. Disconnect the connectors, remove the 3 screws and combination switch with spiral cable.

➡**Do not disassemble the spiral cable or apply oil to it.**

   o. Rotate the stoppers of the glove compartment door to 90 degrees and pull them out inward.

   p. Remove the 2 screws and glove compartment door.

   q. Remove the 2 clips and, using a screwdriver, remove the lower center cluster finish panel. Disconnect the connector.

   r. Remove glove compartment door reinforcement, if equipped.

   s. Disconnect the passenger airbag connector and disengage the airbag connector clamp from the center bracket.

✳✳ CAUTION

When disconnecting the airbag connector, take care not to damage the airbag wire harness.

   t. Remove the 2 bolts and nuts and passenger airbag assembly.

✳✳ CAUTION

Do not store the passenger airbag assembly with the airbag deployment side facing down. Never disassemble the passenger airbag assembly. Take care not to damage the wire harness.

   u. Remove hazard warning switch, using a screwdriver and disconnect the connectors. Tape the screwdriver tip before use.

   v. Remove radio or radio tuner opening cover.

   w. Remove the 2 screws, and using a screwdriver, remove the cluster finish panel assembly. Disconnect the connectors.

   x. Remove the 3 screws and combination meter. Disconnect the connector.

   y. Remove the No. 1 and No. 2 side defroster nozzles, using a screwdriver. Tape the screwdriver tip before use.

   z. Remove the No. 1 and No. 3 registers, using a screwdriver. Tape the screwdriver tip before use.

   aa. Remove fasteners and lower steering column from under the instrument panel

   bb. Disconnect the connectors, wire harness clamps and remove the 4 bolts, 3 screws and instrument panel.

   cc. Remove No. 1 brace.

   dd. Remove the instrument panel reinforcement.

   ee. Remove the rear console box.

9. Remove the blower unit.

10. Remove A/C unit as follows:

   a. Remove the 2 screws and defroster duct.

   b. Remove the foot air duct.

   c. Disconnect the connectors.

   d. Remove the 2 nuts and A/C unit.

11. Remove or disconnect the following:
- Thermistor
- Expansion valve
- PTC heater
- Drain hose
- Aspirator and hose

12. Remove heater pipe cover as follows:

   a. Disconnect the connector.

   b. Remove the screw.

   c. Release the 2 claws and remove the heater pipe cover.

13. Remove the heater radiator (heater core) as follows:

   a. Remove the 3 screws and clamp.

   b. Remove the 3 screws and 2 clamps.

   c. Pull out the heater radiator.

To install:

14. Install the heater radiator (heater core) into the A/C unit and tighten the screws and clamps.

15. Install the heater pipe cover, engaging the 2 claws, tightening the screw and engage the connector.

16. Install or connect the following:
- Aspirator and hose
- Drain hose

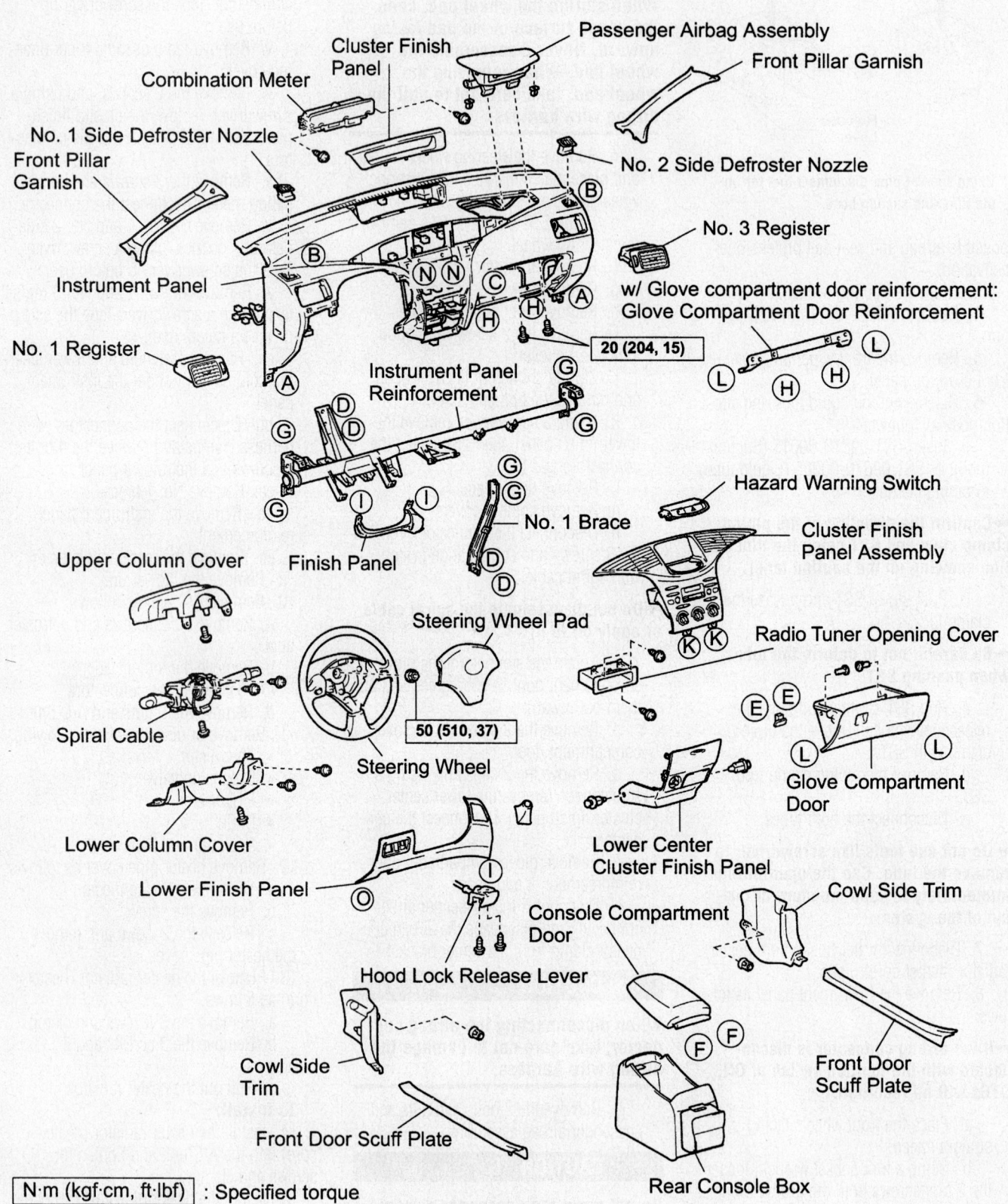

Combination Meter

Cluster Finish Panel

Passenger Airbag Assembly

Front Pillar Garnish

No. 1 Side Defroster Nozzle

Front Pillar Garnish

No. 2 Side Defroster Nozzle

Instrument Panel

No. 3 Register

w/ Glove compartment door reinforcement:
Glove Compartment Door Reinforcement

No. 1 Register

Instrument Panel Reinforcement

20 (204, 15)

Upper Column Cover

Finish Panel

No. 1 Brace

Hazard Warning Switch

Cluster Finish Panel Assembly

Radio Tuner Opening Cover

Spiral Cable

Steering Wheel Pad

50 (510, 37)

Steering Wheel

Lower Column Cover

Lower Finish Panel

Glove Compartment Door

Lower Center Cluster Finish Panel

Console Compartment Door

Cowl Side Trim

Hood Lock Release Lever

Cowl Side Trim

Front Door Scuff Plate

Front Door Scuff Plate

Rear Console Box

N·m (kgf·cm, ft·lbf) : Specified torque

09490_TOYP_G0011

Exploded view of instrument panel assembly and related components—2002–03 Prius

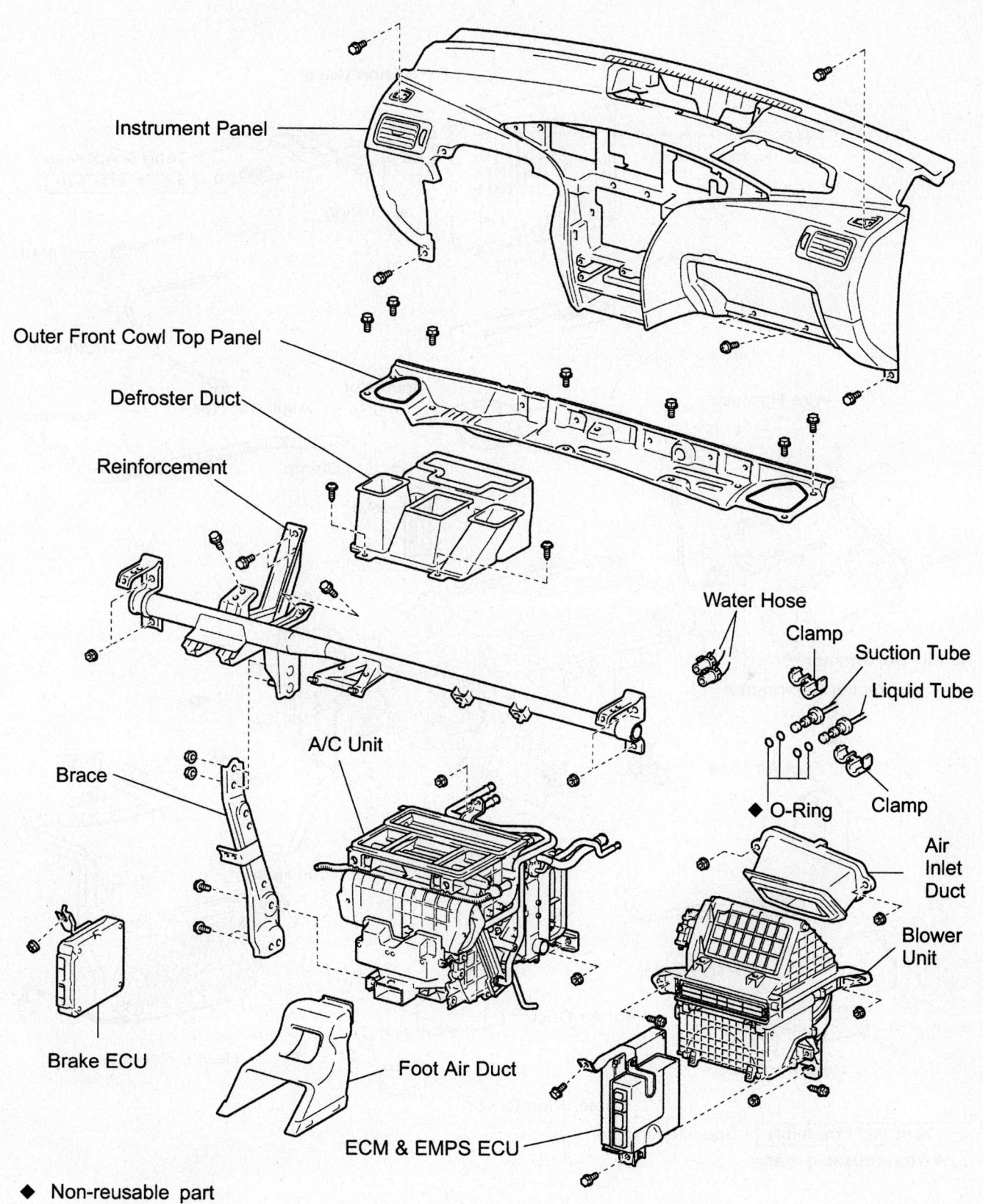

Instrument Panel

Outer Front Cowl Top Panel

Defroster Duct

Reinforcement

Water Hose

Clamp

Suction Tube

Liquid Tube

A/C Unit

◆ O-Ring

Clamp

Air Inlet Duct

Brace

Blower Unit

Brake ECU

Foot Air Duct

ECM & EMPS ECU

◆ Non-reusable part

09490_TOYP_G0012

Exploded view of A/C unit and related components—2002–03 Prius

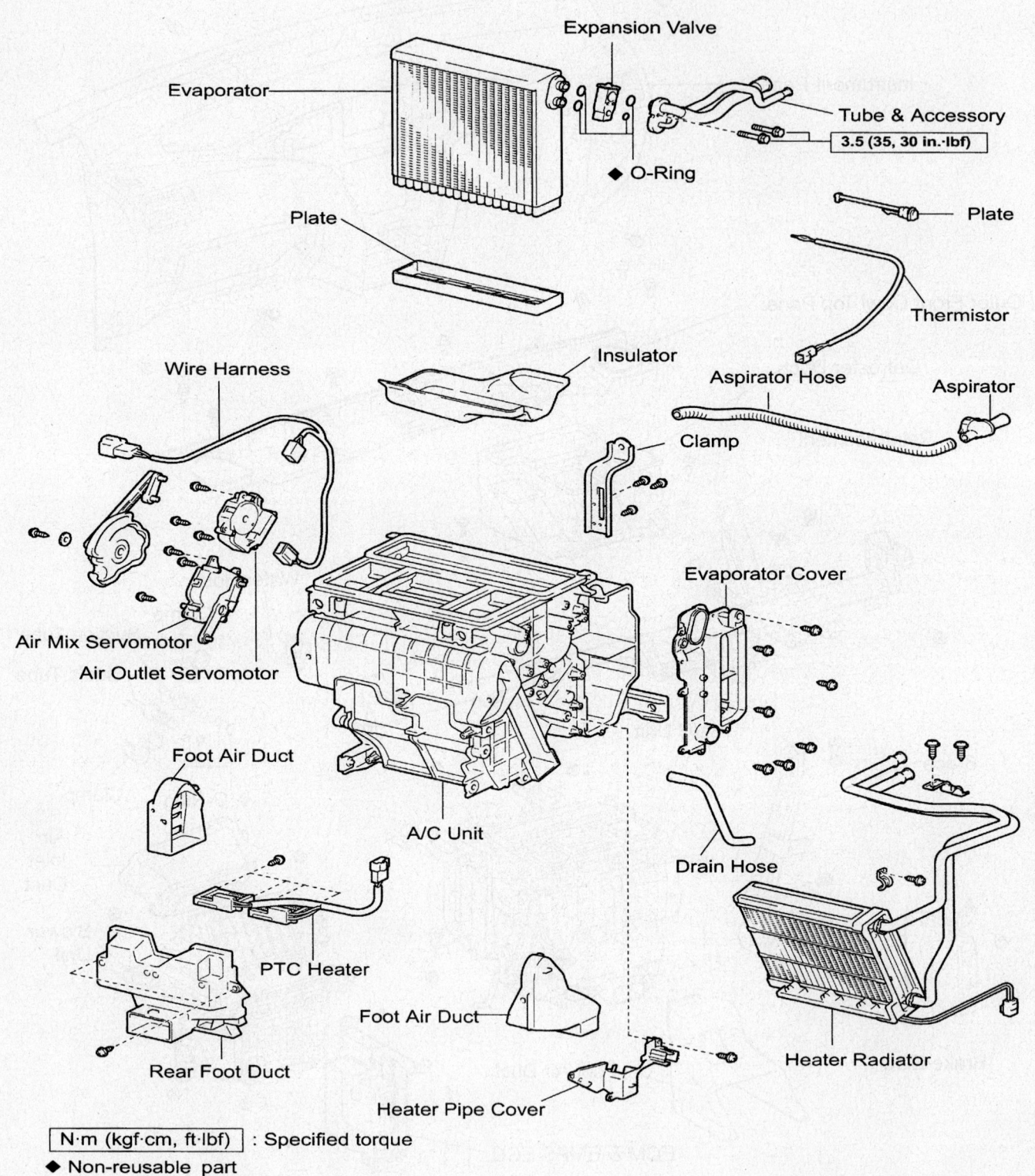

Expansion Valve

Evaporator

Tube & Accessory
3.5 (35, 30 in.·lbf)

◆ O-Ring

Plate

Plate

Thermistor

Wire Harness

Insulator

Aspirator Hose

Aspirator

Clamp

Evaporator Cover

Air Mix Servomotor

Air Outlet Servomotor

Foot Air Duct

A/C Unit

Drain Hose

PTC Heater

Foot Air Duct

Heater Radiator

Rear Foot Duct

Heater Pipe Cover

N·m (kgf·cm, ft·lbf) : Specified torque

◆ Non-reusable part

09490_TOYP_G0013

Heater radiator (heater core) removal from A/C unit—2002–03 Prius

- PTC heater
- Expansion valve
- Thermistor

17. Install the A/C unit into the vehicle as follows:

a. Tighten the 2 mounting nuts to the A/C unit.

b. Connect the connectors.

c. Install the foot air duct.

d. Install the defroster duct and tighten the 2 screws.

18. Install the blower unit.

19. Install the instrument panel as follows:

a. Install the rear console box.

b. Install the instrument panel reinforcement.

c. Install the No. 1 brace.

d. Connect the connectors, wire harness clamps and install the instrument panel. Install and tighten the 4 bolts and 3 screws.

e. Position steering column in place under the instrument panel and tighten the fasteners to 19 ft. lbs. (25 Nm).

f. Install the No. 1 and No. 3 registers.

g. Install the No. 1 and No. 2 side defroster nozzles.

h. Install the combination meter and 3 mounting screws. Engage the connectors.

i. Install the cluster finish panel and 2 screws.

j. Install radio or radio tuner opening cover.

20. Fill the radiator with engine coolant until it reaches the rim of the radiator filler. If the engine coolant level of the radiator drops when grasping the radiator inlet hose outlet hose several times by hand, add more engine coolant.

21. Fill the radiator reservoir tank with engine coolant to the maximum level.

22. Install the radiator cap and start the engine. After repeatedly idling and racing the engine several times for approx. 2 minutes, stop the engine.

23. Remove the radiator cap and fill the radiator with engine coolant until it reaches the rim of the radiator filler. If the engine coolant level of the radiator drops when grasping the radiator inlet hose and outlet hose several times by hand, add more coolant.

24. Install the radiator cap.

25. Perform the activation inspection mode (within 60 seconds) as follows:

a. Turn the ignition switch ON from OFF.

b. With the shift lever in P position, fully depress the accelerator pedal 2 times.

c. With the shift lever in N position, fully depress the accelerator pedal 2 times.

d. With the shift lever in P position, fully depress the accelerator pedal 2 times.

e. Activate the inspecting mode and check that the hybrid system error warning light on the multi-center display flashes.

f. Turning the ignition switch to START starts the engine's continuous operation.

26. Start the engine and warm it up until the radiator fan starts to turn with the engine speed at less than 2,500 rpm. Stop the engine immediately after the radiator fan starts to turn.

27. Stop the engine and cool it down. Cool down the engine until its temperature becomes below 50°C (122°F).

28. Remove the radiator cap and check the engine coolant level of the radiator. If it has dropped, repeat the previous 4 steps.

29. Bleed air from water pump by performing the following:

a. Set the vehicle in the following conditions:

- Ignition switch ON
- Blower speed control dial to LOW
- Temperature control dial to MAX. HOT

b. Operate the water pump in the following conditions:

- The engine stopped
- The blower switch ON
- Temperature control dial at MAX. HOT

30. Operate the water pump until a sound of the air-containing engine coolant can not be heard from the heater core.

31. When the engine coolant remains full, fill the radiator reservoir tank with engine coolant to the maximum level.

32. Install the instrument panel as follows:

a. Install hazard warning switch and connect the connectors.

b. Install the passenger airbag assembly.

c. Engage the airbag connector clamp to the center bracket and connect the passenger airbag connector.

d. Install glove compartment door reinforcement, if equipped.

e. Install the lower center cluster finish panel. Engage the connector and 2 clips.

f. Install the glove compartment door.

g. Install the combination switch with spiral cable and connect the connectors.

h. Install column covers.

i. Install finish panel.

j. Connect the connectors and DLC3 and install the lower finish panel.

k. Install the bolt and screw and close the shifting hole cover.

l. Install the hood lock release lever.

m. Install the front pillar garnishes.

n. Install the cowl side trims.

o. Install front door scuff plates.

p. Install the steering wheel. Torque the set nut to 37 ft. lbs. (50 Nm).

q. Connect the airbag connector. Install the wheel pad to the steering wheel and tighten the 2 screws to 78 inch lbs. (8.8 Nm).

33. Connect water hoses to the heater radiator (heater core) pipes.

34. Connect the liquid tube and suction hose.

35. Install the front wiper and outer front cowl top panel

36. Evacuate and recharge the refrigerant.

37. Connect the negative battery cable.

38. Operate the engine to normal operating temperatures; then, check the climate control operation and check for leaks.

2004–06 Models

1. Before servicing the vehicle, refer to the Precautions Section.

2. Disconnect the negative battery cable. Wait at least 90 seconds after disconnecting the negative battery cable to prevent possible airbag and seat belt pretensioner activation.

3. Place front wheels in the straight ahead position.

4. Drain the cooling system.

5. Recover refrigerant from A/C system.

6. Disconnect the suction hose subassembly as follows:

a. Check SST 09870-00015 installation direction. Set SST so that the stopper side is on the piping clamp lock side.

b. Install SST on the piping clamp.

c. Push down on SST with your thumb while holding the pipe with both hands. Be careful not to bend the pipe.

d. Pull SST until the stopper touches the pipe.

e. Raise SST stopper and remove the piping clamp with SST from the pipe.

f. Remove the piping clamp from SST.

g. Disconnect the suction hose by hand or using a screwdriver.

h. Remove the 2 O-rings from the suction hose. Do not apply excessive force to the suction hose. Seal the opening of the disconnected part using vinyl

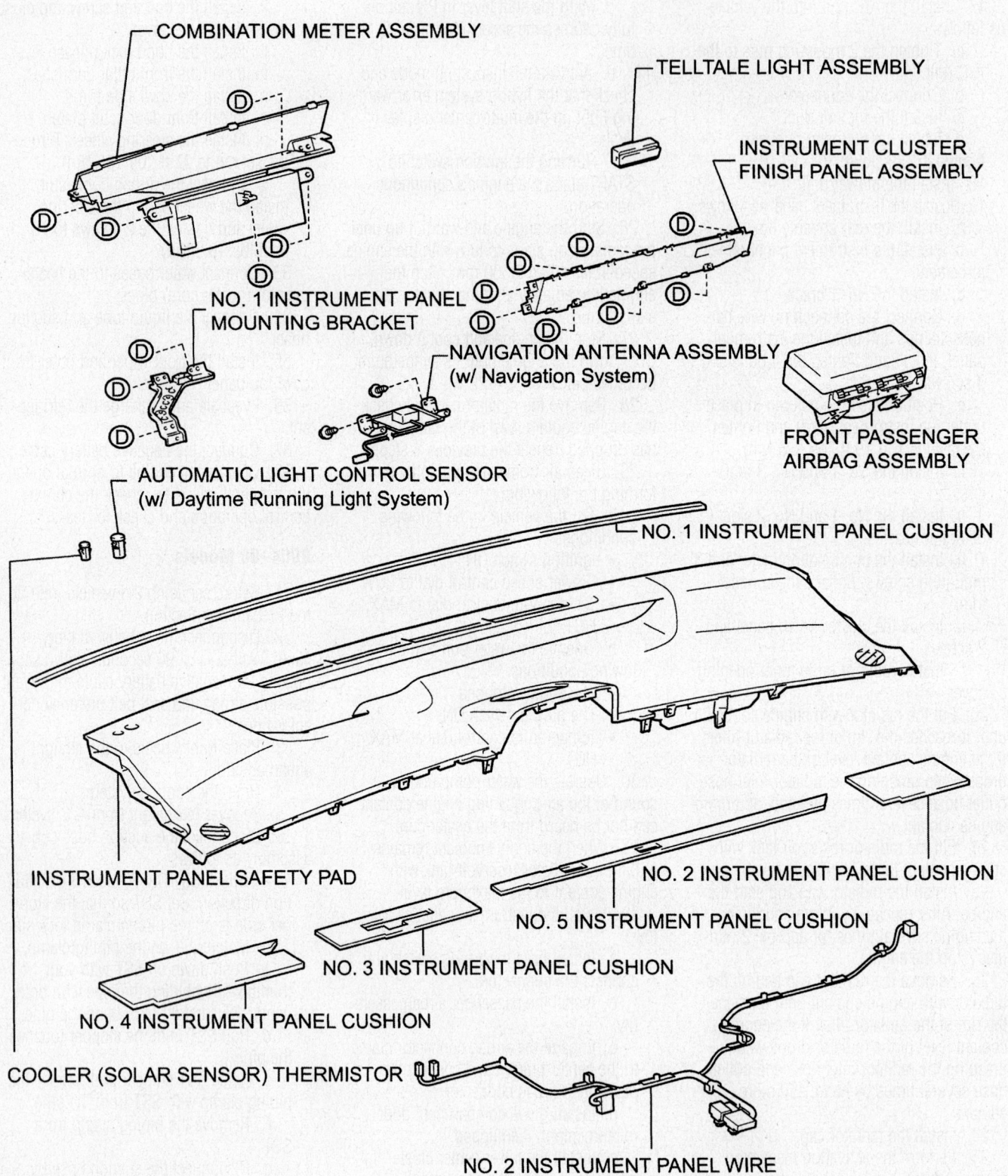

COMBINATION METER ASSEMBLY

TELLTALE LIGHT ASSEMBLY

INSTRUMENT CLUSTER
FINISH PANEL ASSEMBLY

NO. 1 INSTRUMENT PANEL
MOUNTING BRACKET

NAVIGATION ANTENNA ASSEMBLY
(w/ Navigation System)

FRONT PASSENGER
AIRBAG ASSEMBLY

AUTOMATIC LIGHT CONTROL SENSOR
(w/ Daytime Running Light System)

NO. 1 INSTRUMENT PANEL CUSHION

INSTRUMENT PANEL SAFETY PAD

NO. 2 INSTRUMENT PANEL CUSHION

NO. 5 INSTRUMENT PANEL CUSHION

NO. 3 INSTRUMENT PANEL CUSHION

NO. 4 INSTRUMENT PANEL CUSHION

COOLER (SOLAR SENSOR) THERMISTOR

NO. 2 INSTRUMENT PANEL WIRE

09490_TOYP_G0014

Exploded view of instrument panel safety pad and related components—2004–06 Prius

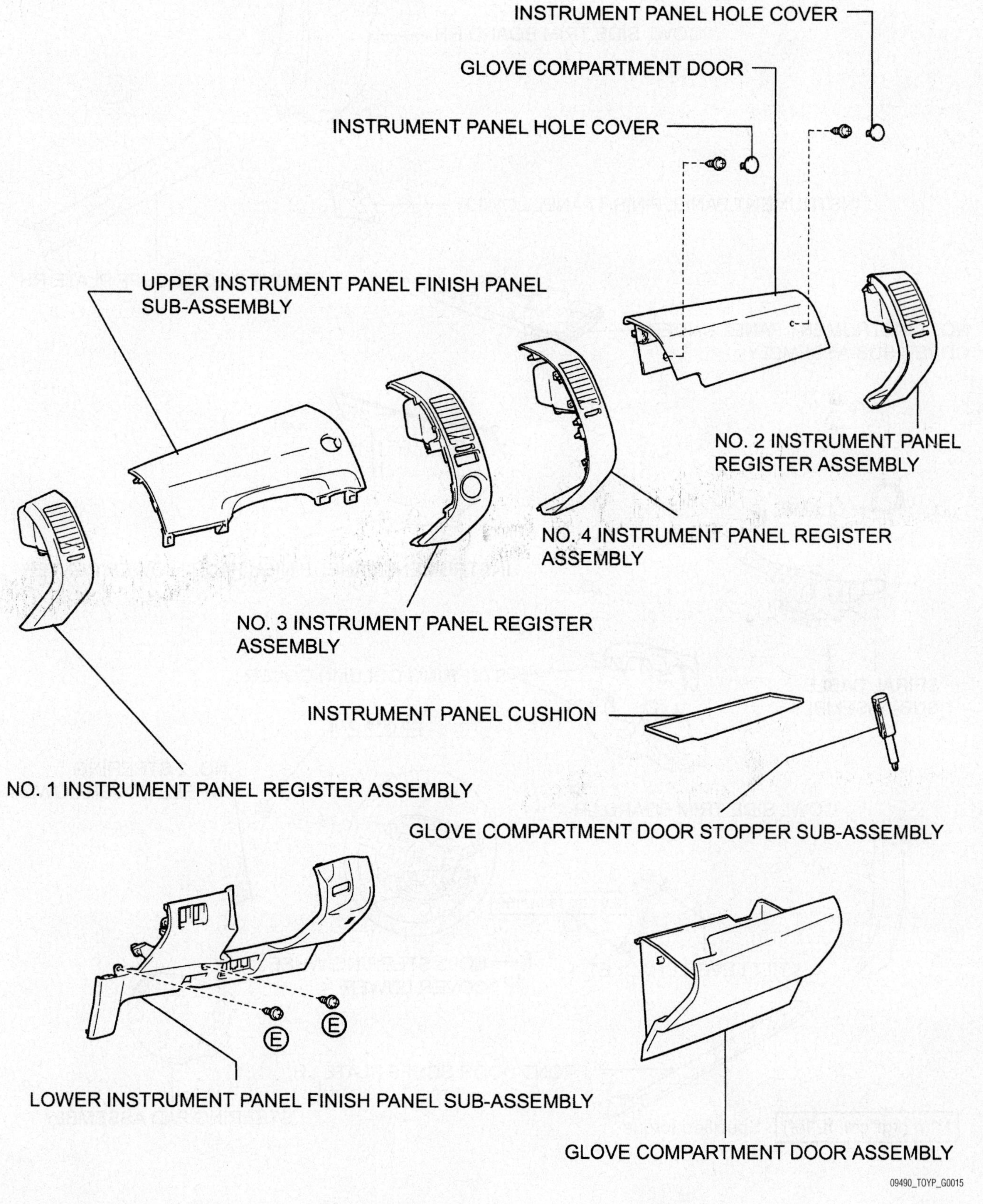

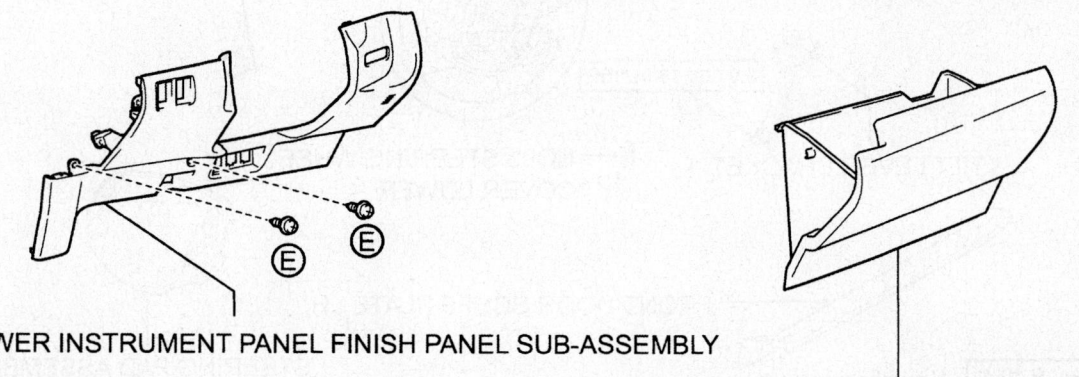

INSTRUMENT PANEL HOLE COVER

GLOVE COMPARTMENT DOOR

INSTRUMENT PANEL HOLE COVER

UPPER INSTRUMENT PANEL FINISH PANEL
SUB-ASSEMBLY

NO. 2 INSTRUMENT PANEL
REGISTER ASSEMBLY

NO. 4 INSTRUMENT PANEL REGISTER
ASSEMBLY

NO. 3 INSTRUMENT PANEL REGISTER
ASSEMBLY

NO. 1 INSTRUMENT PANEL REGISTER ASSEMBLY

INSTRUMENT PANEL CUSHION

GLOVE COMPARTMENT DOOR STOPPER SUB-ASSEMBLY

LOWER INSTRUMENT PANEL FINISH PANEL SUB-ASSEMBLY

GLOVE COMPARTMENT DOOR ASSEMBLY

09490_TOYP_G0015

Exploded view of instrument registers and related components—2004–06 Prius

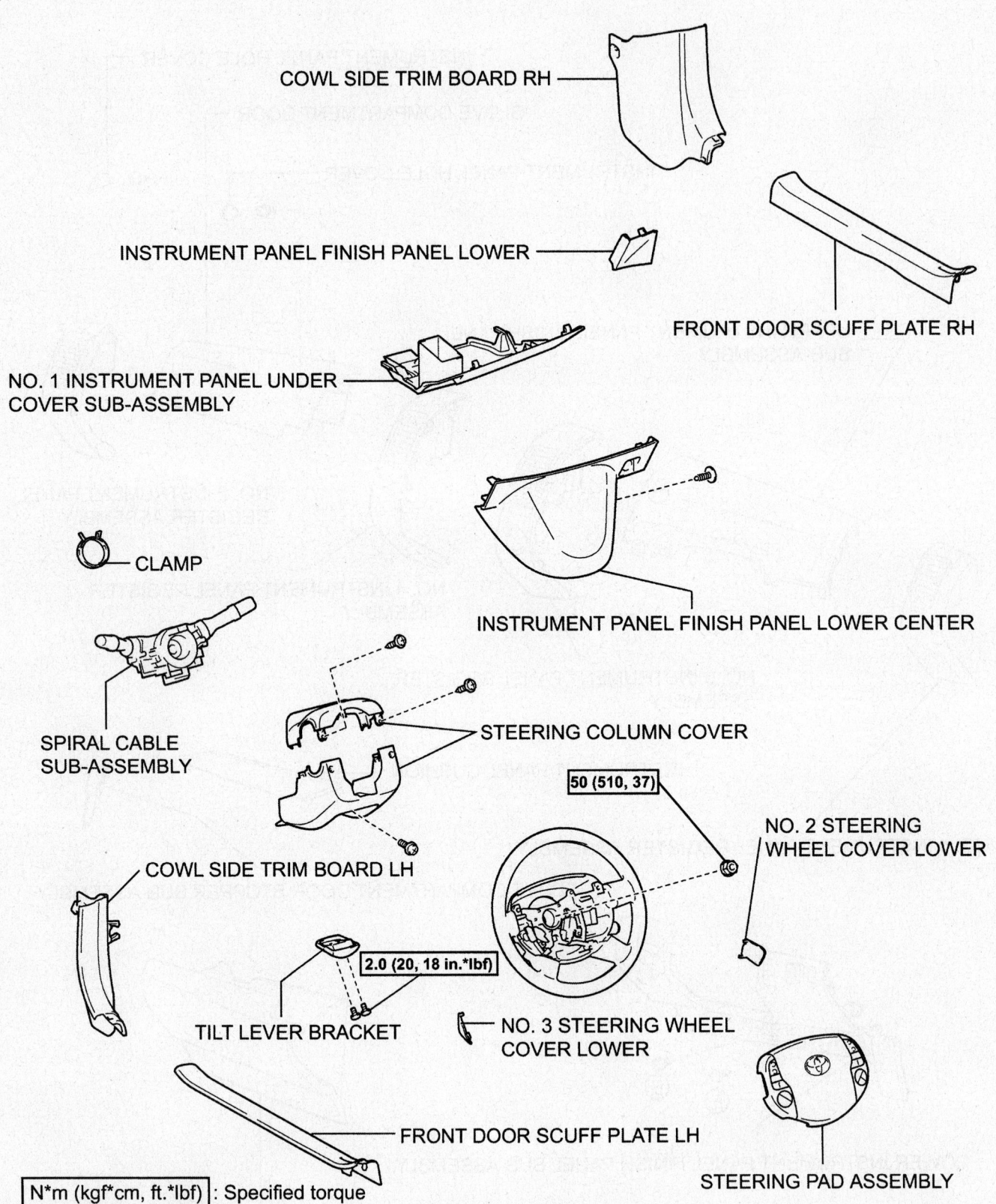

COWL SIDE TRIM BOARD RH

INSTRUMENT PANEL FINISH PANEL LOWER

FRONT DOOR SCUFF PLATE RH

NO. 1 INSTRUMENT PANEL UNDER COVER SUB-ASSEMBLY

CLAMP

INSTRUMENT PANEL FINISH PANEL LOWER CENTER

SPIRAL CABLE SUB-ASSEMBLY

STEERING COLUMN COVER

50 (510, 37)

NO. 2 STEERING WHEEL COVER LOWER

COWL SIDE TRIM BOARD LH

2.0 (20, 18 in.*lbf)

TILT LEVER BRACKET

NO. 3 STEERING WHEEL COVER LOWER

FRONT DOOR SCUFF PLATE LH

STEERING PAD ASSEMBLY

N*m (kgf*cm, ft.*lbf) : Specified torque

09490_TOYP_G0016

Exploded view of lower instrument panel, steering wheel and related components—2004–06 Prius

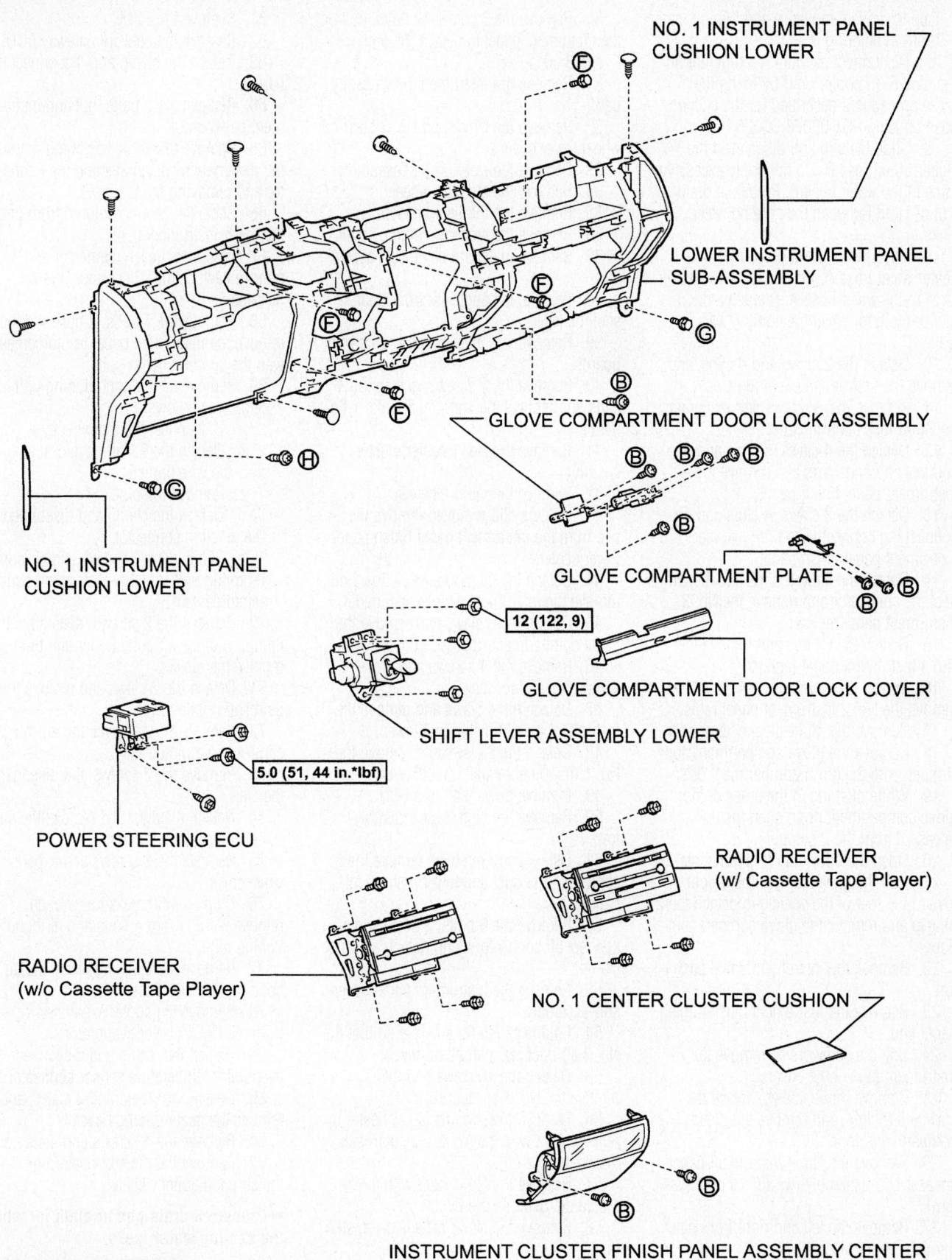

NO. 1 INSTRUMENT PANEL
CUSHION LOWER

LOWER INSTRUMENT PANEL
SUB-ASSEMBLY

GLOVE COMPARTMENT DOOR LOCK ASSEMBLY

NO. 1 INSTRUMENT PANEL
CUSHION LOWER

GLOVE COMPARTMENT PLATE

12 (122, 9)

GLOVE COMPARTMENT DOOR LOCK COVER

SHIFT LEVER ASSEMBLY LOWER

5.0 (51, 44 in.*lbf)

POWER STEERING ECU

RADIO RECEIVER
(w/ Cassette Tape Player)

RADIO RECEIVER
(w/o Cassette Tape Player)

NO. 1 CENTER CLUSTER CUSHION

INSTRUMENT CLUSTER FINISH PANEL ASSEMBLY CENTER

N*m (kgf*cm, ft.*lbf) : Specified torque

09490_TOYP_G0017

Exploded view of lower instrument panel assembly—2004–06 Prius

tape to prevent moisture and foreign matter from entering it.

7. Disconnect the cooler refrigerant liquid pipe E (to cooler unit) by using the same procedures described for the suction hose utilizing SST 09870-00025.

8. Slide the clip and disconnect the heater water hose B. Do not apply excessive force to the water hose B. Prepare a drain pan or cloth for when the cooling water leaks.

9. Slide the clip and disconnect the heater water hose A. Do not apply excessive force to the water hose A. Prepare a drain pan or cloth for when the cooling water leaks.

10. Detach the 2 claws and 4 clips, and remove the instrument panel register.

11. Remove the 2 screws and disconnect the hood lock control cable.

12. Detach the 4 claws, 5 clips and disconnect all connectors and remove the instrument panel finish panel.

13. Detach the 3 claws, 4 clips and disconnect the connector and remove the instrument panel finish panel.

14. Detach the claw, 5 clips and disconnect the connector and remove the No. 3 instrument panel register.

15. Detach the 6 clips and remove the No. 4 instrument panel register.

16. Detach the 2 claws and 4 clips, and remove the No. 2 instrument panel register.

17. Remove the multi-display assembly.

18. Remove the glove compartment door stopper from the glove compartment door.

19. While pushing in the sides of the glove compartment door, open the door to release it from the 2 stoppers.

20. Open the door until it is horizontal.

21. Pull the glove compartment door toward the rear of the vehicle to detach the 2 hinges and remove the glove compartment door.

22. Remove the instrument panel cushion.

23. Remove the instrument cluster finish panel end.

24. Using a screwdriver, remove the 2 instrument panel hole covers.

25. Remove the 2 screws, detach the claw and 2 clips, and remove the glove compartment door.

26. Remove the No. 1 instrument panel speaker sub-assembly (w/ JBL Sound System).

27. Remove the left and right front pillar garnish corner pieces.

28. Remove the left and right front pillar garnishes.

29. Disconnect the passenger airbag connector.

30. Remove the 3 screws, 2 bolts, pull up the instrument panel to detach the 6 claws and 3 clips.

31. Remove the instrument panel safety pad.

32. Remove the No. 2 and No. 3 steering wheel lower covers.

33. Remove the steering pad assembly.

34. Remove the steering wheel.

35. Remove the tilt lever bracket.

36. Remove the steering column cover.

37. Remove the spiral cable sub-assembly.

38. Remove left and right front door scuff plates.

39. Remove left and right cowl side trim boards.

40. Remove the 2 screws, detach the 4 clips and remove the instrument cluster finish panel.

41. Remove the No. 1 center cluster cushion.

42. Remove the radio receiver.

43. Using a clip remover, remove the clip from the instrument panel finish panel lower center.

44. Detach the 4 claws and 2 clips, and remove the instrument panel finish panel.

45. Detach the 6 claws and remove the glove compartment door lock cover.

46. Remove the 4 screws and glove compartment door lock.

47. Detach the 4 claws and remove the instrument panel finish panel lower.

48. Detach the 2 claws and remove the No. 1 instrument panel under cover.

49. Remove power steering ECU.

50. Remove the shift lever assembly lower.

51. Using a clip remover, remove the 8 clips from the duct and lower instrument panel.

52. Remove the 6 bolts, 2 screws, disconnect all connectors, and detach all clamps.

53. Remove the instrument panel lower sub-assembly.

54. Fold back the floor carpet so that the No. 3 air duct rear can be removed.

55. Detach the 10 claws and then remove the No. 3 air duct rear.

56. Remove the clip and No. 3 heater to register duct with the No. 2 side defroster nozzle duct.

57. Remove the No. 1 duct with the No. 1 side defroster nozzle duct.

58. Remove the No. 2 heater-to-register duct.

59. Remove the 2 clips, detach the 3 claws and then remove the defroster nozzle.

60. Remove the transaxle control ECU assembly.

61. Remove the ECM.

62. Remove the network gateway ECU.

63. Detach the clamp and disconnect the harness.

64. Remove the 2 bolts, nut and instrument panel brace.

65. Remove the air conditioning amplifier assembly by disconnecting the connector and removing the 2 screws.

66. Separate steering column from under the lower instrument panel.

67. Disconnect each connector and remove each clamp. Disconnect the wire harness.

68. Remove the 7 bolts, 2 nuts and then remove the instrument panel reinforcement with the air conditioner unit.

69. Remove the air conditioning unit assembly as follows:
    a. Disconnect the 5 connectors.
    b. Detach the 9 clamps and disconnect the wire harness.
    c. Remove the bolt.
    d. Detach the clamp and disconnect the junction connector.
    e. Remove the 2 screws and air conditioning unit from the instrument panel reinforcement.

70. Remove the 2 screws, disengage the fittings with the air conditioner and then remove the blower.

71. Detach the 2 claws and remove the defroster nozzle lower.

72. Remove the 2 screws and air mix control servo motor.

73. Remove the 2 screws and disengage the claw.

74. Detach the claw and remove the air outlet control servo motor.

75. Remove the 2 screws and expansion valve cover.

76. Using a 4mm hexagon wrench, remove the 2 hexagon bolts and air conditioning tube.

77. Remove the 2 O-rings from the air conditioning tube assembly.

78. Remove the cooler expansion valve from the No. 1 cooler evaporator.

79. Detach the clamp and disconnect the evaporator temperature sensor connector.

80. Detach the clamp and 2 claws, and remove the heater piping cover.

81. Remove the 4 screws and 4 clamps.

82. Remove the radiator heater unit from the air conditioner radiator.

➡**Prepare a drain pan or cloth for when the cooling water leaks.**

83. Remove the 2 screws and No. 1 air duct.

84. Remove the 4 screws and quick heater assembly.

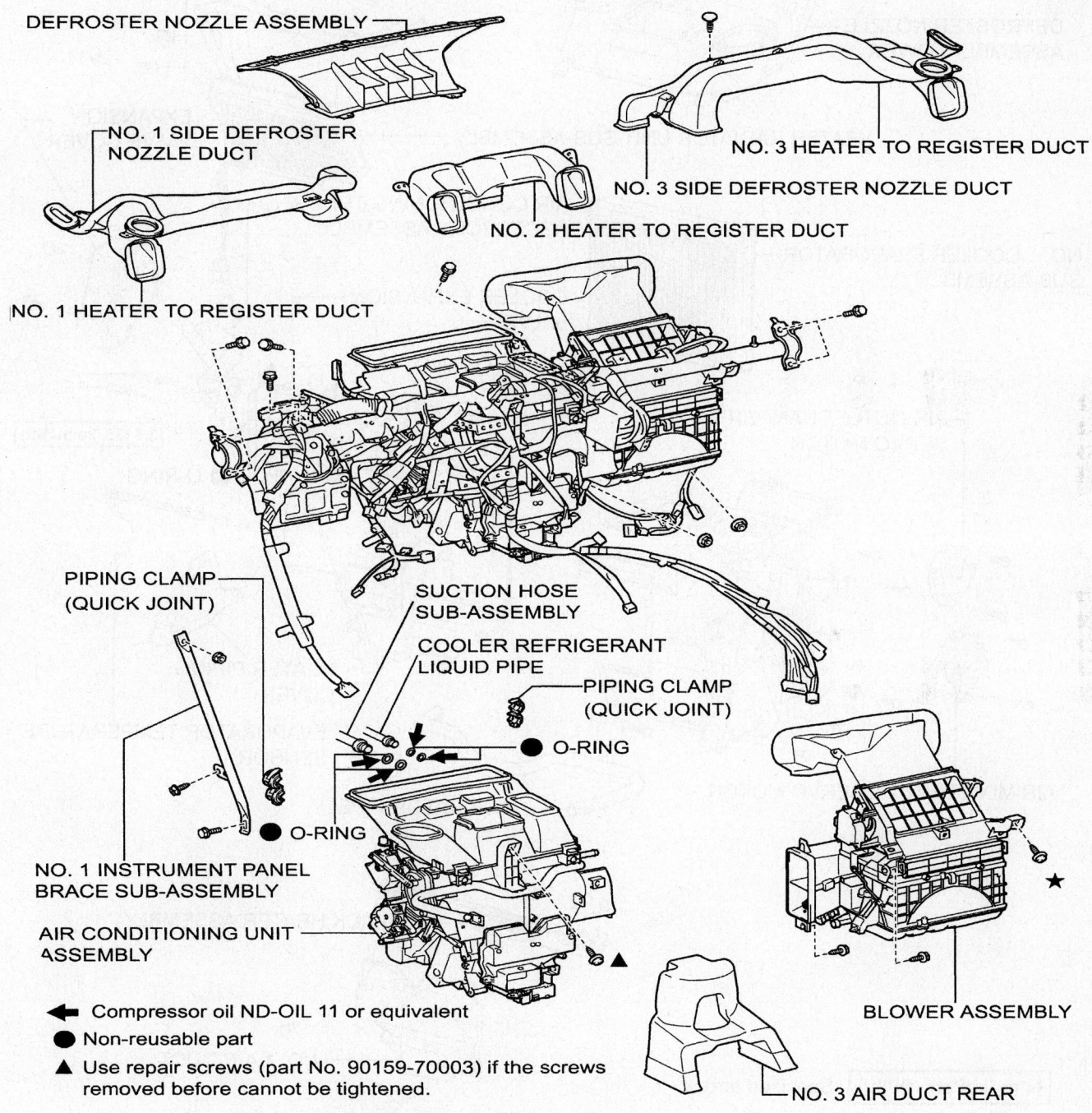

DEFROSTER NOZZLE ASSEMBLY

NO. 1 SIDE DEFROSTER NOZZLE DUCT

NO. 3 HEATER TO REGISTER DUCT

NO. 3 SIDE DEFROSTER NOZZLE DUCT

NO. 2 HEATER TO REGISTER DUCT

NO. 1 HEATER TO REGISTER DUCT

PIPING CLAMP (QUICK JOINT)

SUCTION HOSE SUB-ASSEMBLY

COOLER REFRIGERANT LIQUID PIPE

PIPING CLAMP (QUICK JOINT)

O-RING

O-RING

NO. 1 INSTRUMENT PANEL BRACE SUB-ASSEMBLY

AIR CONDITIONING UNIT ASSEMBLY

BLOWER ASSEMBLY

NO. 3 AIR DUCT REAR

◀ Compressor oil ND-OIL 11 or equivalent

● Non-reusable part

▲ Use repair screws (part No. 90159-70003) if the screws removed before cannot be tightened.

09490_TOYP_G0018

Exploded view of A/C unit and related components—2004–06 Prius

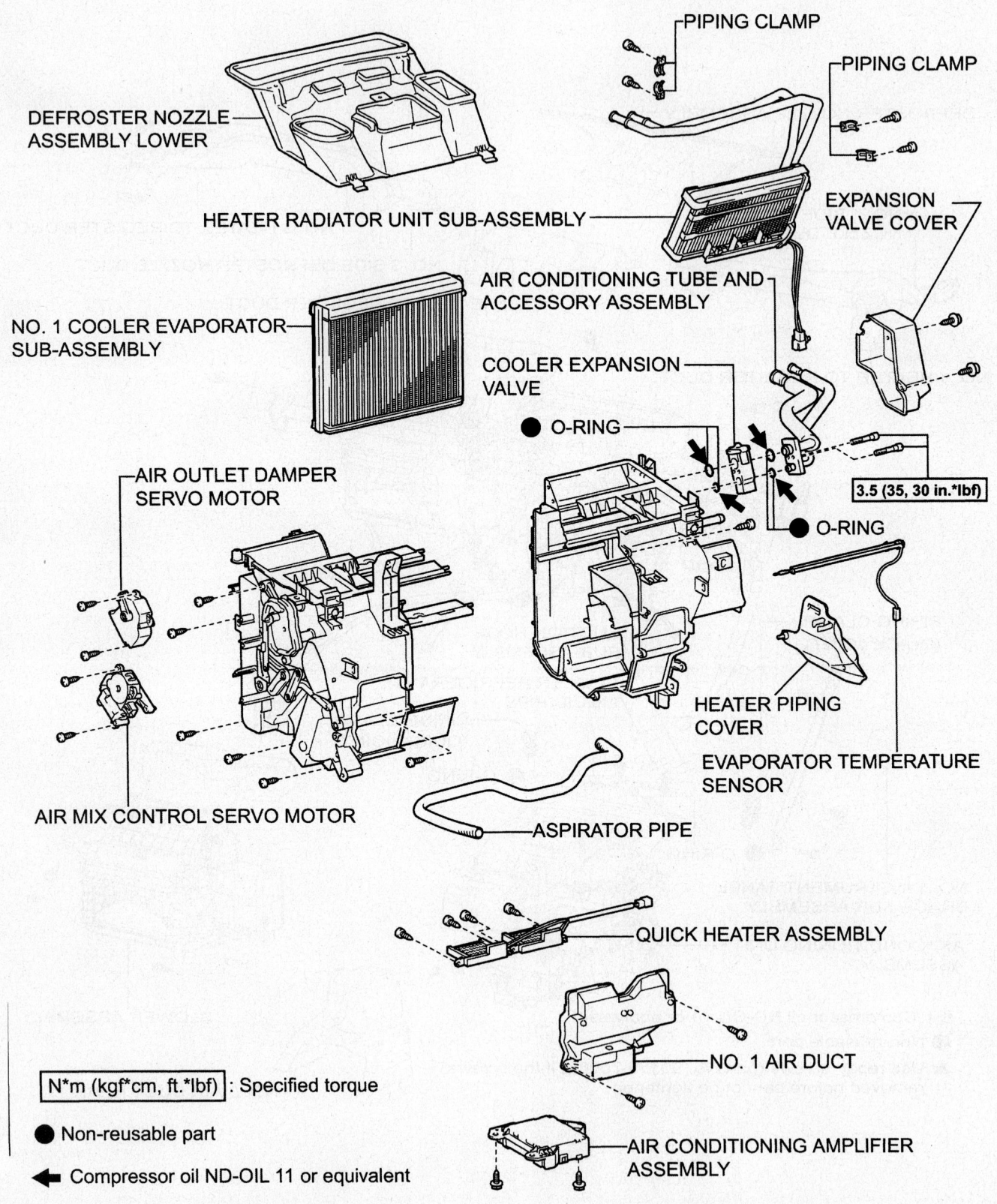

PIPING CLAMP

PIPING CLAMP

DEFROSTER NOZZLE
ASSEMBLY LOWER

HEATER RADIATOR UNIT SUB-ASSEMBLY

EXPANSION
VALVE COVER

AIR CONDITIONING TUBE AND
ACCESSORY ASSEMBLY

NO. 1 COOLER EVAPORATOR
SUB-ASSEMBLY

COOLER EXPANSION
VALVE

● O-RING

3.5 (35, 30 in.*lbf)

● O-RING

AIR OUTLET DAMPER
SERVO MOTOR

HEATER PIPING
COVER

EVAPORATOR TEMPERATURE
SENSOR

AIR MIX CONTROL SERVO MOTOR

ASPIRATOR PIPE

QUICK HEATER ASSEMBLY

NO. 1 AIR DUCT

N*m (kgf*cm, ft.*lbf) : Specified torque

● Non-reusable part

◄ Compressor oil ND-OIL 11 or equivalent

AIR CONDITIONING AMPLIFIER
ASSEMBLY

09490_TOYP_G0019

Heater radiator (heater core) removal from A/C unit—2004–06 Prius

85. Remove the 7 screws and No. 1 cooler evaporator from the heater case.

86. Remove the 2 O-rings from the No. 1 cooler evaporator.

87. Detach the 2 claws and remove the evaporator temperature sensor.

To install:

88. Install the evaporator temperature sensor and attach the 2 claws.

89. Install the No. 1 cooler evaporator sub-assembly as follows:

a. Sufficiently apply compressor oil (ND-OIL 11) to 2 new O-rings and the fitting surface. Install the 2 O-rings to the No. 1 cooler evaporator.

❋❋ CAUTION

Do not use any compressor oil other than ND-OIL 11. If any compressor oil other than ND-OIL 11 is used, compressor motor insulation performance may decrease, resulting in a leakage of electric power.

b. Install the No. 1 cooler evaporator with the 7 screws.

90. Install the quick heater with the 4 screws.

91. Install the No. 1 air duct with the 2 screws.

92. Install the air conditioning radiator to the radiator heater unit and install the 4 clamps and 4 screws.

93. Attach the 2 claws and clamp to install the heater piping cover.

94. Attach the clamp and connect the evaporator temperature sensor connector.

95. Install the cooler expansion valve to the No. 1 cooler evaporator.

96. Install air conditioning tube and accessory assembly as follows:

a. Sufficiently apply compressor oil (ND-OIL 11) to 2 new O-rings and the fitting surface. Install the 2 O-rings to the air conditioning tube assembly.

❋❋ CAUTION

Do not use any compressor oil other than ND-OIL 11. If any compressor oil other than ND-OIL 11 is used, compressor motor insulation performance may decrease, resulting in a leakage of electric power.

b. Install the air conditioning tube to the No. 1 cooler evaporator, placing the cooler expansion valve between them. Using a 0.16 inch (4mm) hexagon wrench, install the 2 hexagon bolts and tighten to 30 inch lbs. (3.5 Nm).

c. Install the expansion cover with the 2 screws.

97. Install the air outlet control servo motor and attach the claw.

98. Attach the claw and install the 2 screws.

99. Install the air mix control servo motor with the 2 screws.

100. Install the defroster nozzle lower and attach the 2 claws.

101. Install the blower and then attach the fittings with the air conditioner and install the 2 screws.

102. Install the air conditioning unit assembly as follows:

a. Install the air conditioning unit to the instrument panel reinforcement with the 2 screws. Use repair screws (part No. 90159-70003) if the screws removed before cannot be tightened.

b. Attach the clamp to connect the junction connector.

c. Install the bolt.

d. Attach the 9 clamps to connect the wire harness.

e. Connect the 5 connectors.

103. Install the instrument panel reinforcement with the 7 bolts.

104. Install the 2 nuts to the air conditioning unit and temporarily tighten them.

105. Connect each connector and install each clamp. Then connect the wire harness.

106. Install the air conditioning amplifier assembly with the 2 screws and engage the connector.

107. Place steering column into position underneath the lower instrument panel and tighten the mounting bolts.

108. Install the instrument panel brace with the 2 bolts and nut.

109. Attach the clamp to connect the harness.

110. Fully tighten the air conditioning unit with the 2 nuts. Tighten the left nut first, then the right nut second.

111. Install the network gateway ECU.

112. Install the ECM.

113. Install the transaxle control ECU assembly.

114. Install the defroster nozzle and attach the 3 claws.

115. Install the 2 clips.

116. Install the No. 2 heater-to-register duct.

117. Install the duct with the No. 1 side defroster nozzle duct.

118. Install the clip and then the duct with the No. 2 side defroster nozzle duct.

119. Attach the 10 claws to install the No. 3 air duct rear, then return the carpet to its original position.

120. Install the lower instrument panel sub-assembly.

121. Connect all the connectors and attach all the clamps.

122. Install the 2 screws, 6 bolts, 8 clips to the duct and lower instrument panel.

123. Install the lower shift lever assembly.

124. Install the power steering ECU.

125. Attach the 2 claws to install the No. 1 instrument panel under cover.

126. Attach the 4 claws to install the instrument panel finish panel.

127. Install the glove compartment door lock with the 4 screws.

128. Attach the 6 claws to install the glove compartment door lock cover.

129. Attach the 4 claws and 2 clips to install the instrument panel finish panel lower center. Install the clip.

130. Install the radio receiver.

131. Install the No. 1 center cluster cushion.

132. Attach the 4 claws and 2 clips to install the instrument cluster finish panel assembly center. Install the 2 screws.

133. Install left and right cowl side trim boards.

134. Install left and right front door scuff plates.

135. Install the spiral cable sub-assembly.

136. Install the steering column cover.

137. Install the tilt lever bracket.

138. Install the steering wheel and tighten the center nut to 37 ft. lbs. (50 Nm).

139. Install the steering pad assembly.

140. Install the No. 2 and No. 3 steering wheel lower covers.

141. Attach the 6 claws and 3 clips to install the instrument panel safety pad. Install the 2 bolts and 3 screws.

142. Connect the passenger airbag connector.

143. Install left and right front pillar garnishes.

144. Install left and right front pillar garnish corner pieces.

145. Install the No. 1 instrument panel speaker sub-assembly (w/ JBL Sound System).

146. Attach the claw and 2 clips to install the glove compartment door. Install the 2 screws.

147. Install the 2 instrument panel hole covers.

148. Install the instrument cluster finish panel end.

149. Install the instrument panel cushion.

150. Attach the 2 hinges to install the glove compartment door.

151. While pushing in the sides of the glove compartment door, engage it to the 2 stoppers.

152. Install glove compartment door stopper sub-assembly.

153. Install the multi-display assembly.

154. Attach the 2 claws and 4 clips to install the No. 2 instrument panel register.

155. Attach the 6 clips to install the No. 4 instrument panel register.

156. Connect the connector, attach the claw and 5 clips to install the No. 3 instrument panel register.

157. Connect the connector, attach the 3 claws and 4 clips to install the upper instrument panel finish panel.

158. Connect the connectors, attach the 4 claws and 5 clips to install the lower instrument panel finish panel.

159. Connect the hood lock control cable and install the 2 screws.

160. Attach the 2 claws and 4 clips to install the instrument panel register.

161. Connect the heater water hose A and slide the clip.

162. Connect the heater water hose B and slide the clip.

163. Connect the cooler refrigerant liquid pipe E (to cooler unit) as follows:

a. Remove the attached vinyl tape from the pipe's disconnected part.

b. Sufficiently apply compressor oil (ND-OIL 11) to 2 new O-rings and the pipe's connecting part.

CAUTION

Do not use any compressor oil other than ND-OIL 11. If any compressor oil other than ND-OIL 11 is used, compressor motor insulation performance may decrease, resulting in a leakage of electric power.

c. Install the O-rings to the pipe.

d. Insert the pipe joint into the cooler unit fitting hole securely.

e. Using the piping clamp, connect the cooler unit refrigerant liquid pipe E. Ensure that the piping clamp is securely engaged.

164. Connect the suction hose sub-assembly by using the same procedures described for the cooler unit refrigerant liquid pipe E.

165. Refill the cooling system.

166. Evacuate and recharge the refrigerant.

167. Connect the negative battery cable.

168. Operate the engine to normal operating temperatures; then, check the climate control operation and check for coolant and refrigerant leaks.

169. Perform calibration.

170. Perform initialization.

171. Check the SRS warning light.

Cylinder Head

REMOVAL & INSTALLATION

1. Before servicing the vehicle, refer to the Precautions Section.

2. Properly relieve the fuel system pressure.

3. Drain engine oil.

4. Drain transaxle and engine coolant.

5. Disconnect the negative battery cable. Wait at least 90 seconds after disconnecting the negative battery cable to prevent possible airbag and seat belt pretensioner activation.

6. While wearing insulating gloves, slide up the lever of the service plug grip and remove while turning the lever to the left. Be sure to insulate the service plug with insulating tape.

CAUTION

Do not touch the high voltage connectors and terminals for 5 minutes after removing the service plug grip.

7. Remove the 2 bolts and 2 compression rings, then disconnect the exhaust pipe assembly front from the exhaust manifold.

8. On 2002–03 Prius models, while wearing insulating gloves, remove the converter and inverter assembly as follows:

a. Disconnect the battery power cable connector and insulate it with packaging tape.

b. Remove the inverter terminal cover.

c. Disconnect the circuit breaker sensor by first moving the outer cover section away toward the wire side.

d. Remove the circuit breaker sensor.

e. Verify that there is 0V by measuring the voltage between the terminals of the three phase connector (U-V, V-W, U-W), using a voltmeter at each terminal and body ground.

f. Remove the mounting bolts and disconnect the MG1 and MG2 power cables. Protect the connector parts with insulating tape.

g. Disconnect the ground cable.

h. Disconnect the 3 water hoses from the converter and inverter assembly.

i. Remove the 4 mounting bolts and the converter and inverter assembly.

9. On 2004–06 Prius models, while wearing insulating gloves, remove the inverter with converter assembly as follows:

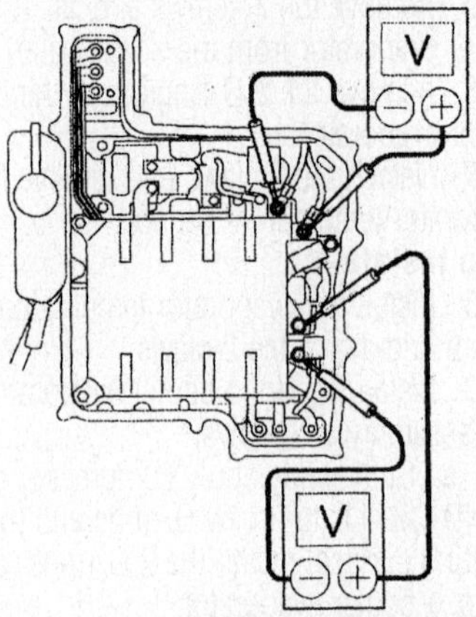

09490_TOYP_G0007

Verifying that there is 0V at the inverter with converter—Prius

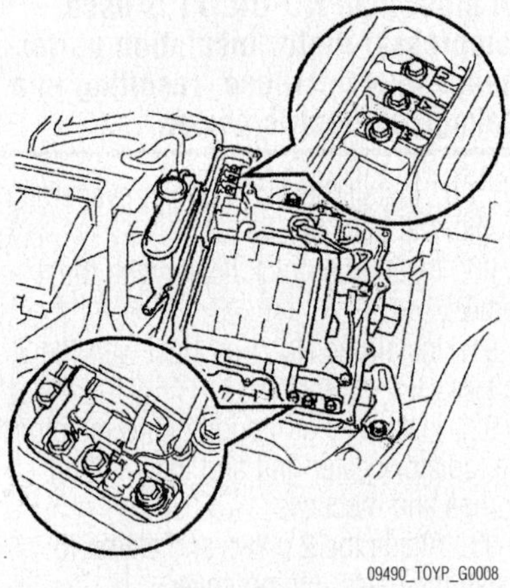

09490_TOYP_G0008

Measuring voltage between the terminals of the three phase connector—Prius

a. Remove the left and right engine under covers.

b. Drain the HV coolant.

c. Remove the radiator support opening cover.

d. Remove the inverter cover.

e. Verify that there is 0V at the inverter with converter, using a voltmeter utilizing a measuring range of DC 400V or more.

f. Again, verify that there is 0V by measuring the voltage between the terminals of the three phase connector (U-V, V-W, U-W), using a voltmeter utilizing a measuring range of DC 400V or more.

g. Disconnect the No. 1, No. 2 and No. 6 inverter cooling hoses.

h. Disconnect the No. 1 circuit breaker sensor by first moving the outer section away toward the wire side.

i. Disconnect the 2 frame wire connectors from the inverter with converter assembly and protect the electrode and connector parts with insulating tape.

j. Use a small screwdriver to lift up the green lock pin, then disconnect the connector for the air conditioning inverter.

k. Disconnect any remaining wiring connectors including the engine main wiring harness.

l. Remove the mounting bolts and disconnect the MG1 and MG2 power cables. Protect the connector parts with insulating tape.

m. Remove the 3 mounting bolts and the inverter with converter assembly.

10. Remove timing chain.

11. Disconnect associated connectors and wire harnesses.

12. Remove the fuel pipe clamp.

13. Disconnect the fuel tube from the fuel delivery pipe. Even if the fuel tube is stuck and cannot be disconnected, do not use any tools. Push and pull the parts with the quick connector pinched to disconnect the tube.

14. Cover the disconnected fuel tube and fuel delivery pipe with a plastic bag in order to prevent foreign objects from entering them.

15. Disconnect the radiator inlet hose from the cylinder head.

16. Disconnect the hoses from the cylinder head as shown in the illustration.

17. Remove the bolt and disconnect the No. 1 water by-pass pipe.

18. Disconnect the hoses, remove the bolt and disconnect the oil dipstick guide.

19. Remove the camshafts.

20. Using an 8mm bi-hexagon wrench, loosen the cylinder head bolts in several steps in the proper sequence. Then remove the cylinder head bolts and washer.

✳✳ CAUTION

When removing the bolt, do not drop the washer into the engine. Removing the cylinder head bolts in the wrong order may cause damage to the cylinder head.

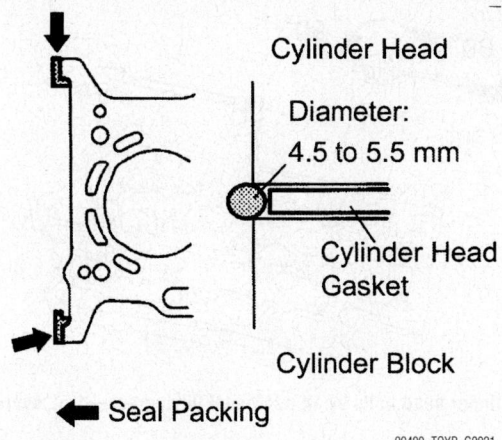

Seal packing correctly applied for cylinder head installation—1.5L hybrid engine

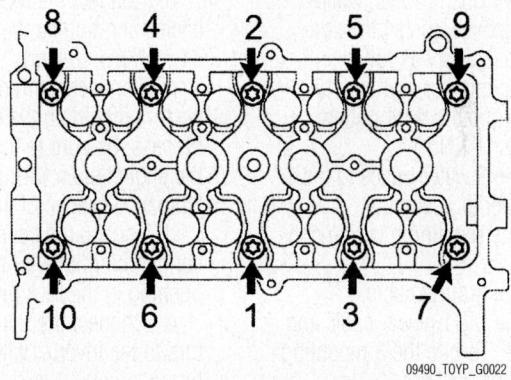

Cylinder head bolt tightening sequence—1.5L hybrid engine

21. Remove the cylinder head and gasket.

To install:

22. Install the cylinder head along with a new gasket.

23. Apply seal packing with a 0.177–0.217 inch (4.5–5.5mm) thickness. Install the cylinder head within 3 minutes of applying seal packing.

24. Apply a light coat of engine oil to the threads of the cylinder head bolts.

25. Using several steps, install and tighten the 10 cylinder head bolts and plate washers uniformly with an 8mm bi-hexagon wrench in the proper sequence to 21 ft. lbs. (29 Nm).

26. Mark the front of the cylinder head bolt with paint.

27. Retighten the cylinder head bolts by an additional 90 degrees and then another 90 degrees. Check that the paint mark is now 180 degrees opposite to the front.

28. Install the camshafts.

29. Install the oil dipstick guide with the bolt and tighten to 80 inch lbs. (9 Nm). Connect the hose.

30. Connect the water by-pass pipe with the bolt and tighten to 80 inch lbs. (9 Nm).

31. Connect the hoses.

32. Connect the radiator inlet hose.

33. Push the fuel main tube into the fuel delivery pipe until it makes a "click" sound. If the fuel tube is connected too tightly, apply a light coat of engine oil to the tip of the fuel delivery pipe. After connecting, check that the fuel tube is securely connected by pulling it.

34. Install the fuel pipe clamp.

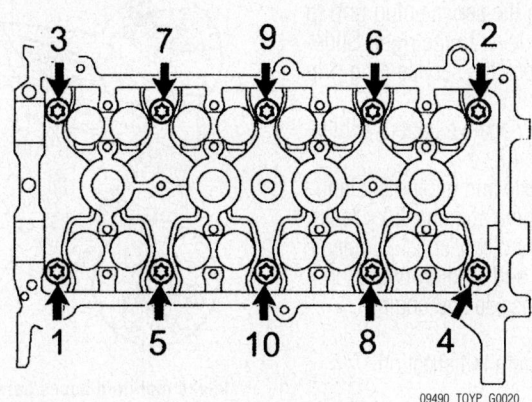

Cylinder head bolt loosening sequence—1.5L hybrid engine

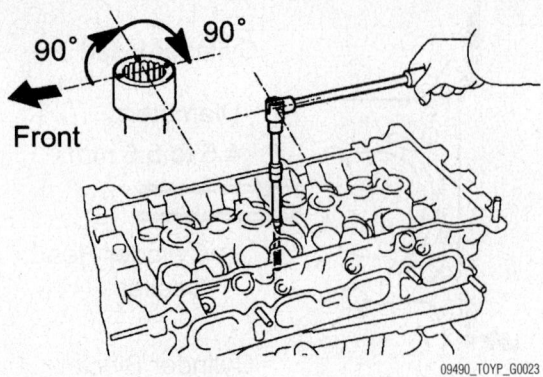

Retightening the cylinder head bolts by an additional 90 degrees—1.5L hybrid engine

35. Connect all associated connectors and wire harnesses.

36. Install the timing chain.

37. On 2002–03 Prius models, while wearing insulating gloves, install the converter and inverter assembly as follows:

   a. Install the converter and inverter assembly and torque the 4 mounting bolts to 15 ft. lbs. (21 Nm).

   b. Connect the 3 water hoses to the converter and inverter assembly.

   c. Connect any remaining wiring connectors.

   d. Connect the ground cable.

   e. Connect the MG1 power cable and 3 mounting bolts. Torque the 3 mounting bolts to 62 inch lbs. (7 Nm).

   f. Connect the 3 MG2 power cables and 6 mounting bolts. Torque the **A** mounting bolts to 14 ft. lbs. (19.5 Nm), and the **B** mounting bolts to 71 inch lbs. (8 Nm).

   g. Install the circuit breaker sensor and connector cover. Torque the 2 mounting screws to 15 ft. lbs. (20 Nm).

   h. Connect the circuit breaker sensor.

   i. Install the inverter terminal cover and gasket. Tighten the 4 screws to 71 inch lbs. (8 Nm).

38. On 2004–06 Prius models, while wearing insulating gloves, install the inverter with converter assembly as follows:

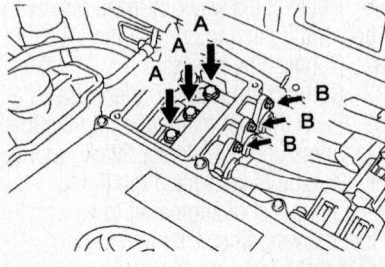

MG2 power cables and mounting bolts—2002–03 Prius

   a. Install the inverter with converter assembly and tighten the three mounting bolts to 16 ft. lbs. (21 Nm).

   b. Connect the MG1 and MG2 power cables and tighten the mounting bolts to 71 inch lbs. (8 Nm).

   c. Connect any remaining wiring connectors including the engine main wiring harness. Be sure to insert the grommet of the engine main wiring harness into the U-shaped groove of the inverter case.

   d. Engage the connector for the air conditioning inverter and secure by pushing in the lock pin..

   e. Connect the 2 frame wire connectors to the inverter with converter assembly.

   f. Connect the No. 1 circuit breaker sensor.

   g. Connect the No. 6, No. 2 and No. 1 inverter cooling hoses.

   h. Install the inverter cover and tighten the mounting fasteners to 8 ft. lbs. (11 Nm).

   i. Install the radiator support opening cover.

   j. Install the left and right engine under covers.

39. Install the front exhaust assembly.

40. While wearing insulating gloves, joint the service plug grip with the HV battery. While pushing the service plug grip to the right, rotate the lever to the right. Slide the lever down to lock the service plug grip in place.

41. Connect the negative battery cable.

42. Refill engine oil.

43. Refill transaxle and engine coolant.

44. Start the engine and check for leaks, check for abnormal noises, shock slippage, correct shift points and smooth operation.

45. Recheck transaxle and engine coolant.

46. Perform system initialization (2004–06 models).

Rocker Arms/Shafts

The 1.5L hybrid engine does not utilize rocker arms/shafts. On the 1.5L hybrid engine, the camshaft directly actuates the valves.

Intake Manifold

REMOVAL & INSTALLATION

1. Before servicing the vehicle, refer to the Precautions Section.

2. Disconnect the negative battery cable. Wait at least 90 seconds after disconnecting the negative battery cable to prevent possible airbag and seat belt pretensioner activation.

3. Drain the engine coolant.

4. Remove the air cleaner assembly.

5. Remove the cylinder head.

6. Remove the oil dipstick guide.

7. On 2002–03 Prius models, remove the throttle body assembly as follows:

   a. Disconnect the throttle position sensor connector.

   b. Disconnect the throttle control motor connector.

   c. Disconnect the 2 PCV hoses from throttle body.

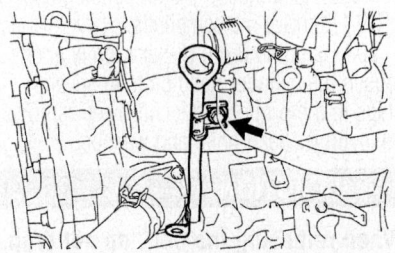

Remove/install oil dipstick guide—1.5L hybrid engine

Intake manifold hoses/harnesses to disconnect/connect—1.5L hybrid engine

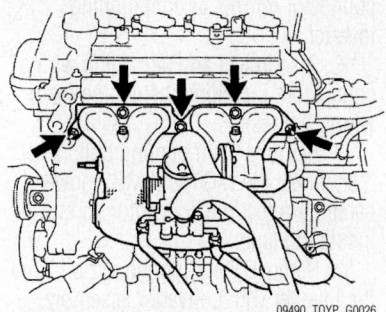

09490_TOYP_G0026

Location of intake manifold mounting bolts and nuts—1.5L hybrid engine

d. Disconnect the 2 water by-pass hoses.

e. Remove the bolt, 2 nuts, accelerator bracket and throttle body from the intake manifold.

f. Remove the throttle body gasket.

8. Remove the bolt and knock control sensor with bracket.

9. Disconnect the wiring harness from the bracket.

10. Disconnect the ventilation hose.

11. Disconnect the water by-pass hose.

12. Remove the 3 bolts and 2 nuts, then remove the intake manifold and gasket.

To install:

13. Install a new gasket, then install the intake manifold with the bolts, nuts and brackets. Uniformly tighten the bolts and nuts in several passes to 15 ft. lbs. (20 Nm).

14. Connect the water by-pass hose.

15. Connect the ventilation hose.

16. Install the knock control sensor with bracket with the bolt and tighten to 80 inch lbs. (9 Nm).

17. On 2002–03 Prius models, install the throttle body assembly and tighten the mounting bolt and nuts to 15 ft. lbs. (20 Nm).

18. Apply engine oil to a new O-ring, then install it to the dipstick guide.

19. Install the dipstick guide with the bolt and tighten to 80 inch lbs. (9 Nm).

20. Install the cylinder head.

21. Install the dipstick.

22. Install the air cleaner assembly.

23. Refill the engine coolant.

24. Connect the negative battery cable.

25. Start the engine and check for coolant leaks.

Exhaust Manifold

REMOVAL & INSTALLATION

✳✳ CAUTION

To avoid the danger of being burned, do not service the exhaust system while it is hot. Service should be performed only after the system cools down.

1. Before servicing the vehicle, refer to the Precautions Section.

2. Properly relieve the fuel system pressure.

3. Disconnect the negative battery cable. Wait at least 90 seconds after disconnecting the negative battery cable to prevent possible airbag and seat belt pretensioner activation.

4. While wearing insulating gloves, slide up the lever of the service plug grip and remove while turning the lever to the left. Be sure to insulate the service plug with insulating tape.

✳✳ CAUTION

Do not touch the high voltage connectors and terminals for 5 minutes after removing the service plug grip.

5. Remove the cylinder head.

6. Remove the 4 bolts and exhaust manifold insulator.

7. Remove the 3 bolts and 2 nuts, then remove the exhaust manifold.

8. Clean the sealing surfaces of the exhaust manifold and the cylinder head.

To install:

9. Install a new gasket, then install the exhaust manifold.

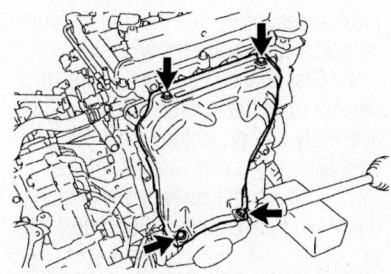

09490_TOYP_G0027

Removal/installation of exhaust manifold insulator—1.5L hybrid engine

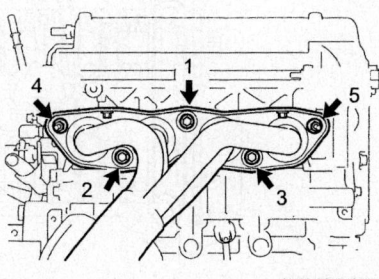

09490_TOYP_G0028

Exhaust manifold nut and bolt tightening sequence—1.5L hybrid engine

10. Tighten the 3 bolts and 2 nuts in the proper sequence to 20 ft. lbs. (27 Nm).

11. Install the exhaust manifold insulator with the 4 bolts and tighten to 71 inch lbs. (8 Nm).

12. Install the cylinder head.

13. Install the service plug.

14. Connect the negative battery cable.

15. Run the engine and check for exhaust leaks.

Front Crankshaft Seal

REMOVAL & INSTALLATION

There are 2 ways to replace the oil seal: remove it with the timing chain cover removed, or remove it with the timing chain cover installed.

1. If the timing chain cover is removed, perform the following:

a. Using a screwdriver, remove the oil seal. Tape the screwdriver tip before use.

b. Using SST 09950-60010, 09950-70010 and a hammer, tap in a new oil seal until its surface is flush with the timing chain cover edge. Be careful not to tap the oil seal at an angle. Keep the lip free of foreign objects.

c. Apply multi-purpose grease to the lip of the oil seal.

2. If the timing chain cover is installed, perform the following:

a. Using a knife, cut off the lip of the oil seal.

b. Using a screwdriver with the tip wrapped in tape, pry out the oil seal.

c. After removal, check if the crankshaft is not damaged. If it is damaged, smooth the surface with 400-grit sandpaper.

d. Apply multi-purpose grease to the lip of a new oil seal. Keep the lip free of foreign objects.

e. Using SST 09223-22010 and a hammer, tap in the oil seal until its surface is flush with the timing chain cover edge. Be careful not to tap the oil seal at an angle. Wipe any extra grease off the crankshaft.

Camshaft and Valve Lifters

REMOVAL & INSTALLATION

1. Before servicing the vehicle, refer to the Precautions Section.

2. Properly relieve the fuel system pressure.

3. Drain engine oil.

4. Disconnect the negative battery

cable. Wait at least 90 seconds after disconnecting the negative battery cable to prevent possible airbag and seat belt pretensioner activation.

5. While wearing insulating gloves, slide up the lever of the service plug grip and remove while turning the lever to the left. Be sure to insulate the service plug with insulating tape.

✳✳ CAUTION

Do not touch the high voltage connectors and terminals for 5 minutes after removing the service plug grip.

6. On 2002–03 Prius models, while wearing insulating gloves, remove the converter and inverter assembly as follows:

a. Disconnect the battery power cable connector and insulate it with packaging tape.

b. Remove the inverter terminal cover.

c. Disconnect the circuit breaker sensor by first moving the outer cover section away toward the wire side.

d. Remove the circuit breaker sensor.

e. Verify that there is 0V by measuring the voltage between the terminals of the three phase connector (U-V, V-W, U-W), using a voltmeter at each terminal and body ground.

f. Remove the mounting bolts and disconnect the MG1 and MG2 power cables. Protect the connector parts with insulating tape.

g. Disconnect the ground cable.

h. Disconnect the 3 water hoses from the converter and inverter assembly.

i. Remove the 4 mounting bolts and the converter and inverter assembly.

7. On 2004–06 Prius models, while wearing insulating gloves, remove the inverter with converter assembly as follows:

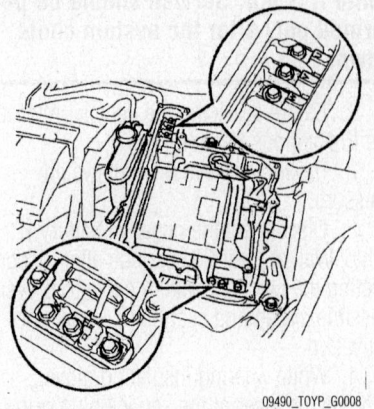

Measuring voltage between the terminals of the three phase connector—Prius

a. Remove the left and right engine under covers.

b. Drain the HV coolant.

c. Remove the radiator support opening cover.

d. Remove the inverter cover.

e. Verify that there is 0V at the inverter with converter, using a voltmeter utilizing a measuring range of DC 400V or more.

f. Again, verify that there is 0V by measuring the voltage between the terminals of the three phase connector (U-V, V-W, U-W), using a voltmeter utilizing a measuring range of DC 400V or more.

g. Disconnect the No. 1, No. 2 and No. 6 inverter cooling hoses.

h. Disconnect the No. 1 circuit breaker sensor by first moving the outer section away toward the wire side.

i. Disconnect the 2 frame wire connectors from the inverter with converter assembly and protect the electrode and connector parts with insulating tape.

j. Use a small screwdriver to lift up the green lock pin, then disconnect the

connector for the air conditioning inverter.

k. Disconnect any remaining wiring connectors including the engine main wiring harness.

l. Remove the mounting bolts and disconnect the MG1 and MG2 power cables. Protect the connector parts with insulating tape.

m. Remove the 3 mounting bolts and the inverter with converter assembly.

8. Remove timing chain.

9. Disconnect associated connectors and wire harnesses.

10. Remove the fuel pipe clamp.

11. Disconnect the fuel tube from the fuel delivery pipe. Even if the fuel tube is stuck and cannot be disconnected, do not use any tools. Push and pull the parts with the quick connector pinched to disconnect the tube.

12. Cover the disconnected fuel tube and fuel delivery pipe with a plastic bag in order to prevent foreign objects from entering them.

13. Disconnect the radiator inlet hose from the cylinder head.

14. Disconnect the hoses from the cylinder head.

15. Remove the bolt and disconnect the No. 1 water by-pass pipe.

16. Disconnect the hoses, remove the bolt and disconnect the oil dipstick guide.

17. Remove the No. 1 and No. 2 camshaft bearing caps in the proper sequence, then remove the camshaft and No. 2 camshaft.

➡ **Uniformly loosen the bolts, keeping the camshaft level.**

18. Remove the valve lifters. Keep the valve lifters in the correct order so that they can be returned to their original locations when reassembling.

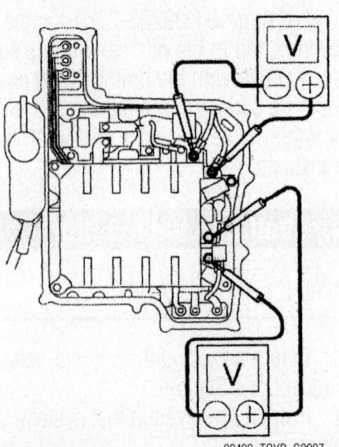

Verifying that there is 0V at the inverter with converter—Prius

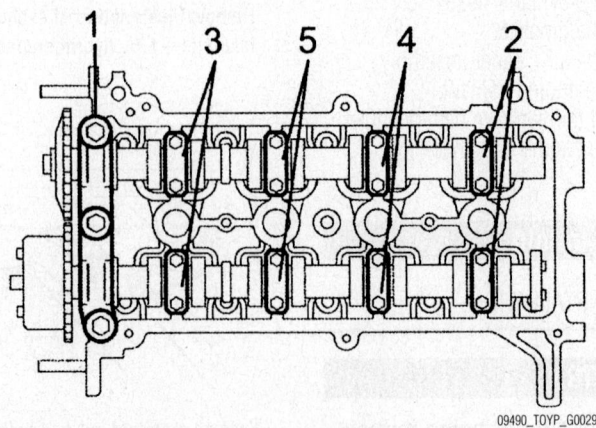

Camshaft cap bolt loosening sequence—1.5L hybrid engine

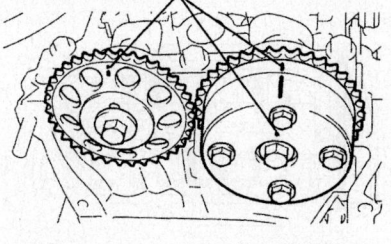

Timing Mark

09490_TOYP_G0030

Timing mark on the camshaft timing gear facing upward—1.5L hybrid engine

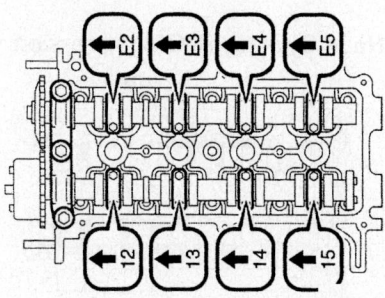

09490_TOYP_G0031

Check the front marks and numbers on the No. 1 and No. 2 camshaft bearing caps—1.5L hybrid engine

To install:

19. Apply a light coat of engine oil to the valve lifter and install the valve lifter. Check that the valve lifter rotates smoothly by hand.

➡**If turning the camshaft with the chain removed, turn the crankshaft counterclockwise by 40 degrees from TDC/compression.**

20. Apply engine oil to the cam and cylinder head journals.

21. Place the camshaft and No. 2 camshaft on the cylinder head with the timing mark on the camshaft timing gear facing upward.

22. Check the front marks and numbers on the No. 1 and No. 2 camshaft bearing caps, then temporarily install them.

23. Uniformly tighten the No. 2 camshaft bearing caps in several steps in the proper sequence to 9.6 ft. lbs. (13 Nm).

24. Uniformly loosen the bolts, keeping the camshaft level.

25. Install the No. 1 camshaft bearing cap. Torque to 17 ft. lbs. (23 Nm).

26. Connect the water by-pass pipe with the bolt and tighten to 80 inch lbs. (9 Nm).

27. Connect the hoses.

28. Connect the radiator inlet hose.

29. Push the fuel main tube into the fuel delivery pipe until it makes a "click" sound.

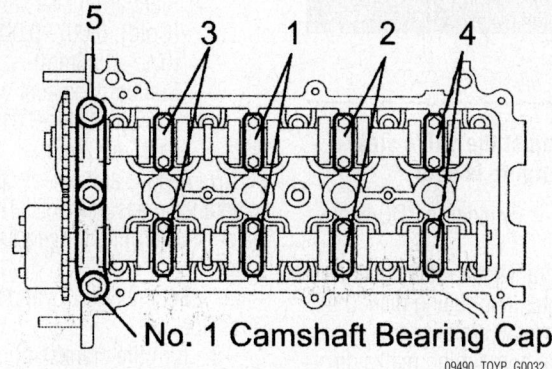

No. 1 Camshaft Bearing Cap

09490_TOYP_G0032

Camshaft cap bolt tightening sequence—1.5L hybrid engine

If the fuel tube is connected too tightly, apply a light coat of engine oil to the tip of the fuel delivery pipe. After connecting, check that the fuel tube is securely connected by pulling it.

30. Install the fuel pipe clamp.

31. Connect all associated connectors and wire harnesses.

32. Install the timing chain.

33. On 2002–03 Prius models, while wearing insulating gloves, install the converter and inverter assembly as follows:

    a. Install the converter and inverter assembly and torque the 4 mounting bolts to 15 ft. lbs. (21 Nm).

    b. Connect the 3 water hoses to the converter and inverter assembly.

    c. Connect any remaining wiring connectors.

    d. Connect the ground cable.

    e. Connect the MG1 power cable and 3 mounting bolts. Torque the 3 mounting bolts to 62 inch lbs. (7 Nm).

    f. Connect the 3 MG2 power cables and 6 mounting bolts. Torque the **A** mounting bolts to 14 ft. lbs. (19.5 Nm), and the **B** mounting bolts to 71 inch lbs. (8 Nm).

    g. Install the circuit breaker sensor and connector cover. Torque the 2 mounting screws to 15 ft. lbs. (20 Nm).

    h. Connect the circuit breaker sensor.

    i. Install the inverter terminal cover and gasket. Tighten the 4 screws to 71 inch lbs. (8 Nm).

34. On 2004–06 Prius models, while wearing insulating gloves, install the inverter with converter assembly as follows:

    a. Install the inverter with converter assembly and tighten the three mounting bolts to 16 ft. lbs. (21 Nm).

    b. Connect the MG1 and MG2 power cables and tighten the mounting bolts to 71 inch lbs. (8 Nm).

    c. Connect any remaining wiring connectors including the engine main wiring

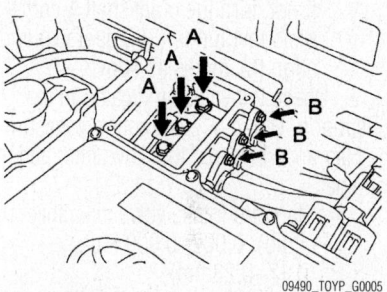

09490_TOYP_G0005

MG2 power cables and mounting bolts—2002–03 Prius

harness. Be sure to insert the grommet of the engine main wiring harness into the U-shaped groove of the inverter case.

    d. Engage the connector for the air conditioning inverter and secure by pushing in the lock pin..

    e. Connect the 2 frame wire connectors to the inverter with converter assembly.

    f. Connect the No. 1 circuit breaker sensor.

    g. Connect the No. 6, No. 2 and No. 1 inverter cooling hoses.

    h. Install the inverter cover and tighten the mounting fasteners to 8 ft. lbs. (11 Nm).

    i. Install the radiator support opening cover.

    j. Install the left and right engine under covers.

35. While wearing insulating gloves, joint the service plug grip with the HV battery. While pushing the service plug grip to the right, rotate the lever to the right. Slide the lever down to lock the service plug grip in place.

36. Connect the negative battery cable.

37. Refill engine oil.

38. Start the engine and check for leaks, check for abnormal noises, shock slippage, correct shift points and smooth operation.

39. Perform system initialization (2004–06 models).

Valve Lash

ADJUSTMENT

→**Inspect and adjust the valve clearance when the engine is cold.**

1. Set the No. 1 cylinder to TDC/compression.
   - Turn the crankshaft pulley until its timing notch and timing mark 0 of the chain cover are aligned.
   - Check that both timing marks on the camshaft timing sprocket and camshaft timing gear are facing upward as shown in the illustration. If not, turn the crankshaft 1 complete revolution (360 degrees) and align the marks as above.

2. Check the valves indicated in the illustration. Using a feeler gauge, measure the clearance between the valve lifter and camshaft.
   - Standard intake valve clearance (Cold): 0.007–0.009 inch (0.17–0.23mm)
   - Standard exhaust valve clearance (Cold): 0.011–0.013 inch (0.27–0.33mm)

3. Record any out-of-specification valve clearance measurements. They will be used later to determine the required replacement lifter.

4. Turn the crankshaft 1 complete revolution until its timing notch and timing mark 0 of the chain cover are aligned.

5. Check the valves indicated in the illustration. Using a feeler gauge, measure the clearance between the valve lifter and camshaft.

- Standard intake valve clearance (Cold): 0.007–0.009 inch (0.17–0.23mm)
- Standard exhaust valve clearance (Cold): 0.011–0.013 inch (0.27–0.33mm)

6. Record any out-of-specification valve clearance measurements. They will be used later to determine the required replacement lifter.

7. Set the No. 1 cylinder to TDC/compression.

8. Turn the crankshaft pulley until its timing notch and timing mark 0 of the chain cover are aligned.

9. Check that both timing marks on the camshaft timing sprocket and valve timing controller assembly are facing upward. If not, turn the crankshaft 1 complete revolution (360 degrees) and align the marks as above.

10. Put paint marks on the timing chain where the timing marks of the camshaft timing sprocket and the camshaft timing gear are located.

11. Using an 8mm hexagon wrench, remove the screw plug.

12. Insert a screwdriver into the service hole of the chain tensioner to hold the stopper plate of the chain tensioner at an upward position. Lifting up the stopper plate of the chain tensioner unlocks the plunger.

13. Keeping the stopper plate of the chain tensioner lifted, slightly rotate the hexagonal lobe of the No. 2 camshaft to the right with an adjustable wrench so the plunger of the chain tensioner is pushed. When the camshaft No. 2 is slightly rotated to the right, the plunger is pushed.

14. Keeping the adjustable wrench

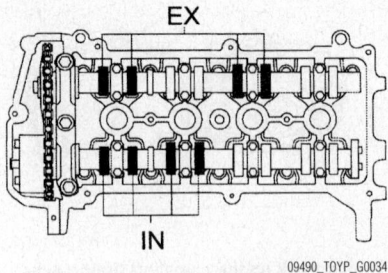

No. 1 Cylinder TDC/Compression

Check the clearance of the valves indicated (No. 1 cylinder)—1.5L hybrid

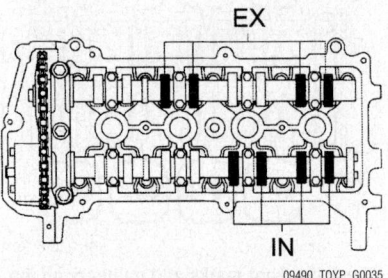

No. 4 Cylinder TDC/Compression

Check the clearance of the following valves (No. 4 cylinder)—1.5L hybrid

installed, remove the screwdriver with the plunger pushed. Do not move the adjustable wrench.

→**Removing the screwdriver lowers the stopper plate and locks the plunger.**

15. Insert a 0.118 inch (3.0mm) diameter bar into the hole of the stopper plate with the stopper plate of the chain tensioner lowered and locked. If the bar cannot be inserted into the hole of the stopper plate, rotate the No. 2 camshaft slightly to the left and right. Then that bar can be inserted easily.

16. Secure the bar with tape.

17. Hold the hexagonal lobe of the camshaft No. 2 with the adjustable wrench.

18. Using SST 09023-38400, loosen the bolt.

19. Using several steps, uniformly loosen and remove the 11 bearing cap bolts in the sequence shown in the illustration. Then remove the 5 bearing caps. Loosen each bolt uniformly, keeping the camshaft level.

20. Remove the flange bolt with the No. 2 camshaft lifted up. Then detach the No. 2 camshaft and the camshaft timing sprocket.

21. Using several steps, uniformly loosen and remove the 8 bearing cap bolts in the sequence shown in the illustration. Then remove the 4 bearing caps. Loosen

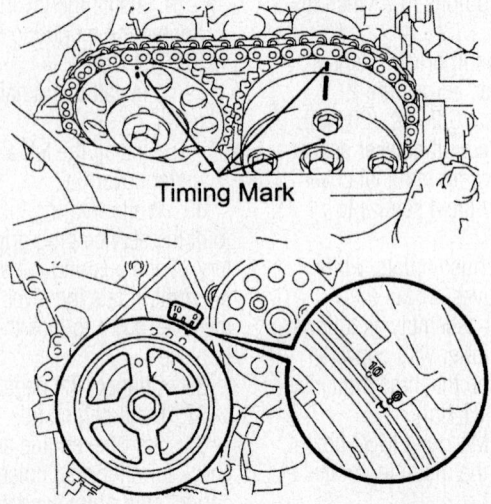

Timing Mark

Check that both timing marks on the camshaft timing sprocket and camshaft timing gear are facing upward—1.5L hybrid engine

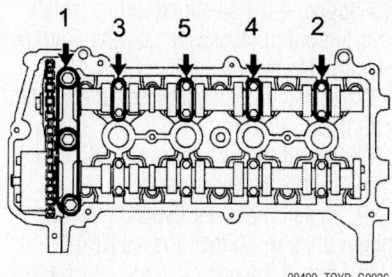

Remove the 11 bearing cap bolts in the sequence—1.5L hybrid engine

09490_TOYP_G0036

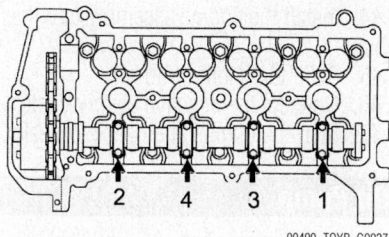

Remove the 8 bearing cap bolts in the sequence—1.5L hybrid engine

09490_TOYP_G0037

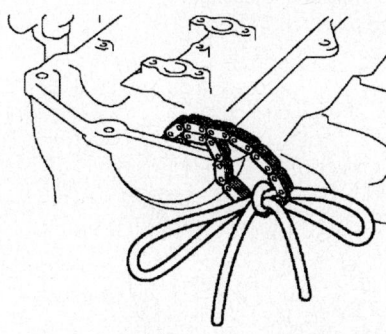

09490_TOYP_G0038

Tie the timing chain with a string—1.5L hybrid engine

each bolt uniformly, keeping the camshaft level.

22. Hold the timing chain with one hand, and remove the camshaft and the camshaft timing gear assembly.

23. Tie the timing chain with a string as shown in the illustration.

✳✳ CAUTION

Be careful not to drop anything inside the timing chain cover.

24. Remove the valve lifters.

25. Using a micrometer, measure the thickness of the removed lifter.

26. Calculate the thickness of a new lifter so that the valve clearance comes within the specified value.

27. Select a new lifter with the thickness as close to the calculated values as possible.

- EXAMPLE: (Intake) Measured valve clearance = 0.0158 inch (0.40mm)
- 0.0158 inch (0.40mm)—0.0079 inch (0.20mm) = 0.0079 inch (0.20mm) (Measured—Specification = Excess clearance)
- Used lifter measurement = 0.2067 inch (5.25mm)
- 0.0079 inch (0.20mm) + 0.2067 inch (5.25mm) = 0.2146 inch (5.45mm) (Excess clearance + Used lifter = Ideal new lifter)
- Closest new lifter = 5.45 mm (0.2146 in.); select lifter (0.2150 inch (5.46mm))

➡**Lifters are available in 35 sizes in increments of 0.0008 inch (0.020mm), from 0.1992 inch (5.060mm) to 0.2260 inch (5.740mm).**

Starter Motor

The Toyota Prius, being a sophisticated, hybrid vehicle that utilizes both electric and gasoline (internal combustion) engine power for mobility does not require (or come equipped with) a starter motor as a part of its starting system. The function of the starter motor is performed by a pair of electrical motor-generators, a computerized shunt system to control them, a mechanical power splitter that acts as a second differential, and a battery pack that serves as an energy reservoir.

Oil Pan

REMOVAL & INSTALLATION

1. Before servicing the vehicle, refer to the Precautions Section.

2. Disconnect the negative battery cable. Wait at least 90 seconds after disconnecting the negative battery cable to prevent possible airbag and seat belt pretensioner activation.

3. Drain engine oil.

4. Remove engine assembly.

5. Install engine to engine stand.

6. Remove the timing chain.

7. Remove the cylinder head.

8. Remove engine wire.

9. Remove the 2 nuts, bolt and water bypass pipe.

10. Remove the thermostat.

11. Remove the knock sensor.

12. Remove the oil pressure switch.

13. Remove the engine coolant drain union.

14. Remove the oil filter.

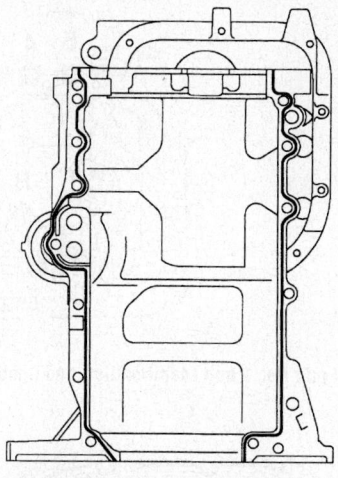

Seal Width 2 - 3 mm

09490_TOYP_G0039

Oil pan No. 1 seal width—1.5L hybrid engine

15. Using a 12mm hexagon wrench, remove the oil filter union.

16. Remove the 9 bolts and 2 nuts of the No. 2 oil pan.

17. Insert the blade of SST 09032-00100 between the oil pan No. 1 and oil pan No. 2, and cut off applied sealer and remove the oil pan. Be careful not to the damage the oil pan contact surface of the oil pan No. 1. or the oil pan No. 2 flange.

18. Remove the bolt and 2 nuts, oil strainer and gasket.

19. Uniformly loosen and remove the 13 bolts, in several passes.

20. Using screwdriver remove the oil pan No. 1 by prying the portions between the cylinder block and oil pan No. 1.

21. Remove the 2 O-rings from the cylinder block.

To install:

22. Remove any old packing (FIPG) material and be careful not to drop any oil on the contact surface of the oil pan No. 1 and cylinder block. Using a razor blade and gasket scraper, remove all the old packing (FIPG) material from the gasket surfaces and sealing grooves. Thoroughly clean all components to remove all the loose material. Using a non-residue solvent, clean both sealing surfaces.

23. Apply seal packing to the oil pan No. 1 with a seal width of 0.08–0.12 inch (2–3mm). Avoid applying an excessive amount to the surface. Parts must be assembled within 3 minutes of application. Otherwise the material must be removed and reapplied. Immediately remove nozzle from the tube and reinstall cap.

24. Install new O-rings to the cylinder block.

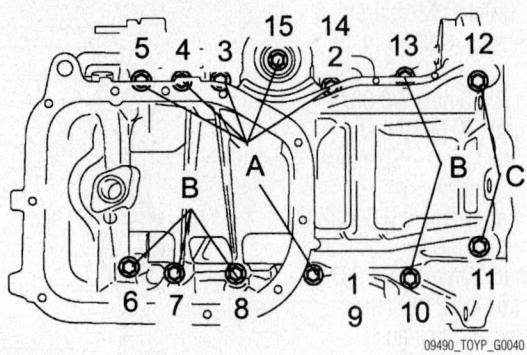

Oil pan No. 1 bolt identification and tightening sequence—1.5L hybrid engine

09490_TOYP_G0040

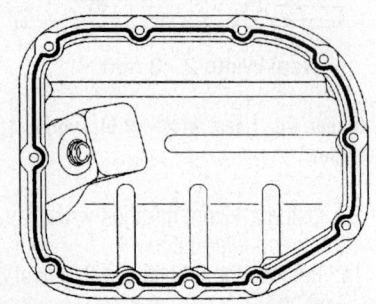

Seal Width 2.5 - 3.5 mm

09490_TOYP_G0041

Correct oil pan seal width—1.5L hybrid engine

25. Using a plastic-faced hammer, lightly tap the oil pan No. 1 to ensure a proper fit.

26. Install and uniformly tighten the 13 bolts, in several passes, in sequence to 18 ft. lbs. (24 Nm).

27. Each bolt indicated in the illustration shown is the following length:
Bolt A—1.929 inches (49mm)
Bolt B—3.465 inches (88mm)
Bolt C—5.669 inches (144mm)

28. Install rear crankshaft oil seal. Wipe seal packing away from the contact surface of the cylinder block assembly and oil seal.

29. Install a new gasket, and oil strainer with the bolt and 2 nuts. Tighten the bolt and 2 nuts to 8 ft. lbs. (11 Nm).

30. Remove any old FIPG material and be careful not to drop any oil on the contact surface of the main bearing cap and oil pan. Using a razor blade and gasket scraper, remove all the old FIPG material from the gasket surfaces and sealing grooves. Thoroughly clean all components to remove all the loose material. Using a non-residue solvent, clean both sealing surfaces.

31. Apply seal packing to the oil pan with a seal width of 0.098–0.138 inch (2.5–3.5mm). Avoid applying an excessive amount to the surface. Parts must be assembled within 3 minutes of application. Otherwise the material must be removed

and reapplied. Immediately remove nozzle from the tube and reinstall cap.

32. Install the oil pan with the 9 bolts and 2 nuts. Uniformly tighten the bolts and nuts in several passes to 80 inch lbs. (9 Nm).

33. Install the oil filter union and tighten to 21 ft. lbs. (30 Nm).

34. Install the oil filter.

35. Apply adhesive to 2 or 3 threads and install the engine coolant drain union. Torque the union to 25 ft. lbs. (35 Nm) and

after applying the specified torque, rotate the drain union clockwise until its drain port is facing downward.

36. Install the knock sensor and tighten to 29 ft. lbs. (39 Nm).

37. Install the oil pressure switch.

38. Install the thermostat.

39. Install the water bypass pipe and tighten bolts to 80 inch lbs. (9 Nm)

40. Install the engine wire.

41. Install the cylinder head.

42. Install the timing chain.

43. Remove the engine assembly from the engine stand.

44. Install the engine assembly into the vehicle.

45. Refill the engine with oil.

46. Connect the negative battery cable.

47. Start the engine and check for leaks.

Oil Pump

REMOVAL & INSTALLATION

1. Before servicing the vehicle, refer to the Precautions Section.

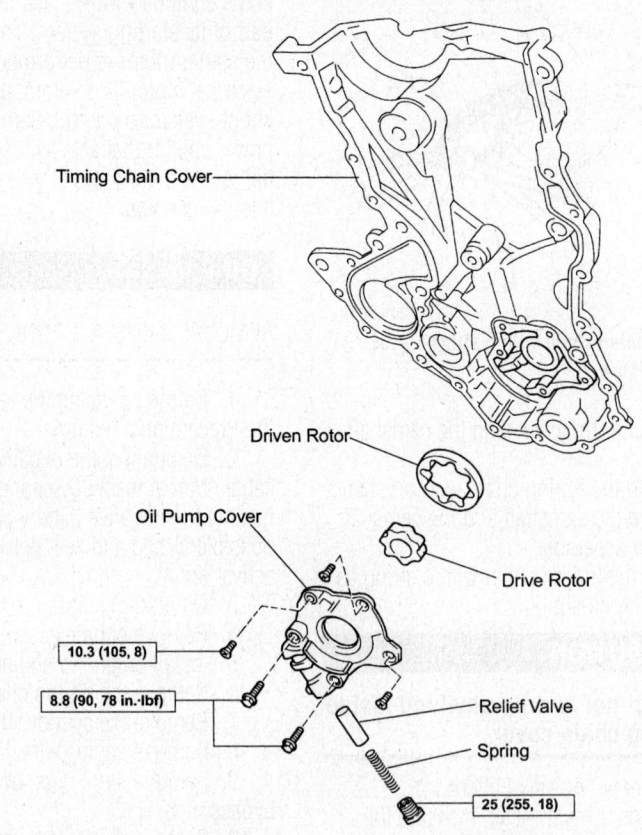

N·m (kgf·cm, ft·lbf) : Specified torque

Exploded view of the oil pump assembly—1.5L hybrid engine

09490_TOYP_G0042

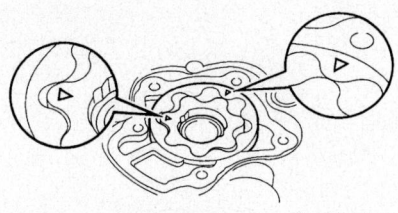

09490_TOYP_G0043

Install the rotors into timing chain cover with the marks facing the oil pump cover side—1.5L hybrid engine

2. Disconnect the negative battery cable. Wait at least 90 seconds after disconnecting the negative battery cable to prevent possible airbag and seat belt pretensioner activation.

3. Drain the engine oil.

4. Remove the timing chain cover (refer to the timing chain procedure).

5. Remove the 2 bolts, 3 screws and oil pump cover from the timing chain cover.

6. Remove the drive and driven rotors.

7. Remove the plug, spring and relief valve.

To install:

8. Insert the relief valve and spring into the pump body hole, and install the plug. Torque the plug to 18 ft. lbs. (25 Nm).

9. Place the drive and driven rotors into timing chain cover with the marks facing the oil pump cover side.

10. Install the oil pump cover to the timing chain cover with the 2 bolts and 3 screws. Torque the bolts to 78 inch lbs. (8.8 Nm), and the screws to 8 ft. lbs. (10.3 Nm).

11. Install the timing chain cover (refer to the timing chain procedure).

12. Refill engine with engine oil.

13. Connect the negative battery cable.

14. Start the engine and check the engine oil pressure.

15. Check that no leaks are present.

Rear Main Seal

REMOVAL & INSTALLATION

1. Before servicing the vehicle, refer to the Precautions Section.

2. Remove or disconnect the following:
   - Engine assembly
   - Flywheel from the crankshaft

3. Using a knife, cut off the lip of the oil seal.

4. Using a screwdriver with the tip wrapped in tape, carefully pry out the oil seal without scratching the sealing surface of the crankshaft. If it is damaged, smooth the surface with 400-grit sandpaper.

To install:

5. Apply multi-purpose grease to the lip of the new seal. Keep the lip free of foreign materials.

6. Install or connect the following:
   - Seal in the retainer using a suitable seal driver. Wipe any extra grease off the crankshaft.
   - Flexplate/flywheel
   - Engine assembly

Timing Chain

REMOVAL & INSTALLATION

❄❄ CAUTION

The hybrid system uses high voltage circuits, so improper handling could cause electric shock or leakage. During service, be sure to follow procedures.

1. Before servicing the vehicle, refer to the Precautions Section.

2. Properly relieve the fuel system pressure.

3. Disconnect the negative battery cable. Wait at least 90 seconds after disconnecting the negative battery cable to prevent possible airbag and seat belt pretensioner activation.

4. While wearing insulating gloves, slide up the lever of the service plug grip and remove while turning the lever to the left. Be sure to insulate the service plug with insulating tape.

❄❄ CAUTION

Do not touch the high voltage connectors and terminals for 5 minutes after removing the service plug grip.

5. Remove the outer front cowl top panel.

6. Remove right engine under cover.

7. Drain the engine coolant.

8. Remove or disconnect the following:
   - Air cleaner assembly
   - Brake fluid level sensor connector
   - Brake fluid reservoir tank mounting bolts (2) and suspend it with rope
   - Brake fluid reservoir tank bracket
   - Ignition connectors (4)
   - Fuel injector connectors (4)
   - Vacuum Switching Valve (VSV) connectors (2)
   - Camshaft Position (CMP) sensor connector
   - Water temperature connector
   - Camshaft timing oil control valve connector

   - Air cleaner inlet hose
   - Engine coolant reservoir tank
   - VSV from engine mounting insulator
   - Accessory drive belt

9. Place a floor jack under the engine to support it, with a block of wood between the engine and the jack.

10. Remove or disconnect the following:
   - Right engine mounting insulator
   - Engine wiring from cylinder head cover
   - Ignition coils (4)
   - PCV hoses (2)
   - 7 bolts, 2 seal washers, 2 nuts, cylinder head cover and gasket

11. Set the No. 1 cylinder to Top Dead Center/compression by turning the crankshaft pulley and aligning its groove with timing mark **"0"** of the timing chain cover.

12. Check that both timing marks on the camshaft timing sprocket and valve timing controller assembly are facing right up. If not, turn the crankshaft 1 revolution (360 degrees) and align the marks.

13. Remove the crankshaft pulley bolt and the crankshaft pulley.

14. Remove or disconnect the following:
   - Crankshaft Position (CKP) sensor
   - Right engine mounting bracket
   - Water pump
   - Oil control valve
   - Timing chain cover
   - Chain tensioner
   - Chain tensioner slipper
   - Chain vibration damper
   - Timing chain

To install:

15. Install the timing chain.

16. After setting the crankshaft at ATDC 40–140 degrees, set cams of intake and exhaust timing sprockets at ATDC 20 degrees and then the reset the crankshaft at ATDC 20 degrees.

17. Install the chain vibration damper with the 2 bolts and tighten to 80 inch lbs. (9 Nm).

18. Align the match marks of timing chain mark plate (Yellow), camshaft timing sprocket, camshaft timing gear and crankshaft timing sprocket to install the timing chain. To prevent the exhaust camshaft from spring back, turn it using a wrench and set it at the mark on a chain.

19. Install the chain tensioner slipper.

20. While rotating the lock plate of the tensioner up-ward, push in the plunger of the tensioner.

21. While rotating the lock plate of the tensioner down-ward, insert a bar of 0.098 inch (2.5mm) into the holes in the lock plate.

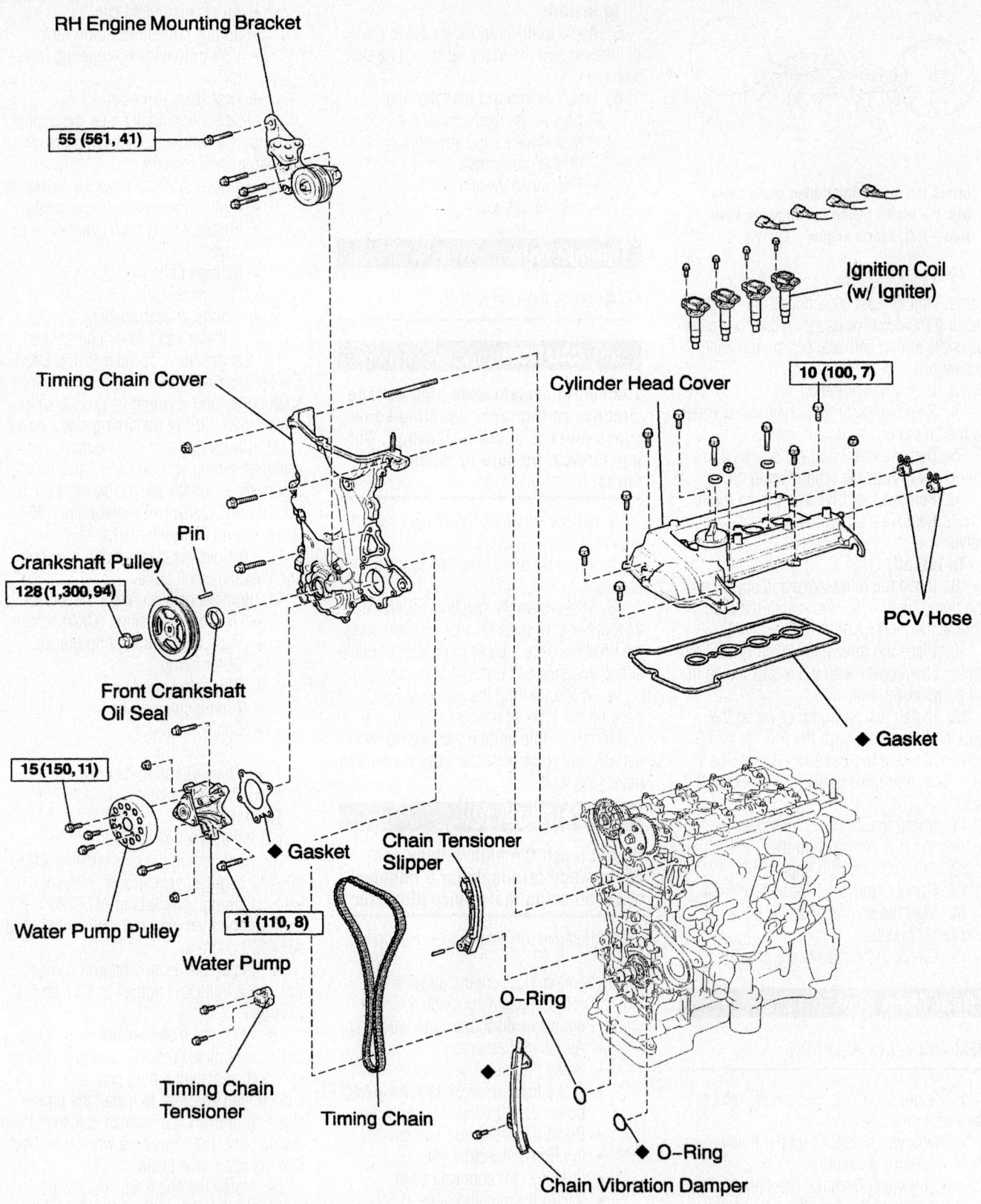

RH Engine Mounting Bracket

55 (561, 41)

Ignition Coil (w/ Igniter)

Timing Chain Cover

Cylinder Head Cover

10 (100, 7)

Pin

Crankshaft Pulley

128 (1,300, 94)

Front Crankshaft Oil Seal

PCV Hose

◆ Gasket

15 (150, 11)

Chain Tensioner Slipper

◆ Gasket

Water Pump Pulley

11 (110, 8)

Water Pump

O-Ring

Timing Chain Tensioner

Timing Chain

◆

O-Ring

Chain Vibration Damper

N·m (kgf·cm, ft·lbf) : Specified torque
◆ Non-reusable part

Exploded view of timing cover, timing chain assembly and related components—1.5L hybrid engine

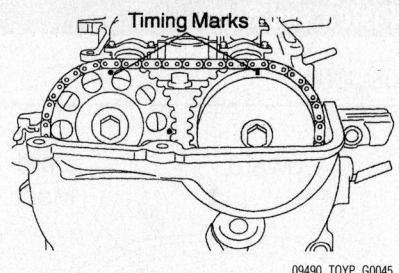

Aligning both timing marks on the camshaft timing sprocket and valve timing controller assembly—1.5L hybrid engine

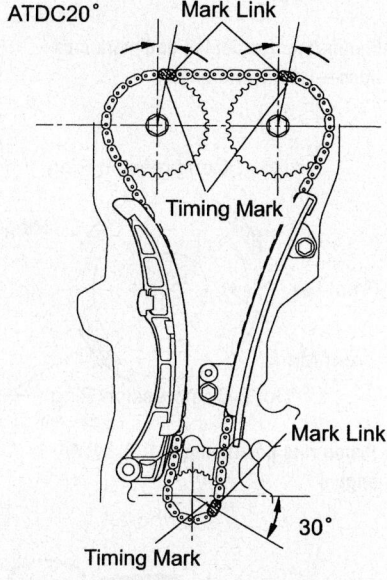

Align the match marks of timing chain mark plate (Yellow), camshaft timing sprocket, camshaft timing gear and crankshaft timing sprocket—1.5L hybrid engine

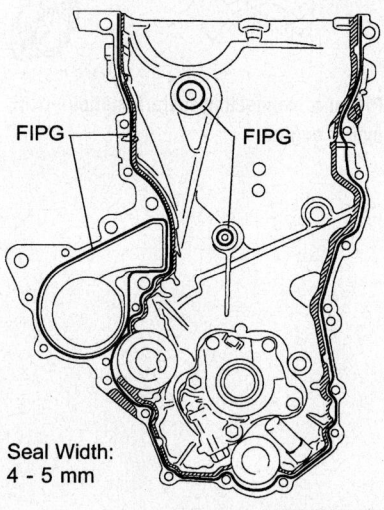

Apply seal packing to the timing chain cover—1.5L hybrid engine

22. Install the chain tensioner with the 2 bolts and tighten to 80 inch lbs. (9 Nm).

23. Remove the bar from the chain tensioner.

24. Check that the tension between the intake and exhaust camshaft timing sprocket.

25. Remove any old packing (FIPG) material and be careful not to drop any oil on the contact surfaces of the timing chain cover, cylinder head and cylinder block. Using a razor blade and a gasket scraper, remove all the old packing (FIPG) material from the gasket surfaces and sealing grooves. Thoroughly clean all components to remove all the loose material. Using a non-residue solvent, clean both sealing surfaces.

26. Apply seal packing to the timing chain cover. Install a nozzle that has been cut to a 0.16–0.20 inch (4–5mm) opening. FIPG shall be accumulated in the groove for FIPG to a depth of 0.10 inch (2.5mm) or more. Avoid applying an excessive amount

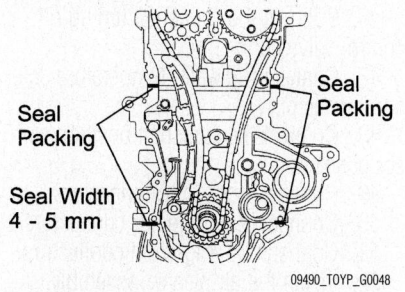

Apply seal packing to 4 locations—1.5L hybrid engine

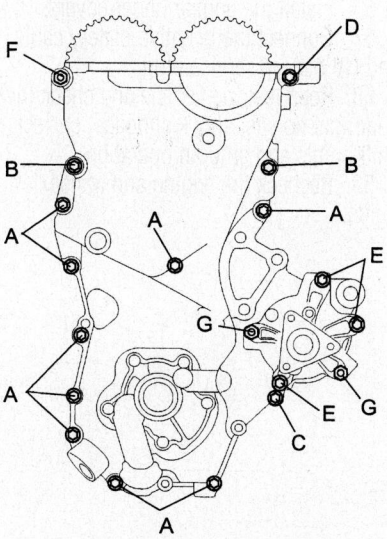

Correct installation of the timing chain cover and water pump bolts and nuts—1.5L hybrid engine

to the surface. Parts must be assembled within 3 minutes of application. Otherwise the material must be removed and reapplied. Immediately remove nozzle from the tube and reinstall cap.

27. Apply seal packing to 4 locations. Install a nozzle that has been cut to a 0.16–0.20 inch (4–5mm) opening. Avoid applying an excessive amount to the surface. Parts must be assembled within 3 minutes of application. Otherwise the material must be removed and reapplied. Immediately remove nozzle from the tube and reinstall cap.

28. Install 2 new O-rings to the cylinder block and oil pan No. 1.

29. Install the timing chain cover, new O-ring and water pump with the 16 bolts and 3 nuts. Uniformly tighten the bolts and nut in several passes to the following torque values:

- Bolt A, 0.787 inch (20mm)—8 ft. lbs. (11 Nm)
- Bolt B, 1.181 inch (30mm)—18 ft. lbs. (24 Nm)
- Bolt C, 1.378 inch (35mm)—8 ft. lbs. (11 Nm)
- Bolt D, 0.787 inch (20mm)—18 ft. lbs. (24 Nm)
- Bolt E, 1.378 inch (35mm)—8 ft. lbs. (11 Nm)
- Nut F—18 ft. lbs. (24 Nm)
- Nut G—8 ft. lbs. (11 Nm)

➡**Pay attention not to wrap the chain and slipper over the chain cover seal line. After installing the chain cover, must install the mounting bracket and water pump within 15 minutes.**

30. Apply seal packing to threads of the engine mounting bracket mounting bolt, but do not apply seal packing to 2 or 3 threads of the bolt end.

31. Install the right mounting bracket with the 4 bolts and tighten to 41 ft. lbs. (55 Nm).

32. Install the crankshaft position sensor and tighten the bolt at the sensor to 66 inch lbs. (7.5 Nm), and the other bolts to 8 ft. lbs. (11 Nm).

33. Install the oil control valve and tighten to 66 inch lbs. (7.5 Nm).

34. Install the crankshaft pulley as follows:

    a. Clean the crankshaft pulley inside.

    b. Install the pin to the crankshaft.

    c. Align the hole in the crank pulley with the pin position and install the crank pulley.

    d. Using SST 09213-70011 and 09330-00021, install the pulley bolt and tighten to 94 ft. lbs. (128 Nm).

Seal Packing

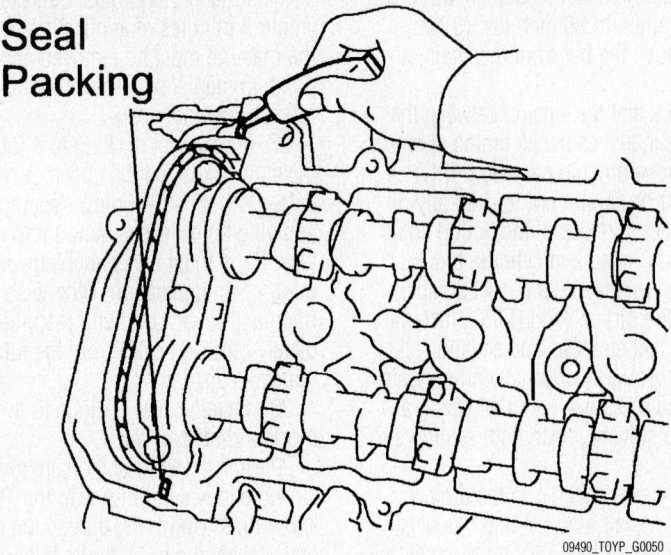

Apply seal packing to 2 locations as shown—1.5L hybrid engine

35. Remove any old packing (FIPG) material and apply seal packing to 2 locations as shown in the illustration.

36. Install the gasket to the cylinder head cover.

➡**Part must be assembled within 3 minutes of seal packing application. Otherwise the material must be remove and reapplied.**

37. Install the cylinder head cover and cable bracket with the 7 bolts, 2 seal washers and 2 nuts. Uniformly tighten the bolts and nuts, in the several passes, in the sequence to 7 ft. lbs. (10 Nm).

38. Connect the 2 PCV hoses to the cylinder head cover.

39. Connect the engine wire to cylinder head cover.

40. Install the ignition coils.

41. Install the RH engine mounting insulator with the 5 bolts and 2 nuts.

42. Install VSV to right engine mounting insulator

43. Install the drive belt.

44. Install the engine coolant reservoir tank.

45. Install the air inlet.

46. Connect the Camshaft timing oil control valve connector.

47. Connect the water temperature sensor connector.

48. Connect the camshaft position sensor connector.

49. Connect the 2 VSV connectors.

50. Connect the 4 injector connectors.

51. Connect the 4 ignition connectors.

52. Install the air cleaner assembly.

53. Install brake fluid reservoir tank.

54. Install the outer front cowl top panel assembly.

55. Fill the engine with coolant.

56. Install the engine under covers.

57. Connect the negative battery cable and HV battery service plug.

58. Road test the vehicle and check for abnormal noises, shock slippage, correct shift points and smooth operation.

59. Recheck the engine and transaxle fluids.

Piston and Ring

POSITIONING

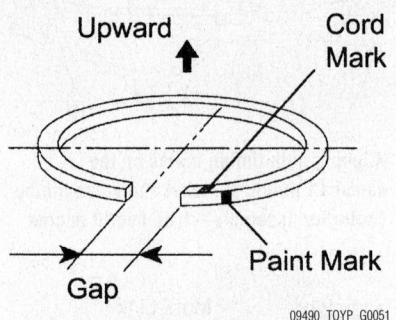

Piston ring positioning and mark locations—1.5L hybrid engine

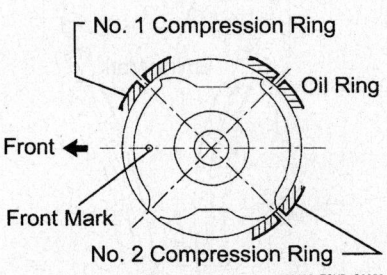

Piston ring positioning—1.5L hybrid engine

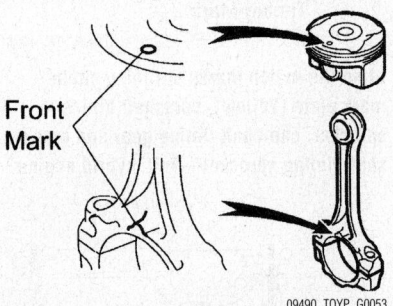

Piston-to-connecting rod orientation—1.5L hybrid engine

FUEL SYSTEM

Fuel System Service Precautions

Safety is the most important factor when performing not only fuel system maintenance, but any type of maintenance. Failure to conduct maintenance and repairs in a safe manner may result in serious personal injury or death. Maintenance and testing of the vehicle's fuel system components can be accomplished safely and effectively by adhering to the following rules and guidelines.

• To avoid the possibility of fire and personal injury, always disconnect the negative battery cable unless the repair or test procedure requires that battery voltage be applied.

• Always relieve the fuel system pressure prior to disconnecting any fuel system component (injector, fuel rail, pressure regulator, etc.), fitting or fuel line connection. Exercise extreme caution whenever relieving the fuel system pressure, to avoid exposing your skin, face and eyes to fuel spray. Please be advised that fuel under pressure may penetrate the skin or any part of the body that it contacts.

• Always place a shop towel or cloth around the fitting or connection prior to loosening to absorb any excess fuel due to spillage. Ensure that all fuel spillage (should it occur) is quickly removed from the engine surfaces. Ensure that all fuel soaked cloths or towels are deposited into a suitable waste container.

• Always keep a dry chemical (Class B) fire extinguisher near the work area.

• Do not allow fuel spray or fuel vapors to come into contact with a spark or open flame.

• Always use a back-up wrench when loosening and tightening fuel line connection fittings. This will prevent unnecessary stress and torsion on fuel line piping. Always follow the proper torque specifications.

• Always replace worn fuel fitting O-rings with new. Do not substitute fuel hose or equivalent, where fuel pipe is installed.

Fuel System Pressure

RELIEVING

✳✳ CAUTION

The fuel system pressure relief procedure must be performed before disconnecting any part of the fuel system. After performing this procedure, pressure will remain in the fuel line. When disconnecting the fuel line, place a cloth or equivalent over fittings to reduce the risk of fuel spray.

1. Before servicing the vehicle, refer to the Precautions Section.
2. Remove the integration relay (unit C: C/OPN relay) from the engine room junction block.
3. Start the vehicle and allow the engine to run until it stops, then turn the power switch **OFF**. This may set off a trouble code (DTC P0171: system too lean).
4. Check that the engine does not start.
5. Remove the fuel filler cap from the filler neck to release the fuel vapor pressure in the fuel tank.
6. Disconnect the negative battery cable. Wait at least 90 seconds after disconnecting the negative battery cable to prevent possible airbag and seat belt pretensioner activation.
7. Install the integration relay (unit C: C/OPN relay) to the engine room junction block.
8. After servicing the fuel system, connect the negative battery cable.
9. Start the engine and check for leaks in the system.

Fuel Filter

REMOVAL & INSTALLATION

The fuel filter for Prius models is an integral component of the tank-mounted fuel pump assembly. Refer to the Fuel Pump Removal procedure later in this section.

Fuel Pump

REMOVAL & INSTALLATION

1. Before servicing the vehicle, refer to the Precautions Section.
2. Relieve the pressure from the fuel system.
3. Disconnect the negative battery cable.
4. Remove instrument panel finish panel lower center.
5. Remove the front floor panel brace and front exhaust pipe assembly.
6. Remove the rear seat cushion as follows:
   a. Detach the seat cushion's 2 front hooks from the vehicle body by choosing a hook to detach first. Place your hands near one of the hooks, then lift the seat cushion to detach the hook.
   b. Repeat for the other hook.
   c. Detach the seat cushion's rear hook.
   d. Remove the seat cushion.
7. Remove the butyl tape and rear floor service hole cover.

✳✳ CAUTION

Remove dirt or foreign objects on the fuel line connectors before any disconnecting procedures. Do not allow any scratches or foreign objects on the parts when disconnecting them as the fuel connectors have O-rings that seal the pipes. Perform such work by hand. Do not use any tools. Do not forcibly bend, twist or turn the nylon tube. Protect the connecting part by covering it with a plastic bag after disconnecting the tube. If the connector and pipe are stuck, push and pull them to release them.

8. Disconnect the fuel pump connector.
9. Disconnect the wire-to-wire connector.
10. Pinch the retainer of the fuel tube connector, then pull out the fuel tube connector to disconnect the fuel tank to canister tube from the pipe.
11. Remove the checker of the fuel tube connector from the pipe.
12. Pinch the retainer of the fuel tube connector, and then pull out the fuel tube connector to disconnect the No. 2 fuel tank main tube from the pipe.
13. Pinch the retainer and pull out the fuel tank vent hose connector with the fuel tank vent hose connector pushed to the pipe side to disconnect the fuel tank vent hose from the canister filter.
14. Pinch the retainer and pull out the suction tube connector with the suction tube connector pushed to the pipe side to disconnect the fuel suction tube from the fuel tank to filler pipe.
15. Pinch the retainer and pull out the No. 1 canister tube connector with the No. 1 canister tube connector pushed to the pipe side to disconnect the No. 1 canister tube from the fuel tank to filler pipe.
16. Set a transmission jack to the fuel tank.
17. Remove the fuel filler pipe clamp and fuel tube connector from the fuel tank inlet pipe.

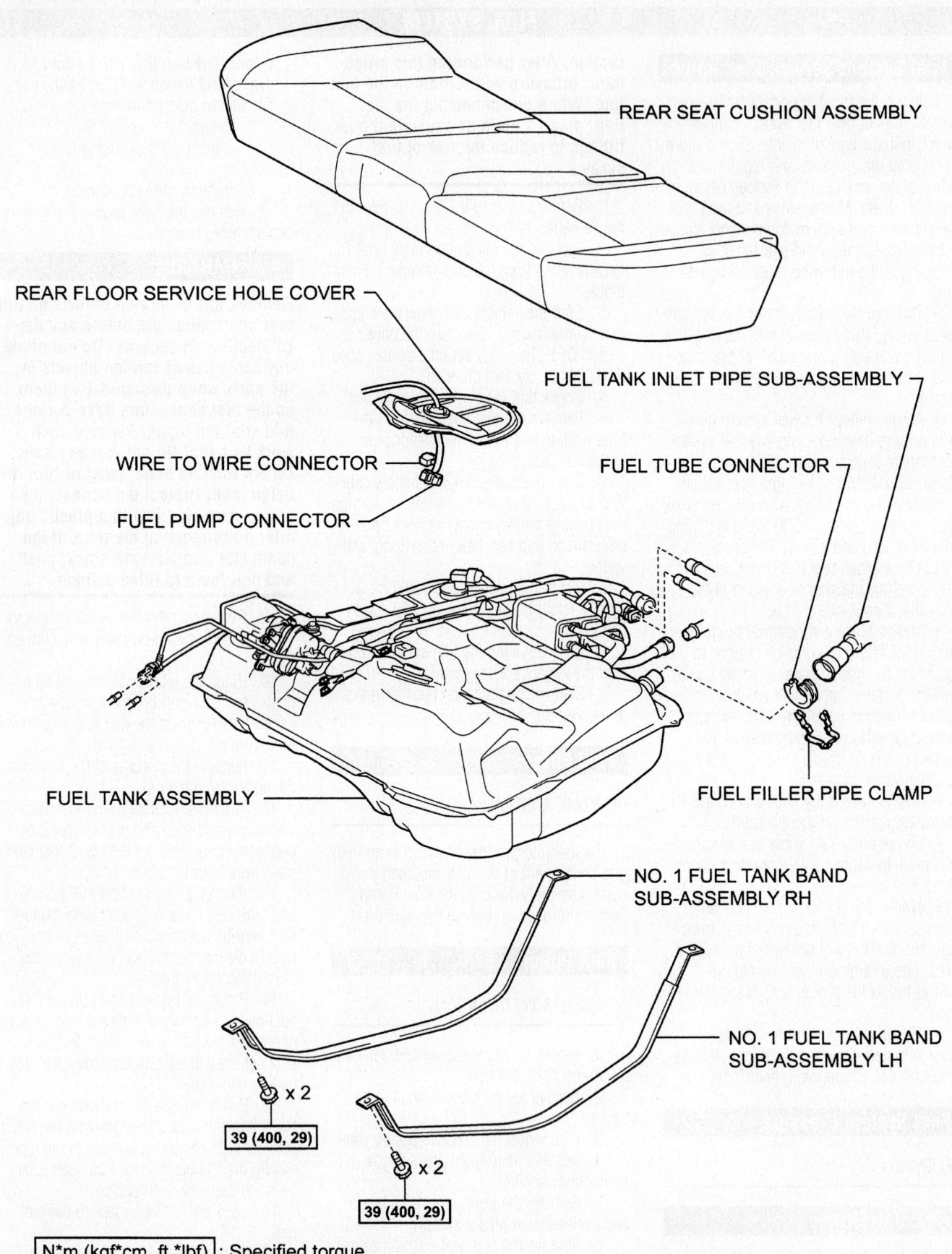

REAR SEAT CUSHION ASSEMBLY

REAR FLOOR SERVICE HOLE COVER

FUEL TANK INLET PIPE SUB-ASSEMBLY

WIRE TO WIRE CONNECTOR

FUEL TUBE CONNECTOR

FUEL PUMP CONNECTOR

FUEL TANK ASSEMBLY

FUEL FILLER PIPE CLAMP

NO. 1 FUEL TANK BAND SUB-ASSEMBLY RH

NO. 1 FUEL TANK BAND SUB-ASSEMBLY LH

× 2

39 (400, 29)

× 2

39 (400, 29)

N*m (kgf*cm, ft.*lbf) : Specified torque

Fuel tank access below the rear seat cushion

09490_TOYP_G0054

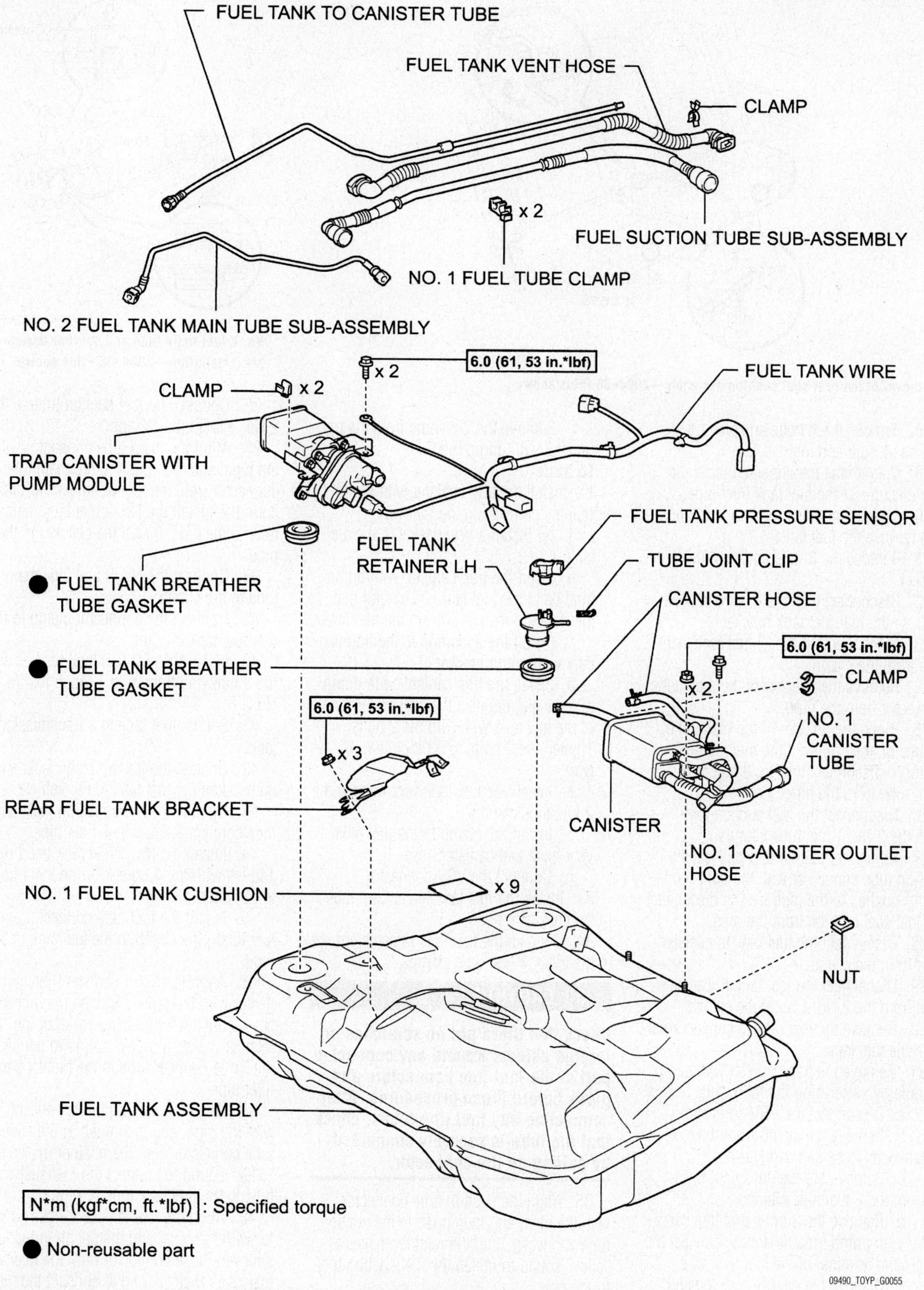

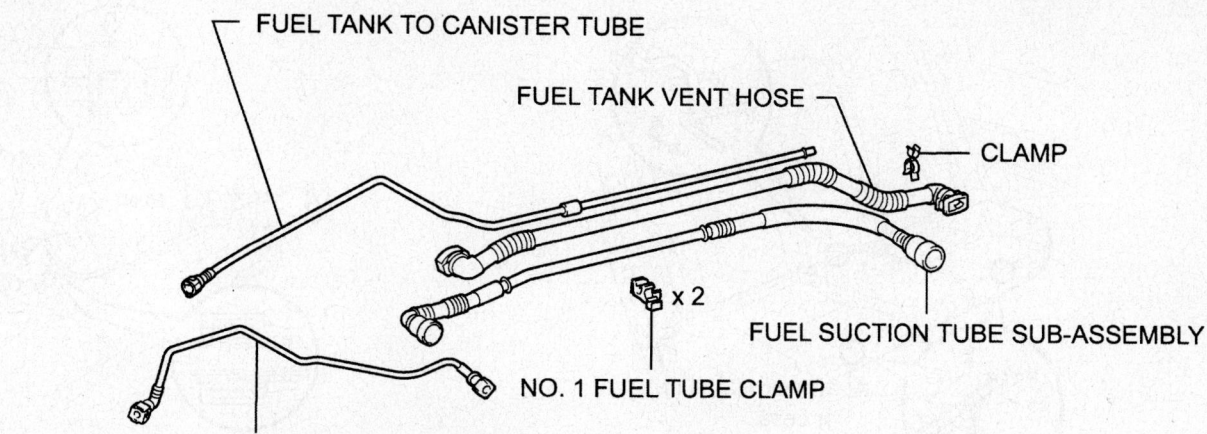

FUEL TANK TO CANISTER TUBE

FUEL TANK VENT HOSE

CLAMP

x 2

FUEL SUCTION TUBE SUB-ASSEMBLY

NO. 1 FUEL TUBE CLAMP

NO. 2 FUEL TANK MAIN TUBE SUB-ASSEMBLY

x 2   6.0 (61, 53 in.*lbf)   FUEL TANK WIRE

CLAMP   x 2

TRAP CANISTER WITH
PUMP MODULE

FUEL TANK PRESSURE SENSOR

TUBE JOINT CLIP

FUEL TANK
RETAINER LH   CANISTER HOSE

● FUEL TANK BREATHER
TUBE GASKET

6.0 (61, 53 in.*lbf)

x 2   CLAMP

● FUEL TANK BREATHER
TUBE GASKET

NO. 1
CANISTER TUBE

6.0 (61, 53 in.*lbf)

x 3

REAR FUEL TANK BRACKET

CANISTER

NO. 1 CANISTER OUTLET
HOSE

NO. 1 FUEL TANK CUSHION   x 9

NUT

FUEL TANK ASSEMBLY

N*m (kgf*cm, ft.*lbf) : Specified torque

● Non-reusable part

09490_TOYP_G0055

Exploded view of the fuel tank and related components—2004–06 Prius shown

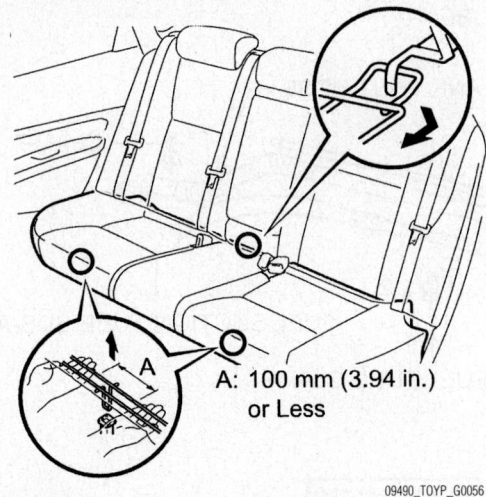

A: 100 mm (3.94 in.) or Less

09490_TOYP_G0056

Removal of the rear seat cushion assembly—2004–06 Prius shown

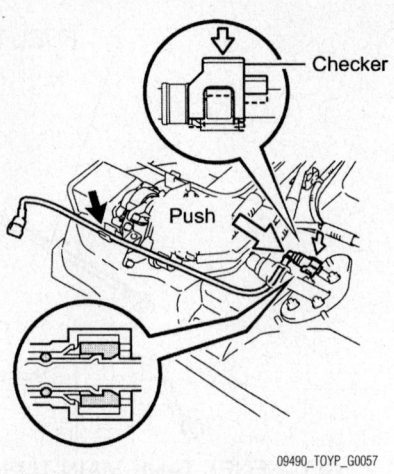

Checker

Push

09490_TOYP_G0057

No. 2 fuel main tube and checker assembly installation—2004–06 Prius shown

18. Remove the 4 bolts and No. 1 fuel tank band right and left.

19. Operate the transmission jack, and then disconnect the fuel tank inlet pipe.

20. Operate the transmission jack, and then remove the fuel tank.

21. Remove the 3 nuts and rear fuel tank bracket.

22. Disconnect the wire to wire connector from the rear fuel tank bracket.

23. Disconnect the No. 2 fuel tank main tube from the clamp.

24. Remove the checker of the main tube connector from the pipe.

25. Pinch the retainer of the main tube connector, then pull out the fuel tube connector to disconnect the No. 2 fuel tank main tube from the pipe.

26. Disconnect the fuel suction tube from the 2 No. 1 fuel tube clamps.

27. Pinch the retainer and pull out the suction tube connector with the quick connector pushed to the pipe side to disconnect the fuel suction tube from the pipe.

28. Disconnect the fuel tank to canister tube from the clamp.

29. Disconnect the fuel tank to canister tube from the 2 No. 1 fuel tube clamps.

30. Remove the fuel tank to canister tube from the fuel tank.

31. Remove the trap canister with pump module by performing the following:

a. Disconnect the VSV connector.

b. Remove the clamp from the fuel tank vent hose and canister hose.

c. Remove the fuel tank vent hose from the 2 fuel tube clamps.

d. Remove the 2 bolts and trap canister with pump module and disconnect the ground terminal of the fuel tank wire.

e. Remove the gasket from the fuel tank.

f. Remove the 2 clamps from the trap canister with pump module.

To install:

32. Install the trap canister with pump module by performing the following:

a. (a) Install a new gasket to the fuel tank.

b. Insert the trap canister with pump module to the fuel tank. Be careful that the gasket does not drop in the fuel tank.

c. Install the 2 clamps to the trap canister with pump module.

d. Install the trap canister with pump module and connect the ground terminal of the fuel tank wire with the 2 bolts. Tighten the 2 bolts to 53 inch lbs. (6 Nm).

e. Install the fuel tank vent hose to the 2 fuel tube clamps.

f. Install the clamp to the fuel tank vent hose and canister hose.

g. Connect the VSV connector.

33. Install the fuel tank to canister tube to the canister's hose.

34. Connect the fuel tank to canister tube to the 2 No. 1 fuel tube clamps.

✢✢ CAUTION

Check that there are no scratches or foreign objects around any connected part of the fuel line connectors and pipes before these procedures. After connecting any fuel line tubes, check that the tube is securely connected by pulling on the connector.

35. Align the suction tube connector with the pipe, and then push in the suction tube connector until the retainer makes a "click" sound to install the fuel suction tube to the pipe.

36. Connect the fuel suction tube to the 2 No. 1 fuel tube clamps.

37. Align the main tube connector with the pipe, and then push in the main tube connector until the retainer makes a "click" sound to install the No. 2 fuel tank main tube to the pipe. Install the checker to the pipe.

38. Connect the No. 2 fuel tank main tube to the clamp.

39. Connect the connector clamp to the rear fuel tank bracket.

40. Install the rear fuel tank bracket with the 3 nuts and tighten to 53 inch lbs. (6 Nm).

41. Set the fuel tank to a transmission jack.

42. Operate the transmission jack, and then install the fuel tank to the vehicle.

43. Operate the transmission jack, and then connect the fuel tank inlet pipe.

44. Install the No. 1 fuel tank band right and left with the 4 bolts. Torque the 4 bolts to 29 ft. lbs. (39 Nm).

45. Install the fuel tube connector and fuel filler pipe clamp to the fuel tank inlet pipe.

46. Align the No. 1 canister tube connector with the pipe, and then push in the No. 1 canister tube connector until the retainer makes a "click" sound to connect the No. 1 canister tube to the fuel tank to filler pipe.

47. Align the suction tube connector with the pipe, and then push in the suction tube connector until the retainer makes a "click" sound to connect the fuel suction tube to the fuel tank to filler pipe.

48. Align the fuel tank vent hose connector with the pipe, and then push in the fuel tank vent hose connector until the retainer makes a "click" sound to connect the fuel tank vent hose to the canister filter.

49. Align the fuel tube connector with the pipe, and then push in the fuel tube connector until the retainer makes a "click" sound to connect the No. 2 fuel tank main tube to the pipe. Install the checker to the pipe.

50. Align the fuel tank to canister tube connector with the pipe, and then push in the fuel tank to canister tube connector until the retainer makes a "click" sound to connect the fuel tank to canister tube to the pipe.

51. Install the front exhaust pipe

52. Connect the negative battery cable.

53. Check for fuel and exhaust leaks.

54. Install the front floor panel brace.

55. Install the instrument panel finish panel lower center.

56. Attach new butyl tape to the rear floor service hole cover.

57. Connect the wire-to-wire connector.

58. Connect the fuel pump connector.

59. Install the rear floor service hole cover while adjusting it to the 3 convex parts of the floor panel.

✳✳ WARNING

Be careful that the rear floor service hole cover does not overlap the convex parts of the floor panel when installing.

60. Install the rear seat cushion by engaging the 3 seat hooks.

61. Perform initialization.

Fuel Injector

REMOVAL & INSTALLATION

1. Before servicing the vehicle, refer to the Precautions Section.

2. Relieve the pressure from the fuel system.

3. Disconnect the negative battery cable.

4. Remove the windshield wipers and

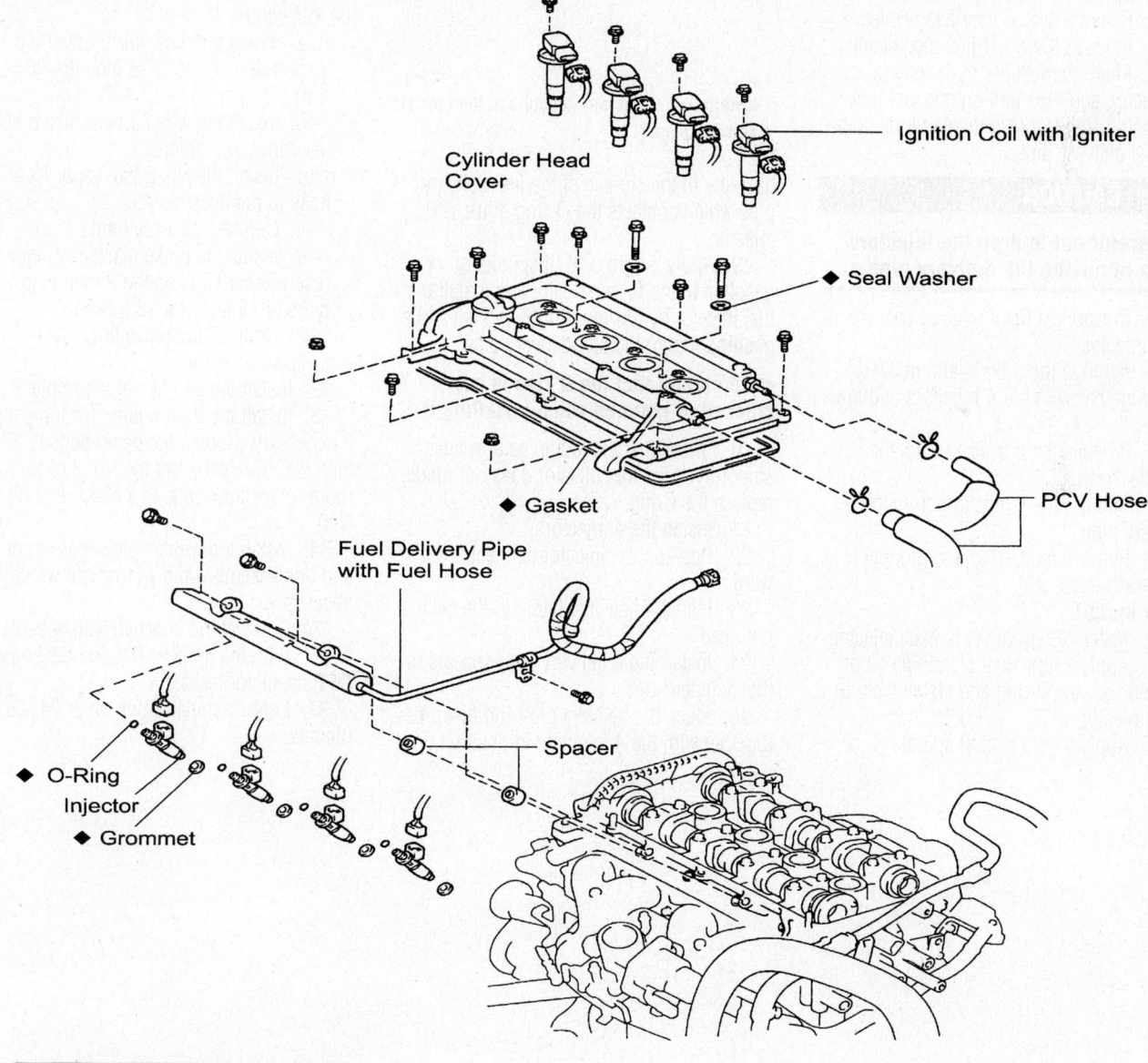

N·m (kgf·cm, ft·lbf) : Specified torque

◆ Non-reusable part

Fuel injectors and cylinder head cover.

windshield wiper motor and link assembly, if necessary.

5. Remove the front outer cowl top panel, if necessary.

6. Remove the air cleaner assembly.

7. On 2004–06 Prius models, perform the following:

a. Disconnect the brake fluid level switch connector.

b. Remove the 2 bolts.

c. Disconnect the claw fitting, then remove the brake master cylinder reservoir.

d. Remove the No. 2 fuel vapor feed hose from the hose clamp.

e. Remove the connector clamp.

f. Remove the wire harness clamp, the 3 bolts and reservoir bracket.

8. Remove the cylinder head cover.

9. Remove the No. 1 fuel pipe clamp.

10. Pinch the retainer of the fuel tube connector, and then pull out the fuel tube connector to disconnect the fuel tube from the fuel delivery pipe.

✳✳ CAUTION

Be careful not to drop the injectors when removing the delivery pipe.

11. Disconnect the 4 injector connectors from injector.

12. Remove the 3 bolts and delivery pipe together with the 4 injectors and fuel pipe.

13. Remove the 2 spacers from the cylinder head.

14. Pull out the 4 injectors from the delivery pipe.

15. Remove the O-ring and grommet from each injector.

To install:

16. Install the grommet to each injector.

17. Apply a light coat of spindle oil or gasoline to new O-ring and install them to each injector.

18. Apply a light coat of spindle oil or

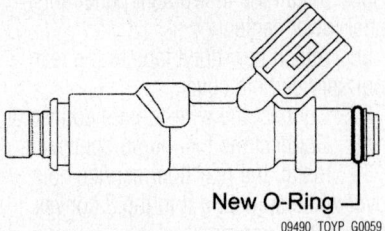

New O-Ring

09490_TOYP_G0059

Installation of new O-ring to the injectors.

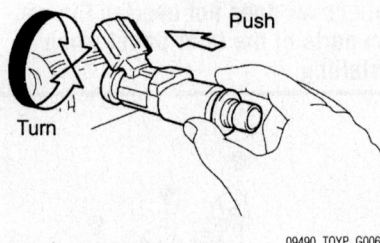

Push

Turn

09490_TOYP_G0060

Installation of the fuel injectors to the fuel delivery pipe.

gasoline to the surface of the fuel delivery pipe which contacts the O-ring of the fuel injector.

19. Apply a light coat of spindle oil or gasoline to the O-ring again, and install the fuel injector by turning it right and left while pushing it onto the fuel delivery pipe.

➡**Be careful that the O-ring is not cracked or jammed when installing it.**

20. Check that the fuel injector rotates smoothly. If the fuel injector does not rotate, replace the Oring.

21. Install the 4 injectors.

22. Position the injector connector outward.

23. Install 4 new insulators to the cylinder head.

24. Install the 2 delivery pipe spacers to the cylinder head.

25. Place the delivery pipe and fuel pipe together with the 4 injectors in position on

the cylinder head and then temporarily tighten the 3 bolts.

26. Check that the injectors rotate smoothly. If the fuel injectors do not rotate, replace the O-ring.

27. Tighten the 2 bolts holding the delivery pipe to the cylinder head and tighten to 14 ft. lbs. (19 Nm). Tighten the bolt holding the fuel pipe to the cylinder head to 80 inch lbs. (9 Nm).

28. Align the fuel tube connector with the pipe, then push in the fuel tube connector until the retainer makes a "click" sound to connect the fuel tube to the fuel delivery pipe.

29. Install the No. 1 fuel pipe clamp.

30. Install the cylinder head cover.

31. On 2004–06 Prius models, perform the following:

a. Install the reservoir bracket and tighten the 3 bolts to 75 inch lbs. (8.5 Nm)..

b. Install the wire harness clamp and the connector clamp.

c. Install the No. 2 fuel vapor feed hose to the hose clamp.

d. Connect the claw fitting.

e. Install the brake master cylinder reservoir and tighten the 2 mounting bolts to 75 inch lbs. (8.5 Nm).

f. Connect the brake fluid level switch connector.

32. Install the air cleaner assembly.

33. Install the front outer cowl top panel, if necessary. Tighten the panel bolts to 57 inch lbs. (6.4 Nm), and the No. 2 engine room relay block bolts to 74 inch lbs. (8.4 Nm).

34. Install the windshield wiper motor and link assembly and windshield wipers, if necessary.

35. Connect the negative battery cable.

36. With the ignition **ON** and the engine **OFF** check for leaks.

37. Perform initialization on 2004–06 models.

DRIVE TRAIN

Transaxle Assembly

REMOVAL & INSTALLATION

2002–03 Models

1. Before servicing the vehicle, refer to the Precautions Section.

2. Disconnect the negative battery cable. Wait at least 90 seconds after disconnecting the negative battery cable to prevent possible airbag and seat belt pretensioner activation.

3. Drain the transaxle fluid.

4. Remove the engine and transaxle as an assembly.

5. Remove the 2 bolts and dust cover.

6. Remove the 6 bolts and separate the transaxle from the engine.

7. Using SST 09213-58012 and 09330-00021, hold the crankshaft pulley, remove the 6 bolts and damper disc and spring.

8. Using SST 09213-58012 and 09330-00021, hold the crankshaft pulley, remove the 6 bolts and flywheel.

To install:

9. Apply adhesive to 2 or 3 threads of the bolt end.

10. Install and uniformly tighten the new 6 bolts in several passes, in a star sequence to 62 ft. lbs. (84 Nm).

11. Retighten the flywheel bolts by 90 degrees in the same star sequence.

12. Insert SST 09301-001 10 in the damper disc, then insert them in the flywheel. Take care not to insert damper disc in the wrong direction.

13. Set the damper spring on the damper disc.

14. Install and tighten the 6 bolts in a star sequence, starting the bolt locating near the knock pin on the top. Tighten the bolts to 12 ft. lbs. (16 Nm).

15. Following the star sequence, tighten the bolts at a time evenly. Move SST up and down, right and left lightly, after checking that the disc in the center, tighten the bolts.

16. Attach the transaxle to the engine, and install the 6 bolts. Torque the bolts to 24 ft. lbs. (33 Nm).

17. Install the dust cover.

18. Install the left engine mounting bracket. Torque the bolts to 38 ft. lbs. (52 Nm).

19. Fill the transaxle with the correct amount and type of fluid.

20. Connect the negative battery cable.

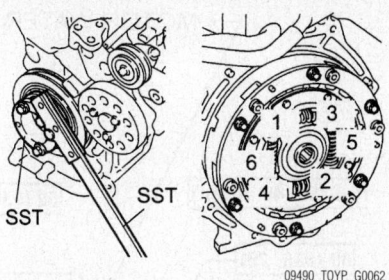

09490_TOYP_G0062

Flywheel bolt tightening sequence— 2002–03 models

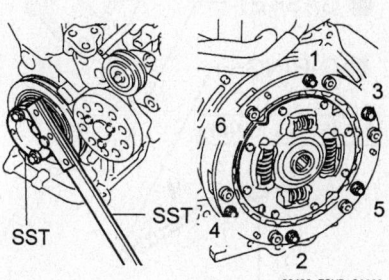

09490_TOYP_G0063

Damper disc and spring bolt tightening sequence—2002–03 models

21. Perform the test drive of the vehicle and check the function of the engine and transaxle.

2004–06 Models

1. Before servicing the vehicle, refer to the Precautions Section.

2. Disconnect the negative battery cable. Wait at least 90 seconds after disconnecting the cable from the negative (-) battery terminal to prevent airbag and seat belt pretensioner activation.

3. While wearing insulating gloves, slide up the lever of the service plug grip and remove while turning the lever to the left. Be sure to insulate the service plug with insulating tape.

✳✳ CAUTION

Do not touch the high voltage connectors and terminals for 5 minutes after removing the service plug grip.

4. Drain engine coolant and transaxle fluid.

5. Remove or disconnect the following:
- Front wheels
- Engine under covers
- Hood
- Left and right front wiper arms
- Cowl top front panel

6. While wearing insulating gloves,

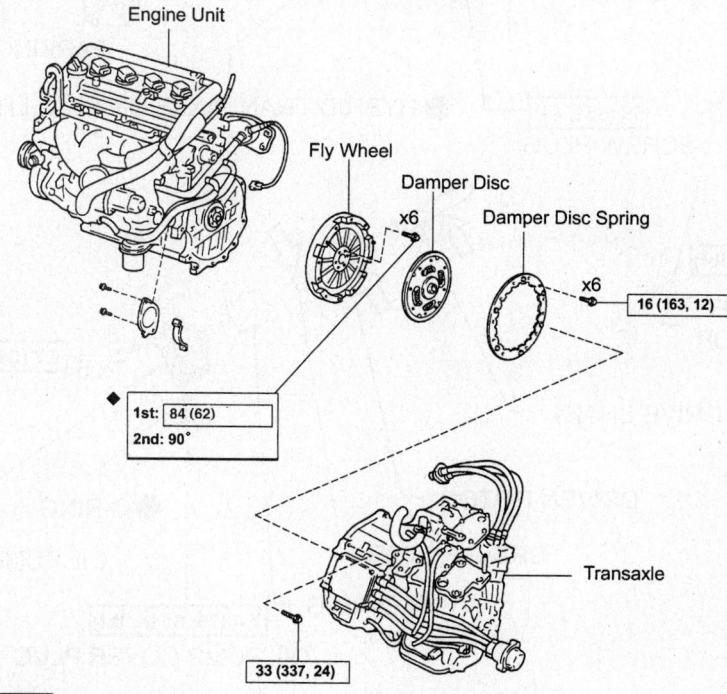

Engine Unit

Fly Wheel

Damper Disc

Damper Disc Spring

x6

x6

16 (163, 12)

◆ 1st: 84 (62)
2nd: 90°

Transaxle

33 (337, 24)

N·m (kgf·cm, ft·lbf) : Specified torque
◆ Non-reusable part

09490_TOYP_G0061

Transaxle unit, engine unit and related components—2002–03 models

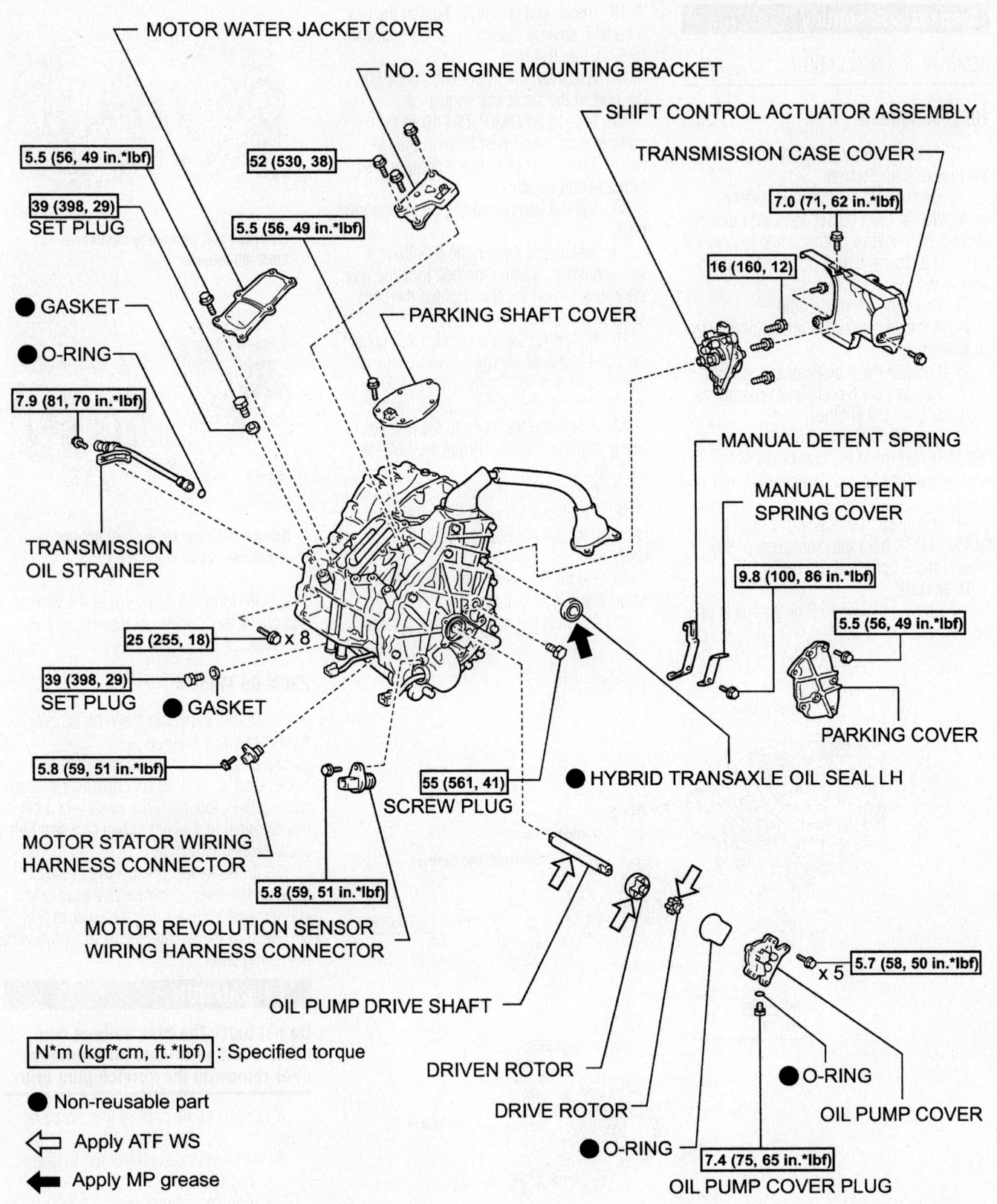

MOTOR WATER JACKET COVER

NO. 3 ENGINE MOUNTING BRACKET

SHIFT CONTROL ACTUATOR ASSEMBLY

TRANSMISSION CASE COVER

5.5 (56, 49 in.*lbf)

39 (398, 29)
SET PLUG

52 (530, 38)

5.5 (56, 49 in.*lbf)

7.0 (71, 62 in.*lbf)

16 (160, 12)

● GASKET

● O-RING

PARKING SHAFT COVER

7.9 (81, 70 in.*lbf)

TRANSMISSION
OIL STRAINER

MANUAL DETENT SPRING

MANUAL DETENT
SPRING COVER

9.8 (100, 86 in.*lbf)

5.5 (56, 49 in.*lbf)

25 (255, 18)   x 8

39 (398, 29)
SET PLUG   ● GASKET

PARKING COVER

5.8 (59, 51 in.*lbf)

55 (561, 41)

● HYBRID TRANSAXLE OIL SEAL LH

SCREW PLUG

MOTOR STATOR WIRING
HARNESS CONNECTOR

5.8 (59, 51 in.*lbf)

MOTOR REVOLUTION SENSOR
WIRING HARNESS CONNECTOR

5.7 (58, 50 in.*lbf)   x 5

OIL PUMP DRIVE SHAFT

N*m (kgf*cm, ft.*lbf) : Specified torque

DRIVEN ROTOR

DRIVE ROTOR

● O-RING

OIL PUMP COVER

● Non-reusable part

⇦ Apply ATF WS

⬅ Apply MP grease

● O-RING

7.4 (75, 65 in.*lbf)
OIL PUMP COVER PLUG

Transaxle unit and related components—2004–06 Prius models

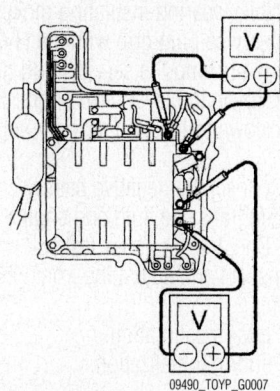

Verifying that there is 0V at the inverter with converter—Prius

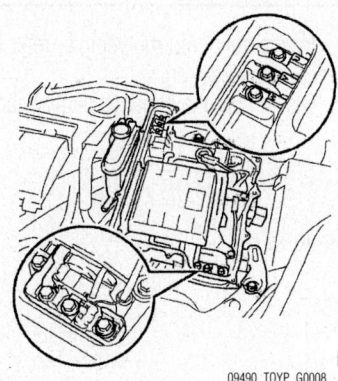

Measuring voltage between the terminals of the three phase connector—Prius

remove the inverter with converter assembly as follows:

    a. Remove the left and right engine under covers.

    b. Drain the HV coolant.

    c. Remove the radiator support opening cover.

    d. Remove the inverter cover.

    e. Verify that there is 0V at the inverter with converter, using a voltmeter utilizing a measuring range of DC 400V or more.

    f. Again, verify that there is 0V by measuring the voltage between the terminals of the three phase connector (U-V, V-W, U-W), using a voltmeter utilizing a measuring range of DC 400V or more.

    g. Disconnect the No. 1, No. 2 and No. 6 inverter cooling hoses.

    h. Disconnect the No. 1 circuit breaker sensor by first moving the outer section away toward the wire side.

    i. Disconnect the 2 frame wire connectors from the inverter with converter assembly and protect the electrode and connector parts with insulating tape.

    j. Use a small screwdriver to lift up the green lock pin, then disconnect the

connector for the air conditioning inverter.

    k. Disconnect any remaining wiring connectors including the engine main wiring harness.

    l. Remove the mounting bolts and disconnect the MG1 and MG2 power cables. Protect the connector parts with insulating tape.

    m. Remove the 3 mounting bolts and the inverter with converter assembly.

  7. Remove or disconnect the following:

- Air cleaner assembly
- Associated wiring harnesses and clamps
- 3 bolts and transmission case cover from the transaxle
- Bolt and ground wire
- 4 clamps and 3 cooling hoses from inverter
- Front exhaust pipe
- Left and right front axle hub nuts
- Left front stabilizer link
- Left and right tie rod ends
- Left and right front lower control arms
- Left and right front axle hub/bearing assemblies
- Left and right drive shafts
- Engine torque rod
- Front suspension crossmember
- 2 PCV hoses

  8. Install the No. 1 (12281-22021) and No. 2 (12281-15040) engine hangers in the correct direction. Torque the hanger bolt (91512-B1016) to 28 ft. lbs. (38 Nm).

  9. Attach an engine chain hoist to the engine hangers to keep the engine suspended.

✻✻ WARNING

Do not attempt to hang the engine by hooking the chain to any other parts.

  10. Using a transmission jack, support the hybrid vehicle transaxle.

  11. Remove the nut and disconnect the No. 3 engine mounting bracket from the engine mounting insulator.

  12. Remove the 2 bolts, starter cover and housing side cover.

  13. Remove the 6 bolts and transaxle unit.

➡**To avoid damage to the knock pin, do not pry the connecting portion of the HV transaxle and the engine.**

  14. Remove the No. 6 clamp and inverter cooling hose.

  15. Remove the 3 bolts and No. 3 engine mounting bracket.

  16. Remove the 6 bolts and clamp.

To install:

  17. Install the clamp in position with the 6 bolts and tighten to 80 inch lbs. (9 Nm).

  18. Install the No. 3 engine mounting bracket with the 3 bolts and tighten to 38 ft. lbs. (52 Nm).

  19. Connect the No. 6 inverter cooling hose and install the clamp.

  20. Install the transaxle to the engine in the vehicle. Tighten the 6 bolts to 24 ft. lbs. (33 Nm).

➡**Ensure that the knock pin is installed on the engine side. Place the transaxle in a horizontal position and align the knock pin to its hole. Then tighten the 6 bolts in the correct positions.**

  21. Install the housing side cover and starter cover and tighten the 2 bolts to 23 ft. lbs. (32 Nm).

  22. Install the No. 3 engine mounting bracket to the engine mounting insulator with the nut. Torque the nut to 59 ft. lbs. (80 Nm).

  23. Remove the engine hangers.

  24. Install or connect the following:

- 2 PCV hoses
- Front suspension crossmember. Torque the left and right front corner bolts to 83 ft. lbs. (113 Nm). Torque the left and right rear corner bolts to 116 ft. lbs. (157 Nm).
- Engine torque rod. Torque the 2 nuts and 2 bolts, at the engine, to 74 ft. lbs. (100 Nm). Torque the 2 nuts and 2 bolts, at the body side, to 41 ft. lbs. (56 Nm).
- Left and right drive shafts
- Left and right front axle hub/bearing assemblies
- Left and right front lower control arms
- Left and right tie rod ends
- Left front stabilizer link
- Left and right front axle hub nuts
- Front exhaust pipe
- 3 cooling hoses to the inverter with the 4 clamps

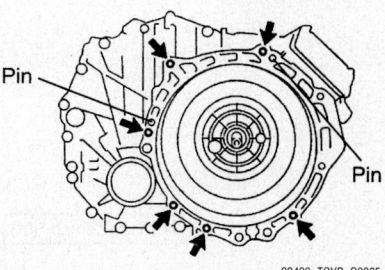

Correct positions of the transaxle mounting bolts—2004–06 Prius

- Ground wire and bolt tightened to 80 inch lbs. (9 Nm)
- Transmission case cover to the transaxle with 3 bolts tightened to 62 inch lbs. (7 Nm)
- Associated wiring harnesses and clamps
- Air cleaner assembly

25. While wearing insulating gloves, install the inverter with converter assembly as follows:

a. Install the inverter with converter assembly and tighten the three mounting bolts to 16 ft. lbs. (21 Nm).

b. Connect the MG1 and MG2 power cables and tighten the mounting bolts to 71 inch lbs. (8 Nm).

c. Connect any remaining wiring connectors including the engine main wiring harness. Be sure to insert the grommet of the engine main wiring harness into the U-shaped groove of the inverter case.

d. Engage the connector for the air conditioning inverter and secure by pushing in the lock pin..

e. Connect the 2 frame wire connectors to the inverter with converter assembly.

f. Connect the No. 1 circuit breaker sensor.

g. Connect the No. 6, No. 2 and No. 1 inverter cooling hoses.

h. Install the inverter cover and tighten the mounting fasteners to 8 ft. lbs. (11 Nm).

i. Install the radiator support opening cover.

j. Install the left and right engine under covers.

26. Install or connect the following:
- Cowl top front panel
- Left and right front wiper arms
- Hood
- Engine under covers
- Front wheels

27. While wearing insulating gloves, joint the service plug grip with the HV battery. While pushing the service plug grip to the right, rotate the lever to the right. Slide the lever down to lock the service plug grip in place.

28. Connect the negative battery cable.

29. Add transaxle fluid and engine coolant. Check systems for leaks.

30. Inspect and adjust the front wheel alignment.

31. Perform calibration.

32. Perform initialization.

Halfshaft

REMOVAL & INSTALLATION

1. Before servicing the vehicle, refer to the Precautions Section.

2. Disconnect the negative battery cable.

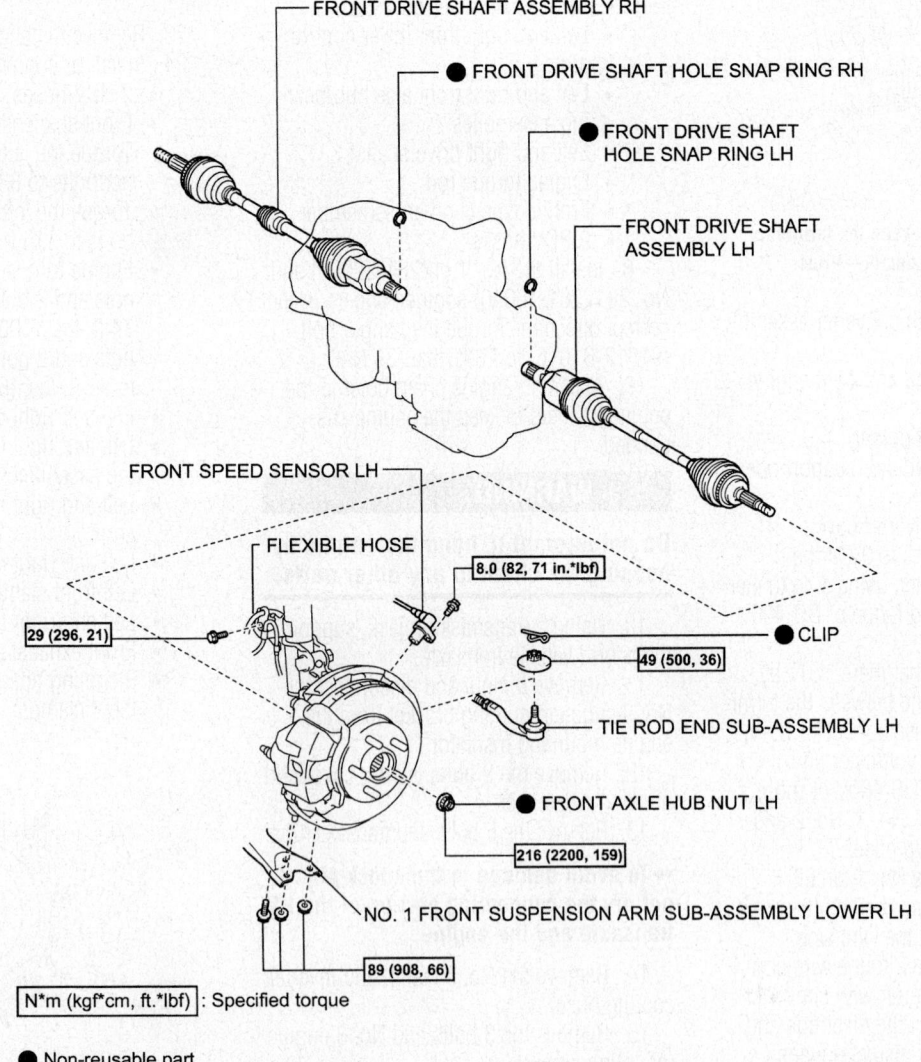

FRONT DRIVE SHAFT ASSEMBLY RH

● FRONT DRIVE SHAFT HOLE SNAP RING RH

● FRONT DRIVE SHAFT HOLE SNAP RING LH

FRONT DRIVE SHAFT ASSEMBLY LH

FRONT SPEED SENSOR LH

FLEXIBLE HOSE

8.0 (82, 71 in.*lbf)

29 (296, 21)

● CLIP

49 (500, 36)

TIE ROD END SUB-ASSEMBLY LH

● FRONT AXLE HUB NUT LH

216 (2200, 159)

NO. 1 FRONT SUSPENSION ARM SUB-ASSEMBLY LOWER LH

89 (908, 66)

N*m (kgf*cm, ft.*lbf) : Specified torque

● Non-reusable part

Front drive shafts and related components

3. Drain transaxle fluid.

4. After draining the fluid, tighten the drain plug, along with a new gasket to 29 ft. lbs. (39 Nm).

5. Raise and safely support the vehicle.

6. Remove the front wheel.

7. Unstake the axle hub nut and using a 30mm socket, remove the axle hub nut.

8. Remove the bolt and disconnect the speed sensor wire and flexible hose clamp from the strut.

9. Remove the bolt and front speed sensor from the steering knuckle.

➡**Keep both the tip and installation part of the speed sensor free of foreign matter.**

10. Remove the clip, castle nut and disconnect the tie rod end from the steering knuckle.

11. Remove the bolt, 2 nuts and disconnect the front lower ball joint from the front lower suspension arm.

12. Using a plastic-faced hammer, tap the end of the front drive shaft and disengage the fitting between the front drive shaft and front axle. If it is difficult to disengage, tap the end of the front drive shaft LH with a brass bar and hammer.

13. Push the front axle outward from the vehicle to remove the front drive shaft from the front axle. Be careful not to push the front axle outward from the vehicle more than necessary to remove it.

➡**Be careful not to damage the rubber boots. Hang the drive shaft down with a string or equivalent.**

14. Remove the front fender apron seal.

15. Hook the SST 09520-01010 and 09520-24010 claw in position to remove the front drive shaft.

❋❋ CAUTION

Be careful not to damage the oil seal. Be careful not to damage the front drive shaft boot. Be careful not to drop the front drive shaft.

16. Check for noticeable looseness when turning the joint up and down, left and right, and in the thrust direction.

17. Check for cracks, damage or grease leaks on the joint boot.

➡**Carry the drive shaft levelly.**

To install:

18. Apply ATF to the spline of the inboard joint.

19. Align the spline of the front drive shaft and insert the front drive shaft using a brass bar and hammer.

➡**Face the snap ring cut area downward. Be careful not to damage the oil seal. Be careful not to damage the front drive shaft boot.**

20. Install the front fender apron seal.

21. Install the front drive shaft dust cover (right drive shaft).

22. Push the front axle outward from the vehicle to align the spline of the front drive shaft with the front axle and insert.

23. Connect the front suspension lower arm to the front lower ball joint and tighten the bolt and 2 nuts to 105 ft. lbs. (142 Nm) for 2002–03 models, and 66 ft. lbs. (89 Nm) for 2004–06 models.

24. Connect the tie rod end to the steering knuckle and install it with the castle nut. Torque the castle nut to 36 ft. lbs. (49 Nm) and install a new cotter pin.

➡**The cotter pin hole alignment should be done after tightening the castle nut up to 60 degrees beyond the torque specification.**

25. Connect the front speed sensor wire and flexible hose clamp to the strut with the bolt.

26. Install the front speed sensor to the steering knuckle and tighten the bolt to 71 inch lbs. (8 Nm).

27. Using a 30mm socket wrench, install a new hub nut and tighten to 159 ft. lbs. (216 Nm).

28. Using a chisel and hammer, stake the hub nut.

29. Install the front wheel.

30. Refill the transaxle fluid to the proper level.

31. Connect the negative battery cable.

32. Check the front wheel alignment.

33. Check the ABS speed sensor signal.

CV-Joints

OVERHAUL

1. Before servicing the vehicle, refer to the Precautions Section.

2. Remove the drive shaft.

3. Remove the large seal retaining clamp from the inboard joint. Discard the clamp.

4. Remove the small seal retaining clamp from the inboard joint. Discard the clamp.

5. Separate the joint housing from the boot.

6. Remove grease from the inboard joint.

7. Put match marks on the inboard joint and outboard joint, but do not make the match marks with a punch and hammer.

8. Remove the inboard joint from the outboard joint.

9. Hold the outboard joint with vise with aluminum plates in between. Do not overtighten the vise.

10. Using a snap ring expander, remove the front drive inner shaft snap ring.

11. Put match marks on the tripod joint and outboard joint. Do not make match marks with a punch and hammer.

12. Using a brass bar and hammer, tap out the tripod joint but do not hit the roller position.

13. Remove the No. 2 inboard joint boot clamp, inboard joint boot and inboard joint boot clamp.

14. Remove the drive shaft damper setting clamp (right drive shaft) and remove the drive shaft damper from the right outboard joint.

15. Using a flat-head screwdriver, remove the No. 2 outboard joint boot clamp.

16. Using a flat-head screwdriver, remove the outboard joint boot clamp.

17. Remove the outboard joint boot from the outboard joint.

18. Remove grease from the outboard joint.

19. Using a flat-head screwdriver, remove the front drive shaft hole snap ring.

20. Using SST 09950-00020 and a press, press out the front drive shaft dust cover. Be careful not to drop the inboard joint.

To install:

21. When installing new boot and damper clamps, observe the following:

- When using a one touch clamp: After installing the new clamp, stake the clamp using a flat-head screwdriver.
- When using a hook type: Using needle-nose pliers, align the concave part and protrusion of the new clamp in order to lock it.

➡**Do not scratch the boots. Do not deform the claw of the hook.**

22. Using a press, install a new dust cover to the inboard joint shaft.

23. Install a new front drive shaft hole snap ring.

24. Wrap protective tape around the spline of the outboard joint.

25. Install a new No. 2 outboard joint boot clamp.

26. Install the outboard joint boot.

27. Install a new outboard joint boot clamp.

28. Apply 4.9–5.3 oz. (125–135g) grease to the joint of the outboard joint and outboard joint boot.

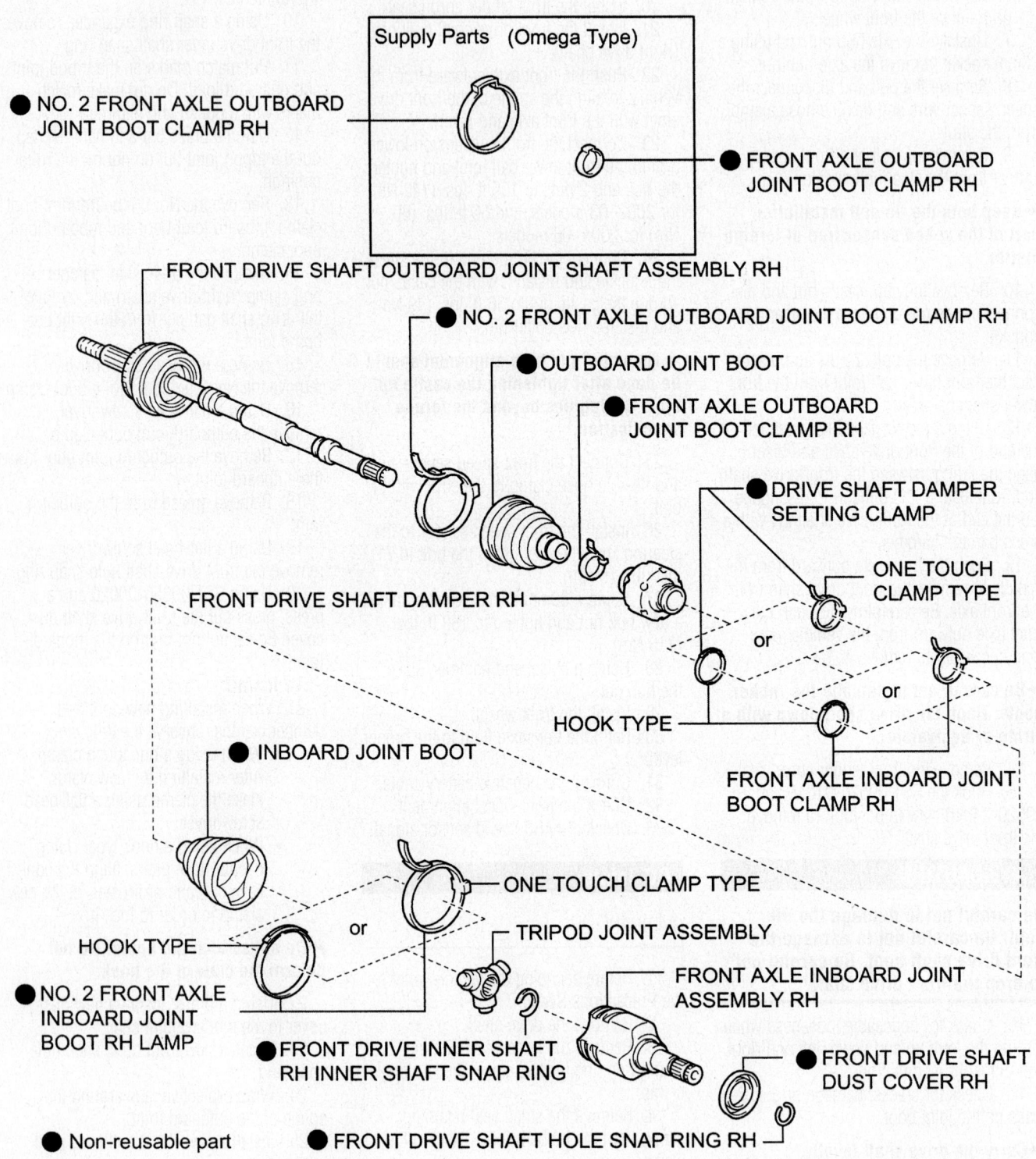

Supply Parts   (Omega Type)

● NO. 2 FRONT AXLE OUTBOARD
JOINT BOOT CLAMP RH

● FRONT AXLE OUTBOARD
JOINT BOOT CLAMP RH

FRONT DRIVE SHAFT OUTBOARD JOINT SHAFT ASSEMBLY RH

● NO. 2 FRONT AXLE OUTBOARD JOINT BOOT CLAMP RH

● OUTBOARD JOINT BOOT

● FRONT AXLE OUTBOARD
JOINT BOOT CLAMP RH

● DRIVE SHAFT DAMPER
SETTING CLAMP

ONE TOUCH
CLAMP TYPE

FRONT DRIVE SHAFT DAMPER RH

HOOK TYPE

or

or

FRONT AXLE INBOARD JOINT
BOOT CLAMP RH

● INBOARD JOINT BOOT

ONE TOUCH CLAMP TYPE

TRIPOD JOINT ASSEMBLY

FRONT AXLE INBOARD JOINT
ASSEMBLY RH

HOOK TYPE

or

● NO. 2 FRONT AXLE
INBOARD JOINT
BOOT RH LAMP

● FRONT DRIVE INNER SHAFT
RH INNER SHAFT SNAP RING

● FRONT DRIVE SHAFT
DUST COVER RH

● Non-reusable part

● FRONT DRIVE SHAFT HOLE SNAP RING RH

Exploded view of tripod joint and drive shaft assembly—Right side shown

09490_TOYP_G0067

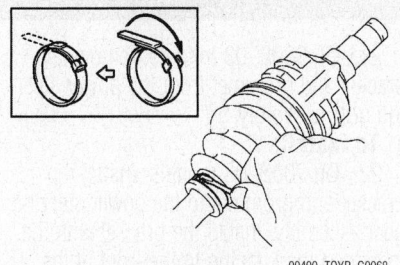

Staking the one touch clamp—Tripod joint

09490_TOYP_G0068

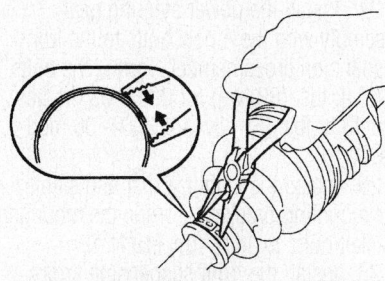

Locking the hook type clamp using needle-nose pliers—Tripod joint

09490_TOYP_G0069

➡ **Do not apply grease in the groove.**

29. Install the outboard joint boot into the outboard joint groove.

30. Hold the front drive shaft in a vise between 2 aluminum plates. Do not overtighten the vise.

31. Set SST 09521-24010 to the No. 2 outboard joint boot clamp and slightly tighten the SST bolt while pushing the outboard joint on.

32. Hold SST and tighten the SST bolt so that the clearance is within 0.031 in. (0.8mm) or less. Do not damage the outboard joint.

33. Remove SST.

34. Using SST 09240-00020, measure the clearance of the No. 2 outboard joint boot clamp. The clearance should measure within 0.031 in. (0.8mm) or less.

➡ **If the clearance exceeds the maximum, retighten it.**

35. Set SST 09521-24010 to the outboard joint boot clamp and slightly tighten the SST bolt while pushing the outboard joint on.

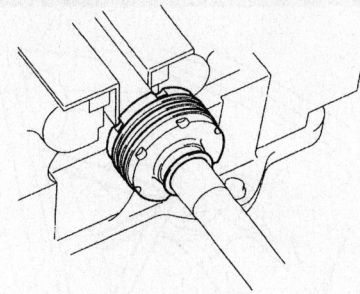

427 +-2.0 mm

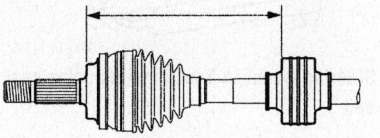

09490_TOYP_G0070

Standard distance measurement for drive shaft damper (right drive shaft)—Tripod joint

36. Hold SST and tighten the SST bolt so that the clearance is within 0.031 in. (0.8mm) or less. Do not damage the outboard joint.

37. Remove SST.

38. Using SST 09240-00020, measure the clearance of the outboard joint boot clamp. The clearance should measure within 0.031 in. (0.8mm) or less.

➡ **If the clearance exceeds the maximum, retighten it.**

39. Install the drive shaft damper to the right drive shaft outboard joint. The standard distance should measure 16.81+/-0.079 inches (427+/-2.0mm). Be sure to install in the correct direction.

40. Install a new drive shaft damper clamp.

41. Install a new inboard joint boot clamp

42. Install inboard joint boot

43. Install a new No. 2 inboard joint boot clamp

44. Hold the drive shaft in a vise between 2 aluminum plates. Do not overtighten the vise.

45. Remove the protective tape.

46. Align the match marks and install the

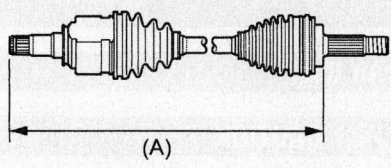

(A)

09490_TOYP_G0071

Measuring the drive shaft dimension—Tripod joint

tripod joint onto the outboard joint. Face the serration side of the tripod joint outward and install it to the outboard joint end.

47. Using a brass bar and hammer, drive the tripod joint in.

✳ CAUTION

Do not hit the roller portion. Do not attach any foreign matter on the tripod joint.

48. Using a snap ring expander, install a new front drive inner shaft snap ring.

49. Apply 4.9–5.3 oz. (125–135g) grease in the joint of the inboard joint and outboard joint boot.

50. Align the match marks and install the inboard joint onto the outboard joint.

51. Install the inboard joint boot into the inboard joint and outboard joint grooves. Do not apply grease in the groove.

52. Install a new No. 2 front axle inboard joint boot clamp.

➡ **Do not scratch the boots.**

53. Install a new front axle inboard joint boot clamp.

➡ **Do not scratch the boots.**

54. Check if the front drive shaft is within the following standard dimension:
  • Left drive shaft: 22.216 inches (564.3mm)
  • Right drive shaft: 33.319 inches (846.3mm)

55. Check for noticeable looseness when turning the joint up and down, left and right, and in the thrust direction.

56. Check for cracks, damage or grease leaks on the joint boot.

➡ **Always carry the drive shaft levelly.**

57. Install the drive shaft.

STEERING

Air Bag

✳✳ CAUTION

Most vehicles are equipped with an air bag system. The system must be disabled before performing service on or around system components, steering column, instrument panel components, wiring and sensors. Failure to follow safety and disabling procedures could result in accidental air bag deployment, possible personal injury and unnecessary system repairs.

PRECAUTIONS

Several precautions must be observed when handling the inflator module to avoid accidental deployment and possible personal injury.

• Never carry the inflator module by the wires or connector on the underside of the module.

• When carrying a live inflator module, hold securely with both hands, and ensure that the bag and trim cover are pointed away.

• Place the inflator module on a bench or other surface with the bag and trim cover facing up.

• With the inflator module on the bench, never place anything on or close to the module that may be thrown in the event of an accidental deployment.

DISARMING

To avoid personal injury when working on vehicles equipped with an air bag, the negative battery cable must be disconnected and at least 90 seconds must elapse before working on the system. Failure to do so may result in deployment of the air bag.

REARMING

To rearm the air bag system, simply reconnect the battery cable(s).

Rack and Pinion Steering Gear

REMOVAL & INSTALLATION

1. Before servicing the vehicle, refer to the Precautions Section.
2. Place the front wheels in the straight ahead position.

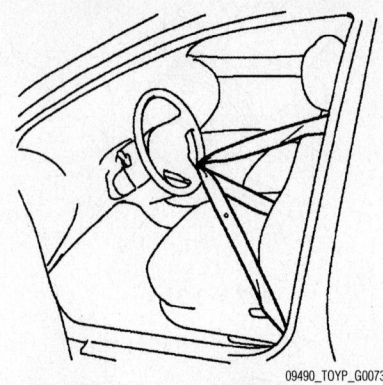
09490_TOYP_G0073
Steering wheel in fixed position using seat belt

3. Disconnect the negative battery cable and wait at least 90 seconds before working on the vehicle to disarm the air bag.
4. Fix the steering wheel with the seat belt in order to prevent rotation and damage to the spiral cable.
5. Remove engine under covers.
6. If equipped, remove the steering column hole cover sheet.
7. On 2002–03 models, perform the following:
   a. Disconnect the 2 connectors.
   b. Disconnect the 2 clamps.
   c. Remove the bolt and disconnect the ground wire.
   d. Remove the 2 bolts and EMPS bracket.
8. Put match marks in the sliding yoke and intermediate shaft.
9. Loosen the top bolt and remove the lower bolt to separate the sliding yoke.
10. If necessary, remove the steering column hole cover from the body. Be careful not to damage the clips.
11. Remove the front wheels.
12. Disconnect the tie rod ends from the steering knuckles.
13. Disconnect the stabilizer links from the struts.
14. Disconnect the lower suspension arm from the lower ball joint.
15. Place match marks on the intermediate shaft assembly and power steering gear.
16. Remove the bolt and disconnect the intermediate shaft from the control valve shaft.
17. Disconnect the torque rod.
18. Remove the front suspension crossmember assembly.
19. Remove the stabilizer bar mounting brackets and the stabilizer bar.
20. Remove the 4 bolts and power steering gear assembly from the front suspension crossmember.

21. On 2002–03 models, remove the bracket and grommet from the power steering gear assembly.

To install:

22. On 2002–03 models, install the bracket and grommet to the power steering gear assembly. Install the bracket with the inscribed mark facing to the front of the vehicle.
23. On 2004–06 models, install the steering column hole cover to the steering gear.
24. Install the power steering gear assembly with the 4 new bolts to the front suspension crossmember. Torque the bolts to 61 ft. lbs. (83 Nm) for 2002–03 modles, and 43 ft. lbs. (58 Nm) for 2004–06 models.
25. Install the stabilizer bar and stabilizer bar mounting brackets. Torque the mounting bracket bolts to 14 ft. lbs. (19 Nm).
26. Install the front suspension crossmember. Torque the left and right front corner bolts to 83 ft. lbs. (113 Nm). Torque the left and right rear corner bolts to 116 ft. lbs. (157 Nm).
27. Connect the torque rod. Tighten the through bolt to 74 ft. lbs. (100 Nm).
28. Align the match marks and connect the intermediate shaft to the control valve shaft. Torque the bolt to 26 ft. lbs. (35 Nm).
29. On 2002–03 models, perform the following:
   a. Install the EMPS bracket and tighten the 2 bolts to 52 inch lbs. (5.5 Nm).
   b. Connect the ground wire and tighten the bolt to 52 inch lbs. (5.5 Nm).
   c. Connect the 2 clamps.
   d. Connect the 2 connectors.
30. Connect the left and right ball joints to the front lower control arms. Torque to 105 ft. lbs. (142 Nm) for 2002–03 models, and 66 ft. lbs. (89 Nm) for 2004–06 models.
31. Connect the stabilizer bar link to the strut. If the ball joint turns together with the nut, use a hexagon wrench to hold the stud. Torque to 55 ft. lbs. (74 Nm).
32. Connect the tie rod end to the steering knuckle. Torque the tie rod end nut to 36 ft. lbs. (49 Nm) and install a new cotter pin.
33. Install the front wheels.
34. Put the dust seal back to the engine compartment side.
35. Align the match marks on the intermediate shaft assembly and control valve shaft. Install and tighten the bolts to 26 ft. lbs. (35 Nm)

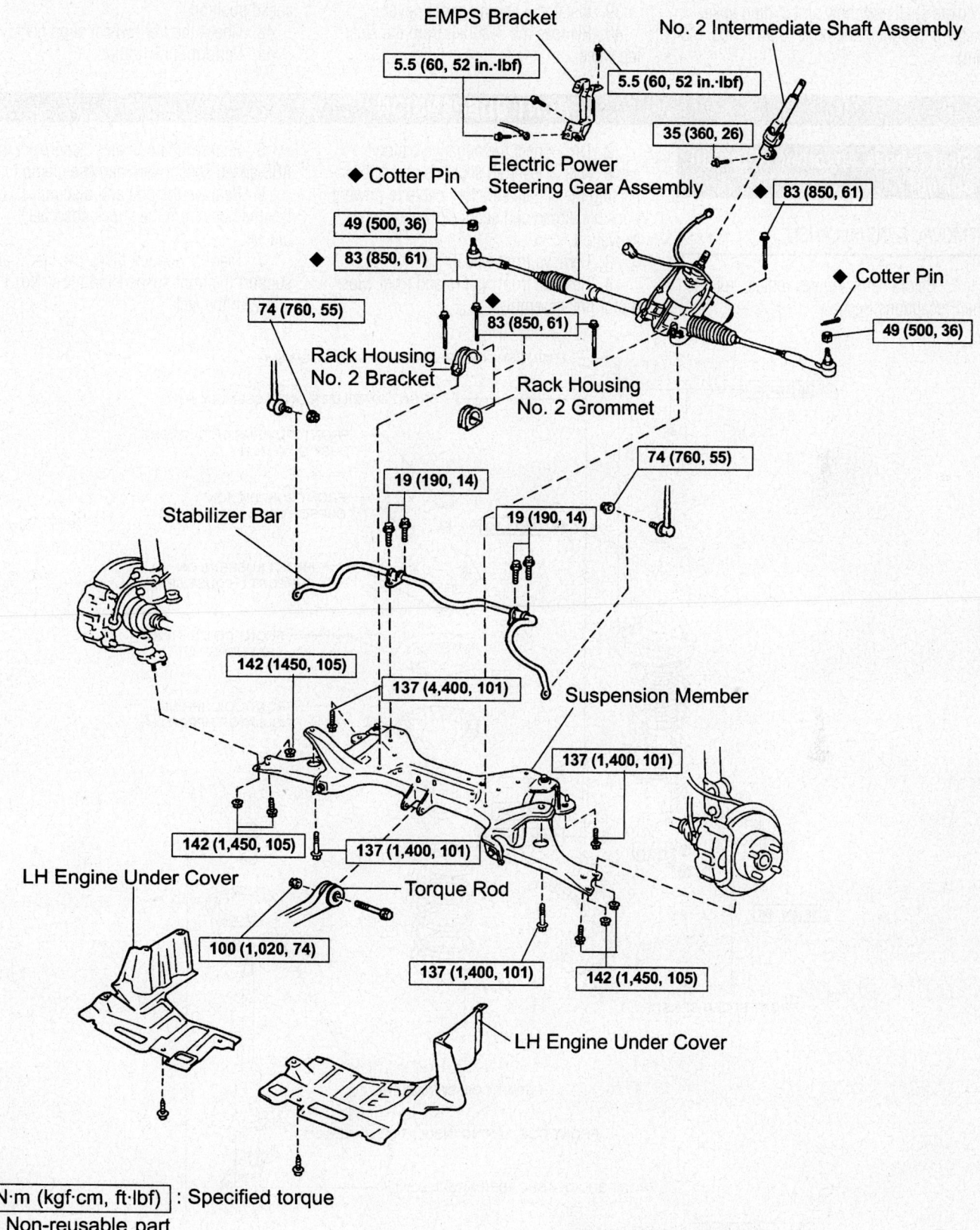

EMPS Bracket

No. 2 Intermediate Shaft Assembly

5.5 (60, 52 in.·lbf)

5.5 (60, 52 in.·lbf)

35 (360, 26)

◆ Cotter Pin

Electric Power
Steering Gear Assembly

◆ 83 (850, 61)

49 (500, 36)

◆ 83 (850, 61)

◆ Cotter Pin

74 (760, 55)

◆ 83 (850, 61)

49 (500, 36)

Rack Housing
No. 2 Bracket

Rack Housing
No. 2 Grommet

74 (760, 55)

19 (190, 14)

19 (190, 14)

Stabilizer Bar

142 (1450, 105)

137 (4,400, 101)

Suspension Member

137 (1,400, 101)

142 (1,450, 105)

137 (1,400, 101)

LH Engine Under Cover

Torque Rod

100 (1,020, 74)

137 (1,400, 101)

142 (1,450, 105)

LH Engine Under Cover

N·m (kgf·cm, ft·lbf) : Specified torque
◆ Non-reusable part

09490_TOYP_G0072

Exploded view of the front power steering gear and related components—2002–03 model shown

36. Install the steering column hole cover.
37. Align the match marks on the intermediate shaft assembly and sliding yoke. Install and tighten the bolts to 26 ft. lbs. (35 Nm)

38. If equipped, install the steering column hole cover sheet.
39. Install the engine under covers.
40. Remove the seat belt from the steering wheel.

41. Connect the negative battery cable.
42. Place the front wheels in the straight ahead position.
43. Check the front wheel alignment.
44. Perform initialization.

FRONT SUSPENSION

Strut (Shock Absorber With Coil Spring)

REMOVAL & INSTALLATION

1. Before servicing the vehicle, refer to the Precautions Section.

2. Disconnect the negative battery cable. Wait at least 90 seconds after disconnecting the negative battery cable to prevent possible airbag and seat belt pretensioner activation.
3. Remove front wheel.
4. Remove front wipers and front wiper motor link assembly.

5. Remove the bolt and disconnect the ABS speed sensor wire harness clamp.
6. Remove the bolt and disconnect the flexible hose from the shock absorber bracket.
7. Place a wooden block on a jack, and support the front suspension lower No. 1 arm with the jack.

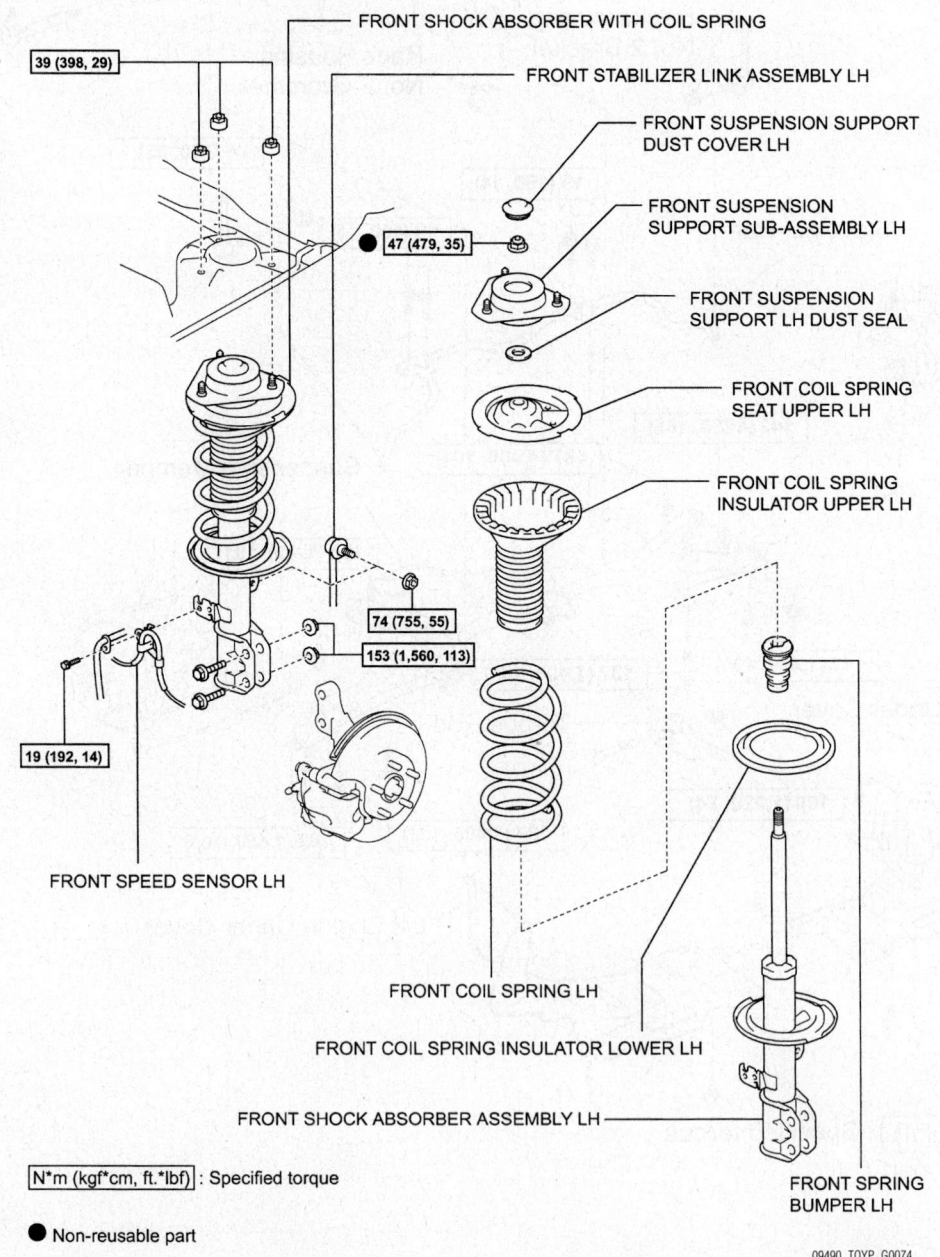

FRONT SHOCK ABSORBER WITH COIL SPRING
FRONT STABILIZER LINK ASSEMBLY LH
FRONT SUSPENSION SUPPORT DUST COVER LH
FRONT SUSPENSION SUPPORT SUB-ASSEMBLY LH
FRONT SUSPENSION SUPPORT LH DUST SEAL
FRONT COIL SPRING SEAT UPPER LH
FRONT COIL SPRING INSULATOR UPPER LH

39 (398, 29)
47 (479, 35)
74 (755, 55)
153 (1,560, 113)
19 (192, 14)

FRONT SPEED SENSOR LH

FRONT COIL SPRING LH
FRONT COIL SPRING INSULATOR LOWER LH
FRONT SHOCK ABSORBER ASSEMBLY LH
FRONT SPRING BUMPER LH

N*m (kgf*cm, ft.*lbf) : Specified torque
● Non-reusable part

Exploded view of the front strut assembly and related components

09490_TOYP_G0074

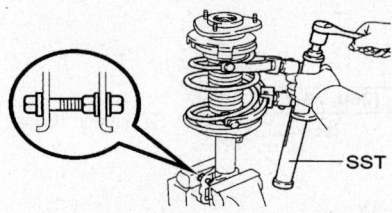

09490_TOYP_G0075

Securing the strut assembly in a vise while installing the spring compressor

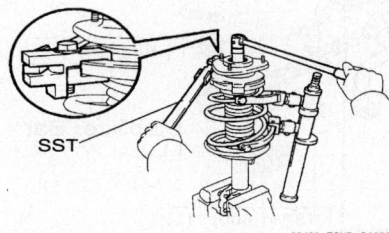

09490_TOYP_G0076

Removing the nut while holding the spring seat.

8. Remove the nut and separate the front stabilizer link from the strut assembly.

➡**Use a hexagon wrench to hold the stud if the ball joint turns together with the nut.**

9. Remove the 2 nuts on the lower side of the front strut assembly, but keep the bolts inserted.

10. Remove the 3 nuts from the top of the strut assembly.

11. Lower the jack slowly. Remove the 2 bolts on the lower side and the front strut assembly.

➡**Ensure that the speed sensor front LH is completely disconnected from the front shock absorber with coil spring.**

12. Separate the coil spring from the strut assembly as follows:

    a. Install 2 nuts and a bolt to the bracket at the lower side of the shock absorber and secure it in a vise.

    b. Using SST 09727-30021, compress the coil spring.

    c. Remove the cap from the suspension support.

    d. Using SST 09729-22031 to hold the spring seat, remove the nut.

    e. Remove the suspension support, dust seal, spring seat, upper insulator, coil spring, spring bumper and lower insulator.

To install:

13. Install the coil spring by assembling the strut assembly as follows:

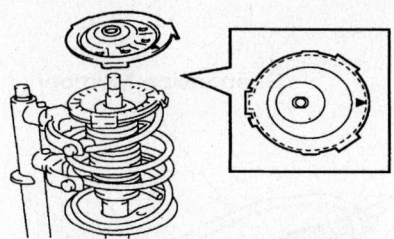

09490_TOYP_G0077

Install the spring seat to the shock absorber with the arrow mark facing to the outside of the vehicle.

    a. Install the lower insulator onto the strut.

    b. Install the spring bumper to piston rod.

    c. Using SST 09727-30021, compress the coil spring. Do not use an impact wrench. It will damage the SST.

    d. Install the coil spring to the shock absorber, fitting the lower end of the coil spring into the gap of the spring lower seat.

    e. Install the upper insulator.

    f. Install the spring seat to the shock absorber with the arrow mark facing to the outside of the vehicle.

    g. Install the dust seal and suspension support.

    h. Using SST 09729-22031 to hold the suspension support, install a new nut and torque to 34 ft. lbs. (47 Nm).

    i. Remove the tool.

    j. Apply MP grease No.2 into the suspension support. Do not touch grease on rubber surface of upper support.

    k. Install the cap.

14. Install the strut assembly to the steering knuckle and tighten the steering knuckle-to-strut assembly nuts and the bolts to 113 ft. lbs. (153 Nm)

15. Connect the stabilizer shaft link to strut assembly and tighten the stabilizer link-to-strut assembly nut to 55 ft. lbs. (74 Nm)

16. Install the flexible brake hose to the shock absorber bracket.

17. Connect the ABS speed sensor wire harness clamp.

18. Install the front wiper motor link assembly and the front wipers.

19. Install the front wheel.

20. Lower the vehicle.

21. Install the nuts securing the strut assembly to the body of the vehicle and tighten to 29 ft. lbs. (39 Nm).

22. Connect the negative battery cable.

Coil Spring

REMOVAL & INSTALLATION

For coil spring service on the Prius, refer to the strut removal and installation procedure.

Lower Ball Joint

REMOVAL & INSTALLATION

1. Before servicing the vehicle, refer to the Precautions Section.

2. Remove the steering knuckle with the hub/bearing assembly (refer to the wheel bearings procedure later in this section).

3. Mount the steering knuckle in a vise.

4. Remove the cotter pin and nut.

5. Using SST 09628-62011, remove the lower ball joint.

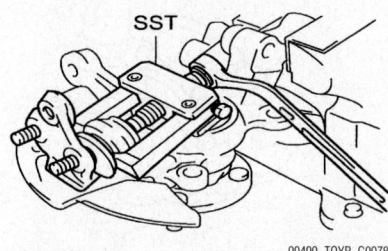

09490_TOYP_G0078

Separate the ball joint from the steering knuckle.

To install:

6. Install the lower ball joint and tighten the nut to 76 ft. lbs. (103 Nm).

7. Install a new cotter pin. If the holes for the cotter pin are not aligned, tighten the nut further up to 60 degrees.

8. Install the steering knuckle with the hub/bearing assembly (refer to the wheel bearings procedure later in this section).

9. Check the ABS speed sensor signal.

10. Check the front wheel alignment.

Lower Control Arm

REMOVAL & INSTALLATION

1. Before servicing the vehicle, refer to the Precautions Section.

2. Place front wheels facing straight ahead.

3. Disconnect the negative battery cable.

4. Raise and safely support the front of the vehicle. Let the control arms hang free.

5. Remove or disconnect the following:

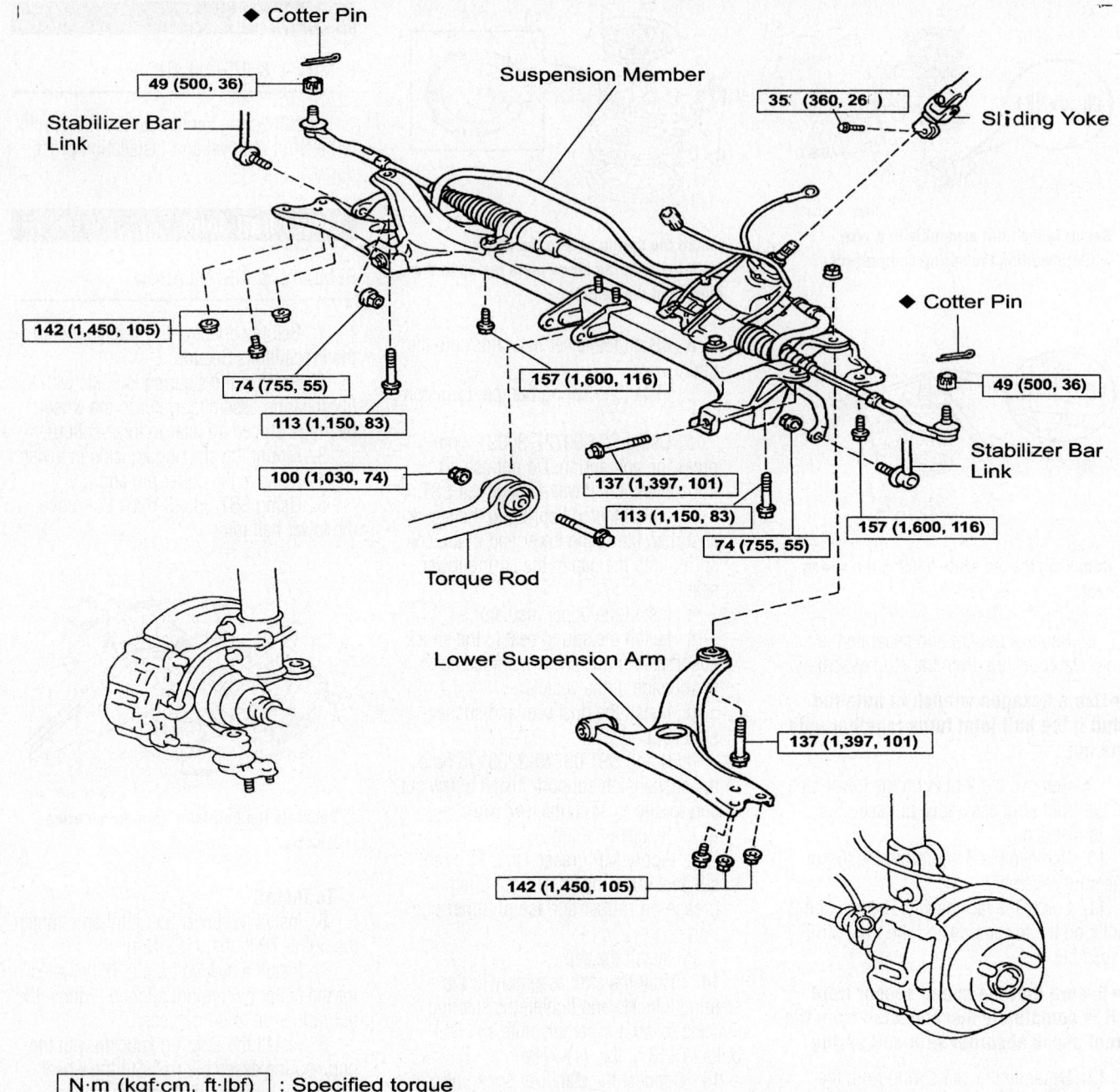

N·m (kgf·cm, ft·lbf) : Specified torque
◆ Non-reusable part

09490_TOYP_G0079

Exploded view of the front lower control arm and related components—2002–03 model shown

- Front wheels
- Steering column hole cover silencer, if necessary
- Engine under cover
- Front exhaust pipe assembly, if necessary
- Tie rod end cotter pin and nut
- Tie rod end from the steering knuckle
- Stabilizer bar link from the strut. If the ball joint turns together with the nut, use a hexagon wrench to hold the stud.

- Bolt and 2 nuts, and disconnect the lower suspension arm from the lower ball joint
- Loosen the 2 lower suspension arm set bolts
- Steering sliding yoke
- Drive shafts, if necessary

6. Support the suspension member with a transmission jack.

7. Remove the bolt and nut, disconnect the torque rod from the suspension member.

8. Remove the 4 bolts and disconnect the suspension member from the body.

9. Remove the 2 lower suspension arm set bolts and disconnect the lower suspension arm from suspension crossmember.

To install:

10. Install the front lower control arm to the suspension crossmember and temporarily tighten the front suspension lower No. 1 arm with the 2 bolts and nut.

11. Install the front suspension crossmember. Torque the left and right front corner bolts to 83 ft. lbs. (113 Nm). Torque the left and right rear corner bolts to 116 ft. lbs. (157 Nm).

12. Install the torque rod. Tighten the 2 nuts and 2 bolts, at the engine, to 74 ft. lbs. (100 Nm). Torque the 2 bolts, at the body side, to 44 ft. lbs. (60 Nm) for 2002–03 models, and 41 ft. lbs. (56 Nm) for 2004–06 models.

13. Install or connect the following:
- Left and right drive shafts.
- Steering sliding yoke. Torque the pinch bolt to 26 ft. lbs. (35 Nm).
- Left and right ball joints to the front lower control arms. Torque to 105 ft. lbs. (142 Nm) for 2002–03 models, and 66 ft. lbs. (89 Nm) for 2004–06 models.
- Stabilizer bar link to the strut. If the ball joint turns together with the nut, use a hexagon wrench to hold the stud. Torque to 55 ft. lbs. (74 Nm).
- Tie rod end to the steering knuckle. Torque the tie rod end nut to 36 ft. lbs. (49 Nm) and install a new clip.
- Front wheels
- Front exhaust pipe assembly, if necessary
- Steering column hole cover silencer, if necessary

14. Lower the vehicle and bounce it up and down several times to stabilize the front suspension.

15. Fully tighten the 2 front lower control arm bolts. Torque to 101 ft. lbs. (137 Nm). Keep the nut from rotating while tightening the rear-side bolt. Lower the tires to the ground using a 4-post lift.

16. Connect the negative battery cable.

17. Check the front wheel alignment.

CONTROL ARM BUSHING REPLACEMENT

The control arm bushings are serviced with the control arm as an assembly.

Wheel Bearings

ADJUSTMENT

The front wheel bearings are a cartridge type design and cannot be adjusted.

REMOVAL & INSTALLATION

2002–03 Models

1. Before servicing the vehicle, refer to the Precautions Section.
2. Raise and safely support the vehicle.
3. Remove the front wheel.
4. Remove the front axle hub nut.
5. Remove the 2 caliper mounting bolts

and separate the caliper from the mounting and support from the vehicle with wire. Do not allow the caliper to hang low enough as to put tension on the brake hose.

6. Remove the front disc brake rotor.

7. Disconnect the front ABS speed sensor connector.

8. Loosen, but do not remove the nuts connecting the steering knuckle to the strut assembly.

9. Disconnect the tie rod end from the steering knuckle.

10. Remove the bolt and 2 nuts and disconnect the front lower suspension control arm from the lower ball joint.

11. Separate the front drive shaft from the hub/bearing and steering knuckle assembly.

12. Separate the steering knuckle and hub/bearing from the strut assembly.

13. Mount the steering knuckle and hub/bearing assembly securely in a soft vice.

14. Using SST 09520-00031, remove the hub/bearing assembly from the steering knuckle.

15. Using SST 09950-40011 and SST 09950-60010, remove the inner race (outside) from the hub/bearing assembly.

16. Remove the dust cover.

17. Remove the snap ring using snap ring pliers.

18. Place the inner race on the outside of the bearing and using SST 09527-17011, SST 09950-60010, SST 09950-70010 and a press, remove the bearing.

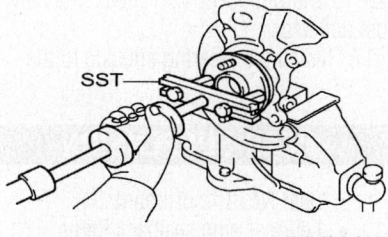

09490_TOYP_G0080

Removal of axle hub from steering knuckle—2002–03 Prius

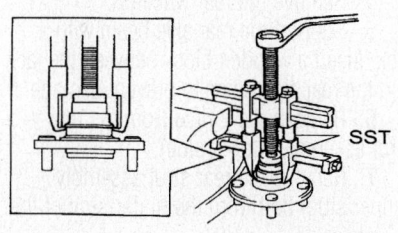

09490_TOYP_G0081

Removal of the inner race (outside) from the hub/bearing assembly—2002–03 Prius

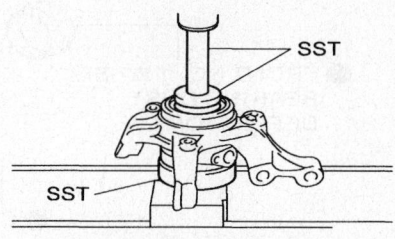

09490_TOYP_G0082

Place inner race on outside of bearing and remove the bearing—2002–03 Prius

To install

19. Using SST 09950-60010, SST 09950-70010 and a press, install a new bearing into the steering knuckle.

20. Install a new snap ring using snap ring pliers.

21. Install the dust cover. Torque the 3 bolts to 74 inch lbs. (8 Nm).

22. Using SST 09608-32010, SST 09950-60010, SST 09950-70010 and a press, install the axle hub.

23. Install the steering knuckle to the strut assembly. Torque the nuts and bolts to 113 ft. lbs. (153 Nm)

24. Install the front drive shaft.

25. Install the lower ball joint to the front lower suspension control arm. Torque the bolt and 2 nuts to 105 ft. lbs. (142 Nm).

26. Connect the tie rod end to the steering knuckle. Torque the tie rod end nut to 36 ft. lbs. (49 Nm) and install a new cotter pin.

27. Connect the front ABS speed sensor connector.

28. Install the front disc brake rotor.

29. Install the front disc brake caliper.

30. Install the front axle hub nut. Torque the nut to 159 ft. lbs. (216 Nm), then stake the hub nut, using a chisel and hammer.

31. Install the front wheel.

32. Inspect and adjust the front wheel alignment.

33. Check ABS speed sensor signal.

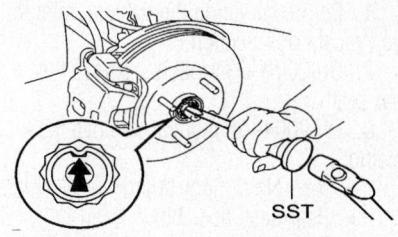

09490_TOYP_G0083

Staking the hub nut after installation

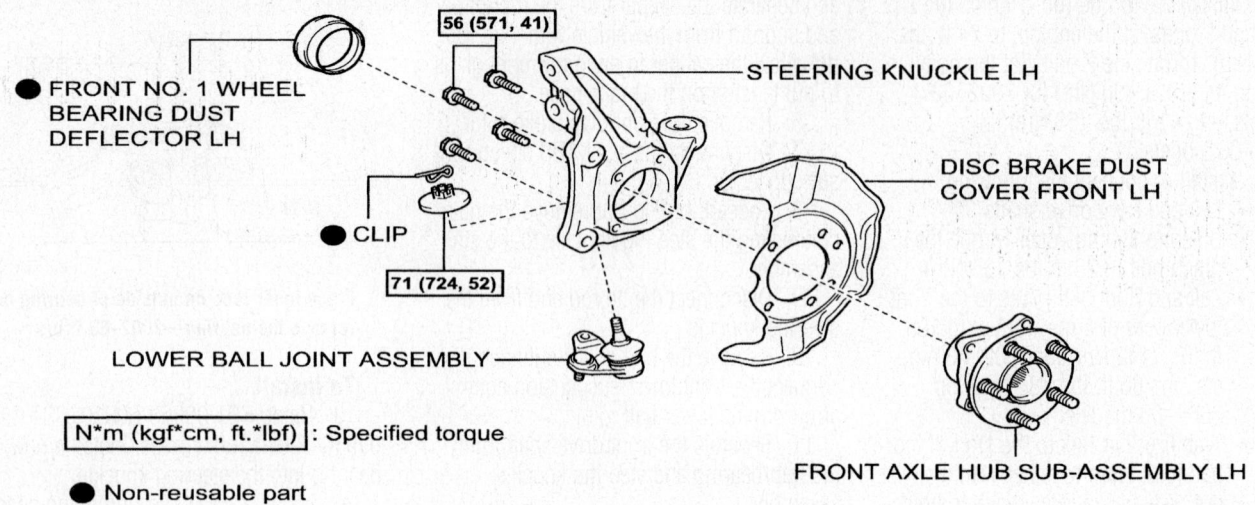

● FRONT NO. 1 WHEEL BEARING DUST DEFLECTOR LH

56 (571, 41)

STEERING KNUCKLE LH

● CLIP

71 (724, 52)

DISC BRAKE DUST COVER FRONT LH

LOWER BALL JOINT ASSEMBLY

N*m (kgf*cm, ft.*lbf) : Specified torque

● Non-reusable part

FRONT AXLE HUB SUB-ASSEMBLY LH

09490_TOYP_G0084

Exploded view of front hub/bearing assembly and related components

2004–06 Models

1. Before servicing the vehicle, refer to the Precautions Section.

2. Raise and safely support the vehicle.

3. Remove the front wheel.

4. Remove the front axle hub nut.

5. Disconnect the front ABS speed sensor connector.

6. Remove the 2 caliper mounting bolts and separate the caliper from the mounting and support from the vehicle with wire. Do not allow the caliper to hang low enough as to put tension on the brake hose.

7. Remove the front disc brake rotor.

8. Disconnect the tie rod from the steering knuckle.

9. Remove the bolt and 2 nuts and disconnect the front lower suspension control arm from the lower ball joint.

10. Separate the front drive shaft from the hub/bearing and steering knuckle assembly.

11. Separate the steering knuckle from the strut assembly.

12. Mount the steering knuckle and hub/bearing assembly securely in a soft vice. Using a screwdriver, remove the dust deflector from the steering knuckle.

13. Remove the 4 mounting bolts and hub/bearing assembly along with the dust cover.

To install:

14. Install the hub/bearing assembly and dust cover with the 4 mounting bolts.

15. Tighten the hub/bearing assembly mounting bolts to 41 ft. lbs. (56 Nm).

16. Using SST 09950-70010, 09608-320-10 and 09950-60020, press in a new dust deflector.

17. Install the steering knuckle to the strut assembly. Torque the nuts and bolts to 113 ft. lbs. (153 Nm)

18. Install the front drive shaft.

19. Install the lower ball joint to the front lower suspension control arm. Torque the bolt and 2 nuts to 66 ft. lbs. (89 Nm).

20. Connect the tie rod end to the steering knuckle. Torque the tie rod end nut to 36 ft. lbs. (49 Nm) and install a new clip.

21. Install the front disc brake rotor.

22. Install the front disc brake caliper.

23. Connect the front ABS speed sensor connector.

24. Install the front axle hub nut. Torque the nut to 159 ft. lbs. (216 Nm), then stake the hub nut, using a chisel and hammer.

25. Install the front wheel.

26. Inspect and adjust the front wheel alignment.

27. Check ABS speed sensor signal.

REAR SUSPENSION

Strut (Shock Absorber and Coil Spring)

REMOVAL & INSTALLATION

1. Before servicing the vehicle, refer to the Precautions Section.

2. On 2002–03 models, remove the rear seat.

3. On 2004–06 models, remove the following:

- Rear No. 2 floor board
- Rear deck floor box
- Rear deck trim cover
- Tonneau cover assembly
- Left rear seatback assembly
- Rear No. 1 floor board
- Left rear side seatback frame
- Rear No. 4 floor board
- Left deck floor box
- Left deck trim side panel assembly
- Battery carrier bracket

4. Remove the rear wheel(s).

5. Support the rear axle beam with a jack. Insert a wooden block between the jack and the rear axle beam to prevent damage.

6. Remove the 2 nuts from the rear strut assembly (upper side).

7. Remove the rear strut assembly (upper side) bolt from the under-side of the vehicle.

8. Remove the nut and spacer from the rear strut assembly (lower side).

9. Remove the rear strut assembly while slowly lowering the jack.

➡ **Seat the jack so that no extra load is placed on the strut assembly on the opposite side of the vehicle.**

10. Separate the coil spring from the strut assembly as follows:

   a. Use a 6mm socket hexagon wrench to secure the piston rod of the shock absorber and loosen the nut.

➡ **Do not remove the nut. Sufficiently insert the hexagon wrench.**

   b. Attach SST 09727-30021 to the coil spring so that the upper and lower hooks of the installed area are as wide as possible.

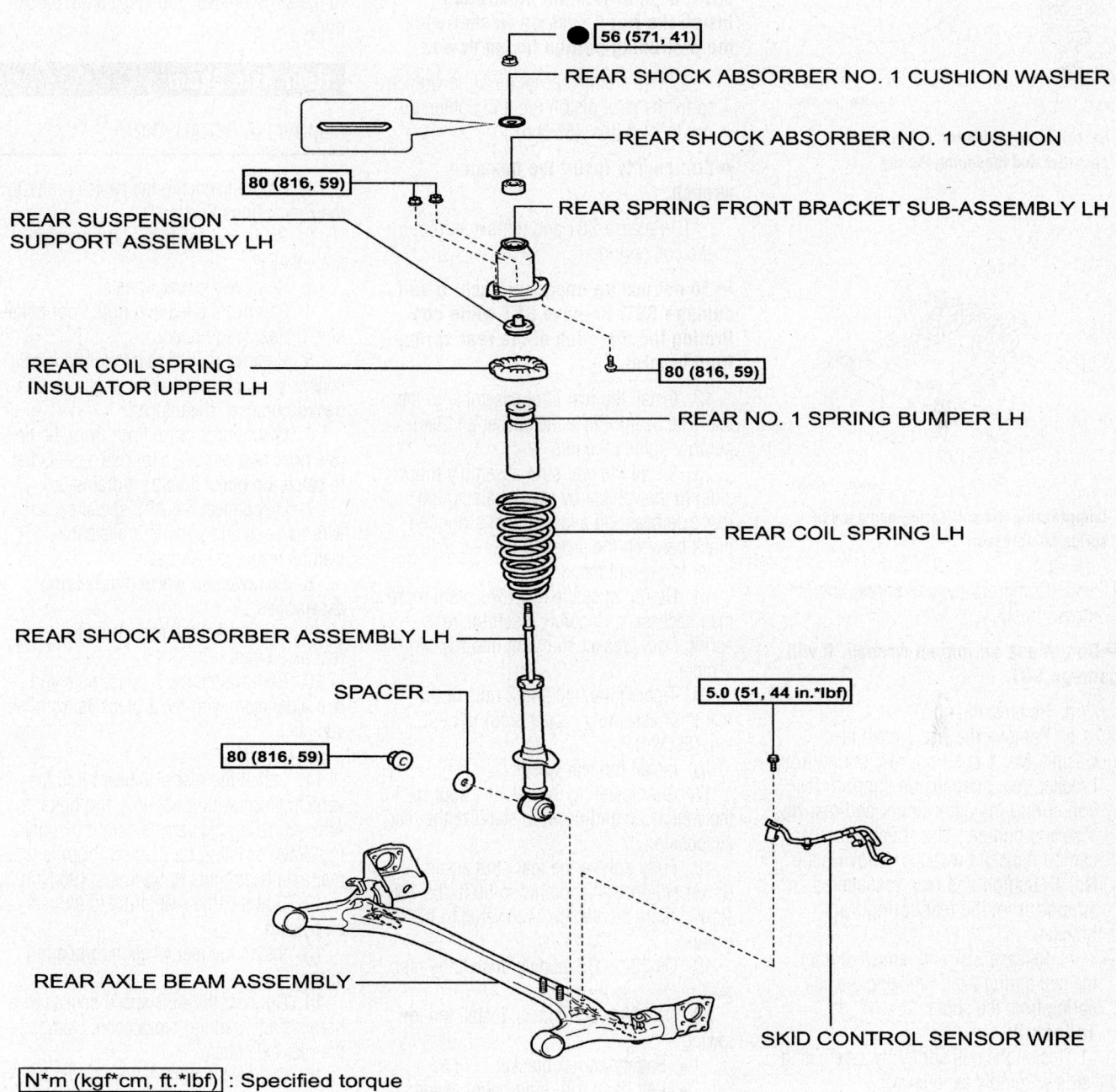

56 (571, 41) ●

REAR SHOCK ABSORBER NO. 1 CUSHION WASHER

REAR SHOCK ABSORBER NO. 1 CUSHION

80 (816, 59)

REAR SUSPENSION
SUPPORT ASSEMBLY LH

REAR SPRING FRONT BRACKET SUB-ASSEMBLY LH

80 (816, 59)

REAR COIL SPRING
INSULATOR UPPER LH

REAR NO. 1 SPRING BUMPER LH

REAR COIL SPRING LH

REAR SHOCK ABSORBER ASSEMBLY LH

SPACER

5.0 (51, 44 in.*lbf)

80 (816, 59)

REAR AXLE BEAM ASSEMBLY

SKID CONTROL SENSOR WIRE

N*m (kgf*cm, ft.*lbf) : Specified torque

● Non-reusable part

09490_TOYP_G0085

Exploded view of the rear strut assembly and related components

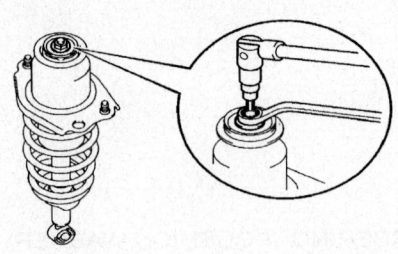

09490_TOYP_G0086

Securing the piston rod of the shock absorber and loosening the nut.

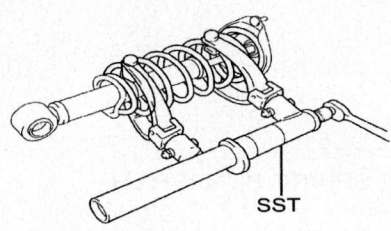

SST

09490_TOYP_G0087

Compressing the coil spring using a coil spring compressor.

    c. Compress the coil spring until it moves freely.

➡ **Do not use an impact wrench. It will damage SST.**

    d. Remove the nut.
    e. Remove the No. 1 cushion washer, No. 1 cushion, rear spring front bracket, rear suspension support, rear coil spring insulator upper and rear No. 1 spring bumper. The shock absorber can be replaced without removing the No. 1 cushion and rear suspension support from the rear spring front bracket.
    f. Release SST and remove it from the coil spring after removing the coil spring from the strut.

To install:

11. Install the coil spring by assembling the strut assembly as follows:
    a. Using SST 09727-30021, compress the coil spring.

➡ **Do not use an impact wrench. It will damage SST.**

    b. Fit the coil spring end into the recessed part of the strut assembly lower seat.
    c. Fit the rear coil spring insulator upper to the rear spring front bracket.
    d. Install the rear No. 1 spring

bumper, rear suspension support, rear spring front bracket, No. 1 cushion and No. 1 cushion washer.

➡ **Install the rear spring front bracket so that it is aligned with the strut lower bush, as shown in the illustration. Install the No. 1 cushion washer with the protruding portion facing down.**

    e. Use a 6mm socket hexagon wrench to fix the strut piston rod and tighten the nut to 41 ft. lbs. (56 Nm).

➡ **Sufficiently insert the hexagon wrench.**

    f. Release SST and remove it from the coil spring.

➡ **Do not use an impact wrench. It will damage SST. Remove SST while confirming the direction of the rear spring front bracket.**

12. Install the rear strut assembly to the rear axle beam. Place the spacer and temporarily tighten the nut.
13. Install the rear strut assembly (upper side) to the vehicle by slowly raising the rear axle beam on a jack. Insert a wooden block between the jack and the rear axle beam to prevent damage.
14. Do not raise the rear axle beam more than necessary. Securely insert the rear spring front bracket stud bolt into the vehicle.
15. Tighten the bolt and 2 nuts of the rear strut assembly (upper side) to 59 ft. lbs. (80 Nm).
16. Install the rear wheel.
17. After lowering the vehicle, bounce the vehicle up and down to stabilize the rear suspension.
18. Fully tighten the rear strut assembly (lower side) installation nut to 59 ft. lbs. (80 Nm). Ensure the vehicle is lowered to the ground.
19. On 2002–03 models, install the rear seat.
20. On 2004–06 models, install the following:
- Battery carrier bracket
- Left deck trim side panel assembly
- Left deck floor box
- Rear No. 4 floor board
- Left rear side seatback frame
- Rear No. 1 floor board
- Left rear seatback assembly
- Tonneau cover assembly
- Rear deck trim cover
- Rear deck floor box
- Rear No. 2 floor board

21. Inspect the rear wheel alignment.

Coil Spring

REMOVAL & INSTALLATION

    For coil spring service on the Prius, refer to the strut removal and installation procedure.

Rear Axle Beam

REMOVAL & INSTALLATION

1. Before servicing the vehicle, refer to the Precautions Section.
2. Raise and safely support the rear of the vehicle.
3. Remove the rear wheels.
4. Remove the left and right floor panel side plates, if equipped.
5. Remove the left rear height control sensor sub-assembly and rear sensor connecting bracket, if equipped.
6. Disconnect brake lines from the flexible hose and remove clip. Use a container to catch the brake fluid as it drains out.
7. Disconnect the ABS speed sensor wire harness and parking brake cable clamps from the axle beam.
8. Remove rear wheel hub/bearing assemblies.
9. Disconnect the rear struts from the rear axle beam.
10. Remove the the 2 bolts, nuts and rear axle beam with the 2 brackets from the vehicle.

    To install:

11. Install the rear axle beam into the vehicle along with the 2 nuts and bolts. After adjusting the vehicle height by pushing down or lifting up the body, torque the rear axle beam nuts to 66 ft. lbs. (90 Nm).
12. Connect the rear struts to the rear axle beam.
13. Install the rear wheel hub/bearing assemblies.
14. Connect the ABS speed sensor wire harness and parking brake cable clamps to the rear axle beam.
15. Connect the brake lines to the flexible hose and install clip. Tighten the line to 11 ft. lbs. (15 Nm).
16. Install the left rear height control sensor sub-assembly and rear sensor connecting bracket, if equipped.
17. Install the left and right floor panel side plates, if equipped.
18. Bleed the brake system.
19. Install the rear wheels and lower the vehicle.
20. Check the ABS sensor signal.

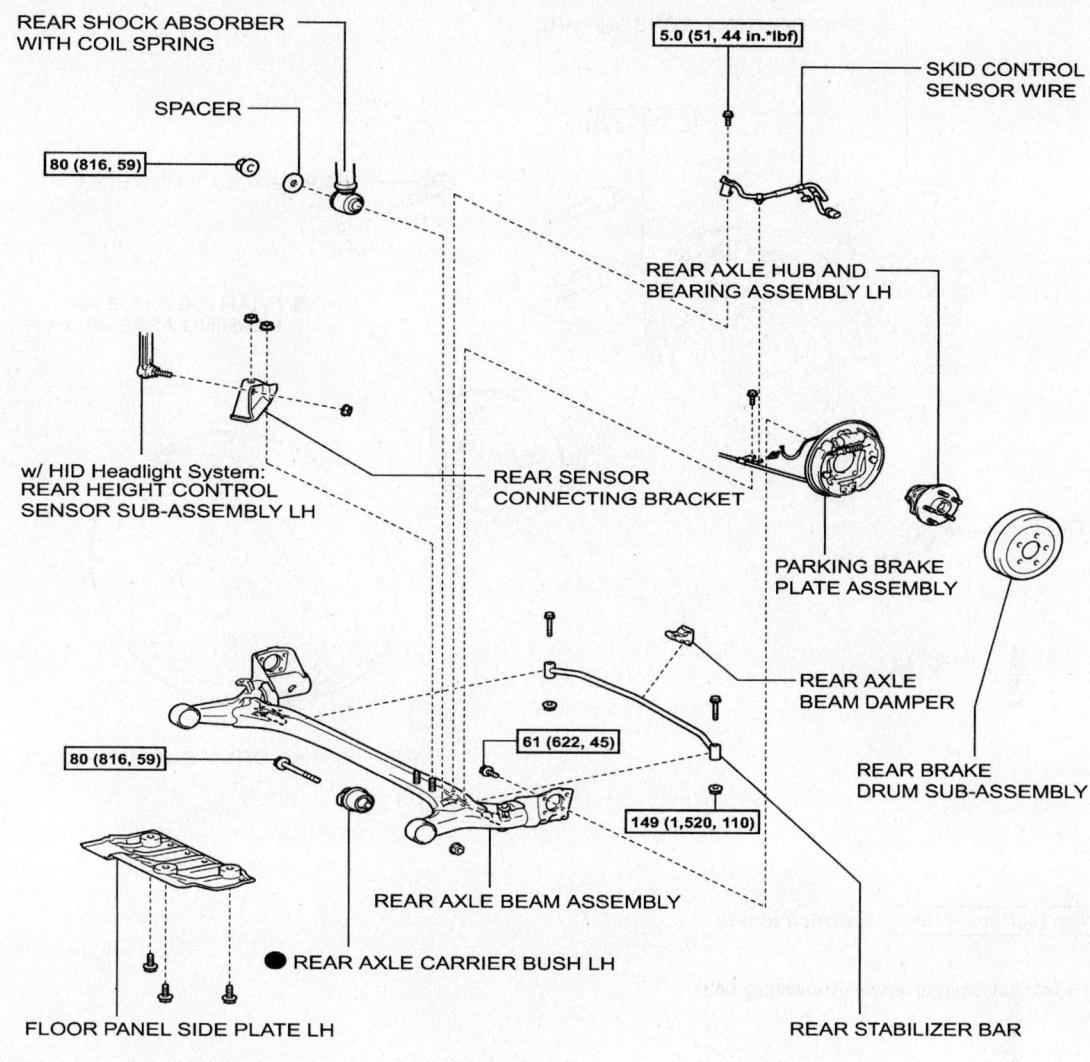

REAR SHOCK ABSORBER WITH COIL SPRING

SPACER

80 (816, 59)

5.0 (51, 44 in.*lbf)

SKID CONTROL SENSOR WIRE

REAR AXLE HUB AND BEARING ASSEMBLY LH

w/ HID Headlight System: REAR HEIGHT CONTROL SENSOR SUB-ASSEMBLY LH

REAR SENSOR CONNECTING BRACKET

PARKING BRAKE PLATE ASSEMBLY

REAR AXLE BEAM DAMPER

REAR BRAKE DRUM SUB-ASSEMBLY

80 (816, 59)

61 (622, 45)

149 (1,520, 110)

REAR AXLE BEAM ASSEMBLY

● REAR AXLE CARRIER BUSH LH

FLOOR PANEL SIDE PLATE LH

REAR STABILIZER BAR

N*m (kgf*cm, ft.*lbf) : Specified torque ● Non-reusable part

09490_TOYP_G0088

Exploded view of rear suspension components—2004–06 model shown

Rear Stabilizer Bar

1. Before servicing the vehicle, refer to the Precautions Section.

2. Raise and safely support the rear of the vehicle.

3. Remove the rear axle beam damper from the rear axle beam (2004–06 models).

4. Remove the 2 bolts, 2 nuts and stabilizer bar.

To install:

5. Install the rear stabilizer bar. On 2004–06 models, be sure that the mark position faces towards the vehicle's right rear.

6. Hold the bolts and tighten the nuts to 110 ft. lbs. (149 Nm).

7. Install the rear axle beam damper to the center of the rear axle beam (2004–06 models).

8. Lower the vehicle.

Wheel Bearings

ADJUSTMENT

The rear wheel bearings are a cartridge type design and cannot be adjusted.

REMOVAL & INSTALLATION

1. Before servicing the vehicle, refer to the Precautions Section.

2. Raise and safely support the vehicle.

3. Remove the rear wheel.

4. Remove the rear brake drum.

5. Disconnect the rear ABS speed sensor connector.

6. Remove automatic adjusting lever, shoe adjuster and adjusting lever spring from the brake shoes assembly, if necessary.

7. Remove the hub bolts and hub/bearing assembly.

To install:

8. Install the hub/bearing assembly.

9. Tighten the hub assembly bolts to 38 ft. lbs (52 Nm) on 2002–03 models, or 45 ft. lbs. (61 Nm) on 2004–06 models.

10. Install adjusting lever spring, shoe adjuster and automatic adjusting lever from the brake shoes assembly, if removed earlier.

11. Connect the rear ABS speed sensor connector.

12. Install the rear brake drum.

13. Install the rear wheel.

14. Lower the vehicle.

15. Inspect the rear wheel alignment.

16. Check to ensure that the brakes are free from dragging and that proper braking is obtained.

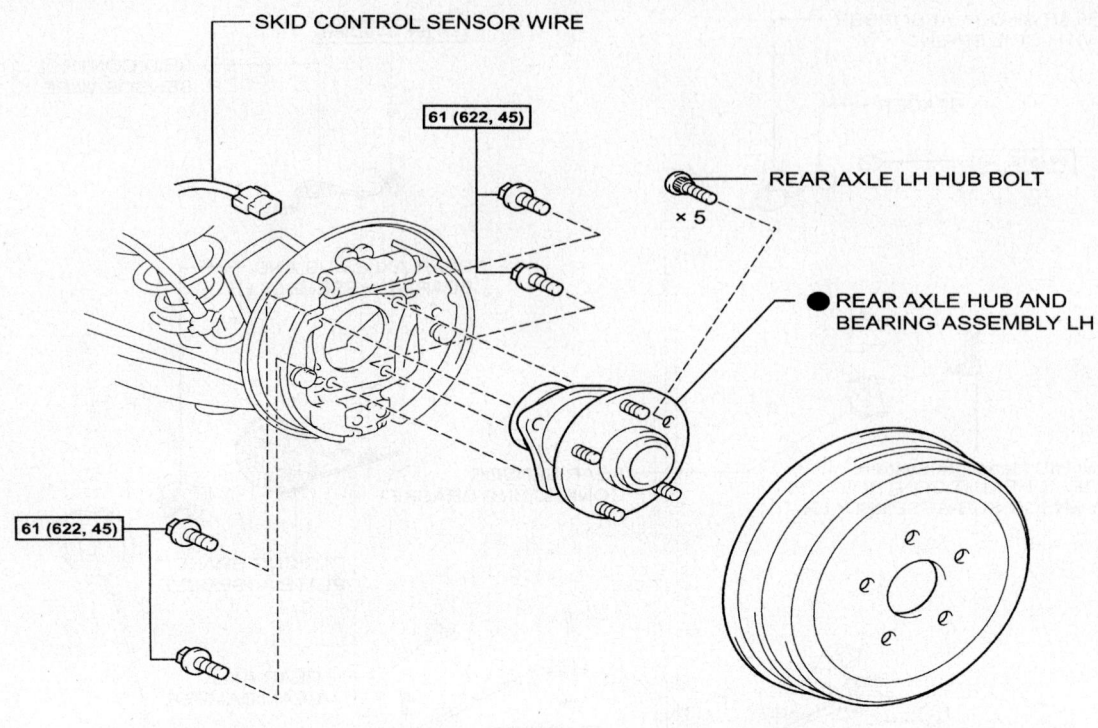

SKID CONTROL SENSOR WIRE

61 (622, 45)

REAR AXLE LH HUB BOLT

× 5

● REAR AXLE HUB AND BEARING ASSEMBLY LH

61 (622, 45)

REAR BRAKE DRUM SUB-ASSEMBLY

N*m (kgf*cm, ft.*lbf) : Specified torque

09490_TOYP_G0089

Remove/install 4 rear hub/bearing assembly mounting bolts

FRONT BRAKES

Brake Caliper

REMOVAL & INSTALLATION

1. Before servicing the vehicle, refer to the Precautions Section.

2. Set brake control **OFF**, if necessary.

3. Raise and safely support the front of the vehicle.

4. Remove the front wheel.

5. Remove the bolt and the gasket attaching the brake hose to the caliper.

6. Disconnect the brake hose, and plug the openings in the caliper and the brake hose to prevent fluid loss and contamination.

7. Hold the front disc brake cylinder slide pin using a wrench.

8. Remove the 2 caliper mounting bolts and remove the caliper from the vehicle.

9. Remove the 2 disc brake pads from the front disc brake cylinder mounting.

10. Remove the anti-squeal shims from the disc brake pads.

11. Remove the front disc brake pad support plates.

12. Remove the slide pins and bush dust boots from the front disc brake cylinder mounting, if necessary.

13. Remove the front disc brake cylinder mounting from the steering knuckle, if necessary.

To install:

14. Install the front disc brake cylinder mounting to the steering knuckle, if removed. Torque the 2 mounting bolts to 79 ft. lbs (107 Nm) on 2002–03 models, or 81 ft. lbs (109 Nm) on 2004–06 models.

15. Install new bush dust boots and slide pins into the front disc brake cylinder mounting, if necessary. Apply lithium soap base glycol grease to the sealing, sliding and fitting areas before installation.

16. Install the front disc brake pad support plates.

17. Install the anti-squeal shims onto the disc brake pads.

18. Install the disc brake pads onto the front disc brake cylinder mounting.

19. Install the brake caliper onto the mounting along with the 2 caliper mounting bolts. Torque the 2 mounting bolts to 25 ft. lbs (34 Nm).

20. Connect the brake hose to the caliper with bolt and a new gasket. Tighten the caliper brake hose bolt to 24 ft. lbs. (33 Nm).

21. Install the wheel.

22. Lower the vehicle.

23. Fill the master cylinder to the proper level with clean brake fluid.

24. Bleed the air out of the brake system.

25. Recheck the fluid level.

26. Repeatedly press the brake pedal to bring the pads in contact with the rotor.

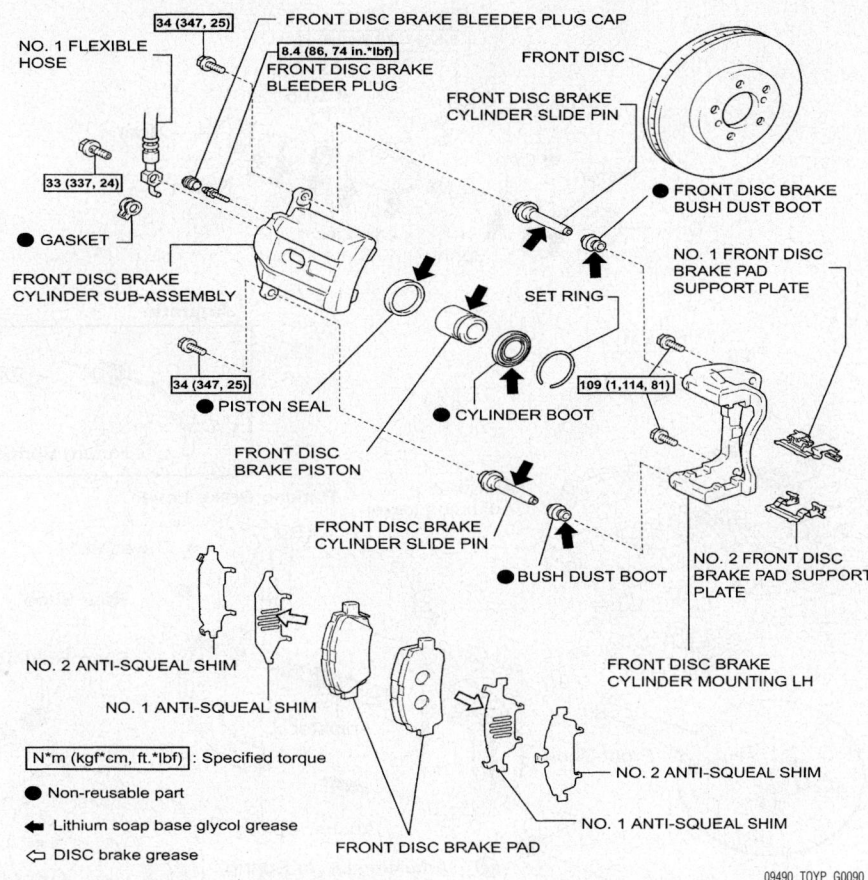

NO. 1 FLEXIBLE HOSE

34 (347, 25)

FRONT DISC BRAKE BLEEDER PLUG CAP

8.4 (86, 74 in.*lbf)
FRONT DISC BRAKE BLEEDER PLUG

FRONT DISC

FRONT DISC BRAKE CYLINDER SLIDE PIN

33 (337, 24)

● GASKET

● FRONT DISC BRAKE BUSH DUST BOOT

FRONT DISC BRAKE CYLINDER SUB-ASSEMBLY

SET RING

NO. 1 FRONT DISC BRAKE PAD SUPPORT PLATE

34 (347, 25)

● PISTON SEAL

● CYLINDER BOOT

109 (1,114, 81)

FRONT DISC BRAKE PISTON

FRONT DISC BRAKE CYLINDER SLIDE PIN

● BUSH DUST BOOT

NO. 2 FRONT DISC BRAKE PAD SUPPORT PLATE

NO. 2 ANTI-SQUEAL SHIM

NO. 1 ANTI-SQUEAL SHIM

FRONT DISC BRAKE CYLINDER MOUNTING LH

N*m (kgf*cm, ft.*lbf) : Specified torque

● Non-reusable part

◀ Lithium soap base glycol grease

◁ DISC brake grease

NO. 2 ANTI-SQUEAL SHIM

NO. 1 ANTI-SQUEAL SHIM

FRONT DISC BRAKE PAD

09490_TOYP_G0090

Exploded view of front disc brake assembly components—2004–06 Prius

Disc Brake Pads

REMOVAL & INSTALLATION

1. Before servicing the vehicle, refer to the Precautions Section.
2. Raise and safely support the front of the vehicle.
3. Remove the front wheel.
4. Hold the front disc brake cylinder slide pin using a wrench.

➡**Caliper removal is not necessary to service the brake pads.**

5. Remove the 2 caliper mounting bolts and separate the caliper from the mounting and support from the vehicle with wire. Do not allow the caliper to hang low enough so to put tension on the brake hose.
6. Remove the brake pads.

To install:

7. Install the anti-squeal shims onto the disc brake pads.

8. Install the disc brake pads.
9. Install the brake caliper onto the mounting along with the 2 caliper mounting bolts. Torque the 2 mounting bolts to 25 ft. lbs (34 Nm).
10. Install the wheel.
11. Lower the vehicle.
12. Repeatedly press the brake pedal to bring the pads in contact with the rotor.

REAR BRAKES

Brake Drums

REMOVAL & INSTALLATION

1. Before servicing the vehicle, refer to the Precautions Section.
2. Raise and safely support the rear of the vehicle.
3. Remove the rear wheels.
4. Put match marks on the rear brake drum and the axle hub.
5. Release the parking brake and remove the brake drum.

6. If the brake drum cannot be removed easily, do the following steps.
   a. Remove the plug and insert a screwdriver through the hole in the backing plate.
   b. Using another screwdriver, reduce the brake shoe adjuster by turning the adjusting wheel.

To install:

7. Aligning the match marks, install brake drum.
8. Adjust the drum brake shoe clearance as follows:
   a. Temporarily install 2 hub nuts.

   b. Remove the hole plug, and turn the adjuster to expand the shoe until the drum locks up.
   c. Rotate the adjuster back by 8 notches.
   d. Install the hole plug.
   e. Remove the 2 hub nuts.
9. Install the wheel. Torque the wheel lugs to 76 ft. lbs. (103 Nm).
10. Lower the vehicle.
11. Check brake fluid level in the reservoir and add to proper level with clean brake fluid, if necessary.

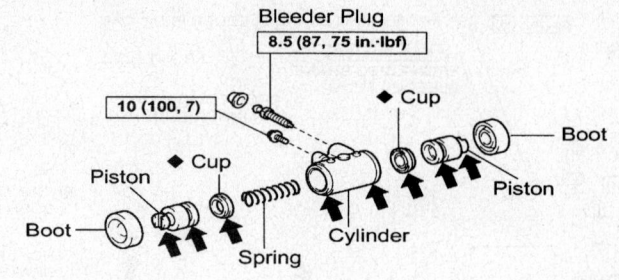

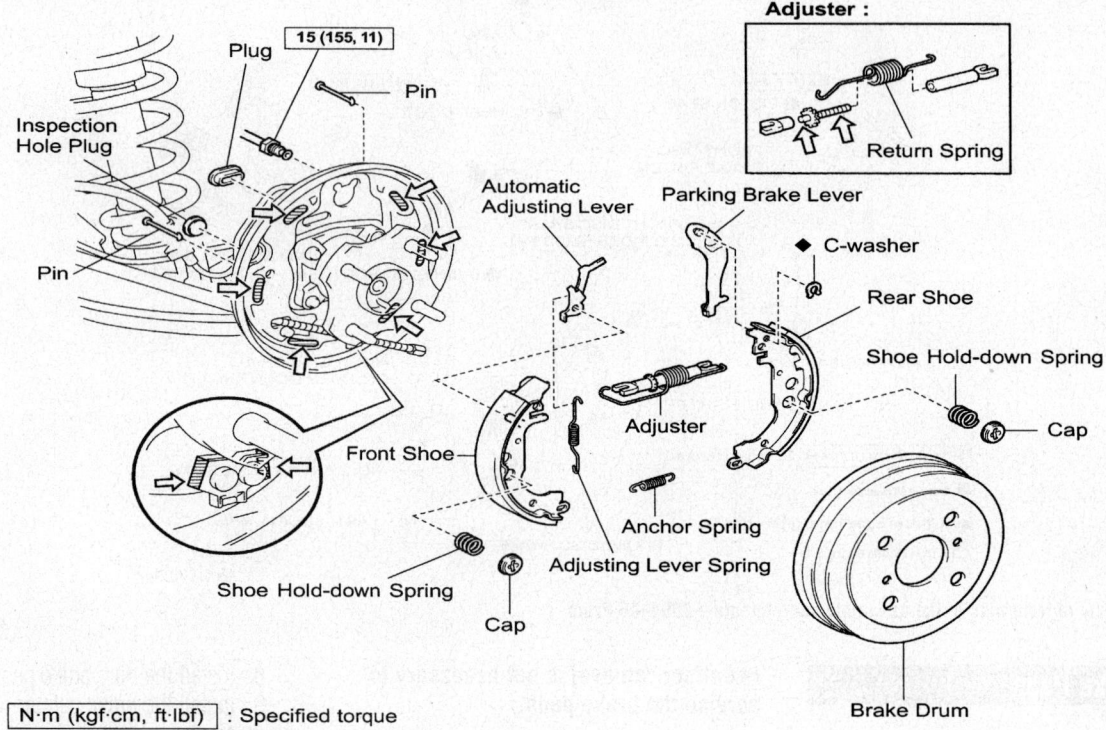

N·m (kgf·cm, ft·lbf) : Specified torque
◆ Non-reusable part
⇐ High temperature grease
◀ Lithium soap base glycol grease

09490_TOYP_G0091

Exploded view of rear brake drum assembly components

Brake Shoes

REMOVAL & INSTALLATION

1. Before servicing the vehicle, refer to the Precautions Section.
2. Remove the rear wheels.
3. Remove the rear brake drums.
4. Disconnect the spring and parking brake shoe adjuster set from the brake shoes.
5. Remove the cup, shoe hold-down spring and pin.
6. Remove front brake shoe.
7. Remove the parking brake shoe adjuster set from the rear brake shoe.
8. Remove the automatic adjusting lever spring and automatic adjusting lever.
9. Remove anchor spring from the rear brake shoe.

10. Remove the cup, shoe hold-down spring and pin.
11. Using needle-nose pliers, disconnect the parking brake cable, and remove the rear brake shoe.
12. Using a screwdriver, remove the C-washer and parking brake lever.
To install:
13. Apply high-temperature grease to the shoe attached surface of the backing plate.
14. Using needle-nose pliers, install the parking brake lever to the rear brake shoe with a new C-washer.
15. Using needle-nose pliers, connect the parking brake cable to the parking brake lever.
16. Install the rear brake shoe, pin, shoe hold-down spring and cup.
17. Install the automatic adjusting lever and automatic adjusting lever spring to the front brake shoe.

18. Apply high-temperature grease to the adjusting bolt and assemble the parking brake shoe adjuster set.
19. Install the rear brake shoe return spring to the adjuster set and install set.
20. Connect the anchor spring to the front and rear brake shoes.
21. Install the front brake shoe, pin, shoe hold-down spring and cup.
22. Connect the shoe return spring to the front and rear brake shoes.
23. Check to make sure that all the components of the brake shoe assembly are installed properly.
24. Install the brake drums.
25. Install the wheels.
26. Lower the vehicle.
27. Press the brake pedal 3–5 times to adjust the brake shoe clearance.

BRAKES **17-113**
DRIVE TRAIN **17-70**
ENGINE REPAIR...............**17-13**
FRONT SUSPENSION **17-97**
FUEL SYSTEM **17-64**
REAR SUSPENSION **17-105**
SPECIFICATIONS AND
   MAINTENANCE CHARTS......**17-3**
Engine and Vehicle Identification...17-3
General Engine Specifications 17-3
Engine Tune-Up Specifications 17-3
Firing Order 17-4
Accessory Drive Belt Routing 17-4
Capacities 17-5
Valve Specifications......................17-5
Camshaft Specifications 17-6
Crankshaft and Connecting Rod
   Specifications 17-7
Piston and Ring Specifications......17-7
Torque Specifications 17-8
Wheel Alignment 17-10
Tire, Wheel and Ball Joint
   Specifications 17-10
Brake Specifications 17-11
Scheduled Maintenance
   Intervals.................................17-12
STEERING **17-94**
A
Air Bag......................................17-94
   Disarming 17-94
   Precautions............................17-94
Alternator 17-13
   Removal & Installation.............17-13
Automatic Transaxle....................17-76
   Removal & Installation.............17-76
B
Brake Caliper 17-113
   Removal & Installation.............17-113
Brake Shoes...............................17-118
   Removal & Installation..........17-118
C
Camshaft and Valve Lifters 17-40
   Inspection 17-45
   Removal & Installation...........17-40

Clutch......................................17-74
   Adjustment.............................17-75
   Removal & Installation............17-74
Coil Spring 17-106
   Removal & Installation..........17-106
CV-Joints...................................17-90
   Overhaul 17-90
Cylinder Head 17-31
   Removal & Installation............17-31
D
Disc Brake Pads.........................17-117
   Removal & Installation..........17-117
Distributor..................................17-13
E
Engine Assembly 17-13
   Removal & Installation.............17-13
Exhaust Manifold 17-39
   Removal & Installation.............17-39
F
Fuel Filter 17-65
   Removal & Installation............17-65
Fuel Injector..............................17-68
   Removal & Installation............17-68
Fuel Pump 17-65
   Removal & Installation............17-65
Fuel System Service
   Precautions.............................17-64
H
Halfshaft....................................17-83
   Removal & Installation.............17-83
Heater Core................................17-19
   Removal & Installation.............17-19
Hydraulic Clutch System 17-76
   Bleeding.................................17-76
I
Ignition Timing 17-13
   Adjustment.............................17-13
Intake Manifold 17-38
   Removal & Installation.............17-38
L
Lower Ball Joint..........................17-100
   Removal & Installation..........17-100

Lower Control Arm (Front) 17-101
   Removal & Installation..........17-101
   Control Arm Bushing
      Replacement 17-102
Lower Control Arm (Rear)..........17-111
   Removal & Installation..........17-111
M
Manual Transaxle Assembly 17-70
   Removal & Installation.............17-70
O
Oil Pan......................................17-49
   Removal & Installation.............17-49
Oil Pump 17-51
   Removal & Installation.............17-51
P
Piston and Ring 17-64
   Positioning 17-64
Power Steering Gear 17-94
   Removal & Installation.............17-94
R
Rear Main Seal 17-54
   Removal & Installation.............17-54
Reliving Fuel System Pressure 17-64
   Relieving................................17-64
S
Shock Absorber 17-105
   Removal & Installation..........17-105
Stabilizer Bar (Front)...................17-99
   Removal & Installation.............17-99
Stabilizer Bar (Rear)...................17-109
   Removal & Installation..........17-109
Starter Motor 17-49
   Removal & Installation.............17-49
Strut...17-97
   Removal & Installation.............17-97
   Strut Overhaul 17-98
T
Timing Belt, Cover and
   Crankshaft Seal 17-55
   Removal & Installation.............17-55
Timing Chain, Sprockets, Front
   Cover and Seal 17-56
   Removal & Installation.............17-56
Transfer Case Assembly...............17-83
   Removal & Installation.............17-83

U

Upper Control Arm 17-111
    Removal & Installation..........17-111

V

Valve Lash 17-45
    Adjustment..............................17-45

W

Water Pump....................................17-27
    Removal & Installation.............17-27
Wheel Bearing 17-102
    Adjustment...............................17-102
    Removal & Installation..........17-102

Wheel Bearing 17-112
    Adjustment...............................17-112
    Removal & Installation..........17-112

SPECIFICATIONS AND MAINTENANCE CHARTS

ENGINE AND VEHICLE IDENTIFICATION

	Engine							Model Year	
Code/ID ①	Liters (cc)	Cu. In.	Cyl.	Fuel Sys.	Engine Type	Eng. Mfg.		Code ②	Year
1AZ-FE/H	2.0 (1998)	122	4	SFI	DOHC	Toyota		2	2002
③	2.4 (2362)	144	4	SFI	DOHC	Toyota		3	2003
2GE-FE/K	3.5 (3498)	213	6	SFI	DOHC	Toyota		4	2004
								5	2005
								6	2006

SFI: Sequential Fuel Injection

DOHC: Double Overhead Camshaft

① 5th digit of the vehicle identification number (VIN)

② 10th digit of the Vehicle Identification Number (VIN)

③ 2AZ-FE/D and/or 2AZ-FE/E

09490_RAV4_C0001

GENERAL ENGINE SPECIFICATIONS

Year	Model	Engine Displacement Liters	Engine Series Code/ID	Net Horsepower @ rpm	Net Torque @ rpm (ft. lbs.)	Bore x Stroke (in.)	Com-pression Ratio	Oil Pressure @ rpm
2002	RAV4	2.0	1AZ-FE/H	127@5400	132@4600	3.40x3.40	9.5:1	NA
2003	RAV4	2.0	1AZ-FE/H	127@5400	132@4600	3.40x3.40	9.5:1	NA
2004	RAV4	2.4	①	155@5600	163@4000	3.48x3.78	NA	36@3000
2005	RAV4	2.4	①	166@6000	165@4000	3.48x3.78	NA	②
2006	RAV4	2.4	①	166@6000	165@4000	3.48x3.78	NA	②
	RAV4	3.5	2GR-FE/K	269@6200	246@4700	NA	NA	③

NA: Not Available

① 2AZ-FE/D and/or 2AZ-FE/E

② 4.3 psi or more at idle

③ 11.6 psi or more at idle

09490_RAV4_C0002

ENGINE TUNE-UP SPECIFICATIONS

Year	Engine Displacement Liters	Engine Code/ID	Spark Plug Gap (in.)	Ignition Timing (deg.)	Fuel Pump (psi)	Idle Speed (rpm)		Valve Clearance	
						MT	AT	Intake	Exhaust
2002	2.0	1AZ-FE/H	0.043	①	44-50	700-800	700-800	0.0080-0.0110	0.0120-0.0160
2003	2.0	1AZ-FE/H	0.043	①	44-50	700-800	700-800	0.0080-0.0110	0.0120-0.0160
2004	2.4	②	0.043	①	44-50	600-700	600-700	0.0080-0.0110	0.0120-0.0160
2005	2.4	②	0.043	①	44-50	600-700	600-700	0.0080-0.0110	0.0120-0.0160
2006	2.4	②	0.043	③	44-50	—	600-700	0.0075-0.0114	0.0150-0.0189
	3.5	2GR-FE/K	0.043	④	44-50	—	600-700	NA	NA

NOTE: The Vehicle Emission Control Information label often reflects specification changes made during production.

The label figures must be used if they differ from those in this chart.

NA: Not Available

① 8-12 degrees BTDC with terminals TC and CG connected to DLC3

② 2AZ-FE/D and/or 2AZ-FE/E

③ 8-12 degrees BTDC at idle. Connect SST when using intelligent tester tool.
  15-12 degrees BTDC at idle. Disconnect SST when not using intelligent tester tool.

④ 8-12 degrees BTDC when using intelligent tester tool.
  8-12 degrees BTDC at idle. Connect SST when not using intelligent tester tool.
  5-15 degrees BTDC at idle. Disconnect SST when not using intelligent tester tool.

09490_RAV4_C0003

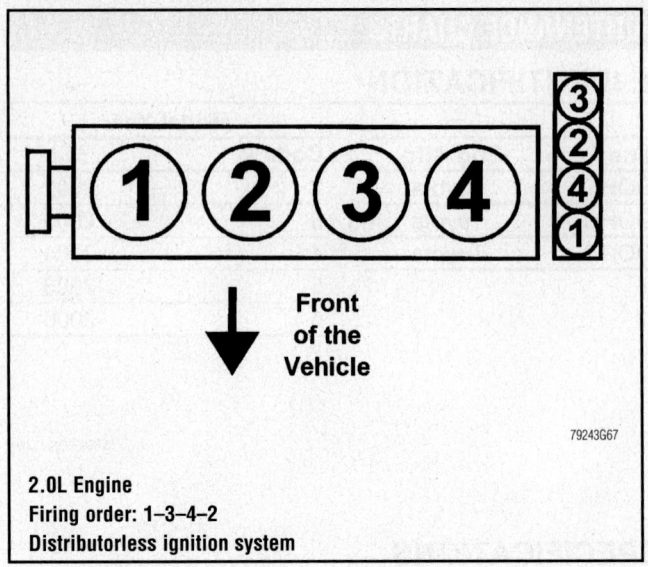

2.0L Engine
Firing order: 1–3–4–2
Distributorless ignition system

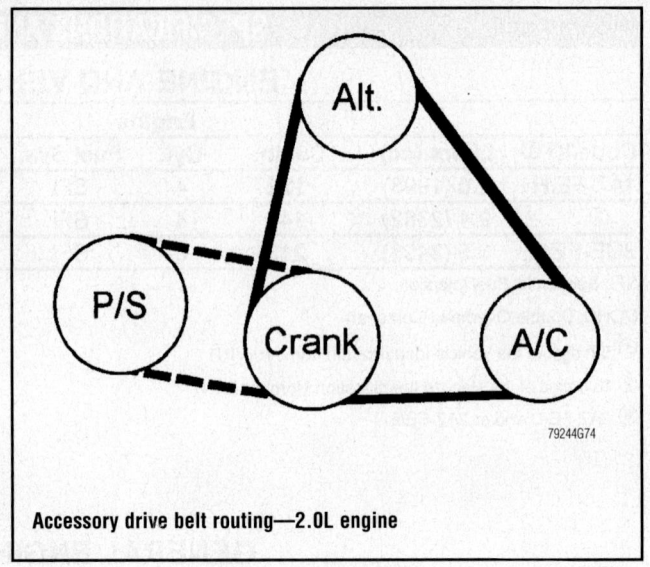

Accessory drive belt routing—2.0L engine

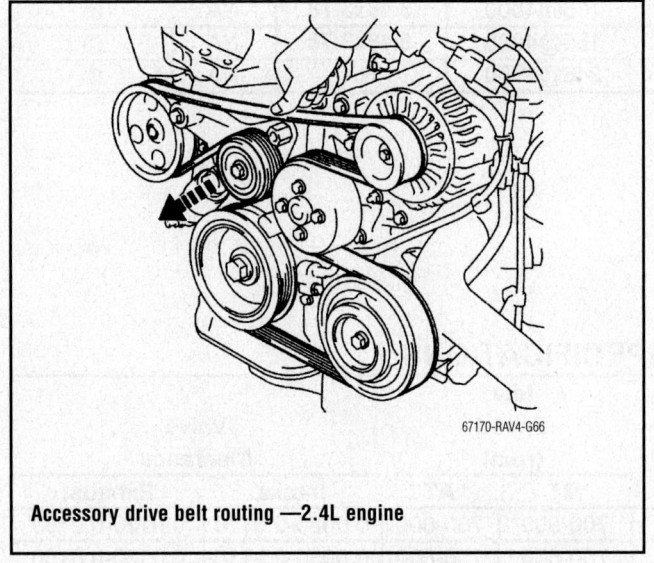

Accessory drive belt routing —2.4L engine

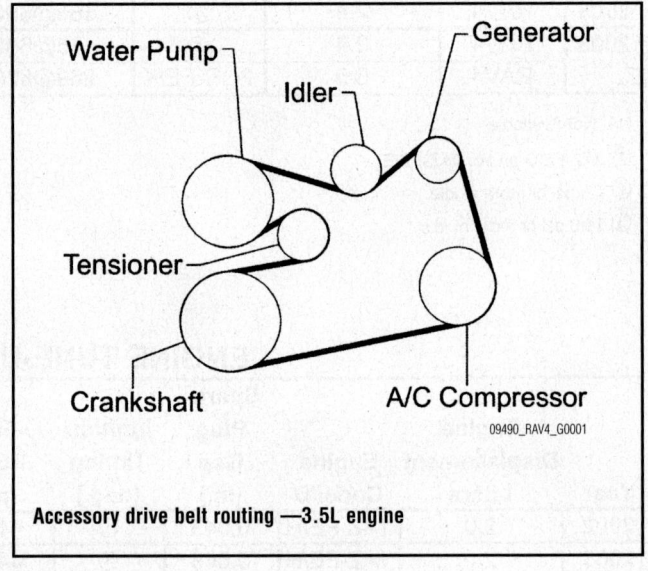

Accessory drive belt routing —3.5L engine

CAPACITIES

Year	Model	Engine Displacement Liters	Engine Code/ID	Engine Oil with Filter (qts.)	Transmission (pts.) 5-Spd	Transmission (pts.) Auto.*	Transfer Case (pts.)	Drive Axle Front (pts.)	Drive Axle Rear (pts.)	Fuel Tank (gal.)	Cooling System (qts.)
2002	RAV4	2.0	1AZ-FE/H	4.6	①	②	2.0	—	2.0	14.8	③
2003	RAV4	2.0	1AZ-FE/H	4.6	①	②	2.0	—	2.0	14.8	③
2004	RAV4	2.4	④	4.0	①	②	2.0	—	2.0	14.8	③
2005	RAV4	2.4	④	4.0	①	7.4	2.0	—	2.0	14.8	⑤
2006	RAV4	2.4	④	4.5	—	7.4	1.0	—	1.0	15.9	7.2
	RAV4	3.5	2GR-FE/K	6.4	—	NA	1.0	—	1.0	15.9	⑥

*After draining, add the following amounts, then fill to the cold full line.

NA: Not Available

① 2wd: 5.2
   4wd: 7.2

② 2wd: 7.0
   4wd: 8.2

③ MT: 6.7
   AT: 6.6

④ 2AZ-FE/D and/or 2AZ-FE/E

⑤ MT: 7.0
   AT: 7.2

⑥ STD: 9.4
   TWG: 9.8

09490_RAV4_C0004

VALVE SPECIFICATIONS

Year	Engine Displacement Liters	Engine Code/ID	Seat Angle (deg.)	Face Angle (deg.)	Spring Test Pressure (lbs. @ in.)	Spring Installed Height (in.)	Stem-to-Guide Clearance (in.) Intake	Stem-to-Guide Clearance (in.) Exhaust	Stem Diameter (in.) Intake	Stem Diameter (in.) Exhaust
2002	2.0	1AZ-FE/H	45	44.5	41.4-45.9@ 1.339	1.3390	0.0010-0.0024	0.0012-0.0026	0.2154-0.2159	0.2152-0.2157
2003	2.0	1AZ-FE/H	45	44.5	41.4-45.9@ 1.339	1.3390	0.0010-0.0024	0.0012-0.0026	0.2154-0.2159	0.2152-0.2157
2004	2.4	①	45	44.5	77-86@ 0.969	1.3390	0.0010-0.0024	0.0012-0.0026	0.2154-0.2159	0.2152-0.2157
2005	2.4	①	NA	NA	NA	1.3390	0.0010-0.0024	0.0012-0.0026	0.2154-0.2159	0.2152-0.2157
2006	2.4	①	NA	NA	NA	1.8670	0.0010-0.0024	0.0012-0.0026	0.2154-0.2159	0.2152-0.2157
	3.5	2GR-FE/K	NA	NA	NA	1.7898	0.0010-0.0024	0.0012-0.0026	0.2154-0.2159	0.2152-0.2157

NA: Not Available

① 2AZ-FE/D and/or 2AZ-FE/E

09490_RAV4_C0005

CAMSHAFT SPECIFICATIONS
All measurements in inches unless noted

Year	Engine Displacement Liters	Engine Code/ID	Journal Dia.	Brg. Oil Clearance	Shaft End-play ①	Circle Runout	Lobe Height	
							Intake	Exhaust
2002	2.0	1AZ-FE/H	②	③	④	0.0012	1.8305-1.8345	1.8106-1.8143
2003	2.0	1AZ-FE/H	②	③	④	0.0012	1.8305-1.8345	1.8106-1.8143
2004	2.4	⑤	②	⑥	④	0.0012	1.8305-1.8345	1.8106-1.8143
2005	2.4	⑤	②	⑥	④	0.0012	1.8305-1.8345	1.8106-1.8143
2006	2.4	⑤	⑦	⑧	④	0.0012	1.8624-1.8664	1.8104-1.8143
	3.5	2GR-FE/K	⑨	⑩	0.0031-0.0051	0.0016	1.7447-1.7487	1.7426-1.7465

① Thrust clearance

② No1: 1.4162-1.4167
   all others: 0.9040-0.9045

③ intake No1: 0.0003-0.0015
   all others: 0.0010-0.0024
   exhaust No1: 0.0006-0.0021
   all others: 0.0010-0.0024

④ intake: 0.0016-0.0037
   exhaust: 0.0032-0.0053

⑤ 2AZ-FE/D and/or 2AZ-FE/E

⑥ intake No1: 0.0003-0.0015
   all others: 0.0010-0.0024
   exhaust No1: 0.0006-0.0021
   all others: 0.0010-0.0031

⑦ No1: 1.4162-1.4167
   all others: 0.9039-0.9045

⑧ intake No1: 0.0003-0.0015
   all others: 0.0010-0.0620
   exhaust No1: 0.0016-0.0031
   all others: 0.0010-0.0620

⑨ No1: 1.4152-1.4157
   all others: 1.0220-1.0226

⑩ No1: 0.0016-0.0031
   all others: 0.0010-0.0024

09490_RAV4_C0006

CRANKSHAFT AND CONNECTING ROD SPECIFICATIONS

All measurements are given in inches.

Year	Engine Displ. Liters	Engine Code/ID	Crankshaft Main Brg. Journal Dia.	Crankshaft Main Brg. Oil Clearance	Crankshaft Shaft End-play	Crankshaft Thrust on No.	Connecting Rod Journal Diameter	Connecting Rod Oil Clearance	Connecting Rod Side Clearance
2002	2.0	1AZ-FE/H	2.1649-2.1655	0.0010-0.0016	0.0008-0.0087	3	1.8894-1.8898	0.0009-0.0019	0.0063-0.0143
2003	2.0	1AZ-FE/H	2.1649-2.1655	0.0010-0.0016	0.0008-0.0087	3	1.8894-1.8898	0.0009-0.0019	0.0063-0.0143
2004	2.4	①	2.0654-2.1648	0.0009-0.0019	0.0016-0.0094	2	1.8894-1.8898	0.0009-0.0019	0.0063-0.0143
2005	2.4	①	2.1649-2.1654	0.0007-0.00016	0.0016-0.0094	NA	1.8894-1.8898	0.0009-0.0019	0.0063-0.0143
2006	2.4	①	2.1830-2.1654	0.0003-0.0010	NA	NA	1.8894-1.8898	0.0009-0.0019	0.0063-0.0143
	3.5	2GR-FE/K	2.4011-2.4016	NA	0.0016-0.0094	NA	NA	0.0018-0.0026	NA

NA: Not Available

① 2AZ-FE/D and/or 2AZ-FE/E

09490_RAV4_C0007

PISTON AND RING SPECIFICATIONS

All measurements are given in inches.

Year	Engine Displ. Liters	Engine Code/ID	Piston Clearance	Ring Gap Top Comp.	Ring Gap Bottom Comp.	Ring Gap Oil Control	Ring Side Clearance Top Comp.	Ring Side Clearance Bottom Comp.	Ring Side Clearance Oil Control
2002	2.0	1AZ-FE/H	0.0025-0.0034	0.0118-0.0157	0.0185-0.0244	0.0039-0.0138	0.0008-0.0028	0.0008-0.0024	0.0028-0.0059
2003	2.0	1AZ-FE/H	0.0025-0.0034	0.0118-0.0157	0.0185-0.0244	0.0039-0.0138	0.0008-0.0028	0.0008-0.0024	0.0028-0.0059
2004	2.4	①	0.0020-0.0029	0.0087-0.0126	0.0197-0.0236	0.0039-0.0138	0.0066-0.0086	0.0066-0.0082	0.0082-0.0114
2005	2.4	①	0.0020-0.0029	0.0087-0.0126	0.0197-0.0236	0.0039-0.0138	0.0066-0.0086	0.0066-0.0082	0.0082-0.0114
2006	2.4	①	NA	0.0094-0.0122 ②	0.0130-0.0169 ②	0.0040-0.0119 ②	0.0008-0.0028	0.0008-0.0024	0.0008-0.0028
	3.5	2GR-FE/K	NA	0.0098-0.0138	0.0197-0.0236	0.0039-0.0157	0.0008-0.0028	0.0008-0.0024	0.0028-0.0059

NA: Not Available

① 2AZ-FE/D and/or 2AZ-FE/E

② measure the piston ring a little beyond the bottom of the ring travel, 4.33 inch from the top of the cylinder block

09490_RAV4_C0008

TORQUE SPECIFICATIONS

All readings in ft. lbs.

Year	Engine Displacement Liters	Engine Code/ID	Cylinder Head Bolts	Main Bearing Bolts	Rod Bearing Bolts	Crankshaft Damper Bolts	Flywheel Bolts	Manifold Intake	Manifold Exhaust	Spark Plugs	Oil Pan Drain Plug
2002	2.0	1AZ-FE/H	①	②	③	125	④	22	25	15	18
2003	2.0	1AZ-FE/H	①	②	③	125	④	22	25	15	18
2004	2.4	⑤	①	②	③	132	④	22	27	14	18
2005	2.4	⑤	①	②	③	132	④	22	27	14	18
2006	2.4	⑤	⑥	⑦	③	133	72	22	27	18	30
	3.5	2GR-FE/K	⑧	⑨	③	184	132	15	15	13	30

① Step 1: 58
  Step 2: plus 90 degrees

② Step 1: 15 ft. lbs.
  Step 2: 30 ft. lbs.
  Step 3: plus 90 degrees

③ Step 1: 18 ft. lbs.
  Step 2: plus 90 degrees

④ MT: 96 ft. lbs.
  AT: 72 ft. lbs.

⑤ 2AZ-FE/D and/or 2AZ-FE/E

⑥ Step 1: 52
  Step 2: plus 90 degrees

⑦ Step 1: 30
  Step 2: plus 90 degrees

⑧ Step 1: 27 ft. lbs.
  Step 2: plus 90 degrees
  Step 3: plus 90 degrees

⑨ Step 1: 45 ft. lbs.
  Step 2: plus 90 degrees
  main bearing cap bolt: 38

09490_RAV4_C0009

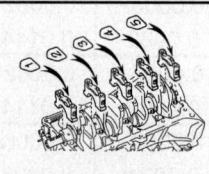
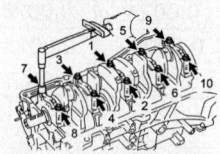
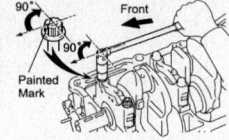

(a) Install the 5 main bearing caps in their proper locations.
(b) Install the main bearing cap bolts.
• The main bearing cap bolts are tightened in 2 progressive steps (steps (2), (3) and (5)).
• If any one of the main bearing cap bolts is broken or deformed, replace it.
  (1) Apply a light coat of engine oil on the threads and under the main bearing cap bolts.
  (2) Install and uniformly tighten the 10 main bearing cap bolts in several passes, in the sequence shown.
  Torque: 20 N·m (205 kgf·cm, 15 ft·lbf)
  (3) Retighten the 10 main bearing cap bolts in several passes, in the sequence shown.
  Torque: 40 N·m (410 kgf·cm, 30 ft·lbf)
  If any one of the main bearing cap bolts does not meet the torque specification, replace the main bearing cap bolt.
  (4) Mark the front of the main bearing cap bolt with paint.
  (5) Retighten the main bearing cap bolts by 90° in the numerical order shown.
  (6) Check that the painted mark is now at a 90° angle to the front.
(c) Check that the crankshaft turns smoothly.

09490_RAV4_G0002

Main bearing bolt torque sequence—2002–2003 2.0L engine and 2004–2005 2.4L engine

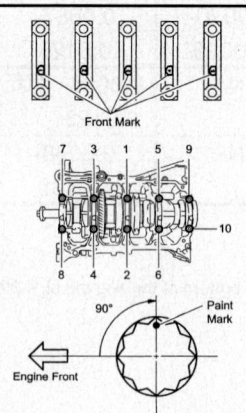

(a) Apply engine oil to the upper bearing, and install the crankshaft onto the cylinder block.
(b) Examine the front marks and numbers, and install the bearing caps onto the cylinder block in the order shown in the illustration.
(c) Apply a light coat of engine oil to the threads and under the heads of the bearing cap bolts.
(d) Using several steps, uniformly install and tighten the 10 bearing cap bolts in the sequence shown in the illustration.
  Torque: 40 N*m (408 kgf*cm, 30 ft.*lbf)
(e) Mark the front of the bearing cap bolts with paint.
(f) Retighten the 10 bearing cap bolts by 90° in the same sequence.
(g) Check that the paint marks are now at a 90° angle to the front.
(h) Check that the crankshaft turns smoothly.

09490_RAV4_G0003

Main bearing bolt torque sequence—2006 2.4L engine

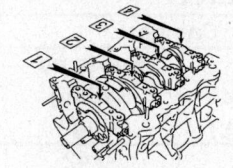

(a) Install the 2 thrust washers under the No. 2 journal position of the cylinder block with the oil grooves facing outward.

(b) Apply engine oil to the upper bearing, then place the crankshaft on the cylinder block.

(c) Confirm the front marks and numbers of the main bearing caps and install the bearing caps on the cylinder block.

A number is marked on each main bearing cap to indicate the installation position.

(d) Apply a light coat of engine oil on the threads and under the heads of the bearing cap bolts.

(e) Temporarily install the 8 main bearing cap bolts to the inside positions.

(f) Insert the main bearing cap by hand until the clearance between the main bearing cap and the cylinder block is less than 6 mm (0.23 in.) by marking the 2 internal bearing cap bolts as a guide.
Bolt length:
    100.0 to 102.0 mm (3.937 to 4.016 in.)

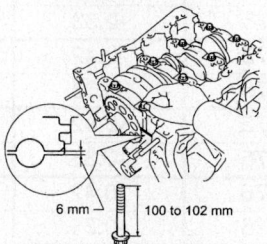

6 mm   100 to 102 mm

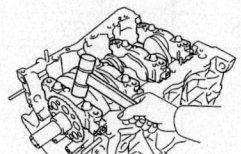

(g) Using a plastic-faced hammer, lightly tap the bearing cap to ensure a proper fit.

(h) Apply a light coat of engine oil to the threads and under the heads of the 8 main bearing cap bolts.

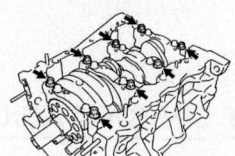

(i) Install the 8 main bearing cap bolts to the outside positions.
Bolt length:
    105.5 to 107.5 mm (4.154 to 4.232 in.)

(j) Install the crankshaft bearing cap bolts.
HINT:
The main bearing cap bolts are tightened in 2 progressive steps.

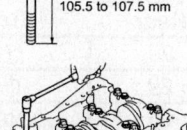

105.5 to 107.5 mm

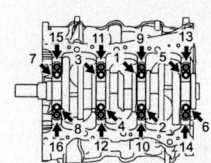

(k) Step 1
(1) Install and uniformly tighten the 16 main bearing cap bolts in the sequence shown in the illustration.
Torque: 61 N*m (622 kgf*cm, 45 ft.*lbf)
If any of the main bearing cap bolts does not meet the specified torque, replace it.

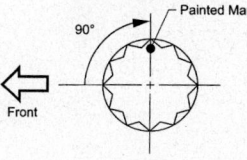

(l) Step 2
(1) Mark the front of the bearing cap bolts with paint.
(2) Tighten the bearing cap bolts another 90° in the order shown in step 1.
(3) Check that the painted mark is now at a 90° angle to the front.

(m) Check that the crankshaft turns smoothly.

(n) Install and uniformly tighten the 8 main bearing cap bolts in several steps, in the sequence shown in the illustration.
Torque: 52 N*m (530 kgf*cm, 38 ft.*lbf)
Bolt length:
    45 mm (1.77 in.) for bolt A
    30 mm (1.18 in.) for except bolt A

(o) Check that the crankshaft turns smoothly.

Painted Mark
90°
Front

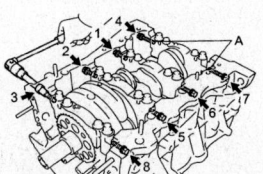

7. **INSTALL CONNECTING ROD BEARING**
(a) Install the connecting rod bearing to the connecting rod and bearing cap.

09490_RAV4_G0004

Main bearing bolt torque sequence—2006 3.5L engine

WHEEL ALIGNMENT

Year	Model		Caster Range (+/-Deg.)	Caster Preferred Setting (Deg.)	Camber Range (+/-Deg.)	Camber Preferred Setting (Deg.)	Toe-in (in.)	Steering Axis Inclination (Deg.)
2002	RAV4	2WD	0.75	+2.00	0.75	-0.42	0+/-0.08	11+/-0.75
		4WD	0.75	+1.92	0.75	-0.33	0.04+/-0.08	10.75+/-0.75
2003	RAV4	2WD	0.75	+2.00	0.75	-0.42	0+/-0.08	11+/-0.75
		4WD	0.75	+1.92	0.75	-0.33	0.04+/-0.08	10.75+/-0.75
2004	RAV4	2WD	0.75	+2.00	0.75	-0.42	0+/-0.08	11+/-0.75
		4WD	0.75	+1.92	0.75	-0.33	0.04+/-0.08	10.75+/-0.75
2005	RAV4	2WD	0.75	+2.00	0.75	-0.42	0+/-0.08	11+/-0.75
		4WD	0.75	+1.92	0.75	-0.33	0.04+/-0.08	10.75+/-0.75
2006	RAV4	2WD	0.75	+2.00	0.75	-0.42	0+/-0.08	11+/-0.75
		4WD	0.75	+1.92	0.75	-0.33	0.04+/-0.08	10.75+/-0.75

09490_RAV4_C0010

TIRE, WHEEL AND BALL JOINT SPECIFICATIONS

Year	Model	OEM Tires Standard	OEM Tires Optional	Tire Pressures (psi) Front	Tire Pressures (psi) Rear	Wheel Size	Ball Joint Inspection	Lugnut Torque (ft. lbs.)
2002	RAV4	P215/70R16	P235/60HR16	29	29	①	9-43 ②	76
2003	RAV4	P215/70R16	P235/60HR16	29	29	③	9-43 ②	76
2004	RAV4	P215/70R16	None	④	④	NA	9-43 ②	76
2005	RAV4	⑤	⑤	④	④	NA	NA	76
2006	RAV4	⑤	⑤	④	④	NA	NA	76

NA: Not Available

OEM: Original Equipment Manufacturer

PSI: Pounds Per Square Inch

① Replace if any measurable movement is found.

② Torque required in inch lbs. to rotate ball joint when removed from the knuckle

③ Steel wheel: 6.5J; aluminum wheel: 7JJ

④ See placard on vehicle

⑤ base model: P215/70R16, P225/65R17. Optional P235/55R18

   sport model: P235/55R18

   limited model: P225/65R17

09490_RAV4_C0011

BRAKE SPECIFICATIONS
All measurements in inches unless noted

Year	Model		Brake Disc			Brake Drum Diameter			Minimum Lining Thickness	Brake Caliper	
			Original Thickness	Minimum Thickness	Maximum Runout	Original Inside Diameter	Max. Wear Limit	Maximum Machine Diameter		Bracket Bolts (ft. lbs.)	Mounting Bolts (ft. lbs.)
2002	RAV4		0.984	0.906	0.0020	9.000	—	9.079	0.039	78	20
2003	RAV4		0.984	0.906	0.0020	9.000	—	9.079	0.039	78	20
2004	RAV4	F	0.984	0.906	0.0020	—	—	—	0.039	78	20
		R	0.354	0.317	0.0039	9.000	—	9.079	0.039	—	33
2005	RAV4	F	0.984	0.906	0.0020	—	—	—	0.039	78	20
		R	0.354	0.317	0.0039	9.000	—	9.079	0.039	—	33
2006	RAV4	F	①	②	0.0020	—	—	—	0.039	72	25
		R	0.472	0.413	0.0059	—	—	—	0.039	65	20

F: Front

R: Rear

① 275 disc: 0.984
　 296 disc: 1.102

② 275 disc: 0.886
　 296 disc: 0.984

09490_RAV4_C0012

SCHEDULED MAINTENANCE INTERVALS
TOYOTA—RAV4

TO BE SERVICED	TYPE OF SERVICE	VEHICLE MILEAGE INTERVAL (x1000)																		
		5	10	15	20	25	30	35	40	45	50	55	60	65	70	75	80	85	90	95
Automatic transmission and differential fluid	S/I			✓			✓			✓			✓			✓			✓	
Ball joints and boots	S/I			✓			✓			✓			✓			✓			✓	
Brake system	S/I			✓			✓			✓			✓			✓			✓	
Charcoal canister	S/I												✓							
Drive belts	S/I						✓						✓						✓	
Driveshaft bushing	L						✓						✓						✓	
Engine coolant	R						✓						✓						✓	
Engine oil & filter	R	✓	✓	✓	✓	✓	✓	✓	✓	✓	✓	✓	✓	✓	✓	✓	✓	✓	✓	✓
Exhaust pipes & mounts	S/I			✓			✓			✓			✓			✓			✓	
Fuel tank cap gasket	S/I						✓						✓						✓	
Halfshaft boots & flange bolts	S/I			✓			✓			✓			✓			✓			✓	
Limited slip differential fluid	R						✓						✓						✓	
Manual transmission and differential fluid	S/I						✓						✓						✓	
Platinum spark plugs	R												✓							
Propeller shaft bolts	S/I			✓			✓			✓			✓			✓			✓	
Steering linkage	S/I			✓			✓			✓			✓			✓			✓	
Tires (rotate)	S/I	✓	✓	✓	✓	✓	✓	✓	✓	✓	✓	✓	✓	✓	✓	✓	✓	✓	✓	✓
Transfer case and differential fluid	S/I			✓			✓			✓			✓			✓			✓	
Valves	S/I												✓							

R: Replace   S/I: Service or Inspect   L: Lubricate

FREQUENT OPERATION MAINTENANCE (SEVERE SERVICE)

If a vehicle is operated under any of the following conditions it is considered severe service:

- Towing a trailer or using a camper or car-top carrier.

- Repeated short trips of less than 5 miles in temperatures below freezing.

- Excessive idling or low-speed driving for long distances as in heavy commercial use, such as delivery, taxi or police cars.

- Operating on rough, muddy or salt-covered roads.

- Operating on unpaved or dusty roads.

Oil filter: service or inspect every 5000 miles or 4 months, whichever occurs first.

Brake linings and discs or drums: service or inspect every 5000 miles or 4 months, whichever occurs first.

Steering linkage: service or inspect every 5000 miles or 4 months, whichever occurs first.

Ball joints and boots: service or inspect every 5000 miles or 4 months, whichever occurs first.

Brake discs & pads (front): service or inspect every 6000 miles.

Halfshaft boots: service or inspect every 5000 miles or 4 months. Retighten the flange bolts, whichever occurs first.

Body chassis bolts and nuts: service or inspect every 5000 miles or 4 months, whichever occurs first.

Transmission and differential fluid: replace every 15,000 miles or 12 months, whichever occurs first.

Transfer case and differential fluid: replace every 15,000 miles or 12 months, whichever occurs first.

Timing belt: replace every 60,000 miles or 48 months, whichever occurs first.

ENGINE REPAIR

➡**Disconnecting the negative battery cable on some vehicles may interfere with the functions of the on board computer system. The computer may undergo a relearning process once the negative battery cable is reconnected.**

Distributor

All vehicles are equipped with a distibutorless ignition system.

Alternator

REMOVAL & INSTALLATION

2.0L Engine

1. Before servicing the vehicle, refer to the Precautions Section.
2. Disconnect the negative battery cable.
3. Remove or disconnect the following:
   • Electrical wiring from the alternator
   • Loosen the adjusting lockbolt and the pivot bolt.
   • Loosen the adjusting bolt to relieve tension on the drive belt, if equipped with air conditioning
   • Drive belt

➡**It may be necessary to remove other belts for access.**

   • Pivot bolt and the adjusting lockbolt
   • Alternator

To install:

4. Install or connect the following:
   • Alternator
   • Adjusting lockbolt and the pivot bolt
   • Drive belt
5. Adjust the drive belt tension to:
   • New belt with A/C: 140–190 lbs.
   • Used belt with A/C: 100–120 lbs.
   • New belt without A/C: 100–150 lbs.
   • Used belt without A/C: 75–115 lbs.
6. Install or connect the following:
   • Tighten the pivot bolt to 38 ft. lbs. (52 Nm).
   • Tighten the adjusting lockbolt to 13 ft. lbs. (18 Nm).
   • Electrical wiring to the alternator

2.4L Engine

1. Before servicing the vehicle, refer to the Precautions Section.
2. Disconnect the negative battery cable.

➡**Wait at least 90 seconds after disconnecting the negative battery cable before starting any repair work to prevent air bag and seat belt pretensioner activation.**

3. Remove or disconnect the following:
   • Electrical wiring from the alternator
   • Drive belt
   • 1 adjusting and 2 mounting bolts
   • Alternator

To install:

4. Installation is the reverse of removal. Observe the following torques:
   • M8 bolts: 15 ft. lbs. (21Nm)
   • M10 bolts: 38 ft. lbs. (52Nm)

3.5L Engine

1. Before servicing the vehicle, refer to the Precautions Section.
2. Disconnect the negative battery cable.

➡**Wait at least 90 seconds after disconnecting the negative battery cable before starting any repair work to prevent air bag and seat belt pretensioner activation.**

3. Drain the cooling system. Remove the radiator.
4. Remove the right front tire and wheel assembly.
5. Remove the rear engine undercover assembly. Remove the front suspension member reinforcement, right side.
6. Remove the radiator reservoir tank.
7. Remove the drive belt. Disconnect the alternator wiring harness and cable.
8. Remove the bolt from the cylinder block. Remove the two alternator retaining bolts. Remove the alternator from the vehicle.

To install:

9. Installation is the reverse of the removal procedure.
10. Tighten the two alternator mounting bolts to 32 ft. lbs. (43 Nm).
11. Tighten the bolt to the cylinder block to 15 ft. lbs. (20 NM).
12. Be sure to fill the cooling system with the proper grade and type engine coolant.

Ignition Timing

ADJUSTMENT

All engines are equipped with a Distibutorless Ignition System (DIS). No timing adjustment is possible.

Engine Assembly

REMOVAL & INSTALLATION

2002–2004

1. Before servicing the vehicle, refer to the Precautions Section.
2. Relieve the fuel system pressure.
3. Remove or disconnect the following:
   • Negative battery cable
   • Battery
   • Hood
   • Engine undercover
4. Drain the engine coolant and oil.
5. Drain the transaxle assembly.
6. Remove or disconnect the following:
   • Air cleaner and case
   • Accelerator cable from the throttle body, bracket and clamps
7. Disconnect and remove the engine wire from the No. 2 relay block, as follows:
   • No. 2 relay block from the body by removing the 2 bolts
   • Upper cover to the relay block
   • Electrical connectors
   • Engine wire, by removing the 2 nuts
   • Charcoal canister
   • Alternator
   • Upper and lower radiator hoses
   • Water inlet from the engine by removing the 2 nuts
   • Heater hoses
   • Fuel hose, by placing a rag under the fuel inlet hose
   • Starter by disconnecting the electrical connectors and 2 bolts, if equipped with manual transaxle
   • Ground cable from the transaxle by removing the bolt
   • Clutch release cylinder from the transaxle, if equipped with manual transaxle
   • Transaxle control cables (2 cables for manual transaxle or 1 for automatic transaxle) from the transaxle.
   • Transaxle cable from the front suspension crossmember and engine mounting centermember by removing the 2 bolts, if equipped with an automatic transaxle
   • Transaxle oil cooler hoses, if equipped with an automatic transaxle or 4WD with manual transaxle
8. Detach the following:
   • Vapor pressure sensor connector
   • Igniter connector

- Ignition coil connector
- Noise filter connector
- Ignition coil wire
- Manifold Absolute Pressure (MAP) sensor connector
- MAP sensor vacuum hose from the gas filter on the intake manifold
- Brake booster hose from the intake manifold
- Differential lock control solenoid connector, if equipped with a 4WD manual transaxle
- Ground strap from cowl

9. Detach the engine wire from the passenger compartment, as follows:
   - Right side scuff plate
   - Right side trim
   - Right side carpet center cover
   - 2 ECM connectors
   - 2 connectors from the bracket connectors
   - No. 4 junction block connector
   - Wire clamp from the bracket
   - Engine wire from the passenger compartment

10. Remove the front exhaust pipe, as follows:
    - 3 nuts and the front exhaust pipe from the exhaust manifold, using a 14mm deep socket wrench; discard the gasket
    - 2 bolts and 2 nuts holding the front exhaust pipe to the catalytic converter
    - Front exhaust pipe and 2 gaskets

11. Remove the compressor from the engine and suspend the compressor securely.

➡ **It is not necessary to remove the air conditioning compressor lines in order to remove the engine.**

12. Remove or disconnect the following:
    - Halfshaft, if equipped with 4WD
    - Halfshaft
    - Sway bar

13. Remove the front suspension crossmember assembly, as follows:
    - 2 centermember set nuts holding the centermember to the middle of the crossmember.
    - 2 rack and pinion assembly set bolts/nuts from the crossmember. Securely suspend the steering gear assembly.
    - Catalytic converter with pipe from the ring
    - Support the suspension crossmember with a jack
    - 6 bolts from the suspension crossmember

- Suspension crossmember with the lower suspension arms

14. Remove the engine mounting centermember, as follows:
    - 2 bolts holding the centermember to the front engine mounting insulator
    - 2 bolts holding the centermember to the body
    - Centermember

15. Disconnect the power steering pump from the engine, as follows:
    - 2 vacuum hoses from the steering pump
    - Adjusting bolt for the power steering unit. Loosen the pivot bolt to the power steering pump and remove the drive belt. Use Torque Wrench Adapter tool 09249-63010 and a deep socket to loosen the pivot bolt.
    - Power steering pump from the engine by removing the 3 bracket bolts

16. Install a engine hanger to the engine.
17. Attach the engine sling device to the engine hangers.
18. Remove or disconnect the following:
    - Left side engine mounting bracket from the mounting insulator by removing the 2 nuts and 2 bolts
    - Ground connector next to the right side engine mount
    - Right side engine mounting bracket from the mounting insulator by removing the bolt and 2 nuts

19. Lower the engine and transaxle and at the same time, raise the vehicle to gain clearance to the remove the engine.
20. Place the assembly on a stand and separate the engine from the transaxle.

To install:
21. Install or connect the following:
    - Engine and transaxle assembly
    - Left side engine mounting bracket to the mounting insulator. Tighten both nuts/bolts to 47 ft. lbs. (64 Nm).
    - Bolt and 2 nuts to hold the right side engine mounting bracket to the mounting insulator. Tighten the bolt to 27 ft. lbs. (37 Nm) and both nuts to 38 ft. lbs. (52 Nm).
    - Ground connector next to the right side engine mount
    - Engine sling and hanger

22. Install the power steering pump, as follows:
    - Pump with the bracket. Tighten the 3 bolts to 32 ft. lbs. (43 Nm).
    - Pivot and adjusting bolts. Tighten the pivot bolt to 32 ft. lbs. (43 Nm)

and the adjusting bolt to 29 ft. lbs. (39 Nm).
    - Drive belt. Adjust the tension.
    - Both air hoses to the power steering pump

23. Install or connect the following:
    - Engine mounting centermember to the body; install the 4 bolts but do not tighten the bolts at this time

24. Install or connect the front crossmember, as follows:
    - Suspension crossmember with the lower control arms. Torque both bolts crossmember-to-chassis bolts to 152 ft. lbs. (206 Nm).
    - Rack and pinion. Tighten both nuts/bolts to 83 ft. lbs. (113 Nm).
    - Centermember to the crossmember. Tighten both nuts to 82 ft. lbs. (112 Nm).
    - Tighten the lower control arm rear brackets to 101 ft. lbs. (137 Nm).
    - Tighten both engine mounting centermember-to-front engine mounting insulator bolts to 59 ft. lbs. (80 Nm).
    - Tighten both engine mounting centermember-to-body bolts to 26 ft. lbs. (35 Nm).

25. Install or connect the following:
    - Sway bar
    - Halfshafts
    - Driveshaft, if equipped with 4WD

26. Install the air conditioning compressor. Tighten nut/bolts, as follows:
    - Stud bolt to 34 ft. lbs. (47 Nm)
    - Bolt to 27 ft. lbs. (37 Nm)
    - Nut to 20 ft. lbs. (27 Nm)

27. Install or connect the following:
    - Air conditioning compressor connector
    - Front exhaust pipe with new gaskets. Tighten the 3 nuts to 46 ft. lbs. (62 Nm) and 2 bolts to 35 ft. lbs. (48 Nm).

28. Attach the engine wire to the passenger compartment, as follows:
    - Engine wire through the cowl panel
    - Wire clamp to the bracket
    - Both ECM connectors
    - Both connectors on the bracket
    - No. 4 junction block
    - Right side floor carpet center cover
    - Right side cowl side trim
    - Right side scuff plate

29. Connect the following:
    - Vapor pressure sensor connector
    - Igniter connector
    - Ignition coil connector
    - Noise filter connector
    - Ignition coil wire
    - MAP sensor connector

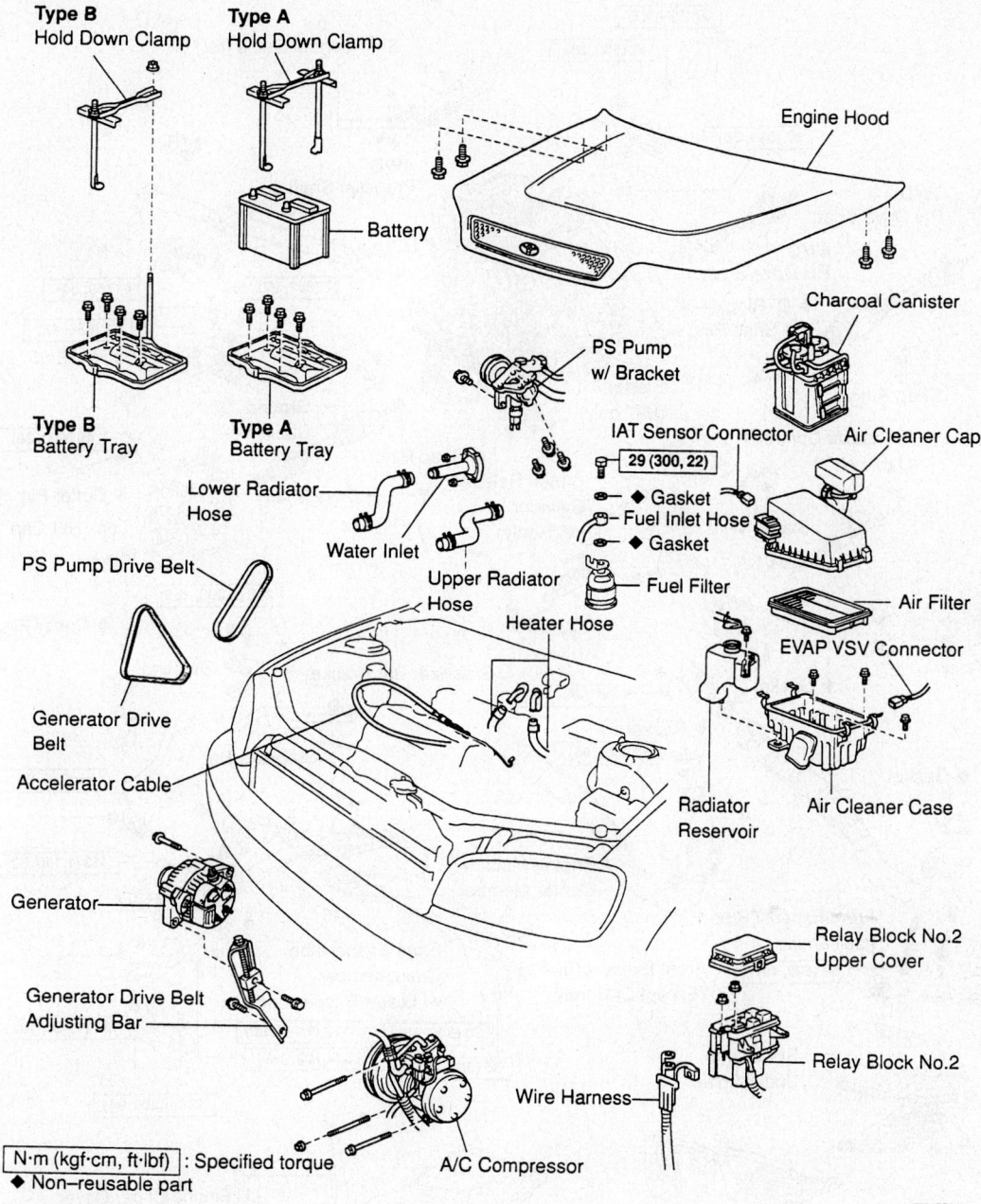

Engine accessory removal components—2.0L

- MAP sensor vacuum hose to the gas filter on the intake manifold
- Brake booster hose to the intake manifold
- Differential lock control solenoid connector, if equipped with 4WD manual transaxle
- Ground strap to cowl
30. Install or connect the following:
    - Transaxle oil cooler hoses, if equipped with an automatic transaxle or 4WD with manual transaxle
    - Transaxle control cable(s) to the transaxle

- Transaxle control cable to the front crossmember and engine mounting centermember, for an automatic transaxle
- Clutch release cylinder, for manual transaxle Tighten both bolts to 108 inch lbs. (12 Nm).
- Ground cable to the transaxle
- Starter, for a manual transaxle
- Fuel inlet hose to the fuel filter, using new gaskets. Tighten the union bolt to 22 ft. lbs. (29 Nm).
- Heater hoses
- Water inlet to the engine. Tighten both nuts to 78 inch lbs. (8.8 Nm).

- Upper and lower radiator hoses
- Alternator
- Charcoal canister
31. Connect the engine wire to the No. 2 relay box, as follows:
    - Engine wire to the No. 2 relay block with the 2 nuts
    - Connector
    - Upper cover
    - No. 2 relay block to the body with the 2 bolts
32. Install or connect the following:
    - Accelerator cable to the throttle body, cable bracket and clamps
    - Air cleaner case and cap

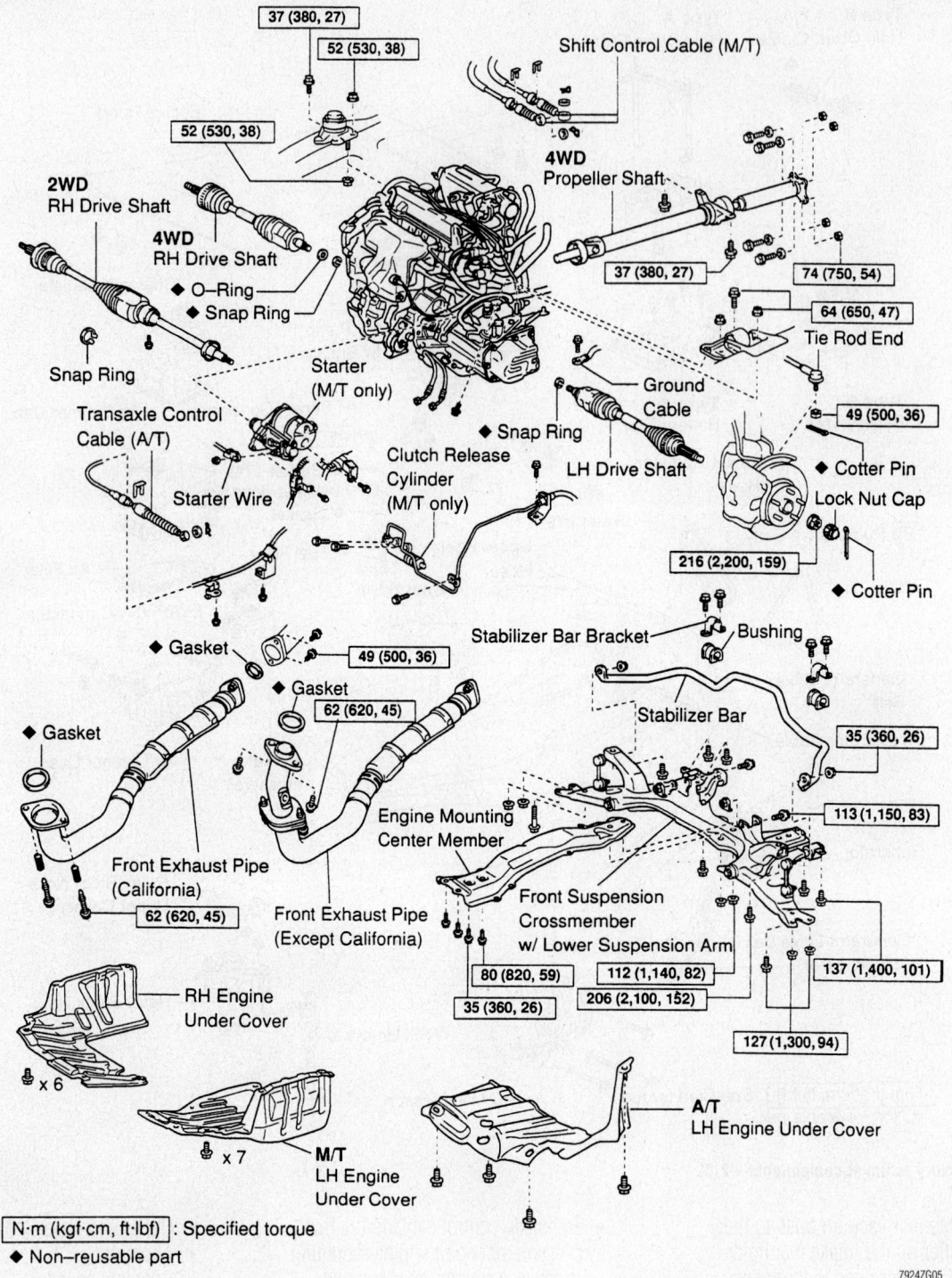

37 (380, 27)
52 (530, 38)
52 (530, 38)

Shift Control Cable (M/T)

4WD
Propeller Shaft

2WD
RH Drive Shaft

4WD
RH Drive Shaft

◆ O–Ring
◆ Snap Ring

37 (380, 27)

74 (750, 54)

64 (650, 47)

Tie Rod End

Snap Ring

Starter
(M/T only)

Ground
Cable

49 (500, 36)

Transaxle Control
Cable (A/T)

◆ Snap Ring

◆ Cotter Pin

Lock Nut Cap

Starter Wire

Clutch Release
Cylinder
(M/T only)

LH Drive Shaft

216 (2,200, 159)

◆ Cotter Pin

◆ Gasket

Stabilizer Bar Bracket

Bushing

49 (500, 36)

◆ Gasket

62 (620, 45)

Stabilizer Bar

35 (360, 26)

◆ Gasket

113 (1,150, 83)

Engine Mounting
Center Member

Front Exhaust Pipe
(California)

62 (620, 45)

Front Exhaust Pipe
(Except California)

Front Suspension
Crossmember
w/ Lower Suspension Arm

137 (1,400, 101)

80 (820, 59)

112 (1,140, 82)

35 (360, 26)

206 (2,100, 152)

127 (1,300, 94)

RH Engine
Under Cover

x 6

x 7

M/T
LH Engine
Under Cover

A/T
LH Engine Under Cover

N·m (kgf·cm, ft·lbf) : Specified torque
◆ Non–reusable part

7924ZG05

Engine assembly and related components—2.0L

- Battery
- Negative battery cables
33. Fill the transaxle with oil.
34. Fill the engine with oil.
35. Fill the engine coolant.
36. Start the engine and check for leaks.
37. Install or connect the following:
- Engine undercovers
- Hood

38. Recheck all fluid levels.
39. Check and/or adjust the front wheel alignment.

2005

1. Before servicing the vehicle, refer to the Precautions Section.
2. Properly discharge the fuel system. Discharge the air conditioning system.

3. Disconnect the negative battery cable.

➡**Wait at least 90 seconds after disconnecting the negative battery cable before starting any repair work to prevent air bag and seat belt pretensioner activation.**

4. Remove the engine undercovers.
5. Drain the transaxle fluid. Drain the

transfer case fluid, if equipped. Drain the engine oil. Drain the engine coolant.

6. Matchmark and remove the hood.

7. Remove the radiator reservoir. Remove the cowl top ventilation louver.

8. Disconnect the positive battery cable. Remove the battery. Remove the battery tray. Remove the five bolts and the battery bracket.

9. Remove the air cleaner assembly.

10. Remove the J/B cover, located inside the engine compartment. Loosen the bolt. Remove the assembly. Pull out the wire connector. Disconnect the connector. Disconnect the wire clamp.

11. Remove the two screws and the glove box door. Disconnect the six engine wire connectors from the ECM and the instrument panel wire. Pull out the engine wire from the cabin.

12. Disconnect the accelerator cable. Disconnect the ground wire from the cylinder head.

13. Remove the air conditioning hoses from the compressor. Disconnect the air conditioning hoses from the power steering pump.

14. Remove the radiator hoses. Disconnect the heater hoses.

15. Drain the power steering fluid. Disconnect and plug the power steering pump hoses.

16. Remove the fuel tube connector. Disconnect the connector from the fusible link block.

17. If equipped with manual transaxle, remove the two clips and washers and disconnect the two control cables.

18. Remove the starter.

19. If equipped with manual transaxle, remove the heat insulator and disconnect the clutch release cylinder and line from the transaxle.

20. If equipped with automatic transaxle, remove the nut holding the control shift lever to the control cable. Remove the clip and disconnect the control cable. Disconnect the control cable from the cable clamp.

21. Disconnect the ground cable from the transaxle.

22. If equipped with automatic transaxle, disconnect and plug the transaxle cooler lines.

23. If equipped with 4WD, remove the driveshaft.

24. Remove the front halfshafts.

25. Remove the four bolts, two compression rings, two gaskets and the front exhaust pipe.

26. Disconnect the front stabilizer bar links.

27. Remove the two bolts and disconnect the power steering gear assembly.

28. Attach a suitable engine removal fixture to the engine using engine hanger tools 12281-28010 and 12281-28020 or equivalent.

➡ **Do not attempt to hang the engine by hooking the chain to any other part.**

29. Remove the two nuts and bolt holding the right side engine mounting insulator to the timing chain cover.

30. Remove the bolt holding the left side engine mounting insulator to the mounting bracket.

31. Support the suspension crossmember with a jack. Remove the six bolts holding the suspension crossmember engine mounting member to the body.

32. Carefully lower the engine assembly out of the vehicle.

➡ **Be sure the assembly is clear of all wiring, hoses and cables before removing it.**

33. Separate the engine, transaxle, front suspension crossmember and center engine mounting member, as required.

To install:

34. Installation is the reverse of the removal procedure.

35. Tighten the six upper engine to transaxle bolts to 34 ft. lbs. for bolt "A" and 47 ft. lbs. for bolt "B".

36. Tighten the four lower engine to transaxle bolts to 32 ft. lbs.

37. If equipped with automatic transaxle, tighten the torque converter retaining bolts to 30 ft. lbs.

➡ **Install the green colored bolt first.**

38. Tighten the six suspension crossmember engine mounting bolts to 29 ft. lbs for bolt "A", 83 ft. lbs. for bolt "B" and 116 ft. lbs. for bolt "C".

39. Be sure to fill the engine with the proper grade and type engine oil.

40. Be sure to fill the transaxle with the proper grade and type transaxle fluid.

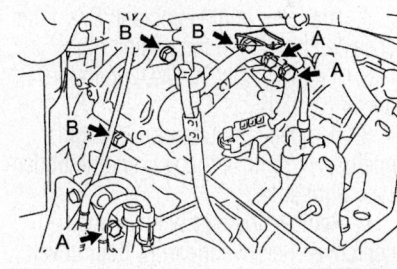

Engine to transaxle bolt locations—2005 2.4L engine

09490_RAV4_G0005

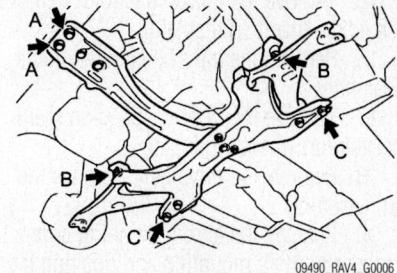

Suspension crossmember bolt locations— 2005 2.4L engine

09490_RAV4_G0006

41. Be sure to fill the cooling system with the proper grade and type engine coolant.

42. Recharge the air conditioning system.

43. Start the engine and check for leaks, correct as required.

2006

2.4L ENGINE

1. Before servicing the vehicle, refer to the Precautions Section.

2. Properly discharge the fuel system.

3. Disconnect the negative battery cable.

➡ **Wait at least 90 seconds after disconnecting the negative battery cable before starting any repair work to prevent air bag and seat belt pretensioner activation.**

4. Drain the engine coolant. Drain the transaxle fluid. Drain the engine oil.

5. Matchmark and remove the hood. Remove the radiator support opening cover.

6. Disconnect the positive battery cable. Remove the battery. Remove the battery tray.

7. Remove the front tire and wheel assemblies.

8. Remove the number one and number two engine undercovers.

9. Remove the front fender apron, right side.

10. Remove the radiator. Remove the radiator reservoir.

11. Remove the purge VSV assembly. Remove the air cleaner assembly.

12. Remove the front battery carrier, the battery bracket reinforcement and the carrier bracket.

13. Disconnect the hoses from the booster vacuum tube. Disconnect the heater hoses.

14. Disconnect the fuel tube. Remove the throttle body. Disconnect the number two ventilation hose. Remove the fuel delivery pipe sub assembly.

15. Remove the intake manifold. Remove the intake manifold insulator.

16. Remove the transaxle control cable assembly.

17. Remove the front suspension member reinforcement, right side.

18. Remove the drive belt. Remove the fan assembly.

19. Remove the air conditioning compressor from its mounting and position it to the side. Do not disconnect the refrigerant lines.

20. Remove the ECM.

21. Remove the engine room relay block cover. Remove the nut and disconnect the two engine wire connectors. Remove the bolt and engine wire cover.

22. Disconnect the engine wire from the engine wire cover. Disconnect the ground cable. Remove the nut from the positive battery terminal to disconnect the engine wire.

23. Disconnect the ground cable from the clamp located near the starter. Remove the bolt and ground cable.

24. Remove the front exhaust pipe assembly. Remove the front halfshaft assembly.

25. Remove the driveshaft, if equipped with 4WD.

26. Remove the front and rear engine mounting insulators.

27. Attach a suitable engine removal fixture to the engine using engine hanger tools 12281-28010 and 12282-28010 or equivalent.

➡ Do not attempt to hang the engine by hooking the chain to any other part.

28. Carefully remove the engine assembly out of the vehicle.

➡ Be sure the assembly is clear of all wiring, hoses and cables before removing it.

29. Separate the engine and transaxle, as required.

To install:

30. Installation is the reverse of the removal procedure.

31. Be sure to fill the engine with the proper grade and type engine oil.

32. Be sure to fill the transaxle with the proper grade and type transaxle fluid.

33. Be sure to fill the cooling system with the proper grade and type engine coolant.

34. Check and adjust the front end alignment, as required.

➡ If the vehicle has both cruise control and VSC, when replacing the ECM or the engine perform VSC recognition.

35. Turn the ignition switch "ON".

36. After waiting five seconds, turn the cruise control main switch "ON".

37. Keep the cruise control main switch "ON" for approximately five seconds or more.

➡ The VSC recognition will not be reflected in the cruise control system until the ignition switch is turned "OFF" and "ON".

➡ Perform the automatic transaxle initialization procedure when replacing the automatic transaxle, engine or ECM.

38. Turn the ignition switch OFF.

39. Connect the intelligent tester tool together with the controller area network vehicle interface module (CAN VIN) to the DLC3.

40. Turn the ignition switch to the ON position.

41. Push the intelligent tester tool main switch to the ON position.

42. Select the items, DIAGNOSIS/ ENHANCED OBD II.

43. Perform the reset memory procedure from the ENGINE menu.

➡ After performing the reset memory, be sure to perform the road test procedure. For roadtest procedure information, refer to the intelligent tester instruction manual.

44. Start the engine and check for leaks, correct as required.

3.5L ENGINE

1. Before servicing the vehicle, refer to the Precautions Section.

2. Properly discharge the fuel system. Discharge the air conditioning system.

3. Disconnect the negative battery cable.

➡ Wait at least 90 seconds after disconnecting the negative battery cable before starting any repair work to prevent air bag and seat belt pretensioner activation.

4. Drain the engine coolant. Drain the transaxle fluid. Drain the engine oil.

5. If equipped with 4WD, drain the transfer case oil.

6. Remove the number one engine undercover. Remove the rear engine undercover, right side.

7. Remove the number two engine undercover. Remove the front floor cover.

8. Matchmark and remove the hood. Remove the V bank cover subassembly.

9. Remove the nine clips and the radiator support opening cover.

10. Disconnect the positive battery cable. Remove the battery. Remove the battery tray. Remove the bolts and the battery bracket. Remove the battery bracket reinforcement.

11. Remove the air cleaner assembly.

12. Remove the engine room upper relay block cover. Remove the nut and disconnect the three engine wire connectors.

13. Remove the nut and disconnect the starter wire.

14. Disconnect the fuel hoses. Disconnect the neater hoses.

15. Disconnect the ECM connectors. Remove the ECM.

16. Disconnect and plug the air conditioning compressor lines.

17. Disconnect and plug the transaxle fluid cooler hoses. Disconnect the transaxle control cable assembly.

18. Disconnect the front exhaust pipe assembly. Disconnect the center exhaust pipe assembly.

19. Remove the driveshaft along with the center bearing shaft assembly.

20. Remove the front tire and wheel assemblies.

21. Remove the front axle hub nuts.

22. Disconnect both stabilizer link assemblies.

23. Disconnect the steering intermediate shaft.

24. Disconnect the tie rod end sub-assemblies.

25. Remove the halfshafts.

26. Attach a suitable engine removal fixture to the engine using engine hanger tools 12281-31120 and 12282-31100 or equivalent.

➡ Do not raise the engine more than necessary. If the engine is raised excessively, the vehicle may also be lifted up.

27. Position an engine lifter underneath the engine.

28. Remove the column hole cover. Loosen the bolt holding the intermediate shaft and slide the intermediate shaft. Matchmark the intermediate shaft and pinion.

29. Remove the two bolts and nuts, and disconnect the engine mounting insulator, right side.

30. Remove the two bolts and nuts, and disconnect the engine mounting insulator, left side.

31. Remove the six bolts and the front suspension member brace, right and left sides.

32. Remove the six bolts, crossmember and suspension member.

33. Using a chain block, slowly remove the engine assembly from the vehicle and the intermediate shaft from the pinion. Position the assembly in a suitable holding fixture.

➡**Make sure that the engine is clear of all wiring, hoses and cables before removing it. While lowering the assembly do not allow it to come in contact with the vehicle.**

34. Separate the engine, transaxle, front suspension crossmember assembly (including steering gear) and transfer case, as required.

To install:

35. Installation is the reverse of the removal procedure.

36. Tighten the suspension member and crossmember bolts to 71 ft. lbs. for bolt "A", 107 ft. lbs. for bolt "B".

37. Tighten the member brace rear bolts to 107 ft. lbs. for bolt "C" and 69 ft. lbs. for bolt "D".

38. Be sure to fill the engine with the proper grade and type engine oil.

39. Be sure to fill the transaxle with the proper grade and type transaxle fluid.

40. Be sure to fill the cooling system with the proper grade and type engine coolant.

41. Recharge the air conditioning system.

42. Check and adjust the front end alignment, as required.

➡**If the vehicle has both cruise control and VSC, when replacing the ECM or the engine perform VSC recognition.**

Suspension crossmember bolt locations— 3.5L engine

43. Turn the ignition switch "ON".

44. After waiting five seconds, turn the cruise control main switch "ON".

45. Keep the cruise control main switch "ON" for approximately five seconds or more.

➡**The VSC recognition will not be reflected in the cruise control system until the ignition switch is turned "OFF" and "ON".**

➡**Perform the automatic transaxle initialization procedure when replacing the automatic transaxle, engine or ECM.**

46. Turn the ignition switch OFF.

47. Connect the intelligent tester tool together with the controller area network vehicle interface module (CAN VIN) to the DLC3.

48. Turn the ignition switch to the ON position.

49. Push the intelligent tester tool main switch to the ON position.

50. Select the items, DIAGNOSIS/ ENHANCED OBD II.

51. Perform the reset memory procedure from the ENGINE menu.

➡**After performing the reset memory, be sure to perform the road test procedure. For roadtest procedure information, refer to the intelligent tester instruction manual.**

52. Start the engine and check for leaks, correct as required.

Heater Core

REMOVAL & INSTALLATION

2002–2005

1. Before servicing the vehicle, refer to the Precautions Section.

2. Position the front wheels in the straight ahead position.

3. Discharge the air conditioning system. Drain the engine coolant.

4. Disconnect the negative battery cable.

➡**Wait at least 90 seconds after disconnecting the negative battery cable before starting any repair work to prevent air bag and seat belt pretensioner activation.**

5. Disconnect and plug the refrigerant lines at the evaporator. Disconnect the heater hoses at the heater core.

6. Using a torx socket, loosen the two

torx screws at the wheel pad. Pull out the wheel pad from the steering wheel and disconnect the air bag connector. Remove the wheel pad.

➡**If the air bag connector is disconnected with the ignition switch "ON", DTC's will be recorded. When storing the wheel pad, keep the upper surface of the pad facing upward. Never disassemble the wheel pad. When removing the wheel pad, take care not to pull the air bag wire harness.**

7. Disconnect the connector. Match-mark the steering wheel. Remove the steering wheel retaining nut. Using the proper removal tool, remove the steering wheel.

8. Remove the door scuff plate. Remove the cowl side trim. Remove the number one switch hole base.

9. Remove the three screws and disconnect the connector. Using a pry tool, pry out the lower finish panel.

10. Remove the four bolts and then remove the lower panel insert.

11. Remove the three screws and the lower column cover. Disconnect the column upper cover from the column tube and remove it.

12. Disconnect the airbag connector from the spiral cable. Disconnect the three connectors from the spiral cable, light control switch and headlight dimmer/wiper/washer switch. Release the three claws and remove the spiral cable.

➡**Do not disassemble the spiral cable or apply oil to it.**

13. Push the claw and pull out the light control switch and the headlight dimmer

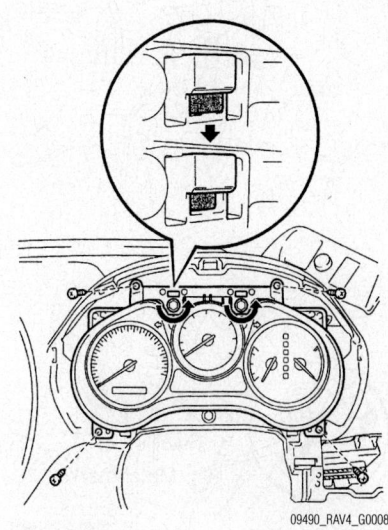

Combination meter removal locating points—2002–2005

switch. Push the claw and pull out the wiper/washer switch.

14. Using a pry tool, remove the cluster finish panel.

15. To remove the combination meter, push out the connector by loosening the two bolts. Check that the red colored indicator goes back as indicated in the illustration. Remove the four screws.

➡**The operation is complete when the bolt fails to catch.**

16. Using a pry tool, remove the instrument cluster finish panel center number one and number two.

17. Remove the two screws. Using a pry tool, pry out the finish panel. Remove the four bolts and move the radio forward. Disconnect the radio connectors and remove the radio from its mounting.

18. Remove the number one instrument panel undercover subassembly. Disconnect the heater control cables.

➡**When the cables are disconnected, be careful not to bend the cables.**

19. Remove the two screws. Using a pry tool, pry out the heater control assembly from its mounting.

20. Remove the two screws. Using a pry tool, pry out the instrument panel cluster finish center panel then disconnect the connector.

21. If equipped with manual transaxle, remove the shift lever knob. Using a pry tool, pry out the center console panel.

22. Remove the two bolts, two screws and rear console box. Disconnect the connector.

23. Remove the six screws and the front console box. Using a pry tool, carefully pry out the console box, toward the passenger's side of the vehicle.

24. Remove the number one and number two instrument panel undercovers. Remove the glove box. Remove the instrument side panel.

25. Disconnect the hood lock control cable. Remove the four bolts and the left instrument panel lower insert.

26. Remove the two bolts and disconnect the main shaft lower dust cover.

27. Disconnect the number two intermediate shaft assembly.

28. Place matchmarks on the number two intermediate shaft assembly and the steering intermediate extension.

29. Loosen the bolt "A" and remove the bolt "B". Disconnect the number two intermediate shaft assembly. Remove the main shaft lower dust cover.

30. Disconnect the connectors and the wire harness clamps from the steering column assembly.

31. Remove the three steering column retaining bolts.

32. Place matchmarks on the sliding yoke and main shaft assembly. Place matchmarks on the sliding yoke and number two intermediate shaft.

33. Remove bolt "A" and the number two intermediate shaft assembly. Remove bolt "B" and the sliding yoke.

34. Remove the front pillar garnish.

➡**If the vehicle is equipped with curtain shield airbag, cover the airbag with cloth or a piece of nylon approximately 27.56 inch by 4.72 inch. Install the ends of the cover with tape, as shown in the illustration. Cover the curtain shield airbag with the protective cover as soon as the pillar garnish is removed.**

35. Disconnect the instrument panel electrical connectors. Disconnect the wire harness clamps. Remove the four screws, two nuts and eight bolts. Remove the instrument panel from the vehicle.

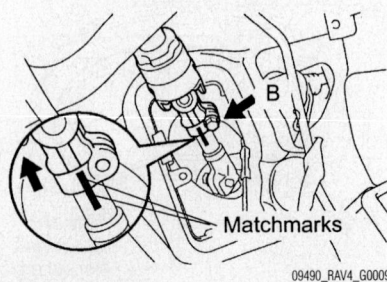

Number two intermediate shaft bolt location—2002–2005

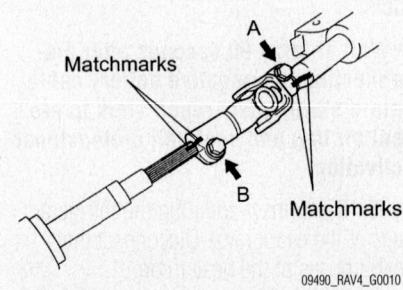

09490_RAV4_G0010

Sliding yoke bolt location—2002–2005

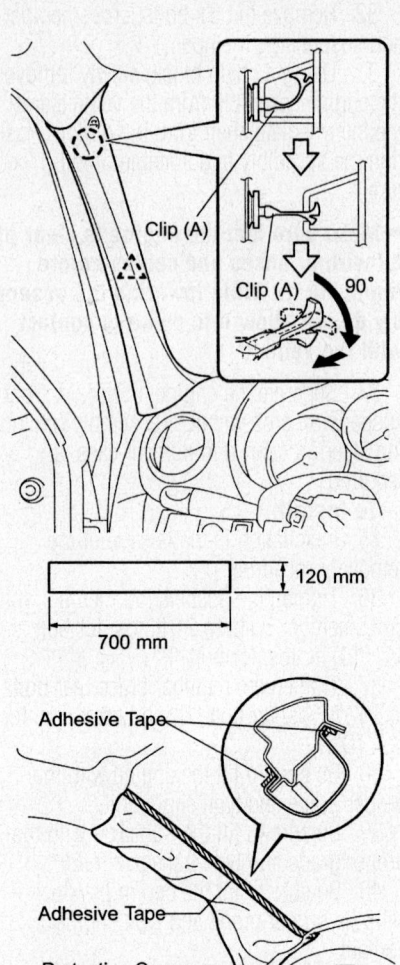

09490_RAV4_G0011

Side pillar garnish removal—2002–2005

36. Remove the three bolts, three nuts and then remove the number one and number two braces.

37. Remove the brake pedal spring. Remove the five bolts, nut and the reinforcement bar assembly.

38. Disconnect the blower resistor, blower motor connectors and harness clamps at the blower unit.

39. Disconnect the air inlet servomotor connector.

40. Remove the two screws, two nuts and the blower unit.

41. Remove the rear duct at the air conditioning unit, if equipped.

42. Remove the drain hose. Remove the two nuts and the air conditioning unit.

43. Remove the thermistor. Remove the evaporator. Remove the expansion valve. Remove the heater core.

To install:

44. Installation is the reverse of the removal procedure.

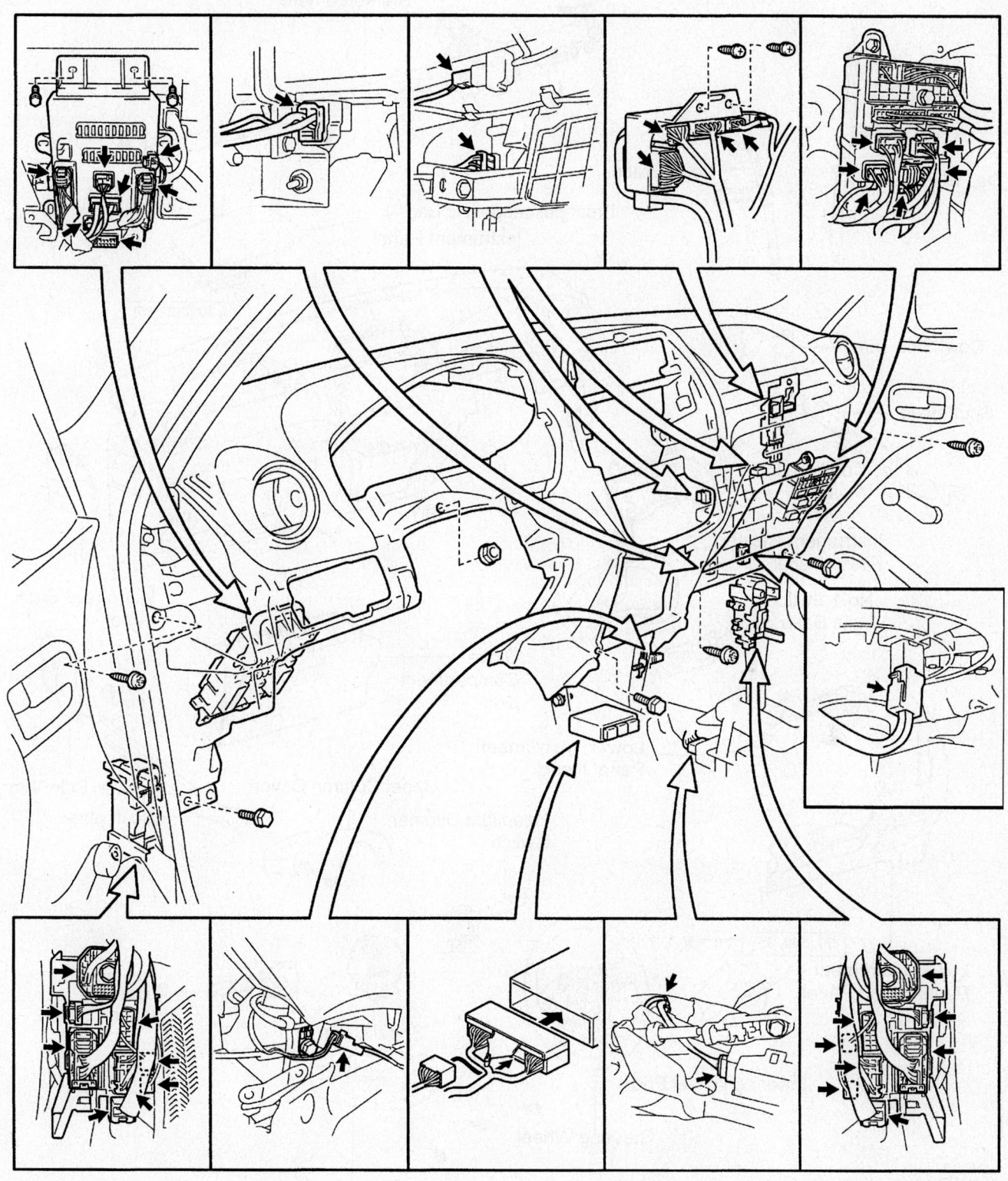

09490_RAV4_G0012

Instrument panel removal points—2002–2005

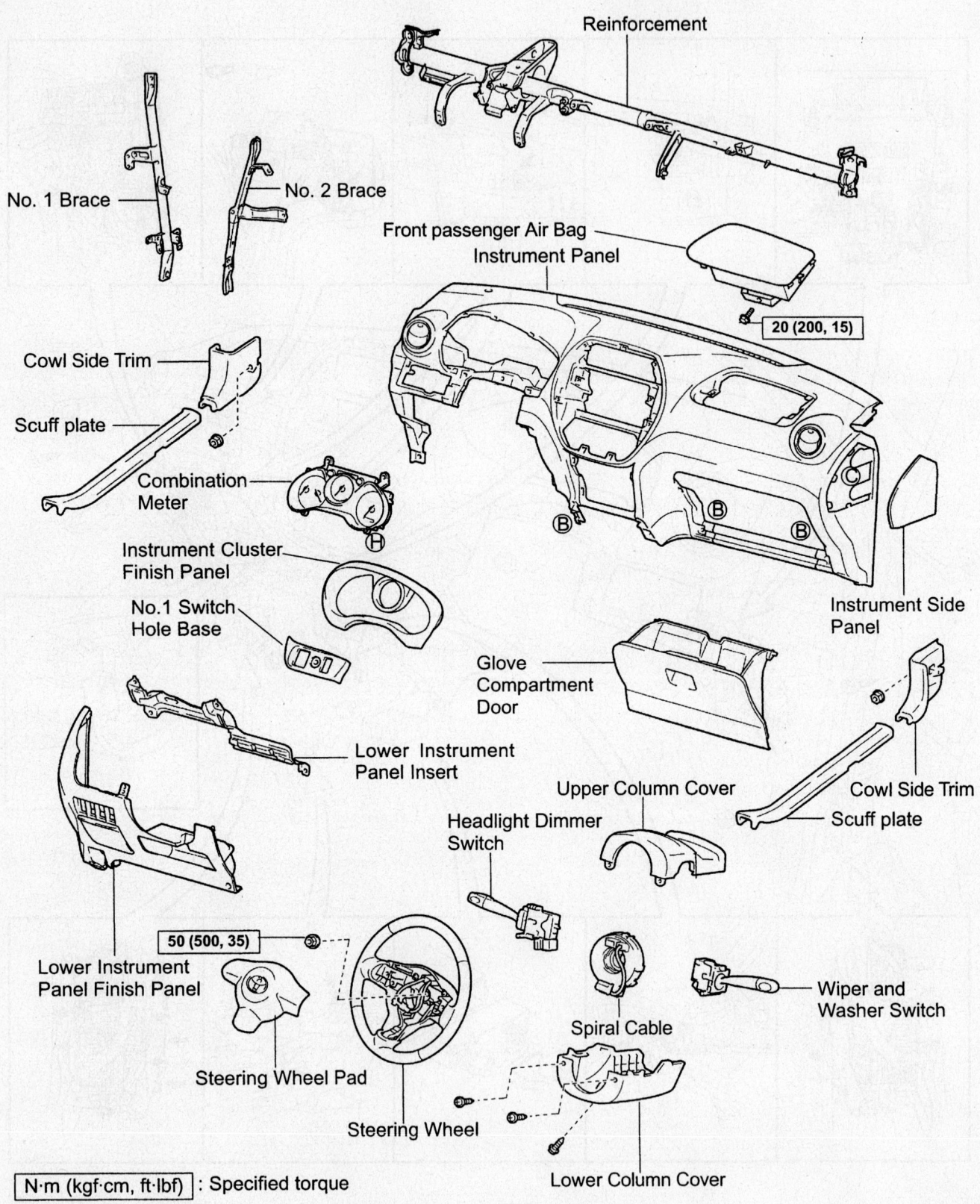

Reinforcement

No. 1 Brace

No. 2 Brace

Front passenger Air Bag
Instrument Panel

20 (200, 15)

Cowl Side Trim

Scuff plate

Combination
Meter

Instrument Cluster
Finish Panel

No.1 Switch
Hole Base

Instrument Side
Panel

Glove
Compartment
Door

Cowl Side Trim

Scuff plate

Lower Instrument
Panel Insert

Upper Column Cover

Headlight Dimmer
Switch

50 (500, 35)

Lower Instrument
Panel Finish Panel

Steering Wheel Pad

Steering Wheel

Spiral Cable

Wiper and
Washer Switch

Lower Column Cover

N·m (kgf·cm, ft·lbf) : Specified torque

09490_RAV4_G0013

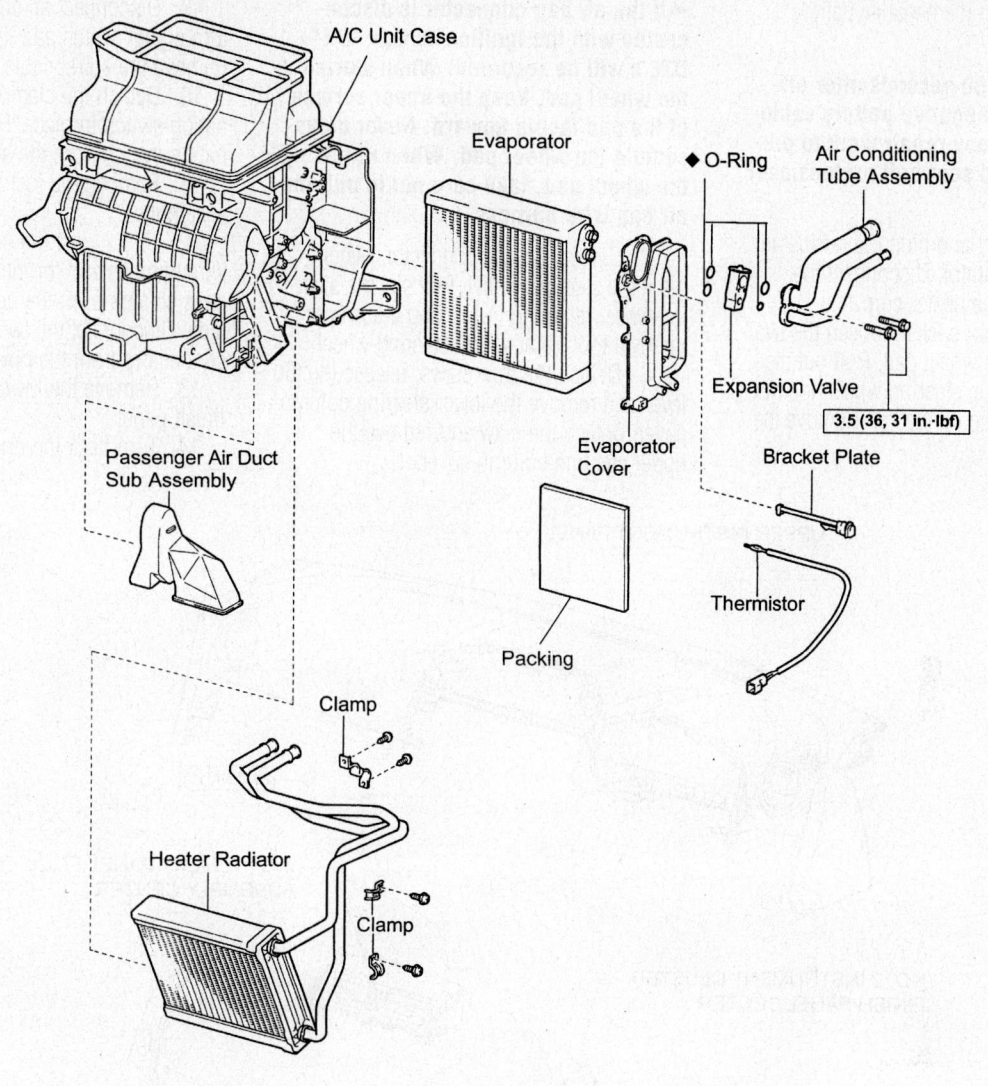

N·m (kgf·cm, ft·lbf) : Specified torque
◆ Non-reusable part

09490_RAV4_G0014

Heater core and related components—2002–2005

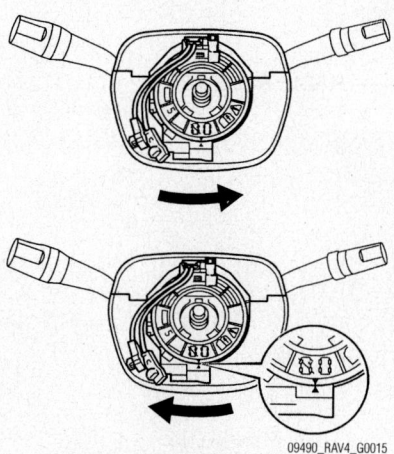

09490_RAV4_G0015

Spiral cable alignment

45. Tighten the sliding yoke to number intermediate shaft retaining bolt to 26 ft. lbs.

46. Tighten the three steering column retaining bolts to 15 ft. lbs.

47. Tighten the number two intermediate shaft assembly retaining bolt to 26 ft. lbs.

48. To center the spiral cable, check that the front wheels are in the straight ahead position. Turn the cable counterclockwise by hand until it becomes hard to turn. Then rotate the cable clockwise about 2.5 turns to align the marks.

➡**The cable will rotate about 2.5 turns to either left or right of the center.**

49. Tighten the steering wheel locknut to 35 ft. lbs.

50. Tighten the wheel pad retaining screws to 78 inch lbs.

51. Check the steering wheel center point.

52. Recharge the air conditioning system.

53. Fill the cooling system with the proper grade and type engine coolant.

54. Start the engine and check for leaks, correct as required.

2006

1. Before servicing the vehicle, refer to the Precautions Section.

2. Position the front wheels in the straight ahead position.

3. Discharge the air conditioning system. Drain the engine coolant.

4. Disconnect the negative battery cable.

➡ **Wait at least 90 seconds after disconnecting the negative battery cable before starting any repair work to prevent air bag and seat belt pretensioner activation.**

5. Disconnect and plug the refrigerant lines at the evaporator. Disconnect the heater hoses at the heater core.

6. Using a torx socket, loosen the two torx screws at the wheel pad. Pull out the wheel pad from the steering wheel and disconnect the air bag connector. Remove the wheel pad.

➡ **If the air bag connector is disconnected with the ignition switch "ON", DTC's will be recorded. When storing the wheel pad, keep the upper surface of the pad facing upward. Never disassemble the wheel pad. When removing the wheel pad, take care not to pull the air bag wire harness.**

7. Disconnect the connector. Matchmark the steering wheel. Remove the steering wheel retaining nut. Using the proper removal tool, remove the steering wheel.

8. Detach the four claws, release the tilt lever and remove the lower steering column cover. Detach the claw and remove the upper steering column cover.

9. Disconnect all connectors from the turn signal switch and the spiral cable. Remove the spiral cable.

10. Detach the clamp holding the combination switch in place. Remove the combination switch from the steering column.

11. Remove the instrument panel sub-assembly.

12. Disconnect the power steering motor wire harness and torque sensor wire harness clamps from the power steering ECU side. Disconnect the two steering column connectors from the power steering ECU.

13. Remove the lower instrument panel finish panel.

14. Turn back the drivers side carpet.

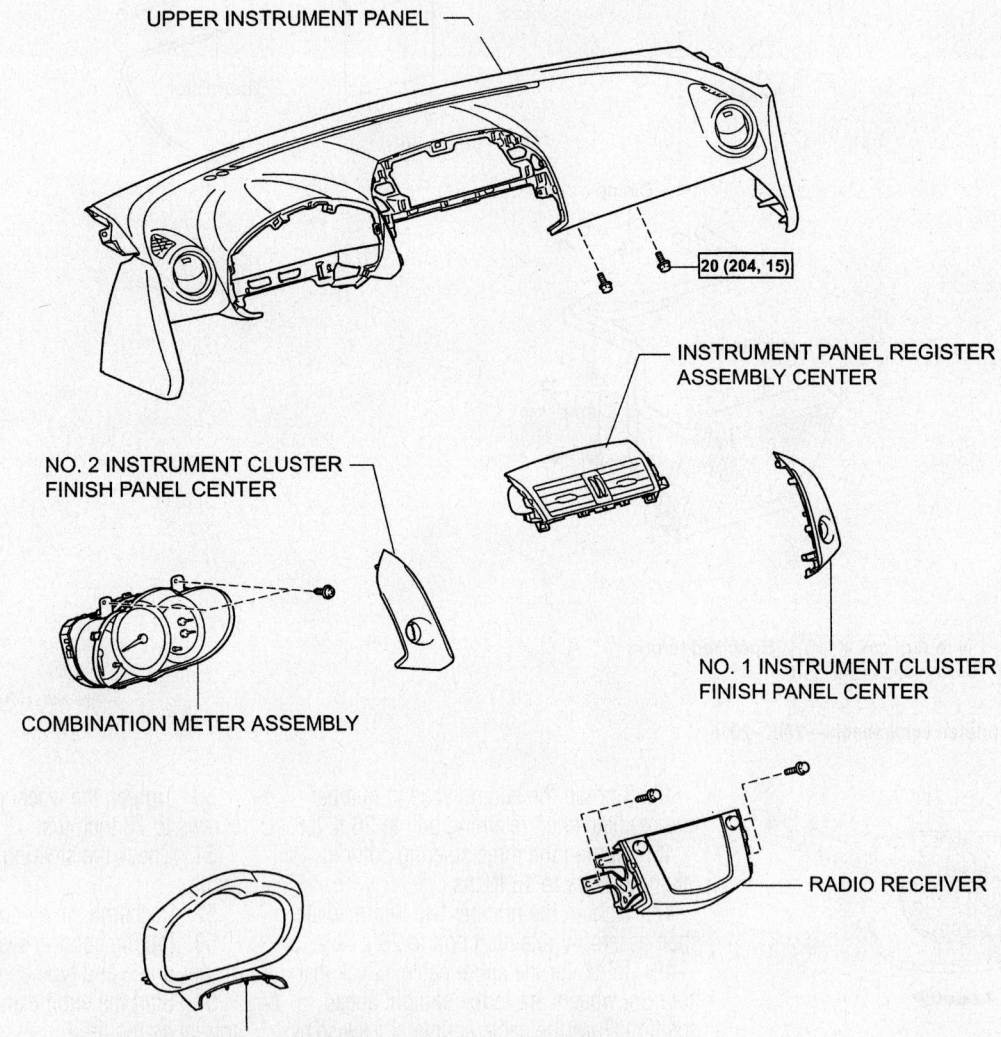

UPPER INSTRUMENT PANEL

20 (204, 15)

INSTRUMENT PANEL REGISTER ASSEMBLY CENTER

NO. 2 INSTRUMENT CLUSTER FINISH PANEL CENTER

NO. 1 INSTRUMENT CLUSTER FINISH PANEL CENTER

COMBINATION METER ASSEMBLY

RADIO RECEIVER

INSTRUMENT CLUSTER FINISH PANEL SUB-ASSEMBLY

N*m (kgf*cm, ft.*lbf) : Specified torque

09490_RAV4_G0016

Upper instrument panel and related components—2006

Remove the two clips and remove the steering column hole cover silencer sheet.

15. Place matchmarks on the steering intermediate shaft and steering gear. Remove the bolt and detach the steering gear.

16. Place matchmarks on the steering intermediate shaft and the steering column. Remove the bolt and detach the steering intermediate shaft from the steering column.

17. Remove the brake pedal support assembly.

18. Disconnect the connectors and the wire harness clamps from the steering column assembly.

19. Remove the bolts and nuts and remove the steering column from the instrument panel reinforcement.

20. Remove the headlight dimmer switch assembly. Remove the windshield wiper switch assembly.

21. Remove the number one and the number two instrument cluster finish panel center assemblies.

22. Remove the radio.

23. Remove the two screws. Disconnect the connectors and remove the air conditioning control.

24. If equipped with automatic climate control, remove the air mix control switch, the blower control switch and the air vent mode control switch.

25. Remove the number one console upper panel garnish. Remove the number two console upper panel garnish.

26. Remove the shift lever knob. Using a pry tool, detach the two clips, four claws

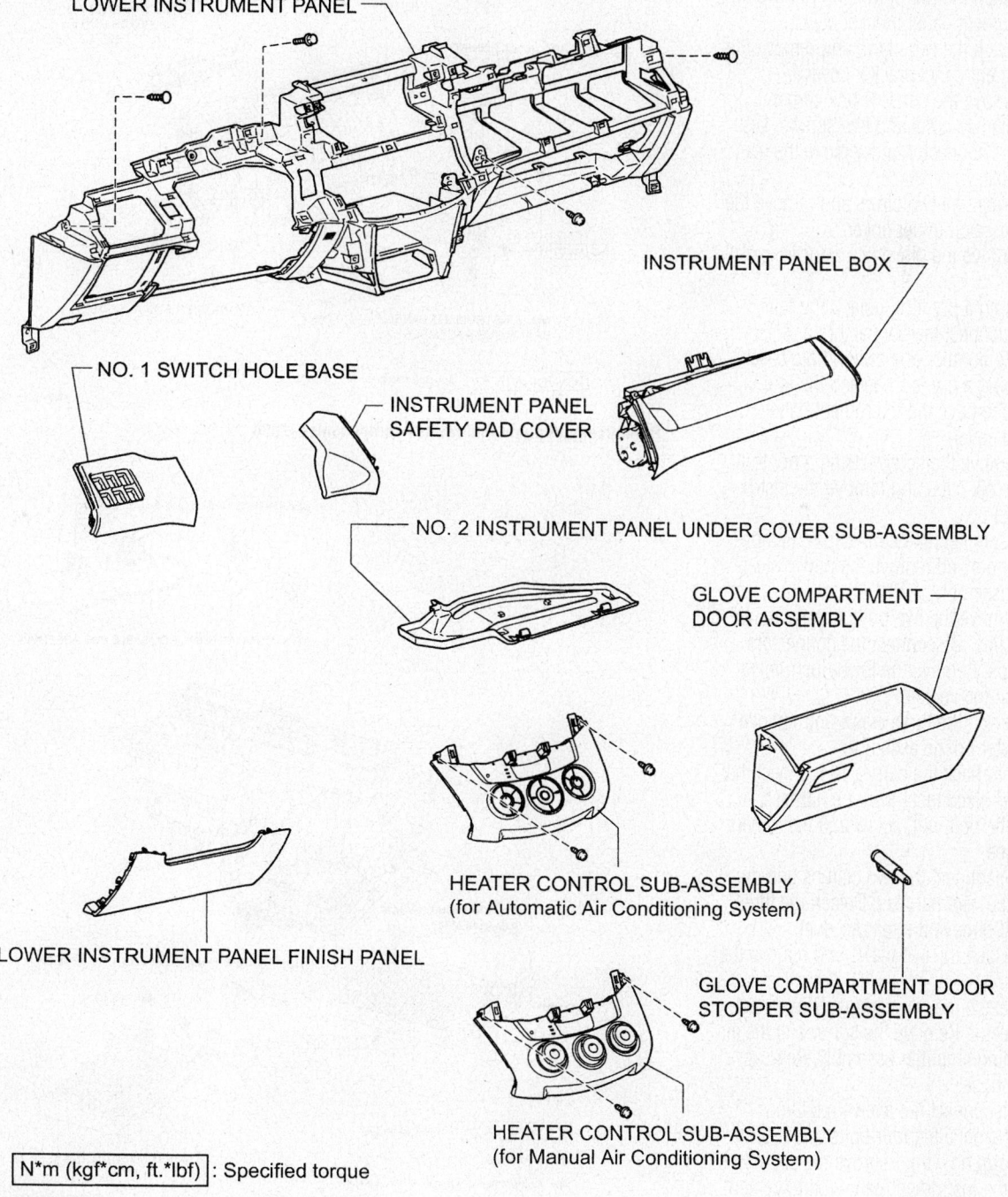

LOWER INSTRUMENT PANEL

INSTRUMENT PANEL BOX

NO. 1 SWITCH HOLE BASE

INSTRUMENT PANEL SAFETY PAD COVER

NO. 2 INSTRUMENT PANEL UNDER COVER SUB-ASSEMBLY

GLOVE COMPARTMENT DOOR ASSEMBLY

HEATER CONTROL SUB-ASSEMBLY
(for Automatic Air Conditioning System)

LOWER INSTRUMENT PANEL FINISH PANEL

GLOVE COMPARTMENT DOOR STOPPER SUB-ASSEMBLY

HEATER CONTROL SUB-ASSEMBLY
(for Manual Air Conditioning System)

N*m (kgf*cm, ft.*lbf) : Specified torque

Lower instrument panel and related components—2006

and remove the upper console panel. Disconnect the connector.

27. Detach the two clips and two claws. Disconnect the connectors and remove the switch base.

28. Remove the two screws. Detach the two clips and remove the cup holder box.

29. Using a pry tool, detach the four claws and two clips. Disconnect the connector and remove the upper rear console.

30. Using a pry tool, detach the six claws and remove the console rear end panel.

31. Detach the two claws and remove the right instrument panel bracket cover.

32. Detach the two claws and remove the left instrument panel bracket cover.

33. Remove the console box carpet. Remove the two bolts and two screws. Disconnect the connector and remove the rear console box.

34. Detach the two claws and remove the instrument panel under cover.

35. Remove the glove box door assembly.

36. Using a pry tool, detach the four clips. Disconnect the connectors and remove the number one switch hole base.

37. Using a pry tool, detach the four claws and remove the instrument panel safety pad cover.

38. Remove the screw. Using a pry tool, detach the six clips and remove the instrument panel box.

39. Disconnect the connectors. Remove the three nuts and remove the power steering ECU assembly.

40. Remove the two bolts, three screws and two clips. Disconnect the connectors and clamps. Remove the lower instrument panel from the vehicle.

41. Detach the three claws and remove the defroster nozzle assembly.

42. Fold back the carpet. Disconnect the clamp and disconnect the wire harness. Remove the bolt, nut, screw and instrument panel brace.

43. Disconnect the two clamps and disconnect the wire harness. Detach the three claws and remove the rear air duct.

44. Detach the two claws and remove the air duct.

45. Detach the two clamps. Disconnect the connector. Remove the screw and the air conditioning amplifier assembly. Remove the drain hose.

46. Disconnect the twelve retaining clamps. Remove the four bolts and disconnect the ground wire. Remove the six bolts. Remove the instrument panel reinforcement.

47. Remove the bolt, nut and the air conditioner unit assembly.

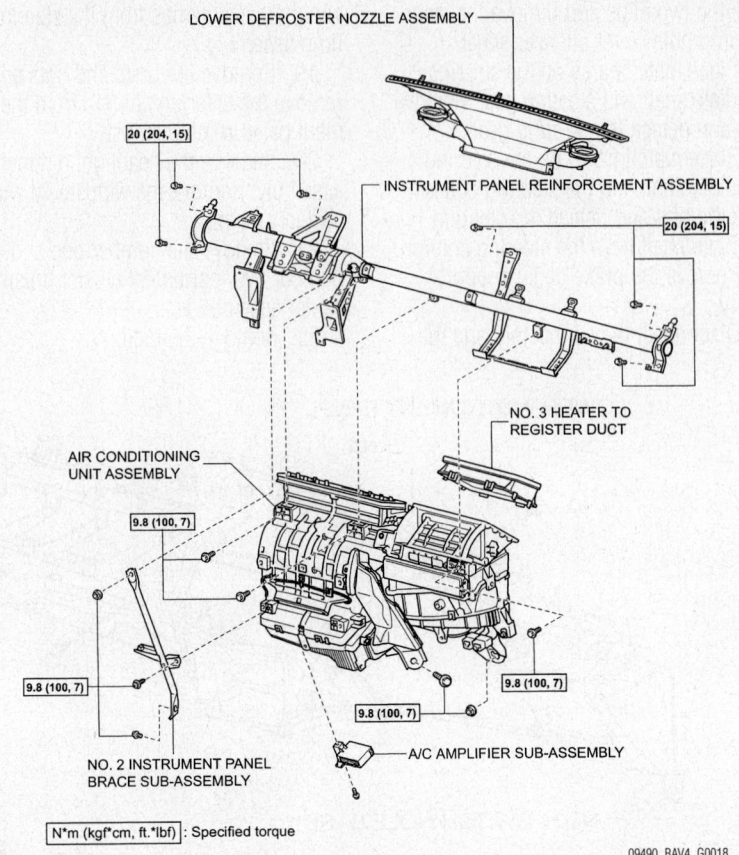

Reinforcement panel and related components—2006

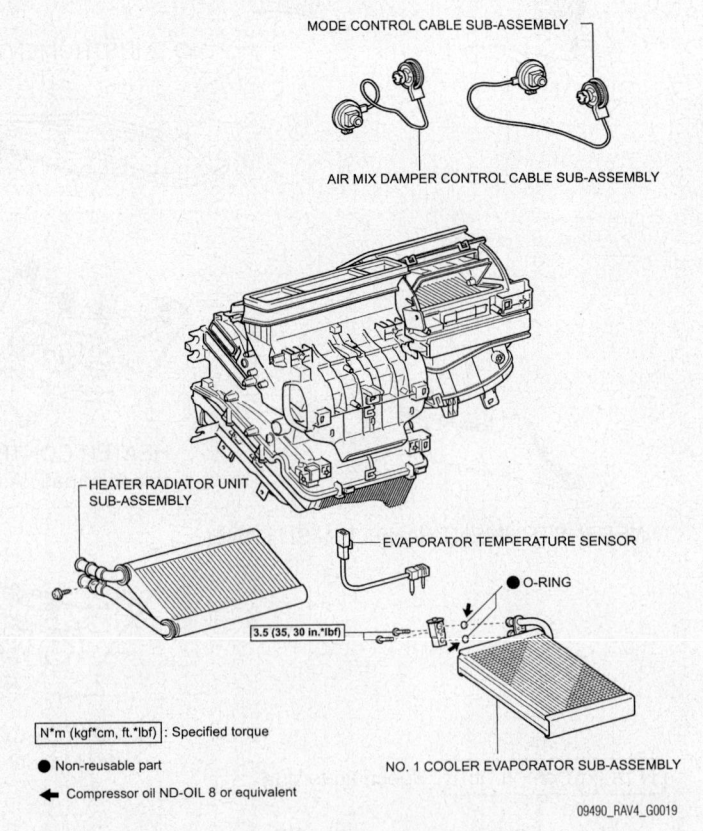

Manual air conditioning heater core and related components—2006

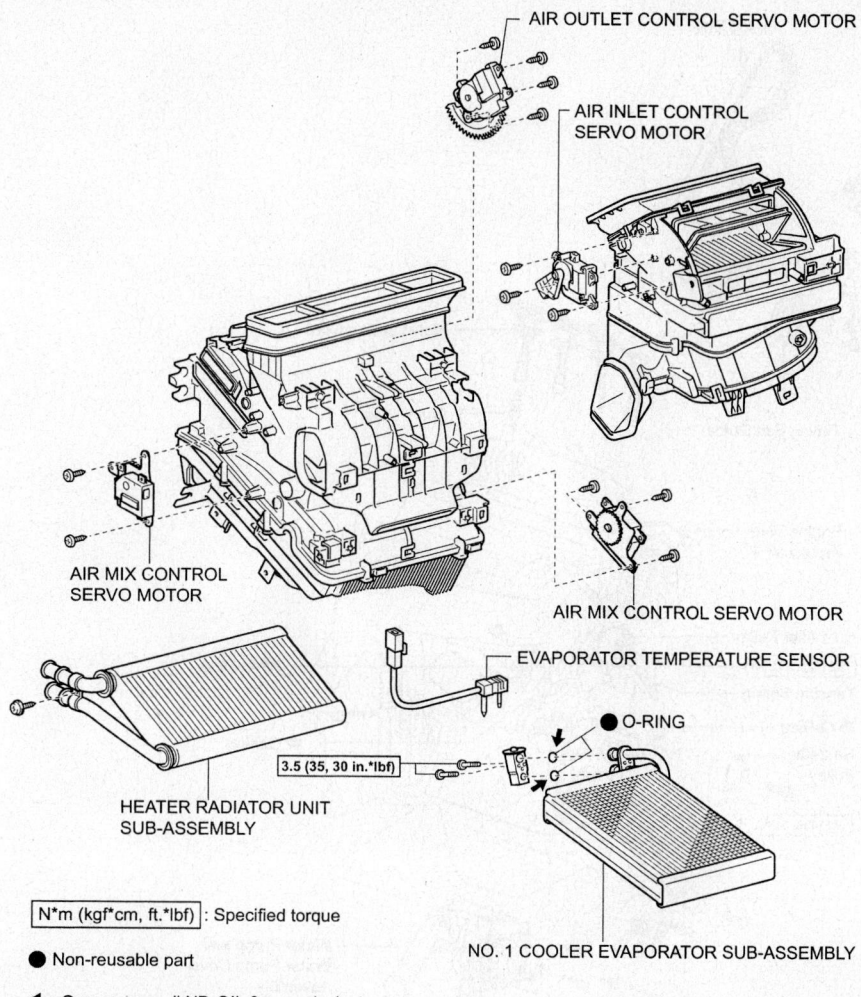

AIR OUTLET CONTROL SERVO MOTOR

AIR INLET CONTROL SERVO MOTOR

AIR MIX CONTROL SERVO MOTOR

AIR MIX CONTROL SERVO MOTOR

EVAPORATOR TEMPERATURE SENSOR

● O-RING

3.5 (35, 30 in.*lbf)

HEATER RADIATOR UNIT SUB-ASSEMBLY

NO. 1 COOLER EVAPORATOR SUB-ASSEMBLY

N*m (kgf*cm, ft.*lbf) : Specified torque

● Non-reusable part

◀ Compressor oil ND-OIL 8 or equivalent

09490_RAV4_G0020

Automatic air conditioning heater core and related components—2006

48. Remove the number three heater register duct. Remove the air duct. Remove the air outlet control servo motor. Remove the air mix control servo motor.

49. If equipped with manual air conditioning remove the mode control cable and the airmix damper control cable.

50. Remove the heater core from its mounting.

To install:

51. Installation is the reverse of the removal procedure.

52. Tighten the steering column retaining bolts to 18 ft. lbs.

53. Tighten the sliding yoke to number intermediate shaft retaining bolt to 26 ft. lbs.

54. Tighten the sliding yoke to steering gear retaining bolt to 26 ft. lbs.

55. To center the spiral cable, check that the front wheels are in the straight ahead position. Turn the cable counterclockwise by hand until it becomes hard to turn. Then rotate the cable clockwise about 2.5 turns to align the marks.

➡**The cable will rotate about 2.5 turns to either left or right of the center.**

56. Tighten the steering wheel locknut to 37 ft. lbs.

57. Tighten the wheel pad retaining screws to 78 inch lbs.

58. Check the steering wheel center point.

59. Recharge the air conditioning system.

60. Fill the cooling system with the proper grade and type engine coolant.

61. Start the engine and check for leaks, correct as required.

Water Pump

REMOVAL & INSTALLATION

2.0L Engine

1. Before servicing the vehicle, refer to the Precautions Section.

2. Remove or disconnect the following:
- Negative battery cable
- Right side engine undercover

3. Drain the engine coolant from the radiator and engine.

4. Remove or disconnect the following:

- Drive belt
- Lower radiator hose from the water inlet
- Drive belt tension spring and the No. 2 idler pulley
- Crankshaft Position (CKP) sensor connector clamp
- Alternator drive belt adjusting bar
- 2 water pump-to-water bypass pipe nuts
- 3 water pump bolts in the sequence
- Water pump cover from the water bypass pipe
- Water pump and water pump cover assembly
- Gasket and 2 O-rings from the water pump and water bypass pipe
- 3 bolts, water pump and gasket, from the water pump cover

To install:

5. Install or connect the following:
- Water pump to the water pump cover, using a new gasket. Tighten the 3 bolts to 78 inch lbs. (9 Nm).
- New O-ring and gasket to the water pump cover
- New O-ring to the water bypass pipe, by applying soapy water to the O-ring
- Water pump cover to the water bypass pipe; do not install the nuts at this time
- Water pump. Tighten the 3 bolts, in sequence, to 78 inch lbs. (9 Nm).
- Water pump cover to the water pump pipe. Tighten the 2 bolts to 82 inch lbs. (9 Nm).
- Alternator drive belt adjusting bar. Tighten the bolt to 20 ft. lbs. (27 Nm).
- CKP connector clamp
- No. 2 idler pulley and drive belt tension spring
- Lower radiator hose
- Drive belt
- Negative battery cable

6. Fill the engine and radiator with engine coolant.

7. Start the engine and check for leaks.

8. Install the right side engine undercover.

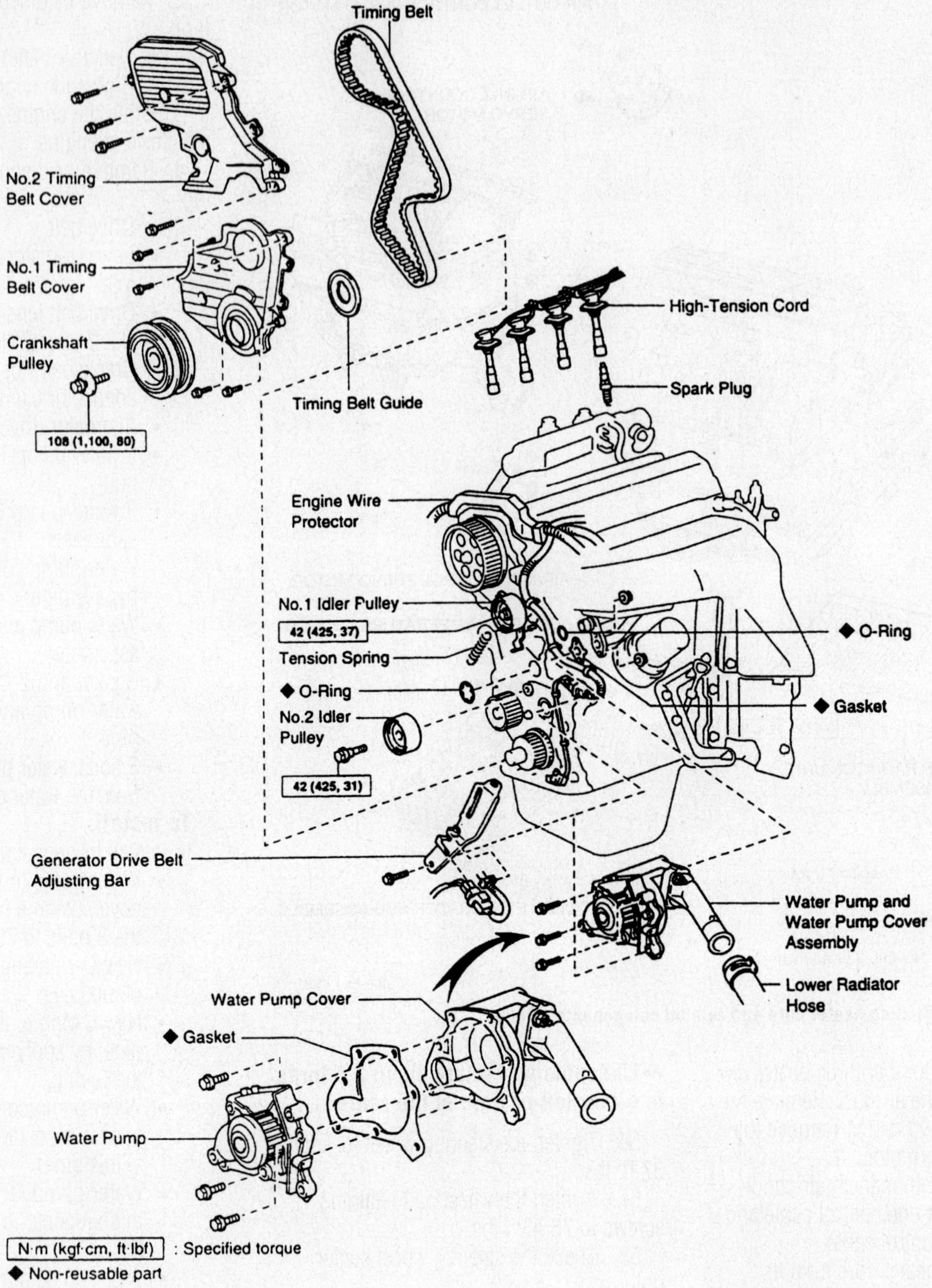

Timing Belt

No.2 Timing Belt Cover

No.1 Timing Belt Cover

Crankshaft Pulley

Timing Belt Guide

108 (1,100, 80)

High-Tension Cord

Spark Plug

Engine Wire Protector

No.1 Idler Pulley

42 (425, 37)

Tension Spring

◆ O-Ring

No.2 Idler Pulley

42 (425, 31)

◆ O-Ring

◆ Gasket

Generator Drive Belt Adjusting Bar

Water Pump and Water Pump Cover Assembly

Lower Radiator Hose

Water Pump Cover

◆ Gasket

Water Pump

N·m (kgf·cm, ft·lbf) : Specified torque
◆ Non-reusable part

7924ZG10

Water pump and related components—2.0L

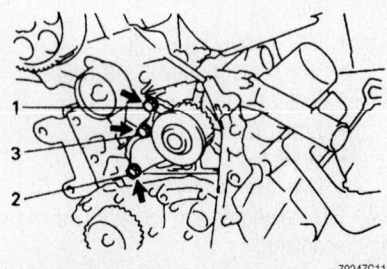

7924ZG11

Water pump bolt loosening sequences—2.0L engine

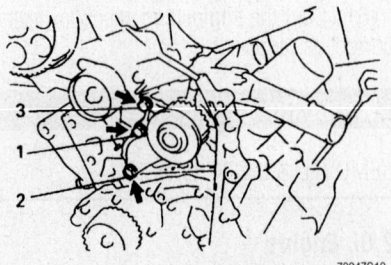

7924ZG12

Water pump bolt tightening sequence—2.0L engine

2.4L Engine

2002–2005

1. Before servicing the vehicle, refer to the Precautions Section.

2. Disconnect the negative battery cable.

➡**Wait at least 90 seconds after disconnecting the negative battery cable before starting any repair work to prevent air bag and seat belt pretensioner activation.**

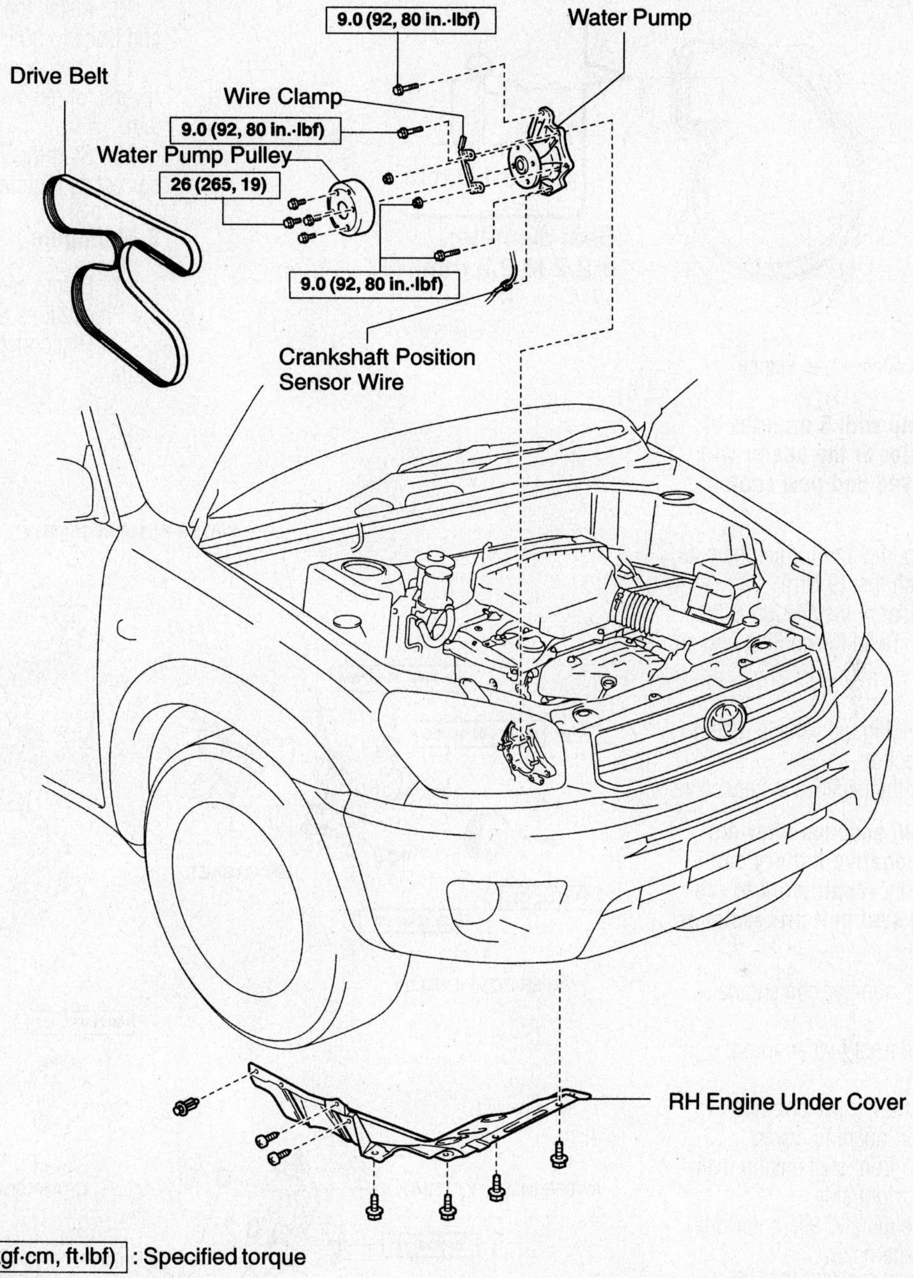

9.0 (92, 80 in.·lbf)   Water Pump

Drive Belt

Wire Clamp

9.0 (92, 80 in.·lbf)

Water Pump Pulley

26 (265, 19)

9.0 (92, 80 in.·lbf)

Crankshaft Position
Sensor Wire

RH Engine Under Cover

N·m (kgf·cm, ft·lbf) : Specified torque

67170-RAV4-G31

Water pump and related components—2.4L engine

3. Drain the cooling system.
4. Remove the accessory drive belt.
5. Remove the water pump pulley.
6. Remove the crankshaft position sensor wire and clamp.
7. Remove the 4 bolts and 2 nuts from the pump.
8. Using a small prybar, remove the water pump.
9. Remove all traces of the gasket material from the pump and block.

To install:

10. Apply a 2.5mm wide bead of RTV gasket material to the pump sealing surface as shown.

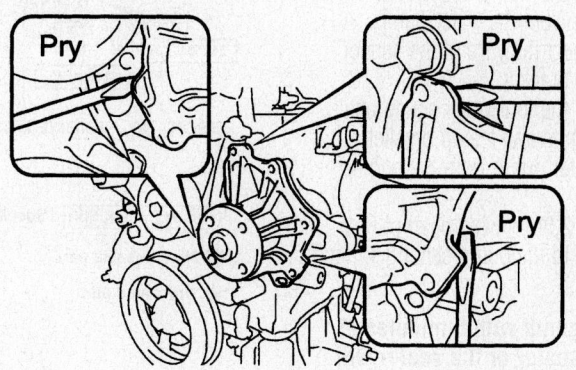

67170-RAV4-G32

Water pump removal points—2.4L engine

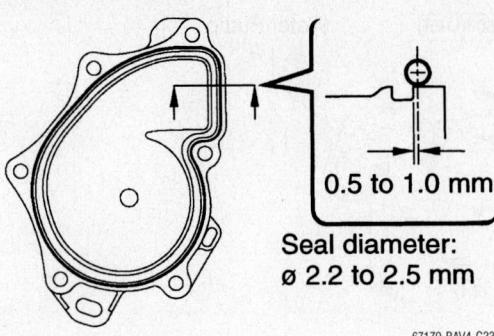

RTV gasket application—2.4L engine

0.5 to 1.0 mm

Seal diameter:
ø 2.2 to 2.5 mm

67170-RAV4-G33

➡ **Install the pump with 5 minutes of applying the sealer or the sealer will have to be removed and new sealer applied.**

11. Install the pump and torque the nuts and bolts to 80 inch lbs. (9 Nm).

12. The remainder of installation is the reverse of removal. Refill the cooling system.

2006

1. Before servicing the vehicle, refer to the Precautions Section.

2. Disconnect the negative battery cable.

➡ **Wait at least 90 seconds after disconnecting the negative battery cable before starting any repair work to prevent air bag and seat belt pretensioner activation.**

3. Remove the number one engine undercover.

4. Remove the front fender apron, right side.

5. Drain the cooling system. Remove the radiator support opening cover.

6. Remove the front suspension member reinforcement, right side.

7. Remove the fan and alternator drive belt. Remove the alternator.

8. Using tool SST09960-10010 remove the four retaining bolts and the water pump pulley.

9. Remove the clamp of the crankshaft position sensor from the water pump.

10. Disconnect the wire of the sensor from the clamp bracket.

11. Remove the four water pump retaining bolts, two nuts and clamp bracket. Remove the water pump from the engine.

To install:

12. Apply a 2.5mm wide bead of RTV gasket material to the pump sealing surface as shown.

➡ **Install the pump with 5 minutes of applying the sealer or the sealer will have to be removed and new sealer applied.**

13. Install the pump and torque the nuts and bolts to 80 inch lbs. (9 Nm).

14. The remainder of installation is the reverse of removal. Refill the cooling system.

15. Start the vehicle and check for leaks, correct as required.

3.5L Engine

1. Before servicing the vehicle, refer to the Precautions Section.

2. Disconnect the negative battery cable.

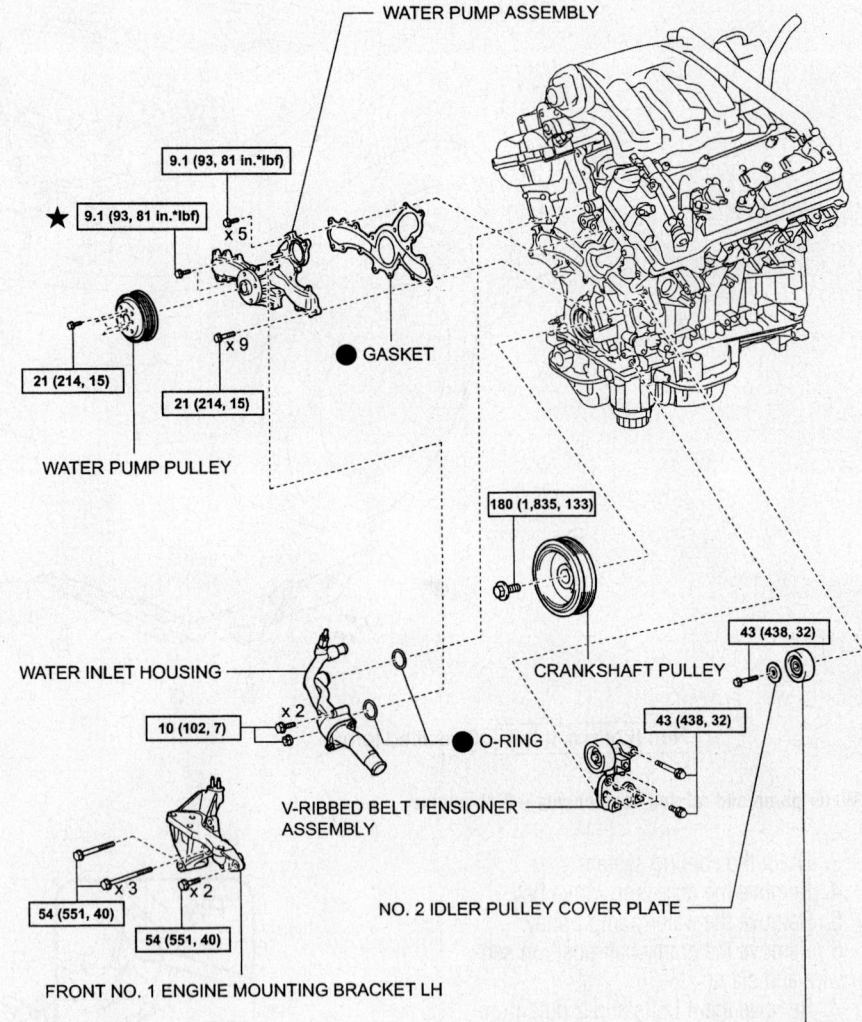

WATER PUMP ASSEMBLY

9.1 (93, 81 in.*lbf)

★ 9.1 (93, 81 in.*lbf)

x 5

GASKET

x 9

21 (214, 15)

21 (214, 15)

WATER PUMP PULLEY

180 (1,835, 133)

WATER INLET HOUSING

43 (438, 32)

CRANKSHAFT PULLEY

10 (102, 7)

x 2

O-RING

43 (438, 32)

V-RIBBED BELT TENSIONER ASSEMBLY

x 3

x 2

54 (551, 40)

54 (551, 40)

NO. 2 IDLER PULLEY COVER PLATE

FRONT NO. 1 ENGINE MOUNTING BRACKET LH

NO. 2 IDLER PULLEY SUB-ASSEMBLY

N*m (kgf*cm, ft.*lbf) : Specified torque

● Non-reusable part

★ Precoated part

Water pump and related components—3.5L engine

09490_RAV4_G0021

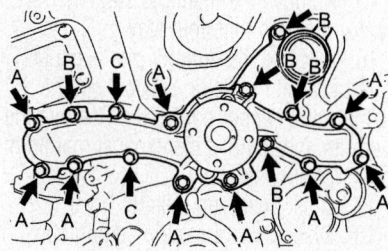

09490_RAV4_G0022

Water pump bolt tightening sequence and bolt markings—3.5L engine

➡**Wait at least 90 seconds after disconnecting the negative battery cable before starting any repair work to prevent air bag and seat belt pretensioner activation.**

3. Remove the engine from the vehicle and position it in a suitable holding fixture.

4. Remove the number one engine mounting bracket, left side.

5. Remove the water inlet housing. Remove the crankshaft pulley.

6. Remove the water pump pulley. Remove the number two idler pulley sub-assembly. Remove the belt tensioner assembly.

7. Remove the 16 water pump retaining bolts. Remove the water pump from its mounting.

To install:

8. Install the water pump to the engine using a new gasket.

9. Tighten the retaining bolts to 15 ft. lbs. for bolts marked "A" and 81 inch lbs for bolts marked "B" and "C" and in the proper sequence.

10. Continue the installation in the reverse order of the removal procedure.

Cylinder Head

REMOVAL & INSTALLATION

2.0L Engine

1. Before servicing the vehicle, refer to the Precautions Section.

2. Release the fuel system pressure.

3. Remove or disconnect the following:
- Negative battery cable
- Right side engine undercover

4. Drain the engine coolant.

5. Remove or disconnect the following:
- Camshafts
- Cylinder head bolts in several passes
- Cylinder head with the intake manifold

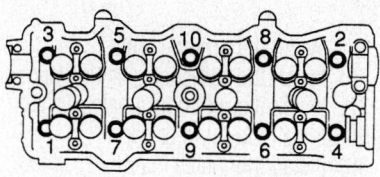

7924ZG17

Cylinder head bolt torque sequence—2.0L engine

- Air hose from the intake manifold
- 2 bolts and the air tube
- Intake manifold and gasket
- Air hose from the cylinder head port
- Air hose
- Fuel delivery pipe and the injectors
- Oil pressure switch

To install:

6. Install or connect the following:
- Oil pressure switch
- Fuel injectors and the delivery pipe
- Air hose to the cylinder head port
- Intake manifold with new gaskets. Tighten the nut/bolts to 14 ft. lbs. (19 Nm).
- Air tube with the 2 bolts
- Air hose to the intake manifold

7. Clean the gasket mating surfaces using care not to damage the aluminum components, replace the gasket; then, lower the cylinder head onto the engine. Be sure the dowel pins are aligned and no hoses or wires are between the head and cylinder block.

8. Tighten the cylinder head bolts in 3 progressive steps, as follows:

a. Apply a light coat of engine oil to the cylinder head bolts.

b. Tighten the cylinder head bolts, in sequence, to 15 ft. lbs. (26 Nm).

c. Tighten the cylinder head bolts, in sequence, to 30 ft. lbs. (52 Nm).

d. Mark the front of the cylinder head bolt with paint.

e. Retighten the cylinder head bolts by 90 degrees in sequence.

9. Install or connect the following:
- Intake and exhaust camshafts
- Negative battery cable

10. Refill the engine with coolant, start the engine, warm up and check for leaks.

11. Bleed the cooling system and top off coolant as necessary.

12. Install the right side engine undercover.

13. Check ignition timing and road test the vehicle for proper operation.

2.4L Engine

2002–2005

1. Before servicing the vehicle, refer to the Precautions Section.

2. Properly relieve the fuel system pressure.

3. Disconnect the negative battery cable.

➡**Wait at least 90 seconds after disconnecting the negative battery cable before starting any repair work to prevent air bag and seat belt pretensioner activation.**

4. Remove the right engine undercover.

5. Drain the coolant.

6. Drain the oil.

7. Remove the air cleaner assembly.

8. Remove the drive belt.

9. Remove the alternator.

10. Remove the power steering pump.

11. Remove the ignition coils.

12. Remove the spark plugs.

13. Remove the injectors.

14. Remove the exhaust manifold.

15. Remove the oil filler cap.

16. Remove the PCV hoses and valve.

17. Remove the intake manifold.

18. Remove all wires and harnesses connected to the head.

19. Remove the timing chain.

20. Remove the camshaft sprocket.

21. Remove the VVT sprocket.

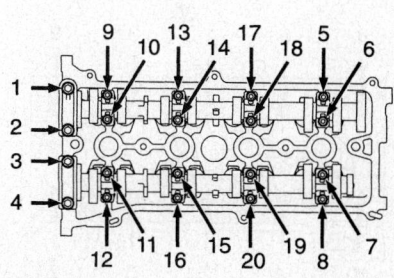

67170-RAV4-G21

Camshaft bolt removal sequence— 2002–2005 2.4L engine

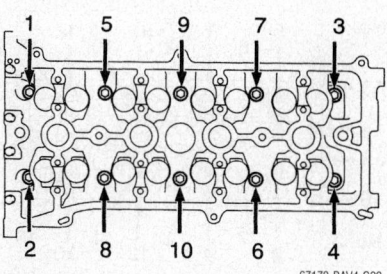

67170-RAV4-G22

Cylinder head bolt removal sequence— 2.4L engine

22. Remove the camshaft timing oil control valve.

23. Remove the camshafts.

24. Loosen the 10 head bolts, evenly and in several passes, in the sequence shown.

25. Remove the plate washers.

26. Remove the cylinder head. It may be necessary to pry it loose. Discard the gasket.

To install:

27. Thoroughly clean all gasket surfaces.

28. Install a new gasket with the identification number upwards.

29. Carefully install the head.

30. Apply a light coating of engine oil to the threads and under the head of each bolt. Install the bolts and tighten them evenly and in several passes, in the sequence shown, to 58 ft. lbs. (79 Nm).

31. Matchmark the head of each bolt and the cylinder head. Tighten each bolt in sequence an additional 90 degrees.

32. Install the camshafts.

33. Check and adjust the valve clearance.

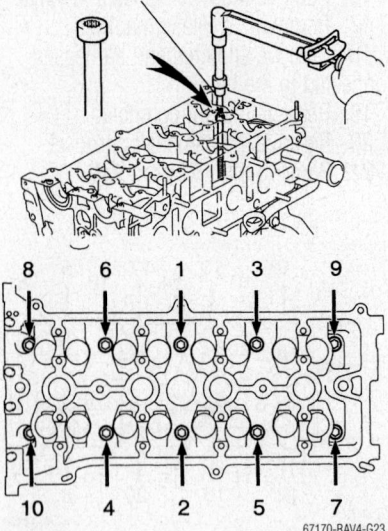

Cylinder head bolt torque sequence—2.4L engine

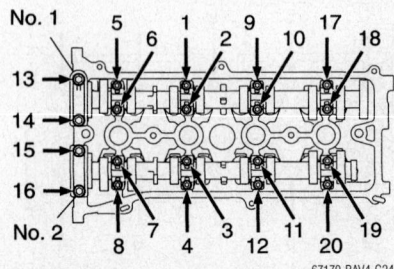

Camshaft bolt torque sequence—2.4L engine

34. Install the camshaft and VVT sprockets.

35. Install the timing chain.

36. Install the camshaft position sensor.

37. Install the oil control valve.

38. Install the oil filler cap.

39. Install the intake manifold.

40. Install the PCV valve and hoses.

41. Install the engine wiring.

42. Install the exhaust manifold.

43. Install the injectors.

44. Install the spark plugs.

45. Install the ignition coils.

46. Install the power steering pump.

47. Install the alternator.

48. Install the drive belt.

49. Install the air cleaner assembly.

50. Refill all fluids.

51. Install the undercover.

2006

1. Before servicing the vehicle, refer to the Precautions Section.

2. Properly relieve the fuel system pressure.

3. Disconnect the negative battery cable.

➡**Wait at least 90 seconds after disconnecting the negative battery cable before starting any repair work to prevent air bag and seat belt pretensioner activation.**

4. Remove the radiator support opening cover.

5. Remove the right front tire and wheel assembly.

6. Remove the number one engine undercover. Remove the front fender apron, right side. Remove the number one engine cover.

7. Drain the engine coolant. Drain the engine oil.

8. Remove the air cleaner assembly.

9. Disconnect the purge line hose from the throttle body. Disconnect the water bypass hoses from the throttle body. Disconnect the number one throttle body hose from the throttle body.

10. Disconnect the throttle position sensor and control motor connector. Disconnect the wire harness clamp. Disconnect the fuel tube from the clamp.

11. Remove the four throttle body retaining bolts and remove the throttle body assembly. Remove the gasket.

12. Remove the fuel delivery pipe subassembly.

13. Remove the intake manifold. Remove the intake manifold insulator.

14. Remove the front exhaust pipe. Remove the oil dipstick. Remove the oil dipstick guide.

15. Remove the manifold stay. Remove the number two manifold stay.

16. Remove the number one exhaust manifold heat insulator. Disconnect the air/fuel ratio sensor connector. Remove the five nuts and remove the exhaust manifold converter subassembly.

17. Disconnect the radiator hose.

18. Disconnect the radio setting condenser connector. Disconnect the engine oil pressure switch connector. Disconnect the engine coolant temperature sensor connector.

19. Disconnect the camshaft position sensor connector. Remove the bolt and ground cable.

20. Remove the front suspension member reinforcement, right side.

21. Remove the alternator. Remove the right side engine mounting insulator. Remove the idler pulley.

22. Remove the ignition coil assembly. Remove the spark plugs.

23. Remove the cylinder head cover retaining bolts. Remove the cylinder head cover from the engine.

24. To remove the belt tensioner assembly, lift the engine upward, using a transmission jack. Remove the bolt, nut and the belt tensioner assembly.

➡**Do not lift the engine more than necessary.**

25. Remove the crankshaft position sensor.

26. Remove the oil pan.

27. Turn the crankshaft pulley until its groove and the timing mark "0" of the timing chain cover are aligned. Check that each timing mark of the camshaft timing gear and sprocket is aligned with each timing mark located on the NO. 1 and No. 2 bearing caps, as shown in the illustration.

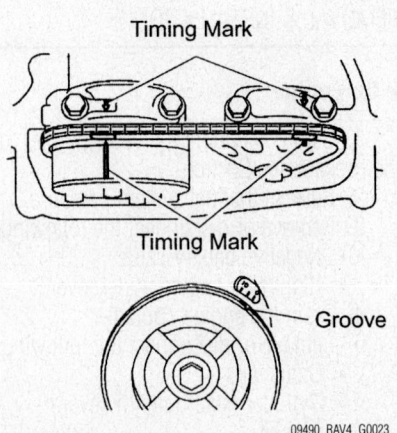

Camshaft gear and sprocket alignment—2006 2.4L engine

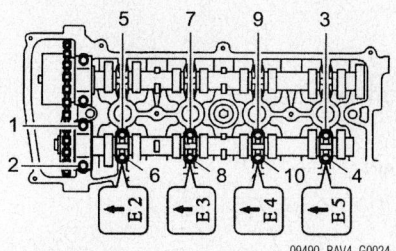

Camshaft number two bolt removal sequence—2006 2.4L engine

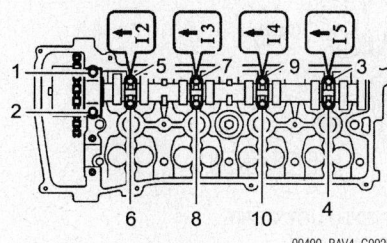

Camshaft number one bolt removal sequence—2006 2.4L engine

➡**If not rotate the engine 360 degrees to align the timing marks.**

28. To remove the number two camshaft, paint matchmarks on the chain in alignment with the timing marks on the camshaft timing gear and camshaft timing sprocket.

29. Remove the two nuts, tensioner and gasket. While holding the camshaft with a wrench, loosen the camshaft timing set bolt.

30. Using several steps, uniformly loosen and remove the ten bearing cap bolts in the proper sequence. Remove the five bearing caps.

31. While holding the number two camshaft, remove the camshaft timing sprocket set bolt.

➡**Remove the camshaft timing sprocket from the camshaft with the timing chain wrapped on the sprocket.**

32. Remove the camshaft timing sprocket from the timing chain.

33. To remove the number one camshaft, using several steps uniformly loosen and remove the ten bearing cap bolts in the proper sequence. Remove the five bearing caps.

34. Remove the camshaft and camshaft timing gear while holding the timing chain with your hand.

35. Tie the timing chain with a string to the side of the engine block.

➡**Be careful not to drop anything inside the timing chain cover.**

36. Remove the crankshaft pulley. Remove the engine mounting bracket, right side.

37. Remove the timing chain cover. Remove the number one crankshaft position sensor plate. Remove the timing chain guide.

38. Remove the chain tensioner slipper. Remove the number one chain vibration damper. Remove the chain subassembly.

39. Remove the camshaft timing oil control valve assembly.

40. Using several steps, uniformly loosen and remove the ten cylinder head bolts in the proper sequence.

➡**Head warpage or cracking could result from removing the bolts in the wrong order.**

41. Carefully remove the cylinder head from the engine. Remove and discard the head gasket.

To install:

42. Thoroughly clean all gasket surfaces.

43. Install a new gasket with the identification lot number upwards.

44. Carefully install the head.

45. Apply a light coating of engine oil to the threads and under the head of each bolt. Install the bolts and tighten them evenly and in several passes, in the sequence shown, to 52 ft. lbs. (70 Nm).

46. Matchmark the head of each bolt and the cylinder head. Tighten each bolt in sequence an additional 90 degrees. Check that the paint mark is now at a 90 degree angle to the front.

47. Install the camshafts. Tighten the retaining bolts to 22 ft. lbs. for No.1 and No.2 bearing cap, and to 80 inch lbs. for the No. 3 bearing cap.

➡**Be sure to install the camshafts with the timing mark of the camshaft timing gear on top.**

48. When installing the throttle body be sure to use a new gasket. Tighten the retaining bolts to 22 ft. lbs.

49. When installing the cylinder head cover, apply seal packing (three bond 1207B, or equivalent) to the two locations, shown in the illustration. Tighten the retaining bolts to 8 ft. lbs. for bolt "A", 10 ft. lbs. for bolt "B" and 8 ft. lbs. for the nut.

50. Continue the installation in the reverse order of the removal procedure.

51. Be sure to fill the cooling system with the proper grade and type engine coolant.

52. Be sure to fill the engine with the proper grade and type engine oil.

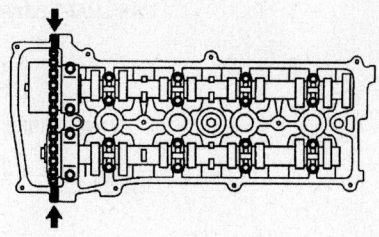

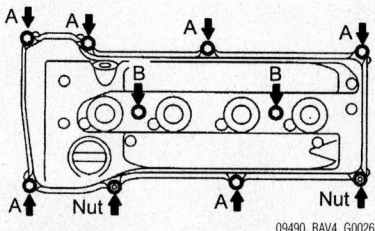

Cylinder head cover sealant location and bolt identification—2006 2.4L engine

53. Start the engine and check for leaks, correct as required.

3.5L Engine

➡**The following procedure is performed with the engine removed from the vehicle.**

1. Before servicing the vehicle, refer to the Precautions Section.

2. Properly relieve the fuel system pressure.

3. Disconnect the negative battery cable.

➡**Wait at least 90 seconds after disconnecting the negative battery cable before starting any repair work to prevent air bag and seat belt pretensioner activation.**

4. Remove the engine and position it in a suitable holding fixture.

5. Remove the oil filler cap subassembly. Remove the spark plugs.

6. Remove the camshaft timing control valve assembly.

7. Remove the VVT sensor. Remove the water inlet. Remove the oil pipe.

8. Remove the twelve retaining bolts and remove the cylinder head cover for bank one.

➡**Upon installation be sure that removed components are reinstalled in their original locations.**

9. Remove the number one oil pipe.

10. Remove the twelve retaining bolts and remove the cylinder head cover for bank two.

➡**Upon installation be sure that removed components are reinstalled in their original locations.**

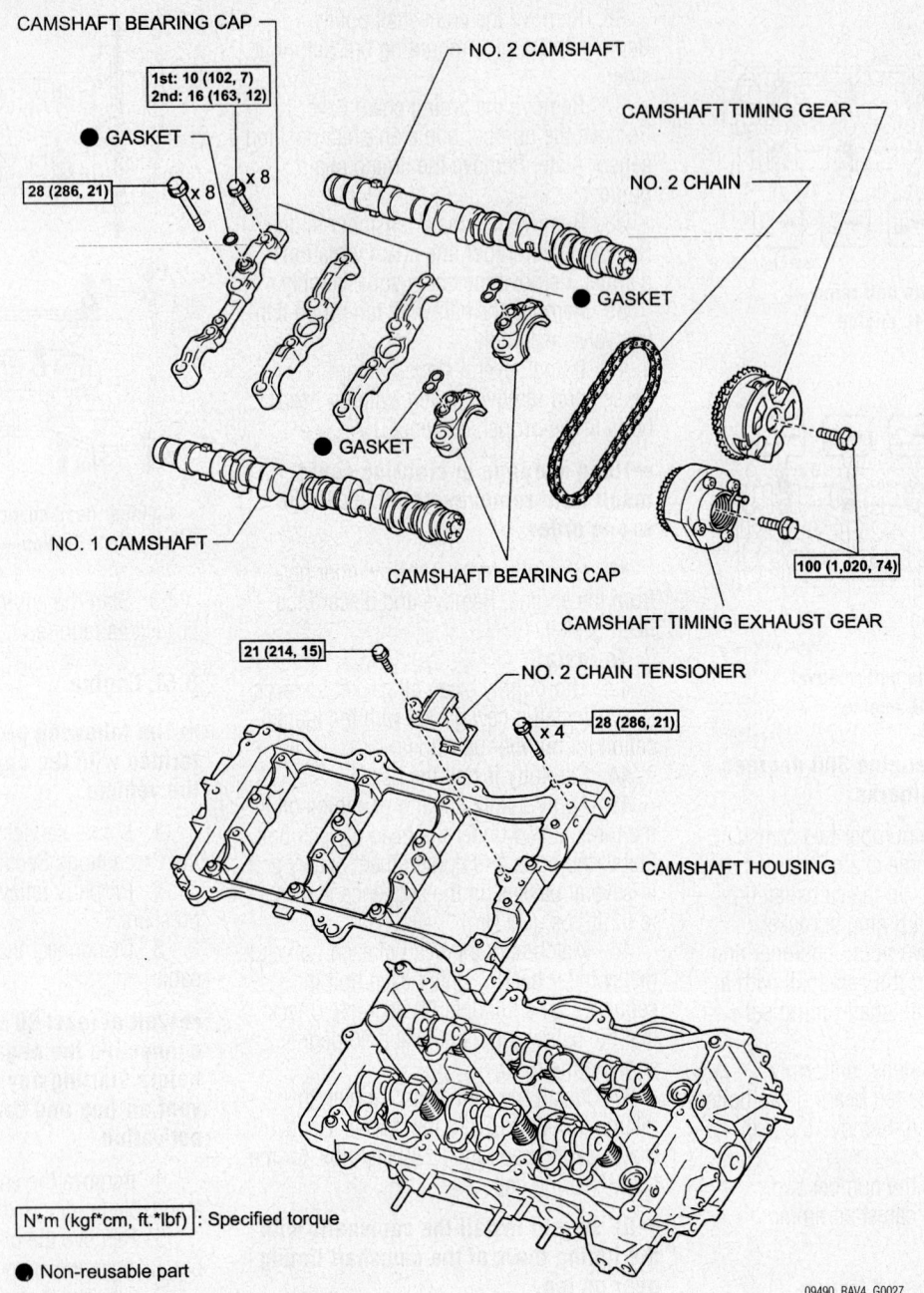

CAMSHAFT BEARING CAP

1st: 10 (102, 7)
2nd: 16 (163, 12)

● GASKET

28 (286, 21) x 8

x 8

NO. 2 CAMSHAFT

CAMSHAFT TIMING GEAR

NO. 2 CHAIN

● GASKET

NO. 1 CAMSHAFT

CAMSHAFT BEARING CAP

● GASKET

CAMSHAFT TIMING EXHAUST GEAR

100 (1,020, 74)

21 (214, 15)

NO. 2 CHAIN TENSIONER

28 (286, 21) x 4

CAMSHAFT HOUSING

N*m (kgf*cm, ft.*lbf): Specified torque

● Non-reusable part

09490_RAV4_G0027

Cylinder head and related components (bank one)—3.5L engine

11. Remove the number one engine mounting bracket. Remove the water inlet housing.

12. Remove the oil filter cap assembly. Remove the oil filter.

13. Remove the number two oil pan. Remove the oil strainer assembly. Remove the number one oil pan. Remove the seven retaining bolts and remove the oil pan baffle plate.

14. Remove the crankshaft pulley. Remove the water pump.

15. Remove the timing chain.

16. Remove the camshafts.

17. Remove the camshaft housing sub-assembly by prying between the cylinder head and camshaft housing with a suitable pry tool.

➡**Be careful not to damage the contact surfaces of the cylinder head and the camshaft housing.**

18. Remove the valve rocker arms. Remove the valve lash adjusters.

➡**Be sure to keep parts arranged in a logical order for reinstallation in their exact locations.**

19. Uniformly loosen the eight cylinder head retaining bolts for bank one in the proper sequence. Remove the bolts and washers. Remove the cylinder head from the engine and discard the gasket.

➡**Be careful not to drop the washers into the cylinder head. Head warpage or cracking could occur from removing the bolts in an incorrect order. Be sure to keep all components labeled and in the correct order for reinstallation**

20. To remove the cylinder head for bank two, first, uniformly loosen and remove the two bolts, as shown in the illustration. Than, uniformly loosen the eight cylinder head retaining bolts in the proper sequence.

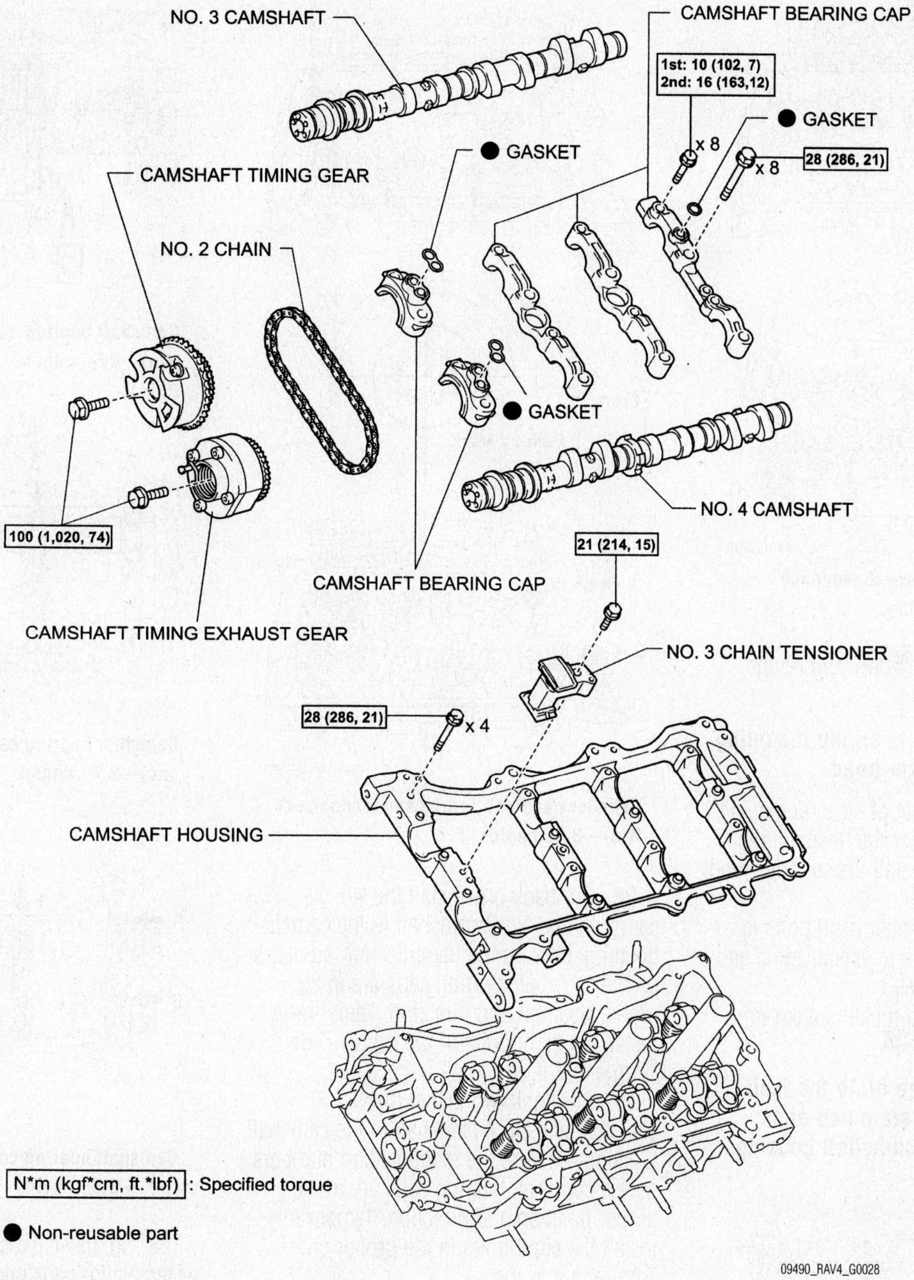

NO. 3 CAMSHAFT

CAMSHAFT BEARING CAP

1st: 10 (102, 7)
2nd: 16 (163,12)

× 8

● GASKET

28 (286, 21)

× 8

CAMSHAFT TIMING GEAR

● GASKET

NO. 2 CHAIN

100 (1,020, 74)

● GASKET

NO. 4 CAMSHAFT

CAMSHAFT TIMING EXHAUST GEAR

CAMSHAFT BEARING CAP

21 (214, 15)

NO. 3 CHAIN TENSIONER

28 (286, 21)   × 4

CAMSHAFT HOUSING

N*m (kgf*cm, ft.*lbf) : Specified torque

● Non-reusable part

09490_RAV4_G0028

Cylinder head and related components (bank two)—3.5L engine

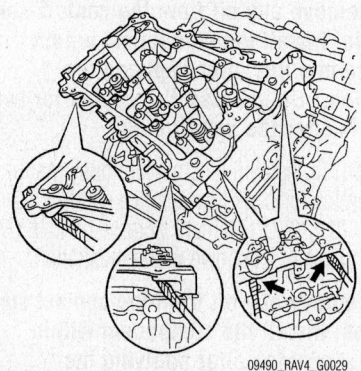

09490_RAV4_G0029

Camshaft housing subassembly removal points—3.5L engine

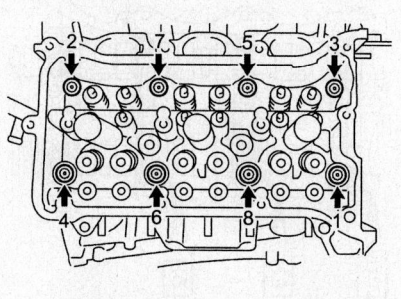

09490_RAV4_G0030

Cylinder head bolt removal sequence (bank one)—3.5L engine

Remove the bolts and washers. Remove the cylinder head from the engine and discard the gasket.

➡ Be careful not to drop the washers into the cylinder head. Head warpage or cracking could occur from removing the bolts in an incorrect order. Be sure to keep all components labeled and in the correct order for reinstallation

To install:

21. Position a new cylinder head gasket on the cylinder block. Be sure that the front face of the lot number, stamped on the gasket, is facing upward.

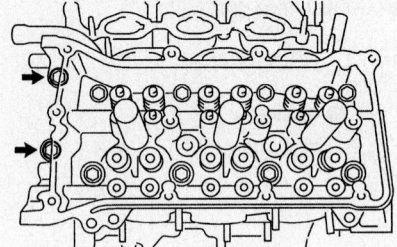

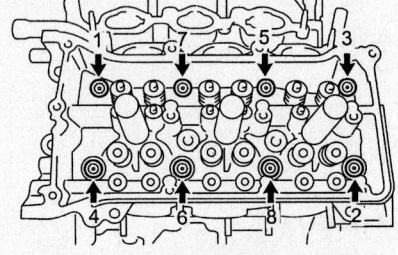

Cylinder head bolt removal sequence (bank two)—3.5L engine

22. Position the cylinder head on the cylinder block.

→**Ensure that no oil is on the mounting surface of the cylinder head.**

23. Apply a thin coat of clean engine oil to the threads and under the heads of the cylinder head bolts. Install the cylinder head bolts.

24. Tighten the cylinder head bolts in three progressive steps to specification and in the proper sequence.

25. Install the valve rocker subassemblies.

→**Apply clean engine oil to the lash adjuster tips, valve stem cap ends, camshaft journals, camshaft housings and bearing caps.**

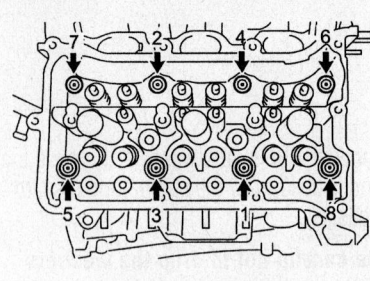

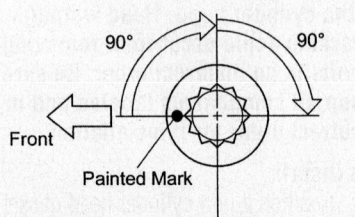

Cylinder head bolt torque sequence (bank one)—3.5L engine

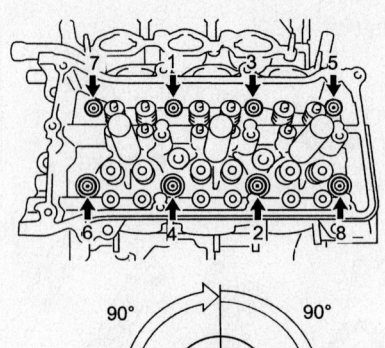

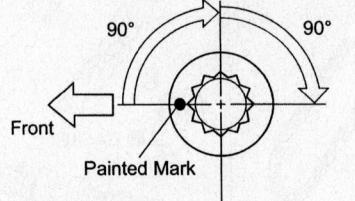

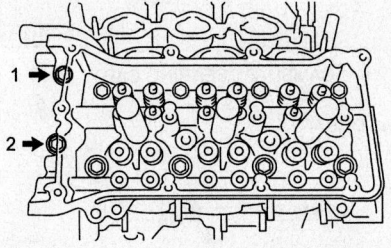

Cylinder head bolt torque sequence (bank two)—3.5L engine

26. For bank one install the No. 1 camshaft and No 2 camshaft to the camshaft housing. Be sure the identification numbers on the camshaft bearing caps are in the proper position and direction. Temporarily install the eight bolts in the proper order. Torque to 7 ft. lbs.

27. For bank two install the No. 3 camshaft and No 4 camshaft to the camshaft housing. Be sure the identification numbers on the camshaft bearing caps are in the proper position and direction. Temporarily install the eight bolts in the proper order. Torque to 7 ft. lbs.

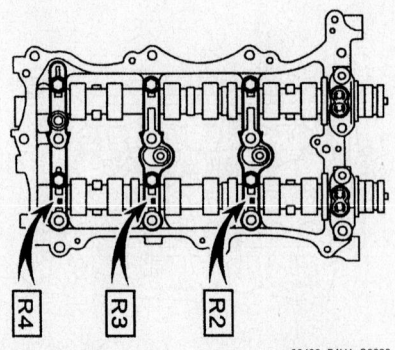

Camshaft bearing cap identification (bank one)—3.5L engine

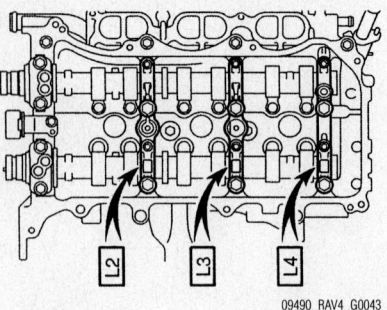

Camshaft bearing cap identification (bank two)—3.5L engine

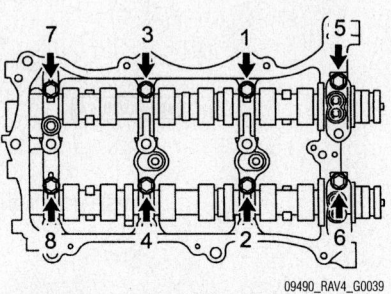

Camshaft bearing cap bolt location (bank one)—3.5L engine

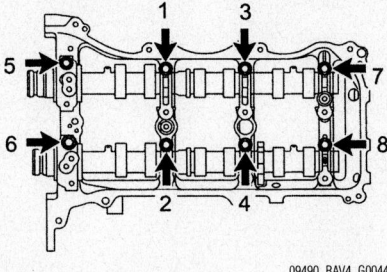

Camshaft bearing cap bolt location (bank two)—3.5L engine

28. To install the camshaft housing subassembly for bank one, apply a continuous bead (0.138–0.177 inch) of three bond 1207B or equivalent as shown in the illustration.

→**Remove any oil from the contact surfaces. Install the component within three minutes after applying the sealant. Do not start the engine for two hours after the installation.**

29. To install the camshaft housing subassembly for bank two, apply a continuous bead (0.138–0.177 inch) of three bond 1207B or equivalent as shown in the illustration.

→**Remove any oil from the contact surfaces. Install the component within three minutes after applying the sealant. Do not start the engine for two hours after the installation.**

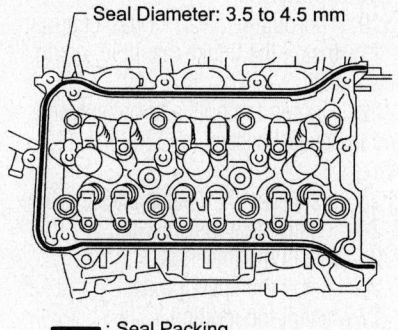

— : Seal Packing

09490_RAV4_G0040

Sealant application (bank one)—3.5L engine

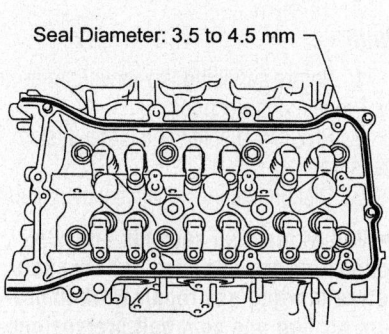

— : Seal Packing

09490_RAV4_G0045

Sealant application (bank two)—3.5L engine

30. Install the camshaft housing and tighten the bolts to 21 ft. lbs and in the proper sequence.

➡**Make sure that the knock pin of the camshaft is positioned as shown in the illustration, before installing the housing.**

31. Tighten the eight bolts to 12 ft. lbs. and in the proper sequence. Be sure to clean any seal packing. Install three new gaskets.

32. Continue the installation in the reverse order of the removal procedure.

33. When installing the cylinder head cover, apply seal packing, three bond 1207B or equivalent, as shown in the illustration.

➡**Remove any oil from the contact surfaces. Install the component within three minutes after applying the sealant. Do not start the engine for two hours after the installation.**

34. Install the cover retaining bolts and torque to 15 ft. lbs for bolt "A" and to 7 ft. lbs. for all other bolts and in the proper sequence.

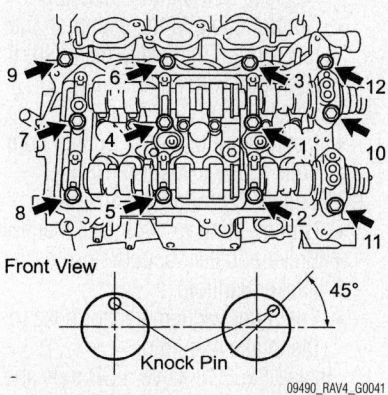

Front View

09490_RAV4_G0041

Camshaft housing bolt torque sequence (bank one)—3.5L engine

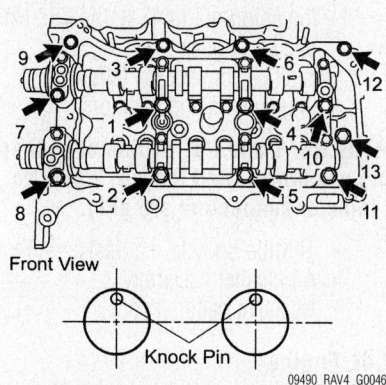

Front View

09490_RAV4_G0046

Camshaft housing bolt torque sequence (bank two)—3.5L engine

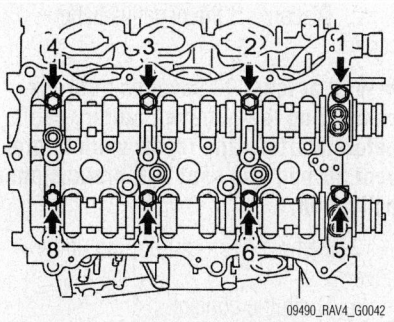

09490_RAV4_G0042

Camshaft bearing cap bolt location (bank one)—3.5L engine

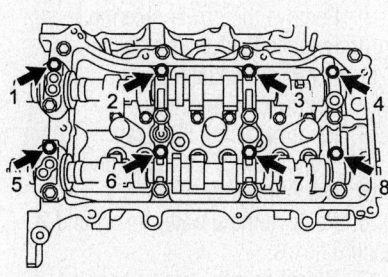

09490_RAV4_G0047

Camshaft bearing cap bolt location (bank two)—3.5L engine

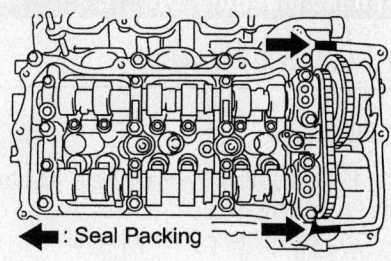

: Seal Packing

09490_RAV4_G0048

Cylinder head cover seal packing application (bank one)—3.5L engine

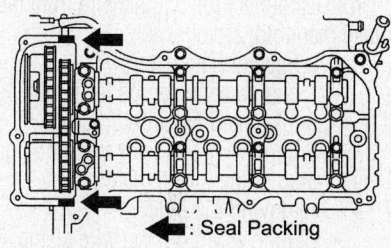

: Seal Packing

09490_RAV4_G0049

Cylinder head cover seal packing application (bank two)—3.5L engine

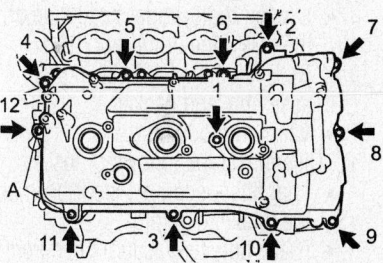

09490_RAV4_G0050

Cylinder head cover bolt torque sequence (bank one)—3.5L engine

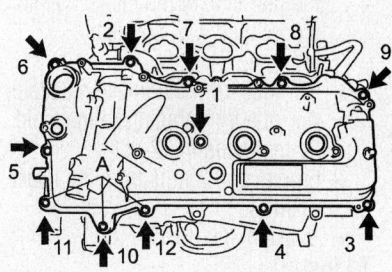

09490_RAV4_G0051

Cylinder head cover bolt torque sequence (bank two)—3.5L engine

Intake Manifold

REMOVAL & INSTALLATION

2.0L Engine

1. Before servicing the vehicle, refer to the Precautions Section.
2. Properly relieve the fuel system pressure.
3. Remove or disconnect the following:
   - Negative battery cable
   - Air cleaner assembly
   - Throttle body from the intake manifold
4. Disconnect the engine wire from the intake manifold, as follows:
   - 4 injector connectors
   - 2 engine wire clamps from the intake manifold wire brackets
   - Engine wire protector from the right side of the intake manifold by removing the bolt
   - Engine wire from the wire clamp
5. Remove the EGR valve, EGR pipe and modulator, as follows:
   - Both vacuum hoses from the Exhaust Gas Recirculation (EGR) Vacuum Switching Valve (VSV)
   - Vacuum modulator from the clamp on the intake manifold
   - Loosen the cylinder head side of the EGR pipe union nut
   - Both nuts, the EGR valve, pipe assembly and gasket
   - Vacuum modulator
6. Disconnect the following hoses:
   - Fuel filter vacuum sensor hose on the intake manifold
   - Brake booster vacuum hose from the intake manifold
   - Ground strap from the intake manifold
7. Remove or disconnect the following:
   - Intake manifold stay by removing the 2 bolts
   - Control cable from the clamp on the rear side of the intake manifold, if equipped with automatic transaxle
   - Air hose from the intake manifold
   - Air tube from the intake manifold, by removing the 2 bolts
   - 6 bolts and 2 nuts from the intake manifold
   - Intake manifold

To install:

8. Install or connect the following:
   - Intake manifold. Tighten the 6 bolts and 2 nuts to 14 ft. lbs. (19 Nm).
   - Air tube with the 2 bolts

   - Air hose to the intake manifold
   - Control cable to the clamp on the rear side of the intake manifold, if equipped with an automatic transaxle
   - Intake manifold stay. Tighten both bolts to 31 ft. lbs. (42 Nm).
9. Connect the following hoses:
   - Ground strap to the intake manifold
   - Brake booster vacuum hose to the intake manifold
   - Fuel filter vacuum sensor hose to the intake manifold
10. Install the EGR valve, EGR pipe and the vacuum modulator, as follows:
    - Vacuum modulator
    - EGR valve and pipe. Tighten both nuts to 108 inch lbs. (13 Nm) and the union nut to 43 ft. lbs. (59 Nm).
    - Vacuum hoses
11. Install or connect the following:
    - Engine wire and injectors

➡ **The No. 1 and No. 3 injector connectors are brown, and the No. 2 and No. 4 injector connectors are gray.**

   - Throttle body to the intake manifold
   - Air cleaner assembly
   - Negative battery cable

2.4L Engine

2002–2005

1. Before servicing the vehicle, refer to the Precautions Section.
2. Disconnect the negative battery cable.

➡ **Wait at least 90 seconds after disconnecting the negative battery cable before starting any repair work to prevent air bag and seat belt pretensioner activation.**

3. Remove the right engine undercover.
4. Drain the coolant.
5. Drain the oil.
6. Remove the air cleaner assembly.
7. Remove the drive belt.
8. Remove the alternator.
9. Remove the power steering pump.
10. Remove the ignition coils.
11. Remove the spark plugs.
12. Remove the injectors.
13. Remove the oil filler cap.
14. Remove the PCV hoses and valve.
15. Disconnect the TPS.
16. Remove the 2 water hoses and 2 vacuum hoses.
17. Disconnect the wiring harness.
18. Remove the 5 bolts and 2 nuts.
19. Remove the intake manifold.

To install:

20. Thoroughly clean all gasket surfaces.
21. Install the intake manifold, using a new gasket.
22. Tighten the nuts and bolts evenly and in several passes, to 22 ft. lbs. (30 Nm).
23. Install the PCV valve and hoses.
24. Install the engine wiring.
25. Install the injectors.
26. Install the spark plugs.
27. Install the ignition coils.
28. Install the power steering pump.
29. Install the alternator.
30. Install the drive belt.
31. Install the air cleaner assembly.
32. Refill all fluids.
33. Install the under-cover.

2006

1. Before servicing the vehicle, refer to the Precautions Section.
2. Properly discharge the fuel system pressure.
3. Disconnect the negative battery cable.

➡ **Wait at least 90 seconds after disconnecting the negative battery cable before starting any repair work to prevent air bag and seat belt pretensioner activation.**

4. Remove the right engine undercover.
5. Drain the engine coolant.
6. Remove the air cleaner cap.
7. Remove the throttle body assembly.
8. Remove the fuel delivery pipe. Remove the fuel tube.
9. Remove the heater water inlet hose. Remove the heater water outlet hose.
10. Disconnect the union to check valve hose from the power brake booster.
11. Disconnect the camshaft timing oil control valve connector. Remove the wire harness clamp.
12. Remove the union to check valve hose from the vacuum hose clamp.
13. Remove the five intake manifold retaining bolts. Remove the intake manifold from the engine. Discard the gasket.

To install:

14. Thoroughly clean all gasket surfaces.
15. Install the intake manifold, using a new gasket.
16. Tighten the nuts and bolts evenly and in several passes, to 22 ft. lbs. (30 Nm).
17. Continue the installation in the reverse order of the removal procedure.
18. Be sure to fill the engine with the proper grade and type engine coolant.
19. Start the engine and check for leaks, correct as required.

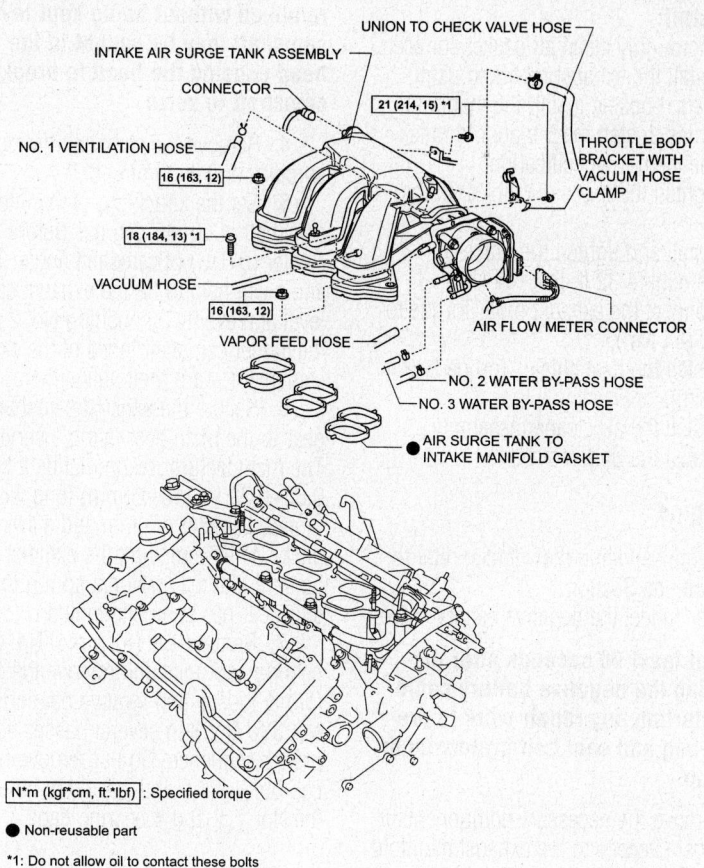

Intake manifold and related components—3.5L engine

3.5L Engine

➡**The following procedure is performed with the engine removed from the vehicle.**

1. Before servicing the vehicle, refer to the Precautions Section.

2. Disconnect the negative battery cable.

➡**Wait at least 90 seconds after disconnecting the negative battery cable before starting any repair work to prevent air bag and seat belt pretensioner activation.**

3. Remove the engine from the vehicle and position it in a suitable holding fixture.

4. Disconnect the two water by pass hoses from the throttle body.

5. Disconnect the vapor feed hose. Disconnect the throttle body connector and clamp.

6. Disconnect the number two ventilation hose. Disconnect the union to check valve hose.

7. Remove the bolt and the vacuum hose clamp. Disconnect the connector.

8. Remove the intake air surge tank

retaining bolts. Remove the two nuts, two bolts and the assembly from the engine. Discard the gaskets.

9. Remove the fuel rail assembly.

10. Remove the intake manifold retaining bolts. Remove the intake manifold from the engine.

To install:

11. Thoroughly clean all gasket surfaces.

12. Install the intake manifold, using a new gasket.

13. Tighten the nuts and bolts evenly and in several passes, to specification.

14. Install the intake air surge tank assembly, using new gaskets. Tighten the retaining bolts to 15 ft. lbs. and the nuts to 12 ft. lbs.

➡**Do not allow oil to contact the retaining bolts and nuts.**

15. Continue the installation in the reverse order of the removal procedure.

Exhaust Manifold

REMOVAL & INSTALLATION

2.0L Engine

1. Before servicing the vehicle, refer to the Precautions Section.

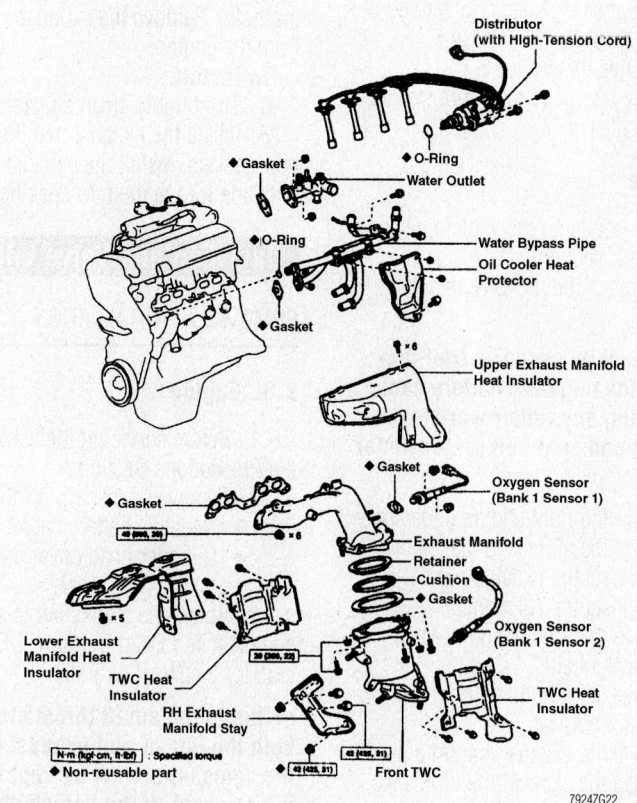

Exhaust manifold and components—2.0L engine

2. Remove or disconnect the following:
- Negative battery cable
- Front exhaust pipe from the exhaust manifold, using a 14mm deep socket wrench; discard the gasket
- Main Oxygen (O₂) sensor and the sub Oxygen (O₂) sensor connectors
- 6 bolts and the upper manifold heat insulator
- 2 right side exhaust manifold stay-to-cylinder block bolts
- 6 nuts, the exhaust manifold and the Three-Way Catalytic (TWC) converter assembly
- Exhaust manifold and front catalytic converter

To install:

3. Install or connect the following:
- Catalytic converter to the exhaust manifold. Tighten the nuts/bolts to 22 ft. lbs. (29 Nm).
- Exhaust manifold and the front TWC assembly. Tighten the 6 nuts, in several passes to 36 ft. lbs. (49 Nm).
- Right side manifold stay. Tighten both bolts to 31 ft. lbs. (42 Nm).
- Manifold upper heat insulator with the 6 bolts and attach the main Oxygen (O₂) and the sub Oxygen (O₂) sensor connectors
- Front exhaust pipe to the TWC, using a new gasket. Tighten the 3 nuts to 46 ft. lbs. (62 Nm).
- Negative battery cable

4. Start the engine and be sure that there are no exhaust leaks.

2.4L Engine

1. Before servicing the vehicle, refer to the Precautions Section.
2. Disconnect the negative battery cable.

➡**Wait at least 90 seconds after disconnecting the negative battery cable before starting any repair work to prevent air bag and seat belt pretensioner activation.**

3. Remove the right engine undercover.
4. Remove the air cleaner assembly.
5. Disconnect the exhaust pipe.
6. Disconnect the A/F sensor.
7. Remove the 3 bolts and nut and remove the heat shield.
8. Remove the 2 bolts holding the exhaust manifold braces.
9. Remove the exhaust manifold nuts.
10. Remove the exhaust manifold. Discard the gaskets.

To install:
11. Thoroughly clean all gasket surfaces.
12. Install the exhaust manifold using new gaskets. Loosely install the manifold-to-brace nuts. Install the 5 manifold nuts and torque them to specification.
13. Tighten the brace nuts to 32 ft. lbs. (44 Nm).
14. Install and tighten the brace-to-crankcase nuts to 32 ft. lbs. (44 Nm).
15. Connect the exhaust pipe. Torque to 32 ft. lbs. (44 Nm).
16. Install the heat shield. Torque to 9 ft. lbs. (12 Nm).
17. Install the air cleaner assembly.
18. Install the undercover.

3.5L Engine

1. Before servicing the vehicle, refer to the Precautions Section.
2. Disconnect the negative battery cable.

➡**Wait at least 90 seconds after disconnecting the negative battery cable before starting any repair work to prevent air bag and seat belt pretensioner activation.**

3. Remove the necessary components in order to gain access to the exhaust manifold retaining bolts.
4. Disconnect the front exhaust pipe retaining bolts.
5. Remove the exhaust manifold retaining bolts. Remove the exhaust manifold from the engine.

To install:
6. Thoroughly clean all gasket surfaces.
7. Install the exhaust manifold using new gaskets. Install the exhaust manifold nuts and torque them to specification.

Camshaft and Valve Lifters

REMOVAL & INSTALLATION

2.0L Engine

1. Before servicing the vehicle, refer to the Precautions Section.
2. Remove or disconnect the following:
- Negative battery cable
- Cylinder head cover and the upper timing belt cover

3. Rotate the crankshaft to set the engine at Top Dead Center (TDC)/compression for the No. 1 cylinder.

➡**Due to the small thrust clearance on both the intake and exhaust camshafts, the camshafts must be kept level during removal. If the camshafts are**

removed without being kept level, the camshaft may be caught in the cylinder head causing the head to break or the camshaft to seize.

4. Remove the camshaft timing sprocket and the timing belt.
5. Set the knock pin of the intake camshaft at 10–45 degrees Before Top Dead Center (BTDC) of camshaft angle. This angle will help to lift the exhaust camshaft level and evenly by pushing No. 2 and No. 4 cylinder camshaft lobes of the exhaust camshaft toward their valve lifters.
6. Secure the exhaust camshaft sub-gear to the main gear using a service bolt. The manufacturer recommends a bolt 0.63–0.79 in. (16–20mm) long with a thread diameter of 6mm and a 1mm thread pitch. When removing the exhaust camshaft, be sure that the torsional spring force of the sub-gear has been eliminated.
7. Remove the No. 1 and No. 2 rear bearing cap bolts and remove the cap. Uniformly loosen and remove bearing cap bolts No. 3 to No. 8 in several passes and in the proper sequence. Do not remove bearing cap bolts No. 9 and 10 at this time. Remove the No. 1, 2 and 4 bearing caps.

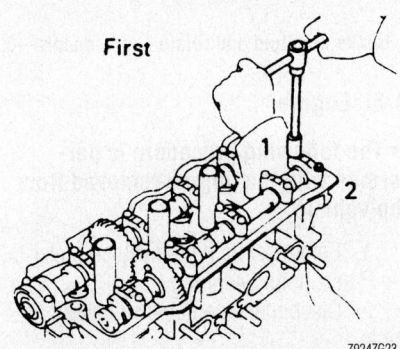

Exhaust camshaft bolt removal sequence: Step 1—2.0L engine

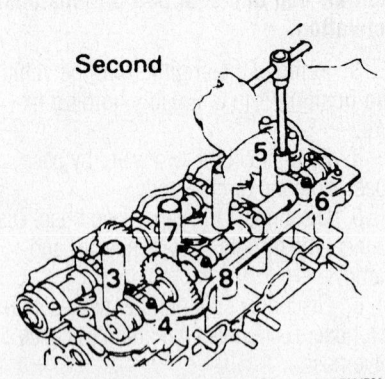

Exhaust camshaft bolt removal sequence: Step 2—2.0L engine

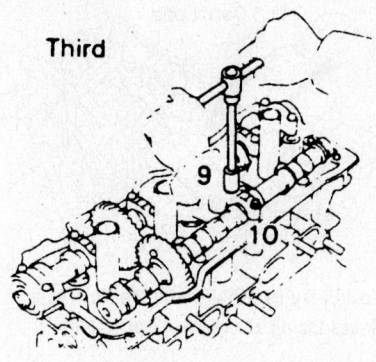

Exhaust camshaft bolt removal sequence: Step 3—2.0L engine

8. Alternately loosen and remove bearing cap bolts No. 9 and 10. As these bolts are loosened check to see that the camshaft is being lifted out straight and level.

➡**If the camshaft is not lifting out straight and level retighten No. 9 and 10 bearing cap bolts. Reverse the order of Steps 5 through 7 and reset the intake camshaft knock pin to 10–45 degrees BTDC and repeat Steps 5 through 7 again. Do not attempt to pry the camshaft from its mounting.**

9. Remove the No. 3 bearing cap and exhaust camshaft from the engine.

10. Set the knock pin of the intake camshaft at 80–115 degrees BTDC of camshaft angle. This angle will help to lift the intake camshaft level and evenly by pushing No. 1 and No. 3 cylinder camshaft lobes of the intake camshaft toward their valve lifters.

11. Remove the No. 1 and No. 2 front bearing cap bolts and remove the front bearing cap and oil seal. If the cap will not come apart easily, leave it in place without the bolts.

12. Uniformly loosen and remove bearing cap bolts No. 3 to No. 8 in several phases and in the proper sequence. Do not remove bearing cap bolts No. 9 and 10 at this time. Remove No. 1, 3 and 4 bearing caps.

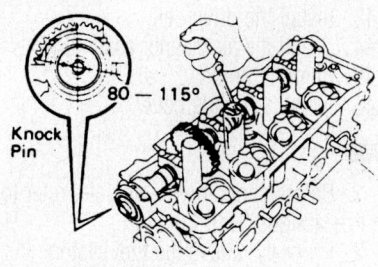

Intake camshaft knock pin alignment— 2.0L engine

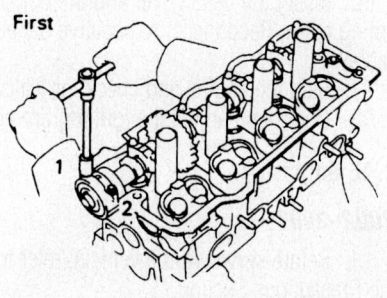

Intake camshaft bolt removal sequence: Step 1—2.0L engine

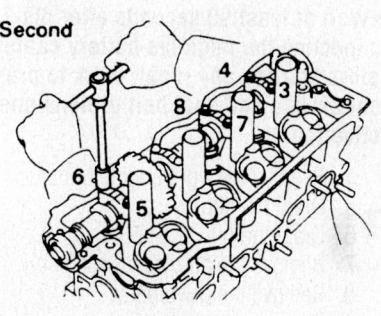

Intake camshaft bolt removal sequence: Step 2—2.0L engine

13. Alternately loosen and remove bearing cap bolts No. 9 and 10. As these bolts are loosened and after breaking the adhesion on the front bearing cap, check to see that the camshaft is being lifted out straight and level.

➡**If the camshaft is not lifting out straight and level retighten No. 9 and 10 bearing cap bolts. Reverse steps 10 through 12, than start over from Step 10. Do not attempt to pry the camshaft from its mounting.**

14. Remove the No. 2 bearing cap with the intake camshaft from the engine.

15. Remove the valve adjusting shims from the engine. Be sure to replace the shims to their original location.

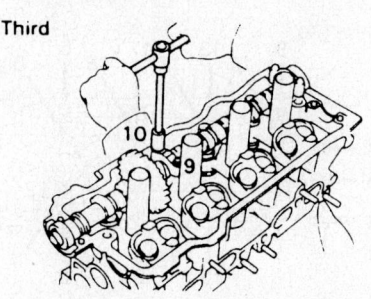

Intake camshaft bolt removal sequence: Step 3—2.0L engine

To install:

16. Install the valve adjusting shims to the engine.

17. Before installing the intake camshaft, apply multi-purpose grease to the thrust portion of the camshaft.

18. Position the camshaft at 80–115 degrees BTDC of camshaft angle on the cylinder head.

19. Apply sealant to the front bearing cap.

20. Coat the bearing cap bolts with clean engine oil.

21. Tighten the camshaft bearing caps evenly and in several passes to 14 ft. lbs. (19 Nm) in the proper sequence.

22. Set the knock pin of the camshaft at 10–45 degrees BTDC of camshaft angle.

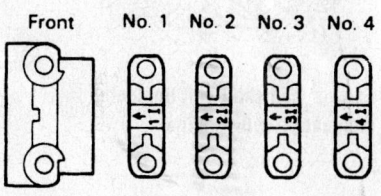

Intake camshaft bearing cap positioning— 2.0L engine

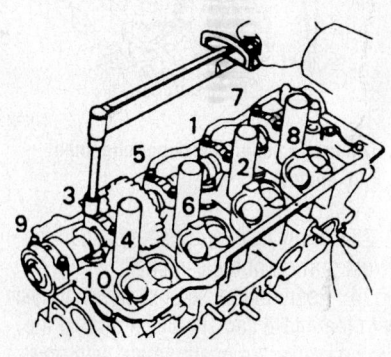

Intake camshaft bolt tightening sequence—2.0L engine

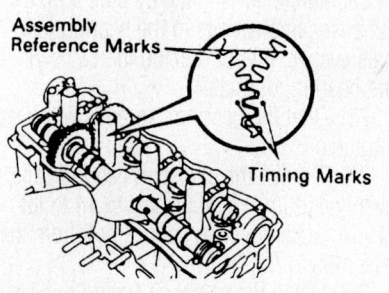

Camshaft timing mark alignment—2.0L engine

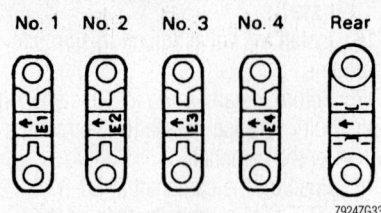

Exhaust camshaft bearing cap positioning—2.0L engine

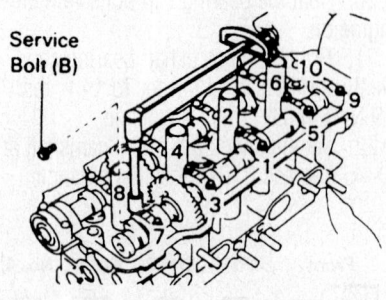

Exhaust camshaft bolt tightening sequence—2.0L engine

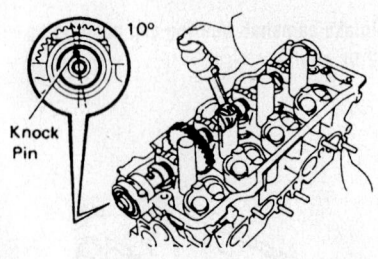

Exhaust camshaft knock pin alignment—2.0L engine

23. Apply multipurpose grease to the thrust portion of the camshaft.

24. Position the exhaust camshaft gear with the intake camshaft gear so that the timing marks are in alignment with one another. Be sure to use the proper alignment marks on the gears. Do not use the assembly reference marks.

25. Turn the intake camshaft clockwise or counterclockwise little by little until the exhaust camshaft sits in the bearing journals evenly without rocking the camshaft on the bearing journals.

26. Coat the bearing cap bolts with clean engine oil.

27. Tighten the camshaft bearing caps evenly and in several passes to 14 ft. lbs. (19 Nm). Remove the service bolt from the assembly.

28. Install the camshaft timing pulleys and the timing belt.

29. Adjust the valve clearance.

30. Install the head cover and the upper timing cover. Reconnect the negative battery cable.

31. Start the engine and check for leaks.

32. Check and adjust the ignition timing.

2.4L Engine

2002–2005

1. Before servicing the vehicle, refer to the Precautions Section.

2. Properly relieve the fuel system pressure.

3. Disconnect the negative battery cable.

➡ **Wait at least 90 seconds after disconnecting the negative battery cable before starting any repair work to prevent air bag and seat belt pretensioner activation.**

4. Remove the right engine undercover.
5. Drain the coolant.
6. Drain the oil.
7. Remove the air cleaner assembly.
8. Remove the drive belt.
9. Remove the alternator.
10. Remove the power steering pump.
11. Remove the ignition coils.
12. Remove the spark plugs.
13. Remove the injectors.
14. Remove the exhaust manifold.
15. Remove the oil filler cap.
16. Remove the PCV hoses and valve.
17. Remove the intake manifold.
18. Remove all wires and harnesses connected to the head.
19. Remove the timing chain.
20. Remove the camshaft sprocket.
21. Remove the VVT sprocket.
22. Remove the camshaft timing oil control valve.
23. Remove the camshafts.

To install:

24. Coat the camshafts and bearings with clean engine oil.

25. Install the camshafts with the No.1 cam lobes facing as shown.

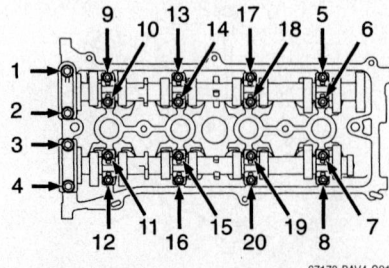

Camshaft bolt removal sequence—2002–2005 2.4L engine

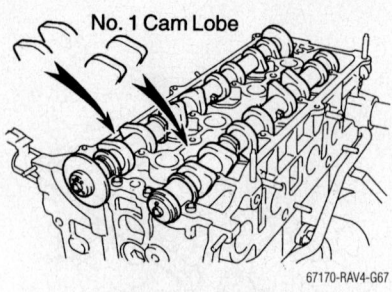

Install the camshafts with the No.1 cam lobes facing as shown—2.4L engine

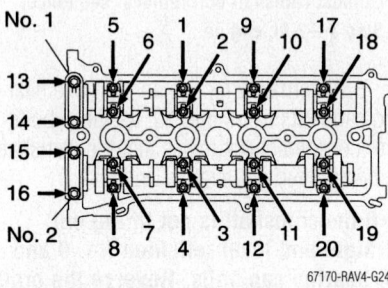

Camshaft bolt torque sequence—2.4L engine

26. Coat the threads and heads of the bearing cap bolts with clean engine oil.

27. Torque the bolts evenly, in several passes, in the sequence shown, to 22 ft. lbs. (29.5 Nm) for Nos. 1 and 2; 80 inch lbs. (9 Nm) for all the others.

28. Check and adjust the valve clearance.

29. Install the camshaft and VVT sprockets.

30. Install the timing chain.
31. Install the camshaft position sensor.
32. Install the oil control valve.
33. Install the oil filler cap.
34. Install the intake manifold.
35. Install the PCV valve and hoses.
36. Install the engine wiring.
37. Install the exhaust manifold.
38. Install the injectors.
39. Install the spark plugs.
40. Install the ignition coils.
41. Install the power steering pump.
42. Install the alternator.
43. Install the drive belt.
44. Install the air cleaner assembly.
45. Refill all fluids.
46. Install the undercover.

2006

1. Before servicing the vehicle, refer to the Precautions Section.

2. Properly relieve the fuel system pressure.

3. Disconnect the negative battery cable.

→**Wait at least 90 seconds after disconnecting the negative battery cable before starting any repair work to prevent air bag and seat belt pretensioner activation.**

4. Remove the radiator support opening cover.

5. Remove the right front tire and wheel assembly.

6. Remove the number one engine undercover. Remove the front fender apron, right side. Remove the number one engine cover.

7. Drain the engine coolant. Drain the engine oil.

8. Remove the air cleaner assembly.

9. Disconnect the purge line hose from the throttle body. Disconnect the water bypass hoses from the throttle body. Disconnect the number one throttle body hose from the throttle body.

10. Disconnect the throttle position sensor and control motor connector. Disconnect the wire harness clamp. Disconnect the fuel tube from the clamp.

11. Remove the four throttle body retaining bolts and remove the throttle body assembly. Remove the gasket.

12. Remove the fuel delivery pipe sub-assembly.

13. Remove the intake manifold. Remove the intake manifold insulator.

14. Remove the front exhaust pipe. Remove the oil dipstick. Remove the oil dipstick guide.

15. Remove the manifold stay. Remove the number two manifold stay.

16. Remove the number one exhaust manifold heat insulator. Disconnect the air/fuel ratio sensor connector. Remove the five nuts and remove the exhaust manifold converter subassembly.

17. Disconnect the radiator hose.

18. Disconnect the radio setting condenser connector. Disconnect the engine oil pressure switch connector. Disconnect the engine coolant temperature sensor connector.

19. Disconnect the camshaft position sensor connector. Remove the bolt and ground cable.

20. Remove the front suspension member reinforcement, right side.

21. Remove the alternator. Remove the right side engine mounting insulator. Remove the idler pulley.

22. Remove the ignition coil assembly. Remove the spark plugs.

23. Remove the cylinder head cover retaining bolts. Remove the cylinder head cover from the engine.

24. To remove the belt tensioner assembly, lift the engine upward, using a transmission jack. Remove the bolt, nut and the belt tensioner assembly.

→**Do not lift the engine more than necessary.**

25. Remove the crankshaft position sensor.

26. Remove the oil pan.

27. Turn the crankshaft pulley until its groove and the timing mark "0" of the timing chain cover are aligned. Check that each timing mark of the camshaft timing gear and sprocket is aligned with each timing mark located on the NO. 1 and No. 2 bearing caps, as shown in the illustration.

→**If not rotate the engine 360 degrees to align the timing marks.**

28. To remove the number two camshaft, paint matchmarks on the chain in alignment with the timing marks on the camshaft timing gear and camshaft timing sprocket.

29. Remove the two nuts, tensioner and gasket. While holding the camshaft with a wrench, loosen the camshaft timing set bolt.

30. Using several steps, uniformly loosen and remove the ten bearing cap bolts in the proper sequence. Remove the five bearing caps.

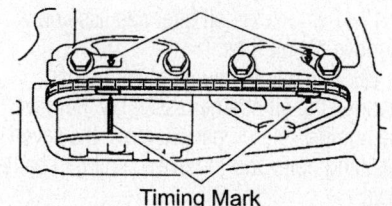

Timing Mark

Timing Mark

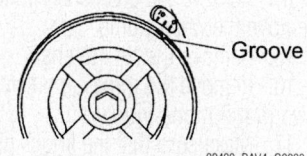

Groove

Camshaft gear and sprocket alignment—2006 2.4L engine

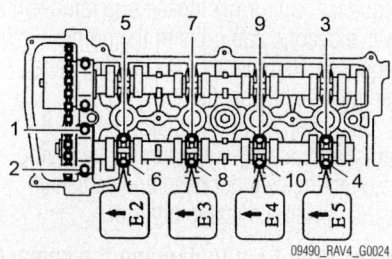

Camshaft number two bolt removal sequence—2006 2.4L engine

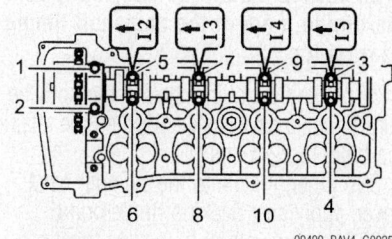

Camshaft number one bolt removal sequence—2006 2.4L engine

31. While holding the number two camshaft, remove the camshaft timing sprocket set bolt.

→**Remove the camshaft timing sprocket from the camshaft with the timing chain wrapped on the sprocket.**

32. Remove the camshaft timing sprocket from the timing chain.

33. To remove the number one camshaft, using several steps uniformly loosen and remove the ten bearing cap bolts in the proper sequence. Remove the five bearing caps.

34. Remove the camshaft and camshaft timing gear while holding the timing chain with your hand.

35. Tie the timing chain with a string to the side of the engine block.

→**Be careful not to drop anything inside the timing chain cover.**

To install:

36. Installation is the reverse of the removal procedure.

37. Tighten the camshaft retaining bolts to 22 ft. lbs. for No.1 and No.2 bearing cap, and to 80 inch lbs. for the No. 3 bearing cap.

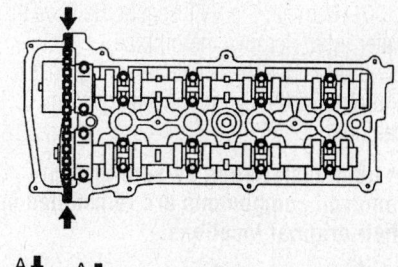

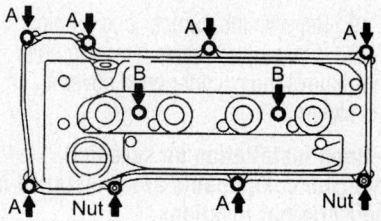

Cylinder head cover sealant location and bolt identification—2006 2.4L engine

➡️**Be sure to install the camshafts with the timing mark of the camshaft timing gear on top.**

38. When installing the throttle body be sure to use a new gasket. Tighten the retaining bolts to 22 ft. lbs.

39. When installing the cylinder head cover, apply seal packing (three bond 1207B, or equivalent) to the two locations, shown in the illustration. Tighten the retaining bolts to 8 ft. lbs. for bolt "A", 10 ft. lbs. for bolt "B" and 8 ft. lbs. for the nut.

40. Be sure to fill the cooling system with the proper grade and type engine coolant.

41. Be sure to fill the engine with the proper grade and type engine oil.

42. Start the engine and check for leaks, correct as required.

3.5L Engine

➡️**The following procedure is performed with the engine removed from the vehicle.**

1. Before servicing the vehicle, refer to the Precautions Section.

2. Properly relieve the fuel system pressure.

3. Disconnect the negative battery cable.

➡️**Wait at least 90 seconds after disconnecting the negative battery cable before starting any repair work to prevent air bag and seat belt pretensioner activation.**

4. Remove the engine and position it in a suitable holding fixture.

5. Remove the oil filler cap subassembly. Remove the spark plugs.

6. Remove the camshaft timing control valve assembly.

7. Remove the VVT sensor. Remove the water inlet. Remove the oil pipe.

8. Remove the twelve retaining bolts and remove the cylinder head cover for bank one.

➡️**Upon installation be sure that removed components are reinstalled in their original locations.**

9. Remove the number one oil pipe.

10. Remove the twelve retaining bolts and remove the cylinder head cover for bank two.

➡️**Upon installation be sure that removed components are reinstalled in their original locations.**

11. Remove the number one engine mounting bracket. Remove the water inlet housing.

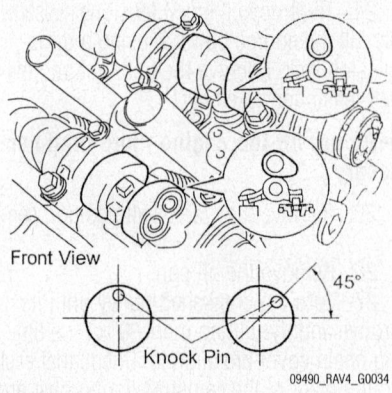

Front View

Knock pin alignment (bank one)—3.5L engine

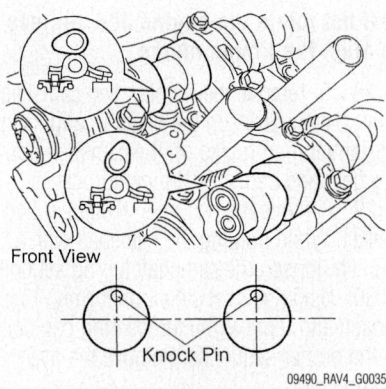

Front View

Knock pin alignment (bank two)—3.5L engine

12. Remove the oil filter cap assembly. Remove the oil filter.

13. Remove the number two oil pan. Remove the oil strainer assembly. Remove the number one oil pan. Remove the seven retaining bolts and remove the oil pan baffle plate.

14. Remove the crankshaft pulley. Remove the water pump.

15. Remove the timing chain.

16. Remove the three camshaft gaskets (two in front, one in rear).

17. Make sure that the knock pin of the camshaft is positioned, as shown in the illustration.

18. Uniformly loosen and remove the eight bearing cap bolts in the proper sequence. Uniformly loosen and remove the twelve bearing cap bolts in the proper sequence. Remove the five bearing caps. Remove the camshafts.

19. If removing the lifters, remove the camshaft housing subassembly by prying between the cylinder head and camshaft housing with a suitable pry tool.

➡️**Be careful not to damage the contact surfaces of the cylinder head and the camshaft housing.**

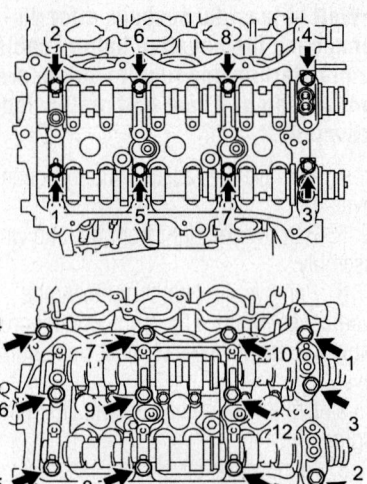

Camshaft bolt removal sequence (bank one)—3.5L engine

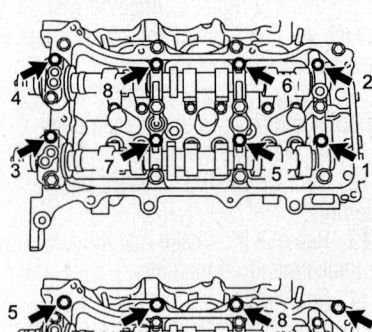

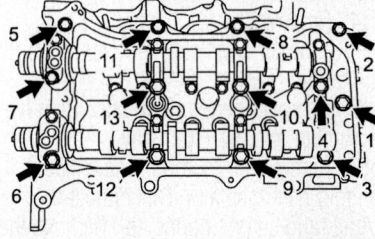

Camshaft bolt removal sequence (bank two)—3.5L engine

20. Remove the valve rocker arms. Remove the valve lash adjusters.

➡️**Be sure to keep parts arranged in a logical order for reinstallation in their exact locations.**

To install:

21. Installation is the reverse of the removal procedure.

22. When installing the cylinder head cover, apply seal packing, three bond 1207B or equivalent, as shown in the illustration.

➡️**Remove any oil from the contact surfaces. Install the component within three minutes after applying the sealant. Do not start the engine for two hours after the installation.**

23. Install the cover retaining bolts and torque to 15 ft. lbs for bolt "A" and to 7 ft. lbs. for all other bolts and in the proper sequence.

INSPECTION

1. Before servicing the vehicle, refer to the Precautions Section.
2. Remove the camshaft from the engine.
3. Check the camshaft bearing journals for damage and binding.
4. If the journals are binding, check the cylinder head for damage.
5. Check the cylinder head for clogged oil holes.
6. Check the camshaft surface for abnormal wear and damage. Replace the camshaft, as required.
7. Measure the camshaft lobe surface and replace the camshaft if not within specification.
8. Measure the camshaft journal diameter and replace the camshaft if not within specification.
9. Measure the camshaft run out and replace the camshaft if not within specification.

Valve Lash

ADJUSTMENT

2.0L Engine

1. Before servicing the vehicle, refer to the Precautions Section.
2. Remove the cylinder head cover.
3. Use a wrench to turn the crankshaft until the notch in the pulley aligns with timing mark **0** of the No. 1 timing belt cover. This will ensure that the No. 1 piston is at Top Dead Center (TDC) of the compression stroke.

➡**Check that the valve lifters on the No. 1 cylinder are loose and those on the No. 4 cylinder are tight. If not, rotate the crankshaft 1 complete revolution (360 degrees) and then realign the marks.**

4. Using a flat feeler gauge measure the clearance between the camshaft lobe and the valve lifter on the first set of valves shown. This measurement should correspond to specifications.

➡**If the measurement is within specifications, go on to the next step. If not, record the measurement taken for each individual valve.**

5. Rotate the crankshaft 1 complete revolution and realign the timing marks.
6. Measure the clearance of the second set of valves.

➡**If the measurement for this set of valves (and also the previous one) is within specifications, go no further, the procedure is finished. If not, record the measurements and proceed to the next step.**

7. Rotate the crankshaft to position the intake camshaft lobe of the cylinder to be adjusted, facing upward.

➡**Both intake and exhaust valve clearance may be adjusted at the same time, if required.**

8. Using a suitable tool, turn the valve lifter so the notch is easily accessible.
9. Install tool 09248-55010 between both camshaft lobes and turn the handle so the tool presses down both intake and exhaust valve lifters evenly.
10. Using a suitable tool and a magnet, remove the valve shims.
11. Measure the thickness of the old shim with a micrometer. Using this measurement and the clearance made earlier (from Step 3 or 5), determine what size replacement shim will be required in order to bring the valve clearance into specification.

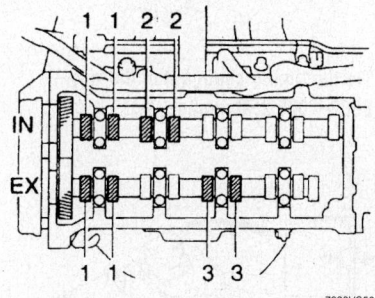

Adjust these valve first—2.0L engine

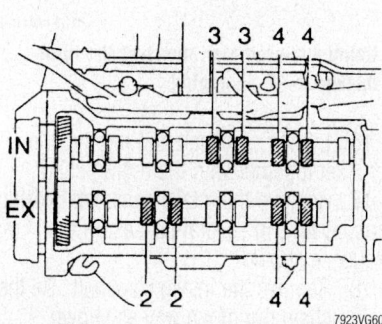

Adjust these valve second—2.0L engine

➡**Replacement shims are available in 27 sizes, in increments of 0.0020 in. (0.05mm). Shim sizes are 0.0787– 0.1299 in. (2.00–3.30mm).**

12. Install the new shim, remove the special tool; then, recheck the valve clearances.
13. Install the cylinder head covers.

2.4L Engine

2002–2005

Perform this procedure on a cold engine only!
1. Before servicing the vehicle, refer to the Precautions Section.
2. Disconnect the negative battery cable.

➡**Wait at least 90 seconds after disconnecting the negative battery cable before starting any repair work to prevent air bag and seat belt pretensioner activation.**

3. Remove the right side engine undercover.
4. Remove the air cleaner assembly.
5. Remove the cylinder head cover.
6. Turn the crankshaft pulley and align its groove with the "0" mark on the timing cover. This sets the engine to No. 1 TDC compression.
7. Check that the timing marks on the camshaft sprocket and VVT sprocket are aligned with the timing marks on the camshaft No.1 and 2 bearing caps.

➡**Valve clearance (cold) should be 0.008-0.011 inch for intake; 0.012-0.016 inch for exhaust.**

8. Check the clearance on the Nos. 1 and 2 intake valves and the Nos. 1 and 3 exhaust valves, as shown.
9. Turn the crankshaft 1 full revolution clockwise (360 degrees).
10. Check the clearance on Nos. 3 and 4 intake, and Nos. 2 and 4 exhaust.

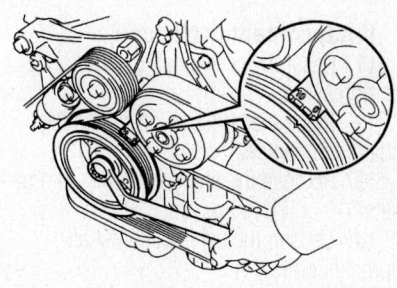

Turn the crankshaft pulley and align its groove with the "0" mark on the timing cover—2.4L engine

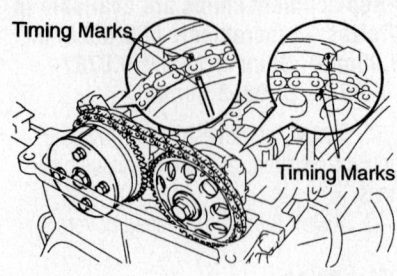

67170-RAV4-G02

Check that the timing marks on the camshaft sprocket and VVT sprocket are aligned with the timing marks on the camshaft No.1 and 2 bearing caps—2.4L engine

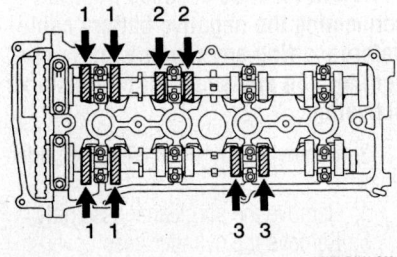

67170-RAV4-G03

Check the clearance on the Nos. 1 and 2 intake valves and the Nos. 1 and 3 exhaust valves—2.4L engine

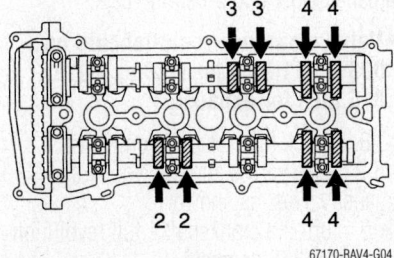

67170-RAV4-G04

Check the clearance on Nos. 3 and 4 intake, and Nos. 2 and 4 exhaust—2.4L engine

If Adjustment Is Needed

11. Reset the crankshaft to No. 1 TDC compression.

12. Place matchmarks on the timing chain and camshaft sprockets.

13. Remove the chain tensioner and gaskets.

14. Loosen the exhaust camshaft sprocket bolt.

15. Remove the exhaust camshaft bearing caps, evenly, in several passes, in the sequence shown.

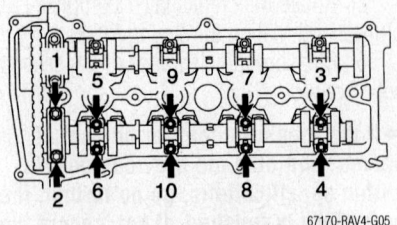

67170-RAV4-G05

Remove the exhaust camshaft bearing caps, evenly, in several passes, in the sequence shown—2002–2005 2.4L engine

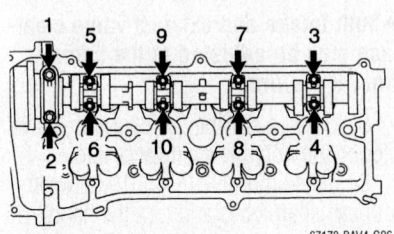

67170-RAV4-G06

Remove the intake camshaft bearing caps, evenly, in several passes, in the sequence shown—2002–2005 2.4L engine

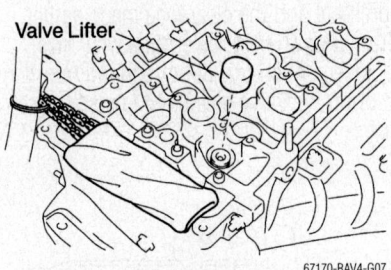

67170-RAV4-G07

Tie the timing chain out of the way—2002–2005 2.4L engine

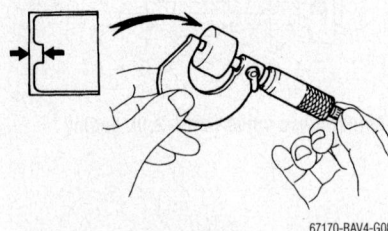

67170-RAV4-G08

Using a micrometer, measure the lifter thickness—2.4L engine

16. Lift the camshaft and remove the sprocket together with the timing chain.

17. Remove the intake camshaft bearing caps, evenly, in several passes, in the sequence shown.

18. Remove the intake camshaft. Tie the timing chain out of the way as shown.

19. For any valve needing adjustment,

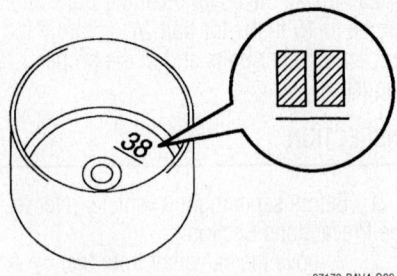

67170-RAV4-G09

An ID number inside the lifter shows the 2 decimal place size—2.4L engine

remove the lifter. Determine replacement lifter size.

   a. Using a micrometer, measure the lifter thickness.

   b. Calculate the thickness of a new liter to bring the valve clearance into the proper range.

   c. Select a lifter with a thickness as close as possible to correct the specified value, from the accompanying charts. Lifters are available in 35 sizes in 0.0008 inch (0.020mm) increments from 5.060mm to 5.740mm.

➡An ID number inside the lifter shows the 2 decimal place size. So, a 38 mark would indicate a lifter that is 5.38mm thick.

   d. Coat the replacement lifter with clean engine oil and install it.

20. When all new lifters are installed, align the crankshaft timing mark with the "0" mark on the timing cover.

21. Hold the chain and install the intake camshaft, aligning all marks.

22. Coat the threads and heads of the bearing cap bolts with clean engine oil.

23. Torque the bolts evenly, in several passes, in the sequence shown, to 22 ft. lbs. (29.5 Nm) for Nos. 1 and 2; 80 inch lbs. (9 Nm) for all the others.

24. Install the exhaust camshaft, aligning all marks and install the bearing caps in the same manner as you did with the intake caps.

25. Recheck all timing marks.

26. Install the tensioner. Torque to 80 inch lbs. (9 Nm).

27. Recheck the valve timing by setting the crankshaft timing mark to align with the "0" mark on the timing cover. All timing marks should align

28. The remainder of installation is the reverse of removal.

2006

1. Before servicing the vehicle, refer to the Precautions Section.

Valve Lifter Selection Chart (Intake)

New Lifter Thickness — mm (in.)

Lifter No.	Thickness	Lifter No.	Thickness	Lifter No.	Thickness
06	5.060 (0.1992)	30	5.300 (0.2087)	54	5.540 (0.2181)
08	5.080 (0.2000)	32	5.320 (0.2094)	56	5.560 (0.2189)
10	5.100 (0.2008)	34	5.340 (0.2102)	58	5.580 (0.2197)
12	5.120 (0.2016)	36	5.360 (0.2110)	60	5.600 (0.2205)
14	5.140 (0.2024)	38	5.380 (0.2118)	62	5.620 (0.2213)
16	5.160 (0.2031)	40	5.400 (0.2126)	64	5.640 (0.2220)
18	5.180 (0.2039)	42	5.420 (0.2134)	66	5.660 (0.2228)
20	5.200 (0.2047)	44	5.440 (0.2142)	68	5.680 (0.2236)
22	5.220 (0.2055)	46	5.460 (0.2150)	70	5.700 (0.2244)
24	5.240 (0.2063)	48	5.480 (0.2157)	72	5.720 (0.2252)
26	5.260 (0.2071)	50	5.500 (0.2165)	74	5.740 (0.2260)
28	5.280 (0.2079)	52	5.520 (0.2173)		

Intake valve clearance (Cold):
0.19 to 0.29 mm (0.008 to 0.011 in.)

EXAMPLE:
The 5.250 mm (0.2067 in.) lifter is installed, and the measured clearance is 0.400 mm (0.0157 in.).
Replace the 5.250 mm (0.2067 in.) lifter with a new No. 42 lifter.

67170-RAV4-G10

Intake valve lifter size selection chart—2.4L engine

Valve Lifter Selection Chart (Intake) — Installed lifter thickness mm (in.) across top row (5.060 (0.1992) through 5.740 (0.2260)); Measure clearance mm (in.) down left column (0.000–0.030 (0.0000–0.0012) through 0.911–0.930 (0.0359–0.0366)), with new lifter numbers at each intersection.

Valve Lifter Selection Chart (Exhaust)

The main chart is a large triangular lookup matrix. Its axes are:

Column headers — Installed lifter thickness mm (in.):
5.060 (0.1992), 5.080 (0.2000), 5.100 (0.2008), 5.120 (0.2016), 5.140 (0.2024), 5.160 (0.2031), 5.180 (0.2039), 5.200 (0.2047), 5.210 (0.2051), 5.220 (0.2055), 5.230 (0.2059), 5.240 (0.2063), 5.250 (0.2067), 5.260 (0.2071), 5.270 (0.2075), 5.280 (0.2079), 5.290 (0.2083), 5.300 (0.2087), 5.310 (0.2091), 5.320 (0.2094), 5.330 (0.2098), 5.340 (0.2102), 5.350 (0.2106), 5.360 (0.2110), 5.370 (0.2114), 5.380 (0.2118), 5.390 (0.2122), 5.400 (0.2126), 5.410 (0.2130), 5.420 (0.2134), 5.430 (0.2138), 5.440 (0.2142), 5.450 (0.2146), 5.460 (0.2150), 5.470 (0.2154), 5.480 (0.2157), 5.490 (0.2161), 5.500 (0.2165), 5.510 (0.2169), 5.520 (0.2173), 5.530 (0.2177), 5.540 (0.2181), 5.550 (0.2185), 5.560 (0.2189), 5.570 (0.2193), 5.580 (0.2197), 5.590 (0.2201), 5.600 (0.2205), 5.620 (0.2213), 5.640 (0.2220), 5.660 (0.2228), 5.680 (0.2236), 5.700 (0.2244), 5.720 (0.2252), 5.740 (0.2260)

Row headers — Measure clearance mm (in.):
0.000–0.030 (0.0000–0.0012), 0.031–0.050 (0.0012–0.0020), 0.051–0.070 (0.0020–0.0028), 0.071–0.090 (0.0028–0.0035), 0.091–0.110 (0.0036–0.0043), 0.111–0.130 (0.0044–0.0051), 0.131–0.150 (0.0052–0.0059), 0.151–0.170 (0.0059–0.0067), 0.171–0.190 (0.0067–0.0075), 0.191–0.210 (0.0075–0.0083), 0.211–0.230 (0.0083–0.0091), 0.231–0.250 (0.0091–0.0098), 0.251–0.270 (0.0099–0.0106), 0.271–0.290 (0.0107–0.0114), 0.291–0.299 (0.0115–0.0118), 0.300–0.400 (0.0118–0.0157), 0.401–0.420 (0.0158–0.0165), 0.421–0.440 (0.0166–0.0173), 0.441–0.460 (0.0174–0.0181), 0.461–0.480 (0.0181–0.0189), 0.481–0.500 (0.0189–0.0197), 0.501–0.520 (0.0197–0.0205), 0.521–0.540 (0.0205–0.0213), 0.541–0.560 (0.0213–0.0220), 0.561–0.580 (0.0220–0.0228), 0.581–0.600 (0.0229–0.0236), 0.601–0.620 (0.0237–0.0244), 0.621–0.640 (0.0244–0.0252), 0.641–0.660 (0.0252–0.0260), 0.661–0.680 (0.0260–0.0268), 0.681–0.700 (0.0268–0.0276), 0.701–0.720 (0.0276–0.0283), 0.721–0.740 (0.0284–0.0291), 0.741–0.760 (0.0292–0.0299), 0.761–0.780 (0.0300–0.0307), 0.781–0.800 (0.0307–0.0315), 0.801–0.820 (0.0315–0.0323), 0.821–0.840 (0.0323–0.0331), 0.841–0.860 (0.0331–0.0339), 0.861–0.880 (0.0339–0.0346), 0.881–0.900 (0.0347–0.0354), 0.901–0.920 (0.0355–0.0362), 0.921–0.940 (0.0363–0.0370), 0.941–0.960 (0.0370–0.0378), 0.961–0.980 (0.0378–0.0386), 0.981–1.000 (0.0386–0.0394), 1.001–1.020 (0.0394–0.0402), 1.021–1.040 (0.0402–0.0409), 1.041–1.060 (0.0410–0.0417), 1.061–1.080 (0.0418–0.0425)

The body cells contain lifter numbers (06 through 74) indicating the replacement lifter for each intersection of installed thickness and measured clearance.

New Lifter Thickness mm (in.)

Lifter No.	Thickness	Lifter No.	Thickness	Lifter No.	Thickness
06	5.060 (0.1992)	30	5.300 (0.2087)	54	5.540 (0.2181)
08	5.080 (0.2000)	32	5.320 (0.2094)	56	5.560 (0.2189)
10	5.100 (0.2008)	34	5.340 (0.2102)	58	5.580 (0.2197)
12	5.120 (0.2016)	36	5.360 (0.2110)	60	5.600 (0.2205)
14	5.140 (0.2024)	38	5.380 (0.2118)	62	5.620 (0.2213)
16	5.160 (0.2031)	40	5.400 (0.2126)	64	5.640 (0.2220)
18	5.180 (0.2039)	42	5.420 (0.2134)	66	5.660 (0.2228)
20	5.200 (0.2047)	44	5.440 (0.2142)	68	5.680 (0.2236)
22	5.220 (0.2055)	46	5.460 (0.2150)	70	5.700 (0.2244)
24	5.240 (0.2063)	48	5.480 (0.2157)	72	5.720 (0.2252)
26	5.260 (0.2071)	50	5.500 (0.2165)	74	5.740 (0.2260)
28	5.280 (0.2079)	52	5.520 (0.2173)		

Exhaust valve clearance (Cold):
0.30 to 0.40 mm (0.012 to 0.016 in.)

EXAMPLE:
The 5.340 mm (0.2102 in.) lifter is installed, and the measured clearance is 0.430 mm (0.0169 in.).
Replace the 5.340 mm (0.2102 in.) lifter with a new No. 42 lifter.

Exhaust valve lifter size selection chart—2.4L engine

67170-RAV4-G11

2. Disconnect the negative battery cable.

➡**Wait at least 90 seconds after disconnecting the negative battery cable before starting any repair work to prevent air bag and seat belt pretensioner activation.**

3. Remove the right front wheel and tire assembly. Remove the engine undercover. Remove the front fender apron, right side.

4. Remove the number one engine cover.

5. Remove the ignition coil. Remove the spark plugs.

6. Remove the cylinder head cover.

7. Position the number one cylinder to TDC on the compression stroke. Check and record the clearance on the number one and two intake valves and on the number one and three exhaust valves.

8. Rotate the engine 360 degrees and set the number four cylinder to TDC on the compression stroke. Check and record the clearance on the number three and four intake valves and on the number two and four exhaust valves.

9. If adjustment is necessary, remove the camshafts. Remove the valve lifters.

10. Using a micrometer, measure the lifter thickness.

11. Calculate the thickness of a new liter to bring the valve clearance into the proper range.

12. Select a lifter with a thickness as close as possible to correct the specified value, from the accompanying charts. Lifters are available in 35 sizes in 0.0008 inch (0.020mm) increments from 5.060mm to 5.740mm.

13. Reinstall removed components in the reverse order of the removal procedure.

14. When installing the cylinder head cover, apply seal packing (three bond 1207B, or equivalent) to the two locations, shown in the illustration. Tighten the retaining bolts to 8 ft. lbs. for bolt "A", 10 ft. lbs. for bolt "B" and 8 ft. lbs. for the nut.

Starter Motor

REMOVAL & INSTALLATION

2.0L Engine

1. Before servicing the vehicle, refer to the Precautions Section.

2. Disconnect the negative battery cable.

3. Remove the engine coolant reservoir.

4. Remove the air cleaner cap assembly, as follows:

a. Disconnect the following:
- Skid control relay connectors
- High tension cord from the air cleaner hose and resonator
- Intake Air Temperature (IAT) sensor connector
- Positive Crankcase Ventilation (PCV) hose from the air cleaner hose
- Air hose from the air cleaner cap assembly

b. The air cleaner cap assembly from the air cleaner case assembly by removing the 4 clamps.

c. Loosen the hose clamp and disconnect the air cleaner hose from the throttle body.

d. Remove the air cleaner cap assembly.

5. Remove or disconnect the following:
- Vacuum Switching Valve (VSV) from the air cleaner case assembly
- 3 bolts and the air cleaner case assembly
- Starter electrical connectors
- Both bolts and the starter

To install:

6. Install or connect the following:
- Starter. Tighten the bolts to 29 ft. lbs. (38 Nm).
- Starter electrical connectors
- Air cleaner case assembly
- VSV to the air cleaner case assembly

7. Install or connect the air cleaner cap assembly, as follows:

a. Install the air cleaner cap assembly.

b. Connect the air cleaner hose to the throttle body and tighten the hose clamp.

c. Install the air cleaner cap assembly to the air cleaner case assembly by installing the 4 clamps.

d. Connect the following:
- Air hose to the air cleaner cap assembly
- PCV hose to the air cleaner hose
- IAT sensor connector
- High tension cord to the air cleaner hose and resonator
- Skid control relay connectors

8. Install the engine coolant reservoir.

2.4L Engine

2002–2005

1. Before servicing the vehicle, refer to the Precautions Section.

2. Disconnect the negative battery cable.

3. Disconnect the hose from the transaxle level gauge.

4. Disconnect the wires from the starter.

5. Remove the 2 bolts and lift out the starter.

6. Installation is the reverse of removal. Torque the bolts to 27 ft. lbs. (37 Nm).

2006

1. Before servicing the vehicle, refer to the Precautions Section.

2. Disconnect the negative battery cable.

➡**Wait at least 90 seconds after disconnecting the negative battery cable before starting any repair work to prevent air bag and seat belt pretensioner activation.**

3. Disconnect the positive battery cable. Remove the battery clamp. Remove the battery insulator. Remove the battery.

4. Remove the front battery bracket. Remove the battery bracket reinforcement.

5. Disconnect the starter connector.

6. Open the terminal cap. Remove the nut and disconnect the starter wire.

7. Remove the starter retaining bolts. Remove the starter from the engine.

To install:

8. Installation is the reverse of the removal procedure.

9. Tighten the retaining bolts to 27 ft. lbs.

3.5L Engine

1. Before servicing the vehicle, refer to the Precautions Section.

2. Disconnect the negative battery cable.

➡**Wait at least 90 seconds after disconnecting the negative battery cable before starting any repair work to prevent air bag and seat belt pretensioner activation.**

3. Disconnect the positive battery cable. Remove the battery clamp. Remove the battery insulator. Remove the battery.

4. Remove the front battery bracket. Remove the battery bracket reinforcement.

5. Disconnect the starter connector.

6. Open the terminal cap. Remove the nut and disconnect the starter wire.

7. Remove the starter retaining bolts. Remove the starter from the engine.

To install:

8. Installation is the reverse of the removal procedure.

9. Tighten the retaining bolts to 27 ft. lbs.

Oil Pan

REMOVAL & INSTALLATION

2.0L Engine

1. Before servicing the vehicle, refer to the Precautions Section.

2. Disconnect the negative battery cable.

3. Remove the right side engine undercover.

4. Drain the crankcase oil.

5. Remove or disconnect the following:
- Dipstick
- Front exhaust pipe
- Stiffener plate from the engine by removing the 2 (manual transaxle) or 3 (automatic transaxle) bolts.
- 2 nuts and 17 bolts from the oil pan
- Oil pan and discard the gasket

To install:

6. Clean all gasket surfaces completely.

7. Apply a thin bead of sealer to the oil pan mounting surfaces.

8. Install or connect the following:
- Oil pan. Tighten the nuts/bolts to 48 inch lbs. (5 Nm).
- Stiffener plate. Tighten the bolts to 27 ft. lbs. (37 Nm).
- Front exhaust pipe

9. Fill the engine with oil to the proper level.

10. Start the engine and check for leaks. Recheck the engine oil level.

11. Install the right side engine cover.

2.4L Engine

2002–2005

1. Before servicing the vehicle, refer to the Precautions Section.

2. Disconnect the negative battery cable.

➡**Wait at least 90 seconds after disconnecting the negative battery cable before starting any repair work to prevent air bag and seat belt pretensioner activation.**

3. Remove the engine undercover.

4. Drain the engine oil.

5. Disconnect the exhaust pipe.

6. Remove the oil pan bolts, nuts and pan.

To install:

7. Clean all gasket surfaces completely.

8. Installation is the reverse of removal. Clean the gasket mating surfaces. Always use new gasket material. The new RTV material should be a bead about 4mm in diameter. Parts must be assembled within 5 minutes. Torque the bolts and nuts to 80 inch lbs. (9Nm).

2006

1. Before servicing the vehicle, refer to the Precautions Section.

2. Disconnect the negative battery cable.

➡**Wait at least 90 seconds after disconnecting the negative battery cable before starting any repair work to prevent air bag and seat belt pretensioner activation.**

3. Remove the engine undercover. Drain the engine oil. Remove the oil filter.

4. Install engine hanger tool 12281-28010 and 12282-28010. Install the engine lifting fixture to the engine hanger tools.

5. Remove the twelve retaining bolts and two nuts.

6. Insert the blade of tool SST09032-00100 between the crankcase, chain cover and the oil pan, cut off the applied sealer and remove the oil pan from the engine.

To install:

7. Clean all gasket surfaces completely.

8. Apply a continuous bead (0.118–0.157 inch) of seal packing (three bond 1207B or equivalent).

To install:

9. Clean all gasket surfaces completely.

10. Install the retaining bolts and nuts.

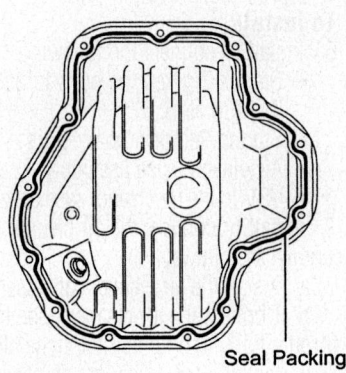

Seal Packing

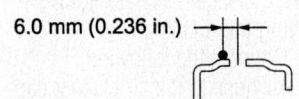

6.0 mm (0.236 in.)

Seal Diameter: 3.0 to 4.0 mm

09490_RAV4_G0052

Oil pan sealant application—2006 2.4L engine

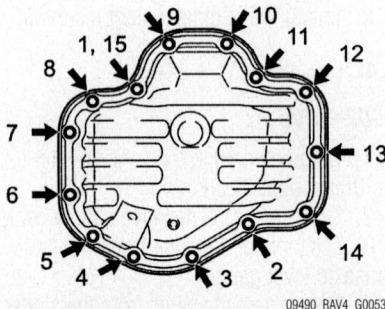

09490_RAV4_G0053

Oil pan bolt torque sequence—2006 2.4L engine

Tighten the retaining bolts to 80 inch lbs. and in the proper sequence.

11. Continue the installation in the reverse order of the removal procedure.

3.5L Engine

➡**The following procedure is performed with the engine removed from the vehicle.**

1. Before servicing the vehicle, refer to the Precautions Section.

2. Properly relieve the fuel system pressure.

3. Disconnect the negative battery cable.

➡**Wait at least 90 seconds after disconnecting the negative battery cable before starting any repair work to prevent air bag and seat belt pretensioner activation.**

4. Remove the engine and position it in a suitable holding fixture.

5. Remove the oil dipstick and dipstick tube. Remove the oil filler cap subassembly.

6. Remove the oil filter.

7. Remove the lower oil pan retaining bolts and nuts.

8. Insert the blade of tool SST09032-00100 at the oil pan, cut off the applied sealer and remove the lower oil pan from the engine.

9. Remove the bolt, two nuts, oil strainer and gasket.

10. Remove the upper oil pan retaining bolts and nuts. Remove the upper oil pan from the engine.

To install:

11. Installation is the reverse of the removal procedure.

12. When installing the upper oil pan, apply a continuous bead (0.118–0.156 inch) of seal packing (three bond 1207B or equivalent).

➡**Remove any oil from the contact surfaces. Install the component within three minutes after applying the sealant. Tighten the bolts within fifteen minutes after applying the seal packing. Do not start the engine for two hours after the installation.**

13. Install the retaining bolts and nuts. Tighten to 7 ft. lbs. for bolt "A" and to 15 ft. lbs for all others.

14. Install the oil strainer subassembly. Tighten the retaining bolt to 7 ft. lbs.

15. When installing the lower oil pan, apply a continuous bead (0.118–0.156 inch) of seal packing (three bond 1207B or equivalent).

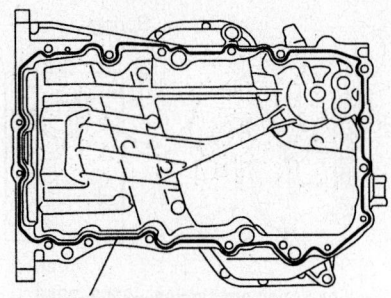

Seal Diameter:
3.0 to 4.0 mm (0.118 to 0.156 in.)

09490_RAV4_G0054

Upper oil pan sealant application—3.5L engine

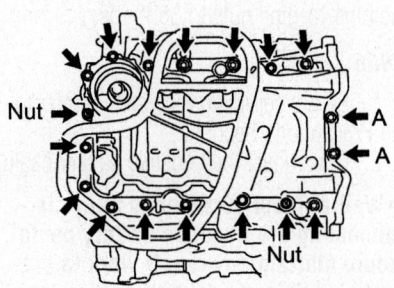

09490_RAV4_G0055

Upper oil pan bolt torque sequence—3.5L engine

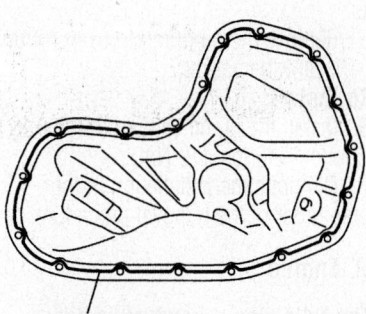

Seal Diameter:
3.0 to 4.0 mm (0.118 to 0.156 in.)

09490_RAV4_G0056

Lower oil pan sealant application—3.5L engine

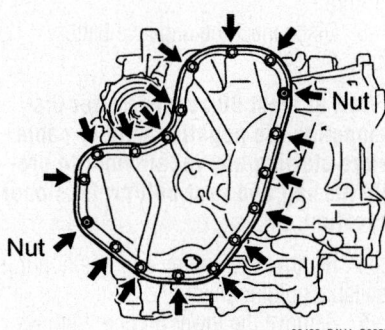

09490_RAV4_G0057

Lower oil pan bolt torque sequence—3.5L engine

➡**Remove any oil from the contact surfaces. Install the component within three minutes after applying the sealant. Tighten the bolts within fifteen minutes after applying the seal packing. Do not start the engine for two hours after the installation.**

16. Continue the installation in the reverse order of the removal procedure.

Oil Pump

REMOVAL & INSTALLATION

2.0L Engine

1. Before servicing the vehicle, refer to the Precautions Section.
2. Remove or disconnect the following:
   • Negative battery cable
   • Hood
   • Right side engine undercover
3. Drain the engine oil.
4. Remove or disconnect the following:
   • Front exhaust pipe
   • Rear end stiffener plate
   • Oil dipstick
   • 17 bolts and 2 nuts from the oil pan
5. Insert the blade of the Oil Pan Seal Cutting tool 09032-00100 between the oil pan and the cylinder block; then, cut off the applied sealer and remove the oil pan.

➡**Do not use the tool for the oil pump body side and rear oil seal retainer.**

6. Remove the bolts, nuts, oil strainer and gasket.
7. Carefully suspend the engine with a sling device.
8. Remove or disconnect the following:
   • Timing belt
   • No. 2 idler pulley and crankshaft timing pulley
   • Oil pump's pulley, using the Variable Pin Wrench Set 09960-10010
   • Crankshaft Position (CKP) sensor
   • Oil pump, by discarding the gasket

To install:
9. Install or connect the following:
   • Oil pump, using a new gasket. Tighten the 12 bolts to 82 inch lbs. (9 Nm).

➡**The long bolts are 35mm and all the others are 25mm.**

   • CKP sensor
   • Oil pump pulley. Tighten the nut to 18 ft. lbs. (24 Nm).
   • Crankshaft timing pulley and No. 2 idler pulley
   • Timing belt

10. Remove the engine sling.
11. Install the oil strainer with a new gasket. Tighten the nuts/bolts to 48 inch lbs. (5 Nm).
12. Remove any old sealant from the oil pan flange and thoroughly clean both sealing surfaces.
13. Apply a 3–5mm bead of sealant to the oil pan flange.

➡**The pan must be installed within 5 minutes of sealant application or the procedure will have to be repeated.**

14. Install or connect the following:
   • Oil pan. Tighten the 17 bolts and 2 nuts to 48 inch lbs. (5 Nm).
   • Dipstick
   • Rear end stiffener plate. Tighten the bolts to 27 ft. lbs. (37 Nm).
   • Front exhaust pipe
   • Negative battery cable
   • Hood
15. Refill the engine with oil.

❊❊ WARNING

Be sure to prime the oil pump prior to initial engine start-up or engine damage may occur because of low oil pressure.

16. Start the engine and check for leaks.
17. Recheck the engine oil level.
18. Install the right side engine undercover.

2.4L Engine

2002–2005

1. Before servicing the vehicle, refer to the Precautions Section.
2. Disconnect the negative battery cable.

➡**Wait at least 90 seconds after disconnecting the negative battery cable before starting any repair work to prevent air bag and seat belt pretensioner activation.**

3. Drain the oil.
4. Remove the air cleaner assembly.
5. Remove the ABS actuator.
6. Remove the right engine mount insulator.
7. Remove the timing chain.
8. Remove the crankshaft sprocket.
9. Remove the oil pump drive chain and sprockets.
10. Remove the 3 bolts and the oil pump.
11. If disassembling, remove the relief valve spring. Remove the pump body cover.
12. Check the clearance between the tip

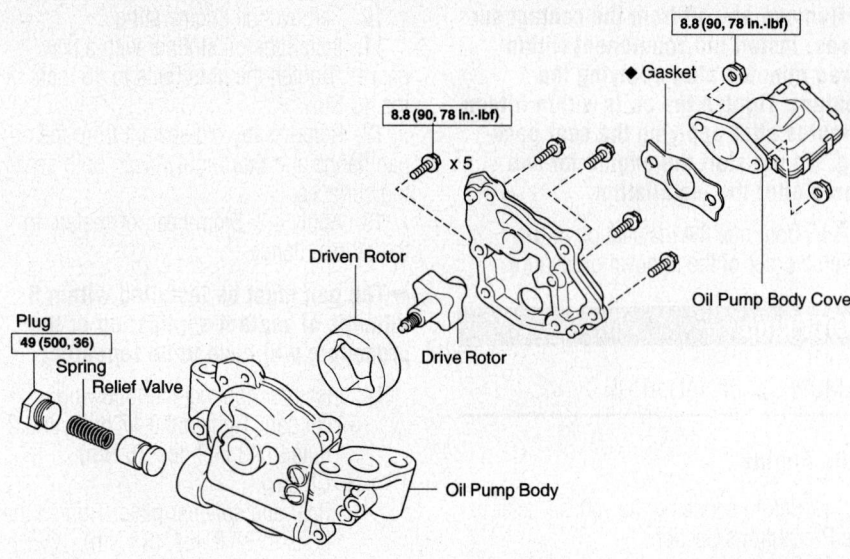

8.8 (90, 78 in.-lbf)

◆ Gasket

8.8 (90, 78 in.-lbf)

x 5

Driven Rotor

Drive Rotor

Oil Pump Body Cover

Plug
49 (500, 36)

Spring
Relief Valve

Oil Pump Body

N·m (kgf·cm, ft·lbf) : Specified torque
◆ Non–reusable part

Oil pump—2.4L engine

67170-RAV4-G34

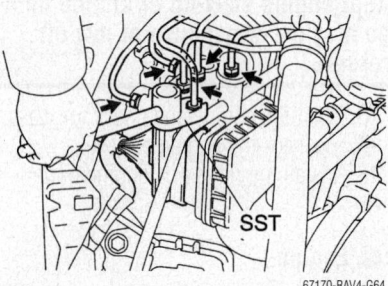

SST

67170-RAV4-G64

ABS actuator removal—2002–2005 2.4L engine

of the drive rotor and the driven rotor. Clearance should be 0.0138 inch max.

13. Place a straight-edge across the rotors and pump body. Check the side clearance. Clearance should be 0.0063 inch max.

14. Check the clearance between the driven rotor and pump body. Clearance should be 0.0128 inch max.

15. Install the cover. Torque to 78 inch lbs. (8.8 Nm).

16. Install the relief valve. Torque to 36 ft. lbs. (49 Nm).

17. Install the strainer assembly. Torque to 78 inch lbs. (8.8 Nm).

To install:

18. Install the pump with a new gasket. Torque to 14 ft. lbs. (19 Nm).

19. Install the drive chain and sprockets. The crankshaft key should be at the 9:00 o'clock position and the cutout on the oil pump shaft should be at the 12:00 o'clock

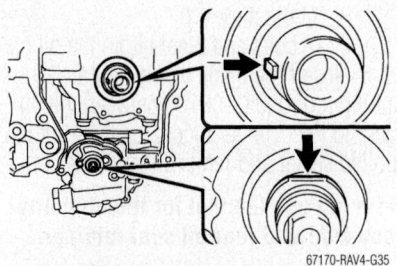

67170-RAV4-G35

The crankshaft key should be at the 9:00 o'clock position and the cutout on the oil pump shaft should be at the 12:00 o'clock position—2002–2005 2.4L engine

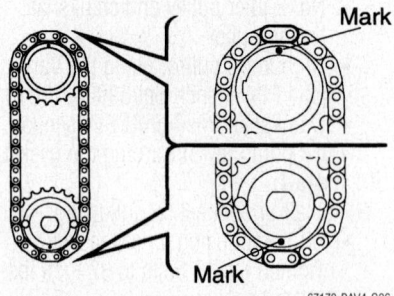

Mark

Mark

67170-RAV4-G36

The sprocket timing marks should align with the colored chain links—2002–2005 2.4L engine

position. The sprocket timing marks should align with the colored chain links. Torque the sprocket nut to 22 ft. lbs. (29 Nm).

20. Install the tensioner. Torque to 9 ft. lbs. (12 Nm).

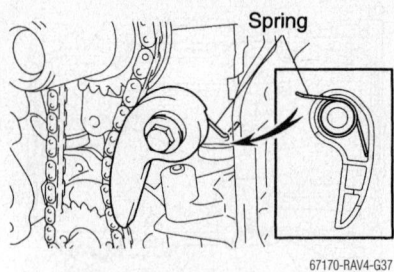

Spring

67170-RAV4-G37

Tension spring positioning—2002–2005 2.4L engine

21. Install the oil pan.

22. The remainder of installation is the reverse of removal. Torque the exhaust pipe-to-manifold nuts to 32 ft. lbs. (43 Nm); the pipe-to-pipe nuts to 36 ft. lbs. (49 Nm).

2006

1. Before servicing the vehicle, refer to the Precautions Section.

2. Disconnect the negative battery cable.

➡**Wait at least 90 seconds after disconnecting the negative battery cable before starting any repair work to prevent air bag and seat belt pretensioner activation.**

3. Remove the timing chain subassembly.

4. Remove the three oil pump retaining bolts.

5. Remove the oil pump from its mounting. Discard the gasket.

To install:

6. Install the pump with a new gasket. Torque to 14 ft. lbs. (19 Nm).

7. Continue the installation in the reverse order of the removal procedure.

3.5L Engine

➡**The following procedure is performed with the engine removed from the vehicle.**

1. Before servicing the vehicle, refer to the Precautions Section.

2. Properly relieve the fuel system pressure.

3. Disconnect the negative battery cable.

➡**Wait at least 90 seconds after disconnecting the negative battery cable before starting any repair work to prevent air bag and seat belt pretensioner activation.**

4. Remove the engine and position it in a suitable holding fixture.

5. Remove the lower oil pan. Remove the oil strainer sub assembly. Remove the upper oil pan.

6. Remove the two O-rings from the oil pump. Remove the water inlet housing. Remove the oil pipe.

7. Remove the cylinder head cover retaining bolts. Remove the cylinder head cover.

8. Remove the crankshaft pulley.

9. Remove the twenty three bolts and two nuts. Remove the timing chain cover with the oil pump.

➡**When removing the timing chain cover be careful when prying between the timing chain cover and cylinder head or cylinder block. Do not damage the contact surfaces.**

10. Remove the oil pump relief valve. Remove the eight bolts and the oil pump cover, drive rotor and driven rotor.

11. Repair or replace defective parts as required.

12. The clearance between the drive rotor and driven rotor should be 0.0024–0.0063 inch.

13. The clearance between the timing chain cover and driven rotor should be 0.0098–0.0128 inch.

14. The clearance between the rotors and the precision measuring straightedge should be 0.0012–0.0035 inch.

To install:

15. Replace the timing chain cover oil seal.

16. Install the oil pump cover. Tighten the retaining bolts to 80 inch lbs. in an alternating sequence.

➡**Be sure to install the right bolts in the right holes. Bolts are two sizes, 0.087 inch and 1.58 inch in length.**

17. Install the oil pump relief valve. Tighten the plug to 36 ft. lbs.

18. Install the timing chain cover sub-assembly, with the oil pump.

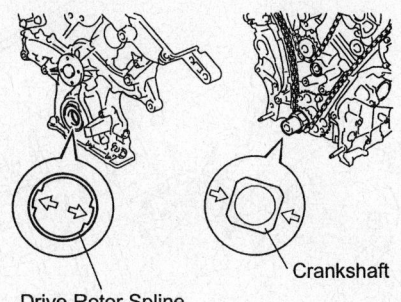

Drive Rotor Spline   Crankshaft

09490_RAV4_G0059

Oil pump alignment—3.5L engine

➡**Be sure to align the oil pump's drive rotor spline and the crankshaft, as shown in the illustration.**

19. Apply a continuous bead (0.118 inch or more) of seal packing (three bond 1207B or equivalent) to the engine.

➡**Be sure to clean and degrease the contact surfaces. Make sure these surfaces are free of oil and dirt before applying the seal packing. Install the component within three minutes and tighten the bolts within fifteen minutes after applying the seal packing. Do not start the engine for at least two hours.**

20. Apply a continuous bead of seal packing (three bond 1207B or equivalent) to the timing chain cover. See illustration for location and amount of seal packing to apply.

➡**Be sure to clean and degrease the contact surfaces. Make sure these surfaces are free of oil and dirt before applying the seal packing. Install the component within three minutes and tighten the bolts within fifteen minutes after applying the seal packing. Do not start the engine for at least two hours.**

21. Install the timing chain cover. Tighten the retaining bolts and nuts in the following order, area "1", area "2", area "3" and area "4". Tighten the bolts and nuts to 15 ft. lbs for area "1" area "2" and area "3". Tighten bolt "A" in area "4" to 32 ft. lbs. Tighten all other bolts in area "4" to 15 ft. lbs.

➡**Be sure that there is no oil on the bolts. Bolt "A" is 1.57 inch long, bolt "B" is 2.17 inch long and bolt "C" is 0.98 inch long.**

22. When installing the cylinder head cover, apply seal packing, three bond 1207B or equivalent.

➡**Remove any oil from the contact surfaces. Install the component within three minutes after applying the sealant. Do not start the engine for two hours after the installation.**

23. Install the cover retaining bolts and torque to 15 ft. lbs for bolt "A" and to 7 ft. lbs. for all other bolts and in the proper sequence.

24. Continue the installation is the reverse order of the removal procedure.

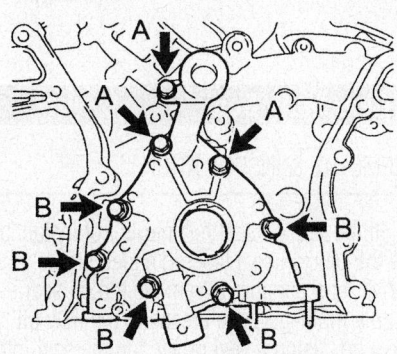

09490_RAV4_G0058

Oil pump cover bolt identification—3.5L engine

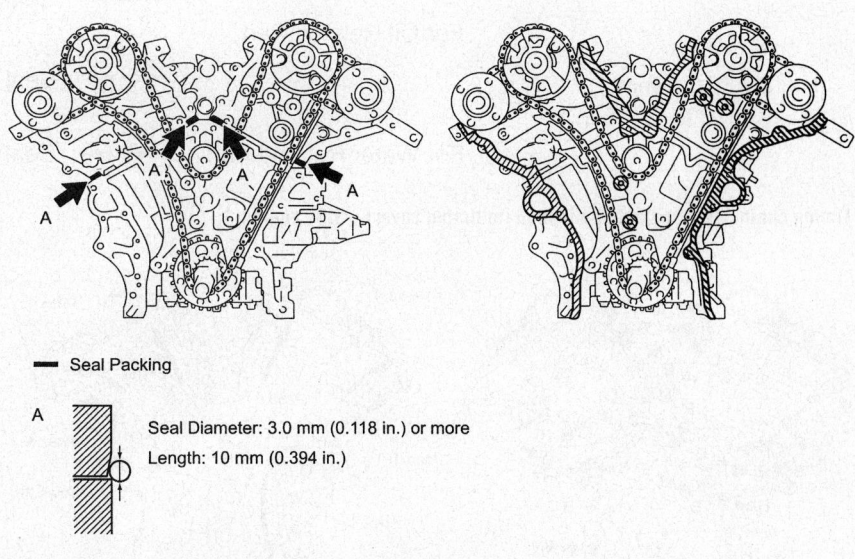

━ Seal Packing

A ▨ Seal Diameter: 3.0 mm (0.118 in.) or more
Length: 10 mm (0.394 in.)

09490_RAV4_G0060

Timing chain cover sealant application (to engine block)—3.5L engine

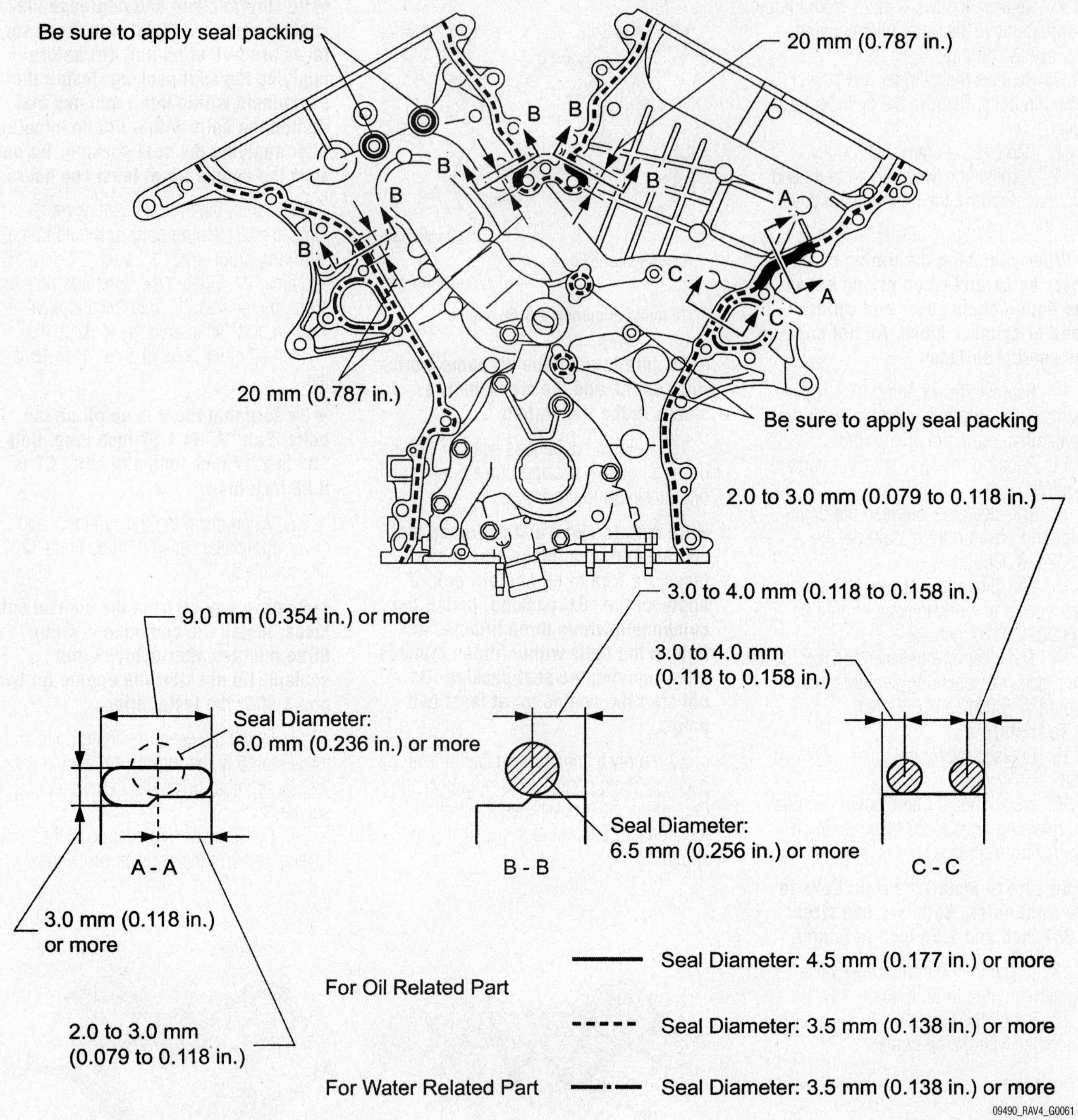

Be sure to apply seal packing

20 mm (0.787 in.)

20 mm (0.787 in.)

Be sure to apply seal packing

2.0 to 3.0 mm (0.079 to 0.118 in.)

3.0 to 4.0 mm (0.118 to 0.158 in.)

3.0 to 4.0 mm (0.118 to 0.158 in.)

9.0 mm (0.354 in.) or more

Seal Diameter: 6.0 mm (0.236 in.) or more

A - A

3.0 mm (0.118 in.) or more

2.0 to 3.0 mm (0.079 to 0.118 in.)

B - B

Seal Diameter: 6.5 mm (0.256 in.) or more

C - C

For Oil Related Part

——————— Seal Diameter: 4.5 mm (0.177 in.) or more

- - - - - - - Seal Diameter: 3.5 mm (0.138 in.) or more

For Water Related Part

—·—·—·— Seal Diameter: 3.5 mm (0.138 in.) or more

09490_RAV4_G0061

Timing chain cover sealant application (to timing cover)—3.5L engine

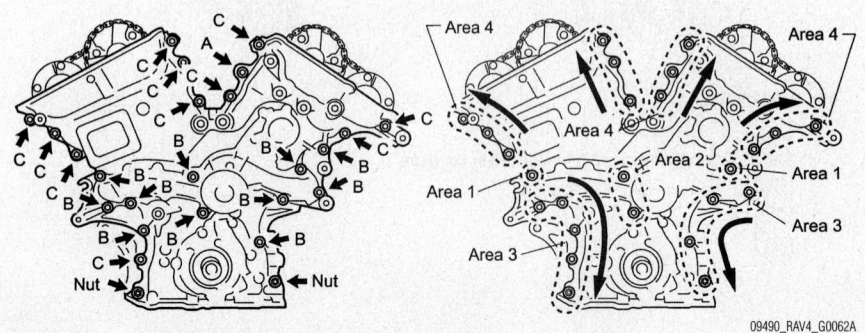

09490_RAV4_G0062A

Timing chain cover bolt location and torque sequence—3.5L engine

Rear Main Seal

REMOVAL & INSTALLATION

If the rear oil seal retainer is not installed to the block, use a tapered ended screwdriver and hammer to remove the oil seal. Apply multi-purpose grease to the new oil seal lip. Using a seal driver, tap the seal into place. Be careful not to install it slantwise.

1. Before servicing the vehicle, refer to the Precautions Section.

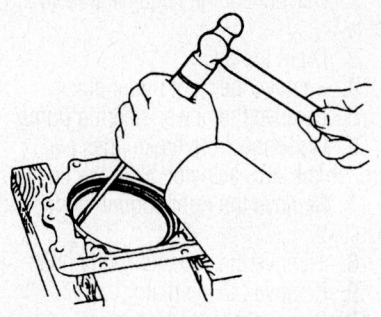

Carefully tap the old seal from the
retainer—2.0L Engine

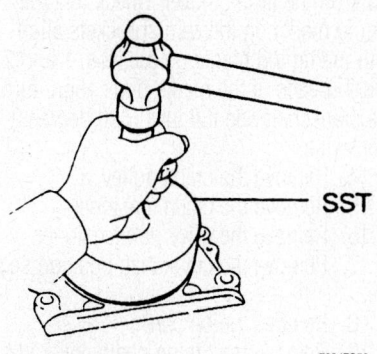

Use the proper sized driver to seat the
seal—2.0L Engine

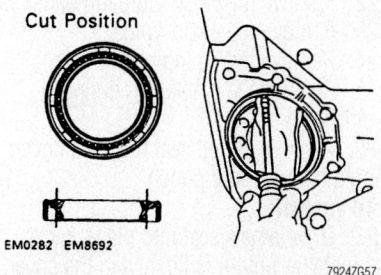

Cut off the oil seal lip, then pry the seal
out of the retaining plate—2.0L Engine

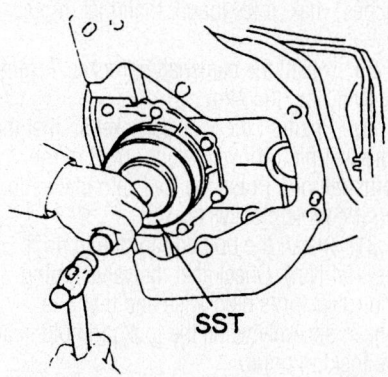

Tap a new seal into place—2.0L Engine

If the rear oil seal retainer is installed on
the cylinder block, using a knife, cut off the
lip of the seal. Using a taped ended prytool,
pry the old seal out of the retainer. Inspect
the oil seal lip contacting surface of the
crankshaft for cracks or damage. Apply multipurpose grease to the new oil seal, then
tap the seal in place with a seal installer. Be
careful not to install the seal slantwise.

Timing Belt, Cover and Crankshaft Seal

REMOVAL & INSTALLATION

2.0L Engine

The timing belt is not adjustable.
1. Before servicing the vehicle, refer to
the Precautions Section.
2. Disconnect the negative battery cable.

✴✴ CAUTION

**To avoid air bag deployment, if
equipped, work must be started after
approximately 90 seconds or longer
from the time the ignition switch is
turned to the LOCK position and the
negative battery cable is disconnected from the battery.**

3. Disconnect the power steering reservoir tank and remove the reservoir bracket.
4. Detach the wiring harness bracket for
the Data Link Connector 1 (DLC1).
5. Remove the alternator and alternator
bracket.
6. If equipped with ABS brakes, remove
the ABS actuator.
7. Remove the right front wheel and the
fender apron seal.
8. Remove the power steering drive
belt.
9. Slightly raise the engine using a
block of wood and floor jack under the oil
pan to prevent damage.
10. Remove the 4 bolts, 2 nuts, and right
side mounting bracket.
11. Remove the spark plugs.
12. Using SST 09213-54015, or equivalent, loosen the crankshaft pulley bolt and
remove it by pulling it straight off the crankshaft.
13. Using SST 09249-63010, or equivalent, loosen the retaining bolts and remove
the right engine mounting bracket.
14. Remove the upper (No. 2) timing belt
cover.
15. Install the crankshaft pulley to the
crankshaft and temporarily install the retaining bolt.

16. Turn the crankshaft pulley and align
its groove with the timing mark **0** of the
No. 1 timing belt cover. Check that the hole
of the camshaft timing pulley is aligned with
the timing mark of the bearing cap. If not,
turn the crankshaft 360 degrees and align
the marks.

➡**If the timing belt is to be reused,
matchmark the timing belt to the timing
pulleys and timing belt covers so the
belt can be reinstalled in its original
position. Also, be sure to mark an
arrow on the belt to indicate which
direction it was turning.**

17. Remove the timing belt from the
camshaft timing pulley.
18. Hold the camshaft sprocket with a
spanner wrench and remove the mounting
bolt. Remove the camshaft pulley.
19. Remove the crankshaft pulley bolt
and remove the crankshaft pulley.
20. Remove the No. 1 timing belt cover.
21. Remove the timing belt guide and
the timing belt.
22. Remove the No. 1 idler pulley and
tension spring.
23. Remove the No. 2 idler pulley.
24. Remove the crankshaft timing pulley.
25. Support the oil pump sprocket with a
spanner wrench, then remove the mounting
bolt and remove the sprocket.

To install:

26. Install the oil pump pulley. Tighten
the nut to 18 ft. lbs. (24 Nm).
27. Install the crankshaft timing pulley.
Align the pulley set key with the key groove
of the pulley. Slide on the pulley facing the
flange side inward.
28. Install the No. 2 idler pulley and
tighten the mounting bolt to 31 ft. lbs. (42
Nm). Be sure that the pulley moves
smoothly.
29. Install the No. 1 idler pulley with the
bolt and the tension spring. Pry the pulley

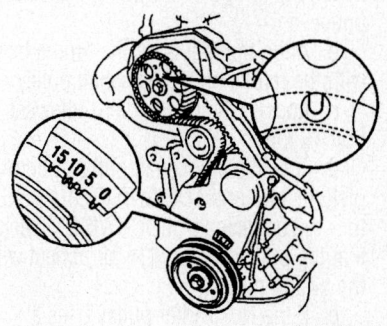

It is necessary to align the timing reference indicators prior to removing the timing belt—2.0L engine

toward the left as far as it will go and tighten the bolt. Make sure that the pulley moves smoothly.

30. Temporarily install the timing belt. Using the crankshaft pulley bolt, turn the crankshaft and position the key groove of the crankshaft timing pulley upward. If reusing the timing belt, align the points marked during removal.

31. Install the timing belt on the crankshaft timing pulley, oil pump pulley, No. 1 idler pulley, water pump pulley and the No. 2 idler pulley.

32. Install the timing belt guide.

➡ **If the old timing belt is being reinstalled, be sure the directional arrow is facing in the original direction and that the belt and sprocket/cover matchmarks are properly aligned.**

33. Install the lower (No. 1) timing belt cover and new gasket with the 4 bolts.

34. Align the crankshaft pulley set key with the pulley key groove. Temporarily install the crankshaft pulley and bolt.

35. Align the camshaft knock pin with the groove of the pulley, and slide the timing pulley onto the camshaft with the plate washer and set bolt.

36. Tighten the pulley set bolt to 40 ft. lbs. (54 Nm).

❄❄ WARNING

If any binding is felt when adjusting the timing belt tension by turning the crankshaft, STOP turning the engine, because the pistons may be hitting the valves.

37. Turn the crankshaft pulley and align the **0** mark on the lower (No. 1) timing belt cover.

38. Finish installing the timing belt and check the valve timing, as follows:

a. If reusing the old timing belt, align the matchmarks made previously and install the timing belt onto the camshaft pulley.

b. Align the marks on the timing belt with the marks on the camshaft pulley.

c. Loosen the No. 1 idler pulley set bolt ½ turn.

d. Turn the crankshaft pulley 2 complete revolutions TDC to TDC. ALWAYS turn the crankshaft CLOCKWISE. Check that the pulleys are still in alignment with the timing marks.

e. If the No. 1 idler pulley uses a green tension spring, slowly turn the crankshaft pulley 1⅞ revolutions, and align its groove with the mark at 45

degrees BTDC (for the No. 1 cylinder) of the No. 1 timing belt cover.

f. Tighten the No. 1 idler pulley set bolt to 31 ft. lbs. (42 Nm).

g. Be sure there is belt tension between the crankshaft and camshaft timing pulleys.

39. Place the right side engine mounting bracket in position but do not install the bolts.

40. Install the upper (No. 2) timing cover with a new gasket(s).

41. Remove the engine crankshaft pulley bolt and pulley.

42. Using SST 09249-63010, or equivalent, install the mounting bolts for the right side mounting bracket. Tighten the mounting bolts to 38 ft. lbs. (52 Nm).

43. Align the crankshaft pulley set key with the pulley key groove. Install the pulley. Tighten the pulley bolt to 80 ft. lbs. (108 Nm).

44. Install the spark plugs.

45. Install the right side mounting insulator, as follows:

a. Attach the mounting insulator to the body and mounting bracket with the 4 bolts and 2 nuts.

b. Tighten the 3 bolts to hold the mounting insulator to the body. Tighten the bolts to 47 ft. lbs. (64 Nm).

c. Tighten the 2 nuts and bolt to hold the mounting insulator to the mounting bracket. Tighten the bolt to 27 ft. lbs. (37 Nm) and the nut to 38 ft. lbs. (52 Nm).

46. Install and adjust the power steering pump drive belt.

47. Install the right side engine undercover.

48. Install the right front wheel.

49. Lower the engine.

50. If equipped, install the ABS actuator.

51. Install the alternator and alternator bracket.

52. Install the wiring harness bracket for the DLC1.

53. Install the power steering reservoir bracket and reservoir.

54. Connect the negative battery cable.

55. Start the engine and check the timing.

Timing Chain, Sprockets, Front Cover and Seal

REMOVAL & INSTALLATION

2.4L Engine

2002–2005

1. Before servicing the vehicle, refer to the Precautions Section.

2. Disconnect the negative battery cable.

3. Drain the oil.

4. Remove the right under cover.

5. Remove the power steering pump.

6. Disconnect the brake lines and remove the ABS actuator. Plug the lines.

7. Remove the right engine mount insulator.

8. Remove the accessory drive belt.

9. Remove the alternator.

10. Remove the air cleaner assembly.

11. Remove the ignition coils.

12. Remove the cylinder head cover.

13. Turn the crankshaft clockwise and align the groove in the pulley with the "0" mark on the timing cover. Check that the timing marks on the cam sprockets align with the timing marks on the Nos. 1 and 2 bearing caps. If the marks don't align, turn the crankshaft one full turn (360 degrees) clockwise.

14. Remove the crank pulley.

15. Remove the chain tensioner.

16. Remove the drive belt tensioner.

17. Remove the crankshaft position sensor.

18. Remove the oil pan.

19. Remove the timing chain cover (14 bolts and 2 nuts)

20. Remove the crank angle sensor plate.

21. Remove the chain tension slipper.

22. Remove the chain vibration damper.

23. Remove the chain guide.

24. Remove the timing chain.

25. Remove the crankshaft timing sprocket.

26. Remove the oil seal from the cover, using a hammer and punch.

To install:

27. Drive a new seal into place with a seal installer until it is flush with the cover. Coat the seal lip with multi-purpose grease.

28. Check the timing chain with the chain full stretched. The length of 16 consecutive links should be 122.6mm (4.827 inches) max. If it's longer than that, replace it.

29. Install the camshaft sprocket. Torque to 40 ft. lbs. (54 Nm).

30. Position the VVT sprocket so that the sprocket pin groove is slight right of the camshaft pin. Press the VVT into place turning it counterclockwise.

31. Install the bolt and torque to 40 ft. lbs. (54 Nm). Check that the valve timing controller turns clockwise and that it is locked securely when the lock pin hole is at the locking point.

32. Turn the camshafts so that the timing marks on the cam sprockets align with the

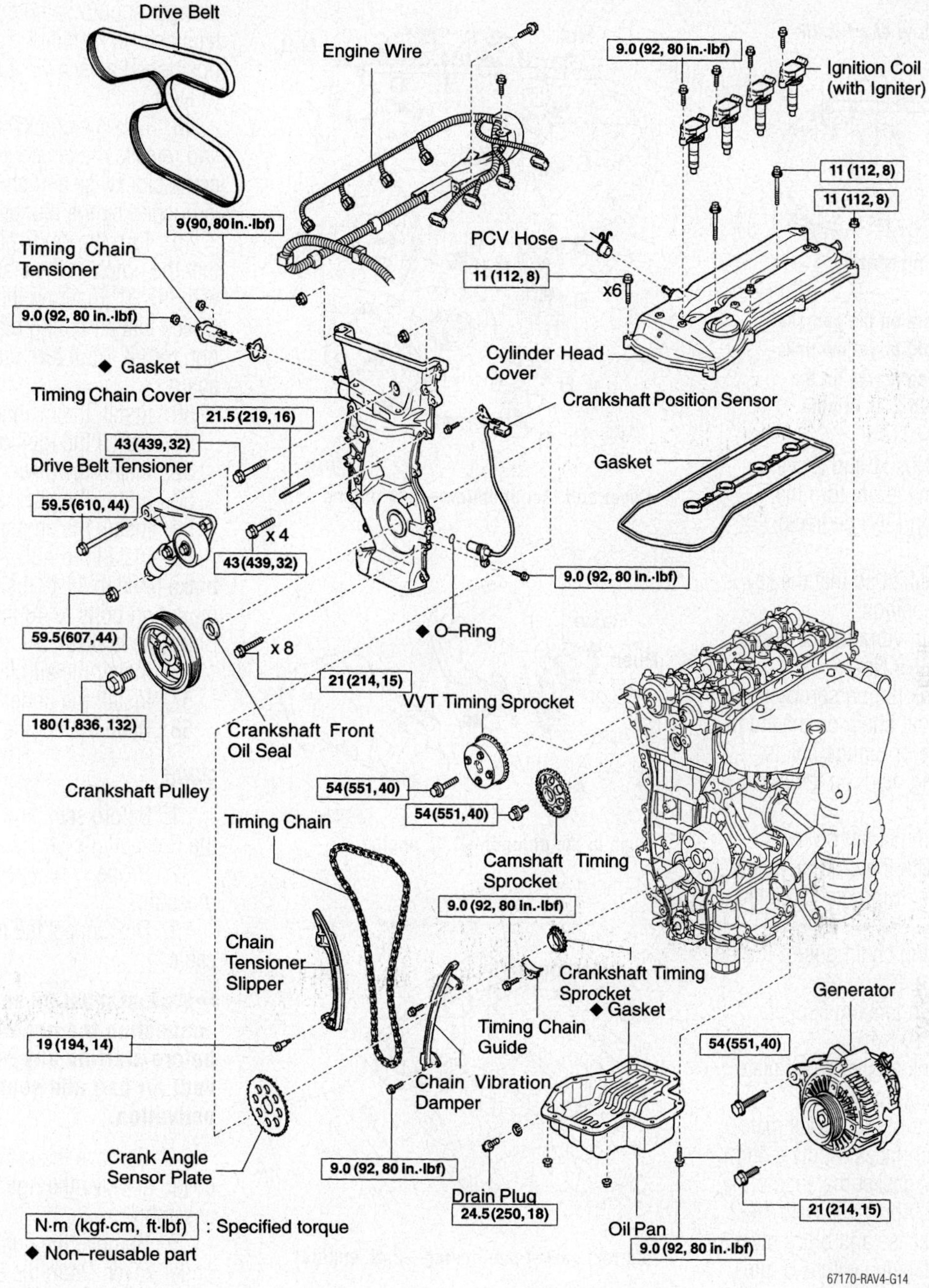

Drive Belt
Engine Wire
9.0 (92, 80 in.·lbf)
Ignition Coil (with Igniter)
11 (112, 8)
11 (112, 8)
9 (90, 80 in.·lbf)
PCV Hose
Timing Chain Tensioner
11 (112, 8)
x6
9.0 (92, 80 in.·lbf)
◆ Gasket
Cylinder Head Cover
Timing Chain Cover
Crankshaft Position Sensor
21.5 (219, 16)
43 (439, 32)
Drive Belt Tensioner
Gasket
59.5 (610, 44)
x 4
43 (439, 32)
9.0 (92, 80 in.·lbf)
59.5 (607, 44)
◆ O–Ring
x 8
21 (214, 15)
180 (1,836, 132)
VVT Timing Sprocket
Crankshaft Front Oil Seal
54 (551, 40)
Crankshaft Pulley
54 (551, 40)
Timing Chain
Camshaft Timing Sprocket
9.0 (92, 80 in.·lbf)
Chain Tensioner Slipper
Crankshaft Timing Sprocket
◆ Gasket
Generator
Timing Chain Guide
19 (194, 14)
54 (551, 40)
Chain Vibration Damper
Crank Angle Sensor Plate
9.0 (92, 80 in.·lbf)
Drain Plug
24.5 (250, 18)
21 (214, 15)
N·m (kgf·cm, ft·lbf) : Specified torque
◆ Non–reusable part
Oil Pan
9.0 (92, 80 in.·lbf)
67170-RAV4-G14

Timing chain and related components—2.4L engine

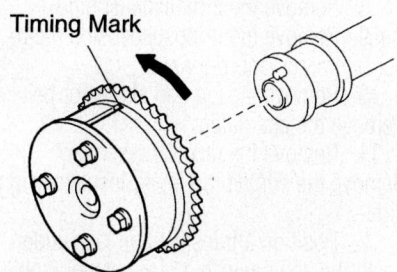

Timing Mark
67170-RAV4-G12

Position the VVT sprocket so that the sprocket pin groove is slight right of the camshaft pin—2002–2005 2.4L engine

Timing Marks
Timing Marks
67170-RAV4-G13

Turn the camshafts so that the timing marks on the cam sprockets align with the timing marks on the Nos. 1 and 2 bearing caps—2002–2005 2.4L engine

Blue or Orange Mark Link
Timing Mark
67170-RAV4-G15

Install the timing chain on the crank sprocket with the blue or orange link is aligned with the timing mark on the crank sprocket—2002–2005 2.4L engine

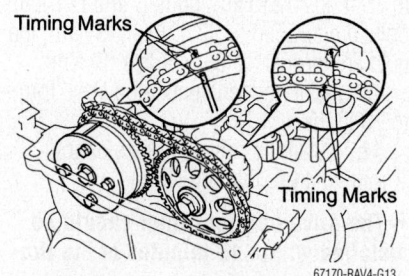

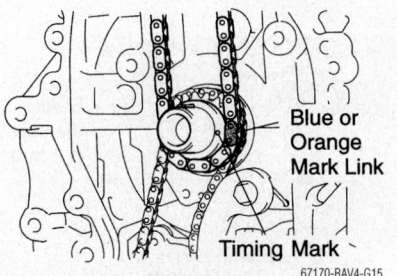

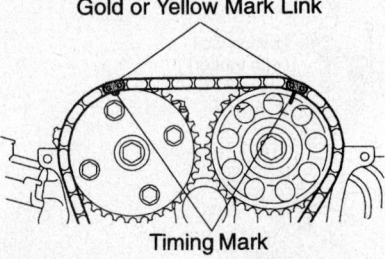

67170-RAV4-G16

Install the timing chain on the camshaft sprockets with the gold or yellow links aligned with the timing marks on the sprockets—2002–2005 2.4L engine

timing marks on the Nos. 1 and 2 bearing caps. If the marks don't align, turn the crankshaft one full turn (360 degrees) clockwise.

33. Turn the crankshaft so that the key is at the 12:00 o'clock position.

34. Install the chain vibration damper. Torque to 80 inch lbs. (9 Nm).

35. Install the crank timing sprocket.

36. Install the timing chain on the crank sprocket with the blue or orange link is aligned with the timing mark on the crank sprocket.

37. Install the timing chain on the camshaft sprockets with the gold or yellow links aligned with the timing marks on the sprockets.

38. Install the timing chain guide. Torque to 80 inch lbs. (9 Nm).

39. Install the chain tension slipper. Torque to 14 ft. lbs. (19 Nm).

40. Install the crank angle sensor plate with the "F" mark facing forward.

41. Clean all traces of old gasket material from the mating surfaces. Apply a 4mm diameter bead of RTV gasket material to the cover mating surface. Apply sealer to the 2 areas where the crankcase and block meet. The cover must be installed with 3-5 minutes of applying the sealer.

42. Install the cover and torque the bolts as shown in the accompanying illustration.

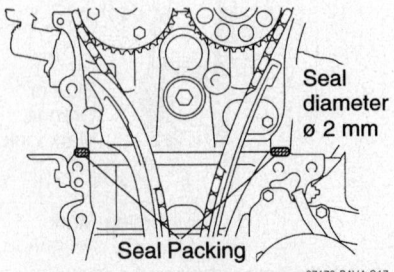

67170-RAV4-G17

Apply sealer to the 2 areas where the crankcase and block meet—2002–2005 2.4L engine

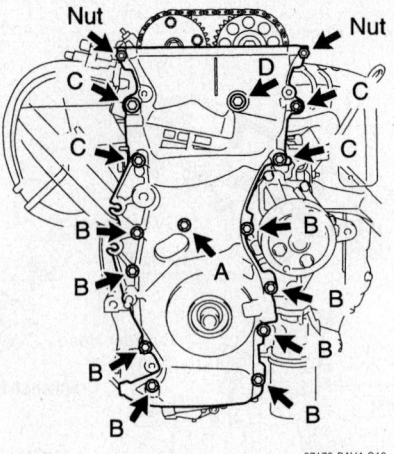

67170-RAV4-G18

Cover bolt identification—2.4L engine

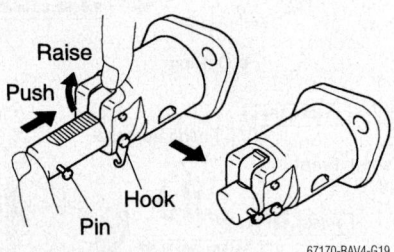

67170-RAV4-G19

Push in the plunger—2.4L engine

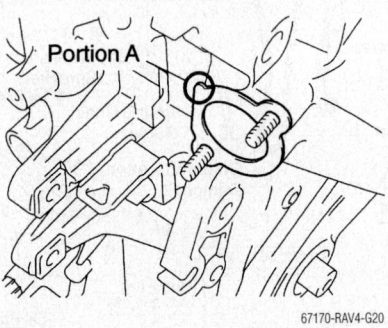

67170-RAV4-G20

Correct gasket positioning—2.4L engine

Bolt A 10 mm bolts are torqued to 80 inch lbs. (9 Nm); bolt B 12mm bolts are torqued to 15 ft. lbs. (21 Nm); bolts C and D 14mm bolts are torqued to 32 ft. lbs. (43 Nm); the nuts are torqued to 80 inch lbs. (9 Nm).

43. Install the tensioner stud bolt. Torque to 16 ft. lbs. (21 Nm).

44. Install the drive belt tensioner. Torque to 44 ft. lbs. (60 Nm).

➡ **The drive belt tensioner should be installed within 15 minutes of the timing cover.**

45. Install the crankshaft position sensor.

46. Install the oil pan.

47. Install the chain tensioner. The

plunger should be held retracted until the tensioner is installed. The gasket should be positioned as shown. Torque to 80 inch lbs. (9 Nm).

48. Turn the crankshaft counterclockwise and release the tensioner plunger. Turn the crank clockwise and check that the slipper is pushed by the plunger.

49. Turn the crank puller clockwise so that the notch in the crank pulley is aligned with the "0" mark on the timing cover. Check that all timing marks are aligned. If not, turn it 1 full turn clockwise and check again.

50. Install the cylinder head cover.

51. Install the ignition coils.

52. Install the power steering pump.

53. Install the alternator.

54. Install the engine mount insulator.

55. Install the ABS actuator. Torque the brake lines to 11 ft. lbs. (15 Nm); the mounting bolts to 48 inch lbs. (5 Nm). Bleed the brakes.

56. Install the air cleaner.

57. Install the under cover.

58. Refill the engine oil.

2006

1. Before servicing the vehicle, refer to the Precautions Section.

2. Properly relieve the fuel system pressure.

3. Disconnect the negative battery cable.

➡ **Wait at least 90 seconds after disconnecting the negative battery cable before starting any repair work to prevent air bag and seat belt pretensioner activation.**

4. Remove the radiator support opening cover. Remove the right tire and wheel assembly.

5. Remove the number one engine under cover. Drain the engine oil.

6. Remove the right front fender apron.

7. Remove the number one engine cover.

8. Remove the front exhaust pipe.

9. Remove the front suspension member reinforcement, right side.

10. Remove the fan and alternator belt. Remove the alternator.

11. Remove the radiator reservoir. Remove the engine mounting insulator, right side.

12. Position a transmission jack underneath the engine, with a piece of wood on the jack. Remove the four bolts and two nuts. Remove the right engine mounting insulator.

➡**Do not apply excessive force to the return tube when removing the engine mounting insulator.**

13. Remove the front engine mounting insulator.

14. Remove the idler pulley. Remove the ignition coil assembly. Remove the spark plugs.

15. Remove the cylinder head cover retaining bolts. Remove the cylinder head cover.

16. Remove the oil pan.

17. Position the engine number one cylinder at TDC on the compression stroke.

➡**Check that the timing mark of the camshaft timing gear and sprocket is aligned with the timing mark located on the number one and number two bearing caps. If not rotate the engine 360 degrees.**

18. Remove the crankshaft pulley.

19. To remove the number one chain tensioner assembly, remove the two nuts, chain tensioner and gasket.

➡**Do not turn the crankshaft without the chain tensioner in place.**

20. Remove the engine mounting bracket, right side.

21. Remove the belt tensioner assembly.

➡**Using the proper equipment, lift the engine upward to remove the retaining bolt. Do not raise the engine more than necessary.**

22. Remove the crankshaft position sensor.

23. Remove the timing chain cover retaining bolts and nuts. Remove the timing chain cover from the engine.

➡**Remove the cover by prying the portions between the timing chain cover, cylinder head and cylinder block with a pry tool. Be careful not to damage the contact surfaces.**

24. Remove the number one crankshaft position sensor plate. Remove the bolt and the timing chain guide.

25. Remove the bolt and the chain tensioner slipper.

26. Remove the two bolts and the number one chain vibration damper. Remove the timing chain.

27. Remove the crankshaft timing sprocket.

28. To remove the number two timing chain, turn the crankshaft ninety degrees counterclockwise to align the adjusting hole of the oil pump driveshaft sprocket with the groove of the oil pump.

29. Insert a 4mm diameter bar into the adjusting hole of the oil pump driveshaft sprocket to lock the gear in position. Remove the nut.

30. Remove the bolt, chain tensioner plate and spring. Remove the oil pump drive sprocket, oil pump driveshaft and number two timing chain.

To install:

31. Drive a new seal into place with a seal installer until it is flush with the cover. Coat the seal lip with multi-purpose grease.

32. Check the timing chain with the chain full stretched. The length of 16 consecutive links should be 122.6mm (4.827 inches) max. If it's longer than that, replace it.

33. To install the number two timing chain, position the crankshaft key in the left horizontal position. Turn the cutout of the driveshaft so that it faces upward.

34. Align the yellow links with the timing marks of each gear. Install the sprockets onto the crankshaft and oil pump shaft with the chain wrapped on the gears.

35. Temporarily tighten the oil pump driveshaft sprocket with the nut.

36. Insert the damper spring into the adjusting hole, and then install the chain tensioner plate with the bolt. Tighten the bolt to 9 ft. lbs.

37. Align the adjusting hole of the oil pump driveshaft sprocket with the groove of the oil pump.

38. Insert a 4mm diameter bar into the adjusting hole of the oil pump driveshaft gear to lock the gear in position. Tighten the nut to 22 ft. lbs.

39. Install the crankshaft timing sprocket. Install the number one chain vibration damper. Tighten the bolts to 80 inch lbs.

40. To install the number one timing chain, be sure the number one cylinder is at TDC on the compression stroke. Position the key on the crankshaft upward.

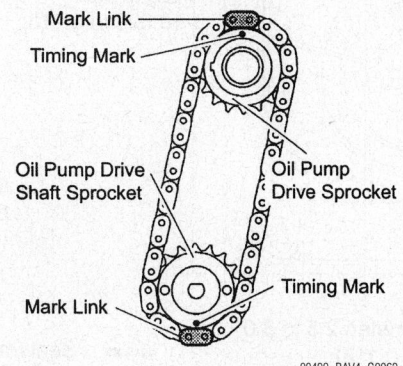

Number two timing chain alignment—2006 2.4L engine

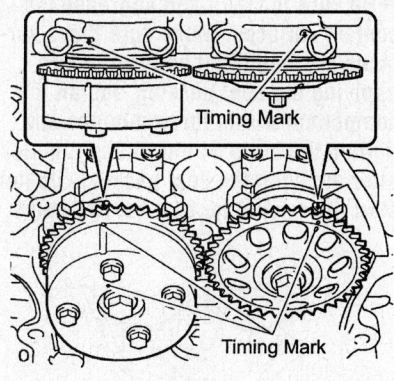

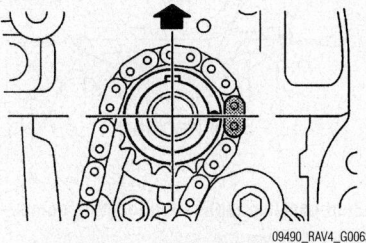

TDC alignment—2006 2.4L engine

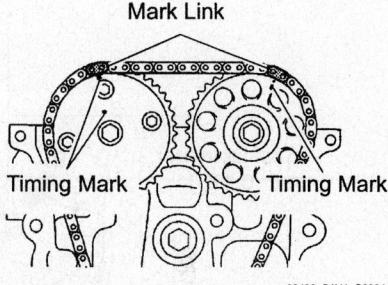

Number one timing chain alignment—2006 2.4L engine

41. Install the chain on the crankshaft sprocket with the gold or orange mark aligned with the timing mark on the crankshaft.

42. Align the gold or yellow links with each timing mark located on the camshaft timing gear and sprocket. Install the chain.

43. Install the chain tensioner slipper. Tighten the bolt to 14 ft. lbs.

44. Install the timing chain guide and bolt. Tighten to 80 inch lbs.

45. Install the number one crankshaft position sensor plate. Be sure the "F" mark is facing forward.

46. To install the timing chain cover, remove any old packing material. Be careful not to drop oil on the contact surfaces.

47. Apply a continuous bead (0157–0.177 inch) of seal packing (three bond 1207B or equivalent) as shown in the illustration.

➡️**Be sure to clean and degrease the contact surfaces. Make sure these surfaces are free of oil and dirt before applying the seal packing. Install the component within three minutes and tighten the bolts within fifteen minutes after applying the seal packing. Do not start the engine for at least two hours.**

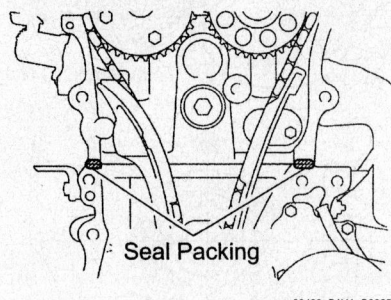

Seal Packing

09490_RAV4_G0065

Seal packing application locating points— 2006 2.4L engine

48. Apply a continuous bead of seal packing (three bond 1207B or equivalent) as shown in the illustration.

➡️**Be sure to clean and degrease the contact surfaces. Make sure these surfaces are free of oil and dirt before applying the seal packing. Install the component within three minutes and tighten the bolts within fifteen minutes after applying the seal packing. Do not start the engine for at least two hours.**

49. Install the timing chain cover retaining bolts and nuts. Tighten bolt "A" to 80 inch lbs. Tighten bolt "B" to 18 ft. lbs. Tighten bolt "C" to 41 ft. lbs. Tighten the nuts to 8 ft. lbs.

50. Release the ratchet pawl of the number one chain tensioner assembly. Push in the plunger and hook the hook to the pin so that the plunger is pushed in.

51. Install the chain tensioner using a new gasket. Tighten the nuts to 80 inch lbs.

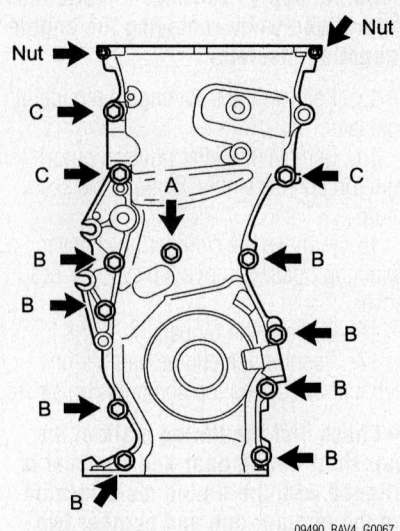

09490_RAV4_G0067

Timing chain cover bolt location and identification—2006 2.4L engine

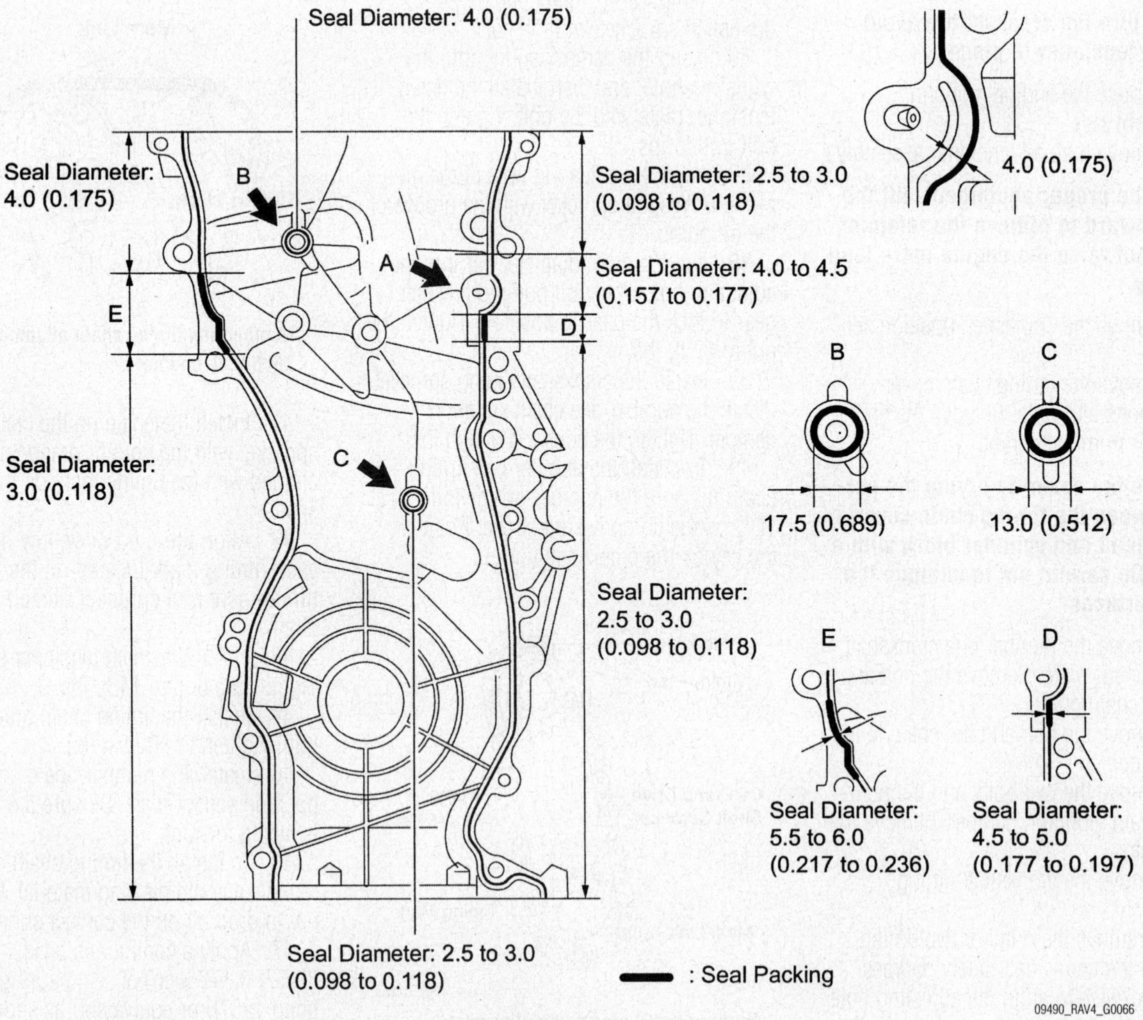

Seal Diameter: 4.0 (0.175)

Seal Diameter: 4.0 (0.175)

B

A

Seal Diameter: 2.5 to 3.0 (0.098 to 0.118)

Seal Diameter: 4.0 to 4.5 (0.157 to 0.177)

E

D

Seal Diameter: 3.0 (0.118)

C

Seal Diameter: 2.5 to 3.0 (0.098 to 0.118)

Seal Diameter: 2.5 to 3.0 (0.098 to 0.118)

A

4.0 (0.175)

B

17.5 (0.689)

C

13.0 (0.512)

E

Seal Diameter: 5.5 to 6.0 (0.217 to 0.236)

D

Seal Diameter: 4.5 to 5.0 (0.177 to 0.197)

▬▬ : Seal Packing

09490_RAV4_G0066

Timing chain cover seal packing application locating points—2006 2.4L engine

➡**When installing the tensioner, set the hook again if the hook releases the plunger.**

52. Continue the installation in the reverse order of the removal procedure.

3.5L Engine

➡**The following procedure is performed with the engine removed from the vehicle.**

1. Before servicing the vehicle, refer to the Precautions Section.

2. Properly relieve the fuel system pressure.

3. Disconnect the negative battery cable.

➡**Wait at least 90 seconds after disconnecting the negative battery cable before starting any repair work to prevent air bag and seat belt pretensioner activation.**

4. Remove the engine and position it in a suitable holding fixture.

5. Remove the oil filler cap subassembly. Remove the spark plugs.

6. Remove the camshaft timing control valve assembly.

7. Remove the VVT sensor. Remove the water inlet. Remove the oil pipe.

8. Remove the twelve retaining bolts and remove the cylinder head cover for bank one.

➡**Upon installation be sure that removed components are reinstalled in their original locations.**

9. Remove the number one oil pipe.

10. Remove the twelve retaining bolts and remove the cylinder head cover for bank two.

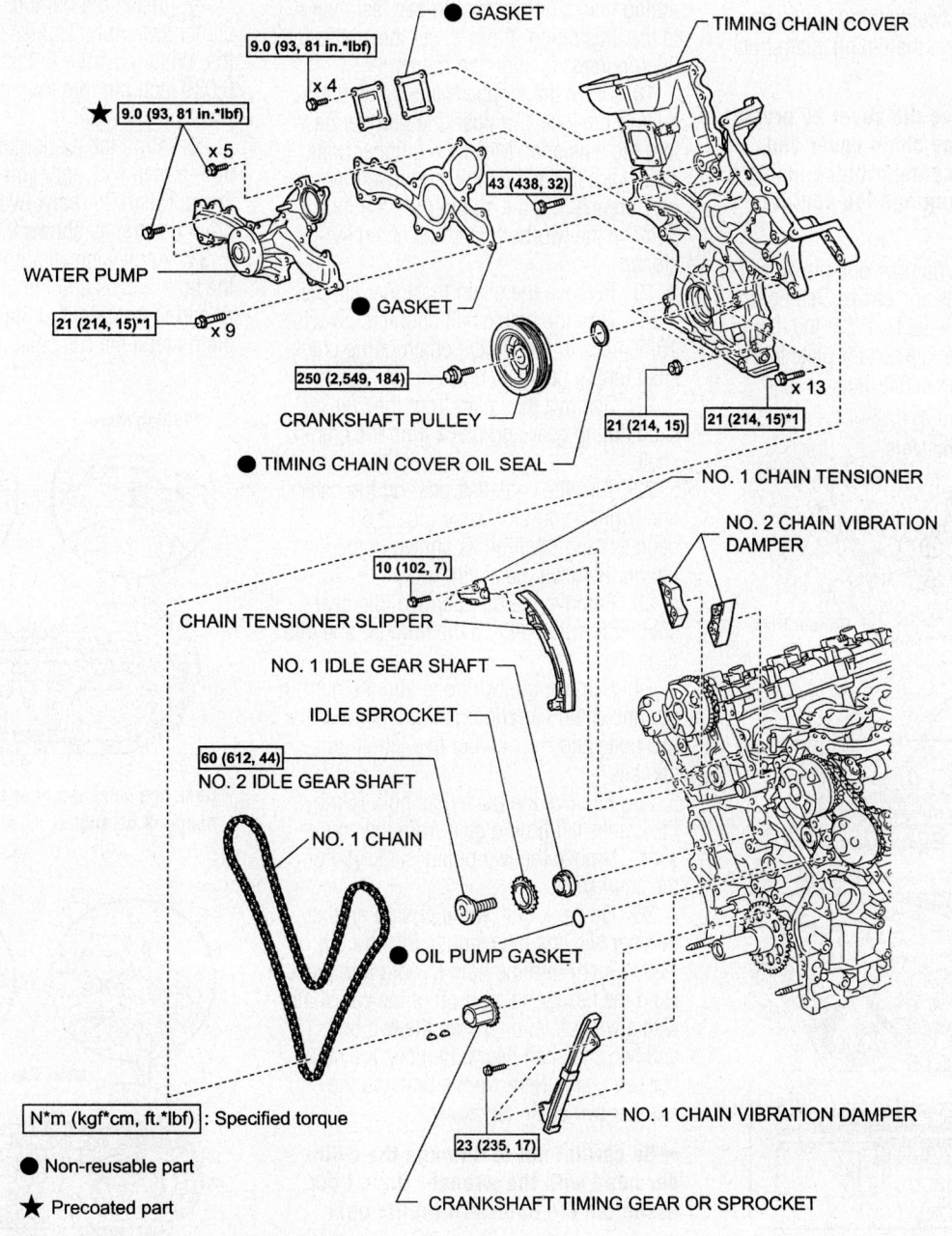

● GASKET — TIMING CHAIN COVER

9.0 (93, 81 in.*lbf) x 4

★ 9.0 (93, 81 in.*lbf) x 5

43 (438, 32)

WATER PUMP

21 (214, 15)*1 x 9

● GASKET

250 (2,549, 184)
CRANKSHAFT PULLEY

● TIMING CHAIN COVER OIL SEAL

21 (214, 15) 21 (214, 15)*1 x 13

NO. 1 CHAIN TENSIONER

NO. 2 CHAIN VIBRATION DAMPER

10 (102, 7)

CHAIN TENSIONER SLIPPER

NO. 1 IDLE GEAR SHAFT

IDLE SPROCKET

60 (612, 44)
NO. 2 IDLE GEAR SHAFT

NO. 1 CHAIN

● OIL PUMP GASKET

N*m (kgf*cm, ft.*lbf) : Specified torque

● Non-reusable part

★ Precoated part

*1: Do not allow oil to contact these bolts

NO. 1 CHAIN VIBRATION DAMPER

23 (235, 17)

CRANKSHAFT TIMING GEAR OR SPROCKET

09490_RAV4_G0068

Timing chain and related components—3.5L engine

➡ **Upon installation be sure that removed components are reinstalled in their original locations.**

11. Remove the number one engine mounting bracket. Remove the water inlet housing.

12. Remove the oil filter cap assembly. Remove the oil filter.

13. Remove the number two oil pan. Remove the oil strainer assembly. Remove the number one oil pan. Remove the seven retaining bolts and remove the oil pan baffle plate.

14. Remove the crankshaft pulley. Remove the water pump.

15. Remove the twenty three bolts and two nuts, and remove the timing chain front cover.

➡ **Carefully remove the cover by prying between the timing chain cover and cylinder head or cylinder block, using a pry tool. Do not damage the contact surfaces.**

16. Position the number one piston at TDC on the compression stroke. Temporarily tighten the pulley set bolt. Set the timing mark on the crank angle sensor plate to the right side block bore center line.

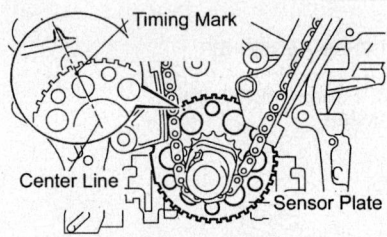

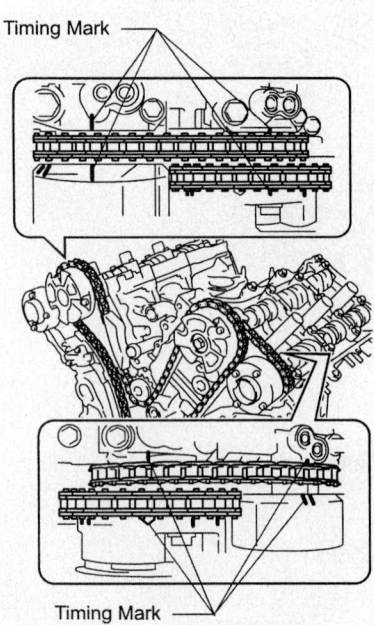

Timing chain alignment—3.5L engine

09490_RAV4_G0069

for Bank 1:

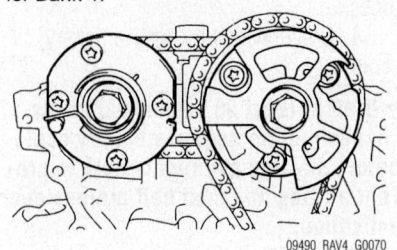

09490_RAV4_G0070

Bank one camshaft gear positioning—3.5L engine

17. Check that the timing marks of the camshaft timing gears are aligned with the timing marks of the bearing cap, as shown in the illustration. If not rotate the crankshaft 360 degrees, to align the marks.

18. Move the stopper plate upward to release the lock and push the plunger deep into the tensioner. Move the stopper plate downward to set the lock, and insert a hexagon wrench into the stopper plate hole. Remove the two bolts and the chain tensioner.

19. Remove the chain tensioner slipper.

20. Turn the crankshaft counterclockwise 10 degrees to loosen the chain of the crankshaft timing gear.

21. Remove the chain from the crankshaft timing gear and place it on the crankshaft.

22. Turn the camshaft gear on the bank one (right bank) clockwise (about 60 degrees) and position as shown in the illustration. Remove the chain.

23. Remove the number two idle gear shaft, idle sprocket and the number one idle gear shaft.

24. Remove the two bolts and the number one chain vibration damper. Remove the two bolts and the number two vibration damper.

25. Remove the pulley set bolt. Remove the crankshaft timing gear from the crankshaft. Remove the two pulley set keys from the crankshaft.

26. On bank one, while raising up the number two chain tensioner, insert a pin (0.039 inch) into the hole to hold it in place. Hold the hexagonal portion of the camshaft with a wrench, and remove the two bolts and two camshaft gears. Remove the number two chain. Remove the bolt and the number two chain tensioner.

➡ **Be careful not to damage the cylinder head with the wrench. Do not disassemble the camshaft timing gear.**

27. On bank two, while raising up the number two chain tensioner, insert a pin (0.039 inch) into the hole to hold it in place.

Hold the hexagonal portion of the camshaft with a wrench, and remove the two bolts and two camshaft gears. Remove the number two chain. Remove the bolt and the number three chain tensioner.

➡ **Be careful not to damage the cylinder head with the wrench. Do not disassemble the camshaft timing gear.**

To install:

28. Install the number two chain tensioner with the bolt. Tighten the bolt to 15 ft. lbs. While pushing in the tensioner, insert a 0.039 inch pin into the hole to hold it in place.

29. Install the number three chain tensioner assembly. Tighten the bolt to 15 ft. lbs. While pushing in the tensioner, insert a 0.039 inch pin into the hole to hold it in place.

30. Align the mark plate (yellow) with the timing marks (1 dot mark for bank one and two dot mark for bank two) of the camshaft timing gears, as shown in the illustration.

31. Apply a small amount of engine oil to the bolt threads and the bolt sealing surface. Align the knock pin of the camshaft with the pin hole on the camshaft timing gear.

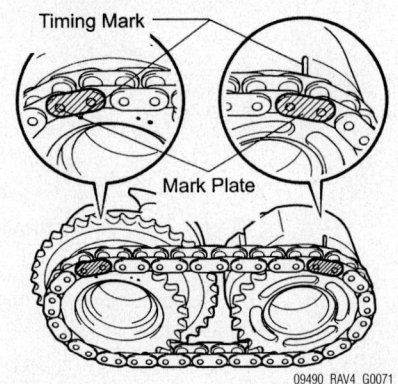

09490_RAV4_G0071

Bank one camshaft gear and chain alignment—3.5L engine

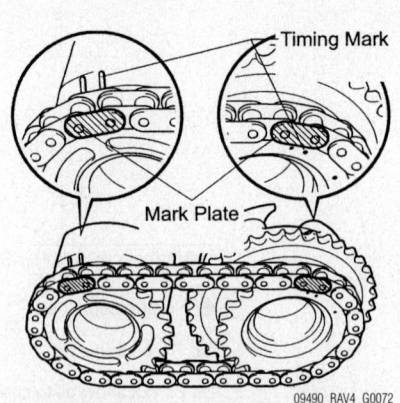

09490_RAV4_G0072

Bank two camshaft gear and chain alignment—3.5L engine

32. Install the camshaft timing gear and the camshaft timing exhaust gear with the number two chain installed.

33. Hold the hexagonal portion of the camshaft with a wrench, and tighten the bolts to 74 ft. lbs. Remove the pin.

34. Install the crankshaft gear or sprocket. Install the vibration damper. Tighten the bolts to 17 ft. lbs.

35. Install the number two chain vibration damper.

36. Apply a light coat of clean engine oil to the rotating surface of the number one idle gear shaft.

37. Temporarily install the number one idle gear shaft and idle sprocket with the number two idle gear shaft while aligning the knock pin of the number one idle gear shaft with the knock pin groove of the cylinder block. Tighten the bolt to 44 ft. lbs.

➡ **Be careful of the idle gear direction.**

38. Align the mark plate and the timing mark of the number one timing chain.

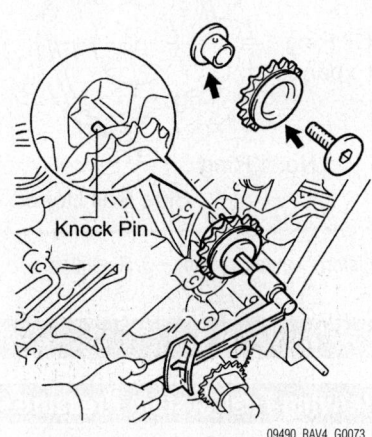

Knock pin location and installation—3.5L engine

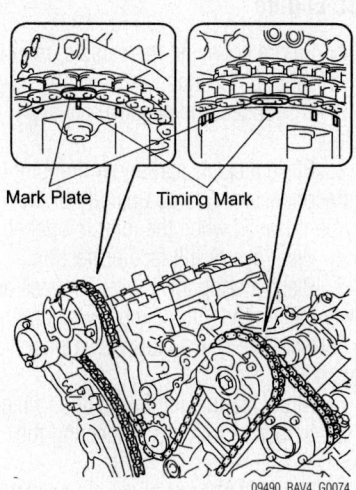

Number one timing chain alignment—3.5L engine

➡ **Do not pass the chain over the crankshaft, just put it on. The chain mark plate is orange.**

39. Turn the camshaft timing gear on bank one counterclockwise to tighten the chain between both banks.

➡ **If reusing the idle sprocket, align one of the idle sprocket's chain plate marks with one of the chain's chain plates when installing the idle sprocket.**

40. Align the mark plate and timing mark. Install the chain on to the crankshaft timing gear. Note that the chain mark plate is yellow. Temporarily tighten the pulley set bolt.

41. Turn the crankshaft clockwise to set it to the bank one bore center line (TDC on the compression stroke).

42. Install the chain tensioner slipper.

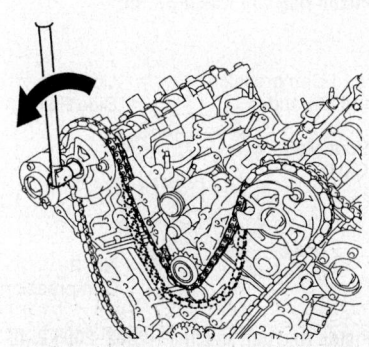

When Reusing Idle Sprocket

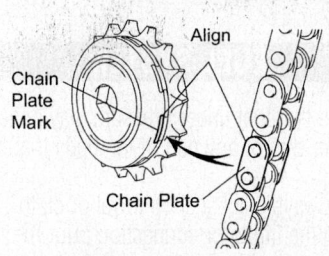

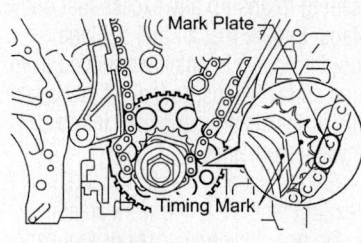

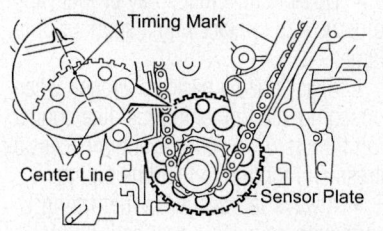

Idle sprocket and mark plate alignment—3.5L engine

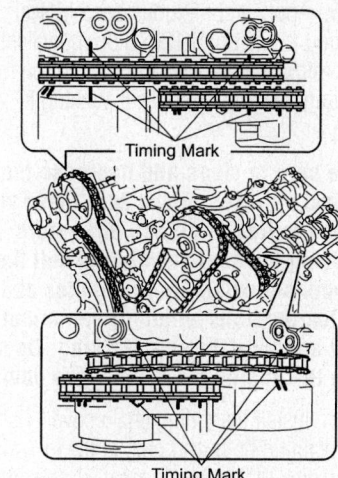

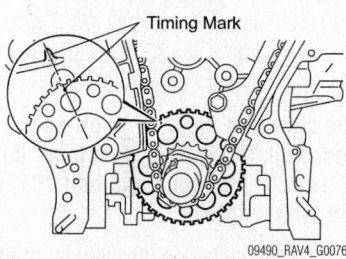

Timing chain number one and timing chain number two positioning—3.5L engine

43. To install the number one chain tensioner assembly, move the stopper plate upward to release the lock. Push the plunger deep into the tensioner.

44. Move the stopper plate downward to set the lock. Insert a hexagon wrench into the hole of the stopper plate. Install the chain tensioner bolts and tighten to 7 ft. lbs.

45. Remove the lock pin of the chain tensioner. Check that each timing mark is aligned with the crankshaft at TDC of the compression stroke.

46. Replace the timing chain cover oil seal.

47. Install the timing chain cover sub-assembly, with the oil pump.

➡ **Be sure to align the oil pump's drive rotor spline and the crankshaft, as shown in the illustration.**

48. Apply a continuous bead (0.118 inch or more) of seal packing (three bond 1207B or equivalent) to the engine.

➡ **Be sure to clean and degrease the contact surfaces. Make sure these surfaces are free of oil and dirt before applying the seal packing. Install the component within three minutes and tighten the bolts within fifteen minutes after applying the seal packing. Do not start the engine for at least two hours.**

49. Apply a continuous bead of seal packing (three bond 1207B or equivalent) to the timing chain cover. See illustration for location and amount of seal packing to apply.

➡ **Be sure to clean and degrease the contact surfaces. Make sure these surfaces are free of oil and dirt before applying the seal packing. Install the component within three minutes and tighten the bolts within fifteen minutes after applying the seal packing. Do not start the engine for at least two hours.**

50. Install the timing chain cover. Tighten the retaining bolts and nuts in the following order, area "1", area "2", area "3" and area "4". Tighten the bolts and nuts to 15 ft. lbs for area "1" area "2" and area "3". Tighten bolt "A" in area "4" to 32 ft. lbs. Tighten all other bolts in area "4" to 15 ft. lbs.

➡ **Be sure that there is no oil on the bolts. Bolt "A" is 1.57 inch long, bolt "B" is 2.17 inch long and bolt "C" is 0.98 inch long.**

51. When installing the cylinder head cover, apply seal packing, three bond 1207B or equivalent.

➡ **Remove any oil from the contact surfaces. Install the component within three minutes after applying the sealant. Do not start the engine for two hours after the installation.**

52. Install the cover retaining bolts and torque to 15 ft. lbs for bolt "A" and to 7 ft. lbs. for all other bolts and in the proper sequence.

53. Continue the installation is the reverse order of the removal procedure.

Piston and Ring

POSITIONING

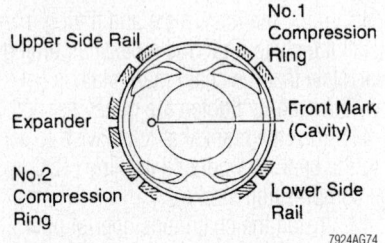

Piston ring gap spacing—2.0L

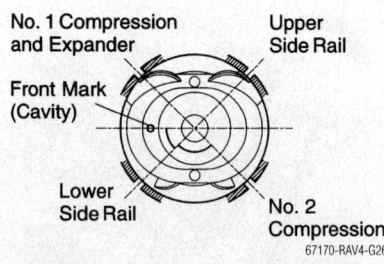

Piston ring gap spacing—2002–2005 2.4L engine

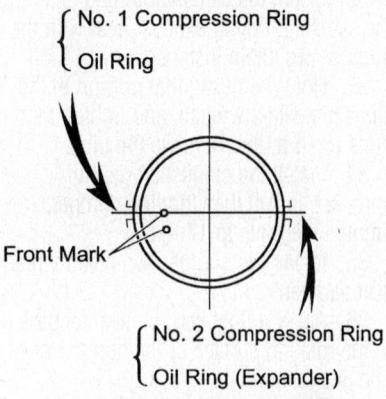

Piston ring gap spacing—2006 2.4L engine

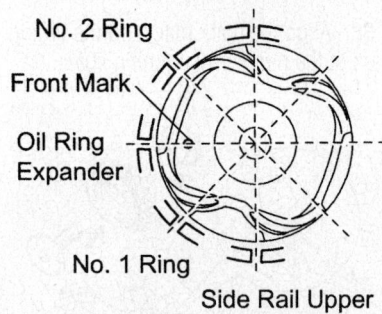

Piston ring gap spacing—3.5L engine

FUEL SYSTEM

Fuel System Service Precautions

Safety is the most important factor when performing not only fuel system maintenance but any type of maintenance. Failure to conduct maintenance and repairs in a safe manner may result in serious personal injury or death. Maintenance and testing of the vehicle's fuel system components can be accomplished safely and effectively by adhering to the following rules and guidelines.

• To avoid the possibility of fire and personal injury, always disconnect the negative battery cable unless the repair or test procedure requires that battery voltage be applied.

• Always relieve the fuel system pressure prior to disconnecting any fuel system component (injector, fuel rail, pressure regulator, etc.), fitting or fuel line connection. Exercise extreme caution whenever relieving fuel system pressure, to avoid exposing skin, face and eyes to fuel spray. Please be advised that fuel under pressure may penetrate the skin or any part of the body that it contacts.

• Always place a shop towel or cloth around the fitting or connection prior to loosening to absorb any excess fuel due to spillage. Ensure that all fuel spillage (should it occur) is quickly removed from engine surfaces. Ensure that all fuel soaked cloths or towels are deposited into a suitable waste container.

• Always keep a dry chemical (Class B) fire extinguisher near the work area.

• Do not allow fuel spray or fuel vapors to come into contact with a spark or open flame.

• Always use a back-up wrench when loosening and tightening fuel line connection fittings. This will prevent unnecessary stress and torsion to fuel line piping.

• Always replace worn fuel fitting O-rings with new. Do not substitute fuel hose or equivalent, where fuel pipe is installed.

Reliving Fuel System Pressure

RELIEVING

2.0L Engine

1. Before servicing the vehicle, refer to the Precautions Section.

2. Disconnect the negative battery cable.

3. Place a catch-pan under the joint to be disconnected. A large quantity of fuel may be released when the joint is opened.

4. Wear eye or full face protection.

5. Place a shop towel over the area and slowly loosen the joint using a wrench of the correct size. Use a back-up wrench if needed.

6. Allow the fuel left in the line to bleed off slowly before fully disconnecting the joint.

7. Plug the opened lines immediately to prevent fuel spillage or the entry of dirt.

8. Dispose of the released fuel properly.

9. After connecting fuel lines, connect the negative battery cable and start the engine.

10. Check for leaks and repair as needed.

2.4L Engine

2002–2005

1. Before servicing the vehicle, refer to the Precautions Section.

2. Disconnect the wire at the fuel pump.

3. Start the engine and run it until it shuts off.

4. Disconnect the negative battery cable.

5. Place a catch-pan under the joint to be disconnected. A large quantity of fuel may be released when the joint is opened.

6. Disconnect the high pressure fuel line.

7. After connecting fuel lines, reconnect the wire and start the engine.

8. Check for leaks and repair as needed.

2006

➡**After performing this procedure, pressure will remain in the fuel line. Use care when disconnecting any fuel lines.**

1. Before servicing the vehicle, refer to the Precautions Section.

2. Remove the console box.

3. Disconnect the connector.

4. Start the engine. After the engine has stopped, turn the ignition switch to the OFF position.

➡**DTC P0171 (system lean) may set.**

5. Check that the engine does not start.

6. Disconnect the negative battery cable.

7. Remove the fuel tank cap.

8. Connect the electrical connector. Install the console box.

3.5L Engine

➡**After performing this procedure, pressure will remain in the fuel line. Use care when disconnecting any fuel lines.**

1. Before servicing the vehicle, refer to the Precautions Section.

2. Remove the console box.

3. Disconnect the connector.

4. Start the engine. After the engine has stopped, turn the ignition switch to the OFF position.

➡**DTC P0171/P0172 (system lean) may set.**

5. Check that the engine does not start.

6. Disconnect the negative battery cable.

7. Remove the fuel tank cap.

8. Connect the electrical connector. Install the console box.

Fuel Filter

REMOVAL & INSTALLATION

2.0L Engine

1. Before servicing the vehicle, refer to the Precautions Section.

2. Properly release fuel system pressure.

3. Remove or disconnect the following:
   - Negative battery cable
   - Fuel filter's protective shield

4. Place a pan under the delivery pipe to catch the dripping fuel and slowly loosen the union bolt or flare nut to bleed off the fuel pressure.

5. Drain the remaining fuel.

6. Remove or disconnect the following:
   - Inlet and outlet lines
   - Fuel filter

To install:

7. Coat the flare nut, union nut and bolt threads with engine oil.

8. Hand tighten the inlet line to the fuel filter.

➡**When tightening the fuel line bolts to the fuel filter, use a torque wrench. The tightening torque is very important, as under or over tightening may cause fuel leakage. Insure that there is no fuel line interference and that there is sufficient clearance between it and any other parts.**

9. Install or connect the following:
   - Fuel filter. Tighten the inlet bolts to 22 ft. lbs. (30 Nm).
   - Delivery pipe using new gaskets. Tighten the union bolt to 22 ft. lbs. (30 Nm).

10. Run the engine for a few minutes and check for any fuel leaks.

11. Install the protective shield.

2.4L and 3.5L Engines

The filter is part of the fuel pump module and is not normally serviced.

Fuel Pump

REMOVAL & INSTALLATION

2.0L Engine

1. Before servicing the vehicle, refer to the Precautions Section.

2. Relieve the fuel system pressure.

3. Remove or disconnect the following:
   - Negative battery cable
   - Left side rear seat assembly
   - Floor service hole by pulling back the carpet; then, remove the 4 screws
   - Fuel pump and sender gauge connector

➡**Loosen the fuel cap to relieve any fuel pressure within the tank.**

   - Fuel pipe union bolt and both gaskets
   - Fuel pump outlet pipe
   - Return vent hose from the fuel pump
   - 8 fuel pump bolts and the pump assembly from the tank

To install:

4. Install or connect the following:
   - Fuel pump to the fuel tank. Tighten the 8 bolts to 31 inch lbs. (3.5 Nm).
   - Return vent hose to the fuel pump
   - Outlet pipe to the fuel pump, using new gaskets. Tighten the union bolts to 22 ft. lbs. (29 Nm).
   - Fuel pump and sender gauge connector
   - Floor hole cover with the 4 screws
   - Carpet
   - Left rear seat assembly
   - Negative battery cable
   - Fuel cap

5. Start the vehicle and check for leaks.

2.4L Engine

2002–2005

1. Before servicing the vehicle, refer to the Precautions Section.

2. Relieve the fuel system pressure.

3. Remove or disconnect the following:
   - Negative battery cable
   - Left side rear seat assembly
   - Floor service hole by pulling back the carpet; then, remove the 4 screws
   - Fuel pump and sender gauge connector

➡**Loosen the fuel cap to relieve any fuel pressure within the tank.**

   - Fuel pipe
   - Fuel pump outlet pipe
   - Return vent hose from the fuel pump
   - 8 fuel pump bolts and the pump assembly from the tank

To install:

4. Install or connect the following:

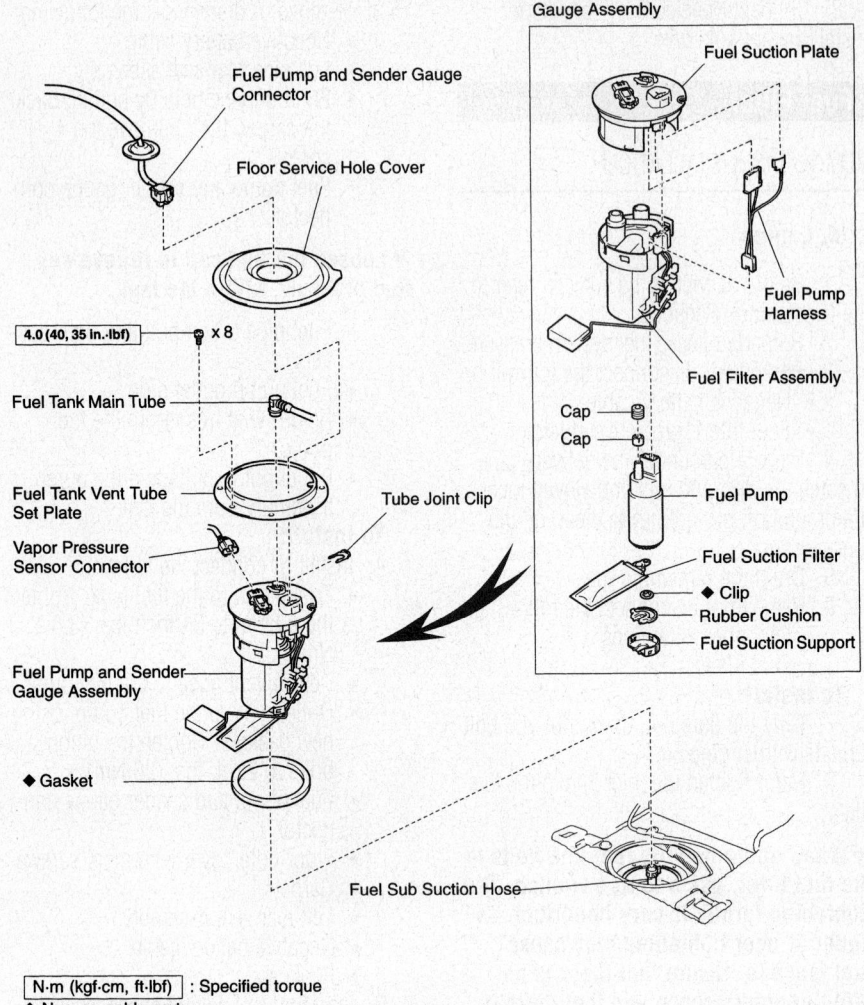

4.0 (40, 35 in.·lbf) ♦ × 8

Fuel Pump and Sender Gauge Connector

Floor Service Hole Cover

Fuel Tank Main Tube

Fuel Tank Vent Tube Set Plate

Vapor Pressure Sensor Connector

Fuel Pump and Sender Gauge Assembly

♦ Gasket

Tube Joint Clip

Fuel Sub Suction Hose

Fuel Pump and Sender Gauge Assembly

Fuel Suction Plate

Fuel Pump Harness

Fuel Filter Assembly

Cap
Cap

Fuel Pump

Fuel Suction Filter

♦ Clip
Rubber Cushion
Fuel Suction Support

N·m (kgf·cm, ft·lbf) : Specified torque
♦ Non–reusable part

67170-RAV4-G28

Fuel pump and related components—2003–2005 2.4L engine

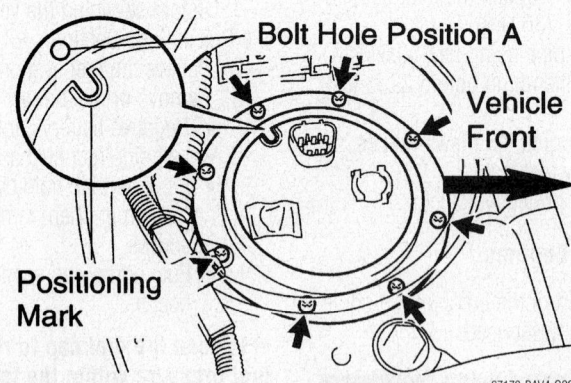

Bolt Hole Position A

Vehicle Front

Positioning Mark

67170-RAV4-G29

Proper fuel pump installation—2002–2005 2.4L engine

- Fuel pump to the fuel tank. Position the unit as shown. Tighten the 8 bolts to 35 inch lbs. (4 Nm).
- Return vent hose to the fuel pump
- Fuel pump and sender gauge connector

- Floor hole cover with the 4 screws
- Carpet
- Left rear seat assembly
- Negative battery cable
- Fuel cap
5. Start the vehicle and check for leaks.

2006

➡The fuel tank must first be removed from the vehicle. Be sure to check and adjust the fuel level as required, before removing the fuel tank. Take all the necessary precautions to avoid safety and fuel disposal problems.

1. Before servicing the vehicle, refer to the Precautions Section.
2. Properly relieve the fuel system pressure.
3. Disconnect the negative battery cable.

➡Wait at least 90 seconds after disconnecting the negative battery cable before starting any repair work to prevent air bag and seat belt pretensioner activation.

4. Remove the front floor carpet.
5. Disconnect the number two parking brake cable assembly.
6. Disconnect the fuel tank main tube assembly.
7. Disconnect the fuel tank to filler pipe hose.
8. Disconnect the fuel tank breather hose.
9. Remove the fuel tank filler pipe.
10. Position a suitable jack under the fuel tank. Remove the six bolts and three fuel tank bands.
11. Slightly lower the suitable jack.

➡Be careful not to cut the wing nuts.

12. Fold back about half of each cushion rubber so that the wire harness can be detached.
13. Disconnect the fuel pump connector and sender gauge connector.
14. Detach the wire harness from the four clamps. Carefully remove the fuel tank from the vehicle.
15. Remove the joint clip and fuel tank main tube.
16. Remove the eight bolts and the fuel tank vent tube set plate.
17. Disconnect the fuel hose and remove the fuel pump assembly from the fuel tank. Discard the gasket.

To install:

18. Installation is the reverse of the removal procedure.
19. Tighten the eight retaining bolts to 35 inch lbs, and in an alternating sequence.
20. Tighten the fuel tank retaining bolts to 30 ft. lbs.
21. Tighten the fuel tank filler pipe bolts to 17 ft. lbs.
22. Start the engine and check for leaks. Correct as required.

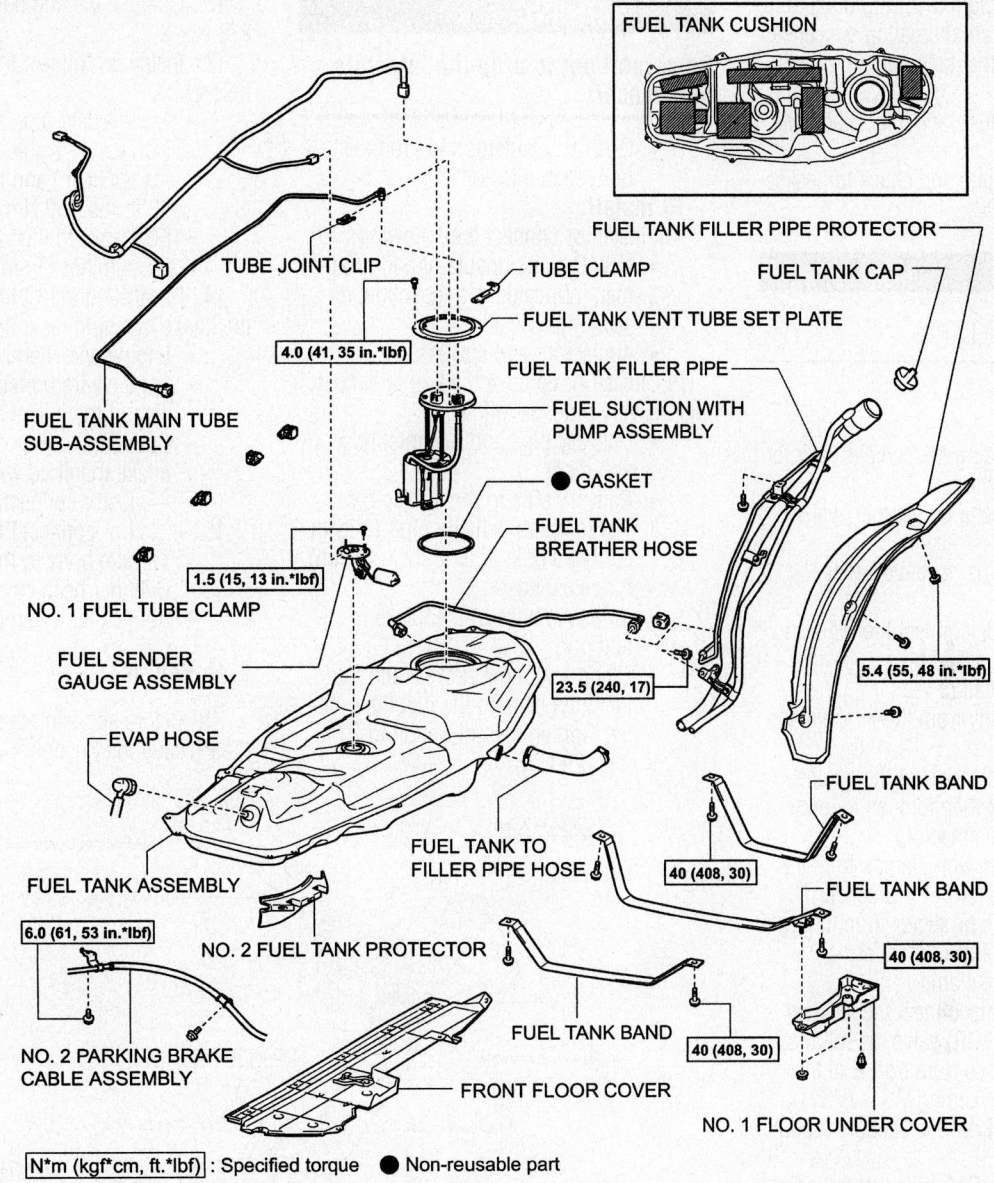

FUEL TANK CUSHION

TUBE JOINT CLIP

TUBE CLAMP

FUEL TANK FILLER PIPE PROTECTOR

FUEL TANK CAP

FUEL TANK VENT TUBE SET PLATE

4.0 (41, 35 in.*lbf)

FUEL TANK FILLER PIPE

FUEL TANK MAIN TUBE SUB-ASSEMBLY

FUEL SUCTION WITH PUMP ASSEMBLY

● GASKET

1.5 (15, 13 in.*lbf)

FUEL TANK BREATHER HOSE

NO. 1 FUEL TUBE CLAMP

FUEL SENDER GAUGE ASSEMBLY

23.5 (240, 17)

5.4 (55, 48 in.*lbf)

EVAP HOSE

FUEL TANK BAND

FUEL TANK ASSEMBLY

40 (408, 30)

FUEL TANK BAND

6.0 (61, 53 in.*lbf)

FUEL TANK TO FILLER PIPE HOSE

40 (408, 30)

NO. 2 FUEL TANK PROTECTOR

FUEL TANK BAND

NO. 2 PARKING BRAKE CABLE ASSEMBLY

40 (408, 30)

FRONT FLOOR COVER

NO. 1 FLOOR UNDER COVER

N*m (kgf*cm, ft.*lbf) : Specified torque   ● Non-reusable part

09490_RAV4_G0079

Fuel pump and related components—2006 2.4L engine and 3.5L engine

3.5L Engine

➡The fuel tank must first be removed from the vehicle. Be sure to check and adjust the fuel level as required, before removing the fuel tank. Take all the necessary precautions to avoid safety and fuel disposal problems.

1. Before servicing the vehicle, refer to the Precautions Section.
2. Properly relieve the fuel system pressure.
3. Disconnect the negative battery cable.

➡Wait at least 90 seconds after disconnecting the negative battery cable before starting any repair work to prevent air bag and seat belt pretensioner activation.

4. Remove the front floor carpet.
5. Disconnect the number two parking brake cable assembly.
6. Disconnect the fuel tank main tube assembly.
7. Disconnect the fuel tank to filler pipe hose.
8. Disconnect the fuel tank breather hose.
9. Remove the fuel tank filler pipe.
10. Position a suitable jack under the fuel tank. Remove the six bolts and three fuel tank bands.
11. Slightly lower the suitable jack.

➡Be careful not to cut the wing nuts.

12. Fold back about half of each cushion rubber so that the wire harness can be detached.
13. Disconnect the fuel pump connector and sender gauge connector.
14. Detach the wire harness from the four clamps. Carefully remove the fuel tank from the vehicle.
15. Remove the joint clip and fuel tank main tube.
16. Remove the eight bolts and the fuel tank vent tube set plate.
17. Disconnect the fuel hose and remove the fuel pump assembly from the fuel tank. Discard the gasket.

To install:

18. Installation is the reverse of the removal procedure.

19. Tighten the eight retaining bolts to 35 inch lbs, and in an alternating sequence.

20. Tighten the fuel tank retaining bolts to 30 ft. lbs.

21. Tighten the fuel tank filler pipe bolts to 17 ft. lbs.

22. Start the engine and check for leaks. Correct as required.

Fuel Injector

REMOVAL & INSTALLATION

2.0L Engine

1. Before servicing the vehicle, refer to the Precautions Section.

2. Properly discharge the fuel system pressure.

3. Disconnect the negative battery cable.

4. Remove or disconnect the following:
- Air cleaner assembly
- Cylinder head cover
- Throttle body from the intake manifold

5. Remove or disconnect the engine wire from the intake manifold, as follows:
- 4 injector connectors
- Both engine wire clamps from the intake manifold wire brackets
- Engine wire protector from the right side of the intake manifold
- Engine wire clamp

6. Remove or disconnect the Exhaust Gas Recirculation (EGR) valve, as follows:
- Vacuum hose from port **E** of the Vacuum Switching Valve (VSV)
- EGR hose from the vacuum modulator
- Loosen the EGR pipe nut from the cylinder head
- Both nuts, EGR valve, pipe assembly and gasket

7. Disconnect the engine compartment R/B No. 2

8. Remove or disconnect the fuel inlet hose and delivery pipe, as follows:
- Union bolt, both gaskets and the fuel inlet hose from the fuel filter outlet
- Air assist hose from the intake manifold port
- Air assist hose
- Loosen both delivery pipe-to-cylinder head bolts
- Delivery pipe from the 4 injectors
- Delivery pipe and fuel inlet hose assembly

9. Remove or disconnect the following:
- 4 injectors and spacers

✳✳ WARNING

Be careful not to drop the injectors and spacers.

- O-rings, insulator and grommet from each injector

To install:

10. Install or connect the following:
- New O-rings, insulator and grommet, lubricated with gasoline, to each injector
- 4 injectors and spacers

11. Install or connect the fuel inlet hose and delivery pipe, as follows:
- Delivery pipe and fuel inlet hose assembly
- Delivery pipe to the 4 injectors. Tighten both delivery pipe-to-cylinder head bolts to 9 ft. lbs. (13 Nm).
- Air assist hose
- Air assist hose to the intake manifold port
- Union bolt, new gaskets and the fuel inlet hose to the fuel filter outlet. Tighten the union bolt to 22 ft. lbs. (29 Nm).

12. Connect the engine compartment R/B No. 2

13. Install or connect the EGR valve, as follows:
- New gasket, pipe assembly and EGR valve. Tighten the nut to 9 ft. lbs. (13 Nm) and the union nut to 43 ft. lbs. (59 Nm).
- EGR hose to the vacuum modulator
- Vacuum hose from port **E** of the VSV

14. Install or connect the engine wire to the intake manifold, as follows:
- Engine wire clamp
- Engine wire protector to the right side of the intake manifold
- Both engine wire clamps to the intake manifold wire brackets
- 4 injector connectors

15. Install or connect the following:
- Throttle body to the intake manifold
- Cylinder head cover
- Air cleaner assembly

2.4L Engine

1. Before servicing the vehicle, refer to the Precautions Section.

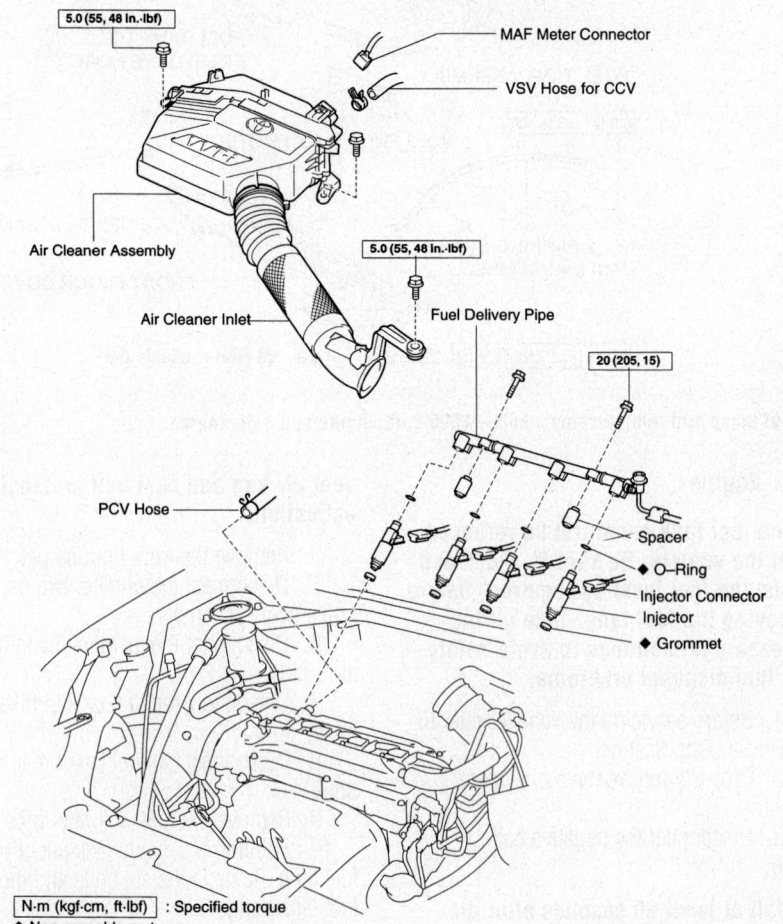

Fuel injectors and related parts—2.4L engine

MAF Meter Connector
VSV Hose for CCV
5.0 (55, 48 in.-lbf)
Air Cleaner Assembly
5.0 (55, 48 in.-lbf)
Air Cleaner Inlet
Fuel Delivery Pipe
20 (205, 15)
PCV Hose
Spacer
◆ O-Ring
Injector Connector
Injector
◆ Grommet

N·m (kgf·cm, ft·lbf) : Specified torque
◆ Non-reusable part

67170-RAV4-G30

2. Properly relieve the fuel system pressure.

3. Disconnect the negative battery cable.

➡**Wait at least 90 seconds after disconnecting the negative battery cable before starting any repair work to prevent air bag and seat belt pretensioner activation.**

4. Disconnect the PCV hose.

5. Remove the air cleaner assembly with the MAF sensor.

6. Disconnect the injector connectors.

7. Remove the bolts holding the delivery pipe to the cylinder head.

8. Remove the delivery pipe with the injectors attached.

9. Remove the 2 spacers and 4 grommets.

10. Pull the injectors from the pipe. Discard the O-rings.

To install:

11. Installation is the reverse of removal.

12. Coat the new O-rings with gasoline. Push the injectors onto the pipes and make sure they rotate freely.

13. Position the assembly onto the head and install the bolts finger tight. Make sure that the injectors still rotate freely. If not, replace the O-rings. Torque the bolts to 15 ft. lbs. (20 Nm).

14. Start the engine and check for leaks, correct as required.

3.5L Engine

1. Before servicing the vehicle, refer to the Precautions Section.

2. Properly relieve the fuel system pressure.

3. Disconnect the negative battery cable.

➡**Wait at least 90 seconds after disconnecting the negative battery cable before starting any repair work to prevent air bag and seat belt pretensioner activation.**

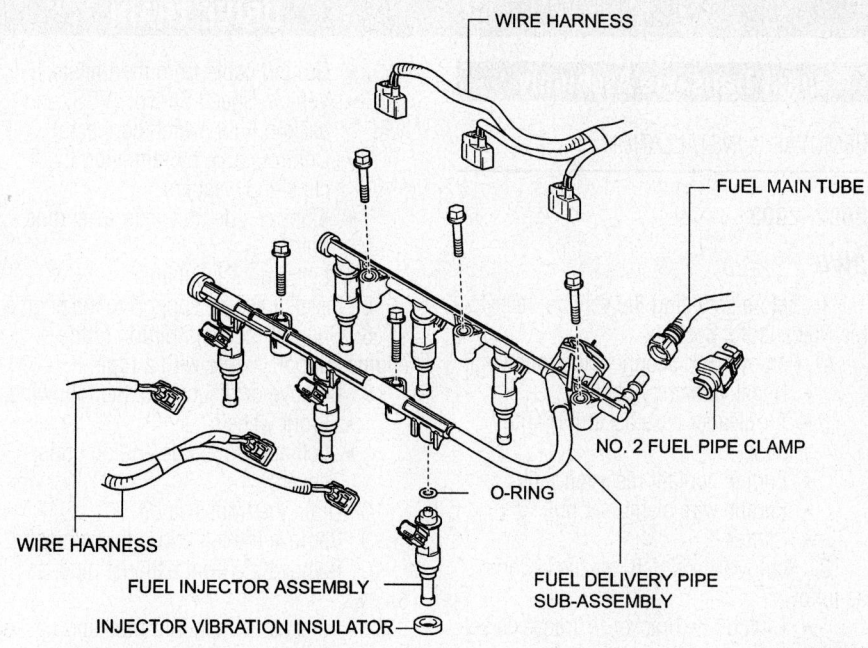

Fuel injectors and related components—3.5L engine

4. Remove the engine from the vehicle and position it in a suitable holding fixture.

5. Remove the intake air surge tank assembly.

6. Disconnect the number two main tube. Pinch the tube connector and then pull out the pipe.

7. Disconnect the six injector connectors. Remove the five bolts and the fuel delivery rail together with the injectors.

8. Remove the six insulators from the intake manifold.

9. Pull out the injectors from the rail. Remove and discard the O-rings.

To install:

10. Installation is the reverse of removal.

11. Coat the new O-rings with gasoline. Push the injectors onto the pipes and make sure they rotate freely.

12. Position the assembly onto the head and install the bolts finger tight. Make sure that the injectors still rotate freely. If not, replace the O-rings. Torque the bolts to 15 ft. lbs. (20 Nm).

13. Start the engine and check for leaks, correct as required.

To install:

14. Installation is the reverse of removal.

15. Coat the new O-rings with gasoline. Push the injectors onto the pipes and make sure they rotate freely.

16. Position the assembly onto the head and install the bolts finger tight. Make sure that the injectors still rotate freely. If not, replace the O-rings. Torque the bolts to 15 ft. lbs. (20 Nm).

17. Start the engine and check for leaks, correct as required.

DRIVE TRAIN

Manual Transaxle Assembly

REMOVAL & INSTALLATION

2002–2003

2WD

1. Before servicing the vehicle, refer to the Precautions Section.
2. Remove or disconnect the following:
   - Negative battery cable
   - Air cleaner case assembly with hose
   - Engine coolant reservoir tank
   - Engine wire clamp set nut
   - Starter
3. Remove the clutch release cylinder, as follows:
   - Clutch line bracket-to-transaxle set bolts
   - Release cylinder and line
4. Remove or disconnect the following:
   - Ground cable from the transaxle
   - Vehicle Speed Sensor (VSS) and backup light switch connector
   - Control cable by removing the 4 clips and washers
   - 4 upper side transaxle-to-engine bolts
   - Left mount insulator
5. Install a engine support to the engine.
6. Support rack and pinion to the engine support fixture with a rope.
7. Remove or disconnect the following:
   - Front wheels
   - Left and right side engine under-covers
8. Drain the transaxle oil.
9. Remove the left and right halfshafts.
10. Remove the front exhaust pipe, as follows:
    - 3 exhaust manifold nuts and gasket
    - Both exhaust pipe-to-center exhaust pipe bolts
    - Exhaust pipe
11. Remove the front suspension cross-member assembly with the sway bar, as follows:
    a. Support the front suspension crossmember with a jack.
    b. Disconnect the ring from the center exhaust pipe.
    c. Remove the 2 set bolts and nuts of the power steering rack and pinion assembly.
    d. Remove the suspension cross-member assembly with the sway bar by removing the 2 nuts and 6 bolts.
12. Remove the engine mounting center-member by removing the 4 bolts.
13. Jack up the transaxle slightly.
14. Remove or disconnect the following:
    - Left mounting bracket from the mounting insulator by removing the set bolt
    - Stiffener plate, No. 2 rear endplate and transaxle lower side mounting bolt
15. Lower the engine left side.
16. Remove or disconnect the following:
    - Transaxle
    - Transaxle case protector by removing both bolts

To install:

17. Install or connect the following:
    - Transaxle case protector. Tighten both bolts to 18 ft. lbs. (25 Nm).
    - Transaxle
18. Install the No. 2 rear endplate and transaxle bolts. Tighten the bolts, as follows:
    - Bolt C: 22 ft. lbs. (29 Nm)
    - Bolt D: 34 ft. lbs. (46 Nm)
    - Bolt E: 18 ft. lbs. (25 Nm)
    - Bolt F: 78 inch lbs. (9.0 Nm)
19. Install or connect the following:
    - Stiffener plate. Tighten both bolts to 27 ft. lbs. (37 Nm).
    - Engine left mounting insulator to the left mounting bracket. Tighten the bolt to 47 ft. lbs. (64 Nm).
    - Engine mount centermember. Tighten the radiator support bolts to 26 ft. lbs. (35 Nm) and the mount insulator to 59 ft. lbs. (80 Nm).
20. Install the front suspension cross-member with the sway bar, as follows:
    a. Install the sway bar and suspension crossmember. Tighten the nuts/bolts, as follows:
    - Vehicle bolt A: 152 ft. lbs. (206 Nm)
    - Lower control arm bracket bolt B: 101 ft. lbs. (137 Nm)

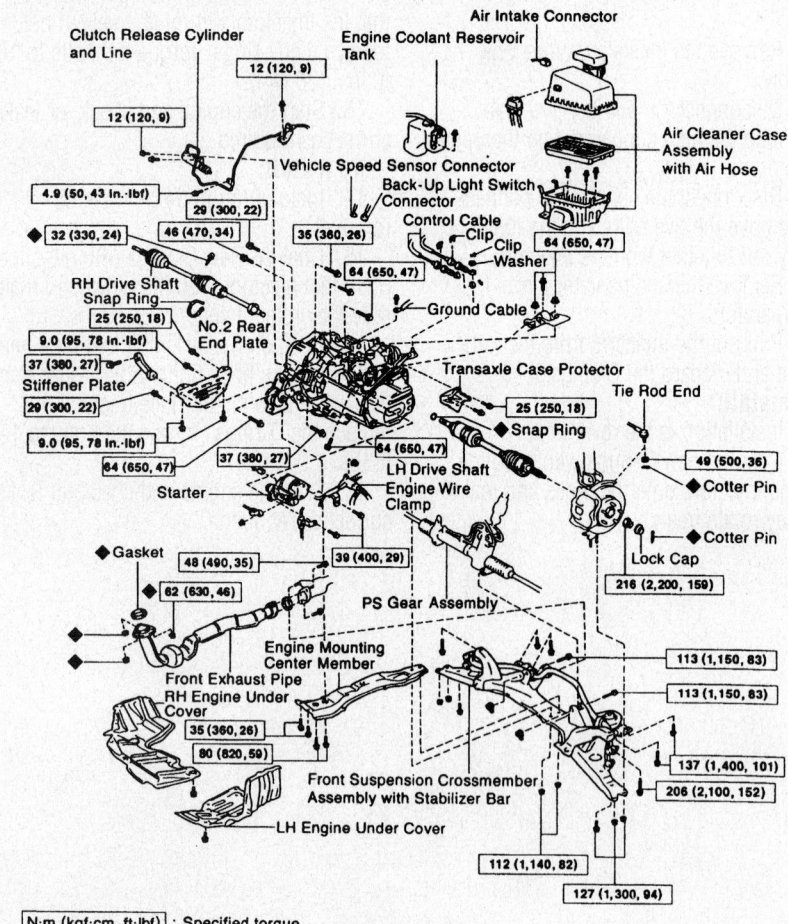

N·m (kgf·cm, ft·lbf) : Specified torque
◆ **Non-reusable part**

7924ZG61

Manual transaxle and related components—2002–2003 2WD

- Rear mounting bracket bolt C: 82 ft. lbs. (112 Nm)

b. Connect the rack and pinion to the crossmember. Tighten both nuts/bolts to 83 ft. lbs. (113 Nm).

c. Connect the ring for the center exhaust pipe.

21. Install the front exhaust pipe, as follow:

- Pipe with new gaskets
- Front pipe to the center exhaust pipe. Tighten both bolts to 35 ft. lbs. (48 Nm).
- Front exhaust pipe to the exhaust manifold. Tighten the 3 nuts to 46 ft. lbs. (62 Nm).

22. Install or connect the following:

- Left and right halfshafts
- Front wheels
- Engine left mounting insulator. Tighten the fasteners to 47 ft. lbs. (64 Nm).

23. Remove the engine support fixture.

24. Install or connect the following:

- 4 transaxle upper side mount bolts. Tighten bolt A to 47 ft. lbs. (64 Nm) and bolt B to 26 ft. lbs. (35 Nm).
- Ground cable with the clips and washers
- VSS and backup light switch connectors
- Ground cable to the transaxle
- Clutch release cylinder and line
- Starter. Tighten both bolts to 29 ft. lbs. (39 Nm).
- Engine wire clamp with the nut
- Engine coolant reservoir tank
- Air cleaner case assembly with the air hose
- Negative battery cable

25. Fill the transaxle with fluid. Check all fluids.

4WD

1. Before servicing the vehicle, refer to the Precautions Section.

2. Remove or disconnect the following:

- Transaxle/engine assembly
- Transaxle case protector, by removing the 2 bolts
- Starter
- Transfer vacuum actuator bracket, by removing the 4 bolts

3. Remove the transfer vacuum actuator assembly, as follows:

- 4 solenoid hoses from the transfer vacuum actuator assembly
- Transfer vacuum actuator assembly, by removing the 2 bolts

4. Remove or disconnect the following:

- Right transfer stiffener plate, by removing the 5 bolts

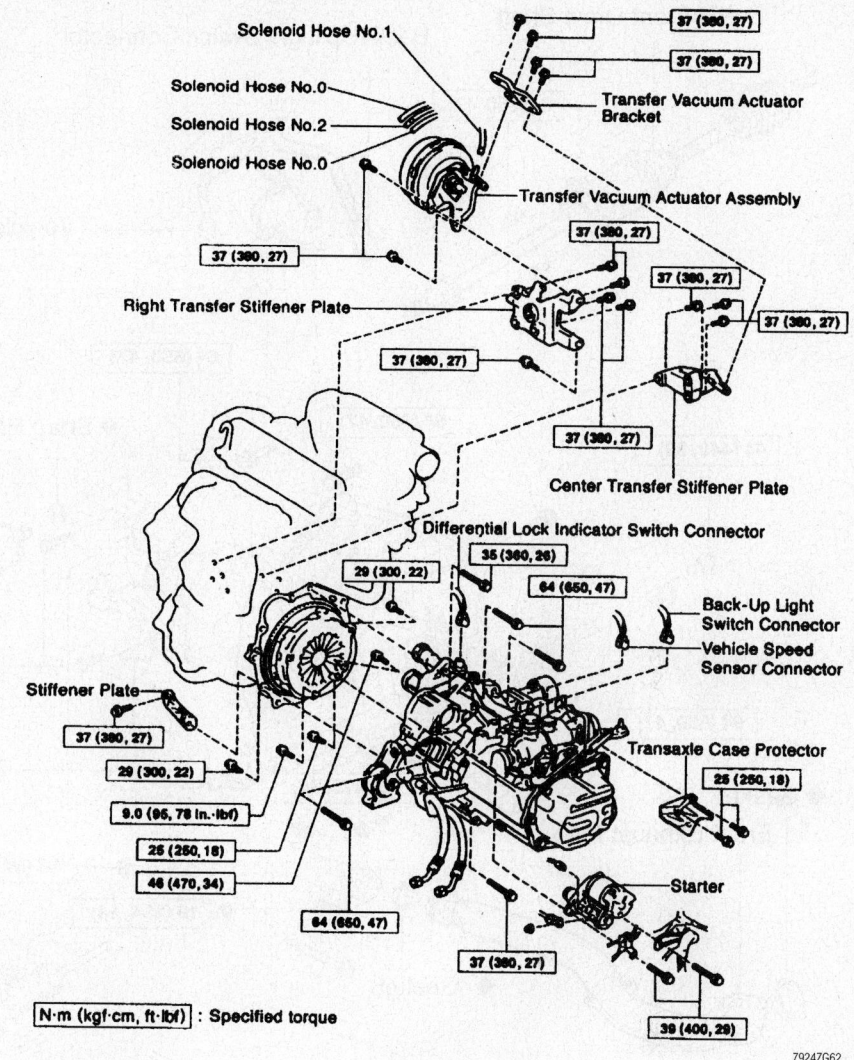

Manual transaxle and related components—2002–2003 4WD

- Center transfer stiffener plate, by removing the 3 bolts
- Stiffener plate by removing the 2 bolts
- Transaxle from the engine, by removing the 9 transaxle mount bolts

To install:

5. Connect the transaxle to the engine. Tighten the 9 bolts, as follows:

- Bolt A: 47 ft. lbs. (64 Nm)
- Bolt B: 26 ft. lbs. (35 Nm)
- Bolt C: 22 ft. lbs. (29 Nm)
- Bolt D: 34 ft. lbs. (46 Nm)
- Bolt E: 18 ft. lbs. (25 Nm)
- Bolt F: 78 inch lbs. (9.0 Nm)

6. Install or connect the following:

- Stiffener plate. Tighten both bolts to 27 ft. lbs. (37 Nm).
- Center transfer stiffener plate. Tighten the 3 bolts to 27 ft. lbs. (37 Nm).
- Right transfer stiffener plate. Tighten the 5 bolts to 27 ft. lbs. (37 Nm).

7. Install the transfer vacuum actuator assembly, as follows:

- Transfer vacuum actuator assembly. Tighten both bolts to 27 ft. lbs. (37 Nm).
- 4 solenoid hoses to the transfer vacuum actuator assembly

8. Install or connect the following:

- Transfer vacuum actuator bracket. Tighten the 4 bolts to 27 ft. lbs. (37 Nm).
- Starter. Tighten both bolts to 29 ft. lbs. (39 Nm).
- Transaxle case protector. Tighten both bolts to 18 ft. lbs. (25 Nm).
- Transaxle/engine assembly

2004–2005

2WD

1. Before servicing the vehicle, refer to the Precautions Section.

2. Disconnect the negative battery cable.

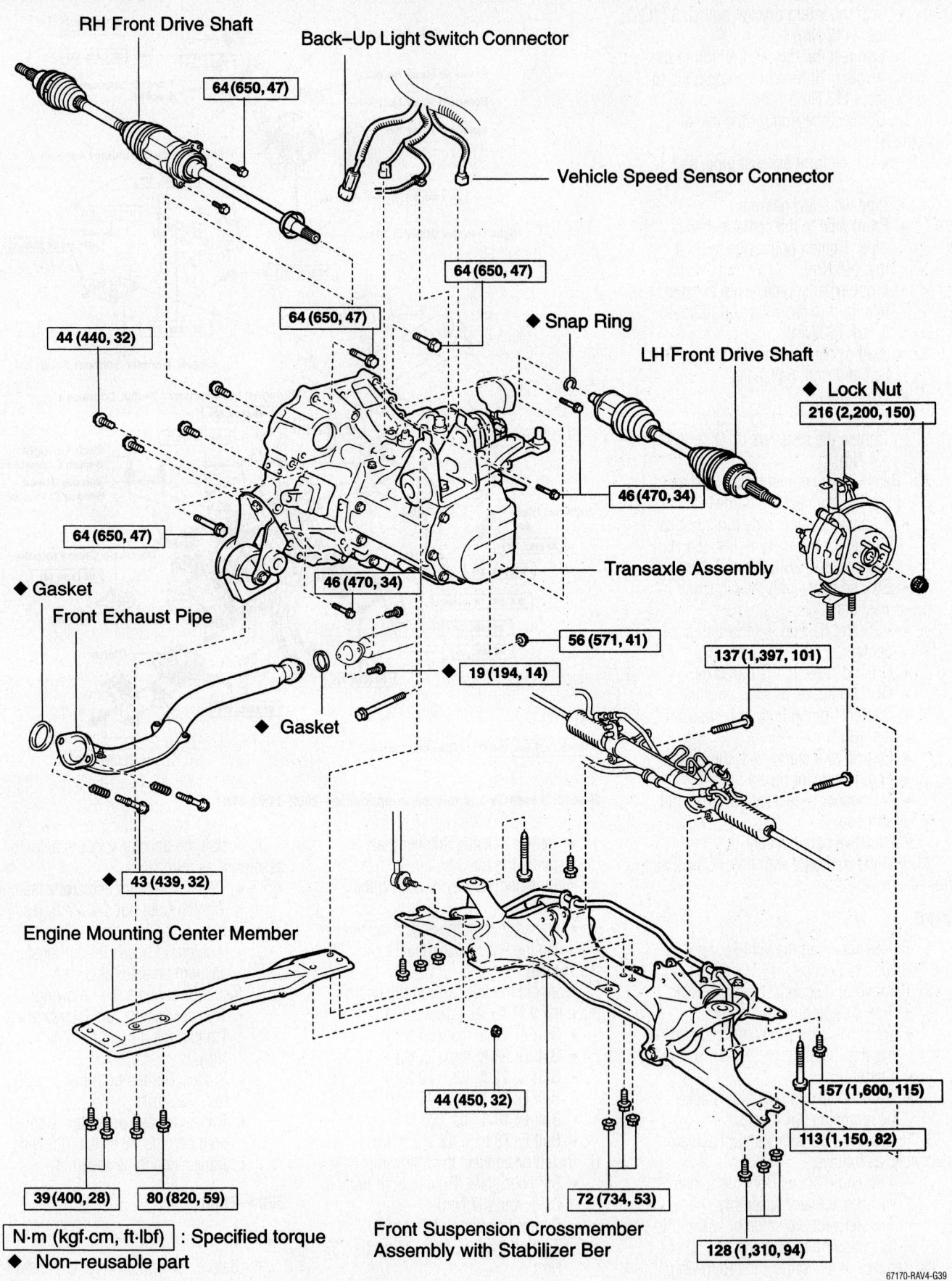

RH Front Drive Shaft

Back–Up Light Switch Connector

64 (650, 47)

Vehicle Speed Sensor Connector

◆ Snap Ring

44 (440, 32)

64 (650, 47)

64 (650, 47)

64 (650, 47)

LH Front Drive Shaft

◆ Lock Nut

216 (2,200, 150)

46 (470, 34)

64 (650, 47)

Transaxle Assembly

◆ Gasket

Front Exhaust Pipe

46 (470, 34)

56 (571, 41)

137 (1,397, 101)

◆ 19 (194, 14)

◆ Gasket

◆ 43 (439, 32)

Engine Mounting Center Member

44 (450, 32)

157 (1,600, 115)

113 (1,150, 82)

39 (400, 28)   80 (820, 59)

72 (734, 53)

128 (1,310, 94)

N·m (kgf·cm, ft·lbf) : Specified torque
◆ Non–reusable part

Front Suspension Crossmember
Assembly with Stabilizer Ber

67170-RAV4-G39

Manual transaxle and related components—2004–2005 2WD

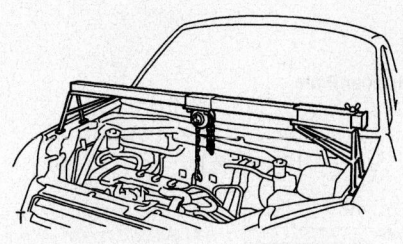

67170-RAV4-G41

Attach an engine support fixture, such as 12281-28010, or equivalent—2004–2005 2WD

3. Remove the hood.
4. Remove the air cleaner assembly.
5. Remove the coolant reservoir.
6. Remove the control cables from the transaxle.
7. Remove the wiring from the speed sensor and back-up light switch.
8. Remove the ground cable from the transaxle.
9. Remove the starter.
10. Remove the clutch release cylinder.
11. Remove the 5 upper transaxle-to-engine bolts.
12. Attach an engine support fixture, such as 12281-28010, or equivalent.
13. Disconnect the left engine mount from the transaxle.
14. Remove the engine under-covers.
15. Drain the transaxle.
16. Remove the front exhaust pipe.
17. Remove the halfshafts.
18. Remove the stabilizer bar.
19. Remove the power steering gear mounting bolts. Support the gear.
20. Remove the front suspension cross-member.
21. Remove the engine mount center member.
22. Using a transaxle jack, slightly raise the transaxle.
23. Remove the lower transaxle-to-engine bolts and lower the transaxle from the vehicle.
24. Installation is the reverse of removal. Observe the following torques:
   • Lower transaxle-to-engine bolts (see the illustration): bolts "A" 34 ft. lbs. (46 Nm); bolt "B" 32 ft. lbs. (44 Nm)
   • Engine mount center member (see the illustration): bolt "E" 59 ft. lb. (80 Nm); bolts "F" 28 ft. lbs. (39 Nm)
   • Front suspension crossmember (see the illustration): bolt "A" 82 ft. lbs. (113 Nm); bolt "B" 115 ft. lbs. (157 Nm); bolts "C" and "D" 53 ft. lbs. (72 Nm)

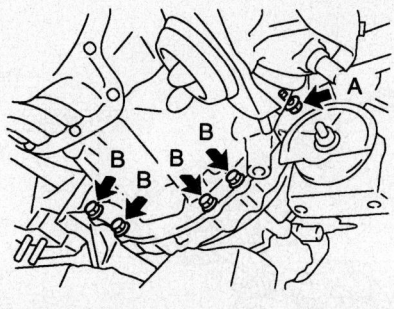

67170-RAV4-G43

Lower transaxle-to-engine bolts—2004–2005 2WD

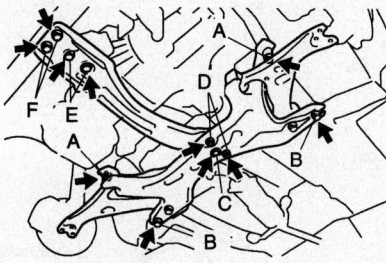

67170-RAV4-G42

Center member and front member bolts—2004–2005 2WD

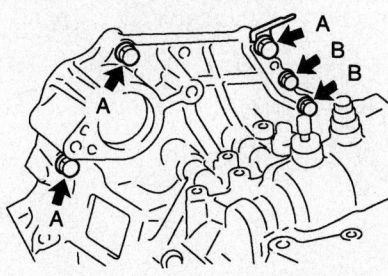

67170-RAV4-G40

Upper transaxle-to-engine bolts—2004–2005 2WD

• Power steering gear set bolts: 101 ft. lbs. (137 Nm)
• Stabilizer bar links: 32 ft. lbs. (44 Nm)
• Exhaust pipe-to-manifold: 32 ft. lbs. (43 Nm)
• Exhaust pipe-to-pipe: 14 ft. lbs. (19 Nm)
• Left engine mount through-bolt: 41 ft. lbs. (56 Nm)
• Upper transaxle-to-engine bolts (see the illustration): bolts "A" 47 ft. lbs. (64 Nm); bolts "B" 34 ft. lbs. (46 Nm)
25. Clutch release cylinder (see the illustration): bolt "A" 18 ft. lbs. (25 Nm); bolt "B" 9 ft. lbs. (12 Nm); bolt "C" 44 inch lbs. (5 Nm)
   • Starter: 28 ft. lbs. (37 Nm)
   • Hood: 10 ft. lbs. (13 Nm)

4WD

1. Before servicing the vehicle, refer to the Precautions Section.
2. Disconnect the negative battery cable.
3. Remove the engine and transaxle as a unit, from the vehicle.
4. Remove the transaxle case protector.
5. Remove the engine-to-transaxle wiring.
6. Remove the stiffener plate.
7. Remove the 10 bolts and separate the transaxle from the engine.
8. Installation is the reverse of removal. Observe the following torques:
   • Transaxle-to-engine (see the illustration): bolt "A" 47 ft. lbs. (64 Nm); bolt "B" 34 ft. lbs. (46 Nm); bolt "C" 32 ft. lbs. (44 Nm)
   • Stiffener plate: 25 ft. lbs. (34 Nm)
   • Case protector: 13 ft. lbs. (18 Nm)

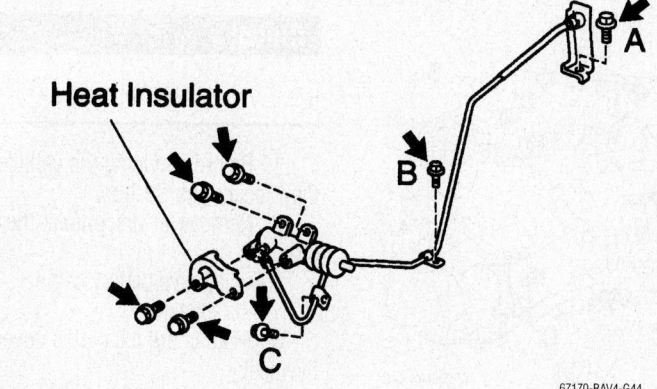

Heat Insulator

67170-RAV4-G44

Clutch release cylinder bolts—2004 2WD

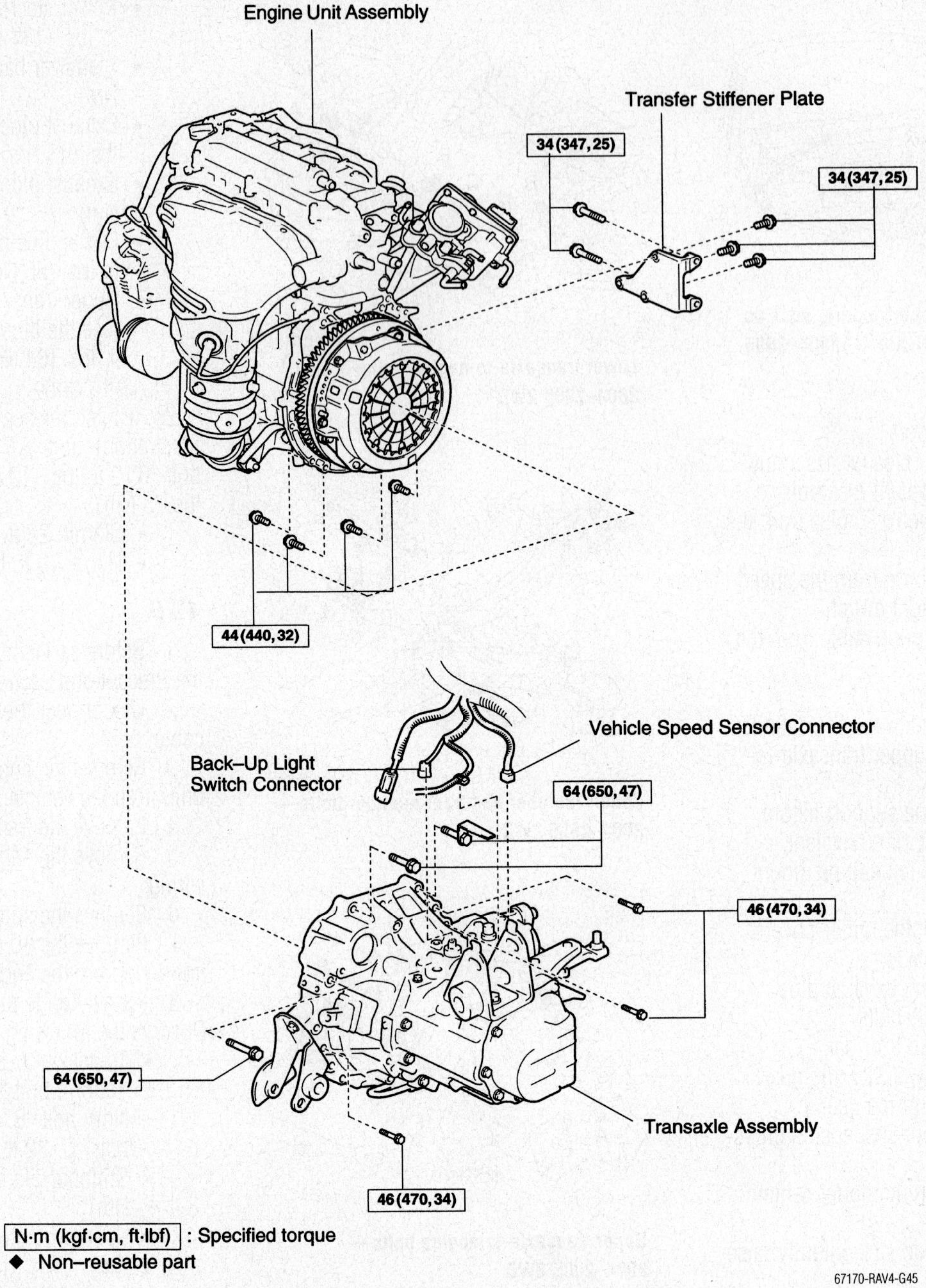

Engine Unit Assembly

Transfer Stiffener Plate

34 (347, 25)

34 (347, 25)

44 (440, 32)

Back–Up Light
Switch Connector

Vehicle Speed Sensor Connector

64 (650, 47)

46 (470, 34)

64 (650, 47)

Transaxle Assembly

46 (470, 34)

N·m (kgf·cm, ft·lbf) : Specified torque
◆ Non–reusable part

67170-RAV4-G45

Manual transaxle and related components—2004–2005 4WD

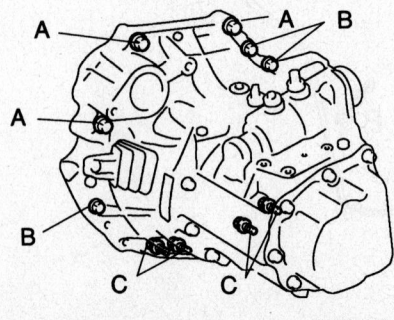

A   A   B

A

B

C   C

67170-RAV4-G46

**Manual transaxle bolt identification—
2004–2005 4WD**

Clutch

REMOVAL & INSTALLATION

1. Before servicing the vehicle, refer to the Precautions Section.

2. Remove or disconnect the following:

- Negative battery cable
- Transaxle

3. Matchmark the clutch cover to the flywheel.

4. Remove the clutch pressure plate retaining bolts in small amounts and in a crisscross pattern to relieve the clutch disc spring tension.

5. At the clutch cover, loosen each bolt 1 turn until spring tension is released.

6. Remove or disconnect the following:

- Clutch cover set bolts and pull off the clutch cover with the clutch disc
- Clutch cover-to-flywheel bolts

7. If the clutch release bearing is to be replaced, perform the following:

a. Remove the bearing retaining clip(s), the bearing and hub.

b. Remove the release fork and the boot.

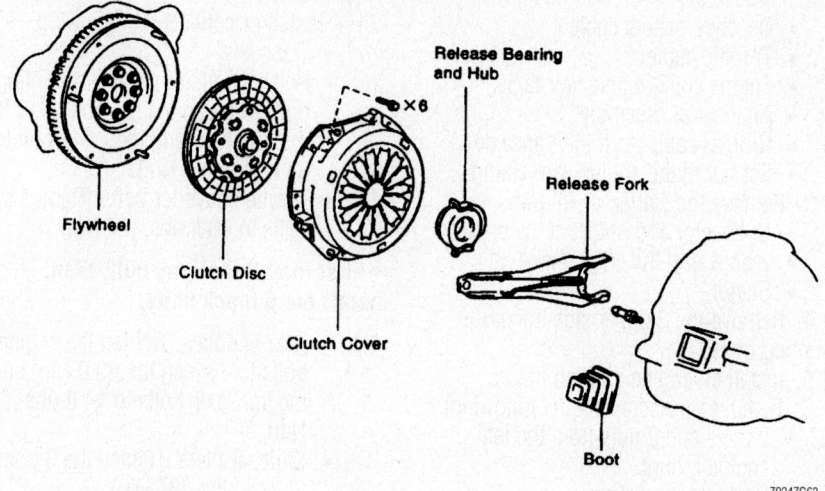

Clutch and related components—2002–2003

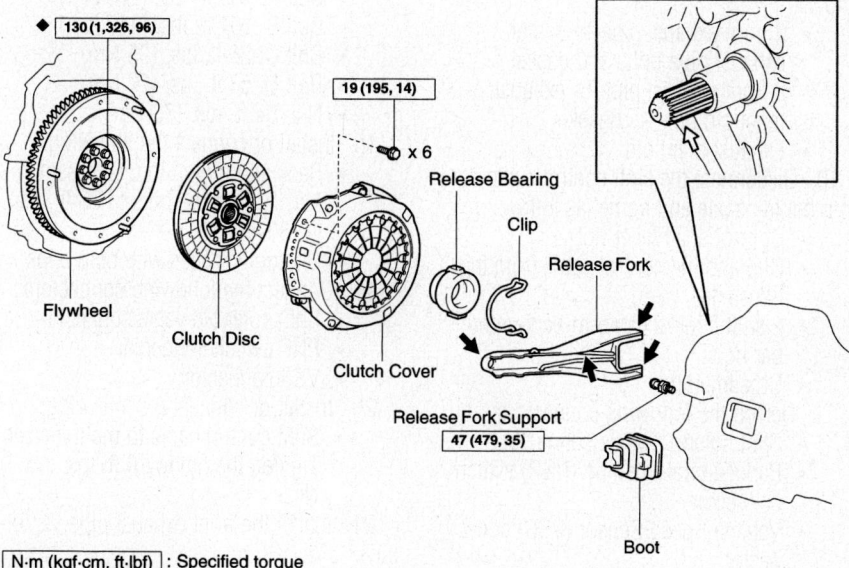

N·m (kgf·cm, ft·lbf) : Specified torque
◆ Non–reusable part
⇐ Clutch spline grease
⬅ Release hub grease

Clutch and related components–2004–2005

c. The bearing is press fitted to the hub.

d. Clean all parts and lightly grease the input shaft splines and all of the contact points.

e. Install the bearing/hub assembly, the fork, the boot and the retaining clip(s) in their original locations.

To install:

8. Inspect the flywheel surface for cracks, heat scoring (blue marks) and warpage. Replace or resurface the flywheel, if any damage is present.

➡ **Before installing any new parts, be sure they are clean. During installation, do not get grease or oil on any of the components, as this will shorten clutch life considerably.**

9. Using a clutch alignment tool, position the clutch disc against the flywheel. The raised center section of the disc faces the transaxle.

10. Install or connect the following:
- Clutch cover onto the flywheel by aligning the matchmarks
- Clutch cover. Tighten the bolts in a crisscross pattern to 14 ft. lbs. (19 Nm).

11. Lubricate the release fork pivot and contact points, release bearing, bearing hub and input shaft spline surfaces with a suitable molybdenum disulfide lithium based or multi-purpose grease.

12. Install or connect the following:
- Boot, release fork, hub and the bearing assemblies
- Transaxle
- Negative battery cable

ADJUSTMENT

1. Before servicing the vehicle, refer to the Precautions Section.

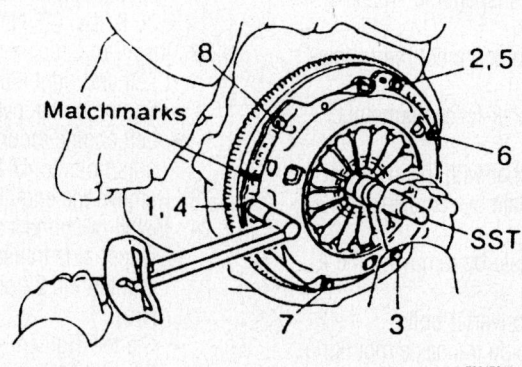

Clutch bolt torque sequence

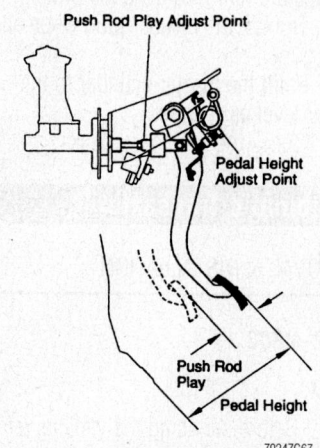

Clutch pedal height measurement location

2. Check that the pedal height is correct. Pedal height from the floor panel should be: 6.889–7.283 in. (175–185mm) for 2002–2003; 6.653–7.047 in. (169–179mm) for 2004–2005.

3. If necessary to adjust the pedal height, loosen the locknut and turn the stopper bolt until the height is correct. Tighten the locknut.

4. Push the pedal inward until the beginning of the clutch resistance is felt. Free-play should be 0.197–0.591 in. (5–15mm).

5. Gently push on the pedal until the resistance begins to increase a little. Pushrod play at the pedal top should be 0.039–0.197 in. (1–5mm).

6. If necessary, adjust the pedal free-play and the pushrod play, as follows:

   a. Loosen the locknut and turn the push the rod until the free-play and pushrod play are correct.

   b. Tighten the locknut.

Hydraulic Clutch System

BLEEDING

1. Before servicing the vehicle, refer to the Precautions Section.

2. Fill the clutch reservoir with brake fluid. Check the reservoir level frequently and add fluid as needed.

3. Connect one end of a vinyl tube to the bleeder plug on the slave cylinder and submerge the other end into a clear container half-filled with brake fluid.

4. Slowly pump the clutch pedal several times.

5. Have an assistant hold the clutch pedal down and loosen the bleeder plug until fluid and/or air starts to run out of the bleeder plug. Close the bleeder plug while the pedal is held to the floor.

6. Repeat Steps 2 and 3 until all the air bubbles are removed from the system.

7. Tighten the bleeder plug when all the air is gone.

8. Refill the master cylinder to the proper level as required.

9. Check the system for leaks.

Automatic Transaxle

REMOVAL & INSTALLATION

2002–2003

2WD

1. Before servicing the vehicle, refer to the Precautions Section.

2. Remove or disconnect the following:
- Negative battery cable
- Throttle cable
- Engine coolant reservoir tank
- Air cleaner assembly
- Ground cable from the transaxle
- Set nut of the engine wire clamp

3. Remove the starter, as follows:
- Connector and nut from the starter
- 2 bolts and the engine wire
- Starter

4. Remove the 3 upper side transaxle mounting bolts.

5. Install an engine support fixture.

6. Remove or disconnect the following:
- 2 bolts and 2 nuts from the left engine mount
- Engine undercovers

7. Drain the fluid from the transaxle.

8. Remove the left and right halfshafts.

9. Remove the front exhaust pipe, as follows:
- 2 front exhaust pipe-to-center exhaust pipe bolts and gasket
- 3 front exhaust pipe-to-exhaust manifold nuts and gasket
- Exhaust manifold

10. Disconnect the shift control cable from the transaxle and frame, as follows:
- Control shaft lever nut
- Clip and the control cable from the transaxle
- 2 shift control cable-to-centermember bolts
- Crossmember

11. Detach the following connectors:
- Shift solenoid valve connector
- Park/Neutral Position (PNP) switch connector
- Vehicle Speed Sensor (VSS) connector

12. Remove or disconnect the following:
- Oil cooler hoses from the transaxle
- Rack and pinion from the crossmember by removing both nuts/bolts

13. Support the rack and pinion.

14. Support the suspension crossmember with a floor jack.

15. Remove or disconnect the following:
- Crossmember-to-centermember fasteners
- Crossmember with the sway bar
- Stiffener plate by removing the 3 bolts
- Rear endplate by removing the 4 bolts
- 6 torque converter bolts
- Both rear side transaxle mounting bolts
- Transaxle

To install:

16. Install or connect the following:
- Transaxle
- Both rear side transaxle mounting bolts. Tighten the top bolt to 18 ft. lbs. (25 Nm) and the lower bolt to 34 ft. lbs. (46 Nm).
- Torque converter bolts. Tighten the bolts to 20 ft. lbs. (27 Nm).

➡ **First install the gray bolt; then, install the 5 black bolts.**

- Rear endplate. Tighten the engine bolts to 78 inch lbs. (9.0 Nm) and the transaxle bolts to 14 ft. lbs. (19 Nm).
- Stiffener plate. Tighten the 3 bolts to 27 ft. lbs. (37 Nm).

17. Install the front suspension crossmember and centermember with the sway bar. Tighten the nuts/bolts, as follows:
- Bolt A: 152 ft. lbs. (206 Nm)
- Bolt B: 101 ft. lbs. (137 Nm)
- Bolt C: 26 ft. lbs. (35 Nm)
- Bolt D: 53 ft. lbs. (72 Nm)
- Nut: 54 ft. lbs. (73 Nm)

18. Install or connect the following:
- Rack and pinion to the crossmember. Tighten the nuts to 83 ft. lbs. (113 Nm).
- Oil cooler hoses with both clips

19. Connect the following connectors:
- Shift solenoid valve connector
- PNP switch connector
- VSS connector

20. Install or connect the following:
- Shift control cable to the transaxle. Tighten the nut to 10 ft. lbs. (13 Nm).

21. Install the front exhaust pipe, as follows:
- Front exhaust pipe, using new gaskets
- Front exhaust pipe to the exhaust manifold. Tighten the 3 nuts to 46 ft. lbs. (62 Nm).
- Front exhaust pipe to the center exhaust pipe. Tighten both bolts to 35 ft. lbs. (48 Nm).

22. Install or connect the following:
- Left and right halfshafts
- Engine undercovers
- Left engine mount. Tighten the both nuts/bolts to 47 ft. lbs. (64 Nm).

23. Remove the engine fixture.

24. Install or connect the following:
- Upper side transaxle mount. Tighten the 3 bolts to 47 ft. lbs. (64 Nm).
- Starter. Tighten both bolts to 29 ft. lbs. (39 Nm).
- Engine wire

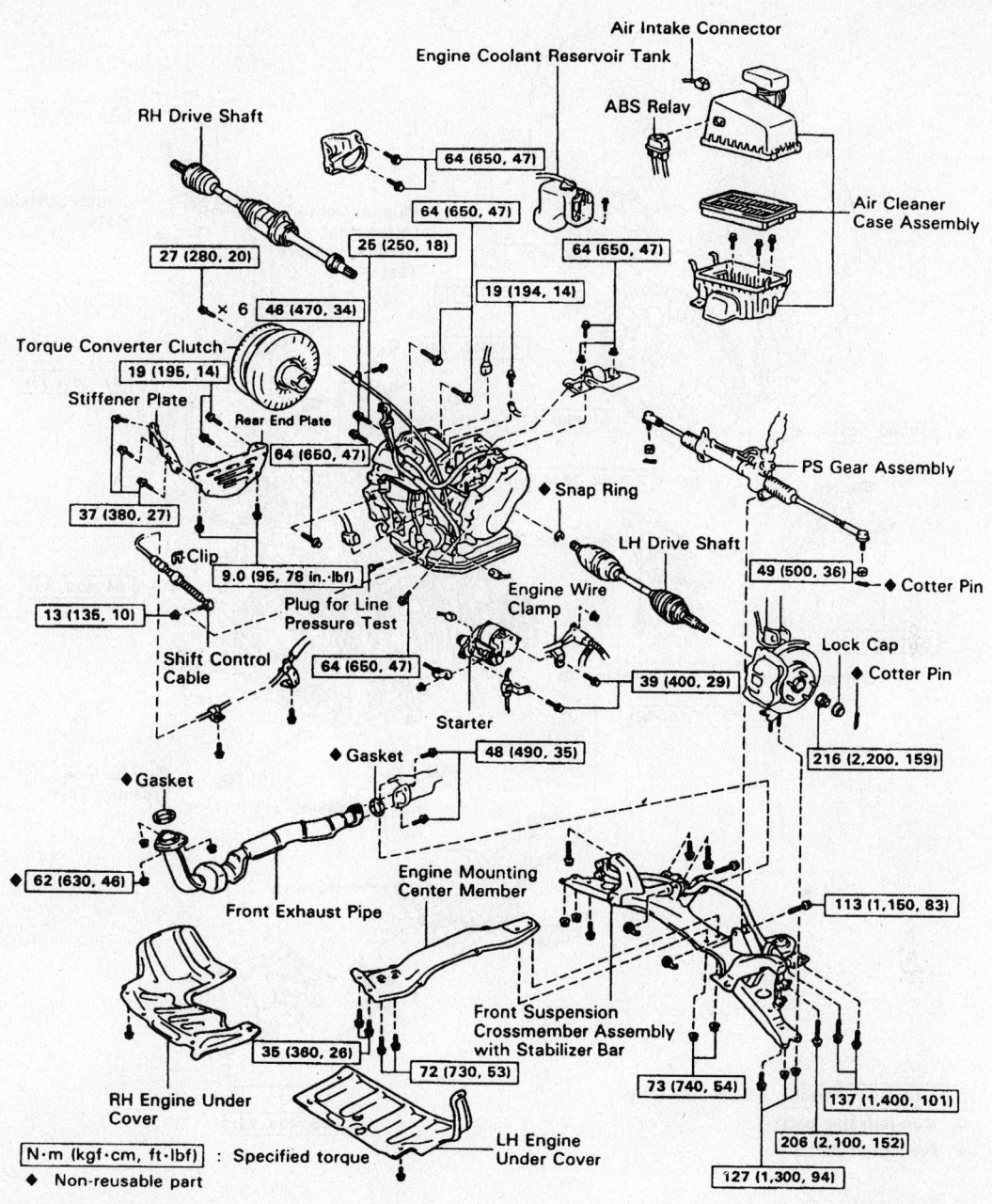

Air Intake Connector
Engine Coolant Reservoir Tank
ABS Relay
RH Drive Shaft
64 (650, 47)
26 (250, 18)
64 (650, 47)
Air Cleaner Case Assembly
27 (280, 20)
19 (194, 14)
× 6 48 (470, 34)
Torque Converter Clutch
19 (195, 14)
Stiffener Plate
Rear End Plate
64 (650, 47)
PS Gear Assembly
37 (380, 27)
♦ Snap Ring
LH Drive Shaft
Clip
49 (500, 36) ♦ Cotter Pin
9.0 (95, 78 in.·lbf)
Engine Wire Clamp
Plug for Line Pressure Test
13 (135, 10)
Lock Cap
Shift Control Cable
♦ Cotter Pin
64 (650, 47)
39 (400, 29)
Starter
216 (2,200, 159)
48 (490, 35)
♦ Gasket
♦ Gasket
♦ Gasket
Engine Mounting Center Member
113 (1,150, 83)
62 (630, 46)
Front Exhaust Pipe
Front Suspension Crossmember Assembly with Stabilizer Bar
35 (360, 26)
72 (730, 53)
73 (740, 54)
137 (1,400, 101)
RH Engine Under Cover
206 (2,100, 152)
N·m (kgf·cm, ft·lbf) : Specified torque
♦ Non-reusable part
LH Engine Under Cover
127 (1,300, 94)

7924ZG64

Automatic transaxle and related components—2002–2003 2WD

- Starter wire with the nut
- Engine wire clamp set nut
- Ground cable to the transaxle. Tighten the bolt to 14 ft. lbs. (19 Nm).
- Air cleaner assembly
- Engine coolant reservoir tank
- Throttle cable
- Negative battery cable
25. Check all fluids.

4WD

1. Before servicing the vehicle, refer to the Precautions Section.

2. Remove or disconnect the following:
- Negative battery cable
- Engine/transaxle assembly
- Starter
- Stiffener plate, by removing the 3 bolts
- Rear endplate, by removing the 4 bolts
- 6 torque converter clutch mounting bolts
- Connectors and wiring harness, from the transaxle
- Center stiffener plate, by removing the 4 bolts

3. Remove the transaxle with the transfer assembly, as follows:
- 2 bolts
- 5 transaxle mounting bolts
- Transaxle from the engine

To install:

4. Install the transaxle. Tighten the 14mm head bolts to 47 ft. lbs. (64 Nm) and the 12mm head bolts to 34 ft. lbs. (46 Nm).

5. Install or connect the following:
- Tighten both bolts to 27 ft. lbs. (37 Nm).
- Center stiffener plate. Tighten the 4 bolts to 27 ft. lbs. (37 Nm).

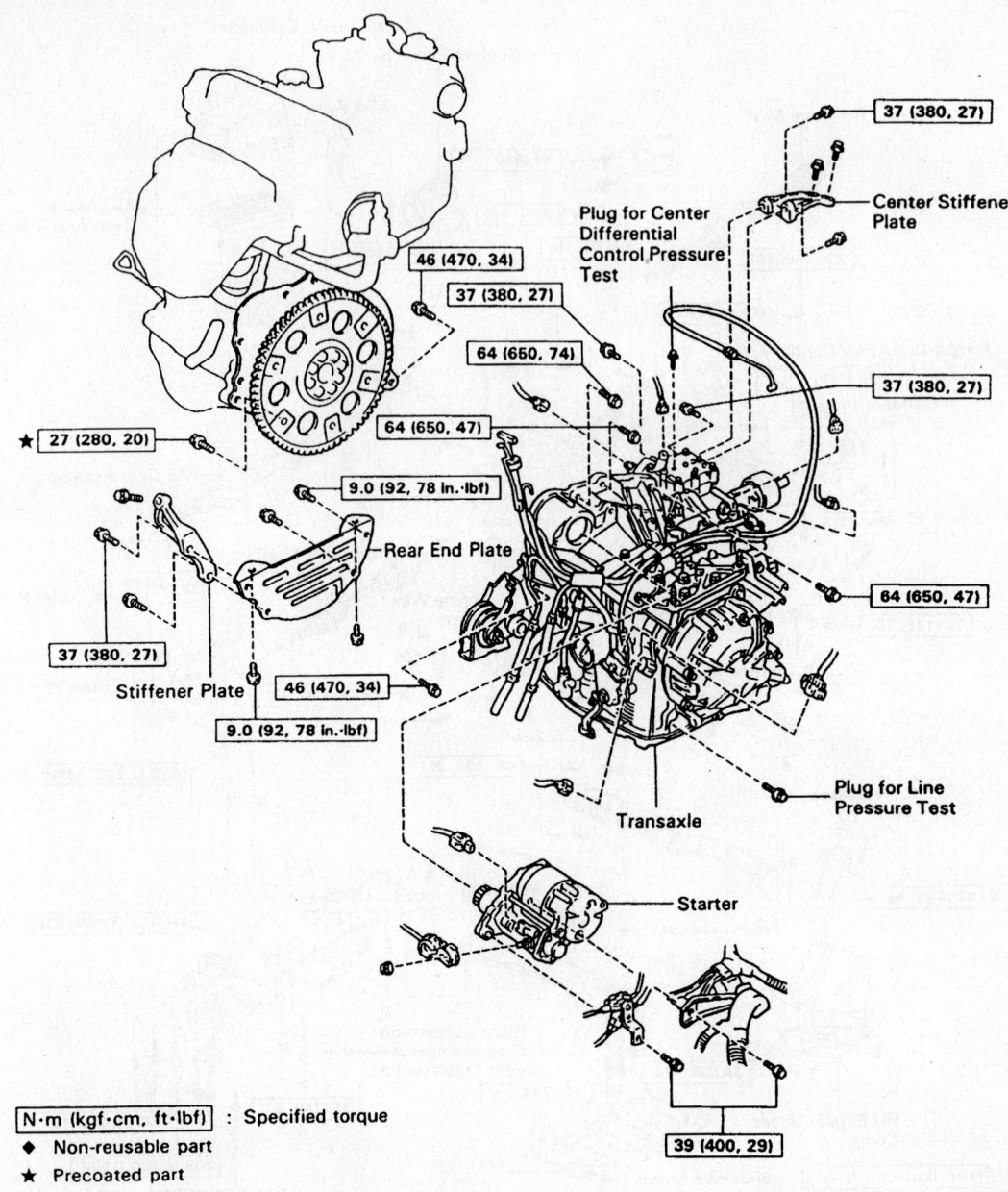

$\boxed{\text{N·m (kgf·cm, ft·lbf)}}$: Specified torque

◆ Non-reusable part

★ Precoated part

7924ZG62A

Automatic transaxle and related components—2002–2003 4WD

- Connectors and the wiring harness to the transaxle
- Torque converter clutch mounting bolts. Tighten each bolt to 20 ft. lbs. (27 Nm).

➡**Coat the threads of the bolts with an approved locking compound.**

- Rear endplate. Tighten the 4 bolts to 80 inch lbs. (9.0 Nm).
- Stiffener plate. Tighten the 3 bolts to 27 ft. lbs. (37 Nm).
- Starter. Tighten both bolts to 29 ft. lbs. (39 Nm).
- Engine/transaxle assembly
- Negative battery cable

2004–2005

2WD

1. Before servicing the vehicle, refer to the Precautions Section.
2. Remove the hood.
3. Remove the air cleaner assembly.
4. Remove the coolant reservoir.
5. Remove the control cables from the transaxle.
6. Remove the wiring from the speed sensor and back-up light switch.
7. Remove the ground cable from the transaxle.
8. Remove the starter.

9. Remove the clutch release cylinder.
10. Remove the 3 upper transaxle-to-engine bolts.
11. Attach an engine support fixture, such as 12281-28010, or equivalent.
12. Disconnect the left engine mount from the transaxle.
13. Remove the engine under-covers.
14. Drain the transaxle.
15. Remove the front exhaust pipe.
16. Remove the halfshafts.
17. Remove the stabilizer bar.
18. Disconnect the oil cooler lines.
19. Remove the power steering gear mounting bolts. Support the gear.

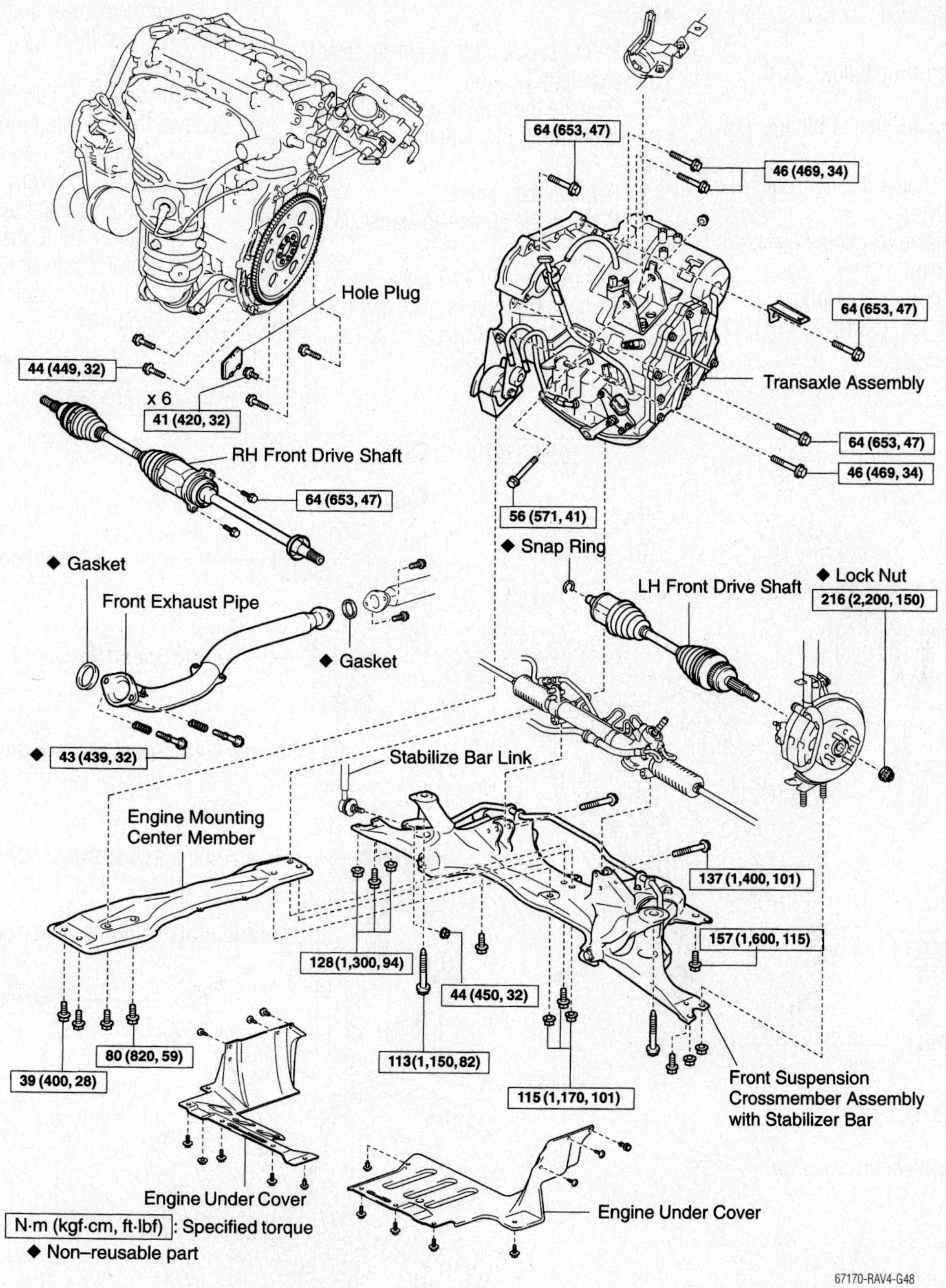

Hole Plug

44 (449, 32)

x 6
41 (420, 32)

RH Front Drive Shaft

64 (653, 47)

64 (653, 47)

46 (469, 34)

64 (653, 47)

Transaxle Assembly

64 (653, 47)

46 (469, 34)

56 (571, 41)

◆ Snap Ring

LH Front Drive Shaft

◆ Lock Nut
216 (2,200, 150)

◆ Gasket

Front Exhaust Pipe

◆ Gasket

◆ 43 (439, 32)

Stabilize Bar Link

Engine Mounting
Center Member

137 (1,400, 101)

157 (1,600, 115)

128 (1,300, 94)

44 (450, 32)

80 (820, 59)

113 (1,150, 82)

39 (400, 28)

115 (1,170, 101)

Front Suspension
Crossmember Assembly
with Stabilizer Bar

Engine Under Cover

N·m (kgf·cm, ft·lbf) : Specified torque
◆ Non–reusable part

Engine Under Cover

67170-RAV4-G48

Automatic transaxle and related components—2004–2005 2WD

20. Remove the front suspension cross-member.

21. Using a transmission jack, slightly raise the transaxle.

22. Remove the 4 lower transaxle-to-engine bolts.

23. Remove the front side mounting bolt, the 2 rear side mounting bolts and the transaxle.

To install:

24. Installation is the reverse of removal. Observe the following torques:

- Lower transaxle-to-engine bolts: 32 ft. lbs. (44 Nm)

- Front side mounting bolt, the 2 rear side mounting bolts: 34 ft. lbs. (46 Nm)
- Torque converter bolts: 32 ft. lbs. (41 Nm). Torque the green bolt first.
- Front suspension crossmember (see the illustration): bolt A 28 ft. lbs. (39 Nm); bolt B 59 ft. lbs. (80 Nm); bolts C 80 ft. lbs. (113 Nm); bolt D 85 ft. lbs. (115 Nm); bolt E 115 ft. lbs. (157 Nm); nuts 85 ft. lbs. (115 Nm)
- Power steering gear bolts: 101 ft. lbs. (137 Nm)

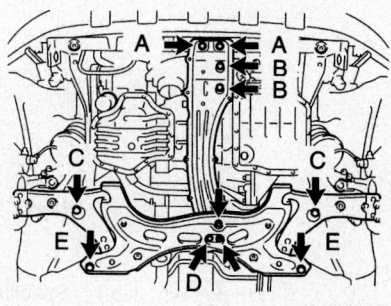

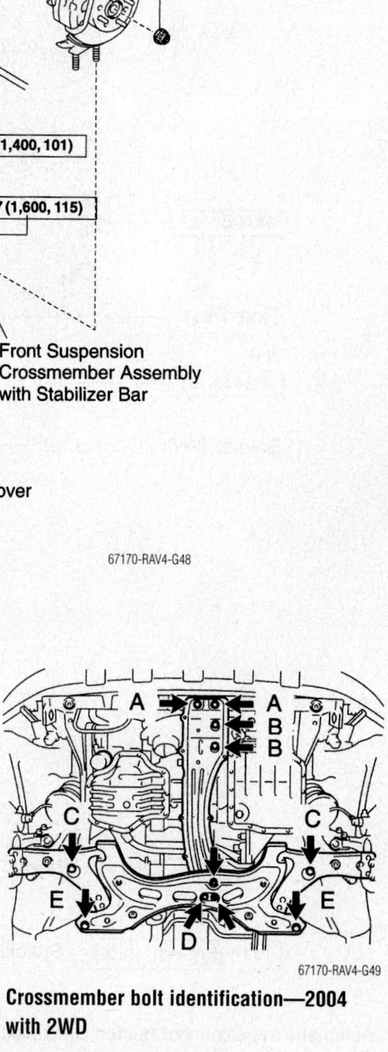

67170-RAV4-G49

Crossmember bolt identification—2004 with 2WD

- Stabilizer bar links: 32 ft. lbs. (44 Nm)
- Exhaust pipe-to-manifold: 32 ft. lbs. (43 Nm)
- Exhaust pipe-to-pipe: 14 ft. lbs. (19 Nm)
- Left engine mount through-bolt: 41 ft. lbs. (56 Nm)
- Upper transaxle-to-engine bolts: 47 ft. lbs. (64 Nm)
- Starter: 28 ft. lbs. (37 Nm)
- Hood: 10 ft. lbs. (13 Nm)

4WD

1. Before servicing the vehicle, refer to the Precautions Section.
2. Remove the engine and transaxle as a unit. Position it in a suitable holding fixture.
3. Remove the starter.
4. Remove the engine-to-transaxle wiring.
5. Remove the stiffener plate.
6. Remove the access plate and remove the 6 torque converter bolts.

7. Remove lower bolts and separate the transaxle from the engine.

To install:

8. Installation is the reverse of removal. Observe the following torques:
- Transaxle-to-engine lower bolts: 32 ft. lbs. (44 Nm)
- Stiffener plate: 25 ft. lbs. (34 Nm)
- Starter: 27 ft. lbs. (37 Nm)
- Torque converter bolts: 30 ft. lbs. (41 Nm). Install the green colored bolt first.

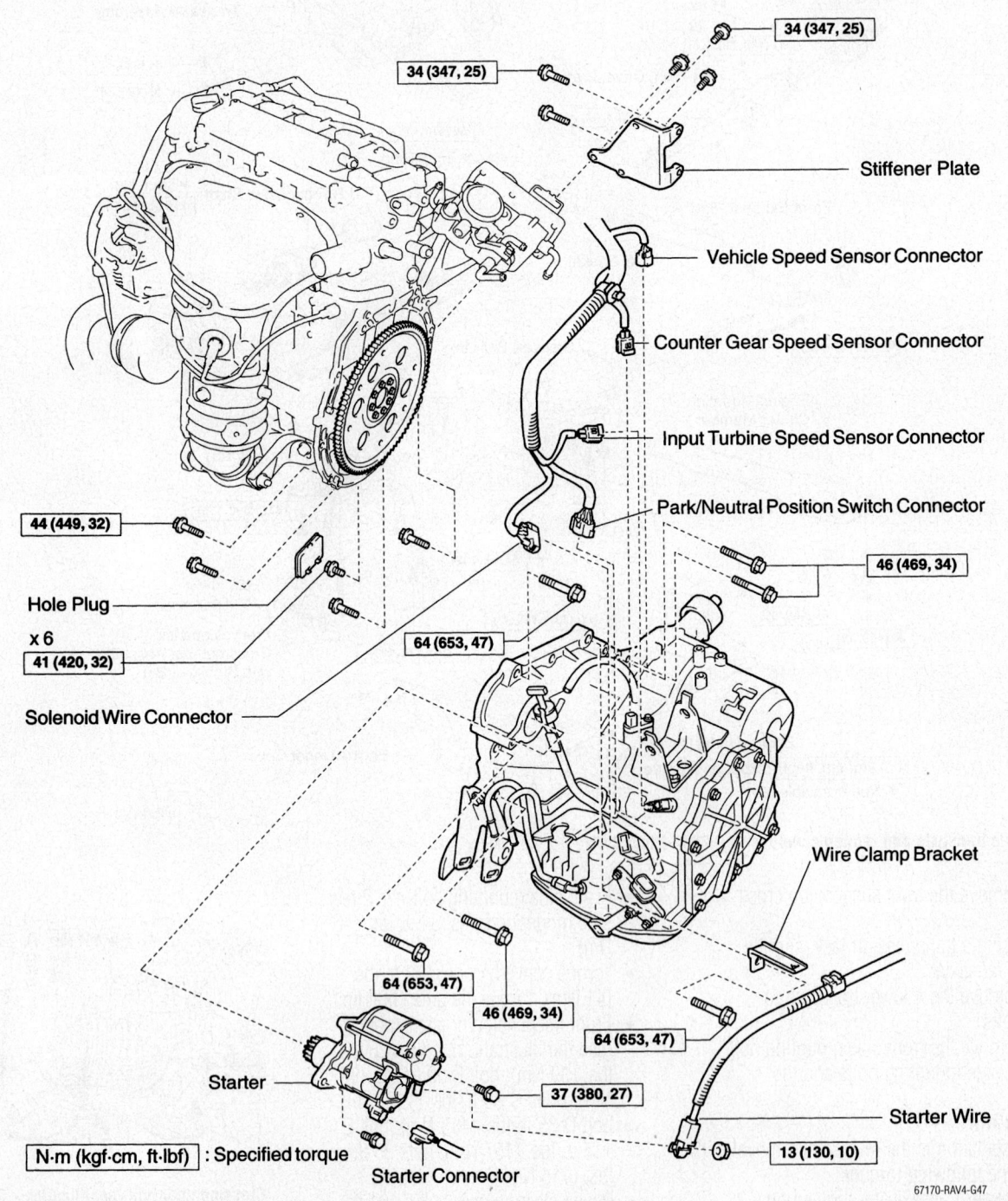

34 (347, 25)

34 (347, 25)

Stiffener Plate

Vehicle Speed Sensor Connector

Counter Gear Speed Sensor Connector

Input Turbine Speed Sensor Connector

Park/Neutral Position Switch Connector

44 (449, 32)

46 (469, 34)

Hole Plug

x 6

41 (420, 32)

64 (653, 47)

Solenoid Wire Connector

Wire Clamp Bracket

64 (653, 47)

46 (469, 34)

64 (653, 47)

Starter

37 (380, 27)

Starter Wire

N·m (kgf·cm, ft·lbf) : Specified torque

Starter Connector

13 (130, 10)

67170-RAV4-G47

2006

2WD

1. Before servicing the vehicle, refer to the Precautions Section.

2. Properly relieve the fuel system pressure.

3. Disconnect the negative battery cable.

➡ **Wait at least 90 seconds after disconnecting the negative battery cable before starting any repair work to prevent air bag and seat belt pretensioner activation.**

4. Drain the transaxle fluid.

5. Remove the engine and transaxle assembly from the vehicle and position it in a suitable holding fixture.

6. Remove the starter. Disconnect the wiring harnesses.

7. Remove the transaxle oil cooler.

8. Remove the rear engine mounting bracket.

9. Remove the front engine mounting bracket.

10. Remove the engine mounting bracket, left side.

11. Remove the flywheel housing inspection cover.

12. Remove the six torque converter to flywheel retaining bolts.

13. Remove the lower side transaxle to engine retaining bolts. Remove the upper side transaxle to engine retaining bolts.

14. Separate the transaxle from the engine.

15. Remove the transaxle dipstick tube assembly. Remove the number one transaxle control cable bracket.

16. Remove the torque converter from the transaxle.

To install:

17. When installing the torque converter use a caliper to measure dimension "A", between the transaxle and the end surface of the drive plate.

18. Using the caliper and a straight edge, measure the dimension "B" and check that "B" is greater than "A". Measurement should be 0.039 inch or more.

19. On U241E transaxle, tighten the lower side transaxle to engine retaining bolts to 27 ft. lbs. Tighten the upper side transaxle to engine retaining bolts to 47 ft.

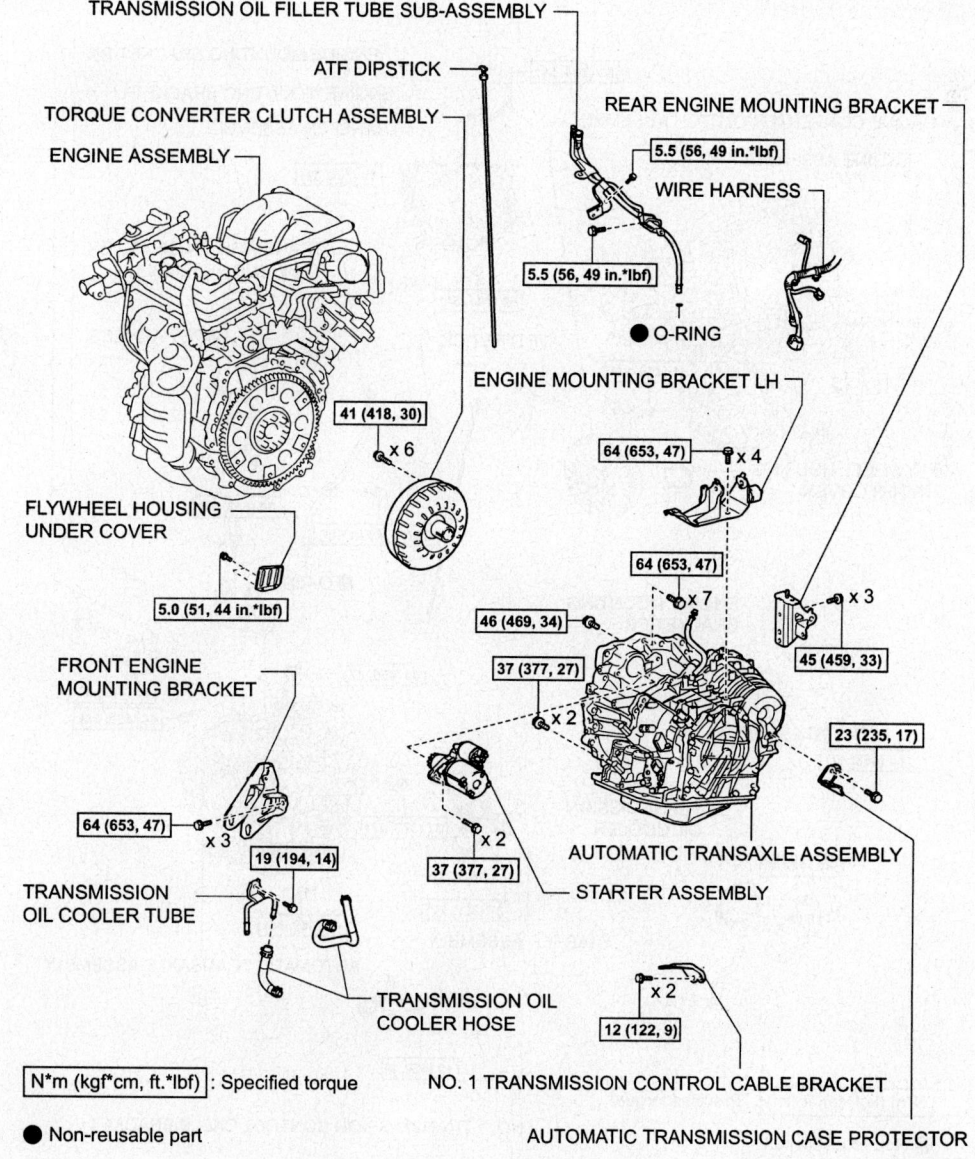

TRANSMISSION OIL FILLER TUBE SUB-ASSEMBLY

ATF DIPSTICK

TORQUE CONVERTER CLUTCH ASSEMBLY

REAR ENGINE MOUNTING BRACKET

ENGINE ASSEMBLY

5.5 (56, 49 in.*lbf)

WIRE HARNESS

5.5 (56, 49 in.*lbf)

● O-RING

ENGINE MOUNTING BRACKET LH

41 (418, 30) x 6

64 (653, 47) x 4

FLYWHEEL HOUSING UNDER COVER

64 (653, 47)

46 (469, 34) x 7

x 3

45 (459, 33)

5.0 (51, 44 in.*lbf)

37 (377, 27)

x 2

23 (235, 17)

FRONT ENGINE MOUNTING BRACKET

64 (653, 47) x 3

19 (194, 14)

x 2

37 (377, 27)

AUTOMATIC TRANSAXLE ASSEMBLY

STARTER ASSEMBLY

TRANSMISSION OIL COOLER TUBE

TRANSMISSION OIL COOLER HOSE

x 2

12 (122, 9)

N*m (kgf*cm, ft.*lbf) : Specified torque

NO. 1 TRANSMISSION CONTROL CABLE BRACKET

● Non-reusable part

AUTOMATIC TRANSMISSION CASE PROTECTOR

09490_RAV4_G0082

Automatic transaxle and related components—2006 2WD

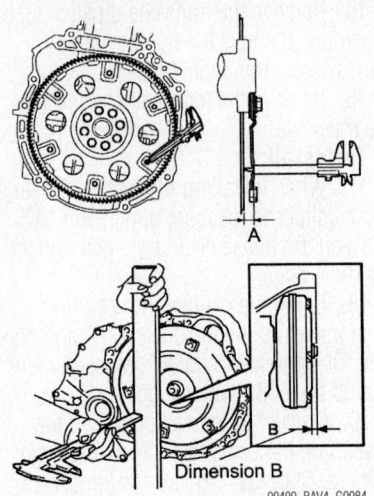

Torque converter installation measurement points

Dimension B

09490_RAV4_G0084

lbs. (for the top three bolts) and 34 ft. lbs. (for the bottom lower bolts, on each side).

20. On U151E transaxle, tighten the lower side transaxle to engine retaining bolts to 27 ft. lbs. (for the bottom two bolts). And 34 ft. lbs. (for the upper bolt). Tighten the upper side transaxle to engine retaining bolts to 47 ft. lbs.

21. Tighten the torque converter to flexplate retaining bolts to 30 ft. lbs.

22. Continue the installation in the reverse order of the removal procedure.

➡**Perform the automatic transmission initialization procedure when replacing the automatic transmission, engine or ECM.**

23. Turn the ignition switch OFF.

24. Connect the intelligent tester tool together with the controller area network vehicle interface module (CAN VIN) to the DLC3.

25. Turn the ignition switch to the ON position.

26. Push the intelligent tester tool main switch to the ON position.

27. Select the items, DIAGNOSIS/ ENHANCED OBD II.

28. Perform the reset memory procedure from the ENGINE menu.

➡**After performing the reset memory, be sure to perform the road test procedure. For roadtest procedure information, refer to the intelligent tester instruction manual.**

4WD

1. Before servicing the vehicle, refer to the Precautions Section.

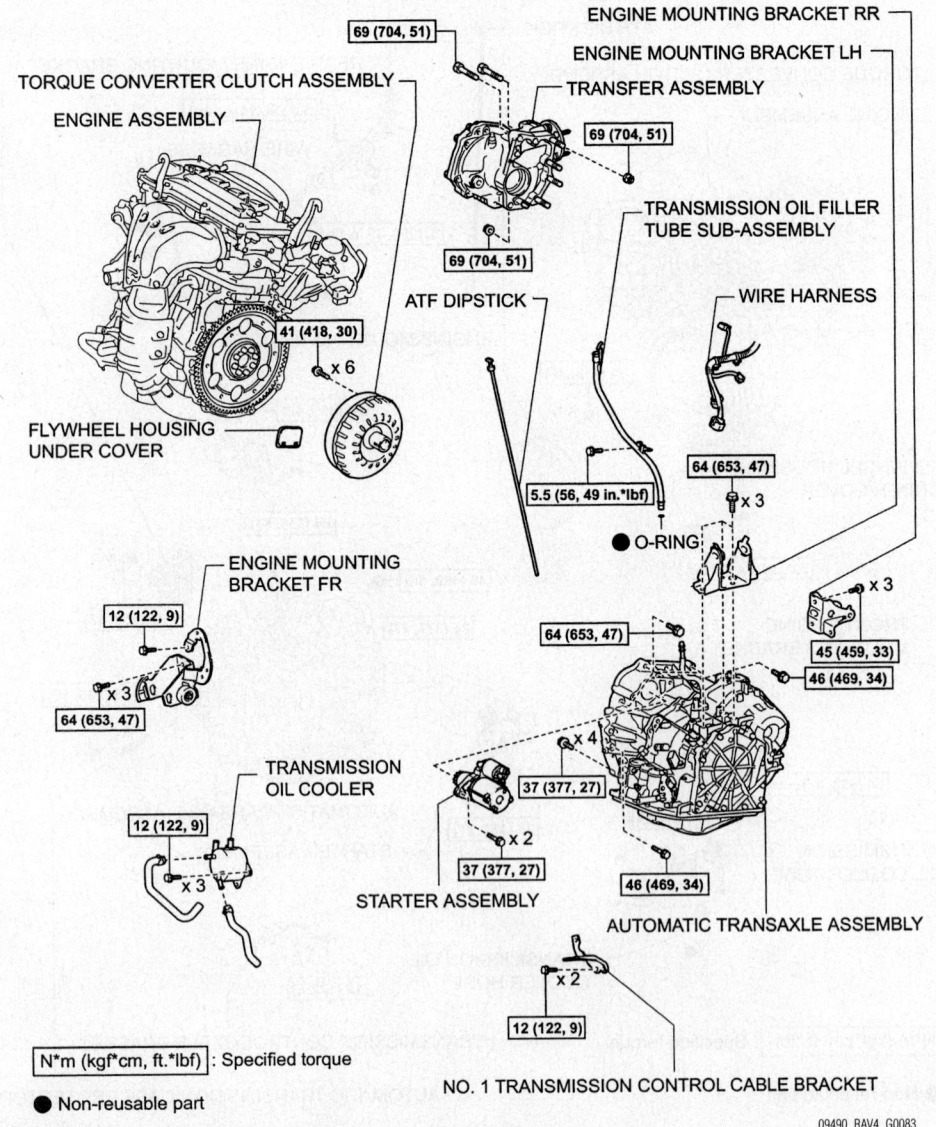

N*m (kgf*cm, ft.*lbf) : Specified torque

● Non-reusable part

Automatic transaxle and related components—2006 4WD

09490_RAV4_G0083

2. Properly relieve the fuel system pressure.

3. Disconnect the negative battery cable.

➡**Wait at least 90 seconds after disconnecting the negative battery cable before starting any repair work to prevent air bag and seat belt pretensioner activation.**

4. Drain the transaxle fluid. Drain the transfer case fluid.

5. Remove the engine and transaxle assembly from the vehicle and position it in a suitable holding fixture.

6. Remove the starter. Disconnect the wiring harnesses.

7. Remove the transaxle oil cooler.

8. Remove the right rear engine mounting bracket.

9. Remove the front rear engine mounting bracket.

10. Remove the left engine mounting bracket.

11. Remove the transfer case retaining bolts. Separate the transfer case for the transaxle.

12. Remove the flywheel housing inspection cover.

13. Remove the six torque converter to flywheel retaining bolts.

14. Remove the lower side transaxle to engine retaining bolts. Remove the upper side transaxle to engine retaining bolts.

15. Separate the transaxle from the engine.

16. Remove the transaxle dipstick tube assembly. Remove the number one transaxle control cable bracket.

17. Remove the torque converter from the transaxle.

To install:

18. When installing the torque converter use a caliper to measure dimension "A", between the transaxle and the end surface of the drive plate.

19. Using the caliper and a straight edge, measure the dimension "B" and check that "B" is greater than "A". Measurement should be 0.039 inch or more.

20. On U140F transaxle, tighten the lower side transaxle to engine retaining bolts to 27 ft. lbs. Tighten the upper side transaxle to engine retaining bolts to 47 ft. lbs. (for the top three bolts) and 34 ft. lbs. (for the bottom lower bolts, on each side).

21. On U151F transaxle, tighten the lower side transaxle to engine retaining bolts to 27 ft. lbs. (for the bottom two bolts). And 34 ft. lbs. (for the upper bolt). Tighten the upper side transaxle to engine retaining bolts to 47 ft. lbs.

22. Tighten the torque converter to flex-plate retaining bolts to 30 ft. lbs.

23. Tighten the transfer case to transaxle retaining bolts to 51 ft. lbs.

24. Continue the installation in the reverse order of the removal procedure.

➡**Perform the automatic transmission initialization procedure when replacing the automatic transmission, engine or ECM.**

25. Turn the ignition switch OFF.

26. Connect the intelligent tester tool together with the controller area network vehicle interface module (CAN VIN) to the DLC3.

27. Turn the ignition switch to the ON position.

28. Push the intelligent tester tool main switch to the ON position.

29. Select the items, DIAGNOSIS/ENHANCED OBD II.

30. Perform the reset memory procedure from the ENGINE menu.

➡**After performing the reset memory, be sure to perform the road test procedure. For roadtest procedure information, refer to the intelligent tester instruction manual.**

Transfer Case Assembly

REMOVAL & INSTALLATION

The transfer case is part of the transaxle/transaxle assembly and is removed with those units.

Halfshaft

REMOVAL & INSTALLATION

Front

2002–2003

1. Before servicing the vehicle, refer to the Precautions Section.

2. Remove or disconnect the following:
   - Negative battery cable
   - Engine undercover

3. Drain the transaxle.

4. Remove or disconnect the following:
   - Anti-lock Brake System (ABS) sensor by removing the bolt, if equipped
   - Cotter pin, lock cap and the locknut holding the halfshaft to the steering knuckle
   - Tie rod ends, from the steering knuckle

   - Sway bar link, from the lower control arm
   - Lower ball joint, from the lower control arm
   - Halfshaft from the axle hub, using a plastic hammer

5. If working on a 2WD right side half-shaft and the vehicle is equipped with a manual transaxle, perform the following to remove the halfshaft:
   - Snapring from the center bearing bracket, using a brass bar and hammer
   - Bolt and the center bearing bracket
   - Halfshaft with the center halfshaft
   - 2 bolts and the center bearing bracket

6. If working on a 2WD right side half-shaft and the vehicle is equipped with an automatic transaxle, perform the following to remove the halfshaft:
   - 2 bolts of the center bearing bracket and pull out the halfshaft together with the center bearing case and center halfshaft
   - 3 bolts and the center bearing bracket

7. If working on a 2WD left side, perform the following:
   - Halfshaft, using a brass bar and hammer
   - Snapring from the transaxle

8. If working of a 4WD right side half-shaft, perform the following:
   - Halfshaft, using a brass bar and hammer
   - Snapring from the transaxle
   - O-ring

9. If working on a 4WD left side, perform the following:
   - Air cleaner
   - Transaxle case protector
   - Halfshaft, by prying it out using a hub wrench
   - Snapring

To install:

10. If working on a 4WD left side, perform the following:
   - Snapring
   - Halfshaft to the transaxle
   - Transaxle case protector
   - Air cleaner

11. If working of a 4WD right side half-shaft, perform the following:
   - Snapring to the transaxle
   - New O-ring
   - Halfshaft to the transaxle

12. If working on a 2WD left side, perform the following:
   - Snapring
   - Halfshaft to the transaxle

13. If working on a 2WD right side half-

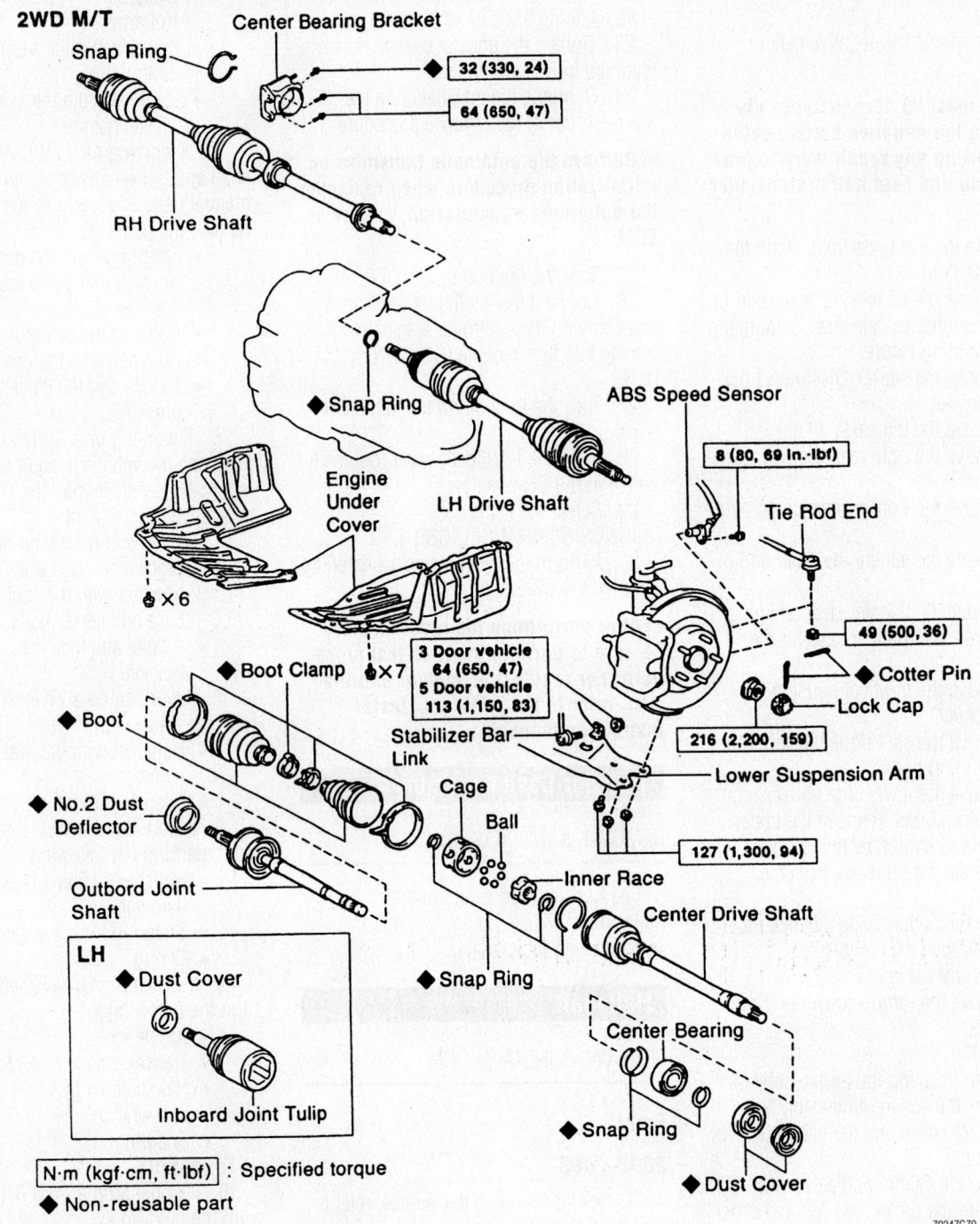

2WD M/T

Center Bearing Bracket

Snap Ring

◆ 32 (330, 24)

64 (650, 47)

RH Drive Shaft

◆ Snap Ring

Engine Under Cover

LH Drive Shaft

ABS Speed Sensor

8 (80, 69 In.·lbf)

Tie Rod End

49 (500, 36)

◆ Cotter Pin

Lock Cap

◆ x 6

◆ Boot Clamp ◆ x 7

3 Door vehicle
64 (650, 47)
5 Door vehicle
113 (1,150, 83)

216 (2,200, 159)

Stabilizer Bar Link

Lower Suspension Arm

◆ Boot

◆ No.2 Dust Deflector

Cage

Ball

127 (1,300, 94)

Outbord Joint Shaft

Inner Race

Center Drive Shaft

LH

◆ Dust Cover

◆ Snap Ring

◆ Snap Ring

Center Bearing

Inboard Joint Tulip

◆ Snap Ring

N·m (kgf·cm, ft·lbf) : Specified torque

◆ Non-reusable part

◆ Dust Cover

7924ZG70

Front halfshaft and related components (2WD with manual transaxle)—2002–2003

shaft and the vehicle is equipped with an automatic transaxle, perform the following to remove the halfshaft:

- Center bearing bracket. Tighten the 3 bolts to 47 ft. lbs. (64 Nm).
- Halfshaft together with the center bearing case and center halfshaft. Tighten both bolts to 47 ft. lbs. (64 Nm).

14. If working on a 2WD right side half-shaft and the vehicle is equipped with a

manual transaxle, perform the following to remove the halfshaft:

- Center bearing bracket
- Halfshaft with the center halfshaft
- Center bearing bracket. Tighten the bolt to 24 ft. lbs. (32 Nm).
- Snapring to the center bearing bracket

15. Install or connect the following:
- Halfshaft to the axle hub
- Lower ball joint to the lower control

arm. Tighten the nuts/bolt to 94 ft. lbs. (127 Nm).

- Sway bar link to the lower control arm. Tighten the nut to 47 ft. lbs. (64 Nm) for 3-door or to 83 ft. lbs. (113 Nm) for 5-door.
- Tie rod end to the steering knuckle. Tighten the nut to 36 ft. lbs. (49 Nm).
- New tie rod end cotter pin
- Halfshaft to the axle hub. Tighten

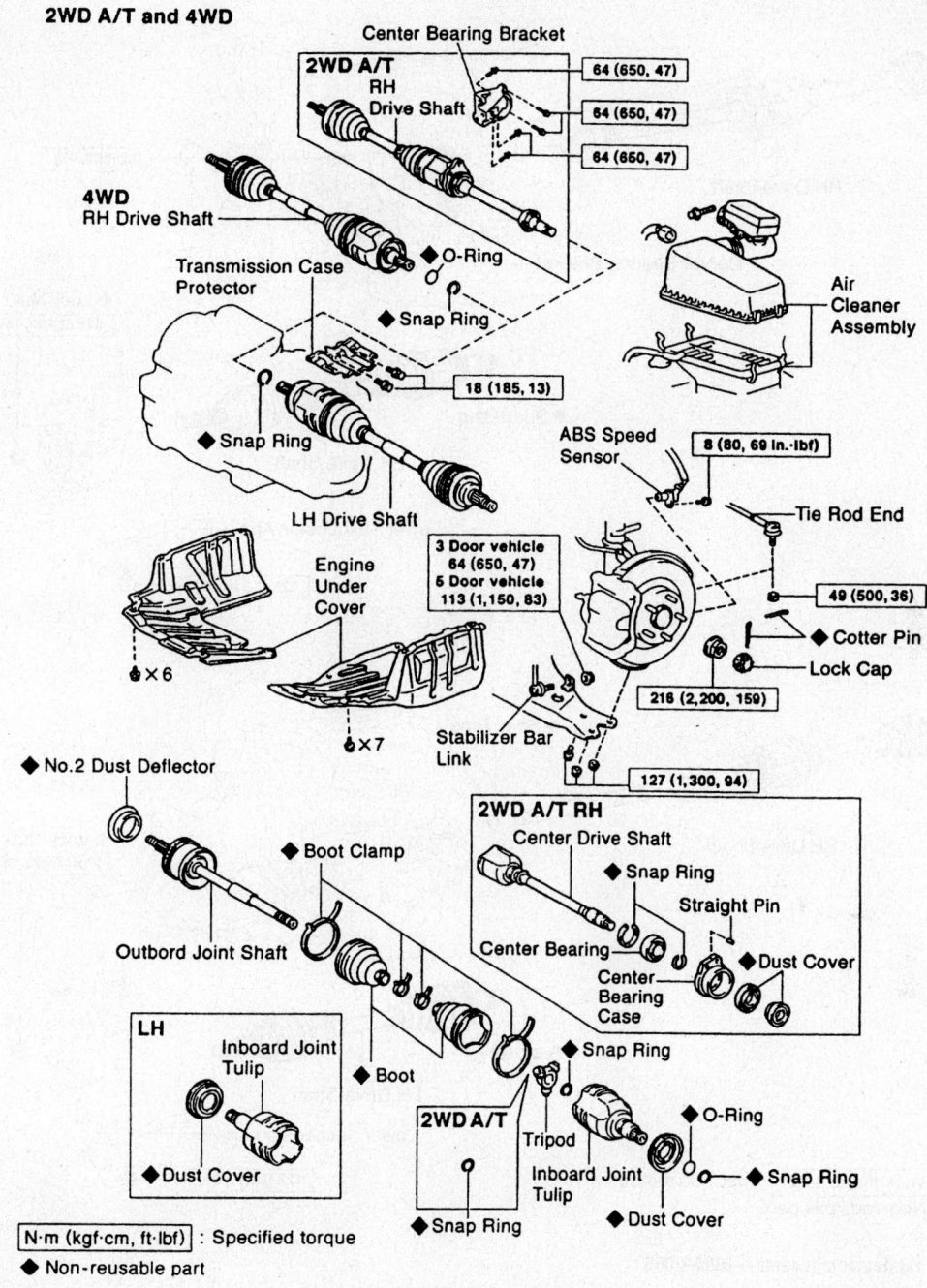

2WD A/T and 4WD

Front halfshaft and related components (2WD and 4WD with automatic transaxle)—2002–2003

the locknut to 159 ft. lbs. (216 Nm).
• Lock cap and cotter pin
• ABS speed sensor with the bolt, if equipped
16. Fill the transaxle with gear oil (manual transaxle) or ATF (automatic transaxle).
17. Install or connect the following:
• Engine undercover
• Wheels
• Negative battery cable
18. Check the ABS sensor signal.

2004–2005

1. Before servicing the vehicle, refer to the Precautions Section.
2. Remove the wheel.
3. Drain the transaxle. With 4WD, drain the transfer case.
4. Unstake and remove the locknut.
5. Unbolt the lower arm from the ball joint.
6. Remove the halfshaft from the hub, using a plastic hammer.

7. Remove the halfshaft. On 2WD, the right side shaft is retained by 2 bolts. All others are removed with a slidehammer and special tool 09520-01010, or equivalent.
8. Remove the snapring.
To install:
9. Installation is the reverse of removal. The new snapring is installed with the split facing downward. After installation, check that there is 2–3mm of axial play. Observe the following torques:

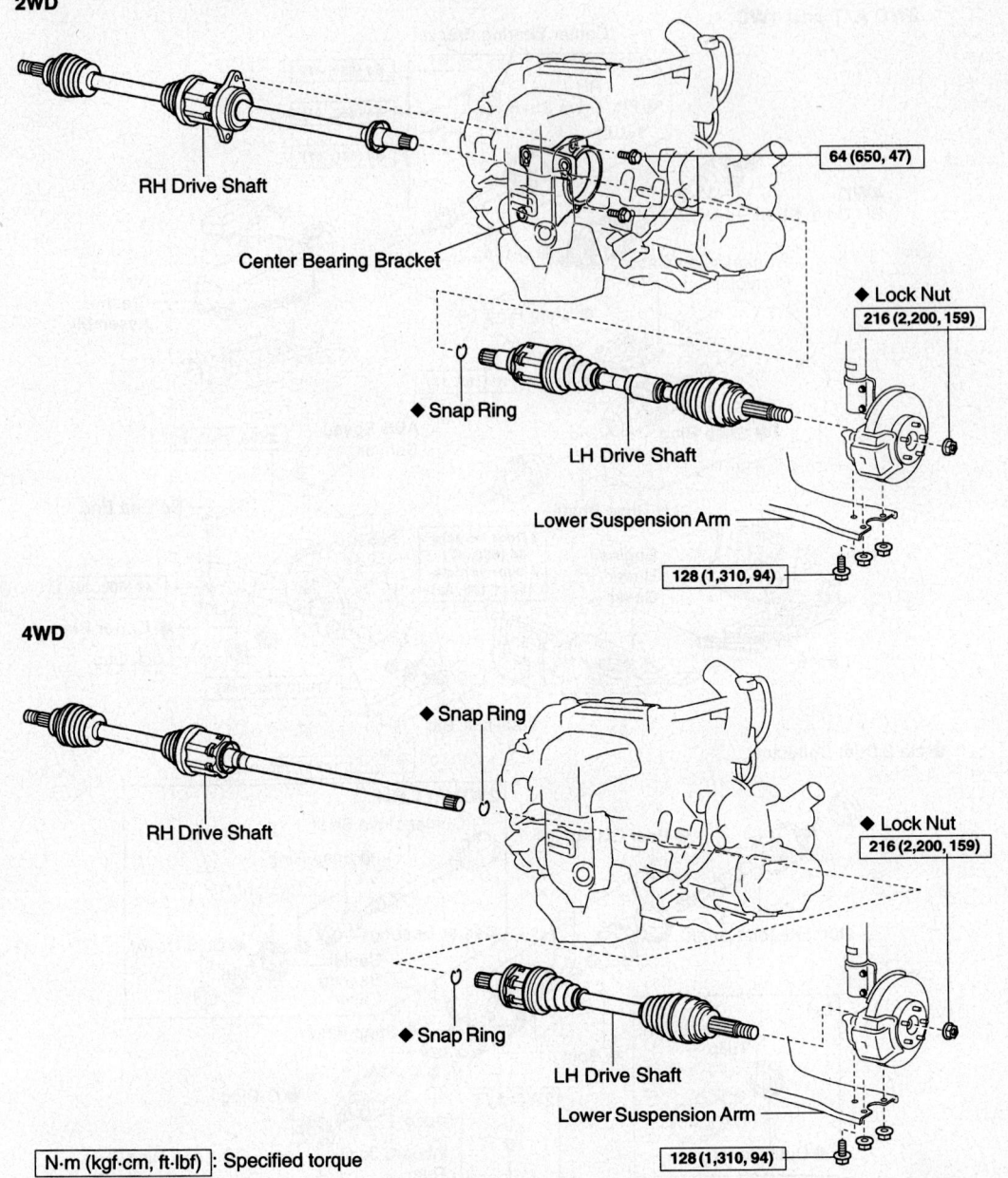

2WD

RH Drive Shaft

Center Bearing Bracket

64 (650, 47)

◆ Lock Nut
216 (2,200, 159)

◆ Snap Ring

LH Drive Shaft

Lower Suspension Arm

128 (1,310, 94)

4WD

◆ Snap Ring

RH Drive Shaft

◆ Lock Nut
216 (2,200, 159)

◆ Snap Ring

LH Drive Shaft

Lower Suspension Arm

128 (1,310, 94)

N·m (kgf·cm, ft·lbf) : Specified torque
◆ Non–reusable part

67170-RAV4-G51

Front halfshafts and related components—2004–2005

- Right halfshaft bolts: 47 ft. lbs. (64 Nm)
- Lower arm-to-ball joint: 94 ft. lbs. (128 Nm)
- Hub nut: 159 ft. lbs. (216 Nm)

2006

1. Before servicing the vehicle, refer to the Precautions Section.
2. Disconnect the negative battery cable.

➡**Wait at least 90 seconds after disconnecting the negative battery cable before starting any repair work to pre-** vent air bag and seat belt pretensioner activation.

3. Raise and support the vehicle safely.
4. Remove the tire and wheel assembly. Remove the front axle hub nut.
5. Drain the transaxle fluid.
6. Disconnect the left and right speed sensors.
7. Remove the left and right brake calipers.
8. Disconnect the left and right front stabilizer link assemblies.
9. Disconnect the left and right front lower number one arm subassemblies.

10. Matchmark the halfshaft and the axle hub, left and right side.

➡**Do not punch the marks.**

11. Using a plastic hammer, disconnect the steering knuckle with the axle hub, left and right side.

➡**Be careful not to damage the boot and speed sensor rotor. Do not push out excessively the halfshaft from the axle assembly.**

12. Disconnect the left and right tie rod subassemblies.
13. On the left side, using tool

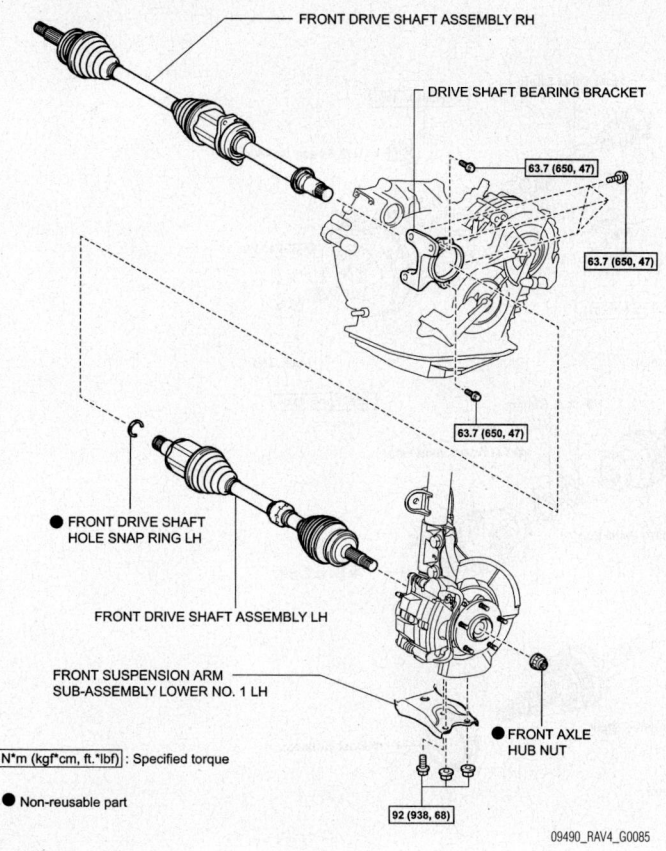

Front halfshaft and related components—2006 2WD

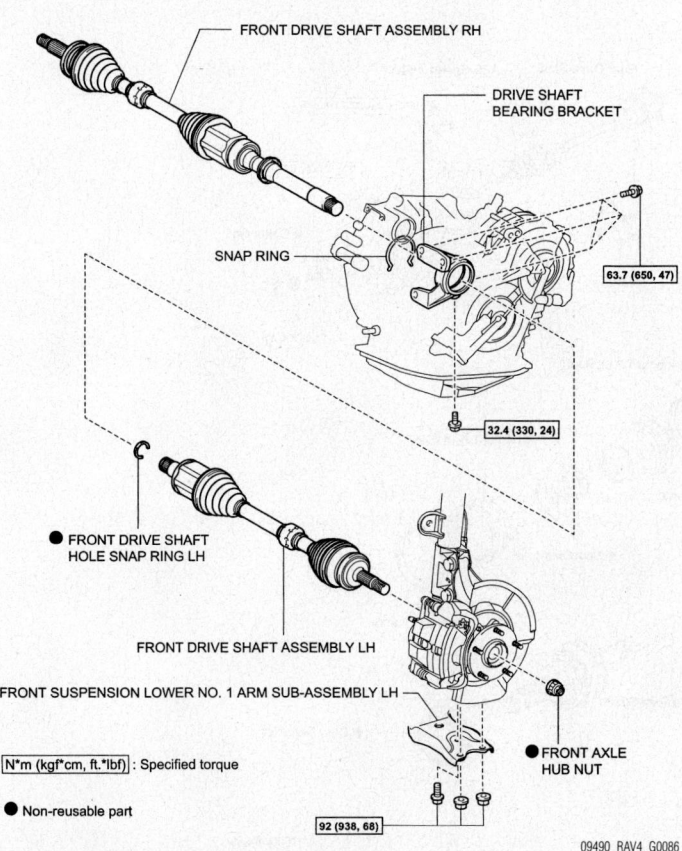

Front halfshaft and related components—2006 4WD

SST09520-01010, or equivalent remove the front halfshaft.

14. On 2WD, to remove the right half-shaft remove the two bolts and pull out the halfshaft together with the halfshaft bearing case. Remove the halfshaft from the transaxle.

➡ **Be careful not to damage the boot and speed sensor rotor. Do not push out excessively the halfshaft from the axle assembly.**

15. On 4WD, to remove the right half-shaft use a brass bar and hammer to remove the halfshaft.

➡ **Do not damage the oil seal, boot or allow the halfshaft to fall out.**

16. Support the front axle assembly.

➡ **The hub bearing could be damaged if it is subjected to the vehicle weight. If it is necessary to place weight on the hub bearing, such as moving it when the halfshaft is removed, support it using too SST09608-16042, or equivalent.**

To install:

17. Installation is the reverse of the removal procedure.

18. On vehicles manufactured from 11/2005 thru 01/2006 tighten the front axle hub nut to 159 ft. lbs. On vehicles manufactured after 01/2006 and equipped with the 2.4L engine, tighten the axle hub nut to 159 ft. lbs. On vehicles manufactured after 01/2006 and equipped with the 3.5L engine, tighten the axle hub nut to 215 ft. lbs.

19. On 2WD, tighten the right side bearing bracket bolts to 47 ft. lbs.

20. On 4WD, tighten the right side bearing bracket bolts to 24 ft. lbs.

21. Check and adjust the wheel alignment, as required.

22. Be sure to fill the transaxle with the proper grade and type transaxle fluid.

23. Start the engine and check for leaks.

Rear

2002–2005

1. Before servicing the vehicle, refer to the Precautions Section.

2. Remove or disconnect the following:
   - Negative battery cable
   - Rear wheels
   - Anti-lock Brake System (ABS) speed sensor from the axle assembly by removing the bolt, if equipped
   - Cotter pin, lock cap and the nut

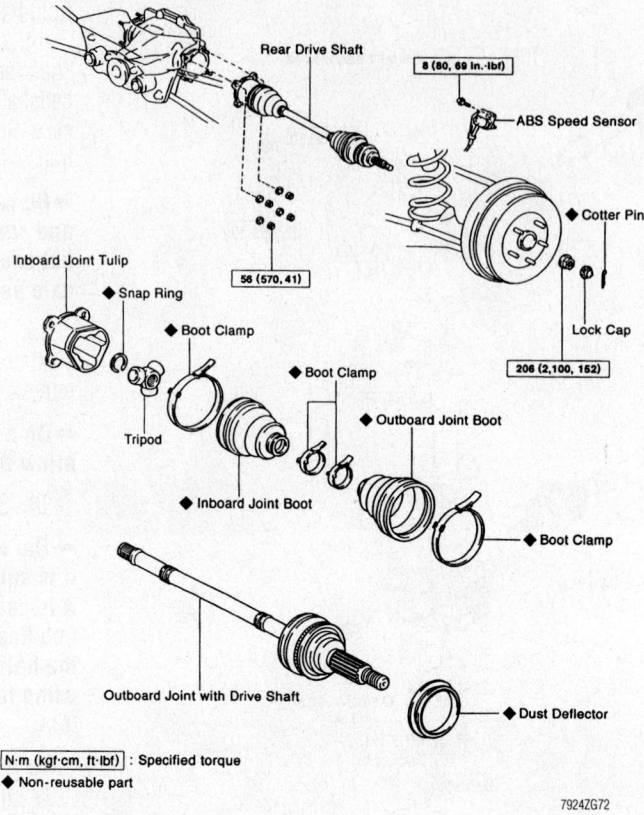

Rear Drive Shaft

8 (80, 69 in.-lbf)

ABS Speed Sensor

Cotter Pin

58 (570, 41)

Inboard Joint Tulip

◆ Snap Ring

◆ Boot Clamp

Lock Cap

Tripod

◆ Boot Clamp

206 (2,100, 152)

◆ Outboard Joint Boot

◆ Inboard Joint Boot

◆ Boot Clamp

Outboard Joint with Drive Shaft

◆ Dust Deflector

N·m (kgf·cm, ft·lbf) : Specified torque
◆ Non-reusable part

7924ZG72

Rear halfshaft and related components—2002–2003

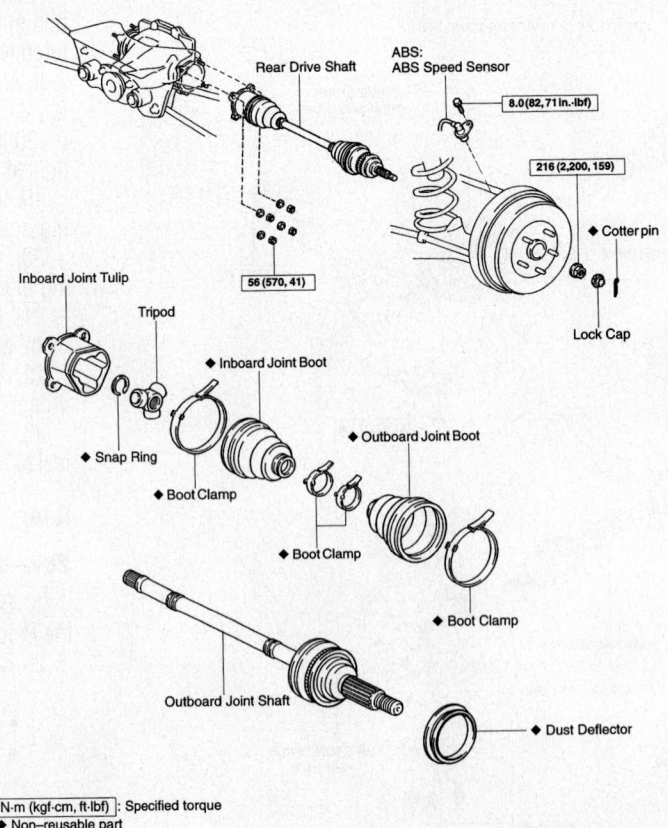

Rear Drive Shaft

ABS:
ABS Speed Sensor

8.0 (82, 71 in.-lbf)

216 (2,200, 159)

56 (570, 41)

Inboard Joint Tulip

Tripod

◆ Cotter pin

◆ Inboard Joint Boot

Lock Cap

◆ Snap Ring

◆ Outboard Joint Boot

◆ Boot Clamp

◆ Boot Clamp

◆ Boot Clamp

Outboard Joint Shaft

◆ Dust Deflector

N·m (kgf·cm, ft·lbf) : Specified torque
◆ Non-reusable part

67170-RAV4-G56

Rear halfshaft and related components—2004–2005

holding the halfshaft to the axle carrier

3. Place matchmarks on the halfshaft and side gear shaft.

4. Remove or disconnect the following:
- Halfshaft from the differential side gear shaft, by removing the 4 nuts and washers
- Halfshaft from the axle carrier, using a plastic hammer

To install:

5. Install or connect the following:
- Halfshaft to the axle carrier
- Halfshaft to the differential side gear shaft, by aligning the marks. Tighten the 4 nuts to 41 ft. lbs. (56 Nm).
- Nut, lock cap and the cotter pin to hold the halfshaft to the axle carrier. Tighten the nut to 152 ft. lbs. (206 Nm) for 2002–2003; 159 ft. lbs. (216 Nm) for 2004–2005.
- ABS sensor. Tighten the bolt to 69 inch lbs. (8 Nm).
- Rear wheels
- Negative battery cable

6. Check the ABS sensor signal.

2006

1. Before servicing the vehicle, refer to the Precautions Section.

2. Disconnect the negative battery cable.

➡**Wait at least 90 seconds after disconnecting the negative battery cable before starting any repair work to prevent air bag and seat belt pretensioner activation.**

3. Raise and support the vehicle safely.

4. Remove the tire and wheel assembly.

5. Drain the differential oil.

6. Remove the tailpipe assembly.

7. Remove the driveshaft with the center bearing.

8. Remove the rear axle shaft nut.

9. Support the rear differential using a suitable jack.

10. Fix the nuts in place and remove bolt "A" "B" and "C". Do not loosen the nuts, loosen the bolts. Slowly lower the jack and tilt the rear differential carrier, as shown in the illustration.

11. Using a suitable tool disconnect the left and right rear halfshafts from the differential carrier.

12. On 2WD, disconnect the skid control sensor wire.

13. On 4WD, disconnect the rear speed sensor, left side.

14. Put matchmarks on the halfshaft and the axle hub. Do not punch the marks.

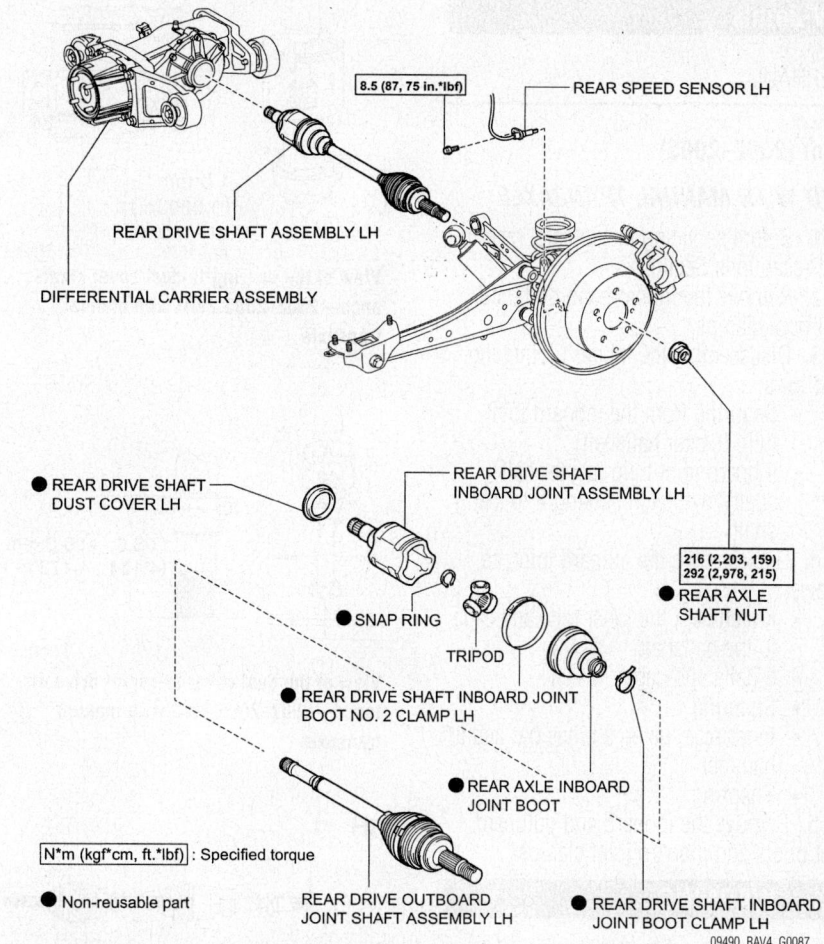

Rear halfshaft and related components—2006

15. Using a plastic faced hammer, separate the halfshaft from the axle hub.

➡**Be careful not to damage the boot and speed sensor rotor. Do not excessively push out the halfshaft from the axle.**

16. Support the rear halfshaft assembly.

➡**The hub bearing could be damaged if it is subjected to the vehicle weight. If it is necessary to place weight on the hub bearing, such as moving it when the halfshaft is removed, support it using too SST09608-16042, or equivalent.**

To install:

17. Installation is the reverse of the removal procedure.

18. Tighten the differential carrier bolts to 63 ft. lbs. for bolt "A", 103 ft. lbs. for bolt "B".

19. Check and adjust the wheel alignment, as required.

20. Be sure to fill the differential with the proper grade and type fluid.

21. Start the engine and check for leaks.

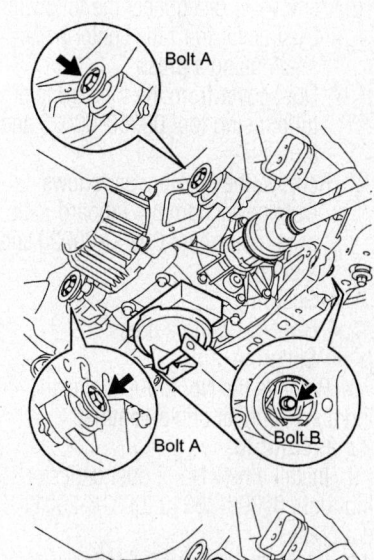

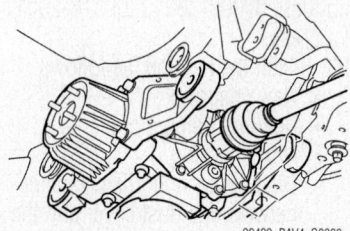

Rear differential bolt locations—2006

CV-Joints

OVERHAUL

Front (2002–2005)

2WD WITH MANUAL TRANSAXLE

1. Before servicing the vehicle, refer to the Precautions Section.
2. Remove the inboard and outboard joint boot clamps.
3. Disassemble the inboard joint tulip, as follows:
   - Snapring from the inboard joint tulip (center halfshaft)
   - Inboard joint tulip (center halfshaft), by matchmarking it to the shaft
4. Disassemble the inboard joint, as follows:
   - Matchmark the inner race and cage to the halfshaft
   - 6 balls and cage
   - Snapring
   - Inner race, using a brass bar and a hammer
   - Snapring
5. Remove the inboard and outboard joint boots and inboard joint clamps.

❄❄ WARNING

Do not disassemble the outboard joint.

6. Remove or disconnect the following:
   - Dust cover from the center half-shaft, using a press
   - Dust cover from the inboard joint tulip, using tool 09950-00020 and a press
7. Remove the bearing, as follows:
   - Dust cover from the inboard joint tulip, using tool 09950-00020 and a press
   - Snapring
   - Bearing, using a press
   - Snapring
8. Remove the No. 2 dust deflector, using a screwdriver and a hammer.
   To assemble:
9. Install a new No. 2 dust deflector, using tools 09309-36010, 09316-20011 and a press.
10. Install the bearing, as follows:
   - New snapring
   - Bearing, using a press
   - New snapring
   - Dust cover, until the clearance between the dust cover and the bearing is 0.039 in. (1.0mm)

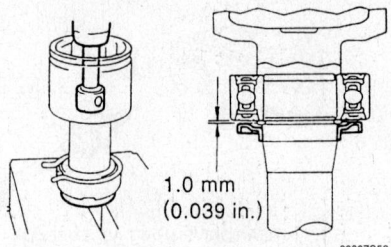

View of the bearing-to-dust cover clearance—2002–2005 2WD with manual transaxle

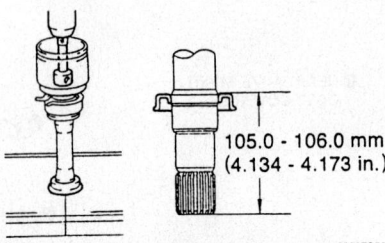

View of the dust cover-to-center drive distance—2002–2005 2WD with manual transaxle

RH

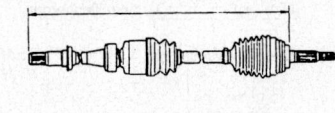

LH

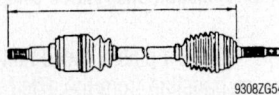

Measuring the front halfshaft lengths—2002–2005 2WD with manual transaxle

11. Install or connect the following:
   - Right dust cover, until the distance from the tip of the center drive is 4.134–4.173 in. (105.0–106.0mm) to the inner edge of the dust cover
   - Left side dust cover, using a press
12. Temporarily install new outboard/inboard joint boots using new clamps, as follows:
   a. Warp tape around the halfshaft splines.
   b. Install the new outboard joint boot onto the halfshaft with both new clamps.
   c. Install the new inboard joint boot onto the halfshaft.
13. Assemble the inboard joint, as follows:
   - New snapring
   - Cage

➡**The smaller diameter side must face outboard.**

   - Inner race, using a brass bar and hammer by aligning the match-marks

❄❄ WARNING

Be careful not to damage the inner race.

   - New snapring
14. Install the outboard joint boot packed with grease from the boot kit.
15. Install the inboard joint tulip, as follows:
   - Cage to the inner race by aligning the matchmarks
   - 6 cage balls

➡**Lubricate the balls with grease to keep them from falling.**

   - Inboard joint tulip, by aligning the matchmarks
   - New snapring
   - Temporarily, install the inboard joint boot packed with grease from the kit
16. Install the boot clamps to both boots, as follows:
   - Both boots to the shaft grooves
   - Halfshaft length should be 32.988–33.382 in. (837.9–847.9mm) for the right side or 21.165–21.559 in. (537.6–547.6mm) for the left side
   - Both new clamps on the inboard joint boot
   - Crimp the new clamps using tool 09521-24010
   - Adjust the crimp clearance to 0.047–0.157 in. (1.2–4.0mm)

2WD WITH AUTOMATIC TRANSAXLE AND 4WD

1. Before servicing the vehicle, refer to the Precautions Section.
2. Remove the inboard and outboard joint boot clamps.
3. Disassemble the inboard joint tulip, as follows:
   - Matchmark the tri-pot, inboard joint tulip or center halfshaft to the half-shaft

❄❄ WARNING

Do not use punch marks.

   - Inboard joint tulip from the half-shaft
4. Remove the inboard and outboard joint clamps.
5. Remove the tri-pot joint, as follows:
   - Snapring

- Matchmark the tri-pot joint to the halfshaft
- Tri-pot joint, using a brass bar and hammer

6. Remove or disconnect the following:
- Inboard and outboard joint boots

➡ **Do not disassemble the outboard joint.**

- Dust cover from the center half-shaft, using a press, for 2WD on the right side
- Dust cover from the inboard joint tulip, using tool 09950-00020 and a press, for 2WD on the left side and 4WD

7. Disassemble the center halfshaft, as follows:
- Snapring
- Bearing case, using a press
- Straight pin from the bearing case, using a pin punch and hammer
- Dust cover, using tool 09950-00020 and a press
- Snapring
- Bearing, using a press

8. Remove the No. 2 dust deflector, using a screwdriver and hammer.

To assemble:

9. Install a new No. 2 dust deflector, using a press.

10. Assemble the center halfshaft, as follows:
- Straight pin into the bearing case, using a pin punch and hammer
- New bearing, using tools 09959-60010, 09950-70010 and a press
- New snapring
- Bearing with the bearing case assembly to the center halfshaft,

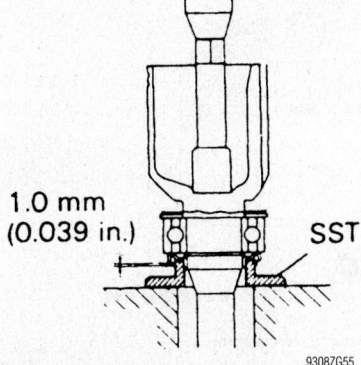

1.0 mm (0.039 in.) **SST**

9308ZG55

View of the bearing-to-dust cover clearance—2002–2005 2WD with automatic transaxle and 4WD

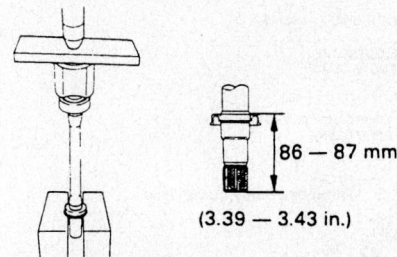

86 — 87 mm

(3.39 — 3.43 in.)

9308ZG56

View of the dust cover-to-center drive distance—2002–2005 2WD with automatic transaxle and 4WD

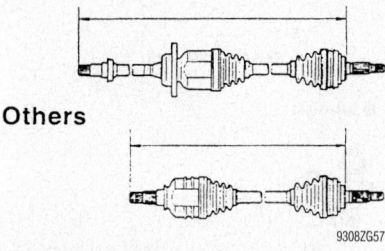

2WD A/T RH

Others

9308ZG57

Measuring the front halfshaft lengths—2002–2005 2WD with automatic transaxle and 4WD

using tool 09710-30021 and a press
- New snapring
- New dust cover, until the clearance between the dust cover and the bearing is 0.039 in. (1.0mm)

11. Install or connect the following:
- Right dust cover (2WD), until the distance from the tip of the center drive is 3.39–3.34 in. (86–87mm) to the inner edge of the dust cover
- Left side dust cover (2WD and 4WD), using a press

12. Temporarily install new outboard/inboard joint boots using new clamps, as follows:
  a. Warp tape around the halfshaft splines.
  b. Install the new outboard joint boot onto the halfshaft.
  c. Install the new inboard joint boot onto the halfshaft.

13. Install the tri-pot joint, as follows:
- Tri-pot joint, face the beveled side toward the outboard joint and align the matchmarks
- Tri-pot joint onto the halfshaft, using a press

- New snapring

14. Install the outboard joint boot packed with grease from the boot kit.

15. Install the inboard joint tulip, as follows:
- Pack the inboard joint boot with grease from the boot kit
- Inboard joint tulip, by aligning the matchmarks
- Temporarily, install the inboard joint boot packed with grease from the kit

16. Install the boot clamps to both boots, as follows:
- Both boots to the shaft grooves
- Halfshaft length should be 33.055–33.449 in. (839.6–849.6mm) for the right side on 2WD with automatic transaxle, 21.397–21.791 in. (543.5–553.5mm) for the left side on 2WD with automatic transaxle, 19.929–20.323 in. (506.2–516.2mm) for the right side on 4WD or 19.803–20.197 in. (503–511mm) for the left side on 4WD
- Both new boot clamps boot
- Bend the band and lock it using a screwdriver

Front (2006)

2WD

1. Before servicing the vehicle, refer to the Precautions Section.

2. Remove the front axle inboard joint boot number two clamp.

3. Remove the front axle inboard joint boot clamp. Remove the front axle inboard joint boot.

4. Matchmark and remove the front inboard joint assembly, left side.

5. Matchmark and remove the front inboard joint assembly, right side.

6. Remove the front halfshaft damper clamp, left side. Remove the front halfshaft damper.

7. Remove the front axle outboard joint boot number two clamp.

8. Remove the front axle outboard joint boot clamp. Remove the front axle outboard joint boot.

9. Remove the front halfshaft hole snapring.

10. On the left side, using tool SST09950-00020 and a press, press out the shaft dust cover. Be careful not to damage the inboard joint.

11. On the right side, using tool SST09950-00020 and a press, press out the shaft dust cover. Be careful not to damage the inboard joint.

12. Remove the halfshaft bearing case,

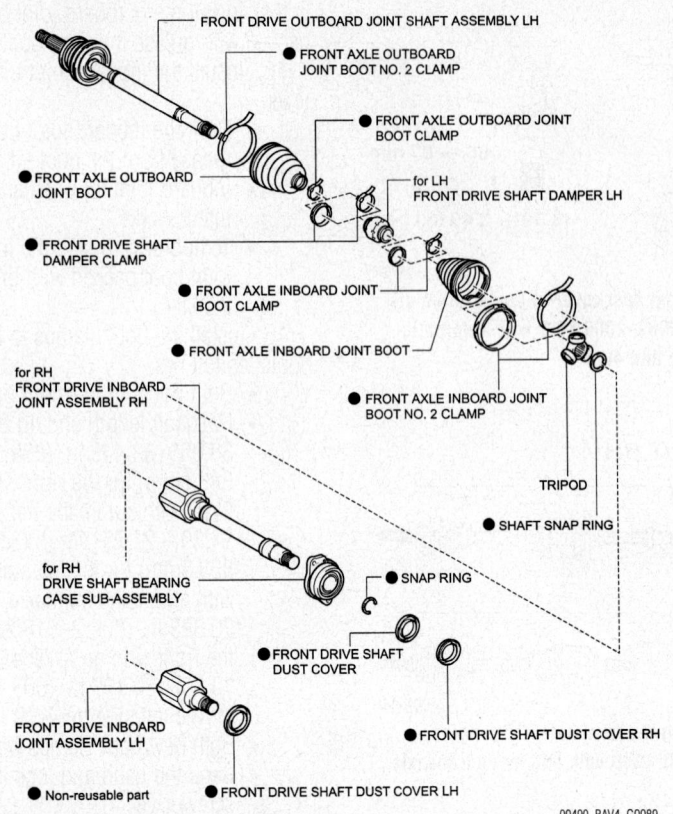

Front halfshaft exploded view—2006 2WD

09490_RAV4_G0089

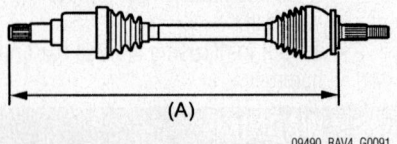

(A)

09490_RAV4_G0091

Front and rear halfshaft measurement locating points—2006

inches for the left shaft and 36.000 inches for the right shaft on vehicles equipped with the 3.5L engine.

4WD

1. Before servicing the vehicle, refer to the Precautions Section.

2. Remove the front axle inboard joint boot number two clamp.

3. Remove the front axle inboard joint boot clamp. Remove the front axle inboard joint boot.

4. Matchmark and remove the front inboard joint assembly, left side.

5. Matchmark and remove the front inboard joint assembly, right side.

6. Remove the front halfshaft damper clamp, left and right sides. Remove the front halfshaft damper, left and right sides.

7. Remove the front axle outboard joint boot number two clamp.

right side. Remove the front halfshaft dust cover, using tool SST09950-00020 and a press.

13. Remove the front halfshaft bearing, using tool SST09527-10011 and a press.

To assemble:

14. Assembly is the reverse of the disassembly procedure.

15. On the right side, when installing the front halfshaft dust cover, the cover depth is 0.04 inch. The distance from the tip of the center halfshaft to the halfshaft dust cover is 3.583–3.622 inch for the 2.4L engine and 4.330–4.370 for the 3.5L engine.

16. Pack the outboard joint shaft and boot with grease. Grease capacity is 6.7–7.1 ounces for the 2.4L engine and 7.2–7.6 ounces for the 3.5L engine.

17. When installing the damper clamp to the outboard joint side the distance should be 6.339–6.496 inches.

18. Pack the inboard joint shaft and boot with grease. Grease capacity is 6.2–6.5 ounces.

19. Check and measure the assembled halfshaft, as shown in the illustration. Dimension "A" should be 23.039 inches for the left shaft and 35.512 inches for the right shaft on vehicles equipped with the 2.4L engine. Dimension "A" should be 22.476

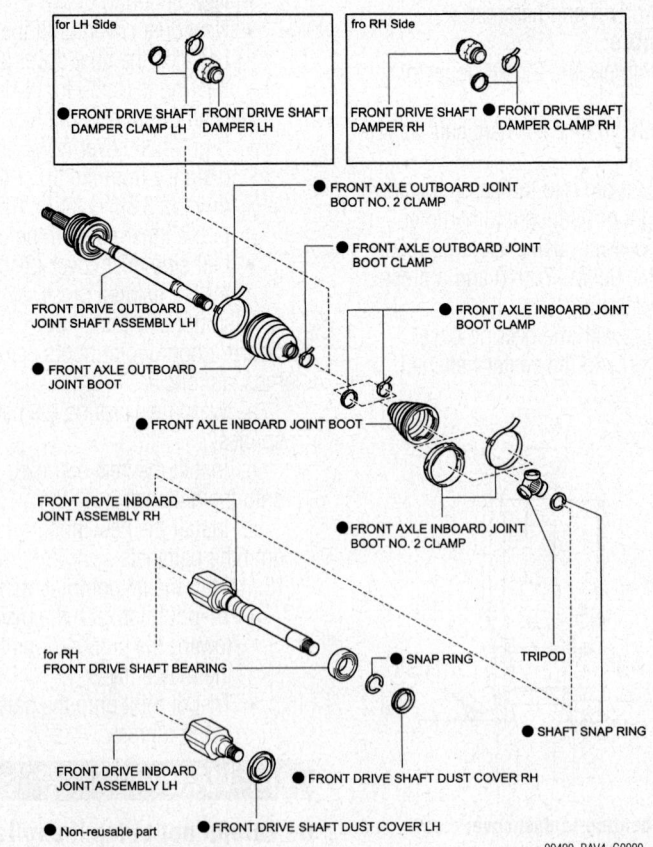

Front halfshaft exploded view—2006 4WD

09490_RAV4_G0090

8. Remove the front axle outboard joint boot clamp. Remove the front axle outboard joint boot.

9. Remove the front halfshaft hole snapring.

10. On the left side, using tool SST09950-00020 and a press, press out the shaft dust cover. Be careful not to damage the inboard joint.

11. On the right side, using tool SST09950-00020 and a press, press out the shaft dust cover. Be careful not to damage the inboard joint.

12. Remove the halfshaft bearing case, right side. Remove the front halfshaft dust cover, using tool SST09950-00020 and a press.

13. Remove the front halfshaft bearing, using tool SST09527-10011 and a press.

To assemble:

14. Assembly is the reverse of the disassembly procedure.

15. On the right side, when installing the front halfshaft dust cover, the cover depth is 4.09 plus/minus 0.02 inches.

16. Pack the outboard joint shaft and boot with grease. Grease capacity is 6.7–7.1 ounces for the 2.4L engine and 7.2–7.6 ounces for the 3.5L engine.

17. When installing the left side damper clamp to the outboard joint side the distance should be 6.339–6.496 inches.

18. Pack the inboard joint shaft and boot with grease. Grease capacity is 6.2–6.5 ounces.

19. Check and measure the assembled halfshaft, as shown in the illustration. Dimension "A" should be 23.039 inches for the left shaft and 35.193 inches for the right shaft on vehicles equipped with the 2.4L engine. Dimension "A" should be 22.476 inches for the left shaft and 36.315 inches for the right shaft on vehicles equipped with the 3.5L engine.

Rear

2002–2005

1. Before servicing the vehicle, refer to the Precautions Section.

2. Remove the inboard and outboard joint boot clamps.

3. Disassemble the inboard joint tulip, as follows:

- Matchmark the tri-pot, inboard joint tulip or center halfshaft to the halfshaft

✳✳ WARNING

Do not use punch marks.

- Inboard joint tulip from the halfshaft
4. Remove the tri-pot joint, as follows:
- Snapring
- Matchmark the tri-pot joint to the halfshaft

✳✳ WARNING

Do not use punch marks.

- Tri-pot joint, using a brass bar and hammer

✳✳ WARNING

Do not tap the roller.

5. Remove or disconnect the following:
- Inboard and outboard joint boots

➡**Do not disassemble the outboard joint.**

- No. 2 dust deflector from the center halfshaft, using a screwdriver and hammer

To assemble:

6. Install a new No. 2 dust deflector, using tools 09309-36010, 09316-20011 and a press.

7. Temporarily install new outboard/inboard joint boots using new clamps, as follows:

   a. Warp tape around the halfshaft splines.

   b. Install the new outboard joint boot onto the halfshaft.

   c. Install the new inboard joint boot onto the halfshaft.

8. Install the tri-pot joint, as follows:
- Tri-pot joint, face the beveled side toward the outboard joint and align the matchmarks
- Tri-pot joint onto the halfshaft, using a brass bar and hammer

✳✳ WARNING

Be careful not to tap the roller.

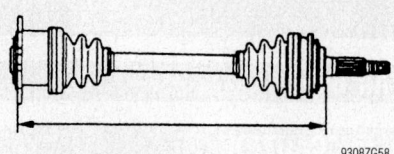

9308ZG58

Rear halfshaft measurement locating points—2002–2005

- New snapring
9. Install the outboard joint boot packed with grease from the boot kit.
10. Install the inboard joint tulip, as follows:

- Pack the inboard joint boot with grease from the boot kit
- Inboard joint tulip, by aligning the matchmarks
- Inboard joint boot packed with grease from the kit

11. Install the boot clamps to both boots, as follows:

- Both boots to the shaft grooves
- Halfshaft length should be 23.392–23.795 in. (594.4–604.4mm) for the right side or 21.590–21.984 in. (548.4–558.4mm) for the left side
- New boot clamps boot
- Bend the band and lock it using a screwdriver

2006

1. Before servicing the vehicle, refer to the Precautions Section.

2. Remove the rear halfshaft inboard joint boot number two clamp.

3. Remove the rear halfshaft inboard joint boot clamp. Remove the boot.

4. Remove the rear halfshaft inboard joint assembly.

5. Remove the halfshaft inboard joint shaft snapring.

6. Remove the rear halfshaft dust cover, using tool SST09950-00020 and a press.

To assemble:

7. Assembly is the reverse of the disassembly procedure.

8. Pack the inboard joint shaft and boot with grease. Grease capacity is 3.0–3.4

9. Check and measure the assembled halfshaft, as shown in the illustration. Dimension "A" should be 28.890 inches.

Air Bag

✳✳ CAUTION

These vehicles are equipped with an air bag system. The system must be disarmed before performing service on, or around, system components, the steering column, instrument panel components, wiring and sensors. Failure to follow the safety precautions and the disarming procedure could result in accidental air bag deployment, possible injury and unnecessary system repairs.

PRECAUTIONS

Several precautions must be observed when handling the inflator module to avoid accidental deployment and possible personal injury.

• Never carry the inflator module by the wires or connector on the underside of the module.

• When carrying a live inflator module, hold securely with both hands and ensure that the bag and trim cover are pointed away.

• Place the inflator module on a bench or other surface with the bag and trim cover facing up.

• With the inflator module on the bench, never place anything on or close to the module which may be thrown in the event of an accidental deployment.

DISARMING

To avoid personal injury when working on vehicles equipped with an air bag, the negative battery cable must be disconnected and at least 90 seconds must elapse before working on the system. Failure to do so may result in deployment of the air bag.

Power Steering Gear

REMOVAL & INSTALLATION

2002–2003

1. Before servicing the vehicle, refer to the Precautions Section.
2. Disconnect the negative battery cable.

✳✳ CAUTION

To avoid personal injury when working on air bag equipped vehicles, work must be started after 90 seconds or longer from the time the ignition switch is turned to the LOCK position and the negative battery terminal is disconnected. If the air bag system is disconnected with the ignition switch at the ON or ACC, diagnostic codes will be set. When removing the air bag, take care not to pull the air bag wiring harness. When carrying the wheel pad, carry it with the upper surface facing away. When storing it, keep the upper surface of the pad facing upward.

3. Turn the key to the **LOCK** position and lock the steering wheel in place.
4. Place a drain pan under the steering rack.
5. Remove or disconnect the following:

• Front wheels
• Right and left side engine undercovers
• Right and left side tie rod ends from the steering knuckle
• Front exhaust pipe
• Sway bar with the links

6. Disconnect the No. 2 intermediate shaft from the rack and pinion, as follows:

a. Loosen the top bolt.
b. Remove the lower bolt holding the No. 2 intermediate shaft to the rack and pinion.
c. Shift the No. 2 intermediate shaft and place matchmarks on the control valve shaft and the No. 2 intermediate shaft.
d. Disconnect the No. 2 shaft from the rack and pinion.

7. Install or connect the following:

• Pressure feed and return tubes from the rack and pinion, using a line wrench
• Pressure feed and return tube clamps, by removing the bolt
• Right and left lower control arms, from the steering knuckle

8. Remove the front suspension crossmember assembly, as follows:

• Both centermember set nuts, holding the centermember to the middle of the crossmember.

• Both rack and pinion assembly set bolts and nuts from the crossmember.
• Securely suspend the steering gear assembly.
• Support the suspension crossmember with a jack.
• Both bolts from the suspension crossmember
• Suspension crossmember with the lower suspension arms

9. Remove the rack and pinion.

To install:

10. Install or connect the following:
• Rack and pinion

11. Install the crossmember to the vehicle, as follows:

• Suspension crossmember with the lower control arms. Tighten both bolts to 152 ft. lbs. (206 Nm).
• Rack and pinion. Tighten the nuts/bolts to 83 ft. lbs. (113 Nm).
• Centermember to the crossmember. Tighten both set nuts to 82 ft. lbs. (112 Nm).

12. Install or connect the following:

• Right and left lower control arms
• Pressure feed and return tubes clamps
• Pressure feed and return tubes to the rack and pinion. Tighten the tubes to 26 ft. lbs. (36 Nm), using a torque wrench with a fulcrum length of 11.81 inches (300mm).
• Steering column No. 2 intermediate shaft to the rack and pinion. Align the marks and tighten the upper and lower pinch bolts to 26 ft. lbs. (35 Nm).
• Stabilizer bar links. Tighten the nuts to 22 ft. lbs. (29 Nm).
• Front exhaust pipe with new gaskets. Tighten the bolts to 35 ft. lbs. (48 Nm) and the nuts to 46 ft. lbs. (62 Nm).
• Right and left side tie rod ends to the steering knuckle. Tighten the nuts to 36 ft. lbs. (49 Nm) and install new cotter pins.
• Right and left side engine undercovers
• Front wheels

13. Fill the power steering unit and bleed the system. Check for leaks.
14. Check and/or adjust the front wheel alignment.

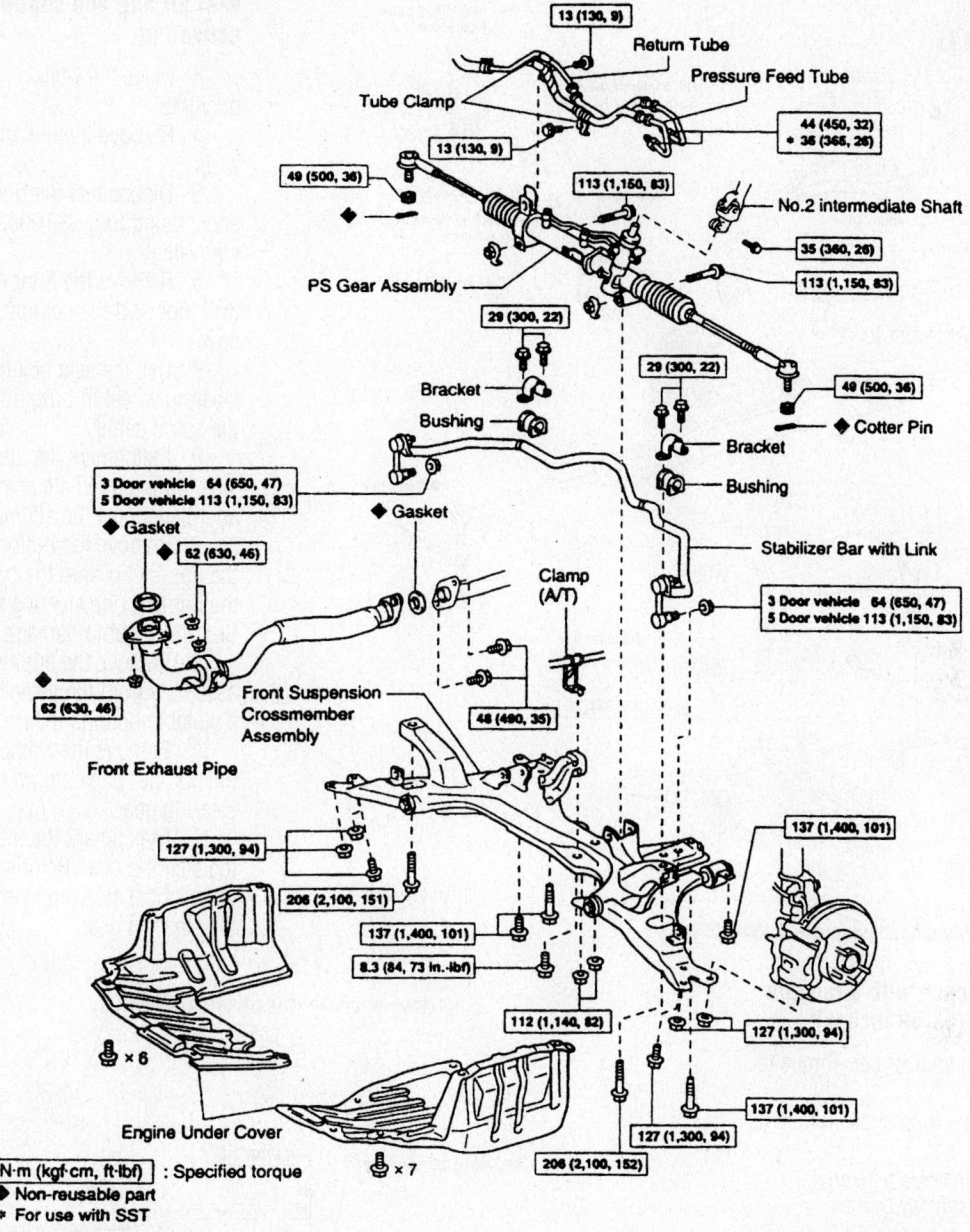

13 (130, 9)
Return Tube
Pressure Feed Tube
Tube Clamp
13 (130, 9)
44 (450, 32)
* 36 (365, 26)
49 (500, 36)
113 (1,150, 83)
No.2 intermediate Shaft
35 (360, 26)
PS Gear Assembly
113 (1,150, 83)
29 (300, 22)
Bracket
29 (300, 22)
Bushing
40 (500, 36)
3 Door vehicle 64 (650, 47)
5 Door vehicle 113 (1,150, 83)
Cotter Pin
◆ Gasket
Bracket
◆ 62 (630, 46)
Bushing
◆ Gasket
Stabilizer Bar with Link
Clamp
(A/T)
3 Door vehicle 64 (650, 47)
5 Door vehicle 113 (1,150, 83)
62 (630, 46)
Front Suspension
Crossmember
Assembly
48 (490, 35)
Front Exhaust Pipe
137 (1,400, 101)
127 (1,300, 94)
206 (2,100, 151)
137 (1,400, 101)
8.3 (84, 73 in.·lbf)
112 (1,140, 82)
127 (1,300, 94)
137 (1,400, 101)
Engine Under Cover
× 6
127 (1,300, 94)
N·m (kgf·cm, ft·lbf) : Specified torque
◆ Non-reusable part
* For use with SST
× 7
206 (2,100, 152)

7924ZG75

Steering gear and related components—2002–2003

2004–2005

1. Before servicing the vehicle, refer to the Precautions Section.
2. Disconnect the negative battery cable.

➡**Wait at least 90 seconds after disconnecting the negative battery cable before starting any repair work to prevent air bag and seat belt pretensioner activation.**

3. Place the wheels in a straight-ahead position.
4. Remove the steering wheel pad.

5. Remove the steering wheel.
6. Remove the intermediate shaft from the gear.
7. Remove the steering column hole cover from the dash panel.
8. Disconnect the tie rod ends.
9. Disconnect the stabilizer bar.
10. Disconnect the pressure and return lines.
11. Remove the 2 gear mounting bolts.
12. Matchmark the intermediate shaft extension and control valve shaft.
13. Remove the intermediate shaft extension bolt and remove the gear.

To install:

14. Place the gear in position. Align the matchmarks and connect the intermediate extension shaft. Torque to 26 ft. lbs. (35 Nm).
15. Install the column cover plate and clamp.
16. Install the gear mounting bolts. Torque to 101 ft. lbs. (137 Nm).
17. Connect the pressure and return lines to the gear with the 2 bolts. Torque to 9 ft. lbs. (12.5 Nm).
18. Connect the pressure and return tube fittings. Torque to 29 ft. lbs. (40 Nm).

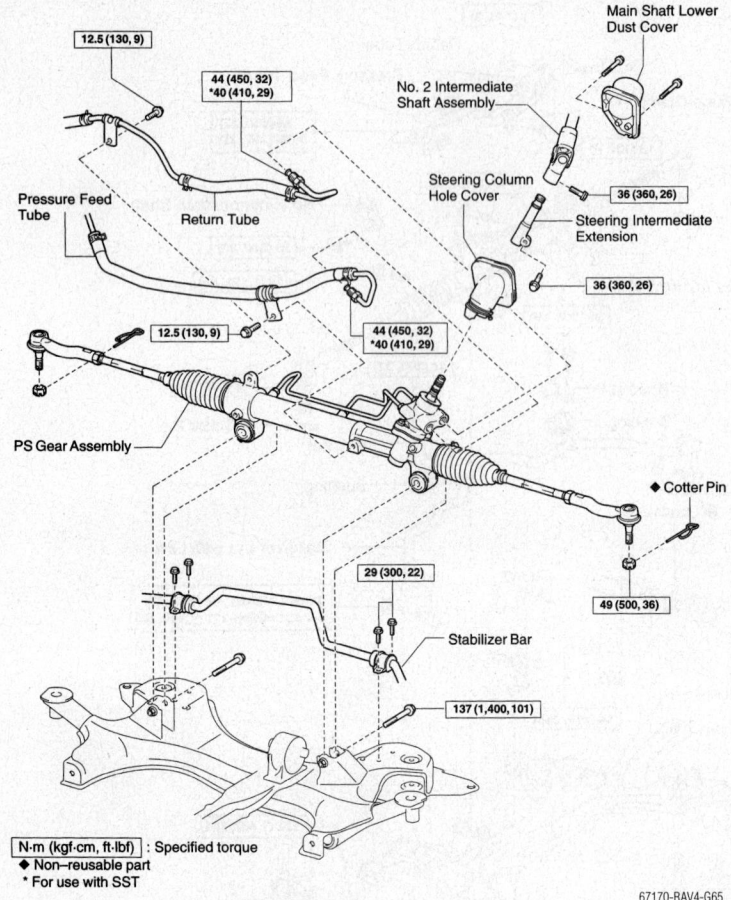

Steering gear and related components—2004–2005

➡Use a torque wrench with a fulcrum length of 345mm. (13.58 inches).

19. Connect the stabilizer bar. Torque to 22 ft. lbs. (29 Nm).

20. Connect the tie rods ends. Torque to 36 ft. lbs. (49 Nm).

21. Connect the intermediate shaft. Torque to 26 ft. lbs. (36 Nm).

22. Install the center spiral cable.

23. Install the power steering pump. Bleed the system.

24. Install the steering wheel. Make sure that everything is still in the straight-ahead position.

25. Torque the steering wheel nut to 37 ft. lbs. (50 Nm).

26. Install the pad.

27. Check the alignment.

2006

1. Before servicing the vehicle, refer to the Precautions Section.

2. Disconnect the negative battery cable.

➡Wait at least 90 seconds after disconnecting the negative battery cable before starting any repair work to pre-

vent air bag and seat belt pretensioner activation.

3. Place the wheels in a straight-ahead position.

4. Remove the tire and wheel assemblies.

5. Disconnect the right and left tie rod ends, using tool SST09628-62011 or equivalent.

6. Remove the floor carpet. Remove the two clips and the column hole silencer cover.

7. Use the seat belt to position the steering wheel in order to avoid breakage of the spiral cable.

8. Matchmark the sliding yoke of the steering intermediate shaft. Remove the bolt and disconnect the sliding yoke.

9. Remove the bottom clip and detach the upper clip from the body and disconnect the number one steering column hole cover. Be careful not to damage the clips.

10. Remove the engine and transaxle assembly from the vehicle and position it in a suitable holding fixture.

11. Remove the clamp and disconnect the number one column hole cover from the steering gear.

12. Matchmark the intermediate shaft of the steering gear. Remove the bolt and disconnect the steering intermediate shaft from the steering gear.

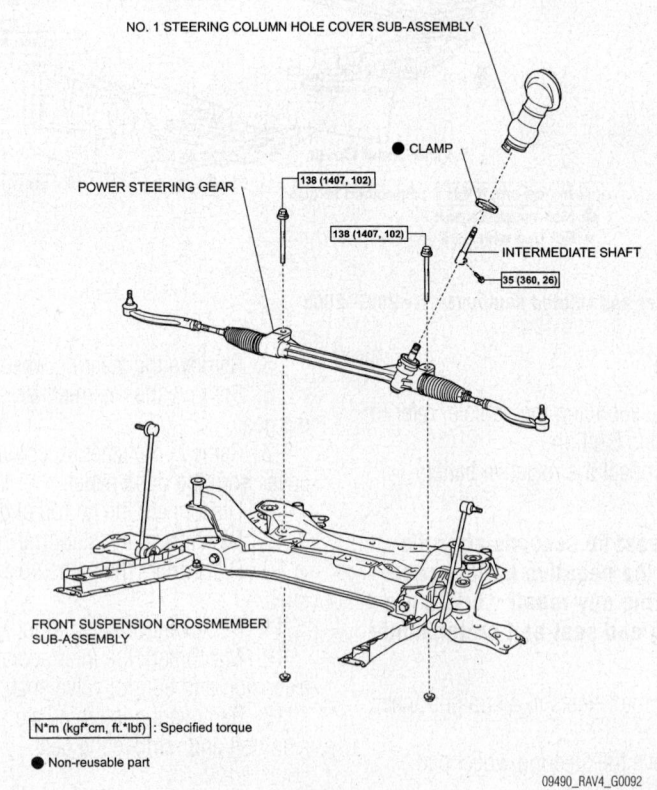

Steering gear and related components—2006

13. Remove the two bolts, two nuts and the steering gear from the crossmember. Be sure to keep the nut from rotating while turning the bolt.

To install:

14. Installation is the reverse of the removal procedure.

15. Tighten the steering gear retaining bolts to 102 ft. lbs. Be sure to keep the nut from rotating while turning the bolt.

16. Tighten the intermediate shaft retaining bolt to 26 ft. lbs.

17. Tighten the sliding yoke retaining bolt to 26 ft. lbs.

18. Tighten the tie rod end castle nut to 36 ft. lbs. If the holes for the clip are not aligned, tighten the nut an additional 60 degrees. Be sure to use a new cotter pin.

19. Check and adjust the alignment, as required.

FRONT SUSPENSION

Strut

REMOVAL & INSTALLATION

2002–2003

1. Before servicing the vehicle, refer to the Precautions Section.

2. Remove or disconnect the following:

- Negative battery cable
- Wheel

➡**Do not support the weight of the vehicle on the suspension arm.**

- Brake hose from the strut
- Anti-lock Brake System (ABS) electrical connection to the strut bolt, if equipped

➡**It is not necessary to disconnect the brake hose from the brake caliper.**

- Strut from the steering knuckle
- Suspension support bracket from the top of the strut tower
- Strut

To install:

3. Install or connect the following:

- Suspension support bracket to the top of the strut tower
- Strut to the strut tower. Tighten the 3 nuts to 59 ft. lbs. (80 Nm).
- Steering knuckle to the strut lower bracket. Tighten the nuts to 117 ft. lbs. (158 Nm).
- ABS electrical connector to the strut. Tighten the bolt to 48 inch lbs. (5.4 Nm).
- Brake line to the strut. Tighten the bolt to 14 ft. lbs. (19 Nm).

4. If the brake lines were opened, add brake fluid and bleed the brake system.

5. Install or connect the following:

- Wheel
- Negative battery cable

6. Check and/or adjust the front wheel alignment.

2004–2005

1. Before servicing the vehicle, refer to the Precautions Section.

2. Raise and support the vehicle safely.

3. Remove the tire and wheel.

4. Disconnect the stabilizer bar link from the strut.

5. Remove the brake hose and wire harness from the strut.

6. Loosen, don't remove, the 2 lower strut nuts

7. Remove the 3 upper end nuts.

8. Remove the 2 lower bolts and nuts.

To install:

9. Installation is the reverse of removal. Observe the following torques:

- Upper end nuts: 59 ft. lbs. (80 Nm)
- Lower end bolts/nuts: 105 ft. lbs. (143 Nm)
- Stabilizer bar links: 32 ft. lbs. (44 Nm)

2006

1. Before servicing the vehicle, refer to the Precautions Section.

2. Raise and support the vehicle safely.

3. Remove the tire and wheel assembly.

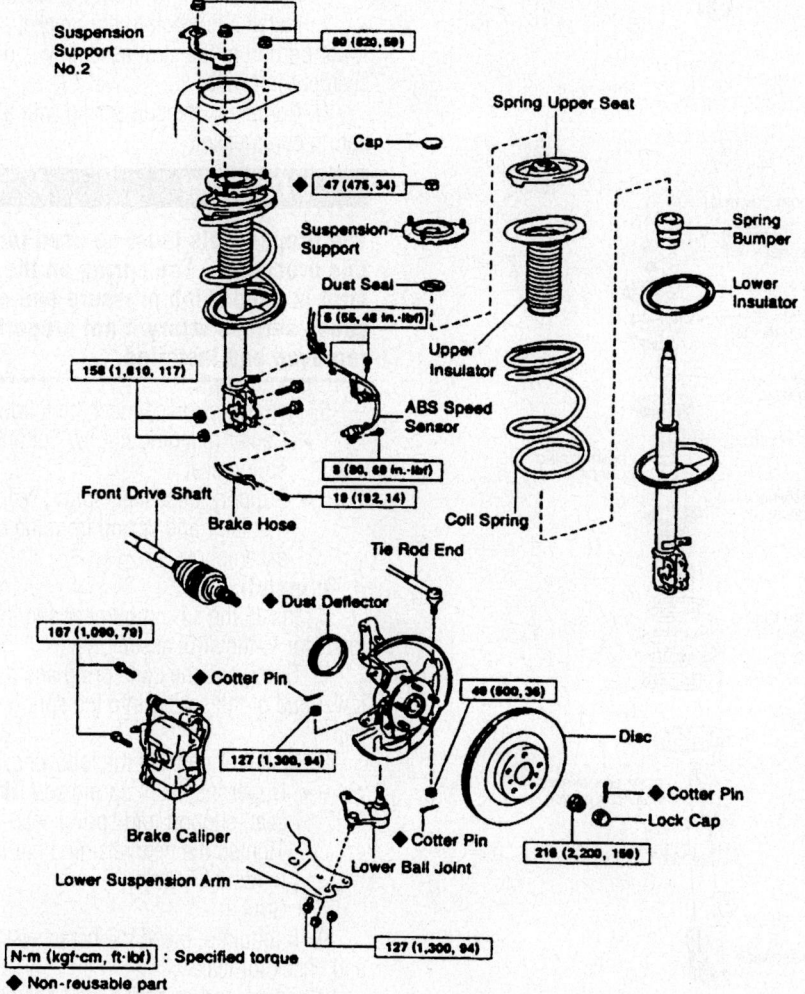

Suspension Support No.2
88 (820, 59)
Spring Upper Seat
Cap
47 (476, 34)
Suspension Support
Spring Bumper
Dust Seal
Lower Insulator
6 (55, 48 in.-lbf)
Upper Insulator
158 (1,610, 117)
ABS Speed Sensor
6 (80, 69 in.-lbf)
Front Drive Shaft
19 (192, 14)
Brake Hose
Coil Spring
Tie Rod End
167 (1,060, 79)
Dust Deflector
Cotter Pin
49 (500, 36)
Disc
127 (1,300, 94)
Cotter Pin
Lock Cap
Brake Caliper
216 (2,200, 159)
Cotter Pin
Lower Suspension Arm
Lower Ball Joint
127 (1,300, 94)
N·m (kgf·cm, ft·lbf) : Specified torque
◆ Non-reusable part
7924ZG78

Strut and related components—2002–2003

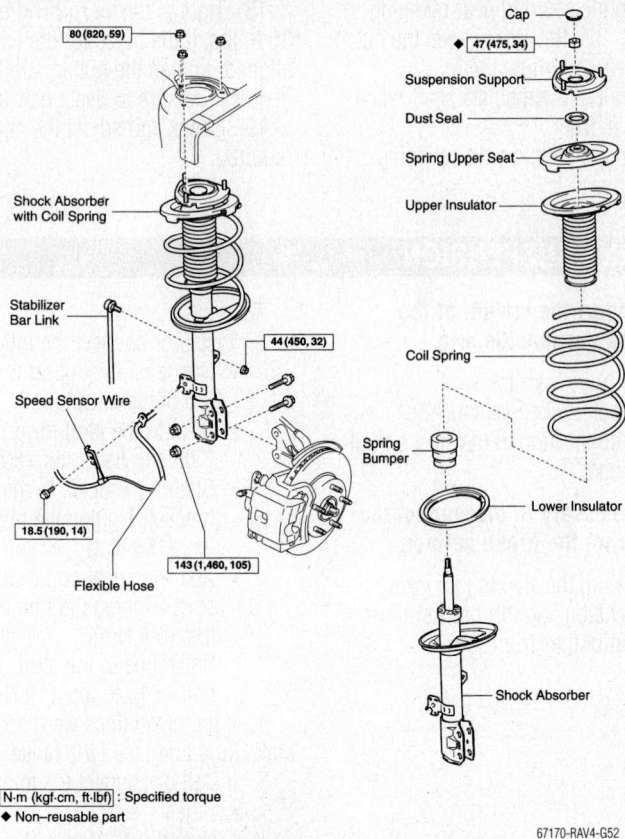

80 (820, 59)

Cap

◆ **47 (475, 34)**

Suspension Support

Dust Seal

Spring Upper Seat

Upper Insulator

Shock Absorber with Coil Spring

Stabilizer Bar Link

44 (450, 32)

Speed Sensor Wire

Coil Spring

Spring Bumper

Lower Insulator

18.5 (190, 14)

143 (1,460, 105)

Flexible Hose

Shock Absorber

N·m (kgf·cm, ft·lbf) : Specified torque
◆ Non–reusable part

67170-RAV4-G52

Front strut and related components—2004–2005

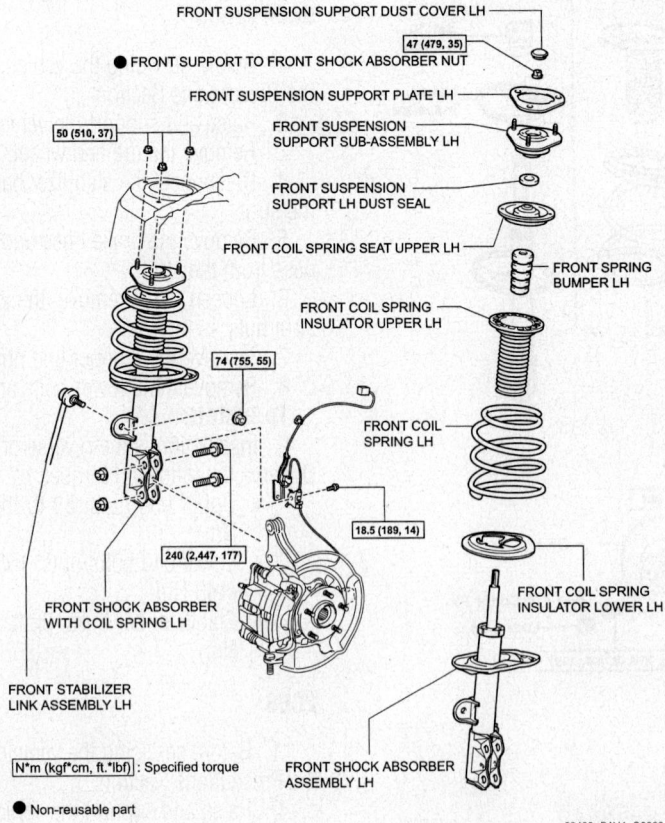

FRONT SUSPENSION SUPPORT DUST COVER LH

47 (479, 35)

● FRONT SUPPORT TO FRONT SHOCK ABSORBER NUT

FRONT SUSPENSION SUPPORT PLATE LH

50 (510, 37)

FRONT SUSPENSION SUPPORT SUB-ASSEMBLY LH

FRONT SUSPENSION SUPPORT LH DUST SEAL

FRONT COIL SPRING SEAT UPPER LH

FRONT SPRING BUMPER LH

FRONT COIL SPRING INSULATOR UPPER LH

74 (755, 55)

FRONT COIL SPRING LH

18.5 (189, 14)

240 (2,447, 177)

FRONT SHOCK ABSORBER WITH COIL SPRING LH

FRONT COIL SPRING INSULATOR LOWER LH

FRONT STABILIZER LINK ASSEMBLY LH

N·m (kgf·cm, ft·lbf) : Specified torque

FRONT SHOCK ABSORBER ASSEMBLY LH

● Non-reusable part

09490_RAV4_G0093

Front strut and related components (without Sport Package)—2006

4. Remove the front speed sensor.

5. Remove the stabilizer link assembly.

6. Remove the two bolts and disconnect the strut from the steering knuckle.

7. Remove the three strut upper retaining bolts.

8. Remove the strut from the vehicle.

To install:

9. Installation is the reverse of the removal procedure.

10. Tighten the three upper strut retaining nuts to 37 ft. lbs.

11. Tighten the lower strut retaining bolts to 177 ft. lbs.

12. To stabilize the suspension, lower the vehicle to ground height. Press down on the vehicle several times to stabilize the suspension.

13. Check and adjust the alignment, as required.

STRUT OVERHAUL

1. Before servicing the vehicle, refer to the Precautions Section.

2. Remove the strut from the vehicle.

3. Install a nut/bolt to the bracket at the lower portion of the strut assembly and secure it in a vise.

4. Compress the coil spring with a spring compressor.

✳✳ CAUTION

The proper tools must be used for this procedure. The spring on the strut is under high pressure and can cause serious injury if not properly removed and installed.

5. Remove or disconnect the following:
- Center retaining nut, by holding the spring seat
- Support, dust seal, spring seat, insulator and spring from the strut assembly

To install:

6. Install the spring bumper and lower insulator to the strut assembly.

7. Compress the coil spring and fit the lower end of the spring into the spring seat gap.

8. Install or connect the following:
- Upper insulator, spring seat, dust seal, support and spring seat. Tighten the new retaining nut to 34 ft. lbs. (47 Nm).
- Strut

9. If required, bleed the brake system and check for leaks.

10. Check and/or adjust the front wheel alignment.

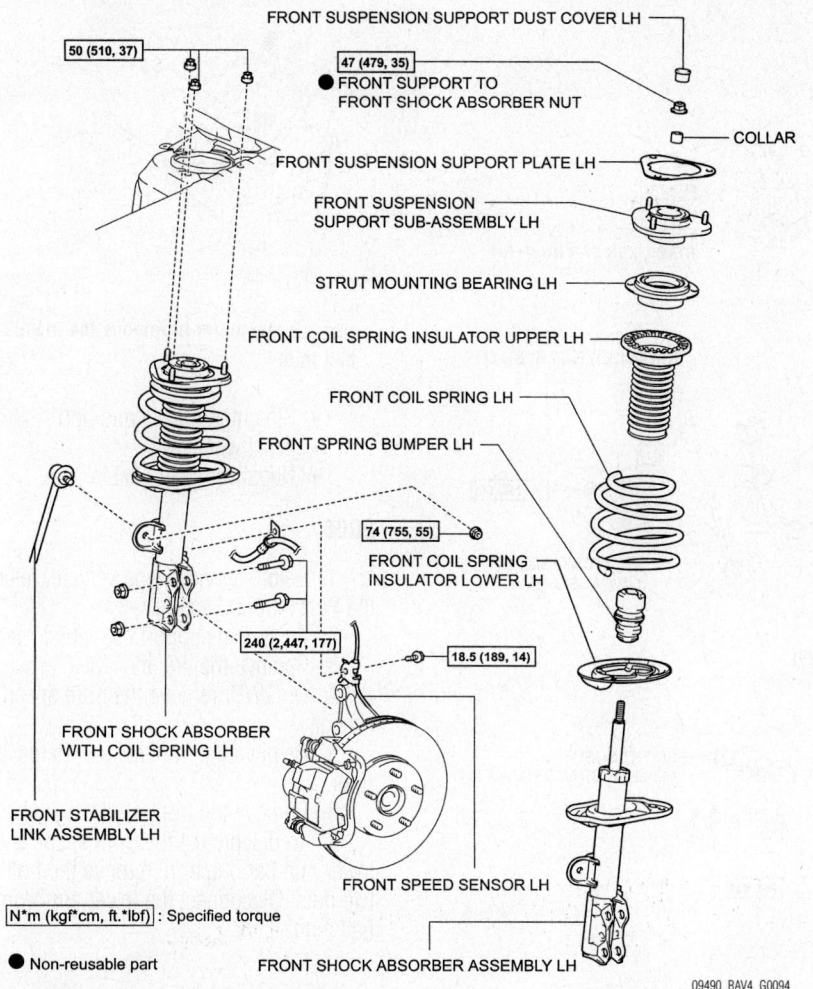

Front strut and related components (Sport Package)—2006

Stabilizer Bar

REMOVAL & INSTALLATION

2002–2005

1. Before servicing the vehicle, refer to the Precautions Section.
2. Raise and support the vehicle safely.
3. Remove the tire and wheel assembly.
4. Remove the stabilizer bar link retaining nuts.
5. Remove the stabilizer bar bracket bolts, bushings and brackets.
6. Remove the stabilizer bar from the vehicle.

To install:

7. Installation is the reverse order of the removal procedure.
8. Be sure to install the bushings so that the cutout will face rearward. Install the bushings to the outside of the bushing stopper.
9. Tighten the stabilizer bar link nuts to 32 ft. lbs.

10. Tighten the stabilizer bracket bolts to 22 ft. lbs.

2006

1. Before servicing the vehicle, refer to the Precautions Section.
2. Raise and support the vehicle safely.
3. Remove the tire and wheel assembly.
4. Remove the stabilizer bar link retaining nuts.
5. Remove the four bolts and remove the left front suspension member brace.
6. Remove the four bolts and remove the right front suspension member brace.
7. Remove the stabilizer bar from the crossmember.
8. Remove the bushings from the stabilizer bar.

To install:

9. Installation is the reverse order of the removal procedure.
10. Install the bushings to the inner side of each bushing stopper on the stabilizer bar.
11. Install the bushing with its slit facing the vehicle rear side.
12. Tighten the stabilizer bar link nuts to 55 ft. lbs.
13. Tighten the front suspension member brace bolts to 64 ft. lbs.
14. To stabilize the suspension, lower the vehicle to ground height. Press down on the vehicle several times to stabilize the suspension.

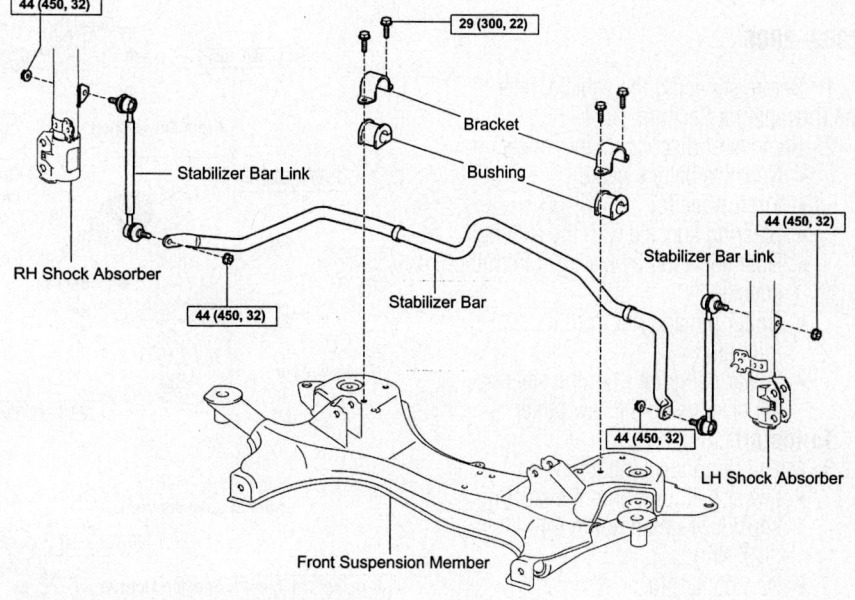

Front stabilizer bar and related components—2002–2005

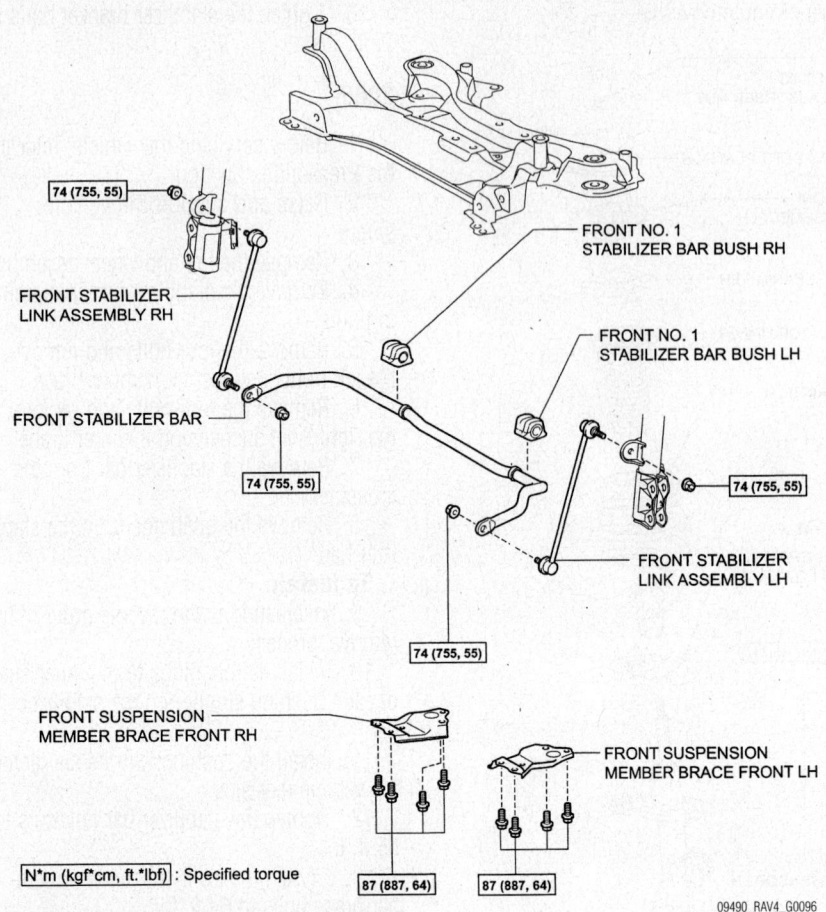

Front stabilizer bar and related components—2006

09490_RAV4_G0096

N*m (kgf*cm, ft.*lbf) : Specified torque

74 (755, 55)
87 (887, 64)
87 (887, 64)

FRONT NO. 1 STABILIZER BAR BUSH RH
FRONT NO. 1 STABILIZER BAR BUSH LH
FRONT STABILIZER LINK ASSEMBLY RH
FRONT STABILIZER BAR
FRONT STABILIZER LINK ASSEMBLY LH
FRONT SUSPENSION MEMBER BRACE FRONT RH
FRONT SUSPENSION MEMBER BRACE FRONT LH

Lower Ball Joint

REMOVAL & INSTALLATION

2002–2005

1. Before servicing the vehicle, refer to the Precautions Section.
2. Remove or disconnect the following:
   - Negative battery cable
   - Front wheel(s)
   - Steering knuckle with the axle hub
   - Dust deflector, by prying it from the knuckle
   - Cotter pin and nut from the ball joint stud
   - Lower ball joint from the steering knuckle, using a 2-jaw puller

To install:

3. Install or connect the following:
   - Lower ball joint onto the steering knuckle. Tighten nut to 94 ft. lbs. (127 Nm).
   - New cotter pin
   - ABS speed sensor, by aligning it the dust deflector hole
   - New dust deflector, using a driver

7924ZG81

Use a 2-jaw puller to remove the lower ball joint

- Steering knuckle and hub
- Front wheel(s)
- Negative battery cable

2006

1. Before servicing the vehicle, refer to the Precautions Section.
2. Raise and support the vehicle safely.
3. Remove the tire and wheel assembly.
4. On 2WD, remove the front speed sensor.
5. Remove the front caliper. Remove the rotor.
6. Remove the front axle hub nut.
7. To disconnect the front suspension lower number one arm, remove the bolt and two nuts. Disconnect the lower arm from the ball joint.

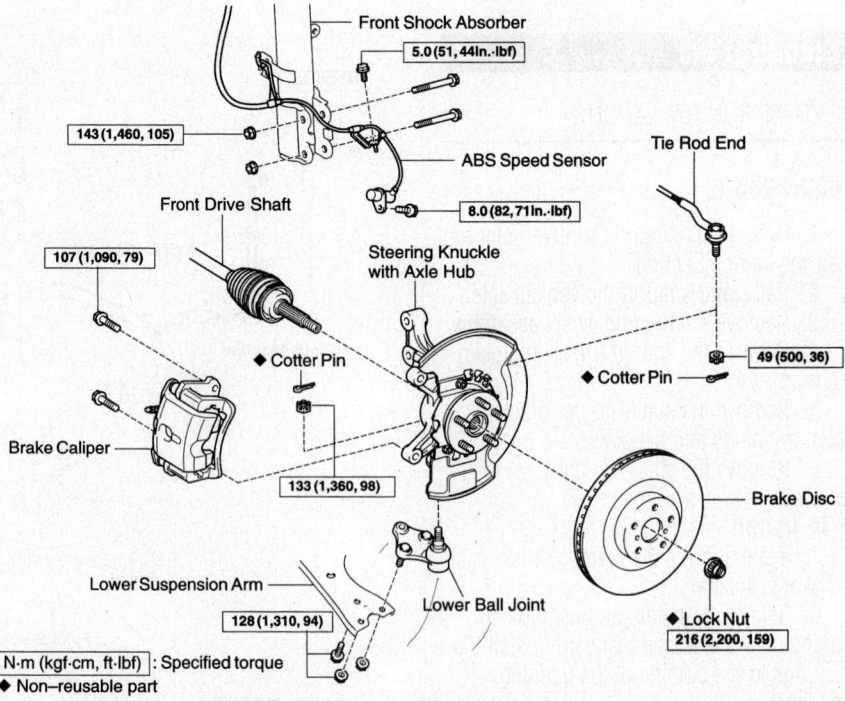

Lower ball joint and related components—2004–2005

N·m (kgf·cm, ft·lbf) : Specified torque
◆ Non–reusable part

67170-RAV4-G54

Front Shock Absorber
5.0 (51, 44In.·lbf)
143 (1,460, 105)
ABS Speed Sensor
Tie Rod End
Front Drive Shaft
8.0 (82, 71In.·lbf)
107 (1,090, 79)
Steering Knuckle with Axle Hub
◆ Cotter Pin
◆ Cotter Pin
49 (500, 36)
Brake Caliper
133 (1,360, 98)
Brake Disc
Lower Suspension Arm
Lower Ball Joint
◆ Lock Nut
216 (2,200, 159)
128 (1,310, 94)

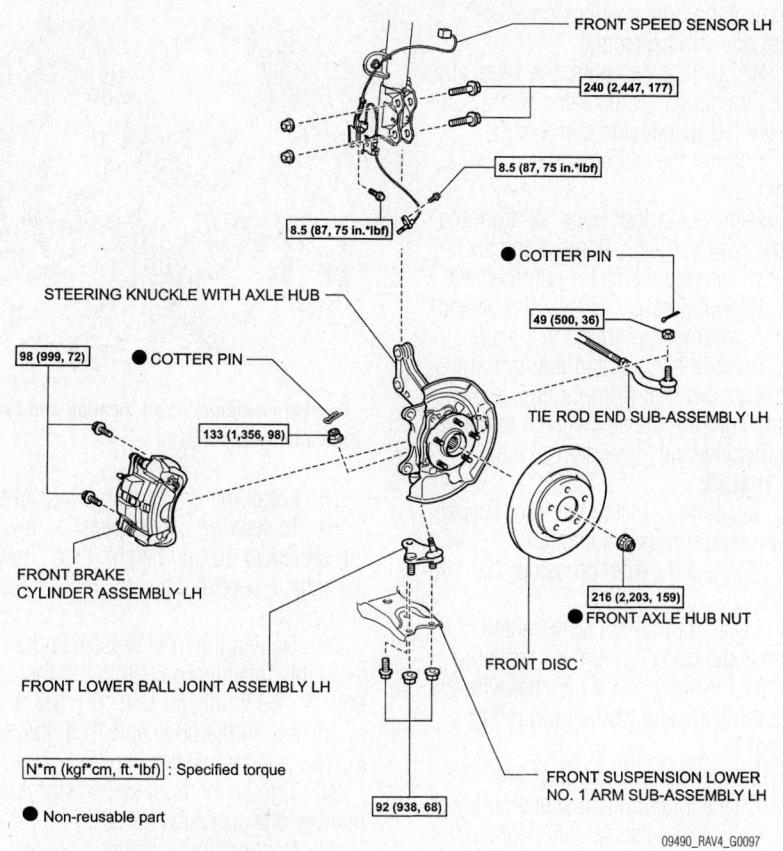

Lower ball joint and related components—2006

8. Using the proper tools, disconnect the tie rod end.

9. Remove the steering knuckle with the axle hub, using the proper removal tools.

10. Remove the cotter pin and nut. Using tool SST09628-62011, remove the lower ball joint.

To install:

11. Installation is the reverse of the removal procedure.

12. Be sure to check and adjust the alignment, as required.

Lower Control Arm

REMOVAL & INSTALLATION

2002–2005

1. Before servicing the vehicle, refer to the Precautions Section.

2. Remove or disconnect the following:
   - Wheel
   - Sway bar link
   - Control arm-to-ball joint bolts
   - Steering rack mount bolts
   - Suspension member subassembly (5 bolts and 2 nuts)
   - Control arm from subassembly (2 bolts, 1 nut)

To install:

3. Installation is the reverse of removal. Observe the following torques:
- Control arm-to-subassembly: 101 ft. lbs. (137 Nm)
- Steering rack mount bolts: 101 ft. lbs. (137 Nm)

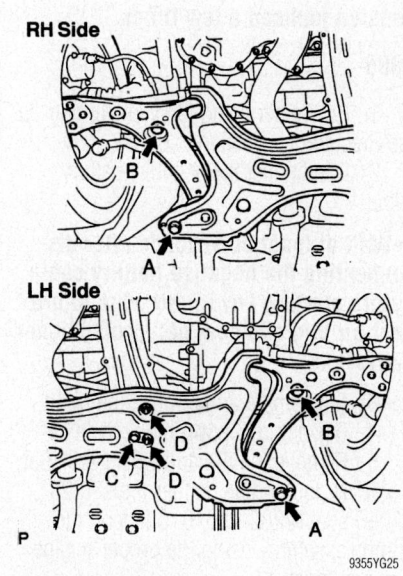

Front suspension subassembly bolt torques—2002–2005

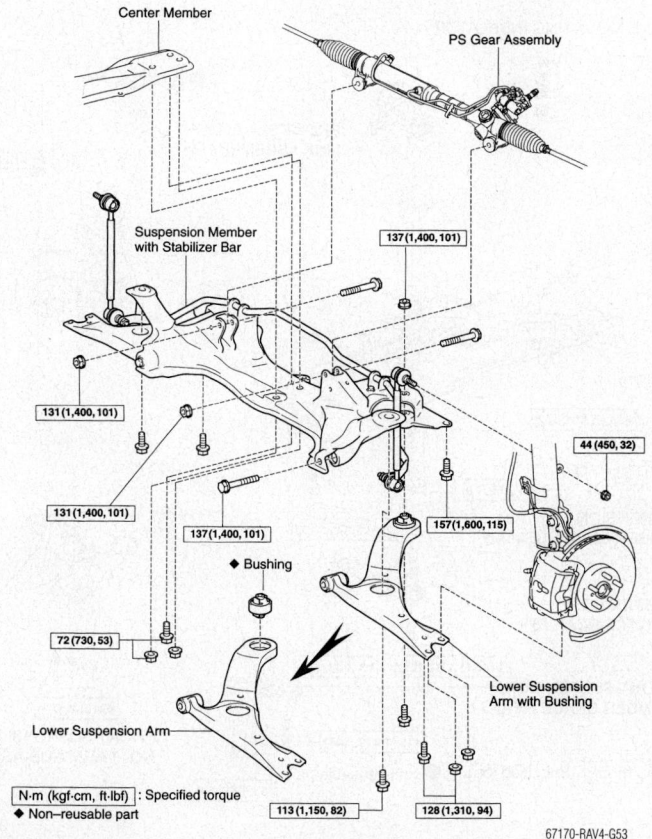

Lower control arm and related components—2004–2005

4. For subassembly torques, see the accompanying illustration.
- A: 115 ft. lbs. (157 Nm)
- B: 82 ft. lbs. (113 Nm)
- C: 53 ft. lbs. (72Nm)
- D: 53 ft. lbs. (72Nm)

➡**Fully tighten the lower arm bolts with the wheels on the ground and the suspension jounced a few times.**

2006

1. Before servicing the vehicle, refer to the Precautions Section.
2. Disconnect the negative battery cable.

➡**Wait at least 90 seconds after disconnecting the negative battery cable before starting any repair work to prevent air bag and seat belt pretensioner activation.**

3. Matchmark and remove the hood.
4. Raise and support the vehicle safely.
5. Remove the tire and wheel assembly.
6. Install engine hanger tools 12281-28010 and 12282-28010. Suspend the engine assembly using the proper engine removal tools.
7. Disconnect the front stabilizer links.

Disconnect the front suspension lower number one arm assembly.

8. Remove the two nuts, two bolts and engine mounting rear insulator. Remove the bolt from the suspension member.
9. Support the crossmember, using a suitable jack.
10. Remove the four bolts "A" from the member reinforcement. Remove the six bolts "B" from the member reinforcement.
11. Carefully lower the jack and disconnect the crossmember from the vehicle.
12. Remove the bolt and nut from the suspension member (front). Remove the bolt and nut from the suspension member (rear). Remove the lower control arm.

To install:
13. Temporarily install the front suspension lower arm in its mounting.
14. Connect the front crossmember subassembly.
15. Install, but do not fully tighten, the four retaining bolts "A", the six retaining bolts "B", the suspension member with the bolt to the body and the rear mounting insulator bolts.
16. Install the front stabilizer link.
17. Connect the front suspension lower number one arm subassembly. Tighten the two bolts and nut to 68 ft. lbs.

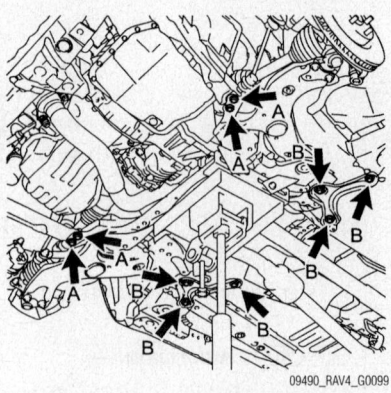

09490_RAV4_G0099

Front crossmember bolt location and identification—2006

18. Install the tire and wheel assembly.
19. To stabilize the suspension, lower the vehicle to ground height. Press down on the vehicle several times to stabilize the suspension.
20. Tighten the front crossmember subassembly retaining bolts to 64 ft. lbs. for bolt "A", 69 ft. lbs. for bolt "B", 107 ft. lbs. for the bolt to the body and 70 ft. lbs. for the rear engine insulator bolts.
21. Tighten the front suspension lower number one arm subassembly.
22. Continue the installation in the reverse order of the removal procedure.
23. Be sure to check and adjust the alignment, as required.

CONTROL ARM BUSHING REPLACEMENT

1. Before servicing the vehicle, refer to the Precautions Section.
2. Remove the component from the vehicle and position it in a suitable holding fixture.
3. Matchmark the control arm with the triangle mark on the bushing.
4. Press out the old bushing.
5. Press in the new bushing, aligning the matchmarks.

Wheel Bearing

ADJUSTMENT

The wheel bearing is not adjustable.

REMOVAL & INSTALLATION

2002–2005

1. Before servicing the vehicle, refer to the Precautions Section.
2. Remove or disconnect the following:
- Negative battery cable
- Front wheels

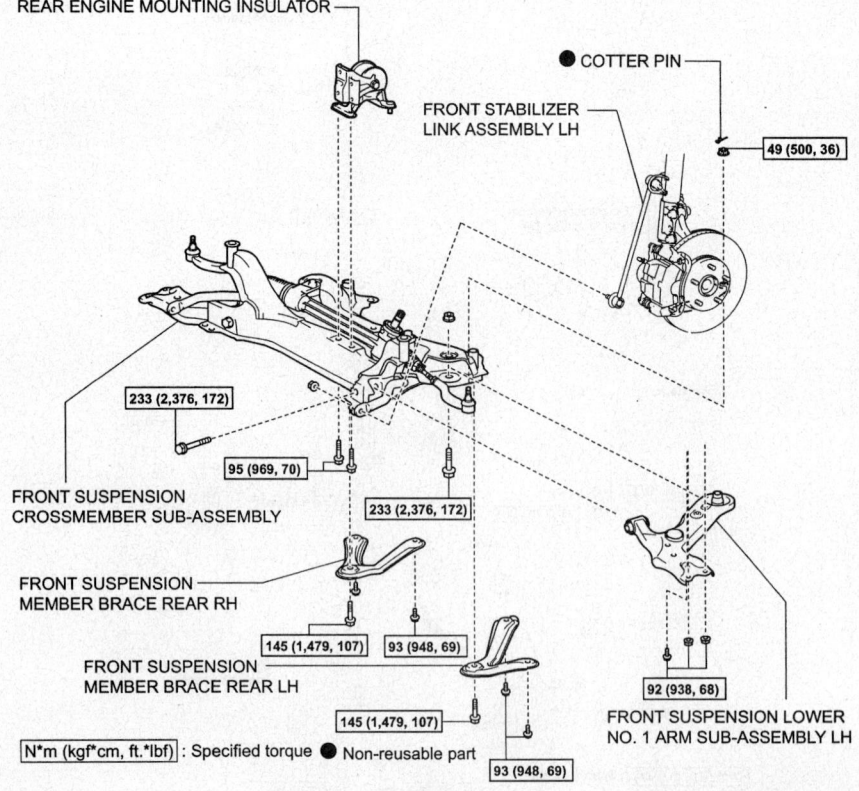

REAR ENGINE MOUNTING INSULATOR

COTTER PIN

FRONT STABILIZER LINK ASSEMBLY LH

49 (500, 36)

233 (2,376, 172)

95 (969, 70)

FRONT SUSPENSION CROSSMEMBER SUB-ASSEMBLY

233 (2,376, 172)

FRONT SUSPENSION MEMBER BRACE REAR RH

145 (1,479, 107)   93 (948, 69)

FRONT SUSPENSION MEMBER BRACE REAR LH

145 (1,479, 107)

92 (938, 68)

FRONT SUSPENSION LOWER NO. 1 ARM SUB-ASSEMBLY LH

N*m (kgf*cm, ft.*lbf) : Specified torque ● Non-reusable part

93 (948, 69)

09490_RAV4_G0098

Lower control arm and related components—2006

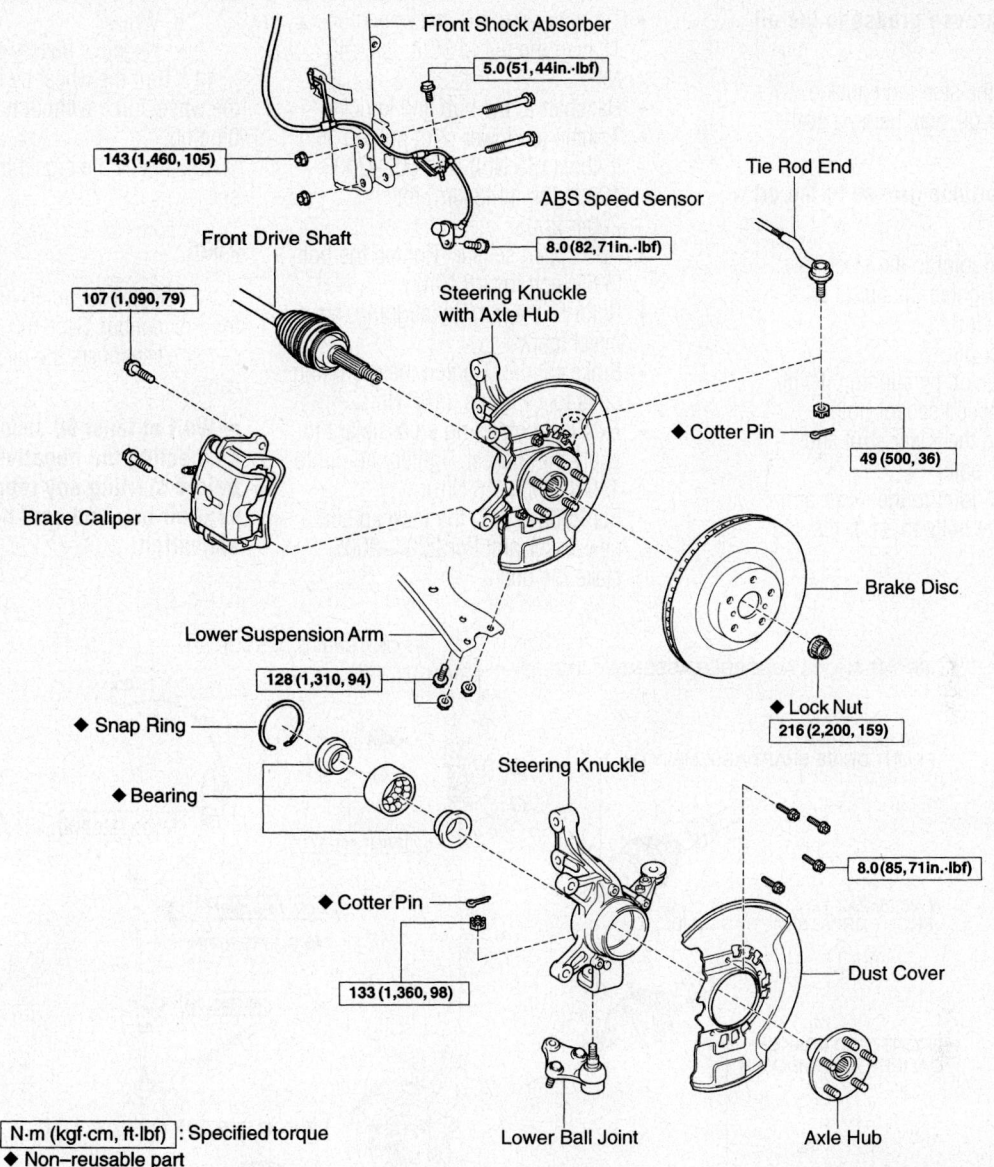

Front Shock Absorber

5.0 (51, 44 in.·lbf)

143 (1,460, 105)

ABS Speed Sensor

Tie Rod End

Front Drive Shaft

107 (1,090, 79)

Steering Knuckle with Axle Hub

8.0 (82, 71 in.·lbf)

◆ Cotter Pin

49 (500, 36)

Brake Caliper

Brake Disc

Lower Suspension Arm

128 (1,310, 94)

◆ Lock Nut
216 (2,200, 159)

◆ Snap Ring

◆ Bearing

Steering Knuckle

◆ Cotter Pin

8.0 (85, 71 in.·lbf)

133 (1,360, 98)

Dust Cover

N·m (kgf·cm, ft·lbf) : Specified torque
◆ Non–reusable part

Lower Ball Joint

Axle Hub

67170-RAV4-G50

Front hub and related components—2004–2005

- For 2002–2003, the cotter pin and lockcap from the halfshaft end. For 2004–2005, unstake the nut.
- Halfshaft locknut, by applying the front brakes
- Brake caliper and support it on a wire
3. Matchmark the rotor to the hub.
4. Remove or disconnect the following:
- Rotor
- Anti-lock Brake System (ABS) speed sensor from the steering knuckle, if equipped
- Loosen the strut's lower end nuts
- Tie rod end from the steering knuckle
- Lower control arm from the ball joint, by removing the bolt and 2 nuts
- Halfshaft from the axle hub

➡**Secure the halfshaft aside using a wire. Be careful not to damage the shaft boot or ABS sensor rotor.**

- Both strut's lower end nuts
- Steering knuckle
5. Clamp the steering knuckle in a vise with soft jaws to protect the knuckle.
6. Remove or disconnect the following:
- Dust deflector, by prying it from the hub
- Ball joint from the steering knuckle
- Hub from the knuckle, using slide hammer
- Inner race from the hub, using press and arbor tool
- 4 bolts and the dust cover
- Inner oil seal, using Seal Removal tool 09308-00010

- Outer oil seal, using Seal Removal tool 09308-00010
- Snapring
7. Install inner race (removed from the hub) on the outside of the bearing
8. Remove the steering knuckle bearing, using a bearing driver
To install:
9. Clean bearing seating surfaces with a clean, dry rag.
10. Install or connect the following:
- Bearing into the knuckle, using a press and Bearing Installer tool 09608-32010
- Snapring
- Dust cover. Tighten the 4 bolts to 74 inch lbs. (8 Nm).
- New outer oil seal, using a seal driver

➡**Apply multi-purpose grease to the oil seal lip.**

- Hub into the steering knuckle
- New inner oil seal, using a seal driver

➡**Apply multi-purpose grease to the oil seal lip.**

- Lower ball joint to the steering knuckle. Tighten the nut to 94 ft. lbs. (127 Nm).
- New cotter pin
- Dust deflector, by aligning it with the ABS speed sensor hole
- Knuckle to the lower strut and install the bolts
- Lower ball joint to the lower arm. Tighten the bolts to 94 ft. lbs. (127 Nm).

- Tie rod end to the steering knuckle. Tighten the nut to 36 ft. lbs. (49 Nm).
- Halfshaft to the hub and knuckle
- Tighten the lower strut nuts to 117 ft. lbs. (158 Nm) for 2002–2003; 105 ft. lbs. (143 Nm) for 2004–2005.
- ABS speed sensor. Tighten the bolt to 69 inch lbs. (8 Nm).
- Rotor to the hub, by aligning the matchmark
- Brake caliper. Tighten the mounting bolts to 79 ft. lbs. (107 Nm).
- Axle locknut, using an assistant to apply the brakes. Tighten the nut to 159 ft. lbs. (216 Nm).
- For 2002–2003, the lockcap and a new cotter pin. For 2004–2005, stake the nut.

- Wheel
- Negative battery cable

11. Turn the wheel by hand, verify that the wheel turns without noise and without binding.

12. Check the signal from the ABS sensor.

2006

1. Before servicing the vehicle, refer to the Precautions Section.

2. Disconnect the negative battery cable.

➡**Wait at least 90 seconds after disconnecting the negative battery cable before starting any repair work to prevent air bag and seat belt pretensioner activation.**

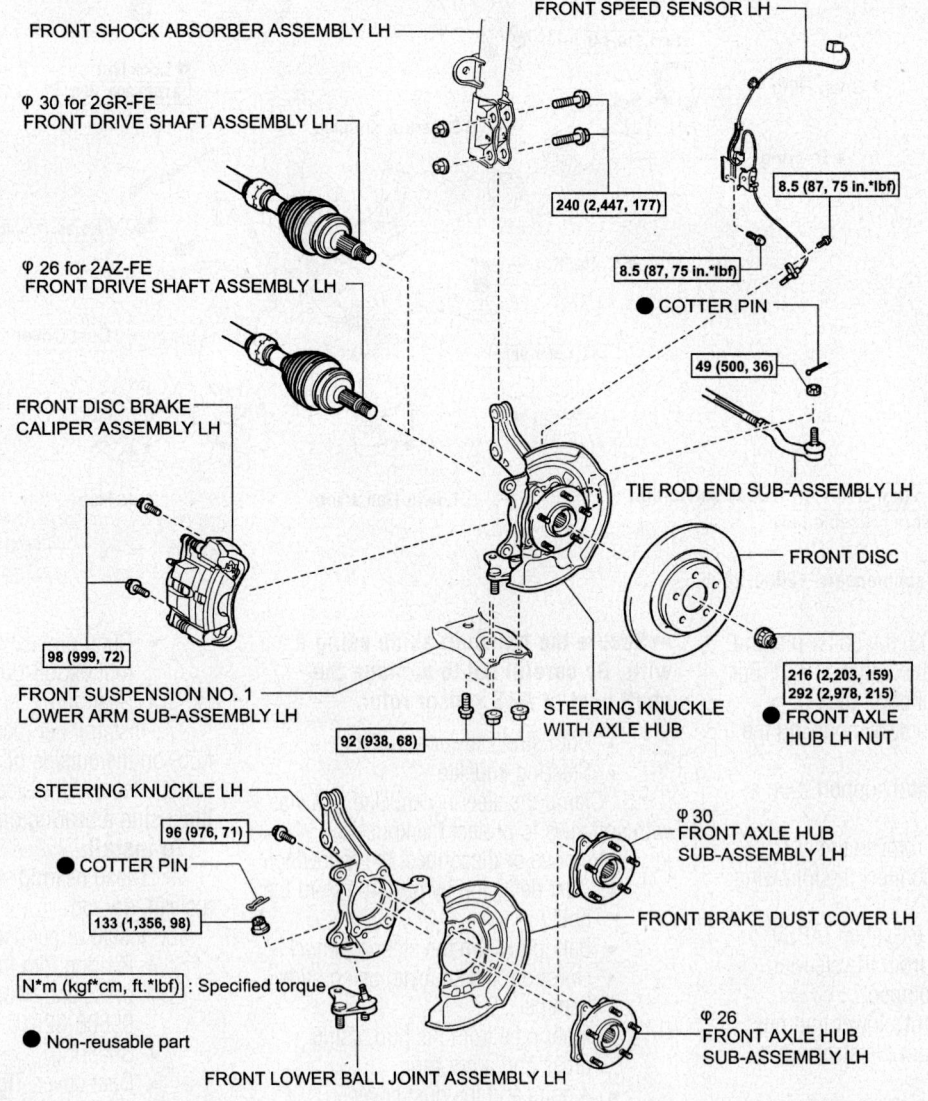

Front hub and related components—2006

09490_RAV4_G0100

3. Raise and support the vehicle safely.

4. Drain the transaxle fluid.

5. Remove the tire and wheel assembly. Remove the front axle hub nut.

6. Remove the front speed sensor. Remove the brake caliper. Remove the rotor.

7. Disconnect the tie rod end, using the proper tools.

8. Disconnect the front suspension number one lower arm subassembly.

9. Remove the two bolts and two nuts. Disconnect the strut from the steering knuckle.

10. Matchmark the halfshaft and the axle hub.

11. Remove the steering knuckle with the axle hub.

➡**Be careful not to damage the boot and the speed sensor rotor. Do not excessively push out the halfshaft from the axle assembly.**

12. Remove the four bolts and the axle hub from the steering knuckle. Remove the dust cover from the steering knuckle.

➡**Do not place the hub and bearing's magnet rotor side so that it is facing downward. Do not allow the magnet rotor side to become damaged or contact foreign matter.**

To install:

13. Installation is the reverse of the removal procedure.

14. Tighten the four axle hub bolts to 71 ft. lbs.

15. Tighten the two steering knuckle to axle hub bolts to 177 ft. lbs.

16. Tighten the front axle hub nut to 159 ft. lbs. on vehicles manufactured from 11/2005 to 01/2006.

17. Tighten the front axle hub nut to 159 ft. lbs. for vehicles manufactured after 01/2006 and equipped with the 2.4L engine, for vehicles equipped with the 3.5L engine tighten the nut to 215 ft. lbs.

18. Be sure to refill the transaxle with the proper grade and type transaxle fluid.

19. Start the engine and check for leaks, correct as required.

20. Check and adjust the alignment, as required.

REAR SUSPENSION

Shock Absorber

REMOVAL & INSTALLATION

2002–2005

1. Before servicing the vehicle, refer to the Precautions Section.

2. Remove the rear wheel.

3. Support the No. 1 control arm with a floor jack.

4. Remove or disconnect the following:
   • Suspension cap from inside the vehicle

• Both upper shock absorber nuts, retainers and cushion
• Shock absorber from the lower control arm by removing the bolt and 2 retainers
• Shock absorber

To install:

5. Install or connect the following:
   • Shock absorber
   • Shock absorber to the lower control arm retainers. Tighten the bolt to 27 ft. lbs. (37 Nm).
   • Shock absorber to the chassis cushion and retainers. Tighten both nuts to 18 ft. lbs. (25 Nm) for

2002–2003; 11 ft. lbs. (14.5 Nm) for 2004–2005.
• Suspension cap
• Wheel

2006

1. Before servicing the vehicle, refer to the Precautions Section.

2. Disconnect the negative battery cable.

➡**Wait at least 90 seconds after disconnecting the negative battery cable before starting any repair work to prevent air bag and seat belt pretensioner activation.**

3. Raise and support the vehicle safely.

4. Remove the tire and wheel assembly.

5. Support the number two suspension arm, using a suitable jack.

6. Remove the bolt and two nuts from the suspension member and axle carrier.

7. Remove the two bolts and disconnect the chock absorber with the bracket.

8. Remove the nut and bolt from the shock absorber upper side.

9. Remove the shock absorber from the vehicle.

To install:

10. Installation is the reverse of the removal procedure.

11. Do not apply final tightening torque until the suspension is stabilized.

12. To stabilize the suspension, lower the vehicle to ground height. Press down on the vehicle several times to stabilize the suspension.

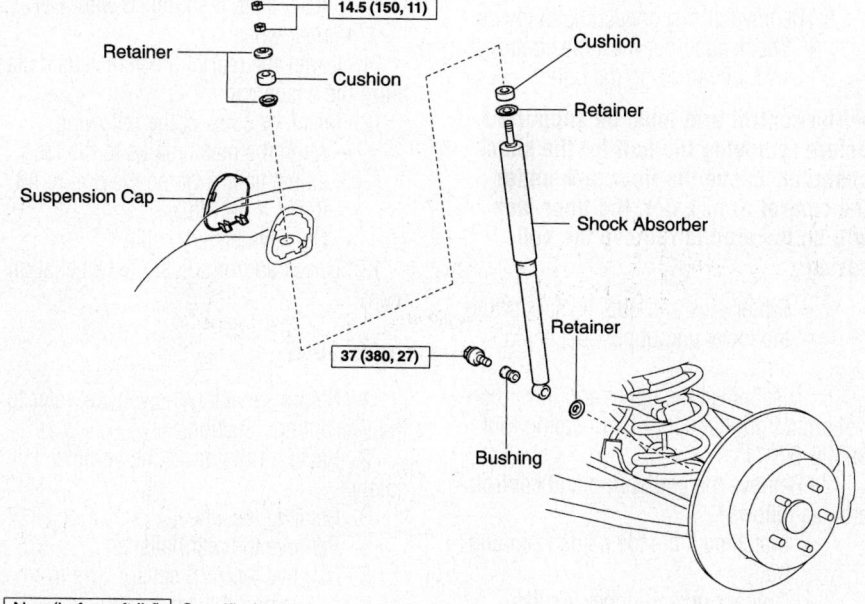

N·m (kgf·cm, ft·lbf) :Specified torque

67170-RAV4-G57

Rear shock absorber and related components—2004–2005

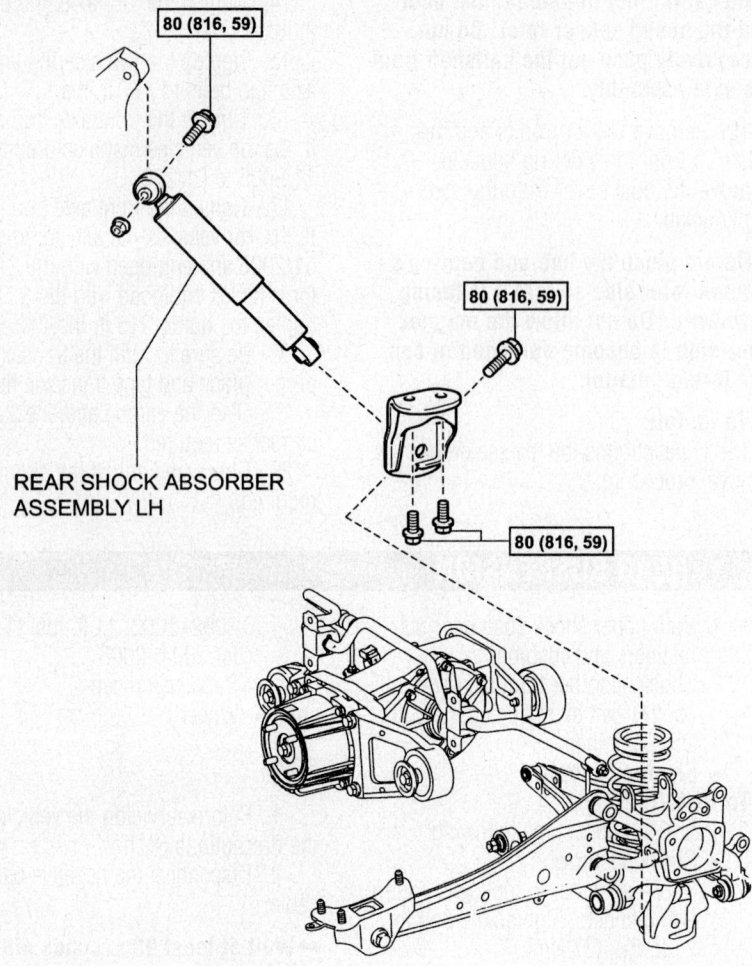

80 (816, 59)

80 (816, 59)

80 (816, 59)

REAR SHOCK ABSORBER
ASSEMBLY LH

N*m (kgf*cm, ft.*lbf) : Specified torque

09490_RAV4_G0101

Rear shock absorber and related components—2006

Coil Spring

REMOVAL & INSTALLATION

2002–2003

1. Before servicing the vehicle, refer to the Precautions Section.
2. Remove or disconnect the following:
   - Negative battery cable
   - Axle shaft, if equipped with 2WD
   - Halfshaft, if equipped with 4WD
   - Brake drum
   - Both brake line clamp bolts
   - Parking brake cable clamp bolt
   - Anti-lock Brake System (ABS) speed sensor and wiring harness, if equipped with ABS
   - Rear axle hub with the brake, by removing the 4 bolts
3. Support the hub securely.
4. Support the control arm with a floor jack.

5. Remove or disconnect the following:
   - Shock absorber from the control arm, by removing the bolt

➡**The control arm must be supported before removing the bolt for the shock absorber. Leave the floor jack under the control arm. Later, the floor jack will be lowered to remove the coil spring.**

   - Cotter pins and nuts by supporting the lower and upper suspension arms
6. Disconnect the upper and lower control arms from the control arm, using tool 09628-62011
7. Remove the coil spring and control arm, as follows:
   - Matchmark the toe adjust cam and body.
   - Coil spring and upper insulator, by loosening the bolt and lowering the control arm.

   - Bolt, toe-adjust cam, 2 attachments, nut and control arm
   - Bolt and spring bumper

To install:

8. Install or connect the following:
   - Spring bumper. Tighten the bolt to 108 inch lbs. (13 Nm).
   - Control arm, 2 attachments, toe-adjust cam, bolt and nut; do not tighten the bolt at this time
   - Spring and upper insulator
9. Raise the control arm with a floor jack.
10. Install or connect the following:
   - Upper and lower suspension arms to the control arm. Tighten the nuts to 76 ft. lbs. (103 Nm).
   - New cotter pins
   - Sock absorber to the control arm. Tighten the bolt to 27 ft. lbs. (37 Nm).
   - Rear axle hub with the brake. Tighten the 4 bolts to 59 ft. lbs. (80 Nm).
   - ABS speed sensor and wiring harness, if equipped. Tighten the ABS speed sensor to 69 inch lbs. (8 Nm) and the wiring harness to 108 inch lbs. (13 Nm).
   - Parking brake cable clamp. Tighten the bolt to 14 ft. lbs. (19 Nm).
   - Both brake line cable clamps. Tighten the bracket bolt to 13 ft. lbs. (18 Nm) and the clamp bolt to 108 inch lbs. (13 Nm).
   - Brake drum
   - Rear halfshaft, if equipped with 4WD
   - Axle shaft, if equipped with 2WD
   - Rear wheel
11. Lower the rear of the vehicle and stabilize the suspension.
12. Install or connect the following:
   - Align the matchmarks to the toe-adjust cam. Tighten the bolt to 98 ft. lbs. (132 Nm).
   - Negative battery cable
13. Check and/or adjust the wheel alignment

2002–2005

1. Before servicing the vehicle, refer to the Precautions Section.
2. Raise and support the vehicle safely.
3. Remove the wheel.
4. Remove the rear halfshaft.
5. Remove the ABS sensor wire from the lower arm and the sensor.
6. Remove the brake drum or caliper and rotor.

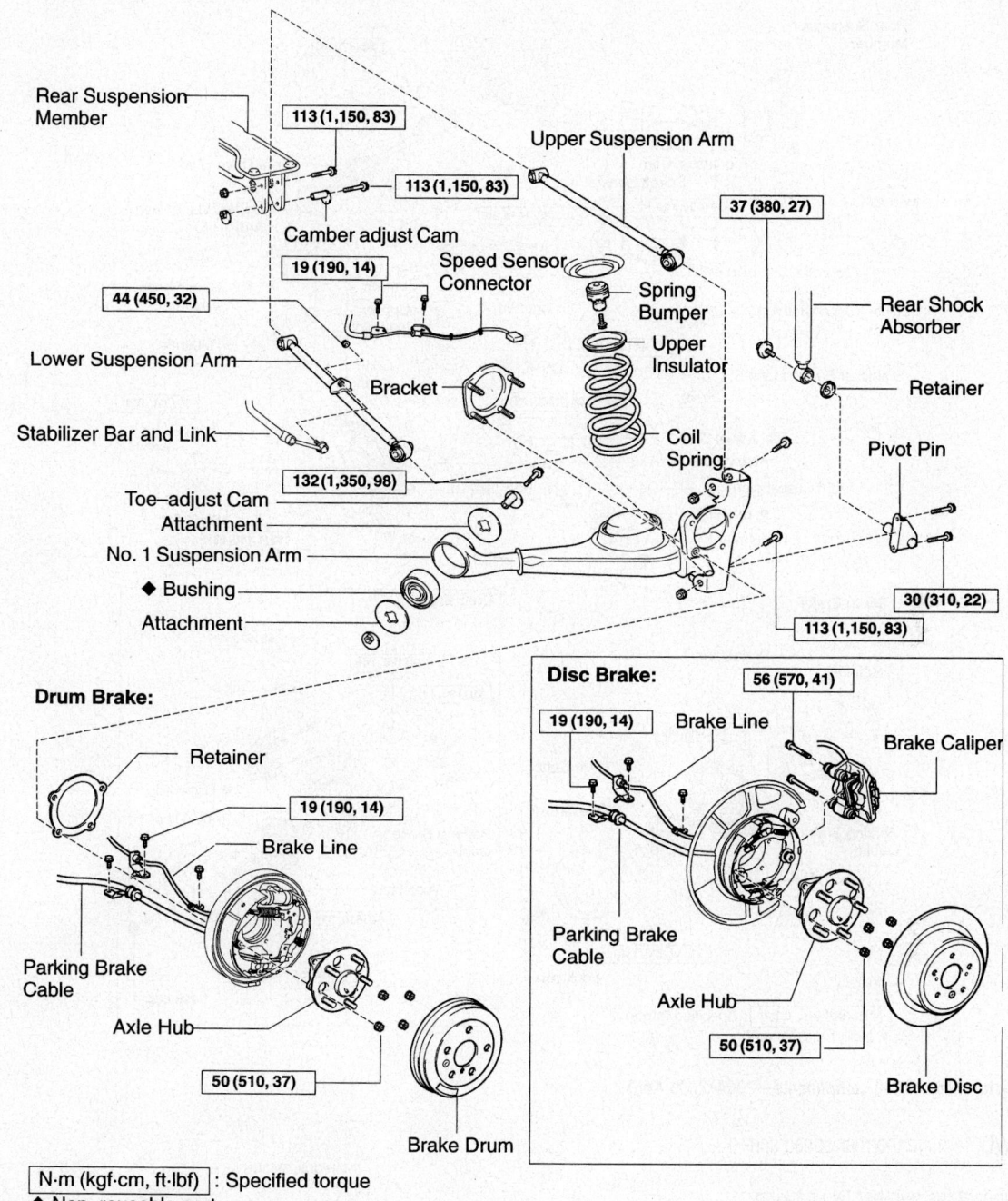

Drum Brake:

Disc Brake:

N·m (kgf·cm, ft·lbf) : Specified torque
◆ Non–reusable part

67170-RAV4-G59

Rear suspension and related components—2004–2005 2WD

7. Remove the brake line from the control arm.

8. Remove the parking brake cable from the control arm.

9. Remove the rear hub.

10. Support the lower arm with a jack.

11. Remove the shock absorber.

12. Remove the stabilizer bar from the lower arm.

13. Lower the jack to remove tension from the spring.

14. Remove the 2 bolts and nuts and remove the upper arm.

15. Matchmark the camber adjusting cam and the suspension member.

16. Remove the 2 bolts and nuts and remove the lower arm.

17. Matchmark the toe adjusting cam and body.

18. Loosen the bolt and remove the spring and insulator.

To install:

19. Installation is the reverse of removal. Observe the following torques:
- Coil spring bolt: 98 ft. lbs. (132 Nm)

- Lower arm bolts: 83 ft. lbs. (113 Nm)
- Upper arm bolts: 83 ft. lbs. (113 Nm)

2006

1. Before servicing the vehicle, refer to the Precautions Section.

2. Raise and support the vehicle safely.

3. Remove the tire and wheel assembly.

4. On 2WD, remove the skid control sensor wire.

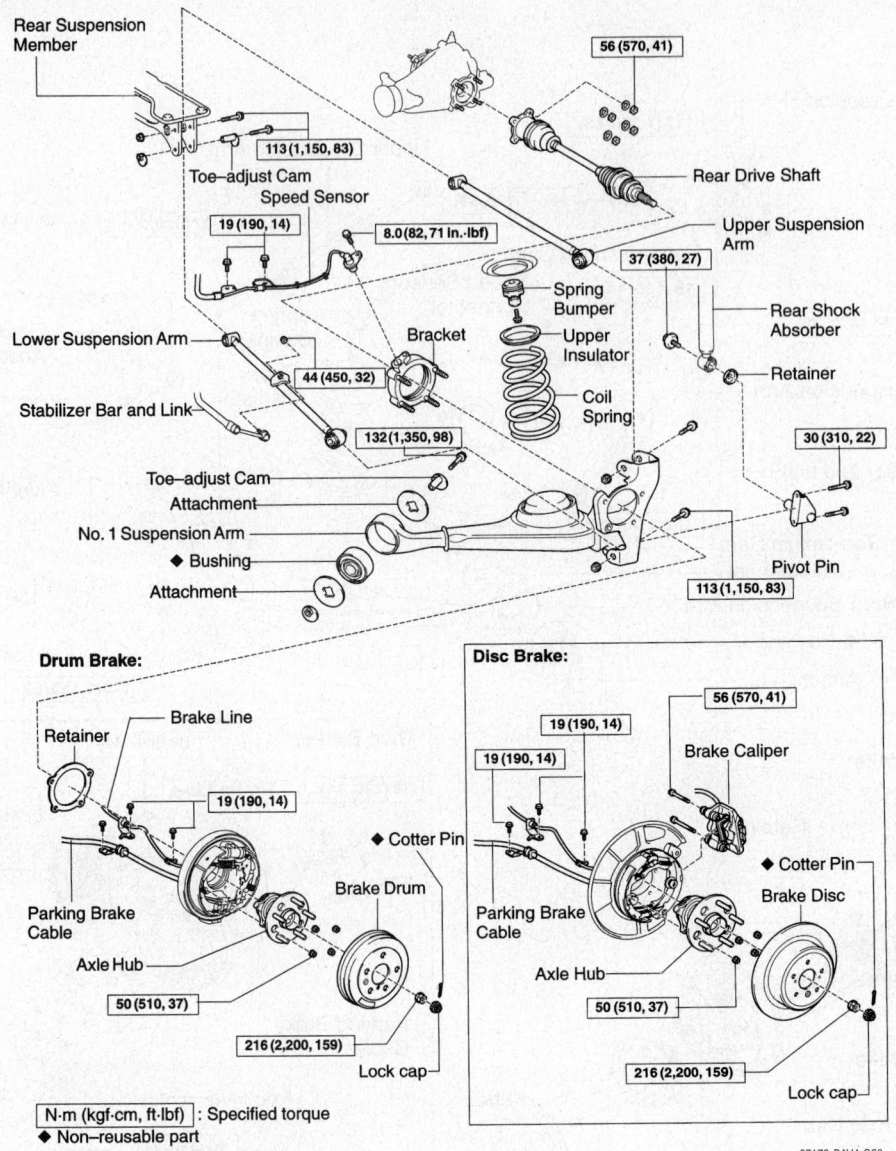

Rear Suspension Member

56 (570, 41)

113 (1,150, 83)

Toe–adjust Cam
Speed Sensor

Rear Drive Shaft

19 (190, 14)

8.0 (82, 71 in.·lbf)

Upper Suspension Arm

37 (380, 27)

Rear Shock Absorber

Spring Bumper

Lower Suspension Arm

Bracket

Upper Insulator

44 (450, 32)

Retainer

Stabilizer Bar and Link

Coil Spring

132 (1,350, 98)

30 (310, 22)

Toe–adjust Cam Attachment

No. 1 Suspension Arm

◆ Bushing

Attachment

Pivot Pin

113 (1,150, 83)

Drum Brake:

Disc Brake:

Retainer

Brake Line

56 (570, 41)

19 (190, 14)

19 (190, 14)

Brake Caliper

19 (190, 14)

◆ Cotter Pin

Parking Brake Cable

Brake Drum

◆ Cotter Pin

Parking Brake Cable

Brake Disc

Axle Hub

Axle Hub

50 (510, 37)

50 (510, 37)

216 (2,200, 159)

216 (2,200, 159)

Lock cap

Lock cap

N·m (kgf·cm, ft·lbf) : Specified torque
◆ Non–reusable part

67170-RAV4-G60

Rear suspension and related components—2004–2005 4WD

5. On 4WD, remove the rear speed sensor wire.

6. Disconnect the number two parking brake cable assembly.

7. Disconnect the rear stabilizer link assembly.

8. To disconnect the rear suspension number two arm assembly, loosen the bolt from the suspension member side. Support the number two suspension arm with a suitable jack.

➡**Do not remove the bolt, only loosen it.**

9. Remove the bolt and nut from the axle carrier side. Slowly lower the jack and disconnect the number two suspension arm from the axle carrier.

10. Remove the upper spring insulator.

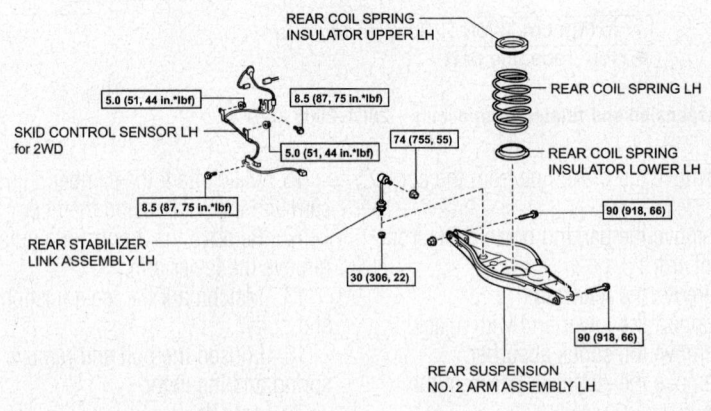

REAR COIL SPRING INSULATOR UPPER LH

5.0 (51, 44 in.*lbf)

8.5 (87, 75 in.*lbf)

REAR COIL SPRING LH

SKID CONTROL SENSOR LH
for 2WD

5.0 (51, 44 in.*lbf)

74 (755, 55)

REAR COIL SPRING INSULATOR LOWER LH

8.5 (87, 75 in.*lbf)

90 (918, 66)

REAR STABILIZER LINK ASSEMBLY LH

30 (306, 22)

90 (918, 66)

REAR SUSPENSION NO. 2 ARM ASSEMBLY LH

N*m (kgf*cm, ft.*lbf) : Specified torque

09490_RAV4_G0102

Rear coil spring and related components—2006

Remove the spring. Remove the lower insulator.

To install:

11. Installation is the reverse of the removal procedure.

12. Do not apply final tightening torque to the rear suspension number two arm assembly until the suspension is stabilized.

13. To stabilize the suspension, lower the vehicle to ground height. Press down on the vehicle several times to stabilize the suspension.

14. Check and adjust the alignment, as required.

Stabilizer Bar

REMOVAL & INSTALLATION

2002–2005

→**The fuel tank must first be removed from the vehicle. Be sure to check and adjust the fuel level as required, before removing the fuel tank. Take all the necessary precautions to avoid safety and fuel disposal problems.**

1. Before servicing the vehicle, refer to the Precautions Section.

2. Properly relieve the fuel system pressure.

3. Disconnect the negative battery cable.

→**Wait at least 90 seconds after disconnecting the negative battery cable before starting any repair work to prevent air bag and seat belt pretensioner activation.**

4. Remove the front floor carpet.

5. Disconnect the number two parking brake cable assembly.

6. Disconnect the fuel tank main tube assembly.

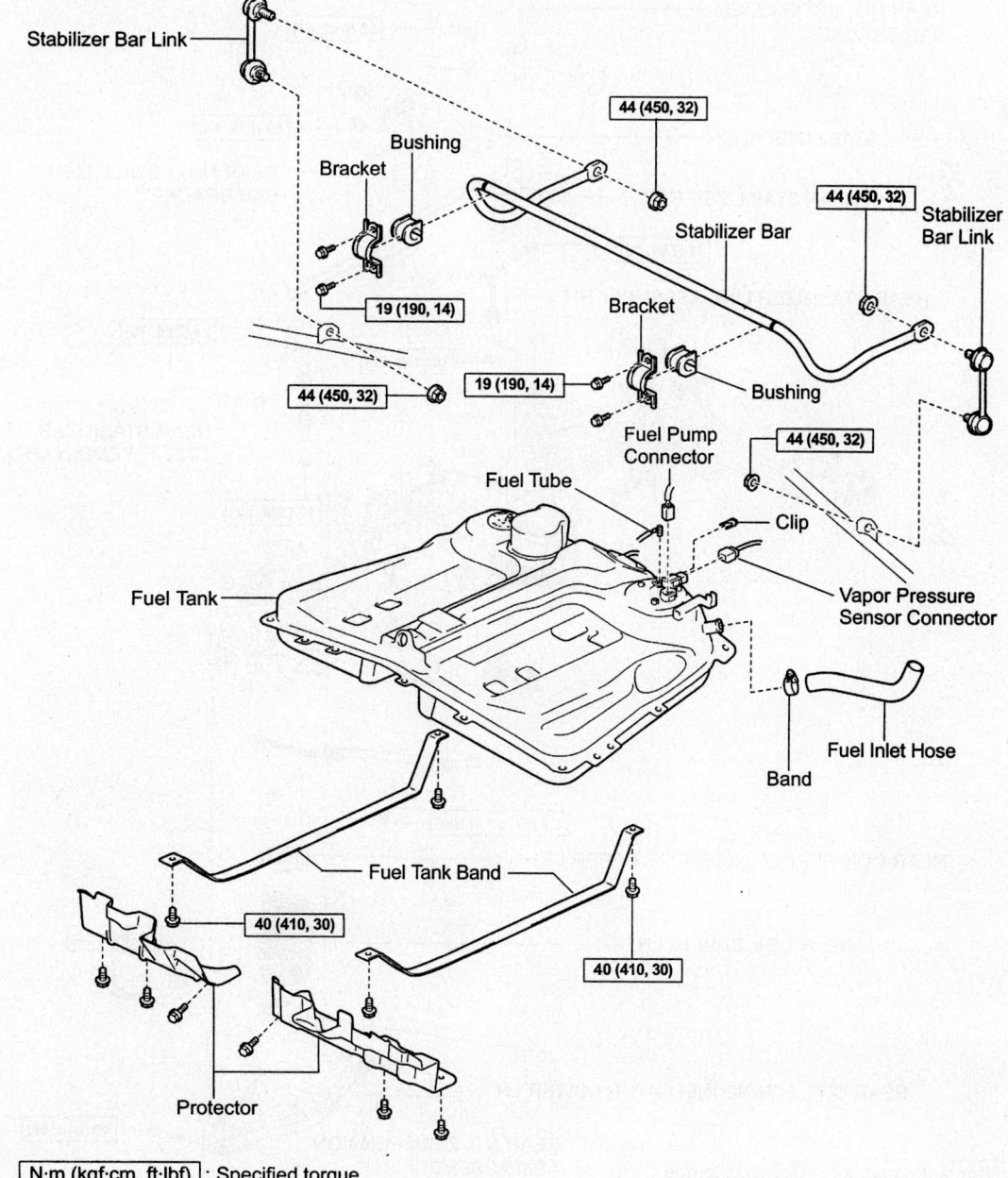

N·m (kgf·cm, ft·lbf) : Specified torque

Rear stabilizer and related components—2002–2005

7. Disconnect the fuel tank to filler pipe hose.

8. Disconnect the fuel tank breather hose.

9. Remove the fuel tank filler pipe.

10. Position a suitable jack under the fuel tank. Remove the six bolts and three fuel tank bands.

11. Slightly lower the suitable jack.

➡**Be careful not to cut the wing nuts.**

12. Fold back about half of each cushion rubber so that the wire harness can be detached.

13. Disconnect the fuel pump connector and sender gauge connector.

14. Detach the wire harness from the four clamps. Carefully remove the fuel tank from the vehicle.

15. Remove the stabilizer bar links. Remove the stabilizer bracket bolt, rubber insulators and brackets.

16. Remove the stabilizer bar from the vehicle.

To install:

17. Installation is the reverse of the removal procedure.

18. Tighten the stabilizer bar links to 32 ft. lbs.

19. Tighten the stabilizer bracket and bushing bolts to 14 ft. lbs.

20. Be sure to install the bushing so that the cutoff will face the lower point. Install the bushing to the inside of the paint line.

21. Tighten the fuel tank retaining bolts to 30 ft. lbs.

22. Tighten the fuel tank filler pipe bolts to 17 ft. lbs.

23. Start the engine and check for leaks. Correct as required.

2006

1. Before servicing the vehicle, refer to the Precautions Section.

2. Raise and support the vehicle safely.

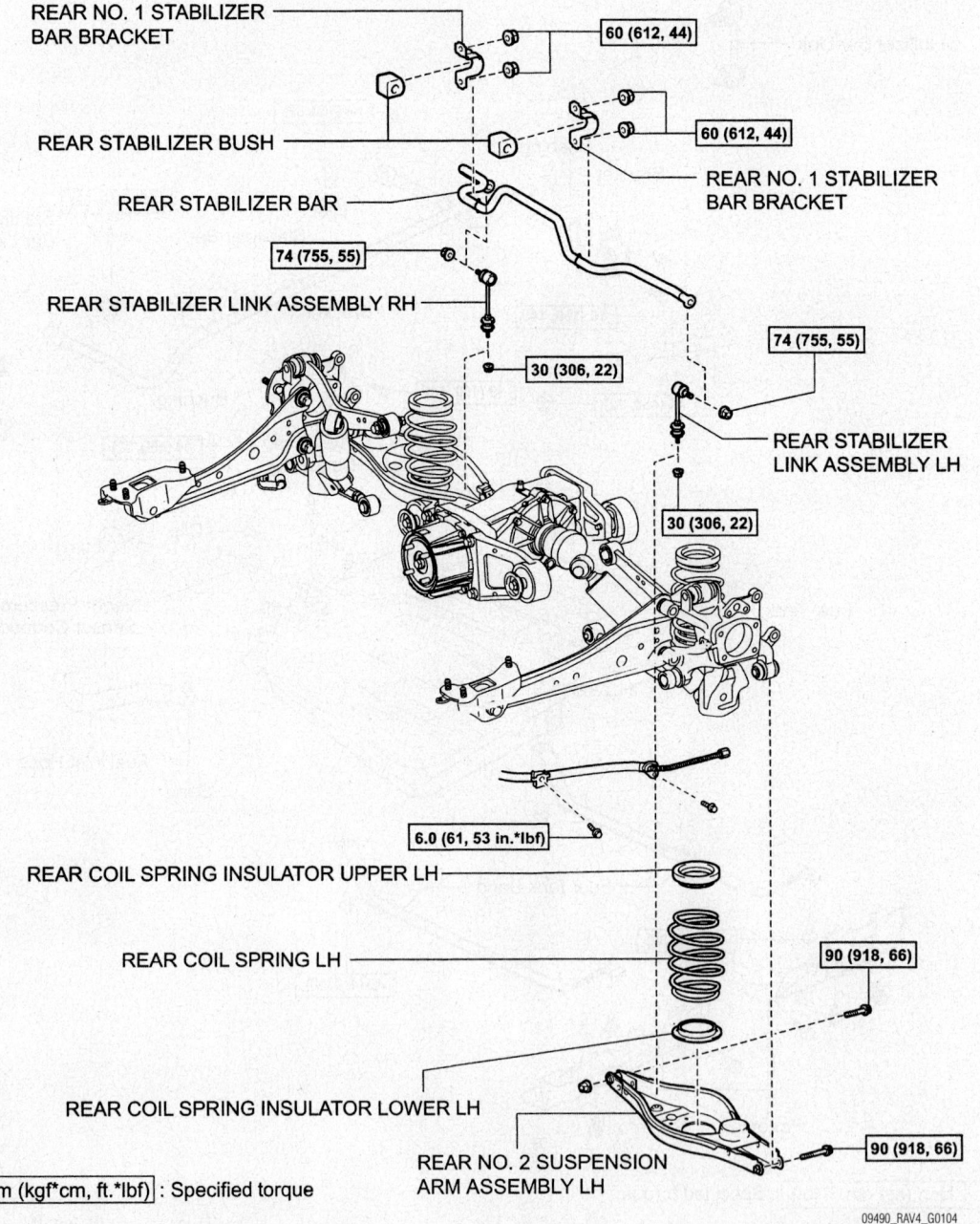

REAR NO. 1 STABILIZER BAR BRACKET — 60 (612, 44)

REAR STABILIZER BUSH

REAR STABILIZER BAR

REAR NO. 1 STABILIZER BAR BRACKET — 60 (612, 44)

74 (755, 55)

REAR STABILIZER LINK ASSEMBLY RH

30 (306, 22)

74 (755, 55)

REAR STABILIZER LINK ASSEMBLY LH

30 (306, 22)

6.0 (61, 53 in.*lbf)

REAR COIL SPRING INSULATOR UPPER LH

REAR COIL SPRING LH

90 (918, 66)

REAR COIL SPRING INSULATOR LOWER LH

REAR NO. 2 SUSPENSION ARM ASSEMBLY LH

90 (918, 66)

N*m (kgf*cm, ft.*lbf) : Specified torque

09490_RAV4_G0104

Rear stabilizer and related components—2006

3. Remove the tire and wheel assembly.

4. Remove the nut and disconnect the link from the suspension number two arm. Remove the nut and the link from the stabilizer bar.

5. Remove the rear number two suspension arms (lower control arm).

6. Remove the coil springs.

7. Remove the stabilizer bracket retaining bolts. Remove the stabilizer bar from the vehicle. Remove the bushings from the bar.

➡**When removing the bar be sure not to damage the sensor wire, brake hose etc.**

To install:

8. Installation is the reverse of the removal procedure.

9. Tighten the stabilizer bar links to 55 ft. lbs.

10. Tighten the stabilizer bracket and bushing bolts to 44 ft. lbs.

11. Install each bushing to the outer side of the bushing stopper on each stabilizer bar. Install each bushing with its slit facing the vehicle front side.

12. Do not apply final tightening torque until the suspension is stabilized.

13. To stabilize the suspension, lower the vehicle to ground height. Press down on the vehicle several times to stabilize the suspension.

14. Check and adjust the alignment, as required.

Lower Control Arm

REMOVAL & INSTALLATION

2004–2005

1. Before servicing the vehicle, refer to the Precautions Section.

2. Raise and support the vehicle safely.

3. Remove the wheel.

4. Remove the rear halfshaft.

5. Remove the ABS sensor wire from the lower arm and the sensor.

6. Remove the brake drum or caliper and rotor.

7. Remove the brake line from the control arm.

8. Remove the parking brake cable from the control arm.

9. Remove the rear hub.

10. Support the lower arm with a jack.

11. Remove the shock absorber.

12. Remove the stabilizer bar from the lower arm.

13. Lower the jack to remove tension from the spring.

14. Remove the 2 bolts and nuts and remove the upper arm.

15. Matchmark the camber adjusting cam and the suspension member.

16. Remove the 2 bolts and nuts and remove the lower arm.

17. Matchmark the toe adjusting cam and body.

18. Loosen the bolt and remove the spring and insulator.

19. Remove the lower control arm from the vehicle.

To install:

20. Installation is the reverse of removal. Observe the following torques:
- Coil spring bolt: 98 ft. lbs. (132 Nm)
- Lower arm bolts: 83 ft. lbs. (113 Nm)
- Upper arm bolts: 83 ft. lbs. (113 Nm)

2006

1. Before servicing the vehicle, refer to the Precautions Section.

2. Raise and support the vehicle safely.

3. Remove the tire and wheel assembly.

4. On 2WD, remove the skid control sensor wire.

5. On 4WD, remove the rear speed sensor wire.

6. Disconnect the number two parking brake cable assembly.

7. Disconnect the rear stabilizer link assembly.

8. To disconnect the rear suspension number two arm assembly, loosen the bolt from the suspension member side. Support the number two suspension arm with a suitable jack.

➡**Do not remove the bolt, only loosen it.**

9. Remove the bolt and nut from the axle carrier side. Slowly lower the jack and disconnect the number two suspension arm from the axle carrier.

10. Remove the upper spring insulator. Remove the spring. Remove the lower insulator.

11. Remove the bolt, nut and suspension arm from the suspension member.

To install:

12. Installation is the reverse of the removal procedure.

13. Do not apply final tightening torque to the rear suspension number two arm assembly until the suspension is stabilized.

14. To stabilize the suspension, lower the vehicle to ground height. Press down on the vehicle several times to stabilize the suspension.

15. Check and adjust the alignment, as required.

Upper Control Arm

REMOVAL & INSTALLATION

2006

1. Before servicing the vehicle, refer to the Precautions Section.

2. Raise and support the vehicle safely.

3. Remove the tire and wheel assembly.

4. On 2WD, remove the skid control sensor wire.

5. On 4WD, remove the rear speed sensor wire.

6. Remove the upper control arm retaining bolts. Remove the upper control arm from the vehicle.

To install:

7. Installation is the reverse of the removal procedure.

8. Do not apply final tightening torque to the component until the suspension is stabilized.

9. To stabilize the suspension, lower the vehicle to ground height. Press down on the vehicle several times to stabilize the suspension.

10. Check and adjust the alignment, as required.

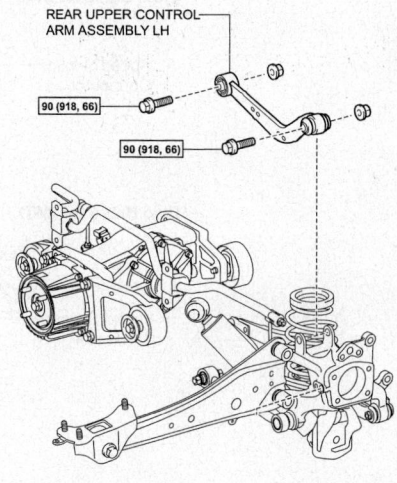

REAR UPPER CONTROL ARM ASSEMBLY LH

90 (918, 66)

90 (918, 66)

09490_RAV4_G0105

Rear upper control arm and related components—2006

Wheel Bearing

ADJUSTMENT

The wheel bearing is not adjustable.

REMOVAL & INSTALLATION

2002–2005

1. Before servicing the vehicle, refer to the Precautions Section.
2. Raise and support the vehicle safely.
3. Remove the tire and wheel assembly.
4. With 4WD, remove the halfshaft.
5. With disc brakes, remove the caliper and rotor.
6. With drum brakes, remove the brake drum.
7. With 2WD, disconnect the speed sensor.

8. Remove the 4 nuts and the hub/bearing assembly.

To install:

9. Installation is the reverse of removal.
10. Torque the hub bolts to 37 ft. lbs. (50 Nm).

2006

1. Before servicing the vehicle, refer to the Precautions Section.
2. Properly relieve the fuel system pressure.
3. Disconnect the negative battery cable.

➥**Wait at least 90 seconds after disconnecting the negative battery cable before starting any repair work to prevent air bag and seat belt pretensioner activation.**

4. Raise and support the vehicle safely.
5. Remove the tire and wheel assembly.
6. On 4WD, remove the rear axle shaft nut.
7. Remove the caliper. Remove the rotor.
8. On 2WD remove the skid control sensor wire.
9. On 4WD, remove the rear speed sensor.
10. Remove the rear suspension number one arm.
11. Disconnect the shock absorber from the axle carrier.
12. On 2WD, remove the four bolts, and the axle hub and bearing from the axle carrier.
13. On 4WD, matchmark the halfshaft and the axle hub and bearing. Remove the

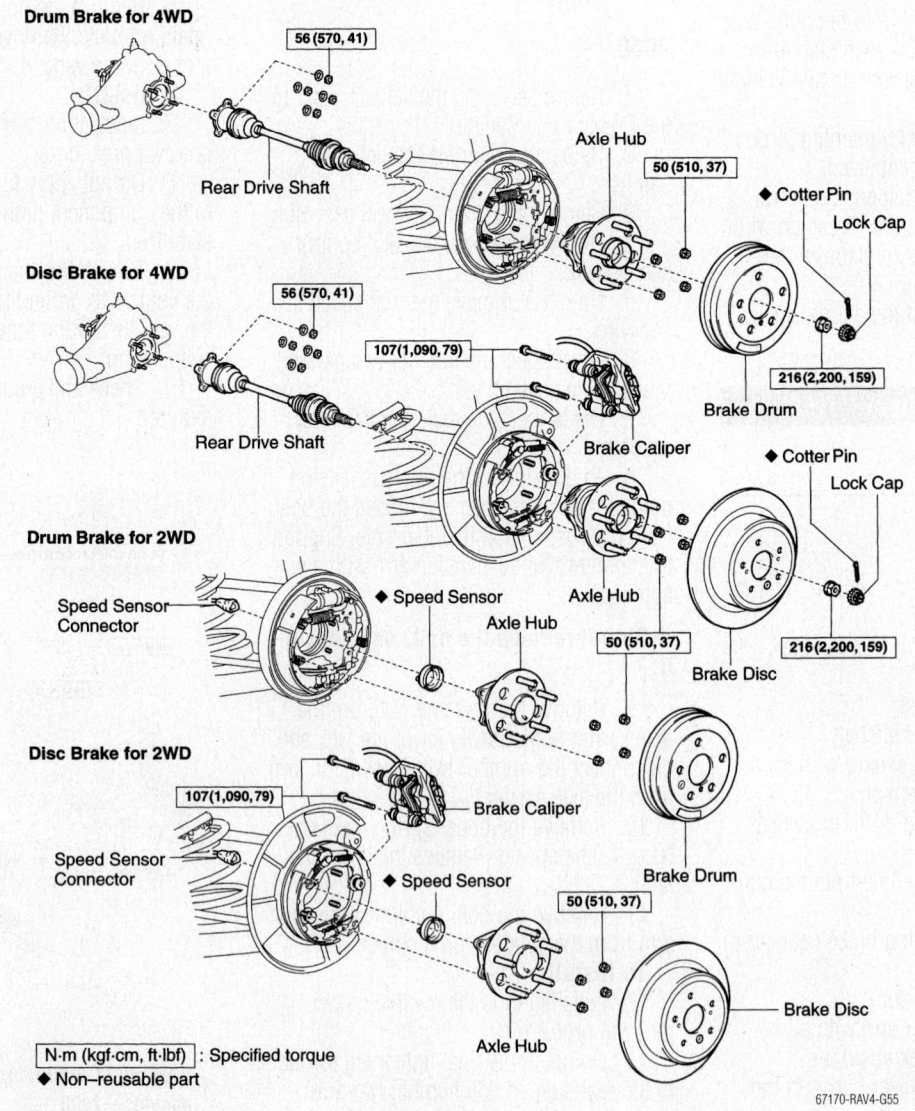

N·m (kgf·cm, ft·lbf) : Specified torque
◆ Non–reusable part

67170-RAV4-G55

Rear hub and related components—2004–2005

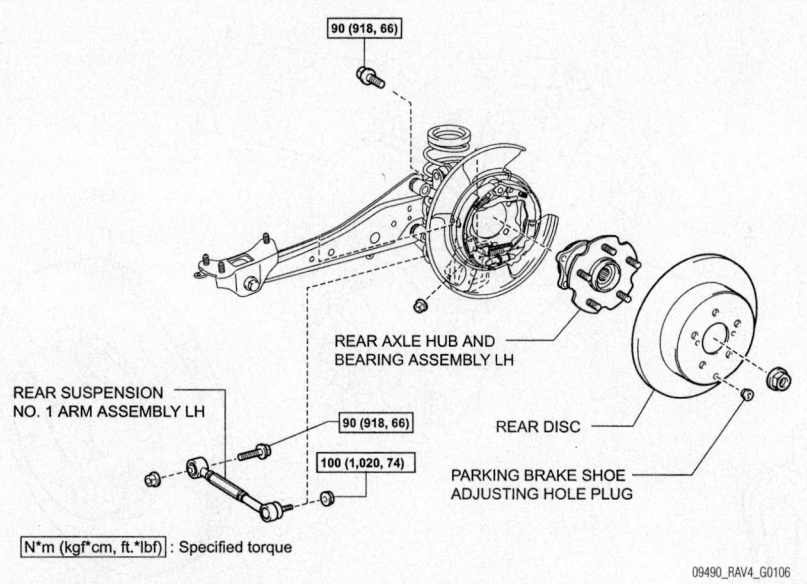

Rear hub and related components—2006

four bolts, and the axle hub and bearing from the axle carrier.

➡**Do not place the hub and bearing's magnet rotor side so that it is facing downward. Do not allow the magnet rotor side to become damaged or contact foreign matter.**

To install:

14. Installation is the reverse of the removal procedure.

15. Tighten the four hub and bearing bolts to 68 ft. lbs.

16. Do not apply final tightening torque to the component until the suspension is stabilized.

17. To stabilize the suspension, lower the vehicle to ground height. Press down on the vehicle several times to stabilize the suspension.

18. Check and adjust the alignment, as required.

BRAKES

Brake Caliper

REMOVAL & INSTALLATION

Front

1. Before servicing the vehicle, refer to the Precautions Section.

2. Raise and safely support the vehicle.

3. Remove the wheel(s).

4. If the caliper is being replaced, remove the union bolt and 2 washers and remove the flexible brake hose from the caliper. Use a suitable container to catch the brake fluid as it drains out. Discard the washers.

5. Hold the sliding pin and loosen the 2 caliper mounting bolts. Remove the bolts and remove the caliper from the torque plate.

6. Remove the brake pads and brake hardware.

To install:

7. Install the brake pads and brake hardware.

8. Install the caliper to the torque plate with the 2 mounting bolts.

9. Reconnect the flexible brake hose to the caliper with 2 new washers and the union bolt. Torque the union bolt to 22 ft. lbs. (30 Nm).

10. Refill the master cylinder with brake fluid and bleed the brake system.

11. Check for proper operation and make sure there are no leaks.

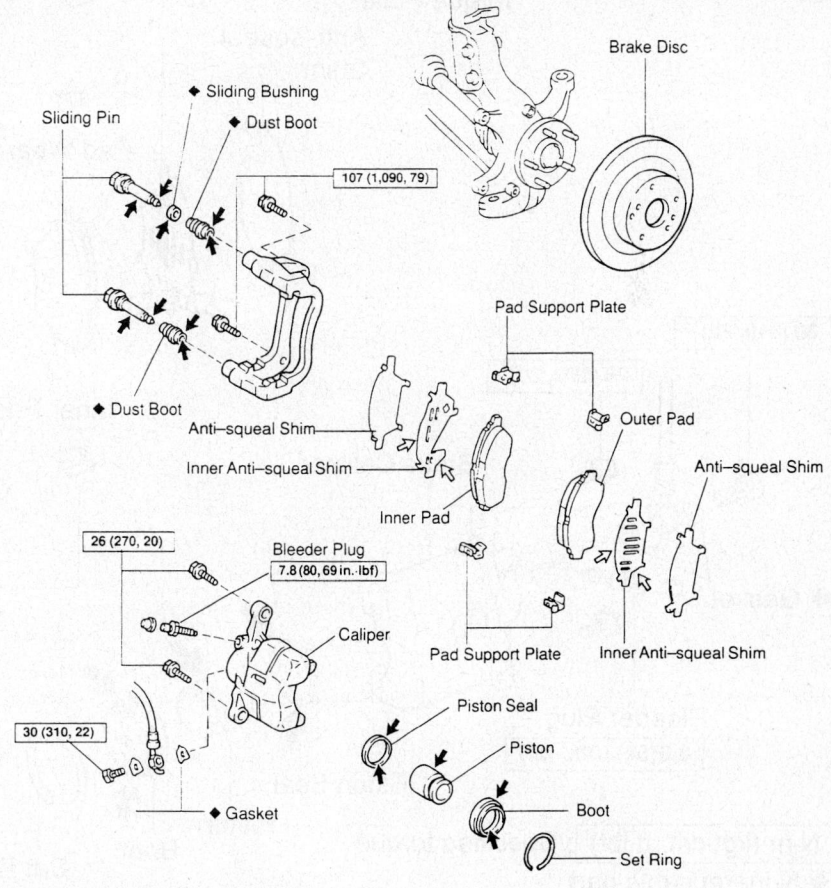

Front brake caliper and related components—2002–2003

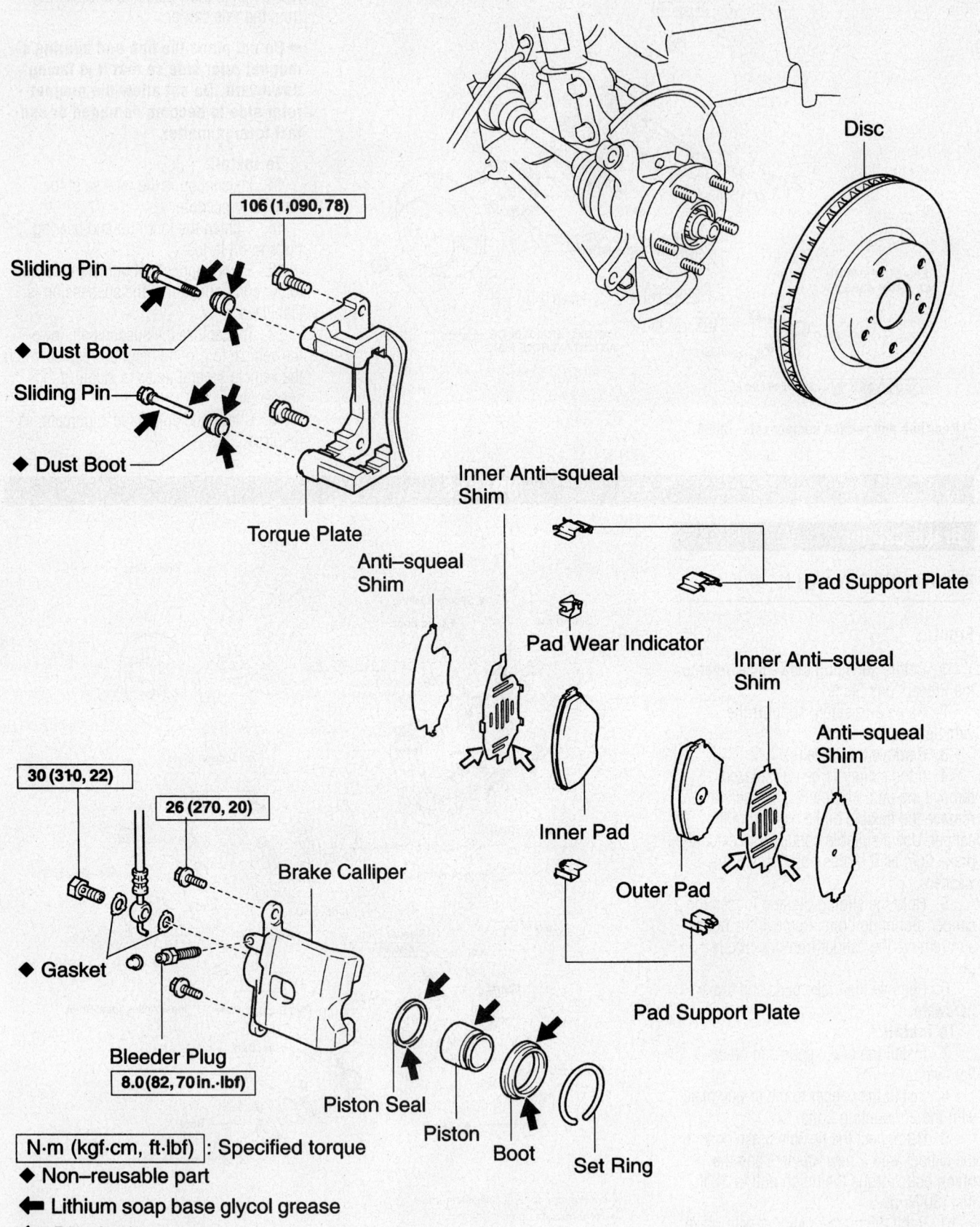

Disc

106 (1,090, 78)

Sliding Pin

◆ Dust Boot

Sliding Pin

◆ Dust Boot

Torque Plate

Inner Anti–squeal Shim

Anti–squeal Shim

Pad Wear Indicator

Pad Support Plate

Inner Anti–squeal Shim

Anti–squeal Shim

Inner Pad

Outer Pad

Pad Support Plate

30 (310, 22)

26 (270, 20)

Brake Calliper

◆ Gasket

Bleeder Plug

8.0 (82, 70 in.·lbf)

Piston Seal

Piston

Boot

Set Ring

N·m (kgf·cm, ft·lbf) : Specified torque

◆ Non–reusable part

◀ Lithium soap base glycol grease

◁ Disc brake grease

67170-RAV4-G61

Front brake caliper and related components—2004–2005

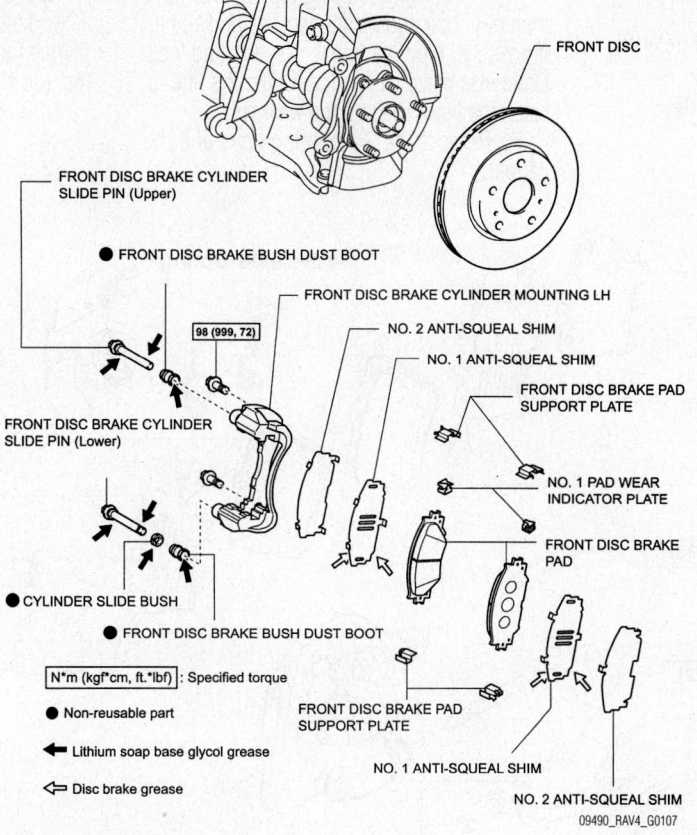

Front brake caliper and related components (275 disc)—2006

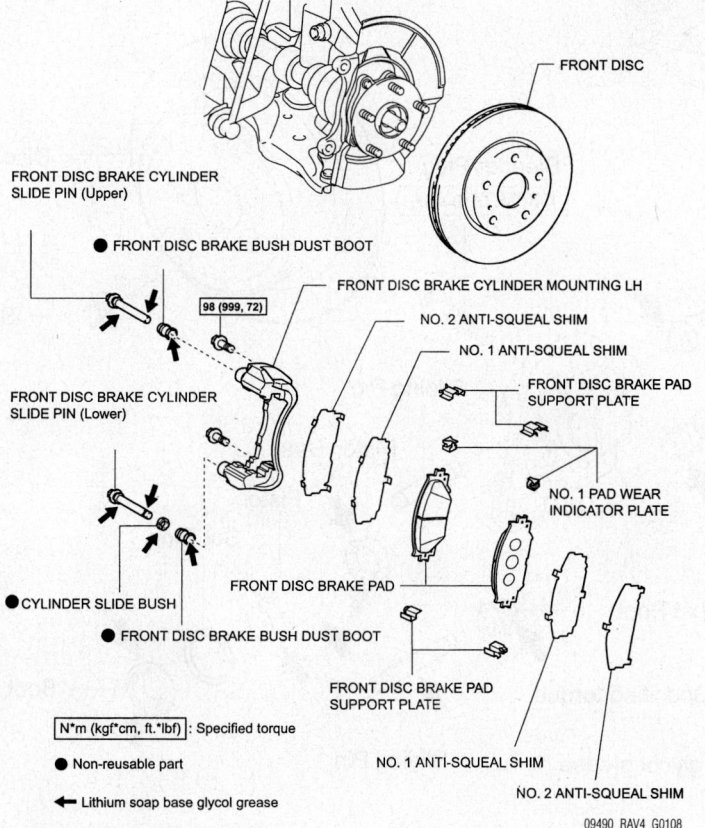

Front brake caliper and related components (296 disc)—2006

Rear

1. Before servicing the vehicle, refer to the Precautions Section.
2. Raise and safely support the vehicle.
3. Remove the wheel.

4. If the caliper is being replaced, remove the union bolt and disconnect the brake line. If possible, plug the line to prevent fluid loss. Otherwise place a container under the line to catch the fluid. Discard the washers.
5. Remove the 2 mounting bolts and the caliper.

To install:

6. Installation is the reverse of removal. Torque the brake line union bolt to 22 ft. lbs. (30 Nm). Use new washers.
7. Bleed the brakes.

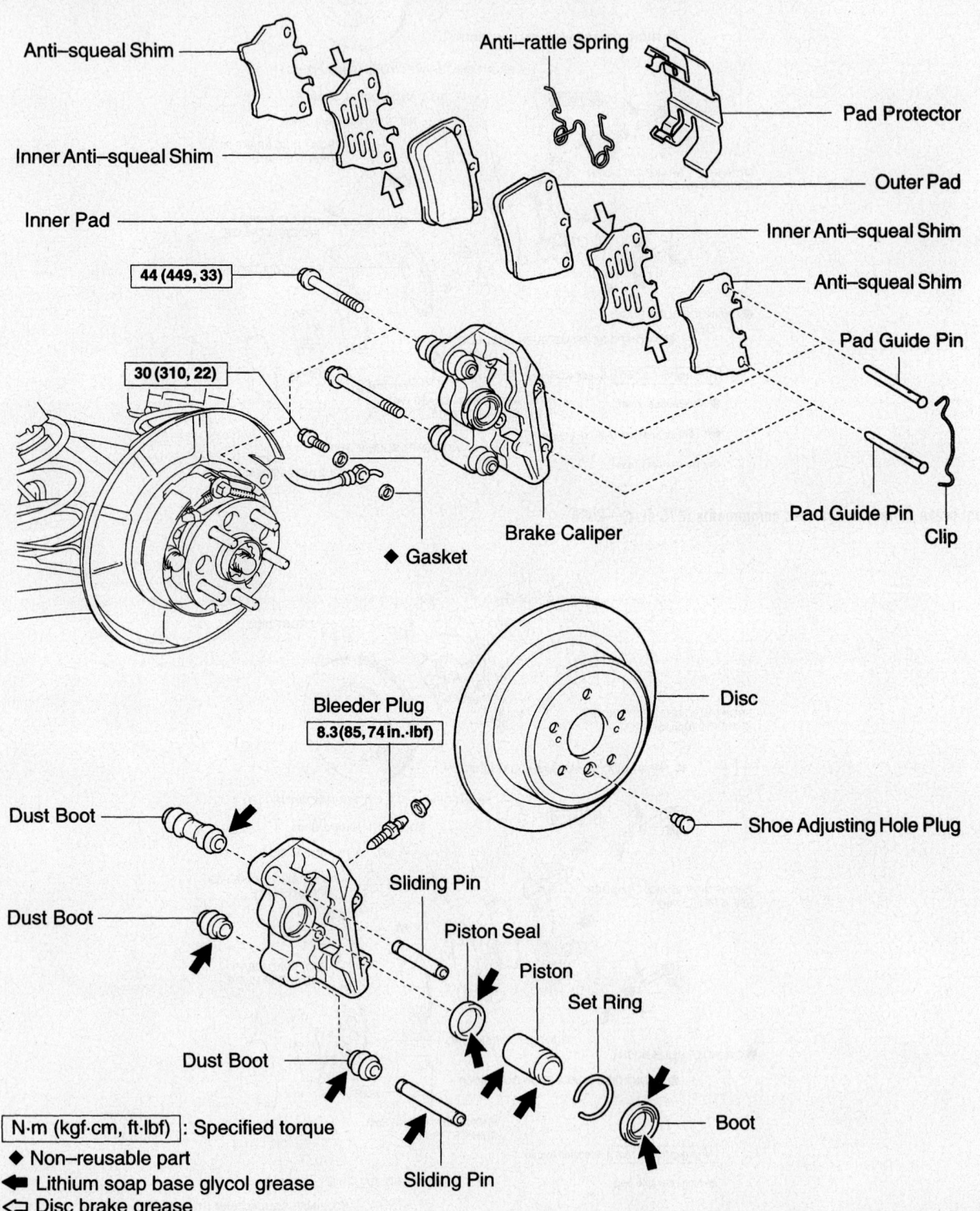

Rear caliper and related components—2002-2005

67170-RAV4-G63

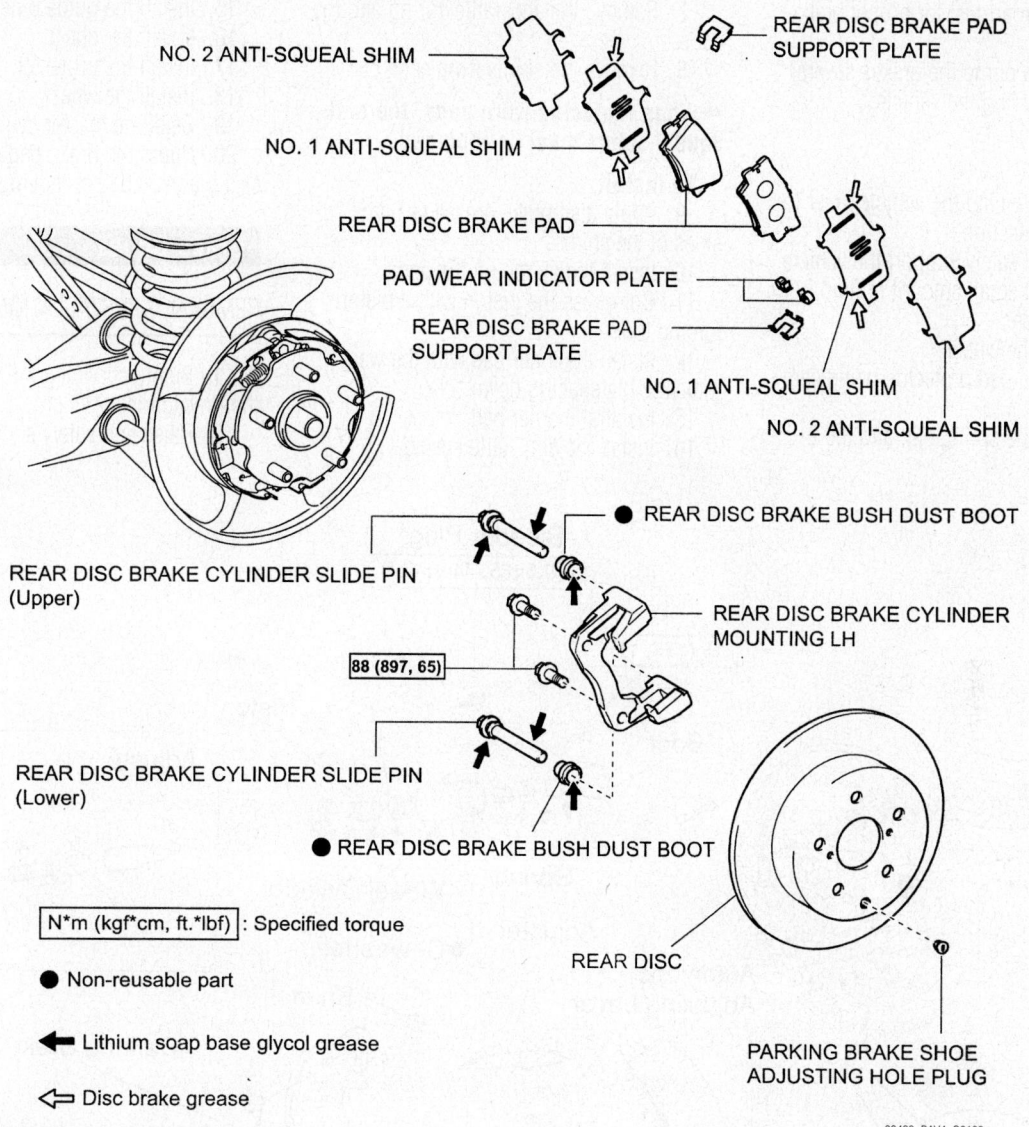

NO. 2 ANTI-SQUEAL SHIM

REAR DISC BRAKE PAD SUPPORT PLATE

NO. 1 ANTI-SQUEAL SHIM

REAR DISC BRAKE PAD

PAD WEAR INDICATOR PLATE

REAR DISC BRAKE PAD SUPPORT PLATE

NO. 1 ANTI-SQUEAL SHIM

NO. 2 ANTI-SQUEAL SHIM

REAR DISC BRAKE CYLINDER SLIDE PIN (Upper)

● REAR DISC BRAKE BUSH DUST BOOT

REAR DISC BRAKE CYLINDER MOUNTING LH

88 (897, 65)

REAR DISC BRAKE CYLINDER SLIDE PIN (Lower)

● REAR DISC BRAKE BUSH DUST BOOT

REAR DISC

PARKING BRAKE SHOE ADJUSTING HOLE PLUG

N*m (kgf*cm, ft.*lbf) : Specified torque

● Non-reusable part

◄ Lithium soap base glycol grease

⇐ Disc brake grease

09490_RAV4_G0109

Rear brake caliper and related components—2006

Disc Brake Pads

REMOVAL & INSTALLATION

Front

1. Before servicing the vehicle, refer to the Precautions Section.
2. Raise and safely support the vehicle.
3. Remove the wheel(s).
4. Temporarily install 2 wheel stud nuts to hold the brake rotor in place.
5. If necessary, siphon a sufficient quantity of brake fluid from the master cylinder reservoir to prevent any brake fluid from overflowing the master cylinder when removing or installing new pads. This may be necessary, as the piston must be forced into the caliper bore to provide sufficient clearance when installing the pads.

6. Grasp the caliper from behind and carefully pull it towards you. This will start to seat the piston(s) in its bore. Using a C-clamp or other suitable tool, press the piston the remaining way into the caliper. Be careful not to cock the piston in the bore. Also, do not force the piston or the caliper and piston may be damaged.
7. Hold the sliding pin and loosen the 2 caliper mounting bolts. Remove the bolts and remove the caliper from the torque plate.
8. Secure the caliper assembly out of the way with a wire; so as not to stress the flexible hose.
9. Slide out the old brake pads along with any anti-squeal shims, springs, pad wear indicators and pad support plates. Make sure to note the position of all assorted pad hardware.

To install:
10. Check the brake disc (rotor) for thickness and run-out. Inspect the caliper and piston assembly for breaks, cracks, fluid seepage or other damage. Overhaul or replace as necessary.
11. Install the pad support plates into the torque plate.
12. Install the pad wear indicators onto the pads. Be sure the arrow on the indicator plate is pointing in the direction of rotation.
13. Install the anti-squeal shims on the outside of each pad and then install the pad assemblies into the torque plate.
14. Install the caliper to the torque plate with the 2 mounting bolts. Torque the bolts to 20 ft. lbs. (26 Nm).
15. Remove the 2 temporary wheel stud nuts and check that the rotor turns freely.
16. Reinstall the wheel(s). Safely lower

the vehicle, and road-test for proper brake operation.

17. Be sure to pump the brakes several times prior to moving the vehicle.

Rear

1. Before servicing the vehicle, refer to the Precautions Section.
2. Raise and safely support the vehicle.
3. Remove a small amount of fluid from the master cylinder.
4. Remove the wheel.
5. Pry off the pad protector, being careful not to bend it.
6. Remove the spring clip and the 2 pad guide pins.

7. Remove the anti-rattle spring and the pads.
8. Remove the shims from each pad.

➡️**When replacing worn pads, the anti-squeal shims must be replaced.**

To install:

9. Apply disc brake grease to both sides of the shims.
10. Install the shims on the pads.
11. Compress the piston with a piston forcing tool.
12. Install the inner pad with the wear indicator plate facing down.
13. Install the outer pad.
14. Install the anti-rattle spring.

15. Install the guide pins.
16. Install the clip.
17. Install the protector.
18. Install the wheel.
19. Refill the master cylinder.
20. Pump the brake pedal a few times to seat the pads before moving the vehicle.

Brake Shoes

REMOVAL & INSTALLATION

1. Before servicing the vehicle, refer to the Precautions Section.
2. Raise and safely support the vehicle.

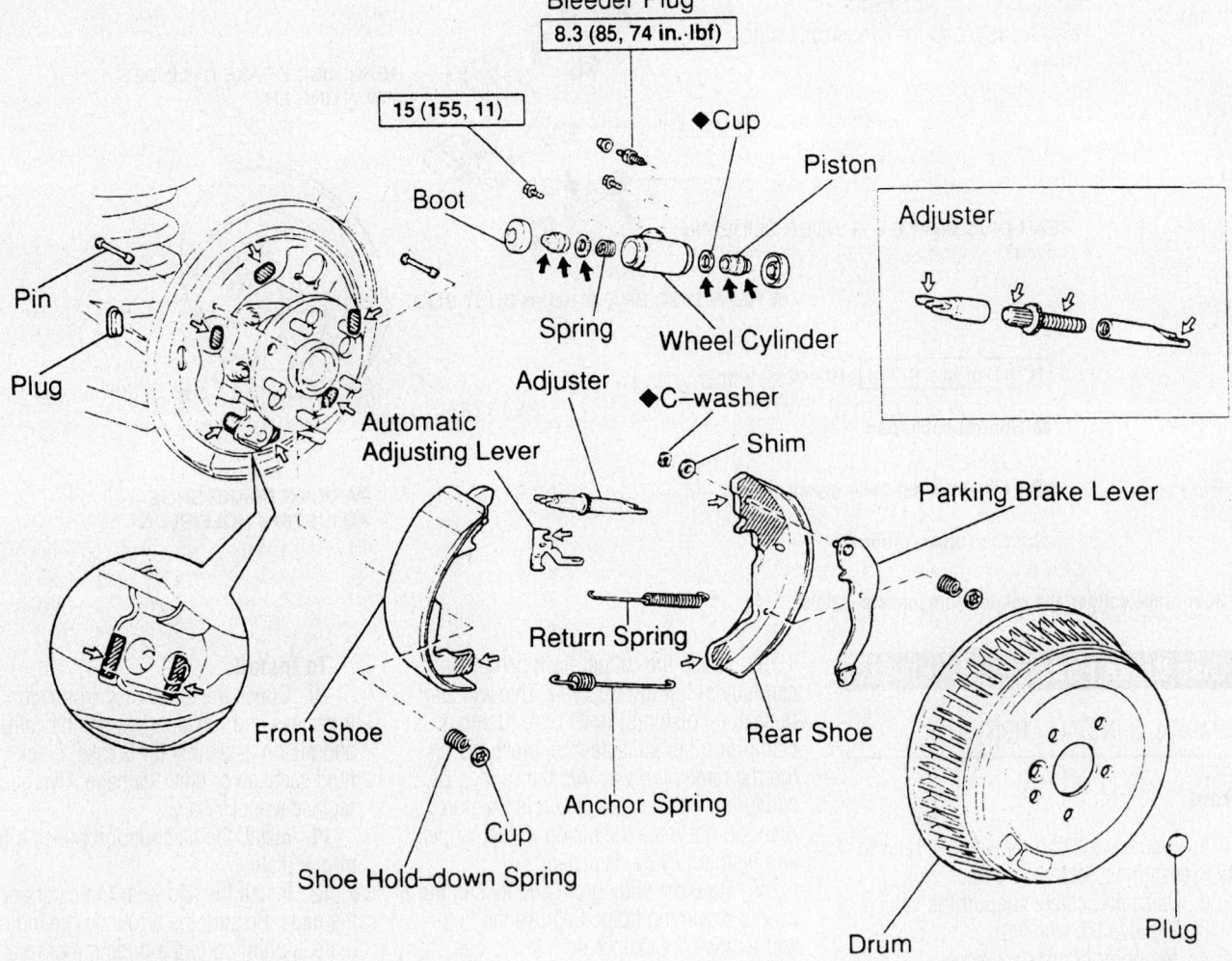

N·m (kgf·cm, ft·lbf) : Specified torque

◆ Non-reusable part
➡️ Lithium soap base glycol grease
⇨ High temperature grease

93026G80

Rear drum brake and related components—2002–2003

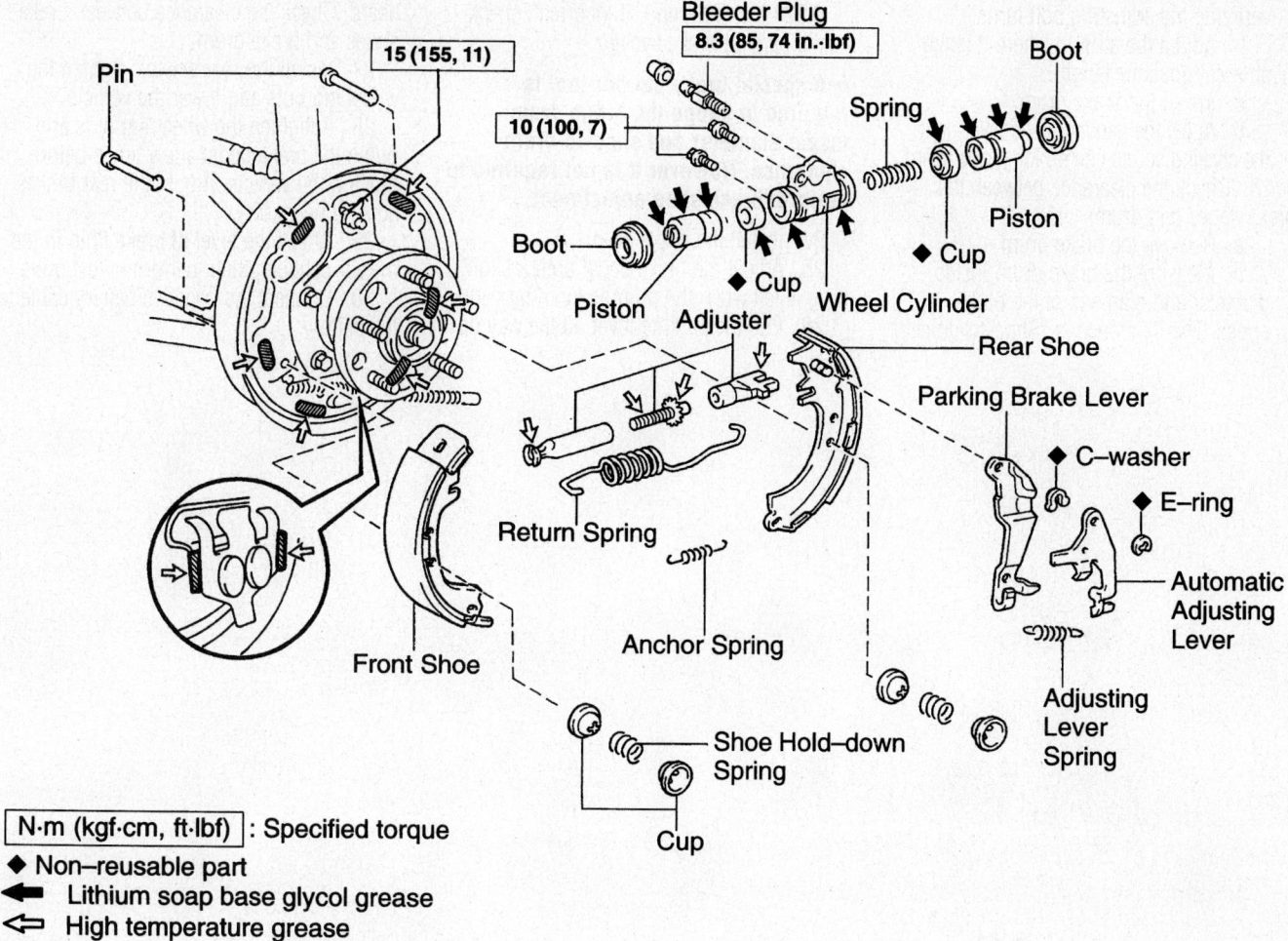

Bleeder Plug
8.3 (85, 74 in.·lbf)

15 (155, 11)

10 (100, 7)

Pin

Boot

Spring

Boot

Piston

Cup

Boot

Piston

Cup

Wheel Cylinder

Adjuster

Rear Shoe

Parking Brake Lever

C–washer

E–ring

Automatic Adjusting Lever

Return Spring

Anchor Spring

Adjusting Lever Spring

Front Shoe

Shoe Hold–down Spring

Cup

N·m (kgf·cm, ft·lbf) : Specified torque

◆ Non–reusable part

◀ Lithium soap base glycol grease

◁ High temperature grease

67170-RAV4-G62

Rear drum brake and related components—2004–2005

3. Loosen the rear wheel lug nuts slightly. Release the parking brake.

4. Block the front wheels, raise the rear of the vehicle, and safely support it with jackstands.

5. Remove the wheel lug nuts and the wheel.

6. Remove the brake drum retaining screws, if equipped. Remove the brake drum.

7. If the drum is difficult to remove, perform the following:

   a. Insert the end of a bent wire (a coat hanger will do nicely) through the hole in the brake drum and hold the automatic adjusting lever away from the adjuster.

   b. Reduce the brake shoe adjustment by turning the adjuster bolt with a brake tool.

   c. The drum should now be loose enough to remove without much effort.

8. Carefully unhook the return spring from the leading (front) brake shoe.

9. Press the hold down spring retainer in and turn the pin on the front brake shoe.

10. Remove the hold down spring, retainers and the pin for the front brake shoe.

11. Pull out the brake shoe and unhook the anchor spring from the lower edge.

12. Remove the hold down spring from the trailing (rear) shoe. Pull the shoe out with the adjuster, automatic adjuster assembly and springs attached. Disconnect the parking brake cable. Remove the tension/return and anchor springs from the rear shoe.

13. Unhook the adjusting lever spring from the rear shoe and then remove the automatic adjuster assembly.

To install:

14. Inspect the shoes for signs of unusual wear or scoring.

15. Check the wheel cylinder for any sign of fluid seepage or frozen pistons.

16. Clean and inspect the brake backing plate and all other components. Check that the brake drum inner diameter is within specified limits. Lubricate the backing plate at the positions the brakes come in contact

with the backing plate. Also lubricate the anchor plate.

17. Mount the automatic adjuster assembly onto a new rear brake shoe.

18. Connect the parking brake cable to the rear shoe and then install the automatic adjusting lever, spring and E-ring. Position the rear shoe so the lower end rides in the anchor plate and the upper end is against the boot of the wheel cylinder.

19. Install the pin and the hold down spring. Press the retainer down over the pin and rotate the pin so the crimped edge is held by the retainer.

20. Place the front brake into position and install the anchor spring between the front and rear shoes. Stretch the spring enough so the front shoe will fit as the rear did. Install the hold down spring, pin and retainer to the front brake shoe.

21. Connect the return spring to the front brake shoe.

22. Check the operation of the automatic adjuster mechanism:

a. Apply the parking brake lever and verifying the adjusting bolt turns.

b. Adjust the strut to where it is the shortest possible length.

c. Install the brake drum.

d. Apply the parking brake lever until the clicking sound can no longer be heard.

23. Check the clearance between the brake shoes and drum:

a. Remove the brake drum.

b. Measure the brake drum inside diameter and diameter of the brake shoes. The difference is "Shoe-to-drum clearance" and should be approximately 0.024 inch (0.6mm). If incorrect, check the parking brake system.

➡**A special brake caliper tool is required to gauge the brake drum inside diameter and shoe-to-drum clearance. However it is not required to perform brake shoe adjustment.**

24. Install the brake drum.

25. Adjust the brake pedal until a slight drag is felt when the drum is spun by hand.

26. Pull the parking lever all the way up until a clicking sound can no longer be heard. Check the clearance between brake shoes and brake drum.

27. Install the rear wheels, tighten the wheel lug nuts and lower the vehicle.

28. Retighten the wheel lug nuts and pump the brake pedal a few times before moving the vehicle. Adjust the rear brakes again if necessary.

29. Check the level of brake fluid in the master cylinder, then, perform a test drive.

30. Connect the negative battery cable to the battery.

BRAKES **18-82**
DRIVE TRAIN **18-60**
ENGINE REPAIR **18-8**
FUEL SYSTEM **18-53**
SPECIFICATIONS AND
  MAINTENANCE CHARTS......**18-2**
Engine and Vehicle
  Identification 18-2
General Engine Specifications 18-2
Engine Tune-Up Specifications 18-2
Firing Order 18-3
Accessory Drive Belt Routing 18-3
Capacities 18-3
Valve Specifications 18-4
Crankshaft and Connecting Rod
  Specifications 18-4
Piston and Ring Specifications 18-4
Torque Specifications 18-5
Wheel Alignment 18-5
Tire, Wheel and Ball Joint
  Specifications 18-6
Brake Specifications 18-6
Scheduled Maintenance
  Intervals..........................18-7
STEERING &
  SUSPENSION.................**18-68**
A
Air Bag..............................18-68
  Disarming..........................18-68
  Precautions..........................18-68
Alternator...........................18-8
  Removal & Installation..............18-8
B
Brake Shoes..........................18-85
  Removal & Installation............18-85
C
Camshaft and Valve Lifters 18-36
  Removal & Installation..............18-36
Coil Spring 18-76
  Removal & Installation..............18-76
CV-Joints............................18-66
  Overhaul 18-66

Cylinder Head.......................18-31
  Removal & Installation............18-31
D
Distributor..........................18-8
E
Engine Assembly 18-8
  Removal & Installation..............18-8
Exhaust Manifold 18-35
  Removal & Installation..............18-35
F
Front Brake Caliper 18-82
  Removal & Installation............18-82
Front Brake Pads.......................18-84
  Removal & Installation............18-84
Front Crankshaft Seal 18-36
  Removal & Installation............18-36
Front Halfshaft 18-64
  Removal & Installation............18-64
Fuel Filter 18-54
  Removal & Installation............18-54
Fuel Injector..........................18-57
  Removal & Installation............18-57
Fuel Pump 18-54
  Removal & Installation............18-54
Fuel System Pressure 18-53
  Relieving..........................18-53
Fuel System Service
  Precautions..........................18-53
H
Heater Core..........................18-21
  Removal & Installation............18-21
I
Ignition Timing 18-8
  Adjustment........................18-8
Intake Manifold 18-34
  Removal & Installation............18-34
L
Lower Ball Joint.......................18-76
  Removal & Installation............18-76
O
Oil Pan.............................18-50
  Removal & Installation............18-50

Oil Pump 18-52
  Removal & Installation............18-52
P
Piston and Rings 18-53
  Positioning 18-53
Power Rack and Pinion Steering
  Gear..............................18-69
  Removal & Installation............18-69
R
Rear Brake Caliper 18-84
  Removal & Installation............18-84
Rear Brake Pads.......................18-84
  Removal & Installation............18-84
Rear Halfshaft 18-66
  Removal & Installation............18-66
Rear Main Seal 18-52
  Removal & Installation............18-52
S
Shock Absorber 18-73
  Removal & Installation............18-73
Starter Motor 18-45
  Removal & Installation............18-45
Strut................................18-71
  Removal & Installation............18-71
T
Timing Belt 18-46
  Removal & Installation............18-46
Transfer Case 18-64
  Removal & Installation............18-64
Transmission Assembly 18-60
  Removal & Installation............18-60
V
Valve Lash 18-44
  Adjustment........................18-44
W
Water Pump 18-28
  Removal & Installation............18-28
Wheel Bearings.......................18-76
  Removal & Installation............18-76

SPECIFICATIONS AND MAINTENANCE CHARTS

ENGINE AND VEHICLE IDENTIFICATION

Engine							Model Year	
Code ①	Liters (cc)	Cu. In.	Cyl.	Fuel Sys.	Engine Type	Eng. Mfg.	Code ②	Year
1MZ-FE	3.0 (2995)	183	6	MFI	DOHC	Toyota	2	2002
3MZ-FE	3.3 (NA)	NA	6	SFI	DOHC	Toyota	3	2003
							4	2004
							5	2005
							6	2006

MFI: Multi-port Fuel Injection

DOHC: Double Overhead Camshaft

① Stamped on the left side of the engine block

② 10th digit of the Vehicle Identification Number (VIN)

09490-SIEN-C0001

GENERAL ENGINE SPECIFICATIONS

Year	Model	Engine Displacement Liters	Engine Series ID	Net Horsepower @ rpm	Net Torque @ rpm (ft. lbs.)	Bore x Stroke (in.)	Compression Ratio	Oil Pressure @ rpm
2002	Sienna	3.0	1MZ-FE	210@5800	220@4400	3.44x3.27	10.5:1	43-78@3000
2003	Sienna	3.0	1MZ-FE	210@5800	220@4400	3.44x3.27	10.5:1	43-78@3000
2004	Sienna	3.3	3MZ-FE	215@5600	222@3600	3.62x3.27	NA	43-78@3000
2005-06	Sienna	3.3	3MZ-FE	215@5600	222@3600	3.62x3.27	NA	43-78@3000

09490-SIEN-C0002

ENGINE TUNE-UP SPECIFICATIONS

Year	Engine Displacement Liters	Engine ID	Spark Plug Gap (in.)	Ignition Timing (deg.)	Fuel Pump (psi)	Idle Speed (rpm)	Valve Clearance Intake	Valve Clearance Exhaust
2002	3.0	1MZ-FE	0.043	10B	38-44	650-750	0.006-0.010	0.010-0.014
2003	3.0	1MZ-FE	0.043	10B	38-44	650-750	0.006-0.010	0.010-0.014
2004	3.3	3MZ-FE	0.039-0.043	①	44-50	550-650	0.006-0.010	0.010-0.014
2005-06	3.3	3MZ-FE	0.039-0.043	①	44-50	550-650	0.006-0.010	0.010-0.014

NOTE: The Vehicle Emission Control Information label often reflects specification changes made during production.

The label figures must be used if they differ from those in this chart.

B: Before top dead center

① With terminal TC and CG of DLC3 connected: 8-12 degrees BTDC

With terminal TC and CG of DLC3 disconnected: 7-24 degrees BTDC

09490-SIEN-C0003

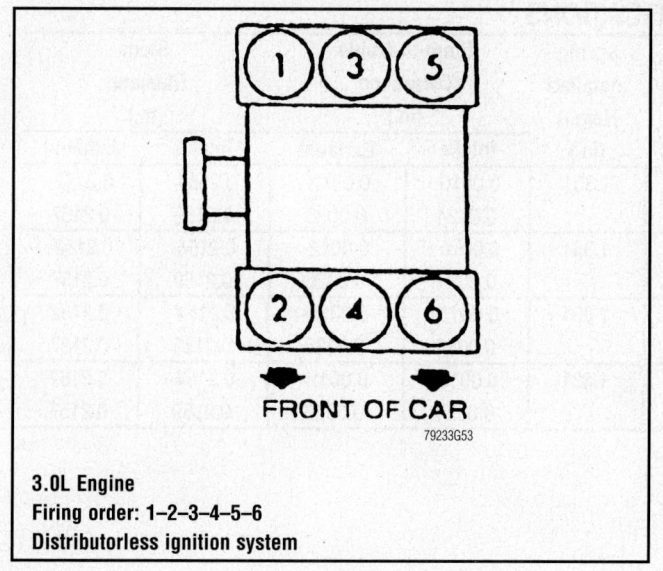

3.0L Engine
Firing order: 1–2–3–4–5–6
Distributorless ignition system

79233G53

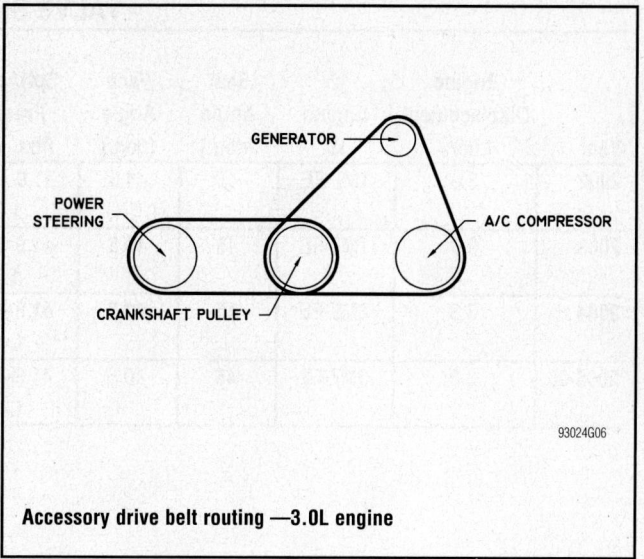

Accessory drive belt routing —3.0L engine

93024G06

CAPACITIES

Year	Model	Engine Displacement Liters	Engine ID	Engine Oil with Filter (qts.)	Transmission (pts.) 5-Spd	Transmission (pts.) Auto.	Transfer Case (pts.)	Drive Axle Front (pts.)	Drive Axle Rear (pts.)	Fuel Tank (gal.)	Cooling System (qts.)
2002	Sienna	3.0	1MZ-FE	5.0	—	10.0	—	—	—	20.9	①
2003	Sienna	3.0	1MZ-FE	5.0	—	10.0	—	—	—	20.9	①
2004	Sienna	3.3	3MZ-FE	5.0	—	②	2.0	—	2.0	20.0	12.4
2005-06	Sienna	3.3	3MZ-FE	5.0	—	②	2.0	—	2.0	20.0	12.4

① w/o rear heater: 10.0
   With rear heater: 11.0
② 2wd: 7.4 pts.
   4wd: 7.6 pts.

09490-SIEN-C0004

VALVE SPECIFICATIONS

Year	Engine Displacement Liters	Engine ID	Seat Angle (deg.)	Face Angle (deg.)	Spring Test Pressure (lbs. @ in.)	Spring Installed Height (in.)	Stem-to-Guide Clearance (in.)		Stem Diameter (in.)	
							Intake	Exhaust	Intake	Exhaust
2002	3.0	1MZ-FE	45	44.5	41.9-46.3@ 1.33	1.331	0.0010-0.0024	0.0012-0.0026	0.2154-0.2159	0.2152-0.2157
2003	3.0	1MZ-FE	45	44.5	41.9-46.3@ 1.33	1.331	0.0010-0.0024	0.0012-0.0026	0.2154-0.2159	0.2152-0.2157
2004	3.3	3MZ-FE	45	40.5	41.9-46.3@ 1.437	1.331	0.0010-0.0024	0.0012-0.0026	0.2154-0.2159	0.2152-0.2157
2005-06	3.3	3MZ-FE	45	40.5	41.9-46.3@ 1.437	1.331	0.0010-0.0024	0.0012-0.0026	0.2154-0.2159	0.2152-0.2157

09490-SIEN-C0005

CRANKSHAFT AND CONNECTING ROD SPECIFICATIONS
All measurements are given in inches.

Year	Engine Displacement Liters	Engine ID	Crankshaft				Connecting Rod		
			Main Brg. Journal Dia.	Main Brg. Oil Clearance	Shaft End-play	Thrust on No.	Journal Diameter	Oil Clearance	Side Clearance
2002	3.0	1MZ-FE	2.4011-2.4016	①	0.0016-0.0095	2	2.0863-2.0866	0.0015-0.0025	0.0059-0.0118
2003	3.0	1MZ-FE	2.4011-2.4016	①	0.0016-0.0095	2	2.0863-2.0866	0.0015-0.0025	0.0059-0.0118
2004	3.3	3MZ-FE	2.4011-2.4016	①	0.0016-0.0095	2	2.0863-2.0866	0.0015-0.0026	0.0059-0.0118
2005-06	3.3	3MZ-FE	2.4011-2.4016	①	0.0016-0.0095	2	2.0863-2.0866	0.0015-0.0026	0.0059-0.0118

① Journals 1 and 4: 0.0006 - 0.0013 in.
  Journals 2 and 3: 0.0010 - 0.0018 in.

09490-SIEN-C0006

PISTON AND RING SPECIFICATIONS
All measurements are given in inches.

Year	Engine Displ. Liters	Engine ID	Piston Clearance	Ring Gap			Ring Side Clearance		
				Top Comp.	Bottom Comp.	Oil Control	Top Comp.	Bottom Comp.	Oil Control
2002	3.0	1MZ-FE	①	0.0098-0.0138	0.0138-0.0177	0.0059-0.0157	0.0008-0.0028	0.0008-0.0024	SNUG
2003	3.0	1MZ-FE	①	0.0098-0.0138	0.0138-0.0177	0.0059-0.0157	0.0008-0.0028	0.0008-0.0024	SNUG
2004	3.3	3MZ-FE	0.0013-0.0023	0.0118-0.0138	0.0197-0.0236	0.0059-0.0157	0.0012-0.0031	0.0008-0.0024	0.0012-0.0043
2005-06	3.3	3MZ-FE	0.0013-0.0023	0.0118-0.0157	0.0197-0.0236	0.0059-0.0157	0.0012-0.0031	0.0008-0.0024	0.0012-0.0043

① AISIN piston: 0.0033-0.0042
  MAHLE piston: 0.0013-0.0023

09490-SIEN-C0007

TORQUE SPECIFICATIONS
All readings in ft. lbs.

Year	Engine Displacement Liters	Engine ID	Cylinder Head Bolts	Main Bearing Bolts	Rod Bearing Bolts	Crankshaft Damper Bolts	Flywheel Bolts	Manifold Intake	Manifold Exhaust	Spark Plugs	Oil Pan Drain Plug
2002	3.0	1MZ-FE	①	②	④	159	61	11	36	13	33
2003	3.0	1MZ-FE	①	②	④	159	61	11	36	13	33
2004	3.3	3MZ-FE	①	②	④	162	61	11	36	18	33
2005-06	3.3	3MZ-FE	①	②	④	162	61	11	36	18	33

① Step 1: 40 ft. lbs.
Step 2: Plus 90 degrees
Recessed bolt: 13 ft. lbs.

② 6-point bolts: 20 ft. lbs.
12-point bolts:
Step 1: 16 ft. lbs.
Step 2: Plus 90 degrees

③ Step 1: 18 ft. lbs.
Step 2: Plus 90 degrees

④ Step 1: 18 ft. lbs.
Step 2: +90 degrees

09490-SIEN-C0008

WHEEL ALIGNMENT

Year	Model		Caster Range (+/-Deg.)	Caster Preferred Setting (Deg.)	Camber Range (+/-Deg.)	Camber Preferred Setting (Deg.)	Toe-in (in.)	Inside Wheel Angle (Deg.)
2002	Sienna	F	0.75	1.53	0.75	-0.50	0.10+/-0.08	34.32
		R	—	—	0.75	-0.92	0.09+/-0.12	—
2003	Sienna	F	0.75	1.53	0.75	-0.50	0.10+/-0.08	34.32
		R	—	—	0.75	-0.92	0.09+/-0.12	—
2004	Sienna FWD	F	0.75	2.80	0.75	-0.28	0+/-0.08	42.75
		R	—	—	0.50	-1.37	0.11+/-0.12	—
	4WD	F	0.75	2.67	0.75	-0.28	0+/-0.08	42.7
		R	—	—	0.50	-1.42	0.05+/-0.12	—
2005-06	Sienna FWD	F	0.75	2.80	0.75	-0.28	0+/-0.08	42.75
		R	—	—	0.50	-1.37	0.11+/-0.12	—
	4WD	F	0.75	2.67	0.75	-0.28	0+/-0.08	42.7
		R	—	—	0.50	-1.42	0.05+/-0.12	—

09490-SIEN-C0009

TIRE, WHEEL AND BALL JOINT SPECIFICATIONS

Year	Model	OEM Tires		Tire Pressures (psi)		Wheel Size	Ball Joint Inspection	Lugnut Torque (ft. lbs.)
		Standard	Optional	Front	Rear			
2002	Sienna CE	P205/70R15 95S	none	35	35	6.5	①	76
	Sienna LE	P205/70R15 95S	P215/65R15 95S	35②	35②	6.5	①	76
	Sienna XLE	P215/65R15 95S	none	35②	35②	6.5	①	76
2003	Sienna CE	P205/70R15 95S	none	35	35	6.5	①	76
	Sienna LE	P205/70R15 95S	P215/65R15 95S	35②	35②	6.5	①	76
	Sienna XLE	P215/65R15 95S	none	35②	35②	6.5	①	76
2004	Sienna	P215/65R16 96T	P225/60R17 98T	35	35	6.5	③	76
2005-06	Sienna	P215/65R16 96T	P225/60R17 98T	35	35	6.5	③	76

OEM: Original Equipment Manufacturer

NA: Information not available

PSI: Pounds Per Square Inch

① Ball joint turning torque should be 30 inch lbs.

② P215/65R15
   Up to 6 passengers: 32
   Trailer towing or up to vehicle capacity weight: 35

③ Ball joint turning torque should be 8.7-30 inch lbs.

09490-SIEN-C0010

BRAKE SPECIFICATIONS
All measurements in inches unless noted

Year	Model		Brake Disc			Brake Drum Diameter		Minimum Lining Thickness	Brake Caliper	
			Original Thickness	Minimum Thickness	Maximum Runout	Original Inside Diameter	Maximum Machine Diameter		Bracket Bolts (ft. lbs.)	Mounting Bolts (ft. lbs.)
2002	Sienna	F	1.102	1.024	0.0020	—	—	0.039	79	25
		R	—	—	—	9.84	9.921	0.039	—	—
2003	Sienna	F	1.102	1.024	0.0020	—	—	0.039	79	25
		R	—	—	—	9.84	9.921	0.039	—	—
2004	Sienna	F	1.102	1.024	0.0020	—	—	0.039	79	25
		R	0.472	0.413	0.0039	10.00	10.08	0.039	65	25
2005-06	Sienna	F	1.102	1.024	0.0020	—	—	0.039	79	25
		R	0.472	0.413	0.0039	10.00	10.08	0.039	65	25

F: Front

R: Rear

09490-SIEN-C0011

SCHEDULED MAINTENANCE INTERVALS
TOYOTA—SIENNA

TO BE SERVICED	TYPE OF SERVICE	VEHICLE MILEAGE INTERVAL (x1000)																		
		5	10	15	20	25	30	35	40	45	50	55	60	65	70	75	80	85	90	95
Automatic transmission and differential fluid	S/I			✓			✓			✓			✓			✓			✓	
Ball joints and boots	S/I			✓			✓			✓			✓			✓			✓	
Brake linings, discs/drums, lines & hoses	S/I			✓			✓			✓			✓			✓			✓	
Charcoal canister	S/I												✓							
Drive belts	S/I						✓						✓						✓	
Engine coolant	R						✓						✓						✓	
Engine oil & filter	R	✓	✓	✓	✓	✓	✓	✓	✓	✓	✓	✓	✓	✓	✓	✓	✓	✓	✓	✓
Exhaust pipes & mounts	S/I			✓			✓			✓			✓			✓			✓	
Fuel lines & connections, fuel tank vapor vent system hoses, fuel tank band	S/I						✓						✓						✓	
Fuel tank cap gasket	S/I						✓						✓						✓	
Halfshaft boots & flange bolts	S/I			✓			✓			✓			✓			✓			✓	
Non-platinum spark plugs	R						✓						✓						✓	
Platinum spark plugs	R												✓							
Rack and pinion assembly	S/I			✓			✓			✓			✓			✓			✓	
Steering linkage	S/I			✓			✓			✓			✓			✓			✓	
Valves	S/I												✓							

R: Replace   S/I: Service or Inspect   L: Lubricate

FREQUENT OPERATION MAINTENANCE (SEVERE SERVICE)

If a vehicle is operated under any of the following conditions it is considered severe service:

- Towing a trailer or using a camper or car-top carrier.

- Repeated short trips of less than 5 miles in temperatures below freezing.

- Excessive idling or low-speed driving for long distances as in heavy commercial use, such as delivery, taxi or police cars.

- Operating on rough, muddy or salt-covered roads.

- Operating on unpaved or dusty roads.

Oil filter: service or inspect every 5000 miles or 4 months, whichever occurs first.

Brake linings and discs or drums: service or inspect every 5000 miles or 4 months, whichever occurs first.

Steering linkage: service or inspect every 5000 miles or 4 months, whichever occurs first.

Ball joints and boots: service or inspect every 5000 miles or 4 months, whichever occurs first.

Brake discs & pads (front): service or inspect every 6000 miles.

Halfshaft boots: service or inspect every 5000 miles or 4 months. Retighten the flange bolts, whichever occurs first.

Body chassis bolts and nuts: service or inspect every 5000 miles or 4 months, whichever occurs first.

Transmission and differential fluid: replace every 15,000 miles or 12 months, whichever occurs first.

Timing belt: replace every 60,000 miles or 48 months, whichever occurs first.

09490-SIEN-C0012

ENGINE REPAIR

Distributor

Sienna models are equipped with a distibutorless ignition system.

Alternator

REMOVAL AND INSTALLATION

3.0L Engine

1. Before servicing the vehicle, refer to the precautions section.
2. Remove or disconnect the following:
   - Alternator electrical connectors
   - Wiring harness from the clip
   - Pivot bolt
   - Plate washer
   - Adjusting lockbolt
   - Drive belt
   - Alternator

To install:

3. Install or connect the following:
   - Alternator
   - Drive belt. Tension the belt to 170–180 lbs. for a new belt or 95–135 lbs. for a used belt.
   - Adjusting lockbolt. Tighten the bolt to 13 ft. lbs. (18 Nm).
   - Plate washer
   - Pivot bolt. Tighten the bolt to 41 ft. lbs. (56 Nm).
   - Wiring harness from the clip
   - Alternator electrical connectors

3.3L Engine

1. Before servicing the vehicle, refer to the precautions section.
2. Remove or disconnect the following:
   - Alternator electrical connectors
   - Wiring harness from the clip
   - Pivot bolt
   - Plate washer
   - Adjusting lockbolt
   - Drive belt
   - Alternator

To install:

3. Install or connect the following:
   - Alternator
   - Drive belt. Tension the belt to 170–180 lbs. for a new belt or 95–135 lbs. for a used belt.
   - Adjusting lockbolt. Tighten the bolt to 13 ft. lbs. (18 Nm).
   - Plate washer
   - Pivot bolt. Tighten the bolt to 43 ft. lbs. (58 Nm).
   - Wiring harness from the clip
   - Alternator electrical connectors

Ignition Timing

ADJUSTMENT

Ignition timing is controlled by the ECM and is not adjustable.

Engine Assembly

REMOVAL & INSTALLATION

3.0L Engine

1. Before servicing the vehicle, refer to the precautions section.
2. Matchmark the hood position.
3. Remove or disconnect the following:
   - Hood
   - Wiper and blade assembly
   - Top cowl seal and panel
   - Window washer hoses from the ventilator louvers
   - Left and right ventilator louvers
   - Heater air duct
4. Properly relieve the fuel system pressure.
5. Remove or disconnect the following:
   - Both battery cables
   - Battery and tray
6. Drain the engine coolant.
7. Drain the engine oil.
8. Remove or disconnect the following:
   - Intake air cleaner and case assembly
   - Cruise control actuator, if equipped
   - Upper and lower radiator hoses
   - Radiator
   - Automatic transmission oil cooler lines
   - Any connectors, hoses and sensors that would interfere with engine removal
   - Engine Control Module (ECM) engine wiring harness from inside the glove box; then, pull the harness into the engine compartment
   - Compressor

➡ **It may be necessary to remove the air conditioning compressor lines in order to remove the engine.**

   - Automatic transmission shifter cable from the transaxle
   - Header pipes from the exhaust manifolds
   - Left and right fender apron seals
   - Halfshafts
   - Stabilizer links and the steering intermediate shaft

   - Power steering pump
   - Engine undercover
   - Engine hanger to the engine
   - Engine sling device to the engine hangers
   - Right-hand motor mount and moving control rod
   - Front suspension lower braces
9. Lower the engine, transaxle and front suspension member as an assembly from the vehicle.

To install:

10. Raise the engine, transaxle and front suspension member as an assembly into the vehicle.
11. Install the front suspension lower braces, and tighten the fasteners, as follows:
    - Bolt A: 134 ft. lbs. (181 Nm)
    - Bolt B: 24 ft. lbs. (32 Nm)
    - Nut C: 27 ft. lbs. (36 Nm)
12. Install or connect the following:
    - Moving control rod. Tighten the bolts to 47 ft. lbs. (64 Nm).
    - Right-hand motor mount. Tighten the bolts to 23 ft. lbs. (32 Nm).
    - Engine sling device from the engine hangers
    - Engine undercover
    - Power steering pump hoses
    - Stabilizer links and the steering intermediate shaft
    - Halfshafts
    - Left and right fender apron seals
    - Header pipes to the exhaust manifolds
    - Automatic transmission shifter cable to the transaxle
    - Air conditioning compressor to the engine
13. Push the wiring harness into the glove box.
14. Install or connect the following:
    - ECM
    - Any connectors, hoses and sensors that were removed
    - Automatic transmission oil cooler lines
    - Upper and lower radiator hoses and fit the radiator
    - Cruise control actuator, if removed
    - Intake air cleaner and case assembly
15. Fill the engine oil to proper level.
16. Fill the engine with coolant.
17. Install or connect the following:
    - Battery tray and battery
    - Battery cables
    - Heater air duct
    - Left and right ventilator louvers

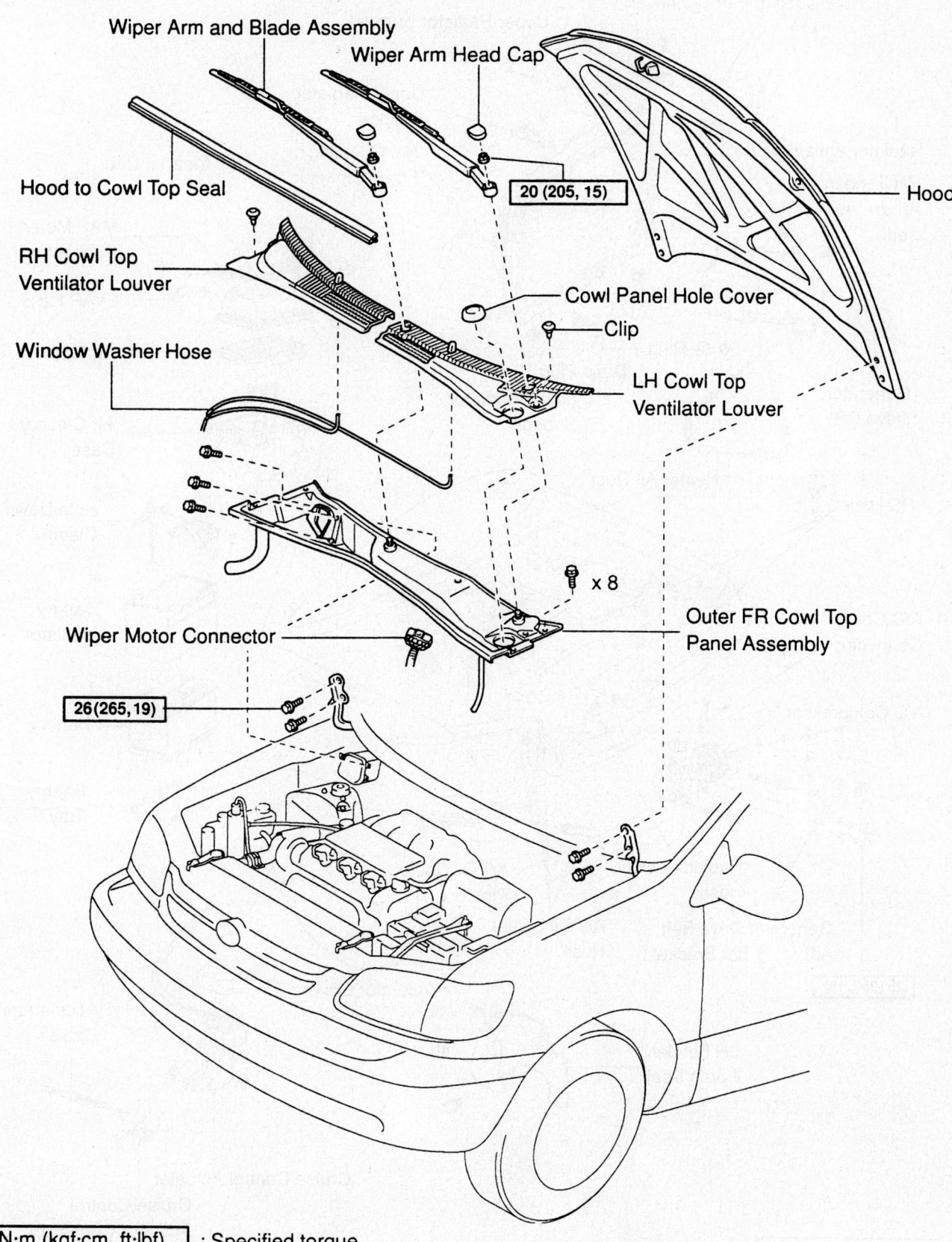

Wiper Arm and Blade Assembly

Wiper Arm Head Cap

Hood to Cowl Top Seal

20 (205, 15)

Hood

RH Cowl Top Ventilator Louver

Cowl Panel Hole Cover

Clip

Window Washer Hose

LH Cowl Top Ventilator Louver

x 8

Wiper Motor Connector

Outer FR Cowl Top Panel Assembly

26 (265, 19)

N·m (kgf·cm, ft·lbf) : Specified torque

7924ZG06

Exploded view of the top cowl and related components—2002–03

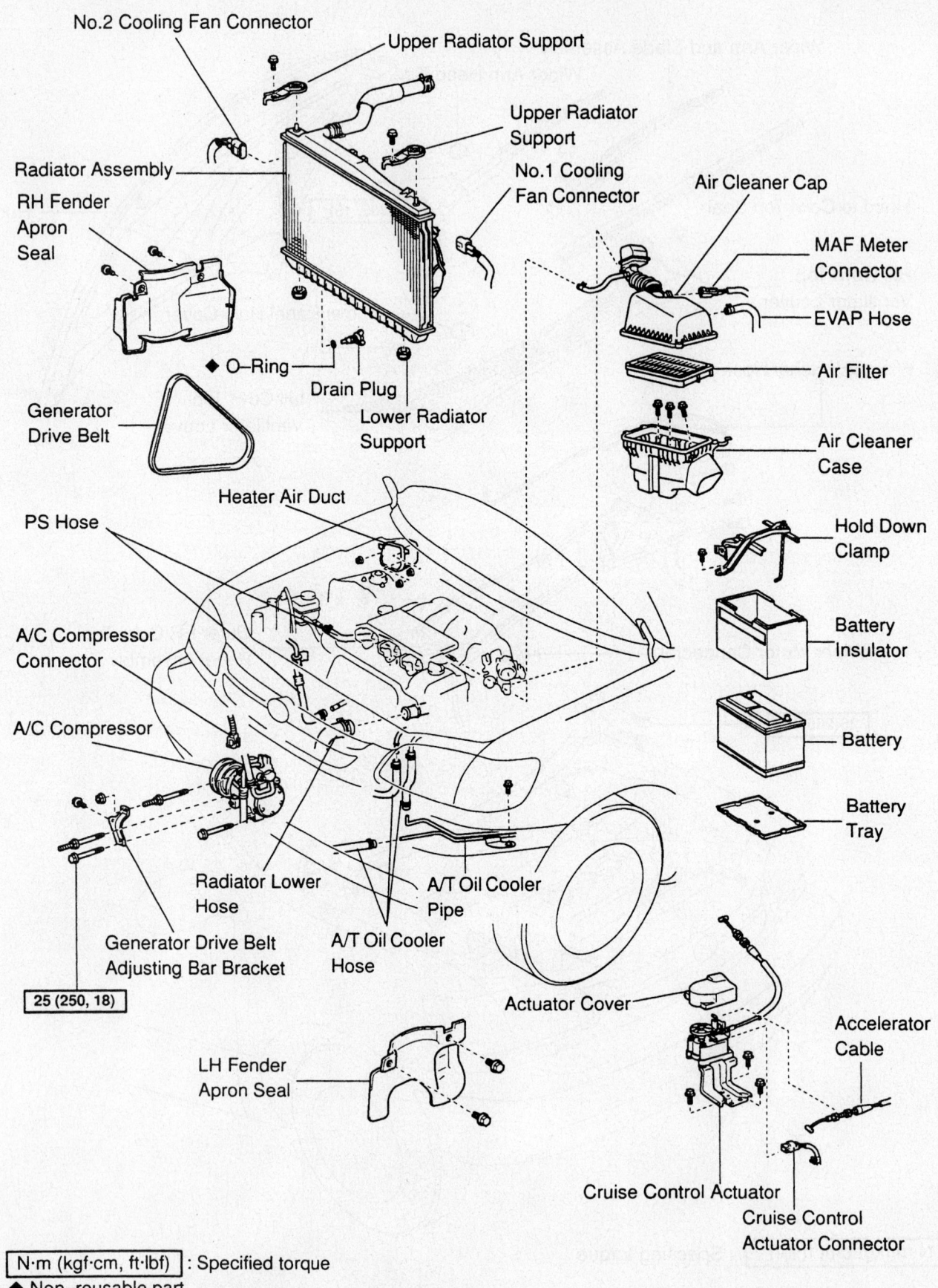

No.2 Cooling Fan Connector

Upper Radiator Support

Upper Radiator Support

No.1 Cooling Fan Connector

Air Cleaner Cap

MAF Meter Connector

EVAP Hose

Air Filter

Air Cleaner Case

Radiator Assembly

RH Fender Apron Seal

◆ O-Ring

Drain Plug

Lower Radiator Support

Generator Drive Belt

Heater Air Duct

PS Hose

A/C Compressor Connector

A/C Compressor

Radiator Lower Hose

Generator Drive Belt Adjusting Bar Bracket

A/T Oil Cooler Pipe

A/T Oil Cooler Hose

25 (250, 18)

LH Fender Apron Seal

Hold Down Clamp

Battery Insulator

Battery

Battery Tray

Actuator Cover

Accelerator Cable

Cruise Control Actuator

Cruise Control Actuator Connector

N·m (kgf·cm, ft·lbf) : Specified torque

◆ Non-reusable part

Exploded view of engine pre-removal components—3.0L engine

7924ZG07

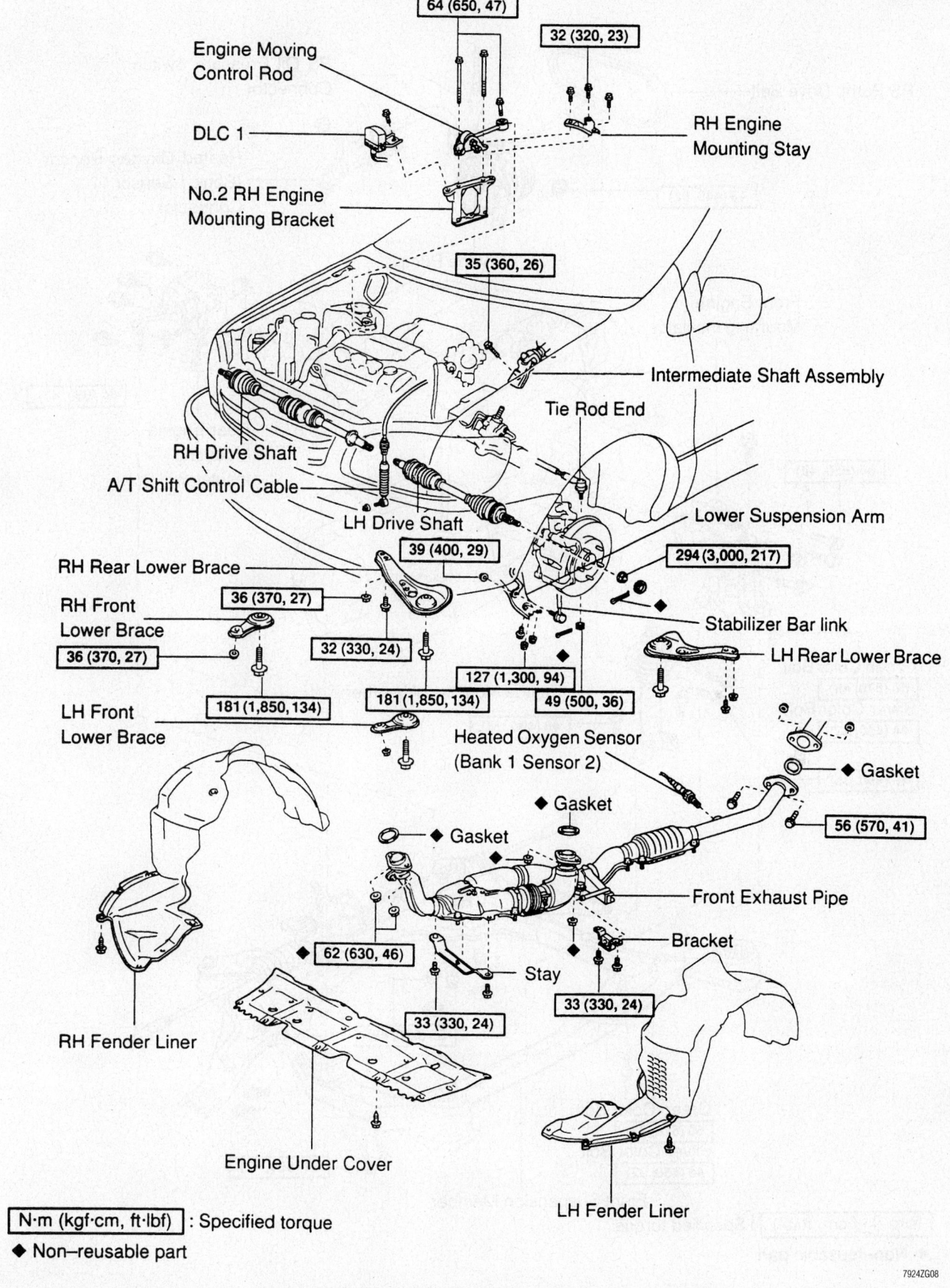

64 (650, 47)

32 (320, 23)

Engine Moving
Control Rod

DLC 1

RH Engine
Mounting Stay

No.2 RH Engine
Mounting Bracket

35 (360, 26)

Intermediate Shaft Assembly

Tie Rod End

RH Drive Shaft

A/T Shift Control Cable

LH Drive Shaft

Lower Suspension Arm

294 (3,000, 217)

RH Rear Lower Brace

39 (400, 29)

RH Front
Lower Brace

36 (370, 27)

Stabilizer Bar link

36 (370, 27)

32 (330, 24)

LH Rear Lower Brace

127 (1,300, 94)

LH Front
Lower Brace

181 (1,850, 134)

181 (1,850, 134)

49 (500, 36)

Heated Oxygen Sensor
(Bank 1 Sensor 2)

◆ Gasket

56 (570, 41)

◆ Gasket

◆ Gasket

Front Exhaust Pipe

RH Fender Liner

◆ **62 (630, 46)**

Bracket

Stay

33 (330, 24)

33 (330, 24)

Engine Under Cover

LH Fender Liner

N·m (kgf·cm, ft·lbf) : Specified torque

◆ Non-reusable part

7924ZG08

Exploded view of engine removal and installation tightening specifications of the related components—3.0L engine

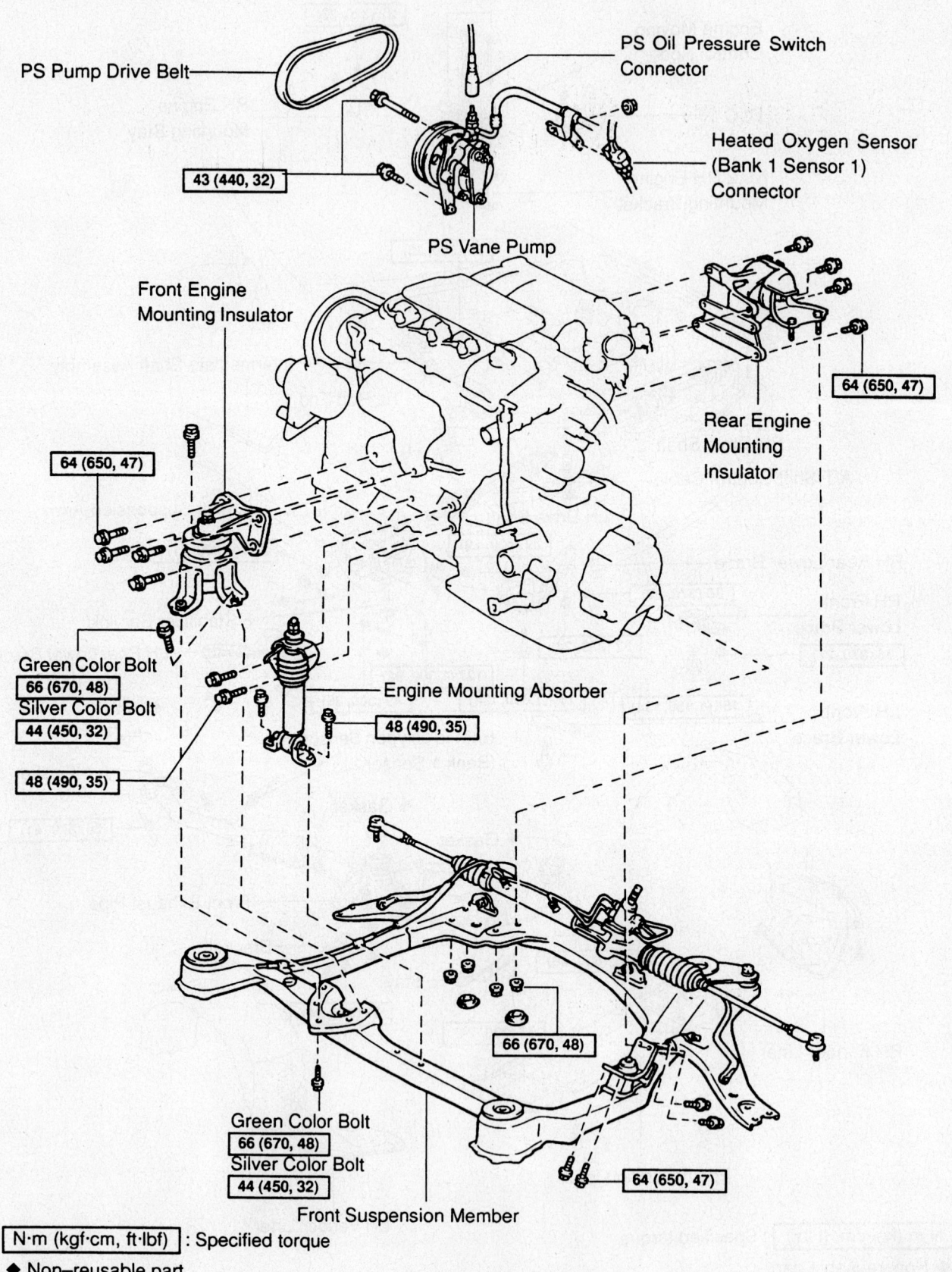

PS Pump Drive Belt

PS Oil Pressure Switch
Connector

Heated Oxygen Sensor
(Bank 1 Sensor 1)
Connector

43 (440, 32)

PS Vane Pump

Front Engine
Mounting Insulator

Rear Engine
Mounting
Insulator

64 (650, 47)

64 (650, 47)

Green Color Bolt
66 (670, 48)
Silver Color Bolt
44 (450, 32)

48 (490, 35)

Engine Mounting Absorber

48 (490, 35)

66 (670, 48)

Green Color Bolt
66 (670, 48)
Silver Color Bolt
44 (450, 32)

64 (650, 47)

Front Suspension Member

N·m (kgf·cm, ft·lbf) : Specified torque

◆ Non–reusable part

7924ZG09

Exploded view of the suspension component removal and installation for engine removal—3.0L engine

- Window washer hoses from the ventilator louvers
- Top cowl seal and panel
- Wiper and blade assembly
- Hood
- New oil filter
18. Refill the engine with oil.

19. Refill the engine with engine coolant.
20. Install the engine undercovers.
21. Start the engine and check for leaks.

3.3L Engine

1. Before servicing the vehicle, refer to the precautions section.

2. Properly relieve the fuel system pressure.
3. Disconnect the negative battery cable.
4. Remove the right hand front wheel.
5. Remove the engine under cover.
6. Remove the fender liners.

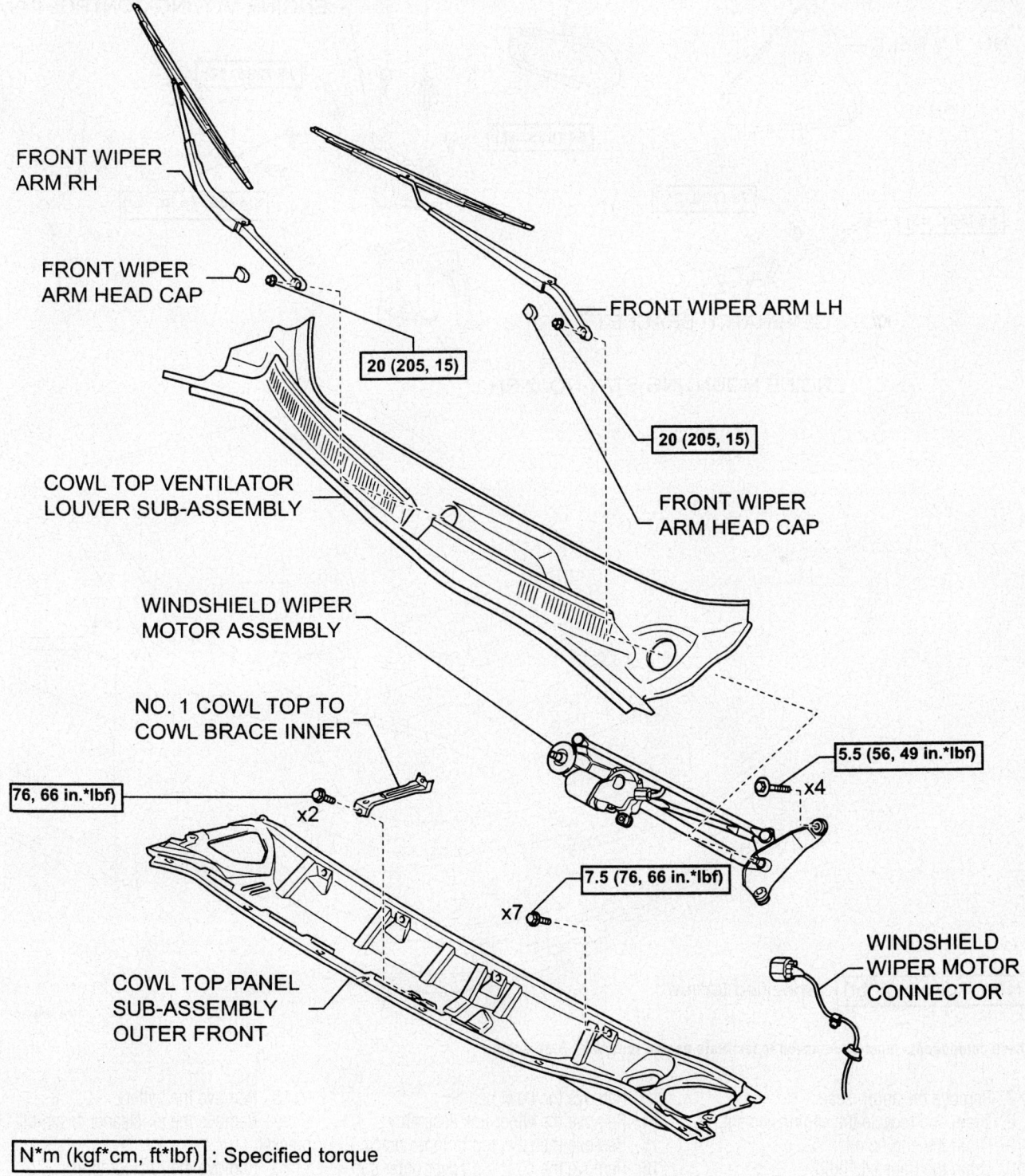

FRONT WIPER ARM RH

FRONT WIPER ARM HEAD CAP

FRONT WIPER ARM LH

20 (205, 15)

20 (205, 15)

FRONT WIPER ARM HEAD CAP

COWL TOP VENTILATOR LOUVER SUB-ASSEMBLY

WINDSHIELD WIPER MOTOR ASSEMBLY

5.5 (56, 49 in.*lbf)

NO. 1 COWL TOP TO COWL BRACE INNER

76, 66 in.*lbf

x2

x4

7.5 (76, 66 in.*lbf)

x7

WINDSHIELD WIPER MOTOR CONNECTOR

COWL TOP PANEL SUB-ASSEMBLY OUTER FRONT

N*m (kgf*cm, ft*lbf) : Specified torque

09490-SIENA-G0008

Locations of cowl panel and wiper assembly components—3.3L engine

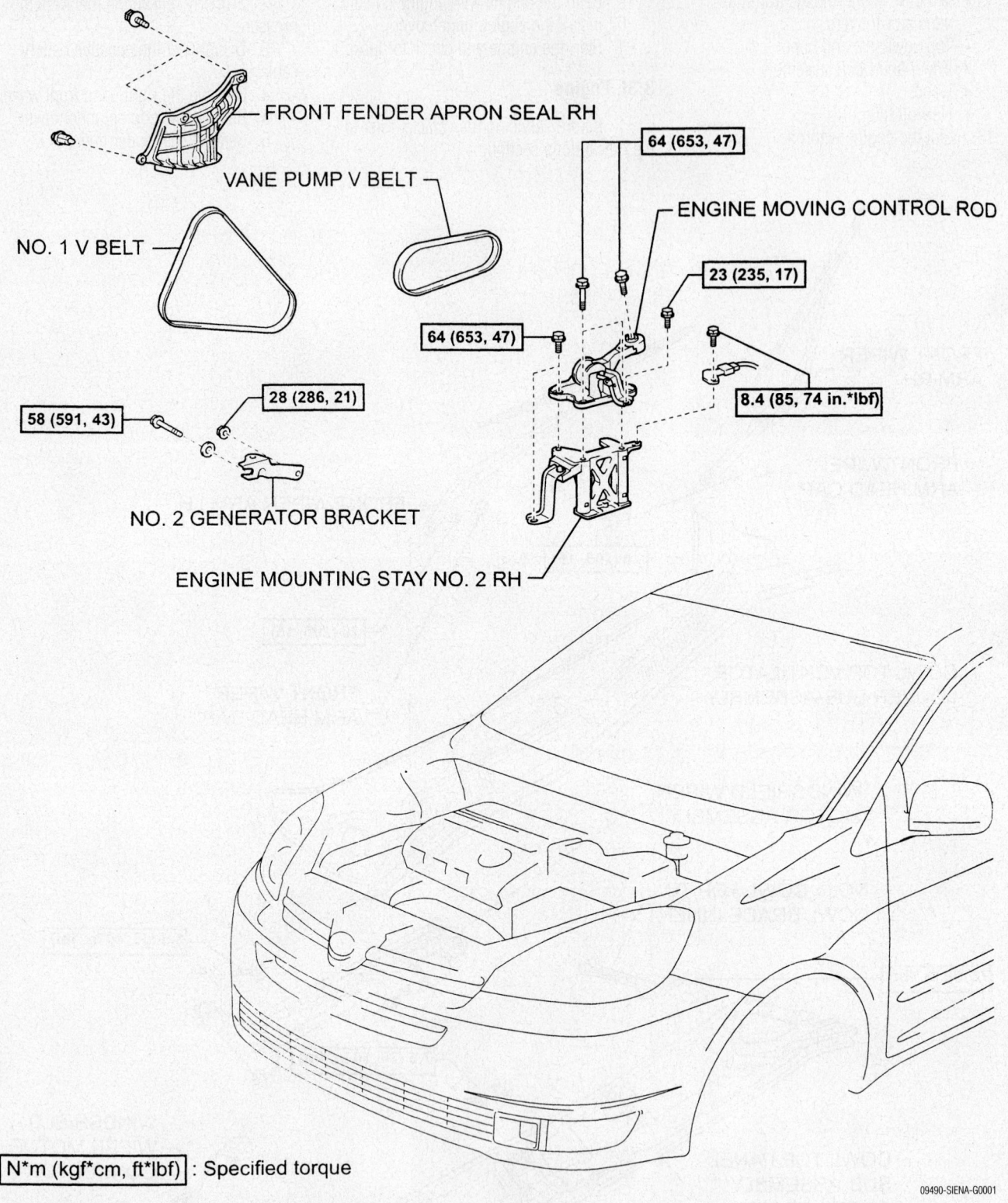

FRONT FENDER APRON SEAL RH

64 (653, 47)

VANE PUMP V BELT

ENGINE MOVING CONTROL ROD

NO. 1 V BELT

23 (235, 17)

64 (653, 47)

8.4 (85, 74 in.*lbf)

58 (591, 43)

28 (286, 21)

NO. 2 GENERATOR BRACKET

ENGINE MOUNTING STAY NO. 2 RH

N*m (kgf*cm, ft*lbf) : Specified torque

09490-SIENA-G0001

These components must be removed to facilitate engine removal—3.3L engine

7. Remove the apron seals.
8. Drain and recycle the engine coolant.
9. Drain the engine oil.
10. Drain the transaxle fluid.
11. If equipped, drain the transfer oil.
12. Remove the wiper arms.

13. Remove the cowl panel.
14. Remove the wiper link assembly.
15. Remove the cowl top to inner brace.
16. Remove the cowl top cover outer sub assembly.
17. Remove the engine cover.

18. Remove the battery.
19. Remove the air cleaner assembly and bracket.
20. Remove the A/C compressor to crankshaft pulley belt.
21. Remove the alternator.

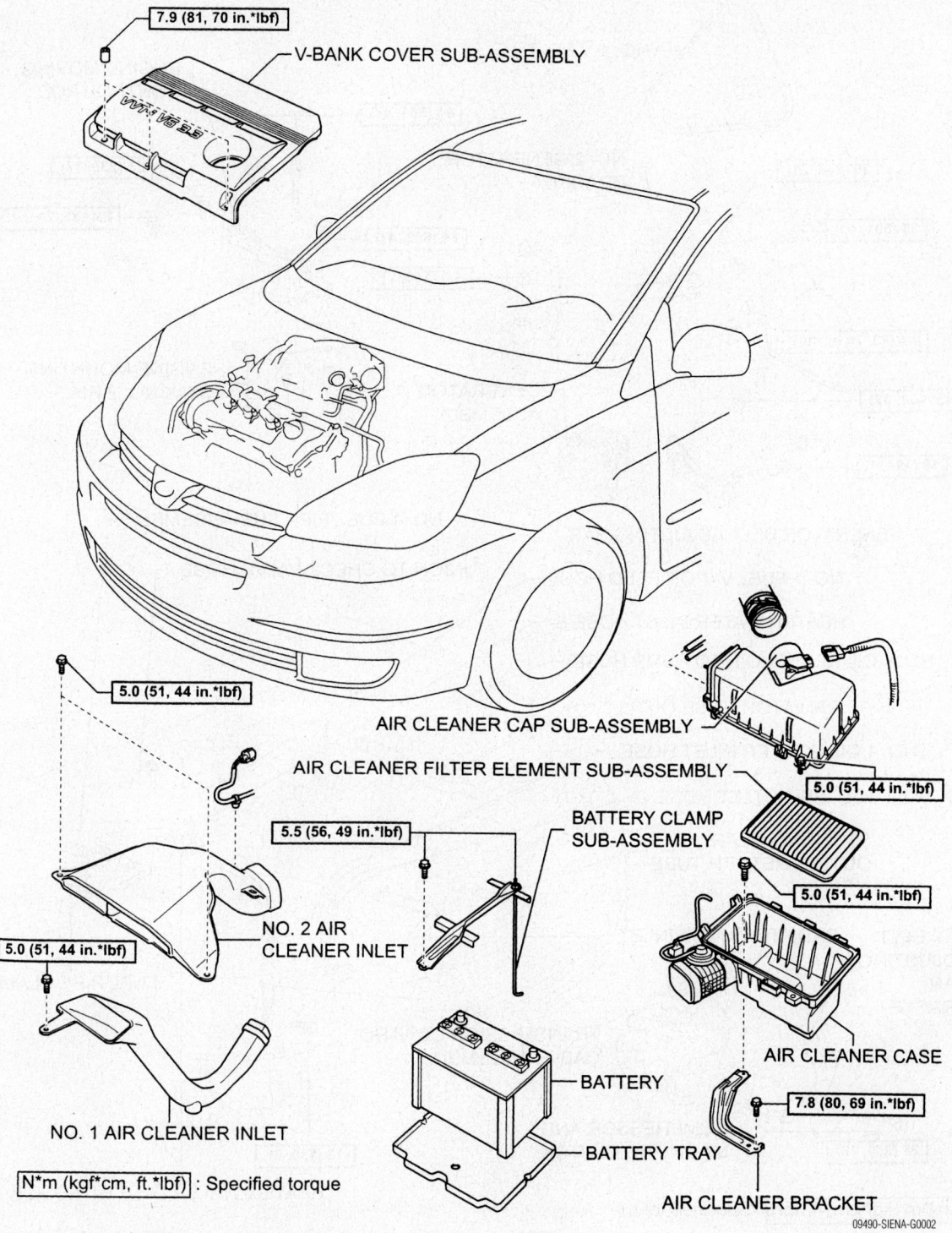

7.9 (81, 70 in.*lbf)

V-BANK COVER SUB-ASSEMBLY

5.0 (51, 44 in.*lbf)

AIR CLEANER CAP SUB-ASSEMBLY

AIR CLEANER FILTER ELEMENT SUB-ASSEMBLY

5.0 (51, 44 in.*lbf)

5.5 (56, 49 in.*lbf)

BATTERY CLAMP SUB-ASSEMBLY

5.0 (51, 44 in.*lbf)

NO. 2 AIR CLEANER INLET

5.0 (51, 44 in.*lbf)

AIR CLEANER CASE

7.8 (80, 69 in.*lbf)

NO. 1 AIR CLEANER INLET

BATTERY

BATTERY TRAY

AIR CLEANER BRACKET

N*m (kgf*cm, ft.*lbf) : Specified torque

09490-SIENA-G0002

Location of the air cleaner and battery assembly components—3.3L engine

22. Remove the engine moving control rod bolts and rod.

23. Remove the number 2 engine mounting stay as follows:

   a. Remove the bolt and wire harness bracket.

   b. Remove the bolt, stay and bracket.

24. Remove the number 2 alternator bracket and the alternator belt adjusting bar.

25. Separate the compressor and magnetic clutch.

26. Separate the transmission control assembly as follows:

   a. Remove the nut from the control shaft lever and disconnect the control cable assembly from the lever.

   b. Remove the clip and disconnect the cable from the bracket.

   c. Disconnect the cable assembly from the transmission bracket.

   d. Remove the nuts and remove the cable control assembly.

27. Disconnect the following hoses:
- Union to check valve hose
- Number 1 fuel vapor feed hose
- Heater hoses
- Radiator hoses

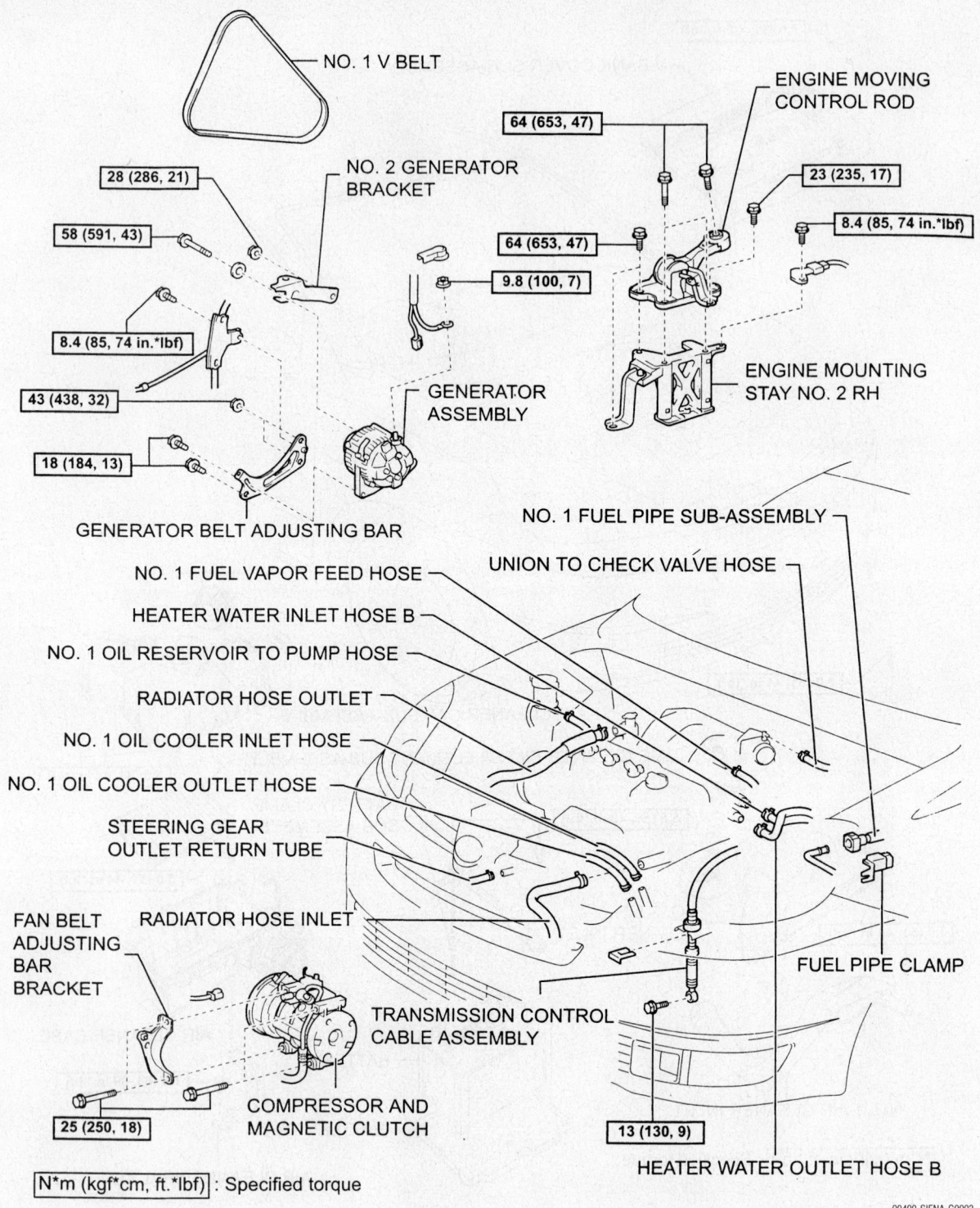

NO. 1 V BELT

ENGINE MOVING CONTROL ROD

64 (653, 47)

NO. 2 GENERATOR BRACKET

28 (286, 21)

23 (235, 17)

58 (591, 43)

8.4 (85, 74 in.*lbf)

64 (653, 47)

9.8 (100, 7)

8.4 (85, 74 in.*lbf)

43 (438, 32)

GENERATOR ASSEMBLY

ENGINE MOUNTING STAY NO. 2 RH

18 (184, 13)

GENERATOR BELT ADJUSTING BAR

NO. 1 FUEL PIPE SUB-ASSEMBLY

NO. 1 FUEL VAPOR FEED HOSE

UNION TO CHECK VALVE HOSE

HEATER WATER INLET HOSE B

NO. 1 OIL RESERVOIR TO PUMP HOSE

RADIATOR HOSE OUTLET

NO. 1 OIL COOLER INLET HOSE

NO. 1 OIL COOLER OUTLET HOSE

STEERING GEAR OUTLET RETURN TUBE

FAN BELT ADJUSTING BAR BRACKET

RADIATOR HOSE INLET

FUEL PIPE CLAMP

TRANSMISSION CONTROL CABLE ASSEMBLY

COMPRESSOR AND MAGNETIC CLUTCH

25 (250, 18)

13 (130, 9)

HEATER WATER OUTLET HOSE B

N*m (kgf*cm, ft.*lbf) : Specified torque

09490-SIENA-G0003

These hoses, brackets and related components must be removed to facilitate engine removal—3.3L engine

- Oil cooler hoses
- Oil reservoir to pump hose

28. Disconnect the steering gear outlet return tube.

29. Remove the glove box door.

30. Disconnect the harness from the PCM.

31. Disconnect the harness from the junction block by removing the nut, separat-ing the harness, then using a screwdriver release the junction block and pull the wire upwards. Pull out the harness and remove the body ground.

32. On 4 wheel drive models, remove the propeller shaft.

33. On 4 wheel drive models, remove the center exhaust pipe assembly.

34. On 2 wheel drive models, remove the front exhaust pipe assembly.

35. Disconnect the front stabilizer link on both sides.

36. Remove the axle hub nut from both sides.

37. Disconnect the front speed sensor wire.

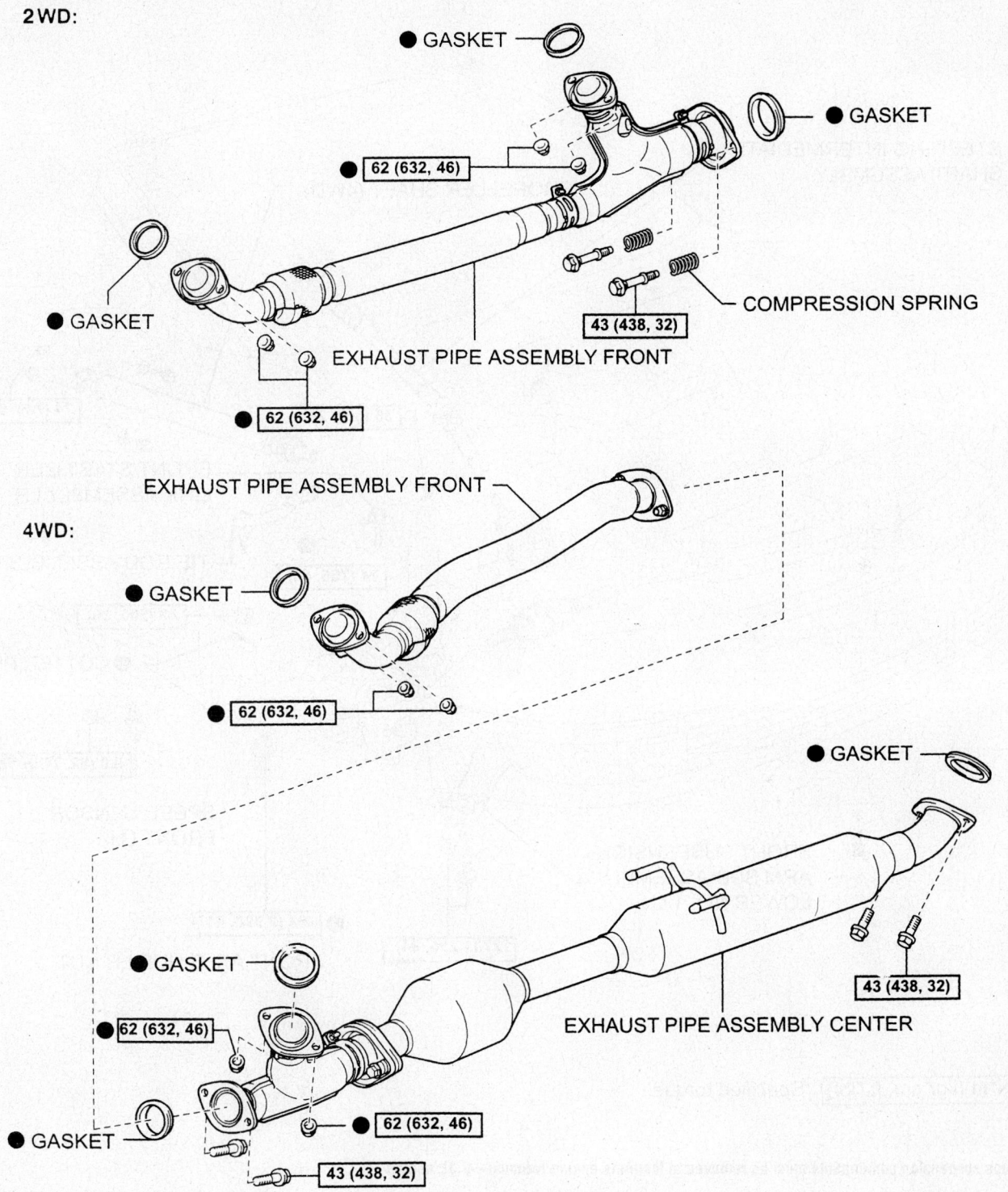

2WD:

● GASKET

● GASKET

62 (632, 46)

● GASKET

COMPRESSION SPRING

43 (438, 32)

EXHAUST PIPE ASSEMBLY FRONT

● 62 (632, 46)

EXHAUST PIPE ASSEMBLY FRONT

4WD:

● GASKET

● 62 (632, 46)

● GASKET

● GASKET

● 62 (632, 46)

EXHAUST PIPE ASSEMBLY CENTER

43 (438, 32)

● 62 (632, 46)

● GASKET

43 (438, 32)

09490-SIENA-G0004

Location of 2 wheel drive model front exhaust assembly and 4 wheel drive model center exhaust assembly—3.3L engine

38. Disconnect the tie rod end on both sides.

39. Disconnect the front suspension control arm assembly on both sides.

40. Remove the front half shafts.

41. Position the wheels in the straight ahead position, remove the dust cover from the steering intermediate shaft. make match marks on the shaft and steering gear assembly and remove the intermediate shaft bolt and separate the shaft from the gear.

42. Install a dolly or suitable tool to support the engine.

43. Remove the bolts and nuts from the frame side plate on both sides.

44. Remove the bolts and nuts from the front suspension member rear brace on both sides.

45. Lower the dolly slowly to remove the engine and transmission assembly from the vehicle.

To install:

46. Installation is the reverse of removal, please note the following:

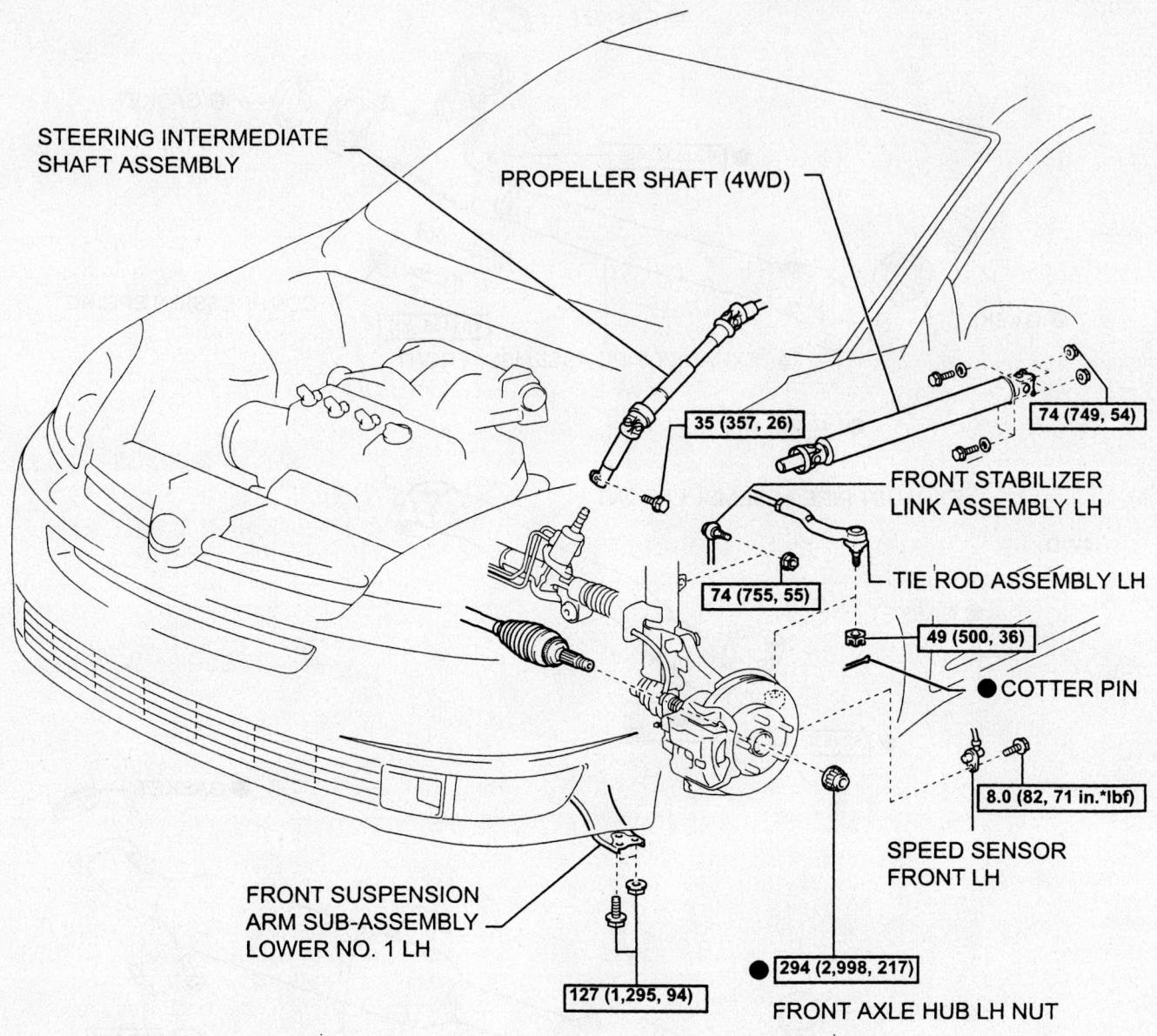

STEERING INTERMEDIATE SHAFT ASSEMBLY

PROPELLER SHAFT (4WD)

35 (357, 26)

74 (749, 54)

FRONT STABILIZER LINK ASSEMBLY LH

TIE ROD ASSEMBLY LH

74 (755, 55)

49 (500, 36)

● COTTER PIN

8.0 (82, 71 in.*lbf)

SPEED SENSOR FRONT LH

FRONT SUSPENSION ARM SUB-ASSEMBLY LOWER NO. 1 LH

● 294 (2,998, 217)

FRONT AXLE HUB LH NUT

127 (1,295, 94)

N*m (kgf*cm, ft.*lbf) : Specified torque

09490-SIENA-G0005

These suspension components must be removed to facilitate engine removal—3.3L engine

a. If removed after engine removal, tighten the alternator bracket to 43 ft. lbs. (58 Nm).

b. If removed after engine removal, tighten the pump bracket to 24 ft. lbs. (32 Nm).

c. If removed after engine removal, tighten the right hand mounting bracket to 40 ft. lbs. (54 Nm).

d. If removed after engine removal, tighten the number 2 manifold stay to 36 ft. lbs. (49 Nm).

e. If removed after engine removal, tighten the wheel shaft bearing bracket to 47 ft. lbs. (64 Nm).

f. If removed after engine removal, tighten the rear engine mounting bracket on 4 wheel drive models to 47 ft. lbs. (64 Nm).

g. Attach the transmission to the engine, refer to the transmission removal and installation procedure for bolt location and torque specifications.

h. If removed after engine removal, tighten the front engine mounting bracket to 47 ft. lbs. (64 Nm).

i. If removed after engine removal, tighten the front frame assembly mounting insulator to 70 ft. lbs. (95 Nm) and the insulator front nut to 64 ft. lbs. (87

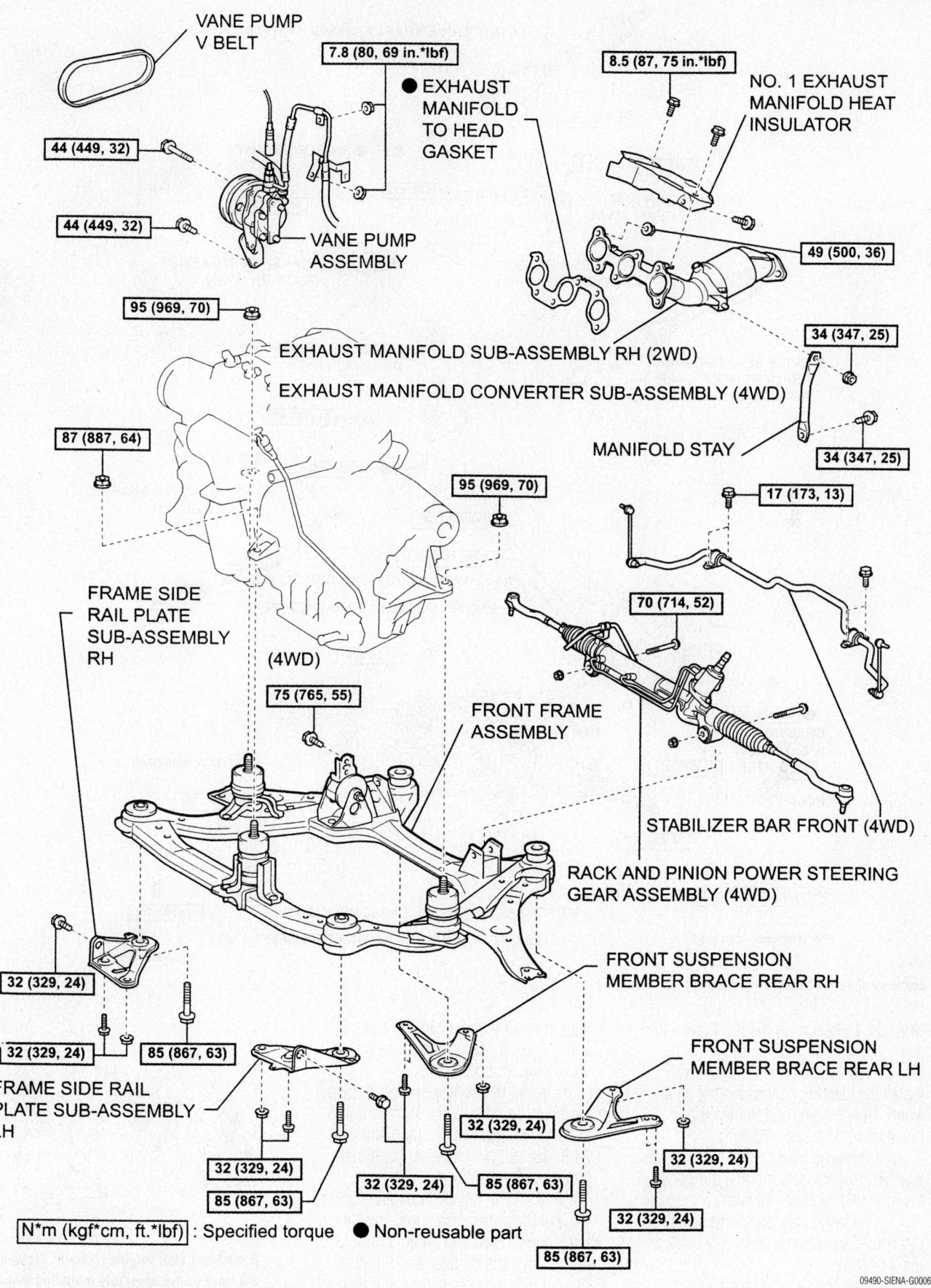

VANE PUMP
V BELT

7.8 (80, 69 in.*lbf)

● EXHAUST
MANIFOLD
TO HEAD
GASKET

8.5 (87, 75 in.*lbf)

NO. 1 EXHAUST
MANIFOLD HEAT
INSULATOR

44 (449, 32)

44 (449, 32)

VANE PUMP
ASSEMBLY

49 (500, 36)

95 (969, 70)

EXHAUST MANIFOLD SUB-ASSEMBLY RH (2WD)

EXHAUST MANIFOLD CONVERTER SUB-ASSEMBLY (4WD)

34 (347, 25)

87 (887, 64)

MANIFOLD STAY

34 (347, 25)

95 (969, 70)

17 (173, 13)

FRAME SIDE
RAIL PLATE
SUB-ASSEMBLY
RH

70 (714, 52)

(4WD)

75 (765, 55)

FRONT FRAME
ASSEMBLY

STABILIZER BAR FRONT (4WD)

RACK AND PINION POWER STEERING
GEAR ASSEMBLY (4WD)

32 (329, 24)

FRONT SUSPENSION
MEMBER BRACE REAR RH

32 (329, 24)

85 (867, 63)

FRONT SUSPENSION
MEMBER BRACE REAR LH

FRAME SIDE RAIL
PLATE SUB-ASSEMBLY
LH

32 (329, 24)

32 (329, 24)

85 (867, 63)

32 (329, 24)

85 (867, 63)

32 (329, 24)

N*m (kgf*cm, ft.*lbf) : Specified torque   ● Non-reusable part

85 (867, 63)

Subframe and related components—3.3L engine

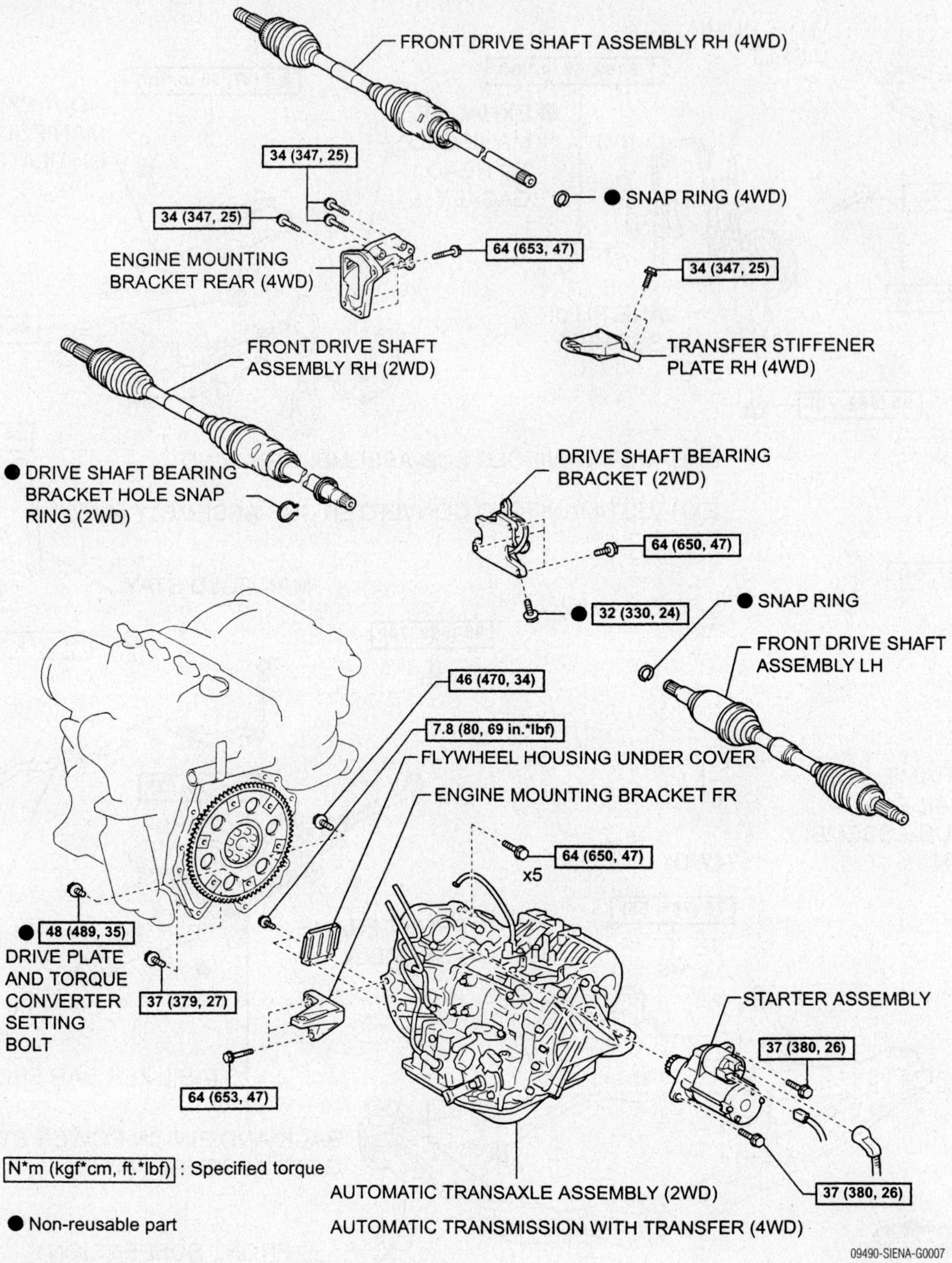

FRONT DRIVE SHAFT ASSEMBLY RH (4WD)

34 (347, 25)

● SNAP RING (4WD)

34 (347, 25)

64 (653, 47)

34 (347, 25)

ENGINE MOUNTING BRACKET REAR (4WD)

TRANSFER STIFFENER PLATE RH (4WD)

FRONT DRIVE SHAFT ASSEMBLY RH (2WD)

DRIVE SHAFT BEARING BRACKET (2WD)

● DRIVE SHAFT BEARING BRACKET HOLE SNAP RING (2WD)

64 (650, 47)

32 (330, 24)

● SNAP RING

FRONT DRIVE SHAFT ASSEMBLY LH

46 (470, 34)

7.8 (80, 69 in.*lbf)

FLYWHEEL HOUSING UNDER COVER

ENGINE MOUNTING BRACKET FR

64 (650, 47)

x5

● 48 (489, 35)

DRIVE PLATE AND TORQUE CONVERTER SETTING BOLT

37 (379, 27)

STARTER ASSEMBLY

37 (380, 26)

64 (653, 47)

N*m (kgf*cm, ft.*lbf) : Specified torque

● Non-reusable part

AUTOMATIC TRANSAXLE ASSEMBLY (2WD)

AUTOMATIC TRANSMISSION WITH TRANSFER (4WD)

37 (380, 26)

09490-SIENA-G0007

Locations of various mounting brackets—3.3L engine

Nm). On 4 wheel drive models tighten the rear insulator nut to 55 ft. lbs. (75 Nm).

j. If removed after engine removal, install the steering gear assembly on 4 wheel drive models and tighten the retainers to 52 ft. lbs. (70 Nm).

k. If removed after engine removal, on 4 wheel drive models, tighten the stabilizer bar to 13 ft. lbs. (44 Nm).

l. If removed after engine removal, tighten the vane pump assembly to 32 ft. lbs. (44 Nm).

m. Install the engine/transmission assembly. Tighten the frame side rail

bolts A to 63 ft. lbs. (85 Nm), bolts B and C to 24 ft. lbs. (32 Nm). refer to the illustration for bolt location.

n. Install the front suspension member brace on both sides. Tighten bolt A to 63 ft. lbs. (85 Nm), bolts B and C to 24 ft. lbs. (32 Nm). refer to the illustration for bolt location.

47. Refill the engine with engine coolant.

48. Fill the transmission and transfer with the correct type and amount of fluid.

49. Change the oil filter and fill the crankcase with the correct type and amount of oil.

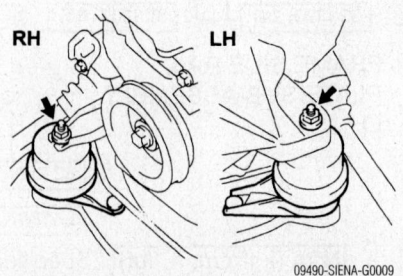

RH   LH

09490-SIENA-G0009

If removed after engine removal, tighten the front frame assembly mounting insulator nuts—3.3L engine and . . .

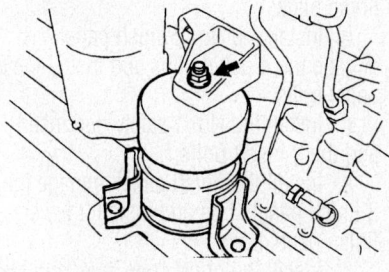

09490-SIENA-G0010

. . . the insulator front nut—3.3L engine

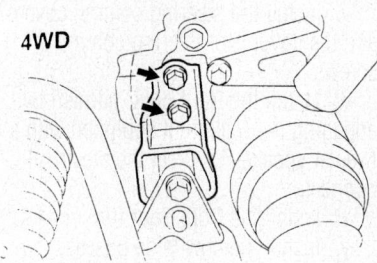

09490-SIENA-G0011

4 wheel drive models rear insulator nut location—3.3L engine

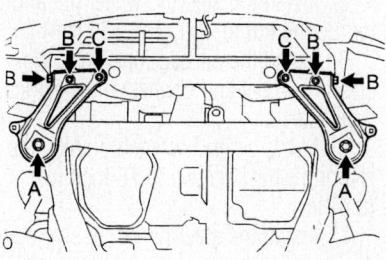

09490-SIENA-G0012

Location of the frame side rail bolts—3.3L engine

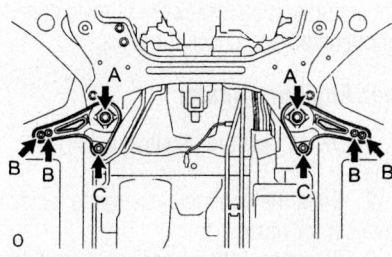

09490-SIENA-G0013

Front suspension member brace bolt locations—3.3L engine

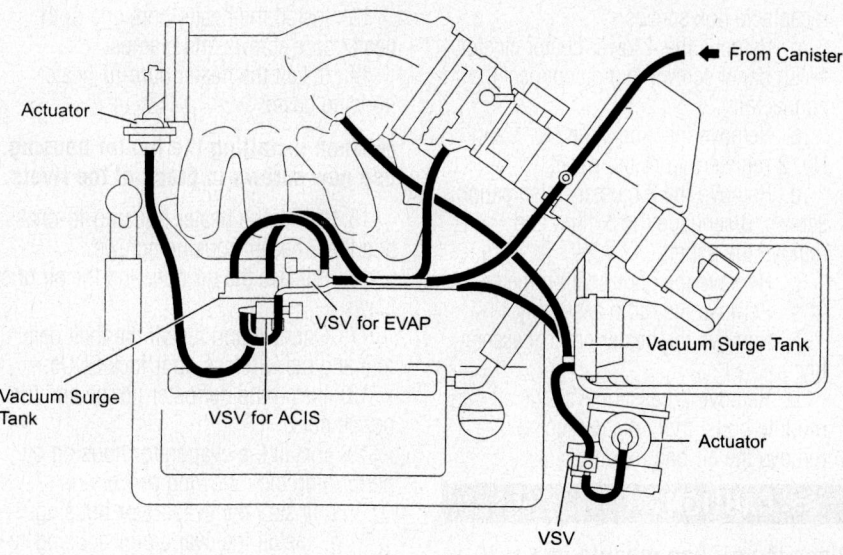

09490-SIENA-G0014

Vacuum hose routing–3.3L engine

50. Install the engine undercovers.
51. Start the engine and check for leaks.

Heater Core

REMOVAL & INSTALLATION

Front Heater

1. Disconnect the negative battery cable.
2. Drain the cooling system into a clean container for reuse.
3. Disconnect the heater hoses from the heater core.
4. Remove the steering wheel by performing the following procedure:
    a. Position the front wheels facing straight-ahead.
    b. Remove the steering wheel side covers.
    c. Using a Torx® wrench, loosen the 2 screws located at each side of the steering wheel until the screw's circumference groove catches on the screw case.
    d. Pull the air bag module from the steering wheel and disconnect the electrical connector.

✳✳ CAUTION

Place the air bag module in a safe place with the front side facing upward.

    e. Remove the steering wheel nut.
    f. Place alignment marks on the steering wheel and the main shaft.
    g. Using a steering wheel puller, press the steering wheel from the steering column.

5. Remove the instrument panel and reinforcement by performing the following procedure:
    a. Remove the front door scuff plates.
    b. Remove the cowl side boards.
    c. Remove the front door trim covers.
    d. Remove the front pillar garnish by disengaging the 5 clips. If equipped with a tweeter speaker, disconnect the electrical connector.
    e. Remove the steering column covers-to-steering column screws and the covers.
    f. Remove the combination switch-to-steering column screws, disconnect the electrical connector(s) and remove the combination switch.
    g. Remove the 2 hood open lever screws and the hood open lever.
    h. Remove the 2 lower finish panel bolts and disengage the panel from the 3 clips.
    i. Remove the 2 No. 1 safety pad insert bolts and the insert.
    j. Remove the 2 No. 2 finish panel bolts and disengage the panel from the 4 clips.
    k. In the left side of the glove compartment, pry out the glove box door finish plate and disconnect the air bag module connector.
    l. Remove the glove box 3 nuts and 2 screws and the glove box.
    m. Remove the center cluster finish panel by disengaging the claw (bottom center) and 4 clips (one at each corner).

n. Remove the ashtray, the 2 ashtray receptacle box screws.

o. Remove the 4 lower center cluster finish panel screws and disconnect the connector.

p. Remove the clock, the No. 1 and No. 2 registers from the panel.

q. Remove the 3 cluster finish panel screws, disengage the 8 clips and remove the panel.

r. Remove the combination meter.

s. Remove the radio assembly.

t. Remove the heater control assembly.

u. Remove 2 passenger's side air bag module bolts; then, disconnect and remove the air bag module.

❈❈ CAUTION

Place the air bag module in a safe place with the front side facing upward.

v. Remove the instrument panel-to-chassis 5 bolts and nut.

w. Remove the audio amplifier.

x. Remove the No. 1 and No. 2 braces.

y. Remove the No. 2 cowl brace.

z. Remove the instrument panel reinforcement.

6. Remove the evaporator housing by performing the following procedure:

a. Discharge and recover the air conditioning system refrigerant.

b. In the engine compartment, remove the refrigerant lines-to-cowl connector bolts; then, disconnect the lines and discard the O-rings.

c. Disconnect the electrical connector at the evaporator housing.

d. Disconnect the wiring harness clamp.

e. Remove the evaporator housing-to-chassis 2 rivets, 3 bolts and nut.

f. Remove the evaporator housing.

7. Remove the 4 defroster nozzle nuts and the nozzle.

8. Disconnect and remove the theft deterrent and the wireless door lock ECUs.

9. Release the 2 air duct claws and the air duct.

10. Remove the 2 heater housing-to-chassis rivets and the heater housing.

➡ **When installing the heater housing, use new screws in place of the rivets.**

11. Remove the heater core-to-heater housing cover.

12. Remove both heater core screws and clamps; then, remove the heater core.

To install:

13. Install the heater core and both heater core screws and clamps.

14. Install the heater core-to-heater housing cover.

➡ **When installing the heater housing, use new screws in place of the rivets.**

15. Install the heater housing-to-chassis and the 2 heater housing screws.

16. Release the air duct and the air duct claws.

17. Connect and install the theft deterrent and the wireless door lock ECUs.

18. Install the defroster nozzle and the 4 nozzle nuts.

19. Install the evaporator housing by performing the following procedure:

a. Install the evaporator housing.

b. Install the evaporator housing-to-chassis 2 rivets, 3 bolts and nut.

c. Connect the wiring harness clamp.

d. Connect the electrical connector at the evaporator housing.

e. In the engine compartment, use new O-rings and install the refrigerant lines-to-cowl connector and install the bolts.

20. Install the instrument panel and reinforcement by performing the following procedure:

a. Install the instrument panel reinforcement.

b. Install the No. 2 cowl brace.

c. Install the No. 1 and No. 2 braces.

d. Install the audio amplifier.

e. Install the instrument panel-to-chassis 5 bolts and nut.

f. Connect and install the air bag module and the 2 passenger's side air bag module bolts.

g. Install the heater control assembly.

h. Install the radio assembly.

i. Install the combination meter.

j. Install the cluster finish panel, engage the 8 clips and install the panel screws.

k. Install the No. 1 and No. 2 registers and the clock to the panel.

l. Connect the lower center cluster finish panel connector and install the 4 lower center cluster finish panel screws.

m. Install the 2 ashtray receptacle box screws and the ashtray.

n. Install the center cluster finish panel by engaging the 4 clips (1 at each corner) and the claw (bottom center).

o. Install the glove box and the glove box 3 nuts and 2 screws.

p. In the left side of the glove compartment, connect the air bag module connector and install the glove box door finish plate.

q. Install the No. 2 finish panel, engage the 4 panel clips and install the 3 panel bolts.

r. Install the No. 1 safety pad insert and the 2 insert bolts.

s. Install the finish panel, engage the 3 finish panel clips and install 2 lower finish panel bolts.

t. Install the hood open lever and the 2 hood open lever screws.

u. Install the combination switch, connect the electrical connector(s) and install the combination switch-to-steering column screws.

v. Install the steering column covers and the covers-to-steering column screws.

w. Install the front pillar garnish by engaging the 5 clips. If equipped with a tweeter speaker, connect the electrical connector.

x. Install the front door trim covers.

y. Install the cowl side boards.

z. Install the front door scuff plates.

21. Install the steering wheel by performing the following procedure:

a. Install the steering wheel to the steering column.

b. Align the steering wheel-to-main shaft marks.

c. Install the steering wheel nut and torque the nut to 25 ft. lbs. (34 Nm).

d. Install the air bag module to the steering wheel and connect the electrical connector.

e. Using a Torx® wrench, tighten the steering wheel screws to 78 inch lbs. (8.8 Nm).

f. Install the steering wheel side covers.

22. Connect the heater hoses to the heater core.

23. Refill the cooling system.

24. Connect the negative battery cable.

25. Evacuate and charge the air conditioning system.

26. Run the engine to normal operating temperatures; then, check the climate control operation and check for leaks.

Rear Auxiliary Heater

1. Disconnect the negative battery cable.

2. Drain the cooling system into a clean container for reuse.

3. Disconnect the heater hoses from the rear heater core.

4. Remove the front seats.

5. Remove the front door scuff plates.

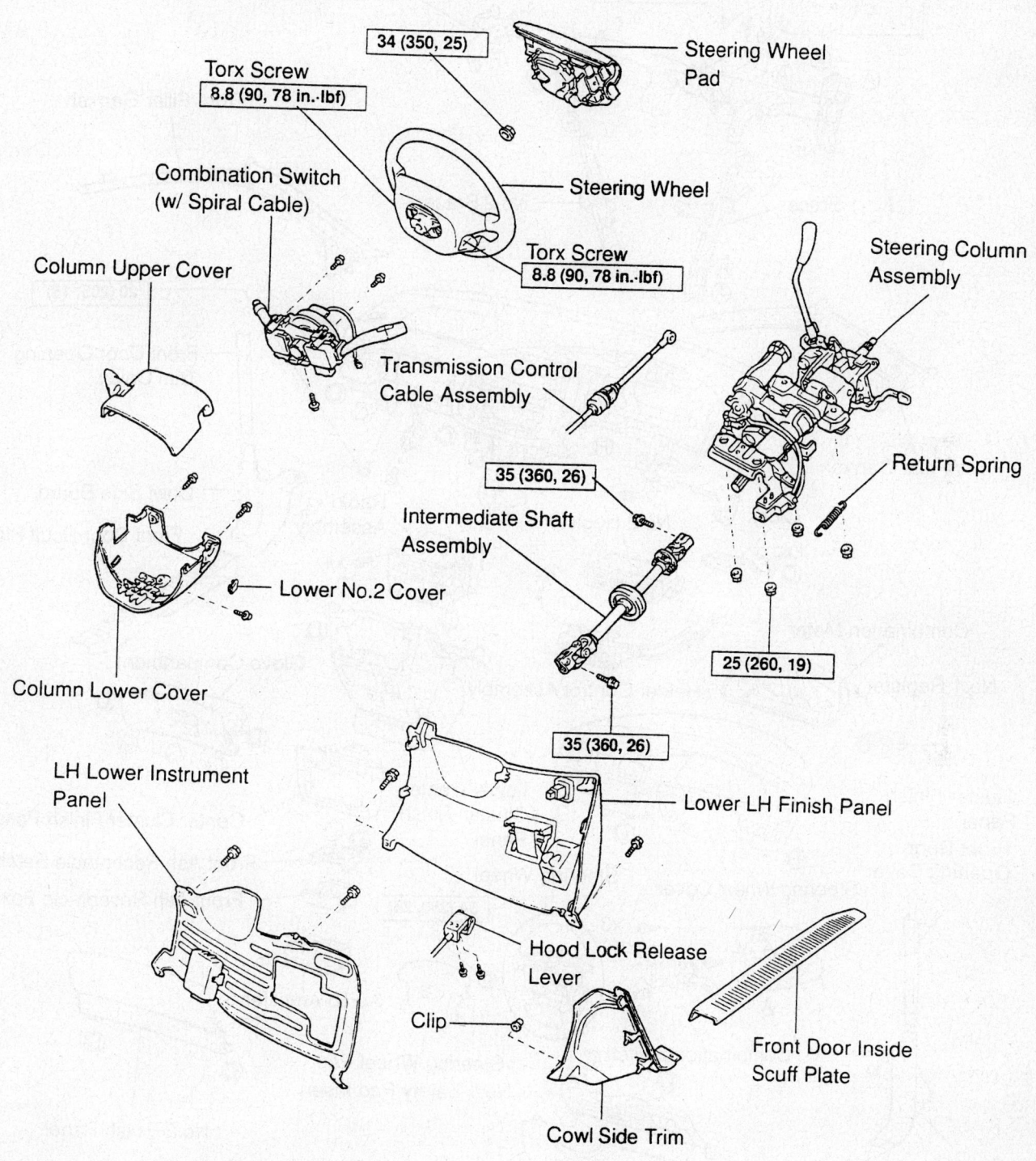

34 (350, 25)

Steering Wheel Pad

Torx Screw
8.8 (90, 78 in.·lbf)

Combination Switch
(w/ Spiral Cable)

Steering Wheel

Column Upper Cover

Torx Screw
8.8 (90, 78 in.·lbf)

Steering Column Assembly

Transmission Control Cable Assembly

Return Spring

35 (360, 26)

Intermediate Shaft Assembly

Lower No.2 Cover

25 (260, 19)

Column Lower Cover

35 (360, 26)

LH Lower Instrument Panel

Lower LH Finish Panel

Hood Lock Release Lever

Clip

Front Door Inside Scuff Plate

Cowl Side Trim

N·m (kgf·cm, ft·lbf) : Specified torque

93113GH3

Exploded view of the steering wheel, steering column and related components

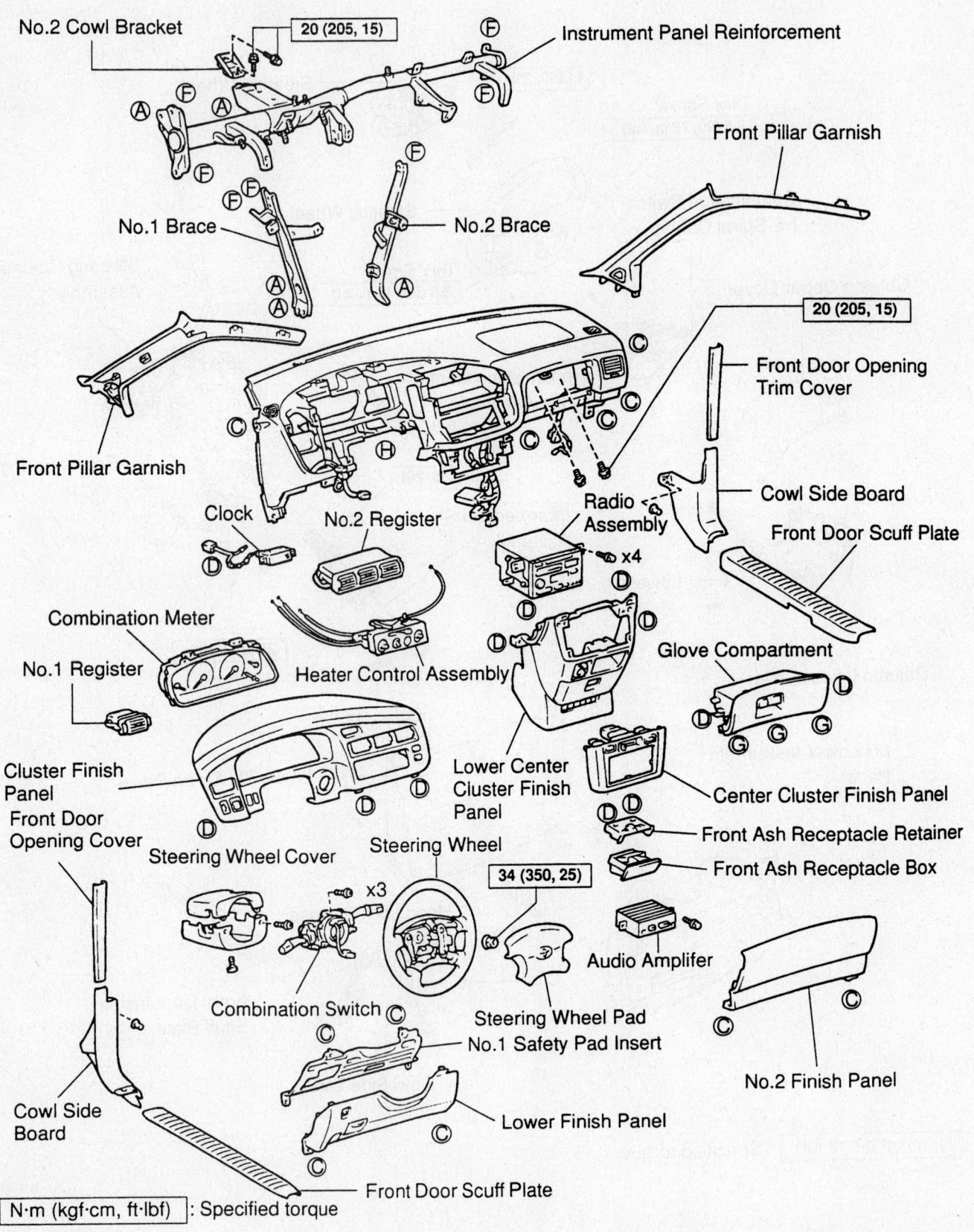

No.2 Cowl Bracket

20 (205, 15)

Instrument Panel Reinforcement

Front Pillar Garnish

No.1 Brace

No.2 Brace

20 (205, 15)

Front Door Opening Trim Cover

Front Pillar Garnish

Cowl Side Board

Front Door Scuff Plate

Clock

No.2 Register

Radio Assembly

x4

Combination Meter

Heater Control Assembly

Glove Compartment

No.1 Register

Cluster Finish Panel

Front Door Opening Cover

Lower Center Cluster Finish Panel

Center Cluster Finish Panel

Front Ash Receptacle Retainer

Front Ash Receptacle Box

Steering Wheel Cover

Steering Wheel

34 (350, 25)

Audio Amplifer

x3

Steering Wheel Pad

No.1 Safety Pad Insert

Combination Switch

No.2 Finish Panel

Cowl Side Board

Lower Finish Panel

Front Door Scuff Plate

N·m (kgf·cm, ft·lbf) : Specified torque

93113GH4

Exploded view of the instrument panel and related components

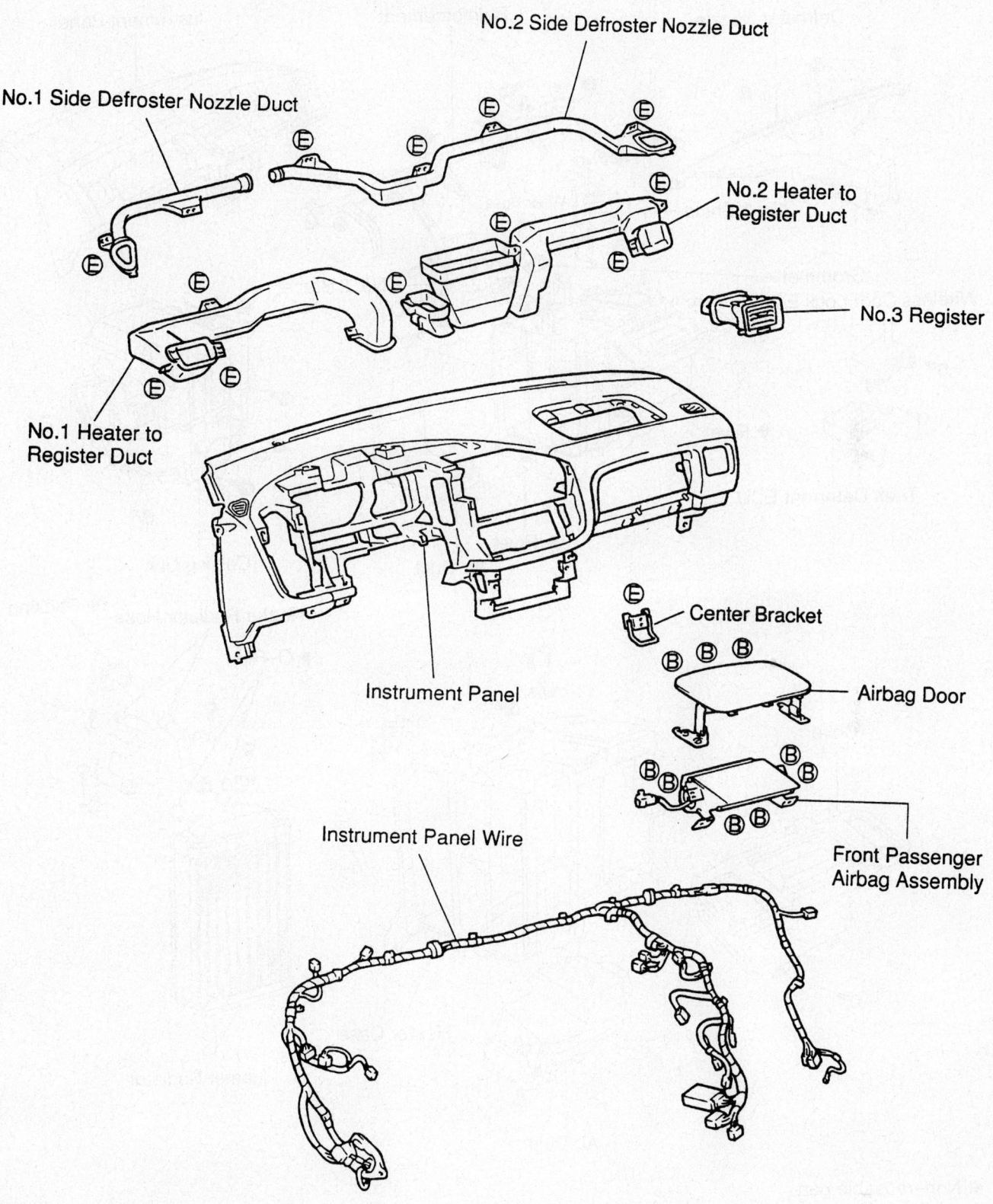

No.2 Side Defroster Nozzle Duct

No.1 Side Defroster Nozzle Duct

No.2 Heater to Register Duct

No.3 Register

No.1 Heater to Register Duct

Center Bracket

Airbag Door

Instrument Panel

Instrument Panel Wire

Front Passenger Airbag Assembly

93113GH5

Exploded view of the ventilation system and related components

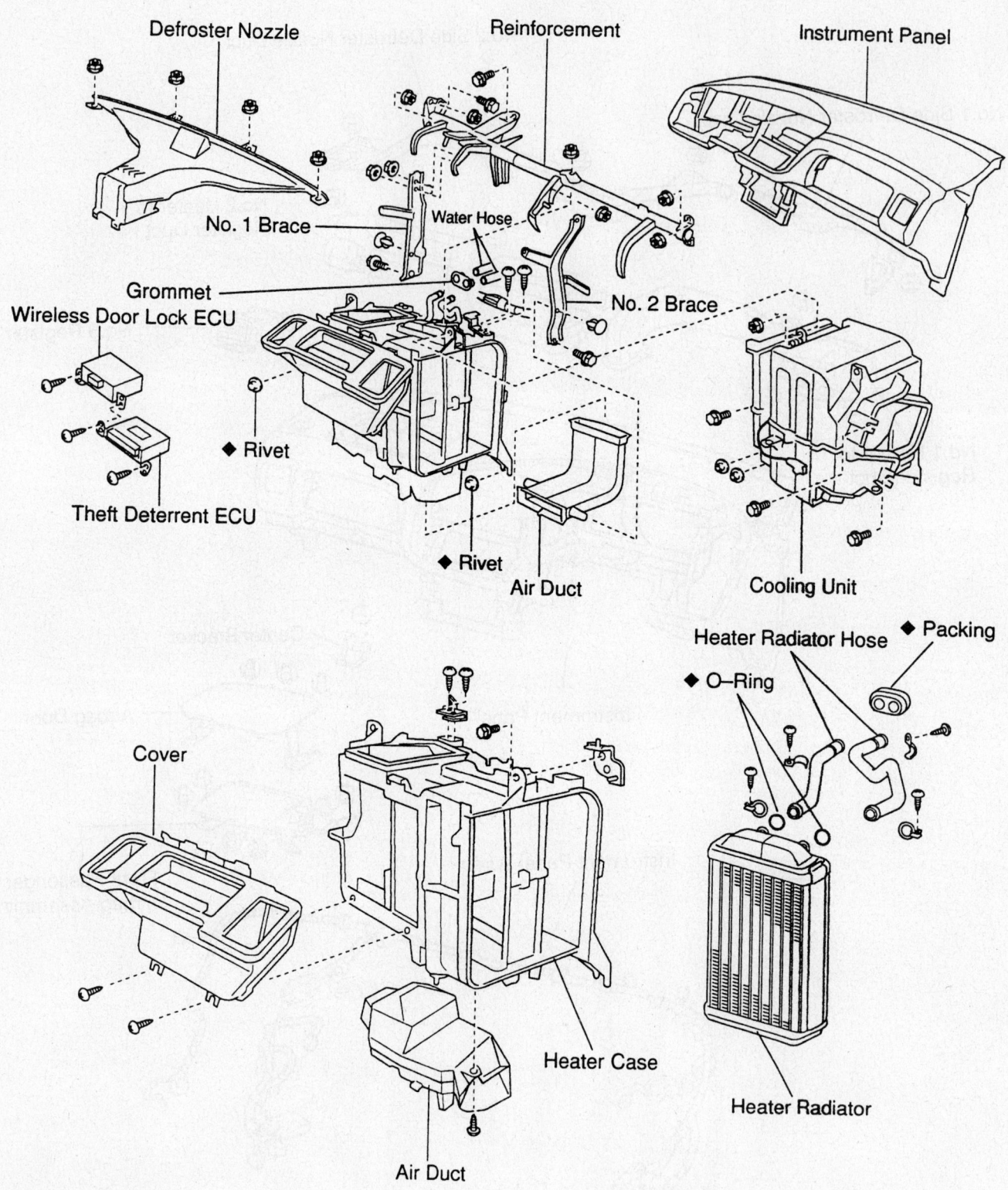

Defroster Nozzle

Reinforcement

Instrument Panel

No. 1 Brace

Water Hose

No. 2 Brace

Grommet

Wireless Door Lock ECU

◆ Rivet

Theft Deterrent ECU

◆ Rivet

Air Duct

Cooling Unit

Heater Radiator Hose

◆ Packing

◆ O–Ring

Cover

Heater Case

Air Duct

Heater Radiator

◆ Non–reusable part

Exploded view of the heater core, heater housing, evaporator housing and related components

93113GH6

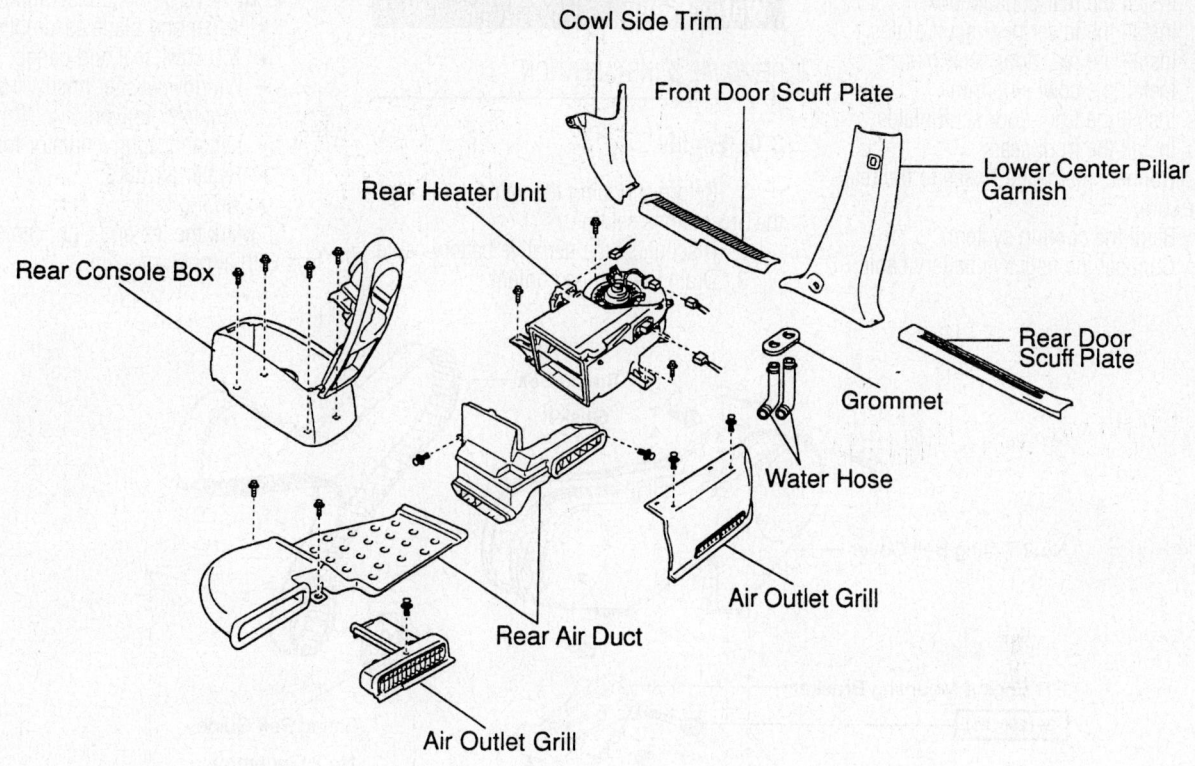

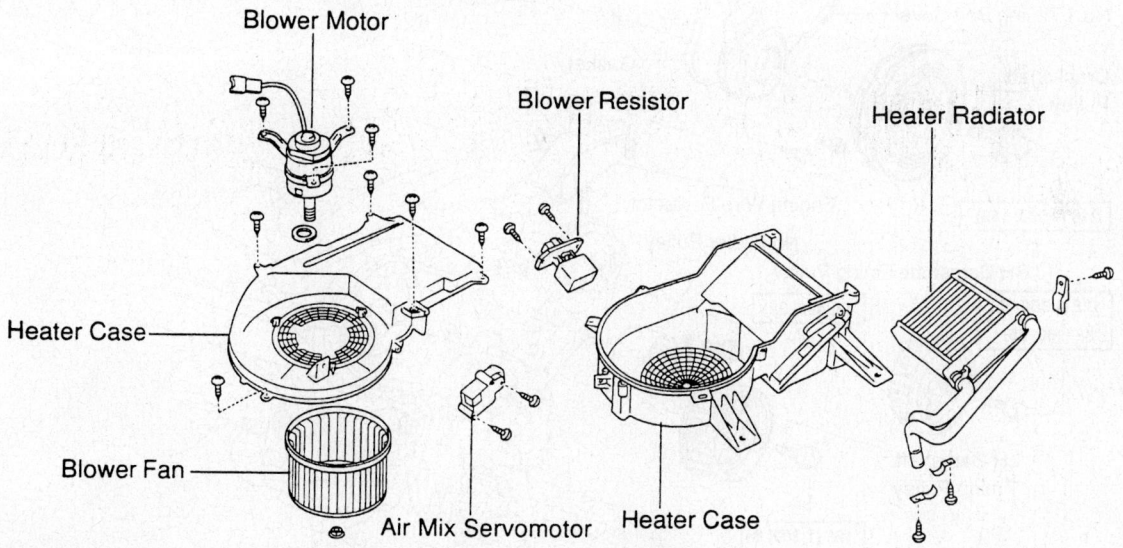

Exploded view of the rear heater core, the rear heater housing and related components

6. Remove the cowl side trim.
7. Remove the rear door scuff plates.
8. Remove the lower door scuff plates.
9. Remove the rear console box.
10. Remove the left side air outlet grille.
11. Pull the carpet rearward.
12. Remove the 3 clips and the air outlet grille.
13. Remove the rear air duct 2 bolts, 2 clips and the duct.
14. Disconnect the electrical connectors.

15. Remove the 3 rear heater housing bolts and the housing.
16. Remove both heater core-to-heater housing screws and clamps.
17. Remove the heater core-to-heater housing screw and plate.
18. Remove the heater core.
To install:
19. Install the heater core.
20. Install the heater core-to-heater housing screw and plate.

21. Install both heater core-to-heater housing screws and clamps.
22. Install the rear heater housing and the 3 housing bolts.
23. Connect the electrical connectors.
24. Install the rear air duct and the 2 bolts and 2 clips.
25. Install the 3 clips and the air outlet grille.
26. Move the carpet forward.
27. Install the left side air outlet grille.

28. Install the rear console box.
29. Install the lower door scuff plates.
30. Install the rear door scuff plates.
31. Install the cowl side trim.
32. Install the front door scuff plates.
33. Install the front seats.
34. Connect the heater hoses to the rear heater core.
35. Refill the cooling system.
36. Connect the negative battery cable.

Water Pump

REMOVAL & INSTALLATION

3.0L Engine

1. Before servicing the vehicle, refer to the precautions section.
2. Disconnect the negative battery cable.
3. Drain the engine coolant.

4. Remove or disconnect the following:
- Wiper and blade assembly
- Top cowl seal and panel
- Window washer hoses, from the ventilator louvers
- Left and right ventilator louvers
- Heater air duct
- Timing belt
5. Mark the left and right camshaft pulleys with a touch of paint.

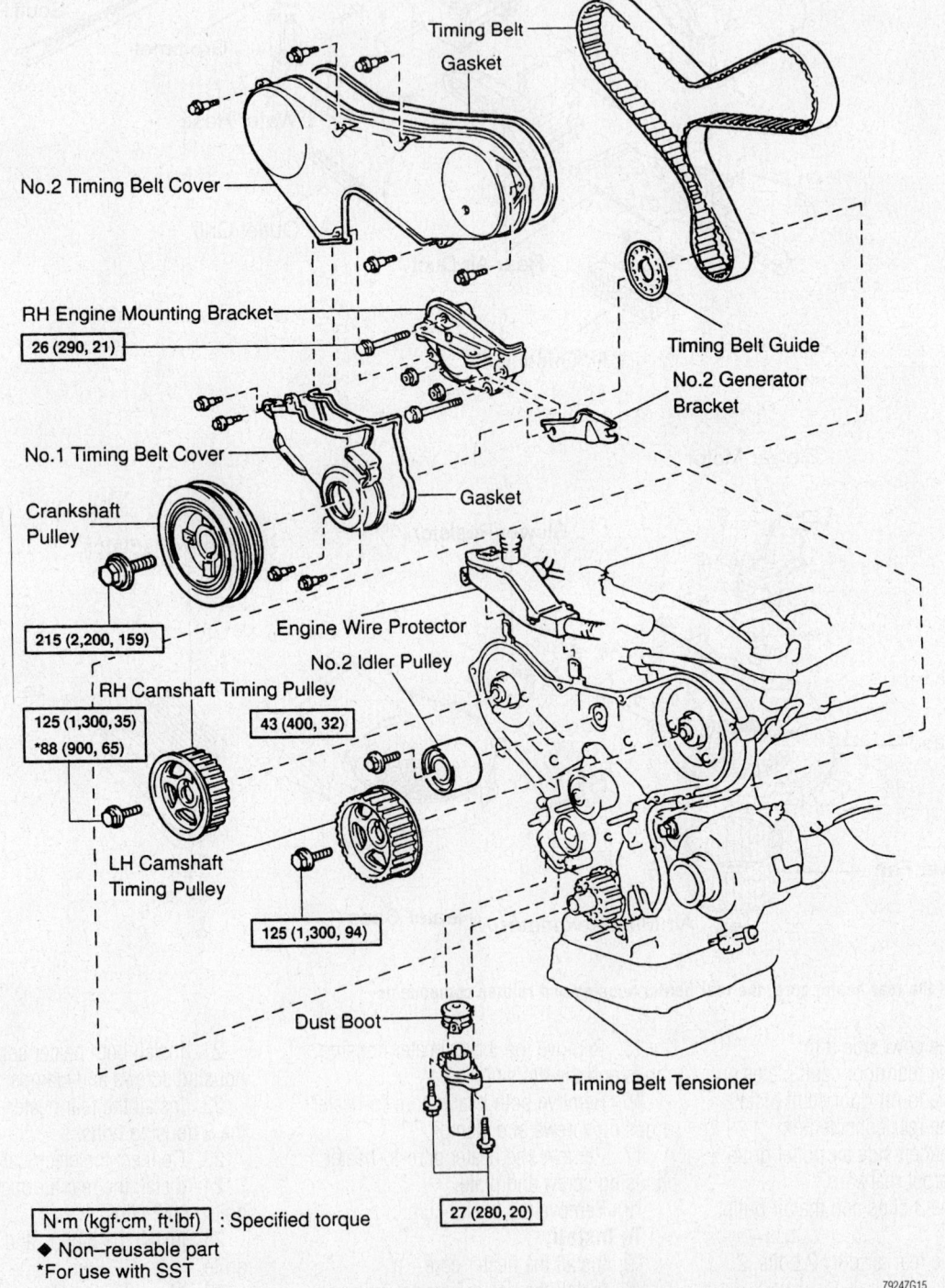

N·m (kgf·cm, ft·lbf) : Specified torque
◆ Non-reusable part
*For use with SST

Exploded view of the components to gain access to the water pump—3.0L engine

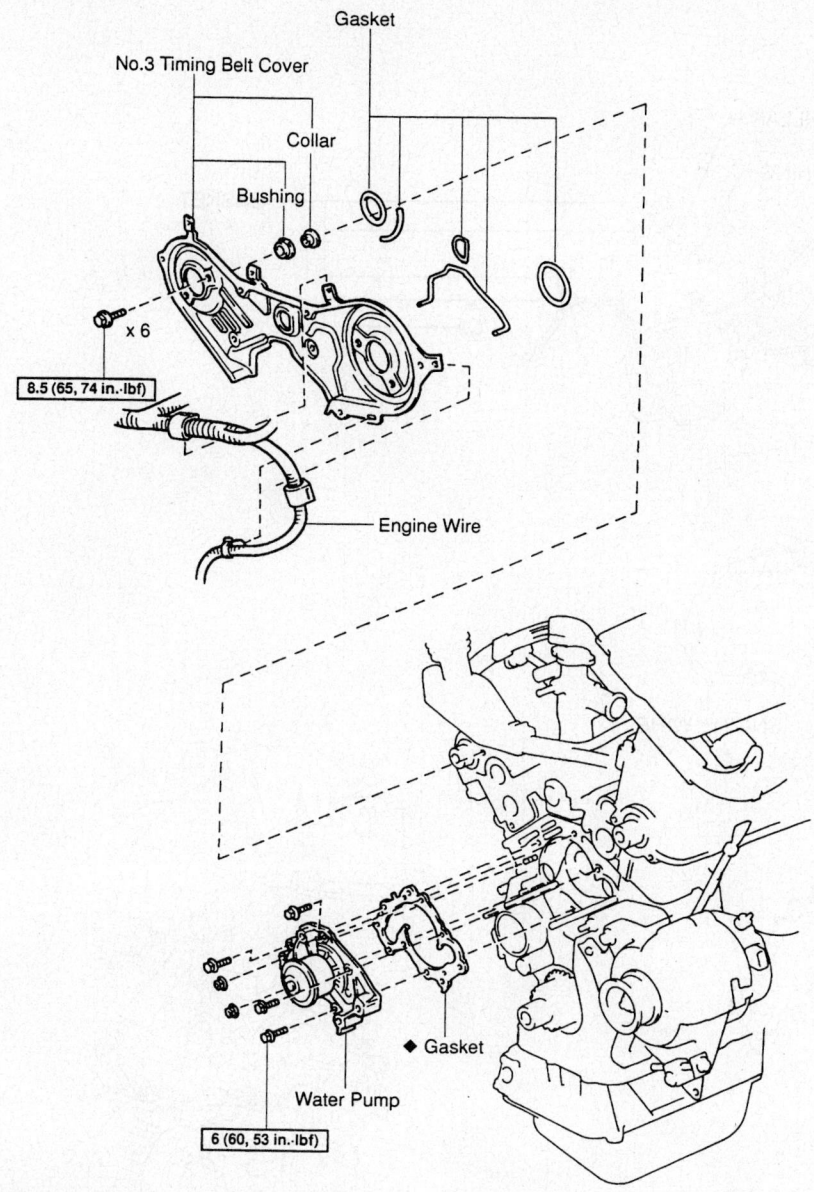

Exploded view of the water pump and related components—3.0L engine

N·m (kgf·cm, ft·lbf) : Specified torque
◆ Non—reusable part

6. Remove or disconnect the following:
- Right and left camshaft pulleys bolts
- Pulleys from the engine

➡**Be sure not to mix up the pulleys.**

- No. 2 idler pulley by removing the bolt
- 3 clamps and engine wire from the rear timing belt cover
- 6 No. 3 timing belt cover-to-engine bolts
- Water pump nuts/bolts
- Water pump and gasket from the engine

To install:

7. Check that the water pump turns smoothly. Also check the air hole for coolant leakage.

8. Apply liquid sealer to the gasket, water pump and engine block.

9. Install or connect the following:
- Water pump, using a new gasket. Tighten the nuts/bolts to 53 inch lbs. (6 Nm).
- Rear timing belt cover. Tighten the 6 bolts to 74 inch lbs. (9 Nm).
- Engine wire with the 3 clamps to the rear timing belt cover

- No. 2 idler pulley. Tighten the bolt to 32 ft. lbs. (43 Nm).

➡**After tightening the bolt, be sure the idler pulley moves smoothly.**

- Right-hand camshaft pulley, with the flange side **outward**.

➡**Be sure to align the knock pin hole on the camshaft pulley with the knock pin on the camshaft.**

- Tighten the camshaft bolt to 65 ft. lbs. (88 Nm), using the removal tools
- Left-hand camshaft pulley, with the flange side **inward**.

➡**Be sure to align the knock pin hole on the camshaft pulley with the knock pin on the camshaft.**

- Tighten the camshaft bolt to 94 ft. lbs. (125 Nm), using the removal tools
- Timing belt
10. Fill the engine coolant.
11. Install or connect the following:
- Heater air duct
- Left and right ventilator louvers
- Window washer hoses to the ventilator louvers
- Top cowl seal and panel
- Wiper and blade assembly
- Negative battery cable
12. Start the engine.
13. Top off the engine coolant and check for leaks.

3.3L Engine

1. Before servicing the vehicle, refer to the precautions section.
2. Properly relieve the fuel system pressure.
3. Disconnect the negative battery cable.
4. Drain and recycle the engine coolant.
5. Remove the right hand front wheel.
6. Remove the wiper arms.
7. Remove the cowl panel.
8. Remove the wiper link assembly.
9. Remove the cowl top to inner brace.
10. Remove the cowl top cover outer sub assembly.
11. Remove the right hand apron seal.
12. Remove the A/C compressor to crankshaft pulley belt.
13. Remove the vane pump belt.
14. Remove the engine moving control rod bolts and rod.
15. Remove the number 2 engine mounting stay as follows:

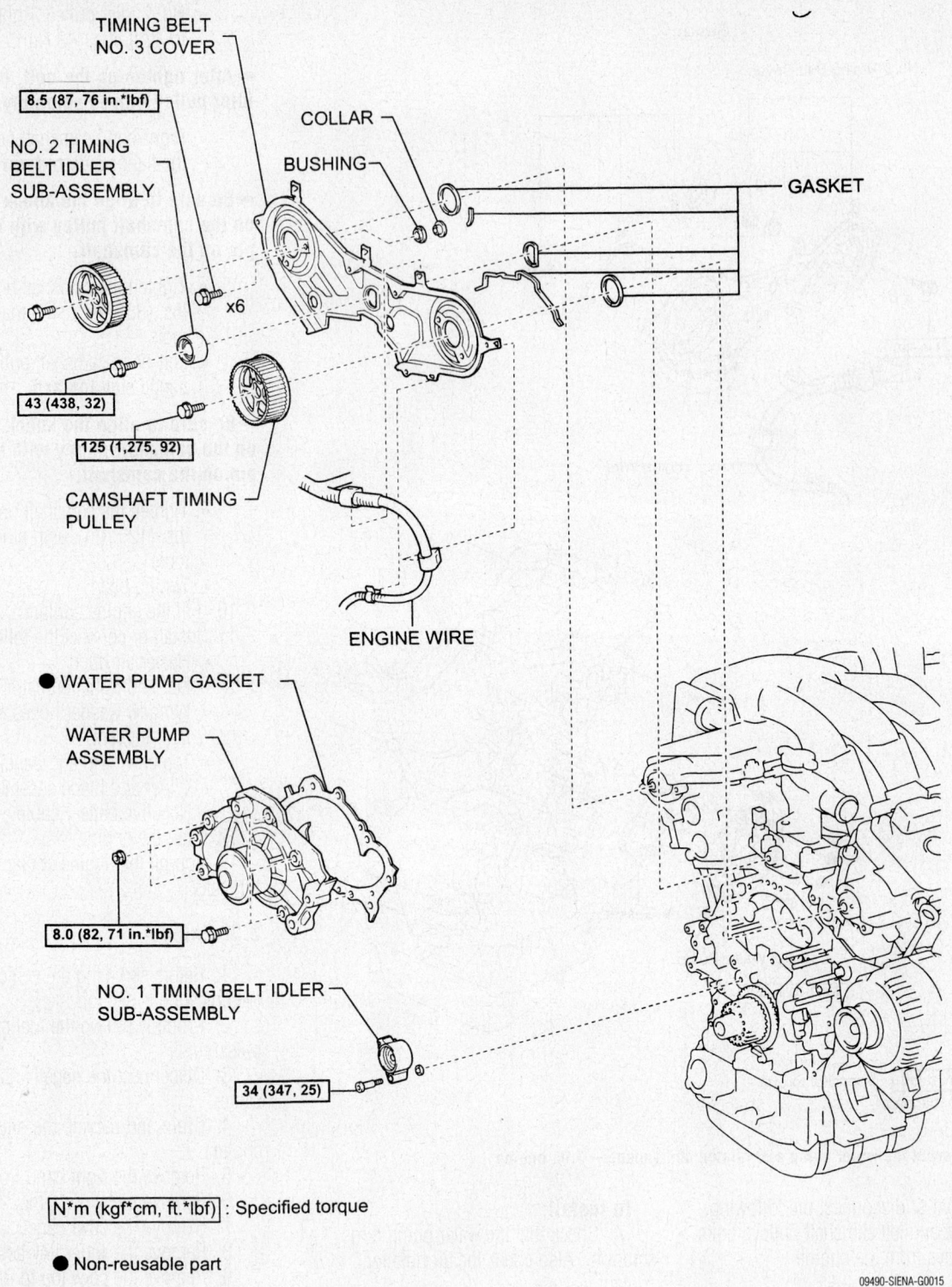

TIMING BELT NO. 3 COVER

8.5 (87, 76 in.*lbf)

NO. 2 TIMING BELT IDLER SUB-ASSEMBLY

COLLAR

BUSHING

GASKET

x6

43 (438, 32)

125 (1,275, 92)

CAMSHAFT TIMING PULLEY

ENGINE WIRE

● WATER PUMP GASKET

WATER PUMP ASSEMBLY

8.0 (82, 71 in.*lbf)

NO. 1 TIMING BELT IDLER SUB-ASSEMBLY

34 (347, 25)

N*m (kgf*cm, ft.*lbf) : Specified torque

● Non-reusable part

09490-SIENA-G0015

Exploded view of the water pump and related components—3.3L engine

a. Remove the bolt and wire harness bracket.

b. Remove the bolt, stay and bracket.

16. Remove the timing belt covers 1 and 2. Refer to the timing belt removal and installation procedure in this manual.

17. Remove the right hand engine mount bracket.

18. Remove the timing belt guide and timing belt. Refer to the timing belt removal and installation procedure in this manual.

19. Remove the timing belt guide number 2. Refer to the timing belt removal and installation procedure in this manual.

20. Remove the camshaft pulley and lower timing cover. Refer to the timing belt removal and installation procedure in this manual.

21. Remove the timing belt idler sub assembly 1. Refer to the timing belt removal and installation procedure in this manual.

22. Loosen the bolts and remove the water pump and gasket.

To install:

23. Clean all gasket mating surfaces and install a new gasket.

24. Install the water pump and tighten the bolts to 71 inch lbs. (8 Nm).

25. Install the remaining components in the reverse order of removal. Refer to the timing belt removal and installation procedure in this manual for all timing belt related component installation.

26. Fill the cooling system and check for leaks.

Cylinder Head

REMOVAL & INSTALLATION

3.0L Engine

1. Before servicing the vehicle, refer to the precautions section.

2. Remove or disconnect the following:
   - Wiper and blade assembly
   - Top cowl seal and panel
   - Window washer hoses from the ventilator louvers
   - Left and right ventilator louvers
   - Heater air duct

3. Relieve the fuel pressure.
   - Turn the ignition key to the **OFF** position
   - Negative battery cable

➡ **Wait at least 90 seconds from the time the negative battery was disconnected to start work.**

4. Drain the cooling system.

5. Remove or disconnect the following:
   - Accelerator and throttle cables, if equipped with an automatic transaxle
   - Air cleaner cover, air flow meter and the air duct
   - Cruise control actuator and bracket, if equipped
   - 2 engine ground straps
   - Right engine mounting support
   - Radiator hoses
   - 2 heater hoses
   - Fuel feed and return lines from the fuel rail assembly
   - Pressure hose from the hydraulic motor
   - V-bank cover

6. Disconnect the following vacuum hoses:
   - Fuel pressure control Vacuum Switching Valve (VSV)
   - Fuel pressure regulator
   - Cylinder head rear plate
   - Intake air control valve VSV
   - Exhaust Gas Recirculation (EGR) vacuum modulator
   - EGR valve

7. Disconnect the following wiring and hoses:
   - Intake air control valve
   - Fuel pressure regulator
   - EGR VSV

8. Remove the 2 nuts and the emission control valve set.

9. Disconnect the following hoses:
   - Brake booster vacuum hose
   - PCV hose
   - Intake air control valve vacuum hose

10. Remove or disconnect the following:
    - Data Link Connector (DLC) from the mounting bracket
    - 2 ground straps from the intake chamber
    - Hydraulic motor pressure hose from the intake chamber
    - Right Oxygen (O₂) sensor connector from the power steering pressure tube
    - 2 nuts and the power steering pressure tube from the intake chamber
    - Both power steering air hoses
    - Engine hanger and the intake chamber support
    - EGR pipe and gaskets

11. Disconnect the following wiring:
    - Throttle Position (TP) sensor connector
    - Idle Air Control (IAC) valve connector
    - EGR gas temperature connector
    - Air conditioning idle up connector

12. Disconnect the following vacuum hoses:
    - 2 vacuum hoses from the Thermal Vacuum Valve (TVV)
    - Vacuum hose from the cylinder head rear plate
    - Vacuum hose from the charcoal canister

13. Remove or disconnect the following:
    - Air assist hose and the 2 water bypass hoses
    - Air intake chamber
    - Left engine wiring harness and move it aside
    - Wiring harness from the rear of the engine
    - Right engine wiring harness and move it aside
    - Ignition coils and move them aside
    - Timing belt
    - Camshaft pulleys and the timing belt rear cover
    - Cylinder head rear plate
    - Water inlet pipe
    - Air assist hose and vacuum hose
    - Intake manifold and fuel rail assembly
    - Water outlet
    - EGR pipe from the right exhaust manifold
    - Front exhaust pipe and exhaust manifolds
    - Dipstick assembly and the power steering pump bracket
    - Valve covers and the Camshaft Position (CMP) sensor
    - Camshafts

14. Be sure the engine is at/or near ambient temperature and remove the 2 (1 on each head) 8mm recessed hex bolts. Loosen and remove the 8 head bolts evenly, in 3 passes, in the reverse order of the

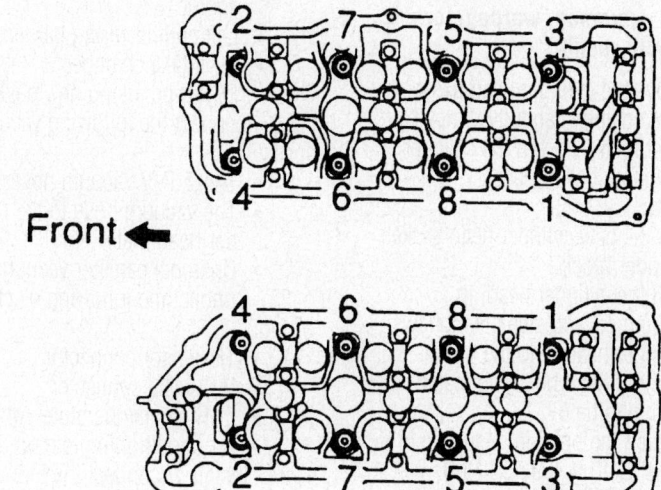

Front ←

Cylinder head bolt loosening sequence

7924ZG19

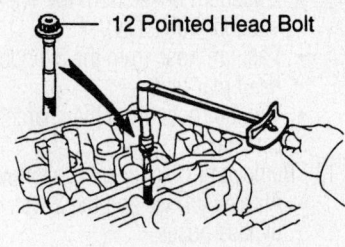

12 Pointed Head Bolt

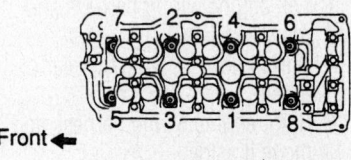

Front ◄

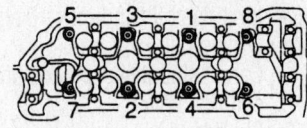

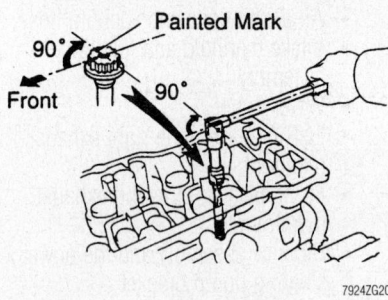

Painted Mark

90° Front 90°

7924ZG20

Cylinder head bolt tightening sequence—
3.0L and 3.3L engines engine

installation sequence. Carefully lift the head from the engine; if necessary to pry the head loose, take great care not to damage the mating surfaces. Place the head on wood blocks in a clean work area.

➡ **If the cylinder head bolts are loosened out of sequence, warpage or cracking could result.**

15. Remove the cylinder head gasket. With a gasket scraper, carefully remove all the old gasket material from the cylinder head and engine block surfaces.

To install:

16. Place the new cylinder head gasket onto the cylinder block.

17. Install the cylinder head, in sequence, using several steps, as follows:
- Cylinder head onto the gasket
- Cylinder head bolts lubricated with clean engine oil
- Tighten the bolts in sequence in 3 steps to 40 ft. lbs. (54 Nm).

➡ **If any bolt does not meet the torque, replace it.**

- Mark the forward edge of each bolt with paint, then tighten each bolt, in proper sequence, an additional 90 degrees.
- Check that each painted mark is now at a 90 degrees angle to the front

➡ **The paint mark applied to the bolt in the 9 o'clock position and should now be in the 12 o'clock position.**

- Remaining 8mm bolts, lubricated with engine oil. Tighten both bolts to 13 ft. lbs. (18 Nm).

18. Install the camshafts.

19. Check and adjust the valves.

20. Apply sealant to the cylinder heads where the camshaft supports meet the cylinder heads.

21. Install or connect the following:
- Cylinder head covers, using new gaskets
- Dipstick and power steering pump bracket
- Exhaust manifolds. Tighten the nuts to 36 ft. lbs. (49 Nm).
- EGR pipe to the right exhaust manifold
- Water outlet
- Intake manifold and the fuel rail assembly. Tighten the intake manifold nuts/bolts to 11 ft. lbs. (15 Nm).
- Air assist hose and the 2 water bypass hoses
- Water inlet pipe and cylinder head rear plate
- Timing belt rear cover and camshaft pulleys
- Timing belt
- Spark plugs and ignition coils
- Right engine wiring harness
- Wiring harness to the rear of the engine
- Left engine wiring harness
- Air intake chamber
- EGR pipe, using new gaskets

22. Connect the following vacuum hoses:
- The 2 TVV vacuum hoses
- The vacuum hose to the rear cylinder head plate
- Charcoal canister vacuum hose

23. Connect the following electrical wiring:
- TP sensor connector
- IAC valve connector
- EGR gas temperature connector
- Air conditioning idle up connector

24. Install or connect the following:
- Engine hanger and the intake chamber support
- Both power steering air hoses

- Power steering pressure tube to the intake chamber
- O2 sensor connector to the pressure tube.
- Both ground straps, to the intake chamber
- DLC to the bracket

25. Connect the following hoses:
- Power brake booster vacuum hose
- PCV hose
- IAC valve vacuum hose

26. Install or connect the following:
- Emission control valve set and related vacuum hoses and connectors
- V-bank cover
- Pressure hose to the hydraulic motor
- Fuel lines to the fuel rail assembly
- Heater and radiator hoses
- Right engine mounting support
- Both engine ground straps
- Upper front suspension brace, if removed. Tighten the nuts to 59 ft. lbs. (80 Nm).
- Cruise control actuator and bracket
- Air cleaner, air flow meter and air duct assembly
- Accelerator and throttle cables, if equipped with an automatic transaxle

27. Fill the cooling system.

28. Install or connect the following:
- Negative battery cable
- Heater air duct
- Left and right ventilator louvers
- Window washer hoses from the ventilator louvers
- Top cowl seal and panel
- Wiper and blade assembly

29. Start the engine and check for leaks.

30. Bleed the air from the cooling system.

31. Road test the vehicle and check for unusual noise, shock, slippage, correct shift points and smooth operation.

32. Recheck the coolant and engine oil levels.

3.3L Engine

1. Before servicing the vehicle, refer to the precautions section.

2. Properly relieve the fuel system pressure.

3. Disconnect the negative battery cable.

4. Drain the engine oil.

5. Drain and recycle the engine coolant.

6. Remove the right hand front wheel.

7. Remove the wiper arms.

8. Remove the cowl panel.

9. Remove the wiper link assembly.

10. Remove the cowl top to inner brace.

11. Remove the cowl top cover outer sub assembly.

12. Remove the engine cover.

13. Remove the air cleaner assembly.

14. Remove the air surge tank.

15. Disconnect the fuel pipe sub assembly.

16. Disconnect the heater hose inlet pipe.

17. Remove the intake manifold.

18. Disconnect the radiator inlet hose.

19. Disconnect the water outlet as follows:

    a. Disconnect the engine coolant temperature sensor.

    b. Remove the clamp, bolts nuts and washers.

    c. Lock the clamp open and remove the water outlet with the by pass hose and remove the gaskets.

20. Remove the right hand front fender apron seal.

21. Remove the compressor to crankshaft pulley belt.

22. Remove the vane pump belt.

23. Remove the engine moving control rod.

24. Remove the number 2 engine mounting stay as follows:

    a. Remove the bolt and wire harness bracket.

    b. Remove the bolt, stay and bracket.

25. Remove the number 2 alternator bracket and the alternator belt adjusting bar.

26. Remove the crankshaft pulley.

27. Remove the timing belt covers 1 and 2. Refer to the timing belt removal and installation procedure in this manual.

28. Remove the right hand engine mount bracket.

29. Remove the timing belt guide and timing belt. Refer to the timing belt removal and installation procedure in this manual.

30. Remove the timing belt guide number 2. Refer to the timing belt removal and installation procedure in this manual.

31. Remove the camshaft pulley and lower timing cover. Refer to the timing belt removal and installation procedure in this manual.

32. Remove the vane pump.

33. Remove the propeller shaft on 4 wheel drive models.

34. Remove the front exhaust pipe on 2 wheel drive models, or center exhaust pipe on 4 wheel drive models.

35. Remove the number 1 exhaust manifold heat insulator.

36. Remove the manifold stay.

37. Disconnect the oxygen sensor and remove the right hand exhaust manifold.

38. Disconnect the front exhaust pipe assembly.

39. Remove the number 3 manifold converter and number e exhaust manifold hear insulator.

40. Disconnect the oxygen sensor and remove the left hand exhaust manifold.

41. Remove the dipstick tube.

42. Remove the ignition coil.

43. Remove the cylinder head covers.

44. Remove the camshafts and their sub assemblies.

45. Disconnect the VVT sensor connector.

46. Disconnect the camshaft timing oil control valve connector and wire harness clamp.

47. Remove the hexagon bolt on the right hand side head.

48. On the left hand side, make sure to disconnect the ground cable, wire harness clamp bracket and water inlet. Remove the hexagon bolt.

49. Be sure the engine is at/or near ambient temperature and remove the 2 (1 on each head) 8mm recessed hex bolts. Loosen and remove the 8 head bolts evenly, in 3 passes, in the reverse order of the installation sequence. Carefully lift the head

from the engine; if necessary to pry the head loose, take great care not to damage the mating surfaces. Place the head on wood blocks in a clean work area.

➡**If the cylinder head bolts are loosened out of sequence, warpage or cracking could result.**

50. Remove the cylinder head gasket. With a gasket scraper, carefully remove all the old gasket material from the cylinder head and engine block surfaces.

To install:

51. Place the new cylinder head gasket onto the cylinder block.

52. Install the cylinder head, in sequence, using several steps, as follows:

- Cylinder head onto the gasket
- Cylinder head bolts lubricated with clean engine oil
- Tighten the bolts in sequence in 3 steps to 40 ft. lbs. (54 Nm).

➡**If any bolt does not meet the torque, replace it.**

- Mark the forward edge of each bolt with paint, then tighten each bolt, in proper sequence, an additional 90 degrees.

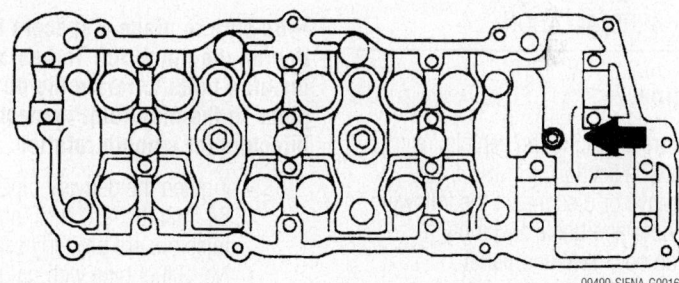

Remove the hexagon bolt on the right hand side head—3.3L engine

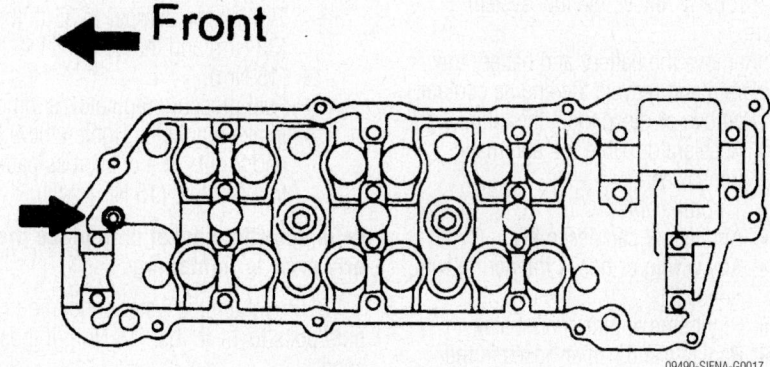

Remove the hexagon bolt on the left hand side head—3.3L engine

- Check that each painted mark is now at a 90 degrees angle to the front

➡**The paint mark applied to the bolt in the 9 o'clock position and should now be in the 12 o'clock position.**

- Tighten the hexagon bolts, lubricated with engine oil. Tighten to 14 ft. lbs. (18 Nm).

53. Tighten the water outlet bolt to 14 ft. lbs. (18 Nm).
54. Install the camshafts.
55. Check and adjust the valves.
56. Apply sealant to the cylinder heads where the camshaft supports meet the cylinder heads. Install the cylinder head covers, using new gaskets.
57. Install the remaining components in the reverse order of removal.
58. Fill the engine with correct grade and amount of oil.
59. Fill the cooling system.
60. Start the engine and check for leaks.
61. Bleed the air from the cooling system.
62. Road test the vehicle and check for unusual noise, shock, slippage, correct shift points and smooth operation.
63. Recheck the coolant and engine oil levels.

Intake Manifold

REMOVAL & INSTALLATION

3.0L Engine

1. Before servicing the vehicle, refer to the precautions section.
2. Remove or disconnect the following:
   - Wiper and blade assembly
   - Top cowl seal and panel
   - Window washer hoses from the ventilator louvers
   - Left and right ventilator louvers
   - Heater air duct
3. Properly relieve the fuel system pressure.
4. Remove the battery and battery tray.
5. Drain and recycle the engine coolant.
6. Remove or disconnect the following:
   - Accelerator cable, on automatic transaxles
   - Throttle cable
   - Air cleaner cap assembly
   - Any wiring or hoses interfering with removal
   - Right side engine mount stay
   - Radiator and heater hoses in the way of the intake manifold removal

- V-bank cover
- All the vacuum hose and wiring for the emission control valve set
- Air intake chamber and discard the gasket
- Exhaust Gas Recirculation (EGR) pipe and discard the gaskets
- Hydraulic motor pressure hose from the air intake chamber
- Engine wiring harnesses from the left side, right side, rear and No. 3 timing belt cover
- Front exhaust pipe, if necessary
- Timing belt, camshaft timing pulleys, No. 2 idler pulley and No. 3 timing belt cover
- Cylinder head rear plate
- 2 bolts, nuts and plate washers with the intake manifold.

➡**The delivery pipes with injectors will be attached to the manifold.**

- Other fuel related components such as the No. 2 fuel pipe and pulsation damper, if needed
- Delivery pipes from the intake manifold

7. Clean and inspect the intake manifold mating surfaces. Scrape all old gasket martial off.

To install:

8. Install or connect the following:
   - Delivery pipes with injectors to the intake manifold.

➡**Be sure to place 4 spacers in position on the manifold. Temporarily install 4 bolts to retain the delivery pipes to the manifold. Inspect the injectors for smooth rotation.**

- Tighten the delivery pipes bolts to 84 inch lbs. (10 Nm), once the injectors are properly seated
- No. 2 fuel pipe with union bolts and gaskets. Tighten the bolts to 24 ft. lbs. (32 Nm).
- No. 1 fuel pipe with pulsation damper, using 4 new gaskets. Tighten the damper to 35 ft. lbs. (32 Nm) and the bolt to 11 ft. lbs. (15 Nm).
- Fuel pressure regulator, if removed
- Intake manifold. Tighten the 9 bolts and 2 nuts in a crisscross pattern to 11 ft. lbs. (15 Nm).

➡**Be sure the gasket is in place properly prior to tightening.**

9. Retighten the water outlet mounting nuts/bolts to 11 ft. lbs. (15 Nm), if loosened.
10. Install or connect the following:

- Air assist hose and water inlet pipe, using a new O-ring, by applying a small amount of soapy water. Tighten the fastener(s) to 14 ft. lbs. (20 Nm).
- Ground strap
- Vacuum hoses removed to the air intake chamber and vacuum tank
- Any remaining components, using new gaskets. Tighten the air intake chamber nuts/bolts to 32 ft. lbs. (43 Nm), the EGR pipe nuts to 108 inch lbs. (12 Nm) and the emission control valve set to 69 inch lbs. (8 Nm).
- Air cleaner assembly
- Heater hoses
- Battery and tray
- Throttle cable with bracket onto the throttle body
- Accelerator cable, by adjusting it, if equipped with an automatic transaxle

11. Refill the cooling system
12. Install or connect the following:
    - Negative battery cable
    - Heater air duct
    - Left and right ventilator louvers
    - Window washer hoses from the ventilator louvers
    - Top cowl seal and panel
    - Wiper and blade assembly
13. Start the engine and inspect for leaks.

3.3L Engine

1. Before servicing the vehicle, refer to the precautions section.
2. Properly relieve the fuel system pressure.
3. Disconnect the negative battery cable.
4. Drain the engine oil.
5. Drain and recycle the engine coolant.
6. Remove the right hand front wheel.
7. Remove the wiper arms.
8. Remove the cowl panel.
9. Remove the wiper link assembly.
10. Remove the cowl top to inner brace.
11. Remove the cowl top cover outer sub assembly.
12. Remove the engine cover.
13. Remove the air cleaner assembly.
14. Remove the emission control valve set.
15. Remove the air surge tank.
16. Disconnect the fuel pipe sub assembly.
17. Disconnect the heater hose inlet pipe.
18. Remove the nut and ground cable.

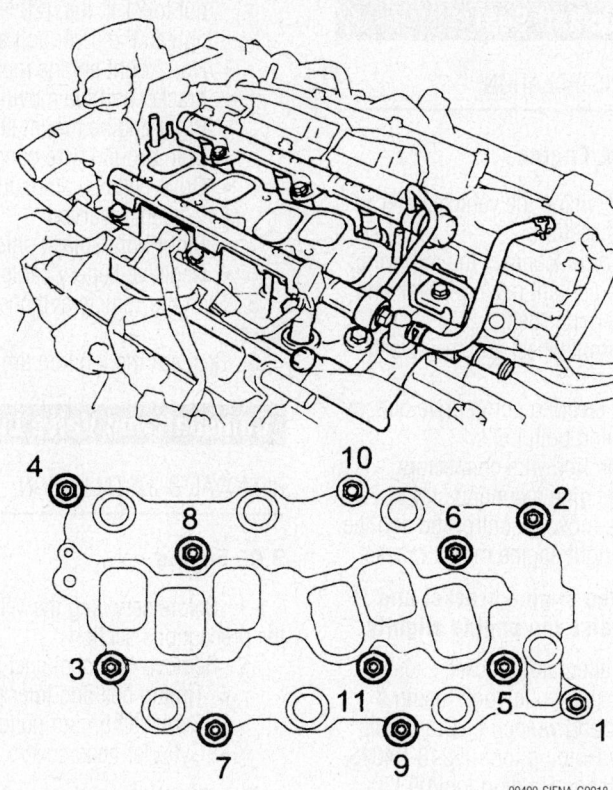

Remove the intake manifold nuts and bolts in the sequence shown—3.3L engine

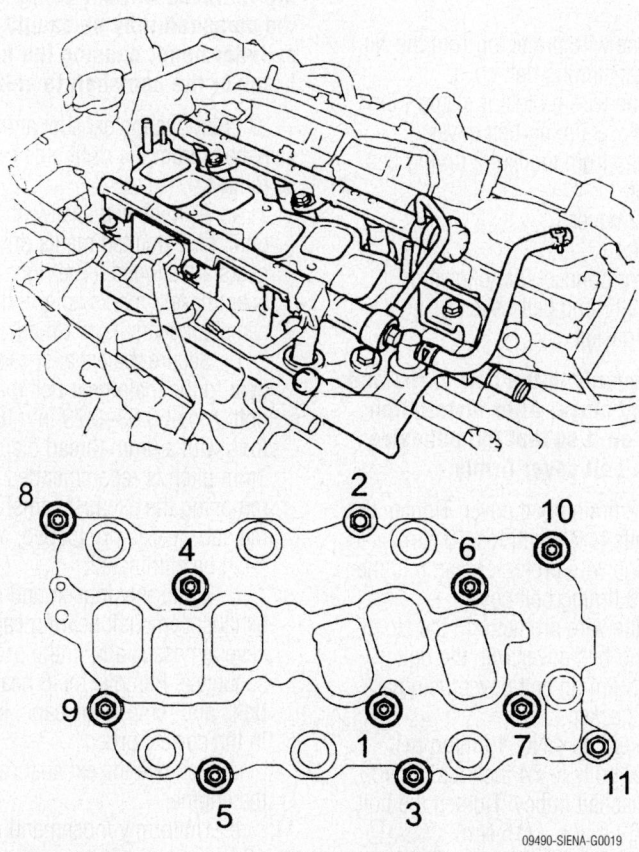

Tighten the intake manifold nuts and bolts in the sequence shown—3.3L engine

19. Disconnect the fuel injector connectors.

20. Remove the intake manifold nuts and bolts in the sequence shown using several passes.

21. Remove the intake manifold.

To install:

22. Install the intake manifold. Tighten the retainers using several passes to 11 ft. lbs. (15 Nm).

23. Install the remaining components in the reverse order of removal.

Exhaust Manifold

REMOVAL & INSTALLATION

3.0L and 3.3L Engine

FRONT MANIFOLD

➡**Removing the oil filter helps gain access to a lower bolt in the front exhaust manifold.**

1. Before servicing the vehicle, refer to the precautions section.

2. Remove or disconnect the following:
   - Negative battery cable
   - Engine undercovers
   - Front exhaust pipe from the exhaust manifolds, by removing the nuts

➡**Check for access to some of the manifold lower bolts, if so remove any possible.**

   - Heated Oxygen (HO$_2$) sensor
   - Exhaust manifold stay, by removing the bolt and nut
   - Remaining exhaust manifold nuts; then, separate the exhaust manifold from the engine

To install:

3. Install or connect the following:
   - Exhaust manifold, using a new gasket. Uniformly, tighten the bolts to 36 ft. lbs. (49 Nm).

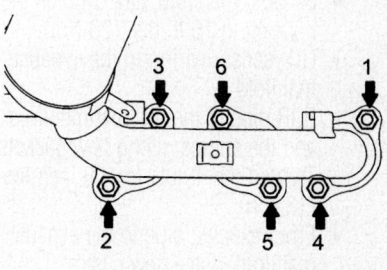

Remove the exhaust manifold nuts and bolts in the sequence shown—3.3L engine

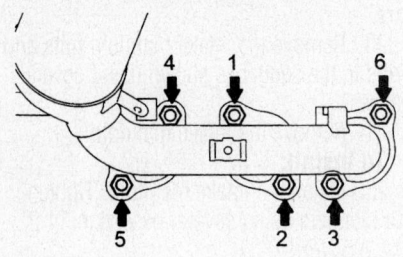

09490-SIENA-G0021

Exhaust manifold torque sequence—3.3L engine

- Exhaust manifold stay. Tighten the nut/bolt to 15 ft. lbs. (20 Nm).
- Heated Oxygen (HO2) sensor to the exhaust manifold
- Front exhaust pipe to the exhaust manifold, using a new gasket. Tighten both nuts to 46 ft. lbs. (62 Nm).
- Engine undercovers
- Negative battery cable

REAR MANIFOLD

1. Before servicing the vehicle, refer to the precautions section.
2. Remove or disconnect the following:
   - Negative battery cable
   - Engine undercovers
   - Front exhaust pipe from both exhaust manifolds, from below the engine
   - Exhaust Gas Recirculation (EGR) pipe from the rear exhaust manifold, by removing the 4 nuts
   - Heated Oxygen (HO2) sensor wiring, from the right exhaust manifold
   - Exhaust manifold stay
   - 6 exhaust manifold nuts and the exhaust manifold

To install:
3. Install or connect the following:
   - Exhaust manifold to the engine, using a new gasket. Tighten the 6 nuts to 36 ft. lbs. (49 Nm).
   - Exhaust manifold stay. Tighten the nut/bolt to 15 ft. lbs. (20 Nm).
   - HO2 sensor wiring to the exhaust manifold
   - EGR pipe to the exhaust manifold and the engine, using new gaskets. Tighten the 4 nuts to 108 inch lbs. (12 Nm).
   - Front exhaust pipe to the exhaust manifold, use a new gasket. Tighten both nuts to 46 ft. lbs. (62 Nm).
   - Engine undercovers
   - Negative battery cable

Front Crankshaft Seal

REMOVAL & INSTALLATION

3.0L and 3.3L Engine

1. Before servicing the vehicle, refer to the precautions section.
2. Remove or disconnect the following:
   - Engine coolant reservoir tank and the alternator belt
   - Right front wheel and the splash shield
   - Power steering pump drive belt, by loosening both bolts
   - Both ground wire connectors
   - Right engine mounting stay
   - Engine moving control rod and the No. 2 right engine mount bracket

➡ **To extract the engine bracket and control rod, raise the engine slightly.**

   - No. 2 alternator bracket
   - Crankshaft pulley bolt, using a prybar and wrench or Crankshaft Pulley Holding tool 09213-54015 and Flange Holding tool 09330-00021
   - Crankshaft pulley, using a puller
   - No. 1 timing belt cover
3. Remove the No. 2 timing belt cover, as follows:
   - Engine wire protector from the No. 3 (rear) timing belt cover
   - Engine wire protector clamp from the No. 3 timing belt cover
   - 5 bolts from the No. 2 timing belt cover
   - No. 2 cover

To install:
4. Install or connect the following:
   - No. 2 timing belt cover, using a new gasket

➡ **Install it evenly to the part of the belt cover shaded black. After installation, press down on it so that the adhesive sticks to the belt cover firmly.**

   - No. 2 timing belt cover. Tighten the 5 bolts to 74 inch lbs. (8 Nm).
   - Engine wire protector clamp to the No. 3 timing belt cover
   - Engine wire protector to the No. 3 timing belt cover with the bolt
   - No. 3 timing belt cover, using a new gasket
   - Tighten the 4 No. 1 timing belt cover bolts to 74 inch lbs. (8 Nm).
   - Crankshaft pulley. Tighten the bolt to 159 ft. lbs. (215 Nm).
   - No. 2 alternator bracket. Tighten the

nut to 21 ft. lbs. (28 Nm). Do not tighten the pivot bolt at this time.
   - No. 2 right engine mounting bracket and the moving control rod
   - Right engine mount stay
   - Both ground wire connectors
   - Drive belts by adjusting them
   - Coolant reservoir
   - Right front splash shield and wheel
   - Negative battery cable
5. Start the vehicle and check for any leaks.
6. Recheck the ignition timing.

Camshaft and Valve Lifters

REMOVAL & INSTALLATION

3.0L Engine

1. Before servicing the vehicle, refer to the precautions section.
2. Remove or disconnect the following:
   - Timing belt and idler pulley
   - Camshaft timing pulleys
   - Cylinder head covers

➡ **The thrust clearance on both the intake and exhaust camshafts is very small; the camshafts must be kept level during removal. If the camshafts are removed without being kept level, the camshaft may be caught in the cylinder head, causing the head to break or the camshaft to seize.**

3. Remove the exhaust and intake camshafts from the right side cylinder head, as follows:
   a. Turn the camshaft with a wrench until the 2 pointed marks drive and driven gears are aligned. (The right camshaft gears have 2 marks apiece; the left side camshaft gears have 1 mark each.)
   b. Secure the exhaust camshaft sub-gear to the main gear using a service bolt. A bolt 0.63–0.79 in. (16–20mm) long with a 6mm thread diameter and a 1mm pitch is recommended. When removing the exhaust camshaft be sure the sub-gear is not loaded; all the force must be eliminated.
   c. Uniformly loosen and remove the exhaust camshaft bearing cap bolts in several passes and in the proper sequence. Remove the 8 bearing cap bolts and remove the caps, keeping them in the correct order.
   d. Remove the exhaust camshaft from the engine.
   e. Uniformly loosen and remove the 10 bearing cap bolts in several passes,

Intake

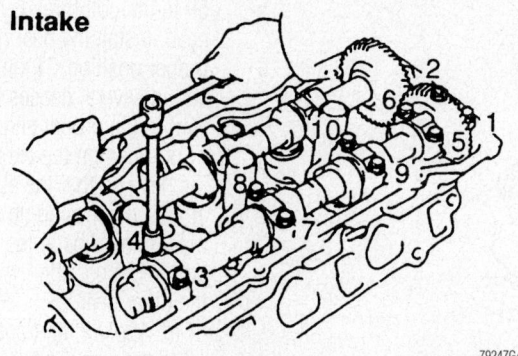

7924ZG44

Right intake camshaft bearing cap bolt loosening sequence—3.0L engine

Exhaust

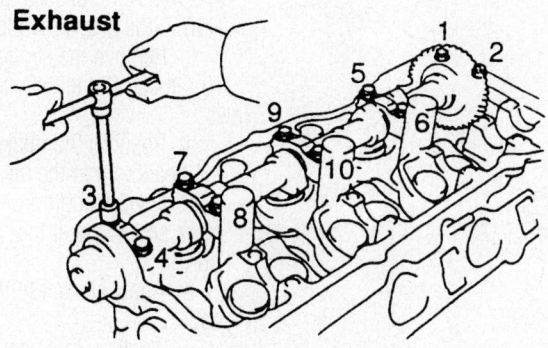

7924ZG45

Right side exhaust camshaft bearing cap bolt loosening sequence—3.0L engine

Intake

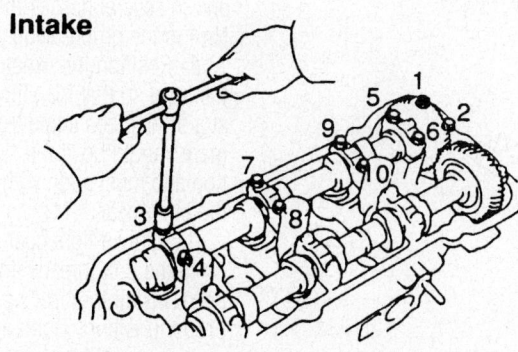

7924ZG46

Left intake camshaft bearing cap bolt loosening sequence—3.0L engine

Exhaust

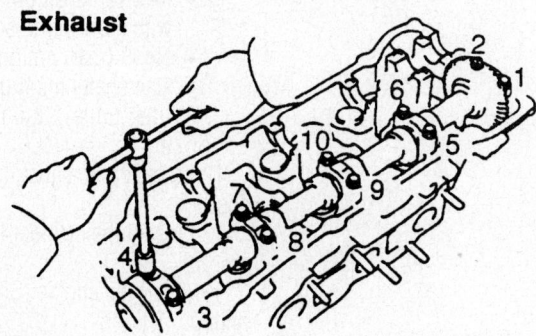

7924ZG47

Left side exhaust camshaft bearing cap bolt loosening sequence—3.0L engine

in the proper sequence. Remove the bearing caps, keeping them in order, remove the oil seal, then lift out the intake camshaft.

4. Remove the exhaust and intake camshafts from the left side cylinder head, as follows:

a. Turn the camshaft with a wrench until the pointed marks on the drive and driven gears are aligned. (The right camshaft gears have 2 marks apiece; the left side camshaft gears have 1 mark each.)

b. Secure the exhaust camshaft sub-gear to the main gear using a service bolt. A bolt 16–20mm long with a 6mm thread diameter and a 1mm pitch is recommended. When removing the exhaust camshaft be sure the sub-gear is not loaded; all the force must be eliminated.

c. Uniformly loosen and remove the exhaust camshaft bearing cap bolts in several passes and in the proper sequence. Remove the 8 bearing cap bolts and remove the caps. Keep the caps in the correct order.

d. Remove the exhaust camshaft from the engine.

e. Uniformly loosen and remove the 10 bearing cap bolts in several passes, in the reverse order of the installation sequence. Remove the bearing caps, keeping them in order, remove the oil seal, then lift out the intake camshaft.

5. Remove the valve lifter shims and hydraulic lifters. Identify each lifter and shim as it is removed so it can be reinstalled in the same position. If the lifters are to be reused, store them upside down in a sealed container.

To install:

6. Install the valve lifters into their original positions and install the shims. Check valve clearance and replace the shims as necessary.

7. When reinstalling, remember that the camshafts must be handled carefully and kept straight and level to avoid damage.

8. Before installing the camshafts in either cylinder head, apply multi-purpose grease to each camshaft.

9. Install the right camshafts, as follows:

a. Position the intake camshaft on the head so that the alignment marks are at a 90 degrees angle from vertical. The mark should be at the "3 o'clock" position.

b. Apply sealant to the No. 1 bearing cap.

c. Apply a light coat of clean engine

Exhaust

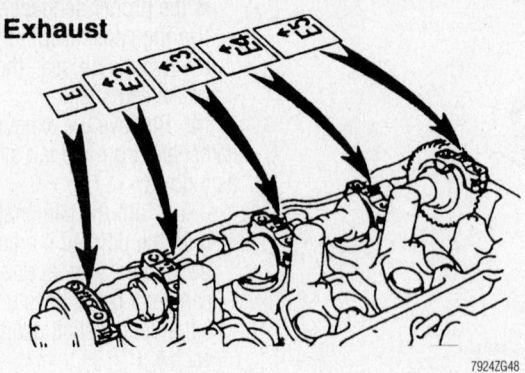

7924ZG48

Right exhaust bearing caps must be placed in their proper locations—3.0L engine

Exhaust

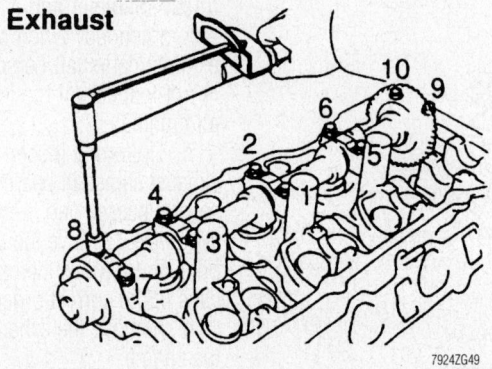

7924ZG49

Right exhaust camshaft bearing cap bolt tightening sequence—3.0L engine

Intake

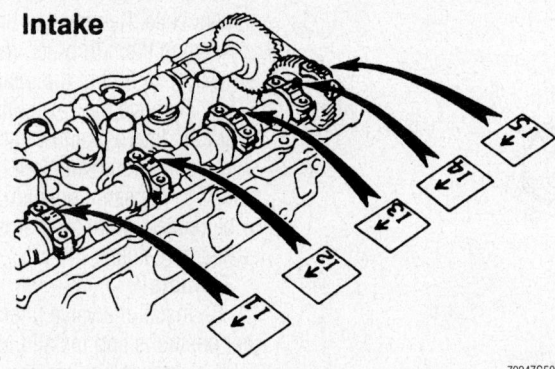

7924ZG50

Right intake bearing caps must be placed in their proper locations—3.0L engine

Intake

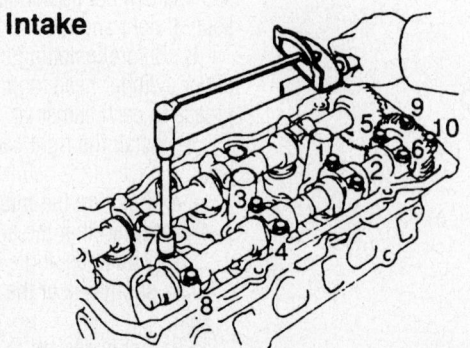

7924ZG51

Right intake camshaft bearing cap bolt tightening sequence—3.0L engine

oil to the bolt threads and under the bolt head. Install the bearing caps to their proper position. Tighten the bolts evenly and in several passes to 12 ft. lbs. (16 Nm) in the proper sequence.

d. Position the exhaust camshaft on the head so that the alignment marks are at a 90 degrees angle from vertical. The mark should be at the "9 o'clock" position and must align with the marks on the other gear.

e. Apply a light coat of clean engine oil to the bolt threads and under the bolt head. Install the bearing caps to their proper position. Tighten the bolts evenly and in several passes to 12 ft. lbs. (16 Nm) in the proper sequence.

f. Remove the service bolt.

10. Install the left camshafts, as follows:

a. Position the intake camshaft on the head so that the alignment mark is at a 90 degrees angle from vertical. The mark should be at the "9 o'clock" position.

b. Apply sealant to the No. 1 bearing cap.

c. Apply a light coat of clean engine oil to the bolt threads and under the bolt head. Install the bearing caps to their proper position. Tighten the bolts evenly and in several passes to 12 ft. lbs. (16 Nm) in the proper sequence.

d. Position the exhaust camshaft on the head so that the alignment marks are at a 90 degrees angle from vertical. The mark should be at the "3 o'clock" position and must align with the marks on the other gear.

e. Apply a light coat of clean engine oil to the bolt threads and under the bolt head. Install the bearing caps to their proper position. Tighten the bolts evenly and in several passes to 12 ft. lbs. (16 Nm) in the proper sequence.

f. Remove the service bolt.

11. Install or connect the following:
- New camshaft oil seals, lubricated with multi-purpose grease
- No. 3 (rear) timing belt cover
- Camshaft timing gears
- Idler pulley, timing belt and covers

12. Check and adjust the valve clearance.

13. Install the cylinder head (valve) covers.

14. Start the engine. Check the ignition timing.

15. Test drive the vehicle.

16. Check all fluid levels.

Exhaust

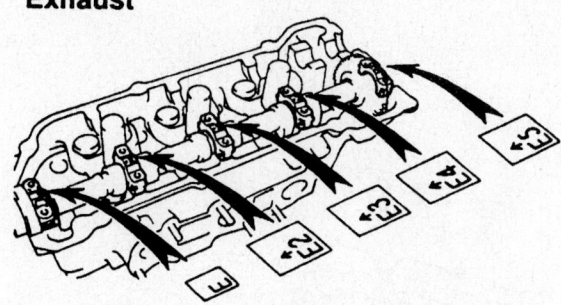

Exhaust

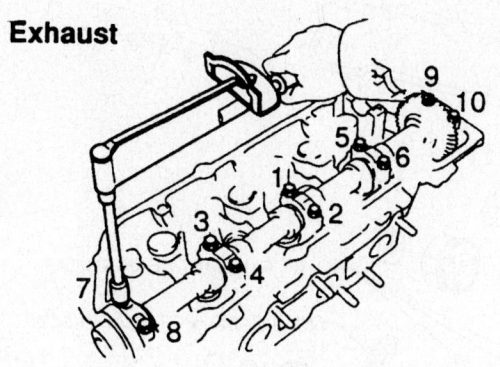

Left exhaust bearing caps locations and bolt tightening sequence—3.0L engine

Intake

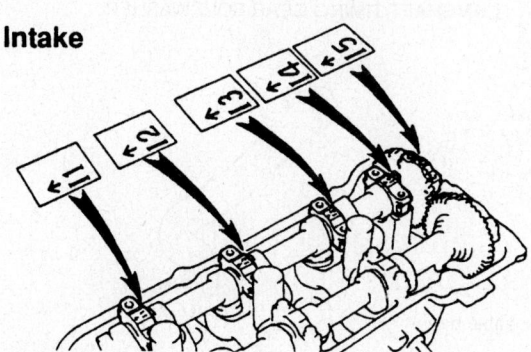

Intake

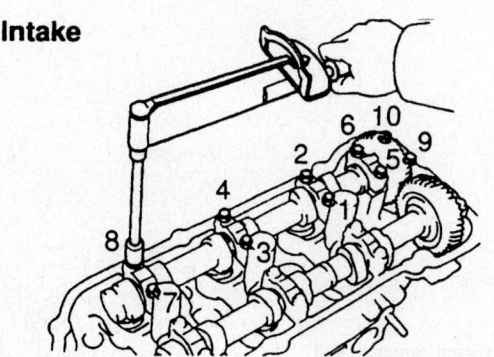

Left intake camshaft bearing cap locations and bolt tightening sequence—3.0L engine

3.3L Engine

LEFT SIDE

1. Before servicing the vehicle, refer to the precautions section.
2. Properly relieve the fuel system pressure.
3. Drain and recycle the engine coolant.
4. Disconnect the negative battery cable.
5. Remove the right hand front wheel.
6. Remove the wiper arms.
7. Remove the cowl panel.
8. Remove the wiper link assembly.
9. Remove the cowl top to inner brace.
10. Remove the cowl top cover outer sub assembly.
11. Remove the radiator inlet hose.
12. Remove the ignition coil.
13. Remove the cylinder head cover.
14. Remove the apron seal.
15. Remove the A/C compressor to crankshaft pulley belt.
16. Remove the vane pump belt.
17. Remove the engine moving control rod bolts and rod.
18. Remove the number 2 engine mounting stay.
19. Remove the number 2 alternator bracket and the alternator belt adjusting bar.
20. Remove the crankshaft pulley.
21. Remove the timing belt covers 1 and 2. Refer to the timing belt removal and installation procedure in this manual.
22. Remove the right hand engine mount bracket.
23. Remove the timing belt guide number 2. Refer to the timing belt removal and installation procedure in this manual.
24. Remove the timing belt.
25. Remove the timing belt guide number 2 idler sub assembly. Refer to the timing belt removal and installation procedure in this manual.
26. Remove the camshaft pulley and lower timing cover. Refer to the timing belt removal and installation procedure in this manual.
27. Remove the timing belt number 3 cover. Refer to the timing belt removal and installation procedure in this manual.
28. Align the timing marks of the camshaft drive and driven gears by using a wrench to turn the camshaft.
29. Attach the exhaust camshaft sub gear to the main gear with a 16–20mm long x 6mm thread diameter bolt and tighten to 48 inch lbs. (5 Nm).

➡When removing the camshaft, be sure the torsional spring force of the sub gear has been taken up by the service bolt.

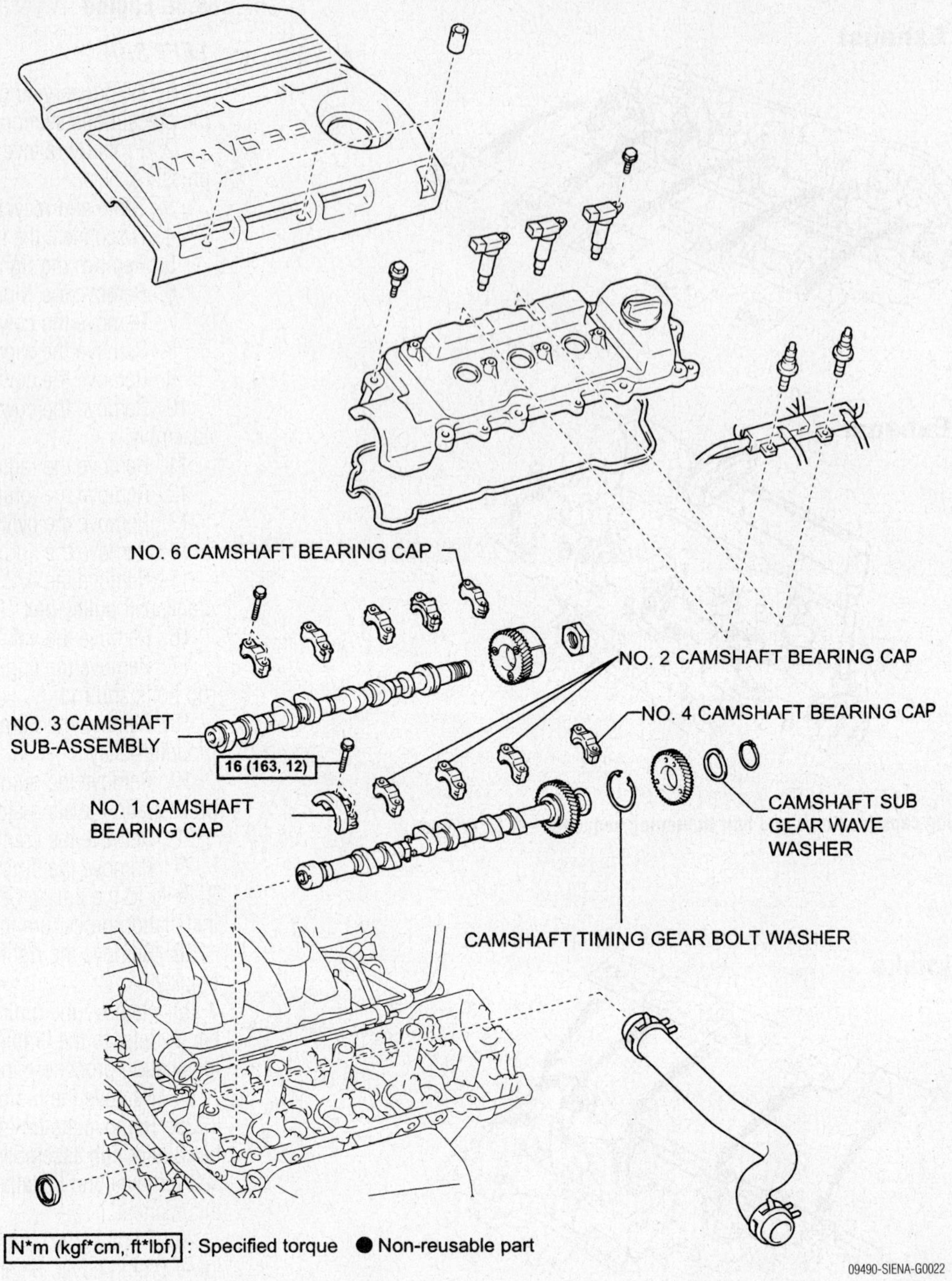

NO. 6 CAMSHAFT BEARING CAP

NO. 3 CAMSHAFT SUB-ASSEMBLY

16 (163, 12)

NO. 1 CAMSHAFT BEARING CAP

NO. 2 CAMSHAFT BEARING CAP

NO. 4 CAMSHAFT BEARING CAP

CAMSHAFT SUB GEAR WAVE WASHER

CAMSHAFT TIMING GEAR BOLT WASHER

N*m (kgf*cm, ft*lbf) : Specified torque ● Non-reusable part

09490-SIENA-G0022

Exploded view of the left side camshaft assemblies—3.3L engine

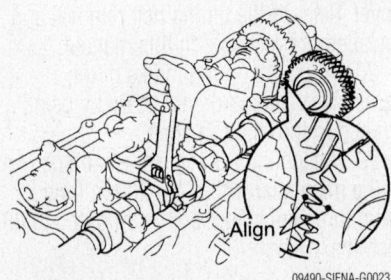

Align

09490-SIENA-G0023

Align the timing marks of the left side camshaft drive and driven gears by using a wrench to turn the camshaft—3.3L engine

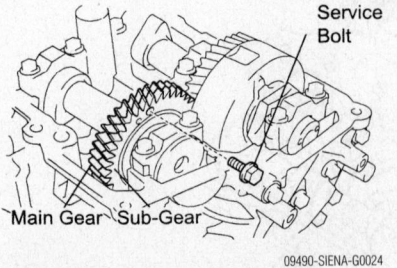

Service Bolt

Main Gear   Sub-Gear

09490-SIENA-G0024

Attach the left side exhaust camshaft sub gear to the main gear with a 16–20mm long x 6mm thread diameter bolt—3.3L engine

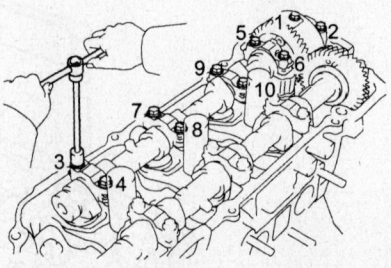

09490-SIENA-G0025

Loosen the number 3 camshaft cap bolts in sequence—3.3L engine

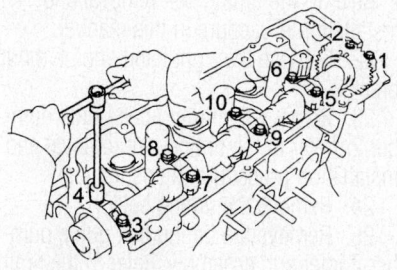

Loosen the number 4 camshaft cap bolts
in sequence—3.3L engine

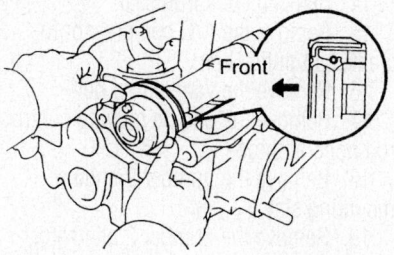

Make sure to not turn the seal lip over and
insert the seal until it stops—3.3L engine

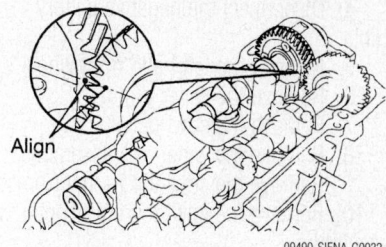

Align the timing marks on the number 3
camshaft as shown—3.3L engine

30. Mark the bearing caps prior to
removal so they can be installed in their
original positions.

31. Loosen the number 3 camshaft cap
bolts in sequence using several steps.

32. Loosen the number 4 camshaft cap
bolts in sequence using several steps.

To install:

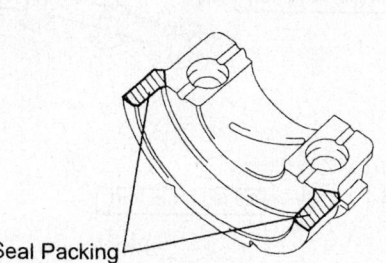

Apply seal packing to the number 1 bearing cap at the locations shown—3.3L
engine

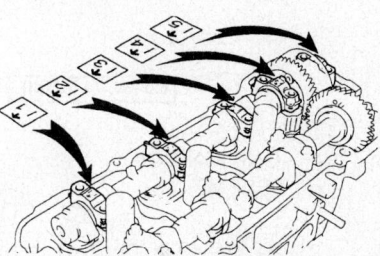

Install the number 3 camshaft bearing
caps in their original locations—3.3L

⁂ WARNING

**The camshaft has a small thrust
clearance, make sure to keep the
camshaft level during installation to
avoid damage to the camshaft and
cylinder head.**

33. Apply clean engine oil to the
camshaft thrust and journal locations.

34. Place the number 4 camshaft at a 90
degree angle of timing on the cylinder head
as illustrated.

35. Apply a multi purpose grease to a
new oil seal lip and install the seal. Make
sure to not turn the seal lip over and insert
the seal until it stops. Remove any packing
material from the seal surface.

36. Apply seal packing to the number 1
bearing cap as illustrated. Install the cap
within 5 minutes of applying the packing.
Do not let the seal come into contact with
engine oil until at least two hours after it
has been installed.

37. Install the number 4 camshaft bearing caps in their original locations, apply a

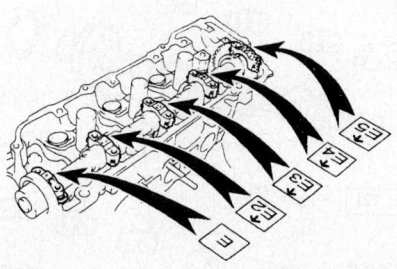

Install the number 4 camshaft bearing
caps in their original locations—3.3L
engine

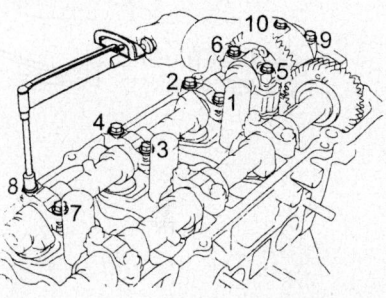

Tighten the number 3 camshaft bearing
caps in this sequence—3.3L engine

light coat of oil to the cap bolt thread and
using several passes; tighten the bolts to 12
ft. lbs. (16 Nm) in the sequence shown.

38. Align the timing marks on the number 3 camshaft as shown.

39. Install the number 3 camshaft bearing caps in their original locations, apply a
light coat of oil to the cap bolt thread and
using several passes; tighten the bolts to 12
ft. lbs. (16 Nm) in the sequence shown.

40. Install the remaining components in
the reverse order of removal.

RIGHT SIDE

1. Before servicing the vehicle, refer to
the precautions section.

2. Properly relieve the fuel system
pressure.

3. Drain and recycle the engine coolant.

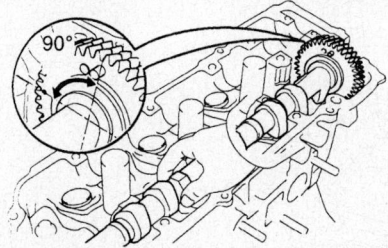

Place the number 4 camshaft at a 90
degree angle of timing on the cylinder
head—3.3L engine

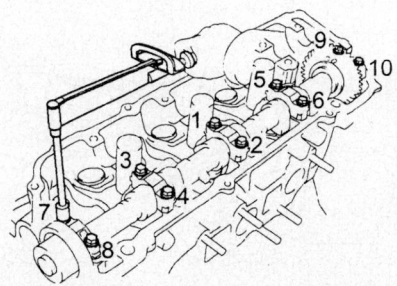

Tighten the number 4 camshaft bearing
caps in this sequence—3.3L engine

4. Disconnect the negative battery cable.

5. Remove the left hand front wheel.

6. Remove the wiper arms.

7. Remove the cowl panel.

8. Remove the wiper link assembly.

9. Remove the cowl top to inner brace.

10. Remove the cowl top cover outer sub assembly.

11. Remove the radiator inlet hose.

12. Remove the ignition coil.

13. Remove the cylinder head cover.

14. Remove the apron seal.

15. Remove the A/C compressor to crankshaft pulley belt.

16. Remove the vane pump belt.

17. Remove the engine moving control rod bolts and rod.

18. Remove the number 2 engine mounting stay.

19. Remove the number 2 alternator bracket and the alternator belt adjusting bar.

20. Remove the crankshaft pulley.

21. Remove the timing belt covers 1 and

2. Refer to the timing belt removal and installation procedure in this manual.

22. Remove the right hand engine mount bracket.

23. Remove the timing belt guide number 2. Refer to the timing belt removal and installation procedure in this manual.

24. Remove the timing belt.

25. Remove the timing belt guide number 2 idler sub assembly. Refer to the timing belt removal and installation procedure in this manual.

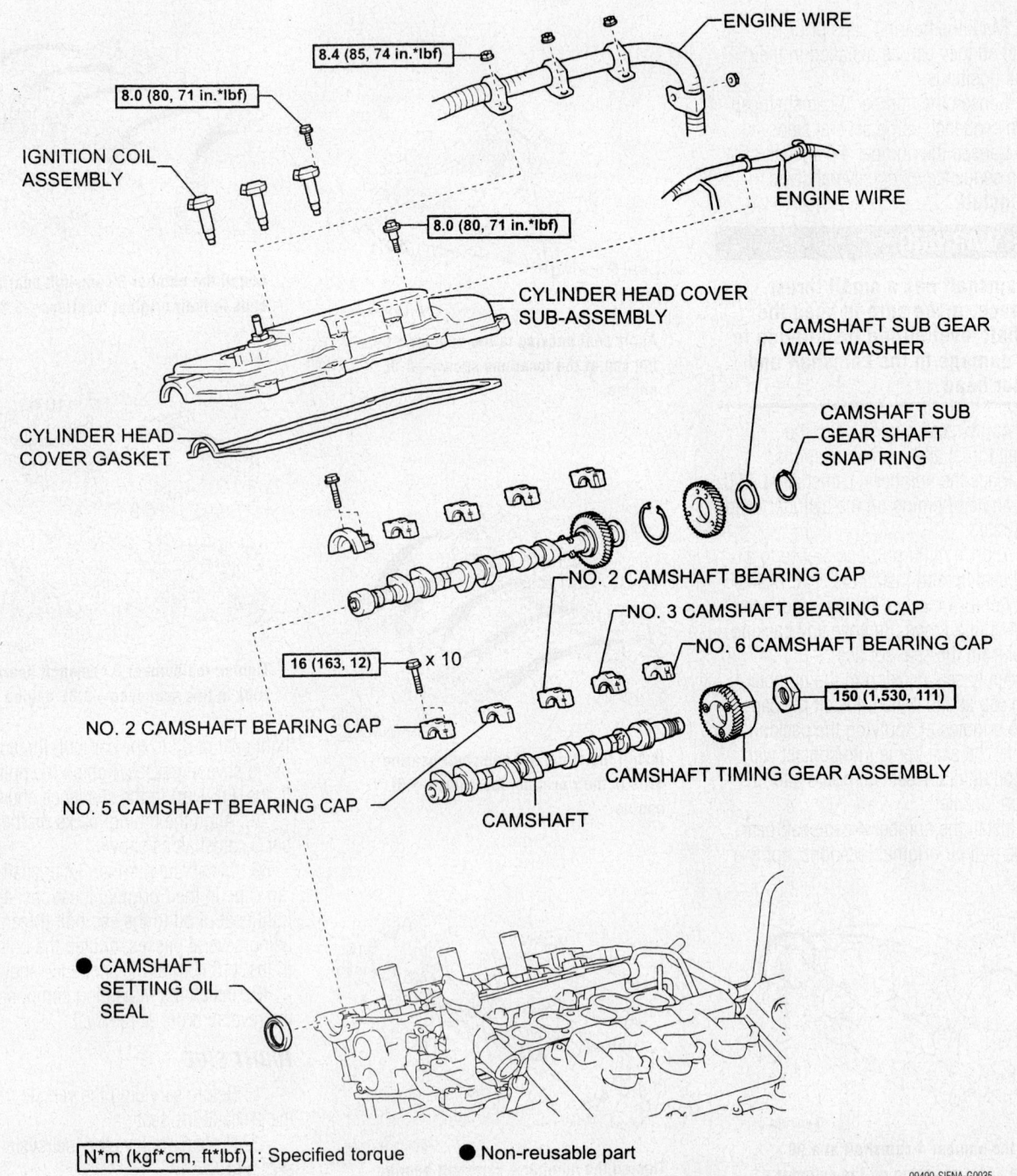

Exploded view of the right side camshaft assemblies—3.3L engine

09490-SIENA-G0035

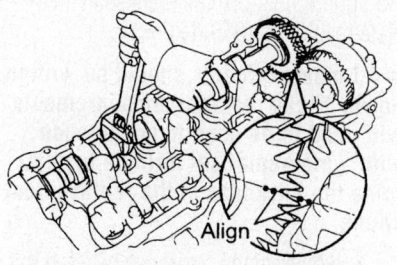

Align the timing marks of the right side camshaft drive and driven gears by using a wrench to turn the camshaft—3.3L engine

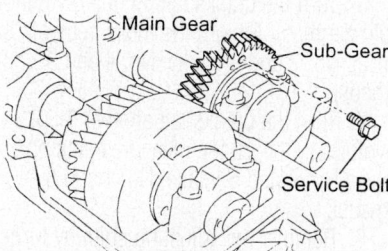

Attach the right side exhaust camshaft sub gear to the main gear with a 16–20mm long x 6mm thread diameter bolt—3.3L engine

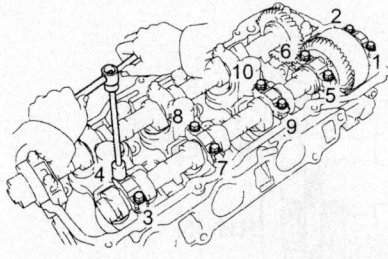

Loosen the number 1 camshaft cap bolts in sequence—3.3L engine

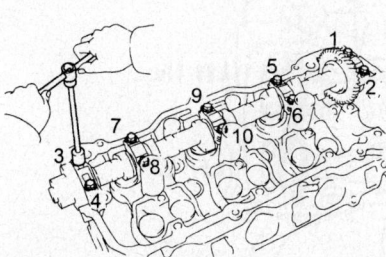

Loosen the number 2 camshaft cap bolts in sequence—3.3L engine

26. Remove the camshaft pulley and lower timing cover. Refer to the timing belt removal and installation procedure in this manual.

27. Remove the timing belt number 3 cover. Refer to the timing belt removal and installation procedure in this manual.

28. Align the timing marks of the camshaft drive and driven gears by using a wrench to turn the camshaft.

29. Attach the exhaust camshaft sub gear to the main gear with a 16–20mm long x 6mm thread diameter bolt and tighten to 48 inch lbs. (5 Nm).

➡ When removing the camshaft, be sure the torsional spring force of the sub gear has been taken up by the service bolt.

30. Mark the bearing caps prior to removal so they can be installed in their original positions.

31. Loosen the number 1 camshaft cap bolts in sequence using several steps.

32. Loosen the number 2 camshaft cap bolts in sequence using several steps.

To install:

✳✳ WARNING

The camshaft has a small thrust clearance, make sure to keep the camshaft level during installation to avoid damage to the camshaft and cylinder head.

33. Apply clean engine oil to the camshaft thrust and journal locations.

34. Place the number 2 camshaft at a 90 degree angle of timing on the cylinder head as illustrated.

35. Apply a multi purpose grease to a new oil seal lip and install the seal. Make sure to not turn the seal lip over and insert the seal until it stops. Remove any packing material from the seal surface.

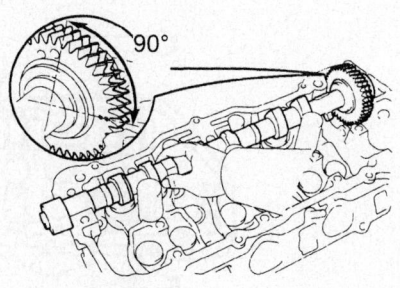

Place the number 2 camshaft at a 90 degree angle of timing on the cylinder head—3.3L engine

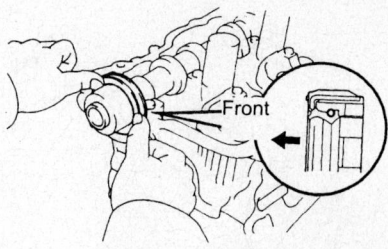

Make sure to not turn the seal lip over and insert the seal until it stops—3.3L engine

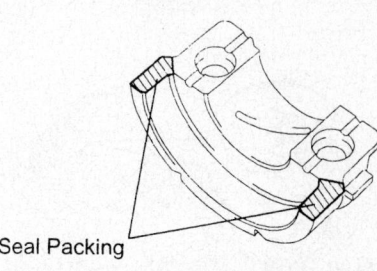

Apply seal packing to the number 1 bearing cap as illustrated—3.3L engine

36. Apply seal packing to the number 1 bearing cap as illustrated. Install the cap within 5 minutes of applying the packing. Do not let the seal come into contact with engine oil until at least two hours after it has been installed.

37. Install the number 2 camshaft bearing caps in their original locations, apply a light coat of oil to the cap bolt thread and using several passes; tighten the bolts to 12 ft. lbs. (16 Nm) in the sequence shown.

38. Align the timing marks on the number 1 camshaft as shown.

39. Install the number 1 camshaft bearing caps in their original locations, apply a light coat of oil to the cap bolt thread and using several passes; tighten the bolts to 12 ft. lbs. (16 Nm) in the sequence shown.

40. Install the remaining components in the reverse order of removal.

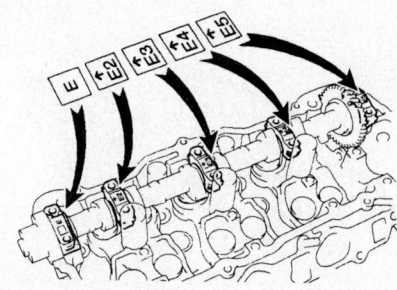

Install the number 2 camshaft bearing caps in their original locations—3.3L

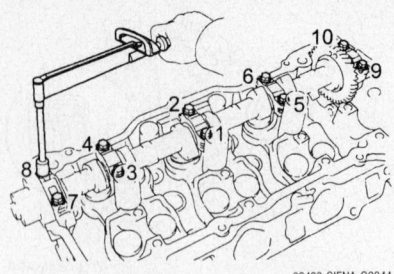

09490-SIENA-G0044

Tighten the number 2 camshaft bearing caps in this sequence—3.3L engine

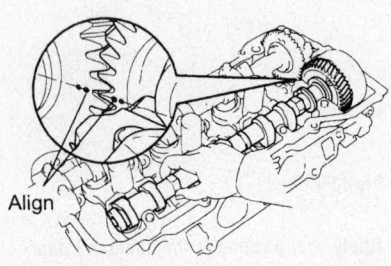

Align

09490-SIENA-G0045

Align the timing marks on the number 1 camshaft as shown—3.3L engine

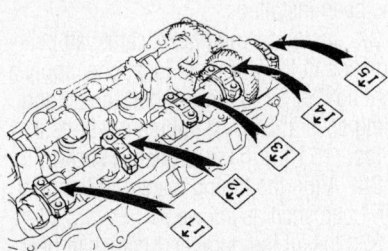

09490-SIENA-G0046

Install the number 1 camshaft bearing caps in their original locations—3.3L engine

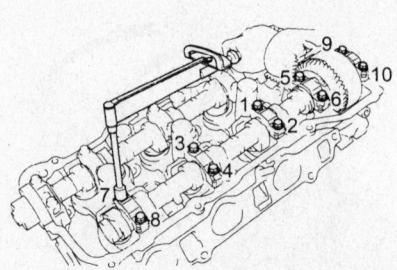

09490-SIENA-G0047

Tighten the number 1 camshaft bearing caps in this sequence—3.3L engine

Valve Lash

ADJUSTMENT

3.0L and 3.3L Engine

➡**Adjust the valve clearance when the engine is cold.**

1. Before servicing the vehicle, refer to the precautions section.

2. Remove or disconnect the following:
- Negative battery cable. If equipped with an air bag, wait at least 90 seconds before proceeding.
- Accelerator/throttle cable from the throttle linkage
- Air cleaner cover, air flow meter and air duct assembly
- V-bank cover
- Emission control valve set
- Air intake chamber
- Engine harness from the injectors and the ignition coils
- Ignition coils and keep them in order for reassembly
- Spark plugs
- Cylinder head covers

3. Turn the crankshaft pulley and align its groove with the timing mark **0** of the No. 1 timing cover.

4. Check that the valve lifters on the No. 1 intake are loose and the No. 1 exhaust are tight. If not, turn the crankshaft 1 complete revolution (360 degrees).

➡**All measurements should be written down. These recorded measurements will need to be used in conjunction with a mathematical formula to determine the thickness of the replacement shims.**

5. Measure the clearance between the valve lifters and the camshaft. Record the measurements on valves No. 1 and 6 intake; No. 2 and 3 exhaust.

   a. The intake valve clearance cold is 0.006–0.010 in. (0.15–0.25mm).

   b. The exhaust valve clearance cold is 0.010–0.014 in. (0.25–0.35mm).

6. Turn the crankshaft ⅔ of a revolution (240 degrees). Record the measurements on valves No. 2 and 3 intake; No. 4 and 5 exhaust.

7. Turn the crankshaft another ⅔ of a revolution. Record the measurements on valves No. 4 and 5 intake; No. 1 and 6 exhaust.

8. Remove the adjusting shim by turning the crankshaft to position the cam lobe of the camshaft in the up position on the valve to be adjusted. Using a small thin flat bladed tool, turn the valve lifter so that the notches are perpendicular to the camshaft. Press down the valve lifter with tool 09248-55010 part A. Place too 09248-55010 part

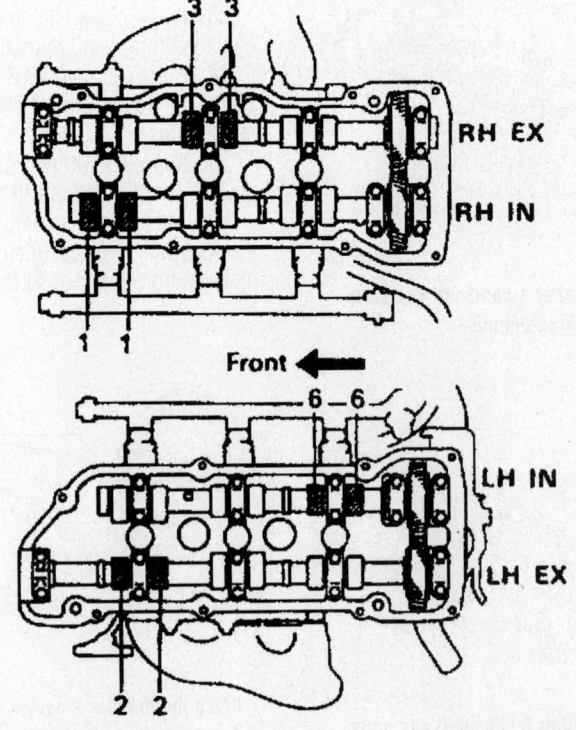

7923VG65

Adjust these valves during the 1st step

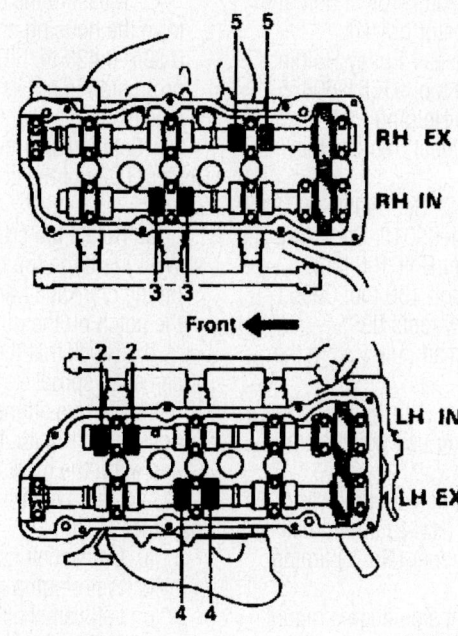

Adjust these valves during the 2nd step

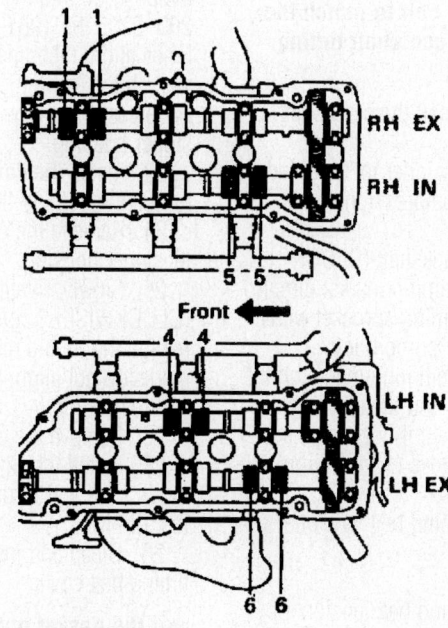

Adjust these valves during the 3rd step

B between the camshaft and the valve lifter; remove part A.

9. Remove the adjusting shim with a magnet and a small screwdriver.

10. Determine the replacement adjusting shim size by either using the charts or the following formulas:

- Intake: N = T + (A—0.008 in./0.020mm)
- Exhaust: N = T + (A—0.012 in./0.30mm)

- T = Thickness of removed shim
- A = Measured valve clearance
- N = Thickness of new shim

11. Select a new shim with a thickness as close as possible to the calculated value. Install the new replacement shim.

➡**Shims are available in 17 sizes in increments of 0.0020 in. (0.050mm), from 0.0984 in. (2.500mm) to 0.1299 in. (3.300mm).**

12. Recheck the valve clearance.
13. Install or connect the following:
- Cylinder head covers
- Spark plugs and the ignition coils
- Engine wiring harness to the injectors and the coils
- Intake chamber
- Emission control valve set
- V-bank cover
- Air flow meter, air duct and air cleaner cover
- Negative battery cable

Starter Motor

REMOVAL & INSTALLATION

3.0L Engine

1. Before servicing the vehicle, refer to the precautions section.
2. Remove or disconnect the following:
- Battery
- Battery tray
3. Remove or disconnect the cruise control actuator, if equipped, as follows:
- Actuator connector and clamp
- 3 bolts and the actuator with the bracket
4. Remove or disconnect the following:
- Automatic transaxle shift control cable
- Engine wiring
- Starter electrical connectors
- Both bolts, shift control cable clamp and the starter

To install:
5. Install or connect the following:
- Starter and the shift control cable clamp. Tighten the bolts to 27 ft. lbs. (37 Nm).
- Starter electrical connectors
- Engine wiring
- Automatic transaxle shift control cable
6. Install or connect the following, if equipped with cruise control:
- 3 bolts and the actuator with the bracket
- Actuator connector and clamp
7. Install or connect the following:
- Battery tray
- Battery

3.3L Engine

1. Before servicing the vehicle, refer to the precautions section.
2. Remove the battery and tray.
3. Remove the air cleaner assembly and inlet tubes.
4. Remove the air cleaner bracket.

5. Remove the wiring from the starter.

6. Remove the 2 bolts and lower the starter.

7. Installation is the reverse of removal. Torque the starter bolts to 26 ft. lbs. (37 Nm).

Timing Belt

REMOVAL & INSTALLATION

3.0L Engine

1. Disconnect the negative battery cable.

2. Remove the outer front cowl top panel assembly by performing the following procedure:

a. Remove the wiper arm/blade assemblies head caps, nuts and assemblies.

b. Remove the head-to-cowl seal and the cowl panel hole cover.

c. Disconnect the windshield washer clip and hose.

d. Remove both (right and left) cowl top ventilator louvers.

e. Disconnect the electrical connector from the windshield wiper motor.

f. Remove the outer front cowl top panel assembly-to-cowl bolts and the panel.

3. Raise and safely support the vehicle.

4. Remove the right front wheel assembly and apron seal.

5. Remove the alternator by performing the following procedure:

a. Loosen the pivot bolt, adjusting lockbolt and adjusting bolt, then, remove the drive belt.

b. Disconnect the alternator's electrical connector.

c. Remove the nut and the alternator wire.

d. Disconnect the wiring harness from the clip.

e. Remove the alternator-to-bracket pivot bolt, the washer, adjusting lockbolt and alternator.

6. Loosen the power steering pump's mount and adjusting bolt, then, remove the drive belt.

7. Disconnect the hose from the engine coolant reservoir.

8. Disconnect the Diagnostic Link Connector 1 (DLC1) from the No. 2 right side engine mounting bracket.

9. Remove the right side engine mounting stay, the engine moving control rod and the No. 2 right side engine mounting bracket.

10. Loosen the alternator's pivot bolt, the nut and the No. 2 alternator bracket.

11. Using the Crankshaft Pulley Holding tool 09213-54015, Bolt tool 91651-60855 and Companion Flange Holding tool 09330-00021, or equivalent, remove the crankshaft pulley bolt.

12. Using a Puller "C" Set 09950-50011 (Hanger 150 tool 09951-05010, Slide Arm tool 09952-5010, Center Bolt 100 tool 09953-05010, Center Bolt 150 tool 09953-05020 and 2 No. 2 Claw tools 09954-05020), pull the crankshaft pulley from the crankshaft.

13. Remove the lower (No. 1) timing belt cover. Remove the timing belt guide from the crankshaft.

14. Remove the engine wire protector clamps from the upper (No. 2) timing belt cover and remove the upper (No. 2) timing belt cover.

15. Remove the right side engine mounting brace

➡If reusing the timing belt, be sure that you can still read the installation marks. If not, place new installation marks on the timing belt to match the timing marks of the camshaft timing pulleys.

16. Temporarily install the crankshaft pulley bolt.

17. Set the No. 1 cylinder to Top Dead Center (TDC) of the compression stroke, as follows:

a. Rotate the crankshaft (CLOCKWISE) to align the timing marks: dimple on the crankshaft timing sprocket with the notch on the oil pump body.

b. Check that the timing marks on the camshaft sprockets and the rear timing belt cover are aligned; if not, rotate the crankshaft 360 degrees (1 revolution) and align the marks.

18. Remove the timing belt tensioner and the timing belt.

To install:

19. Inspect the timing belt tensioner by performing the following procedures:

a. Inspect the seal for leakage; if leakage is suspected, replace the tensioner.

b. Using both hands to hold the tensioner facing upward, strongly press the pushrod against a solid surface. If the pushrod moves, replace the tensioner.

❊❊ WARNING

Never hold the tensioner with the pushrod facing downward.

c. Measure the pushrod protrusion from the housing end, it should be 0.394–0.425 in. (10.0–10.8mm); if the protrusion is not as specified, replace the tensioner.

20. Set the No. 1 cylinder to Top Dead Center (TDC) of the compression stroke, as follows:

a. Rotate the crankshaft (CLOCKWISE) to align the timing marks: dimple on the crankshaft timing sprocket with the notch on the oil pump body.

b. Check that the timing marks on the camshaft sprockets and the rear timing belt cover are aligned; if not, rotate the crankshaft 1 revolution (360 degrees) and align the marks.

21. Install the timing belt in the following order:

a. Crankshaft timing sprocket

b. Water pump pulley

c. Left camshaft timing sprocket

d. No. 2 idler pulley

e. Right camshaft sprocket

f. No. 1 idler pulley

22. Using a vertical press, slowly press the pushrod into the housing using 200–2205 lbs. (981–9807 N) until the holes align, then, install a 1.27mm Allen® wrench to secure the pushrod and release the press. Install the dust boot on the tensioner housing.

23. Install the timing belt tensioner and torque the bolts to 20 ft. lbs. (27 Nm).

24. Remove the Allen® wrench from the tensioner housing.

25. Slowly, rotate the crankshaft (CLOCKWISE) 2 complete revolutions and realign the timing marks. If the timing marks do not align, remove the timing belt and reinstall it.

26. Remove the crankshaft pulley bolt.

27. Install the right side engine mounting bracket and torque the bolts to 21 ft. lbs. (28 Nm).

28. Clean and install the upper (No. 2) timing belt cover.

➡If the gasket material on the timing belt covers is cracked, peeling or etc., replace it.

29. Install the timing belt guide on the crankshaft with the cup side facing outward.

30. Clean and install the lower (No. 1) timing belt cover.

➡If the gasket material on the timing belt covers is cracked, peeling or etc., replace it.

31. Install the crankshaft pulley.

32. Using the Crankshaft Pulley Holding tool 09213-54015, Bolt tool 91651-60855

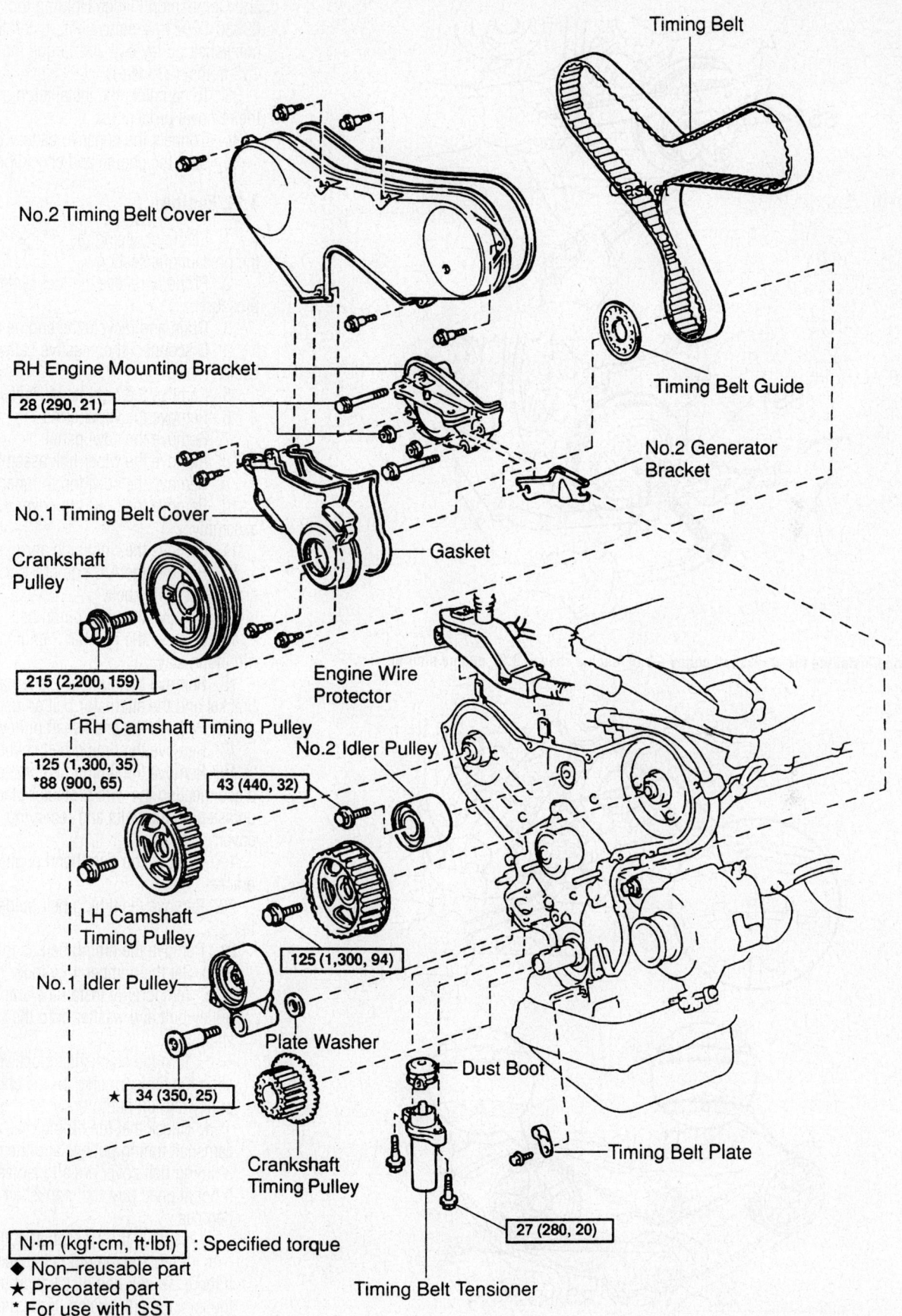

Timing Belt

No.2 Timing Belt Cover

Gasket

Timing Belt Guide

RH Engine Mounting Bracket

28 (290, 21)

No.2 Generator Bracket

No.1 Timing Belt Cover

Crankshaft Pulley

Gasket

215 (2,200, 159)

Engine Wire Protector

RH Camshaft Timing Pulley

No.2 Idler Pulley

125 (1,300, 35)
*88 (900, 65)

43 (440, 32)

LH Camshaft Timing Pulley

125 (1,300, 94)

No.1 Idler Pulley

Plate Washer

Dust Boot

★ 34 (350, 25)

Crankshaft Timing Pulley

Timing Belt Plate

27 (280, 20)

N·m (kgf·cm, ft·lbf) : Specified torque
◆ Non–reusable part
★ Precoated part
* For use with SST

Timing Belt Tensioner

93025G19

Exploded view of the timing belt assembly—3.0L engine shown, 3.3L engine similar

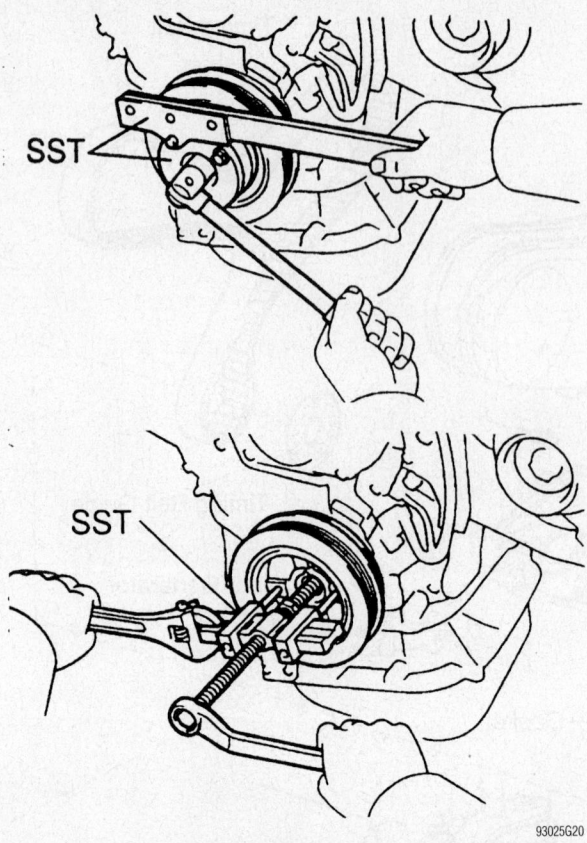

Removing/installing the crankshaft pulley—3.0L engine shown, 3.3L engine similar

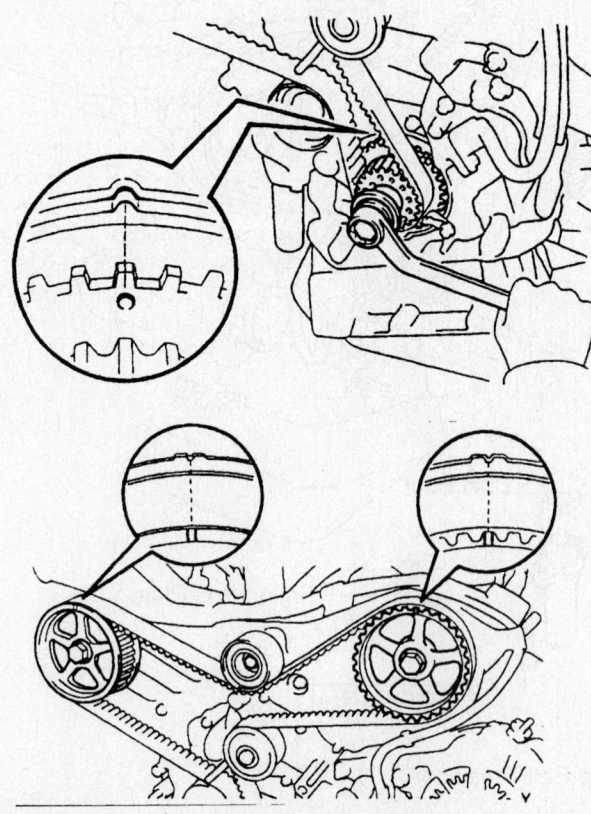

View of the timing mark locations—3.0L engine shown, 3.3L engine similar

and Companion Flange Holding tool 09330-00021, or equivalent, install the crankshaft pulley bolt and torque the bolt to 159 ft. lbs. (215 Nm).

33. To complete the installation, reverse the removal procedures.

34. Connect the negative battery cable.

35. Start the engine and check for leaks.

3.3L Engine

1. Before servicing the vehicle, refer to the precautions section.

2. Properly relieve the fuel system pressure.

3. Drain and recycle the engine coolant.

4. Disconnect the negative battery cable.

5. Remove the right hand front wheel.

6. Remove the wiper arms.

7. Remove the cowl panel.

8. Remove the wiper link assembly.

9. Remove the cowl top to inner brace.

10. Remove the cowl top cover outer sub assembly.

11. Remove the right side apron seal.

12. Remove the A/C compressor to crankshaft pulley belt.

13. Remove the vane pump belt.

14. Remove the number 2 engine mounting stay.

15. Remove the number 2 alternator bracket and the alternator belt adjusting bar.

16. Remove the crankshaft pulley.

17. Remove the timing belt cover 1.

18. Remove the timing belt cover 2 by disconnecting the wire protector clamps, unfastening the bolts and removing the cover.

19. Remove the right hand engine mount bracket.

20. Remove the timing belt guide number 2.

21. Remove the timing belt as follows:

   a. Set the number 1 cylinder to TDC.

   b. Temporarily install the crankshaft pulley bolt and washer onto the crankshaft.

   c. Turn the crankshaft clockwise and align the timing marks on the crankshaft pulley and oil pump body as illustrated.

   d. Check that the timing marks on the camshaft timing pulleys and the number 3 timing belt cover are aligned as shown. If not aligned turn the crankshaft 360 degrees.

   e. Remove the crankshaft pulley bolt.

   f. If reusing the old belt, check to see if there are still belt installation marks at the locations illustrated. If the marks are gone, put new marks on the old belt.

   g. Set the number 1 cylinder to 60

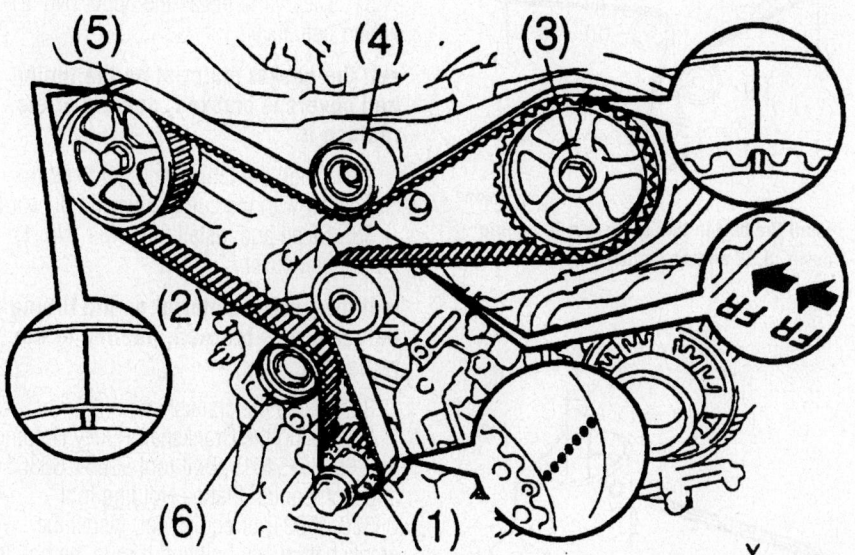

Timing belt alignment—3.0L engine shown, 3.3L engine similar

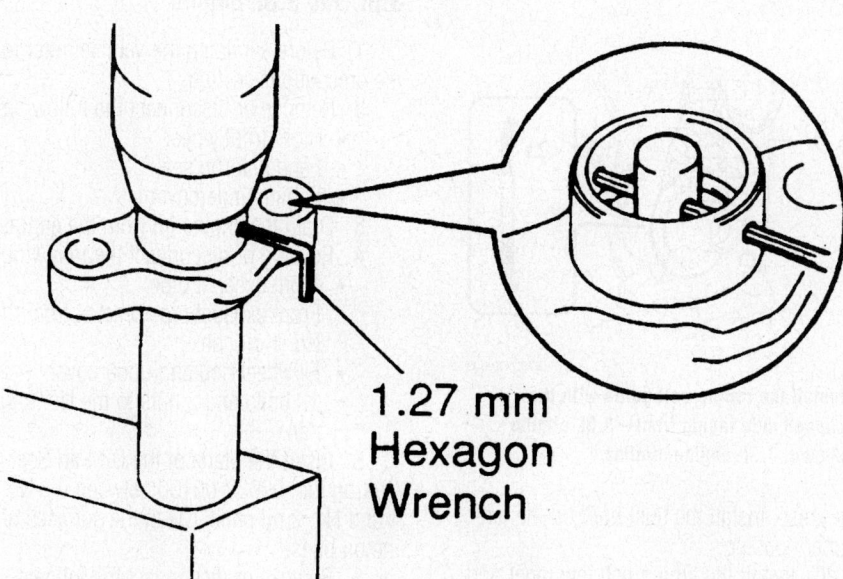

1.27 mm Hexagon Wrench

Timing belt tensioner installation preparation—3.0L engine shown, 3.3L engine similar

degrees BTDC by turning the crankshaft counterclockwise 60 degrees.

➡ **If the belt is disengaged, having the crankshaft at the wrong angle can cause the piston to strike the cylinder head and cause major engine damage. Always make sure to set the crankshaft pulley at the correct angle.**

    h. Remove the timing belt tensioner.
    i. Remove the belt in this order:
- No. 1 idler pulley
- Right camshaft sprocket
- No. 2 idler pulley
- Left camshaft timing sprocket
- Water pump pulley
- Crankshaft timing sprocket

To install:

22. Inspect the timing belt tensioner by performing the following procedures:
    a. Inspect the seal for leakage; if leakage is suspected, replace the tensioner.
    b. Using both hands to hold the tensioner facing upward, strongly press the pushrod against a solid surface. If the pushrod moves, replace the tensioner.

✳✳ WARNING

Never hold the tensioner with the pushrod facing downward.

    c. Measure the pushrod protrusion from the housing end, it should be 0.394–0.425 in. (10.0–10.8mm); if the protrusion is not as specified, replace the tensioner.

23. Set the No. 1 cylinder to Top Dead Center (TDC) of the compression stroke, as follows:
    a. Rotate the crankshaft 60 degrees clockwise to align the timing marks: dimple on the crankshaft timing sprocket with the notch on the oil pump body.
    b. Check that the timing marks on the camshaft pulley and the number 3 timing belt cover are aligned; if not, rotate the crankshaft 1 revolution (360 degrees) and align the marks.

24. Install the timing belt in the following order:
    a. Crankshaft timing sprocket
    b. Water pump pulley
    c. Left camshaft timing sprocket
    d. No. 2 idler pulley
    e. Right camshaft sprocket
    f. No. 1 idler pulley

25. Using a vertical press, slowly press the pushrod into the housing using 200–2205 lbs. (981–9807 N) until the holes align, then install a 1.5mm Allen® wrench to secure the pushrod and release

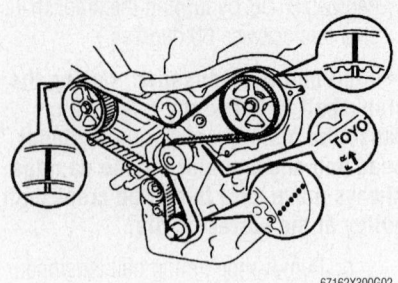

If the timing belt is re-used, check that the 3 original installation marks are visible on the belt as shown

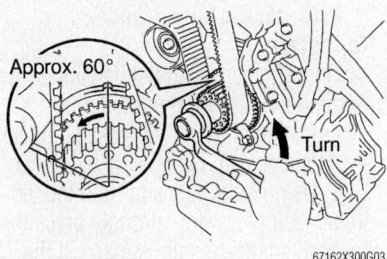

Turn the crankshaft counterclockwise by 60 degrees—3.0L engine shown, 3.3L engine similar

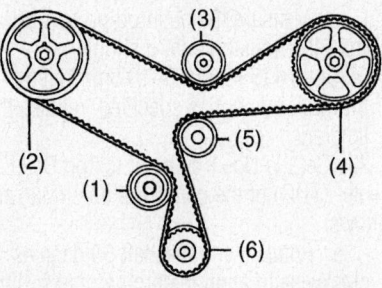

Remove the belt from the pulleys in this order—3.0L engine shown, 3.3L engine similar

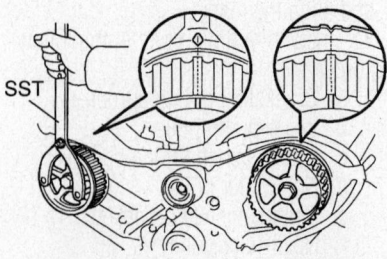

Turn the camshaft pulleys back into alignment so the marks align with the notches on the inner cover—3.0L engine shown, 3.3L engine similar

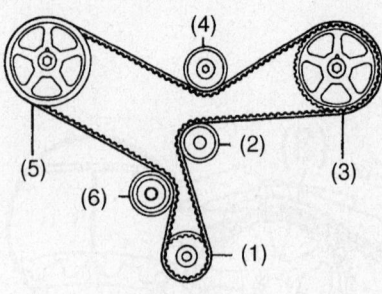

Install the belt in this order—3.0L engine shown, 3.3L engine similar

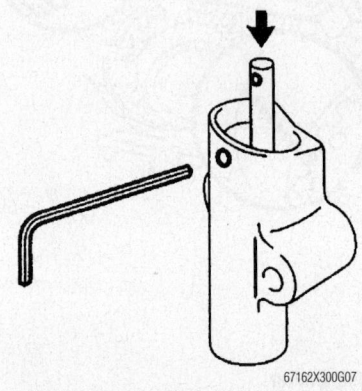

Set the tensioner in a press and collapse the plunger. Do not apply more that 2,205 lbs (9.8 kN) of force. Insert a suitable metal rod through the holes to hold the plunger in position—3.0L engine shown, 3.3L engine similar

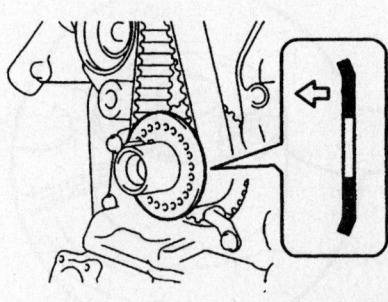

Install the timing belt guide with the cupped side facing front—3.0L engine shown, 3.3L engine similar

the press. Install the dust boot on the tensioner housing.

26. Install the timing belt tensioner and torque the bolts to 20 ft. lbs. (27 Nm).

27. Remove the Allen® wrench from the tensioner housing.

28. Slowly, rotate the crankshaft clockwise 2 complete revolutions and realign the timing marks. If the timing marks do not align, remove the timing belt and reinstall it.

29. Remove the crankshaft pulley bolt.

30. Install the right side engine mounting bracket and torque the bolts to 21 ft. lbs. (28 Nm).

31. Clean and install the upper (No. 2) timing belt cover.

➡ **If the gasket material on the timing belt covers is cracked, peeling or etc., replace it.**

32. Install the timing belt guide on the crankshaft with the cup side facing outward.

33. Clean and install the lower (No. 1) timing belt cover.

➡ **If the gasket material on the timing belt covers is cracked, peeling or etc., replace it.**

34. Install the crankshaft pulley.

35. Using the Crankshaft Pulley Holding tool 09213-54015, Bolt tool 91651-60855 and Companion Flange Holding tool 09330-00021, or equivalent, install the crankshaft pulley bolt and torque the bolt to 162 ft. lbs. (220 Nm).

36. To complete the installation, reverse the removal procedures.

37. Connect the negative battery cable.

38. Start the engine and check for leaks.

Oil Pan

REMOVAL & INSTALLATION

3.0L and 3.3L Engine

1. Before servicing the vehicle, refer to the precautions section.

2. Remove or disconnect the following:
   - Right front wheel
   - Fender apron seal
   - Engine undercover

3. Drain the engine oil from the engine.

4. Remove or disconnect the following:
   - Front exhaust pipe
   - Front exhaust pipe bracket from the No. 1 oil pan
   - Flywheel housing undercover
   - 10 bolts and 2 nuts to the No. 2 oil pan

5. Insert the blade of the Oil Pan Seal Cutting tool 09032-00100 between the No. 1 and No. 2 oil pans. Clean the surfaces of the oil pans.

6. Remove or disconnect the following:
   - 3 oil strainer nuts and gasket

7. Remove the No. 1 oil pan, as follows:
   - 2 bolts and the flywheel housing undercover
   - 17 bolts and 2 nuts to the No. 1 oil pan

➡**Make a note of the position of the each bolt. When replacing the bolts into the oil pan, place each bolt in the position from which it was removed.**

- Oil pan, by prying the portions between the cylinder block and the oil pan

➡**Be careful not to damage the contact surfaces.**

- Baffle plate from the No. 1 oil pan

To install:

8. Clean all mating surfaces of the oil pans.

9. Install the baffle plate to the No. 1 oil pan and tighten to 69 inch lbs. (8 Nm).

10. Install the No. 1 oil pan, as follows:

a. Using a non residue solvent, clean both sealing surfaces to the oil pan.

b. Apply liquid sealant to the oil pan and engine block.

c. Install the oil pan with the 17 bolts and 2 nuts. Uniformly tighten the bolts and nuts in several passes.

d. Tighten the No. 1 oil pan bolts, as follows:

- 10mm head bolt: 69 inch lbs. (8 Nm)
- 12mm head bolt: 14 ft. lbs. (20 Nm)
- 14mm head bolt: 27 ft. lbs. (37 Nm)

e. Install the flywheel housing undercover with the 2 bolts. Tighten the bolts to 69 inch lbs. (8 Nm).

11. Install the oil strainer with the 3 nuts. Tighten the nuts to 69 inch lbs. (8 Nm).

12. Install the No. 2 oil pan, as follows:

a. Using a non residue solvent, clean both sealing surfaces to the oil pan.

b. Apply liquid sealant to the oil pan and engine block.

c. Install the No. 2 oil pan with the 10 bolts and 2 nuts. Uniformly tighten the bolts and nuts in several passes. Tighten the bolts to 69 inch lbs. (8 Nm).

13. Install or connect the following:

- Flywheel housing undercover
- Front exhaust pipe bracket to the No. 1 oil pan. Tighten the bolts to 15 ft. lbs. (21 Nm).

14. Install the front exhaust pipe, as follows:

- Temporarily install the 3 new gaskets and the front exhaust pipe with the 2 bolts and 6 nuts
- Tighten the 4 exhaust manifolds-to-front exhaust pipe nuts to 46 ft. lbs. (62 Nm).
- Tighten the both front exhaust pipe-to-center exhaust pipe nuts/bolts to 41 ft. lbs. (56 Nm).
- Bracket. Tighten both bolts to 14 ft. lbs. (19 Nm).
- Support stay. Tighten both bolts to 22 ft. lbs. (29 Nm).

15. Install or connect the following:

- Engine undercover
- Right fender apron seal
- Right front wheel

16. Fill the engine with oil.

17. Start the engine and check for leaks.

3.3L Engine

1. Before servicing the vehicle, refer to the precautions section.

2. Properly relieve the fuel system pressure.

3. Drain the engine oil.

4. Disconnect the negative battery cable.

5. Remove the right hand front wheel.

6. Remove the wiper arms.

7. Remove the cowl panel.

8. Remove the wiper link assembly.

9. Remove the cowl top to inner brace.

10. Remove the cowl top cover outer sub assembly.

11. Remove the right side apron seal.

12. Remove the A/C compressor to crankshaft pulley belt.

13. Remove the vane pump belt.

14. Remove the number 2 engine mounting stay.

15. Remove the number 2 alternator bracket and the alternator belt adjusting bar.

16. Remove the crankshaft pulley.

17. Remove the timing belt cover 1.

18. Remove the timing belt cover 2 by disconnecting the wire protector clamps, unfastening the bolts and removing the cover.

19. Remove the right hand engine mount bracket.

20. Remove the timing belt guide number 2.

21. Remove the timing belt.

22. Remove the number 2 timing belt idler sub assembly, camshaft timing pulley and the number 3 timing belt cover.

23. Remove the number 1 timing belt idler sub assembly and the crankshaft timing pulley.

24. Remove the center exhaust pipe assembly on 4WD models.

25. Remove the front exhaust pipe assembly.

26. Remove the number 3 manifold converter and number 2 exhaust manifold heat shield and then remove the number 2 exhaust manifold converter sub assembly.

27. Disconnect the A/C compressor electrical connector, remove the bolts, adjusting bar bracket and the compressor. Set aside with the hoses still attached.

28. Remove the A/C compressor bracket.

29. On 4 WD model, disconnect the rear engine mounting insulator.

30. Remove the nuts and separate but do not remove the front engine mount insulator.

31. Disconnect the power steering hose from the frame, remove the front right hand engine mounting insulator nuts.

32. Support the engine with a block of wood and a floor jack, raise the engine slightly and remove the front right hand engine mounting insulator.

33. Remove the right hand engine mount bracket.

34. Remove the 10 bolts and 2 nuts to the No. 2 oil pan.

35. Insert the blade of the Oil Pan Seal Cutting tool 09032-00100 between the No. 1 and No. 2 oil pans.

36. Clean the surfaces of the oil pans.

37. Remove the oil strainer nuts and gasket

38. Remove the No. 1 oil pan, as follows:

a. Remove the 2 bolts and the flywheel housing undercover

b. Remove the 17 bolts and 2 nuts to the No. 1 oil pan.

➡**Make a note of the position of the each bolt. When replacing the bolts into the oil pan, place each bolt in the position from which it was removed.**

c. Remove the oil pan, by prying the portions between the cylinder block and the oil pan.

➡**Be careful not to damage the contact surfaces.**

d. Remove the baffle plate from the No. 1 oil pan.

39. Remove the oil pump bolts and the pump.

To install:

40. Clean all mating surfaces of the oil pans.

41. Install the baffle plate to the No. 1 oil pan and tighten to 69 inch lbs. (8 Nm).

42. Install the No. 1 oil pan, as follows:

a. Using a non residue solvent, clean both sealing surfaces to the oil pan.

b. Apply liquid sealant to the oil pan and engine block.

c. Install the oil pan with the 17 bolts and 2 nuts. Uniformly tighten the bolts and nuts in several passes.

d. Tighten the No. 1 oil pan bolts, as follows:

- 10mm head bolt: 71 inch lbs. (8 Nm)
- 12mm head bolt: 14 ft. lbs. (20 Nm)
- 14mm head bolt: 27 ft. lbs. (37 Nm)

e. Install the flywheel housing undercover with the 2 bolts. Tighten the bolts to 69 inch lbs. (8 Nm).

43. Install the oil strainer with the 3 nuts. Tighten the nuts to 69 inch lbs. (8 Nm).

44. Install the No. 2 oil pan, as follows:

a. Using a non residue solvent, clean both sealing surfaces to the oil pan.

b. Apply liquid sealant to the oil pan and engine block.

c. Install the No. 2 oil pan with the 10 bolts and 2 nuts. Uniformly tighten the bolts and nuts in several passes. Tighten the bolts to 71 inch lbs. (8 Nm).

45. Install the remaining components in the reverse order of removal.

Oil Pump

REMOVAL & INSTALLATION

3.0L and 3.3L Engine

1. Before servicing the vehicle, refer to the precautions section.

2. Remove or disconnect the following:
- Oil pan
- Crankshaft Position (CKP) sensor
- 9 oil pump bolts

➡**Make a note of the position of the each bolt. When replacing the bolts into the oil pump body, place each bolt in the position from which it was removed.**

- Oil pump body, by prying between the oil pump and main bearing cap
- O-ring from the cylinder block
- Plug, gasket, spring and relief valve from the oil pump body
- 9 screws, pump body cover, drive and driven rotors

To install:

3. Install or connect the following:
- Driven rotors, drive, pump body cover, using the 9 screws
- Oil pump relief valve, spring, gasket and the plug to the oil pump body
- New O-ring on the cylinder block

4. Using a non residue solvent, clean both sealing surfaces to the oil pump.

5. Apply liquid sealant to the oil pump and engine block.

6. Install or connect the following:
- Oil pump

➡**Be sure to engage the spline teeth of the oil pump drive gear with the large teeth of the crankshaft.**

- 9 oil pump bolts. On 3.0L engines, tighten the bolts in several passes to 69 inch lbs. (8 Nm), for 10mm or to 14 ft. lbs. (20 Nm), for 12mm. On 3.3L engines, tighten the bolts in several passes. Tighten bolt A to 71 inch lbs. (8 Nm), bolt B to 14 ft. lbs. (20 Nm) and bolt C to 32 ft. lbs. (43 Nm). refer to the illustration for bolt locations.
- CKP sensor. Tighten the bolt to 69 inch lbs. (8 Nm).
- Baffle plate to the No. oil pan. Tighten to 69 inch lbs. (8 Nm).
- No. 1 oil pan, oil strainer and No. 2 oil pan

7. Refill the engine with oil.
8. Start the engine and inspect for leaks.
9. Recheck the engine oil level.

Rear Main Seal

REMOVAL & INSTALLATION

3.0L and 3.3L Engine

If the rear oil seal retainer is not installed to the block, use a tapered ended screwdriver and hammer to remove the oil seal. Apply multi-purpose grease to the new oil seal lip. Using a seal driver, tap the

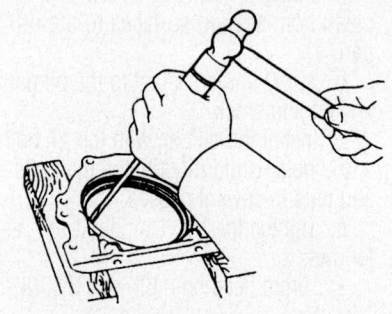

Carefully tap the old seal from the retainer

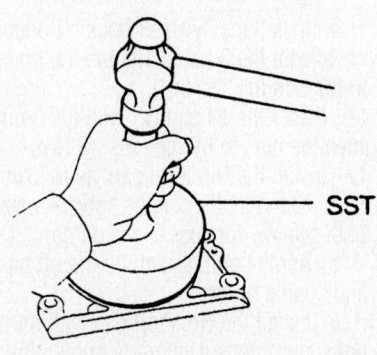

Use the proper sized driver to seat the seal

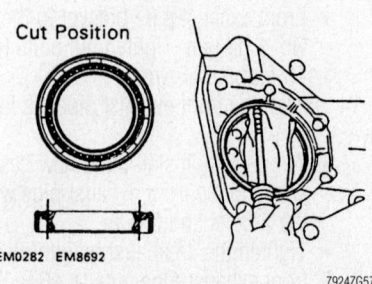

Cut off the oil seal lip, then pry the seal out of the retaining plate

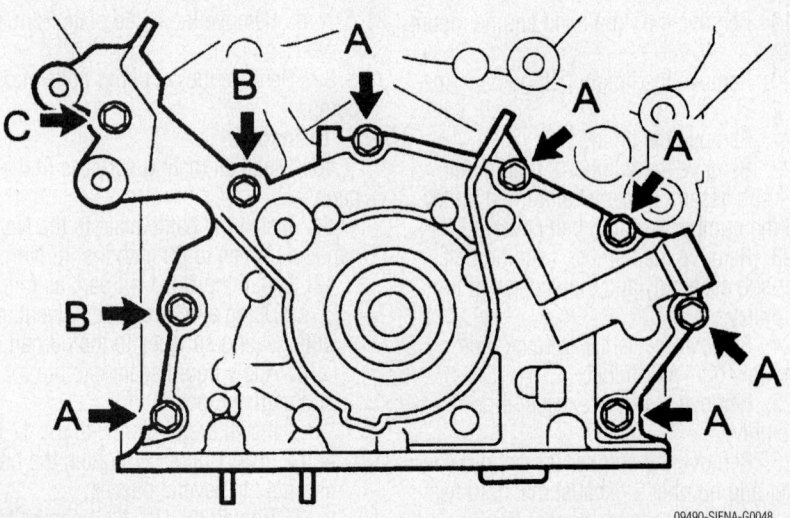

Oil pump bolt locations—3.3L engine

09490-SIENA-G0048

SST
7924ZG58

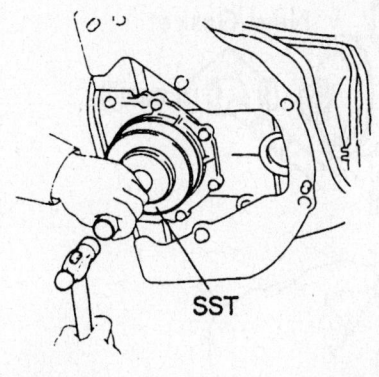

Tap a new seal into place

seal into place. Be careful not to install it slantwise.

1. Before servicing the vehicle, refer to the precautions section.

If the rear oil seal retainer is installed on the cylinder block, using a knife, cut off the lip of the seal. Using a taped ended prytool, pry the old seal out of the retainer. Inspect the oil seal lip contacting surface of the crankshaft for cracks or damage. Apply multipurpose grease to the new oil seal, then tap the seal in place with a seal installer. Be careful not to install the seal slantwise.

Piston and Rings

POSITIONING

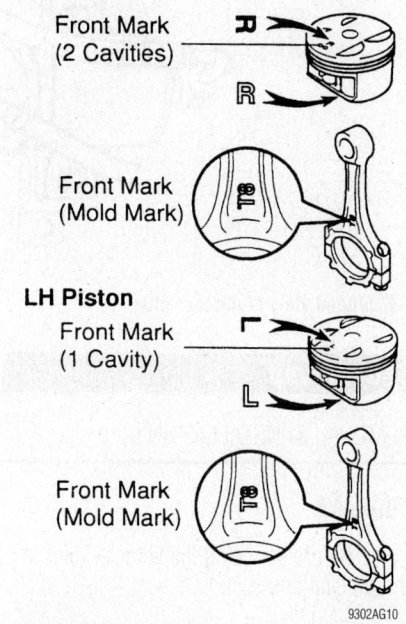

RH Piston
Front Mark
(2 Cavities)
R
R
Front Mark
(Mold Mark)
LH Piston
Front Mark
(1 Cavity)
L
L
Front Mark
(Mold Mark)
9302AG10

Piston/connecting rod-to-engine positioning

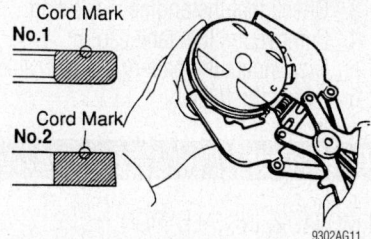

Cord Mark
No.1
Cord Mark
No.2
9302AG11

Piston ring positioning

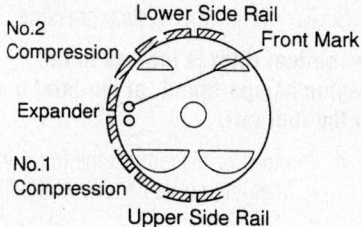

RH Piston
Lower Side Rail
No.2
Compression
Front Mark
Expander
No.1
Compression
Upper Side Rail
LH Piston
Lower Side Rail
No.2
Compression
Front Mark
Expander
No.1
Compression
Upper Side Rail
9302AG12

Piston ring identification

FUEL SYSTEM

Fuel System Service Precautions

Safety is the most important factor when performing not only fuel system maintenance but any type of maintenance. Failure to conduct maintenance and repairs in a safe manner may result in serious personal injury or death. Work on a vehicle's fuel system components can be accomplished safely and effectively by adhering to the following rules and guidelines.

- To avoid the possibility of fire and personal injury, always disconnect the negative battery cable unless the repair or test procedure requires that battery voltage be applied.
- Always relieve the fuel system pressure prior to disconnecting any fuel system component (injector, fuel rail, pressure regulator, etc.) fitting or fuel line connection. Exercise extreme caution whenever relieving fuel system pressure, to avoid exposing skin, face and eyes to fuel spray. Please be advised that fuel under pressure may penetrate the skin or any part of the body that it contacts.
- Always place a shop towel or cloth around the fitting or connection prior to loosening to absorb any excess fuel due to spillage. Ensure that all fuel spillage is quickly remove from engine surfaces. Ensure that all fuel-soaked cloths or towels are deposited into a flame-proof waste container with a lid.
- Always keep a dry chemical (Class B) fire extinguisher near the work area.
- Do not allow fuel spray or fuel vapors to come into contact with a light bulb, spark or open flame.
- Always use a second wrench when loosening or tightening fuel line connection fittings. This will prevent unnecessary stress and torsion to fuel piping. Always follow the proper torque specifications.
- Always replace worn fuel fitting O-rings with new ones. Do not substitute fuel hose where rigid pipe is installed.

Fuel System Pressure

RELIEVING

2002–03

1. Before servicing the vehicle, refer to the precautions section.
2. Disconnect the negative battery terminal. Wait at least 90 seconds prior to working on models equipped with an airbag.
3. Place a catch-pan under the joint to be disconnected. A large quantity of fuel may be released when the joint is opened.

➡Wear eye or full-face protection.

4. Place a shop towel over the area and slowly loosen the joint using a wrench of the correct size. Use a back-up wrench if needed.
5. Allow the fuel left in the line to bleed off slowly before fully disconnecting the joint.
6. Plug the opened lines immediately to prevent fuel spillage or the entry of dirt.
7. Dispose of the released fuel properly.
8. After adjoining fuel lines, connect the negative battery cable and start the engine.
9. Check for leaks and repair as needed.

2004–06

1. Remove the fuel circuit opening relay from the engine compartment relay block.
2. Start the engine. After the engine stops, turn the ignition to OFF.

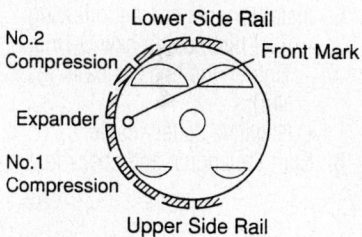

3. Check that the engine won't start.
4. Remove the fuel tank cap.
5. Disconnect the battery ground cable.
6. Install the relay.

Fuel Filter

REMOVAL & INSTALLATION

1. Before servicing the vehicle, refer to the precautions section.
2. Disconnect the negative battery cable.
3. Relieve the fuel system pressure.

➡**The fuel filter is located in the engine compartment, at the inlet line to the fuel rail.**

4. Remove or disconnect the following:
   • Inlet and outlet lines from the filter
   • Fuel filter

To install:

5. Install or connect the following:
   • Fuel filter, using new O-rings. Tighten the lines to 22 ft. lbs. (29 Nm).
   • Negative battery cable
6. Start the engine and check for leaks.

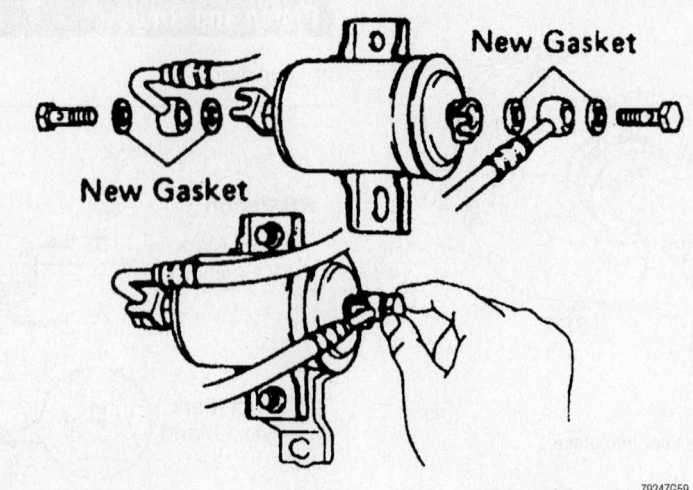

Exploded view of the fuel filter

Fuel Pump

REMOVAL & INSTALLATION

2002–03

1. Before servicing the vehicle, refer to the precautions section.

2. Relieve the fuel pressure.
3. Disconnect the negative battery cable. Wait at least 90 seconds before proceeding on models with an airbag.
4. Drain the fuel tank.
5. Remove or disconnect the following:
   • Fuel tank
   • Access plate bolts

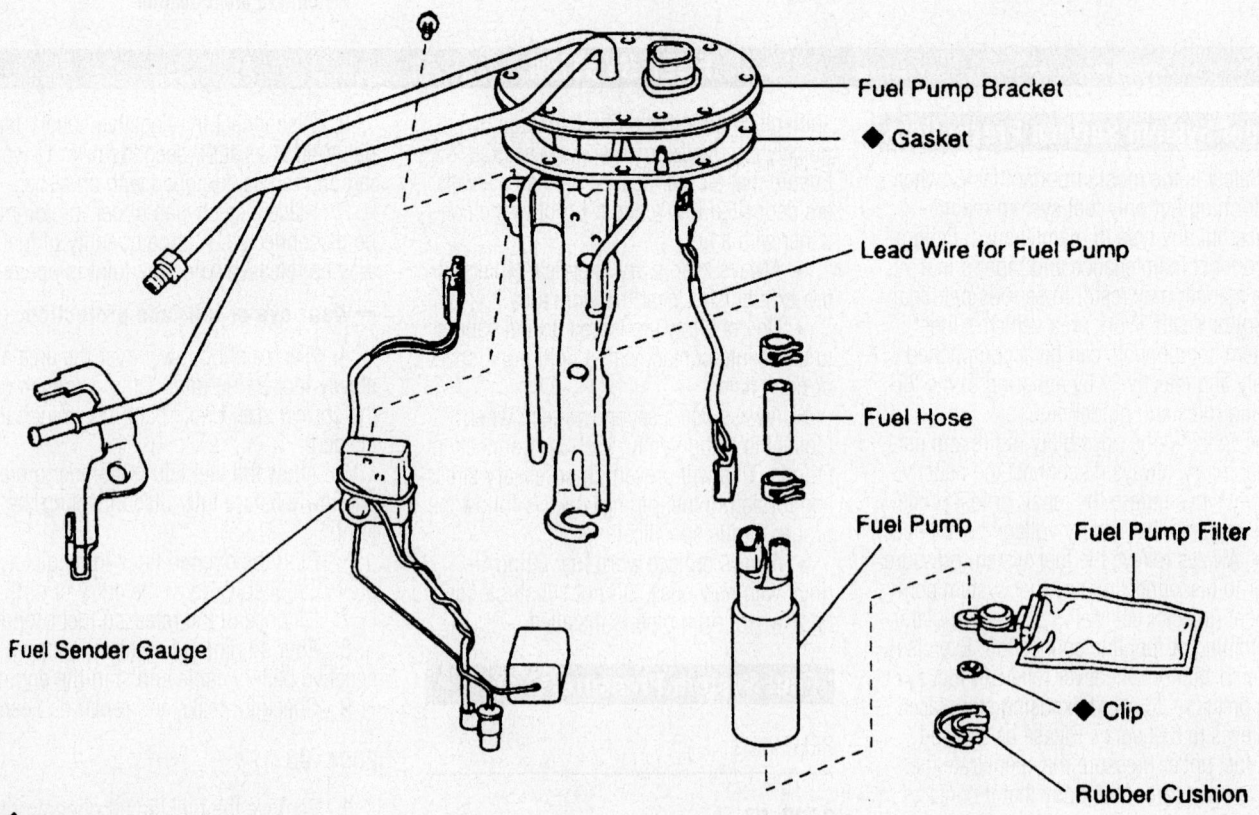

◆ **Non-reusable part**

Exploded view of the fuel pump, bracket and related components—2002

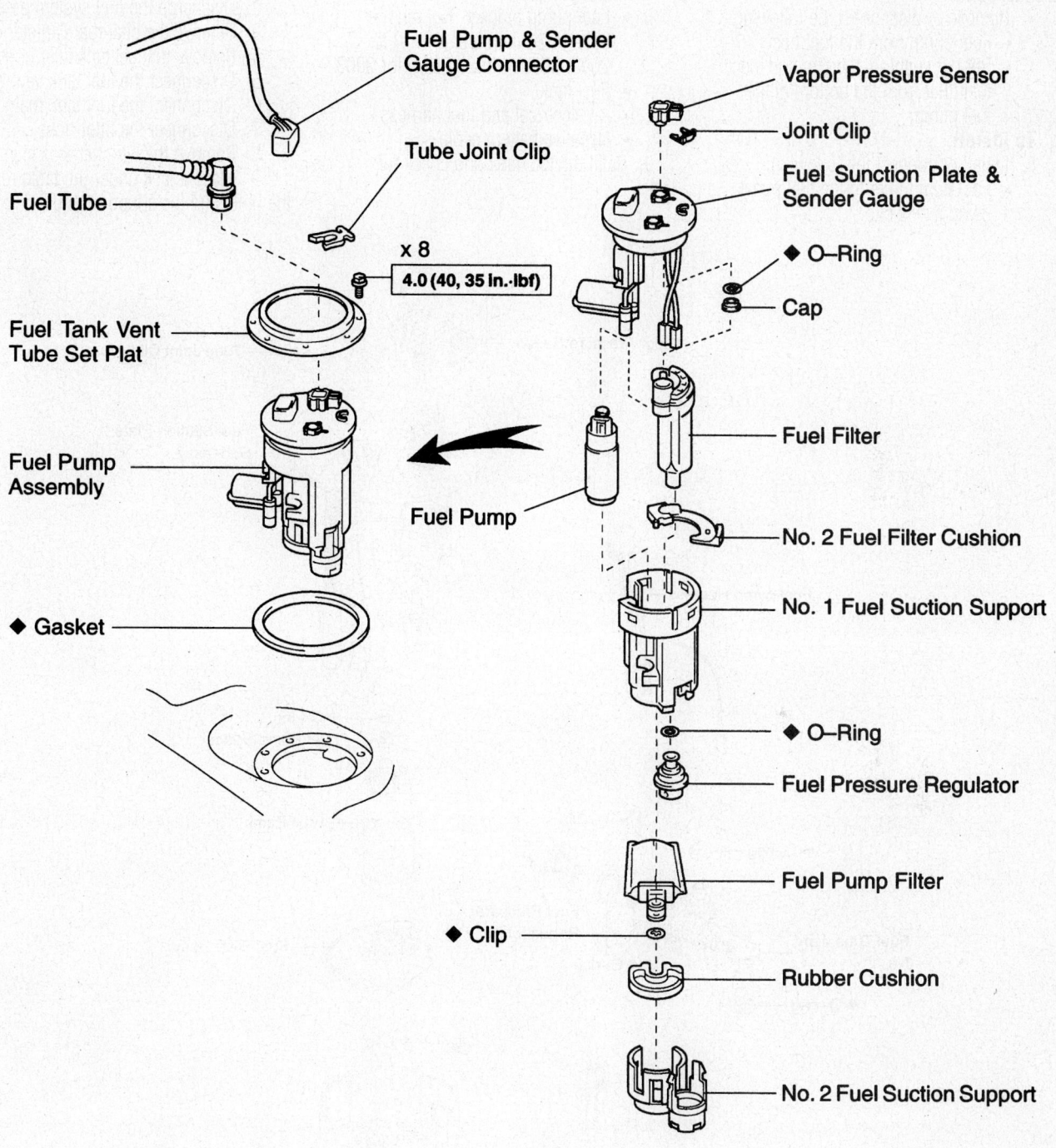

Fuel Pump & Sender
Gauge Connector

Vapor Pressure Sensor

Joint Clip

Tube Joint Clip

Fuel Sunction Plate &
Sender Gauge

Fuel Tube

◆ O–Ring

x 8

4.0 (40, 35 In.·lbf)

Cap

Fuel Tank Vent
Tube Set Plat

Fuel Filter

Fuel Pump
Assembly

Fuel Pump

No. 2 Fuel Filter Cushion

◆ Gasket

No. 1 Fuel Suction Support

◆ O–Ring

Fuel Pressure Regulator

Fuel Pump Filter

◆ Clip

Rubber Cushion

No. 2 Fuel Suction Support

N·m (kgf·cm, ft·lbf) : Specified torque

◆ Non–reusable part

67170-SIEN-G01

Exploded view of the fuel pump and related components—2003

- Fuel pump assembly
- Fuel pump electrical connectors.

6. Pull the bracket from the lower side of the fuel pump.

7. Remove or disconnect the following:
- Fuel pump from the fuel hose
- Rubber cushion, the clip and the fuel filter from the bottom of the fuel pump.

To install:

8. Install or connect the following:
- Fuel pump filter to the fuel pump using a new clip

- Fuel pump to the bracket using new gaskets
- Fuel hose to the outlet port of the fuel pump
- Fuel pump bracket. Tighten the bolts to 26 inch lbs. (3 Nm) for 2002; 35 inch lbs. (4 Nm) for 2003.
- Fuel tank
- All electrical and fuel harness
- Negative battery cable

9. Refill the fuel tank and check for leaks.

2004–06

1. Before servicing the vehicle, refer to the precautions section.
2. Discharge the fuel system pressure.
3. Remove the charcoal canister cover.
4. Remove the fuel tank filler hose cover.
5. Disconnect the fuel tank vent hose.
6. Disconnect the fuel tank main tube.
7. Disconnect the filler hose.
8. Remove the wire harness clamps.
9. Place a jack under the tank, remove the bolts and the support bands.

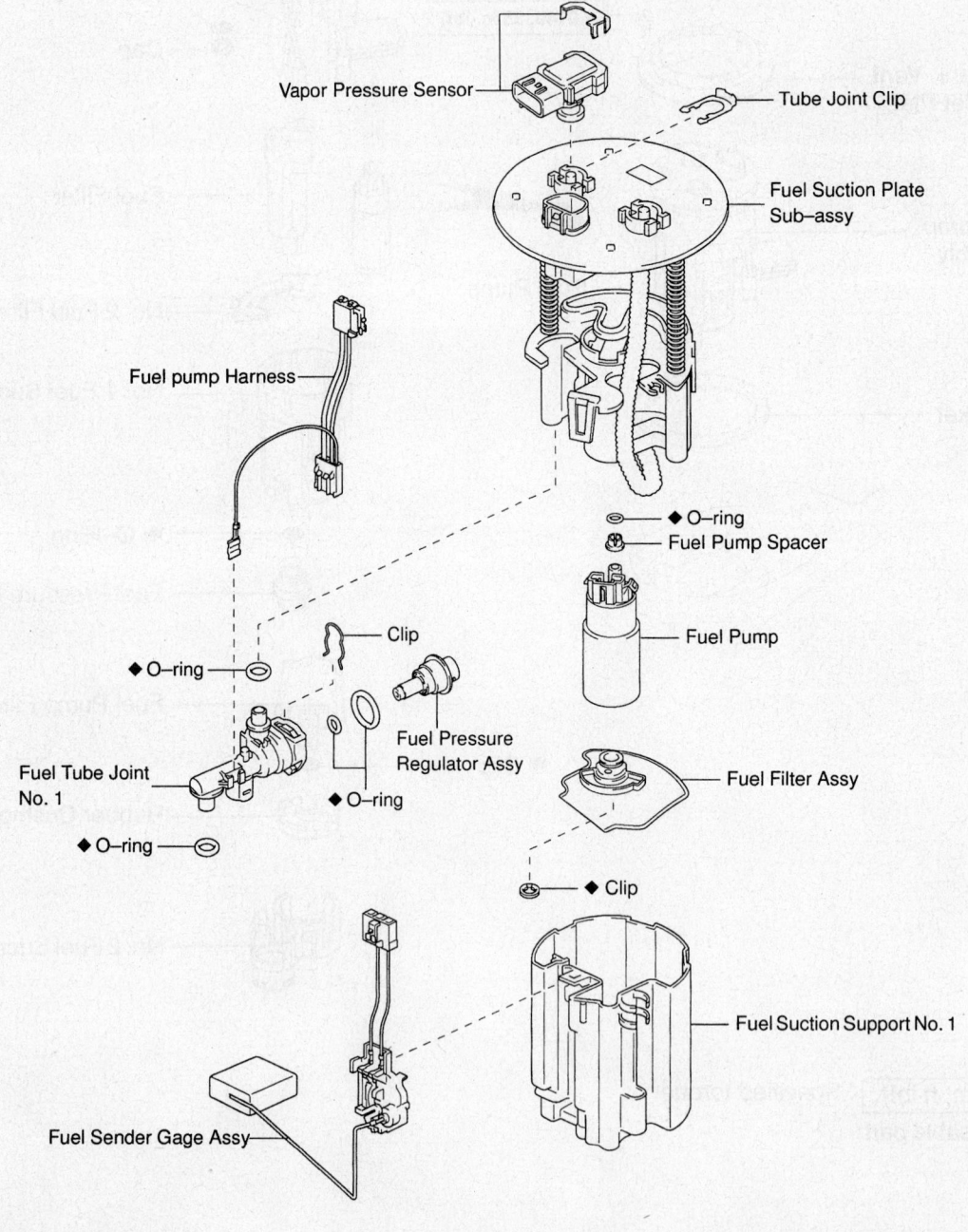

◆ Non–reusable part

Fuel pump module exploded view—2004–06

67170-SIEN-G02

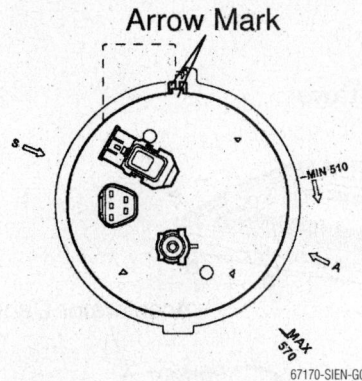

Module alignment marks—2004–06

67170-SIEN-G03

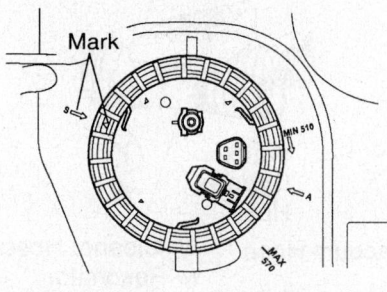

Retainer alignment marks—2004–06

67170-SIEN-G04

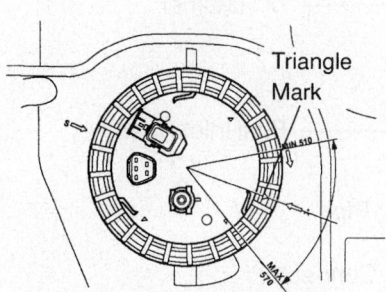

Tightening reference—2004–06

67170-SIEN-G05

10. Disconnect all remaining wiring and lower the tank.

11. Remove the tube joint clip and pull out the fuel main tube from the fuel pump module.

12. Remove the main tube from the tank.

13. Using tool 09808-14020, or equivalent lock ring tool, remove the lockring from the fuel pump module.

14. Remove the module from the tank.

15. Remove the joint clip and remove the pressure sensor.

16. Remove the sender assembly.

17. Wrap the tip of a small screwdriver with tape and disconnect the 4 snap retainers, and remove the fuel suction plate.

18. Disconnect the snap retainers, disconnect the connector, and remove the fuel pump.

19. Remove the O-ring and spacer from the pump.

20. Installation is the reverse of removal. Use a new O-ring coated with clean gasoline.

➡**Prior to assembly, all new parts must be stored at room temperature for a minimum of 12 hours.**

21. Make sure that the gasket groove is clean. Use a new gasket. Align the arrow on the fuel suction tube and the tank suction support.

22. Align the marks on the fuel pump module retainer and the fuel tank.

23. Position the retainer on the module and push down. Hold the module and turn the retainer by hand, one complete turn.

➡**Make sure that the anti-rotation tab is in the groove during tightening. The "S" arrow on the fuel tank indicates "0" degrees position. Make sure that the retainer isn't cross-threaded.**

24. Using the special tool, torque the retainer to 59–67 ft. lbs. (80–90 Nm). The triangle mark on the retainer should be about 1½ turn from the start.

Fuel Injector

REMOVAL & INSTALLATION

3.0L Engine

1. Before servicing the vehicle, refer to the precautions section.

2. Remove or disconnect the following:
   - Outer front cowl top panel assembly
   - Air cleaner cap with hose
   - Negative battery cable. Work must be started approximately 90 seconds or longer after the negative battery cable has been disconnected, if equipped with an air bag.
   - Coolant
   - Accelerator and throttle cables
   - V-bank cover
   - Emission valve control set
   - No. 2 EGR pipe
   - Hydraulic motor pressure pipe from the water inlet and air inlet chamber
   - Air intake chamber assembly
   - Injector wiring
   - Air assist pipe from the bracket on the No. 1 fuel pipe
   - Air assist hoses from the intake manifold

- Fuel return hose from the No. 1 fuel pipe
- Fuel inlet hose for the fuel filter
- 2 union bolts holding the No. 2 fuel pipe to the delivery pipes
- Fuel return hose from the fuel pressure regulator
- Union bolt for the right hand delivery pipe, 2 gaskets, 2 bolts, left hand delivery pipe together with the 3 injectors and the No. 2 fuel pipe
- Union bolt for the delivery pipe and 2 gaskets from the No. 2 fuel pipe
- The 3 bolts, right hand delivery pipe together with the 3 injectors and the No. 1 fuel pipe
- The 4 spacers from the intake manifold
- The 6 injectors from the delivery pipes
- The two O-rings and two grommets from each injector

To install:

3. Install or connect the following:
   - 2 new grommets to each injector
   - New O-rings, with a light coat of fuel, to each injector
   - Injectors
   - The 4 spacers on the intake manifold
   - Right hand delivery pipe and the No. 1 fuel pipe together with the 3 injectors in position on the intake manifold
   - Bolt holding the right side delivery pipe, temporarily, to the intake manifold
   - Left hand delivery pipe and the No. 2 fuel pipe together with the 3 injectors in position on the intake manifold
   - Fuel return hose to the fuel pressure regulator

4. Temporarily install the 2 bolts holding the left hand delivery pipe to the intake manifold.

5. Temporarily install the No. 2 fuel pipe to the left side delivery pipe with the union bolt and 2 new gaskets.

6. Check that the injectors rotate smoothly. If they do not, Replace the O-rings.

7. Position the injector connector outward. Tighten the 4 bolts holding the delivery pipes to the intake manifold and tighten to 7 ft. lbs. (10 Nm). Tighten the bolt holding the No. 1 fuel pipe to the intake manifold to 14 ft. lbs. (20 Nm). Tighten the 2 union bolts holding the no. 2 fuel pipe to the delivery pipes to 24 ft. lbs. (32 Nm).

8. Install or connect the following:

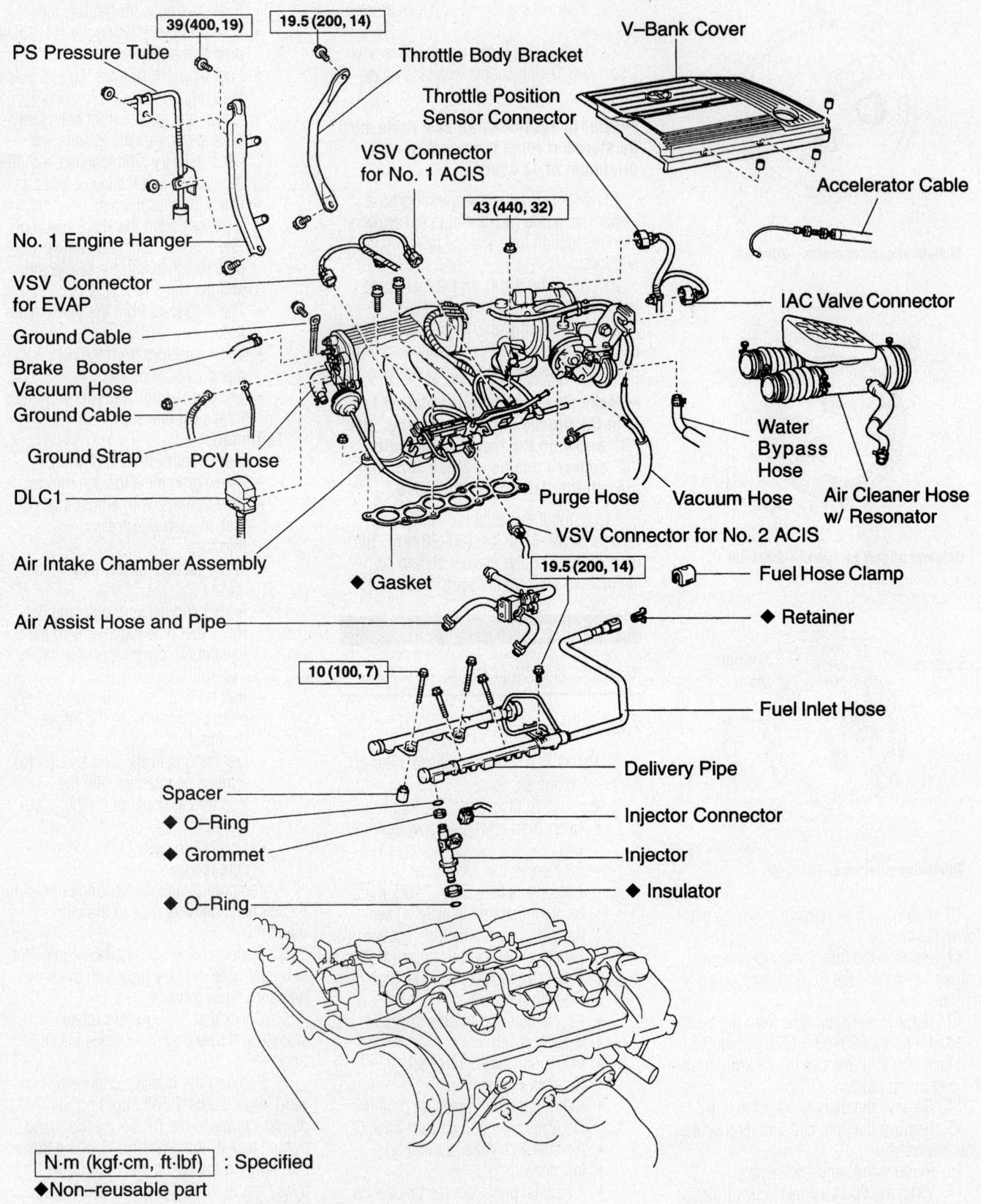

PS Pressure Tube

39 (400, 19)

19.5 (200, 14)

Throttle Body Bracket

Throttle Position Sensor Connector

VSV Connector for No. 1 ACIS

V–Bank Cover

43 (440, 32)

Accelerator Cable

No. 1 Engine Hanger

VSV Connector for EVAP

Ground Cable

Brake Booster Vacuum Hose

Ground Cable

Ground Strap

PCV Hose

IAC Valve Connector

Water Bypass Hose

Air Cleaner Hose w/ Resonator

DLC1

Air Intake Chamber Assembly

Air Assist Hose and Pipe

◆ **Gasket**

Purge Hose

Vacuum Hose

VSV Connector for No. 2 ACIS

Fuel Hose Clamp

19.5 (200, 14)

◆ **Retainer**

10 (100, 7)

Fuel Inlet Hose

Delivery Pipe

Spacer

◆ **O–Ring**

◆ **Grommet**

◆ **O–Ring**

Injector Connector

Injector

◆ **Insulator**

N·m (kgf·cm, ft·lbf) : Specified

◆Non–reusable part

67170-SIEN-G06

Fuel injectors and related parts—3.0L engine

- Fuel inlet and return hoses. Union bolt: 22 ft. lbs. (30 Nm)
- Fuel return hose to the No. 1 fuel pipe. Pass the fuel return hose under the heater hoses.
- Air assist hoses to the intake manifold
- Air assist pipe to the bracket on the No. 1 fuel pipe
- Fuel injector wiring connectors

- Air intake chamber assembly
- Hydraulic motor pressure pipe to the intake chamber. Bolts: 69 inch lbs. (8 Nm)
- No. 2 EGR pipe with new gaskets, tighten to 9 ft. lbs. (12 Nm)
- Emission control valve set
- V-bank cover
- Air cleaner hose
- Throttle and accelerator cables

- Coolant
- Air cleaner cap with hose
- Outer front cowl top panel assembly
- Negative battery cable

3.3L Engine

1. Before servicing the vehicle, refer to the precautions section.
2. Relieve the fuel system pressure.

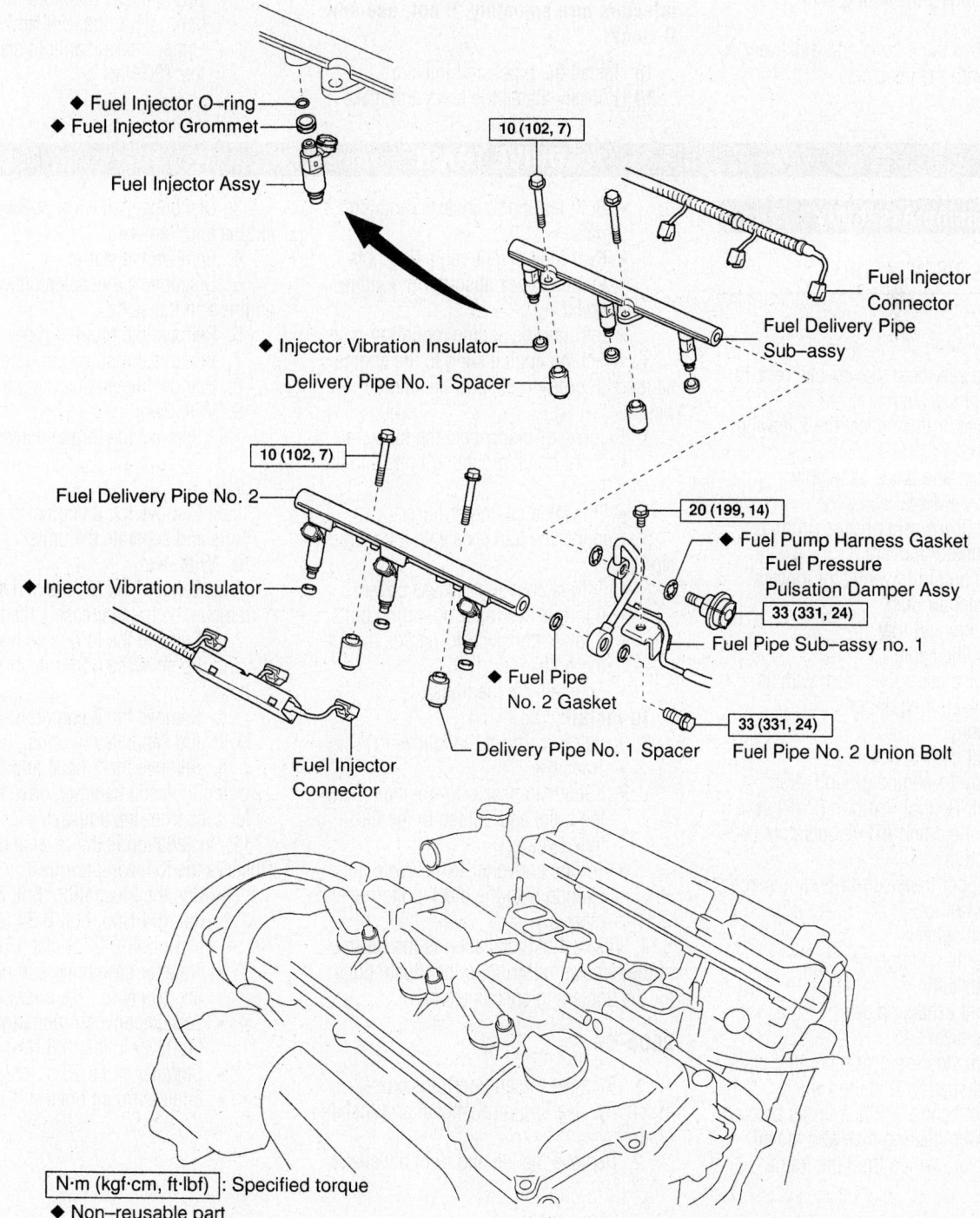

N·m (kgf·cm, ft·lbf) : Specified torque
◆ Non-reusable part

67170-SIEN-G07

Fuel injectors and related parts—3.3L engine

3. Drain the coolant.

4. Remove the wiper arms.

5. Remove the wiper motor.

6. Remove the cowl tops.

7. Remove the V-bank cover.

8. Remove the air cleaner assembly.

9. Remove the emission control valve set.

10. Remove the upper intake manifold (intake air surge tank). Discard the gasket.

11. Remove the fuel pipe sub-assembly.

12. Disconnect the wiring at the injectors.

13. Remove the 4 bolts and 2 delivery pipe along with the injectors.

14. Remove the delivery pipe spacers and insulators from the manifold.

15. Pull each injector from the pipe.

To install:

16. Install new O-rings on each injector. Apply a light coating of gasoline to the O-rings and mating points on the pipes.

17. Using a twisting motion, install the injectors on the pipes.

➡**Be careful to avoid twisting the O-rings. After installation, check that the injectors turn smoothly. If not, use new O-rings.**

18. Install the pipes and injectors.

19. Loosely install the bolts and make

sure that the injectors still turn freely. If not, replace the O-rings.

20. Torque the bolts to 84 inch lbs. (10 Nm).

21. The remainder of installation is the reverse of removal. Observe the following torques:

- Fuel line union bolt: 24 ft. lbs. (33 Nm)
- Pulsation damper: 24 ft. lbs. (33 Nm)
- Fuel feed pipe: 14 ft. lbs. (20 Nm)
- Upper intake manifold (air surge tank): 21 ft. lbs. (28 Nm)
- Upper intake manifold stays: 14 ft. lbs. (20 Nm)

DRIVE TRAIN

Transmission Assembly

REMOVAL & INSTALLATION

2002–03

1. Before servicing the vehicle, refer to the precautions section.

2. Remove or disconnect the following:

- Hood
- Wiper and blade assembly
- Top cowl seal and panel
- Window washer hoses, from the ventilator louvers
- Left and right ventilator louvers
- Heater air duct
- Battery and tray
- Throttle cable
- Cruise control actuator with its bracket, if equipped
- Starter
- Shift control cable
- Body-to-engine ground strap
- Park/Neutral Position (PNP) switch, solenoid and ATF temperature connectors
- 5 upper transaxle-to-engine mounting bolts
- Front wheel
- Engine undercover
- Halfshafts
- Front exhaust pipe
- Stabilizer bar
- Both steering gear mounting bolts and support it in the vehicle
- Shift control cable from its bracket
- Power steering pipe and the oil cooler clamps from the frame

- Both left-side transaxle mounting nuts
- Rear-side engine mounting nuts
- Engine shock absorber mounting bolts
- 3 front-side engine mounting bolts

3. Attach an engine sling to the engine hangers in order to support the engine weight.

4. Remove or disconnect the following:

- Front frame mounting bolts and the frame
- Transaxle oil cooler lines

5. Support the transaxle with a transmission jack.

- Torque converter access cover
- 6 torque converter mounting bolts
- 3 lower transaxle-to-engine mounting bolts
- Engine from the transaxle

To install:

6. Install or connect the following:

- Transaxle
- 3 lower transaxle-to-engine mounting bolts and tighten to the illustrated value.
- Torque converter-to-flexplate bolts, starting with the black bolt, then the other 5.

7. The rest of installation is the reverse of the removal referring to the illustrations for the tightening specifications.

2004–06

1. Remove the engine/transaxle assembly. See Engine Removal and Installation.

2. Remove the left and right halfshafts.

3. Disconnect all wiring between the engine and transaxle.

4. Remove the starter.

5. Disconnect all cables between the engine and transaxle.

6. Remove the oil filler tube.

7. Remove the oil cooler tubes.

8. Remove the engine mount bracket.

9. With 2wd:

a. Remove the torque converter cover.

b. Remove the 6 torque converter bolts.

c. Remove the 8 engine-to-transaxle bolts and separate the units.

10. With 4wd:

a. Remove the 5 bolts and 1 nut and remove the transfer case stiffener plate.

b. Remove the torque converter cover.

c. Remove the 6 torque converter bolts.

d. Remove the 8 engine-to-transaxle bolts and separate the units.

e. Remove the 2 bolts and 6 nuts. Using a plastic hammer, drive the transfer case from the transaxle.

11. Installation is the reverse of removal. Observe the following torques:

- See the illustration: bolt A 47 ft. lbs. (64 Nm); bolt B 34 ft. lbs. (46 Nm); bolt C 27 ft. lbs. (37 Nm)
- Transfer case bolts and nuts: 51 ft. lbs. (69 Nm). Use a new gasket.
- Torque converter bolts (green bolt first): 35 ft. lbs. (48 Nm)
- Stiffener plate: 25 ft. lbs. (34 Nm)
- Engine mount bracket: 47 ft. lbs. (64 Nm)

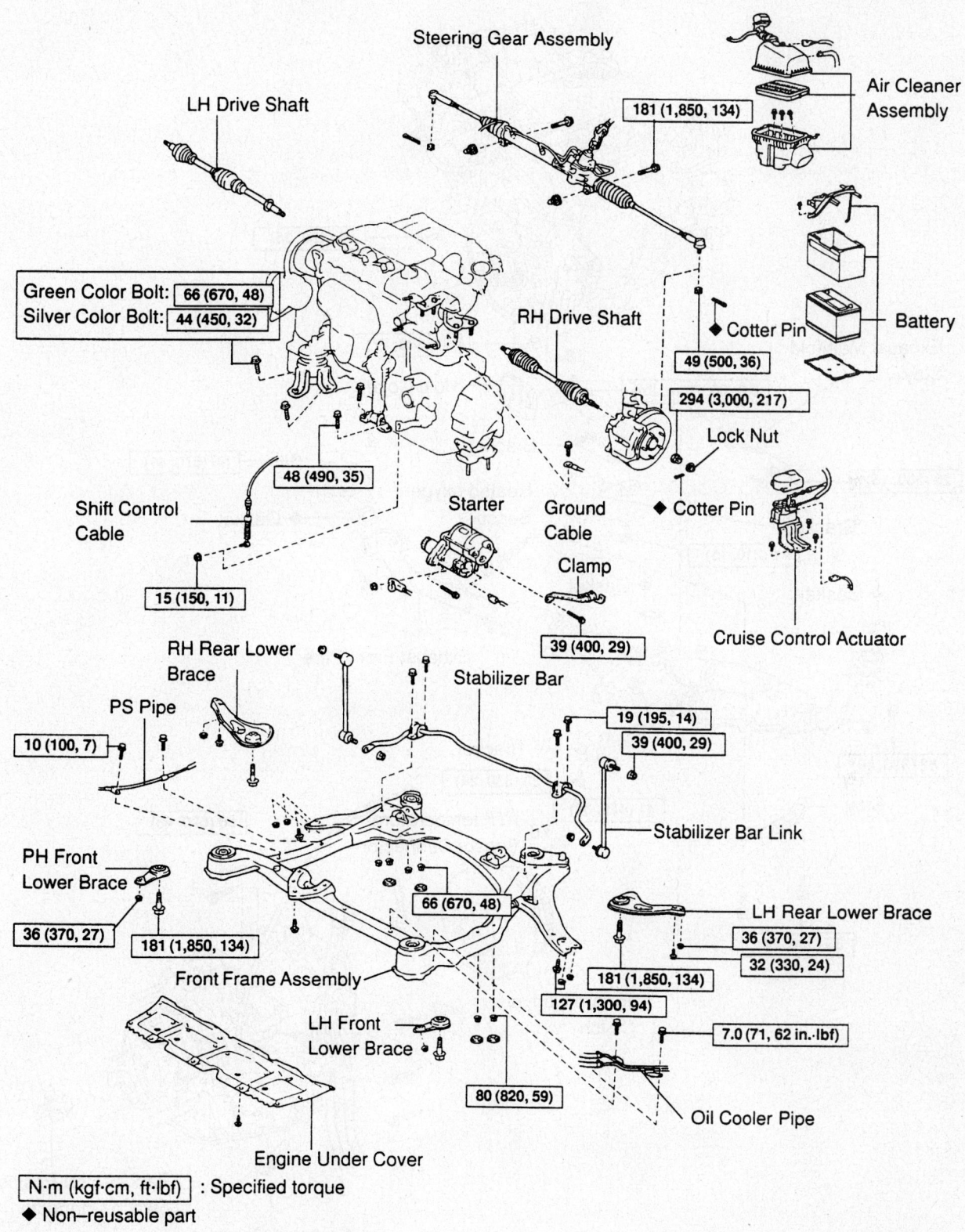

Steering Gear Assembly

LH Drive Shaft

Air Cleaner Assembly

181 (1,850, 134)

Green Color Bolt: 66 (670, 48)
Silver Color Bolt: 44 (450, 32)

RH Drive Shaft

◆ Cotter Pin

Battery

49 (500, 36)

294 (3,000, 217)

Lock Nut

◆ Cotter Pin

48 (490, 35)

Shift Control Cable

Starter

Ground Cable

Clamp

Cruise Control Actuator

15 (150, 11)

39 (400, 29)

RH Rear Lower Brace

Stabilizer Bar

19 (195, 14)

39 (400, 29)

PS Pipe

10 (100, 7)

Stabilizer Bar Link

PH Front Lower Brace

LH Rear Lower Brace

36 (370, 27)

36 (370, 27)

181 (1,850, 134)

66 (670, 48)

32 (330, 24)

Front Frame Assembly

181 (1,850, 134)

LH Front Lower Brace

127 (1,300, 94)

7.0 (71, 62 in.·lbf)

80 (820, 59)

Oil Cooler Pipe

Engine Under Cover

N·m (kgf·cm, ft·lbf) : Specified torque
◆ Non–reusable part

7924ZG65

Exploded view of the transaxle removal and installation components—2002–03

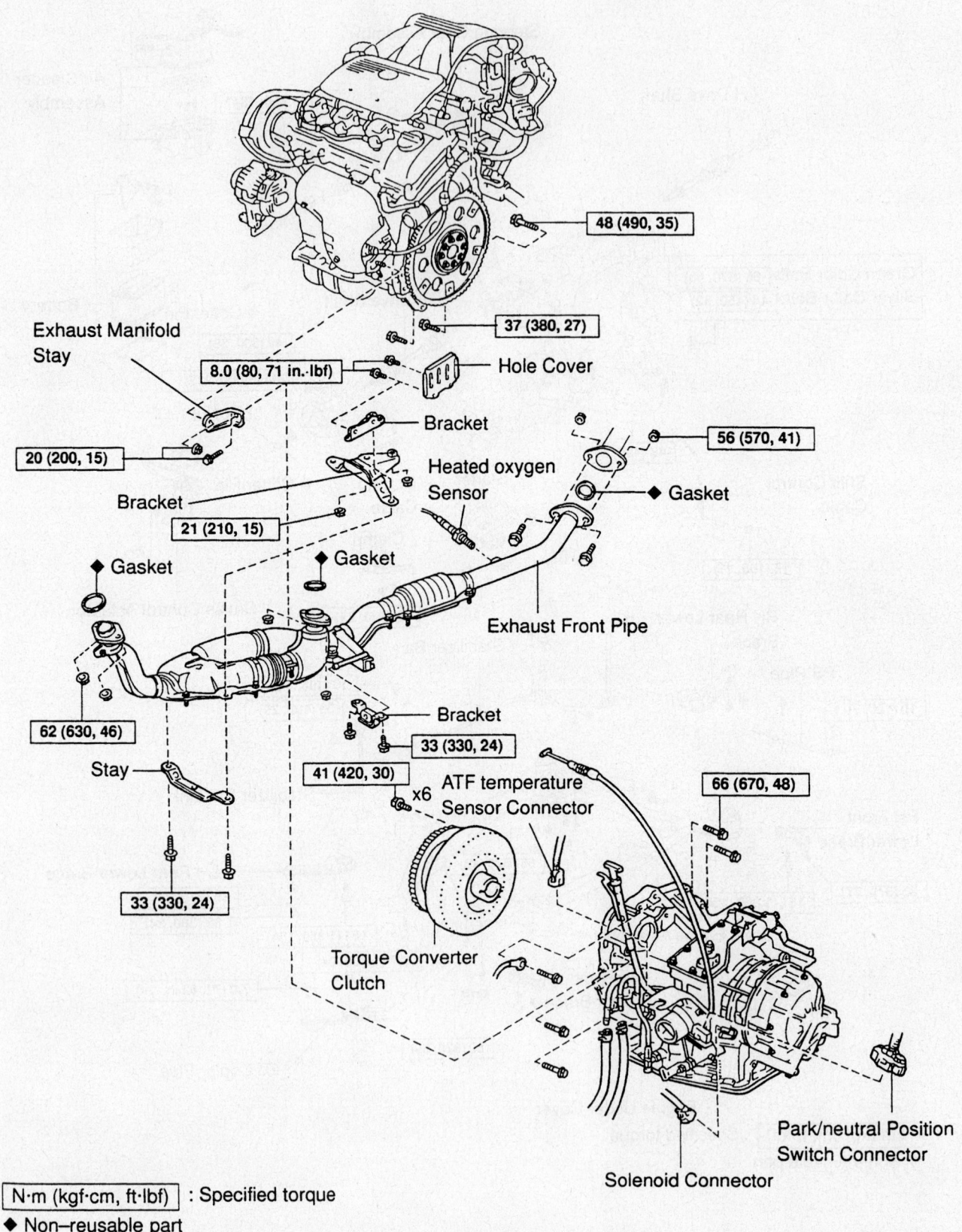

48 (490, 35)

37 (380, 27)

Exhaust Manifold Stay

8.0 (80, 71 in.·lbf)

Hole Cover

Bracket

56 (570, 41)

20 (200, 15)

Bracket

Heated oxygen Sensor

◆ Gasket

21 (210, 15)

◆ Gasket

◆ Gasket

Exhaust Front Pipe

62 (630, 46)

Bracket

Stay

33 (330, 24)

41 (420, 30)

ATF temperature Sensor Connector

66 (670, 48)

×6

33 (330, 24)

Torque Converter Clutch

Park/neutral Position Switch Connector

Solenoid Connector

N·m (kgf·cm, ft·lbf) : Specified torque

◆ Non-reusable part

7924ZG66

Exploded view of the transaxle removal and installation components—2002–03

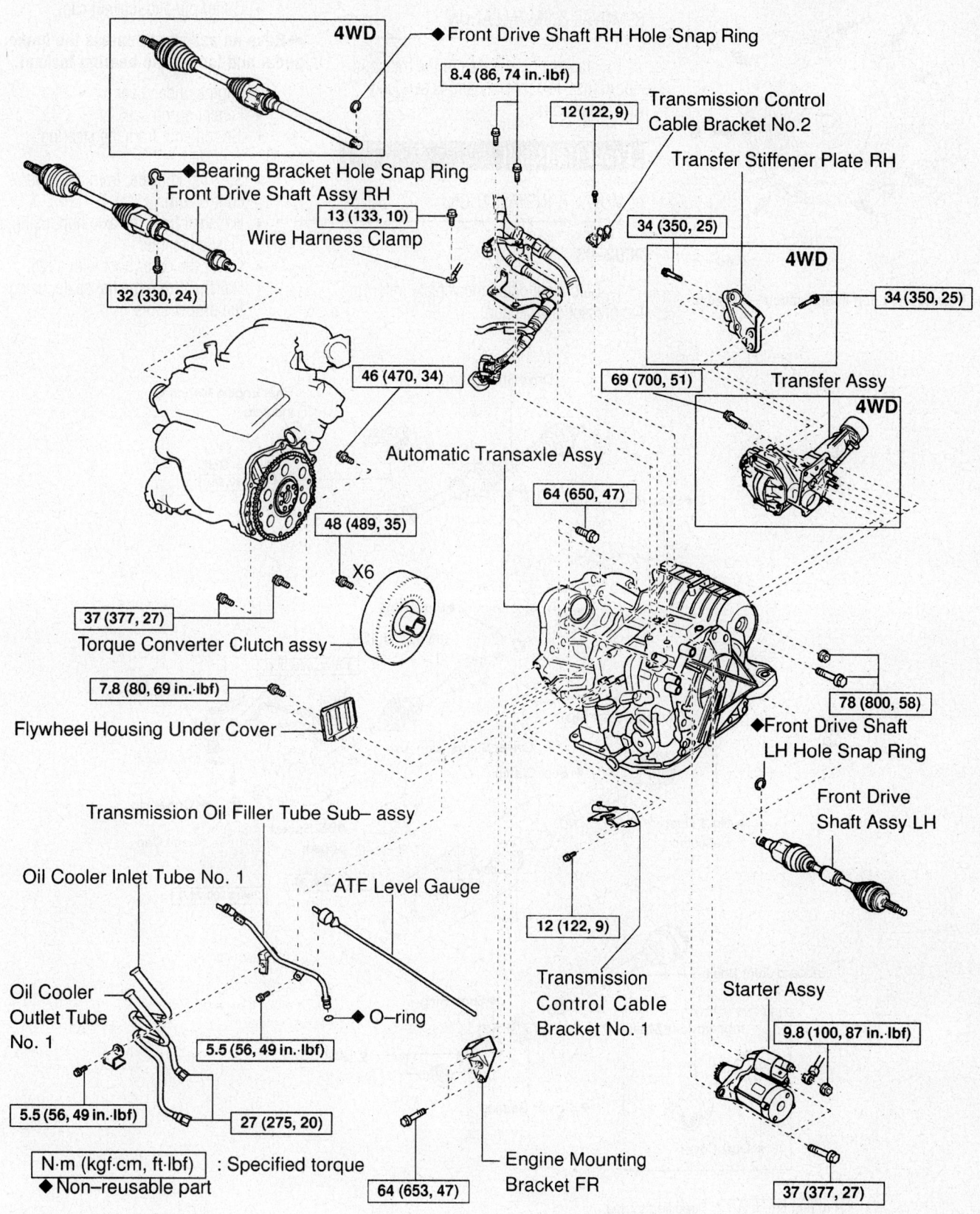

4WD

◆Front Drive Shaft RH Hole Snap Ring

8.4 (86, 74 in.·lbf)

12 (122, 9)

Transmission Control
Cable Bracket No.2

Transfer Stiffener Plate RH

◆Bearing Bracket Hole Snap Ring
Front Drive Shaft Assy RH

Wire Harness Clamp

13 (133, 10)

34 (350, 25)

4WD

34 (350, 25)

32 (330, 24)

46 (470, 34)

69 (700, 51)

Transfer Assy

4WD

Automatic Transaxle Assy

64 (650, 47)

48 (489, 35)

X6

37 (377, 27)

Torque Converter Clutch assy

78 (800, 58)

◆Front Drive Shaft
LH Hole Snap Ring

Front Drive
Shaft Assy LH

7.8 (80, 69 in.·lbf)

Flywheel Housing Under Cover

Transmission Oil Filler Tube Sub−assy

Oil Cooler Inlet Tube No. 1

ATF Level Gauge

12 (122, 9)

Transmission
Control Cable
Bracket No. 1

Starter Assy

Oil Cooler
Outlet Tube
No. 1

◆O−ring

5.5 (56, 49 in.·lbf)

9.8 (100, 87 in.·lbf)

5.5 (56, 49 in.·lbf)

27 (275, 20)

N·m (kgf·cm, ft·lbf) : Specified torque
◆Non−reusable part

64 (653, 47)

Engine Mounting
Bracket FR

37 (377, 27)

67170-SIEN-G08

Exploded view of the transaxle removal and installation components—2004–06

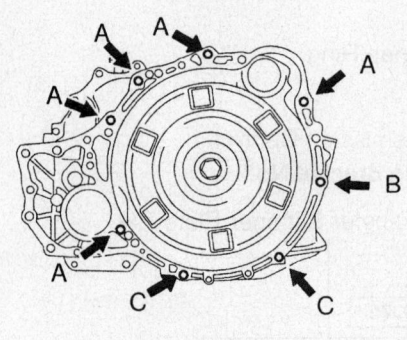

Transmission bolt identification—2004-06

67170-SIEN-G09

Transfer Case

REMOVAL & INSTALLATION

The transfer case is part of the transmission/transaxle assembly and is serviced with those units.

Front Halfshaft

REMOVAL & INSTALLATION

2002–03

1. Before servicing the vehicle, refer to the precautions section.

2. Remove or disconnect the following:
   • Front wheels
   • Cotter pin and locknut cap

➡ **Have an assistant depress the brake pedal and loosen the bearing locknut.**

   • Engine undercover
   • Fender apron seal
   • Tie rod end, from the steering knuckle
   • Steering knuckle, from the lower control arm
   • Halfshaft from the axle hub, using a plastic hammer
   • Cover the outer boot with a rag
   • Halfshaft from the transaxle, using the proper tools

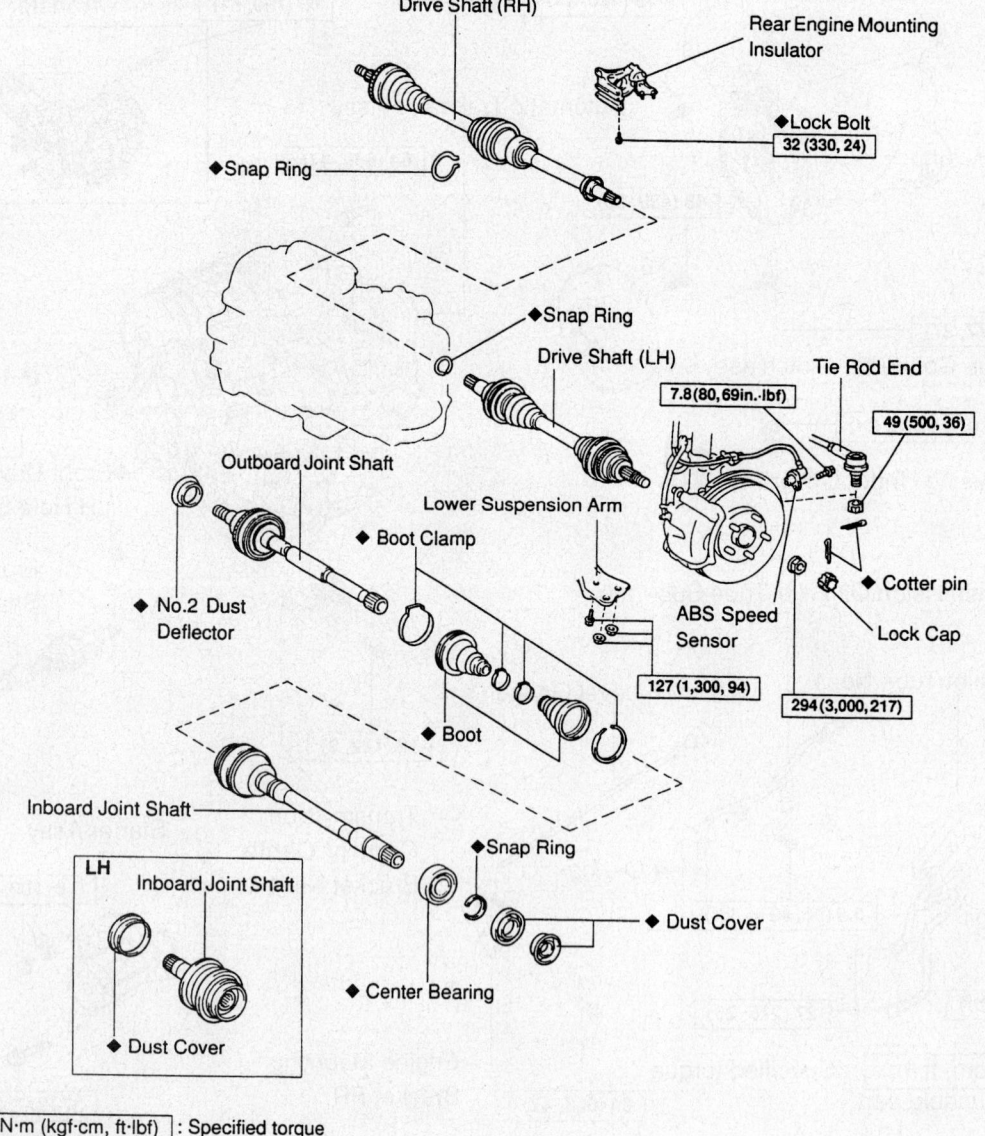

N·m (kgf·cm, ft·lbf) : Specified torque
◆ Non-reusable part

Exploded view of halfshaft—2002–03

7924ZG73

To install:

3. Reverse the removal procedures to complete installation, tightening fasteners to specifications.

4. Fill the transaxle with gear oil, install the fender apron, check front end alignment and test drive.

➡**If the cotter pin holes do not align, always correct by tightening the nut until the next hole aligns.**

5. Install a new cotter pin.

2004–06

1. Drain the transaxle fluid.
2. With 4wd, drain the transfer case.
3. Remove the wheel.
4. Unstake the hub nut, and, with the brake applied, remove the hub nut.
5. Disconnect the stabilizer link.
6. Remove the speed sensor.

7. Disconnect the tie rod end from the knuckle.
8. Disconnect the lower arm from the ball joint.
9. Using a plastic hammer, drive the halfshaft from the hub.
10. On the left side with 2wd and both sides with 4wd, using a slidehammer with adapter, pull the halfshaft from the transaxle.

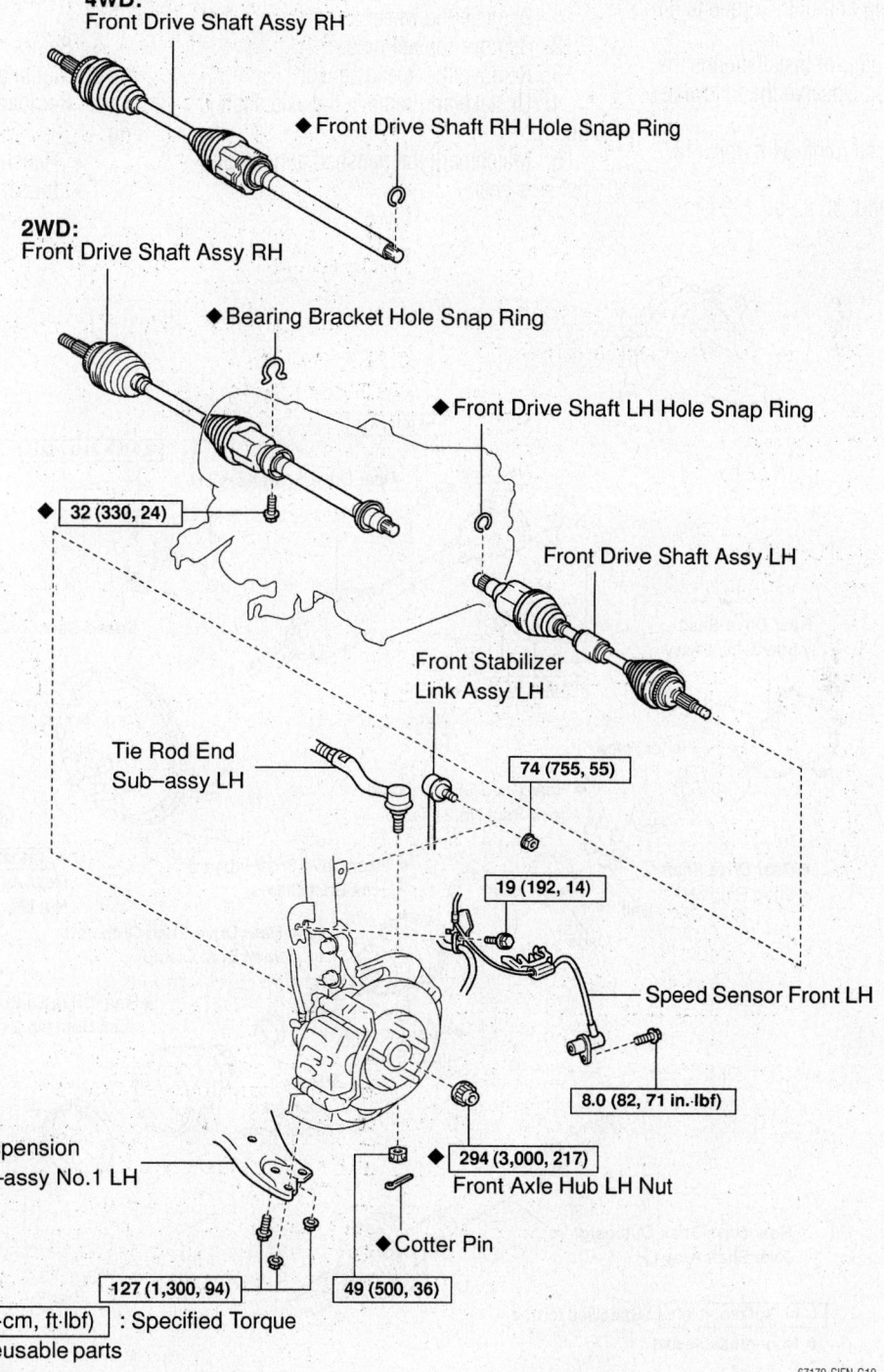

4WD:
Front Drive Shaft Assy RH

◆ Front Drive Shaft RH Hole Snap Ring

2WD:
Front Drive Shaft Assy RH

◆ Bearing Bracket Hole Snap Ring

◆ Front Drive Shaft LH Hole Snap Ring

◆ 32 (330, 24)

Front Drive Shaft Assy LH

Front Stabilizer
Link Assy LH

Tie Rod End
Sub–assy LH

74 (755, 55)

19 (192, 14)

Speed Sensor Front LH

8.0 (82, 71 in.·lbf)

Front Suspension
Arm Sub–assy No.1 LH

◆ 294 (3,000, 217)
Front Axle Hub LH Nut

◆Cotter Pin

127 (1,300, 94)   49 (500, 36)

N·m (kgf·cm, ft·lbf) : Specified Torque
◆Non–reusable parts

67170-SIEN-G10

Front halfshaft exploded view—2004–06 models

11. On the right side, with 2wd, remove the halfshaft bearing bracket snapring. Remove the bolt and the halfshaft from the bearing bracket.

To install:

12. Coat the splines of the inboard end with clean ATF.

13. Drive the left shaft (2wd and both shafts 4wd) into place with a hammer and brass drift. Install the snapring with the opening downward.

14. Install the right side (2wd) shaft. Install the snapring and bolt. Torque to 24 ft. lbs. (32 Nm).

15. The remainder of installation is the reverse of removal. Observe the following torques:

- Arm-to-ball joint: 94 ft. lbs. (127 Nm)
- Tie rod end: 36 ft. lbs. (49 Nm).

Advance the nut no more than 60 degrees to align the hole.
- Stabilizer link: 55 ft. lbs. (74 Nm)
- New hub nut: 217 ft. lbs. (294 Nm). Stake the nut.

Rear Halfshaft

REMOVAL & INSTALLATION

2004–06

1. Remove the wheel.
2. Remove the tail pipe.
3. Remove the speed sensor.
4. Unstake and remove the axle shaft nut.
5. Matchmark the halfshaft and differential side gear.

6. Remove the 4 nuts and washers and remove the shaft.

7. Installation is the reverse of removal. Torque the 4 nuts t 41 ft. lbs. (56 Nm). Torque the axle shaft nut to 159 ft. lbs. (216 Nm). Stake the nut.

CV-Joints

OVERHAUL

2002

1. Before servicing the vehicle, refer to the precautions section.
2. Remove or disconnect the following:

- Halfshaft
- Inboard joint boot clamps

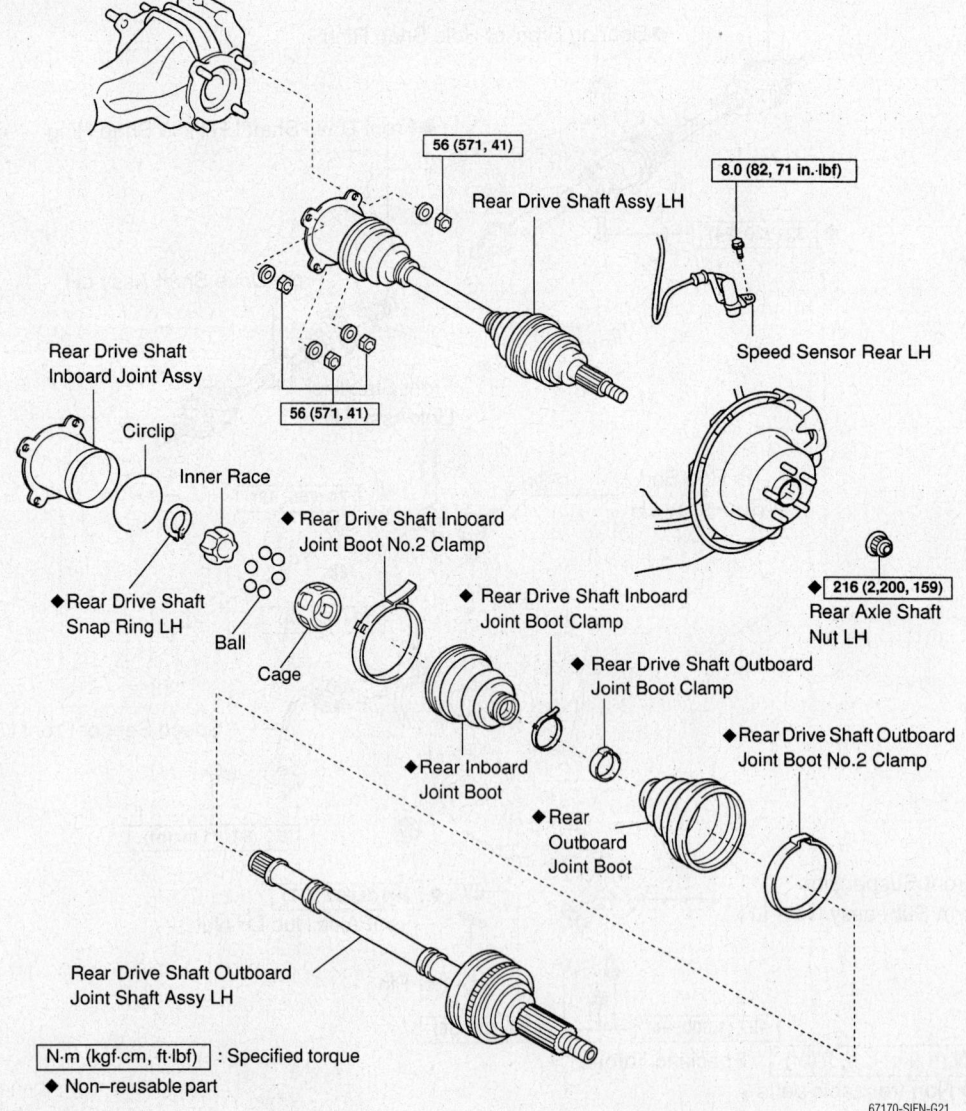

Rear Drive Shaft Assy LH

56 (571, 41)

8.0 (82, 71 in.·lbf)

Speed Sensor Rear LH

Rear Drive Shaft Inboard Joint Assy

Circlip

Inner Race

◆Rear Drive Shaft Snap Ring LH

Ball

Cage

◆ Rear Drive Shaft Inboard Joint Boot No.2 Clamp

56 (571, 41)

◆ Rear Drive Shaft Inboard Joint Boot Clamp

◆Rear Inboard Joint Boot

◆ Rear Drive Shaft Outboard Joint Boot Clamp

◆Rear Outboard Joint Boot

216 (2,200, 159)
Rear Axle Shaft Nut LH

◆Rear Drive Shaft Outboard Joint Boot No.2 Clamp

Rear Drive Shaft Outboard Joint Shaft Assy LH

N·m (kgf·cm, ft·lbf) : Specified torque
◆ Non–reusable part

67170-SIEN-G21

Rear halfshaft and related parts—2004–06 models

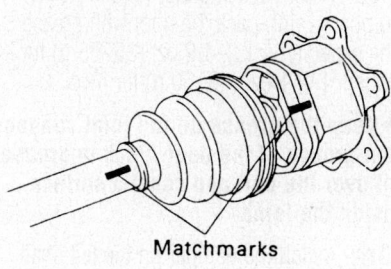

Place matchmarks on the inboard joint outer race and shaft

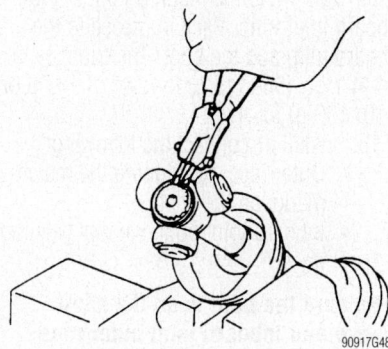

Use snapring expanders to extract the snapring from the end

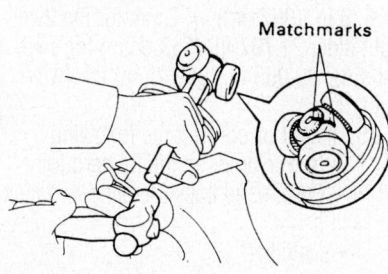

With a brass bar and hammer, tap the joint hard enough to remove

3. Clean the joint before removing the boot.

4. Slide the inboard joint boot toward the outboard joint.

5. Place matchmarks on the inboard joint tulip and the shaft.

6. Remove the inboard joint tulip from the driveshaft.

7. Clamp the halfshaft in a vise.

8. Remove or disconnect the following:
- Snapring and disassemble the tri-pot joint
- Tri-pot joint from the halfshaft, using a brass bar and hammer

❋❋ WARNING

Be careful not to punch the roller.

- Inboard joint boot
- Outboard joint boot clamps and boot

❋❋ WARNING

Do not disassemble the outboard joint.

To assemble:

9. Temporarily, install the new boot and new boot clamps to the outboard joint.

➡**Before installing the boot, wrap vinyl tape around the spline of the shaft to prevent damaging the boot.**

10. Temporarily, install the new boot and the new boot clamps for the inboard joint to the halfshaft.

11. Assemble the tri-pot joint, as follows:
   a. Place the beveled side of the tri-pot axial spline toward the outboard joint.
   b. Align the matchmarks placed before disassembly.
   c. Using a brass bar and hammer, tap in the tri-pot joint onto the driveshaft. Do not punch the roller.

12. Install a new snapring.

13. Before assembling the boot to the outboard joint, pack the boot with grease. The capacity is 4.2–4.6 oz. (120–130 g).

➡**Keep the grease off the joint connection groove of the boot. Pack in grease all over the ball and contact surface inside the joint.**

14. Assemble the inboard joint to the inboard joint tulip. Pack in grease to the inboard tulip and the boot. The capacity is 7.6–7.9 oz. (215–225 g).

15. Install or connect the following:
- Outer race, by aligning the matchmarks on the shaft
- Inboard joint boot, without twisting it

➡**Be sure the boot is on the shaft groove and inboard joint outer race groove.**

16. Set the length of the shaft to 19.146–19.546 in. (486.4–496.41mm).

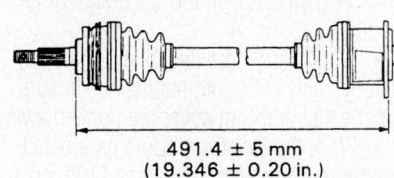

**491.4 ± 5 mm
(19.346 ± 0.20 in.)**

Set the length of the halfshaft at the points shown here

17. Install or connect the following:
- Both clamps to the inboard joint boot; bend back the band and lock it
- Halfshaft

2003

1. Before servicing the vehicle, refer to the precautions section.

2. Remove or disconnect the following:
- Halfshaft
- Inboard joint boot clamps

3. Clean the joint before removing the boot.

4. Slide the inboard joint boot toward the outboard joint.

5. Place matchmarks on the inboard joint tulip and the shaft.

6. Remove the inboard joint tulip from the driveshaft.

7. Clamp the halfshaft in a vise.

8. Remove or disconnect the following:
- Snapring and disassemble the tri-pot joint
- Tri-pot joint from the halfshaft, using a brass bar and hammer

❋❋ WARNING

Be careful not to punch the roller.

- Inboard joint boot
- Outboard joint boot clamps and boot

❋❋ WARNING

Do not disassemble the outboard joint.

To assemble:

9. Temporarily, install the new boot and new boot clamps to the outboard joint.

➡**Before installing the boot, wrap vinyl tape around the spline of the shaft to prevent damaging the boot.**

10. Temporarily, install the new boot and the new boot clamps for the inboard joint to the halfshaft.

11. Assemble the tri-pot joint, as follows:
   a. Place the beveled side of the tri-pot axial spline toward the outboard joint.
   b. Align the matchmarks placed before disassembly.
   c. Using a brass bar and hammer, tap in the tri-pot joint onto the driveshaft. Do not punch the roller.

12. Install a new snapring.

13. Before assembling the boot to the outboard joint, pack the boot with grease. The capacity is 3.7–4.4 oz. (105–125 g).

➡**Keep the grease off the joint connection groove of the boot. Pack in grease all over the ball and contact surface inside the joint.**

14. Assemble the inboard joint to the inboard joint tulip. Pack in grease to the inboard tulip and the boot. The capacity is 4.2–4.6 oz. (120–130 g) for the joint side; 2.1–2.3 oz. (52.5–57.5 g) for the boot side.

15. Install or connect the following:
- Outer race, by aligning the matchmarks on the shaft
- Inboard joint boot, without twisting it

➡**Be sure the boot is on the shaft groove and inboard joint outer race groove.**

16. Set the length of the shaft (inboard end to outboard joint flange) to 23.071 in. (586mm) +/- 2mm for left side; 34.709 in. (881.6mm) +/- 2mm for the right side.

17. Install or connect the following:
- Both clamps to the inboard joint boot; bend back the band and lock it
- Halfshaft

2004–06

1. Before servicing the vehicle, refer to the precautions section.

2. Remove or disconnect the following:
- Halfshaft
- Inboard joint boot clamps

3. Clean the joint before removing the boot.

4. Slide the inboard joint boot toward the outboard joint.

5. Place matchmarks on the inboard joint tulip and the shaft.

6. Remove the inboard joint tulip from the driveshaft.

7. Clamp the halfshaft in a vise.

8. Remove or disconnect the following:
- Snapring and disassemble the tri-pot joint
- Tri-pot joint from the halfshaft, using a brass bar and hammer

✳✳ WARNING

Be careful not to punch the roller.

- Inboard joint boot
- Outboard joint boot clamps and boot

✳✳ WARNING

Do not disassemble the outboard joint.

To assemble:

9. Temporarily, install the new boot and new boot clamps to the outboard joint.

➡**Before installing the boot, wrap vinyl tape around the spline of the shaft to prevent damaging the boot.**

10. Temporarily, install the new boot and the new boot clamps for the inboard joint to the halfshaft.

11. Assemble the tri-pot joint, as follows:

a. Place the beveled side of the tri-pot axial spline toward the outboard joint.

b. Align the matchmarks placed before disassembly.

c. Using a brass bar and hammer, tap in the tri-pot joint onto the driveshaft. Do not punch the roller.

12. Install a new snapring.

13. Before assembling the boot to the outboard joint, pack the boot with grease. The capacity is 1.2–1.9 oz. (35–55 g) for 2wd; 1.4–2.1 oz. (40–60 g) for 4wd.

➡**Keep the grease off the joint connection groove of the boot. Pack in grease all over the ball and contact surface inside the joint.**

14. Install the damper on the left shaft. The distance from the outer end flange of the damper to the outer end flange of the outer joint should be 8.7 inches (221mm) +/- 2mm.

15. Assemble the inboard joint to the inboard joint tulip. Pack in grease to the inboard tulip and the boot. The capacity is 3.4–4.1 oz. (95–115 g) for 2wd; 3.9–4.6 oz. (110–130 g) for 4wd.

16. Install or connect the following:
- Outer race, by aligning the matchmarks on the shaft
- Inboard joint boot, without twisting it

➡**Be sure the boot is on the shaft groove and inboard joint outer race groove.**

17. Set the length of the shaft (inboard end to outboard joint flange) to 25.134 in. (638.4mm) +/- 5mm for 2wd left side; 37.678 in. (957mm) +/- 5mm for the 2wd right side; 21.797 in. (553.6mm) for 4wd left side; 39.065 in. (992.2mm) for 4wd right side.

18. Install or connect the following:
- Both clamps to the inboard joint boot; bend back the band and lock it
- Halfshaft

STEERING & SUSPENSION

Air Bag

✳✳ CAUTION

Some vehicles are equipped with an air bag system. The system must be disabled before performing service on or around system components, steering column, instrument panel components, wiring and sensors. Failure to follow safety and disabling procedures could result in accidental air bag deployment, possible personal injury and unnecessary system repairs.

PRECAUTIONS

Several precautions must be observed when handling the inflator module to avoid accidental deployment and possible personal injury.

- Never carry the inflator module by the wires or connector on the underside of the module.
- When carrying a live inflator module, hold securely with both hands, and ensure that the bag and trim cover are pointed away.
- Place the inflator module on a bench or other surface with the bag and trim cover facing up.

- With the inflator module on the bench, never place anything on or close to the module, which may be thrown in the event of an accidental deployment.

DISARMING

To avoid personal injury when working on vehicles equipped with an air bag, the negative battery cable must be disconnected and at least 90 seconds must elapse before working on the system. Failure to do so may result in deployment of the air bag.

Power Rack and Pinion Steering Gear

REMOVAL & INSTALLATION

2002–03

1. Before servicing the vehicle, refer to the precautions section.
2. Remove or disconnect the following:
   • Negative battery cable

→**Wait at least 90 seconds before working on the vehicle to allow the Supplemental Restraint System (SRS) system to disarm.**

   • Right and left side fender apron seals
   • Right and left tie rod ends
3. Place matchmarks on the intermediate shaft.
4. Remove or disconnect the following:
   • Pinch bolt and the intermediate shaft out from under the vehicle
   • Power steering line clamp

• Pressure and feed lines
• Stabilizer bar, unbolt it but do not remove it
• Heated Oxygen (HO$_2$) sensor
• Both gear assembly set bolts and nuts, by lifting the stabilizer bar
• Gear assembly from the left side of the vehicle

To install:
5. Install or connect the following:
   • Gear assembly to the left side of the vehicle

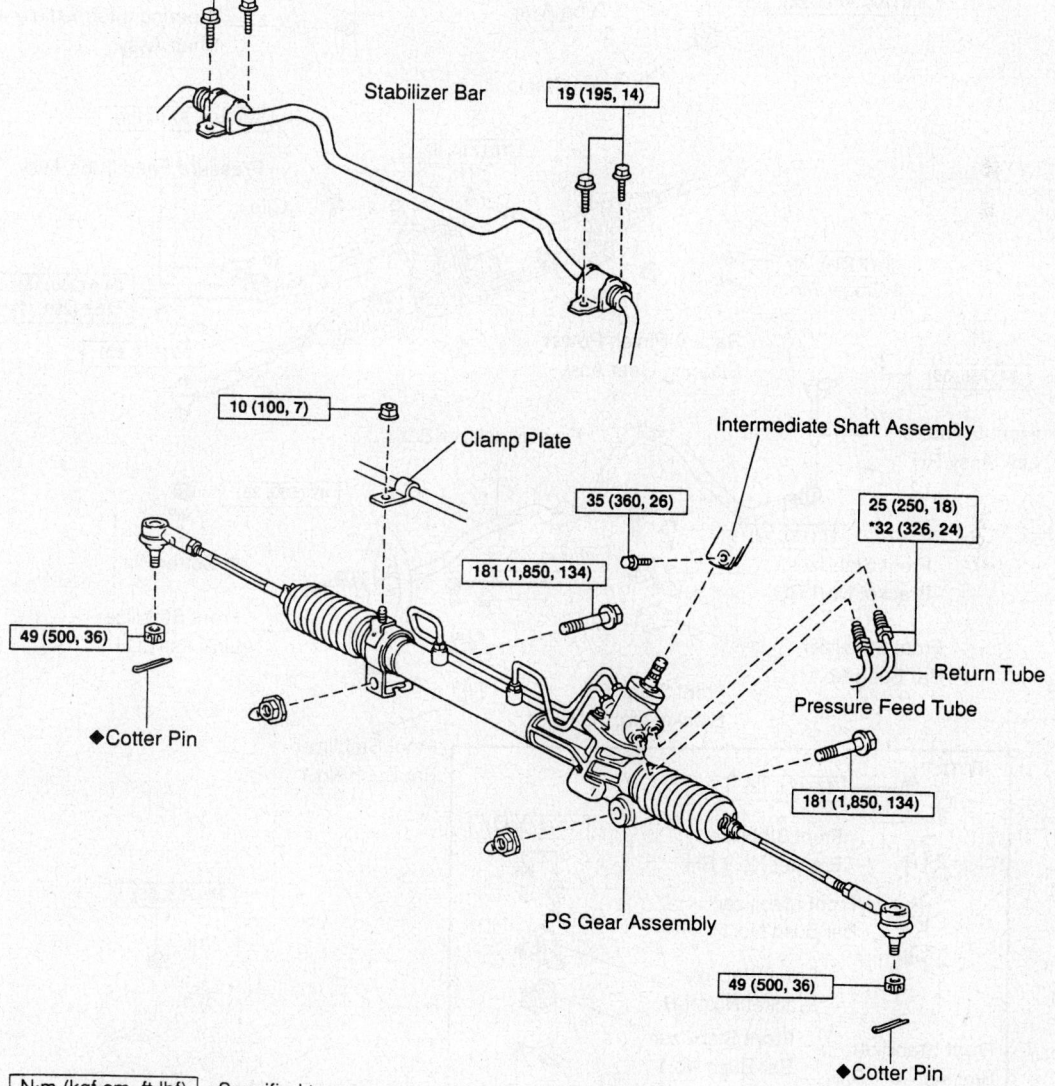

19 (195, 14)
Stabilizer Bar
19 (195, 14)
10 (100, 7)
Clamp Plate
Intermediate Shaft Assembly
35 (360, 26)
181 (1,850, 134)
25 (250, 18) *32 (326, 24)
49 (500, 36)
Return Tube
Pressure Feed Tube
◆Cotter Pin
181 (1,850, 134)
PS Gear Assembly
49 (500, 36)
◆Cotter Pin

N·m (kgf·cm, ft·lbf) : Specified torque
◆ Non–reusable part
* For use with SST

7924ZG76

Exploded view of the power steering gear and related components—2002–03

✳✳ WARNING

Be careful not to damage the power steering lines.

- Tighten the gear assembly set bolts and nuts to 134 ft. lbs. (181 Nm), by lifting the stabilizer bar
- HO$_2$ sensor
- Stabilizer bar. Tighten the bolt to 14 ft. lbs. (19 Nm) and the nut to 29 ft. lbs. (39 Nm).
- Pressure and feed return lines. Tighten them to 18 ft. lbs. (25 Nm).

- Line clamps. Tighten the nut to 84 inch lbs. (10 Nm).
- Intermediate shaft, by aligning the joint and main shaft matchmarks. Tighten to 26 ft. lbs. (35 Nm).
- Tie rod ends
- Fender apron seals. Securely tighten the bolts.

6. Remove or disconnect the following:
- Steering wheel pad
- Steering wheel

7. Position the front wheels facing straight-ahead. Do this with the front of the vehicle on jackstands.

8. Center the spiral cable.

9. Install the steering wheel at the straight-ahead position. Temporarily tighten the wheel set nut. Attach the wiring.

10. Bleed the power steering system.

11. Check the steering wheel center point. Tighten the steering nut to 26 ft. lbs. (35 Nm).

12. Check and/or adjust the front wheel alignment.

2004–06

1. Place the wheels in a straight-ahead position.

2. Remove the wheels.

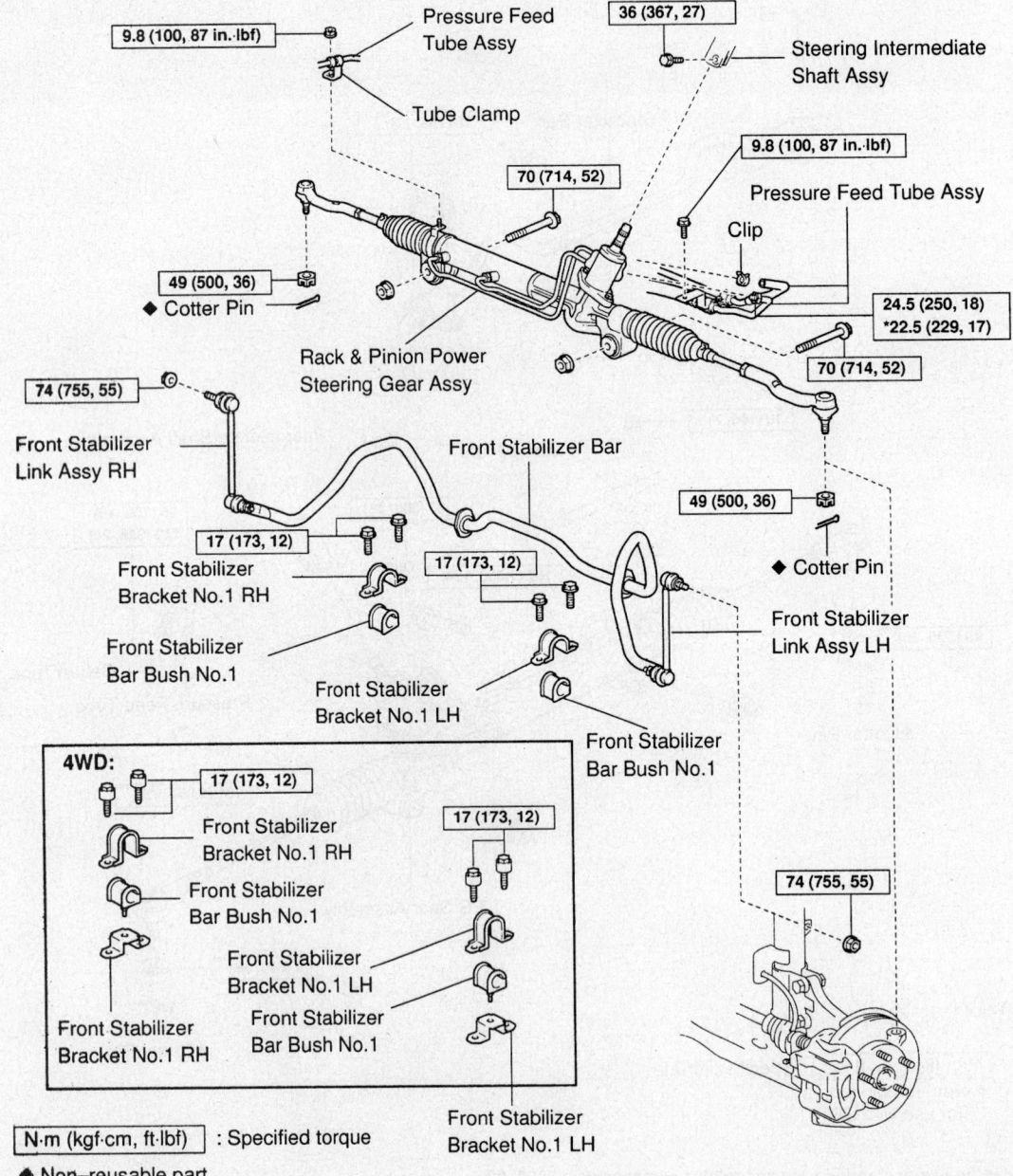

9.8 (100, 87 in.·lbf) — Pressure Feed Tube Assy
Tube Clamp
36 (367, 27) — Steering Intermediate Shaft Assy
9.8 (100, 87 in.·lbf) — Pressure Feed Tube Assy
Clip
70 (714, 52)
49 (500, 36)
◆ Cotter Pin
Rack & Pinion Power Steering Gear Assy
24.5 (250, 18)
*22.5 (229, 17)
70 (714, 52)
74 (755, 55)
Front Stabilizer Link Assy RH
Front Stabilizer Bar
17 (173, 12)
Front Stabilizer Bracket No.1 RH
17 (173, 12)
49 (500, 36)
◆ Cotter Pin
Front Stabilizer Link Assy LH
Front Stabilizer Bar Bush No.1
Front Stabilizer Bracket No.1 LH
Front Stabilizer Bar Bush No.1

4WD:
17 (173, 12)
Front Stabilizer Bracket No.1 RH
Front Stabilizer Bar Bush No.1
17 (173, 12)
Front Stabilizer Bracket No.1 LH
Front Stabilizer Bar Bush No.1
Front Stabilizer Bracket No.1 RH
74 (755, 55)
Front Stabilizer Bracket No.1 LH

N·m (kgf·cm, ft·lbf) : Specified torque
◆ Non–reusable part
* For use with SST

67170-SIEN-G11

3. Matchmark and remove the intermediate shaft from the gear.

4. Remove the steering column hole cover from the dash panel.

5. Disconnect the tie rod ends.

6. Disconnect the stabilizer bar.

7. Disconnect the pressure and return lines.

8. Remove the 2 gear mounting bolts.

9. Matchmark the intermediate shaft extension and control valve shaft.

10. Remove the intermediate shaft extension bolt and remove the gear.

To install:

11. Place the gear in position. Align the matchmarks and connect the intermediate extension shaft. Torque to 27 ft. lbs. (36 Nm).

12. Install the column cover plate and clamp.

13. Install the gear mounting bolts. Torque to 52 ft. lbs. (70 Nm).

14. Connect the stabilizer bar. Torque to 12 ft. lbs. (17 Nm).

15. Connect the tie rods ends. Torque to 36 ft. lbs. (49 Nm).

16. Connect the intermediate shaft. Torque to 26 ft. lbs. (36 Nm).

17. Check the alignment.

Strut

REMOVAL & INSTALLATION

2002–03

1. Before servicing the vehicle, refer to the precautions section.

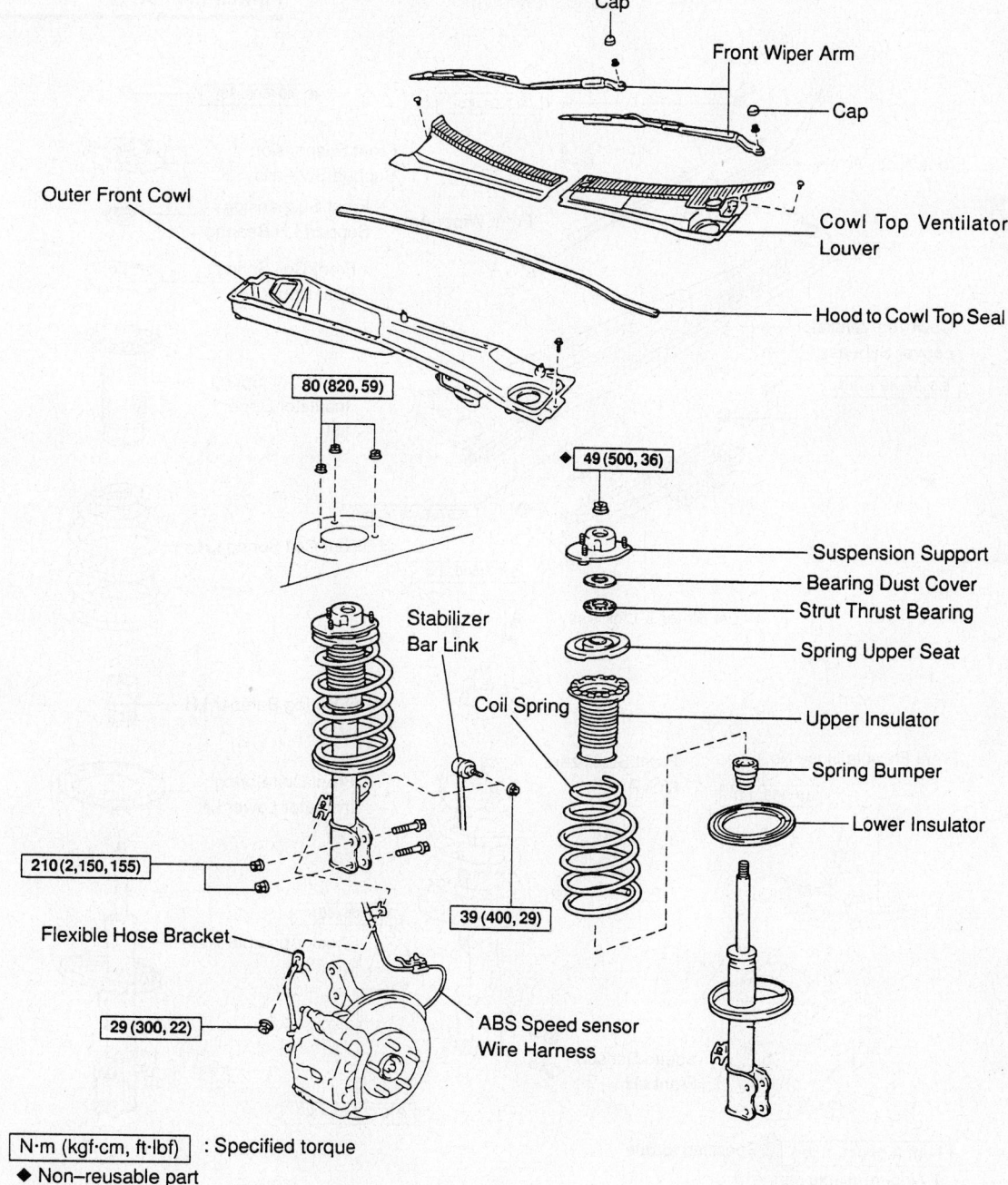

View of the front strut assembly and related components—2002–03

7924ZG79

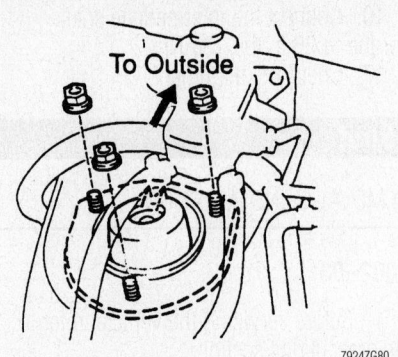

To Outside
7924ZG80

Cutaway view of the upper strut bearing position for installation—2002-03

➡**Do not support the weight of the vehicle on the suspension arm; the arm will deform under its weight.**

2. Remove or disconnect the following:
   - Wheel
   - Brake hose and the Anti-lock Brake System (ABS) speed sensor wire from the strut
   - Sway bar link from the strut
   - Outer front cowl top panel
3. Matchmark the strut lower bracket and camber adjust cam, if equipped.
4. Remove or disconnect the following:
   - Lower strut end from the steering knuckle's lower arm
   - 3 upper strut mounting plate to the upper wheel arch nuts
   - Strut

To install:

5. Align the upper suspension support hole with the strut piston or end, so they fit properly.
6. Install or connect the following:
   - Strut piston rod end to the upper suspension support. Tighten the new nut to 29–40 ft. lbs. (39–54 Nm).

⁂ WARNING

Do not use an impact wrench to tighten the nut.

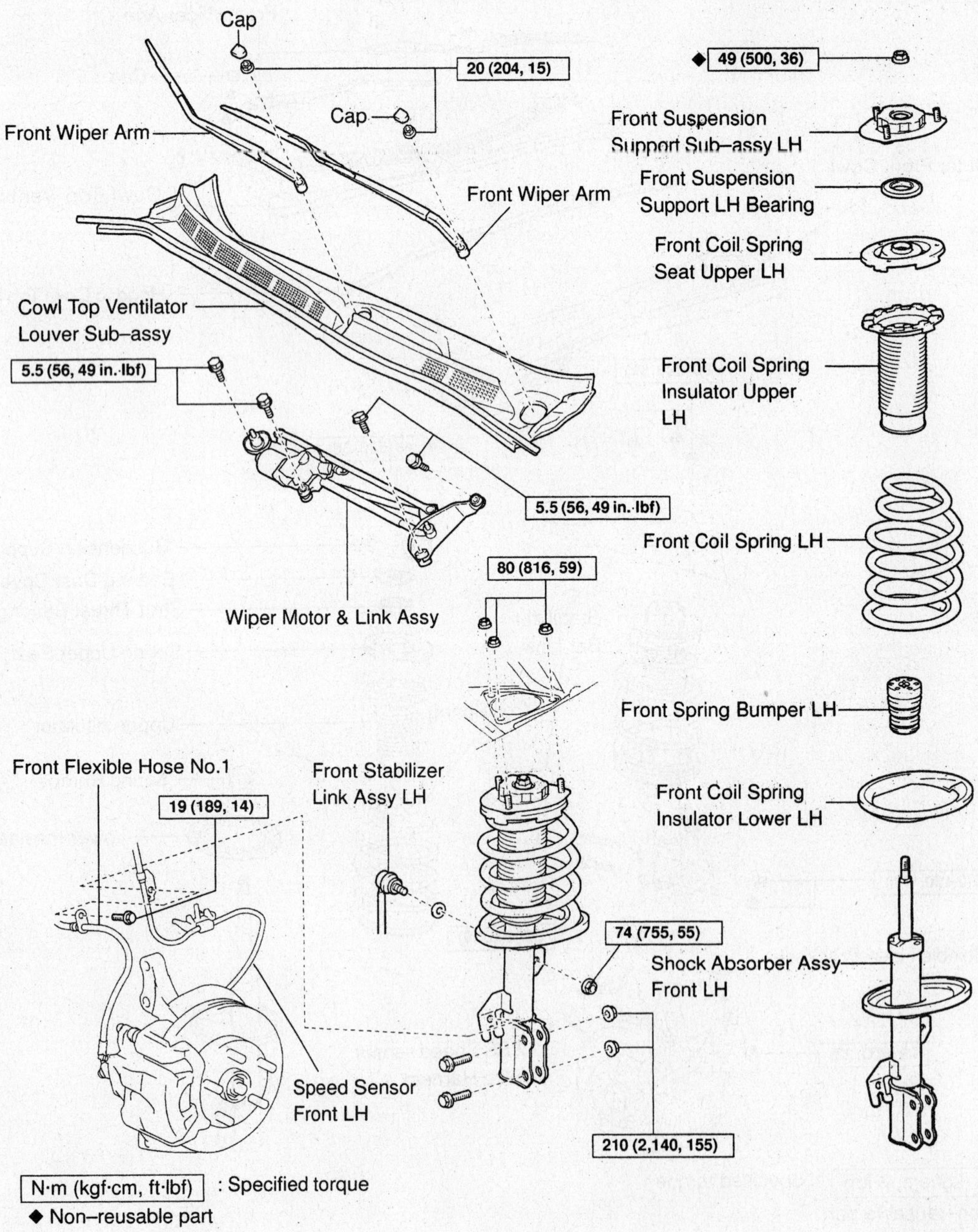

Cap
20 (204, 15)
◆ 49 (500, 36)
Cap
Front Wiper Arm
Front Suspension Support Sub–assy LH
Front Suspension Support LH Bearing
Front Wiper Arm
Front Coil Spring Seat Upper LH
Cowl Top Ventilator Louver Sub–assy
5.5 (56, 49 in.·lbf)
Front Coil Spring Insulator Upper LH
5.5 (56, 49 in.·lbf)
Front Coil Spring LH
80 (816, 59)
Wiper Motor & Link Assy
Front Spring Bumper LH
Front Flexible Hose No.1
Front Stabilizer Link Assy LH
19 (189, 14)
Front Coil Spring Insulator Lower LH
74 (755, 55)
Shock Absorber Assy Front LH
Speed Sensor Front LH
210 (2,140, 155)
N·m (kgf·cm, ft·lbf) : Specified torque
◆ Non–reusable part
67170-SIEN-G12

Front suspension and related components—2004-06

- Lubricate the suspension support bearing with multi-purpose grease. Pack the upper support space with multi-purpose grease, after installation.
- Tighten the 3 suspension support-to-wheel arch nuts to 47 ft. lbs. (64 Nm).
- Tighten the strut-to-steering knuckle arm bolts to 156 ft. lbs. (211 Nm).
- Sway bar link to the strut. Tighten the nut to 29 ft. lbs. (39 Nm).
- Outer front cowl top panel
- ABS speed sensor and the brake hose to the strut, if equipped.
- Wheel

2004–06

1. Remove the wheel.
2. Remove the wiper arms.
3. Remove the wiper motor.
4. Remove the top cowl.
5. Remove the stabilizer link from the strut.
6. Loosen the strut rod locknut. Don't remove it.
7. Remove the brake hose bracket from the strut.
8. Remove the lower strut bolts.
9. Remove the 3 upper strut nuts.
10. Remove the strut.
11. Installation is the reverse of removal. Observe the following torques:

- Upper nuts: 59 ft. lbs. (80 Nm)
- Lower nuts/bolts: 155 ft. lbs. (210 Nm)
- Strut rod locknut: 36 ft. lbs. (49 Nm)
- Stabilizer link: 55 ft. lbs. (74 Nm)

Shock Absorber

REMOVAL & INSTALLATION

2002–03

1. Before servicing the vehicle, refer to the precautions section.
2. Support the axle beam with jacks.
3. Remove or disconnect the following:

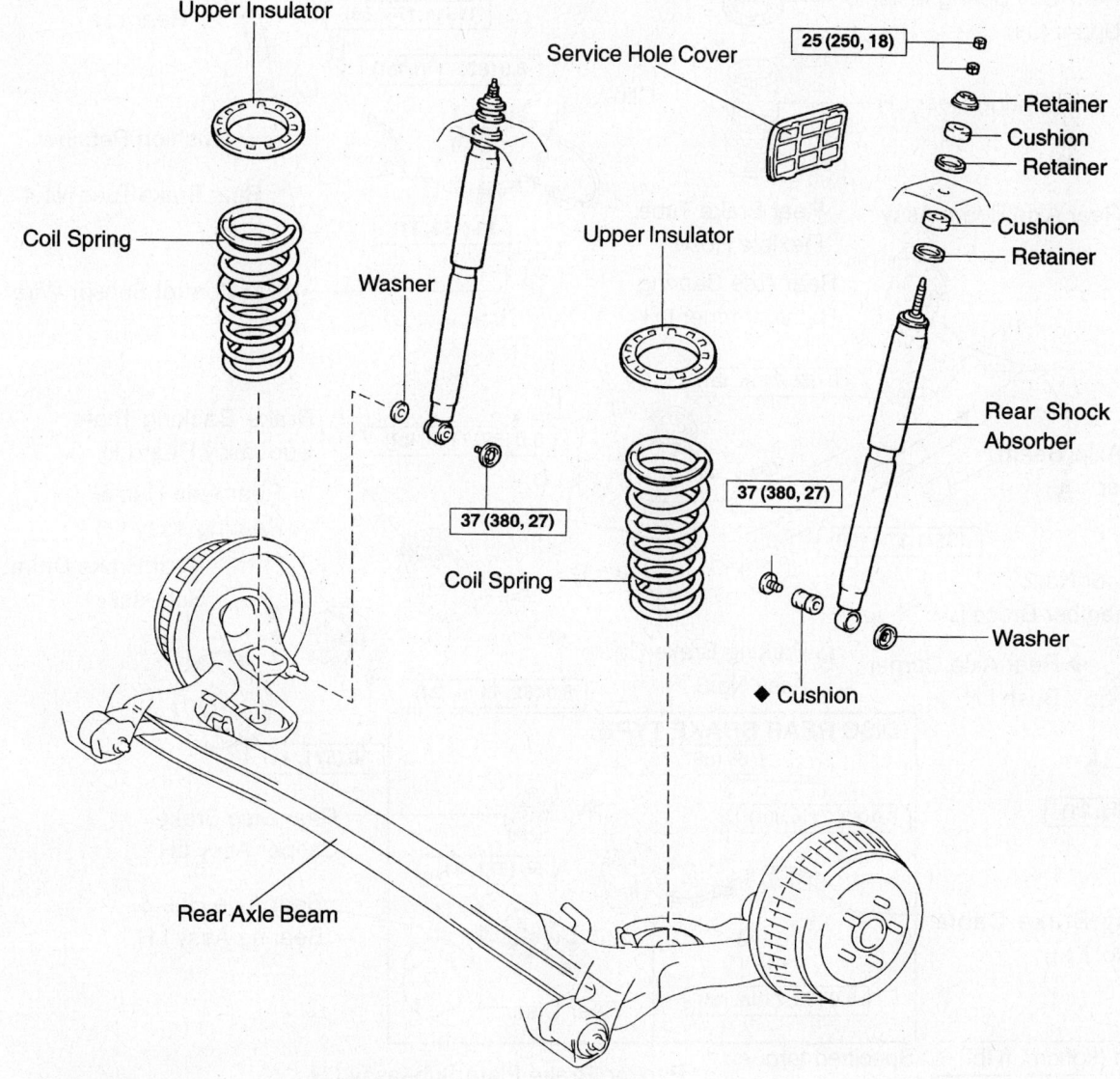

N·m (kgf·cm, ft·lbf) : Specified torque
◆ Non–reusable part

Rear suspension components—2002–03 models

2WD DRIVE TYPE:

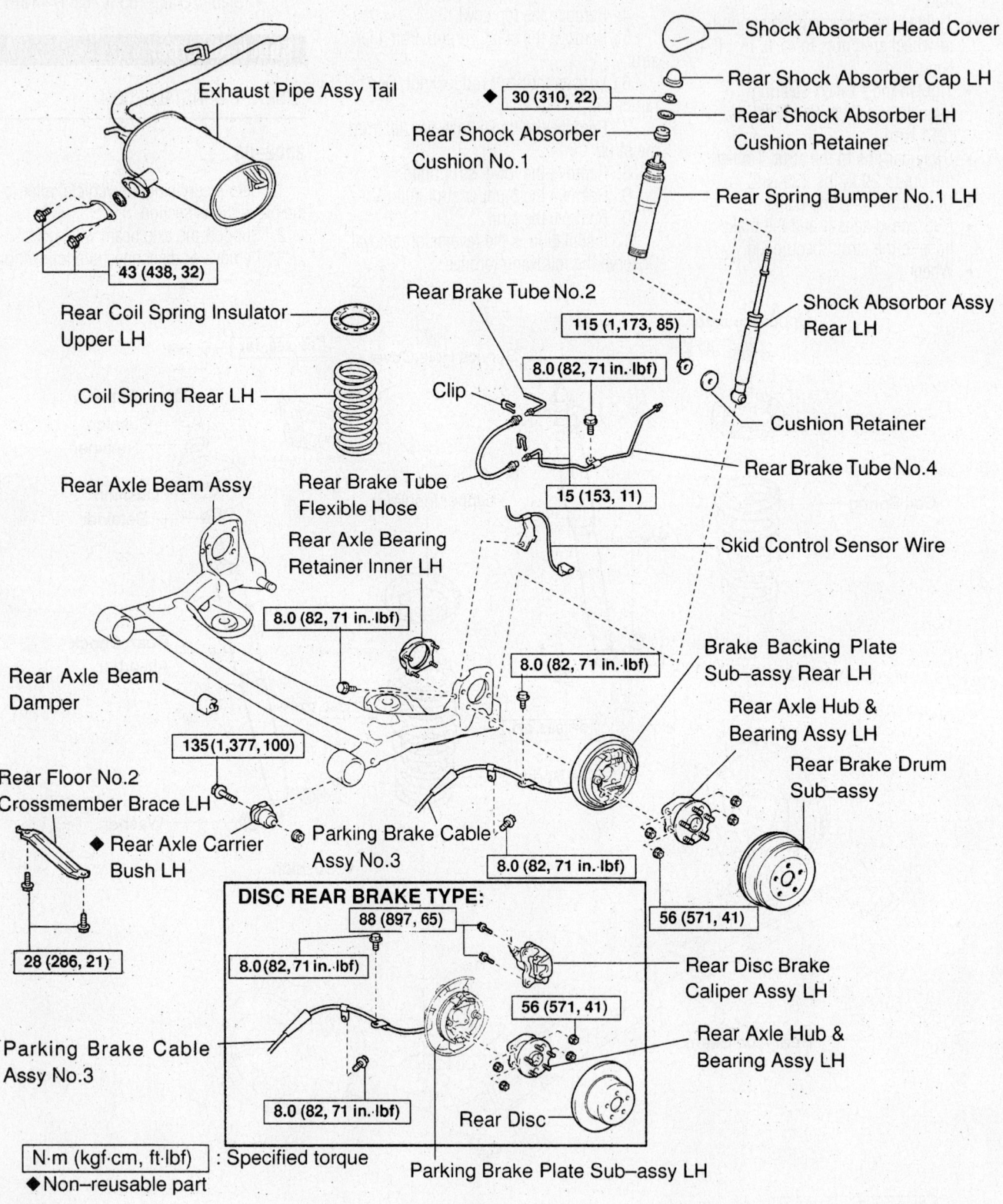

Exhaust Pipe Assy Tail

Shock Absorber Head Cover

Rear Shock Absorber Cap LH

◆ 30 (310, 22)

Rear Shock Absorber LH Cushion Retainer

Rear Shock Absorber Cushion No.1

Rear Spring Bumper No.1 LH

43 (438, 32)

Rear Coil Spring Insulator Upper LH

Rear Brake Tube No.2

115 (1,173, 85)

Shock Absorbor Assy Rear LH

8.0 (82, 71 in·lbf)

Coil Spring Rear LH

Clip

Cushion Retainer

Rear Brake Tube No.4

Rear Axle Beam Assy

Rear Brake Tube Flexible Hose

15 (153, 11)

Skid Control Sensor Wire

Rear Axle Bearing Retainer Inner LH

8.0 (82, 71 in·lbf)

8.0 (82, 71 in·lbf)

Brake Backing Plate Sub–assy Rear LH

Rear Axle Beam Damper

Rear Axle Hub & Bearing Assy LH

Rear Brake Drum Sub–assy

135 (1,377, 100)

Rear Floor No.2 Crossmember Brace LH

◆ Rear Axle Carrier Bush LH

Parking Brake Cable Assy No.3

8.0 (82, 71 in·lbf)

28 (286, 21)

56 (571, 41)

DISC REAR BRAKE TYPE:

88 (897, 65)

8.0 (82, 71 in·lbf)

Rear Disc Brake Caliper Assy LH

56 (571, 41)

Parking Brake Cable Assy No.3

Rear Axle Hub & Bearing Assy LH

8.0 (82, 71 in·lbf)

N·m (kgf·cm, ft·lbf) : Specified torque

◆ Non–reusable part

Rear Disc

Parking Brake Plate Sub–assy LH

67170-SIEN-G13

Rear suspension components—2004–06 2wd models

4WD DRIVE TYPE:

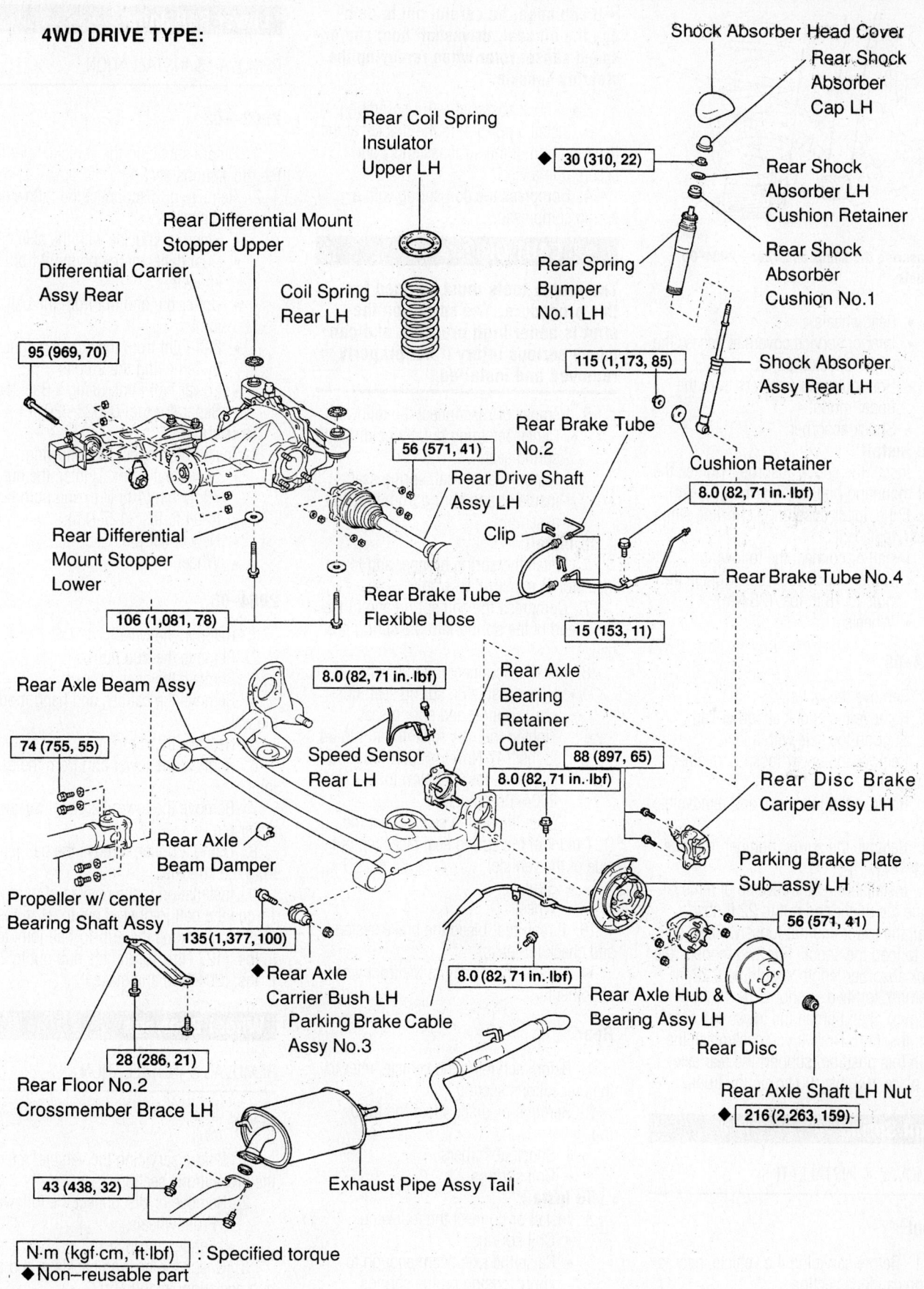

Shock Absorber Head Cover

Rear Shock Absorber Cap LH

Rear Coil Spring Insulator Upper LH

Rear Differential Mount Stopper Upper

Differential Carrier Assy Rear

Coil Spring Rear LH

95 (969, 70)

30 (310, 22)

Rear Shock Absorber LH Cushion Retainer

Rear Shock Absorber Cushion No.1

Rear Spring Bumper No.1 LH

115 (1,173, 85)

Shock Absorber Assy Rear LH

Rear Brake Tube No.2

56 (571, 41)

Rear Drive Shaft Assy LH

Cushion Retainer

8.0 (82, 71 in.·lbf)

Rear Differential Mount Stopper Lower

Clip

Rear Brake Tube Flexible Hose

Rear Brake Tube No.4

106 (1,081, 78)

15 (153, 11)

Rear Axle Beam Assy

8.0 (82, 71 in.·lbf)

Rear Axle Bearing Retainer Outer

Speed Sensor Rear LH

88 (897, 65)

74 (755, 55)

8.0 (82, 71 in.·lbf)

Rear Disc Brake Cariper Assy LH

Rear Axle Beam Damper

Parking Brake Plate Sub-assy LH

Propeller w/ center Bearing Shaft Assy

135(1,377,100)

Rear Axle Carrier Bush LH

56 (571, 41)

Parking Brake Cable Assy No.3

8.0 (82, 71 in.·lbf)

28 (286, 21)

Rear Axle Hub & Bearing Assy LH

Rear Disc

Rear Floor No.2 Crossmember Brace LH

Rear Axle Shaft LH Nut

216 (2,263, 159)

43 (438, 32)

Exhaust Pipe Assy Tail

N·m (kgf·cm, ft·lbf) : Specified torque
◆ Non–reusable part

67170-SIEN-G14

Rear suspension components—2004–06 4wd models

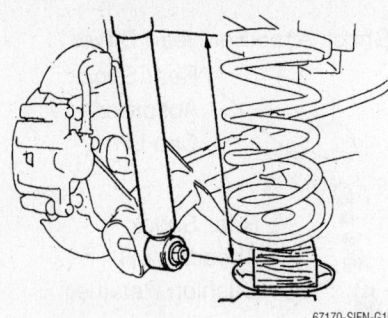

Measuring the shock absorber—2004–06 models

- Rear wheels
- Interior service covers to access the upper shock mounts
- Both nuts and retainers from the upper mount
- Shock absorber

To install:

4. Install the shock absorber. Tighten the lower mounting bolt to 27 ft. lbs. (37 Nm).

5. If the upper cushion is showing signs of wear, replace it.

6. Install or connect the following:
- Upper shock absorber. Tighten the nuts to 18 ft. lbs. (25 Nm).
- Wheels

2004–06

1. Remove the wheel.
2. Remove the shock absorber cap.
3. Support the axle with a jack.
4. Remove the upper locknut, retainer and bushing.
5. Remove the lower nut and remove the shock.
6. Remove the spring bumper from the shock.
7. Installation is the reverse of removal. Torque the upper end nut to 22 ft. lbs. Install the lower end nut loosely. Raise the axle to load the shock. For 2wd models, the shock absorber length should be 9.22 in. (234mm); for 4wd, it should be 10.16 in. (258mm), then tighten the lower end nut to 85 ft. lbs. (115 Nm). If you can't reach the nut in this position, support the rear axle and place 198 lbs. (90 kg) in the trunk.

Coil Spring

REMOVAL & INSTALLATION

Front

1. Before servicing the vehicle, refer to the precautions section.
2. Remove or disconnect the following:
- Wheel

➡️ **If equipped, be careful not to damage the oil seal, driveshaft boot and/or speed sensor rotor when removing the steering knuckle.**

- Shock absorber (strut assembly)

3. Install a nut/bolt to the bracket at the lower portion of the strut assembly and secure it in a vise.

4. Compress the coil spring with a spring compressor.

✸✸ CAUTION

The proper tools must be used for this procedure. The spring on the strut is under high pressure and can cause serious injury if not properly removed and installed.

5. Remove or disconnect the following:
- Center retaining nut, by holding the spring seat
- Support, dust seal, spring seat, insulator and spring from the strut assembly

To install:

6. Install the spring bumper and lower insulator to the strut assembly.

7. Compress the coil spring and fit the lower end of the spring into the spring seat gap.

8. Install or connect the following:
- Upper insulator, spring seat, dust seal, support and spring seat. Tighten the new retaining nut to 34 ft. lbs. (47 Nm) for 2002–03 models; 36 ft. lbs. (49 Nm) for 2004–06 models.

9. Rotate the spring seat so that the OUT mark of the spring seat faces the outside of the vehicle.
- Strut
- Wheel

10. If required, bleed the brake system and check for leaks.

11. Check and/or adjust the front wheel alignment.

Rear

1. Before servicing the vehicle, refer to the precautions section.
2. Remove or disconnect the following:
- Shock absorbers
- Coil springs

To install:

3. Install or connect the following:
- Coil springs
- Raise the axle beam enough to apply tension on the springs
- Shock absorbers

Lower Ball Joint

REMOVAL & INSTALLATION

2002—03

1. Before servicing the vehicle, refer to the precautions section.
2. Remove or disconnect the following:
- Wheel
- Steering knuckle with the axle hub
- Dust deflector, by prying it from the knuckle
- Cotter pin and nut from the ball joint
- Ball joint from the steering knuckle, by removing the 2 bolts
- Lower ball joint, using a Ball Joint Separator tool 09628-62011

To install:

3. Install or connect the following:
- Lower ball joint. Tighten the nut to 76 ft. lbs. (103 Nm) and both bolts to 94 ft. lbs. (127 Nm).
- New cotter pin
- Wheel

2004–06

1. Remove the wheel.
2. Remove the hub nut.
3. Remove the speed sensor.
4. Remove the caliper, and hang it out of the way.
5. Remove the rotor.
6. Remove the lower arm from the ball joint.
7. Remove the lower ball joint nut and cotter pin.
8. Using a puller, remove the ball joint from the knuckle.
9. Installation is the reverse of removal. Torque the ball joint stud nut to 91 ft. lbs. (123 Nm). Torque the arm-to-ball joint to 94 ft. lbs. (127 Nm). Torque the hub nut to 217 ft. lbs. (294 Nm) and stake it.

Wheel Bearings

REMOVAL & INSTALLATION

Front

1. Before servicing the vehicle, refer to the precautions section.
2. Remove or disconnect the following:
- Front wheels
- Fender apron seal

3. Check the bearing backlash and axle hub deviation, as follows:

a. Remove the 2 brake caliper set bolts.

b. Hang the caliper using stiff wire on the shock absorber assembly.

c. Remove the rotor.

d. Place a dial indicator near the center of the axle hub and check the backlash in the bearing shaft direction.

e. Backlash maximum should read 0.0020 inch (0.05mm). If greater than specified, replace the bearing.

f. Using the dial indicator, check the deviation at the surface of the axle hub outside and hub bolt. Maximum is 0.0020 inch (0.05mm). If greater than specified, replace the axle hub.

4. Install the rotor and caliper assembly.
5. Remove or disconnect the following:
- Cotter pin (discard it) and lockcap off the center hub nut
- Driveshaft locknut, by applying the front brakes
- Tie rod end, from the steering knuckle
- Left and right stabilizer end brackets, from the lower arms
- Both nuts and the lower arm from the ball joint
- Driveshaft from the axle hub. Secure the shaft aside using wire.

✷✷ WARNING

Be careful not to damage the shaft boot or Anti-lock Brake System (ABS) sensor rotor.

- Both brake caliper mounting bolts and the caliper.

➡**Support caliper from the vehicle using wire.**

- Brake rotor
- Sensor from the steering knuckle, if equipped with ABS

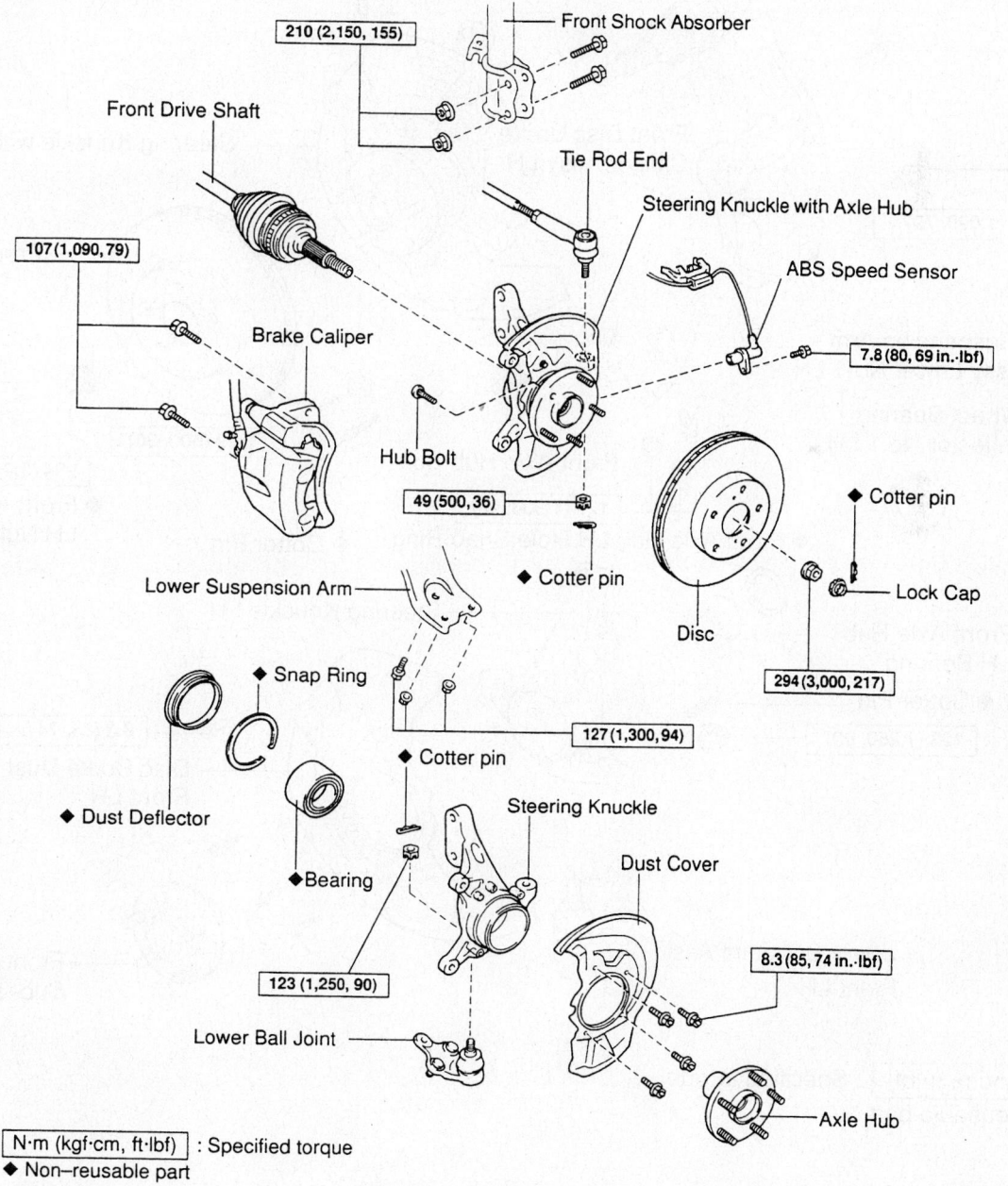

N·m (kgf·cm, ft·lbf) : Specified torque
◆ Non–reusable part

7924ZG82

Exploded view of the front hub, bearing and steering knuckle assembly—2002–03 models

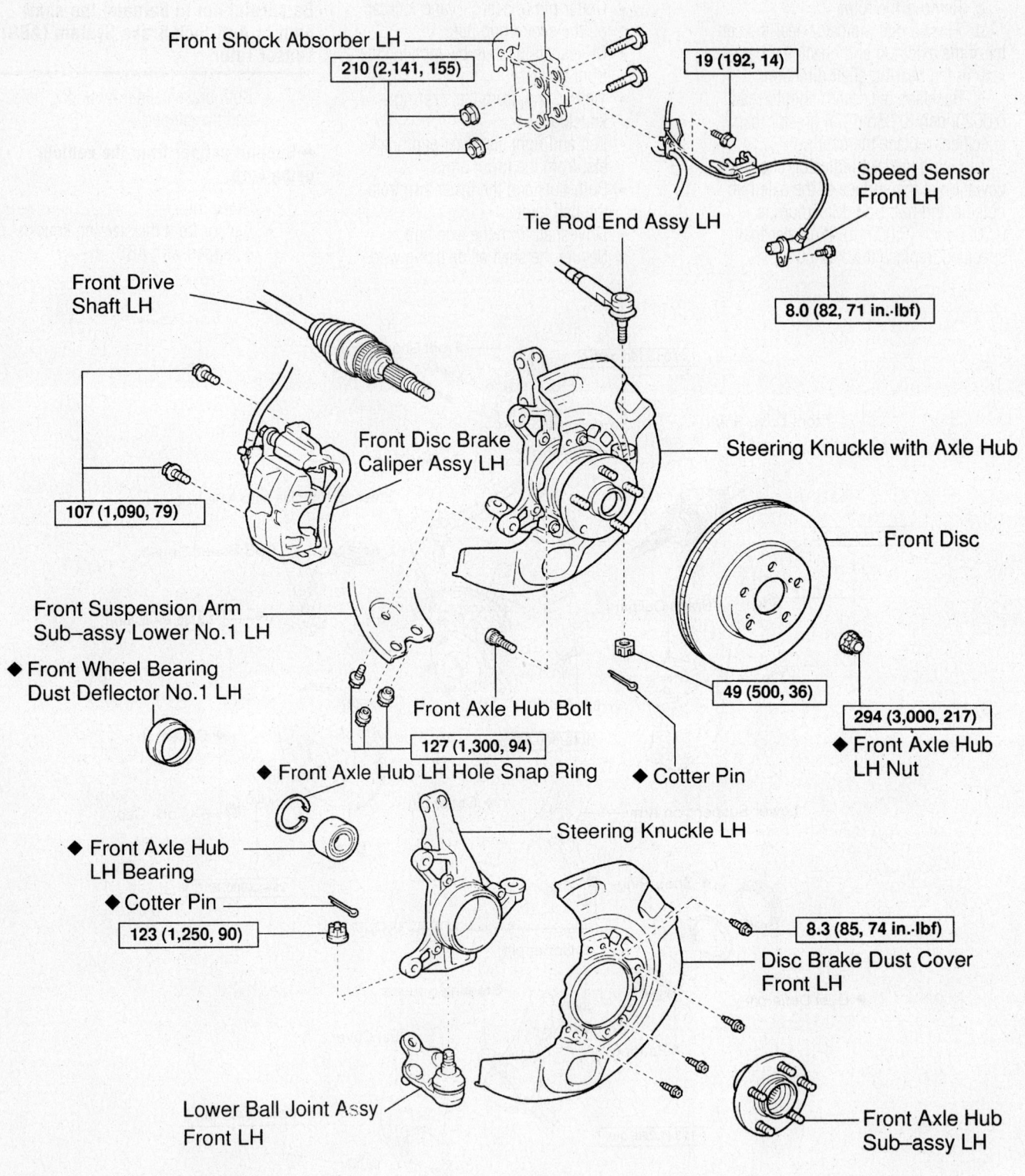

Front Shock Absorber LH

210 (2,141, 155)

19 (192, 14)

Tie Rod End Assy LH

Speed Sensor
Front LH

8.0 (82, 71 in.·lbf)

Front Drive
Shaft LH

Front Disc Brake
Caliper Assy LH

Steering Knuckle with Axle Hub

107 (1,090, 79)

Front Disc

Front Suspension Arm
Sub–assy Lower No.1 LH

◆ Front Wheel Bearing
Dust Deflector No.1 LH

Front Axle Hub Bolt

49 (500, 36)

127 (1,300, 94)

294 (3,000, 217)

◆ Front Axle Hub LH Hole Snap Ring

◆ Cotter Pin

◆ Front Axle Hub
LH Nut

◆ Front Axle Hub
LH Bearing

Steering Knuckle LH

◆ Cotter Pin

8.3 (85, 74 in.·lbf)

123 (1,250, 90)

Disc Brake Dust Cover
Front LH

Lower Ball Joint Assy
Front LH

Front Axle Hub
Sub–assy LH

N·m (kgf·cm, ft·lbf) : Specified torque
◆ Non–reusable part

Exploded view of the front hub, bearing and steering knuckle assembly—2004–06 models

- Both nuts from the lower end of the shock
- Steering knuckle and hub assembly

6. Clamp the steering knuckle in a vise with soft jaws to protect the knuckle.

7. Remove or disconnect the following:
- Dust deflector from the hub, using a screwdriver
- Bearing inner oil seal, by prying it from the knuckle
- Snapring from the knuckle bore
- Dust deflector from the steering knuckle
- Axle hub, by pulling it from the dust deflector, using a 2-armed mechanical puller
- Inner (inside) bearing race from the bearing, using the puller
- Sensor control rotor from the axle hub, using Torx® wrench
- Outer bearing race, using the puller
- Outer bearing seal, using the puller

8. Position the inner (outside) race inside the bearing.

9. Using a brass rod, tap the bearing from the steering knuckle.

To install:

10. Clean all the oil seal and bearing seating surfaces with a clean, dry rag.

11. Install or connect the following:

- Bearing into the bore, using a Bearing Driver tool 09608-32010 and a press
- New outer oil seal, driving it into the steering knuckle, by inserting the seal side lip into the factory tool
- Brake disc cover to the steering knuckle with the bolts

12. Apply multi-purpose grease between the oil seal lip, oil seal and bearing.

13. Install or connect the following:
- Hub, by pressing it into the knuckle
- New snapring into the knuckle
- New oil seal, by pressing it into the knuckle once lubricated with multi-purpose grease
- Dust deflector, by pressing it into the knuckle.

✴✴ WARNING

Align the speed sensor holes in the dust deflector and steering knuckle, if equipped with ABS.

- Ball joint to the steering knuckle. Tighten the bolts to 94 ft. lbs. (127 Nm).
- Steering knuckle/hub assembly and

temporarily install the lower shock bolts
- Lower ball joint to the lower arm. Tighten the bolt and nuts to 94 ft. lbs. (127 Nm).
- Tie rod to the knuckle. Tighten the nut to 36 ft. lbs. (49 Nm).
- New cotter pin
- Tighten the lower shock nuts to 156 ft. lbs. (211 Nm).
- Both side stabilizer end brackets to the lower arm. Tighten the fasteners to 43 ft. lbs. (58 Nm) for 2002–03; 55 ft. lbs. (74 Nm).
- Front ABS sensor. Tighten it to 69 inch lbs. (8 Nm).
- Front brake rotor and caliper.
- Driveshaft locknut, by applying the brakes. Tighten it to 217 ft. lbs. (294 Nm).
- Lockcap and new cotter pin
- Front fender apron seal
- Front wheel. Tighten the lug nuts to 76 ft. lbs. (103 Nm).

Rear

2002–03

1. Before servicing the vehicle, refer to the precautions section.

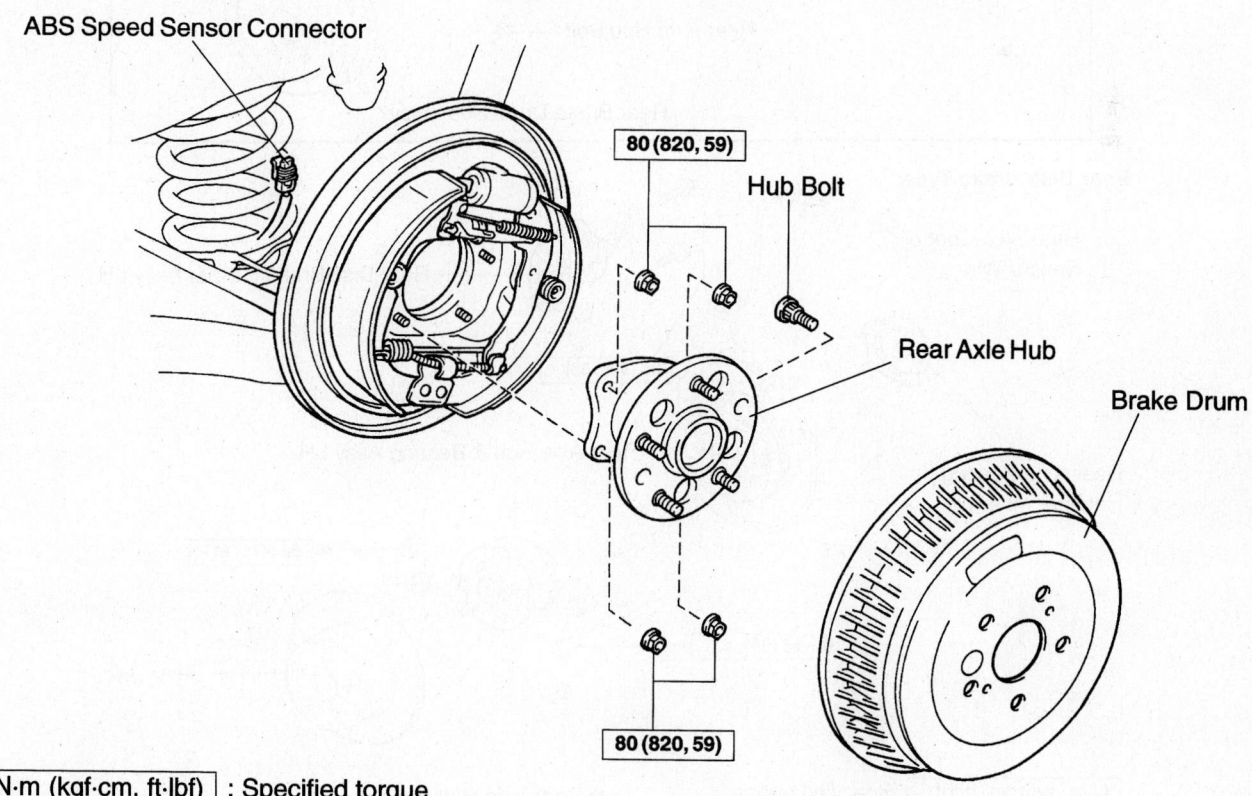

ABS Speed Sensor Connector

80 (820, 59)

Hub Bolt

Rear Axle Hub

Brake Drum

N·m (kgf·cm, ft·lbf) : Specified torque

◆ Non–reusable part

Rear hub assembly—2002–03 models

67170-SIEN-G18

2. Remove or disconnect the following:
- Rear wheel
- Brake drum
- Anti-lock Brake System (ABS) speed sensor connector
- 4 rear axle hub assembly nuts
- Hub assembly

To install:

3. Install or connect the following:
- New hub assembly. Tighten the nuts to 59 ft. lbs. (80 Nm).
- ABS speed sensor

4. Install or connect the following:
- Brake drum
- Wheels
5. Test drive the vehicle.

2004–06 W/2WD

1. Before servicing the vehicle, refer to the precautions section.
2. Remove the wheel.
3. Remove the brake drum (if equipped) or caliper and rotor. Hang the caliper out of the way.

a. Place a dial indicator near the center of the axle hub and check the backlash in the bearing shaft direction.

b. Backlash maximum should read 0.0020 inch (0.05mm). If greater than specified, replace the bearing.

c. Using the dial indicator, check the deviation at the surface of the axle hub outside and hub bolt. Maximum is 0.0020 inch (0.05mm). If greater than specified, replace the axle hub.

4. Remove the ABS sensor wire.

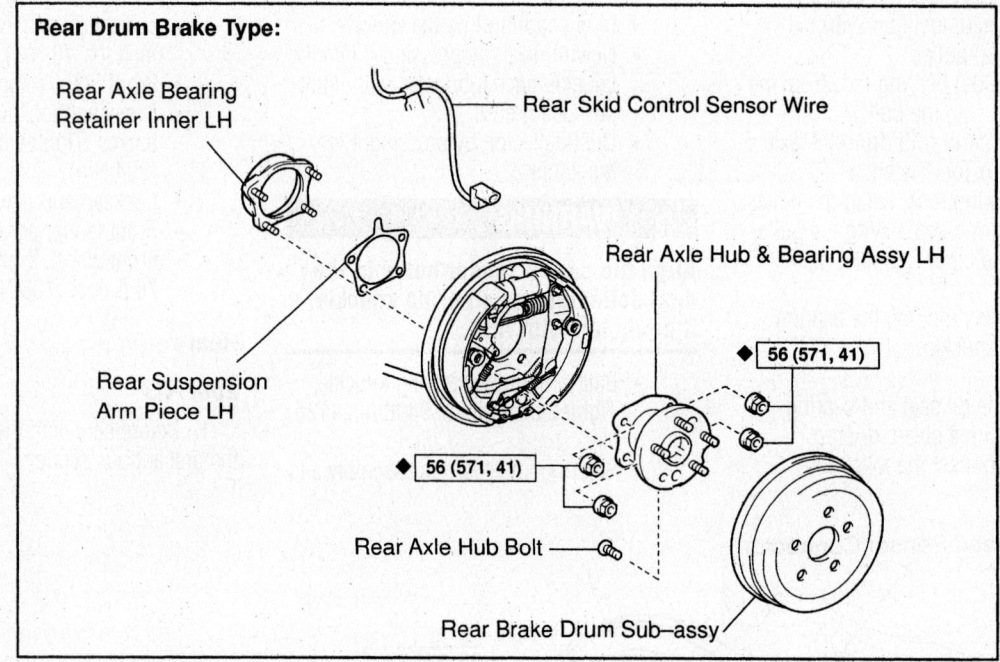

Rear Drum Brake Type:

Rear Axle Bearing Retainer Inner LH

Rear Skid Control Sensor Wire

Rear Axle Hub & Bearing Assy LH

56 (571, 41)

Rear Suspension Arm Piece LH

56 (571, 41)

Rear Axle Hub Bolt

Rear Brake Drum Sub–assy

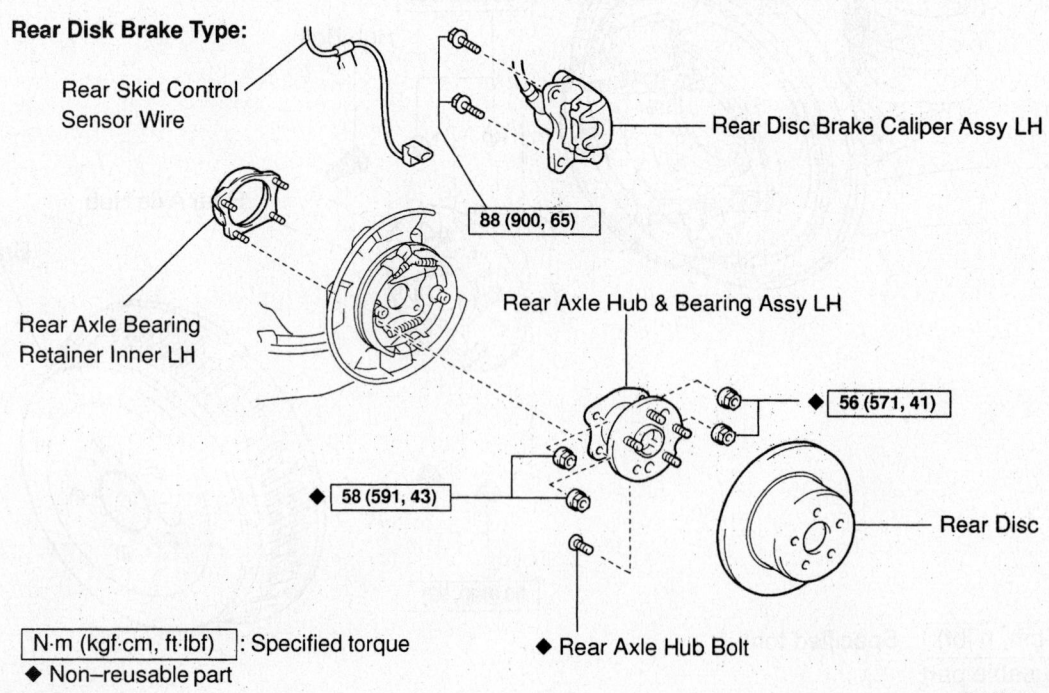

Rear Disk Brake Type:

Rear Skid Control Sensor Wire

Rear Disc Brake Caliper Assy LH

88 (900, 65)

Rear Axle Hub & Bearing Assy LH

Rear Axle Bearing Retainer Inner LH

56 (571, 41)

58 (591, 43)

Rear Disc

N·m (kgf·cm, ft·lbf) : Specified torque

◆ Rear Axle Hub Bolt

◆ Non-reusable part

67170-SIEN-G19

5. Remove the 4 bolts and the hub/bearing assembly.

6. Installation is the reverse of removal. Torque the hub bolts to 41 ft. lbs. (56 Nm).

2004–06 W/4WD

1. Before servicing the vehicle, refer to the precautions section.
2. Remove the wheel.
3. Unstake and remove the axle shaft nut.

4. Remove the caliper and rotor. Hang the caliper out of the way.

a. Place a dial indicator near the center of the axle hub and check the backlash in the bearing shaft direction.

b. Backlash maximum should read 0.0020 inch (0.05mm). If greater than specified, replace the bearing.

c. Using the dial indicator, check the deviation at the surface of the axle hub

outside and hub bolt. Maximum is 0.0020 inch (0.05mm). If greater than specified, replace the axle hub.

5. Remove the halfshaft.
6. Remove the ABS sensor wire.
7. Remove the 4 bolts and the hub/bearing assembly.

8. Installation is the reverse of removal. Torque the hub bolts to 41 ft. lbs. (56 Nm).

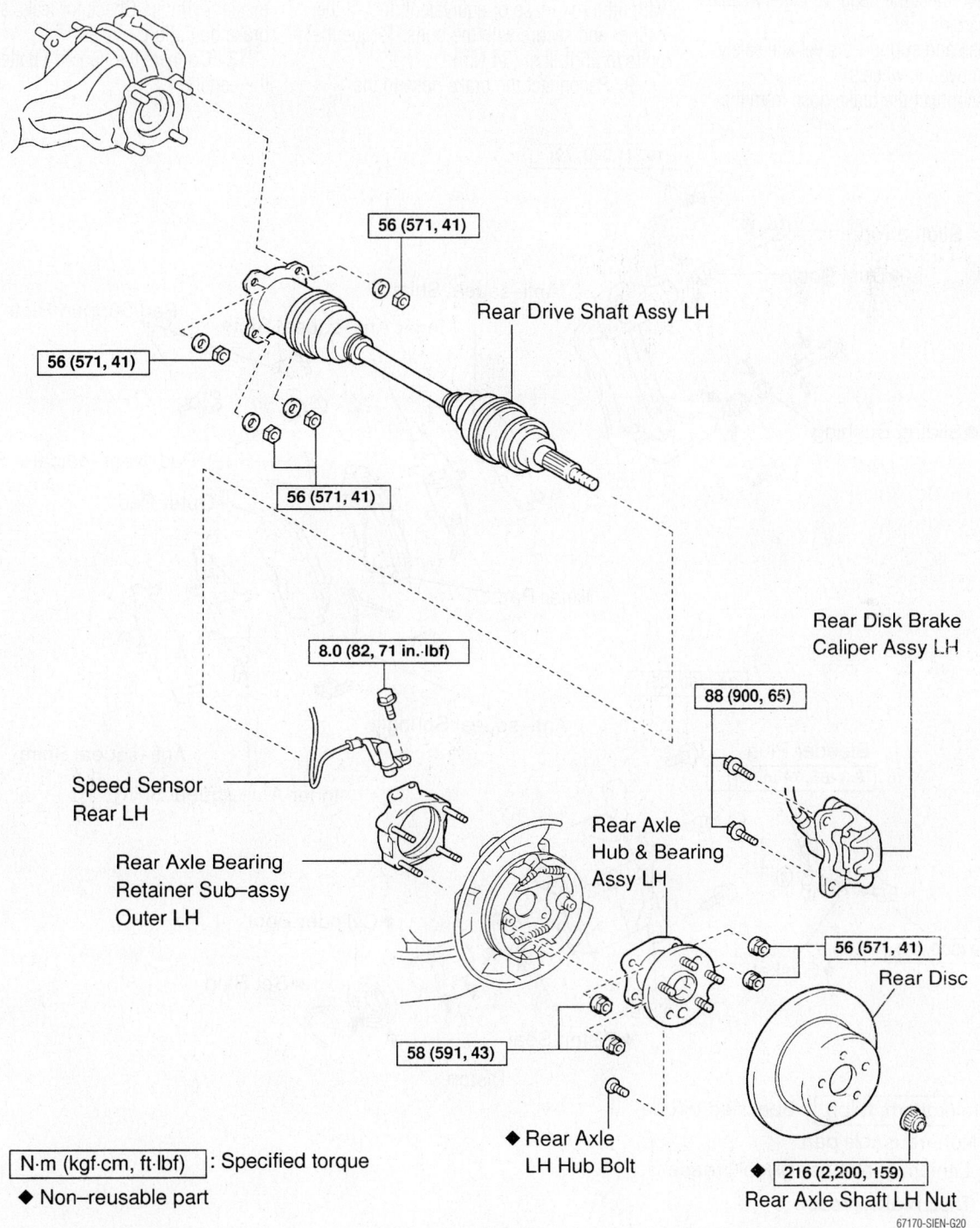

56 (571, 41)

Rear Drive Shaft Assy LH

56 (571, 41)

56 (571, 41)

Rear Disk Brake
Caliper Assy LH

8.0 (82, 71 in. lbf)

88 (900, 65)

Speed Sensor
Rear LH

Rear Axle
Hub & Bearing
Assy LH

Rear Axle Bearing
Retainer Sub–assy
Outer LH

56 (571, 41)

Rear Disc

58 (591, 43)

◆ Rear Axle
LH Hub Bolt

◆ 216 (2,200, 159)
Rear Axle Shaft LH Nut

N·m (kgf·cm, ft·lbf) : Specified torque

◆ Non–reusable part

67170-SIEN-G20

Rear hub assembly—2004–06 models with 4wd

BRAKES

Front Brake Caliper

REMOVAL & INSTALLATION

2002–03

1. Before servicing the vehicle, refer to the precautions section.
2. Disconnect the negative battery cable from the battery.
3. Raise and support the vehicle safely.
4. Remove the wheels.
5. Disconnect the brake hose from the caliper by removing the union bolt and 2 gaskets. Plug the end of the hose to prevent loss of fluid.
6. Remove the bolts that attach the caliper to the torque plate.
7. Lift the bottom of the caliper up and remove the caliper assembly.

To install:

8. Grease the caliper slides and bolts with lithium grease or equivalent. Install the caliper and secure with the bolts. Torque the bolts to 25 ft. lbs. (34 Nm).
9. Reconnect the brake hose to the caliper, using 2 new washers. Make sure the flexible hose lock is securely in the lock hole of the caliper. Torque the union bolt to 21 ft. lbs. (29 Nm). Also, verify that the brake hose is not twisted.
10. Fill the brake system to the proper level and bleed the brake system.
11. Install the tire and wheel assembly.
12. Top off the brake fluid level in the master cylinder. Check for leaks and proper brake operation.
13. Connect the negative battery cable to the battery.

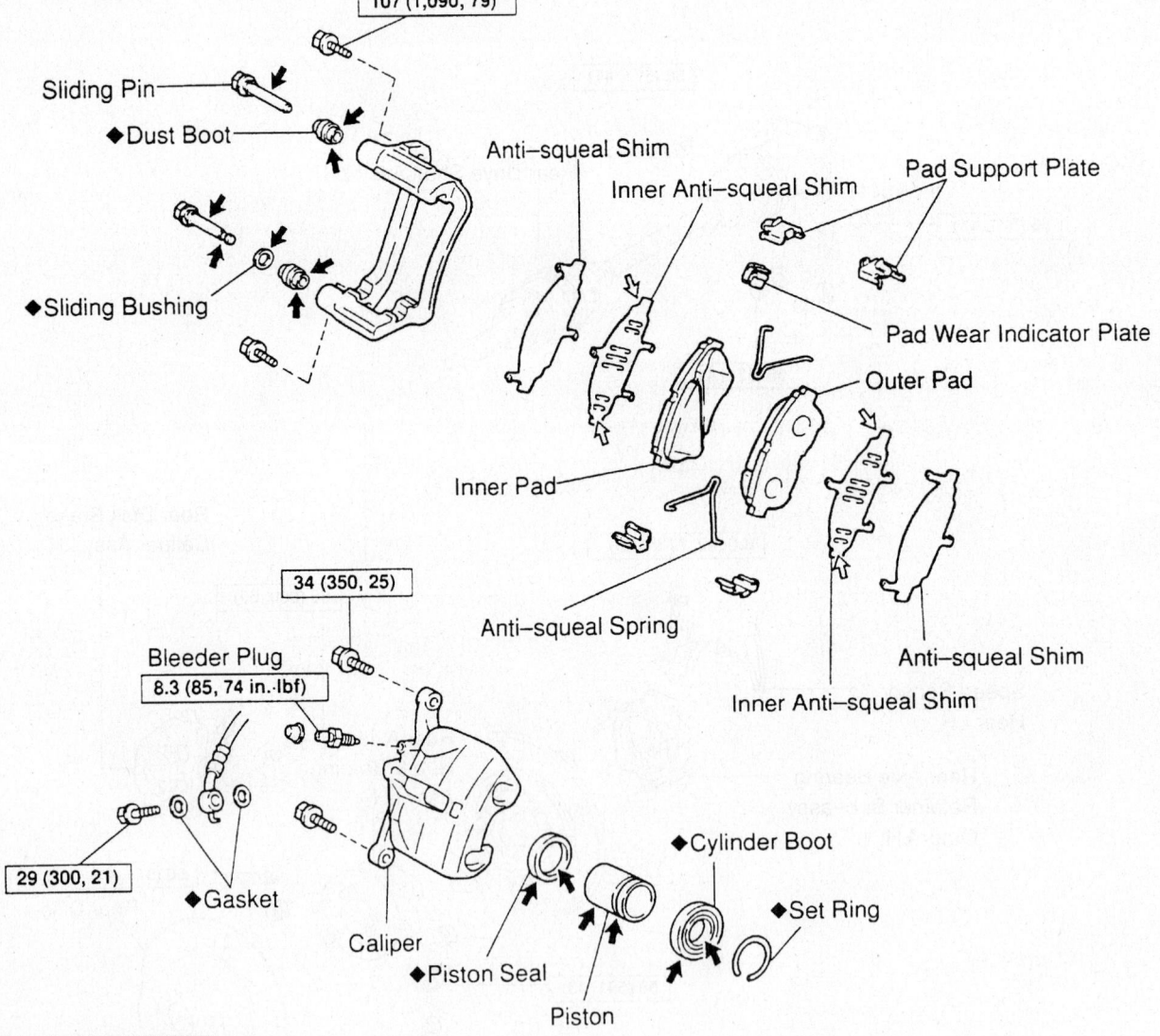

N·m (kgf·cm, ft·lbf) : Specified torque
◆ Non–reusable part
➡ Lithium soap base glycol grease
⇨ Disc brake grease

93026G72

Exploded view of the front disc brake caliper assembly—2002–03 models

2004–06

1. Before servicing the vehicle, refer to the precautions section.
2. Disconnect the negative battery cable from the battery.
3. Raise and support the vehicle safely.
4. Remove the wheels.

5. Disconnect the brake hose from the caliper by removing the union bolt and 2 gaskets. Plug the end of the hose to prevent loss of fluid.
6. Remove the bolts that attach the caliper to the torque plate.
7. Lift the bottom of the caliper up and remove the caliper assembly.

To install:
8. Grease the caliper slides and bolts with lithium grease or equivalent. Install the caliper and secure with the bolts. Torque the bolts to 25 ft. lbs. (34 Nm).
9. Reconnect the brake hose to the caliper, using 2 new washers. Make sure the flexible hose lock is securely in the lock

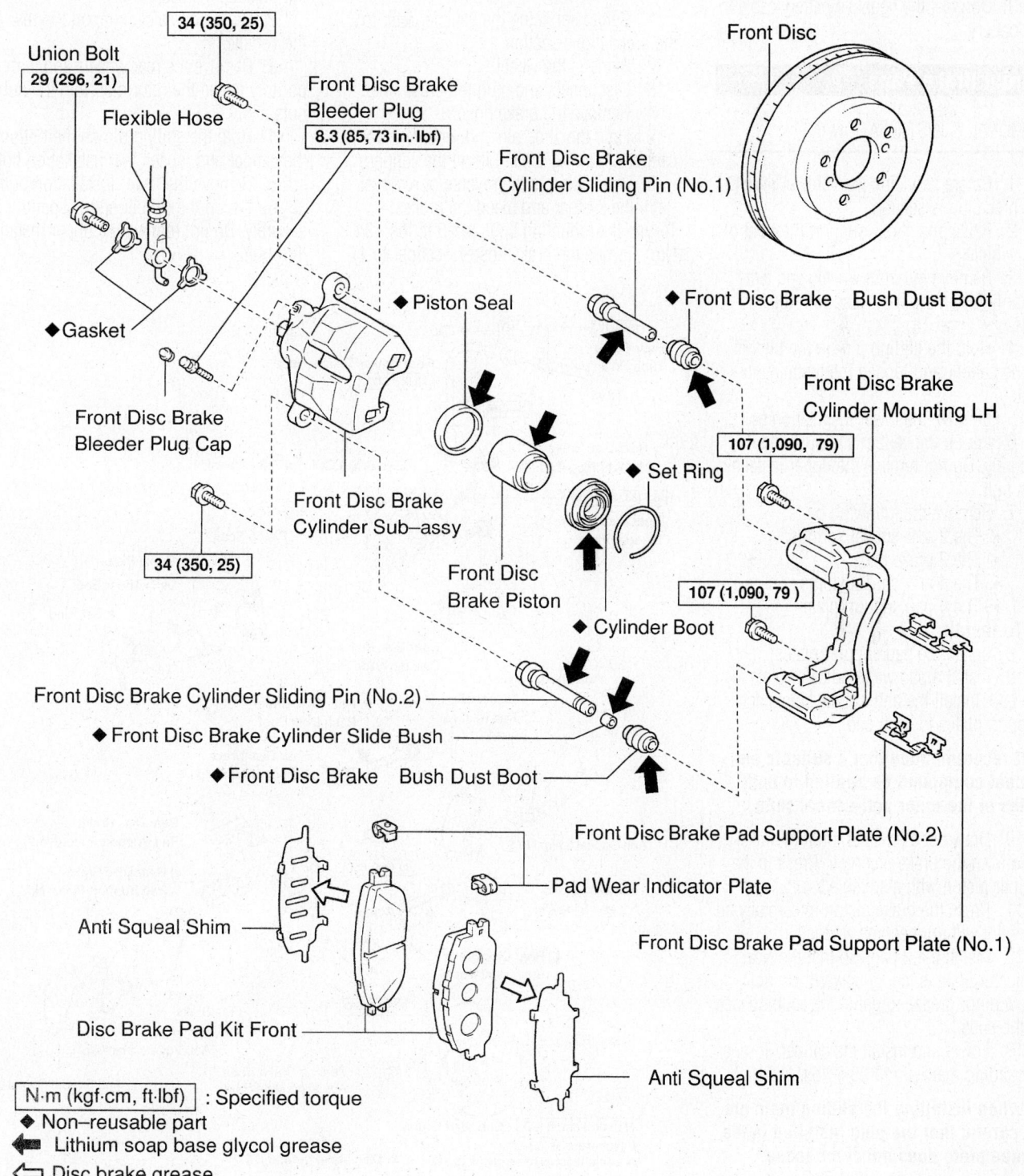

Union Bolt
34 (350, 25)
29 (296, 21)
Flexible Hose
Front Disc Brake Bleeder Plug
8.3 (85, 73 in.·lbf)
Front Disc
Front Disc Brake Cylinder Sliding Pin (No.1)
◆Gasket
◆ Piston Seal
◆ Front Disc Brake   Bush Dust Boot
Front Disc Brake Bleeder Plug Cap
Front Disc Brake Cylinder Mounting LH
Front Disc Brake Cylinder Sub–assy
◆ Set Ring
107 (1,090, 79)
Front Disc Brake Piston
34 (350, 25)
107 (1,090, 79)
◆ Cylinder Boot
Front Disc Brake Cylinder Sliding Pin (No.2)
◆ Front Disc Brake Cylinder Slide Bush
◆Front Disc Brake   Bush Dust Boot
Front Disc Brake Pad Support Plate (No.2)
Pad Wear Indicator Plate
Anti Squeal Shim
Front Disc Brake Pad Support Plate (No.1)
Disc Brake Pad Kit Front
Anti Squeal Shim

N·m (kgf·cm, ft·lbf) : Specified torque
◆ Non–reusable part
⬅ Lithium soap base glycol grease
⇐ Disc brake grease

67170-SIEN-G22

Front disc brake components—2004–06

hole of the caliper. Torque the union bolt to 21 ft. lbs. (29 Nm). Also, verify that the brake hose is not twisted.

10. Fill the brake system to the proper level and bleed the brake system.

11. Install the tire and wheel assembly.

12. Top off the brake fluid level in the master cylinder. Check for leaks and proper brake operation.

13. Connect the negative battery cable to the battery.

Front Brake Pads

REMOVAL & INSTALLATION

1. Before servicing the vehicle, refer to the precautions section.

2. Raise and safely support the front of the vehicle.

3. Remove the front wheels and temporarily fasten the rotor disc with the hub nuts.

4. Hold the sliding pin on the bottom of the caliper and loosen the installation bolt.

5. Remove the lower installation bolt.

6. Lift up the caliper and suspend it securely. Do not remove the upper installation bolt.

7. Remove the following parts:
   • The 2 anti-squeal springs.
   • The 2 brake pads.
   • The 4 anti-squeal shims.
   • The 4 pad support plates.

To install:

8. Install the pad support plates.

9. Install a pad wear indicator plate to the pad. Install the anti-squeal shims and support plates to each pad.

➡ **It recommended that a suitable anti-squeal compound be applied to both sides of the inner anti-squeal shim.**

10. Draw out a small amount of brake fluid from the brake reservoir. Press in the caliper piston with a suitable tool.

11. Press the brake piston in carefully so the boot will not become wedged.

12. Install the 2 pads so that the wear indicator plate is facing upward. Do not allow oil or grease to get in the rubbing face of the pads.

13. Lower and install the caliper. Torque the sliding main pin to 25 ft. lbs. (34 Nm).

➡ **When installing the sliding main pin, be careful that the plug installed in the torque plate does not come loose.**

14. Install the front wheels and lower the vehicle.

15. Check the fluid level in the master cylinder and add as necessary. Be sure to pump the brake pedal a few times before road-testing the vehicle.

Rear Brake Caliper

REMOVAL & INSTALLATION

1. Before servicing the vehicle, refer to the precautions section.

2. Remove the wheel.

3. Disconnect and plug the brake line.

4. Remove the brake hose.

5. Hold the slide pin and remove the 2 caliper mounting bolts. Lift off the caliper.

6. Installation is the reverse of removal. Refill the system and bleed the brakes. Torque the mounting bolts to 25 ft. lbs. (34 Nm). Torque the brake hose-to-caliper to 17 ft. lbs. (23 Nm). Torque the steel brake line to 11 ft. lbs. (15 Nm).

Rear Brake Pads

REMOVAL & INSTALLATION

1. Before servicing the vehicle, refer to the precautions section.

2. Raise and safely support the rear of the vehicle.

3. Remove the rear wheels and temporarily fasten the rotor disc with the hub nuts.

4. Hold the sliding pin on the bottom of the caliper and loosen the installation bolt.

5. Remove the lower installation bolt.

6. Lift up the caliper and suspend it securely. Do not remove the upper installation bolt.

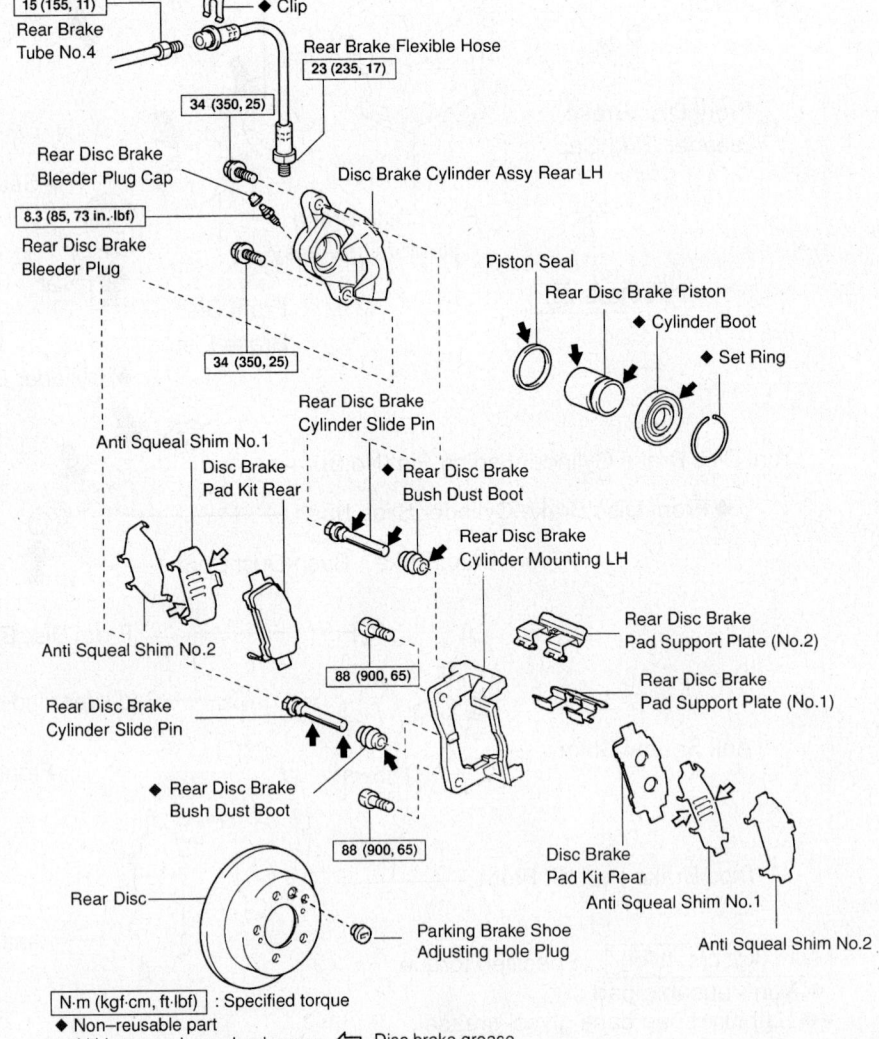

15 (155, 11)
Rear Brake Tube No.4
◆ Clip
Rear Brake Flexible Hose
23 (235, 17)
34 (350, 25)
Rear Disc Brake Bleeder Plug Cap
Disc Brake Cylinder Assy Rear LH
8.3 (85, 73 in. lbf)
Rear Disc Brake Bleeder Plug
Piston Seal
Rear Disc Brake Piston
◆ Cylinder Boot
◆ Set Ring
34 (350, 25)
Rear Disc Brake Cylinder Slide Pin
Anti Squeal Shim No.1
Disc Brake Pad Kit Rear
◆ Rear Disc Brake Bush Dust Boot
Rear Disc Brake Cylinder Mounting LH
Anti Squeal Shim No.2
88 (900, 65)
Rear Disc Brake Pad Support Plate (No.2)
Rear Disc Brake Pad Support Plate (No.1)
Rear Disc Brake Cylinder Slide Pin
◆ Rear Disc Brake Bush Dust Boot
88 (900, 65)
Rear Disc
Parking Brake Shoe Adjusting Hole Plug
Disc Brake Pad Kit Rear
Anti Squeal Shim No.1
Anti Squeal Shim No.2
N·m (kgf·cm, ft·lbf) : Specified torque
◆ Non–reusable part
⬅ Lithium soap base glycol grease ⬅ Disc brake grease

67170-SIEN-G23

Rear disc brake components

7. Remove the following parts:
- The 2 anti-squeal springs.
- The 2 brake pads.
- The 4 anti-squeal shims.
- The 4 pad support plates.

To install:

8. Install the pad support plates.

9. Install a pad wear indicator plate to the pad. Install the anti-squeal shims and support plates to each pad.

➡**It recommended that a suitable anti-squeal compound (available at your local parts house) be applied to both sides of the inner anti-squeal shim.**

10. Draw out a small amount of brake fluid from the brake reservoir. Press in the caliper piston with a suitable tool.

11. Press the brake piston in carefully so the boot will not become wedged.

12. Install the 2 pads so that the wear indicator plate is facing upward. Do not allow oil or grease to get in the rubbing face of the pads.

13. Lower and install the caliper. Torque the sliding main pin to 25 ft. lbs. (34 Nm).

➡**When installing the sliding main pin, be careful that the plug installed in the torque plate does not come loose.**

14. Install the rear wheels and lower the vehicle.

15. Check the fluid level in the master cylinder and add as necessary. Be sure to pump the brake pedal a few times before road-testing the vehicle.

Brake Shoes

REMOVAL AND INSTALLATION

1. Before servicing the vehicle, refer to the precautions section.

2. Disconnect the negative battery cable from the battery.

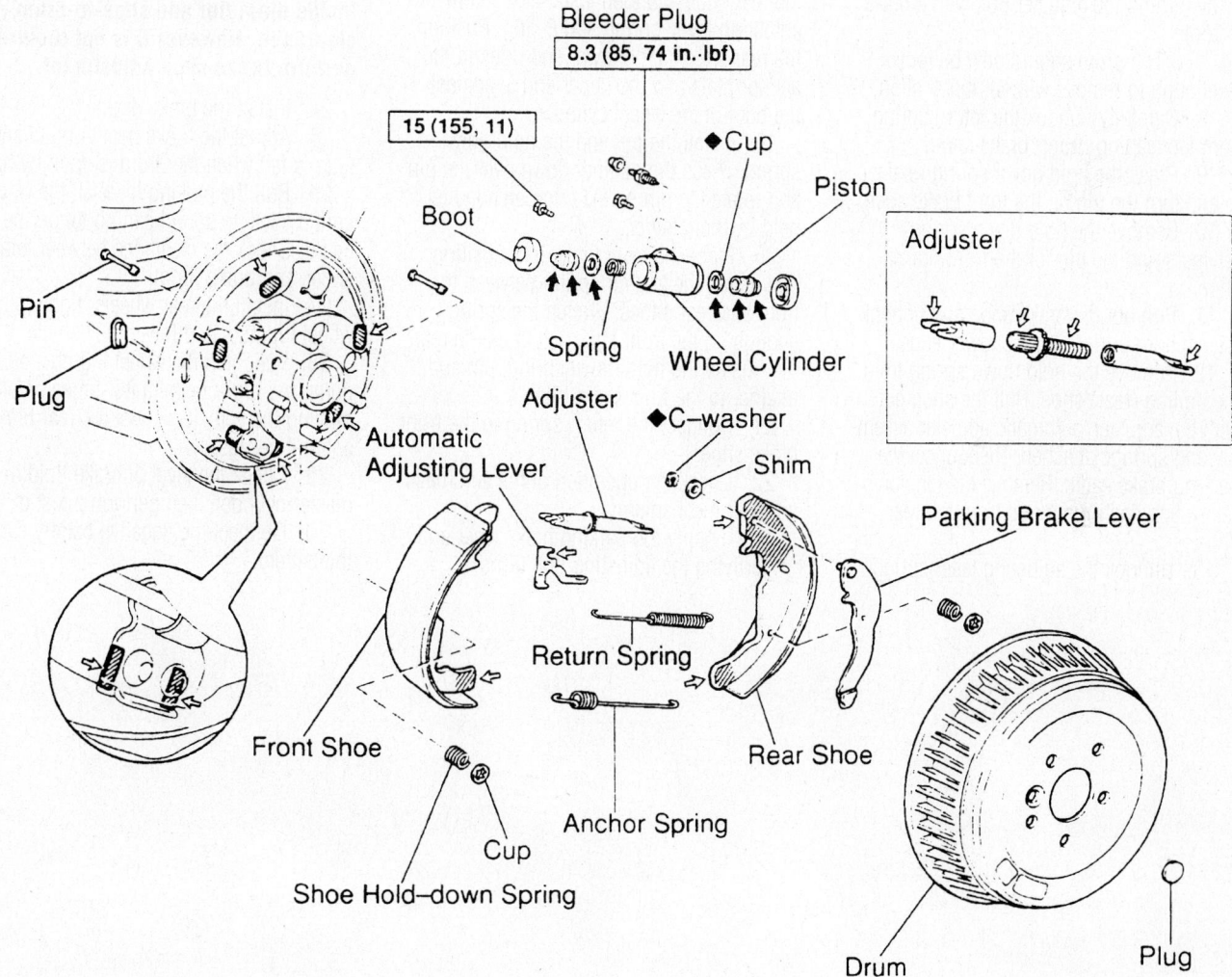

N·m (kgf·cm, ft·lbf) : Specified torque

◆ Non–reusable part

➡ Lithium soap base glycol grease

⇨ High temperature grease

Exploded view of the rear drum brake components—Sienna

93026G80

3. Loosen the rear wheel lug nuts slightly. Release the parking brake.

4. Block the front wheels, raise the rear of the vehicle, and safely support it with jackstands.

5. Remove the wheel lug nuts and the wheel.

6. Remove the brake drum retaining screws, if equipped. Remove the brake drum.

7. If the drum is difficult to remove, perform the following:

a. Insert the end of a bent wire (a coat hanger will do nicely) through the hole in the brake drum and hold the automatic adjusting lever away from the adjuster.

b. Reduce the brake shoe adjustment by turning the adjuster bolt with a brake tool.

c. The drum should now be loose enough to remove without much effort.

8. Carefully unhook the return spring from the leading (front) brake shoe.

9. Press the hold down spring retainer in and turn the pin on the front brake shoe.

10. Remove the hold down spring, retainers and the pin for the front brake shoe.

11. Pull out the brake shoe and unhook the anchor spring from the lower edge.

12. Remove the hold down spring from the trailing (rear) shoe. Pull the shoe out with the adjuster, automatic adjuster assembly and springs attached. Disconnect the parking brake cable. Remove the tension/return and anchor springs from the rear shoe.

13. Unhook the adjusting lever spring from the rear shoe and then remove the automatic adjuster assembly.

To install:

14. Inspect the shoes for signs of unusual wear or scoring.

15. Check the wheel cylinder for any sign of fluid seepage or frozen pistons.

16. Clean and inspect the brake backing plate and all other components. Check that the brake drum inner diameter is within specified limits. Lubricate the backing plate at the positions the brakes come in contact with the backing plate. Also lubricate the anchor plate.

17. Mount the automatic adjuster assembly onto a new rear brake shoe.

18. Connect the parking brake cable to the rear shoe and then install the automatic adjusting lever, spring and E-ring. Position the rear shoe so the lower end rides in the anchor plate and the upper end is against the boot of the wheel cylinder.

19. Install the pin and the hold down spring. Press the retainer down over the pin and rotate the pin so the crimped edge is held by the retainer.

20. Place the front brake into position and install the anchor spring between the front and rear shoes. Stretch the spring enough so the front shoe will fit as the rear did. Install the hold down spring, pin and retainer to the front brake shoe.

21. Connect the return spring to the front brake shoe.

22. Check the operation of the automatic adjuster mechanism:

a. Apply the parking brake lever and verifying the adjusting bolt turns.

b. Adjust the strut to where it is the shortest possible length.

c. Install the brake drum.

d. Apply the parking brake lever until the clicking sound can no longer be heard.

23. Check the clearance between the brake shoes and drum:

a. Remove the brake drum.

b. Measure the brake drum inside diameter and diameter of the brake shoes. The difference is "Shoe-to-drum clearance" and should be approximately 0.024 inch (0.6mm). If incorrect, check the parking brake system.

➡ **A special brake caliper tool is required to gauge the brake drum inside diameter and shoe-to-drum clearance. However it is not required to perform brake shoe adjustment.**

24. Install the brake drum.

25. Adjust the brake pedal until a slight drag is felt when the drum is spun by hand.

26. Pull the parking lever all the way up until a clicking sound can no longer be heard. Check the clearance between brake shoes and brake drum.

27. Install the rear wheels, tighten the wheel lug nuts and lower the vehicle.

28. Retighten the wheel lug nuts and pump the brake pedal a few times before moving the vehicle. Adjust the rear brakes again if necessary.

29. Check the level of brake fluid in the master cylinder, then perform a test drive.

30. Connect the negative battery cable to the battery.

TOYOTA

Tacoma

BRAKES **19-138**
DRIVE TRAIN **19-100**
ENGINE REPAIR...............**19-19**
FRONT SUSPENSION **19-115**
FUEL SYSTEM **19-95**
REAR SUSPENSION **19-135**
SPECIFICATIONS AND
  MAINTENANCE CHARTS......**19-3**
Engine and Vehicle Identification ..19-3
General Engine Specifications 19-3
Engine Tune-Up Specifications 19-4
Firing Order 19-5
Accessory Drive Belt Routing 19-5
Capacities 19-7
Valve Specifications.....................19-8
Camshaft Specifications 19-9
Crankshaft and Connecting Rod
  Specifications 19-10
Piston and Ring Specifications 19-11
Torque Specifications 19-12
Wheel Alignment 19-15
Tire, Wheel and Ball Joint
  Specifications 19-16
Brake Specifications 19-17
Scheduled Maintenance
  Intervals..................................19-18
STEERING **19-112**
A
Air Bag......................................19-112
  Disarming 19-112
  Precautions............................19-112
Alternator 19-19
  Removal & Installation............19-19
Automatic Transmission
  Assembly.................................19-102
  Removal & Installation............19-102
B
Brake Caliper 19-138
  Removal & Installation............19-138
Brake Drums..............................19-140
  Removal & Installation..........19-140
Brake Shoes..............................19-142
  Removal & Installation..........19-142
C
Camshaft and Valve Lifters 19-52
  Inspection...............................19-66
  Removal & Installation............19-52

Clutch Assembly.....................19-101
  Removal & Installation..........19-101
Coil Spring 19-117
  Removal & Installation..........19-117
CV-Joints.................................19-108
  Overhaul 19-108
Cylinder Head 19-38
  Removal & Installation............19-38
D
Disc Brake Pads.......................19-140
  Removal & Installation..........19-140
Distributor................................19-19
E
Engine Assembly 19-20
  Removal & Installation............19-20
Exhaust Manifold 19-51
  Removal & Installation............19-51
F
Front Axle Shaft, Bearing and
  Seal 19-108
  Removal & Installation..........19-108
Front Wheel Bearings 19-130
  Adjustment..............................19-130
  Removal & Installation............19-130
Fuel Filter 19-96
  Removal & Installation............19-96
Fuel Injector.............................19-98
  Removal & Installation............19-98
Fuel Pump 19-96
  Removal & Installation............19-96
Fuel System Pressure 19-95
  Relieving.................................19-95
Fuel System Service
  Precautions............................19-95
H
Halfshaft...................................19-104
  Removal & Installation..........19-104
Heater Core..............................19-25
  Removal & Installation............19-25
Hydraulic Clutch System 19-102
  Bleeding.................................19-102
I
Ignition Timing 19-19
  Adjustment..............................19-19
Intake Manifold.........................19-46
  Removal & Installation............19-46

L
Leaf Springs 19-135
  Removal & Installation..........19-135
Lower Ball Joint........................19-123
  Removal & Installation..........19-123
Lower Control Arm 19-126
  Lower Control Arm Bushing
    Replacement 19-129
  Removal & Installation..........19-126
M
Manual Transmission
  Assembly................................19-100
  Removal & Installation..........19-100
O
Oil Pan.....................................19-76
  Removal & Installation..........19-76
Oil Pump 19-78
  Removal & Installation..........19-78
P
Pinion Seal 19-111
  Removal & Installation..........19-111
Piston and Ring 19-95
  Positioning 19-95
Power Steering Gear 19-113
  Removal & Installation..........19-113
R
Rear Axle Shaft, Bearing and
  Seal 19-109
  Removal & Installation..........19-109
Rear Main Seal 19-83
  Removal & Installation..........19-83
S
Shock Absorbers (Front)...........19-115
  Assembly & Disassembly 19-117
  Removal & Installation..........19-115
Shock Absorbers (Rear)............19-135
  Removal & Installation..........19-135
Spindle Bearings 19-108
  Removal, Packing &
    Installation..........................19-108
Stabilizer Bar (Front).................19-118
  Removal & Installation..........19-118
Stabilizer Bar (Rear).................19-138
  Removal & Installation..........19-138
Starter 19-76
  Removal & Installation..........19-76

T

Timing Belt, Cover and
  Crankshaft Seal 19-84
    Removal & Installation............19-84
Timing Chain Cover, Seal and
  Timing Chain.............................19-86
    Removal & Installation...........19-86
    Seal Replacement 19-89

Transfer Case Assembly.............19-104
    Removal & Installation..........19-104

U

Upper Ball Joint.......................19-119
    Removal & Installation..........19-119
Upper Control Arm 19-124
    Removal & Installation..........19-124
Upper Control Arm Bushing
  Replacement.....................19-125

V

Valve Lash 19-66
    Adjustment.............................19-66

W

Water Pump 19-36
    Removal & Installation............19-36

SPECIFICATIONS AND MAINTENANCE CHARTS

ENGINE AND VEHICLE IDENTIFICATION

Engine							Model Year	
Code/ID ① ②	Liters (cc)	Cu. In.	Cyl.	Fuel Sys.	Engine Type	Eng. Mfg.	Code ③	Year
2RZ-FE/L	2.4 (2438)	149	4	MFI	DOHC	Toyota	2	2002
3RZ-FE/M	2.7 (2693)	164	4	MFI	DOHC	Toyota	3	2003
2TR-FE/X	2.7 (2693)	164	4	MFI	DOHC	Toyota	4	2004
5VZ-FE/N	3.4 (3378)	206	6	MFI	DOHC	Toyota	5	2005
1GR-FE/U	4.0 (3956)	241	6	MFI	DOHC	Toyota	6	2006

MFI: Multi-port Fuel Injection

DOHC: Double Overhead Camshaft

① Except 1GR-FE/U engine: stamped on the left side of the engine block. Engine ID is fifth character of the VIN number.

② 1GR-FE/U engine: stamped on the right side of the engine block. Ingine ID is the fifth character of the VIN number.

③ 10th digit of the VIN number

09490_TACO_C0001

GENERAL ENGINE SPECIFICATIONS

Year	Model	Engine Displacement Liters	Engine Series Code/ID	Net Horsepower @ rpm	Net Torque @ rpm (ft. lbs.)	Bore x Stroke (in.)	Com-pression Ratio	Oil Pressure @ rpm
2002	Tacoma	2.4	2RZ-FE/L	142@5000	160@4000	3.74x3.38	9.5:1	36-71@3000
		2.7	3RZ-FE/M	150@4800	177@4000	3.74x3.74	9.5:1	36-71@3000
		3.4	5VZ-FE/N	190@4800	220@3600	3.68x3.23	9.6:1	NA
2003	Tacoma	2.4	2RZ-FE/L	142@5000	160@4000	3.74x3.38	9.5:1	36-71@3000
		2.7	3RZ-FE/M	150@4800	177@4000	3.74x3.74	9.5:1	36-71@3000
		3.4	5VZ-FE/N	190@4800	220@3600	3.68x3.23	9.6:1	NA
2004	Tacoma	2.4	2RZ-FE/L	142@5000	160@4000	3.74x3.38	9.5:1	36-71@3000
		2.7	3RZ-FE/M	150@4800	177@4000	3.74x3.74	9.5:1	36-71@3000
		3.4	5VZ-FE/N	190@4800	220@3600	3.68x3.23	9.6:1	NA
2005	Tacoma	2.7	2TR-FE/X	159@5200	236@5200	3.74x3.74	NA	43-85@3000
		4.0	1GR-FE/U	180@3800	266@4000	3.70x3.74	NA	23-75@3000
2006	Tacoma	2.7	2TR-FE/X	159@5200	236@5200	3.74x3.74	NA	43-85@3000
		4.0	1GR-FE/U	180@3800	266@4000	3.70x3.74	NA	23-75@3000

NA: Not Available

09490_TACO_C0002

ENGINE TUNE-UP SPECIFICATIONS

Year	Engine Displacement Liters	Engine Code/ID	Spark Plug Gap (in.)	Ignition Timing (deg.)		Fuel Pump (psi)	Idle Speed (rpm) MT	Idle Speed (rpm) AT	Valve Clearance Intake	Valve Clearance Exhaust
2002	2.4	2RZ-FE/L	0.043	5B	①	38-44	650-750	—	0.006-0.010	0.010-0.014
	2.7	3RZ-FE/M	0.043	5B	①	38-44	650-750	650-750	0.006-0.010	0.010-0.014
	3.4	5VZ-FE/N	0.043	10B	①	38-44	650-750	650-750	0.006-0.010	0.010-0.014
2003	2.4	2RZ-FE/L	0.043	5B	①	38-44	650-750	—	0.006-0.010	0.010-0.014
	2.7	3RZ-FE/M	0.043	5B	①	38-44	650-750	650-750	0.006-0.010	0.010-0.014
	3.4	5VZ-FE/N	0.043	10B	①	38-44	650-750	650-750	0.006-0.010	0.010-0.014
2004	2.4	2RZ-FE/L	0.043	5B	①	38-44	650-750	—	0.006-0.010	0.010-0.014
	2.7	3RZ-FE/M	0.043	5B	①	38-44	650-750	650-750	0.006-0.010	0.010-0.014
	3.4	5VZ-FE/N	0.043	10B	①	38-44	650-750	650-750	0.006-0.010	0.010-0.014
2005	2.7	2TR-FE/X	0.039-0.043	3-7B	②	40.8-41.7	600-700	600-700	③	③
	4.0	1GR-FE/U	0.039-0.043	7-24B	②	40.8-41.7	650-750	650-750	0.006-0.010	0.011-0.015
2006	2.7	2TR-FE/X	0.039-0.043	3-7B	②	40.8-41.7	600-700	600-700	③	③
	4.0	1GR-FE/U	0.039-0.043	7-24B	②	40.8-41.7	650-750	650-750	0.006-0.010	0.011-0.015

NOTE: The Vehicle Emission Control Information label often reflects specification changes made during production.

The label figures must be used if they differ from those in this chart.

B: Before top dead center

① With terminals TE1 and E1 connected to DLC1

② With terminals TC and CG of the DLC3 connected

③ Automatic adjustment

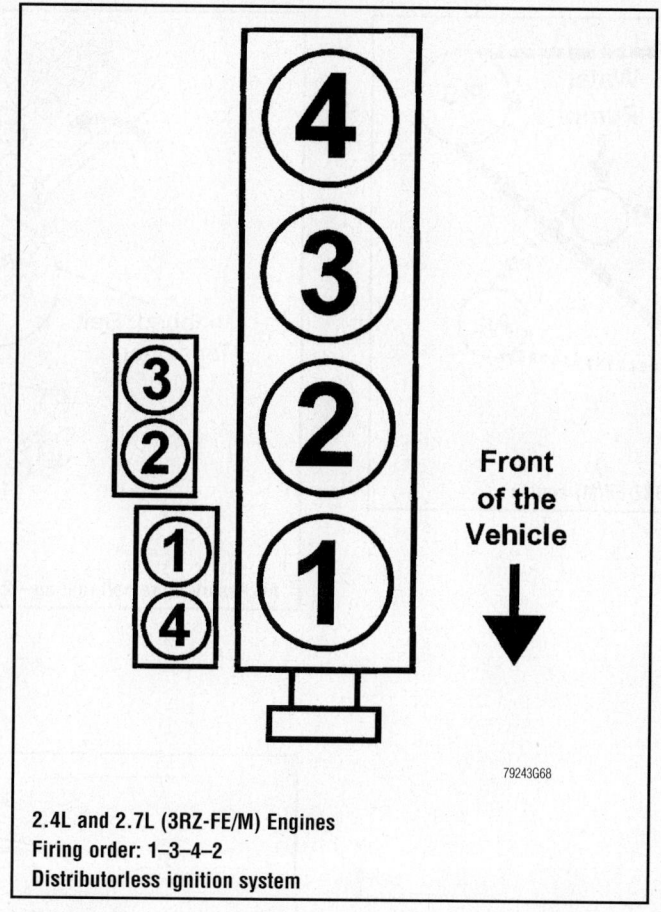

2.4L and 2.7L (3RZ-FE/M) Engines
Firing order: 1–3–4–2
Distributorless ignition system

79243G68

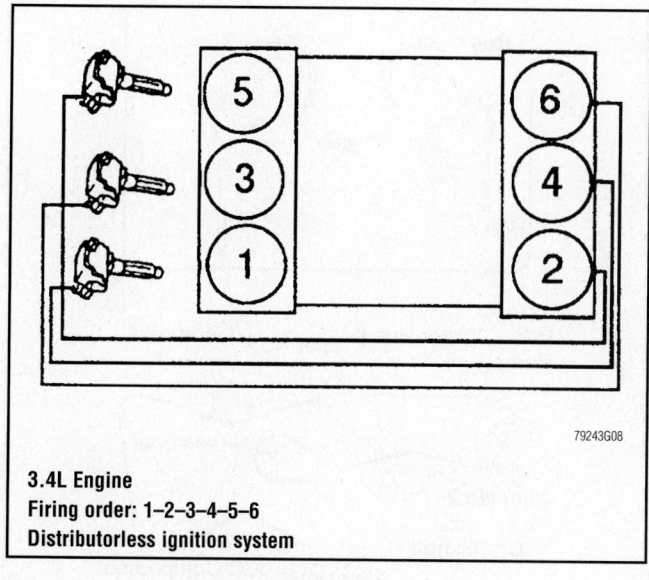

3.4L Engine
Firing order: 1–2–3–4–5–6
Distributorless ignition system

79243G08

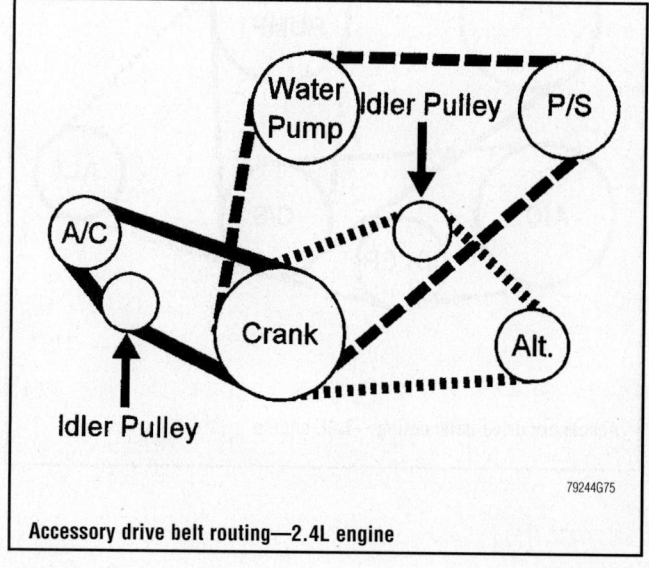

Accessory drive belt routing—2.4L engine

79244G75

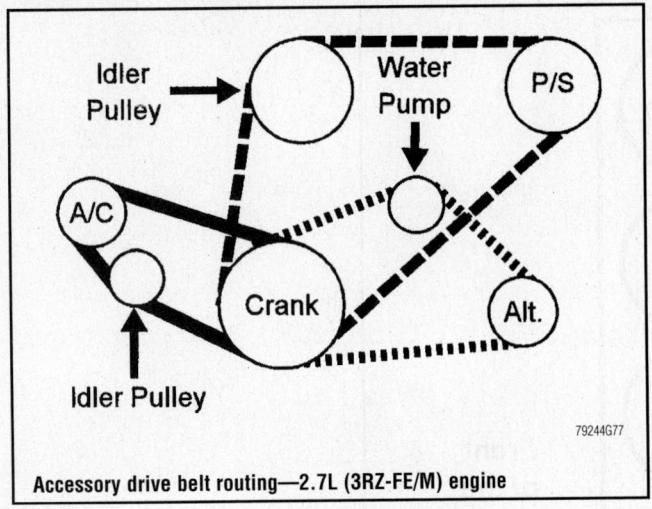

Accessory drive belt routing—2.7L (3RZ-FE/M) engine

79244G77

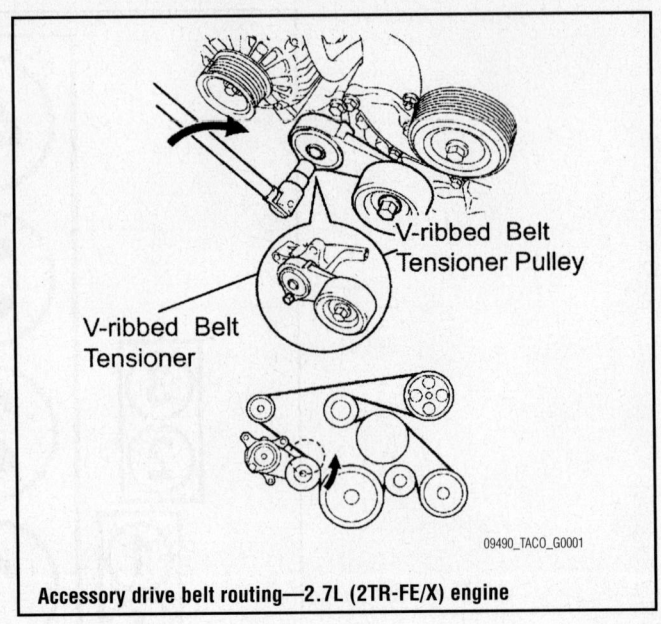

Accessory drive belt routing—2.7L (2TR-FE/X) engine

09490_TACO_G0001

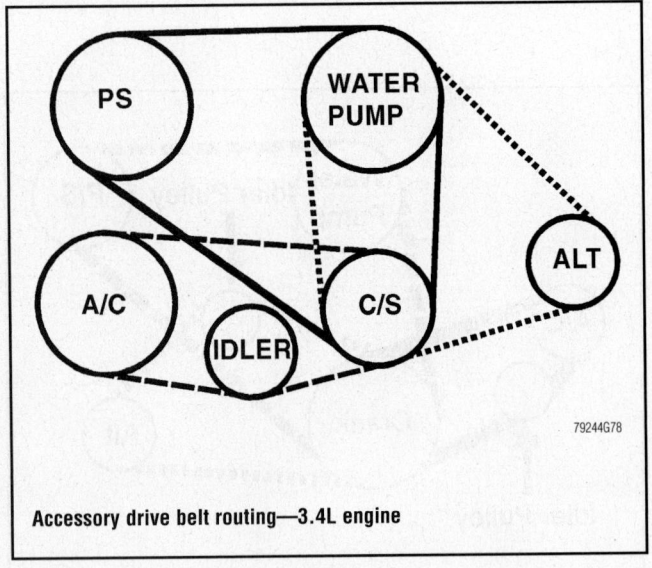

Accessory drive belt routing—3.4L engine

79244G78

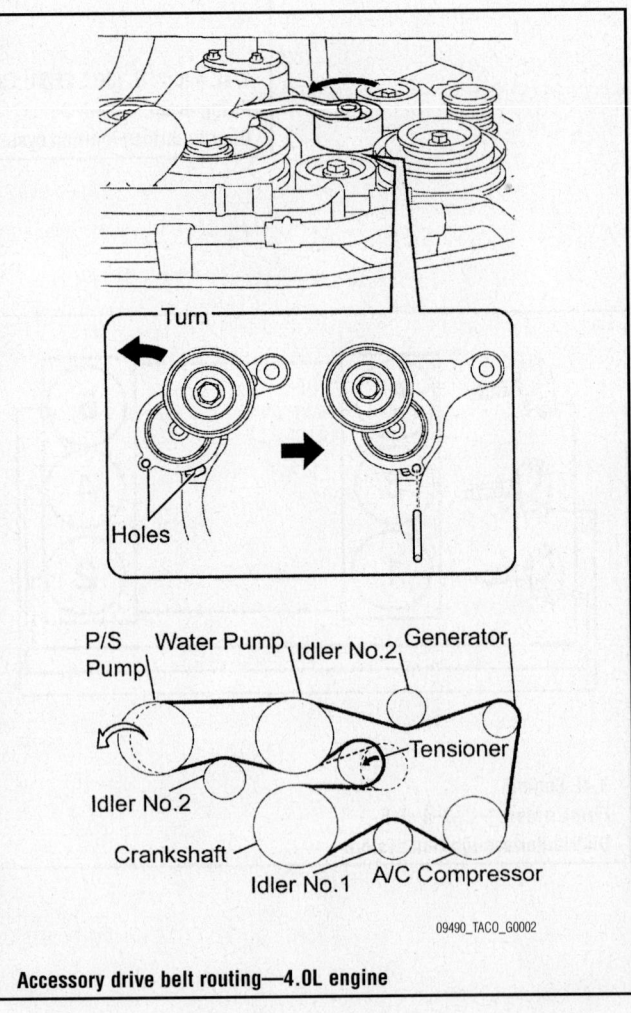

Accessory drive belt routing—4.0L engine

09490_TACO_G0002

CAPACITIES

Year	Model	Engine Displacement Liters	Engine Code/ID	Engine Oil with Filter (qts.)	Transmission (pts.)		Transfer Case (pts.)	Drive Axle		Fuel Tank (gal.)	Cooling System (qts.)
					5-Spd	Auto.		Front (pts.)	Rear (pts.)		
2002	Tacoma	2.4	2RZ-FE/L	5.8	①	②	—	—	2.8	18.5	③
		2.7	3RZ-FE/M	5.8	①	②	2.2	2.5	④	18.5	③
		3.4	5VZ-FE/N	5.5	⑤	②	2.2	2.5	④	18.5	⑥
2003	Tacoma	2.4	2RZ-FE/L	5.8	①	②	—	—	2.8	18.5	③
		2.7	3RZ-FE/M	5.8	①	②	2.2	2.5	④	18.5	③
		3.4	5VZ-FE/N	5.5	⑤	②	2.2	2.5	④	18.5	⑥
2004	Tacoma	2.4	2RZ-FE/L	5.8	①	②	—	—	2.8	18.5	③
		2.7	3RZ-FE/M	5.8	①	②	2.2	2.5	④	18.5	③
		3.4	5VZ-FE/N	5.5	⑤	②	2.2	2.5	④	18.5	⑥
2005	Tacoma	2.7	2TR-FE/X	5.8	①	⑦	2.2	3.2	⑧	21.1	9.1
		4.0	1GR-FE/U	⑨	3.6	⑦	2.2	3.2	⑧	21.1	⑩
2006	Tacoma	2.7	2TR-FE/X	5.8	①	⑦	2.2	3.2	⑧	21.1	9.1
		4.0	1GR-FE/U	⑨	3.6	⑦	2.2	3.2	⑧	21.1	⑩

① 2WD: 5.4
   4WD: 4.6

② A44D: 5.0
   A340E: 3.4
   A340F: 4.2

③ 2WD M/T: 8.5
   2WD A/T: 8.2
   4WD M/T: 8.8
   4WD A/T: 8.7

④ 4wd extra long: 5.16
   Short models w/diff. Lock: 5.60
   Extra long models w/diff. Lock: 6.36
   w/o diff. Lock: 5.38

⑤ W59:
   2WD: 5.4
   4WD: 5.2
   R150, R150F:
   2WD: 5.4
   4WD: 4.6

⑥ w/rear heater: 9.5
   w/o rear heater: 8.5

⑦ 4 speed: 4.0
   5 speed: 6.0

⑧ 2WD: 7.0
   4WD: 6.0

⑨ 2WD, except PreRunner: 4.5
   4WD and PreRunner: 5.2

⑩ MT: 10.3
   AT: 10.1

09490_TACO_C0004

VALVE SPECIFICATIONS

Year	Engine Displacement Liters	Engine Code/ID	Seat Angle (deg.)	Face Angle (deg.)	Spring Test Pressure (lbs. @ in.)	Spring Installed Height (in.)	Stem-to-Guide Clearance (in.)		Stem Diameter (in.)	
							Intake	Exhaust	Intake	Exhaust
2002	2.4	2RZ-FE/L	45	44.5	40.0-46.0@ 1.406	1.4060	0.0010- 0.0024	0.0012- 0.0026	0.2350- 0.2356	0.2348- 0.2354
	2.7	3RZ-FE/M	45	44.5	40.0-46.0@ 1.406	1.4060	0.0010- 0.0024	0.0012- 0.0026	0.2350- 0.2356	0.2348- 0.2354
	3.4	5VZ-FE/N	45	44.5	41.9-46.3@ 1.311	1.3110	0.0010- 0.0024	0.0012- 0.0026	0.2350- 0.2356	0.2348- 0.2354
2003	2.4	2RZ-FE/L	45	44.5	40.0-46.0@ 1.406	1.4060	0.0010- 0.0024	0.0012- 0.0026	0.2350- 0.2356	0.2348- 0.2354
	2.7	3RZ-FE/M	45	44.5	40.0-46.0@ 1.406	1.4060	0.0010- 0.0024	0.0012- 0.0026	0.2350- 0.2356	0.2348- 0.2354
	3.4	5VZ-FE/N	45	44.5	41.9-46.3@ 1.311	1.3110	0.0010- 0.0024	0.0012- 0.0026	0.2350- 0.2356	0.2348- 0.2354
2004	2.4	2RZ-FE/L	45	44.5	40.0-46.0@ 1.406	1.4060	0.0010- 0.0024	0.0012- 0.0026	0.2350- 0.2356	0.2348- 0.2354
	2.7	3RZ-FE/M	45	44.5	40.0-46.0@ 1.406	1.4060	0.0010- 0.0024	0.0012- 0.0026	0.2350- 0.2356	0.2348- 0.2354
	3.4	5VZ-FE/N	45	44.5	41.9-46.3@ 1.311	1.3110	0.0010- 0.0024	0.0012- 0.0026	0.2350- 0.2356	0.2348- 0.2354
2005	2.7	2TR-FE/X	45	NA	NA	1.9106	0.0010- 0.0024	0.0012- 0.0026	0.2154- 0.2159	0.2151- 0.2157
	4.0	1GR-FE/U	NA	44.5	41.9-46.3@ 1.311	1.882	0.0010- 0.0024	0.0012- 0.0026	0.2154- 0.2159	0.2152- 0.2158
2006	2.7	2TR-FE/X	45	NA	NA	1.9106	0.0010- 0.0024	0.0012- 0.0026	0.2154- 0.2159	0.2151- 0.2157
	4.0	1GR-FE/U	NA	44.5	41.9-46.3@ 1.311	1.882	0.0010- 0.0024	0.0012- 0.0026	0.2154- 0.2159	0.2152- 0.2158

NA: Not Available

09490_TACO_C0005

CAMSHAFT SPECIFICATIONS
All measurements in inches unless noted

Year	Engine Displacement Liters	Engine Code/ID	Journal Dia.	Brg. Oil Clearance	Shaft End-play ①	Circle Runout	Lobe Height Intake	Lobe Height Exhaust
2002	2.4	2RZ-FE/L	1.0614-1.0620	0.0010-0.0024	0.0016-0.0037	0.0024	1.7839-1.7878	1.7740 1.7779
	2.7	3RZ-FE/M	1.0614-1.0620	0.0010-0.0024	0.0016-0.0037	0.0024	1.7839-1.7878	1.7740 1.7779
	3.4	5VZ-FE/N	1.0610-1.0616	0.0014-0.0028	0.0013-0.0031	0.0024	1.6657-1.6697	1.6520-1.6559
2003	2.4	2RZ-FE/L	1.0614-1.0620	0.0010-0.0024	0.0016-0.0037	0.0024	1.7839-1.7878	1.7740 1.7779
	2.7	3RZ-FE/M	1.0614-1.0620	0.0010-0.0024	0.0016-0.0037	0.0024	1.7839-1.7878	1.7740 1.7779
	3.4	5VZ-FE/N	1.0610-1.0616	0.0014-0.0028	0.0013-0.0031	0.0024	1.6657-1.6697	1.6520-1.6559
2004	2.4	2RZ-FE/L	1.0614-1.0620	0.0010-0.0024	0.0016-0.0037	0.0024	1.7839-1.7878	1.7740 1.7779
	2.7	3RZ-FE/M	1.0614-1.0620	0.0010-0.0024	0.0016-0.0037	0.0024	1.7839-1.7878	1.7740 1.7779
	3.4	5VZ-FE/N	1.0610-1.0616	0.0014-0.0028	0.0013-0.0031	0.0024	1.6657-1.6697	1.6520-1.6559
2005	2.7	2TR-FE/X	②	③	0.0039-0.0090	0.0012	1.6872-1.6911	1.6872 1.6911
	4.0	1GR-FE/U	④	⑤	0.0160-0.0350	0.0024	⑥	⑥
2006	2.7	2TR-FE/X	②	③	0.0039-0.0090	0.0012	1.6872-1.6911	1.6872 1.6911
	4.0	1GR-FE/U	④	⑤	0.0160-0.0350	0.0024	⑥	⑥

NA: Not Available

① Thrust clearance

② No. 1: 1.4153-1.4159
   All others: 1.0614-1.0620

③ No. 1: 0.0014-0.0029
   All others: 0.0010-0.0024

④ No. 1: 1.4162-1.4167
   All others: 0.9039-0.9045

⑤ No. 1: 0.0016-0.0031
   All others: 0.0010-0.0024

⑥ No. 1 camshaft: 1.7389-1.7428
   No. 2 camshaft: 1.7551-1.7591
   No. 3 camshaft (sub assembly): 1.7389-1.7428
   No. 2 camshaft (sub assembly): 1.7551-1.7591

09490_TACO_C0006

CRANKSHAFT AND CONNECTING ROD SPECIFICATIONS

All measurements are given in inches.

Year	Engine Displacement Liters	Engine Code/ID	Crankshaft				Connecting Rod		
			Main Brg. Journal Dia.	Main Brg. Oil Clearance	Shaft End-play	Thrust on No.	Journal Diameter	Oil Clearance	Side Clearance
2002	2.4	2RZ-FE/L	2.3617-2.3622	0.0009-0.0022	0.0008-0.0087	2	2.0861-2.0866	0.0012-0.0022	0.0063-0.0123
	2.7	3RZ-FE/M	2.2615-2.3620	0.0012-0.0022	0.0008-0.0087	3	2.0861-2.0866	0.0009-0.0022	0.0063-0.0123
	3.4	5VZ-FE/N	2.5191-2.5197	0.0008-0.0015	0.0008-0.0087	2	2.1648-2.1654	0.0009-0.0021	0.0059-0.0130
2003	2.4	2RZ-FE/L	2.3617-2.3622	0.0009-0.0022	0.0008-0.0087	2	2.0861-2.0866	0.0012-0.0022	0.0063-0.0123
	2.7	3RZ-FE/M	2.2615-2.3620	0.0012-0.0022	0.0008-0.0087	3	2.0861-2.0866	0.0009-0.0022	0.0063-0.0123
	3.4	5VZ-FE/N	2.5191-2.5197	0.0008-0.0015	0.0008-0.0087	2	2.1648-2.1654	0.0009-0.0021	0.0059-0.0130
2004	2.4	2RZ-FE/L	2.3617-2.3622	0.0009-0.0022	0.0008-0.0087	2	2.0861-2.0866	0.0012-0.0022	0.0063-0.0123
	2.7	3RZ-FE/M	2.2615-2.3620	0.0012-0.0022	0.0008-0.0087	3	2.0861-2.0866	0.0009-0.0022	0.0063-0.0123
	3.4	5VZ-FE/N	2.5191-2.5197	0.0008-0.0015	0.0008-0.0087	2	2.1648-2.1654	0.0009-0.0021	0.0059-0.0130
2005	2.7	2TR-FE/X	①	②	0.0008-0.0087	NA	NA	0.0009-0.0019	0.0059-0.0138
	4.0	1GR-FE/U	2.8342-2.8346	0.0007-0.0012	NA	NA	NA	0.0010-0.0018	0.0059-0.0118
2006	2.7	2TR-FE/X	①	②	0.0008-0.0087	NA	NA	0.0009-0.0019	0.0059-0.0138
	4.0	1GR-FE/U	2.8342-2.8346	0.0007-0.0012	NA	NA	NA	0.0010-0.0018	0.0059-0.0118

NA: Not Available

① No. 3: 2.3615-2.3620
  All others: 2.3619-2.3622

② No. 3: 0.0012-0.0022
  All others: 0.0009-0.0019

09490_TACO_C0007

PISTON AND RING SPECIFICATIONS
All measurements are given in inches.

Year	Engine Displacement Liters	Engine Code/ID	Piston Clearance	Ring Gap			Ring Side Clearance		
				Top Compression	Bottom Compression	Oil Control	Top Compression	Bottom Compression	Oil Control
2002	2.4	2RZ-FE/L	0.0012-0.0020	0.0118-0.0169	0.0177-0.0236	0.0051-0.0150	0.0008-0.0028	0.0012-0.0028	SNUG
	2.7	3RZ-FE/M	0.0019-0.0028	0.0118-0.0157	0.0157-0.0194	0.0051-0.0150	0.0008-0.0028	0.0012-0.0028	SNUG
	3.4	5VZ-FE/N	0.0053-0.0060	0.0118-0.0197	0.0157-0.0236	0.0059-0.0217	0.0016-0.0031	0.0012-0.0028	SNUG
2003	2.4	2RZ-FE/L	0.0012-0.0020	0.0118-0.0169	0.0177-0.0236	0.0051-0.0150	0.0008-0.0028	0.0012-0.0028	SNUG
	2.7	3RZ-FE/M	0.0019-0.0028	0.0118-0.0157	0.0157-0.0194	0.0051-0.0150	0.0008-0.0028	0.0012-0.0028	SNUG
	3.4	5VZ-FE/N	0.0053-0.0060	0.0118-0.0197	0.0157-0.0236	0.0059-0.0217	0.0016-0.0031	0.0012-0.0028	SNUG
2004	2.4	2RZ-FE/L	0.0012-0.0020	0.0118-0.0169	0.0177-0.0236	0.0051-0.0150	0.0008-0.0028	0.0012-0.0028	SNUG
	2.7	3RZ-FE/M	0.0019-0.0028	0.0118-0.0157	0.0157-0.0194	0.0051-0.0150	0.0008-0.0028	0.0012-0.0028	SNUG
	3.4	5VZ-FE/N	0.0053-0.0060	0.0118-0.0197	0.0157-0.0236	0.0059-0.0217	0.0016-0.0031	0.0012-0.0028	SNUG
2005	2.7	2TR-FE/X	0.0007-0.0020	0.0087-0.0134	0.0177-0.0224	0.0039-0.0157	0.0008-0.0030	0.0008-0.0026	0.0008-0.0028
	4.0	1GR-FE/U	0.0031-0.0040	0.0118-0.0157	0.0157-0.0197	0.0039-0.0157	0.0008-0.0028	0.0008-0.0024	0.0028-0.0060
2006	2.7	2TR-FE/X	0.0007-0.0020	0.0087-0.0134	0.0177-0.0224	0.0039-0.0157	0.0008-0.0030	0.0008-0.0026	0.0008-0.0028
	4.0	1GR-FE/U	0.0031-0.0040	0.0118-0.0157	0.0157-0.0197	0.0039-0.0157	0.0008-0.0028	0.0008-0.0024	0.0028-0.0060

09490_TACO_C0008

TORQUE SPECIFICATIONS
All readings in ft. lbs.

Year	Engine Displacement Liters	Engine Code/ID	Cylinder Head Bolts	Main Bearing Bolts	Rod Bearing Bolts	Crankshaft Damper Bolts	Flywheel Bolts	Manifold		Spark Plugs	Oil Pan Drain Plug
								Intake	Exhaust		
2002	2.4	2RZ-FE/L	①	②	③	193	④	22	36	14	27
	2.7	3RZ-FE/M	①	②	③	193	⑤	22	36	14	27
	3.4	5VZ-FE/N	⑥	⑦	⑧	184	63	13	30	13	NA
2003	2.4	2RZ-FE/L	①	②	③	193	④	22	36	14	27
	2.7	3RZ-FE/M	①	②	③	193	⑤	22	36	14	27
	3.4	5VZ-FE/N	⑥	⑦	⑧	184	63	13	30	13	NA
2004	2.4	2RZ-FE/L	①	②	③	193	④	22	36	14	27
	2.7	3RZ-FE/M	①	②	③	193	⑤	22	36	14	27
	3.4	5VZ-FE/N	⑥	⑦	⑧	184	⑨	13	30	13	NA
2005	2.7	2TR-FE/X	①	②	⑧	192	⑩	18	27	13	28
	4.0	1GR-FE/U	⑪	⑫	⑧	185	61	19	16	13	30
2006	2.7	2TR-FE/X	①	②	⑧	192	⑩	18	27	13	28
	4.0	1GR-FE/U	⑪	⑫	⑧	185	61	19	16	13	30

NA: Information not available

① Step 1: 29 ft. lbs.
   Step 2: Plus 90 degrees
   Step 3: Plus 90 degrees

② Step 1: 29 ft. lbs.
   Step 2: Plus 90 degrees

③ Step 1: 33 ft. lbs.
   Step 2: Plus 90 degrees

④ MT: 65 ft. lbs.
   AT: 54 ft. lbs.

⑤ MT: 19 ft. lbs. +90 degrees
   AT: 54 ft. lbs.

⑥ Step 1: 25 ft. lbs.
   Step 2: Plus 90 degrees
   Step 3: Plus 90 degrees
   Recessed head: 13 ft. lbs.

⑦ Step 1: 45 ft. lbs.
   Step 2: Plus 90 degrees

⑧ Step 1: 18 ft. lbs.
   Step 2: Plus 90 degrees

⑨ AT: 61 ft. lbs.
   MT: 63 ft. lbs.

⑩ AT: 55 ft. lbs.
   MT: 20 ft. lbs. plus 90 degrees

⑪ right side: 27 ft. lbs. Then + 90 degrees. Then + 180 degrees
   left side (recessed head): 27 ft. lbs. Then plus 180 degrees
   left side (0.55 inch head): 22 ft. lbs.

⑫ 12 pointed head: 45 ft. lbs. Then plus 90 degrees
   12mm head: 18 ft. lbs.

09490_TACO_C0009

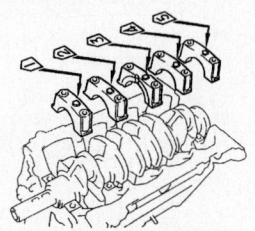

Install the 5 main bearing caps in their proper locations.

Apply a light coat of engine oil on the threads and under the heads of the main bearing cap bolts.

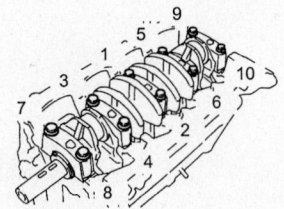

Install and uniformly tighten the 10 bolts of the main bearing caps in several passes, in the sequence shown.
Torque: 39 N·m (400 kgf·cm, 29 ft·lbf)
If any one of the main bearing cap bolts does not meet the torque specification, replace the main bearing cap bolt.

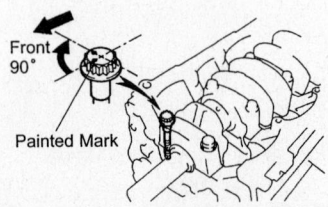

Mark the front of the main bearing cap bolt with paint.
Retighten the main bearing cap bolts by 90° in the numerical order shown above.
Check that the painted mark is now at a 90° angle to the front.
Check that the crankshaft turns smoothly.
Check the crankshaft thrust clearance.

09490_TACO_G0003

Main bearing torque sequence—2.4L and 2.7L (3RZ-FE/M) engines

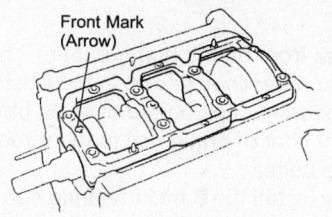

Front Mark
(Arrow)

Install the main bearing cap with the front mark facing forward.

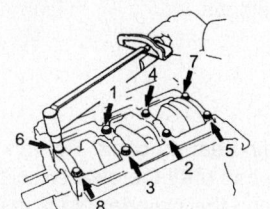

Apply a light coat of engine oil on the threads and under the heads of the main bearing cap bolts.
Install and uniformly tighten the 8 main bearing cap bolts in several passes, in the sequence shown.
Torque: 61 N·m (625 kgf·cm, 45 ft·lbf)

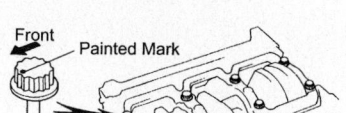

Front   Painted Mark

Mark the front of the main bearing cap bolt with paint.

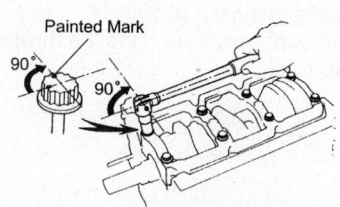

Painted Mark

90°   90°

Retighten the main bearing cap bolts by 90° in the numerical order shown.
Check that the painted mark is now at a 90° angle to the front.
Check that the crankshaft turns smoothly.
Check the crankshaft thrust clearance.

09490_TACO_G0004

Main bearing torque sequence—3.4L engine

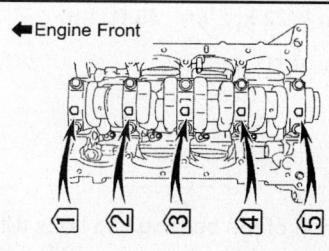

Engine Front

Install the 5 crankshaft bearing caps in their proper locations.

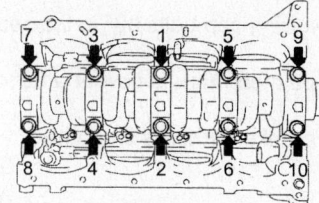

Step 1
(1) Install and uniformly tighten the 10 main bearing cap bolts in the sequence shown in the illustration.
Torque: 39 N m (398 kgf cm, 29 ft. lbf)

Step 2
(1) Mark the front of the bearing cap bolts with paint.
(2) Retighten the bearing cap bolts by 90° in the order above.
(3) Check that the painted mark is now at a 90° angle to the front.
Check that the crankshaft turns smoothly.
Check the crankshaft thrust clearance.

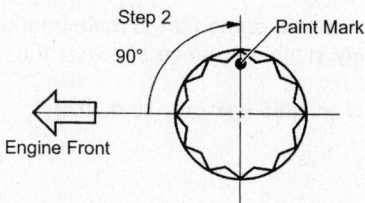

Step 2   Paint Mark

90°

Engine Front

09490_TACO_G0005

Main bearing torque sequence—2.7L (2TR-FE/X) engine

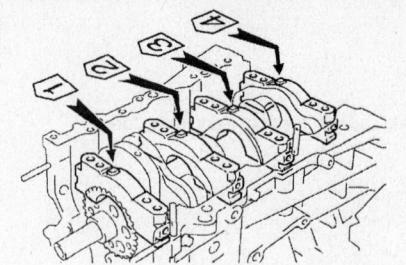

Examine the front marks and numbers, check the sequence number is as shown in the illustration and install the bearing caps on the cylinder block.
Apply a light coat of engine oil to the threads of bearing cap bolts.
Temporarily install the 8 main bearing cap bolts in the inside positions.

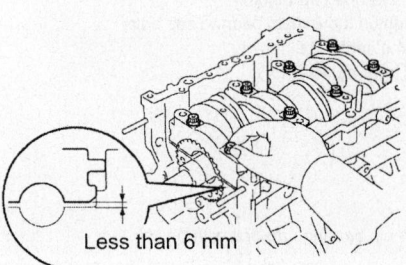

Less than 6 mm

Install the main bearing caps. Tighten the 2 bolts for each bearing cap until the clearance between the bearing cap and the cylinder block is under 6 mm (0.23 in.).

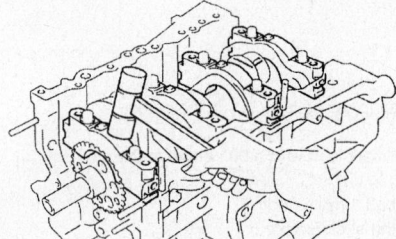

Using a plastic-faced hammer, lightly tap the bearing cap to ensure a proper fit.
Apply a light coat of engine oil to the threads of main bearing cap bolts.

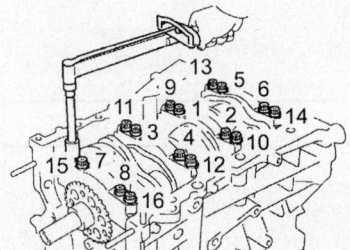

Install the 16 main bearing cap bolts. Using several steps, tighten the bolts uniformly in the sequence shown in the illustration.
Torque: 61 N m (622 kgf cm, 45 ft. lbf)

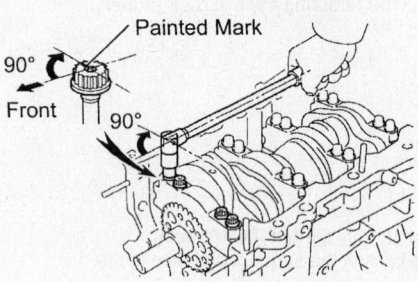

Painted Mark
90°
Front
90°

Mark the front side of the bearing cap bolts with paint.
Retighten the bearing cap bolts 90° in the sequence as shown.
Check that the painted mark is now at a 90° angle from the front.
Check that the crankshaft turns smoothly.

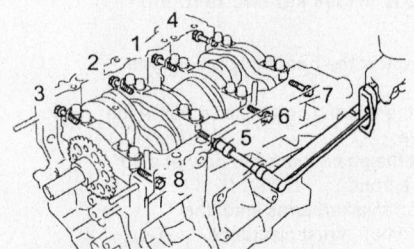

Using several steps, tighten the 8 main bearing cap bolts uniformly in the sequence shown in the illustration.
Torque: 25 N m (255 kgf cm, 18 ft. lbf)

09490_TACO_G0006

Main bearing torque sequence—4.0L engine

WHEEL ALIGNMENT

Year	Model	Caster Range (+/-Deg.)	Caster Preferred Setting (Deg.)	Camber Range (+/-Deg.)	Camber Preferred Setting (Deg.)	Toe-in (in.)	Steering Axis Inclination (Deg.)
2002	2WD	0.75	+0.67	0.75	0	0.06+/-0.08	10.00
	4WD	0.75	0.30	0.75	+1.62	0.06+/-0.08	10.40
2003	2WD	0.75	+0.67	0.75	0	0.06+/-0.08	10.00
	4WD	0.75	0.30	0.75	+1.62	0.06+/-0.08	10.40
2004	2WD exc. PreRunner	0.75	+0.67	0.75	0	0.06+/-0.08	10.00
	4WD & PreRunner	0.75	0.30	0.75	+1.62	0.06+/-0.08	10.40
2005	2WD exc. PreRunner	0.75	+0.67	0.75	0	0.06+/-0.08	10.00
	4WD & PreRunner	0.75	0.30	0.75	+1.62	0.06+/-0.08	10.40
2006	2WD exc. PreRunner	0.75	+0.67	0.75	0	0.06+/-0.08	10.00
	4WD & PreRunner	0.75	0.30	0.75	+1.62	0.06+/-0.08	10.40

All alignment figures based on nominal ride height and standard tires

09490_TACO_C0010

TIRE, WHEEL AND BALL JOINT SPECIFICATIONS

Year	Model	OEM Tires Standard	OEM Tires Optional	Tire Pressures (psi) Front	Tire Pressures (psi) Rear	Wheel Size	Ball Joint Inspection ①	Lugnut Torque (ft. lbs.)
2002	2wd Reg. Cab	P205/75R15	P235/55R16	Std: 29 Opt: 29	Std: 29 Opt: 32	6-JJ Opt: 6.5J	Upper 4-30 Lower 0.8-30	83
	4wd Reg. Cab	P225/75SR15	P265/70R16	Std: 26 Opt: 26	Std: 29 Opt: 26	6-JJ Opt.: 7-JJ	Upper 6-39 Lower 1-22	83
	4wd xtracab	P225/75SR15	P265/70R16	26	29	7-JJ	Upper 6-39 Lower 1-22	83
	2wd xtracab S-Runner	P235/55R16	None	29	32	7-JJ	Upper 4-30 Lower 0.8-30	83
	2wd xtracab PreRunner	P225/75R15	P265/70R16	Std: 26 Opt: 26	Std: 29 Opt: 26	7-JJ	Upper 6-39 Lower 1-22	83
	4wd Crew Cab	P225/75SR15	P265/70R16	Std: 26 Opt: 26	Std: 29 Opt: 26	7-JJ	Upper 6-39 Lower 1-22	83
	2wd Crew Cab PreRunner	P225/75R15	P265/70R16	Std: 26 Opt: 26	Std: 29 Opt: 26	7-JJ	Upper 6-39 Lower 1-22	83
2003	2wd Reg. Cab	P205/75R15	P235/55R16	Std: 29 Opt: 29	Std: 29 Opt: 32	6-JJ Opt: 6.5J	Upper 4-30 Lower 0.8-30	83
	4wd Reg. Cab	P225/75SR15	P265/70R16	Std: 26 Opt: 26	Std: 29 Opt: 26	6-JJ Opt.: 7-JJ	Upper 6-39 Lower 1-22	83
	4wd xtracab	P225/75SR15	P265/70R16	26	29	7-JJ	Upper 6-39 Lower 1-22	83
	2wd xtracab S-Runner	P235/55R16	None	29	32	7-JJ	Upper 4-30 Lower 0.8-30	83
	2wd xtracab PreRunner	P225/75R15	P265/70R16	Std: 26 Opt: 26	Std: 29 Opt: 26	7-JJ	Upper 6-39 Lower 1-22	83
	4wd Crew Cab	P225/75SR15	P265/70R16	Std: 26 Opt: 26	Std: 29 Opt: 26	7-JJ	Upper 6-39 Lower 1-22	83
	2wd Crew Cab PreRunner	P225/75R15	P265/70R16	Std: 26 Opt: 26	Std: 29 Opt: 26	7-JJ	Upper 6-39 Lower 1-22	83
2004	Reg. Cab 2wd	P205/75R15	P235/55R16	Std: 29 Opt: 29	Std: 29 Opt: 32	std. 6-JJ opt. 6.5	②	83
	Reg. Cab 4wd	P225/75SR15	P265/70R16	Std: 26 ③	Std: 29 ③	7-JJ	②	83
	Reg. Cab PreRunner	P225/75SR15	P265/70R16	③	③	7-JJ	②	83
	Extended Cab 2wd	P205/75R15	P235/55R16	③	③	std. 6-JJ opt. 6.5	②	83
	Extended Cab 4wd	P225/75SR15	P265/70R16	③	③	7-JJ	②	83
	Extended Cab PreRunner	P225/75SR15	P265/70R16	③	③	7-JJ	②	83
	S-Runner	P235/55R16	None	③	③	6.5	②	83
2005	Reg. Cab 2wd	P215/70R15	None	29	32	NA	②	85
	Reg. Cab 4wd	P245/75R16	None	29	29	NA	②	85
	Reg. Cab PreRunner	P245/75R16	None	29	29	NA	②	85
	Extended Cab 2wd	P215/70R15	None	③	③	NA	②	85
	Extended Cab 4wd	P245/75R16	None	③	③	NA	②	85
	Extended Cab PreRunner	P245/75R16	None	③	③	NA	②	85
	X-Runner	P255/45R18	None	29	29	NA	②	85
	Double Cab	P245/75R16	None	29	29	NA	②	85
2006	Reg. Cab 2wd	P215/70R15	None	29	32	NA	②	85
	Reg. Cab 4wd	P245/75R16	None	29	29	NA	②	85
	Reg. Cab PreRunner	P245/75R16	None	29	29	NA	②	85
	Extended Cab 2wd	P215/70R15	None	③	③	NA	②	85

TIRE, WHEEL AND BALL JOINT SPECIFICATIONS

Year	Model	OEM Tires Standard	OEM Tires Optional	Tire Pressures (psi) Front	Tire Pressures (psi) Rear	Wheel Size	Ball Joint Inspection ①	Lugnut Torque (ft. lbs.)
2006 cont.	Extended Cab 4wd	P245/75R16	None	③	③	NA	②	85
	Extended Cab PreRunner	P245/75R16	None	③	③	NA	②	85
	X-Runner	P255/45R18	None	29	29	NA	②	85
	Double Cab	P245/75R16	None	29	29	NA	②	85

OEM: Original Equipment Manufacturer

PSI: Pounds Per Square Inch

NA: Information not available

STD: Standard

OPT: Optional

① Torque required in inch lbs. to rotate ball joint when removed from the knuckle

② 2wd exc. PreRunner: Upper 4-30 inch lbs.; lower 0.8-30 inch lbs.

    4wd and PreRunner: Upper 6-39 inch lbs.; lower 1-22 inch lbs.

    Lower ball joint excessive play, all models: 0.020 inch

③ See driver's side door placard

09490_TACO_C0011A

BRAKE SPECIFICATIONS
All measurements in inches unless noted

Year	Model	Brake Disc Original Thickness	Brake Disc Minimum Thickness	Brake Disc Maximum Runout	Brake Drum Diameter Original Inside Diameter	Brake Drum Diameter Maximum Machine Diameter	Minimum Lining Thickness Front	Minimum Lining Thickness Rear	Brake Caliper Bracket Bolts (ft. lbs.)	Brake Caliper Mounting Bolts (ft. lbs.)
2002	2WD	0.866	0.787	0.0028	10.00	10.08	0.039	—	80	65
	4WD	0.866	0.787	0.0028	11.61	11.69	—	0.039	—	90
2003	2WD	0.866	0.787	0.0028	10.00	10.08	0.039	—	80	65
	4WD	0.866	0.787	0.0028	11.61	11.69	—	0.039	—	90
2004	2WD	0.866	0.787	0.0028	10.00	10.08	0.039	—	80	65
	4WD	0.866	0.787	0.0028	11.61	11.69	—	0.039	—	90
2005	2WD	0.984	0.906	0.0020	10.00	10.08	0.039	—	80	27
	①	1.102	1.024	0.0020	10.00	10.08	—	0.039	91	—
2006	2WD	0.984	0.906	0.0020	10.00	10.08	0.039	—	80	27
	①	1.102	1.024	0.0020	10.00	10.08	—	0.039	91	—

① 4WD and PreRunner

09490_TACO_C0012

SCHEDULED MAINTENANCE INTERVALS
TOYOTA—TACOMA

TO BE SERVICED	TYPE OF	VEHICLE MILEAGE INTERVAL (x1000)																		
		5	10	15	20	25	30	35	40	45	50	55	60	65	70	75	80	85	90	95
Automatic transmission and differential fluid	S/I			✓			✓			✓			✓			✓			✓	
Ball joints and boots	S/I			✓			✓			✓			✓			✓			✓	
Brake linings, discs/drums, lines & hoses	S/I			✓			✓			✓			✓			✓			✓	
Charcoal canister	S/I												✓							
Drive belts	S/I						✓						✓						✓	
Driveshaft bushing (4WD)	L						✓						✓						✓	
Engine coolant	R						✓						✓						✓	
Engine oil & filter	R	✓	✓	✓	✓	✓	✓	✓	✓	✓	✓	✓	✓	✓	✓	✓	✓	✓	✓	✓
Exhaust pipes & mounts	S/I			✓			✓			✓			✓			✓			✓	
Fuel lines & connections, fuel tank vapor vent system hoses, fuel tank band	S/I						✓						✓						✓	
Fuel tank cap gasket	S/I						✓						✓						✓	
Halfshaft boots & flange bolts	S/I			✓			✓			✓			✓			✓			✓	
Limited slip differential fluid	R						✓						✓						✓	
Manual transmission and differential fluid	S/I						✓						✓						✓	
Non-platinum spark plugs	R						✓						✓						✓	
Platinum spark plugs	R												✓							
Propeller shaft (4WD)	L			✓			✓			✓			✓			✓			✓	
Propeller shaft bolts	S/I			✓			✓			✓			✓			✓			✓	
Rack and pinion assembly	S/I			✓			✓			✓			✓			✓			✓	
Rear wheel bearing	L						✓						✓						✓	
Rotate tires	S/I	✓	✓	✓	✓	✓	✓	✓	✓	✓	✓	✓	✓	✓	✓	✓	✓	✓	✓	✓
Steering linkage	S/I			✓			✓			✓			✓			✓			✓	
Valves	S/I												✓							

R: Replace   S/I: Service or Inspect   L: Lubricate

FREQUENT OPERATION MAINTENANCE (SEVERE SERVICE)

If a vehicle is operated under any of the following conditions it is considered severe service:

- Towing a trailer or using a camper or car-top carrier.
- Repeated short trips of less than 5 miles in temperatures below freezing.
- Excessive idling or low-speed driving for long distances as in heavy commercial use, such as delivery, taxi or police cars.
- Operating on rough, muddy or salt-covered roads.
- Operating on unpaved or dusty roads.

Oil filter: service or inspect every 5000 miles or 4 months, whichever occurs first.

Brake linings and discs or drums: service or inspect every 5000 miles or 4 months, whichever occurs first.

Steering linkage: service or inspect every 5000 miles or 4 months, whichever occurs first.

Ball joints and boots: service or inspect every 5000 miles or 4 months, whichever occurs first.

Brake discs & pads (front): service or inspect every 6000 miles.

Halfshaft boots: service or inspect every 5000 miles or 4 months. Retighten the flange bolts, whichever occurs first.

Body chassis bolts and nuts: service or inspect every 5000 miles or 4 months, whichever occurs first.

Transmission and differential fluid: replace every 15,000 miles or 12 months, whichever occurs first.

Transfer case and differential fluid: replace every 15,000 miles or 12 months, whichever occurs first.

Timing belt: replace every 60,000 miles or 48 months, whichever occurs first.

ENGINE REPAIR

Distributor

All engines are equipped with distributorless ignition systems.

Alternator

REMOVAL & INSTALLATION

On some vehicles, the alternator is mounted very low on the engine. It may be necessary to remove the gravel shield and work from beneath the vehicle in order to gain access to the alternator. Replacing the alternator while the engine is cold is recommended.

2002–2004

1. Before servicing the vehicle, refer to the Precautions Section.
2. Disconnect the negative battery cable.
3. Disconnect the alternator wiring.
4. Remove the drive belt.
5. Remove the alternator retaining bolts.
6. Remove the alternator from the vehicle.
   To install:
7. Installation is the reverse of the removal procedure.
8. On the 2.4L and 2.7L engines, tighten

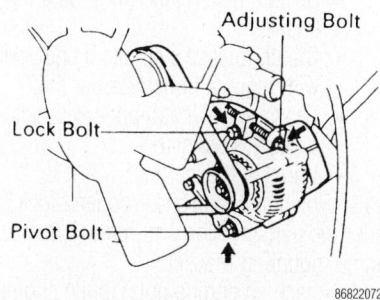

Locations of the adjusting, pivot and lock-bolts—2.4L and 2.7L (3RZ-FE/M) engines

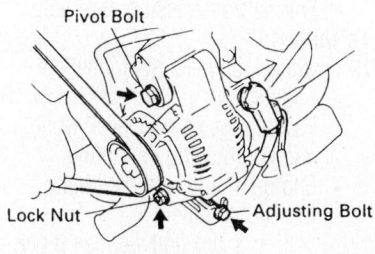

Locations of the adjusting and pivot bolts and the locknut—3.4L Engine

the locknut to 21 ft. lbs. (29 Nm) and the pivot bolt to 43 ft. lbs. (59 Nm).

9. On the 3.4L engine, tighten the locknut 25 ft. lbs. (33 Nm) and the pivot bolt 38 ft. lbs. (51 Nm).

2005–2006

1. Before servicing the vehicle, refer to the Precautions Section.
2. Disconnect the negative battery cable.
3. Remove the V bank cover.
4. Remove the radiator support to frame seal, left side.
5. Remove the radiator fan shroud.
6. Disconnect the alternator harness wiring.
7. Remove the drive belt.
8. Remove the alternator retaining bolts.
9. Remove the alternator from the vehicle.
   To install:
10. Installation is the reverse of the removal procedure.
11. Tighten the alternator retaining bolts to 32 ft. lbs. (43 Nm.).

Ignition Timing

ADJUSTMENT

All engines use a distributorless ignition system referred to as Direct Ignition System (DIS). All spark advance is permanently set by the PCM.

2.4L Engine

➡ **The ignition timing is not adjustable but can be checked.**

1. Before servicing the vehicle, refer to the Precautions Section.
2. Warm the engine to normal operating temperature.
3. Attach a hand-held tester to the Data Link Connector 3 (DLC3) under the dashboard on the driver's side.
4. Jumper terminals T_{E1} and E_1 of the DLC1.
5. Check the idle speed.
6. Aim the timing light at the timing indicator and check the ignition timing. Timing should be between 3–7 degrees BTDC at idle.
7. For a further check on ignition timing, disconnect the hand-held tester from the DLC3 and disconnect the jumper wire from the DLC1.

8. Point the timing light at the crankshaft pulley and read the timing. Timing should be between 7–18 degrees BTDC at idle.
9. Remove timing light from the engine.

2.7L (2TR-FE/X) Engine

➡ **The ignition timing is not adjustable but can be checked. Be sure that all electrical systems are OFF.**

WITH INTELEGENT TESTER

1. Connect the tool to the DLC3.
2. Turn the ignition switch ON.
3. Turn the tester ON.
4. Start the engine and allow it to warm up.
5. Select the following menu items: DIAGNOSIS/ENHANCED OBDII/DATA LIST/PRIMARY/IGN ADVANCE. Check that the ignition timing advances immediately when the engine speed is increased.

WITHOUT INTELEGENT TESTER

1. Turn the ignition switch ON.
2. Start the engine and allow it to warm up.
3. Install the timing light.

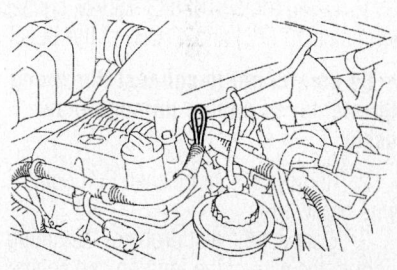

Wiring harness location—2.7L (2TR-FE/X) engine

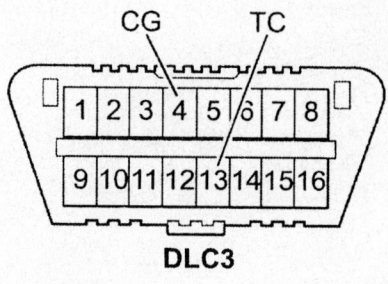

DLC3 terminal pin location points—2005–2006

➡Be sure to use a timing light that detects the first signal. After checking be sure to wrap the wire harness with tape.

4. Using too SST09843-18040, connect terminals 13 (TC) and 4 (CG) of the DLC3.

➡Be careful not to connect the wrong terminals, as engine damage may occur.

5. Check that the ignition timing is within specification. Be sure all electrical systems are OFF.

4.0L Engine

➡The ignition timing is not adjustable but can be checked. Be sure that all electrical systems are OFF. Be sure that the cooling fan motor is OFF.

WITH INTELEGENT TESTER

1. Warm up the engine.
2. Connect the tool to the DLC3.
3. Select the following menu items: DIAGNOSIS/ENHANCED OBDII/DATA LIST/PRIMARY/IGN ADVANCE.
4. Inspect the ignition timing during idling.
5. Check that the ignition timing advances immediately when the engine speed is increased.

WITHOUT INTELEGENT TESTER

1. Using too SST09843-18040, connect terminals 13 (TC) and 4 (CG) of the DLC3.

➡Be careful not to connect the wrong terminals, as engine damage may occur.

2. Remove the air cleaner. Pull out the wire harness.
3. Connect the test probe of the timing light to the wire of the ignition coil connector for the No. 1 cylinder.

➡Be sure to use a timing light that detects the first signal. After checking

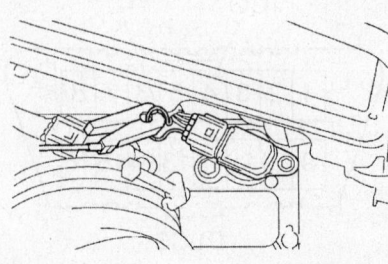

09490_TACO_G0009

Wiring harness location—4.0L engine

be sure to wrap the wire harness with tape.

4. Check the ignition timing during idling. Check that the ignition timing is within specification. Be sure all electrical systems are OFF.

Engine Assembly

REMOVAL & INSTALLATION

2.4L Engine

1. Before servicing the vehicle, refer to the Precautions Section.
2. Properly relieve the fuel system pressure.
3. Turn the ignition switch **OFF**.
4. Remove or disconnect the following:
   - Battery cables; negative cable first
   - Hood by matchmarking the hood hinges
   - Engine undercover
5. Drain the engine oil, transmission oil and cooling system.
6. Remove or disconnect the following:
   - Radiator
   - Drive belts
   - Loosen the lockbolt and adjusting bolt to the idler pulley, if equipped with power steering
   - Loosen the idler pulley nut and adjusting bolt, if equipped with air conditioning
   - Fan (with fan clutch), water pump pulley and fan shroud
   - Accelerator cable from the throttle body, if equipped with a manual transaxle
   - Accelerator and throttle cables from the throttle body, if equipped with an automatic transaxle
   - Actuator cover and cruise control cable from the actuator, if equipped with cruise control
   - Air cleaner cap, Mass Air Flow (MAF) and resonator
   - Air cleaner case
   - Intake air connector
   - Air conditioning compressor and bracket, if equipped with air conditioning
   - Alternator wires from the alternator
   - Heater hoses at the cowl panel
   - Brake booster vacuum hose
   - EVAP hose
   - Vacuum hose, if equipped with 4WD with Automatic Disconnecting Differential (ADD)
   - Both power steering hoses, if equipped with power steering

   - Fuel return hose
   - Fuel inlet hose
7. Remove the power steering pump as follows:
   - Nut and power steering pulley
   - Both bolts and the power steering pump
8. Disconnect the Engine Control Module (ECM) wiring from the ECM as follows:
   - Right front door scuff plate
   - Cowl panel side trim by removing the clip
   - 4 ECM electrical connectors.
9. Detach the engine wiring harness and connectors as follows:
   - Igniter connector
   - Ground strap from the cowl top panel
   - Both engine wiring harness clamps
   - Engine wiring harness retainer to the cowl panel nuts and pull out the engine wiring harness from the vehicle
10. Disconnect the front exhaust pipe from the exhaust manifold and catalytic converter.
11. If equipped with manual transmission, remove the shift lever assembly as follows:
    - Shift lever knob
    - 4 screws and shift lever boot
    - 6 bolts, shift lever assembly and baffle
12. Remove or disconnect the following:
    - Driveshaft
    - Speedometer cable from the transmission
    - Clutch release cylinder, if equipped with manual transmission
    - Cross-shaft, if equipped with automatic transmission
    - Wires at the starter
13. Position a jack and wooden block under the transmission and remove the rear engine mounting bracket.
14. Attach an engine hoist to the engine hangers.
15. Remove or disconnect the following:
    - Nuts and bolts from the engine mounts
    - Engine/transmission assembly
To install:
16. Install or connect the following:
    - Engine/transmission assembly, by keeping the engine level, while aligning the engine mounts
    - Engine mount fasteners but do not fully tighten them
17. Position a jack and wooden block under the transmission.
18. Install the rear engine mounting bracket. Tighten the frame bolts to 43 ft. lbs.

(58 Nm) and the mount bolts to 13 ft. lbs. (18 Nm).

19. Remove the jack and engine hoist.
20. Install or connect the following:
    - Tighten the engine mounts to 28 ft. lbs. (38 Nm).
    - Starter wires to the starter
    - Clutch release cylinder, if equipped with manual transmission
    - Cross-shaft, if equipped with automatic transmission
    - Speedometer to the transmission
    - Driveshaft
21. If equipped with manual transmission, install the shift lever assembly as follows:
    - Baffle and shift lever assembly with the 6 bolts
    - Shift lever boot with the 4 screws
    - Shift lever knob
    - Front exhaust pipe to the exhaust manifold
    - All wires and connectors
    - Cowl side trim and clip
    - Front door scuff plate
    - Alternator wires to the alternator
22. Install the power steering pump as follows:
    - Power steering pump to the bracket with the 2 bolts. Tighten the bolts to 43 ft. lbs. (58 Nm).
    - Power steering pulley with the nut. Tighten the nut to 32 ft. lbs. (43 Nm).
23. Install or connect the following:
    - All hoses
    - Compressor, if equipped with air conditioning
    - Intake air connector. Tighten the 2 bolts to 13 ft. lbs. (18 Nm).
    - Water pump pulley, fan shroud, fan (with fan clutch) and alternator drive belt.
    - Drive belt, if equipped with air conditioning
    - Power steering drive belt
    - Accelerator cable to the throttle body, if equipped with manual transmission
    - Accelerator and throttle cables to the throttle body, if equipped with automatic transmission
    - Air cleaner case
    - MAF meter, resonator and air cleaner cap
    - Radiator with the support tabs through the radiator service holes. Tighten the bolts to 108 inch lbs. (13 Nm).
    - Lower radiator hose to the radiator
    - Oil cooler hoses to the radiator, if equipped with an automatic transmission

- No. 2 fan shroud.
- Radiator reservoir hose to the radiator
- Upper radiator hose to the radiator
- Air pipe
- Radiator grille to the vehicle with the 11 clips
- Both fillers
- Clearance lights to the grille with the 4 bolts and 2 clips
- Both battery cables

24. Fill the engine oil, engine coolant and transmission oil.
25. Start the engine and check for leaks.
26. Check ignition timing.
27. Install or connect the following:
    - Engine undercover
    - Hood
28. Roadtest the vehicle and check all fluids.

2.7L (3RZ-FE/M) Engine

1. Before servicing the vehicle, refer to the Precautions Section.
2. Properly relieve the fuel system pressure.
3. Turn the ignition switch **OFF**.
4. Remove or disconnect the following:
    - Battery cables; negative cable first
    - Hood
    - Engine undercover
5. Drain the engine oil, transmission oil and cooling system.
6. Remove or disconnect the following:
    - Radiator
    - Idler pulley and drive belt, if equipped with power steering
    - Idler pulley nut/bolt and drive belt, if equipped with air conditioning
    - Alternator drive belt, fan (with fan clutch), water pump pulley and fan shroud
    - Accelerator cable from the throttle body, if equipped with a manual transaxle
    - Accelerator and throttle cables from the throttle body, if equipped with a automatic transaxle
    - Actuator cover and the cruise control cable from the actuator, if equipped with cruise control
    - Air cleaner cap
    - Mass Air Flow (MAF) meter and resonator
    - Air cleaner case
    - Intake air connector
    - Air conditioning compressor and bracket, if equipped with air conditioning
    - Alternator wires
    - Heater hoses at the cowl panel

7. Disconnect the following hoses:
    - Brake booster vacuum hose
    - Evaporative Emissions (EVAP) hose
    - Vacuum hose, if equipped with 4WD with Automatic Disconnecting Differential (ADD)
    - Both power steering hoses, if equipped with power steering
    - Fuel return hose
    - Fuel inlet hose
8. Remove the power steering pump as follows:
    - Nut and power steering pulley
    - Power steering pump
9. Disconnect the Engine Control Module (ECM) wiring from the ECM as follows:
    - 4 screws to the right front door scuff plate
    - Scuff plate
    - Cowl panel side trim by removing the clip
    - 4 ECM electrical connectors
10. Detach the engine wiring harness and connectors from the vehicle as follows:
    - Igniter
    - Ground strap from the cowl top panel
    - 2 engine wiring harness clamps
    - Engine wiring harness retainer to cowl panel nut and wiring harness
11. Disconnect the front exhaust pipe from the exhaust manifold and catalytic converter.
12. If equipped with manual transmission, remove the shift lever assembly as follows:
13. Remove or disconnect the following:
    - Shift lever knob
    - 4 screws and shift lever boot
    - 6 bolts, shift lever assembly and baffle
14. Remove or disconnect the following:
    - Driveshaft
    - Speedometer cable from the transmission
    - Clutch release cylinder, if equipped with manual transmission
    - Cross-shaft, If equipped with automatic transmission
    - Wires at the starter
15. Position a jack and wooden block under the transmission.
16. Remove the rear engine mounting bracket.
17. Attach an engine hoist to the engine hangers.
18. Remove or disconnect the following:
    - Nuts and bolts from the engine mounts
    - Engine/transmission

To install:

19. Attach the engine hoist to the engine hangers.

20. Install or connect the following:
- Engine/transmission assembly

➡**Keep the engine level, while aligning the engine mounts.**

- Engine mount fasteners but do not fully tighten them
- Position a jack and wooden block under the transmission
- Rear engine mounting bracket. Tighten the frame bolts to 19 ft. lbs. (26 Nm) and the mount bolts to 13 ft. lbs. (18 Nm).

21. Remove the jack and engine hoist.

22. Install or connect the following:
- Tighten the engine mounts to 28 ft. lbs. (38 Nm).
- Starter wires
- Clutch release cylinder, if equipped with manual transmission
- Cross-shaft, if equipped with automatic transmission
- Speedometer to the transmission
- Driveshaft

23. If equipped with manual transmission, install the shift lever assembly as follows:
- Baffle and shift lever assembly with the 6 bolts
- Shift lever boot with the 4 screws
- Shift lever knob

24. Install or connect the following:
- Front exhaust pipe to the exhaust manifold
- Engine wiring harness
- Cowl side trim and clip
- Front door scuff plate
- Alternator wires

25. Install the power steering pump as follows:
- Power steering pump to the bracket. Tighten the 2 bolts to 43 ft. lbs. (58 Nm).
- Power steering pulley. Tighten the nut to 32 ft. lbs. (43 Nm).

26. Install or connect the following:
- All hoses
- Heater hoses at the cowl panel
- Compressor, if equipped with air conditioning
- Intake air connector. Tighten the 2 bolts to 13 ft. lbs. (18 Nm).

27. Install the water pump pulley, fan shroud, fan (with fan clutch) and alternator drive belt as follows:
- Fan (with the fan clutch), water pump pulley and fan shroud in position
- Water pump pulley but do not tighten the nuts

- Alternator drive belt
- Stretch the alternator belt tight. Tighten the fan nuts to 16 ft. lbs. (21 Nm).
- Adjust the alternator drive belt

28. Install or connect the following:
- Adjust the drive belt, if equipped with air conditioning
- Adjust the power steering drive belt
- Accelerator cable to the throttle body, if equipped with manual transmission
- Accelerator and throttle cables to the throttle body, if equipped with automatic transmission
- Air cleaner case
- MAF meter, resonator and air cleaner cap
- Radiator with the tabs on the supports through the radiator service holes. Tighten the bolts to 108 inch lbs. (13 Nm).
- Lower radiator hose to the radiator
- Oil cooler hoses to the radiator, if equipped with automatic transmission
- No. 2 fan shroud
- Radiator reservoir hose to the radiator
- Upper radiator hose to the radiator
- Air pipe with the 2 bolts, if removed
- Radiator grille with the 11 clips
- 2 fillers
- Clearance lights to the grille with the 4 bolts and 2 clips
- Negative and positive cables to the battery

29. Fill the engine oil, engine coolant and transmission oil.

30. Start the engine and check for leaks.

31. Check ignition timing.

32. Install or connect the following:
- Engine undercover
- Hood

33. Road test the vehicle and check all fluids.

2.7L (2TR-FE/X) Engine

1. Before servicing the vehicle, refer to the Precautions Section.

2. Properly disable the SRS system.

3. Properly relieve the fuel system pressure.

4. Disconnect the negative battery cable. Disconnect the positive battery cable.

5. Matchmark and remove the hood.

6. On 4WD and PreRunner, remove the number one engine cover.

7. On 4WD and PreRunner (regular cab) remove the number two engine cover.

8. Drain the engine coolant. Remove the battery. Remove the battery tray. Drain the engine oil.

9. Remove the radiator support to frame seal, left side.

10. Remove the fan shroud. Remove the air cleaner cap sub assembly. Remove the air cleaner filter element subassembly. Remove the air cleaner case.

11. Separate the vane pump. Remove the radiator hoses. Separate the air conditioning compressor from the engine and position it to the side.

12. Disconnect the heater hoses from the heater core.

13. Disconnect the fuel line hoses. Separate the fuel vapor feed hose assembly. Disconnect the number one air injection hose.

14. Disconnect all engine wiring harnesses and connectors in the engine compartment.

➡**Disconnect the connector from the ECM, then pull the engine wire harness to the compartment side. Do not forcibly pull the wire harness to the engine compartment side.**

15. Remove the exhaust pipe assembly. Remove the front exhaust pipe assembly.

16. Remove the transmission.

17. Install engine hanger, tool 12281-75040 and retainer bolts 91552-A1020. Attach a suitable engine lifting fixture.

18. Remove the engine retaining bolts and mounts.

➡**Check to be sure nothing will interfere with the removal of the engine. Check to be sure all required wires, hoses and connectors are disconnected.**

19. Remove the engine from the vehicle.

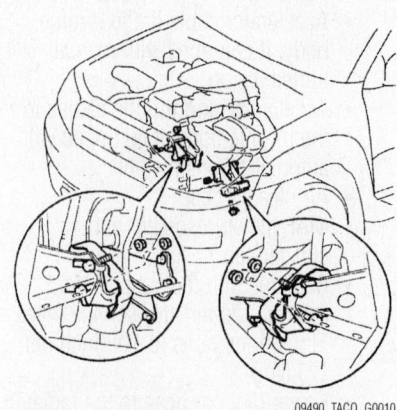

09490_TACO_G0010

Engine retaining bolt locations—2.7L (2TR-FE/X) engine

To install:

20. Installation is the reverse of the removal procedure.

21. Tighten the four mounting insulator bolts to 28 ft. lbs.

22. Be sure to fill the cooling system with the proper grade and type engine coolant.

23. Be sure to fill the engine with the proper grade and type engine oil.

24. Check ignition timing. Check idle speed.

25. Start the engine and check for leaks, correct as required.

3.4L Engine

2WD

1. Before servicing the vehicle, refer to the Precautions Section.

2. Properly relieve the fuel system pressure.

3. Remove or disconnect the following:
   - Hood
   - Battery
   - Engine undercovers

4. Drain the engine coolant.

5. Drain the engine oil.

6. Remove or disconnect the following:
   - Radiator

7. Remove the power steering drive belt as follows:
   - Stretch the belt and loosen the fan pulley mounting nuts
   - Loosen the lockbolt, pivot bolt and adjusting bolt and remove the drive belt

8. Remove or disconnect the following:
   - Air conditioning drive belt by loosening the idle pulley nut and adjusting bolt, if equipped with air conditioning
   - Loosen the lockbolt, pivot bolt and adjusting bolt and the alternator drive belt
   - Fan with the fluid coupling and fan pulleys
   - Power steering pump, do not disconnect the lines from the pump
   - Compressor, if equipped with air conditioning; do not disconnect the lines from the compressor
   - Air cleaner cap
   - Mass Air Flow (MAF) meter and resonator
   - Air cleaner case and filter

9. Disconnect the following cables:
   - Actuator cable with the bracket, if equipped with cruise control
   - Accelerator cable
   - Throttle cable, if equipped with an automatic transmission

10. Disconnect the following hoses:
    - Heater hoses
    - Brake booster vacuum hose
    - Evaporative Emission (EVAP) hose
    - Fuel return hose
    - Fuel inlet hose

11. Detach the starter wire and connectors as follows:
    - Ground strap by removing the bolt
    - Positive cable from the battery
    - 3 starter wire clamps and connector

12. Detach the alternator connector and wire.

13. Detach the engine wiring harness and connectors as follows:
    - Right front door scuff plate.
    - Cowl panel side trim, by removing the clip
    - Engine Control Module (ECM)
    - 2 connectors from the cowl wire
    - Igniter
    - Ground strap
    - 6 engine wiring harness clamps
    - Engine wiring harness

14. If equipped with manual transmission, remove the shift lever assembly as follows:
    - Shift lever knob
    - 4 screws and the shift lever boot
    - Shift lever assembly and gasket, by removing the 6 bolts

15. Remove or disconnect the following:
    - Stabilizer bar
    - Driveshaft from the transmission
    - Speedometer cable
    - Front exhaust pipe
    - Clutch release cylinder, if equipped with a manual transmission
    - Cross-shaft, if equipped with an automatic transmission

16. Place a jack under the transmission.

17. Remove or disconnect the following:
    - Transmission rear mounting bracket, by removing the 8 bolts
    - Air conditioning compressor wire clamp, if equipped with air conditioning

18. If necessary, install a No. 2 engine hanger with 2 bolts. Tighten the 2 bolts to 30 ft. lbs. (40 Nm).

19. Attach the engine hoist chain to the 2 engine hangers.

20. Remove or disconnect the following:
    - 4 bolts/nuts holding the engine front mounting insulators to the frame
    - Engine/transmission assembly

To install:

21. Install or connect the following:
    - Engine
    - Engine mounts to the body mountings. Install the bolts and nuts but do not tighten at this time.

22. Remove the engine chain hoist the No. 2 engine hanger.

23. Install or connect the following:
    - Air conditioning wire, if equipped with air conditioning
    - Transmission mounting bracket. Tighten the frame bolts to 43 ft. lbs. (58 Nm) and the mounting insulator bolts to 13 ft. lbs. (18 Nm).
    - Tighten the engine mounting nuts and bolts to 28 ft. lbs. (38 Nm).
    - Cross-shaft, if equipped with an automatic transmission
    - Clutch release cylinder, if equipped with a manual transmission. Tighten the bolts to 108 inch lbs. (13 Nm).
    - Front exhaust pipe
    - Speedometer cable
    - Driveshaft
    - Stabilizer bar

24. Install the shift lever assembly, as follows:

25. Install or connect the following:
    - New gasket and shift lever assembly with the 6 bolts
    - Shift lever boot with the 4 screws
    - Shift lever knob

26. Install or connect the following:
    - All engine wiring harness, hoses and cables
    - Air cleaner case and air filter
    - MAF meter, resonator and air cleaner cap
    - Air conditioning compressor, if equipped
    - Remaining components

27. Fill the engine with oil.

28. Fill the engine and radiator with coolant.

29. Install the engine undercover.

30. Start the engine and check for leaks.

4WD

1. Before servicing the vehicle, refer to the Precautions Section.

2. Properly relieve the fuel system pressure.

3. Remove or disconnect the following:
   - Transmission
   - Hood
   - Battery from the vehicle
   - Engine undercovers

4. Drain the engine coolant.

5. Drain the engine oil.

6. Remove or disconnect the following:
   - Radiator

7. Remove the power steering drive belt, as follows:
   - Stretch the belt and loosen the fan pulley mounting nuts

- Loosen the lockbolt, pivot bolt and adjusting bolt
- Drive belt from the engine

8. Remove or disconnect the following:
- Air conditioning drive belt by loosening the idle pulley nut and adjusting bolt, if equipped with air conditioning
- Alternator drive belt
- Fan with the fluid coupling and fan pulleys
- Power steering pump; do not disconnect the lines from the pump
- Compressor, if equipped with air conditioning. Do not disconnect the lines from the compressor.
- Air cleaner cap, Mass Air Flow Meter (MAF) meter and resonator
- Air cleaner case and filter

9. Disconnect the following cables:
- Actuator cable with the bracket, if equipped with cruise control
- Accelerator cable
- Throttle cable, if equipped with an automatic transmission

10. Disconnect the following hoses:
- Heater hoses
- Brake booster vacuum hose
- Evaporative Emissions (EVAP) hose
- Automatic Disconnecting Differential (ADD) vacuum hose
- Fuel return hose
- Fuel inlet hose

11. Detach the starter wire and connectors as follows:

12. Remove or disconnect the following:
- Ground strap by removing the bolt

13. Disconnect the positive cable from the battery, as follows:
- 3 starter wire clamps and connector
- Automatic Disconnecting Differential (ADD) indicator switch connector

14. Detach the alternator connector and wire.

15. Detach the engine wiring harness and connectors, as follows:
- Right front door scuff plate
- Cowl panel side trim, by removing the clip
- Engine Control Module (ECM)
- 2 connectors from the cowl wire
- Igniter connector
- Ground strap
- 6 engine wiring harness clamps
- Engine wiring harness

16. Remove or disconnect the following:
- Air conditioning compressor wire clamp, if equipped with air conditioning

17. If necessary, install a No. 2 engine hanger with 2 bolts. Tighten the 2 bolts to 30 ft. lbs. (40 Nm).

18. Attach the engine hoist chain to the 2 engine hangers.

19. Remove or disconnect the following:
- 4 Engine front mounting insulators-to-frame bolts/nuts
- Engine

To install:
20. Install or connect the following:
- Engine
- Engine mounts to the body mountings. Install the bolts and nuts but do not tighten at this time.

21. Remove the engine chain hoist the No. 2 engine hanger.

22. Install or connect the following:
- Air conditioning wire with the bolt, if equipped with air conditioning
- Tighten the engine mounting nuts and bolts to 28 ft. lbs. (38 Nm).
- All engine wiring harness, hoses and cables
- Air cleaner case and air filter
- MAF meter, resonator and air cleaner cap
- Air conditioning compressor, if equipped
- Fan with the fluid coupling and fan pulleys. Tighten the nuts to 48 inch lbs. (5.4 Nm).
- Alternator drive belt
- Adjust the air conditioning drive belt, if equipped
- Power steering pump, pump pulley and the drive belt
- Radiator

23. Fill the engine with oil.
24. Fill the engine and radiator with coolant.
25. Install or connect the following:
- Hood
- Engine undercover
- Transmission

26. Start the engine and check for leaks.

4.0L Engine

1. Before servicing the vehicle, refer to the Precautions Section.
2. Properly disable the SRS system.
3. Properly relieve the fuel system pressure.
4. Disconnect the negative battery cable. Disconnect the positive battery cable.
5. Disconnect the windshield washer hose. Matchmark and remove the hood.
6. Drain the engine coolant. Remove the battery. Remove the battery tray. Drain the engine oil.
7. Remove the radiator support to frame seal, left side.
8. Remove the fan shroud. Remove the radiator hoses. Remove the radiator.

9. Remove the V bank cover.
10. Remove the transmission. If equipped with manual transmission, remove the clutch cover and clutch.
11. Remove the fan pulley. Remove the vane pump assembly. Remove the alternator.
12. Separate the air conditioning compressor from the engine and position it to the side.
13. Disconnect the heater hoses from the heater core. Disconnect the fuel line hoses. Remove the intake air surge tank.
14. To separate the main engine wire harness, remove the glove box door. Remove the instrument panel finish panel subassembly, lower right side. Disconnect the ECM connectors. Disconnect the connectors from the ECU, if equipped with 4WD. Pull the harness into the engine compartment. Disconnect the front differential connector, if equipped with 4WD. Disconnect the three connectors from the engine compartment relay block. Separate the engine wire from the engine compartment relay block.
15. Disconnect the ground cable from the cylinder head.
16. Remove the front exhaust pipe assembly.
17. Install engine hanger, tool 12281-31070 and 12282-31050 and retainer bolts 90119-08177. Attach a suitable engine lifting fixture.
18. Remove the engine retaining bolts and mounts.

➡**Check to be sure nothing will interfere with the removal of the engine. Check to be sure all required wires, hoses and connectors are disconnected.**

19. Remove the engine from the vehicle.
To install:
20. Installation is the reverse of the removal procedure.

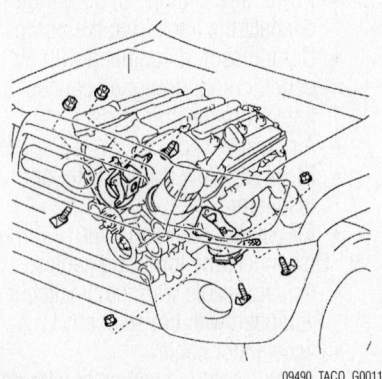

09490_TACO_G0011

Engine retaining bolt locations—4.0L engine

21. Tighten the four mounting insulator bolts to 28 ft. lbs.

22. If equipped with automatic transmission, perform the automatic transmission initialization procedure listed below.

➡**Perform the automatic transmission initialization procedure when replacing the automatic transmission, engine or ECM.**

23. On the A340E automatic transmission, initialization is completed by connecting and disconnecting the negative battery cable.

24. On the A750E automatic transmission, initialization is performed using the intelligent tester tool.

25. Turn the ignition switch OFF.

26. Connect the intelligent tester tool together with the controller area network vehicle interface module (CAN VIN) to the DLC3.

27. Turn the ignition switch to the ON position.

28. Push the intelligent tester tool main switch to the ON position.

29. Select the items, DIAGNOSIS/ENHANCED OBD II.

30. Perform the reset memory procedure from the ENGINE menu.

➡**After performing the reset memory, be sure to perform the roadtest procedure. For roadtest procedure information, refer to the intelligent tester instruction manual.**

31. Be sure to fill the cooling system with the proper grade and type engine coolant.

32. Be sure to fill the engine with the proper grade and type engine oil.

33. Check ignition timing. Check idle speed.

34. Start the engine and check for leaks, correct as required.

Heater Core

REMOVAL & INSTALLATION

2002–2004

1. Before servicing the vehicle, refer to the Precautions Section.

2. Disconnect the negative battery cable.

✽✽ CAUTION

After the negative battery cable has been disconnected, wait at least 1½ minutes for the air bag module to deplete its energy.

3. Drain the cooling system into a clean container for reuse.

4. Disconnect the heater hoses from the heater core.

5. Remove the steering wheel by performing the following procedure:

a. Position the front wheels in the straight-ahead position.

b. At both sides of the steering wheel, remove the side covers.

c. Using a Torx® wrench, loosen the steering wheel Torx® screws until the screw's circumference ring catches on the screw case.

d. Carefully, lift the air bag module, disconnect the electrical connector and remove the air bag.

✽✽ CAUTION

Place the air bag module in a safe location with the front facing upward.

e. Remove the steering wheel nut.

f. Using a steering wheel puller, press the steering wheel from the steering column.

6. Remove the instrument panel and reinforcement by performing the following procedure:

a. Remove the steering column cover screws and the covers.

b. Remove the combination switch-to-steering column screws and the combination switch.

c. Remove the 2 hood lock release lever screws and the hood lock release lever.

d. Remove the fuse box opening cover.

e. Remove the 4 lower left side finish panel bolts, the screw and the panel.

f. If equipped, remove the 2 rear console box bolts/screws and the rear console box.

g. Remove the front console box.

h. Remove the 2 upper console box mounting bracket screws and the bracket.

i. Remove the 2 lower center cover clips and the cover.

j. Remove the No. 2 heater-to-register duct screw and the duct.

k. Remove the heater control knobs.

l. Remove the heater control panel.

m. Remove the 2 center cluster finish panel screws, disengage the 5 clips, disconnect the electrical connectors and remove the panel.

n. Pry out the cigar lighter hole bezel.

o. Remove the front ashtray and the center cluster finish panel retainer.

p. Pry out the starter switch bezel.

q. Remove the 3 cluster finish panel screws and the panel.

r. Remove the radio and stereo opening cover.

s. Remove the combination meter and disconnect the electrical connectors.

t. Remove the 2 No. 1 register screws and the register.

u. Remove the No. 1 heater-to-register duct screw and the duct.

v. If equipped with a column shifter, disconnect the transmission control cable from the steering column.

w. Remove the steering column-to-instrument panel nuts/bolts and the lower steering column joint bolt; then, carefully, remove the steering column.

x. Remove the 2 glove compartment door screws and the door.

y. Remove the 3 glove compartment reinforcement screws and the reinforcement.

z. Remove the No. 4 heater-to-register duct.

aa. Disconnect the No. 1 undercover and disconnect the passenger's side air bag module electrical connector.

bb. Remove the 3 lower No. 2 finish panel screws and the panel.

cc. Remove the heater control screws.

dd. Remove the lower center finish panel.

ee. Remove the instrument panel-to-chassis nuts/bolts and the instrument panel.

ff. Remove the 3 No. 1 brace bolts and the brace.

gg. Remove the 2 No. 2 brace bolts and the brace.

hh. Remove the center heater-to-register duct.

ii. Remove the defroster nozzle.

jj. Remove the reinforcement-to-chassis 3 bolts, 4 nuts and the reinforcement.

7. Remove the defroster nozzle and the heater-to-register duct.

8. Remove the evaporator housing by performing the following procedure:

a. Discharge and recover the air conditioning system refrigerant.

b. Disconnect the refrigerant lines from the evaporator core. Discard the O-rings and plug the openings to prevent contamination.

c. Remove the 2 grommets and the drain pipe grommet.

d. Disconnect the electrical connectors.

e. Remove the 3 evaporator housing-to-chassis screws, the bolt and the housing.

9. Remove the 2 heater housing-to-

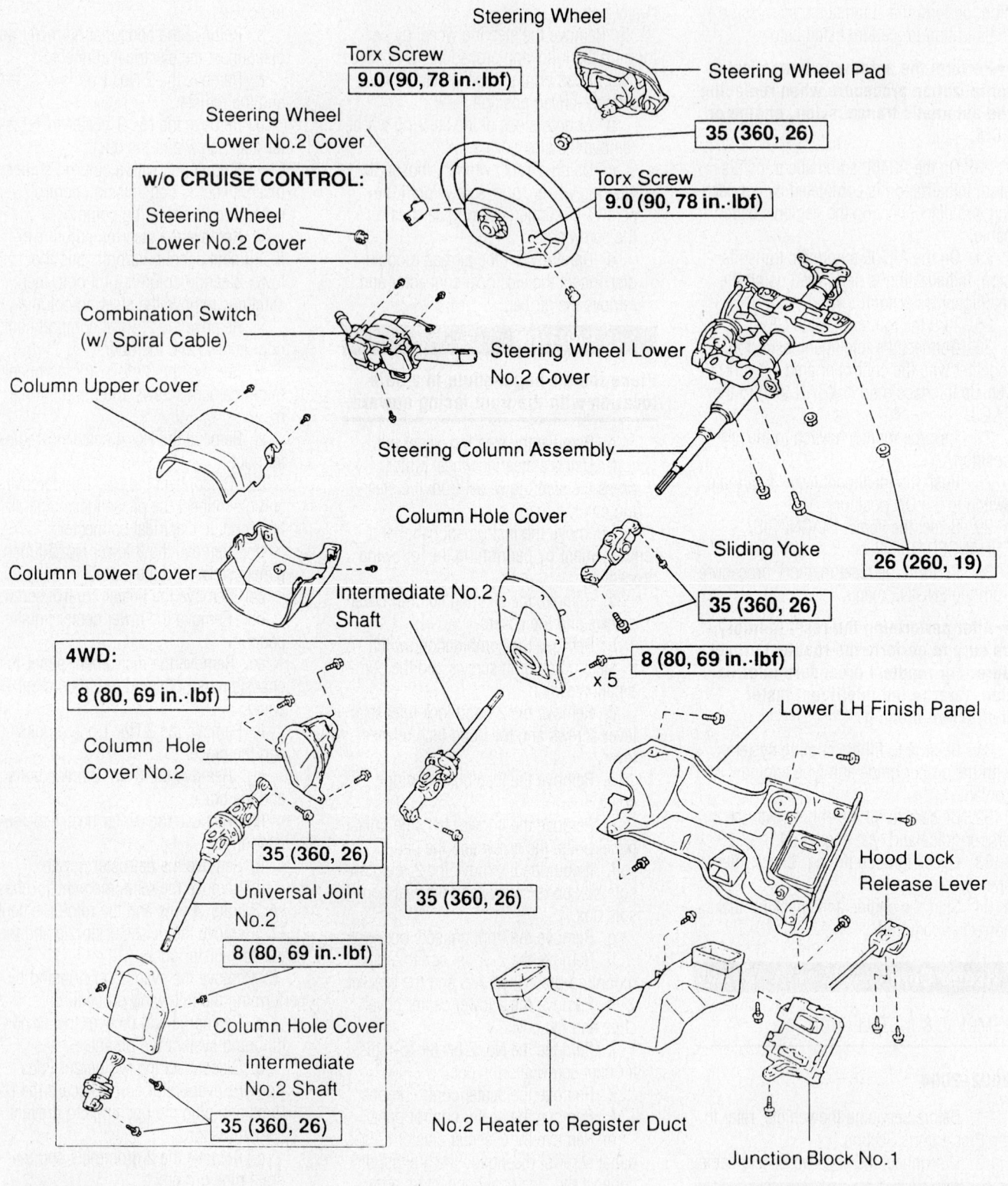

Steering Wheel

Torx Screw
9.0 (90, 78 in.·lbf)

Steering Wheel Pad

35 (360, 26)

Steering Wheel
Lower No.2 Cover

Torx Screw
9.0 (90, 78 in.·lbf)

w/o CRUISE CONTROL:

Steering Wheel
Lower No.2 Cover

Combination Switch
(w/ Spiral Cable)

Steering Wheel Lower
No.2 Cover

Column Upper Cover

Steering Column Assembly

Column Hole Cover

Sliding Yoke

26 (260, 19)

Column Lower Cover

Intermediate No.2
Shaft

35 (360, 26)

8 (80, 69 in.·lbf)

x 5

Lower LH Finish Panel

4WD:

8 (80, 69 in.·lbf)

Column Hole
Cover No.2

35 (360, 26)

Universal Joint
No.2

8 (80, 69 in.·lbf)

35 (360, 26)

Hood Lock
Release Lever

Column Hole Cover

Intermediate
No.2 Shaft

35 (360, 26)

No.2 Heater to Register Duct

Junction Block No.1

N·m (kgf·cm, ft·lbf) : Specified torque

Floor shift steering column and related components—2002–2004

93113GI9

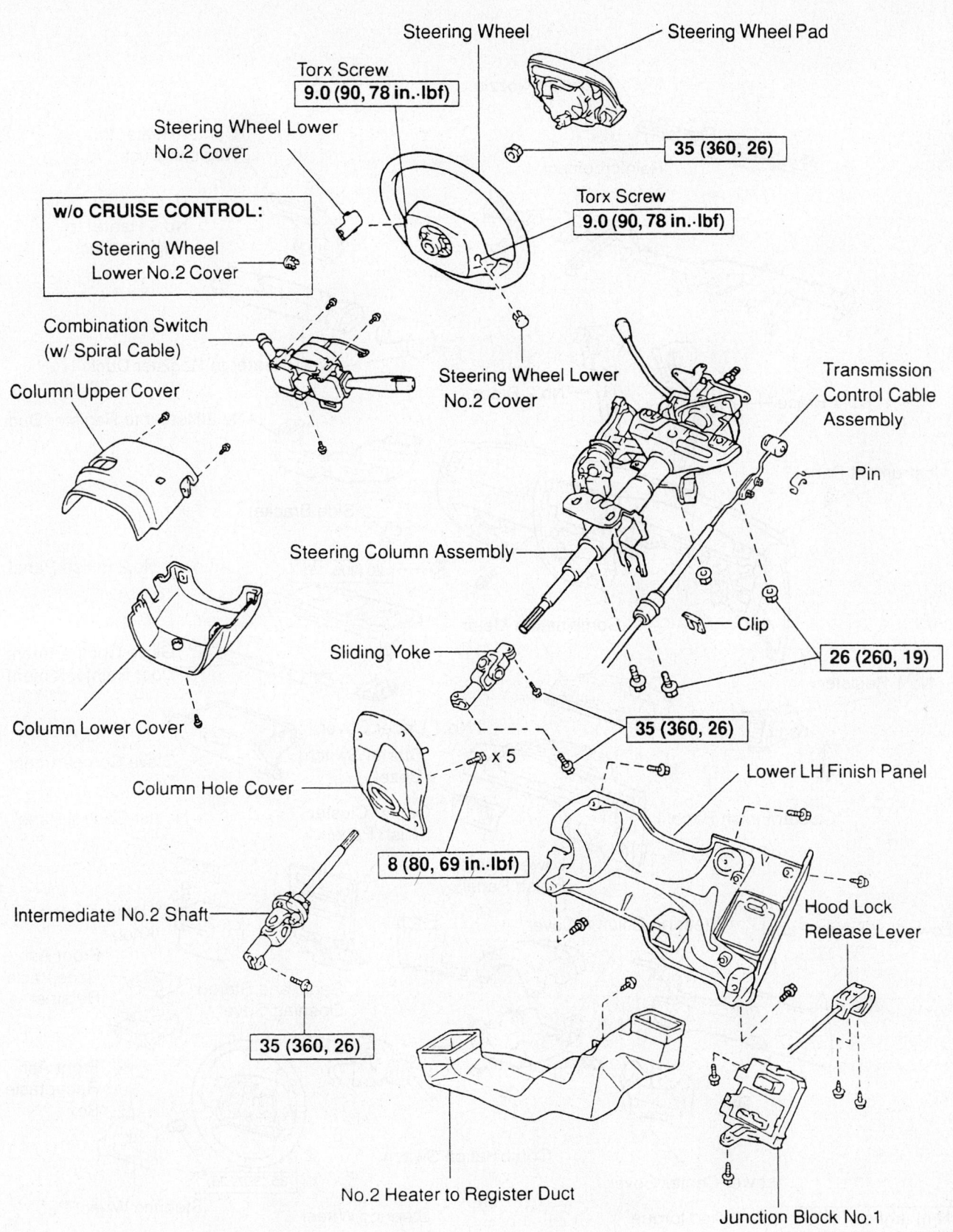

Steering Wheel

Steering Wheel Pad

Torx Screw
9.0 (90, 78 in.·lbf)

Steering Wheel Lower
No.2 Cover

35 (360, 26)

w/o CRUISE CONTROL:

Steering Wheel
Lower No.2 Cover

Torx Screw
9.0 (90, 78 in.·lbf)

Combination Switch
(w/ Spiral Cable)

Steering Wheel Lower
No.2 Cover

Transmission
Control Cable
Assembly

Column Upper Cover

Pin

Steering Column Assembly

Clip

Column Lower Cover

Sliding Yoke

26 (260, 19)

Column Hole Cover

35 (360, 26)

Lower LH Finish Panel

× 5

8 (80, 69 in.·lbf)

Intermediate No.2 Shaft

Hood Lock
Release Lever

35 (360, 26)

No.2 Heater to Register Duct

Junction Block No.1

N·m (kgf·cm, ft·lbf) : Specified torque

Column shift steering column and related components—2002–2004

93113GI0

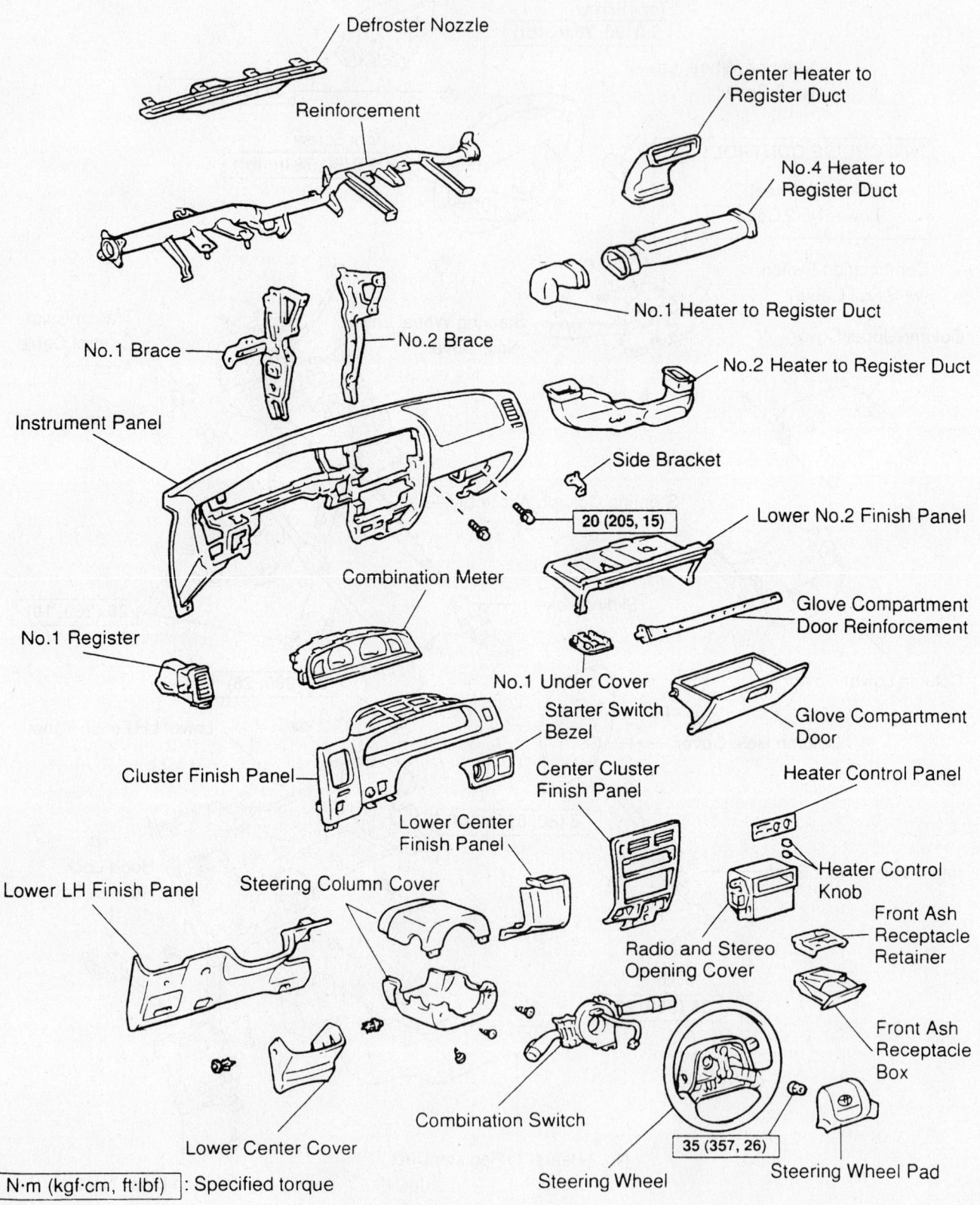

Defroster Nozzle

Center Heater to Register Duct

Reinforcement

No.4 Heater to Register Duct

No.1 Heater to Register Duct

No.1 Brace

No.2 Brace

No.2 Heater to Register Duct

Instrument Panel

Side Bracket

20 (205, 15)

Lower No.2 Finish Panel

Combination Meter

Glove Compartment Door Reinforcement

No.1 Register

No.1 Under Cover

Starter Switch Bezel

Glove Compartment Door

Cluster Finish Panel

Center Cluster Finish Panel

Heater Control Panel

Lower Center Finish Panel

Heater Control Knob

Lower LH Finish Panel

Steering Column Cover

Front Ash Receptacle Retainer

Radio and Stereo Opening Cover

Front Ash Receptacle Box

Combination Switch

Lower Center Cover

35 (357, 26)

Steering Wheel

Steering Wheel Pad

N·m (kgf·cm, ft·lbf) : Specified torque

Instrument panel and related components—2002–2004

93113GJ1A

chassis bolts, the nut and the heater housing.

10. Remove the 3 heater core-to-heater housing screws, the 2 plates and clamp.

11. Remove the heater core from the heater housing.

To install:

12. Install the heater core to the heater housing.

13. Install the 3 heater core-to-heater housing screws, the 2 plates and clamp.

14. Install the heater housing, the nut and the 2 heater housing to-chassis bolts.

15. Install the evaporator housing by performing the following procedure:

a. Install the evaporator housing, the bolt and the 3 housing-to-chassis screws.

b. Connect the electrical connectors.

c. Install the 2 grommets and the drain pipe grommet.

d. Using new O-rings, connect the refrigerant lines to the evaporator core.

16. Install the defroster nozzle and the heater-to-register duct.

17. Install the instrument panel and reinforcement by performing the following procedure:

a. Install the reinforcement-to-chassis 3 bolts, 4 nuts and the reinforcement.

b. Install the defroster nozzle.

c. Install the center heater-to-register duct.

d. Install the No. 2 brace and the 2 brace bolts.

e. Install the No. 1 brace and the 3 brace bolts.

f. Install the instrument panel and the instrument panel-to-chassis nuts/bolts.

g. Install the lower center finish panel.

h. Install the heater control screws.

i. Install the lower No. 2 finish panel and the 3 panel screws.

j. Connect the passenger's side air

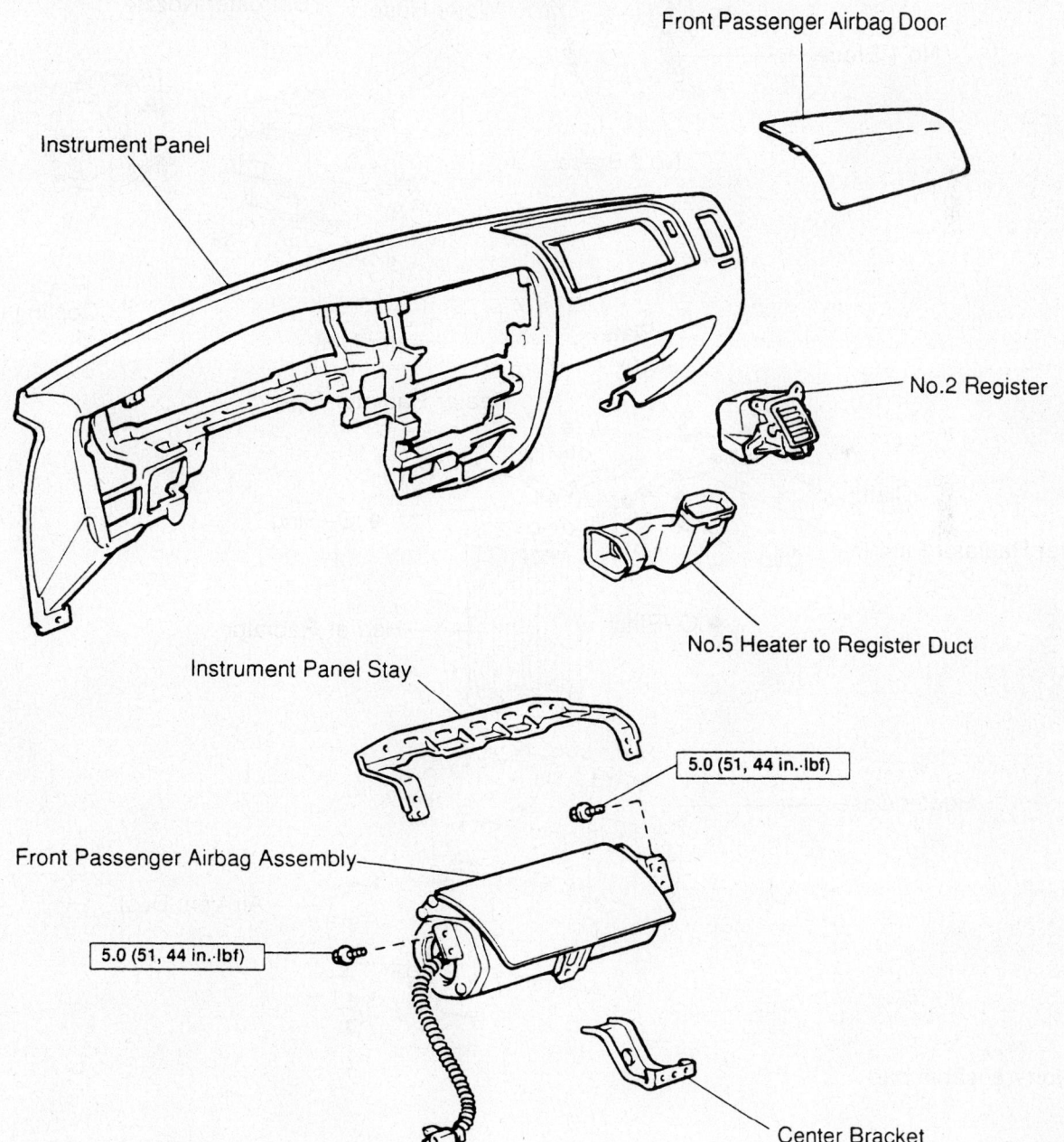

Front Passenger Airbag Door

Instrument Panel

No.2 Register

No.5 Heater to Register Duct

Instrument Panel Stay

5.0 (51, 44 in.·lbf)

Front Passenger Airbag Assembly

5.0 (51, 44 in.·lbf)

Center Bracket

93113GJ2A

Passenger side air bag module and related components—2002–2004

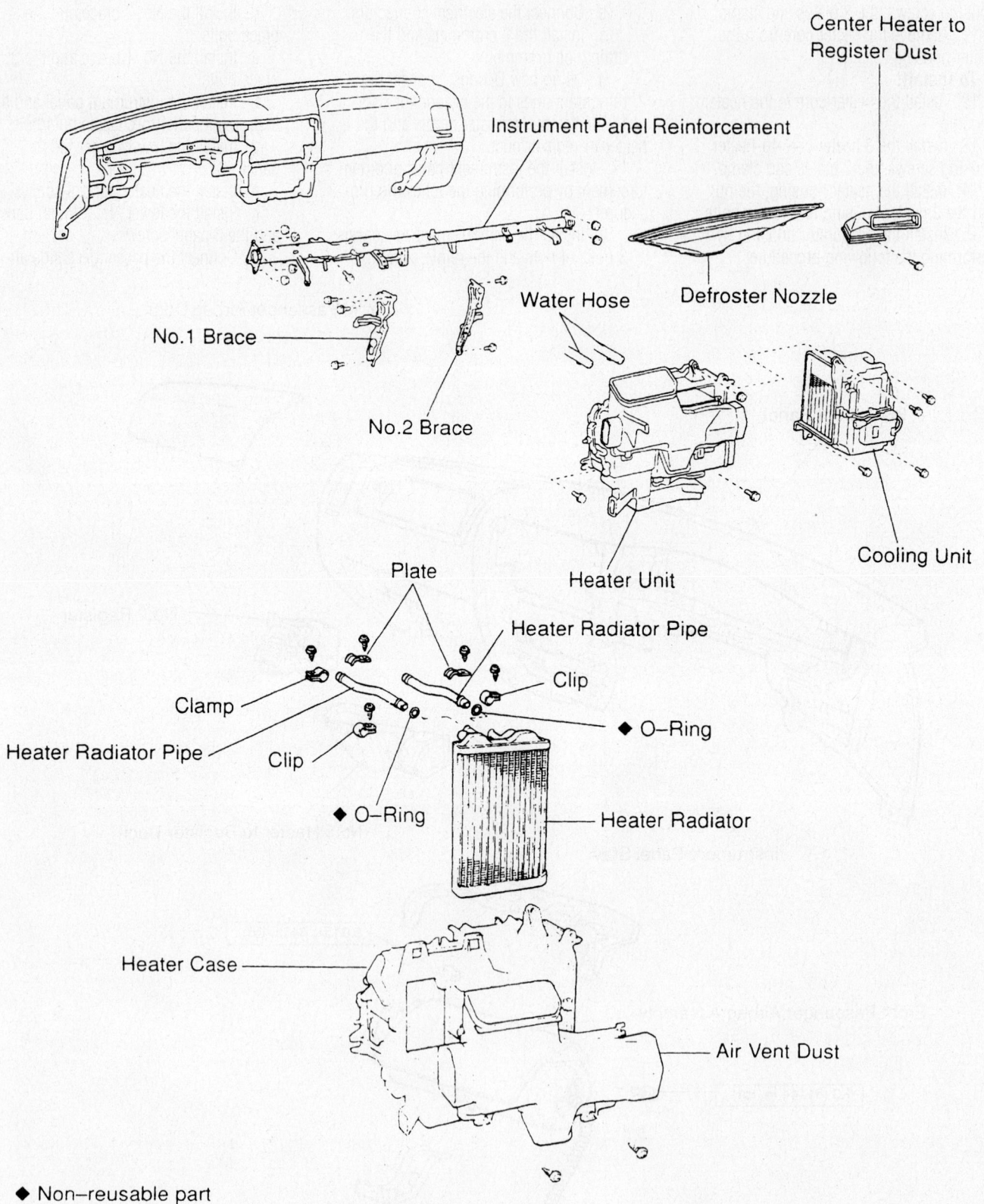

Instrument Panel Reinforcement

Center Heater to Register Dust

No.1 Brace

No.2 Brace

Water Hose

Defroster Nozzle

Heater Unit

Cooling Unit

Plate

Heater Radiator Pipe

Clip

Clamp

♦ O–Ring

Heater Radiator Pipe

Clip

♦ O–Ring

Heater Radiator

Heater Case

Air Vent Dust

♦ Non–reusable part

Heater core and related components—2002–2004

93113GJ3

bag module electrical connector and the No. 1 undercover.

k. Install the No. 4 heater-to-register duct.

l. Install the glove compartment reinforcement and the 3 reinforcement screws.

m. Install the glove compartment door and the 2 door screws.

n. Install the steering column and torque the steering column-to-instrument panel nuts/bolts to 19 ft. lbs. (26 Nm) and the lower steering column joint bolt to 26 ft. lbs. (35 Nm).

o. If equipped with a column shifter, connect the transmission control cable to the steering column.

p. Install the No. 1 heater-to-register duct and the duct screw.

q. Install the No. 1 register and the 2 register screws.

r. Install the combination meter and connect the electrical connectors.

s. Install the radio and stereo opening cover.

t. Install the cluster finish panel and the 3 panel screws.

u. Pry out the starter switch bezel.

v. Install the front ashtray and the center cluster finish panel retainer.

w. Install the cigar lighter hole bezel.

x. Install the center cluster finish panel, engage the 5 clips, connect the electrical connectors and install the 2 panel screws.

y. Install the heater control panel.

z. Install the heater control knobs.

aa. Install the No. 2 heater-to-register duct and the duct screw.

bb. Install the lower center cover and engage the 2 clips.

cc. Install the upper console box mounting bracket and the 2 bracket screws.

dd. Install the front console box.

ee. If equipped, install the rear console box and the 2 rear console box bolts/screws.

ff. Install the lower left side finish panel, the 4 bolts and the screw.

gg. Install the fuse box opening cover.

hh. Install the hood lock release lever and the 2 hood lock release lever screws.

ii. Install the combination switch-to-steering column and the combination switch screws.

jj. Install the steering column cover and the covers screws.

18. Install the steering wheel by performing the following procedure:

a. Install the steering wheel from the steering column.

b. Install the steering wheel nut and torque to 26 ft. lbs. (35 Nm).

c. Carefully, install the air bag module and connect the electrical connector.

d. Using a Torx® wrench, tighten the steering wheel screws to 78 inch lbs. (8.8 Nm).

e. At both sides of the steering wheel, install the side covers.

19. Connect the heater hoses to the heater core.

20. Refill the cooling system.

21. Connect the negative battery cable.

22. Evacuate and charge the air conditioning system refrigerant.

23. Run the engine to normal operating temperatures; then, check the climate control operation and check for leaks.

2005–2006

1. Before servicing the vehicle, refer to the Precautions Section.

2. Properly disarm the SRS system.

3. Disconnect the negative battery cable.

> ✳✳ **CAUTION**
>
> **After the negative battery cable has been disconnected, wait at least 1½ minutes for the air bag module to deplete its energy.**

4. Properly discharge the air conditioning system. Drain the engine coolant.

5. Disconnect and plug the air condition compressor refrigerant lines.

6. Disconnect the heater hoses at the heater core.

7. Remove the front windshield wiper arm head cap. Remove the wiper arms.

8. Remove the right and left front fender to cowl side seals. Remove the cowl top ventilator louver subassembly.

9. Position the front wheels in the straight ahead position.

10. Remove the lower number two steering wheel cover after disengaging the two claws using the proper tool.

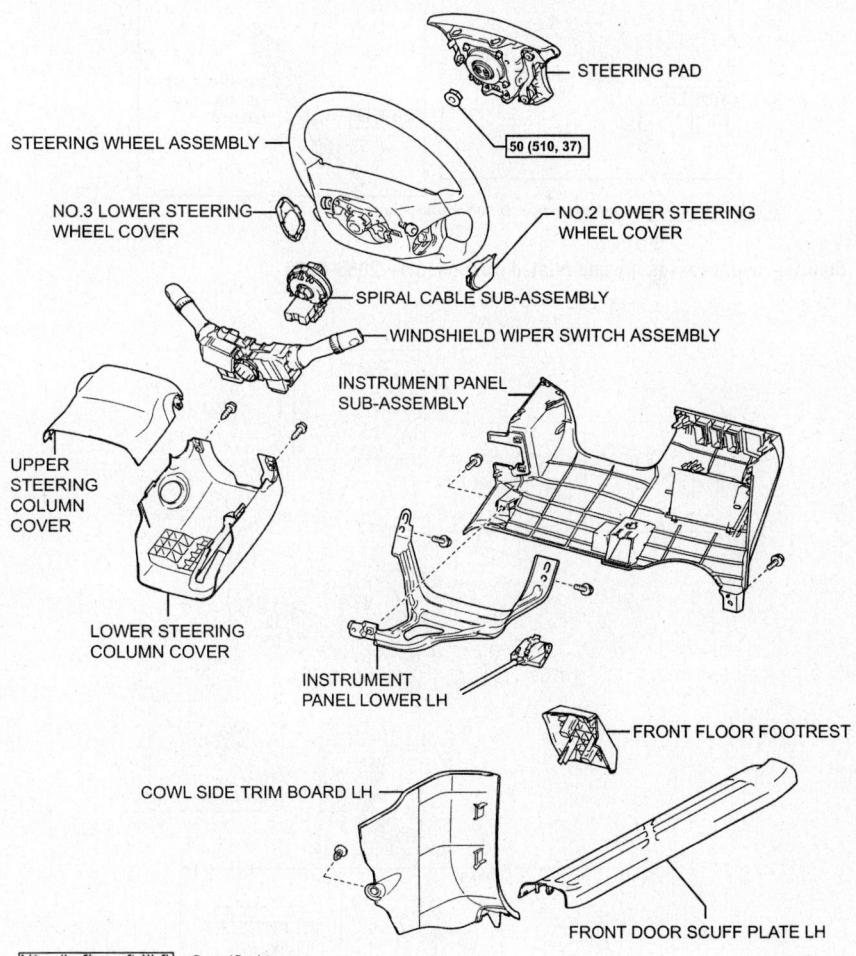

N*m (kgf*cm, ft.*lbf) : Specified torque

09490_TACO_G0014

Steering wheel and related components—2005–2006

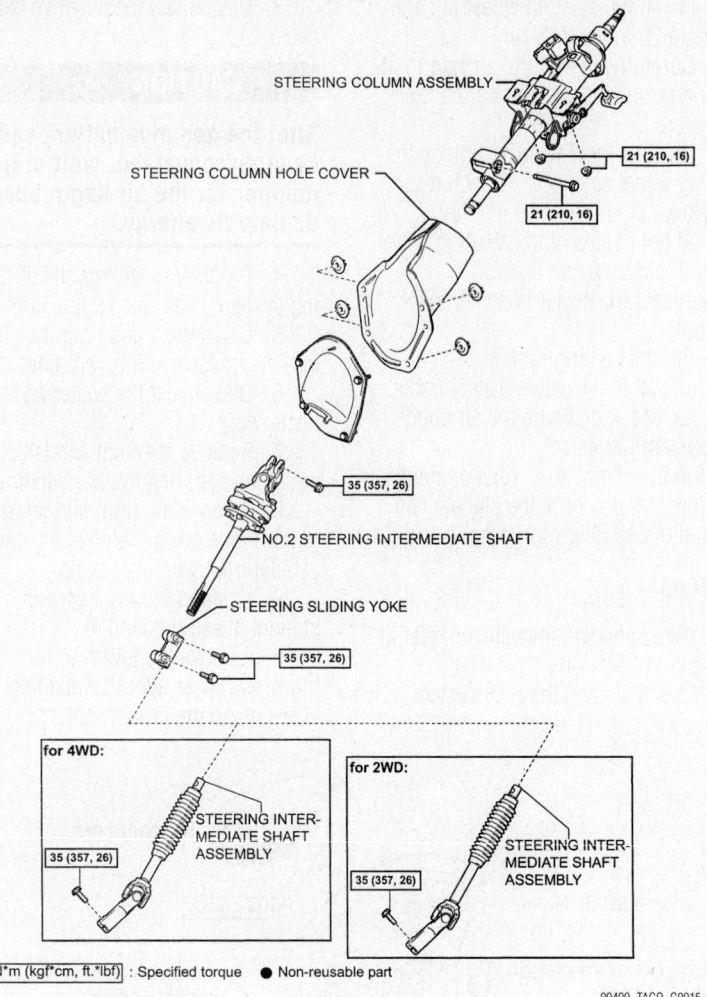

STEERING COLUMN ASSEMBLY

STEERING COLUMN HOLE COVER

21 (210, 16)

21 (210, 16)

35 (357, 26)

NO.2 STEERING INTERMEDIATE SHAFT

STEERING SLIDING YOKE

35 (357, 26)

for 4WD:

STEERING INTERMEDIATE SHAFT ASSEMBLY

35 (357, 26)

for 2WD:

STEERING INTERMEDIATE SHAFT ASSEMBLY

35 (357, 26)

N*m (kgf*cm, ft.*lbf) : Specified torque ● Non-reusable part

09490_TACO_G0015

Steering column assembly and related components—2005–2006

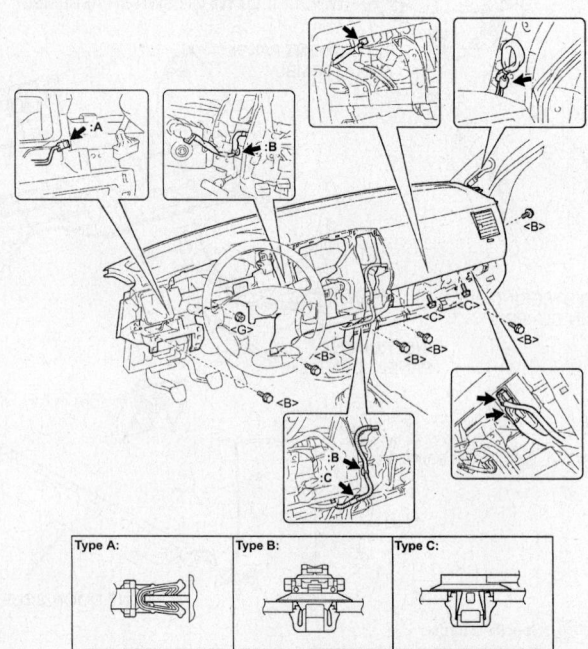

:A

:B

<B>

<B>

<B>

<C>

<B>

<B>

<B>

:B

:C

Type A:	Type B:	Type C:

09490_TACO_G0012

Instrument panel retaining clip locations—2005–2006

11. Remove the lower number three steering wheel cover after disengaging the two claws using the proper tool.

12. Using a torx socket wrench loosen the two retaining screws until the groove along the screw circumference fits in the screw case.

13. Remove the lower number two steering wheel cover after disengaging the two claws using the proper tool.

14. Pull the steering pad out of the steering wheel assembly and support the steering pad using your hand.

➡**Take care as not to pull the air bag wire harness.**

15. Disconnect the horn ground harness from the steering pad. Disconnect the air bag connectors. Remove the steering pad.

16. Remove the steering wheel retainer nut. Using the proper removal tool, remove the steering wheel from the steering column.

17. Disengage the two claws, using the proper tool, and remove the lower steering column cover.

18. Disengage the claw, using the proper tool, and remove the upper steering column cover.

19. To remove the spiral cable assembly, disconnect the air bag connector and the connector from the spiral cable subassembly. Take care as not to damage the air bag with the harness.

20. Disengage the three claws and remove the spiral cable subassembly.

21. Remove the windshield wiper switch assembly.

22. Remove the front floor footrest. Remove the right and left front door scuff plates. Remove the right and left cowl side trim boards.

23. Separate the right and left front door opening trim weatherstrips. Remove the right and left front pillar garnish.

24. Disengage the four clips and two claws and remove the instrument panel under tray.

25. Disengage the four clips and remove the instrument panel access hole cover.

26. If equipped with automatic transmission, remove the shift knob. Disengage the six clips and one claw. Remove the console upper rear panel subassembly.

27. If equipped with manual transmission, disengage the four clips and one claw. Remove the console upper rear panel subassembly.

28. Remove the console box carpet. Remove the two screws. Disengage the four claws and remove the rear console box assembly.

finish panel sub assembly. Disconnect the electrical connectors.

37. Remove the four screws and the combination meter assembly. Disconnect the two electrical connectors.

38. Disengage the ten clips and remove the instrument cluster center finish panel subassembly.

39. Disengage the claw and separate the glove box door stopper from the glove box door assembly.

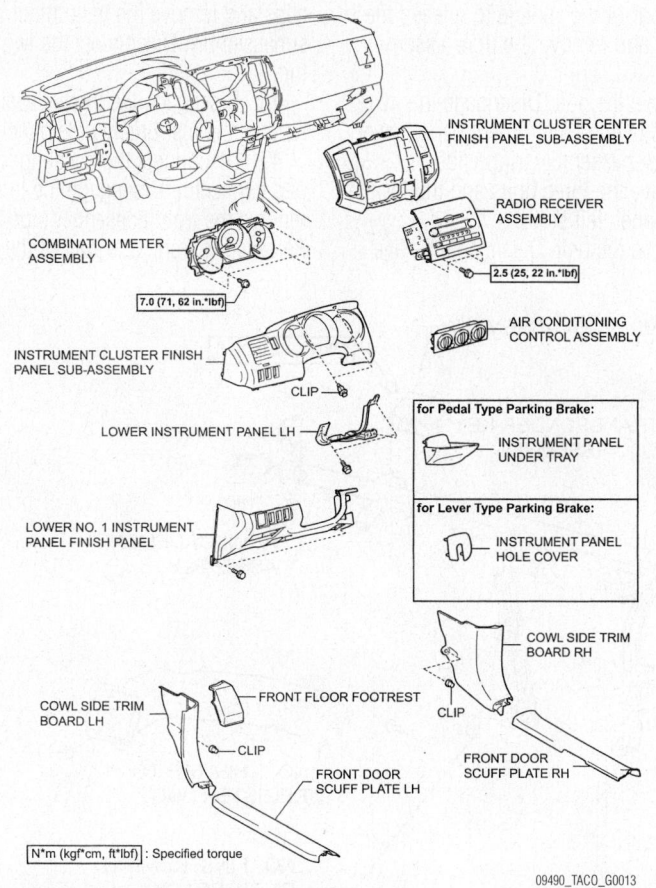

Instrument panel and related components—2005–2006

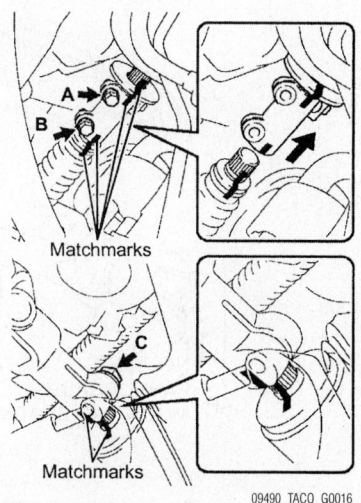

Matchmarking the steering gear assembly—2005–2006

29. If equipped with automatic transmission, disengage the four claws and two clips and remove the instrument panel cup holder tray. Remove the two screws. Disengage the two clips and one claw and remove the front console box.

30. If equipped with manual transmission, remove the shift lever knob. Remove the two screws. Disengage the claw and two clips. Remove the front console box.

31. If the vehicle is equipped with a bench seat, remove the shift knob. Remove the box bottom mat. Remove the instrument panel cup holder. Remove the number two box bottom mat. Remove the clip and two bolts. Disengage the claw and two clips and remove the front console box.

32. Disengage the four clips and remove the air conditioning control assembly. Disconnect the two electrical connectors.

33. Remove the radio.

34. Separate the hood lock control lever subassembly.

35. Remove the two bolts. Disengage the three clips and remove the lower number one instrument panel finish panel. Disconnect the electrical connectors.

36. Remove the two clips. Disengage the six clips and remove the instrument cluster

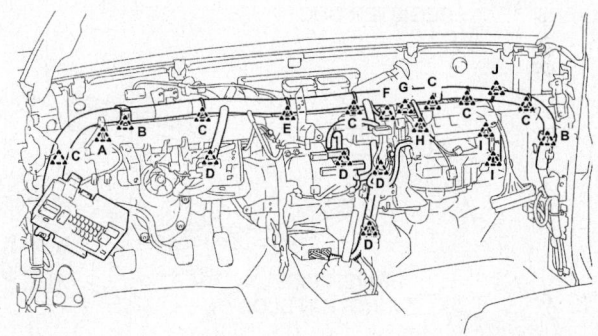

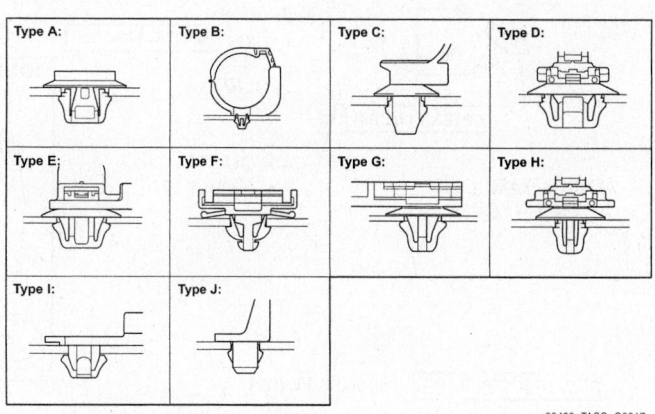

Air conditioning unit clip location—2005–2006

40. Slightly deform the upper part of the glove box door assembly and release the two stoppers and open the door assembly until it becomes horizontal.

41. Pull the glove box door assembly toward the rear of the vehicle and release the two stoppers and open the glove box door until it becomes horizontal.

42. Pull the glove box door assembly toward the rear of the vehicle to release the three hinges and remove the door assembly.

43. Remove the bolt. Disengage the two clips and remove the instrument panel lower finish panel subassembly, right side.

44. Remove the three bolts and the lower instrument panel, left side.

45. Remove the bolt. Disengage the five clips and remove the instrument lower cover subassembly. Disconnect the two connectors.

46. Disengage the three claws and remove the instrument side panel, right side.

47. Disconnect the passenger's side air bag connector. Disengage the ten hooks and remove the front passenger side air bag. Release the front side wall of the air bag

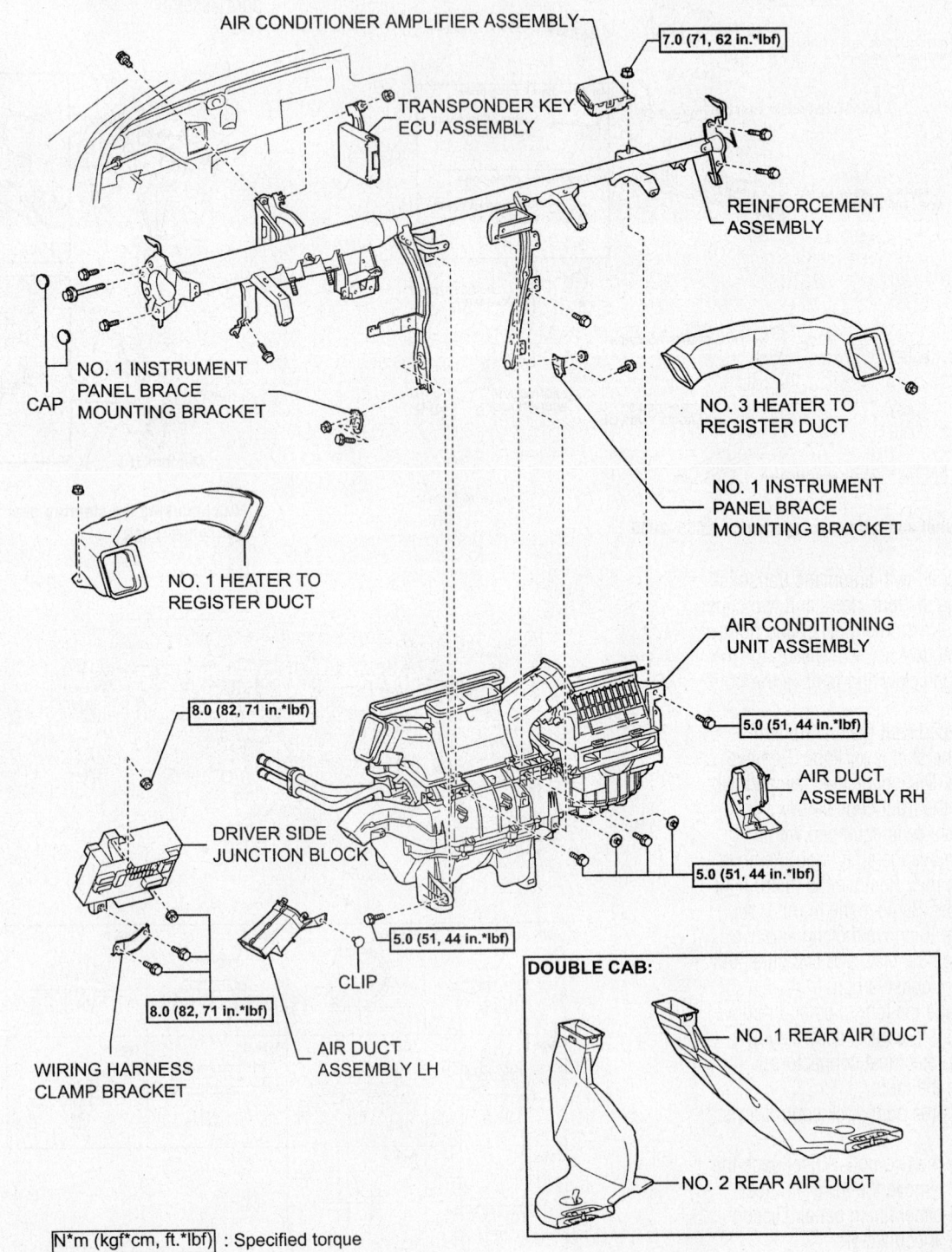

AIR CONDITIONER AMPLIFIER ASSEMBLY

7.0 (71, 62 in.*lbf)

TRANSPONDER KEY ECU ASSEMBLY

REINFORCEMENT ASSEMBLY

CAP

NO. 1 INSTRUMENT PANEL BRACE MOUNTING BRACKET

NO. 3 HEATER TO REGISTER DUCT

NO. 1 INSTRUMENT PANEL BRACE MOUNTING BRACKET

NO. 1 HEATER TO REGISTER DUCT

AIR CONDITIONING UNIT ASSEMBLY

8.0 (82, 71 in.*lbf)

5.0 (51, 44 in.*lbf)

AIR DUCT ASSEMBLY RH

DRIVER SIDE JUNCTION BLOCK

5.0 (51, 44 in.*lbf)

5.0 (51, 44 in.*lbf)

CLIP

8.0 (82, 71 in.*lbf)

WIRING HARNESS CLAMP BRACKET

AIR DUCT ASSEMBLY LH

DOUBLE CAB:

NO. 1 REAR AIR DUCT

NO. 2 REAR AIR DUCT

N*m (kgf*cm, ft.*lbf) : Specified torque

09490_TACO_G0018

Reinforcement assembly and related components—2005–2006

door from the other hook and remove the front passenger side air bag assembly.

48. Tape up the steering column cover and upper steering wheel subassembly using protective tape. Disengage the four clamps and disconnect the four connectors. Remove the eight bolts, nut and instrument subassembly.

49. Remove the number one heater to register duct. Remove the number three heater to register duct.

50. On double cab vehicles, remove the number two rear duct and the number one rear duct.

51. Remove the clip. Disengage the four claws and remove the air duct assembly heater, left side.

52. Remove the clip. Disengage the three claws and remove the air duct assembly heater, right side.

53. Remove the bolt, nut and instrument panel brace mounting bracket, left side.

54. Remove the bolt, nut and instrument panel brace mounting bracket, right side.

55. Disengage the four clips and remove the steering column hole cover.

56. Matchmark the steering gear sliding yoke, steering gear intermediate shaft subassembly number two and the steering gear intermediate shaft assembly.

57. Remove bolts "A" and "B" from the steering sliding yoke.

58. Slide the steering yoke up and separate it from the steering intermediate shaft subassembly number two.

59. Pull down the steering sliding yoke from the steering intermediate shaft assembly to remove it.

60. Disconnect the steering column electrical connectors. Remove the steering column retaining bolts. Remove the steering column from the vehicle.

61. Disconnect the connector. Remove the nut and the transponder key ECU assembly.

62. Disconnect the connector. Remove the nut and the air conditioner amplifier assembly.

63. Disengage the four clamps and disconnect the six connectors from the air conditioning unit assembly. Remove the three bolts. Disconnect the connector. Disconnect the two air bag connectors. Disengage the twenty clamps.

64. Remove the five bolts and two nuts. Remove the two caps and seven bolts. Disengage the reinforcement hook of the air conditioning unit. Remove the reinforcement.

65. Remove the air conditioning unit assembly from the vehicle.

66. Remove the heater to register center

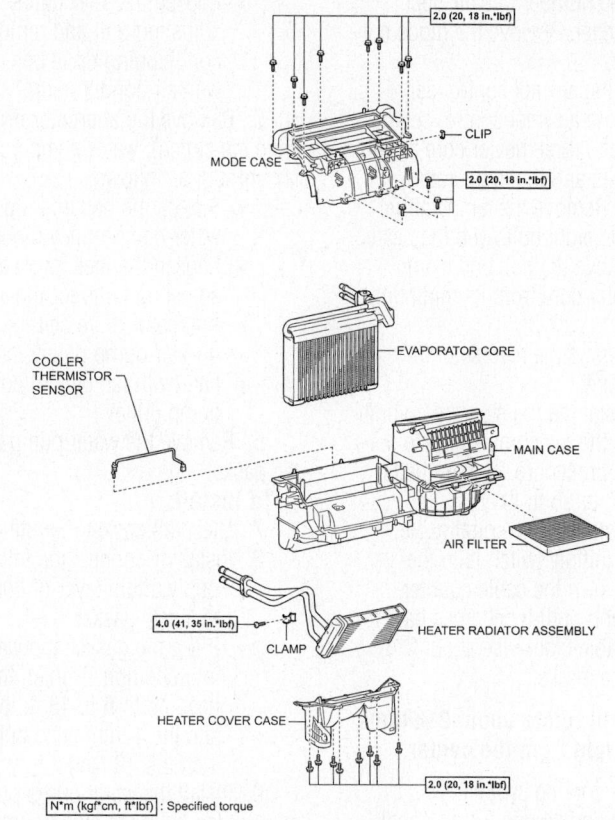

Air conditioning unit and related components—2005–2006

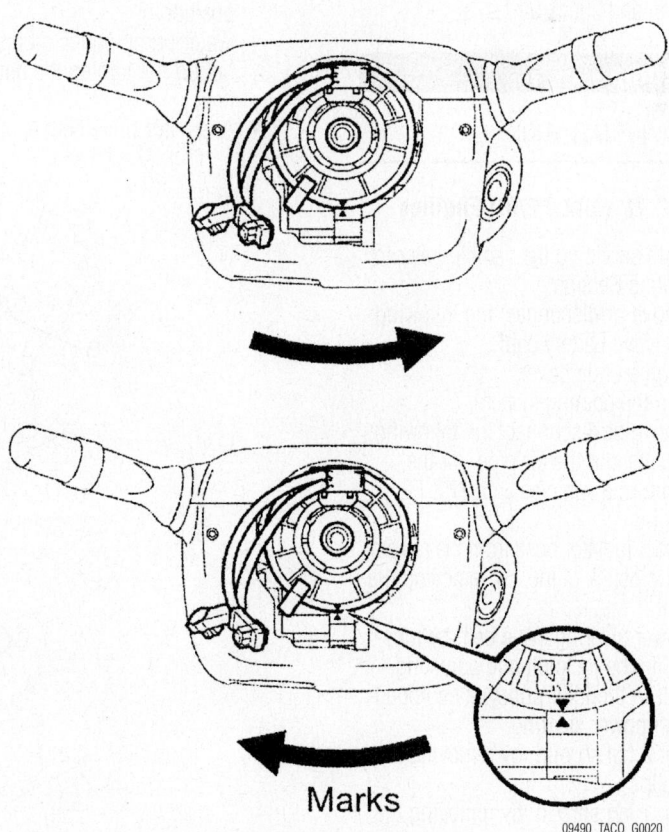

Spiral cable alignment marks—2005–2006

duct. Remove the number one air duct. Remove the air filter. Remove the mode control servo motor.

67. Remove the air mix control servo motor. Remove the air inlet control servo motor. Remove the three heater core assembly retaining bolts and the temperature mounting plate. Remove the temperature cam. Remove the eight bolts and the heater cover case. Remove the bolt and clamp. Remove the heater core from its mounting.

To install:

68. Installation is the reverse of the removal procedure.

69. When installing the steering column be sure to align the matchmarks made during the removal procedure. Tighten bolts "A", "B" and "C" to 26 ft. lbs.

70. When installing the spiral cable, check that the ignition switch is in the "OFF" position. Turn the cable counterclockwise by hand until it becomes hard to turn. Turn the cable clockwise about 2½ turns to align the marks.

➡**The cable will rotate about 2½ turns both left and right from the center.**

71. Refill the cooling system.

72. Evacuate and charge the air conditioning system refrigerant.

73. Run the engine to normal operating temperatures; then, check the climate control operation and check for leaks.

Water Pump

REMOVAL & INSTALLATION

2.4L and 2.7L (3RZ-FE/M) Engines

1. Before servicing the vehicle, refer to the Precautions Section.
2. Remove or disconnect the following:
   - Negative battery cable
   - Engine undercover
3. Drain the cooling system.
4. Remove or disconnect the following:
   - 2 bolts and the air pipe, for the California vehicles with 3RZ-FE engine
   - Upper radiator hose from the radiator
   - Oil dipstick guide, by removing the bolt
   - Power steering drive belt, by loosening the lockbolt and adjusting bolt to the idler pulley, if equipped with power steering
   - No. 2 fan shroud, by removing the 2 clips
   - No. 1 fan shroud, by removing the 4 bolts

   - Loosen the idler pulley nut and adjusting bolt and remove the air conditioning drive belt, if equipped with air conditioning
5. Remove the alternator drive belt, fan (with fan clutch), water pump pulley and the fan shroud, as follows:
   - Stretch the belt and loosen the water pump pulley mounting nuts
   - Loosen the lock, pivot and the adjusting bolts for the alternator
   - Alternator drive belt
   - 4 water pump pulley mounting nuts
   - Fan (with fan clutch) and the water pump pulley
6. Remove the water pump and discard the gasket.

To install:

7. Clean all gasket mounting surfaces.
8. Install or connect the following:
   - Apply a thin layer of liquid sealant to a new gasket
   - Place the gasket and water pump into position. Tighten the 14mm head bolts **A** to 18 ft. lbs. (25 Nm) and the 12mm head bolts to 78 inch lbs. (9 Nm)
9. Install the water pump pulley, fan shroud, fan (with fan clutch) and the alternator drive belt, as follows:
   - Fan (with the fan clutch), water pump pulley and the fan shroud in position
   - Water pump pulley mounting nuts but do not tighten the nuts at this time
   - Alternator drive belt

   - Stretch the alternator belt tight. Tighten the fan nuts to 16 ft. lbs. (21 Nm).
   - Adjust the alternator drive belt
10. Install or connect the following:
    - Adjust the drive belt, if equipped with air conditioning
    - No. 1 fan shroud, by installing the 4 bolts
    - No. 2 fan shroud, with the 2 clips
    - Adjust the power steering drive belt
    - Oil dipstick guide, with the bolt
    - Upper radiator hose to the radiator
    - Air pipe, If removed
    - Negative battery cable
11. Fill and bleed the cooling system.
12. Start the engine and check for leaks.
13. Install the engine undercover.

2.7L (2TR-FE/X) Engine

1. Before servicing the vehicle, refer to the Precautions Section.

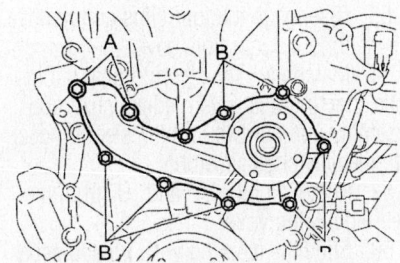

09490_TACO_G0021

Water pump bolt identification—2.7L (2TR-FE/X) engine

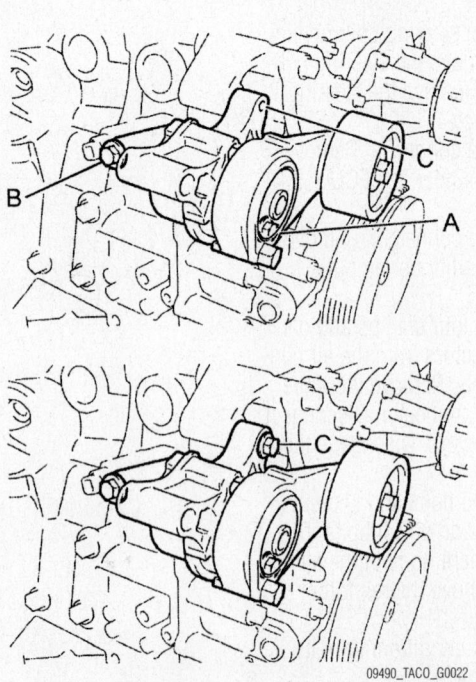

09490_TACO_G0022

Belt tensioner bolt identification and location—2.7L (2TR-FE/X) engine

2. Disconnect the negative battery cable.

3. On 4WD and PreRunner, remove the four retaining bolts and remove the number one engine undercover subassembly.

4. Drain the engine coolant.

5. Remove the radiator support to frame seal, left side.

6. Remove the fan shroud. Remove the alternator.

7. Remove the three bolts, and remove the belt tensioner assembly.

8. Remove the water pump retaining bolts. Remove the water pump from the engine.

To install:

9. Clean all gasket mounting surfaces.

10. Using a new gasket install the water pump to the engine. Tighten bolts "A" to 15 ft. lbs. Tighten bolts "B" to 80 inch lbs.

11. Install the belt tensioner assembly. Tighten bolt "B" to 30 ft. lbs. Tighten bolt "A" to 16 ft. lbs. Tighten bolt "C" to 32 ft. lbs.

➡**Check that the bolt holes on the belt tensioner and timing chain cover are aligned, prior to installing bolt "C".**

12. Continue the installation in the reverse order of the removal procedure.

13. Fill the cooling system with the proper grade and type engine coolant.

14. Start the engine and check for leaks.

3.4L Engine

1. Before servicing the vehicle, refer to the Precautions Section.

2. Remove or disconnect the following:
- Negative battery cable
- Engine undercover

3. Drain the engine coolant.

4. Remove the upper radiator hose.

5. Remove the power steering drive belt, as follows:
- Stretch the belt and loosen the fan pulley mounting nuts
- Loosen the lockbolt, pivot bolt and the adjusting bolt
- Drive belt

6. Remove or disconnect the following:
- Air conditioning drive belt, by loosening the idler pulley nut and adjusting bolt
- Lockbolt, pivot bolt and the adjusting bolt
- Alternator drive belt
- No. 2 fan shroud, by removing the 2 clips
- Fan with the fluid coupling and fan pulleys
- Power steering pump and move it

aside without disconnecting the lines from the pump
- Compressor from the engine and move it aside without disconnecting the compressor lines, if equipped with air conditioning
- Air conditioning bracket, if equipped with air conditioning

7. Remove the No. 2 timing belt cover, as follows:
- Camshaft Position (CMP) sensor connector from the No. 2 timing belt cover
- 3 spark plug wire clamps from the No. 2 timing belt cover
- 6 bolts and the timing belt cover

8. Remove the fan bracket, as follows:
- Power steering adjusting strut, by removing the nut
- Fan bracket, by removing the bolt and nut

9. Set the No. 1 cylinder to Top Dead Center (TDC) of the compression stroke, as follows:

a. Turn the crankshaft pulley and align its groove with the timing mark **0** of the No. 1 timing belt cover.

b. Check that the timing marks of the camshaft timing pulleys and the No. 3 timing belt cover are aligned. If not, turn the crankshaft pulley 1 revolution (360 degrees).

10. Remove the camshaft timing pulleys, as follows:

a. Remove the timing belt tensioner by alternately loosening the 2 bolts.

b. Using Variable Wrench Set No. 09960-10010, remove the pulley bolt, the timing pulley and the knock pin.

c. Remove the 2 timing pulleys with the timing belt.

11. Remove or disconnect the following:
- Thermostat
- No. 2 oil cooler hose, from the water pump
- Water pump, by removing the 7 bolts

12. Thoroughly clean the mating surfaces.

To install:

13. Install or connect the following:
- Apply sealant (PN 08826-00100) to the water pump

✳✳ WARNING

Parts must be assembled within 5 minutes of application. Otherwise the material must be removed and reapplied.

- Water pump. Tighten the bolts to 14 ft. lbs. (20 Nm).
- No. 2 oil cooler hose
- Thermostat
- Left camshaft timing pulley. Tighten the pulley bolt to 81 ft. lbs. (110 Nm).

14. Set the No. 1 cylinder to TDC of the compression stroke.

15. Connect the timing belt to the left camshaft timing pulley. Check that the installation mark on the timing belt is aligned with the end of the No. 1 timing belt cover, as follows:

a. Using Variable Pin Wrench Set 09960-01000, slightly turn the left camshaft timing pulley clockwise. Align the installation mark on the timing belt with the timing mark of the camshaft timing pulley and hang the timing belt on the left camshaft timing pulley.

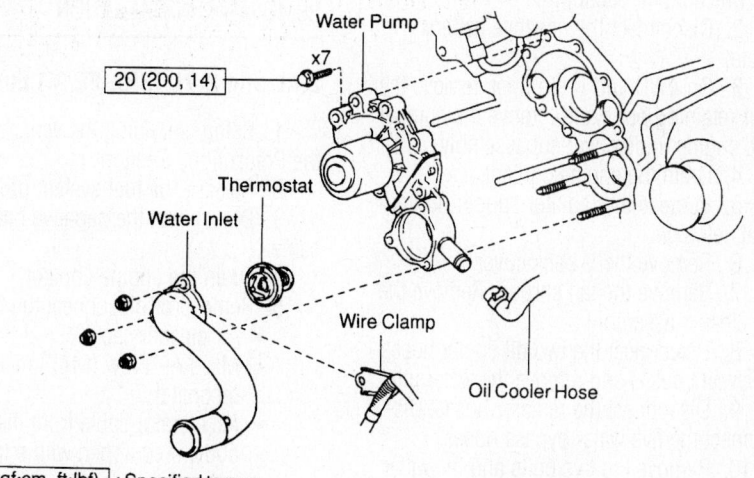

N·m(kgf·cm, ft·lbf) : Specified torque
◆ Non–Reusable part

Water pump and related components—3.4L engine

7924YG08

b. Align the timing marks of the left camshaft pulley and the No. 3 timing belt cover.

c. Check that the timing belt has tension between the crankshaft timing pulley and the left camshaft timing pulley.

16. Install the right camshaft timing pulley and the timing belt.

17. Set the timing belt tensioner, as follows:

a. Using a press, slowly press in the pushrod using 220–2,205 lbs. (981–9,807 N) of force.

b. Align the holes of the pushrod and housing, pass a 1.5mm hexagon wrench through the holes to keep the setting position of the pushrod.

c. Release the press and install the dust boot to the tensioner.

d. Install the timing belt tensioner and alternately tighten the bolts to 20 ft. lbs. (28 Nm).

e. Using pliers, remove the 1.5mm hexagon wrench from the belt tensioner.

18. Check the valve timing, as follows:

a. Slowly turn the crankshaft pulley 2 revolutions from the TDC-to-TDC; always turn the crankshaft pulley clockwise.

b. Check that each pulley aligns with the timing marks. If the timing marks do not align, remove the timing belt and reinstall it.

19. Install or connect the following:
- Fan bracket, with the bolt and nut
- Remaining components
- Negative battery cable

20. Fill with engine coolant.

21. Start the engine and check for leaks.

4.0L Engine

1. Before servicing the vehicle, refer to the Precautions Section.

2. Disconnect the negative battery cable.

3. On 4WD and PreRunner, remove the four retaining bolts and remove the number one engine undercover subassembly.

4. Drain the engine coolant.

5. Remove the radiator support to frame seal, left side.

6. Remove the V bank cover.

7. Remove the fan shroud. Remove the air cleaner assembly.

8. Disconnect the two oil cooler hoses (with oil cooler) and remove the water inlet.

9. Disconnect the radiator hoses. Disconnect the five water bypass hoses.

10. Remove the five bolts and the water inlet. Remove the O-ring from the water outlet pipe. Remove the gasket from the water pump.

Water pump bolt identification—4.0L engine

11. Remove the two bolts and remove the number two idler pulley subassembly.

12. Remove the alternator.

13. Remove the mounting bolt and separate the air conditioning compressor suction hose subassembly. Disconnect the air condition compressor connector. Remove the four bolts and separate the compressor from the belt tensioner assembly.

14. Remove belt tensioner assembly.

15. Remove the water pump retaining bolts. Remove the water pump from the engine.

To install:

16. Clean all gasket mounting surfaces.

17. Using a new gasket install the water pump to the engine. Tighten bolts "A" to 80 inch lbs. Tighten bolts "B" to 17 ft. lbs.

18. Install the belt tensioner assembly.

19. Continue the installation in the reverse order of the removal procedure.

20. Fill the cooling system with the proper grade and type engine coolant.

21. Start the engine and check for leaks.

Cylinder Head

REMOVAL & INSTALLATION

2.4L and 2.7L (3RZ-FE/M) Engines

1. Before servicing the vehicle, refer to the Precautions Section.

2. Release the fuel system pressure.

3. Disconnect the negative battery cable.

4. Drain the engine coolant.

5. Remove or disconnect the following:
- Air cleaner cap
- Mass Air Flow (MAF) meter and resonator
- Accelerator cable from the throttle body, if equipped with a manual transmission
- Accelerator and throttle cables from the throttle body, if equipped with an automatic transmission
- Cruise control cable from the actuator, if equipped with cruise control
- Intake air connector
- Air hose for Idle Air Control (IAC)
- Vacuum sensing hose
- Wire clamp for the engine wiring harness
- Oil dipstick guide
- Power steering belt
- Power steering pulley pump and bracket
- Positive Crankcase Ventilation (PCV) hoses
- Distributor
- Spark plug wires from the spark plugs
- Engine wiring harness
- Air conditioning compressor, if equipped with air conditioning
- Oil pressure sensor
- Engine Coolant Temperature (ECT) sensor connector
- ECT sender gauge connector
- Exhaust Gas Recirculation (EGR) gas temperature sensor connector
- Vacuum Switching Valve (VSV) connector
- 2 vacuum hose from the VSV
- Ground strap from the cowl top panel
- Engine wiring harness from the air intake chamber
- Throttle Position (TP) sensor connector
- IAC valve connector
- Crankshaft Position (CKP) sensor connector
- Knock Sensor (KS) connector
- Data Link Connector 1 (DLC1) from the bracket
- Engine wiring harness clamp
- EGR pipe
- Intake chamber stay
- Air intake chamber assembly

6. Disconnect the following hoses:
- Evaporative Emissions (EVAP) hose from the throttle body
- Brake booster vacuum hose from the union
- Water bypass hose from the water bypass pipe
- Water bypass hose from the cylinder head rear cover

7. Remove or disconnect the following:
- Injector connectors
- Fuel inlet pipe
- Hoses and the fuel return pipe
- Delivery pipe and injectors
- Intake manifold
- Front exhaust pipe
- Exhaust manifold and gasket
- Water outlet

- Cylinder head rear cover
- Spark plugs
- Front engine hanger
- Engine wiring harness brackets
- Cylinder head cover

8. Set No. 1 cylinder to Top Dead Center (TDC) of the compression stroke. The groove on the crankshaft pulley should align with the **0** mark on the timing chain cover and the timing marks (1 and 2 dots) of the camshaft gears should form a straight line in respect to the cylinder head surface. If not, turn the crankshaft 1 revolution (360 degrees).

9. Remove or disconnect the following:
- Chain tensioner and gasket
- Camshaft timing gear
- Exhaust camshafts

10. Remove the intake camshaft, as follows:

a. Uniformly, loosen and remove the bearing cap bolts in the reverse order of the tightening in several passes, in sequence.

b. Remove the bearing caps and camshaft. Make a note of the bearing cap positions for proper installation.

➡**If the camshaft is not being lifted out straight and level, reinstall the No. 3 bearing cap with the 2 bolts. Then, alternately loosen and remove the 2 bearing cap bolts with the camshaft gear pulled up.**

11. Remove or disconnect the following:
- Valve lifters and shims

➡**Arrange the valve lifters and shims in correct order.**

- Cylinder head, by uniformly loosen and remove the cylinder head bolts in the reverse order of the tightening, in sequence, using several passes

To install:

12. Before installing, thoroughly clean the gasket mating surfaces and check for warpage.

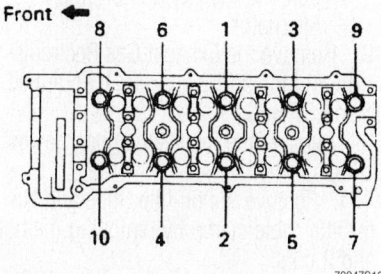

Cylinder head bolt torque sequence—2.4L and 2.7L (3RZ-FE/M) engines

7924ZG18

13. Apply sealant (PN 08826-00080) to the 2 locations. Place a new head gasket on the block and install the cylinder head.

14. Install the cylinder head as follows:

a. Lightly coat the cylinder head bolts with engine oil.

b. Install the bolts and tighten, in several passes, in the sequence. Tighten all bolts to 29 ft. lbs. (39 Nm).

c. Mark the front of the bolt with paint and retighten bolts 90 degrees in the proper sequence.

d. Retighten an additional 90 degrees. Check that the painted mark is now facing rearward.

15. Install or connect the following:
- Tighten the 2 front mounting bolts to 15 ft. lbs. (21 Nm).
- Valve lifters and shims in their proper locations. Check that the valve lifter rotates smoothly by hand.
- Intake and exhaust camshafts

16. Set No. 1 cylinder to TDC compression stroke. The groove on the crankshaft pulley should align with the **0** mark on the timing chain cover and the timing marks (1 and 2 dots) of the camshaft gears should form a straight line in respect to the cylinder head surface. If not, turn the crankshaft 1 revolution (360 degrees).

17. Install the timing gear, as follows:

a. Place the gear over the straight pin of the intake camshaft.

b. Hold the intake camshaft with a wrench. Install and tighten the bolt to 54 ft. lbs. (74 Nm).

c. Hold the exhaust camshaft and install the bolt and distributor gear. Tighten the bolt to 34 ft. lbs. (46 Nm).

18. Install or connect the following:
- Chain tensioner, using a new gasket (mark toward the front)
- Recheck the valve timing
- Check and adjust the valve clearance
- Spark plugs
- Semi-circular plug

19. Recheck the engine for proper valve timing.

20. Install or connect the following:
- Cylinder head cover, using a new gasket
- Engine wiring harness brackets
- Front engine hanger. Tighten the bolts to 30 ft. lbs. (42 Nm).
- Cylinder head rear cover. Tighten the bolts to 10 ft. lbs. (13 Nm).
- Water outlet, using a new gasket. Tighten the bolts to 14 ft. lbs. (20 Nm).
- Upper radiator hose

- Exhaust manifold. Tighten the bolts to 36 ft. lbs. (49 Nm).
- Remaining components
- Negative battery cable

21. Fill the engine and radiator with engine coolant.

22. Start the engine and check for leaks.

23. Check the ignition timing. Road test the vehicle for proper operation.

24. Recheck all fluid levels.

2.7L (2TR-FE/X) Engine

➡**The engine must first be removed from the vehicle.**

1. Before servicing the vehicle, refer to the Precautions Section.

2. Properly discharge the fuel system pressure.

3. Properly disarm the SRS system.

4. Disconnect the negative battery cable.

5. Remove the hood subassembly.

6. Remove the engine from the vehicle and position it in a suitable holding fixture.

7. Remove the intake air connector. Remove the alternator.

8. Remove the number one exhaust manifold heat insulator. Remove the air switching valve assembly.

9. Remove the exhaust manifold.

10. Remove the belt tensioner assembly. Remove the number one idler pulley subassembly. Remove the idle pulley assembly and bracket.

11. Remove the crankshaft position sensor. Remove the camshaft position sensor.

12. Remove the number one intake manifold.

13. Remove the cylinder head cover. Remove the crankshaft pulley. Remove the oil gauge subassembly.

14. Remove the number two oil pan subassembly. Remove the oil strainer. Remove the oil pap subassembly.

15. Remove the timing chain cover. Remove the timing chain guide. Remove the number one chain tensioner assembly.

16. Remove the chain tensioner slipper. Remove the number one chain vibration damper. Remove the chain subassembly.

17. Remove the camshaft. Remove the valve rocker arms.

18. Remove the valve lash adjuster assembly.

19. Disconnect the water hoses. Disconnect the engine coolant temperature sensor wire. Remove the bolts and separate the wire from the harnesses. Remove the bolt, and then separate the wire harness bracket.

20. Loosen the cylinder head retaining bolts in several steps and in the proper removal sequence. Remove the bolts.

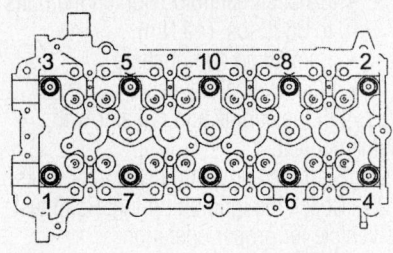

Cylinder head bolt loosening sequence—2.7L (2TR-FE/X) engine

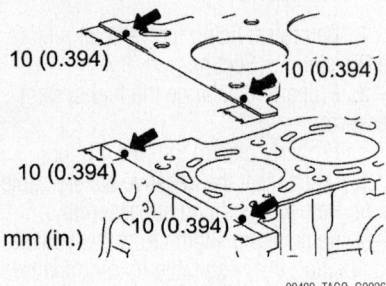

Cylinder head gasket sealant application—2.7L (2TR-FE/X) engine

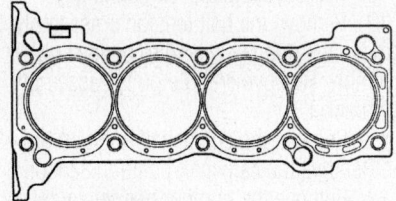

Cylinder head gasket positioning—2.7L (2TR-FE/X) engine

➡Be careful not to drop the washers into the cylinder head. Head warpage and cracking could result in removing the bolts in the wrong order.

To install:

21. Before installing, thoroughly clean the gasket mating surfaces and check for warpage.

22. Apply continuous beads of seal packing, part number 08826-00080 or equivalent, to the cylinder block upper side and cylinder head gasket upper side. Bead width should be 0.15–0.28 inch.

➡Remove any oil from the contact surface. Install the cylinder head gasket within three minutes after applying the seal packing. Install the cylinder head bolts within fifteen minutes after applying the seal packing. Do not fill the engine with oil for at least four hours.

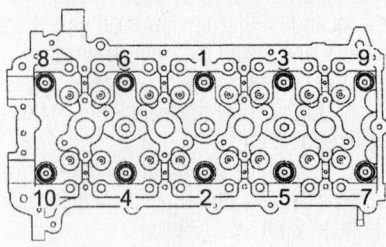

Cylinder head bolt torque sequence—2.7L (2TR-FE/X) engine

23. Be sure to position the cylinder head gasket on the engine block with the lot number stamp facing upward.

➡Be sure that the head gasket is installed in the correct direction. Position the gasket gently in order to avoid damaging the gasket with the bottom part of the cylinder head.

24. Apply a light coat of clean engine oil to the cylinder head bolts. Tighten the cylinder head bolts, in two successive steps, to specification and in the proper sequence.

25. Continue the installation in the reverse of the removal procedure.

26. Be sure to fill the cooling system with the proper grade and type engine coolant.

27. Be sure to fill the engine oil with the proper grade and type oil.

➡Do not fill the engine with oil for at least four hours after the cylinder head has been installed.

28. Start the engine and check for leaks. Correct as required.

3.4L Engine

1. Before servicing the vehicle, refer to the Precautions Section.

2. Disconnect the negative battery cable.

3. Relieve the fuel system pressure.

4. Remove the engine undercover.

5. Drain the cooling system.

6. Remove or disconnect the following:
- Front exhaust pipe
- Air cleaner cap
- Mass Air Flow (MAF) meter and resonator

7. Disconnect the following cables:
- Actuator cable from the bracket, if equipped with cruise control
- Accelerator cable
- Throttle cable, if equipped with an automatic transmission
- Heater hose
- Upper radiator hose

- Power steering drive belt
- Air conditioning drive belt, by loosening the idle pulley nut and adjusting bolt
- Loosen the lockbolt, pivot bolt and adjusting bolt and the alternator drive belt
- No. 2 fan shroud by removing the 2 clips
- Fan with the fluid coupling and fan pulleys
- Power steering pump and move it aside without disconnecting the pump lines
- Compressor and move it aside without disconnecting the compressor lines, if equipped with air conditioning
- Air conditioning bracket, if equipped with air conditioning
- Spark plug wires with the ignition coils
- Spark plugs
- No. 2 timing belt cover

8. Remove the fan bracket, as follows:

a. Remove the power steering adjusting strut by removing the nut.

b. Remove the fan bracket by removing the bolt and nut.

9. Set the No. 1 cylinder at Top Dead Center (TDC) of the compression stroke.

a. Turn the crankshaft pulley and align its groove with the timing mark **0** on the No. 1 timing belt cover.

b. Check that the timing marks of the camshaft timing pulleys and the No. 3 timing belt cover are aligned. If not, turn the crankshaft pulley 1 revolution (360 degrees).

10. Remove the timing belt tensioner by alternately loosening the 2 bolts.

11. Remove the camshaft timing pulleys, as follows:

a. Using Variable Pin Wrench Set tool 09960-10010, remove the pulley bolt, the timing pulley and the knock pin.

b. Remove the 2 timing pulleys with the timing belt.

12. Remove or disconnect the following:
- Bolt and the No. 2 idler pulley
- Alternator

13. Remove the Exhaust Gas Recirculation (EGR) pipe and 2 gaskets, if equipped with an EGR valve

14. Remove the intake chamber stay as follows:

a. Remove the oil filler tube and No. 1 throttle cable clamp by removing the bolt and 2 nuts.

b. Remove the intake chamber stay by removing the 2 bolts.

15. Remove the following connectors:

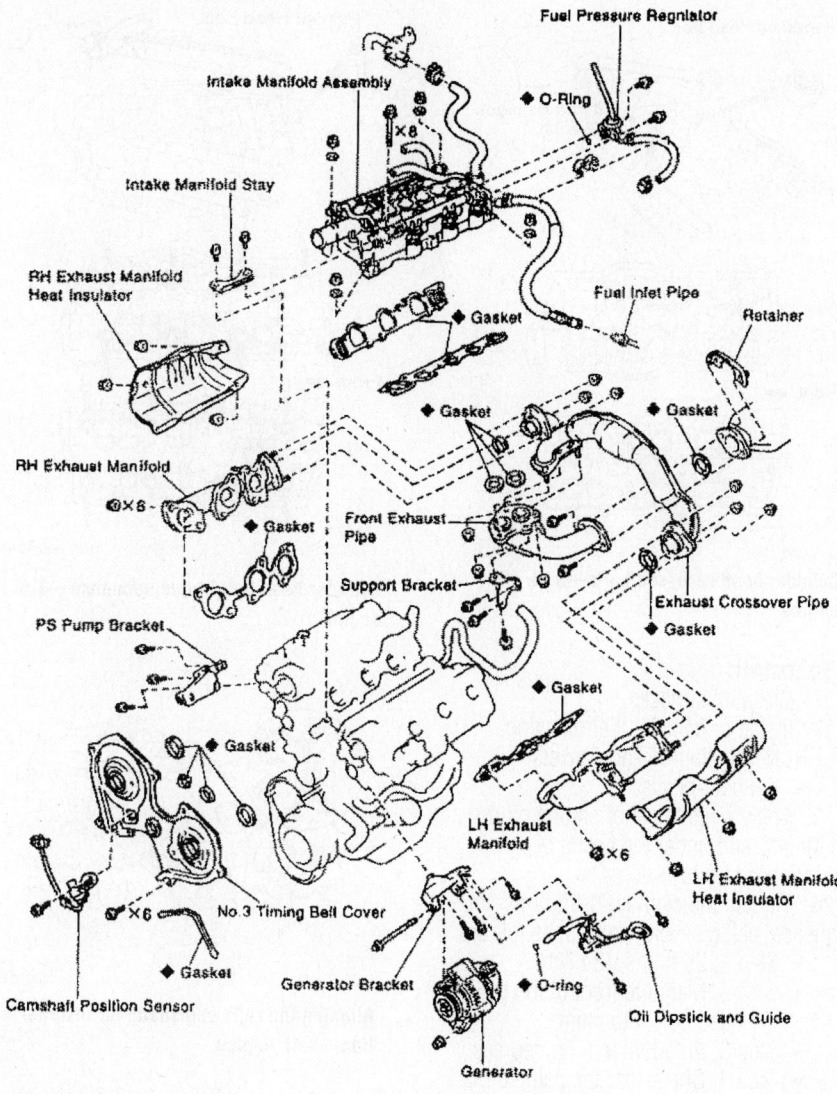

Cylinder head and related components—3.4L engine

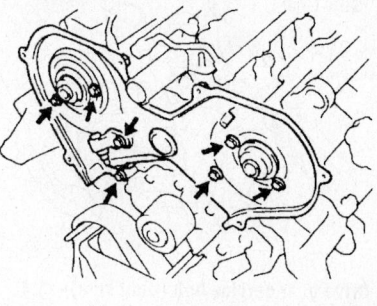

Rear timing belt cover bolt locations—3.4L engine

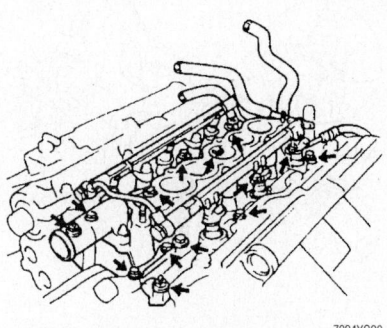

Intake manifold bolt locations—3.4L engine

◆ Non-reusable part

- VSV connector for the fuel pressure control.
- Throttle position sensor
- IAC valve connector
- EGR valve gas temperature sensor, if equipped
- Vacuum Switching Valve (VSV) connector for the EGR valve, if equipped
16. Disconnect the following hoses:
- Positive Crankcase Ventilation (PCV) hoses
- Water bypass hoses.
- Air assist hose from the intake air connector
- 2 vacuum sensing hoses from the VSV
- Evaporative Emissions (EVAP) hose
- Air hose, from the power steering

- Air hose from the air conditioning idle up valve, if equipped with air conditioning
17. Remove or disconnect the following:
- 4 bolts, 2 nuts and the air intake chamber assembly
- Intake air connector
18. Disconnect the engine wiring harness from the intake manifold, as follows:
- Oil pressure sensor connector
- Crankshaft position sensor connector
- 6 injector connectors
- Engine Coolant Temperature (ECT) sender gauge connector
- ECT sensor connector
- Knock (KS) sensor connector
- Camshaft Position (CMP) sensor connector

- 3 engine wiring harness clamps
- 3 bolts and the engine wiring harness from the cylinder head
19. Remove or disconnect the following:
- CMP sensor
- No. 3 (rear) timing belt cover, by removing the 6 bolts
- Fuel pressure regulator
- Intake manifold assembly
- Power steering pump bracket
- Oil dipstick and guide
- Exhaust crossover pipe and gaskets, by removing the 6 nuts
- Left-hand exhaust manifold, by removing the heat insulator and 6 nuts
- Right-hand exhaust manifold, by removing the heat insulator and 6 nuts
- 8 bolts, seal washers, cylinder head cover and gasket
- Both cylinder head covers
- Semi-circular plugs
- Right exhaust camshafts
- Right-hand intake camshaft
- Left exhaust camshafts
- Left-hand intake camshaft
- Valve lifters and shims from the cylinder head; arrange the valve lifters and shims in correct order

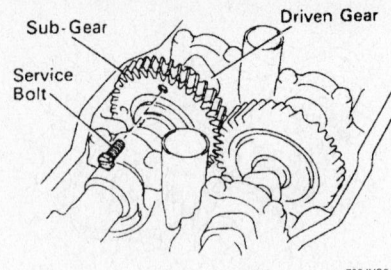

Drive gear service bolt (right side)—3.4L engine

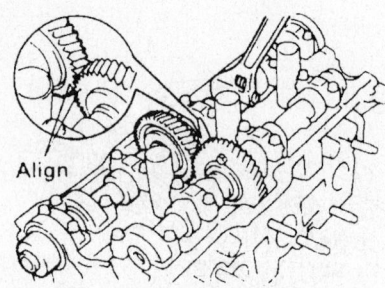

Aligning the timing mark (1 dot mark) of the left camshafts—3.4L engine

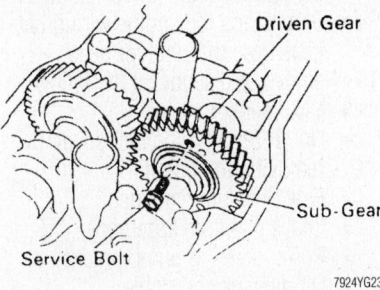

Drive gear service bolt (left side)—3.4L engine

20. Remove the cylinder heads, as follows:
- Bolt and disconnect the ground strap
- Cylinder head (recessed head) bolt on each cylinder head, using an 8mm hexagon wrench; then repeat for the other side
- 8 cylinder head (12-pointed head) bolts on each cylinder head, by loosening the bolts in several passes and in the reverse order of the tightening sequence
- 16 cylinder head bolts and plate washers.
- Cylinder head

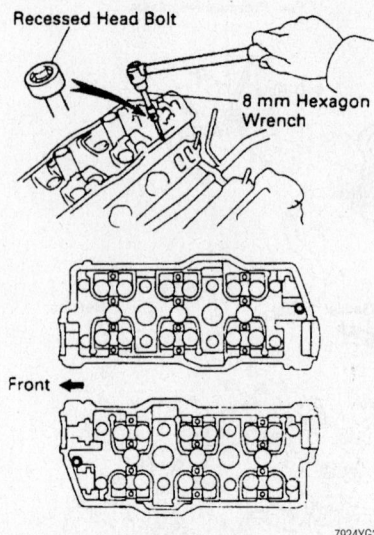

Cylinder head recessed bolts—3.4L engine

To install:
21. Clean all surfaces.
22. Install or connect the following:
- New cylinder head gaskets
- Cylinder heads
23. Apply a light coat of engine oil on the threads and under the heads of the cylinder head bolts.
24. Tighten the cylinder head bolts, in sequence, using several passes, as follows:
- Step 1: 25 ft. lbs. (34 Nm)
- Step 2: Mark the front of the cylinder head bolt with paint
- Step 3: An additional 90 degrees
- Step 4: Check that the painted mark is now at a 90 degrees angle to the front
25. Install the recessed head cylinder head bolts, as follows:
- Apply a light coat of engine oil on the threads and under the heads of the cylinder head bolts
- Cylinder head bolt on each cylinder head, using a 8mm hexagon wrench; then, repeat for the other side, as shown. Tighten the bolts to 13 ft. lbs. (18 Nm).
- Bolt and the ground strap
26. Install or connect the following:
- Valve lifters and shims

➡Check that the valve lifter rotates smoothly by hand.
- Camshafts
- Check and adjust the valve clearance
- Semi-circular plugs
- Cylinder head covers. Tighten the bolts, in several passes, to 53 inch lbs. (6 Nm).

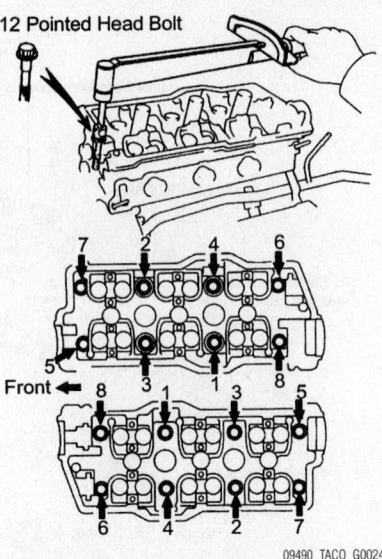

Cylinder head bolt torque sequence—3.4L engine

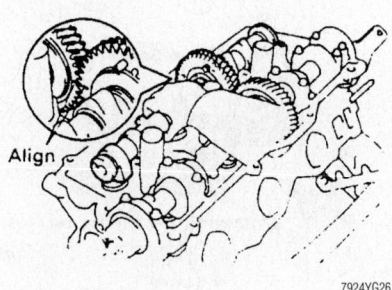

Aligning the right camshafts for installation—3.4L engine

- Exhaust manifolds, with new gaskets. Tighten the nuts to 30 ft. lbs. (40 Nm).
- Exhaust manifold heat insulators. Tighten the nuts to 71 inch lbs. (8 Nm).
- Exhaust crossover pipe. Tighten the nuts to 33 ft. lbs. (45 Nm).
- Alternator bracket. Tighten to 14 ft. lbs. (18 Nm).
- Oil dipstick and guide, using a new O-ring
- Power steering bracket. Tighten the fasteners to 14 ft. lbs. (18 Nm).
- New gaskets and the intake manifold assembly. Tighten the bolts and nuts to 13 ft. lbs. (18 Nm).
- Intake manifold stay with the 2 bolts. Tighten the bolts to 14 ft. lbs. (18 Nm).
- Fuel inlet hose
- Fuel pressure regulator
- No. 3 timing belt cover. Tighten the bolts to 80 inch lbs. (9 Nm).

- CMP sensor. Tighten to 71 inch lbs. (8 Nm).
- Engine wiring harness

27. Install or connect the intake air connector, as follows:
- Intake manifold. Tighten the bolts and nuts to 14 ft. lbs. (19 Nm).
- DLC1 to the bracket on the intake manifold
- Ground strap to the intake manifold
- Brake booster vacuum hose to the intake air connector
- 2 fuel return hoses
- Engine wiring harness to the intake manifold
- Idle up valve connector, if equipped with air conditioning

28. Install or connect the following:
- Air intake chamber assembly. Tighten the bolts and nuts to 14 ft. lbs. (18.5 Nm).
- Hoses

29. Attach the following connectors:
- VSV connector for the fuel pressure control
- TP sensor connector
- IAC valve connector
- EGR gas temperature connector, if equipped with an EGR valve
- VSV connector, if equipped with an EGR valve
- Intake chamber stay
- New gaskets and the EGR pipe. Tighten the clamp nuts to 71 inch lbs. (8 Nm) and the EGR pipe nuts to 14 ft. lbs. (18 Nm).
- Alternator but do not tighten the bolts and nuts at this time.
- No. 2 timing belt idler. Tighten the bolt to 30 ft. lbs. (40 Nm).

➡**Check that the pulley bracket moves smoothly.**

- Left camshaft timing pulley

30. Set the No. 1 cylinder to TDC of the compression stroke, as follows:
a. Turn the crankshaft pulley and align its groove with the timing mark **0** on the No. 1 timing belt cover.
b. Turn the camshaft, align the knock pin hole of the camshaft with the timing mark of the No. 3 timing belt cover.
c. Turn the camshaft timing pulley, align the timing marks of the camshaft timing pulley and the No. 3 timing belt cover.

31. Connect the timing belt to the left camshaft timing pulley, as follows:
a. Check that the installation mark on the timing belt is aligned with the end of the No. 1 timing belt cover.
b. Using Variable Pin Wrench Set

09960-01000, slightly turn the left camshaft timing pulley clockwise. Align the installation mark on the timing belt with the timing mark of the camshaft timing pulley and hang the timing belt on the left camshaft timing pulley.
c. Align the timing marks of the left camshaft pulley and the No. 3 timing belt cover.
d. Check that the timing belt has tension between the crankshaft timing pulley and the left camshaft timing pulley.

32. Install the right camshaft timing pulley and the timing belt, as follows:
a. Align the installation mark on the timing belt with the timing mark of the right camshaft timing pulley, and hang the timing belt on the right camshaft timing pulley with the flange side facing inward.
b. Slide the right camshaft timing pulley on the camshaft. Align the timing marks on the right camshaft timing pulley and the No. 3 timing belt cover.
c. Align the knock pin hole of the camshaft with the knock pin groove of the pulley and install the knock pin. Install the bolt and tighten to 81 ft. lbs. (110 Nm).

33. Set the timing belt tensioner as follows:
a. Using a press, slowly press in the pushrod using 220–2,205 lbs. (981–9,807 N) of force.
b. Align the holes of the pushrod and housing, pass a 1.5mm hexagon wrench through the holes to keep the setting position of the pushrod.
c. Release the press and install the dust boot to the tensioner.

34. Install the timing belt tensioner. Tighten the bolts alternately to 20 ft. lbs. (28 Nm).

35. Using pliers, remove the 1.5mm hexagon wrench from the belt tensioner.

36. Check the valve timing.

37. Install or connect the following:
- Remaining components
- Negative battery cable

38. Fill the radiator with engine coolant.
39. Start the engine and check for leaks.
40. Check the ignition timing.
41. Install the engine undercover.
42. Road test the vehicle.
43. Recheck all fluid levels.

4.0L Engine

2WD–BANK 1

➡**The engine must first be removed from the vehicle.**

1. Before servicing the vehicle, refer to the Precautions Section.
2. Properly discharge the fuel system pressure.
3. Properly disarm the SRS system.
4. Disconnect the negative battery cable. Disconnect the positive battery cable. Remove the battery.
5. Remove the hood subassembly.
6. Remove the engine from the vehicle and position it in a suitable holding fixture.
7. Remove the timing chain or belt cover subassembly.
8. Remove the chain subassembly.
9. Remove the number one cool inlet. Remove the front exhaust pipe assembly. Remove the three bolts and remove the exhaust manifold stay.
10. Disconnect the air fuel ratio sensor connector. Remove the six nuts; remove the exhaust manifold and gasket.
11. Disconnect the number one fuel pipe subassembly. Disconnect the number two fuel pipe subassembly.
12. Disconnect the fuel injector connectors. Remove the ten intake manifold retaining bolts. Remove the intake manifold from the engine.
13. Disconnect the engine coolant temperature sensor connector. Disconnect the heater hose. Remove the two bolts and four nuts, and then remove the water bypass joint RR and two gaskets. Remove the O-ring from the water outlet hose.
14. While raising the chain tensioner number two, insert a 0.039 inch diameter pin, into the hole. Hold the hexagonal portion of the camshaft in place, using a wrench.

➡**Be careful not to damage the cylinder head and valve lifter with the wrench.**

15. Remove the two bolts. Remove the camshaft timing gear, camshaft timing gear assembly and timing chain number two.

➡**Do not disassemble the camshaft timing gear assembly.**

16. Remove the bolt and then remove the chain tensioner number two.
17. Remove the camshafts.
18. Remove the two bolts. Separate the two ground cables.
19. Loosen the cylinder head retaining bolts in several steps and in the proper removal sequence. Remove the bolts.

➡**Be careful not to drop the washers into the cylinder head. Head warpage and cracking could result in removing the bolts in the wrong order.**

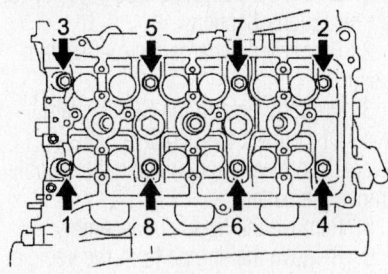

Bank 1 cylinder head bolt loosening sequence—4.0L engine

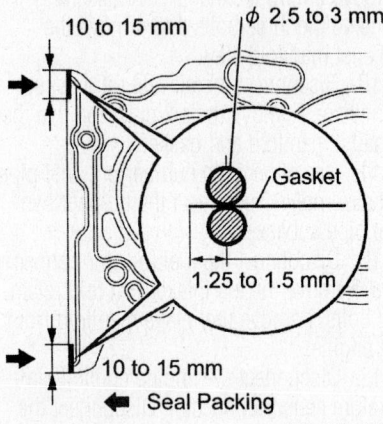

Cylinder head gasket sealant application—4.0L engine

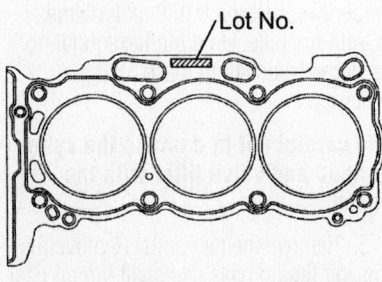

Bank 1 cylinder head gasket positioning—4.0L engine

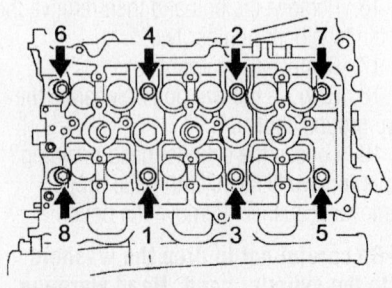

Bank 1 cylinder head bolt torque sequence—4.0L engine

20. Lift the cylinder head from the dowels on the cylinder block and remove it from the engine.

To install:

21. Before installing, thoroughly clean the gasket mating surfaces and check for warpage. Remove any old packing material. Be careful not to drop any oil on the contact surfaces of the cylinder head or engine block.

22. Apply continuous beads of seal packing, part number 08826-00080 or equivalent, to the cylinder head gasket. Bead width should be 0.098–0.118 inch.

➡**Install the cylinder head gasket within three minutes after applying the seal packing. Install the cylinder head bolts within fifteen minutes after applying the seal packing.**

23. Be sure to position the cylinder head gasket on the engine block with the lot number stamp facing upward.

➡**Be sure that the head gasket is installed in the correct direction. Position the gasket gently in order to avoid damaging the gasket with the bottom part of the cylinder head.**

24. Apply a light coat of clean engine oil to the cylinder head bolts. Tighten the cylinder head bolts, in two successive steps, to specification and in the proper sequence.

➡**Replace defective bolts, as required.**

25. Continue the installation in the reverse of the removal procedure.

26. Be sure to fill the cooling system with the proper grade and type engine coolant.

27. Be sure to fill the engine oil with the proper grade and type oil.

➡**Do not fill the engine with oil for at least four hours after the cylinder head has been installed.**

28. Start the engine and check for leaks. Correct as required.

2WD–BANK 2

➡**The engine must first be removed from the vehicle.**

1. Before servicing the vehicle, refer to the Precautions Section.

2. Properly discharge the fuel system pressure.

3. Properly disarm the SRS system.

4. Disconnect the negative battery cable. Disconnect the positive battery cable. Remove the battery.

5. Remove the hood subassembly.

6. Remove the engine from the vehicle and position it in a suitable holding fixture.

7. Remove the timing chain or belt cover subassembly.

8. Remove the chain subassembly.

9. Remove the number one cool inlet. Remove the number two front exhaust pipe assembly. Remove the three bolts and remove the exhaust manifold stay.

10. Disconnect the air fuel ratio sensor connector. Remove the six nuts; remove the exhaust manifold and gasket.

11. Disconnect the fuel injector connectors. Remove the ten intake manifold retaining bolts. Remove the intake manifold from the engine.

12. Disconnect the engine coolant temperature sensor connector. Disconnect the heater hose. Remove the two bolts and four nuts, and then remove the water bypass joint RR and two gaskets. Remove the O-ring from the water outlet hose.

13. Remove the two bolts and then remove the chain vibration damper number one.

14. While pushing down on chain tensioner number two, insert a 0.039 inch diameter pin, into the hole. Hold the hexagonal portion of the camshaft in place, using a wrench.

➡**Be careful not to damage the cylinder head and valve lifter with the wrench.**

15. Remove the two bolts. Remove the camshaft timing gear, camshaft timing gear assembly and timing chain number two.

➡**Do not disassemble the camshaft timing gear assembly.**

16. Remove the bolt and then remove the chain tensioner number three.

17. Remove the camshafts.

18. Remove the two bolts. Separate the two ground cables.

19. Loosen the cylinder head retaining bolts in several steps and in the proper removal sequence. Remove the bolts.

➡**Be careful not to drop the washers into the cylinder head. Head warpage and cracking could result in removing the bolts in the wrong order.**

20. Lift the cylinder head from the dowels on the cylinder block and remove it from the engine.

To install:

21. Before installing, thoroughly clean the gasket mating surfaces and check for warpage. Remove any old packing material. Be careful not to drop any oil on the contact surfaces of the cylinder head or engine block.

22. Apply continuous beads of seal packing, part number 08826-00080 or equivalent, to the cylinder head gasket. Bead width should be 0.098–0.118 inch.

➡ **Install the cylinder head gasket within three minutes after applying the seal packing. Install the cylinder head bolts within fifteen minutes after applying the seal packing.**

23. Be sure to position the cylinder head gasket on the engine block with the lot number stamp facing upward.

➡ **Be sure that the head gasket is installed in the correct direction. Position the gasket gently in order to avoid damaging the gasket with the bottom part of the cylinder head.**

24. Apply a light coat of clean engine oil to the cylinder head bolts. Tighten the cylinder head bolts, in two successive steps, to specification and in the proper sequence.

➡ **Replace defective bolts, as required.**

25. Continue the installation in the reverse of the removal procedure.
26. Be sure to fill the cooling system with the proper grade and type engine coolant.
27. Be sure to fill the engine oil with the proper grade and type oil.

➡ **Do not fill the engine with oil for at least four hours after the cylinder head has been installed.**

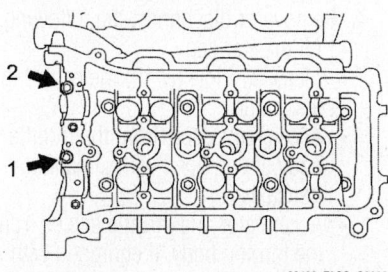

Bank 2 recessed cylinder head bolt loosening sequence—4.0L engine

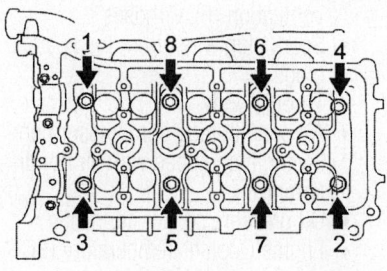

Bank 2 cylinder head bolt loosening sequence—4.0L engine

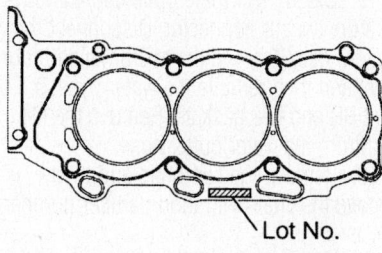

Bank 2 cylinder head gasket positioning—4.0L engine

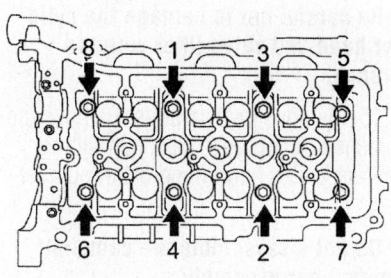

Bank 2 Cylinder head bolt torque sequence—4.0L engine

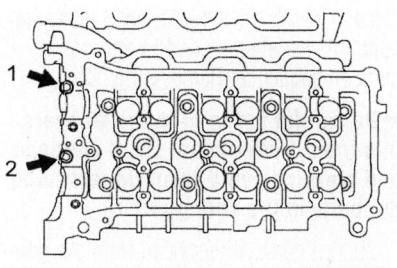

Bank 2 recessed cylinder head bolt torque sequence—4.0L engine

28. Start the engine and check for leaks. Correct as required.

4WD AND PRERUNNER–BANK 1

1. Before servicing the vehicle, refer to the Precautions Section.
2. Properly discharge the fuel system pressure.
3. Properly disarm the SRS system.
4. Disconnect the negative battery cable. Disconnect the positive battery cable. Remove the battery.
5. Remove the power steering gear assembly.
6. Remove the front differential carrier assembly.
7. Remove the timing chain or belt cover subassembly.
8. Remove the chain subassembly.

9. Remove the number one cool inlet. Remove the front exhaust pipe assembly. Remove the three bolts and remove the exhaust manifold stay.
10. Disconnect the air fuel ratio sensor connector. Remove the six nuts; remove the exhaust manifold and gasket.
11. Disconnect the number one fuel pipe subassembly. Disconnect the number two fuel pipe subassembly.
12. Disconnect the fuel injector connectors. Remove the ten intake manifold retaining bolts. Remove the intake manifold from the engine.
13. Disconnect the engine coolant temperature sensor connector. Disconnect the heater hose. Remove the two bolts and four nuts, and then remove the water bypass joint RR and two gaskets. Remove the O-ring from the water outlet hose.
14. While raising the chain tensioner number two, insert a 0.039 inch diameter pin, into the hole. Hold the hexagonal portion of the camshaft in place, using a wrench.

➡ **Be careful not to damage the cylinder head and valve lifter with the wrench.**

15. Remove the two bolts. Remove the camshaft timing gear, camshaft timing gear assembly and timing chain number two.

➡ **Do not disassemble the camshaft timing gear assembly.**

16. Remove the bolt and then remove the chain tensioner number two.
17. Remove the camshafts.
18. Remove the two bolts. Separate the two ground cables.
19. Loosen the cylinder head retaining bolts in several steps and in the proper removal sequence. Remove the bolts.

➡ **Be careful not to drop the washers into the cylinder head. Head warpage and cracking could result in removing the bolts in the wrong order.**

20. Lift the cylinder head from the dowels on the cylinder block and remove it from the engine.

To install:
21. Before installing, thoroughly clean the gasket mating surfaces and check for warpage. Remove any old packing material. Be careful not to drop any oil on the contact surfaces of the cylinder head or engine block.
22. Apply continuous beads of seal packing, part number 08826-00080 or equivalent, to the cylinder head gasket. Bead width should be 0.098–0.118 inch.

➡**Install the cylinder head gasket within three minutes after applying the seal packing. Install the cylinder head bolts within fifteen minutes after applying the seal packing.**

23. Be sure to position the cylinder head gasket on the engine block with the lot number stamp facing upward.

➡**Be sure that the head gasket is installed in the correct direction. Position the gasket gently in order to avoid damaging the gasket with the bottom part of the cylinder head.**

24. Apply a light coat of clean engine oil to the cylinder head bolts. Tighten the cylinder head bolts, in two successive steps, to specification and in the proper sequence.

➡**Replace defective bolts, as required.**

25. Continue the installation in the reverse of the removal procedure.
26. Be sure to fill the cooling system with the proper grade and type engine coolant.
27. Be sure to fill the engine oil with the proper grade and type oil.

➡**Do not fill the engine with oil for at least four hours after the cylinder head has been installed.**

28. Start the engine and check for leaks. Correct as required.

4WD AND PRERUNNER–BANK 2

1. Before servicing the vehicle, refer to the Precautions Section.
2. Properly discharge the fuel system pressure.
3. Properly disarm the SRS system.
4. Disconnect the negative battery cable. Disconnect the positive battery cable. Remove the battery.
5. Remove the power steering gear assembly.
6. Remove the front differential carrier assembly.
7. Remove the timing chain or belt cover subassembly.
8. Remove the chain subassembly.
9. Remove the number one cool inlet. Remove the number two front exhaust pipe assembly. Remove the three bolts and remove the exhaust manifold stay.
10. Disconnect the air fuel ratio sensor connector. Remove the six nuts; remove the exhaust manifold and gasket.
11. Disconnect the fuel injector connectors. Remove the ten intake manifold retaining bolts. Remove the intake manifold from the engine.

12. Disconnect the engine coolant temperature sensor connector. Disconnect the heater hose. Remove the two bolts and four nuts, and then remove the water bypass joint RR and two gaskets. Remove the O-ring from the water outlet hose.
13. Remove the two bolts and then remove the chain vibration damper number one.
14. While pushing down on chain tensioner number two, insert a 0.039 inch diameter pin, into the hole. Hold the hexagonal portion of the camshaft in place, using a wrench.

➡**Be careful not to damage the cylinder head and valve lifter with the wrench.**

15. Remove the two bolts. Remove the camshaft timing gear, camshaft timing gear assembly and timing chain number two.

➡**Do not disassemble the camshaft timing gear assembly.**

16. Remove the bolt and then remove the chain tensioner number three.
17. Remove the camshafts.
18. Remove the two bolts. Separate the two ground cables.
19. Loosen the cylinder head retaining bolts in several steps and in the proper removal sequence. Remove the bolts.

➡**Be careful not to drop the washers into the cylinder head. Head warpage and cracking could result in removing the bolts in the wrong order.**

20. Lift the cylinder head from the dowels on the cylinder block and remove it from the engine.
To install:
21. Before installing, thoroughly clean the gasket mating surfaces and check for warpage. Remove any old packing material. Be careful not to drop any oil on the contact surfaces of the cylinder head or engine block.
22. Apply continuous beads of seal packing, part number 08826-00080 or equivalent, to the cylinder head gasket. Bead width should be 0.098–0.118 inch.

➡**Install the cylinder head gasket within three minutes after applying the seal packing. Install the cylinder head bolts within fifteen minutes after applying the seal packing.**

23. Be sure to position the cylinder head gasket on the engine block with the lot number stamp facing upward.

➡**Be sure that the head gasket is installed in the correct direction. Position the gasket gently in order to avoid damaging the gasket with the bottom part of the cylinder head.**

24. Apply a light coat of clean engine oil to the cylinder head bolts. Tighten the cylinder head bolts, in two successive steps, to specification and in the proper sequence.

➡**Replace defective bolts, as required.**

25. Continue the installation in the reverse of the removal procedure.
26. Be sure to fill the cooling system with the proper grade and type engine coolant.
27. Be sure to fill the engine oil with the proper grade and type oil.

➡**Do not fill the engine with oil for at least four hours after the cylinder head has been installed.**

28. Start the engine and check for leaks. Correct as required.

Intake Manifold

REMOVAL & INSTALLATION

2.4L Engine

1. Relieve the fuel system pressure.
2. Disconnect the negative battery cable.
3. Drain the engine coolant.
4. Remove or disconnect the following:
   - Air cleaner cap
   - Mass Air Flow (MAF) meter and the resonator
   - Accelerator cable from the throttle body, if equipped with a manual transaxle
   - Accelerator and throttle cables from the throttle body, if equipped with an automatic transaxle
   - Intake air connector
   - Air conditioning idle-up valve, if equipped with air conditioning
   - No. 1 and No. 2 Positive Crankcase Ventilation (PCV) hoses
   - Spark plug wires from the spark plugs
   - Throttle body
   - Air conditioning compressor connector, if equipped with air conditioning
   - Oil pressure sensor connector
   - Engine Coolant Temperature (ECT) sensor connector
   - Exhaust Gas Recirculation (EGR) gas temperature sensor connector

- EGR Vacuum Switching Valve (VSV) connector
5. Disconnect the engine wiring harness, as follows:
  - 2 bolts and the harness from the intake chamber
  - 5 engine harness clamps and harness
6. Remove or disconnect the following:
  - Knock (KS) sensor connector
  - Crankshaft Position (CKP) sensor connector
  - Fuel pressure control VSV connector
  - Data Link Connector 1 (DLC1) from the bracket
  - 2 engine wiring harness clamps
  - Engine wiring harness
  - Fuel injectors
  - EGR valve and vacuum modulator
  - Intake chamber stay, by removing the 2 bolts
  - Fuel return pipe, by removing the hoses and 2 bolts
7. Remove the intake chamber, as follows:
  - Vacuum hose, from the gas filter
  - Brake booster vacuum hose, from the intake chamber
  - 3 bolts, 2 nuts, air intake chamber and gasket
8. Remove the fuel inlet tube, by removing the union bolts
9. Remove the delivery pipe and injectors, as follows:
  - Vacuum hose, from the fuel pressure regulator
  - Bolts and delivery pipe together with the 4 injectors
  - 4 insulators, from the 4 spacers
  - Injectors from the delivery pipe
  - O-ring and grommet, from each injector
10. Remove or disconnect the following:
  - Intake manifold, by removing the 3 bolts and 2 nuts
  - Gasket

To install:
11. Clean the intake manifold surfaces.
12. Install or connect the following:
  - New gasket
  - Intake manifold. Tighten the bolts and nuts to 22 ft. lbs. (30 Nm).
13. Install the injectors to the delivery pipe, as follows:
  - New grommet to the injector
  - New O-ring onto the injector, by lubricating it with gasoline
  - Injectors to the delivery pipe
  - Injector with the connector facing upward
  - Injectors and delivery pipe. Tighten the bolts to 15 ft. lbs. (21 Nm).

➡ **Check that the injectors rotate smoothly.**

  - Fuel tube, with new gaskets. Tighten the union bolts to 22 ft. lbs. (30 Nm).
14. Install the air intake chamber, as follows:
  - New gasket
  - Air intake chamber. Tighten the bolts and nuts to 15 ft. lbs. (21 Nm).
  - Vacuum hose, to the gas filter
  - Brake booster vacuum hose, to the intake chamber
15. Install or connect the following:
  - Fuel return pipe
  - Intake chamber stay. Tighten the bolts to 14 ft. lbs. (20 Nm).
  - EGR valve and vacuum modulator
  - Injector connectors
16. Connect the engine wiring harness to the engine, as follows:
  - Engine wiring harness to the intake manifold

- Engine wiring harness clamps
- DLC1 to the bracket
- Fuel pressure control VSV connector
- KS sensor connector
- CKP sensor connector
- 5 engine wiring harness clamps
- Engine wiring harness, to the intake chamber
17. Install or connect the following:
- EGR VSV connector
- EGR gas temperature sensor connector
- ECT sensor connector
- Oil pressure sensor connector
- Compressor connector, if equipped with air conditioning
- Throttle body
- Spark plug wires, to the spark plugs
- No. 1 and No. 2 PCV hoses
- Air conditioning idle up valve, if equipped with air conditioning
- Air intake connector

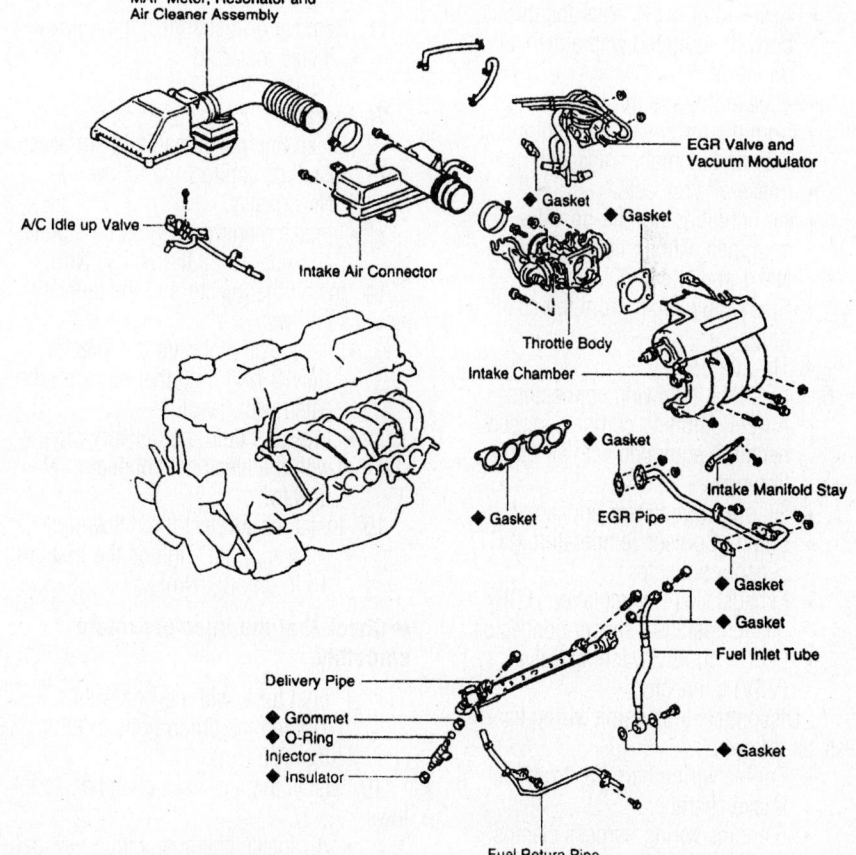

◆ Non-reusable part

7924YG32

Intake manifold and related components—2.4L and 2.7L (3RZ-FE/M) engines

- Accelerator cable, to the throttle body, if equipped with a manual transaxle
- Throttle and accelerator cables to the throttle body, if equipped with automatic transaxle
- MAF meter, resonator and the air cleaner cap
- Negative battery cable

18. Refill the cooling system.
19. Start the engine and check for leaks.
20. Check the ignition timing. Road test the vehicle for proper operation.
21. Recheck all fluid levels.

2.7L (3RZ-FE/M) Engine

1. Before servicing the vehicle, refer to the Precautions Section.
2. Relieve the fuel system pressure.
3. Disconnect the negative battery cable.
4. Drain the engine coolant.
5. Remove or disconnect the following:
- Air cleaner cap
- Mass Air Flow (MAF) meter and resonator
- Accelerator cable, from the throttle body, if equipped with a manual transaxle
- Accelerator and throttle cables, from the throttle body, if equipped with an automatic transaxle
- Intake air connector
- Air conditioning idle-up valve, if equipped with air conditioning
- No. 1 and No. 2 PCV hoses
- Spark plug wires, from the spark plugs
- Throttle body

6. Detach the following connectors:
- Air conditioning compressor connector, if equipped with air conditioning
- Oil pressure sensor connector
- Engine Coolant Temperature (ECT) sensor connector
- Exhaust Gas Recirculation (EGR) gas temperature sensor connector
- EGR Vacuum Switching Valve (VSV) connector

7. Disconnect the engine wiring harness, as follows:
- Engine wiring harness, from the intake chamber
- 5 engine wiring harness clamps and engine wiring harness
- Knock (KS) sensor connector
- Crankshaft Position (CKP) sensor connector
- Fuel pressure control VSV connector

- Data Link Connector 1 (DLC1), from the bracket
- 2 engine wiring harness clamps.
- Engine wiring harness from the engine

8. Remove or disconnect the following:
- Fuel injectors
- EGR valve and vacuum modulator
- Intake chamber stay
- Fuel return pipe

9. Remove the intake chamber, as follows:
- Vacuum hose, from the gas filter
- Brake booster vacuum hose, from the intake chamber
- Intake chamber and gasket
- Fuel inlet tube, by removing the union bolts

10. Remove the delivery pipe and injectors, as follows:
- Vacuum hose, from the fuel pressure regulator
- Delivery pipe, with the 4 injectors
- 4 insulators, from the 4 spacers
- 4 injectors, from the delivery pipe
- O-ring and grommet, from each injector

11. Remove or disconnect the following:
- Intake manifold
- Gasket

To install:

12. Clean the intake manifold surfaces.
13. Install or connect the following:
- New gasket
- Intake manifold. Tighten the bolts and nuts to 22 ft. lbs. (30 Nm).

14. Install the injectors to the delivery pipe, as follows:
- New grommet onto the injector
- New O-ring onto the injector, lubricated with gasoline
- Injectors onto the delivery pipe, with the electrical connector upward

15. Install or connect the following:
- Delivery pipe. Tighten the bolts to 15 ft. lbs. (21 Nm).

➡**Check that the injectors rotate smoothly.**

- Fuel tube, with new gaskets. Tighten the union bolts to 22 ft. lbs. (30 Nm).

16. Install the air intake chamber, as follows:
- Air intake chamber with a new gasket. Tighten the bolts and nuts to 15 ft. lbs. (21 Nm).
- Vacuum hose, to the gas filter
- Brake booster vacuum hose, to the intake chamber

17. Install or connect the following:

- Fuel return pipe
- Intake chamber stay. Tighten the bolts to 14 ft. lbs. (20 Nm).

18. Install the EGR valve and vacuum modulator, as follows:
- New gasket, EGR valve and vacuum modulator. Tighten the bolt to 74 inch lbs. (9 Nm) and the nuts to 14 ft. lbs. (19 Nm).
- Vacuum hoses, to the EGR VSV
- Water bypass hose
- EGR pipe, using new gaskets. Tighten the bolts to 14 ft. lbs. (18 Nm), intake manifold nuts to 14 ft. lbs. (19 Nm) and the cylinder head nuts to 15 ft. lbs. (20 Nm).
- Injector connectors

19. Connect the engine wiring harness to the engine, as follows:
- Engine wiring harness, to the intake manifold
- 2 engine wiring harness clamps
- DLC1, to the bracket
- Fuel pressure control VSV connector
- KS sensor connector
- CKP sensor connector
- 5 engine wiring harness clamps.
- Engine wiring harness, to the intake chamber

20. Attach the following connectors to the engine:
- EGR VSV connector
- EGR gas temperature sensor connector
- ECT sensor connector
- Oil pressure sensor connector
- Compressor connector, if equipped with air conditioning

21. Install or connect the following:
- Throttle body
- Spark plug wires, to the spark plugs
- No. 1 and No. 2 PCV hoses
- Air conditioning idle up valve, if equipped with air conditioning
- Air intake connector, by installing the 2 bolts, hose clamp and 2 air hoses
- Accelerator cable, if equipped with a manual transaxle
- Throttle and accelerator cables, if equipped with an automatic transaxle
- MAF meter, resonator and the air cleaner cap
- Negative battery cable

22. Refill the cooling system.
23. Start the engine and check for leaks.
24. Check the ignition timing. Road test the vehicle for proper operation.
25. Recheck all fluid levels.

2.7L (2TR-FE/X) Engine

1. Before servicing the vehicle, refer to the Precautions Section.
2. Properly relieve the fuel system pressure.
3. Properly disarm the SRS system
4. Disconnect the negative battery cable.
5. On 4WD and PreRunner, remove the engine undercover subassembly.
6. Drain the engine coolant. Remove the air intake connector.
7. Remove the throttle body and motor assembly. Disconnect the fuel hoses.
8. Disconnect the fuel vapor feed hose from the VSR. Disconnect the vacuum hose. Remove the bolt, and then remove the clamp bracket.
9. Disconnect the number two water bypass hose. Disconnect the number three ventilation hose. Disconnect the VSV connector.
10. Disengage the engine wire harness clamp. Disconnect the air conditioning compressor magnetic clutch connector.

11. Disengage the wire harness clamp. Remove the bolt and harness clamp bracket. Disconnect the three connectors.
12. Remove the retaining nut inside the relay block. Disconnect the engine wire harness from the relay block.
13. Remove the five bolts and two nuts retaining the intake manifold in place. Remove the intake manifold from the engine.

To install:

14. Clean all surfaces.
15. Install a new gasket onto the intake manifold. Position the intake manifold to the engine.
16. Install the retaining bolts and tighten to specification, in an alternating sequence.
17. Continue the installation in the reverse order of the removal procedure.
18. Be sure to fill the cooling system with the proper grade and type engine coolant.
19. Start the engine and check for leaks. Correct as required.

3.4L Engine

1. Before servicing the vehicle, refer to the Precautions Section.
2. Disconnect the negative battery cable.
3. Relieve the fuel system pressure.
4. Drain the engine coolant.
5. Remove or disconnect the following:
   - Spark plug wires from the spark plugs
   - Air cleaner cap
   - Mass Air Flow (MAF) meter and resonator
   - Actuator cable with the bracket, if equipped with cruise control
   - Accelerator cable
   - Throttle cable, if equipped with an automatic transmission
   - Exhaust Gas Recirculation (EGR) pipe and gaskets, if equipped with an EGR valve
   - Oil filler tube and No. 1 throttle cable clamp, by removing the bolt and 2 nuts
   - Intake chamber stay, by removing the 2 bolts
   - Vacuum Switching Valve (VSV) connector, for the fuel pressure control
   - Throttle Position (TP) sensor connector
   - Idle Air Control (IAC) valve connector
   - EGR gas temperature connector, if equipped with an EGR valve
   - VSV connector for the EGR valve, if equipped with an EGR valve
   - Disconnect the Positive Crankcase Ventilation (PCV) hoses
   - Water bypass hoses
   - Air assist hose from the intake air connector
   - 2 vacuum sensing hoses from the VSV
   - Evaporative Emission (EVAP) hose
   - Air hose from the power steering
   - Air hose from the air conditioning idle up valve, if equipped with air conditioning
   - Air intake chamber assembly
   - Engine wiring harness from the intake air connector
   - 2 fuel return hoses
   - Brake booster vacuum hose, from the intake air connector
   - Ground strap, from the intake air connector
   - Data Link Connector 1 (DLC1) from the intake air connector bracket
   - Idle up valve connector, if equipped with air conditioning

FUEL HOSE

NO. 2 FUEL HOSE

NO. 3 VENTILATION HOSE

● GASKET

25 (255, 18)

25 (255, 18)

25 (255, 18)

25 (255, 18)

INTAKE MANIFOLD

25 (255, 18)

N*m (kgf*cm, ft*lbf) : Specified torque

VACUUM HOSE

● Non-reusable part

09490_TACO_G0038

Intake manifold and related components—2.7L (2TR-FE/X) engine

- Intake air connector
- Upper radiator hose, from the engine
- Oil pressure sensor connector
- Crankshaft Position (CKP) sensor connector
- 6 injector connectors
- Engine Coolant Temperature (ECT) sender gauge connector
- ECT sensor connector
- Knock (KS) sensor connector
- Camshaft Position (CMP) sensor connector
- 3 engine wiring harness clamps
- Engine wiring harness, from the cylinder head
- Fuel pressure regulator
- Heater hose
- Camshaft Position sensor.
- Fuel inlet hose
- Intake manifold stay
- Intake manifold assembly

To install:

6. Clean all surfaces.
7. Install or connect the following:
- New gaskets
- Intake manifold assembly. Tighten the bolts and nuts to 13 ft. lbs. (18 Nm).
- Intake manifold stay. Tighten the bolts to 13 ft. lbs. (18 Nm).
- Fuel inlet hose
- CMP sensor. Tighten it to 71 inch lbs. (8 Nm).
- Engine wiring harness to the cylinder head
- 3 engine wiring harness clamps
- Oil pressure sensor connector
- CKP sensor connector
- 6 injector connectors
- ECT sender gauge connector
- ECT sensor connector
- KS sensor connector
- CMP sensor connector
- Heater hose
- Intake manifold. Tighten the bolts and nuts to 14 ft. lbs. (18.5 Nm).
- DLC1 to the bracket on the intake manifold
- Ground strap to the intake manifold
- Brake booster vacuum hose, to the intake air connector
- 2 fuel return hoses
- Engine wiring harness to the intake manifold
- Idle up valve connector, if equipped with air conditioning
- Air intake chamber assembly to the engine. Tighten the bolts and nuts to 14 ft. lbs. (18.5 Nm).
- PCV hoses
- Water bypass hoses

- Air assist hose to the intake manifold
- 2 vacuum sensing hoses to the VSV
- EVAP hose
- Air hose to the power steering
- Air hose to the air conditioning idle up valve, if equipped with air conditioning
- VSV connector for the fuel pressure control
- TP sensor connector
- IAC valve connector
- EGR gas temperature connector, if equipped with an EGR valve
- VSV connector for the EGR valve, if equipped with an EGR valve
- Intake chamber stay. Tighten the bolts to 30 ft. lbs. (40 Nm).
- New O-ring to the oil filler tube
- Oil filler tube end into the tube hole in the oil pan
- Oil filler tube and No. 1 throttle cable clamp
- New gaskets and the EGR pipe.

Tighten the clamp nuts to 71 inch lbs. (8 Nm) and the EGR pipe nuts to 14 ft. lbs. (18 Nm).
- Fuel pressure regulator
- 3 clamps for the spark plug wires, to the No. 2 timing belt cover
- CMP connector to the No. 2 timing belt cover
- Upper radiator hose
8. Fill with engine coolant.
9. Install or connect the following:
- Spark plug wires to the spark plugs
- Actuator cable with the bracket, if equipped with cruise control
- Accelerator cable
- Throttle cable, if equipped with an automatic transmission
- Air cleaner hose
- Negative battery cable
10. Fill the radiator with engine coolant.
11. Start the engine and check for leaks.

4.0L Engine

1. Before servicing the vehicle, refer to the Precautions Section.

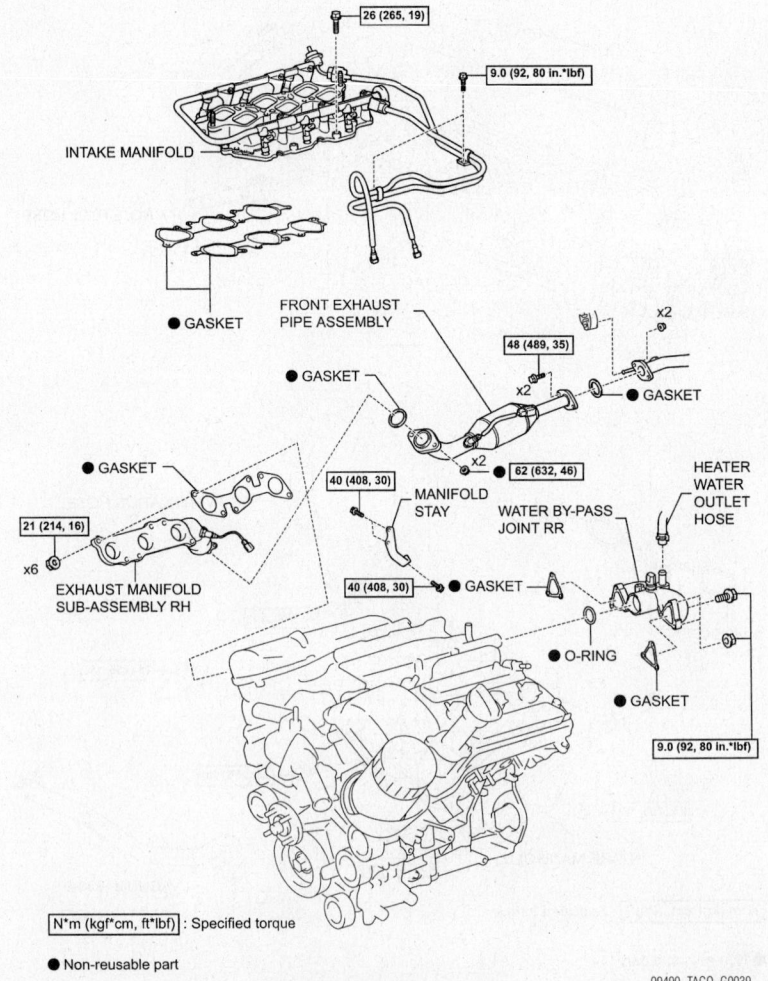

Intake manifold and related components—4.0L engine

2. Properly relieve the fuel system pressure.

3. Properly disarm the SRS system

4. Disconnect the negative battery cable.

5. Drain the engine coolant. Remove the air cleaner assembly.

6. Disconnect the fuel injector wiring connectors.

7. Remove the necessary components in order to gain access to the intake manifold retaining bolts.

8. Remove the intake manifold retaining bolts. Remove the intake manifold from the engine.

To install:

9. Clean all surfaces.

10. Install a new gasket on each cylinder head.

➡**Align the ports of the gasket and the cylinder head. Be careful of the installation direction. Position the intake manifold to the engine.**

11. Install the retaining bolts and tighten to specification, in an alternating sequence.

12. Continue the installation in the reverse order of the removal procedure.

13. Be sure to fill the cooling system with the proper grade and type engine coolant.

14. Start the engine and check for leaks. Correct as required.

Exhaust Manifold

REMOVAL & INSTALLATION

2.4L and 2.7L (3RZ-FE/M) Engines

1. Before servicing the vehicle, refer to the Precautions Section.

2. Remove or disconnect the following:
- Clamp from the support bracket
- Support bracket
- Front exhaust pipe and gaskets from the exhaust manifold
- Heat insulator
- Exhaust manifold and gasket

Front exhaust pipe to exhaust manifold nut and bolt locations—2.4L and 2.7L(3RZ-FE/M) engines

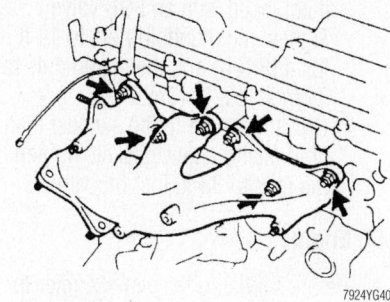

Exhaust manifold bolt locations—2.4L and 2.7L (3RZ-FE/M) engines

To install:

3. Install or connect the following:
- Exhaust manifold and gasket. Tighten the nuts to 36 ft. lbs. (49 Nm).
- Heat insulator. Tighten the bolts and nuts to 48 inch lbs. (5.5 Nm).
- Front exhaust pipe assembly to the exhaust manifold. Tighten the nuts to 46 ft. lbs. (62 Nm).

- Support bracket. Tighten the bolts to 29 ft. lbs. (39 Nm).
- Clamp. Tighten the bolt to 14 ft. lbs. (19 Nm).

4. Start the engine.

5. Check for exhaust leaks.

2.7L (2TR-FE/X) Engine

1. Before servicing the vehicle, refer to the Precautions Section.

2. Disconnect the negative battery cable.

3. Disconnect the exhaust manifold to exhaust flange nuts.

4. Remove the necessary components to gain access to the exhaust manifold retaining bolts.

5. Remove the exhaust manifold retaining nuts. Discard the nuts. Remove the exhaust manifold from the engine.

To install:

6. Installation is the reverse of the removal procedure.

7. Be sure to use new gaskets. Be sure to use new retaining nuts.

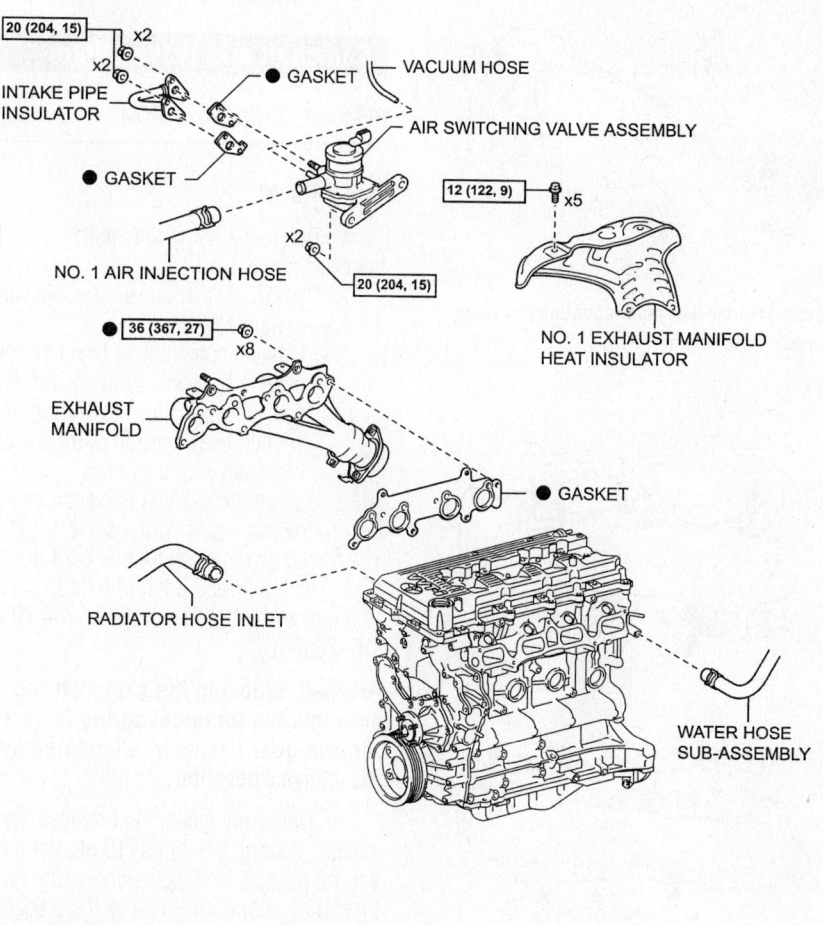

N*m (kgf*cm, ft*lbf) : Specified torque

● Non-reusable part

Exhaust manifold and related components—2.7L (2TR-FE/X) engine

3.4L Engine

1. Before servicing the vehicle, refer to the Precautions Section.
2. Remove or disconnect the following:
   - Exhaust crossover pipe, from the exhaust manifold by removing the 3 nuts
   - Exhaust Gas Recirculation (EGR) pipe, from the exhaust manifold, on the left manifold equipped with an EGR valve
   - Exhaust manifold heat insulator, by removing the 3 nuts
   - Exhaust manifold

To install:

3. Install or connect the following:
   - Exhaust manifold, using a new gasket. Tighten the nuts to 30 ft. lbs. (40 Nm).
   - Exhaust heat insulator. Tighten the nuts to 71 inch lbs. (8 Nm).
   - EGR pipe to the exhaust manifold,

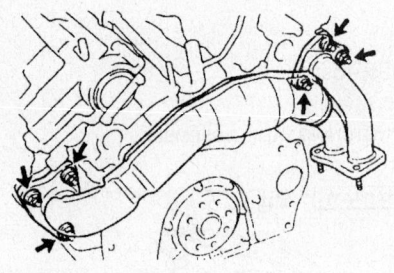

Exhaust crossover pipe mounting nut locations—3.4L engine

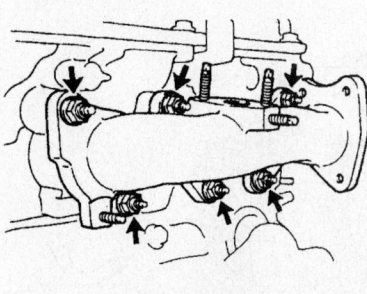

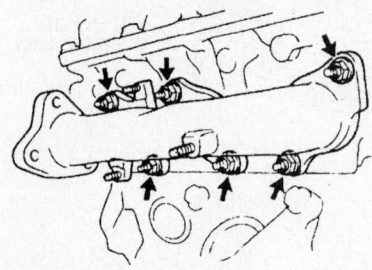

Exhaust manifold bolt locations—3.4L engine

if equipped with an EGR valve. Tighten the manifold nuts to 14 ft. lbs. (18 Nm) and the clamp nuts to 71 inch lbs. (8 Nm).
   - Crossover pipe to the exhaust manifold, using a new gasket. Tighten the nuts to 33 ft. lbs. (45 Nm).

4.0L Engine

1. Before servicing the vehicle, refer to the Precautions Section.
2. Disconnect the negative battery cable.
3. Disconnect the exhaust manifold to exhaust flange nuts.
4. Remove the necessary components to gain access to the exhaust manifold retaining bolts.
5. Remove the exhaust manifold retaining nuts. Discard the nuts. Remove the exhaust manifold from the engine.

To install:

6. Installation is the reverse of the removal procedure.
7. Be sure to use new gaskets. Be sure to use new retaining nuts.

Camshaft and Valve Lifters

REMOVAL & INSTALLATION

2.4L Engine

1. Before servicing the vehicle, refer to the Precautions Section.
2. Remove or disconnect the following:
   - Timing chain
   - Exhaust camshaft by bringing the service bolt hole of the driven sub-gear upwards. Turn the hexagon wrench head portion of the exhaust camshaft with a wrench.
3. Secure the exhaust camshaft sub-gear to the main gear with a service bolt. The thread diameter should be 0.23 in. (6mm) with a thread pitch of 0.04 in. (1.0mm) and a bolt length of 0.63–0.79 in. (16–20mm).

➡**When removing the camshaft, be sure that the torsional spring force of the sub-gear has been eliminated by the above operation.**

4. Uniformly loosen and remove the exhaust bearing cap bolts (10 of them), in several passes. Use the reverse order of the tightening sequence. Remove the 5 bearing caps and the camshaft. Do the same for the intake camshafts.

➡**If the camshaft is not being lifted out straight and level, reinstall the No. 3**

cap with the 2 bolts. Alternately loosen, then remove the bearing cap bolts with the camshaft pulled up. Do not pry on or force the camshaft.

5. Inspect the camshafts for excessive runout. Inspect the cam lobes and journals. The bearings are part of the cam and should be inspected for flaking or scoring. If the bearings are damaged, replace the caps and the cylinder head as a set. The camshaft journal oil and thrust clearances should be checked.

To install:

6. Install the intake camshaft, as follows:
   a. Apply multi-purpose grease to the thrust portion of the intake camshaft.

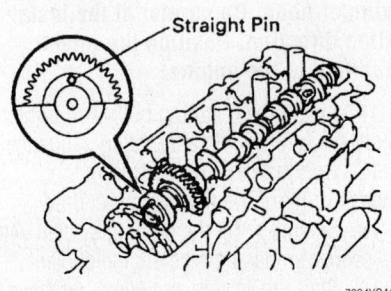

Install the camshaft with the pin facing upwards—2.4L engine

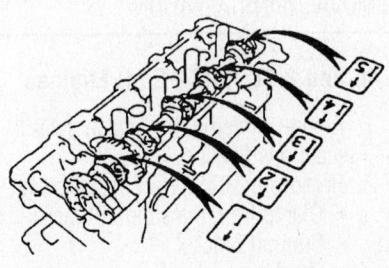

Intake camshaft bearing cap locations—2.4L engine

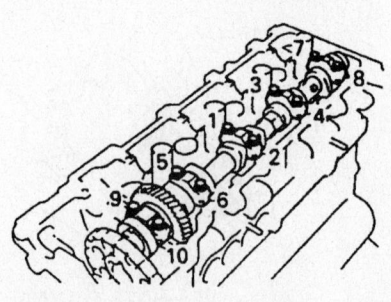

Intake camshaft bearing cap torque sequence—2.4L engine

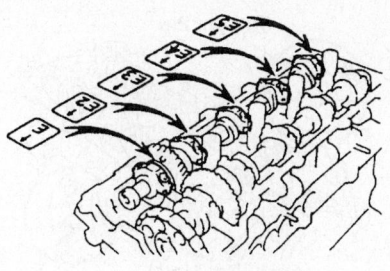

Exhaust camshaft bearing cap locations—2.4L engine

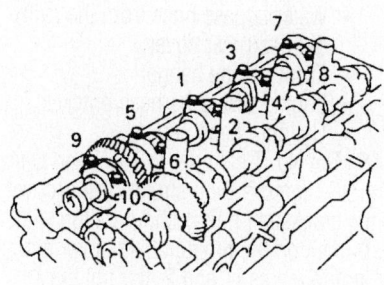

Exhaust camshaft bearing cap torque sequence—2.4L engine

b. Position the intake camshaft with the pin facing upward.

c. Install the bearing caps in their proper locations. Apply a light coat of engine oil to the threads and install the cap bolts. Uniformly tighten the cap bolts in the sequence shown to 12 ft. lbs. (16 Nm).

7. Install the exhaust camshaft, as follows:

a. Apply engine oil to the thrust portion of the intake camshaft.

b. Engage the exhaust camshaft gear to the intake camshaft gear by matching the timing marks (1 and 2 dots) on each other.

c. Roll down the exhaust camshaft onto the bearing journals while engaging the gears with each other. Install the bearing caps in their proper locations.

d. Apply a light coat of engine oil to the threads and install the cap bolts. Uniformly tighten the cap bolts in the sequence shown to 12 ft. lbs. (16 Nm).

e. Remove the service bolt from the driven sub-gear. Check that the intake and exhaust camshafts turn smoothly.

8. Set No. 1 cylinder to Top Dead Center (TDC) of the compression stroke. The crankshaft pulley groove aligns with the **0** mark on timing cover and camshaft timing marks with 1 dot and 2 dots will be in a straight line on the cylinder head surface.

9. Install the timing gear, as follows:

a. Place the gear over the straight pin of the intake camshaft.

b. Hold the intake camshaft with a wrench. Install and tighten the bolt to 54 ft. lbs. (74 Nm).

c. Hold the exhaust camshaft and install the bolt and distributor gear. Tighten the bolt to 34 ft. lbs. (46 Nm).

10. Install the chain tensioner, using a new gasket (mark toward the front), as follows:

a. Release the ratchet pawl, fully push in the plunger and apply the hook to the pin so that the plunger cannot spring out.

b. Turn the crankshaft pulley clockwise to provide some slack for the chain on the tensioner side.

c. Push the tensioner by hand until it touches the head installation surface, then install the 2 nuts. Tighten the nuts to 13 ft. lbs. (18 Nm). Check that the hook of the tensioner is not released.

d. Turn the crankshaft to the left so that the hook of the chain tensioner is released from the pin of the plunger, allowing the plunger to spring out and the slipper to be pushed into the chain.

11. Check and adjust the valve clearance. Intake valve clearance is 0.006–0.010 inch (0.15–0.25mm) and exhaust valve clearance is 0.010–0.014 inch (0.25–0.35mm).

12. Recheck the engine for proper valve timing. Check and adjust the valve clearance.

13. Install the spark plugs and the semi-circular plug.

14. Recheck the engine for proper valve timing. Install the valve cover and engine hangers. Tighten the engine hanger bolts to 30 ft. lbs. (42 Nm).

15. Reinstall all other parts from the timing chain removal. Fill any fluids, start the engine, top off the fluids.

2.7L (3RZ-FE/M) Engine

1. Before servicing the vehicle, refer to the Precautions Section.

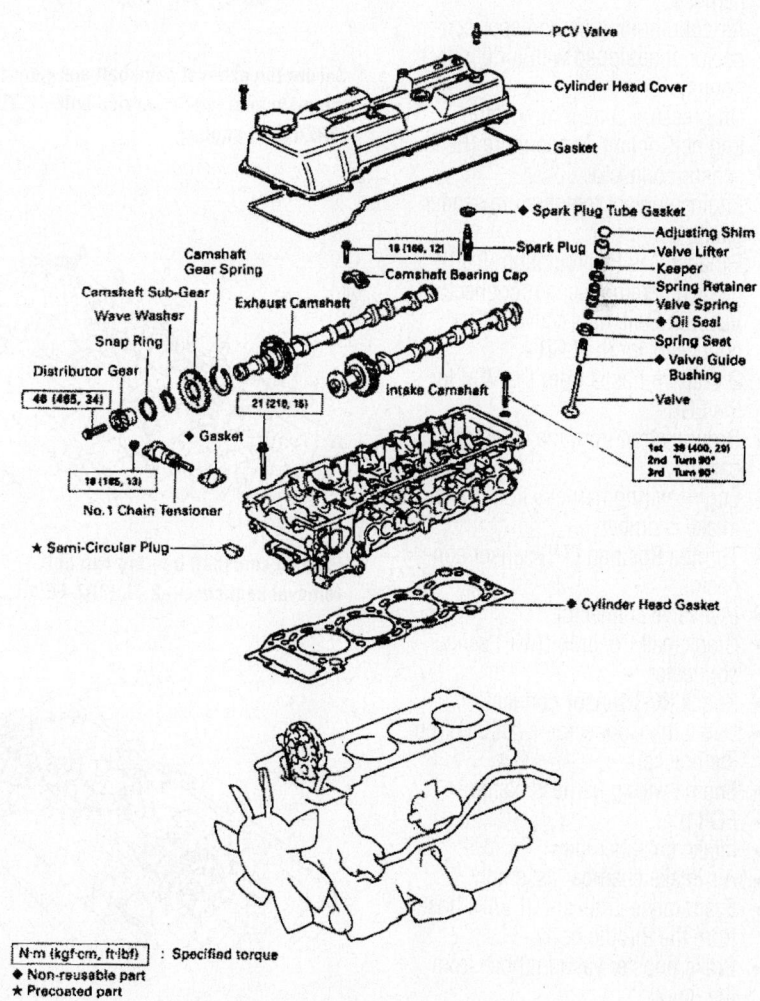

Camshafts and related components—2.7L (3RZ-FE/M) engine

2. Disconnect the negative battery cable.

3. Drain the engine coolant.

4. Remove or disconnect the following:
- Air cleaner cap
- Mass Air Flow (MAF) meter and the resonator
- Accelerator cable from the throttle body, if equipped with a manual transmission
- Accelerator and throttle cables from the throttle body, if equipped with an automatic transmission
- Cruise control cable from the actuator, if equipped with cruise control
- Intake air connector
- Air hose for Idle Air Control (IAC)
- Vacuum sensing hose
- Wire clamp for the engine wiring harness
- Positive Crankcase Ventilation (PCV) hoses
- Spark plug wires from the spark plugs
- Engine wiring harness clamps and harness
- Air conditioning compressor connector, if equipped with air conditioning
- Oil pressure sensor connector
- Engine Coolant Temperature (ECT) sensor connector
- Engine coolant temperature sender gauge connector
- Exhaust Gas Recirculation (EGR) gas temperature sensor connector
- Vacuum Switching Valve (VSV) connector for the EGR
- 2 vacuum hoses from the VSV for the EGR
- Ground strap from the cowl top panel
- Engine wiring harness from the air intake chamber
- Throttle Position (TP) sensor connector
- IAC valve connector
- Crankshaft Position (CKP) sensor connector
- Knock (KS) sensor connector
- Data Link Connector 1 (DLC1) from the bracket
- Engine wiring harness clamp
- EGR pipe
- Intake chamber stay
- Air intake chamber assembly.
- Evaporative Emission (EVAP) hose from the throttle body
- Brake booster vacuum hose from the union
- Water bypass hose from the water bypass pipe

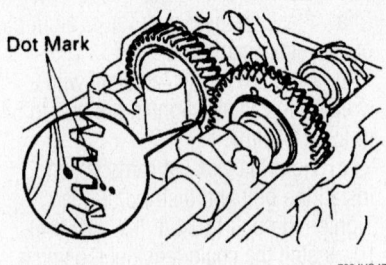

Camshafts TDC/compression timing marks. Marks with 1 and 2 dots will be in straight line on cylinder head surface—2.7L (3RZ-FE/M) engine

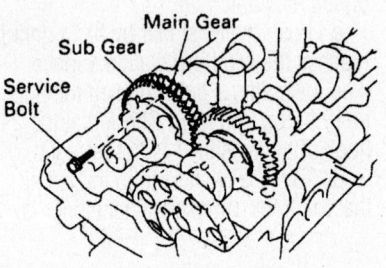

Secure the exhaust camshaft sub-gear to the main gear with a service bolt—2.7L (3RZ-FE/M) engine

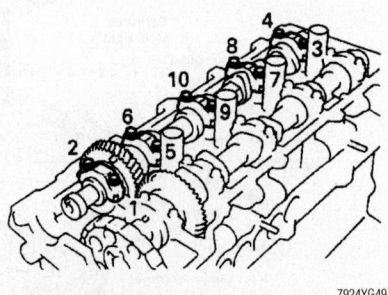

Exhaust camshaft bearing cap bolt removal sequence—2.7L (3RZ-FE/M)

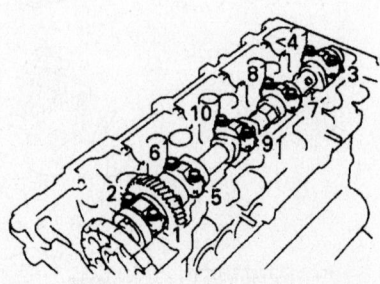

Intake camshaft bearing cap bolt removal sequence—2.7L (3RZ-FE/M) engine

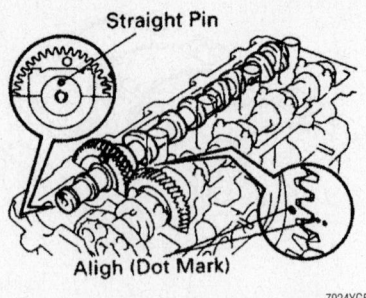

Engage both camshaft gears while matching the timing marks—2.7L (3RZ-FE/M) engine

- Water bypass hose from the cylinder head rear cover
- Front engine hanger
- Engine wiring harness brackets
- Cylinder head cover

5. Set No. 1 cylinder to Top Dead Center (TDC) compression stroke. The groove on the crankshaft pulley should align with the **0** mark on the timing chain cover and the timing marks (1 and 2 dots) of the camshaft gears should form a straight line in respect to the cylinder head surface. If not, turn the crankshaft 1 revolution (360 degrees).

6. Remove the chain tensioner and gasket.

7. Remove the camshaft timing gear as follows:

a. Remove the 2 semi-circular plugs.

b. Place matchmarks on the camshaft timing gear and No. 1 timing chain.

c. Hold the hexagon head portion of the exhaust camshaft with a wrench and remove the fastener and distributor gear.

d. Hold the hexagon head portion of the intake camshaft with a wrench and remove the bolt.

e. Remove the camshaft timing gear and chain from the intake camshaft and leave on the slipper and damper.

8. Remove exhaust camshafts:

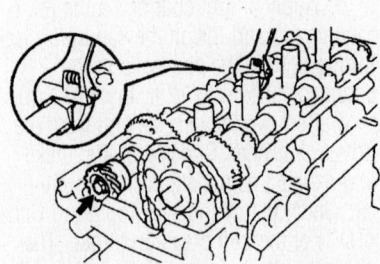

Using a wrench to hold the camshaft—2.7L (3RZ-FE/M) engine

a. Bring the service bolt hole of the driven sub-gear upward by turning the hexagon head portion of the exhaust camshaft with a wrench.

b. Secure the exhaust camshaft sub-gear to the driven gear with a service bolt (6mm diameter, 0.63–0.79 inches in length and 1.0mm in thread pitch).

➡ **When removing the camshaft, be sure that the torsional spring force of the sub-gear has been eliminated by the above operation.**

c. Uniformly loosen and remove the bearing cap bolts in several passes, in the sequence shown.

d. Remove the bearing caps and camshaft. Make a note of the bearing cap positions for proper installation.

9. Remove or disconnect the following:
- Intake camshaft bearing cap bolts in several passes, in the sequence shown
- Bearing caps and camshaft. Make a note of the bearing cap positions for proper installation.

➡ **If the camshaft is not being lifted out straight and level, reinstall the No. 3 bearing cap with the 2 bolts. Then, alternately loosen and remove the 2 bearing cap bolts with the camshaft gear pulled up.**

- Valve lifters and shims from the cylinder head. Arrange the valve lifters and shims in correct order.

To install:

10. Install the valve lifters and shims in their proper locations. Check that the valve lifter rotates smoothly by hand.

11. Install the intake camshaft, as follows:

a. Apply engine oil to the thrust portion of the intake camshaft.

b. Position the intake camshaft with the knock pin facing upward.

c. Install the bearing caps in their proper locations. Apply a light coat of engine oil to the threads and install the cap bolts. Uniformly tighten the cap bolts, in sequence, to 12 ft. lbs. (16 Nm).

12. Install the exhaust camshaft, as shown:

a. Apply engine oil to the thrust portion of the intake camshaft.

b. Engage the exhaust camshaft gear to the intake camshaft gear by matching the timing marks (1 and 2 dots) on each other.

c. Roll down the exhaust camshaft onto the bearing journals while engaging the gears with each other. Install the bearing caps in their proper locations.

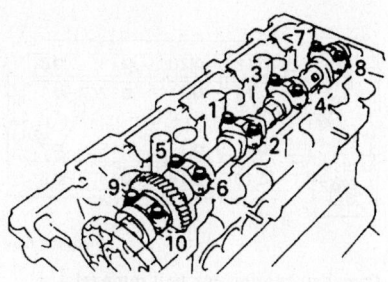

Intake camshaft bearing cap bolt torque—2.7L (3RZ-FE/M) engine

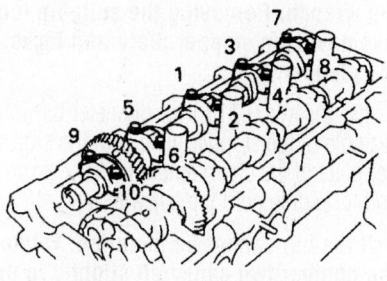

Exhaust camshaft bearing cap bolt torque sequence—2.7L (3RZ-FE/M) engine

d. Apply a light coat of engine oil to the threads and install the cap bolts. Uniformly tighten the cap bolts in the sequence shown to 12 ft. lbs. (16 Nm).

e. Remove the service bolt from the driven sub-gear. Check that the intake and exhaust camshafts turns smoothly.

13. Set No. 1 cylinder to Top Dead Center (TDC) of the compression stroke: Crankshaft pulley groove align with **0** mark on timing cover and camshafts timing marks with 1 dot and 2 dots will be straight line on the cylinder head surface.

14. Install the timing gear. Place the gear over the straight pin of the intake camshaft.

a. Hold the intake camshaft with a wrench. Install and tighten the bolt to 54 ft. lbs. (74 Nm).

b. Hold the exhaust camshaft and install the bolt and distributor gear. Tighten the bolt to 34 ft. lbs. (46 Nm).

15. Install or connect the following:
- Chain tensioner, using a new gasket (mark toward the front)
- Recheck the engine for proper valve timing. Check and adjust the valve clearance.
- Semi-circular plug
- Recheck the engine for proper valve timing.
- Cylinder head cover with a new gasket

- Engine wiring harness brackets
- Front engine hanger. Tighten the bolts to 30 ft. lbs. (42 Nm).
- Air intake chamber assembly. Tighten the bolts to 15 ft. lbs. (20 Nm).
- Hoses
- Intake chamber stay
- Air intake chamber stay. Tighten the bolts to 15 ft. lbs. (20 Nm).
- EGR pipe. Tighten the bolts to 13 ft. lbs. (18 Nm), nut "A" to 14 ft. lbs. (19 Nm) and nut "B" to 15 ft. lbs. (20 Nm).
- Engine wiring harness
- Spark plug wires to the spark plugs
- PCV hoses
- Intake air connector. Tighten the bolts to 13 ft. lbs. (18 Nm).
- Air hose for the IAC
- Vacuum sensing hose
- Wire clamp for the engine wiring harness
- Cruise control cable to the actuator, if equipped with cruise control
- Accelerator cable to the throttle body, if equipped with a manual transmission
- Accelerator and throttle cables to the throttle body, if equipped with an automatic transmission
- Air cleaner cap, MAF meter and the resonator assembly
- Negative battery cable

16. Refill the cooling system.

17. Start the engine and check for leaks.

18. Check the ignition timing. Road test the vehicle for proper operation.

19. Recheck all fluid levels.

2.7L (2TR-FE/X) Engine

1. Before servicing the vehicle, refer to the Precautions Section.

2. Disconnect the negative battery cable.

3. On vehicles equipped with 4WD, remove the engine undercover subassembly.

4. Drain the engine coolant. Remove the radiator support to frame seal, left side. Remove the fan shroud.

5. Remove the air cleaner cap subassembly. Remove the intake air connector.

6. Disconnect the ignition coil connectors. Disconnect the throttle body motor connector. Disconnect the VSV connector.

7. Disconnect the camshaft position sensor connector. Disconnect the engine wire harness clamps. Remove the ignition coils. Disconnect the PCV hose.

8. Remove the cylinder head cover retaining bolts. Remove the cylinder head cover.

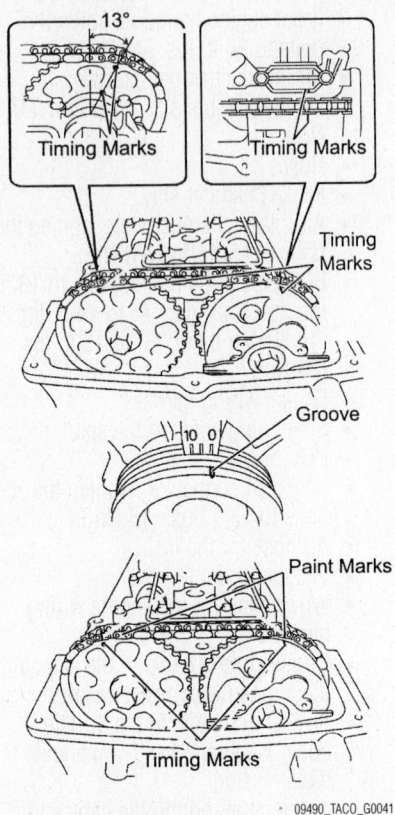

Camshaft positioning—2.7L (2TR-FE/X) engine

9. Remove the two timing chain guide bolts. Remove the timing chain guide. Remove the O-ring.

10. Position the number one cylinder at TDC on the compression stroke.

➡Turn the crankshaft pulley clockwise to align the timing mark notch with the timing mark "0". Paint marks on the timing chain plates that align with the timing marks on the camshaft timing gear.

11. Hold the hexagonal lobe of the number two camshaft, with a suitable tool. Loosen the bolt. Remove the head straight screw plug. Insert a suitable tool into the service hole of the chain tensioner to hold the stopper plate of the chain tensioner lifted up.

➡Lifting up the stopper plate of the chain tensioner unlocks the plunger.

12. While keeping the stopper plate of the chain tensioner lifted up, slightly rotate the hexagonal lobe of the number two camshaft clockwise so that the plunger of the chain tensioner is pushed. Be careful not to damage the camshaft oil delivery pipe.

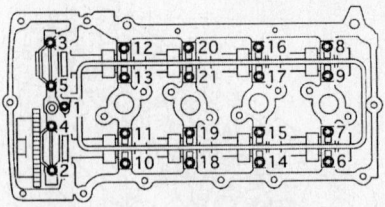

Camshaft bearing cap bolt removal sequence—2.7L (2TR-FE/X) engine

➡With the wrench still installed, remove the suitable tool with the plunger still pushed in. Do not remove the wrench. Removing the suitable tool lifts down the stopper plate and locks the plunger.

13. Insert a 0.118 inch diameter bar into the hole of the stopper plate with the stopper plate of the chain tensioner lifter down and locked. Secure the bar with tape.

➡If the bar cannot be installed, rotate the number two camshaft slightly to the left and right. Then insert the bar.

14. Remove the camshaft timing gear bolt. Remove the gear.

15. Using several steps, uniformly loosen and remove the camshaft bearing cap bolts in the proper sequence. Remove the camshaft oil delivery pipe and O-ring. Remove the camshaft bearing cap number one and eight camshaft bearing caps number two. Remove the number one camshaft and the number two camshaft.

16. Tie the timing chain with a piece of wire.

17. Clamp the camshaft in a soft jaw vise, be sure that the camshaft timing gear does not rotate. Do not clamp the camshaft too tightly in the vise.

18. Cover the four oil path holes of the cam journal with vinyl tape.

➡One of the two grooves on the cam journal is for retarding cam timing (upper) the other is for advancing cam timing (lower). Each groove has two oil paths. Plug one of the two paths for each groove with a piece of rubber before wrapping the cam journal with the tape.

19. Puncture the tape covering the advance side path and the retard side path on the opposite side.

20. Apply about 29 psi air pressure into the two paths, from the two punctures. When applying air pressure, cover the paths with a shop rag to prevent oil splashes.

21. Confirm that the camshaft timing

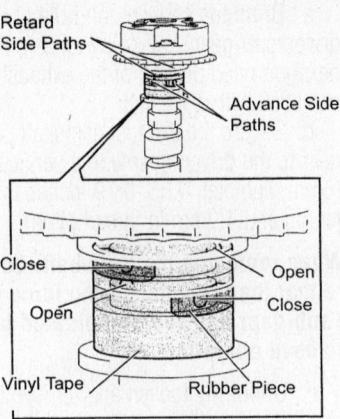

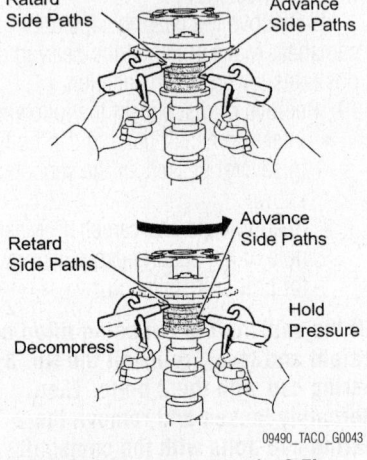

Camshaft timing gear removal—2.7L (2TR-FE/X) engine

gear revolves in the advance direction, when reducing the air pressure on the retard side.

➡The lock pin is released and the camshaft timing groove revolves in the advance direction.

22. When the camshaft timing gear reaches the most advanced position, release the air pressure on the retard side path, then release the air pressure on the advance side path.

➡If the air pressure on the advance path is released first, the camshaft timing gear assembly occasionally shifts in the retard direction abruptly. This may damage the lock pin. Be sure to release the air pressure on the retard side first.

23. Remove the fringe bolt of the camshaft timing gear.

➡Do not remove the other three bolts.

To install:
24. Clean all surfaces.
25. Put the camshaft timing gear and the camshaft together by aligning the key groove and the straight pin.

26. Gently press the gear against the camshaft and turn the gear. Push further at the position where the pin fits the groove.

➡ **Be sure not to turn the camshaft timing gear to the retard angle side (to the right angle).**

27. Check that there is no clearance between the gear's fringe and the camshaft. Tighten the fringe bolt to 58 ft. lbs.

28. Check that the camshaft timing gear can move to the retard side (the right angle) and is locked in the extreme retard position.

29. Check that the valve rocker arm is correctly installed. Apply clean engine oil to the camshaft's cam portion and the cylinder head journals.

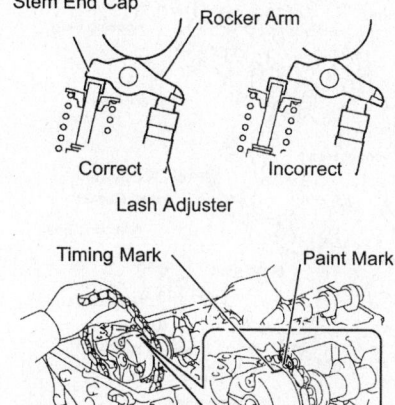

Valve rocker arm installation—2.7L (2TR-FE/X) engine

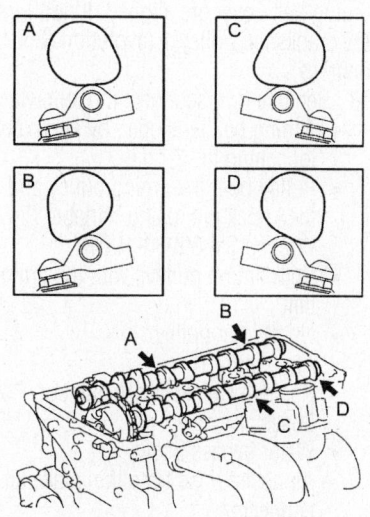

Camshaft positioning—2.7L (2TR-FE/X) engine

30. Install the chain onto the camshaft timing gear, with the painted mark of the link aligned with the timing mark of the camshaft timing gear.

31. Position the two camshafts in there mounting on the engine, see illustration.

➡ **Align the paint mark with the timing mark before installing the camshaft.**

32. Provisionally install the number one camshaft bearing cap. Check the proper location of each of the number two camshaft bearing caps and install them.

33. Install a new O-ring onto the number one camshaft bearing cap. Provisionally install the camshaft oil delivery pipe.

34. Tighten the camshaft bearing cap bolts to specification and in the proper sequence.

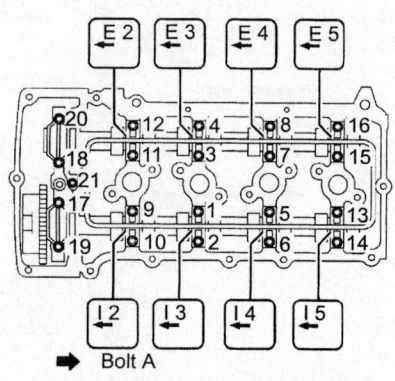

Camshaft bearing cap location, identification and torque sequence—2.7L (2TR-FE/X) engine

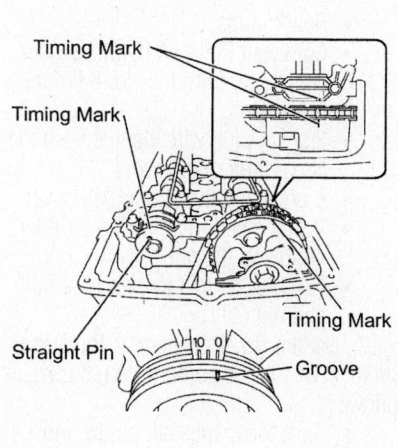

Camshaft timing mark alignment—2.7L (2TR-FE/X) engine

➡ **Bolt "A" is tightened to 9 ft. lbs. All other bolts are tightened to 11 ft. lbs.**

35. Check that each timing mark is set as indicated in the illustration.

36. Install the timing chain onto the camshaft timing gear, with the paint mark aligned with the timing mark on the camshaft timing gear.

37. Align the number two camshaft straight pin and timing gear straight pin hole. Install the camshaft timing gear onto the number two camshaft.

➡ **If the straight pin and straight pin hole are difficult to align, slightly rotate the number two camshaft to the left and right, then attempt to align them.**

38. Hold the hexagonal lobe of the number two camshaft with a wrench. Tighten the bolt to 58 ft. lbs.

39. Remove the 0.118 inch diameter bar from the chain tensioner. Apply adhesive, part number 08833-00070 or equivalent, to two or three threads of the timing gear case with head straight screw plug. Install the timing gear case with head straight screw plug and tighten to 12 ft. lbs.

40. Install a new O-ring onto the camshaft bearing cap. Install the two timing chain guide bolts. Tighten to 7 ft. lbs.

41. Apply seal packing, part number 08826-00080 or equivalent, to the cylinder head as indicated in the illustration. Provisionally install the cylinder head cover bolts and nuts. Tighten bolts "A" to 80 inch lbs. Tighten bolts "B" to 80 inch lbs. Retighten bolts "A" to 80 inch lbs.

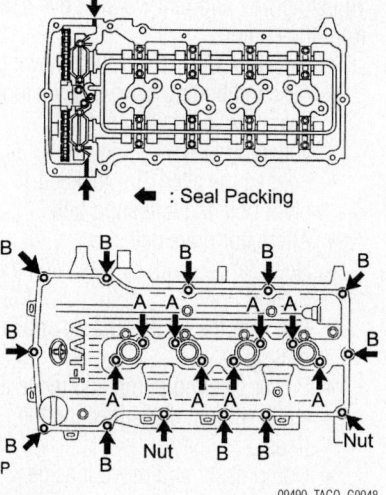

Cylinder head cover bolt sealant application and bolt identification—2.7L (2TR-FE/X) engine

➡**Be sure to remove any oil from the contact surfaces of the cylinder head cover and the cylinder head. Install the cover within three minutes after applying the seal packing. Do not add engine oil for at least two hours after installing the cover.**

42. Continue the installation in the reverse order of the removal procedure.

43. Be sure to fill the engine with the proper grade and type engine coolant.

44. Be sure to fill the engine with the proper grade and type engine oil.

45. Start the engine and check for leaks. Correct as required.

3.4L Engine

1. Before servicing the vehicle, refer to the Precautions Section.

2. Remove or disconnect the following:
- Negative battery cable
- Engine undercover

3. Drain the cooling system.

4. Remove or disconnect the following:
- Air cleaner cap
- Mass Air Flow (MAF) meter and the resonator
- Actuator cable with the bracket, if equipped with cruise control
- Accelerator cable
- Throttle cable, if equipped with an automatic transmission
- Heater hose
- Upper radiator hose from the engine

5. Remove the power steering drive belt, as follows:

   a. Stretch the belt and loosen the fan pulley mounting nuts.

   b. Loosen the lockbolt, pivot bolt and adjusting bolt and remove the drive belt from the engine.

6. Remove or disconnect the following:
- Air conditioning drive belt, by loosening the idle pulley nut and adjusting bolt
- Loosen the alternator lockbolt, pivot bolt and adjusting bolt
- Alternator drive belt
- No. 2 fan shroud, by removing the 2 clips
- Fan with the fluid coupling and fan pulleys
- Power steering pump and move it aside without disconnecting the lines
- Compressor and move it aside without disconnecting the lines, if equipped with air conditioning
- Air conditioning bracket, if equipped with air conditioning

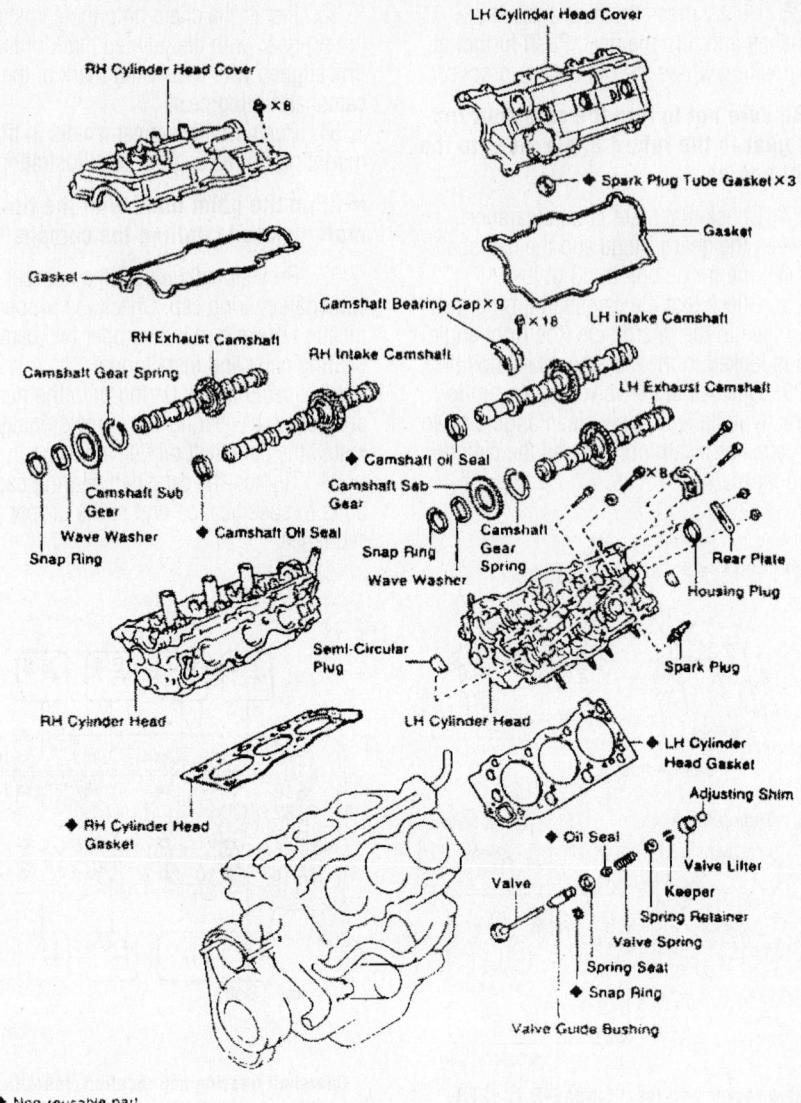

Camshafts and related components—3.4L engine

◆ Non-reusable part

7924YG75

- Spark plug wires with the ignition coils
- Spark plugs
- Camshaft Position (CMP) sensor connector, from the No. 2 timing belt cover
- 3 spark plug wire clamps from the No. 2 timing belt cover
- 6 bolts and the timing belt cover
- Power steering adjusting strut by removing the nut
- Fan bracket by removing the bolt and nut

7. Set the No. 1 cylinder at Top Dead Center (TDC) of the compression stroke, as follows:

   a. Turn the crankshaft pulley and align its groove with the timing mark **0** on the No. 1 timing belt cover.

   b. Check that the timing marks of the camshaft timing pulleys and the No. 3

timing belt cover are aligned. If not, turn the crankshaft pulley 1 revolution (360 degrees).

8. Remove or disconnect the following:
- Timing belt tensioner, by alternately loosening the 2 bolts
- Pulley bolt, the timing pulley and the knock pin, using Variable Pin Wrench Set 09960-10010
- Both timing pulleys with the timing belt
- No. 2 idler pulley
- Alternator
- Positive Crankcase Ventilation PCV hoses
- Water bypass hoses
- Air assist hose from the intake air connector
- 2 vacuum sensing hoses from the Vacuum Switching Valve (VSV)
- Evaporative Emissions (EVAP) hose

- Air hose from the power steering
- Air hose from the air conditioning idle up valve, if equipped with air conditioning
- 4 bolts, 2 nuts and the air intake chamber assembly
- Intake air connector
- Camshaft Position (CMP) sensor
- No. 3 (rear) timing belt cover by removing the 6 bolts
- 8 bolts, seal washers, both cylinder head cover and gaskets
- Semi-circular plugs

9. Remove the right exhaust camshafts, as follows:

a. Bring the service bolt hole of the driven sub-gear upward by turning the hexagon head portion of the exhaust camshaft with a wrench.

b. Align the timing mark (2 dot marks) of the camshaft drive and driven gears by turning the camshaft with a wrench.

c. Secure the exhaust camshaft sub-gear to the driven gear with a service bolt (6mm diameter, 16–20mm bolt length and 1.0mm in thread pitch).

➡When removing the camshaft, be sure the torsional spring force of the sub-gear has been eliminated by the above operation.

d. Uniformly loosen and remove the bearing cap bolts in several passes, in the sequence shown.

e. Remove the bearing caps and camshaft. Make a note of the bearing cap positions for proper installation.

➡Do not pry on or attempt to force the camshaft with a tool or other object.

10. Remove the right-hand intake camshaft, as follows:

a. Uniformly loosen and remove the bearing cap bolts in several passes, in the sequence shown.

b. Remove the bearing caps, oil seal and camshaft. Make a note of the bearing cap positions for proper installation.

11. Remove the left exhaust camshafts, as follows:

a. Align the timing mark (1 dot mark) of the camshaft drive and driven gears by turning the camshaft with a wrench.

b. Secure the exhaust camshaft sub-gear to the driven gear with a service bolt (6mm diameter, 16–20mm bolt length and 1.0mm in thread pitch).

➡When removing the camshaft, be sure that the torsional spring force of the sub-gear has been eliminated by the above operation.

c. Uniformly loosen and remove the bearing cap bolts in several passes, in the sequence shown.

d. Remove the bearing caps and camshaft. Make a note of the bearing cap positions for proper installation.

12. Remove the left-hand intake camshaft, as follows:

a. Uniformly loosen and remove the bearing cap bolts in several passes, in the sequence shown.

b. Remove the bearing caps, oil seal and camshaft. Make a note of the bearing cap positions for proper installation.

13. Remove the valve lifters and shims from the cylinder head. Arrange the valve lifters and shims in correct order.

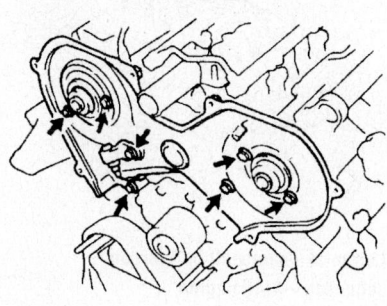

Rear timing belt cover bolt locations—3.4L engine

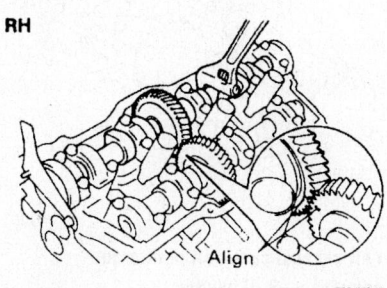

Aligning the timing marks (2 dot marks) of the right camshafts—3.4L engine

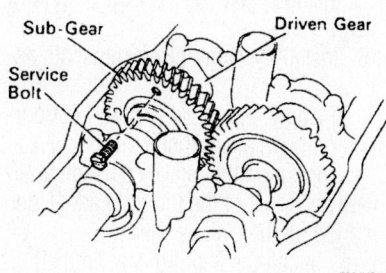

Drive gear service bolt (right side)—3.4L engine

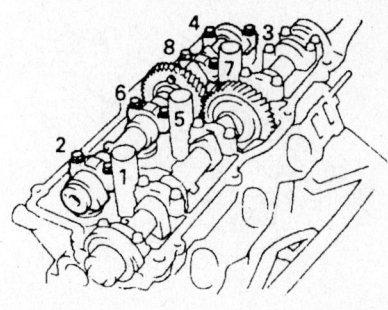

Right exhaust camshaft bolt removal sequence—3.4L engine

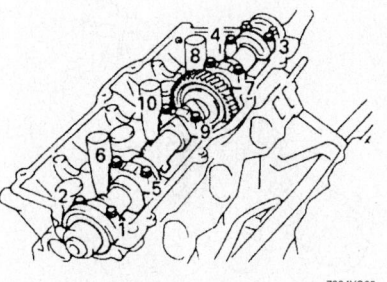

Right intake camshaft bolt removal sequence—3.4L engine

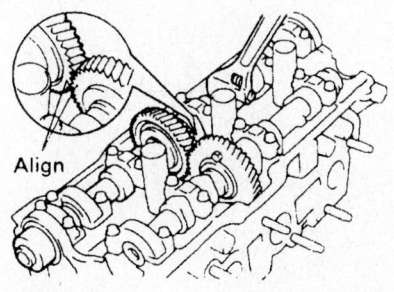

Aligning the timing mark (1 dot mark) of the left camshafts—3.4L engine

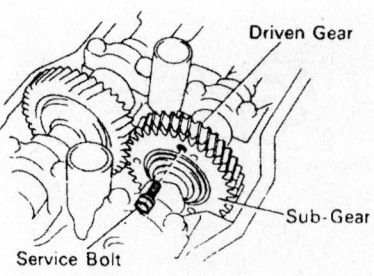

Drive gear service bolt (left side)—3.4L engine

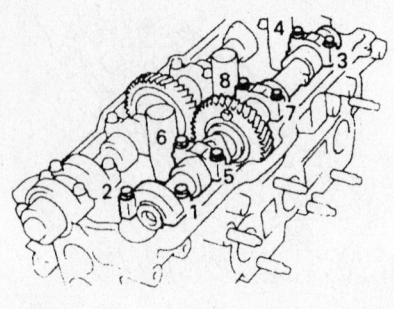

7924YG68

Left exhaust camshaft bolt removal sequence—3.4L engine

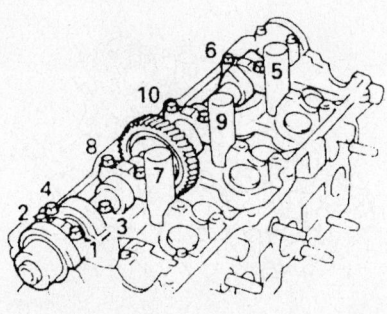

7924YG69

Left intake camshaft bolt removal sequence—3.4L engine

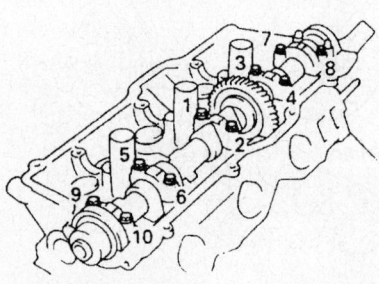

7924YG70

Right intake camshaft bolt torque sequence—3.4L engine

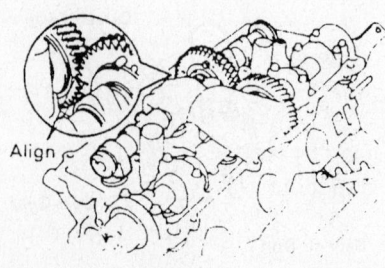

7924YG71

Aligning the right camshafts for installation—3.4L engine

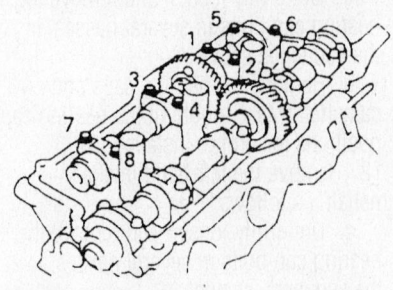

7924YG72

Right exhaust camshaft bolt torque sequence—3.4L engine

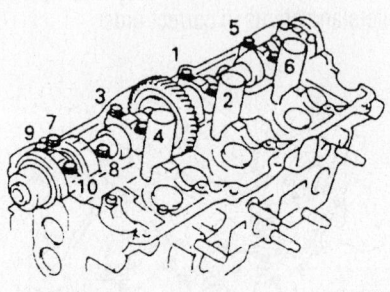

7924YG73

Left intake camshaft bolt torque sequence—3.4L engine

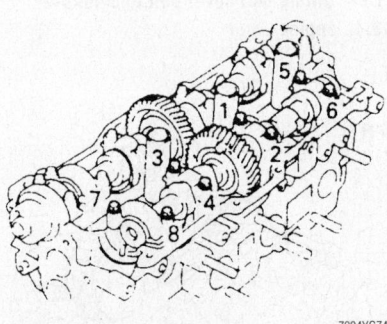

7924YG74

Left exhaust camshaft bolt torque sequence—3.4L engine

To install:

14. Clean all surfaces.

15. Install the valve lifters and shims. Check that the valve lifter rotates smoothly by hand.

16. Install the right intake camshaft, as follows:

   a. Apply engine oil to the thrust portion of the intake camshaft.

   b. Position the intake camshaft at 90 degrees angle of the timing mark (2 dot marks) on the cylinder head.

   c. Install the bearing caps in their proper locations. Apply a light coat of engine oil to the threads and install the cap bolts.

   d. Apply a light coat of engine oil on the threads and under the heads of the bearing cap bolts.

   e. Uniformly tighten the cap bolts in the sequence shown to 12 ft. lbs. (16 Nm).

17. Install the right exhaust camshaft, as follows:

   a. Apply engine oil to the thrust portion of the intake camshaft.

   b. Align the timing marks (2 dot marks) of the camshaft drive and driven gears.

   c. Roll down the exhaust camshaft onto the bearing journals while engaging the gears with each other. Install the bearing caps in their proper locations.

   d. Apply a light coat of engine oil to the threads and install the cap bolts.

   e. Apply a light coat of engine oil on the threads and under the heads of the bearing cap bolts.

   f. Uniformly tighten the cap bolts in the sequence shown to 12 ft. lbs. (16 Nm).

   g. Remove the service bolt from the driven sub-gear. Check that the intake and exhaust camshafts turns smoothly.

   h. Align the timing marks (2 dot marks) of the camshaft drive and driven gears by turning the camshaft with a wrench.

18. Install the left intake camshaft, as follows:

   a. Apply engine oil to the thrust portion of the intake camshaft.

   b. Position the intake camshaft at 90 degrees angle of the timing mark (1 dot mark) on the cylinder head.

   c. Install the bearing caps in their proper locations. Apply a light coat of engine oil to the threads and install the cap bolts.

   d. Apply a light coat of engine oil on the threads and under the heads of the bearing cap bolts.

   e. Uniformly tighten the cap bolts in the sequence shown to 12 ft. lbs. (16 Nm).

19. Install the left exhaust camshaft, as follows:

   a. Apply engine oil to the thrust portion of the intake camshaft.

   b. Align the timing marks (1 dot mark) of the camshaft drive and driven gears.

   c. Roll down the exhaust camshaft onto the bearing journals while engaging the gears with each other. Install the bearing caps in their proper locations.

   d. Apply a light coat of engine oil to the threads and install the cap bolts.

e. Apply a light coat of engine oil on the threads and under the heads of the bearing cap bolts.

f. Uniformly tighten the cap bolts in the sequence shown to 12 ft. lbs. (16 Nm).

g. Remove the service bolt.

20. Check and adjust the valve clearance.

21. Install or connect the following:
- Semi-circular plugs
- Cylinder head covers. Tighten the bolts, in several passes, to 53 inch lbs. (6 Nm).
- Alternator bracket. Tighten the bolts to 14 ft. lbs. (18 Nm).
- No. 3 timing belt cover. Tighten the 6 bolts to 80 inch lbs. (9 Nm).
- CMP sensor. Tighten it to 71 inch lbs. (8 Nm).
- Intake air connector
- Hoses
- Alternator but do not tighten the bolts and nuts at this time
- No. 2 timing belt idler. Tighten the bolt to 30 ft. lbs. (40 Nm).

➡ **Check that the pulley bracket moves smoothly.**

22. Install the left camshaft timing pulley, as follows:

a. Install the knock pin to the camshaft.

b. Align the knock pin hose of the camshaft with the knock pin groove of the timing pulley.

c. Slide the timing pulley on the camshaft with the flange side facing outward. Tighten the pulley bolt to 81 ft. lbs. (110 Nm).

23. Set the No. 1 cylinder to TDC of the compression stroke.

a. Turn the crankshaft pulley and align its groove with the timing mark **0** on the No. 1 timing belt cover.

b. Turn the camshaft, align the knock pin hole of the camshaft with the timing mark of the No. 3 timing belt cover.

c. Turn the camshaft timing pulley, align the timing marks of the camshaft timing pulley and the No. 3 timing belt cover.

24. Connect the timing belt to the left camshaft timing pulley, as follows:

a. Check that the installation mark on the timing belt is aligned with the end of the No. 1 timing belt cover.

b. Using Variable Pin Wrench Set 09960-01000 or equivalent, slightly turn the left camshaft timing pulley clockwise. Align the installation mark on the timing belt with the timing mark of the camshaft

timing pulley, and hang the timing belt on the left camshaft timing pulley.

c. Align the timing marks of the left camshaft pulley and the No. 3 timing belt cover.

d. Check that the timing belt has tension between the crankshaft timing pulley and the left camshaft timing pulley.

25. Install the right camshaft timing pulley and the timing belt, as follows:

a. Align the installation mark on the timing belt with the timing mark of the right camshaft timing pulley, and hang the timing belt on the right camshaft timing pulley with the flange side facing inward.

b. Slide the right camshaft timing pulley on the camshaft. Align the timing marks on the right camshaft timing pulley and the No. 3 timing belt cover.

c. Align the knock pin hole of the camshaft with the knock pin groove of the pulley.

d. Install the knock pin. Tighten the bolt to 81 ft. lbs. (110 Nm).

26. Set the timing belt tensioner, as follows:

a. Using a press, slowly press in the pushrod using 220–2,205 lbs. (981–9,807 N) of force.

b. Align the holes of the pushrod and housing, pass a 1.5mm hexagon wrench through the holes to keep the setting position of the pushrod.

c. Release the press and install the dust boot to the tensioner.

27. Install the timing belt tensioner and alternately tighten the bolts to 20 ft. lbs. (28 Nm). Using pliers, remove the 1.5mm hexagon wrench from the belt tensioner.

28. Check the valve timing, as follows:

a. Slowly turn the crankshaft pulley 2 revolutions from the TDC-to-TDC. Always turn the crankshaft pulley clockwise.

b. Check that each pulley aligns with the timing marks. If the timing marks do not align, remove the timing belt and reinstall it.

29. Install or connect the following:
- Fan bracket with the bolt and nut
- Power steering adjusting strut with the nut
- No. 2 timing belt cover. Tighten the bolts to 80 inch lbs. (9 Nm).
- Negative battery cable

30. Fill the radiator with engine coolant.

31. Start the engine and check for leaks.

32. Check the ignition timing.

33. Install the engine undercover.

34. Road test the vehicle.

35. Recheck all fluid levels.

4.0L Engine

BANK 1

1. Before servicing the vehicle, refer to the Precautions Section.

2. Disconnect the negative battery cable.

3. Drain the engine coolant. Remove the V bank cover.

4. Remove the air cleaner assembly.

5. Disconnect the two water bypass hoses. Disconnect the fuel vapor feed hose. Disconnect the ventilation hose. Disconnect the VSV connectors.

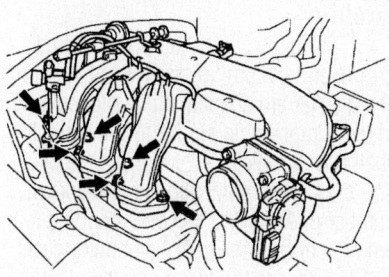

09490_TACO_G0049

Intake surge tank bolt locations—4.0L engine

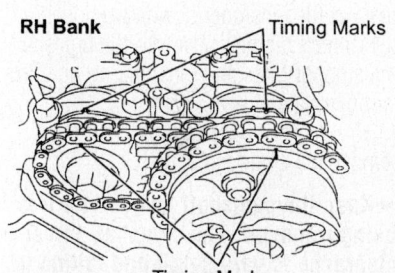

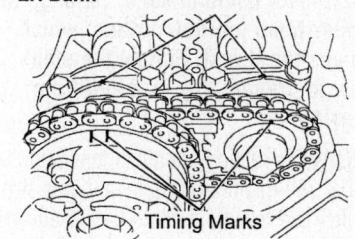

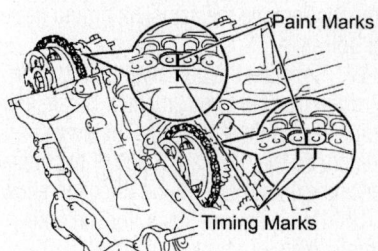

09490_TACO_G0050

Bank 1 camshaft timing mark alignment—4.0L engine

6. Disconnect the throttle body motor connector. Separate the three wire harness clamps and hose clamp.

7. If equipped with manual transmission, remove the nut, then separate the clutch flexible hose bracket from the surge tank stay.

8. Remove the two bolts and the throttle body bracket. Remove the bolt and the oil baffle plate. Remove the four bolts and the two serge tank stays.

9. Remove the two nuts. Remove the four bolts, intake air surge tank and gasket.

10. Remove the ignition coil assembly.

11. Remove the cylinder head cover retaining bolts. Remove the cylinder head cover.

12. Turn the crankshaft pulley until its groove and the timing mark "0" of the timing chain cover are aligned. If not aligned at TDC of the compression stroke, turn the crankshaft one complete revolution, in the direction of rotation. Paint alignment marks on the number one chain links corresponding to the timing marks of the camshaft timing gears.

13. Remove the four bolts, and then remove the timing chain cover plate and gasket.

14. While turning the stopper plate of the tensioner upward, push the plunger of the chain tensioner. While turning the stopper plate of the tensioner downward, insert a 0.118 inch diameter bar into the holes in the stopper plate and tensioner to hold the stopper plate.

15. Remove the two bolts, and then remove the chain tensioner.

➡Keep the camshaft level while it is being removed. The camshaft thrust clearance is very small and failing to keep it level could crack or damage the cylinder head journal surface, which receives the thrust. Follow the steps below to prevent this problem from occurring.

16. While raising the chain tensioner number two insert a 0.039 inch diameter pin into the hole to hold it. Hold the hexagonal portion of the number two camshaft with a wrench. Remove the camshaft timing gear set bolt.

17. Separate the camshaft timing gear from the number two camshaft. Rotate the camshafts counterclockwise, using a wrench, so that the cam lobes of the number one cylinder face in the direction shown.

18. Using several steps, loosen and remove the eight bearing cap bolts uniformly and in the proper removal sequence. Remove the four bearing caps and the number two camshaft.

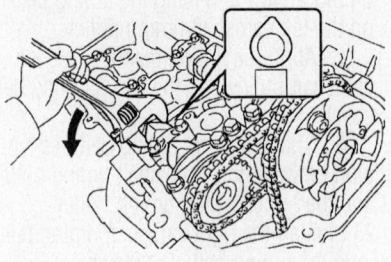

Bank 1 camshaft lobe removal positioning—4.0L engine

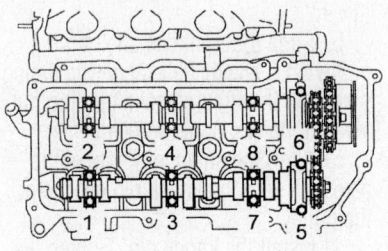

Bank 1 camshaft number two bearing cap bolt removal sequence—4.0L engine

19. Remove the number two chain tensioner bolt. Remove the number chain tensioner and camshaft timing gear.

➡Keep the camshaft level while it is being removed. The camshaft thrust clearance is very small and failing to keep it level could crack or damage the cylinder head journal surface, which receives the thrust. Follow the steps below to prevent this problem from occurring.

20. Hold the hexagonal portion of the number one camshaft, with a wrench. Loosen the camshaft timing gear set bolt.

➡Do not disassemble the camshaft timing gear assembly.

21. Slide the camshaft timing gear and separate the number one chain from the camshaft timing gear.

22. Rotate the number one camshaft counterclockwise, using the wrench so that the cam lobes on the number one cylinder face downward.

23. Using several steps, loosen and remove the eight bearing cap bolts uniformly and in the proper removal sequence. Remove the four bearing caps.

24. Remove the camshaft timing gear set bolt with the number one camshaft lifted up. Remove the number one camshaft and camshaft timing gear with the number two chain.

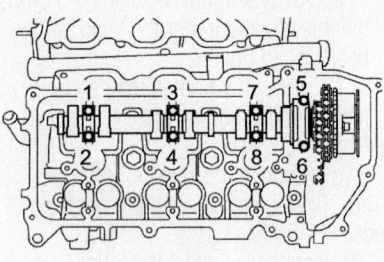

Bank 1 camshaft number one bearing cap bolt removal sequence—4.0L engine

25. Tie the number one chain to the side. Be careful not to drop anything inside the timing chain cover.

To install:

➡Keep the camshaft level while it is being installed. The camshaft thrust clearance is very small and failing to keep it level could crack or damage the cylinder head journal surface, which receives the thrust.

26. Align the yellow mark link with the timing mark (1 dot mark) of the camshaft timing gear. Apply new engine oil to the thrust portion and journal of the camshafts.

27. Temporarily install the number one chain onto the number two chain of the camshaft timing gear.

28. Align the knock pin hole of the camshaft timing gear with the knock pin of

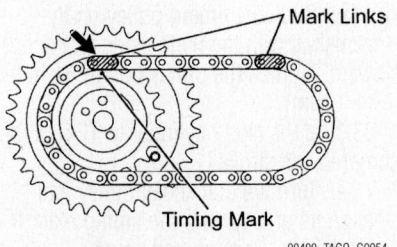

Mark Links

Timing Mark

Bank 1 camshaft number one alignment marks—4.0L engine

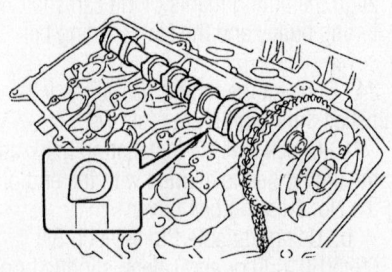

Bank 1 camshaft number one lobe installation positioning—4.0L engine

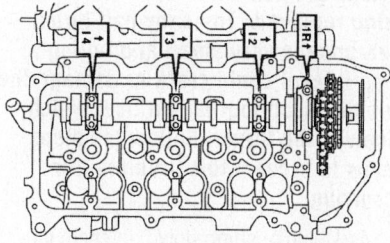

Bank 1 camshaft number one bearing cap bolt installation sequence—4.0L engine

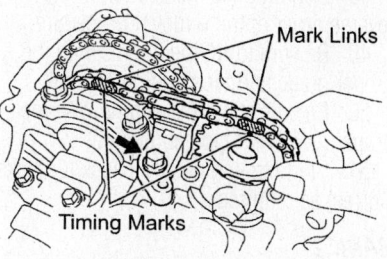

Mark Links

Timing Marks

Bank 1 camshaft number two alignment marks—4.0L engine

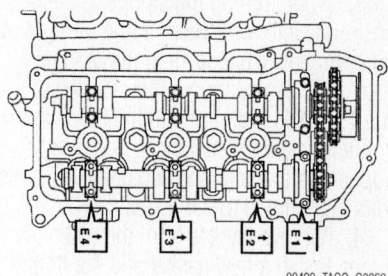

Bank 1 camshaft number two bearing cap bolt installation sequence—4.0L engine

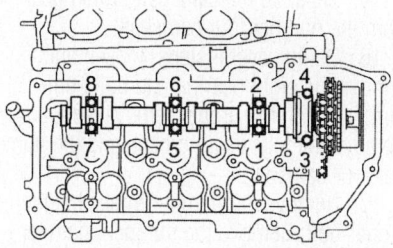

Bank 1 camshaft number one bearing cap bolt torque sequence—4.0L engine

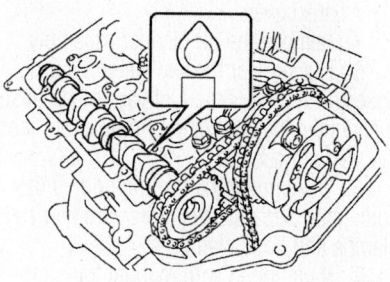

Bank 1 camshaft number two lobe installation positioning—4.0L engine

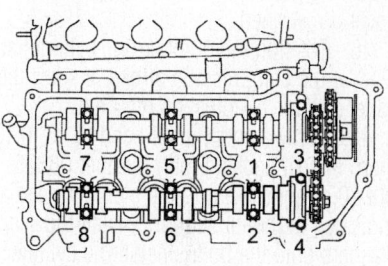

Bank 1 camshaft number two bearing cap bolt torque sequence—4.0L engine

the number one camshaft. Insert the number one camshaft into the camshaft timing gear.

29. Temporarily install the camshaft timing gear set bolt. Install the number one camshaft onto the right cylinder head with the cam lobes of the number one cylinder facing downward, as indicated in the illustration.

30. Install the four bearing caps, in the proper location. Apply a light coat of engine oil to the threads and under the heads of the cap bolts.

31. Using several steps, uniformly install and tighten the bearing cap bolts in the proper sequence to 80 inch lbs. for the 10mm bolts and 18 ft. lbs. for the 12mm bolts.

32. Rotate the number one camshaft clockwise, using a wrench so that the timing mark of the camshaft timing gear is aligned with the timing mark of the camshaft bearing cap.

33. Align the paint mark of the number one chain with the timing mark of the camshaft timing gear.

34. Hold the hexagonal portion of the number one camshaft with a wrench, and tighten the camshaft timing gear set bolt to 74 ft. lbs.

35. While pushing in on the number two chain tensioner, insert a 0.039 inch pin into the hole to hold it.

36. Temporarily install the camshaft timing gear and chain tensioner number two and align the yellow mark links with the

timing marks (1 dot mark) on the camshaft timing gears. Tighten the bolt to 14 ft. lbs.

➡**Keep the camshaft level while it is being installed. The camshaft thrust clearance is very small and failing to keep it level could crack or damage the cylinder head journal surface, which receives the thrust.**

37. Install the number two camshaft onto the right cylinder head with the cam lobes of the number one cylinder facing upward, as indicated in the illustration.

38. Install the four bearing caps in the proper location. Apply a light coat of clean engine oil to the threads and under the heads of the bolts.

39. Using several steps, uniformly install and tighten the eight bearing cap bolts in sequence to 80 inch lbs, for the 10mm head bolts and 18 ft. lbs. for the 12mm head bolts.

40. Rotate the number two camshaft clockwise, using a wrench, so that the lock pin of the number two camshaft is aligned with the knock pin hole of the camshaft timing gear.

41. Hold the hexagonal portion of the number two camshaft, with a wrench, and install the camshaft timing gear set bolt and tighten it to 74 ft. lbs.

42. Remove the pin from the number two chain tensioner.

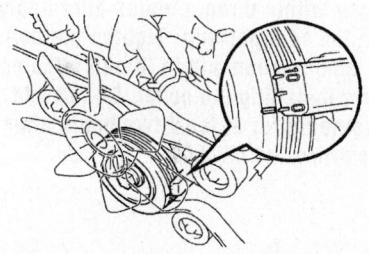

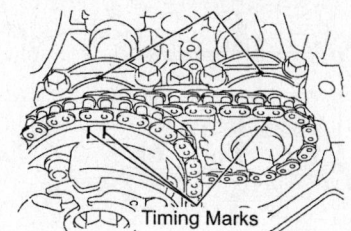

RH Bank — Timing Marks

LH Bank — Timing Marks

Timing Marks

Bank 1 and Bank 2 timing mark alignment—4.0L engine

43. While turning the stopper plate of the tensioner clockwise, push in the plunger of the tensioner. While turning the stopper plate of the tensioner counterclockwise, insert a 0.138 inch bar into the holes in the stopper plate and tensioner to hold the stopper plate. Install the two chain tensioner bolts and tighten to 7.4 ft. lbs.

44. Remove the bar from the chain tensioner. Install a new gasket and the timing chain cover plate. Torque the bolts to 80 inch lbs.

45. Turn the crankshaft pulley two complete revolutions slowly until its groove and the timing mark "0" of the timing chain cover are aligned.

46. Position the number one cylinder at TDC of the compression stroke. Inspect the valve clearance, adjust as required.

47. Apply a continuous bead (0.08–0.12 inch) of seal packing, part number 08826-00080 or equivalent, to the cylinder head as indicated in the illustration. Install the seal washers onto the bolts. Install the cylinder head cover bolts and nuts. Tighten bolts "A" to 7.4 ft. lbs. Tighten bolts "B" to 80 inch lbs. Tighten nuts to 80 inch lbs.

➡ **Be sure to remove any oil from the contact surfaces of the cylinder head cover and the cylinder head. Install the cover within three minutes after applying the seal packing. Tighten the bolts to specification within fifteen minutes after installing the cover. Do not add engine oil for at least two hours after installing the cover.**

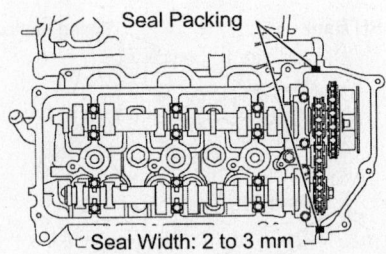

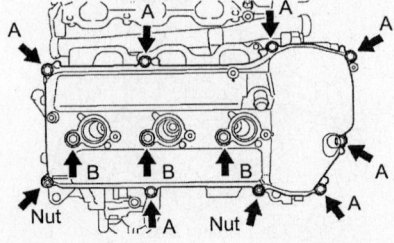

Bank 1 cylinder head cover bolt sealant application and bolt identification—4.0L engine

48. Continue the installation in the reverse order of the removal procedure.

49. Be sure to fill the engine with the proper grade and type engine coolant.

50. Be sure to fill the engine with the proper grade and type engine oil.

51. Start the engine and check for leaks. Correct as required.

BANK 2

1. Before servicing the vehicle, refer to the Precautions Section.

2. Disconnect the negative battery cable.

3. Drain the engine coolant. Remove the V bank cover.

4. Remove the air cleaner assembly.

5. Disconnect the two water bypass hoses. Disconnect the fuel vapor feed hose. Disconnect the ventilation hose. Disconnect the VSV connectors.

6. Disconnect the throttle body motor connector. Separate the three wire harness clamps and hose clamp.

7. If equipped with manual transmission, remove the nut, then separate the clutch flexible hose bracket from the surge tank stay.

8. Remove the two bolts and the throttle body bracket. Remove the bolt and the oil baffle plate. Remove the four bolts and the two serge tank stays.

9. Remove the two nuts. Remove the four bolts, intake air surge tank and gasket.

10. Remove the ignition coil assembly.

11. Remove the cylinder head cover retaining bolts. Remove the cylinder head cover.

12. Turn the crankshaft pulley until its groove and the timing mark "0" of the timing chain cover are aligned. If not aligned at TDC of the compression stroke, turn the crankshaft one complete revolution, in the direction of rotation. Paint alignment marks on the number one chain links corresponding to the timing marks of the camshaft timing gears.

13. While turning the stopper plate of the tensioner upward, push in the plunger of the chain tensioner. While turning the stopper plate of the tensioner downward, insert a 0.138 inch bar into the holes in the stopper plate and tensioner to hold the stopper plate. Remove the two bolts and then remove the number one chain tensioner assembly.

➡ **Never rotate the crankshaft with the chain tensioner removed. When rotating the camshaft with the tensioner removed, rotate the crankshaft counterclockwise forty degrees from TDC, first.**

➡ **Keep the camshaft level while it is being removed. The camshaft thrust clearance is very small and failing to keep it level could crack or damage the cylinder head journal surface, which receives the thrust. Follow the steps below to prevent this problem from occurring.**

14. While pushing down on chain tensioner number three insert a 0.039 inch diameter pin into the hole to hold it. Hold the hexagonal portion of the number four camshaft with a wrench. Remove the camshaft timing gear set bolt.

15. Separate the camshaft timing gear from the number four camshaft.

16. Using several steps, loosen and remove the eight bearing cap bolts uniformly and in the proper removal sequence. Remove the four bearing caps and the number four camshaft.

17. Remove the number three chain tensioner bolt. Remove the number chain tensioner and camshaft timing gear.

➡ **Keep the camshaft level while it is being removed. The camshaft thrust clearance is very small and failing to keep it level could crack or damage the cylinder head journal surface, which receives the thrust. Follow the steps below to prevent this problem from occurring.**

18. Release the chain tension between the camshaft gear (LH bank) and the crankshaft timing gear by turning the crankshaft pulley counterclockwise slightly.

19. Hold the hexagonal portion of the number three camshaft, with a wrench. Loosen the camshaft timing gear set bolt.

➡ **Do not disassemble the camshaft timing gear assembly.**

20. Slide the camshaft timing gear and separate the number one chain from the camshaft timing gear.

21. Using several steps, loosen and

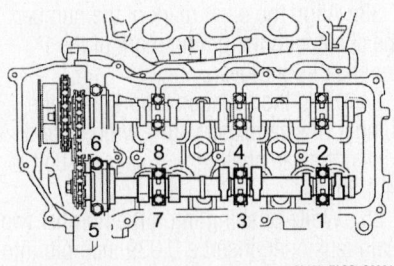

Bank 2 camshaft number four bearing cap bolt removal sequence—4.0L engine

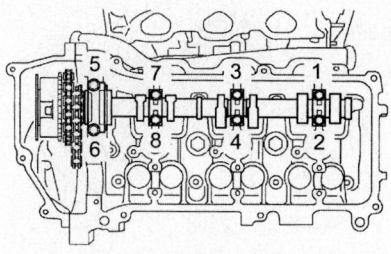

Bank 2 camshaft number three bearing
cap bolt removal sequence—4.0L engine

remove the eight bearing cap bolts uniformly and in the proper removal sequence.
Remove the four bearing caps.

22. Remove the camshaft timing gear set
bolt with the number three camshaft lifted
up. Remove the number three camshaft and
camshaft timing gear with the number two
chain.

23. Tie the number one chain to the side.
Be careful not to drop anything inside the
timing chain cover.

To install:

➡Keep the camshaft level while it is
being installed. The camshaft thrust
clearance is very small and failing to
keep it level could crack or damage the
cylinder head journal surface, which
receives the thrust.

24. Align the yellow mark link with the
timing mark (2 dot mark) of the camshaft
timing gear. Apply new engine oil to the
thrust portion and journal of the camshafts.

25. Temporarily install the number one
chain onto the number two chain of the
camshaft timing gear.

26. Align the knock pin hole of the
camshaft timing gear with the knock pin of
the number three camshaft. Insert the number three camshaft into the camshaft timing
gear.

27. Temporarily install the camshaft timing gear set bolt. Install the number three
camshaft onto the left cylinder head with the

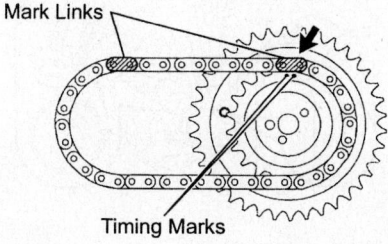

Bank 2 camshaft number three alignment
marks—4.0L engine

Bank 2 camshaft number three lobe installation positioning—4.0L engine

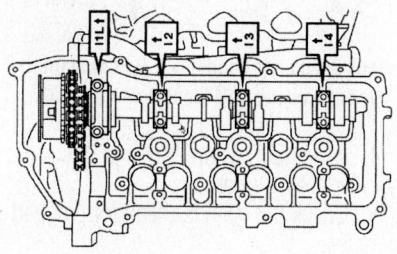

Bank 2 camshaft number three bearing
cap bolt installation sequence—4.0L
engine

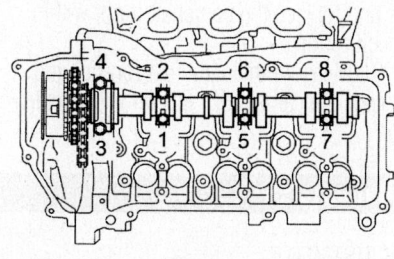

Bank 2 camshaft number three bearing
cap bolt torque sequence—4.0L engine

cam lobes of the number two cylinder facing downward, as indicated in the illustration.

28. Install the four bearing caps, in the
proper location. Apply a light coat of engine
oil to the threads and under the heads of the
cap bolts.

29. Using several steps, uniformly install
and tighten the bearing cap bolts in the
proper sequence to 80 inch lbs. for the
10mm bolts and 18 ft. lbs. for the 12mm
bolts.

30. Rotate the number one camshaft
clockwise, using a wrench so that the timing mark of the camshaft timing gear is
aligned with the timing mark of the
camshaft bearing cap.

31. Align the paint mark of the number
one chain with the timing mark of the
camshaft timing gear.

32. Hold the hexagonal portion of the
number three camshaft with a wrench, and
tighten the camshaft timing gear set bolt to
74 ft. lbs.

33. While pushing in on the number
three chain tensioner, insert a 0.039 inch
pin into the hole to hold it.

34. Temporarily install the camshaft timing gear and chain tensioner number three
and align the yellow mark links with the
timing marks (1 dot mark and 2 dot marks)
on the camshaft timing gears. Tighten the
bolt to 14 ft. lbs.

➡Keep the camshaft level while it is
being installed. The camshaft thrust
clearance is very small and failing to
keep it level could crack or damage the
cylinder head journal surface, which
receives the thrust.

35. Align the knock pin hole in the
camshaft timing gear with the knock pin of
the number four camshaft, and insert the
number four camshaft into the camshaft
timing gear.

36. Temporarily install the camshaft timing gear set bolt.

37. Install the four bearing caps in the
proper location. Apply a light coat of clean
engine oil to the threads and under the
heads of the bolts.

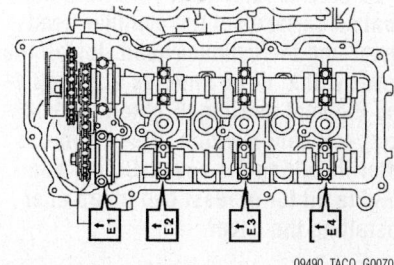

Bank 2 camshaft number four bearing cap
bolt installation sequence—4.0L engine

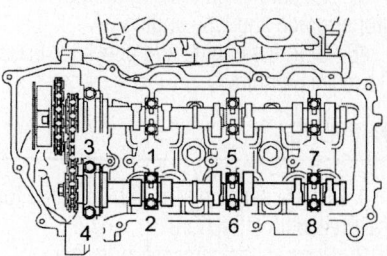

Bank 2 camshaft number four bearing cap
bolt torque sequence—4.0L engine

38. Using several steps, uniformly install and tighten the eight bearing cap bolts in sequence to 80 inch lbs, for the 10mm head bolts and 18 ft. lbs. for the 12mm head bolts.

39. Hold the hexagonal portion of the number four camshaft, with a wrench, and install the camshaft timing gear set bolt and tighten it to 74 ft. lbs.

40. Remove the pin from the number three chain tensioner.

41. Release the chain tension between the camshaft timing gear (RH bank) and the crankshaft timing gear by turning the crankshaft pulley clockwise slightly.

42. While turning the stopper plate of the tensioner clockwise, push in the plunger of the tensioner. While turning the stopper plate of the tensioner counterclockwise, insert a 0.138 inch bar into the holes in the stopper plate and tensioner to hold the stopper plate. Install the two chain tensioner bolts and tighten to 7.4 ft. lbs. Remove the bar from the chain tensioner.

43. Position the number one cylinder at TDC of the compression stroke. Inspect the valve clearance, adjust as required.

44. Apply a continuous bead (0.08–0.12 inch) of seal packing, part number 08826-00080 or equivalent, to the cylinder head as indicated in the illustration. Install the seal washers onto the bolts. Install the cylinder head cover bolts and nuts. Tighten bolts "A" to 7.4 ft. lbs. Tighten bolts "B" to 80 inch lbs. Tighten nuts to 80 inch lbs.

➡ **Be sure to remove any oil from the contact surfaces of the cylinder head cover and the cylinder head. Install the cover within three minutes after applying the seal packing. Tighten the bolts to specification within fifteen minutes after installing the cover. Do not add engine oil for at least two hours after installing the cover.**

45. Continue the installation in the reverse order of the removal procedure.

46. Be sure to fill the engine with the proper grade and type engine coolant.

47. Be sure to fill the engine with the proper grade and type engine oil.

48. Start the engine and check for leaks. Correct as required.

INSPECTION

1. Before servicing the vehicle, refer to the Precautions Section.

2. Remove the camshaft from the engine.

3. Check the camshaft bearing journals for damage and binding.

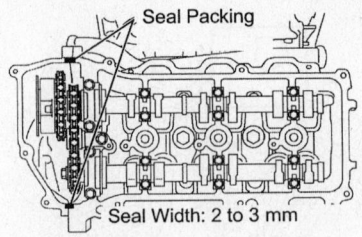

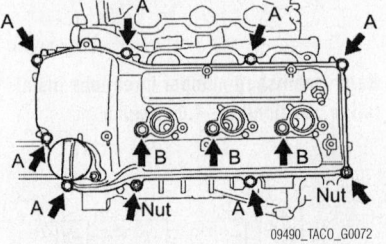

Bank 2 cylinder head cover bolt sealant application and bolt identification—4.0L

4. If the journals are binding, check the cylinder head for damage.

5. Check the cylinder head for clogged oil holes.

6. Check the camshaft surface for abnormal wear and damage. Replace the camshaft, as required.

7. Measure the camshaft lobe surface and replace the camshaft if not within specification.

8. Measure the camshaft journal diameter and replace the camshaft if not within specification.

9. Measure the camshaft run out and replace the camshaft if not within specification.

Valve Lash

ADJUSTMENT

2.4L Engine

1. Before servicing the vehicle, refer to the Precautions Section.

2. Remove or disconnect the following:
- Negative battery cable
- Air intake connector
- Positive Crankcase Ventilation (PCV) hoses
- Spark plug wires
- 4 clamps and the engine wiring harness
- Air conditioning compressor connector, if equipped with air conditioning
- Oil pressure sensor connector
- Engine Coolant Temperature (ECT) sensor connector
- Distributor connector
- Cylinder head cover

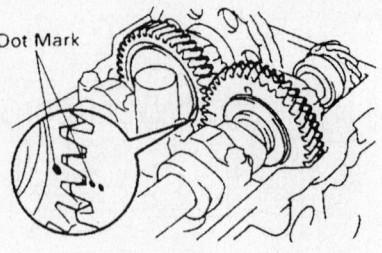

Aligning the timing marks—2.4L and 2.7L (3RZ-FE/M) engines

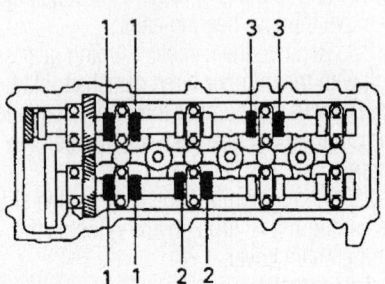

First valve adjustment—2.4L and 2.7L (3RZ-FE/M) engines

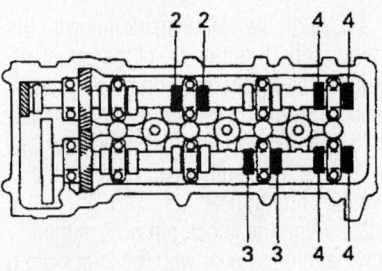

Second valve adjustment—2.4L and 2.7L (3RZ-FE/M) engines

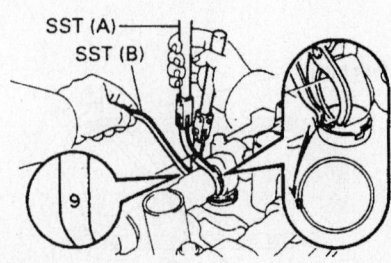

Removing adjusting shim using the special tools shown above—2.4L and 2.7L (3RZ-FE/M) engines

Adjusting Shim Selection Chart (Intake)

Intake valve shim selection chart—2.4L and 2.7L (3RZ-FE/M) engines

New shim thickness

Shim No.	Thickness mm (in.)	Shim No.	Thickness mm (in.)
1	2.500 (0.0984)	10	2.950 (0.1161)
2	2.550 (0.1004)	11	3.000 (0.1181)
3	2.600 (0.1024)	12	3.050 (0.1201)
4	2.650 (0.1043)	13	3.100 (0.1220)
5	2.700 (0.1063)	14	3.150 (0.1240)
6	2.750 (0.1083)	15	3.200 (0.1260)
7	2.800 (0.1102)	16	3.250 (0.1280)
8	2.850 (0.1122)	17	3.300 (0.1299)
9	2.900 (0.1142)		

HINT:
New shims have the thickness in millimeters imprinted on the face.

Intake valve clearance (Cold):
0.15 – 0.25 mm (0.006 – 0.010 in.)
EXAMPLE:
The 2.800 mm (0.1102 in.) shim is installed, and the measured clearance is 0.440 mm (0.0173 in.). Replace the 2.800 mm (0.1102 in.) shim with a new No.12 shim.

67170-TACO-G01

Adjusting Shim Selection Chart (Exhaust)

New shim thickness — mm (in.)

Shim No.	Thickness	Shim No.	Thickness
1	2.500 (0.0984)	10	2.950 (0.1161)
2	2.550 (0.1004)	11	3.000 (0.1181)
3	2.600 (0.1024)	12	3.050 (0.1201)
4	2.650 (0.1043)	13	3.100 (0.1220)
5	2.700 (0.1063)	14	3.150 (0.1240)
6	2.750 (0.1083)	15	3.200 (0.1260)
7	2.800 (0.1102)	16	3.250 (0.1280)
8	2.850 (0.1122)	17	3.300 (0.1299)
9	2.900 (0.1142)		

HINT:
New shims have the thickness in millimeters imprinted on the face.

Exhaust valve clearance (Cold):
0.25 – 0.35 mm (0.010 – 0.014 in.)

EXAMPLE:
The 2.800 mm (0.1102 in.) shim is installed, and the measured clearance is 0.440 mm (0.0173 in.). Replace the 2.800 mm (0.1102 in.) shim with a new No.10 shim.

Installed shim thickness mm (in.)

New shim thickness columns (mm): 2.500 (0.0984), 2.520 (0.0992), 2.540 (0.1000), 2.550 (0.1004), 2.560 (0.1008), 2.580 (0.1016), 2.600 (0.1024), 2.620 (0.1031), 2.640 (0.1039), 2.650 (0.1043), 2.660 (0.1047), 2.680 (0.1055), 2.700 (0.1063), 2.710 (0.1067), 2.720 (0.1071), 2.730 (0.1075), 2.740 (0.1079), 2.750 (0.1083), 2.760 (0.1087), 2.770 (0.1091), 2.780 (0.1094), 2.790 (0.1098), 2.800 (0.1102), 2.810 (0.1106), 2.820 (0.1110), 2.830 (0.1114), 2.840 (0.1118), 2.860 (0.1122), 2.870 (0.1126), 2.880 (0.1130), 2.890 (0.1134), 2.900 (0.1142), 2.910 (0.1146), 2.920 (0.1150), 2.930 (0.1154), 2.940 (0.1157), 2.950 (0.1161), 2.960 (0.1165), 2.970 (0.1169), 2.980 (0.1173), 2.990 (0.1177), 3.000 (0.1181), 3.010 (0.1185), 3.020 (0.1189), 3.030 (0.1193), 3.040 (0.1197), 3.050 (0.1201), 3.060 (0.1205), 3.070 (0.1209), 3.080 (0.1213), 3.090 (0.1217), 3.100 (0.1220), 3.120 (0.1228), 3.140 (0.1236), 3.150 (0.1240), 3.160 (0.1244), 3.180 (0.1252), 3.200 (0.1260), 3.220 (0.1268), 3.240 (0.1276), 3.250 (0.1280), 3.260 (0.1283), 3.280 (0.1291), 3.300 (0.1299)

Measured clearance mm (in.):
0.000 – 0.030 (0.0000 – 0.0012)
0.031 – 0.050 (0.0012 – 0.0020)
0.051 – 0.070 (0.0020 – 0.0028)
0.071 – 0.090 (0.0028 – 0.0035)
0.091 – 0.110 (0.0036 – 0.0043)
0.111 – 0.130 (0.0044 – 0.0051)
0.131 – 0.150 (0.0052 – 0.0059)
0.151 – 0.170 (0.0059 – 0.0067)
0.171 – 0.190 (0.0067 – 0.0075)
0.191 – 0.210 (0.0075 – 0.0083)
0.211 – 0.230 (0.0083 – 0.0091)
0.231 – 0.249 (0.0091 – 0.0098)
0.250 – 0.350 (0.0098 – 0.0138)
0.351 – 0.370 (0.0138 – 0.0146)
0.371 – 0.390 (0.0146 – 0.0154)
0.391 – 0.410 (0.0154 – 0.0161)
0.411 – 0.430 (0.0162 – 0.0169)
0.431 – 0.450 (0.0170 – 0.0177)
0.451 – 0.470 (0.0178 – 0.0185)
0.471 – 0.490 (0.0185 – 0.0193)
0.491 – 0.510 (0.0193 – 0.0201)
0.511 – 0.530 (0.0201 – 0.0209)
0.531 – 0.550 (0.0209 – 0.0217)
0.551 – 0.570 (0.0217 – 0.0224)
0.571 – 0.590 (0.0225 – 0.0232)
0.591 – 0.610 (0.0233 – 0.0240)
0.611 – 0.630 (0.0241 – 0.0248)
0.631 – 0.650 (0.0248 – 0.0256)
0.651 – 0.670 (0.0256 – 0.0264)
0.671 – 0.690 (0.0264 – 0.0272)
0.691 – 0.710 (0.0272 – 0.0280)
0.711 – 0.730 (0.0280 – 0.0287)
0.731 – 0.750 (0.0288 – 0.0295)
0.751 – 0.770 (0.0296 – 0.0303)
0.771 – 0.790 (0.0304 – 0.0311)
0.791 – 0.810 (0.0311 – 0.0319)
0.811 – 0.830 (0.0319 – 0.0327)
0.831 – 0.850 (0.0327 – 0.0335)
0.851 – 0.870 (0.0335 – 0.0343)
0.871 – 0.890 (0.0343 – 0.0350)
0.891 – 0.910 (0.0351 – 0.0358)
0.911 – 0.930 (0.0359 – 0.0366)
0.931 – 0.950 (0.0367 – 0.0374)
0.951 – 0.970 (0.0374 – 0.0382)
0.971 – 0.990 (0.0382 – 0.0390)
0.991 – 1.010 (0.0390 – 0.0398)
1.011 – 1.030 (0.0398 – 0.0406)
1.031 – 1.050 (0.0406 – 0.0413)
1.051 – 1.070 (0.0414 – 0.0421)
1.071 – 1.090 (0.0422 – 0.0429)
1.091 – 1.110 (0.0430 – 0.0437)
1.111 – 1.130 (0.0437 – 0.0445)
1.131 – 1.150 (0.0445 – 0.0413)

67170-TACO-G02

Exhaust valve shim selection chart—2.4L and 2.7L (3RZ-FE/M) engines

3. Set the No. 1 cylinder to Top Dead Center (TDC) of the compression stroke, as follows:

a. Turn the crankshaft pulley clockwise and align its groove with the **0** mark on the timing chain cover.

b. Check that the timing marks (1 and 2 dots) of the camshaft drive and driven gears are in a straight line on the cylinder head surface. If not, turn the crankshaft 1 revolution (360 degrees) and align the marks.

4. Inspect the valve clearance, as follows:

a. Measure the clearance between the valve lifter and the camshaft. Measure the 1st and 2nd intake and the 1st and 3rd exhaust valves.

b. Turn the crankshaft pulley 1 revolution (360 degrees) and align the marks as above. Measure the 3rd and 4th intake and the 2nd and 4th exhaust valves.

5. Valve clearance "cold" should be:
- Intake: 0.006–0.010 in. (0.15–0.25mm)
- Exhaust: 0.010–0.014 in. (0.25–0.35mm)

6. Adjust the valve clearance by using adjusting shims, as follows:

a. Turn the equipment driveshaft so that the cam lobe for the valve to be adjusted faces up.

b. Using SST 09248-55040, press down the valve lifter and place SST 09248-05420, between the camshaft and the valve lifter.

c. Remove SST 09248-55040.

d. Remove the adjusting shim with a small flat prying tool and a magnetic finger.

e. Determine the replacement adjusting shim size according to the following formula or use the adjusting shim charts.

f. Using a micrometer, measure the thickness of the removed shim. Calculate the thickness of a new shim so that the valve clearance comes within the specified value:
- T: Thickness of the removed shim
- A: Measured valve clearance
- N: Thickness of the new shim

g. Intake: $N = T + (A−0.008$ in. $(0.20mm))$

h. Exhaust: $N = T + (A−0.012$ in. $(0.30mm))$

i. Install a new adjusting shim. Place it on the valve lifter. Using the SST 09248-55040, press down the valve lifter.

j. Remove the SST 09248-05420.

k. Recheck the valve clearance.

7. Install or connect the following:
- Cylinder head cover
- Engine wiring harness and clamps
- Distributor connector
- ECT sensor connector
- Oil pressure sensor connector
- Air conditioning compressor connector, if disconnected
- Spark plug wires
- PCV hoses
- Air intake connector
- Negative battery cable

8. Check the ignition timing.

2.7L (3RZ-FE/M) Engine

1. Before servicing the vehicle, refer to the Precautions Section.

2. Disconnect the negative battery cable.

3. Drain the engine coolant.

4. Remove or disconnect the following:
- Intake air connector
- Positive Crankcase Ventilation (PCV) hoses
- Spark plug wires
- Engine wiring harness clamps and harness
- Air conditioning compressor connector, if equipped with air conditioning
- Oil pressure sensor connector
- Engine Coolant Temperature (ECT) sensor connector
- Distributor connector
- Cylinder head cover

5. Set the No. 1 cylinder to Top Dead Center (TDC) of the compression stroke, as follows:

a. Turn the crankshaft pulley clockwise and align its groove with the **0** mark on the timing chain cover.

b. Check that the timing marks (1 and 2 dots) of the camshaft drive and driven gears are in a straight line on the cylinder head surface. If not, turn the crankshaft 1 revolution (360 degrees) and align the marks.

6. Inspect the valve clearance, as follows:

a. Measure the clearance between the valve lifter and the camshaft. Measure the 1st and 2nd intake and the 1st and 3rd exhaust valves.

b. Turn the crankshaft pulley 1 revolution (360 degrees) and align the marks as above. Measure the 3rd and 4th intake and the 2nd and 4th exhaust valves.

7. Valve clearance cold should be:
- Intake: 0.006–0.010 in. (0.15–0.25mm)
- Exhaust: 0.010–0.014 in. (0.25–0.35mm)

8. Adjust the valve clearance by using adjusting shims, as follows:

a. Turn the camshaft so the cam lobe for the valve to be adjusted faces up.

b. Using SST 09248-55040, press down the valve lifter and place SST 09248-05420, between the camshaft and the valve lifter. Remove SST 09248-55040.

c. Remove the adjusting shim with a small flat prying tool and a magnetic finger.

d. Determine the replacement adjusting shim size according to the following formula or use the adjusting shim charts.

e. Using a micrometer, measure the thickness of the removed shim. Calculate the thickness of a new shim so the valve clearance comes within the specified value.
- T: Thickness of the removed shim
- A: Measured valve clearance
- N: Thickness of the new shim

f. Intake: $N = T + (A−0.008$ in. $(0.20mm))$

g. Exhaust: $N = T + (A−0.012$ in. $(0.30mm))$

h. Install a new adjusting shim. Place it on the valve lifter. Using the SST 09248-55040, press down the valve lifter and remove SST 09248-05420.

i. Recheck the valve clearance.

9. Install or connect the following:
- Cylinder head cover
- Engine wiring harness and clamps
- Distributor connector
- ECT sensor connector
- Oil pressure sensor connector
- Air conditioning compressor connector, if disconnected
- Spark plug wires
- PCV hoses
- Intake air connector
- Negative battery cable

10. Refill with engine coolant.

11. Check the ignition timing.

3.4L Engine

1. Before servicing the vehicle, refer to the Precautions Section.

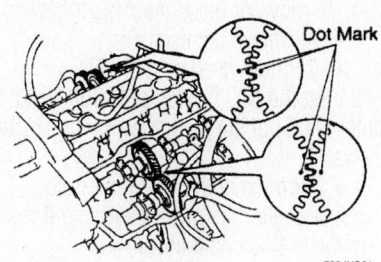

7924YG81

Aligning the timing marks—3.4L engine

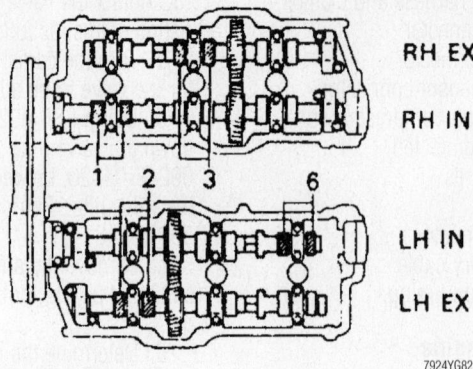

First valve adjustment—3.4L engine

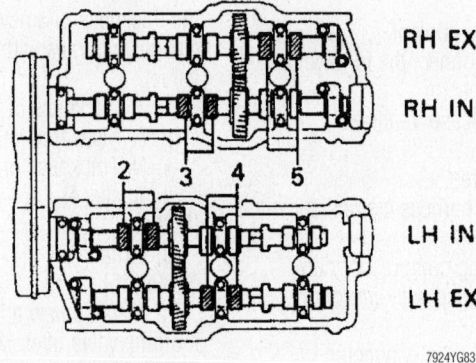

Second valve adjustment—3.4L engine

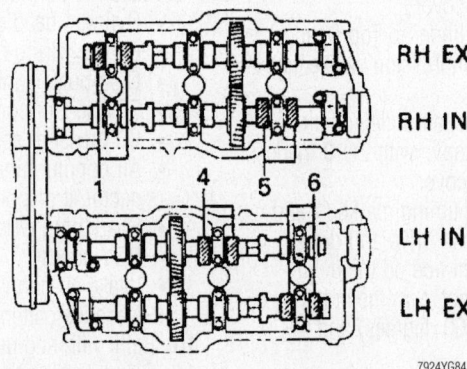

Third valve adjustment—3.4L engine

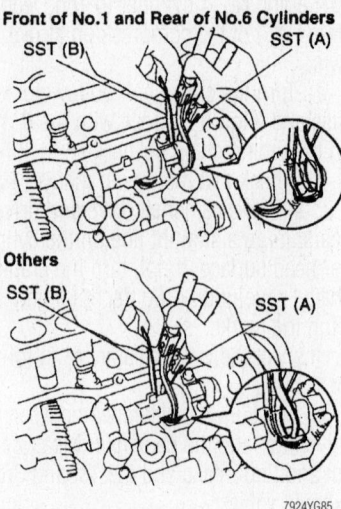

Removing the adjusting shim—3.4L engine

2. Disconnect the negative battery cable.

3. Drain the engine coolant.

4. Remove or disconnect the following:
- Air intake connector
- Cylinder head cover

5. Set the No. 1 cylinder to Top Dead Center (TDC) of the compression stroke, as follows:

a. Turn the crankshaft pulley clockwise and align its groove with the **0** mark on the timing chain cover.

b. Check that the timing marks (1 and 2 dots) of the camshaft drive and driven gears are in a straight line on the cylinder head surface. If not, turn the crankshaft 1 revolution (360 degrees) and align the marks.

6. Inspect the valve clearance, as follows:

a. Measure the clearance between the valve lifter and the camshaft. Measure the 1st intake and the 3rd exhaust valves on the right head and the 6th intake and the 2nd exhaust valves on the left head.

b. Turn the crankshaft ⅔ of a revolution (240 degrees) and adjust the 3rd intake and the 5th exhaust valves on the right head and the 2nd intake and the 4th exhaust valves on the left head.

c. Turn the crankshaft ⅔ of a revolution (240 degrees) and adjust the 5th intake and the 1st exhaust valves on the right head and the 4th intake and the 6th exhaust valves on the left head.

7. Valve clearance cold should be:
- Intake: 0.006–0.009 in. (0.13–0.23mm)
- Exhaust: 0.011–0.014 in. (0.27–0.37mm)

8. Adjust the valve clearance by using adjusting shims, as follows:

a. Turn the equipment camshaft so that the cam lobe for the valve to be adjusted faces up.

b. Turn the valve lifter so that the notches are perpendicular to the camshaft.

c. Using SST 09248-55040, press down the valve lifter and place SST 09248-05420, between the camshaft and the valve lifter. Remove SST 09248-55040.

d. Remove the adjusting shim with a small flat prying tool and a magnetic finger.

e. Determine the replacement adjusting shim size according to the following formula or use the adjusting shim charts.

f. Using a micrometer, measure the thickness of the removed shim. Calculate the thickness of a new shim so that the valve clearance comes within the specified value.
- T: Thickness of the removed shim
- A: Measured valve clearance
- N: Thickness of the new shim

g. Intake: N = T + (A−0.007 in. (0.18mm))

Adjusting Shim Selection Chart (Intake)

Intake valve clearance (Cold):
0.13 – 0.23 mm (0.006 – 0.009 in.)

EXAMPLE: The 2.800 mm (0.1102 in.) shim is installed, and the measured clearance is 0.450 mm (0.0177 in.). Replace the 2.800 mm (0.1102 in.) shim with a new No.10 shim.

New shim thickness — mm (in.)

Shim No.	Thickness	Shim No.	Thickness
1	2.500 (0.0984)	10	2.950 (0.1161)
2	2.550 (0.1004)	11	3.000 (0.1181)
3	2.600 (0.1024)	12	3.050 (0.1201)
4	2.650 (0.1043)	13	3.100 (0.1220)
5	2.700 (0.1063)	14	3.150 (0.1240)
6	2.750 (0.1083)	15	3.200 (0.1260)
7	2.800 (0.1102)	16	3.250 (0.1280)
8	2.850 (0.1122)	17	3.300 (0.1299)
9	2.900 (0.1142)		

HINT: New shims have the thickness in millimeters imprinted on the face.

Installed shim thickness — mm (in.) (column headers across top):
2.500 (0.0984), 2.520 (0.0992), 2.540 (0.1000), 2.550 (0.1004), 2.560 (0.1008), 2.580 (0.1016), 2.600 (0.1024), 2.620 (0.1031), 2.640 (0.1039), 2.660 (0.1047), 2.670 (0.1051), 2.680 (0.1055), 2.690 (0.1059), 2.710 (0.1067), 2.720 (0.1071), 2.730 (0.1075), 2.740 (0.1079), 2.760 (0.1087), 2.770 (0.1091), 2.780 (0.1094), 2.790 (0.1098), 2.810 (0.1106), 2.820 (0.1110), 2.830 (0.1114), 2.840 (0.1118), 2.860 (0.1126), 2.870 (0.1130), 2.880 (0.1134), 2.890 (0.1138), 2.900 (0.1142), 2.910 (0.1146), 2.920 (0.1150), 2.930 (0.1154), 2.940 (0.1157), 2.950 (0.1161), 2.960 (0.1165), 2.970 (0.1169), 2.980 (0.1173), 2.990 (0.1177), 3.010 (0.1185), 3.020 (0.1189), 3.030 (0.1193), 3.040 (0.1197), 3.050 (0.1201), 3.060 (0.1205), 3.080 (0.1213), 3.100 (0.1220), 3.120 (0.1228), 3.140 (0.1236), 3.150 (0.1240), 3.160 (0.1244), 3.180 (0.1252), 3.200 (0.1260), 3.220 (0.1268), 3.240 (0.1276), 3.250 (0.1280), 3.260 (0.1283), 3.280 (0.1291), 3.300 (0.1299)

Measured clearance — mm (in.) (row labels down left side):
0.000 – 0.020 (0.0000 – 0.0008)
0.021 – 0.040 (0.0008 – 0.0016)
0.041 – 0.060 (0.0016 – 0.0024)
0.061 – 0.080 (0.0024 – 0.0031)
0.081 – 0.100 (0.0032 – 0.0039)
0.101 – 0.120 (0.0040 – 0.0047)
0.121 – 0.129 (0.0048 – 0.0051)
0.130 – 0.230 (0.0051 – 0.0091)
0.231 – 0.240 (0.0091 – 0.0094)
0.241 – 0.260 (0.0095 – 0.0102)
0.261 – 0.280 (0.0103 – 0.0110)
0.281 – 0.300 (0.0111 – 0.0118)
0.301 – 0.320 (0.0119 – 0.0126)
0.321 – 0.340 (0.0126 – 0.0134)
0.341 – 0.360 (0.0134 – 0.0142)
0.361 – 0.380 (0.0142 – 0.0150)
0.381 – 0.400 (0.0150 – 0.0157)
0.401 – 0.420 (0.0158 – 0.0165)
0.421 – 0.440 (0.0166 – 0.0173)
0.441 – 0.460 (0.0174 – 0.0181)
0.461 – 0.480 (0.0181 – 0.0189)
0.481 – 0.500 (0.0189 – 0.0197)
0.501 – 0.520 (0.0197 – 0.0205)
0.521 – 0.540 (0.0205 – 0.0213)
0.541 – 0.560 (0.0213 – 0.0220)
0.561 – 0.580 (0.0221 – 0.0228)
0.581 – 0.600 (0.0229 – 0.0236)
0.601 – 0.620 (0.0237 – 0.0244)
0.621 – 0.640 (0.0244 – 0.0252)
0.641 – 0.660 (0.0252 – 0.0260)
0.661 – 0.680 (0.0260 – 0.0268)
0.681 – 0.700 (0.0268 – 0.0276)
0.701 – 0.720 (0.0276 – 0.0283)
0.721 – 0.740 (0.0284 – 0.0291)
0.741 – 0.760 (0.0292 – 0.0299)
0.761 – 0.780 (0.0300 – 0.0307)
0.781 – 0.800 (0.0307 – 0.0315)
0.801 – 0.820 (0.0315 – 0.0323)
0.821 – 0.840 (0.0323 – 0.0331)
0.841 – 0.860 (0.0331 – 0.0339)
0.861 – 0.880 (0.0339 – 0.0346)
0.881 – 0.900 (0.0347 – 0.0354)
0.901 – 0.920 (0.0355 – 0.0362)
0.921 – 0.940 (0.0363 – 0.0370)
0.941 – 0.960 (0.0370 – 0.0378)
0.961 – 0.980 (0.0378 – 0.0386)
0.981 – 1.000 (0.0386 – 0.0394)
1.001 – 1.020 (0.0394 – 0.0402)
1.021 – 1.030 (0.0402 – 0.0406)

(The chart interior contains a matrix of shim numbers 1–17 cross-referencing installed shim thickness and measured clearance.)

Intake valve shim selection chart—3.4L engines

67170-TACO-G03

Adjusting Shim Selection Chart (Exhaust)

The installed shim thickness columns (mm (in.)) run across the top:
2.500 (0.0984), 2.520 (0.0992), 2.540 (0.1000), 2.550 (0.1004), 2.560 (0.1008), 2.580 (0.1016), 2.600 (0.1024), 2.620 (0.1031), 2.640 (0.1039), 2.650 (0.1043), 2.660 (0.1047), 2.670 (0.1051), 2.680 (0.1055), 2.690 (0.1059), 2.700 (0.1063), 2.710 (0.1067), 2.720 (0.1071), 2.730 (0.1075), 2.740 (0.1079), 2.750 (0.1083), 2.760 (0.1087), 2.770 (0.1091), 2.780 (0.1094), 2.790 (0.1098), 2.800 (0.1102), 2.810 (0.1106), 2.820 (0.1110), 2.830 (0.1114), 2.840 (0.1118), 2.850 (0.1122), 2.860 (0.1126), 2.870 (0.1130), 2.880 (0.1134), 2.890 (0.1138), 2.900 (0.1142), 2.910 (0.1146), 2.920 (0.1150), 2.930 (0.1154), 2.940 (0.1157), 2.950 (0.1161), 2.960 (0.1165), 2.970 (0.1169), 2.980 (0.1173), 2.990 (0.1177), 3.000 (0.1181), 3.010 (0.1185), 3.020 (0.1189), 3.030 (0.1193), 3.040 (0.1197), 3.050 (0.1201), 3.060 (0.1205), 3.080 (0.1213), 3.100 (0.1220), 3.120 (0.1228), 3.140 (0.1236), 3.150 (0.1240), 3.160 (0.1244), 3.180 (0.1252), 3.200 (0.1260), 3.220 (0.1268), 3.240 (0.1276), 3.250 (0.1280), 3.260 (0.1283), 3.280 (0.1291), 3.300 (0.1299)

The measured clearance rows (mm (in.)) run down the left side:
0.000 – 0.020 (0.0000 – 0.0008), 0.021 – 0.040 (0.0008 – 0.0016), 0.041 – 0.060 (0.0016 – 0.0024), 0.061 – 0.080 (0.0024 – 0.0031), 0.081 – 0.100 (0.0032 – 0.0039), 0.101 – 0.120 (0.0040 – 0.0047), 0.121 – 0.140 (0.0048 – 0.0055), 0.141 – 0.160 (0.0056 – 0.0063), 0.161 – 0.180 (0.0063 – 0.0071), 0.181 – 0.200 (0.0071 – 0.0079), 0.201 – 0.220 (0.0079 – 0.0087), 0.221 – 0.240 (0.0087 – 0.0094), 0.241 – 0.260 (0.0095 – 0.0102), 0.261 – 0.269 (0.0103 – 0.0106), 0.270 – 0.370 (0.0106 – 0.0146), 0.371 – 0.380 (0.0146 – 0.0150), 0.381 – 0.400 (0.0150 – 0.0157), 0.401 – 0.420 (0.0158 – 0.0165), 0.421 – 0.440 (0.0166 – 0.0173), 0.441 – 0.460 (0.0174 – 0.0181), 0.461 – 0.480 (0.0181 – 0.0189), 0.481 – 0.500 (0.0189 – 0.0197), 0.501 – 0.520 (0.0197 – 0.0205), 0.521 – 0.540 (0.0205 – 0.0213), 0.541 – 0.560 (0.0213 – 0.0220), 0.561 – 0.580 (0.0221 – 0.0228), 0.581 – 0.600 (0.0229 – 0.0236), 0.601 – 0.620 (0.0237 – 0.0244), 0.621 – 0.640 (0.0244 – 0.0252), 0.641 – 0.660 (0.0252 – 0.0260), 0.661 – 0.680 (0.0260 – 0.0268), 0.681 – 0.700 (0.0268 – 0.0276), 0.701 – 0.720 (0.0276 – 0.0283), 0.721 – 0.740 (0.0284 – 0.0291), 0.741 – 0.760 (0.0292 – 0.0299), 0.761 – 0.780 (0.0300 – 0.0307), 0.781 – 0.800 (0.0307 – 0.0315), 0.801 – 0.820 (0.0315 – 0.0323), 0.821 – 0.840 (0.0323 – 0.0331), 0.841 – 0.860 (0.0331 – 0.0339), 0.861 – 0.880 (0.0339 – 0.0346), 0.881 – 0.900 (0.0347 – 0.0354), 0.901 – 0.920 (0.0355 – 0.0362), 0.921 – 0.940 (0.0363 – 0.0370), 0.941 – 0.960 (0.0370 – 0.0378), 0.961 – 0.980 (0.0378 – 0.0386), 0.981 – 1.000 (0.0386 – 0.0394), 1.001 – 1.020 (0.0394 – 0.0402), 1.021 – 1.040 (0.0402 – 0.0409), 1.041 – 1.060 (0.0410 – 0.0417), 1.061 – 1.080 (0.0418 – 0.0425), 1.081 – 1.100 (0.0426 – 0.0433), 1.101 – 1.120 (0.0433 – 0.0441), 1.121 – 1.140 (0.0441 – 0.0449), 1.141 – 1.160 (0.0449 – 0.0457), 1.161 – 1.170 (0.0457 – 0.0461)

The body of the chart lists the replacement shim number (1–17) for each combination of installed shim thickness and measured clearance.

New shim thickness

Shim No.	Thickness mm (in.)	Shim No.	Thickness mm (in.)
1	2.500 (0.0984)	10	2.950 (0.1161)
2	2.550 (0.1004)	11	3.000 (0.1181)
3	2.600 (0.1024)	12	3.050 (0.1201)
4	2.650 (0.1043)	13	3.100 (0.1220)
5	2.700 (0.1063)	14	3.150 (0.1240)
6	2.750 (0.1083)	15	3.200 (0.1260)
7	2.800 (0.1102)	16	3.250 (0.1280)
8	2.850 (0.1122)	17	3.300 (0.1299)
9	2.900 (0.1142)		

HINT: New shims have the thickness in millimeters imprinted on the face.

Exhaust valve clearance (Cold):
0.27 – 0.37 mm (0.011 – 0.014 in.)

EXAMPLE: The 2.800 mm (0.1102 in.) shim is installed, and the measured clearance is 0.450 mm (0.0177 in.). Replace the 2.800 mm (0.1102 in.) shim with a new No.10 shim.

Exhaust valve shim selection chart—3.4L engines

6770-TACO-G04

h. Exhaust: N = T + (A–0.013 in. (0.32mm))

i. Install a new adjusting shim. Place it on the valve lifter. Using the SST 09248-55040, press down the valve lifter and remove SST 09248-05420.

j. Recheck the valve clearance.

9. Install or connect the following:
- Cylinder head cover
- Intake air connector
- Negative battery cable

10. Refill with engine coolant.

11. Start the engine and check for leaks.

4.0L Engine

1. Before servicing the vehicle, refer to the Precautions Section.

2. Disconnect the negative battery cable.

3. Drain the engine coolant. Remove the V bank cover.

4. Remove the air cleaner assembly.

5. Disconnect the two water bypass hoses. Disconnect the fuel vapor feed hose. Disconnect the ventilation hose. Disconnect the VSV connectors.

6. Disconnect the throttle body motor connector. Separate the three wire harness clamps and hose clamp.

7. If equipped with manual transmission, remove the nut, then separate the clutch flexible hose bracket from the surge tank stay.

8. Remove the two bolts and the throttle body bracket. Remove the bolt and the oil baffle plate. Remove the four bolts and the two serge tank stays.

9. Remove the two nuts. Remove the four bolts, intake air surge tank and gasket.

10. Remove the ignition coil assembly.

11. Remove the cylinder head cover retaining bolts. Remove the cylinder head cover.

12. Turn the crankshaft pulley until its groove and the timing mark "0" of the timing chain cover are aligned. If not aligned at TDC of the compression stroke, turn the crankshaft one complete revolution, in the direction of rotation.

13. Using a feeler gauge, check and record the valve clearance on the following valves: right bank, exhaust number three and intake number one, left bank exhaust number two and intake number six.

14. Rotate the crankshaft 240 degrees clockwise and using a feeler gauge, check and record the following valves: right bank, exhaust number five and intake number three, left bank exhaust number four and intake number two.

15. Rotate the crankshaft 240 degrees clockwise and using a feeler gauge, check and record the following valves: right bank, exhaust number one and intake number five, left bank exhaust number six and intake number four.

16. If adjustment is required, position the engine at TDC on the compression stroke.

17. Place paint marks on the number one chain links corresponding to the timing marks of the camshaft timing gears.

18. Remove the chain tensioner assembly number one. Remove the number two camshaft.

19. Remove the chain tensioner assembly number two. Remove the camshaft.

20. Remove the number four camshaft subassembly. Remove the chain tensioner assembly number three.

21. Remove the number three camshaft subassembly.

22. Remove the valve lifters.

23. Determine the replacement adjusting shim size according to the following formula or use the adjusting shim charts.

24. Using a micrometer, measure the thickness of the removed shim. Calculate the thickness of a new shim so that the valve clearance comes within the specified value.
- T: Thickness of the removed shim
- A: Measured valve clearance
- N: Thickness of the new shim
  a. Intake: N = T + A
  b. Exhaust: N = T + A

25. Select a new lifter with a thickness as close as possible to the calculated value.

26. Install removed components in the reverse order of the removal procedure.

27. When installing the cylinder head cover, apply a continuous bead (0.08–0.12 inch) of seal packing, part number 08826-00080 or equivalent, to the cylinder head as indicated in the illustration. Install the seal washers onto the bolts. Install the cylinder head cover bolts and nuts. Tighten bolts "A" to 7.4 ft. lbs. Tighten bolts "B" to 80 inch lbs. Tighten nuts to 80 inch lbs.

➡ **Be sure to remove any oil from the contact surfaces of the cylinder head cover and the cylinder head. Install the cover within three minutes after applying the seal packing. Tighten the bolts to specification within fifteen minutes after installing the cover. Do not add engine oil for at least two hours after installing the cover.**

28. Continue the installation in the reverse order of the removal procedure.

29. Be sure to fill the engine with the proper grade and type engine coolant.

30. Be sure to fill the engine with the proper grade and type engine oil.

31. Start the engine and check for leaks. Correct as required.

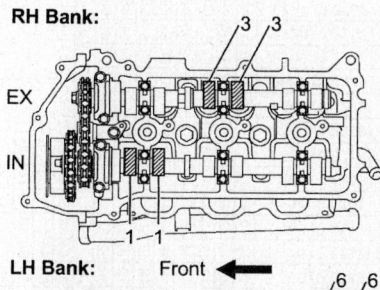

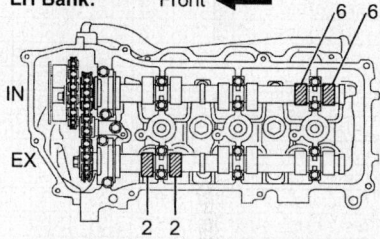

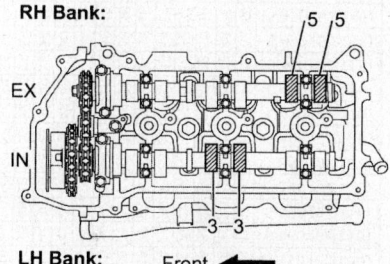

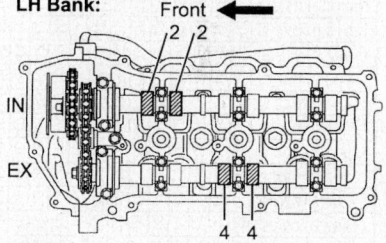

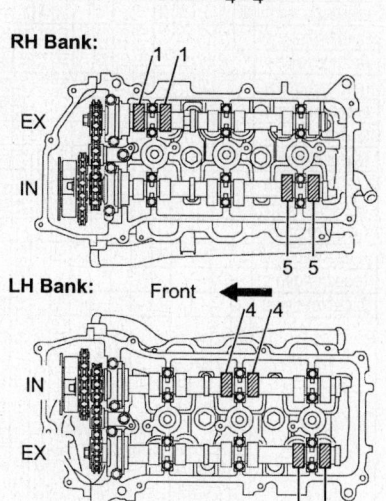

09490_TACO_G0073

Valve clearance location—4.0L engine

06	5.060 (0.1992)	30	5.300 (0.2087)	54	5.540 (0.2181)
08	5.080 (0.2000)	32	5.320 (0.2094)	56	5.560 (0.2189)
10	5.100 (0.2008)	34	5.340 (0.2102)	58	5.580 (0.2197)
12	5.120 (0.2016)	36	5.360 (0.2110)	60	5.600 (0.2205)
14	5.140 (0.2024)	38	5.380 (0.2118)	62	5.620 (0.2213)
16	5.160 (0.2031)	40	5.400 (0.2126)	64	5.640 (0.2220)
18	5.180 (0.2039)	42	5.420 (0.2134)	66	5.660 (0.2228)
20	5.200 (0.2047)	44	5.440 (0.2142)	68	5.680 (0.2236)
22	5.220 (0.2055)	46	5.460 (0.2150)	70	5.700 (0.2244)
24	5.240 (0.2063)	48	5.480 (0.2157)	72	5.720 (0.2252)
26	5.260 (0.2071)	50	5.500 (0.2165)	74	5.740 (0.2260)
28	5.280 (0.2079)	52	5.520 (0.2173)		

09490_TACO_G0074

Shim selection chart, part one—4.0L engine

09490_TACO_G0075

Shim number / thickness legend:

No.	mm (in.)	No.	mm (in.)	No.	mm (in.)
06	5.060 (0.1992)	30	5.300 (0.2087)	54	5.540 (0.2181)
08	5.080 (0.2000)	32	5.320 (0.2094)	56	5.560 (0.2189)
10	5.100 (0.2008)	34	5.340 (0.2102)	58	5.580 (0.2197)
12	5.120 (0.2016)	36	5.360 (0.2110)	60	5.600 (0.2205)
14	5.140 (0.2024)	38	5.380 (0.2118)	62	5.620 (0.2213)
16	5.160 (0.2031)	40	5.400 (0.2126)	64	5.640 (0.2220)
18	5.180 (0.2039)	42	5.420 (0.2134)	66	5.660 (0.2228)
20	5.200 (0.2047)	44	5.440 (0.2142)	68	5.680 (0.2236)
22	5.220 (0.2055)	46	5.460 (0.2150)	70	5.700 (0.2244)
24	5.240 (0.2063)	48	5.480 (0.2157)	72	5.720 (0.2252)
26	5.260 (0.2071)	50	5.500 (0.2165)	74	5.740 (0.2260)
28	5.280 (0.2079)	52	5.520 (0.2173)		

Shim selection chart, part two—4.0L engine

Starter

REMOVAL & INSTALLATION

2002–2004

2.4L ENGINE

1. Before servicing the vehicle, refer to the Precautions Section.
2. Remove or disconnect the following:
   - Negative battery cable
   - Engine cover
   - Accelerator cable and intake air connector
   - Intake manifold assembly
   - Starter motor
   - Starter electrical connectors

To install:

3. Install or connect the following:
   - Starter. Tighten both bolts to 29 ft. lbs. (39 Nm).
   - Intake manifold
   - Accelerator cable and intake air connector
   - Engine cover
   - Negative battery cable

EXCEPT 2.4L ENGINE

1. Before servicing the vehicle, refer to the Precautions Section.
2. Remove or disconnect the following:
   - Negative battery cable
   - Starter electrical connectors
   - Starter

To install:

3. Install or connect the following:
   - Starter. Tighten the fasteners to 29 ft. lbs. (39 Nm).
   - Electrical connections
   - Negative battery cable

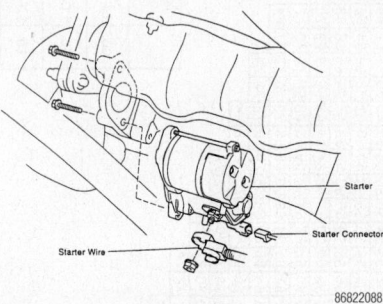

86822088

Starter motor location—except 2.4L, 2.7L (TR-FE/X) and 4.0L engines

2005–2006

2.7L (TR-FE/X) ENGINE

1. Before servicing the vehicle, refer to the Precautions Section.
2. Disconnect the negative battery cable.
3. Remove the terminal cap. Disconnect the electrical connections.
4. Disconnect the positive battery cable.
5. Remove the starter retaining bolts. Remove the starter from the vehicle.

To install:

6. Installation is the reverse of the removal procedure.
7. Tighten the retaining bolts to 27 ft. lbs.

4.0L ENGINE

1. Before servicing the vehicle, refer to the Precautions Section.
2. Disconnect the negative battery cable.
3. Remove the engine undercover assembly.
4. On 2WD and PreRunner, remove the number two manifold stay.
5. On 4WD vehicles, remove the number two exhaust front pipe assembly. Remove the five clips and then remove the front fender splash shield, left side. Remove the number two steering intermediate shaft.
6. Disconnect the starter electrical connectors.
7. Disconnect the positive battery cable.
8. Remove the starter retaining bolts. Remove the starter from the vehicle.

To install:

9. Installation is the reverse of the removal procedure.
10. Tighten the retaining bolts to 27 ft. lbs.

Oil Pan

REMOVAL & INSTALLATION

2002–2004

1. Before servicing the vehicle, refer to the Precautions Section.
2. Disconnect the negative battery cable.
3. Drain the engine oil.
4. Remove or disconnect the following:
   - Engine undercover
   - Front differential, if equipped with 4WD
   - Oil pan, separate it from the engine using SST 09032-00100 and a brass bar

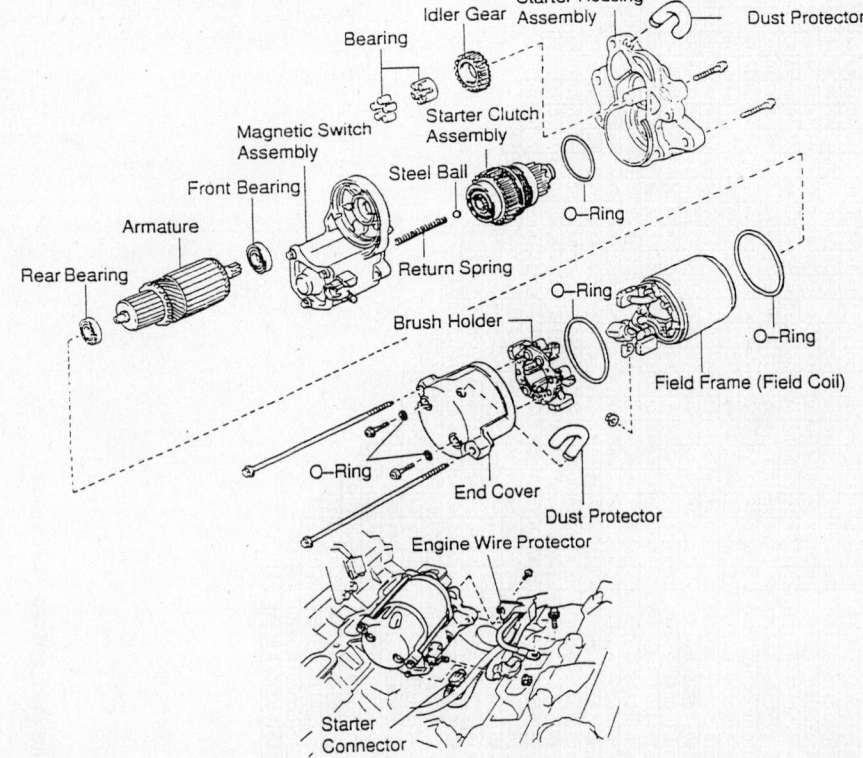

93162G18

Starter motor location—2.4L engine

To install:

5. Apply seal packing to the oil pan.

6. Install the oil pan to the cylinder block. Tighten the nuts and bolts to 67 inch lbs. (8 Nm)

☀☀ WARNING

If parts are not assembled within 5 minutes of applying time, the effectiveness of the seal packing is lost and must be removed and reapplied.

7. Install or connect the following:
   • Front differential, if removed
   • Engine undercover
   • Negative battery cable
8. Fill with engine oil.
9. Start the engine and check for leaks.

2005–2006

2.7L (2TR-FE/X) ENGINE

1. Before servicing the vehicle, refer to the Precautions Section.
2. Disconnect the negative battery cable.
3. Raise and support the vehicle safely.
4. Remove the engine undercover. Drain the engine oil.
5. Remove the necessary components in order to gain access to the lower oil pan retaining bolts.
6. Remove the eighteen bolts and two nuts. Insert the blade of tool SST09032-00100 between the pans. Cut through the sealer and remove the lower oil pan from the engine.

➡**Be careful not to damage the contact surface of the oil pans.**

7. Remove the two bolts and nuts. Remove the oil strainer. Discard the gasket.
8. To remove the upper oil pan, remove the sixteen bolts and two nuts. Remove the upper oil pan from the engine, by prying it apart using a suitable tool.

➡**Be careful not to damage the sealing surface between the upper oil pan and the cylinder block.**

To install:

9. Apply a continuous bead (0.079–0.118 inch in diameter) of seal packing, part number 08826-00080 or equivalent, to the sealing surface of the oil pan.

➡**Remove any oil from the contact surface. Install the upper oil pan within three minutes of applying the seal packing. Do not start the engine for at least two hours after the installation of the oil pan.**

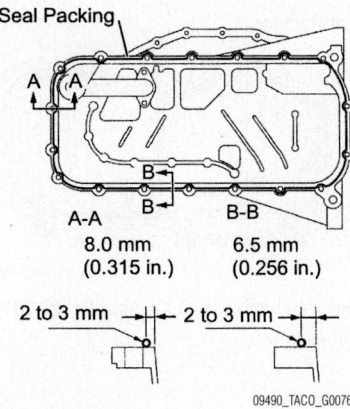

Upper oil pan sealant application—2.7L (2TR-FE/X) engine

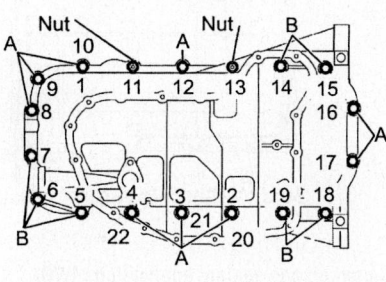

Upper oil pan bolt torque sequence—2.7L (2TR-FE/X) engine

10. Loosely install the upper oil pan bolts and nuts. Bolt "A" is 0.79 inch long and bolt "B" is 1.57 inch long. Uniformly tighten the bolts to 19 ft. lbs, in the proper sequence.
11. Install the oil strainer assembly. Torque the bolts to 19 ft. lbs.
12. Apply a continuous bead (0.118–0.157 inch in diameter) of seal packing, part number 08826-00080 or equivalent, to the sealing surface of the oil pan.

➡**Remove any oil from the contact surface. Install the lower oil pan within three minutes of applying the seal packing. Do not start the engine for at least two hours after the installation of the oil pan.**

13. Loosely install the lower oil pan bolts and nuts. Uniformly tighten the bolts to 80 inch lbs, in the proper sequence.
14. Continue the installation in the reverse order of the removal procedure.
15. Be sure to fill the engine with the proper grade and type engine coolant.
16. Be sure to fill the engine with the proper grade and type engine oil.
17. Start the engine and check for leaks. Correct as required.

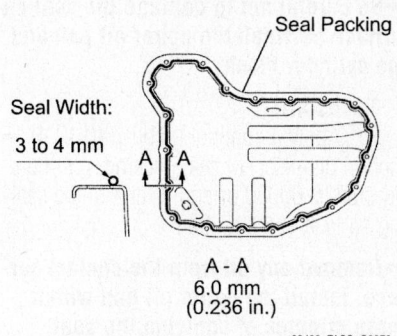

Lower oil pan sealant application—2.7L (2TR-FE/X) engine

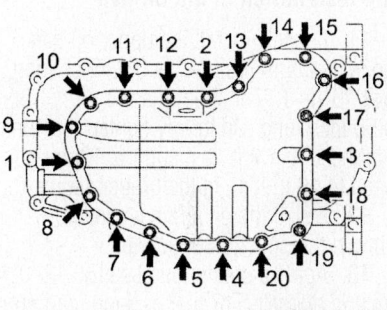

Lower oil pan bolt torque sequence—2.7L (2TR-FE/X) engine

4.0L ENGINE

1. Before servicing the vehicle, refer to the Precautions Section.
2. Disconnect the negative battery cable.
3. Raise and support the vehicle safely.
4. Remove the engine undercover. Drain the engine oil.
5. Remove the necessary components in order to gain access to the lower oil pan retaining bolts.
6. Remove the fifteen bolts and two nuts (2WD vehicles) and ten bolts and two nuts (4WD and PreRunner). Insert the blade of tool SST09032-00100 between the pans. Cut through the sealer and remove the lower oil pan from the engine.

➡**Be careful not to damage the contact surface of the oil pans.**

7. Remove the two bolts and nuts. Remove the oil strainer. Discard the gasket.
8. On 4WD and PreRunner, remove the four housing bolts. Remove the flywheel housing undercover.
9. To remove the upper oil pan, remove the seventeen bolts and two nuts. Remove the upper oil pan from the engine, by prying it apart using a suitable tool.

➡️Be careful not to damage the sealing surface between the upper oil pan and the cylinder block.

To install:

10. Apply a continuous bead (0.12–0.16 inch in diameter) of seal packing, part number 08826-00080 or equivalent, to the sealing surface of the oil pan.

➡️Remove any oil from the contact surface. Install the upper oil pan within three minutes of applying the seal packing. Tighten the pan bolts to specification within fifteen minutes after applying the seal packing. Do not start the engine for at least two hours after the installation of the oil pan.

11. Loosely install the upper oil pan bolts and nuts. Bolt "A" is 0.98 inch long, bolt "B" is 1.77 inch long and bolt "C" is 0.55 inch long. Uniformly tighten the 10mm bolt head to 7.4 ft. lbs, and the 12mm bolt head to 16 ft. lbs., in the proper sequence.

12. Install the oil strainer assembly. Torque the bolts to 80 inch lbs.

13. Apply a continuous bead (0.12–0.16 inch in diameter) of seal packing, part number 08826-00080 or equivalent, to the sealing surface of the oil pan.

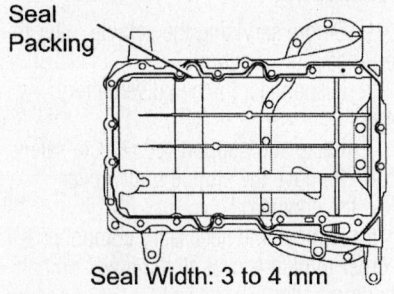

Upper oil pan sealant application (2WD)—4.0L engine

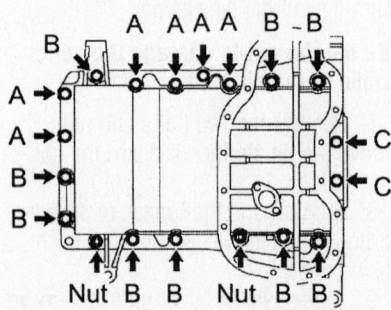

Upper oil pan bolt torque sequence (2WD)—4.0L engine

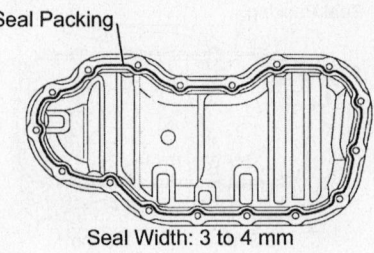

Lower oil pan sealant application (2WD)—4.0L engine

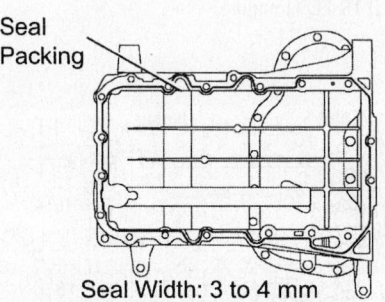

Upper oil pan sealant application (4WD and PreRunner)—4.0L engine

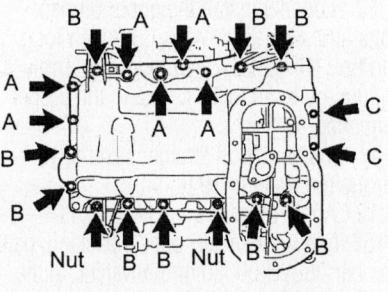

Upper oil pan bolt torque sequence (4WD and PreRunner)—4.0L engine

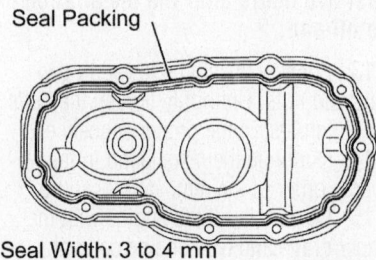

Lower oil pan sealant application (4WD and PreRunner)—4.0L engine

➡️Remove any oil from the contact surface. Install the lower oil pan within three minutes of applying the seal packing. Tighten the pan bolts to specification within fifteen minutes after applying the seal packing. Do not start the engine for at least two hours after the installation of the oil pan.

14. Loosely install the lower oil pan bolts and nuts. Uniformly tighten the bolts to 80 inch lbs, and the nuts to 7.4 ft. lbs., in several steps.

15. Continue the installation in the reverse order of the removal procedure.

16. Be sure to fill the engine with the proper grade and type engine coolant.

17. Be sure to fill the engine with the proper grade and type engine oil.

18. Start the engine and check for leaks. Correct as required.

Oil Pump

REMOVAL & INSTALLATION

2.4L Engine

➡️The oil pump assembly is mounted in the timing chain cover. To properly service the oil pump, the timing chain cover should be removed from the cylinder block.

1. Before servicing the vehicle, refer to the Precautions Section.

2. Disconnect the negative battery cable.

3. Drain the oil and the cooling system.

4. Remove or disconnect the following:

- Engine undercover
- Front differential and halfshaft assembly, if equipped with 4WD
- Upper radiator hose from the radiator
- Oil dipstick guide, by removing the bolt
- Power steering drive belt, by loosening the lockbolt and adjusting bolt, if equipped with power steering
- No. 2 fan shroud, by removing the 2 clips
- No. 1 fan shroud, by removing the 4 bolts
- Drive belt, if equipped with air conditioning
- Alternator drive belt, fan (with fan clutch), water pump pulley and the fan shroud
- Cylinder head
- Air conditioning compressor and

bracket, if equipped with air conditioning
- Alternator, adjusting bar and bracket
- Crankshaft Position (CKP) sensor, by removing the 2 bolts
- Stiffener plates by removing the 8 bolts, if equipped with 2WD
- Flywheel housing undercover and dust seal
- Oil pan.
- Oil strainer and gasket
- Crankshaft pulley
- Timing chain cover
- Oil pump from the front cover, by removing the 9 screws
- Oil pump cover, drive rotor, driven rotor and O-ring

5. Remove the relief valve, as follows:

a. Using snapring pliers, remove the snapring for the relief valve.

b. Remove the retainer, spring(s) and relief valve from the front cover.

To install:

6. Install the relief valve, as follows:

a. Install the relief valve, spring(s) and retainer to the valve cover.

b. Using snapring pliers, install the snapring to hold the relief valve.

7. Install the drive and driven rotors as follows:

a. Place the drive and driven rotors into the pump body.

b. Place a new O-ring to the pump body.

c. Install the pump cover with the 9 screws.

8. Install the remaining components in the reverse order of removal.

9. Fill the cooling system and fill the engine with oil.

10. Connect the negative battery cable.

11. Start the engine and check for leaks.

12. Adjust ignition timing. Road test the vehicle for proper operation.

13. Recheck all fluid levels.

2.7L (3RZ-FE/M) Engine

➡**The oil pump assembly is mounted in the timing chain cover. To properly service the oil pump, the timing chain cover should be removed from the cylinder block.**

1. Before servicing the vehicle, refer to the Precautions Section.

2. Disconnect the negative battery cable.

3. Drain the oil and the cooling system.

4. Remove or disconnect the following:
- Engine undercover
- Front differential and halfshaft assembly, if equipped with 4WD

- 2 bolts and the air pipe, for California vehicles with 3RZ-FE engine
- Upper radiator hose from the radiator
- Oil dipstick guide
- Power steering drive belt, by loosening the lockbolt and adjusting bolt, if equipped with power steering
- No. 2 fan shroud by removing the 2 clips
- No. 1 fan shroud by removing the 4 bolts
- Drive belt, by loosening the idler pulley nut and adjusting bolt, if equipped with air conditioning
- Alternator drive belt, fan (with fan clutch), water pump pulley and the fan shroud
- Cylinder head
- Air conditioning compressor and bracket with the lines attached, if equipped with air conditioning
- Alternator, adjusting bar and bracket
- Crankshaft Position (CKP) sensor by removing the 2 bolts
- Stiffener plates by removing the 8 bolts, if equipped with 2WD
- Flywheel housing undercover and dust seal
- Oil pan
- Oil strainer and gasket
- Crankshaft pulley
- Timing chain cover
- Oil pump from the front cover by removing the 9 screws
- Oil pump cover, drive rotor, driven rotor and O-ring

5. Remove the relief valve, as follows:

a. Using snapring pliers, remove the snapring for the relief valve.

b. Remove the retainer, spring(s) and relief valve from the front cover.

To install:

6. Install the relief valve, as follows:

a. Install the relief valve, spring(s) and retainer to the valve cover.

b. Using snapring pliers, install the snapring to hold the relief valve.

7. Install the drive and driven rotors, as follows:

a. Place the drive and driven rotors into the pump body.

b. Place a new O-ring to the pump body.

c. Install the pump cover with the 9 screws.

8. Install the remaining components in the reverse order of removal.

9. Fill the cooling system and fill the engine with oil.

10. Connect the negative battery cable.

11. Start the engine and check for leaks.

12. Adjust ignition timing. Road test the vehicle for proper operation.

13. Recheck all fluid levels.

2.7L (2TR-FE/X) Engine

1. Before servicing the vehicle, refer to the Precautions Section.

2. Disconnect the negative battery cable.

3. Remove the engine and position it in a suitable holding fixture.

4. Remove the air intake connector. Remove the alternator.

5. Remove the belt tensioner assembly. Remove the idler pulley subassembly. Remove the air conditioning idler pulley assembly and bracket.

6. Remove the crankshaft position sensor. Remove the camshaft position sensor.

7. Remove the intake manifold. Remove the cylinder head cover.

8. Remove the crankshaft pulley. Remove the oil level gauge subassembly.

9. Remove the lower oil pan. Remove the oil strainer assembly. Remove the upper oil pan.

10. Remove the two nuts and separate the water bypass pipe number one.

11. Remove the nineteen bolts and two nuts retaining the timing chain case cover to its mounting. Remove the timing chain case cover from the engine.

➡**Carefully remove the cover by prying between the cover and the cylinder head or block with a suitable tool. Be sure to cover the tip of the suitable tool prior to usage. Be careful not to damage the contact surfaces of the cylinder block, cylinder head and timing chain cover.**

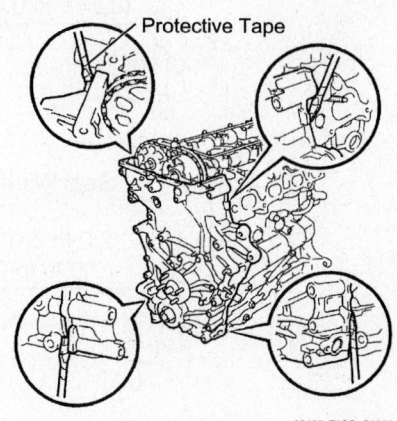

09490_TACO_G0086

Timing chain cover removal points—2.7L (2TR-FE/X) engine

➡ **The oil pump gears are located inside the timing case cover.**

12. Remove and discard the O-rings. Remove the head straight screw plug. Remove the water inlet. Remove the thermostat. Remove the oil seal.

13. Remove the oil pump relief valve. Remove the seven oil pump cover bolts. Remove the gears from the timing chain case cover.

To install:

14. Coat the oil pump gears with clean engine oil. Position the gears in the timing chain case cover with the identification marks facing outward. Check that the rotors revolve smoothly. Install the cover. Alternately tighten the bolts to 80 inch lbs.

15. Coat the oil pump relief valve with clean engine oil. Install the plug, using a new gasket and tighten to 36 ft. lbs.

16. Install a new front case oil seal. Install the thermostat. Install the water inlet.

17. Apply adhesive, part number 08833-00070 or equivalent to the head straight screw plug. Install the plug and tighten to 12 ft. lbs. Install four new O-rings onto the timing chain case cover.

18. Apply continuous beads of seal packing, part number 08826-00080 or equivalent as shown in the illustration.

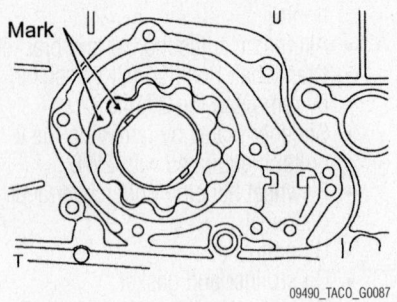

09490_TACO_G0087

Oil pump gear alignment marks—2.7L (2TR-FE/X) engine

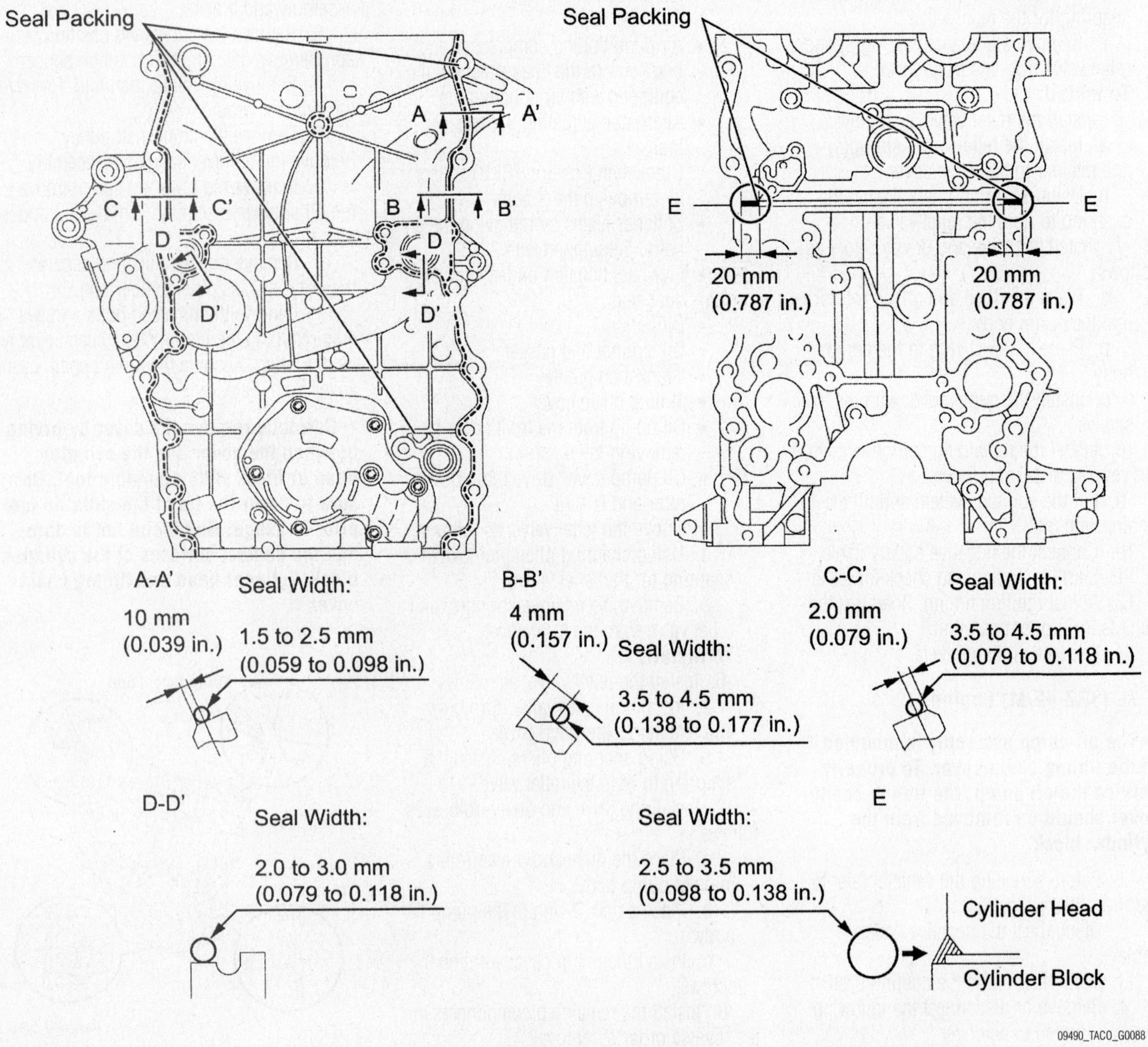

Seal Packing

Seal Packing

A-A'
10 mm
(0.039 in.)

Seal Width:
1.5 to 2.5 mm
(0.059 to 0.098 in.)

B-B'
4 mm
(0.157 in.)

Seal Width:
3.5 to 4.5 mm
(0.138 to 0.177 in.)

C-C'
2.0 mm
(0.079 in.)

Seal Width:
3.5 to 4.5 mm
(0.079 to 0.118 in.)

D-D'

Seal Width:
2.0 to 3.0 mm
(0.079 to 0.118 in.)

E

Seal Width:
2.5 to 3.5 mm
(0.098 to 0.138 in.)

20 mm
(0.787 in.)

20 mm
(0.787 in.)

Cylinder Head

Cylinder Block

09490_TACO_G0088

Timing chain case cover sealant application—2.7L (2TR-FE/X) engine

➡**Remove any oil from the contact sur-faces. Install the timing chain case cover within three minutes and tighten the bolts within fifteen minutes of applying the seal packing. Do not start the engine for at least four hours after installation of the cover.**

19. Align the oil pump drive rotor spline and the crankshaft, as indicated in the illustration. Install the spline and timing chain case cover onto the crankshaft.

20. Loosely install the timing chain case cover retaining bolts and nuts.

➡**If the vehicle is equipped with air conditioning install the bolts that hold the idle pulley bracket in place when installing the idle pulley, as they are for this purpose.**

21. Fully tighten the bolts and nuts, except bolts "A" in the following order: Area 1, Area 3 and then Area 2 to 15 ft. lbs.

22. Fully tighten the bolts "A" in the following order: Area 2 and then Area 3 to 34 ft. lbs.

23. Fully tighten the bolts "E" in Area 4 to 15 ft. lbs.

24. Continue the installation in the reverse order of the removal procedure.

25. When installing the cylinder head cover, apply seal packing, part number

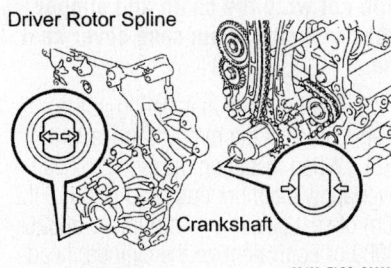

Oil pump drive rotor spline alignment—2.7L (2TR-FE/X) engine

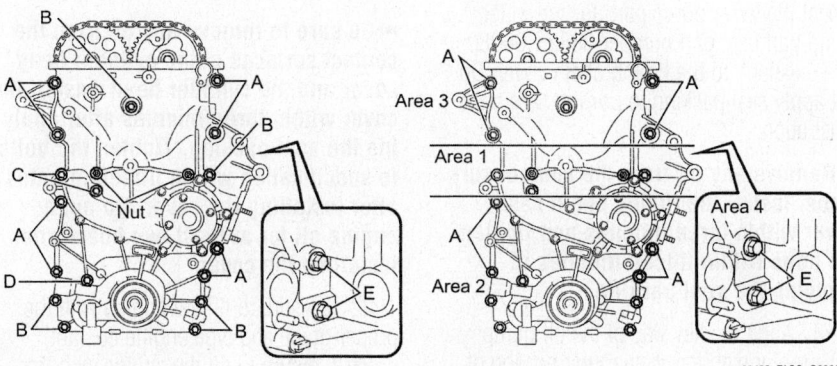

Timing chain case cover bolt location and torque sequence—2.7L (2TR-FE/X) engine

08826-00080 or equivalent, to the cylinder head. Provisionally install the cylinder head cover bolts and nuts. Tighten bolts "A" to 80 inch lbs. Tighten bolts "B" to 80 inch lbs. Retighten bolts "A" to 80 inch lbs.

➡**Be sure to remove any oil from the contact surfaces of the cylinder head cover and the cylinder head. Install the cover within three minutes after apply-ing the seal packing. Do not add engine oil for at least two hours after installing the cover.**

26. Be sure to fill the engine with the proper grade and type engine coolant.

27. Be sure to fill the engine with the proper grade and type engine oil.

28. Start the engine and check for leaks, correct as required.

3.4L Engine

1. Before servicing the vehicle, refer to the Precautions Section.

2. Remove or disconnect the following:
   - Negative battery cable
   - Engine undercover
   - Crankshaft timing pulley
   - Front differential, if equipped with 4WD

3. Drain the engine oil from the engine.

4. Remove or disconnect the following:
   - Timing belt and crankshaft gear
   - Oil cooler tube and clamp, if equipped with automatic transmission
   - Stiffener plate
   - Flywheel housing undercover and dust cover
   - Rear end cover and dust cover
   - Starter wire clamp
   - Crankshaft Position (CKP) sensor
   - Oil pan

➡**Be careful not to damage the baffle plate flange.**

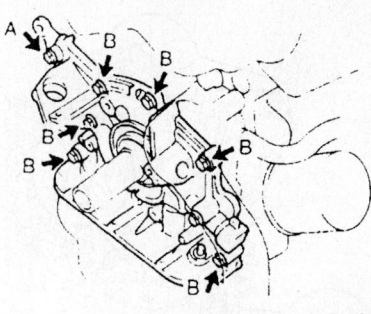

Oil pump bolt identification—3.4L engine

   - Oil strainer
   - Oil baffle plate
   - Oil pump body by removing the 8 bolts.
   - O-ring from the cylinder block

To install:

5. Install or connect the following:
   - Apply Seal Packing PN 08826-00080 to the oil pump
   - New O-ring into the groove of the cylinder block
   - Oil pump to the crankshaft with the spline teeth of the drive rotor engaged with the large teeth of the crankshaft. Tighten the oil pump bolts "A" 15 ft. lbs. (20 Nm) and bolts "B" 31 ft. lbs. (42 Nm)
   - CKP
   - Oil pan baffle plate
   - Oil strainer with a new gasket. Tighten the bolts to 13 ft. lbs. (18 Nm).
   - Remaining components
   - Negative battery cable

6. Fill with engine oil.

7. Start the engine and check for leaks.

4.0L Engine

2WD

1. Before servicing the vehicle, refer to the Precautions Section.

2. Properly relieve the fuel system pressure.

3. Disconnect the negative battery cable. Disconnect the positive battery cable. Remove the battery.

4. Drain the engine coolant. Drain the engine oil.

5. Remove the engine from the vehicle and position it in a suitable holding fixture.

6. Remove the oil level gauge guide. Remove the water inlet. Remove the belt tensioner.

7. Remove the idler pulley number two subassembly. Remove the idler pulley number one sub assembly. Remove the crank-shaft pulley.

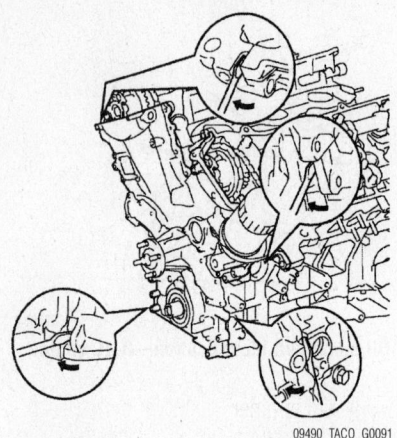

Timing chain case cover removal points—4.0L engine

8. Remove the lower oil pan. Remove the strainer and pickup tube. Remove the upper oil pan.

9. Remove the ignition coil assembly. Remove the cylinder head cover. Remove the camshaft timing oil control valve assembly.

10. Remove the VVT sensor. Remove the oil filter bracket subassembly.

11. Remove the timing chain case cover retaining bolts. Remove the cover from the engine. Remove the O-ring from the left cylinder head.

➡Carefully remove the cover by prying between the cover and the cylinder head or block with a suitable tool. Be sure to cover the tip of the suitable tool prior to usage. Be careful not to damage the contact surfaces of the cylinder block, cylinder head and timing chain cover.

➡The oil pump gears are located inside the timing case cover.

12. Remove the three bolts and remove the oil pipe. Remove the two O-rings. Remove the seven oil pump cover bolts. Remove the gears from the timing chain case cover. Remove the oil pump relief valve.

To install:

13. Coat the oil pump relief valve with clean engine oil. Install the plug, using a new gasket and tighten to 36 ft. lbs.

14. Coat the oil pump gears with clean engine oil. Position the gears in the timing chain case cover with the identification marks facing oil pump cover side. Install the cover. Alternately tighten the bolts to 80 inch lbs. Install the oil pipe, tighten the bolts to 80 inch lbs.

15. Install a new front case oil seal.

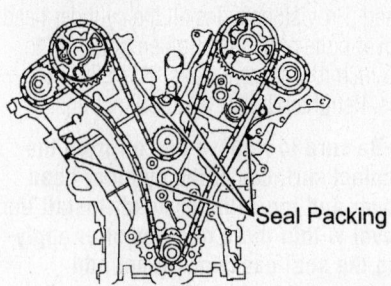

Timing chain case cover seal packing locating points—4.0L engine

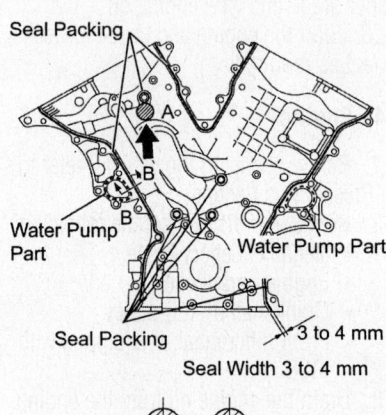

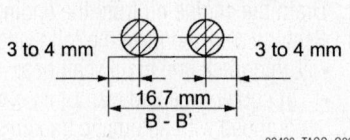

Timing chain case cover sealant application—4.0L engine

Install a new O-ring onto the left cylinder head.

16. Apply continuous beads (0.12–0.16 inch in diameter) of seal packing, part number 08826-00080 or equivalent to the four locations shown in the illustration.

17. Apply continuous beads (0.12–0.16 inch in diameter) of seal packing, part number 08826-00080 or equivalent to all parts except the water pump part: for the water pump part use, part number 08826-00080 or equivalent, to the timing chain cover. Do not apply seal packing to portion "A" in the illustration.

➡Remove any oil from the contact surfaces. Install the timing chain case cover within three minutes and tighten the bolts within fifteen minutes of applying the seal packing.

18. Align the key way of the oil pump drive motor with the rectangular portion of the crankshaft timing gear and slide the timing chain case cover into place.

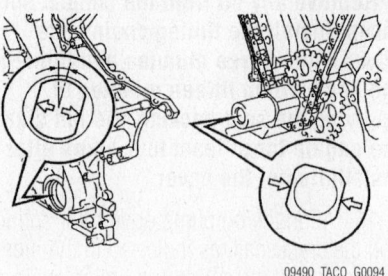

Oil pump drive rotor spline alignment—4.0L engine

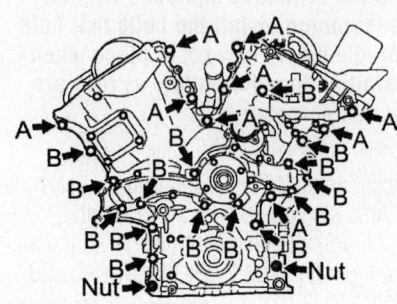

Timing chain case cover bolt location and torque sequence—4.0L engine

19. Install the timing chain case cover bolts. Tighten the bolts and nuts uniformly in several steps to 17 ft. lbs.

➡Do not wrap the chain and slipper over the timing chain case cover seal line.

20. Continue the installation in the reverse order of the removal procedure.

21. When installing the cylinder head cover apply a continuous bead (0.08–0.12 inch) of seal packing, part number 08826-00080 or equivalent, to the cylinder head. Install the seal washers onto the bolts. Install the cylinder head cover bolts and nuts. Tighten bolts "A" to 7.4 ft. lbs. Tighten bolts "B" to 80 inch lbs. Tighten nuts to 80 inch lbs.

➡Be sure to remove any oil from the contact surfaces of the cylinder head cover and the cylinder head. Install the cover within three minutes after applying the seal packing. Tighten the bolts to specification within fifteen minutes after installing the cover. Do not add engine oil for at least two hours after installing the cover.

22. Be sure to fill the engine with the proper grade and type engine coolant.

23. Be sure to fill the engine with the proper grade and type engine oil.

24. Start the engine and check for leaks, correct as required.

4WD AND PRERUNNER

1. Before servicing the vehicle, refer to the Precautions Section.

2. Properly relieve the fuel system pressure.

3. Disconnect the negative battery cable. Disconnect the positive battery cable. Remove the battery.

4. Drain the engine coolant. Drain the engine oil.

5. Remove the power steering gear assembly.

6. If equipped with 4WD, remove the front differential carrier assembly.

7. Remove the V bank cover. Remove the radiator support to frame seal, left side. Remove the fan shroud.

8. Remove the air cleaner assembly. Remove the oil level gauge. Remove the water inlet.

9. Separate the vane pump assembly. Remove the alternator. Remove the air conditioning compressor and position it to the side.

10. Remove the belt tensioner assembly. Remove the idler pulley number two subassembly. Remove the idler pulley number one subassembly. Remove the crankshaft pulley.

11. Remove the lower oil pan. Remove the oil strainer and pickup tube assembly. Remove the upper oil pan.

12. Remove the intake manifold. Remove the cylinder head cover assembly.

13. Remove the camshaft timing oil control valve assembly. Remove the VVT sensor. Remove the oil filter bracket subassembly.

➡**Carefully remove the cover by prying between the cover and the cylinder head or block with a suitable tool. Be sure to cover the tip of the suitable tool prior to usage. Be careful not to damage the contact surfaces of the cylinder block, cylinder head and timing chain cover.**

➡**The oil pump gears are located inside the timing case cover.**

14. Remove the three bolts and remove the oil pipe. Remove the two O-rings. Remove the seven oil pump cover bolts. Remove the gears from the timing chain case cover. Remove the oil pump relief valve.

To install:

15. Coat the oil pump relief valve with clean engine oil. Install the plug, using a new gasket and tighten to 36 ft. lbs.

16. Coat the oil pump gears with clean engine oil. Position the gears in the timing chain case cover with the identification marks facing oil pump cover side. Install the cover. Alternately tighten the bolts to 80 inch lbs. Install the oil pipe, tighten the bolts to 80 inch lbs.

17. Install a new front case oil seal. Install a new O-ring onto the left cylinder head.

18. Apply continuous beads (0.12–0.16 inch in diameter) of seal packing, part number 08826-00080 or equivalent to the four locations shown in the illustration.

19. Apply continuous beads (0.12–0.16 inch in diameter) of seal packing, part number 08826-00080 or equivalent to all parts except the water pump part: for the water pump part use, part number 08826-00080 or equivalent, to the timing chain cover. Do not apply seal packing to portion "A" in the illustration.

➡**Remove any oil from the contact surfaces. Install the timing chain case cover within three minutes and tighten the bolts within fifteen minutes of applying the seal packing.**

20. Align the key way of the oil pump drive motor with the rectangular portion of the crankshaft timing gear and slide the timing chain case cover into place.

21. Install the timing chain case cover bolts. Tighten the bolts and nuts uniformly in several steps to 17 ft. lbs.

➡**Do not wrap the chain and slipper over the timing chain case cover seal line.**

22. Continue the installation in the reverse order of the removal procedure.

23. When installing the cylinder head cover apply a continuous bead (0.08–0.12 inch) of seal packing, part number 08826-00080 or equivalent, to the cylinder head. Install the seal washers onto the bolts. Install the cylinder head cover bolts and nuts. Tighten bolts "A" to 7.4 ft. lbs. Tighten bolts "B" to 80 inch lbs. Tighten nuts to 80 inch lbs.

➡**Be sure to remove any oil from the contact surfaces of the cylinder head cover and the cylinder head. Install the cover within three minutes after applying the seal packing. Tighten the bolts to specification within fifteen minutes after installing the cover. Do not add engine oil for at least two hours after installing the cover.**

24. Be sure to fill the engine with the proper grade and type engine coolant.

25. Be sure to fill the engine with the proper grade and type engine oil.

26. Start the engine and check for leaks, correct as required.

Rear Main Seal

REMOVAL & INSTALLATION

Seal Retainer On Engine

2002–2006

1. Before servicing the vehicle, refer to the Precautions Section.

2. Disconnect the negative battery cable.

3. Remove the transmission.

4. Remove the clutch cover assembly and flywheel (manual transmission) or the flexplate (automatic transmission).

5. Use a small, sharp knife to cut off the lip of the oil seal. Take great care not to score any metal with the knife.

6. Use a small prybar to pry the old seal from the retaining plate. Be careful not to damage the plate. Protect the tip of the tool with tape and pad the fulcrum point with cloth.

7. Inspect the crankshaft and seal lip contact surfaces for any sign of damage.

To install:

8. Apply a light coat of multi-purpose grease to the lip of a new oil seal. Loosely fit the seal into place by hand, making sure it is not crooked.

9. Use a seal driver such as (SST 09223–15030 and 09950–70010) of the correct size to install the seal. Tap it into place until the surface of the seal is flush with the edge of the housing.

➡**Use the correct tools. Homemade substitutes may install the seal crooked, resulting in oil leaks and premature seal failure.**

Seal Retainer Removed

2002–2004

1. Support the retainer on two thin pieces of wood.

2. Use a small prybar to pry the old seal from the retaining plate. Be careful not to damage the plate. Protect the tip of the tool with tape and pad the fulcrum point with cloth.

To install:

3. Apply a light coat of multi-purpose grease to the lip of a new oil seal. Loosely fit the seal into place by hand, making sure it is not crooked.

4. Use a seal driver such as (SST 09223–15030 and 09950–70010) of the

correct size to install the seal. Tap it into place until the surface of the seal is flush with the edge of the housing.

Timing Belt, Cover and Crankshaft Seal

REMOVAL & INSTALLATION

3.4L Engine

1. Before servicing the vehicle, refer to the Precautions Section.

2. Disconnect the negative battery cable.

✳ CAUTION

Work must be started after 90 seconds from the time the ignition switch is turned to the LOCK position and the negative battery cable is disconnected.

3. Raise and safely support the vehicle.
4. Remove the engine undercover.
5. Drain the engine coolant.

6. Disconnect the upper radiator hose from the engine.
7. Remove the power steering drive belt.
8. Remove the air conditioning drive belt by loosening the idler pulley nut and the adjusting bolt.
9. Loosen the lockbolt, pivot bolt, and the adjusting bolt and the alternator drive belt.
10. Remove the No. 2 fan shroud by removing the 2 clips.
11. Remove the fan with the fluid coupling and fan pulleys.

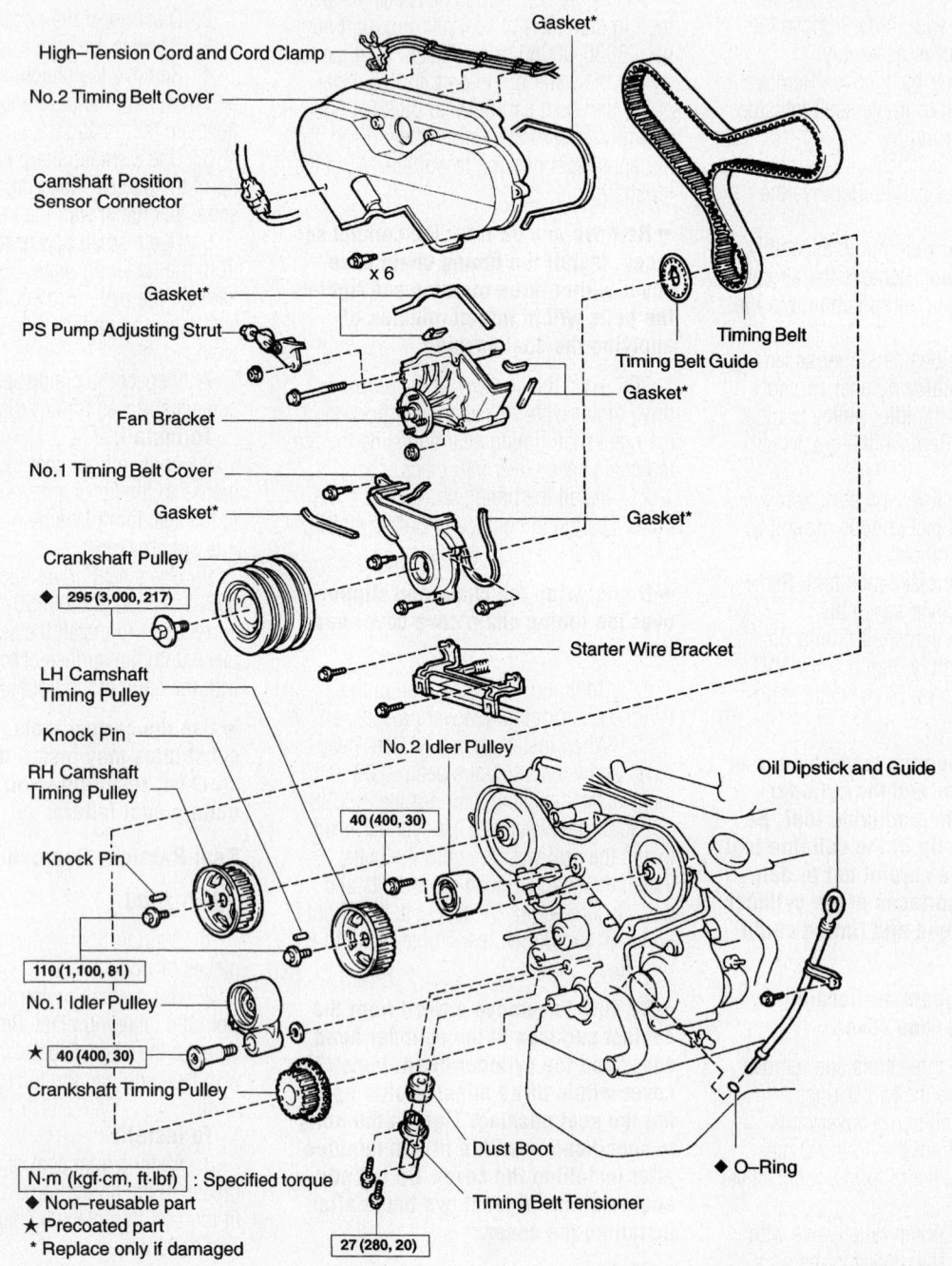

Timing belt and related components—3.4L engine

67170-TACO-G05

12. Disconnect the power steering pump from the engine and set aside. Do not disconnect the lines from the pump.

13. If equipped with air conditioning, disconnect the compressor from the engine and set aside. Do not disconnect the lines from the compressor.

14. If equipped with air conditioning, disconnect the air conditioning bracket.

15. Remove the No. 2 timing belt cover, as follows:

    a. Detach the camshaft position sensor connector from the No. 2 timing belt cover.

    b. Disconnect the 3 spark plug wire clamps from the No. 2 timing belt cover.

    c. Remove the 6 bolts and remove the timing belt cover.

16. Remove the fan bracket, as follows:

    a. Remove the power steering adjusting strut by removing the nut.

    b. Remove the fan bracket by removing the bolt and nut.

17. Set the No. 1 cylinder at Top Dead Center (TDC) of the compression stroke, as follows:

    a. Turn the crankshaft pulley and align its groove with the timing mark **0** of the No. 1 timing belt cover.

    b. Check that the timing marks of the camshaft timing pulleys and the No. 3 timing belt cover are aligned. If not, turn the crankshaft pulley one revolution (360 degrees).

➡️**If reusing the timing belt, be sure that you can still read the installation marks. If not, place new installation marks on the timing belt to match the timing marks of the camshaft timing pulleys.**

18. Remove the timing belt tensioner by alternately loosening the 2 bolts.

19. Remove the camshaft timing pulleys, as follows:

    a. Using SST 09960-10010, or equivalent, remove the pulley bolt, the timing pulley and the knock pin. Remove the 2 timing pulleys with the timing belt.

20. Remove the crankshaft pulley, as follows:

    a. Using SST 09213-54015 and 09330-00021 or their equivalents, loosen the pulley bolt.

    b. Remove the SST tool, the pulley bolt and the pulley.

21. Remove the starter wire bracket and the No. 1 timing belt cover.

22. Remove the timing belt guide and remove the timing belt.

23. Remove the bolt and the No. 2 idler pulley.

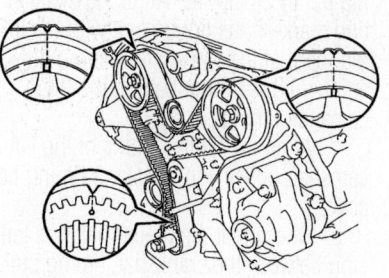

Crankshaft alignment before removing timing belt—3.4L engine

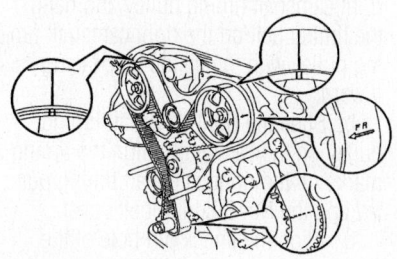

67170-TACO-G07

Check the installation marks on the timing belt—3.4L engine

24. Remove the pivot bolt, the No. 1 idler pulley and the plate washer.

25. Remove the crankshaft gear.

To install:

26. Install the crankshaft timing gear.

    a. Align the timing pulley set key with the key groove of the gear.

    b. Using SST 09214-60010, or equivalent, and a hammer, tap in the timing gear with the flange side facing inward.

27. Install the plate washer and the No. 1 idler pulley with the pivot bolt and tighten it to 26 ft. lbs. (35 Nm). Check that the pulley bracket moves smoothly.

28. Install the No. 2 timing belt idler with the bolt. Tighten the bolt to 30 ft. lbs. (40 Nm). Check that the pulley bracket moves smoothly.

29. Temporarily install the timing belt, as follows:

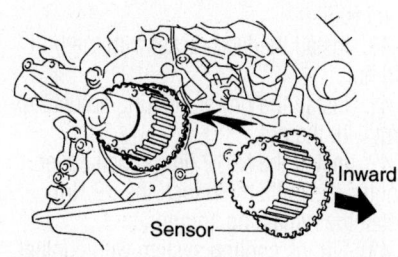

67170-TACO-G08

Tap in the timing gear with the flange side facing inward—3.4L engine

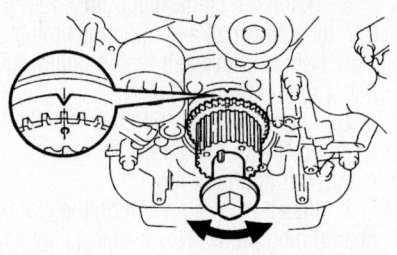

67170-TACO-G09

Turn the crankshaft and align the timing marks of the crankshaft timing pulley and the oil pump body, this sets the No.1 piston at TDC compression—3.4L engine

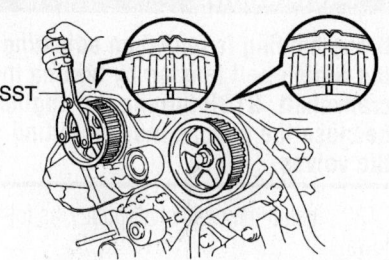

67170-TACO-G10

Camshaft sprocket timing mark alignment—3.4L engine

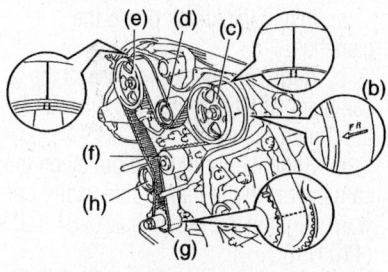

67170-TACO-G11

Timing belt installation sequence: b–timing belt front mark; c–left camshaft timing pulley; d–no.2 idler pulley; e–right camshaft timing pulley; f–water pump pulley; g–crankshaft; h–no.1 idler pulley—3.4L engine

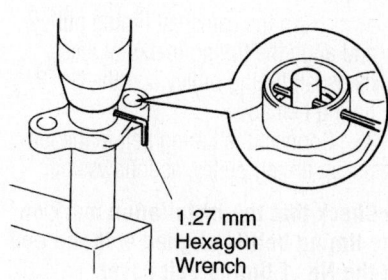

67170-TACO-G12

Compressing the tensioner plunger—3.4L engine

a. Using the crankshaft pulley bolt, turn the crankshaft and align the timing marks of the crankshaft timing pulley and the oil pump body.

b. Align the installation mark on the timing belt with the dot mark of the crankshaft timing pulley.

c. Install the timing belt on the crankshaft timing pulley, No. 1 idler pulley and the water pump pulleys.

30. Install the timing belt guide with the cup side facing outward.

31. Install the No. 1 timing belt cover and starter wire bracket. Tighten the timing belt cover bolts to 80 inch lbs. (9 Nm).

✳✳ WARNING

If any binding is felt when adjusting the timing belt tension by turning the crankshaft, STOP turning the engine, because the pistons may be hitting the valves.

32. Install the crankshaft pulley, as follows:

a. Align the pulley set key with the key groove of the crankshaft pulley.

b. Install the pulley bolt and tighten it to 184 ft. lbs. (250 Nm).

33. Install the left camshaft timing pulley.

a. Install the knock pin to the camshaft.

b. Align the knock pin hose of the camshaft with the knock pin groove of the timing pulley.

c. Slide the timing belt pulley on the camshaft with the flange side facing outward. Tighten the pulley bolt to 81 ft. lbs. (110 Nm).

34. Set the No. 1 cylinder to TDC of the compression stroke, as follows:

a. Turn the crankshaft pulley, and align its groove with the timing mark **0** of the No. 1 timing belt cover.

b. Turn the camshaft to align the knock pin hole of the camshaft with the timing mark of the No. 3 timing belt cover.

c. Turn the camshaft timing pulley, and align the timing marks of the camshaft timing pulley and the No. 3 timing belt cover.

35. Connect the timing belt to the left camshaft timing pulley, as follows:

➡**Check that the installation mark on the timing belt is aligned with the end of the No. 1 timing belt cover.**

a. Using SST 09960-01000, or equivalent, slightly turn the left camshaft timing pulley clockwise. Align the installation mark on the timing belt with the timing mark of the camshaft timing pulley and hang the timing belt on the left camshaft timing pulley.

b. Align the timing marks of the left camshaft pulley and the No. 3 timing belt cover.

c. Check that the timing belt has tension between the crankshaft timing pulley and the left camshaft timing pulley.

36. Install the right camshaft timing pulley and the timing belt, as follows:

a. Align the installation mark on the timing belt with the timing mark of the right camshaft timing pulley and hang the timing belt on the right camshaft timing pulley with the flange side facing inward.

b. Slide the right camshaft timing pulley on the camshaft. Align the timing marks on the right camshaft timing pulley and the No. 3 timing belt cover.

c. Align the knock pin hole of the camshaft with the knock pin groove of the pulley and install the knock pin. Install the bolt and tighten it to 81 ft. lbs. (110 Nm).

37. Set the timing belt tensioner, as follows:

a. Using a press, slowly press in the pushrod using 220–2205 lbs. (981–9807 N) of force.

b. Align the holes of the pushrod and housing, pass a 1.5mm hex wrench through the holes to keep the setting position of the pushrod.

c. Release the press and install the dust boot on the tensioner.

38. Install the timing belt tensioner and alternately tighten the bolts to 20 ft. lbs. (28 Nm). Using pliers, remove the 1.5mm hex wrench from the belt tensioner.

39. Check the valve timing, as follows:

a. Slowly turn the crankshaft pulley 2 revolutions from TDC to TDC. Always turn the crankshaft pulley clockwise.

b. Check that each pulley aligns with the timing marks. If the timing marks do not align, remove the timing belt and reinstall it.

40. Install the fan bracket with the bolt and nut.

41. Install the power steering adjusting strut with the nut.

42. Install the No. 2 timing belt cover. Tighten the bolts to 80 inch lbs. (9 Nm). Install the remaining components.

43. Fill the cooling system with coolant.

44. Connect the negative battery cable.

45. Start the engine and check for leaks.

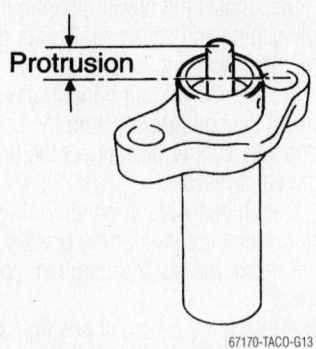

67170-TACO-G13

Measure the plunger pushrod protrusion as shown If the protrusion is not within 0.394–0.425 in. (10-10.8mm), replace the tensioner—3.4L engine

Timing Chain Cover, Seal and Timing Chain

REMOVAL & INSTALLATION

2.4L and 2.7L (3RZ-FE/M) Engines

1. Before servicing the vehicle, refer to the Precautions Section.

2. Disconnect the negative battery cable.

3. Drain the engine coolant.

4. Remove or disconnect the following:

- Engine undercover
- Engine oil
- On 4WD vehicles, the front differential
- Alternator belt, fan with coupling and the water pump pulley
- Cylinder head
- A/C belt, compressor, and the bracket
- Alternator adjusting bar and bracket
- Crankshaft position sensor and O-ring
- On 2WD vehicles, the stiffener plates
- Flywheel housing undercover and dust seal
- Oil pan
- Oil strainer and gasket
- Crankshaft pulley
- Water bypass pipe
- Chain cover assembly. Remove the bolts shown by the arrows.
- No. 1 timing chain and camshaft gear
- Crankshaft timing gear
- No. 1 timing chain tensioner slipper and No. 1 vibration damper
- No. 1 damper

5. On the 2.4L remove the crankshaft position sensor rotor and the timing chain oil jet.

for 2RZ–FE

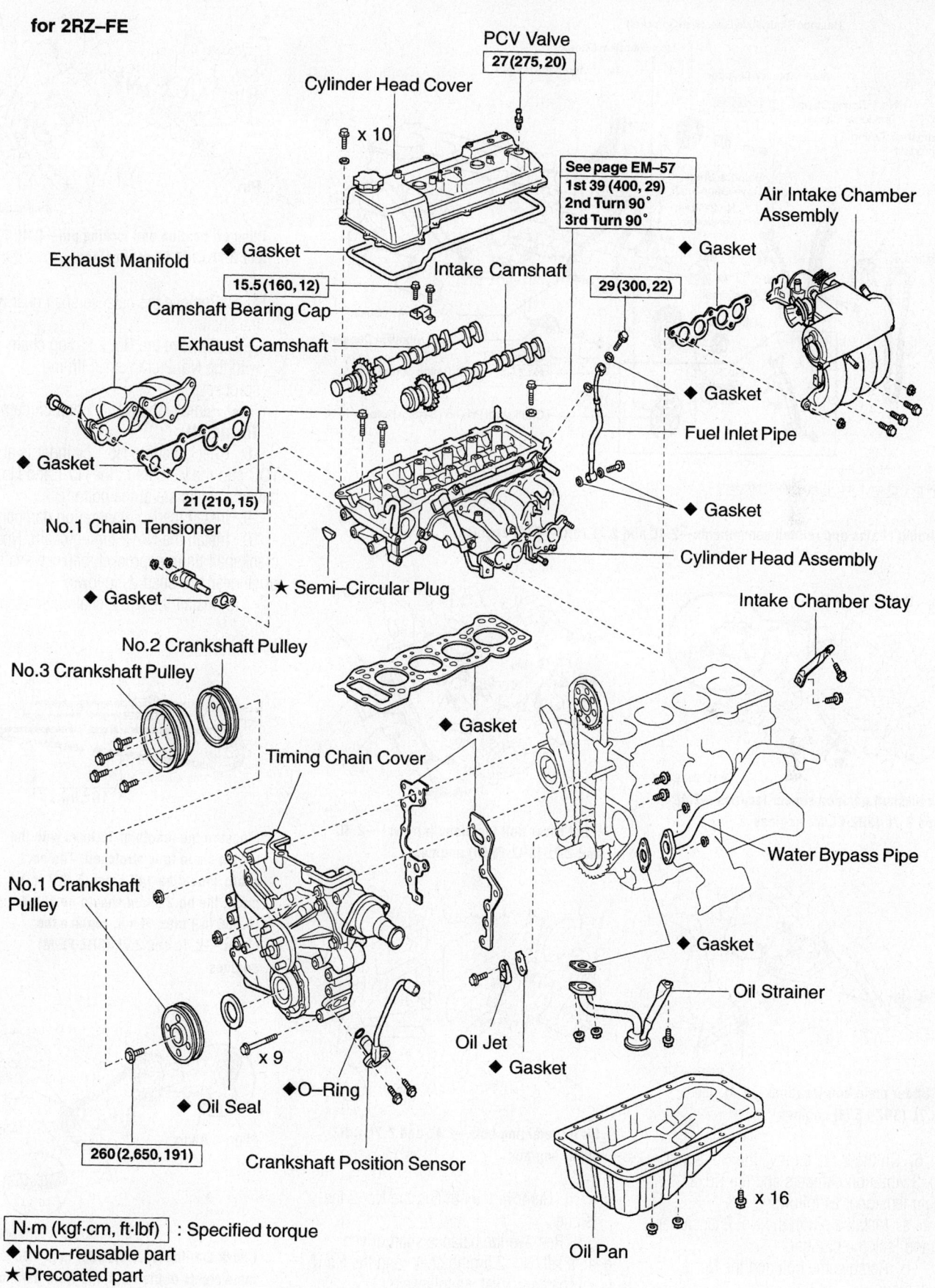

PCV Valve
27 (275, 20)

Cylinder Head Cover
x 10

See page EM–57
1st 39 (400, 29)
2nd Turn 90°
3rd Turn 90°

Air Intake Chamber Assembly

◆ Gasket

◆ Gasket

Exhaust Manifold

Intake Camshaft

15.5 (160, 12)

Camshaft Bearing Cap

Exhaust Camshaft

29 (300, 22)

◆ Gasket

Fuel Inlet Pipe

◆ Gasket

◆ Gasket

21 (210, 15)

No.1 Chain Tensioner

Cylinder Head Assembly

◆ Gasket

★ Semi–Circular Plug

Intake Chamber Stay

No.2 Crankshaft Pulley

No.3 Crankshaft Pulley

◆ Gasket

Timing Chain Cover

Water Bypass Pipe

No.1 Crankshaft Pulley

◆ Gasket

Oil Strainer

x 9

Oil Jet

◆ Gasket

◆ Oil Seal

◆ O–Ring

260 (2,650, 191)

Crankshaft Position Sensor

x 16

N·m (kgf·cm, ft·lbf) : Specified torque
◆ Non–reusable part
★ Precoated part

Oil Pan

67170-TACO-G15

Front cover and related components—2.4L and 2.7L (3RZ-FE/M) engines

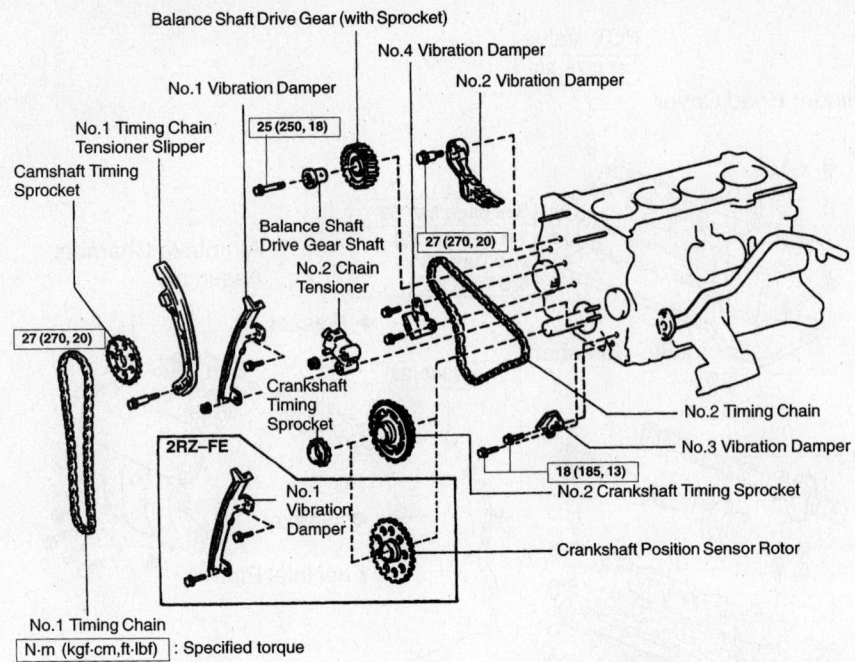

Timing chains and related components—2.4L and 2.7L (3RZ-FE/M) engines

Camshaft Timing Sprocket

No.1 Timing Chain Tensioner Slipper

No.1 Vibration Damper

Balance Shaft Drive Gear (with Sprocket)

No.4 Vibration Damper

No.2 Vibration Damper

25 (250, 18)

Balance Shaft Drive Gear Shaft

No.2 Chain Tensioner

27 (270, 20)

27 (270, 20)

Crankshaft Timing Sprocket

2RZ-FE

No.1 Vibration Damper

No.1 Timing Chain

N·m (kgf·cm,ft-lbf) : Specified torque

No.2 Timing Chain

No.3 Vibration Damper

18 (185, 13)

No.2 Crankshaft Timing Sprocket

Crankshaft Position Sensor Rotor

67170-TACO-G14

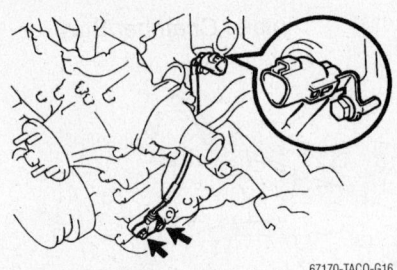

Crankshaft position sensor location—2.4L and 2.7L (3RZ-FE/M) engines

67170-TACO-G16

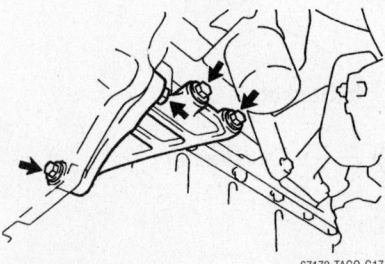

Stiffner plate bolt locations—2.4L and 2.7L (3RZ-FE/M) engines

67170-TACO-G17

Front cover bolt locations (arrows)—2.4L and 2.7L (3RZ-FE/M) engines

67170-TACO-G18

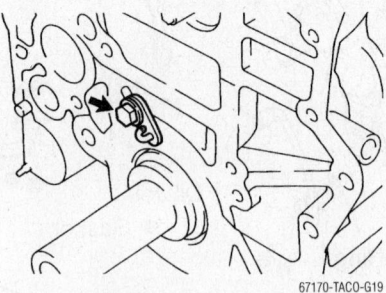

Oil jet retaining bolt—2.4L and 2.7L (3RZ-FE/M) engines

67170-TACO-G19

6. On the 2.7L, remove the No. 2 and No. 3 vibration dampers and the No. 2 chain tensioner as follows:

   a. Install a pin in the No. 2 tensioner and lock the plunger.

   b. Remove the bolt and the No. 2 damper.

   c. Remove the 2 bolts and the No. 3 damper.

   d. Remove the nut and the No. 2 tensioner.

7. Remove the balance shaft driven gear, shaft, No. 2 timing chain and the No. 2 crankshaft sprocket, as follows:

   a. Unbolt the balance shaft driven gear.

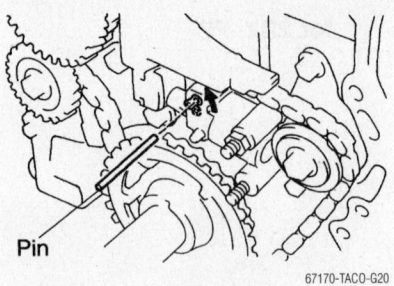

Plunger location and locking pin—2.4L and 2.7L (3RZ-FE/M) engines

67170-TACO-G20

   b. Remove the balance shaft gear with the shaft.

   c. Remove the No. 2 timing chain with the No. 2 crankshaft timing sprocket.

   • Remove the No. 4 vibration damper.

To install:

8. Check that the No.1 cylinder is at TDC and the weights of the NO.1 and No.2 balance shafts are at the bottom.

9. Install the No. 4 vibration dampener.

10. Install the No. 2 timing chain, No. 2 crankshaft timing sprocket, balance shaft drive gear and shaft as follows:

   a. Install the No. 2 chain by matching

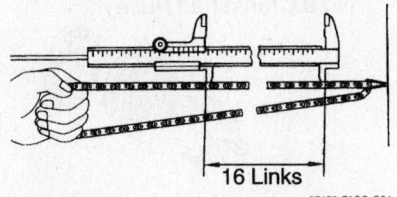

16 Links

67170-TACO-G21

Measure the length of 16 links with the timing chain fully stretched. The no.1 chain should be 147.5mm (5.807 in.) max.; the no.2 chain should be 123.6mm (4.866 in.) max. If not, replace the chain.—2.4L and 2.7L (3RZ-FE/M) engines

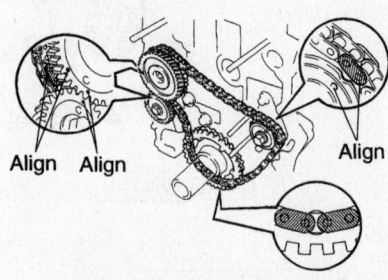

Align Align Align

67170-TACO-G22

Check that the No.1 cylinder is at TDC and the weights of the No.1 and No.2 balance shafts are at the bottom—2.4L and 2.7L (3RZ-FE/M) engines

the marked links with the timing marks on the crankshaft sprocket and balance shaft timing sprocket.

b. Fit the other marked link of the No. 2 chain onto the sprocket behind the large timing mark of the balance shaft gear.

c. Insert the balance shaft gear shaft through the balance shaft drive gear so that it fits into the thrust plate hole. Align the small timing mark of the balance shaft drive gear with the timing mark of the balance shaft timing gear.

d. Install the bolt to the balance shaft gear and tighten to 18 ft. lbs. (25 Nm).

e. Check each timing mark is matched with the corresponding mark link.

11. Install the No. 2, No. 3 vibration dampers and the No. 2 chain tensioner, as follows:

➡**Assemble the chain tensioner with the pin installed, then remove the pin after assembly.**

a. Install the No. 2 chain tensioner with the nut, tighten to 13 ft. lbs. (18 Nm).

b. Install the No. 3 damper with the bolts, tighten to 13 ft. lbs. (18 Nm).

c. Install the No. 2 damper, tighten to 20 ft. lbs. (27 Nm).

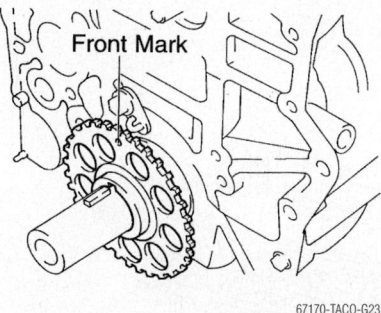

Front Mark

67170-TACO-G23

Crankshaft position sensor rotor location—2.4L and 2.7L (3RZ-FE/M) engines

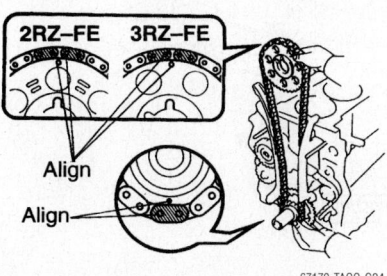

67170-TACO-G24

No.1 timing chain and camshaft timing sprocket identification and alignment—2.4L and 2.7L (3RZ-FE/M) engines

d. Remove the pin from the No. 2 chain tensioner and free the plunger.

12. On the 2.4L, install the crankshaft position sensor rotor and the timing chain oil jet.

13. Install or connect the following:
- No. 1 timing chain tensioner slipper and the No. 1 vibration damper. Torque the No. 1 damper to 22 ft. lbs. (29 Nm); torque the slipper to 20 ft. lbs. (27 Nm). Check that the slipper moves smoothly.
- Crankshaft timing gear.
- No. 1 timing chain and camshaft timing gear.

14. Align the timing mark between the marked link of the No. 1 timing chain, and install the No. 1 timing chain to the gear.

15. Align the timing mark of the crankshaft timing gear with the mark of the No. 1 timing chain, then install the No. 1 timing chain.

16. Tie the No. 1 chain with a wire or cord, make sure it does not come loose.

17. Install or connect the following:
- Timing chain cover assembly

18. Tighten the following:
- 12mm **A** bolts—14 ft. lbs. (20 Nm)
- 12mm **B** bolts—18 ft. lbs. (25 Nm)
- 14mm bolts—32 ft. lbs. (44 Nm)
- 14mm nut—14 ft. lbs. (20 Nm)

19. Attach the water bypass pipe.

20. Remove the cord or wire from the chain.

21. Install or connect the following:
- Crankshaft pulley, tighten the bolt to 193 ft. lbs. (260 Nm). On A/C vehicles, install the crankshaft pulleys with bolts and tighten to 18 ft. lbs. (25 Nm).
- Oil strainer, tighten to 13 ft. lbs. (18 Nm)
- Oil pan, tighten the mounting bolts to 108 inch lbs. (13 Nm)
- Flywheel housing undercover and dust seal

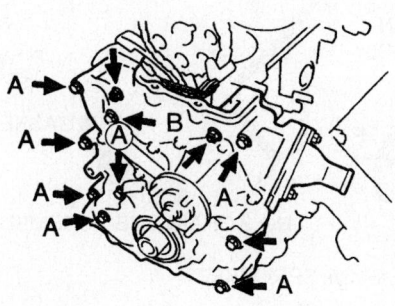

67170-TACO-G25

Timing chain case cover bolt identification—2.4L and 2.7L (3RZ-FE/M) engines

- Stiffener plates on 2WD vehicles, tighten to 27 ft. lbs. (37 Nm)
- Crankshaft position sensor with a new O-ring
- Alternator, adjusting bar and bracket
- A/C compressor and bracket
- Cylinder head
- Water pump pulley and the fluid coupling with the fan
- On 4WD vehicles, the front differential and driveshaft assemblies

22. Adjust the drive belt tension.

23. Fill with engine coolant and engine oil.

24. Install the engine undercover.

25. Connect the negative battery cable.

SEAL REPLACEMENT

Cover Removed

1. Unbolt the timing chain cover assembly. Be careful to loosen only the correct bolts.

2. Pry out the seal from the cover with a flat-bladed tool.

3. It is a good idea to remove the oil pump from the timing cover and replace the O-ring.

To install:

4. Clean and inspect the timing cover area. Install new gaskets around the dowel areas and pump spline.

5. Apply multi-purpose grease to the new oil seal lip.

6. Tap the seal into place with SST 09223–50010/60010 or equivalent, and a hammer. Do this until the seal surface is flush with the cover edge.

7. Install the cover, tighten the bolts as specified for your engine.

8. If the oil pump was removed, install a new O-ring behind the pump prior to installation.

Cover Installed

1. Unbolt and remove the oil pump.

2. Using a knife, carefully cut off the oil seal lip. With a flat-bladed tool, (preferably with tape around it) pry the seal from the cover.

To install:

3. Apply multi-purpose grease to the new oil seal lip.

4. Tap the seal into place with SST 09223–50010/60011 or equivalent seal driver, and a hammer. Do this until the seal surface is flush with the cover edge.

5. Install the oil pump with a new O-ring.

2.7L (2TR-FE/X) Engine

1. Before servicing the vehicle, refer to the Precautions Section.

2. Disconnect the negative battery cable.

3. Remove the engine and position it in a suitable holding fixture.

4. Remove the air intake connector. Remove the alternator.

5. Remove the belt tensioner assembly. Remove the number one idler pulley sub-assembly. Remove the air conditioning idler pulley assembly and bracket.

6. Remove the crankshaft position sensor. Remove the camshaft position sensor.

7. Remove the intake manifold. Remove the cylinder head cover.

8. Position the engine at TDC of the compression stroke. Remove the crankshaft pulley.

9. Remove the oil level gauge sub-assembly. Remove the lower oil pan. Remove the oil strainer assembly. Remove the upper oil pan.

10. Remove the two nuts and separate the water bypass pipe number one.

11. Remove the nineteen bolts and two

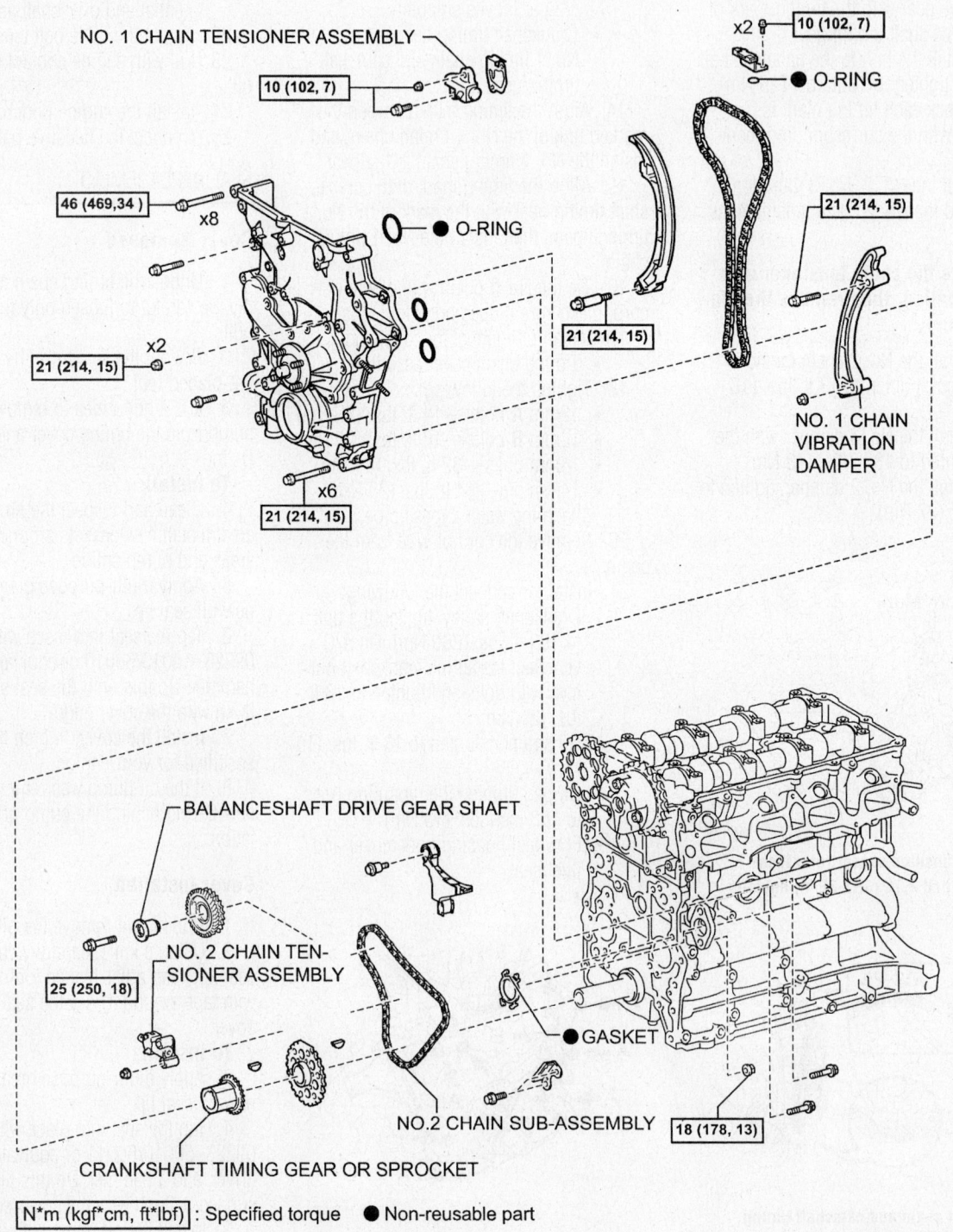

NO. 1 CHAIN TENSIONER ASSEMBLY

10 (102, 7)

×2 | 10 (102, 7)

● O-RING

46 (469,34) ×8

● O-RING

21 (214, 15)

21 (214, 15) ×2

21 (214, 15)

21 (214, 15) ×6

NO. 1 CHAIN VIBRATION DAMPER

BALANCESHAFT DRIVE GEAR SHAFT

25 (250, 18)

NO. 2 CHAIN TEN-SIONER ASSEMBLY

● GASKET

NO.2 CHAIN SUB-ASSEMBLY

18 (178, 13)

CRANKSHAFT TIMING GEAR OR SPROCKET

N*m (kgf*cm, ft*lbf) : Specified torque   ● Non-reusable part

Timing chain and related components—2.7L (2TR-FE/X) engine

09490_TACO_G0096

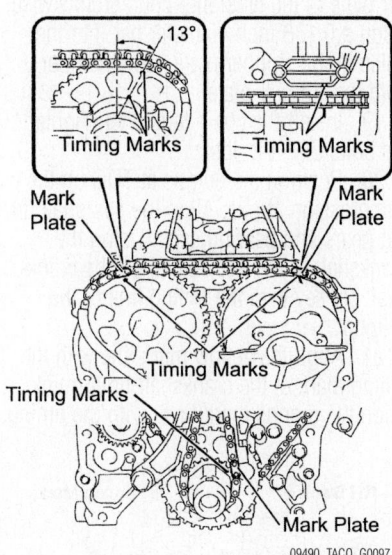

Primary timing chain alignment removal points—2.7L (2TR-FE/X) engine

nuts retaining the timing chain case cover to its mounting. Remove the timing chain case cover from the engine.

➡**Carefully remove the cover by prying between the cover and the cylinder head or block with a suitable tool. Be sure to cover the tip of the suitable tool prior to usage. Be careful not to damage the contact surfaces of the cylinder block, cylinder head and timing chain cover.**

12. Make sure that each matchmark is in the same position as shown in the illustration. Remove the two bolts, timing chain guide and O-ring.

13. Move the stopper plate upward to release the lock, and push the plunger deep into the tensioner.

14. Move the stopper plate downward to set the lock. Insert a 0.118 inch diameter bar into the stopper plate hole. Remove the bolt, nut, number one chain tensioner and gasket.

➡**When the number one chain tensioner is removed do not rotate the crankshaft. When the chain is removed and the camshaft needs to be rotated, rotate the crankshaft 90 degrees to the right.**

15. Remove the bolt and chain tensioner slipper. Remove the two bolts and remove the number one chain vibration damper. Remove the primary timing chain sub-assembly.

16. Remove the crankshaft timing gear or sprocket. Remove the bolt and remove the number two chain vibration damper.

17. Remove the two bolts and remove the number three chain vibration damper.

18. Remove the nut and the number two chain tensioner assembly. Remove the bolt, balance shaft drive gear shaft and balance shaft drive gear. Remove the crankshaft timing sprocket number two and chain.

To install:

19. Install the chain with its marks aligned with the timing marks on the crankshaft timing sprocket and balance shaft timing sprocket.

20. Bring the other mark link of the crankshaft timing sprocket behind the large timing mark of the balance shaft drive gear.

21. Insert the balance shaft drive gear shaft through the balance shaft drive gear so that it fits into the thrust plate hole.

22. Align the small timing mark of the balance shaft drive gear with the timing mark of the balance shaft timing gear.

23. Install the bolt onto the balance shaft drive gear and tighten it to 18 ft. lbs.

24. Check that the timing mark is aligned with the corresponding mark link.

25. Install the number two chain tensioner assembly. Tighten the nut to 13 ft. lbs.

➡**Assemble the chain tensioner with the 0.118 inch diameter bar installed, then remove the bar after assembly. When doing this avoid pushing the vibration damper against the chain.**

26. Install the number three chain vibration damper with the two bolts. Tighten the bolts to 13 ft. lbs.

27. Install the chain vibration damper number two bolt and tighten it to 20 ft. lbs. Remove the pin from the chain tensioner and release the plunger.

28. Install the crankshaft timing gear or sprocket.

29. Install the number one chain vibration damper bolt and nut. Tighten to 15 ft. lbs.

30. Install the primary timing chain onto the sprocket and gear with the painted marks aligned with the timing marks on the sprocket and gear.

➡**The camshaft mark plate is orange. The crankshaft mark plate is yellow.**

31. Use a rope to tie the chain of the crankshaft timing sprocket. Tie the rope near the sprocket.

➡**After the chain tensioner has been installed, remove the rope. The rope is used to prevent gear jumping.**

32. Install the tensioner slipper and tighten the bolt to 15 ft. lbs.

33. Install the number one chain tensioner assembly, using a new gasket. Tighten the bolts to 7 ft. lbs.

34. Install a new front case oil seal. Install the thermostat. Install the water inlet.

35. Apply adhesive, part number 08833-00070 or equivalent to the head straight screw plug. Install the plug and tighten to 12 ft. lbs. Install four new O-rings onto the timing chain case cover.

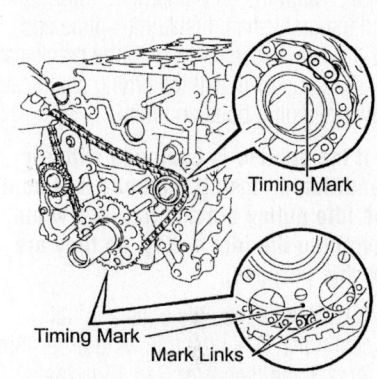

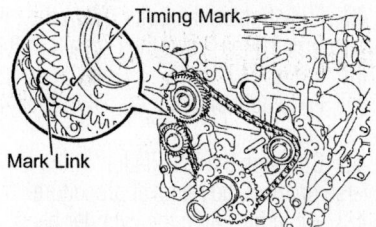

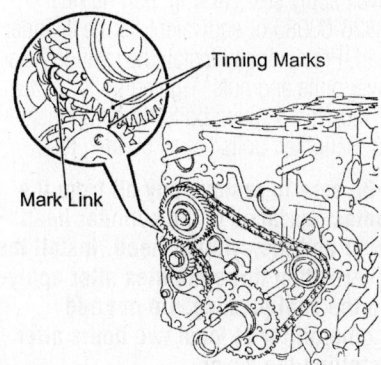

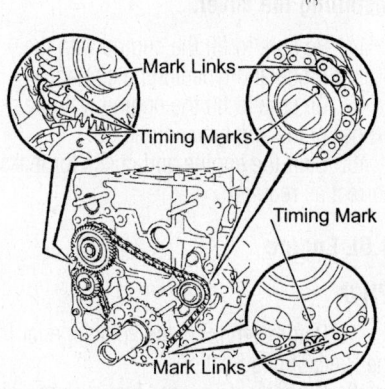

Number two timing chain alignment marks—2.7L (2TR-FE/X) engine

36. Apply continuous beads of seal packing, part number 08826-00080 or equivalent.

➡ **Remove any oil from the contact surfaces. Install the timing chain case cover within three minutes and tighten the bolts within fifteen minutes of applying the seal packing. Do not start the engine for at least four hours after installation of the cover.**

37. Align the oil pump drive rotor spline and the crankshaft. Install the spline and timing chain case cover onto the crankshaft.

38. Loosely install the timing chain case cover retaining bolts and nuts.

➡ **If the vehicle is equipped with air conditioning install the bolts that hold the idle pulley bracket in place when installing the idle pulley, as they are for this purpose.**

39. Fully tighten the bolts and nuts, except bolts "A" in the following order: Area 1, Area 3 and then Area 2 to 15 ft. lbs.

40. Fully tighten the bolts "A" in the following order: Area 2 and then Area 3 to 34 ft. lbs.

41. Fully tighten the bolts "E" in Area 4 to 15 ft. lbs.

42. Continue the installation in the reverse order of the removal procedure.

43. When installing the cylinder head cover, apply seal packing, part number 08826-00080 or equivalent, to the cylinder head. Provisionally install the cylinder head cover bolts and nuts. Tighten bolts "A" to 80 inch lbs. Tighten bolts "B" to 80 inch lbs. Retighten bolts "A" to 80 inch lbs.

➡ **Be sure to remove any oil from the contact surfaces of the cylinder head cover and the cylinder head. Install the cover within three minutes after applying the seal packing. Do not add engine oil for at least two hours after installing the cover.**

44. Be sure to fill the engine with the proper grade and type engine coolant.

45. Be sure to fill the engine with the proper grade and type engine oil.

46. Start the engine and check for leaks, correct as required.

4.0L Engine

2WD

1. Before servicing the vehicle, refer to the Precautions Section.
2. Properly relieve the fuel system pressure.
3. Disconnect the negative battery cable. Disconnect the positive battery cable. Remove the battery.
4. Drain the engine coolant. Drain the engine oil.
5. Remove the engine from the vehicle and position it in a suitable holding fixture.
6. Remove the oil level gauge guide. Remove the water inlet. Remove the belt tensioner.
7. Remove the idler pulley number two subassembly. Remove the idler pulley number one sub assembly. Remove the crankshaft pulley.
8. Remove the lower oil pan. Remove the strainer and pickup tube. Remove the upper oil pan.
9. Remove the intake manifold. Remove the ignition coil assembly. Remove the cylinder head cover. Remove the camshaft timing oil control valve assembly.
10. Remove the VVT sensor. Remove the oil filter bracket subassembly.
11. Remove the timing chain case cover retaining bolts. Remove the cover from the engine. Remove the O-ring from the left cylinder head.

➡ **Carefully remove the cover by prying between the cover and the cylinder head or block with a suitable tool. Be sure to cover the tip of the suitable tool prior to usage. Be careful not to damage the contact surfaces of the cylinder block, cylinder head and timing chain cover.**

12. Using the crankshaft pulley set bolt, turn the crankshaft to align the crankshaft set key with the timing line of the cylinder block. If not aligned at TDC of the compression stroke, turn the crankshaft one complete revolution, in the direction of rotation.
13. While turning the stopper plate of the tensioner upward, push the plunger of the chain tensioner. While turning the stopper plate of the tensioner downward, insert a 0.138 inch diameter bar into the holes in the stopper plate and tensioner to hold the stopper plate.
14. Remove the two bolts, and then remove the chain tensioner.
15. Remove the chain tensioner slipper. Remove the idle gear shaft number two, idle gear number one and idle gear shaft number one.
16. Remove the number two chain vibration damper. Remove the timing chain subassembly.

To install:

17. Install the chain tensioner slipper.
18. While turning the stopper plate of the tensioner clockwise, push in the plunger of the chain tensioner. While turning the stopper plate of the tensioner counterclockwise, insert a 0.138 inch diameter bar into the holes in the stopper plate and tensioner to hold the stopper plate.
19. Install the chain tensioner. Tighten the bolts to 7.4 ft. lbs.
20. Position the engine at TDC on the compression stroke. Align the camshaft timing gears and bearing caps. Using the crankshaft pulley set bolt, align the crankshaft set key with the timing line of the cylinder.
21. Align the yellow mark line with the timing mark of the crankshaft timing link. Align the orange mark links with the timing

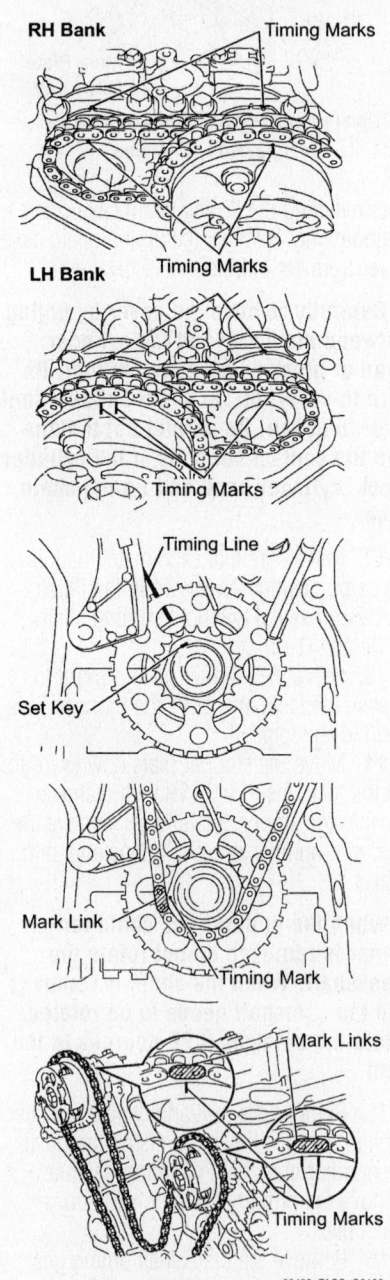

Timing chain alignment—4.0L engine

marks of the camshaft timing gears, and install the chain.

22. Install the number two chain vibration damper.

23. Apply a light coat of clean engine oil to the rotating surface of the idle gear shaft number one.

24. Temporarily install the idle gear shaft number one together with idle gear shaft number two, while aligning the knock pin of idle gear shaft number one with the knock pin groove of the cylinder block.

➡**Be care of the idle gear direction.**

25. Tighten the idle gear shaft number two to 44 ft. lbs. Remove the bar from the chain tensioner.

26. Install a new front case oil seal.

Install a new O-ring onto the left cylinder head.

27. Apply continuous beads (0.12–0.16 inch in diameter) of seal packing, part number 08826-00080 or equivalent to the four locations shown in the illustration.

28. Apply continuous beads (0.12–0.16 inch in diameter) of seal packing, part num-

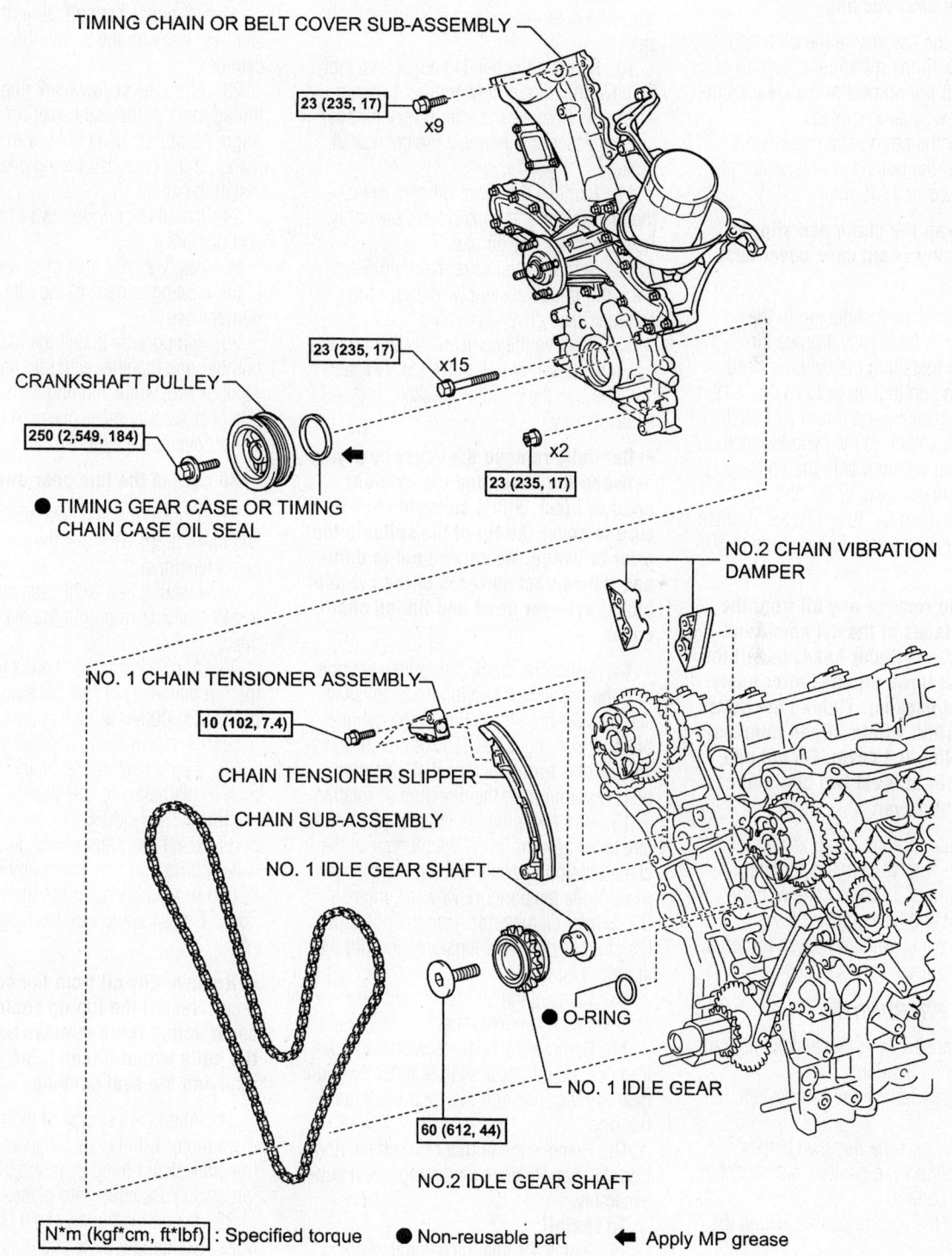

TIMING CHAIN OR BELT COVER SUB-ASSEMBLY

23 (235, 17) ×9

CRANKSHAFT PULLEY

250 (2,549, 184)

23 (235, 17) ×15

23 (235, 17) ×2

● TIMING GEAR CASE OR TIMING CHAIN CASE OIL SEAL

NO.2 CHAIN VIBRATION DAMPER

NO. 1 CHAIN TENSIONER ASSEMBLY

10 (102, 7.4)

CHAIN TENSIONER SLIPPER

CHAIN SUB-ASSEMBLY

NO. 1 IDLE GEAR SHAFT

● O-RING

● NO. 1 IDLE GEAR

60 (612, 44)

NO.2 IDLE GEAR SHAFT

N*m (kgf*cm, ft*lbf) : Specified torque   ● Non-reusable part   ◀ Apply MP grease

Timing chain and related components—4.0L engine

09490_TACO_G0099

ber 08826-00080 or equivalent to all parts except the water pump part: for the water pump part use, part number 08826-00080 or equivalent, to the timing chain cover. Do not apply seal packing to portion "A".

➡ **Remove any oil from the contact surfaces. Install the timing chain case cover within three minutes and tighten the bolts within fifteen minutes of applying the seal packing.**

29. Align the key way of the oil pump drive motor with the rectangular portion of the crankshaft timing gear and slide the timing chain case cover into place.

30. Install the timing chain case cover bolts. Tighten the bolts and nuts uniformly in several steps to 17 ft. lbs.

➡ **Do not wrap the chain and slipper over the timing chain case cover seal line.**

31. Continue the installation in the reverse order of the removal procedure.

32. When installing the cylinder head cover apply a continuous bead (0.08–0.12 inch) of seal packing, part number 08826-00080 or equivalent, to the cylinder head. Install the seal washers onto the bolts. Install the cylinder head cover bolts and nuts. Tighten bolts "A" to 7.4 ft. lbs. Tighten bolts "B" to 80 inch lbs. Tighten nuts to 80 inch lbs.

➡ **Be sure to remove any oil from the contact surfaces of the cylinder head cover and the cylinder head. Install the cover within three minutes after applying the seal packing. Tighten the bolts to specification within fifteen minutes after installing the cover. Do not add engine oil for at least two hours after installing the cover.**

33. Be sure to fill the engine with the proper grade and type engine coolant.

34. Be sure to fill the engine with the proper grade and type engine oil.

35. Start the engine and check for leaks, correct as required.

4WD AND PRERUNNER

1. Before servicing the vehicle, refer to the Precautions Section.

2. Properly relieve the fuel system pressure.

3. Disconnect the negative battery cable. Disconnect the positive battery cable. Remove the battery.

4. Drain the engine coolant. Drain the engine oil.

5. Remove the power steering gear assembly.

6. If equipped with 4WD, remove the front differential carrier assembly.

7. Remove the V bank cover. Remove the radiator support to frame seal, left side. Remove the fan shroud.

8. Remove the air cleaner assembly. Remove the oil level gauge. Remove the water inlet.

9. Separate the vane pump assembly. Remove the alternator. Remove the air conditioning compressor and position it to the side.

10. Remove the belt tensioner assembly. Remove the idler pulley number two subassembly. Remove the idler pulley number one subassembly. Remove the crankshaft pulley.

11. Remove the lower oil pan. Remove the oil strainer and pickup tube assembly. Remove the upper oil pan.

12. Remove the intake manifold. Remove the ignition coil assembly. Remove the cylinder head cover assembly.

13. Remove the camshaft timing oil control valve assembly. Remove the VVT sensor. Remove the oil filter bracket subassembly.

➡ **Carefully remove the cover by prying between the cover and the cylinder head or block with a suitable tool. Be sure to cover the tip of the suitable tool prior to usage. Be careful not to damage the contact surfaces of the cylinder block, cylinder head and timing chain cover.**

14. Using the crankshaft pulley set bolt, turn the crankshaft to align the crankshaft set key with the timing line of the cylinder block. If not aligned at TDC of the compression stroke, turn the crankshaft one complete revolution, in the direction of rotation.

15. While turning the stopper plate of the tensioner upward, push the plunger of the chain tensioner. While turning the stopper plate of the tensioner downward, insert a 0.138 inch diameter bar into the holes in the stopper plate and tensioner to hold the stopper plate.

16. Remove the two bolts, and then remove the chain tensioner.

17. Remove the chain tensioner slipper. Remove the idle gear shaft number two, idle gear number one and idle gear shaft number one.

18. Remove the number two chain vibration damper. Remove the timing chain subassembly.

To install:

19. Install the chain tensioner slipper.

20. While turning the stopper plate of the tensioner clockwise, push in the plunger of the chain tensioner. While turning the stopper plate of the tensioner counterclockwise, insert a 0.138 inch diameter bar into the holes in the stopper plate and tensioner to hold the stopper plate.

21. Install the chain tensioner. Tighten the bolts to 7.4 ft. lbs.

22. Position the engine at TDC on the compression stroke. Align the camshaft timing gears and bearing caps. Using the crankshaft pulley set bolt, align the crankshaft set key with the timing line of the cylinder.

23. Align the yellow mark line with the timing mark of the crankshaft timing link. Align the orange mark links with the timing marks of the camshaft timing gears, and install the chain.

24. Install the number two chain vibration damper.

25. Apply a light coat of clean engine oil to the rotating surface of the idle gear shaft number one.

26. Temporarily install the idle gear shaft number one together with idle gear shaft number two, while aligning the knock pin of idle gear shaft number one with the knock pin groove of the cylinder block.

➡ **Be care of the idle gear direction.**

27. Tighten the idle gear shaft number two to 44 ft. lbs. Remove the bar from the chain tensioner.

28. Install a new front case oil seal. Install a new O-ring onto the left cylinder head.

29. Apply continuous beads (0.12–0.16 inch in diameter) of seal packing, part number 08826-00080 or equivalent to the four locations shown in the illustration.

30. Apply continuous beads (0.12–0.16 inch in diameter) of seal packing, part number 08826-00080 or equivalent to all parts except the water pump part: for the water pump part use, part number 08826-00080 or equivalent, to the timing chain cover. Do not apply seal packing to portion "A".

➡ **Remove any oil from the contact surfaces. Install the timing chain case cover within three minutes and tighten the bolts within fifteen minutes of applying the seal packing.**

31. Align the key way of the oil pump drive motor with the rectangular portion of the crankshaft timing gear and slide the timing chain case cover into place.

32. Install the timing chain case cover bolts. Tighten the bolts and nuts uniformly in several steps to 17 ft. lbs.

➡**Do not wrap the chain and slipper over the timing chain case cover seal line.**

33. Continue the installation in the reverse order of the removal procedure.

34. When installing the cylinder head cover apply a continuous bead (0.08–0.12 inch) of seal packing, part number 08826-00080 or equivalent, to the cylinder head. Install the seal washers onto the bolts. Install the cylinder head cover bolts and nuts. Tighten bolts "A" to 7.4 ft. lbs. Tighten bolts "B" to 80 inch lbs. Tighten nuts to 80 inch lbs.

➡**Be sure to remove any oil from the contact surfaces of the cylinder head cover and the cylinder head. Install the cover within three minutes after applying the seal packing. Tighten the bolts to specification within fifteen minutes after installing the cover. Do not add engine oil for at least two hours after installing the cover.**

35. Be sure to fill the engine with the proper grade and type engine coolant.
36. Be sure to fill the engine with the proper grade and type engine oil.
37. Start the engine and check for leaks, correct as required.

Piston and Ring

POSITIONING

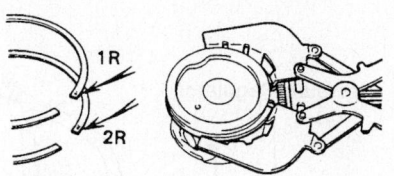

Compression ring identification mark locations—2.7L (3RZ-FE/M) engine

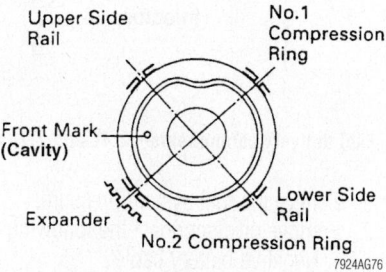

Piston ring end-gap spacing—2.4L and 2.7L (3RZ-FE/M) engines

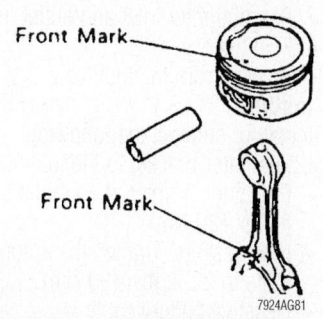

Piston to connecting rod assembly

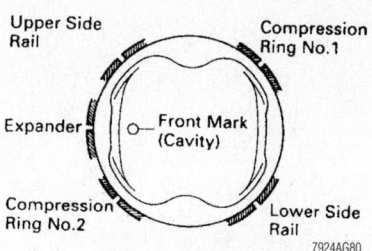

Piston ring end-gap spacing—3.4L engine

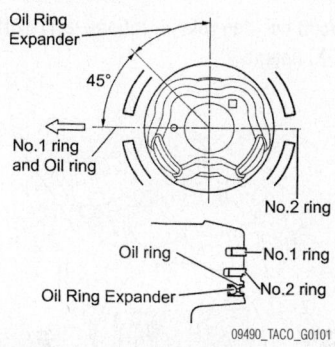

Piston ring positioning—2.7L (2TR-FE/X) engine

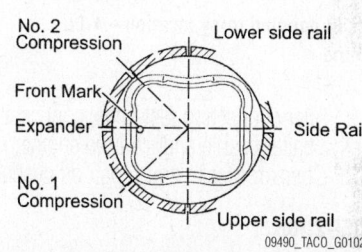

Piston ring positioning—4.0L engine

FUEL SYSTEM

Fuel System Service Precautions

Safety is the most important factor when performing not only fuel system maintenance, but any type of maintenance. Failure to conduct maintenance and repairs in a safe manner may result in serious personal injury or death. Work on a vehicle's fuel system components can be accomplished safely and effectively by adhering to the following rules and guidelines.

• To avoid the possibility of fire and personal injury, always disconnect the negative battery cable unless the repair or test procedure requires that battery voltage be applied.

• Always relieve the fuel system pressure prior to disconnecting any fuel system component (injector, fuel rail, pressure regulator, etc.) fitting or fuel line connection. Exercise extreme caution whenever relieving fuel system pressure, to avoid exposing skin, face and eyes to fuel spray. Please be advised that fuel under pressure may penetrate the skin or any part of the body that it contacts.

• Always place a shop towel or cloth around the fitting or connection prior to loosening to absorb any excess fuel due to spillage. Ensure that all fuel spillage is quickly removed from engine surfaces. Ensure that all fuel-soaked cloths or towels are deposited into a flame-proof waste container with a lid.

• Always keep a dry chemical (Class B) fire extinguisher near the work area.

• Do not allow fuel spray or fuel vapors to come into contact with a light bulb, spark or open flame.

• Always use a second wrench when loosening or tightening fuel line connection fittings. This will prevent unnecessary stress and torsion to fuel piping. Always follow the proper torque specifications.

• Always replace worn fuel fitting O-rings with new ones. Do not substitute fuel hose where rigid pipe is installed.

Fuel System Pressure

RELIEVING

2002–2004

1. Before servicing the vehicle, refer to the Precautions Section.
2. Disconnect the negative battery terminal.
3. Place a catch-pan under the joint to be disconnected. A large quantity of fuel may be released when the joint is opened.
4. Wear eye or full face protection.
5. Place a shop towel over the area and slowly loosen the joint using a wrench of the correct size. Use a back-up wrench if needed.
6. Allow the fuel left in the line to bleed off slowly before fully disconnecting the joint.
7. Plug the opened lines immediately to prevent fuel spillage or the entry of dirt.
8. Dispose of the released fuel properly.

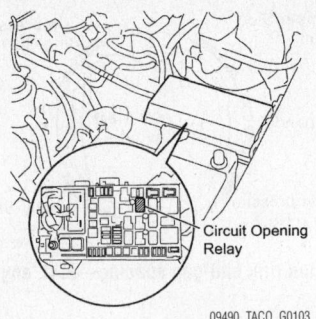

Circuit opening relay location—2.7L (2TR-FE/X) engine

09490_TACO_G0103

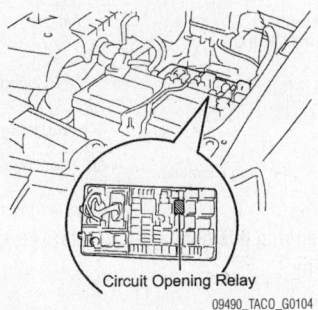

Circuit opening relay location—4.0L engine

09490_TACO_G0104

9. After connecting fuel lines, connect the negative battery cable and start the engine.

10. Check for leaks and repair as needed.

2005–2006

1. Before servicing the vehicle, refer to the Precautions Section.
2. Disconnect the negative battery cable.
3. Remove the engine relay block cover.
4. Remove the circuit opening relay.
5. Connect the negative battery.
6. Start the engine.
7. Turn the ignition switch "ON" after the engine stops.

➡ **Code DTC P0171 (system lean) may be present.**

8. Crank the engine again. Check that the engine stops.
9. Remove the fuel tank cap and completely discharge the pressure in the fuel tank. Install the circuit opening relay.
10. Disconnect the negative battery cable.

Fuel Filter

REMOVAL & INSTALLATION

2002–2004

1. Before servicing the vehicle, refer to the Precautions Section.

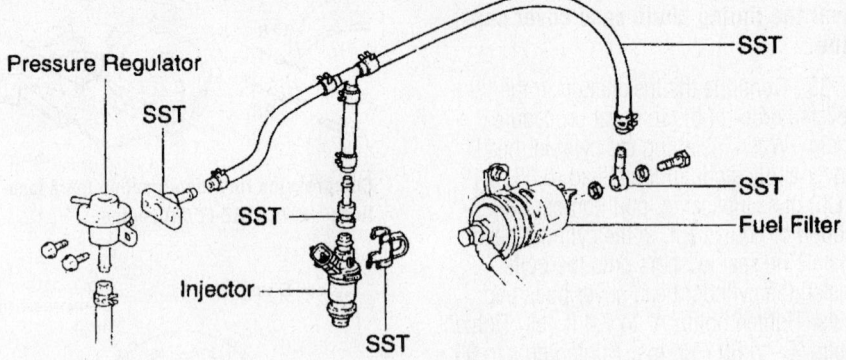

Fuel delivery components—2.4L and 2.7L (3RZ-FE/M) engines

7924YG90

2. Relieve the fuel system pressure.
3. Remove or disconnect the following:
   • Negative battery cable

➡ **The fuel filter is located in the engine compartment, at the inlet line to the fuel rail.**

   • Plug the filter inlet and outlet lines
   • Fuel filter
   • Bracket from the fuel filter

To install:

4. Install or connect the following:
   • Fuel filter bracket to the fuel filter
   • Fuel filter. Tighten the 2 bolts to 14 ft. lbs. (20 Nm).
   • New gaskets. Tighten the union bolts to 22 ft. lbs. (30 Nm).
   • Negative battery cable
5. Start the engine and check for leaks.

Fuel Pump

REMOVAL & INSTALLATION

2002–2004

1. Before servicing the vehicle, refer to the Precautions Section.
2. Relieve the fuel pressure.
3. Disconnect the negative battery cable.
4. Drain the fuel from the fuel tank.
5. Remove or disconnect the following:
   • Fuel tank
   • Fuel pump connector from the clamp
   • Access plate bolts, then pull out the fuel pump assembly from the fuel tank
   • Gasket(s) from the pump bracket

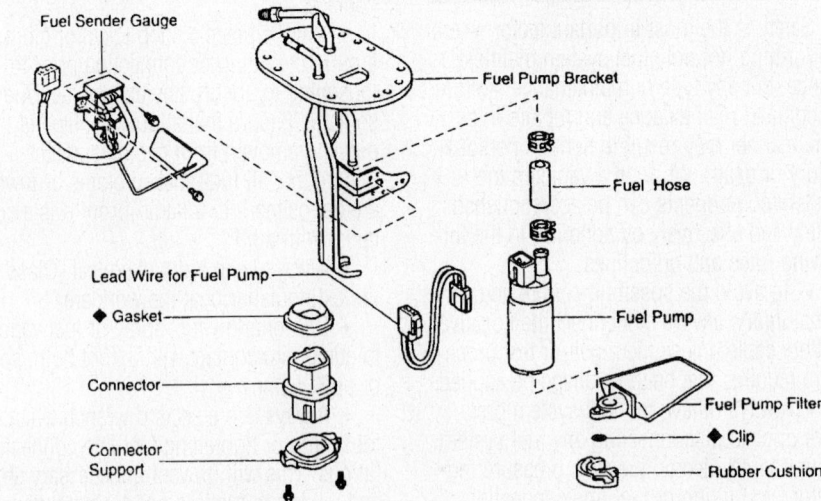

Reference (2WD)

◆ Non-reusable part

Fuel pump and related components—2002–2004

7924YG91

- Fuel pump connector
- Bracket from the lower side of the fuel pump
- Fuel pump from the fuel hose
- Rubber cushion, the clip and the fuel filter at the bottom of the fuel pump

To install:

6. Install or connect the following:
   - Fuel pump filter to the fuel pump with a new clip
   - Fuel pump to the fuel pump bracket

- Fuel hose to the outlet port of the fuel pump
- Fuel pump connector
- Fuel pump assembly with a new gasket(s). Tighten the bolts to 31 inch lbs. (4 Nm).
- Fuel pump connector to the clamp
- Fuel tank
- All electrical and fuel connections
- Negative battery cable

7. Refill the fuel tank and check for leaks.

2005–2006

1. Before servicing the vehicle, refer to the Precautions Section.
2. Relieve the fuel pressure.
3. Disconnect the negative battery cable.
4. Drain the fuel from the fuel tank.
5. On vehicles equipped with off road package and 4.0L engine, remove the number one fuel tank protector subassembly.
6. Remove the fuel tank.
7. Remove the fuel tank main hose and the return hose.

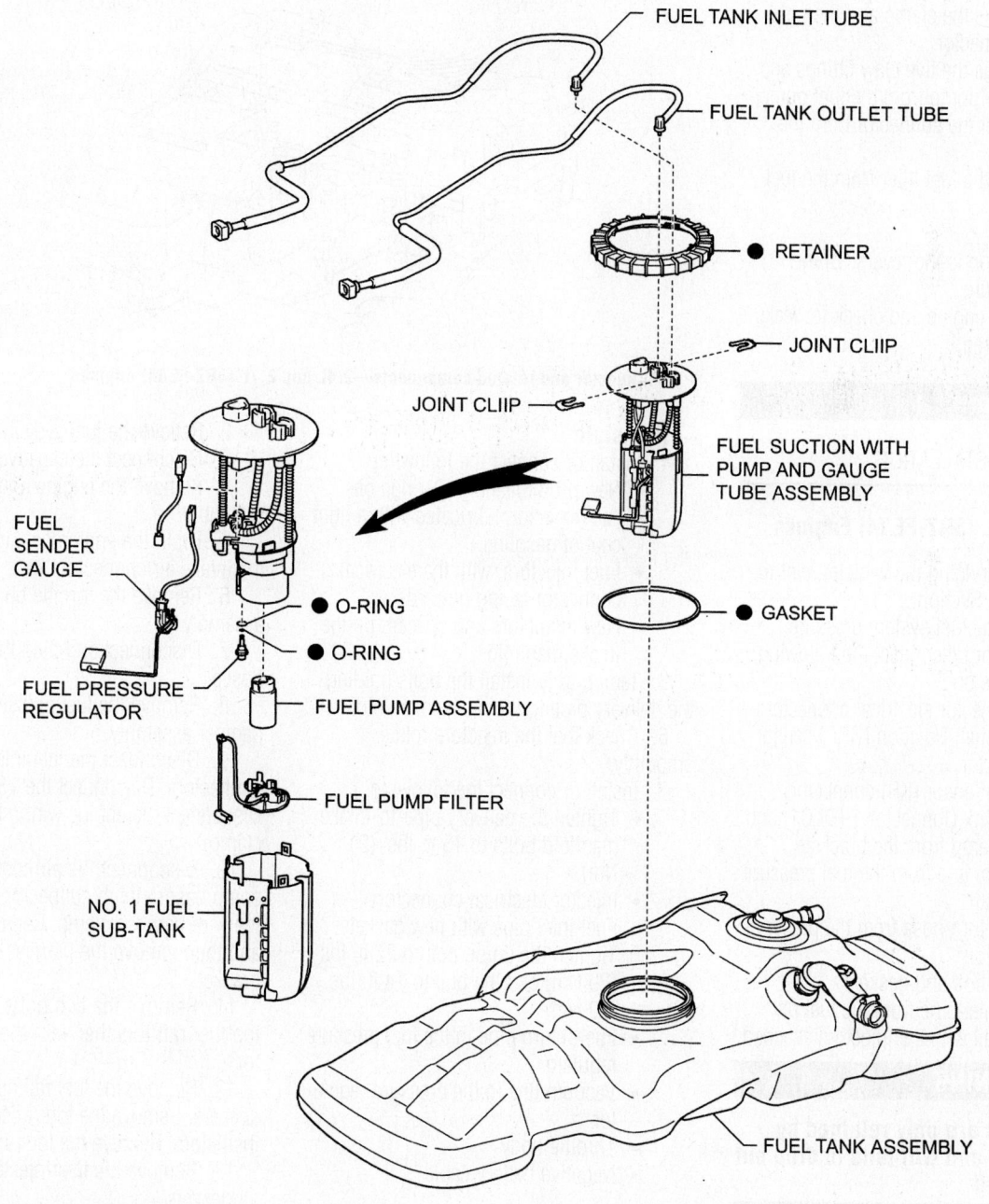

● Non-reusable part

Fuel pump and related components—2005–2006

09490_TACO_G0105

8. Remove the fuel pump and gauge assembly retainer from its mounting, using tool SST09808-14020 or equivalent.

9. Pull the fuel pump and gauge assembly out of the fuel tank. Be careful not to bend the arm of the sender. Remove and discard the gasket.

10. Disconnect the connector from the assembly.

11. Disengage the claw fitting and remove the sender gauge by sliding it forward.

12. Disengage the five claw fittings and remove the fuel pump tank. Separate the connector and disengage the clamp.

13. Disengage the clamp and then disconnect the connector.

14. Disengage the five claw fittings and separate the fuel pump from the fuel pump case. Disconnect the connector from the fuel pump.

15. Remove the fuel filter from the fuel pump.

To install:

16. Installation is the reverse of the removal procedure.

17. Start the engine and check for leaks, correct as required.

Fuel Injector

REMOVAL & INSTALLATION

2.4L and 2.7L (3RZ-FE/M) Engines

1. Before servicing the vehicle, refer to the Precautions Section.

2. Relieve the fuel system pressure.

3. Remove or disconnect the following:
- Throttle body
- Fuel injector electrical connectors
- Crankshaft Position (CKP) sensor connector
- Knock Sensor (KS) connector
- Data Link Connector 1 (DLC1) and wire clamp from the brackets
- Vacuum line from the fuel pressure regulator
- Fuel return hose from the pressure regulator
- Union bolt and gaskets
- Fuel inlet pipe from the fuel rail
- Fuel rail with the injectors attached

✱✱ WARNING

The injectors are only retained by their O-rings and will tend to drop out of the fuel rail.

- 4 insulators from the four spacers
- Fuel injectors from the fuel rail
- O-ring and grommet, discard them

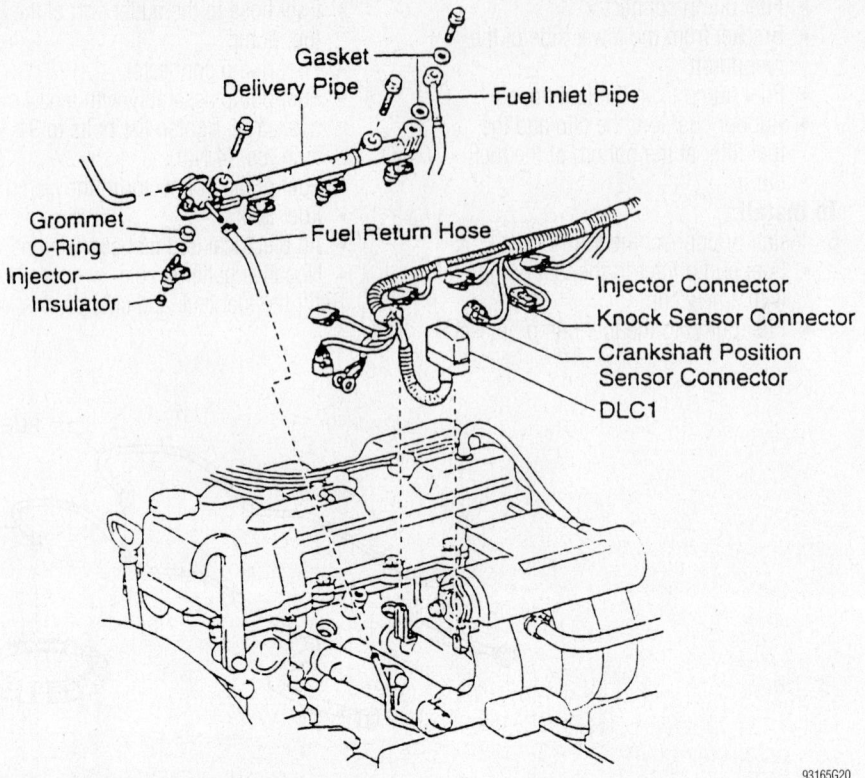

Fuel injector and related components—2.4L and 2.7L (3RZ-FE/M) engines

93165G20

To install:

4. Install or connect the following:
- New grommets and O-rings on each injector, lubricated with a light coat of gasoline
- Fuel injectors, with the electrical connector facing upwards
- New insulators and spacers on the intake manifold

5. Temporarily install the bolts holding the delivery pipe to the intake manifold.

6. Check that the injectors rotate smoothly.

7. Install or connect the following:
- Tighten the delivery pipe-to-intake manifold bolts to 15 ft. lbs. (21 Nm).
- Injector electrical connectors
- Fuel inlet pipe with new gaskets. Tighten the union bolt to 22 ft. lbs. (29 Nm) and the bolt to 14 ft. lbs. (20 Nm).
- Fuel return pipe to the fuel pressure regulator
- Vacuum line to the pressure regulator
- Throttle body
- Negative battery cable

2.7L (2TR-FE/X) Engine

1. Before servicing the vehicle, refer to the Precautions Section.

2. Relieve the fuel system pressure.

3. Disconnect the negative battery cable.

4. Remove the engine undercover sub-assembly.

5. Drain the engine coolant. Remove the intake air connector.

6. Remove the throttle body motor assembly.

7. Disconnect and plug the fuel line hoses.

8. Remove the fuel pressure pulsation damper assembly.

9. Disconnect the fuel injector electrical connectors. Disconnect the VSV connector. Disconnect the engine wiring harness clamp.

10. Disconnect the air conditioning compressor clutch connector. Disconnect the wire harness clamp. Remove the bolt and then remove the harness clamp bracket.

11. Remove the two bolts and remove the fuel rail together with the fuel injectors.

12. Remove the fuel rail number one spacers. Remove the four injector vibration insulators. Remove the four spacers.

13. Remove the fuel injectors. Discard all gaskets.

To install:

14. Installation is the reverse of the removal procedure.

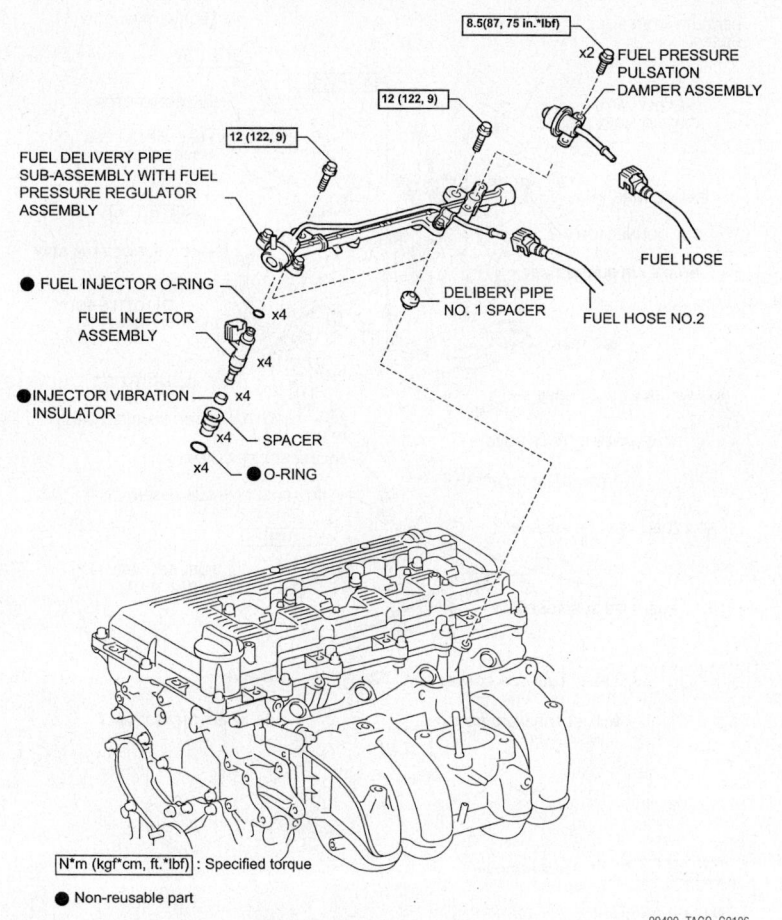

Fuel injector and related components—2.7L (2TR-FE/X) engine

N*m (kgf*cm, ft.*lbf) : Specified torque

● Non-reusable part

09490_TACO_G0106

15. Be sure to use new gaskets and O-rings, as required.

16. Be sure to fill the cooling system with the proper grade and type engine coolant.

17. Start the engine and check for leaks, correct as required.

3.4L Engine

1. Before servicing the vehicle, refer to the Precautions Section.

2. Depressurize the fuel system.

3. Remove or disconnect the following:
- Air cleaner hose
- Upper half of the intake manifold
- Fuel pressure regulator
- Fuel inlet pipe
- Fuel injector electrical connections
- Fuel rail with the injectors
- Spacers from the intake manifold
- Injectors from the delivery pipes
- O-rings and grommets, discard them

To install:

4. Install or connect the following:
- New grommets and O-rings on each injector, lubricated with a light coat of gasoline

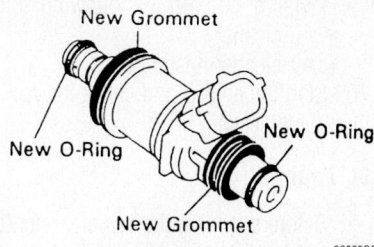

86825GG8

Install new O-rings and grommets on each injector—3.4L engine

- Fuel injector with the electrical connector facing outward
- Spacers on the intake manifold

5. Temporarily install the bolts to hold the delivery pipes to the intake manifold.

6. Check that the injectors rotate smoothly. If they do not, the O-rings have probably been installed incorrectly.

7. Install or connect the following:
- Fuel injector electrical connectors
- Fuel pipe with new gaskets. Tighten the bolts to 25 ft. lbs. (34 Nm) and the delivery pipes-to-intake manifold bolts to 10 ft. lbs. (13 Nm).
- Fuel pipe union with new gaskets. Tighten the clamp bolt to 71 inch lbs. (8 Nm).
- Fuel pressure regulator

8. Inspect the vacuum lines and connections. Look for any loose connections, sharp bends or damage.

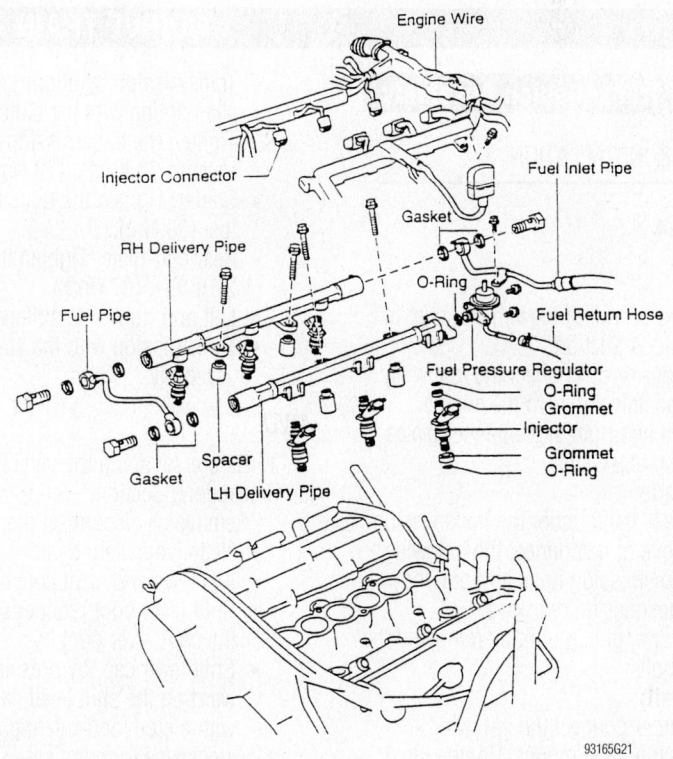

93165G21

Fuel injector and related components—3.4L engine

9. Install or connect the following:
- Air cleaner
- Air cleaner hose

10. Start the engine and check for vacuum and fuel leaks.

4.0L Engine

1. Before servicing the vehicle, refer to the Precautions Section.

2. Relieve the fuel system pressure.

3. Disconnect the negative battery cable.

4. Drain the engine coolant. Remove the V bank cover.

5. Remove the air cleaner assembly.

6. Remove the intake manifold.

7. Disconnect and plug the fuel line hoses.

8. Disconnect the fuel injector connectors. Remove the six bolts and the fuel rail together with the injectors.

9. Pull the injectors out of the fuel rail.

To install:

10. Installation is the reverse of the removal procedure.

11. Be sure to use new gaskets and O-rings, as required.

12. Be sure to fill the cooling system with the proper grade and type engine coolant.

13. Start the engine and check for leaks, correct as required.

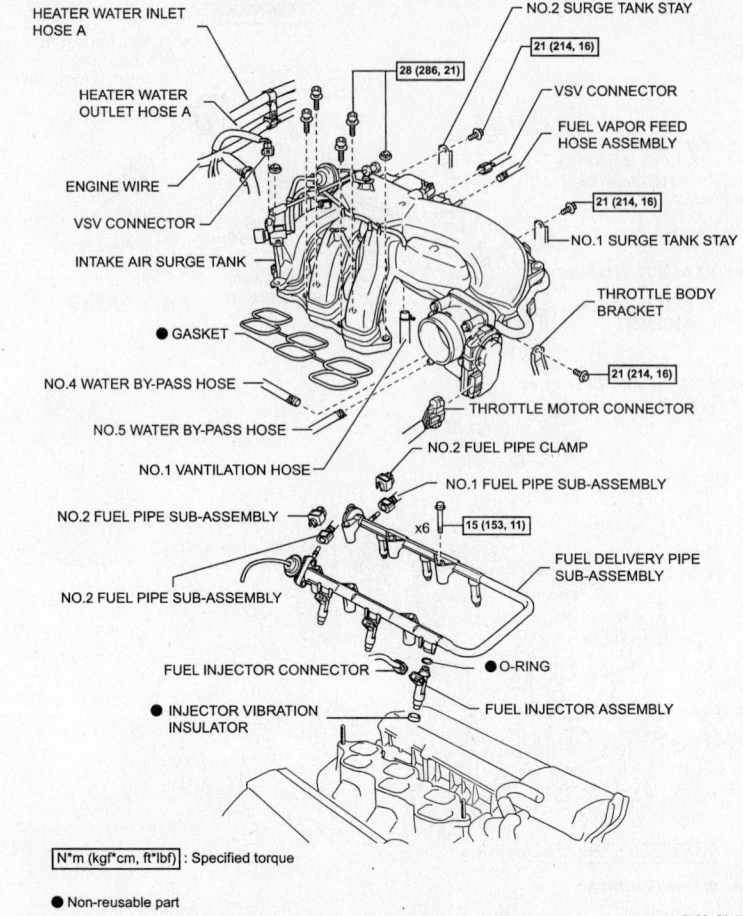

HEATER WATER INLET HOSE A

HEATER WATER OUTLET HOSE A

ENGINE WIRE

VSV CONNECTOR

INTAKE AIR SURGE TANK

● GASKET

NO.4 WATER BY-PASS HOSE

NO.5 WATER BY-PASS HOSE

NO.1 VANTILATION HOSE

NO.2 FUEL PIPE SUB-ASSEMBLY

NO.2 FUEL PIPE SUB-ASSEMBLY

FUEL INJECTOR CONNECTOR

● INJECTOR VIBRATION INSULATOR

NO.2 SURGE TANK STAY

28 (286, 21)

21 (214, 16)

VSV CONNECTOR

FUEL VAPOR FEED HOSE ASSEMBLY

21 (214, 16)

NO.1 SURGE TANK STAY

THROTTLE BODY BRACKET

21 (214, 16)

THROTTLE MOTOR CONNECTOR

NO.2 FUEL PIPE CLAMP

NO.1 FUEL PIPE SUB-ASSEMBLY

x6   15 (153, 11)

FUEL DELIVERY PIPE SUB-ASSEMBLY

● O-RING

FUEL INJECTOR ASSEMBLY

| N*m (kgf*cm, ft*lbf) | : Specified torque

● Non-reusable part

09490_TACO_G0107

Fuel injector and related components—4.0L engine

DRIVE TRAIN

Manual Transmission Assembly

REMOVAL & INSTALLATION

2002–2004

2WD

1. Before servicing the vehicle, refer to the Precautions Section.

2. Remove or disconnect the following:
- Transmission with the engine
- Left and right side stiffener plates
- Rear end-plate
- Starter

3. Place a stand under the transmission.

4. Remove or disconnect the following:
- Transmission bolts and pull the transmission rearward
- Rear engine mount by removing the 4 bolts

To install:

5. Install or connect the following:
- Rear engine mount. Tighten the 4 bolts to 48 ft. lbs. (65 Nm).

- Transmission by aligning the input shaft spline with the clutch disc. Tighten the transmission-to-engine bolts to 53 ft. lbs. (72 Nm).
- Starter. Tighten the bolts to 29 ft. lbs. (39 Nm).
- Rear end-plate. Tighten the bolts to 27 ft. lbs. (37 Nm).
- Left and right side stiffener plates
- Transmission with the engine assembly.

4WD

1. Before servicing the vehicle, refer to the Precautions Section.

2. Remove or disconnect the following:
- Negative battery cable
- 4 screws and front console box
- Shift lever boot retainer screws and the shift lever boot
- Shift lever cap, by pressing downward on the shift lever cap covered with a cloth and rotating it counterclockwise to remove it
- Transfer shift lever, using snapring

pliers to pull it from the transfer case

3. Drain the transmission and the transfer oil.

4. Remove or disconnect the following:
- Front and rear driveshafts
- Speedometer cable and the back-up light switch connector
- 4WD position switch connector, on the Standard cab
- L4 position switch connector, on the Extra cab
- Clutch release cylinder, by moving it aside without disconnecting the clutch line
- Exhaust pipe bracket
- Starter
- Rear end plate by removing the nuts and 2 bolts

5. Support the transmission rear side.

6. Remove or disconnect the following:
- 4 engine rear mount bolts
- O-ring and the crossmember

7. Using a transmission jack, support the transmission.

8. Remove or disconnect the following:
- 6 transmission-to-engine bolts
- 3 wire clamps from the transmission
- Transmission with the transfer case
- Engine rear mounting from the transfer case
- Transfer adapter rear mount bolts
- Transfer case from the transmission

To install:

9. Apply MP grease to the adapter oil seal and shift the 2 shift fork shafts to the high 4 position.

10. Install or connect the following:
- Transfer to the transmission. Tighten the bolts to 17 ft. lbs. (24 Nm).

➡**Be careful not to damage the oil seal by the input gear spline when installing the transfer.**

- Transmission/transfer case assembly, by aligning the input shaft spline with the clutch disc.

11. Support the transmission with a jack.

12. Install or connect the following:
- Tighten the engine to transmission bolts to 53 ft. lbs. (72 Nm)
- Engine rear mount. Tighten the 4 bolts to 48 ft. lbs. (65 Nm).

13. Raise the transmission slightly with a jack.

14. Install or connect the following:
- Crossmember. Tighten the 4 bolts to 48 ft. lbs. (65 Nm).
- Engine rear mount. Tighten the bolts to 14 ft. lbs. (19 Nm).
- Rear end-plate. Tighten the 4 bolts and nuts to 13 ft. lbs. (18 Nm) on R150 and R150F transmissions or to 27 ft. lbs. (37 Nm) on W59 transmissions.
- Starter. Tighten both bolts to 29 ft. lbs. (39 Nm).
- Front exhaust pipe. Tighten the exhaust pipe-to-manifold bolts to 46 ft. lbs. (62 Nm), the exhaust bracket bolts to 33 ft. lbs. (44 Nm) and the exhaust pipe-to-catalytic converter bolts to 35 ft. lbs. (48 Nm).
- Clutch release cylinder. Tighten the 2 bolts to 108 inch lbs. (13 Nm).
- L4 position switch connector on the extra cab or the 4WD position switch connector on the standard cab.
- VSS and the back-up light switch connector.
- Front and rear driveshafts.

15. Refill the transmission to the correct level.

16. Apply MP grease to the transfer shift lever.

17. Install or connect the following:
- Transfer shift lever.
- Snapring, using pliers

18. Install the transmission shift lever, as follows:

a. Apply MP grease to the transmission shift lever.

b. Align the groove of the shift lever cap and the pin par of the case cover. Cover the shift lever cap with a cloth. Pressing down on the shift lever cap, rotate it clockwise to install.

c. Install shift lever boot retainer with the 4 screws.

d. Install the front console box with the 4 screws.

19. Connect the negative battery cable. Start the engine and check for leaks.

20. Road test the vehicle for proper operation. Recheck all fluid levels.

2005–2006

1. Before servicing the vehicle, refer to the Precautions Section.

2. Disconnect the negative battery cable.

3. Remove the upper console rear panel subassembly.

4. Remove the rear console box assembly. Remove the front console box. Remove the shift lever boot assembly. Remove the floor shift lever assembly.

5. Remove the number two engine undercover subassembly.

6. Drain the transmission oil.

7. Remove the front exhaust pipe assembly.

8. Remove the front driveshaft, if equipped.

9. Remove the driveshaft.

➡**Some vehicles are also equipped with a center bearing assembly, remove if equipped.**

10. Remove the starter. Remove the manifold stay.

11. On five speed, separate the clutch release cylinder assembly by removing the two bolts.

12. On six speed, remove the number one clutch housing cover. Remove the clutch release cylinder assembly. Remove the clutch accumulator assembly.

13. Using the proper support equipment, support the transmission (2WD) or transmission/transfer case (4WD).

14. Remove the bolts from the engine rear mounting. Remove the bolts, nuts and frame crossmember number three sub assembly.

15. Remove the bolts and the engine mounting insulator rear number one.

➡**Be sure that the transmission (2WD) or transmission/transfer case (4WD) is properly supported.**

16. Tilt the transmission assembly downward. Disconnect all electrical wires and hoses that will interfere with the removal of the transmission (2WD) or transmission/transfer case (4WD).

➡**When tilting the transmission assembly downward, make sure that the cooling fan does not hit the radiator fan shroud.**

17. Remove the transmission assembly retaining bolts. Remove the transmission (2WD) or transmission/transfer case (4WD) from the vehicle.

➡**If equipped with 4WD, remove the transmission and the transfer case as one unit.**

18. On 4WD separate the transmission from the transfer case as required.

To install:

19. Installation is the reverse of the removal procedure.

20. On five speed, tighten the three upper transmission to engine retaining bolts to 53 ft. lbs. and the four lower bolts to 27 ft. lbs.

21. On five speed, tighten the four upper transmission to engine retaining bolts to 53 ft. lbs. and the five lower bolts to 28 ft. lbs.

22. Be sure to fill the transmission with the proper grade and type oil.

23. Start the engine and check for leaks, correct as required.

Clutch Assembly

REMOVAL & INSTALLATION

1. Before servicing the vehicle, refer to the Precautions Section.

2. Remove or disconnect the following:
- Negative battery cable
- Transmission assembly

3. Matchmark the clutch cover to the flywheel.

4. At the clutch cover, loosen each bolt 1 turn until spring tension is released.

5. Remove or disconnect the following:
- Clutch cover set bolts and the clutch cover with the clutch disc.
- Release bearing retaining clip and withdraw it
- Release fork and boot assembly

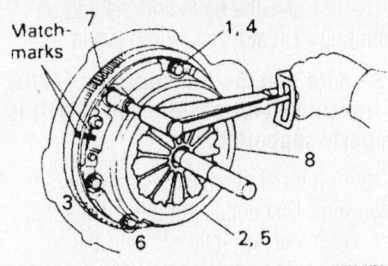

Clutch cover bolt tightening sequence—
2002–2004

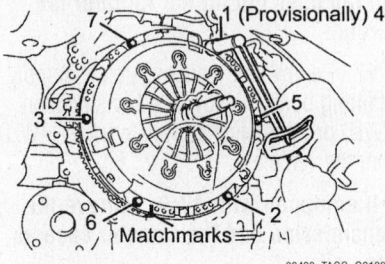

Clutch cover bolt tightening sequence—
2005–2006

To install:

6. Install or connect the following:
- Clutch disc onto the flywheel, using a clutch disc alignment tool
- Clutch cover, position it onto the flywheel and if reusing the old pressure plate, align the match-marks.
- Clutch cover. Tighten the bolts in a crisscross pattern to 14 ft. lbs. (19 Nm).

7. Lubricate the release fork pivot and contact points, the release bearing, bearing hub and input shaft spline surfaces with a suitable molybdenum disulfide lithium based or multi-purpose grease.

8. Install or connect the following:
- Boot, release fork, hub and the bearing assemblies
- Transmission
- Negative battery cable

Hydraulic Clutch System

BLEEDING

1. Before servicing the vehicle, refer to the Precautions Section.

2. Fill the clutch reservoir with brake fluid. Check the reservoir level frequently and add fluid as needed.

3. Connect one end of a vinyl tube to the bleeder plug on the slave cylinder and submerge the other end into a clear container half-filled with brake fluid.

4. Slowly pump the clutch pedal several times.

5. Have an assistant hold the clutch pedal down and loosen the bleeder plug until fluid and/or air starts to run out of the bleeder plug. Close the bleeder plug while the pedal is held to the floor.

6. Repeat Steps 2 and 3 until all the air bubbles are removed from the system.

7. Tighten the bleeder plug when all the air is gone.

8. Refill the master cylinder to the proper level as required.

9. Check the system for leaks.

Automatic Transmission Assembly

REMOVAL & INSTALLATION

2002–2004

A340F

1. Before servicing the vehicle, refer to the Precautions Section.

2. Remove or disconnect the following:
- Automatic Transmission Fluid (ATF) level gauge
- Engine undercover

3. Drain the transmission fluid.

4. Remove or disconnect the following:
- Throttle cable
- No. 1 fan shroud

5. Remove the transmission shift lever assembly and the transfer shift lever, as follows:
- Rear console box
- Front console box with the transfer shift lever knob
- Connectors
- Shift control rod
- Transmission shift lever assembly
- Snapring and pull it from the transfer case
- Oil filler pipe, with the O-ring
- Front and rear driveshaft
- Exhaust pipe
- Speedometer cable
- No. 2 Vehicle Speed Sensor (VSS) connector
- Solenoid connector
- Transfer case neutral position switch connector
- Transfer case L4 position switch connector
- Transfer indicator switch
- Oil cooler pipe
- Automatic Transmission Fluid (ATF) temperature sensor connector
- Park/Neutral Position (PNP) switch connector

- Starter
- 4 stabilizer bar bracket mounting bolts

6. Remove the torque converter bolts, as follows:
- Flywheel housing undercover
- Torque converter clutch mounting bolts, while turning the crankshaft to gain access

7. Remove the front differential rear mounting cushion, as follows:
- Nut, using a hexagon wrench
- Front differential by lifting it

➡ **Be careful not to touch the torque converter clutch housing and the front differential companion flange**

- 2 rear mount cushion bolts

8. Remove or disconnect the following:
- Support the transmission's rear side
- 4 engine rear mount bolts
- 4 nuts, bolts and the crossmember, by supporting the transmission
- Transmission

To install:

9. Install or connect the following:
- Transmission. Tighten the engine-to-transmission bolts to 53 ft. lbs. (71 Nm).
- Crossmember. Tighten the bolts to 48 ft. lbs. (65 Nm).
- Engine rear mount. Tighten the bolts to 14 ft. lbs. (19 Nm).
- Front differential rear mount cushion. Tighten the nut to 64 ft. lbs. (41 Nm).
- Torque converter clutch mount bolt

➡ **Install the green colored bolt, then the 5 others. Tighten the bolts to 30 ft. lbs. (41 Nm).**

- Flywheel housing undercover. Tighten the bolts to 13 ft. lbs. (18 Nm) for 3.4L engine or to 27 ft. lbs. (37 Nm) for 2.7L (3RZ-FE/M) engine.
- Stabilizer bar bracket bolts. Tighten the 4 bolts to 19 ft. lbs. (25 Nm).
- Starter. Tighten the bolts to 29 ft. lbs. (39 Nm).
- Remaining components
- ATF level gauge

10. Fill and check the fluid level.

11. Test drive and check for proper shifting.

A340D AND A340E

1. Before servicing the vehicle, refer to the Precautions Section.

2. Remove or disconnect the following:
- Transmission with the engine and place it on a stand

- Bolts, 2 stiffener plates and rear endplate
3. Turn the crankshaft to gain access to the torque converter bolts.
4. Remove or disconnect the following:
- Torque converter bolts
- Starter
- Transmission-to-engine bolts
- Transmission

To install:

5. Install or connect the following:
- Transmission to the engine. Tighten the bolts to 53 ft. lbs. (71 Nm).
- Starter. Tighten both bolts to 29 ft. lbs. (39 Nm).
- Torque converter. Tighten the bolts to 30 ft. lbs. (41 Nm).
- Stiffener plate and rear endplate. Tighten the bolts to 27 ft. lbs. (37 Nm).
- Starter wires
- Transmission with the engine

2005–2006

A340E

1. Before servicing the vehicle, refer to the Precautions Section.
2. Disconnect the negative battery cable.
3. Remove the driveshaft.

➡**Some vehicles are also equipped with a center bearing assembly, remove if equipped.**

4. Drain the transmission fluid.
5. Remove the transmission dipstick. Remove the two bolts and remove the transmission oil filler tube assembly.
6. Remove the three bolts and clamps. Using tool SST09023-12701, disconnect the oil cooler outlet tube number one.
7. Remove the three bolts and clamps. Using tool SST09023-12701, disconnect the oil cooler inlet tube number one.
8. Remove the manifold stay.
9. On cable type, disconnect the transmission control cable assembly.
10. On floor shift (rod type), remove the clip and separate the floor shift gear shifting rod from the transmission.
11. Support the transmission assembly, with a transmission jack.
12. Remove the eight bolts and two front suspension member brackets.
13. Disconnect the heated oxygen sensor connector. Remove the four bolts from the frame crossmember subassembly number three.
14. Remove the four nuts, four bolts and frame crossmember number three.
15. Remove the starter.

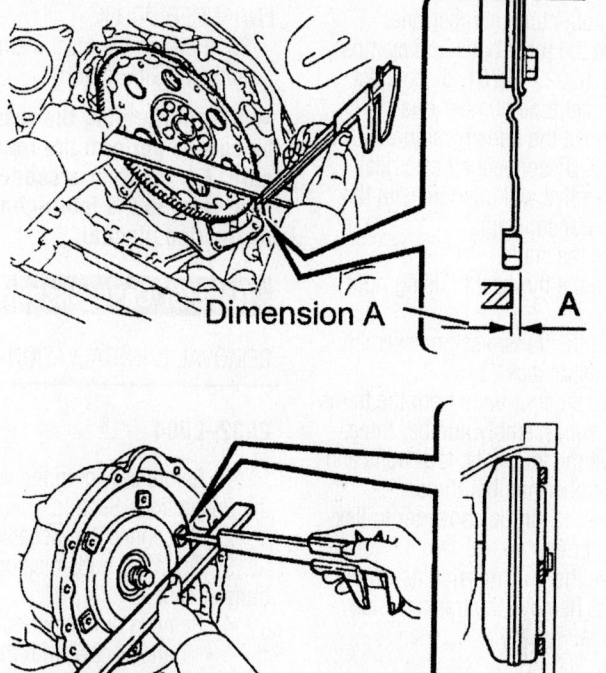

Torque converter installation—2005–2006

09490_TACO_G0109

16. Tilt the transmission assembly downward. Disconnect all electrical wires and hoses that will interfere with the removal of the transmission.

➡**When tilting the transmission assembly downward, make sure that the cooling fan does not hit the radiator fan shroud.**

17. Remove the flywheel housing dust shield. Remove the torque converter to flexplate retaining bolts.
18. Remove the transmission assembly retaining bolts. Remove the transmission from the vehicle.

To install:

19. Installation is the reverse of the removal procedure.
20. When installing the torque converter on to the transmission mainshaft, use a calipers and straight edge and measure between the transmission and the end of the surface of the driveplate. Specification should be 0.8772 inch.
21. Use a calipers and straight edge and measure the dimension "B" shown in the illustration and check that "B" is greater than "A". Specification should be 0.9165 inch or more.
22. Tighten the three upper transmission to engine retaining bolts to 52 ft. lbs. and the four lower bolts to 27 ft. lbs.

23. Tighten the torque converter to flexplate retaining bolts to 35 ft. lbs.
24. Be sure to fill the transmission with the proper grade and type transmission fluid.
25. Start the engine and check for leaks, correct as required.

➡**Perform the automatic transmission initialization procedure when replacing the automatic transmission, engine or ECM.**

26. On the A340E automatic transmission, initialization is completed by connecting and disconnecting the negative battery cable.

A750E

1. Before servicing the vehicle, refer to the Precautions Section.
2. Disconnect the negative battery cable.
3. Remove the number one engine undercover assembly. Remove the number two engine undercover assembly.
4. Disconnect the oxygen sensor. Remove the exhaust pipe assembly.
5. Remove the driveshaft and center bearing assembly.
6. Drain the transmission fluid. Remove the number one manifold stay. Remove the number two manifold stay.
7. Remove the three bolts and clamps.

Using tool SST09023-12701, disconnect the oil cooler outlet tube number one.

8. Remove the three bolts and clamps. Using tool SST09023-12701, disconnect the oil cooler inlet tube number one.

9. Disconnect the transmission control cable assembly. Disconnect all electrical wires and hoses that will interfere with the removal of the transmission.

10. Remove the starter.

11. Remove the flywheel housing dust shield cover.

12. Support the transmission assembly, with a transmission jack.

13. Remove the four bolts from the frame crossmember subassembly number three.

14. Remove the four nuts, four bolts and frame crossmember number three.

15. Remove the torque converter to flex-plate retaining bolts.

16. Remove the transmission assembly retaining bolts. Remove the transmission from the vehicle.

To install:

17. Installation is the reverse of the removal procedure.

18. When installing the torque converter on to the transmission mainshaft, use a calipers and straight edge and measure between the transmission and the end of the surface of the driveplate, dimension "A".

19. Use a calipers and straight edge and measure the dimension "B" shown in the illustration and check that "B" is greater than "A". Specification should be plus 1mm or more.

20. Tighten the five upper transmission to engine retaining bolts to 53 ft. lbs. and the four lower bolts to 27 ft. lbs.

21. Tighten the torque converter to flex-plate retaining bolts to 35 ft. lbs.

22. Be sure to fill the transmission with the proper grade and type transmission fluid.

23. Start the engine and check for leaks, correct as required.

➡**Perform the automatic transmission initialization procedure when replacing the automatic transmission, engine or ECM.**

24. On the A750E automatic transmission, initialization is performed using the intelligent tester tool.

25. Turn the ignition switch OFF.

26. Connect the intelligent tester tool together with the controller area network vehicle interface module (CAN VIN) to the DLC3.

27. Turn the ignition switch to the ON position.

28. Push the intelligent tester tool main switch to the ON position.

29. Select the items, DIAGNOSIS/ENHANCED OBD II.

30. Perform the reset memory procedure from the ENGINE menu.

➡**After performing the reset memory, be sure to perform the road test procedure. For roadtest procedure information, refer to the intelligent tester instruction manual.**

Transfer Case Assembly

REMOVAL & INSTALLATION

2002–2004

1. Before servicing the vehicle, refer to the Precautions Section.

2. Disconnect the negative battery cable.

3. Drain the transmission and the transfer case.

4. Remove or disconnect the following:
   - Transfer case with the transmission
   - Breather hose from the transfer upper cover and the transmission control retainer, if equipped with an automatic transmission
   - Rear engine mounting
   - Dynamic damper

5. Remove the driveshaft upper dust cover and the transfer from the transmissions, as follows:
   - Dust cover bolt from the bracket
   - Transfer case adapter rear mounting bolts
   - Transfer case, by pulling it straight up and away from the transmission.

✳✳ WARNING

Be careful not to damage the adapter rear oil seal with the transfer input gear spline.

To install:

6. Install the transfer case and the driveshaft upper dust cover to the transmission with a new gasket, as follows:
   - Shift the 2 shift fork shafts to the high 4 position
   - Apply MP grease to the adapter oil seal
   - New gasket to the transfer adapter
   - Transfer case to the transmission.

✳✳ WARNING

Take care not to damage the oil seal by the input gear spline.

   - Transfer case adapter. Tighten the rear bolts to 27 ft. lbs. (37 Nm).

   - Dust cover to the bracket. Tighten the bolt to 17 ft. lbs. (23 Nm).

7. Install or connect the following:
   - Engine rear mount. Tighten the bolts to 19 ft. lbs. (25 Nm).
   - Dynamic damper. Tighten the bolts to 27 ft. lbs. (37 Nm).
   - Breather hose, if equipped with an automatic transmission
   - Transfer case with the transmission to the engine

8. Fill the transmission and the transfer case with oil.

9. Test drive the vehicle and check the abnormal noise and smooth operation.

10. Recheck the fluid levels.

2005–2006

1. Before servicing the vehicle, refer to the Precautions Section.

2. Disconnect the negative battery cable.

3. Drain the transfer case oil.

4. Remove the four bolts and remove the transfer case lower protector.

5. Remove the transmission assembly.

6. Remove the eight transfer adaptor rear mounting bolts. Pull the transfer case straight up and remove it from the transmission.

➡**Take care not to damage the adaptor rear oil seal with the transfer case input gear spline.**

To install:

7. Installation is the reverse of the removal procedure.

8. Tighten the eight transfer case mounting bolts to 17 ft. lbs.

9. Be sure to fill the transfer case with the proper grade and type transmission fluid.

10. Start the engine and check for leaks, correct as required.

Halfshaft

REMOVAL & INSTALLATION

2002–2004

1. Before servicing the vehicle, refer to the Precautions Section.

2. Drain the differential oil from the differential.

3. If not equipped with a free-wheeling hub, disconnect the halfshaft from the steering knuckle, as follows:
   - Grease cap
   - Cotter pin and lockcap, from the halfshaft

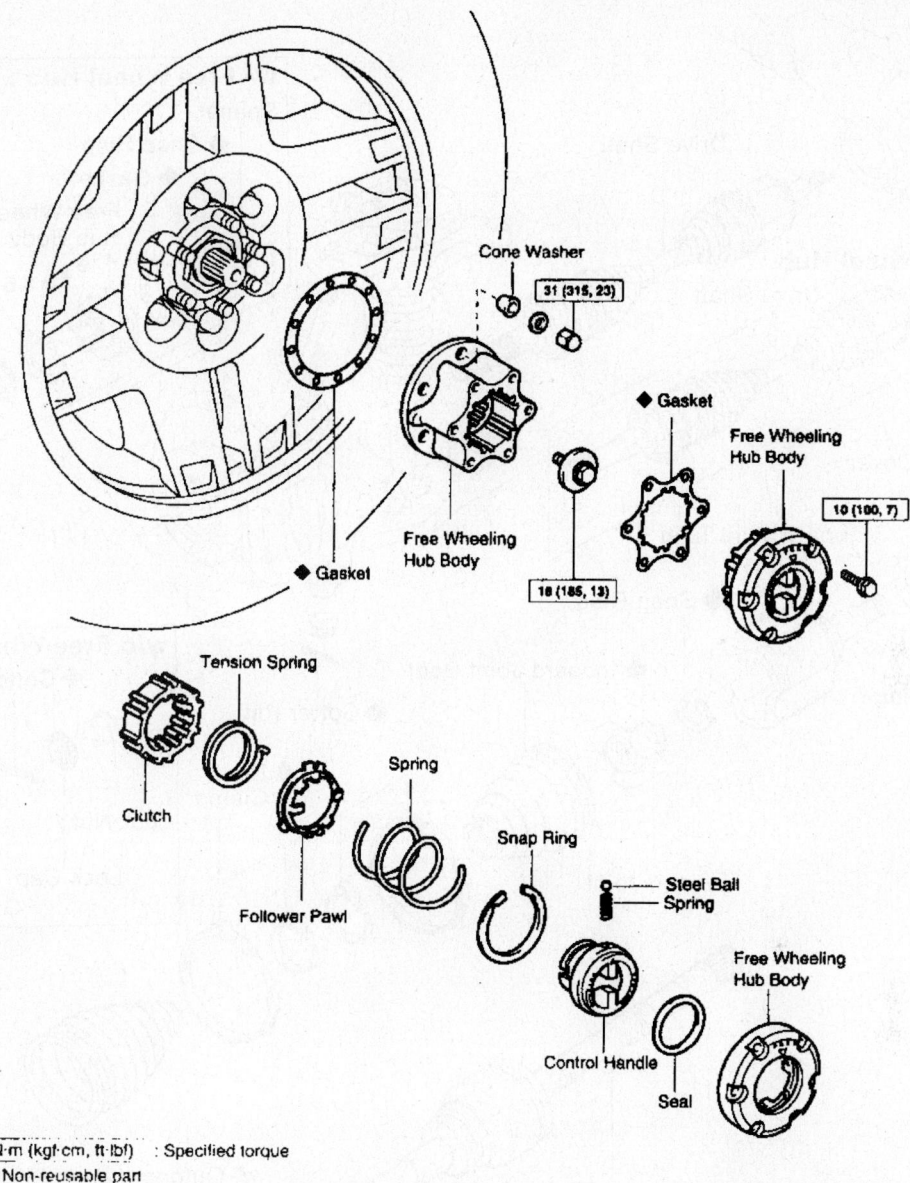

Cone Washer

31 (315, 23)

◆ Gasket

Free Wheeling Hub Body

Free Wheeling Hub Body

10 (100, 7)

◆ Gasket

18 (185, 13)

Tension Spring

Clutch

Spring

Follower Pawl

Snap Ring

Steel Ball

Spring

Free Wheeling Hub Body

Control Handle

Seal

N·m (kgf·cm, ft·lbf) : Specified torque
◆ Non-reusable part

7924YG96

Free wheeling hub assembly and related components—2002–2004

- Locknut from the halfshaft, while having an assistant apply the brakes
4. If equipped with free-wheeling hub, remove the free wheel hub, as follows:
   - Set the control handle to FREE
   - Cover bolts and pull off the cover
   - Center bolt with washer
   - Mounting nuts and washer to the hub body
   - Cone washer, using a brass bar and hammer to tap on the bolt heads
   - Free wheel hub body and gasket
   - Snapring from the end of the half-shaft, using a snapring expander
5. Remove or disconnect the follow-ing:

- Halfshaft from the differential, using a brass bar and hammer
- Cotter pin and nut, from the lower ball joint
- Lower control arm from the lower ball joint
- Halfshaft

➡**If it is difficult to remove the half-shaft from the steering knuckle, use a rubber hammer and tap the halfshaft from the steering knuckle.**

- Snapring from the inboard shaft
To install:
6. Install or connect the following:
   - New snapring to the inboard shaft
   - Halfshaft to the steering knuckle

➡**Push the steering knuckle inwards and at the same time, push the half-shaft into the differential with the snapring opening facing downward. Be sure the halfshaft is fully installed to the differential by checking that it can-not be pulled out by hand.**

- Lower control arm to the lower ball joint. Tighten the nut to 112 ft. lbs. (152 Nm).
- New cotter pin
7. If equipped with a free wheeling hub, install the hub, as follows:
   - Spacer
   - Snapring to the halfshaft, using a snapring expander
   - New gasket on the front axle hub

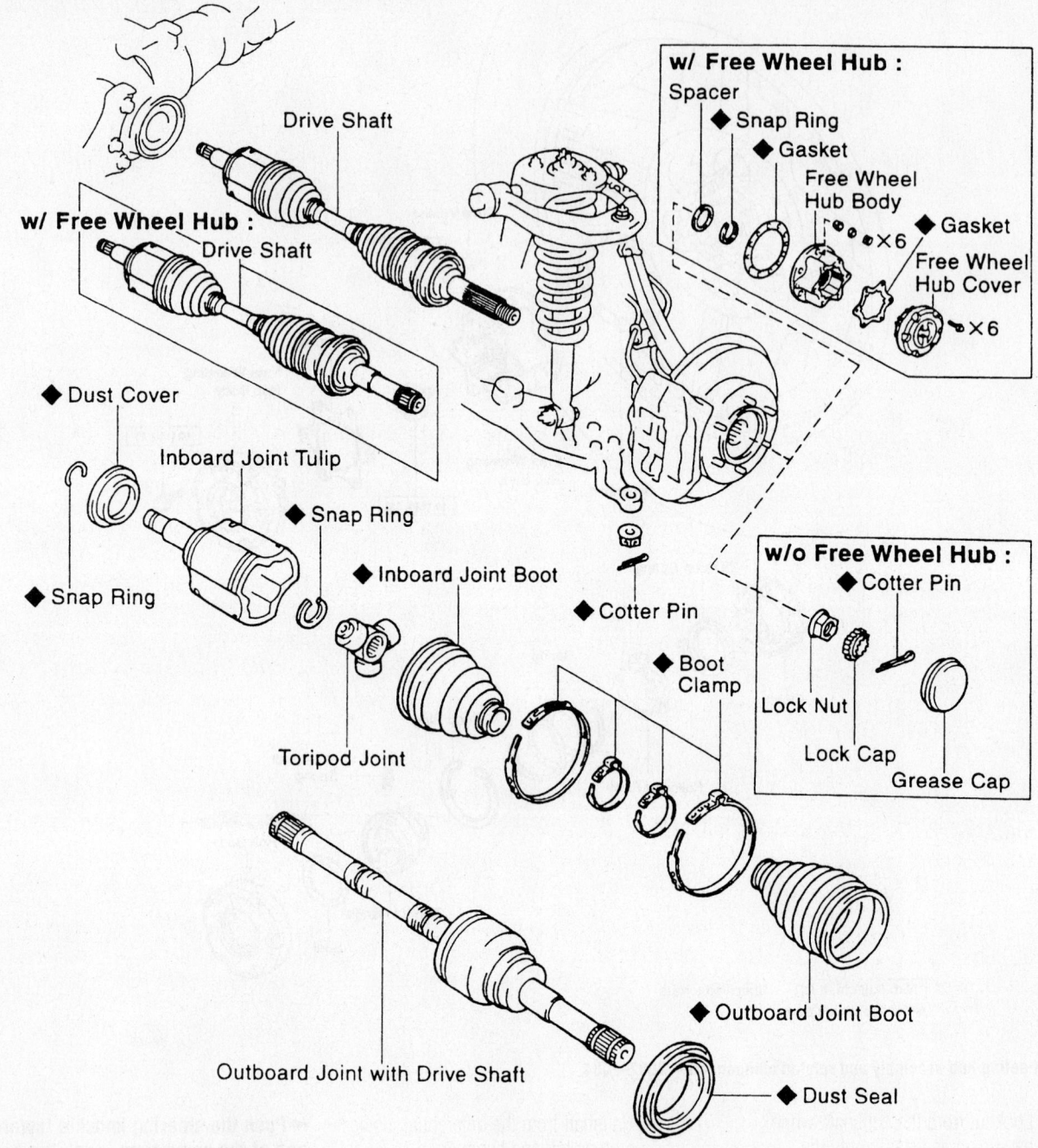

Halfshaft and related components—4WD 2002–2004

- Free wheeling hub body, with the 6 cone washers and nuts. Tighten the 6 nuts to 23 ft. lbs. (31 Nm).
- Bolt with the washer. Tighten the bolt to 13 ft. lbs. (18 Nm).
- Apply multi purpose grease to the inner hub splines
- Set the control handle and clutch to the FREE position

- New gasket on the cover
- Cover to the hub body, with the follower pawl tabs aligned with the non-toothed portions of the hub body.
- Tighten the cover bolts to 84 inch lbs. (10 Nm).

8. If equipped without a free wheeling hub, install the halfshaft to the steering knuckle, as follows:

- Locknut to the halfshaft. Tighten the locknut to 174 ft. lbs. (235 Nm).
- Lockcap and cotter pin to the halfshaft
- Grease cap
- Wheels

9. Fill the differential with gear oil.

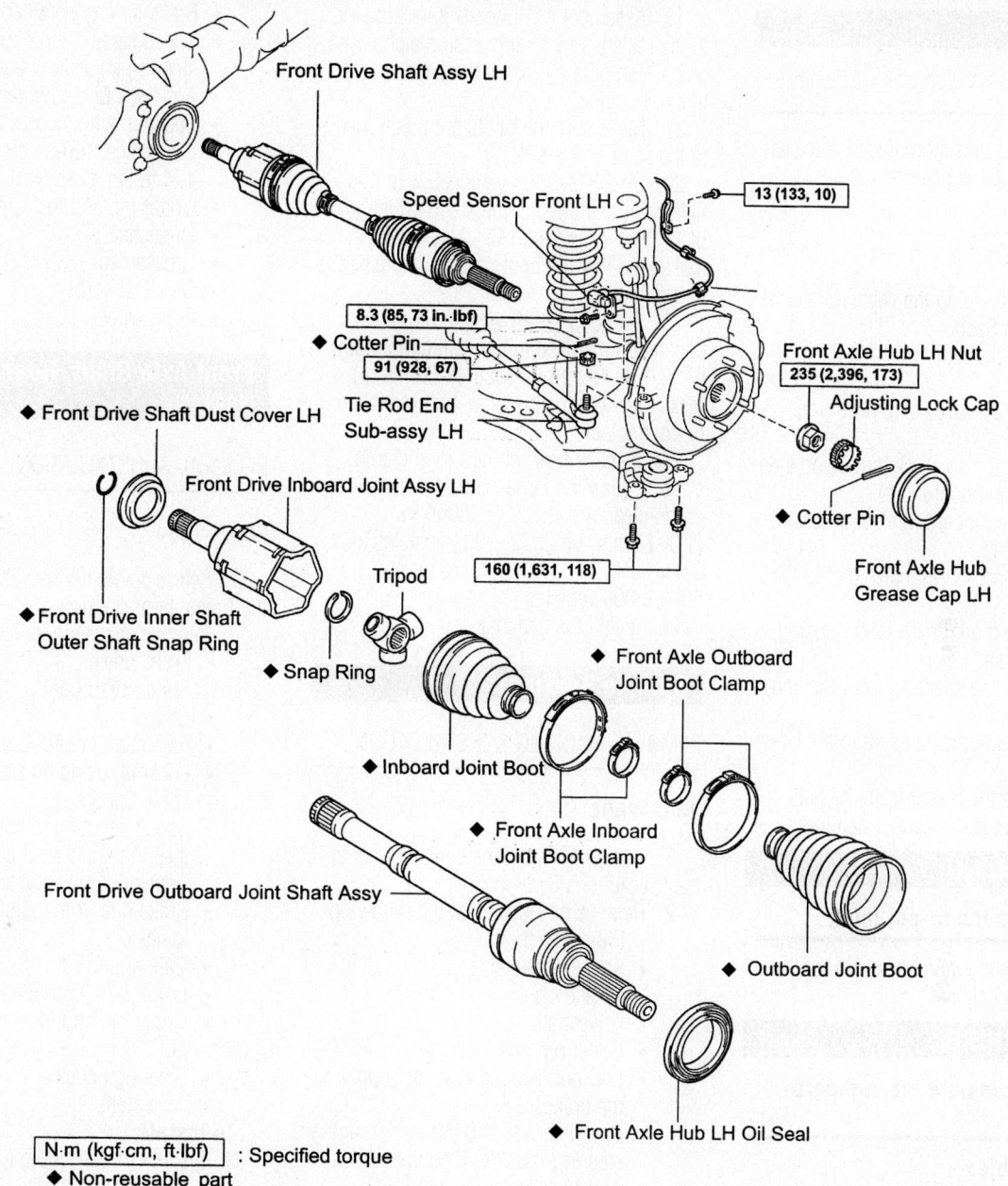

Front Drive Shaft Assy LH

Speed Sensor Front LH

13 (133, 10)

8.3 (85, 73 in.·lbf)

◆ Cotter Pin

91 (928, 67)

Front Axle Hub LH Nut

235 (2,396, 173)

Adjusting Lock Cap

◆ Front Drive Shaft Dust Cover LH

Tie Rod End
Sub-assy LH

Front Drive Inboard Joint Assy LH

◆ Cotter Pin

◆ Front Drive Inner Shaft
Outer Shaft Snap Ring

Front Axle Hub
Grease Cap LH

Tripod

160 (1,631, 118)

◆ Snap Ring

◆ Inboard Joint Boot

◆ Front Axle Outboard
Joint Boot Clamp

◆ Front Axle Inboard
Joint Boot Clamp

Front Drive Outboard Joint Shaft Assy

◆ Outboard Joint Boot

◆ Front Axle Hub LH Oil Seal

N·m (kgf·cm, ft·lbf) : Specified torque
◆ Non-reusable part

09490_TACO_G0110

Halfshaft and related components—4WD 2005–2006

2005–2006

1. Before servicing the vehicle, refer to the Precautions Section.

2. Disconnect the negative battery cable.

3. Raise and support the vehicle safely. Remove the tire and wheel assembly.

4. Drain the differential.

5. Remove the bolt and separate the front speed sensor. Disengage the two clamps. Remove the bolt and separate the speed sensor wire harness from the steering knuckle.

6. Remove the cotter pin and nut. Using tool SST09628-62011 or equivalent, sepa-rate the tie rod end from the steering knuckle.

7. Using a suitable tool and hammer, remove the front axle hub grease cap. Remove the cotter pin and adjusting cap. Remove the front axle hub nut.

8. Remove the two bolts and separate the front lower ball joint attachment front from the steering knuckle.

9. Using tool SST09520-01010, and SST09520-24010 remove the halfshaft. Be careful not to damage the oil seal.

To install:

10. Coat the spline of the inboard joint shaft with clean ATF.

11. Align the shaft splines and install the halfshaft.

➡**Set the snapring with the opening side facing downward. Be careful not to damage the oil seal.**

12. Continue the installation in the reverse order of the removal procedure.

13. Tighten the hub nut to 173 ft. lbs.

14. Be sure to fill the differential with the proper type and grade lubricant.

15. Check and correct leaks, as required.

16. Check and adjust the alignment, as required.

CV-Joints

OVERHAUL

The outboard joint is replaced with half-shaft; no overhaul is possible or necessary.

Inboard Joint

1. Before servicing the vehicle, refer to the Precautions Section.
2. Remove the halfshaft from the vehicle.
3. Remove the large clamp from the inboard joint.
4. Remove the small clamp, using side cutters, from the inboard joint.
5. Slide the inboard joint boot toward the outboard joint.
6. Matchmark the inboard joint to the halfshaft.
7. Remove the inboard joint housing from the halfshaft.
8. Remove the snapring from the end of the halfshaft.
9. Matchmark the halfshaft to the tri-pot joint.
10. Remove the tri-pot from the half-shaft, using a brass bar and a hammer.

✳✳ WARNING

Do not tap on the tri-pot joint.

11. Remove the inboard and outboard boots from the halfshaft.

✳✳ WARNING

Do not disassemble the outboard joint.

To assemble:

12. Wrap vinyl tape around the halfshaft splines to prevent damaging the boots.
13. Install the outboard and inboard boots to the halfshaft with the small end clamps.
14. Assemble the tri-pot joint to the halfshaft with the beveled side facing the outboard joint and align the matchmarks.
15. Install the tri-pot joint, using a brass bar and a hammer.

✳✳ WARNING

Do not tap on the roller.

16. Install the snapring.
17. Lubricate the outboard joint with ½ of the grease supplied with the kit.
18. Assemble the boot to the outboard joint

19. Assemble the inboard joint housing to the halfshaft by aligning the matchmarks.
20. Temporarily install the boot onto the tri-pot housing.
21. Make sure the boots are positioned in the shaft grooves.
22. On 2002–2004 vehicles, with the halfshaft positioned at the standard length of 17.094–17.252 in. (434.2–438.2mm), make sure that the boots are not stretched or contracted.
23. On 2005–2006 vehicles, with the halfshaft positioned at the standard length of 19.071–19.226 in. (484.4–488.4mm), make sure that the boots are not stretched or contracted.
24. Install a new inboard joint clamp.
25. Crimp the large clamp so that the crimp clearance is 0.039–0.059 in. (1.0–1.5mm) on 2002–2004 vehicles and 0.039–0.197 in. (1.0–5.0mm) on 2005–2006 vehicles.
26. Install the halfshaft.

Spindle Bearings

REMOVAL, PACKING & INSTALLATION

2002–2004

1. Before servicing the vehicle, refer to the Precautions Section.
2. Remove or disconnect the following:
   - Front wheel
   - Shock absorber
   - Grease cap
   - Driveshaft
   - Cotter pin and lockcap
   - Locknut, with an assistant applying the brakes
   - Speed sensor and harness from the steering knuckle, if equipped with Anti-lock Brake System (ABS)
   - Brake line from the steering knuckle
   - Caliper and rotor
   - Lower ball joint bolts and the joint from the steering knuckle
   - Cotter pin and axle hub nut
   - Steering knuckle
   - Bearings from the steering knuckle

To install:

3. Install or connect the following:
   - Bearings to the steering knuckle
   - Steering knuckle
   - Cotter pin and axle hub nut. Tighten the nut to 80 ft. lbs. (108 Nm).
   - Lower ball joint to the steering knuckle
   - Caliper and rotor

   - Brake line to the steering knuckle
   - Speed sensor and harness to the steering knuckle, if equipped with Anti-lock Brake System (ABS)
   - Locknut, with an assistant applying the brakes. Torque the locknut to 174 ft. lbs. (235 Nm).
   - Cotter pin and lockcap
   - Driveshaft
   - Grease cap
   - Shock absorber
   - Front wheel

Front Axle Shaft, Bearing and Seal

REMOVAL & INSTALLATION

2002–2004

1. Before servicing the vehicle, refer to the Precautions Section.
2. Remove or disconnect the following:
   - Front wheel
   - Shock absorber
   - Grease cap
   - Axle shaft's cotter pin and lock cap
   - Locknut, using an assistant to apply the brakes
   - Speed sensor and harness from the steering knuckle, if equipped with Anti-lock Brake System (ABS)
   - Brake line from the steering knuckle
   - Caliper and rotor
   - Lower ball joint bolts
   - Cotter pin and loosen the axle hub nut
   - Steering knuckle
   - Axle shaft

To install:

3. Install or connect the following:
   - Axle shaft
   - Steering knuckle
   - Tighten the axle hub nut to 80 ft. lbs. (108 Nm) and the locknut to 174 ft. lbs. (235 Nm).
   - Cotter pin
   - Lower ball joint bolts
   - Caliper and rotor
   - Brake line to the steering knuckle
   - Speed sensor and harness to the steering knuckle, if equipped with Anti-lock Brake System (ABS)
   - Locknut, using an assistant to apply the brakes
   - Axle shaft's cotter pin and lock cap
   - Grease cap
   - Shock absorber
   - Front wheel

Rear Axle Shaft, Bearing and Seal

REMOVAL & INSTALLATION

2002–2004

1. Before servicing the vehicle, refer to the Precautions Section.

2. Remove or disconnect the following:
   - Rear wheel
   - Brake drum

3. Check the bearing backlash and axle shaft deviation, as follows:

   a. Using a dial indicator, check that the backlash in the bearing shaft direction. The maximum is 0.027 in. (0.7mm).

   b. If the backlash exceeds the maximum, replace the bearing.

   c. Using a dial indicator, check the deviation at the surface of the axle shaft outside the hub bolt. Maximum is 0.0039 in. (0.1mm).

   d. If the deviation exceeds the maximum, replace the axle shaft.

4. Remove or disconnect the following:
   - Anti-lock Brake System (ABS) speed sensor from the axle housing, if equipped

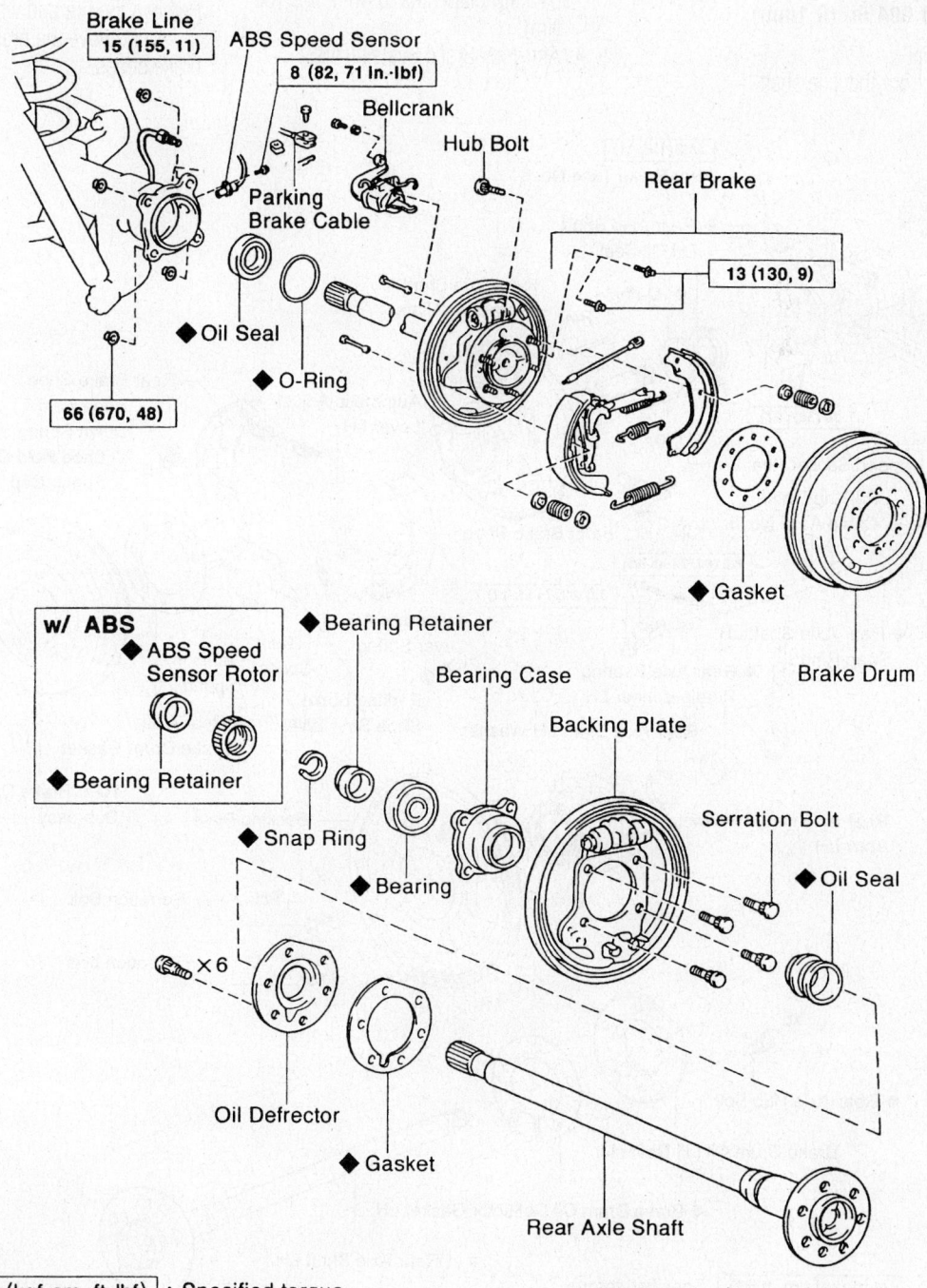

N·m (kgf·cm, ft·lbf) : Specified torque
◆ Non-reusable part

Rear axle shaft and related components—2002–2004

86827G97

- Axle shaft assembly by removing the 4 nuts from the backing plate
- O-ring from the axle housing
- Bearing and retainer (differential side) and ABS speed sensor rotor, if equipped
- Snapring from the axle shaft
- Axle shaft

➡**Inspect the axle shaft and flange run-outs. The axle shaft run-out should be 0.079 in. (2.0mm) and the flange run-out should be 0.004 in. (0.1mm).**

- Outer seal
- Bearing from the axle shaft

To install:

5. Install or connect the following:
- Bearing from the axle shaft
- Outer seal
- Axle shaft
- Snapring to the axle shaft
- Bearing and retainer (differential side) and ABS speed sensor rotor, if equipped
- O-ring to the axle housing
- Axle shaft assembly. Tighten the 4 backing plate nuts to 48 ft. lbs. (66 Nm).
- Anti-lock Brake System (ABS)

speed sensor to the axle housing, if equipped
- Brake drum
- Rear wheel

2005–2006

1. Before servicing the vehicle, refer to the Precautions Section.

2. Disconnect the negative battery cable. Drain the brake fluid.

3. Raise and support the vehicle safely. Remove the tire and wheel assembly.

4. Remove the brake drum. Remove the brake shoes.

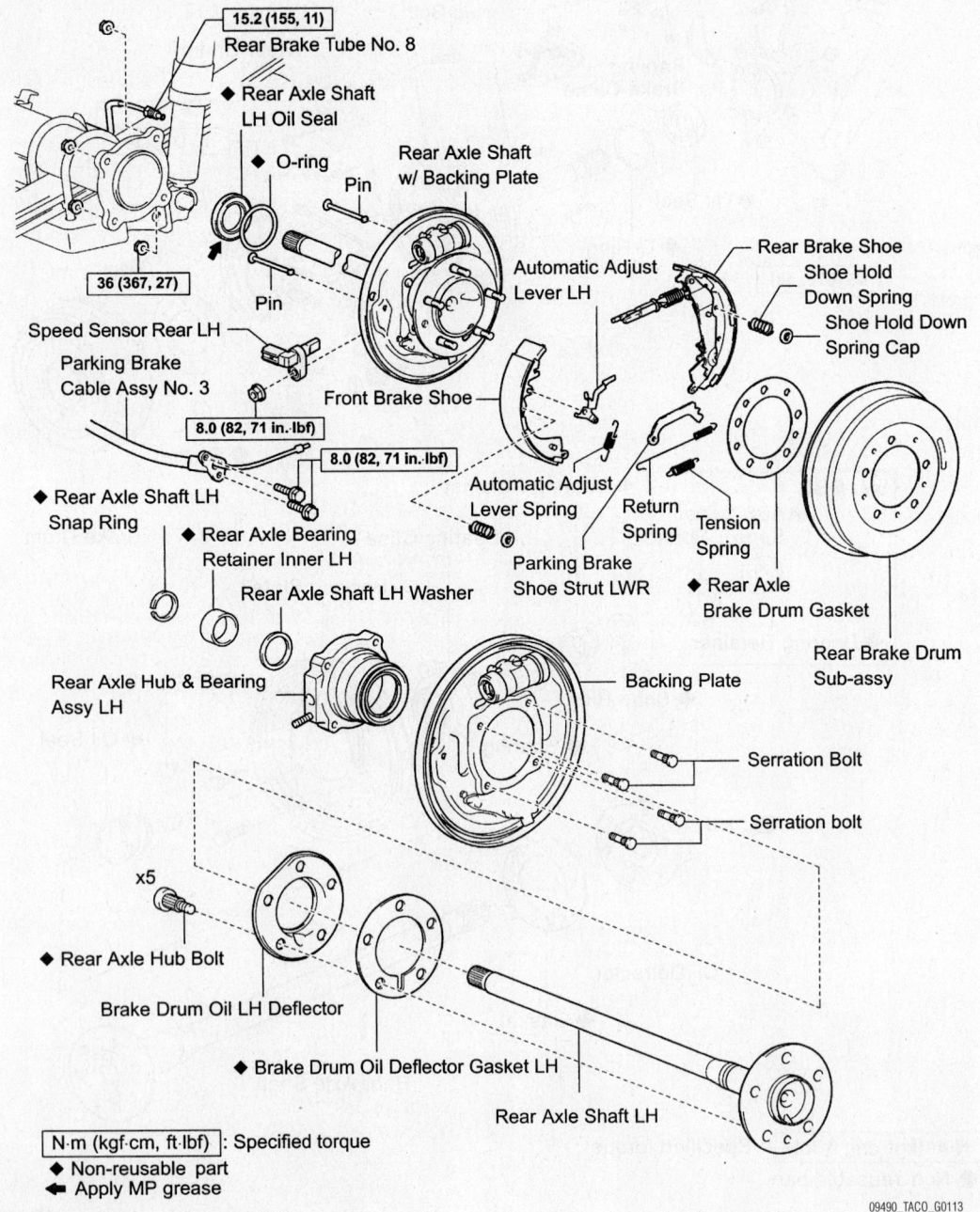

Rear axle shaft and related components—2WD 2005–2006

09490_TACO_G0113

5. Remove the rear speed sensor.

6. Remove the two bolts and disconnect the parking brake cable from the backing plate. Disconnect the brake line at the brake backing plate.

7. Remove the four nuts and the rear axle shaft and backing plate. Remove the O-ring.

8. Remove the rear axle shaft oil seal using tool SST09308-00010.

To install:

9. Installation is the reverse of the removal procedure.

10. Be sure to fill the master cylinder with the proper grade and type brake fluid.

11. Bleed the brakes, as required.

Pinion Seal

REMOVAL & INSTALLATION

Front

1. Before servicing the vehicle, refer to the Precautions Section.

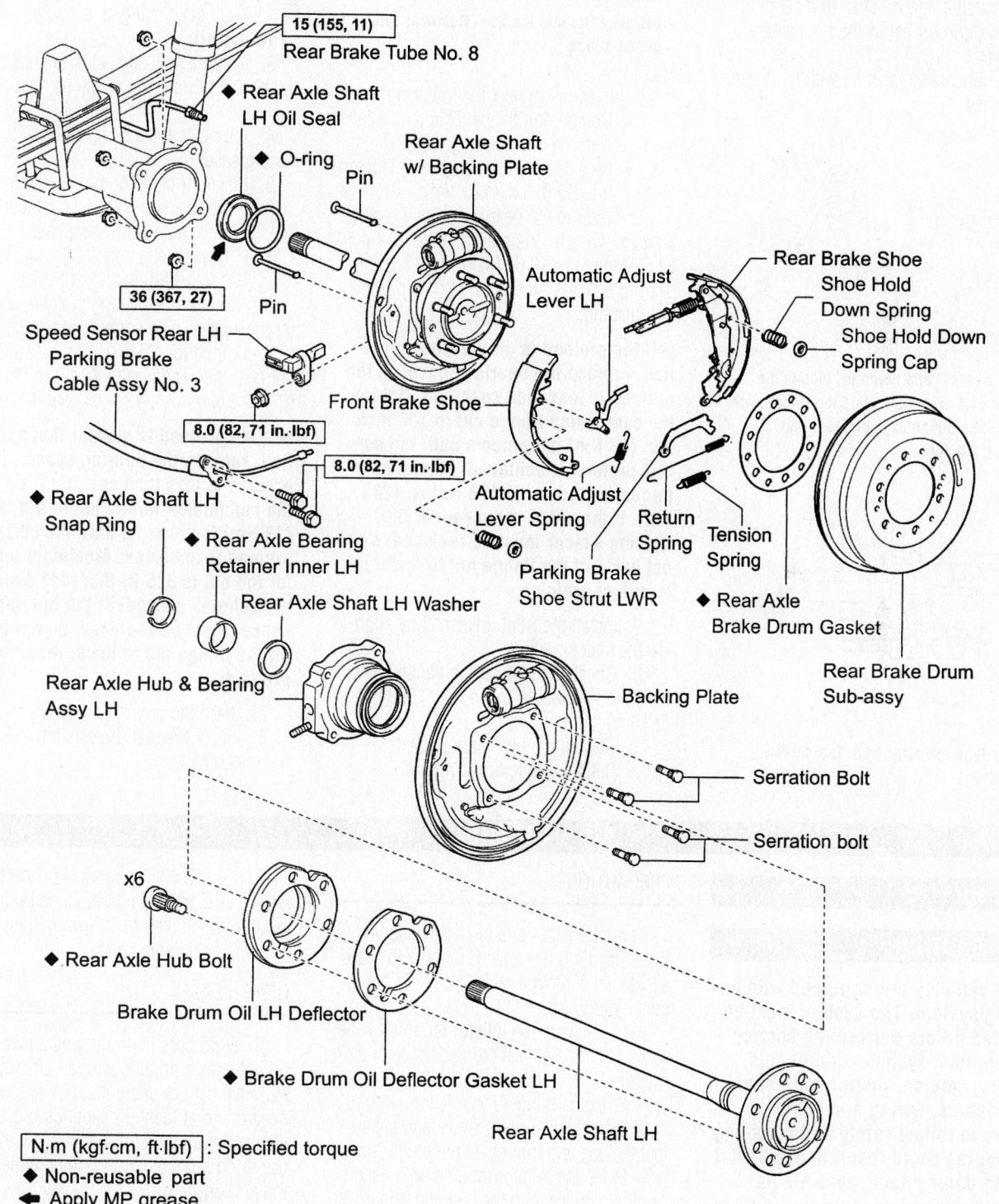

15 (155, 11)
Rear Brake Tube No. 8

◆ Rear Axle Shaft LH Oil Seal

◆ O-ring

Pin

Rear Axle Shaft w/ Backing Plate

Automatic Adjust Lever LH

Rear Brake Shoe
Shoe Hold Down Spring
Shoe Hold Down Spring Cap

36 (367, 27)

Pin

Speed Sensor Rear LH
Parking Brake Cable Assy No. 3

Front Brake Shoe

8.0 (82, 71 in.·lbf)

8.0 (82, 71 in.·lbf)

Automatic Adjust Lever Spring

Automatic Adjust Lever LWR

Parking Brake Shoe Strut LWR

Return Spring

Tension Spring

◆ Rear Axle Brake Drum Gasket

◆ Rear Axle Shaft LH Snap Ring

◆ Rear Axle Bearing Retainer Inner LH

Rear Axle Shaft LH Washer

Rear Axle Hub & Bearing Assy LH

Backing Plate

Rear Brake Drum Sub-assy

Serration Bolt

Serration bolt

x6

◆ Rear Axle Hub Bolt

Brake Drum Oil LH Deflector

◆ Brake Drum Oil Deflector Gasket LH

Rear Axle Shaft LH

N·m (kgf·cm, ft·lbf) : Specified torque
◆ Non-reusable part
◄ Apply MP grease

Rear axle shaft and related components—4WD and PreRunner 2005–2006

09490_TACO_G0114

2. Remove the engine undercover.
3. Drain the differential oil.
4. Remove or disconnect the following:
   - Front driveshaft
   - Companion flange nut, by unstaking it
   - Companion flange
   - Pinion seal, using an extractor

To install:

5. Install a new oil seal, to a depth of 0.059 in. (1.5mm) below the lip, using a seal driver.
6. Lubricate the seal lip with multi-purpose grease.

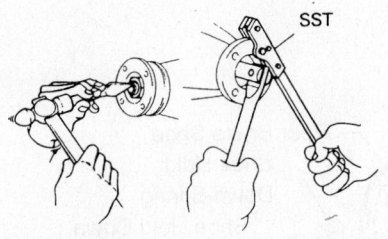

Using a chisel and hammer, loosen the staked part of the nut. Hold the flange with SST 09950-30010 or equivalent and remove the nut

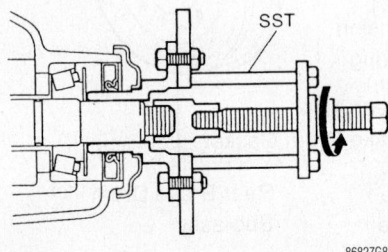

Screw-type extractor from Tool 09950-30010

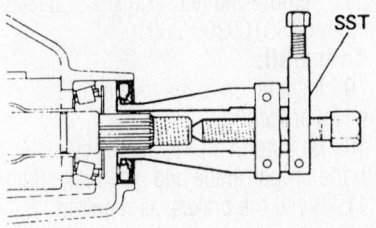

Extractor fits into the Seal Removal Tool 09308-10010

7. Install or connect the following:
   - Companion flange, coat the threads with multi-purpose grease
   - New companion flange nut. Tighten it to 89 ft. lbs. (120 Nm).
8. Measure the bearing preload, using a torque wrench. The correct preload should be 5–9 inch lbs. (0.6–1.0 Nm) for a used bearing or 10–17 inch lbs. (1–2 Nm) for a new bearing.

➡If the preload is greater that specified, replace the bearing spacer. If the preload is less than specified, tighten the companion flange nut in 108 inch lbs. (13 Nm) increments until the correct preload is achieved. Maximum torque for the nut is 173 ft. lbs. (235 Nm). If the value is exceeded, the bearing spacer must be replaced; do not back off the flange nut to lower the torque or preload.

9. Install the front driveshaft by aligning the matchmarks.
10. Check the companion flange run-out; maximum allowable run-out is 0.003 in. (0.10mm).
11. Stake the pinion flange nut.
12. Refill the differential with oil.

Rear

1. Before servicing the vehicle, refer to the Precautions Section.
2. Remove or disconnect the following:
   - Rear driveshaft by matchmarking it
   - Companion flange nut, by loosen the staked portion
   - Companion flange, using a screw-type extractor
   - Oil seal, using an extractor

To install:

3. Install a new oil seal, to a depth of 0.039 in. (1.0mm) below the lip, using a seal driver.
4. Lubricate the seal lip with multi-purpose grease.
5. Install or connect the following:
   - Companion flange, coat the threads with multi-purpose grease
   - New companion flange nut. Tighten it to 109 ft. lbs. (147 Nm).
6. Measure the bearing preload, using a torque wrench. The correct preload should be 8–11 inch lbs. (0.9–1.2 Nm) for a 2 spider gear differential or to 4–7 inch lbs. (0.4–0.8 Nm) for a 4 spider gear differential.

➡If the preload is greater that specified, replace the bearing spacer. If the preload is less than specified, tighten the companion flange nut in 9 ft. lbs. (13 Nm) increments until the correct preload is achieved. Maximum torque for the nut is 325 ft. lbs. (441 Nm). If the value is exceeded, the bearing spacer must be replaced; do not back off the flange nut to lower the torque or preload.

7. Stake the pinion flange nut.
8. Install the rear driveshaft by aligning the matchmarks.

STEERING

Air Bag

❊❊ CAUTION

These vehicles are equipped with an air bag system. The system must be disabled before performing service on or around system components, steering column, instrument panel components, wiring and sensors. Failure to follow safety and disabling procedures could result in accidental air bag deployment, possible personal injury and unnecessary system repairs.

PRECAUTIONS

Several precautions must be observed when handling the inflator module to avoid accidental deployment and possible personal injury.

- Never carry the inflator module by the wires or connector on the underside of the module.
- When carrying a live inflator module, hold securely with both hands, and ensure that the bag and trim cover are pointed away.
- Place the inflator module on a bench or other surface with the bag and trim cover facing up.

- With the inflator module on the bench, never place anything on or close to the module which may be thrown in the event of an accidental deployment.

DISARMING

To avoid personal injury when working on vehicles equipped with an air bag, the negative battery cable must be disconnected and at least 90 seconds must elapse before working on the system. Failure to do so may result in deployment of the air bag.

Power Steering Gear

REMOVAL & INSTALLATION

2002–2004

1. Before servicing the vehicle, refer to the Precautions Section.
2. Remove or disconnect the following:
   - Negative battery cable
   - Right and left tie rod ends from the knuckle
   - Intermediate No. 2 shaft from the steering rack
   - Pressure feed and the return tubes, using SST 09631-22020
   - Power steering rack

To install:

3. Install the power steering rack. Tighten the bolts to 148 ft. lbs. (201 Nm) for 2WD or to the following values for 4WD:
   - Rack assembly bolt: 123 ft. lbs. (167 Nm)
   - Rack assembly nut: 141 ft. lbs. (191 Nm)
   - Bracket nut and bolt: 123 ft. lbs. (167 Nm)
4. Install or connect the following:
   - New O-ring
   - Pressure feed tube. Tighten it to 33 ft. lbs. (45 Nm).
   - Return tube. Tighten it to 36 ft. lbs. (49 Nm) for 2WD or to 29 ft. lbs. (40 Nm) for 4WD

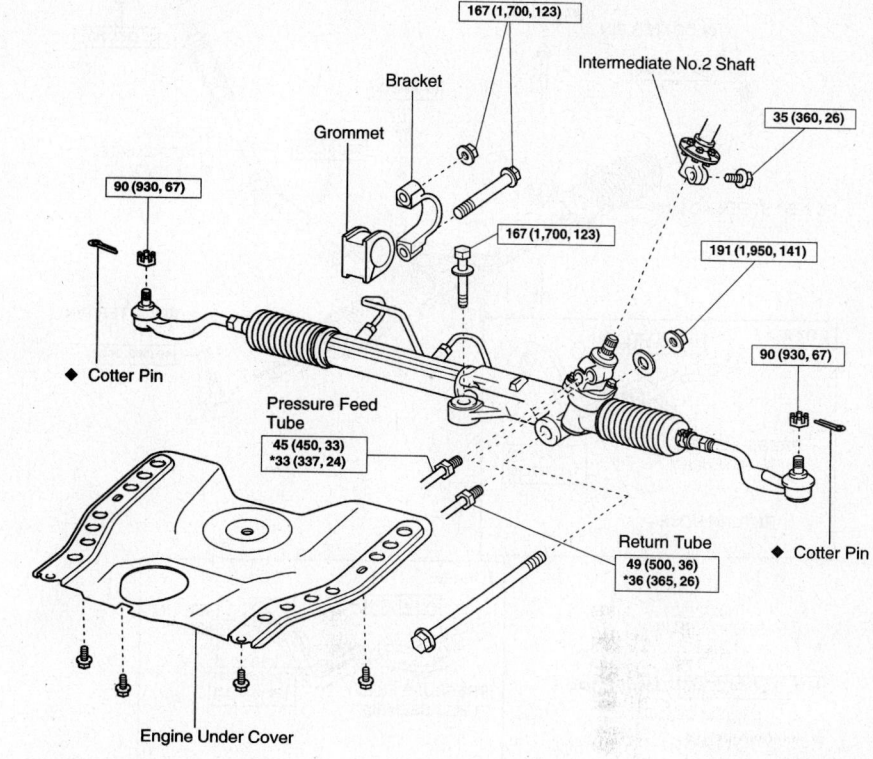

N·m (kgf·cm, ft·lbf) : Specified torque
◆ Non–reusable part
* For use with SST

67170-TACO-G29

Power steering gear and related components—4WD 2002–2004

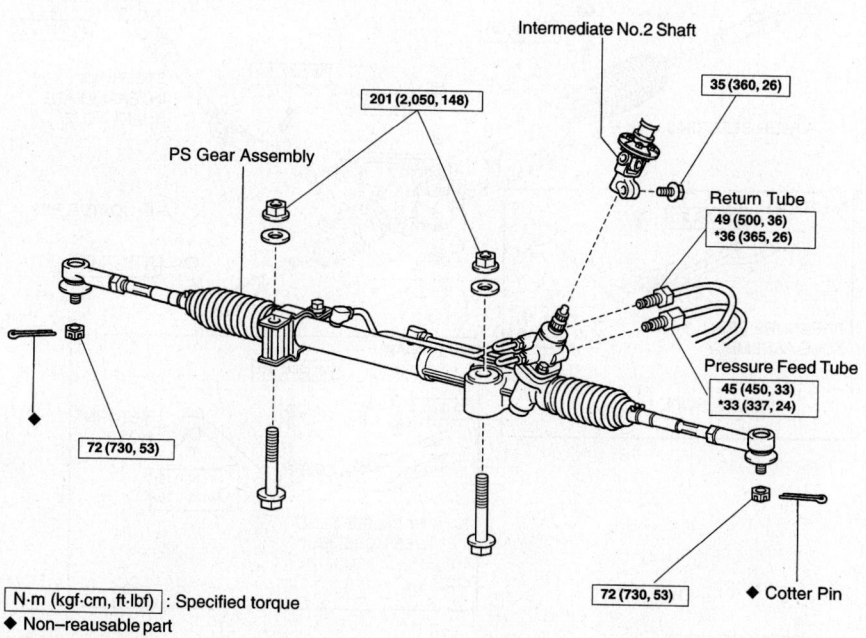

N·m (kgf·cm, ft·lbf) : Specified torque
◆ Non–reusable part
* For use with SST

67170-TACO-G28

Power steering gear and related components—2WD 2002–2004

- Intermediate No. 2 shaft to the steering rack
- Right and left tie rod ends to the steering knuckle
- Tighten the castle nuts to specification
- New cotter pins
- Negative battery cable

5. Check the steering wheel center point.
6. Bleed the power steering system.
7. Check the front wheel alignment. Tighten the tie rod end locknuts to 67 ft. lbs. (90 Nm)

2005–2006

2WD

1. Before servicing the vehicle, refer to the Precautions Section.
2. Position the front wheels in the straight ahead position.
3. Disconnect the negative battery cable. Drain the power steering fluid.
4. Raise and support the vehicle safely. Remove the tire and wheel assemblies.
5. Lock the steering wheel to prevent it from turning.

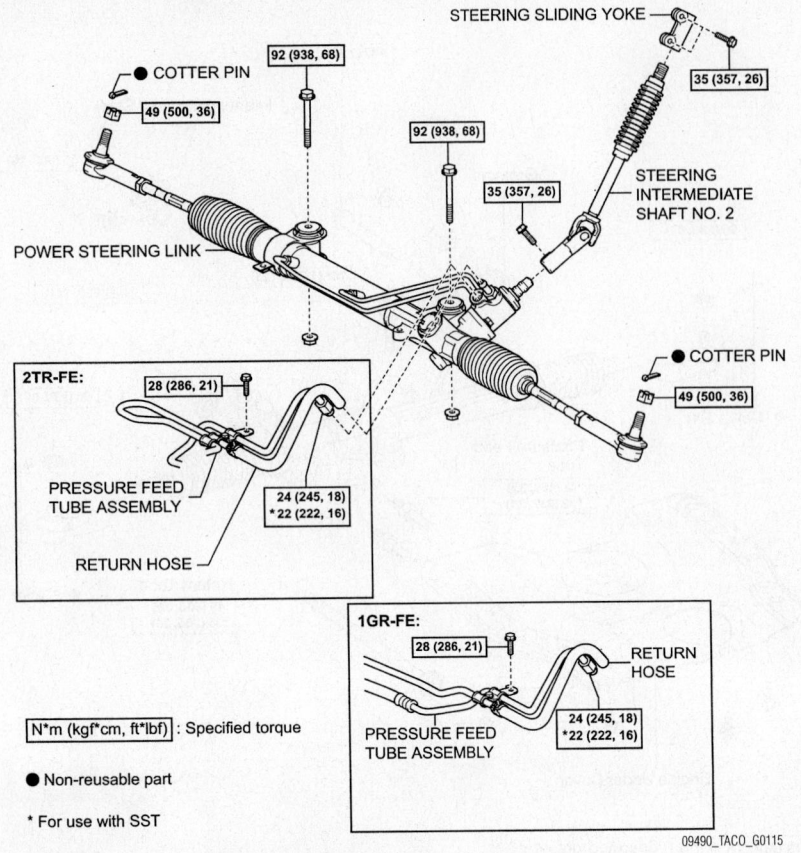

2TR-FE:

PRESSURE FEED TUBE ASSEMBLY

RETURN HOSE

28 (286, 21)

24 (245, 18)
*22 (222, 16)

1GR-FE:

28 (286, 21)

RETURN HOSE

PRESSURE FEED TUBE ASSEMBLY

24 (245, 18)
*22 (222, 16)

COTTER PIN

49 (500, 36)

92 (938, 68)

POWER STEERING LINK

STEERING SLIDING YOKE

35 (357, 26)

92 (938, 68)

35 (357, 26)

STEERING INTERMEDIATE SHAFT NO. 2

COTTER PIN

49 (500, 36)

N*m (kgf*cm, ft*lbf) : Specified torque

● Non-reusable part

* For use with SST

09490_TACO_G0115

Power steering gear and related components—2WD 2005–2006

➡ **The seat belt can be used to prevent rotation.**

6. Place matchmarks on the steering slider yoke, the steering intermediate shaft number two and the steering intermediate shaft.

7. Remove the steering slider yoke bolts. Slide the steering sliding yoke up and separate it from the steering intermediate shaft number two.

8. Pull down the steering sliding yoke from the steering intermediate shaft and remove it.

9. Place matchmarks on the steering intermediate shaft number two and the power steering gear.

10. Remove the bolt from the steering intermediate shaft number two.

11. Slide the steering intermediate shaft number two up and remove it from the power steering gear.

12. Remove the cotter pin and nut. Using tool SST09610-20012 or equivalent, separate the left tie rod end from the left steering knuckle arm.

13. Remove the cotter pin and nut. Using tool SST09610-20012 or equivalent, separate the right tie rod end from the right steering knuckle arm.

14. Remove the bolt and separate the

tube support bracket. Separate the pressure line. Disengage the clip and disconnect the return hose.

15. Remove the power steering gear retaining bolts. Remove the steering gear from the vehicle.

➡ **The nut has a detent, so never turn it. Always turn the bolt.**

To install:

16. Install the power steering gear. Tighten the retaining bolts to 68 ft. lbs.

➡ **The nut has a detent, so never turn it. Always turn the bolt.**

17. Continue the installation in the reverse order of the removal procedure.

18. Be sure to fill the power steering system with the proper grade and type power steering fluid.

19. Bleed the system, as required.

20. Start the engine and check for leaks, correct as required.

21. Check and adjust the alignment, as required.

4WD

1. Before servicing the vehicle, refer to the Precautions Section.

2. Position the front wheels in the straight ahead position.

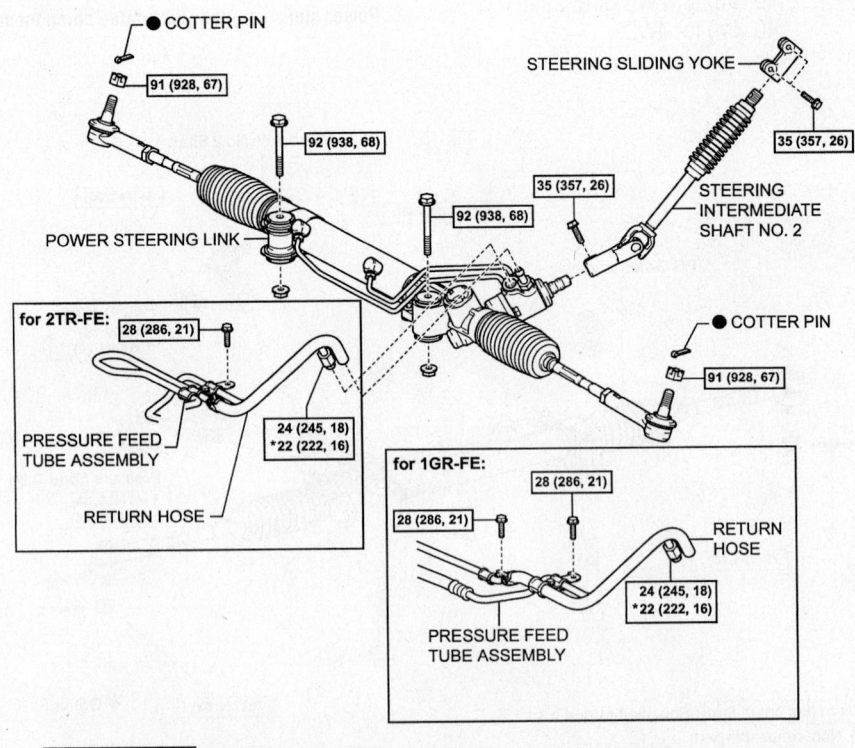

COTTER PIN

91 (928, 67)

STEERING SLIDING YOKE

35 (357, 26)

92 (938, 68)

35 (357, 26)

POWER STEERING LINK

92 (938, 68)

STEERING INTERMEDIATE SHAFT NO. 2

COTTER PIN

91 (928, 67)

for 2TR-FE:

28 (286, 21)

PRESSURE FEED TUBE ASSEMBLY

RETURN HOSE

24 (245, 18)
*22 (222, 16)

for 1GR-FE:

28 (286, 21)

28 (286, 21)

RETURN HOSE

PRESSURE FEED TUBE ASSEMBLY

24 (245, 18)
*22 (222, 16)

N*m (kgf*cm, ft*lbf) : Specified torque   ● Non-reusable part   * For use with SST

09490_TACO_G0116

Power steering gear and related components—4WD and PreRunner 2005–2006

3. Disconnect the negative battery cable. Drain the power steering fluid.

4. Raise and support the vehicle safely. Remove the tire and wheel assemblies.

5. Remove the number one engine undercover subassembly.

6. Remove the front exhaust pipe assembly. On the 4.0L engine, remove the number two exhaust pipe assembly.

7. Remove the driveshaft. Some vehicles also use a center bearing assembly, remove that too.

8. Remove the frame crossmember subassembly.

9. Remove the stabilizer bar.

10. Lock the steering wheel to prevent it from turning.

➡**The seat belt can be used to prevent rotation.**

11. Place matchmarks on the steering slider yoke, the steering intermediate shaft number two and the steering intermediate shaft.

12. Remove the steering slider yoke bolts. Slide the steering sliding yoke up and separate it from the steering intermediate shaft number two.

13. Pull down the steering sliding yoke from the steering intermediate shaft and remove it.

14. Place matchmarks on the steering intermediate shaft number two and the power steering gear.

15. Remove the bolt from the steering intermediate shaft number two.

16. Slide the steering intermediate shaft number two up and remove it from the power steering gear.

17. Remove the cotter pin and nut. Using tool SST09610-20011 or equivalent, separate the left tie rod end from the left steering knuckle arm.

18. Remove the cotter pin and nut. Using tool SST09610-20011 or equivalent, separate the right tie rod end from the right steering knuckle arm.

19. Remove the bolt and separate the

tube support bracket. Separate the pressure line. Disengage the clip and disconnect the return hose.

20. Remove the power steering gear retaining bolts. Tilt the transmission and remove the steering gear from the vehicle.

➡**The nut has a detent, so never turn it. Always turn the bolt.**

To install:

21. Install the power steering gear. Tighten the retaining bolts to 68 ft. lbs.

➡**The nut has a detent, so never turn it. Always turn the bolt.**

22. Continue the installation in the reverse order of the removal procedure.

23. Be sure to fill the power steering system with the proper grade and type power steering fluid.

24. Bleed the system, as required.

25. Start the engine and check for leaks, correct as required.

26. Check and adjust the alignment, as required.

FRONT SUSPENSION

Shock Absorbers

REMOVAL & INSTALLATION

2002–2004

2WD

1. Before servicing the vehicle, refer to the Precautions Section.

2. Remove or disconnect the following:

- Front wheel
- Shock absorber from the lower control arm
- Nut, retainers and the cushion from the top of the shock absorber
- Shock absorber
- Retainers and cushion from the shock absorber

To install:

3. Install or connect the following:

- Retainers and cushion to the shock absorber
- Shock absorber
- Retainers, cushion and nut to the top of the shock absorber. Tighten the nut to 18 ft. lbs. (25 Nm).
- Lower shock absorber-to-lower control arm. Tighten the bolts to 29 ft. lbs. (39 Nm)
- Wheels

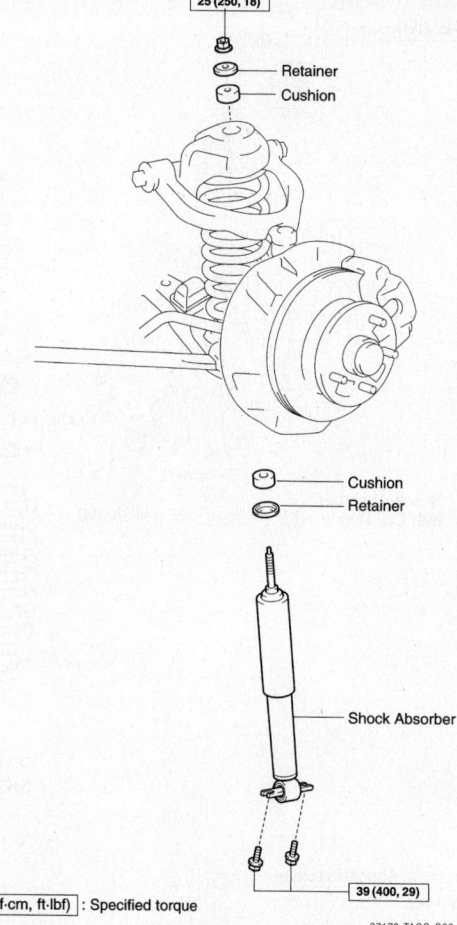

25 (250, 18)

Retainer
Cushion

Cushion
Retainer

Shock Absorber

N·m (kgf·cm, ft·lbf) : Specified torque

39 (400, 29)

67170-TACO-G30

Front shock absorber and related components—2WD 2002–2004

4WD

1. Before servicing the vehicle, refer to the Precautions Section.
2. Remove the wheel.
3. Remove the front wheel.
4. Remove the shock absorber nut and washer from the lower control arm.

❊❊ WARNING

Do not remove the bolt at this time.

5. While slowly lowering the front suspension, remove the lower bolt from the shock absorber assembly.
6. Support the shock absorber.
7. Remove the three nuts from the top of the shock absorber tower.
8. Lower the shock absorber out of the wheel well.

To install:

9. Installation is the reverse of the removal procedure. Please note the following torque specifications:

- 3 upper strut to body nuts: 47 ft. lbs. (64 Nm)
- Lower strut absorber mounting bolt: 101 ft. lbs. (135 Nm)

2005–2006

2WD

1. Before servicing the vehicle, refer to the Precautions Section.
2. Disconnect the negative battery cable.
3. Raise and support the vehicle safely. Remove the tire and wheel assembly.
4. Remove the speed sensor connector. Remove the two bolts. Separate the skid control sensor wire.
5. Remove the bolt and separate the front flexible hose.
6. Remove the upper control arm.
7. Remove the shock absorber lower bolt and nut. Remove the three nuts on the upper side of the shock absorber.
8. Remove the shock absorber and coil spring from the vehicle.

To install:

➡**Always fully tighten rubber bushings when the wheels are in full contact with the ground and the vehicle is at curb height. Bounce the vehicle up and down several times to stabilize the suspension, prior to final tightening of these components.**

9. Position the shock absorber to its mounting on the vehicle.
10. On the left side, install the coil spring onto the body with the lower end of the coil spring facing the outer side of the vehicle.
11. On the right side, install the coil spring onto the body with the lower end of the coil spring facing the inner side of the vehicle.
12. Install the upper retaining nuts. Torque to 47 ft. lbs.
13. Temporarily tighten the lower shock retaining bolt.
14. Continue the installation in the reverse order of the removal procedure.
15. Final tightening torque for the lower shock absorber mounting bolt is 61 ft. lbs.
16. Check and adjust the alignment, as required.

4WD AND PRERUNNER

1. Before servicing the vehicle, refer to the Precautions Section.
2. Disconnect the negative battery cable.
3. Raise and support the vehicle safely. Remove the tire and wheel assembly.
4. Remove the engine undercover assembly.
5. Remove the stabilizer bar.
6. Remove the cotter pin and nut. Using the proper tool, separate the tie rod end from the steering knuckle arm.
7. Remove the shock absorber lower bolt and nut. Remove the three nuts on the upper side of the shock absorber.
8. Remove the shock absorber and coil spring from the vehicle.

To install:

➡**Always fully tighten rubber bushings when the wheels are in full contact with the ground and the vehicle is at curb height. Bounce the vehicle up and down several times to stabilize the suspension, prior to final tightening of these components.**

9. Position the shock absorber to its mounting on the vehicle.
10. On the left side, install the coil spring onto the body with the lower end of the coil spring facing the outer side of the vehicle.
11. On the right side, install the coil spring onto the body with the lower end of the coil spring facing the inner side of the vehicle.
12. Install the upper retaining nuts. Torque to 47 ft. lbs.
13. Temporarily tighten the lower shock retaining bolt.

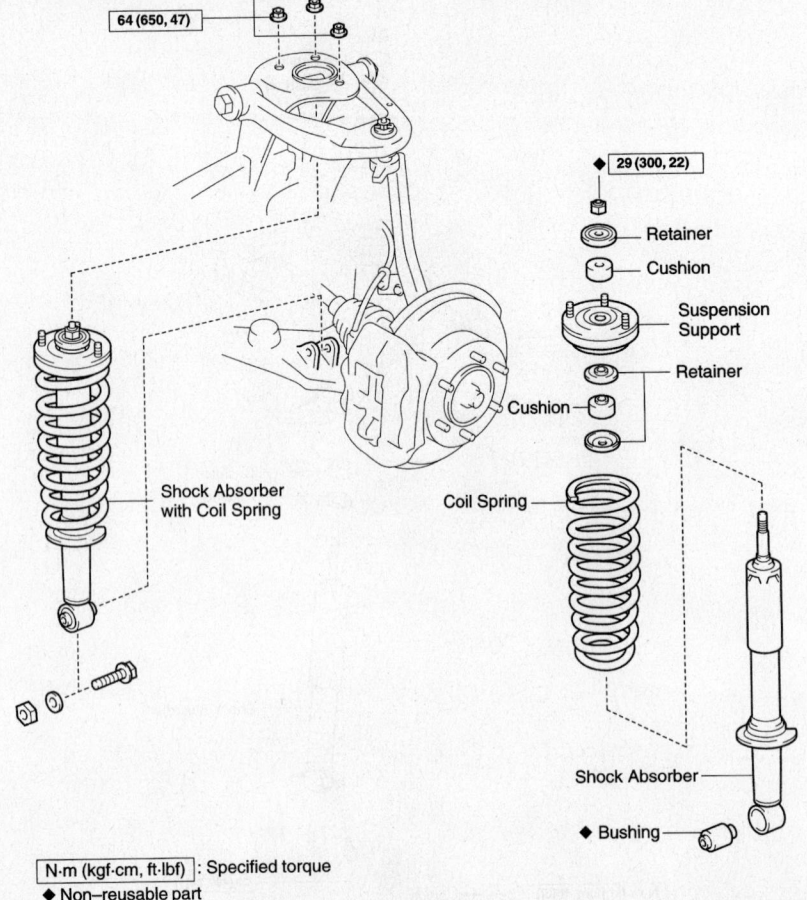

64 (650, 47)

29 (300, 22)

Retainer

Cushion

Suspension Support

Retainer

Cushion

Shock Absorber with Coil Spring

Coil Spring

Shock Absorber

◆ Bushing

N·m (kgf·cm, ft·lbf) : Specified torque
◆ Non–reusable part

67170-TACO-G31A

Front shock absorber and related components—4WD

14. Continue the installation in the reverse order of the removal procedure.

15. Final tightening torque for the lower shock absorber mounting bolt is 61 ft. lbs.

16. Check and adjust the alignment, as required.

ASSEMBLY & DISASSEMBLY

1. Before servicing the vehicle, refer to the Precautions Section.

2. Remove the shock absorber from the vehicle.

3. Position the shock absorber in a suitable holding fixture. Using tool 09727-30021, compress the coil spring.

➡**Do not use an impact wrench. It will damage the tool.**

4. While holding the shock absorber rod, remove the nut.

➡**Do not use an impact wrench. It will damage the shock absorber rod.**

5. Remove the cushion retainer. Remove the number one cushion. Remove the cushion retainer.

6. Remove the coil spring upper insulator. Remove the cushion retainer.

7. Remove the coil spring.

8. Inspect all components for wear and damage. Replace as required. Use new locknuts, as required.

9. Assembly is the reverse of the disassembly procedure.

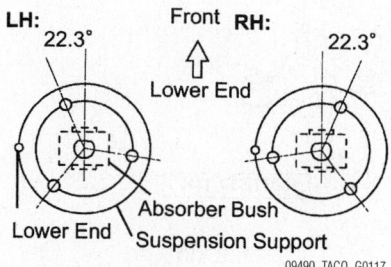

Front shock absorber coil spring alignment—2WD 2005–2006

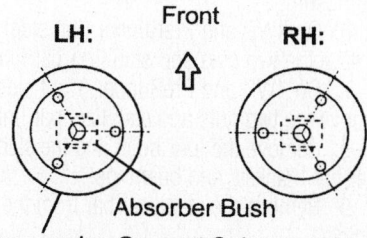

Front shock absorber coil spring alignment—4WD and PreRunner 2005–2006

10. Align the suspension support and the absorber brush, as shown in the illustration.

Coil Spring

REMOVAL & INSTALLATION

2002–2004

2WD

1. Before servicing the vehicle, refer to the Precautions Section.

2. Remove or disconnect the following:
- Shock absorber from the suspension, by removing the 2 bottom bolts and top nut
- Compress the coil spring, using a Spring Compressor
- Nut and sway bar link from the lower control arm
- 2 sway bar bracket bolts on the side of the suspension that the lower control arm is being removed.

➡**This will allow access to the lower control arm through-bolt.**

3. Support the steering knuckle and upper control arm.

4. Remove or disconnect the following:
- Cotter pin and nut from the lower ball joint
- Lower ball joint from the lower control arm, using SST 09628-62011
- Nut from the lower control arm set bolt
- Nut from the strut bar front set bolt
- Lower control arm and strut bar as an assembly, pulling out the 2 bolts

➡**When the lower control arm is removed, set the coil spring aside.**

To install:

5. Install or connect the following:
- Place the end of the coil spring in contact with the lower control arm seat
- Lower control arm, spring, and strut arm to the suspension.
- Strut arm bolt and lower control arm bolt
- Nuts for the strut arm bolt and lower control arm bolt; do not tighten the bolts at this time
- Lower control arm to the lower ball joint. Tighten the nut to 80 ft. lbs. (110 Nm).
- New cotter pin

6. Remove the support from the upper control arm and steering knuckle.

7. Install or connect the following:
- Sway bar bracket to the suspension. Tighten the bolts to 22 ft. lbs. (29 Nm).
- Sway bar link to the lower control arm. Tighten the nut to 22 ft. lbs. (29 Nm).

8. Making sure the coil spring is in its correct position, slowly remove the spring compressor from the coil.

9. Install or connect the following:
- Shock absorber. Tighten the top nut to 18 ft. lbs. (25 Nm) and the bottom 2 bolts to 29 ft. lbs. (39 Nm).
- Wheel
- Stabilize the suspension by pushing up and down on the vehicle
- Tighten the strut bar nut/bolt to 221 ft. lbs. (300 Nm) and the lower control arm bolt/nut to 148 ft. lbs. (200 Nm).

10. Check the front wheel alignment.

4WD

1. Before servicing the vehicle, refer to the Precautions Section.

2. Remove or disconnect the following:
- Strut to the lower control arm nut/bolt.
- 3 strut-to-strut tower nuts/bolts
- Strut

3. Compress the coil spring until there is a clearance on both ends, using SST 09727-30030

4. Remove or disconnect the following:
- Strut center nut
- Suspension support and coil spring
- Insulator from the suspension support

To install:

5. Install or connect the following:
- Insulator to the suspension support

➡**Match the bolt of the suspension support with the cut out part of the insulator.**

- Coil spring to the strut, by compressing it with a coil spring compressor

➡**Fit the lower end of the coil spring into the gap of the spring seat of the strut.**

- Suspension support to the strut rod
- Temporarily tighten a new suspension support center nut

6. Position the suspension support so that a line drawn between the 2 bolts would

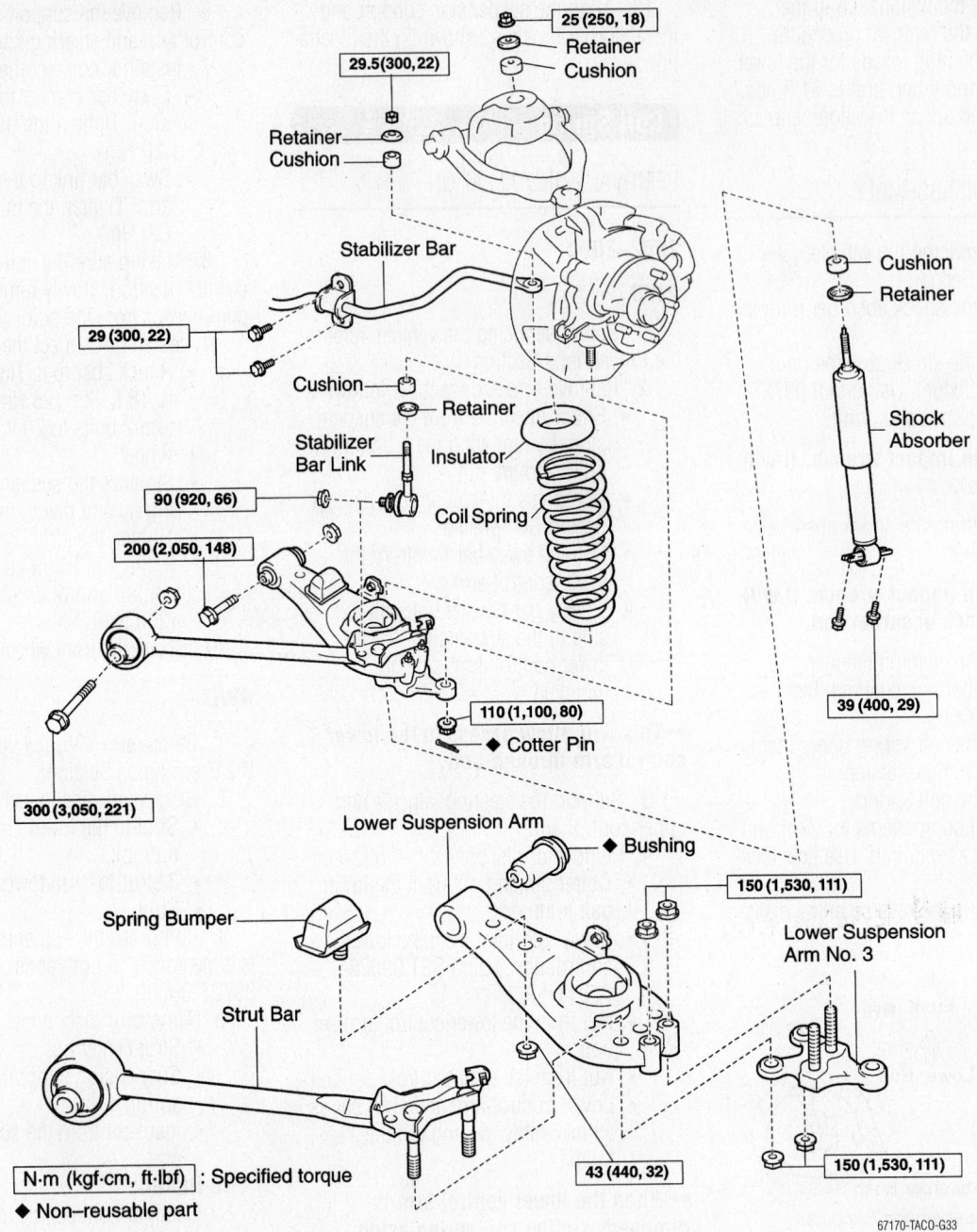

25 (250, 18)
Retainer
Cushion

29.5 (300, 22)

Retainer
Cushion

Stabilizer Bar

29 (300, 22)

Cushion
Retainer

Stabilizer
Bar Link

90 (920, 66)

200 (2,050, 148)

Insulator

Coil Spring

300 (3,050, 221)

110 (1,100, 80)
◆ Cotter Pin

Cushion
Retainer

Shock
Absorber

39 (400, 29)

Lower Suspension Arm

◆ Bushing

150 (1,530, 111)

Spring Bumper

Lower Suspension
Arm No. 3

Strut Bar

N·m (kgf·cm, ft·lbf) : Specified torque

◆ Non−reusable part

43 (440, 32)

150 (1,530, 111)

67170-TACO-G33

Coil spring and related components—2WD 2002–2004

be parallel to the direction of the lower bushing.

7. Remove the compressor from the spring.

8. Install or connect the following:
- Tighten the strut center nut to 22 ft. lbs. (29 Nm).
- Strut
- Strut-to-strut tower. Tighten the 3 nuts to 47 ft. lbs. (64 Nm).
- Strut-to-lower control arm. Tighten the nut/bolt to 101 ft. lbs. (135 Nm).
- Front wheels

Stabilizer Bar

REMOVAL & INSTALLATION

2002–2004

1. Before servicing the vehicle, refer to the Precautions Section.

2. Raise and support the vehicle safely.

3. Remove the tire and wheel assemblies.

4. On 2WD (left side), remove the two nuts, stabilizer bar link, two retainers and two bushings.

5. On 2WD (right side), remove the two nuts, stabilizer bar link, two retainers and two bushings.

6. On 4WD and PreRunner (left side), remove the two nuts and stabilizer bar link.

7. On 4WD and PreRunner (right side), remove the two nuts and stabilizer bar link.

8. Remove the four bolts and the stabilizer bar brackets and bushings.

9. Remove the stabilizer bar from the vehicle.

To install:

10. Installation is the reverse of the removal procedure.

11. On 2WD, tighten the stabilizer bar side bushing bolts to 22 ft. lbs. Tighten the lower arm side bushing bolts to 66 ft. lbs. Tighten the bushing bracket bolts to 22 ft. lbs. Be sure to install the cushion to the portion inside the paint line.

12. On 4WD and PreRunner, tighten the bar links to 66 ft. lbs. Tighten the bushing bracket bolts to 19 ft. lbs.

2005–2006

1. Before servicing the vehicle, refer to the Precautions Section.

2. Raise and support the vehicle safely.

3. Remove the tire and wheel assemblies.

4. On 4WD and PreRunner, remove the engine undercover subassembly.

5. On 2WD (left side), remove the two nuts, stabilizer bar link, two retainers and two bushings.

6. On 2WD (right side), remove the two

nuts, stabilizer bar link, two retainers and two bushings.

7. On 4WD and PreRunner (left side), remove the two nuts and stabilizer bar link.

8. On 4WD and PreRunner (right side), remove the two nuts and stabilizer bar link.

9. Remove the four bolts and the stabilizer bar brackets and bushings.

10. Remove the stabilizer bar from the vehicle.

To install:

11. Installation is the reverse of the removal procedure.

12. On 2WD, tighten the stabilizer bar side bushing bolts to 14 ft. lbs. Tighten the lower arm side bushing bolts to 51 ft. lbs. Tighten the bushing bracket bolts to 16 ft. lbs.

13. On 4WD and PreRunner, tighten the bar links to 52 ft. lbs. Tighten the bushing bracket bolts to 30 ft. lbs.

14. Be sure to install the cushion onto the stabilizer bar with its cut line facing the front.

Upper Ball Joint

REMOVAL & INSTALLATION

2002–2004

2WD

1. Before servicing the vehicle, refer to the Precautions Section.

2. Remove the wheels.

3. Support the lower control arm with a floor jack.

4. Remove or disconnect the following:

- Anti-lock Brake System (ABS) speed sensor wire from the upper control arm
- 2 bolts and camber adjusting shims from the upper control arm

➡**Before removing the shims from the upper control arm, make a note of each shim size and position.**

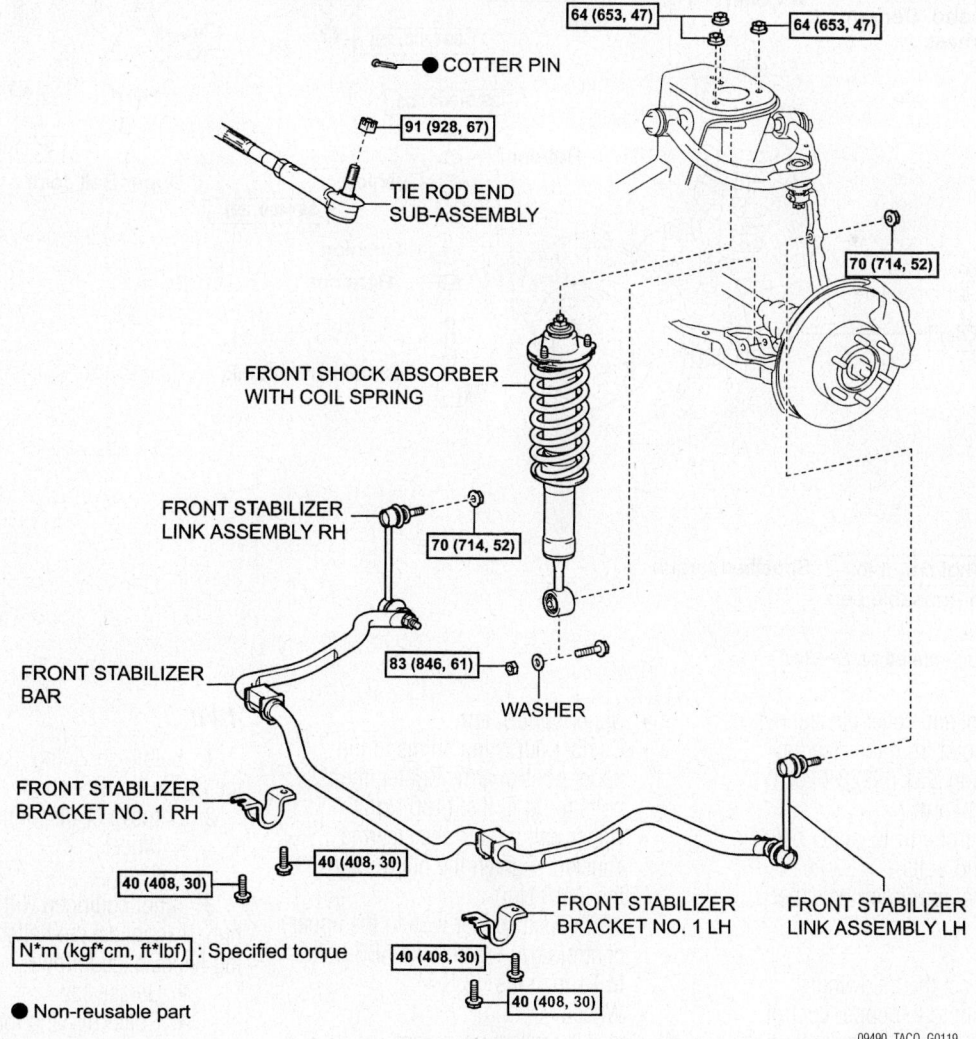

COTTER PIN

64 (653, 47)   64 (653, 47)

91 (928, 67)

TIE ROD END SUB-ASSEMBLY

70 (714, 52)

FRONT SHOCK ABSORBER WITH COIL SPRING

FRONT STABILIZER LINK ASSEMBLY RH

70 (714, 52)

FRONT STABILIZER BAR

83 (846, 61)   WASHER

FRONT STABILIZER BRACKET NO. 1 RH

40 (408, 30)

40 (408, 30)

FRONT STABILIZER BRACKET NO. 1 LH

FRONT STABILIZER LINK ASSEMBLY LH

N*m (kgf*cm, ft*lbf) : Specified torque

40 (408, 30)

40 (408, 30)

● Non-reusable part

Front stabilizer bar and related components—4WD and PreRunner 2005–2006

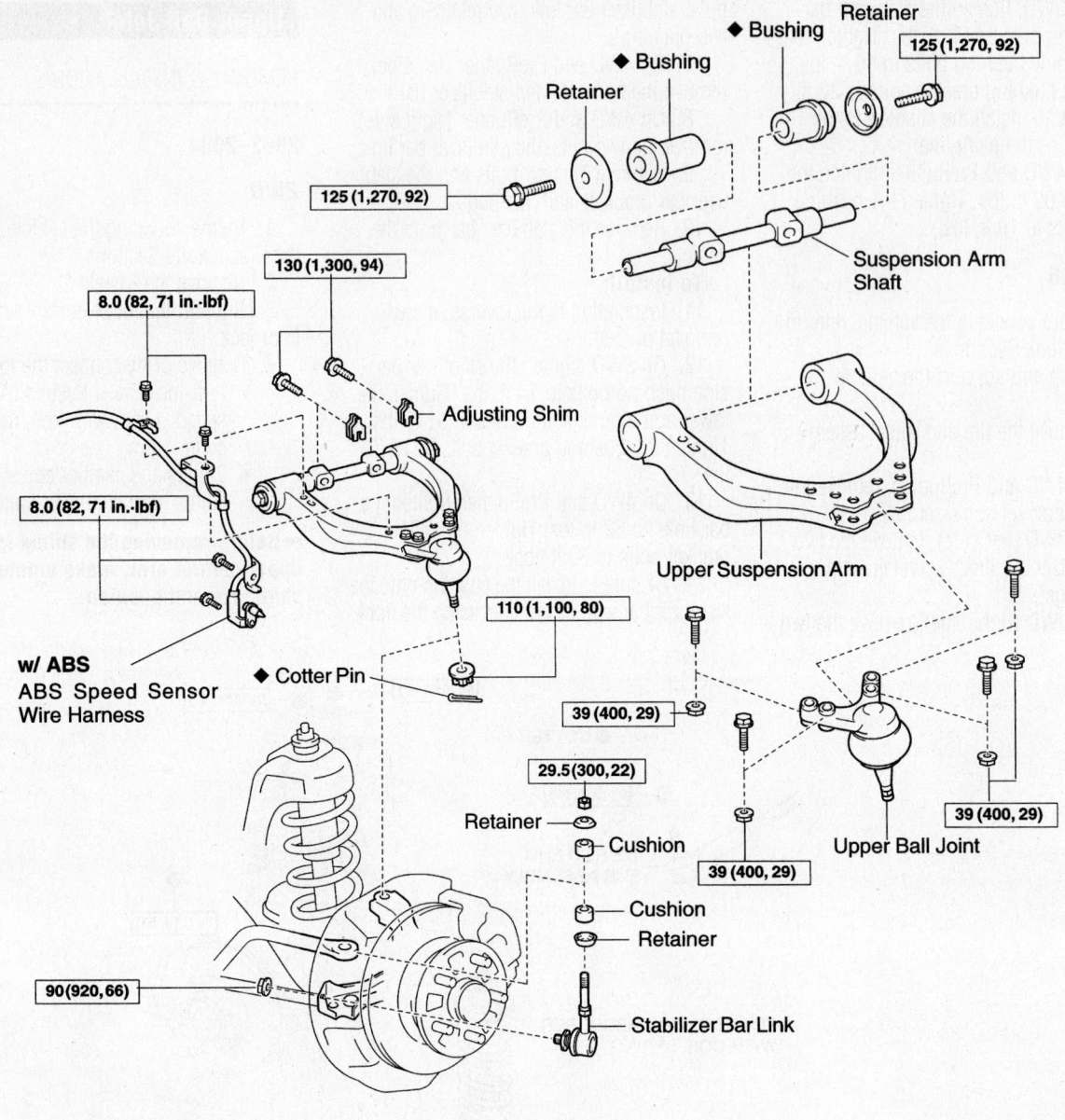

◆ Bushing
Retainer
125 (1,270, 92)

Bushing
Retainer
125 (1,270, 92)

Suspension Arm Shaft

130 (1,300, 94)

8.0 (82, 71 in.·lbf)

Adjusting Shim

8.0 (82, 71 in.·lbf)

Upper Suspension Arm

w/ ABS
ABS Speed Sensor Wire Harness

110 (1,100, 80)

◆ Cotter Pin

39 (400, 29)

29.5(300, 22)

Retainer
Cushion

39 (400, 29)

Upper Ball Joint

39 (400, 29)

Cushion
Retainer

90(920,66)

Stabilizer Bar Link

N·m (kgf·cm, ft·lbf) : Specified torque
◆ Non–reusable part

67170-TACO-G35

Upper control arm and related parts—2wd

- Upper control arm cotter pin and nut.
- Upper ball joint from the steering knuckle, using SST 09628-62011
- Upper control arm
- 4 upper control arm-to-upper ball joint nuts and bolts.
- Upper control arm from the upper ball joint

To install:

5. Install or connect the following:
   - New ball joint to the upper control arm. Tighten the 4 nuts/bolts to 29 ft. lbs. (39 Nm).

- Upper control arm
- Camber adjusting shims to the upper control arm. Tighten the 2 bolts to 94 ft. lbs. (130 Nm).
- Upper ball joint to the steering knuckle. Tighten the nut to 80 ft. lbs. (110 Nm).
- ABS speed sensor wire to the upper control arm. Tighten the ABS bolt to 71 inch lbs. (8 Nm).
- Wheels

6. Check the wheel alignment.

4WD

1. Before servicing the vehicle, refer to the Precautions Section.

2. Remove or disconnect the following:
   - Wheel
   - Strut

3. If not equipped with a FREE wheeling hub, disconnect the halfshaft from the steering knuckle, as follows:
   - Grease cap
   - Cotter pin and lockcap from the halfshaft

4WD

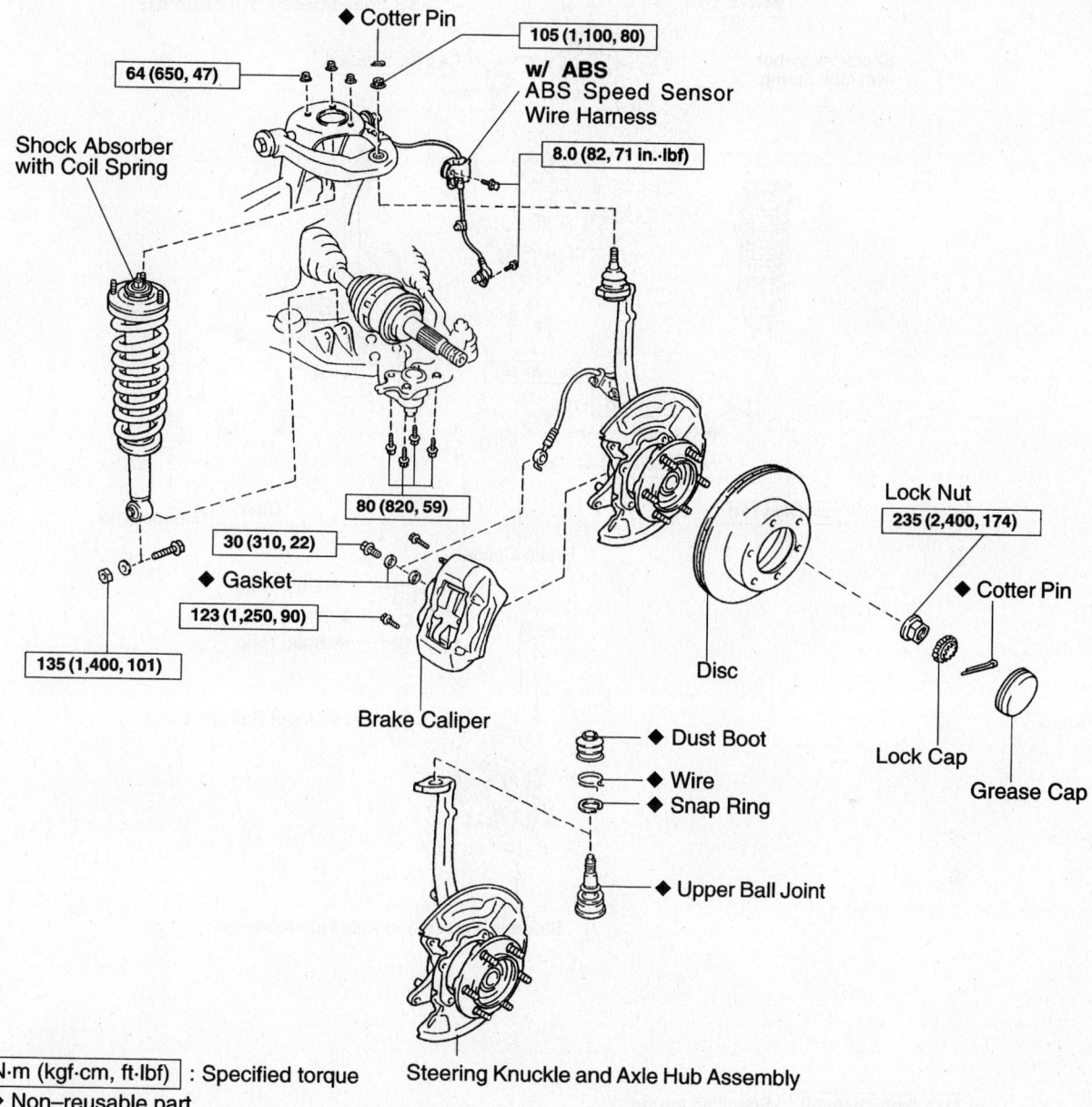

◆ Cotter Pin

105 (1,100, 80)

64 (650, 47)

w/ ABS
ABS Speed Sensor
Wire Harness

Shock Absorber
with Coil Spring

8.0 (82, 71 in.·lbf)

80 (820, 59)

Lock Nut
235 (2,400, 174)

30 (310, 22)

◆ Gasket

◆ Cotter Pin

123 (1,250, 90)

135 (1,400, 101)

Disc

Lock Cap

Grease Cap

Brake Caliper

◆ Dust Boot

◆ Wire

◆ Snap Ring

◆ Upper Ball Joint

N·m (kgf·cm, ft·lbf) : Specified torque

Steering Knuckle and Axle Hub Assembly

◆ Non–reusable part

67170-TACO-G36

Upper ball joint and related components—4WD 2002–2004 except PreRunner

- Locknut from the halfshaft, while having an assistant apply the brakes
4. If equipped with free wheeling hub, remove the free wheel hub, as follows:
  - Set the control handle to FREE
  - Cover bolts and pull off the cover
  - Center bolt with washer
  - Hub body nuts and washer
  - Cone washer, using a brass bar and hammer to tap on the bolt heads
  - Free wheel hub body and gasket
  - Snaping and spacer from the half-

shaft end, using a snapring expander
- Anti-lock Brake System (ABS) speed sensor from the steering knuckle, if equipped with ABS
- Brake hose from the steering knuckle
- Brake caliper support bracket and support it on a wire

✳✳ WARNING

Do not allow the caliper to hang from the brake hose.

5. Remove or disconnect the following:
  - Rotor
  - Lower ball joint from the steering knuckle
  - Upper control arm cotter pin and nut
  - Steering knuckle from the upper control arm, using SST 09950-40010
  - Steering knuckle from the vehicle

➡ **If it is difficult to remove the half-shaft from the steering knuckle, use a rubber hammer to tap the halfshaft from the steering knuckle.**

Pre runner

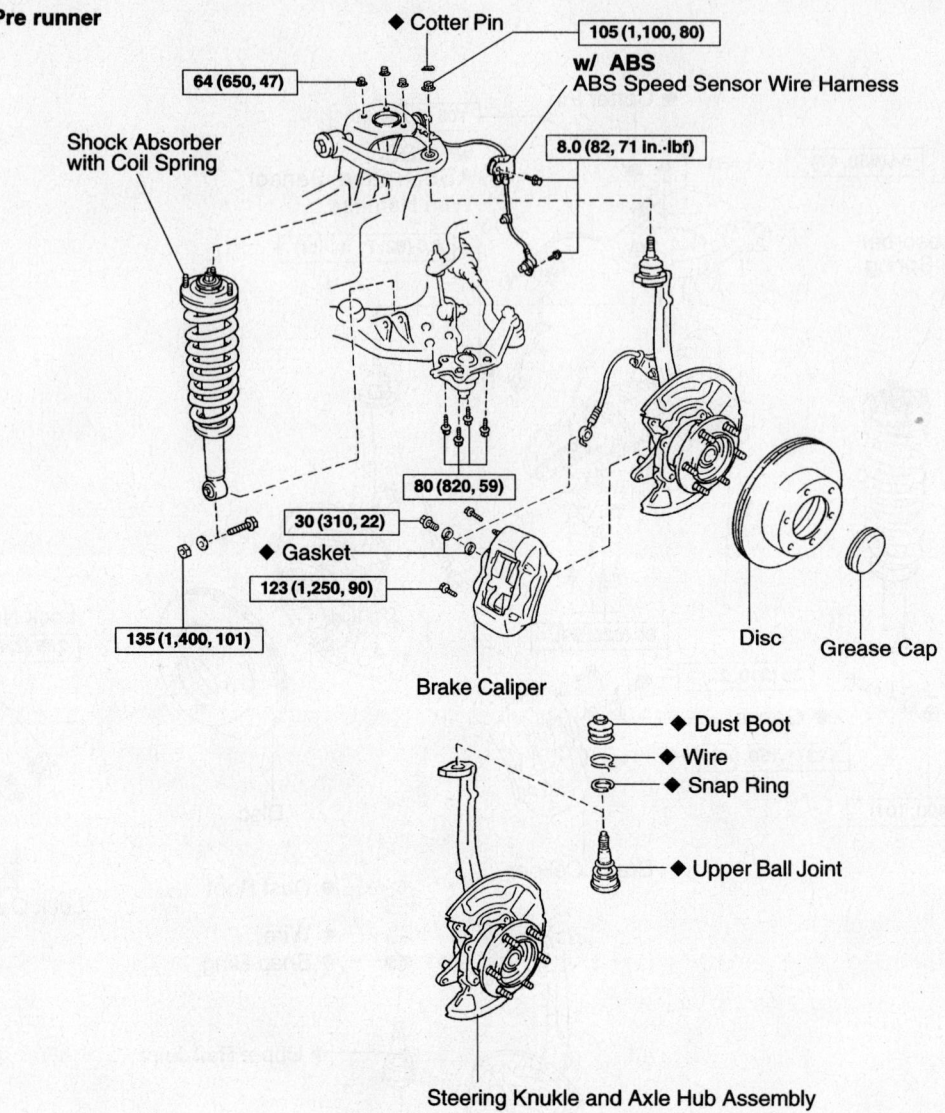

◆ Cotter Pin
105 (1,100, 80)
w/ ABS
ABS Speed Sensor Wire Harness

64 (650, 47)

8.0 (82, 71 in.·lbf)

Shock Absorber
with Coil Spring

80 (820, 59)

30 (310, 22)

◆ Gasket

123 (1,250, 90)

135 (1,400, 101)

Brake Caliper

Disc   Grease Cap

◆ Dust Boot
◆ Wire
◆ Snap Ring

◆ Upper Ball Joint

Steering Knukle and Axle Hub Assembly

N·m (kgf·cm, ft·lbf) : Specified torque
◆ Non–reusable part

67170-TACO-G37

Upper ball joint and related components—2002–2004 PreRunner

- Wire and boot from the upper ball joint
- Snapring from the ball joint, using a snapring expander
- Upper ball joint from the steering knuckle, using SST 09950-40010 (puller set) and a deep socket wrench

To install:

6. Install or connect the following:
- Press in a new upper ball joint, using SST 09309-37010 and a socket wrench

- New snapring, using a snapring expander
- New boot, secured with a new piece of wire
- Steering knuckle to the halfshaft
- Steering knuckle to the lower ball joint by installing the 4 bolts; do not tighten the bolts at this time.
- Upper control arm
- Upper ball joint to the arm. Tighten the nut to 80 ft. lbs. (105 Nm).
- New cotter pin

- Tighten the lower ball joint-to-steering knuckle bolts to 59 ft. lbs. (80 Nm).
- Brake rotor
- Caliper support bracket to the steering knuckle. Tighten both bolts to 90 ft. lbs. (123 Nm).
- Brake hose clamp to the steering knuckle. Tighten the bolt to 13 ft. lbs. (18 Nm).
- ABS speed sensor and wiring harness to the steering knuckle, if equipped with ABS

- Spacer and snapring to the half-shaft, using a snapring expander

7. If equipped with a free wheeling hub, install the hub, as follows:
- New front axle hub gasket
- Free wheeling hub body with the 6 cone washers and nuts. Tighten the 6 nuts to 23 ft. lbs. (31 Nm).
- Bolt with the washer. Tighten the bolt to 13 ft. lbs. (18 Nm).
- Apply multi-purpose grease to the inner hub splines
- Set the control handle and clutch to the FREE position
- New gasket on the cover
- Cover to the hub body with the follower pawl tabs aligned with the non-toothed portions of the hub body
- Tighten the cover bolts to 84 inch lbs. (10 Nm).

8. If equipped without a free wheeling hub, install the halfshaft to the steering knuckle, as follows:
- Locknut to the halfshaft. Tighten the nut to 174 ft. lbs. (235 Nm).
- Halfshaft lockcap and cotter pin
- Grease cab
- Strut. Tighten the strut-to-lower control arm nut to 101 ft. lbs. (135 Nm) and the upper 3 nuts to 47 ft. lbs. (64 Nm).
- Front wheels

9. Check the wheel alignment.

2005–2006

➡️**The upper ball joint is removed with the upper control arm.**

Lower Ball Joint

REMOVAL & INSTALLATION

2002–2004

2WD

1. Before servicing the vehicle, refer to the Precautions Section.
2. Remove the wheel.
3. Support the lower control with a floor jack.
4. Remove or disconnect the following:
- Loosen the 2 lower ball joint set bolts
- Cotter pin and nut from the tie rod
- Tie rod from the ball joint bracket
- Cotter pin and nut from the lower ball joint
- Lower ball joint from the lower control arm

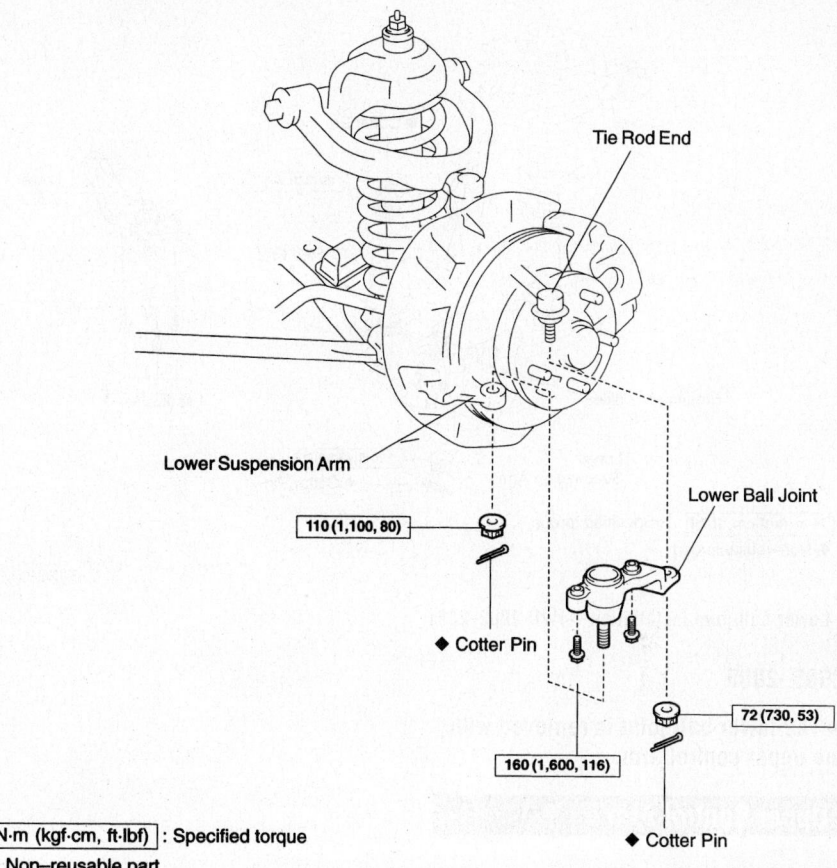

Tie Rod End
Lower Suspension Arm
Lower Ball Joint
110 (1,100, 80)
◆ Cotter Pin
72 (730, 53)
160 (1,600, 116)
◆ Cotter Pin

N·m (kgf·cm, ft·lbf) : Specified torque
◆ Non–reusable part

67160-TACO-G38

Lower ball joint installation—2WD 2002–2004

- Both lower ball joint set bolts
- Ball joint from the suspension

To install:
5. Install or connect the following:
- Lower ball joint to the steering knuckle and lower control arm
- Both lower ball joint set bolts; do not tighten the bolts at this time
- Lower ball joint nut to hold the lower ball joint to the lower control arm. Tighten the nut to 80 ft. lbs. (110 Nm).
- New cotter pin to the lower ball joint
- Tie rod end to the ball joint bracket. Tighten the nut to 53 ft. lbs. (72 Nm).
- New cotter pin to the tie rod end
- Tighten the both lower ball joint set bolts to 116 ft. lbs. (160 Nm).
- Wheel

6. Check the wheel alignment.

4WD

1. Before servicing the vehicle, refer to the Precautions Section.
2. Remove or disconnect the following:
- Wheel
- Loosen the 4 lower ball joint set bolts

- Cotter pin and nut from the tie rod
- Tie rod from the ball joint bracket
- Cotter pin and nut from the lower ball joint
- Lower ball joint from the lower control arm
- 4 lower ball joint set bolts
- Ball joint from the suspension

To install:
3. Install or connect the following:
- Lower ball joint to the steering knuckle and lower control arm
- 4 lower ball joint set bolts; do not tighten the bolts at this time
- Lower ball joint-to-lower control arm nut. Tighten the nut to 103 ft. lbs. (140 Nm).
- New cotter pin to the lower ball joint
- Tie rod end to the ball joint bracket. Tighten the nut to 67 ft. lbs. (90 Nm).
- New cotter pin to the tie rod end
- Tighten the lower ball joint set bolts to 59 ft. lbs. (80 Nm).
- Wheel

4. Check the wheel alignment.

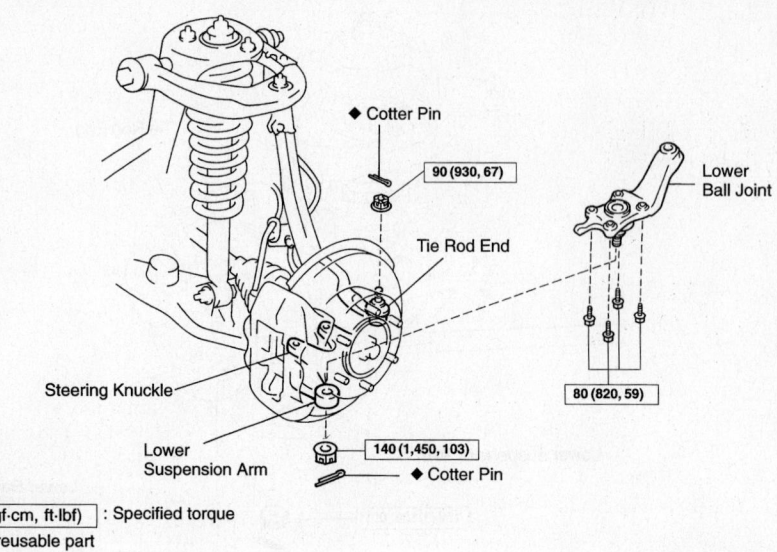

Lower ball joint installation—4WD 2002–2004

◆ Cotter Pin
90 (930, 67)
Tie Rod End
Lower Ball Joint
80 (820, 59)
Steering Knuckle
Lower Suspension Arm
140 (1,450, 103)
◆ Cotter Pin

N·m (kgf·cm, ft·lbf) : Specified torque
◆ Non–reusable part

67160-TACO-G39

2005–2006

➡ **The lower ball joint is removed with the upper control arm.**

Upper Control Arm

REMOVAL & INSTALLATION

2002–2004

2WD

1. Before servicing the vehicle, refer to the Precautions Section.
2. Remove or disconnect the following:
   - Front wheel
   - Anti-lock Brake System (ABS) speed sensor and wire harness
   - Stabilizer bar link
   - Steering knuckle from the upper bar joint
3. Loosen the 2 bolts; then, remove the front and rear alignment adjusting shims.
4. Make note of the number and thickness of the front and rear shims.
5. Remove or disconnect the following:
   - Upper control arm
   - Upper ball joint from the arm

To install:

6. Install or connect the following:
   - Upper ball joint to the arm. Tighten the fasteners to 29 ft. lbs. (39 Nm).

➡ **Do not lose the camber adjusting shims. Record the position and thickness of the camber shims so that these can be reinstalled to there original locations. Install the equal number and thickness of shims into there locations.**

- Upper control arm with the shims. Tighten the mounting bolts to 94 ft. lbs. (130 Nm).
- Steering knuckle to the upper ball joint
- Stabilizer bar link
- ABS speed sensor and wire harness
- Front wheel. Tighten the lug nuts to 83 ft. lbs. (110 Nm).

4WD AND PRERUNNER

1. Before servicing the vehicle, refer to the Precautions Section.
2. Remove or disconnect the following:
   - Front wheel
   - Shock and coil spring assembly
   - Anti-lock Brake System (ABS) speed sensor wire harness clamp
3. Upper ball joint, as follows:
   - Cotter pins and loosen the nut
   - Upper ball joint from the control arm, using a ball joint separator

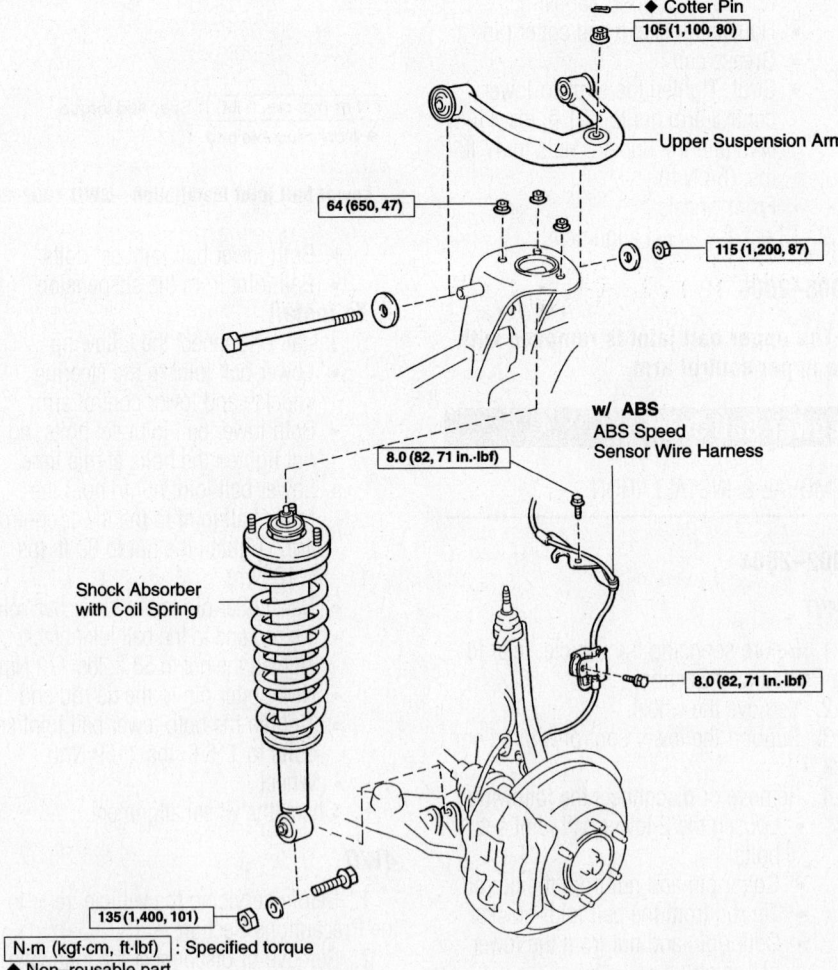

◆ Cotter Pin
105 (1,100, 80)
Upper Suspension Arm
64 (650, 47)
115 (1,200, 87)
w/ ABS
ABS Speed Sensor Wire Harness
8.0 (82, 71 in.-lbf)
Shock Absorber with Coil Spring
8.0 (82, 71 in.-lbf)
135 (1,400, 101)

N·m (kgf·cm, ft·lbf) : Specified torque
◆ Non–reusable part

67170-TACO-G40

Upper control arm and related components—4WD and PreRunner 2002–2004

- Support the steering knuckle
- Nut

4. Detach the control arm, by removing the nut, bolt, washers and lowering the arm.

To install:

5. Install or connect the following:
- Upper control arm with the washer, bolt and nut. Tighten the nut to 87 ft. lbs. (115 Nm).
- Upper ball joint to the control arm. Tighten the mounting nut to 80 ft. lbs. (105 Nm).
- New cotter pin
- ABS speed sensor wire harness clamp. Tighten it to 71 inch lbs. (8 Nm).
- Shock and coil spring assembly

6. Check and/or adjust the alignment.

2005–2006

1. Before servicing the vehicle, refer to the Precautions Section.

2. Disconnect the negative battery cable.

3. Raise and support the vehicle safely. Remove the tire and wheel assemblies.

4. Remove the speed sensor connector. Remove the two bolts and separate the skid control sensor wire.

5. On 2WD vehicles, remove the bolt and separate the front flexible hose.

6. Support the front suspension lower control arm using a jack.

7. Remove the clip and nut. Using the tool SST09628-62011, or equivalent separate the upper ball joint from the steering knuckle.

8. On 4WD and PreRunner, remove the bolt and bracket.

9. Remove the two upper control arm retaining bolts.

10. Remove the upper control arm from the vehicle.

To install:

➡**Always fully tighten rubber bushings when the wheels are in full contact with the ground and the vehicle is at curb height. Bounce the vehicle up and down several times to stabilize the suspension, prior to final tightening of these components.**

11. Position the upper control arm on its mounting in the vehicle.

12. Temporarily tighten the upper control arm mounting bolts.

13. Continue the installation in the reverse order of the removal procedure.

14. Final tightening torque for the upper control arm bushings is 60 ft. lbs. on 2WD vehicles and 85 ft. lbs. on 4WD vehicles and PreRunner.

15. Check and adjust the alignment, as required.

UPPER CONTROL ARM BUSHING REPLACEMENT

2002–2004

2WD

1. Before servicing the vehicle, refer to the Precautions Section.

2. Remove the upper control arm.

3. Cut off the outer edges of the bushing so it is flush with arm tube and the shaft.

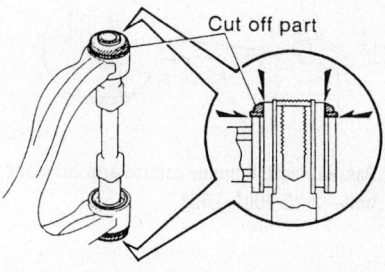

9308YG01

Cut the bushing flush with the upper control arm tube and shaft—2WD 2002–2004

➡**Be careful not to damage the edge of the arm tubes.**

4. Using a shop press with tool 09710-03031, press down the suspension arm tube until it touches tool 09710-03141.

➡**Do not press the tube excessively.**

5. Temporarily install a 1.8–2.0 in. (45–50mm) bolt to the arm shaft on the other side.

6. Using a shop press with tool 09710-03141, remove the bushing from the arm shaft.

7. Repeat this procedure for the other bushing.

To install:

8. Using a shop press and tool 09710-03101, install the new bushing.

9. Place the arm shaft to the bushing.

10. Using a shop press and tool 09710-03101, install the other new bushing.

➡**Pass the arm shaft through the bushing to make sure that the shaft turns freely and there is no axial play**

11. Install the upper control arm.

12. Check and/or adjust the alignment.

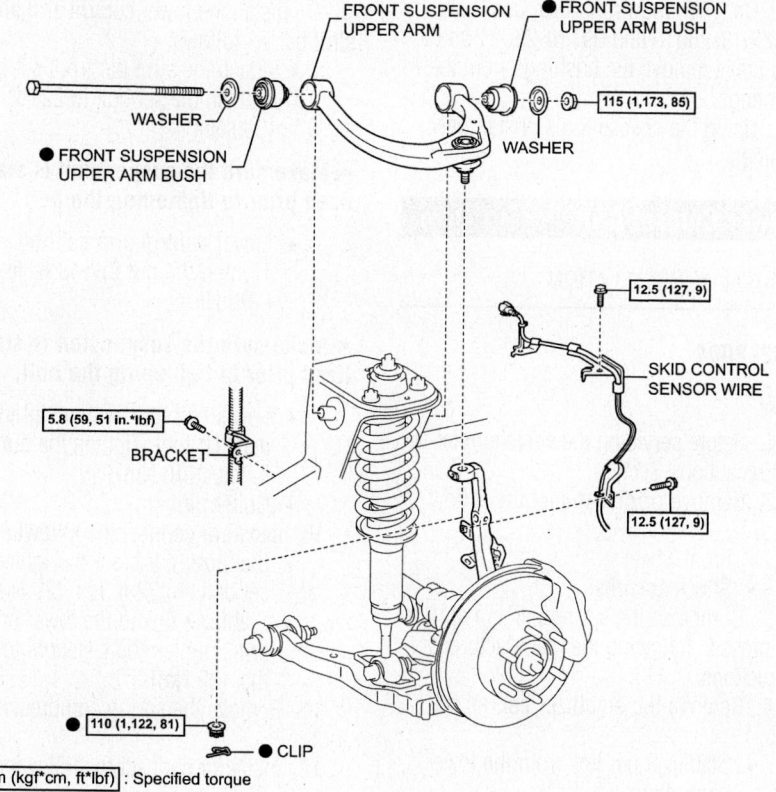

Upper control arm and related components—4WD and PreRunner 2005–2006

09490_TACO_G0120

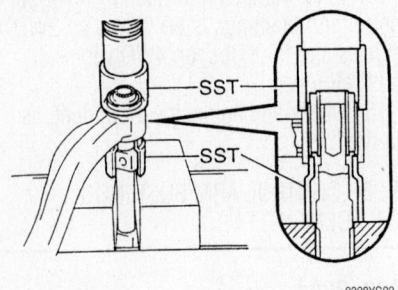

Position the upper control arm bushings with the tools—2WD 2002–2004

9308YG02

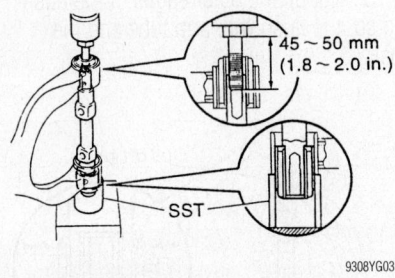

Positioning the upper control arm tube and bolt—2WD 2002–2004

9308YG03

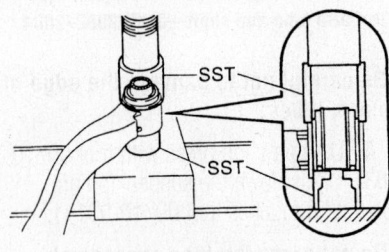

Removing the bushings from the upper control arm—2WD 2002–2004

9308YG04

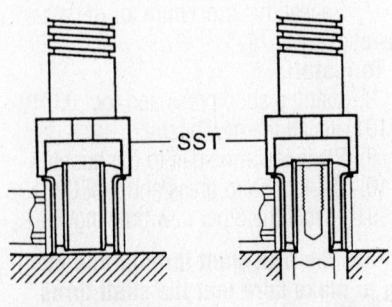

Installing the bushings to the upper control arm—2WD 2002–2004

9308YG05

4WD

1. Before servicing the vehicle, refer to the Precautions Section.

2. Remove the upper control arm from the vehicle.

3. Pry up the bushing flange, using a chisel and a hammer.

4. Using tools 09613-26010, 09613-20060 and 09950-00020 and a shop press, remove the bushing.

To install:

5. Using tools 09223-00010, 09506-35010 and a shop press, press the new bushing into the upper control arm.

6. Install the upper control arm to the chassis.

7. Check and/or adjust the alignment.

2005–2006

1. Before servicing the vehicle, refer to the Precautions Section.

2. Remove the upper control arm from the vehicle.

3. Position the assembly in a suitable holding fixture.

4. Using a hammer and a chisel, raise the flange of the bushing diagonally.

5. On 2WD vehicles, use tool SST09950-40011 remove the bushings from their mountings.

6. On 4WD and PreRunner, use tools SST09613-26010 and 09710-22021 and a shop press remove the bushings from their mountings.

7. Using the special tools, install new bushings.

Lower Control Arm

REMOVAL & INSTALLATION

2002–2004

2WD

1. Before servicing the vehicle, refer to the Precautions Section.

2. Remove or disconnect the following:
- Front wheel
- Shock absorber

3. Compress the spring using a spring compressor, following the manufacturer's instructions.

4. Remove the stabilizer bar, as follows:
- Stabilizer bar link from the lower control arm
- 2 stabilizer bar bracket set bolts

5. Remove the lower control arm and strut bar, as follows:

- Support the upper control arm and steering knuckle assembly
- Cotter pin and nut
- Lower ball joint from the lower control arm
- Loosen the lower control arm set bolt and remove the nut
- Loosen the strut bar front set bolt and remove the nut
- Pull out the bolts and remove the lower control arm along with the strut bar

6. Remove or disconnect the following:
- Coil spring compressor tool and coil
- Strut bar from the lower control arm. Separate the nut and spring bumper.
- Lower suspension arm No. 3

To install:

7. Install or connect the following:
- Lower suspension arm No. 3. Tighten the bolts to 111 ft. lbs. (150 Nm).
- Spring bumper. Tighten it to 32 ft. lbs. (43 Nm) and the strut bar-to-lower control arm to 111 ft. lbs. (150 Nm).

8. Place each end of the coil spring and lower control arm seat in contact when applying the coil spring expander.

9. Install the lower control arm and strut bar, as follows:
- Attach the strut bar front set bolt. Tighten the set bolt to 221 ft. lbs. (300 Nm).

➡**Make sure the suspension is stabilized prior to tightening the bolt.**

- Lower control arm set bolt. Tighten the nut to 148 ft. lbs. (200 Nm).

➡**Make sure the suspension is stabilized prior to tightening the bolt.**

- Lower ball joint with a ball joint installer tool. Tighten the nut to 80 ft. lbs. (110 Nm).
- Cotter pin.

10. Install or connect the following:
- Stabilizer bar bracket. Tighten the set bolts to 22 ft. lbs. (29 Nm).
- Stabilizer link to the lower control arm. Tighten the fasteners to 29 ft. lbs. (39 Nm).

11. Remove the spring compressing tool.

12. Install or connect the following:
- Shock absorber
- Wheel. Tighten the lug nuts.

13. Check and/or adjust the alignment.

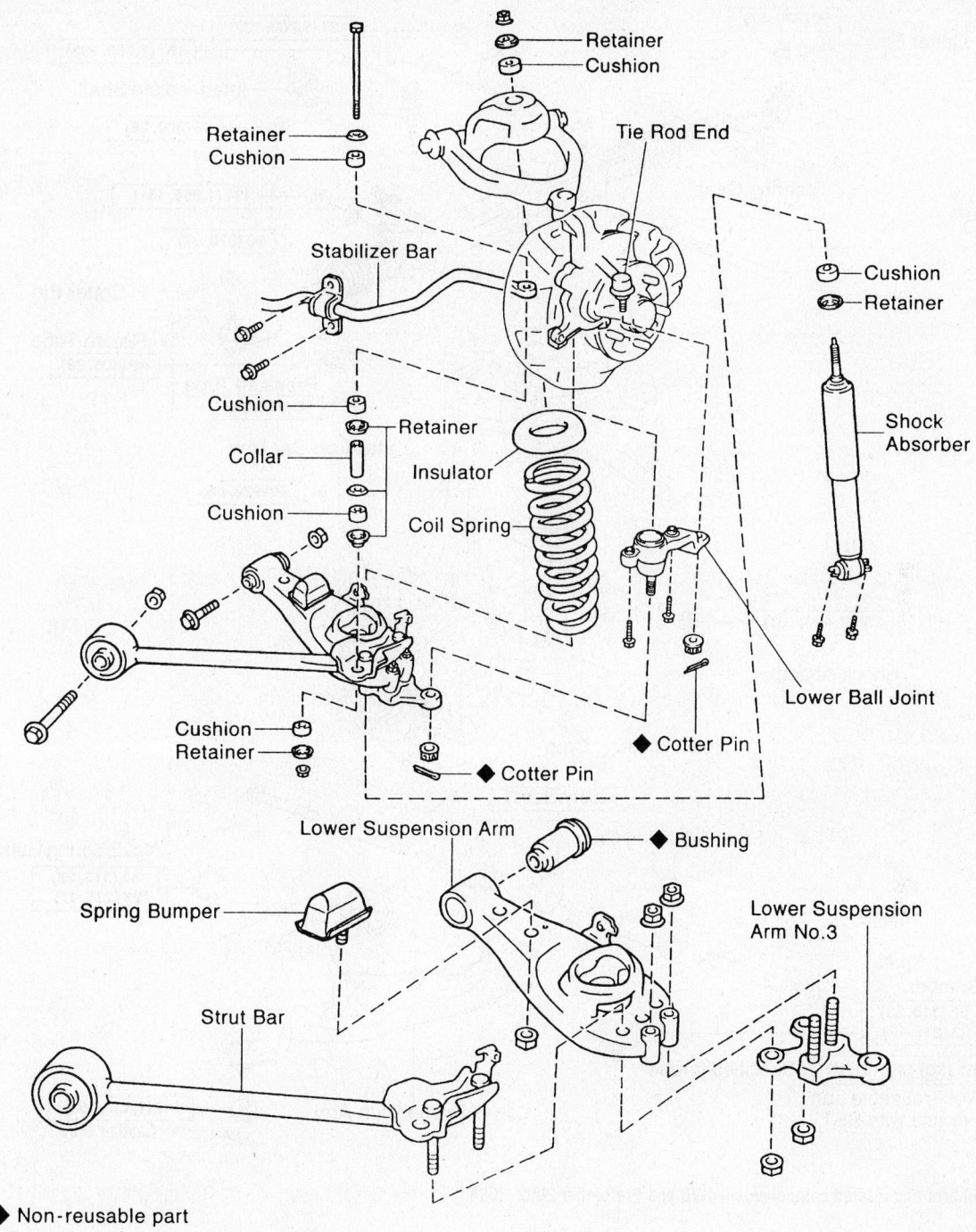

Retainer
Cushion
Tie Rod End
Retainer
Cushion
Stabilizer Bar
Cushion
Retainer
Shock Absorber
Cushion
Retainer
Collar
Insulator
Cushion
Coil Spring
Lower Ball Joint
Cushion
Retainer
◆ Cotter Pin
◆ Cotter Pin
Lower Suspension Arm
◆ Bushing
Spring Bumper
Lower Suspension Arm No.3
Strut Bar

◆ Non-reusable part

86828G01

Lower control arm and related components—2WD 2002–2004

4WD

1. Before servicing the vehicle, refer to the Precautions Section.

2. Remove or disconnect the following:
- Front wheel
- Steering gear assembly
- Stabilizer bar link
- Shock absorber from lower control arm

3. Support the upper control and steering knuckle securely.

4. Remove or disconnect the following:
- Cotter pin and nut from the lower ball joint
- Lower ball joint from the lower control arm

5. Place matchmarks on the front and rear adjusting cams.

6. Remove or disconnect the following:
- 2 bolts, nuts, adjusting cams and lower control arm

- Spring bumpers, with a special tool 09922-10010

To install:

7. Install or connect the following:
- Spring bumpers. Tighten to 17 ft. lbs. (23 Nm).
- Lower control arm, placing it in the appropriate position with the matchmarks. Tighten the arm to 96 ft. lbs. (130 Nm).

8. Install or connect the following:

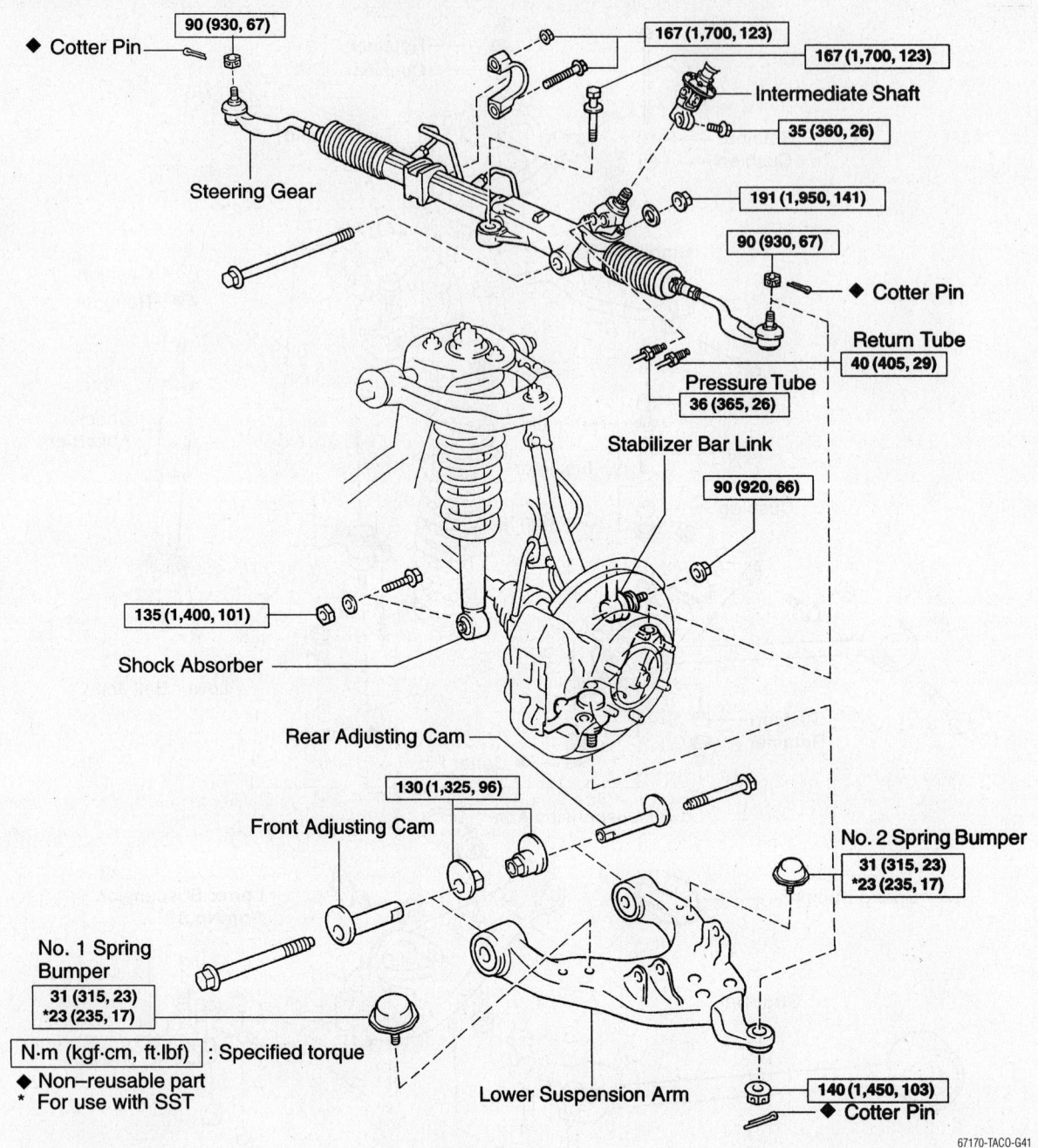

90 (930, 67)

◆ Cotter Pin

167 (1,700, 123)

167 (1,700, 123)

Intermediate Shaft

35 (360, 26)

Steering Gear

191 (1,950, 141)

90 (930, 67)

◆ Cotter Pin

Return Tube

40 (405, 29)

Pressure Tube

36 (365, 26)

Stabilizer Bar Link

90 (920, 66)

135 (1,400, 101)

Shock Absorber

Rear Adjusting Cam

130 (1,325, 96)

Front Adjusting Cam

No. 2 Spring Bumper

31 (315, 23)
*23 (235, 17)

No. 1 Spring Bumper

31 (315, 23)
*23 (235, 17)

N·m (kgf·cm, ft·lbf) : Specified torque

◆ Non–reusable part
* For use with SST

Lower Suspension Arm

140 (1,450, 103)

◆ Cotter Pin

67170-TACO-G41

Lower control arm and related components—4WD and PreRunner 2002–2004

- Lower ball joint. Tighten the nut to 103 ft. lbs. (140 Nm).
- Shock absorber to the lower control arm
- Stabilizer bar link
- Steering gear assembly
- Front wheel. Tighten the lug nuts.
9. Check and/or adjust the alignment.

2005–2006

1. Before servicing the vehicle, refer to the Precautions Section.
2. Raise and support the vehicle safely.

3. Remove the tire and wheel assemblies.
4. To check the lower ball joint, install the hub nuts. Using a dial indicator gauge push the hub nut up and down with a force of 66 ft. lbs. Specification should be 0.020 inch.

➡**If not within specification, replace the lower control arm.**

5. On 2WD vehicles, remove the stabilizer link assembly.
6. Properly support the lower control arm assembly, as required.
7. Remove the bolt, nut and washer.

Separate the front shock absorber with the coil spring from the lower control arm.
8. Remove the two bolts and separate the front lower ball joint attachment from the front axle.
9. On 2WD vehicles, place matchmarks on the camber adjusting cam number two. Remove the two nuts, the two number two camber adjusting cams, the two number one camber adjusting cams.
10. On 4WD and PreRunner, place matchmarks on the camber adjusting cam number two. Remove the nut, the camber adjusting cam number two, the camber adjusting cam

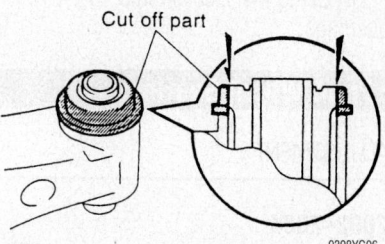

Cutting off part of the bushing—2WD 2002–2004

1. Before servicing the vehicle, refer to the Precautions Section.

2. Remove the lower control arm from the vehicle.

3. Cut off a portion of the bushing to expose the edge of the arm tube.

4. Position the lower control arm on a shop press with the cut side facing downward, resting on tool 09710-30021; then, press the bushing from the arm.

To install:

5. Position a new bushing onto the lower control arm.

6. Position the lower control arm on a shop press, resting on tool 09710-30021.

7. Press the bushing into the lower control arm.

8. Install the lower control arm.

9. Check and/or adjust the alignment.

4WD

1. Before servicing the vehicle, refer to the Precautions Section.

2. Remove the lower control arm from the vehicle.

3. Pry up the bushing flange, using a chisel and a hammer.

4. Using tools 09613-26010, 09632-36010 and 09950-00020 and a shop press, remove the bushing.

To install:

5. Using tools 09316-20011, 09710-30021 and a shop press, press the new bushing into the lower control arm.

6. Install the lower control arm to the chassis.

7. Check and/or adjust the alignment.

2005–2006

1. Before servicing the vehicle, refer to the Precautions Section.

2. Remove the upper control arm from the vehicle.

3. Position the assembly in a suitable holding fixture.

4. Using a hammer and a chisel, raise the flange of the bushing diagonally.

5. Using tool SST09950-40011 remove the bushings from their mountings.

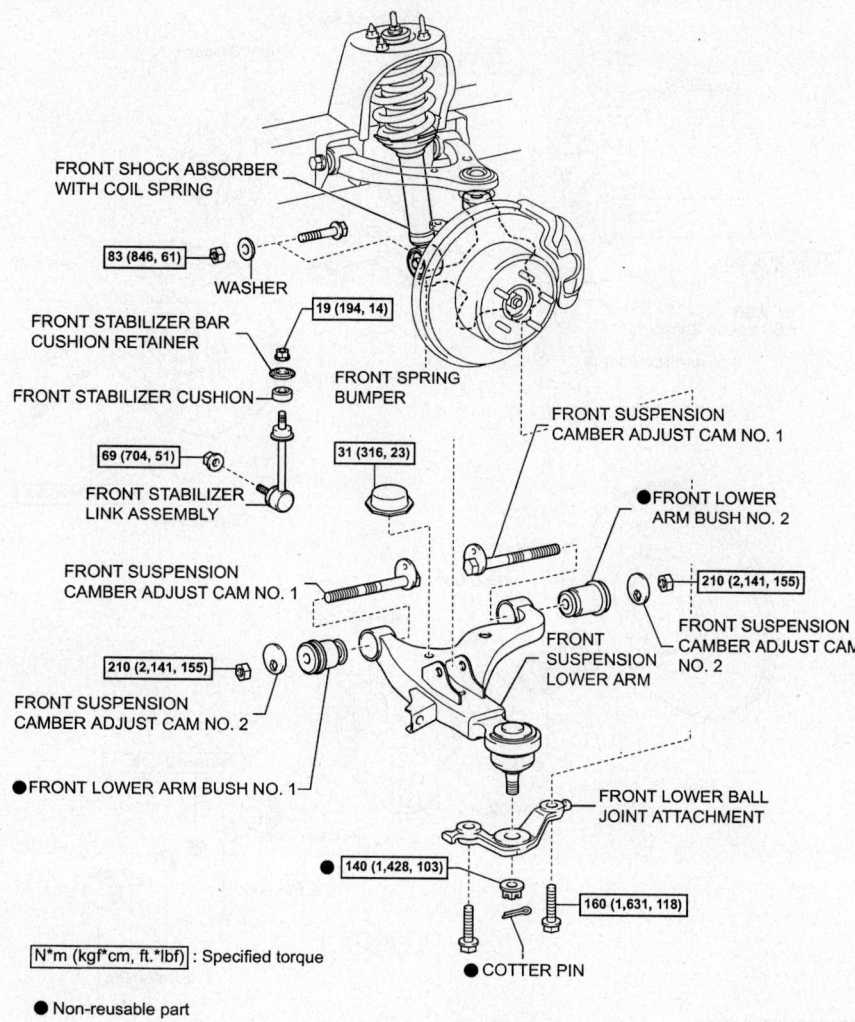

FRONT SHOCK ABSORBER WITH COIL SPRING

83 (846, 61)

WASHER

FRONT STABILIZER BAR CUSHION RETAINER

19 (194, 14)

FRONT STABILIZER CUSHION

FRONT SPRING BUMPER

69 (704, 51)

FRONT STABILIZER LINK ASSEMBLY

31 (316, 23)

FRONT SUSPENSION CAMBER ADJUST CAM NO. 1

FRONT SUSPENSION CAMBER ADJUST CAM NO. 1

●FRONT LOWER ARM BUSH NO. 2

210 (2,141, 155)

FRONT SUSPENSION CAMBER ADJUST CAM NO. 2

FRONT SUSPENSION LOWER ARM

210 (2,141, 155)

FRONT SUSPENSION CAMBER ADJUST CAM NO. 2

●FRONT LOWER ARM BUSH NO. 1

FRONT LOWER BALL JOINT ATTACHMENT

140 (1,428, 103)

160 (1,631, 118)

N*m (kgf*cm, ft.*lbf) : Specified torque

●Non-reusable part

●COTTER PIN

Lower control arm and related components—2WD 2005–2006

number one, the bolt, the toe adjust cam and the toe adjust plate number two.

11. Remove the lower control arm from the vehicle.

12. To remove the ball joint, position the assembly in a vise and using tool SST09628-00011, remove the ball joint from its mounting.

To install:

➡ **Always fully tighten rubber bushings when the wheels are in full contact with the ground and the vehicle is at curb height. Bounce the vehicle up and down several times to stabilize the suspension, prior to final tightening of these components.**

13. Position the lower control arm on its mounting in the vehicle.

14. Align the mating marks made during the removal procedure and temporarily tighten the lower control arm mounting bolts.

15. Install the lower ball joint attachment. Be sure to use a new nut and cotter pin. Tighten the nut to 103 ft. lbs.

16. Install the front lower ball joint attachment with the two bolts. Tighten the bolts to 118 ft. lbs.

17. Continue the installation in the reverse order of the removal procedure.

18. Final tightening torque for the lower control arm is 155 ft. lbs. on 2WD vehicles and 100 ft. lbs. on 4WD vehicles and Pre-Runner.

19. Check and adjust the alignment, as required.

LOWER CONTROL ARM BUSHING REPLACEMENT

2002–2004

2WD

The lower control arm is equipped with a single bushing.

6. Using the special tools, install new bushings.

Front Wheel Bearings

ADJUSTMENT

2002–2004

2WD

1. Tighten the adjusting nut to 26 ft. lbs. (35 Nm).
2. Turn the disc/hub assembly 2–3 times, from the left to the right.
3. Loosen the adjusting nut until it can be turned by hand.
4. Attach a spring tension gauge to 1 lug on the hub assembly. Pull on the gauge and measure the frictional force. Frictional force should be 1–3 lbs. (5.0–14.0 N).
5. Adjust the preload by tightening the nut.
6. Measure the hub axial play. The limit is 0.0020 in. (0.05mm).

REMOVAL & INSTALLATION

2002–2004

2WD

1. Before servicing the vehicle, refer to the Precautions Section.
2. Remove or disconnect the following:
   - Brake caliper support bracket, by removing the 2 bolts. Support the brake caliper with a piece of wire. Do not allow the caliper to hang from the brake hose.
   - Cotter pin, lockcap, nut and the claw washer from the axle hub and disc
   - Axle hub with the disc from the steering knuckle.

✳✳ WARNING

Do not drop the outer bearing when removing the hub.

   - Inner oil seal
   - Inner bearing
   - Bearing outer races, using SST 09527-17011, a brass bar and a hammer
3. If it is necessary to separate the hub and rotor, place matchmarks on the hub and rotor.
4. Remove the 5 bolts and remove the hub from the rotor.

To install:

5. Install or connect the following:
   - New bearing races, using SST 09527-17011 and a press.

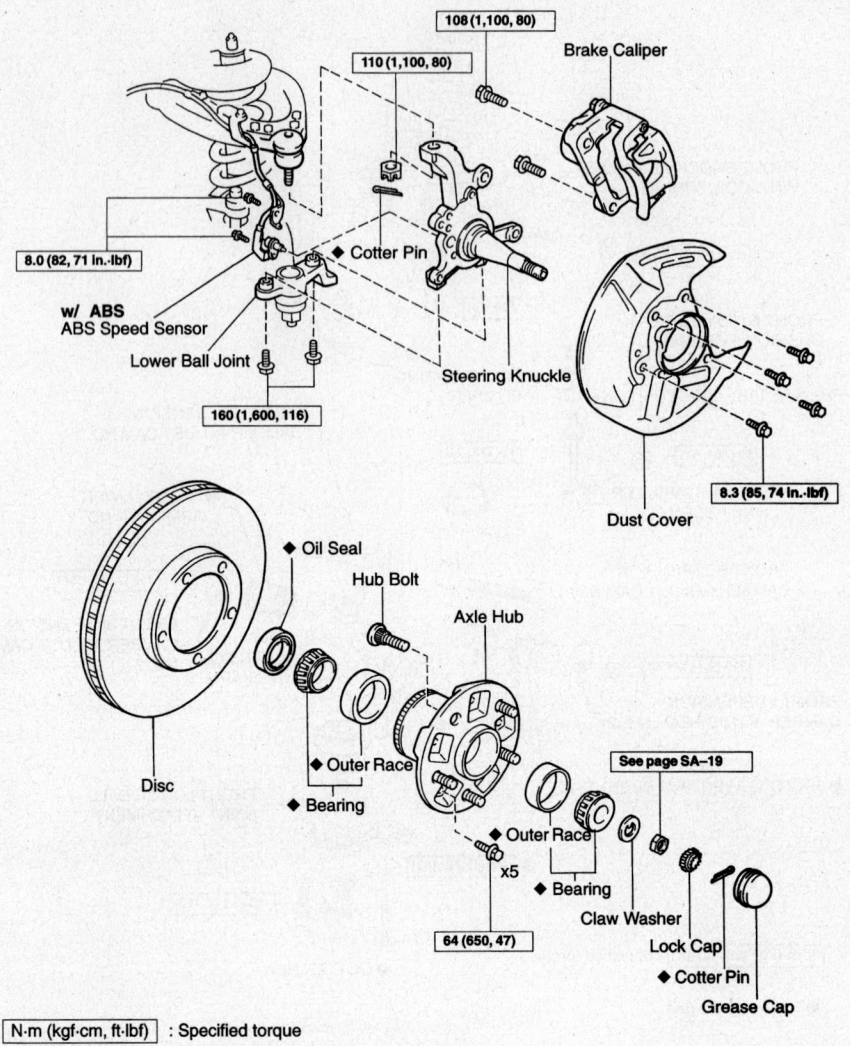

N·m (kgf·cm, ft·lbf) : Specified torque
◆ Non–reusable part

67170-TACO-G42

Front hub and related components—2WD 2002–2004

   - Hub to the rotor. Tighten the 5 bolts to 47 ft. lbs. (64 Nm).
6. Clean all parts.
7. Repack the bearings with multi purpose grease and apply the same grease to the outer bearings.
8. Install or connect the following:
   - Inner bearing and seal to the hub. Coat the inner seal with multi purpose grease.
   - Outer bearing to the hub
   - Hub to the steering knuckle
   - Axle hub to the steering knuckle claw washer and nut
9. Adjust the bearing preload as described above.
10. Install or connect the following:
   - Locknut, cotter pin and the grease cap
   - Disc brake caliper. Tighten the 2 bolts to 80 ft. lbs. (108 Nm).
   - Wheel

4WD

1. Before servicing the vehicle, refer to the Precautions Section.
2. Remove the front wheel.
3. Detach the shock absorber.
4. Disconnect the driveshaft by removing the grease cap and pulling out the cotter pin and lock cap.
5. Apply the brakes to hold the axle from spinning and remove the lock nut.
6. If the vehicle is equipped with antilock brakes, detach the speed sensor and wiring harness clamp from the steering knuckle.
7. Remove or disconnect the following:
   - Banjo bolt and 2 gaskets from the caliper
   - Flexible brake hose from the caliper
   - Brake caliper and then the rotor
   - Lower ball joint
   - Steering knuckle

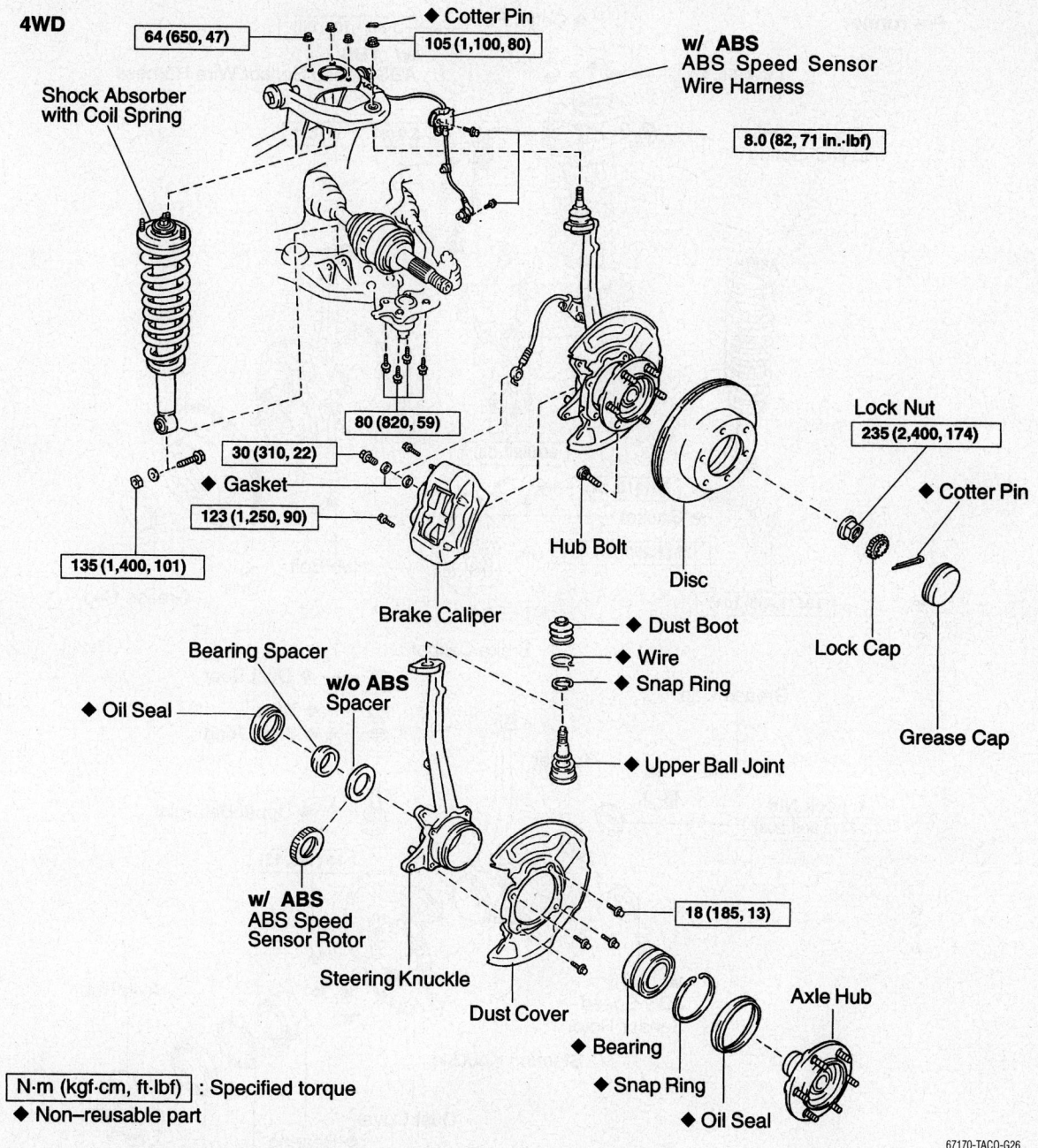

4WD

◆ Cotter Pin
64 (650, 47)
105 (1,100, 80)

w/ ABS
ABS Speed Sensor
Wire Harness

Shock Absorber
with Coil Spring

8.0 (82, 71 in.·lbf)

80 (820, 59)

30 (310, 22)
◆ Gasket
123 (1,250, 90)

135 (1,400, 101)

Brake Caliper

Hub Bolt

Disc

Lock Nut
235 (2,400, 174)

◆ Cotter Pin

Lock Cap

Grease Cap

◆ Dust Boot
◆ Wire
◆ Snap Ring

◆ Upper Ball Joint

Bearing Spacer

◆ Oil Seal

w/o ABS
Spacer

w/ ABS
ABS Speed
Sensor Rotor

Steering Knuckle

Dust Cover

18 (185, 13)

Axle Hub

◆ Bearing
◆ Snap Ring
◆ Oil Seal

N·m (kgf·cm, ft·lbf) : Specified torque
◆ Non–reusable part

67170-TACO-G26

Front hub and related components—4WD except PreRunner 2002–2004

8. Clamp the axle hub in a soft jaw vise.

➡**Close the vise until it holds the hub bolts.**

9. Using the proper seal puller or prytool, remove the oil seal.

10. On vehicles equipped with a free wheel hub and on the Pre Runner, use a chisel and hammer to loosen the staked part of the lock nut.

11. Remove the lock nut. A special service tool may be required.

12. Remove the Antilock Brake System (ABS) speed sensor rotor/spacer.

➡**Do not scratch the speed sensor rotor.**

13. Detach the bolts to the dust shield and shift the shield towards the outside of the hub.

14. Remove the axle from the steering knuckle, a special service tool may be required.

15. Remove the dust cover from the steering knuckle.

16. On vehicles without a free wheeling hub, that are 4WD only, remove the bearing spacer and ABS speed sensor rotor spacer.

17. Remove the outside seal by prying it out with a seal puller.

18. Remove the bearing from the steering knuckle by removing the snapring with a pair of snapring pliers.

19. Press the bearing from the steering knuckle.

To install:

20. Install or connect the following:
- New bearing
- New oil seal
- Axle hub to the steering knuckle. Torque the bolts to 13 ft. lbs. (18 Nm).
- ABS speed sensor rotor

21. On vehicles that are equipped with a free wheel hub, install the bearing spacer.

Pre runner

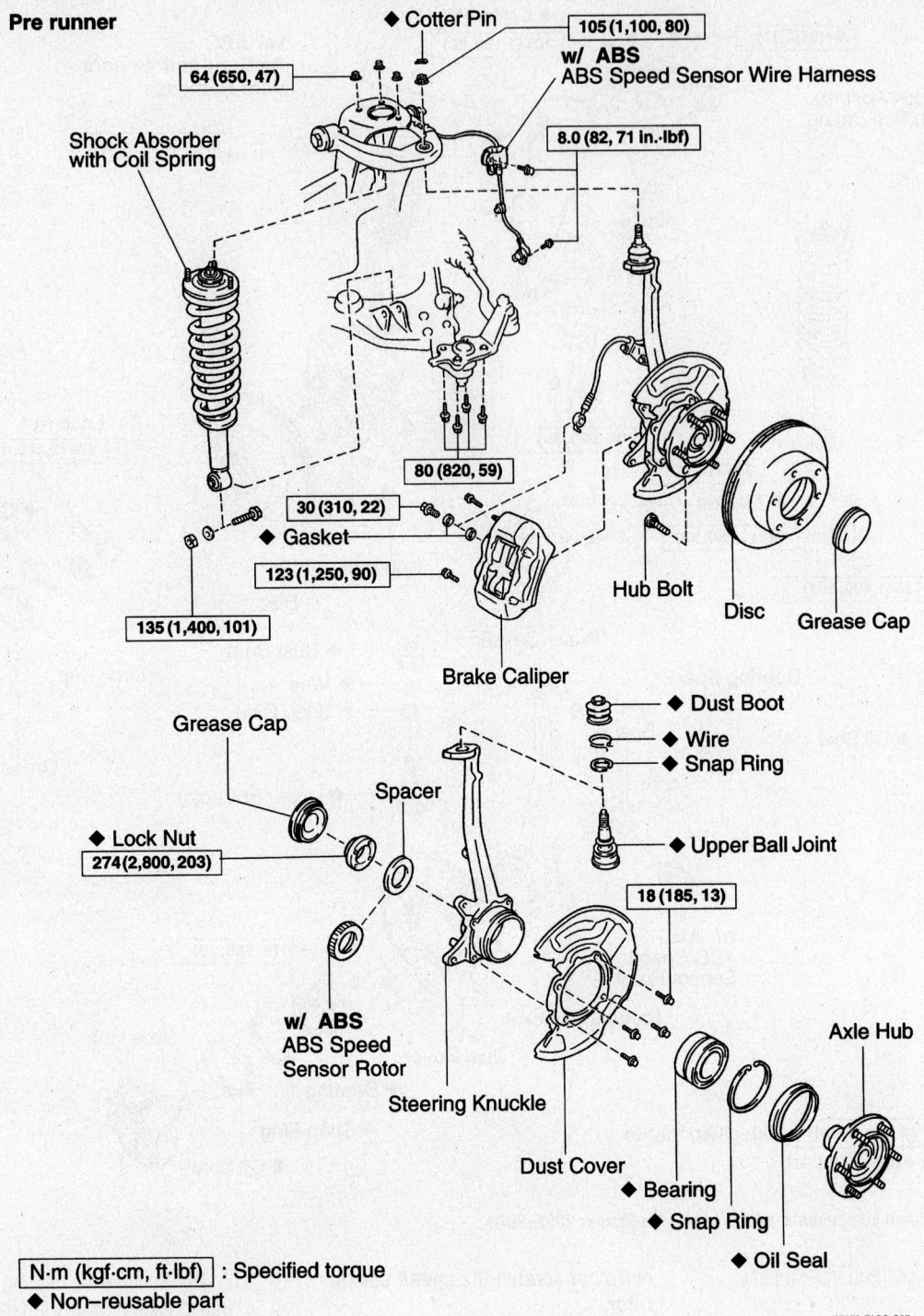

◆ Cotter Pin

105 (1,100, 80)

64 (650, 47)

w/ ABS
ABS Speed Sensor Wire Harness

8.0 (82, 71 in.·lbf)

Shock Absorber
with Coil Spring

80 (820, 59)

30 (310, 22)

◆ **Gasket**

123 (1,250, 90)

135 (1,400, 101)

Hub Bolt

Disc   **Grease Cap**

Brake Caliper

◆ **Dust Boot**

◆ **Wire**

◆ **Snap Ring**

Grease Cap

Spacer

◆ **Lock Nut**
274 (2,800, 203)

◆ **Upper Ball Joint**

18 (185, 13)

w/ ABS
ABS Speed
Sensor Rotor

Steering Knuckle

Axle Hub

Dust Cover

◆ **Bearing**

◆ **Snap Ring**

◆ **Oil Seal**

N·m (kgf·cm, ft·lbf) : Specified torque
◆ Non–reusable part

67170-TACO-G27

Front hub and related components—PreRunner 2002–2004

22. Except Pre-Runner install a new inside oil seal.

23. On the Pre-Runner, install the grease cap.

24. The remainder of the installation procedure is the reverse of removal. New lock nut torque is 203 ft. lbs. (274 Nm) for the PreRunner; 174 ft. lbs. (235 Nm) for all others.

2005–2006

2WD

1. Before servicing the vehicle, refer to the Precautions Section.

2. Disconnect the negative battery cable. Drain the brake fluid.

3. Raise and support the vehicle safely. Remove the tire and wheel assembly.

4. Remove the speed sensor connector.

Remove the two bolts and separate the skid control sensor wire.

5. Remove the bolt and separate the front flexible hose.

6. Remove the front brake caliper. Remove the rotor.

7. Remove the cotter pin and nut. Using the proper tool separate the tie rod end from the steering knuckle.

8. Remove the two bolts and separate

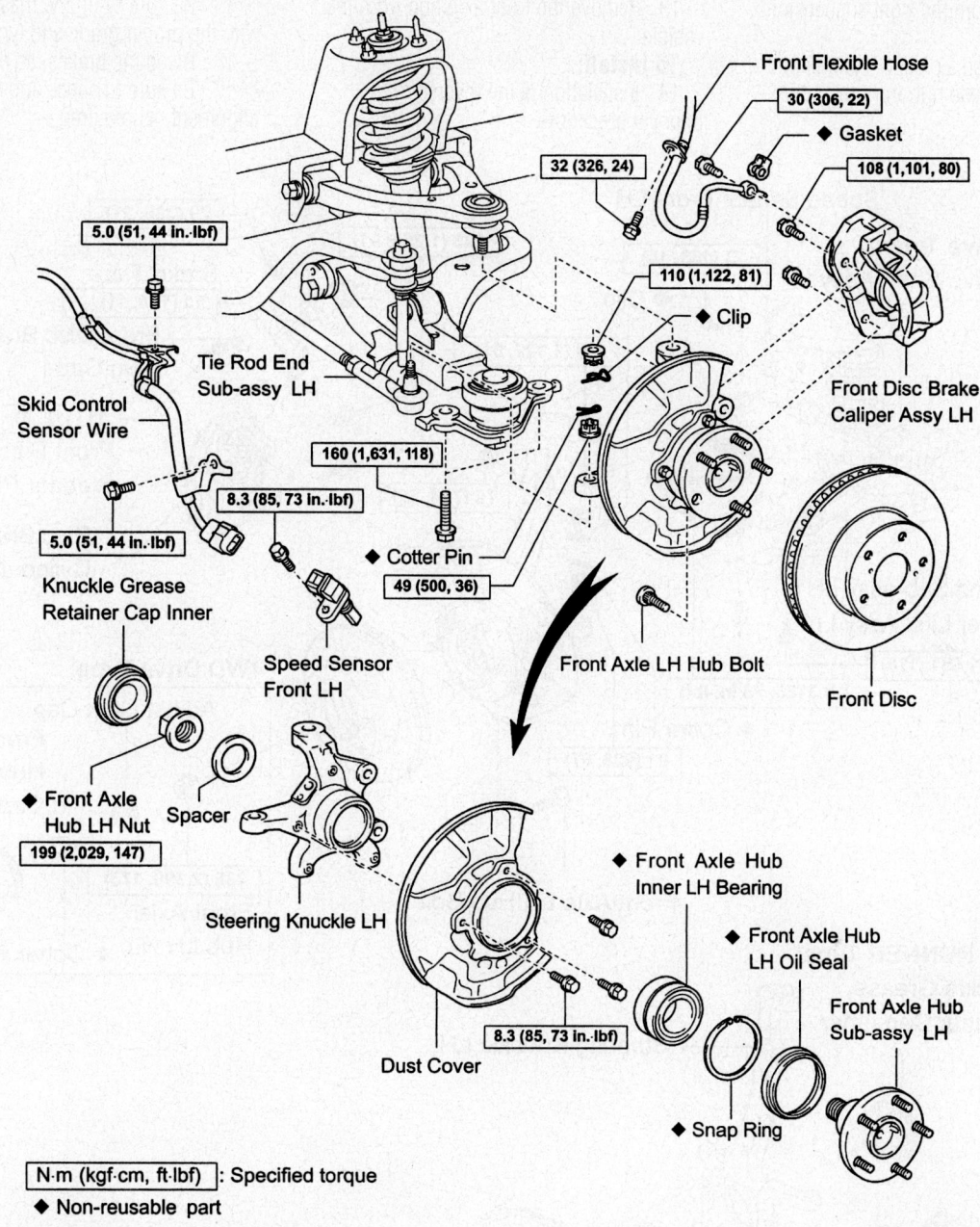

Front Flexible Hose

30 (306, 22)

◆ Gasket

32 (326, 24)

108 (1,101, 80)

5.0 (51, 44 in.·lbf)

110 (1,122, 81)

◆ Clip

Tie Rod End
Sub-assy LH

Skid Control
Sensor Wire

Front Disc Brake
Caliper Assy LH

160 (1,631, 118)

8.3 (85, 73 in.·lbf)

5.0 (51, 44 in.·lbf)

◆ Cotter Pin

Knuckle Grease
Retainer Cap Inner

49 (500, 36)

Front Axle LH Hub Bolt

Speed Sensor
Front LH

Front Disc

◆ Front Axle
Hub LH Nut

Spacer

199 (2,029, 147)

Steering Knuckle LH

◆ Front Axle Hub
Inner LH Bearing

◆ Front Axle Hub
LH Oil Seal

Front Axle Hub
Sub-assy LH

8.3 (85, 73 in.·lbf)

Dust Cover

◆ Snap Ring

N·m (kgf·cm, ft·lbf): Specified torque
◆ Non-reusable part

09490_TACO_G0111

Front hub and related components—2WD 2005–2006

the front lower ball joint attachment from the front axle.

9. Remove the clip and nut. Using the proper tool, separate the upper ball joint from the steering knuckle.

10. Remove the front axle hub from the vehicle.

To install:

11. Installation is the reverse of the removal procedure.

12. Be sure to fill the master cylinder with the proper grade and type brake fluid.

13. Bleed the brakes, as required.

14. Be sure to check and adjust the alignment, as required.

4WD AND PRERUNNER

1. Before servicing the vehicle, refer to the Precautions Section.

2. Disconnect the negative battery cable. Drain the brake fluid.

3. Raise and support the vehicle safely. Remove the tire and wheel assembly.

4. Remove the bolt and separate the speed sensor. Disengage the clamps. Remove the bolt and separate the speed sensor wire harness from the steering knuckle.

5. Remove the front brake caliper. Remove the rotor.

6. Remove the axle hub grease cap. Remove the cotter pin and lock cap. Remove the front axle hub nut.

7. Remove the nut and separate the stabilizer link from the steering knuckle.

8. Remove the cotter pin and nut. Using the proper tool separate the tie rod end from the steering knuckle.

9. Remove the two bolts and separate the front suspension lower arm from the front axle.

10. Properly support the front suspension lower arm with a jack. Remove the clip and nut.

11. Using the proper tool separate the

steering knuckle from the front suspension upper control arm.

12. On 4WD, use a plastic hammer and separate the front axle hub from the front driveshaft.

13. Remove the front axle hub from the vehicle.

To install:

14. Installation is the reverse of the removal procedure.

15. Be sure to fill the master cylinder with the proper grade and type brake fluid.

16. Bleed the brakes, as required.

17. Be sure to check and adjust the alignment, as required.

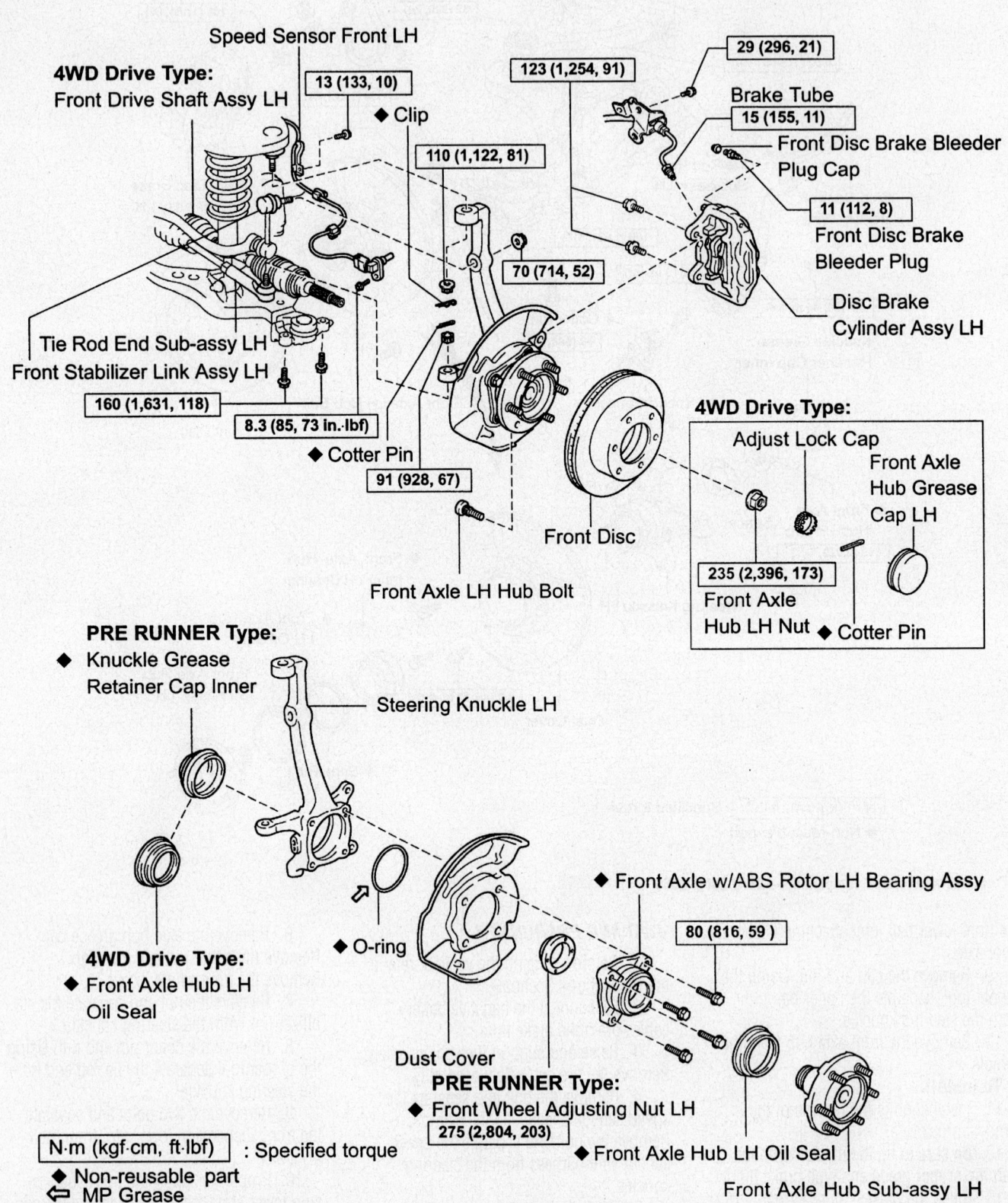

Speed Sensor Front LH

4WD Drive Type:
Front Drive Shaft Assy LH

13 (133, 10)

◆ Clip

123 (1,254, 91)

29 (296, 21)

Brake Tube

15 (155, 11)

Front Disc Brake Bleeder Plug Cap

110 (1,122, 81)

11 (112, 8)

Front Disc Brake Bleeder Plug

70 (714, 52)

Disc Brake Cylinder Assy LH

Tie Rod End Sub-assy LH

Front Stabilizer Link Assy LH

160 (1,631, 118)

8.3 (85, 73 in.·lbf)

◆ Cotter Pin

91 (928, 67)

Front Axle LH Hub Bolt

Front Disc

4WD Drive Type:
Adjust Lock Cap

Front Axle Hub Grease Cap LH

235 (2,396, 173)

Front Axle Hub LH Nut

◆ Cotter Pin

PRE RUNNER Type:
◆ Knuckle Grease Retainer Cap Inner

Steering Knuckle LH

4WD Drive Type:
◆ Front Axle Hub LH Oil Seal

◆ O-ring

Dust Cover

PRE RUNNER Type:
◆ Front Wheel Adjusting Nut LH

275 (2,804, 203)

◆ Front Axle w/ABS Rotor LH Bearing Assy

80 (816, 59)

◆ Front Axle Hub LH Oil Seal

Front Axle Hub Sub-assy LH

N·m (kgf·cm, ft·lbf) : Specified torque

◆ Non-reusable part

⇐ MP Grease

Front hub and related components—4WD and PreRunner 2005–2006

09490_TACO_G0112

REAR SUSPENSION

Shock Absorbers

REMOVAL & INSTALLATION

1. Before servicing the vehicle, refer to the Precautions Section.
2. Raise and support the vehicle safely.
3. Remove the tire and wheel assembly.
4. Lower the floor jack to take tension off of the spring.
5. Remove or disconnect the following:
   - Shock absorber from the rear axle housing
   - Nut, retainers and the cushions holding the shock absorber to the frame
   - Shock absorber with the washers and bushings

To install:

➡**Always fully tighten rubber bushings when the wheels are in full contact with the ground and the vehicle is at curb height. Bounce the vehicle up and down several times to stabilize the suspension, prior to final tightening of these components.**

6. Install the shock absorber to the frame with the washers and bushings.
7. On 2002–2004 vehicles tighten the shock absorber-to-frame nut on 2WD to 19 ft. lbs. (25 Nm) and on 4WD and PreRunner to 53 ft. lbs. (72 Nm).
8. On 2005–2006 vehicles tighten the shock absorber-to-upper frame nut on to 15 ft. lbs. (20 Nm).
9. Connect the shock absorber to the rear axle housing.
10. On 2002–2004 vehicles tighten the bolt on 2WD to 19 ft. lbs. (25 Nm) and on 4WD and PreRunner to 53 ft. lbs. (72 Nm).
11. On 2005–2006 vehicles tighten the bolt to 74 ft. lbs. (100 Nm).
12. Install the wheels.

Leaf Springs

REMOVAL & INSTALLATION

2002–2004

2WD

1. Before servicing the vehicle, refer to the Precautions Section.

2. Loosen the rear wheel lug nuts.
3. Raise the rear of the vehicle. Support the frame and rear axle housing with stands.
4. Remove the lug nuts and the wheel.
5. Remove the cotter pin, nut, and washer from the lower end of the shock absorber.
6. Detach the shock absorber from the spring seat.
7. Remove the parking brake cable clamp.

➡**Remove the parking brake equalizer, if necessary.**

8. Unfasten the U-bolt nuts and remove the spring seat assemblies.
9. Adjust the height of the rear axle housing so that the weight of the rear axle is removed from the rear springs.
10. Unfasten the spring shackle retaining nuts. Withdraw the spring shackle inner plate. Carefully pry out the spring shackle with a bar.
11. Remove the spring bracket pin from the front end of the spring hanger and remove the rubber bushing.
12. Remove the spring. Use care not to damage the hydraulic brake line or the parking brake cable.

To install:

13. Install the rubber bushing in the eye of the spring.
14. Align the eye of the spring with the spring hanger bracket and drive the pin through the bracket holes and rubber bushings.

➡**Use soapy water or glass cleaner as a lubricant, if necessary, to aid in pin installation. Never use oil or grease.**

15. Finger-tighten the spring hanger nuts and/or bolts.
16. Install the rubber bushing in the spring eye at the opposite end of the spring.
17. Raise the free end of the spring. Install the spring shackle through the bushing and the bracket.
18. Install the shackle inner plate and finger-tighten the retaining nuts.
19. Center the bolt head in the hole which is provided in the spring seat on the axle housing.
20. Fit the U-bolts over the axle housing. Install the lower spring seat.
21. Tighten the U-bolt nuts to 90 ft. lbs. (120 Nm)
22. Install the parking brake cable and clamp. Install the equalizer, if removed.
23. Tighten the hanger pin and shackle

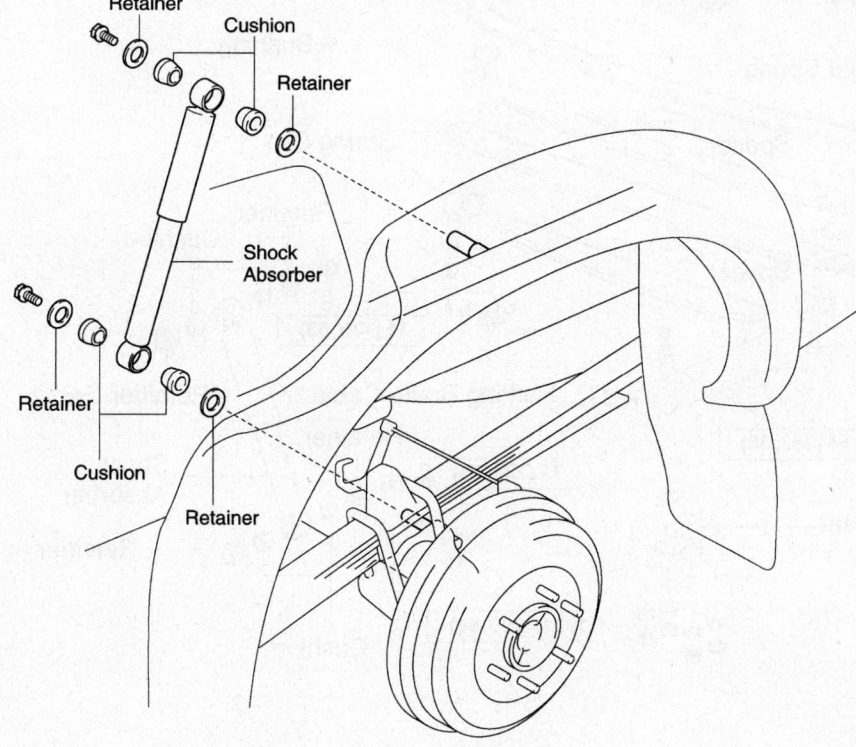

67170-TACO-G32A

Rear shock absorber and related components

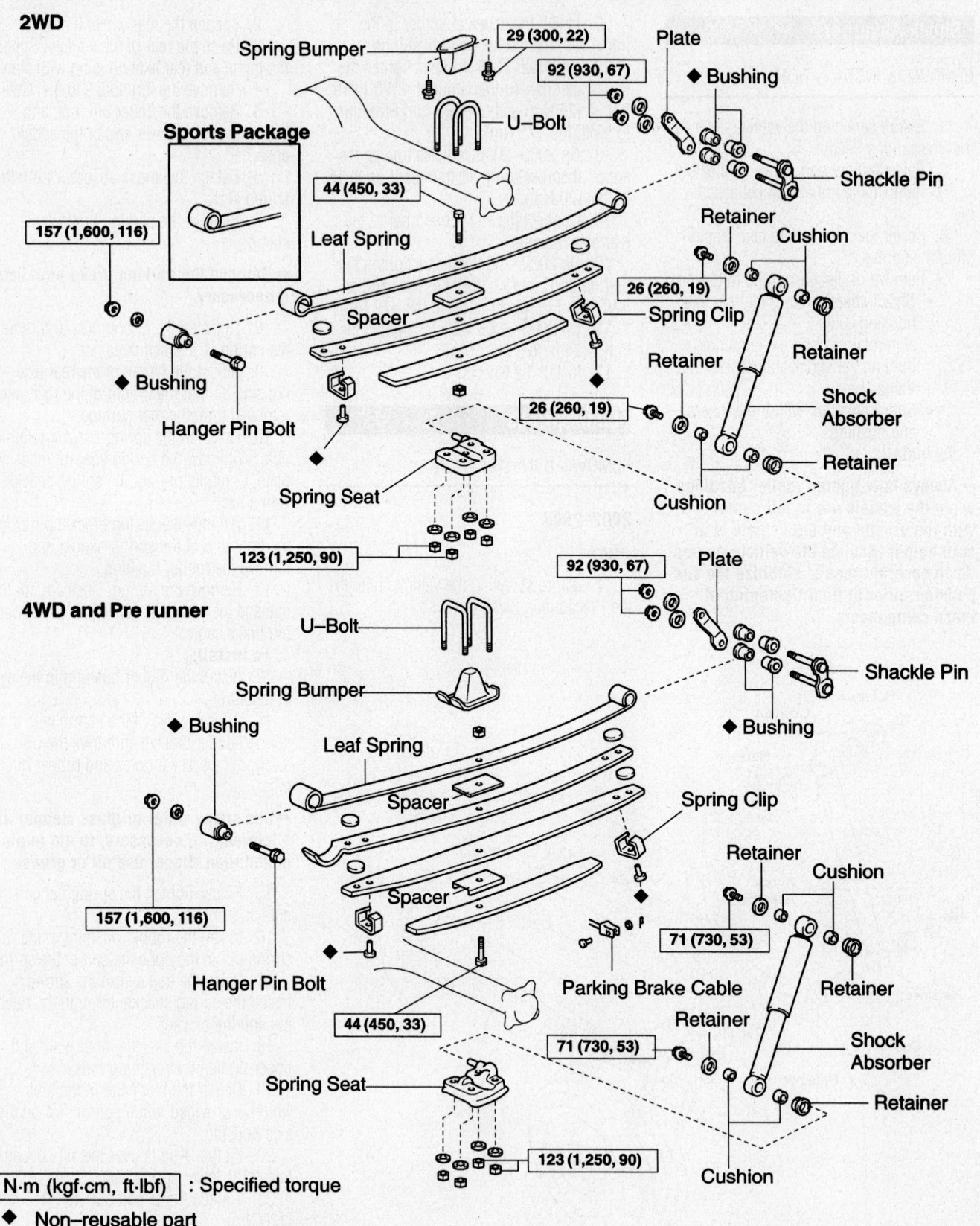

2WD

Spring Bumper

29 (300, 22)

Plate

Bushing

92 (930, 67)

U–Bolt

Shackle Pin

Sports Package

44 (450, 33)

Retainer

Cushion

157 (1,600, 116)

Leaf Spring

26 (260, 19)

Spring Clip

Retainer

Spacer

Retainer

Shock Absorber

◆ Bushing

Hanger Pin Bolt

26 (260, 19)

Retainer

Cushion

Spring Seat

Cushion

123 (1,250, 90)

92 (930, 67)

Plate

4WD and Pre runner

U–Bolt

Shackle Pin

Spring Bumper

◆ Bushing

◆ Bushing

Leaf Spring

Retainer

Cushion

Spacer

Spring Clip

157 (1,600, 116)

Spacer

71 (730, 53)

Retainer

Hanger Pin Bolt

Parking Brake Cable

44 (450, 33)

71 (730, 53)

Retainer

Shock Absorber

Spring Seat

123 (1,250, 90)

Cushion

Retainer

N·m (kgf·cm, ft·lbf) : Specified torque

◆ Non–reusable part

67170-TACO-G34

Rear leaf spring and related components—2002–2004

nuts. Install the shock absorber bushings and washers. Tighten and install the cotter pins.

24. Install the stabilizer link and hand-tighten its retaining nuts.

25. Install the wheels. Lower the vehicle.

26. Bounce the truck several times to set the suspension and then tighten the shock absorber bolt. Tighten the hanger pin nut or bolt to 115 ft. lbs. (120 Nm).

27. Tighten the shackle pin to 67 ft. lbs. (91 Nm).

4WD

1. Before servicing the vehicle, refer to the Precautions Section.

2. Loosen the rear wheel lug nuts.

3. Raise the rear of the vehicle. Support the frame and rear axle housing with stands.

4. Remove the lug nuts and the wheel.

5. Remove the cotter pin, nut and washer from the lower end of the shock absorber.

6. Detach the shock absorber from the spring seat.

7. Remove the parking brake cable clamp.

➡**Remove the parking brake equalizer, if necessary.**

8. Unfasten the U-bolt and nuts, then remove the spring seat assemblies.

9. Adjust the height of the rear axle housing so that the weight of the rear axle is removed from the rear springs.

10. Unfasten the spring shackle retaining nuts. Withdraw the spring shackle inner plate. Carefully pry out the spring shackle with a bar.

11. Remove the spring bracket pin from the front end of the spring hanger and remove the rubber bushing.

12. Remove the spring. Use care not to damage the hydraulic brake line or the parking brake cable.

To install:

13. Install the rubber bushing in the eye of the spring.

14. Align the eye of the spring with the spring hanger bracket and drive the pin through the bracket holes and rubber bushings.

➡**Use soapy water or glass cleaner as a lubricant, if necessary, to aid in pin installation. Never use oil or grease.**

15. Finger-tighten the spring hanger nuts and/or bolts.

16. Install the rubber bushing in the spring eye at the opposite end of the spring.

17. Raise the free end of the spring.

Install the spring shackle through the bushing and the bracket.

18. Install the shackle inner plate and finger-tighten the retaining nuts.

19. Center the bolt head in the hole which is provided in the spring seat on the axle housing.

20. Fit the U-bolts over the axle housing. Install the lower spring seat. Install the spring bumper, if equipped.

21. Tighten the U-bolt nuts to 90 ft. lbs. (120 Nm)

22. Install the parking brake cable and clamp. Install the equalizer, if removed.

23. Tighten the hanger pin and shackle nuts. Install the shock absorber bushings and washers. Tighten and install the cotter pins.

24. Install the stabilizer link and hand-tighten its retaining nuts.

25. Install the wheels and remove the stands. Lower the vehicle.

26. Bounce the vehicle several times to set the suspension and then tighten the shock absorber bolt. Tighten the hanger pin nut or bolt to 115 ft. lbs. (120 Nm).

27. Tighten the shackle pin to 67 ft. lbs. (91 Nm).

2005–2006

1. Before servicing the vehicle, refer to the Precautions Section.

2. Raise and support the vehicle safely.

3. Remove the tire and wheel assemblies.

4. On 4WD and PreRunner, remove the spare tire.

5. Support the rear axle housing. Remove the bolt nut and washer.

6. On 2WD separate the chock absorber from the rear spring seat.

7. On 4WD and PreRunner, separate the shock absorber from the rear axle housing.

8. Remove the bolt and then separate the parking brake cable.

9. On 2WD remove the two nuts and spring bumper. Remove the four nuts and four washers. Remove the spring seat and two U-bolts.

10. On 4WD and PreRunner, remove the four nuts and four washers. Remove the spring seat and two U-bolts. Remove the spring bumper.

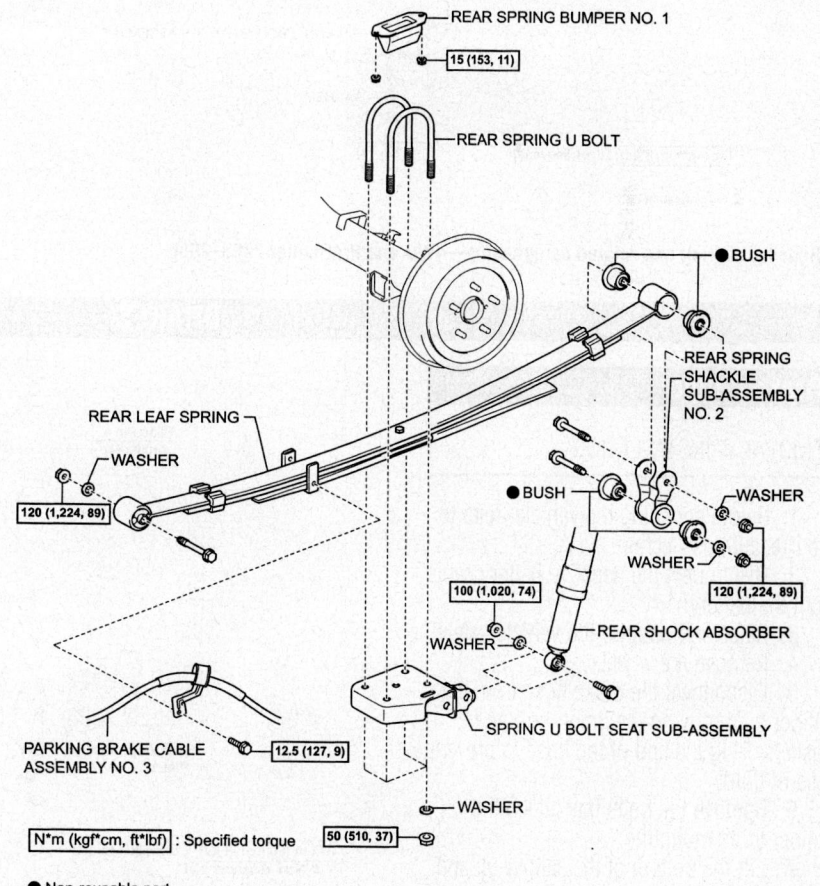

REAR SPRING BUMPER NO. 1

15 (153, 11)

REAR SPRING U BOLT

● BUSH

REAR SPRING SHACKLE SUB-ASSEMBLY NO. 2

REAR LEAF SPRING

WASHER

120 (1,224, 89)

● BUSH

WASHER

WASHER

100 (1,020, 74)

120 (1,224, 89)

REAR SHOCK ABSORBER

WASHER

SPRING U BOLT SEAT SUB-ASSEMBLY

PARKING BRAKE CABLE ASSEMBLY NO. 3

12.5 (127, 9)

N*m (kgf*cm, ft*lbf) : Specified torque

50 (510, 37)

WASHER

● Non-reusable part

09490_TACO_G0122

Rear leaf spring and related components—2WD 2005–2006

11. Remove the nut, washer and through bolt. Remove the spring from the vehicle.

➡Be careful not to drop the spring when removing the through bolt.

To install:

➡Always fully tighten rubber bushings when the wheels are in full contact with the ground and the vehicle is at curb height. Bounce the vehicle up and down several times to stabilize the suspension, prior to final tightening of these components.

12. Installation is the reverse of the removal procedure.

13. Tighten the rear spring U-bolts to 37 ft. lbs. Be sure that the lengths of all of the U-bolts under the spring seat are the same.

Stabilizer Bar

REMOVAL & INSTALLATION

1. Before servicing the vehicle, refer to the Precautions Section.
2. Raise and support the vehicle safely.
3. Remove the tire and wheel assemblies.
4. On the left side, remove the two nuts and stabilizer bar link.
5. On the right side, remove the two nuts and stabilizer bar link.
6. Remove the four bolts and the stabilizer bar brackets and bushings.
7. Remove the stabilizer bar from the vehicle.

To install:

8. Installation is the reverse of the removal procedure.
9. Tighten the stabilizer bar and bushing bracket bolts to 20 ft. lbs.
10. Tighten the stabilizer link assembly bolts to 51 ft. lbs
11. Be sure to install the bushing onto the outer side of the mark on the stabilizer bar.

REAR SPRING U BOLT

REAR SPRING BUMPER NO. 1

REAR LEAF SPRING

BUSH

REAR SPRING SHACKLE SUB-ASSEMBLY NO. 2

120 (1,224, 89)

WASHER

WASHER

BUSH

WASHER

120 (1,224, 89)

PARKING BRAKE CABLE ASSEMBLY NO. 3

12.5 (127, 9)

100 (1,020, 74)

WASHER

REAR SHOCK ABSORBER

SPRING U BOLT SEAT SUB-ASSEMBLY

WASHER

N·m (kgf·cm, ft·lbf) : Specified torque

● Non-reusable part

50 (510, 37)

09490_TACO_G0123

Rear leaf spring and related components—4WD and PreRunner 2005–2006

BRAKES

Brake Caliper

REMOVAL & INSTALLATION

1. Before servicing the vehicle, refer to the Precautions Section.
2. Disconnect the negative battery cable from the battery.
3. Raise and support the vehicle safely.
4. Remove the wheels.
5. Disconnect the brake hose from the caliper by removing the union bolt and 2 gaskets. Plug the end of the hose to prevent loss of fluid.
6. Remove the bolts that attach the caliper to its mounting.
7. Lift the bottom of the caliper up and remove the caliper assembly.

To install:

8. Grease the caliper slides and bolts

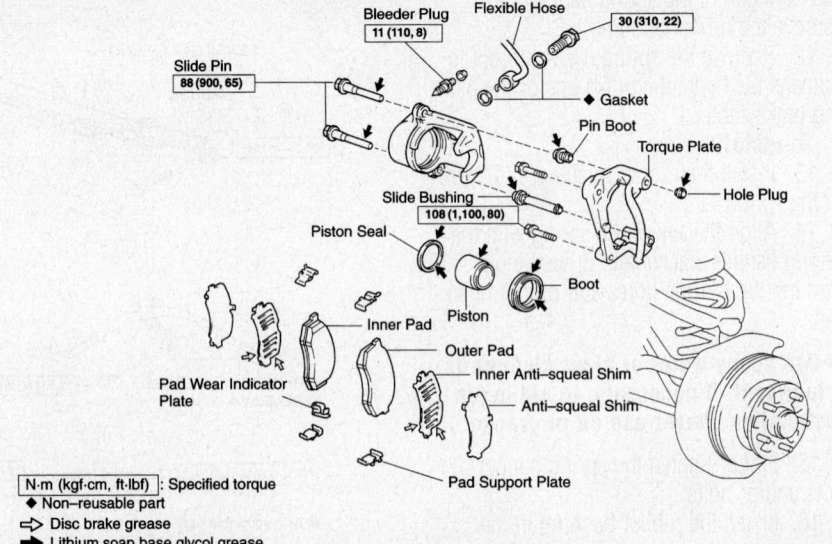

Bleeder Plug 11 (110, 8)
Flexible Hose
30 (310, 22)
Slide Pin 88 (900, 65)
◆ Gasket
Pin Boot
Torque Plate
Slide Bushing 108 (1,100, 80)
Hole Plug
Piston Seal
Boot
Piston
Inner Pad
Outer Pad
Inner Anti-squeal Shim
Anti-squeal Shim
Pad Wear Indicator Plate
Pad Support Plate

N·m (kgf·cm, ft·lbf) : Specified torque
◆ Non-reusable part
⇨ Disc brake grease
➡ Lithium soap base glycol grease

Brake caliper and related components—2WD 2002–2004

67170-TACO-G43

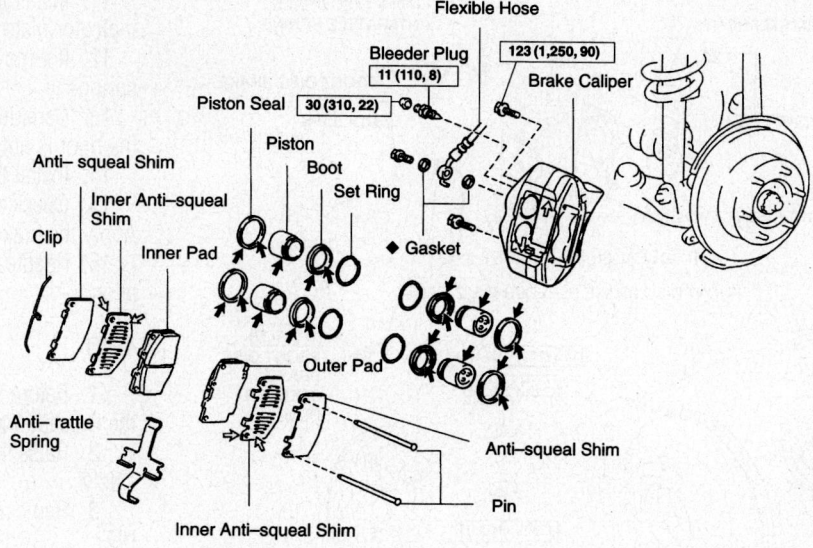

Flexible Hose

Bleeder Plug
`11 (110, 8)`
Piston Seal `30 (310, 22)`
`123 (1,250, 90)`
Brake Caliper

Piston
Boot
Set Ring

Anti– squeal Shim
Inner Anti–squeal Shim

Clip
Inner Pad

◆ Gasket

Anti– rattle Spring

Outer Pad

Anti–squeal Shim

Pin

Inner Anti–squeal Shim

`N·m (kgf·cm, ft·lbf)` : Specified torque
◆ Non–reusable part
➡ Lithium soap base glycol grease
⇨ Disc brake grease

67170-TACO-G44

Brake caliper and related components—4WD 2002–2004

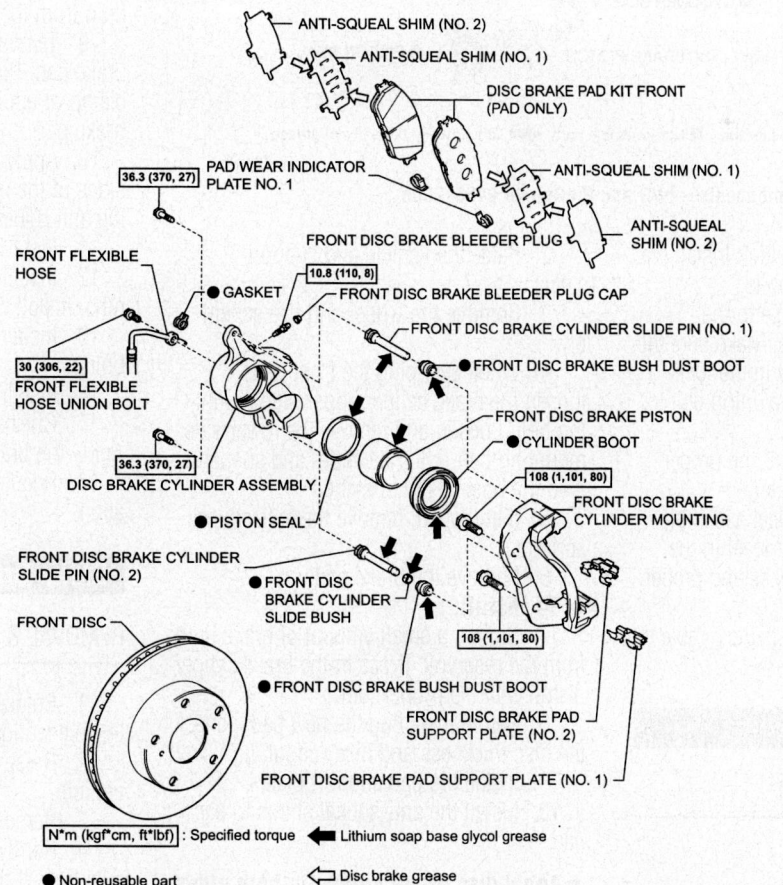

ANTI-SQUEAL SHIM (NO. 2)

ANTI-SQUEAL SHIM (NO. 1)

DISC BRAKE PAD KIT FRONT (PAD ONLY)

ANTI-SQUEAL SHIM (NO. 1)

`36.3 (370, 27)`
PAD WEAR INDICATOR PLATE NO. 1

ANTI-SQUEAL SHIM (NO. 2)

FRONT FLEXIBLE HOSE

FRONT DISC BRAKE BLEEDER PLUG

● GASKET `10.8 (110, 8)`
FRONT DISC BRAKE BLEEDER PLUG CAP

FRONT DISC BRAKE CYLINDER SLIDE PIN (NO. 1)

`30 (306, 22)`
FRONT FLEXIBLE HOSE UNION BOLT

● FRONT DISC BRAKE BUSH DUST BOOT

FRONT DISC BRAKE PISTON

`36.3 (370, 27)`
DISC BRAKE CYLINDER ASSEMBLY

● CYLINDER BOOT

`108 (1,101, 80)`
FRONT DISC BRAKE CYLINDER MOUNTING

● PISTON SEAL

FRONT DISC BRAKE CYLINDER SLIDE PIN (NO. 2)

● FRONT DISC BRAKE CYLINDER SLIDE BUSH

FRONT DISC

● FRONT DISC BRAKE BUSH DUST BOOT

`108 (1,101, 80)`

FRONT DISC BRAKE PAD SUPPORT PLATE (NO. 2)

FRONT DISC BRAKE PAD SUPPORT PLATE (NO. 1)

`N*m (kgf*cm, ft*lbf)` : Specified torque
◀ Lithium soap base glycol grease
● Non-reusable part
◁ Disc brake grease

09490_TACO_G0124

Brake caliper and related components—2WD 2005–2006

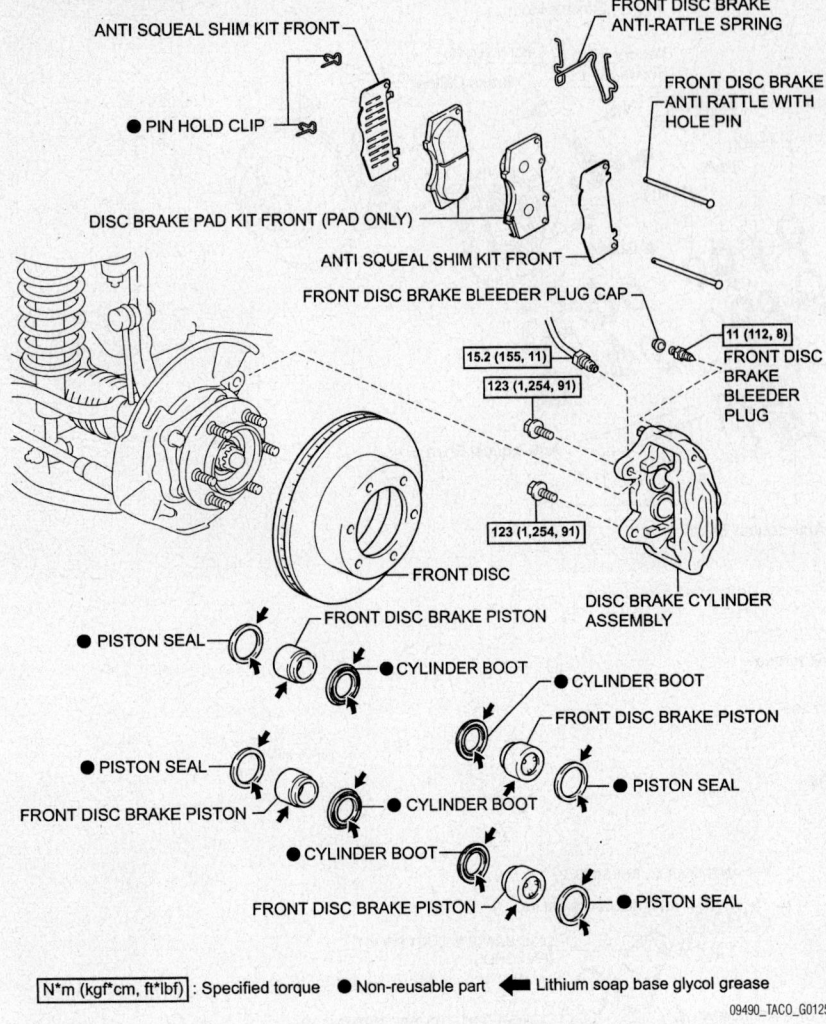

ANTI SQUEAL SHIM KIT FRONT

● PIN HOLD CLIP

DISC BRAKE PAD KIT FRONT (PAD ONLY)

FRONT DISC BRAKE ANTI-RATTLE SPRING

FRONT DISC BRAKE ANTI RATTLE WITH HOLE PIN

ANTI SQUEAL SHIM KIT FRONT

FRONT DISC BRAKE BLEEDER PLUG CAP

15.2 (155, 11)

123 (1,254, 91)

11 (112, 8) FRONT DISC BRAKE BLEEDER PLUG

123 (1,254, 91)

FRONT DISC

FRONT DISC BRAKE PISTON

DISC BRAKE CYLINDER ASSEMBLY

● PISTON SEAL

● CYLINDER BOOT

● CYLINDER BOOT

● PISTON SEAL

FRONT DISC BRAKE PISTON

● CYLINDER BOOT

FRONT DISC BRAKE PISTON

● CYLINDER BOOT

FRONT DISC BRAKE PISTON

● PISTON SEAL

● PISTON SEAL

N*m (kgf*cm, ft*lbf) : Specified torque ● Non-reusable part ◄ Lithium soap base glycol grease

09490_TACO_G0125

Brake caliper and related components—4WD and PreRunner 2005–2006

with lithium grease or equivalent. Install the caliper and secure with the bolts.

9. Connect the brake hose to the caliper, using 2 new washers. Make sure the flexible hose lock is securely in the lock hole of the caliper. Torque the union bolt to 22 ft. lbs. (30 Nm).

10. Fill the brake system to the proper level and bleed the brake system.

11. Install the tire and wheel assembly.

12. Top off the brake fluid level in the master cylinder. Check for leaks and proper brake operation.

13. Connect the negative battery cable to the battery.

Disc Brake Pads

REMOVAL & INSTALLATION

2WD

1. Before servicing the vehicle, refer to the Precautions Section.

2. Raise the vehicle and support it safely.

3. Remove the wheel and tire assembly.

4. When servicing the front pads, loosen the brake caliper upper side mounting bolt. Loosen and remove the lower side mounting bolt. Lift the caliper and suspend it so the hose is not stretched.

5. If equipped, remove the anti-squeal spring.

6. Remove the brake pads.

To install:

7. Siphon a small amount of brake fluid from the reservoir. Press in the brake caliper piston with the proper tool.

8. Before installing the new pads, check the disc thickness and disc runout.

9. Install the pad support plates.

10. Install the anti-squeal shims to each pad.

➡**Apply disc brake grease to both sides of the inner anti-squeal shims.**

11. Install the disc pads so the wear indicator plate is facing downward.

12. If removed, install the anti-squeal springs.

13. Carefully install the brake caliper so the boot is not wedged.

14. Install the wheel and tire assembly.

15. Check and adjust the fluid level. Apply the brake pedal several times.

16. Roadtest the vehicle for proper operation.

4WD

1. Before servicing the vehicle, refer to the Precautions Section.

2. Raise the vehicle and support it safely.

3. Remove the wheel and tire assembly.

4. Remove the clip, pins, and the anti-rattle spring.

5. Remove the pads and the anti-squeal shims.

6. Remove the caliper, but do not disconnect the brake hose.

To install:

7. Before installing the new pads, check the disc thickness and disc runout.

8. Siphon out a small amount of brake fluid from the reservoir.

9. Temporarily install the old inner brake pad. Press in the pistons with a C-clamp or equivalent. Remove the old inner brake pad.

10. Apply disc brake grease to both sides of the inner anti-squeal shim. Install the anti-squeal shims to the new pads.

11. Install the pads.

12. Install the anti-rattle springs and pins. Install the clip.

13. Install the caliper and the mounting bolts.

14. Install the wheel and tire assembly.

15. Check and adjust the fluid level. Apply the brake pedal several times.

16. Roadtest the vehicle for proper operation.

Brake Drums

REMOVAL & INSTALLATION

1. Before servicing the vehicle, refer to the Precautions Section.

2. Raise and safely support the vehicle.

3. Remove the rear wheel(s).

4. Remove the brake drum from the axle hub. If there is difficulty in removing the drum, insert a suitable tool through the hole in the rear of the backing plate, and

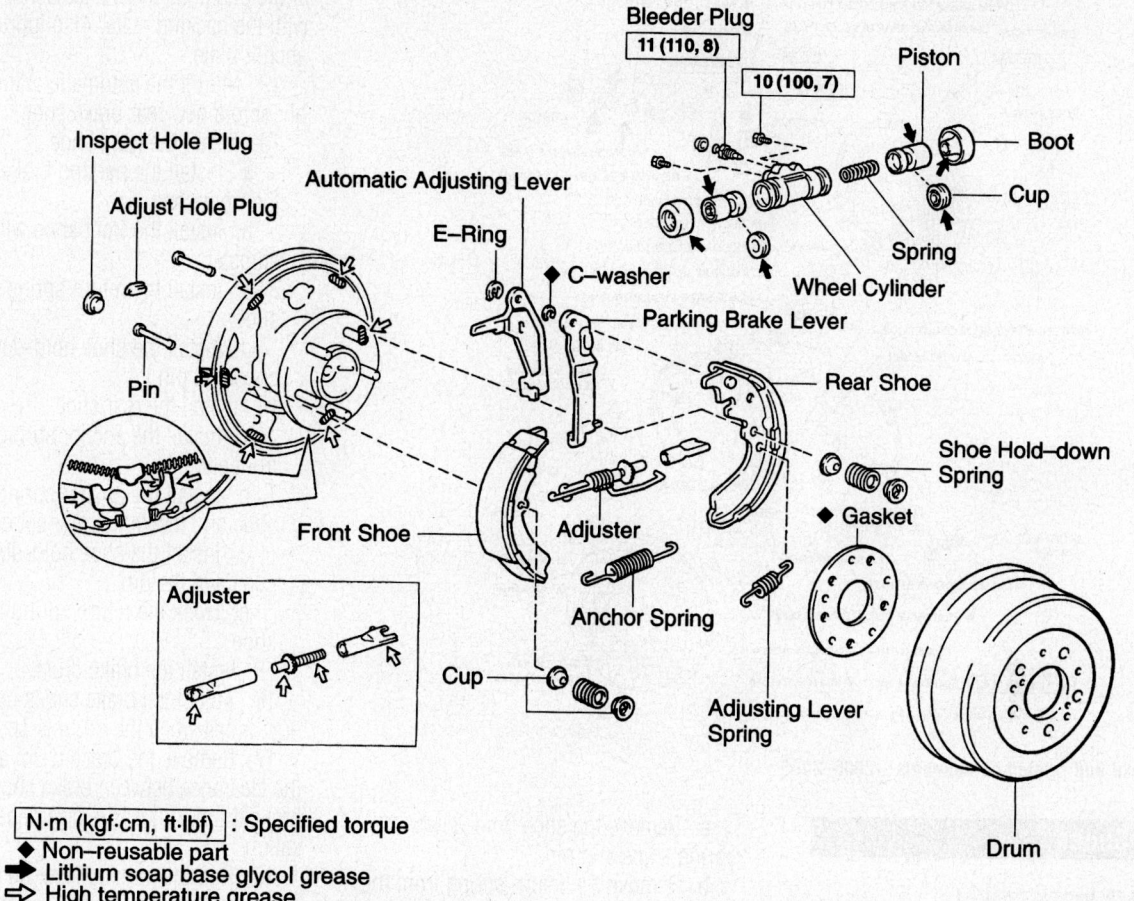

Rear brake and related components—2WD 2002–2004

N·m (kgf·cm, ft·lbf) : Specified torque
◆ Non–reusable part
➡ Lithium soap base glycol grease
⇨ High temperature grease

67170-TACO-G47

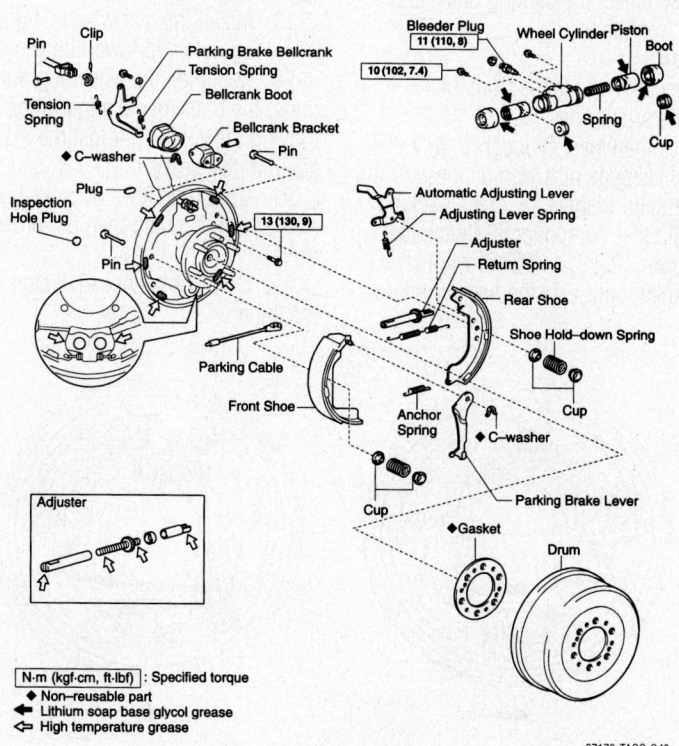

N·m (kgf·cm, ft·lbf) : Specified torque
◆ Non–reusable part
⬅ Lithium soap base glycol grease
⇦ High temperature grease

Rear brake and related components—4WD 2002–2004

67170-TACO-G48

hold the automatic adjusting lever away from the adjuster. Using another suitable tool at the same time, reduce the brake shoe adjuster by turning the adjusting wheel.

To install:

5. Install the brake drum and pull the parking brake lever all the way up until a clicking sound can no longer be heard.

6. Verify that the rear wheels will not turn. If the rear wheels turn, adjust the parking brake cable as necessary.

7. Release the parking brake and remove the brake drum. Measure the brake drum inside diameter and diameter of the brake shoes. Check that the difference between the diameters is the correct shoe clearance. Clearance is 0.024 inch (6mm) for 2002–2004 vehicles and 0.020 inch (5mm) for 2005–2006 vehicles.

8. If the brake shoe clearance is not correct, adjust the brake shoes until the clearance is correct.

9. Install the brake drum, replace the wheel(s), and safely lower the vehicle.

10. Road-test the vehicle for proper brake operation.

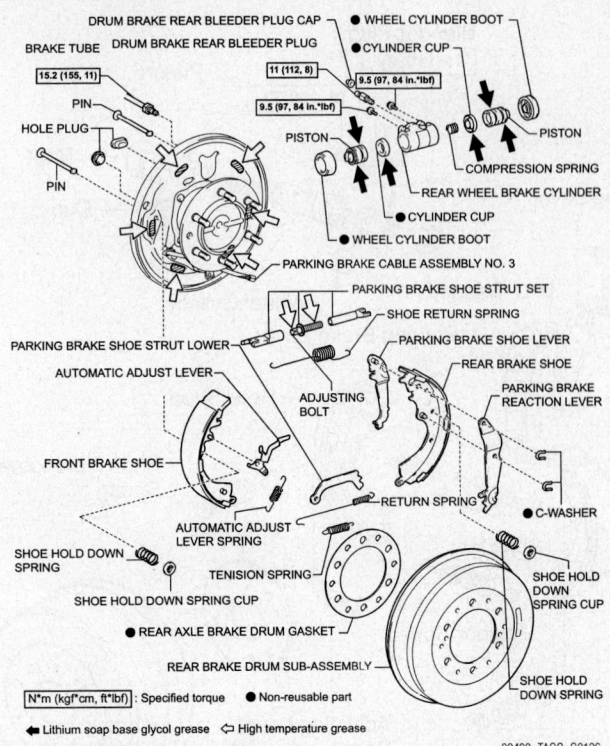

DRUM BRAKE REAR BLEEDER PLUG CAP
BRAKE TUBE — DRUM BRAKE REAR BLEEDER PLUG
● WHEEL CYLINDER BOOT
● CYLINDER CUP
15.2 (155, 11)
11 (112, 8)
9.5 (97, 84 in.*lbf)
PIN
HOLE PLUG
9.5 (97, 84 in.*lbf)
PISTON
● PISTON
PIN
COMPRESSION SPRING
REAR WHEEL BRAKE CYLINDER
● CYLINDER CUP
● WHEEL CYLINDER BOOT
PARKING BRAKE CABLE ASSEMBLY NO. 3
PARKING BRAKE SHOE STRUT SET
SHOE RETURN SPRING
PARKING BRAKE SHOE STRUT LOWER
PARKING BRAKE SHOE LEVER
AUTOMATIC ADJUST LEVER
REAR BRAKE SHOE
ADJUSTING BOLT
PARKING BRAKE REACTION LEVER
FRONT BRAKE SHOE
AUTOMATIC ADJUST LEVER SPRING
RETURN SPRING
● C-WASHER
SHOE HOLD DOWN SPRING
TENISION SPRING
SHOE HOLD DOWN SPRING CUP
SHOE HOLD DOWN SPRING CUP
● REAR AXLE BRAKE DRUM GASKET
REAR BRAKE DRUM SUB-ASSEMBLY
SHOE HOLD DOWN SPRING
N*m (kgf*cm, ft*lbf) : Specified torque   ● Non-reusable part
◀ Lithium soap base glycol grease   ◁ High temperature grease

09490_TACO_G0126

Rear brake and related components—2005–2006

Brake Shoes

REMOVAL & INSTALLATION

1. Before servicing the vehicle, refer to the Precautions Section.
2. Loosen the rear wheel lug nuts slightly.
3. Raise and support the vehicle safely.
4. Remove the wheel lug nuts and the wheel.
5. Remove the brake drum.
6. If the drum is difficult to remove, perform the following:
   a. Insert a flat prying tool through the hole in the brake drum and hold the automatic adjusting lever away from the adjuster.
   b. Reduce the brake shoe adjustment by turning the adjuster bolt with a brake tool.
   c. The drum should now be loose enough to remove without much effort.
7. Remove the rear shoe.
   a. Carefully unhook the return spring from the brake shoe.
   b. Remove the shoe hold-down spring, cups and the pin.
   c. Disconnect the anchor spring from the rear shoe and remove the rear shoe.
   d. Disconnect the anchor spring from the front shoe.
8. Remove the front shoe.

a. Remove the shoe hold-down spring, cups and pin.
b. Remove the return spring from the front shoe.
c. Remove the front shoe with the adjuster.
d. Disconnect the parking brake cable from the front shoe.

To install:

9. Inspect the shoes for signs of unusual wear or scoring.
10. Check the wheel cylinder for any sign of fluid seepage or frozen pistons.
11. Clean and inspect the brake backing plate and all other components. Check that the brake drum inner diameter is within specified limits. Lubricate the backing plate

at the positions the brakes come in contact with the backing plate. Also lubricate the anchor plate.

12. Mount the automatic adjuster assembly onto a new rear brake shoe.
13. Install the front shoe.
   a. Install the parking brake cable to the front shoe.
   b. Install the front shoe with the adjuster.
   c. Install the return spring to the front shoe.
   d. Install the shoe hold-down spring, cups and pin.
14. Install the rear shoe.
   a. Install the anchor spring to the front shoe.
   b. Install the anchor spring to the rear shoe and install the rear shoe.
   c. Install the shoe hold-down spring, cups and the pin.
   d. Hook the return spring to the brake shoe.
15. Install the brake drum.
16. Adjust the brake shoes until a slight drag is felt when the drum is spun by hand.
17. Remove the brake drum and check the clearance between brake shoes and brake drum. Adjust the clearance to specification.
18. Pull the parking lever all the way up until a clicking sound can no longer be heard. Verify that the drum doesn't turn. If the drum turns, adjust the parking brake cable.
19. Install the rear wheels, tighten the wheel lug nuts and lower the vehicle.
20. Retighten the wheel lug nuts and pump the brake pedal a few times before moving the vehicle. Adjust the rear brakes again if necessary.
21. Check the level of brake fluid in the master cylinder, and then perform a test drive.
22. Connect the negative battery cable to the battery.

LH:
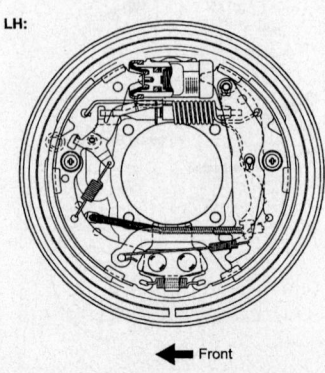
← Front

RH:
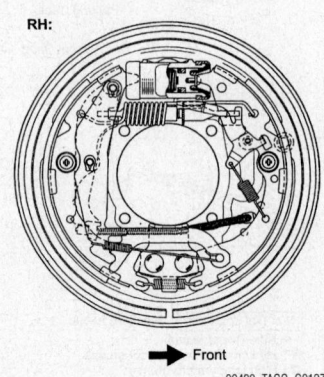
→ Front

09490_TACO_G0127

Rear brake shoes—assembled view

TOYOTA

Tundra

BRAKES **20-84**
DRIVE TRAIN **20-71**
ENGINE REPAIR **20-14**
FRONT SUSPENSION **20-79**
FUEL SYSTEM **20-67**
REAR SUSPENSION **20-83**
SPECIFICATIONS AND
  MAINTENANCE CHARTS **20-3**
Engine and Vehicle
  Identification 20-3
General Engine Specifications 20-3
Engine Tune-Up Specifications 20-4
Firing Order 20-4
Accessory Drive Belt Routing 20-5
Capacities 20-6
Valve Specifications 20-6
Camshaft Specifications 20-7
Crankshaft and Connecting Rod
  Specifications 20-8
Piston and Ring Specifications 20-8
Torque Specifications 20-9
Wheel Alignment 20-11
Tire, Wheel and Ball Joint
  Specifications 20-12
Brake Specifications 20-12
Scheduled Maintenance
  Intervals 20-13
STEERING **20-78**

A
Air Bag 20-78
  Disarming 20-78
  Precautions 20-78
Alternator 20-14
  Removal & Installation 20-14
Automatic Transmission
  Assembly 20-73
  Removal & Installation 20-73

B
Brake Caliper 20-84
  Removal & Installation 20-84
Brake Drums 20-85
  Removal & Installation 20-85
Brake Shoes 20-86
  Removal & Installation 20-86

C
Camshaft and Valve Lifters 20-37
  Inspection 20-49
  Removal & Installation 20-37
Clutch Assembly 20-72
  Removal & Installation 20-72
CV-Joints 20-75
  Overhaul 20-75
Cylinder Head 20-27
  Removal & Installation 20-27

D
Disc Brake Pads 20-85
  Removal & Installation 20-85

E
Engine Assembly 20-15
  Removal & Installation 20-15
Exhaust Manifold 20-37
  Removal & Installation 20-37

F
Front Wheel Bearing 20-82
  Removal & Installation 20-82
Fuel Filter 20-67
  Removal & Installation 20-67
Fuel Injector 20-69
  Removal & Installation 20-69
Fuel Pump 20-68
  Removal & Installation 20-68
Fuel System Pressure 20-67
  Relieving 20-67
Fuel System Service
  Precautions 20-67

H
Halfshaft 20-75
  Removal & Installation 20-75
Heater Core 20-20
  Removal & Installation 20-20
Hydraulic Clutch System 20-72
  Bleeding 20-72

I
Ignition Timing 20-14
  Adjustment 20-14
Intake Manifold 20-34
  Removal & Installation 20-34

L
Leaf Spring 20-83
  Removal & Installation 20-83
Lower Ball Joint 20-80
  Removal & Installation 20-80
Lower Control Arm 20-81
  Lower Control Arm Bushing
    Replacement 20-81
  Removal & Installation 20-81

M
Manual Transmission 20-71
  Removal & Installation 20-71

O
Oil Pan 20-55
  Removal & Installation 20-55
Oil Pump 20-56
  Removal & Installation 20-56

P
Pinion Seal 20-77
  Removal & Installation 20-77
Piston and Ring 20-66
  Positioning 20-66
Power Steering Gear 20-78
  Removal & Installation 20-78

R
Rear Axle Shaft, Bearing and
  Seal 20-76
  Removal & Installation 20-76
Rear Main Seal 20-66
  Removal & Installation 20-66

S
Shock Absorber (Rear) 20-83
  Removal & Installation 20-83
Shock Absorber (Front) 20-79
  Assembly & Disassembly 20-79
  Removal & Installation 20-79
Stabilizer Bar (Rear) 20-84
  Removal & Installation 20-84
Stabilizer Bar (Front) 20-80
  Removal & Installation 20-80
Starter Motor 20-58
  Removal & Installation 20-58

T

Timing Belt, Cover and
Crankshaft Seal 20-59
Removal & Installation...........20-59
Timing Chain, Sprockets, Front
Cover and Seal 20-64
Removal & Installation...........20-64
Transfer Case Assembly...............20-74
Removal & Installation...........20-74

U

Upper Ball Joint.........................20-80
Removal & Installation...........20-80
Upper Control Arm 20-81
Removal & Installation...........20-81
Upper Control Arm Bushing
Replacement........................20-81

V

Valve Lash 20-49
Adjustment.............................20-49

W

Water Pump................................20-18
Removal & Installation...........20-18

SPECIFICATIONS AND MAINTENANCE CHARTS

ENGINE AND VEHICLE IDENTIFICATION

		Engine					Model Year	
Code ① ②	Liters (cc)	Cu. In.	Cyl.	Fuel Sys.	Engine Type	Eng. Mfg.	Code ③	Year
5VZ-FE/N	3.4 (3378)	206	6	MFI	DOHC	Toyota	2	2002
1GR-FE/U	4.0 (3956)	241	6	MFI	DOHC	Toyota	3	2003
2UZ-FE/T	4.7 (4664)	285	8	SFI	DOHC	Toyota	4	2004
							5	2005
							6	2006

SFI: Sequential Fuel Injection

MFI: Multi-port Fuel Injection

DOHC: Double Overhead Camshaft

① Except 1GR-FE/U engine: stamped on the left side of the engine block. Engine ID is fifth character of the VIN number.

② 1GR-FE/U engine: stamped on the right side of the engine block. Ingine ID is the fifth character of the VIN number.

③ 10th digit of the VIN number

09490_TUND_C0001

GENERAL ENGINE SPECIFICATIONS

Year	Model	Engine Displacement Liters	Engine Series Code/ID	Net Horsepower @ rpm	Net Torque @ rpm (ft. lbs.)	Bore x Stroke (in.)	Compression Ratio	Oil Pressure @ rpm
2002	Tundra	3.4	5VZ-FE/N	190@4800	220@3600	3.68x3.23	9.6:1	36-75@3000
		4.7	2UZ-FE/T	245@4800	315@3400	3.70x3.70	9.6:1	45-65@3000
2003	Tundra	3.4	5VZ-FE/N	190@4800	220@3600	3.68x3.23	9.6:1	36-75@3000
		4.7	2UZ-FE/T	245@4800	315@3400	3.70x3.70	9.6:1	45-65@3000
2004	Tundra	3.4	5VZ-FE/N	190@4800	220@3600	3.68x3.23	9.6:1	36-75@3000
		4.7	2UZ-FE/T	245@4800	315@3400	3.70x3.70	9.6:1	45-65@3000
2005	Tundra	4.0	1GR-FE/U	236@5200	266@4000	3.70x3.74	NA	43-85@3000
		4.7	2UZ-FE/T	271@5400	313@3400	3.70x3.70	9.6:1	43-85@3000
2006	Tundra	4.0	1GR-FE/U	236@5200	266@4000	3.70x3.74	NA	43-85@3000
		4.7	2UZ-FE/T	271@5400	313@3400	3.70x3.70	9.6:1	43-85@3000

NA: Not Available

09490_TUND_C0002

ENGINE TUNE-UP SPECIFICATIONS

Year	Engine Displacement Liters	Engine Code/ID	Spark Plug Gap (in.)	Ignition Timing (deg.)	Fuel Pump (psi)	Idle Speed (rpm) MT	Idle Speed (rpm) AT	Valve Clearance Intake	Valve Clearance Exhaust
2002	3.4	5VZ-FE/N	0.043	8-12 ①	38-44	650-750	650-750	0.006-0.010	0.010-0.014
	4.7	2UZ-FE/T	0.043	8-12 ①	38-44	650-750	650-750	0.006-0.009	0.011-0.014
2003	3.4	5VZ-FE/N	0.043	8-12 ①	38-44	650-750	650-750	0.006-0.010	0.010-0.014
	4.7	2UZ-FE/T	0.043	8-12 ①	38-44	650-750	650-750	0.006-0.009	0.011-0.014
2004	3.4	5VZ-FE/N	0.043	8-12 ①	38-44	650-750	650-750	0.006-0.010	0.010-0.014
	4.7	2UZ-FE/T	0.043	8-12 ①	38-44	650-750	650-750	0.006-0.009	0.011-0.014
2005	4.0	1GR-FE/U	0.039-0.043	7-24B ①	40.8-41.7	650-750	650-750	0.006-0.010	0.011-0.015
	4.7	2UZ-FE/T	0.043	8-12 ①	38-44	650-750	650-750	0.006-0.009	0.011-0.014
2006	4.0	1GR-FE/U	0.039-0.043	7-24B ①	40.8-41.7	650-750	650-750	0.006-0.010	0.011-0.015
	4.7	2UZ-FE/T	0.043	8-12 ①	38-44	650-750	650-750	0.006-0.009	0.011-0.014

NOTE: The Vehicle Emission Control Information label often reflects specification changes made during production.

The label figures must be used if they differ from those in this chart.

B: Before top dead center

① With terminals TE1 and E1 connected to DLC1

09490_TUND_C0003

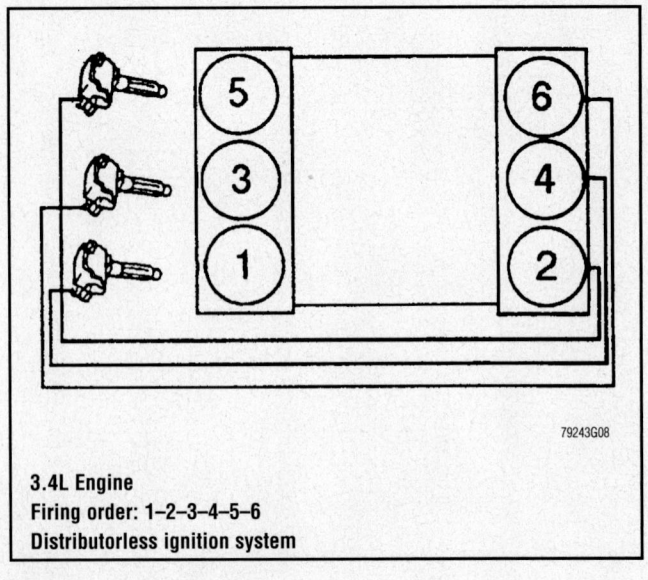

3.4L Engine
Firing order: 1–2–3–4–5–6
Distributorless ignition system

79243G08

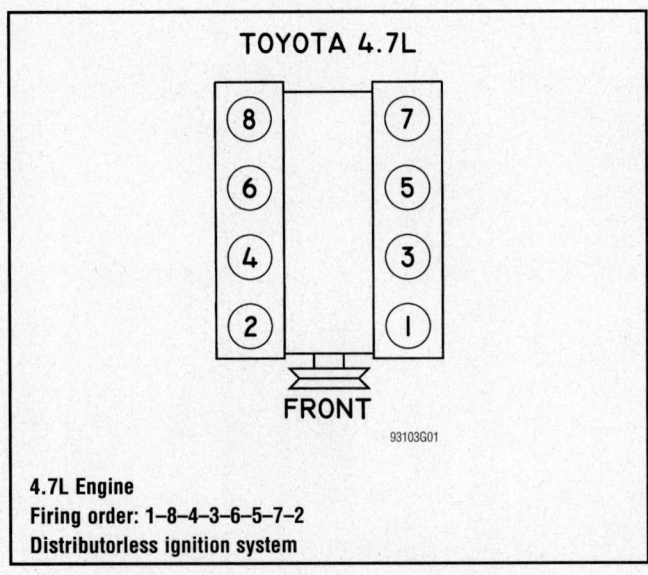

TOYOTA 4.7L

FRONT

93103G01

4.7L Engine
Firing order: 1–8–4–3–6–5–7–2
Distributorless ignition system

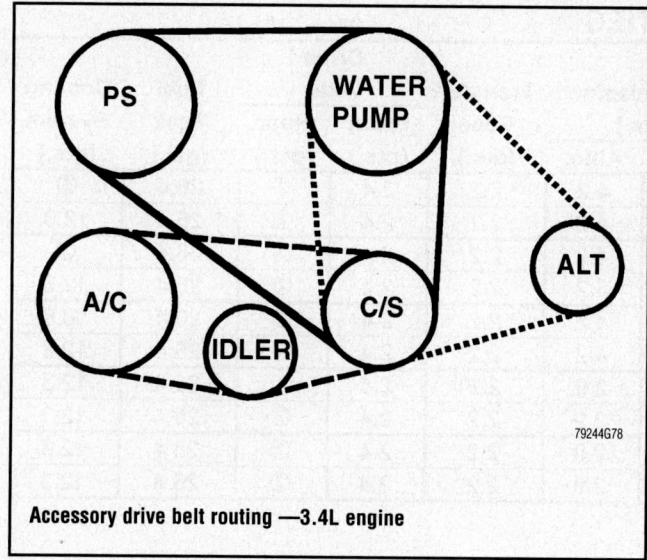

Accessory drive belt routing —3.4L engine

79244G78

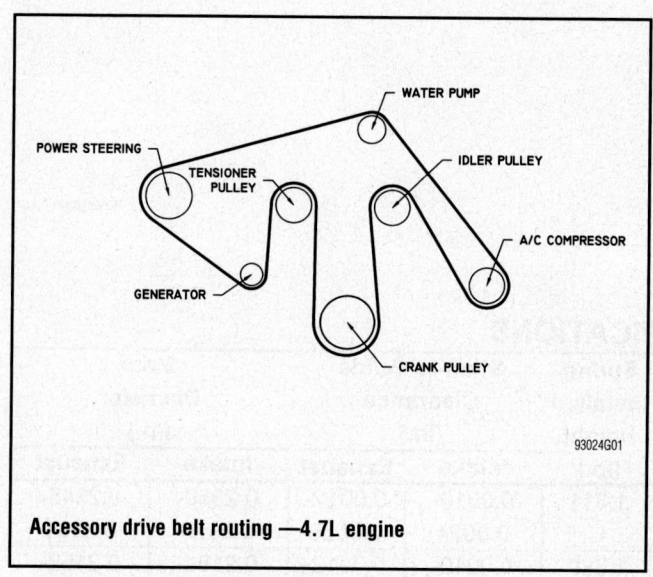

Accessory drive belt routing —4.7L engine

93024G01

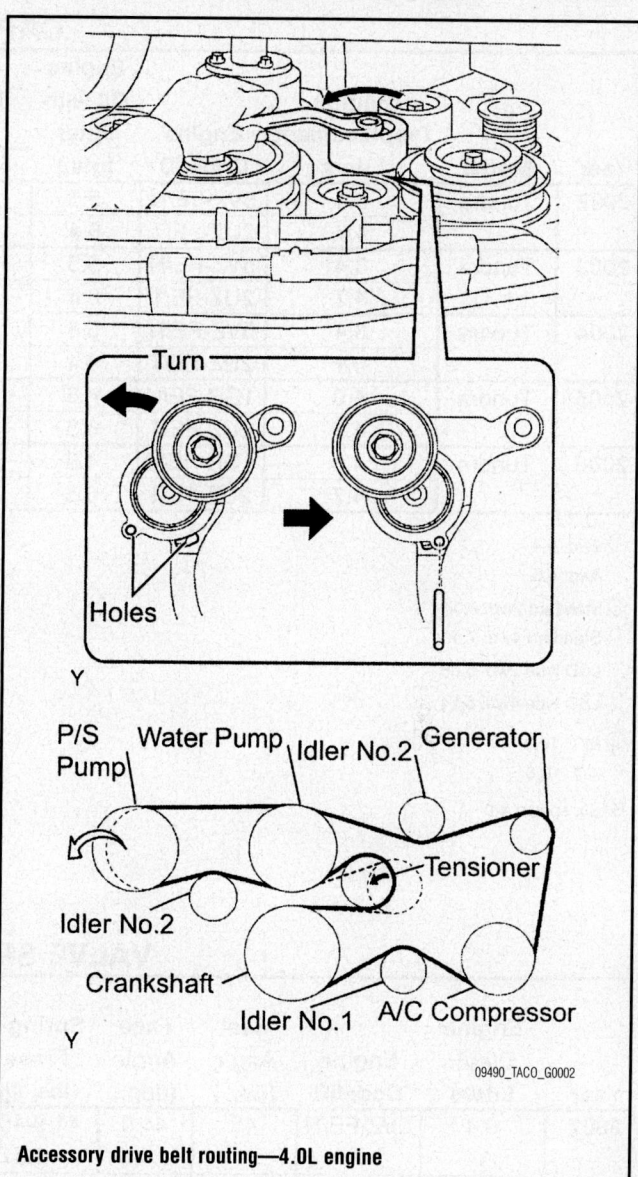

Accessory drive belt routing—4.0L engine

09490_TACO_G0002

CAPACITIES

Year	Model	Engine Displacement Liters	Engine Code/ID	Engine Oil with Filter (qts.)	Transmission (pts.) 5-Spd	Transmission (pts.) Auto.	Transfer Case (pts.)	Drive Axle Front (pts.)	Drive Axle Rear (pts.)	Fuel Tank (gal.)	Cooling System (qts.)
2002	Tundra	3.4	5VZ-FE/N	5.5	①	4.2	2.2	2.4	②	26.4	③
		4.7	2UZ-FE/T	6.4	①	4.2	2.2	2.4	②	26.4	12.3
2003	Tundra	3.4	5VZ-FE/N	5.5	①	4.2	2.2	2.4	②	26.4	③
		4.7	2UZ-FE/T	6.4	①	4.2	2.2	2.4	②	26.4	12.3
2004	Tundra	3.4	5VZ-FE/N	5.5	①	4.2	2.2	2.4	②	26.4	10.5
		4.7	2UZ-FE/T	6.4	①	4.2	2.2	2.4	②	26.4	12.3
2005	Tundra	4.0	1GR-FE/U	4.8	④	2.9	2.2	2.4	②	26.4	12.3
		4.7	2UZ-FE/T	6.5	④	2.9	2.2	2.4	②	26.4	12.3
2006	Tundra	4.0	1GR-FE/U	4.8	④	2.9	2.2	2.4	②	26.4	12.3
		4.7	2UZ-FE/T	6.5	④	2.9	2.2	2.4	②	26.4	12.3

① 2wd: 5.4
   4wd: 4.6

② Standard 2wd: 8.04
   Standard 4wd: 7.40
   LSD type 2wd: 6.66
   LSD type 4wd: 6.24

③ M/T: 10.3
   A/T: 10.0

④ Six speed: 4.0

09490_TUND_C0004

VALVE SPECIFICATIONS

Year	Engine Displ. Liters	Engine Code/ID	Seat Angle (deg.)	Face Angle (deg.)	Spring Test Pressure (lbs. @ in.)	Spring Installed Height (in.)	Stem-to-Guide Clearance (in.) Intake	Stem-to-Guide Clearance (in.) Exhaust	Stem Diameter (in.) Intake	Stem Diameter (in.) Exhaust
2002	3.4	5VZ-FE/N	45	44.5	41.9-46.3@ 1.311	1.311	0.0010-0.0024	0.0012-0.0026	0.2350-0.2356	0.2348-0.2354
	4.7	2UZ-FE/T	45	44.5	45.9-50.7@ 1.378	1.380	0.0010-0.0024	0.0012-0.0026	0.2154-0.2159	0.2152-0.2157
2003	3.4	5VZ-FE/N	45	44.5	41.9-46.3@ 1.311	1.311	0.0010-0.0024	0.0012-0.0026	0.2350-0.2356	0.2348-0.2354
	4.7	2UZ-FE/T	45	44.5	45.9-50.7@ 1.378	1.380	0.0010-0.0024	0.0012-0.0026	0.2154-0.2159	0.2152-0.2157
2004	3.4	5VZ-FE/N	45	44.5	41.9-46.3@ 1.311	1.311	0.0010-0.0024	0.0012-0.0026	0.2350-0.2356	0.2348-0.2354
	4.7	2UZ-FE/T	45	44.5	45.9-50.7@ 1.378	1.380	0.0010-0.0024	0.0012-0.0026	0.2154-0.2159	0.2152-0.2157
2005	4.0	1GR-FE/U	NA	44.5	41.9-46.3@ 1.311	1.882	0.0010-0.0024	0.0012-0.0026	0.2154 0.2159	0.2152 0.2157
	4.7	2UZ-FE/T	45	44.5	45.9-50.7@ 1.378	1.380	0.0010-0.0024	0.0012-0.0026	0.2154-0.2159	0.2152-0.2157
2006	4.0	1GR-FE/U	NA	44.5	41.9-46.3@ 1.311	1.882	0.0010-0.0024	0.0012-0.0026	0.2154 0.2159	0.2152 0.2157
	4.7	2UZ-FE/T	45	44.5	45.9-50.7@ 1.378	1.380	0.0010-0.0024	0.0012-0.0026	0.2154-0.2159	0.2152-0.2157

NA: Not Available

09490_TUND_C0005

CAMSHAFT SPECIFICATIONS
All measurements in inches unless noted

Year	Engine Displacement Liters	Engine Code/ID	Journal Dia.	Brg. Oil Clearance	Shaft End-play ①	Circle Runout	Lobe Height	
							Intake	Exhaust
2002	3.4	5VZ-FE/N	1.0610-1.0616	0.0014-0.0028	0.0013-0.0031	0.0024	1.6657-1.6697	1.6520-1.6559
	4.7	2UZ-FE/T	1.0612-1.0618	0.0012-0.0026	②	0.0031	1.6512-1.6551	1.6520-1.6559
2003	3.4	5VZ-FE/N	1.0610-1.0616	0.0014-0.0028	0.0013-0.0031	0.0024	1.6657-1.6697	1.6520-1.6559
	4.7	2UZ-FE/T	1.0612-1.0618	0.0012-0.0026	②	0.0031	1.6512-1.6551	1.6520-1.6559
2004	3.4	5VZ-FE/N	1.0610-1.0616	0.0014-0.0028	0.0013-0.0031	0.0024	1.6657-1.6697	1.6520-1.6559
	4.7	2UZ-FE/T	1.0612-1.0618	0.0012-0.0026	②	0.0031	1.6512-1.6551	1.6520-1.6559
2005	4.0	1GR-FE/U	③	④	0.0160-0.0350	0.0024	⑤	⑤
	4.7	2UZ-FE/T	1.0612-1.0618	0.0012-0.0026	②	0.0031	1.6512-1.6551	1.6520-1.6559
2006	4.0	1GR-FE/U	③	④	0.0160-0.0350	0.0024	⑤	⑤
	4.7	2UZ-FE/T	1.0612-1.0618	0.0012-0.0026	②	0.0031	1.6512-1.6551	1.6520-1.6559

① Thrust clearance

② Intake: 0.0016-0.0033
   Exhaust: 0.0011-0.0030

③ No. 1: 1.4162-1.4167
   All others: 0.9039-0.9045

④ No. 1: 0.0016-0.0031
   All others: 0.0010-0.0024

⑤ No. 1 camshaft: 1.7389-1.7428
   No. 2 camshaft: 1.7551-1.7591
   No. 3 camshaft (sub assembly): 1.7389-1.7428
   No. 2 camshaft (sub assembly): 1.7551-1.7591

09490_TUND_C0006

CRANKSHAFT AND CONNECTING ROD SPECIFICATIONS

All measurements are given in inches.

Year	Engine Displ. Liters	Engine Code/ID	Crankshaft				Connecting Rod		
			Main Brg. Journal Dia.	Main Brg. Oil Clearance	Shaft End-play	Thrust on No.	Journal Diameter	Oil Clearance	Side Clearance
2002	3.4	5VZ-FE/N	2.5191-2.5197	0.0008-0.0015	0.0008-0.0087	2	2.1648-2.1654	0.0009-0.0021	0.0059-0.0130
	4.7	2UZ-FE/T	2.6373-2.6378	0.0016-0.0023	0.0008-0.0087	3	2.0465-2.0472	0.0011-0.0021	0.0063-0.0138
2003	3.4	5VZ-FE/N	2.5191-2.5197	0.0008-0.0015	0.0008-0.0087	2	2.1648-2.1654	0.0009-0.0021	0.0059-0.0130
	4.7	2UZ-FE/T	2.6373-2.6378	0.0016-0.0023	0.0008-0.0087	3	2.0465-2.0472	0.0011-0.0021	0.0063-0.0138
2004	3.4	5VZ-FE/N	2.5191-2.5197	0.0008-0.0015	0.0008-0.0087	2	2.1648-2.1654	0.0009-0.0021	0.0059-0.0130
	4.7	2UZ-FE/T	2.6373-2.6378	0.0016-0.0023	0.0008-0.0087	3	2.0465-2.0472	0.0011-0.0021	0.0063-0.0138
2005	4.0	1GR-FE/U	2.8342-2.8346	0.0007-0.0012	NA	NA	NA	0.0010-0.0018	0.0059-0.0118
	4.7	2UZ-FE/T	2.6373-2.6378	0.0016-0.0023	0.0008-0.0087	NA	2.0465-2.0472	0.0011-0.0021	0.0063-0.0138
2006	4.0	1GR-FE/U	2.8342-2.8346	0.0007-0.0012	NA	NA	NA	0.0010-0.0018	0.0059-0.0118
	4.7	2UZ-FE/T	2.6373-2.6378	0.0016-0.0023	0.0008-0.0087	NA	2.0465-2.0472	0.0011-0.0021	0.0063-0.0138

NA: Not Available

09490_TUND_C0007

PISTON AND RING SPECIFICATIONS

All measurements are given in inches.

Year	Engine Displ. Liters	Engine Code/ID	Piston Clearance	Ring Gap			Ring Side Clearance		
				Top Compression	Bottom Compression	Oil Control	Top Compression	Bottom Compression	Oil Control
2002	3.4	5VZ-FE/N	0.0053-0.0060	0.0118-0.0197	0.0157-0.0236	0.0059-0.0217	0.0016-0.0031	0.0012-0.0028	SNUG
	4.7	2UZ-FE/T	0.0035-0.0044	0.0118-0.0197	0.0157-0.0256	0.0051-0.0189	0.0012-0.0031	0.0012-0.0028	SNUG
2003	3.4	5VZ-FE/N	0.0053-0.0060	0.0118-0.0197	0.0157-0.0236	0.0059-0.0217	0.0016-0.0031	0.0012-0.0028	SNUG
	4.7	2UZ-FE/T	0.0035-0.0044	0.0118-0.0197	0.0157-0.0256	0.0051-0.0189	0.0012-0.0031	0.0012-0.0028	SNUG
2004	3.4	5VZ-FE/N	0.0053-0.0060	0.0118-0.0197	0.0157-0.0236	0.0059-0.0217	0.0016-0.0031	0.0012-0.0028	SNUG
	4.7	2UZ-FE/T	0.0035-0.0044	0.0118-0.0197	0.0157-0.0256	0.0051-0.0189	0.0012-0.0031	0.0012-0.0028	SNUG
2005	4.0	1GR-FE/U	0.0031-0.0040	0.0118-0.0157	0.0157-0.0197	0.0039-0.0157	0.0008-0.0028	0.0008-0.0024	0.0028-0.0060
	4.7	2UZ-FE/T	0.0035-0.0044	0.0118-0.0197	0.0157-0.0256	0.0051-0.0189	0.0012-0.0031	0.0012-0.0028	SNUG
2006	4.0	1GR-FE/U	0.0031-0.0040	0.0118-0.0157	0.0157-0.0197	0.0039-0.0157	0.0008-0.0028	0.0008-0.0024	0.0028-0.0060
	4.7	2UZ-FE/T	0.0035-0.0044	0.0118-0.0197	0.0157-0.0256	0.0051-0.0189	0.0012-0.0031	0.0012-0.0028	SNUG

09490_TUND_C0008

TORQUE SPECIFICATIONS
All readings in ft. lbs.

Year	Engine Displacement Liters	Engine Code/ID	Cylinder Head Bolts	Main Bearing Bolts	Rod Bearing Bolts	Crankshaft Damper Bolts	Flywheel Bolts	Manifold Intake	Manifold Exhaust	Spark Plugs	Oil Pan Drain Plug
2002	3.4	5VZ-FE/N	①	②	③	217	63	13	30	13	28
	4.7	2UZ-FE/T	④	⑤	③	181	⑥	13	33	13	29
2003	3.4	5VZ-FE/N	①	②	③	217	63	13	30	13	28
	4.7	2UZ-FE/T	④	⑤	③	181	⑥	13	33	13	29
2004	3.4	5VZ-FE/N	①	②	③	217	63	13	30	13	28
	4.7	2UZ-FE/T	④	⑤	③	181	⑥	13	33	13	29
2005	4.0	1GR-FE/U	⑦	⑧	⑨	185	61	19	16	13	30
	4.7	2UZ-FE/T	⑩	⑤	⑨	181	⑥	13	32	13	29
2006	4.0	1GR-FE/U	⑦	⑧	⑨	185	61	19	16	13	30
	4.7	2UZ-FE/T	⑩	⑤	⑨	181	⑥	13	32	13	29

① Step 1: 25 ft. lbs.
　Step 2: Plus 90 degrees
　Step 3: Plus 90 degrees
　Recessed head: 13 ft. lbs.

② Step 1: 45 ft. lbs.
　Step 2: Plus 90 degrees

③ Step 1: 18 ft. lbs.
　Step 2: Plus 90 degrees

④ Step 1: 24 ft. lbs.
　Step 2: Plus 180 degrees

⑤ Step 1: 20 ft. lbs.
　Step 2: Plus 90 degrees

⑥ Step 1: 35 ft. lbs.
　Step 2: Plus 90 degrees

⑦ right side: 27 ft. lbs. Then + 90 degrees. Then + 180 degrees.
　left side (recessed head): 27 ft. lbs. Then plus 180 degrees
　left side (0.55 inch head): 22 ft. lbs.

⑧ 12 pointed head: 45 ft. lbs. Then plus 90 degrees
　12mm head: 18 ft. lbs.

⑨ Step 1: 18 ft. lbs.
　Step 2: Plus 90 degrees

⑩ Step 1: 30 ft. lbs.
　Step 2: Plus 90 degrees
　Step 3: Plus 90 degrees

09490_TUND_C0009

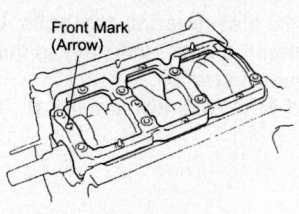

Install the main bearing cap with the front mark facing forward.

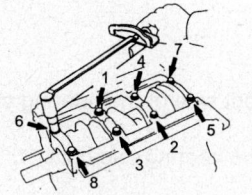

Apply a light coat of engine oil on the threads and under the heads of the main bearing cap bolts.
Install and uniformly tighten the 8 main bearing cap bolts in several passes, in the sequence shown.
Torque: 61 N·m (625 kgf·cm, 45 ft·lbf)

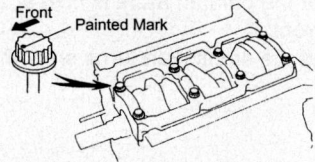

Mark the front of the main bearing cap bolt with paint.

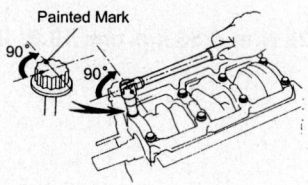

Retighten the main bearing cap bolts by 90° in the numerical order shown.
Check that the painted mark is now at a 90° angle to the front.
Check that the crankshaft turns smoothly.
Check the crankshaft thrust clearance.

09490_TACO_G0004

Main bearing torque sequence—3.4L engine

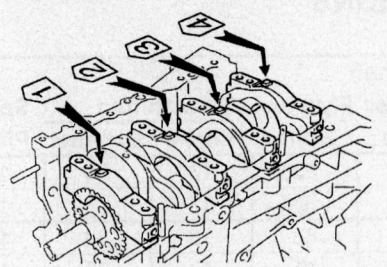

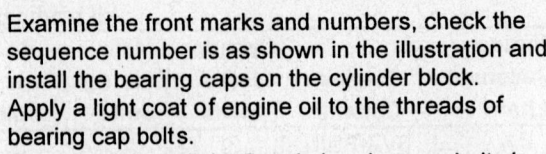

Examine the front marks and numbers, check the sequence number is as shown in the illustration and install the bearing caps on the cylinder block.
Apply a light coat of engine oil to the threads of bearing cap bolts.
Temporarily install the 8 main bearing cap bolts in the inside positions.

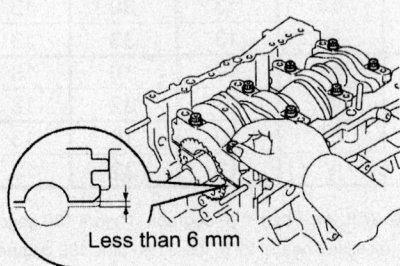

Less than 6 mm

Install the main bearing caps. Tighten the 2 bolts for each bearing cap until the clearance between the bearing cap and the cylinder block is under 6 mm (0.23 in.).

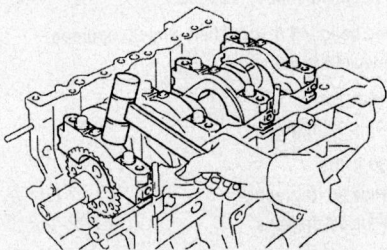

Using a plastic-faced hammer, lightly tap the bearing cap to ensure a proper fit.
Apply a light coat of engine oil to the threads of main bearing cap bolts.

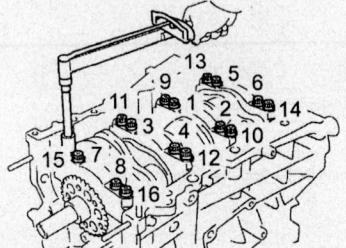

Install the 16 main bearing cap bolts. Using several steps, tighten the bolts uniformly in the sequence shown in the illustration.
Torque: 61 N m (622 kgf cm, 45 ft. lbf)

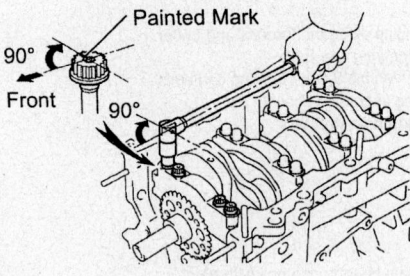

Painted Mark

90°
Front
90°

Mark the front side of the bearing cap bolts with paint.
Retighten the bearing cap bolts 90° in the sequence as shown.
Check that the painted mark is now at a 90° angle from the front.
Check that the crankshaft turns smoothly.

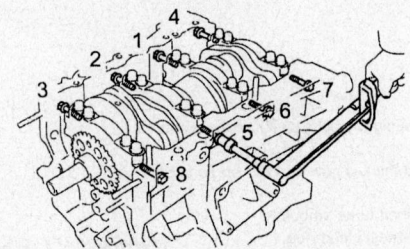

Using several steps, tighten the 8 main bearing cap bolts uniformly in the sequence shown in the illustration.
Torque: 25 N m (255 kgf cm, 18 ft. lbf)

09490_TACO_G0006

Main bearing torque sequence—4.0L engine

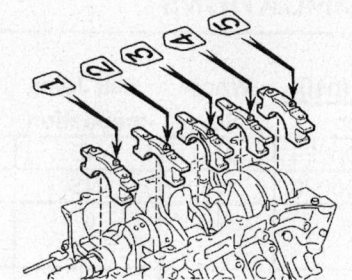

Install the 5 main bearing caps in their proper locations.

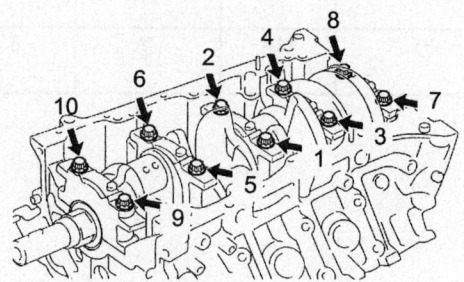

Apply a light coat of engine oil on the threads and under the main bearing cap bolts.

Install and uniformly tighten the 10 main bearing cap bolts in several passes, in the sequence shown.

Torque: 27 N·m (275 kgf·cm, 20 ft·lbf)

If any one of the main bearing cap bolts does not meet the torque specification, replace the main bearing cap bolt.

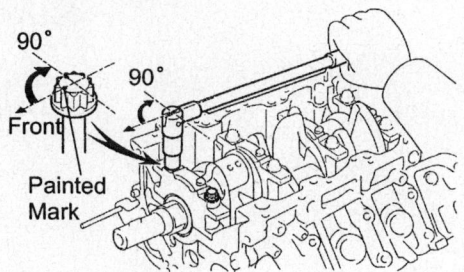

Mark the front of the main bearing cap bolt with paint.

Retighten the main bearing cap bolts by 90° in the numerical order shown.

Check that the painted mark is now at a 90° angle to the front.

Check that the crankshaft turns smoothly.

Check crankshaft thrust clearance

09490_TUND_G0001

Main bearing torque sequence—4.7L engine

WHEEL ALIGNMENT

| Year | Model | Caster | | Camber | | Toe-in (in.) | Steering Axis Inclination (Deg.) |
		Range (+/-Deg.)	Preferred Setting (Deg.)	Range (+/-Deg.)	Preferred Setting (Deg.)		
2002	2WD	0.75	2.05	0.75	-0.12	0.06+/-0.08	10.87
	4WD	0.75	1.07	0.75	0.33	0.07+/-0.08	10.41
2003	2WD	0.75	2.05	0.75	-0.12	0.06+/-0.08	10.87
	4WD	0.75	1.07	0.75	0.33	0.07+/-0.08	10.41
2004	2WD	0.75	2.05	0.75	-0.12	0.06+/-0.08	10.87
	4WD	0.75	1.07	0.75	0.33	0.07+/-0.08	10.41
2005	2WD	0.75	2.05	0.75	-0.12	0.06+/-0.08	10.87
	4WD	0.75	1.07	0.75	0.33	0.07+/-0.08	10.41
2006	2WD	0.75	2.05	0.75	-0.12	0.06+/-0.08	10.87
	4WD	0.75	1.07	0.75	0.33	0.07+/-0.08	10.41

All alignment figures based on nominal ride height and standard tires

09490_TUND_C0011

TIRE, WHEEL AND BALL JOINT SPECIFICATIONS

| Year | Model | OEM Tires | | Tire Pressures (psi) | | Wheel Size | Ball Joint Inspection | Lugnut Torque (ft. lbs.) |
		Standard	Optional	Front	Rear			
2002	Tundra	P245/70R16	P265/70R16	26	Std:35/Opt:29	NA	NS	83
2003	Tundra	P245/70R16	P265/70R16	26	Std:35/Opt:29	NA	NS	83
2004	Tundra	P245/70R16	P265/70R16 P265/65R17	①	①	NA	②	83
2005	Tundra	P245/70R16	P265/65R17	①	①	NA	②	83
2006	Tundra	P245/70R16	P265/65R17	①	①	NA	②	83

OEM: Original Equipment Manufacturer

PSI: Pounds Per Square Inch

STD: Standard

OPT: Optional

NS: Not specified by manufacturer

NA: Not available

① See placard on vehicle

② Upper: turning torque within 6-39 inch lbs.

   Lower: turning torque within 1-22 inch lbs.

09490_TUND_C0012

BRAKE SPECIFICATIONS
All measurements in inches unless noted

| Year | Brake Disc | | | Brake Drum | | Minimum Lining Thickness | | Caliper Mounting Bolts (ft. lbs.) |
	Original Thickness	Minimum Thickness	Maximum Runout	Original Inside Diameter	Maximum Machine Diameter	Front	Rear	
2002	1.102	1.024	0.0028	11.61	11.69	0.039	0.039	90
2003	1.102	1.024	0.0028	11.61	11.69	0.039	0.039	90
2004	1.102	1.024	0.0028	11.61	11.69	0.039	0.039	90
2005	1.102	1.024	0.0028	11.61	11.69	0.039	0.039	90
2006	1.102	1.024	0.0028	11.61	11.69	0.039	0.039	90

09490_TUND_C0010

SCHEDULED MAINTENANCE INTERVALS
TOYOTA—TUNDRA

TO BE SERVICED	TYPE OF SERVICE	VEHICLE MILEAGE INTERVAL (x1000)																		
		5	10	15	20	25	30	35	40	45	50	55	60	65	70	75	80	85	90	95
Automatic transmission and differential fluid	S/I			✓			✓			✓			✓			✓			✓	
Ball joints and boots	S/I			✓			✓			✓			✓			✓			✓	
Brake system	S/I			✓			✓			✓			✓			✓			✓	
Charcoal canister	S/I												✓							
Drive belts	S/I						✓						✓						✓	
Driveshaft bushing	L						✓						✓						✓	
Engine coolant	R						✓						✓						✓	
Engine oil & filter	R	✓	✓	✓	✓	✓	✓	✓	✓	✓	✓	✓	✓	✓	✓	✓	✓	✓	✓	✓
Exhaust system	S/I			✓			✓			✓			✓			✓			✓	
Fuel lines	S/I						✓						✓						✓	
Fuel tank cap gasket	S/I						✓						✓						✓	
Halfshaft boots & flange bolts	S/I			✓			✓			✓			✓			✓			✓	
Limited slip differential fluid	R						✓						✓						✓	
Manual transmission and differential fluid	S/I						✓						✓						✓	
Non-platinum spark plugs	R						✓						✓						✓	
Platinum spark plugs	R												✓							
Propeller shaft (4WD)	L			✓			✓			✓			✓			✓			✓	
Propeller shaft bolts	S/I			✓			✓			✓			✓			✓			✓	
Steering gear	S/I			✓			✓			✓			✓			✓			✓	
Steering linkage	S/I			✓			✓			✓			✓			✓			✓	
Tires (rotate)	S/I	✓	✓	✓	✓	✓	✓	✓	✓	✓	✓	✓	✓	✓	✓	✓	✓	✓	✓	✓
Valves	S/I												✓							

R: Replace   S/I: Service or Inspect   L: Lubricate

FREQUENT OPERATION MAINTENANCE (SEVERE SERVICE)

If a vehicle is operated under any of the following conditions it is considered severe service:

- Towing a trailer or using a camper or car-top carrier.

- Repeated short trips of less than 5 miles in temperatures below freezing.

- Excessive idling or low-speed driving for long distances as in heavy commercial use, such as delivery, taxi or police cars.

- Operating on rough, muddy or salt-covered roads.

- Operating on unpaved or dusty roads.

Oil filter: service or inspect every 5000 miles or 4 months, whichever occurs first.

Brake linings and discs or drums: service or inspect every 5000 miles or 4 months, whichever occurs first.

Steering linkage: service or inspect every 5000 miles or 4 months, whichever occurs first.

Ball joints and boots: service or inspect every 5000 miles or 4 months, whichever occurs first.

Brake discs & pads (front): service or inspect every 6000 miles.

Halfshaft boots: service or inspect every 5000 miles or 4 months. Retighten the flange bolts, whichever occurs first.

Body chassis bolts and nuts: service or inspect every 5000 miles or 4 months, whichever occurs first.

Transmission and differential fluid: replace every 15,000 miles or 12 months, whichever occurs first.

Transfer case and differential fluid: replace every 15,000 miles or 12 months, whichever occurs first.

ENGINE REPAIR

➡️**Disconnecting the negative battery cable on some vehicles may interfere with the functions of the on board computer system. The computer may undergo a relearning process once the negative battery cable is reconnected.**

Alternator

REMOVAL & INSTALLATION

3.4L Engine

1. Before servicing the vehicle, refer to the Precautions Section.
2. Disconnect the negative battery cable.
3. Remove or disconnect the following:
   - Negative battery cable
   - Alternator wiring
   - Alternator locknut, pivot bolt, nut and adjusting bolt
   - Drive belt
   - Alternator

To install:
4. Install or connect the following:
   - Alternator
   - Drive belt. Tighten the locknut 25 ft. lbs. (33 Nm) and the pivot bolt 38 ft. lbs. (51 Nm).
   - Alternator wiring
   - Negative battery cable

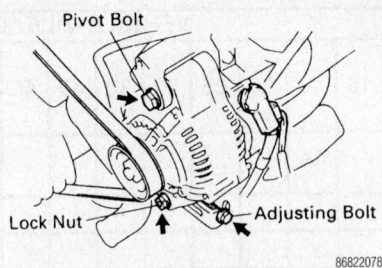

Adjusting location points—3.4L Engine

4.0L Engine

1. Before servicing the vehicle, refer to the Precautions Section.
2. Disconnect the negative battery cable.
3. Remove the V bank cover.
4. Remove the engine number one under cover assembly.
5. Remove the drive belt.
6. Remove the battery.
7. Disconnect the alternator electrical wires.
8. Remove the alternator retaining bolts. Remove the alternator from the vehicle.

To install:
9. Installation is the reverse of the removal procedure.
10. Tighten the alternator retaining bolts to 32 ft. lbs.

4.7L Engine

2002–2004

1. Before servicing the vehicle, refer to the Precautions Section.
2. Drain the cooling system.
3. Remove or disconnect the following:
   - Negative battery cable
   - Accessory drive belt
   - Engine under cover
   - Radiator
   - Power steering pump pulley
   - Alternator harness connectors
   - Alternator

To install:
4. Install or connect the following:
   - Alternator. Tighten the fasteners to 29 ft. lbs. (39 Nm).
   - Alternator harness connectors
   - Power steering pump pulley
   - Radiator
   - Engine under cover
   - Accessory drive belt
   - Negative battery cable
5. Fill the cooling system.
6. Start the engine and check for leaks.

2005–2006

1. Before servicing the vehicle, refer to the Precautions Section.
2. Disconnect the negative battery cable.
3. Remove the engine number one under cover assembly.
4. Remove the throttle body cover sub-assembly.
5. Disconnect the air cleaner hose from the throttle body.
6. Loosen the belt tension, by turning the belt tensioner counterclockwise. Remove the drive belt.
7. Remove the power steering pump assembly.
8. Disconnect the alternator electrical wires.
9. Remove the alternator retaining bolts. Remove the alternator from the vehicle.

To install:
10. Installation is the reverse of the removal procedure.
11. Tighten the alternator retaining bolts to 29 ft. lbs.

Ignition Timing

ADJUSTMENT

The engines are equipped with a Distributorless Ignition System (DIS). No timing adjustment is possible.

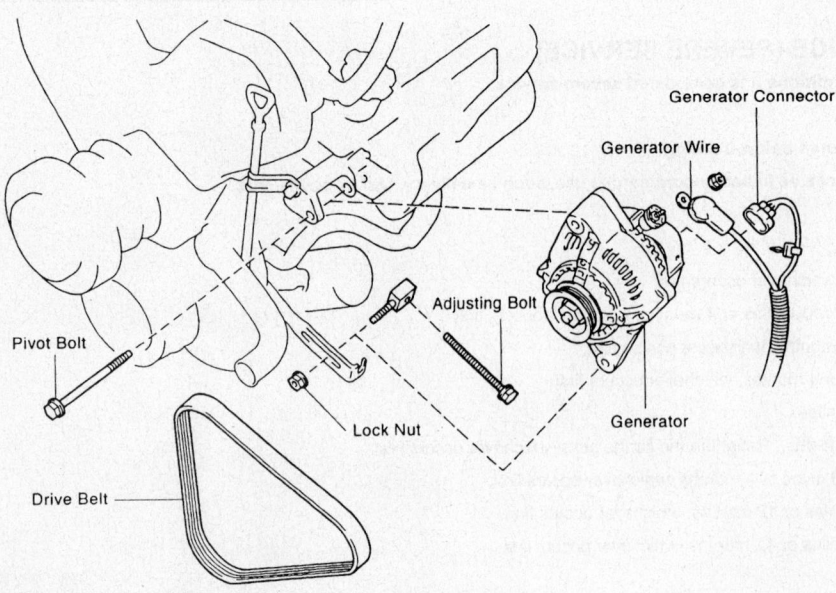

Alternator and related components—3.4L Engine

4.0L Engine

➡The ignition timing is not adjustable but can be checked. Be sure that all electrical systems are OFF. Be sure that the cooling fan motor is OFF.

WITH INTELEGENT TESTER

1. Warm up the engine.
2. Connect the tool to the DLC3.
3. Select the following menu items: DIAGNOSIS/ENHANCED OBDII/DATA LIST/PRIMARY/IGN ADVANCE.
4. Inspect the ignition timing during idling.
5. Check that the ignition timing advances immediately when the engine speed is increased.

WITHOUT INTELEGENT TESTER

1. Using too SST09843-18040, connect terminals 13 (TC) and 4 (CG) of the DLC3.

➡Be careful not to connect the wrong terminals, as engine damage may occur.

2. Remove the air cleaner. Pull out the wire harness.

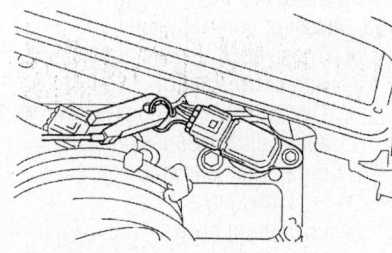

09490_TACO_G0009

Wiring harness location—4.0L engine

3. Connect the test probe of the timing light to the wire of the ignition coil connector for the No. 1 cylinder.

➡Be sure to use a timing light that detects the first signal. After checking be sure to wrap the wire harness with tape.

4. Check the ignition timing during idling. Check that the ignition timing is within specification. Be sure all electrical systems are OFF.

4.7L Engine (2005–2006)

➡The ignition timing is not adjustable but can be checked. Be sure that all electrical systems are OFF. Be sure that the cooling fan motor is OFF.

WITH INTELEGENT TESTER

1. Warm up the engine.
2. Connect the tool to the DLC3.
3. Refer to the tester operating manual for additional details.
4. Inspect the ignition timing during idling.
5. Check that the ignition timing advances immediately when the engine speed is increased.

WITHOUT INTELEGENT TESTER

1. Remove the throttle body cover.
2. Using too SST09843-18040, connect terminals TC and CG of the DLC3.

➡Be careful not to connect the wrong terminals, as engine damage may occur.

3. Connect the test probe of the timing light to the wire of the ignition coil connector for the No. 1 cylinder.
4. Check the ignition timing.

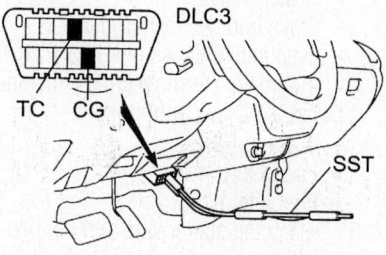

09490_TUND_G0002

Wiring harness location—2005–2006 4.7L engine

Engine Assembly

REMOVAL & INSTALLATION

3.4L Engine

2WD

1. Before servicing the vehicle, refer to the Precautions Section.
2. Properly relieve the fuel system pressure.
3. Remove or disconnect the following:
   • Hood
   • Battery
   • Engine under covers
4. Drain the engine coolant.
5. Drain the engine oil.
   • Radiator
   • Fan with the fluid coupling and fan pulleys
   • Air cleaner cap
   • Air cleaner case and filter

6. Disconnect the following hoses:
   • Heater hoses
   • Brake booster vacuum hose
   • Evaporative Emissions (EVAP) hose
   • Vacuum hose
   • Fuel return hose
   • Fuel inlet hose
7. Detach the starter wire and connectors, as follows:
   • Ground strap, by removing the bolt
   • Starter wires
8. Remove or disconnect the following:
   • Alternator connector and wire
   • Throttle cable, if equipped with an automatic transmission
   • Cruise control cable, if equipped with cruise control
9. Disconnect the engine wiring harness, as follows:
   • Glove box door
   • Lower the finish No. 2 panel
   • Heater to register duct
   • 3 Engine Control Module (ECM) connectors
   • 2 cassette connectors and the 2 wire clamps from the lower finish panel
   • Engine wiring harness clamp
10. Remove or disconnect the following:
    • Igniter connector
    • Ground strap
    • 2 engine wiring harness retainer-to-cowl panel nuts and pull out the engine wiring harness
11. If equipped with a manual transmission, remove or disconnect the following:
    • Shift lever knob
    • 4 shift lever boot screws
    • 6 shift lever assembly bolts, the assembly and gasket
12. Remove or disconnect the following:
    • Driveshaft from the transmission
    • Speedometer cable

➡Do not lose the felt protector and washers.

    • Front exhaust pipe
    • Clutch release cylinder, if equipped with a manual transmission
    • Nut and the control cable
13. Place a jack under the transmission.
14. Remove or disconnect the following:
    • Transmission rear mounting bracket by removing the 8 bolts
    • Bolt and the air conditioning compressor wire clamp, if equipped with air conditioning
15. If necessary, install a No. 2 engine hanger with 2 bolts. Tighten the 2 bolts to 30 ft. lbs. (40 Nm).
16. Attach the engine hoist chain to the 2 engine hangers.

17. Remove or disconnect the following:
- 4 engine front mounting insulators-to-frame bolts and nuts
- Engine from the transmission

To install:

18. Install or connect the following:
- Engine to the transmission
- Engine mounts to the body mountings. Install the bolts and nuts but do not tighten at this time.

19. Remove the engine chain hoist the No. 2 engine hanger.

20. Install or connect the following:
- Air conditioning wire with the bolt, if equipped with air conditioning
- Transmission mounting bracket. Tighten the frame bolts to 43 ft. lbs. (58 Nm) and the mounting insulator bolts to 13 ft. lbs. (18 Nm).
- Tighten the engine mounting nuts and bolts to 28 ft. lbs. (38 Nm).
- Control cable
- Clutch release cylinder, if equipped with a manual transmission. Torque the bolts to 9 ft. lbs. (12 Nm).

21. Install or connect the following:
- Front exhaust pipe
- Speedometer cable
- Driveshaft

22. If equipped with a manual transmission, install or connect the following:
- 6 shift lever assembly bolts, the assembly and gasket
- 4 shift lever boot screws
- Shift lever knob

23. Install or connect the following:
- All engine wiring harness, hoses and cables
- Fan with the fluid coupling and fan pulleys. Tighten the nuts to 48 inch lbs. (5.4 Nm).
- Air cleaner case and air filter
- Radiator

24. Install or connect the following hoses:
- Fuel inlet hose
- Fuel return hose
- Vacuum hose
- Evaporative Emissions (EVAP) hose
- Brake booster vacuum hose
- Heater hoses

25. Fill the engine with oil.

26. Fill the engine and radiator with coolant.

27. Install or connect the following:
- Engine undercover
- Battery
- Hood

28. Start the engine and check for leaks.

29. Make any necessary adjustments and road test the vehicle.

4WD

1. Before servicing the vehicle, refer to the Precautions Section.

2. Remove or disconnect the following:
- Transmission
- Hood

3. Release the fuel system pressure.

4. Remove or disconnect the following:
- Battery
- Engine undercovers

5. Drain the engine coolant.

6. Drain the engine oil.

7. Remove or disconnect the following:
- Radiator
- Fan with the fluid coupling and fan pulleys
- Air cleaner cap
- Mass Air Flow (MAF) meter and the resonator
- Cruise control cable, if equipped with cruise control
- Throttle cable, if equipped with an automatic transmission

8. Disconnect the following hoses:
- Heater hoses
- Brake booster vacuum hose
- Evaporative Emissions (EVAP) hose
- Automatic Disconnecting Differential (ADD) vacuum hose
- Vacuum hose
- Fuel return hose
- Fuel inlet hose

9. Detach the starter wire and connectors, as follows:
- Ground strap, by removing the bolt
- 3 starter wire clamps and connector

10. Detach the alternator connector and wire.

11. Disconnect the engine wiring harness, as follows:
- Glove box door
- Lower the finish No. 2 panel
- 3 Engine Control Module (ECM) connectors
- 2 cassette connectors and the 2 wire clamps from the lower finish panel
- Igniter connector
- Ground strap
- Engine wiring harness clamp

12. Remove or disconnect the following:
- 2 engine wiring harness retainer-to-cowl panel nuts and wiring harness
- Air conditioning compressor wire clamp and compressor bracket, if equipped with air conditioning

13. If necessary, install a No. 2 engine hanger with 2 bolts. Tighten the 2 bolts to 30 ft. lbs. (40 Nm).

14. Attach the engine hoist chain to the 2 engine hangers.

15. Remove or disconnect the following:
- 4 engine front mounting insulators-to-frame bolts and nuts
- Engine

To install:

16. Install or connect the following:
- Engine
- Engine mounts-to-body mountings. Install the bolts and nuts but do not tighten at this time.

17. Remove the engine chain hoist the No. 2 engine hanger.

18. Install or connect the following:
- Air conditioning wire with the bolt and the compressor bracket, if equipped with air conditioning
- Tighten the engine mounting nuts and bolts to 28 ft. lbs. (38 Nm).

19. Install the engine wiring harness, as follows:
- Engine wiring harness clamp
- Ground strap
- Igniter connector
- 2 cassette connectors and the 2 wire clamps from the lower finish panel
- 3 Engine Control Module (ECM) connectors
- Lower the finish No. 2 panel
- Glove box door

20. Install or connect the following:
- Cruise control cable, if equipped with cruise control
- Throttle cable, if equipped with an automatic transmission

21. Connect the following hoses:
- Fuel inlet hose
- Fuel return hose
- Vacuum hose
- Automatic Disconnecting Differential (ADD) vacuum hose
- Evaporative Emissions (EVAP) hose
- Brake booster vacuum hose
- Heater hoses

22. Install or connect the following:
- All wires, hoses and cables
- Fan with the fluid coupling and fan pulleys. Tighten the nuts to 48 inch lbs. (5.4 Nm).
- Air cleaner case and air filter
- MAF meter, resonator and the air cleaner cap
- Radiator

23. Fill the engine with oil.

24. Fill the engine and radiator with coolant.

25. Install or connect the following:
- Transmission and refill it with transmission oil
- Engine undercover
- Battery
- Hood

26. Start the engine, make any necessary adjustments and check for leaks.

4.0L Engine

1. Before servicing the vehicle, refer to the Precautions Section.
2. Properly disable the SRS system.
3. Properly relieve the fuel system pressure.
4. Disconnect the negative battery cable.
5. Matchmark and remove the hood. Remove the V bank cover.
6. Remove the transmission.
7. Remove the front exhaust pipe assembly.
8. Drain the engine coolant. Remove the radiator.
9. Drain the engine oil.
10. Disconnect the positive battery cable. Remove the battery.
11. Remove the air cleaner assembly. Remove the drive belt. Remove the engine fan. Remove the fan pulley.
12. Remove the power steering pump. Position the unit to the side.
13. Remove the alternator. Remove the air conditioning compressor. Position it to the side.
14. Disconnect the heater hoses from the heater core. Disconnect the fuel line hoses. Remove the intake air surge tank.
15. To separate the main engine wire harness, remove the glove box door. Remove the instrument panel finish panel subassembly, lower right side. Remove the heater to register duct. Disconnect the ECM connectors. Disconnect the engine wire from the engine wire bracket and remove the two nuts, bolt and bracket. Pull the engine wire from the cowl panel.
16. Install engine hanger, tool 12281-31070 and 12282-31050 and retainer bolts 90119-08177. Attach a suitable engine lifting fixture.

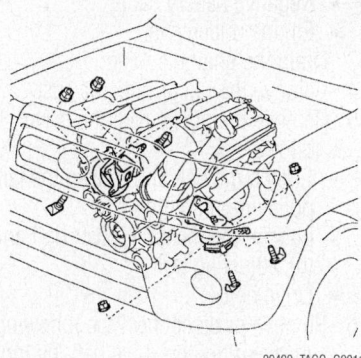

Engine retaining bolt locations—4.0L engine

09490_TACO_G0011

17. Remove the engine retaining bolts and mounts.

➡**Check to be sure nothing will interfere with the removal of the engine. Check to be sure all required wires, hoses and connectors are disconnected.**

18. Remove the engine from the vehicle.
To install:
19. Installation is the reverse of the removal procedure.
20. Tighten the four mounting insulator bolts to 28 ft. lbs.
21. Be sure to fill the cooling system with the proper grade and type engine coolant.
22. Be sure to fill the engine with the proper grade and type engine oil.
23. Check ignition timing. Check idle speed.

➡**On 2005–2006 vehicles perform the automatic transmission initialization procedure when replacing the automatic transmission, engine or ECM.**

24. Turn the ignition switch OFF.
25. Connect the intelligent tester tool to the DLC3.
26. Turn the ignition switch to the ON position.
27. Push the intelligent tester tool main switch to the ON position.
28. Select the items, DIAGNOSIS/ ENHANCED OBD II.
29. Perform the reset memory procedure from the ENGINE menu.

➡**After performing the reset memory, be sure to perform the road test procedure. For roadtest procedure information, refer to the intelligent tester instruction manual.**

30. Start the engine and check for leaks, correct as required.

4.7L Engine

2002–2004

1. Before servicing the vehicle, refer to the Precautions Section.
2. Relieve the fuel system pressure.
3. Drain the cooling system.
4. Drain the engine oil.
5. Remove or disconnect the following:
   - Battery and tray
   - Hood
   - Engine appearance cover
   - Air intake pipe
   - Engine under covers
   - Coolant recovery tank
   - Radiator hoses

- Radiator and fan shroud
- Accessory drive belt
- Cooling fan and pulley
- Powertrain Control Module (PCM) harness connectors and pass the wiring harness through the firewall
- Accelerator cable
- Power steering vacuum hoses
- Alternator harness connectors
- Heater hoses
- Engine control wiring harness and grommet at the firewall
- Ground cable connector
- Fuel lines
- Evaporative Emissions (EVAP) canister hoses
- Wire clamp at right inner fender
- Negative battery cable at the relay box and right inner fender
- Positive battery cable
- Center console
- Transmission shift lever assembly
- Transfer case shift lever and rod
- Exhaust front pipes
- Stabilizer bar
- Front and rear driveshafts
- A/C compressor
- Power steering pump

6. Attach a hoist to the engine lifting eyes.
7. Remove or disconnect the following:
   - Transfer case skid plate
   - Left and right motor mounts
   - Transmission mount crossmember
8. Attach a hoist to the engine lifting eyes and raise the powertrain out of the vehicle.
To install:
9. Lower the powertrain into the vehicle.
10. Install or connect the following:
   - Transmission mount crossmember. Tighten the bolts to 37 ft. lbs. (50 Nm) and the nuts to 55 ft. lbs. (74 Nm).
   - Transfer case skid plate
   - Left and right motor mounts. Tighten the fasteners to 22 ft. lbs. (30 Nm).
   - Power steering pump. Tighten the bolts to 13 ft. lbs. (17 Nm).
   - A/C compressor. Tighten the bolts to 36 ft. lbs. (49 Nm).
   - Front driveshaft. Tighten the fasteners to 59 ft. lbs. (80 Nm).
   - Rear driveshaft. Tighten the fasteners to 78 ft. lbs. (106 Nm).
   - Stabilizer bar. Tighten the bracket bolts to 13 ft. lbs. (18 Nm) and the link nuts to 18 ft. lbs. (25 Nm).
   - Exhaust front pipes
   - Transfer case shift lever and rod

- Transmission shift lever assembly
- Center console
- Positive battery cable
- Negative battery cable at the relay box and right inner fender
- Wire clamp at right inner fender
- EVAP canister hoses
- Fuel lines
- Ground cable connector
- Engine control wiring harness and grommet at the firewall
- Heater hoses
- Alternator harness connectors
- Power steering vacuum hoses
- Accelerator cable
- PCM harness connectors
- Cooling fan and pulley
- Accessory drive belt
- Radiator and fan shroud
- Radiator hoses
- Coolant recovery tank
- Engine under covers
- Air intake pipe
- Engine appearance cover
- Hood
- Battery and tray

11. Fill the crankcase to the correct level.
12. Fill the cooling system.
13. Start the engine and check for leaks.

2005-2006

➡ **On 2WD the transmission is removed along with the engine. On 4WD the automatic transmission must be removed before removing the engine.**

1. Before servicing the vehicle, refer to the Precautions Section.
2. Properly disable the SRS system.
3. Properly relieve the fuel system pressure.
4. Disconnect the negative battery cable. Disconnect the positive battery cable.
5. Matchmark and remove the hood.
6. On 4WD remove the front driveshaft assembly. Remove the automatic transmission.
7. Drain the cooling system. Remove the radiator.
8. Drain the engine oil.
9. Remove the air cleaner hose. Remove the air cleaner cap. Remove the air cleaner filter element subassembly. Remove the air cleaner case.
10. Loosen the fan assembly. Remove the drive belt. Remove the engine fan.
11. To separate the main engine wire harness, remove the glove box door. Remove the instrument panel finish panel subassembly. Disconnect the ECM connectors. Disconnect the two wire harness con-

nector. Disconnect the engine wire from the engine wire bracket and remove the two nuts, bolt and bracket. Pull the engine wire from the cowl panel. Disconnect the ground strap.
12. Disconnect the two power steering air hoses from the hose clamp on the number three timing belt cover, right side.
13. Disconnect the hose clamp for the power steering air hose.
14. Disconnect the power steering air hose from the upper intake manifold.
15. Disconnect and plug the heater hoses.
16. Disconnect the air inlet hose from the charcoal canister.
17. Disconnect the EVAP hose from the VSV.
18. Disconnect the brake booster hose. Disconnect the alternator wiring. Disconnect the fuel lines.
19. Disconnect the heated oxygen sensor connectors. Remove the two bolts, three nuts, gasket, ring and front right exhaust pipe from the exhaust manifold.
20. Disconnect the heated oxygen sensor connectors. Remove the two bolts, three nuts, gasket, ring and front left exhaust pipe from the exhaust manifold.
21. Remove the driveshaft, with the center bearing assembly.
22. Remove the front stabilizer bar assembly.
23. Disconnect the power steering hoses.
24. Remove the transmission control cable assembly.
25. Remove the air conditioning compressor and position it to the side.
26. Remove the power steering pump and position it to the side.
27. Install the engine hangers and retainer bolts. Attach a suitable engine lifting fixture.
28. Remove the engine retaining bolts.
29. On 2WD, remove the eight bolts, two nuts and frame crossmember. Remove the four bolts and engine rear mounting bracket from the transmission.
30. Remove the engine assembly from the vehicle.

➡ **Check to be sure nothing will interfere with the removal of the engine. Check to be sure all required wires, hoses and connectors are disconnected.**

To install:
31. Installation is the reverse of the removal procedure.
32. Tighten the four mounting insulator bolts to 27 ft. lbs.

33. On 2WD tighten the number one engine rear mounting insulator bolts to 48 ft. lbs.
34. On 2WD tighten the four frame crossmember bolts to 53 ft. lbs.
35. On 2WD tighten the four crossmember to transmission nuts to 13 ft. lbs.
36. Be sure to fill the cooling system with the proper grade and type engine coolant.
37. Be sure to fill the engine with the proper grade and type engine oil.

➡ **On 2005–2006 vehicles perform the automatic transmission initialization procedure when replacing the automatic transmission, engine or ECM.**

38. Turn the ignition switch OFF.
39. Connect the intelligent tester tool to the DLC3.
40. Turn the ignition switch to the ON position.
41. Push the intelligent tester tool main switch to the ON position.
42. Select the items, DIAGNOSIS/ENHANCED OBD II.
43. Perform the reset memory procedure from the ENGINE menu.

➡ **After performing the reset memory, be sure to perform the road test procedure. For roadtest procedure information, refer to the intelligent tester instruction manual.**

44. Start the engine and check for leaks, correct as required.

Water Pump

REMOVAL & INSTALLATION

3.4L Engine

1. Before servicing the vehicle, refer to the Precautions Section.
2. Remove or disconnect the following:
   - Negative battery cable
   - Engine undercover
3. Drain the engine coolant.
4. Remove the upper radiator hose.
5. Remove the power steering drive belt, as follows:
   - Stretch the belt and loosen the fan pulley mounting nuts
   - Loosen the lockbolt, pivot bolt and the adjusting bolt
   - Drive belt
6. Remove or disconnect the following:
   - Air conditioning drive belt, by loosening the idler pulley nut and adjusting bolt

- Lockbolt, pivot bolt and the adjusting bolt
- Alternator drive belt
- No. 2 fan shroud, by removing the 2 clips
- Fan with the fluid coupling and fan pulleys
- Power steering pump and move it aside without disconnecting the lines from the pump
- Compressor from the engine and move it aside without disconnecting the compressor lines, if equipped with air conditioning
- Air conditioning bracket, if equipped with air conditioning

7. Remove the No. 2 timing belt cover, as follows:
- Camshaft Position (CMP) sensor connector from the No. 2 timing belt cover
- 3 spark plug wire clamps from the No. 2 timing belt cover
- 6 bolts and the timing belt cover

8. Remove the fan bracket, as follows:
- Power steering adjusting strut, by removing the nut
- Fan bracket, by removing the bolt and nut

9. Set the No. 1 cylinder to Top Dead Center (TDC) of the compression stroke, as follows:

a. Turn the crankshaft pulley and align its groove with the timing mark **O** of the No. 1 timing belt cover.

b. Check that the timing marks of the camshaft timing pulleys and the No. 3 timing belt cover are aligned. If not, turn the crankshaft pulley 1 revolution (360 degrees).

10. Remove the camshaft timing pulleys, as follows:

a. Remove the timing belt tensioner by alternately loosening the 2 bolts.

b. Using Variable Wrench Set No. 09960-10010, remove the pulley bolt, the timing pulley and the knock pin.

c. Remove the 2 timing pulleys with the timing belt.

11. Remove or disconnect the following:
- Thermostat
- No. 2 oil cooler hose, from the water pump
- Water pump, by removing the 7 bolts

12. Thoroughly clean the mating surfaces.

To install:

13. Install or connect the following:
- Apply sealant (PN 08826-00100) to the water pump

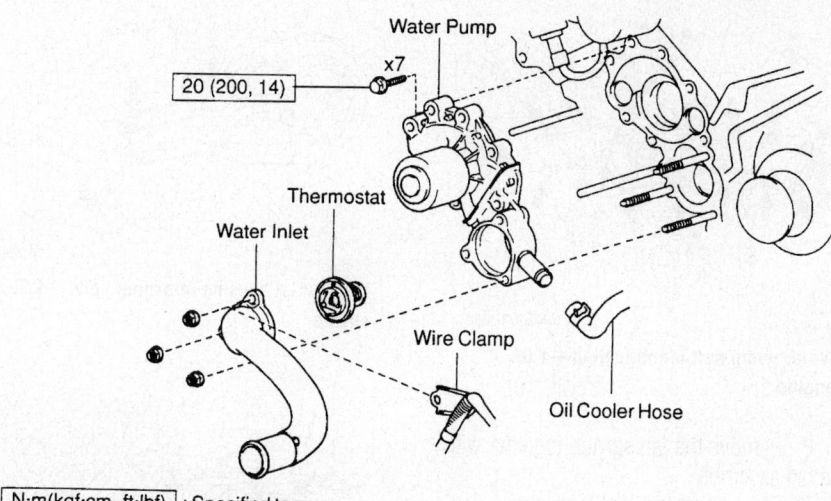

20 (200, 14)

Water Pump

x7

Thermostat

Water Inlet

Wire Clamp

Oil Cooler Hose

N·m(kgf·cm, ft·lbf) : Specified torque
◆ Non–Reusable part

7924YG08

Water pump and related components—3.4L engine

❋❋ **WARNING**

Parts must be assembled within 5 minutes of application. Otherwise the material must be removed and reapplied.

- Water pump. Tighten the bolts to 14 ft. lbs. (20 Nm).
- No. 2 oil cooler hose
- Thermostat
- Left camshaft timing pulley. Tighten the pulley bolt to 81 ft. lbs. (110 Nm).

14. Set the No. 1 cylinder to TDC of the compression stroke.

15. Connect the timing belt to the left camshaft timing pulley. Check that the installation mark on the timing belt is aligned with the end of the No. 1 timing belt cover, as follows:

a. Using Variable Pin Wrench Set 09960-01000, slightly turn the left camshaft timing pulley clockwise. Align the installation mark on the timing belt with the timing mark of the camshaft timing pulley and hang the timing belt on the left camshaft timing pulley.

b. Align the timing marks of the left camshaft pulley and the No. 3 timing belt cover.

c. Check that the timing belt has tension between the crankshaft timing pulley and the left camshaft timing pulley.

16. Install the right camshaft timing pulley and the timing belt.

17. Set the timing belt tensioner, as follows:

a. Using a press, slowly press in the

pushrod using 220–2,205 lbs. (981–9,807 N) of force.

b. Align the holes of the pushrod and housing, pass a 1.5mm hexagon wrench through the holes to keep the setting position of the pushrod.

c. Release the press and install the dust boot to the tensioner.

d. Install the timing belt tensioner and alternately tighten the bolts to 20 ft. lbs. (28 Nm).

e. Using pliers, remove the 1.5mm hexagon wrench from the belt tensioner.

18. Check the valve timing, as follows:

a. Slowly turn the crankshaft pulley 2 revolutions from the TDC-to-TDC; always turn the crankshaft pulley clockwise.

b. Check that each pulley aligns with the timing marks. If the timing marks do not align, remove the timing belt and reinstall it.

19. Install or connect the following:
- Fan bracket, with the bolt and nut
- Remaining components
- Negative battery cable

20. Fill with engine coolant.

21. Start the engine and check for leaks.

4.0L Engine

1. Before servicing the vehicle, refer to the Precautions Section.

2. Disconnect the negative battery cable.

3. Remove the five retaining bolts and remove the number one engine undercover subassembly.

4. Drain the engine coolant.

5. Remove the V bank cover. Remove the drive belt.

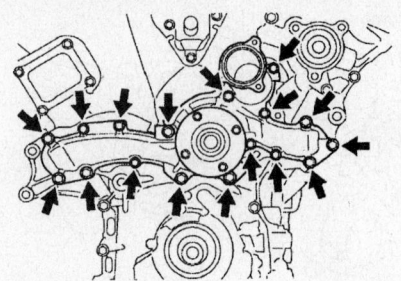

Water pump bolt identification—4.0L engine

6. Remove the fan shroud together with the fan assembly.

7. Disconnect the ventilation hose number two. Remove the air cleaner assembly.

8. Disconnect the two oil cooler hoses (with oil cooler) and remove the water inlet.

9. Disconnect the radiator hoses. Disconnect the five water bypass hoses.

10. Remove the five bolts and the water inlet. Remove the O-ring from the water outlet pipe. Remove the gasket from the water pump.

11. Remove the two bolts and remove the number two idler pulley subassembly.

12. Remove the alternator.

13. Remove the mounting bolt and separate the air conditioning compressor suction hose subassembly. Disconnect the air condition compressor connector. Remove the four bolts and separate the compressor from the belt tensioner assembly.

14. Remove belt tensioner assembly.

15. Remove the water pump retaining bolts. Remove the water pump from the engine.

To install:

16. Clean all gasket mounting surfaces.

17. Using a new gasket install the water pump to the engine. Tighten bolts "A" to 80 inch lbs. Tighten bolts "B" to 17 ft. lbs.

18. Install the belt tensioner assembly.

19. Continue the installation in the reverse order of the removal procedure.

20. Fill the cooling system with the proper grade and type engine coolant.

21. Start the engine and check for leaks.

4.7L Engine

1. Before servicing the vehicle, refer to the Precautions Section.

2. Drain the cooling system.

3. Remove or disconnect the following:
- Negative battery cable
- Timing belt

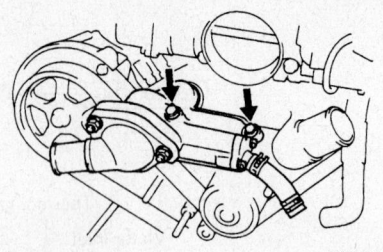

Water inlet housing attaching bolts—4.7L engine

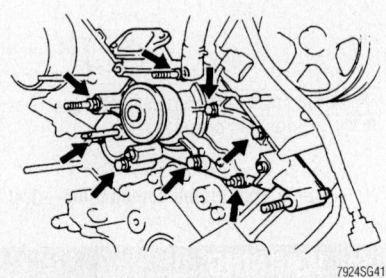

Water pump bolt locations—4.7L engine

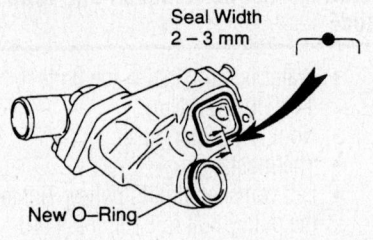

Seal Width 2 – 3 mm

New O-Ring

Water inlet housing sealant application—4.7L engine

- No. 2 idler pulley
- Radiator hose
- Bypass hose
- Water inlet housing assembly
- Water pump

To install:

4. Install or connect the following:
- Water pump. Use a new gasket and tighten the bolts to 15 ft. lbs. (21 Nm). Tighten the stud bolt and nut to 13 ft. lbs. (18 Nm).
- Water inlet housing assembly. Use a new O-ring and apply sealant as shown. Tighten the bolts to 13 ft. lbs. (18 Nm).
- Bypass hose
- Radiator hose
- No. 2 idler pulley
- Timing belt
- Negative battery cable

5. Fill the cooling system.

6. Start the engine and check for leaks.

Heater Core

REMOVAL & INSTALLATION

2002–2004

1. Before servicing the vehicle, refer to the Precautions Section.

2. Disconnect the negative battery cable.

3. Drain the cooling system into a clean container for reuse.

4. Disconnect the heater hoses from the heater core.

5. Remove the steering wheel by performing the following procedure:

a. Position the front wheels facing straight-ahead.

b. Remove the steering wheel side covers.

c. Using a Torx® wrench, loosen the 2 screws located at each side of the steering wheel until the screw's circumference groove catches on the screw case.

d. Pull the air bag module from the steering wheel and disconnect the electrical connector.

❋❋ CAUTION

Place the air bag module in a safe place with the front side facing upward.

e. Remove the steering wheel nut.

f. Place alignment marks on the steering wheel and the main shaft.

g. Using a steering wheel puller, press the steering wheel from the steering column.

6. Remove the instrument panel and reinforcement by performing the following procedure:

a. Remove the front door scuff plates, the cowl side trim and the front door opening trim.

b. At the driver's side, remove the 2 assist grip plugs, the 2 screws and assist grip and the front pillar garnish.

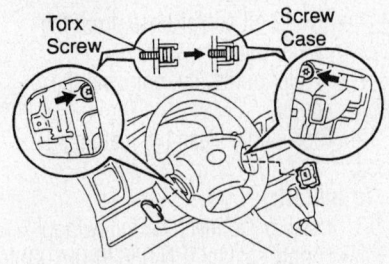

Torx Screw — Screw Case

Steering pad Torx bolt locating points

c. At the passenger's side, remove the 4 assist grip plugs, the 4 screws, the 2 assist grips and the front pillar garnish.

d. Remove the instrument cluster finish panel.

e. Remove the 2 screws and the hood lock control cable.

f. Remove the 2 screws and the fuel lid control cable lever.

g. Remove the lower No. 1 panel screw and the panel.

h. Remove the lower left side panel.

i. Remove the 3 steering column cover screws and the covers.

j. At the steering column, disconnect the electrical connectors; then, remove the clamp, the 3 screws and the combination switch.

k. Remove the No. 2 heater-to-register duct screw and the duct.

l. Remove the steering column-to-instrument panel bolts and the steering column.

m. At the combination meter, disconnect the electrical connectors; then, remove the 4 screws and the combination meter.

n. Remove the glove compartment door stoppers, the 2 screws and the glove box door.

o. At the passenger's side air bag module, remove the No. 1 undercover, pull the air bag connector up from the undercover and disconnect it; then, remove the air bag.

✳✳ CAUTION

Place the air bag module in a safe place with the front side facing upward.

p. Remove the 3 lower No. 2 panel screws and the panel.

q. Remove the center cluster; then,

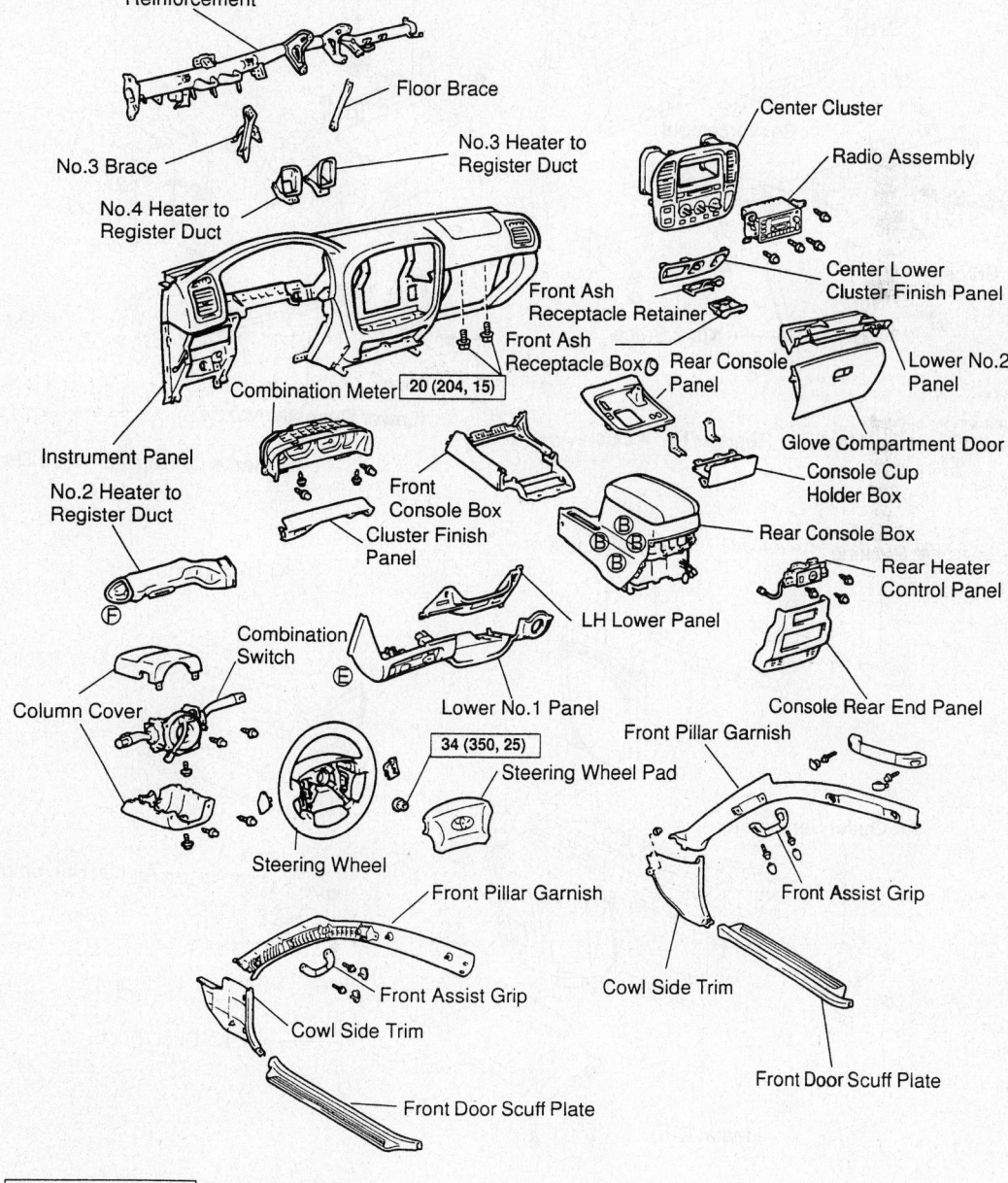

N·m (kgf·cm, ft·lbf) : Specified torque

Instrument panel and related components—2002–2004

93113GG7A

pry the center cluster from the dash by prying the 8 clips in the following order:
- Left side
- Right side
- Top left side
- Top right side

r. Remove the 4 radio screws, pull the radio outward, disconnect the electrical connectors and remove the radio.

s. At the rear console panel, remove the transfer shift lever knob; then, pry the panel upward disengaging the 4 clips (2 on each side) and remove the panel.

t. At the rear of the console, remove the 2 rear end panel-to-console screws; then, pry the end panel rearward disengaging the 2 clips and remove the panel.

u. If not equipped with a rear air conditioning system, disconnect the connector and control cable; then, remove the 3 rear heater control panel screws and the panel.

v. Remove the 4 rear console box-to-chassis screws/bolts and the console box.

w. Remove the center lower cluster finish panel by prying panel rearward disengaging the 5 clips; then, disconnect the electrical connector.

x. Remove the 2 front console-to-chassis bolts/screws, disengage the 2 clips and remove the console.

y. At the instrument panel, disconnect

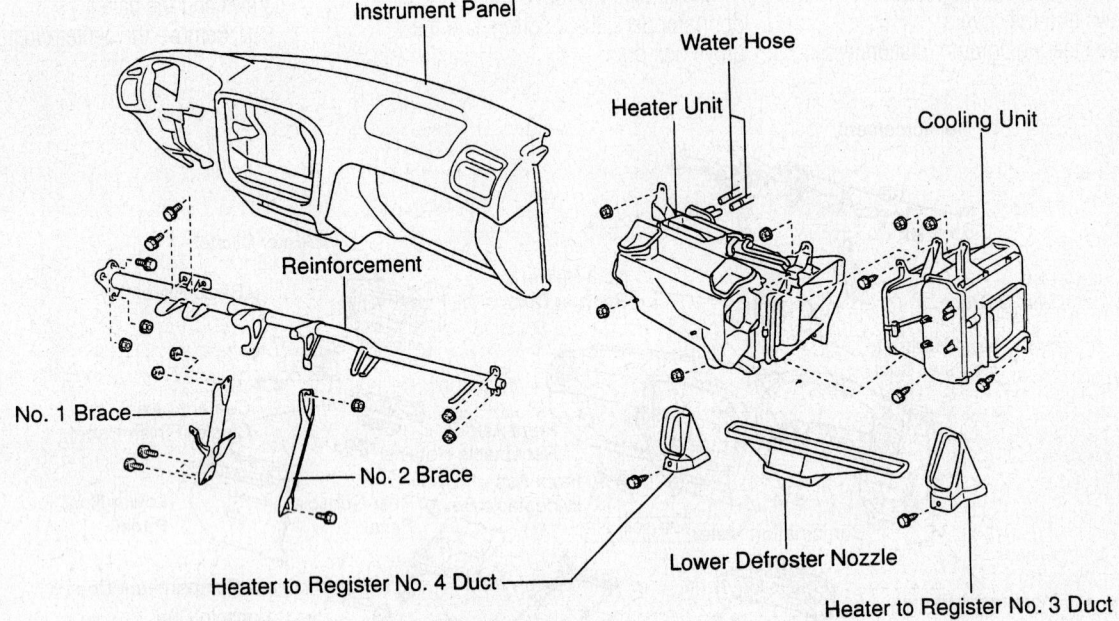

Instrument Panel — Water Hose — Heater Unit — Cooling Unit — Reinforcement — No. 1 Brace — No. 2 Brace — Heater to Register No. 4 Duct — Lower Defroster Nozzle — Heater to Register No. 3 Duct

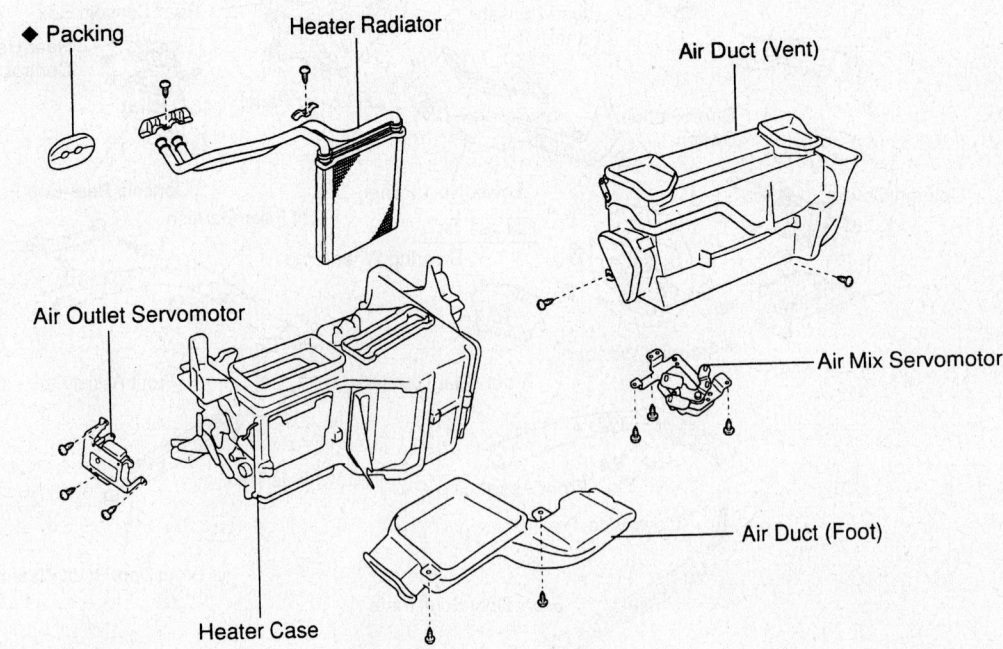

◆ Packing — Heater Radiator — Air Duct (Vent) — Air Outlet Servomotor — Air Mix Servomotor — Heater Case — Air Duct (Foot)

◆ Non-reusable part

93113GG0

Heater core housing and related components—2002–2004

the junction connectors (the connectors can be disconnected by loosening the bolts), the instrument panel-to-chassis 8 bolts and 2 nuts. Using an assistant, remove the instrument panel.

z. Disconnect the electrical connector and remove the ECM.

aa. Remove the No. 3 and No. 4 heater-to-register ducts.

bb. Remove the floor brace, the No. 1 brace and the reinforcement.

7. Remove the evaporator housing by performing the following procedure:

a. Discharge and recover the air conditioning system refrigerant.

b. Remove the air conditioning liquid line clamp.

c. Remove the air conditioning suction line clamp.

d. Disconnect both air conditioning lines and plug the openings to prevent contamination. Discard the 4 O-rings.

e. Remove the antenna relay electrical connector, the 2 screws and the relay.

f. Remove the evaporator housing-to-chassis 4 screws/2 nuts and the housing.

8. Remove the heater housing by performing the following procedure:

a. Remove the defroster nozzle.

b. Disconnect the electrical connector.

c. Remove the 4 nuts and the heater housing.

9. Remove the heater core-to-heater housing packing, the screw, the bracket, the clamp and the heater core.

To install:

10. Install the heater core, the clamp, the bracket, the screw and the heater core-to-heater housing packing.

11. Install the heater housing by performing the following procedure:

a. Install the heater housing and the 4 nuts.

b. Connect the electrical connector.

c. Install the defroster nozzle.

12. Install the evaporator housing by performing the following procedure:

a. Install the evaporator housing and the housing-to-chassis 4 screws and 2 nuts.

b. Install the antenna relay, the 2 screws and the electrical connector.

c. Using new O-rings, connect both air conditioning lines.

d. Install the air conditioning liquid line and suction line clamp.

13. Install the instrument panel and reinforcement by performing the following procedure:

a. Install the reinforcement, the No. 1 brace and the floor brace.

b. Install the No. 3 and No. 4 heater-to-register ducts.

c. Install the ECM and connect the electrical connector.

d. Using an assistant, install the instrument panel, connect the junction connectors, the instrument panel-to-chassis 8 bolts and 2 nuts.

e. Install the front the console, engage the 2 clips and install the 2 console-to-chassis bolts/screws.

f. Connect the electrical connector; then, install the center lower cluster finish panel by engaging the 5 clips.

g. Install the console box and the 4 rear console box-to-chassis screws/bolts.

h. If not equipped with a rear air conditioning system, install rear heater control panel, the 3 panel screws; then, connect the connector and control cable.

i. Install the rear of the console and engage the 2 clips; then, install the 2 rear end panel-to-console screws.

j. Install the rear console panel and engage the 4 clips (2 on each side); then, install the transfer shift lever knob.

k. Install the radio, connect the electrical connectors and the 4 radio screws.

l. Install the center cluster and engage the 8 center cluster clips.

m. Install the lower No. 2 panel and the 3 panel screws.

n. Install the passenger's side air bag module, connect it and install the No. 1 undercover.

o. Install the glove box door, the 2 screws and the glove compartment door stoppers.

p. Install the combination meter and the 4 screws; then, connect the electrical connectors.

q. Install the steering column and the steering column-to-instrument panel bolts.

r. Install the No. 2 heater-to-register duct and the duct screw.

s. At the steering column, install the combination switch, the 3 screws and the clamp; then, connect the electrical connectors.

t. Install the steering column covers and the 3 covers screws.

u. Install the lower left side panel.

v. Install the lower No. 1 panel and the panel screw.

w. Install the fuel lid control cable lever and the 2 screws.

x. Install the hood lock control cable and the 2 screws.

y. Install the instrument cluster finish panel.

z. At the passenger's side, install the front pillar garnish, the 2 assist grips, the 4 screws and the 4 assist grip plugs.

aa. At the driver's side, install the front pillar garnish, assist grip, the 2 screws and the 2 assist grip plugs.

bb. Install the front door scuff plates, the cowl side trim and the front door opening trim.

14. Install the steering wheel by performing the following procedure:

a. Install the steering wheel to the steering column.

b. Align the steering wheel-to-main shaft marks.

c. Install the steering wheel nut and torque to 25 ft. lbs. (34 Nm).

d. Install the air bag module to the steering wheel and connect the electrical connector.

e. Using a Torx® wrench, tighten the 2 screws located at each side of the steering wheel to 78 inch lbs. (8.8 Nm).

f. Install the steering wheel side covers.

15. Connect the heater hoses to the heater core.

16. Refill the cooling system.

17. Connect the negative battery cable.

a. Evacuate and charge the air conditioning system refrigerant.

18. Run the engine to normal operating temperatures; then, check the climate control operation and check for leaks.

2005–2006

1. Before servicing the vehicle, refer to the Precautions Section.

2. Disarm the SRS system.

➡**Wait at least 90 seconds before starting any repair work.**

3. Position the front wheels in the straight ahead position.

4. Disconnect the negative battery cable.

5. Discharge the air conditioning system.

6. Drain the engine coolant.

7. Disconnect and plug the air conditioning lines at the evaporator core, using tool SST09870-00025, or equivalent.

8. Remove the glove compartment. Remove the lower number two finish panel. Remove the lower center cover.

9. Remove the lower left finish panel. Remove the instrument panel lower cover.

10. Remove the number four heater to register duct.

11. Disconnect the electrical connectors. Remove the three retaining screws. Remove the cooling unit from the vehicle.

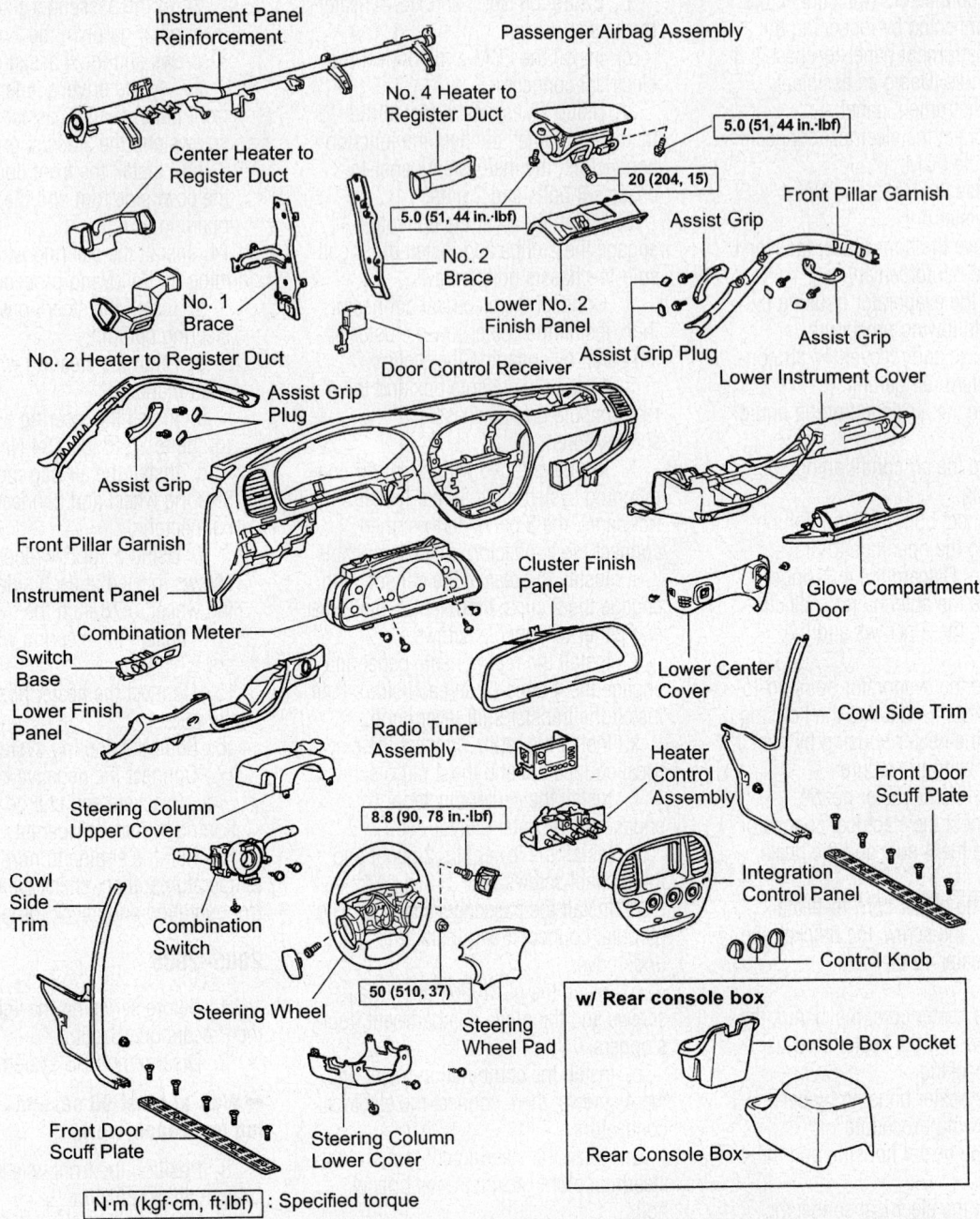

Instrument Panel Reinforcement

Passenger Airbag Assembly

No. 4 Heater to Register Duct

5.0 (51, 44 in.·lbf)

Center Heater to Register Duct

20 (204, 15)

5.0 (51, 44 in.·lbf)

Front Pillar Garnish

No. 2 Brace

Assist Grip

No. 1 Brace

Lower No. 2 Finish Panel

Assist Grip

No. 2 Heater to Register Duct

Door Control Receiver

Assist Grip Plug

Lower Instrument Cover

Assist Grip Plug

Assist Grip

Front Pillar Garnish

Glove Compartment Door

Instrument Panel

Cluster Finish Panel

Combination Meter

Switch Base

Cowl Side Trim

Lower Finish Panel

Radio Tuner Assembly

Lower Center Cover

Front Door Scuff Plate

Steering Column Upper Cover

8.8 (90, 78 in.·lbf)

Heater Control Assembly

Cowl Side Trim

Combination Switch

Integration Control Panel

Steering Wheel

50 (510, 37)

Control Knob

w/ Rear console box

Console Box Pocket

Steering Wheel Pad

Front Door Scuff Plate

Steering Column Lower Cover

Rear Console Box

N·m (kgf·cm, ft·lbf) : Specified torque

09490_TUND_G0003

Instrument panel and related components—2005–2006 Standard and Access Cab

12. Disconnect and plug the heater hoses at the heater core.

13. Remove the front door scuff plate. Remove the cowl side trim.

14. On Double Cab, remove the right and left side trim board. Remove the side panel.

15. Remove the assist grips. Remove the front pillar garnish.

16. Remove the steering wheel lower cover number two and three. Carefully pull the steering wheel pad from its mounting and disconnect the airbag connectors and horn terminal.

➡**If the airbag connector is disconnected with the ignition switch in the ON or ACC position, DTC codes will be set.**

17. Remove the steering wheel pad. When storing the pad assembly, keep the upper surface of the pad facing upward. Never disassemble the steering wheel pad.

➡**When removing the pad, take care not to pull on the airbag wire harness.**

18. Remove the steering wheel locknut. Using a steering wheel removal tool,

remove the steering wheel from the vehicle. Disconnect the connector.

19. Remove the three screws and the upper and lower steering column covers.

20. Disconnect the connectors and screws. Remove the combination switch. Remove the spiral cable assembly.

21. Remove the two screws and hood lock release lever. Remove the four bolts and lower finish panel. Disconnect the connectors.

22. Remove the switch base assembly. Remove the number two heater to register duct. Remove the brake pedal return spring.

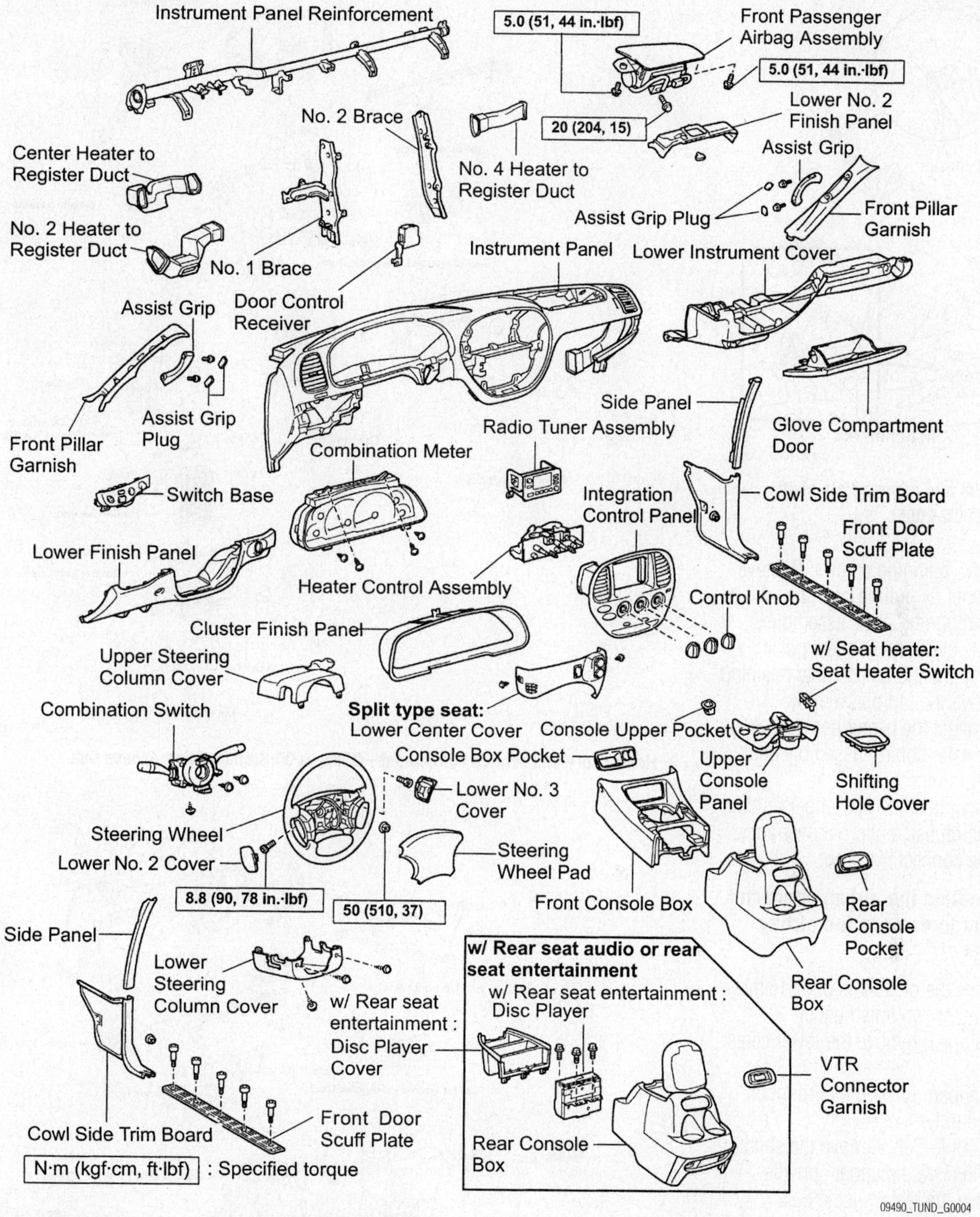

Instrument Panel Reinforcement

5.0 (51, 44 in.·lbf)

Front Passenger Airbag Assembly

5.0 (51, 44 in.·lbf)

No. 2 Brace

20 (204, 15)

Lower No. 2 Finish Panel

Center Heater to Register Duct

No. 4 Heater to Register Duct

Assist Grip

No. 2 Heater to Register Duct

No. 1 Brace

Assist Grip Plug

Front Pillar Garnish

Assist Grip

Door Control Receiver

Instrument Panel

Lower Instrument Cover

Front Pillar Garnish

Assist Grip Plug

Switch Base

Combination Meter

Side Panel

Radio Tuner Assembly

Glove Compartment Door

Lower Finish Panel

Integration Control Panel

Cowl Side Trim Board

Front Door Scuff Plate

Heater Control Assembly

Control Knob

w/ Seat heater: Seat Heater Switch

Cluster Finish Panel

Upper Steering Column Cover

Combination Switch

Split type seat:
Lower Center Cover

Console Box Pocket

Console Upper Pocket

Upper Console Panel

Shifting Hole Cover

Steering Wheel

Lower No. 2 Cover

Lower No. 3 Cover

Steering Wheel Pad

Front Console Box

Rear Console Pocket

8.8 (90, 78 in.·lbf)

50 (510, 37)

Side Panel

Lower Steering Column Cover

w/ Rear seat audio or rear seat entertainment

w/ Rear seat entertainment: Disc Player

Rear Console Box

w/ Rear seat entertainment: Disc Player Cover

Cowl Side Trim Board

Front Door Scuff Plate

Rear Console Box

VTR Connector Garnish

N·m (kgf·cm, ft·lbf) : Specified torque

09490_TUND_G0004

Instrument panel and related components—2005–2006 Double Cab

23. Matchmark the sliding yoke and number two intermediate shaft assembly. Remove bolt "A" and bolt "B". Slide the sliding yoke and remove it.

24. Remove the three bolts and the column hole cover number two.

25. If equipped with automatic transmission, disconnect the control cable from the column shift lever assembly.

26. Disconnect the electrical connectors. Remove the four steering column set nuts. Pull out the steering column assembly with the number two universal joint assembly.

27. Matchmark the steering column assembly and the number two universal joint assembly. Remove the bolt and the number two universal joint assembly.

28. Matchmark the number two intermediate shaft assembly and the control valve shaft. Remove the bolt and the number two intermediate shaft assembly.

29. Remove the cluster finish panel.

30. Remove the combination meter retaining screws. Remove the combination meter. Disconnect the electrical connectors.

31. Remove the three control knobs.

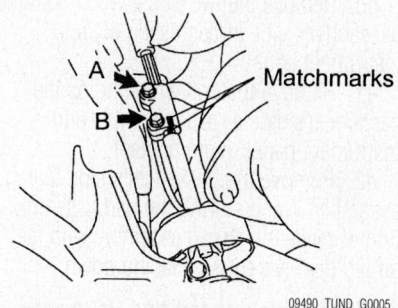

Matchmarks

09490_TUND_G0005

Sliding yoke bolt location—2005–2006

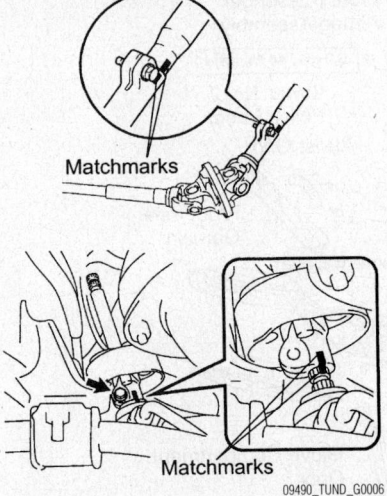

Matchmarks

Matchmarks

09490_TUND_G0006

Universal joint and intermediate shaft markings—2005–2006

Remove the five retaining screws. Remove the control panel by pulling carefully at the inside air conditioning grille assemblies. Disconnect the electrical connectors.

32. Remove the radio assembly retaining screws. Remove the radio assembly.

33. Disconnect the heater control cables. Remove the heater control assembly and disconnect the connectors.

34. To disconnect the passenger's side airbag connector, use a clip remover and disengage the connector clamp.

➡**When handling the airbag connector, take care not to damage the airbag wire harness.**

35. Remove the glove box door. Remove the lower number two finish panel.

36. If equipped, remove the lower center cover.

37. If equipped, remove the front console box assembly.

38. On Double Cab, remove the shifting hole cover. Remove the upper console panel. Remove the rear console box. Remove the console box pocket.

39. Remove the front console box. Remove the lower instrument cover.

40. Remove the two bolts which hold the passenger's side airbag assembly and instrument panel.

41. Remove the bolt which holds the passenger's side airbag assembly and instrument panel reinforcement.

42. Remove the passenger's side airbag assembly. When storing the pad assembly, do not place the airbag assembly with the airbag deployment side facing down.

➡**When removing the pad, take care not to pull on the airbag wire harness.**

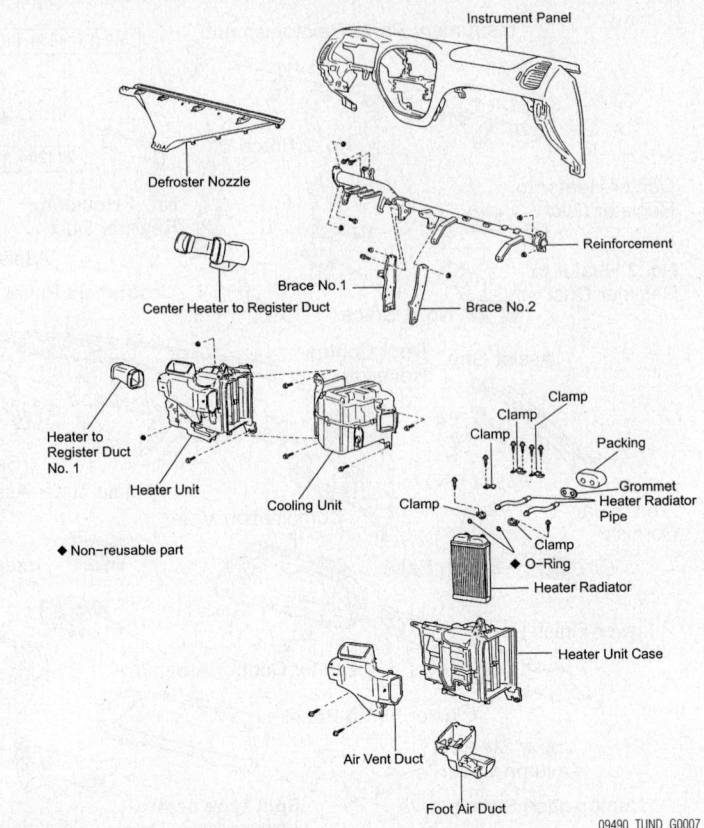

Instrument Panel

Defroster Nozzle

Reinforcement

Brace No.1

Center Heater to Register Duct

Brace No.2

Heater to Register Duct No. 1

Heater Unit

Cooling Unit

Clamp

Clamp

Clamp

Clamp

Clamp

Packing

Grommet

Heater Radiator Pipe

Clamp

O–Ring

Heater Radiator

Heater Unit Case

Air Vent Duct

Foot Air Duct

◆ Non–reusable part

09490_TUND_G0007

Heater core and related component—2005–2006 Standard and Access Cab

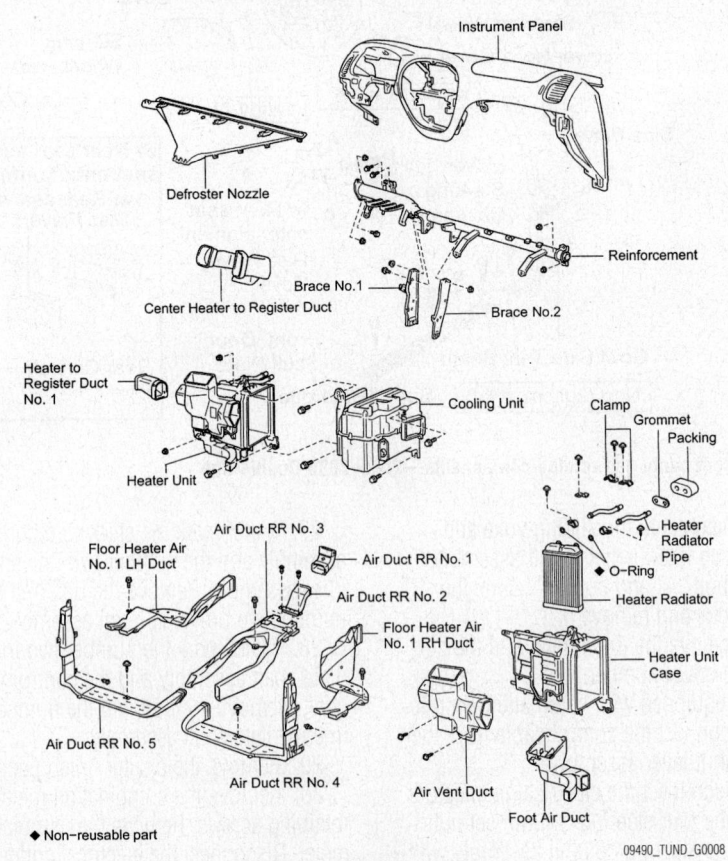

Instrument Panel

Defroster Nozzle

Reinforcement

Brace No.1

Center Heater to Register Duct

Brace No.2

Heater to Register Duct No. 1

Cooling Unit

Clamp

Grommet

Packing

Heater Unit

Air Duct RR No. 3

Floor Heater Air No. 1 LH Duct

Air Duct RR No. 1

Air Duct RR No. 2

Floor Heater Air No. 1 RH Duct

Heater Radiator Pipe

O–Ring

Heater Radiator

Heater Unit Case

Air Duct RR No. 5

Air Duct RR No. 4

Air Vent Duct

Foot Air Duct

◆ Non–reusable part

09490_TUND_G0008

Heater core and related components—2005–2006 Double Cab

43. If equipped, remove the rear console box. Remove the number four heater to register duct.

44. Remove the retaining nut located in the center of the combination meter housing. Disconnect the connectors. Remove the instrument panel from the vehicle.

45. Remove the center heater to register duct.

46. Disconnect the connectors. Remove the bolt and door control receiver. Remove the three bolts, nut and number one brace.

47. Remove the bolt, nut and number two brace.

48. Remove the three bolts, four nuts and instrument panel reinforcement bar assembly.

49. Remove the defroster nozzle and heater to register duct.

50. Remove the three heater unit retaining screws. Remove the heater unit from the vehicle.

51. Remove the heater core from the heater nit.

To install:

52. Installation is the reverse of the removal procedure.

53. Tighten the number two intermediate shaft assembly bolt to 26 ft. lbs.

54. Tighten the number two universal joint assembly bolt to 26 ft. lbs.

55. Tighten the steering column set nuts to 19 ft. lbs.

56. Tighten bolts "A" and "B" of the sliding yoke to 26 ft. lbs.

57. When installing the center spiral cable, check that the front tires are in the straight ahead position. Turn the cable counterclockwise by hand until it feels firm. Rotate the cable clockwise about 2½ turns to align the marks.

➡**The cable will rotate about 2½ turns to both the right and left from the center.**

58. Tighten the steering wheel retaining bolt to 37 ft. lbs.

59. Fill the cooling system with the proper grade and type engine coolant.

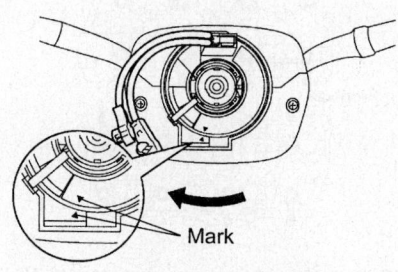

09490_TUND_G0009

Spiral cable alignment—2005–2006

60. Properly recharge the air condition system.

61. Start the engine and check for leaks, correct as required.

Cylinder Head

REMOVAL & INSTALLATION

3.4L Engine

1. Before servicing the vehicle, refer to the Precautions Section.
2. Disconnect the negative battery cable.
3. Relieve the fuel system pressure.
4. Remove the engine undercover.
5. Drain the cooling system.
6. Remove or disconnect the following:
   - Front exhaust pipe
   - Air cleaner cap
   - Mass Air Flow (MAF) meter and resonator
7. Disconnect the following cables:
   - Actuator cable from the bracket, if equipped with cruise control
   - Accelerator cable
   - Throttle cable, if equipped with an automatic transmission
   - Heater hose
   - Upper radiator hose
   - Power steering drive belt
   - Air conditioning drive belt, by loosening the idle pulley nut and adjusting bolt
   - Loosen the lockbolt, pivot bolt and adjusting bolt and the alternator drive belt
   - No. 2 fan shroud by removing the 2 clips
   - Fan with the fluid coupling and fan pulleys
   - Power steering pump and move it aside without disconnecting the pump lines
   - Compressor and move it aside without disconnecting the compressor lines, if equipped with air conditioning
   - Air conditioning bracket, if equipped with air conditioning
   - Spark plug wires with the ignition coils
   - Spark plugs
   - No. 2 timing belt cover
8. Remove the fan bracket, as follows:
   a. Remove the power steering adjusting strut by removing the nut.
   b. Remove the fan bracket by removing the bolt and nut.
9. Set the No. 1 cylinder at Top Dead Center (TDC) of the compression stroke.

a. Turn the crankshaft pulley and align its groove with the timing mark **0** on the No. 1 timing belt cover.

b. Check that the timing marks of the camshaft timing pulleys and the No. 3 timing belt cover are aligned. If not, turn the crankshaft pulley 1 revolution (360 degrees).

10. Remove the timing belt tensioner by alternately loosening the 2 bolts.

11. Remove the camshaft timing pulleys, as follows:
   a. Using Variable Pin Wrench Set tool 09960-10010, remove the pulley bolt, the timing pulley and the knock pin.
   b. Remove the 2 timing pulleys with the timing belt.

12. Remove or disconnect the following:
   - Bolt and the No. 2 idler pulley
   - Alternator

13. Remove the Exhaust Gas Recirculation (EGR) pipe and 2 gaskets, if equipped with an EGR valve

14. Remove the intake chamber stay as follows:
   a. Remove the oil filler tube and No. 1 throttle cable clamp by removing the bolt and 2 nuts.
   b. Remove the intake chamber stay by removing the 2 bolts.

15. Remove the following connectors:
   - VSV connector for the fuel pressure control.
   - Throttle position sensor
   - IAC valve connector
   - EGR valve gas temperature sensor, if equipped
   - Vacuum Switching Valve (VSV) connector for the EGR valve, if equipped

16. Disconnect the following hoses:
   - Positive Crankcase Ventilation (PCV) hoses
   - Water bypass hoses.
   - Air assist hose from the intake air connector
   - 2 vacuum sensing hoses from the VSV
   - Evaporative Emissions (EVAP) hose
   - Air hose, from the power steering
   - Air hose from the air conditioning idle up valve, if equipped with air conditioning

17. Remove or disconnect the following:
   - 4 bolts, 2 nuts and the air intake chamber assembly
   - Intake air connector

18. Disconnect the engine wiring harness from the intake manifold, as follows:
   - Oil pressure sensor connector
   - Crankshaft position sensor connector

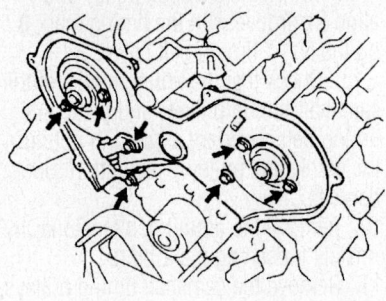

Rear timing belt cover bolt locations—3.4L engine

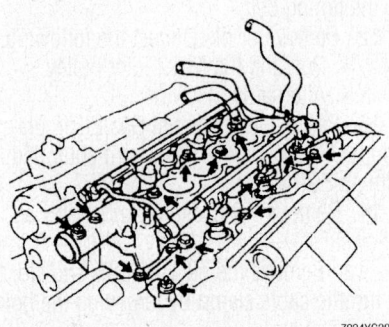

Intake manifold bolts and nuts locations—3.4L engine

- 6 injector connectors
- Engine Coolant Temperature (ECT) sender gauge connector
- ECT sensor connector
- Knock (KS) sensor connector
- Camshaft Position (CMP) sensor connector
- 3 engine wiring harness clamps
- 3 bolts and the engine wiring harness from the cylinder head

19. Remove or disconnect the following:
- CMP sensor
- No. 3 (rear) timing belt cover, by removing the 6 bolts
- Fuel pressure regulator
- Intake manifold assembly
- Power steering pump bracket
- Oil dipstick and guide
- Exhaust crossover pipe and gaskets, by removing the 6 nuts
- Left-hand exhaust manifold, by removing the heat insulator and 6 nuts
- Right-hand exhaust manifold, by removing the heat insulator and 6 nuts
- 8 bolts, seal washers, cylinder head cover and gasket
- Both cylinder head covers
- Semi-circular plugs

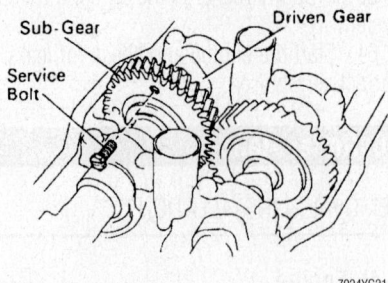

Drive gear service bolt (right side)—3.4L engine

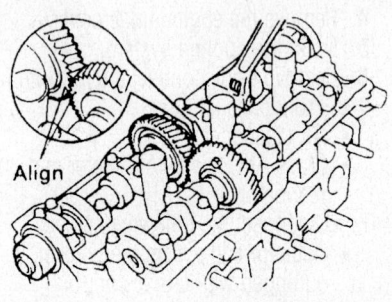

Aligning the timing mark (1 dot mark) of the left camshafts—3.4L engine

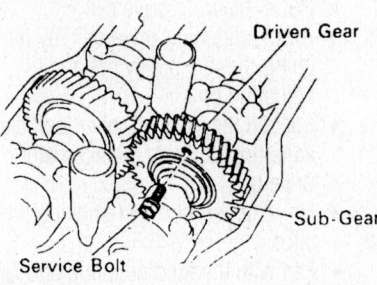

Drive gear service bolt (left side)—3.4L engine

- Right exhaust camshafts
- Right-hand intake camshaft
- Left exhaust camshafts
- Left-hand intake camshaft
- Valve lifters and shims from the cylinder head; arrange the valve lifters and shims in correct order

20. Remove the cylinder heads, as follows:
- Bolt and disconnect the ground strap
- Cylinder head (recessed head) bolt on each cylinder head, using an 8mm hexagon wrench; then repeat for the other side
- 8 cylinder head (12-pointed head) bolts on each cylinder head, by

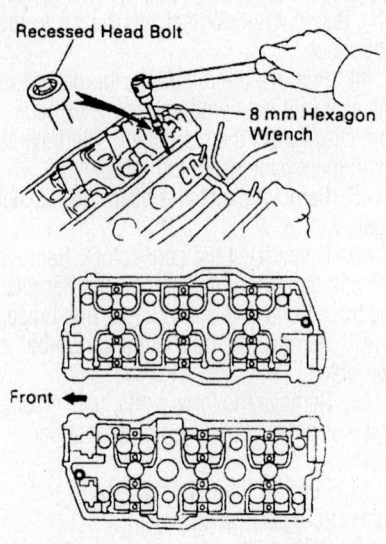

Cylinder head recessed bolt location—3.4L engine

loosening the bolts in several passes and in the reverse order of the tightening sequence
- 16 cylinder head bolts and plate washers.
- Cylinder head

To install:
21. Clean all surfaces.
22. Install or connect the following:
- New cylinder head gaskets
- Cylinder heads
23. Apply a light coat of engine oil on the threads and under the heads of the cylinder head bolts.
24. Tighten the cylinder head bolts, in sequence, using several passes, as follows:

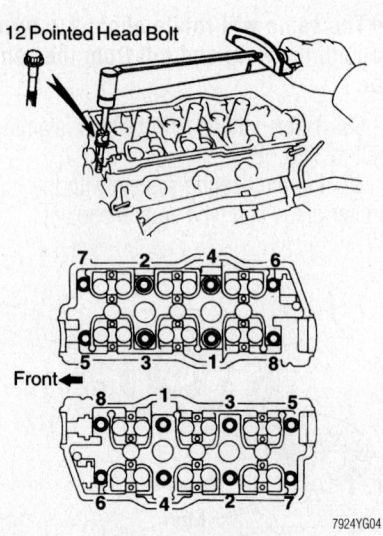

Cylinder head bolt torque sequence—3.4L engine

- Step 1: 25 ft. lbs. (34 Nm)
- Step 2: Mark the front of the cylinder head bolt with paint
- Step 3: An additional 90 degrees
- Step 4: Check that the painted mark is now at a 90 degrees angle to the front

25. Install the recessed head cylinder head bolts, as follows:
- Apply a light coat of engine oil on the threads and under the heads of the cylinder head bolts
- Cylinder head bolt on each cylinder head, using a 8mm hexagon wrench; then, repeat for the other side, as shown. Tighten the bolts to 13 ft. lbs. (18 Nm).
- Bolt and the ground strap

26. Install or connect the following:
- Valve lifters and shims

➡**Check that the valve lifter rotates smoothly by hand.**

- Camshafts
- Check and adjust the valve clearance
- Semi-circular plugs
- Cylinder head covers. Tighten the bolts, in several passes, to 53 inch lbs. (6 Nm).
- Exhaust manifolds, with new gaskets. Tighten the nuts to 30 ft. lbs. (40 Nm).
- Exhaust manifold heat insulators. Tighten the nuts to 71 inch lbs. (8 Nm).
- Exhaust crossover pipe. Tighten the nuts to 33 ft. lbs. (45 Nm).
- Alternator bracket. Tighten to 14 ft. lbs. (18 Nm).
- Oil dipstick and guide, using a new O-ring
- Power steering bracket. Tighten the fasteners to 14 ft. lbs. (18 Nm).
- New gaskets and the intake manifold assembly. Tighten the bolts and nuts to 13 ft. lbs. (18 Nm).
- Intake manifold stay with the 2

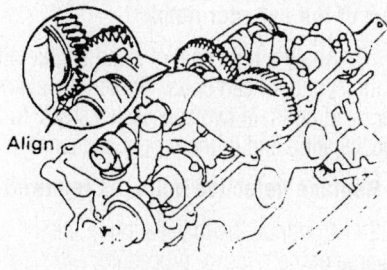

Aligning the right camshafts for installation—3.4L engine

7924YG26

bolts. Tighten the bolts to 14 ft. lbs. (18 Nm).
- Fuel inlet hose
- Fuel pressure regulator
- No. 3 timing belt cover. Tighten the bolts to 80 inch lbs. (9 Nm).
- CMP sensor. Tighten to 71 inch lbs. (8 Nm).
- Engine wiring harness

27. Install or connect the intake air connector, as follows:
- Intake manifold. Tighten the bolts and nuts to 14 ft. lbs. (19 Nm).
- DLC1 to the bracket on the intake manifold
- Ground strap to the intake manifold
- Brake booster vacuum hose to the intake air connector
- 2 fuel return hoses
- Engine wiring harness to the intake manifold
- Idle up valve connector, if equipped with air conditioning

28. Install or connect the following:
- Air intake chamber assembly. Tighten the bolts and nuts to 14 ft. lbs. (18.5 Nm).
- Hoses

29. Attach the following connectors:
- VSV connector for the fuel pressure control
- TP sensor connector
- IAC valve connector
- EGR gas temperature connector, if equipped with an EGR valve
- VSV connector, if equipped with an EGR valve
- Intake chamber stay
- New gaskets and the EGR pipe. Tighten the clamp nuts to 71 inch lbs. (8 Nm) and the EGR pipe nuts to 14 ft. lbs. (18 Nm).
- Alternator but do not tighten the bolts and nuts at this time.
- No. 2 timing belt idler. Tighten the bolt to 30 ft. lbs. (40 Nm).

➡**Check that the pulley bracket moves smoothly.**

- Left camshaft timing pulley

30. Set the No. 1 cylinder to TDC of the compression stroke, as follows:
a. Turn the crankshaft pulley and align its groove with the timing mark **0** on the No. 1 timing belt cover.
b. Turn the camshaft, align the knock pin hole of the camshaft with the timing mark of the No. 3 timing belt cover.
c. Turn the camshaft timing pulley, align the timing marks of the camshaft timing pulley and the No. 3 timing belt cover.

31. Connect the timing belt to the left camshaft timing pulley, as follows:
a. Check that the installation mark on the timing belt is aligned with the end of the No. 1 timing belt cover.
b. Using Variable Pin Wrench Set 09960-01000, slightly turn the left camshaft timing pulley clockwise. Align the installation mark on the timing belt with the timing mark of the camshaft timing pulley and hang the timing belt on the left camshaft timing pulley.
c. Align the timing marks of the left camshaft pulley and the No. 3 timing belt cover.
d. Check that the timing belt has tension between the crankshaft timing pulley and the left camshaft timing pulley.

32. Install the right camshaft timing pulley and the timing belt, as follows:
a. Align the installation mark on the timing belt with the timing mark of the right camshaft timing pulley, and hang the timing belt on the right camshaft timing pulley with the flange side facing inward.
b. Slide the right camshaft timing pulley on the camshaft. Align the timing marks on the right camshaft timing pulley and the No. 3 timing belt cover.
c. Align the knock pin hole of the camshaft with the knock pin groove of the pulley and install the knock pin. Install the bolt and tighten to 81 ft. lbs. (110 Nm).

33. Set the timing belt tensioner as follows:
a. Using a press, slowly press in the pushrod using 220–2,205 lbs. (981–9,807 N) of force.
b. Align the holes of the pushrod and housing, pass a 1.5mm hexagon wrench through the holes to keep the setting position of the pushrod.
c. Release the press and install the dust boot to the tensioner.

34. Install the timing belt tensioner. Tighten the bolts alternately to 20 ft. lbs. (28 Nm).

35. Using pliers, remove the 1.5mm hexagon wrench from the belt tensioner.

36. Check the valve timing.

37. Install or connect the following:
- Remaining components
- Negative battery cable

38. Fill the radiator with engine coolant.

39. Start the engine and check for leaks.

40. Check the ignition timing.

41. Install the engine undercover.

42. Road test the vehicle.

43. Recheck all fluid levels.

4.0L Engine

BANK 1

1. Before servicing the vehicle, refer to the Precautions Section.

2. Properly discharge the fuel system pressure.

3. Properly disarm the SRS system.

4. Disconnect the negative battery cable. Disconnect the positive battery cable. Remove the battery.

5. Remove the V bank cover. Remove the number one engine undercover assembly.

6. Disconnect the number one fuel pipe subassembly. Disconnect the number two fuel pipe subassembly.

7. Remove the drive belt. Remove the radiator shroud and fan.

8. Drain the cooling system. Disconnect the upper and lower radiator hoses. As required, remove the radiator from the vehicle.

9. Remove the alternator. Disconnect the power steering pump and position it to the side. Disconnect the air conditioning compressor and position it to the side.

10. Remove the intake air surge tank. Remove the ignition coil assembly. Remove the VVT sensor.

11. Remove the camshaft timing oil control valve assembly. Remove the cylinder head covers. Remove the oil level gauge guide.

12. Disconnect the fuel injector connectors. Remove the ten intake manifold retaining bolts. Remove the intake manifold from the engine.

13. Disconnect the engine coolant temperature sensor connector. Disconnect the heater hose. Remove the two bolts and four nuts, and then remove the water bypass joint RR and two gaskets. Remove the O-ring from the water outlet hose.

14. Remove the oil control valve filter from the cylinder head.

15. Disconnect the two heated oxygen sensor connectors. Remove the front exhaust pipe retaining bolts. Remove the front exhaust pipe. Remove the front exhaust pipe subassembly. Remove the three bolts and remove the exhaust manifold stay.

16. Disconnect the air fuel ratio sensor connector. Remove the six nuts; remove the exhaust manifold and gasket.

17. Remove the belt tensioner assembly. Remove the timing chain.

18. While raising the chain tensioner number two, insert a 0.039 inch diameter pin, into the hole. Hold the hexagonal portion of the camshaft in place, using a wrench.

➡ **Be careful not to damage the cylinder head and valve lifter with the wrench.**

19. Remove the two bolts. Remove the camshaft timing gear, camshaft timing gear assembly and timing chain number two.

➡ **Do not disassemble the camshaft timing gear assembly.**

20. Remove the bolt and then remove the chain tensioner number two.

21. Remove the camshafts.

22. Remove the two bolts. Separate the two ground cables.

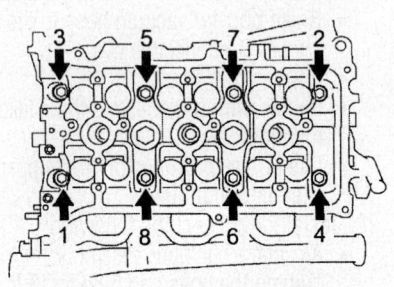

09490_TACO_G0029

Bank 1 cylinder head bolt loosening sequence—4.0L engine

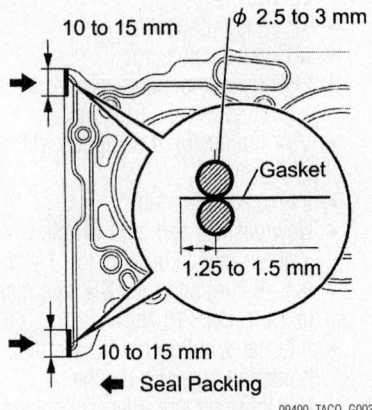

09490_TACO_G0030

Cylinder head gasket sealant application—4.0L engine

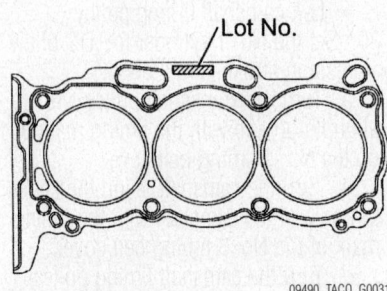

09490_TACO_G0031

Bank 1 cylinder head gasket positioning—4.0L engine

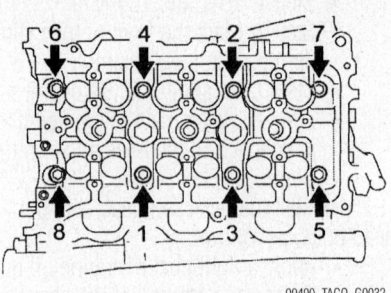

09490_TACO_G0032

Bank 1 cylinder head bolt torque sequence—4.0L engine

23. Loosen the cylinder head retaining bolts in several steps and in the proper removal sequence. Remove the bolts.

➡ **Be careful not to drop the washers into the cylinder head. Head warpage and cracking could result in removing the bolts in the wrong order.**

24. Lift the cylinder head from the dowels on the cylinder block and remove it from the engine.

To install:

25. Before installing, thoroughly clean the gasket mating surfaces and check for warpage. Remove any old packing material. Be careful not to drop any oil on the contact surfaces of the cylinder head or engine block.

26. Apply continuous beads of seal packing, part number 08826-00080 or equivalent, to the cylinder head gasket. Bead width should be 0.098–0.118 inch.

➡ **Install the cylinder head gasket within three minutes after applying the seal packing. Install the cylinder head bolts within fifteen minutes after applying the seal packing.**

27. Be sure to position the cylinder head gasket on the engine block with the lot number stamp facing upward.

➡ **Be sure that the head gasket is installed in the correct direction. Position the gasket gently in order to avoid damaging the gasket with the bottom part of the cylinder head.**

28. Apply a light coat of clean engine oil to the cylinder head bolts. Tighten the cylinder head bolts, in two successive steps, to specification and in the proper sequence.

➡ **Replace defective bolts, as required.**

29. Continue the installation in the reverse of the removal procedure.

30. Be sure to fill the cooling system with the proper grade and type engine coolant.

31. Be sure to fill the engine oil with the proper grade and type oil.

➡ **Do not fill the engine with oil for at least four hours after the cylinder head has been installed.**

32. Start the engine and check for leaks. Correct as required.

BANK 2

1. Before servicing the vehicle, refer to the Precautions Section.

2. Properly discharge the fuel system pressure.

3. Properly disarm the SRS system.

4. Disconnect the negative battery cable. Disconnect the positive battery cable. Remove the battery.

5. Remove the V bank cover. Remove the number one engine undercover assembly.

6. Disconnect the number one fuel pipe subassembly. Disconnect the number two fuel pipe subassembly.

7. Remove the drive belt. Remove the radiator shroud and fan.

8. Drain the cooling system. Disconnect the upper and lower radiator hoses. As required, remove the radiator from the vehicle.

9. Remove the alternator. Disconnect the power steering pump and position it to the side. Disconnect the air conditioning compressor and position it to the side.

10. Remove the intake air surge tank. Remove the ignition coil assembly. Remove the VVT sensor.

11. Remove the camshaft timing oil control valve assembly. Remove the cylinder head covers. Remove the oil level gauge guide.

12. Disconnect the fuel injector connectors. Remove the ten intake manifold retaining bolts. Remove the intake manifold from the engine.

13. Disconnect the engine coolant temperature sensor connector. Disconnect the heater hose. Remove the two bolts and four nuts, and then remove the water bypass joint RR and two gaskets. Remove the O-ring from the water outlet hose.

14. Remove the oil control valve filter from the cylinder head.

15. Disconnect the two heated oxygen sensor connectors. Remove the front exhaust pipe retaining bolts. Remove the front exhaust pipe. Remove the front exhaust pipe subassembly. Remove the three bolts and remove the exhaust manifold stay.

16. Disconnect the air fuel ratio sensor connector. Remove the six nuts; remove the exhaust manifold and gasket.

17. Remove the belt tensioner assembly. Remove the timing chain.

18. Remove the two bolts and then remove the chain vibration damper number one.

19. While pushing down on chain tensioner number two, insert a 0.039 inch diameter pin, into the hole. Hold the hexagonal portion of the camshaft in place, using a wrench.

➡ **Be careful not to damage the cylinder head and valve lifter with the wrench.**

20. Remove the two bolts. Remove the camshaft timing gear, camshaft timing gear assembly and timing chain number two.

➡ **Do not disassemble the camshaft timing gear assembly.**

21. Remove the bolt and then remove the chain tensioner number three.

22. Remove the camshafts.

23. Remove the two bolts. Separate the two ground cables.

24. Loosen the cylinder head retaining bolts in several steps and in the proper removal sequence. Remove the bolts.

➡ **Be careful not to drop the washers into the cylinder head. Head warpage and cracking could result in removing the bolts in the wrong order.**

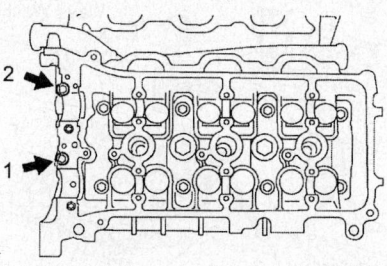

Bank 2 recessed cylinder head bolt loosening sequence—4.0L engine

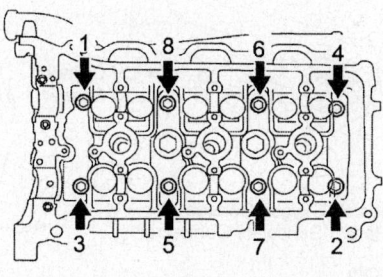

Bank 2 cylinder head bolt loosening sequence—4.0L engine

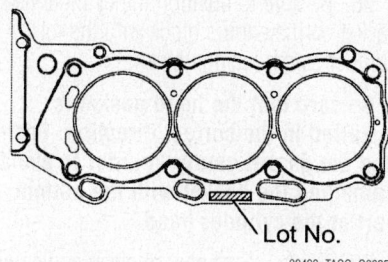

Bank 2 cylinder head gasket positioning—4.0L engine

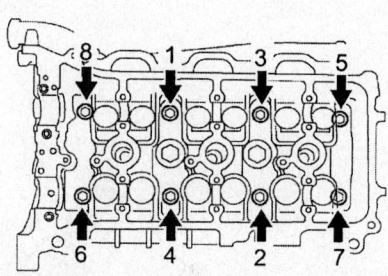

Bank 2 Cylinder head bolt torque sequence—4.0L engine

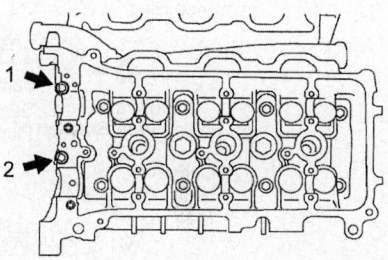

Bank 2 recessed cylinder head bolt torque sequence—4.0L engine

25. Lift the cylinder head from the dowels on the cylinder block and remove it from the engine.

To install:

26. Before installing, thoroughly clean the gasket mating surfaces and check for warpage. Remove any old packing material. Be careful not to drop any oil on the contact surfaces of the cylinder head or engine block.

27. Apply continuous beads of seal packing, part number 08826-00080 or equivalent, to the cylinder head gasket. Bead width should be 0.098–0.118 inch.

➡ **Install the cylinder head gasket within three minutes after applying the seal packing. Install the cylinder head bolts within fifteen minutes after applying the seal packing.**

28. Be sure to position the cylinder head gasket on the engine block with the lot number stamp facing upward.

➡ **Be sure that the head gasket is installed in the correct direction. Position the gasket gently in order to avoid damaging the gasket with the bottom part of the cylinder head.**

29. Apply a light coat of clean engine oil to the cylinder head bolts. Tighten the cylinder head bolts, in two successive steps, to specification and in the proper sequence.

➡ **Replace defective bolts, as required.**

30. Continue the installation in the reverse of the removal procedure.

31. Be sure to fill the cooling system with the proper grade and type engine coolant.

32. Be sure to fill the engine oil with the proper grade and type oil.

➡ **Do not fill the engine with oil for at least four hours after the cylinder head has been installed.**

33. Start the engine and check for leaks. Correct as required.

4.7L Engine

2002–2004

1. Before servicing the vehicle, refer to the Precautions Section.

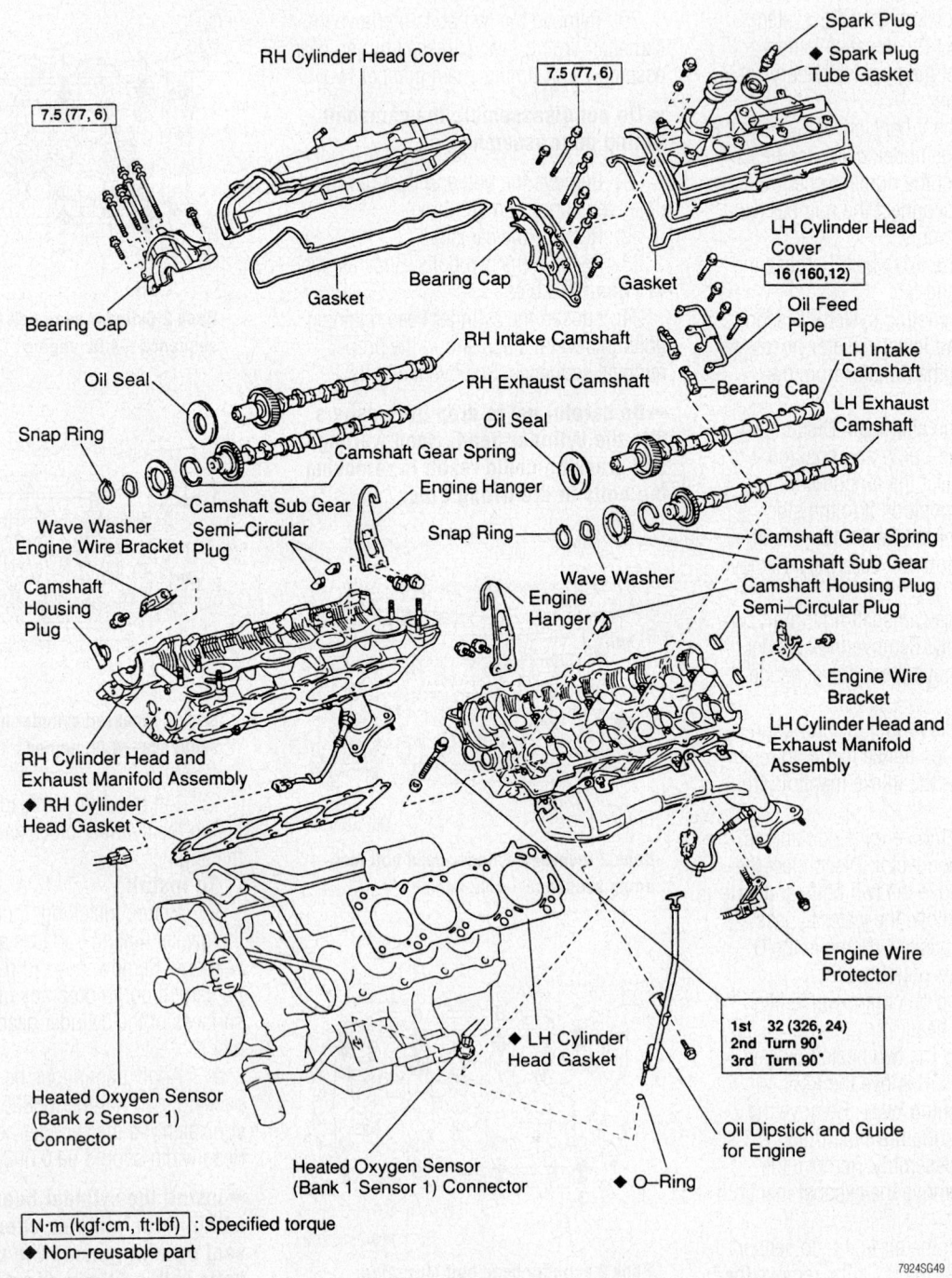

Cylinder head and related components—2002–2004 4.7L engine

N·m (kgf·cm, ft·lbf) : Specified torque
◆ Non-reusable part

7924SG49

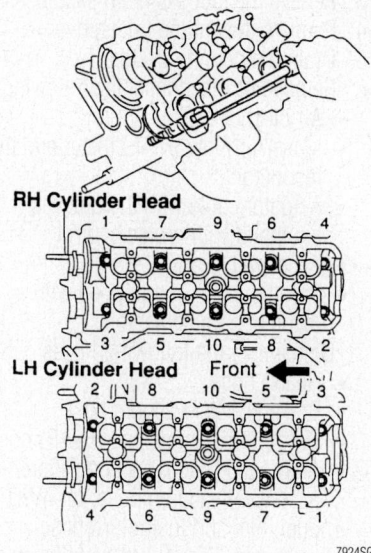

Cylinder head bolt loosening sequence— 2002–2004 4.7L engine

2. Drain the cooling system.

3. Relieve the fuel system pressure.

4. Remove or disconnect the following:
- Battery and tray
- Engine appearance cover
- Engine under covers
- Air intake assembly
- Accessory drive belt
- A/C compressor and bracket
- Cooling fan and bracket
- Radiator
- Idler pulley
- Front covers
- Timing belt. Refer to the Timing Belt unit repair section.
- Camshaft sprockets
- Camshaft Position (CMP) sensor
- Power steering pump
- Exhaust front pipes
- Transmission dipstick tube
- Ignition coils
- Rear timing belt covers
- Fuel lines
- Intake manifold
- Water inlet housing assembly
- Front and rear water bypass joints
- Engine lifting eyes
- Oil dipstick tube
- Valve covers
- Camshafts
- Cylinder heads with the exhaust manifolds attached. Loosen the bolts in the sequence shown.

To install:

5. Install the cylinder heads with new gaskets. Tighten the bolts in sequence as follows:
  a. Step 1: 24 ft. lbs. (32 Nm)
  b. Step 2: Plus 180 degrees

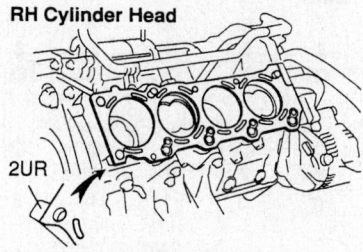

Cylinder head gasket identification—4.7L engine

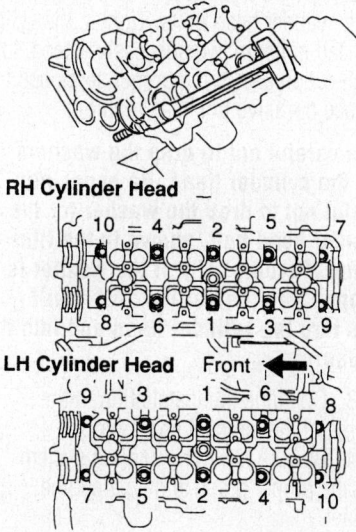

Cylinder head bolt torque sequence— 2002–2004 4.7L engine

6. Install or connect the following:
- Camshafts
- Valve covers
- Oil dipstick tube
- Engine lifting eyes
- Front and rear water bypass joints
- Water inlet housing assembly
- Intake manifold
- Fuel lines
- Rear timing belt covers
- Ignition coils
- Transmission dipstick tube
- Exhaust front pipes
- Power steering pump

- CMP sensor
- Camshaft sprockets
- Timing belt
- Front covers
- Idler pulley
- Radiator
- Cooling fan and bracket
- A/C compressor and bracket
- Accessory drive belt
- Air intake assembly
- Engine under covers
- Engine appearance cover
- Battery and tray

7. Fill the cooling system.

8. Start the engine and check for leaks.

2005–2006

1. Before servicing the vehicle, refer to the Precautions Section.

2. Properly discharge the fuel system pressure.

3. Properly disarm the SRS system.

4. Disconnect the negative battery cable.

5. Remove the timing belt.

6. Remove the camshaft.

7. Remove the transmission oil filler tube subassembly.

8. Remove the intake manifold.

9. Remove the water inlet housing bolts. Remove the water inlet housing and O-ring.

10. Remove the air pump assembly.

11. Remove the water bypass joint nuts. Remove the water bypass joint and gaskets. Remove the water bypass pipe subassembly.

12. Remove the number two air switching valve assembly.

13. Remove the rear water bypass joint retaining nuts. Remove the rear water bypass joint and gaskets. Remove the engine wire.

14. Remove the bolt, stud bolt and rear timing plate, right side.

15. Remove the two bolts and the number two rear timing plate, left side.

16. Remove the bolt and remove the rear timing plate, left side.

17. Remove the front exhaust pipe.

18. Loosen the cylinder head retaining bolts in several steps and in the proper removal sequence. Remove the bolts.

➡ **Be careful not to drop the washers into the cylinder head. Be especially careful not to drop the washer for the cylinder head bolt, shown in the illustration (Portion "A"). If the washer is dropped into the portion "A" it will pass thru the cylinder head and into the oil pan. Head warpage and cracking could result in removing the bolts in the wrong order.**

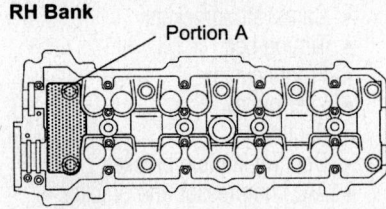

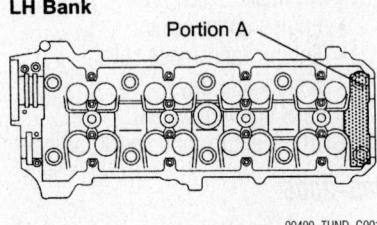

Cylinder head bolt washer (portion "A") location—2005–2006 4.7L engine

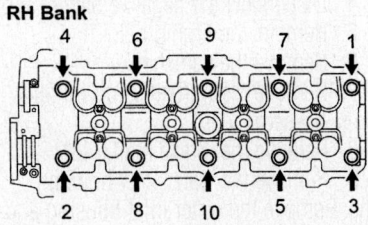

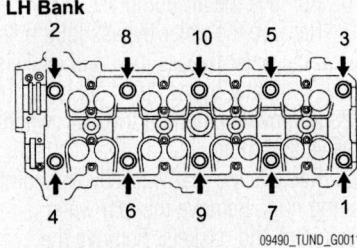

Cylinder head bolt loosening sequence—2005–2006 4.7L engine

19. Lift the cylinder head from the dowels on the cylinder block and remove it from the engine.

➡The cylinder head should not be tilted so as to secure the valve lifter. If the cylinder head is tilted, remove the valve lifter and check that the adjusting shim is set correctly.

To install:

20. Before installing, thoroughly clean the gasket mating surfaces and check for warpage. Be careful not to drop any oil on the contact surfaces of the cylinder head or engine block.

21. Position the new cylinder head gasket on the engine block. Note the identification marks on the engine block, to ensure

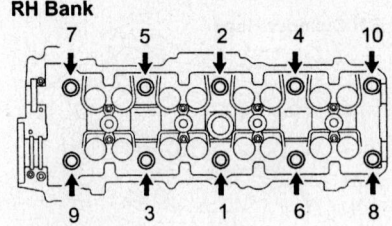

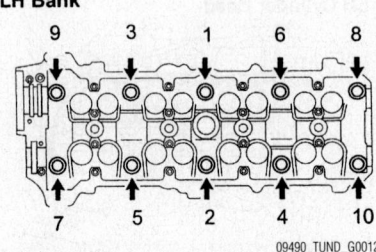

Cylinder head bolt tightening sequence—2005–2006 4.7L engine

proper positioning. "2R" for the right bank and "2L" for the left bank.

22. Apply a light coat of clean engine oil to the cylinder head bolts. Tighten the cylinder head bolts, in two successive steps, to specification and in the proper sequence. Replace defective bolts, as required.

➡Be careful not to drop the washers into the cylinder head. Be especially careful not to drop the washer for the cylinder head bolt, shown in the illustration (Portion "A"). If the washer is dropped into the portion "A" it will pass thru the cylinder head and into the oil pan.

23. Continue the installation in the reverse of the removal procedure.

24. Be sure to fill the cooling system with the proper grade and type engine coolant.

25. Be sure to fill the engine oil with the proper grade and type oil.

➡Do not fill the engine with oil for at least four hours after the cylinder head has been installed.

26. Start the engine and check for leaks. Correct as required.

Intake Manifold

REMOVAL & INSTALLATION

3.4L Engine

1. Before servicing the vehicle, refer to the Precautions Section.
2. Disconnect the negative battery cable.

3. Relieve the fuel system pressure.
4. Remove the engine undercover.
5. Drain the cooling system.
6. Remove or disconnect the following:
   - Air cleaner cap
   - Mass Air Flow (MAF) meter and the resonator
   - Actuator cable with the bracket, if equipped with cruise control
   - Accelerator cable
   - Throttle cable, if equipped with automatic transmission
7. Disconnect the following hoses:
   - Heater hose
   - Brake booster vacuum hose
   - Evaporative Emissions (EVAP) hose
   - Automatic Disconnecting Differential (ADD) vacuum hose, for 4WD
   - Fuel inlet and fuel return hose
8. Remove or disconnect the following:
   - Spark plug wires, with the ignition coils
   - Intake chamber stay
   - No. 2 timing belt cover
   - Air intake chamber assembly
   - Throttle Position (TP) sensor connector
   - Idle Air Control (IAC) valve connector
   - Positive Crankcase Ventilation (PCV) hoses
   - Water bypass hoses
   - Air assist hose from the throttle body
   - Intake air connector
   - Engine wiring harness
   - Fuel return hose
   - Vacuum hose, from the fuel pressure regulator
   - Ground strap, from the intake air connector
   - Data Link Connector 1 (DLC1), from the bracket
   - 6 injector connectors
   - Engine Coolant Temperature (ECT) sensor and sender gauge connectors
   - Engine wiring harness protector from the cylinder head
   - Fuel pressure regulator

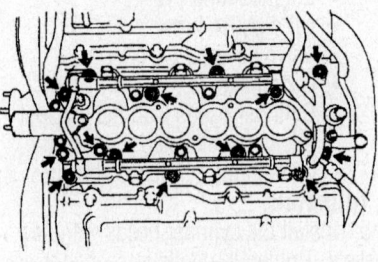

Intake manifold bolts and nuts—3.4L engine

- Intake manifold assembly
- Intake manifold stay
- Intake manifold, delivery pipes and the injectors assembly with the gaskets

To install:

9. Install or connect the following:
   - New gaskets
   - Intake manifold assembly. Tighten the bolts and nuts to 13 ft. lbs. (18 Nm).
   - Intake manifold stay. Tighten the bolts to 14 ft. lbs. (18 Nm).
   - Fuel pressure regulator
   - Engine wiring harness to the cylinder head, by installing the 3 bolts
   - 3 engine wiring harness clamps
   - 6 injector connectors
   - ECT sender gauge connector
   - ECT sensor connector
   - Intake manifold. Tighten the bolts and nuts to 14 ft. lbs. (18.5 Nm).
   - DLC1 to the bracket on the intake manifold
   - Ground strap to the intake manifold, by installing the bolt
   - Brake booster vacuum hose, to the intake air connector
   - 2 fuel return hoses
   - Engine wiring harness to the intake manifold
   - Air intake chamber assembly to the engine. Tighten the bolts and nuts to 14 ft. lbs. (18.5 Nm).
   - Intake chamber stay. Tighten the bolts to 30 ft. lbs. (40 Nm).
   - New O-ring to the oil filler tube
   - Oil filler tube end into the tube hole in the oil pan
   - Oil filler tube and No. 1 throttle cable clamp
   - No. 2 timing belt cover. Tighten the bolts to 80 inch lbs. (9 Nm).
   - PCV hoses
   - Water bypass hoses
   - Air assist hose to the throttle body
   - IAC valve connector
   - TP sensor connector
   - Brake booster vacuum hose
   - EVAP hose
   - Automatic Disconnecting Differential (ADD) vacuum hose, for 4WD
   - Fuel inlet and fuel return hose
   - 3 spark plug wire clamps to the No. 2 timing belt cover
   - CMP connector to the No. 2 timing belt cover
   - Spark plug wires with the ignition coils
   - Heater hose
   - Actuator cable with the bracket, if equipped with cruise control
   - Accelerator cable
   - Throttle cable, if equipped with automatic transmission
   - MAF meter, resonator and the air cleaner cap
   - Negative battery cable

10. Fill the radiator with engine coolant.
11. Start the engine and check for leaks.
12. Install the engine undercover.
13. Road test the vehicle.
14. Recheck all fluid levels.

4.0L Engine

1. Before servicing the vehicle, refer to the Precautions Section.
2. Properly relieve the fuel system pressure.
3. Properly disarm the SRS system.
4. Disconnect the negative battery cable.
5. Drain the engine coolant. Remove the air cleaner assembly.
6. Disconnect the fuel injector wiring connectors.
7. Remove the necessary components in order to gain access to the intake manifold retaining bolts.
8. Remove the intake manifold retaining bolts. Remove the intake manifold from the engine.

To install:

9. Clean all surfaces.
10. Install a new gasket on each cylinder head.

➡ **Align the ports of the gasket and the cylinder head. Be careful of the installation direction. Position the intake manifold to the engine.**

11. Install the retaining bolts and tighten to specification, in an alternating sequence.
12. Continue the installation in the reverse order of the removal procedure.
13. Be sure to fill the cooling system with the proper grade and type engine coolant.
14. Start the engine and check for leaks. Correct as required.

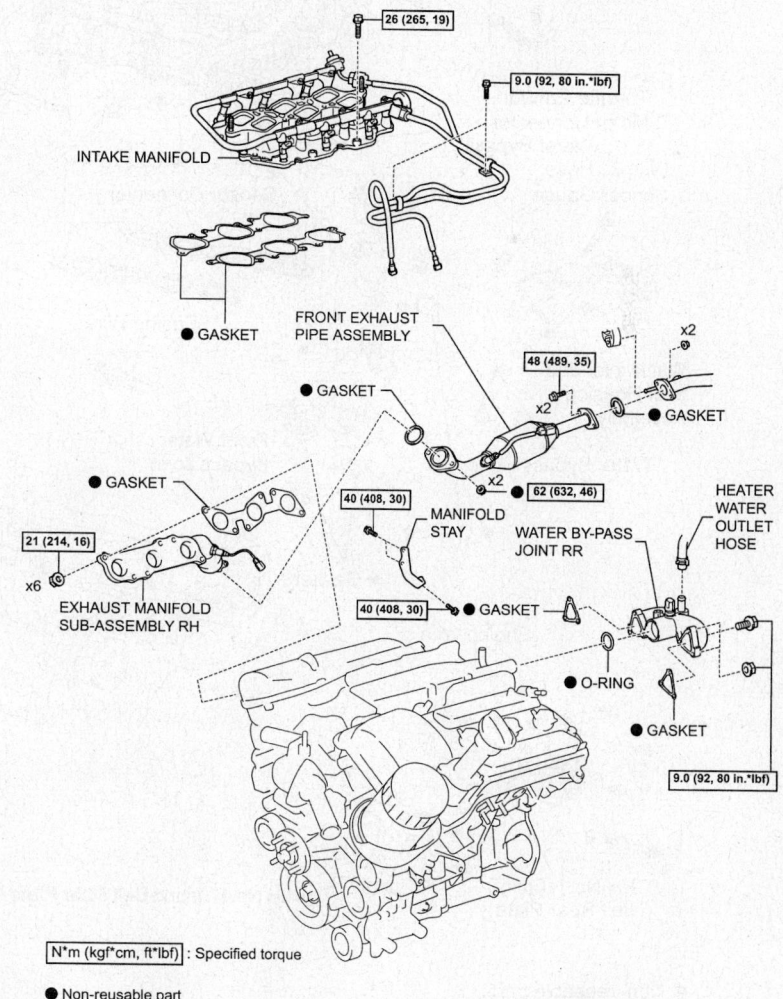

Intake manifold and related components—4.0L engine

4.7L Engine

1. Before servicing the vehicle, refer to the Precautions Section.
2. Drain the cooling system.
3. Relieve the fuel system pressure.
4. Remove or disconnect the following:
   - Negative battery cable
   - Engine appearance cover
   - Accelerator cable
   - Throttle Position (TP) sensor connector
   - Accelerator pedal position sensor
   - Throttle motor connector
   - Evaporative Emissions (EVAP) vacuum switching valve connector
   - Fuel injector connectors
   - Engine Coolant Temperature (ECT) sensor connector
   - ETC gauge sender connector
   - Heated Oxygen (HO2S) sensor connectors
   - Fuel pressure regulator vacuum hose
   - Positive Crankcase Ventilation (PCV) valve and hose

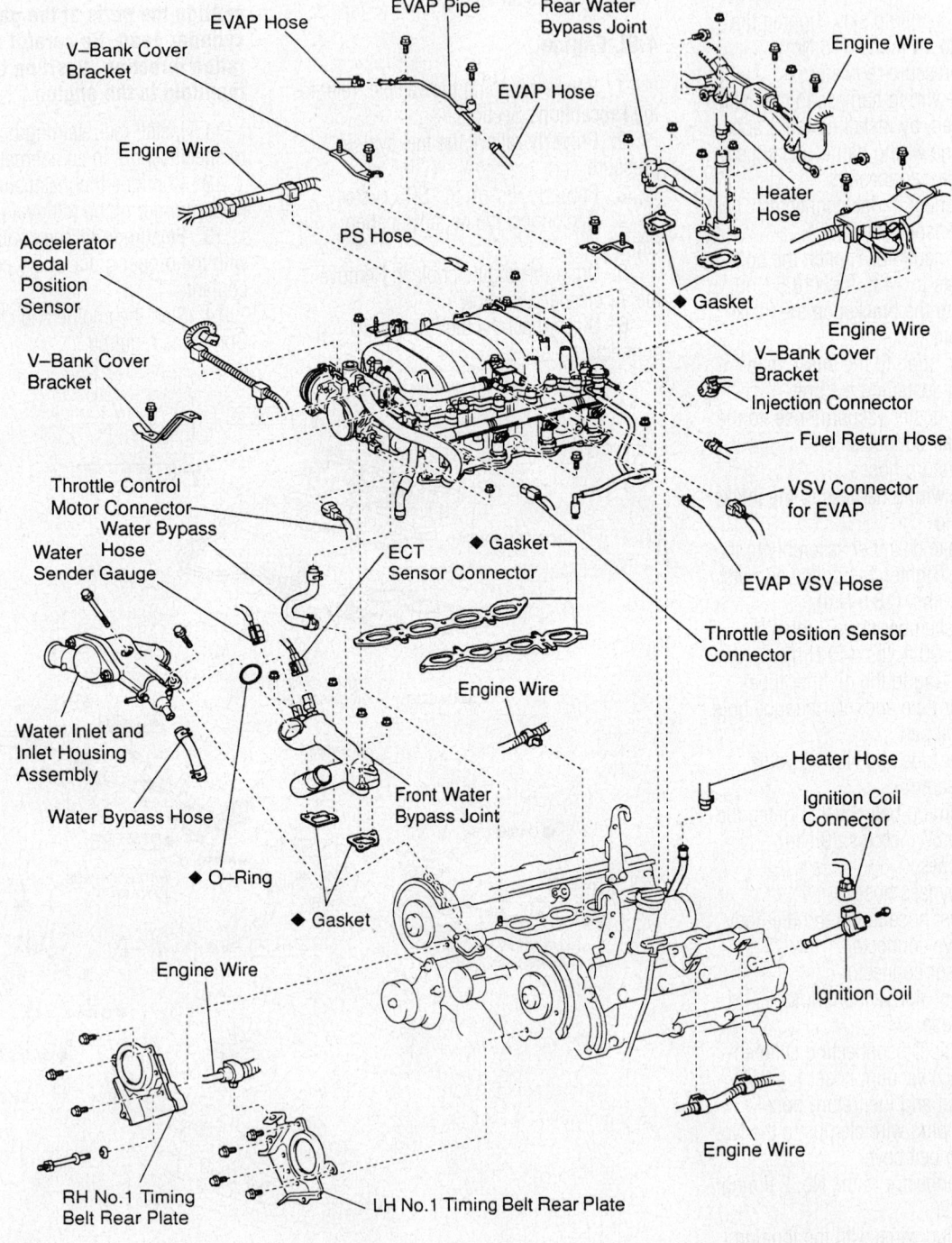

◆ Non-reusable part

Intake manifold and related components—4.7L engine

7924SG50

- EVAP hoses
- Power steering vacuum hoses
- Water bypass hose
- Engine control wiring harness clamps
- Cylinder head ground cables
- Intake manifold wire harness protector
- EVAP pipe
- Engine appearance cover brackets
- Intake manifold

To install:

5. Install or connect the following:
- Intake manifold. Tighten the fasteners to 13 ft. lbs. (18 Nm).
- Engine appearance cover brackets
- EVAP pipe
- Intake manifold wire harness protector
- Cylinder head ground cables
- Engine control wiring harness clamps
- Water bypass hose
- Power steering vacuum hoses
- EVAP hoses
- PCV valve and hose
- Fuel pressure regulator vacuum hose
- HO2S sensor connectors
- ETC gauge sender connector
- ECT sensor connector
- Fuel injector connectors
- EVAP vacuum switching valve connector
- Throttle motor connector
- Accelerator pedal position sensor
- TP sensor connector
- Accelerator cable
- Engine appearance cover
- Negative battery cable

6. Fill the cooling system.
7. Start the engine and check for leaks.

Exhaust Manifold

REMOVAL & INSTALLATION

3.4L Engine

1. Before servicing the vehicle, refer to the Precautions Section.
2. Remove or disconnect the following:
- Exhaust crossover pipe, from the exhaust manifold by removing the 3 nuts
- Exhaust Gas Recirculation (EGR) pipe, from the exhaust manifold, on the left manifold equipped with an EGR valve
- Exhaust manifold heat insulator, by removing the 3 nuts
- Exhaust manifold

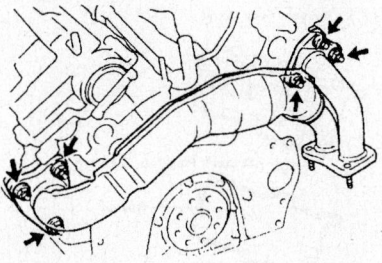

Exhaust crossover pipe mounting nut locations—3.4L engine

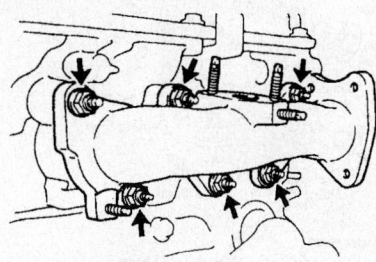

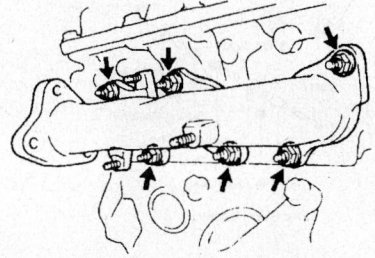

Exhaust manifold nuts—3.4L engine

To install:

3. Install or connect the following:
- Exhaust manifold, using a new gasket. Tighten the nuts to 30 ft. lbs. (40 Nm).
- Exhaust heat insulator. Tighten the nuts to 71 inch lbs. (8 Nm).
- EGR pipe to the exhaust manifold, if equipped with an EGR valve. Tighten the manifold nuts to 14 ft. lbs. (18 Nm) and the clamp nuts to 71 inch lbs. (8 Nm).
- Crossover pipe to the exhaust manifold, using a new gasket. Tighten the nuts to 33 ft. lbs. (45 Nm).

4.0L Engine

1. Before servicing the vehicle, refer to the Precautions Section.
2. Disconnect the negative battery cable.
3. Disconnect the exhaust manifold to exhaust flange nuts.
4. Remove the necessary components to gain access to the exhaust manifold retaining bolts.

5. Remove the exhaust manifold retaining nuts. Discard the nuts. Remove the exhaust manifold from the engine.

To install:

6. Installation is the reverse of the removal procedure.
7. Be sure to use new gaskets. Be sure to use new retaining nuts.

4.7L Engine

1. Before servicing the vehicle, refer to the Precautions Section.
2. Attach a hoist to the engine lifting eyes.
3. Remove or disconnect the following:
- Negative battery cable
- Heated Oxygen (HO2S) sensor connectors
- Exhaust manifold heat shield
- Exhaust front pipe
- Motor mount
- Motor mount bracket
- Exhaust manifold

To install:

➡**Use new exhaust manifold nuts for assembly.**

4. Install or connect the following:
- Exhaust manifold. Tighten the nuts to 32 ft. lbs. (44 Nm).
- Motor mount bracket. Tighten the bolts to 27 ft. lbs. (36 Nm).
- Motor mount. Tighten the fasteners to 22 ft. lbs. (30 Nm).
- Exhaust front pipe. Tighten the nuts to 46 ft. lbs. (62 Nm).
- Exhaust manifold heat shield
- HO2S sensor connectors
- Negative battery cable

5. Start the engine and check for leaks.

Camshaft and Valve Lifters

REMOVAL & INSTALLATION

3.4L Engine

1. Before servicing the vehicle, refer to the Precautions Section.
2. Remove or disconnect the following:
- Negative battery cable
- Engine undercover

3. Drain the cooling system.
4. Remove or disconnect the following:
- Air cleaner cap
- Mass Air Flow (MAF) meter and the resonator
- Actuator cable with the bracket, if equipped with cruise control
- Accelerator cable
- Throttle cable, if equipped with an automatic transmission

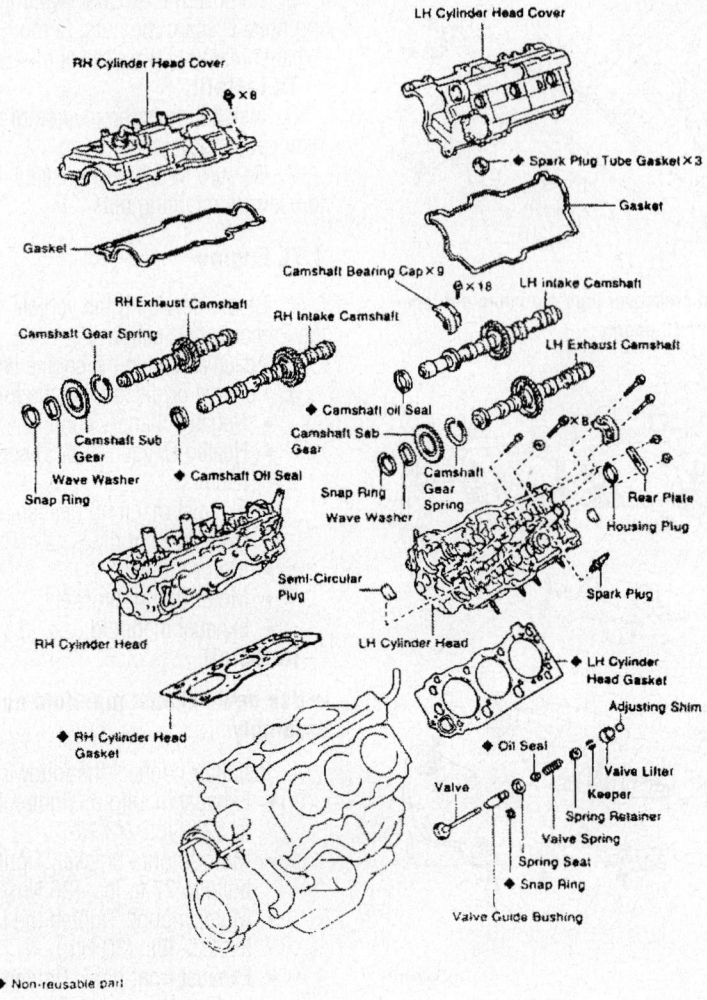

Camshafts and related components—3.4L engine

- Heater hose
- Upper radiator hose from the engine

5. Remove the power steering drive belt, as follows:

 a. Stretch the belt and loosen the fan pulley mounting nuts.

 b. Loosen the lockbolt, pivot bolt and adjusting bolt and remove the drive belt from the engine.

6. Remove or disconnect the following:

- Air conditioning drive belt, by loosening the idle pulley nut and adjusting bolt
- Loosen the alternator lockbolt, pivot bolt and adjusting bolt
- Alternator drive belt
- No. 2 fan shroud, by removing the 2 clips
- Fan with the fluid coupling and fan pulleys
- Power steering pump and move it aside without disconnecting the lines

- Compressor and move it aside without disconnecting the lines, if equipped with air conditioning
- Air conditioning bracket, if equipped with air conditioning
- Spark plug wires with the ignition coils
- Spark plugs
- Camshaft Position (CMP) sensor connector, from the No. 2 timing belt cover
- 3 spark plug wire clamps from the No. 2 timing belt cover
- 6 bolts and the timing belt cover
- Power steering adjusting strut by removing the nut
- Fan bracket by removing the bolt and nut

7. Set the No. 1 cylinder at Top Dead Center (TDC) of the compression stroke, as follows:

 a. Turn the crankshaft pulley and align its groove with the timing mark **0** on the No. 1 timing belt cover.

 b. Check that the timing marks of the camshaft timing pulleys and the No. 3 timing belt cover are aligned. If not, turn the crankshaft pulley 1 revolution (360 degrees).

8. Remove or disconnect the following:

- Timing belt tensioner, by alternately loosening the 2 bolts
- Pulley bolt, the timing pulley and the knock pin, using Variable Pin Wrench Set 09960-10010
- Both timing pulleys with the timing belt
- No. 2 idler pulley
- Alternator
- Positive Crankcase Ventilation PCV hoses
- Water bypass hoses
- Air assist hose from the intake air connector
- 2 vacuum sensing hoses from the Vacuum Switching Valve (VSV)
- Evaporative Emissions (EVAP) hose
- Air hose from the power steering
- Air hose from the air conditioning idle up valve, if equipped with air conditioning
- 4 bolts, 2 nuts and the air intake chamber assembly
- Intake air connector
- Camshaft Position (CMP) sensor
- No. 3 (rear) timing belt cover by removing the 6 bolts
- 8 bolts, seal washers, both cylinder head cover and gaskets
- Semi-circular plugs

9. Remove the right exhaust camshafts, as follows:

 a. Bring the service bolt hole of the driven sub-gear upward by turning the hexagon head portion of the exhaust camshaft with a wrench.

 b. Align the timing mark (2 dot marks) of the camshaft drive and driven gears by turning the camshaft with a wrench.

 c. Secure the exhaust camshaft sub-gear to the driven gear with a service bolt (6mm diameter, 16–20mm bolt length and 1.0mm in thread pitch).

➡ **When removing the camshaft, be sure the torsional spring force of the sub-gear has been eliminated by the above operation.**

 d. Uniformly loosen and remove the bearing cap bolts in several passes, in the sequence shown.

 e. Remove the bearing caps and camshaft. Make a note of the bearing cap positions for proper installation.

➡ **Do not pry on or attempt to force the camshaft with a tool or other object.**

10. Remove the right-hand intake camshaft, as follows:

a. Uniformly loosen and remove the bearing cap bolts in several passes, in the sequence shown.

b. Remove the bearing caps, oil seal and camshaft. Make a note of the bearing cap positions for proper installation.

11. Remove the left exhaust camshafts, as follows:

a. Align the timing mark (1 dot mark) of the camshaft drive and driven gears by turning the camshaft with a wrench.

b. Secure the exhaust camshaft sub-gear to the driven gear with a service bolt (6mm diameter, 16–20mm bolt length and 1.0mm in thread pitch).

➡ **When removing the camshaft, be sure that the torsional spring force of the sub-gear has been eliminated by the above operation.**

c. Uniformly loosen and remove the bearing cap bolts in several passes, in the sequence shown.

d. Remove the bearing caps and camshaft. Make a note of the bearing cap positions for proper installation.

12. Remove the left-hand intake camshaft, as follows:

a. Uniformly loosen and remove the bearing cap bolts in several passes, in the sequence shown.

b. Remove the bearing caps, oil seal and camshaft. Make a note of the bearing cap positions for proper installation.

13. Remove the valve lifters and shims from the cylinder head. Arrange the valve lifters and shims in correct order.

To install:

14. Clean all surfaces.

15. Install the valve lifters and shims. Check that the valve lifter rotates smoothly by hand.

16. Install the right intake camshaft, as follows:

a. Apply engine oil to the thrust portion of the intake camshaft.

b. Position the intake camshaft at 90 degrees angle of the timing mark (2 dot marks) on the cylinder head.

c. Install the bearing caps in their proper locations. Apply a light coat of engine oil to the threads and install the cap bolts.

d. Apply a light coat of engine oil on the threads and under the heads of the bearing cap bolts.

e. Uniformly tighten the cap bolts in the sequence shown to 12 ft. lbs. (16 Nm).

17. Install the right exhaust camshaft, as follows:

a. Apply engine oil to the thrust portion of the intake camshaft.

b. Align the timing marks (2 dot marks) of the camshaft drive and driven gears.

c. Roll down the exhaust camshaft onto the bearing journals while engaging the gears with each other. Install the bearing caps in their proper locations.

d. Apply a light coat of engine oil to the threads and install the cap bolts.

e. Apply a light coat of engine oil on the threads and under the heads of the bearing cap bolts.

f. Uniformly tighten the cap bolts in the sequence shown to 12 ft. lbs. (16 Nm).

g. Remove the service bolt from the driven sub-gear. Check that the intake and exhaust camshafts turns smoothly.

h. Align the timing marks (2 dot marks) of the camshaft drive and driven gears by turning the camshaft with a wrench.

18. Install the left intake camshaft, as follows:

a. Apply engine oil to the thrust portion of the intake camshaft.

b. Position the intake camshaft at 90 degrees angle of the timing mark (1 dot mark) on the cylinder head.

c. Install the bearing caps in their proper locations. Apply a light coat of engine oil to the threads and install the cap bolts.

d. Apply a light coat of engine oil on the threads and under the heads of the bearing cap bolts.

e. Uniformly tighten the cap bolts in the sequence shown to 12 ft. lbs. (16 Nm).

19. Install the left exhaust camshaft, as follows:

a. Apply engine oil to the thrust portion of the intake camshaft.

b. Align the timing marks (1 dot mark) of the camshaft drive and driven gears.

c. Roll down the exhaust camshaft onto the bearing journals while engaging the gears with each other. Install the bearing caps in their proper locations.

d. Apply a light coat of engine oil to the threads and install the cap bolts.

e. Apply a light coat of engine oil on the threads and under the heads of the bearing cap bolts.

f. Uniformly tighten the cap bolts in the sequence shown to 12 ft. lbs. (16 Nm).

g. Remove the service bolt.

20. Check and adjust the valve clearance.

21. Install or connect the following:

- Semi-circular plugs
- Cylinder head covers. Tighten the bolts, in several passes, to 53 inch lbs. (6 Nm).
- Alternator bracket. Tighten the bolts to 14 ft. lbs. (18 Nm).
- No. 3 timing belt cover. Tighten the 6 bolts to 80 inch lbs. (9 Nm).
- CMP sensor. Tighten it to 71 inch lbs. (8 Nm).
- Intake air connector
- Hoses
- Alternator but do not tighten the bolts and nuts at this time
- No. 2 timing belt idler. Tighten the bolt to 30 ft. lbs. (40 Nm).

➡ **Check that the pulley bracket moves smoothly.**

22. Install the left camshaft timing pulley, as follows:

a. Install the knock pin to the camshaft.

b. Align the knock pin hose of the camshaft with the knock pin groove of the timing pulley.

c. Slide the timing pulley on the camshaft with the flange side facing outward. Tighten the pulley bolt to 81 ft. lbs. (110 Nm).

23. Set the No. 1 cylinder to TDC of the compression stroke.

a. Turn the crankshaft pulley and align its groove with the timing mark **0** on the No. 1 timing belt cover.

b. Turn the camshaft, align the knock pin hole of the camshaft with the timing mark of the No. 3 timing belt cover.

c. Turn the camshaft timing pulley, align the timing marks of the camshaft timing pulley and the No. 3 timing belt cover.

24. Connect the timing belt to the left camshaft timing pulley, as follows:

a. Check that the installation mark on the timing belt is aligned with the end of the No. 1 timing belt cover.

b. Using Variable Pin Wrench Set 09960-01000 or equivalent, slightly turn the left camshaft timing pulley clockwise. Align the installation mark on the timing belt with the timing mark of the camshaft timing pulley, and hang the timing belt on the left camshaft timing pulley.

c. Align the timing marks of the left camshaft pulley and the No. 3 timing belt cover.

d. Check that the timing belt has tension between the crankshaft timing pulley and the left camshaft timing pulley.

25. Install the right camshaft timing pulley and the timing belt, as follows:

a. Align the installation mark on the timing belt with the timing mark of the right camshaft timing pulley, and hang the timing belt on the right camshaft timing pulley with the flange side facing inward.

b. Slide the right camshaft timing pulley on the camshaft. Align the timing marks on the right camshaft timing pulley and the No. 3 timing belt cover.

c. Align the knock pin hole of the camshaft with the knock pin groove of the pulley.

d. Install the knock pin. Tighten the bolt to 81 ft. lbs. (110 Nm).

26. Set the timing belt tensioner, as follows:

a. Using a press, slowly press in the pushrod using 220–2,205 lbs. (981–9,807 N) of force.

b. Align the holes of the pushrod and housing, pass a 1.5mm hexagon wrench through the holes to keep the setting position of the pushrod.

c. Release the press and install the dust boot to the tensioner.

27. Install the timing belt tensioner and alternately tighten the bolts to 20 ft. lbs. (28 Nm). Using pliers, remove the 1.5mm hexagon wrench from the belt tensioner.

28. Check the valve timing, as follows:

a. Slowly turn the crankshaft pulley 2 revolutions from the TDC-to-TDC. Always turn the crankshaft pulley clockwise.

b. Check that each pulley aligns with the timing marks. If the timing marks do not align, remove the timing belt and reinstall it.

29. Install or connect the following:

- Fan bracket with the bolt and nut
- Power steering adjusting strut with the nut

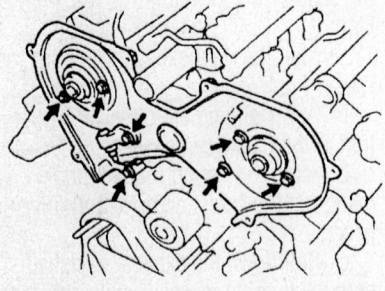

Rear timing belt cover mounting bolt locations—3.4L engine

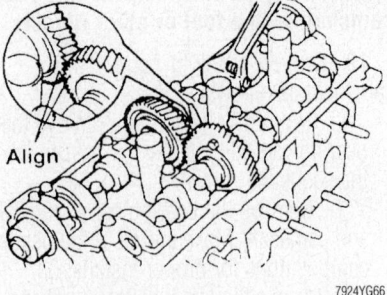

Aligning the timing mark (1 dot mark) of the left camshafts—3.4L engine

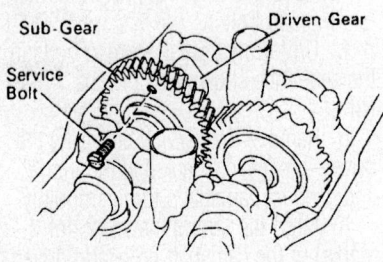

Drive gear service bolt (right side)—3.4L engine

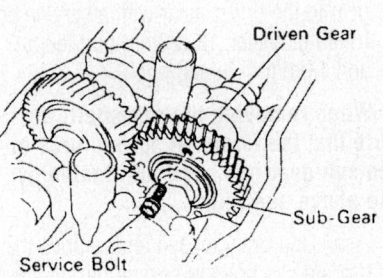

Drive gear service bolt (left side)—3.4L engine

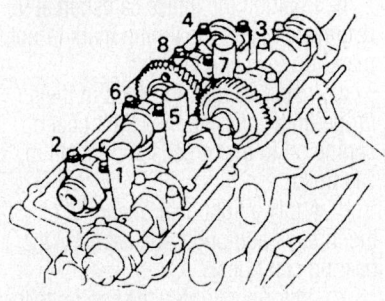

Right exhaust camshaft bolt removal sequence—3.4L engine

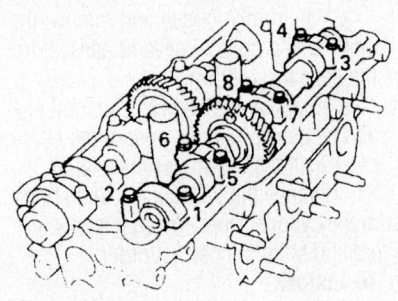

Left exhaust camshaft bolt removal sequence—3.4L engine

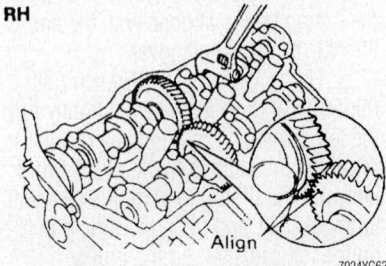

Aligning the timing marks (2 dot marks) of the right camshafts—3.4L engine

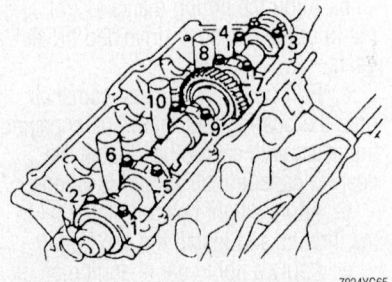

Right intake camshaft bolt removal sequence—3.4L engine

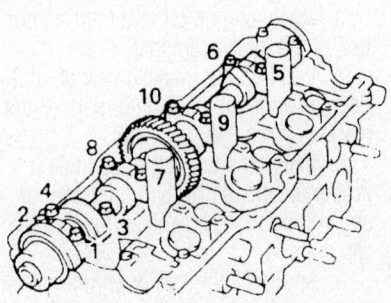

Left intake camshaft bolt removal sequence—3.4L engine

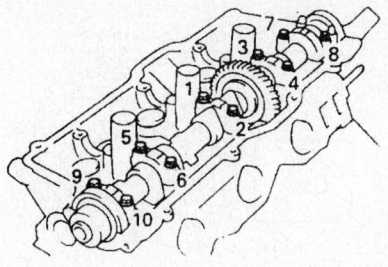

Right intake camshaft bolt tightening sequence—3.4L engine

7924YG70

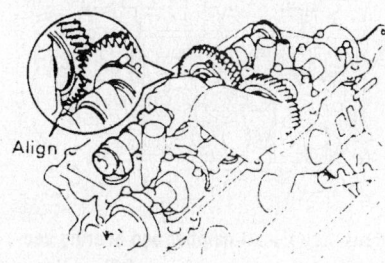

Aligning the right camshafts for installation—3.4L engine

7924YG71

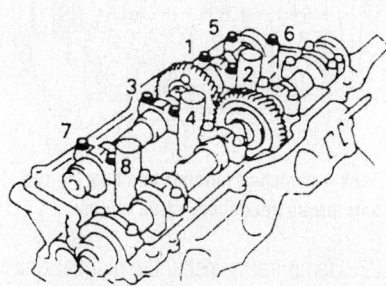

Right exhaust camshaft bolt tightening sequence—3.4L engine

7924YG72

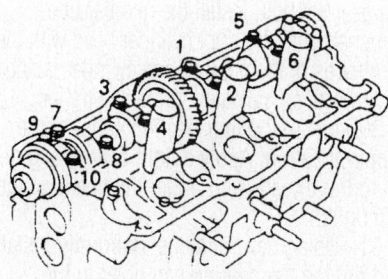

Left intake camshaft bolt tightening sequence—3.4L engine

7924YG73

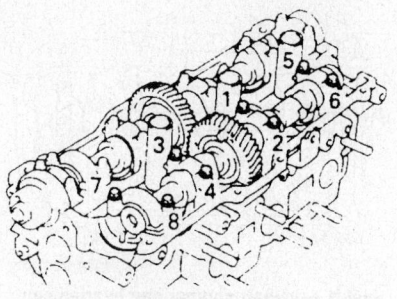

Left exhaust camshaft bolt tightening sequence—3.4L engine

7924YG74

- No. 2 timing belt cover. Tighten the bolts to 80 inch lbs. (9 Nm).
- Negative battery cable
30. Fill the radiator with engine coolant.
31. Start the engine and check for leaks.
32. Check the ignition timing.
33. Install the engine undercover.
34. Road test the vehicle.
35. Recheck all fluid levels.

4.0L Engine

BANK 1

1. Before servicing the vehicle, refer to the Precautions Section.
2. Properly disarm the SRS system.
3. Disconnect the negative battery cable.
4. Drain the engine coolant. Remove the V bank cover.
5. Remove the air cleaner assembly.
6. Disconnect the two water bypass hoses. Disconnect the fuel vapor feed hose. Disconnect the ventilation hose. Disconnect the VSV connectors.
7. Disconnect the throttle body motor connector. Separate the three wire harness clamps and hose clamp.
8. Remove the two bolts and the throttle body bracket. Remove the bolt and the oil baffle plate. Remove the four bolts and the two serge tank stays.
9. Remove the two nuts. Remove the four bolts, intake air surge tank and gasket.
10. Remove the ignition coil assembly.
11. Remove the cylinder head cover retaining bolts. Remove the cylinder head cover.
12. Turn the crankshaft pulley until its groove and the timing mark "0" of the timing chain cover are aligned. If not aligned at TDC of the compression stroke, turn the crankshaft one complete revolution, in the direction of rotation. Paint alignment marks on the number one chain links corresponding to the timing marks of the camshaft timing gears.

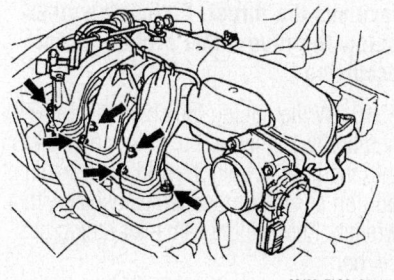

Intake surge tank bolt locations—4.0L engine

09490_TACO_G0049

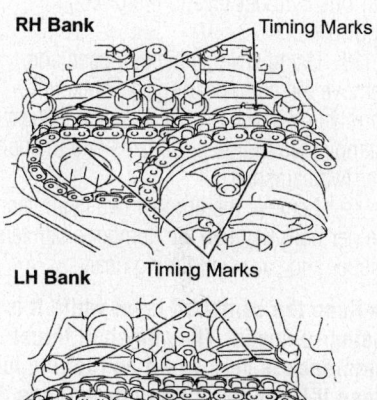

09490_TACO_G0050

Bank 1 camshaft timing mark alignment—4.0L engine

13. Remove the four bolts, and then remove the timing chain cover plate and gasket.
14. While turning the stopper plate of the tensioner upward, push the plunger of the chain tensioner. While turning the stopper plate of the tensioner downward, insert a 0.118 inch diameter bar into the holes in the stopper plate and tensioner to hold the stopper plate.
15. Remove the two bolts, and then remove the chain tensioner.

➡**Keep the camshaft level while it is being removed. The camshaft thrust clearance is very small and failing to keep it level could crack or damage the** cylinder head journal surface, which receives the thrust. Follow the steps below to prevent this problem from occurring.

16. While raising the chain tensioner number two insert a 0.039 inch diameter pin into the hole to hold it. Hold the hexagonal portion of the number two camshaft with a wrench. Remove the camshaft timing gear set bolt.

17. Separate the camshaft timing gear from the number two camshaft. Rotate the camshafts counterclockwise, using a wrench, so that the cam lobes of the number one cylinder face in the direction shown.

18. Using several steps, loosen and remove the eight bearing cap bolts uniformly and in the proper removal sequence. Remove the four bearing caps and the number two camshaft.

19. Remove the number two chain tensioner bolt. Remove the number chain tensioner and camshaft timing gear.

➡ Keep the camshaft level while it is being removed. The camshaft thrust clearance is very small and failing to keep it level could crack or damage the cylinder head journal surface, which receives the thrust. Follow the steps below to prevent this problem from occurring.

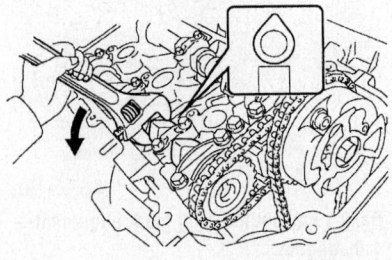

Bank 1 camshaft lobe removal positioning—4.0L engine

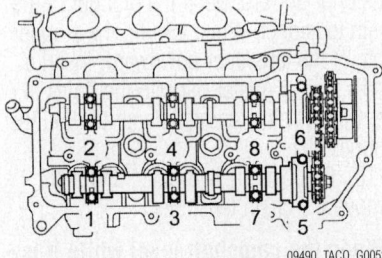

Bank 1 camshaft number two bearing cap bolt removal sequence—4.0L engine

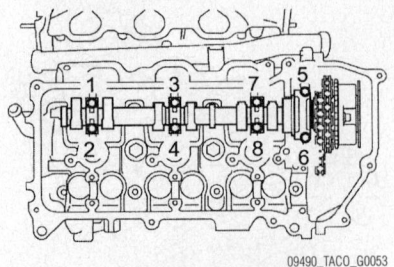

Bank 1 camshaft number one bearing cap bolt removal sequence—4.0L engine

20. Hold the hexagonal portion of the number one camshaft, with a wrench. Loosen the camshaft timing gear set bolt.

➡ Do not disassemble the camshaft timing gear assembly.

21. Slide the camshaft timing gear and separate the number one chain from the camshaft timing gear.

22. Rotate the number one camshaft counterclockwise, using the wrench so that the cam lobes on the number one cylinder face downward.

23. Using several steps, loosen and remove the eight bearing cap bolts uniformly and in the proper removal sequence. Remove the four bearing caps.

24. Remove the camshaft timing gear set bolt with the number one camshaft lifted up. Remove the number one camshaft and camshaft timing gear with the number two chain.

25. Tie the number one chain to the side. Be careful not to drop anything inside the timing chain cover.

To install:

➡ Keep the camshaft level while it is being installed. The camshaft thrust clearance is very small and failing to keep it level could crack or damage the cylinder head journal surface, which receives the thrust.

26. Align the yellow mark link with the timing mark (1 dot mark) of the camshaft timing gear. Apply new engine oil to the thrust portion and journal of the camshafts.

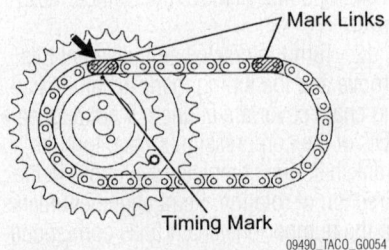

Bank 1 camshaft number one alignment marks—4.0L engine

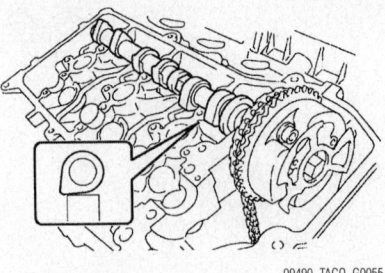

Bank 1 camshaft number one lobe installation positioning—4.0L engine

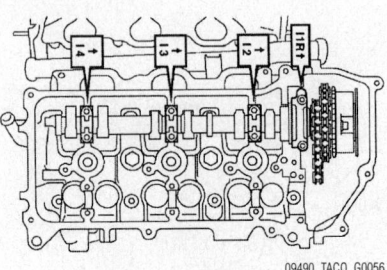

Bank 1 camshaft number one bearing cap bolt installation sequence—4.0L engine

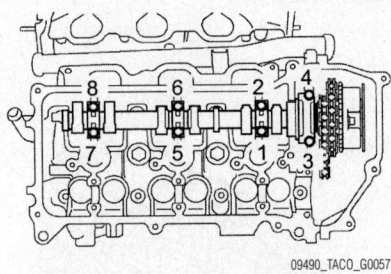

Bank 1 camshaft number one bearing cap bolt torque sequence—4.0L engine

27. Temporarily install the number one chain onto the number two chain of the camshaft timing gear.

28. Align the knock pin hole of the camshaft timing gear with the knock pin of the number one camshaft. Insert the number one camshaft into the camshaft timing gear.

29. Temporarily install the camshaft timing gear set bolt. Install the number one camshaft onto the right cylinder head with the cam lobes of the number one cylinder facing downward, as indicated in the illustration.

30. Install the four bearing caps, in the proper location. Apply a light coat of engine oil to the threads and under the heads of the cap bolts.

31. Using several steps, uniformly install and tighten the bearing cap bolts in the proper sequence to 80 inch lbs. for the 10mm bolts and 18 ft. lbs. for the 12mm bolts.

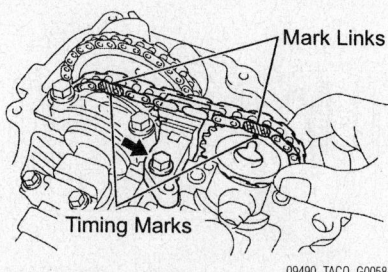

Bank 1 camshaft number two alignment marks—4.0L engine

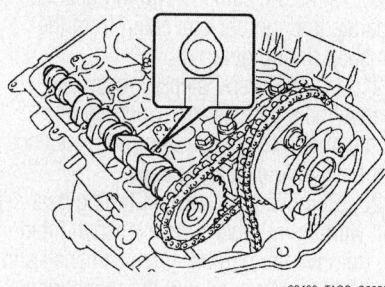

Bank 1 camshaft number two lobe installation positioning—4.0L engine

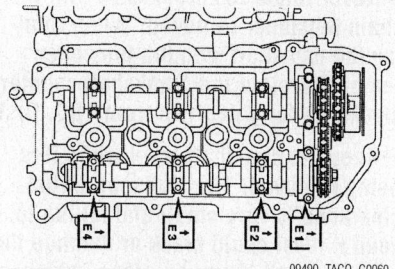

Bank 1 camshaft number two bearing cap bolt installation sequence—4.0L engine

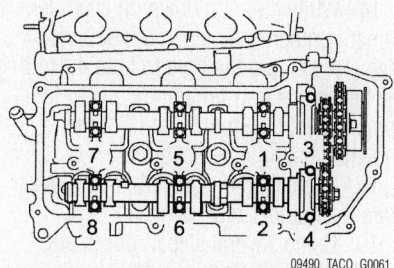

Bank 1 camshaft number two bearing cap bolt torque sequence—4.0L engine

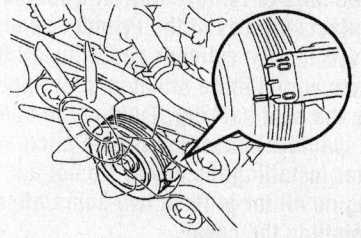

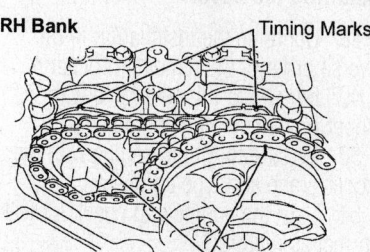

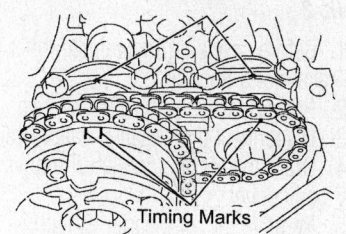

Bank 1 and Bank 2 timing mark alignment—4.0L engine

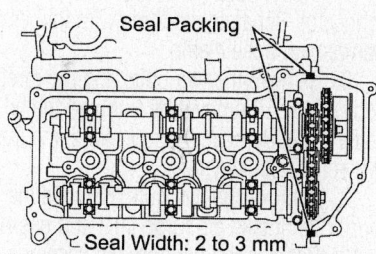

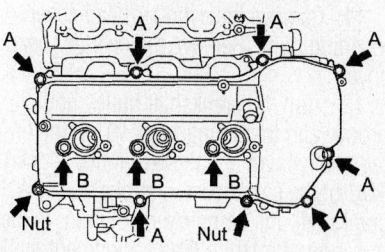

Bank 1 cylinder head cover bolt sealant application and bolt identification—4.0L engine

32. Rotate the number one camshaft clockwise, using a wrench so that the timing mark of the camshaft timing gear is aligned with the timing mark of the camshaft bearing cap.

33. Align the paint mark of the number one chain with the timing mark of the camshaft timing gear.

34. Hold the hexagonal portion of the number one camshaft with a wrench, and tighten the camshaft timing gear set bolt to 74 ft. lbs.

35. While pushing in on the number two chain tensioner, insert a 0.039 inch pin into the hole to hold it.

36. Temporarily install the camshaft timing gear and chain tensioner number two and align the yellow mark links with the timing marks (1 dot mark) on the camshaft timing gears. Tighten the bolt to 14 ft. lbs.

➡**Keep the camshaft level while it is being installed. The camshaft thrust clearance is very small and failing to keep it level could crack or damage the cylinder head journal surface, which receives the thrust.**

37. Install the number two camshaft onto the right cylinder head with the cam lobes of the number one cylinder facing upward, as indicated in the illustration.

38. Install the four bearing caps in the proper location. Apply a light coat of clean engine oil to the threads and under the heads of the bolts.

39. Using several steps, uniformly install and tighten the eight bearing cap bolts in sequence to 80 inch lbs, for the 10mm head bolts and 18 ft. lbs. for the 12mm head bolts.

40. Rotate the number two camshaft clockwise, using a wrench, so that the lock pin of the number two camshaft is aligned with the knock pin hole of the camshaft timing gear.

41. Hold the hexagonal portion of the number two camshaft, with a wrench, and install the camshaft timing gear set bolt and tighten it to 74 ft. lbs.

42. Remove the pin from the number two chain tensioner.

43. While turning the stopper plate of the tensioner clockwise, push in the plunger of the tensioner. While turning the stopper plate of the tensioner counterclockwise, insert a 0.138 inch bar into the holes in the stopper plate and tensioner to hold the stopper plate. Install the two chain tensioner bolts and tighten to 7.4 ft. lbs.

44. Remove the bar from the chain tensioner. Install a new gasket and the timing chain cover plate. Torque the bolts to 80 inch lbs.

45. Turn the crankshaft pulley two complete revolutions slowly until its groove and the timing mark "0" of the timing chain cover are aligned.

46. Position the number one cylinder at TDC of the compression stroke. Inspect the valve clearance, adjust as required.

47. Apply a continuous bead (0.08–0.12 inch) of seal packing, part number 08826-00080 or equivalent, to the cylinder head as indicated in the illustration. Install the seal washers onto the bolts. Install the cylinder head cover bolts and nuts. Tighten bolts "A" to 7.4 ft. lbs. Tighten bolts "B" to 80 inch lbs. Tighten nuts to 80 inch lbs.

➡Be sure to remove any oil from the contact surfaces of the cylinder head cover and the cylinder head. Install the cover within three minutes after applying the seal packing. Tighten the bolts to specification within fifteen minutes after installing the cover. Do not add engine oil for at least two hours after installing the cover.

48. Continue the installation in the reverse order of the removal procedure.

49. Be sure to fill the engine with the proper grade and type engine coolant.

50. Be sure to fill the engine with the proper grade and type engine oil.

51. Start the engine and check for leaks. Correct as required.

BANK 2

1. Before servicing the vehicle, refer to the Precautions Section.

2. Properly disarm the SRS system.

3. Disconnect the negative battery cable.

4. Drain the engine coolant. Remove the V bank cover.

5. Remove the air cleaner assembly.

6. Disconnect the two water bypass hoses. Disconnect the fuel vapor feed hose. Disconnect the ventilation hose. Disconnect the VSV connectors.

7. Disconnect the throttle body motor connector. Separate the three wire harness clamps and hose clamp.

8. Remove the two bolts and the throttle body bracket. Remove the bolt and the oil baffle plate. Remove the four bolts and the two serge tank stays.

9. Remove the two nuts. Remove the four bolts, intake air surge tank and gasket.

10. Remove the ignition coil assembly.

11. Remove the cylinder head cover retaining bolts. Remove the cylinder head cover.

12. Turn the crankshaft pulley until its groove and the timing mark "0" of the timing chain cover are aligned. If not aligned at TDC of the compression stroke, turn the crankshaft one complete revolution, in the direction of rotation. Paint alignment marks on the number one chain links corresponding to the timing marks of the camshaft timing gears.

13. While turning the stopper plate of the tensioner upward, push in the plunger of the chain tensioner. While turning the stopper plate of the tensioner downward, insert a 0.138 inch bar into the holes in the stopper plate and tensioner to hold the stopper plate. Remove the two bolts and then remove the number one chain tensioner assembly.

➡Never rotate the crankshaft with the chain tensioner removed. When rotating the camshaft with the tensioner removed, rotate the crankshaft counterclockwise forty degrees from TDC, first.

➡Keep the camshaft level while it is being removed. The camshaft thrust clearance is very small and failing to keep it level could crack or damage the cylinder head journal surface, which receives the thrust. Follow the steps below to prevent this problem from occurring.

14. While pushing down on chain tensioner number three insert a 0.039 inch diameter pin into the hole to hold it. Hold the hexagonal portion of the number four camshaft with a wrench. Remove the camshaft timing gear set bolt.

15. Separate the camshaft timing gear from the number four camshaft.

16. Using several steps, loosen and remove the eight bearing cap bolts uniformly and in the proper removal sequence. Remove the four bearing caps and the number four camshaft.

17. Remove the number three chain tensioner bolt. Remove the number chain tensioner and camshaft timing gear.

➡Keep the camshaft level while it is being removed. The camshaft thrust clearance is very small and failing to keep it level could crack or damage the cylinder head journal surface, which receives the thrust. Follow the steps below to prevent this problem from occurring.

18. Release the chain tension between the camshaft gear (LH bank) and the crankshaft timing gear by turning the crankshaft pulley counterclockwise slightly.

19. Hold the hexagonal portion of the number three camshaft, with a wrench. Loosen the camshaft timing gear set bolt.

➡Do not disassemble the camshaft timing gear assembly.

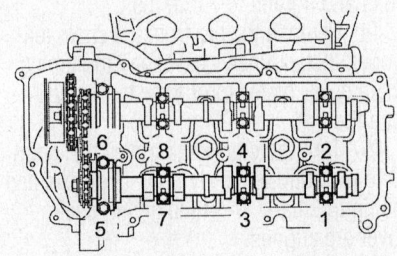

Bank 2 camshaft number four bearing cap bolt removal sequence—4.0L engine

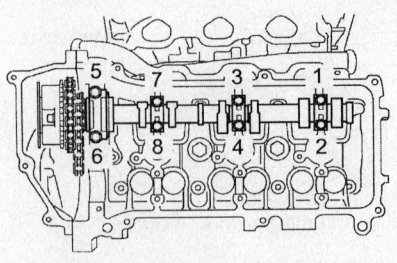

Bank 2 camshaft number three bearing cap bolt removal sequence—4.0L engine

20. Slide the camshaft timing gear and separate the number one chain from the camshaft timing gear.

21. Using several steps, loosen and remove the eight bearing cap bolts uniformly and in the proper removal sequence. Remove the four bearing caps.

22. Remove the camshaft timing gear set bolt with the number three camshaft lifted up. Remove the number three camshaft and camshaft timing gear with the number two chain.

23. Tie the number one chain to the side. Be careful not to drop anything inside the timing chain cover.

To install:

➡Keep the camshaft level while it is being installed. The camshaft thrust clearance is very small and failing to keep it level could crack or damage the cylinder head journal surface, which receives the thrust.

24. Align the yellow mark link with the timing mark (2 dot mark) of the camshaft timing gear. Apply new engine oil to the thrust portion and journal of the camshafts.

25. Temporarily install the number one chain onto the number two chain of the camshaft timing gear.

26. Align the knock pin hole of the camshaft timing gear with the knock pin of the number three camshaft. Insert the number three camshaft into the camshaft timing gear.

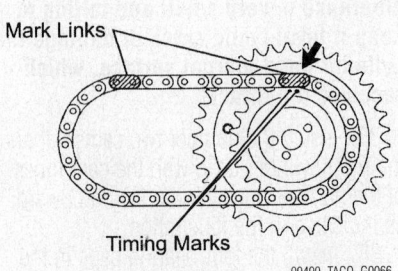

Bank 2 camshaft number three alignment marks—4.0L engine

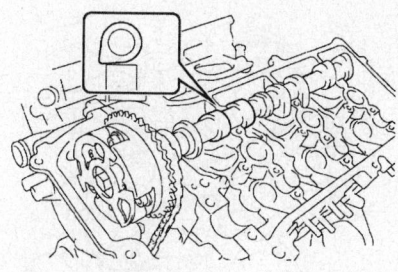

Bank 2 camshaft number three lobe installation positioning—4.0L engine

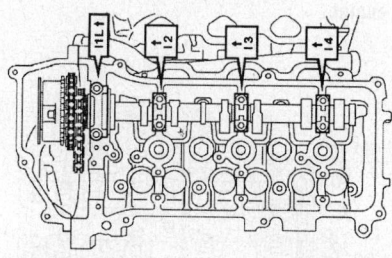

Bank 2 camshaft number three bearing cap bolt installation sequence—4.0L

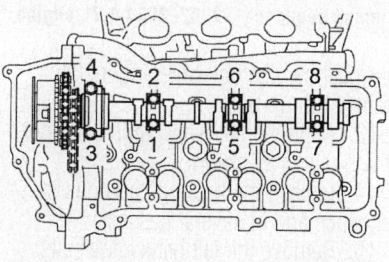

Bank 2 camshaft number three bearing cap bolt torque sequence—4.0L engine

27. Temporarily install the camshaft timing gear set bolt. Install the number three camshaft onto the left cylinder head with the cam lobes of the number two cylinder facing downward, as indicated in the illustration.

28. Install the four bearing caps, in the proper location. Apply a light coat of engine oil to the threads and under the heads of the cap bolts.

29. Using several steps, uniformly install and tighten the bearing cap bolts in the proper sequence to 80 inch lbs. for the 10mm bolts and 18 ft. lbs. for the 12mm bolts.

30. Rotate the number one camshaft clockwise, using a wrench so that the timing mark of the camshaft timing gear is aligned with the timing mark of the camshaft bearing cap.

31. Align the paint mark of the number

one chain with the timing mark of the camshaft timing gear.

32. Hold the hexagonal portion of the number three camshaft with a wrench, and tighten the camshaft timing gear set bolt to 74 ft. lbs.

33. While pushing in on the number three chain tensioner, insert a 0.039 inch pin into the hole to hold it.

34. Temporarily install the camshaft timing gear and chain tensioner number three and align the yellow mark links with the timing marks (1 dot mark and 2 dot marks) on the camshaft timing gears. Tighten the bolt to 14 ft. lbs.

➡ **Keep the camshaft level while it is being installed. The camshaft thrust clearance is very small and failing to keep it level could crack or damage the cylinder head journal surface, which receives the thrust.**

35. Align the knock pin hole in the camshaft timing gear with the knock pin of the number four camshaft, and insert the number four camshaft into the camshaft timing gear.

36. Temporarily install the camshaft timing gear set bolt.

37. Install the four bearing caps in the proper location. Apply a light coat of clean engine oil to the threads and under the heads of the bolts.

38. Using several steps, uniformly install and tighten the eight bearing cap bolts in

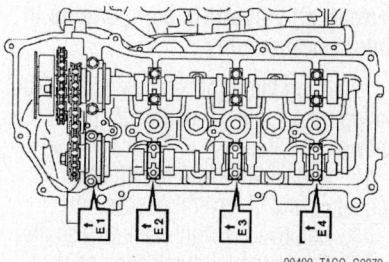

Bank 2 camshaft number four bearing cap bolt installation sequence—4.0L engine

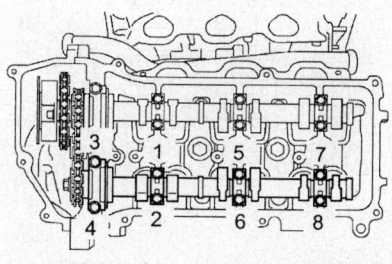

Bank 2 camshaft number four bearing cap bolt torque sequence—4.0L engine

sequence to 80 inch lbs, for the 10mm head bolts and 18 ft. lbs. for the 12mm head bolts.

39. Hold the hexagonal portion of the number four camshaft, with a wrench, and install the camshaft timing gear set bolt and tighten it to 74 ft. lbs.

40. Remove the pin from the number three chain tensioner.

41. Release the chain tension between the camshaft timing gear (RH bank) and the crankshaft timing gear by turning the crankshaft pulley clockwise slightly.

42. While turning the stopper plate of the tensioner clockwise, push in the plunger of the tensioner. While turning the stopper plate of the tensioner counterclockwise, insert a 0.138 inch bar into the holes in the stopper plate and tensioner to hold the stopper plate. Install the two chain tensioner bolts and tighten to 7.4 ft. lbs. Remove the bar from the chain tensioner.

43. Position the number one cylinder at TDC of the compression stroke. Inspect the valve clearance, adjust as required.

44. Apply a continuous bead (0.08–0.12 inch) of seal packing, part number 08826-00080 or equivalent, to the cylinder head as indicated in the illustration. Install the seal washers onto the bolts. Install the cylinder head cover bolts and nuts. Tighten bolts "A" to 7.4 ft. lbs. Tighten bolts "B" to 80 inch lbs. Tighten nuts to 80 inch lbs.

➡ **Be sure to remove any oil from the contact surfaces of the cylinder head cover and the cylinder head. Install the cover within three minutes after applying the seal packing. Tighten the bolts to specification within fifteen minutes**

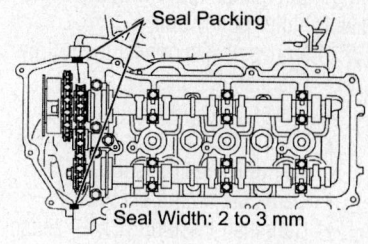

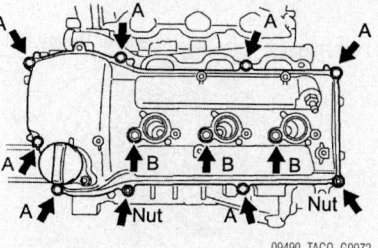

Bank 2 cylinder head cover bolt sealant application and bolt identification—4.0L engine

after installing the cover. Do not add engine oil for at least two hours after installing the cover.

45. Continue the installation in the reverse order of the removal procedure.

46. Be sure to fill the engine with the proper grade and type engine coolant.

47. Be sure to fill the engine with the proper grade and type engine oil.

48. Start the engine and check for leaks. Correct as required.

4.7L Engine

2002–2004

1. Before servicing the vehicle, refer to the Precautions Section.

2. Drain the cooling system.

3. Relieve the fuel system pressure.

4. Remove or disconnect the following:
- Negative battery cable
- Engine under covers
- Engine appearance cover
- Air intake hose
- Accessory drive belt
- Cooling fan
- Radiator
- Idler pulley
- Upper and middle timing belt covers
- A/C compressor
- Cooling fan bracket
- Alternator
- Accessory drive belt tensioner

5. Set the engine to Top Dead Center (TDC) with the camshaft sprocket timing marks aligned with the rear cover timing marks.

6. Rotate the crankshaft to 50 degrees After TDC as shown. The crankshaft pulley timing mark should align with the center of the No. 2 idler pulley bolt.

7. Remove or disconnect the following:
- Crankshaft pulley
- Lower timing cover
- Timing belt. Refer to the Timing Belt unit repair section.
- Camshaft timing sprockets
- Camshaft Position (CMP) sensor

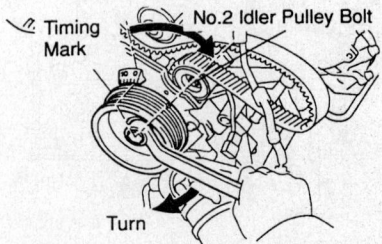

Setting the crankshaft to 50 degrees ATDC—2002–2004 4.7L engine

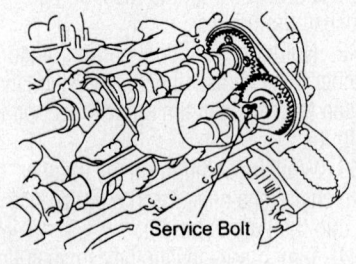

Camshaft service bolt installation—2002–2004 4.7L engine

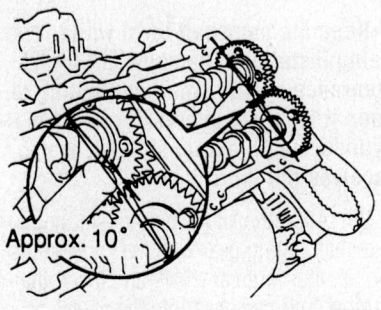

Right bank camshaft timing mark (1 dot marks) alignment—2002–2004 4.7L

- Ignition coils
- Valve cover
- Timing belt rear covers

8. Rotate the right bank camshafts as necessary to access the exhaust camshaft sub-gear service bolt hole and install a 6mm x 1.0mm bolt.

➡ **Keep all valvetrain components in order for assembly.**

9. Align the right bank camshaft 1 dot timing marks to a **10** degree angle as shown.

10. Loosen the bearing cap bolts in sequence and in several passes.

11. Remove the right bank camshafts.

12. Rotate the left bank camshafts as necessary to access the exhaust camshaft

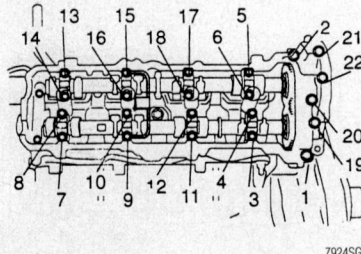

Right bank camshaft bearing cap bolt loosening sequence—2002–2004 4.7L engine

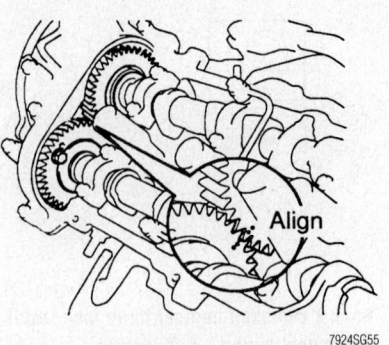

Left bank camshaft timing mark (2 dot marks) alignment—2002–2004 4.7L engine

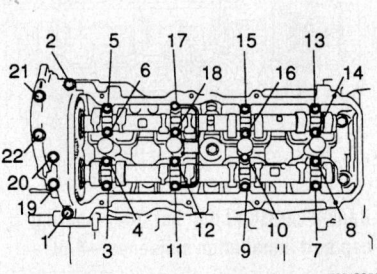

Left bank camshaft bearing cap bolt loosening sequence—2002–2004 4.7L engine

sub-gear service bolt hole and install a 6mm x 1.0mm bolt.

13. Align the left bank camshaft 2 dot timing marks as shown.

14. Loosen the bearing cap bolts in sequence and in several passes.

15. Remove the left bank camshafts.

16. Remove the valve lifters and shims.

To install:

17. Ensure that the crankshaft is at 50 degrees After TDC.

18. Install or connect the following:
- Valve lifters and shims in their original positions
- Right bank camshafts with the 1 dot timing marks at 10 degrees
- Left bank camshafts with the 2 dot timing marks aligned
- Left and right bank camshaft bear-

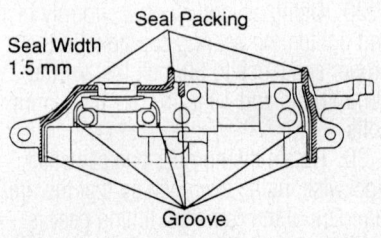

Apply a 1.5mm bead of sealant to the front bearing caps—4.7L engine

ing caps in their original positions. Apply sealant to the front bearing caps as shown.

• Camshaft oil seals

19. The bearing cap bolts vary in length and are identified as follows:

- A: 3.70 inches (94mm)
- B: 2.83 inches (72mm)
- C: 0.98 inches (25mm)
- D: 2.05 inches (52mm)
- E: 1.50 inches (38mm)

20. Bolts in positions **A**, **B** and **C** are installed dry.

21. Lubricate the threads and under the contact flange for bolts in positions **D** and **E**.

22. Install oil feed pipes and the bearing cap bolts according to position in the illustrations.

23. Tighten the camshaft bearing bolts in sequence and in several passes to the following specifications:

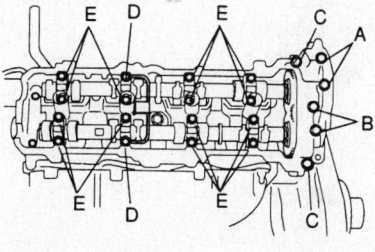

Right bank bearing cap bolt location— 4.7L engine

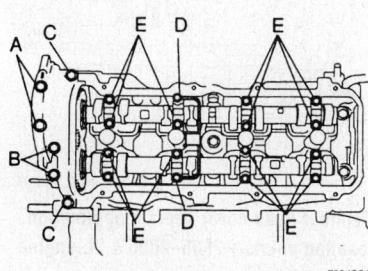

Left camshaft bearing cap bolt locations— 4.7L engine

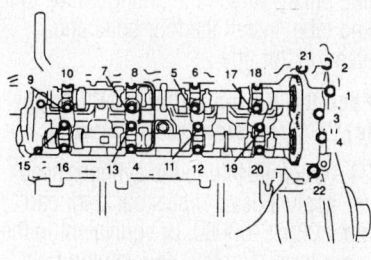

Right bank camshaft bearing cap bolt torque sequence—2002–2004 4.7L engine

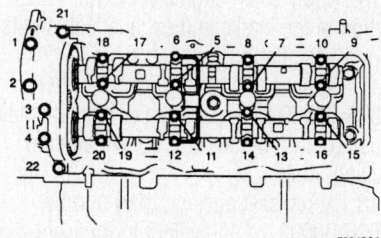

Left bank camshaft bearing cap bolt torque sequence—2002–2004 4.7L engine

- Bolt C: 66 inch lbs. (7.5 Nm)
- All others: 12 ft. lbs. (16 Nm)

24. Remove the service bolts from the exhaust camshaft gears.

25. Install or connect the following:

- Timing belt rear covers
- Valve cover
- Ignition coils
- CMP sensor
- Camshaft timing sprockets. Tighten the bolts to 80 ft. lbs. (108 Nm).
- Timing belt
- Lower timing cover
- Crankshaft pulley. Tighten the bolt to 181 ft. lbs. (245 Nm).
- Accessory drive belt tensioner
- Alternator
- Cooling fan bracket
- A/C compressor
- Upper and middle timing belt covers
- Idler pulley. Tighten the bolt to 27 ft. lbs. (37 Nm).
- Radiator
- Cooling fan
- Accessory drive belt
- Air intake hose
- Engine appearance cover
- Engine under covers
- Negative battery cable

26. Fill the cooling system.

27. Start the engine and check for leaks.

2005–2006

1. Before servicing the vehicle, refer to the Precautions Section.

2. Properly relieve the fuel system pressure.

3. Properly disarm the SRS system.

4. Disconnect the negative battery cable.

5. Remove the timing belt.

6. Remove the ignition coil assembly.

7. Remove the cylinder head cover retaining bolts. Remove the cylinder head covers.

8. Hold the camshaft in place, using the proper wrench. Remove the four bolts and the timing pulley

➡Do not remove the four bolts shown in the illustration. If any of these bolts are loosened or removed, the backlash of the gear in the timing tube will go out of adjustment. If this is the case, replace the timing tube assembly with a new one.

9. Remove the camshaft position sensor retaining bolt, stud bolt and sensor.

➡The thrust clearance of the camshaft is small; the camshaft must be kept level while it is being removed. If it is not kept level, the portion of the cylinder head which receives the shaft thrust make crack or become damaged. To avoid this follow the steps below.

10. Position the crankshaft pulley as shown in the illustration.

11. Bring the service bolt hole of the subgear upward by turning the hexagonal head portion of the exhaust camshaft with a wrench. Secure the subgear to the main gear with a service bolt.

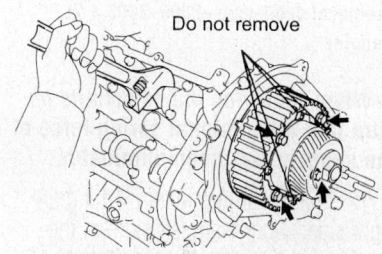

Timing pulley bolt locations—2005–2006 4.7L engine

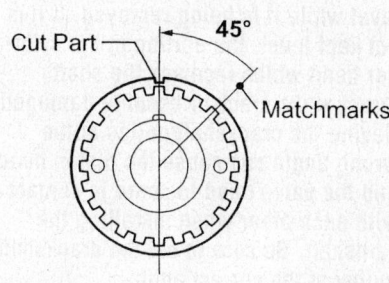

Crankshaft pulley alignment—2005–2006 4.7L engine

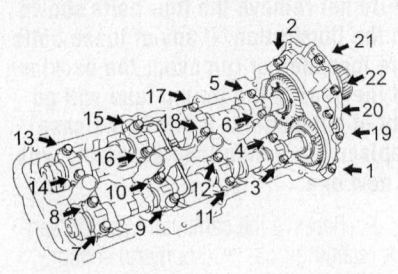

Right bank camshaft bearing cap bolt removal sequence—2005–2006 4.7L engine

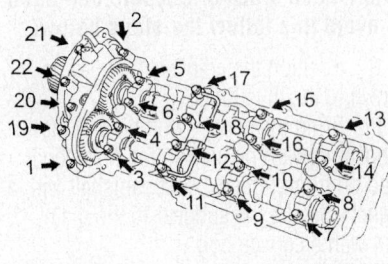

Left bank camshaft bearing cap bolt removal sequence—2005–2006 4.7L engine

➡**When removing the camshafts be sure that the torsional spring force of the subgear has been eliminated.**

12. Set the timing mark (1-dot mark right side. 2-dot mark left side) of the camshaft main gear at approximately 10 degrees, by turning the hexagonal portion of the exhaust camshaft with a wrench.

13. Uniformly loosen and remove the bearing cap bolts in several steps and in the proper sequence.

14. Remove the bolts, oil feed pipe, the nine bearing caps, camshaft housing plug, oil control valve filter and camshafts. Be sure to keep the removed components in order.

To install:

➡**The thrust clearance of the camshaft is small; the camshaft must be kept level while it is being removed. If it is not kept level, the portion of the cylinder head which receives the shaft thrust make crack or become damaged. Having the crankshaft pulley at the wrong angle can cause the piston head and the valve head to come in contact with each other when installing the camshaft. Be sure to set the crankshaft pulley at the correct angle.**

15. Installation is the reverse of the removal procedure.

16. Apply clean engine oil to the thrust portion of the intake and exhaust camshafts, prior to installation.

17. Apply seal packing, part number 08826-00080, or equivalent to the camshaft housing plug. Be sure to remove any old packing, first.

18. Apply seal packing, part number 08826-00080, or equivalent to the front bearing cap. Be sure to remove any old packing.

➡**Parts must be assembled within five minutes after applying the seal packing; otherwise the material must be removed and reapplied.**

19. When installing the new seal washers to the bearing cap bolts, apply a light coat of clean engine oil to the threads and under the heads of the bolts ("D" and "E") see illustration. Do not apply engine oil under the bolt heads of the other bolts.

20. The bearing cap bolts vary in length and are identified as follows:
- A: 3.70 inches (94mm)
- B: 2.83 inches (72mm)
- C: 0.98 inches (25mm)
- D: 2.05 inches (52mm)
- E: 1.50 inches (38mm)

21. Tighten the bearing cap bolts in several steps and in the proper sequence. Tighten all bolts to 12 ft. lbs. except bolt "C". Tighten bolt "C" to 66 inch. lbs.

22. Check the valve clearance.

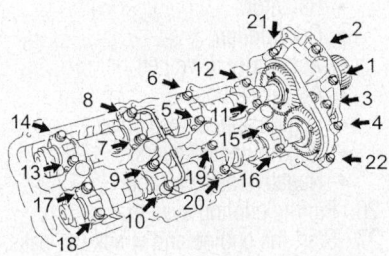

Right bank camshaft bearing cap bolt torque sequence—2005–2006 4.7L engine

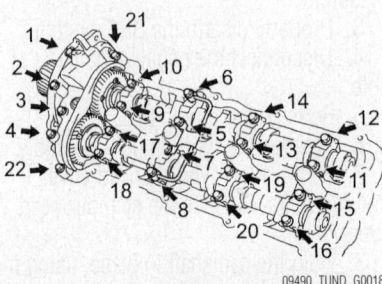

Left bank camshaft bearing cap bolt torque sequence—2005–2006 4.7L engine

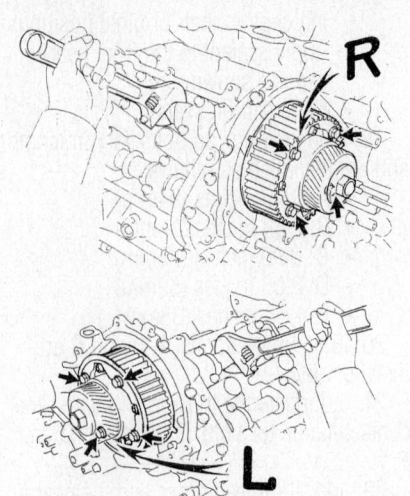

Timing pulley identification and bolt locations—2005–2006 4.7L engine

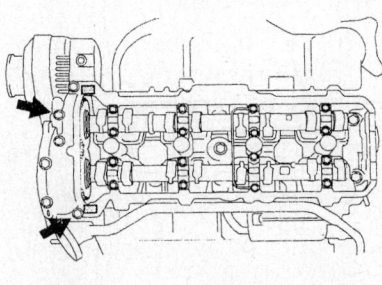

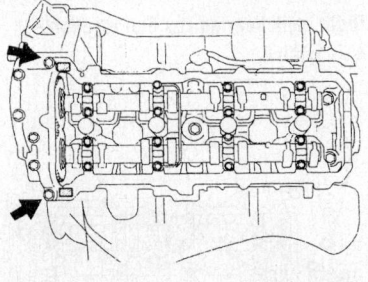

Cylinder head cover sealant application location points—2005–2006 4.7L engine

23. When installing the camshaft timing pulley subassembly, align the timing tube knock pin with the knock pin groove of the timing pulley. Attach the timing pulley to the timing tube. Install the four bolts and tighten to 72 ft. lbs.

➡**Face the timing pulley's "R" (right side) "L" (left side) mark forward.**

24. When installing the cylinder head cover, apply a bead of seal packing, part number 08836-00080, or equivalent to the cylinder head. Tighten the retaining bolts to 53 inch lbs in several steps and in an alternating sequence.

INSPECTION

1. Before servicing the vehicle, refer to the Precautions Section.

2. Remove the camshaft from the engine.

3. Check the camshaft bearing journals for damage and binding.

4. If the journals are binding, check the cylinder head for damage.

5. Check the cylinder head for clogged oil holes.

6. Check the camshaft surface for abnormal wear and damage. Replace the camshaft, as required.

7. Measure the camshaft lobe surface and replace the camshaft if not within specification.

8. Measure the camshaft journal diameter and replace the camshaft if not within specification.

9. Measure the camshaft run out and replace the camshaft if not within specification.

Valve Lash

ADJUSTMENT

3.4L Engine

1. Before servicing the vehicle, refer to the Precautions Section.

2. Disconnect the negative battery cable.

3. Drain the engine coolant.

4. Remove or disconnect the following:

- Air intake connector
- Cylinder head cover

5. Set the No. 1 cylinder to Top Dead Center (TDC) of the compression stroke, as follows:

a. Turn the crankshaft pulley clockwise and align its groove with the **0** mark on the timing chain cover.

b. Check that the timing marks (1 and 2 dots) of the camshaft drive and driven

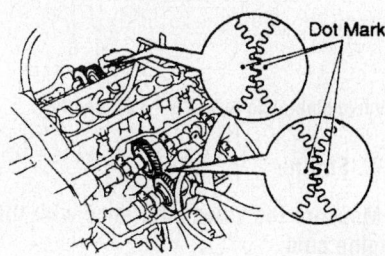

Aligning the timing marks—3.4L engine

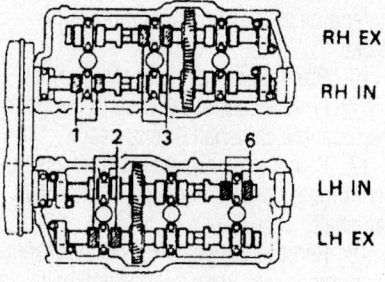

First valve adjustment—3.4L engine

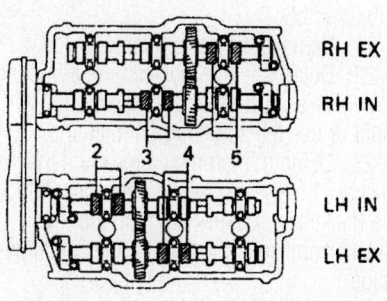

Second valve adjustment—3.4L engine

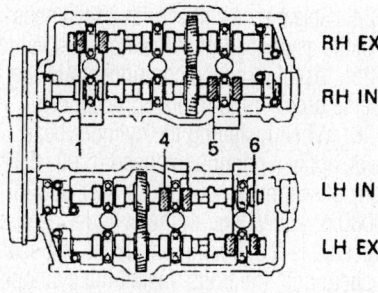

Third valve adjustment—3.4L engine

gears are in a straight line on the cylinder head surface. If not, turn the crankshaft 1 revolution (360 degrees) and align the marks.

6. Inspect the valve clearance, as follows:

a. Measure the clearance between the valve lifter and the camshaft. Measure the 1st intake and the 3rd exhaust valves on the right head and the 6th intake and the 2nd exhaust valves on the left head.

b. Turn the crankshaft ⅔ of a revolution (240 degrees) and adjust the 3rd intake and the 5th exhaust valves on the right head and the 2nd intake and the 4th exhaust valves on the left head.

c. Turn the crankshaft ⅔ of a revolution (240 degrees) and adjust the 5th

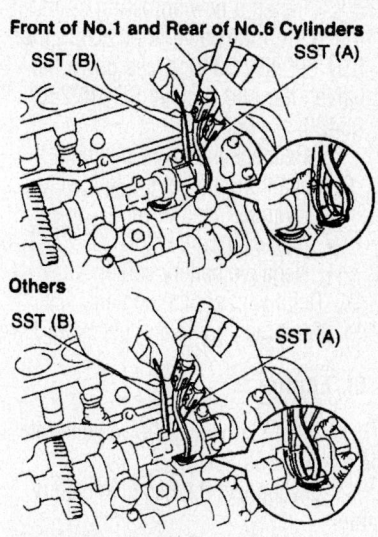

Removing the adjusting shim—3.4L engine

intake and the 1st exhaust valves on the right head and the 4th intake and the 6th exhaust valves on the left head.

7. Valve clearance cold should be:

- Intake: 0.006–0.009 in. (0.13–0.23mm)
- Exhaust: 0.011–0.014 in. (0.27–0.37mm)

8. Adjust the valve clearance by using adjusting shims, as follows:

a. Turn the equipment camshaft so that the cam lobe for the valve to be adjusted faces up.

b. Turn the valve lifter so that the notches are perpendicular to the camshaft.

c. Using SST 09248-55040, press down the valve lifter and place SST 09248-05420, between the camshaft and the valve lifter. Remove SST 09248-55040.

d. Remove the adjusting shim with a small flat prying tool and a magnetic finger.

e. Determine the replacement adjusting shim size according to the following formula or use the adjusting shim charts.

f. Using a micrometer, measure the thickness of the removed shim. Calculate the thickness of a new shim so that the valve clearance comes within the specified value.

- T: Thickness of the removed shim
- A: Measured valve clearance
- N: Thickness of the new shim

g. Intake: $N = T + (A - 0.007$ in. (0.18mm))

h. Exhaust: $N = T + (A - 0.013$ in. (0.32mm))

i. Install a new adjusting shim. Place it on the valve lifter. Using the SST 09248-55040, press down the valve lifter and remove SST 09248-05420.

j. Recheck the valve clearance.

9. Install or connect the following:
- Cylinder head cover
- Intake air connector
- Negative battery cable

10. Refill with engine coolant.

11. Start the engine and check for leaks.

4.0L Engine

1. Before servicing the vehicle, refer to the Precautions Section.

2. Disconnect the negative battery cable.

3. Drain the engine coolant. Remove the V bank cover.

4. Remove the air cleaner assembly.

5. Disconnect the two water bypass hoses. Disconnect the fuel vapor feed hose. Disconnect the ventilation hose. Disconnect the VSV connectors.

6. Disconnect the throttle body motor connector. Separate the three wire harness clamps and hose clamp.

7. Remove the two bolts and the throttle body bracket. Remove the bolt and the oil baffle plate. Remove the four bolts and the two surge tank stays.

8. Remove the two nuts. Remove the four bolts, intake air surge tank and gasket.

9. Remove the ignition coil assembly.

10. Remove the cylinder head cover retaining bolts. Remove the cylinder head cover.

11. Turn the crankshaft pulley until its groove and the timing mark "0" of the timing chain cover are aligned. If not aligned at TDC of the compression stroke, turn the crankshaft one complete revolution, in the direction of rotation.

12. Using a feeler gauge, check and record the valve clearance on the following valves: right bank, exhaust number three and intake number one, left bank exhaust number two and intake number six.

13. Rotate the crankshaft 240 degrees clockwise and using a feeler gauge, check and record the following valves: right bank, exhaust number five and intake number three, left bank exhaust number four and intake number two.

14. Rotate the crankshaft 240 degrees clockwise and using a feeler gauge, check and record the following valves: right bank, exhaust number one and intake number five, left bank exhaust number six and intake number four.

15. If adjustment is required, position the engine at TDC on the compression stroke.

16. Place paint marks on the number one chain links corresponding to the timing marks of the camshaft timing gears.

17. Remove the chain tensioner assembly number one. Remove the number two camshaft.

18. Remove the chain tensioner assembly number two. Remove the camshaft.

19. Remove the number four camshaft subassembly. Remove the chain tensioner assembly number three.

20. Remove the number three camshaft subassembly.

21. Remove the valve lifters.

22. Determine the replacement adjusting shim size according to the following formula or use the adjusting shim charts.

23. Using a micrometer, measure the thickness of the removed shim. Calculate the thickness of a new shim so that the valve clearance comes within the specified value.

- T: Thickness of the removed shim
- A: Measured valve clearance
- N: Thickness of the new shim

a. Intake: $N = T + A$

b. Exhaust: $N = T + A$

24. Select a new lifter with a thickness as close as possible to the calculated value.

25. Install removed components in the reverse order of the removal procedure.

26. When installing the cylinder head cover, apply a continuous bead (0.08–0.12 inch) of seal packing, part number 08826-00080 or equivalent, to the cylinder head as indicated in the illustration. Install the seal washers onto the bolts. Install the cylinder head cover bolts and nuts. Tighten bolts "A" to 7.4 ft. lbs. Tighten bolts "B" to 80 inch lbs. Tighten nuts to 80 inch lbs.

➡**Be sure to remove any oil from the contact surfaces of the cylinder head cover and the cylinder head. Install the cover within three minutes after applying the seal packing. Tighten the bolts to specification within fifteen minutes after installing the cover. Do not add engine oil for at least two hours after installing the cover.**

27. Continue the installation in the reverse order of the removal procedure.

28. Be sure to fill the engine with the proper grade and type engine coolant.

29. Be sure to fill the engine with the proper grade and type engine oil.

30. Start the engine and check for leaks. Correct as required.

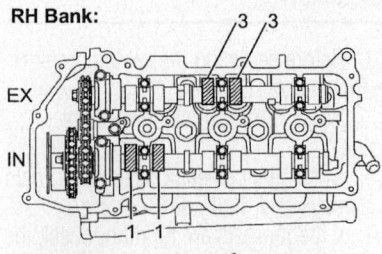

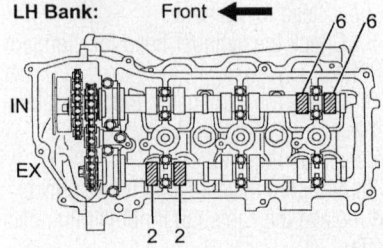

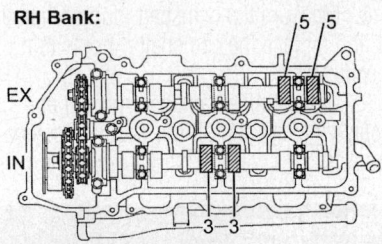

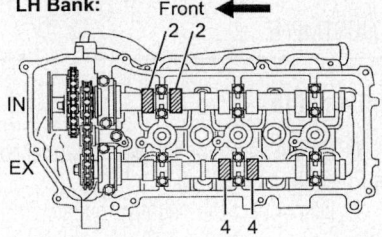

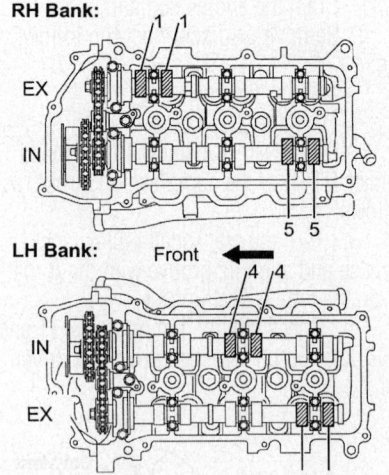

09490_TACO_G0073

Valve clearance location—4.0L engine

4.7L Engine

➡**Measure the valve clearance with the engine cold.**

1. Before servicing the vehicle, refer to the Precautions Section.

06	5.060 (0.1992)	30	5.300 (0.2087)	54	5.540 (0.2181)
08	5.080 (0.2000)	32	5.320 (0.2094)	56	5.560 (0.2189)
10	5.100 (0.2008)	34	5.340 (0.2102)	58	5.580 (0.2197)
12	5.120 (0.2016)	36	5.360 (0.2110)	60	5.600 (0.2205)
14	5.140 (0.2024)	38	5.380 (0.2118)	62	5.620 (0.2213)
16	5.160 (0.2031)	40	5.400 (0.2126)	64	5.640 (0.2220)
18	5.180 (0.2039)	42	5.420 (0.2134)	66	5.660 (0.2228)
20	5.200 (0.2047)	44	5.440 (0.2142)	68	5.680 (0.2236)
22	5.220 (0.2055)	46	5.460 (0.2150)	70	5.700 (0.2244)
24	5.240 (0.2063)	48	5.480 (0.2157)	72	5.720 (0.2252)
26	5.260 (0.2071)	50	5.500 (0.2165)	74	5.740 (0.2260)
28	5.280 (0.2079)	52	5.520 (0.2173)		

09490_TACO_G0074

Shim selection chart, part one—4.0L engine

Shim thickness reference (mm (in.) / shim number):

mm (in.)	No.	mm (in.)	No.	mm (in.)	No.
5.060 (0.1992)	06	5.300 (0.2087)	30	5.540 (0.2181)	54
5.080 (0.2000)	08	5.320 (0.2094)	32	5.560 (0.2189)	56
5.100 (0.2008)	10	5.340 (0.2102)	34	5.580 (0.2197)	58
5.120 (0.2016)	12	5.360 (0.2110)	36	5.600 (0.2205)	60
5.140 (0.2024)	14	5.380 (0.2118)	38	5.620 (0.2213)	62
5.160 (0.2031)	16	5.400 (0.2126)	40	5.640 (0.2220)	64
5.180 (0.2039)	18	5.420 (0.2134)	42	5.660 (0.2228)	66
5.200 (0.2047)	20	5.440 (0.2142)	44	5.680 (0.2236)	68
5.220 (0.2055)	22	5.460 (0.2150)	46	5.700 (0.2244)	70
5.240 (0.2063)	24	5.480 (0.2157)	48	5.720 (0.2252)	72
5.260 (0.2071)	26	5.500 (0.2165)	50	5.740 (0.2260)	74
5.280 (0.2079)	28	5.520 (0.2173)	52		

Measured clearance (mm (in.)) — left axis of shim selection chart:

Clearance mm	Clearance in.
0.000 - 0.020	(0.0000 - 0.0008)
0.021 - 0.040	(0.0008 - 0.0016)
0.041 - 0.060	(0.0016 - 0.0024)
0.061 - 0.080	(0.0024 - 0.0031)
0.081 - 0.100	(0.0032 - 0.0039)
0.101 - 0.120	(0.0040 - 0.0047)
0.121 - 0.140	(0.0048 - 0.0055)
0.141 - 0.160	(0.0056 - 0.0063)
0.161 - 0.180	(0.0063 - 0.0071)
0.181 - 0.200	(0.0071 - 0.0079)
0.201 - 0.220	(0.0079 - 0.0087)
0.221 - 0.240	(0.0087 - 0.0094)
0.241 - 0.260	(0.0095 - 0.0102)
0.261 - 0.280	(0.0103 - 0.0110)
0.281 - 0.289	(0.0111 - 0.0114)
0.391 - 0.410	(0.0154 - 0.0161)
0.411 - 0.430	(0.0162 - 0.0169)
0.431 - 0.450	(0.0170 - 0.0177)
0.451 - 0.470	(0.0178 - 0.0185)
0.471 - 0.490	(0.0185 - 0.0193)
0.491 - 0.510	(0.0193 - 0.0201)
0.511 - 0.530	(0.0201 - 0.0209)
0.531 - 0.550	(0.0209 - 0.0217)
0.551 - 0.570	(0.0217 - 0.0224)
0.571 - 0.590	(0.0225 - 0.0232)
0.591 - 0.610	(0.0233 - 0.0240)
0.611 - 0.630	(0.0241 - 0.0248)
0.631 - 0.650	(0.0248 - 0.0256)
0.651 - 0.670	(0.0256 - 0.0264)
0.671 - 0.690	(0.0264 - 0.0272)
0.691 - 0.710	(0.0272 - 0.0280)
0.711 - 0.730	(0.0280 - 0.0287)
0.731 - 0.750	(0.0288 - 0.0295)
0.751 - 0.770	(0.0296 - 0.0303)
0.771 - 0.790	(0.0304 - 0.0311)
0.791 - 0.810	(0.0311 - 0.0319)
0.811 - 0.830	(0.0319 - 0.0327)
0.831 - 0.850	(0.0327 - 0.0335)
0.851 - 0.870	(0.0335 - 0.0343)
0.871 - 0.890	(0.0343 - 0.0350)
0.891 - 0.910	(0.0351 - 0.0358)
0.911 - 0.930	(0.0359 - 0.0366)
0.931 - 0.950	(0.0367 - 0.0374)
0.951 - 0.970	(0.0374 - 0.0382)
0.971 - 0.990	(0.0382 - 0.0390)
0.991 - 1.010	(0.0390 - 0.0398)

Installed shim thickness (top axis of chart): 5.060 (0.1992) through 5.740 (0.2260).

09490_TACO_G0075

Shim selection chart, part two—4.0L engine

Intake valve clearance (Cold):
0.15 – 0.25 mm (0.006 – 0.010 in.)

EXAMPLE:
The 2.300 mm (0.0906 in.) shim is installed, and the measured clearance is 0.440 mm (0.0173 in.). Replace the 2.300 mm (0.0906 in.) shim with a No. 54 shim.

New shim thickness

mm (in.)

Shim No.	Thickness	Shim No.	Thickness	Shim No.	Thickness
00	2.000 (0.0787)	28	2.280 (0.0898)	56	2.560 (0.1008)
02	2.020 (0.0795)	30	2.300 (0.0906)	58	2.580 (0.1016)
04	2.040 (0.0803)	32	2.320 (0.0913)	60	2.600 (0.1024)
06	2.060 (0.0811)	34	2.340 (0.0921)	62	2.620 (0.1031)
08	2.080 (0.0819)	36	2.360 (0.0929)	64	2.640 (0.1039)
10	2.100 (0.0827)	38	2.380 (0.0937)	66	2.660 (0.1047)
12	2.120 (0.0835)	40	2.400 (0.0945)	68	2.680 (0.1055)
14	2.140 (0.0843)	42	2.420 (0.0953)	70	2.700 (0.1063)
16	2.160 (0.0850)	44	2.440 (0.0961)	72	2.720 (0.1071)
18	2.180 (0.0858)	46	2.460 (0.0969)	74	2.740 (0.1079)
20	2.200 (0.0866)	48	2.480 (0.0976)	76	2.760 (0.1087)
22	2.220 (0.0874)	50	2.500 (0.0984)	78	2.780 (0.1094)
24	2.240 (0.0882)	52	2.520 (0.0992)	80	2.800 (0.1102)
26	2.260 (0.0890)	54	2.540 (0.1000)		

Intake valve clearance shim selection chart—4.7L engine

792ASG71

New shim thickness

Shim No.	Thickness	Shim No.	Thickness	Shim No.	Thickness
00	2.000 (0.0787)	28	2.280 (0.0898)	56	2.560 (0.1008)
02	2.020 (0.0795)	30	2.300 (0.0906)	58	2.580 (0.1016)
04	2.040 (0.0803)	32	2.320 (0.0913)	60	2.600 (0.1024)
06	2.060 (0.0811)	34	2.340 (0.0921)	62	2.620 (0.1031)
08	2.080 (0.0819)	36	2.360 (0.0929)	64	2.640 (0.1039)
10	2.100 (0.0827)	38	2.380 (0.0937)	66	2.660 (0.1047)
12	2.120 (0.0835)	40	2.400 (0.0945)	68	2.680 (0.1055)
14	2.140 (0.0843)	42	2.420 (0.0953)	70	2.700 (0.1063)
16	2.160 (0.0850)	44	2.440 (0.0961)	72	2.720 (0.1071)
18	2.180 (0.0858)	46	2.460 (0.0969)	74	2.740 (0.1079)
20	2.200 (0.0866)	48	2.480 (0.0976)	76	2.760 (0.1087)
22	2.220 (0.0874)	50	2.500 (0.0984)	78	2.780 (0.1094)
24	2.240 (0.0882)	52	2.520 (0.0992)	80	2.800 (0.1102)
28	2.280 (0.0890)	54	2.540 (0.1000)		

mm (in.)

Exhaust valve clearance (Cold):
0.25 – 0.35 mm (0.010 – 0.014 in.)

EXAMPLE:
The 2.300 mm (0.0906 in.) shim is installed, and the measured clearance is 0.440 mm (0.0173 in.). Replace the 2.300 mm (0.0906 in.) shim with a No. 44 shim.

Exhaust valve clearance shim selection chart—4.7L engine

7924SG72

2. Drain the cooling system.

3. Remove or disconnect the following:
- Negative battery cable
- Ignition coils
- Valve covers

4. Set the engine to the top of the compression stroke with the valves closed for the cylinder to be measured.

5. Check the valve clearance. The valve clearance specifications are as follows:
- Intake: 0.006–0.010 in. (0.15–0.25mm)
- Exhaust: 0.010–0.014 in. (0.25–0.35mm)

6. Record the measurements for each valve.

7. When all valve clearances have been measured, remove the camshafts.

8. Remove the valve shims and measure them. Note this measurement along with the clearance measurement recorded earlier.

9. Using the valve clearance and shim thickness measurements, find replacement shims in the Adjusting Shim Selection charts.

10. Install or connect the following:
- Replacement valve shims
- Camshafts
- Valve covers
- Ignition coils
- Negative battery cable

11. Fill the cooling system.

12. Start the engine and check for leaks.

Oil Pan

REMOVAL & INSTALLATION

3.4L Engine

1. Before servicing the vehicle, refer to the Precautions Section.

2. Disconnect the negative battery cable.

3. Drain the engine oil.

4. Remove or disconnect the following:
- Engine undercover
- Front differential, if equipped with 4WD
- Oil pan, separate it from the engine using SST 09032-00100 and a brass bar

To install:

5. Apply seal packing to the oil pan.

6. Install the oil pan to the cylinder block. Tighten the nuts and bolts to: 67 inch lbs. (8 Nm)

✸✸ WARNING

If parts are not assembled within 5 minutes of applying time, the effec-

tiveness of the seal packing is lost and must be removed and reapplied.

7. Install or connect the following:
- Front differential, if removed
- Engine undercover
- Negative battery cable

8. Fill with engine oil.

9. Start the engine and check for leaks.

4.0L Engine

1. Before servicing the vehicle, refer to the Precautions Section.

2. Disconnect the negative battery cable.

3. Raise and support the vehicle safely.

4. Remove the engine undercover. Drain the engine oil.

5. Remove the necessary components in order to gain access to the lower oil pan retaining bolts.

6. Remove the fifteen bolts and two nuts that retain the oil pan to the engine. Insert the blade of tool SST09032-00100 between the pans. Cut through the sealer and remove the lower oil pan from the engine.

➡**Be careful not to damage the contact surface of the oil pans.**

7. Remove the two bolts and nuts. Remove the oil strainer. Discard the gasket.

8. Remove the four housing bolts. Remove the flywheel housing undercover.

9. To remove the upper oil pan, remove the seventeen bolts and two nuts. Remove the upper oil pan from the engine, by prying it apart using a suitable tool.

➡**Be careful not to damage the sealing surface between the upper oil pan and the cylinder block.**

To install:

10. Apply a continuous bead (0.12–0.16 inch in diameter) of seal packing, part number 08826-00080 or equivalent, to the sealing surface of the oil pan.

➡**Remove any oil from the contact surface. Install the upper oil pan within three minutes of applying the seal packing. Tighten the pan bolts to specification within fifteen minutes after applying the seal packing. Do not start the engine for at least two hours after the installation of the oil pan.**

11. Loosely install the upper oil pan bolts and nuts. Bolt "A" is 0.98 inch long, bolt "B" is 1.77 inch long and bolt "C" is 0.55 inch long. Uniformly tighten the 14mm bolt to 7.0 ft. lbs, and the other bolts and nuts to 17 ft. lbs., in the proper sequence.

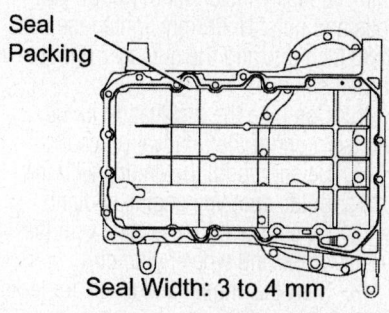

Upper oil pan sealant application—4.0L engine

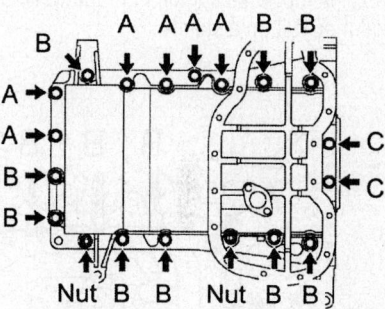

Upper oil pan bolt torque sequence—4.0L engine

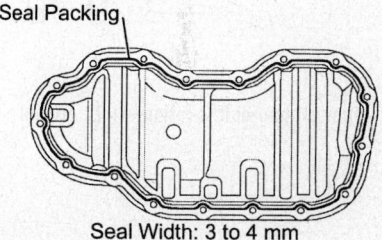

Lower oil pan sealant application—4.0L engine

12. Install the oil strainer assembly. Torque the bolts to 80 inch lbs.

13. Apply a continuous bead (0.12–0.16 inch in diameter) of seal packing, part number 08826-00080 or equivalent, to the sealing surface of the oil pan.

➡**Remove any oil from the contact surface. Install the lower oil pan within three minutes of applying the seal packing. Tighten the pan bolts to specification within fifteen minutes after applying the seal packing. Do not start the engine for at least two hours after the installation of the oil pan.**

14. Loosely install the lower oil pan bolts and nuts. Uniformly tighten the bolts to 80 inch lbs, and the nuts to 7.0 ft. lbs., in several steps.

15. Continue the installation in the reverse order of the removal procedure.

16. Be sure to fill the engine with the proper grade and type engine coolant.

17. Be sure to fill the engine with the proper grade and type engine oil.

18. Start the engine and check for leaks. Correct as required.

4.7L Engine

1. Before servicing the vehicle, refer to the Precautions Section.

2. Remove the engine from the vehicle and mount it on a stand.

3. Remove or disconnect the following:
- Oil dipstick tube
- Lower oil pan
- Oil pan baffle
- Upper oil pan

To install:

4. The upper oil pan bolts are different lengths and are identified as follows:
- A: 0.79 inch (20mm) w/10mm head
- B: 0.98 inch (25mm) w/12mm head
- C: 2.36 inch (60mm) w/12mm head
- D: 1.38 inch (35mm) w/10mm head

5. Apply silicone sealant to the upper oil pan as shown.

6. Install the upper oil pan and tighten the fasteners in several passes to the following specifications:

- 10mm: 66 inch lbs. (7.5 Nm)
- 12mm: 21 ft. lbs. (28 Nm)

7. Install or connect the following:
- Oil pan baffle. Tighten the fasteners to 66 inch lbs. (7.5 Nm).
- Lower oil pan. Tighten the fasteners in several passes to 66 inch lbs. (7.5 Nm).
- Oil dipstick tube

8. Install the engine.

Oil Pump

REMOVAL & INSTALLATION

3.4L Engine

1. Before servicing the vehicle, refer to the Precautions Section.

2. Remove or disconnect the following:
- Negative battery cable
- Engine undercover
- Crankshaft timing pulley
- Front differential, if equipped with 4WD

3. Drain the engine oil from the engine.

4. Remove or disconnect the following:
- Timing belt and crankshaft gear
- Oil cooler tube and clamp, if equipped with automatic transmission
- Stiffener plate
- Flywheel housing undercover and dust cover
- Rear end cover and dust cover
- Starter wire clamp
- Crankshaft Position (CKP) sensor
- Oil pan

➡ **Be careful not to damage the baffle plate flange.**

- Oil strainer
- Oil baffle plate
- Oil pump body by removing the 8 bolts.
- O-ring from the cylinder block

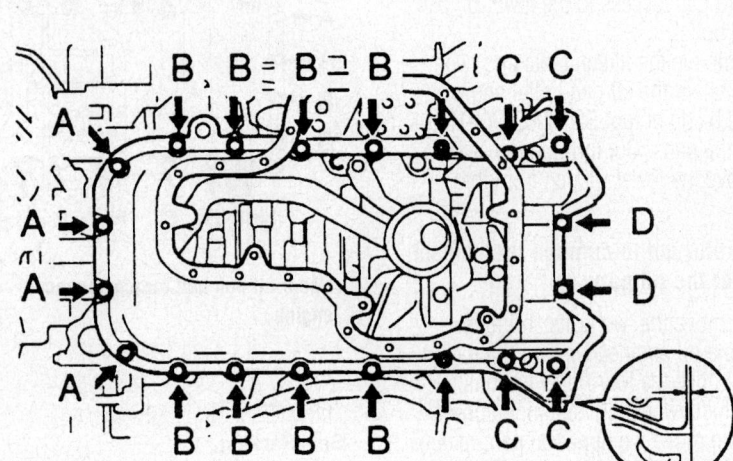

Upper oil pan bolt location—4.7L engine

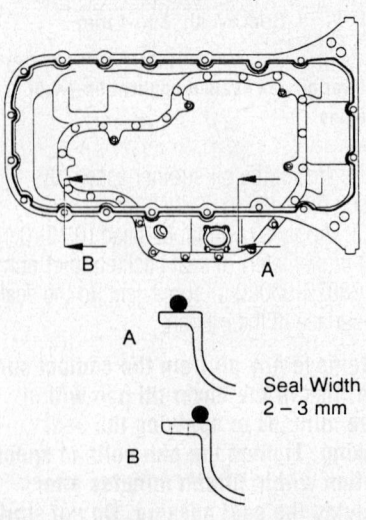

Upper oil pan sealant application—4.7L engine

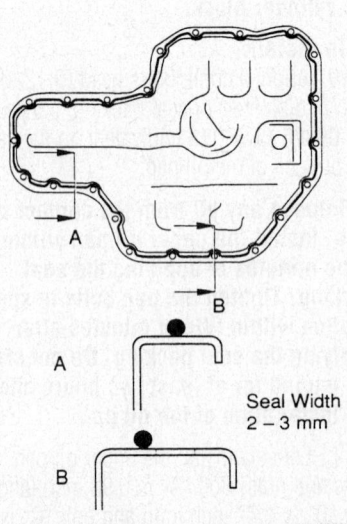

Lower oil pan sealant application—4.7L engine

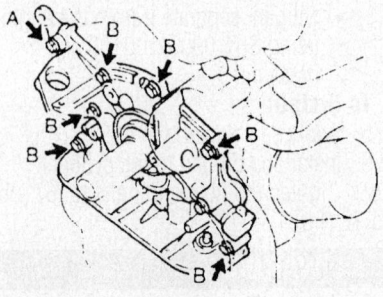

Oil pump bolt identification—3.4L engine

To install:

5. Install or connect the following:
- Apply Seal Packing PN 08826-00080 to the oil pump
- New O-ring into the groove of the cylinder block
- Oil pump to the crankshaft with the spline teeth of the drive rotor engaged with the large teeth of the crankshaft. Tighten the oil pump bolts "A" 15 ft. lbs. (20 Nm) and bolts "B" 31 ft. lbs. (42 Nm)
- CKP
- Oil pan baffle plate
- Oil strainer with a new gasket. Tighten the bolts to 13 ft. lbs. (18 Nm).
- Remaining components
- Negative battery cable

6. Fill with engine oil.
7. Start the engine and check for leaks.

4.0L Engine

1. Before servicing the vehicle, refer to the Precautions Section.
2. Properly relieve the fuel system pressure.
3. Disconnect the negative battery cable. Disconnect the positive battery cable. Remove the battery.
4. Drain the engine coolant. Remove the radiator.
5. Remove the air cleaner assembly.
6. Separate the vane pump assembly. Remove the alternator. Remove the air conditioning compressor and position it to the side.
7. Remove the belt tensioner assembly. Remove the idler pulley number two subassembly. Remove the idler pulley number one subassembly. Remove the crankshaft pulley.
8. Remove the lower oil pan. Remove the oil strainer and pickup tube assembly. Remove the upper oil pan.
9. Remove the intake manifold. Remove the cylinder head cover assembly.
10. Remove the camshaft timing oil control valve assembly. Remove the VVT sensor. Remove the oil filter bracket subassembly.
11. Remove the timing chain case cover retaining bolts. Remove the cover from the engine. Remove the O-ring from the left cylinder head.

➡Carefully remove the cover by prying between the cover and the cylinder head or block with a suitable tool. Be sure to cover the tip of the suitable tool prior to usage. Be careful not to damage the contact surfaces of the cylinder

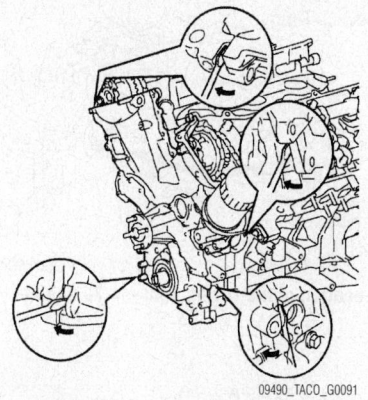

Timing chain case cover removal points— 4.0L engine

block, cylinder head and timing chain cover.

➡The oil pump gears are located inside the timing case cover.

12. Remove the three bolts and remove the oil pipe. Remove the two O-rings. Remove the seven oil pump cover bolts. Remove the gears from the timing chain case cover. Remove the oil pump relief valve.

To install:

13. Coat the oil pump relief valve with clean engine oil. Install the plug, using a new gasket and tighten to 36 ft. lbs.
14. Coat the oil pump gears with clean engine oil. Position the gears in the timing chain case cover with the identification marks facing oil pump cover side. Install the cover. Alternately tighten the bolts to 80 inch lbs. Install the oil pipe, tighten the bolts to 80 inch lbs.
15. Install a new front case oil seal. Install a new O-ring onto the left cylinder head.
16. Apply continuous beads (0.12–0.16 inch in diameter) of seal packing, part number 08826-00080 or equivalent to the four locations shown in the illustration.
17. Apply continuous beads (0.12–0.16 inch in diameter) of seal packing, part number 08826-00080 or equivalent to all parts except the water pump part: for the water pump part use, part number 08826-00080 or equivalent, to the timing chain cover. Do not apply seal packing to portion "A" in the illustration.

➡Remove any oil from the contact surfaces. Install the timing chain case cover within three minutes and tighten the bolts within fifteen minutes of applying the seal packing.

18. Align the key way of the oil pump drive motor with the rectangular portion of

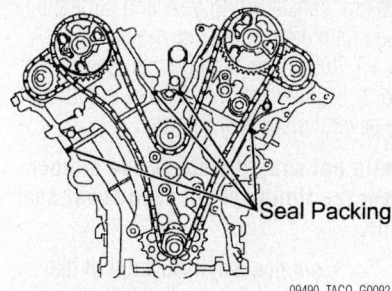

Timing chain case cover seal packing locating points—4.0L engine

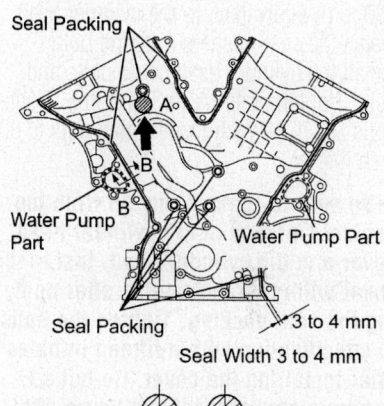

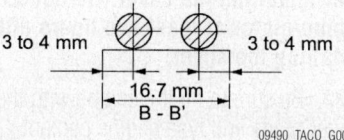

Timing chain case cover sealant application—4.0L engine

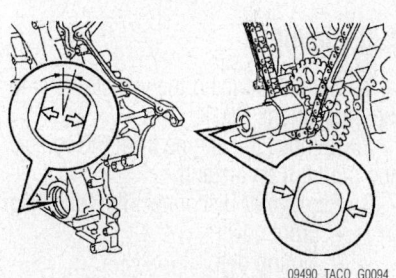

Oil pump drive rotor spline alignment— 4.0L engine

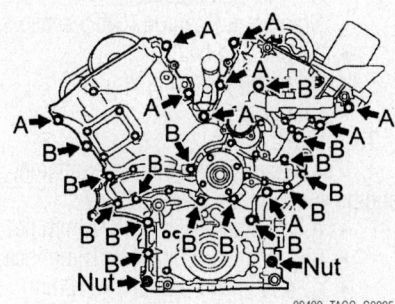

Timing chain case cover bolt location and torque sequence—4.0L engine

the crankshaft timing gear and slide the timing chain case cover into place.

19. Install the timing chain case cover bolts. Tighten the bolts and nuts uniformly in several steps to 17 ft. lbs.

➡**Do not wrap the chain and slipper over the timing chain case cover seal line.**

20. Continue the installation in the reverse order of the removal procedure.

21. When installing the cylinder head cover apply a continuous bead (0.08–0.12 inch) of seal packing, part number 08826-00080 or equivalent, to the cylinder head. Install the seal washers onto the bolts. Install the cylinder head cover bolts and nuts. Tighten bolts "A" to 7.4 ft. lbs. Tighten bolts "B" to 80 inch lbs. Tighten nuts to 80 inch lbs.

➡**Be sure to remove any oil from the contact surfaces of the cylinder head cover and the cylinder head. Install the cover within three minutes after applying the seal packing. Tighten the bolts to specification within fifteen minutes after installing the cover. Do not add engine oil for at least two hours after installing the cover.**

22. Be sure to fill the engine with the proper grade and type engine coolant.

23. Be sure to fill the engine with the proper grade and type engine oil.

24. Start the engine and check for leaks, correct as required.

4.7L Engine

1. Before servicing the vehicle, refer to the Precautions Section.

2. Remove the engine from the vehicle and mount it on a stand.

3. Remove or disconnect the following:
   - Front cover
   - Timing belt.
   - Timing belt idler pulleys
   - Crankshaft timing sprocket
   - Oil dipstick tube
   - Oil filter and bracket
   - Crankshaft Position (CKP) sensor
   - Oil pan and baffle
   - Oil pump pickup tube
   - Oil pump

To install:

4. The upper oil pan bolts are different lengths and are identified as follows:
   - A: 1.38 inch (35mm) w/12mm head
   - B: 1.97 inch (50mm) w/12mm head
   - C: 4.17 inch (106mm) w/12mm head
   - D: 1.57 inch (40mm) w/14mm head

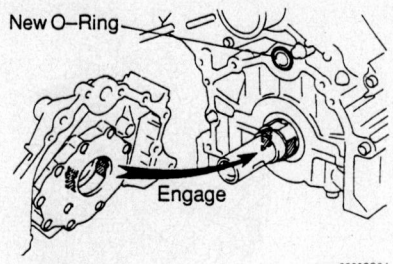

Location of the O-ring seal—4.7L engine

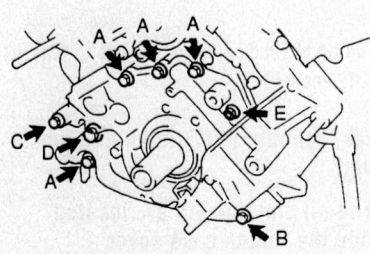

Oil pump bolt location—4.7L engine

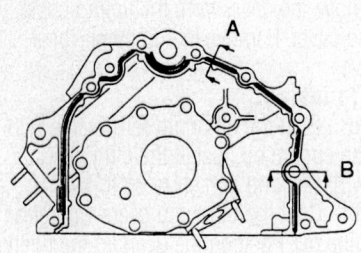

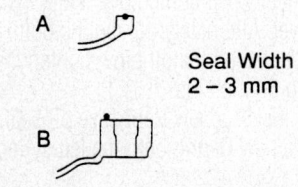

Oil pump housing sealant application—4.7L engine

   - E: 1.18 inch (30mm) w/6mm hex head

5. Install a new O-ring on the engine block.

6. Apply silicone sealant to the oil pump housing as shown.

7. Install the oil pump. Tighten the bolts in several passes to the following specifications:
   - 12mm: 11 ft. lbs. (15.5 Nm)
   - 14mm: 22 ft. lbs. (30.5 Nm)
   - 6mm Hex: 11 ft. lbs. (15.5 Nm)

8. Install or connect the following:
   - Oil pump pickup tube. Tighten the bolts to 66 inch lbs. (7.5 Nm).

   - Oil pan and baffle
   - CKP sensor
   - Oil filter and bracket. Tighten the bolts to 13 ft. lbs. (18 Nm).
   - Oil dipstick tube
   - Crankshaft timing sprocket
   - Timing belt idler pulleys
   - Timing belt
   - Front cover

9. Install the engine.

Starter Motor

REMOVAL & INSTALLATION

3.4L Engine

1. Before servicing the vehicle, refer to the Precautions Section.

2. Remove or disconnect the following:
   - Negative battery cable
   - Starter electrical connectors
   - Starter

To install:

3. Install or connect the following:
   - Starter. Tighten the fasteners to 29 ft. lbs. (39 Nm).
   - Electrical connections
   - Negative battery cable

4.0L Engine

1. Before servicing the vehicle, refer to the Precautions Section.

2. Disconnect the negative battery cable.

3. Remove the engine undercover assembly.

4. Disconnect the starter electrical connectors.

5. Disconnect the positive battery cable.

6. Remove the starter retaining bolts. Remove the starter from the vehicle.

To install:

7. Installation is the reverse of the removal procedure.

8. Tighten the retaining bolts to 27 ft. lbs.

4.7L Engine

1. Before servicing the vehicle, refer to the Precautions Section.

2. Drain the cooling system.

3. Relieve the fuel system pressure.

4. Remove or disconnect the following:
   - Negative battery cable
   - Engine appearance cover
   - Air intake tube
   - Intake manifold
   - Starter motor mounting bolts
   - Starter wiring connectors
   - Starter motor

To install:

5. Install or connect the following:
- Starter motor
- Starter wiring connectors. Tighten the cable nut to 86 inch lbs. (10 Nm).
- Starter motor mounting bolts. Tighten the bolts to 29 ft. lbs. (39 Nm).
- Intake manifold
- Air intake tube
- Engine appearance cover
- Negative battery cable
6. Fill the cooling system.
7. Start the engine and check for leaks.

Timing Belt, Cover and Crankshaft Seal

REMOVAL & INSTALLATION

3.4L Engine

1. Disconnect the negative battery cable.

✳✳ CAUTION

Work must be started after 90 seconds from the time the ignition switch is turned to the LOCK position and the negative battery cable is disconnected.

2. Raise and safely support the vehicle.
3. Remove the engine undercover.
4. Drain the engine coolant.
5. Disconnect the upper radiator hose from the engine.
6. Remove the power steering drive belt.
7. Remove the air conditioning drive belt by loosening the idler pulley nut and the adjusting bolt.
8. Loosen the lockbolt, pivot bolt, and the adjusting bolt and the alternator drive belt.
9. Remove the No. 2 fan shroud by removing the 2 clips.
10. Remove the fan with the fluid coupling and fan pulleys.
11. Disconnect the power steering pump from the engine and set aside. Do not disconnect the lines from the pump.
12. If equipped with air conditioning, disconnect the compressor from the engine and set aside. Do not disconnect the lines from the compressor.
13. If equipped with air conditioning, disconnect the air conditioning bracket.
14. Remove the No. 2 timing belt cover, as follows:

a. Detach the camshaft position sensor connector from the No. 2 timing belt cover.

b. Disconnect the 3 spark plug wire clamps from the No. 2 timing belt cover.

c. Remove the 6 bolts and remove the timing belt cover.

15. Remove the fan bracket, as follows:

a. Remove the power steering adjusting strut by removing the nut.

b. Remove the fan bracket by removing the bolt and nut.

16. Set the No. 1 cylinder at Top Dead Center (TDC) of the compression stroke, as follows:

a. Turn the crankshaft pulley and align its groove with the timing mark **0** of the No. 1 timing belt cover.

b. Check that the timing marks of the camshaft timing pulleys and the No. 3 timing belt cover are aligned. If not, turn the crankshaft pulley one revolution (360 degrees).

➡**If reusing the timing belt, be sure that you can still read the installation marks. If not, place new installation marks on the timing belt to match the timing marks of the camshaft timing pulleys.**

17. Remove the timing belt tensioner by alternately loosening the 2 bolts.

18. Remove the camshaft timing pulleys, as follows:

a. Using SST 09960-10010, or equivalent, remove the pulley bolt, the timing pulley and the knock pin. Remove the 2 timing pulleys with the timing belt.

19. Remove the crankshaft pulley, as follows:

a. Using SST 09213-54015 and 09330-00021 or their equivalents, loosen the pulley bolt.

b. Remove the SST tool, the pulley bolt and the pulley.

20. Remove the starter wire bracket and the No. 1 timing belt cover.

21. Remove the timing belt guide and remove the timing belt.

22. Remove the bolt and the No. 2 idler pulley.

23. Remove the pivot bolt, the No. 1 idler pulley and the plate washer.

24. Remove the crankshaft gear.

To install:

25. Install the crankshaft timing gear.

a. Align the timing pulley set key with the key groove of the gear.

b. Using SST 09214-60010, or equivalent, and a hammer, tap in the timing gear with the flange side facing inward.

26. Install the plate washer and the No. 1 idler pulley with the pivot bolt and tighten it to 26 ft. lbs. (35 Nm). Check that the pulley bracket moves smoothly.

27. Install the No. 2 timing belt idler with the bolt. Tighten the bolt to 30 ft. lbs. (40 Nm). Check that the pulley bracket moves smoothly.

28. Temporarily install the timing belt, as follows:

a. Using the crankshaft pulley bolt, turn the crankshaft and align the timing marks of the crankshaft timing pulley and the oil pump body.

b. Align the installation mark on the timing belt with the dot mark of the crankshaft timing pulley.

c. Install the timing belt on the crankshaft timing pulley, No. 1 idler pulley and the water pump pulleys.

29. Install the timing belt guide with the cup side facing outward.

30. Install the No. 1 timing belt cover and starter wire bracket. Tighten the timing belt cover bolts to 80 inch lbs. (9 Nm).

Turn the crankshaft clockwise to align the timing marks before removing the timing belt—3.4L engine

79245G37

❊❊ WARNING

If any binding is felt when adjusting the timing belt tension by turning the crankshaft, STOP turning the engine, because the pistons may be hitting the valves.

31. Install the crankshaft pulley, as follows:

a. Align the pulley set key with the key groove of the crankshaft pulley.

b. Install the pulley bolt and tighten it to 184 ft. lbs. (250 Nm).

32. Install the left camshaft timing pulley.

a. Install the knock pin to the camshaft.

b. Align the knock pin hose of the camshaft with the knock pin groove of the timing pulley.

c. Slide the timing belt pulley on the camshaft with the flange side facing outward. Tighten the pulley bolt to 81 ft. lbs. (110 Nm).

33. Set the No. 1 cylinder to TDC of the compression stroke, as follows:

a. Turn the crankshaft pulley, and align its groove with the timing mark **0** of the No. 1 timing belt cover.

b. Turn the camshaft to align the knock pin hole of the camshaft with the timing mark of the No. 3 timing belt cover.

c. Turn the camshaft timing pulley, and align the timing marks of the camshaft timing pulley and the No. 3 timing belt cover.

34. Connect the timing belt to the left camshaft timing pulley, as follows:

➡**Check that the installation mark on the timing belt is aligned with the end of the No. 1 timing belt cover.**

a. Using SST 09960-01000, or equivalent, slightly turn the left camshaft timing pulley clockwise. Align the installation mark on the timing belt with the timing mark of the camshaft timing pulley and hang the timing belt on the left camshaft timing pulley.

b. Align the timing marks of the left camshaft pulley and the No. 3 timing belt cover.

c. Check that the timing belt has tension between the crankshaft timing pulley and the left camshaft timing pulley.

35. Install the right camshaft timing pulley and the timing belt, as follows:

a. Align the installation mark on the timing belt with the timing mark of the right camshaft timing pulley and hang

the timing belt on the right camshaft timing pulley with the flange side facing inward.

b. Slide the right camshaft timing pulley on the camshaft. Align the timing marks on the right camshaft timing pulley and the No. 3 timing belt cover.

c. Align the knock pin hole of the camshaft with the knock pin groove of the pulley and install the knock pin. Install the bolt and tighten it to 81 ft. lbs. (110 Nm).

36. Set the timing belt tensioner, as follows:

a. Using a press, slowly press in the pushrod using 220–2205 lbs. (981–9807 N) of force.

b. Align the holes of the pushrod and housing, pass a 1.5mm hex wrench through the holes to keep the setting position of the pushrod.

c. Release the press and install the dust boot on the tensioner.

37. Install the timing belt tensioner and alternately tighten the bolts to 20 ft. lbs. (28 Nm). Using pliers, remove the 1.5mm hex wrench from the belt tensioner.

38. Check the valve timing, as follows:

a. Slowly turn the crankshaft pulley 2 revolutions from TDC to TDC. Always turn the crankshaft pulley clockwise.

b. Check that each pulley aligns with the timing marks. If the timing marks do not align, remove the timing belt and reinstall it.

39. Install the fan bracket with the bolt and nut.

40. Install the power steering adjusting strut with the nut.

41. Install the No. 2 timing belt cover. Tighten the bolts to 80 inch lbs. (9 Nm). Install the remaining components.

42. Fill the cooling system with coolant.

43. Connect the negative battery cable.

44. Start the engine and check for leaks.

4.7L Engine

1. Disconnect the negative battery cable.

2. Raise and safely support the vehicle.

3. Remove the oil pan protector and the engine under cover.

4. Drain the cooling system.

5. Lower the vehicle and remove the battery clamp cover.

6. From the top of the engine, remove the fuel return hose, the engine cover nuts/bolts and the cover.

7. Remove the air cleaner and the intake air connector assembly.

8. Remove the cooling fan pulley by performing the following procedures:

a. Loosen the 4 fan clutch-to-fan pulley nuts.

b. Using a box-end wrench on the serpentine drive belt tensioner bolt, rotate the tensioner counterclockwise and remove the drive belt.

➡**The serpentine drive belt tensioner bolt is a left-hand thread.**

c. Remove the fan clutch-to-fan pulley nuts, the fan, the clutch assembly and the fan pulley.

9. Remove the radiator by performing the following procedures:

a. Disconnect the upper, lower and reservoir hoses from the radiator.

b. Disconnect and plug the automatic transmission oil cooler at the radiator. Disconnect the automatic transmission oil cooler hoses from the fan shroud clamp.

c. Remove the radiator reservoir tank.

d. Remove the fan shroud-to-radiator bolts and the shroud.

e. Remove the 2 upper radiator-to-chassis nuts.

f. Remove the middle radiator-to-chassis nut/bolts and brackets.

g. Carefully, lift the radiator from the vehicle.

10. Remove the power steering pump.

11. Remove the serpentine drive belt idler pulley bolt, cover plate and pulley.

12. Remove the right side (No. 3) timing belt cover.

13. Remove the left side (No. 3) timing belt cover by performing the following procedures:

a. Disconnect the engine wire from both wire clamps.

b. Disconnect the camshaft position sensor wire from the wire clamp on the left-side (No.3) timing belt cover.

c. Disconnect the sensor connector from the connector bracket.

d. Disconnect the sensor connector.

e. Remove the wire grommet from the left-side (No. 3) timing belt cover.

f. Remove the oil cooler tube bolts and tube.

14. Remove the middle (No. 2) timing belt cover bolts and cover.

15. Remove the air conditioning compressor and position it to the side.

16. Remove the cooling fan bracket nuts/bolts and bracket.

➡**If reusing the timing belt, make sure that there are 3 installation marks on the belt; if there are none, install them.**

17. Using the Crankshaft Pulley Holding tool 09213-70010, Bolt tool 90105-08076

and Companion Flange Holding tool 09330-00021, or equivalent, loosen the crankshaft pulley bolt.

18. Position the No. 1 cylinder to approximately 50 degrees After Top Dead Center (ATDC) of the compression stroke by performing the following procedures:

a. Rotate the crankshaft pulley (CLOCKWISE) to align its groove with the timing mark "0" on the lower (No. 1) timing belt cover.

b. Check that the camshaft sprocket timing marks are aligned with the rear timing belt plate marks; if not, rotate the crankshaft 1 revolution (360 degrees).

c. Rotate the crankshaft pulley approximately 50 degrees (CLOCKWISE) and align the crankshaft pulley timing mark between the centers of the crankshaft pulley bolt and the idler pulley bolt.

✴✴ WARNING

If the timing belt is disengaged, having the crankshaft pulley in the wrong angle can cause the valve to come into contact with the piston when removing the camshaft pulley.

19. Remove the crankshaft pulley bolt.

➡️**If reusing the timing belt and the installation marks have disappeared, place new installation marks on the timing belt to match the camshaft timing sprocket marks.**

➡️**To avoid meshing the timing sprocket and the timing belt, secure one with a string; then, place matchmarks on the timing belt and the right-side camshaft timing sprocket.**

20. Remove the timing belt tensioner bolts and the tensioner.

21. Using the Camshaft Holding tool 09960-10010, or equivalent, slightly turn the left-side camshaft sprocket clockwise to loosen the tension spring. Then, disconnect the timing belt from the camshaft sprockets.

22. Remove the alternator by performing the following procedures:

a. Disconnect the electrical connector from the alternator.

b. Remove the rubber cap/nut and disconnect the battery wire from the alternator.

c. Disconnect the wire clamp from the alternator cord clip.

d. Remove the alternator-to-engine nuts/bolts and the alternator.

23. Remove the serpentine drive belt tensioner nuts/bolts and the tensioner.

24. Using the Crankshaft Puller Assembly tool 09950-50012, or equivalent, press the crankshaft pulley from the crankshaft.

✴✴ WARNING

DO NOT rotate the crankshaft pulley.

25. Remove the lower (No. 1) timing belt cover bolts and the cover.

26. Remove the timing belt guide, spacer and the timing belt.

To install:

➡️**With the timing belt removed, this is a perfect opportunity to inspect and/or replace the water pump.**

27. Inspect the timing belt tensioner by performing the following procedures:

a. Inspect the seal for leakage; if leakage is suspected, replace the tensioner.

b. Using both hands to hold the tensioner facing upward, strongly press the

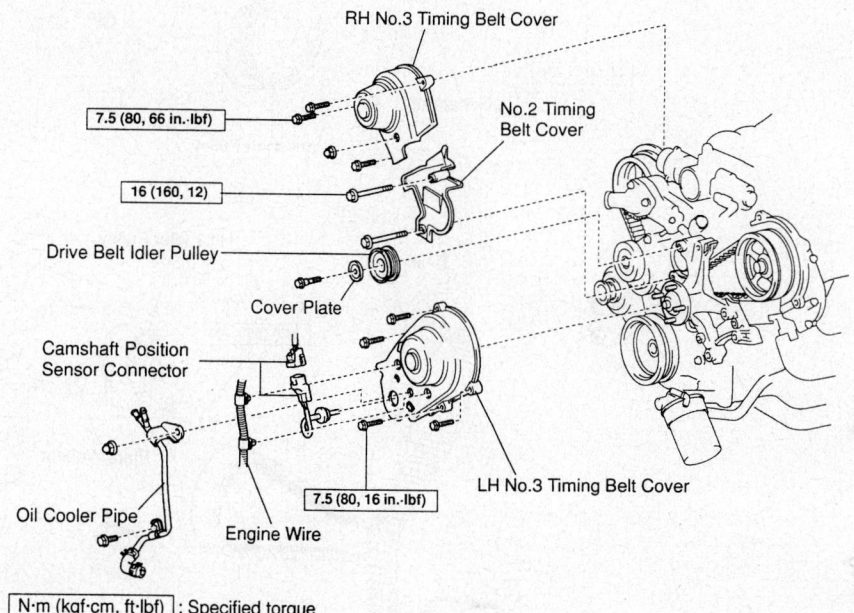

Upper timing belt and related components—4.7L engine

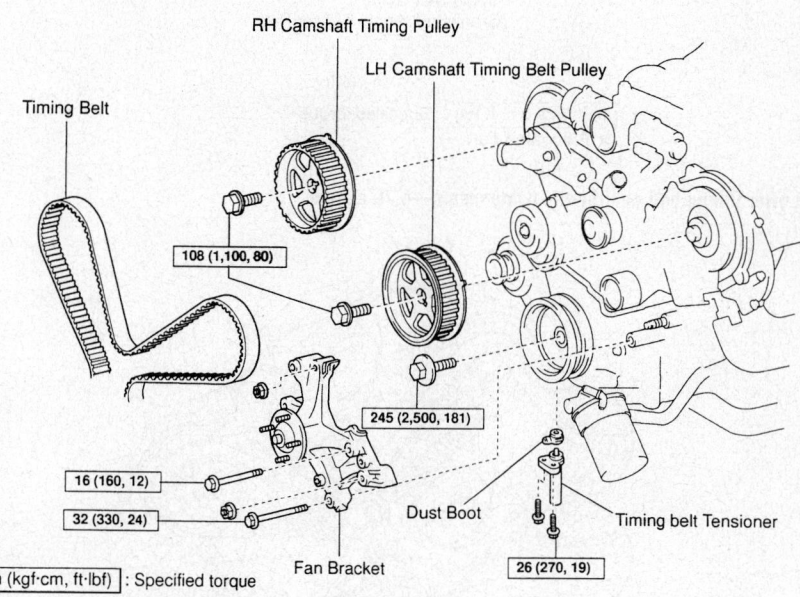

Upper timing sprockets and related components—4.7L engine

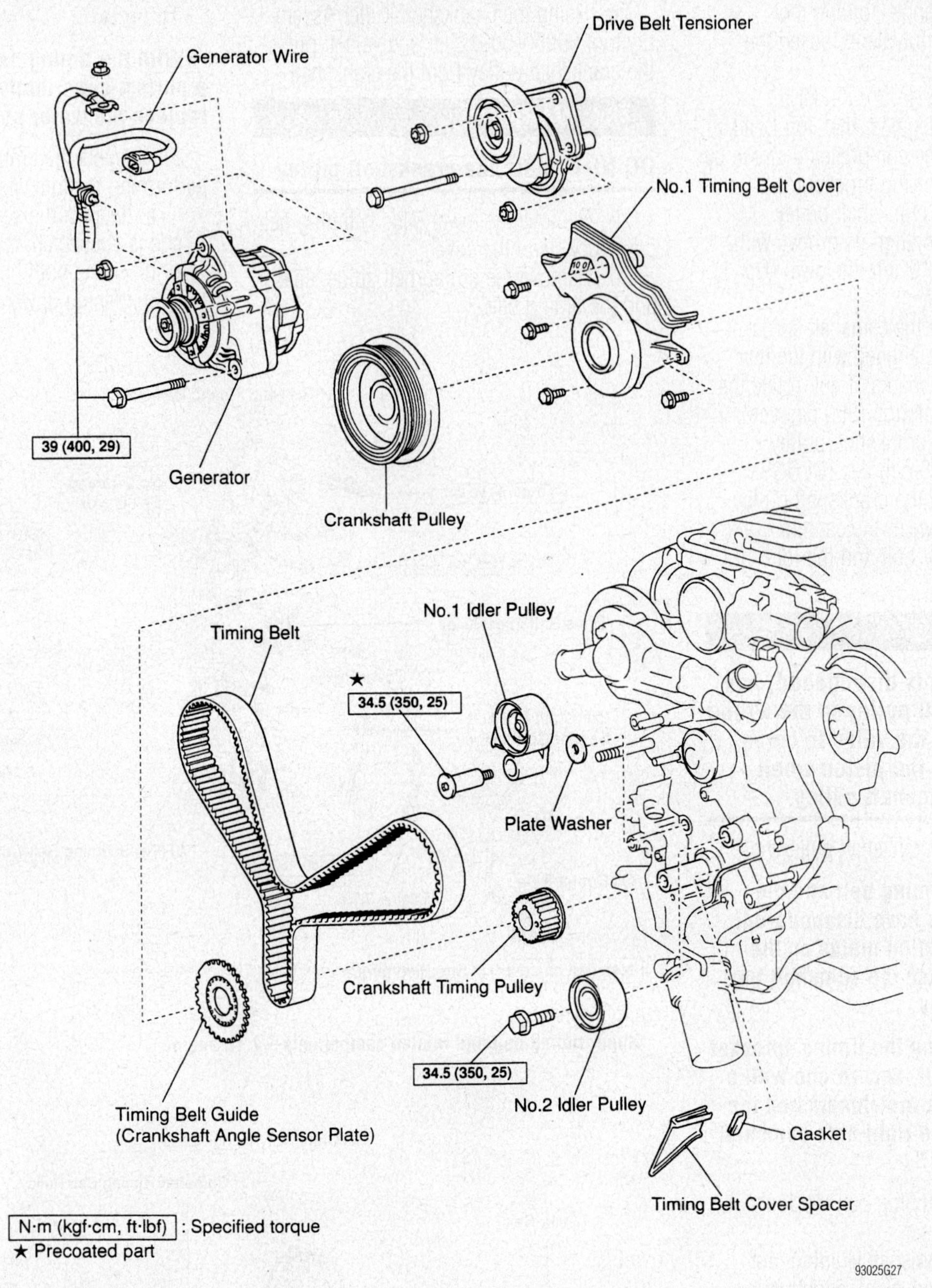

Generator Wire

Drive Belt Tensioner

No.1 Timing Belt Cover

39 (400, 29)

Generator

Crankshaft Pulley

Timing Belt

No.1 Idler Pulley

★ 34.5 (350, 25)

Plate Washer

Crankshaft Timing Pulley

Timing Belt Guide
(Crankshaft Angle Sensor Plate)

34.5 (350, 25)

No.2 Idler Pulley

Gasket

Timing Belt Cover Spacer

N·m (kgf·cm, ft·lbf) : Specified torque
★ Precoated part

93025G27

Lower timing belt and related components—4.7L engine

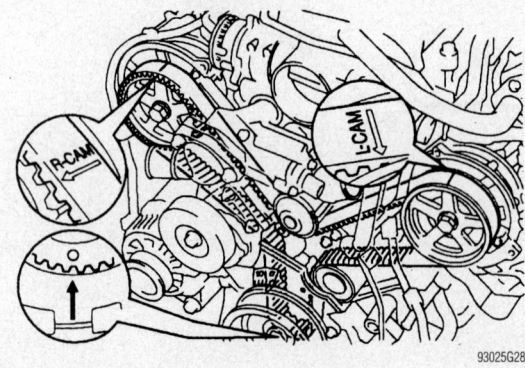

R-CAM

L-CAM

Timing belt and sprocket alignment—4.7L engine

93025G28

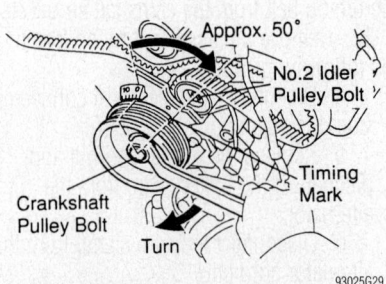

Approx. 50°

No.2 Idler
Pulley Bolt

Timing
Mark

Crankshaft
Pulley Bolt

Turn

93025G29

Aligning of crankshaft pulley timing mark with the center line of the crankshaft pulley bolt and the idler pulley bolt—4.7L engine

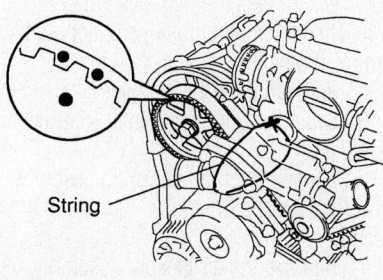

Securing the timing belt with string and matchmarking the camshaft with the timing belt—4.7L engine

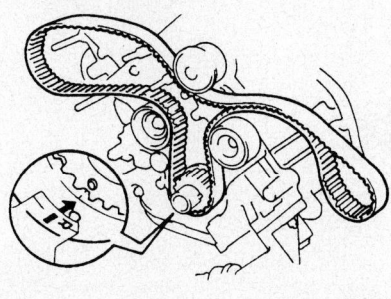

Installing the timing belt on the crankshaft sprocket—4.7L engine

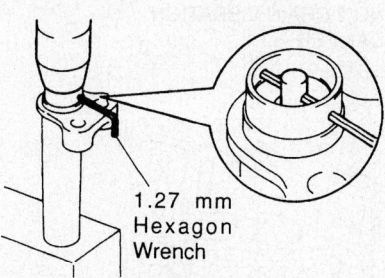

Securing the timing belt tensioner pushrod—4.7L engine

Checking the TDC alignment marks after rotating the crankshaft 2 revolutions—4.7L engine

pushrod against a solid surface. If the pushrod moves, replace the tensioner.

✳✳ WARNING

Never hold the tensioner with the pushrod facing downward.

c. Measure the pushrod protrusion from the housing end, it should be 0.413–0.453 in. (10.5–11.5mm). If the protrusion is not as specified, replace the tensioner.

28. Temporarily install the timing belt by performing the following procedures:

a. Align the timing belt's installation mark with the crankshaft timing sprocket.

b. Install the timing belt on the crankshaft timing sprocket, the No. 1 idler pulley and the No. 2 idler pulley.

29. Install the gasket to the timing belt cover spacer and install the cover spacer.

30. Install the timing belt guide with the cup side facing outward.

31. Install the lower (No. 1) timing belt cover.

32. Install the crankshaft pulley by performing the following procedures:

a. Align the crankshaft pulley with the crankshaft key.

b. Using the Crankshaft Installer tool 09223-46011, or equivalent, and a hammer, tap the crankshaft pulley into position.

33. Install the serpentine drive belt tensioner and torque the tensioner-to-engine bolts to 12 ft. lbs. (16 Nm) for 2002–2004 vehicles and 19 ft. lbs. (26 Nm) for 2005–2006 vehicles.

➡ To install the serpentine drive belt tensioner, use a bolt 4.18 in. (106mm) in length.

34. Check that the crankshaft pulley's timing mark is aligned with the centers of the idler pulley and crankshaft pulley bolts.

35. Install the alternator and torque the alternator-to-engine nuts/bolts to 29 ft. lbs. (39 Nm). Connect the alternator's electrical connectors and clip.

36. Install the timing belt to the left-side camshaft by performing the following procedures:

a. Rotate the left-side camshaft pulley to align the timing belt installation mark with the camshaft sprocket's timing mark and slide the belt onto the camshaft timing sprocket.

b. Using the Camshaft Holding tool 09960-10010, or equivalent, slightly turn the left-side camshaft sprocket counterclockwise to place tension on the timing

belt between the crankshaft sprocket and the camshaft sprocket.

37. Rotate the right-side camshaft pulley to align the timing belt installation mark with the camshaft sprocket's timing mark and slide the belt onto the camshaft timing sprocket.

38. Using a vertical press, slowly press the pushrod into the housing using 200–2205 lbs. (981–9807 N) until the holes align, then, install a 1.27mm Allen® wrench to secure the pushrod and release the press. Install the dust boot on the tensioner housing.

39. Install the timing belt tensioner and torque the bolts to 19 ft. lbs. (26 Nm).

40. Using a pair of pliers, remove the Allen® wrench from the tensioner housing.

41. Check the valve timing by performing the following procedure:

a. Temporarily install the crankshaft pulley bolt.

b. Slowly, rotate the crankshaft pulley 2 revolutions (CLOCKWISE) and realign the TDC marks.

➡ If the pulley/sprocket timing marks do not realign, remove the timing belt and reinstall it.

42. Using the Crankshaft Pulley Holding tool 09213-70010, Bolt tool 90105-08076 and Companion Flange Holding tool 09330-00021, or equivalent, torque the crankshaft pulley bolt to 181 ft. lbs. (245 Nm).

43. Install the cooling fan bracket and torque the 12mm (head size) bolt to 12 ft. lbs. (16 Nm) and the 14mm (head size) bolt to 24 ft. lbs. (32 Nm).

44. Install the air conditioning compressor.

45. Install the middle (No. 2) timing belt cover and torque the bolts to 12 ft. lbs. (16 Nm).

46. Install the upper right-side (No. 3) timing belt cover and torque the bolts to 66 inch lbs. (7.5 Nm).

47. Install the upper left-side (No. 3) timing belt cover by performing the following procedures:

a. Install the oil cooler tube and bolt.

b. Feed the Camshaft Position Sensor (CPS) through the left-side (No. 3) timing belt cover hole.

c. Install the left-side (No. 3) timing belt cover and torque the bolts to 66 inch lbs. (7.5 Nm).

d. Install the wire grommet to the left-side (No. 3) timing belt cover.

e. Install the sensor connector to the connector bracket and connect the sensor connector.

f. Install the sensor wire and the engine wire to the clamps on the left-side (No. 3) timing belt cover.

48. Install the drive belt idler pulley and cover plate; then, torque the pulley bolt to 27 ft. lbs. (37 Nm) on 2002–2004 vehicles and to 29 ft. lbs. (39 Nm) on 2005–2006 vehicles.

49. To complete the installation, reverse the removal procedures.

50. Refill the cooling system and connect the negative battery cable.

Timing Chain, Sprockets, Front Cover and Seal

REMOVAL & INSTALLATION

4.0L Engine

1. Before servicing the vehicle, refer to the Precautions Section.
2. Properly relieve the fuel system pressure.

3. Disconnect the negative battery cable. Disconnect the positive battery cable. Remove the battery.
4. Remove the V bank cover.
5. Drain the engine coolant. Remove the radiator.
6. Loosen the fluid coupling assembly (fan). Remove the drive belt. Remove the fan assembly.
7. Remove the air cleaner assembly. Remove the oil level gauge. Remove the water inlet.

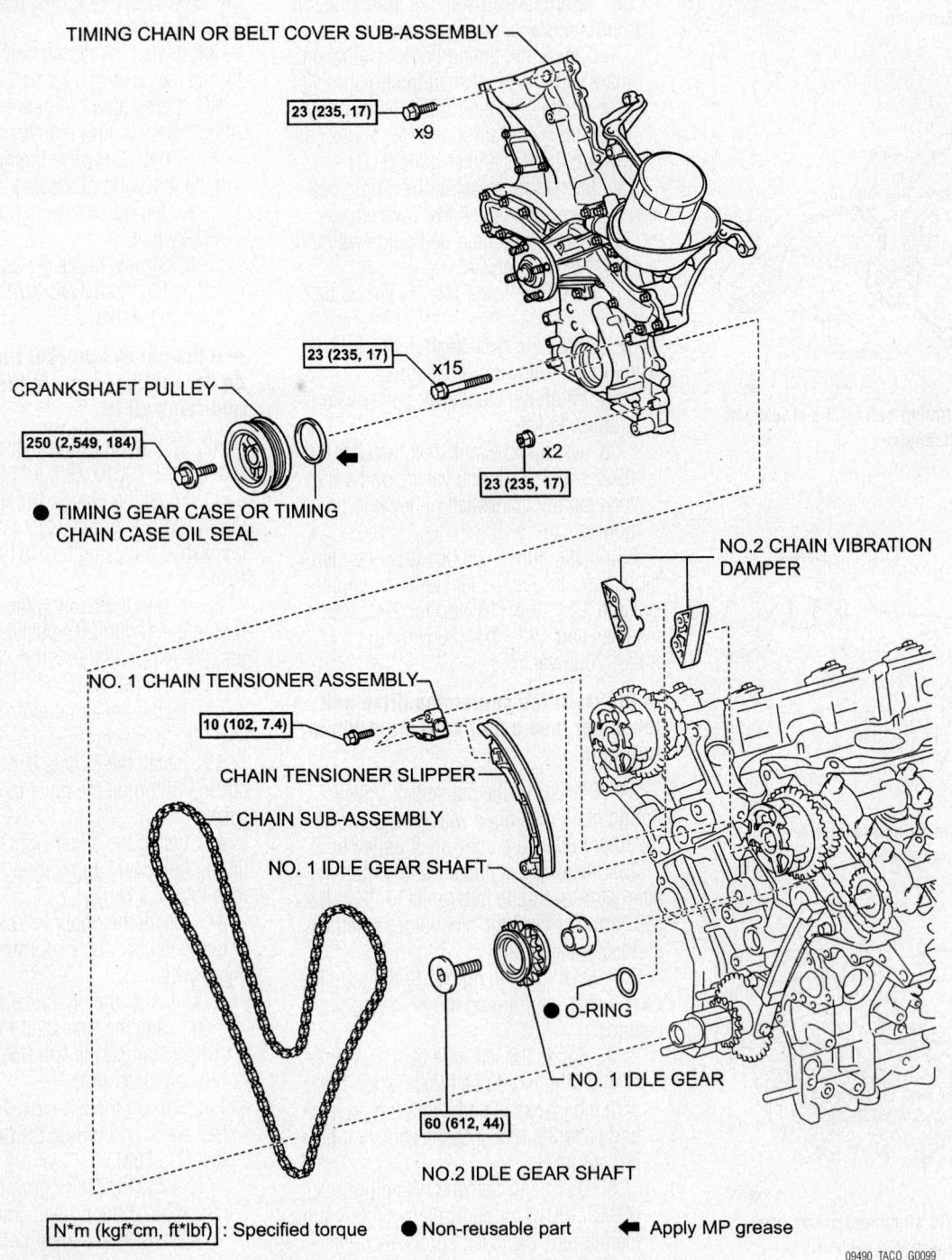

TIMING CHAIN OR BELT COVER SUB-ASSEMBLY

23 (235, 17)   x9

23 (235, 17)   x15

CRANKSHAFT PULLEY

250 (2,549, 184)

x2

23 (235, 17)

● TIMING GEAR CASE OR TIMING CHAIN CASE OIL SEAL

NO.2 CHAIN VIBRATION DAMPER

NO. 1 CHAIN TENSIONER ASSEMBLY

10 (102, 7.4)

CHAIN TENSIONER SLIPPER

CHAIN SUB-ASSEMBLY

NO. 1 IDLE GEAR SHAFT

● O-RING

NO. 1 IDLE GEAR

60 (612, 44)

NO.2 IDLE GEAR SHAFT

N*m (kgf*cm, ft*lbf) : Specified torque   ● Non-reusable part   ◀ Apply MP grease

Timing chain and related components—4.0L engine

09490_TACO_G0099

8. Separate the vane pump assembly (power steering pump). Remove the alternator. Remove the air conditioning compressor and position it to the side.

9. Remove the belt tensioner assembly. Remove the idler pulley number two subassembly. Remove the idler pulley number one subassembly. Remove the crankshaft pulley.

10. Remove the lower oil pan. Remove the oil strainer and pickup tube assembly. Remove the upper oil pan.

11. Remove the intake manifold. Remove the ignition coil assembly. Remove the cylinder head cover assembly.

12. Remove the camshaft timing oil control valve assembly. Remove the VVT sensor. Remove the oil filter bracket subassembly.

➡**Carefully remove the cover by prying between the cover and the cylinder head or block with a suitable tool. Be sure to cover the tip of the suitable tool prior to usage. Be careful not to damage the contact surfaces of the cylinder block, cylinder head and timing chain cover.**

13. Using the crankshaft pulley set bolt, turn the crankshaft to align the crankshaft set key with the timing line of the cylinder block. If not aligned at TDC of the compression stroke, turn the crankshaft one complete revolution, in the direction of rotation

14. While turning the stopper plate of the tensioner upward, push the plunger of the chain tensioner. While turning the stopper plate of the tensioner downward, insert a 0.138 inch diameter bar into the holes in the stopper plate and tensioner to hold the stopper plate.

15. Remove the two bolts, and then remove the chain tensioner.

16. Remove the chain tensioner slipper. Remove the idle gear shaft number two, idle gear number one and idle gear shaft number one.

17. Remove the number two chain vibration damper. Remove the timing chain subassembly.

To install:

18. Install the chain tensioner slipper.

19. While turning the stopper plate of the tensioner clockwise, push in the plunger of the chain tensioner. While turning the stopper plate of the tensioner counterclockwise, insert a 0.138 inch diameter bar into the holes in the stopper plate and tensioner to hold the stopper plate.

20. Install the chain tensioner. Tighten the bolts to 7.4 ft. lbs.

21. Position the engine at TDC on the compression stroke. Align the camshaft timing gears and bearing caps. Using the crankshaft pulley set bolt, align the crankshaft set key with the timing line of the cylinder.

22. Align the yellow mark line with the timing mark of the crankshaft timing link. Align the orange mark links with the timing marks of the camshaft timing gears, and install the chain.

23. Install the number two chain vibration damper.

24. Apply a light coat of clean engine oil to the rotating surface of the idle gear shaft number one.

25. Temporarily install the idle gear shaft number one together with idle gear shaft number two, while aligning the knock pin of idle gear shaft number one with the knock pin groove of the cylinder block.

➡**Be care of the idle gear direction.**

26. Tighten the idle gear shaft number two to 44 ft. lbs. Remove the bar from the chain tensioner.

27. Install a new front case oil seal. Install a new O-ring onto the left cylinder head.

28. Apply continuous beads (0.12–0.16 inch in diameter) of seal packing, part number 08826-00080 or equivalent to the four locations shown in the illustration.

29. Apply continuous beads (0.12–0.16 inch in diameter) of seal packing, part number 08826-00080 or equivalent to all parts except the water pump part: for the water pump part use, part number 08826-00080 or equivalent, to the timing chain cover. Do not apply seal packing to portion "A".

➡**Remove any oil from the contact surfaces. Install the timing chain case cover within three minutes and tighten the bolts within fifteen minutes of applying the seal packing.**

30. Align the key way of the oil pump drive motor with the rectangular portion of the crankshaft timing gear and slide the timing chain case cover into place.

31. Install the timing chain case cover bolts. Tighten the bolts and nuts uniformly in several steps to 17 ft. lbs.

➡**Do not wrap the chain and slipper over the timing chain case cover seal line.**

32. Continue the installation in the reverse order of the removal procedure.

33. When installing the cylinder head cover apply a continuous bead (0.08–0.12 inch) of seal packing, part number 08826-00080 or equivalent, to the cylinder head. Install the seal washers onto the bolts. Install the cylinder head cover bolts and nuts. Tighten bolts "A" to 7.4 ft. lbs. Tighten bolts "B" to 80 inch lbs. Tighten nuts to 80 inch lbs.

➡**Be sure to remove any oil from the contact surfaces of the cylinder head cover and the cylinder head. Install the cover within three minutes after applying the seal packing. Tighten the bolts to specification within fifteen minutes**

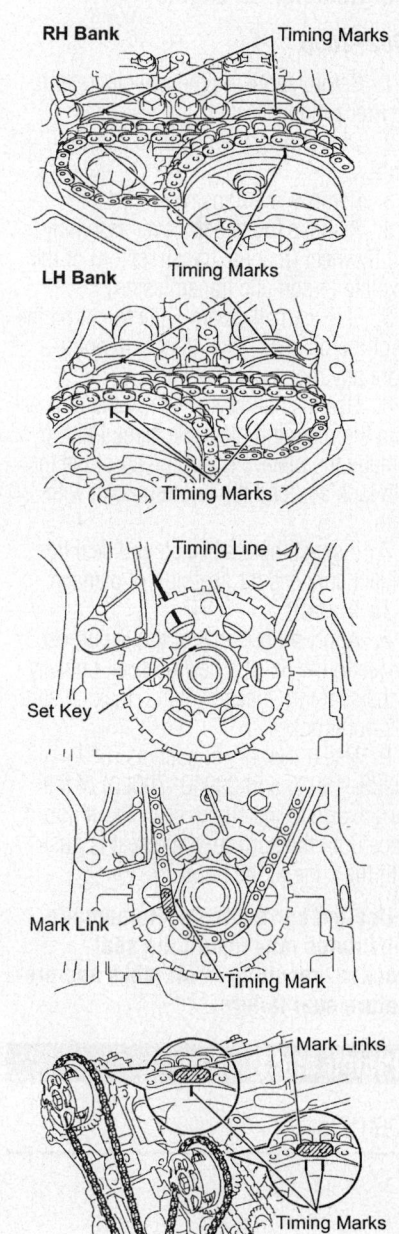

Timing chain alignment—4.0L engine

after installing the cover. **Do not add engine oil for at least two hours after installing the cover.**

34. Be sure to fill the engine with the proper grade and type engine coolant.

35. Be sure to fill the engine with the proper grade and type engine oil.

36. Start the engine and check for leaks, correct as required.

Rear Main Seal

REMOVAL & INSTALLATION

Seal Retainer On Engine

2002–2006

1. Before servicing the vehicle, refer to the Precautions Section.

2. Disconnect the negative battery cable.

3. Remove the transmission.

4. Remove the clutch cover assembly and flywheel (manual transmission) or the flexplate (automatic transmission).

5. Use a small, sharp knife to cut off the lip of the oil seal. Take great care not to score any metal with the knife.

6. Use a small prybar to pry the old seal from the retaining plate. Be careful not to damage the plate. Protect the tip of the tool with tape and pad the fulcrum point with cloth.

7. Inspect the crankshaft and seal lip contact surfaces for any sign of damage.

To install:

8. Apply a light coat of multi-purpose grease to the lip of a new oil seal. Loosely fit the seal into place by hand, making sure it is not crooked.

9. Use a seal driver such as (SST 09223–15030 and 09950–70010) of the correct size to install the seal. Tap it into place until the surface of the seal is flush with the edge of the housing.

➡ **Use the correct tools. Homemade substitutes may install the seal crooked, resulting in oil leaks and premature seal failure.**

Piston and Ring

POSITIONING

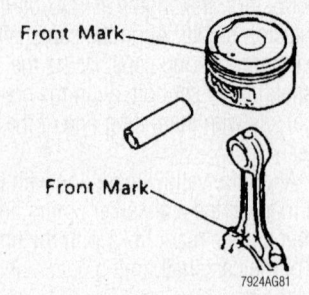

Piston to connecting rod assembly—3.4L engine

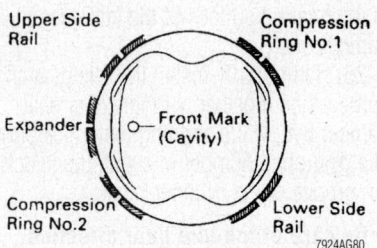

Piston ring end-gap spacing—3.4L engine

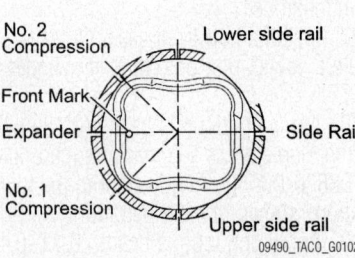

Piston ring positioning—4.0L engine

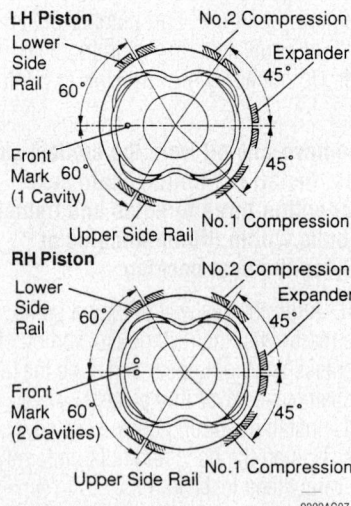

Piston ring positioning—4.7L engine

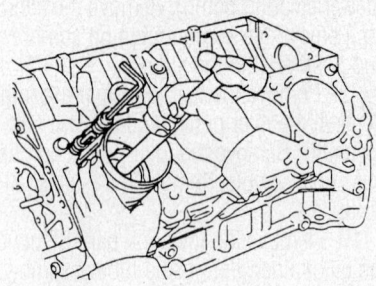

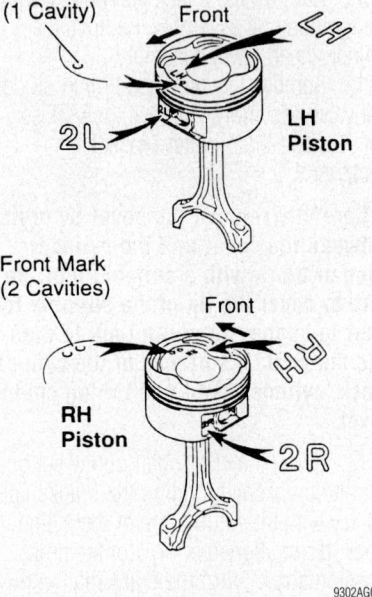

Piston positioning—4.7L engine

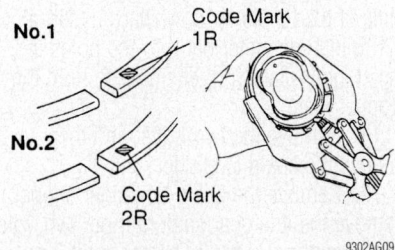

Piston ring identification—4.7L engine

FUEL SYSTEM

Fuel System Service Precautions

Safety is the most important factor when performing not only fuel system maintenance but any type of maintenance. Failure to conduct maintenance and repairs in a safe manner may result in serious personal injury or death. Maintenance and testing of the vehicle's fuel system components can be accomplished safely and effectively by adhering to the following rules and guidelines.

• To avoid the possibility of fire and personal injury, always disconnect the negative battery cable unless the repair or test procedure requires that battery voltage be applied.

• Always relieve the fuel system pressure prior to disconnecting any fuel system component (injector, fuel rail, pressure regulator, etc.), fitting or fuel line connection. Exercise extreme caution whenever relieving fuel system pressure, to avoid exposing skin, face and eyes to fuel spray. Please be advised that fuel under pressure may penetrate the skin or any part of the body that it contacts.

• Always place a shop towel or cloth around the fitting or connection prior to loosening to absorb any excess fuel due to spillage. Ensure that all fuel spillage (should it occur) is quickly removed from engine surfaces. Ensure that all fuel soaked cloths or towels are deposited into a suitable waste container.

• Always keep a dry chemical (Class B) fire extinguisher near the work area.

• Do not allow fuel spray or fuel vapors to come into contact with a spark or open flame.

• Always use a back-up wrench when loosening and tightening fuel line connection fittings. This will prevent unnecessary stress and torsion to fuel line piping.

• Always replace worn fuel fitting O-rings with new. Do not substitute fuel hose or equivalent, where fuel pipe is installed.

Fuel System Pressure

RELIEVING

2002–2004

1. Before servicing the vehicle, refer to the Precautions Section.
2. Disconnect the fuel pump connector near the fuel tank.
3. Start the engine and allow it to run

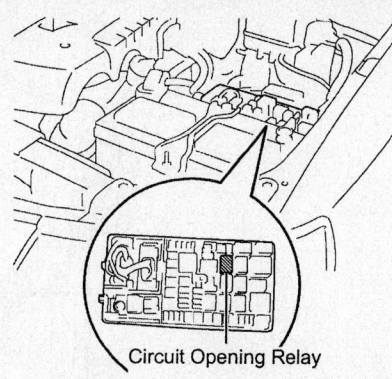

Circuit Opening Relay

09490_TACO_G0104

Circuit opening relay location—2005–2006

until it stalls. Crank the engine for a few seconds to relieve additional fuel pressure.
4. Disconnect the negative battery cable.
5. When repairs are complete, connect the negative battery cable.

2005–2006

1. Before servicing the vehicle, refer to the Precautions Section.
2. Disconnect the negative battery cable.
3. Remove the engine relay block cover.
4. Remove the circuit opening relay.
5. Connect the negative battery.
6. Start the engine.
7. Turn the ignition switch "ON" after the engine stops.

➡Code DTC P0171 (system lean) may be present.

8. Crank the engine again. Check that the engine stops.
9. Remove the fuel tank cap and completely discharge the pressure in the fuel tank. Install the circuit opening relay.
10. Disconnect the negative battery cable.

Fuel Filter

REMOVAL & INSTALLATION

2002–2004

1. Before servicing the vehicle, refer to the Precautions Section.
2. Relieve the fuel system pressure.
3. Remove or disconnect the following:

• Negative battery cable
• Fuel lines
• Fuel filter

To install:
4. Install the fuel filter.
5. Use new washers and tighten the fuel line bolts to the following specifications:

• Banjo bolt fittings: 21 ft. lbs. (29 Nm)
• Flare nut fitting: 28 ft. lbs. (38 Nm)
6. Connect the negative battery cable.
7. Start the engine and check for leaks.

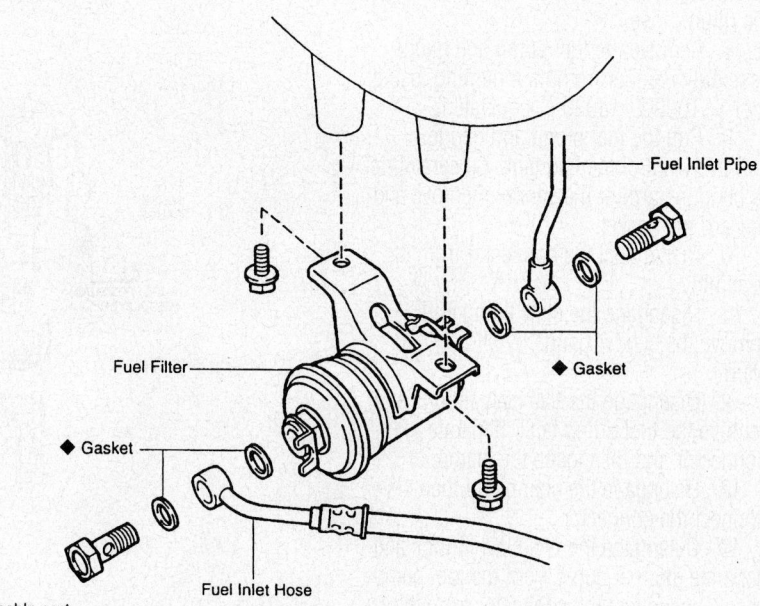

Fuel Inlet Pipe

◆ Gasket

Fuel Filter

◆ Gasket

Fuel Inlet Hose

◆ Non-reusable part

7924SG28

Always use new gaskets when replacing the fuel filter

Fuel Pump

REMOVAL & INSTALLATION

2002–2004

1. Before servicing the vehicle, refer to the Precautions Section.
2. Relieve the fuel system pressure.
3. Remove or disconnect the following:
   - Negative battery cable
   - Fuel tank
   - Fuel pump harness connector
   - Fuel lines
   - Fuel pump module

To install:

4. Install or connect the following:
   - Fuel pump module. Tighten the bolts to 35 inch lbs. (4 Nm).
   - Fuel lines
   - Fuel pump harness connector
   - Fuel tank
   - Negative battery cable
5. Start the engine and check for leaks.

2005–2006

1. Before servicing the vehicle, refer to the Precautions Section.
2. Relieve the fuel pressure.
3. Disconnect the negative battery cable.
4. Drain the fuel from the fuel tank.
5. If equipped, remove the number one fuel tank protector subassembly.
6. Remove the fuel tank.
7. Remove the fuel tank main hose and the return hose.
8. Remove the fuel pump and gauge assembly retainer from its mounting, using tool SST09808-14020 or equivalent.
9. Pull the fuel pump and gauge assembly out of the fuel tank. Be careful not to bend the arm of the sender. Remove and discard the gasket.
10. Disconnect the connector from the assembly.
11. Disengage the claw fitting and remove the sender gauge by sliding it forward.
12. Disengage the five claw fittings and remove the fuel pump tank. Separate the connector and disengage the clamp.
13. Disengage the clamp and then disconnect the connector.
14. Disengage the five claw fittings and separate the fuel pump from the fuel pump case. Disconnect the connector from the fuel pump.
15. Remove the fuel filter from the fuel pump.

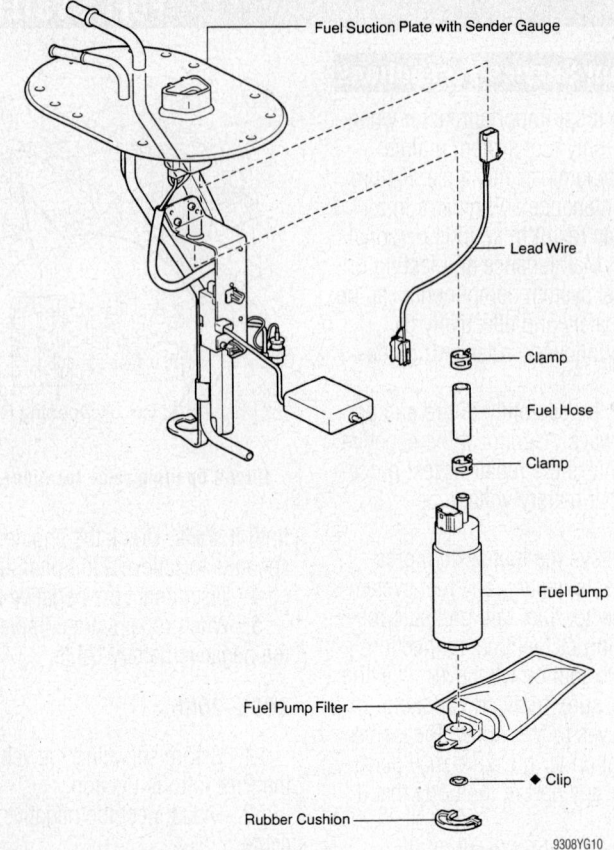

Fuel pump and related components—2002–2004

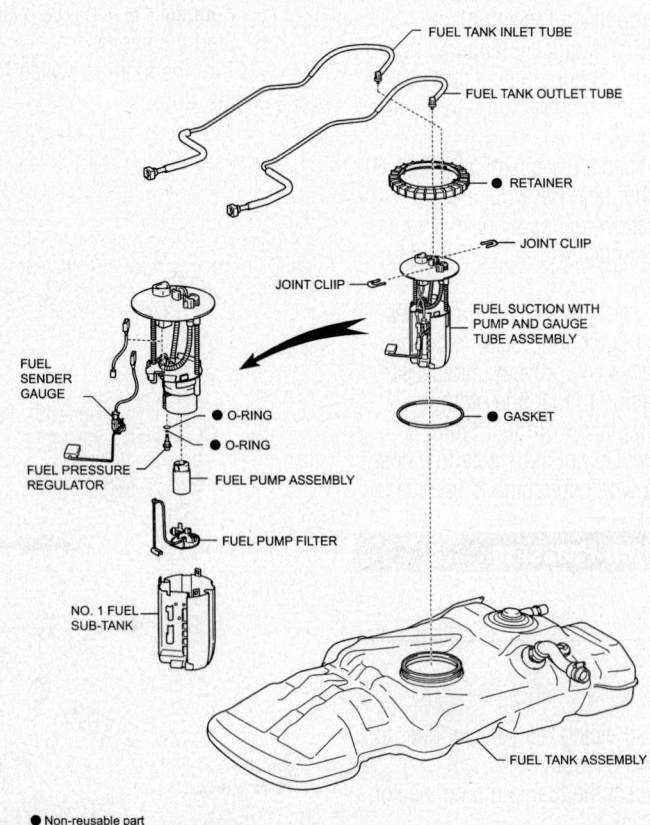

● Non-reusable part

Fuel pump and related components—2005–2006

To install:

16. Installation is the reverse of the removal procedure.

17. Start the engine and check for leaks, correct as required.

Fuel Injector

REMOVAL & INSTALLATION

3.4L Engine

1. Before servicing the vehicle, refer to the Precautions Section.
2. Depressurize the fuel system.
3. Remove or disconnect the following:
   - Air cleaner hose
   - Upper half of the intake manifold
   - Fuel pressure regulator
   - Fuel inlet pipe
   - Fuel injector electrical connections
   - Fuel rail with the injectors
   - Spacers from the intake manifold
   - Injectors from the delivery pipes
   - O-rings and grommets, discard them

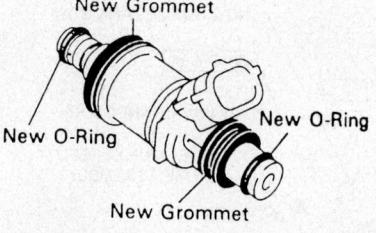

Install new O-rings and grommets on each injector—3.4L engine

To install:

4. Install or connect the following:
   - New grommets and O-rings on each injector, lubricated with a light coat of gasoline
   - Fuel injector with the electrical connector facing outward
   - Spacers on the intake manifold
5. Temporarily install the bolts to hold the delivery pipes to the intake manifold.
6. Check that the injectors rotate smoothly. If they do not, the O-rings have probably been installed incorrectly.
7. Install or connect the following:

- Fuel injector electrical connectors
- Fuel pipe with new gaskets. Tighten the bolts to 25 ft. lbs. (34 Nm) and the delivery pipes-to-intake manifold bolts to 10 ft. lbs. (13 Nm).
- Fuel pipe union with new gaskets. Tighten the clamp bolt to 71 inch lbs. (8 Nm).
- Fuel pressure regulator

8. Inspect the vacuum lines and connections. Look for any loose connections, sharp bends or damage.
9. Install or connect the following:
   - Air cleaner
   - Air cleaner hose
10. Start the engine and check for vacuum and fuel leaks.

4.0L Engine

1. Before servicing the vehicle, refer to the Precautions Section.
2. Relieve the fuel system pressure.
3. Disconnect the negative battery cable.
4. Drain the engine coolant. Remove the V bank cover.
5. Remove the air cleaner assembly.
6. Remove the intake manifold.
7. Disconnect and plug the fuel line hoses.
8. Disconnect the fuel injector connectors. Remove the six bolts and the fuel rail together with the injectors.
9. Pull the injectors out of the fuel rail.

To install:

10. Installation is the reverse of the removal procedure.
11. Be sure to use new gaskets and O-rings, as required.
12. Be sure to fill the cooling system with the proper grade and type engine coolant.
13. Start the engine and check for leaks, correct as required.

4.7L Engine

2002–2004

1. Before servicing the vehicle, refer to the Precautions Section.
2. Relieve the fuel system pressure.
3. Remove or disconnect the following:
   - Negative battery cable
   - Engine appearance cover
   - Air intake tube
   - Fuel lines
   - Fuel pulsation damper
   - Fuel pressure regulator vacuum line
   - Accelerator cable and bracket
   - Positive Crankcase Ventilation (PCV) valve and hose

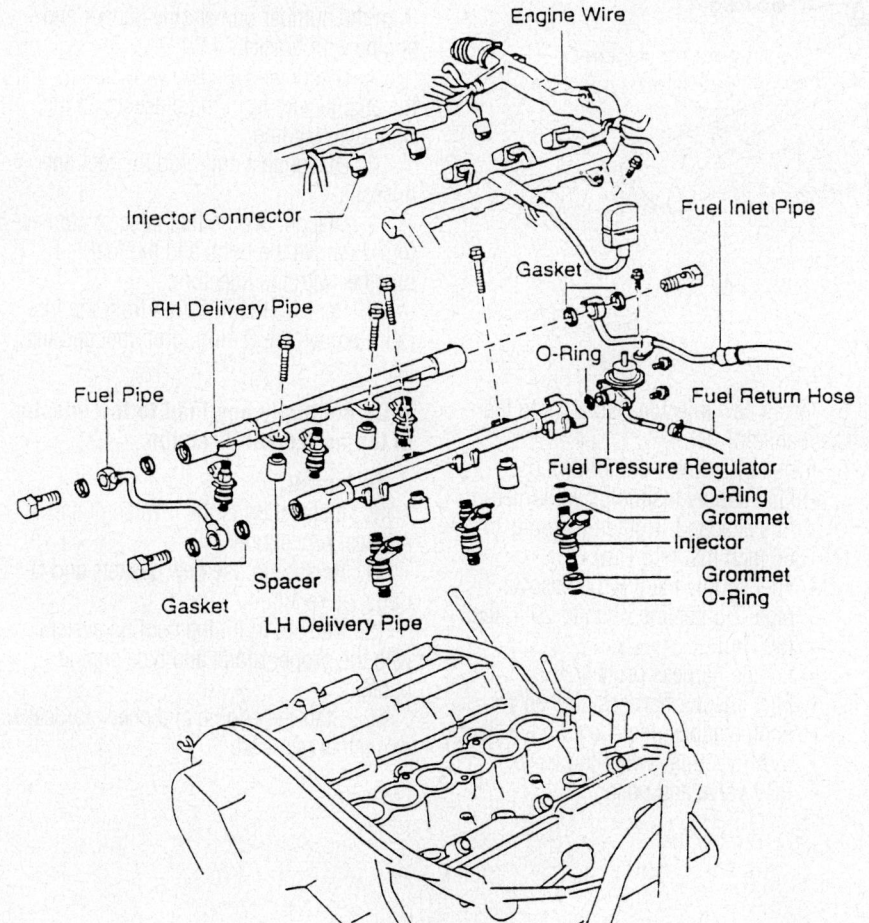

Fuel injector arrangement and related components—3.4L engine

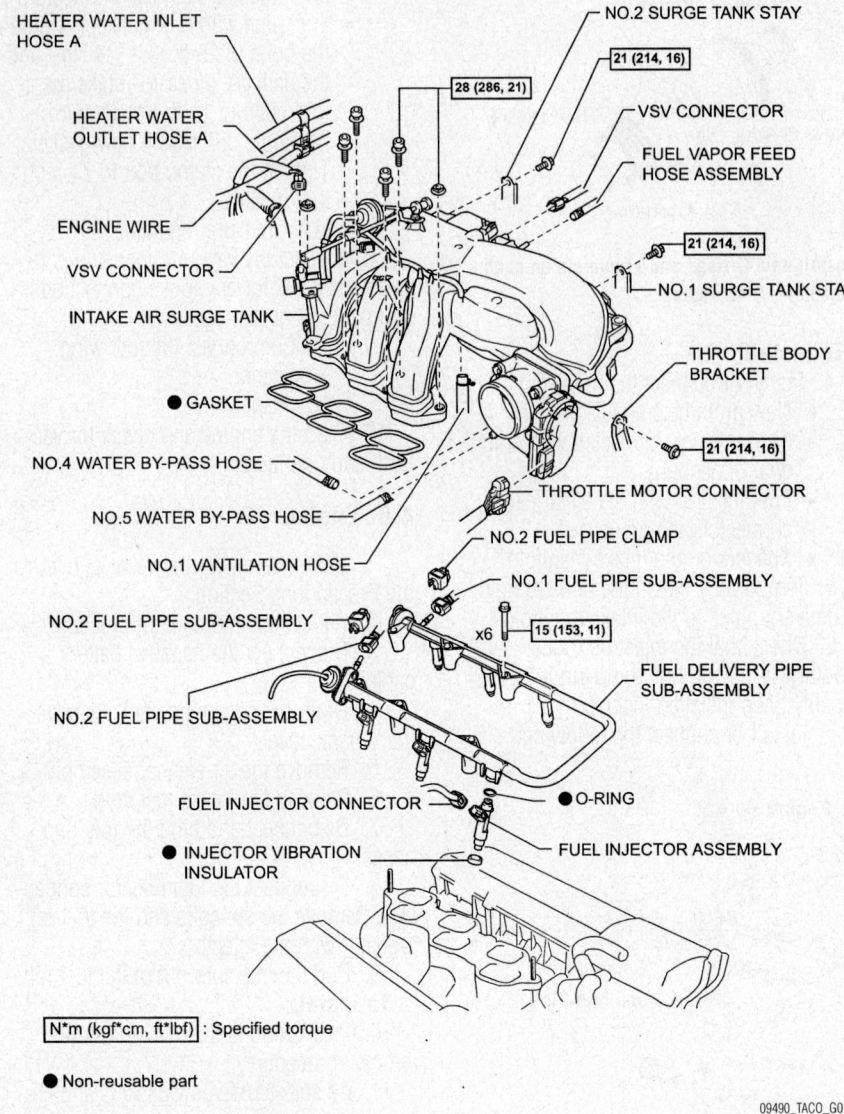

HEATER WATER INLET HOSE A

HEATER WATER OUTLET HOSE A

ENGINE WIRE

VSV CONNECTOR

INTAKE AIR SURGE TANK

● **GASKET**

NO.4 WATER BY-PASS HOSE

NO.5 WATER BY-PASS HOSE

NO.1 VANTILATION HOSE

NO.2 FUEL PIPE SUB-ASSEMBLY

NO.2 FUEL PIPE SUB-ASSEMBLY

FUEL INJECTOR CONNECTOR

● **INJECTOR VIBRATION INSULATOR**

NO.2 SURGE TANK STAY

21 (214, 16)

28 (286, 21)

VSV CONNECTOR

FUEL VAPOR FEED HOSE ASSEMBLY

21 (214, 16)

NO.1 SURGE TANK STAY

THROTTLE BODY BRACKET

21 (214, 16)

THROTTLE MOTOR CONNECTOR

NO.2 FUEL PIPE CLAMP

NO.1 FUEL PIPE SUB-ASSEMBLY

x6   **15 (153, 11)**

FUEL DELIVERY PIPE SUB-ASSEMBLY

● **O-RING**

FUEL INJECTOR ASSEMBLY

N*m (kgf*cm, ft*lbf) : Specified torque

● Non-reusable part

09490_TACO_G0107

Fuel injector and related components—4.0L engine

- Evaporative Emissions (EVAP) vacuum switching valve
- Engine appearance cover brackets
- Fuel injector harness connectors
- Engine harness protector
- Fuel supply manifold crossover pipe
- Fuel supply manifolds with injectors attached
- Fuel injectors

To install:

4. Install the fuel injectors to the supply manifold with new O-ring seals and new grommets.

5. Install new injector insulators to the intake manifold.

6. Install or connect the following:
   - Fuel supply manifolds with injectors attached. Tighten the bolts to 66 inch lbs. (7.5 Nm).
   - Fuel supply manifold crossover pipe. Tighten the bolts to 29 ft. lbs. (39 Nm).
   - Engine harness protector
   - Fuel injector harness connectors
   - Engine appearance cover brackets
   - EVAP vacuum switching valve
   - PCV valve and hose

- Accelerator cable and bracket
- Fuel pressure regulator vacuum line
- Fuel pulsation damper
- Fuel lines
- Air intake tube
- Engine appearance cover
- Negative battery cable

7. Start the engine and check for leaks.

2005–2006

1. Before servicing the vehicle, refer to the Precautions Section.

2. Relieve the fuel system pressure.

3. Disconnect the negative battery cable.

4. Remove the throttle body cover sub-assembly. Remove the air cleaner assembly.

5. Remove the fuel pressure pulsation damper, upper gasket, fuel main hose and lower gasket.

6. Disconnect the PCV hose. Disconnect the VSV connector at the EVAP. Disconnect the EVAP hose. Remove the VSV for the EVAP from the intake manifold.

7. Remove the throttle body cover bracket.

8. Disconnect the engine wire clamps from the number one engine hanger and engine wire bracket.

9. Disconnect the two wire clamps on the engine wire from the brackets on the right delivery line.

10. Disconnect and plug the fuel line hoses.

11. Disconnect the fuel injector connectors. Remove the bolts and the fuel rail together with the injectors.

12. Remove the injectors from the fuel rail. Remove the O-ring, grommet and insulator.

→**Do not apply any load to the injector in the horizontal direction.**

To install:

13. Installation is the reverse of the removal procedure.

14. Be sure to use new gaskets and O-rings, as required.

15. Be sure to fill the cooling system with the proper grade and type engine coolant.

16. Start the engine and check for leaks, correct as required.

DRIVE TRAIN

Manual Transmission

REMOVAL & INSTALLATION

2002–2004

2WD

1. Before servicing the vehicle, refer to the Precautions Section.
2. Remove or disconnect the following:
   - Negative battery cable
3. Drain the transmission oil.
4. Remove the shift lever assembly, as follows:
   - Shift lever knob
   - 4 screws, shift lever boot retainer and shift lever boot
   - 6 bolts, shift lever assembly and baffle
   - Turn over the dust boot
   - Shift lever cap, cover it with a cloth
   - Shift lever cap by pressing downward and rotating it counterclockwise
   - Shift lever
5. Remove or disconnect the following:
   - Driveshaft
   - Vehicle Speed Sensor (VSS) and the back-up light switch connectors
   - Oxygen (O_2) sensor connector
   - Front exhaust pipe from the exhaust manifold and catalytic converter
   - Clutch release cylinder
   - Starter wires
   - Starter
6. Position a jack and wooden block under the transmission.
7. Remove or disconnect the following:
   - Rear endplate
   - Rear engine mount bracket
   - Crossmember
8. Attach an engine hoist to the engine hangers.
9. Remove or disconnect the following:
   - Engine mounts
   - Engine/transmission assembly out of the vehicle
10. Safely support the engine/transmission assembly.
11. Remove or disconnect the following:
    - Transmission-to-engine bolts
    - Transmission mount
To install:
12. Install or connect the following:
    - Transmission
    - Transmission mount. Tighten the 4 bolts to 48 ft. lbs. (65 Nm).

- Tighten the 6 transmission-to-engine bolts to 53 ft. lbs. (72 Nm).
- Crossmember. Torque the bolts to 53 ft. lbs. (72 Nm).
- Rear engine mount-to-crossmember bolts to 13 ft. lbs. (18 Nm).
- Starter. Tighten the 2 bolts to 29 ft. lbs. (39 Nm).
- Rear endplate. Tighten the 4 bolts to 27 ft. lbs. (37 Nm).
13. Install or connect the following:
    - Tighten the engine mounts to 28 ft. lbs. (38 Nm).
    - Starter wires
    - Clutch release cylinder
    - Driveshaft
    - Front exhaust pipe. Tighten the exhaust pipe-to-manifold bolts to 46 ft. lbs. (62 Nm) and the exhaust pipe-to-catalytic converter bolts to 35 ft. lbs. (48 Nm).
    - Remaining components
    - Negative battery cable
14. Install the shift lever assembly, as follows:
    - Shift lever
    - Shift lever cap, cover it with a cloth
    - Shift lever cap by pressing downward and rotating it clockwise
    - Turn over the dust boot
    - 6 bolts, shift lever assembly and baffle
    - 4 screws, shift lever boot retainer and shift lever boot
    - Shift lever knob
15. Fill the transmission with oil.
16. Start the engine and check for leaks.
17. Install the engine undercover.
18. Roadtest the vehicle and check all fluids.

4WD

1. Before servicing the vehicle, refer to the Precautions Section.
2. Disconnect negative battery cable.
3. Remove the transmission shift lever assembly, as follows:
   - Shift lever knob
   - 4 screws, shift lever boot retainer and shift lever boot
   - 6 bolts, shift lever assembly and baffle
   - Turn over the dust boot
   - Shift lever cap, cover it with a cloth
   - Shift lever cap by pressing downward and rotating it counterclockwise
   - Shift lever
   - Transfer shift lever, using snapring

pliers to pull it from the transfer case
4. Drain the transmission and the transfer oil.
5. Remove or disconnect the following:
   - Driveshafts
   - Vehicle Speed Sensor (VSS), back-up light switch connector and the transfer indicator switch connector
   - Clutch release cylinder, move it aside without disconnecting the clutch line
   - Oxygen (O_2) sensor
   - Front exhaust pipe bracket
   - Starter
   - Rear endplate by removing the nuts and 2 bolts
6. Using a transmission jack, support the transmission.
7. Remove or disconnect the following:
   - 4 engine rear mount bolts
   - 8 bolts and the frame crossmember from the side frame
   - 6 transmission-to-engine bolts
   - 3 wire clamps from the transmission
   - Transmission with the transfer case
   - 4 engine rear mount bolts from the transfer case
   - Transfer adapter rear mount bolts
   - Transfer case from the transmission
To install:
8. Apply MP grease to the adapter oil seal and shift the 2 shift fork shafts to the high 4 position.
9. Install or connect the following:
   - Transfer case to the transmission. Tighten the bolts to 17 ft. lbs. (24 Nm).

✵✵ WARNING

Be careful not to damage the oil seal by the input gear spline when installing the transfer.

- Engine rear mounting. Tighten the 4 bolts to 48 ft. lbs. (65 Nm).
- Transmission/transfer case assembly
10. Support the transmission with a jack. Align the input shaft spline with the clutch disc and push the transmission with the transfer fully into position.
11. Install or connect the following:
    - Tighten the engine-to-transmission bolts to 53 ft. lbs. (72 Nm).
    - Crossmember. Tighten the 4 bolts to 53 ft. lbs. (72 Nm) and the 4 engine rear mount bolts to 13 ft. lbs. (18 Nm).

- Stabilizer bar
- Rear endplate Tighten the 2 bolts and nuts to 27 ft. lbs. (37 Nm).
- Starter. Tighten the bolts to 29 ft. lbs. (39 Nm).
- Front exhaust pipe. Tighten the manifold bolts to 46 ft. lbs. (62 Nm) and the converter bolts to 35 ft. lbs. (48 Nm).
- Clutch release cylinder. Tighten the bolts to 9 ft. lbs. (12 Nm).
- Driveshafts
- Remaining components

12. Install the transmission shift lever assembly, as follows:
- Apply MP grease to the shift lever
- Shift lever
- Shift lever cap, cover it with a cloth
- Shift lever cap by pressing downward and rotating it clockwise
- Turn over the dust boot
- 6 bolts, shift lever assembly and baffle
- 4 screws, shift lever boot retainer and shift lever boot
- Shift lever knob

13. Connect the negative battery cable.
14. Start the engine and check for leaks.
15. Roadtest the vehicle for proper operation. Recheck all fluid levels.

2005–2005

1. Before servicing the vehicle, refer to the Precautions Section.
2. Disconnect negative battery cable.
3. Remove the shift lever knob subassembly.
4. Remove the four bolts and the shift lever lock boot retainer. Remove the shift lever boot.
5. To remove the floor shift lever assembly, turn over the dust boot. Press down on the shift lever cap and turn it counterclockwise. Pull out the shift lever.
6. Remove the number one engine under cover subassembly. Drain the transmission.
7. Remove the front exhaust pipe. Remove the driveshaft along with the center bearing assembly.
8. Remove the manifold stay. Remove the starter.
9. Remove the number one clutch housing cover. Remove the clutch release cylinder assembly. Remove the clutch accumulator assembly.
10. Properly support the transmission assembly, using the proper transmission jack.
11. Remove the four set bolts of the engine rear mounting on the crossmember.

Remove the four nuts, washers, bolts and crossmember.
12. Remove the four bolts and the engine rear mounting insulator.
13. Remove the two bolts, nuts and stabilizer bar brackets.
14. Tilt the transmission assembly downward. Disconnect the speedometer sensor connector. Disconnect the backup light switch connector. Disconnect the wiring harness.
15. Remove the nine transmission retaining bolts. Separate and remove the transmission from the engine.

To install:
16. Installation is the reverse of the removal procedure.
17. Tighten the five upper transmission to engine retaining bolts to 53 ft. lbs. Tighten the four lower transmission to engine retaining bolts to 27 ft. lbs.

➡**Fifth upper bolt is located on the lower side of the transmission that has three bolt holes.**

18. Tighten the crossmember bolts to 53 ft. lbs. Tighten the engine rear mounting set bolts to 13 ft. lbs.
19. Be sure to fill the transmission with the proper grade and type fluid.
20. Start the engine and check for leaks.
21. Roadtest the vehicle.

Clutch Assembly

REMOVAL & INSTALLATION

1. Before servicing the vehicle, refer to the Precautions Section.
2. Remove or disconnect the following:
- Negative battery cable
- Transmission assembly

3. Matchmark the clutch cover to the flywheel.
4. At the clutch cover, loosen each bolt 1 turn until spring tension is released.
5. Remove or disconnect the following:
- Clutch cover set bolts and the clutch cover with the clutch disc.
- Release bearing retaining clip and withdraw the it
- Release fork and boot assembly

To install:
6. Install or connect the following:
- Clutch disc onto the flywheel, using a clutch disc alignment tool
- Clutch cover, position it onto the flywheel and if reusing the old pressure plate, align the matchmarks.
- Clutch cover. Tighten the bolts in a

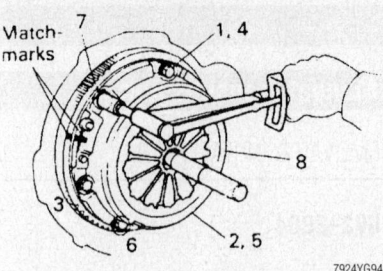

Clutch bolt tightening sequence—2002–2004

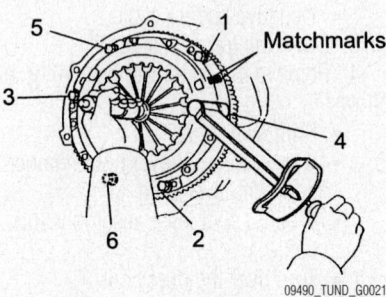

Clutch bolt tightening sequence—2005–2006

crisscross pattern to 14 ft. lbs. (19 Nm).
7. Lubricate the release fork pivot and contact points, the release bearing, bearing hub and input shaft spline surfaces with a suitable molybdenum disulfide lithium based or multi-purpose grease.
8. Install or connect the following:
- Boot, release fork, hub and the bearing assemblies
- Transmission
- Negative battery cable

Hydraulic Clutch System

BLEEDING

1. Before servicing the vehicle, refer to the Precautions Section.
2. Fill the clutch reservoir with brake fluid. Check the reservoir level frequently and add fluid as needed.
3. Connect one end of a vinyl tube to the bleeder plug on the slave cylinder and submerge the other end into a clear container half-filled with brake fluid.
4. Slowly pump the clutch pedal several times.
5. Have an assistant hold the clutch pedal down and loosen the bleeder plug until fluid and/or air starts to run out of the bleeder plug. Close the bleeder plug while the pedal is held to the floor.
6. Repeat Steps 2 and 3 until all the air bubbles are removed from the system.

7. Tighten the bleeder plug when all the air is gone.

8. Refill the master cylinder to the proper level as required.

9. Check the system for leaks.

Automatic Transmission Assembly

REMOVAL & INSTALLATION

3.4L Engine

1. Before servicing the vehicle, refer to the Precautions Section.

2. Remove or disconnect the following:
   - Negative battery cable
   - Throttle cable
   - Transmission dipstick
   - Oil filler tube and discard the O-ring

3. Shift the transfer case into **H4** position.

4. Remove or disconnect the following:
   - Transfer case shift lever knob
   - No. 1 engine under cover
   - Front and center exhaust pipes
   - Driveshaft(s)
   - No. 1 and 2 Vehicle Speed Sensor (VSS) connectors
   - Shift control cable
   - Oil cooler pipe
   - Automatic Transmission Fluid (ATF) sensor connector
   - Park/Neutral Position (PNP) switch
   - Rear endplate
   - Torque converter bolts
   - Crossmember
   - Engine rear mounting insulator
   - Starter
   - Transmission

To install:

5. Install or connect the following:
   - Transmission. Torque the bolts to 53 ft. lbs. (71 Nm).
   - Starter. Torque the bolts to 29 ft. lbs. (39 Nm).
   - Engine rear mounting insulator. Torque the bolts to 48 ft. lbs. (65 Nm).
   - Crossmember. Torque the nuts to 53 ft. lbs. (72 Nm) and the bolts to 13 ft. lbs. (18 Nm)..
   - Torque converter. Torque the bolts to 30 ft. lbs. (41 Nm).
   - Rear endplate. Torque the bolts to 13 ft. lbs. (18 Nm).
   - Park/Neutral Position (PNP) switch
   - Automatic Transmission Fluid (ATF) sensor connector
   - Oil cooler pipe. Torque the bolts to 25 ft. lbs. (34 Nm).

- Shift control cable. Torque the bolts to 9 ft. lbs. (12 Nm).
- No. 1 and 2 Vehicle Speed Sensor (VSS) connectors
- Driveshaft(s)
- Front and center exhaust pipes
- No. 1 engine under cover
- Transfer case shift lever knob
- Oil filler tube using a new O-ring
- Transmission dipstick
- Throttle cable
- Negative battery cable

4.0L and 4.7L Engines

1. Before servicing the vehicle, refer to the Precautions Section.

2. Disconnect the negative battery cable.

3. Raise and support the vehicle safely.

4. Remove the engine number one under cover assembly. Drain the transmission fluid.

5. Remove the front exhaust pipe assembly.

6. Matchmark and remove the driveshaft assembly.

7. Disconnect the NT and SP2 speed sensor connectors.

8. Disconnect the transmission wire connector. Disconnect the neutral safety switch connector.

9. Remove the nut and disconnect the shift control cable. Remove the two bolts and the shift control cable bracket from the transmission.

10. Remove the bolts and clamps retaining the fluid cooler lines. Using tool SST09023-12701 or equivalent, disconnect and plug the lines.

11. Remove the two bolts, nuts and stabilizer bar bushings and brackets. Disconnect the stabilizer bar.

12. On 4.0L engine, remove the starter.

13. On 4.0L engine, remove the flywheel housing side cover. On 4.7L engine remove the rear end plate cover.

14. Remove the torque converter retaining bolts.

15. Properly support the transmission assembly, using the proper transmission jack.

16. Remove the four set bolts of the engine rear mounting on the crossmember. Remove the four nuts, washers, bolts and crossmember.

17. Remove the number one engine rear mounting insulator bolts and the engine rear mounting insulator.

18. Disconnect the wiring harness from the transmission.

19. Lower the rear end of the transmission. Remove the nine transmission retain-

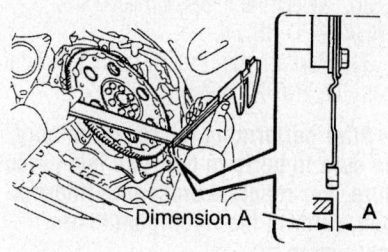

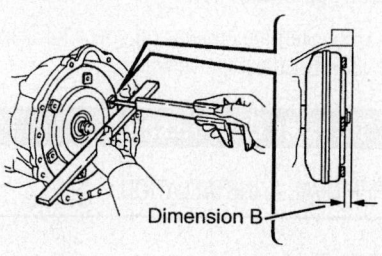

Torque converter installation—2005–2006

ing bolts. Separate and remove the transmission from the engine.

To install:

20. Installation is the reverse of the removal procedure.

21. When installing the torque converter on to the transmission housing, use a calipers and straight edge and measure between the transmission and the end of the surface of the driveplate.

22. Use a calipers and straight edge and measure the dimension "B" shown in the illustration and check that "B" is greater than "A". Specification should be 0.04 inch or more.

23. Tighten the 17mm head transmission to engine retaining bolts to 52 ft. lbs. and the 14mm head bolts to 27 ft. lbs.

24. Tighten the torque converter to flexplate retaining bolts to 35 ft. lbs.

➡**Install the green colored bolt first, then the remaining five bolts.**

25. Be sure to fill the transmission with the proper grade and type fluid.

➡**On 2005–2006 vehicles equipped with the A750E and A750F automatic transmissions, initialization is performed using the intelligent tester tool. Perform the automatic transmission initialization procedure when replacing the automatic transmission, engine or ECM.**

26. Turn the ignition switch OFF.

27. Connect the intelligent tester tool to the DLC3.

28. Turn the ignition switch to the ON position.

29. Push the intelligent tester tool main switch to the ON position.

30. Select the items, DIAGNOSIS/ENHANCED OBD II.

31. Perform the reset memory procedure from the ENGINE menu.

➡**After performing the reset memory, be sure to perform the road test procedure. For roadtest procedure information, refer to the intelligent tester instruction manual.**

32. Start the engine and check for leaks.
33. Roadtest the vehicle.

Transfer Case Assembly

REMOVAL & INSTALLATION

3.4L Engine

1. Before servicing the vehicle, refer to the Precautions Section.
2. Drain the transfer case oil.
3. Shift the transfer shift lever into the **H4** position.
4. Remove or disconnect the following:
   - Transfer case shift lever knob
   - 4 screws, transfer case shift lever boot retainer and the shift lever boot
   - Snapring from the transfer case shift lever and the lever
   - Breather hose from the transfer case
   - Front and rear driveshafts
   - Dynamic damper from the transfer case
   - Crossmember from the rear of the transmission
   - 4 bolts and the engine rear mount from the transfer case adapter
   - Vehicle Speed Sensor (VSS) connector
   - Transfer Detection Switch (TDS) connector
5. Support the transfer case
6. Remove or disconnect the following:
   - 8 transfer case-to-transfer adapter bolts
   - Transfer case

To install:

7. Install or connect the following:
   - Transfer case
   - Transfer case-to-transfer adapter bolts. Torque the 8 bolts to 17 ft. lbs. (24 Nm).
   - Transfer Detection Switch (TDS) connector
   - Vehicle Speed Sensor (VSS) connector
   - Engine rear mount to the transfer case adapter. Torque the 4 bolts to 48 ft. lbs. (65 Nm).

   - Crossmember to the rear of the transmission. Torque the 4 nuts/bolts to 53 ft. lbs. (72 Nm).
   - Crossmember to the chassis. Torque the 4 bolts to 13 ft. lbs. (18 Nm).
   - Dynamic damper to the transfer case. Torque the 2 bolts to 28 ft. lbs. (38 Nm).
   - Front and rear driveshafts
   - Breather hose to the transfer case to a depth of 0.51 in. (13mm) or more
   - Snapring to the transfer case's shift lever
   - 4 screws, transfer case shift lever boot retainer and the shift lever boot
   - Transfer case shift lever knob
8. Refill the transfer case to the correct level.
9. Test drive the vehicle.

4.0L Engine

1. Before servicing the vehicle, refer to the Precautions Section.
2. Shift the transfer case into 2WD.
3. Disconnect the negative battery cable. Drain the transfer case oil.
4. Disconnect the breather hose from the transfer case.
5. Remove the front exhaust pipe assembly.
6. Remove the front driveshaft assembly.
7. Remove the number three crossmember subassembly.
8. Disconnect the transfer case detection switches and motor actuator connectors.
9. Properly support the transfer case assembly, using the proper transmission jack.
10. Remove the eight transfer case mounting bolts and clamps.
11. Carefully pull the transfer case out from the transfer adaptor down and toward the rear.

To install:

12. Installation is the reverse of the removal procedure.
13. Tighten the transfer case retaining bolts to 17 ft. lbs.
14. Be sure to fill the transfer case with the proper grade and type oil.
15. Start the engine and check for leaks.
16. Roadtest the vehicle.

4.7L Engine

2002–2004

1. Before servicing the vehicle, refer to the Precautions Section.

2. Drain the transfer case oil.
3. Turn the touch select 2–4 switch **ON**.
4. Remove or disconnect the following:
   - Breather hose from the transfer case
   - Left and right exhaust pipes
   - Front and rear driveshafts
   - Crossmember from the rear of the transmission
   - 4 bolts and the engine rear mount from the transfer case adapter
   - Vehicle Speed Sensor (VSS) connector
   - Transfer Detection Switch (TDS) connectors
   - Motor actuator connectors
5. Support the transfer case
6. Remove or disconnect the following:
   - 8 transfer case-to-transfer adapter bolts
   - Transfer case

To install:

7. Install or connect the following:
   - Transfer case
   - Transfer case-to-transfer adapter bolts. Torque the 8 bolts to 17 ft. lbs. (24 Nm).
   - Motor actuator connectors
   - Transfer Detection Switch (TDS) connector
   - Vehicle Speed Sensor (VSS) connector
   - Engine rear mount to the transfer case adapter. Torque the 4 bolts to 48 ft. lbs. (65 Nm).
   - Crossmember to the rear of the transmission. Torque the 4 nuts/bolts to 53 ft. lbs. (72 Nm).
   - Crossmember to the chassis. Torque the 4 bolts to 13 ft. lbs. (18 Nm).
   - Dynamic damper to the transfer case. Torque the 2 bolts to 28 ft. lbs. (38 Nm).
   - Front and rear driveshafts
   - Left and right exhaust pipes
   - Breather hose to the transfer case to a depth of 0.51 in. (13mm) or more
8. Refill the transfer case to the correct level.
9. Test drive the vehicle.

2005–2006

1. Before servicing the vehicle, refer to the Precautions Section.
2. Shift the transfer case into 2WD.
3. Disconnect the negative battery cable. Drain the transfer case oil.
4. Disconnect the breather hose from the transfer case.
5. Remove the front exhaust pipe assembly.

6. Remove the front driveshaft assembly.

7. Remove the number three crossmember subassembly.

8. Disconnect the transfer case detection switches and motor actuator connectors.

9. Properly support the transfer case assembly, using the proper transmission jack.

10. Remove the eight transfer case mounting bolts and clamps.

11. Carefully pull the transfer case out from the transfer adaptor down and toward the rear.

To install:

12. Installation is the reverse of the removal procedure.

13. Tighten the transfer case retaining bolts to 17 ft. lbs.

14. Be sure to fill the transfer case with the proper grade and type oil.

15. Start the engine and check for leaks.

16. Roadtest the vehicle.

Halfshaft

REMOVAL & INSTALLATION

1. Before servicing the vehicle, refer to the Precautions Section.
2. Remove or disconnect the following:
   - Front wheel
   - Under cover
3. Drain the differential oil.
4. Remove or disconnect the following:
   - Grease cap
   - Cotter pin and lock cap
   - Halfshaft locknut by applying the brakes
   - Lower control arm from the lower ball joint
   - Halfshaft from the steering knuckle, using a plastic hammer
   - Left shock absorber, for the left halfshaft
   - Right halfshaft, using a brass bar and a hammer
   - Left halfshaft, using tools 09520-01010 and 09520-24010
   - Snapring from the inboard joint shaft

To install:

5. Install or connect the following:
   - New snapring, onto the inboard joint shaft with the opening facing downward
   - Halfshafts to the differential using a brass bar and a hammer
   - Halfshafts to the steering knuckles

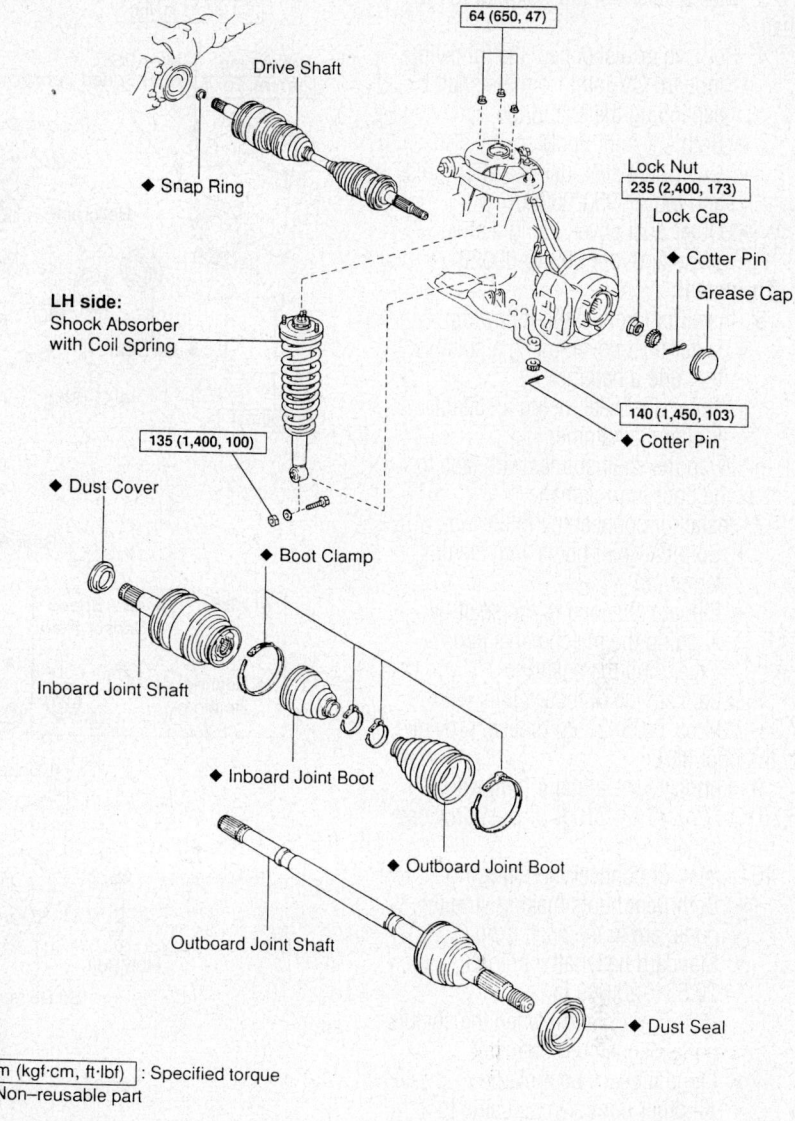

Lock Nut
235 (2,400, 173)

64 (650, 47)

Drive Shaft

♦ Snap Ring

Lock Cap

♦ Cotter Pin

Grease Cap

♦ Cotter Pin

140 (1,450, 103)

LH side:
Shock Absorber with Coil Spring

135 (1,400, 100)

♦ Dust Cover

♦ Boot Clamp

Inboard Joint Shaft

♦ Inboard Joint Boot

♦ Outboard Joint Boot

Outboard Joint Shaft

♦ Dust Seal

N·m (kgf·cm, ft·lbf) : Specified torque
N ♦ Non–reusable part

9308YG12

Front halfshaft and related components

❊❊ WARNING

Be careful not to damage the oil seal, boot or dust seal.

- Lower control arm to the lower ball joint using a new cotter pin. Torque the ball joint nut to 103 ft. lbs. (140 Nm).
- Left shock absorber, for the left halfshaft
- Halfshaft locknut by applying the brakes. Torque the nut to 173 ft. lbs. (235 Nm).
- Lock cap and a new cotter pin
- Grease cap
6. Refill the differential with oil.
7. Install or connect the following:
   - Under cover
   - Front wheel

CV-Joints

OVERHAUL

Outer CV-joint

The outer CV-joint is serviced with the axle shaft as an assembly. The outer CV-joint boot can be serviced by removing the inner CV-joint.

Inner CV-joint

1. Before servicing the vehicle, refer to the Precautions Section.
2. Remove or disconnect the following:
   - Halfshaft from the vehicle
   - Large boot clamps
   - Small boot clamps

3. Matchmark inboard CV-joint to the shaft

4. Remove or disconnect the following:
- Inboard CV-joint from the shaft by expanding the snapring
- Both CV-joint boots
- Outer dust seal, using a shop press and tool 09950-00020
- Outer dust cover, using a shop press and tool 09950-00020

To install:

5. Install or connect the following:
- Outer dust cover, using a suitable tool and a hammer
- Outer dust seal, using a suitable tool and a hammer

6. Wrap the shaft splines with tape to protect the boot from damage.

7. Install or connect the following:
- Both CV-joint boots with clamps, temporarily
- Inboard CV-joint to the shaft by aligning the matchmarks and expanding the snapring

8. Lubricate the outboard joint with 7.23–7.94 oz. (205–225g) grease, provided in the boot kit.

9. Lubricate the inboard joint with 6.70–7.41 oz. (190–210g) grease, provided in the boot kit.

10. Install or connect the following:
- Both joint boots making sure the boots are in the shaft groove
- Standard halfshaft length is 20.531–20.689 in. (521.5–525.5mm) when the shaft is not expanded or contracted
- Large inboard boot clamp
- All other boot clamps using tool 09521-24010. Tighten the crimping tool until the clamp clearance is 0.039–0.059 in. (1.0–1.5mm)
- Halfshaft

Rear Axle Shaft, Bearing and Seal

REMOVAL & INSTALLATION

1. Before servicing the vehicle, refer to the Precautions Section.

2. Remove or disconnect the following:
- Rear wheel
- Brake drum and gasket

3. Remove or disconnect the following:
- Anti-lock Brake System (ABS) speed sensor from the rear axle housing, if equipped
- Brake line from the wheel cylinder, using tool 09023-00100

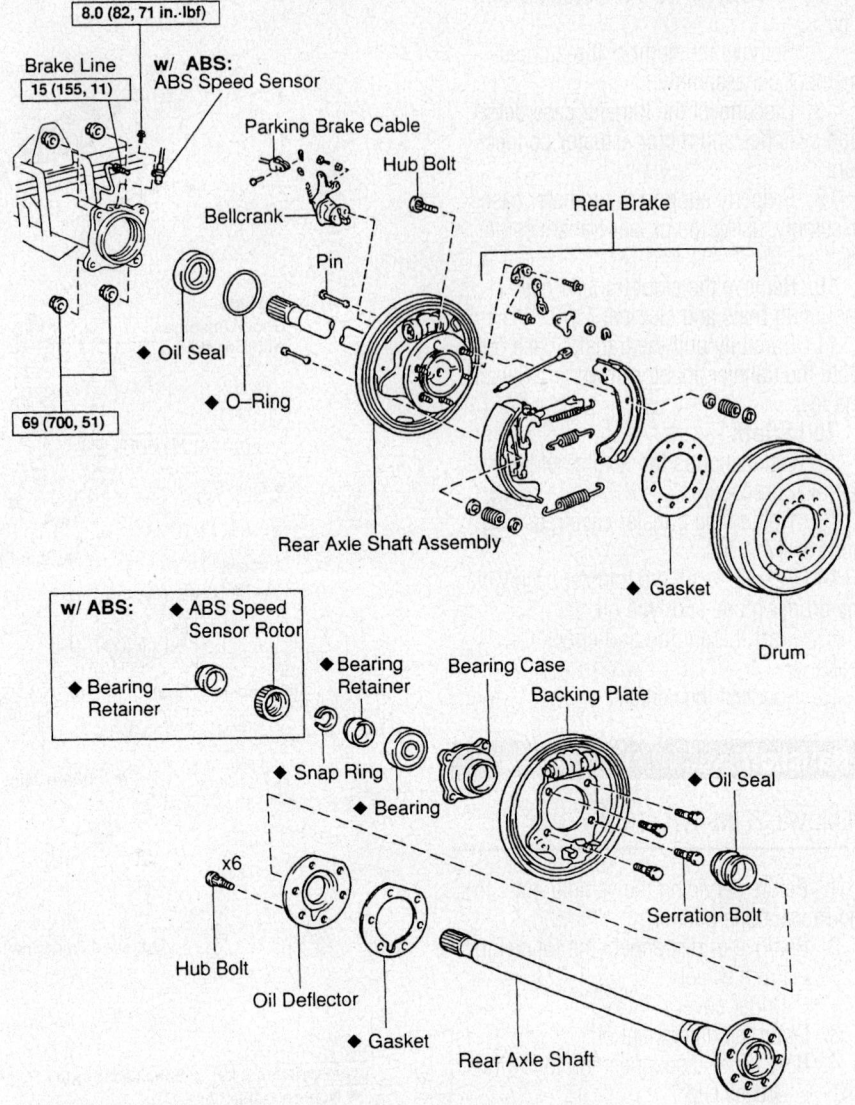

Rear axle and related components

- Parking brake cable
- 4 backing plate nuts
- Axle shaft assembly, by pulling it from the axle housing

> ☆☆ **WARNING**
>
> **Be careful not to damage the oil seal.**

- O-ring from the rear axle housing
- Inner side oil seal using tool 09308-00010

4. If equipped with ABS, perform the following:

a. Remove and discard the 4 serration bolt nuts; then, using a hammer, drive the bolts from the backing plate.

b. Using a grinder, grind the retainer and sensor rotor surfaces; then, chisel them out.

5. Remove the snapring from the axle shaft.

6. Remove the axle shaft from the backing plate, as follows:

a. Position tool 09521-25011 onto the backing plate with the 4 nuts.

b. Using a shop press, remove the axle shaft and bearing retainer from the backing plate.

7. Using tool 09308-00010, pull the oil seal from the backing plate.

8. Using a shop press and tools 09223-56010 and 09950-60010, press the bearing from the backing plate.

To install:

9. Install or connect the following:
   - Bearing into the backing plate, using a shop press and tools 09223-56010 and 09950-60010
   - New O-ring to the rear axle housing
   - New oil seal into the backing plate, using a hammer and tools 09950-70010 and 09950-60010

10. Install the axle shaft to the backing plate, as follows:
    - New outer side seal, lubricate the oil seal lip with multi-purpose grease
    - Backing plate and bearing retainer onto the rear axle shaft
    - Axle shaft onto the backing plate, by pressing it using a shop press and tool 09316-60011
    - New snapring

✳✳ WARNING

Be careful not to damage the oil seal.

11. Install or connect the following:
    - New sensor rotor and new bearing retainer onto the axle shaft, using a shop press and tool 09316-60011 to a standard length of 4.77–4.85 in. (121.2–123.2mm), if equipped with ABS
    - New inner side oil seal, using a hammer and tools 09950-60020 and 09950-70010
    - Axle shaft assembly. Torque the bolts to 51 ft. lbs. (69 Nm).

✳✳ WARNING

Be careful not to damage the oil seal.

- Parking brake cable
- Brake line to the wheel cylinder, using tool 09023-00100. Torque the brake line to 11 ft. lbs. (15 Nm).
- Rear brake assembly
- ABS speed sensor to the rear axle housing. Torque it to 7.1 ft. lbs. (8.0 Nm).

12. Using a dial indicator, check the bearing backlash and the axle shaft deviation. If the bearing backlash exceeds a maximum or 0.028 in. (0.7mm), replace it. If the axle shaft deviation exceeds the maximum of 0.004 in. (0.1mm), replace it.

13. Install or connect the following:
    - New gasket and brake drum
    - Rear wheel. Torque the lug nuts to 81 ft. lbs. (110 Nm) on 2002–2004 vehicles and to 83 ft. lbs. (112 Nm) on 2005–2006 vehicles.

14. Bleed the brake system.

15. Check the ABS speed sensor signal.

Pinion Seal

REMOVAL & INSTALLATION

Front

1. Before servicing the vehicle, refer to the Precautions Section.

2. Remove the under cover.

3. Drain the differential housing oil.

4. Remove the front driveshaft.

5. Remove the companion flange, as follows:
   - Loosen the staked part of the nut, using a chisel and a hammer
   - Companion flange nut, using tool 09330-00021
   - Companion flange, using tools 09950-30011 and 09954-03010

6. Remove the oil seal and slinger, as follows:
   - Oil seal, using tool 09308-10010
   - Oil slinger

To install:

7. Install or connect the following:
   - Oil slinger
   - New oil seal, using a hammer and tool 09554-22010 to a depth of 0.165–0.189 in. (4.2–4.8mm).

8. Install the companion flange, as follows:
   - Companion flange
   - New nut, lubricated with hypoid gear oil
   - Torque the nut to 80 ft. lbs. (108 Nm), using tool 09330-00021.

9. Adjust the drive pinion preload

10. Rotate the drive pinion, using a torque wrench while tightening the flange

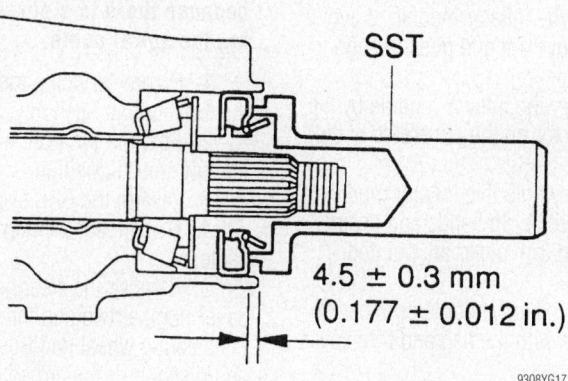

| N·m (kgf·cm, ft·lbf) : Specified torque
◆ Non–reusable part

9308YG16

Front differential and related components

SST

4.5 ± 0.3 mm
(0.177 ± 0.012 in.)

9308YG17

Positioning the front pinion seal in the differential housing

nut to make sure the bearing preload is 10.4–16.5 inch lbs. (1.2–1.9 Nm) for a new bearing or 5.2–8.7 inch lbs. (0.6–1.0 Nm) for a used bearing. Tighten the flange nut to achieve the preload torque readings originally recorded.

※ CAUTION

Never loosen the pinion nut to reduce bearing preload.

11. Install or connect the following:
   • Drive pinion nut, stake it
   • Front driveshaft. Tighten the fasteners to 54 ft. lbs. (74 Nm).
   • Under cover
12. Fill the differential with gear lubricant and check for leaks.

Rear

1. Before servicing the vehicle, refer to the Precautions Section.
2. Drain the differential housing oil.
3. Remove the rear driveshaft.

4. Remove the companion flange, as follows:
   • Loosen the staked part of the nut, using a chisel and a hammer
   • Companion flange nut, using tool 09330-00021
   • Companion flange, using tools 09950-30011 and 09954-03010
   • Oil seal, using tool 09308-10010

To install:

5. Install the new oil seal until it is flush with the housing, using a plastic hammer and tools 09316-12010 and 09649-17010

➡ **Use vinyl tape to connect both oil seal installation tools.**

6. Install the companion flange, as follows:
   • Companion flange
   • New nut, lubricated with hypoid gear oil
   • Torque the nut to 109 ft. lbs. (147 Nm), using tool 09330-00021.

7. Adjust the drive pinion preload
8. Rotate the drive pinion, using a torque wrench while tightening the flange nut to make sure the bearing preload is 11.4–16.7 inch lbs. (1.3–1.9 Nm) for a new bearing or 4.3–6.9 inch lbs. (0.5–0.8 Nm) for a used bearing. On 2005–2006 vehicles with LSD, 5.3–8.0 inch lbs. (0.6–0.9 Nm) for a used bearing. Tighten the flange nut to achieve the preload torque readings originally recorded.

※※ CAUTION

Never loosen the pinion nut to reduce bearing preload.

9. Install or connect the following:
   • Drive pinion nut, stake it
   • Rear driveshaft. Tighten the fasteners to 54 ft. lbs. (74 Nm) on 2002–2004 vehicles and to 65 ft. lbs. (88 Nm).
10. Refill the differential with gear lubricant and check for leaks.

STEERING

Air Bag

※※ CAUTION

These vehicles are equipped with an air bag system. The system must be disarmed before performing service on, or around, system components, the steering column, instrument panel components, wiring and sensors. Failure to follow the safety precautions and the disarming procedure could result in accidental air bag deployment, possible injury and unnecessary system repairs.

PRECAUTIONS

Several precautions must be observed when handling the inflator module to avoid accidental deployment and possible personal injury.
   • Never carry the inflator module by the wires or connector on the underside of the module.
   • When carrying a live inflator module, hold securely with both hands and ensure that the bag and trim cover are pointed away.
   • Place the inflator module on a bench or other surface with the bag and trim cover facing up.
   • With the inflator module on the bench,

never place anything on or close to the module which may be thrown in the event of an accidental deployment.

DISARMING

To avoid personal injury when working on vehicles equipped with an air bag, the negative battery cable must be disconnected and at least 90 seconds must elapse before working on the system. Failure to do so may result in deployment of the air bag.

Power Steering Gear

REMOVAL & INSTALLATION

➡ **Remove the steering wheel assembly before removing the steering gear, because there is a possibility of breaking the spiral cable.**

1. Before servicing the vehicle, refer to the Precautions Section.
2. Position the front wheels in the straight-ahead position.
3. Disarm the SRS system.
4. Disconnect the negative battery cable.
5. Remove the steering wheel lower cover number two and three. Carefully pull the steering wheel pad from its mounting and disconnect the airbag connectors and horn terminal.

➡ **If the airbag connector is disconnected with the ignition switch in the ON or ACC position, DTC codes will be set.**

6. Remove the steering wheel pad. When storing the pad assembly, keep the upper surface of the pad facing upward. Never disassemble the steering wheel pad.

➡ **When removing the pad, take care not to pull on the airbag wire harness.**

7. Remove the steering wheel locknut. Using a steering wheel removal tool, remove the steering wheel from the vehicle. Disconnect the connector.
8. Remove the left and right outer tie-rod ends from the steering knuckles.
9. Matchmark the No. 2 intermediate shaft to the steering gear input shaft.
10. Remove or disconnect the following:
   • Clamp plate
   • Pressure feed and return tubes from the power steering gear, using tool 09631-22020
   • Power steering gear assembly

To install:

11. Install or connect the following:
   • Power steering gear assembly. Torque the set bolt to 123 ft. lbs. (165 Nm) and the set nut/bolt to 96 ft. lbs. (91 Nm).
   • Pressure feed and return tubes to the power steering gear, using tool 09631-22020

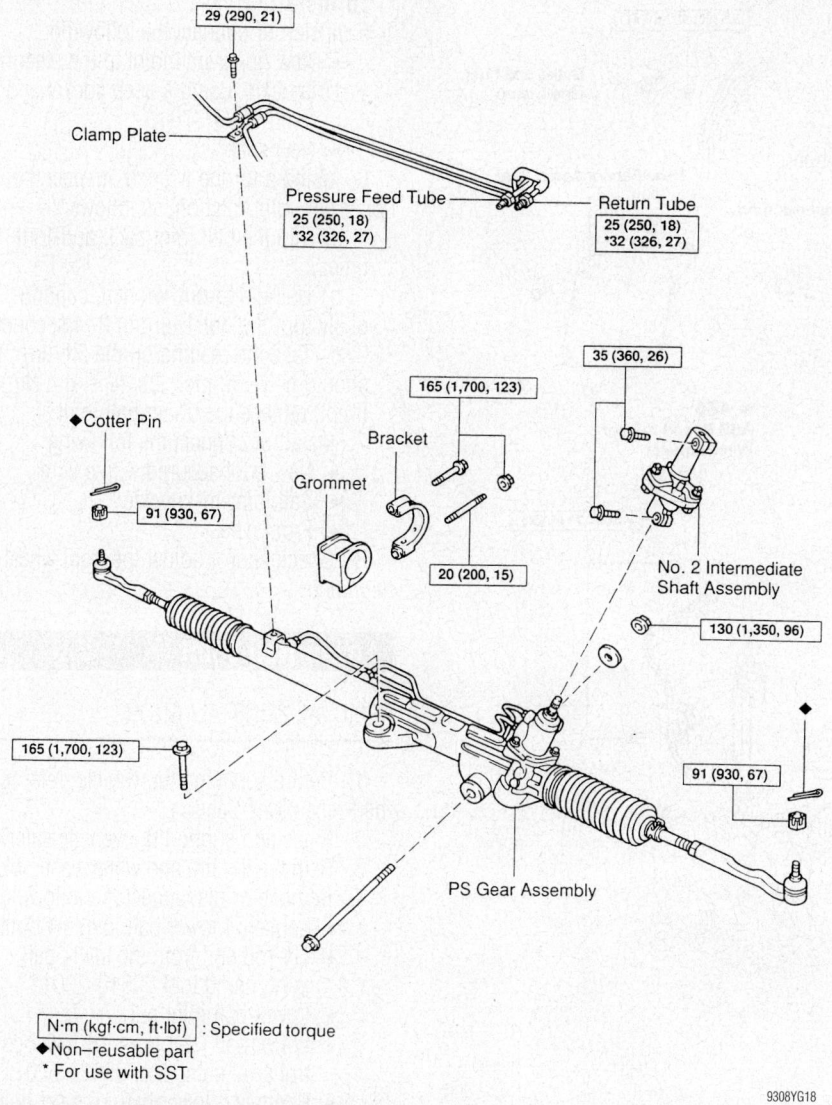

29 (290, 21)

Clamp Plate

Pressure Feed Tube
25 (250, 18)
*32 (326, 27)

Return Tube
25 (250, 18)
*32 (326, 27)

35 (360, 26)

165 (1,700, 123)

◆Cotter Pin

Bracket

Grommet

91 (930, 67)

20 (200, 15)

No. 2 Intermediate
Shaft Assembly

130 (1,350, 96)

165 (1,700, 123)

91 (930, 67)

◆

PS Gear Assembly

N·m (kgf·cm, ft·lbf) : Specified torque
◆Non–reusable part
* For use with SST

9308YG18

Power steering gear and related components

- Clamp plate. Torque the bolt to 21 ft. lbs. (29 Nm).
- No. 2 intermediate shaft to the steering gear input shaft
- Left and right outer tie-rod ends to the steering knuckles. Torque the nuts to 67 ft. lbs. (91 Nm).

12. To center the spiral cable, check that the front tires are in the straight ahead position. Turn the cable counterclockwise by hand until it feels firm. Rotate the cable clockwise about 2½ turns to align the marks.

➡**The cable will rotate about 2½ turns to both the right and left from the center.**

13. Continue the installation in the reverse order of the removal procedure.

14. Tighten the steering wheel retaining locknut to 26 ft. lbs. on 2002–2004 vehicles and to 37 ft. lbs. on 2005–2006 vehicles.

15. Fill and bleed the power steering system.

16. Check and/or adjust the wheel alignment, as necessary.

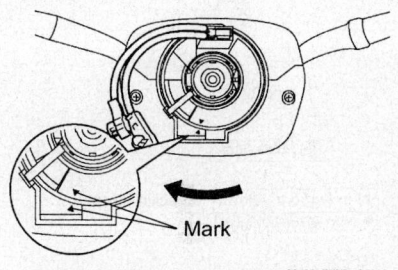

09490_TUND_G0009

Spiral cable alignment—2005–2006

Mark

FRONT SUSPENSION

Shock Absorber

REMOVAL & INSTALLATION

1. Before servicing the vehicle, refer to the Precautions Section.
2. Raise and support the vehicle safely.
3. Remove the tire and wheel assembly.
4. Remove the shock absorber lower retaining nut.

➡**Wrap the bolt head with tape, to prevent the bolt head from damaging the halfshaft boot.**

5. Pry down on the suspension arm and remove the retaining bolt.
6. Remove the three upper shock absorber retaining nuts.

7. Remove the shock absorber from the vehicle.

To install:

8. Installation is the reverse of the removal procedure.
9. Tighten the upper nuts to 47 ft. lbs. (64 Nm).
10. Tighten the lower bolt to 100 ft. lbs. (135 Nm).

ASSEMBLY & DISASSEMBLY

1. Before servicing the vehicle, refer to the Precautions Section.
2. Remove the shock absorber from the vehicle.
3. Position the shock absorber in a suitable holding fixture. Using tool 09727-30021, compress the coil spring.

➡**Do not use an impact wrench. It will damage the tool.**

4. While holding the shock absorber rod, remove the nut.

➡**Do not use an impact wrench. It will damage the shock absorber rod.**

5. Remove the cushion retainer. Remove the number one cushion. Remove the cushion retainer.
6. Remove the coil spring upper insulator. Remove the cushion retainer.
7. Remove the coil spring.
8. Inspect all components for wear and damage. Replace as required. Use new locknuts, as required.
9. Assembly is the reverse of the disassembly procedure.

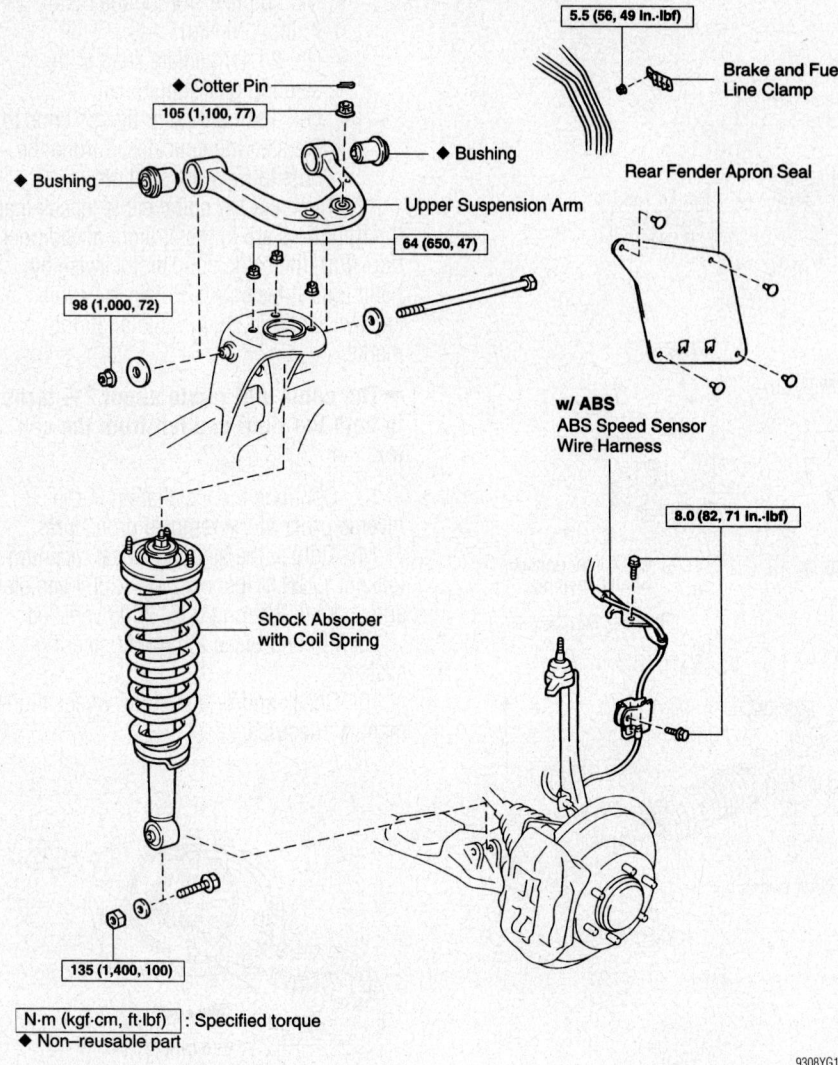

5.5 (56, 49 in.-lbf)

Brake and Fuel Line Clamp

◆ Cotter Pin

105 (1,100, 77)

◆ Bushing

◆ Bushing

Upper Suspension Arm

64 (650, 47)

98 (1,000, 72)

Rear Fender Apron Seal

w/ ABS
ABS Speed Sensor Wire Harness

8.0 (82, 71 in.-lbf)

Shock Absorber with Coil Spring

135 (1,400, 100)

N·m (kgf·cm, ft·lbf) : Specified torque
◆ Non–reusable part

9308YG19

Front suspension and related components

10. Align the suspension support and the absorber brush, as shown in the illustration.

Stabilizer Bar

REMOVAL & INSTALLATION

1. Before servicing the vehicle, refer to the Precautions Section.
2. Raise and support the vehicle safely.
3. Remove the tire and wheel assembly.
4. Remove the two nuts and disconnect the stabilizer bar links from the lower suspension arms.
5. Remove the two bolts, nuts, and stabilizer bar with the cushions and brackets.
6. Remove the two brackets and cushions from the stabilizer bar.

To install:
7. Installation is the reverse of the removal procedure.

8. Tighten the upper stabilizer bar link nuts to 14 ft. lbs. (19 Nm).
9. Tighten the bushing retainer nuts to 27 ft. lbs. (37 Nm).
10. Tighten the lower stabilizer bar link nuts to 51 ft. lbs. (69 Nm).

Upper Ball Joint

REMOVAL & INSTALLATION

1. Before servicing the vehicle, refer to the Precautions Section.
2. Raise and support the vehicle safely.
3. Remove the tire and wheel assembly.
4. Remove or disconnect the following:
   - Steering knuckle with the axle hub
   - Wire and boot
   - Snapring
   - Upper ball joint from the steering knuckle, using a deep socket wrench and tool 09050-40011

To install:
5. Install or connect the following:
   - New upper ball joint to the steering knuckle, using a deep socket and tool 09309-37010
   - New snapring
6. Using a torque wrench, inspect the upper ball joint rotation, as follows:
   a. Flip the ball joint back-and-forth 5 times.
   b. Using a torque wrench, continuously turn the nut 1 turn in 2–4 seconds.
   c. Take the reading on the 5th turn; it should be 6–39 inch lbs. (0.7–4.4 Nm). If not, replace the upper ball joint.
7. Install or connect the following:
   - New boot secured with a wire
   - Ball joint-to-knuckle
   - Front wheel
8. Check and/or adjust the front wheel alignment.

Lower Ball Joint

REMOVAL & INSTALLATION

1. Before servicing the vehicle, refer to the Precautions Section.
2. Raise and support the vehicle safely.
3. Remove the tire and wheel assembly.
4. Remove or disconnect the following:
   - Loosen 4 lower ball joint set bolts
   - Tie-rod end from the lower ball joint, using tool 09610-20012
   - Lower ball joint nut.
   - Lower ball joint from the lower control arm, using tool 09628-62011
   - Remove 4 lower ball joint set bolts

➡ **Be sure to properly support the upper suspension arm and steering knuckle assembly.**

To install:
5. Install or connect the following:
   - New lower ball joint to the lower control. Torque the bolts to 103 ft. lbs. (140 Nm) on 2002–2004 vehicles and to 117 ft. lbs. (159 Nm) on 2005–2006 vehicles.
   - New cotter pin
   - Tie-rod end to the lower ball joint. Torque the nut to 67 ft. lbs. (91 Nm).
   - Lower ball joint set bolts. Torque the 4 bolts to 59 ft. lbs. (80 Nm) on 2002–2004 vehicles and to 48 ft. lbs. (65 Nm) on 2005–2006 vehicles.
   - Front wheel
6. Check and/or adjust the front wheel alignment.

Upper Control Arm

REMOVAL & INSTALLATION

1. Before servicing the vehicle, refer to the Precautions Section.
2. Raise and support the vehicle safely.
3. Remove the tire and wheel assembly.
4. Remove or disconnect the following:
   - Shock absorber
   - Wheel speed sensor harness, if equipped with Anti-lock Brake System (ABS)
5. Upper ball joint, as follows:
   - Cotter pin and loosen the nut
   - Upper ball joint from the upper control arm, using tool 09950-40011
   - Steering knuckle, support it securely
   - Upper ball joint nut
6. Remove or disconnect the following:
   - 4 clips and the fender apron seal
   - Brake/fuel line clamp nut and clamp
   - Both upper control arm-to-chassis nuts/bolts
   - Upper control arm

To install:

7. Install or connect the following:
   - Upper control arm. Torque both upper control arm-to-chassis nuts/bolts to 72 ft. lbs. (98 Nm).
   - Brake/fuel line clamp nut and clamp. Torque the clamp nut to 49 inch lbs. (5.5 Nm).
   - Fender apron seal
   - Upper ball joint. Torque the nut to 77 ft. lbs. (105 Nm).
   - New cotter pin
   - Steering knuckle
   - Wheel speed sensor harness, if equipped with Anti-lock Brake System (ABS). Torque it to 71 inch lbs. (8.0 Nm).
   - Shock absorber
   - Front wheel
8. Check and/or adjust the wheel alignment.

UPPER CONTROL ARM BUSHING REPLACEMENT

1. Before servicing the vehicle, refer to the Precautions Section.
2. Remove the upper control arm from the vehicle.
3. Remove the control arm bushings, as follows:
   - Pry up the bushing flange, using a chisel and a hammer
   - Press the bushing(s) from the upper control arm, using a shop press and tools 09613-26010, 09631-20060 and 09950-00020

To install:

4. Lubricate the new control arm bushings with liquid soap.
5. Press the bushings into the control arm until the bushing flange contacts the housing edge of the control arm, using a shop press, a steel plate and tools 09631-12090 and 09710-30021
6. Install the upper control arm to the vehicle.
7. Check and/or adjust the wheel alignment.

Lower Control Arm

REMOVAL & INSTALLATION

1. Before servicing the vehicle, refer to the Precautions Section.
2. Raise and support the vehicle safely.
3. Remove the tire and wheel assembly.
4. Disconnect the tie-rod end, as follows:
   - Cotter pin and nut
   - Tie-rod end from the lower ball joint, using tool 09610-20012
5. Remove or disconnect the following:
   - Power steering gear set bolts and nuts
   - Stabilizer bar link from the lower control arm
   - Shock absorber from the lower control arm.
6. Disconnect the lower ball joint, as follows:
   - Cotter pin and nut
   - Lower ball joint from the lower control arm
7. Matchmark both front and rear cam plates and chassis frame.
8. Remove the lower control arm while slightly shifting the power steering gear rearward.

To install:

9. Install or connect the following:
   - Lower control arm while slightly shifting the power steering gear rearward
   - Align both front and rear cam plates and chassis frame matchmarks. Torque both bolts to 96 ft. lbs. (130 Nm).
10. Connect the lower ball joint, as follows:
    - Lower ball joint to the lower control arm. Torque the nut to 103 ft. lbs. (140 Nm).
    - New cotter pin
11. Install or connect the following:
    - Shock absorber to the lower control arm. Torque the nut/bolt to 100 ft. lbs. (135 Nm).
    - Stabilizer bar link to the lower control arm. Torque the nut to 51 ft. lbs. (69 Nm).
    - Power steering gear set bolts and nuts. Torque the set bolt and clamp nut/bolt to 122 ft. lbs. and the set nut/bolt to 96 ft. lbs. (130 Nm)
    - Tie-rod end to the lower ball joint. Torque the nut to 67 ft. lbs. (91 Nm).
    - New cotter pin
    - Front wheel
12. Check and/or adjust the wheel alignment.

LOWER CONTROL ARM BUSHING REPLACEMENT

1. Before servicing the vehicle, refer to the Precautions Section.
2. Remove the lower control arm from the vehicle.

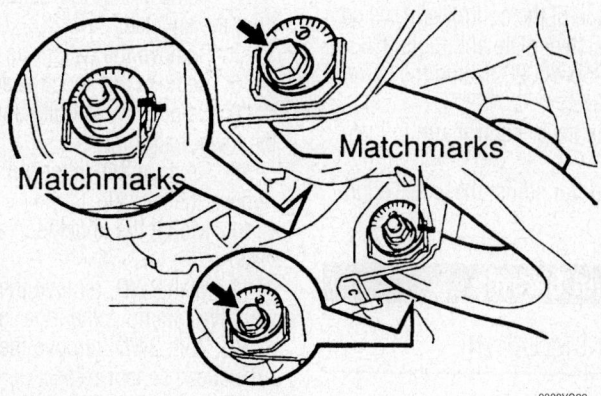

9308YG22

Lower control arm cam plate alignment

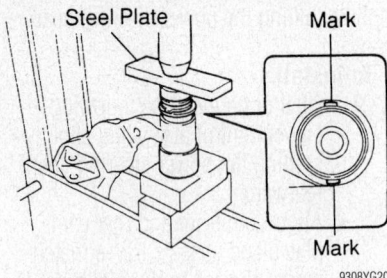

Lower control arm number one bushing installed direction

9308YG20

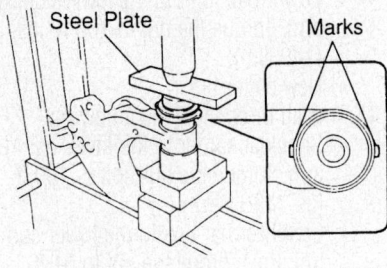

Lower control arm number two bushing installed direction

9308YG21

3. Remove the control arm bushings, as follows:
- Pry up the bushing flange, using a chisel and a hammer
- Press the bushing(s) from the upper control arm, using a shop press and tools 09613-26010, 09632-36010 and 09950-00020

To install:

4. Lubricate the new control arm bushings with liquid soap.

5. Press the No. 1 bushing into the control arm until the bushing flange contacts the housing edge of the control arm, using a shop press, a steel plate and tools 09631-12090 and 09502-12010, facing the correct direction.

6. Press the No. 2 bushing into the control arm until the bushing flange contacts the housing edge of the control arm, using a shop press, a steel plate and tools 09631-12090 and 09950-60020, facing the correct direction.

7. Install the lower control arm to the vehicle.

8. Check and/or adjust the wheel alignment.

Front Wheel Bearing

REMOVAL & INSTALLATION

1. Before servicing the vehicle, refer to the Precautions Section.

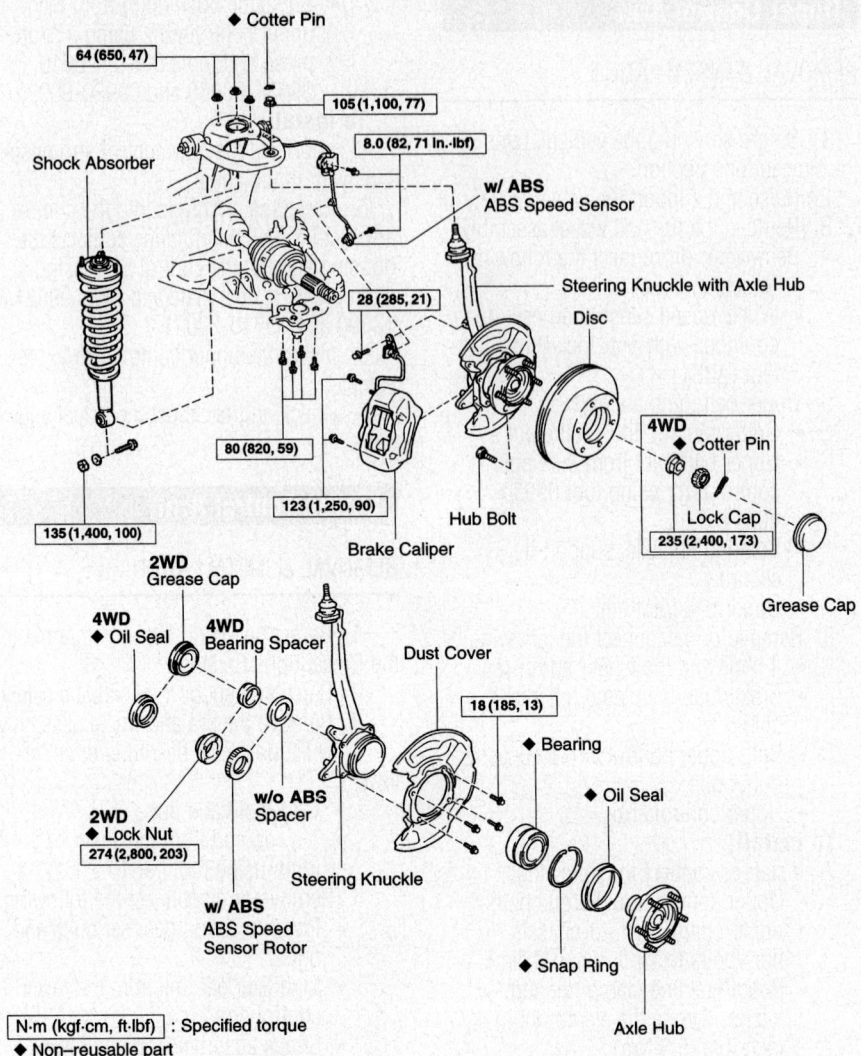

Front hub and related components

2. Raise and support the vehicle safely.
3. Remove the tire and wheel assembly.
4. Remove the grease cap.
5. On 4WD remove the front driveshaft assembly. Remove the cotter pin and lock cap. Apply the brakes and remove the locknut.
6. Remove the ABS sensor harness from the knuckle.
7. Remove the caliper and rotor.
8. Remove the shock absorber.
9. Remove the 4 bolts and disconnect the lower ball joint.
10. Remove the cotter pin and nut and remove the knuckle.
11. Mount the knuckle in a soft-jawed vise.
12. With 2WD, remove the grease cap; with 4WD, remove the inner oil seal.
13. With 2WD, remove the locknut and ABS speed sensor rotor.
14. Remove the 4 bolts and shift the dust cover towards the hub.

15. Remove the hub from the knuckle with a puller.
16. Remove the outer seal.
17. Remove the snapring and press out the bearing.

To install:

18. Press a new bearing into place.
19. Install a new snapring.
20. Drive a new outer seal into place. Coat the seal lip with MP grease.
21. Press the hub onto the knuckle. Torque the bolts to 13 ft. lbs. (18 Nm).
22. Install the speed sensor rotor or spacer.
23. With 2WD, install a new locknut. Torque to 203 ft. lbs. (Nm). Stake the nut.
24. With 4WD, install the bearing spacer using a driver.
25. With 2WD, install the grease cap.
26. With 4WD, install a new inner oil seal using a seal driver. Coat the seal lip with MP grease.

27. With 4WD, insert the halfshaft into the hub and temporarily tighten the nut.

28. Connect the steering knuckle to the upper arm.

29. Install the nut, torque it to 77 ft. lbs. (105 Nm) and install a new cotter pin. If the hole doesn't line up, tighten the nut up to 60 degrees more.

30. Connect the lower ball joint to the knuckle. Torque the four bolts to 59 ft. lbs. (80 Nm) on 2002–2004 vehicles and bolts to 48 ft. lbs. (65 Nm) on 2005–2006 vehicles.

31. Install the shock absorber.

32. Install the caliper.

33. Attach the brake line clamp to the knuckle.

34. Connect the ABS wiring.

35. With 4WD, install the driveshaft locknut. Torque to 173 ft. lbs. (235 Nm). Install the lock cap and a new cotter pin. If the hole doesn't align, tighten the nut up to an additional 60 degrees.

36. Install the grease cap.

37. Install the wheel.

38. Pump the brake a few times before driving.

39. Check the alignment.

REAR SUSPENSION

Shock Absorber

REMOVAL & INSTALLATION

1. Before servicing the vehicle, refer to the Precautions Section.

2. Raise and support the vehicle safely.

3. Remove the tire and wheel assembly.

4. Lower the floor jack to take tension off of the spring.

5. Remove or disconnect the following:
- Shock absorber from the rear axle housing
- Nut, retainers and the cushions holding the shock absorber to the frame
- Shock absorber with the washers and bushings

To install:

➡ Always fully tighten rubber bushings when the wheels are in full contact with the ground and the vehicle is at curb height. Bounce the vehicle up and down several times to stabilize the suspension, prior to final tightening of these components.

6. Install the shock absorber to the frame with the washers and bushings. Tighten the shock absorber-to-frame nut to 64 ft. lbs. (87 Nm).

7. Connect the shock absorber to the rear axle housing. Tighten the bolt to 15 ft. lbs. (20 Nm).

8. Install the tire and wheel assembly.

Leaf Spring

REMOVAL & INSTALLATION

1. Before servicing the vehicle, refer to the Precautions Section.

2. Raise and support the vehicle safely.

3. Remove the tire and wheel assembly.

4. Lower the floor jack to take tension off of the spring.

5. Support the axle with a floor jack.

6. Remove or disconnect the following:
- 4 spring seat nuts and seat
- Both leaf spring-to-chassis nuts/bolts
- Leaf spring

To install:

7. Install or connect the following:
- Leaf spring
- Both leaf spring-to-chassis nuts/bolts. Torque both nuts/bolts to 125 ft. lbs. (170 Nm).
- Spring seat. Torque the 4 nuts to 98 ft. lbs. (133 Nm).
- Rear wheel

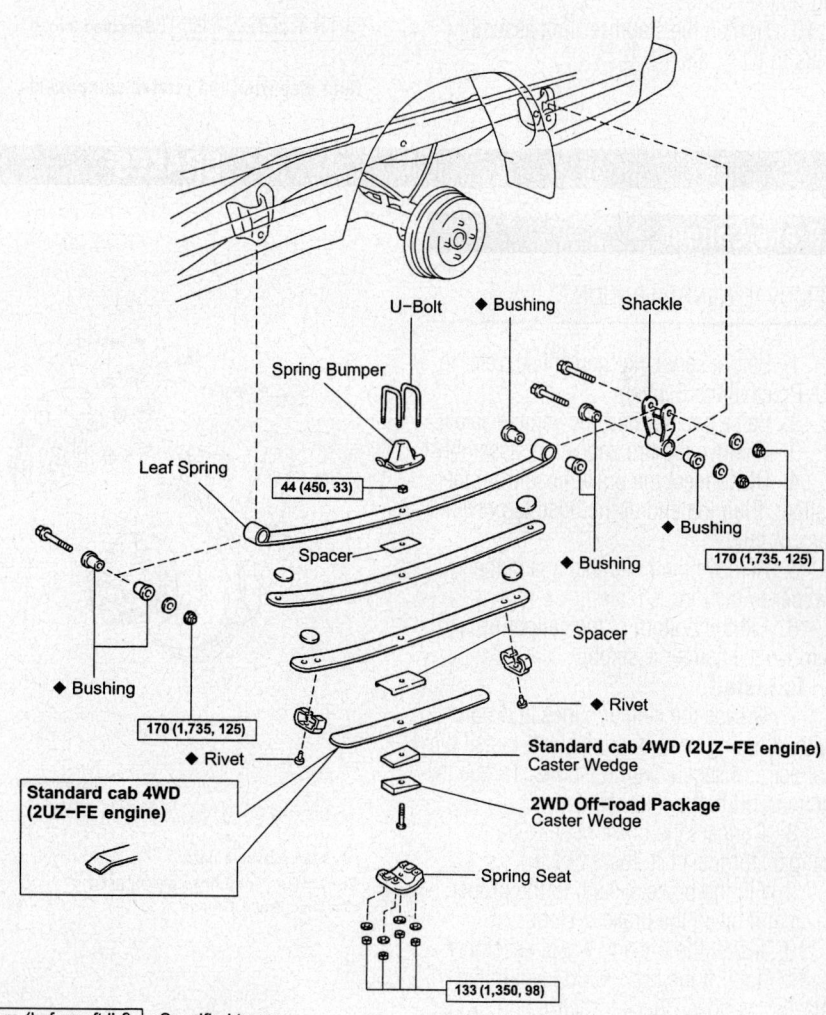

N·m (kgf·cm, ft·lbf) : Specified torque
◆ Non-reusable part

Rear spring and related components

Stabilizer Bar

REMOVAL & INSTALLATION

1. Before servicing the vehicle, refer to the Precautions Section.

2. Raise and support the vehicle safely.

3. Remove the tire and wheel assemblies.

4. On the left side, remove the two nuts and stabilizer bar link.

5. On the right side, remove the two nuts and stabilizer bar link.

6. Remove the four bolts and the stabilizer bar brackets and bushings.

7. Remove the stabilizer bar from the vehicle.

To install:

8. Installation is the reverse of the removal procedure.

9. Tighten the stabilizer bar and bushing bracket bolts to 21 ft. lbs.

10. Tighten the stabilizer link assembly bolts to 51 ft. lbs

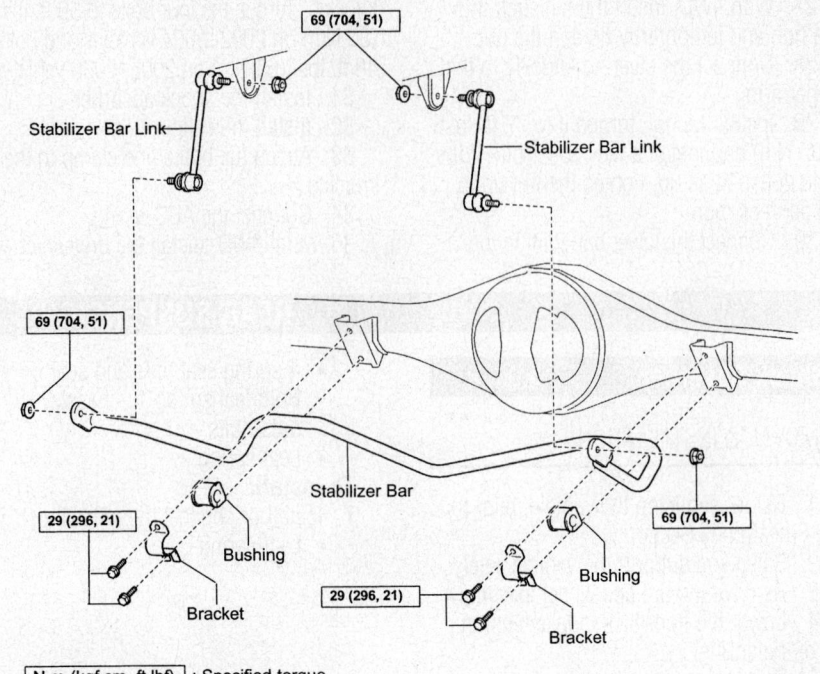

Rear stabilizer and related components

BRAKES

Brake Caliper

REMOVAL & INSTALLATION

1. Before servicing the vehicle, refer to the Precautions Section.

2. Raise and support the vehicle safely.

3. Remove the tire and wheel assembly.

4. Disconnect the brake hose from the caliper. Plug the end of the hose to prevent loss of fluid.

5. Remove the bolts that attach the caliper to the torque plate.

6. Lift the bottom of the caliper up and remove the caliper assembly.

To install:

7. Grease the caliper slides and bolts with lithium grease or equivalent. Install the caliper and secure with the bolts. Torque the bolts to 90 ft. lbs. (123 Nm).

8. Connect the brake hose to the caliper. Torque 11 ft. lbs. (15 Nm).

9. Fill the brake system to the proper level and bleed the brake system.

10. Install the tire and wheel assembly.

11. Top off the brake fluid level in the master cylinder. Check for leaks and proper brake operation.

12. Connect the negative battery cable to the battery.

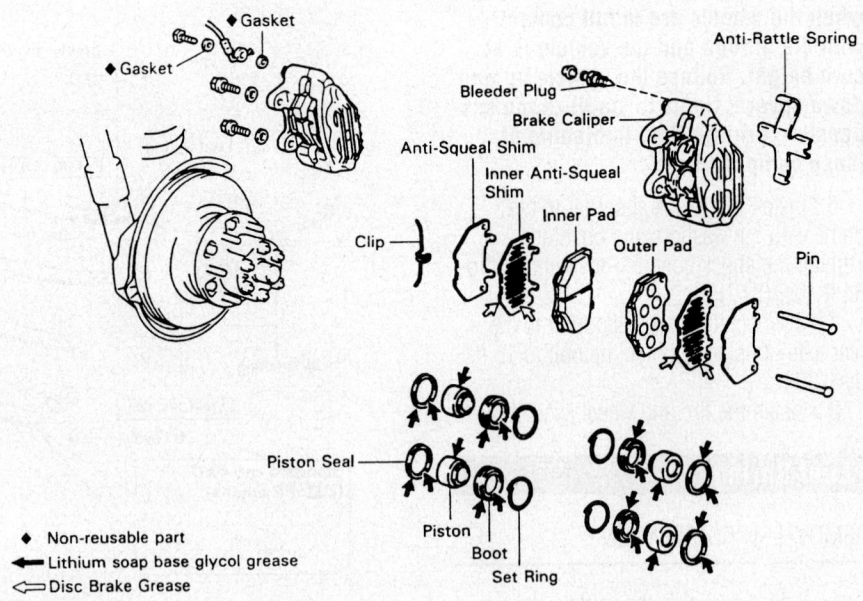

Front disc brake and related components—2002–2003

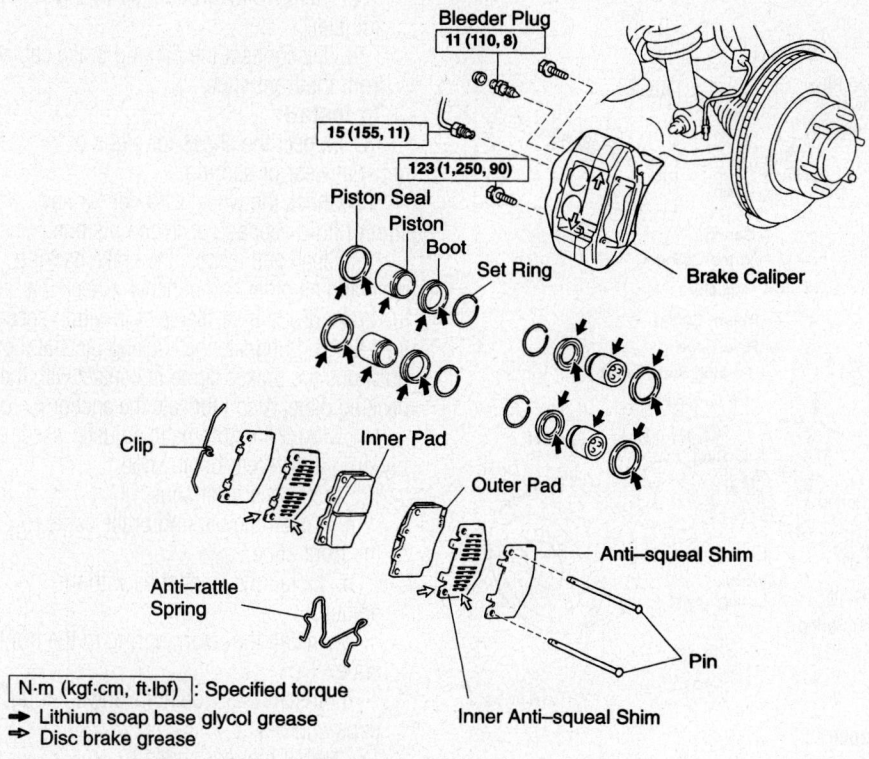

Front disc brake and related components—2004–2006

67170-TUND-G01

Disc Brake Pads

REMOVAL & INSTALLATION

1. Before servicing the vehicle, refer to the Precautions Section.

2. Raise the vehicle and support it safely.

3. Remove the tire and wheel assembly.

4. Remove the clip, pins and anti-rattle spring.

5. Withdraw the pads and remove the anti-squeal shims.

To install:

6. Before installing the new pads, check the disc thickness and disc runout.

7. Siphon out a small amount of brake fluid from the reservoir.

8. Press in the pistons with a hammer handle or equivalent.

9. Apply disc brake grease to both sides of the inner anti-squeal shim. Install the anti-squeal shims to the new pads.

10. Install the pads.

11. Install the anti-rattle springs and pins. Install the clip.

12. Install the wheels.

13. Check and adjust the fluid level. Apply the brake pedal several times.

14. Road-test the vehicle for proper operation.

Brake Drums

REMOVAL & INSTALLATION

1. Before servicing the vehicle, refer to the Precautions Section.

2. Raise and safely support the vehicle.

3. Remove the tire and wheel assembly.

4. Remove the brake drum from the axle hub. If there is difficulty in removing the drum, insert a suitable tool through the hole in the rear of the backing plate, and hold the automatic adjusting lever away from the adjuster. Using another suitable tool at the same time, reduce the brake shoe adjuster by turning the adjusting wheel.

To install:

5. Install the brake drum and pull the parking brake lever all the way up until a clicking sound can no longer be heard.

6. Verify that the rear wheels will not turn. If the rear wheels turn, adjust the parking brake cable as necessary.

7. Release the parking brake and remove the brake drum. Measure the brake drum inside diameter and diameter of the brake shoes. Replace as required.

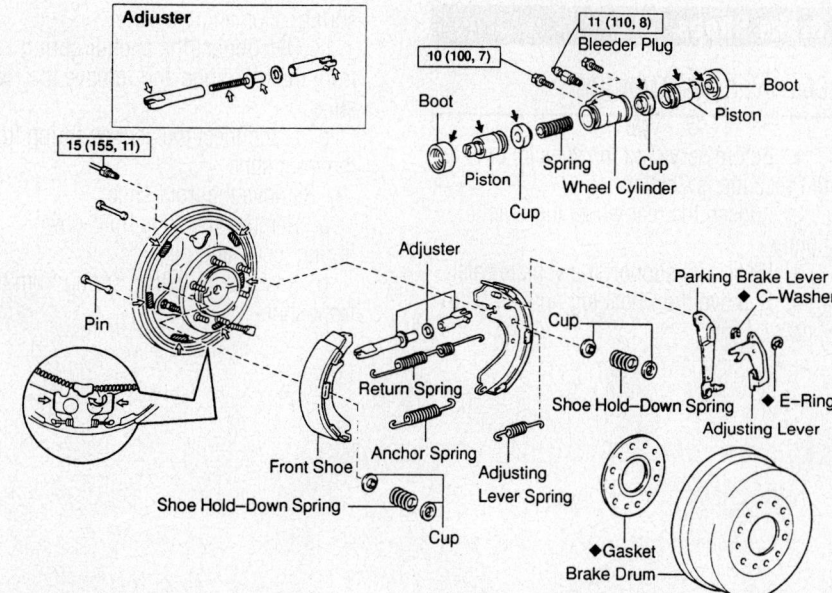

Rear brake shoes and related components—2002–2003

93026G77

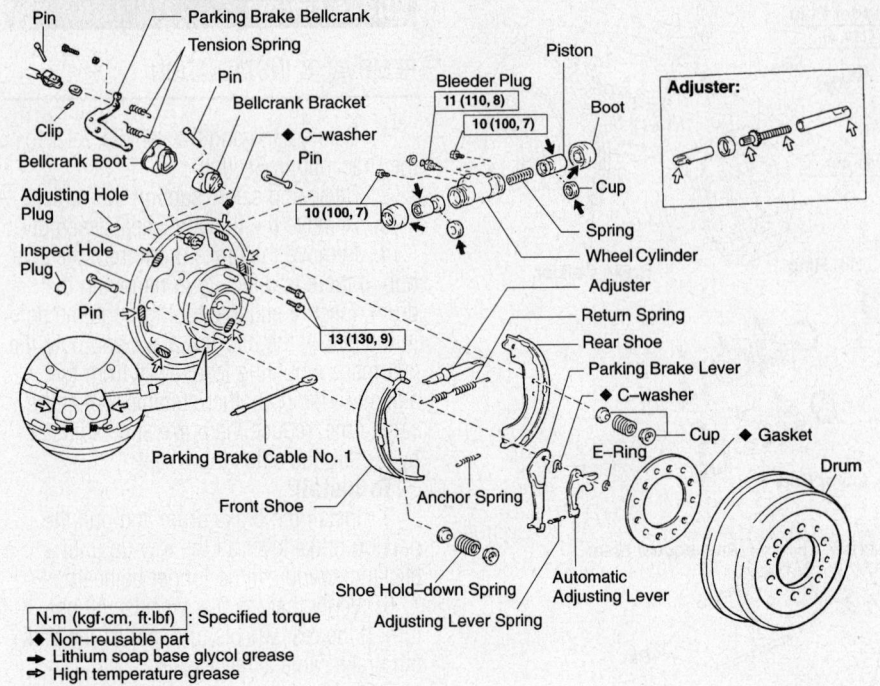

Pin
Parking Brake Bellcrank
Tension Spring
Pin
Bellcrank Bracket
Clip
◆ C-washer
Bellcrank Boot
Pin
Adjusting Hole Plug
10 (100, 7)
Inspect Hole Plug
Pin
13 (130, 9)

Piston
Bleeder Plug
11 (110, 8)
10 (100, 7)
Boot
Cup
Spring
Wheel Cylinder
Adjuster
Return Spring
Rear Shoe
Parking Brake Lever
◆ C-washer
Cup   ◆ Gasket
E-Ring

Adjuster:

Parking Brake Cable No. 1
Front Shoe
Anchor Spring
Shoe Hold-down Spring
Adjusting Lever Spring
Automatic Adjusting Lever
Drum

N·m (kgf·cm, ft·lbf) : Specified torque
◆ Non-reusable part
➡ Lithium soap base glycol grease
➡ High temperature grease

67170-TUND-G03

Rear brake shoes and related components—2004–2006

8. Install the brake drum, replace the wheel(s), and safely lower the vehicle.
9. Road-test the vehicle for proper brake operation.

Brake Shoes

REMOVAL & INSTALLATION

1. Before servicing the vehicle, refer to the Precautions Section.
2. Loosen the rear wheel lug nuts slightly.
3. Raise and support the vehicle safely.
4. Remove the wheel lug nuts and the wheel.

5. Remove the brake drum.
6. Remove the rear shoe.
   a. Carefully unhook the return spring from the brake shoe.
   b. Remove the shoe hold-down spring, cups and the pin.
   c. Disconnect the anchor spring from the rear shoe and remove the rear shoe.
   d. Disconnect the anchor spring from the front shoe.
7. Remove the front shoe.
   a. Remove the shoe hold-down spring, cups and pin.
   b. Remove the return spring from the front shoe.

   c. Remove the front shoe with the adjuster.
   d. Disconnect the parking brake cable from the front shoe.

To install:
8. Inspect the shoes for signs of unusual wear or scoring.
9. Check the wheel cylinder for any sign of fluid seepage or frozen pistons.
10. Clean and inspect the brake backing plate and all other components. Check that the brake drum inner diameter is within specified limits. Lubricate the backing plate at the positions the brakes come in contact with the backing plate. Also lubricate the anchor plate.
11. Mount the automatic adjuster assembly onto a new rear brake shoe.
12. Install the front shoe.
    a. Install the parking brake cable to the front shoe.
    b. Install the front shoe with the adjuster.
    c. Install the return spring to the front shoe.
    d. Install the shoe hold-down spring, cups and pin.
13. Install the rear shoe.
    a. Install the anchor spring to the front shoe.
    b. Install the anchor spring to the rear shoe and install the rear shoe.
    c. Install the shoe hold-down spring, cups and the pin.
    d. Hook the return spring to the brake shoe.
14. Install the brake drum.
15. Adjust the brake shoes until a slight drag is felt when the drum is spun by hand.
16. Check the level of brake fluid in the master cylinder, and then perform a test drive.
17. Connect the negative battery cable to the battery.

GLOSSARY

ABS: Anti-lock braking system. An electro-mechanical braking system which is designed to minimize or prevent wheel lock-up during braking.

ABSOLUTE PRESSURE: Atmospheric (barometric) pressure plus the pressure gauge reading.

ACCELERATOR PUMP: A small pump located in the carburetor that feeds fuel into the air/fuel mixture during acceleration.

ACCUMULATOR: A device that controls shift quality by cushioning the shock of hydraulic oil pressure being applied to a clutch or band.

ACTUATING MECHANISM: The mechanical output devices of a hydraulic system, for example, clutch pistons and band servos.

ACTUATOR: The output component of a hydraulic or electronic system.

ADVANCE: Setting the ignition timing so that spark occurs earlier before the piston reaches top dead center (TDC).

ADAPTIVE MEMORY (ADAPTIVE STRATEGY): The learning ability of the TCM or PCM to redefine its decision-making process to provide optimum shift quality.

AFTER TOP DEAD CENTER (ATDC): The point after the piston reaches the top of its travel on the compression stroke.

AIR BAG: Device on the inside of the car designed to inflate on impact of crash, protecting the occupants of the car.

AIR CHARGE TEMPERATURE (ACT) SENSOR: The temperature of the airflow into the engine is measured by an ACT sensor, usually located in the lower intake manifold or air cleaner.

AIR CLEANER: An assembly consisting of a housing, filter and any connecting ductwork. The filter element is made up of a porous paper, sometimes with a wire mesh screening, and is designed to prevent airborne particles from entering the engine through the carburetor or throttle body.

AIR INJECTION: One method of reducing harmful exhaust emissions by injecting air into each of the exhaust ports of an engine. The fresh air entering the hot exhaust manifold causes any remaining fuel to be burned before it can exit the tailpipe.

AIR PUMP: An emission control device that supplies fresh air to the exhaust manifold to aid in more completely burning exhaust gases.

AIR/FUEL RATIO: The ratio of air-to-gasoline by weight in the fuel mixture drawn into the engine.

ALDL (assembly line diagnostic link): Electrical connector for scanning ECM/PCM/TCM input and output devices.

ALIGNMENT RACK: A special drive-on vehicle lift apparatus/measuring device used to adjust a vehicle's toe, caster and camber angles.

ALL WHEEL DRIVE: Term used to describe a full time four wheel drive system or any other vehicle drive system that continuously delivers power to all four wheels. This system is found primarily on station wagon vehicles and SUVs not utilized for significant off road use.

ALTERNATING CURRENT (AC): Electric current that flows first in one direction, then in the opposite direction, continually reversing flow.

ALTERNATOR: A device which produces AC (alternating current) which is converted to DC (direct current) to charge the car battery.

AMMETER: An instrument, calibrated in amperes, used to measure the flow of an electrical current in a circuit. Ammeters are always connected in series with the circuit being tested.

AMPERAGE: The total amount of current (amperes) flowing in a circuit.

AMPLIFIER: A device used in an electrical circuit to increase the voltage of an output signal.

AMP/HR. RATING (BATTERY): Measurement of the ability of a battery to deliver a stated amount of current for a stated period of time. The higher the amp/hr. rating, the better the battery.

AMPERE: The rate of flow of electrical current present when one volt of electrical pressure is applied against one ohm of electrical resistance.

ANALOG COMPUTER: Any microprocessor that uses similar (analogous) electrical signals to make its calculations.

ANODIZED: A special coating applied to the surface of aluminum valves for extended service life.

ANTIFREEZE: A substance (ethylene or propylene glycol) added to the coolant to prevent freezing in cold weather.

ANTI-FOAM AGENTS: Minimize fluid foaming from the whipping action encountered in the converter and planetary action.

ANTI-WEAR AGENTS: Zinc agents that control wear on the gears, bushings, and thrust washers.

ANTI-LOCK BRAKING SYSTEM: A supplementary system to the base hydraulic system that prevents sustained lock-up of the wheels during braking as well as automatically controlling wheel slip.

ANTI-ROLL BAR: See stabilizer bar.

ARC: A flow of electricity through the air between two electrodes or contact points that produces a spark.

ARMATURE: A laminated, soft iron core wrapped by a wire that converts electrical energy to mechanical energy as in a motor or relay. When rotated in a magnetic field, it changes mechanical energy into electrical energy as in a generator.

ATDC: After Top Dead Center.

ATF: Automatic transmission fluid.

ATMOSPHERIC PRESSURE: The pressure on the Earth's surface caused by the weight of the air in the atmosphere. At sea level, this pressure is 14.7 psi at 32°F (101 kPa at 0°C).

ATOMIZATION: The breaking down of a liquid into a fine mist that can be suspended in air.

AUXILIARY ADD-ON COOLER: A supplemental transmission fluid cooling device that is installed in series with the heat exchanger (cooler), located inside the radiator, to provide additional support to cool the hot fluid leaving the torque converter.

AUXILIARY PRESSURE: An added fluid pressure that is introduced into a regulator or balanced valve system to control valve movement. The auxiliary pressure itself can be either a fixed or a variable value. (See balanced valve; regulator valve.)

AWD: All wheel drive.

AXIAL FORCE: A side or end thrust force acting in or along the same plane as the power flow.

AXIAL PLAY: Movement parallel to a shaft or bearing bore.

AXLE CAPACITY: The maximum load-carrying capacity of the axle itself, as specified by the manufacturer. This is usually a higher number than the GAWR.

AXLE RATIO: This is a number (3.07:1, 4.56:1, for example) expressing the ratio between driveshaft revolutions and wheel revolutions. A low numerical ratio allows the engine to work easier because it doesn't have to turn as fast. A high numerical ratio means that the engine has to turn more rpm's to move the wheels through the same number of turns.

BACKFIRE: The sudden combustion of gases in the intake or exhaust system that results in a loud explosion.

BACKLASH: The clearance or play between two parts, such as meshed gears.

BACKPRESSURE: Restrictions in the exhaust system that slow the exit of exhaust gases from the combustion chamber.

BAKELITE®: A heat resistant, plastic insulator material commonly used in printed circuit boards and transistorized components.

BALANCED VALVE: A valve that is positioned by opposing auxiliary hydraulic pressures and/or spring force. Examples include mainline regulator, throttle, and governor valves. (See regulator valve.)

BAND: A flexible ring of steel with an inner lining of friction material. When tightened around the outside of a drum, a planetary member is held stationary to the transmission/transaxle case.

BALL BEARING: A bearing made up of hardened inner and outer races between which hardened steel balls roll.

BALL JOINT: A ball and matching socket connecting suspension components (steering knuckle to lower control arms). It permits rotating movement in any direction between the components that are joined.

BARO (BAROMETRIC PRESSURE SENSOR): Measures the change in the intake manifold pressure caused by changes in altitude.

BAROMETRIC MANIFOLD ABSOLUTE PRESSURE (BMAP) SENSOR: Operates similarly to a conventional MAP sensor; reads intake manifold pressure and is also responsible for determining altitude and barometric pressure prior to engine operation.

BAROMETRIC PRESSURE: (See atmospheric pressure.)

BALLAST RESISTOR: A resistor in the primary ignition circuit that lowers voltage after the engine is started to reduce wear on ignition components.

BATTERY: A direct current electrical storage unit, consisting of the basic active materials of lead and sulfuric acid, which converts chemical energy into electrical energy. Used to provide current for the operation of the starter as well as other equipment, such as the radio, lighting, etc.

BEAD: The portion of a tire that holds it on the rim.

BEARING: A friction reducing, supportive device usually located between a stationary part and a moving part.

BEFORE TOP DEAD CENTER (BTDC): The point just before the piston reaches the top of its travel on the compression stroke.

BELTED TIRE: Tire construction similar to bias-ply tires, but using two or more layers of reinforced belts between body plies and the tread.

BEZEL: Piece of metal surrounding radio, headlights, gauges or similar components; sometimes used to hold the glass face of a gauge in the dash.

BIAS-PLY TIRE: Tire construction, using body ply reinforcing cords which run at alternating angles to the center line of the tread.

BI-METAL TEMPERATURE SENSOR: Any sensor or switch made of two dissimilar types of metal that bend when heated or cooled due to the different expansion rates of the alloys. These types of sensors usually function as an on/off switch.

BLOCK: See Engine Block.

BLOW-BY: Combustion gases, composed of water vapor and unburned fuel, that leak past the piston rings into the crankcase during normal engine operation. These gases are removed by the PCV system to prevent the buildup of harmful acids in the crankcase.

BOOK TIME: See Labor Time.

BOOK VALUE: The average value of a car, widely used to determine trade-in and resale value.

BOOST VALVE: Used at the base of the regulator valve to increase mainline pressure.

BORE: Diameter of a cylinder.

BRAKE CALIPER: The housing that fits over the brake disc. The caliper holds the brake pads, which are pressed against the discs by the caliper pistons when the brake pedal is depressed.

BRAKE HORSEPOWER (BHP): The actual horsepower available at the engine flywheel as measured by a dynamometer.

BRAKE FADE: Loss of braking power, usually caused by excessive heat after repeated brake applications.

BRAKE HORSEPOWER: Usable horsepower of an engine measured at the crankshaft.

BRAKE PAD: A brake shoe and lining assembly used with disc brakes.

BRAKE PROPORTIONING VALVE: A valve on the master cylinder which restricts hydraulic brake pressure to the wheels to a specified amount, preventing wheel lock-up.

BREAKAWAY: Often used by Chrysler to identify first-gear operation in D and 2 ranges. In these ranges, first-gear operation depends on a one-way roller clutch that holds on acceleration and releases (breaks away) on deceleration, resulting in a freewheeling coast-down condition.

BRAKE SHOE: The backing for the brake lining. The term is, however, usually applied to the assembly of the brake backing and lining.

BREAKER POINTS: A set of points inside the distributor, operated by a cam, which make and break the ignition circuit.

BRINNELLING: A wear pattern identified by a series of indentations at regular intervals. This condition is caused by a lack of lube, overload situations, and/or vibrations.

BTDC: Before Top Dead Center.

BUMP: Sudden and forceful apply of a clutch or band.

BUSHING: A liner, usually removable, for a bearing; an anti-friction liner used in place of a bearing.

CALIFORNIA ENGINE: An engine certified by the EPA for use in California only; conforms to more stringent emission regulations than Federal engine.

CALIPER: A hydraulically activated device in a disc brake system,

which is mounted straddling the brake rotor (disc). The caliper contains at least one piston and two brake pads. Hydraulic pressure on the piston(s) forces the pads against the rotor.

CAPACITY: The quantity of electricity that can be delivered from a unit, as from a battery in ampere-hours, or output, as from a generator.

CAMBER: One of the factors of wheel alignment. Viewed from the front of the car, it is the inward or outward tilt of the wheel. The top of the tire will lean outward (positive camber) or inward (negative camber).

CAMSHAFT: A shaft in the engine on which are the lobes (cams) which operate the valves. The camshaft is driven by the crankshaft, via a belt, chain or gears, at one half the crankshaft speed.

CAPACITOR: A device which stores an electrical charge.

CARBON MONOXIDE (CO): A colorless, odorless gas given off as a normal byproduct of combustion. It is poisonous and extremely dangerous in confined areas, building up slowly to toxic levels without warning if adequate ventilation is not available.

CARBURETOR: A device, usually mounted on the intake manifold of an engine, which mixes the air and fuel in the proper proportion to allow even combustion.

CASTER: The forward or rearward tilt of an imaginary line drawn through the upper ball joint and the center of the wheel. Viewed from the sides, positive caster (forward tilt) lends directional stability, while negative caster (rearward tilt) produces instability.

CATALYTIC CONVERTER: A device installed in the exhaust system, like a muffler, that converts harmful byproducts of combustion into carbon dioxide and water vapor by means of a heat-producing chemical reaction.

CENTRIFUGAL ADVANCE: A mechanical method of advancing the spark timing by using flyweights in the distributor that react to centrifugal force generated by the distributor shaft rotation.

CENTRIFUGAL FORCE: The outward pull of a revolving object, away from the center of revolution. Centrifugal force increases with the speed of rotation.

CETANE RATING: A measure of the ignition value of diesel fuel. The higher the cetane rating, the better the fuel. Diesel fuel cetane rating is roughly comparable to gasoline octane rating.

CHECK VALVE: Any one-way valve installed to permit the flow of air, fuel or vacuum in one direction only.

CHOKE: The valve/plate that restricts the amount of air entering an engine on the induction stroke, thereby enriching the air/fuel ratio.

CHUGGLE: Bucking or jerking condition that may be engine related and may be most noticeable when converter clutch is engaged; similar to the feel of towing a trailer.

CIRCLIP: A split steel snapring that fits into a groove to hold various parts in place.

CIRCUIT BREAKER: A switch which protects an electrical circuit from overload by opening the circuit when the current flow exceeds a pre-determined level. Some circuit breakers must be reset manually, while most reset automatically.

CIRCUIT: Any unbroken path through which an electrical current can flow. Also used to describe fuel flow in some instances.

CIRCUIT, BYPASS: Another circuit in parallel with the major circuit through which power is diverted.

CIRCUIT, CLOSED: An electrical circuit in which there is no interruption of current flow.

CIRCUIT, GROUND: The non-insulated portion of a complete circuit used as a common potential point. In automotive circuits, the ground is composed of metal parts, such as the engine, body sheet metal, and frame and is usually a negative potential.

CIRCUIT, HOT: That portion of a circuit not at ground potential. The hot circuit is usually insulated and is connected to the positive side of the battery.

CIRCUIT, OPEN: A break or lack of contact in an electrical circuit, either intentional (switch) or unintentional (bad connection or broken wire).

CIRCUIT, PARALLEL: A circuit having two or more paths for current flow with common positive and negative tie points. The same voltage is applied to each load device or parallel branch.

CIRCUIT, SERIES: An electrical system in which separate parts are connected end to end, using one wire, to form a single path for current to flow.

CIRCUIT, SHORT: A circuit that is accidentally completed in an electrical path for which it was not intended.

CLAMPING (ISOLATION) DIODES: Diodes positioned in a circuit to prevent self-induction from damaging electronic components.

CLEARCOAT: A transparent layer which, when sprayed over a vehicle's paint job, adds gloss and depth as well as an additional protective coating to the finish.

CLUTCH: Part of the power train used to connect/disconnect power to the rear wheels.

CLUTCH, FLUID: The same as a fluid coupling. A fluid clutch or coupling performs the same function as a friction clutch by utilizing fluid friction and inertia as opposed to solid friction used by a friction clutch. (See fluid coupling.)

CLUTCH, FRICTION: A coupling device that provides a means of smooth and positive engagement and disengagement of engine torque to the vehicle powertrain. Transmission of power through the clutch is accomplished by bringing one or more rotating drive members into contact with complementing driven members.

COAST: Vehicle deceleration caused by engine braking conditions.

COEFFICIENT OF FRICTION: The amount of surface tension between two contacting surfaces; identified by a scientifically calculated number.

COIL: Part of the ignition system that boosts the relatively low voltage supplied by the car's electrical system to the high voltage required to fire the spark plugs.

COMBINATION MANIFOLD: An assembly which includes both the intake and exhaust manifolds in one casting.

COMBINATION VALVE: A device used in some fuel systems that routes fuel vapors to a charcoal storage canister instead of venting them into the atmosphere. The valve relieves fuel tank pressure and allows fresh air into the tank as the fuel level drops to prevent a vapor lock situation.

COMBUSTION CHAMBER: The part of the engine in the cylinder head where combustion takes place.

COMPOUND GEAR: A gear consisting of two or more simple gears with a common shaft.

COMPOUND PLANETARY: A gearset that has more than the three elements found in a simple gearset and is constructed by combining members of two planetary gearsets to create additional gear ratio possibilities.

COMPRESSION CHECK: A test involving removing each spark plug and inserting a gauge. When the engine is cranked, the gauge will record a pressure reading in the individual cylinder. General operating condition can be determined from a compression check.

COMPRESSION RATIO: The ratio of the volume between the piston and cylinder head when the piston is at the bottom of its stroke (bottom dead center) and when the piston is at the top of its stroke (top dead center).

COMPUTER: An electronic control module that correlates input data according to prearranged engineered instructions; used for the management of an actuator system or systems.

CONDENSER: An electrical device which acts to store an electrical charge, preventing voltage surges.

2. A radiator-like device in the air conditioning system in which refrigerant gas condenses into a liquid, giving off heat.

CONDUCTOR: Any material through which an electrical current can be transmitted easily.

CONNECTING ROD: The connecting link between the crankshaft and piston.

CONSTANT VELOCITY JOINT: Type of universal joint in a halfshaft assembly in which the output shaft turns at a constant angular velocity without variation, provided that the speed of the input shaft is constant.

CONTINUITY: Continuous or complete circuit. Can be checked with an ohmmeter.

CONTROL ARM: The upper or lower suspension components which are mounted on the frame and support the ball joints and steering knuckles.

CONVENTIONAL IGNITION: Ignition system which uses breaker points.

CONVERTER: (See torque converter.)

CONVERTER LOCKUP: The switching from hydrodynamic to direct mechanical drive, usually through the application of a friction element called the converter clutch.

COOLANT: Mixture of water and anti-freeze circulated through the engine to carry off heat produced by the engine.

CORROSION INHIBITOR: An inhibitor in ATF that prevents corrosion of bushings, thrust washers, and oil cooler brazed joints.

COUNTERSHAFT: An intermediate shaft which is rotated by a mainshaft and transmits, in turn, that rotation to a working part.

COUPLING PHASE: Occurs when the torque converter is operating at its greatest hydraulic efficiency. The speed differential between the impeller and the turbine is at its minimum. At this point, the stator freewheels, and there is no torque multiplication.

CRANKCASE: The lower part of an engine in which the crankshaft and related parts operate.

CRANKSHAFT: Engine component (connected to pistons by connecting rods) which converts the reciprocating (up and down) motion of pistons to rotary motion used to turn the driveshaft.

CURB WEIGHT: The weight of a vehicle without passengers or payload, but including all fluids (oil, gas, coolant, etc.) and other equipment specified as standard.

CURRENT: The flow (or rate) of electrons moving through a circuit. Current is measured in amperes (amp).

CURRENT FLOW CONVENTIONAL: Current flows through a circuit from the positive terminal of the source to the negative terminal (plus to minus).

CURRENT FLOW, ELECTRON: Current or electrons flow from the negative terminal of the source, through the circuit, to the positive terminal (minus to plus).

CV-JOINT: Constant velocity joint.

CYCLIC VIBRATIONS: The off-center movement of a rotating object that is affected by its initial balance, speed of rotation, and working angles.

CYLINDER BLOCK: See engine block.

CYLINDER HEAD: The detachable portion of the engine, usually fastened to the top of the cylinder block and containing all or most of the combustion chambers. On overhead valve engines, it contains the valves and their operating parts. On overhead cam engines, it contains the camshaft as well.

CYLINDER: In an engine, the round hole in the engine block in which the piston(s) ride.

DATA LINK CONNECTOR (DLC): Current acronym/term applied to the federally mandated, diagnostic junction connector that is used to monitor ECM/PC/TCM inputs, processing strategies, and outputs including diagnostic trouble codes (DTCs).

DEAD CENTER: The extreme top or bottom of the piston stroke.

DECELERATION BUMP: When referring to a torque converter clutch in the applied position, a sudden release of the accelerator pedal causes a forceful reversal of power through the drivetrain (engine braking), just prior to the apply plate actually being released.

DELAYED (LATE OR EXTENDED): Condition where shift is expected but does not occur for a period of time, for example, where clutch or band engagement does not occur as quickly as expected during part throttle or wide open throttle apply of accelerator or when manually downshifting to a lower range.

DETENT: A spring-loaded plunger, pin, ball, or pawl used as a holding device on a ratchet wheel or shaft. In automatic transmissions, a detent mechanism is used for locking the manual valve in place.

DETENT DOWNSHIFT: (See kickdown.)

DETERGENT: An additive in engine oil to improve its operating characteristics.

DETONATION: An unwanted explosion of the air/fuel mixture in the combustion chamber caused by excess heat and compression, advanced timing, or an overly lean mixture. Also referred to as "ping".

DEXRON®: A brand of automatic transmission fluid.

DIAGNOSTIC TROUBLE CODES (DTCs): A digital display from the control module memory that identifies the input, processor, or output device circuit that is related to the powertrain emission/driveability malfunction detected. Diagnostic trouble codes can be read by the MIL to flash any codes or by using a handheld scanner.

DIAPHRAGM: A thin, flexible wall separating two cavities, such as in a vacuum advance unit.

DIESELING: The engine continues to run after the car is shut off; caused by fuel continuing to be burned in the combustion chamber.

DIFFERENTIAL: A geared assembly which allows the transmission of motion between drive axles, giving one axle the ability to rotate faster than the other, as in cornering.

DIFFERENTIAL AREAS: When opposing faces of a spool valve are acted upon by the same pressure but their areas differ in size, the face with the larger area produces the differential force and valve movement. (See spool valve.)

DIFFERENTIAL FORCE: (See differential areas)

DIGITAL READOUT: A display of numbers or a combination of numbers and letters.

DIGITAL VOLT OHMMETER: An electronic diagnostic tool used to measure voltage, ohms and amps as well as several other functions, with the readings displayed on a digital screen in tenths, hundredths and thousandths.

DIODE: An electrical device that will allow current to flow in one direction only.

DIRECT CURRENT (DC): Electrical current that flows in one direction only.

DIRECT DRIVE: The gear ratio is 1:1, with no change occurring in the torque and speed input/output relationship.

DISC BRAKE: A hydraulic braking assembly consisting of a brake disc, or rotor, mounted on an axle shaft, and a caliper assembly containing, usually two brake pads which are activated by hydraulic pressure. The pads are forced against the sides of the disc, creating friction which slows the vehicle.

DISPERSANTS: Suspend dirt and prevent sludge buildup in a liquid, such as engine oil.

DOUBLE BUMP (DOUBLE FEEL): Two sudden and forceful applies of a clutch or band.

DISPLACEMENT: The total volume of air that is displaced by all pistons as the engine turns through one complete revolution.

DISTRIBUTOR: A mechanically driven device on an engine which is responsible for electrically firing the spark plug at a pre-determined point of the piston stroke.

DOHC: Double overhead camshaft.

DOUBLE OVERHEAD CAMSHAFT: The engine utilizes two camshafts mounted in one cylinder head. One camshaft operates the exhaust valves, while the other operates the intake valves.

DOWEL PIN: A pin, inserted in mating holes in two different parts allowing those parts to maintain a fixed relationship.

DRIVELINE: The drive connection between the transmission and the drive wheels.

DRIVE TRAIN: The components that transmit the flow of power from the engine to the wheels. The components include the clutch, transmission, driveshafts (or axle shafts in front wheel drive), U-joints and differential.

DRUM BRAKE: A braking system which consists of two brake shoes and one or two wheel cylinders, mounted on a fixed backing plate, and a brake drum, mounted on an axle, which revolves around the assembly.

DRY CHARGED BATTERY: Battery to which electrolyte is added when the battery is placed in service.

DVOM: Digital volt ohmmeter

DWELL: The rate, measured in degrees of shaft rotation, at which an electrical circuit cycles on and off.

DYNAMIC: An application in which there is rotating or reciprocating motion between the parts.

EARLY: Condition where shift occurs before vehicle has reached proper speed, which tends to labor engine after upshift.

EBCM: See Electronic Control Unit (ECU).

ECM: See Electronic Control Unit (ECU).

ECU: Electronic control unit.

ELECTRODE: Conductor (positive or negative) of electric current.

ELECTROLYSIS: A surface etching or bonding of current conducting transmission/transaxle components that may occur when grounding straps are missing or in poor condition.

ELECTROLYTE: A solution of water and sulfuric acid used to activate the battery. Electrolyte is extremely corrosive.

ELECTROMAGNET: A coil that produces a magnetic field when current flows through its windings.

ELECTROMAGNETIC INDUCTION: A method to create (generate) current flow through the use of magnetism.

ELECTROMAGNETISM: The effects surrounding the relationship between electricity and magnetism.

ELECTROMOTIVE FORCE (EMF): The force or pressure (voltage) that causes current movement in an electrical circuit.

ELECTRONIC CONTROL UNIT: A digital computer that controls engine (and sometimes transmission, brake or other vehicle system) functions based on data received from various sensors. Examples used by some manufacturers include Electronic Brake Control Module (EBCM), Engine Control Module (ECM), Powertrain Control Module (PCM) or Vehicle Control Module (VCM).

ELECTRONIC IGNITION: A system in which the timing and firing of the spark plugs is controlled by an electronic control unit, usually called a module. These systems have no points or condenser.

ELECTRONIC PRESSURE CONTROL (EPC) SOLENOID: A specially designed solenoid containing a spool valve and spring assembly to control fluid mainline pressure. A variable current flow, controlled by the ECM/PCM, varies the internal force of the solenoid on the spool valve and resulting mainline pressure. (See variable force solenoid.)

ELECTRONICS: Miniaturized electrical circuits utilizing semiconductors, solid-state devices, and printed circuits. Electronic circuits utilize small amounts of power.

ELECTRONIFICATION: The application of electronic circuitry to a mechanical device. Regarding automatic transmissions, electrification is incorporated into converter clutch lockup, shift scheduling, and line pressure control systems.

ELECTROSTATIC DISCHARGE (ESD): An unwanted, high-voltage electrical current released by an individual who has taken on a static charge of electricity. Electronic components can be easily damaged by ESD.

ELEMENT: A device within a hydrodynamic drive unit designed with a set of blades to direct fluid flow.

ENAMEL: Type of paint that dries to a smooth, glossy finish.

END BUMP (END FEEL OR SLIP BUMP): Firmer feel at end of shift when compared with feel at start of shift.

END-PLAY: The clearance/gap between two components that allows for expansion of the parts as they warm up, to prevent binding and to allow space for lubrication.

ENERGY: The ability or capacity to do work.

ENGINE: The primary motor or power apparatus of a vehicle, which converts liquid or gas fuel into mechanical energy.

ENGINE BLOCK: The basic engine casting containing the cylinders, the crankshaft main bearings, as well as machined surfaces for the mounting of other components such as the cylinder head, oil pan, transmission, etc.

ENGINE BRAKING: Use of engine to slow vehicle by manually downshifting during zero-throttle coast down.

ENGINE CONTROL MODULE (ECM): Manages the engine and incorporates output control over the torque converter clutch solenoid. (Note: Current designation for the ECM in late model vehicles is PCM.)

ENGINE COOLANT TEMPERATURE (ECT) SENSOR: Prevents converter clutch engagement with a cold engine; also used for shift timing and shift quality.

EP LUBRICANT: EP (extreme pressure) lubricants are specially formulated for use with gears involving heavy loads (transmissions, differentials, etc.).

ETHYL: A substance added to gasoline to improve its resistance to knock, by slowing down the rate of combustion.

ETHYLENE GLYCOL: The base substance of antifreeze.

EXHAUST MANIFOLD: A set of cast passages or pipes which conduct exhaust gases from the engine.

FAIL-SAFE (BACKUP) CONTROL: A substitute value used by the PCM/TCM to replace a faulty signal from an input sensor. The temporary value allows the vehicle to continue to be operated.

FAST IDLE: The speed of the engine when the choke is on. Fast idle speeds engine warm-up.

FEDERAL ENGINE: An engine certified by the EPA for use in any of the 49 states (except California).

FEEDBACK: A circuit malfunction whereby current can find another path to feed load devices.

FEELER GAUGE: A blade, usually metal, of precisely predetermined thickness, used to measure the clearance between two parts.

FILAMENT: The part of a bulb that glows; the filament creates high resistance to current flow and actually glows from the resulting heat.

FINAL DRIVE: An essential part of the axle drive assembly where final gear reduction takes place in the powertrain. In RWD applications and north-south FWD applications, it must also change the power flow direction to the axle shaft by ninety degrees. (Also see axle ratio).

FIRING ORDER: The order in which combustion occurs in the cylinders of an engine. Also the order in which spark is distributed to the plugs by the distributor.

FIRM: A noticeable quick apply of a clutch or band that is considered normal with medium to heavy throttle shift; should not be confused with harsh or rough.

FLAME FRONT: The term used to describe certain aspects of the fuel explosion in the cylinders. The flame front should move in a controlled pattern across the cylinder, rather than simply exploding immediately.

FLARE (SLIPPING): A quick increase in engine rpm accompanied by momentary loss of torque; generally occurs during shift.

FLAT ENGINE: Engine design in which the pistons are horizontally opposed. Porsche, Subaru and some old VW are common examples of flat engines.

FLAT RATE: A dealership term referring to the amount of money paid to a technician for a repair or diagnostic service based on that particular service versus dealership's labor time (NOT based on the actual time the technician spent on the job).

FLAT SPOT: A point during acceleration when the engine seems to lose power for an instant.

FLOODING: The presence of too much fuel in the intake manifold and combustion chamber which prevents the air/fuel mixture from firing, thereby causing a no-start situation.

FLUID: A fluid can be either liquid or gas. In hydraulics, a liquid is used for transmitting force or motion.

FLUID COUPLING: The simplest form of hydrodynamic drive, the fluid coupling consists of two look-alike members with straight radial varies referred to as the impeller (pump) and the turbine. Input torque is always equal to the output torque.

FLUID DRIVE: Either a fluid coupling or a fluid torque converter. (See hydrodynamic drive units.)

FLUID TORQUE CONVERTER: A hydrodynamic drive that has the ability to act both as a torque multiplier and fluid coupling. (See hydrodynamic drive units; torque converter.)

FLUID VISCOSITY: The resistance of a liquid to flow. A cold fluid (oil) has greater viscosity and flows more slowly than a hot fluid (oil).

FLYWHEEL: A heavy disc of metal attached to the rear of the crankshaft. It smoothes the firing impulses of the engine and keeps the crankshaft turning during periods when no firing takes place. The starter also engages the flywheel to start the engine.

FOOT POUND (ft. lbs., lbs. ft. or sometimes, ft. lb.): The amount of energy or work needed to raise an item weighing one pound, a distance of one foot.

FREEZE PLUG: A plug in the engine block which will be pushed out if the coolant freezes. Sometimes called expansion plugs, they protect the block from cracking should the coolant freeze.

FRICTION: The resistance that occurs between contacting surfaces. This relationship is expressed by a ratio called the coefficient of friction (CL).

FRICTION, COEFFICIENT OF: The amount of surface tension between two contacting surfaces; expressed by a scientifically calculated number.

FRONT END ALIGNMENT: A service to set caster, camber and toe-in to the correct specifications. This will ensure that the car steers and handles properly and that the tires wear properly.

FRICTION MODIFIER: Changes the coefficient of friction of the fluid between the mating steel and composition clutch/band surfaces during the engagement process and allows for a certain amount of intentional slipping for a good "shift-feel".

FRONTAL AREA: The total frontal area of a vehicle exposed to air flow.

FUEL FILTER: A component of the fuel system containing a porous paper element used to prevent any impurities from entering the engine through the fuel system. It usually takes the form of a canister-like housing, mounted in-line with the fuel hose, located anywhere on a vehicle between the fuel tank and engine.

FUEL INJECTION: A system replacing the carburetor that sprays fuel into the cylinder through nozzles. The amount of fuel can be more precisely controlled with fuel injection.

FULL FLOATING AXLE: An axle in which the axle housing extends through the wheel giving bearing support on the outside of the housing. The front axle of a four-wheel drive vehicle is usually a full floating axle, as are the rear axles of many larger (1 ton and over) pick-ups and vans.

FULL-TIME FOUR-WHEEL DRIVE: A four-wheel drive system that continuously delivers power to all four wheels. A differential between the front and rear driveshafts permits variations in axle speeds to control gear wind-up without damage.

FULL THROTTLE DETENT DOWNSHIFT: A quick apply of accelerator pedal to its full travel, forcing a downshift.

FUSE: A protective device in a circuit which prevents circuit overload by breaking the circuit when a specific amperage is present. The device is constructed around a strip or wire of a lower amperage rating than the circuit it is designed to protect. When an amperage higher than that stamped on the fuse is present in the circuit, the strip or wire melts, opening the circuit.

FUSIBLE LINK: A piece of wire in a wiring harness that performs the same job as a fuse. If overloaded, the fusible link will melt and interrupt the circuit.

FWD: Front wheel drive.

GAWR: (Gross axle weight rating) the total maximum weight an axle is designed to carry.

GCW: (Gross combined weight) total combined weight of a tow vehicle and trailer.

GARAGE SHIFT: initial engagement feel of transmission, neutral to reverse or neutral to a forward drive.

GARAGE SHIFT FEEL: A quick check of the engagement quality and responsiveness of reverse and forward gears. This test is done with the vehicle stationary.

GEAR: A toothed mechanical device that acts as a rotating lever to transmit power or turning effort from one shaft to another. (See gear ratio.)

GEAR RATIO: A ratio expressing the number of turns a smaller gear will make to turn a larger gear through one revolution. The ratio is found by dividing the number of teeth on the smaller gear into the number of teeth on the larger gear.

GEARBOX: Transmission

GEAR REDUCTION: Torque is multiplied and speed decreased by the factor of the gear ratio. For example, a 3:1 gear ratio changes an input torque of 180 ft. lbs. and an input speed of 2700 rpm to 540 Ft. lbs. and 900 rpm, respectively. (No account is taken of frictional losses, which are always present.)

GEARTRAIN: A succession of intermeshing gears that form an assembly and provide for one or more torque changes as the power input is transmitted to the power output.

GEL COAT: A thin coat of plastic resin covering fiberglass body panels.

GENERATOR: A device which produces direct current (DC) necessary to charge the battery.

GOVERNOR: A device that senses vehicle speed and generates a hydraulic oil pressure. As vehicle speed increases, governor oil pressure rises.

GROUND CIRCUIT: (See circuit, ground.)

GROUND SIDE SWITCHING: The electrical/electronic circuit control switch is located after the circuit load.

GVWR: (Gross vehicle weight rating) total maximum weight a vehicle is designed to carry including the weight of the vehicle, passengers, equipment, gas, oil, etc.

HALOGEN: A special type of lamp known for its quality of brilliant white light. Originally used for fog lights and driving lights.

HARD CODES: DTCs that are present at the time of testing; also called continuous or current codes.

HARSH(ROUGH): An apply of a clutch or band that is more noticeable than a firm one; considered undesirable at any throttle position.

HEADER TANK: An expansion tank for the radiator coolant. It can be located remotely or built into the radiator.

HEAT RANGE: A term used to describe the ability of a spark plug to carry away heat. Plugs with longer nosed insulators take longer to carry heat off effectively.

HEAT RISER: A flapper in the exhaust manifold that is closed when the engine is cold, causing hot exhaust gases to heat the intake manifold providing better cold engine operation. A thermostatic spring opens the flapper when the engine warms up.

HEAVY THROTTLE: Approximately three-fourths of accelerator pedal travel.

HEMI: A name given an engine using hemispherical combustion chambers.

HERTZ (HZ): The international unit of frequency equal to one cycle per second (10,000 Hertz equals 10,000 cycles per second).

HIGH-IMPEDANCE DVOM (DIGITAL VOLT-OHMMETER): This styled device provides a built-in resistance value and is capable of limiting circuit current flow to safe milliamp levels.

HIGH RESISTANCE: Often refers to a circuit where there is an excessive amount of opposition to normal current flow.

HORSEPOWER: A measurement of the amount of work; one horsepower is the amount of work necessary to lift 33,000 lbs. one foot in one minute. Brake horsepower (bhp) is the horsepower delivered by an engine on a dynamometer. Net horsepower is the power remaining (measured at the flywheel of the engine) that can be used to turn the wheels after power is consumed through friction and running the engine accessories (water pump, alternator, air pump, fan etc.)

HOT CIRCUIT: (See circuit, hot; hot lead.)

HOT LEAD: A wire or conductor in the power side of the circuit. (See circuit, hot.)

HOT SIDE SWITCHING: The electrical/electronic circuit control switch is located before the circuit load.

HUB: The center part of a wheel or gear.

HUNTING (BUSYNESS): Repeating quick series of up-shifts and downshifts that causes noticeable change in engine rpm, for example, as in a 4-3-4 shift pattern.

HYDRAULICS: The use of liquid under pressure to transfer force of motion.

HYDROCARBON (HC): Any chemical compound made up of hydrogen and carbon. A major pollutant formed by the engine as a by-product of combustion.

HYDRODYNAMIC DRIVE UNITS: Devices that transmit power solely by the action of a kinetic fluid flow in a closed recirculating path. An impeller energizes the fluid and discharges the high-speed jet stream into the turbine for power output.

HYDROMETER: An instrument used to measure the specific gravity of a solution.

HYDROPLANING: A phenomenon of driving when water builds up under the tire tread, causing it to lose contact with the road. Slowing down will usually restore normal tire contact with the road.

HYPOID GEARSET: The drive pinion gear may be placed below or above the centerline of the driven gear; often used as a final drive gearset.

IDLE MIXTURE: The mixture of air and fuel (usually about 14:1) being fed to the cylinders. The idle mixture screw(s) are sometimes adjusted as part of a tune-up.

IDLER ARM: Component of the steering linkage which is a geometric duplicate of the steering gear arm. It supports the right side of the center steering link.

IMPELLER: Often called a pump, the impeller is the power input (drive) member of a hydrodynamic drive. As part of the torque converter cover, it acts as a centrifugal pump and puts the fluid in motion.

INCH POUND (inch lbs.; sometimes in. lb. or in. lbs.): One twelfth of a foot pound.

INDUCTANCE: The force that produces voltage when a conductor is passed through a magnetic field.

INDUCTION: A means of transferring electrical energy in the form of a magnetic field. Principle used in the ignition coil to increase voltage.

INITIAL FEEL: A distinct firmer feel at start of shift when compared with feel at finish of shift.

INJECTOR: A device which receives metered fuel under relatively low pressure and is activated to inject the fuel into the engine under relatively high pressure at a predetermined time.

INPUT: In an automatic transmission, the source of power from the engine is absorbed by the torque converter, which provides the power input into the transmission. The turbine drives the input(turbine)shaft.

INPUT SHAFT: The shaft to which torque is applied, usually carrying the driving gear or gears.

INTAKE MANIFOLD: A casting of passages or pipes used to conduct air or a fuel/air mixture to the cylinders.

INTERNAL GEAR: The ring-like outer gear of a planetary gearset with the gear teeth cut on the inside of the ring to provide a mesh with the planet pinions.

ISOLATION (CLAMPING) DIODES: Diodes positioned in a circuit to prevent self-induction from damaging electronic components.

IX ROTARY GEAR PUMP: Contains two rotating members, one shaped with internal gear teeth and the other with external gear teeth. As the gears separate, the fluid fills the gaps between gear teeth, is pulled across a crescent-shaped divider, and then is forced to flow through the outlet as the gears mesh.

IX ROTARY LOBE PUMP: Sometimes referred to as a gerotor type pump. Two rotating members, one shaped with internal lobes and the other with external lobes, separate and then mesh to cause fluid to flow.

JOURNAL: The bearing surface within which a shaft operates.

JUMPER CABLES: Two heavy duty wires with large alligator clips used to provide power from a charged battery to a discharged battery mounted in a vehicle.

JUMPSTART: Utilizing the sufficiently charged battery of one vehicle to start the engine of another vehicle with a discharged battery by the use of jumper cables.

KEY: A small block usually fitted in a notch between a shaft and a hub to prevent slippage of the two parts.

KICKDOWN: Detent downshift system; either linkage, cable, or electrically controlled.

KILO: A prefix used in the metric system to indicate one thousand.

KNOCK: Noise which results from the spontaneous ignition of a portion of the air-fuel mixture in the engine cylinder caused by overly advanced ignition timing or use of incorrectly low octane fuel for that engine.

KNOCK SENSOR: An input device that responds to spark knock, caused by over advanced ignition timing.

LABOR TIME: A specific amount of time required to perform a certain repair or diagnostic service as defined by a vehicle or after-market manufacturer .

LACQUER: A quick-drying automotive paint.

LATE: Shift that occurs when engine is at higher than normal rpm for given amount of throttle.

LIGHT-EMITTING DIODE (LED): A semiconductor diode that emits light as electrical current flows through it; used in some electronic display devices to emit a red or other color light.

LIGHT THROTTLE: Approximately one-fourth of accelerator pedal travel.

LIMITED SLIP: A type of differential which transfers driving force to the wheel with the best traction.

LIMP-IN MODE: Electrical shutdown of the transmission/ transaxle output solenoids, allowing only forward and reverse gears that are hydraulically energized by the manual valve. This permits the vehicle to be driven to a service facility for repair.

LIP SEAL: Molded synthetic rubber seal designed with an outer sealing edge (lip) that points into the fluid containing area to be sealed. This type of seal is used where rotational and axial forces are present.

LITHIUM-BASE GREASE: Chassis and wheel bearing grease using lithium as a base. Not compatible with sodium-base grease.

LOAD DEVICE: A circuit's resistance that converts the electrical energy into light, sound, heat, or mechanical movement.

LOAD RANGE: Indicates the number of plies at which a tire is rated. Load range B equals four-ply rating; C equals six-ply rating; and, D equals an eight-ply rating.

LOAD TORQUE: The amount of output torque needed from the transmission/transaxle to overcome the vehicle load.

LOCKING HUBS: Accessories used on part-time four-wheel drive systems that allow the front wheels to be disengaged from the drive train when four-wheel drive is not being used. When four-wheel drive is desired, the hubs are engaged, locking the wheels to the drive train.

LOCKUP CONVERTER: A torque converter that operates hydraulically and mechanically. When an internal apply plate (lockup plate) clamps to the torque converter cover, hydraulic slippage is eliminated.

LOCK RING: See Circlip or Snapring

MAGNET: Any body with the property of attracting iron or steel.

MAGNETIC FIELD: The area surrounding the poles of a magnet that is affected by its attraction or repulsion forces.

MAIN LINE PRESSURE: Often called control pressure or line pressure, it refers to the pressure of the oil leaving the pump and is controlled by the pressure regulator valve.

MALFUNCTION INDICATOR LAMP (MIL): Previously known as a check engine light, the dash-mounted MIL illuminates and signals the driver that an emission or driveability problem with the powertrain has been detected by the ECM/PCM. When this occurs, at least one diagnostic trouble code (DTC) has been stored into the control module memory.

MANIFOLD ABSOLUTE PRESSURE (MAP) SENSOR: Reads the amount of air pressure (vacuum) in the engine's intake manifold system; its signal is used to analyze engine load conditions.

MANIFOLD VACUUM: Low pressure in an engine intake manifold formed just below the throttle plates. Manifold vacuum is highest at idle and drops under acceleration.

MANIFOLD: A casting of passages or set of pipes which connect the cylinders to an inlet or outlet source.

MANUAL LEVER POSITION SWITCH (MLPS): A mechanical switching unit that is typically mounted externally to the transmission/transaxle to inform the PCM/ECM which gear range the driver has selected.

MANUAL VALVE: Located inside the transmission/transaxle, it is directly connected to the driver's shift lever. The position of the manual valve determines which hydraulic circuits will be charged with oil pressure and the operating mode of the transmission.

MANUAL VALVE LEVER POSITION SENSOR (MVLPS): The input from this device tells the TCM what gear range was selected.

MASS AIR FLOW (MAF) SENSOR: Measures the airflow into the engine.

MASTER CYLINDER: The primary fluid pressurizing device in a hydraulic system. In automotive use, it is found in brake and hydraulic clutch systems and is pedal activated, either directly or, in a power brake system, through the power booster.

MacPherson STRUT: A suspension component combining a shock absorber and spring in one unit.

MEDIUM THROTTLE: Approximately one-half of accelerator pedal travel.

MEGA: A metric prefix indicating one million.

MEMBER: An independent component of a hydrodynamic unit such as an impeller, a stator, or a turbine. It may have one or more elements.

MERCON: A fluid developed by Ford Motor Company in 1988. It contains a friction modifier and closely resembles operating characteristics of Dexron.

METAL SEALING RINGS: Made from cast iron or aluminum, their primary application is with dynamic components involving pressure sealing circuits of rotating members. These rings are designed with either butt or hook lock end joints.

METER (ANALOG): A linear-style meter representing data as lengths; a needle-style instrument interfacing with logical numerical increments. This style of electrical meter uses relatively low impedance internal resistance and cannot be used for testing electronic circuitry.

METER (DIGITAL): Uses numbers as a direct readout to show values. Most meters of this style use high impedance internal resistance and must be used for testing low current electronic circuitry.

MICRO: A metric prefix indicating one-millionth (0.000001).

MILLI: A metric prefix indicating one-thousandth (0.001).

MINIMUM THROTTLE: The least amount of throttle opening required for upshift; normally close to zero throttle.

MISFIRE: Condition occurring when the fuel mixture in a cylinder fails to ignite, causing the engine to run roughly.

MODULE: Electronic control unit, amplifier or igniter of solid state or integrated design which controls the current flow in the ignition primary circuit based on input from the pick-up coil. When the module opens the primary circuit, high secondary voltage is induced in the coil.

MODULATED: In an electronic-hydraulic converter clutch system (or shift valve system), the term modulated refers to the pulsing of a solenoid, at a variable rate. This action controls the buildup of oil pressure in the hydraulic circuit to allow a controlled amount of clutch slippage.

MODULATED CONVERTER CLUTCH CONTROL (MCCC): A pulse width duty cycle valve that controls the converter lockup apply pressure and maximizes smoother transitions between lock and unlock conditions.

MODULATOR PRESSURE (THROTTLE PRESSURE): A hydraulic signal oil pressure relating to the amount of engine load, based on either the amount of throttle plate opening or engine vacuum.

MODULATOR VALVE: A regulator valve that is controlled by engine vacuum, providing a hydraulic pressure that varies in relation to engine torque. The hydraulic torque signal functions to delay the shift pattern and provide a line pressure boost. (See throttle valve.)

MOTOR: An electromagnetic device used to convert electrical energy into mechanical energy.

MULTIPLE-DISC CLUTCH: A grouping of steel and friction lined plates that, when compressed together by hydraulic pressure acting upon a piston, lock or unlock a planetary member.

MULTI-WEIGHT: Type of oil that provides adequate lubrication at both high and low temperatures.

needed to move one amp through a resistance of one ohm.

MUSHY: Same as soft; slow and drawn out clutch apply with very little shift feel.

MUTUAL INDUCTION: The generation of current from one wire circuit to another by movement of the magnetic field surrounding a current-carrying circuit as its ampere flow increases or decreases.

NEEDLE BEARING: A bearing which consists of a number (usually a large number) of long, thin rollers.

NITROGEN OXIDE (NOx): One of the three basic pollutants found in the exhaust emission of an internal combustion engine. The amount of NOx usually varies in an inverse proportion to the amount of HC and CO.

NONPOSITIVE SEALING: A sealing method that allows some minor leakage, which normally assists in lubrication.

O2 SENSOR: Located in the engine's exhaust system, it is an input device to the ECM/PCM for managing the fuel delivery and ignition system. A scanner can be used to observe the fluctuating voltage readings produced by an O2 sensor as the oxygen content of the exhaust is analyzed.

O-RING SEAL: Molded synthetic rubber seal designed with a circular cross-section. This type of seal is used primarily in static applications.

OBD II (ON-BOARD DIAGNOSTICS, SECOND GENERATION): Refers to the federal law mandating tighter control of 1996 and newer vehicle emissions, active monitoring of related devices, and standardization of terminology, data link connectors, and other technician concerns.

OCTANE RATING: A number, indicating the quality of gasoline based on its ability to resist knock. The higher the number, the better the quality. Higher compression engines require higher octane gas.

OEM: Original Equipment Manufactured. OEM equipment is that furnished standard by the manufacturer.

OFFSET: The distance between the vertical center of the wheel and the mounting surface at the lugs. Offset is positive if the center is outside the lug circle; negative offset puts the center line inside the lug circle.

OHM'S LAW: A law of electricity that states the relationship between voltage, current, and resistance. Volts = amperes x ohms

OHM: The unit used to measure the resistance of conductor-to-electrical flow. One ohm is the amount of resistance that limits current flow to one ampere in a circuit with one volt of pressure.

OHMMETER: An instrument used for measuring the resistance, in ohms, in an electrical circuit.

ONE-WAY CLUTCH: A mechanical clutch of roller or sprag design that resists torque or transmits power in one direction only. It is used to either hold or drive a planetary member.

ONE-WAY ROLLER CLUTCH: A mechanical device that transmits or holds torque in one direction only.

OPEN CIRCUIT: A break or lack of contact in an electrical circuit, either intentional (switch) or unintentional (bad connection or broken wire).

ORIFICE: Located in hydraulic oil circuits, it acts as a restriction. It slows down fluid flow to either create back pressure or delay pressure buildup downstream.

OSCILLOSCOPE: A piece of test equipment that shows electric impulses as a pattern on a screen. Engine performance can be analyzed by interpreting these patterns.

OUTPUT SHAFT: The shaft which transmits torque from a device, such as a transmission.

OUTPUT SPEED SENSOR (OSS): Identifies transmission/transaxle output shaft speed for shift timing and may be used to calculate TCC slip; often functions as the VSS (vehicle speed sensor).

OVERDRIVE: (1.) A device attached to or incorporated in a transmission/transaxle that allows the engine to turn less than one full revolution for every complete revolution of the wheels. The net effect is to reduce engine rpm, thereby using less fuel. A typical overdrive gear ratio would be .87:1, instead of the normal 1:1 in high gear. (2.) A gear assembly which produces more shaft revolutions than that transmitted to it.

OVERDRIVE PLANETARY GEARSET: A single planetary gearset designed to provide a direct drive and overdrive ratio. When coupled to a three-speed transmission/transaxle configuration, a four-speed/overdrive unit is present.

OVERHEAD CAMSHAFT (OHC): An engine configuration in which the camshaft is mounted on top of the cylinder head and operates the valve either directly or by means of rocker arms.

OVERHEAD VALVE (OHV): An engine configuration in which all of the valves are located in the cylinder head and the camshaft is located in the cylinder block. The camshaft operates the valves via lifters and pushrods.

OVERRUNCLUTCH: Another name for a one-way mechanical clutch. Applies to both roller and sprag designs.

OVERSTEER: The tendency of some vehicles, when steering into a turn, to over-respond or steer more than required, which could result in excessive slip of the rear wheels. Opposite of under-steer.

OXIDATION STABILIZERS: Absorb and dissipate heat. Automatic transmission fluid has high resistance to varnish and sludge buildup that occurs from excessive heat that is generated primarily in the torque converter. Local temperatures as high as 6000F (3150C) can occur at the clutch plates during engagement, and this heat must be absorbed and dissipated. If the fluid cannot withstand the heat, it burns or oxidizes, resulting in an almost immediate destruction of friction materials, clogged filter screen and hydraulic passages, and sticky valves.

OXIDES OF NITROGEN: See nitrogen oxide (NOx).

OXYGEN SENSOR: Used with a feedback system to sense the presence of oxygen in the exhaust gas and signal the computer which can use the voltage signal to determine engine operating efficiency and adjust the air/fuel ratio.

PARALLEL CIRCUIT: (See circuit, parallel.)

PARTS WASHER: A basin or tub, usually with a built-in pump mechanism and hose used for circulating chemical solvent for the purpose of cleaning greasy, oily and dirty components.

PART-TIME FOUR WHEEL DRIVE: A system that is normally in the two wheel drive mode and only runs in four-wheel drive when the system is manually engaged because more traction is desired. Two or four wheel drive is normally selected by a lever to engage the front axle, but if locking hubs are used, these must also be manually engaged in the Lock position. Otherwise, the front axle will not drive the front wheels.

PASSIVE RESTRAINT: Safety systems such as air bags or automatic seat belts which operate with no action required on the part of the driver or passenger. Mandated by Federal regulations on all vehicles sold in the U.S. after 1990.

PAYLOAD: The weight the vehicle is capable of carrying in addition to its own weight. Payload includes weight of the driver, passengers and cargo, but not coolant, fuel, lubricant, spare tire, etc.

PCM: Powertrain control module.

PCV VALVE: A valve usually located in the rocker cover that vents crankcase vapors back into the engine to be reburned.

PERCOLATION: A condition in which the fuel actually "boils," due to excessive heat. Percolation prevents proper atomization of the fuel causing rough running.

PICK-UP COIL: The coil in which voltage is induced in an electronic ignition.

PING: A metallic rattling sound produced by the engine during acceleration. It is usually due to incorrect ignition timing or a poor grade of gasoline.

PINION: The smaller of two gears. The rear axle pinion drives the ring gear which transmits motion to the axle shafts.

PINION GEAR: The smallest gear in a drive gear assembly.

PISTON: A disc or cup that fits in a cylinder bore and is free to move. In hydraulics, it provides the means of converting hydraulic pressure into a usable force. Examples of piston applications are found in servo, clutch, and accumulator units.

PISTON RING: An open-ended ring which fits into a groove on the outer diameter of the piston. Its chief function is to form a seal between the piston and cylinder wall. Most automotive pistons have three rings: two for compression sealing; one for oil sealing.

PITMAN ARM: A lever which transmits steering force from the steering gear to the steering linkage.

PLANET CARRIER: A basic member of a planetary gear assembly that carries the pinion gears.

PLANET PINIONS: Gears housed in a planet carrier that are in constant mesh with the sun gear and internal gear. Because they have their own independent rotating centers, the pinions are capable of rotating around the sun gear or the inside of the internal gear.

PLANETARY GEAR RATIO: The reduction or overdrive ratio developed by a planetary gearset.

PLANETARY GEARSET: In its simplest form, it is made up of a basic assembly group containing a sun gear, internal gear, and planet carrier. The gears are always in constant mesh and offer a wide range of gear ratio possibilities.

PLANETARY GEARSET (COMPOUND): Two planetary gearsets combined together.

PLANETARY GEARSET (SIMPLE): An assembly of gears in constant mesh consisting of a sun gear, several pinion gears mounted in a carrier, and a ring gear. It provides gear ratio and direction changes, in addition to a direct drive and a neutral.

PLY RATING: A. rating given a tire which indicates strength (but not necessarily actual plies). A two-ply/four-ply rating has only two plies, but the strength of a four-ply tire.

POLARITY: Indication (positive or negative) of the two poles of a battery.

PORT: An opening for fluid intake or exhaust.

POSITIVE SEALING: A sealing method that completely prevents leakage.

POTENTIAL: Electrical force measured in volts; sometimes used interchangeably with voltage.

POWER: The ability to do work per unit of time, as expressed in horsepower; one horsepower equals 33,000 ft. lbs. of work per minute, or 550 ft. lbs. of work per second.

POWER FLOW: The systematic flow or transmission of power through the gears, from the input shaft to the output shaft.

POWER-TO-WEIGHT RATIO: Ratio of horsepower to weight of car.

POWERTRAIN: See Drivetrain.

POWERTRAIN CONTROL MODULE (PCM): Current designation for the engine control module (ECM). In many cases, late model vehicle control units manage the engine as well as the transmission. In other settings, the PCM controls the engine and is interfaced with a TCM to control transmission functions.

Ppm: Parts per million; unit used to measure exhaust emissions.

PREIGNITION: Early ignition of fuel in the cylinder, sometimes due to glowing carbon deposits in the combustion chamber. Preignition can be damaging since combustion takes place prematurely.

PRELOAD: A predetermined load placed on a bearing during assembly or by adjustment.

PRESS FIT: The mating of two parts under pressure, due to the inner diameter of one being smaller than the outer diameter of the other, or vice versa; an interference fit.

PRESSURE: The amount of force exerted upon a surface area.

PRESSURE CONTROL SOLENOID (PCS): An output device that provides a boost oil pressure to the mainline regulator valve to control line pressure. Its operation is determined by the amount of current sent from the PCM.

PRESSURE GAUGE: An instrument used for measuring the fluid pressure in a hydraulic circuit.

PRESSURE REGULATOR VALVE: In automatic transmissions, its purpose is to regulate the pressure of the pump output and supply the basic fluid pressure necessary to operate the transmission. The regulated fluid pressure may be referred to as mainline pressure, line pressure, or control pressure.

PRESSURE SWITCH ASSEMBLY (PSA): Mounted inside the transmission, it is a grouping of oil pressure switches that inputs to the PCM when certain hydraulic passages are charged with oil pressure.

PRESSURE PLATE: A spring-loaded plate (part of the clutch) that transmits power to the driven (friction) plate when the clutch is engaged.

PRIMARY CIRCUIT: The low voltage side of the ignition system which consists of the ignition switch, ballast resistor or resistance wire, bypass, coil, electronic control unit and pick-up coil as well as the connecting wires and harnesses.

PROFILE: Term used for tire measurement (tire series), which is the ratio of tire height to tread width.

PROM (PROGRAMMABLE READ-ONLY MEMORY): The heart of the computer that compares input data and makes the engineered program or strategy decisions about when to trigger the appropriate output based on stored computer instructions.

PULSE GENERATOR: A two-wire pickup sensor used to produce a fluctuating electrical signal. This changing signal is read by the controller to determine the speed of the object and can be used to measure transmission/transaxle input speed, output speed, and vehicle speed.

PSI: Pounds per square inch; a measurement of pressure.

PULSE WIDTH DUTY CYCLE SOLENOID (PULSE WIDTH MODULATED SOLENOID): A computer-controlled solenoid that turns on and off at a variable rate producing a modulated oil pressure; often referred to as a pulse width modulated (PWM) solenoid. Employed in many electronic automatic transmissions and transaxles, these solenoids are used to manage shift control and converter clutch hydraulic circuits.

PUSHROD: A steel rod between the hydraulic valve lifter and the valve rocker arm in overhead valve (OHV) engines.

PUMP: A mechanical device designed to create fluid flow and pressure buildup in a hydraulic system.

QUARTER PANEL: General term used to refer to a rear fender. Quarter panel is the area from the rear door opening to the tail light area and from rear wheel well to the base of the trunk and roof-line.

RACE: The surface on the inner or outer ring of a bearing on which the balls, needles or rollers move.

RACK AND PINION: A type of automotive steering system using a pinion gear attached to the end of the steering shaft. The pinion meshes with a long rack attached to the steering linkage.

RADIAL TIRE: Tire design which uses body cords running at right angles to the center line of the tire. Two or more belts are used to give tread strength. Radials can be identified by their characteristic sidewall bulge.

RADIATOR: Part of the cooling system for a water-cooled engine, mounted in the front of the vehicle and connected to the engine with rubber hoses. Through the radiator, excess combustion heat is dissipated into the atmosphere through forced convection using a water and glycol based mixture that circulates through, and cools, the engine.

RANGE REFERENCE AND CLUTCH/BAND APPLY CHART: A guide that shows the application of clutches and bands for each gear, within the selector range positions. These charts are extremely useful for understanding how the unit operates and for diagnosing malfunctions.

RAVIGNEAUX GEARSET: A compound planetary gearset that features matched dual planetary pinions (sets of two) mounted in a single planet carrier. Two sun gears and one ring mesh with the carrier pinions.

REACTION MEMBER: The stationary planetary member, in a planetary gearset, that is grounded to the transmission/transaxle case through the use of friction and wedging devices known as bands, disc clutches, and one-way clutches.

REACTION PRESSURE: The fluid pressure that moves a spool valve against an opposing force or forces; the area on which the opposing force acts. The opposing force can be a spring or a combination of spring force and auxiliary hydraulic force.

REACTOR, TORQUE CONVERTER: The reaction member of a fluid torque converter, more commonly called a stator. (See stator.)

REAR MAIN OIL SEAL: A synthetic or rope-type seal that prevents oil from leaking out of the engine past the rear main crankshaft bearing.

RECIRCULATING BALL: Type of steering system in which recirculating steel balls occupy the area between the nut and worm wheel, causing a reduction in friction.

RECTIFIER: A device (used primarily in alternators) that permits electrical current to flow in one direction only.

REDUCTION: (See gear reduction.)

REGULATOR VALVE: A valve that changes the pressure of the oil in a hydraulic circuit as the oil passes through the valve by bleeding off (or exhausting) some of the volume of oil supplied to the valve.

REFRIGERANT 12 (R-12) or 134 (R-134): The generic name of the refrigerant used in automotive air conditioning systems.

REGULATOR: A device which maintains the amperage and/or voltage levels of a circuit at predetermined values.

RELAY: A switch which automatically opens and/or closes a circuit.

RELAY VALVE: A valve that directs flow and pressure. Relay valves simply connect or disconnect interrelated passages without restricting the fluid flow or changing the pressure.

RELIEF VALVE: A spring-loaded, pressure-operated valve that limits oil pressure buildup in a hydraulic circuit to a predetermined maximum value.

RELUCTOR: A wheel that rotates inside the distributor and triggers the release of voltage in an electronic ignition.

RESERVOIR: The storage area for fluid in a hydraulic system; often called a sump.

RESIN: A liquid plastic used in body work.

RESIDUAL MAGNETISM: The magnetic strength stored in a material after a magnetizing field has been removed.

RESISTANCE: The opposition to the flow of current through a circuit or electrical device, and is measured in ohms. Resistance is equal to the voltage divided by the amperage.

RESISTOR SPARK PLUG: A spark plug using a resistor to shorten the spark duration. This suppresses radio interference and lengthens plug life.

RESISTOR: A device, usually made of wire, which offers a preset amount of resistance in an electrical circuit.

RESULTANT FORCE: The single effective directional thrust of the fluid force on the turbine produced by the vortex and rotary forces acting in different planes.

RETARD: Set the ignition timing so that spark occurs later (fewer degrees before TDC).

RHEOSTAT: A device for regulating a current by means of a variable resistance.

RING GEAR: The name given to a ring-shaped gear attached to a differential case, or affixed to a flywheel or as part of a planetary gear set.

ROADLOAD: grade.

ROCKER ARM: A lever which rotates around a shaft pushing down (opening) the valve with an end when the other end is pushed up by the pushrod. Spring pressure will later close the valve.

ROCKER PANEL: The body panel below the doors between the wheel opening.

ROLLER BEARING: A bearing made up of hardened inner and outer races between which hardened steel rollers move.

ROLLER CLUTCH: A type of one-way clutch design using rollers and springs mounted within an inner and outer cam race assembly.

ROTARY FLOW: The path of the fluid trapped between the blades of the members as they revolve with the rotation of the torque converter cover (rotational inertia).

ROTOR: (1.) The disc-shaped part of a disc brake assembly, upon which the brake pads bear; also called, brake disc. (2.) The device mounted atop the distributor shaft, which passes current to the distributor cap tower contacts.

ROTARY ENGINE: See Wankel engine.

RPM: Revolutions per minute (usually indicates engine speed).

RTV: A gasket making compound that cures as it is exposed to the atmosphere. It is used between surfaces that are not perfectly machined to one another, leaving a slight gap that the RTV fills and in which it hardens. The letters RTV represent room temperature vulcanizing.

RUN-ON: Condition when the engine continues to run, even when the key is turned off. See dieseling.

SEALED BEAM: A automotive headlight. The lens, reflector and filament from a single unit.

SEATBELT INTERLOCK: A system whereby the car cannot be started unless the seatbelt is buckled.

SECONDARY CIRCUIT: The high voltage side of the ignition system, usually above 20,000 volts. The secondary includes the ignition coil, coil wire, distributor cap and rotor, spark plug wires and spark plugs.

SELF-INDUCTION: The generation of voltage in a current-carrying wire by changing the amount of current flowing within that wire.

SEMI-CONDUCTOR: A material (silicon or germanium) that is neither a good conductor nor an insulator; used in diodes and transistors.

SEMI-FLOATING AXLE: In this design, a wheel is attached to the axle shaft, which takes both drive and cornering loads. Almost all solid axle passenger cars and light trucks use this design.

SENDING UNIT: A mechanical, electrical, hydraulic or electromagnetic device which transmits information to a gauge.

SENSOR: Any device designed to measure engine operating conditions or ambient pressures and temperatures. Usually electronic in nature and designed to send a voltage signal to an on-board computer, some sensors may operate as a simple on/off switch or they may provide a variable voltage signal (like a potentiometer) as conditions or measured parameters change.

SERIES CIRCUIT: (See circuit, series.)

SERPENTINE BELT: An accessory drive belt, with small multiple v-ribs, routed around most or all of the engine-powered accessories such as the alternator and power steering pump. Usually both the front and the back side of the belt comes into contact with various pulleys.

SERVO: In an automatic transmission, it is a piston in a cylinder assembly that converts hydraulic pressure into mechanical force and movement; used for the application of the bands and clutches.

SHIFT BUSYNESS: When referring to a torque converter clutch, it is the frequent apply and release of the clutch plate due to uncommon driving conditions.

SHIFT VALVE: Classified as a relay valve, it triggers the automatic shift in response to a governor and a throttle signal by directing fluid to the appropriate band and clutch apply combination to cause the shift to occur.

SHIM: Spacers of precise, predetermined thickness used between parts to establish a proper working relationship.

SHIMMY: Vibration (sometimes violent) in the front end caused by misaligned front end, out of balance tires or worn suspension components.

SHORT CIRCUIT: An electrical malfunction where current takes the path of least resistance to ground (usually through damaged insulation). Current flow is excessive from low resistance resulting in a blown fuse.

SHUDDER: Repeated jerking or stick-slip sensation, similar to chuggle but more severe and rapid in nature, that may be most noticeable during certain ranges of vehicle speed; also used to define condition after converter clutch engagement.

SIMPSON GEARSET: A compound planetary gear train that integrates two simple planetary gearsets referred to as the front planetary and the rear planetary.

SINGLE OVERHEAD CAMSHAFT: See overhead camshaft.

SKIDPLATE: A metal plate attached to the underside of the body to protect the fuel tank, transfer case or other vulnerable parts from damage.

SLAVE CYLINDER: In automotive use, a device in the hydraulic clutch system which is activated by hydraulic force, disengaging the clutch.

SLIPPING: Noticeable increase in engine rpm without vehicle speed increase; usually occurs during or after initial clutch or band engagement.

SLUDGE: Thick, black deposits in engine formed from dirt, oil, water, etc. It is usually formed in engines when oil changes are neglected.

SNAP RING: A circular retaining clip used inside or outside a shaft or part to secure a shaft, such as a floating wrist pin.

SOFT: Slow, almost unnoticeable clutch apply with very little shift feel.

SOFTCODES: DTCs that have been set into the PCM memory but are not present at the time of testing; often referred to as history or intermittent codes.

SOHC: Single overhead camshaft.

SOLENOID: An electrically operated, magnetic switching device.

SPALLING: A wear pattern identified by metal chips flaking off the hardened surface. This condition is caused by foreign particles, overloading situations, and/or normal wear.

SPARK PLUG: A device screwed into the combustion chamber of a spark ignition engine. The basic construction is a conductive core inside of a ceramic insulator, mounted in an outer conductive base. An electrical charge from the spark plug wire travels along the conductive core and jumps a preset air gap to a grounding point or points at the end of the conductive base. The resultant spark ignites the fuel/air mixture in the combustion chamber.

SPECIFIC GRAVITY (BATTERY): The relative weight of liquid (battery electrolyte) as compared to the weight of an equal volume of water.

SPLINES: Ridges machined or cast onto the outer diameter of a shaft or inner diameter of a bore to enable parts to mate without rotation.

SPLIT TORQUE DRIVE: In a torque converter, it refers to parallel paths of torque transmission, one of which is mechanical and the other hydraulic.

SPONGY PEDAL: A soft or spongy feeling when the brake pedal is depressed. It is usually due to air in the brake lines.

SPOOLVALVE: A precision-machined, cylindrically shaped valve made up of lands and grooves. Depending on its position in the valve bore, various interconnecting hydraulic circuit passages are either opened or closed.

SPRAG CLUTCH: A type of one-way clutch design using cams or contoured-shaped sprags between inner and outer races. (See one-way clutch.)

SPRUNG WEIGHT: The weight of a car supported by the springs.

SQUARE-CUT SEAL: Molded synthetic rubber seal designed with a square- or rectangular-shaped cross-section. This type of seal is used for both dynamic and static applications.

SRS: Supplemental restraint system

STABILIZER (SWAY) BAR: A bar linking both sides of the suspension. It resists sway on turns by taking some of added load from one wheel and putting it on the other.

STAGE: The number of turbine sets separated by a stator. A turbine set may be made up of one or more turbine members. A three-element converter is classified as a single stage.

STALL: In fluid drive transmission/transaxle applications, stall refers to engine rpm with the transmission/transaxle engaged and the vehicle stationary; throttle valve can be in any position between closed and wide open.

STALL SPEED: In fluid drive transmission/transaxle applications, stall speed refers to the maximum engine rpm with the transmission/transaxle engaged and vehicle stationary, when the throttle valve is wide open. (See stall; stall test.)

STALL TEST: A procedure recommended by many manufacturers to help determine the integrity of an engine, the torque converter stator, and certain clutch and band combinations. With the shift lever in each of the forward and reverse positions and with the brakes firmly applied, the accelerator pedal is momentarily pressed to the wide open throttle (WOT) position. The engine rpm reading at full throttle can provide clues for diagnosing the condition of the items listed above.

STALL TORQUE: The maximum design or engineered torque ratio of a fluid torque converter, produced under stall speed conditions. (See stall speed.)

STARTER: A high-torque electric motor used for the purpose of starting the engine, typically through a high ratio geared drive connected to the flywheel ring gear.

STATIC: A sealing application in which the parts being sealed do not move in relation to each other.

STATOR (REACTOR): The reaction member of a fluid torque converter that changes the direction of the fluid as it leaves the turbine to enter the impeller vanes. During the torque multiplication phase, this action assists the impeller's rotary force and results in an increase in torque.

STEERING GEOMETRY: Combination of various angles of suspension components (caster, camber, toe-in); roughly equivalent to front end alignment.

STRAIGHT WEIGHT: Term designating motor oil as suitable for use within a narrow range of temperatures. Outside the narrow temperature range its flow characteristics will not adequately lubricate.

STROKE: The distance the piston travels from bottom dead center to top dead center.

SUBSTITUTION: Replacing one part suspected of a defect with a like part of known quality.

SUMP: The storage vessel or reservoir that provides a ready source of fluid to the pump. In an automatic transmission, the sump is the oil pan. All fluid eventually returns to the sump for recycling into the hydraulic system.

SUN GEAR: In a planetary gearset, it is the center gear that meshes with a cluster of planet pinions.

SUPERCHARGER: An air pump driven mechanically by the engine through belts, chains, shafts or gears from the crankshaft. Two general types of supercharger are the positive displacement and centrifugal type, which pump air in direct relationship to the speed of the engine.

SUPPLEMENTAL RESTRAINT SYSTEM: See air bag.

SURGE: Repeating engine-related feeling of acceleration and deceleration that is less intense than chuggle.

SWITCH: A device used to open, close, or redirect the current in an electrical circuit.

SYNCHROMESH: A manual transmission/transaxle that is equipped with devices (synchronizers) that match the gear speeds so that the transmission/transaxle can be downshifted without clashing gears.

SYNTHETIC OIL: Non-petroleum based oil.

TACHOMETER: A device used to measure the rotary speed of an engine, shaft, gear, etc., usually in rotations per minute.

TDC: Top dead center. The exact top of the piston's stroke.

TEFLON SEALING RINGS: Teflon is a soft, durable, plastic-like material that is resistant to heat and provides excellent sealing. These rings are designed with either scarf-cut joints or as one-piece rings. Teflon sealing rings have replaced many metal ring applications.

TERMINAL: A device attached to the end of a wire or cable to make an electrical connection.

TEST LIGHT, CIRCUIT-POWERED: Uses available circuit voltage to test circuit continuity.

TEST LIGHT, SELF-POWERED: Uses its own battery source to test circuit continuity.

THERMISTOR: A special resistor used to measure fluid temperature; it decreases its resistance with increases in temperature.

THERMOSTAT: A valve, located in the cooling system of an engine, which is closed when cold and opens gradually in response to engine heating, controlling the temperature of the coolant and rate of coolant flow.

THERMOSTATIC ELEMENT: A heat-sensitive, spring-type device that controls a drain port from the upper sump area to the lower sump. When the transaxle fluid reaches operating temperature, the port is closed and the upper sump fills, thus reducing the fluid level in the lower sump.

THROTTLE POSITION (TP) SENSOR: Reads the degree of throttle opening; its signal is used to analyze engine load conditions. The ECM/PCM decides to apply the TCC, or to disengage it for coast or load conditions that need a converter torque boost.

THROTTLE PRESSURE/MODULATOR PRESSURE: A hydraulic signal oil pressure relating to the amount of engine load, based on either the amount of throttle plate opening or engine vacuum.

THROTTLE VALVE: A regulating or balanced valve that is controlled mechanically by throttle linkage or engine vacuum. It sends a hydraulic signal to the shift valve body to control shift timing and shift quality. (See balanced valve; modulator valve.)

THROW-OUT BEARING: As the clutch pedal is depressed, the

throwout bearing moves against the spring fingers of the pressure plate, forcing the pressure plate to disengage from the driven disc.

TIE ROD: A rod connecting the steering arms. Tie rods have threaded ends that are used to adjust toe-in.

TIE-UP: Condition where two opposing clutches are attempting to apply at same time, causing engine to labor with noticeable loss of engine rpm.

TIMING BELT: A square-toothed, reinforced rubber belt that is driven by the crankshaft and operates the camshaft.

TIMING CHAIN: A roller chain that is driven by the crankshaft and operates the camshaft.

TIRE ROTATION: Moving the tires from one position to another to make the tires wear evenly.

TOE-IN (OUT): A term comparing the extreme front and rear of the front tires. Closer together at the front is toe-in; farther apart at the front is toe-out.

TOP DEAD CENTER (TDC): The point at which the piston reaches the top of its travel on the compression stroke.

TORQUE: Measurement of turning or twisting force, expressed as foot-pounds or inch-pounds.

TORQUE CONVERTER: A turbine used to transmit power from a driving member to a driven member via hydraulic action, providing changes in drive ratio and torque. In automotive use, it links the driveplate at the rear of the engine to the automatic transmission.

TORQUE CONVERTER CLUTCH: The apply plate (lockup plate) assembly used for mechanical power flow through the converter.

TORQUE PHASE: Sometimes referred to as slip phase or stall phase, torque multiplication occurs when the turbine is turning at a slower speed than the impeller, and the stator is reactionary (stationary). This sequence generates a boost in output torque.

TORQUE RATING (STALL TORQUE): The maximum torque multiplication that occurs during stall conditions, with the engine at wide open throttle (WOT) and zero turbine speed.

TORQUE RATIO: An expression of the gear ratio factor on torque effect. A 3:1 gear ratio or 3:1 torque ratio increases the torque input by the ratio factor of 3. Input torque (100 ft. lbs.) x 3 = output torque (300 ft. lbs.)

TRACTION: The amount of usable tractive effort before the drive wheels slip on the road contact surface.

TORSION BAR SUSPENSION: Long rods of spring steel which take the place of springs. One end of the bar is anchored and the other arm (attached to the suspension) is free to twist. The bars' resistance to twisting causes springing action.

TRACK: Distance between the centers of the tires where they contact the ground.

TRACTION CONTROL: A control system that prevents the spinning of a vehicle's drive wheels when excess power is applied.

TRACTIVE EFFORT: The amount of force available to the drive wheels, to move the vehicle.

TRANSAXLE: A single housing containing the transmission and differential. Transaxles are usually found on front engine/front wheel drive or rear engine/rear wheel drive cars.

TRANSDUCER: A device that changes energy from one form to another. For example, a transducer in a microphone changes sound energy to electrical energy. In automotive air-conditioning controls used in automatic temperature systems, a transducer changes an electrical signal to a vacuum signal, which operates mechanical doors.

TRANSMISSION: A powertrain component designed to modify torque and speed developed by the engine; also provides direct drive, reverse, and neutral.

TRANSMISSION CONTROL MODULE (TCM): Manages transmission functions. These vary according to the manufacturer's product design but may include converter clutch operation, electronic shift scheduling, and mainline pressure.

TRANSMISSION FLUID TEMPERATURE (TFT) SENSOR: Originally called a transmission oil temperature (TOT) sensor, this input device to the ECM/PCM senses the fluid temperature and provides a resistance value. It operates on the thermistor principle.

TRANSMISSION INPUT SPEED (TIS) SENSOR: Measures turbine shaft (input shaft) rpm's and compares to engine rpm's to determine torque converter slip. When compared to the transmission output speed sensor or VSS, gear ratio and clutch engagement timing can be determined.

TRANSMISSION OIL TEMPERATURE (TOT) SENSOR: (See transmission fluid temperature (TFT) sensor.)

TRANSMISSION RANGE SELECTOR (TRS) SWITCH: Tells the module which gear shift position the driver has chosen.

TRANSFER CASE: A gearbox driven from the transmission that delivers power to both front and rear driveshafts in a four-wheel drive system. Transfer cases usually have a high and low range set of gears, used depending on how much pulling power is needed.

TRANSISTOR: A semi-conductor component which can be actuated by a small voltage to perform an electrical switching function.

TREAD WEAR INDICATOR: Bars molded into the tire at right angles to the tread that appear as horizontal bars when 1/16 in. of tread remains.

TREAD WEAR PATTERN: The pattern of wear on tires which can be "read" to diagnose problems in the front suspension.

TUNE-UP: A regular maintenance function, usually associated with the replacement and adjustment of parts and components in the electrical and fuel systems of a vehicle for the purpose of attaining optimum performance.

TURBINE: The output (driven) member of a fluid coupling or fluid torque converter. It is splined to the input (turbine) shaft of the transmission.

TURBOCHARGER: An exhaust driven pump which compresses intake air and forces it into the combustion chambers at higher than atmospheric pressures. The increased air pressure allows more fuel to be burned and results in increased horsepower being produced.

TURBULENCE: The interference of molecules of a fluid (or vapor) with each other in a fluid flow.

TYPE F: Transmission fluid developed and used by Ford Motor Company up to 1982. This fluid type provides a high coefficient of friction.

TYPE 7176: The preferred choice of transmission fluid for Chrysler automatic transmissions and transaxles. Developed in 1986, it closely resembles Dexron and Mercon. Type 7176 is the recommended service fill fluid for all Chrysler products utilizing a lockup torque converter dating back to 1978.

U-JOINT (UNIVERSAL JOINT): A flexible coupling in the drive train that allows the driveshafts or axle shafts to operate at different angles and still transmit rotary power.

UNDERSTEER: The tendency of a car to continue straight ahead while negotiating a turn.

UNIT BODY: Design in which the car body acts as the frame.

UNLEADED FUEL: Fuel which contains no lead (a common gasoline additive). The presence of lead in fuel will destroy the functioning elements of a catalytic converter, making it useless.

UNSPRUNG WEIGHT: The weight of car components not supported by the springs (wheels, tires, brakes, rear axle, control arms, etc.).

UPSHIFT: A shift that results in a decrease in torque ratio and an increase in speed.

VACUUM: A negative pressure; any pressure less than atmospheric pressure.

VACUUM ADVANCE: A device which advances the ignition timing in response to increased engine vacuum.

VACUUM GAUGE: An instrument used for measuring the existing vacuum in a vacuum circuit or chamber. The unit of measure is inches (of mercury in a barometer).

VACUUM MODULATOR: Generates a hydraulic oil pressure in response to the amount of engine vacuum.

VALVES: Devices that can open or close fluid passages in a hydraulic system and are used for directing fluid flow and controlling pressure.

VALVE BODY ASSEMBLY: The main hydraulic control assembly of the transmission/transaxle that contains numerous valves, check balls, and other components to control the distribution of pressurized oil throughout the transmission.

VALVE CLEARANCE: The measured gap between the end of the valve stem and the rocker arm, cam lobe or follower that activates the valve.

VALVE GUIDES: The guide through which the stem of the valve passes.

The guide is designed to keep the valve in proper alignment.

VALVE LASH (clearance): The operating clearance in the valve train.

VALVE TRAIN: The system that operates intake and exhaust valves, consisting of camshaft, valves and springs, lifters, pushrods and rocker arms.

VAPOR LOCK: Boiling of the fuel in the fuel lines due to excess heat. This will interfere with the flow of fuel in the lines and can completely stop the flow. Vapor lock normally only occurs in hot weather.

VARIABLE DISPLACEMENT (VARIABLE CAPACITY) VANE PUMP: Slipper-type vanes, mounted in a revolving rotor and contained within the bore of a movable slide, capture and then force fluid to flow. Movement of the slide to various positions changes the size of the vane chambers and the amount of fluid flow. **Note:** GM refers to this pump design as variable displacement, and Ford terms it variable capacity.

VARIABLE FORCE SOLENOID (VFS): Commonly referred to as the electronic pressure control (EPC) solenoid, it replaces the cable/linkage style of TV system control and is integrated with a spool valve and spring assembly to control pressure. A variable computer-controlled current flow varies the internal force of the solenoid on the spool valve and resulting control pressure.

VARIABLE ORIFICE THERMAL VALVE: Temperature-sensitive hydraulic oil control device that adjusts the size of a circuit path opening. By altering the size of the opening, the oil flow rate is adapted for cold to hot oil viscosity changes.

VARNISH: Term applied to the residue formed when gasoline gets old and stale.

VCM: See Electronic Control Unit (ECU).

VEHICLE SPEED SENSOR (VSS): Provides an electrical signal to the computer module, measuring vehicle speed, and affects the torque converter clutch engagement and release.

VESPEL SEALING RINGS: Hard plastic material that produces excellent sealing in dynamic settings. These rings are found in late versions of the 4T60 and in all 4T60-E and 4T80-E transaxles.

VISCOSITY: The ability of a fluid to flow. The lower the viscosity rating, the easier the fluid will flow. 10 weight motor oil will flow much easier than 40 weight motor oil.

VISCOSITY INDEX IMPROVERS: Keeps the viscosity nearly constant with changes in temperature. This is especially important at low temperatures, when the oil needs to be thin to aid in shifting and for cold-weather starting. Yet it must not be so thin that at high temperatures it will cause excessive hydraulic leakage so that pumps are unable to maintain the proper pressures.

VISCOUS CLUTCH: A specially designed torque converter clutch apply plate that, through the use of a silicon fluid, clamps smoothly and absorbs torsional vibrations.

VOLT: Unit used to measure the force or pressure of electricity. It is defined as the pressure needed to move one amp through the resistance of one ohm.

VOLTAGE: The electrical pressure that causes current to flow. Voltage is measured in volts (V).

VOLTAGE, APPLIED: The actual voltage read at a given point in a circuit. It equals the available voltage of the power supply minus the losses in the circuit up to that point.

VOLTAGE DROP: The voltage lost or used in a circuit by normal loads such as a motor or lamp or by abnormal loads such as a poor (high-resistance) lead or terminal connection.

VOLTAGE REGULATOR: A device that controls the current output of the alternator or generator.

VOLTMETER: An instrument used for measuring electrical force in units called volts. Voltmeters are always connected parallel with the circuit being tested.

VORTEX FLOW: The crosswise or circulatory flow of oil between the blades of the members caused by the centrifugal pumping action of the impeller.

WANKEL ENGINE: An engine which uses no pistons. In place of pistons, triangular-shaped rotors revolve in specially shaped housings.

WATER PUMP: A belt driven component of the cooling system that mounts on the engine, circulating the coolant under pressure.

WATT: The unit for measuring electrical power. One watt is the product of one ampere and one volt (watts equals amps times volts). Wattage is the horsepower of electricity (746 watts equal one horsepower).

WHEEL ALIGNMENT: Inclusive term to describe the front end geometry (caster, camber, toe-in/out).

WHEEL CYLINDER: Found in the automotive drum brake assembly, it is a device, actuated by hydraulic pressure, which, through internal pistons, pushes the brake shoes outward against the drums.

WHEEL WEIGHT: Small weights attached to the wheel to balance the wheel and tire assembly. Out-of-balance tires quickly wear out and also give erratic handling when installed on the front.

WHEELBASE: Distance between the center of front wheels and the center of rear wheels.

WIDE OPEN THROTTLE (WOT): Full travel of accelerator pedal.

WORK: The force exerted to move a mass or object. Work involves motion; if a force is exerted and no motion takes place, no work is done. Work per unit of time is called power. Work = force x distance = ft. lbs. 33,000 ft. lbs. in one minute = 1 horsepower

ZERO-THROTTLE COAST DOWN: A full release of accelerator pedal while vehicle is in motion and in drive range.

Commonly Used Abbreviations

2

2WD	Two Wheel Drive

4

4WD	Four Wheel Drive

A

A/C	Air Conditioning
ABDC	After Bottom Dead Center
ABS	Anti-lock Brakes
AC	Alternating Current
ACL	Air cleaner
ACT	Air Charge Temperature
AIR	Secondary Air Injection
ALCL	Assembly Line Communications Link
ALDL	Assembly Line Diagnostic Link
AT	Automatic Transaxle/Transmission
ATDC	After Top Dead Center
ATF	Automatic Transmission Fluid
ATS	Air Temperature Sensor
AWD	All Wheel Drive

B

BAP	Barometric Absolute Pressure
BARO	Barometric Pressure
BBDC	Before Bottom Dead Center
BCM	Body Control Module
BDC	Bottom Dead Center
BPT	Backpressure Transducer
BTDC	Before Top Dead Center
BVSV	Bimetallic Vacuum Switching Valve

C

CAC	Charge Air Cooler
CARB	California Air Resources Board
CAT	Catalytic Converter
CCC	Computer Command Control
CCCC	Computer Controlled Catalytic Converter
CCCI	Computer Controlled Coil Ignition
CCD	Computer Controlled Dwell
CDI	Capacitor Discharge Ignition
CEC	Computerized Engine Control
CFI	Continuous Fuel Injection
CIS	Continuous Injection System
CIS-E	Continuous Injection System - Electronic
CKP	Crankshaft Position
CL	Closed Loop
CMP	Camshaft Position
CPP	Clutch Pedal Position
CTOX	Continuous Trap Oxidizer System
CTP	Closed Throttle Position
CVC	Constant Vacuum Control
CYL	Cylinder

D

DBC	Dual Bed Catalyst
DC	Direct Current
DFI	Direct Fuel Injection
DIS	Distributorless Ignition System
DLC	Data Link Connector
DMM	Digital Multimeter
DOHC	Double Overhead Camshaft
DRB	Diagnostic Readout Box
DTC	Diagnostic Trouble Code
DTM	Diagnostic Test Mode
DVOM	Digital Volt/Ohmmeter

E

EBCM	Electronic Brake Control Module
ECM	Engine Control Module
ECT	Engine Coolant Temperature
ECU	Engine Control Unit or Electronic Control Unit
EDIS	Electronic Distributorless Ignition System
EEC	Electronic Engine Control
EEPROM	Electrically Erasable Programmable Read Only Memory
EFE	Early Fuel Evaporation
EGR	Exhaust Gas Recirculation
EGRT	Exhaust Gas Recirculation Temperature
EGRVC	EGR Valve Control
EPROM	Erasable Programmable Read Only Memory
EVAP	Evaporative Emissions
EVP	EGR Valve Position

F

FBC	Feedback Carburetor
FEEPROM	Flash Electrically Erasable Programmable Read Only Memory
FF	Flexible Fuel
FI	Fuel Injection
FT	Fuel Trim
FWD	Front Wheel Drive

G

GND	Ground

H

HAC	High Altitude Compensation
HEGO	Heated Exhaust Gas Oxygen sensor
HEI	High Energy Ignition
HO2 Sensor	Heated Oxygen Sensor

I

IAC	Idle Air Control
IAT	Intake Air Temperature
ICM	Ignition Control Module
IFI	Indirect Fuel Injection
IFS	Inertia Fuel Shutoff
ISC	Idle Speed Control
IVSV	Idle Vacuum Switching Valve

Commonly Used Abbreviations

K

KOEO	Key On, Engine Off
KOER	Key ON, Engine Running
KS	Knock Sensor

M

MAF	Mass Air Flow
MAP	Manifold Absolute Pressure
MAT	Manifold Air Temperature
MC	Mixture Control
MDP	Manifold Differential Pressure
MFI	Multiport Fuel Injection
MIL	Malfunction Indicator Lamp or Maintenance
MST	Manifold Surface Temperature
MVZ	Manifold Vacuum Zone

N

NVRAM	Nonvolatile Random Access Memory

O

O2 Sensor	Oxygen Sensor
OBD	On-Board Diagnostic
OC	Oxidation Catalyst
OHC	Overhead Camshaft
OL	Open Loop

P

P/S	Power Steering
PAIR	Pulsed Secondary Air Injection
PCM	Powertrain Control Module
PCS	Purge Control Solenoid
PCV	Positive Crankcase Ventilation
PIP	Profile Ignition Pick-up
PNP	Park/Neutral Position
PROM	Programmable Read Only Memory
PSP	Power Steering Pressure
PTO	Power Take-Off
PTOX	Periodic Trap Oxidizer System

R

RABS	Rear Anti-lock Brake System
RAM	Random Access Memory
ROM	Read Only Memory
RPM	Revolutions Per Minute
RWAL	Rear Wheel Anti-lock Brakes
RWD	Rear Wheel Drive

S

SBC	Single Bed Converter
SBEC	Single Board Engine Controller
SC	Supercharger
SCB	Supercharger Bypass
SFI	Sequential Multiport Fuel Injection
SIR	Supplemental Inflatible Restraint
SOHC	Single Overhead Camshaft
SPL	Smoke Puff Limiter
SPOUT	Spark Output
SRI	Service Reminder Indicator
SRS	Supplemental Restraint System
SRT	System Readiness Test
SSI	Solid State Ignition
ST	Scan Tool
STO	Self-Test Output

T

TAC	Thermostatic Air Cleaner
TBI	Throttle Body Fuel Injection
TC	Turbocharger
TCC	Torque Converter Clutch
TCM	Transmission Control Module
TDC	Top Dead Center
TFI	Thick Film Ignition
TP	Throttle Position
TR Sensor	Transaxle/Transmission Range Sensor
TVV	Thermal Vacuum Valve
TWC	Three-way Catalytic Converter

V

VAF	Volume Air Flow, or Vane Air Flow
VAPS	Variable Assist Power Steering
VRV	Vacuum Regulator Valve
VSS	Vehicle Speed Sensor
VSV	Vacuum Switching Valve

W

WOT	Wide Open Throttle
WU-TWC	Warm Up Three-way Catalytic Converter

ENGLISH TO METRIC CONVERSION: TORQUE

To convert foot-pounds (ft. lbs.) to Newton-meters (Nm), multiply the number of ft. lbs. by 1.36

To convert Newton-meters (Nm) to foot-pounds (ft. lbs.), multiply the number of Nm by 0.7376

ft. lbs.	Nm	ft. lbs.	Nm	ft. lbs.	Nm	ft. lbs.	Nm
0.1	0.1	34	46.2	76	103.4	118	160.5
0.2	0.3	35	47.6	77	104.7	119	161.8
0.3	0.4	36	49.0	78	106.1	120	163.2
0.4	0.5	37	50.3	79	107.4	121	164.6
0.5	0.7	38	51.7	80	108.8	122	165.9
0.6	0.8	39	53.0	81	110.2	123	167.3
0.7	1.0	40	54.4	82	111.5	124	168.6
0.8	1.1	41	55.8	83	112.9	125	170.0
0.9	1.2	42	57.1	84	114.2	126	171.4
1	1.4	43	58.5	85	115.6	127	172.7
2	2.7	44	59.8	86	117.0	128	174.1
3	4.1	45	61.2	87	118.3	129	175.4
4	5.4	46	62.6	88	119.7	130	176.8
5	6.8	47	63.9	89	121.0	131	178.2
6	8.2	48	65.3	90	122.4	132	179.5
7	9.5	49	66.6	91	123.8	133	180.9
8	10.9	50	68.0	92	125.1	134	182.2
9	12.2	51	69.4	93	126.5	135	183.6
10	13.6	52	70.7	94	127.8	136	185.0
11	15.0	53	72.1	95	129.2	137	186.3
12	16.3	54	73.4	96	130.6	138	187.7
13	17.7	55	74.8	97	131.9	139	189.0
14	19.0	56	76.2	98	133.3	140	190.4
15	20.4	57	77.5	99	134.6	141	191.8
16	21.8	58	78.9	100	136.0	142	193.1
17	23.1	59	80.2	101	137.4	143	194.5
18	24.5	60	81.6	102	138.7	144	195.8
19	25.8	61	83.0	103	140.1	145	197.2
20	27.2	62	84.3	104	141.4	146	198.6
21	28.6	63	85.7	105	142.8	147	199.9
22	29.9	64	87.0	106	144.2	148	201.3
23	31.3	65	88.4	107	145.5	149	202.6
24	32.6	66	89.8	108	146.9	150	204.0
25	34.0	67	91.1	109	148.2	151	205.4
26	35.4	68	92.5	110	149.6	152	206.7
27	36.7	69	93.8	111	151.0	153	208.1
28	38.1	70	95.2	112	152.3	154	209.4
29	39.4	71	96.6	113	153.7	155	210.8
30	40.8	72	97.9	114	155.0	156	212.2
31	42.2	73	99.3	115	156.4	157	213.5
32	43.5	74	100.6	116	157.8	158	214.9
33	44.9	75	102.0	117	159.1	159	216.2

METRIC TO ENGLISH CONVERSION: TORQUE

To convert foot-pounds (ft. lbs.) to Newton-meters (Nm), multiply the number of ft. lbs. by 1.36

To convert Newton-meters (Nm) to foot-pounds (ft. lbs.), multiply the number of Nm by 0.7376

Nm	ft. lbs.	Nm	ft. lbs.	Nm	ft. lbs.	Nm	ft. lbs.	Nm	ft. lbs.
0.1	0.1	34	25.0	76	55.9	118	86.8	160	117.6
0.2	0.1	35	25.7	77	56.6	119	87.5	161	118.4
0.3	0.2	36	26.5	78	57.4	120	88.2	162	119.1
0.4	0.3	37	27.2	79	58.1	121	89.0	163	119.9
0.5	0.4	38	27.9	80	58.8	122	89.7	164	120.6
0.6	0.4	39	28.7	81	59.6	123	90.4	165	121.3
0.7	0.5	40	29.4	82	60.3	124	91.2	166	122.1
0.8	0.6	41	30.1	83	61.0	125	91.9	167	122.8
0.9	0.7	42	30.9	84	61.8	126	92.6	168	123.5
1	0.7	43	31.6	85	62.5	127	93.4	169	124.3
2	1.5	44	32.4	86	63.2	128	94.1	170	125.0
3	2.2	45	33.1	87	64.0	129	94.9	171	125.7
4	2.9	46	33.8	88	64.7	130	95.6	172	126.5
5	3.7	47	34.6	89	65.4	131	96.3	173	127.2
6	4.4	48	35.3	90	66.2	132	97.1	174	127.9
7	5.1	49	36.0	91	66.9	133	97.8	175	128.7
8	5.9	50	36.8	92	67.6	134	98.5	176	129.4
9	6.6	51	37.5	93	68.4	135	99.3	177	130.1
10	7.4	52	38.2	94	69.1	136	100.0	178	130.9
11	8.1	53	39.0	95	69.9	137	100.7	179	131.6
12	8.8	54	39.7	96	70.6	138	101.5	180	132.4
13	9.6	55	40.4	97	71.3	139	102.2	181	133.1
14	10.3	56	41.2	98	72.1	140	102.9	182	133.8
15	11.0	57	41.9	99	72.8	141	103.7	183	134.6
16	11.8	58	42.6	100	73.5	142	104.4	184	135.3
17	12.5	59	43.4	101	74.3	143	105.1	185	136.0
18	13.2	60	44.1	102	75.0	144	105.9	186	136.8
19	14.0	61	44.9	103	75.7	145	106.6	187	137.5
20	14.7	62	45.6	104	76.5	146	107.4	188	138.2
21	15.4	63	46.3	105	77.2	147	108.1	189	139.0
22	16.2	64	47.1	106	77.9	148	108.8	190	139.7
23	16.9	65	47.8	107	78.7	149	109.6	191	140.4
24	17.6	66	48.5	108	79.4	150	110.3	192	141.2
25	18.4	67	49.3	109	80.1	151	111.0	193	141.9
26	19.1	68	50.0	110	80.9	152	111.8	194	142.6
27	19.9	69	50.7	111	81.6	153	112.5	195	143.4
28	20.6	70	51.5	112	82.4	154	113.2	196	144.1
29	21.3	71	52.2	113	83.1	155	114.0	197	144.9
30	22.1	72	52.9	114	83.8	156	114.7	198	145.6
31	22.8	73	53.7	115	84.6	157	115.4	199	146.3
32	23.5	74	54.4	116	85.3	158	116.2	200	147.1
33	24.3	75	55.1	117	86.0	159	116.9	201	147.8

ENGLISH/METRIC CONVERSION: TEMPERATURE

To convert Fahrenheit (F°) to Celsius (C°), take F° temperature and subtract 32, multiply the result by 5 and divide the result by 9
To convert Celsius (C°) to Fahrenheit (F°), take C° temperature and multiply it by 9, divide the result by 5 and add 32

F°	C°	F°	C°	C°	F°	C°	F°
-40	-40.0	150	65.6	-38	-36.4	46	114.8
-35	-37.2	155	68.3	-36	-32.8	48	118.4
-30	-34.4	160	71.1	-34	-29.2	50	122
-25	-31.7	165	73.9	-32	-25.6	52	125.6
-20	-28.9	170	76.7	-30	-22	54	129.2
-15	-26.1	175	79.4	-28	-18.4	56	132.8
-10	-23.3	180	82.2	-26	-14.8	58	136.4
-5	-20.6	185	85.0	-24	-11.2	60	140
0	-17.8	190	87.8	-22	-7.6	62	143.6
1	-17.2	195	90.6	-20	-4	64	147.2
2	-16.7	200	93.3	-18	-0.4	66	150.8
3	-16.1	205	96.1	-16	3.2	68	154.4
4	-15.6	210	98.9	-14	6.8	70	158
5	-15.0	212	100.0	-12	10.4	72	161.6
10	-12.2	215	101.7	-10	14	74	165.2
15	-9.4	220	104.4	-8	17.6	76	168.8
20	-6.7	225	107.2	-6	21.2	78	172.4
25	-3.9	230	110.0	-4	24.8	80	176
30	-1.1	235	112.8	-2	28.4	82	179.6
35	1.7	240	115.6	0	32	84	183.2
40	4.4	245	118.3	2	35.6	86	186.8
45	7.2	250	121.1	4	39.2	88	190.4
50	10.0	255	123.9	6	42.8	90	194
55	12.8	260	126.7	8	46.4	92	197.6
60	15.6	265	129.4	10	50	94	201.2
65	18.3	270	132.2	12	53.6	96	204.8
70	21.1	275	135.0	14	57.2	98	208.4
75	23.9	280	137.8	16	60.8	100	212
80	26.7	285	140.6	18	64.4	102	215.6
85	29.4	290	143.3	20	68	104	219.2
90	32.2	295	146.1	22	71.6	106	222.8
95	35.0	300	148.9	24	75.2	108	226.4
100	37.8	305	151.7	26	78.8	110	230
105	40.6	310	154.4	28	82.4	112	233.6
110	43.3	315	157.2	30	86	114	237.2
115	46.1	320	160.0	32	89.6	116	240.8
120	48.9	325	162.8	34	93.2	118	244.4
125	51.7	330	165.6	36	96.8	120	248
130	54.4	335	168.3	38	100.4	122	251.6
135	57.2	340	171.1	40	104	124	255.2
140	60.0	345	173.9	42	107.6	126	258.8
145	62.8	350	176.7	44	111.2	128	262.4

LENGTH CONVERSION

To convert inches (in.) to millimeters (mm), multiply the number of inches by 25.4
To convert millimeters (mm) to inches (in.), multiply the number of millimeters by 0.04

Inches	Millimeters	Inches	Millimeters	Inches	Millimeters	Inches	Millimeters
0.0001	0.00254	0.005	0.1270	0.09	2.286	4	101.6
0.0002	0.00508	0.006	0.1524	0.1	2.54	5	127.0
0.0003	0.00762	0.007	0.1778	0.2	5.08	6	152.4
0.0004	0.01016	0.008	0.2032	0.3	7.62	7	177.8
0.0005	0.01270	0.009	0.2286	0.4	10.16	8	203.2
0.0006	0.01524	0.01	0.254	0.5	12.70	9	228.6
0.0007	0.01778	0.02	0.508	0.6	15.24	10	254.0
0.0008	0.02032	0.03	0.762	0.7	17.78	11	279.4
0.0009	0.02286	0.04	1.016	0.8	20.32	12	304.8
0.001	0.0254	0.05	1.270	0.9	22.86	13	330.2
0.002	0.0508	0.06	1.524	1	25.4	14	355.6
0.003	0.0762	0.07	1.778	2	50.8	15	381.0
0.004	0.1016	0.08	2.032	3	76.2	16	406.4

ENGLISH/METRIC CONVERSION: LENGTH

To convert inches (in.) to millimeters (mm), multiply the number of inches by 25.4
To convert millimeters (mm) to inches (in.), multiply the number of millimeters by 0.04

Inches		Millimeters	Inches		Millimeters	Inches		Millimeters
Fraction	Decimal	Decimal	Fraction	Decimal	Decimal	Fraction	Decimal	Decimal
1/64	0.016	0.397	11/32	0.344	8.731	11/16	0.688	17.463
1/32	0.031	0.794	23/64	0.359	9.128	45/64	0.703	17.859
3/64	0.047	1.191	3/8	0.375	9.525	23/32	0.719	18.256
1/16	0.063	1.588	25/64	0.391	9.922	47/64	0.734	18.653
5/64	0.078	1.984	13/32	0.406	10.319	3/4	0.750	19.050
3/32	0.094	2.381	27/64	0.422	10.716	49/64	0.766	19.447
7/64	0.109	2.778	7/16	0.438	11.113	25/32	0.781	19.844
1/8	0.125	3.175	29/64	0.453	11.509	51/64	0.797	20.241
9/64	0.141	3.572	15/32	0.469	11.906	13/16	0.813	20.638
5/32	0.156	3.969	31/64	0.484	12.303	53/64	0.828	21.034
11/64	0.172	4.366	1/2	0.500	12.700	27/32	0.844	21.431
3/16	0.188	4.763	33/64	0.516	13.097	55/64	0.859	21.828
13/64	0.203	5.159	17/32	0.531	13.494	7/8	0.875	22.225
7/32	0.219	5.556	35/64	0.547	13.891	57/64	0.891	22.622
15/64	0.234	5.953	9/16	0.563	14.288	29/32	0.906	23.019
1/4	0.250	6.350	37/64	0.578	14.684	59/64	0.922	23.416
17/64	0.266	6.747	19/32	0.594	15.081	15/16	0.938	23.813
9/32	0.281	7.144	39/64	0.609	15.478	61/64	0.953	24.209
19/64	0.297	7.541	5/8	0.625	15.875	31/32	0.969	24.606
5/16	0.313	7.938	41/64	0.641	16.272	63/64	0.984	25.003
21/64	0.328	8.334	21/32	0.656	16.669	1/1	1.000	25.400
			43/64	0.672	17.066			